Marks'
Standard Handbook
for Mechanical Engineers

Other McGraw-Hill Handbooks of Interest

Marks'
Standard Handbook for Mechanical Engineers

Revised by a staff of specialists

EUGENE A. AVALLONE   *Editor*

Consulting Engineer; Professor of Mechanical Engineering, Emeritus
The City College of the City University of New York

THEODORE BAUMEISTER III   *Editor*

Retired Consultant, Information Systems Department
E. I. du Pont de Nemours & Co.

Tenth Edition

McGRAW-HILL

New York   San Francisco   Washington, D.C.   Auckland   Bogotá
Caracas   Lisbon   London   Madrid   Mexico City   Milan
Montreal   New Delhi   San Juan   Singapore
Sydney   Tokyo   Toronto

Library of Congress Cataloged The First Issue
   of this title as follows:

Standard handbook for mechanical engineers. 1st-ed.;
   1916–

   New York, McGraw-Hill.

      v. Illus. 18–24 cm.

   Title varies: 1916–58; Mechanical engineers' handbook.
   Editors: 1916–51, L. S. Marks.—1958–   T. Baumeister.
   Includes bibliographies.

      1. Mechanical engineering—Handbooks, manuals, etc.   I. Marks,
   Lionel Simeon, 1871– ed.   II. Baumeister, Theodore, 1897–
   ed.   III. Title; Mechanical engineers' handbook.
   TJ151.S82   502′.4′621   16–12915

Library of Congress Catalog Card Number: 87-641192

MARKS' STANDARD HANDBOOK FOR MECHANICAL ENGINEERS

McGraw-Hill

A Division of The McGraw·Hill Companies

1 2 3 4 5 6 7 8 9 0   DOW/DOW   9 0 1 0 9 8 7 6

ISBN 0-07-004997-1

*The sponsoring editors for this book were Robert W. Hauserman and Robert Esposito, the editing
supervisor was David E. Fogarty, and the production supervisor was Suzanne W. B. Rapcavage.
It was set in Times Roman by Progressive Information Technologies.*

Printed and bound by R. R. Donnelley & Sons Company.

This book is printed on acid-free paper.

*The editors and the publishers will be grateful to readers who notify them of any inaccuracy or
important omission in this book.*

Contents

For the detailed contents of any section consult the title page of that section.

Contributors

Abraham Abramowitz *Consulting Engineer; Professor of Electrical Engineering, Emeritus, The City College, The City University of New York* (ILLUMINATION)

Vincent M. Altamuro *President, VMA, Inc., Toms River, NJ* (MATERIAL HOLDING AND FEEDING. CONVEYOR MOVING AND HANDLING. AUTOMATED GUIDED VEHICLES AND ROBOTS. MATERIAL STORAGE AND WAREHOUSING. METHODS ENGINEERING. AUTOMATED MANUFACTURING. INDUSTRIAL PLANTS)

Alger Anderson *Vice President, Engineering, Research & Product Development, Lift-Tech International, Inc.* (OVERHEAD TRAVELING CRANES)

William Antis* *Technical Director, Maynard Research Council, Inc., Pittsburgh, PA* (METHODS ENGINEERING)

Dennis N. Assanis *Professor of Mechanical Engineering, University of Michigan* (INTERNAL COMBUSTION ENGINES)

Klemens C. Baczewski *Consulting Engineer* (CARBONIZATION OF COAL AND GAS MAKING)

Glenn W. Baggley *Manager, Regenerative Systems, Bloom Engineering Co., Inc.* (COMBUSTION FURNACES)

Frederick G. Bailey *Consulting Engineer; formerly Technical Coordinator, Thermodynamics and Applications Engineering, General Electric Co.* (STEAM TURBINES)

Antonio F. Baldo *Professor of Mechanical Engineering, Emeritus, The City College, The City University of New York* (NONMETALLIC MATERIALS. MACHINE ELEMENTS)

Robert D. Bartholomew *Engineer, Powell Labs., Ltd.* (CORROSION)

George F. Baumeister *President, EMC Process Corp., Newport, DE* (MATHEMATICAL TABLES)

Heard K. Baumeister *Senior Engineer, Retired, International Business Machines Corp.* (MECHANISM)

Howard S. Bean* *Late Physicist, National Bureau of Standards* (GENERAL PROPERTIES OF MATERIALS)

E. R. Behnke* *Product Manager, CM Chain Division, Columbus, McKinnon Corp.* (CHAINS)

John T. Benedict *Retired Standards Engineer and Consultant, Society of Automotive Engineers* (AUTOMOTIVE ENGINEERING)

C. H. Berry* *Late Gordon McKay Professor of Mechanical Engineering, Harvard University; Late Professor of Mechanical Engineering, Northeastern University* (PREFERRED NUMBERS)

Louis Bialy *Director, Codes & Product Safety, Otis Elevator Company* (ELEVATORS, DUMBWAITERS, AND ESCALATORS)

Malcolm Blair *Technical and Research Director, Steel Founders Society of America* (IRON AND STEEL CASTINGS)

Omer W. Blodgett *Senior Design Consultant, Lincoln Electric Co.* (WELDING AND CUTTING)

Donald E. Bolt *Engineering Manager, Heat Transfer Products Dept., Foster Wheeler Energy Corp.* (POWER PLANT HEAT EXCHANGERS)

Claus Borgnakke *Associate Professor of Mechanical Engineering, University of Michigan* (INTERNAL COMBUSTION ENGINES)

G. David Bounds *Senior Engineer, PanEnergy Corp.* (PIPELINE TRANSMISSION)

William J. Bow *Director, Retired, Heat Transfer Products Department, Foster Wheeler Energy Corp.* (POWER PLANT HEAT EXCHANGERS)

James L. Bowman *Senior Engineering Consultant, Rotary-Reciprocating Compressor Division, Ingersoll-Rand Co.* (COMPRESSORS)

Aine Brazil *Vice President, Thornton-Tomasetti/Engineers* (STRUCTURAL DESIGN OF BUILDINGS)

Frederic W. Buse* *Chief Engineer, Standard Pump Division, Ingersoll-Rand Co.* (DISPLACEMENT PUMPS)

C. P. Butterfield *Chief Engineer, Wind Technology Division, National Renewable Energy Laboratory* (WIND POWER)

Benson Carlin* *President, O.E.M. Medical, Inc.* (SOUND, NOISE, AND ULTRASONICS)

C. L. Carlson* *Late Fellow Engineer, Research Labs., Westinghouse Electric Corp.* (NONFERROUS METALS)

Vittorio (Rino) Castelli *Senior Research Fellow, Xerox Corp.* (FRICTION, FLUID FILM BEARINGS)

Michael J. Clark *Manager, Optical Tool Engineering and Manufacturing, Bausch & Lomb, Rochester, NY* (OPTICS)

Ashley C. Cockerill *Staff Engineer, Motorola Corp.* (ENGINEERING STATISTICS AND QUALITY CONTROL)

Aaron Cohen *Retired Center Director, Lyndon B. Johnson Space Center, NASA and Zachry Professor, Texas A&M University* (ASTRONAUTICS)

Arthur Cohen *Manager, Standards and Safety Engineering, Copper Development Assn.* (COPPER AND COPPER ALLOYS)

D. E. Cole *Director, Office for Study of Automotive Transportation, Transportation Research Institute, University of Michigan* (INTERNAL COMBUSTION ENGINES)

James M. Connolly *Section Head, Projects Department, Jacksonville Electric Authority* (COST OF ELECTRIC POWER)

Robert T. Corry* *Retired Associate Professor of Mechanical and Aerospace Engineering, Polytechnic University* (INSTRUMENTS)

Paul E. Crawford *Partner; Connolly, Bove, Lodge & Hutz; Wilmington, DE* (PATENTS, TRADEMARKS, AND COPYRIGHTS)

M. R. M. Crespo da Silva* *University of Cincinnati* (ATTITUDE DYNAMICS, STABILIZATION, AND CONTROL OF SPACECRAFT)

Julian H. Dancy *Consulting Engineer, Formerly Senior Technologist, Technology Division, Fuels and Lubricants Technology Department, Texaco, Inc.* (LUBRICANTS AND LUBRICATION)

Benjamin B. Dayton *Consulting Physicist, East Flat Rock, NC* (HIGH-VACUUM PUMPS)

Rodney C. DeGroot *Research Plant Pathologist, Forest Products Lab., USDA* (WOOD)

Joseph C. Delibert *Retired Executive, The Babcock and Wilcox Co.* (STEAM BOILERS)

Donald D. Dodge *Supervisor, Retired, Product Quality and Inspection Technology, Manufacturing Development, Ford Motor Co.* (NONDESTRUCTIVE TESTING)

Joseph S. Dorson *Senior Engineer, Columbus McKinnon Corp.* (CHAIN)

Michael B. Duke *Chief, Solar Systems Exploration, Johnson Space Center, NASA* (ASTRONOMICAL CONSTANTS OF THE SOLAR SYSTEM. DYNAMIC ENVIRONMENTS. SPACE ENVIRONMENT)

F. J. Edeskuty *Retired Associate, Los Alamos National Laboratory* (CRYOGENICS)

O. Elnan* *University of Cincinnati* (SPACE-VEHICLE TRAJECTORIES, FLIGHT MECHANICS, AND PERFORMANCE. ORBITAL MECHANICS)

Robert E. Eppich *Vice President, Technology, American Foundrymen's Society* (IRON AND STEEL CASTINGS)

C. James Erickson* *Principal Consultant, Engineering Department. E. I. du Pont de Nemours & Co.* (ELECTRICAL ENGINEERING)

George H. Ewing* *Retired President and Chief Executive Officer, Texas Eastern Gas Pipeline Co. and Transwestern Pipeline Co.* (PIPELINE TRANSMISSION)

Erich A. Farber *Distinguished Service Professor Emeritus; Director, Emeritus, Solar Energy and Energy Conversion Lab., University of Florida* (HOT AIR ENGINES. SOLAR ENERGY. DIRECT ENERGY CONVERSION)

D. W. Fellenz* *University of Cincinnati* (SPACE-VEHICLE TRAJECTORIES, FLIGHT MECHANICS, AND PERFORMANCE. ATMOSPHERIC ENTRY)

Arthur J. Fiehn* *Late Retired Vice President, Project Operations Division, Burns & Roe, Inc.* (COST OF ELECTRIC POWER)

Sanford Fleeter *Professor of Mechanical Engineering and Director, Thermal Sciences and Propulsion Center, School of Mechanical Engineering, Purdue University* (JET PROPULSION AND AIRCRAFT PROPELLERS)

William L. Gamble *Professor of Civil Engineering, University of Illinois at Urbana-Champaign* (CEMENT, MORTAR, AND CONCRETE. REINFORCED CONCRETE DESIGN AND CONSTRUCTION)

*Contributions by authors whose names are marked with an asterisk were made for the previous edition and have been revised or rewritten by others for this edition. The stated professional position in these cases is that held by the author at the time of his or her contribution.

Daniel G. Garner* Senior Program Manager, Institute of Nuclear Power Operations, Atlanta, GA (NUCLEAR POWER)

Burt Garofab Senior Engineer, Pittston Corp. (MINES, HOISTS, AND SKIPS. LOCOMOTIVE HAULAGE, COAL MINES)

Siamak Ghofranian Senior Engineer, Rockwell Aerospace (DOCKING OF TWO FREE-FLYING SPACECRAFT)

Samuel V. Glorioso Section Chief, Metallic Materials, Johnson Space Center, NASA (STRESS CORROSION CRACKING)

Norman Goldberg Consulting Engineer (HEATING, VENTILATION, AND AIR CONDITIONING)

David T. Goldman Deputy Manager, U.S. Department of Energy, Chicago Operations Office (MEASURING UNITS)

Frank E. Goodwin Vice President, Materials Science, ILZRO, Inc. (BEARING METALS. LOW-MELTING-POINT METALS AND ALLOYS. ZINC AND ZINC ALLOYS)

Don Graham Manager, Turning Programs, Carboloy, Inc. (CEMENTED CARBIDES)

John E. Gray* ERCI, Intl. (NUCLEAR POWER)

David W. Green Supervisory Research General Engineer, Forest Products Lab., USDA (WOOD)

Walter W. Guy Chief, Crew and Thermal Systems Division, Johnson Space Center, NASA (SPACECRAFT LIFE SUPPORT AND THERMAL MANAGEMENT)

Harold V. Hawkins* Late Manager, Product Standards and Services, Columbus McKinnon Corp. (DRAGGING, PULLING, AND PUSHING. PIPELINE FLEXURE STRESSES)

Keith L. Hawthorne Senior Assistant Vice President, Transportation Technology Center, Association of American Railroads (RAILWAY ENGINEERING)

V. T. Hawthorne Vice President, Engineering and Technical Services, American Steel Foundries (RAILWAY ENGINEERING)

J. Edmund Hay U.S. Department of the Interior (EXPLOSIVES)

Roger S. Hecklinger Project Director, Roy F. Weston of New York, Inc. (INCINERATION)

Terry L. Henshaw Consulting Engineer, Battle Creek, MI (DISPLACEMENT PUMPS)

Roland Hernandez Research General Engineer, Forest Products Lab., USDA (WOOD)

Hoyt C. Hottel Professor Emeritus, Massachusetts Institute of Technology (RADIANT HEAT TRANSFER)

R. Eric Hutz Associate; Connolly, Bove, Lodge, & Hutz; Wilmington, DE (PATENTS, TRADEMARKS, AND COPYRIGHTS)

Michael W. M. Jenkins Professor, Aerospace Design, Georgia Institute of Technology (AERONAUTICS)

Peter K. Johnson Director, Marketing and Public Relations, Metal Powder Industries Federation (POWDERED METALS)

Randolph T. Johnson Naval Surface Warfare Center (ROCKET FUELS)

Robert L. Johnston Branch Chief, Materials, Johnson Space Center, NASA (METALLIC MATERIALS FOR AEROSPACE APPLICATIONS. MATERIALS FOR USE IN HIGH-PRESSURE OXYGEN SYSTEMS)

Byron M. Jones Retired Associate Professor of Electrical Engineering, School of Engineering, University of Tennessee at Chattanooga (ELECTRONICS)

Scott K. Jones Associate Professor, Department of Accounting, University of Delaware (COST ACCOUNTING)

Robert Jorgensen Engineering Consultant (FANS)

Serope Kalpakjian Professor of Mechanical and Materials Engineering, Illinois Institute of Technology (METAL REMOVAL PROCESSES AND MACHINE TOOLS)

Igor J. Karassik Late Senior Consulting Engineer, Ingersoll-Dresser Pump Co. (CENTRIFUGAL AND AXIAL FLOW PUMPS)

Robert W. Kennard* Lake-Sumter Community College, Leesburg, FL (ENGINEERING STATISTICS AND QUALITY CONTROL)

Edwin E. Kintner* Executive Vice President, GPU Nuclear Corp., Parsippany, NJ (NUCLEAR POWER)

J. Randolph Kissell Partner, The TGB Partnership (ALUMINUM AND ITS ALLOYS)

Andrew C. Klein Associate Professor, Nuclear Engineering, Oregon State University (ENVIRONMENTAL CONTROL. OCCUPATIONAL SAFETY AND HEALTH. FIRE PROTECTION)

Ezra S. Krendel Emeritus Professor of Operations Research and Statistics, Wharton School, University of Pennsylvania (HUMAN FACTORS AND ERGONOMICS. MUSCLE GENERATED POWER)

A. G. Kromis* University of Cincinnati (SPACE-VEHICLE TRAJECTORIES, FLIGHT MECHANICS, AND PERFORMANCE)

P. G. Kuchuris, Jr.* Market Planning Manager, International Harvester Co. (OFF-HIGHWAY VEHICLES AND EARTHMOVING EQUIPMENT)

L. D. Kunsman* Late Fellow Engineer, Research Labs., Westinghouse Electric Corp. (NONFERROUS METALS)

Colin K. Larsen Vice President, Blue Giant U.S.A. Corp. (SURFACE HANDLING)

Lubert J. Leger Deputy Branch Chief, Materials, Johnson Space Center, NASA (SPACE ENVIRONMENT)

John H. Lewis Technical Staff, Pratt & Whitney, Division of United Technologies Corp.; Adjunct Associate Professor, Hartford Graduate Center, Renssealear Polytechnic Institute (GAS TURBINES)

Peter E. Liley Professor, School of Mechanical Engineering, Purdue University (THERMODYNAMICS, THERMODYNAMIC PROPERTIES OF SUBSTANCES)

Michael K. Madsen Manager, Industrial Products Engineering, Neenah Foundry Co. (FOUNDRY PRACTICE AND EQUIPMENT)

C. J. Manney* Consultant, Columbus McKinnon Corp. (HOISTS)

Ernst K. H. Marburg Manager, Product Standards and Service, Columbus McKinnon Corp. (LIFTING, HOISTING, AND ELEVATING. DRAGGING, PULLING, AND PUSHING. LOADING, CARRYING, AND EXCAVATING)

Adolph Matz* Late Professor Emeritus of Accounting, The Wharton School, University of Pennsylvania (COST ACCOUNTING)

Leonard Meirovitch University Distinguished Professor, Department of Engineering Science and Mechanics, Virginia Polytechnic Institute and State University (VIBRATION)

Sherwood B. Menkes Professor of Mechanical Engineering, Emeritus, The City College, The City University of New York (FLYWHEEL ENERGY STORAGE)

George W. Michalec Consulting Engineer, Formerly Professor and Dean of Engineering and Science, Stevens Institute of Technology (GEARING)

Duane K. Miller Welding Design Engineer, Lincoln Electric Co. (WELDING AND CUTTING)

Russell C. Moody Supervisory Research General Engineer, Forest Products Lab., USDA (WOOD)

Ralph L. Moore* Retired Systems Consultant, E. I. du Pont de Nemours & Co. (AUTOMATIC CONTROLS)

Thomas L. Moser Deputy Associate Administrator, Office of Space Flight, NASA Headquarters, NASA (SPACE-VEHICLE STRUCTURES)

George J. Moshos Professor Emeritus of Computer and Information Science, New Jersey Institute of Technology (COMPUTERS)

Otto Muller-Girard Consulting Engineer (INSTRUMENTS)

James W. Murdock Late Consulting Engineer (MECHANICS OF FLUIDS)

Gregory V. Murphy Process Control Consultant, DuPont Co. (AUTOMATIC CONTROLS)

Joseph F. Murphy Supervisory General Engineer, Forest Products Lab., USDA (WOOD)

John Nagy Retired Supervisory Physical Scientist, U.S. Department of Labor, Mine Safety and Health Administration (DUST EXPLOSIONS)

B. W. Niebel Professor Emeritus of Industrial Engineering, The Pennsylvania State University (INDUSTRIAL ECONOMICS AND MANAGEMENT)

Paul E. Norian Special Assistant, Regulatory Applications, Office of Nuclear Regulatory Research, U.S. Nuclear Regulatory Commission (NUCLEAR POWER)

Nunzio J. Palladino* Dean Emeritus, College of Engineering, Pennsylvania State University (NUCLEAR POWER)

D. J. Patterson Professor of Mechanical Engineering, Emeritus, University of Michigan (INTERNAL COMBUSTION ENGINES)

Harold W. Paxton United States Steel Professor Emeritus, Carnegie Mellon University (IRON AND STEEL)

Richard W. Perkins Professor of Mechanical, Aerospace, and Manufacturing Engineering, Syracuse University (WOODCUTTING TOOLS AND MACHINES)

W. R. Perry* University of Cincinnati (ORBITAL MECHANICS. SPACE-VEHICLE TRAJECTORIES, FLIGHT MECHANICS, AND PERFORMANCE)

Kenneth A. Phair Senior Mechanical Engineer, Stone and Webster Engineering Corp. (GEOTHERMAL POWER)

Orvis E. Pigg Section Head, Structural Analysis, Johnson Space Center, NASA (SPACE-VEHICLE STRUCTURES)

Henry O. Pohl Chief, Propulsion and Power Division, Johnson Space Center, NASA (SPACE PROPULSION)

Charles D. Potts Retired Project Engineer, Engineering Department, E. I. du Pont de Nemours & Co. (ELECTRICAL ENGINEERING)

R. Ramakumar Professor of Electrical Engineering, Oklahoma State University (WIND POWER)

Pascal M. Rapier Scientist III, Retired, Lawrence Berkeley Laboratory (ENVIRONMENTAL CONTROL. OCCUPATIONAL SAFETY AND HEALTH. FIRE PROTECTION)

James D. Redmond President, Technical Marketing Services, Inc. (STAINLESS STEEL)

Albert H. Reinhardt Technical Staff, Pratt & Whitney, Division of United Technologies Corp. (GAS TURBINES)

Warren W. Rice Senior Project Engineer, Piedmont Engineering Corp. (MECHANICAL REFRIGERATION)

George J. Roddam Sales Engineer, Lectromelt Furnace Division, Salem Furnace Co. (ELECTRIC FURNACES AND OVENS)

Louis H. Roddis* Late Consulting Engineer, Charleston, SC (NUCLEAR POWER)

Darrold E. Roen Late Manager, Sales & Special Engineering & Government Products, John Deere (OFF-HIGHWAY VEHICLES)

Ivan L. Ross* International Manager, Chain Conveyor Division, ACCO (OVERHEAD CONVEYORS)

Robert J. Ross Supervisory Research General Engineer, Forest Products Lab., USDA (WOOD)

J. W. Russell* University of Cincinnati (SPACE-VEHICLE TRAJECTORIES, FLIGHT MECHANICS, AND PERFORMANCE. LUNAR- AND INTERPLANETARY-FLIGHT MECHANICS)

A. J. Rydzewski Project Engineer, Engineering Department, E. I. du Pont de Nemours & Co. (MECHANICAL REFRIGERATION)

C. Edward Sandifer   *Professor, Western Connecticut State University, Danbury, CT* (MATHEMATICS)

Adel F. Sarofim   *Lammot du Pont Professor of Chemical Engineering, Massachusetts Institute of Technology* (RADIANT HEAT TRANSFER)

Martin D. Schlesinger   *Wallingford Group, Ltd.* (FUELS)

John R. Schley   *Manager, Technical Marketing, RMI Titanium Co.* (TITANIUM AND ZIRCONIUM)

Matthew S. Schmidt   *Senior Engineer, Rockwell Aerospace* (DOCKING OF TWO FREE-FLYING SPACECRAFT)

George Sege   *Technical Assistant to the Director, Office of Nuclear Regulatory Research, U.S. Nuclear Regulatory Commission* (NUCLEAR POWER)

James D. Shearouse, III   *Senior Development Engineer, The Dow Chemical Co.* (MAGNESIUM AND MAGNESIUM ALLOYS)

David A. Shifler   *Metallurgist, Naval Surface Warfare Center* (CORROSION)

Rajiv Shivpuri   *Professor of Industrial, Welding, and Systems Engineering, Ohio State University* (PLASTIC WORKING OF METALS)

William T. Simpson   *Research Forest Products Technologist, Forest Products Lab., USDA* (WOOD)

Kenneth A. Smith   *Edward R. Gilliland Professor of Chemical Engineering, Massachusetts Institute of Technology* (TRANSMISSION OF HEAT BY CONDUCTION AND CONVECTION)

Lawrence H. Sobel*   *University of Cincinnati* (VIBRATION OF STRUCTURES)

James G. Speight   *Western Research Institute* (FUELS)

Ivan K. Spiker   *NASA, Retired* (STRUCTURAL COMPOSITES)

Robert D. Steele   *Manager, Turbine and Rehabilitation Design, Voith Hydro, Inc.* (HYDRAULIC TURBINES)

Robert F. Steidel, Jr.   *Professor of Mechanical Engineering, Retired, University of California, Berkeley* (MECHANICS OF SOLIDS)

Stephen R. Swanson   *Professor of Mechanical Engineering, University of Utah* (FIBER COMPOSITE MATERIALS)

John Symonds   *Fellow Engineer, Retired, Oceanic Division, Westinghouse Electric Corp.* (MECHANICAL PROPERTIES OF MATERIALS)

Anton TenWolde   *Research Physicist, Forest Products Lab., USDA* (WOOD)

W. David Teter   *Professor of Civil Engineering, University of Delaware* (SURVEYING)

Helmut Thielsch*   *President, Thielsch Engineering Associates* (PIPE, PIPE FITTINGS, AND VALVES)

Michael C. Tracy   *Captain, U.S. Navy* (MARINE ENGINEERING)

John H. Tundermann   *Vice President, Research and Technology, INCO Alloys Intl., Inc.* (METALS AND ALLOYS FOR USE AT ELEVATED TEMPERATURES. NICKEL AND NICKEL ALLOYS)

Charles O. Velzy   *Consultant* (INCINERATION)

Harry C. Verakis   *Supervisory Physical Scientist, U.S. Department of Labor, Mine Safety and Health Administration* (DUST EXPLOSIONS)

Arnold S. Vernick   *Associate, Geraghty & Miller, Inc.* (WATER)

J. P. Vidosic   *Regents' Professor Emeritus of Mechanical Engineering, Georgia Institute of Technology* (MECHANICS OF MATERIALS)

Robert J. Vondrasek   *Assistant Vice President of Engineering, National Fire Protection Assoc.* (COST OF ELECTRIC POWER)

Michael W. Washo   *Engineering Associate, Eastman Kodak Co.* (BEARINGS WITH ROLLING CONTACT)

Harold M. Werner*   *Consultant* (PAINTS AND PROTECTIVE COATINGS)

Robert H. White   *Supervisory Wood Scientist, Forest Products Lab., USDA* (WOOD)

Thomas W. Wolff   *Instructor, Retired, Mechanical Engineering Dept., The City College, The City University of New York* (SURFACE TEXTURE DESIGNATION, PRODUCTION, AND CONTROL)

John W. Wood, Jr.   *Applications Specialist, Fluidtec Engineered Products, Coltec Industries* (PACKINGS AND SEALS)

Dedication

On the occasion of the publication of the tenth edition of *Marks' Standard Handbook for Mechanical Engineers,* we note that this is also the eightieth anniversary of the publication of the first edition. The Editors and publisher proffer this brief dedication to all those who have been instrumental in the realization of the goals set forth by Lionel S. Marks in the preface to the first edition.

First, we honor the memory of the deceased Editors, Lionel S. Marks and Theodore Baumeister. Lionel S. Marks' concept of a *Mechanical Engineers' Handbook* came to fruition with the publication of the first edition in 1916; Theodore Baumeister followed as Editor with the publication of the sixth edition in 1958.

Second, we are indebted to our contributors, past and present, who so willingly mined their expertise to gather material for inclusion in the Handbook, thereby sharing it with others, far and wide.

Third, we acknowledge our wide circle of readers—engineers and others—who have used the Handbook in the conduct of their work and, from time to time, have provided cogent commentary, suggestions, and expressions of loyalty.

Preface to the Tenth Edition

In the preparation of the tenth edition of "Marks," the Editors had two major continuing objectives. First, to modernize and update the contents as required, and second, to hold to the high standard maintained for eighty years by the previous Editors, Lionel S. Marks and Theodore Baumeister.

The Editors have found it instructive to leaf through the first edition of *Marks' Handbook* and to peruse its contents. Some topics still have currency as we approach the end of the twentieth century; others are of historical interest only. Certainly, the passage of 80 years since the publication of the first edition sends a clear message that "things change"!

The replacement of the U.S. Customary System (USCS) of units by the International System (SI) is still far from complete, and proceeds at different rates not only in the engineering professions, but also in our society in general. Accordingly, duality of units has been retained, as appropriate.

Established practice combined with new concepts and developments are the underpinnings of our profession. Among the most significant and far-reaching changes are the incorporation of microprocessors into many tools and devices, both new and old. An ever-increasing number of production processes are being automated with robots performing dull or dangerous jobs.

Workstations consisting of personal computers and a selection of software seemingly without limits are almost universal. Not only does the engineer have powerful computational and analytical tools at hand, but also those same tools have been applied in diverse areas which appear to have no bounds. A modern business or manufacturing entity without a keyboard and a screen is an anomaly.

The Editors are cognizant of the competing requirements to offer the user a broad spectrum of information that has been the hallmark of the Marks' Handbook since its inception, and yet to keep the size of the one volume within reason. This has been achieved through the diligent efforts and cooperation of contributors, reviewers, and the publisher.

Last, the Handbook is ultimately the responsibility of the Editors. Meticulous care has been exercised to avoid errors, but if any are inadvertently included, the Editors will appreciate being so informed so that corrections can be incorporated in subsequent printings of this edition.

Ardsley, NY   EUGENE A. AVALLONE
Newark, DE   THEODORE BAUMEISTER III

Preface to the First Edition*

This Handbook is intended to supply both the practicing engineer and the student with a reference work which is authoritative in character and which covers the field of mechanical engineering in a comprehensive manner. It is no longer possible for a single individual or a small group of individuals to have so intimate an acquaintance with any major division of engineering as is necessary if critical judgment is to be exercised in the statement of current practice and the selection of engineering data. Only by the cooperation of a considerable number of specialists is it possible to obtain the desirable degree of reliability. This Handbook represents the work of fifty specialists.

Each contributor is to be regarded as responsible for the accuracy of his section. The number of contributors required to ensure sufficiently specialized knowledge for all the topics treated is necessarily large. It was found desirable to enlist the services of thirteen specialists for an adequate handling of the "Properties of Engineering Materials." Such topics as "Automobiles," "Aeronautics," "Illumination," "Patent Law," "Cost Accounting," "Industrial Buildings," "Corrosion," "Air Conditioning," "Fire Protection," "Prevention of Accidents," etc., though occupying relatively small spaces in the book, demanded each a separate writer.

A number of the contributions which deal with engineering practice, after examination by the Editor-in-Chief, were submitted by him to one or more specialists for criticism and suggestions. Their cooperation has proved of great value in securing greater accuracy and in ensuring that the subject matter does not embody solely the practice of one individual but is truly representative.

An accuracy of four significant figures has been assumed as the desirable limit; figures in excess of this number have been deleted, except in special cases. In the mathematical tables only four significant figures have been kept.

The Editor-in-Chief desires to express here his appreciation of the spirit of cooperation shown by the Contributors and of their patience in submitting to modifications of their sections. He wishes also to thank the Publishers for giving him complete freedom and hearty assistance in all matters relating to the book from the choice of contributors to the details of typography.

Cambridge, Mass.
April 23, 1916

LIONEL S. MARKS

* Excerpt.

Symbols and Abbreviations

For symbols of chemical elements, see Sec. 6; for abbreviations applying to metric weights and measures and SI units, Sec. 1; SI unit prefixes are listed on p. 1-19.

Pairs of parentheses, brackets, etc., are frequently used in this work to indicate corresponding values. For example, the statement that "the cost per kW of a 30,000-kW plant is $86; of a 15,000-kW plant, $98; and of an 8,000-kW plant, $112," is condensed as follows: The cost per kW of a 30,000 (15,000) [8,000]-kW plant is $86 (98) [112].

In the citation of references readers should always attempt to consult the latest edition of referenced publications.

A or Å	Angstrom unit $= 10^{-10}$ m; 3.937×10^{-11} in
A	mass number $= N + Z$; ampere
AA	arithmetical average
AAA	Am. Automobile Assoc.
AAMA	American Automobile Manufacturers' Assoc.
AAR	Assoc. of Am. Railroads
AAS	Am. Astronautical Soc.
ABAI	Am. Boiler & Affiliated Industries
abs	absolute
a.c.	aerodynamic center
a-c, ac	alternating current
ACI	Am. Concrete Inst.
ACM	Assoc. for Computing Machinery
ACRMA	Air Conditioning and Refrigerating Manufacturers Assoc.
ACS	Am. Chemical Soc.
ACSR	aluminum cable steel-reinforced
ACV	air cushion vehicle
A.D.	anno Domini (in the year of our Lord)
AEC	Atomic Energy Commission (U.S.)
a-f, af	audio frequency
AFBMA	Anti-friction Bearings Manufacturers' Assoc.
AFS	Am. Foundrymen's Soc.
AGA	Am. Gas Assoc.
AGMA	Am. Gear Manufacturers' Assoc.
ahp	air horsepower
AIChE	Am. Inst. of Chemical Engineers
AIEE	Am. Inst. of Electrical Engineers (see IEEE)
AIME	Am. Inst. of Mining Engineers
AIP	Am. Inst. of Physics
AISC	American Institute of Steel Construction, Inc.
AISE	Am. Iron & Steel Engineers
AISI	Am. Iron and Steel Inst.
a.m.	ante meridiem (before noon)
a-m, am	amplitude modulation
Am. Mach.	Am. Machinist (New York)
AMA	Acoustical Materials Assoc.
AMCA	Air Moving & Conditioning Assoc., Inc.
amu	atomic mass unit
AN	ammonium nitrate (explosive); Army-Navy Specification
AN-FO	ammonium nitrate-fuel oil (explosive)
ANC	Army-Navy Civil Aeronautics Committee
ANS	Am. Nuclear Soc.
ANSI	American National Standards Institute
antilog	antilogarithm of
API	Am. Petroleum Inst.
approx	approximately
APWA	Am. Public Works Assoc.
AREA	Am. Railroad Eng. Assoc.
ARI	Air Conditioning and Refrigeration Inst.
ARS	Am. Rocket Soc.
ASCE	Am. Soc. of Civil Engineers
ASHRAE	Am. Soc. of Heating, Refrigerating, and Air Conditioning Engineers
ASLE	Am. Soc. of Lubricating Engineers
ASM	Am. Soc. of Metals
ASME	Am. Soc. of Mechanical Engineers
ASST	Am. Soc. of Steel Treating
ASTM	Am. Soc. for Testing and Materials
ASTME	Am. Soc. of Tool & Manufacturing Engineers
atm	atmosphere
Auto. Ind.	Automotive Industries (New York)
avdp	avoirdupois
avg, ave	average
AWG	Am. Wire Gage
AWPA	Am. Wood Preservation Assoc.
AWS	American Welding Soc.
AWWA	American Water Works Assoc.
b	barns
bar	barometer
B&S	Brown & Sharp (gage); Beams and Stringers
bbl	barrels
B.C.	before Christ
B.C.C.	body centered cubic
Bé	Baumé (degrees)
B.G.	Birmingham gage (hoop and sheet)
bgd	billions of gallons per day
BHN	Brinell Hardness Number
bhp	brake horsepower
BLC	boundary layer control
B.M.	board measure; bench mark
bmep	brake mean effective pressure
B of M, BuMines	Bureau of Mines
BOD	biochemical oxygen demand

bp	boiling point	d-c, dc	direct current
Bq	bequerel	def	definition
bsfc	brake specific fuel consumption	deg	degrees
BSI	British Standards Inst.	diam. (dia)	diameter
Btu	British thermal units	DO	dissolved oxygen
Btuh, Btu/h	Btu per hr	D_2O	deuterium (heavy water)
bu	bushels	d.p.	double pole
Bull.	Bulletin	DP	Diametral pitch
Buweaps	Bureau of Weapons, U.S. Navy	DPH	diamond pyramid hardness
BWG	Birmingham wire gage	DST	daylight saving time
c	velocity of light	d^2 tons	breaking strength, d = chain wire diam, in.
°C	degrees Celsius (centigrade)	DX	direct expansion
C	coulomb	e	base of Napierian logarithmic system (= 2.7182 +)
CAB	Civil Aeronautics Board	EAP	equivalent air pressure
CAGI	Compressed Air & Gas Inst.	EDR	equivalent direct radiation
cal	calories	EEI	Edison Electric Inst.
C-B-R	chemical, biological & radiological (filters)	eff	efficiency
CBS	Columbia Broadcasting System	e.g.	exempli gratia (for example)
cc, cm³	cubic centimeters	ehp	effective horsepower
CCR	critical compression ratio	EHV	extra high voltage
c to c	center to center	*El. Wld.*	Electrical World (New York)
cd	candela	elec	electric
c.f.	centrifugal force	elong	elongation
cf.	confer (compare)	emf	electromotive force
cfh, ft³/h	cubic feet per hour	*Engg.*	Engineering (London)
cfm, ft³/min	cubic feet per minute	*Engr.*	The Engineer (London)
C.F.R.	Cooperative Fuel Research	ENT	emergency negative thrust
cfs, ft³/s	cubic feet per second	EP	extreme pressure (lubricant)
cg	center of gravity	ERDA	Energy Research & Development Administration (successor to AEC; see also NRC)
cgs	centimeter-gram-second		
Chm. Eng.	Chemical Eng'g (New York)	Eq.	equation
chu	centigrade heat unit	est	estimated
C.I.	cast iron	etc.	et cetera (and so forth)
cir	circular	et seq.	et sequens (and the following)
cir mil	circular mils	eV	electron volts
cm	centimeters	evap	evaporation
CME	Chartered Mech. Engr. (IMechE)	exp	exponential function of
C.N.	cetane number	exsec	exterior secant of
coef	coefficient	ext	external
COESA	U.S. Committee on Extension to the Standard Atmosphere	°F	degrees Fahrenheit
col	column	F	farad
colog	cologarithm of	FAA	Federal Aviation Agency
const	constant	F.C.	fixed carbon, %
cos	cosine of	FCC	Federal Communications Commission; Federal Constructive Council
cos⁻¹	angle whose cosine is, inverse cosine of		
cosh	hyperbolic cosine of	F.C.C.	face-centered-cubic (alloys)
cosh⁻¹	inverse hyperbolic cosine of	ff.	following (pages)
cot	cotangent of	fhp	friction horsepower
cot⁻¹	angle whose cotangent is (see cos⁻¹)	Fig.	figure
coth	hyperbolic cotangent of	F.I.T.	Federal income tax
coth⁻¹	inverse hyperbolic cotangent of	f-m, fm	frequency modulation
covers	coversed sine of	F.O.B.	free on board (cars)
c.p.	circular pitch; center of pressure	FP	fore perpendicular
cp	candle power	FPC	Federal Power Commission
cp	coef of performance	fpm, ft/min	feet per minute
CP	chemically pure	fps	foot-pound-second system
CPH	close packed hexagonal	ft/s	feet per second
cpm,	cycles per minute	F.S.	Federal Specifications
cycles/min		FSB	Federal Specifications Board
cps, cycles/s	cycles per second	fsp	fiber saturation point
CSA	Canadian Standards Assoc.	ft	feet
csc	cosecant of	fc	foot candles
csc⁻¹	angle whose cosecant is (see cos⁻¹)	fL	foot lamberts
csch	hyperbolic cosecant of	ft · lb	foot-pounds
csch⁻¹	inverse hyperbolic cosecant of	g	acceleration due to gravity
cu	cubic	g	grams
cyl	cylinder	gal	gallons
db, dB	decibel	gc	gigacycles per sec

GCA	ground-controlled approach		J&P	joists and planks
g · cal	gram-calories		*Jour.*	Journal
gd	Gudermannian of		JP	jet propulsion fuel
G.E.	General Electric Co.		*k*	isentropic exponent; conductivity
GEM	ground effect machine		K	degrees Kelvin (Celsius abs)
GFI	gullet feed index		K	Knudsen number
G.M.	General Motors Co.		kB	kilo Btu (1000 Btu)
GMT	Greenwich Mean Time		kc	kilocycles
GNP	gross national product		kcps	kilocycles per sec
gpcd	gallons per capita day		kg	kilograms
gpd	gallons per day; grams per denier		kg · cal	kilogram-calories
gpm, gal/min	gallons per minute		kg · m	kilogram-meters
gps, gal/s	gallons per second		kip	1000 lb or 1 kilo-pound
gpt	grams per tex		kips	thousands of pounds
H	henry		km	kilometers
h	Planck's constant = 6.624×10^{-27} erg-sec		kmc	kilomegacycles per sec
$\hbar$	Planck's constant, $\hbar = h/2\pi$		kmcps	kilomegacycles per sec
HEPA	high efficiency particulate matter		kpsi	thousands of pounds per sq in
h-f, hf	high frequency		ksi	one kip per sq in, 1000 psi (lb/in²)
hhv	high heat value		kts	knots
horiz	horizontal		kVA	kilovolt-amperes
hp	horsepower		kW	kilowatts
h-p	high-pressure		kWh	kilowatt-hours
HPAC	Heating, Piping, & Air Conditioning (Chicago)		L	lamberts
hp · hr	horsepower-hour		l, L	litres
hr, h	hours		£	Laplace operational symbol
HSS	high speed steel		lb	pounds
H.T.	heat-treated		L.B.P.	length between perpendiculars
HTHW	high temperature hot water		lhv	low heat value
Hz	hertz = 1 cycle/s (cps)		lim	limit
IACS	International Annealed Copper Standard		lin	linear
IAeS	Institute of Aerospace Sciences		ln	Napierian logarithm of
ibid.	ibidem (in the same place)		loc. cit.	loco citato (place already cited)
ICAO	International Civil Aviation Organization		log	common logarithm of
ICC	Interstate Commerce Commission		LOX	liquid oxygen explosive
ICE	Inst. of Civil Engineers		l-p, lp	low pressure
ICI	International Commission on Illumination		LPG	liquified petroleum gas
I.C.T.	International Critical Tables		lpw, lm/W	lumens per watt
I.D., ID	inside diameter		lx	lux
i.e.	id est (that is)		L.W.L.	load water line
IEC	International Electrotechnical Commission		lm	lumen
IEEE	Inst. of Electrical & Electronics Engineers (successor to AIEE, *q.v.*)		m	metres
			M	thousand; Mach number; moisture, %
IES	Illuminating Engineering Soc.		mA	milliamperes
i-f, if	intermediate frequency		*Machy.*	Machinery (New York)
IGT	Inst. of Gas Technology		max	maximum
ihp	indicated horsepower		MBh	thousands of Btu per hr
IMechE	Inst. of Mechanical Engineers		mc	megacycles per sec
imep	indicated mean effective pressure		m.c.	moisture content
Imp	Imperial		Mcf	thousand cubic feet
in., in	inches		mcps	megacycles per sec
in. · lb	inch-pounds		*Mech. Eng.*	Mechanical Eng'g (ASME)
in · lb			mep	mean effective pressure
INA	Inst. of Naval Architects		METO	maximum, except during take-off
Ind. & Eng.	Industrial & Eng'g Chemistry (Easton, PA)		me V	million electron volts
Chem.			MF	maintenance factor
int	internal		mhc	mean horizontal candles
i-p, ip	intermediate pressure		mi	mile
ipm, in/min	inches per minute		MIL-STD	U.S. Military Standard
ipr	inches per revolution		min	minutes; minimum
IPS	iron pipe size		mip	mean indicated pressure
IRE	Inst. of Radio Engineers (see IEEE)		MKS	meter-kilogram-second system
IRS	Internal Revenue Service		MKSA	meter-kilogram-second-ampere system
ISO	International Organization for Standardization		mL	millilamberts
isoth	isothermal		ml, mL	millilitre = 1.000027 cm³
ISTM	International Soc. for Testing Materials		mlhc	mean lower hemispherical candles
IUPAC	International Union of Pure & Applied Chemistry		mm	millimetres
J	joule		mm-free	mineral matter free

mmf	magnetomotive force	psi, lb/in^2	lb per sq in
mol	mole	psia	lb per sq in. abs
mp	melting point	psig	lb per sq in. gage
MPC	maximum permissible concentration	pt	point; pint
mph, mi/h	miles per hour	PVC	polyvinyl chloride
MRT	mean radiant temperature	Q	10^{18} Btu
ms	manuscript; milliseconds	qt	quarts
msc	mean spherical candles	q.v.	quod vide (which see)
MSS	Manufacturers Standardization Soc. of the Valve & Fittings Industry	r	roentgens
		R	gas constant
Mu	micron, micro	R	deg Rankine (Fahrenheit abs); Reynolds number
MW	megawatts	rad	radius; radiation absorbed dose; radian
MW day	megawatt day	RBE	see rem
MWT	mean water temperature	R-C	resistor-capacitor
n	polytropic exponent	RCA	Radio Corporation of America
N	number (in mathematical tables)	R&D	research & development
N	number of neutrons; newton	RDX	cyclonite, a military explosive
N_s	specific speed	rem	Roentgen equivalent man (formerly RBE)
NA	not available	rev	revolutions
NAA	National Assoc. of Accountants	r-f, rf	radio frequency
NACA	National Advisory Committee on Aeronautics (see NASA)	RMA	Rubber Manufacturers Assoc.
NACM	National Assoc. of Chain Manufacturers	rms	square root of mean square
NASA	National Aeronautics and Space Administration	rpm, r/min	revolutions per minute
nat.	natural	rps, r/s	revolutions per second
NBC	National Broadcasting Company	RSHF	room sensible heat factor
NBFU	National Board of Fire Underwriters	ry.	railway
NBS	National Bureau of Standards	s	entropy
NCN	nitrocarbonitrate (explosive)	s	seconds
NDHA	National District Hearing Assoc.	S	sulfur, %; siemens
NEC®	National Electric Code® (National Electrical Code® and NEC® are registered trademarks of the National Fire Protection Association, Inc., Quincy, MA.)	SAE	Soc. of Automotive Engineers
		sat	saturated
		SBI	steel Boiler Inst.
NEMA	National Electrical Manufacturers Assoc.	scfm	standard cu ft per min
NFPA	National Fire Protection Assoc.	SCR	silicon controlled rectifier
NLGI	National Lubricating Grease Institute	sec	secant of
nm	nautical miles	sec^{-1}	angle whose secant is (see cos^{-1})
No. (Nos.)	number(s)	Sec.	Section
NPSH	net positive suction head	sech	hyperbolic secant of
NRC	Nuclear Regulator Commission (successor to AEC; see also ERDA)	sech^{-1}	inverse hyperbolic secant of
		segm	segment
NTP	normal temperature and pressure	SE No.	steam emulsion number
O.D., OD	outside diameter (pipes)	sfc	specific fuel consumption, lb per hphr
O.H.	open-hearth (steel)	sfm, sfpm	surface feet per minute
O.N.	octane number	shp	shaft horsepower
op. cit.	opere citato (work already cited)	SI	International System of Units (Le Système International d'Unites)
OSHA	Occupational Safety & Health Administration		
OSW	Office of Saline Water	sin	sine of
OTS	Office of Technical Services, U.S. Dept. of Commerce	sin^{-1}	angle whose sine is (see cos^{-1})
oz	ounces	sinh	hyperbolic sine of
p. (pp.)	page (pages)	sinh^{-1}	inverse hyperbolic sine of
Pa	pascal	SME	Society of Manufacturing Engineers (successor to ASTME)
P.C.	propulsive coefficient		
PE	polyethylene	SNAME	Soc. of Naval Architects and Marine Engineers
PEG	polyethylene glycol	SP	static pressure
P.E.L.	proportional elastic limit	sp	specific
PETN	an explosive	specif	specification
pf	power factor	sp gr	specific gravity
PFI	Pipe Fabrication Inst.	sp ht	specific heat
PIV	peak inverse voltage	spp	species unspecified (botanical)
p.m.	post meridiem (after noon)	SPS	standard pipe size
PM	preventive maintenance	sq	square
P.N.	performance number	sr	steradian
ppb	parts per billion	SSF	sec Saybolt Furol
PPI	plan position indicator	SSU	seconds Saybolt Universal (same as SUS)
ppm	parts per million	std	standard
press	pressure	SUS	Saybolt Universal seconds (same as SSU)
$Proc.$	Proceedings	SWG	Standard (British) wire gage
PSD	power spectral density, g^2/cps	T	tesla

TAC	Technical Advisory Committee on Weather Design Conditions (ASHRAE)	USS	United States Standard
tan	tangent of	USSG	U.S. Standard Gage
$\tan^{-1}$	angle whose tangent is (see $\cos^{-1}$)	UTC	Coordinated Universal Time
tanh	hyperbolic tangent of	V	volt
$\tanh^{-1}$	inverse hyperbolic tangent of	VCF	visual comfort factor
TDH	total dynamic head	VCI	visual comfort index
TEL	tetraethyl lead	VDI	Verein Deutscher Ingenieure
temp	temperature	vel	velocity
THI	temperature-humidity (discomfort) index	vers	versed sine of
thp	thrust horsepower	vert	vertical
TNT	trinitrotoluol (explosive)	VHF	very high frequency
torr	$= 1$ mm Hg $= 1.332$ millibars $(1/760)$ atm $= (1.013250/760)$ dynes per cm^2	VI	viscosity index
		viz.	videlicet (namely)
TP	total pressure	V.M.	volatile matter, %
tph	tons per hour	vol	volume
tpi	turns per in	VP	velocity pressure
TR	transmitter-receiver	vs.	versus
Trans.	Transactions	W	watt
		Wb	weber
T.S.	tensile strength; tensile stress	W&M	Washburn & Moen wire gage
tsi	tons per sq in	w.g.	water gage
ttd	terminal temperature difference	WHO	World Health Organization
UHF	ultra high frequency	W.I.	wrought iron
UKAEA	United Kingdom Atomic Energy Authority	W.P.A.	Western Pine Assoc.
UL	Underwriters' Laboratory	wt	weight
ult	ultimate	yd	yards
UMS	universal maintenance standards	Y.P.	yield point
USAF	U.S. Air Force	yr	year(s)
USCG	U.S. Coast Guard	Y.S.	yield strength; yield stress
USCS	U.S. Commercial Standard; U.S. Customary System	*z*	atomic number; figure of merit
USDA	U.S. Dept. of Agriculture	*Zeit.*	Zeitschrift
USFPL	U.S. Forest Products Laboratory	Δ	mass defect
USGS	U.S. Geologic Survey	μc	microcurie
USHEW	U.S. Dept. of Health, Education & Welfare	σ, s	Boltzmann constant
USN	U.S. Navy	μ	micro $(= 10^{-6})$, as in μs
USP	U.S. Pharmacopoeia	μm	micrometer (micron) $= 10^{-6}$ m $(10^{-3}$ mm)
USPHS	U.S. Public Health Service	Ω	ohm

MATHEMATICAL SIGNS AND SYMBOLS

$+$	plus (sign of addition)	$\neq$	not equal to
$+$	positive	$\rightarrow \doteq$	approaches
$-$	minus (sign of subtraction)	$\propto$	varies as
$-$	negative	∞	infinity
$\pm (\mp)$	plus or minus (minus or plus)	$\sqrt{\ }$	square root of
$\times$	times, by (multiplication sign)	$\sqrt[3]{\ }$	cube root of
$\cdot$	multiplied by	$\therefore$	therefore
$\div$	sign of division	$\parallel$	parallel to
$/$	divided by	$(\ )[\ ]\{\ \}$	parentheses, brackets and braces; quantities enclosed by them to be taken together in multiplying, dividing, etc.
$:$	ratio sign, divided by, is to		
$::$	equals, as (proportion)	$\overline{AB}$	length of line from A to B
$<$	less than	π	pi $(= 3.14159^+)$
$>$	greater than	$\circ$	degrees
$\ll$	much less than	$'$	minutes
$\gg$	much greater than	$''$	seconds
$=$	equals	$\angle$	angle
$\equiv$	identical with	dx	differential of x
$\sim$	similar to	Δ	(delta) difference
$\approx$	approximately equals	Δx	increment of x
$\cong$	approximately equals, congruent	$\partial u/\partial x$	partial derivative of u with respect to x
$\leq$	equal to or less than	$\int$	integral of
$\geq$	equal to or greater than		

$\int_b^a$ integral of, between limits a and b

$\oint$ line integral around a closed path

Σ (sigma) summation of

$f(x), F(x)$ functions of x

$\exp x = e^x$ [$e = 2.71828$ (base of natural, or Napierian, logarithms)]

∇ del or nabla, vector differential operator

∇^2 Laplacian operator

$\pounds$ Laplace operational symbol

$4!$ factorial $4 = 4 \times 3 \times 2 \times 1$

$|x|$ absolute value of x

$\dot{x}$ first derivative of x with respect to time

$\ddot{x}$ second derivative of x with respect to time

$\mathbf{A} \times \mathbf{B}$ vector product; magnitude of $\mathbf{A}$ times magnitude of $\mathbf{B}$ times sine of the angle from $\mathbf{A}$ to $\mathbf{B}$; $AB \sin \overline{AB}$

$\mathbf{A} \cdot \mathbf{B}$ scalar product; magnitude of $\mathbf{A}$ times magnitude of $\mathbf{B}$ times cosine of the angle from $\mathbf{A}$ to $\mathbf{B}$; $AB \cos \overline{AB}$

Mathematical Tables and Measuring Units

BY

GEORGE F. BAUMEISTER *President, EMC Process Corp., Newport, DE*
DAVID T. GOLDMAN *Deputy Manager, U.S. Department of Energy, Chicago Operations Office*

1.1 MATHEMATICAL TABLES
by George F. Baumeister

REFERENCES FOR MATHEMATICAL TABLES: Dwight, ''Mathematical Tables of Elementary and Some Higher Mathematical Functions,'' McGraw-Hill. Dwight, ''Tables of Integrals and Other Mathematical Data,'' Macmillan. Jahnke and Emde, ''Tables of Functions,'' B. G. Teubner, Leipzig, or Dover. Pierce-Foster, ''A Short Table of Integrals,'' Ginn. ''Mathematical Tables from Handbook of Chemistry and Physics,'' Chemical Rubber Co. ''Handbook of Mathematical Functions,'' NBS.

Table 1.1.1 Segments of Circles, Given h/c

Given: h = height; c = chord. To find the diameter of the circle, the length of arc, or the area of the segment, form the ratio h/c, and find from the table the value of (diam/c), (arc/c); then, by a simple multiplication,

$$\text{diam} = c \times (\text{diam}/c)$$
$$\text{arc} = c \times (\text{arc}/c)$$
$$\text{area} = h \times c \times (\text{area}/h \times c)$$

The table gives also the angle subtended at the center, and the ratio of h to D.

$\dfrac{h}{c}$	$\dfrac{\text{Diam}}{c}$	Diff	$\dfrac{\text{Arc}}{c}$	Diff	$\dfrac{\text{Area}}{h \times c}$	Diff	Central angle, v	Diff	$\dfrac{h}{\text{Diam}}$	Diff
.00			1.000	0	.6667	0	0.00°	458	.0000	4
1	25.010	12490	1.000	1	.6667	2	4.58	458	.0004	12
2	12.520	*4157	1.001	1	.6669	2	9.16	457	.0016	20
3	8.363	*2073	1.002	2	.6671	4	13.73	457	.0036	28
4	6.290	*1240	1.004	3	.6675	5	18.30	454	.0064	35
.05	5.050	*823	1.007	3	.6680	6	22.84°	453	.0099	43
6	4.227	*586	1.010	3	.6686	7	27.37	451	.0142	50
7	3.641	*436	1.013	4	.6693	8	31.88	448	.0192	58
8	3.205	*337	1.017	4	.6701	9	36.36	446	.0250	64
9	2.868	*268	1.021	5	.6710	10	40.82	442	.0314	71
.10	2.600	*217	1.026	6	.6720	11	45.24°	439	.0385	77
1	2.383	*180	1.032	6	.6731	12	49.63	435	.0462	83
2	2.203	*150	1.038	6	.6743	13	53.98	432	.0545	88
3	2.053	*127	1.044	7	.6756	14	58.30	427	.0633	94
4	1.926	*109	1.051	8	.6770	15	62.57	423	.0727	99
.15	1.817	*94	1.059	8	.6785	16	66.80°	418	.0826	103
6	1.723	*82	1.067	8	.6801	17	70.98	413	.0929	107
7	1.641	*72	1.075	9	.6818	18	75.11	409	.1036	111
8	1.569	*63	1.084	10	.6836	19	79.20	403	.1147	116
9	1.506	56	1.094	9	.6855	20	83.23	399	.1263	116
.20	1.450	50	1.103	11	.6875	21	87.21°	392	.1379	120
1	1.400	44	1.114	10	.6896	22	91.13	387	.1499	123
2	1.356	39	1.124	12	.6918	23	95.00	381	.1622	124
3	1.317	35	1.136	11	.6941	24	98.81	375	.1746	127
4	1.282	32	1.147	12	.6965	24	102.56	370	.1873	127
.25	1.250	28	1.159	12	.6989	25	106.26°	364	.2000	128
6	1.222	26	1.171	13	.7014	27	109.90	358	.2128	130
7	1.196	23	1.184	13	.7041	27	113.48	352	.2258	129
8	1.173	21	1.197	14	.7068	28	117.00	345	.2387	130
9	1.152	19	1.211	14	.7096	29	120.45	341	.2517	130
.30	1.133	17	1.225	14	.7125	29	123.86°	334	.2647	130
1	1.116	15	1.239	15	.7154	31	127.20	328	.2777	129
2	1.101	13	1.254	15	.7185	31	130.48	322	.2906	128
3	1.088	13	1.269	15	.7216	32	133.70	316	.3034	128
4	1.075	11	1.284	16	.7248	32	136.86	311	.3162	127
.35	1.064	10	1.300	16	.7280	34	139.97°	305	.3289	125
6	1.054	8	1.316	16	.7314	34	143.02	299	.3414	124
7	1.046	8	1.332	17	.7348	35	146.01	293	.3538	123
8	1.038	7	1.349	17	.7383	36	148.94	288	.3661	122
9	1.031	6	1.366	17	.7419	36	151.82	282	.3783	119
.40	1.025	5	1.383	18	.7455	37	154.64°	277	.3902	119
1	1.020	5	1.401	18	.7492	38	157.41	271	.4021	116
2	1.015	4	1.419	18	.7530	38	160.12	266	.4137	115
3	1.011	3	1.437	18	.7568	39	162.78	261	.4252	112
4	1.008	2	1.455	19	.7607	40	165.39	256	.4364	111
.45	1.006	3	1.474	19	.7647	40	167.95°	251	.4475	109
6	1.003	1	1.493	19	.7687	41	170.46	245	.4584	107
7	1.002	1	1.512	19	.7728	41	172.91	241	.4691	105
8	1.001	1	1.531	20	.7769	42	175.32	237	.4796	103
· 9	1.000	0	1.551	20	.7811	43	177.69	231	.4899	101
.50	1.000		1.571		.7854		180.00°		.5000	

* Interpolation may be inaccurate at these points.

Table 1.1.2   Segments of Circles, Given h/D

Given: h = height; D = diameter of circle. To find the chord, the length of arc, or the area of the segment, form the ratio h/D, and find from the table the value of (chord/D), (arc/D), or (area/D^2); then by a simple multiplication,

$$\text{chord} = D \times (\text{chord}/D)$$
$$\text{arc} = D \times (\text{arc}/D)$$
$$\text{area} = D^2 \times (\text{area}/D^2)$$

This table gives also the angle subtended at the center, the ratio of the arc of the segment to the whole circumference, and the ratio of the area of the segment to the area of the whole circle.

$\dfrac{h}{D}$	$\dfrac{\text{Arc}}{D}$	Diff	$\dfrac{\text{Area}}{D^2}$	Diff	Central angle, v	Diff	$\dfrac{\text{Chord}}{D}$	Diff	$\dfrac{\text{Arc}}{\text{Circum}}$	Diff	$\dfrac{\text{Area}}{\text{Circle}}$	Diff
.00	0.000	2003	.0000	13	0.00°	2296	.0000	*1990	.0000	*638	.0000	17
1	.2003	*835	.0013	24	22.96	*956	.1990	*810	.0638	*265	.0017	31
2	.2838	*644	.0037	32	32.52	*738	.2800	*612	.0903	*205	.0048	39
3	.3482	*545	.0069	36	39.90	*625	.3412	*507	.1108	*174	.0087	47
4	.4027	*483	.0105	42	46.15	*553	.3919	*440	.1282	*154	.0134	53
.05	.4510	*439	.0147	45	51.68°	*504	.4359	*391	.1436	*139	.0187	58
6	.4949	*406	.0192	50	56.72	*465	.4750	*353	.1575	*130	.0245	63
7	.5355	*380	.0242	52	61.37	*435	.5103	*323	.1705	121	.0308	67
8	.5735	*359	.0294	56	65.72	*411	.5426	*298	.1826	114	.0375	71
9	.6094	*341	.0350	59	69.83	*391	.5724	*276	.1940	108	.0446	74
.10	.6435	*326	.0409	61	73.74°	*374	.6000	*258	.2048	104	.0520	78
1	.6761	*314	.0470	64	77.48	*359	.6258	*241	.2152	100	.0598	82
2	.7075	*302	.0534	66	81.07	*347	.6499	*227	.2252	96	.0680	84
3	.7377	*293	.0600	68	84.54	*335	.6726	*214	.2348	93	.0764	87
4	.7670	*284	.0668	71	87.89	*326	.6940	*201	.2441	91	.0851	90
.15	.7954	276	.0739	72	91.15°	316	.7141	*191	.2532	88	.0941	92
6	.8230	270	.0811	74	94.31	309	.7332	*181	.2620	86	.1033	94
7	.8500	263	.0885	76	97.40	302	.7513	*171	.2706	83	.1127	97
8	.8763	258	.0961	78	100.42	295	.7684	162	.2789	82	.1224	99
9	.9021	252	.1039	79	103.37	289	.7846	154	.2871	81	.1323	101
.20	0.9273	248	.1118	81	106.26°	284	.8000	146	.2952	79	.1424	103
1	0.9521	243	.1199	82	109.10	279	.8146	139	.3031	77	.1527	104
2	0.9764	240	.1281	84	111.89	274	.8285	132	.3108	76	.1631	107
3	1.0004	235	.1365	84	114.63	271	.8417	125	.3184	75	.1738	108
4	1.0239	233	.1449	86	117.34	266	.8542	118	.3259	74	.1846	109
.25	1.0472	229	.1535	88	120.00°	263	.8660	113	.3333	73	.1955	111
6	1.0701	227	.1623	88	122.63	260	.8773	106	.3406	72	.2066	112
7	1.0928	224	.1711	89	125.23	256	.8879	101	.3478	72	.2178	114
8	1.1152	222	.1800	90	127.79	254	.8980	95	.3550	70	.2292	115
9	1.1374	219	.1890	92	130.33	251	.9075	90	.3620	70	.2407	116
.30	1.1593	217	.1982	92	132.84°	249	.9165	85	.3690	69	.2523	117
1	1.1810	215	.2074	93	135.33	247	.9250	80	.3759	69	.2640	119
2	1.2025	214	.2167	93	137.80	245	.9330	74	.3828	68	.2759	119
3	1.2239	212	.2260	95	140.25	242	.9404	70	.3896	67	.2878	120
4	1.2451	210	.2355	95	142.67	241	.9474	65	.3963	67	.2998	121
.35	1.2661	209	.2450	96	145.08°	240	.9539	61	.4030	67	.3119	122
6	1.2870	208	.2546	96	147.48	238	.9600	56	.4097	66	.3241	123
7	1.3078	206	.2642	97	149.86	237	.9656	52	.4163	66	.3364	123
8	1.3284	206	.2739	97	152.23	235	.9708	47	.4229	65	.3487	124
9	1.3490	204	.2836	98	154.58	235	.9755	43	.4294	65	.3611	124
.40	1.3694	204	.2934	98	156.93°	233	.9798	39	.4359	65	.3735	125
1	1.3898	203	.3032	98	159.26	233	.9837	34	.4424	65	.3860	126
2	1.4101	202	.3130	99	161.59	231	.9871	31	.4489	64	.3986	126
3	1.4303	202	.3229	99	163.90	232	.9902	26	.4553	64	.4112	126
4	1.4505	201	.3328	100	166.22	230	.9928	22	.4617	64	.4238	126
.45	1.4706	201	.3428	99	168.52°	230	.9950	18	.4681	64	.4364	127
6	1.4907	201	.3527	100	170.82	230	.9968	14	.4745	64	.4491	127
7	1.5108	200	.3627	100	173.12	229	.9982	10	.4809	64	.4618	127
8	1.5308	200	.3727	100	175.41	230	.9992	6	.4873	63	.4745	128
9	1.5508	200	.3827	100	177.71	229	.9998	2	.4936	64	.4873	127
.50	1.5708		.3927		180.00°		1.0000		.5000		.5000	

* Interpolation may be inaccurate at these points.

Table 1.1.3   Regular Polygons

n = number of sides

$v = 360°/n$ = angle subtended at the center by one side

$$a = \text{length of one side} = R\left(2\sin\frac{v}{2}\right) = r\left(2\tan\frac{v}{2}\right)$$

$$R = \text{radius of circumscribed circle} = a\left(\tfrac{1}{2}\csc\frac{v}{2}\right) = r\left(\sec\frac{v}{2}\right)$$

$$r = \text{radius of inscribed circle} = R\left(\cos\frac{v}{2}\right) = a\left(\tfrac{1}{2}\cot\frac{v}{2}\right)$$

$$\text{Area} = a^2\left(\tfrac{1}{4}n\cot\frac{v}{2}\right) = R^2(\tfrac{1}{2}n\sin v) = r^2\left(n\tan\frac{v}{2}\right)$$

n	v	$\dfrac{\text{Area}}{a^2}$	$\dfrac{\text{Area}}{R^2}$	$\dfrac{\text{Area}}{r^2}$	$\dfrac{R}{a}$	$\dfrac{R}{r}$	$\dfrac{a}{R}$	$\dfrac{a}{r}$	$\dfrac{r}{R}$	$\dfrac{r}{a}$
3	120°	0.4330	1.299	5.196	0.5774	2.000	1.732	3.464	0.5000	0.2887
4	90°	1.000	2.000	4.000	0.7071	1.414	1.414	2.000	0.7071	0.5000
5	72°	1.721	2.378	3.633	0.8507	1.236	1.176	1.453	0.8090	0.6882
6	60°	2.598	2.598	3.464	1.0000	1.155	1.000	1.155	0.8660	0.8660
7	51°.43	3.634	2.736	3.371	1.152	1.110	0.8678	0.9631	0.9010	1.038
8	45°	4.828	2.828	3.314	1.307	1.082	0.7654	0.8284	0.9239	1.207
9	40°	6.182	2.893	3.276	1.462	1.064	0.6840	0.7279	0.9397	1.374
10	36°	7.694	2.939	3.249	1.618	1.052	0.6180	0.6498	0.9511	1.539
12	30°	11.20	3.000	3.215	1.932	1.035	0.5176	0.5359	0.9659	1.866
15	24°	17.64	3.051	3.188	2.405	1.022	0.4158	0.4251	0.9781	2.352
16	22°.50	20.11	3.062	3.183	2.563	1.020	0.3902	0.3978	0.9808	2.514
20	18°	31.57	3.090	3.168	3.196	1.013	0.3129	0.3168	0.9877	3.157
24	15°	45.58	3.106	3.160	3.831	1.009	0.2611	0.2633	0.9914	3.798
32	11°.25	81.23	3.121	3.152	5.101	1.005	0.1960	0.1970	0.9952	5.077
48	7°.50	183.1	3.133	3.146	7.645	1.002	0.1308	0.1311	0.9979	7.629
64	5°.625	325.7	3.137	3.144	10.19	1.001	0.0981	0.0983	0.9968	10.18

Table 1.1.4   Binomial Coefficients

$(n)_0 = 1 \quad (n)_1 = n \quad (n)_2 = \dfrac{n(n-1)}{1\times 2} \quad (n)_3 = \dfrac{n(n-1)(n-2)}{1\times 2 \times 3}$ etc. in general $(n)_r = \dfrac{n(n-1)(n-2)\cdots[n-(r-1)]}{1\times 2 \times 3 \times \cdots \times r}$. Other notations: $nC_r = \begin{pmatrix} n \\ r \end{pmatrix} = (n)_r$

n	$(n)_0$	$(n)_1$	$(n)_2$	$(n)_3$	$(n)_4$	$(n)_5$	$(n)_6$	$(n)_7$	$(n)_8$	$(n)_9$	$(n)_{10}$	$(n)_{11}$	$(n)_{12}$	$(n)_{13}$
1	1	1	......	......	......	......	......	......	......	......	......	......	......	......
2	1	2	1	......	......	......	......	......	......	......	......	......	......	......
3	1	3	3	1	......	......	......	......	......	......	......	......	......	......
4	1	4	6	4	1	......	......	......	......	......	......	......	......	......
5	1	5	10	10	5	1	......	......	......	......	......	......	......	......
6	1	6	15	20	15	6	1	......	......	......	......	......	......	......
7	1	7	21	35	35	21	7	1	......	......	......	......	......	......
8	1	8	28	56	70	56	28	8	1	......	......	......	......	......
9	1	9	36	84	126	126	84	36	9	1	......	......	......	......
10	1	10	45	120	210	252	210	120	45	10	1	......	......	......
11	1	11	55	165	330	462	462	330	165	55	11	1	......	......
12	1	12	66	220	495	792	924	792	495	220	66	12	1	......
13	1	13	78	286	715	1287	1716	1716	1287	715	286	78	13	1
14	1	14	91	364	1001	2002	3003	3432	3003	2002	1001	364	91	14
15	1	15	105	455	1365	3003	5005	6435	6435	5005	3003	1365	455	105

NOTE: For $n = 14$, $(n)_{14} = 1$; for $n = 15$, $(n)_{14} = 15$, and $(n)_{15} = 1$.

Table 1.1.5 Compound Interest. Amount of a Given Principal

The amount A at the end of n years of a given principal P placed at compound interest today is $A = P \times x$ or $A = P \times y$, according as the interest (at the rate of r percent per annum) is compounded annually, or continuously; the factor x or y being taken from the following tables.

Years	$r = 6$	8	10	12	14	16	18	20	22
				Values of x (interest compounded annually: $A = P \times x$)					
1	1.0600	1.0800	1.1000	1.1200	1.1400	1.1600	1.1800	1.2000	1.2200
2	1.1236	1.1664	1.2100	1.2544	1.2996	1.3456	1.3924	1.4400	1.4884
3	1.1910	1.2597	1.3310	1.4049	1.4815	1.5609	1.6430	1.7280	1.8158
4	1.2625	1.3605	1.4641	1.5735	1.6890	1.8106	1.9388	2.0736	2.2153
5	1.3382	1.4693	1.6105	1.7623	1.9254	2.1003	2.2878	2.4883	2.7027
6	1.4185	1.5869	1.7716	1.9738	2.1950	2.4364	2.6996	2.9860	3.2973
7	1.5036	1.7138	1.9487	2.2107	2.5023	2.8262	3.1855	3.5832	4.0227
8	1.5938	1.8509	2.1436	2.4760	2.8526	3.2784	3.7589	4.2998	4.9077
9	1.6895	1.9990	2.3579	2.7731	3.2519	3.8030	4.4355	5.1598	5.9874
10	1.7908	2.1589	2.5937	3.1058	3.7072	4.4114	5.2338	6.1917	7.3046
11	1.8983	2.3316	2.8531	3.4786	4.2262	5.1173	6.1759	7.4301	8.9117
12	2.0122	2.5182	3.1384	3.8960	4.8179	5.9360	7.2876	8.9161	10.872
13	2.1329	2.7196	3.4523	4.3635	5.4924	6.8858	8.5994	10.699	13.264
14	2.2609	2.9372	3.7975	4.8871	6.2613	7.9875	10.147	12.839	16.182
15	2.3966	3.1722	4.1772	5.4736	7.1379	9.2655	11.974	15.407	19.742
16	2.5404	3.4259	4.5950	6.1304	8.1372	10.748	14.129	18.488	24.086
17	2.6928	3.7000	5.0545	6.8660	9.2765	12.468	16.672	22.186	29.384
18	2.8543	3.9960	5.5599	7.6900	10.575	14.463	19.673	26.623	35.849
19	3.0256	4.3157	6.1159	8.6128	12.056	16.777	23.214	31.948	43.736
20	3.2071	4.6610	6.7275	9.6463	13.743	19.461	27.393	38.338	53.358
25	4.2919	6.8485	10.835	17.000	26.462	40.874	62.669	95.396	144.21
30	5.7435	10.063	17.449	29.960	50.950	85.850	143.37	237.38	389.76
40	10.286	21.725	45.259	93.051	188.88	378.72	750.38	1469.8	2847.0
50	18.420	46.902	117.39	289.00	700.23	1670.7	3927.4	9100.4	20796.6
60	32.988	101.26	304.48	897.60	2595.9	7370.2	20555.1	56347.5	151911.2

NOTE: This table is computed from the formula $x = [1 + (r/100)]^n$.

Years	$r = 6$	8	10	12	14	16	18	20	22
				Values of y (interest compounded continuously: $A = P \times y$)					
1	1.0618	1.0833	1.1052	1.1275	1.1503	1.1735	1.1972	1.2214	1.2461
2	1.1275	1.1735	1.2214	1.2712	1.3231	1.3771	1.4333	1.4918	1.5527
3	1.1972	1.2712	1.3499	1.4333	1.5220	1.6161	1.7160	1.8221	1.9348
4	1.2712	1.3771	1.4918	1.6161	1.7507	1.8965	2.0544	2.2255	2.4109
5	1.3499	1.4918	1.6487	1.8221	2.0138	2.2255	2.4596	2.7183	3.0042
6	1.4333	1.6161	1.8221	2.0544	2.3164	2.6117	2.9447	3.3201	3.7434
7	1.5220	1.7507	2.0138	2.3164	2.6645	3.0649	3.5254	4.0552	4.6646
8	1.6161	1.8965	2.2255	2.6117	3.0649	3.5966	4.2207	4.9530	5.8124
9	1.7160	2.0544	2.4596	2.9447	3.5254	4.2207	5.0531	6.0496	7.2427
10	1.8221	2.2255	2.7183	3.3201	4.0552	4.9530	6.0496	7.3891	9.0250
11	1.9348	2.4109	3.0042	3.7434	4.6646	5.8124	7.2427	9.0250	11.246
12	2.0544	2.6117	3.3201	4.2207	5.3656	6.8210	8.6711	11.023	14.013
13	2.1815	2.8292	3.6693	4.7588	6.1719	8.0045	10.381	13.464	17.462
14	2.3164	3.0649	4.0552	5.3656	7.0993	9.3933	12.429	16.445	21.758
15	2.4596	3.3201	4.4817	6.0496	8.1662	11.023	14.880	20.086	27.113
16	2.6117	3.5966	4.9530	6.8210	9.3933	12.936	17.814	24.533	33.784
17	2.7732	3.8962	5.4739	7.6906	10.805	15.180	21.328	29.964	42.098
18	2.9447	4.2207	6.0496	8.6711	12.429	17.814	25.534	36.598	52.457
19	3.1268	4.5722	6.6859	9.7767	14.296	20.905	30.569	44.701	65.366
20	3.3201	4.9530	7.3891	11.023	16.445	24.533	36.598	54.598	81.451
25	4.4817	7.3891	12.182	20.086	33.115	54.598	90.017	148.41	244.69
30	6.0496	11.023	20.086	36.598	66.686	121.51	221.41	403.43	735.10
40	11.023	24.533	54.598	121.51	270.43	601.85	1339.4	2981.0	6634.2
50	20.086	54.598	148.41	403.43	1096.6	2981.0	8103.1	22026.5	59874.1
60	36.598	121.51	403.43	1339.4	4447.1	14764.8	49020.8	162754.8	540364.9

FORMULA: $y = e^{(r/100) \times n}$.

Table 1.1.6 Principal Which Will Amount to a Given Sum

The principal P, which, if placed at compound interest today, will amount to a given sum A at the end of n years $P = A \times x'$ or $P = A \times y'$, according as the interest (at the rate of r percent per annum) is compounded annually, or continuously; the factor x' or y' being taken from the following tables.

Years	r = 6	8	10	12	14	16	18	20	22

Values of x' (interest compounded annually: $P = A \times x'$)

Years	r = 6	8	10	12	14	16	18	20	22
1	.94340	.92593	.90909	.89286	.87719	.86207	.84746	.83333	.81967
2	.89000	.85734	.82645	.79719	.76947	.74316	.71818	.69444	.67186
3	.83962	.79383	.75131	.71178	.67497	.64066	.60863	.57870	.55071
4	.79209	.73503	.68301	.63552	.59208	.55229	.51579	.48225	.45140
5	.74726	.68058	.62092	.56743	.51937	.47611	.43711	.40188	.37000
6	.70496	.63017	.56447	.50663	.45559	.41044	.37043	.33490	.30328
7	.66506	.58349	.51316	.45235	.39964	.35383	.31393	.27908	.24859
8	.62741	.54027	.46651	.40388	.35056	.30503	.26604	.23257	.20376
9	.59190	.50025	.42410	.36061	.30751	.26295	.22546	.19381	.16702
10	.55839	.46319	.38554	.32197	.26974	.22668	.19106	.16151	.13690
11	.52679	.42888	.35049	.28748	.23662	.19542	.16192	.13459	.11221
12	.49697	.39711	.31863	.25668	.20756	.16846	.13722	.11216	.09198
13	.46884	.36770	.28966	.22917	.18207	.14523	.11629	.09346	.07539
14	.44230	.34046	.26333	.20462	.15971	.12520	.09855	.07789	.06180
15	.41727	.31524	.23939	.18270	.14010	.10793	.08352	.06491	.05065
16	.39365	.29189	.21763	.16312	.12289	.09304	.07078	.05409	.04152
17	.37136	.27027	.19784	.14564	.10780	.08021	.05998	.04507	.03403
18	.35034	.25025	.17986	.13004	.09456	.06914	.05083	.03756	.02789
19	.33051	.23171	.16351	.11611	.08295	.05961	.04308	.03130	.02286
20	.31180	.21455	.14864	.10367	.07276	.05139	.03651	.02608	.01874
25	.23300	.14602	.09230	.05882	.03779	.02447	.01596	.01048	.00693
30	.17411	.09938	.05731	.03338	.01963	.01165	.00697	.00421	.00257
40	.09722	.04603	.02209	.01075	.00529	.00264	.00133	.00068	.00035
50	.05429	.02132	.00852	.00346	.00143	.00060	.00025	.00011	.00005
60	.03031	.00988	.00328	.00111	.00039	.00014	.00005	.00002	.00001

FORMULA: $x' = [1 + (r/100)]^{-n} = 1/x$.

Years	r = 6	8	10	12	14	16	18	20	22

Values of y' (interest compounded continuously: $P = A \times y'$)

Years	r = 6	8	10	12	14	16	18	20	22
1	.94176	.92312	.90484	.88692	.86936	.85214	.83527	.81873	.80252
2	.88692	.85214	.81873	.78663	.75578	.72615	.69768	.67032	.64404
3	.83527	.78663	.74082	.69768	.65705	.61878	.58275	.54881	.51685
4	.78663	.72615	.67032	.61878	.57121	.52729	.48675	.44933	.41478
5	.74082	.67032	.60653	.54881	.49659	.44933	.40657	.36788	.33287
6	.69768	.61878	.54881	.48675	.43171	.38289	.33960	.30119	.26714
7	.65705	.57121	.49659	.43171	.37531	.32628	.28365	.24660	.21438
8	.61878	.52729	.44933	.38289	.32628	.27804	.23693	.20190	.17204
9	.58275	.48675	.40657	.33960	.28365	.23693	.19790	.16530	.13807
10	.54881	.44933	.36788	.30119	.24660	.20190	.16530	.13534	.11080
11	.51685	.41478	.33287	.26714	.21438	.17204	.13807	.11080	.08892
12	.48675	.38289	.30119	.23693	.18637	.14661	.11533	.09072	.07136
13	.45841	.35345	.27253	.21014	.16203	.12493	.09633	.07427	.05727
14	.43171	.32628	.24660	.18637	.14086	.10646	.08046	.06081	.04596
15	.40657	.30119	.22313	.16530	.12246	.09072	.06721	.04979	.03688
16	.38289	.27804	.20190	.14661	.10646	.07730	.05613	.04076	.02960
17	.36059	.25666	.18268	.13003	.09255	.06587	.04689	.03337	.02375
18	.33960	.23693	.16530	.11533	.08046	.05613	.03916	.02732	.01906
19	.31982	.21871	.14957	.10228	.06995	.04783	.03271	.02237	.01530
20	.30119	.20190	.13534	.09072	.06081	.04076	.02732	.01832	.01228
25	.22313	.13534	.08208	.04979	.03020	.01832	.01111	.00674	.00409
30	.16530	.09072	.04979	.02732	.01500	.00823	.00452	.00248	.00136
40	.09072	.04076	.01832	.00823	.00370	.00166	.00075	.00034	.00015
50	.04979	.01832	.00674	.00248	.00091	.00034	.00012	.00005	.00002
60	.02732	.00823	.00248	.00075	.00022	.00007	.00002	.00001	.00000

FORMULA: $y' = e^{-(r/100) \times n} = 1/y$.

Table 1.1.7   Amount of an Annuity

The amount S accumulated at the end of n years by a given annual payment Y set aside at the end of each year is $S = Y \times v$, where the factor v is to be taken from the following table (interest at r percent per annum, compounded annually).

Years	r = 6	8	10	12	14	16	18	20	22
					Values of v				
1	1.0000	1.0000	1.0000	1.0000	1.0000	1.0000	1.0000	1.0000	1.0000
2	2.0600	2.0800	2.1000	2.1200	2.1400	2.1600	2.1800	2.2000	2.2200
3	3.1836	3.2464	3.3100	3.3744	3.4396	3.5056	3.5724	3.6400	3.7084
4	4.3746	4.5061	4.6410	4.7793	4.9211	5.0665	5.2154	5.3680	5.5242
5	5.6371	5.8666	6.1051	6.3528	6.6101	6.8771	7.1542	7.4416	7.7396
6	6.9753	7.3359	7.7156	8.1152	8.5355	8.9775	9.4420	9.9299	10.442
7	8.3938	8.9228	9.4872	10.089	10.730	11.414	12.142	12.916	13.740
8	9.8975	10.637	11.436	12.300	13.233	14.240	15.327	16.499	17.762
9	11.491	12.488	13.579	14.776	16.085	17.519	19.086	20.799	22.670
10	13.181	14.487	15.937	17.549	19.337	21.321	23.521	25.959	28.657
11	14.972	16.645	18.531	20.655	23.045	25.733	28.755	32.150	35.962
12	16.870	18.977	21.384	24.133	27.271	30.850	34.931	39.581	44.874
13	18.882	21.495	24.523	28.029	32.089	36.786	42.219	48.497	55.746
14	21.015	24.215	27.975	32.393	37.581	43.672	50.818	59.196	69.010
15	23.276	27.152	31.772	37.280	43.842	51.660	60.965	72.035	85.192
16	25.673	30.324	35.950	42.753	50.980	60.925	72.939	87.442	104.93
17	28.213	33.750	40.545	48.884	59.118	71.673	87.068	105.93	129.02
18	30.906	37.450	45.599	55.750	68.394	84.141	103.74	128.12	158.40
19	33.760	41.446	51.159	63.440	78.969	98.603	123.41	154.74	194.25
20	36.786	45.762	57.275	72.052	91.025	115.38	146.63	186.69	237.99
25	54.865	73.106	98.347	133.33	181.87	249.21	342.60	471.98	650.96
30	79.058	113.28	164.49	241.33	356.79	530.31	790.95	1181.9	1767.1
40	154.76	259.06	422.59	767.09	1342.0	2360.8	4163.2	7343.9	12936.5
50	290.34	573.77	1163.9	2400.0	4994.5	10435.6	21813.1	45497.2	94525.3
60	533.13	1253.2	3034.8	7471.6	18535.1	46057.5	114189.7	281732.6	690501.0

FORMULA: $v \{[1 + (r/100)]^n - 1\} \div (r/100) = (x - 1) \div (r/100)$.

Table 1.1.8   Annuity Which Will Amount to a Given Sum (Sinking Fund)

The annual payment Y which, if set aside at the end of each year, will amount with accumulated interest to a given sum S at the end of n years is $Y = S \times v'$, where the factor v' is given below (interest at r percent per annum, compounded annually).

Years	r = 6	8	10	12	14	16	18	20	22
					Values of v'				
1	1.00000	1.00000	1.00000	1.00000	1.00000	1.00000	1.00000	1.00000	1.00000
2	.48544	.48077	.47619	.47170	.46729	.46296	.45872	.45455	.45045
3	.31411	.30803	.30211	.29635	.29073	.28526	.27992	.27473	.26966
4	.22859	.22192	.21547	.20923	.20320	.19738	.19174	.18629	.18102
5	.17740	.17046	.16380	.15741	.15128	.14541	.13978	.13438	.12921
6	.14336	.13632	.12961	.12323	.11716	.11139	.10591	.10071	.09576
7	.11914	.11207	.10541	.09912	.09319	.08761	.08236	.07742	.07278
8	.10104	.09401	.08744	.08130	.07557	.07022	.06524	.06061	.05630
9	.08702	.08008	.07364	.06768	.06217	.05708	.05239	.04808	.04411
10	.07587	.06903	.06275	.05698	.05171	.04690	.04251	.03852	.03489
11	.06679	.06008	.05396	.04842	.04339	.03886	.03478	.03110	.02781
12	.05928	.05270	.04676	.04144	.03667	.03241	.02863	.02526	.02228
13	.05296	.04652	.04078	.03568	.03116	.02718	.02369	.02062	.01794
14	.04758	.04130	.03575	.03087	.02661	.02290	.01968	.01689	.01449
15	.04296	.03683	.03147	.02682	.02281	.01936	.01640	.01388	.01174
16	.03895	.03298	.02782	.02339	.01962	.01641	.01371	.01144	.00953
17	.03544	.02963	.02466	.02046	.01692	.01395	.01149	.00944	.00775
18	.03236	.02670	.02193	.01794	.01462	.01188	.00964	.00781	.00631
19	.02962	.02413	.01955	.01576	.01266	.01014	.00810	.00646	.00515
20	.02718	.02185	.01746	.01388	.01099	.00867	.00682	.00536	.00420
25	.01823	.01368	.01017	.00750	.00550	.00401	.00292	.00212	.00154
30	.01265	.00883	.00608	.00414	.00280	.00189	.00126	.00085	.00057
40	.00646	.00386	.00226	.00130	.00075	.00042	.00024	.00014	.00008
50	.00344	.00174	.00086	.00042	.00020	.00010	.00005	.00002	.00001
60	.00188	.00080	.00033	.00013	.00005	.00002	.00001	.00000	.00000

FORMULA: $v' = (r/100) \div \{[1 + (r/100)]^n - 1\} = 1/v$.

Table 1.1.9 Present Worth of an Annuity

The capital C which, if placed at interest today, will provide for a given annual payment Y for a term of n years before it is exhausted is $C = Y \times w$, where the factor w is given below (interest at r percent per annum, compounded annually).

Years	$r = 6$	8	10	12	14	16	18	20	22
					Values of w				
1	.94340	.92590	.90910	.89290	.87720	.86210	.84750	.83330	.81970
2	1.8334	1.7833	1.7355	1.6901	1.6467	1.6052	1.5656	1.5278	1.4915
3	2.6730	2.5771	2.4869	2.4018	2.3216	2.2459	2.1743	2.1065	2.0422
4	3.4651	3.3121	3.1699	3.0373	2.9137	2.7982	2.6901	2.5887	2.4936
5	4.2124	3.9927	3.7908	3.6048	3.4331	3.2743	3.1272	2.9906	2.8636
6	4.9173	4.6229	4.3553	4.1114	3.8887	3.6847	3.4976	3.3255	3.1669
7	5.5824	5.2064	4.8684	4.5638	4.2883	4.0386	3.8115	3.6046	3.4155
8	6.2098	5.7466	5.3349	4.9676	4.6389	4.3436	4.0776	3.8372	3.6193
9	6.8017	6.2469	5.7590	5.3282	4.9464	4.6065	4.3030	4.0310	3.7863
10	7.3601	6.7101	6.1446	5.6502	5.2161	4.8332	4.4941	4.1925	3.9232
11	7.8869	7.1390	6.4951	5.9377	5.4527	5.0286	4.6560	4.3271	4.0354
12	8.3838	7.5361	6.8137	6.1944	5.6603	5.1971	4.7932	4.4392	4.1274
13	8.8527	7.9038	7.1034	6.4235	5.8424	5.3423	4.9095	4.5327	4.2028
14	9.2950	8.2442	7.3667	6.6282	6.0021	5.4675	5.0081	4.6106	4.2646
15	9.7122	8.5595	7.6061	6.8109	6.1422	5.5755	5.0916	4.6755	4.3152
16	10.106	8.8514	7.8237	6.9740	6.2651	5.6685	5.1624	4.7296	4.3567
17	10.477	9.1216	8.0216	7.1196	6.3729	5.7487	5.2223	4.7746	4.3908
18	10.828	9.3719	8.2014	7.2497	6.4674	5.8178	5.2732	4.8122	4.4187
19	11.158	9.6036	8.3649	7.3658	6.5504	5.8775	5.3162	4.8435	4.4415
20	11.470	9.8181	8.5136	7.4694	6.6231	5.9288	5.3527	4.8696	4.4603
25	12.783	10.675	9.0770	7.8431	6.8729	6.0971	5.4669	4.9476	4.5139
30	13.765	11.258	9.4269	8.0552	7.0027	6.1772	5.5168	4.9789	4.5338
40	15.046	11.925	9.7791	8.2438	7.1050	6.2335	5.5482	4.9966	4.5439
50	15.762	12.233	9.9148	8.3045	7.1327	6.2463	5.5541	4.9995	4.5452
60	16.161	12.377	9.9672	8.3240	7.1401	6.2492	5.5553	4.9999	4.5454

FORMULA: $w = \{1 - [1 + (r/100)]^{-n}\} \div [r/100] = v/x$.

Table 1.1.10 Annuity Provided for by a Given Capital

The annual payment Y provided for a term of n years by a given capital C placed at interest today is $Y = C \times w'$ (interest at r percent per annum, compounded annually; the fund supposed to be exhausted at the end of the term).

Years	$r = 6$	8	10	12	14	16	18	20	22
					Values of w'				
1	1.0600	1.0800	1.1000	1.1200	1.1400	1.1600	1.1800	1.2000	1.2200
2	.54544	.56077	.57619	.59170	.60729	.62296	.63872	.65455	.67045
3	.37411	.38803	.40211	.41635	.43073	.44526	.45992	.47473	.48966
4	.28859	.30192	.31547	.32923	.34320	.35738	.37174	.38629	.40102
5	.23740	.25046	.26380	.27741	.29128	.30541	.31978	.33438	.34921
6	.20336	.21632	.22961	.24323	.25716	.27139	.28591	.30071	.31576
7	.17914	.19207	.20541	.21912	.23319	.24761	.26236	.27742	.29278
8	.16104	.17401	.18744	.20130	.21557	.23022	.24524	.26061	.27630
9	.14702	.16008	.17364	.18768	.20217	.21708	.23239	.24808	.26411
10	.13587	.14903	.16275	.17698	.19171	.20690	.22251	.23852	.25489
11	.12679	.14008	.15396	.16842	.18339	.19886	.21478	.23110	.24781
12	.11928	.13270	.14676	.16144	.17667	.19241	.20863	.22526	.24228
13	.11296	.12652	.14078	.15568	.17116	.18718	.20369	.22062	.23794
14	.10758	.12130	.13575	.15087	.16661	.18290	.19968	.21689	.23449
15	.10296	.11683	.13147	.14682	.16281	.17936	.19640	.21388	.23174
16	.09895	.11298	.12782	.14339	.15962	.17641	.19371	.21144	.22953
17	.09544	.10963	.12466	.14046	.15692	.17395	.19149	.20944	.22775
18	.09236	.10670	.12193	.13794	.15462	.17188	.18964	.20781	.22631
19	.08962	.10413	.11955	.13576	.15266	.17014	.18810	.20646	.22515
20	.08718	.10185	.11746	.13388	.15099	.16867	.18682	.20536	.22420
25	.07823	.09368	.11017	.12750	.14550	.16401	.18292	.20212	.22154
30	.07265	.08883	.10608	.12414	.14280	.16189	.18126	.20085	.22057
40	.06646	.08386	.10226	.12130	.14075	.16042	.18024	.20014	.22008
50	.06344	.08174	.10086	.12042	.14020	.16010	.18005	.20002	.22001
60	.06188	.08080	.10033	.12013	.14005	.16002	.18001	.20000	.22000

FORMULA: $w' = [r/100] \div \{1 - [1 + (r/100)]^{-n}\} = 1/w = v' + (r/100)$.

Table 1.1.11   Ordinates of the Normal Density Function

$$f(x) = \frac{1}{\sqrt{2\pi}} e^{-x^2/2}$$

x	.00	.01	.02	.03	.04	.05	.06	.07	.08	.09
.0	.3989	.3989	.3989	.3988	.3986	.3984	.3982	.3980	.3977	.3973
.1	.3970	.3965	.3961	.3956	.3951	.3945	.3939	.3932	.3925	.3918
.2	.3910	3902	.3894	.3885	.3876	.3867	.3857	.3847	.3836	.3825
.3	.3814	.3802	.3790	.3778	.3765	.3752	.3739	.3725	.3712	.3697
.4	.3683	.3668	.3653	.3637	.3621	.3605	.3589	.3572	.3555	.3538
.5	.3521	.3503	.3485	.3467	.3448	.3429	.3410	.3391	.3372	.3352
.6	.3332	.3312	.3292	.3271	.3251	.3230	.3209	.3187	.3166	.3144
.7	.3123	.3101	.3079	.3056	.3034	.3011	.2989	.2966	.2943	.2920
.8	.2897	.2874	.2850	.2827	.2803	.2780	.2756	.2732	.2709	.2685
.9	.2661	.2637	.2613	.2589	.2565	.2541	.2516	.2492	.2468	.2444
1.0	.2420	.2396	.2371	.2347	.2323	.2299	.2275	.2251	.2227	.2203
1.1	.2179	.2155	.2131	.2107	.2083	.2059	.2036	.2012	.1989	.1965
1.2	.1942	.1919	.1895	.1872	.1849	.1826	.1804	.1781	.1758	.1736
1.3	.1714	.1691	.1669	.1647	.1626	.1604	.1582	.1561	.1539	.1518
1.4	.1497	.1476	.1456	.1435	.1415	.1394	.1374	.1354	.1334	.1315
1.5	.1295	.1276	.1257	.1238	.1219	.1200	.1182	.1163	.1154	.1127
1.6	.1109	.1092	.1074	.1057	.1040	.1023	.1006	.0989	.0973	.0957
1.7	.0940	.0925	.0909	.0893	.0878	.0863	.0848	.0833	.0818	.0804
1.8	.0790	.0775	.0761	.0748	.0734	.0721	.0707	.0694	.0681	.0669
1.9	.0656	.0644	.0632	.0620	.0608	.0596	.0584	.0573	.0562	.0551
2.0	.0540	.0529	.0519	.0508	.0498	.0488	.0478	.0468	.0459	.0449
2.1	.0440	.0431	.0422	.0413	.0404	.0396	.0387	.0379	.0371	.0363
2.2	.0355	.0347	.0339	.0332	.0325	.0317	.0310	.0303	.0297	.0290
2.3	.0283	.0277	.0270	.0264	.0258	.0252	.0246	.0241	.0235	.0229
2.4	.0224	.0219	.0213	.0208	.0203	.0198	.0194	.0189	.0184	.0180
2.5	.0175	.0171	.0167	.0163	.0158	.0154	.0151	.0147	.0143	.0139
2.6	.0136	.0132	.0129	.0126	.0122	.0119	.0116	.0113	.0110	.0107
2.7	.0104	.0101	.0099	.0096	.0093	.0091	.0088	.0086	.0084	.0081
2.8	.0079	.0077	.0075	.0073	.0071	.0069	.0067	.0065	.0063	.0061
2.9	.0060	.0058	.0056	.0055	.0053	.0051	.0050	.0048	.0047	.0046
3.0	.0044	.0043	.0042	.0040	.0039	.0038	.0037	.0036	.0035	.0034
3.1	.0033	.0032	.0031	.0030	.0029	.0028	.0027	.0026	.0025	.0025
3.2	.0024	.0023	.0022	.0022	.0021	.0020	.0020	.0019	.0018	.0018
3.3	.0017	.0017	.0016	.0016	.0015	.0015	.0014	.0014	.0013	.0013
3.4	.0012	.0012	.0012	.0011	.0011	.0010	.0010	.0010	.0009	.0009
3.5	.0009	.0008	.0008	.0008	.0008	.0007	.0007	.0007	.0007	.0006
3.6	.0006	.0006	.0006	.0005	.0005	.0005	.0005	.0005	.0005	.0004
3.7	.0004	.0004	.0004	.0004	.0004	.0004	.0003	.0003	.0003	.0003
3.8	.0003	.0003	.0003	.0003	.0003	.0002	.0002	.0002	.0002	.0002
3.9	.0002	.0002	.0002	.0002	.0002	.0002	.0002	.0002	.0001	.0001

NOTE: x is the value in left-hand column + the value in top row.
$f(x)$ is the value in the body of the table. *Example:* $x = 2.14$; $f(x) = 0.0404$.

Table 1.1.12   Cumulative Normal Distribution

$$F(x) = \int_{-\infty}^{x} \frac{1}{\sqrt{2\pi}} e^{-t^2/2}\, dt$$

x	.00	.01	.02	.03	.04	.05	.06	.07	.08	.09
.0	.5000	.5040	.5080	.5120	.5160	.5199	.5239	.5279	.5319	.5359
.1	.5398	.5438	.5478	.5517	.5557	.5596	.5636	.5675	.5714	.5735
.2	.5793	.5832	.5871	.5910	.5948	.5987	.6026	.6064	.6103	.6141
.3	.6179	.6217	.6255	.6293	.6331	.6368	.6406	.6443	.6480	.6517
.4	.6554	.6591	.6628	.6664	.6700	.6736	.6772	.6808	.6844	.6879
.5	.6915	.6950	.6985	.7019	.7054	.7088	.7123	.7157	.7190	.7224
.6	.7257	.7291	.7324	.7357	.7389	.7422	.7454	.7486	.7517	.7549
.7	.7580	.7611	.7642	.7673	.7703	.7734	.7764	.7793	.7823	.7852
.8	.7881	.7910	.7939	.7967	.7995	.8023	.8051	.8078	.8106	.8133
.9	.8159	.8186	.8212	.8238	.8264	.8289	.8315	.8340	.8365	.8389
1.0	.8413	.8438	.8461	.8485	.8508	.8531	.8554	.8577	.8599	.8621
1.1	.8643	.8665	.8686	.8708	.8729	.8749	.8770	.8790	.8810	.8830
1.2	.8849	.8869	.8888	.8906	.8925	.8943	.8962	.8980	.8997	.9015
1.3	.9032	.9049	.9066	.9082	.9099	.9115	.9131	.9147	.9162	.9177
1.4	.9192	.9207	.9222	.9236	.9251	.9265	.9279	.9292	.9306	.9319
1.5	.9332	.9345	.9357	.9370	.9382	.9394	.9406	.9418	.9429	.9441
1.6	.9452	.9463	.9474	.9484	.9495	.9505	.9515	.9525	.9535	.9545
1.7	.9554	.9564	.9573	.9582	.9591	.9599	.9608	.9616	.9625	.9633
1.8	.9641	.9649	.9656	.9664	.9671	.9678	.9686	.9693	.9699	.9706
1.9	.9713	.9719	.9726	.9732	.9738	.9744	.9750	.9756	.9761	.9767
2.0	.9772	.9778	.9783	.9788	.9793	.9798	.9803	.9808	.9812	.9817
2.1	.9812	.9826	.9830	.9834	.9838	.9842	.9846	.9850	.9854	.9857
2.2	.9861	.9864	.9868	.9871	.9875	.9878	.9881	.9884	.9887	.9890
2.3	.9893	.9896	.9898	.9901	.9904	.9906	.9909	.9911	.9913	.9916
2.4	.9918	.9920	.9922	.9925	.9927	.9929	.9931	.9932	.9934	.9936
2.5	.9938	.9940	.9941	.9943	.9945	.9946	.9948	.9949	.9951	.9952
2.6	.9953	.9955	.9956	.9957	.9959	.9960	.9961	.9962	.9963	.9964
2.7	.9965	.9966	.9967	.9968	.9969	.9970	.9971	.9972	.9973	.9974
2.8	.9974	.9975	.9976	.9977	.9977	.9978	.9979	.9979	.9980	.9981
2.9	.9981	.9982	.9982	.9983	.9984	.9984	.9985	.9985	.9986	.9986
3.0	.9986	.9987	.9987	.9988	.9988	.9989	.9989	.9989	.9990	.9990
3.1	.9990	.9991	.9991	.9991	.9992	.9992	.9992	.9992	.9993	.9993
3.2	.9993	.9993	.9994	.9994	.9994	.9994	.9994	.9995	.9995	.9995
3.3	.9995	.9995	.9995	.9996	.9996	.9996	.9996	.9996	.9996	.9997
3.4	.9997	.9997	.9997	.9997	.9997	.9997	.9997	.9997	.9997	.9998

NOTE: $x = (a - \mu)/\sigma$ where a is the observed value, μ is the mean, and σ is the standard deviation.
x is the value in the left-hand column + the value in the top row.
$F(x)$ is the probability that a point will be less than or equal to x.
$F(x)$ is the value in the body of the table. *Example:* The probability that an observation will be less than or equal to 1.04 is .8508.
NOTE: $F(-x) = 1 - F(x)$.

Table 1.1.13   Cumulative Chi-Square Distribution

$$F(t) = \int_0^t \frac{x^{(n-2)/2}e^{-x/2}\,dx}{2^{n/2}[(n-2)/2]!}$$

n	.005	.010	.025	.050	.100	.250	.500	.750	.900	.950	.975	.990	.995
1	.000039	.00016	.00098	.0039	.0158	.101	.455	1.32	2.70	3.84	5.02	6.62	7.86
2	.0100	.0201	.0506	.103	.211	.575	1.39	2.77	4.61	5.99	7.38	9.21	10.6
3	.0717	.155	.216	.352	.584	1.21	2.37	4.11	6.25	7.81	9.35	11.3	12.8
4	.207	.297	.484	.711	1.06	1.92	3.36	5.39	7.78	9.49	11.1	13.3	14.9
5	.412	.554	.831	1.15	1.61	2.67	4.35	6.63	9.24	11.1	12.8	15.1	16.7
6	.676	.872	1.24	1.64	2.20	3.45	5.35	7.84	10.6	12.6	14.4	16.8	18.5
7	.989	1.24	1.69	2.17	2.83	4.25	6.35	9.04	12.0	14.1	16.0	18.5	20.3
8	1.34	1.65	2.18	2.73	3.49	5.07	7.34	10.2	13.4	15.5	17.5	20.1	22.0
9	1.73	2.09	2.70	3.33	4.17	5.90	8.34	11.4	14.7	16.9	19.0	21.7	23.6
10	2.16	2.56	3.25	3.94	4.87	6.74	9.34	12.5	16.0	18.3	20.5	23.2	25.2
11	2.60	3.05	3.82	4.57	5.58	7.58	10.3	13.7	17.3	19.7	21.9	24.7	26.8
12	3.07	3.57	4.40	5.23	6.30	8.44	11.3	14.8	18.5	21.0	23.3	26.2	28.3
13	3.57	4.11	5.01	5.89	7.04	9.30	12.3	16.0	19.8	22.4	24.7	27.7	29.8
14	4.07	4.66	5.63	6.57	7.79	10.2	13.3	17.1	21.1	23.7	26.1	29.1	31.3
15	4.60	5.23	6.26	7.26	8.55	11.0	14.3	18.2	22.3	25.0	27.5	30.6	32.8
16	5.14	5.81	6.91	7.96	9.31	11.9	15.3	19.4	23.5	26.3	28.8	32.0	34.3
17	5.70	6.41	7.56	8.67	10.1	12.8	16.3	20.5	24.8	27.6	30.2	33.4	35.7
18	6.26	7.01	8.23	9.39	10.9	13.7	17.3	21.6	26.0	28.9	31.5	34.8	37.2
19	6.84	7.63	8.91	10.1	11.7	14.6	18.3	22.7	27.2	30.1	32.9	36.2	38.6
20	7.43	8.26	9.59	10.9	12.4	15.5	19.3	23.8	28.4	31.4	34.2	37.6	40.0
21	8.03	8.90	10.3	11.6	13.2	16.3	20.3	24.9	29.6	32.7	35.5	38.9	41.4
22	8.64	9.54	11.0	12.3	14.0	17.2	21.3	26.0	30.8	33.9	36.8	40.3	42.8
23	9.26	10.2	11.7	13.1	14.8	18.1	22.3	27.1	32.0	35.2	38.1	41.6	44.2
24	9.89	10.9	12.4	13.8	15.7	19.0	23.3	28.2	33.2	36.4	39.4	43.0	45.6
25	10.5	11.5	13.1	14.6	16.5	19.9	24.3	29.3	34.4	37.7	40.6	44.3	46.9
26	11.2	12.2	13.8	15.4	17.3	20.8	25.3	30.4	35.6	38.9	41.9	45.6	48.3
27	11.8	12.9	14.6	16.2	18.1	21.7	26.3	31.5	36.7	40.1	43.2	47.0	49.6
28	12.5	13.6	15.3	16.9	18.9	22.7	27.3	32.6	37.9	41.3	44.5	48.3	51.0
29	13.1	14.3	16.0	17.7	19.8	23.6	28.3	33.7	39.1	42.6	45.7	49.6	52.3
30	13.8	15.0	16.8	18.5	20.6	24.5	29.3	34.8	40.3	43.8	47.0	50.9	53.7

NOTE: n is the number of degrees of freedom.
Values for t are in the body of the table. *Example:* The probability that, with 16 degrees of freedom, a point will be ≤ 23.5 is .900.

Table 1.1.14 Cumulative "Student's" Distribution

$$F(t) = \int_{-\infty}^{t} \frac{\left(\dfrac{n-1}{2}\right)!}{\left(\dfrac{n-2}{2}\right)! \sqrt{\pi n}\left(1 + \dfrac{x^2}{n}\right)^{(n+1)/2}} dx$$

n \ F	.75	.90	.95	.975	.99	.995	.9995
1	1.000	3.078	6.314	12.70	31.82	63.66	636.3
2	.816	1.886	2.920	4.303	6.965	9.925	31.60
3	.765	1.638	2.353	3.182	4.541	5.841	12.92
4	.741	1.533	2.132	2.776	3.747	4.604	8.610
5	.727	1.476	2.015	2.571	3.365	4.032	6.859
6	.718	1.440	1.943	2.447	3.143	3.707	5.959
7	.711	1.415	1.895	2.365	2.998	3.499	5.408
8	.706	1.397	1.860	2.306	2.896	3.355	5.041
9	.703	1.383	1.833	2.262	2.821	3.250	4.781
10	.700	1.372	1.812	2.228	2.764	3.169	4.587
11	.697	1.363	1.796	2.201	2.718	3.106	4.437
12	.695	1.356	1.782	2.179	2.681	3.055	4.318
13	.694	1.350	1.771	2.160	2.650	3.012	4.221
14	.692	1.345	1.761	2.145	2.624	2.977	4.140
15	.691	1.341	1.753	2.131	2.602	2.947	4.073
16	.690	1.337	1.746	2.120	2.583	2.921	4.015
17	.689	1.333	1.740	2.110	2.567	2.898	3.965
18	.688	1.330	1.734	2.101	2.552	2.878	3.922
19	.688	1.328	1.729	2.093	2.539	2.861	3.883
20	.687	1.325	1.725	2.086	2.528	2.845	3.850
21	.686	1.323	1.721	2.080	2.518	2.831	3.819
22	.686	1.321	1.717	2.074	2.508	2.819	3.792
23	.685	1.319	1.714	2.069	2.500	2.807	3.768
24	.685	1.318	1.711	2.064	2.492	2.797	3.745
25	.684	1.316	1.708	2.060	2.485	2.787	3.725
26	.684	1.315	1.706	2.056	2.479	2.779	3.707
27	.684	1.314	1.703	2.052	2.473	2.771	3.690
28	.683	1.313	1.701	2.048	2.467	2.763	3.674
29	.683	1.311	1.699	2.045	2.462	2.756	3.659
30	.683	1.310	1.697	2.042	2.457	2.750	3.646
40	.681	1.303	1.684	2.021	2.423	2.704	3.551
60	.679	1.296	1.671	2.000	2.390	2.660	3.460
120	.677	1.289	1.658	1.980	2.385	2.617	3.373

NOTE: n is the number of degrees of freedom.

Values for t are in the body of the table. *Example:* The probability that, with 16 degrees of freedom, a point will be ≤ 2.921 is .995.

NOTE: $F(-t) = 1 - F(t)$.

Table 1.1.15 Cumulative *F* Distribution

m degrees of freedom in numerator; *n* in denominator

$$G(F) = \int_0^F \frac{[(m + n - 2)/2]!\, m^{m/2} n^{n/2} x^{(m-2)/2} (n + mx)^{-(m+n)/2}}{[(m - 2)/2]!\,[(n - 2)/2]!}\, dx$$

Upper 5% points ($F_{.95}$)

Degrees of freedom for denominator	Degrees of freedom for numerator																		
	1	2	3	4	5	6	7	8	9	10	12	15	20	24	30	40	60	120	∞
1	161	200	216	225	230	234	237	239	241	242	244	246	248	249	250	251	252	253	254
2	18.5	19.0	19.2	19.2	19.3	19.3	19.4	19.4	19.4	19.4	19.4	19.4	19.4	19.5	19.5	19.5	19.5	19.5	19.5
3	10.1	9.55	9.28	9.12	9.01	8.94	8.89	8.85	8.81	8.79	8.74	8.70	8.66	8.64	8.62	8.59	8.57	8.55	8.53
4	7.71	6.94	6.59	6.39	6.26	6.16	6.09	6.04	6.00	5.96	5.91	5.86	5.80	5.77	5.75	5.72	5.69	5.66	5.63
5	6.61	5.79	5.41	5.19	5.05	4.95	4.88	4.82	4.77	4.74	4.68	4.62	4.56	4.53	4.50	4.46	4.43	4.40	4.37
6	5.99	5.14	4.76	4.53	4.39	4.28	4.21	4.15	4.10	4.06	4.00	3.94	3.87	3.84	3.81	3.77	3.74	3.70	3.67
7	5.59	4.74	4.35	4.12	3.97	3.87	3.79	3.73	3.68	3.64	3.57	3.51	3.44	3.41	3.38	3.34	3.30	3.27	3.23
8	5.32	4.46	4.07	3.84	3.69	3.58	3.50	3.44	3.39	3.35	3.28	3.22	3.15	3.12	3.08	3.04	3.01	2.97	2.93
9	5.12	4.26	3.86	3.63	3.48	3.37	3.29	3.23	3.18	3.14	3.07	3.01	2.94	2.90	2.86	2.83	2.79	2.75	2.71
10	4.96	4.10	3.71	3.48	3.33	3.22	3.14	3.07	3.02	2.98	2.91	2.85	2.77	2.74	2.70	2.66	2.62	2.58	2.54
11	4.84	3.98	3.59	3.36	3.20	3.09	3.01	2.95	2.90	2.85	2.79	2.72	2.65	2.61	2.57	2.53	2.49	2.45	2.40
12	4.75	3.89	3.49	3.26	3.11	3.00	2.91	2.85	2.80	2.75	2.69	2.62	2.54	2.51	2.47	2.43	2.38	2.34	2.30
13	4.67	3.81	3.41	3.18	3.03	2.92	2.83	2.77	2.71	2.67	2.60	2.53	2.46	2.42	2.38	2.34	2.30	2.25	2.21
14	4.60	3.74	3.34	3.11	2.96	2.85	2.76	2.70	2.65	2.60	2.53	2.46	2.39	2.35	2.31	2.27	2.22	2.18	2.13
15	4.54	3.68	3.29	3.06	2.90	2.79	2.71	2.64	2.59	2.54	2.48	2.40	2.33	2.29	2.25	2.20	2.16	2.11	2.07
16	4.49	3.63	3.24	3.01	2.85	2.74	2.66	2.59	2.54	2.49	2.42	2.35	2.28	2.24	2.19	2.15	2.11	2.06	2.01
17	4.45	3.59	3.20	2.96	2.81	2.70	2.61	2.55	2.49	2.45	2.38	2.31	2.23	2.19	2.15	2.10	2.06	2.01	1.96
18	4.41	3.55	3.16	2.93	2.77	2.66	2.58	2.51	2.46	2.41	2.34	2.27	2.19	2.15	2.11	2.06	2.02	1.97	1.92
19	4.38	3.52	3.13	2.90	2.74	2.63	2.54	2.48	2.42	2.38	2.31	2.23	2.16	2.11	2.07	2.03	1.98	1.93	1.88
20	4.35	3.49	3.10	2.87	2.71	2.60	2.51	2.45	2.39	2.35	2.28	2.20	2.12	2.08	2.04	1.99	1.95	1.90	1.84
21	4.32	3.47	3.07	2.84	2.68	2.57	2.49	2.42	2.37	2.32	2.25	2.18	2.10	2.05	2.01	1.96	1.92	1.87	1.81
22	4.30	3.44	3.05	2.82	2.66	2.55	2.46	2.40	2.34	2.30	2.23	2.15	2.07	2.03	1.98	1.94	1.89	1.84	1.78
23	4.28	3.42	3.03	2.80	2.64	2.53	2.44	2.37	2.32	2.27	2.20	2.13	2.05	2.01	1.96	1.91	1.86	1.81	1.76
24	4.26	3.40	3.01	2.78	2.62	2.51	2.42	2.36	2.30	2.25	2.18	2.11	2.03	1.98	1.94	1.89	1.84	1.79	1.73
25	4.24	3.39	2.99	2.76	2.60	2.49	2.40	2.34	2.28	2.24	2.16	2.09	2.01	1.96	1.92	1.87	1.82	1.77	1.71
30	4.17	3.32	2.92	2.69	2.53	2.42	2.33	2.27	2.21	2.16	2.09	2.01	1.93	1.89	1.84	1.79	1.74	1.68	1.62
40	4.08	3.23	2.84	2.61	2.45	2.34	2.25	2.18	2.12	2.08	2.00	1.92	1.84	1.79	1.74	1.69	1.64	1.58	1.51
60	4.00	3.15	2.76	2.53	2.37	2.25	2.17	2.10	2.04	1.99	1.92	1.84	1.75	1.70	1.65	1.59	1.53	1.47	1.39
120	3.92	3.07	2.68	2.45	2.29	2.18	2.09	2.02	1.96	1.91	1.83	1.75	1.66	1.61	1.55	1.50	1.43	1.35	1.25
∞	3.84	3.00	2.60	2.37	2.21	2.10	2.01	1.94	1.88	1.83	1.75	1.67	1.57	1.52	1.46	1.39	1.32	1.22	1.00

Upper 1% points ($F_{.99}$)

Degrees of freedom for denominator	Degrees of freedom for numerator																		
	1	2	3	4	5	6	7	8	9	10	12	15	20	24	30	40	60	120	∞
1	4052	5000	5403	5625	5764	5859	5928	5982	6023	6056	6106	6157	6209	6235	6261	6287	6313	6339	6366
2	98.5	99.0	99.2	99.2	99.3	99.3	99.4	99.4	99.4	99.4	99.4	99.4	99.4	99.5	99.5	99.5	99.5	99.5	99.5
3	34.1	30.8	29.5	28.7	28.2	27.9	27.7	27.5	27.3	27.2	27.1	26.9	26.7	26.6	26.5	26.4	26.3	26.2	26.1
4	21.2	18.0	16.7	16.0	15.5	15.2	15.0	14.8	14.7	14.5	14.4	14.2	14.0	13.9	13.8	13.7	13.7	13.6	13.5
5	16.3	13.3	12.1	11.4	11.0	10.7	10.5	10.3	10.2	10.1	9.89	9.72	9.55	9.47	9.38	9.29	9.20	9.11	9.02
6	13.7	10.9	9.78	9.15	8.75	8.47	8.26	8.10	7.98	7.87	7.72	7.56	7.40	7.31	7.23	7.14	7.06	6.97	6.88
7	12.2	9.55	8.45	7.85	7.46	7.19	6.99	6.84	6.72	6.62	6.47	6.31	6.16	6.07	5.99	5.91	5.82	5.74	5.65
8	11.3	8.65	7.59	7.01	6.63	6.37	6.18	6.03	5.91	5.81	5.67	5.52	5.36	5.28	5.20	5.12	5.03	4.95	4.86
9	10.6	8.02	6.99	6.42	6.06	5.80	5.61	5.47	5.35	5.26	5.11	4.96	4.81	4.73	4.65	4.57	4.48	4.40	4.31
10	10.0	7.56	6.55	5.99	5.64	5.39	5.20	5.06	4.94	4.85	4.71	4.56	4.41	4.33	4.25	4.17	4.08	4.00	3.91
11	9.65	7.21	6.22	5.67	5.32	5.07	4.89	4.74	4.63	4.54	4.40	4.25	4.10	4.02	3.94	3.86	3.78	3.69	3.60
12	9.33	6.93	5.95	5.41	5.06	4.82	4.64	4.50	4.39	4.30	4.16	4.01	3.86	3.78	3.70	3.62	3.54	3.45	3.36
13	9.07	6.70	5.74	5.21	4.86	4.62	4.44	4.30	4.19	4.10	3.96	3.82	3.66	3.59	3.51	3.43	3.34	3.25	3.17
14	8.86	6.51	5.56	5.04	4.70	4.46	4.28	4.14	4.03	3.94	3.80	3.66	3.51	3.43	3.35	3.27	3.18	3.09	3.00
15	8.68	6.36	5.42	4.89	4.56	4.32	4.14	4.00	3.89	3.80	3.67	3.52	3.37	3.29	3.21	3.13	3.05	2.96	2.87
16	8.53	6.23	5.29	4.77	4.44	4.20	4.03	3.89	3.78	3.69	3.55	3.41	3.26	3.18	3.10	3.02	2.93	2.84	2.75
17	8.40	6.11	5.19	4.67	4.34	4.10	3.93	3.79	3.68	3.59	3.46	3.31	3.16	3.08	3.00	2.92	2.83	2.75	2.65
18	8.29	6.01	5.09	4.58	4.25	4.01	3.84	3.71	3.60	3.51	3.37	3.23	3.08	3.00	2.92	2.84	2.75	2.66	2.57
19	8.19	5.93	5.01	4.50	4.17	3.94	3.77	3.63	3.52	3.43	3.30	3.15	3.00	2.92	2.84	2.76	2.67	2.58	2.49
20	8.10	5.85	4.94	4.43	4.10	3.87	3.70	3.56	3.46	3.37	3.23	3.09	2.94	2.86	2.78	2.69	2.61	2.52	2.42
21	8.02	5.78	4.87	4.37	4.04	3.81	3.64	3.51	3.40	3.31	3.17	3.03	2.88	2.80	2.72	2.64	2.55	2.46	2.36
22	7.95	5.72	4.82	4.31	3.99	3.76	3.59	3.45	3.35	3.26	3.12	2.98	2.83	2.75	2.67	2.58	2.50	2.40	2.31
23	7.88	5.66	4.76	4.26	3.94	3.71	3.54	3.41	3.30	3.21	3.07	2.93	2.78	2.70	2.62	2.54	2.45	2.35	2.26
24	7.82	5.61	4.72	4.22	3.90	3.67	3.50	3.36	3.26	3.17	3.03	2.89	2.74	2.66	2.58	2.49	2.40	2.31	2.21
25	7.77	5.57	4.68	4.18	3.86	3.63	3.46	3.32	3.22	3.13	2.99	2.85	2.70	2.62	2.53	2.45	2.36	2.27	2.17
30	7.56	5.39	4.51	4.02	3.70	3.47	3.30	3.17	3.07	2.98	2.84	2.70	2.55	2.47	2.39	2.30	2.21	2.11	2.01
40	7.31	5.18	4.31	3.83	3.51	3.29	3.12	2.99	2.89	2.80	2.66	2.52	2.37	2.29	2.20	2.11	2.02	1.92	1.80
60	7.08	4.98	4.13	3.65	3.34	3.12	2.95	2.82	2.72	2.63	2.50	2.35	2.20	2.12	2.03	1.94	1.84	1.73	1.60
120	6.85	4.79	3.95	3.48	3.17	2.96	2.79	2.66	2.56	2.47	2.34	2.19	2.03	1.95	1.86	1.76	1.66	1.53	1.38
∞	6.63	4.61	3.78	3.32	3.02	2.80	2.64	2.51	2.41	2.32	2.18	2.04	1.88	1.79	1.70	1.59	1.47	1.32	1.00

NOTE: *m* is the number of degrees of freedom in the numerator of *F*; *n* is the number of degrees of freedom in the denominator of *F*.
Values for *F* are in the body of the table.
G is the probability that a point, with *m* and *n* degrees of freedom will be $\leq F$.
Example: With 2 and 5 degrees of freedom, the probability that a point will be ≤ 13.3 is .99.
SOURCE: ''Chemical Engineers' Handbook,'' 5th edition, by R. H. Perry and C. H. Chilton, McGraw-Hill, 1973. Used with permission.

Table 1.1.16   Standard Distribution of Residuals

a = any positive quantity
y = the number of residuals which are numerically $< a$
r = the probable error of a single observation
n = number of observations

$\dfrac{a}{r}$	$\dfrac{y}{n}$	Diff	$\dfrac{a}{r}$	$\dfrac{y}{n}$	Diff
0.0	.000		2.5	.908	
		54			13
1	.054		6	.921	
		53			10
2	.107		7	.931	
		53			10
3	.160		8	.941	
		53			9
4	.213		9	.950	
		51			7
0.5	.264		3.0	.957	
		50			6
6	.314		1	.963	
		49			6
7	.363		2	.969	
		48			5
8	.411		3	.974	
		45			4
9	.456		4	.978	
		44			4
1.0	.500		3.5	.982	
		42			3
1	.542		6	.985	
		40			2
2	.582		7	.987	
		37			3
3	.619		8	.990	
		36			1
4	.655		9	.991	
		33			2
1.5	.688		4.0	.993	
		31			6
6	.719				
		29	5.0	.999	
7	.748				
		27			
8	.775				
		25			
9	.800				
		23			
2.0	.823				
		20			
1	.843				
		19			
2	.862				
		17			
3	.879				
		16			
4	.895				
		13			

Table 1.1.17   Factors for Computing Probable Error

n	Bessel $\dfrac{0.6745}{\sqrt{(n-1)}}$	Bessel $\dfrac{0.6745}{\sqrt{n(n-1)}}$	Peters $\dfrac{0.8453}{\sqrt{n(n-1)}}$	Peters $\dfrac{0.8453}{n\sqrt{n-1}}$	n	Bessel $\dfrac{0.6745}{\sqrt{(n-1)}}$	Bessel $\dfrac{0.6745}{\sqrt{n(n-1)}}$	Peters $\dfrac{0.8453}{\sqrt{n(n-1)}}$	Peters $\dfrac{0.8453}{n\sqrt{n-1}}$
2	.6745	.4769	.5978	.4227	30	.1252	.0229	.0287	.0052
3	.4769	.2754	.3451	.1993	31	.1231	.0221	.0277	.0050
4	.3894	.1947	.2440	.1220	32	.1211	.0214	.0268	.0047
5	.3372	.1508	.1890	.0845	33	.1192	.0208	.0260	.0045
6	.3016	.1231	.1543	.0630	34	.1174	.0201	.0252	.0043
7	.2754	.1041	.1304	.0493	35	.1157	.0196	.0245	.0041
8	.2549	.0901	.1130	.0399	36	.1140	.0190	.0238	.0040
9	.2385	.0795	.0996	.0332	37	.1124	.0185	.0232	.0038
10	.2248	.0711	.0891	.0282	38	.1109	.0180	.0225	.0037
11	.2133	.0643	.0806	.0243	39	.1094	.0175	.0220	.0035
12	.2034	.0587	.0736	.0212	40	.1080	.0171	.0214	.0034
13	.1947	.0540	.0677	.0188	45	.1017	.0152	.0190	.0028
14	.1871	.0500	.0627	.0167	50	.0964	.0136	.0171	.0024
15	.1803	.0465	.0583	.0151	55	.0918	.0124	.0155	.0021
16	.1742	.0435	.0546	.0136	60	.0878	.0113	.0142	.0018
17	.1686	.0409	.0513	.0124	65	.0843	.0105	.0131	.0016
18	.1636	.0386	.0483	.0114	70	.0812	.0097	.0122	.0015
19	.1590	.0365	.0457	.0105	75	.0784	.0091	.0113	.0013
20	.1547	.0346	.0434	.0097	80	.0759	.0085	.0106	.0012
21	.1508	.0329	.0412	.0090	85	.0736	.0080	.0100	.0011
22	.1472	.0314	.0393	.0084	90	.0715	.0075	.0094	.0010
23	.1438	.0300	.0376	.0078	95	.0696	.0071	.0089	.0009
24	.1406	.0287	.0360	.0073	100	.0678	.0068	.0085	.0008
25	.1377	.0275	.0345	.0069					
26	.1349	.0265	.0332	.0065					
27	.1323	.0255	.0319	.0061					
28	.1298	.0245	.0307	.0058					
29	.1275	.0237	.0297	.0055					

Table 1.1.18 Decimal Equivalents

From minutes and seconds into decimal parts of a degree

0'	0°.0000	0"	0°.0000
1	.0167	1	.0003
2	.0333	2	.0006
3	.05	3	.0008
4	.0667	4	.0011
5'	.0833	5"	.0014
6	.10	6	.0017
7	.1167	7	.0019
8	.1333	8	.0022
9	.15	9	.0025
10'	0°.1667	10"	0°.0028
1	.1833	1	.0031
2	.20	2	.0033
3	.2167	3	.0036
4	.2333	4	.0039
15'	.25	15"	.0042
6	.2667	6	.0044
7	.2833	7	.0047
8	.30	8	.005
9	.3167	9	.0053
20'	0°.3333	20"	0°.0056
1	.35	1	.0058
2	.3667	2	.0061
3	.3833	3	.0064
4	.40	4	.0067
25'	.4167	25"	.0069
6	.4333	6	.0072
7	.45	7	.0075
8	.4667	8	.0078
9	.4833	9	.0081
30'	0°.50	30"	0°.0083
1	.5167	1	.0086
2	.5333	2	.0089
3	.55	3	.0092
4	.5667	4	.0094
35'	.5833	35"	.0097
6	.60	6	.01
7	.6167	7	.0103
8	.6333	8	.0106
9	.65	9	.0108
40'	0°.6667	40"	0°.0111
1	.6833	1	.0114
2	.70	2	.0117
3	.7167	3	.0119
4	.7333	4	.0122
45'	.75	45"	.0125
6	.7667	6	.0128
7	.7833	7	.0131
8	.80	8	.0133
9	.8167	9	.0136
50'	0°.8333	50"	0°.0139
1	.85	1	.0142
2	.8667	2	.0144
3	.8833	3	.0147
4	.90	4	.015
55'	.9167	55"	.0153
6	.9333	6	.0156
7	.95	7	.0158
8	.9667	8	.0161
9	.9833	9	.0164
60'	1.00	60"	0°.0167

From decimal parts of a degree into minutes and seconds (exact values)

0°.00	0'	0°.50	30'
1	0' 36"	1	30' 36"
2	1' 12"	2	31' 12"
3	1' 48"	3	31' 48"
4	2' 24"	4	32' 24"
0°.05	3'	0°.55	33'
6	3' 36"	6	33' 36"
7	4' 12"	7	34' 12"
8	4' 48"	8	34' 48"
9	5' 24"	9	35' 24"
0°.10	6'	0°.60	36'
1	6' 36"	1	36' 36"
2	7' 12"	2	37' 12"
3	7' 48"	3	37' 48"
4	8' 24"	4	38' 24"
0°.15	9'	0°.65	39'
6	9' 36"	6	39' 36"
7	10' 12"	7	40' 12"
8	10' 48"	8	40' 48"
9	11' 24"	9	41' 24"
0°.20	12'	0°.70	42'
1	12' 36"	1	42' 36"
2	13' 12"	2	43' 12"
3	13' 48"	3	43' 48"
4	14' 24"	4	44' 24"
0°.25	15'	0°.75	45'
6	15' 36"	6	45' 36"
7	16' 12"	7	46' 12"
8	16' 48"	8	46' 48"
9	17' 24"	9	47' 24"
0°.30	18'	0°.80	48'
1	18' 36"	1	48' 36"
2	19' 12"	2	49' 12"
3	19' 48"	3	49' 48"
4	20' 24"	4	50' 24"
0°.35	21'	0°.85	51'
6	21' 36"	6	51' 36"
7	22' 12"	7	52' 12"
8	22' 48"	8	52' 48"
9	23' 24"	9	53' 24"
0°.40	24'	0°.90	54'
1	24' 36"	1	54' 36"
2	25' 12"	2	55' 12"
3	25' 48"	3	55' 48"
4	26' 24"	4	56' 24"
0°.45	27'	0°.95	57'
6	27' 36"	6	57' 36"
7	28' 12"	7	58' 12"
8	28' 48"	8	58' 48"
9	29' 24"	9	59' 24"
0°.50	30'	1°.00	60'

0°.000	0".0
1	3".6
2	7".2
3	10".8
4	14".4
0°.005	18"
6	21".6
7	25".2
8	28".8
9	32".4
0°.010	36"

Common fractions

8ths	16ths	32nds	64ths	Exact decimal values
			1	.01 5625
		1	2	.03 125
			3	.04 6875
	1	2	4	.06 25
			5	.07 8125
		3	6	.09 375
			7	.10 9375
1	2	4	8	.12 5
			9	.14 0625
		5	10	.15 625
			11	.17 1875
	3	6	12	.18 75
			13	.20 3125
		7	14	.21 875
			15	.23 4375
2	4	8	16	.25
			17	.26 5625
		9	18	.28 125
			19	.29 6875
	5	10	20	.31 25
			21	.32 8125
		11	22	.34 375
			23	.35 9375
3	6	12	24	.37 5
			25	.39 0625
		13	26	.40 625
			27	.42 1875
	7	14	28	.43 75
			29	.45 3125
		15	30	.46 875
			31	.48 4375
4	8	16	32	.50
			33	.51 5625
		17	34	.53 125
			35	.54 6875
	9	18	36	.56 25
			37	.57 8125
		19	38	.59 375
			39	.60 9375
5	10	20	40	.62 5
			41	.64 0625
		21	42	.65 625
			43	.67 1875
	11	22	44	.68 75
			45	.70 3125
		23	46	.71 875
			47	.73 4375
6	12	24	48	.75
			49	.76 5625
		25	50	.78 125
			51	.79 6875
	13	26	52	.81 25
			53	.82 8125
		27	54	.84 375
			55	.85 9375
7	14	28	56	.87 5
			57	.89 0625
		29	58	.90 625
			59	.92 1875
	15	30	60	.93 75
			61	.95 3125
		31	62	.96 875
			63	.98 4375

1.2 MEASURING UNITS

by David T. Goldman

REFERENCES: "International Critical Tables," McGraw-Hill. "Smithsonian Physical Tables," Smithsonian Institution. "Landolt-Börnstein: Zahlenwerte und Funktionen aus Physik, Chemie, Astronomie, Geophysik und Technik," Springer. "Handbook of Chemistry and Physics," Chemical Rubber Co. "Units and Systems of Weights and Measures; Their Origin, Development, and Present Status," NBS LC 1035 (1976). "Weights and Measures Standards of the United States, a Brief History," NBS Spec. Pub. 447 (1976). "Standard Time," Code of Federal Regulations, Title 49. "Fluid Meters, Their Theory and Application," 6th ed., chaps. 1–2, ASME, 1971. H. E. Huntley, "Dimensional Analysis," Richard & Co., New York, 1951. "U.S. Standard Atmosphere, 1962," Government Printing Office. Public Law 89-387, "Uniform Time Act of 1966." Public Law 94-168, "Metric Conversion Act of 1975." ASTM E380-91a, "Use of the International Standards of Units (SI) (the Modernized Metric System)." The International System of Units," NIST Spec. Pub. 330. "Guidelines for Use of the Modernized Metric System," NBS LC 1120. "NBS Time and Frequency Dissemination Services," NBS Spec. Pub. 432. "Factors for High Precision Conversion," NBS LC 1071. American Society of Mechanical Engineers SI Series, ASME SI 1–9. Jespersen and Fitz-Randolph, "From Sundials to Atomic Clocks: Understanding Time and Frequency," NBS, Monograph 155. ANSI/IEEE Std 268-1992, "American National Standard for Metric Practice."

U.S. CUSTOMARY SYSTEM (USCS)

The USCS, often called the "inch-pound system," is the system of units most commonly used for measures of weight and length (Table 1.2.1). The units are identical for practical purposes with the corresponding English units, but the capacity measures differ from those used in the British Commonwealth, the U.S. gallon being defined as 231 cu in and the bushel as 2,150.42 cu in, whereas the corresponding British Imperial units are, respectively, 277.42 cu in and 2,219.36 cu in (1 Imp gal = 1.2 U.S. gal, approx; 1 Imp bu = 1.03 U.S. bu, approx).

Table 1.2.1 U.S. Customary Units

Units of length

12 inches	= 1 foot
3 feet	= 1 yard
5½ yards = 16½ feet	= 1 rod, pole, or perch
40 poles = 220 yards	= 1 furlong
8 furlongs = 1,760 yards = 5,280 feet	= 1 mile
3 miles	= 1 league
4 inches	= 1 hand
9 inches	= 1 span

Nautical units

6,076.11549 feet	= 1 international nautical mile
6 feet	= 1 fathom
120 fathoms	= 1 cable length
1 nautical mile per hr	= 1 knot

Surveyor's or Gunter's units

7.92 inches	= 1 link
100 links = 66 ft = 4 rods	= 1 chain
80 chains	= 1 mile
33⅓ inches	= 1 vara (Texas)

Units of area

144 square inches	= 1 square foot
9 square feet	= 1 square yard
30¼ square yards	= 1 square rod, pole, or perch

160 square rods = 10 square chains = 43,560 square feet = 5,645 sq varas (Texas)	= 1 acre
640 acres = 1 square mile =	1 "section" of U.S. government-surveyed land
1 circular inch = area of circle 1 inch in diameter	= 0.7854 sq in
1 square inch	= 1.2732 circular inches
1 circular mil	= area of circle 0.001 in in diam
1,000,000 cir mils	= 1 circular inch

Units of volume

1,728 cubic inches	= 1 cubic foot
231 cubic inches	= 1 gallon
27 cubic feet	= 1 cubic yard
1 cord of wood	= 128 cubic feet
1 perch of masonry	= 16½ to 25 cu ft

Liquid or fluid measurements

4 gills	= 1 pint
2 pints	= 1 quart
4 quarts	= 1 gallon
7.4805 gallons	= 1 cubic foot

(There is no standard liquid barrel; by trade custom, 1 bbl of petroleum oil, unrefined = 42 gal. The capacity of the common steel barrel used for refined petroleum products and other liquids is 55 gal.)

Apothecaries' liquid measurements

60 minims	= 1 liquid dram or drachm
8 drams	= 1 liquid ounce
16 ounces	= 1 pint

Water measurements

The miner's inch is a unit of water volume flow no longer used by the Bureau of Reclamation. It is used within particular water districts where its value is defined by statute. Specifically, within many of the states of the West the miner's inch is ¹⁄₅₀ cubic foot per second. In others it is equal to ¹⁄₄₀ cubic foot per second, while in the state of Colorado, 38.4 miner's inch is equal to 1 cubic-foot per second. In SI units, these correspond to $.32 \times 10^{-6}$ m³/s, $.409 \times 10^{-6}$ m³/s, and $.427 \times 10^{-6}$ m³/s, respectively.

Dry measures

2 pints	= 1 quart
8 quarts	= 1 peck
4 pecks	= 1 bushel
1 std bbl for fruits and vegetables	= 7,056 cu in or 105 dry qt, struck measure

Shipping measures

1 Register ton	= 100 cu ft
1 U.S. shipping ton	= 40 cu ft
	= 32.14 U.S. bu or 31.14 Imp bu
1 British shipping ton	= 42 cu ft
	= 32.70 Imp bu or 33.75 U.S. bu

Board measurements

(Based on nominal not actual dimensions; see Table 12.2.8)

1 board foot = { 144 cu in = volume of board / 1 ft sq and 1 in thick

The international log rule, based upon ¼ in kerf, is expressed by the formula

$$X = 0.904762(0.22\,D^2 - 0.71\,D)$$

where X is the number of board feet in a 4-ft section of a log and D is the top diam in in. In computing the number of board feet in a log, the taper is taken at ½ in per 4 ft linear, and separate computation is made for each 4-ft section.

Weights
(The grain is the same in all systems.)

Avoirdupois weights

16 drams = 437.5 grains	= 1 ounce
16 ounces = 7,000 grains	= 1 pound
100 pounds	= 1 cental
2,000 pounds	= 1 short ton
2,240 pounds	= 1 long ton
1 std lime bbl, small	= 180 lb net
1 std lime bbl, large	= 280 lb net
Also (in Great Britain):	
14 pounds	= 1 stone
2 stone = 28 pounds	= 1 quarter
4 quarters = 112 pounds	= 1 hundredweight (cwt)
20 hundredweight	= 1 long ton

Troy weights

24 grains	= 1 pennyweight (dwt)
20 pennyweights = 480 grains	= 1 ounce
12 ounces = 5,760 grains	= 1 pound

1 assay ton = 29,167 milligrams, or as many milligrams as there are troy ounces in a ton of 2,000 lb avoirdupois. Consequently, the number of milligrams of precious metal yielded by an assay ton of ore gives directly the number of troy ounces that would be obtained from a ton of 2,000 lb avoirdupois.

Apothecaries' weights

20 grains	= 1 scruple ℈
3 scruples = 60 grains	= 1 dram ʒ
8 drams	= 1 ounce ʒ
12 ounces = 5,760 grains	= 1 pound

Weight for precious stones
1 carat = 200 milligrams
(Used by almost all important nations)

Circular measures

60 seconds	= 1 minute
60 minutes	= 1 degree
90 degrees	= 1 quadrant
360 degrees	= circumference
57.2957795 degrees	= 1 radian (or angle having
(= 57°17′44.806″)	arc of length equal to radius)

METRIC SYSTEM

In the United States the name "metric system" of length and mass units is commonly taken to refer to a system that was developed in France about 1800. The unit of length was equal to 1/10,000,000 of a quarter meridian (north pole to equator) and named the metre. A cube 1/10th metre on a side was the litre, the unit of volume. The mass of water filling this cube was the kilogram, or standard of mass; i.e., 1 litre of water = 1 kilogram of mass. Metal bars and weights were constructed conforming to these prescriptions for the metre and kilogram. One bar and one weight were selected to be the primary representations. The kilogram and the metre are now defined independently, and the litre, although for many years defined as the volume of a kilogram of water at the temperature of its maximum density, 4°C, and under a pressure of 76 cm of mercury, is now equal to 1 cubic decimeter.

In 1866, the U.S. Congress formally recognized metric units as a legal system, thereby making their use permissible in the United States. In 1893, the Office of Weights and Measures (now the National Bureau of Standards), by executive order, fixed the values of the U.S. yard and pound in terms of the meter and kilogram, respectively, as 1 yard = 3,600/3,937 m; and 1 lb = 0.453 592 4277 kg. By agreement in 1959 among the national standards laboratories of the English-speaking nations, the relations in use now are: 1 yd = 0.9144 m, whence 1 in =

25.4 mm exactly; and 1 lb = 0.453 592 37 kg, or 1 lb = 453.59 g (nearly).

THE INTERNATIONAL SYSTEM OF UNITS (SI)

In October 1960, the Eleventh General (International) Conference on Weights and Measures redefined some of the original metric units and expanded the system to include other physical and engineering units. This expanded system is called, in French, Le Système International d'Unités (abbreviated SI), and in English, The International System of Units.

The Metric Conversion Act of 1975 codifies the voluntary conversion of the U.S. to the SI system. It is expected that in time all units in the United States will be in SI form. For this reason, additional tables of units, prefixes, equivalents, and conversion factors are included below (Tables 1.2.2 and 1.2.3).

SI consists of seven base units, two supplementary units, a series of derived units consistent with the base and supplementary units, and a series of approved prefixes for the formation of multiples and submultiples of the various units (see Tables 1.2.2 and 1.2.3). Multiple and submultiple prefixes in steps of 1,000 are recommended. (See ASTM E380-91a for further details.)

Base and supplementary units are defined [NIST Spec. Pub. 330 (1991)] as:

Metre   The metre is defined as the length of path traveled by light in a vacuum during a time interval 1/299 792 458 of a second.

Kilogram   The kilogram is the unit of mass; it is equal to the mass of the international prototype of the kilogram.

Second   The second is the duration of 9,192,631,770 periods of the radiation corresponding to the transition between the two hyperfine levels of the ground state of the cesium 133 atom.

Ampere   The ampere is that constant current which, if maintained in two straight parallel conductors of infinite length, of negligible cross section, and placed 1 metre apart in vacuum, would produce between these conductors a force equal to 2×10^{-7} newton per metre of length.

Kelvin   The kelvin, unit of thermodynamic temperature, is the fraction 1/273.16 of the thermodynamic temperature of the triple point of water.

Mole   The mole is the amount of substance of a system which contains as many elementary entities as there are atoms in 0.012 kilogram of carbon 12. (When the mole is used, the elementary entities must be specified and may be atoms, molecules, ions, electrons, other particles, or specified groups of such particles.)

Candela   The candela is the luminous intensity, in a given direction, of a source that emits monochromatic radiation of frequency 540×10^{12} hertz and that has a radiant intensity in that direction of $1/683$ watt per steradian.

Radian   The unit of measure of a plane angle with its vertex at the center of a circle and subtended by an arc equal in length to the radius.

Steradian   The unit of measure of a solid angle with its vertex at the center of a sphere and enclosing an area of the spherical surface equal to that of a square with sides equal in length to the radius.

SI conversion factors are listed in Table 1.2.4 alphabetically (adapted from ASTM E380-91a, "Standard Practice for Use of the International System of Units (SI) (the Modernized Metric System)." Conversion factors are written as a number greater than one and less than ten with six or fewer decimal places. This number is followed by the letter E (for exponent), a plus or minus symbol, and two digits which indicate the power of 10 by which the number must be multiplied to obtain the correct value. For example:

$$3.523\ 907\ E - 02 \text{ is } 3.523\ 907 \times 10^{-2} \text{ or } 0.035\ 239\ 07$$

An asterisk (*) after the sixth decimal place indicates that the conversion factor is exact and that all subsequent digits are zero. All other conversion factors have been rounded off.

Table 1.2.2 SI Units

Quantity	Unit	SI symbol	Formula
Base units*			
Length	metre	m	
Mass	kilogram	kg	
Time	second	s	
Electric current	ampere	A	
Thermodynamic temperature	kelvin	K	
Amount of substance	mole	mol	
Luminous intensity	candela	cd	
Supplementary units*			
Plane angle	radian	rad	
Solid angle	steradian	sr	
Derived units*			
Acceleration	metre per second squared		m/s^2
Activity (of a radioactive source)	disintegration per second		(disintegration)/s
Angular acceleration	radian per second squared		rad/s^2
Angular velocity	radian per second		rad/s
Area	square metre		m^2
Density	kilogram per cubic metre		kg/m^3
Electric capacitance	farad	F	A · s/V
Electrical conductance	siemens	S	A/V
Electric field strength	volt per metre		V/m
Electric inductance	henry	H	V · s/A
Electric potential difference	volt	V	W/A
Electric resistance	ohm	Ω	V/A
Electromotive force	volt	V	W/A
Energy	joule	J	N · m
Entropy	joule per kelvin		J/K
Force	newton	N	kg · m/s^2
Frequency	hertz	Hz	1/s
Illuminance	lux	lx	lm/m^2
Luminance	candela per square metre		cd/m^2
Luminous flux	lumen	lm	cd:sr
Magnetic field strength	ampere per metre		A/m
Magnetic flux	weber	Wb	V · s
Magnetic flux density	tesla	T	Wb/m^2
Magnetic potential difference	ampere	A	
Power	watt	W	J/s
Pressure	pascal	Pa	N/m^2
Quantity of electricity	coulomb	C	A · s
Quantity of heat	joule	J	N · m
Radiant intensity	watt per steradian		W/sr
Specific heat capacity	joule per kilogram-kelvin		J/(kg · K)
Stress	pascal	Pa	N/m^2
Thermal conductivity	watt per metre-kelvin		W/(m · K)
Velocity	metre per second		m/s
Viscosity, dynamic	pascal-second		Pa · s
Viscosity, kinematic	square metre per second		m^2/s
Voltage	volt	V	W/A
Volume	cubic metre		m^3
Wave number	reciprocal metre		1/m
Work	joule	J	N · m
Units in use with the SI†			
Time	minute	min	1 min = 60 s
	hour	h	1 h = 60 min = 3,600 s
	day	d	1 d = 24 h = 86,400 s
Plane angle	degree	°	1° = π/180 rad
	minute‡	′	1′ = (1/60)° = (π/10,800) rad
	second‡	″	1″ = (1/60)′ = (π/648,000) rad
Volume	litre	L	1 L = 1 dm^3 = 10^{-3} m^3
Mass	metric ton	t	1 t = 10^3 m^3
	unified atomic mass unit§	u	1 u = 1.660 57 × 10^{-27} kg
Energy	electronvolt§	eV	1 eV = 1.602 19 × 10^{-19} J

* ASTM E380-91a.

† These units are not part of SI, but their use is both so widespread and important that the International Committee for Weights and Measures in 1969 recognized their continued use with the SI (see NIST Spec. Pub. 330).

‡ Use discouraged, except for special fields such as cartography.

§ Values in SI units obtained experimentally. These units are to be used in specialized fields only.

Table 1.2.3   SI Prefixes*

Multiplication factors	Prefix	SI symbol
$1\ 000\ 000\ 000\ 000\ 000\ 000\ 000\ 000 = 10^{24}$	yotta	Y
$1\ 000\ 000\ 000\ 000\ 000\ 000\ 000 = 10^{21}$	zeta	Z
$1\ 000\ 000\ 000\ 000\ 000\ 000 = 10^{18}$	exa	E
$1\ 000\ 000\ 000\ 000\ 000 = 10^{15}$	peta	P
$1\ 000\ 000\ 000\ 000 = 10^{12}$	tera	T
$1\ 000\ 000\ 000 = 10^{9}$	giga	G
$1\ 000\ 000 = 10^{6}$	mega	M
$1\ 000 = 10^{3}$	kilo	k
$100 = 10^{2}$	hecto†	h
$10 = 10^{1}$	deka†	da
$0.1 = 10^{-1}$	deci†	d
$0.01 = 10^{-2}$	centi†	c
$0.001 = 10^{-3}$	milli	m
$0.000\ 001 = 10^{-6}$	micro	μ
$0.000\ 000\ 001 = 10^{-9}$	nano	n
$0.000\ 000\ 000\ 001 = 10^{-12}$	pico	p
$0.000\ 000\ 000\ 000\ 001 = 10^{-15}$	femto	f
$0.000\ 000\ 000\ 000\ 000\ 001 = 10^{-18}$	atto	a
$0.000\ 000\ 000\ 000\ 000\ 000\ 001 = 10^{-21}$	zepto	z
$0.000\ 000\ 000\ 000\ 000\ 000\ 000\ 001 = 10^{-24}$	yocto	y

* ANSI/IEEE Std 268-1992.
† To be avoided where practical.

Table 1.2.4   SI Conversion Factors

To convert from	to	Multiply by
abampere	ampere (A)	$1.000\ 000$*E+01
abcoulomb	coulomb (C)	$1.000\ 000$*E+01
abfarad	farad (F)	$1.000\ 000$*E+09
abhenry	henry (H)	$1.000\ 000$*E−09
abmho	siemens (S)	$1.000\ 000$*E+09
abohm	ohm (Ω)	$1.000\ 000$*E−09
abvolt	volt (V)	$1.000\ 000$*E−08
acre-foot (U.S. survey)^a	metre^3 (m^3)	$1.233\ 489$ E+03
acre (U.S. survey)^a	metre^2 (m^2)	$4.046\ 873$ E+03
ampere, international U.S. (A$_{\text{INT−US}}$)^b	ampere (A)	$9.998\ 43$ E−01
ampere, U.S. legal 1948 (A$_{\text{US−48}}$)	ampere (A)	$1.000\ 008$ E+00
ampere-hour	coulomb (C)	$3.600\ 000$*E+03
angstrom	metre (m)	$1.000\ 000$*E−10
are	metre^2 (m^2)	$1.000\ 000$*E+02
astronomical unit	metre (m)	$1.495\ 98$ E+11
atmosphere (normal)	pascal (Pa)	$1.013\ 25$ E+05
atmosphere (technical = 1 kg$_f$/cm^2)	pascal (Pa)	$9.806\ 650$*E+04
bar	pascal (Pa)	$1.000\ 000$*E+05
barn	metre^2 (m^2)	$1.000\ 000$*E−28
barrel (for crude petroleum, 42 gal)	metre^3 (m^3)	$1.589\ 873$ E−01
board foot	metre^3 (m^3)	$2.359\ 737$ E−03
British thermal unit (International Table)^c	joule (J)	$1.055\ 056$ E+03
British thermal unit (mean)	joule (J)	$1.055\ 87$ E+03
British thermal unit (thermochemical)	joule (J)	$1.054\ 350$ E+03
British thermal unit (39°F)	joule (J)	$1.059\ 67$ E+03
British thermal unit (59°F)	joule (J)	$1.054\ 80$ E+03
British thermal unit (60°F)	joule (J)	$1.054\ 68$ E+03
Btu (thermochemical)/foot^2-second	watt/metre^2 (W/m^2)	$1.134\ 893$ E+04
Btu (thermochemical)/foot^2-minute	watt/metre^2 (W/m^2)	$1.891\ 489$ E+02
Btu (thermochemical)/foot^2-hour	watt/metre^2 (W/m^2)	$3.152\ 481$ E+00
Btu (thermochemical)/inch^2-second	watt/metre^2 (W/m^2)	$1.634\ 246$ E+06
Btu (thermochemical)·in/s·ft^2·°F (k, thermal conductivity)	watt/metre-kelvin (W/m·K)	$5.188\ 732$ E+02
Btu (International Table)·in/s·ft^2·°F (k, thermal conductivity)	watt/metre-kelvin (W/m·K)	$5.192\ 204$ E+02
Btu (thermochemical)·in/h·ft^2·°F (k, thermal conductivity)	watt/metre-kelvin (W/m·K)	$1.441\ 314$ E−01
Btu (International Table)·in/h·ft^2·°F (k, thermal conductivity)	watt/metre-kelvin (W/m·K)	$1.442\ 279$ E−01
Btu (International Table)/ft^2	joule/metre^2 (J/m^2)	$1.135\ 653$ E+04
Btu (thermochemical)/ft^2	joule/metre^2 (J/m^2)	$1.134\ 893$ E+04
Btu (International Table)/h·ft^2·°F (C, thermal conductance)	watt/metre^2-kelvin (W/m^2·K)	$5.678\ 263$ E+00
Btu (thermochemical)/h·ft^2·°F (C, thermal conductance)	watt/metre^2-kelvin (W/m^2·K)	$5.674\ 466$ E+00
Btu (International Table)/pound-mass	joule/kilogram (J/kg)	$2.326\ 000$*E+03

Table 1.2.4   SI Conversion Factors   (Continued)

To convert from	to	Multiply by
Btu (thermochemical)/pound-mass	joule/kilogram (J/kg)	2.324 444 E+03
Btu (International Table)/lbm · °F (c, heat capacity)	joule/kilogram-kelvin (J/kg · K)	4.186 800*E+03
Btu (thermochemical)/lbm · °F (c, heat capacity)	joule/kilogram-kelvin (J/kg · K)	4.184 000 E+03
Btu (International Table)/s · ft² · °F	watt/metre²-kelvin (W/m² · K)	2.044 175 E+04
Btu (thermochemical)/s · ft² · °F	watt/metre²-kelvin (W/m² · K)	2.042 808 E+04
Btu (International Table)/hour	watt (W)	2.930 711 E−01
Btu (thermochemical)/second	watt (W)	1.054 350 E+03
Btu (thermochemical)/minute	watt (W)	1.757 250 E+01
Btu (thermochemical)/hour	watt (W)	2.928 751 E−01
bushel (U.S.)	metre³ (m³)	3.523 907 E−02
calorie (International Table)	joule (J)	4.186 800*E+00
calorie (mean)	joule (J)	4.190 02 E+00
calorie (thermochemical)	joule (J)	4.184 000*E+00
calorie (15°C)	joule (J)	4.185 80 E+00
calorie (20°C)	joule (J)	4.181 90 E+00
calorie (kilogram, International Table)	joule (J)	4.186 800*E+03
calorie (kilogram, mean)	joule (J)	4.190 02 E+03
calorie (kilogram, thermochemical)	joule (J)	4.184 000*E+03
calorie (thermochemical)/centimetre²-minute	watt/metre² (W/m²)	6.973 333 E+02
cal (thermochemical)/cm²	joule/metre² (J/m²)	4.184 000*E+04
cal (thermochemical)/cm² · s	watt/metre² (W/m²)	4.184 000*E+04
cal (thermochemical)/cm · s · °C	watt/metre-kelvin (W/m · K)	4.184 000*E+02
cal (International Table)/g	joule/kilogram (J/kg)	4.186 800*E+03
cal (International Table)/g · °C	joule/kilogram-kelvin (J/kg · K)	4.186 800*E+03
cal (thermochemical)/g	joule/kilogram (J/kg)	4.184 000*E+03
cal (thermochemical)/g · °C	joule/kilogram-kelvin (J/kg · K)	4.184 000*E+03
calorie (thermochemical)/second	watt (W)	4.184 000*E+00
calorie (thermochemical)/minute	watt (W)	6.973 333 E−02
carat (metric)	kilogram (kg)	2.000 000*E−04
centimetre of mercury (0°C)	pascal (Pa)	1.333 22 E+03
centimetre of water (4°C)	pascal (Pa)	9.806 38 E+01
centipoise	pascal-second (Pa · s)	1.000 000*E−03
centistokes	metre²/second (m²/s)	1.000 000*E−06
chain (engineer or ramden)	meter (m)	3.048* E+01
chain (surveyor or gunter)	meter (m)	2.011 684 E+01
circular mil	metre² (m²)	5.067 075 E−10
cord	metre³ (m³)	3.624 556 E+00
coulomb, international U.S. (C_{INT-US})[b]	coulomb (C)	9.998 43 E−01
coulomb, U.S. legal 1948 (C_{US-48})	coulomb (C)	1.000 008 E+00
cup	metre³ (m³)	2.365 882 E−04
curie	bequerel (Bq)	3.700 000*E+10
day (mean solar)	second (s)	8.640 000 E+04
day (sidereal)	second (s)	8.616 409 E+04
degree (angle)	radian (rad)	1.745 329 E−02
degree Celsius	kelvin (K)	$t_K = t_{°C} + 273.15$
degree centigrade	kelvin (K)	$t_K = t_{°C} + 273.15$
degree Fahrenheit	degree Celsius	$t_C = (t_{°F} - 32)/1.8$
degree Fahrenheit	kelvin (K)	$t_K = (t_{°F} + 459.67)/1.8$
deg F · h · ft²/Btu (thermochemical) (R, thermal resistance)	kelvin-metre²/watt (K · m²/W)	1.762 280 E−01
deg F · h · ft²/Btu (International Table) (R, thermal resistance)	kelvin-metre²/watt (K · m²/W)	1.761 102 E−01
degree Rankine	kelvin (K)	$t_K = t_{°R}/1.8$
dram (avoirdupois)	kilogram (kg)	1.771 845 E−03
dram (troy or apothecary)	kilogram (kg)	3.887 934 E−03
dram (U.S. fluid)	kilogram (kg)	3.696 691 E−06
dyne	newton (N)	1.000 000*E−05
dyne-centimetre	newton-metre (N · m)	1.000 000*E−07
dyne-centimetre²	pascal (Pa)	1.000 000*E−01
electron volt	joule (J)	1.602 19 E−19
EMU of capacitance	farad (F)	1.000 000*E+09
EMU of current	ampere (A)	1.000 000*E+01
EMU of electric potential	volt (V)	1.000 000*E−08
EMU of inductance	henry (H)	1.000 000*E−09
EMU of resistance	ohm (Ω)	1.000 000*E−09
ESU of capacitance	farad (F)	1.112 650 E−12
ESU of current	ampere (A)	3.335 6 E−10
ESU of electric potential	volt (V)	2.997 9 E+02
ESU of inductance	henry (H)	8.987 554 E+11

Table 1.2.4   SI Conversion Factors   (Continued)

To convert from	to	Multiply by
ESU of resistance	ohm (Ω)	8.987 554 E+11
erg	joule (J)	1.000 000*E−07
erg/centimetre²-second	watt/metre² (W/m²)	1.000 000*E−03
erg/second	watt (W)	1.000 000*E−07
farad, international U.S. (F_{INT-US})	farad (F)	9.995 05 E−01
faraday (based on carbon 12)	coulomb (C)	9.648 70 E+04
faraday (chemical)	coulomb (C)	9.649 57 E+04
faraday (physical)	coulomb (C)	9.652 19 E+04
fathom (U.S. survey)[a]	metre (m)	1.828 804 E+00
fermi (femtometer)	metre (m)	1.000 000*E−15
fluid ounce (U.S.)	metre³ (m³)	2.957 353 E−05
foot	metre (m)	3.048 000*E−01
foot (U.S. survey)[a]	metre (m)	3.048 006 E−01
foot³/minute	metre³/second (m³/s)	4.719 474 E−04
foot³/second	metre³/second (m³/s)	2.831 685 E−02
foot³ (volume and section modulus)	metre³ (m³)	2.831 685 E−02
foot²	metre² (m²)	9.290 304*E−02
foot⁴ (moment of section)[d]	metre⁴ (m⁴)	8.630 975 E−03
foot/hour	metre/second (m/s)	8.466 667 E−05
foot/minute	metre/second (m/s)	5.080 000 E−03
foot/second	metre/second (m/s)	3.048 000*E−01
foot²/second	metre²/second (m²/s)	9.290 304*E−02
foot of water (39.2°F)	pascal (Pa)	2.988 98 E+03
footcandle	lumen/metre² (lm/m²)	1.076 391 E+01
footcandle	lux (lx)	1.076 391 E+01
footlambert	candela/metre² (cd/m²)	3.426 259 E+00
foot-pound-force	joule (J)	1.355 818 E+00
foot-pound-force/hour	watt (W)	3.766 161 E−04
foot-pound-force/minute	watt (W)	2.259 697 E−02
foot-pound-force/second	watt (W)	1.355 818 E+00
foot-poundal	joule (J)	4.214 011 E−02
ft²/h (thermal diffusivity)	metre²/second (m²/s)	2.580 640*E−05
foot/second²	metre/second² (m/s²)	3.048 000*E−01
free fall, standard	metre/second² (m/s²)	9.806 650*E+00
furlong	metre (m)	2.011 68 *E+02
gal	metre/second² (m/s²)	1.000 000*E−02
gallon (Canadian liquid)	metre³ (m³)	4.546 090 E−03
gallon (U.K. liquid)	metre³ (m³)	4.546 092 E−03
gallon (U.S. dry)	metre³ (m³)	4.404 884 E−03
gallon (U.S. liquid)	metre³ (m³)	3.785 412 E−03
gallon (U.S. liquid)/day	metre³/second (m³/s)	4.381 264 E−08
gallon (U.S. liquid)/minute	metre³/second (m³/s)	6.309 020 E−05
gamma	tesla (T)	1.000 000*E−09
gauss	tesla (T)	1.000 000*E−04
gilbert	ampere-turn	7.957 747 E−01
gill (U.K.)	metre³ (m³)	1.420 654 E−04
gill (U.S.)	metre³ (m³)	1.182 941 E−04
grade	degree (angular)	9.000 000*E−01
grade	radian (rad)	1.570 796 E−02
grain (1/7,000 lbm avoirdupois)	kilogram (kg)	6.479 891*E−05
gram	kilogram (kg)	1.000 000*E−03
gram/centimetre³	kilogram/metre³ (kg/m³)	1.000 000*E+03
gram-force/centimetre²	pascal (Pa)	9.806 650*E+01
hectare	metre² (m²)	1.000 000*E+04
henry, international U.S. (H_{INT-US})	henry (H)	1.000 495 E+00
hogshead (U.S.)	metre³ (m³)	2.384 809 E−01
horsepower (550 ft·lbf/s)	watt (W)	7.456 999 E+02
horsepower (boiler)	watt (W)	9.809 50 E+03
horsepower (electric)	watt (W)	7.460 000*E+02
horsepower (metric)	watt (W)	7.354 99 E+02
horsepower (water)	watt (W)	7.460 43 E+02
horsepower (U.K.)	watt (W)	7.457 0 E+02
hour (mean solar)	second (s)	3.600 000 E+03
hour (sidereal)	second (s)	3.590 170 E+03
hundredweight (long)	kilogram (kg)	5.080 235 E+01
hundredweight (short)	kilogram (kg)	4.535 924 E+01
inch	metre (m)	2.540 000*E−02
inch²	metre² (m²)	6.451 600*E−04
inch³ (volume and section modulus)	metre³ (m³)	1.638 706 E−05
inch³/minute	metre³/second (m³/s)	2.731 177 E−07
inch⁴ (moment of section)[d]	metre⁴ (m⁴)	4.162 314 E−07
inch/second	metre/second (m/s)	2.540 000*E−02
inch of mercury (32°F)	pascal (Pa)	3.386 389 E+03

Table 1.2.4   SI Conversion Factors   (*Continued*)

To convert from	to	Multiply by
inch of mercury (60°F)	pascal (Pa)	3.376 85 E+03
inch of water (39.2°F)	pascal (Pa)	2.490 82 E+02
inch of water (60°F)	pascal (Pa)	2.488 4 E+02
inch/second²	metre/second² (m/s²)	2.540 000*E−02
joule, international U.S. (J_{INT-US})[b]	joule (J)	1.000 182 E+00
joule, U.S. legal 1948 (J_{US-48})	joule (J)	1.000 017 E+00
kayser	1/metre (1/m)	1.000 000*E+02
kelvin	degree Celsius	$t_C = t_K - 273.15$
kilocalorie (thermochemical)/minute	watt (W)	6.973 333 E+01
kilocalorie (thermochemical)/second	watt (W)	4.184 000*E+03
kilogram-force (kgf)	newton (N)	9.806 650*E+00
kilogram-force-metre	newton-metre (N·m)	9.806 650*E+00
kilogram-force-second²/metre (mass)	kilogram (kg)	9.806 650*E+00
kilogram-force/centimetre²	pascal (Pa)	9.806 650*E+04
kilogram-force/metre³	pascal (Pa)	9.806 650*E+00
kilogram-force/millimetre²	pascal (Pa)	9.806 650*E+06
kilogram-mass	kilogram (kg)	1.000 000 E+00
kilometre/hour	metre/second (m/s)	2.777 778 E−01
kilopond	newton (N)	9.806 650*E+00
kilowatt hour	joule (J)	3.600 000*E+06
kilowatt hour, international U.S. (kWh$_{INT-US}$)[b]	joule (J)	3.600 655 E+06
kilowatt hour, U.S. legal 1948 (kWh$_{US-48}$)	joule (J)	3.600 061 E+06
kip (1,000 lbf)	newton (N)	4.448 222 E+03
kip/inch² (ksi)	pascal (Pa)	6.894 757 E+06
knot (international)	metre/second (m/s)	5.144 444 E−01
lambert	candela/metre² (cd/m²)	3.183 099 E+03
langley	joule/metre² (J/m²)	4.184 000*E+04
league, nautical (international and U.S.)	metre (m)	5.556 000*E+03
league (U.S. survey)[a]	metre (m)	4.828 042 E+03
league, nautical (U.K.)	metre (m)	5.559 552*E+03
light year	metre (m)	9.460 55 E+15
link (engineer or ramden)	metre (m)	3.048* E−01
link (surveyor or gunter)	metre (m)	2.011 68* E−01
litre[e]	metre³ (m³)	1.000 000*E−03
lux	lumen/metre² (lm/m²)	1.000 000*E+00
maxwell	weber (Wb)	1.000 000*E−08
mho	siemens (S)	1.000 000*E+00
microinch	metre (m)	2.540 000*E−08
micron (micrometre)	metre (m)	1.000 000*E−06
mil	metre (m)	2.540 000*E−05
mile, nautical (international and U.S.)	metre (m)	1.852 000*E+03
mile, nautical (U.K.)	metre (m)	1.853 184*E+03
mile (international)	metre (m)	1.609 344*E+03
mile (U.S. survey)[a]	metre (m)	1.609 347 E+03
mile² (international)	metre² (m²)	2.589 988 E+06
mile² (U.S. survey)[a]	metre² (m²)	2.589 998 E+06
mile/hour (international)	metre/second (m/s)	4.470 400*E−01
mile/hour (international)	kilometre/hour	1.609 344*E+00
millimetre of mercury (0°C)	pascal (Pa)	1.333 224 E+02
minute (angle)	radian (rad)	2.908 882 E−04
minute (mean solar)	second (s)	6.000 000 E+01
minute (sidereal)	second (s)	5.983 617 E+01
month (mean calendar)	second (s)	2.268 000 E+06
oersted	ampere/metre (A/m)	7.957 747 E+01
ohm, international U.S. (Ω_{INT-US})	ohm (Ω)	1.000 495 E+00
ohm-centimetre	ohm-metre (Ω·m)	1.000 000*E−02
ounce-force (avoirdupois)	newton (N)	2.780 139 E−01
ounce-force-inch	newton-metre (N·m)	7.061 552 E−03
ounce-mass (avoirdupois)	kilogram (kg)	2.834 952 E−02
ounce-mass (troy or apothecary)	kilogram (kg)	3.110 348 E−02
ounce-mass/yard²	kilogram/metre² (kg/m²)	3.390 575 E−02
ounce (avoirdupois)(mass)/inch³	kilogram/metre³ (kg/m³)	1.729 994 E+03
ounce (U.K. fluid)	metre³ (m³)	2.841 307 E−05
ounce (U.S. fluid)	metre³ (m³)	2.957 353 E−05
parsec	metre (m)	3.083 74 E+16
peck (U.S.)	metre³ (m³)	8.809 768 E−03
pennyweight	kilogram (kg)	1.555 174 E−03
perm (0°C)	kilogram/pascal-second-metre² (kg/Pa·s·m²)	5.721 35 E−11
perm (23°C)	kilogram/pascal-second-metre² (kg/Pa·s·m²)	5.745 25 E−11

Table 1.2.4   SI Conversion Factors   (Continued)

To convert from	to	Multiply by
perm-inch (0°C)	kilogram/pascal-second-metre (kg/Pa·s·m)	1.453 22 E−12
perm-inch (23°C)	kilogram/pascal-second-metre (kg/Pa·s·m)	1.459 29 E−12
phot	lumen/metre² (lm/m²)	1.000 000*E+04
pica (printer's)	metre (m)	4.217 518 E−03
pint (U.S. dry)	metre³ (m³)	5.506 105 E−04
pint (U.S. liquid)	metre³ (m³)	4.731 765 E−04
point (printer's)	metre	3.514 598*E−04
poise (absolute viscosity)	pascal-second (Pa·s)	1.000 000*E−01
poundal	newton (N)	1.382 550 E−01
poundal-foot²	pascal (Pa)	1.488 164 E+00
poundal-second/foot²	pascal-second (Pa·s)	1.488 164 E+00
pound-force (lbf avoirdupois)	newton (N)	4.448 222 E+00
pound-force-inch	newton-metre (N·m)	1.129 848 E−01
pound-force-foot	newton-metre (N·m)	1.355 818 E+00
pound-force-foot/inch	newton-metre/metre (N·m/m)	5.337 866 E+01
pound-force-inch/inch	newton-metre/metre (N·m/m)	4.448 222 E+00
pound-force/inch	newton-metre (N/m)	1.751 268 E+02
pound-force/foot	newton-metre (N/m)	1.459 390 E+01
pound-force/foot²	pascal (Pa)	4.788 026 E+01
pound-force/inch² (psi)	pascal (Pa)	6.894 757 E+03
pound-force-second/foot²	pascal-second (Pa·s)	4.788 026 E+01
pound-mass (lbm avoirdupois)	kilogram (kg)	4.535 924 E−01
pound-mass (troy or apothecary)	kilogram (kg)	3.732 417 E−01
pound-mass-foot² (moment of inertia)	kilogram-metre² (kg·m²)	4.214 011 E−02
pound-mass-inch² (moment of inertia)	kilogram-metre² (kg·m²)	2.926 397 E−04
pound-mass-foot²	kilogram/metre² (kg/m²)	4.882 428 E+00
pound-mass/second	kilogram/second (kg/s)	4.535 924 E−01
pound-mass/minute	kilogram/second (kg/s)	7.559 873 E−03
pound-mass/foot³	kilogram/metre³ (kg/m³)	1.601 846 E+01
pound-mass/inch³	kilogram/metre³ (kg/m³)	2.767 990 E+04
pound-mass/gallon (U.K. liquid)	kilogram/metre³ (kg/m³)	9.977 633 E+01
pound-mass/gallon (U.S. liquid)	kilogram/metre³ (kg/m³)	1.198 264 E+02
pound-mass/foot-second	pascal-second (Pa·s)	1.488 164 E+00
quart (U.S. dry)	metre³ (m³)	1.101 221 E−03
quart (U.S. liquid)	metre³ (m³)	9.463 529 E−04
rad (radiation dose absorbed)	gray (Gy)	1.000 000*E−02
rem (dose equivalent)	sievert (Sv)	1.000 000*E−02
rhe	metre²/newton-second (m²/N·s)	1.000 000*E+01
rod (U.S. survey)[a]	metre (m)	5.029 210 E+00
roentgen	coulomb/kilogram (C/kg)	2.579 760*E−04
second (angle)	radian (rad)	4.848 137 E−06
second (sidereal)	second (s)	9.972 696 E−01
section (U.S. survey)[a]	metre² (m²)	2.589 998 E+06
shake	second (s)	1.000 000*E−08
slug	kilogram (kg)	1.459 390 E+01
slug/foot³	kilogram/metre³ (kg/m³)	5.153 788 E+02
slug/foot-second	pascal-second (Pa·s)	4.788 026 E+01
statampere	ampere (A)	3.335 640 E−10
statcoulomb	coulomb (C)	3.335 640 E−10
statfarad	farad (F)	1.112 650 E−12
stathenry	henry (H)	8.987 554 E+11
statmho	siemens (S)	1.112 650 E−12
statohm	ohm (Ω)	8.987 554 E+11
statvolt	volt (V)	2.997 925 E+02
stere	metre³ (m³)	1.000 000*E+00
stilb	candela/metre² (cd/m²)	1.000 000*E+04
stokes (kinematic viscosity)	metre²/second (m²/s)	1.000 000*E−04
tablespoon	metre³ (m³)	1.478 676 E−05
teaspoon	metre³ (m³)	4.928 922 E−06
ton (assay)	kilogram (kg)	2.916 667 E−02
ton (long, 2,240 lbm)	kilogram (kg)	1.016 047 E+03
ton (metric)	kilogram (kg)	1.000 000*E+03
ton (nuclear equivalent of TNT)	joule (J)	4.20 E+09
ton (register)	metre³ (m³)	2.831 685 E+00
ton (short, 2,000 lbm)	kilogram (kg)	9.071 847 E+02
ton (short, mass)/hour	kilogram/second (kg/s)	2.519 958 E−01
ton (long, mass)/yard³	kilogram/metre³ (kg/m³)	1.328 939 E+03
tonne	kilogram (kg)	1.000 000*E+03
torr (mm Hg, 0°C)	pascal (Pa)	1.333 22 E+02
township (U.S. survey)[a]	metre² (m²)	9.323 994 E+07
unit pole	weber (Wb)	1.256 637 E−07

Table 1.2.4 SI Conversion Factors *(Continued)*

To convert from	to	Multiply by
volt, international U.S. $(V_{INT-US})^b$	volt (V)	1.000 338 E+00
volt, U.S. legal 1948 (V_{US-48})	volt (V)	1.000 008 E+00
watt, international U.S. $(W_{INT-US})^b$	watt (W)	1.000 182 E+00
watt, U.S. legal 1948 (W_{US-48})	watt (W)	1.000 017 E+00
watt/centimetre^2	watt/metre^2 (W/m^2)	1.000 000*E+04
watt-hour	joule (J)	3.600 000*E+03
watt-second	joule (J)	1.000 000*E+00
yard	metre (m)	9.144 000*E−01
yard^2	metre^2 (m^2)	8.361 274 E−01
yard^3	metre^3 (m^3)	7.645 549 E−01
yard^3/minute	metre^3/second (m^3/s)	1.274 258 E−02
year (calendar)	second (s)	3.153 600 E+07
year (sidereal)	second (s)	3.155 815 E+07
year (tropical)	second (s)	3.155 693 E+07

^{a} Based on the U.S. survey foot (1 ft = 1,200/3,937 m).
^{b} In 1948 a new international agreement was reached on absolute electrical units, which changed the value of the volt used in this country by about 300 parts per million. Again in 1969 a new base of reference was internationally adopted making a further change of 8.4 parts per million. These changes (and also changes in ampere, joule, watt, coulomb) require careful terminology and conversion factors for exact use of old information. Terms used in this guide are:
Volt as used prior to January 1948 — volt, international U.S. (V_{INT-US})
Volt as used between January 1948 and January 1969 — volt, U.S. legal 1948 (V_{INT-48})
Volt as used since January 1969 — volt (V)
Identical treatment is given the ampere, coulomb, watt, and joule.
^{c} This value was adopted in 1956. Some of the older International Tables use the value 1.055 04 E+03. The exact conversion factor is 1.055 055 852 62*E+03.
^{d} Moment of inertia of a plane section about a specified axis.
^{e} In 1964, the General Conference on Weights and Measures adopted the name "litre" as a special name for the cubic decimetre. Prior to this decision the litre differed slightly (previous value, 1.000028 dm^3), and in expression of precision, volume measurement, this fact must be kept in mind.

SYSTEMS OF UNITS

The principal units of interest to mechanical engineers can be derived from three base units which are considered to be dimensionally independent of each other. The British "gravitational system," in common use in the United States, uses units of **length, force,** and **time** as base units and is also called the "foot-pound-second system." The metric system, on the other hand, is based on the meter, kilogram, and second, units of **length, mass,** and **time,** and is often designated as the "MKS system." During the nineteenth century a metric "gravitational system," based on a kilogram-force (also called a "kilopond") came into general use. With the development of the International System of Units (SI), based as it is on the original metric system for mechanical units, and the general requirements by members of the European Community that only SI units be used, it is anticipated that the kilogram-force will fall into disuse to be replaced by the newton, the SI unit of force. Table 1.2.5 gives the base units of four systems with the corresponding derived unit given in parentheses.

In the definitions given below, the "standard kilogram body" refers to the international kilogram prototype, a platinum-iridium cylinder kept in the International Bureau of Weights and Measures in Sèvres, just outside Paris. The "standard pound body" is related to the kilogram by a precise numerical factor: 1 lb = 0.453 592 37 kg. This new "unified" pound has replaced the somewhat smaller Imperial pound of the United Kingdom and the slightly larger pound of the United States (see NBS Spec. Pub. 447). The "standard locality" means sea level, 45° latitude, or more strictly any locality in which the acceleration due to gravity has the value 9.80 665 m/s^2 = 32.1740 ft/s^2, which may be called the **standard acceleration** (Table 1.2.6).

The **pound force** is the force required to support the standard pound body against gravity, *in vacuo,* in the standard locality; or, it is the force which, if applied to the standard pound body, supposed free to move, would give that body the "standard acceleration." The word *pound* is used for the unit of both force and mass and consequently is ambiguous. To avoid uncertainty, it is desirable to call the units "pound force" and "pound mass," respectively. The slug has been defined as that mass which will accelerate at 1 ft/s^2 when acted upon by a one pound force. It is therefore equal to 32.1740 pound-mass.

The **kilogram force** is the force required to support the standard kilogram against gravity, *in vacuo,* in the standard locality; or, it is the force which, if applied to the standard kilogram body, supposed free to move, would give that body the "standard acceleration." The word *kilogram* is used for the unit of both force and mass and consequently is ambiguous. It is for this reason that the General Conference on Weights and Measures declared (in 1901) that the kilogram was the unit of mass, a concept incorporated into SI when it was formally approved in 1960.

The **dyne** is the force which, if applied to the standard gram body, would give that body an acceleration of 1 cm/s^2; i.e., 1 dyne = 1/980.665 of a gram force.

The **newton** is that force which will impart to a 1-kilogram mass an acceleration of 1 m/s^2.

Table 1.2.5 Systems of Units

Quantity	Dimensions of units in terms of L/M/F/T	British "gravitational system"	Metric "gravitational system"	CGS system	SI system
Length	L	1 ft	1 m	1 cm	1 m
Mass	M	(1 slug)		1 g	1 kg
Force	F	1 lb	1 kg	(1 dyne)	(1 N)
Time	T	1 s	1 s	1 s	1 s

Table 1.2.6 Acceleration of Gravity

Latitude, deg	g			Latitude, deg	g		
	m/s²	ft/s²	g/g^0		m/s²	ft/s²	g/g^0
0	9.780	32.088	0.9973	50	9.811	32.187	1.0004
10	9.782	32.093	0.9975	60	9.819	32.215	1.0013
20	9.786	32.108	0.9979	70	9.826	32.238	1.0020
30	9.793	32.130	0.9986	80	9.831	32.253	1.0024
40	9.802	32.158	0.9995	90	9.832	32.258	1.0026

NOTE: Correction for altitude above sea level: -3 mm/s² for each 1,000 m; -0.003 ft/s² for each 1,000 ft.
SOURCE: U.S. Coast and Geodetic Survey, 1912.

TEMPERATURE

The SI unit for thermodynamic temperature is the **kelvin, K,** which is the fraction 1/273.16 of the thermodynamic temperature of the triple point of water. Thus 273.16 K is the **fixed (base) point** on the **kelvin scale.**

Another unit used for the measurement of temperature is degrees **Celsius** (formerly **centigrade**), °C. The relation between a thermodynamic temperature T and a Celsius temperature t is

$$t = T - 273.16 \text{ K}$$

Thus the unit Celsius degree is equal to the unit kelvin, and a difference of temperature would be the same on either scale.

In the USCS temperature is measured in degrees **Fahrenheit, F.** The relation between the Celsius and the Fahrenheit scales is

$$t_{°C} = (t_{°F} - 32)/1.8$$

(For temperature-conversion tables, see Sec. 4.)

TERRESTRIAL GRAVITY

Standard acceleration of gravity is $g^0 = 9.80665$ m per sec per sec, or 32.1740 ft per sec per sec. This value g^0 is assumed to be the value of g at sea level and latitude 45°.

MOHS SCALE OF HARDNESS

This scale is an arbitrary one which is used to describe the hardness of several mineral substances on a scale of 1 through 10 (Table 1.2.7). The given number indicates a higher relative hardness compared with that of substances below it; and a lower relative hardness than those above it. For example, an unknown substance is scratched by quartz, but it, in turn, scratches feldspar. The unknown has a hardness of between 6 and 7 on the Mohs scale.

Table 1.2.7 Mohs Scale of Hardness

1. Talc	5. Apatite	8. Topaz
2. Gypsum	6. Feldspar	9. Sapphire
3. Calc-spar	7. Quartz	10. Diamond
4. Fluorspar		

TIME

Kinds of Time Three kinds of time are recognized by astronomers: sidereal, apparent solar, and mean solar time. The **sidereal day** is the interval between two consecutive transits of some fixed celestial object across any given meridian, or it is the interval required by the earth to make one complete revolution on its axis. The interval is constant, but it is inconvenient as a time unit because the noon of the sidereal day occurs at all hours of the day and night. The **apparent solar day** is the interval between two consecutive transits of the sun across any given meridian. On account of the variable distance between the sun and earth, the variable speed of the earth in its orbit, the effect of the moon, etc., this interval is not constant and consequently cannot be kept by any simple mechanisms, such as clocks or watches. To overcome the objection noted above, the **mean solar day** was devised. The mean solar day is

the length of the average apparent solar day. Like the sidereal day it is constant, and like the apparent solar day its noon always occurs at approximately the same time of day. By international agreement, beginning Jan. 1, 1925, the astronomical day, like the civil day, is from midnight to midnight. The hours of the astronomical day run from 0 to 24, and the hours of the civil day usually run from 0 to 12 A.M. and 0 to 12 P.M. In some countries the hours of the civil day also run from 0 to 24.

The Year Three different kinds of year are used: the sidereal, the tropical, and the anomalistic. The **sidereal year** is the time taken by the earth to complete one revolution around the sun from a given star to the same star again. Its length is 365 days, 6 hours, 9 minutes, and 9 seconds. The **tropical year** is the time included between two successive passages of the vernal equinox by the sun, and since the equinox moves westward 50.2 seconds of arc a year, the tropical year is shorter by 20 minutes 23 seconds in time than the sidereal year. As the seasons depend upon the earth's position with respect to the equinox, the tropical year is the year of civil reckoning. The **anomalistic year** is the interval between two successive passages of the perihelion, viz., the time of the earth's nearest approach to the sun. The anomalistic year is used only in special calculations in astronomy.

The Second Although the second is ordinarily defined as 1/86,400 of the mean solar day, this is not sufficiently precise for many scientific purposes. Scientists have adopted more precise definitions for specific purposes: in 1956, one in terms of the length of the tropical year 1900 and, more recently, in 1967, one in terms of a specific atomic frequency.

Frequency is the reciprocal of time for 1 cycle; the unit of frequency is the **hertz** (Hz), defined as 1 cycle/s.

The Calendar The **Gregorian calendar,** now used in most of the civilized world, was adopted in Catholic countries of Europe in 1582 and in Great Britain and her colonies Jan. 1, 1752. The average length of the Gregorian calendar year is 365¼ − ³⁄₄₀₀ days, or 365.2425 days. This is equivalent to 365 days, 5 hours, 49 minutes, 12 seconds. The length of the tropical year is 365.2422 days, or 365 days, 5 hours, 48 minutes, 46 seconds. Thus the Gregorian calendar year is longer than the tropical year by 0.0003 day, or 26 seconds. This difference amounts to 1 day in slightly more than 3,300 years and can properly be neglected.

Standard Time Prior to 1883, each city of the United States had its own time, which was determined by the time of passage of the sun across the local meridian. A system of standard time had been used since its first adoption by the railroads in 1883 but was first legalized on Mar. 19, 1918, when Congress directed the Interstate Commerce Commission to establish limits of the standard time zones. Congress took no further steps until the **Uniform Time Act of 1966** was enacted, followed with an amendment in 1972. This legislation, referred to as "the Act," transferred the regulation and enforcement of the law to the Department of Transportation.

By the legislation of 1918, with some modifications by the Act, the contiguous United States is divided into four **time zones,** each of which, theoretically, was to span 15 degrees of longitude. The first, the **Eastern** zone, extends from the Atlantic westward to include most of Michigan and Indiana, the eastern parts of Kentucky and Tennessee, Georgia, and Florida, except the west half of the panhandle. **Eastern standard time** is based upon the mean solar time of the 75th meridian west of Greenwich, and is 5 hours slower than **Greenwich Mean Time (GMT).** (See also discussion of UTC below.) The second or **Central zone** extends westward to include most of North Dakota, about half of South Dakota and Nebraska, most of Kansas, Oklahoma, and all but the two most westerly counties of Texas. **Central standard time** is based upon the mean solar time of the 90th meridian west of Greenwich, and is 6 hours slower than GMT. The third or **Mountain zone** extends westward to include Montana, most of Idaho, one county of Oregon, Utah, and Arizona. **Mountain standard time** is based upon the mean solar time of the 105th meridian west of Greenwich, and is 7 hours slower than GMT. The fourth or **Pacific zone** includes all of the remaining 48 contiguous states. **Pacific standard time** is based on the mean solar time of the 120th meridian west of Greenwich, and is 8 hours slower than GMT. Exact locations of boundaries may be obtained from the Department of Transportation.

In addition to the above four zones there are four others that apply to the noncontiguous states and islands. The most easterly is the **Atlantic zone,** which includes Puerto Rico and the Virgin Islands, where the time is 4 hours slower than GMT. Eastern standard time is used in the Panama Canal strip. To the west of the Pacific time zone there are the **Yukon,** the **Alaska-Hawaii,** and **Bering zones** where the times are, respectively, 9, 10, and 11 hours slower than GMT. The system of standard time has been adopted in all civilized countries and is used by ships on the high seas.

The Act directs that from the first Sunday in April to the last Sunday in October, the time in each zone is to be advanced one hour for advanced time or daylight saving time (DST). However, any state-by-state enactment may exempt the entire state from using advanced time. By this provision Arizona and Hawaii do not observe advanced time (as of 1973). By the 1972 amendment to the Act, a state split by a time-zone boundary may exempt from using advanced time all that part which is in one zone without affecting the rest of the state. By this amendment, 80 counties of Indiana in the Eastern zone are exempt from using advanced time, while 6 counties in the northwest corner and 6 counties in the southwest, which are in Central zone, do observe advanced time.

Pursuant to its assignment of carrying out the Act, the Department of Transportation has stipulated that municipalities located on the boundary between the Eastern and Central zones are in the Central zone; those on the boundary between the Central and Mountain zones are in the Mountain zone (except that Murdo, SD, is in the Central zone); those on the boundary between Mountain and Pacific time zones are in the Mountain zone. In such places, when the time is given, it should be specified as Central, Mountain, etc.

Standard Time Signals   The National Institute of Standards and Technology broadcasts time signals from **station WWV,** Ft. Collins, CO, and from **station WWVH,** near Kekaha, Kaui, HI. The broadcasts by WWV are on radio carrier frequencies of 2.5, 5, 10, 15, and 20 MHz, while those by WWVH are on radio carrier frequencies of 2.5, 5, 10, and 15 MHz. Effective Jan. 1, 1975, time announcements by both WWV and WWVH are referred to as **Coordinated Universal Time, UTC,** the international coordinated time scale used around the world for most timekeeping purposes. UTC is generated by reference to International Atomic Time (TAI), which is determined by the Bureau International de l'Heure on the basis of atomic clocks operating in various establishments in accordance with the definition of the second. Since the difference between UTC and TAI is defined to be a whole number of seconds, a "leap second" is periodically added to or subtracted from UTC to take into account variations in the rotation of the earth. Time (i.e., clock time) is given in terms of 0 to 24 hours a day, starting with 0000 at midnight at Greenwich zero longitude. The beginning of each 0.8-second-long audio tone marks the end of an announced time interval. For example, at 2:15 P.M., UTC, the voice announcement would be: "At the tone fourteen hours fifteen minutes Coordinated Universal Time," given during the last 7.5 seconds of each minute. The tone markers from both stations are given simultaneously, but owing to propagation interferences may not be received simultaneously.

Beginning 1 minute after the hour, a 600-Hz signal is broadcast for about 45 s. At 2 min after the hour, the standard musical pitch of 440 Hz is broadcast for about 45 s. For the remaining 57 min of the hour, alternating tones of 600 and 500 Hz are broadcast for the first 45 s of each minute (see NIST Spec. Pub. 432). The time signal can also be received via long-distance telephone service from Ft. Collins. In addition to providing the musical pitch, these tone signals may be of use as markers for automated recorders and other such devices.

DENSITY AND RELATIVE DENSITY

Density of a body is its mass per unit volume. With SI units densities are in kilograms per cubic meter. However, giving densities in grams per cubic centimeter has been common. With the USCS, densities are given in pounds per mass cubic foot.

Table 1.2.8   Relative Densities at 60°/60°F Corresponding to Degrees API and Weights per U.S. Gallon at 60°F

$$\left(\text{Calculated from the formula, relative density} = \frac{141.5}{131.5 + \text{deg API}}\right)$$

Degrees API	Relative density	Lb per U.S. gallon	Degrees API	Relative density	Lb per U.S. gallon
10	1.0000	8.328	56	0.7547	6.283
11	0.9930	8.270	57	0.7507	6.249
12	0.9861	8.212	58	0.7467	6.216
13	0.9792	8.155	59	0.7428	6.184
14	0.9725	8.099	60	0.7389	6.151
15	0.9659	8.044	61	0.7351	6.119
16	0.9593	7.989	62	0.7313	6.087
17	0.9529	7.935	63	0.7275	6.056
18	0.9465	7.882	64	0.7238	6.025
19	0.9402	7.830	65	0.7201	5.994
20	0.9340	7.778	66	0.7165	5.964
21	0.9279	7.727	67	0.7128	5.934
22	0.9218	7.676	68	0.7093	5.904
23	0.9159	7.627	69	0.7057	5.874
24	0.9100	7.578	70	0.7022	5.845
25	0.9042	7.529	71	0.6988	5.817
26	0.8984	7.481	72	0.6953	5.788
27	0.8927	7.434	73	0.6919	5.759
28	0.8871	7.387	74	0.6886	5.731
29	0.8816	7.341	75	0.6852	5.703
30	0.8762	7.296	76	0.6819	5.676
31	0.8708	7.251	77	0.6787	5.649
32	0.8654	7.206	78	0.6754	5.622
33	0.8602	7.163	79	0.6722	5.595
34	0.8550	7.119	80	0.6690	5.568
35	0.8498	7.076	81	0.6659	5.542
36	0.8448	7.034	82	0.6628	5.516
37	0.8398	6.993	83	0.6597	5.491
38	0.8348	6.951	84	0.6566	5.465
39	0.8299	6.910	85	0.6536	5.440
40	0.8251	6.870	86	0.6506	5.415
41	0.8203	6.830	87	0.6476	5.390
42	0.8155	6.790	88	0.6446	5.365
43	0.8109	6.752	89	0.6417	5.341
44	0.8063	6.713	90	0.6388	5.316
45	0.8017	6.675	91	0.6360	5.293
46	0.7972	6.637	92	0.6331	5.269
47	0.7927	6.600	93	0.6303	5.246
48	0.7883	6.563	94	0.6275	5.222
49	0.7839	6.526	95	0.6247	5.199
50	0.7796	6.490	96	0.6220	5.176
51	0.7753	6.455	97	0.6193	5.154
52	0.7711	6.420	98	0.6166	5.131
53	0.7669	6.385	99	0.6139	5.109
54	0.7628	6.350	100	0.6112	5.086
55	0.7587	6.316			

NOTE: The weights in this table are weights in air at 60°F with humidity 50 percent and pressure 760 mm.

Table 1.2.9   Relative Densities at 60°/60°F Corresponding to Degrees Baumé for Liquids Lighter than Water and Weights per U.S. Gallon at 60°F

$$\left(\text{Calculated from the formula, relative density } \frac{60°}{60°}\text{F} = \frac{140}{130 + \text{deg Baumé}}\right)$$

Degrees Baumé	Relative density	Lb per gallon	Degrees Baumé	Relative density	Lb per gallon
10.0	1.0000	8.328	56.0	0.7527	6.266
11.0	0.9929	8.269	57.0	0.7487	6.233
12.0	0.9859	8.211	58.0	0.7447	6.199
13.0	0.9790	8.153	59.0	0.7407	6.166
14.0	0.9722	8.096	60.0	0.7368	6.134
15.0	0.9655	8.041	61.0	0.7330	6.102
16.0	0.9589	7.986	62.0	0.7292	6.070
17.0	0.9524	7.931	63.0	0.7254	6.038
18.0	0.9459	7.877	64.0	0.7216	6.007
19.0	0.9396	7.825	65.0	0.7179	5.976
20.0	0.9333	7.772	66.0	0.7143	5.946
21.0	0.9272	7.721	67.0	0.7107	5.916
22.0	0.9211	7.670	68.0	0.7071	5.886
23.0	0.9150	7.620	69.0	0.7035	5.856
24.0	0.9091	7.570	70.0	0.7000	5.827
25.0	0.9032	7.522	71.0	0.6965	5.798
26.0	0.8974	7.473	72.0	0.6931	5.769
27.0	0.8917	7.425	73.0	0.6897	5.741
28.0	0.8861	7.378	74.0	0.6863	5.712
29.0	0.8805	7.332	75.0	0.6829	5.685
30.0	0.8750	7.286	76.0	0.6796	5.657
31.0	0.8696	7.241	77.0	0.6763	5.629
32.0	0.8642	7.196	78.0	0.6731	5.602
33.0	0.8589	7.152	79.0	0.6699	5.576
34.0	0.8537	7.108	80.0	0.6667	5.549
35.0	0.8485	7.065	81.0	0.6635	5.522
36.0	0.8434	7.022	82.0	0.6604	5.497
37.0	0.8383	6.980	83.0	0.6573	5.471
38.0	0.8333	6.939	84.0	0.6542	5.445
39.0	0.8284	6.898	85.0	0.6512	5.420
40.0	0.8235	6.857	86.0	0.6482	5.395
41.0	0.8187	6.817	87.0	0.6452	5.370
42.0	0.8140	6.777	88.0	0.6422	5.345
43.0	0.8092	6.738	89.0	0.6393	5.320
44.0	0.8046	6.699	90.0	0.6364	5.296
45.0	0.8000	6.661	91.0	0.6335	5.272
46.0	0.7955	6.623	92.0	0.6306	5.248
47.0	0.7910	6.586	93.0	0.6278	5.225
48.0	0.7865	6.548	94.0	0.6250	5.201
49.0	0.7821	6.511	95.0	0.6222	5.178
50.0	0.7778	6.476	96.0	0.6195	5.155
51.0	0.7735	6.440	97.0	0.6167	5.132
52.0	0.7692	6.404	98.0	0.6140	5.100
53.0	0.7650	6.369	99.0	0.6114	5.088
54.0	0.7609	6.334	100.0	0.6087	5.066
55.0	0.7568	6.300			

Relative density is the ratio of the density of one substance to that of a second (or reference) substance, both at some specified temperature. Use of the earlier term *specific gravity* for this quantity is discouraged. For solids and liquids water is almost universally used as the reference substance. Physicists use a reference temperature of 4°C (= 39.2°F); U.S. engineers commonly use 60°F. With the introduction of SI units, it may be found desirable to use 59°F, since 59°F and 15°C are equivalents.

For gases, relative density is generally the ratio of the density of the gas to that of air, both at the same temperature, pressure, and dryness (as regards water vapor). Because equal numbers of moles of gases occupy equal volumes, the ratio of the molecular weight of the gas to that of air may be used as the relative density of the gas. When this is done, the molecular weight of air may be taken as 28.9644.

The relative density of liquids is usually measured by means of a **hydrometer.** In addition to a scale reading in relative density as defined above, other arbitrary scales for hydrometers are used in various trades and industries. The most common of these are the **API** and **Baumé.** The API (American Petroleum Institute) scale is approved by the American Petroleum Institute, the ASTM, the U.S. Bureau of Mines, and the National Bureau of Standards and is recommended for exclusive use in the U.S. petroleum industry, superseding the Baumé scale for liquids lighter than water. The relation between **API degrees** and relative density (see Table 1.2.8) is expressed by the following equation:

$$\text{Degrees API} = \frac{141.5}{\text{rel dens } 60°/60°\text{F}} - 131.5$$

The relative densities corresponding to the indications of the **Baumé hydrometer** are given in Tables 1.2.9 and 1.2.10.

Table 1.2.10   Relative Densities at 60°/60°F Corresponding to Degrees Baumé for Liquids Heavier than Water

$$\left(\text{Calculated from the formula, relative density } \frac{60°}{60°}\text{F} = \frac{145}{145 - \text{deg Baumé}}\right)$$

Degrees Baumé	Relative density	Degrees Baumé	Relative density	Degrees Baumé	Relative density
0	1.0000	24	1.1983	48	1.4948
1	1.0069	25	1.2083	49	1.5104
2	1.0140	26	1.2185	50	1.5263
3	1.0211	27	1.2288	51	1.5426
4	1.0284	28	1.2393	52	1.5591
5	1.0357	29	1.2500	53	1.5761
6	1.0432	30	1.2609	54	1.5934
7	1.0507	31	1.2719	55	1.6111
8	1.0584	32	1.2832	56	1.6292
9	1.0662	33	1.2946	57	1.6477
10	1.0741	34	1.3063	58	1.6667
11	1.0821	35	1.3182	59	1.6860
12	1.0902	36	1.3303	60	1.7059
13	1.0985	37	1.3426	61	1.7262
14	1.1069	38	1.3551	62	1.7470
15	1.1154	39	1.3679	63	1.7683
16	1.1240	40	1.3810	64	1.7901
17	1.1328	41	1.3942	65	1.8125
18	1.1417	42	1.4078	66	1.8354
19	1.1508	43	1.4216	67	1.8590
20	1.1600	44	1.4356	68	1.8831
21	1.1694	45	1.4500	69	1.9079
22	1.1789	46	1.4646	70	1.9333
23	1.1885	47	1.4796		

CONVERSION AND EQUIVALENCY TABLES

Note for Use of Conversion Tables (Tables 1.2.11 through 1.2.34)

Subscripts after any figure, 0_s, 9_s, etc., mean that that figure is to be repeated the indicated number of times.

Table 1.2.11 Length Equivalents

Centimetres	Inches	Feet	Yards	Metres	Chains	Kilometres	Miles
1	0.3937	0.03281	0.01094	0.01	0.0_34971	10^{-5}	0.0_56214
2.540	1	0.08333	0.02778	0.0254	0.001263	0.0_4254	0.0_41578
30.48	12	1	0.3333	0.3048	0.01515	0.0_33048	0.0_31894
91.44	36	3	1	0.9144	0.04545	0.0_39144	0.0_55682
100	39.37	3.281	1.0936	1	0.04971	0.001	0.0_36214
2012	792	66	22	20.12	1	0.02012	0.0125
100000	39370	3281	1093.6	1000	49.71	1	0.6214
160934	63360	5280	1760	1609	80	1.609	1

(As used by metrology laboratories for precise measurements, including measurements of surface texture)*

Angstrom units Å	Surface texture (U.S.), microinch μin	Light bands,† monochromatic helium light count‡	Surface texture foreign, μm	Precision measurements, § 0.0001 in	Close-tolerance measurements, 0.001 in (mils)	Metric unit, mm	USCS unit, in
1	0.003937	0.0003404	0.0001	0.0_43937	0.0_53937	0.0_61	0.0_83937
254	1	0.086	0.0254	0.01	0.001	0.0_4254	0.0_51
2937.5	11.566	1	0.29375	0.11566	0.011566	0.0_329375	0.0_411566
10,000	39.37	3.404	1	0.3937	0.03937	0.001	0.0_43937
25,400	100	8.646	2.54	1	0.1	0.00254	0.0001
254,000	1000	86.46	25.4	10	1	0.0254	0.001
10,000,000	39,370	3404	1000	393.7	39.37	1	0.03937
254,000,000	1,000,000	86,460	25,400	10,000	1000	25.4	1

* Computed by J. A. Broadston.
† One light band equals one-half corresponding wavelength. Visible-light wavelengths range from red at 6,500 Å to violet at 4,100 Å.
‡ One helium light band = 0.000011661 in = 2937.5 Å; one krypton 86 light band = 0.0000119 in = 3,022.5 Å; one mercury 198 light band = 0.00001075 in = 2,730 Å.
§ The designations "precision measurements," etc., are not necessarily used in all metrology laboratories.

Table 1.2.12   Conversion of Lengths*

	Inches to millimetres	Millimetres to inches	Feet to metres	Metres to feet	Yards to metres	Metres to yards	Miles to kilometres	Kilometres to miles
1	25.40	0.03937	0.3048	3.281	0.9144	1.094	1.609	0.6214
2	50.80	0.07874	0.6096	6.562	1.829	2.187	3.219	1.243
3	76.20	0.1181	0.9144	9.843	2.743	3.281	4.828	1.864
4	101.60	0.1575	1.219	13.12	3.658	4.374	6.437	2.485
5	127.00	0.1969	1.524	16.40	4.572	5.468	6.047	3.107
6	152.40	0.2362	1.829	19.69	5.486	6.562	9.656	3.728
7	177.80	0.2756	2.134	22.97	6.401	7.655	11.27	4.350
8	203.20	0.3150	2.438	26.25	7.315	8.749	12.87	4.971
9	228.60	0.3543	2.743	29.53	8.230	9.843	14.48	5.592

*EXAMPLE: 1 in = 25.40 mm.

Common fractions of an inch to millimetres (from 1/64 to 1 in)

64ths	Millimetres	64ths	Millimetres	64th	Millimetres	64ths	Millimetres	64ths	Millimetres	64ths	Millimetres
1	0.397	13	5.159	25	9.922	37	14.684	49	19.447	57	22.622
2	0.794	14	5.556	26	10.319	38	15.081	50	19.844	58	23.019
3	1.191	15	5.953	27	10.716	39	15.478	51	20.241	59	23.416
4	1.588	16	6.350	28	11.112	40	15.875	52	20.638	60	23.812
5	1.984	17	6.747	29	11.509	41	16.272	53	21.034	61	24.209
6	2.381	18	7.144	30	11.906	42	16.669	54	21.431	62	24.606
7	2.778	19	7.541	31	12.303	43	17.066	55	21.828	63	25.003
8	3.175	20	7.938	32	12.700	44	17.462	56	22.225	64	25.400
9	3.572	21	8.334	33	13.097	45	17.859				
10	3.969	22	8.731	34	13.494	46	18.256				
11	4.366	23	9.128	35	13.891	47	18.653				
12	4.762	24	9.525	36	14.288	48	19.050				

Decimals of an inch to millimetres (0.01 to 0.99 in)

	0	1	2	3	4	5	6	7	8	9
.0		0.254	0.508	0.762	1.016	1.270	1.524	1.778	2.032	2.286
.1	2.540	2.794	3.048	3.302	3.556	3.810	4.064	4.318	4.572	4.826
.2	5.080	5.334	5.588	5.842	6.096	6.350	6.604	6.858	7.112	7.366
.3	7.620	7.874	8.128	8.382	8.636	8.890	9.144	9.398	9.652	9.906
.4	10.160	10.414	10.668	10.922	11.176	11.430	11.684	11.938	12.192	12.446
.5	12.700	12.954	13.208	13.462	13.716	13.970	14.224	14.478	14.732	14.986
.6	15.240	15.494	15.748	16.002	16.256	16.510	16.764	17.018	17.272	17.526
.7	17.780	18.034	18.288	18.542	18.796	19.050	19.304	19.558	19.812	20.066
.8	20.320	20.574	20.828	21.082	21.336	21.590	21.844	22.098	22.352	22.606
.9	22.860	23.114	23.368	23.622	23.876	24.130	24.384	24.638	24.892	25.146

Millimetres to decimals of an inch (from 1 to 99 mm)

	0.	1.	2.	3.	4.	5.	6.	7.	8.	9.
0		0.0394	0.0787	0.1181	0.1575	0.1969	0.2362	0.2756	0.3150	0.3543
1	0.3937	0.4331	0.4724	0.5118	0.5512	0.5906	0.6299	0.6693	0.7087	0.7480
2	0.7874	0.8268	0.8661	0.9055	0.9449	0.9843	1.0236	1.0630	1.1024	1.1417
3	1.1811	1.2205	1.2598	1.2992	1.3386	1.3780	1.4173	1.4567	1.4961	1.5354
4	1.5748	1.6142	1.6535	1.6929	1.7323	1.7717	1.8110	1.8504	1.8898	1.9291
5	1.9685	2.0079	2.0472	2.0866	2.1260	2.1654	2.2047	2.2441	2.2835	2.3228
6	2.3622	2.4016	2.4409	2.4803	2.5197	2.5591	2.5984	2.6378	2.6772	2.7165
7	2.7559	2.7953	2.8346	2.8740	2.9134	2.9528	2.9921	3.0315	3.0709	3.1102
8	3.1496	3.1890	3.2283	3.2677	3.3071	3.3465	3.3858	3.4252	3.4646	3.5039
9	3.5433	3.5827	3.6220	3.6614	3.7008	3.7402	3.7795	3.8189	3.8583	3.8976

Table 1.2.13 Area Equivalents
(1 hectare = 100 ares = 10,000 centiares or square metres)

Square metres	Square inches	Square feet	Square yards	Square rods	Square chains	Roods	Acres	Square miles or sections
1	1550	10.76	1.196	0.0395	0.002471	0.0_39884	0.0_22471	0.0_63861
0.0_36452	1	0.006944	0.0_37716	0.0_42551	0.0_51594	0.0_66377	0.0_61594	0.0_92491
0.09290	144	1	0.1111	0.003673	0.0_32296	0.0_49183	0.0_42296	0.0_73587
0.8361	1296	9	1	0.03306	0.002066	0.0_38264	0.0002066	0.0_63228
25.29	39204	272.25	30.25	1	0.0625	0.02500	0.00625	0.0_99766
404.7	627264	4356	484	16	1	0.4	0.1	0.0001562
1012	1568160	10890	1210	40	2.5	1	0.25	0.0_33906
4047	6272640	43560	4840	160	10	4	1	0.001562
2589988		27878400	3097600	102400	6400	2560	640	1

Table 1.2.14 Conversion of Areas*

	Sq in to sq cm	Sq cm to sq in	Sq ft to sq m	Sq m to sq ft	Sq yd to sq m	Sq m to sq yd	Acres to hectares	Hectares to acres	Sq mi to sq km	Sq km to sq mi
1	6.452	0.1550	0.0929	10.76	0.8361	1.196	0.4047	2.471	2.590	0.3861
2	12.90	0.3100	0.1858	21.53	1.672	2.392	0.8094	4.942	5.180	0.7722
3	19.35	0.4650	0.2787	32.29	2.508	3.588	1.214	7.413	7.770	1.158
4	25.81	0.6200	0.3716	43.06	3.345	4.784	1.619	9.884	10.360	1.544
5	32.26	0.7750	0.4645	53.82	4.181	5.980	2.023	12.355	12.950	1.931
6	38.71	0.9300	0.5574	64.58	5.017	7.176	2.428	14.826	15.540	2.317
7	45.16	1.085	0.6503	75.35	5.853	8.372	2.833	17.297	18.130	2.703
8	51.61	1.240	0.7432	86.11	6.689	9.568	3.237	19.768	20.720	3.089
9	58.06	1.395	0.8361	96.88	7.525	10.764	3.642	22.239	23.310	3.475

* EXAMPLE: 1 in^2 = 6.452 cm^2.

Table 1.2.15 Volume and Capacity Equivalents

Cubic inches	Cubic feet	Cubic yards	U.S. Apothecary fluid ounces	U.S. quarts Liquid	U.S. quarts Dry	U.S. gallons	U.S. bushels	Cubic decimetres or litres
1	0.0_35787	0.0_42143	0.5541	0.01732	0.01488	0.0_24329	0.0_34650	0.01639
1728	1	0.03704	957.5	29.92	25.71	7.481	0.8036	28.32
46656	27	1	25853	807.9	694.3	202.2	21.70	764.6
1.805	0.001044	0.0_43868	1	0.03125	0.02686	0.007812	0.0_38392	0.02957
57.75	0.03342	0.001238	32	1	0.8594	0.25	0.02686	0.9464
67.20	0.03889	0.001440	37.24	1.164	1	0.2909	0.03125	1.101
231	0.1337	0.004951	128	4	3.437	1	0.1074	3.785
2150	1.244	0.04609	1192	37.24	32	9.309	1	35.24
61.02	0.03531	0.001308	33.81	1.057	0.9081	0.2642	0.02838	1

Table 1.2.16 Conversion of Volumes or Cubic Measure*

	Cu in to mL	mL to cu in	Cu ft to cu m	Cu m to cu ft	Cu yd to cu m	Cu m to cu yd	Gallons to cu ft	Cu ft to gallons
1	16.39	0.06102	0.02832	35.31	0.7646	1.308	0.1337	7.481
2	32.77	0.1220	0.05663	70.63	1.529	2.616	0.2674	14.96
3	49.16	0.1831	0.08495	105.9	2.294	3.924	0.4010	22.44
4	65.55	0.2441	0.1133	141.3	3.058	5.232	0.5347	29.92
5	81.94	0.3051	0.1416	176.6	3.823	6.540	0.6684	37.40
6	98.32	0.3661	0.1699	211.9	4.587	7.848	0.8021	44.88
7	114.7	0.4272	0.1982	247.2	5.352	9.156	0.9358	52.36
8	131.1	0.4882	0.2265	282.5	6.116	10.46	1.069	59.84
9	147.5	0.5492	0.2549	317.8	6.881	11.77	1.203	67.32

* EXAMPLE: 1 in^3 = 16.39 mL.

Table 1.2.17   Conversion of Volumes or Capacities*

	Fluid ounces to mL	mL to fluid ounces	Liquid pints to litres	Litres to liquid pints	Liquid quarts to litres	Litres to liquid quarts	Gallons to litres	Litres to gallons	Bushels to hectolitres	Hectolitres to bushels
1	29.57	0.03381	0.4732	2.113	0.9463	1.057	3.785	0.2642	0.3524	2.838
2	59.15	0.06763	0.9463	4.227	1.893	2.113	7.571	0.5284	0.7048	5.676
3	88.72	0.1014	1.420	6.340	2.839	3.170	11.36	0.7925	1.057	8.513
4	118.3	0.1353	1.893	8.454	3.785	4.227	15.14	1.057	1.410	11.35
5	147.9	0.1691	2.366	10.57	4.732	5.284	18.93	1.321	1.762	14.19
6	177.4	0.2092	2.839	12.68	5.678	6.340	22.71	1.585	2.114	17.03
7	207.0	0.2367	3.312	14.79	6.624	7.397	26.50	1.849	2.467	19.86
8	236.6	0.2705	3.785	16.91	7.571	8.454	30.28	2.113	2.819	22.70
9	266.2	2.3043	4.259	19.02	8.517	9.510	34.07	2.378	3.171	25.54

* EXAMPLE: 1 fluid oz = 29.57 mL.

Table 1.2.18   Mass Equivalents

Kilograms	Grains	Ounces Troy and apoth	Ounces Avoirdupois	Pounds Troy and apoth	Pounds Avoirdupois	Tons Short	Tons Long	Tons Metric
1	15432	32.15	35.27	2.6792	2.205	0.0_21102	0.0_39842	0.001
0.0_46480	1	0.0_22083	0.0_22286	0.0_31736	0.0_31429	0.0_77143	0.0_66378	0.0_66480
0.03110	480	1	1.09714	0.08333	0.06857	0.0_43429	0.0_43061	0.0_43110
0.02835	437.5	0.9115	1	0.07595	0.0625	0.0_43125	0.0_42790	0.0_42835
0.3732	5760	12	13.17	1	0.8229	0.0_34114	0.0_33673	0.0_33732
0.4536	7000	14.58	16	1.215	1	0.0005	0.0_44464	0.0_44536
907.2	140_6	29167	320_3	2431	2000	1	0.8929	0.9072
1016	15680_4	32667	35840	2722	2240	1.12	1	1.016
1000	15432356	32151	35274	2679	2205	1.102	0.9842	1

Table 1.2.19   Conversion of Masses*

	Grains to grams	Grams to grains	Ounces (avdp) to grams	Grams to ounces (avdp)	Pounds (avdp) to kilograms	Kilograms to pounds (avdp)	Short tons (2000 lb) to metric tons	Metric tons (1000 kg) to short tons	Long tons (2240 lb) to metric tons	Metric tons to long tons
1	0.06480	15.43	28.35	0.03527	0.4536	2.205	0.907	1.102	1.016	0.984
2	0.1296	30.86	56.70	0.07055	0.9072	4.409	1.814	2.205	2.032	1.968
3	0.1944	46.30	85.05	0.1058	1.361	6.614	2.722	3.307	3.048	2.953
4	0.2592	61.73	113.40	0.1411	1.814	8.818	3.629	4.409	4.064	3.937
5	0.3240	77.16	141.75	0.1764	2.268	11.02	4.536	5.512	5.080	4.921
6	0.3888	92.59	170.10	0.2116	2.722	13.23	5.443	6.614	6.096	5.905
7	0.4536	108.03	198.45	0.2469	3.175	15.43	6.350	7.716	7.112	6.889
8	0.5184	123.46	226.80	0.2822	3.629	17.64	7.257	8.818	8.128	7.874
9	0.5832	138.89	255.15	0.3175	4.082	19.84	8.165	9.921	9.144	8.858

* EXAMPLE: 1 grain = 0.06480 grams.

Table 1.2.20   Pressure Equivalents

Pascals N/m^2	Bars 10^5 N/m^2	Poundsf per in^2	Atmospheres	Columns of mercury at temperature 0°C and $g = 9.80665$ m/s² cm	in	Columns of water at temperature 15°C and $g = 9.80665$ m/s² cm	in
1	10^{-5}	0.000145	0.00001	0.00075	0.000295	0.01021	0.00402
100000	1	14.504	0.9869	75.01	29.53	1020.7	401.8
6894.8	0.068948	1	0.06805	5.171	2.036	70.37	27.703
101326	1.0132	14.696	1	76.000	29.92	1034	407.1
1333	0.0133	0.1934	0.01316	1	0.3937	13.61	5.357
3386	0.03386	0.4912	0.03342	2.540	1	34.56	13.61
97.98	0.0009798	0.01421	0.000967	0.07349	0.02893	1	0.3937
248.9	0.002489	0.03609	0.002456	0.1867	0.07349	2.540	1

Table 1.2.21   Conversion of Pressures*

	Lb/in² to bars	Bars to lb/in²	Lb/in² to atmospheres	Atmospheres to lb/in²	Bars to atmospheres	Atmospheres to bars
1	0.06895	14.504	0.06805	14.696	0.98692	1.01325
2	0.13790	29.008	0.13609	29.392	1.9738	2.0265
3	0.20684	43.511	0.20414	44.098	2.9607	3.0397
4	0.27579	58.015	0.27218	58.784	3.9477	4.0530
5	0.34474	72.519	0.34823	73.480	4.9346	5.0663
6	0.41368	87.023	0.40826	88.176	5.9215	6.0795
7	0.48263	101.53	0.47632	102.87	6.9085	7.0927
8	0.55158	116.03	0.54436	117.57	7.8954	8.1060
9	0.62053	130.53	0.61241	132.26	8.8823	9.1192

* EXAMPLE: 1 lb/in² = 0.06895 bar.

Table 1.2.22   Velocity Equivalents

cm/s	m/s	m/min	km/h	ft/s	ft/min	mi/h	Knots
1	0.01	0.6	0.036	0.03281	1.9685	0.02237	0.01944
100	1	60	3.6	3.281	196.85	2.237	1.944
1.667	0.01667	1	0.06	0.05468	3.281	0.03728	0.03240
27.78	0.2778	16.67	1	0.9113	54.68	0.6214	0.53996
30.48	0.3048	18.29	1.097	1	60	0.6818	0.59248
0.5080	0.005080	0.3048	0.01829	0.01667	1	0.01136	0.00987
44.70	0.4470	26.82	1.609	1.467	88	1	0.86898
51.44	0.5144	30.87	1.852	1.688	101.3	1.151	1

Table 1.2.23   Conversion of Linear and Angular Velocities*

	cm/s to ft/min	ft/min to cm/s	cm/s to mi/h	mi/h to cm/s	ft/s to mi/h	mi/h to ft/s	rad/s to r/min	r/min to rad/s
1	1.97	0.508	0.0224	44.70	0.682	1.47	9.55	0.1047
2	3.94	1.016	0.0447	89.41	1.364	2.93	19.10	0.2094
3	5.91	1.524	0.0671	134.1	2.045	4.40	28.65	0.3142
4	7.87	2.032	0.0895	178.8	2.727	5.87	38.20	0.4189
5	9.84	2.540	0.1118	223.5	3.409	7.33	47.75	0.5236
6	11.81	3.048	0.1342	268.2	4.091	8.80	57.30	0.6283
7	13.78	3.556	0.1566	312.9	4.773	10.27	66.84	0.7330
8	15.75	4.064	0.1790	357.6	5.455	11.73	76.39	0.8378
9	17.72	4.572	0.2013	402.3	6.136	13.20	85.94	0.9425

* EXAMPLE: 1 cm/s = 1.97 ft/min.

Table 1.2.24   Acceleration Equivalents

cm/s²	m/s²	m/(h·s)	km/(h·s)	ft/(h·s)	ft/s²	ft/min²	mi/(h·s)	knots/s
1	0.01	36.00	0.036	118.1	0.03281	118.1	0.02237	0.01944
100	1	3600	3.6	11811	3.281	11811	2.237	1.944
0.02778	0.0002778	1	0.001	3.281	0.0009113	3.281	0.0006214	0.0005400
27.78	0.2778	1000	1	3281	0.9113	3281	0.6214	0.5400
0.008467	0.00008467	0.3048	0.0003048	1	0.0002778	1	0.0001894	0.0001646
30.48	0.3048	1097	1.097	3600	1	3600	0.6818	0.4572
0.008467	0.00008467	0.3048	0.0003048	1	0.0002778	1	0.0001894	0.0001646
44.70	0.4470	1609	1.609	5280	1.467	5280	1	0.8690
51.44	0.5144	1852	1.852	6076	1.688	6076	1.151	1

Table 1.2.25   Conversion of Accelerations*

	cm/s² to ft/min²	km/(h·s) to mi/(h·s)	km/(h·s) to knots/s	ft/s² to mi/(h·s)	ft/s² to knots/s	ft/min² to cm/s²	mi/(h·s) to m/(h·s)	mi/(h·s) to knots/s	knots/s to mi/(h·s)	knots/s to km/(h·s)
1	118.1	0.6214	0.5400	0.6818	0.4572	0.008467	1.609	0.8690	1.151	1.852
2	236.2	1.243	1.080	1.364	0.9144	0.01693	3.219	1.738	2.302	3.704
3	354.3	1.864	1.620	2.045	1.372	0.02540	4.828	2.607	3.452	5.556
4	472.4	2.485	2.160	2.727	1.829	0.03387	6.437	3.476	4.603	7.408
5	590.6	3.107	2.700	3.409	2.286	0.04233	8.046	4.345	5.754	9.260
6	708.7	3.728	3.240	4.091	2.743	0.05080	9.656	5.214	6.905	11.11
7	826.8	4.350	3.780	4.772	3.200	0.05927	11.27	6.083	8.056	12.96
8	944.9	4.971	4.320	5.454	3.658	0.06774	12.87	6.952	9.206	14.82
9	1063	5.592	4.860	6.136	4.115	0.07620	14.48	7.821	10.36	16.67

* EXAMPLE: 1 cm/s² = 118.1 ft/min².

Table 1.2.26   Energy or Work Equivalents

Joules or Newton-metre	Kilogramf-metres	Foot-poundsf	Kilowatt hours	Metric horsepower-hours	Horsepower-hours	Litre-atmospheres	Kilocalories	British thermal units
1	0.10197	0.7376	0.0_62778	0.0_63777	0.0_63725	0.009869	0.0_32388	0.0_39478
9.80665	1	7.233	0.0_52724	0.0_537037	0.0_53653	0.09678	0.002342	0.009295
1.356	0.1383	1	0.0_63766	0.0_651206	0.0_650505	0.01338	0.0_33238	0.001285
3.600×10^6	3.671×10^5	2.655×10^6	1	1.3596	1.341	35528	859.9	3412
2.648×10^6	270000	1.9529×10^6	0.7355	1	0.9863	26131	632.4	2510
2.6845×10^6	2.7375×10^5	1.98×10^6	0.7457	1.0139	1	26493	641.2	2544
101.33	10.333	74.74	0.0_42815	0.0_43827	0.0_43775	1	0.02420	0.09604
4186.8	426.9	3088	0.001163	0.001581	0.001560	41.32	1	3.968
1055	107.6	778.2	0.0_32931	0.0_33985	0.0_33930	10.41	0.25200	1

Table 1.2.27   Conversion of Energy, Work, Heat*

	Ft·lbf to joules	Joules to ft·lbf	Ft·lbf to Btu	Btu to ft·lbf	Kilogramf-metres to kilocalories	Kilocalories to kilogramf-metres	Joules to calories	Calories to joules
1	1.3558	0.7376	0.001285	778.2	0.002342	426.9	0.2388	4.187
2	2.7116	1.4751	0.002570	1,556	0.004685	853.9	0.4777	8.374
3	4.0674	2.2127	0.003855	2,334	0.007027	1,281	0.7165	12.56
4	5.4232	2.9503	0.005140	3,113	0.009369	1,708	0.9554	16.75
5	6.7790	3.6879	0.006425	3,891	0.01172	2,135	1.194	20.93
6	8.1348	4.4254	0.007710	4,669	0.01405	2,562	1.433	25.12
7	9.4906	5.1630	0.008995	5,447	0.01640	2,989	1.672	29.31
8	10.8464	5.9006	0.01028	6,225	0.01874	3,415	1.911	33.49
9	12.2022	6.6381	0.01156	7,003	0.02108	3,842	2.150	37.68

* EXAMPLE: 1 ft·lbf = 1.3558 J.

Table 1.2.28   Power Equivalents

Horsepower	Kilowatts	Metric horsepower	Kgf·m per s	Ft·lbf per s	Kilocalories per s	Btu per s
1	0.7457	1.014	76.04	550	0.1781	0.7068
1.341	1	1.360	102.0	737.6	0.2388	0.9478
0.9863	0.7355	1	75	542.5	0.1757	0.6971
0.01315	0.009807	0.01333	1	7.233	0.002342	0.009295
0.00182	0.001356	0.00184	0.1383	1	0.0_33238	0.001285
5.615	4.187	5.692	426.9	3088	1	3.968
1.415	1.055	1.434	107.6	778.2	0.2520	1

Table 1.2.29   Conversion of Power*

	Horsepower to kilowatts	Kilowatts to horsepower	Metric horsepower to kilowatts	Kilowatts to metric horsepower	Horsepower to metric horsepower	Metric horsepower to horsepower
1	0.7457	1.341	0.7355	1.360	1.014	0.9863
2	1.491	2.682	1.471	2.719	2.028	1.973
3	2.237	4.023	2.206	4.079	3.042	2.959
4	2.983	5.364	2.942	5.438	4.055	3.945
5	3.729	6.705	3.677	6.798	5.069	4.932
6	4.474	8.046	4.412	8.158	6.083	5.918
7	5.220	9.387	5.147	9.520	7.097	6.904
8	5.966	10.73	5.883	10.88	8.111	7.891
9	6.711	12.07	6.618	12.24	9.125	8.877

* EXAMPLE: 1 hp = 0.7457 kW.

Table 1.2.30   Density Equivalents*

Grams per mL	Lb per cu in	Lb per cu ft	Short tons (2,000 lb) per cu yd	Lb per U.S. gal
1	0.03613	62.43	0.8428	8.345
27.68	1	1728	23.33	231
0.01602	0.0$_3$5787	1	0.0135	0.1337
1.187	0.04287	74.7	1	9.902
0.1198	0.004329	7.481	0.1010	1

* EXAMPLE: 1 g per mL = 62.43 lb per cu ft.

Table 1.2.31   Conversion of Density

Grams per mL to lb per cu ft	Lb per cu ft to grams per mL	Grams per mL to short tons per cu yd	Short tons per cu yd to grams per mL
62.43	0.01602	0.8428	1.187
187.28	0.04805	2.5283	3.560
312.14	0.08009	4.2139	5.933
437.00	0.11213	5.8995	8.306
561.85	0.14416	7.5850	10.679

Table 1.2.32   Thermal Conductivity

Calories per cm $\cdot$ s $\cdot$ °C	Watts per cm $\cdot$ °C	Calories per cm $\cdot$ h $\cdot$ °C	Btu $\cdot$ ft per ft$^2 \cdot$ h $\cdot$ °F	Btu $\cdot$ in per ft$^2 \cdot$ day $\cdot$ °F
1	4.1868	3,600	241.9	69,670
0.2388	1	860	57.79	16,641
0.0002778	0.001163	1	0.0672	19.35
0.004134	0.01731	14.88	1	288
0.00001435	0.00006009	0.05167	0.00347	1

Table 1.2.33   Thermal Conductance

Calories per cm$^2 \cdot$ s $\cdot$ °C	Watts per cm$^2 \cdot$ °C	Calories per cm$^2 \cdot$ h $\cdot$ °C	Btu per ft$^2 \cdot$ h $\cdot$ °F	Btu per ft$^2 \cdot$ day $\cdot$ °F
1	4.1868	3,600	7,373	176,962
0.2388	1	860	1,761	42,267
0.0002778	0.001163	1	2.048	49.16
0.0001356	0.0005678	0.4882	1	24
0.000005651	0.00002366	0.02034	0.04167	1

Table 1.2.34   Heat Flow

Calories per cm$^2 \cdot$ s	Watts per cm^2	Calories per cm$^2 \cdot$ h	Btu per ft$^2 \cdot$ h	Btu per ft$^2 \cdot$ day
1	4.1868	3,600	13,272	318,531
0.2388	1	860	3,170	76,081
0.0002778	0.001163	1	3.687	88.48
0.00007535	0.0003154	0.2712	1	24
0.000003139	0.00001314	0.01130	0.04167	1

Section **2**

Mathematics

C. EDWARD SANDIFER *Professor, Western Connecticut State University, Danbury, CT.*
GEORGE J. MOSHOS *Professor Emeritus of Computer and Information Science, New Jersey Institute of Technology*

2.1 MATHEMATICS
by C. Edward Sandifer

REFERENCES: Conte and DeBoor, ''Elementary Numerical Analysis: An Algorithmic Approach,'' McGraw-Hill. Boyce and DiPrima, ''Elementary Differential Equations and Boundary Value Problems,'' Wiley. Hamming, ''Numerical Methods for Scientists and Engineers,'' McGraw-Hill. Kreyszig, ''Advanced Engineering Mathematics,'' Wiley.

SETS, NUMBERS, AND ARITHMETIC

Sets and Elements

The concept of a set appears throughout modern mathematics. A **set** is a well-defined list or collection of objects and is generally denoted by capital letters, $A, B, C, \ldots$. The objects composing the set are called **elements** and are denoted by lowercase letters, $a, b, x, y, \ldots$. The notation

$$x \in A$$

is read ''x is an element of A'' and means that x is one of the objects composing the set A.

There are two basic ways to describe a set. The first way is to list the elements of the set.

$$A = \{2, 4, 6, 8, 10\}$$

This often is not practical for very large sets.

The second way is to describe properties which determine the elements of the set.

$$A = \{\text{even numbers from 2 to 10}\}$$

This method is sometimes awkward since a single set may sometimes be described in several different ways.

In describing sets, the symbol : is read ''such that.'' The expression

$$B = \{x : x \text{ is an even integer}, x > 1, x < 11\}$$

is read ''B equals the set of all x such that x is an even integer, x is greater than 1, and x is less than 11.''

Two sets, A and B, are equal, written $A = B$, if they contain exactly the same elements. The sets A and B above are equal. If two sets, X and Y, are not equal, it is written $X \neq Y$.

Subsets A set C is a subset of a set A, written $C \subseteq A$, if each element in C is also an element in A. It is also said that C is contained in A. Any set is a subset of itself. That is, $A \subseteq A$ always. A is said to be an ''improper subset of itself.'' Otherwise, if $C \subseteq A$ and $C \neq A$, then C is a proper subset of A.

Two theorems are important about subsets:
(Fundamental theorem of set equality)

$$\text{If } X \subseteq Y \quad \text{and} \quad Y \subseteq X, \quad \text{then } X = Y \quad (2.1.1)$$

(Transitivity)

$$\text{If } X \subseteq Y \quad \text{and} \quad Y \subseteq Z, \quad \text{then } X \subseteq Z \quad (2.1.2)$$

Universe and Empty Set In an application of set theory, it often happens that all sets being considered are subsets of some fixed set, say integers or vectors. This fixed set is called the **universe** and is sometimes denoted U.

It is possible that a set contains no elements at all. The set with no elements is called the **empty set** or the **null set** and is denoted $\varnothing$.

Set Operations New sets may be built from given sets in several ways. The **union** of two sets, denoted $A \cup B$, is the set of all elements belonging to A or to B, or to both.

$$A \cup B = \{x : x \in A \quad \text{or} \quad x \in B\}$$

The union has the properties:

$$A \subseteq A \cup B \quad \text{and} \quad B \subseteq A \cup B \quad (2.1.3)$$

The **intersection** is denoted $A \cap B$ and consists of all elements, each of which belongs to both A and B.

$$A \cap B = \{x : x \in A \quad \text{and} \quad x \in B\}$$

The intersection has the properties

$$A \cap B \subseteq A \quad \text{and} \quad A \cap B \subseteq B \quad (2.1.4)$$

If $A \cap B = \varnothing$, then A and B are called **disjoint**.

In general, a union makes a larger set and an intersection makes a smaller set.

The **complement** of a set A is the set of all elements in the universe set which are not in A. This is written

$$\sim A = \{x : x \in U, \quad x \notin A\}$$

The difference of two sets, denoted $A - B$, is the set of all elements which belong to A but do not belong to B.

Algebra on Sets The operations of union, intersection, and complement obey certain laws known as **Boolean algebra**. Using these laws, it is possible to convert an expression involving sets into other equivalent expressions. The laws of Boolean algebra are given in Table 2.1.1.

Venn Diagrams To give a pictorial representation of a set, **Venn diagrams** are often used. Regions in the plane are used to correspond to sets, and areas are shaded to indicate unions, intersections, and complements. Examples of Venn diagrams are given in Fig. 2.1.1.

Numbers

Numbers are the basic instruments of computation. It is by operations on numbers that calculations are made. There are several different kinds of numbers.

Natural numbers, or counting numbers, denoted **N**, are the whole numbers greater than zero. Sometimes zero is included as a natural number. Any two natural numbers may be added or multiplied to give

Table 2.1.1 Laws of Boolean Algebra

1. Idempotency	
$A \cup A = A$	$A \cap A = A$
2. Associativity	
$(A \cup B) \cup C = A \cup (B \cup C)$	$(A \cap B) \cap C = A \cap (B \cap C)$
3. Commutativity	
$A \cup B = B \cup A$	$A \cap B = B \cap A$
4. Distributivity	
$A \cup (B \cap C) = (A \cup B) \cap (A \cup C)$	$A \cap (B \cup C) = (A \cap B) \cup (A \cap C)$
5. Identity	
$A \cup \varnothing = A$	$A \cap U = A$
$A \cup U = U$	$A \cap \varnothing = \varnothing$
6. Complement	
$A \cup \sim A = U$	$A \cap \sim A = \varnothing$
$\sim(\sim A) = A$	
$\sim U = \varnothing$	
$\sim \varnothing = U$	
7. DeMorgan's laws	
$\sim(A \cup B) = \sim A \cap \sim B$	$\sim(A \cap B) = \sim A \cup \sim B$

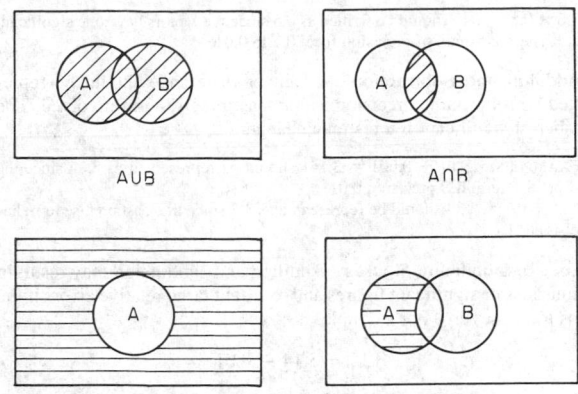

Fig. 2.1.1 Venn diagrams.

another natural number, but subtracting them may produce a negative number, which is not a natural number, and dividing them may produce a fraction, which is not a natural number.

Integers, or whole numbers, are denoted by **Z.** They include both positive and negative numbers and zero. Integers may be added, subtracted, and multiplied, but division might not produce an integer.

Real numbers, denoted **R,** are essentially all values which it is possible for a measurement to take, or all possible lengths for line segments. **Rational numbers** are real numbers that are the quotient of two integers, for example, $^{11}/_{78}$. **Irrational numbers** are not the quotient of two integers, for example, π and $\sqrt{2}$. Within the real numbers, it is always possible to add, subtract, multiply, and divide (except division by zero).

Complex numbers, or **imaginary numbers,** denoted **C,** are an extension of the real numbers that include the square root of -1, denoted i. Within the real numbers, only positive numbers have square roots. Within the complex numbers, all numbers have square roots.

Any complex number z can be written uniquely as $z = x + iy$, where x and y are real. Then x is the real part of z, denoted $\text{Re}(z)$, and y is the imaginary part, denoted $\text{Im}(z)$.

The **complex conjugate,** or simply conjugate of a complex number, z is $\bar{z} = x - iy$.

If $z = x + iy$ and $w = u + iv$, then z and w may be manipulated as follows:

$$z + w = (x + u) + i(y + v)$$
$$z - w = (x - u) + i(y - v)$$
$$zw = xu - yv + i(xv + yu)$$
$$\frac{z}{w} = \frac{xu + yv + i(yu - xv)}{u^2 + v^2}$$

As sets, the following relation exists among these different kinds of numbers:

$$\mathbf{N} \subseteq \mathbf{Z} \subseteq \mathbf{R} \subseteq \mathbf{C}$$

Functions

A **function** f is a rule that relates two sets A and B. Given an element x of the set A, the function assigns a unique element y from the set B. This is written

$$y = f(x)$$

The set A is called the **domain** of the function, and the set B is called the **range.** It is possible for A and B to be the same set.

Functions are usually described by giving the rule. For example,

$$f(x) = 3x + 4$$

is a rule for a function with range and domain both equal to **R.** Given a value, say, 2, from the domain, $f(2) = 3(2) + 4 = 10$.

If two functions f and g have the same range and domain and if the ranges are numbers, then f and g may be added, subtracted, multiplied, or divided according to the rules of the range. If $f(x) = 3x + 4$ and $g(x) = \sin(x)$ and both have range and domain equal to **R,** then

$$f + g(x) = 3x + 4 + \sin(x)$$

and

$$\frac{f}{g}(x) = \frac{3x + 4}{\sin x}$$

Dividing functions occasionally leads to complications when one of the functions assumes a value of zero. In the example f/g above, this occurs when $x = 0$. The quotient cannot be evaluated for $x = 0$ although the quotient function is still meaningful. In this case, the function f/g is said to have a **pole** at $x = 0$.

Polynomial functions are functions of the form

$$f(x) = \sum_{i=0}^{n} a_i x^i$$

where $a_n \neq 0$. The domain and range of polynomial functions are always either **R** or **C.** The number n is the degree of the polynomial.

Polynomials of degree 0 or 1 are called **linear;** of degree 2 they are called **parabolic** or **quadratic;** and of degree 3 they are called **cubic.**

The values of f for which $f(x) = 0$ are called the **roots of** f. A polynomial of degree n has at most n roots. There is exactly one exception to this rule: If $f(x) = 0$ is the constant zero function, the degree of f is zero, but f has infinitely many roots.

Roots of polynomials of degree 1 are found as follows: Suppose the polynomial is $f(x) = ax + b$. Set $f(x) = 0$ and solve for x. Then $x = -b/a$.

Roots of polynomials of degree 2 are often found using the quadratic formula. If $f(x) = ax^2 + bx + c$, then the two roots of f are given by the **quadratic formula:**

$$x_1 = \frac{-b + \sqrt{b^2 - 4ac}}{2a} \quad \text{and} \quad x_2 = \frac{-b - \sqrt{b^2 - 4ac}}{2a}$$

Roots of a polynomial of degree 3 fall into two types.

Equations of the Third Degree with Term in x^2 Absent

Solution: After dividing through by the coefficient of x^3, any equation of this type can be written $x^3 = Ax + B$. Let $p = A/3$ and $q = B/2$. The general solution is as follows:

CASE 1. $q^2 - p^3$ positive. One root is real, viz.,

$$x_1 = \sqrt[3]{q + \sqrt{q^2 - p^3}} + \sqrt[3]{q - \sqrt{q^2 - p^3}}$$

The other two roots are imaginary.

CASE 2. $q^2 - p^3$ zero. Three roots real, but two of them equal.

$$x_1 = 2\sqrt[3]{q} \qquad x_2 = -\sqrt[3]{q} \qquad x_3 = -\sqrt[3]{q}$$

CASE 3. $q^2 - p^3$ negative. All three roots are real and distinct. Determine an angle u between 0 and $180°$, such that $\cos u = q/(p\sqrt{p})$. Then

$$x_1 = 2\sqrt{p}\,\cos(u/3)$$
$$x_2 = 2\sqrt{p}\,\cos(u/3 + 120°)$$
$$x_3 = 2\sqrt{p}\,\cos(u/3 + 240°)$$

Graphical Solution: Plot the curve $y_1 = x^3$, and the straight line $y_2 = Ax + B$. The abscissas of the points of intersection will be the roots of the equation.

Equations of the Third Degree (General Case)

Solution: The general cubic equation, after dividing through by the coefficient of the highest power, may be written $x^3 + ax^2 + bx + c = 0$. To get rid of the term in x^2, let $x = x_1 - a/3$. The equation then becomes $x_1^3 = Ax_1 + B$, where $A = 3(a/3)^2 - b$, and $B = -2(a/3)^3 + b(a/3) - c$. Solve this equation for x_1, by the method above, and then find x itself from $x = x_1 - (a/3)$.

Graphical Solution: Without getting rid of the term in x^2, write the equation in the form $x^3 = -a[x + (b/2a)]^2 + [a(b/2a)^2 - c]$, and solve by the graphical method.

Arithmetic

When numbers, functions, or vectors are manipulated, they always obey certain properties, regardless of the types of the objects involved. Elements may be added or subtracted only if they are in the same universe set. Elements in different universes may sometimes be multiplied or divided, but the result may be in a different universe. Regardless of the universe sets involved, the following properties hold true:

1. Associative laws. $a + (b + c) = (a + b) + c$, $a(bc) = (ab)c$
2. Identity laws. $0 + a = a$, $1a = a$
3. Inverse laws. $a - a = 0$, $a/a = 1$
4. Distributive law. $a(b + c) = ab + ac$
5. Commutative laws. $a + b = b + a$, $ab = ba$

Certain universes, for example, matrices, do not obey the commutative law for multiplication.

SIGNIFICANT FIGURES AND PRECISION

Number of Significant Figures   In engineering computations, the data are ordinarily the result of measurement and are correct only to a limited number of significant figures. Each of the numbers 3.840 and 0.003840 is said to be given ''correct to four figures''; the true value lies in the first case between 0.0038395 and 0.0038405. The **absolute error** is less than 0.001 in the first case, and less than 0.000001 in the second; but the **relative error** is the same in both cases, namely, an error of less than ''one part in 3,840.''

If a number is written as 384,000, the reader is left in doubt whether the number of correct significant figures is 3, 4, 5, or 6. This doubt can be removed by writing the number as 3.84×10^5, or 3.840×10^5, or 3.8400×10^5, or 3.84000×10^5.

In any numerical computation, the possible or desirable degree of accuracy should be decided on and the computation should then be so arranged that the required number of significant figures, and no more, is secured. Carrying out the work to a larger number of places than is justified by the data is to be avoided, (1) because the form of the results leads to an erroneous impression of their accuracy and (2) because time and labor are wasted in superfluous computation.

The unit value of the least significant figure in a number is its **precision.** The number 123.456 has six significant figures and has precision 0.001.

Two ways to represent a real number are as **fixed-point** or as **floating-point,** also known as ''scientific notation.''

In scientific notation, a number is represented as a product of a **mantissa** and a power of 10. The mantissa has its first significant figure either immediately before or immediately after the decimal point, depending on which convention is being used. The power of 10 used is called the **exponent.** The number 123.456 may be represented as either

$$0.123456 \times 10^3 \qquad \text{or} \qquad 1.23456 \times 10^2$$

Fixed-point representations tend to be more convenient when the quantities involved will be added or subtracted or when all measurements are taken to the same precision. Floating-point representations are more convenient for very large or very small numbers or when the quantities involved will be multiplied or divided.

Many different numbers may share the same representation. For example, 0.05 may be used to represent, with precision 0.01, any value between 0.045000 and 0.054999. The largest value a number represents, in this case 0.0549999, is sometimes denoted x^*, and the smallest is denoted x_*.

An awareness of precision and significant figures is necessary so that answers correctly represent their accuracy.

Multiplication and Division   A product or quotient should be written with the smallest number of significant figures of any of the factors involved. The product often has a different precision than the factors, but the significant figures must not increase.

EXAMPLES.   (6.)(8.) = 48 should be written as 50 since the factors have one significant figure. There is a loss of precision from 1 to 10.

$0.6 \times 0.8 = .048$ should be written as 0.5 since the factors have one significant figure. There is a gain of precision from 0.1 to 0.01.

Addition and Subtraction   A sum or difference should be represented with the same precision as the least precise term involved. The number of significant figures may change.

EXAMPLES. $3.14 + 0.001 = 3.141$ should be represented as 3.14 since the least precise term has precision 0.01.

$3.14 + 0.1 = 3.24$ should be represented as 3.2 since the least precise term has precision 0.1.

Loss of Significant Figures   Addition and subtraction may result in serious loss of significant figures and resultant large relative errors if the sums are near zero. For example,

$$3.15 - 3.14 = 0.01$$

shows a loss from three significant figures to just one. Where it is possible, calculations and measurements should be planned so that loss of significant figures can be avoided.

Mixed Calculations   When an expression involves both products and sums, significant figures and precision should be noted for each term or factor as it is calculated, so that correct significant figures and precision for the result are known. The calculation should be performed to as much precision as is available and should be rounded to the correct precision when the calculation is finished. This process is frequently done incorrectly, particularly when calculators or computers provide many decimal places in their result but provide no clue as to how many of those figures are significant.

Significant Figures in Evaluating Functions   If $y = f(x)$, then the correct number of significant figures in y depends on the number of significant figures in x and on the behavior of the function f in the neighborhood of x. In general, y should be represented so that all of $f(x)$, $f(x^*)$, and $f(x_*)$ are between y^* and y_*.

EXAMPLES.

$$
\begin{array}{ll}
\text{sqr (2.0)} & \text{sqr (1.95)} = 1.39642 \\
& \text{sqr (2.00)} = 1.41421 \\
& \text{sqr (2.05)} = 1.43178 \\
& \text{so } y = 1.4 \\
\text{sin (1°)} & \sin (0.5) = 0.00872 \\
& \sin (1.0) = 0.01745 \\
& \sin (1.5) = 0.02617 \\
& \text{so } \sin (1°) = 0.0 \\
\text{sin (90°)} & \sin (89.5) = 0.99996 \\
& \sin (90.0) = 1.00000 \\
& \sin (90.5) = 0.99996 \\
& \text{so } \sin (90°) = 1.0000
\end{array}
$$

Note that in finding sin (90°), there was a gain in significant figures from two to five and also a gain in precision. This tends to happen when $f'(x)$ is close to zero. On the other hand, precision and significant figures are often lost when $f'(x)$ or $f''(x)$ are large.

Rearrangement of Formulas   Often a formula may be rewritten in order to avoid a loss of significant figures.

In using the quadratic formula to find the roots of a polynomial, significant figures may be lost if the $ax^2 + bx + c$ has a root near zero. The quadratic formula may be rearranged as follows:

1. Use the quadratic formula to find the root that is not close to 0. Call this root x_1.
2. Then $x_2 = c/ax_1$.

If $f(x) = \sqrt{x + 1} - \sqrt{x}$, then loss of significant figures occurs if x is large. This can be eliminated by ''rationalizing the numerator'' as follows:

$$\frac{(\sqrt{x+1} - \sqrt{x})(\sqrt{x+1} + \sqrt{x})}{\sqrt{x+1} + \sqrt{x}} = \frac{1}{\sqrt{x+1} + \sqrt{x}}$$

and this has no loss of significant figures.

There is an almost unlimited number of "tricks" for rearranging formulas to avoid loss of significant figures, but many of these are very similar to the tricks used in calculus to evaluate limits.

GEOMETRY, AREAS, AND VOLUMES

Geometrical Theorems

Right Triangles $a^2 + b^2 = c^2$. (See Fig. 2.1.2.) $\angle A + \angle B = 90°$. $p^2 = mn$. $a^2 = mc$. $b^2 = nc$.

Oblique Triangles   Sum of angles = 180°. An exterior angle = sum of the two opposite interior angles (Fig. 2.1.2).

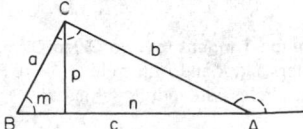

Fig. 2.1.2   Right triangle.

The medians, joining each vertex with the middle point of the opposite side, meet in the center of gravity G (Fig. 2.1.3), which trisects each median.

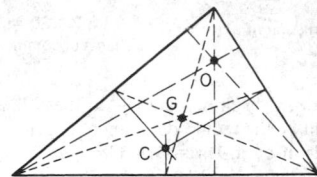

Fig. 2.1.3   Triangle showing medians and center of gravity.

The altitudes meet in a point called the **orthocenter**, O.

The perpendiculars erected at the midpoints of the sides meet in a point C, the center of the circumscribed circle. (In any triangle G, O, and C lie in line, and G is two-thirds of the way from O to C.)

The bisectors of the angles meet in the center of the inscribed circle (Fig. 2.1.4).

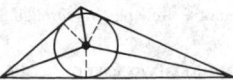

Fig. 2.1.4   Triangle showing bisectors of angles.

The largest side of a triangle is opposite the largest angle; it is less than the sum of the other two sides.

Similar Figures   Any two similar figures, in a plane or in space, can be placed in "perspective," i.e., so that straight lines joining corresponding points of the two figures will pass through a common point (Fig. 2.1.5). That is, of two similar figures, one is merely an enlargement of the other. Assume that each length in one figure is k times the corresponding length in the other; then each area in the first figure is k^2 times the corresponding area in the second, and each volume in the first figure is k^3 times the corresponding volume in the second. If two lines are cut by a set of parallel lines (or parallel planes), the corresponding segments are proportional.

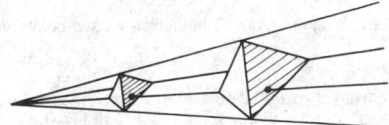

Fig. 2.1.5   Similar figures.

The Circle   An angle that is inscribed in a semicircle is a right angle (Fig. 2.1.6).

A tangent is perpendicular to the radius drawn to the point of contact.

Fig. 2.1.6   Angle inscribed in a semicircle.

Fig. 2.1.7   Dihedral angle.

Dihedral Angles   The dihedral angle between two planes is measured by a plane angle formed by two lines, one in each plane, perpendicular to the edge (Fig. 2.1.7). (For solid angles, see Surfaces and Volumes of Solids.)

In a **tetrahedron,** or triangular pyramid, the four medians, joining each vertex with the center of gravity of the opposite face, meet in a point, the center of gravity of the tetrahedron; this point is ¾ of the way from any vertex to the center of gravity of the opposite face.

The Sphere   (See also Surfaces and Volumes of Solids.) If AB is a diameter, any plane perpendicular to AB cuts the sphere in a circle, of which A and B are called the poles. A great circle on the sphere is formed by a plane passing through the center.

Geometrical Constructions

To Bisect a Line AB (Fig. 2.1.8) (1) From A and B as centers, and with equal radii, describe arcs intersecting at P and Q, and draw PQ, which will bisect AB in M. (2) Lay off $AC = BD =$ approximately half of AB, and then bisect CD.

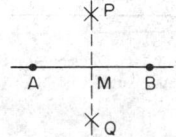

Fig. 2.1.8   Bisectors of a line.

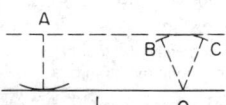

Fig. 2.1.9   Construction of a line parallel to a given line.

To Draw a Parallel to a Given Line l **through a Given Point** A (Fig. 2.1.9) With point A as center draw an arc just touching the line l; with any point O of the line as center, draw an arc BC with the same radius. Then a line through A touching this arc will be the required parallel. Or, use a straightedge and triangle. Or, use a sheet of celluloid with a set of lines parallel to one edge and about ¼ in apart ruled upon it.

To Draw a Perpendicular to a Given Line from a Given Point A **Outside the Line**   (Fig. 2.1.10) (1) With A as center, describe an arc cutting the line at R and S, and bisect RS at M. Then M is the foot of the perpendicular. (2) If A is nearly opposite one end of the line, take any point B of the line and bisect AB in O; then with O as center, and OA or OB as radius, draw an arc cutting the line in M. Or, (3) use a straightedge and triangle.

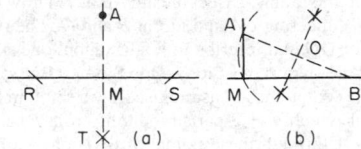

Fig. 2.1.10   Construction of a line perpendicular to a given line from a point not on the line.

To Erect a Perpendicular to a Given Line at a Given Point P (1) Lay off $PR = PS$ (Fig. 2.1.11), and with R and S as centers draw arcs intersecting at *A*. Then *PA* is the required perpendicular. (2) If *P* is near the end of the line, take any convenient point *O* (Fig. 2.1.12) above the line as center, and with radius *OP* draw an arc cutting the line at *Q*. Produce *QO* to meet the arc at *A*; then *PA* is the required perpendicular. (3) Lay off *PB* = 4 units of any scale (Fig. 2.1.13); from *P* and *B* as centers lay off *PA* = 3 and *BA* = 5; then *APB* is a right angle.

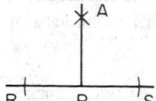

Fig. 2.1.11 Construction of a line perpendicular to a given line from a point on the line.

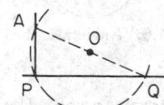

Fig. 2.1.12 Construction of a line perpendicular to a given line from a point on the line.

To Divide a Line *AB* **into** *n* **Equal Parts** (Fig. 2.1.14) Through *A* draw a line *AX* at any angle, and lay off *n* equal steps along this line. Connect the last of these divisions with *B*, and draw parallels through the other divisions. These parallels will divide the given line into *n* equal parts. A similar method may be used to divide a line into parts which shall be proportional to any given numbers.

To Bisect an Angle *AOB* (Fig. 2.1.15) Lay off *OA* = *OB*. From *A* and *B* as centers, with any convenient radius, draw arcs meeting at *M*; then *OM* is the required bisector.

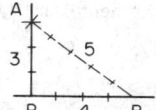

Fig. 2.1.13 Construction of a line perpendicular to a given line from a point on the line.

Fig. 2.1.14 Division of a line into equal parts.

To draw the bisector of an angle when the vertex of the angle is not accessible. Parallel to the given lines *a*, *b*, and equidistant from them, draw two lines *a'*, *b'* which intersect; then bisect the angle between *a'* and *b'*.

To Inscribe a Hexagon in a Circle (Fig. 2.1.16) Step around the circumference with a chord equal to the radius. Or, use a 60° triangle.

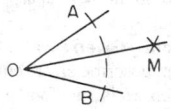

Fig. 2.1.15 Bisection of an angle.

Fig. 2.1.16 Hexagon inscribed in a circle.

To Circumscribe a Hexagon about a Circle (Fig. 2.1.17) Draw a chord *AB* equal to the radius. Bisect the arc *AB* at *T*. Draw the tangent at *T* (parallel to *AB*), meeting *OA* and *OB* at *P* and *Q*. Then draw a circle with radius *OP* or *OQ* and inscribe in it a hexagon, one side being *PQ*.

To Construct a Polygon of *n* **Sides, One Side** *AB* **Being Given** (Fig. 2.1.18) With *A* as center and *AB* as radius, draw a semicircle, and divide it into *n* parts, of which *n* − 2 parts (counting from *B*) are to be used. Draw rays from *A* through these points of division, and complete the construction as in the figure (in which *n* = 7). Note that the center of the polygon must lie in the perpendicular bisector of each side.

To Draw a Tangent to a Circle from an external point *A* (Fig. 2.1.19) Bisect *AC* in *M*; with *M* as center and radius *MC*, draw arc cutting circle in *P*; then *P* is the required point of tangency.

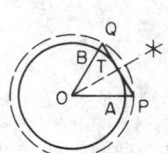

Fig. 2.1.17 Hexagon circumscribed about a circle.

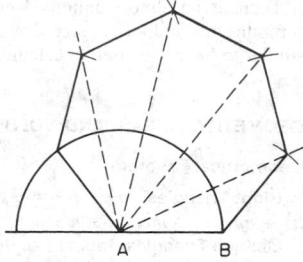

Fig. 2.1.18 Construction of a polygon with a given side.

To Draw a Common Tangent to Two Given Circles (Fig. 2.1.20) Let *C* and *c* be centers and *R* and *r* the radii (*R* > *r*). From *C* as center, draw two concentric circles with radii *R* + *r* and *R* − *r*; draw tangents to

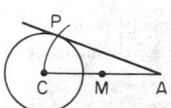

Fig. 2.1.19 Construction of a tangent to a circle.

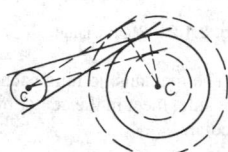

Fig. 2.1.20 Construction of a tangent common to two circles.

these circles from *c*; then draw parallels to these lines at distance *r*. These parallels will be the required common tangents.

To Draw a Circle through Three Given Points *A*, *B*, *C*, or to find the center of a given circular arc (Fig. 2.1.21) Draw the perpendicular bisectors of *AB* and *BC*; these will meet at the center, *O*.

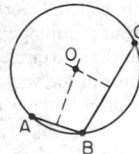

Fig. 2.1.21 Construction of a circle passing through three given points.

To Draw a Circle through Two Given Points *A*, *B*, **and Touching a Given Circle** (Fig. 2.1.22) Draw any circle through *A* and *B*, cutting the given circle at *C* and *D*. Let *AB* and *CD* meet at *E*, and let *ET* be tangent from *E* to the circle just drawn. With *E* as center, and radius *ET*, draw an arc cutting the given circle at *P* and *Q*. Either *P* or *Q* is the required point of contact. (Two solutions.)

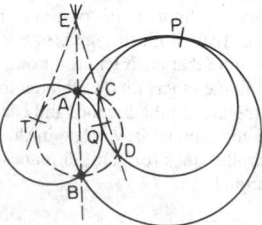

Fig. 2.1.22 Construction of a circle through two given points and touching a given circle.

To Draw a Circle through One Given Point, *A*, **and Touching Two Given Circles** (Fig. 2.1.23) Let *S* be a center of similitude for the two given circles, i.e., the point of intersection of two external (or internal) common tangents. Through S draw any line cutting one circle at two points, the nearer of which shall be called P, and the other at two points, the more remote of which shall be called Q. Through A, P, Q draw a circle cutting SA at B. Then draw a circle through A and B and touching one of the given circles (see preceding construction). This circle will touch the other given circle also. (Four solutions.)

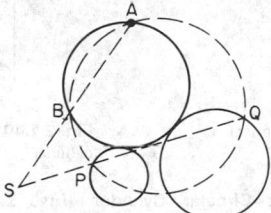

Fig. 2.1.23   Construction of a circle through a given point and touching two given circles.

To Draw an Annulus Which Shall Contain a Given Number of Equal Contiguous Circles   (Fig. 2.1.24) (An annulus is a ring-shaped area enclosed between two concentric circles.) Let $R + r$ and $R - r$ be the inner and outer radii of the annulus, r being the radius of each of the n circles. Then the required relation between these quantities is given by $r = R \sin (180°/n)$, or $r = (R + r) [\sin (180°/n)]/[1 + \sin (180°/n)]$.

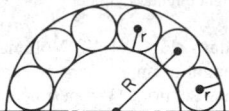

Fig. 2.1.24   Construction of an annulus containing a given number of contiguous circles.

Lengths and Areas of Plane Figures

Right Triangle   (Fig. 2.1.25) $a^2 + b^2 = c^2$. Area $= \frac{1}{2}ab = \frac{1}{2}a^2 \cot A = \frac{1}{2}b^2 \tan A = \frac{1}{4}c^2 \sin 2A$.

Equilateral Triangle   (Fig. 2.1.26) Area $= \frac{1}{4}a^2\sqrt{3} = 0.43301a^2$.

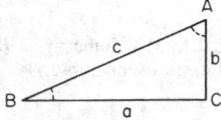

Fig. 2.1.25   Right triangle.

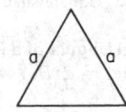

Fig. 2.1.26   Equilateral triangle.

Any Triangle   (Fig. 2.1.27)

$s = \frac{1}{2}(a + b + c)$, $t = \frac{1}{2}(m1 + m2 + m3)$

$r = \sqrt{(s - a)(s - b)(s - c)/s}$ = radius inscribed circle

$R = \frac{1}{2}a/\sin A = \frac{1}{2}b/\sin B = \frac{1}{2}c/\sin C$ = radius circumscribed circle

Area $= \frac{1}{2}$ base $\times$ altitude $= \frac{1}{2}ah = \frac{1}{2}ab \sin C = rs = abc/4R =$ $\pm \frac{1}{2}\{(x_1 y_2 - x_2 y_1) + (x_2 y_3 - x_3 y_2) + (x_3 y_1 - x_1 y_3)\}$, where (x_1, y_1), (x_2, y_2), (x_3, y_3) are coordinates of vertices.

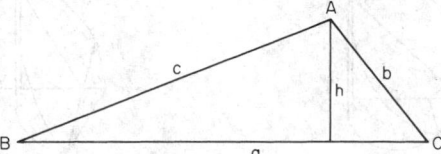

Fig. 2.1.27   Triangle.

Rectangle   (Fig. 2.1.28) Area $= ab = \frac{1}{2}D^2 \sin u$, where u = angle between diagonals D, D.

Rhombus   (Fig. 2.1.29) Area $= a^2 \sin C = \frac{1}{2}D_1D_2$, where C = angle between two adjacent sides; D_1, D_2 = diagonals.

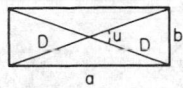

Fig. 2.1.28   Rectangle.

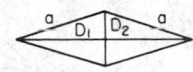

Fig. 2.1.29   Rhombus.

Parallelogram   (Fig. 2.1.30) Area $= bh = ab \sin C = \frac{1}{2}D_1D_2 \sin u$, where u = angle between diagonals D_1 and D_2.

Trapezoid   (Fig. 2.1.31) Area $= \frac{1}{2}(a + b) h$ where bases a and b are parallel.

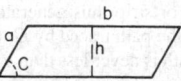

Fig. 2.1.30   Parallelogram.

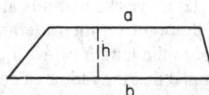

Fig. 2.1.31   Trapezoid.

Any Quadrilateral   (Fig. 2.1.32) Area $= \frac{1}{2}D_1D_2 \sin u$.

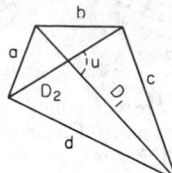

Fig. 2.1.32   Quadrilateral.

Regular Polygons n = number of sides; $v = 360°/n$ = angle subtended at center by one side; a = length of one side $= 2R \sin (v/2) = 2r \tan (v/2)$; R = radius of circumscribed circle $= 0.5 \, a$ csc $(v/2) = r \sec (v/2)$; r = radius of inscribed circle $= R \cos (v/2) = 0.5 \cot (v/2)$; area $= 0.25 \, a^2n \cot (v/2) = 0.5 \, R^2n \sin (v) = r^2n \tan (v/2)$. Areas of regular polygons are tabulated in Table 1.1.3.

Circle   Area $= \pi r^2 = \frac{1}{2}Cr = \frac{1}{4}Cd = \frac{1}{4}\pi d^2 = 0.785398d^2$, where r = radius, d = diameter, C = circumference $= 2 \pi r = \pi d$.

Annulus   (Fig. 2.1.33) Area $= \pi(R^2 - r^2) = \pi(D^2 - d^2)/4 = 2\pi R'b$, where R' = mean radius $= \frac{1}{2}(R + r)$, and $b = R - r$.

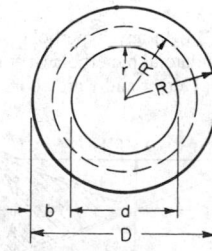

Fig. 2.1.33   Annulus.

Sector   (Fig. 2.1.34) Area $= \frac{1}{2}rs = \pi r^2A/360° = \frac{1}{2} r^2$ rad A, where rad A = radian measure of angle A, and s = length of arc $= r$ rad A.

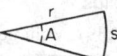

Fig. 2.1.34   Sector.

Segment (Fig. 2.1.35) Area = $\frac{1}{2}r^2(\text{rad } A - \sin A) = \frac{1}{2}[r(s - c) + ch]$, where rad A radian measure of angle A. For small arcs, $s = \frac{1}{3}(8c' - c)$, where c' = chord of half of the arc (Huygens' approximation). Areas of segments are tabulated in Tables 1.1.1 and 1.1.2.

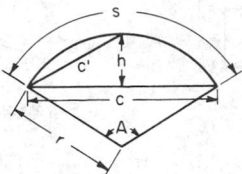

Fig. 2.1.35 Segment.

Ribbon bounded by two parallel curves (Fig. 2.1.36). If a straight line AB moves so that it is always perpendicular to the path traced by its middle point G, then the area of the ribbon or strip thus generated is equal to the length of AB times the length of the path traced by G. (It is assumed that the radius of curvature of G's path is never less than $\frac{1}{2}AB$, so that successive positions of generating line will not intersect.)

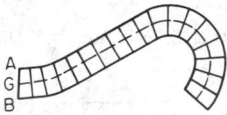

Fig. 2.1.36 Ribbon.

Ellipse (Fig. 2.1.37) Area of ellipse = πab. Area of shaded segment = $xy + ab \sin^{-1}(x/a)$. Length of perimeter of ellipse = $\pi(a + b)K$, where $K = (1 + \frac{1}{4}m^2 + \frac{1}{64}m^4 + \frac{1}{256}m^6 + \ldots)$, $m = (a - b)/(a + b)$.

For $m =$	0.1	0.2	0.3	0.4	0.5
$K =$	1.002	1.010	1.023	1.040	1.064
For $m =$	0.6	0.7	0.8	0.9	1.0
$K =$	1.092	1.127	1.168	1.216	1.273

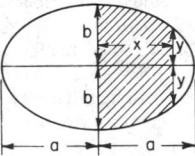

Fig. 2.1.37 Ellipse.

Hyperbola (Fig. 2.1.38) In any hyperbola, shaded area $A = ab \ln[(x/a) + (y/b)]$. In an equilateral hyperbola $(a = b)$, area $A = a^2 \sinh^{-1}(y/a) = a^2 \cosh^{-1}(x/a)$. Here x and y are coordinates of point P.

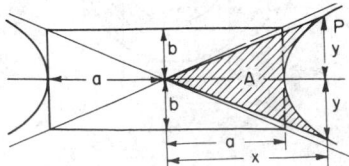

Fig. 2.1.38 Hyperbola.

For lengths and areas of **other curves** see Analytical Geometry.

Surfaces and Volumes of Solids

Regular Prism (Fig. 2.1.39) Volume = $\frac{1}{2}nrah = Bh$. Lateral area = $nah = Ph$. Here n = number of sides; B = area of base; P = perimeter of base.

Right Circular Cylinder (Fig. 2.1.40) Volume = $\pi r^2 h = Bh$. Lateral area = $2\pi rh = Ph$. Here B = area of base; P = perimeter of base.

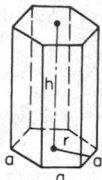

Fig. 2.1.39 Regular prism.

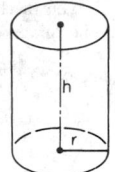

Fig. 2.1.40 Right circular cylinder.

Truncated Right Circular Cylinder (Fig. 2.1.41) Volume = $\pi r^2 h = Bh$. Lateral area = $2\pi rh = Ph$. Here h = mean height = $\frac{1}{2}(h_1 + h_2)$; B = area of base; P = perimeter of base.

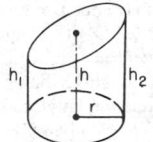

Fig. 2.1.41 Truncated right circular cylinder.

Any Prism or Cylinder (Fig. 2.1.42) Volume = $Bh = Nl$. Lateral area = Ql. Here l = length of an element or lateral edge; B = area of base; N = area of normal section; Q = perimeter of normal section.

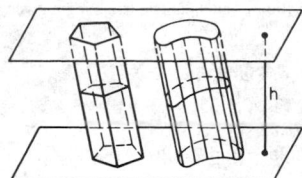

Fig. 2.1.42 Any prism or cylinder.

Special Ungula of a Right Cylinder (Fig. 2.1.43) Volume = $\frac{2}{3}r^2H$. Lateral area = $2rH$. r = radius. (Upper surface is a semiellipse.)

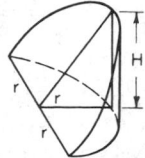

Fig. 2.1.43 Special ungula of a right circular cylinder.

Any Ungula of a right circular cylinder (Figs. 2.1.44 and 2.1.45) Volume = $H(\frac{2}{3}a^3 \pm cB)/(r \pm c) = H[a(r^2 - \frac{1}{3}a^2) \pm r^2c \text{ rad } u]/(r \pm c)$. Lateral area = $H(2ra \pm cs)/(r \pm c) = 2rH(a \pm c \text{ rad } u)/$

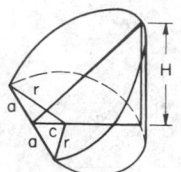

Fig. 2.1.44 Ungula of a right circular cylinder.

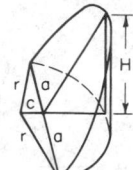

Fig. 2.1.45 Ungula of a right circular cylinder.

$(r \pm c)$. If base is greater (less) than a semicircle, use $+ \; (-)$ sign. $r =$ radius of base; $B =$ area of base; $s =$ arc of base; $u =$ half the angle subtended by arc s at center; rad $u =$ radian measure of angle u.

Regular Pyramid   (Fig. 2.1.46) Volume $= \frac{1}{3}$ altitude $\times$ area of base $= \frac{1}{6}hran$. Lateral area $= \frac{1}{2}$ slant height $\times$ perimeter of base $= \frac{1}{2}san$. Here $r =$ radius of inscribed circle; $a =$ side (of regular polygon); $n =$ number of sides; $s = \sqrt{r^2 + h^2}$. Vertex of pyramid directly above center of base.

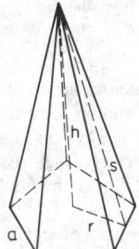

Fig. 2.1.46   Regular pyramid.

Right Circular Cone   Volume $= \frac{1}{3}\pi r^2 h$. Lateral area $= \pi rs$. Here $r =$ radius of base; $h =$ altitude; $s =$ slant height $= \sqrt{r^2 + h^2}$.

Frustum of Regular Pyramid   (Fig. 2.1.47) Volume $= \frac{1}{6}hran[1 + (a'/a) + (a'/a)^2]$. Lateral area $=$ slant height $\times$ half sum of perimeters of bases $=$ slant height $\times$ perimeter of midsection $= \frac{1}{2}sn(r + r')$. Here $r,r' =$ radii of inscribed circles; $s = \sqrt{(r - r')^2 + h^2}$; $a,a' =$ sides of lower and upper bases; $n =$ number of sides.

Frustum of Right Circular Cone   (Fig. 2.1.48) Volume $= \frac{1}{3}\pi r^2 h[1 + (r'/r) + (r'/r)^2] = \frac{1}{3}\pi h(r^2 + rr' + r'^2) = \frac{1}{4}\pi h[r + r')^2 + \frac{1}{3}(r - r')^2]$. Lateral area $= \pi s(r + r'); s = \sqrt{(r - r')^2 + h^2}$.

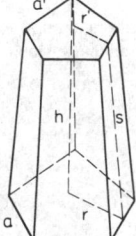   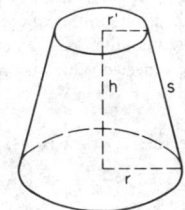

Fig. 2.1.47   Frustum of a regular pyramid.

Fig. 2.1.48   Frustum of a right circular cone.

Any Pyramid or Cone   Volume $= \frac{1}{3}Bh$. $B =$ area of base; $h =$ perpendicular distance from vertex to plane in which base lies.

Any Pyramidal or Conical Frustum   (Fig. 2.1.49) Volume $= \frac{1}{3}h(B + \sqrt{BB'} + B') = \frac{1}{3}hB[1 + (P'/P) + (P'/P)^2]$. Here $B, B' =$ areas of lower and upper bases; $P, P' =$ perimeters of lower and upper bases.

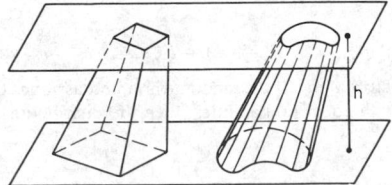

Fig. 2.1.49   Pyramidal frustum and conical frustum.

Sphere   Volume $= V = \frac{4}{3}\pi r^3 = 4.188790 r^3 = \frac{1}{6}\pi d^3 = \frac{2}{3}$ volume of circumscribed cylinder. Area $= A = 4\pi r^2 =$ four great cir-

cles $= \pi d^2 =$ lateral area of circumscribed cylinder. Here $r =$ radius; $d = 2r =$ diameter $= \sqrt[3]{6V/\pi} = \sqrt{A/\pi}$.

Hollow Sphere   or spherical shell. Volume $= \frac{4}{3}\pi(R^3 - r^3) = \frac{1}{6}\pi(D^3 - d^3) = 4\pi R_1^2 t + \frac{1}{3}\pi t^3$. Here $R,r =$ outer and inner radii; $D,d =$ outer and inner diameters; $t =$ thickness $= R - r; R_1 =$ mean radius $= \frac{1}{2}(R + r)$.

Any Spherical Segment. Zone   (Fig. 2.1.50) Volume $= \frac{1}{6}\pi h(3a^2 + 3a_1^2 + h^2)$. Lateral area (zone) $= 2\pi rh$. Here $r =$ radius of sphere. If the inscribed frustum of a cone is removed from the spherical segment, the volume remaining is $\frac{1}{6}\pi hc^2$, where $c =$ slant height of frustum $= \sqrt{h^2 + (a - a_1)^2}$.

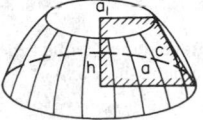

Fig. 2.1.50   Any spherical segment.

Spherical Segment of One Base. Zone   (spherical "cap" of Fig. 2.1.51) Volume $= \frac{1}{6}\pi h(3a^2 + h^2) = \frac{1}{3}\pi h^2(3r - h)$. Lateral area (of zone) $= 2\pi rh = \pi(a^2 + h^2)$.

NOTE. $a^2 = h(2r - h)$, where $r =$ radius of sphere.

Spherical Sector   (Fig. 2.1.51) Volume $= \frac{1}{3}r \times$ area of cap $= \frac{2}{3}\pi r^2 h$. Total area $=$ area of cap $+$ area of cone $= 2\pi rh + \pi ra$.

NOTE. $a^2 = h(2r - h)$.

Spherical Wedge   bounded by two plane semicircles and a **lune** (Fig. 2.1.52). Volume of wedge $\div$ volume of sphere $= u/360°$. Area of lune $\div$ area of sphere $= u/360°$. $u =$ dihedral angle of the wedge.

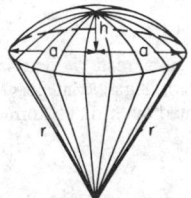   

Fig. 2.1.51   Spherical sector.   **Fig. 2.1.52**   Spherical wedge.

Solid Angles   Any portion of a spherical surface subtends what is called a **solid angle** at the center of the sphere. If the area of the given portion of spherical surface is equal to the square of the radius, the subtended solid angle is called a **steradian,** and this is commonly taken as the unit. The entire solid angle about the center is called a **steregon,** so that 4π steradians $= 1$ steregon. A so-called "solid right angle" is the solid angle subtended by a quadrantal (or trirectangular) spherical triangle, and a "spherical degree" (now little used) is a solid angle equal to $\frac{1}{90}$ of a solid right angle. Hence 720 spherical degrees $= 1$ steregon, or π steradians $= 180$ spherical degrees. If $u =$ the angle which an element of a cone makes with its axis, then the solid angle of the cone contains $2\pi(1 - \cos u)$ steradians.

Regular Polyhedra $A =$ area of surface; $V =$ volume; $a =$ edge.

Name of solid	Bounded by	A/a^2	V/a^3
Tetrahedron	4 triangles	1.7321	0.1179
Cube	6 squares	6.0000	1.0000
Octahedron	8 triangles	3.4641	0.4714
Dodecahedron	12 pentagons	20.6457	7.6631
Icosahedron	20 triangles	8.6603	2.1917

Ellipsoid   (Fig. 2.1.53) Volume $= \frac{4}{3}\pi abc$, where a, b, c = semi-axes.

Torus, or Anchor Ring   (Fig. 2.1.54) Volume $= 2\pi^2 cr^2$. Area $= 4\pi^2 cr$.

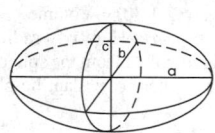

Fig. 2.1.53   Ellipsoid.

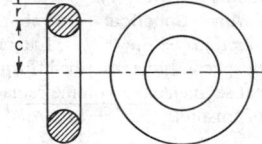

Fig. 2.1.54   Torus.

Volume of a Solid of Revolution   (solid generated by rotating an area bounded above by $f(x)$ around the x axis)

$$V = \pi \int_a^b |f(x)|^2\, dx$$

Area of a Surface of Revolution

$$A = 2\pi \int_a^b y\sqrt{1 + (dy/dx)^2}\, dx$$

Length of Arc of a Plane Curve $y = f(x)$ between values $x = a$ and $x = b$. $s = \int_a^b \sqrt{1 + (dy/dx)^2}\, dx$. If $x = f(t)$ and $y = g(t)$, for $a < t < b$, then

$$s = \int_a^b \sqrt{(dx/dt)^2 + (dy/dt)^2}\, dt$$

PERMUTATIONS AND COMBINATIONS

The product $(1)(2)(3) \ldots (n)$ is written $n!$ and is read "n factorial." By convention, $0! = 1$, and $n!$ is not defined for negative integers.

For large values of n, $n!$ may be approximated by Stirling's formula:

$$n! \approx 2.50663 n^{n+.5} e^{-n}$$

The **binomial coefficient** $C(n, k)$, also written $\binom{n}{k}$, is defined as:

$$C(n, k) = \frac{n!}{k!(n - k)!}$$

$C(n, k)$ is read "n choose k" or as "binomial coefficient n-k." Binomial coefficients have the following properties:
1. $C(n, 0) = C(n, n) = 1$
2. $C(n, 1) = C(n, n - 1) = n$
3. $C(n + 1, k) = C(n, k) + C(n, k - 1)$
4. $C(n, k) = C(n, n - k)$

Binomial coefficients are tabulated in Sec. 1.

Binomial Theorem

If n is a positive integer, then

$$(a + b)^n = \sum_{k=0}^{n} C(n, k) a^k b^{n-k}$$

EXAMPLE.   The third term of $(2x + 3)^7$ is $C(7, 4)(2x)^{7-4}3^4 = [7!/(4!3!)](2x)^3 3^4 = (35)(8\ x^3)(81) = 22680x^3$.

Combinations $C(n, k)$ gives the number of ways k distinct objects can be chosen from a set of n elements. This is the number of k-element subsets of a set of n elements.

EXAMPLE.   The set of four elements $\{a, b, c, d\}$ has $C\{4, 2\} = 6$ two-element subsets, $\{a, b\}$, $\{a, c\}$, $\{a, d\}$, $\{b, c\}$, $\{b, d\}$, and $\{c, d\}$. (Note that $\{a, c\}$ is the same set as $\{c, a\}$.)

Permutations   The number of ways k objects may be arranged from a set of n elements is given by

$$P(n, k) = \frac{n!}{(n - k)!}$$

EXAMPLE.   Two elements from the set $\{a, b, c, d\}$ may be arranged in $C(4, 2) = 12$ ways: ab, ac, ad, ba, bc, bd, ca, cb, cd, da, db, and dc. Note that ac is a different arrangement than ca.

Permutations and combinations are examined in detail in most texts on probability and statistics and on discrete mathematics.

If an event can occur in s ways and can fail to occur in f ways, and if all ways are equally likely, then the probability of the event's occurring is $p = s/(s + f)$, and the probability of failure is $q = f/(s + f) = 1 - p$.

The set of all possible outcomes of an experiment is called the **sample space**, denoted S. Let n be the number of outcomes in the sample set. A subset A of the sample space is called an **event**. The number of outcomes in A is s. Therefore $P(A) = s/n$. The probability that A does not occur is $P(\sim A) = q = 1 - p$.

Always $0 \le p \le 1$ and $P(S) = 1$.

If two events cannot occur simultaneously, then $A \cap B = \varnothing$, and A and B are said to be **mutually exclusive**. Then $P(A \cup B) = P(A) + P(B)$. Otherwise, $P(A \cup B) = P(A) + P(B) - P(A \cap B)$.

Events A and B are **independent** if $P(A \cap B) = P(A)P(B)$.

If E is an event and if $P(E) > 0$, then the probability that A occurs once E has already occurred is called the "conditional probability of A given E," written $P(A|E)$ and defined as

$$P(A|E) = P(A \cap E)/P(E)$$

A and E are independent if $P(A|E) = P(A)$.

If the outcomes in a sample space X are all numbers, then X, together with the probabilities of the outcomes, is called a **random variable**. If x_i is an outcome, then $p_i = P(x_i)$.

The **expected value** of a random variable is

$$E(X) = \sum e_i p_i$$

The **variance** of X is

$$V(X) = \sum [x_i - E(X)]^2 p_i$$

The **standard deviation** is

$$S(X) = \sqrt{[V(X)]}$$

The Binomial, or Bernoulli, Distribution   If an experiment is repeated n times and the probability of a success on any trial is p, then the probability of k successes among those n trials is

$$f(n, k, p) = C(n, k) p^k q^{n-k}$$

Geometric Distribution   If an experiment is repeated until it finally succeeds, let x be the number of failures observed before the first success. Let p be the probability of success on any trial and let $q = 1 - p$. Then

$$P(x = k) = q^k \cdot p$$

Uniform Distribution   If the random variable x assumes the values 1, 2, . . . , n, with equal probabilities, then the distribution is uniform, and

$$P(x = k) = \frac{1}{n}$$

Hypergeometric Distribution—Sampling without Replacement   If a finite population of N elements contains x successes and if n items are selected randomly without replacement, then the probability that k successes will occur among those n samples is

$$h(x; N, n, k) = \frac{C(k, x)C(N - k, n - x)}{C(N, n)}$$

For large values of N, the hypergeometric distribution approaches the binomial distribution, so

$$h(x; N, n, k) \approx f\left(n, k, \frac{x}{N}\right)$$

Poisson Distribution If the average number of successes which occur in a given fixed time interval is m, then let x be the number of successes observed in that time interval. The probability that $x = k$ is

$$p(k, m) = \frac{e^{-m}m^x}{x!} \qquad \text{where } e = 2.71828 \ldots$$

Negative Binomial Distribution If repeated independent trials have probability of success p, then let x be the trial number upon which success number n occurs. Then the probability that $x = k$ is

$$b^*(k; n, p) = C(k - 1, n - 1)p^n q^{k-n}$$

The expected values and variances of these distributions are summarized in the following table:

Distribution	$E(X)$	$V(X)$
Uniform	$(n + 1)/2$	$(n^2 - 1)/12$
Binomial	np	npq
Hypergeometric	nk/N	$[nk(N - n)(1 - k/N)]/[N(N - 1)]$
Poisson	m	m
Geometric	q/p	q/p^2
Negative binomial	nq/p	nq/p^2

LINEAR ALGEBRA

Using linear algebra, it is often possible to express in a single equation a set of relations that would otherwise require several equations. Similarly, it is possible to replace many calculations involving several variables with a few calculations involving vectors and matrices. In general, the equations to which the techniques of linear algebra apply must be linear equations; they can involve no polynomial, exponential, or trigonometric terms.

Vectors

A **row vector** v is a list of numbers written in a row, usually enclosed by parentheses.

$$v = (v_1, v_2, \ldots, v_n)$$

A **column vector** u is a list of numbers written in a column:

$$u = \begin{pmatrix} u_1 \\ u_2 \\ \cdot \\ \cdot \\ \cdot \\ u_n \end{pmatrix}$$

The numbers u_i and v_i may be real or complex, or they may even be variables or functions.

A vector is sometimes called an **ordered n-tuple**. In the case where $n = 2$, it may be called an **ordered pair**.

The numbers v_i are called **components** or **coordinates** of the vector v. The number n is called the **dimension** of v.

Two-dimensional vectors correspond with points in the plane, where v_1 is the x coordinate and v_2 is the y coordinate of the point v. Two-dimensional vectors also correspond with complex numbers, where $z = v_1 + iv_2$.

Three-dimensional vectors correspond to points in space, where v_1, v_2, and v_3 are the x, y, and z coordinates of the point, respectively.

Two- and three-dimensional vectors may be thought of as having a direction and a magnitude. See the section "Analytical Geometry."

Two vectors u and v are equal if:
1. u and v are the same type (either row or column).
2. u and v have the same dimension.
3. Corresponding components are equal; that is, $u_i = v_i$ for $i = 1, 2, \ldots, n$.

Note that the row vectors

$$u = (1, 2, 3) \qquad \text{and} \qquad v = (3, 2, 1)$$

are not equal since the components are not in the same order. Also,

$$u = (1, 2, 3) \qquad \text{and} \qquad v = \begin{pmatrix} 1 \\ 2 \\ 3 \end{pmatrix}$$

are not equal since u is a row vector and v is a column vector.

Vector Transpose If u is a row vector, then the **transpose** of u, written u^T, is the column vector with the same components in the same order as u. Similarly, the transpose of a column vector is the row vector with the same components in the same order. Note that $(u^T)^T = u$.

Vector Addition If u and v are vectors of the same type and the same dimension, then the sum of u and v, written $u + v$, is the vector obtained by adding corresponding components. In the case of row vectors,

$$u + v = (u_1 + v_1, u_2 + v_2, \ldots, u_n + v_n)$$

Scalar Multiplication If a is a number and u is a vector, then the **scalar product** au is the vector obtained by multiplying each component of u by a.

$$au = (au_1, au_2, \ldots, au_n)$$

A number by which a vector is multiplied is called a **scalar**. The **negative** of vector u is written $-u$, and

$$-u = -1u$$

The **zero vector** is the vector with all its components equal to zero.

Arithmetic Properties of Vectors If u, v, and w are vectors of the same type and dimensions, and if a and b are scalars, then vector addition and scalar multiplication obey the following seven rules, known as the *properties of a vector space:*

1. $(u + v) + w = u + (v + w)$ — associative law
2. $u + v = v + u$ — commutative law
3. $u + 0 = u$ — additive identity
4. $u + (-u) = 0$ — additive inverse
5. $a(u + v) = au + av$ — distributive law
6. $(ab)u = a(bu)$ — associative law of multiplication
7. $1u = u$ — multiplicative identity

Inner Product or Dot Product If u and v are vectors of the same type and dimension, then their **inner product** or **dot product**, written uv or $u \cdot v$, is the scalar

$$uv = u_1v_1 + u_2v_2 + \cdots + u_nv_n$$

Vectors u and v are **perpendicular** or **orthogonal** if $uv = 0$.

Magnitude There are two equivalent ways to define the magnitude of a vector u, written $|u|$ or $\|u\|$.

$$|u| = \sqrt{(u \cdot u)}$$

or

$$|u| = \sqrt{u_1^2 + u_2^2 + \cdots + u_n^2}$$

Cross Product or Outer Product If u and v are three-dimensional vectors, then they have a **cross product**, also called **outer product** or **vector product**.

$$u \times v = (u_2v_3 - u_3v_2, v_1u_3 - v_3u_1, u_1v_2 - u_2v_1)$$

The cross product $\mathbf{u} \times \mathbf{v}$ is a three-dimensional vector that is perpendicular to both $\mathbf{u}$ and $\mathbf{v}$. The cross product is not commutative. In fact,

$$\mathbf{u} \times \mathbf{v} = -\mathbf{v} \times \mathbf{u}$$

Cross product and inner product have two properties involving trigonometric functions. If θ is the angle between vectors $\mathbf{u}$ and $\mathbf{v}$, then

$$\mathbf{u}\mathbf{v} = |\mathbf{u}|\,|\mathbf{v}|\cos\theta \quad \text{and} \quad |\mathbf{u} \times \mathbf{v}| = |\mathbf{u}|\,|\mathbf{v}|\sin\theta$$

Matrices

A **matrix** is a rectangular array of numbers. A matrix A with m rows and n columns may be written

$$A = \begin{pmatrix} a_{11} & a_{12} & a_{13} & \cdots & a_{1n} \\ a_{21} & a_{22} & a_{23} & \cdots & a_{2n} \\ a_{31} & a_{32} & a_{33} & \cdots & a_{3n} \\ \cdots & \cdots & \cdots & \cdots & \cdots \\ a_{m1} & a_{m2} & a_{m3} & \cdots & a_{mn} \end{pmatrix}$$

The numbers a_{ij} are called the **entries** of the matrix. The first subscript i identifies the row of the entry, and the second subscript j identifies the column.

Matrices are denoted either by capital letters, A, B, etc., or by writing the general entry in parentheses, (a_{ij}).

The number of rows and the number of columns together define the dimensions of the matrix. The matrix A is an $m \times n$ matrix, read "m by n."

A row vector may be considered to be a $1 \times n$ matrix, and a column vector may be considered as a $n \times 1$ matrix.

The rows of a matrix are sometimes considered as row vectors, and the columns may be considered as column vectors.

If a matrix has the same number of rows as columns, the matrix is called a **square matrix**.

In a square matrix, the entries a_{ii}, where the row index is the same as the column index, are called the diagonal entries.

If a matrix has all its entries equal to zero, it is called a **zero matrix**.

If a square matrix has all its entries equal to zero except its diagonal entries, it is called a **diagonal matrix**.

The diagonal matrix with all its diagonal entries equal to 1 is called the **identity matrix,** and is denoted I, or $I_{n \times n}$ if it is important to emphasize the dimensions of the matrix. The 2×2 and 3×3 identity matrices are:

$$I_{2 \times 2} = \begin{pmatrix} 1 & 0 \\ 0 & 1 \end{pmatrix} \qquad I_{3 \times 3} = \begin{pmatrix} 1 & 0 & 0 \\ 0 & 1 & 0 \\ 0 & 0 & 1 \end{pmatrix}$$

The entries of a square matrix a_{ij} where $i > j$ are said to be below the diagonal. Similarly, those where $i < j$ are said to be above the diagonal.

A square matrix with all entries below (resp. above) the diagonal equal to zero is called *upper-triangular* (resp. *lower-triangular*).

Matrix Addition Matrices A and B may be added only if they have the same dimensions. Then the sum $C = A + B$ is defined by

$$c_{ij} = a_{ij} + b_{ij}$$

That is, corresponding entries of the matrices are added together, just as with vectors. Similarly, matrices may be multiplied by scalars.

Matrix Multiplication Matrices A and B may be multiplied only if the number of columns in A equals the number of rows of B. If A is an $m \times n$ matrix and B is an $n \times p$ matrix, then the product $C = AB$ is an $m \times p$ matrix, defined as follows:

$$c_{ij} = a_{i1}b_{1j} + a_{i2}b_{2j} + \cdots + a_{in}b_{nj}$$
$$= \sum_{k=1}^{n} a_{ik}b_{kj}$$

The entry c_{ij} may also be defined as the dot product of row i of A with the transpose of column j of B.

EXAMPLE. $\begin{pmatrix} 1 & 2 \\ 5 & 6 \end{pmatrix} \begin{pmatrix} 3 & 4 \\ 7 & 8 \end{pmatrix} =$

$$\begin{pmatrix} 1 \times 3 + 2 \times 7 & 1 \times 4 + 2 \times 8 \\ 5 \times 3 + 6 \times 7 & 5 \times 4 + 6 \times 8 \end{pmatrix} = \begin{pmatrix} 17 & 20 \\ 57 & 68 \end{pmatrix}$$

Matrix multiplication is not commutative. Even if A and B are both square, it is hardly ever true that $AB = BA$. Matrix multiplication does have the following properties:

1. $(AB)C = A(BC)$ associative law
2. $A(B + C) = AB + AC$
3. $(B + C)A = BA + CA$ distributive laws

If A is square, then also

4. $AI = IA = A$ multiplicative identity

If A is square, then powers of A, AA, and AAA are denoted A^2 and A^3, respectively.

The transpose of a matrix A, written A^T, is obtained by writing the rows of A as columns. If A is $m \times n$, then A^T is $n \times m$.

EXAMPLE. $\begin{pmatrix} 1 & 2 & 3 \\ 4 & 5 & 6 \end{pmatrix}^T = \begin{pmatrix} 1 & 4 \\ 2 & 5 \\ 3 & 6 \end{pmatrix}$

The transpose has the following properties:

1. $(A^T)^T = A$
2. $(A + B)^T = A^T + B^T$
3. $(AB)^T = B^T A^T$

Note that in property 3, the order of multiplication is reversed.

If $A^T = A$, then A is called *symmetric*.

Linear Equations

A linear equation in two variables is of the form

$$a_1 x_1 + a_2 x_2 = b \quad \text{or} \quad a_1 x + a_2 y = b$$

depending on whether the variables are named x_1 and x_2 or x and y.

In n variables, such an equation has the form

$$a_1 x_1 + a_2 x_2 + \cdots a_n x_n = b$$

Such equations describe lines and planes. Often it is necessary to solve several such equations simultaneously. A set of m linear equations in n variables is called an $m \times n$ system of simultaneous linear equations.

Systems with Two Variables

1×2 Systems An equation of the form

$$a_1 x + a_2 y = b$$

has infinitely many solutions which form a straight line in the xy plane. That line has slope $-a_1/a_2$ and y intercept b/a_2.

2×2 Systems A 2×2 system has the form

$$a_{11}x + a_{12}y = b_1 \qquad a_{21}x + a_{22}y = b_2$$

Solutions to such systems do not always exist.

CASE 1. The system has exactly one solution (Fig. 2.1.55a). The lines corresponding to the equations intersect at a single point. This occurs whenever the two lines have different slopes, so they are not

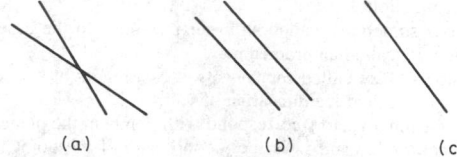

Fig. 2.1.55 Line corresponding to linear equations. (a) One solution; (b) no solutions; (c) infinitely many solutions.

parallel. In this case,

$$\frac{a_{11}}{a_{21}} \neq \frac{a_{12}}{a_{22}} \quad \text{so} \quad a_{11}a_{22} - a_{21}a_{12} \neq 0$$

CASE 2. The system has no solutions (Fig. 2.1.55b). This occurs whenever the two lines have the same slope and different y intercepts, so they are parallel. In this case,

$$\frac{a_{11}}{a_{21}} = \frac{a_{12}}{a_{22}}$$

CASE 3. The system has infinitely many solutions (Fig. 2.1.55c). This occurs whenever the two lines coincide. They have the same slope and y intercept. In this case,

$$\frac{a_{11}}{a_{21}} = \frac{a_{12}}{a_{22}} = \frac{b_1}{b_2}$$

The value $a_{11}a_{22} - a_{21}a_{12}$ is called the **determinant** of the system. A larger $n \times n$ system also has a determinant (see below). A system has exactly one solution when its determinant is not zero.

3 × 2 Systems Any system with more equations than variables is called **overdetermined**. The only case in which a 3 × 2 system has exactly one solution is when one of the equations can be derived from the other two.

One basic way to solve such a system is to treat any two equations as a 2 × 2 system and see if the solution to that subsystem of equations is also a solution to the third equation.

Matrix Form for Systems of Equations The 2 × 2 system of linear equations

$$a_{11}x_1 + a_{12}x_2 = b_1 \quad a_{21}x_1 + a_{22}x_2 = b_2$$

may be written as a matrix equation as follows:

$$\begin{pmatrix} a_{11} & a_{12} \\ a_{21} & a_{22} \end{pmatrix} \begin{pmatrix} x_1 \\ x_2 \end{pmatrix} = \begin{pmatrix} b_1 \\ b_2 \end{pmatrix}$$

or as

$$A\mathbf{x} = \mathbf{b}$$

where A is the 2 × 2 matrix and $\mathbf{x}$ and $\mathbf{b}$ are two-dimensional column vectors. Then, the determinant of A, written det A or $|A|$, is the same as the determinant of the 2 × 2 system:

$$\det A = a_{11}a_{22} - a_{21}a_{12}$$

In general, any $m \times n$ system of simultaneous linear equations may be written as

$$A\mathbf{x} = \mathbf{b}$$

where A is an $m \times n$ matrix, $\mathbf{x}$ is an n-dimensional column vector, and $\mathbf{b}$ is an m-dimensional column vector.

An $n \times n$ (square) system of simultaneous linear equations has exactly one solution whenever its determinant is not zero. Then the system and the matrix A are called **nonsingular**. If the determinant is zero, the system is called **singular**.

Elementary Row Operations on a Matrix There are three operations on a matrix which change the matrix:
1. Multiply each entry in row i by a scalar k (not zero).
2. Interchange row i with row j.
3. Add row i to row j.
Similarly, there are three elementary column operations.
The elementary row operations have the following effects on $|A|$:
1. Multiplying a row (or column) by k multiplies $|A|$ by k.
2. Interchanging two rows (or columns) multiplies $|A|$ by -1.
3. Adding one row (or column) to another does not change $|A|$.

Pivoting, or Reducing, a Column The process of changing the ij entry of a matrix to 1 and changing the rest of column j to zero, by using elementary row operations, is known as **reducing** column j or as **pivoting**

on the ij entry. Combining pivoting, the properties of the elementary row operations, and the fact:

$$|I_{n \times n}| = 1$$

provides a technique for finding the determinant of $n \times n$ matrices.

EXAMPLE. Find $|A|$ where

$$A = \begin{pmatrix} 1 & 2 & -4 \\ 5 & -3 & -7 \\ 3 & -2 & 3 \end{pmatrix}$$

First, pivot on the entry in row 1, column 1, in this case, the 1.
Multiplying row 1 by -5, then adding row 1 to row 2, we first multiply the determinant by -5, then do not change it:

$$-5|A| = \begin{vmatrix} -5 & -10 & 20 \\ 5 & -3 & -7 \\ 3 & -2 & 3 \end{vmatrix} = \begin{vmatrix} -5 & -10 & 20 \\ 0 & -13 & 13 \\ 3 & -2 & 3 \end{vmatrix}$$

Next, multiply row 1 by $\frac{3}{5}$ and add row 1 to row 3:

$$-3|A| = \begin{vmatrix} -3 & -6 & 12 \\ 0 & -13 & 13 \\ 3 & -2 & 3 \end{vmatrix} = \begin{vmatrix} -3 & -6 & 12 \\ 0 & -13 & 13 \\ 0 & -8 & 15 \end{vmatrix}$$

Next, divide row 1 by -3:

$$|A| = \begin{vmatrix} 1 & 2 & -4 \\ 0 & -13 & 13 \\ 0 & -8 & 15 \end{vmatrix}$$

Next, pivot on the entry in row 2, column 2. Multiplying row 2 by $-\frac{8}{13}$ and then adding row 2 to row 3, we get:

$$-\frac{8}{13}|A| = \begin{vmatrix} 1 & 2 & -4 \\ 0 & 8 & -8 \\ 0 & -8 & 15 \end{vmatrix} = \begin{vmatrix} 1 & 2 & -4 \\ 0 & 8 & -8 \\ 0 & 0 & 7 \end{vmatrix}$$

Next, divide row 2 by $-\frac{8}{13}$.

$$|A| = \begin{vmatrix} 1 & 2 & -4 \\ 0 & -13 & 13 \\ 0 & 0 & 7 \end{vmatrix}$$

The determinant of a triangular matrix is the product of its diagonal elements, in this case -91.

Inverses Whenever $|A|$ is not zero, that is, whenever A is nonsingular, then there is another $n \times n$ matrix, denoted A^{-1}, read "A inverse" with the property

$$AA^{-1} = A^{-1}A = I_{n \times n}$$

Then the $n \times n$ system of equations

$$A\mathbf{x} = \mathbf{b}$$

can be solved by multiplying both sides by A^{-1}, so

$$\mathbf{x} = I_{n \times n}\mathbf{x} = A^{-1}A\mathbf{x} = A^{-1}\mathbf{b}$$

so

$$\mathbf{x} = A^{-1}\mathbf{b}$$

The matrix A^{-1} may be found as follows:
1. Make a $n \times 2n$ matrix, with the first n columns the matrix A and the last n columns the identity matrix $I_{n \times n}$.
2. Pivot on each of the diagonal entries of this matrix, one after another, using the elementary row operations.
3. After pivoting n times, the matrix will have in the first n columns the identity matrix, and the last n columns will be the matrix A^{-1}.

EXAMPLE. Solve the system

$$x_1 + 2x_2 - 4x_3 = -4$$
$$5x_1 - 3x_2 - 7x_3 = 6$$
$$3x_1 - 2x_2 + 3x_3 = 11$$

We must invert the matrix

$$A = \begin{pmatrix} 1 & 2 & -4 \\ 5 & -3 & -7 \\ 3 & -2 & 3 \end{pmatrix}$$

This is the same matrix used in the determinant example above. Adjoin the identity matrix to make a 3×6 matrix

$$\begin{pmatrix} 1 & 2 & -4 & 1 & 0 & 0 \\ 5 & -3 & -7 & 0 & 1 & 0 \\ 3 & -2 & 3 & 0 & 0 & 1 \end{pmatrix}$$

Perform the elementary row operations in exactly the same order as in the determinant example.

STEP 1.   Pivot on row 1, column 1.

$$\begin{pmatrix} 1 & 2 & -4 & 1 & 0 & 0 \\ 0 & -13 & 13 & -5 & 1 & 0 \\ 0 & -8 & 15 & -3 & 0 & 1 \end{pmatrix}$$

STEP 2.   Pivot on row 2, column 2.

$$\begin{pmatrix} 1 & 0 & -2 & \frac{3}{13} & \frac{2}{13} & 0 \\ 0 & 1 & -1 & \frac{5}{13} & -\frac{1}{13} & 0 \\ 0 & 0 & 7 & \frac{1}{13} & -\frac{8}{13} & 1 \end{pmatrix}$$

STEP 3.   Pivot on row 3, column 3.

$$\begin{pmatrix} 1 & 0 & 0 & \frac{23}{91} & -\frac{2}{91} & \frac{26}{91} \\ 0 & 1 & 0 & \frac{36}{91} & -\frac{15}{91} & \frac{13}{91} \\ 0 & 0 & 1 & \frac{1}{91} & -\frac{8}{91} & \frac{13}{91} \end{pmatrix}$$

Now, the inverse matrix appears on the right. To solve the equation,

$$\mathbf{x} = A^{-1}\mathbf{b}$$

so,

$$\mathbf{x} = \begin{pmatrix} \frac{23}{91} & -\frac{2}{91} & \frac{26}{91} \\ \frac{36}{91} & -\frac{15}{91} & \frac{13}{91} \\ \frac{1}{91} & -\frac{8}{91} & \frac{13}{91} \end{pmatrix} \begin{pmatrix} -4 \\ 6 \\ 11 \end{pmatrix}$$

$$= \begin{pmatrix} (-4 \times 23 + 6 \times -2 + 11 \times 26)/91 \\ (-4 \times 36 + 6 \times -15 + 11 \times 13)/91 \\ (-4 \times 1 + 6 \times -8 + 11 \times 13)/91 \end{pmatrix} = \begin{pmatrix} 2 \\ -1 \\ 1 \end{pmatrix}$$

The solution to the system is then

$$x_1 = 2 \qquad x_2 = -1 \qquad x_3 = 1$$

Special Matrices   If A is a matrix of complex numbers, then it is possible to take the complex conjugate $a_{ij}*$ of each entry, a_{ij}. This is called the **conjugate** of A and is denoted $A*$.

1. If $a_{ij} = a_{ji}$, then A is **symmetric**.
2. If $a_{ij} = -a_{ji}$, then A is skew or **antisymmetric**.
3. If $A^T = A^{-1}$, then A is **orthogonal**.
4. If $A = A^{-1}$, then A is **involutory**.
5. If $A = A*$, then A is **hermitian**.
6. If $A = -A*$, then A is **skew hermitian**.
7. If $A^{-1} = A*$, then A is **unitary**.

Eigenvalues and Eigenvectors   If A is a square matrix and x is a variable, then the matrix $B = A - xI$ is the **characteristic matrix**, or **eigenmatrix**, of A. The determinant $|A - xI|$ is a polynomial of degree n, called the **characteristic polynomial** of A. The roots of this polynomial, $x_1, x_2, \ldots, x_n$, are the **eigenvalues** of A.

Note that some sources define the characteristic matrix as $xI - A$. If n is odd, then this multiplies the characteristic equation by -1, but the eigenvalues are not changed.

EXAMPLE. $A = \begin{vmatrix} -2 & 5 \\ 2 & 1 \end{vmatrix}$ $B = \begin{vmatrix} -2 - x & 5 \\ 2 & 1 - x \end{vmatrix}$

Then the characteristic polynomial is

$$|B| = (-2 - x)(1 - x) - (2)(5)$$
$$= x^2 + x - 2 - 10$$
$$= x^2 + x - 12$$
$$= (x + 4)(x - 3)$$

The eigenvalues are -4 and $+3$.

A nonzero vector $\mathbf{v}$ satisfying

$$(A - x_i I)\mathbf{v} = \mathbf{0}$$

is called an **eigenvector** of A associated with the eigenvalue x_i. Eigenvectors have the special property

$$A\mathbf{v} = x_i\mathbf{v}$$

Any multiple of an eigenvector is also an eigenvector.

A matrix is nonsingular when none of its eigenvalues are zero.

Rank and Nullity   It is possible that the product of a nonzero matrix A and a nonzero vector $\mathbf{v}$ is zero. This cannot happen if A is nonsingular.

The set of all vectors which become zero when multiplied by A is called the **kernel** of A. The **nullity** of A is the dimension of the kernel. It is a measure of how singular a matrix is.

If A is an $m \times n$ matrix, then the **rank** of A is defined as n − nullity. Rank is at most m.

The technique of pivoting is useful in finding the rank of a matrix. The procedure is as follows:

1. Pivot on each diagonal entry in the matrix, starting with a_{11}.
2. If a row becomes all zero, exchange it with other rows to move it to the bottom of the matrix.
3. If a diagonal entry is zero but the row is not all zero, exchange the column containing the entry with a column to the right not containing a zero in that row.

When the procedure can be carried no further, the nullity is the number of rows of zeros in the matrix.

EXAMPLE.   Find the rank and nullity of the 3×2 matrix:

$$\begin{pmatrix} 1 & 1 \\ 2 & 1 \\ 4 & 1 \end{pmatrix}$$

Pivoting on row 1, column 1, yields

$$\begin{pmatrix} 1 & 0 \\ 0 & -1 \\ 0 & -3 \end{pmatrix}$$

Pivoting on row 2, column 2, yields

$$\begin{pmatrix} 1 & 0 \\ 0 & 1 \\ 0 & 0 \end{pmatrix}$$

Nullity is therefore 1. Rank is $3 - 1 = 2$.

If the rank of a matrix is n, so that

$$\text{Rank} + \text{nullity} = m$$

the matrix is said to be *full rank*.

TRIGONOMETRY

Formal Trigonometry

Angles or Rotations   An **angle** is generated by the rotation of a ray, as Ox, about a fixed point O in the plane. Every angle has an **initial line** (OA) from which the rotation started (Fig. 2.1.56), and a **terminal line** (OB) where it stopped; and the counterclockwise direction of rotation is taken as positive. Since the rotating ray may revolve as often as desired, angles of any magnitude, positive or negative, may be obtained. Two angles are **congruent** if they may be superimposed so that their initial lines coincide and their terminal lines coincide; i.e., two congruent angles are either equal or differ by some multiple of 360°. Two angles are **complementary** if their sum is 90°; **supplementary** if their sum is 180°.

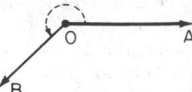

Fig. 2.1.56   Angle.

(The acute angles of a right-angled triangle are complementary.) If the initial line is placed so that it runs horizontally to the right, as in Fig. 2.1.57, then the angle is said to be an angle in the 1st, 2nd, 3rd, or 4th **quadrant** according as the terminal line lies across the region marked I, II, III, or IV.

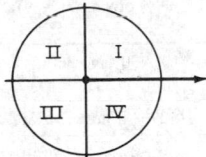

Fig. 2.1.57   Circle showing quadrants.

Units of Angular Measurement

1. *Sexagesimal measure.* (360 degrees = 1 revolution.) Denoted on many calculators by DEG. 1 degree = 1° = $\frac{1}{90}$ of a right angle. The degree is usually divided into 60 equal parts called minutes ('), and each minute into 60 equal parts called seconds ("); while the second is subdivided decimally. But for many purposes it is more convenient to divide the degree itself into decimal parts, thus avoiding the use of minutes and seconds.

2. *Centesimal measure.* Used chiefly in France. Denoted on calculators by GRAD. (400 grades = 1 revolution.) 1 grade − $\frac{1}{100}$ of a right angle. The grade is always divided decimally, the following terms being sometimes used: 1 "centesimal minute" = $\frac{1}{100}$ of a grade; 1 "centesimal second" = $\frac{1}{100}$ of a centesimal minute. In reading Continental books it is important to notice carefully which system is employed.

3. *Radian, or circular, measure.* (π radians = 180 degrees.) Denoted by RAD. 1 radian = the angle subtended by an arc whose length is equal to the length of the radius. The radian is constantly used in higher mathematics and in mechanics, and is always divided decimally. Many theorems in calculus assume that angles are being measured in radians, not degrees, and are not true without that assumption.

$$1 \text{ radian} = 57°.30 - = 57°.2957795131$$
$$= 57°17'44''.806247 = 180°/\pi$$
$$1° = 0.01745 \ldots \text{ radian} = 0.01745\ 32925 \text{ radian}$$
$$1' = 0.00029\ 08882 \text{ radian}$$
$$1'' = 0.00000\ 48481 \text{ radian}$$

Table 2.1.2   Signs of the Trigonometric Functions

If x is in quadrant	I	II	III	IV
sin x and csc x are	+	+	−	−
cos x and sec x are	+	−	−	+
tan x and cot x are	+	−	+	−

Definitions of the Trigonometric Functions   Let x be any angle whose initial line is OA and terminal line OP (see Fig. 2.1.58). Drop a perpendicular from P on OA or OA produced. In the right triangle OMP, the three sides are MP = "side opposite" O (positive if running upward); OM = "side adjacent" to O (positive if running to the right); OP = "hypotenuse" or "radius" (may always be taken as positive); and the six ratios between these sides are the principal trigonometric

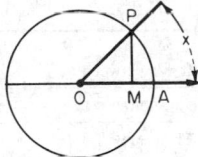

Fig. 2.1.58   Unit circle showing elements used in trigonometric functions.

functions of the angle x; thus:

$$\text{sine of } x = \sin x = \text{opp/hyp} = MP/OP$$
$$\text{cosine of } x = \cos x = \text{adj/hyp} = OM/OP$$
$$\text{tangent of } x = \tan x = \text{opp/adj} = MP/OM$$
$$\text{cotangent of } x = \cot x = \text{adj/opp} = OM/MP$$
$$\text{secant of } x = \sec x = \text{hyp/adj} = OP/OM$$
$$\text{cosecant of } x = \csc x = \text{hyp/opp} = OP/MP$$

The last three are best remembered as the reciprocals of the first three:

$$\cot x = 1/\tan x \qquad \sec x = 1/\cos x \qquad \csc x = 1/\sin x$$

Trigonometric functions, the exponential functions, and complex numbers are all related by the **Euler formula**: $e^{ix} = \cos x + i \sin x$, where $i = \sqrt{-1}$. A special case of this $e^{i\pi} = -1$. Note that here x must be measured in radians.

Variations in the functions as x varies from 0 to 360° are shown in Table 2.1.3. The variations in the sine and cosine are best remembered by noting the changes in the lines MP and OM (Fig. 2.1.59) in the "unit circle" (i.e., a circle with radius = OP = 1), as P moves around the circumference.

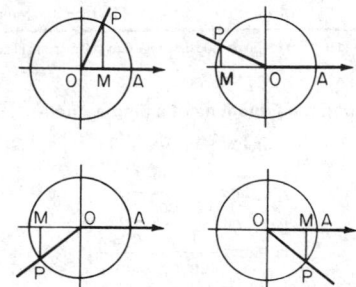

Fig. 2.1.59   Unit circle showing angles in the various quadrants.

Table 2.1.3   Ranges of the Trigonometric Functions

					Values at		
x in DEG x in RAD	0° to 90° (0 to $\pi/2$)	90° to 180° ($\pi/2$ to π)	180° to 270° (π to $3\pi/2$)	270° to 360° ($3\pi/2$ to 2π)	30° ($\pi/6$)	45° ($\pi/4$)	60° ($\pi/3$)
sin x	+0 to +1	+1 to +0	−0 to −1	−1 to −0	$\frac{1}{2}$	$\frac{1}{2}\sqrt{2}$	$\frac{1}{2}\sqrt{3}$
csc x	+∞ to +1	+1 to +∞	−∞ to −1	−1 to −∞	2	$\sqrt{2}$	$\frac{2}{3}\sqrt{3}$
cos x	+1 to +0	−0 to −1	−1 to −0	+0 to +1	$\frac{1}{2}\sqrt{3}$	$\frac{1}{2}\sqrt{2}$	$\frac{1}{2}$
sec x	+1 to +∞	−∞ to −1	−1 to −∞	+∞ to +1	$\frac{2}{3}\sqrt{3}$	$\sqrt{2}$	2
tan x	+0 to +∞	−∞ to −0	+0 to +∞	−∞ to −0	$\frac{1}{2}\sqrt{3}$	1	$\sqrt{3}$
cot x	+∞ to +0	−0 to −∞	+∞ to +0	−0 to −∞	$\sqrt{3}$	1	$\frac{1}{3}\sqrt{3}$

To Find Any Function of a Given Angle (Reduction to the first quadrant.) It is often required to find the functions of any angle x from a table that includes only angles between 0 and 90°. If x is not already between 0 and 360°, first "reduce to the first revolution" by simply adding or subtracting the proper multiple of 360° [for any function of (x) = the same function of $(x \pm n \times 360°)$]. Next **reduce to first quadrant** per table below.

If x is between	90° and 180° ($\pi/2$ and π)	180° and 270° (π and $3\pi/2$)	270° and 360° ($3\pi/2$ and 2π)
Subtract	90° from x ($\pi/2$)	180° from x (π)	270° from x ($3\pi/2$)
Then $\sin x$	$= + \cos (x - 90°)$	$= - \sin (x - 180°)$	$= - \cos (x - 270°)$
$\csc x$	$= + \sec (x - 90°)$	$= - \csc (x - 180°)$	$= - \sec (x - 270°)$
$\cos x$	$= - \sin (x - 90°)$	$= - \cos (x - 180°)$	$= + \sin (x - 270°)$
$\sec x$	$= - \csc (x - 90°)$	$= - \sec (x - 180°)$	$= + \csc (x - 270°)$
$\tan x$	$= - \cot (x - 90°)$	$= + \tan (x - 180°)$	$= - \cot (x - 270°)$
$\cot x$	$= - \tan (x - 90°)$	$= + \cot (x - 180°)$	$= - \tan (x - 270°)$

The "reduced angle" ($x - 90°$, or $x - 180°$, or $x - 270°$) will in each case be an angle between 0 and 90°, whose functions can then be found in the table.

NOTE.   The formulas for sine and cosine are best remembered by aid of the unit circle.

To Find the Angle When One of Its Functions Is Given   In general, there will be two angles between 0 and 360° corresponding to any given function. The rules showing how to find these angles are tabulated below.

Given	First find an *acute* angle x_0 such that	Then the required angles x_1 and x_2 will be*
$\sin x = +a$	$\sin x_0 = a$	x_0 and $180° - x_0$
$\cos x = +a$	$\cos x_0 = a$	x_0 and $[360° - x_0]$
$\tan x = +a$	$\tan x_0 = a$	x_0 and $[180° + x_0]$
$\cot x = +a$	$\cot x_0 = a$	x_0 and $[180° + x_0]$
$\sin x = -a$	$\sin x_0 = a$	$[180° + x_0]$ and $[360° - x_0]$
$\cos x = -a$	$\cos x_0 = a$	$180° - x_0$ and $[180° + x_0]$
$\tan x = -a$	$\tan x_0 = a$	$180° - x_0$ and $[360° - x_0]$
$\cot x = -a$	$\cot x_0 = a$	$180° - x_0$ and $[360° - x_0]$

* The angles enclosed in brackets lie outside the range 0 to 180 deg and hence cannot occur as angles in a triangle.

Relations Among the Functions of a Single Angle

$$\sin^2 x + \cos^2 x = 1$$

$$\tan x = \frac{\sin x}{\cos x}$$

$$\cot x = \frac{1}{\tan x} = \frac{\cos x}{\sin x}$$

$$1 + \tan^2 x = \sec^2 x = \frac{1}{\cos^2 x}$$

$$1 + \cot^2 x = \csc^2 x = \frac{1}{\sin^2 x}$$

$$\sin x = \sqrt{1 - \cos^2 x} = \frac{\tan x}{\sqrt{1 + \tan^2 x}} = \frac{1}{\sqrt{1 + \cot^2 x}}$$

$$\cos x = \sqrt{1 - \sin^2 x} = \frac{1}{\sqrt{1 + \tan^2 x}} = \frac{\cot x}{\sqrt{1 + \cot^2 x}}$$

Functions of Negative Angles $\sin (-x) = -\sin x$; $\cos (-x) = \cos x$; $\tan (-x) = -\tan x$.

Functions of the Sum and Difference of Two Angles

$\sin (x + y) = \sin x \cos y + \cos x \sin y$
$\cos (x + y) = \cos x \cos y - \sin x \sin y$

$\tan (x + y) = (\tan x + \tan y)/(1 - \tan x \tan y)$
$\cot (x + y) = (\cot x \cot y - 1)/(\cot x + \cot y)$
$\sin (x - y) = \sin x \cos y - \cos x \sin y$
$\cos (x - y) = \cos x \cos y + \sin x \sin y$
$\tan (x - y) = (\tan x - \tan y)/(1 + \tan x \tan y)$
$\cot (x - y) = (\cot x \cot y + 1)/(\cot y - \cot x)$
$\sin x + \sin y = 2 \sin \frac{1}{2}(x + y) \cos \frac{1}{2}(x - y)$

$\sin x - \sin y = 2 \cos \frac{1}{2}(x + y) \sin \frac{1}{2}(x - y)$
$\cos x + \cos y = 2 \cos \frac{1}{2}(x + y) \cos \frac{1}{2}(x - y)$
$\cos x - \cos y = - 2 \sin \frac{1}{2}(x + y) \sin \frac{1}{2}(x - y)$

$$\tan x + \tan y = \frac{\sin (x + y)}{\cos x \cos y}; \quad \cot x + \cot y = \frac{\sin (x + y)}{\sin x \sin y}$$

$$\tan x - \tan y = \frac{\sin (x - y)}{\cos x \cos y}; \quad \cot x - \cot y = \frac{\sin (y - x)}{\sin x \sin y}$$

$\sin^2 x - \sin^2 y = \cos^2 y - \cos^2 x = \sin (x + y) \sin (x - y)$
$\cos^2 x - \sin^2 y = \cos^2 y - \sin^2 x = \cos (x + y) \cos (x - y)$
$\sin (45° + x) = \cos (45° - x)$
$\tan (45° + x) = \cot (45° - x)$
$\sin (45° - x) = \cos (45° + x)$
$\tan (45° - x) = \cot (45° + x)$

In the following transformations, a and b are supposed to be positive, $c = \sqrt{a^2 + b^2}$, A = the positive acute angle for which $\tan A = a/b$, and B = the positive acute angle for which $\tan B = b/a$:

$$a \cos x + b \sin x = c \sin (A + x) = c \cos (B - x)$$
$$a \cos x - b \sin x = c \sin (A - x) = c \cos (B + x)$$

Functions of Multiple Angles and Half Angles

$\sin 2x = 2 \sin x \cos x$; $\sin x = 2 \sin \frac{1}{2}x \cos \frac{1}{2}x$
$\cos 2x = \cos^2 x - \sin^2 x = 1 - 2 \sin^2 x = 2 \cos^2 x - 1$

$$\tan 2x = \frac{2 \tan x}{1 - \tan^2 x} \qquad \cot 2x = \frac{\cot^2 x - 1}{2 \cot x}$$

$$\sin 3x = 3 \sin x - 4 \sin^3 x; \quad \tan 3x = \frac{3 \tan x - \tan^3 x}{1 - 3 \tan^2 x}$$

$\cos 3x = 4 \cos^3 x - 3 \cos x$
$\sin (nx) = n \sin x \cos^{n-1} x - (n)_3 \sin^3 x \cos^{n-3} x$
$\qquad\qquad\qquad + (n)_5 \sin^5 x \cos^{n-5} x - \cdots$
$\cos (nx) = \cos^n x - (n)_2 \sin^2 x \cos^{n-2} x + (n)_4 \sin^4 x \cos^{n-4} x - \cdots$

where $(n)_2$, $(n)_3$, . . . , are the binomial coefficients.

$$\sin \tfrac{1}{2}x = \pm \sqrt{\tfrac{1}{2}(1 - \cos x)}. \quad 1 - \cos x = 2 \sin^2 \tfrac{1}{2}x$$

$$\cos \tfrac{1}{2}x = \pm \sqrt{\tfrac{1}{2}(1 + \cos x)}. \quad 1 + \cos x = 2 \cos^2 \tfrac{1}{2}x$$

$$\tan \tfrac{1}{2}x = \pm \sqrt{\frac{1 - \cos x}{1 + \cos x}} = \frac{\sin x}{1 + \cos x} = \frac{1 - \cos x}{\sin x}$$

$$\tan \left(\frac{x}{2} + 45° \right) = \pm \sqrt{\frac{1 + \sin x}{1 - \sin x}}$$

Here the + or − sign is to be used according to the sign of the left-hand side of the equation.

Approximations for $\sin x$, $\cos x$, and $\tan x$ For small values of x,

x measured in radians, the following approximations hold:

$$\sin x \approx x \qquad \tan x \approx x \qquad \cos x \approx 1 - \frac{x^2}{2}$$

The following actually hold:

$$\sin x < x < \tan x \qquad \cos x < \frac{\sin x}{x} < 1$$

As x approaches 0, $\lim [(\sin x)/x] = 1$.

Inverse Trigonometric Functions   The notation $\sin^{-1} x$ (read: antisine of x, or inverse sine of x; sometimes written arc sin x) means the principal angle whose sine is x. Similarly for $\cos^{-1} x$, $\tan^{-1} x$, etc. (The principal angle means an angle between -90 and $+90°$ in case of $\sin^{-1}$ and $\tan^{-1}$, and between 0 and $180°$ in the case of $\cos^{-1}$.)

Solution of Plane Triangles

The "parts" of a plane triangle are its three sides a, b, c, and its three angles A, B, C (A being opposite a). Two triangles are **congruent** if all their corresponding parts are equal. Two triangles are **similar** if their corresponding angles are equal, that is, $A_1 = A_2$, $B_1 = B_2$, and $C_1 = C_2$. Similar triangles may differ in scale, but they satisfy $a_1/a_2 = b_1/b_2 = c_1/c_2$.

Two different triangles may have two corresponding sides and the angle opposite one of those sides equal (Fig. 2.1.60), and still not be congruent. This is the **angle-side-side theorem**.

Otherwise, a triangle is uniquely determined by any three of its parts, as long as those parts are not all angles. To "solve" a triangle means to find the unknown parts from the known. The fundamental formulas are

$$\text{Law of sines: } \frac{a}{b} = \frac{\sin A}{\sin B}$$

$$\text{Law of cosines: } c^2 = a^2 + b^2 - 2ab \cos C$$

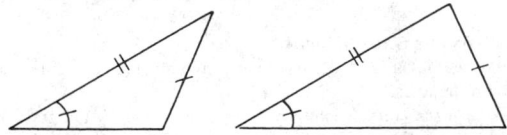

Fig. 2.1.60   Triangles with an angle, an adjacent side, and an opposite side given.

Right Triangles   Use the definitions of the trigonometric functions, selecting for each unknown part a relation which connects that unknown with known quantities; then solve the resulting equations. Thus, in Fig. 2.1.61, if $C = 90°$, then $A + B = 90°$, $c^2 = a^2 + b^2$,

$$\sin A = a/c \qquad \cos A = b/c$$
$$\tan A = a/b \qquad \cot A = b/a$$

If A is very small, use $\tan \frac{1}{2}A = \sqrt{(c - b)/(c + b)}$.

Oblique Triangles   There are four cases. It is highly desirable in all these cases to draw a sketch of the triangle approximately to scale before commencing the computation, so that any large numerical error may be readily detected.

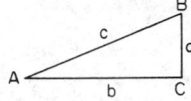

Fig. 2.1.61   Right triangle.

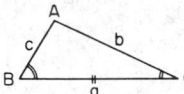

Fig. 2.1.62   Triangle with two angles and the included side given.

CASE 1.   GIVEN TWO ANGLES (provided their sum is $< 180°$) AND ONE SIDE (say a, Fig. 2.1.62). The third angle is known since $A +$

$B + C = 180°$. To find the remaining sides, use

$$b = \frac{a \sin B}{\sin A} \qquad c = \frac{a \sin C}{\sin A}$$

Or, drop a perpendicular from either B or C on the opposite side, and solve by right triangles.

Check: $c \cos B + b \cos C = a$.

CASE 2.   GIVEN TWO SIDES (say a and b) AND THE INCLUDED ANGLE (C); AND SUPPOSE $a > b$ (Fig. 2.1.63).

Method 1: Find c from $c^2 = a^2 + b^2 - 2ab \cos C$; then find the smaller angle, B, from $\sin B = (b/c) \sin C$; and finally, find A from $A = 180° - (B + C)$.

Check: $a \cos B + b \cos A = c$.

Method 2: Find $\frac{1}{2}(A - B)$ from the law of tangents:

$$\tan \tfrac{1}{2}(A - B) = [(a - b)/(a + b)] \cot \tfrac{1}{2}C$$

and $\frac{1}{2}(A + B)$ from $\frac{1}{2}(A + B) = 90° - C/2$; hence $A = \frac{1}{2}(A + B) + \frac{1}{2}(A - B)$ and $B = \frac{1}{2}(A + B) - \frac{1}{2}(A - B)$. Then find c from $c = a \sin C/\sin A$ or $c = b \sin C/\sin B$.

Check: $a \cos B + b \cos A = c$.

Method 3: Drop a perpendicular from A to the opposite side, and solve by right triangles.

CASE 3.   GIVEN THE THREE SIDES (provided the largest is less than the sum of the other two) (Fig. 2.1.64).

Method 1: Find the largest angle A (which may be acute or obtuse) from $\cos A = (b^2 + c^2 - a^2)/2bc$ and then find B and C (which will always be acute) from $\sin B = b \sin A/a$ and $\sin C = c \sin A/a$.

Check: $A + B + C = 180°$.

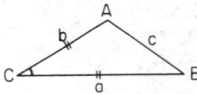

Fig. 2.1.63   Triangle with two sides and the included angle given.

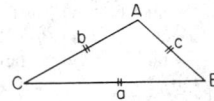

Fig. 2.1.64   Triangle with three sides given.

Method 2: Find A, B, and C from $\tan \frac{1}{2}A = r/(s - a)$, $\tan \frac{1}{2}B = r/(s - b)$, $\tan \frac{1}{2}C = r/(s - c)$, where $s = \frac{1}{2}(a + b + c)$, and $r = \sqrt{(s - a)(s - b)(s - c)/s}$. Check: $A + B + C = 180°$.

Method 3: If only one angle, say A, is required, use

$$\sin \tfrac{1}{2}A = \sqrt{(s - b)(s - c)/bc}$$
$$\text{or} \qquad \cos \tfrac{1}{2}A = \sqrt{s(s - a)/bc}$$

according as $\frac{1}{2}A$ is nearer $0°$ or nearer $90°$.

CASE 4.   GIVEN TWO SIDES (say b and c) AND THE ANGLE OPPOSITE ONE OF THEM (B). This is the "ambiguous case" in which there may be two solutions, or one, or none.

First, try to find $C = c \sin B/b$. If $\sin C > 1$, there is no solution. If $\sin C = 1$, $C = 90°$ and the triangle is a right triangle. If $\sin C < 1$, this determines two angles C, namely, an acute angle C_1, and an obtuse angle $C_2 = 180° - C_1$. Then C_1 will yield a solution when and only when $C_1 + B < 180°$ (see Case 1); and similarly C_2 will yield a solution when and only when $C_2 + B < 180°$ (see Case 1).

Other Properties of Triangles   (See also Geometry, Areas, and Volumes.)

Area $= \frac{1}{2}ab \sin C = \sqrt{s(s - a)(s - b)(s - c)} = rs$ where $s = \frac{1}{2}(a + b + c)$, and $r =$ radius of inscribed circle $= \sqrt{(s - a)(s - b)(s - c)/s}$.

Radius of circumscribed circle $= R$, where

$$2R = a/\sin A = b/\sin B = c/\sin C$$
$$r = 4R \sin \frac{A}{2} \sin \frac{B}{2} \sin \frac{C}{2} = \frac{abc}{4Rs}$$

The length of the bisector of the angle C is

$$z = \frac{2\sqrt{abs(s-c)}}{a+b} = \frac{\sqrt{ab[(a+b)^2 - c^2]}}{a+b}$$

The median from C to the middle point of c is $m = \frac{1}{2}\sqrt{2(a^2 + b^2) - c^2}$.

Hyperbolic Functions

The **hyperbolic sine, hyperbolic cosine,** etc., of any number x, are functions of x which are closely related to the exponential e^x, and which have formal properties very similar to those of the trigonometric functions, sine, cosine, etc. Their definitions and fundamental properties are as follows:

$$\sinh x = \frac{1}{2}(e^x - e^{-x})$$
$$\cosh x = \frac{1}{2}(e^x + e^{-x})$$
$$\tanh x = \sinh x / \cosh x$$

$$\cosh x + \sinh x = e^x$$
$$\cosh x - \sinh x = e^{-x}$$

$$\operatorname{csch} x = 1/\sinh x$$
$$\operatorname{sech} x = 1/\cosh x$$
$$\coth x = 1/\tanh x$$

$$\cosh^2 x - \sinh^2 x = 1$$
$$1 - \tanh^2 x = \operatorname{sech}^2 x$$
$$1 - \coth^2 x = -\operatorname{csch}^2 x$$

$$\sinh(-x) = -\sinh x$$
$$\cosh(-x) = \cosh x$$
$$\tanh(-x) = -\tanh x$$

$$\sinh(x \pm y) = \sinh x \cosh y \pm \cosh x \sinh y$$
$$\cosh(x \pm y) = \cosh x \cosh y \pm \sinh x \sinh y$$
$$\tanh(x \pm y) = (\tanh x \pm \tanh y)/(1 \pm \tanh x \tanh y)$$

$$\sinh 2x = 2\sinh x \cosh x$$
$$\cosh 2x = \cosh^2 x + \sinh^2 x$$
$$\tanh 2x = (2\tanh x)/(1 + \tanh^2 x)$$

$$\sinh \tfrac{1}{2}x = \sqrt{\tfrac{1}{2}(\cosh x - 1)}$$
$$\cosh \tfrac{1}{2}x = \sqrt{\tfrac{1}{2}(\cosh x + 1)}$$
$$\tanh \tfrac{1}{2}x = (\cosh x - 1)/(\sinh x) = (\sinh x)/(\cosh x + 1)$$

The hyperbolic functions are related to the rectangular hyperbola, $x^2 - y^2 = a^2$ (Fig. 2.1.66), in much the same way that the trigonometric functions are related to the circle $x^2 + y^2 = a^2$ (Fig. 2.1.65); the analogy, however, concerns not angles but areas. Thus, in either figure, let A

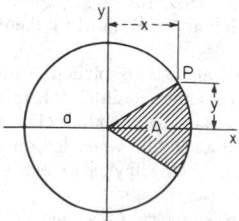

Fig. 2.1.65 Circle.

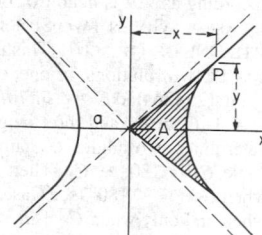

Fig. 2.1.66 Hyperbola.

represent the shaded area, and let $u = A/a^2$ (a pure number). Then for the coordinates of the point P we have, in Fig. 2.1.65, $x = a\cos u$, $y = a\sin u$; and in Fig. 2.1.66, $x = a\cosh u$, $y = a\sinh u$.

The **inverse hyperbolic sine** of y, denoted by $\sinh^{-1} y$, is the number whose hyperbolic sine is y; that is, the notation $x = \sinh^{-1} y$ means $\sinh x = y$. Similarly for $\cosh^{-1} y$, $\tanh^{-1} y$, etc. These functions are closely related to the logarithmic function, and are especially valuable in the integral calculus.

$$\sinh^{-1}(y/a) = \ln(y + \sqrt{y^2 + a^2}) - \ln a$$
$$\cosh^{-1}(y/a) = \ln(y + \sqrt{y^2 - a^2}) - \ln a$$

$$\tanh^{-1}\frac{y}{a} = \tfrac{1}{2}\ln\frac{a+y}{a-y}$$

$$\coth^{-1}\frac{y}{a} = \tfrac{1}{2}\ln\frac{y+a}{y-a}$$

ANALYTICAL GEOMETRY

The Point and the Straight Line

Rectangular Coordinates (Fig. 2.1.67) Let $P_1 = (x_1, y_1)$, $P_2 = (x_2, y_2)$. Then, distance

$$P_1P_2 = \sqrt{(x_2 - x_1)^2 + (y_2 - y_1)^2}$$

slope of $P_1P_2 = m = \tan u = (y_2 - y_1)/(x_2 - x_1)$; coordinates of midpoint are $x = \frac{1}{2}(x_1 + x_2)$, $y = \frac{1}{2}(y_1 + y_2)$; coordinates of point $1/n$th of the way from P_1 to P_2 are $x = x_1 + (1/n)(x_2 - x_1)$, $y = y_1 + (1/n)(y_2 - y_1)$.

Let m_1, m_2 be the slopes of two lines; then, if the lines are parallel, $m_1 = m_2$; if the lines are perpendicular to each other, $m_1 = -1/m_2$.

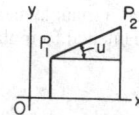

Fig. 2.1.67 Graph of straight line.

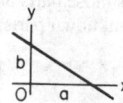

Fig. 2.1.68 Graph of straight line showing intercepts.

Equations of a Straight Line

1. *Intercept form* (Fig. 2.1.68). $x/a + y/b = 1$. (a, b = intercepts of the line on the axes.)
2. *Slope form* (Fig. 2.1.69). $y = mx + b$. ($m = \tan u$ = slope; b = intercept on the y axis.)
3. *Normal form* (Fig. 2.1.70). $x\cos v + y\sin v = p$. (p = perpendicular from origin to line; v = angle from the x axis to p.)

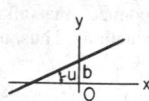

Fig. 2.1.69 Graph of straight line showing slope and vertical intercept.

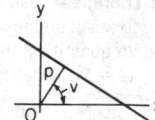

Fig. 2.1.70 Graph of straight line showing perpendicular line from origin.

4. *Parallel-intercept form* (Fig. 2.1.71). $\dfrac{y-b}{c-b} = \dfrac{x}{k}$. ($b$, c = intercepts on two parallels at distance k apart.)

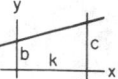

Fig. 2.1.71 Graph of straight line showing intercepts on parallel lines.

5. *General form.* $Ax + By + C = 0$. [Here $a = -C/A$, $b = -C/B$, $m = -A/B$, $\cos v = A/R$, $\sin v = B/R$, $p = -C/R$, where $R = \pm\sqrt{A^2 + B^2}$ (sign to be so chosen that p is positive).]
6. *Line through* (x_1, y_1) *with slope* m. $y - y_1 = m(x - x_1)$.

7. *Line through* (x_1, y_1) *and* (x_2, y_2). $y - y_1 = \dfrac{y_2 - y_1}{x_2 - x_1}(x - x_1)$.

8. *Line parallel to x axis.* $y = a$; to y axis: $x = b$.

Angles and Distances If $u =$ angle from the line with slope m_1 to the line with slope m_2, then

$$\tan u = \frac{m_2 - m_1}{1 + m_2 m_1}$$

If parallel, $m_1 = m_2$.

If perpendicular, $m_1 m_2 = -1$.

If $u =$ angle between the lines $Ax + By + C = 0$ and $A'x + B'y + C' = 0$, then

$$\cos u = \frac{AA' + BB'}{\pm \sqrt{(A^2 + B^2)(A'^2 + B'^2)}}$$

If parallel, $A/A' = B/B'$.

If perpendicular, $AA' + BB' = 0$.

The equation of a line through (x_1, y_1) and meeting a given line $y = mx + b$ at an angle u, is

$$y - y_1 = \frac{m + \tan u}{1 - m \tan u}(x - x_1)$$

The distance from (x_0, y_0) to the line $Ax + By + C = 0$ is

$$D = \left| \frac{Ax_0 + By_0 + C}{\sqrt{A^2 + B^2}} \right|$$

where the vertical bars mean "the absolute value of."

The distance from (x_0, y_0) to a line which passes through (x_1, y_1) and makes an angle u with the x axis is

$$D = (x_0 - x_1) \sin u - (y_0 - y_1) \cos u$$

Polar Coordinates (Fig. 2.1.72) Let (x, y) be the rectangular and (r, θ) the polar coordinates of a given point P. Then $x = r \cos \theta$; $y = r \sin \theta$; $x^2 + y^2 = r^2$.

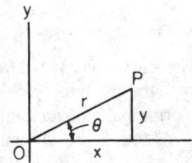

Fig. 2.1.72 Polar coordinates.

Transformation of Coordinates If origin is moved to point (x_0, y_0), the new axes being parallel to the old, $x = x_0 + x'$, $y = y_0 + y'$.

If axes are turned through the angle u, without change of origin,

$$x = x' \cos u - y' \sin u \qquad y = x' \sin u + y' \cos u$$

The Circle

The **equation of a circle** with center (a, b) and radius r is

$$(x - a)^2 + (y - b)^2 = r^2$$

If center is at the origin, the equation becomes $x^2 + y^2 = r^2$. If circle goes through the origin and center is on the x axis at point $(r, 0)$, equation becomes $x^2 + y^2 = 2rx$. The **general equation** of a circle is

$$x^2 + y^2 + Dx + Ey + F = 0$$

It has center at $(-D/2, -E/2)$, and radius $= \sqrt{(D/2)^2 + (E/2)^2 - F}$ (which may be real, null, or imaginary).

Equations of Circle in Parametric Form It is sometimes convenient to express the coordinates x and y of the moving point P (Fig. 2.1.73) in terms of an auxiliary variable, called a **parameter**. Thus, if the parameter be taken as the angle u from the x axis to the radius vector OP, then the equations of the circle in parametric form will be $x = a \cos u$; $y = a$

$\sin u$. For every value of the parameter u, there corresponds a point (x, y) on the circle. The ordinary equation $x^2 + y^2 = a^2$ can be obtained from the parametric equations by eliminating u.

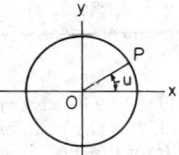

Fig. 2.1.73 Parameters of a circle.

The Parabola

The **parabola** is the locus of a point which moves so that its distance from a fixed line (called the **directrix**) is always equal to its distance from a fixed point F (called the **focus**). See Fig. 2.1.74. The point halfway from focus to directrix is the **vertex**, O. The line through the focus, perpendicular to the directrix, is the **principal axis**. The breadth of the curve at the focus is called the **latus rectum**, or **parameter**, $= 2p$, where p is the distance from focus to directrix.

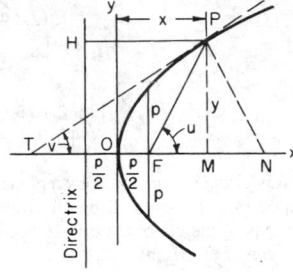

Fig. 2.1.74 Graph of parabola.

NOTE. Any section of a right circular cone made by a plane parallel to a tangent plane of the cone will be a parabola.

Equation of parabola, principal axis along the x axis, origin at vertex (Fig. 2.1.74): $y^2 = 2px$.

Polar equation of parabola, referred to F as origin and Fx as axis (Fig. 2.1.75): $r = p/(1 - \cos \theta)$.

Equation of parabola with principal axis parallel to y axis: $y = ax^2 + bx + c$. This may be rewritten, using a technique called **completing the square**:

$$y = a\left[x^2 + \frac{b}{a}x + \frac{b^2}{4a^2}\right] + c - \frac{b^2}{4a}$$

$$= a\left[x + \frac{b}{2a}\right]^2 + c - \frac{b^2}{4a}$$

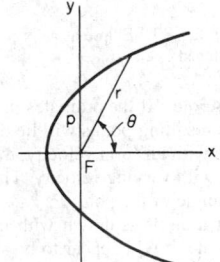

Fig. 2.1.75 Polar plot of parabola.

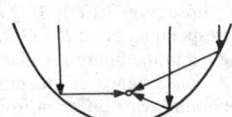

Fig. 2.1.76 Vertical parabola showing rays passing through the focus.

Then: vertex is the point $[-b/2a, c - b^2/4a]$; latus rectum is $p = 1/2a$; and focus is the point $[-b/2a, c - b^2/4a + 1/4a]$.

A parabola has the special property that lines parallel to its principal axis, when reflected off the inside "surface" of the parabola, will all pass through the focus (Fig. 2.1.76). This property makes parabolas useful in designing mirrors and antennas.

The Ellipse

The **ellipse** (as shown in Fig. 2.1.77), has two **foci**, F and F', and two **directrices**, DH and $D'H'$. If P is any point on the curve, $PF + PF'$ is constant, $= 2a$; and PF/PH (or PF'/PH') is also constant, $= e$, where e is the **eccentricity** ($e < 1$). Either of these properties may be taken as the definition of the curve. The relations between e and the semiaxes a and b are as shown in Fig. 2.1.78. Thus, $b^2 = a^2(1 - e^2)$, $ae = \sqrt{a^2 - b^2}$, $e^2 = 1 - (b/a)^2$. The **semilatus rectum** $= p = a(1 - e^2) = b^2/a$. Note that b is always less than a, except in the special case of the circle, in which $b = a$ and $e = 0$.

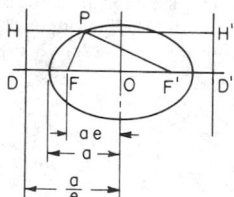

Fig. 2.1.77   Ellipse.

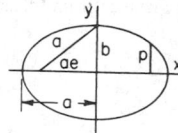

Fig. 2.1.78   Ellipse showing semiaxes.

Any section of a right circular cone made by a plane which cuts all the elements of one nappe of the cone will be an ellipse; if the plane is perpendicular to the axis of the cone, the ellipse becomes a circle.

Equation of ellipse, center at origin:

$$\frac{x^2}{a^2} + \frac{y^2}{b^2} = 1 \qquad \text{or} \qquad y = \pm \frac{b}{a}\sqrt{a^2 - x^2}$$

If $P = (x, y)$ is any point of the curve, $PF = a + ex$, $PF' = a - ex$.

Equations of the ellipse in parametric form: $x = a \cos u$, $y = b \sin u$, where u is the eccentric angle of the point $P = (x, y)$. See Fig. 2.1.81.

Polar equation, focus as origin, axes as in Fig. 2.1.79. $r = p/(1 - e \cos \theta)$.

Equation of the tangent at (x_1, y_1): $b^2 x_1 x + a^2 y_1 y = a^2 b^2$.

The line $y = mx + k$ will be a tangent if $k = \pm\sqrt{a^2 m^2 + b^2}$.

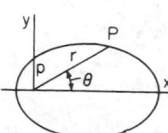

Fig. 2.1.79   Ellipse in polar form.

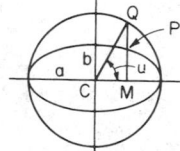

Fig. 2.1.80   Ellipse as a flattened circle.

Ellipse as a Flattened Circle, Eccentric Angle   If the ordinates in a circle are diminished in a constant ratio, the resulting points will lie on an ellipse (Fig. 2.1.80). If Q traces the circle with uniform velocity, the corresponding point P will trace the ellipse, with varying velocity. The angle u in the figure is called the eccentric angle of the point P.

A consequence of this property is that if a circle is drawn with its horizontal scale different from its vertical scale, it will appear to be an ellipse. This phenomenon is common in computer graphics.

The **radius of curvature of an ellipse** at any point $P = (x, y)$ is

$$R = a^2 b^2 (x^2/a^4 + y^2/b^4)^{3/2} = p/\sin^3 v$$

where v is the angle which the tangent at P makes with PF or PF'. At end of major axis, $R = b^2/a = MA$; at end of minor axis, $R = a^2/b = NB$ (see Fig. 2.1.81).

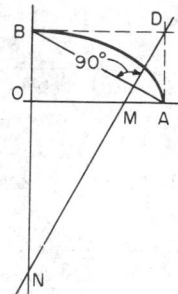

Fig. 2.1.81   Ellipse showing radius of curvature.

The Hyperbola

The **hyperbola** has two **foci**, F and F', at distances $\pm ae$ from the center, and two **directrices**, DH and $D'H'$, at distances $\pm a/e$ from the center (Fig. 2.1.82). If P is any point of the curve, $|PF - PF'|$ is constant, $= 2a$; and PF/PH (or PF'/PH') is also constant, $= e$ (called the **eccentricity**), where $e > 1$. Either of these properties may be taken as the

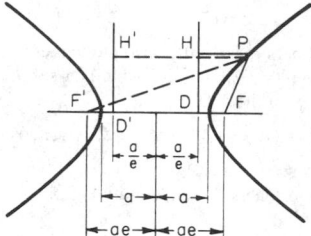

Fig. 2.1.82   Hyperbola.

definition of the curve. The curve has two branches which approach more and more nearly two straight lines called the **asymptotes**. Each asymptote makes with the principal axis an angle whose tangent is b/a. The relations between e, a, and b are shown in Fig. 2.1.83: $b^2 = a^2(e^2 - 1)$, $ae = \sqrt{a^2 + b^2}$, $e^2 = 1 + (b/a)^2$. The semilatus rectum, or ordinate at the focus, is $p = a(e^2 - 1) = b^2/a$.

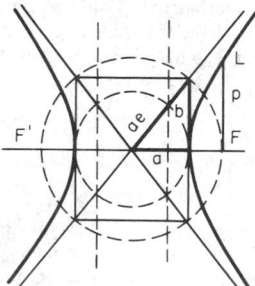

Fig. 2.1.83   Hyperbola showing the asymptotes.

Any section of a right circular cone made by a plane which cuts both nappes of the cone will be a hyperbola.

Equation of the hyperbola, center as origin:

$$\frac{x^2}{a^2} - \frac{y^2}{b^2} = 1 \qquad \text{or} \qquad y = \pm \frac{b}{a}\sqrt{x^2 - a^2}$$

If $P = (x, y)$ is on the right-hand branch, $PF = ex - a$, $PF' = ex + a$. If P is on the left-hand branch, $PF = -ex + a$, $PF' = -ex - a$.

Equations of Hyperbola in Parametric Form   (1) $x = a \cosh u$, $y = b \sinh u$. Here u may be interpreted as A/ab, where A is the area shaded in Fig. 2.1.84. (2) $x = a \sec v$, $y = b \tan v$, where v is an auxiliary angle of no special geometric interest.

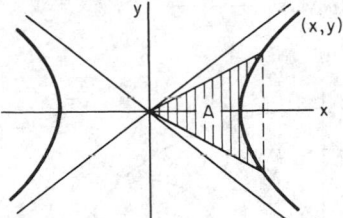

Fig. 2.1.84   Hyperbola showing parametric form.

Polar equation, referred to focus as origin, axes as in Fig. 2.1.85:

$$r = p/(1 - e \cos \theta)$$

Equation of tangent at (x_1, y_1): $b^2 x_1 x - a^2 y_1 y = a^2 b^2$. The line $y = mx + k$ will be a tangent if $k = \pm \sqrt{a^2 m^2 - b^2}$.

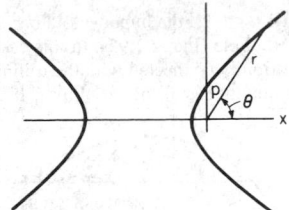

Fig. 2.1.85   Hyperbola in polar form.

The triangle bounded by the asymptotes and a variable tangent is of constant area, $= ab$.

Conjugate hyperbolas are two hyperbolas having the same asymptotes with semiaxes interchanged (Fig. 2.1.86). The equations of the hyperbola conjugate to $x^2/a^2 - y^2/b^2 = 1$ is $x^2/a^2 - y^2/b^2 = -1$.

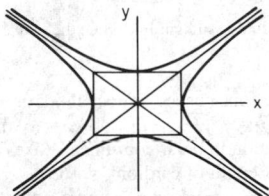

Fig. 2.1.86   Conjugate hyperbolas.

Equilateral Hyperbola $(a = b)$ Equation referred to principal axes (Fig. 2.1.87): $x^2 - y^2 = a^2$.

NOTE. $p = a$ (Fig. 2.1.87). Equation referred to asymptotes as axes (Fig. 2.1.88): $xy = a^2/2$.

Asymptotes are perpendicular. Eccentricity $= \sqrt{2}$. Any diameter is equal in length to its conjugate diameter.

The Catenary

The **catenary** is the curve in which a flexible chain or cord of uniform density will hang when supported by the two ends. Let w = weight of the chain per unit length; T = the tension at any point P; and T_h, T_v = the horizontal and vertical components of T. The horizontal component T_h is the same at all points of the curve.

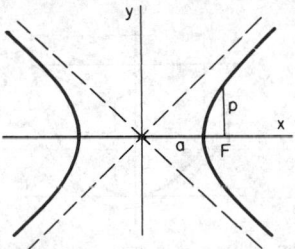

Fig. 2.1.87   Equilateral hyperbola.

The length $a = T_h/w$ is called the **parameter** of the catenary, or the distance from the lowest point O to the **directrix** DQ (Fig. 2.1.89). When a is very large, the curve is very flat.

The rectangular **equation,** referred to the lowest point as origin, is $y = a [\cosh (x/a) - 1]$. In case of very flat arcs (a large), $y = x^2/2a + \cdots$; $s = x + \frac{1}{6}x^3/a^2 + \cdots$, approx, so that in such a case the catenary closely resembles a parabola.

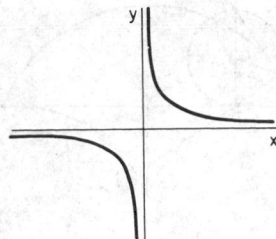

Fig. 2.1.88   Hyperbola with asymptotes as axes.

Calculus properties of the catenary are often discussed in texts on the calculus of variations (Weinstock, "Calculus of Variations," Dover; Ewing, "Calculus of Variations with Applications," Dover).

Problems on the Catenary   (Fig. 2.1.89) When any two of the four quantities, x, y, s, T/w are known, the remaining two, and also the parameter a, can be found, using the following:

$a = x/z$	$s = a \sinh z$
$T = wa \cosh z$	$y/x = (\cosh z - 1)/z$
$s/x = (\sinh z)/z$	$wx/T = z \cosh z$

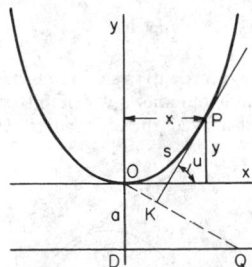

Fig. 2.1.89   Catenary.

NOTE.   If $wx/T < 0.6627$, then there are two values of z, one less than 1.2, and one greater. If $wx/T > 0.6627$, then the problem has no solution.

Given the Length $2L$ **of a Chain Supported at Two Points** A **and** B **Not in the Same Level, to Find** a (See Fig. 2.1.90; b and c are supposed known.) Let $(\sqrt{L^2 - b^2})/c = s/x$; use $s/x = \sinh z/z$ to find z. Then $a = c/z$.

NOTE.   The coordinates of the midpoint M of AB (see Fig. 2.1.90) are $x_0 = a \tanh^{-1} (b/L)$, $y_0 = (L/\tanh z) - a$, so that the position of the lowest point is determined.

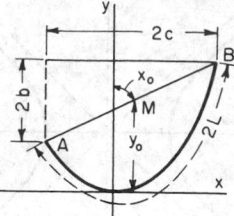

Fig. 2.1.90 Catenary with ends at unequal levels.

Other Useful Curves

The **cycloid** is traced by a point on the circumference of a circle which rolls without slipping along a straight line. Equations of cycloid, in parametric form (axes as in Fig. 2.1.91): $x = a(\text{rad } u - \sin u)$, $y = a(1 - \cos u)$, where a is the radius of the rolling circle, and rad u is the radian measure of the angle u through which it has rolled. The **radius of curvature** at any point P is $PC = 4a \sin (u/2) = 2\sqrt{2ay}$.

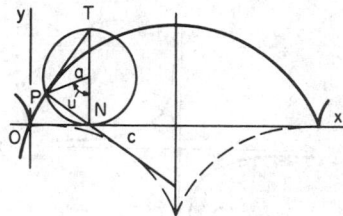

Fig. 2.1.91 Cycloid.

The **trochoid** is a more general curve, traced by any point on a radius of the rolling circle, at distance b from the center (Fig. 2.1.92). It is a prolate trochoid if $b < a$, and a curtate or looped trochoid if $b > a$. The equations in either case are $x = a \text{ rad } u - b \sin u$, $y = a - b \cos u$.

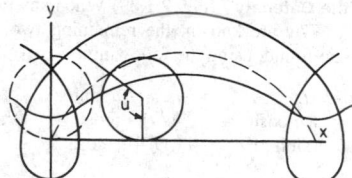

Fig. 2.1.92 Trochoid.

The **epicycloid** (or **hypocycloid**) is a curve generated by a point on the circumference of a circle of radius a which rolls without slipping on the outside (or inside) of a fixed circle of radius c (Fig. 2.1.93 and Fig.

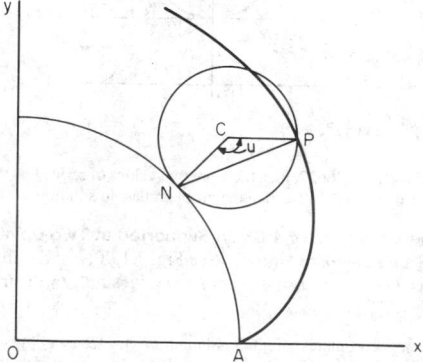

Fig. 2.1.93 Epicycloid.

2.1.94). For the **equations**, put $b = a$ in the equations of the epi- or hypotrochoid, below.

Radius of curvature at any point P is

$$R = \frac{4a(c \pm a)}{c \pm 2a} \times \sin \tfrac{1}{2}u$$

At A, $R = 0$; at D, $R = \dfrac{4a(c \pm a)}{c \pm 2a}$.

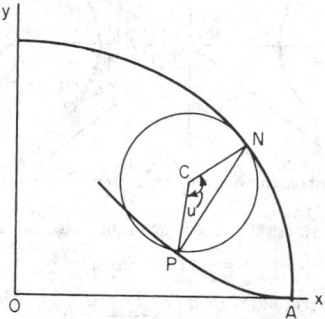

Fig. 2.1.94 Hypocycloid.

Special Cases If $a = \frac{1}{2}c$, the hypocycloid becomes a straight line, diameter of the fixed circle (Fig. 2.1.95). In this case the hypotrochoid traced by any point rigidly connected with the rolling circle (not necessarily on the circumference) will be an ellipse. If $a = \frac{1}{4}c$, the curve

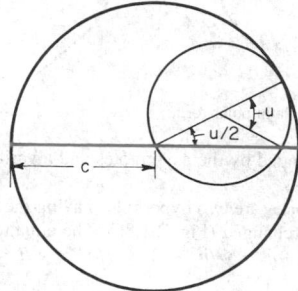

Fig. 2.1.95 Hypocycloid is straight line when the radius of inside circle is half that of the outside circle.

generated will be the four-cusped hypocycloid, or **astroid** (Fig. 2.1.96), whose equation is $x^{2/3} + y^{2/3} = c^{2/3}$. If $a = c$, the epicycloid is the **cardioid**, whose equation in polar coordinates (axes as in Fig. 2.1.97) is $r = 2c(1 + \cos \theta)$. Length of cardioid = $16c$.

The **epitrochoid** (or **hypotrochoid**) is a curve traced by any point rigidly attached to a circle of radius a, at distance b from the center, when this

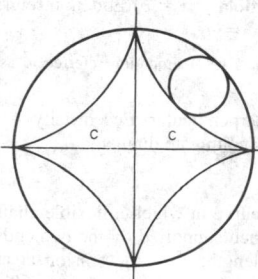

Fig. 2.1.96 Astroid.

circle rolls without slipping on the outside (or inside) of a fixed circle of radius c. The equations are

$$x = (c \pm a) \cos\left(\frac{a}{c}u\right) \pm b \cos\left[\left(1 \pm \frac{a}{c}\right)u\right]$$

$$y = (c \pm a) \sin\left(\frac{a}{c}u\right) - b \sin\left[\left(1 \pm \frac{a}{c}\right)u\right]$$

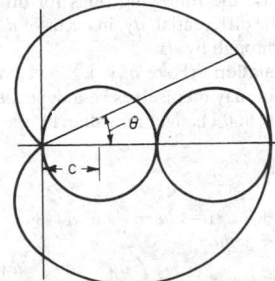

Fig. 2.1.97 Cardioid.

where u = the angle which the moving radius makes with the line of centers; take the upper sign for the epi- and the lower for the hypotrochoid. The curve is called prolate or curtate according as $b < a$ or $b > a$. When $b = a$, the special case of the epi- or hypocycloid arises.

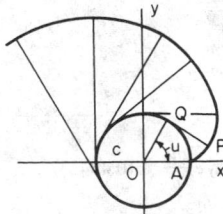

Fig. 2.1.98 Involute of circle.

The **involute of a circle** is the curve traced by the end of a taut string which is unwound from the circumference of a fixed circle, of radius c. If QP is the free portion of the string at any instant (Fig. 2.1.98), QP will be tangent to the circle at Q, and the length of QP = length of arc QA; hence the construction of the curve. The equations of the curve in parametric form (axes as in figure) are $x = c(\cos u + \text{rad } u \sin u)$, $y = c(\sin u - \text{rad } u \cos u)$, where rad u is the radian measure of the angle u which OQ makes with the x axis. Length of arc $AP = \frac{1}{2}c(\text{rad } u)^2$; radius of curvature at P is QP. Polar equations, in terms of parameter

$v(= \text{angle } POQ)$, are $r = c \sec v$, rad $\theta = \tan v - \text{rad } v$. Here, $r = OP$, and rad θ = radian measure of angle, AOP (Fig. 2.1.98).

The **spiral of Archimedes** (Fig. 2.1.99) is traced by a point P which, starting from O, moves with uniform velocity along a ray OP, while the ray itself revolves with uniform angular velocity about O. Polar equation: $r = k \text{ rad } \theta$, or $r = a(\theta°/360°)$. Here $a = 2\pi k$ = the distance measured along a radius, from each coil to the next.

The radius of curvature at P is $R = (k^2 + r^2)^{3/2}/(2k^2 + r^2)$.

The **logarithmic spiral** (Fig. 2.1.100) is a curve which cuts the radii from O at a constant angle v, whose cotangent is m. Polar equation: $r = ae^{m\text{rad}\theta}$. Here a is the value of r when $\theta = 0$. For large negative values of θ, the curve winds around O as an asymptotic point. If PT and PN are the tangent and normal at P, the line TON being perpendicular to OP (not shown in figure), then $ON = rm$, and $PN = r\sqrt{1 + m^2} = r/\sin v$. Radius of curvature at P is PN.

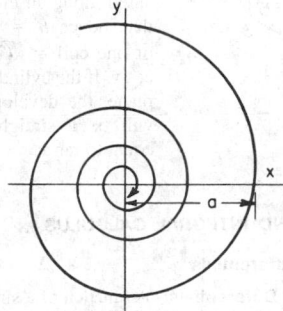

Fig. 2.1.100 Logarithmic spiral.

The **tractrix**, or Schiele's antifriction curve (Fig. 2.1.101), is a curve such that the portion PT of the tangent between the point of contact and the x axis is constant $= a$. Its equation is

$$x = \pm a\left[\cosh^{-1}\frac{a}{y} - \sqrt{1 - \left(\frac{y}{a}\right)^2}\right]$$

or, in parametric form, $x = \pm a(t - \tanh t)$, $y = a/\cosh t$. The x axis is an asymptote of the curve. Length of arc $BP = a \log_e (a/y)$.

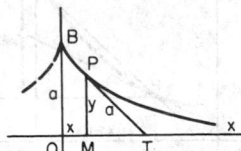

Fig. 2.1.101 Tractrix.

The tractrix describes the path taken by an object being pulled by a string moving along the x axis, where the initial position of the object is B and the opposite end of the string begins at O.

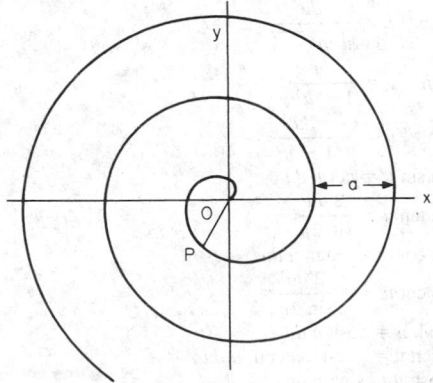

Fig. 2.1.99 Spiral of Archimedes.

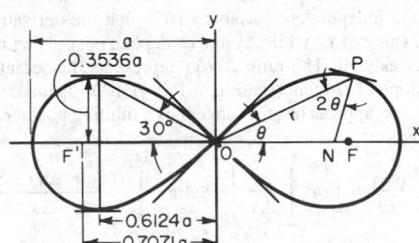

Fig. 2.1.102 Lemniscate.

The **lemniscate** (Fig. 2.1.102) is the locus of a point P the product of whose distances from two fixed points F, F' is constant, equal to $\frac{1}{2}a^2$. The distance $FF' = a\sqrt{2}$. Polar equation is $r = a\sqrt{\cos 2\theta}$. Angle between OP and the normal at P is 2θ. The two branches of the curve cross at right angles at O. Maximum y occurs when $\theta = 30°$ and $r = a/\sqrt{2}$, and is equal to $\frac{1}{4}a\sqrt{2}$. Area of one loop $= a^2/2$.

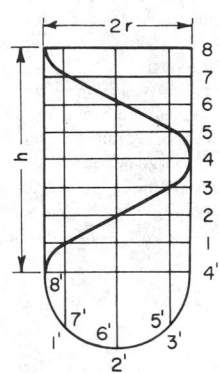

The **helix** (Fig. 2.1.103) is the curve of a screw thread on a cylinder of radius r. The curve crosses the elements of the cylinder at a constant angle, v. The pitch, h, is the distance between two coils of the helix, measured along an element of the cylinder; hence $h = 2\pi r \tan v$. Length of one coil $= \sqrt{(2\pi r)^2 + h^2} = 2\pi r/\cos v$. If the cylinder is rolled out on a plane, the development of the helix will be a straight line, with slope equal to $\tan v$.

Fig. 2.1.103 Helix.

DIFFERENTIAL AND INTEGRAL CALCULUS

Derivatives and Differentials

Derivatives and Differentials A **function** of a single variable x may be denoted by $f(x)$, $F(x)$, etc. The value of the function when x has the value x_0 is then denoted by $f(x_0)$, $F(x_0)$, etc. The **derivative** of a function $y = f(x)$ may be denoted by $f'(x)$, or by dy/dx. The value of the derivative at a given point $x = x_0$ is the **rate of change** of the function at that point; or, if the function is represented by a curve in the usual way (Fig. 2.1.104), the value of the derivative at any point shows the **slope of the curve** (i.e., the slope of the tangent to the curve) at that point (positive if the tangent points upward, and negative if it points downward, moving to the right).

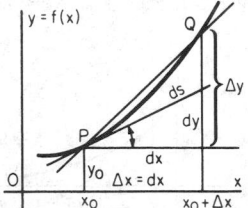

Fig. 2.1.104 Curve showing tangent and derivatives.

The **increment** Δy (read: "delta y") in y is the change produced in y by increasing x from x_0 to $x_0 + \Delta x$; i.e., $\Delta y = f(x_0 + \Delta x) - f(x_0)$. The **differential**, dy, of y is the value which Δy would have if the curve coincided with its tangent. (The differential, dx, of x is the same as Δx when x is the independent variable.) Note that the derivative depends only on the value of x_0, while Δy and dy depend not only on x_0 but on the value of Δx as well. The ratio $\Delta y/\Delta x$ represents the secant slope, and dy/dx the slope of tangent (see Fig. 2.1.104). If Δx is made to approach zero, the secant approaches the tangent as a limiting position, so that the derivative is

$$f'(x) = \frac{dy}{dx} = \lim_{\Delta x \to 0}\left[\frac{\Delta y}{\Delta x}\right] = \lim_{\Delta x \to 0}\left[\frac{f(x_0 + \Delta x) - f(x_0)}{\Delta x}\right]$$

Also, $dy = f'(x)\,dx$.

The symbol "lim" in connection with $\Delta x \to 0$ means "the limit, as Δx approaches 0, of . . . ." (A constant c is said to be the **limit** of a variable u if, whenever any quantity m has been assigned, there is a stage in the variation process beyond which $|c - u|$ is always less than m; or, briefly, c is the limit of u if the difference between c and u can be made to become and remain as small as we please.)

To find the derivative of a given function at a given point: (1) If the function is given only by a curve, measure graphically the slope of the tangent at the point in question; (2) if the function is given by a mathematical expression, use the following rules for differentiation. These rules give, directly, the differential, dy, in terms of dx; to find the derivative, dy/dx, divide through by dx.

Rules for Differentiation (Here u, v, w, . . . represent any functions of a variable x, or may themselves be independent variables. a is a constant which does not change in value in the same discussion; $e = 2.71828$.)

1. $d(a + u) = du$
2. $d(au) = a\,du$
3. $d(u + v + w + \cdots) = du + dv + dw + \cdots$
4. $d(uv) = u\,dv + v\,du$
5. $d(uvw \ldots) = (uvw \ldots)\left(\dfrac{du}{u} + \dfrac{dv}{v} + \dfrac{dw}{w} + \cdots\right)$
6. $d\dfrac{u}{v} = \dfrac{v\,du - u\,dv}{v^2}$
7. $d(u^m) = mu^{m-1}\,du$. Thus, $d(u^2) = 2u\,du$; $d(u^3) = 3u^2\,du$; etc.
8. $d\sqrt{u} = \dfrac{du}{2\sqrt{u}}$
9. $d\left(\dfrac{1}{u}\right) = -\dfrac{du}{u^2}$
10. $d(e^u) = e^u\,du$
11. $d(a^u) = (\ln a)a^u\,du$
12. $d\ln u = \dfrac{du}{u}$
13. $d\log_{10} u = \log_{10} e\,\dfrac{du}{u} = (0.4343 \ldots)\dfrac{du}{u}$
14. $d\sin u = \cos u\,du$
15. $d\csc u = -\cot u\,\csc u\,du$
16. $d\cos u = -\sin u\,du$
17. $d\sec u = \tan u\,\sec u\,du$
18. $d\tan u = \sec^2 u\,du$
19. $d\cot u = -\csc^2 u\,du$
20. $d\sin^{-1} u = \dfrac{du}{\sqrt{1 - u^2}}$
21. $d\csc^{-1} u = -\dfrac{du}{u\sqrt{u^2 - 1}}$
22. $d\cos^{-1} u = \dfrac{du}{\sqrt{1 - u^2}}$
23. $d\sec^{-1} u = \dfrac{du}{u\sqrt{u^2 - 1}}$
24. $d\tan^{-1} u = \dfrac{du}{1 + u^2}$
25. $d\cot^{-1} u = -\dfrac{du}{1 + u^2}$
26. $d\ln\sin u = \cot u\,du$
27. $d\ln\tan u = \dfrac{2\,du}{\sin 2u}$
28. $d\ln\cos u = -\tan u\,du$
29. $d\ln\cot u = -\dfrac{2\,du}{\sin 2u}$
30. $d\sinh u = \cosh u\,du$
31. $d\operatorname{csch} u = -\operatorname{csch} u\,\coth u\,du$
32. $d\cosh u = \sinh u\,du$
33. $d\operatorname{sech} u = -\operatorname{sech} u\,\tanh u\,du$

34. $d \tanh u = \operatorname{sech}^2 u \, du$

35. $d \coth u = - \operatorname{csch}^2 u \, du$

36. $d \sinh^{-1} u = \dfrac{du}{\sqrt{u^2 + 1}}$

37. $d \operatorname{csch}^{-1} u = - \dfrac{du}{u\sqrt{u^2 + 1}}$

38. $d \cosh^{-1} u = \dfrac{du}{\sqrt{u^2 - 1}}$

39. $d \operatorname{sech}^{-1} u = - \dfrac{du}{u\sqrt{1 - u^2}}$

40. $d \tanh^{-1} u = \dfrac{du}{1 - u^2}$

41. $d \coth^{-1} u = \dfrac{du}{1 - u^2}$

42. $d(u^v) = (u^{v-1})(u \ln u \, dv + v \, du)$

Derivatives of Higher Orders The derivative of the derivative is called the second derivative; the derivative of this, the third derivative; and so on. If $y = f(x)$,

$$f'(x) = D_x y = \frac{dy}{dx}$$

$$f''(x) = D_x^2 y = \frac{d^2 y}{dx^2}$$

$$f'''(x) = D_x^3 y = \frac{d^3 y}{dx^3} \quad \text{etc.}$$

NOTE. If the notation $d^2 y/dx^2$ is used, this must not be treated as a fraction, like dy/dx, but as an inseparable symbol, made up of a symbol of operation d^2/dx^2, and an operand y.

The geometric meaning of the second derivative is this: if the original function $y = f(x)$ is represented by a curve in the usual way, then at any point where $f''(x)$ is *positive*, the curve is *concave upward,* and at any point where $f''(x)$ is *negative,* the curve is *concave downward* (Fig. 2.1.105). When $f''(x) = 0$, the curve usually has a **point of inflection.**

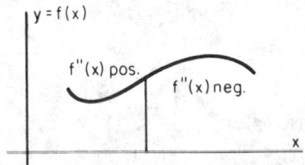

Fig. 2.1.105 Curve showing concavity.

Functions of two or more variables may be denoted by $f(x, y, \ldots)$, $F(x, y, \ldots)$, etc. The derivative of such a function $u = f(x, y, \ldots)$ formed on the assumption that x is the only variable ($y, \ldots$ being regarded for the moment as constants) is called the **partial derivative of u with respect to** x, and is denoted by $f_x(x, y)$ or $D_x u$, or $d_x u/dx$, or $\partial u/\partial x$. Similarly, the partial derivative of u with respect to y is $f_y(x, y)$ or $D_y u$, or $d_y u/dy$, or $\partial u/\partial y$.

NOTE. In the third notation, $d_x u$ denotes the differential of u formed on the assumption that x is the only variable. If the fourth notation, $\partial u/\partial x$, is used, this must not be treated as a fraction like du/dx; the $\partial/\partial x$ is a symbol of operation, operating on u, and the "∂x" must not be separated.

Partial derivatives of the second order are denoted by f_{xx}, f_{xy}, f_{yy}, or by $Du, D_x(D_y u), D_y^2 u$, or by $\partial^2 u/\partial x^2, \partial^2 u/\partial x \, \partial y, \partial^2 u/\partial y^2$, the last symbols being "inseparable." Similarly for higher derivatives. Note that $f_{xy} = f_{yx}$.

If increments $\Delta x, \Delta y$ (or dx, dy) are assigned to the independent variables x, y, the increment, Δu, produced in $u = f(x, y)$ is

$$\Delta u = f(x + \Delta x, y + \Delta y) - f(x, y)$$

while the **differential,** du, i.e., the value which Δu would have if the partial derivatives of u with respect to x and y were constant, is given by

$$du = (f_x) \cdot dx + (f_y) \cdot dy$$

Here the coefficients of dx and dy are the values of the partial derivatives of u at the point in question.

If x and y are functions of a third variable t, then the equation

$$\frac{du}{dt} = (f_x) \frac{dx}{dt} + (f_y) \frac{dy}{dt}$$

expresses the rate of change of u with respect to t, in terms of the separate rate of change of x and y with respect to t.

Implicit Functions If $f(x, y) = 0$, either of the variables x and y is said to be an implicit function of the other. To find dy/dx, either (1) solve for y in terms of x, and then find dy/dx directly; or (2) differentiate the equation through as it stands, remembering that both x and y are variables, and then divide by dx; or (3) use the formula $dy/dx = -(f_x/f_y)$, where f_x and f_y are the partial derivatives of $f(x, y)$ at the point in question.

Maxima and Minima

A **function of one variable,** as $y = f(x)$, is said to have a **maximum** at a point $x = x_0$, if at that point the slope of the curve is zero and the concavity downward (see Fig. 2.1.106); a sufficient condition for a maximum is $f'(x_0) = 0$ and $f''(x_0)$ negative. Similarly, $f(x)$ has a **minimum** if the slope is zero and the concavity upward; a sufficient condition for a minimum is $f'(x_0) = 0$ and $f''(x_0)$ positive. If $f''(x_0) = 0$ and $f'''(x_0) \neq 0$, the point x_0 will be a **point of inflection.** If $f'(x_0) = 0$ and $f''(x_0) = 0$ and $f'''(x_0) = 0$, the point x_0 will be a maximum if $f''''(x_0) < 0$, and a minimum if $f''''(x_0) > 0$. It is usually sufficient, however, in any practical case, to find the values of x which make $f'(x) = 0$, and then decide, from a general knowledge of the curve or the sign of $f'(x)$ to the right and left of x_0, which of these values (if any) give maxima or minima, without investigating the higher derivatives.

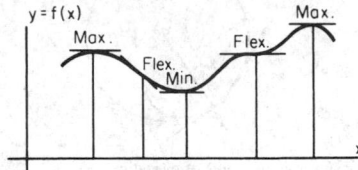

Fig. 2.1.106 Curve showing maxima and minima.

A **function of two variables,** as $u = f(x, y)$, will have a **maximum** at a point (x_0, y_0) if at that point $f_x = 0, f_y = 0$, and $f_{xx} < 0, f_{yy} < 0$; and a **minimum** if at that point $f_x = 0, f_y = 0$, and $f_{xx} > 0, f_{yy} > 0$; provided, in each case, $(f_{xx})(f_{yy}) - (f_{xy})^2$ is positive. If $f_x = 0$ and $f_y = 0$, and f_{xx} and f_{yy} have opposite signs, the point (x_0, y_0) will be a "saddle point" of the surface representing the function.

Indeterminate Forms

In the following paragraphs, $f(x), g(x)$ denote functions which approach 0; $F(x), G(x)$ functions which increase indefinitely; and $U(x)$ a function which approaches 1, when x approaches a definite quantity a. The problem in each case is to find the limit approached by certain combinations of these functions when x approaches a. The symbol $\rightarrow$ is to be read "approaches" or "tends to."

CASE 1. "0/0." To find the limit of $f(x)/g(x)$ when $f(x) \rightarrow 0$ and $g(x) \rightarrow 0$, use the theorem that $\lim [f(x)/g(x)] = \lim [f'(x)/g'(x)]$, where $f'(x)$ and $g'(x)$ are the derivatives of $f(x)$ and $g(x)$. This second limit may be easier to find than the first. If $f'(x) \to 0$ and $g'(x) \to 0$, apply the same theorem a second time: $\lim [f'(x)/g'(x)] = \lim [f''(x)/g''(x)]$, and so on.

CASE 2. "∞/∞." If $F(x) \to \infty$ and $G(x) \to \infty$, then $\lim [F(x)/G(x)] = \lim [F'(x)/G'(x)]$, precisely as in Case 1.

CASE 3. "$0 \cdot \infty$." To find the limit of $f(x) \cdot F(x)$ when $f(x) \to 0$ and $F(x) \to \infty$, write $\lim [f(x) \cdot F(x)] = \lim\{f(x)/[1/F(x)]\}$ or $= \lim \{F(x)/[1/f(x)]\}$, then proceed as in Case 1 or Case 2.

CASE 4. The limit of combinations "0^0" or $[f(x)]^{g(x)}$; "1^∞" or $[U(x)]^{F(x)}$; "∞^0" or $[F(x)]^{b(x)}$ may be found since their logarithms are limits of the type evaluated in Case 3.

CASE 5. "$\infty - \infty$." If $F(x) \to \infty$ and $G(x) \to \infty$, write

$$\lim [F(x) - G(x)] = \lim \frac{1/G(x) - 1/F(x)}{1/[F(x) \cdot G(x)]}$$

then proceed as in Case 1. Sometimes it is shorter to expand the functions in series. It should be carefully noticed that expressions like 0/0, ∞/∞, etc., do not represent mathematical quantities.

Curvature

The **radius of curvature** R of a plane curve at any point P (Fig. 2.1.107) is the distance, measured along the normal, on the concave side of the curve, to the **center of curvature**, C, this point being the limiting position of the point of intersection of the normals at P and a neighboring point Q, as Q is made to approach P along the curve. If the equation of the curve is $y = f(x)$,

$$R = \frac{ds}{du} = \frac{[1 + (y')^2]^{3/2}}{y''}$$

where $ds = \sqrt{dx^2 + dy^2}$ = the differential of arc, $u = \tan^{-1}[f'(x)]$ = the angle which the tangent at P makes with the x axis, and $y' = f'(x)$ and $y'' = f''(x)$ are the first and second derivatives of $f(x)$ at the point P. Note that $dx = ds \cos u$ and $dy = ds \sin u$. The **curvature**, K, at the point P, is $K = 1/R = du/ds$; i.e., the curvature is the rate at which the angle u is changing with respect to the length of arc s. If the slope of the curve is small, $K \approx f''(x)$.

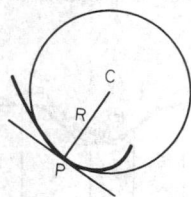

Fig. 2.1.107 Curve showing radius of curvature.

If the equation of the curve in polar coordinates is $r = f(\theta)$, where r = radius vector and θ = polar angle, then

$$R = \frac{[r^2 + (r')^2]^{3/2}}{r^2 - rr'' + 2(r')^2}$$

where $r' = f'(\theta)$ and $r'' = f''(\theta)$.

The **evolute** of a curve is the locus of its centers of curvature. If one curve is the evolute of another, the second is called the **involute** of the first.

Indefinite Integrals

An **integral** of $f(x) \, dx$ is any function whose differential is $f(x) \, dx$, and is denoted by $\int f(x) \, dx$. All the integrals of $f(x) \, dx$ are included in the expression $\int f(x) \, dx + C$, where $\int f(x) \, dx$ is any particular integral, and C is an arbitrary constant. The process of finding (when possible) an integral of a given function consists in recognizing by inspection a function which, when differentiated, will produce the given function; or

in transforming the given function into a form in which such recognition is easy. The most common integrable forms are collected in the following brief table; for a more extended list, see Peirce, "Table of Integrals," Ginn, or Dwight, "Table of Integrals and other Mathematical Data," Macmillan, or "CRC Mathematical Tables."

GENERAL FORMULAS

1. $\int a \, du = a \int du = au + C$

2. $\int (u + v) \, dx = \int u \, dx + \int v \, dx$

3. $\int u \, dv = uv - \int v \, du$ (integration by parts)

4. $\int f(x) \, dx = \int f[F(y)]F'(y) \, dy, \, x = F(y)$

   (change of variables)

5. $\int dy \int f(x, y) \, dx = \int dx \int f(x, y) \, dy$

FUNDAMENTAL INTEGRALS

6. $\int x^n \, dx = \dfrac{x^{n+1}}{n + 1} + C$, when $n \neq -1$

7. $\int \dfrac{dx}{x} = \ln x + C = \ln cx$

8. $\int e^x \, dx = e^x + C$

9. $\int \sin x \, dx = -\cos x + C$

10. $\int \cos x \, dx = \sin x + C$

11. $\int \dfrac{dx}{\sin^2 x} = -\cot x + C$

12. $\int \dfrac{dx}{\cos^2 x} = \tan x + C$

13. $\int \dfrac{dx}{\sqrt{1 - x^2}} = \sin^{-1} x + C = -\cos^{-1} x + C$

14. $\int \dfrac{dx}{1 + x^2} = \tan^{-1} x + C = -\cot^{-1} x + C$

RATIONAL FUNCTIONS

15. $\int (a + bx)^n \, dx = \dfrac{(a + bx)^{n+1}}{(n + 1)b} + C$

16. $\int \dfrac{dx}{a + bx} = \dfrac{1}{b} \ln (a + bx) + C = \dfrac{1}{b} \ln c(a + bx)$

17. $\int \dfrac{dx}{x^n} = -\dfrac{1}{(n - 1)x^{n-1}} + C$ except when $n = 1$

18. $\int \dfrac{dx}{(a + bx)^2} = -\dfrac{1}{b(a + bx)} + C$

19. $\int \dfrac{dx}{1 - x^2} = \frac{1}{2} \ln \dfrac{1 + x}{1 - x} + C = \tanh^{-1} x + C$, when $x < 1$

20. $\int \dfrac{dx}{x^2 - 1} = \frac{1}{2} \ln \dfrac{x - 1}{x + 1} + C = -\coth^{-1} x + C$, when $x > 1$

21. $\int \dfrac{dx}{a + bx^2} = \dfrac{1}{\sqrt{ab}} \tan^{-1} \left(\sqrt{\dfrac{b}{a}} x \right) + C$

22. $\int \dfrac{dx}{a - bx^2} = \dfrac{1}{2\sqrt{ab}} \ln \dfrac{\sqrt{ab} + bx}{\sqrt{ab} - bx} + C$

   $= \dfrac{1}{\sqrt{ab}} \tanh^{-1} \left(\sqrt{\dfrac{b}{a}} x \right) + C$

$[a > 0, b > 0]$

23. $\displaystyle\int \frac{dx}{a + 2bx + cx^2} =$

$$\frac{1}{\sqrt{ac - b^2}} \tan^{-1} \frac{b + cx}{\sqrt{ac - b^2}} + C \qquad [ac - b^2 > 0]$$

$$= \frac{1}{2\sqrt{b^2 - ac}} \ln \frac{\sqrt{b^2 - ac} - b - cx}{\sqrt{b^2 - ac} + b + cx} + C \qquad [b^2 - ac > 0]$$

$$= -\frac{1}{\sqrt{b^2 - ac}} \tanh^{-1} \frac{b + cx}{\sqrt{b^2 - ac}} + C$$

24. $\displaystyle\int \frac{dx}{a + 2bx + cx^2} = -\frac{1}{b + cx} + C$, when $b^2 = ac$

25. $\displaystyle\int \frac{(m + nx)\, dx}{a + 2bx + cx^2} = \frac{n}{2c} \ln (a + 2bx + cx^2)$

$$+ \frac{mc - nb}{c} \int \frac{dx}{a + 2bx + cx^2}$$

26. In $\displaystyle\int \frac{f(x)\, dx}{a + 2bx + cx^2}$, if $f(x)$ is a polynomial of higher than the first degree, divide by the denominator before integrating

27. $\displaystyle\int \frac{dx}{(a + 2bx + cx^2)^p} = \frac{1}{2(ac - b^2)(p - 1)}$

$$\times \frac{b + cx}{(a + 2bx + cx^2)^{p-1}}$$

$$+ \frac{(2p - 3)c}{2(ac - b^2)(p - 1)} \int \frac{dx}{(a + 2bx + cx^2)^{p-1}}$$

28. $\displaystyle\int \frac{(m + nx)\, dx}{(a + 2bx + cx^2)^p} = -\frac{n}{2c(p - 1)} \times$

$$\frac{1}{(a + 2bx + cx^2)^{p-1}} + \frac{mc - nb}{c} \int \frac{dx}{(a + 2bx + cx^2)^p}$$

29. $\displaystyle\int x^{m-1}(a + bx)^n\, dx = \frac{x^{m-1}(a + bx)^{n+1}}{(m + n)b}$

$$- \frac{(m - 1)a}{(m + n)b} \int x^{m-2}(a + bx)^n\, dx$$

$$= \frac{x^m(a + bx)^n}{m + n} + \frac{na}{m + n} \int x^{m-1}(a + bx)^{n-1}\, dx$$

IRRATIONAL FUNCTIONS

30. $\displaystyle\int \sqrt{a + bx}\, dx = \frac{2}{3b} (\sqrt{a + bx})^3 + C$

31. $\displaystyle\int \frac{dx}{\sqrt{a + bx}} = \frac{2}{b} \sqrt{a + bx} + C$

32. $\displaystyle\int \frac{(m + nx)\, dx}{\sqrt{a + bx}} = \frac{2}{3b^2} (3mb - 2an + nbx) \sqrt{a + bx} + C$

33. $\displaystyle\int \frac{dx}{(m + nx)\sqrt{a + bx}}$; substitute $y = \sqrt{a + bx}$, and use 21 and 22

34. $\displaystyle\int \frac{f(x, \sqrt[n]{a + bx})}{F(x, \sqrt[n]{a + bx})}\, dx$; substitute $\sqrt[n]{a + bx} = y$

35. $\displaystyle\int \frac{dx}{\sqrt{a^2 - x^2}} = \sin^{-1} \frac{x}{a} + C = -\cos^{-1} \frac{x}{a} + C$

36. $\displaystyle\int \frac{dx}{\sqrt{a^2 + x^2}} = \ln (x + \sqrt{a^2 + x^2}) + C = \sinh^{-1} \frac{x}{a} + C$

37. $\displaystyle\int \frac{dx}{\sqrt{x^2 - a^2}} = \ln (x + \sqrt{x^2 - a^2}) + C = \cosh^{-1} \frac{x}{a} + C$

38. $\displaystyle\int \frac{dx}{\sqrt{a + 2bx + cx^2}}$

$$= \frac{1}{\sqrt{c}} \ln (b + cx + \sqrt{c}\,\sqrt{a + 2bx + cx^2}) + C, \text{ where } c > 0$$

$$= \frac{1}{\sqrt{c}} \sinh^{-1} \frac{b + cx}{\sqrt{ac - b^2}} + C, \text{ when } ac - b^2 > 0$$

$$= \frac{1}{\sqrt{c}} \cosh^{-1} \frac{b + cx}{\sqrt{b^2 - ac}} + C, \text{ when } b^2 - ac > 0$$

$$= \frac{-1}{\sqrt{-c}} \sin^{-1} \frac{b + cx}{\sqrt{b^2 - ac}} + C, \text{ when } c < 0$$

39. $\displaystyle\int \frac{(m + nx)\, dx}{\sqrt{a + 2bx + cx^2}} = \frac{n}{c} \sqrt{a + 2bx + cx^2}$

$$+ \frac{mc - nb}{c} \int \frac{dx}{\sqrt{a + 2bx + cx^2}}$$

40. $\displaystyle\int \frac{x^m\, dx}{\sqrt{a + 2bx + cx^2}} = \frac{x^{m-1}X}{mc} - \frac{(m - 1)a}{mc} \int \frac{x^{m-2}\, dx}{X}$

$$- \frac{(2m - 1)b}{mc} \int \frac{x^{m-1}}{X}\, dx \text{ when } X = \sqrt{a + 2bx + cx^2}$$

41. $\displaystyle\int \sqrt{a^2 + x^2}\, dx = \frac{x}{2} \sqrt{a^2 + x^2} + \frac{a^2}{2} \ln (x + \sqrt{a^2 + x^2}) + C$

$$= \frac{x}{2} \sqrt{a^2 + x^2} + \frac{a^2}{2} \sinh^{-1} \frac{x}{a} + C$$

42. $\displaystyle\int \sqrt{a^2 - x^2}\, dx = \frac{x}{2} \sqrt{a^2 - x^2} + \frac{a^2}{2} \sin^{-1} \frac{x}{a} + C$

43. $\displaystyle\int \sqrt{x^2 - a^2}\, dx = \frac{x}{2} \sqrt{x^2 - a^2} - \frac{a^2}{2} \ln (x + \sqrt{x^2 - a^2}) + C$

$$= \frac{x}{2} \sqrt{x^2 - a^2} - \frac{a^2}{2} \cosh^{-1} \frac{x}{a} + C$$

44. $\displaystyle\int \sqrt{a + 2bx + cx^2}\, dx = \frac{b + cx}{2c} \sqrt{a + 2bx + cx^2}$

$$+ \frac{ac - b^2}{2c} \int \frac{dx}{\sqrt{a + 2bx + cx^2}} + C$$

TRANSCENDENTAL FUNCTIONS

45. $\displaystyle\int a^x\, dx = \frac{a^x}{\ln a} + C$

46. $\displaystyle\int x^n e^{ax}\, dx$

$$= \frac{x^n e^{ax}}{a} \left[1 - \frac{n}{ax} + \frac{n(n - 1)}{a^2 x^2} - \cdots \pm \frac{n!}{a^n x^n} \right] + C$$

47. $\displaystyle\int \ln x\, dx = x \ln x - x + C$

48. $\displaystyle\int \frac{\ln x}{x^2}\, dx = -\frac{\ln x}{x} - \frac{1}{x} + C$

49. $\displaystyle\int \frac{(\ln x)^n}{x}\, dx = \frac{1}{n + 1} (\ln x))^{n+1} + C$

50. $\displaystyle\int \sin^2 x\, dx = -\tfrac{1}{4} \sin 2x + \tfrac{1}{2} x + C$

$$= -\tfrac{1}{2} \sin x \cos x + \tfrac{1}{2} x + C$$

51. $\displaystyle\int \cos^2 x\, dx = \tfrac{1}{4} \sin 2x + \tfrac{1}{2} x + C$

$$= \tfrac{1}{2} \sin x \cos x + \tfrac{1}{2} x + C$$

52. $\displaystyle\int \sin mx\, dx = -\frac{\cos mx}{m} + C$

53. $\displaystyle\int \cos mx\, dx = \frac{\sin mx}{m} + C$

54. $\displaystyle\int \sin mx \cos nx\, dx = -\frac{\cos (m + n)x}{2(m + n)} - \frac{\cos (m - n)x}{2(m - n)} + C$

55. $\displaystyle\int \sin mx \sin nx\, dx = \frac{\sin (m - n)x}{2(m - n)} - \frac{\sin (m + n)x}{2(m + n)} + C$

56. $\displaystyle\int \cos mx \cos nx\, dx = \frac{\sin (m - n)x}{2(m - n)} + \frac{\sin (m + n)x}{2(m + n)} + C$

57. $\int \tan x \, dx = -\ln \cos x + C$

58. $\int \cot x \, dx = \ln \sin x + C$

59. $\int \dfrac{dx}{\sin x} = \ln \tan \dfrac{x}{2} + C$

60. $\int \dfrac{dx}{\cos x} = \ln \tan \left(\dfrac{\pi}{4} + \dfrac{x}{2} \right) + C$

61. $\int \dfrac{dx}{1 + \cos x} = \tan \dfrac{x}{2} + C$

62. $\int \dfrac{dx}{1 - \cos x} = -\cot \dfrac{x}{2} + C$

63. $\int \sin x \cos x \, dx = \tfrac{1}{2} \sin^2 x + C$

64. $\int \dfrac{dx}{\sin x \cos x} = \ln \tan x + C$

65.* $\int \sin^n x \, dx = -\dfrac{\cos x \sin^{n-1} x}{n} + \dfrac{n-1}{n} \int \sin^{n-2} x \, dx$

66.* $\int \cos^n x \, dx = \dfrac{\sin x \cos^{n-1} x}{n} + \dfrac{n-1}{n} \int \cos^{n-2} x \, dx$

67. $\int \tan^n x \, dx = \dfrac{\tan^{n-1} x}{n-1} - \int \tan^{n-2} x \, dx$

68. $\int \cot^n x \, dx = -\dfrac{\cot^{n-1} x}{n-1} - \int \cot^{n-2} x \, dx$

69. $\int \dfrac{dx}{\sin^n x} = -\dfrac{\cos x}{(n-1) \sin^{n-1} x} + \dfrac{n-2}{n-1} \int \dfrac{dx}{\sin^{n-2} x}$

70. $\int \dfrac{dx}{\cos^n x} = \dfrac{\sin x}{(n-1) \cos^{n-1} x} + \dfrac{n-2}{n-1} \int \dfrac{dx}{\cos^{n-2} x}$

71.† $\int \sin^p x \cos^q x \, dx = \dfrac{\sin^{p+1} x \cos^{q-1} x}{p+q}$

$+ \dfrac{q-1}{p+q} \int \sin^p x \cos^{q-2} x \, dx = -\dfrac{\sin^{p-1} x \cos^{q+1} x}{p+q}$

$+ \dfrac{p-1}{p+q} \int \sin^{p-2} x \cos^q x \, dx$

72.† $\int \sin^{-p} x \cos^q x \, dx = -\dfrac{\sin^{-p+1} x \cos^{q+1} x}{p-1}$

$+ \dfrac{p-q-2}{p-1} \int \sin^{-p+2} x \cos^q x \, dx$

73.† $\int \sin^p x \cos^{-q} x \, dx = \dfrac{\sin^{p+1} x \cos^{-q+1} x}{q-1}$

$+ \dfrac{q-p-2}{q-1} \int \sin^p x \cos^{-q+2} x \, dx$

74. $\int \dfrac{dx}{a + b \cos x} = \dfrac{2}{\sqrt{a^2 - b^2}} \tan^{-1} \left(\sqrt{\dfrac{a-b}{a+b}} \tan \tfrac{1}{2}x \right) + C,$

when $a^2 > b^2$,

$= \dfrac{1}{\sqrt{b^2 - a^2}} \ln \dfrac{b + a \cos x + \sin x \sqrt{b^2 - a^2}}{a + b \cos x} + C,$

when $a^2 < b^2$,

$= \dfrac{2}{\sqrt{b^2 - a^2}} \tanh^{-1} \left(\sqrt{\dfrac{b-a}{b+a}} \tan \tfrac{1}{2}x \right) + C,$ when $a^2 < b^2$

75. $\int \dfrac{\cos x \, dx}{a + b \cos x} = \dfrac{x}{b} - \dfrac{a}{b} \int \dfrac{dx}{a + b \cos x} + C$

76. $\int \dfrac{\sin x \, dx}{a + b \cos x} = -\dfrac{1}{b} \ln (a + b \cos x) + C$

77. $\int \dfrac{A + B \cos x + C \sin x}{a + b \cos x + c \sin x} \, dx = A \int \dfrac{dy}{a + p \cos y}$

$+ (B \cos u + C \sin u) \int \dfrac{\cos y \, dy}{a + p \cos y}$

$- (B \sin u - C \cos u) \int \dfrac{\sin y \, dy}{a + p \cos y}$, where $b = p \cos u$, $c = p$

$\sin u$ and $x - u = y$

78. $\int e^{ax} \sin bx \, dx = \dfrac{a \sin bx - b \cos bx}{a^2 + b^2} e^{ax} + C$

79. $\int e^{ax} \cos bx \, dx = \dfrac{a \cos bx + b \sin bx}{a^2 + b^2} e^{ax} + C$

80. $\int \sin^{-1} x \, dx = x \sin^{-1} x + \sqrt{1 - x^2} + C$

81. $\int \cos^{-1} x \, dx = x \cos^{-1} x - \sqrt{1 - x^2} + C$

82. $\int \tan^{-1} x \, dx = x \tan^{-1} x - \tfrac{1}{2} \ln (1 + x^2) + C$

83. $\int \cot^{-1} x \, dx = x \cot^{-1} x + \tfrac{1}{2} \ln (1 + x^2) + C$

84. $\int \sinh x \, dx = \cosh x + C$

85. $\int \tanh x \, dx = \ln \cosh x + C$

86. $\int \cosh x \, dx = \sinh x + C$

87. $\int \coth x \, dx = \ln \sinh x + C$

88. $\int \operatorname{sech} x \, dx = 2 \tan^{-1}(e^x) + C$

89. $\int \operatorname{csch} x \, dx = \ln \tanh (x/2) + C$

90. $\int \sinh^2 x \, dx = \tfrac{1}{2} \sinh x \cosh x - \tfrac{1}{2}x + C$

91. $\int \cosh^2 x \, dx = \tfrac{1}{2} \sinh x \cosh x + \tfrac{1}{2}x + C$

92. $\int \operatorname{sech}^2 x \, dx = \tanh x + C$

93. $\int \operatorname{csch}^2 x \, dx = -\coth x + C$

Hints on Using Integral Tables It happens with frustrating frequency that no integral table lists the integral that needs to be evaluated. When this happens, one may (a) seek a more complete integral table, (b) appeal to mathematical software, such as Mathematica, Maple, MathCad or Derive, (c) use numerical or approximate methods, such as Simpson's rule (see section "Numerical Methods"), or (d) attempt to transform the integral into one which may be evaluated. Some hints on such transformation follow. For a more complete list and more complete explanations, consult a calculus text, such as Thomas, "Calculus and Analytic Geometry," Addison-Wesley, or Anton, "Calculus with Analytic Geometry," Wiley. One or more of the following "tricks" may be successful.

Trigonometric Substitutions

1. If an integrand contains $\sqrt{(a^2 - x^2)}$, substitute $x = a \sin u$, and $\sqrt{(a^2 - x^2)} = a \cos u$.

2. Substitute $x = a \tan u$ and $\sqrt{(x^2 + a^2)} = a \sec u$.

3. Substitute $x = a \sec u$ and $\sqrt{(x^2 - a^2)} = a \tan u$.

Completing the Square

4. Rewrite $ax^2 + bx + c = a[x + b/(2a)]^2 + (4ac - b^2)/(4a)$; then substitute $u = x + b/(2a)$ and $B = (4ac - b^2)/(4a)$.

---

* If n is an odd number, substitute $\cos x = z$ or $\sin x = z$.

† If p or q is an odd number, substitute $\cos x = z$ or $\sin x = z$.

PARTIAL FRACTIONS

5. For a ratio of polynomials, where the denominator has been completely factored into linear factors $p_i(x)$ and quadratic factors $q_j(x)$, and where the degree of the numerator is less than the degree of the denominator, then rewrite $r(x)/[p_1(x) \ldots p_n(x)q_1(x) \ldots q_m(x)] = A_1/p_1(x) + \cdots + A_n/p_n(x) + (B_1x + C_1)/q_1(x) + \cdots + (B_mx + C_m)/q_m(x)$.

INTEGRATION BY PARTS

6. Change the integral using the formula

$$\int u\,dv = uv - \int v\,du$$

where u and dv are chosen so that (a) v is easy to find from dv, and (b) $v\,du$ is easier to find than $u\,dv$.

Kasube suggests ("A Technique for Integration by Parts," *Am. Math. Month.*, vol. 90, no. 3, Mar. 1983): Choose u in the order of preference LIATE, that is, Logarithmic, Inverse trigonometric, Algebraic, Trigonometric, Exponential.

EXAMPLE. Find $\int x \ln x\,dx$. The logarithmic $\ln x$ has higher priority than does the algebraic x, so let $u = \ln(x)$ and $dv = x\,dx$. Then $du = (1/x)\,dx$; $v = x^2/2$, so $\int x \ln x\,dx = uv - \int v\,du = (x^2/2) \ln x - \int (x^2/2)(1/x)\,dx = (x^2/2) \ln x - \int x/2\,dx = (x^2/2) \ln x - x^2/4 + C$.

Definite Integrals The definite integral of $f(x)\,dx$ from $x = a$ to $x = b$, denoted by $\int_a^b f(x)\,dx$, is the limit (as n increases indefinitely) of a sum of n terms:

$$\int_a^b f(x)\,dx = \lim_{n \to \infty} [f(x_1)\,\Delta x + f(x_2)\,\Delta x + f(x_3)\,\Delta x + \cdots + f(x_n)\,\Delta x]$$

built up as follows: Divide the interval from a to b into n equal parts, and call each part $\Delta x, = (b - a)/n$; in each of these intervals take a value of x (say, $x_1, x_2, \ldots, x_n$), find the value of the function $f(x)$ at each of these points, and multiply it by Δx, the width of the interval; then take the limit of the sum of the terms thus formed, when the number of terms increases indefinitely, while each individual term approaches zero.

Geometrically, $\int_a^b f(x)\,dx$ is the area bounded by the curve $y = f(x)$, the x axis, and the ordinates $x = a$ and $x = b$ (Fig. 2.1.108); i.e., briefly, the "area under the curve, from a to b." The **fundamental theorem** for the evaluation of a definite integral is the following:

$$\int_a^b f(x)\,dx = \left[\int f(x)\,dx\right]_{x=b} - \left[\int f(x)\,dx\right]_{x=a}$$

i.e., the definite integral is equal to the difference between two values of any one of the indefinite integrals of the function in question. In other words, the limit of a sum can be found whenever the function can be integrated.

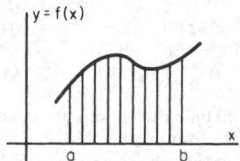

Fig. 2.1.108 Graph showing areas to be summed during integration.

Properties of Definite Integrals

$$\int_a^b = -\int_b^a; \qquad \int_a^c + \int_c^b = \int_a^b$$

MEAN-VALUE THEOREM FOR INTEGRALS

$$\int_a^b F(x)f(x)\,dx = F(X) \int_a^b f(x)\,dx$$

provided $f(x)$ does not change sign from $x = a$ to $x = b$; here X is some (unknown) value of x intermediate between a and b.

MEAN VALUE. The **mean value** of $f(x)$ with respect to x, between a and b, is

$$\bar{f} = \frac{1}{b - a} \int_a^b f(x)\,dx$$

THEOREM ON CHANGE OF VARIABLE. In evaluating $\int_{x=a}^{x=b} f(x)\,dx$, $f(x)\,dx$ may be replaced by its value in terms of a new variable t and dt, and $x = a$ and $x = b$ by the corresponding values of t, provided that throughout the interval the relation between x and t is a one-to-one correspondence (i.e., to each value of x there corresponds one and only one value of t, and to each value of t there corresponds one and only one value of x). So $\int_{x=a}^{x=b} f(x)\,dx = \int_{t=g(a)}^{t=g(b)} f(g(t))\,g'(t)\,dt$.

DIFFERENTIATION WITH RESPECT TO THE UPPER LIMIT. If b is variable, then $\int_a^b f(x)\,dx$ is a function b, whose derivative is

$$\frac{d}{db} \int_a^b f(x)\,dx = f(b)$$

DIFFERENTIATION WITH RESPECT TO A PARAMETER

$$\frac{\partial}{\partial c} \int_a^b f(x, c)\,dx = \int_a^b \frac{\partial f(x, c)}{\partial c}\,dx$$

Functions Defined by Definite Integrals The following definite integrals have received special names:

1. Elliptic integral of the first kind $= F(u, k) = \int_0^u \dfrac{dx}{\sqrt{1 - k^2 \sin^2 x}}$ when $k^2 < 1$.

2. Elliptic integral of the second kind $= E(u, k) = \int_0^u \sqrt{1 - k^2 \sin^2 x}\,dx$, when $k^2 < 1$.

3, 4. Complete elliptic integrals of the first and second kinds; put $u = \pi/2$ in (1) and (2).

5. The probability integral $= \dfrac{2}{\sqrt{\pi}} \int_0^x e^{-x^2}\,dx$.

6. The gamma function $= \Gamma(n) = \int_0^\infty x^{n-1} e^{-x}\,dx$.

Approximate Methods of Integration. Mechanical Quadrature (See also section "Numerical Methods.")

1. Use Simpson's rule (see also Scarborough, "Numerical Mathematical Analyses," Johns Hopkins Press).

2. Expand the function in a converging power series, and integrate term by term.

3. Plot the area under the curve $y = f(x)$ from $x = a$ to $x = b$ on squared paper, and measure this area roughly by "counting squares."

Double Integrals The notation $\iint f(x, y)\,dy\,dx$ means $\int [\int f(x, y)\,dy]\,dx$, the limits of integration in the inner, or first, integral being functions of x (or constants).

EXAMPLE. To find the weight of a plane area whose density, w, is variable, say $w = f(x, y)$. The weight of a typical element, $dx\,dy$, is $f(x, y)\,dx\,dy$. Keeping x and dx constant and summing these elements from, say, $y = F_1(x)$ to $y = F_2(x)$, as determined by the shape of the boundary (Fig. 2.1.109), the weight of a typical strip perpendicular to the x axis is

$$dx \int_{y=F_1(x)}^{y=F_2(x)} f(x, y)\,dy$$

Finally, summing these strips from, say, $x = a$ to $x = b$, the weight of the whole area is

$$\int_{x=a}^{x=b} \left[dx \int_{y=F_1(x)}^{y=F_2(x)} f(x, y)\,dy \right] \text{ or, briefly, } \iint f(x, y)\,dy\,dx$$

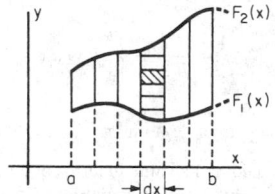

Fig. 2.1.109 Graph showing areas to be summed during double integration.

Triple Integrals The notation $\iiint f(x, y, z)\,dz\,dy\,dx$ means

$$\int \left\{ \int \left[\int f(x, y, z)\,dz \right] dy \right\} dx$$

Such integrals are known as **volume integrals**.

EXAMPLE. To find the mass of a volume which has variable density, say, $w = f(x, y, z)$. If the shape of the volume is described by $a < x < b$, $F_1(x) < y < F_2(x)$, and $G_1(x, y) < z < G_2(x, y)$, then the mass is given by

$$\int_a^b \int_{F_1(x)}^{F_2(x)} \int_{G_1(x,y)}^{G_2(x,y)} f(x, y, z)\,dz\,dy\,dx$$

SERIES AND SEQUENCES

Sequences

A **sequence** is an ordered list of numbers, $x_1, x_2, \ldots, x_n, \ldots$.

An infinite sequence is an infinitely long list. A sequence is often defined by a function $f(n)$, $n = 1, 2, \ldots$. The formula defining $f(n)$ is called the **general term** of the sequence.

The variable n is called the **index** of the sequence. Sometimes the index is taken to start with $n = 0$ instead of $n = 1$.

A sequence **converges** to a limit L if the general term $f(n)$ has limit L as n goes to infinity. If a sequence does not have a unique limit, the sequence is said to "diverge." There are two fundamental ways a function can diverge: (1) It may become infinitely large, in which case the sequence is said to be "unbounded," or (2) it may tend to alternate among two or more values, as in the sequence $x_n = (-1)^n$.

A sequence **alternates** if its odd-numbered terms are positive and its even-numbered terms are negative, or vice versa.

Series

A **series** is a sequence of sums. The terms of the sums are another sequence, $x_1, x_2, \ldots$. Then the series is the sequence defined by $s_n = x_1 + x_2 + \cdots + x_n = \sum_{i=1}^{n} x_i$. The sequence s_n is also called the **sequence of partial sums** of the series.

If the sequence of partial sums converges (resp. diverges), then the series is said to converge (resp. diverge). If the limit of a series is S, then the sequence defined by $r_n = S - s_n$ is called the "error sequence" or the "sequence of truncation errors."

Convergence of Series THEOREM. If a series $s_n = x_1 + x_2 +$ $\cdots + x_n$ converges, then it is necessary (but not sufficient) that the sequence x_n has limit zero. A series of partial sums of an alternating sequence is called an *alternating series*.

THEOREM. An alternating series converges whenever the sequence x_n has limit zero.

A series is a **geometric series** if its terms are of the form ar^n. The value r is called the **ratio** of the series. Usually, for geometric series, the index is taken to start with $n = 0$ instead of $n = 1$.

THEOREM. A geometric series with $x_n = ar^n$, $n = 0, 1, 2, \ldots$, converges if and only if $-1 < r < 1$, and then the limit of the series is $a/(1 - r)$. The partial sums of a geometric series are $s_n = a(1 - r^n)/(1 - r)$.

The series defined by the sequence $x_n = 1/n$, $n = 1, 2, \ldots$, is called the **harmonic series**. The harmonic series diverges.

A series with each term $x_n > 0$ is called a "positive series."

There are a number of tests to determine whether or not a positive series s_n converges.

1. *Comparison test.* If $c_1 + c_2 + \cdots + c_n$ is a positive series that converges, and if $0 < x_n < c_n$, then the series $x_1 + x_2 + \cdots + x_n$ also converges.

If $d_1 + d_2 + \cdots + d_n$ diverges and $x_n > d_n$, then $x_1 + x_2 + \cdots + x_n$ also diverges.

2. *Integral test.* If $f(t)$ is a strictly decreasing function and $f(n) = x_n$, then the series s_n and the integral $\int_1^{\infty} f(t)\,dt$ either both converge or both diverge.

3. *P test.* The series defined by $x_n = 1/n^p$ converges if $p > 1$ and diverges if $p = 1$ or $p < 1$. If $p = 1$, then this is the harmonic series.

4. *Ratio test.* If the limit of the sequence $x_{n+1}/x_n = r$, then the series diverges if $r > 1$, and it converges if $0 < r < 1$. The test is inconclusive if $r = 1$.

5. *Cauchy root test.* If L is the limit of the nth root of the nth term, $\lim x_n^{1/n}$, then the series converges if $L < 1$ and diverges if $L > 1$. If $L = 1$, then the test is inconclusive.

A **power series** is an expression of the form $a_0 + a_1x + a_2x^2 + \cdots + a_nx^n + \cdots$ or $\sum_{i=0}^{\infty} a_ix^i$.

The range of values of x for which a power series converges is the **interval of convergence** of the power series.

General Formulas of Maclaurin and Taylor If $f(x)$ and all its derivatives are continuous in the neighborhood of the point $x = 0$ (or $x = a$), then, for any value of x in this neighborhood, the function $f(x)$ may be expressed as a power series arranged according to ascending powers of x (or of $x - a$), as follows:

$$f(x) = f(0) + \frac{f'(0)}{1!}x + \frac{f''(0)}{2!}x^2 + \frac{f'''(0)}{3!}x^3 + \cdots$$
$$+ \frac{f^{(n-1)}(0)}{(n-1)!}x^{n-1} + (P_n)x^n \qquad \text{(Maclaurin)}$$

$$f(x) = f(a) + \frac{f'(a)}{1!}(x-a) + \frac{f''(a)}{2!}(x-a)^2 + \frac{f'''(a)}{3!}(x-a)^3 +$$
$$\cdots + \frac{f^{(n-1)}(a)}{(n-1)!}(x-a)^{n-1} + (Q_n)(x-a)^n \qquad \text{(Taylor)}$$

Here $(P_n)x^n$, or $(Q_n)(x-a)^n$, is called the **remainder term**; the values of the coefficients P_n and Q_n may be expressed as follows:

$$P_n = [f^{(n)}(sx)]/n! = [(1-t)^{n-1}f^{(n)}(tx)]/(n-1)!$$
$$Q_n = \{f^{(n)}[a + s(x-a)]\}/n!$$
$$= \{(1-t)^{n-1}f^{(n)}[a + t(x-a)]\}/(n-1)!$$

where s and t are certain unknown numbers between 0 and 1; the s form is due to Lagrange, the t form to Cauchy.

The error due to neglecting the remainder term is less than $(\bar{P}_n)x^n$, or $(\bar{Q}_n)(x-a)^n$, where $\bar{P}_n$, or $\bar{Q}_n$, is the largest value taken on by P_n, or

Q_n, when s or t ranges from 0 to 1. If this error, which depends on both n and x, approaches 0 as n increases (for any given value of x), then the general expression with remainder becomes (for that value of x) a convergent infinite series.

The sum of the first few terms of Maclaurin's series gives a good approximation to $f(x)$ for values of x near $x = 0$; Taylor's series gives a similar approximation for values near $x = a$.

The MacLaurin series of some important functions are given below.

Power series may be differentiated term by term, so the derivative of a power series $a_0 + a_1 x + a_2 x^2 + \cdots + a_n x^n$ is $a_1 + 2a_2 x + \cdots + n a_x x^{n-1}$. . . . The power series of the derivative has the same interval of convergence, except that the endpoints may or may not be included in the interval.

Series Expansions of Some Important Functions

The range of values of x for which each of the series is convergent is stated at the right of the series.

Geometrical Series

$$\frac{1}{1 - x} = \sum_{n=0}^{\infty} x^n \qquad\qquad -1 < x < 1$$

$$\frac{1}{(1 - x)^m} = 1 + \sum_{n=1}^{\infty} \frac{(m + n - 1)!}{(m - 1)!\, n!} x^n \qquad -1 < x < 1$$

Exponential and Logarithmic Series

$$e^x = 1 + \frac{x}{1!} + \frac{x^2}{2!} + \frac{x^3}{3!} + \frac{x^4}{4!} + \cdots \qquad [-\infty < x < +\infty]$$

$$a^x = e^{mx} = 1 + \frac{m}{1!} x + \frac{m^2}{2!} x^2 + \frac{m^3}{3!} x^3 + \cdots$$
$$[a > 0,\ -\infty < x < +\infty]$$

where $m = \ln a = (2.3026)(\log_{10} a)$.

$$\ln (1 + x) = x - \frac{x^2}{2} + \frac{x^3}{3} - \frac{x^4}{4} + \frac{x^5}{5} \cdots \qquad [-1 < x < +1]$$

$$\ln (1 - x) = -x - \frac{x^2}{2} - \frac{x^3}{3} - \frac{x^4}{4} - \frac{x^5}{5} - \cdots \qquad [-1 < x < +1]$$

$$\ln \left(\frac{1 + x}{1 - x} \right) = 2 \left(x + \frac{x^3}{3} + \frac{x^5}{5} + \frac{x^7}{7} + \cdots \right) \qquad [-1 < x < +1]$$

$$\ln \left(\frac{x + 1}{x - 1} \right) = 2 \left(\frac{1}{x} + \frac{1}{3x^3} + \frac{1}{5x^5} + \frac{1}{7x^7} + \cdots \right)$$
$$[x < -1 \text{ or } 1 < x]$$

$$\ln x = 2 \left[\frac{x - 1}{x + 1} + \frac{1}{3} \left(\frac{x - 1}{x + 1} \right)^3 + \frac{1}{5} \left(\frac{x - 1}{x + 1} \right)^5 + \cdots \right]$$
$$[0 < x < \infty]$$

$$\ln (a + x) = \ln a + 2 \left[\frac{x}{2a + x} + \frac{1}{3} \left(\frac{x}{2a + x} \right)^3 \right.$$
$$\left. + \frac{1}{5} \left(\frac{x}{2a + x} \right)^5 + \cdots \right]$$
$$[0 < a < +\infty,\ -a < x < +\infty]$$

Series for the Trigonometric Functions
In the following formulas, *all angles must be expressed in radians.* If $D =$ the number of degrees in the angle, and $x =$ its radian measure, then $x = 0.017453D$.

$$\sin x = x - \frac{x^3}{3!} + \frac{x^5}{5!} - \frac{x^7}{7!} + \cdots \qquad [-\infty < x < +\infty]$$

$$\cos x = 1 - \frac{x^2}{2!} + \frac{x^4}{4!} - \frac{x^6}{6!} + \frac{x^8}{8!} - \cdots \qquad [-\infty < x < +\infty]$$

$$\tan x = x + \frac{x^3}{3} + \frac{2x^5}{15} + \frac{17x^7}{315} + \frac{62x^9}{2835} + \cdots$$
$$[-\pi/2 < x < +\pi/2]$$

$$\cot x = \frac{1}{x} - \frac{x}{3} - \frac{x^3}{45} - \frac{2x^5}{945} - \frac{x^7}{4725} - \cdots \qquad [-\pi < x < +\pi]$$

$$\sin^{-1} y = y + \frac{y^3}{6} + \frac{3y^5}{40} + \frac{5y^7}{112} + \cdots \qquad [-1 \le y \le +1]$$

$$\tan^{-1} y = y - \frac{y^3}{3} + \frac{y^5}{5} - \frac{y^7}{7} + \cdots \qquad [-1 \le y \le +1]$$

$$\cos^{-1} y - \tfrac{1}{2}\pi - \sin^{-1} y; \qquad \cot^{-1} y = \tfrac{1}{2}\pi - \tan^{-1} y.$$

Series for the Hyperbolic Functions   (x a pure number)

$$\sinh x = x + \frac{x^3}{3!} + \frac{x^5}{5!} + \frac{x^7}{7!} + \cdots \qquad [-\infty < x < \infty]$$

$$\cosh x = 1 + \frac{x^2}{2!} + \frac{x^4}{4!} + \frac{x^6}{6!} + \cdots \qquad [-\infty < x < \infty]$$

$$\sinh^{-1} y = y - \frac{y^3}{6} + \frac{3y^5}{40} - \frac{5y^7}{112} + \cdots \qquad [-1 < y < +1]$$

$$\tanh^{-1} y = y + \frac{y^3}{3} + \frac{y^5}{5} + \frac{y^7}{7} + \cdots \qquad [-1 < y < +1]$$

ORDINARY DIFFERENTIAL EQUATIONS

An **ordinary differential equation** is one which contains a single independent variable, or argument, and a single dependent variable, or function, with its derivatives of various orders. A **partial differential equation** is one which contains a function of several independent variables, and its partial derivatives of various orders. The order of a differential equation is the order of the highest derivative which occurs in it. A solution of a differential equation is any relation among the variables, involving no derivatives, though possibly involving integrations which, when substituted in the given equation, will satisfy it. The general solution of an ordinary differential equation of the nth order will contain n arbitrary constants.

If specific values of the arbitrary constants are chosen, then a solution is called a **particular solution**. For most problems, all possible particular solutions to a differential equation may be found by choosing values for the constants in a general solution. In some cases, however, other solutions exist. These are called **singular solutions**.

EXAMPLE.   The differential equation $(yy')^2 - a^2 - y^2 = 0$ has general solution $(x - c)^2 + y^2 = a^2$, where c is an arbitrary constant. Additionally, it has the two singular solutions $y = a$ and $y = -a$. The singular solutions form two parallel lines tangent to the family of circles given by the general solution.

The example illustrates a general property of singular solutions; at each point on a singular solution, the singular solution is tangent to some curve given in the general solution.

Methods of Solving Ordinary Differential Equations

DIFFERENTIAL EQUATIONS OF THE FIRST ORDER

1. If possible, separate the variables; i.e., collect all the x's and dx on one side, and all the y's and dy on the other side; then integrate both sides, and add the constant of integration.

2. If the equation is homogeneous in x and y, the value of dy/dx in terms of x and y will be of the form $dy/dx = f(y/x)$. Substituting $y = xt$ will enable the variables to be separated.

Solution: $\log_e x = \displaystyle\int \frac{dt}{f(t) - t} + C.$

3. The expression $f(x, y)\, dx + F(x, y)\, dy$ is an *exact differential* if $\dfrac{\partial f(x, y)}{\partial y} = \dfrac{\partial F(x, y)}{\partial x}$ $(= P$, say). In this case the solution of $f(x, y)\, dx + F(x, y)\, dy = 0$ is

$$\int f(x, y)\, dx + \int [F(x, y) - \int P\, dx]\, dy = C$$

or

$$\int F(x, y)\, dy + \int [f(x, y) - \int P\, dy]\, dx = C$$

4. Linear differential equation of the first order: $\dfrac{dy}{dx} + f(x) \cdot y = F(x)$.

Solution: $y = e^{-P}[\int e^{P}F(x)\, dx + C]$, where $P = \int f(x)\, dx$

5. Bernoulli's equation: $\dfrac{dy}{dx} + f(x) \cdot y = F(x) \cdot y^{n}$. Substituting $y^{1-n} = v$ gives $(dv/dx) + (1 - n)f(x) \cdot v = (1 - n)F(x)$, which is linear in v and x.

6. Clairaut's equation: $y = xp + f(p)$, where $p = dy/dx$. The solution consists of the family of lines given by $y = Cx + f(C)$, where C is any constant, together with the curve obtained by eliminating p between the equations $y = xp + f(p)$ and $x + f'(p) = 0$, where $f'(p)$ is the derivative of $f(p)$.

7. Riccati's equation. $p + ay^{2} + Q(x)y + R(x) = 0$, where $p = dy/dx$ can be reduced to a second-order linear differential equation $(d^{2}u/dx^{2}) + Q(x)(du/dx) + R(x) = 0$ by the substitution $y = du/dx$.

8. Homogeneous equations. A function $f(x, y)$ is **homogeneous** of degree n if $f(rx, ry) = r^{m}f(x, y)$, for all values of r, x, and y. In practice, this means that $f(x, y)$ looks like a polynomial in the two variables x and y, and each term of the polynomial has total degree m. A differential equation is homogeneous if it has the form $f(x, y) = 0$, with f homogeneous. $(xy + x^{2})\, dx + y^{2}\, dy = 0$ is homogeneous. $\cos(xy)\, dx + y^{2}\, dy = 0$ is not.

If an equation is homogeneous, then either of the substitutions $y = vx$ or $x = vy$ will transform the equation into a separable equation.

9. $dy/dx = f[(ax + by + c)/(dx + ey + g)]$ is reduced to a homogeneous equation by substituting $u = ax + by + c$, $v = dx + ey + g$, if $ae - bd = 0$, and $z = ax + by$, $w = dx + ey$ if $ae - bd = 0$.

DIFFERENTIAL EQUATIONS OF THE SECOND ORDER

10. *Dependent variable missing.* If an equation does not involve the variable y, and is of the form $F(x, dy/dx, d^{2}y/dx^{2}) = 0$, then it can be reduced to a first-order equation by substituting $p = dy/dx$ and $dp/dx = d^{2}y/dx^{2}$.

11. *Independent variable missing.* If the equation is of the form $F(y, dy/dx, d^{2}y/dx^{2}) = 0$, and so is missing the variable x, then it can be reduced to a first-order equation by substituting $p = dy/dx$ and $p(dp/dy) = d^{2}y/dx^{2}$.

12. $\dfrac{d^{2}y}{dx^{2}} = -n^{2}y$.

Solution: $y = C_{1} \sin(nx + C_{2})$, or $y = C_{3} \sin nx + C_{4} \cos nx$.

13. $\dfrac{d^{2}y}{dx^{2}} = +n^{2}y$.

Solution: $y = C_{1} \sinh(nx + C_{2})$, or $y = C_{3}e^{nx} + C_{4}e^{-nx}$.

14. $\dfrac{d^{2}y}{dx^{2}} = f(y)$.

Solution: $x = \displaystyle\int \dfrac{dy}{\sqrt{C_{1} + 2P}} + C_{2}$, where $P = \displaystyle\int f(y)\, dy$.

15. $\dfrac{d^{2}y}{dx^{2}} = f(x)$.

Solution: $y = \displaystyle\int P\, dx + C_{1}x + C_{2}$ where $P = \displaystyle\int f(x)\, dx$,

or $y = xP - \displaystyle\int xf(x)\, dx + C_{1}x + C_{2}$.

16. $\dfrac{d^{2}y}{dx^{2}} = f\left(\dfrac{dy}{dx}\right)$. Putting $\dfrac{dy}{dx} = z$, $\dfrac{d^{2}y}{dx^{2}} = \dfrac{dz}{dx}$,

$$x = \int \frac{dz}{f(z)} + C_{1}, \quad \text{and} \quad y = \int \frac{z\,dz}{f(z)} + C_{2}$$

then eliminate z from these two equations.

17. The equation for damped vibrations: $\dfrac{d^{2}y}{dx^{2}} + 2b\dfrac{dy}{dx} + a^{2}y = 0$.

CASE 1. If $a^{2} - b^{2} > 0$, let $m = \sqrt{a^{2} - b^{2}}$.
Solution:

$$y = C_{1}e^{-bx} \sin(mx + C_{2}) \quad \text{or} \quad y = e^{-bx}[C_{3} \sin(mx) + C_{4} \cos(mx)]$$

CASE 2. If $a^{2} - b^{2} = 0$, solution is $y = e^{-bx}(C_{1} + C_{2}x)$.

CASE 3. If $a^{2} - b^{2} < 0$, let $n = \sqrt{b^{2} - a^{2}}$.
Solution:

$$y = C_{1}e^{-bx} \sinh(nx + C_{2}) \quad \text{or} \quad y = C_{3}e^{-(b+n)x} + C_{4}e^{-(b-n)x}$$

18. $\dfrac{d^{2}y}{dx^{2}} + 2b\dfrac{dy}{dx} + a^{2}y = c$.

Solution: $y = \dfrac{c}{a^{2}} + y_{1}$, where $y_{1} =$ the solution of the corresponding equation with second member zero [see type 17 above].

19. $\dfrac{d^{2}y}{dx^{2}} + 2b\dfrac{dy}{dx} + a^{2}y = c \sin(kx)$.

Solution: $y = R \sin(kx - S) + y_{1}$ where $R = c/\sqrt{(a^{2} - k^{2})^{2} + 4b^{2}k^{2}}$, $\tan S = 2bk/(a^{2} - k^{2})$, and $y_{1} =$ the solution of the corresponding equation with second member zero [see type 17 above].

20. $\dfrac{d^{2}y}{dx^{2}} + 2b\dfrac{dy}{dx} + a^{2}y = f(x)$.

Solution: $y = R \sin(kx - S) + y_{1}$ where $R = c/\sqrt{(a^{2} - k^{2})^{2} + 4b^{2}k^{2}}$, $\tan S = 2bk/(a^{2} - k^{2})$, and $y_{1} =$ the solution of the corresponding equation with second member zero [see type 17 above].

If $b^{2} < a^{2}$,

$$y_{0} = \frac{1}{2\sqrt{b^{2} - a^{2}}}\left[e^{m_{1}x} \int e^{-m_{1}x} f(x)\, dx - e^{m_{2}x} \int e^{-m_{2}x}f(x)\, dx \right]$$

where $m_{1} = -b + \sqrt{b^{2} - a^{2}}$ and $m_{2} = -b - \sqrt{b^{2} - a^{2}}$.
If $b^{2} < a^{2}$, let $m = \sqrt{a^{2} - b^{2}}$, then

$$y_{0} = \frac{1}{m}e^{-bx}\left[\sin(mx) \int e^{bx} \cos(mx) \cdot f(x)\, dx \right.$$
$$\left. - \cos(mx) \int e^{bx} \sin(mx) \cdot f(x)\, dx \right]$$

If $b^{2} = a^{2}$, $y_{0} = e^{-bx}\left[x \displaystyle\int e^{bx}f(x)\, dx - \displaystyle\int x \cdot e^{bx} f(x)\, dx \right]$.

Types 17 to 20 are examples of linear differential equations with constant coefficients. The solutions of such equations are often found most simply by the use of Laplace transforms. (See Franklin, "Fourier Methods," pp. 198–229, McGraw-Hill.)

Linear Equations

For the linear equation of the nth order

$$A_{n}(x)\, d^{n}y/dx^{n} + A_{n-1}(x)\, d^{n-1}y/dx^{n-1} + \cdots$$
$$+ A_{1}(x)\, dy/dx + A_{0}(x)y = E(x)$$

the general solution is $y = u + c_{1}u_{1} + c_{2}u_{2} + \cdots + c_{n}u_{n}$. Here u, the particular integral, is any solution of the given equation, and u_{1}, $u_{2}, \ldots, u_{n}$ form a fundamental system of solutions of the homogeneous equation obtained by replacing $E(x)$ by zero. A set of solutions is fundamental, or independent, if its Wronskian determinant $W(x)$ is not zero, where

$$W(x) = \begin{vmatrix} u_1 & u_2 & \cdots & u_n \\ u_1' & u_2' & \cdots & u_n' \\ \cdot & \cdot & & \cdot \\ \cdot & \cdot & & \cdot \\ \cdot & \cdot & & \cdot \\ u_1^{(n-1)} & u_2^{(n-1)} & \cdots & u_n^{(n-1)} \end{vmatrix}$$

For any n functions, $W(x) = 0$ if some one u_i is linearly dependent on the others, as $u_n = k_1 u_1 + k_2 u_2 + \cdots + k_{n-1} u_{n-1}$ with the coefficients k_i constant. And for n solutions of a linear differential equation of the nth order, if $W(x) \neq 0$, the solutions are linearly independent.

Constant Coefficients   To solve the homogeneous equation of the nth order $A_n d^n y/dx^n + A_{n-1} d^{n-1} y/dx^{n-1} + \cdots + A_1 dy/dx + A_0 y = 0$, $A_n \neq 0$, where $A_n, A_{n-1}, \ldots, A_0$ are constants, find the roots of the auxiliary equation

$$A_n p^n + A_{n-1} p^{n-1} + \cdots + A_1 p + A_0 = 0$$

For each simple real root r, there is a term ce^{rx} in the solution. The terms of the solution are to be added together. When r occurs twice among the n roots of the auxiliary equation, the corresponding term is $e^{rx}(c_1 + c_2 x)$. When r occurs three times, the corresponding term is $e^{rx}(c_1 + c_2 x + c_3 x^2)$, and so forth. When there is a pair of conjugate complex roots $a + bi$ and $a - bi$, the real form of the terms in the solution is $e^{ax}(c_1 \cos bx + d_1 \sin bx)$. When the same pair occurs twice, the corresponding term is $e^{ax}[(c_1 + c_2 x) \cos bx + (d_1 + d_2 x) \sin bx]$, and so forth.

Consider next the general nonhomogeneous linear differential equation of order n, with constant coefficients, or

$$A_n d^n y/dx^n + A_{n-1} d^{n-1} y/dx^{n-1} + \cdots + A_1 dy/dx + A_0 y = E(x)$$

We may solve this by adding any particular integral to the complementary function, or general solution, of the homogeneous equation obtained by replacing $E(x)$ by zero. The complementary function may be found from the rules just given. And the particular integral may be found by the methods of the following paragraphs.

Undetermined Coefficients   In the last equation, let the right member $E(x)$ be a sum of terms each of which is of the type k, $k \cos bx$, $k \sin bx$, ke^{ax}, kx, or more generally, $kx^m e^{ax}$, $kx^m e^{ax} \cos bx$, or $kx^m e^{ax} \sin bx$. Here m is zero or a positive integer, and a and b are any real numbers. Then the form of the particular integral I may be predicted by the following rules.

CASE 1. $E(x)$ **is a single term** T. Let D be written for d/dx, so that the given equation is $P(D)y = E(x)$, where $P(D) = A_n D^n + A_{n-1} D^{n-1} + \cdots + A_1 D + A_0 y$. With the term T associate the simplest polynomial $Q(D)$ such that $Q(D)T = 0$. For the particular types k, etc., $Q(D)$ will be $D, D^2 + b^2, D^2 + b^2, D - a, D^2$; and for the general types $kx^m e^{ax}$, etc., $Q(D)$ will be $(D - a)^{m+1}$, $(D^2 - 2aD + a^2 + b^2)^{m+1}$, $(D^2 - 2aD + a^2 + b^2)^{m+1}$. Thus $Q(D)$ will always be some power of a first- or second-degree factor, $Q(D) = F^v$, $F = D - a$, or $F = D^2 - 2aD + a^2 + b^2$.

Use the method described under **Constant Coefficients** to find the terms in the solution of $P(D)y = 0$ and also the terms in the solution of $Q(D)P(D)y = 0$. Then assume the particular integral I is a linear combination with unknown coefficients of those terms in the solution of $Q(D)P(D)y = 0$ which are not in the solution of $P(D)y = 0$. Thus if $Q(D) = F^q$ and F is *not* a factor of $P(D)$, assume $I = (Ax^{q-1} + Bx^{q-2} + \cdots + L)e^{ax}$ when $F = D - a$, and assume $I = (Ax^{q-1} + Bx^{q-2} + \cdots + L)e^{ax} \cos bx + (Mx^{q-1} + Nx^{q-2} + \cdots + R)e^{ax} \sin bx$ when $F = D^2 - 2aD + a^2 + b^2$. When F is a factor of $P(D)$ and the highest power of F which is a divisor of $P(D)$ is F^k, try the I above multiplied by x^k.

CASE 2. $E(x)$ **is a sum of terms**. With each term in $E(x)$, associate a polynomial $Q(D) = F^q$ as before. Arrange in one group all the terms that have the same F. The particular integral of the given equation will be the sum of solutions of equations each of which has one group on the right. For any one such equation, the form of the particular integral is given as for Case 1, with q the highest power of F associated with any term of the group on the right.

After the form has been found in Case 1 or 2, the unknown coefficients follow when we substitute back in the given differential equation, equate coefficients of like terms, and solve the resulting system of simultaneous equations.

Variation of Parameters.   Whenever a fundamental system of solutions $u_1, u_2, \ldots, u_n$ for the homogeneous equation is known, a particular integral of

$$A_n(x)d^n y/dx^n + A_{n-1}(x)d^{n-1}y/dx^{n-1} + \cdots$$
$$+ A_1(x)\, dy/dx + A_0(x)y = E(x)$$

may be found in the form $y = \Sigma v_k u_k$. In this and the next few summations, k runs from 1 to n. The v_k are functions of x, found by integrating their derivatives v_k', and these derivatives are the solutions of the n simultaneous equations $\Sigma v_k' u_k = 0$, $\Sigma v_k' u_k' = 0$, $\Sigma v_k' u_k'' = 0, \cdots$, $\Sigma v_k' u_k^{(n-2)} = 0$, $A_n(x)\Sigma v_k' u_k^{(n-1)} = E(x)$. To find the v_k from $v_k = \int v_k'\, dx + c_k$, any choice of constants will lead to a particular integral.

The special choice $v_k = \displaystyle\int_0^x v_k'\, dx$ leads to the particular integral having $y, y', y'', \ldots, y^{(n-1)}$ each equal to zero when $x = 0$.

The Cauchy-Euler Equidimensional Equation   This has the form

$$k_n x^n d^n y/dx^n + k_{n-1} x^{n-1} d^{n-1}y/dx^{n-1} + \cdots$$
$$+ k_1 x\, dy/dx + k_0 y = F(x)$$

The substitution $x = e^t$, which makes

$$x\, dy/dx = dy/dt$$
$$x^k\, d^k y/dx^k = (d/dt - k + 1) \cdots (d/dt - 2)(d/dt - 1)\, dy/dt$$

transforms this into a linear differential equation with constant coefficients. Its solution $y = g(t)$ leads to $y = g(\ln x)$ as the solution of the given Cauchy-Euler equation.

Bessel's Equation   The general Bessel equation of order n is:

$$x^2 y'' + xy' + (x^2 - n^2)y = 0$$

This equation has general solution

$$y = AJ_n(x) + BJ_{-n}(x)$$

when n is not an integer. Here, $J_n(x)$ and $J_{-n}(x)$ are Bessel functions (see section on Special Functions).

In case $n = 0$, Bessel's equation has solution

$$y = AJ_0(x) + B\left[J_0(x) \ln(x) - \sum_{k=1}^{\infty} \frac{(-1)^k H_k x^{2k}}{2^{2k}(k!)^2} \right]$$

where H_k is the kth partial sum of the harmonic series, $1 + \frac{1}{2} + \frac{1}{3} + \cdots + \frac{1}{k}$.

In case $n = 1$, the solution is

$$y = AJ_1(x) + B\left\{ J_1(x) \ln(x) + 1/x \right.$$
$$\left. - \left[\sum_{k=1}^{\infty} \frac{(-1)^k (H_k + H_{k-1})x^{2k-1}}{2^{2k}k!(k-1)!} \right] \right\}$$

In case $n > 1$, n is an integer, solution is

$$y = AJ_n(x) + B\left\{ J_n(x) \ln(x) + \left[\sum_{k=0}^{\infty} \frac{(-1)^{k+1}(n-1)!x^{2k-n}}{2^{2k+1-n}k!(1-n)^k} \right] \right.$$
$$\left. + \frac{1}{2}\left[\sum_{k=0}^{\infty} \frac{(-1)^{k+1}(H_k + H_{k+1})x^{2k+n}}{2^{2k+n}k!(k+n)!} \right] \right\}$$

Solutions to Bessel's equation may be given in several other forms, often exploiting the relation between H_k and $\ln(k)$ or the so-called *Euler constant*.

General Method of Power Series   Given a general differential equation $F(x, y, y', \ldots) = 0$, the solution may be expanded as a Maclaurin series, so $y = \Sigma_{n=0}^{\infty} a_n x^n$, where $a_n = f^{(n)}(0)/n!$. The power series for y may be differentiated formally, so that $y' = \sum_{n=1}^{\infty} n a_n x^{n-1} = \sum_{n=0}^{\infty} (n+1)a_{n+1}x^n$, and $y'' = \sum_{n=2}^{\infty} n(n-1)a_n x^{n-2} = \sum_{n=0}^{\infty} (n+1)(n+2)a_{n+2}x^n$.

Substituting these series into the equation $F(x, y, y', \ldots) = 0$ often gives useful recursive relationships, giving the value of a_n in terms of previous values. If approximate solutions are useful, then it may be sufficient to take the first few terms of the Maclaurin series as a solution to the equation.

EXAMPLE. Consider $y'' - y' + xy = 0$. The procedure gives $\sum_{n=0}^{\infty} (n+1)(n+2)a_{n+2}x^n - \sum_{n=0}^{\infty}(n+1)a_{n+1}x^n + x\sum_{n=0}^{\infty} a_n x^n = \sum_{n=0}^{\infty} (n+1)(n+2)a_{n+2}x^n - \sum_{n=0}^{\infty}(n+1)a_{n+1}x^n + \sum_{n=1}^{\infty} a_{n-1}x^n = (2a_2 - a_1)x^0 + \sum_{n=1}^{\infty} [(n+1)(n+2)a_{n+2} - (n+1)a_{n+1} + a_{n-1})]\,x^n = 0$. Thus $2a_2 - a_1 = 0$ and, for $n > 0$, $(n+1)(n+2)a_{n+2} - (n+1)a_{n+1} + a_{n-1} = 0$. Thus, a_0 and a_1 may be determined arbitrarily, but thereafter, the values of a_n are determined recursively.

PARTIAL DIFFERENTIAL EQUATIONS

Partial differential equations (PDEs) arise when there are two or more independent variables. Two notations are common for the partial derivatives involved in PDEs, the "del" or fraction notation, where the first partial derivative of f with respect to x would be written $\partial f/\partial x$, and the subscript notation, where it would be written f_x.

In the same way that ordinary differential equations often involve arbitrary constants, solutions to PDEs often involve arbitrary functions.

EXAMPLE. $f_{xy} = 0$ has as its general solution $g(x) + h(y)$. The function g does not depend on y, so $g_y = 0$. Similarly, $f_x = 0$.

PDEs usually involve boundary or initial conditions dictated by the application. These are analogous to initial conditions in ordinary differential equations.

In solving PDEs, it is seldom feasible to find a general solution and then specialize that general solution to satisfy the boundary conditions, as is done with ordinary differential equations. Instead, the boundary conditions usually play a key role in the solution of a problem. A notable exception to this is the case of linear, homogeneous PDEs since they have the property that if f_1 and f_2 are solutions, then $f_1 + f_2$ is also a solution. The wave equation is one such equation, and this property is the key to the solution described in the section "Fourier Series."

Often it is difficult to find exact solutions to PDEs, so it is necessary to resort to approximations or numerical solutions.

Classification of PDEs

Linear A PDE is **linear** if it involves only first derivatives, and then only to the first power. The general form of a linear PDE, in two independent variables, x and y, and the dependent variable z, is $P(x, y, z)f_x + Q(x, y, z)f_y = R(x, y, z)$, and it will have a solution of the form $z = f(x, y)$ if its solution is a function, or $F(x, y, z) = 0$ if the solution is not a function.

Elliptic Laplace's equation $f_{xx} + f_{yy} = 0$ and Poisson's equation $f_{xx} + f_{yy} = g(x, y)$ are the prototypical elliptic equations. They have analogs in more than two variables. They do not explicitly involve the variable time and generally describe steady-state or equilibrium conditions, gravitational potential, where boundary conditions are distributions of mass, electrical potential, where boundary conditions are electrical charges, or equilibrium temperatures, and where boundary conditions are points where the temperature is held constant.

Parabolic $T_t = T_{xx} + T_{yy}$ represents the dynamic condition of diffusion or heat conduction, where $T(x, y, t)$ usually represents the temperature at time t at the point (x, y). Note that when the system reaches steady state, the temperature is no longer changing, so $T_t = 0$, and this becomes Laplace's equation.

Hyperbolic Wave propagation is described by equations of the type $u_{tt} = c^2(u_{xx} + u_{yy})$, where c is the velocity of waves in the medium.

VECTOR CALCULUS

Vector Fields A **vector field** is a function that assigns a vector to each point in a region. If the region is two-dimensional, then the vectors assigned are two-dimensional, and the vector field is a two-dimensional vector field, denoted $\mathbf{F}(x, y)$. In the same way, a three-dimensional vector field is denoted $\mathbf{F}(x, y, z)$.

A three-dimensional vector field can always be written:

$$\mathbf{F}(x, y, z) = f_1(x, y, z)\mathbf{i} + f_2(x, y, z)\mathbf{j} + f_3(x, y, z)\mathbf{k}$$

where $\mathbf{i}$, $\mathbf{j}$, and $\mathbf{k}$ are the basis vectors $(1, 0, 0)$, $(0, 1, 0)$, and $(0, 0, 1)$, respectively. The functions f_1, f_2, and f_3 are called **coordinate functions** of $\mathbf{F}$.

Parameterized Curves If C is a curve from a point A to a point B, either in two dimensions or in three dimensions, then a **parameterization** of C is a vector-valued function $\mathbf{r}(t) = r_1(t)\mathbf{i} + r_2(t)\mathbf{j} + r_3(t)\mathbf{k}$, which satisfies $\mathbf{r}(a) = A$, $\mathbf{r}(b) = B$, and $\mathbf{r}(t)$ is on the curve C, for $a \le t \le b$. It is also necessary that the function $\mathbf{r}(t)$ be continuous and one-to-one. A given curve C has many different parameterizations.

The derivative of a parameterization $\mathbf{r}(t)$ is a vector-valued function $\mathbf{r}'(t) = r_1'(t)\mathbf{i} + r_2'(t)\mathbf{j} + r_3'(t)\mathbf{k}$. The derivative is the velocity function of the parameterization. It is always tangent to the curve C, and the magnitude is the speed of the parameterization.

Line Integrals If $\mathbf{F}$ is a vector field, C is a curve, and $\mathbf{r}(t)$ is a parameterization of C, then the **line integral,** or **work integral,** of $\mathbf{F}$ along C is

$$W = \int_C \mathbf{F} \cdot d\mathbf{r} = \int_a^b \mathbf{F}(\mathbf{r}(t)) \cdot \mathbf{r}'(t)\,dt$$

This is sometimes called the work integral because if $\mathbf{F}$ is a force field, then W is the amount of work necessary to move an object along the curve C from A to B.

Divergence and Curl The **divergence** of a vector field $\mathbf{F}$ is div $\mathbf{F} = f_{1x} + f_{2y} + f_{3z}$. If $\mathbf{F}$ represents the flow of a fluid, then the divergence at a point represents the rate at which the fluid is expanding at that point. Vector fields with div $\mathbf{F} = 0$ are called **incompressible**.

The **curl** of $\mathbf{F}$ is

$$\text{curl } \mathbf{F} = (f_{3y} - f_{2z})\mathbf{i} + (f_{1z} - f_{3x})\mathbf{j} + (f_{2x} - f_{1y})\mathbf{k}$$

If $\mathbf{F}$ is a two-dimensional vector field, then the first two terms of the curl are zero, so the curl is just

$$\text{curl } \mathbf{F} = (f_{2x} - f_{1y})\mathbf{k}$$

If $\mathbf{F}$ represents the flow of a fluid, then the curl represents the rotation of the fluid at a given point. Vector fields with curl $\mathbf{F} = 0$ are called **irrotational**.

Two important facts relate div, grad, and curl:
1. div (curl $\mathbf{F}$) = 0
2. curl (grad f) = $\mathbf{0}$

Conservative Vector Fields A vector field $\mathbf{F} = f_1\mathbf{i} + f_2\mathbf{j} + f_3\mathbf{k}$ is **conservative** if all of the following are satisfied:

$$f_{1y} = f_{2x} \qquad f_{1z} = f_{3x} \qquad \text{and} \qquad f_{2z} = f_{3y}$$

If $\mathbf{F}$ is a two-dimensional vector field, then the second and third conditions are always satisfied, and so only the first condition must be checked. Conservative vector fields have three important properties:

1. $\int_C \mathbf{F} \cdot d\mathbf{r}$ has the same value regardless of what curve C is chosen that connects the points A and B. This property is called **path independence**.

2. $\mathbf{F}$ is the gradient of some function $f(x, y, z)$.

3. Curl $\mathbf{F} = \mathbf{0}$.

In the special case that $\mathbf{F}$ is a conservative vector field, if $\mathbf{F} = $ grad

(f), then

$$\int_C \mathbf{F} \cdot d\mathbf{r} = f(B) - f(A)$$

THEOREMS ABOUT LINE AND SURFACE INTEGRALS

Two important theorems relate line integrals with double integrals. If R is a region in the plane and if C is the curve tracing the boundary of R in the positive (counterclockwise) direction, and if $\mathbf{F}$ is a continuous vector field with continuous first partial derivatives, line integrals on C are related to double integrals on R by Green's theorem and the divergence theorem.

Green's Theorem

$$\int_C \mathbf{F} \cdot d\mathbf{r} = \iint_R \text{curl } (\mathbf{F}) \cdot d\mathbf{S}$$

The right-hand double integral may also be written as $\iint_R |\text{curl } (\mathbf{F})| \, dA$.

Green's theorem describes the total rotation of a vector field in two different ways, on the left in terms of the boundary of the region and on the right in terms of the rotation at each point within the region.

Divergence Theorem

$$\int_C \mathbf{F} \cdot d\mathbf{N} = \iint_R \text{div } (\mathbf{F}) \, dA$$

where $\mathbf{N}$ is the so-called *normal vector field* to the curve C. The divergence theorem describes the expansion of a region in two distinct ways, on the left in terms of the flux across the boundary of the region and on the right in terms of the expansion at each point within the region.

Both Green's theorem and the divergence theorem have corresponding theorems involving surface integrals and volume integrals in three dimensions.

LAPLACE AND FOURIER TRANSFORMS

Laplace Transforms The Laplace transform is used to convert equations involving a time variable t into equations involving a fre-

Table 2.1.4   Laplace Transforms

$f(t)$	$F(s) = \mathscr{L}(f(t))$	Name of function
1. a	a/s	
2. 1	$1/s$	
3. $u_a(t) = \begin{cases} 0 & t < a \\ 1 & t > a \end{cases}$	e^{-as}/s	Heaviside or step function
4. $\delta_a(t) = u_a'(t)$	e^{-as}	Dirac or impulse function
5. e^{at}	$1/(s - a)$	
6. $(1/r)e^{-t/r}$	$1/(rs + 1)$	
7. ke^{-at}	$k/(s + a)$	
8. $\sin at$	$a/(s^2 + a^2)$	
9. $\cos at$	$s/(s^2 + a^2)$	
10. $e^{at} \sin bt$	$b/[(s + a)^2 + b^2]$	
11. $\dfrac{e^{at}}{a + b} - \dfrac{e^{bt}}{a + b}$	$\dfrac{1}{(s - a)(s - b)}$	
12. t	$1/s^2$	
13. t^2	$2/s^3$	
14. t^n	$n!/s^{n+1}$	
15. t^a	$\Gamma(a + 1)/s^{a + 1}$	Gamma function (see "Special Functions")
16. $\sinh at$	$a/(s^2 - a^2)$	
17. $\cosh at$	$s/(s^2 - a^2)$	
18. $t^n e^{at}$	$n!/(s - a)^{n+1}$	
19. $t \cos at$	$(s^2 - a^2)/(s^2 + a^2)^2$	
20. $t \sin at$	$2as/(s^2 + a^2)^2$	
21. $\sin at - at \cos at$	$2a^3/(s^2 + a^2)^2$	
22. $\arctan a/s$	$(\sin at)/t$	

Table 2.1.5   Properties of Laplace Transforms

$f(t)$	$F(s) = \mathscr{L}(f(t))$	Name of rule
1. $f(t)$	$\displaystyle\int_0^\infty e^{-st} f(t) \, dt$	Definition
2. $f(t) + g(t)$	$F(s) + G(s)$	Addition
3. $kf(t)$	$kF(s)$	Scalar multiples
4. $f'(t)$	$sF(s) - f(0^+)$	Derivative laws
5. $f''(t)$	$s^2 F(s) - sf(0^+) - f'(0^+)$	
6. $f'''(t)$	$s^3 F(s) - s^2 f(0^+) - sf'(0^+) - f''(0^+)$	
7. $\displaystyle\int f(t) dt$	$(1/s)F(s) + (1/s)\displaystyle\int f(t) \, dt\vert_{0^+}$	Integral law
8. $f(bt)$	$(1/b)F(s/b)$	Change of scale
9. $e^{at}f(t)$	$F(s - a)$	First shifting
10. $f * g(t)$	$F(s)G(s)$	Convolution
11. $u_a(t)f(t - a)$	$F(s)e^{-at}$	Second shifting
12. $-tf(t)$	$F'(s)$	Derivative in s

quency variable s. There are essentially three reasons for doing this: (1) higher-order differential equations may be converted to purely algebraic equations, which are more easily solved; (2) boundary conditions are easily handled; and (3) the method is well-suited to the theory associated with the **Nyquist stability criteria**.

In Laplace-transformation mathematics the following symbols and equations are used (Tables 2.1.4 and 2.1.5):

$f(t) =$ a function of time

$s =$ a complex variable of the form $(\sigma + j\omega)$

$F(s) =$ an equation expressed in the transform variable s, resulting from operating on a function of time with the Laplace integral

$\mathscr{L} =$ an operational symbol indicating that the quantity which it prefixes is to be transformed into the frequency domain

$f(0^+) =$ the limit from the positive direction of $f(t)$ as t approaches zero

$f(0^-) =$ the limit from the negative direction of $f(t)$ as t approaches zero

Therefore, $F(s) = \mathscr{L}[f(t)]$. The Laplace integral is defined as

$$\mathscr{L} = \int_0^\infty e^{-st} \, dt. \quad \text{Therefore,} \quad \mathscr{L}[f(t)] = \int_0^\infty e^{-st} f(t) \, dt$$

Direct Transforms

EXAMPLE.

$$f(t) = \sin \beta t$$

$$\mathscr{L}[f(t)] = \mathscr{L}(\sin \beta t) = \int_0^\infty \sin \beta t \, e^{-st} \, dt$$

but

$$\sin \beta t = \frac{e^{j\beta t} - e^{-j\beta t}}{2j} \quad \text{where} \quad j^2 = -1$$

$$\mathscr{L}(\sin \beta t) = \frac{1}{2j} \int_0^\infty (e^{j\beta t} - e^{-j\beta t}) e^{-st} \, dt$$

$$= \frac{1}{2j} \left(\frac{-1}{s - j\beta} \right) e^{(-s + j\beta)t} \Big|_0^\infty$$

$$- \frac{1}{2j} \left(\frac{-1}{s + j\beta} \right) e^{(-s - j\beta)t} \Big|_0^\infty$$

$$= \frac{\beta}{s^2 + \beta^2}$$

Table 2.1.4 lists the transforms of common time-variable expressions.

Some special functions are frequently encountered when using Laplace methods.

The **Heavyside,** or **step, function** $u_a(t)$ sometimes written $u(t - a)$, is zero for all $t < a$ and 1 for all $t > a$. Its value at $t = a$ is defined differently in different applications, as 0, ½, or 1, or it is simply left undefined. In Laplace applications, the value of a function at a single point does not matter. The Heavyside function describes a force which is "off" until time $t = a$ and then instantly goes "on."

The **Dirac delta function,** or **impulse function,** $\delta_a(t)$, sometimes written $\delta(t - a)$, is the derivative of the Heavyside function. Its value is always zero, except at $t = a$, where its value is "positive infinity." It is sometimes described as a "point mass function." The delta function describes an impulse or an instantaneous transfer of momentum.

The derivative of the Dirac delta function is called the **dipole function.** It is less frequently encountered.

The **convolution** $f * g(t)$ of two functions $f(t)$ and $g(t)$ is defined as

$$f * g(t) = \int_0^t f(u)g(t - u)\, du$$

Laplace transforms are often used to solve differential equations arising from so-called *linear systems*. Many vibrating systems and electrical circuits are linear systems. If an input function $f_i(t)$ describes the forces exerted upon a system and a response or output function $f_o(t)$ describes the motion of the system, then the **transfer function** $T(s) = F_o(s)/F_i(s)$. Linear systems have the special property that the transfer function is independent of the input function, within the elastic limits of the system. Therefore,

$$\frac{F_o(s)}{F_i(s)} = \frac{G_o(s)}{G_i(s)}$$

This gives a technique for describing the response of a system to a complicated input function if its response to a simple input function is known.

EXAMPLE. Solve $y'' + 2y' - 3y = 8e^t$ subject to initial conditions $y(0) = 2$ and $y'(0) = 0$. Let $y = f(t)$ and $Y = F(s)$. Take Laplace transforms of both sides and substitute for $y(0)$ and $y'(0)$, and get

$$s^2Y - 2s + 2(sY - 2) - 3Y = \frac{8}{s - 1}$$

Solve for Y, apply partial fractions, and get

$$Y = \frac{2s^2 + 2s + 4}{(s + 3)(s - 1)^2}$$
$$= \frac{1}{s + 3} + \frac{s + 1}{(s - 1)^2}$$
$$= \frac{1}{(s + 3)} + \frac{1}{(s - 1)} + \frac{2}{(s - 1)^2}$$

Using the tables of transforms to find what function has Y as its transform, we get

$$y = e^{-3t} + e^t + 2te^t$$

EXAMPLE. A vibrating system responds to an input function $f_i(t) = \sin t$ with a response $f_o(t) = \sin 2t$. Find the system response to the input $g_i(t) = \sin 2t$.

Apply the invariance of the transfer function, and get

$$G_o(s) = \frac{F_o G_i}{F_i}$$
$$= \frac{4(s^2 + 1)}{(s^2 + 4)^2}$$
$$= \frac{2(2)}{s^2 + 2^2} - \frac{12}{16}\left[\frac{16}{(s^2 + 2^2)^2}\right]$$

Applying formulas 8 and 21 from Table 2.1.4 of Laplace transforms,

$$g_o(t) = 2 \sin 2t - \tfrac{3}{4} \sin 2t + \tfrac{1}{2} t \cos 2t$$

Inversion When an equation has been transformed, an explicit solution for the unknown may be directly determined through algebraic manipulation. In automatic-control design, the equation is usually the differential equation describing the system, and the unknown is either the output quantity or the error. The solution gained from the transformed equation is expressed in terms of the complex variable s. For many design or analysis purposes, the solution in s is sufficient, but in some cases it is necessary to retransform the solution in terms of time. The process of passing from the complex-variable (frequency domain) expression to that of time (time domain) is called an **inverse transformation.** It is represented symbolically as

$$\mathscr{L}^{-1}F(s) = f(t)$$

For any $f(t)$ there is only one direct transform, $F(s)$. For any given $F(s)$ there is only one inverse transform $f(t)$. Therefore, tables are generally used for determining inverse transforms. Very complete tables of inverse transforms may be found in Gardner and Barnes, "Transients in Linear Systems." As an example of the inverse procedure consider an equation of the form

$$K = \alpha x(t) + \int \frac{x(t)}{\beta}\, dt$$

It is desired to obtain an expression for $x(t)$ resulting from an instantaneous change in the quantity K. Transforming the last equation yields

$$\frac{K}{s} = X(s)\alpha + \frac{X(s)}{s\beta} + \frac{f^{-1}(0^+)}{s}$$

If $f^{-1}(0)/s = 0$

then

$$X(s) = \frac{K/\alpha}{s + 1/\alpha\beta}$$

$$x(t) = \mathscr{L}^{-1}[X(s)] = \mathscr{L}^{-1}\frac{K/\alpha}{s + 1/\alpha\beta}$$

From Table 2.1.4, $x(t) = \dfrac{K}{\alpha} e^{-t/\alpha\beta}$

Fourier Coefficients Fourier coefficients are used to analyze periodic functions in terms of sines and cosines. If $f(x)$ is a function with period $2L$, then the Fourier coefficients are defined as

$$a_n = \frac{1}{L}\int_{-L}^{L} f(s) \cos \frac{n\pi s}{L}\, ds \qquad n = 0, 1, 2, \ldots$$
$$b_n = \frac{1}{L}\int_{-L}^{L} f(s) \sin \frac{n\pi s}{L}\, ds \qquad n = 1, 2, \ldots$$

Then the Fourier theorem states that

$$f(x) = \frac{a_0}{2} + \sum_{n=1}^{\infty} a_n \cos\left(\frac{n\pi x}{L}\right) + b_n \sin\left(\frac{n\pi x}{L}\right)$$

The series on the right is called the "Fourier series of the function $f(x)$." The convergence of the Fourier series is usually rapid, so that the function $f(x)$ is usually well-approximated by the sum of the first few sums of the series.

Examples of the Fourier Series If $y = f(x)$ is the curve in Figs. 2.1.110 to 2.1.112, then in Fig. 2.1.110,

$$y = \frac{h}{2} - \frac{4h}{\pi^2}\left(\cos \frac{\pi x}{c} + \frac{1}{9}\cos \frac{3\pi x}{c} + \frac{1}{25}\cos \frac{5\pi x}{c} + \cdots\right)$$

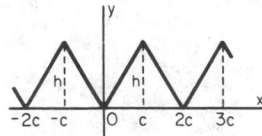

Fig. 2.1.110 Saw-tooth curve.

In Fig. 2.1.111,

$$y = \frac{4h}{\pi}\left(\sin\frac{\pi x}{c} + \frac{1}{3}\sin\frac{3\pi x}{c} + \frac{1}{5}\sin\frac{5\pi x}{c} + \cdots\right)$$

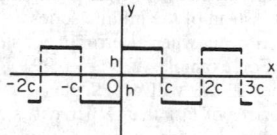

Fig. 2.1.111 Step-function curve.

In Fig. 2.1.112,

$$y = \frac{2h}{\pi}\left(\sin\frac{\pi x}{c} - \frac{1}{2}\sin\frac{2\pi x}{c} + \frac{1}{3}\sin\frac{3\pi x}{c} - \cdots\right)$$

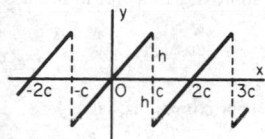

Fig. 2.1.112 Linear-sweep curve.

If the Fourier coefficients of a function $f(x)$ are known, then the coefficients of the derivative $f'(x)$ can be found, when they exist, as follows:

$$a'_n = nb_n \qquad b'_n = -na_n$$

where a'_n and b'_n are the Fourier coefficients of $f'(x)$.

The **complex Fourier coefficients** are defined by:

$$c_n = \frac{1}{2}(a_n - ib_n)$$
$$c_0 = \frac{1}{2}a_0$$
$$c_n = \frac{1}{2}(a_n + ib_n)$$

Then the complex form of the Fourier theorem is

$$f(x) = \sum_{n=-\infty}^{\infty} c_n e^{in\pi x/L}$$

Wave Equation Fourier series are often used in the solution of the wave equation $a^2 u_{xx} = u_{tt}$ where $0 < x < L, t > 0$, and initial conditions are $u(x, 0) = f(x)$ and $u_t(x, 0) = g(x)$. This describes the position of a vibrating string of length L, fixed at both ends, with initial position $f(x)$ and initial velocity $g(x)$. The constant a is the velocity at which waves are propagated along the string, and is given by $a^2 = T/p$, where T is the tension in the string and p is the mass per unit length of the string.

If $f(x)$ is extended to the interval $-L < x < L$ by setting $f(-x) = -f(x)$, then f may be considered periodic of period $2L$. Its Fourier coefficients are

$$a_n = 0 \qquad b_n = \int_{-L}^{L} f(x)\sin\frac{n\pi x}{L}\,\pi\,dx \qquad n = 1, 2, \ldots$$

The solution to the wave equation is

$$u(x, t) = \sum_{n=1}^{\infty} b_n \sin\frac{n\pi x}{L}\cos\frac{n\pi t}{L}$$

Fourier transform A nonperiodic function $f(x)$ requires two func-

tions to describe its Fourier transform:

$$A(w) = \int_{-\infty}^{\infty} f(x)\cos wx\,dx$$

$$B(w) = \int_{-\infty}^{\infty} f(x)\sin wx\,dx$$

Then the **Fourier integral equation** is

$$f(x) = \int_0^{\infty} A(w)\cos wx + B(w)\sin wx\,dw$$

The **complex Fourier transform** of $f(x)$ is defined as

$$F(w) = \int_{-\infty}^{\infty} f(x)\,e^{iwx}\,dx$$

Then the **complex Fourier integral equation** is

$$f(x) = \frac{1}{2\pi}\int_0^{\infty} F(w)e^{-iwx}\,dw$$

Heat Equation The Fourier transform may be used to solve the one-dimensional heat equation $u_t(x, t) = u_{xx}(x, t)$, for $t > 0$, given initial condition $u(x, 0) = f(x)$. Let $F(s)$ be the complex Fourier transform of $f(x)$, and let $U(s, t)$ be the complex Fourier transform of $u(x, t)$. Then the transform of $u_t(x, t)$ is $dU(s, t)/dt$.

Transforming $u_t(x, t) = u_{xx}(s, t)$ yields $dU/dt + s^2 U = 0$ and $U(s, 0) = f(s)$. Solving this using the Laplace transform gives $U(s, t) = F(s)e^{s^2 t}$.

Applying the complex Fourier integral equation, which gives $u(x, t)$ in terms of $U(s, t)$, gives

$$u(x, t) = \frac{1}{2\pi}\int_0^{\infty} U(s, t)e^{-isx}\,ds$$

$$= \frac{1}{2\pi}\int_{-\infty}^{\infty}\int_0^{\infty} f(y)e^{is(y-x)}e^{s^2 t}\,ds\,dy$$

Applying the Euler formula, $e^{ix} = \cos x + i\sin x$,

$$u(x, t) = \frac{1}{2\pi}\int_{-\infty}^{\infty}\int_0^{\infty} f(y)\cos(s(y - x))e^{s^2 t}\,ds\,dy$$

SPECIAL FUNCTIONS

Gamma Function The gamma function is a generalization of the factorial function. It arises in Laplace transforms of polynomials, in continuous probability, and in the solution to certain differential equations. It is defined by the improper integral:

$$\Gamma(x) = \int_0^{\infty} t^{x-1}e^{-t}\,dt$$

The integral converges for $x > 0$ and diverges otherwise. The function is extended to all negative values, except negative integers, by the relation

$$\Gamma(x + 1) = x\Gamma(x)$$

The gamma function is related to the factorial function by

$$\Gamma(n + 1) = n!$$

for all positive integers n.

An important value of the gamma function is

$$\Gamma(0.5) = \pi^{1/2}$$

Other values of the gamma function are found in CRC Standard Mathematical Tables and similar tables.

Beta Function The beta function is a function of two variables and is a generalization of the binomial coefficients. It is closely related to the gamma function. It is defined by the integral:

$$B(x, y) = \int_0^1 t^{x-1}(1 - t)^{y-1}\, dt \qquad \text{for } x, y > 0$$

The beta function can also be represented as a trigonometric integral, by substituting $t = \sin^2 \theta$, as

$$B(x, y) = 2 \int_0^{\pi/2} (\sin \theta)^{2x-1} (\cos \theta)^{2y-1}\, d\theta$$

The beta function is related to the gamma function by the relation

$$B(x, y) = \frac{\Gamma(x)\Gamma(y)}{\Gamma(x + y)}$$

This relation shows that $B(x, y) = B(y, x)$.

Bernoulli Functions The Bernoulli functions are a sequence of periodic functions of period 1 used in approximation theory. Note that for any number x, $[x]$ represents the largest integer less than or equal to x. $[3.14] = 3$ and $[-1.2] = -2$. The Bernoulli functions $B_n(x)$ are defined recursively as follows:

1. $B_0(x) = 1$
2. $B_1(x) = x - [x] - \frac{1}{2}$
3. B_{n+1} is defined so that $B'_{n+1}(x) = B_n(x)$ and so that B_{n+1} is periodic of period 1.

Bessel Functions of the First Kind Bessel functions of the first kind arise in the solution of Bessel's equation of order v:

$$x^2 y'' + xy' + (x^2 - v^2)y = 0$$

When this is solved using series methods, the recursive relations define the Bessel functions of the first kind of order v:

$$J_v(x) = \sum_{k=0}^{\infty} \frac{(-1)^k}{k!(v + k + 1)} \left(\frac{x}{2}\right)^{v+2k}$$

Chebyshev Polynomials The Chebyshev polynomials arise in the solution of PDEs of the form

$$(1 - x^2)y'' - xy' + n^2 y = 0$$

and in approximation theory. They are defined as follows:

$$T_0(x) = 1 \qquad T_2(x) = 2x^2 - 1$$
$$T_1(x) = x \qquad T_3(x) = 4x^3 - 3x$$

For $n > 3$, they are defined recursively by the relation

$$T_{n+1}(x) - 2xT_n(x) + T_{n-1}(x) = 0$$

Chebyshev polynomials are said to be orthogonal because they have the property

$$\int_{-1}^1 \frac{T_n(x)T_m(x)}{(1 - x^2)^{1/2}}\, dx = 0 \qquad \text{for } n \neq m$$

NUMERICAL METHODS

Introduction Classical numerical analysis is based on polynomial approximation of the infinite operations of integration, differentiation, and interpolation. The objective of such analyses is to replace difficult or impossible exact computations with easier approximate computations. The challenge is to make the approximate computations short enough and accurate enough to be useful.

Modern numerical analysis includes Fourier methods, including the *fast Fourier transform* (FFT) and many problems involving the way computers perform calculations. Modern aspects of the theory are changing very rapidly.

Errors Actual value = calculated value + error. There are several sources of errors in a calculation: mistakes, round-off errors, truncation errors, and accumulation errors.

Round-off errors arise from the use of a number not sufficiently accurate to represent the actual value of the number, for example, using 3.14159 to represent the irrational number π, or using 0.56 to represent $\frac{9}{16}$ or 0.5625.

Truncation errors arise when a finite number of steps are used to approximate an infinite number of steps, for example, the first n terms of a series are used instead of the infinite series.

Accumulation errors occur when an error in one step is carried forward into another step. For example, if $x = 0.994$ has been previously rounded to 0.99, then $1 - x$ will be calculated as 0.01, while its true value is 0.006. An error of less than 1 percent is accumulated into an error of over 50 percent in just one step. Accumulation errors are particularly characteristic of methods involving recursion or iteration, where the same step is performed many times, with the results of one iteration used as the input for the next.

Simultaneous Linear Equations The matrix equation $Ax = b$ can be solved directly by finding A^{-1}, or it can be solved iteratively, by the method of **iteration in total steps:**

1. If necessary, rearrange the rows of the equation so that there are no zeros on the diagonal of A.
2. Take as initial approximations for the values of x_i:

$$x_1^{(0)} = \frac{b_1}{a_{11}} \qquad x_2^{(0)} = \frac{b_2}{a_{22}} \qquad \cdots \qquad x_n^{(0)} = \frac{b_n}{a_{nn}}$$

3. For successive approximations, take

$$x_i^{(k+1)} = (b_i - a_{i1}x_1^{(k)} - \cdots - a_{in}x_n^{(k)})/a_{ii}$$

Repeat step 3 until successive approximations for the values of x_i reach the specified tolerance.

A property of iteration by total steps is that it is self-correcting: that is, it can recover both from mistakes and from accumulation errors.

Zeros of Functions An iterative procedure for solving an equation $f(x) = 0$ is the **Newton-Raphson method**. The algorithm is as follows:

1. Choose a first estimate of a root x_0.
2. Let $x_{k+1} = x_k - f(x_k)/f'(x_k)$. Repeat step 2 until the estimate x_k converges to a root r.
3. If there are other roots of $f(x)$, then let $g(x) = f(x)/(x - r)$ and seek roots of $g(x)$.

False Position If two values x_0 and x_1 are known, such that $f(x_0)$ and $f(x_1)$ are opposite signs, then an iterative procedure for finding a root between x_0 and x_1 is the method of false position.

1. Let $m = [f(x_1) - f(x_0)]/(x_1 - x_0)$.
2. Let $x_2 = x_1 - f(x_1)/m$.
3. Find $f(x_2)$.
4. If $f(x_2)$ and $f(x_1)$ have the same sign, then let $x_1 = x_2$. Otherwise, let $x_0 = x_2$.
5. If x_1 is not a good enough estimate of the root, then return to step 1.

Functional Equalities To solve an equation of the form $f(x) = g(x)$, use the methods above to find roots of the equation $f(x) - g(x) = 0$.

Maxima One method for finding the maximum of a function $f(x)$ on an interval $[a, b]$ is to find the roots of the derivative $f'(x)$. The maximum of $f(x)$ occurs at a root or at an endpoint a or b.

Fibonacci Search An iterative procedure for searching for maxima works if $f(x)$ is unimodular on $[a, b]$. That is, f has only one maximum, and no other local maxima, between a and b. This procedure takes advantage of the so-called *golden ratio*, $r = 0.618034 = (\sqrt{5} - 1)/2$, which arises from the Fibonacci sequence.

1. If a is a sufficiently good estimate of the maximum, then stop. Otherwise, proceed to step 2.
2. Let $x_1 = ra + (1 - r)b$, and let $x_2 = (1 - r)a + rb$. Note $x_1 < x_2$. Find $f(x_1)$ and $f(x_2)$.
   a. If $f(x_1) = f(x_2)$, then let $a = x_1$ and $b = x_2$, and go to step 1.
   b. If $f(x_1) < f(x_2)$, then let $a = x_1$, and go to step 1.
   c. If $f(x_1) > f(x_2)$, then let $b = x_2$, and return to step 1.

In cases b and c, computation is saved since the new value of one of x_1 and x_2 will have been used in the previous step. It has been proved that the Fibonacci search is the fastest possible of the general "cutting" type of searches.

Steepest Ascent   If $z = f(x, y)$ is to be maximized, then the method of steepest ascent takes advantage of the fact that the gradient, grad (f) always points in the direction that f is increasing the fastest.

1. Let (x_0, y_0) be an initial guess of the maximum of f.
2. Let e be an initial step size, usually taken to be small.
3. Let $(x_{k+1}, y_{k+1}) = (x_k, y_k) + e$ grad $f(x_k, y_k)/|$grad $f(x_k, y_k)|$.
4. If $f(x_{k+1}, y_{k+1})$ is not greater than $f(x_k, y_k)$, then replace e with $e/2$ (cut the step size in half) and reperform step 3.
5. If (x_k, y_k) is a sufficiently accurate estimate of the maximum, then stop. Otherwise, repeat step 3.

Minimization   The theory of minimization exactly parallels the theory of maximization, since minimizing $z = f(x)$ occurs at the same value of x as maximizing $w = -f(x)$.

Numerical Differentiation   In general, numerical differentiation should be avoided where possible, since differentiation tends to be very sensitive to small errors in the value of the function $f(x)$. There are several approximations to $f'(x)$, involving a "step size" h usually taken to be small:

$$f'(x) = \frac{f(x + h) - f(x)}{h}$$

$$f'(x) = \frac{f(x + h) - f(x - h)}{2h}$$

$$f'(x) = \frac{f(x + 2h) + f(x + h) - f(x - h) - f(x - 2h)}{6h}$$

Other formulas are possible.

If a derivative is to be calculated from an equally spaced sequence of measured data, $y_1, y_2, \ldots, y_n$, then the above formulas may be adapted by taking $y_i = f(x_i)$. Then $h = x_{i+1} - x_i$ is the distance between measurements.

Since there are usually noise or measurement errors in measured data, it is often necessary to smooth the data, expecting that errors will be averaged out. Elementary smoothing is by simple averaging, where a value y_i is replaced by an average before the derivative is calculated. Examples include:

$$y_i \leftarrow \frac{y_{i+1} + y_i + y_{i-1}}{3}$$

$$y_i \leftarrow \frac{y_{i+2} + y_{i+1} + y_i + y_{i-1} + y_{i-2}}{5}$$

More information may be found in the literature under the topics linear filters, digital signal processing, and smoothing techniques.

Numerical Integration

Numerical integration requires a great deal of calculation and is usually done with the aid of a computer. All the methods described here, and many others, are widely available in packaged computer software. There is often a temptation to use whatever software is available without first checking that it really is appropriate. For this reason, it is important that the user be familiar with the methods being used and that he or she ensure that the error terms are tolerably small.

Trapezoid Rule   If an interval $a \leq x \leq b$ is divided into subintervals $x_0, x_1, \ldots, x_n$, then the definite integral

$$\int_a^b f(x)\, dx$$

may be approximated by

$$\sum_{i=1}^n [f(x_i) + f(x_{i-1})] \frac{x_{i+1} - x_i}{2}$$

If the values x_i are equally spaced at distance h and if f_i is written for $f(x_i)$, then the above formula reduces to

$$[f_0 + 2f_1 + 2f_2 + \cdots + 2f_{n-1} + f_n] \frac{h}{2}$$

The error in the trapezoid rule is given by

$$|E_n| \leq \frac{(b - a)^3 |f''(t)|}{12n^2}$$

where t is some value $a \leq t \leq b$.

Simpson's Rule   The most widely used rule for numerical integration approximates the curve with parabolas. The interval $a < x < b$ must be divided into $n/2$ subintervals, each of length $2h$, where n is an even number. Using the notation above, the integral is approximated by

$$[f_0 + 4f_1 + 2f_2 + 4f_3 + \cdots + 4f_{n-1} + f_n] \frac{h}{3}$$

The error term for Simpson's rule is given by $|E_n| < nh^5 |f^{(4)}(t)|/180$, where $a < t < b$.

Simpson's rule is generally more accurate than the trapezoid rule.

Ordinary Differential Equations

Modified Euler Method   Consider a first-order differential equation $dy/dx = f(x, y)$ and initial condition $y = y_0$ and $x = x_0$. Take x_i equally spaced, with $x_{i+1} - x_i = h$. Then the method is:

1. Set $n = 0$.
2. $y_n' = f(x_n, y_n)$ and $y'' = f_x(x_n, y_n) + y_n' f_y(x_n, y_n)$, where f_x and f_y denote partial derivatives.
3. $y_{n+1}' = f(x_{n+1}, y_{n+1})$.

Predictor steps:

4. For $n > 0$, $y_{n+1}^* = y_{n-1} + 2hy_n'$.
5. $y_{n+1}'^* = f(x_{n+1}, y_{n+1}^*)$.

Corrector steps:

6. $y_n^\# + 1 = y_n + [y_{n+1}^* + y_n']h/2$.
7. $y_{n+1}'^\# = f(x_{n+1}, y_{n+1}^\#)$.
8. If required accuracy is not yet obtained for y_{n+1} and y_{n+1}', then substitute $y^\#$ for y^*, in all its forms, and repeat the corrector steps. Otherwise, set $n = n + 1$ and return to step 2.

Other predictor-corrector methods are described in the literature.

Runge-Kutta Methods   These make up a family of widely used methods for ordinary differential equations. Given $dy/dx = f(x, y)$ and $h =$ interval size, third-order method (error proportional to h^4):

$$k_0 = hf(x_n)$$

$$k_1 = hf\left(x_n + \frac{h}{2}, y_n + \frac{k_0}{2}\right)$$

$$k_2 = hf(x_n + h, y_n + 2k_1 - k_0)$$

$$y_{n+1} = y_n + \frac{k_0 + 4k_1 + k_2}{6}$$

Higher-order Runge-Kutta methods are described in the literature. In general, higher-order methods yield smaller error terms.

2.2 COMPUTERS
by George J. Moshos

REFERENCES: Manuals from Computer Manufacturers. Knuth, "The Art of Computer Programming," vols 1, 2, and 3, Addison-Wesley. Yourdon and Constantine, "Structured Design," Prentice-Hall. DeMarco, "Structured Analysis and System Specification," Prentice-Hall. Moshos, "Data Communications," West Publishing. Date, "An Introduction to Database Systems," 4th ed., Addison-Wesley. Wiener and Sincovec, "Software Engineering with Modula-2 and ADA," Wiley. Hamming, "Numerical Methods for Scientists and Engineers," McGraw-Hill. Bowers and Sedore, "SCEPTRE: A Computer Program for Circuit and System Analysis," Prentice-Hall. Tannenbaum, "Operating Systems," Prentice-Hall. Lister, "Fundamentals of Operating Systems, 3d ed., Springer-Verlag. American National Standard Programming Language FORTRAN, ANSI X3.198-1992. Jensen and Wirth, "PASCAL: User Manual and Report," Springer. *Communications, Journal,* and *Computer Surveys,* ACM Computer Society. *Computer, Spectrum,* IEEE.

COMPUTER PROGRAMMING

Machine Types

Computers are machines used for automatically processing information represented by mechanical, electrical, or optical means. They may be classified as analog or digital according to the techniques used to represent and process the information. Analog computers represent information as physically measurable, continuous quantities and process the information by components that have been interconnected to form an analogous model of the problem to be solved. Digital computers, on the other hand, represent information as discrete physical states which have been encoded into symbolic formats, and process the information by sequences of operational steps which have been preplanned to solve the given problem.

When compared to analog computers, digital computers have the advantages of greater versatility in solving scientific, engineering, and commercial problems that involve numerical and nonnumerical information; of an accuracy dictated by significant digits rather than that which can be measured; and of exact reproducibility of results that stay unvitiated by small, random fluctuations in the physical signals. In the past, multiple-purpose analog computers offered advantages of speed and cost in solving a sophisticated class of complex problems dealing with networks of differential equations, but these advantages have disappeared with the advances in solid-state computers. Other than the occasional use of analog techniques for embedding computations as part of a larger system, digital techniques now account almost exclusively for the technology used in computers.

Digital information may be represented as a series of incremental, numerical steps which may be manipulated to position control devices using stepping motors. Digital information may also be encoded into symbolic formats representing digits, alphabetic characters, arithmetic numbers, words, linguistic constructs, points, and pictures which may be processed by a variety of mechanized operators. Machines organized in this manner can handle a more general class of both numerical and nonnumerical problems and so form by far the most common type of digital machines. In fact, the term *computer* has become synonymous with this type of machine.

Digital Machines

Digital machines consist of two kinds of circuits: memory cells, which effectively act to delay signals until needed, and logical units, which perform basic Boolean operations such as AND, OR, NOT, XOR, NAND, and NOR. Memory circuits can be simply defined as units where information can be stored and retrieved on demand. Configurations assembled from the Boolean operators provide the macro opera-tors and functions available to the machine user through encoded instructions. A typical computer might house hundreds of thousands to millions of transistors serving one or the other of these roles.

Both data and the instructions for processing the data can be stored in memory. Each unit of memory has an address at which the contents can be retrieved, or "read." The *read* operation makes the contents at an address available to other parts of the computer without destroying the contents in memory. The contents at an address may be changed by a *write* operation which inserts new information after first nullifying the previous contents. Some types of memory, called read-only memory (**ROM**), can be read from but not written to. They can only be changed at the factory.

Abstractly, the address and the contents at the address serve roles analogous to a variable and the value of the variable. For example, the equation $z = x + y$ specifies that the value of x added to the value of y will produce the value of z. In a similar way, the machine instruction whose format might be:

$$\text{add, address 1, address 2, address 3}$$

will, when executed, add the contents at address 1 to the contents at address 2 and store the result at address 3. As in the equation where the variables remain unaltered while the values of the variables may be changed, the addresses in the instruction remain unaltered while the contents at the address may change.

An essential property of a digital computer is that the sequence of instructions processed to solve a problem is executed without human intervention. When an operator manually controls the sequence of computation, the machine is called a calculator. This distinction between computer and calculator, however, is arbitrary and vague with modern machines. Modern calculators offer opportunity to program a series of operations which can be executed without any required intervention. On the other hand, the computer is often programmed to interrogate the operator for a response before continuing with the solution.

Computers differ from other kinds of mechanical and electrical machines in that computers perform work on information rather than on forces and displacements. A common form of information is numbers. Numbers can be encoded into a mechanized form and processed by the four rules of arithmetic ($+, -, \times, \div$). But numbers are only one kind of information that can be manipulated by the computer. Given an encoded alphabet, words and languages can be formed and the computer can be used to perform such processes as information storage and retrieval, translation, and editing. Given an encoded representation of points and lines, the computer can be used to perform such functions as drawing, recognizing, editing, and displaying graphs, patterns, and pictures.

Because computers have become easily accessible, engineers and scientists from every discipline have reformatted their professional activities to mechanize those aspects which can supplant human thought and decision. In this way, mechanical processes can be viewed as augmenting human physical skills and strength, and information processes can be viewed as augmenting human mental skills and intelligence.

COMPUTER DATA STRUCTURES

Binary Notation

Digital computers represent information by strings of digits which assume one of two values: 0 or 1. These units of information are called **bits,** a word contracted from the term *binary digits*. A string of bits may represent either numerical or nonnumerical information.

In order to achieve efficiency in handling the information, the computer groups the bits together into units containing a fixed number of bits which can be referenced as discrete units. By encoding and formatting these units of information, the computer can act to process them. Units of 8 bits, called **bytes,** are common. A byte can be used to encode the basic symbolic characters which provide the computer with input-output information such as the alphabet, decimal digits, punctuation marks, and special characters.

Bit groups may be organized into larger units of 4 bytes (32 bits) called **words,** or even larger units of 8 bytes called *double words;* and sometimes into smaller units of 2 bytes called *half words.* Besides encoding numerical information and other linguistic constructs, these units are used to encode a repertoire of machine instructions. Older machines and special-purpose machines may have other word sizes.

Computers process numerical information represented as binary numbers. The binary numbering system uses a positional notation similar to the decimal system. For example, the decimal number 596.37 represents the value $5 \times 10^2 + 9 \times 10^1 + 6 \times 10^0 + 3 \times 10^{-1} + 7 \times 10^{-2}$. The value assigned to any of the 10 possible digits in the decimal system depends on its position relative to the decimal point (a weight of 10 to zero or positive exponent is assigned to the digits appearing to the left of the decimal point, and a weight of 10 to a negative exponent is applied to digits to the right of the decimal point). In a similar manner, a binary number uses a radix of 2 and two possible digits: 0 and 1. The radix point in the positional notation separates the whole from the fractional part of the number, just as in the decimal system. The binary number 1011.011 represents a value $1 \times 2^3 + 0 \times 2^2 + 1 \times 2^1 + 1 \times 2^0 + 0 \times 2^{-1} + 1 \times 2^{-2} + 1 \times 2^{-3}$.

The operators available in the computer for setting up the solution of a problem are encoded into the instructions of the machine. The instruction repertoire always includes the usual arithmetic operators to handle numerical calculations. These instructions operate on data encoded in the binary system. However, this is not a serious operational problem, since the user specifies the numbers in the decimal system or by mnemonics, and the computer converts these formats into its own internal binary representation.

On occasions when one must express a number directly in the binary system, the number of digits needed to represent a numerical value becomes a handicap. In these situations, a radix of 8 or 16 (called the **octal** or **hexadecimal** system, respectively) constitutes a more convenient system. Starting with the digit to the left or with the digit to the right of the radix point, groups of 3 or 4 binary digits can be easily converted to equivalent octal or hexadecimal digits, respectively. Appending nonsignificant 0s as needed to the rightmost and leftmost part of the number to complete the set of 3 or 4 binary digits may be necessary. Table 2.2.1 lists the conversions of binary digits to their equivalent octal and hexadecimal representations. In the hexadecimal system, the letters A through F augment the set of decimal digits to represent the digits for 10 through 15. The following examples illustrate the conversion between binary numbers and octal or hexadecimal numbers using the table.

binary number	011	011	110	101	.	001	111	100
octal number	3	3	6	5	.	1	7	4

binary number		0110	1111	0101	.	0011	1110	
hexadecimal number		6	F	5	.	3	E	

Formats for Numerical Data

Three different formats are used to represent numerical information internal to the computer: fixed-point, encoded decimal, and floating-point.

A word or half word in fixed-point format is given as a string of 0s and 1s representing a binary number. The program infers the position of the radix point (immediately to the right of the word representing integers, and immediately to the left of the word representing fractions). Algebraic numbers have several alternate forms: 1's complement, 2's complement, and signed-magnitude. Most often 1's and 2's complement forms are adopted because they lead to a simplification in the

Table 2.2.1   Binary-Hexadecimal and Binary-Octal Conversion

Binary	Hexadecimal	Binary	Octal
0000	0	000	0
0001	1	001	1
0010	2	010	2
0011	3	011	3
0100	4	100	4
0101	5	101	5
0110	6	110	6
0111	7	111	7
1000	8		
1001	9		
1010	A		
1011	B		
1100	C		
1101	D		
1110	E		
1111	F		

hardware needed to perform the arithmetic operations. The sign of a 1's complement number can be changed by replacing the 0s with 1s and the 1s with 0s. To change the sign of a 2's complement number, reverse the digits as with a 1's-complement number and then add a 1 to the resulting binary number. Signed-magnitude numbers use the common representation of an explicit + or − sign by encoding the sign in the leftmost bit as a 0 or 1, respectively.

Many computers provide an encoded-decimal representation as a convenience for applications needing a decimal system. Table 2.2.2 gives three out of over 8000 possible schemes used to encode decimal digits in which 4 bits represent each decade value. Many other codes are possible using more bits per decade, but four bits per decimal digit are common because two decimal digits can then be encoded in one byte. The particular scheme selected depends on the properties needed by the devices in the application.

The floating-point format is a mechanized version of the scientific notation ($\pm M \times 10^{\pm E}$, where $\pm M$ and $\pm E$ represent the signed mantissa and signed exponent of the number). This format makes possible the use of a machine word to encode a large range of numbers. The signed mantissa and signed exponent occupy a portion of the word. The exponent is implied as a power of 2 or 16 rather than of 10, and the radix point is implied to the left of the mantissa. After each operation, the machine adjusts the exponent so that a nonzero digit appears in the most significant digit of the mantissa. That is, the mantissa is **normalized** so that its value lies in the range of $1/b \leq M < 1$ where b is the implied base of the number system (e.g.: $1/2 \leq M < 1$ for a radix of 2, and $1/16 \leq M < 1$ for a radix of 16). Since the zero in this notation has many logical representations, the format uses a standard recognizable form for zero, with a zero mantissa and a zero exponent, in order to avoid any ambiguity.

When calculations need greater precision, floating-point numbers use

Table 2.2.2   Schemes for Encoding Decimal Digits

Decimal digit	BCD	Excess-3	4221 code
0	0000	0011	0000
1	0001	0100	0001
2	0010	0101	0010
3	0011	0110	0011
4	0100	0111	0110
5	0101	1000	1001
6	0110	1001	1100
7	0111	1010	1101
8	1000	1011	1110
9	1001	1100	1111

a two-word representation. The first word contains the exponent and mantissa as in the one-word floating point. Precision is increased by appending the extra word to the mantissa. The terms **single precision** and **double precision** make the distinction between the one- and two-word representations for floating-point numbers, although *extended precision* would be a more accurate term for the two-word form since the added word more than doubles the number of significant digits.

The equivalent decimal precision of a floating-point number depends on the number n of bits used for the unsigned mantissa and on the implied base b (binary, octal, or hexadecimal). This can be simply expressed in equivalent decimal digits p as: $0.0301 (n - \log_2 b) < p < 0.0301 n$. For example, a 32-bit number using 7 bits for the signed exponent of an implied base of 16, 1 bit for the sign of the mantissa, and 24 bits for the value of the mantissa gives a precision of 6.02 to 7.22 equivalent decimal digits. The fractional parts indicate that some 7-digit and some 8-digit numbers cannot be represented with a mantissa of 24 bits. On the other hand, a double-precision number formed by adding another word of 32 bits to the 24-bit mantissa gives a precision of 15.65 to 16.85 equivalent decimal digits.

The range r of possible values in floating-point notation depends on the number of bits used to represent the exponent and the implied radix. For example, for a signed exponent of 7 bits and an implied base of 16, then $16^{-64} \le r \le 16^{63}$.

Formats of Nonnumerical Data

Logical elements, also called **Boolean** elements, have two possible values which simply represent 0 or 1, true or false, yes or no, OFF or ON, etc. These values may be conveniently encoded by a single bit.

A large variety of codes are used to represent the alphabet, digits, punctuation marks, and other special symbols. The most popular ones are the 7-bit ASCII code and the 8-bit EBCDIC code. ASCII and EBCDIC find their genesis in punch-tape and punch-card technologies, respectively, where each character was encoded as a combination of punched holes in a column. Both have now evolved into accepted standards represented by a combination of 0s and 1s in a byte.

Figure 2.2.1 shows the ASCII code. (ASCII stands for American Standard Code for Information Interchange.) The possible 128 bit patterns divide the code into 96 graphic characters (although the codes 0100000 and 1111111 do not represent any printable graphic symbol) and 32 control characters which represent nonprintable characters used in communications, in controlling peripheral machines, or in expanding the code set with other characters or fonts. The graphic codes and the control codes are organized so that subsets of usable codes with fewer bits can be formed and still maintain the pattern.

Data Structure Types

The above types of numerical and nonnumerical data formats are recognized and manipulated by the hardware operations of the computer. Other more complex data structures may be programmed into the computer by building upon these primitive data types. The programmable data structures might include arrays, defined as ordered lists of elements of identical type; **sets**, defined as unordered lists of elements of identical type; **records**, defined as ordered lists of elements that need not be of the same type; **files**, defined as sequential collections of identical records; and **databases**, defined as organized collections of different records or file types.

COMPUTER ORGANIZATION

Principal Components

The principal components of a computer system shown schematically in Fig. 2.2.2 consist of a central processing unit (referred to as the **CPU** or platform), its working memory, an operator's console, file storage, and a collection of add-ons and peripheral devices. A computer system can be viewed as a library of collected data and packages of assembled sequences of instructions that can be executed in the prescribed order by the CPU to solve specific problems or perform utility functions for the users. These sequences are variously called programs, subprograms, routines, subroutines, procedures, functions, etc. Collectively they are called **software** and are directly accessible to the CPU through the working memory. The file devices act analogously to a bookshelf—they store information until it is needed. Only after a program and its data have been transferred from the file devices or from peripheral devices to the working memory can the individual instructions and data be addressed and executed to perform their intended functions.

The CPU functions to monitor the flow of data and instructions into and out of memory during program execution, control the order of instruction execution, decode the operation, locate the operand(s) needed, and perform the operation specified. Two characteristics of the memory and storage components dictate the roles they play in the computer system. They are **access time,** defined as the elapsed time between the instant a read or write operation has been initiated and the instant the

Bits	b7	0	0	0	0	1	1	1	1
	b6	0	0	1	1	0	0	1	1
	b5	0	1	0	1	0	1	0	1

b4	b3	b2	b1								
0	0	0	0	\<NUL\>	\<DLE\>	\<SP\>	0	@	P	`	p
0	0	0	1	\<SOH\>	\<DC1\>	!	1	A	Q	a	q
0	0	1	0	\<STX\>	\<DC2\>	"	2	B	R	b	r
0	0	1	1	\<ETX\>	\<DC3\>	#	3	C	S	c	s
0	1	0	0	\<EOT\>	\<DC4\>	$	4	D	T	d	t
0	1	0	1	\<ENQ\>	\<NAK\>	%	5	E	U	e	u
0	1	1	0	\<ACK\>	\<SYN\>	&	6	F	V	f	v
0	1	1	1	\<BEL\>	\<ETB\>	'	7	G	W	g	w
1	0	0	0	\<BS\>	\<CAN\>	(	8	H	X	h	x
1	0	0	1	\<HT\>	\<EM\>	)	9	I	Y	i	y
1	0	1	0	\<LF\>	\<SUB\>	*	:	J	Z	j	z
1	0	1	1	\<VT\>	\<ESC\>	+	;	K	[	k	{
1	1	0	0	\<FF\>	\<FS\>	,	<	L	\	l	|
1	1	0	1	\<CR\>	\<GS\>	-	=	M	]	m	}
1	1	1	0	\<SO\>	\<RS\>	.	>	N	^	n	~
1	1	1	1	\<SI\>	\<US\>	/	?	O	__	o	\<DEL\>

Fig. 2.2.1 ASCII code set.

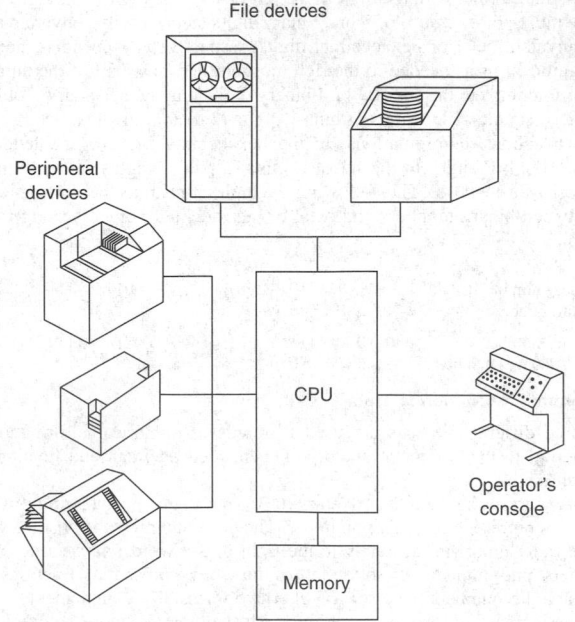

Fig. 2.2.2 Principal components of a computer system.

File devices

Peripheral devices

CPU

Memory

Operator's console

operation is completed, and size, defined by the number of bytes in a module. The faster the access time, the more costly per bit of memory or storage, and the smaller the module. The principal types of memory and storage components from the fastest to the slowest are registers which operate as an integral part of the CPU, cache and main memory which form the working memory, and mass and archival storage which serve for storing files.

The interrelationships among the components in a computer system and their primary performance parameters will be given in context in the following discussion. However, hundreds of manufacturers of computers and computer products have a stake in advancing the technology and adding new functionality to maintain their competitive edge. In such an environment, no performance figures stay current. With this caveat, performance figures given should not be taken as absolutes but only as an indication of how each component contributes to the performance of the total system.

Throughout the discussion (and in the computer world generally), prefixes indicating large numbers are given by the symbols k for kilo (10^3), M for mega (10^6), G for giga (10^9), and T for tera (10^{12}). For memory units, however, these symbols have a slightly altered meaning. Memories are organized in binary units whereby powers of two form the basis for all addressing schemes. According, k refers to a memory size of 1024 (2^{10}) units. Similarly M refers to 1024^2 (1,048,576), G refers to 1024^3, and T refers to 1024^4. For example, 1-Mbyte memory indicates a size of 1,048,576 bytes.

Memory

The main memory, also known as random access memory (**RAM**), is organized into fixed size bit cells (words, bytes, half words, or double words) which can be located by address and whose contents contain the instructions and data currently being executed. Typically RAM modules come in sizes of 1 to 10 Mbytes. The CPU acts to address the individual memory cells during program execution and transfers their contents to and from its internal registers.

Optionally, the working memory may contain an auxiliary memory, called **cache**, which is faster than the main memory. Cache operates on the premise that data and instructions that will shortly be needed are located near those currently being used. If the information is not found in the cache, then it is transferred from the main memory. Transfer rates between the cache and main memory are very fast and are usually made in block sizes of 16 to 64 bytes. Transfers between the cache and the registers are usually made on a word basis. Typically, cache modules come in sizes of a few kbytes to 1 Mbyte. The effective average access times offered by the combined configuration of RAM and cache results in a more powerful (faster) computer.

Central Processing Unit

The CPU makes available a repertoire of instructions which the user uses to set up the problem solutions. Although the specific format for instructions varies among machines, the following illustrates the pattern:

name: operator, operand(s)

The name designates an address whose contents contain the operator and one or more operands. The operator encodes an operation permitted by the hardware of the CPU. The operand(s) refer to the entities used in the operation which may be either data or another instruction specified by address. Some instructions have implied operand(s) and use the bits which would have been used for operand(s) to modify the operator.

To begin execution of a program, the CPU first loads the instructions serially by address into the memory either from a peripheral device, or more frequently, from storage. The internal structure of the CPU contains a number of memory registers whose number, while relatively few, depend on the machine's organization. The CPU acts to transfer the instructions one at a time from memory into a designated register where the individual bits can be interpreted and executed by the hardware. The actions of the following steps in the CPU, known as the *fetch-execute cycle,* control the order of instruction execution.

Step 1: Manually or automatically under program control load the address of the starting instruction into a register called the program register (PR).

Step 2: Fetch and copy the contents at the address in PR into a register called the program content register (PCR).

Step 3: Prepare to fetch the next instruction by augmenting PR to the next address in normal sequence.

Step 4: Interpret the instruction in PCR, retrieve the operands, execute the encoded operation, and then return to step 2. Note that the executed instruction may change the address in PR to start a different instruction sequence.

The speed of machines can be compared by calculating the average execution time of an instruction. Table 2.2.3 illustrates a typical instruction mix used in calculating the average. The instruction mix gives the relative frequency each instruction appears in a compiled list of typical programs and so depends on the types of problems one expects the machine to solve (e.g., scientific, commercial, or combination). The equation

$$t = \sum_i w_i t_i$$

expresses the average instruction execution time t as a function of the execution time t_i for instruction i having a relative frequency w_i in the instruction mix. The reciprocal of t measures the processor's performance as the average number of instructions per second (ips) it can execute.

Table 2.2.3   Instruction Mix

i	Instruction type	Weight w_i
1	Add: Floating point	0.07
2	Fixed point	0.16
3	Multiple: Floating point	0.06
4	Load/store register	0.12
5	Shift: One character	0.11
6	Branch: Conditional	0.21
7	Unconditional	0.17
8	Move 3 words in memory	0.10
	Total	1.00

For machines designed to support scientific and engineering calculations, the floating-point arithmetic operations dominate the time needed to execute an average instruction mix. The speed for such machines is given by the average number of floating-point operations which can be executed per second (flops). This measure, however, can be misleading in comparing different machine models. For example, when the machine has been configured with a cluster of processors cooperating through a shared memory, the rate of the configuration (measured in flops) represents the simple sum of the individual processors' flops rates. This does not reflect the amount of parallelism that can be realized within a given problem.

To compare the performance of different machine models, users often assemble and execute a suite of programs which characterize their particular problem load. This idea has been refined so that in 1992 two suites of **benchmark** programs representing typical scientific, mathematical, and engineering applications were standardized: Specint92 for integer operations, and Specfp92 for floating-point operations. Performance ratings for midsized computers are often reported in units calculated by a weighted average of the processing rates of these programs.

Computer performance depends on a number of interrelated factors used in their design and fabrication, among them compactness, bus size, clock frequency, instruction set size, and number of coprocessors.

The speed that energy can be transmitted through a wire, bounded theoretically at 3×10^{10} cm/s, limits the ultimate speed at which the electronic circuits can operate. The further apart the electronic elements are from each other, the slower the operations. Advances in integrated circuits have produced compact microprocessors operating in the nanosecond range.

The microprocessor's bus size (the width of its data path, or the number of bits that can be sent simultaneously in parallel) affect its performance in two ways: by the number of memory cells that can be directly addressed, and by the number of bits each memory reference can fetch and process at a time. For example, a 16-bit microprocessor can reference 2^{16} 16-bit memory cells and process 16 bits at a time. In order to handle the individual bits, the number of transistors that must be packed into the microprocessor goes up geometrically with the width of the data path. The earliest microprocessors were 8-bit devices, meaning that every memory reference retrieved 8 bits. To retrieve more bits, say 16, 32, or 64 bits, the 8-bit microprocessor had to make multiple references. Microprocessors have become more powerful as the packing technology has improved up to the 32-bit and 64-bit microprocessors currently available.

While normally the circuits operate asynchronously, a computer clock times the sequencing of the instructions. Clock speed is given in hertz (Hz, one cycle per second). Today's clock cycles are in the megahertz (MHz) range. Each instruction takes an integral number of cycles to complete, with one cycle being the minimum. If an instruction completes its operations in the middle of a cycle, the start of the next instruction must wait for the beginning of the next cycle.

Two schemes are used to implement the computer instruction set in the microprocessors. The more traditional complex instruction set computer (CISC) microprocessors implement by hard-wiring some 300 instruction types. Strange to say, the faster alternate-approach reduced instruction set computer (RISC) implements only about 10 to 30 percent of the instruction types by hard wiring, and implements the remaining more-complex instructions by programming them at the factory into read-only memory. Since the fewer hard-wired instructions are more frequently used and can operate faster in the simpler, more-compact RISC environment, the average instruction time is reduced.

To achieve even greater effectiveness and speed calls for more complex coordination in the execution of instructions and data. In one scheme, several microprocessors in a cluster share a common memory to form the machine organization (a multiprocessor or parallel processor). The total work which may come from a single program or independent programs is parceled out to the individual machines which operate independently but are coordinated to work in parallel with the other machines in the cluster. Faster speeds can be achieved when the individual processors can work on different parts of the problem or can be assigned to those parts of the problem solution for which they have been especially designed (e.g., input-output operations or computational operations). Two other schemes, pipelining and array processing, divide an instruction into the separate tasks that must be performed to complete its execution. A pipelining machine executes the tasks concurrently on consecutive pieces of data. An array processor executes the tasks of the different instructions in a sequence simultaneously and coordinates their completion (which might mean abandoning a partially completed instruction if it had been initiated prematurely). These schemes are usually associated with the larger and faster machines.

Operator's Console

The system operator uses the console to initiate or terminate computing tasks, to interrogate the computer to determine the status of the tasks during execution, to give and receive instructions such as mounting a particular file onto a drive or provide operating parameters during operations, and to otherwise monitor the system.

The operator's console consists of a relatively slow-speed keyboard input and a monitor display. The monitor display consists of a video scope which might be simply two-tone or could have a selection of colors or shades (up to 256) to build pictures and icons. Other important scope characteristics are the size of the screen and the resolution measured in points on the screen called **pixels**. The total number of pixels is given by the number of pixels on a horizontal line and the number of pixels on a vertical line (e.g., 1024×768 or 1600×1200). The scope has its own memory which refreshes and controls the display. For convenience and manual speed, a device called a **mouse** can be attached to the console and rolled on a flat surface which in turn moves the **cursor** on the display. This can be used to locate and select options displayed as a menu on the screen. A mouse turned upside down so the ball can be turned by the thumb performs the same function and is called a **trackball**.

File Devices

File devices serve to store libraries of directly accessible programs and data in electronically or optically readable formats. The file devices record the information in large blocks rather than by individual addresses. To be used, the blocks must first be transferred into the working memory. Depending on how selected blocks are located, file devices are categorized as sequential or direct-access. On **sequential** devices the computer locates the information by searching the file from the beginning. Direct-access devices, on the other hand, position the read-write mechanism directly at the location of the needed information. Searching on these devices works concurrently with the CPU and the other devices making up the computer configuration.

Magnetic tapes using arbitrary block sizes form commercial sequential-access products. Besides the disadvantage that the medium must be passed over sequentially to locate the beginning of the needed information, magnetic tape recording does not permit information to be changed in situ. Information can be changed only by reading the information from one tape, making the changes, and writing the changed information onto another tape.

Traditional magnetic tape recorders consist of reels of tape ½ in (12.7 mm) wide, 0.0015 in (0.0381 mm) thick, and 2400 ft (732 m) long. Information is recorded across the tape in 9-bit frames. One bit in each frame, called a **parity bit,** is used for checking purposes and is not transferred into the memory of the computer. The remaining 8 bits record the information using some standard format (e.g.: EBCDIC, modified ASCII, or an internal binary format). Lengthwise the information is recorded using standard densities such as 9600 bits/in, with gaps between blocks sufficient in size to stop the tape transport at the end of a block and before the beginning of the next block. Today's tape units use ½-in, 8-mm, or ¼-in cartridges that have a capacity up to 2.5 Tbytes of uncompacted data or 7.2 Tbytes of compacted data.

Magnetic or **optical disks** that offer a wide choice of options form the commercial direct-access devices. The recording surface consists of a platter (or platters) of recording material mounted on a common spindle rotated at high speed. The read-write heads may be permanently positioned along the radius of the platter or may be mounted on a common arm that can be moved radially to locate any specified track of information. Information is recorded on the **tracks** circumferentially using fixed-size blocks called pages or sectors. **Pages** divide the storage and memory space alike into blocks of 4096 bytes so that program transfers can be made without creating unusable space. **Sectors** nominally describe the physical division of the storage space into equal segments for easier positioning of the read-write heads.

The access time for retrieving information from a disk depends on three separately quoted factors, called seek time, latency time, and transfer time. **Seek time** gives the time needed to position the read-write heads from their current track position to the track containing the information. Average seek time is on the order of 100 ms. Since the faster fixed-head disks require no radial motion, only latency and transfer time need to be factored into the total access time for these devices. **Latency time** is the time needed to locate the start of the information along the circumferential track. This time depends on the speed of revolution of the disk, and, on average, corresponds to the time required to revolve the platter half a turn. The average latency time can be reduced by repeating the information several times around the track. Average latency time is on the order of 2 to 20 ms. **Transfer time,** usually quoted as a rate, gives the rate at which information can be transferred to memory after it has been located. There is a large variation in transfer rates depending on the disk system selected. Typical systems range from 20 kbytes/s to 20 Mbytes/s.

Disk devices are called soft or hard disks, referring to the rigidity of the platter. **Soft disks,** also called **floppy disks,** have a mountable, small, single platter that provides one or two recording surfaces. Soft or floppy might be a misnomer since many systems use diskettes about the size and rigidity of a credit card. Typical floppies have a physical size of 5¼ in or 3½ in and have a capacity of 1.2 Mbytes and 1.44 Mbytes, respectively. **Hard disks** refer to sealed devices whose physical size has been reduced to units of 1.3 to 2.5 in (33 to 63.5 mm), yet their capacity has increased. For example, disk storage of 200 Mbytes is available for small computers, and for more complex systems an array of disks is available having a capacity of from over 500 Mbytes to nearly 2 Gbytes.

Computer architects sometimes refer to file storage as **mass storage** or **archival storage,** depending on whether or not the libraries can be kept off-line from the system and mounted when needed. Disk drives with mountable platter(s) and tape drives constitute the archival storage. Sealed disks that often have fixed heads for faster access are the medium of choice for mass storage.

Peripheral Devices and Add-ons

Peripheral devices function as self-contained external units that work on line to the computer to provide or receive information or to control the flow of information. Add-ons are a special class of units whose circuits can be integrated into the circuitry of the computer hardware to augment the basic functionality of the processors. Section 15 covers the electronic technology associated with these devices.

An input device may be defined as any device that provides a machine-readable source of information. For engineering work, the most common forms of input are punched cards, punched tape, magnetic tape, magnetic ink, touch-tone dials, mark sensing, bar codes, and keyboards (usually in conjunction with a printing mechanism or video scope). Many bench instruments have been reconfigured to include digital devices to provide direct input to computers. Because of the data-handling capabilities of the computer, these instruments can be simpler, smaller, and less expensive than the hand instruments they replace. New devices have also been introduced: devices for visual measurement of distance, area, speed, and coordinate position of an object; or for inspecting color or shades of gray for computer-guided vision. Other methods of input that are finding greater acceptance include handwriting recognition, printed character recognition, voice digitizers, and picture digitizers.

Traditionally, output devices play the role of producing displays for the interpretation of results. A large variety of printers, graphical plotters, video displays, and audio sets have been developed for this purpose. Printers are distinguished by:

Type of print head (letter-quality or dot-matrix)
Type of paper feed (tractor or friction)
Allowable paper sizes
Print control (character, line, or page at a time)
Speed (measured in characters, lines, or pages per minute)
Number of fonts (especially for laser printers)

Graphic plotters and video displays offer variations in size, color capabilities, and quality. The more sophisticated video scopes offer dynamic characteristics capable of animated displays.

A variety of actuators have been developed for driving control mechanisms. Typical developments are in high-precision rack-and-pinion mechanisms and in lead screws that essentially eliminate backlash due to gear trains. For complex numerical control, programmable controllers (called PLCs) can simultaneously control and update data from multiple tasks. These electronically driven mechanisms and controllers, working with input devices, make possible systems for automatic testing of products, real-time control, and robotics with learning and adaptive capabilities.

Computer Sizes

Computer size refers not only to the physical size but also to the number of electronics elements in the system, and so reflects the performance of the system. Between the two ends of the spectrum from the largest and fastest to the smallest and slowest are machines that vary in speed and complexity. Although no nomenclature has been universally adopted that indicates computer size, the following descriptions illustrate a few generally understood terms used for some common configurations.

Personal computers (PCs) have been made possible by the advances in solid-state technology. The name applies to computers that can fit the total complement of hardware on a desktop and operate as stand-alone systems so as to provide immediate dedicated services to an individual user. This popular computer size has been generally credited for spreading computer literacy in today's society. Because of its commercial success, many peripheral devices, add-ons, and software products have been (and are continually being) developed. Laptop PCs are personal computers that have the low weight and size of a briefcase and can easily be transported when peripherals are not immediately needed.

The term **workstation** describes computer systems which have been designed to support complex engineering, scientific, or business applications in a professional environment. Although a top-of-the-line PC or a PC connected as a peripheral to another computer can function like a workstation, one can expect a machine designed as a workstation to offer higher performance than a PC and to support the more specialized peripherals and sophisticated professional software. Nevertheless, the boundary between PCs and workstations changes as the technology advances. Table 2.2.4 lists some published performance values for the spectrum of computers which have been designated as workstations. The spread in speed values represents the statistical average of reported samples distributed over one standard deviation.

Notebook PCs and the smaller sized **palmtop** PCs are portable, battery-operated machines. A typical notebook PC size would be 9 × 11 in (230 × 280 mm) in area, 1 to 2 in (25 to 50 mm) thick, and 2 to 9 lb (1 to 4 kg) in weight. They often have built-in programs stored in ROM. Having 68-pin integrated circuit cards for mass memory that can store as much as some hard disks, and being able to share programs with desktop PCs, these machines find excellent use as portable PCs in some applications and as data acquisition systems. However their undersized keyboards and small scopes limit their usefulness for sustained operations.

Table 2.2.4   Reported Performance Parameters for Workstations

	Workstation range		
	Low	Mid	High
Processor			
Clock speed, MHz	20–33	40–80	100–200
Bus size	16–32	32	64
Number of coprocessors	1–2	1–2	1–4
Instruction set	CISC		RISC
Speed rating			
Specint92	17.1–25.1	32.3–55.7	38.1–77.1
Specfp92	21.2–26.4	43.9–81.9	52.0–120.0
Mips	20.6–36.4	21.9–92.1	86.6–135.4
Mflops	2.6–6.0	4.3–20.9	30.0–50.0
Memory capacity			
Main, Mbytes	2–128		16–128
Cache, kbytes	8–128		64–256
Disk capacity			
Hard, Mbytes	10–80	80–200	200–400
Floppy, Mbytes	1.44		1.44

Computers larger than a PC or a workstation, called **mainframes** (and sometimes minis or maxis, depending on size), serve to support multiusers and multiapplications. A remotely accessible computing center may house several mainframes which either operate alone or cooperate with each other. Their high speed and large memories allow them to handle complex programs. A specific type of mainframe, used to maintain the database of a system, is called a **database machine.** Database machines act in cooperation with a number of user stations in a server-client relationship. In this, the database machine (the server) provides the data and/or the programs and shares the processing with the individual workstations (the clients).

At the upper extreme end of the computer spectrum is the **supercomputer,** the class of the fastest machines that can address large, complex scientific/engineering problems which cannot reasonably be transferred to other machines. Obviously this class of computer must have cache and main memory sizes and speeds commensurate with the speed of the platform. While mass memory sizes must also be large, computers which support large databases may often have larger memories than do supercomputers. Large, complex technical problems must be run with high-precision arithmetic. Because of this, performance is measured in double-precision flops. Supercomputer performance has moved from the current range of 10 Gflops into the Tflops range. To realize these speeds, the designers of supercomputers work at the edge of the available technology, especially in the use of multiple processors operating in parallel. Current clusters of 4 to 16 processors are being expanded to a goal of 100 and more. With multiple processors, however, performance depends as much on the time spent in communication between processors as on the computational speed of the individual processors. In the final analysis, to muster the supercomputer's inherent speed, the development of the software becomes the problem. Some users report that software must often be hand-tailored to the specific problem. The power of the machines, however, can replace years of work in analysis and experimentation.

DISTRIBUTED COMPUTING

Organization of Data Facilities

A distributed computer system can be defined as a collection of computer resources which are remotely located from each other and are interconnected to cooperate in providing their respective services. The resources include both the equipment and the software. Resources distributed to reside near the vicinity where the data is collected or used have an obvious advantage over centralization. But to provide information in a timely and reliable manner, these islands of automation must be integrated.

The size and complexity of an enterprise served by a distributed information system can vary from a single-purpose office to a multiple-plant conglomerate. An enterprise is defined as a system which has been created to accomplish a mission in its environment and whose goals involve risk. Internally it consists of organized functions and facilities which have been prepared to provide its services and accomplish its mission. When stimulated by an external entity, the enterprise acts to produce its planned response. An enterprise must handle both the flow of material (goods) and the flow of information. The information system tracks the material in the material system, but itself handles only the enterprise's information.

The technology for distributing and integrating the total information system comes under the industrial strategy known as computer-integrated business (CIB) or computer-integrated manufacturing (CIM). The following reasons have been cited for developing CIB and CIM:

Most data generated locally has only local significance.
Data integrity resides where it is generated.
The quality and consistency of operational decisions demands not only that all parts of the system work with the same data but that they can receive it in a reliable and timely manner.

If a local processor fails, it may disrupt local operations, but the remaining system should continue to function independently.
Small cohesive processors can be best managed and maintained locally.
Through standards, selection of local processes can be made from the best products in a competitive market that can be integrated into the total system.
Obsolete processors can be replaced by processors implemented by more advance technology that conform to standards without the cost of tailoring the products to the existing system.

Figure 2.2.3 depicts the total information system of an enterprise. The database consists of the organized collection of data the processors use in their operations. Because of differences in their communication requirements, the automated procedures are shown separated into those used in the office and those used on the production floor. In a business environment, the front office operations and back office operations make this separation. While all processes have a critical deadline, the production floor handles real-time operations, defined as processes which must complete their response within a critical deadline or else the results of the operations become moot. This places different constraints on the local-area networks (LANs) that serve the communication needs within the office or within the production floor. To communicate with entities outside the enterprise, the enterprise uses a wide-area network (WAN), normally made up from available public network facilities. For efficient and effective operation, the processes must be interconnected by the communications to share the data in the database and so integrate the services provided.

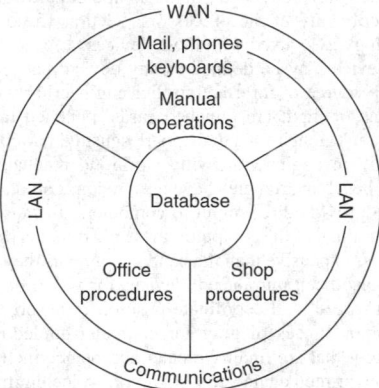

Fig. 2.2.3 Composite view of an enterprise's information system.

Communication Channels

A communication channel provides the connecting path for transmitting signals between a computing system and a remotely located application. Physically the channel may be formed by a wire line using copper, coaxial cable, or optical-fiber cable; or may be formed by a wireless line using radio, microwave, or communication satellites; or may be a combination of these lines.

Capacity, defined as the maximum rate at which information can be transmitted, characterizes a channel independent of the morphic line. Theoretically, an ideal noiseless channel that does not distort the signals has a channel capacity C given by:

$$C = 2W$$

where C is in pulses per second and W is the channel bandwidth. For digital transmission, the Hartley-Shannon theorem sets the capacity of a channel limited by the presence of gaussian noise such as the thermal noise inherent in the components. The formula:

$$C = W \log_2 (1 + S/N)$$

gives the capacity C in bits/s in terms of the signal to noise ratio S/N and the bandwidth W. Since the signal to noise ratio is normally given in decibels divisible by 3 (e.g., 12, 18, 21, 24) the following formula provides a workable approximation to the formula above:

$$C = W(S/N)_{db}/3$$

where $(S/N)_{db}$ is the signal-to-noise ratio expressed in decibels. Other forms of noise, signal distortions, and the methods of signal modulation reduce this theoretical capacity appreciably.

Nominal transmission speeds for electronic channels vary from 1000 bits to almost 20 Mbits per second. Fiber optics, however, form an almost noise-free medium. The transmission speed in fiber optics depends on the amount a signal spreads due to the multiple reflected paths it takes from its source to its destination. Advances in fiber technology have reduced this spread to give unbelievable rates. Effectively, the speeds available in today's optical channels make possible the transmission over a common channel, using digital techniques, of all forms of information: text, voice, and pictures.

Besides agreeing on speed, the transmitter and receiver must agree on the mode of transmission and on the timing of the signals. For stations located remotely from each other, transmission occurs by organizing the bits into groups and transferring them, one bit after another, in a serial mode. One scheme, called **asynchronous** or start-stop transmission, uses separate start and stop signals to frame a small group of bits representing a character. Separate but identical clocks at the transmitter and receiver time the signals. For transmission of larger blocks at faster rates, the stations use **synchronous** transmission which embeds the clock information within the transmitted bits.

Communication Layer Model

Figure 2.2.4 depicts two remotely located stations that must cooperate through communication in accomplishing their respective tasks. The communications substructure provides the communication services needed by the application. The application tasks themselves, however, are outside the scope of the communication substructure. The distinction here is similar to that in a telephone system which is not concerned with the application other than to provide the needed communication service. The figure shows the communication facilities packaged into a hierarchical modular layer architecture in which each node contains identical kinds of functions at the same layer level. The layer functions represent abstractions of real facilities, but need not represent specific hardware or software. The entities at a layer provide communication services to the layer above or can request the services available from the layer below. The services provided or requested are available only at service points which are identified by addresses at the boundaries that interface the adjacent layers.

The top and bottom levels of the layered structure are unique. The topmost layer interfaces and provides the communication services to the noncommunication functions performed at a node dealing with the application task (the user's program). This layer also requests communication services from the layer below. The bottom layer does not have a lower layer through which it can request communication services. This layer acts to create and recognize the physical signals transmitted between the bottom entities of the communicating partners (it arranges the actual transmission). The medium that provides the path for the transfer of signals (a wire, usually) connects the service access points at the bottom layers, but itself lies outside the layer structure.

Virtual communication occurs between peer entities, those at the same level. Peer-to-peer communication must conform to layer protocol, defined as the rules and conventions used to exchange information. Actual physical communication proceeds from the upper layers to the bottom, through the communication medium (wire), and then up through the layer structure of the cooperating node.

Since the entities at each layer both transmit and receive data, the protocol between peer layers controls both input and output data, depending on the direction of transmission. The transmitting entities accomplish this by appending control information to each data unit that they pass to the layer below. This control information is later interpreted and removed by the peer entities receiving the data unit.

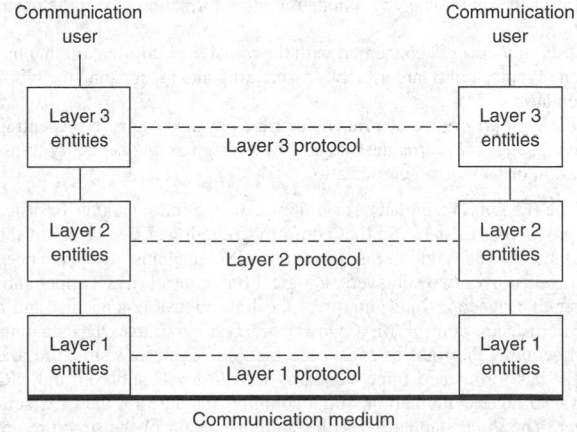

Fig. 2.2.4 Communication layer architecture.

Communication Standards

Table 2.2.5 lists a few of the hundreds of forums seeking to develop and adopt voluntary standards or to coordinate standards activities. Often users establish standards by agreement that fixes some existing practice. The ISO, however, has described a seven-layer model, called the *Reference Model for Open Systems Interconnection* (OSI), for coordinating and expediting the development of new implementation standards. The term *open systems* refers to systems that allow devices to be interconnected and to communicate with each other by conforming to common implementation standards. The ISO model is not of itself an implementation standard nor does it provide a basis for appraising existing implementations, but it partitions the communication facilities into layers of related function which can be independently standardized by different teams of experts. Its importance lies in the fact that both vendors and users have agreed to provide and accept implementation standards that conform to this model.

Table 2.2.5 Some Groups Involved with Communication Standards

CCITT	Comité Consultatif de Télégraphique et Téléphonique
ISO	International Organization for Standardization
ANSI	American National Standards Institute
EIA	Electronic Industries Association
IEEE	Institute of Electrical and Electronics Engineers
MAP/TOP	Manufacturing Automation Protocols and Technical and Office Protocols Users Group
NIST	National Institute of Standards and Technology

The following lists the names the ISO has given the layers in its ISO model together with a brief description of their roles.

Application layer provides no services to the other layers but serves as the interface for the specialized communication that may be required by the actual application, such as file transfer, message handling, virtual terminal, or job transfer.

Presentation layer relieves the node from having to conform to a particular syntactical representation of the data by converting the data formats to those needed by the layer above.

Session layer coordinates the dialogue between nodes including arranging several sessions to use the same transport layer at one time.

Transport layer establishes and releases the connections between peers to provide for data transfer services such as throughput, transit delays, connection setup delays, error rate control, and assessment of resource availability.

Network layer provides for the establishment, maintenance, and release of the route whereby a node directs information toward its destination.

Data link layer is concerned with the transfer of information that has been organized into larger blocks by creating and recognizing the block boundaries.

Physical layer generates and detects the physical signals representing the bits, and safeguards the integrity of the signals against faulty transmission or lack of synchronization.

The IEEE has formulated several implementation standards for office or production floor LANs that conform to the lower two layers of the ISO model. The functions assigned to the ISO data link layer have been distributed over two sublayers, a logical link control (LLC) upper sublayer that generates and interprets the link control commands, and a medium access control (MAC) lower sublayer that frames the data units and acquires the right to access the medium. From this structure, the IEEE has formulated three standards for the MAC sublayer and ISO physical layer combination, and a common standard for the LLC sublayer. The three standards for the bottom portion of the structure are named according to the method used to control the access to the medium: carrier sense multiple access with collision detection (CSMA/CD), token-passing bus access, and token ring access. A wide variety of options have been included for each of these standards which may be selected to tailor specific implementation standards.

CSMA/CD standardized the access method developed by the Xerox Corporation under its trademark Ethernet. The nodes in the network are attached to a common bus, schematically shown in Fig. 2.2.5a. All nodes hear every message transmitted, but accept only those messages addressed to themselves. When a node has a message to transmit, it listens for the line to be free of other traffic before it initiates transmission. However, more than one mode may detect the free line and may

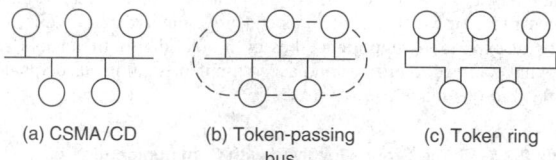

(a) CSMA/CD   (b) Token-passing bus   (c) Token ring

Fig. 2.2.5 LAN structures.

start to transmit. In this situation the signals will collide and produce a detectable change in the energy level present in the line. Even after a station detects a collision it must continue to transmit to make sure that all stations hear the collision (all data frames must be of sufficient length to be present simultaneously on the line as they pass each station). On hearing a collision, all stations that are transmitting wait a random length of time and then attempt to retransmit.

The stations in the **token-passing** bus access method, like the CSMA/CD method, share a common bus and communicate by broadcasting their messages to all stations. Unlike CSMA/CD, token-passing bus stations communicate in an ordered fashion as shown by the dashed line in Fig. 2.2.5b. By using special control frames the stations organize themselves into a logical ring by address (station 40 follows 30 which follows 20 which follows 40). The token is a special control frame which is circulated sequentially from station to station, giving the station that has the token the exclusive right to transmit any message it has

ready for transmission. When a station has no message to transmit, or after it has completed transmission, it passes the token to the next station in the ring. The method features protocol procedures for restructuring the ring when ring membership changes, such as when a station intentionally or through failure leaves the ring, or a new station joins.

The **token-ring** access method connects the stations into a physical ring as shown in Fig. 2.2.5c. A special mechanical connector attaches the station equipment to the medium which when disconnected automatically closes the line to reestablish line continuity. The token has a priority level which may be changed by a station. When a station receives the token, it can start to circulate any data it has ready for transmission at the priority level of the token. As each station receives information from its neighbor, it regenerates the information and continues to circulate it around the ring while retaining a copy of everything destined for itself. The station that had originally sent the information retains the token until the information has been returned uncorrupted. Then it passes the token to the next station. Any station that had changed the priority level of the token has the responsibility for returning it to its previous level in a fair and orderly fashion. Protocol procedures sense failures in a station or faults in the medium.

The MAP/TOP (Manufacturing Automation Protocols and Technical and Office Protocols) Users Group started under the auspices of General Motors and Boeing Information Systems and now has a membership of many thousands of national and international corporations. The corporations in this group have made a commitment to open systems that will allow them to select the best products through standards, agreed to by the group, that will meet their respective requirements. In particular, MAP has standardized options from the IEEE token-passing bus method for production floor LAN implementation, and TOP has standardized options from the IEEE CSMA/CD for office LAN implementations. These standards have also been adopted by NIST for governmentwide use under the title Government Open Systems Interconnections Profile (GOSIP).

The Electronics Industries Association has established three interface standards, RS-232C, RS-422, and RS-423, which are frequently referenced for digital communications. These standards specify the use of multiple lines that interface the equipment at a station and the communication control equipment attached to the medium. **RS-232C** has been the primary standard for several years for low-speed voltage-oriented digital communications. RS-232C uses nonbalanced circuits sharing a common ground wire which, because of their sensitivity to noise, limits the bandwidth and length of the lines. RS-232C specifications call for a maximum line length of about 250 ft at a bandwidth of 10 kHz. **RS-423** also uses nonbalanced circuits but with individual ground wires which allows higher limits to a maximum line length of about 400 ft at a bandwidth of 100 kHz. **RS-422** uses balanced circuits with individual ground wires which allow line lengths up to 4000 ft at bandwidths of 100 kHz.

The common carriers who offer WAN communication services through their public networks have also developed **packet-switching** networks for public use. Packet switching transmits data in a purely digital format, which, when embellished, can replace the common circuit-switching technology used in analog communications such as voice. A packet is a fixed-sized block of digital data with embedded control information. The network serves to deliver the packets to their destination in an efficient and reliable manner.

CCITT has developed a set of standards, called X.25, for the three bottom ISO layers, to interface the public packet-switching networks. One of the set, named the X.21 standard, serves as a replacement to the EIA standards (RS-232C, RS-422, and RS-423) with fewer interconnecting lines whereby an expanded number of functions can be selected by coded digital means. When the equipment at the local site does not support the X.25 protocol, then a protocol converter interface, called a packet assembler/disassembler (PAD), properly structures the data for transmission over public packet-switching networks. While the upper four layers are not addressed by this interface, it is understood that end-to-end communication can take place only when the protocols between the layers at source and destination points agree or are made to conform through protocol converters.

RELATIONAL DATABASE TECHNOLOGY

Design Concepts

As computer hardware has evolved from small working memories and tape storage to large working memories and large disk storage, so has database technology moved from accessing and processing of a single, sequential file to that of multiple, random-access files. A **relational database** can be defined as an organized collection of interconnected tables or records. The records appear like the flat files of older technology. In each record the information is in columns (fields) which identify attributes, and rows (tuples) which list particular instances of the attributes. One column (or more), known as the **primary key,** identifies each row. Obviously, the primary key must be unique for each row.

If the data is to be handled in an efficient and orderly way, the records cannot be organized in a helter-skelter fashion such as simply transporting existing flat files into relational tables. To avoid problems in maintaining and using the database, redundancy should be eliminated by storing each fact at only one place so that, when making additions or deletions, one need not worry about duplicates throughout the database. This goal can be realized by organizing the records into what is known as the *third normal form*. A record is in the third normal form if and only if all nonkey attributes are mutually independent and fully dependent on the primary key.

The advantages of relational databases, assuming proper normalization, are:

Each fact can be stored exactly once.

The integrity of the data resides locally, where it is generated and can best be managed.

The tables can be physically distributed yet interconnected.

Each user can be given his/her own private view of the database without altering its physical structure.

New applications involving only a part of the total database can be developed independently.

The system can be automated to find the best path through the database for the specified data.

Each table can be used in many applications by employing simple operators without having to transfer and manipulate data superfluous to the application.

A large, comprehensive system can evolve from phased design of local systems.

New tables can be added without corrupting everyone's view of the data.

The data in each table can be protected differently for each user (read-only, write-only).

The tables can be made inaccessible to all users who do not have the right to know.

Relational Database Operators

A database system contains the structured collection of data, an on-line catalog and dictionary of data items, and facilities to access and use the data. The system allows users to:

Add new tables
Remove old tables
Insert new data into existing tables
Delete data from existing tables
Retrieve selected data
Manipulate data extracted from several tables
Create specialized reports

As might be expected, these systems include a large collection of operators and built-in functions in addition to those normally used in mathematics. Because of the similarity between database tables and mathematical sets, special set-like operators have been developed to manipulate tables. Table 2.2.6 lists eight typical table operators. The list of functions would normally also include such things as count, sum, average, find the maximum in a column, and find the minimum in a column. A rich collection of report generators offers powerful and flexible capabilities for producing tabular listings, text, graphics (bar charts, pie charts, point plots, and continuous plots), pictorial displays, and voice output.

SOFTWARE ENGINEERING

Programming Goals

Software engineering encompasses the methodologies for analyzing program requirements and for structuring programs to meet the requirements over their life cycle. The objectives are to produce programs that are:

Well documented
Easily read
Proved correct
Bug- (error-) free
Modifiable and maintainable
Implementable in modules

Control-Flow Diagrams

A control-flow diagram, popularly known as a **flowchart,** depicts all possible sequences of a program during execution by representing the control logic as a directed graph with labeled nodes. The theory associated with flowcharts has been refined so that programs can be structured to meet the above objectives. Without loss of generality, the nodes in a flowchart can be limited to the three types shown in Fig. 2.2.6. A **function** may be either a *transformer* which converts input data values into output data values or a *transducer* which converts that data's morphological form. A label placed in the rectangle specifies the function's action. A **predicate** node acts to bifurcate the path through the node. A

Table 2.2.6 Relational Database Operators

Operator	Input	Output
Select	A table and a condition	A table of all tuples that satisfy the given condition
Project	A table and an attribute	A table of all values in the specified attribute
Union	Two tables	A table of all unique tuples appearing in one table or the other
Intersection	Two tables	A table of all tuples the given tables have in common
Difference	Two tables	A table of all tuples appearing in the first and not in the second table
Join	Two tables and a condition	A table concatenating the attributes of the tuples that satisfy the given condition
Divide	A table, two attributes, and list of values	A table of values appearing in one specified attribute of the given table when the table has tuples that satisfies every value in the list in the other given attribute

question labels the diamond representing a predicate node. The answer to the question yields a binary value: 0 or 1, yes or no, ON or OFF. One of the output lines is selected accordingly. A **connector** serves to rejoin separated paths. Normally the circle representing a connector does not contain a label, but when the flowchart is used to document a computer program it may be convenient to label the connector.

Structured programming theory models all programs by their flowcharts by placing minor restrictions on their lines and nodes. Specifically, a flowchart is called a **proper program** if it has precisely one input and one output line, and for every node there exists a path from the input line through the node to the output line. The restriction prohibiting multiple input or output lines can easily be circumvented by funneling the lines through collector nodes. The other restriction simply discards unwanted program structures, since a program with a path that does not reach the output may not terminate.

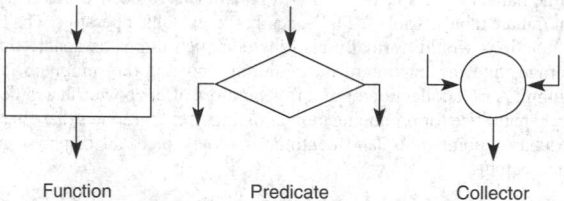

Function   Predicate   Collector

Fig. 2.2.6 Basic flowchart nodes.

Not all proper programs exhibit the desirable properties of meeting the objectives listed above. Figure 2.2.7 lists a group of proper programs whose graphs have been identified as being *well-structured* and useful as basic building blocks for creating other well-structured programs. The name assigned to each of these graph suggests the process each represents. CASE is just a convenient way of showing multiple IFTHENELSEs more compactly.

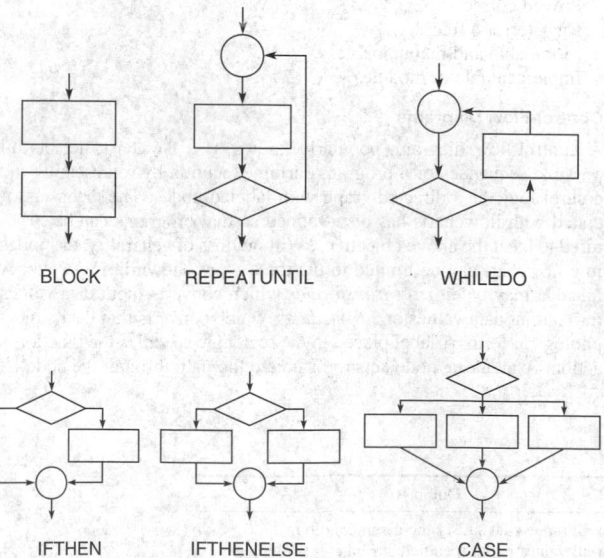

BLOCK   REPEATUNTIL   WHILEDO

IFTHEN   IFTHENELSE   CASE

Fig. 2.2.7 Basic flowchart building blocks.

The **structured programming theorem** states: any proper program can be reconfigured to an equivalent program producing the same transformation of the data by a flowchart containing at most the graphs labeled BLOCK, IFTHENELSE, and REPEATUNTIL.

Every proper program has one input line and one output line like a function block. The synthesis of more complex well-structured pro-

grams is achieved by substituting any of the three building blocks mentioned in the theorem for a function node. In fact, any of the basic building blocks would do just as well. A program so structured will appear as a block of function nodes with a top-down control flow. Because of the top-down structure, the arrow points are not normally shown.

Figure 2.2.8 illustrates the expansion of a program to find the roots of $ax^2 + bx + c = 0$. The flowchart is shown in three levels of detail.

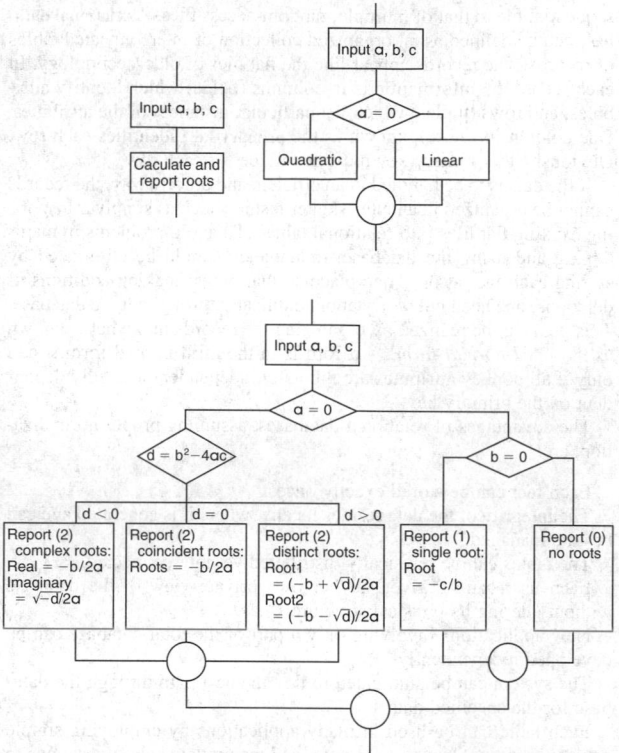

Fig. 2.2.8 Illustration of a control-flow diagram.

Data-Flow Diagrams

Data-flow diagrams structure the actions of a program into a network by tracking the data as it passes through the program. They depict the interworkings of a system by the processes performing the work and the communication between the processes. Data-flow diagrams have proved valuable in analyzing existing or new systems to determine the system requirements and in designing systems to meet those requirements. Figure 2.2.9 shows the four basic elements used to construct a data-flow diagram. The roles each element plays in the system are:

Rectangular boxes lie outside the system and represent the input data sources or output data sinks that communicate with the system. The sources and sinks are also called *terminators*.

Circles (bubbles) represent processes or actions performed by the system in accomplishing its function.

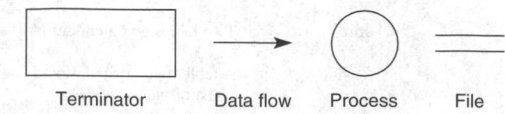

Terminator   Data flow   Process   File

Fig. 2.2.9 Data-flow diagram elements.

Twin parallel lines represent a data file used to collect and store data from among the processes or from a process over time which can be later recalled.

Arcs or vectors connect the other elements and represent data flows.

A label placed with each element makes clear its role in the system. The circles contain verbs and the other elements contain nouns. The arcs tie the system together. An arc between a terminator and a process represents input to or output from the system. An arc between two processes represents output from one process which is input to the other. An arc between a process and a file represents data gathered by the process and stored in the file, or retrieval of data from the file.

Analysis starts with a contextual view of the system studied in its environment. The contextual view gives the name of the system, the collection of terminators, and the data flows that provide the system inputs and outputs; all accompanied by a statement of the system objective. Details on the terminators and data they provide may also be described by text, but often the picture suffices. It is understood that the form of the input and output may not be dictated by the designer since they often involve organizations outside the system. Typical inputs in industrial systems include customer orders, payment checks, purchase orders, requests for quotations, etc. Figure 2.2.10a illustrates a context diagram for a repair shop.

Figure 2.2.10b gives many more operational details showing how the parts of the system interact to accomplish the system's objectives. The

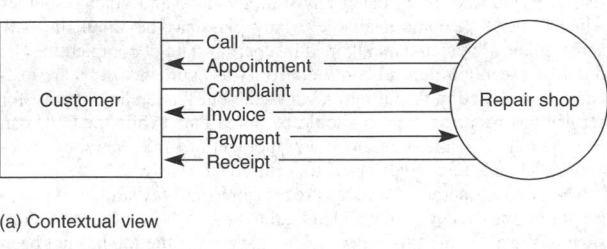

(a) Contextual view

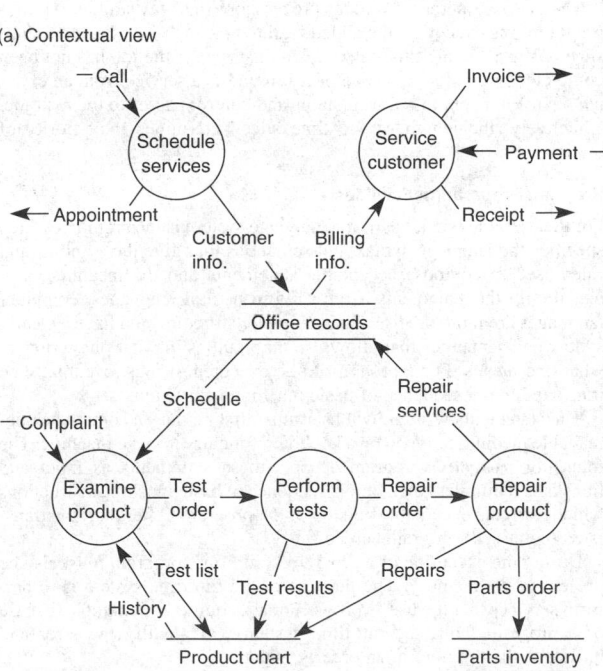

(b) Behavorial view

Fig. 2.2.10   Illustration of a data-flow diagram.

designer can restructure the internal processors and the formats of the data flows. The bubbles in a diagram can be broken down into further details to be shown in another data-flow diagram. This can be repeated level after level until the processes become manageable and understandable. To complete the system description, each bubble in the data-flow charts is accompanied by a control-flow diagram or its equivalent to describe the algorithm used to accomplish the actions and a data dictionary describing the items in the data flows and in the databases.

The techniques of data-flow diagrams lend themselves beautifully to the analysis of existing systems. In a complex system it would be unusual for an individual to know all the details, but all system participants know their respective roles: what they receive, whence they receive it, what they do, what they send, and where they send it. By carefully structuring interviews, the complete system can be synthesized to any desired level of detail. Moreover, each system component can be verified because what is sent from one process must be received by another and what is received by a process must be used by the process. To automate the total system or parts of the system, control bubbles containing transition diagrams can be implemented to control the timing of the processes.

SOFTWARE SYSTEMS

Software Techniques

Two basic operations form the heart of nonnumerical techniques such as those found in handling large database tables. One basic operation, called **sorting**, collates the information in a table by reordering the items by their key into a specified order. The other basic operation, called **searching**, seeks to find items in a table whose keys have the same or related value as a given argument. The search operation may or may not be successful, but in either case further operations follow the search (e.g., retrieve, insert, replace).

One must recognize that computers cannot do mathematics. They can perform a few basic operations such as the four rules of arithmetic, but even in this case the operations are approximations. In fact, computers represent long integers, long rationals, and all the irrational numbers like π and e only as approximations. While computer arithmetic and the computer representation of numbers exceed the precision one commonly uses, the size of problems solved in a computer and the number of operations that are performed can produce misleading results with large computational errors.

Since the computer can handle only the four rules of arithmetic, complex **functions** must be approximated by polynomials or rational fractions. A rational fraction is a polynomial divided by another polynomial. From these curve-fitting techniques, a variety of weighted-average formulas can be developed to approximate the **definite integral** of a function. These formulas are used in the procedures for solving differential and integral equations. While differentiation can also be expressed by these techniques, it is seldom used, since the errors become unacceptable.

Taking advantage of the machine's speed and accuracy, one can solve **nonlinear equations** by trial and error. For example, one can use the Newton-Raphson method to find successive approximations to the roots of an equation. The computer is programmed to perform the calculations needed in each iteration and to terminate the procedure when it has converged on a root. More sophisticated routines can be found in the libraries for finding real, multiple, and complex roots of an equation.

Matrix techniques have been commercially programmed into libraries of prepared modules which can be integrated into programs written in all popular engineering programming languages. These libraries not only contain excellent routines for solving **simultaneous linear equations** and the **eigenvalues** of characteristic matrices, but also embody procedures guarding against ill-conditioned matrices which lead to large computational errors.

Special matrix techniques called **relaxation** are used to solve partial differential equations on the computer. A typical problem requires setting up a grid of hundreds or thousands of points to describe the region and expressing the equation at each point by finite-difference methods. The resulting matrix is very sparse with a regular pattern of nonzero elements. The form of the matrix circumvents the need for handling large arrays of numbers in the computer and avoids problems in computational accuracy normally found in dealing with extremely large matrices.

The computer is an excellent tool for handling **optimization** problems. Mathematically these problems are formulated as problems in finding the maximum or minimum of a nonlinear equation. The excellent techniques that have been developed can deal effectively with the unique complexities these problems have, such as saddle points which represent both a maximum and a minimum.

Another class of problems, called **linear programming** problems, is characterized by the linear constraint of many variables which plot into regions outlined by multidimensional planes (in the two-dimensional case, the region is a plane enclosed by straight lines). Techniques have been developed to find the optimal solution of the variables satisfying some given value or cost objective function. The solution to the problem proceeds by searching the corners of the region defined by the constraining equations to find points which represent minimum points of a cost function or maximum points of a value function.

The best known and most widely used techniques for solving statistical problems are those of linear **statistics.** These involve the techniques of **least squares** (otherwise known as **regression**). For some problems these techniques do not suffice, and more specialized techniques involving nonlinear statistics must be used, albeit a solution may not exist.

Artificial intelligence (AI) is the study and implementation of programs that model knowledge systems and exhibit aspects of intelligence in problem solving. Typical areas of application are in learning, linguistics, pattern recognition, decision making, and theorem proving. In AI, the computer serves to search a collection of heuristic rules to find a match with a current situation and to make inferences or otherwise reorganize knowledge into more useful forms. AI techniques have been utilized to build sophisticated systems, called **expert systems,** to aid in producing a timely response in problems involving a large number of complex conditions.

Operating Systems

The operating system provides the services that support the needs that computer programs have in common during execution. Any list of services would include those needed to configure the resources that will be made available to the users, to attach hardware units (e.g., memory modules, storage devices, coprocessors, and peripheral devices) to the existing configuration, to detach modules, to assign default parameters to the hardware and software units, to set up and schedule users' tasks so as to resolve conflicts and optimize throughput, to control system input and output devices, to protect the system and users' programs from themselves and from each other, to manage storage space in the file devices, to protect file devices from faults and illegal use, to account for the use of the system, and to handle in an orderly way any exception which might be encountered during program execution. A well-designed operating system provides these services in a user-friendly environment and yet makes itself and the computer operating staff transparent to the user.

The design of a computer operating system depends on the number of users which can be expected. The focus of single-user systems relies on the monitor to provide a user-friendly system through dialog menus with icons, mouse operations, and templets. Table 2.2.7 lists some popular operating systems for PCs by their trademark names. The design of a multiuser system attempts to give each user the impression that he/she is the lone user of the system. In addition to providing the accoutrements of a user-friendly system, the design focuses on the order of processing the jobs in an attempt to treat each user in a fair and equitable fashion. The basic issues for determining the order of processing center on the selection of job queues: the number of queues (a simple queue or

a mix of queues), the method used in scheduling the jobs in the queue (first come–first served, shortest job next, or explicit priorities), and the internal handling of the jobs in the queue (batch, multiprogramming, or timesharing).

Table 2.2.7 Some Popular PC Operating Systems

Trademark	Supplier
DOS	Microsoft Corp.
Windows	Microsoft Corp.
OS/2	IBM Corp.
Unix	Unix Systems Laboratory Inc.
Sun/OS	Sun Microsystems Inc.
Macintosh	Apple Computer Inc.

Batch operating systems process jobs in a sequential order. Jobs are collected in batches and entered into the computer with individual job instructions which the operating system interprets to set up the job, to allocate resources needed, to process the job, and to provide the input/output. The operating system processes each job to completion in the order it appears in the batch. In the event a malfunction or fault occurs during execution, the operating system terminates the job currently being executed in an orderly fashion before initiating the next job in sequence.

Multiprogramming operating systems process several jobs concurrently. A job may be initiated any time memory and other resources which it needs become available. Many jobs may be simultaneously active in the system and maintained in a partial state of completion. The order of execution depends on the priority assignments. Jobs are executed to completion or put into a wait state until a pending request for service has been satisfied. It should be noted that, while the CPU can execute only a single program at any moment of time, operations with peripheral and storage devices can occur concurrently.

Timesharing operating systems process jobs in a way similar to multiprogramming except for the added feature that each job is given a short slice of the available time to complete its tasks. If the job has not been completed within its time slice or if it requests a service from an external device, it is put into a wait status and control passes to the next job. Effectively, the length of the time slice determines the priority of the job.

Program Preparation Facilities

For the user, the crucial part of a language system is the grammar which specifies the language syntax and semantics that give the symbols and rules used to compose acceptable statements and the meaning associated with the statements. Compared to natural languages, computer languages are more precise, have a simpler structure, and have a clearer syntax and semantics that allows no ambiguities in what one writes or what one means. For a program to be executed, it must eventually be translated into a sequence of basic machine instructions.

The statements written by a user must first be put on some machine-readable medium or typed on a keyboard for entry into the machine. The translator (**compiler**) program accepts these statements as input and translates (compiles) them into a sequence of basic machine instructions which form the executable version of the program. After that, the translated (compiled) program can be run.

During the execution of a program, a run-time program must also be present in the memory. The purpose of the run-time system is to perform services that the user's program may require. For example, in case of a program fault, the run-time system will identify the error and terminate the program in an orderly manner.

Some language systems do not have a separate compiler to produce machine-executable instructions. Instead the run-time system interprets the statements as written, converts them into a pseudo-code, and executes the coded version.

Commonly needed functions are made available as prepared modules, either as an integral part of the language or from stored libraries. The documentation of these functions must be studied carefully to assure correct selection and utilization.

Languages may be classified as procedure-oriented or problem-oriented. With **procedure-oriented** languages, all the detailed steps must be specified by the user. These languages are usually characterized as being more verbose than problem-oriented languages, but are more flexible and can deal with a wider range of problems. **Problem-oriented** languages deal with more specialized classes of problems. The elements of problem-oriented languages are usually familiar to a knowledgeable professional and so are easier to learn and use than procedure-oriented languages.

The most elementary form of a procedure-oriented language is called an **assembler.** This class of language permits a computer program to be written directly in basic computer instructions using mnemonic operators and symbolic operands. The assembler's translator converts these instructions into machine-usable form.

A further refinement of an assembler permits the use of macros. A **macro** identifies, by an assigned name and a list of formal parameters, a sequence of computer instructions written in the assembler's format and stored in its subroutine library. The macroassembler includes these macro instructions in the translated program along with the instructions written by the programmer.

Besides these basic language systems there exists a large variety of other language systems. These are called higher-level language systems since they permit more complex statements than are permitted by a macroassembler. They can also be used on machines produced by different manufacturers or on machines with different instruction repertoires.

In the field of business programming, **COBOL** (COmmon Business-Oriented Language) is the most popular. This language facilitates the handling of the complex information files found in business and data-processing problems.

Another example of an application area supported by special languages is in the field of problems involving strings of text. SNOBOL and LISP exemplify these string-manipulation or list-processing languages. Applications vary from generating concordances to sophisticated symbolic formula manipulation.

One language of historical value is **ALGOL 60.** It is a landmark in the theoretical development of computer languages. It was designed and standardized by an international committee whose goal was to formulate a language suitable for publishing computer algorithms. Its importance lies in the many language features it introduced which are now common in the more recent languages which succeeded it and in the scientific notation which was used to define it.

FORTRAN (FORmula TRANslator) was one of the first languages catering to the engineering and scientific community where algebraic formulas specify the computations used within the program. It has been standardized several times. The current version is FORTRAN 90 (ANSI X3.198-1992). Each version has expanded the language features and has removed undesirable features which lead to unstructured programs. The new features include new data types like Boolean and character strings, additional operators and functions, and new statements that support programs conforming to the requirements for structured programming.

The **PASCAL** language couples the ideas of ALGOL 60 to those of structured programming. By allowing only appropriate statement types, it guarantees that any program written in the language will be well-structured. In addition, the language introduced new data types and allows programmers to define new complex data structures based on the primitive data types.

The definition of the **Ada** language was sponsored by the Department of Defense as an all-encompassing language for the development and maintenance of very large, software-intensive projects over their life cycle. While it meets software engineering objectives in a manner similar to Pascal, it has many other features not normally found in programming languages. Like other attempts to formulate very large all-inclusive languages, it is difficult to learn and has not found popular favor. Nevertheless, its many unique features make it especially valuable in implementing programs which cannot be easily implemented in other languages (e.g., programs for parallel computations in embedded computers).

By edict, subsets of Ada were forbidden. **Modula-2** was designed to retain the inherent simplicity of PASCAL but include many of the advanced features of Ada. Its advantage lies in implementing large projects involving many programmers. The compilers for this language have rigorous interface cross-checking mechanisms to avoid poor interfaces between components. Another troublesome area is in the implicit use of global data. Modula-2 retains the Ada facilities that allow programmers to share data and avoids incorrectly modifying the data in different program units.

The **C** language was developed by AT&T's Bell Laboratories and subsequently standardized by ANSI. It has a reputation for translating programs into compact and fast code, and for allowing program segments to be precompiled. Its strength rests in the flexibility of the language; for example, it permits statements from other languages to be included in-line in a C program and it offers the largest selection of operators that mirror those available in an assembly language. Because of its flexibility, programs written in C can become unreadable.

Problem-oriented languages have been developed for every discipline. A language might deal with a specialized application within an engineering field, or it might deal with a whole gamut of applications covering one or more fields.

A class of problem-oriented languages that deserves special mention are those for solving problems in **discrete simulation.** GPSS, Simscript, and SIMULA are among the most popular. A simulation (another word for *model*) of a system is used whenever it is desirable to watch a succession of many interrelated events or when there is interplay between the system under study and outside forces. Examples are problems in human-machine interaction and in the modeling of business systems. Typical human-machine problems are the servicing of automatic equipment by a crew of operators (to study crew size and assignments, typically), or responses by shared maintenance crews to equipment subject to unpredictable (random) breakdown. Business models often involve transportation and warehousing studies. A business model could also study the interactions between a business and the rest of the economy such as competitive buying in a raw materials market or competitive marketing of products by manufacturers.

Physical or chemical systems may also be modeled. For example, to study the application of automatic control values in pipelines, the computer model consists of the control system, the valves, the piping system, and the fluid properties. Such a model, when tested, can indicate whether fluid hammer will occur or whether valve action is fast enough. It can also be used to predict pressure and temperature conditions in the fluid when subject to the valve actions.

Another class of problem-oriented languages makes the computer directly accessible to the specialist with little additional training. This is achieved by permitting the user to describe problems to the computer *in terms that are familiar in the discipline of the problem* and for which the language is designed. Two approaches are used. Figures 2.2.11 and 2.2.12 illustrate these.

One approach sets up the computer program directly from the mathematical equations. In fact, problems were formulated in this manner in the past, where analog computers were especially well-suited. Anyone familiar with analog computers finds the transitions to these languages easy. Figure 2.2.11 illustrates this approach using the MIMIC language to write the program for the solution of the initial-value problem:

$$M\ddot{y} + Z\dot{y} + Ky = 1 \qquad \text{and} \qquad \dot{y}(0) = y(0) = 0$$

MIMIC is a digital simulation language used to solve systems of ordinary differential equations. The key step in setting up the solution is to isolate the highest-order derivative on the left-hand side of the equation and equate it to an expression composed of the remaining terms. For the

MIMIC statements	Explanation
$DY2 = (1 - Z * DY1 - K * Y)/M$	Differential equation to be solved. "$*$" is used for multiplication and $DY2$, $DY1$, and Y are defined mnemonics for $\ddot{y}$, $\dot{y}$, and y.
$DY1 = \text{INT}(DY2,0.)$ $Y = \text{INT}(DY1,0.)$	INT(A,B) is used to perform integration. It forms successive values of $B + \int A dt$.
FIN(T,10.)	T is a reserved name representing the independent variable. This statement will terminate execution when $T \geq 10$.
CON(M,K,Z)	Values must be furnished for M, K, and Z. An input with these values must appear after the END card.
PLO($T,DY2$) PLO($T,DY1$) PLO(T,Y)	Three point plots are produced on the line printer; $\ddot{y}$, $\dot{y}$, and y vs. t.
END	Necessary last statement.

Fig. 2.2.11 Illustration of a MIMIC program.

equation above, this results in:

$$\ddot{y} = (1 - Z\dot{y} - Ky)/M$$

The highest-order derivative is derived by equating it to the expression on the right-hand side of the equation. The lower-order derivatives in the expression are generated successively by integrating the highest-order derivative. The MIMIC language permits the user to write these statements in a format closely resembling mathematical notation.

The alternate approach used in problem-oriented languages permits the setup to be described to the computer directly from the block diagram of the physical system. Figure 2.2.12 illustrates this approach

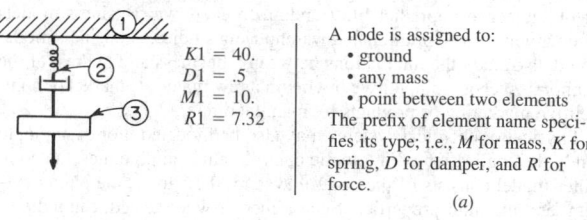

$K1 = 40.$ $D1 = .5$ $M1 = 10.$ $R1 = 7.32$	A node is assigned to: • ground • any mass • point between two elements The prefix of element name specifies its type; i.e., M for mass, K for spring, D for damper, and R for force.

(a)

SCEPTRE statements	Explanation
MECHANICAL DESCRIPTION ELEMENTS $M1, 1 - 3 = 10.$ $K1, 1 - 2 = 40.$ $D1, 2 - 3 = .5$ $R1, 1 - 3 = 7.32$	Specifies the elements and their position in the diagram using the node numbers.
OUTPUT $SM1,VM1$	Results are listed on the line printer. Prefix on the element specifies the quantity to be listed; S for displacement, V for velocity.
RUN CONTROL STOPTIME = 10.	TIME is reserved name for independent variable. Statement will terminate execution of program when TIME is equal to or greater than 10.
END	Necessary statement.

(b)

Fig. 2.2.12 Illustration of SCEPTRE program. (*a*) Problem to be solved; (*b*) SCEPTRE program.

using the SCEPTRE language. SCEPTRE statements are written under headings and subheadings which identify the type of component being described. This language may be applied to network problems of electrical digital-logic elements, mechanical-translation or rotational elements, or transfer-function blocks. The translator for this language develops and sets up the equations directly from this description of the network diagram, and so relieves the user from the mathematical aspects of the problem.

Application Packages

An application package differs from a language in that its components have been organized to solve problems in a particular application rather than to create the components themselves. The user interacts with the package by initiating the operations and providing the data. From an operational view, packages are built to minimize or simplify interactions with the users by using a menu to initiate operations and entering the data through templets.

Perhaps the most widely used application package is the **word processor**. The objective of a word processor is to allow users to compose text in an electronically stored format which can be corrected or modified, and from which a hard copy can be produced on demand. Besides the basic typewriter operations, it contains functions to manipulate text in blocks or columns, to create headers and footers, to number pages, to find and correct words, to format the data in a variety of ways, to create labels, and to merge blocks of text together. The better word processors have an integrated dictionary, a spelling checker to find and correct misspelled words, a grammar checker to find grammatical errors, and a thesaurus. They often have facilities to prepare complex mathematical equations and to include and manipulate graphical artwork, including editing color pictures. When enough page- and document formatting capability has been added, the programs are known as **desktop publishing** programs.

One of the programs that contributed to the early acceptance of personal computers was the **spread sheet** program. These programs simulate the common spread sheet with its columns and rows of interrelated data. The computerized approach has the advantage that the equations are stored so that the results of a change in data can be shown quickly after any change is made in the data. Modern spread sheet programs have many capabilities, including the ability to obtain information from other spread sheets, to produce a variety of reports, and to prepare equations which have complicated logical aspects.

Tools for **project management** have been organized into commercially available application packages. The objectives of these programs are in the planning, scheduling, and controlling the time-oriented activities describing the projects. There are two basically similar techniques used in these packages. One, called **CPM** (critical path method), assumes that the project activities can be estimated deterministically. The other, called **PERT** (project evaluation and review technology), assumes that the activities can be estimated probabilistically. Both take into account such items as the requirement that certain tasks cannot start before the completion of other tasks. The concepts of **critical path** and **float** are crucial, especially in scheduling the large projects that these programs are used for. In both cases tools are included for estimating project schedules, estimating resources needed and their schedules, and representing the project activities in graphical as well as tabular form.

A major use of the digital computer is in **data reduction,** data analysis, and visualization of data. In installations where large amounts of data are recorded and kept, it is often advisable to reduce the amount of data by ganging the data together, by averaging the data with numerical filters to reduce the amount of noise, or by converting the data to a more appropriate form for storage, analysis, visualization, or future processing. This application has been expanded to produce systems for evaluation, automatic testing, and fault diagnosis by coupling the data acquisition equipment to special peripherals that automatically measure and record the data in a digital format and report the data as meaningful, nonphysically measurable parameters associated with a mathematical model.

Computer-aided design/computer-aided manufacturing (**CAD/CAM**) is an integrated collection of software tools which have been designed to make way for innovative methods of fabricating customized products to meet customer demands. The goal of modern manufacturing is to process orders placed for different products sooner and faster, and to fabricate them without retooling. CAD has the tools for prototyping a design and setting up the factory for production. Working within a framework of agile manufacturing facilities that features automated vehicles, handling robots, assembly robots, and welding and painting robots, the factory sets itself up for production under computer control. Production starts with the receipt of an order on which customers may pick options such as color, size, shapes, and features. Manufacturing proceeds with greater flexibility, quality, and efficiency in producing an increased number of products with a reduced workforce. Effectively, CAD/CAM provides for the ultimate just-in-time (JIT) manufacturing.

Two other types of application package illustrate the versatility of data management techniques. One type ties on-line equipment to a computer for collecting real-time data from the production lines. An animated, pictorial display of the production lines forms the heart of the system, allowing supervision in a central control station to continuously track operations. The other type collects time-series data from the various activities in an enterprise. It assists in what is known as **management by exception.** It is especially useful where the detailed data is so voluminous that it is feasible to examine it only in summaries. The data elements are processed and stored in various levels of detail in a seamless fashion. The system stores the reduced data and connects it to the detailed data from which it was derived. The application package allows management, through simple computer operations, to detect a problem at a higher level and to locate and pinpoint its cause through examination of successively lower levels.

Section **3**

Mechanics of Solids and Fluids

BY

ROBERT F. STEIDEL, JR. *Professor of Mechanical Engineering (Retired), University of California, Berkeley*
VITTORIO (RINO) CASTELLI *Senior Research Fellow, Xerox Corp.*
J. W. MURDOCK *Late Consulting Engineer*
LEONARD MEIROVITCH *University Distinguished Professor, Department of Engineering Science and Mechanics, Virginia Polytechnic Institute and State University*

3.1 MECHANICS OF SOLIDS
by Robert F. Steidel, Jr.

REFERENCES: Beer and Johnston, "Mechanics for Engineers," McGraw-Hill. Ginsberg and Genin, "Statics and Dynamics," Wiley. Higdon and Stiles, "Engineering Mechanics," Prentice-Hall. Holowenko, "Dynamics of Machinery," Wiley. Housnor and Hudson, "Applied Mechanics," Van Nostrand. Meriam, "Statics and Dynamics," Wiley. Mabie and Ocvirk, "Mechanisms and Dynamics of Machinery," Wiley. Synge and Griffith, "Principles of Mechanics," McGraw-Hill. Timoshenko and Young, "Advanced Dynamics," McGraw-Hill. Timoshenko and Young, "Engineering Mechanics," McGraw-Hill.

PHYSICAL MECHANICS

Definitions

Force is the action of one body on another which will cause acceleration of the second body unless acted on by an equal and opposite action counteracting the effect of the first body. It is a **vector** quantity.

Time is a measure of the sequence of events. In newtonian mechanics it is an absolute quantity. In relativistic mechanics it is relative to the *frames of reference* in which the sequence of events is observed. The common unit of time is the second.

Inertia is that property of matter which causes a resistance to any change in the motion of a body.

Mass is a quantitative measure of *inertia*.

Acceleration of Gravity Every object which falls in a vacuum at a given position on the earth's surface will have the same acceleration g. Accurate values of the acceleration of gravity as measured *relative* to the earth's surface include the effect of the earth's rotation and flattening at the poles. The international gravity formula for the acceleration of gravity at the earth's surface is $g = 32.0881(1 + 0.005288 \sin^2 \phi - 0.0000059 \sin^2 2\phi)$ ft/s², where ϕ is latitude in degrees. For extreme accuracy, the local acceleration of gravity must also be corrected for the presence of large water or land masses and for height above sea level. The absolute acceleration of gravity for a nonrotating earth discounts the effect of the earth's rotation and is rarely used, except outside the earth's atmosphere. If g_0 represents the absolute acceleration at sea level, the absolute value at an altitude h is $g = g_0 R^2/(R + h)^2$, where R is the radius of the earth, approximately 3,960 mi (6,373 km).

Weight is the resultant force of attraction on the mass of a body due to a gravitational field. On the earth, units of weight are based upon an acceleration of gravity of 32.1740 ft/s² (9.80665 m/s²).

Linear momentum is the product of mass and the linear velocity of a particle and is a vector. The moment of the linear-momentum vector about a fixed axis is the **angular momentum** of the particle about that fixed axis. For a rigid body rotating about a fixed axis, angular momentum is defined as the product of moment of inertia and angular velocity, each measured about the fixed axis.

An increment of **work** is defined as the product of an incremental displacement and the component of the force vector in the direction of the displacement or the component of the displacement vector in the direction of the force. The increment of work done by a couple acting on a body during a rotation of $d\theta$ in the plane of the couple is $dU = M\,d\theta$.

Energy is defined as the capacity of a body to do work by reason of its motion or configuration (see **Work and Energy**).

A **vector** is a directed line segment that has both magnitude and direction. In script or text, a vector is distinguished from a scalar V by a boldface-type **V**. The magnitude of the scalar is the magnitude of the vector, $V = |\mathbf{V}|$.

A **frame of reference** is a specified set of geometric conditions to which other locations, motion, and time are referred. In newtonian mechanics, the fixed stars are referred to as the **primary (inertial) frame of reference**. Relativistic mechanics denies the existence of a primary reference frame and holds that all reference frames must be described relative to each other.

SYSTEMS AND UNITS OF MEASUREMENTS

In *absolute systems,* the units of **length, mass,** and **time** are considered fundamental quantities, and all other units including that of **force** are derived.

In *gravitational systems,* the units of **length, force,** and **time** are considered fundamental qualities, and all other units including that of **mass** are derived.

In the SI system of units, the unit of mass is the kilogram (kg) and the unit of length is the metre (m). A force of one newton (N) is derived as the force that will give 1 kilogram an acceleration of 1 m/s².

In the English engineering system of units, the unit of mass is the pound mass (lbm) and the unit of length is the foot (ft). A force of one pound (1 lbf) is the force that gives a pound mass (1 lbm) an acceleration equal to the standard acceleration of gravity on the earth, 32.1740 ft/s² (9.80665 m/s²). A slug is the mass that will be accelerated 1 ft/s² by a force of 1 lbf. Therefore, 1 slug = 32.1740 lbm. When described in the gravitational system, mass is a derived unit, being the constant of proportionality between force and acceleration, as determined by Newton's second law.

General Laws

NEWTON'S LAWS

I. If a balanced force system acts on a particle at rest, it will remain at rest. If a balanced force system acts on a particle in motion, it will remain in motion in a straight line without acceleration.

II. If an unbalanced force system acts on a particle, it will accelerate in proportion to the magnitude and in the direction of the resultant force.

III. When two particles exert forces on each other, these forces are equal in magnitude, opposite in direction, and collinear.

Fundamental Equation The basic relation between mass, acceleration, and force is contained in Newton's second law of motion. As applied to a particle of mass, $\mathbf{F} = m\mathbf{a}$, force = mass × acceleration. This equation is a vector equation, since the direction of $\mathbf{F}$ must be the direction of $\mathbf{a}$, as well as having $\mathbf{F}$ equal in magnitude to $m\mathbf{a}$. An alternative form of Newton's second law states that the resultant force is equal to the time rate of change of momentum, $\mathbf{F} = d(m\mathbf{v})/dt$.

Law of the Conservation of Mass The mass of a body remains unchanged by any ordinary physical or chemical change to which it may be subjected.

Law of the Conservation of Energy The principle of conservation of energy requires that the total mechanical energy of a system remain unchanged if it is subjected only to forces which depend on position or configuration.

Law of the Conservation of Momentum The linear momentum of a system of bodies is unchanged if there is no resultant external force on the system. The angular momentum of a system of bodies about a fixed axis is unchanged if there is no resultant external moment about this axis.

Law of Mutual Attraction (Gravitation) Two particles attract each other with a force F proportional to their masses m_1 and m_2 and inversely proportional to the square of the distance r between them, or $F = km_1m_2/r^2$, in which k is the gravitational constant. The value of the gravitational constant is $k = 6.673 \times 10^{-11}$ m³/kg · s² in SI or absolute units, or $k = 3.44 \times 10^{-8}$ ft⁴ lb⁻¹ s⁻⁴ in engineering gravitational units.

It should be pointed out that the unit of force F in the SI system is the **newton** and is derived, while the unit force in the gravitational system is the **pound-force** and is a fundamental quantity.

EXAMPLE. Each of two solid steel spheres 6 in in diam will weigh 32.0 lb on the earth's surface. This is the force of attraction between the earth and the steel sphere. The force of mutual attraction between the spheres if they are just touching is 0.000000136 lb.

STATICS OF RIGID BODIES

General Considerations

If the forces acting on a rigid body do not produce any acceleration, they must neutralize each other, i.e., form a **system of forces in equilibrium.** Equilibrium is said to be **stable** when the body with the forces acting upon it returns to its original position after being displaced a very small amount from that position; **unstable** when the body tends to move still farther from its original position than the very small displacement; and **neutral** when the forces retain their equilibrium when the body is in its new position.

External and Internal Forces The forces by which the individual particles of a body act on each other are known as internal forces. All other forces are called external forces. If a body is supported by other bodies while subject to the action of forces, deformations and forces will be produced at the points of support or contact and these internal forces will be distributed throughout the body until equilibrium exists and the body is said to be in a state of tension, compression, or shear. The forces exerted by the body on the supports are known as **reactions.** They are equal in magnitude and opposite in direction to the forces with which the supports act on the body, known as **supporting forces.** The supporting forces are external forces applied to the body.

In considering a body at a definite section, it will be found that all the internal forces act in pairs, the two forces being equal and opposite. The external forces act singly.

General Law When a body is at rest, the forces acting externally to it must form an equilibrium system. This law will hold for any part of the body, in which case the forces acting at any section of the body become external forces when the part on either side of the section is considered alone. In the case of a **rigid body**, any two forces of the same magnitude, but acting in opposite directions in any straight line, may be added or removed without change in the action of the forces acting on the body, provided the strength of the body is not affected.

Composition, Resolution, and Equilibrium of Forces

The **resultant** of several forces acting at a point is a force which will produce the same effect as all the individual forces acting together.

Forces Acting on a Body at the Same Point The resultant R of two forces F_1 and F_2 applied to a rigid body at the same point is represented in magnitude and direction by the diagonal of the parallelogram formed by F_1 and F_2 (see Figs. 3.1.1 and 3.1.2).

$$R = \sqrt{F_1^2 + F_2^2 + 2F_1F_2 \cos a}$$

$$\sin a_1 = (F_2 \sin a)/R \qquad \sin a_2 = (F_1 \sin a)/R$$

When $a = 90°$, $R = \sqrt{F_1^2 + F_2^2}$, $\sin a_1 = F_2/R$, and $\sin a_2 = F_1/R$.

$$\left. \begin{array}{l} \text{When } a = 0°, R = F_1 + F_2 \\ \text{When } a = 180°, R = F_1 - F_2 \end{array} \right\} \quad \begin{array}{l} \text{Forces act in same} \\ \text{straight line.} \end{array}$$

A **force** R **may be resolved into two component forces** intersecting anywhere on R and acting in the same plane as R, by the reverse of the operation shown by Figs. 3.1.1 and 3.1.2; and by repeating the operation with the components, R may be resolved into any number of component forces intersecting R at the same point and in the same plane.

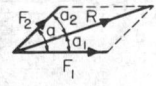

Fig. 3.1.1   **Fig. 3.1.2**

Resultant of Any Number of Forces Applied to a Rigid Body at the Same Point Resolve each of the given forces F into components along three rectangular coordinate axes. If A, B, and C are the angles made with XX, YY, and ZZ, respectively, by any force F, the components will be $F \cos A$ along XX, $F \cos B$ along YY, $F \cos C$ along ZZ; add the components of all the forces along each axis algebraically and obtain $\Sigma F \cos A = \Sigma X$ along XX, $\Sigma F \cos B = \Sigma Y$ along YY, and $\Sigma F \cos C = \Sigma Z$ along ZZ.

The resultant $R = \sqrt{(\Sigma X)^2 + (\Sigma Y)^2 + (\Sigma Z)^2}$. The angles made by the resultant with the three axes are A_r with XX, B_r with YY, C_r with ZZ, where

$$\cos A_r = \Sigma X/R \qquad \cos B_r = \Sigma Y/R \qquad \cos C_r = \Sigma Z/R$$

The **direction** of the **resultant** can be determined by plotting the algebraic sums of the components.

If the forces are all in the same plane, the components of each of the forces along one of the three axes (say ZZ) will be 0; i.e., angle $C_r = 90°$ and $R = \sqrt{(\Sigma X)^2 + (\Sigma Y)^2}$, $\cos A_r = \Sigma X/R$, and $\cos B_r = \Sigma Y/R$.

For equilibrium, it is necessary that $R = 0$; i.e., ΣX, ΣY, and ΣZ must each be equal to zero.

General Law In order that a number of forces acting at the same point shall be in equilibrium, the algebraic sum of their components along any *three* coordinate axes must each be equal to zero. When the forces all act in the same plane, the algebraic sum of their components along any *two* coordinate axes must each equal zero.

When the Forces Form a System in Equilibrium Three unknown forces can be determined if the lines of action of the forces are *all* known and are in different planes. If the forces are all in the same plane, the lines of action being known, only *two* unknown forces can be determined. If the lines of action of the unknown forces are *not* known, only *one* unknown force can be determined in either case.

Couples and Moments

Couple Two parallel forces of equal magnitude (Fig. 3.1.3) which act in opposite directions and are not collinear form a couple. **A couple cannot be reduced to a single force.**

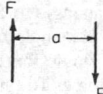

Fig. 3.1.3

Displacement and Change of a Couple The forces forming a couple may be moved about and their magnitude and direction changed, provided they always remain parallel to each other and remain in either the original plane or one parallel to it, and provided the product of one of the forces and the perpendicular distance between the two is constant and the direction of rotation remains the same.

Moment of a Couple The moment of a couple is the product of the magnitude of one of the forces and the perpendicular distance between the lines of action of the forces. Fa = moment of couple; a = arm of couple. If the forces are measured in pounds and the distance a in feet, the **unit of rotation moment** is the foot-pound. If the force is measured in kilograms and the distance in metres, the unit is the metre-kilogram. In the cgs system the unit of rotation moment is 1 cm-dyne.

Rotation moments of couples acting in the same plane are conventionally considered to be positive for counterclockwise moments and negative for clockwise moments, although it is only necessary to be consistent within a given problem. The magnitude, direction, and sense of rotation of a couple are completely determined by its moment axis, or moment vector, which is a line drawn perpendicular to the plane in which the couple acts, with an arrow indicating the direction from which the couple will appear to have right-handed rotation; the length of the line represents the magnitude of the moment of the couple. See

Fig. 3.1.4, in which AB represents the magnitude of the moment of the couple. Looking along the line in the direction of the arrow, the couple will have right-handed rotation in any plane perpendicular to the line.

Composition of Couples   Couples may be combined by adding their moment vectors geometrically, in accordance with the parallelogram rule, in the same manner in which forces are combined.

Couples lying in the same or parallel planes are added algebraically. Let $+28$ lbf · ft ($+38$ N · m), -42 lbf · ft (-57 N · m), and $+70$ lbf · ft (95 N · m) be the moments of three couples in the same or parallel planes; their resultant is a single couple lying in the same or in a parallel plane, whose moment is $\Sigma M = +28 - 42 + 70 = +56$ lbf · ft ($\Sigma M = +38 - 57 + 95 = 76$ N · m).

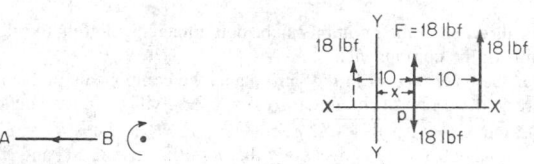

Fig. 3.1.4   **Fig. 3.1.5**

If the polygon formed by the moment vectors of several couples closes itself, the couples form an equilibrium system. Two couples will balance each other when they lie in the same or parallel planes and have the same moment in magnitude, but opposite in sign.

Combination of a Couple and a Single Force in the Same Plane (Fig. 3.1.5)   Given a force $F = 18$ lbf (80 N) acting as shown at distance x from YY, and a couple whose moment is -180 lbf · ft (244 N · m) in the same or parallel plane, to find the resultant. A couple may be changed to any other couple in the same or a parallel plane having the same moment and same sign. Let the couple consist of two forces of 18 lbf (80 N) each and let the arm be 10 ft (3.05 m). Place the couple in such a manner that one of its forces is opposed to the given force at p. This force of the couple and the given force being of the same magnitude and opposite in direction will neutralize each other, leaving the other force of the couple acting at a distance of 10 ft (3.05 m) from p and parallel and equal to the given force 18 lbf (80 N).

General Rule   The resultant of a couple and a single force lying in the same or parallel planes is a single force, equal in magnitude, in the same direction and parallel to the single force, and acting at a distance from the line of action of the single force equal to the moment of the couple divided by the single force. The moment of the resultant force about any point on the line of action of the given single force must be of the same sense as that of the couple, positive if the moment of the couple is positive, and negative if the moment of the couple is negative. If the moment of the couple in Fig. 3.1.5 had been $+$ instead of $-$, the resultant would have been a force of 18 lbf (80 N) acting in the same direction and parallel to F, but at a distance of 10 ft (3.05 m) to the left of it (shown dotted), making the moment of the resultant about any point on F positive.

To effect a parallel displacement of a single force F over a distance a, a couple whose moment is Fa must be added to the system. The sense of the couple will depend upon which way it is desired to displace force F.

The moment of a force with respect to a point is the product of the force F and the perpendicular distance from the point to the line of action of the force.

The Moment of a Force with Respect to a Straight Line   If the force is resolved into components parallel and perpendicular to the given line, the moment of the force with respect to the line is the product of the magnitude of the perpendicular component and the distance from its line of action to the given line.

Forces with Different Points of Application

Composition of Forces   If each force $\mathbf{F}$ is resolved into components parallel to three rectangular coordinate axes XX, YY, and ZZ, the magnitude of the resultant is $R = \sqrt{(\Sigma X)^2 + (\Sigma Y)^2 + (\Sigma Z)^2}$, and its line of action makes angles A_r, B_r, and C_r with axes XX, YY, and ZZ, where cos

$A_r = \Sigma X/R$, cos $B_r = \Sigma Y/R$, and cos $C_r = \Sigma Z/R$; and there are three couples which may be combined by their moment vectors into a single resultant couple having the moment $M_r = \sqrt{(M_x)^2 + (M_y)^2 + (M_z)^2}$, whose moment vector makes angles of A_m, B_m, and C_m with axes XX, YY, and ZZ, such that cos $A_m = M_x/M_r$, cos $B_m = M_y/M_r$, cos $C_m = M_z/M_r$. If this single resulting couple is in the same plane as the single resulting force at the origin or a plane parallel to it, the system may be reduced to a single force $\mathbf{R}$ acting at a distance from $\mathbf{R}$ equal to M_r/R. If the couple and force are not in the same or parallel planes, it is impossible to reduce the system to a single force. If $\mathbf{R} = 0$, i.e., if ΣX, ΣY, and ΣZ all equal zero, the system will reduce to a single couple whose moment is $\mathbf{M}_r$. If $\mathbf{M}_r = 0$, i.e., if M_x, M_y, and M_z all equal zero, the resultant will be a single force $\mathbf{R}$.

When the forces are all in the same plane, the cosine of one of the angles A_r, B_r, or $C_r = 0$, say, $C_r = 90°$. Then $R = \sqrt{(\Sigma X)^2 + (\Sigma Y)^2}$, $M_r = \sqrt{M_x^2 + M_y^2}$, and the final resultant is a force equal and parallel to R, acting at a distance from R equal to M_r/R.

A system of forces in the same plane can always be replaced by either a couple or a single force. If $\mathbf{R} = 0$ and $\mathbf{M}_r \cdot 0$, the resultant is a couple. If $\mathbf{M}_r = 0$ and $R > 0$, the resultant is a single force.

A rigid body is in equilibrium when acted upon by a system of forces whenever $R = 0$ and $M_r = 0$, i.e., when the following six conditions hold true: $\Sigma X = 0$, $\Sigma Y = 0$, $\Sigma Z = 0$, $M_x = 0$, $M_y = 0$, and $M_z = 0$. When the system of forces is in the same plane, equilibrium prevails when the following three conditions hold true: $\Sigma X = 0$, $\Sigma Y = 0$, $\Sigma M = 0$.

Forces Applied to Support Rigid Bodies

The external forces in equilibrium acting upon a body may be statically determinate or indeterminate according to the number of unknown forces existing. When the forces are all in the same plane and act at a common point, two unknown forces may be determined if their lines of action are known, one if unknown.

When the forces are all in the same plane and are parallel, two unknown forces may be determined if the lines of action are known, one if unknown.

When the forces are anywhere in the same plane, three unknown forces may be determined if their lines of action are known, if they are not parallel or do not pass through a common point; if the lines of action are unknown, only one unknown force can be determined.

If the forces all act at a common point but are in different planes, three unknown forces can be determined if the lines of action are known, one if unknown.

If the forces act in different planes but are parallel, three unknown forces can be determined if their lines of action are known, one if unknown.

The first step in the solution of problems in statics is the determination of the supporting forces. The following data are required for the complete knowledge of supporting forces: magnitude, direction, and point of application. According to the nature of the problem, none, one, or two of these quantities are known.

One Fixed Support   The point of application, direction, and magnitude of the load are known. See Fig. 3.1.6. As the body on which the forces act is in equilibrium, the supporting force P must be equal in magnitude and opposite in direction to the resultant of the loads L.

In the case of a **rolling surface**, the point of application of the support is obtained from the center of the connecting bolt A (Fig. 3.1.7), both the direction and magnitude being unknown. The point of application and

Fig. 3.1.6

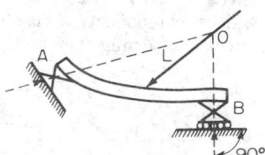

Fig. 3.1.7

line of action of the support at B are known, being determined by the rollers.

When three forces acting in the same plane on the same rigid body are in equilibrium, their lines of action must pass through the same point O. The load L is known in magnitude and direction. The line of action of the support at B is known on account of the rollers. The point of application of the support at A is known. The three forces are in equilibrium and are in the same plane; therefore, the lines of action must meet at the point O.

In the case of the rolling surfaces shown in Fig. 3.1.8, the direction of the support at A is known, the magnitude and point of application unknown. The line of action and point of application of the supporting

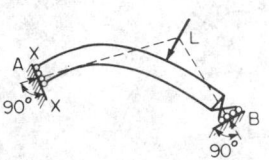

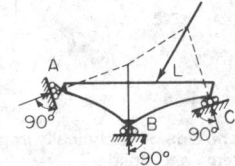

Fig. 3.1.8   **Fig. 3.1.9**

force at B are known, its magnitude unknown. The lines of action of the three forces must meet in a point, and the supporting force at A must be perpendicular to the plane XX. In the case shown in Fig. 3.1.9, the directions and points of application of the supporting forces are known, and the magnitudes unknown. The lines of action of resultant of supports A and B, the support at C and load L must meet at a point. Resolve the resultant of supports at A and B into components at A and B, their direction being determined by the rollers.

If a member of a truss or frame in equilibrium is pinned at two points and loaded at these two points only, the line of action of the forces exerted on the member or by the member at these two points must be along a line connecting the pins.

If the external forces acting upon a rigid body in equilibrium are all in the same plane, the equations $\Sigma X = 0$, $\Sigma Y = 0$, and $\Sigma M = 0$ must be satisfied. When trusses, frames, and other structures are under discussion, these equations are usually used as $\Sigma V = 0$, $\Sigma H = 0$, $\Sigma M = 0$, where V and H represent vertical and horizontal components, respectively.

The **supports** are said to be **statically determinate** when the laws of equilibrium are sufficient for their determination. When the conditions are not sufficient for the determination of the supports or other forces, the structure is said to be **statically indeterminate;** the unknown forces can then be determined from considerations involving the deformation of the material.

When several bodies are so connected to one another as to make up a rigid structure, the forces at the points of connection must be considered as internal forces and are not taken into consideration in the determination of the supporting forces for the structure as a whole.

The distortion of any practically rigid structure under its working loads is so small as to be negligible when determining supporting forces. When the forces acting at the different joints in a built-up structure cannot be determined by dividing the structure up into parts, the structure is said to be **statically indeterminate internally.** A structure may be statically indeterminate internally and still be statically determinate externally.

Fundamental Problems in Graphical Statics

A force may be represented by a straight line in a determined position, and its magnitude by the length of the straight line. The direction in which it acts may be indicated by an arrow.

Polygon of Forces   The parallelogram of two forces intersecting each other (see Figs. 3.1.4 and 3.1.5) leads directly to the graphic composition by means of the triangle of forces. In Fig. 3.1.10, R is called the **closing side,** and represents the resultant of the forces $\mathbf{F}_1$ and $\mathbf{F}_2$ in mag-

nitude and direction. Its position is given by the point of application O. By means of repeated use of the triangle of forces and by omitting the closing sides of the individual triangles, the magnitude and direction of the resultant $\mathbf{R}$ of any number of forces in the same plane and intersect-

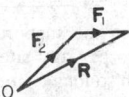

Fig. 3.1.10

ing at a single point can be found. In Fig. 3.1.11 the lines representing the forces start from point O, and in the force polygon (Fig. 3.1.12) they are joined in any order, the arrows showing their directions following around the polygon in the same direction. The magnitude of the resultant at the point of application of the forces is represented by the closing side $\mathbf{R}$ of the force polygon; its direction, as shown by the arrow, is counter to that in the other sides of the polygon.

If the forces are in equilibrium, R must equal zero, i.e., the force polygon must close.

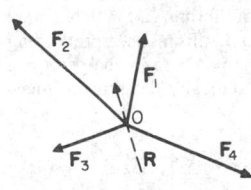

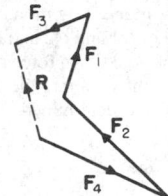

Fig. 3.1.11   **Fig. 3.1.12**

If in a closed polygon one of the forces is reversed in direction, this force becomes the resultant of all the others.

If the **forces** do **not** all lie **in the same plane,** the diagram becomes a polygon in space. The resultant $\mathbf{R}$ of this system may be obtained by adding the forces in space. The resultant is the vector which closes the space polygon. The space polygon may be projected onto three coordinate planes, giving three related plane polygons. Any two of these projections will involve all static equilibrium conditions and will be sufficient for a full description of the force system (see Fig. 3.1.13).

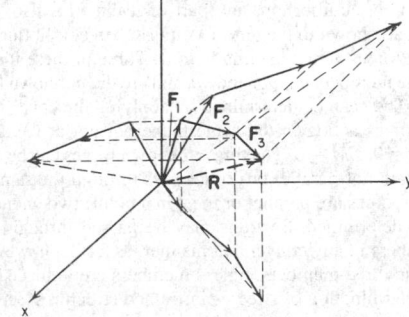

Fig. 3.1.13

Determination of Stresses in Members of a Statically Determinate Plane Structure with Loads at Rest

It will be assumed that the loads are applied at the joints of the structure, i.e., at the points where the different members are connected, and that the connections are pins with no friction. The stresses in the members must then be along lines connecting the pins, unless any member is loaded at more than two points by pin connections. If the members are straight, the forces exerted on them or by them must coincide with the axes of the members, In other words, there shall be no bending stresses in any of the members of the structure.

Equilibrium In order that the whole structure should be in equilibrium, it is necessary that the external forces (loads and supports) shall form a balanced system. Graphical and analytical methods are both of service.

Supporting Forces When the supporting forces are to be determined, it is not necessary to pay any attention to the makeup of the structure under consideration so long as it is practically rigid; the loads may be taken as they occur, or the resultant of the loads may be used instead. When the stresses in the members of the structure are being determined, the loads *must* be distributed at the joints where they belong.

Method of Joints When all the external forces have been determined, any joint at which there are not more than two unknown forces may be taken and these unknown forces determined by the methods of the stress polygon, resolution or moments. In Fig. 3.1.14, let O be the joint of a structure and $\mathbf{F}$ be the only known force; but let $O1$ and $O2$ be two members of the structure joined at O. Then the lines of action of the unknown forces are known and their magnitude may be determined (1) by a **stress polygon** which, for equilibrium, must close; (2) by resolution into H and V components, using the condition of equilibrium $\Sigma H = 0$, $\Sigma V = 0$; or (3) by moments, using any convenient point on the line of action of $O1$ and $O2$ and the condition of equilibrium $\Sigma M = 0$. No more than *two* unknown forces can be determined. In this manner, proceeding from joint to joint, the stresses in all the members of the truss can usually be determined if the structure is statically determinate internally.

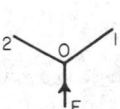

Fig. 3.1.14   **Fig. 3.1.15**

Method of Sections The structure may be divided into parts by passing a section through it cutting some of its members; one part may then be treated as a rigid body and the external forces acting upon it determined. Some of these forces will be the stresses in the members themselves. For example, let xx (Fig. 3.1.15) be a section taken through a truss loaded at P_1, P_2, and P_3, and supported on rollers at S. As the whole truss is in equilibrium, any part of it must be also, and consequently the part shown to the left of xx must be in equilibrium under the action of the forces acting externally to it. Three of these forces are the stresses in the members aa, bb, and bc, and are the unknown forces to be determined. They can be determined by applying the condition of equilibrium of forces acting in the same plane but not at the same point. $\Sigma H = 0$, $\Sigma V = 0$, $\Sigma M = 0$. The three unknown forces can be determined only if they are not parallel or do not pass through the same point; if, however, the forces are parallel or meet in a point, two unknown forces only can be determined. Sections may be passed through a structure cutting members in any convenient manner, as a rule, however, cutting not more than three members, unless members are unloaded.

For the determination of stresses in framed structures, see Sec. 12.2.

CENTER OF GRAVITY

Consider a three-dimensional body of any size, shape, and weight. If it is suspended as in Fig. 3.1.16 by a cord from any point A, it will be in equilibrium under the action of the tension in the cord and the resultant of the gravity or body forces W. If the experiment is repeated by suspending the body from point B, it will again be in equilibrium. If the lines of action of the resultant of the body forces were marked in each case, they would be concurrent at a point G known as the **center of**

gravity or **center of mass.** Whenever the density of the body is uniform, it will be a constant factor and like geometric shapes of different densities will have the same center of gravity. The term **centroid** is used in this case since the location of the center of gravity is of geometric concern only. If densities are nonuniform, like geometric shapes will have the same centroid but different centers of gravity.

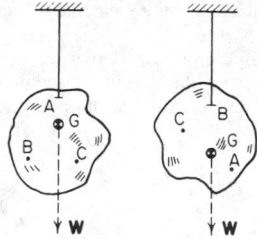

Fig. 3.1.16

Centroids of Technically Important Lines, Areas, and Solids

CENTROIDS OF LINES

Straight Lines The centroid is at its middle point.
Circular Arc AB (Fig. 3.1.17a) $x_0 = r \sin c/\mathrm{rad}\ c$; $y_0 = 2r \sin^2 \frac{1}{2}c/\mathrm{rad}$ c. (rad c = angle c measured in radians.)
Circular Arc AC (Fig. 3.1.17b) $x_0 = r \sin c/\mathrm{rad}\ c$; $y_0 = 0$.

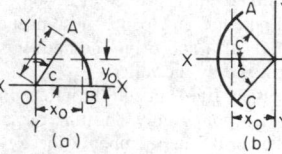

Fig. 3.1.17

Quadrant, AB (Fig. 3.1.18) $x_0 = y_0 = 2r/\pi = 0.6366r$.
Semicircumference, AC (Fig. 3.1.18) $y_0 = 2r/\pi = 0.6366r$; $x_0 = 0$.
Combination of Arcs and Straight Line (Fig. 3.1.19) AD and BC are two quadrants of radius r. $y_0 = \{(AB)r + 2[0.5\pi r(r - 0.6366r)]\} \div \{AB + 2(0.5\pi r)\}$.

Fig. 3.1.18   **Fig. 3.1.19**

CENTROIDS OF PLANE AREAS

Triangle Centroid lies at the intersection of the lines joining the vertices with the midpoints of the sides, and at a distance from any side equal to one-third of the corresponding altitude.
Parallelogram Centroid lies at the point of intersection of the diagonals.
Trapezoid (Fig. 3.1.20) Centroid lies on the line joining the middle points m and n of the parallel sides. The distances h_a and h_b are

$$h_a = h(a + 2b)/3(a + b) \qquad h_b = h(2a + b)/3(a + b)$$

Draw $BE = a$ and $CF = b$; EF will then intersect mn at centroid.
Any Quadrilateral The centroid of any quadrilateral may be determined by the general rule for areas, or graphically by dividing it into two sets of triangles by means of the diagonals. Find the centroid of each of the four triangles and connect the centroids of the triangles belonging to the same set. The intersection of these lines will be cen-

troid of area. Thus, in Fig. 3.1.21, O, O_1, O_2, and O_3 are, respectively, the centroids of the triangles ABD, ABC, BDC, and ACD. The intersection of O_1O_3 with OO_2 gives the centroids.

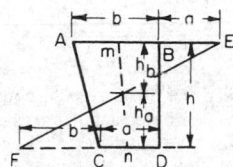

Fig. 3.1.20

Fig. 3.1.21

Segment of a Circle (Fig. 3.1.22) $x_0 = \frac{2}{3}r \sin^3 c/(\text{rad } c - \cos c \sin c)$. A segment may be considered to be a sector from which a triangle is subtracted, and the general rule applied.

Sector of a Circle (Fig. 3.1.23) $x_0 = \frac{2}{3}r \sin c/\text{rad } c$; $y_0 = \frac{4}{3}r \sin^2 \frac{1}{2}c/\text{rad } c$.

Semicircle $x_0 = \frac{4}{3}r/\pi = 0.4244r$; $y_0 = 0$.

Quadrant (90° sector) $x_0 = y_0 = \frac{4}{3}r/\pi = 0.4244r$.

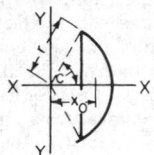

Fig. 3.1.22

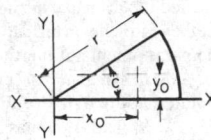

Fig. 3.1.23

Parabolic Half Segment (Fig. 3.1.24)   Area ABO: $x_0 = \frac{3}{5}x_1$; $y_0 = \frac{3}{8}y_1$.

Parabolic Spandrel (Fig. 3.1.24)   Area AOC: $x'_0 = \frac{3}{10}x_1$; $y'_0 = \frac{3}{4}y_1$.

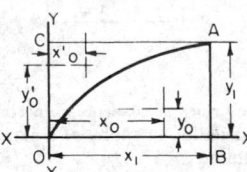

Fig. 3.1.24

Quadrant of an Ellipse (Fig. 3.1.25)   Area OAB: $x_0 = \frac{4}{3}(a/\pi)$; $y_0 = \frac{4}{3}(b/\pi)$.

The centroid of a figure such as that shown in Fig. 3.1.26 may be determined as follows: Divide the area $OABC$ into a number of parts by lines drawn perpendicular to the axis XX, e.g., 11, 22, 33, etc. These parts will be approximately either triangles, rectangles, or trapezoids. The area of each division may be obtained by taking the product of its

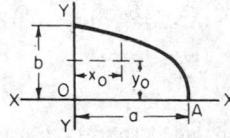

Fig. 3.1.25

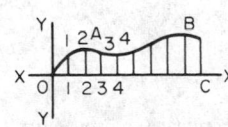

Fig. 3.1.26

mean height and its base. The centroid of each area may be obtained as previously shown. The sum of the moments of all the areas about XX and YY, respectively, divided by the sum of the areas will give approximately the distances from the center of gravity of the whole area to the axes XX and YY. The greater the number of areas taken, the more nearly exact the result.

CENTROIDS OF SOLIDS

Prism or Cylinder with Parallel Bases   The centroid lies in the center of the line connecting the centers of gravity of the bases.

Oblique Frustum of a Right Circular Cylinder (Fig. 3.1.27)   Let 1 2 3 4 be the plane of symmetry. The distance from the base to the centroid is $\frac{1}{2}h + (r^2 \tan^2 c)/8h$, where c is the angle of inclination of the oblique section to the base. The distance of the centroid from the axis of the cylinder is $r^2 \tan c/4h$.

Pyramid or Cone   The centroid lies in the line connecting the centroid of the base with the vertex and at a distance of one-fourth of the altitude above the base.

Truncated Pyramid   If h is the height of the truncated pyramid and A and B the areas of its bases, the distance of its centroid from the surface of A is

$$h(A + 2\sqrt{AB} + 3B)/4(A + \sqrt{AB} + B)$$

Truncated Circular Cone   If h is the height of the frustum and R and r the radii of the bases, the distance from the surface of the base whose radius is R to the centroid is $h(R^2 + 2Rr + 3r^2)/4(R^2 + Rr + r^2)$.

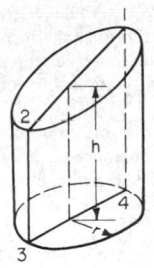

Fig. 3.1.27

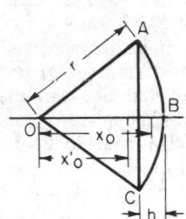

Fig. 3.1.28

Segment of a Sphere (Fig. 3.1.28)   Volume ABC: $x_0 = 3(2r - h)^2/4(3r - h)$.

Hemisphere $x_0 = 3r/8$.

Hollow Hemisphere $x_0 = 3(R^4 - r^4)/8(R^3 - r^3)$, where R and r are, respectively, the outer and inner radii.

Sector of a Sphere (Fig. 3.1.28)   Volume $OABCO$: $x'_0 = \frac{3}{8}(2r - h)$.

Ellipsoid, with Semiaxes a, b, and c   For each octant, distance from center of gravity to each of the bounding planes = $\frac{3}{8} \times$ length of semiaxis perpendicular to the plane considered.

The formulas given for the determination of the centroid of lines and areas can be used to determine the areas and volumes of surfaces and solids of revolution, respectively, by employing the theorems of Pappus, Sec. 2.1.

Determination of Center of Gravity of a Body by Experiment   The center of gravity may be determined by hanging the body up from different points and plumbing down; the point of intersection of the plumb lines will give the center of gravity. It may also be determined as shown in Fig. 3.1.29. The body is placed on knife-edges which rest on platform scales. The sum of the weights registered on the two scales $(w_1 + w_2)$ must equal the weight (w) of the body. Taking a moment axis at either end (say, O), $w_2A/w = x_0 =$ distance from O to plane containing the center of gravity.

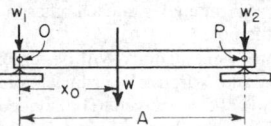

Fig. 3.1.29

Graphical Determination of the Centroids of Plane Areas   See Fig. 3.1.40.

MOMENT OF INERTIA

The **moment of inertia of a solid body** with respect to a given axis is the limit of the sum of the products of the masses of each of the elementary particles into which the body may be conceived to be divided and the square of their distance from the given axis.

If $dm = dw/g$ represents the mass of an elementary particle and y its distance from an axis, the moment of inertia I of the body about this axis will be $I = \int y^2\, dm = \int y^2\, dw/g$.

The moment of inertia may be expressed in weight units ($I_w = \int y^2\, dw$), in which case the moment of inertia in weight units, I_w, is equal to the moment of inertia in mass units, I, multiplied by g.

If $I = k^2 m$, the quantity k is called the **radius of gyration** or the **radius of inertia**.

If a body is considered to be composed of a number of parts, its moment of inertia about an axis is equal to the sum of the moments of inertia of the several parts about the same axis, or $I = I_1 + I_2 + I_3 + \cdots + I_n$.

The **moment of inertia of an area** with respect to a given axis is the limit of the sum of the products of the elementary areas into which the area may be conceived to be divided and the square of their distance (y) from the axis in question. $I = \int y^2\, dA = k^2 A$, where $k =$ **radius of gyration**.

The quantity $\int y^2\, dA$ is more properly referred to as the *second moment of area* since it is not a measure of *inertia* in a true sense.

Formulas for moments of inertia and radii of gyration of various areas follow later in this section.

Relation between the Moments of Inertia of an Area and a Solid The moment of inertia of a solid of elementary thickness about an axis is equal to the moment of inertia of the area of one face of the solid about the same axis multiplied by the mass per unit volume of the solid times the elementary thickness of the solid.

Moments of Inertia about Parallel Axes The moment of inertia of an area or solid about any given axis is equal to the moment of inertia about a parallel axis through the center of gravity plus the square of the distance between the two axes times the area or mass.

In Fig. 3.1.30a, the moment of inertia of the area $ABCD$ about axis YY is equal to I_0 (or the moment of inertia about $Y_0 Y_0$ through the center of gravity of the area and parallel to YY) plus $x_0^2 A$, where $A =$ area of $ABCD$. In Fig. 3.1.30b, the moment of inertia of the mass m about $YY = I_0 + x_0^2 m$. $Y_0 Y_0$ passes through the centroid of the mass and is parallel to YY.

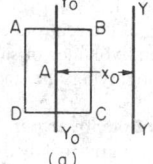 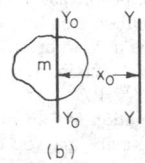

(a) (b)

Fig. 3.1.30

Polar Moment of Inertia The polar moment of inertia (Fig. 3.1.31) is taken about an axis perpendicular to the plane of the area. Referring to Fig. 3.1.31, if I_y and I_x are the moments of inertia of the area A about YY and XX, respectively, then the polar moment of inertia $I_p = I_x + I_y$, or the polar moment of inertia is equal to the sum of the moments of inertia about any two axes at right angles to each other in the plane of the area and intersecting at the pole.

Product of Inertia This quantity will be represented by I_{xy}, and is $\int\int xy\, dy\, dx$, where x and y are the coordinates of any elementary part into which the area may be conceived to be divided. I_{xy} may be positive or negative, depending upon the position of the area with respect to the coordinate axes XX and XY.

Relation between Moments of Inertia about Axes Inclined to Each Other Referring to Fig. 3.1.32, let I_y and I_x be the moments of inertia of the area A about YY and XX, respectively, I'_y and I'_x the moments about $Y'Y'$ and $X'X'$, and I_{xy} and I'_{xy} the products of inertia for XX and YY, and

$X'X'$ and $Y'Y'$, respectively. Also, let c be the angle between the respective pairs of axes, as shown. Then,

$$I'_y = I_y \cos^2 c + I_x \sin^2 c + I_{xy} \sin 2c$$
$$I'_x = I_x \cos^2 c + I_y \sin^2 c - I_{xy} \sin 2c$$

$$I'_{xy} = \frac{I_x - I_y}{2} \sin 2c + I_{xy} \cos 2c$$

Principal Moments of Inertia In every plane area, a given point being taken as the origin, there is at least one pair of rectangular axes in

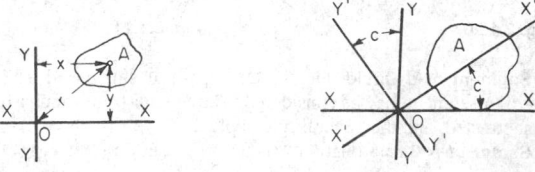

Fig. 3.1.31   **Fig. 3.1.32**

the plane of the area about one of which the moment of inertia is a maximum, and a minimum about the other. These moments of inertia are called the **principal moments of inertia**, and the axes about which they are taken are the **principal axes of inertia**. One of the conditions for principal moments of inertia is that the product of inertia I_{xy} shall equal zero. **Axes of symmetry** of an area are always principal axes of inertia.

Relation between Products of Inertia and Parallel Axes In Fig. 3.1.33, $X_0 X_0$ and $Y_0 Y_0$ pass through the center of gravity of the area parallel to the given axes XX and YY. If I_{xy} is the product of inertia for XX and YY, and $I_{x_0 y_0}$ that for $X_0 X_0$ and $Y_0 Y_0$, then $I_{xy} = I_{x_0 y_0} + abA$.

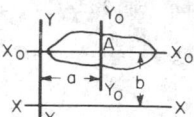

Fig. 3.1.33

Mohr's Circle The **principal moments of inertia** and the location of the **principal axes of inertia** for any point of a plane area may be established graphically as follows.

Given at any point A of a plane area (Fig. 3.1.34), the moments of inertia I_x and I_y about axes X and Y, and the product of inertia I_{xy} relative to X and Y. The graph shown in Fig. 3.1.34b is plotted on rectangular coordinates with moments of inertia as abscissas and products of inertia

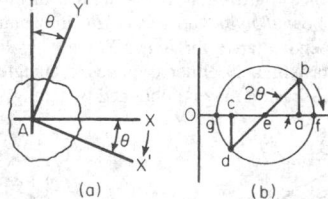

(a)   (b)

Fig. 3.1.34

as ordinates. Lay out $Oa = I_x$ and $ab = I_{xy}$ (upward for positive products of inertia, downward for negative). Lay out $Oc = I_y$ and $cd = negative$ *of* I_{xy}. Draw a circle with bd as diameter. This is **Mohr's circle**. The *maximum* moment of inertia is $I'_x = Of$; the *minimum* moment of inertia is $I'_y = Og$. The principal axes of inertia are located as follows. From axis AX (Fig. 3.1.34a) lay out angular distance $\theta = \frac{1}{2} < bef$. This locates axis AX', one principal axis ($I'_x = Of$). The other principal axis of inertia is AY', perpendicular to AX' ($I'_y = Og$).

The **moment of inertia of any area** may be considered to be made up of the sum or difference of the known moments of inertia of simple figures. For example, the dimensioned figure shown in Fig. 3.1.35 represents the section of a rolled shape with hole *oprs* and may be divided into the semicircle *abc,* rectangle *edkg,* and triangles *mfg* and *hkl,* from which the rectangle *oprs* is to be subtracted. Referring to axis *XX,*

$$I_{xx} = \pi 4^4/8 \text{ for semicircle } abc = (2 \times 11^3)/3 \text{ for rectangle } edkg$$
$$= 2[(5 \times 3^3)/36 + 10^2(5 \times 3)/2] \text{ for the two triangles } mfg$$
$$\text{and } hkl$$

From the sum of these there is to be subtracted $I_{xx} = [(2 \times 3^2)/12 + 4^2(2 \times 3)]$ for the rectangle *oprs.*

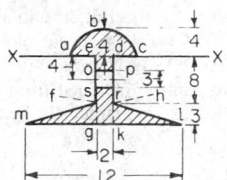

Fig. 3.1.35

If the moment of inertia of the whole area is required about an axis parallel to *XX,* but passing through the center of gravity of the whole area, $I_0 = I_{xx} - x_0^2 A$, where $x_0 =$ distance from *XX* to center of gravity. The **moments of inertia of built-up sections** used in structural work may be found in the same manner, the moments of inertia of the different **rolled sections** being given in Sec. 12.2.

Moments of Inertia of Solids For moments of inertia of solids about parallel axes, $I_x = I_0 + x_0^2 m$.

Moment of Inertia with Reference to Any Axis Let a mass particle *dm* of a body have *x, y,* and *z* as coordinates, *XX, YY,* and *ZZ* being the coordinate axes and *O* the origin. Let *X'X'* be any axis passing through the origin and making angles of *A, B,* and *C* with *XX, YY,* and *ZZ,* respectively. The moment of inertia with respect to this axis then becomes equal to

$$I'_x = \cos^2 A \int (y^2 + z^2) \, dm + \cos^2 B \int (z^2 + x^2) \, dm$$
$$+ \cos^2 C \int (x^2 + y^2) \, dm - 2 \cos B \cos C \int yz \, dm$$
$$- 2 \cos C \cos A \int zx \, dm - 2 \cos A \cos B \int xy \, dm$$

Let the moment of inertia about $XX = I_x = \int (y^2 + z^2) \, dm$, about $YY = I_y = \int (z^2 + x^2) \, dm$, and about $ZZ = I_z = \int (x^2 + y^2) \, dm$. Let the products of inertia about the three coordinate axes be

$$I_{yz} = \int yz \, dm \qquad I_{zx} = \int zx \, dm \qquad I_{xy} = \int xy \, dm$$

Then the moment of inertia I'_x becomes equal to

$$I_x \cos^2 A + I_y \cos^2 B + I_z \cos^2 C - 2I_{yz} \cos B \cos C - 2I_{zx}$$
$$\cos C \cos A - 2I_{xy} \cos A \cos B$$

The moment of inertia of any solid may be considered to be made up of the sum or difference of the moments of inertia of simple solids of which the moments of inertia are known.

Moments of Inertia of Important Solids (Homogeneous)

$m = w/g =$ mass per unit of volume of the body
$M = W/g =$ total mass of body
$r =$ radius
$I =$ moment of inertia (mass units)
$I_w = I \times g =$ moment of inertia (weight units)

Solid circular cylinder about its axis: $I = \pi r^4 ma/2 = Mr^2/2$. ($a =$ length of axis of cylinder.)

Solid circular cylinder about an axis through the center of gravity and perpendicular to axis of cylinder: $I = M[r^2 + (a^2/3)]/4$.

Hollow circular cylinder about its axis: $I = \pi ma(r_1^4 - r_2^4)/2$. ($r_1$ and $r_2 =$ outer and inner radii; $a =$ length.)

Thin hollow circular cylinder about its axis: $I = Mr^2$.

Solid sphere about a diameter: $I = 8m\pi r^5/15 = 2Mr^2/5$.

Thin hollow sphere about a diameter: $I = 2Mr^2/3$.

Thick hollow sphere about a diameter: $I = 8m\pi (r_1^5 - r_2^5)/15$. ($r_1$ and r_2 are outer and inner radii.)

Rectangular prism about an axis through center of gravity and perpendicular to a face whose dimensions are *a* and *b*: $I = M(a^2 + b^2)/12$.

Solid right circular cone about an axis through its apex and perpendicular to its axis: $I = 3M[(r^2/4) + h^2]/5$. ($h =$ altitude of cone, $r =$ radius of base.)

Solid right circular cone about its axis of revolution: $I = 3Mr^2/10$.

Ellipsoid with semiaxes *a, b,* and *c*: *I* about diameter $2c$ (*z* axis) $= 4m\pi abc (a^2 + b^2)/15$. [Equation of ellipsoid: $(x^2/a^2) + (y^2/b^2) + (z^2/c^2) = 1$.]

Ring with Circular Section (Fig. 3.1.36) $I_{yy} = \frac{1}{2}m\pi^2 Ra^2(4R^2 + 3a^2)$; $I_{xx} = m\pi^2 Ra^2[R^2 + (5a^2/4)]$.

Fig. 3.1.36   **Fig. 3.1.37**

Approximate Moments of Inertia of Solids In order to determine the moment of inertia of a solid, it is necessary to know all its dimensions. In the case of a rod of mass *M* (Fig. 3.1.37) and length *l,* with shape and size of the cross section unknown, making the approximation that the weight is all concentrated along the axis of the rod, the moment of inertia about *YY* will be $I_{yy} = \int_0^l (M/l)x^2 \, dx = Ml^2/3$.

A **thin plate** may be treated in the same way (Fig. 3.1.38): $I_{yy} = \int_0^l (M/l)x^2 \, dx$. Here the mass of the plate is assumed concentrated at its middle layer.

Thin Ring, or Cylinder (Fig. 3.1.39) Assume the mass *M* of the ring or cylinder to be concentrated at a distance *r* from *O.* The moment of inertia about an axis through *O* perpendicular to plane of ring or along the axis of the cylinder will be $I = Mr^2$; this will be greater than the exact moment of inertia, and *r* is sometimes taken as the distance from *O* to the center of gravity of the cross section of the rim.

Fig. 3.1.38   **Fig. 3.1.39**

Flywheel Effect The moment of inertia of a solid is often called flywheel effect in the solution of problems dealing with rotating bodies, and is usually expressed in lb · ft² (I_w).

Graphical Determination of the Centroids and Moments of Inertia of Plane Areas Required to find the center of gravity of the area *MNP* (Fig. 3.1.40) and its moment of inertia about any axis *XX.*

Draw any line *SS* parallel to *XX* and at a distance *d* from it. Draw a number of lines such as *AB* and *EF* across the figure parallel to *XX.* From *E* and *F* draw *ER* and *FT* perpendicular to *SS.* Select as a pole any

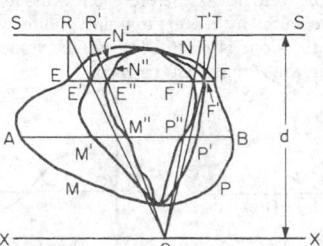

Fig. 3.1.40

point on *XX*, preferably the point nearest the area, and draw *OR* and *OT*, cutting *EF* at *E'* and *F'*. If the same construction is repeated, using other lines parallel to *XX*, a number of points will be obtained, which, if connected by a smooth curve, will give the area *M'N'P'*. Project *E'* and *F'* onto *SS* by lines *E'R'* and *F'T'*. Join *F'* and *T'* with *O*, obtaining *E''* and *F''*; connect the points obtained using other lines parallel to *XX* and obtain an area *M''N''P''*. The area $M'N'P' \times d$ = moment of area *MNP* about the line *XX*, and the distance from *XX* to the centroid *MNP* = area $M'N'P' \times d$/area *MNP*. Also, area $M''N''P'' \times d^2$ = moment of inertia of *MNP* about *XX*. The areas *M'N'P'* and *M''N''P''* can best be obtained by use of a planimeter.

KINEMATICS

Kinematics is the study of the motion of bodies without reference to the forces causing that motion or the mass of the bodies.

The **displacement** of a point is the directed distance that a point has moved on a geometric path from a convenient origin. It is a **vector**, having both magnitude and direction, and is subject to all the laws and characteristics attributed to vectors. In Fig. 3.1.41, the displacement of the point *A* from the origin *O* is the directed distance *O* to *A*, symbolized by the vector **s**.

The **velocity** of a point is the time rate of change of displacement, or $\mathbf{v} = d\mathbf{s}/dt$.

The **acceleration** of a point is the time rate of change of velocity, or $\mathbf{a} = d\mathbf{v}/dt$.

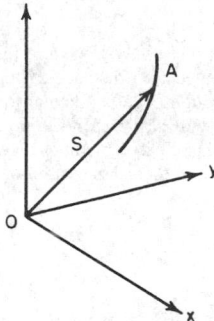

Fig. 3.1.41

The kinematic definitions of velocity and acceleration involve the four variables, *displacement, velocity, acceleration,* and *time*. If we eliminate the variable of time, a third equation of motion is obtained, $ds/v = dt = dv/a$. This differential equation, together with the definitions of velocity and acceleration, make up the *three kinematic equations* of motion, $v = ds/dt$, $a = dv/dt$, and $a\,ds = v\,dv$. These differential equations are usually limited to the scalar form when expressed together, since the last can only be properly expressed in terms of the scalar *dt*. The first two, since they are definitions for velocity and acceleration, are **vector equations**.

A **space-time curve** offers a convenient means for the study of the motion of a point. The **slope of the curve** at any point will represent the **velocity** at that time. In Fig. 3.1.42*a* the slope is constant, as the graph is a straight line; the velocity is therefore uniform. In Fig. 3.1.42*b* the slope of the curve varies from point to point, and the velocity must also vary. At *p* and *q* the slope is zero; therefore, the velocity of the point at the corresponding times must also be zero.

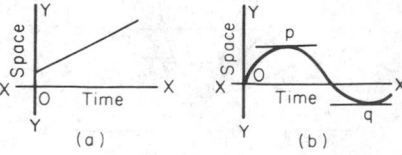

Fig. 3.1.42

A **velocity-time curve** offers a convenient means for the study of acceleration. The **slope of the curve** at any point will represent the **acceleration** at that time. In Fig. 3.1.43*a* the slope is constant; so the acceleration must be constant. In the case represented by the full line, the acceleration is positive; so the velocity is increasing. The dotted line shows a negative acceleration and therefore a decreasing velocity. In Fig. 3.1.43*b* the slope of the curve varies from point to point; so the acceleration must also vary. At *p* and *q* the slope is zero; therefore, the acceleration of the point at the corresponding times must also be zero. The area under the velocity-time curve between any two ordinates such as *NL* and *HT* will represent the distance moved in time interval *LT*. In the case of the uniformly accelerated motion shown by the full line in Fig. 3.1.43*a*, the area *LNHT* is ½(*NL* + *HT*) × (*OT* − *OL*) = mean velocity multiplied by the time interval = space passed over during this time interval. In Fig. 3.1.43*b* the mean velocity can be obtained from the equation of the curve by means of the calculus, or graphically by approximation of the area.

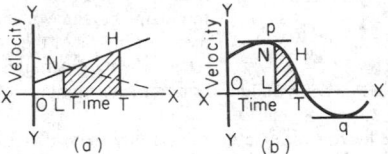

Fig. 3.1.43

An **acceleration-time curve** (Fig. 3.1.44) may be constructed by plotting accelerations as ordinates, and times as abscissas. The area under this curve between any two ordinates will represent the total increase in velocity during the time interval. The area *ABCD* represents the total increase in velocity between time t_1 and time t_2.

General Expressions Showing the Relations between Space, Time, Velocity, and Acceleration for Rectilinear Motion

SPECIAL MOTIONS

Uniform Motion If the **velocity** is **constant**, the **acceleration** must be **zero**, and the point has uniform motion. The space-time curve becomes a straight line inclined toward the time axis (Fig. 3.1.42*a*). The velocity-time curve becomes a straight line parallel to the time axis. For this motion $a = 0$, v = constant, and $s = s_0 + vt$.

Uniformly Accelerated or Retarded Motion If the **velocity** is **not uniform** but the **acceleration** is **constant**, the point has uniformly accelerated motion; the acceleration may be either positive or negative. The space-time curve becomes a parabola and the velocity-time curve becomes a straight line inclined toward the time axis (Fig. 3.1.43*a*). The acceleration-time curve becomes a straight line parallel to the time axis. For this motion a = constant, $v = v_0 + at$, $s = s_0 + v_0 t + \frac{1}{2}at^2$.

If the point starts from rest, $v_0 = 0$. Care should be taken concerning the sign + or − for acceleration.

Composition and Resolution of Velocities and Acceleration

Resultant Velocity A velocity is said to be the resultant of two other velocities when it is represented by a vector that is the geometric sum of the vectors representing the other two velocities. This is the **parallelogram of motion**. In Fig. 3.1.45, **v** is the resultant of $\mathbf{v}_1$ and $\mathbf{v}_2$ and is

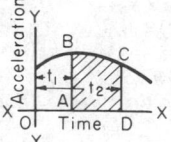

Fig. 3.1.44

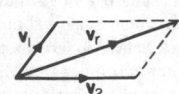

Fig. 3.1.45

represented by the diagonal of a parallelogram of which $\mathbf{v}_1$ and $\mathbf{v}_2$ are the sides; or it is the third side of a triangle of which $\mathbf{v}_1$ and $\mathbf{v}_2$ are the other two sides.

Polygon of Motion The parallelogram of motion may be extended to the polygon of motion. Let $\mathbf{v}_1$, $\mathbf{v}_2$, $\mathbf{v}_3$, $\mathbf{v}_4$ (Fig. 3.1.46a) show the directions of four velocities imparted in the same plane to point O. If the lines $\mathbf{v}_1$, $\mathbf{v}_2$, $\mathbf{v}_3$, $\mathbf{v}_4$ (Fig. 3.1.46b) are drawn parallel to and proportional to the velocities imparted to point O, $\mathbf{v}$ will represent the resultant velocity imparted to O. It will make no difference in what order the velocities are taken in constructing the motion polygon. As long as the arrows showing the direction of the motion follow each other in order about the polygon, the resultant velocity of the point will be represented in magnitude by the closing side of the polygon, but opposite in direction.

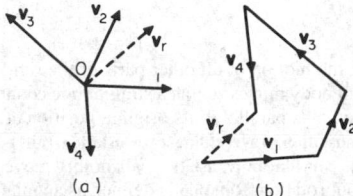

Fig. 3.1.46

Resolution of Velocities Velocities may be resolved into **component velocities** in the same plane, as shown by Fig. 3.1.47. Let the velocity of point O be v_r. In Fig. 3.1.47a this velocity is resolved into two components in the same plane as v_r and at right angles to each other.

$$v_r = \sqrt{(v_1)^2 + (v_2)^2}$$

In Fig. 3.1.47b the components are in the same plane as v_r, but are not at right angles to each other. In this case,

$$v_r = \sqrt{(v_1)^2 + (v_2)^2 + 2v_1 v_2 \cos B}$$

If the components v_1 and v_2 and angle B are known, the direction of v_r can be determined. $\sin bOc = (v_1/v_r) \sin B$. $\sin cOa = (v_2/v_r) \sin B$. Where v_1 and v_2 are at right angles to each other, $\sin B = 1$.

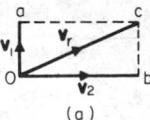

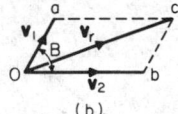

Fig. 3.1.47

Resultant Acceleration Accelerations may be combined and resolved in the same manner as velocities, but in this case the lines or vectors represent accelerations instead of velocities. If the acceleration had components of magnitude a_1 and a_2, the magnitude of the resultant acceleration would be $a = \sqrt{(a_1)^2 + (a_2)^2 + 2a_1 a_2 \cos B}$, where B is the angle between the vectors a_1 and a_2.

Curvilinear Motion in a Plane

The linear velocity $v = ds/dt$ of a point in curvilinear motion is the same as for rectilinear motion. Its direction is tangent to the path of the point. In Fig. 3.1.48a, let $P_1 P_2 P_3$ be the path of a moving point and $\mathbf{V}_1$, $\mathbf{V}_2$, $\mathbf{V}_3$ represent its velocity at points P_1, P_2, P_3, respectively. If O is taken as a pole (Fig. 3.1.48b) and vectors $\mathbf{V}_1$, $\mathbf{V}_2$, $\mathbf{V}_3$ representing the velocities of the point at P_1, P_2, and P_3, are drawn, the curve connecting the terminal points of these vectors is known as the **hodograph** of the motion. This velocity diagram is applicable only to motions all in the same plane.

Acceleration Tangents to the curve (Fig. 3.1.48b) indicate the directions of the **instantaneous velocities**. The direction of the tangents does not, as a rule, coincide with the direction of the accelerations as represented by tangents to the path. If the acceleration a at some point in the

path is resolved by means of a parallelogram into components tangent and normal to the path, the normal acceleration $a_n = v^2/\rho$, where ρ = radius of curvature of the path at the point in question, and the tangential acceleration $a_t = dv/dt$, where v = velocity tangent to the path at the same point. $a = \sqrt{a_n^2 + a_t^2}$. The normal acceleration is constantly directed toward the center of the path.

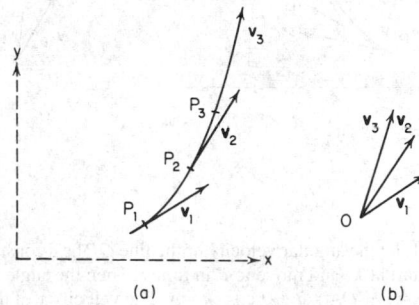

Fig. 3.1.48

EXAMPLE. Figure 3.1.49 shows a point moving in a curvilinear path. At p_1 the velocity is $\mathbf{v}_1$; at p_2 the velocity is $\mathbf{v}_2$. If these velocities are drawn from pole O (Fig. 3.1.49b), $\Delta\mathbf{v}$ will be the difference between $\mathbf{v}_2$ and $\mathbf{v}_1$. The acceleration during travel $p_1 p_2$ will be $\Delta\mathbf{v}/\Delta t$, where Δt is the time interval. The approximation becomes closer to instantaneous acceleration as shorter intervals Δt are employed.

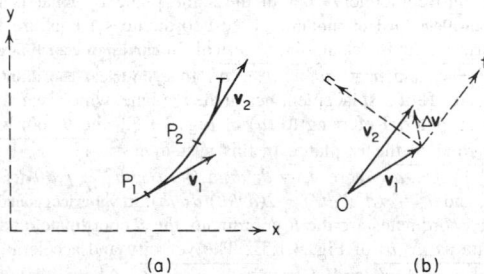

Fig. 3.1.49

The acceleration $\Delta\mathbf{v}/\Delta t$ can be resolved into normal and tangential components leading to $\mathbf{a}_n = \Delta\mathbf{v}_n/\Delta t$, normal to the path, and $\mathbf{a}_r = \Delta\mathbf{v}_p/\Delta t$, tangential to the path.

Velocity and *acceleration* may be expressed in **polar coordinates** such that $v = \sqrt{v_r^2 + v_\theta^2}$ and $a = \sqrt{a_r^2 + a_\theta^2}$. Figure 3.1.50 may be used to explain the r and θ coordinates.

EXAMPLE. At P_1 the velocity is $\mathbf{v}_1$, with components $\mathbf{v}_{1r}$ in the r direction and $\mathbf{v}_{1\theta}$ in the θ direction. At P_2 the velocity is $\mathbf{v}_2$, with components $\mathbf{v}_{2r}$ in the r direction and $\mathbf{v}_{2\theta}$ in the θ direction. It is evident that the difference in velocities $\mathbf{v}_2 - \mathbf{v}_1 = \Delta\mathbf{v}$ will have components $\Delta\mathbf{v}_r$ and $\Delta\mathbf{v}_\theta$, giving rise to accelerations $\mathbf{a}_r$ and $\mathbf{a}_\theta$ in a time interval Δt.

In polar coordinates, $v_r = dr/dt$, $a_r = d^2r/dt^2 - r(d\theta/dt)^2$, $v_\theta = r(d\theta/dt)$, and $a_\theta = r(d^2\theta/dt^2) + 2(dr/dt)(d\theta/dt)$.

If a point P moves on a circular path of radius r with an angular velocity of ω and an angular acceleration of α, the linear velocity of the point P is $v = \omega r$ and the two components of the linear acceleration are $a_n = v^2/r = \omega^2 r = v\omega$ and $a_t = \alpha r$.

If the angular velocity is constant, the point P travels equal circular paths in equal intervals of time. The projected displacement, velocity, and acceleration of the point P on the x and y axes are sinusoidal functions of time, and the motion is said to be **harmonic motion.** Angular velocity is usually expressed in radians per second, and when the number (N) of revolutions traversed per minute (r/min) by the point P is known, the angular velocity of the radius r is $\omega = 2\pi N/60 = 0.10472N$.

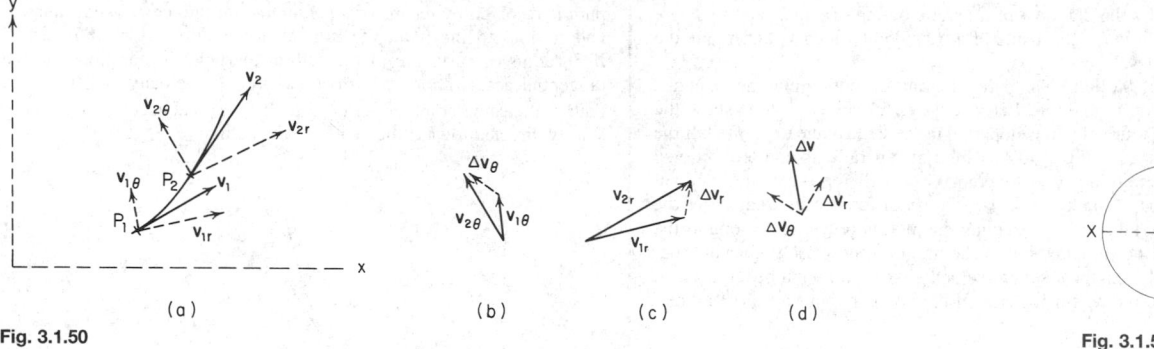

Fig. 3.1.50

(a)   (b)   (c)   (d)

Fig. 3.1.51

In Fig. 3.1.51, let the angular velocity of the line OP be a constant ω. Let the point P start at X' and move to P in time t. Then the angle $\theta = \omega t$. If $OP = r$, $X'A = r - OA = r - r\cos \omega t = s$. The velocity V of the point A on the x axis will equal $ds/dt = \omega r \sin \omega t$, and the acceleration $a = dv/dt = -\omega^2 r \cos \omega t$. The period τ is the time necessary for the point P to complete one cycle of motion $\tau = 2\pi/\omega$, and it is also equal to the time necessary for A to complete a full cycle on the x axis from X' to X and return.

Curvilinear Motion in Space

If **three dimensions** are used, velocities and accelerations may be resolved into components not in the same plane by what is known as the **parallelepiped of motion.** Three coordinate systems are widely used, cartesian, cylindrical, and spherical. In **cartesian coordinates,** $v = \sqrt{v_x^2 + v_y^2 + v_z^2}$ and $a = \sqrt{a_x^2 + a_y^2 + a_z^2}$. In **cylindrical coordinates,** the radius vector $\mathbf{R}$ of displacement lies in the rz plane, which is at an angle with the xz plane. Referring to (a) of Fig. 3.1.52, the θ coordinate is perpendicular to the rz plane. In this system $v = \sqrt{v_r^2 + v_\theta^2 + v_z^2}$ and $a = \sqrt{a_r^2 + a_\theta^2 + a_z^2}$ where $v_r = dr/dt$, $a_r = d^2r/dt^2 - r(d\theta/dt)^2$, $v_\theta = r(d\theta/dt)$, and $a_\theta = r(d^2\theta/dt^2) + 2(dr/dt)(d\theta/dt)$. In **spherical coordinates,** the three coordinates are the R coordinate, the θ coordinate, and the ϕ coordinate as in (b) of Fig. 3.1.52. The velocity and acceleration are $v = \sqrt{v_R^2 + v_\theta^2 + v_\phi^2}$ and $a = \sqrt{a_R^2 + a_\theta^2 + a_\phi^2}$, where $v_R = dR/dt$, $v_\phi = R(d\phi/dt)$, $v_\theta = R \cos \phi(d\theta/dt)$, $a_R = d^2R/dt^2 - R(d\phi/dt)^2 - R \cos^2 \phi(d\theta/dt)^2$, $a_\phi = R(d^2\phi/dt^2) + R \cos \phi \sin \phi (d\theta/dt)^2 + 2(dR/dt)(d\phi/dt)$, and $a_\theta = R \cos \phi (d^2\theta/dt^2) + 2[(dR/dt) \cos \phi - R \sin \phi (d\phi/dt)] d\theta/dt$.

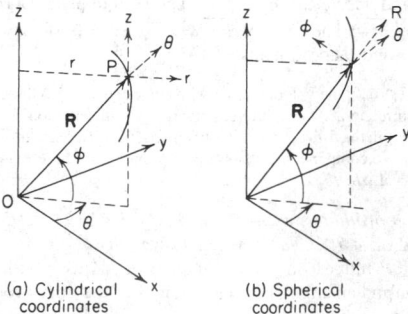

(a) Cylindrical coordinates   (b) Spherical coordinates

Fig. 3.1.52

Motion of Rigid Bodies

A body is said to be **rigid** when the distances between all its particles are invariable. Theoretically, rigid bodies do not exist, but materials used in engineering are rigid under most practical working conditions. The motion of a rigid body can be completely described by knowing the **angular motion** of a line on the rigid body and the **linear motion** of a point on this line and relating the motion of all other parts of the rigid body to these motions. If a rigid body moves so that a straight line connecting any two of its particles remains parallel to its original position at all times, it is said to have **translation.** In **rectilinear translation,** all points move in straight lines. In **curvilinear translation,** all points move on congruent curves but without rotation. **Rotation** is defined as angular motion about an axis, which may or may not be fixed. Rigid body motion in which the paths of all particles lie on parallel planes is called **plane motion.**

Angular Motion

Angular displacement is the change in angular position of a given line as measured from a convenient reference line. In Fig. 3.1.53, consider the motion of the line AB as it moves from its original position $A'B'$. The angle between lines AB and $A'B'$ is the angular displacement of line **AB,** symbolized as θ. It is a directed quantity and is a vector. The usual notation used to designate angular displacement is a vector normal to

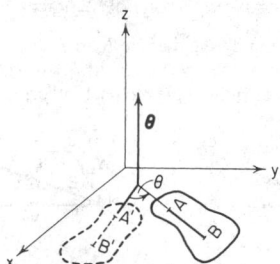

Fig. 3.1.53

the plane in which the angular displacement occurs. The length of the vector is proportional to the magnitude of the angular displacement. For a rigid body moving in three dimensions, the line AB may have angular motion about any three orthogonal axes. For example, the angular displacement can be described in cartesian coordinates as $\theta = \theta_x + \theta_y + \theta_z$, where $\theta = \sqrt{\theta_x^2 + \theta_y^2 + \theta_z^2}$.

Angular velocity is defined as the time rate of change of angular displacement, $\omega = d\theta/dt$. Angular velocity may also have components about any three orthogonal axes.

Angular acceleration is defined as the time rate of change of angular velocity, $\alpha = d\omega/dt = d^2\theta/dt^2$. Angular acceleration may also have components about any three orthogonal axes.

The *kinematic equations* of angular motion of a line are analogous to those for the motion of a point. In referring to Table 3.1.1, $\omega = d\theta/dt$ $\alpha = d\omega/dt$, and $\alpha d\theta = \omega d\omega$. Substitute θ for s, ω for v, and α for a.

Motion of a Rigid Body in a Plane

Plane motion is the motion of a rigid body such that the paths of all particles of that rigid body lie on parallel planes.

Table 3.1.1

Variables	$s = f(t)$	$v = f(t)$	$a = f(t)$	$a = f(s,v)$
Displacement	$s = s_0 + \int_{t_0}^{t} v\,dt$	$s = s_0 + \int_{t_0}^{t}\int_{t_0}^{t} a\,dt\,dt$	$s = s_0 + \int_{v_0}^{v} (v/a)\,dv$
		$v = v_0 + \int_{t_0}^{t} a\,dt$	$\int_{v_0}^{v} v\,dv = \int_{s}^{s_0} a\,ds$
Velocity	$v = ds/dt$		
Acceleration	$a = d^2s/dt^2$	$a = dv/dt$		$a = v\,dv/ds$

Instantaneous Axis When the axis about which any body may be considered to rotate changes its position, any one position is known as an instantaneous axis, and the line through all positions of the instantaneous axis as the **centrode**.

When the velocity of two points in the same plane of a rigid body having plane motion is known, the instantaneous axis for the body will be at the intersection of the lines drawn from each point and perpendicular to its velocity. See Fig. 3.1.54, in which A and B are two points on the rod AB, v_1 and v_2 representing their velocities. O is the instantaneous axis for AB; therefore point C will have velocity shown in a line perpendicular to OC.

Linear velocities of points in a body rotating about an instantaneous axis are proportional to their distances from this axis. In Fig. 3.1.54, $v_1 : v_2 : v_3 = AO : OB : OC$. If the velocities of A and B were parallel, the lines OA and OB would also be parallel and there would be no instantaneous axis. The motion of the rod would be translation, and all points would be moving with the same velocity in parallel straight lines.

If a body has plane motion, the components of the velocities of any two points in the body along the straight line joining them must be equal. Ax must be equal to By and Cz in Fig. 3.1.54.

EXAMPLE. In Fig. 3.1.55a, the velocities of points A and B are known—they are v_1 and v_2, respectively. To find the instantaneous axis of the body, perpendiculars AO and BO are drawn. O, at the intersection of the perpendiculars, is the **instantaneous axis** of the body. To find the velocity of any other point, like C, line OC is drawn and v_3 erected perpendicular to OC with magnitude equal to v_1 (CO/AO). The **angular velocity** of the body will be $\omega = v_1/AO$ or v_2/BO or v_3/CO. The instantaneous axis of a wheel rolling on a rack without slipping (Fig. 3.1.55b) lies at the point of contact O, which has zero linear velocity. All points of the wheel will have velocities perpendicular to radii to O and proportional in magnitudes to their respective distances from O.

Another way to describe the plane motion of a rigid body is with the use of **relative motion**. In Fig. 3.1.56 the velocity of point A is v_1. The angular velocity of the line AB is v_1/r_{AB}. The velocity of B relative to A is $\omega_{AB} \times r_{AB}$. Point B is considered to be moving on a circular path around A as a center. The direction of relative velocity of B to A would be tangent to the circular path in the direction that ω_{AB} would make B move. The velocity of B is the vector sum of the velocity A added to the velocity of B relative to A, $v_B = v_A + v_{B/A}$.

The **acceleration** of B is the vector sum of the acceleration of A added to the acceleration of B relative to A, $a_B = a_A + a_{B/A}$. Care must be taken 0to include the complete relative acceleration of B to A. If B is considered to move on a circular path about A, with a velocity relative to A, it will have an acceleration relative to A that has both normal and tangential components: $a_{B/A} = (a_{B/A})_n + (a_{B/A})_t$.

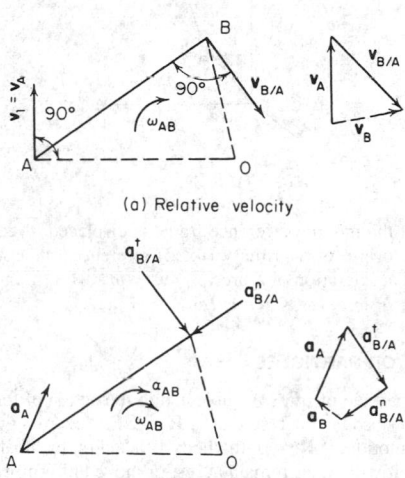

(a) Relative velocity

(b) Relative acceleration

Fig. 3.1.56

If B is a point on a path which lies on the same rigid body as the line AB, a **particle P traveling on the path** will have a velocity v_P at the instant P passes over point B such that $v_P = v_A + v_{B/A} + v_{P/B}$, where the velocity $v_{P/B}$ is the velocity of P relative to path B.

The particle P will have an acceleration a_P at the instant P passes over the point B such that $a_P = a_A + a_{B/A} + a_{P/B} + 2\omega_{AB} \times v_{P/B}$. The term $a_{P/B}$ is the acceleration of P relative to the path at point B. The last term $2\omega_{AB} \, v_{P/B}$ is frequently referred to as the **coriolis acceleration**. The direction is always normal to the path in a sense which would rotate the head of the vector $v_{P/B}$ about its tail in the direction of the angular velocity of the rigid body ω_{AB}.

EXAMPLE. In Fig. 3.1.57, arm AB is rotating counterclockwise about A with a constant angular velocity of 38 r/min or 4 rad/s, and the slider moves outward with a velocity of 10 ft/s (3.05 m/s). At an instant when the slider P is 30 in (0.76 m) from the center A, the acceleration of the slider will have two components. One component is the normal acceleration directed toward the center A. Its magnitude is $\omega^2 r = 4^2 (30/12) = 40$ ft/s² [$\omega^2 r = 4^2 (0.76) = 12.2$ m/s²]. The second is the coriolis acceleration directed normal to the arm AB, upward and to the left. Its magnitude is $2\omega v = 2(4)(10) = 80$ ft/s² [$2\omega v = 2(4)(3.05) = 24.4$ m/s²].

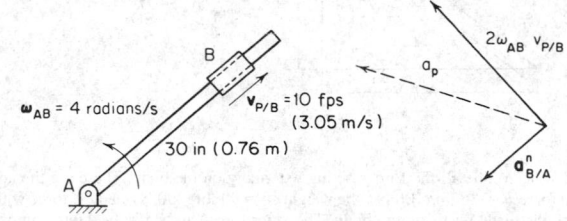

Fig. 3.1.57

General Motion of a Rigid Body

The general motion of a point moving in a coordinate system which is itself in motion is complicated and can best be summarized by using

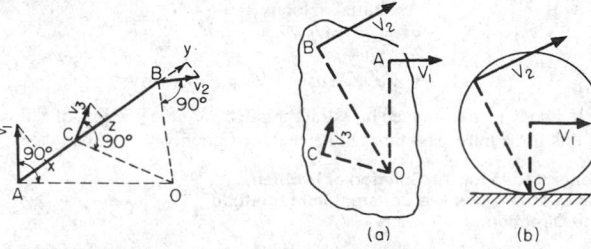

Fig. 3.1.54   **Fig. 3.1.55**

vector notation. Referring to Fig. 3.1.58, let the point P be displaced a vector distance $\mathbf{R}$ from the origin O of a moving reference frame x, y, z which has a velocity $\mathbf{v}_o$ and an acceleration $\mathbf{a}_o$. If point P has a velocity and an acceleration relative to the moving reference plane, let these be $\mathbf{v}_r$ and $\mathbf{a}_r$. The angular velocity of the moving reference fame is ω, and

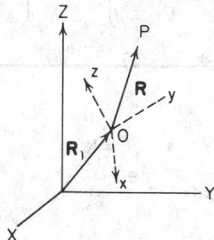

Fig. 3.1.58

the origin of the moving reference frame is displaced a vector distance $\mathbf{R}_1$ from the origin of a primary (fixed) reference frame X, Y, Z. The velocity and acceleration of P are $\mathbf{v}_P = \mathbf{v}_o + \omega \times \mathbf{R} + \mathbf{v}_r$ and $\mathbf{a}_P = \mathbf{a}_o + (d\omega/dt) \times \mathbf{R} + \omega \times (\omega \times \mathbf{R}) + 2\omega \times \mathbf{v}_r + \mathbf{a}_r$.

DYNAMICS OF PARTICLES

Consider a particle of mass m subjected to the action of forces $\mathbf{F}_1$, $\mathbf{F}_2$, $\mathbf{F}_3$, . . . , whose vector resultant is $\mathbf{R} = \Sigma\mathbf{F}$. According to Newton's first law of motion, if $\mathbf{R} = 0$, the body is acted on by a balanced force system, and it will either remain at rest or move uniformly in a straight line. If $\mathbf{R} \neq 0$, Newton's second law of motion states that the body will accelerate in the direction of and proportional to the magnitude of the resultant R. This may be expressed as $\Sigma\mathbf{F} = m\mathbf{a}$. If the resultant of the force system has components in the x, y, and z directions, the resultant acceleration will have proportional components in the x, y, and z direction so that $F_x = ma_x$, $F_y = ma_y$, and $F_z = ma_z$. If the resultant of the force system varies with time, the acceleration will also vary with time.

In **rectilinear motion,** the acceleration and the direction of the unbalanced force must be in the direction of motion. **Forces must be in balance and the acceleration equal to zero in any direction other than the direction of motion.**

EXAMPLE 1. The body in Fig. 3.1.59 has a mass of 90 lbm (40.8 kg) and is subjected to an external horizontal force of 36 lbf (160 N) applied in the direction shown. The coefficient of friction between the body and the inclined plane is 0.1. Required, the velocity of the body at the end of 5 s, if it starts from rest.

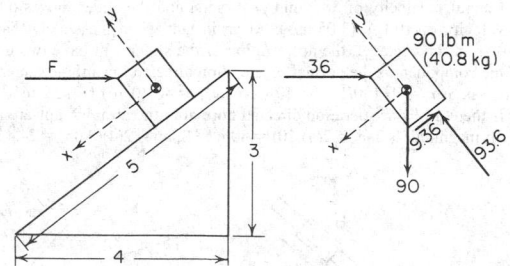

Fig. 3.1.59

First determine all the forces acting externally on the body. These are the applied force $F = 36$ lbf (106 N), the weight $W = 90$ lbf (400 N), and the force with which the plane reacts on the body. The latter force can be resolved into component forces, one normal and one parallel to the surface of the plane. Motion will be downward along the plane since a static analysis will show that the body will slide downward unless the static coefficient of friction is greater than 0.269. In the direction normal to the surface of the plane, the forces must be balanced. The normal force is $(3/5)(36) + (4/5)(90) = 93.6$ lbf (416 N). The frictional force is $93.6 \times 0.1 = 9.36$ lbf (41.6 N). The unbalanced force acting on the body along the

plane is $(3/5)(90) - (4/5)(36) - 9.36 = 15.84$ lbf (70.46 N) downward. $F = (W/9)\,a = (90/g)\,a$; therefore, $a = 0.176\,g = 56.6$ ft/s² (1.725 m/s²). In SI units, $F = ma = 70.46 = 40.8a$; and $a = 1.725$ m/s². The body is acted upon by constant forces and starts from rest; therefore, $v = \int_0^5 a\,dt$, and at the end of 5 s, the velocity would be 28.35 ft/s (8.91 m/s).

EXAMPLE 2. The force with which a rope acts on a body is equal and opposite to the force with which the body acts on the rope, and each is equal to the tension in the rope. In Fig. 3.1.60a, neglecting the weight of the pulley and the rope, the tension in the cord must be the force of 27 lbf. For the 18-lb mass, the unbalanced force is $27 - 18 = 9$ lbf in the upward direction, i.e., $27 - 18 = (18/g)a$, and $a = 16.1$ ft/s² upward. In Fig. 3.1.60b the 27-lb force is replaced by a 27-lb mass. The unbalanced force is still $27 - 18 = 9$ lbf, but it now acts on two masses so that $27 - 18 = (45/g)$ and $a = 6.44$ ft/s². The 18-lb mass is accelerated upward, and the 27-lb mass is accelerated downward. The tension in the rope is equal to 18 lbf plus the unbalanced force necessary to give it an upward acceleration of $g/5$ or $T = 18 + (18/g)(g/5) = 21.6$ lbf. The tension is also equal to 27 lbf less the unbalanced force necessary to give it a downward acceleration of $g/5$ or $T = 27 - (27/g) \times (g/5) = 21.6$ lbf.

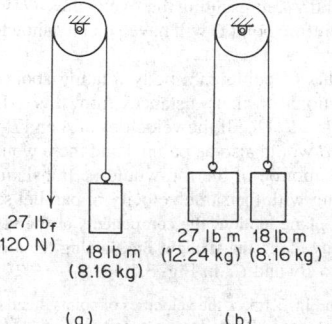

Fig. 3.1.60

In SI units, in Fig. 3.1.60a, the unbalanced force is $120 - 80 = 40$ N, in the upward direction, i.e., $120 - 80 = 8.16a$, and $a = 4.9$ m/s² (16.1 ft/s²). In Fig. 3.1.60b the unbalanced force is still 40 N, but it now acts on the two masses so that $120 - 80 = 20.4a$ and $a = 1.96$ ft/s² (6.44 ft/s²). The tension in the rope is the weight of the 8.16-kg mass in newtons plus the unbalanced force necessary to give it an upward acceleration of 1.96 m/s², $T = 9.807(8.16) + (8.16)(1.96) = 96$ N (21.6 lbf).

General Formulas for the Motion of a Body under the Action of a Constant Unbalanced Force

Let s = space, ft; a = acceleration, ft/s²; v = velocity, ft/s; v_0 = initial velocity, ft/s; h = height, ft; F = force; m = mass; w = weight; g = acceleration due to gravity.

$$\text{Initial velocity} = 0$$
$$F = ma = (w/g)a$$
$$v = at$$
$$s = \tfrac{1}{2}at^2 = \tfrac{1}{2}vt$$
$$v = \sqrt{2as}$$
$$= \sqrt{2gh} \text{ (falling freely from rest)}$$

$$\text{Initial velocity} = v$$
$$F = ma = (w/g)a$$
$$v = v_0 + at$$
$$s = v_0t + \tfrac{1}{2}at^2 = \tfrac{1}{2}v_0t + \tfrac{1}{2}vt$$

If a body is to be moved in a straight line by a force, the line of action of this force must pass through its center of gravity.

General Rule for the Solution of Problems When the Forces Are Constant in Magnitude and Direction

Resolve all the forces acting on the body into two components, one in the direction of the body's motion and one at right angles to it. Add the components in the direction of the body's motion algebraically and find the **unbalanced force,** if any exists.

In **curvilinear motion,** a particle moves along a curved path, and the resultant of the unbalanced force system may have components in directions other than the direction of motion. **The acceleration in any given direction is proportional to the component of the resultant in that direction.** It is common to utilize orthogonal coordinate systems such as **cartesian coordinates, polar coordinates,** and **normal and tangential coordinates** in analyzing forces and accelerations.

EXAMPLE.   A conical pendulum consists of a weight suspended from a cord or light rod and made to rotate in a horizontal circle about a vertical axis with a constant angular velocity of N r/min. For any given constant speed of rotation, the angle θ, the radius r, and the height h will have fixed values. Looking at Fig. 3.1.61, we see that the forces in the vertical direction must be balanced, $T \cos \theta = w$. The forces in the direction normal to the circular path of rotation are unbalanced such that $T \sin \theta = (w/g)a_n = (w/g)\omega^2 r$. Substituting $r = l \sin \theta$ in this last equation gives the value of the tension in the cord $T = (w/g)l\omega^2$. Dividing the second equation by the first and substituting $\tan \theta = r/h$ yields the additional relation that $h = g/\omega^2$.

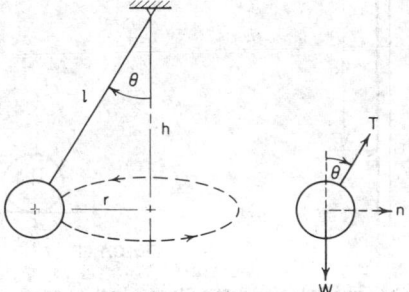

Fig. 3.1.61

An unresisted **projectile** has a motion compounded of the vertical motion of a falling body, and of the horizontal motion due to the horizontal component of the velocity of projection. In Fig. 3.1.62 the only force acting after the projectile starts is gravity, which causes an accelerating downward. The horizontal component of the original velocity v_0 is not changed by gravity. The projectile will rise until the velocity

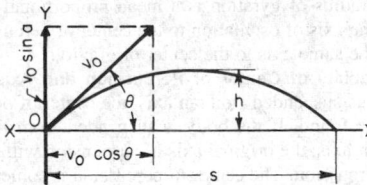

Fig. 3.1.62

given to it by gravity is equal to the vertical component of the starting velocity v_0, and the equation $v_0 \sin \theta = gt$ gives the time t required to reach the highest point in the curve. The same time will be taken in falling if the surface XX is level, and the projectile will therefore be in flight $2t$ s. The distance $s = v_0 \cos \theta \times 2t$, and the maximum height of ascent $h = (v_0 \sin \theta)^2/2g$. The expressions for the coordinates of any point on the path of the projectile are: $x = (v_0 \cos \theta)t$, and $y = (v_0 \sin \theta)t - \frac{1}{2}gt^2$, giving $y = x \tan \theta - (gx^2/2v_0^2 \cos^2 \theta)$ as the equation for the curve of the path. The radius of curvature of the highest point may be found by using the general expression $v^2 = gr$ and solving for r, v being taken equal to $v_0 \cos \theta$.

Simple Pendulum   The period of oscillation $= \tau = 2\pi \sqrt{l/g}$, where l is the length of the pendulum and the length of the swing is not great compared to l.

Centrifugal and Centripetal Forces   When a body revolves about an axis, some connection must exist capable of applying force enough to

the body to constantly deviate it toward the axis. This deviating force is known as **centripetal force.** The equal and opposite resistance offered by the body to the connection is called the **centrifugal force.** The acceleration toward the axis necessary to keep a particle moving in a circle about that axis is v^2/r; therefore, the force necessary is $ma = mv^2/r = wv^2/gr = w\pi^2 N^2 r/900g$, where $N = $ r/min. This force is constantly directed toward the axis.

The centrifugal force of a solid body revolving about an axis is the same as if the whole mass of the body were concentrated at its center of gravity. Centrifugal force $= wv^2/gr = mv^2/r = w\omega^2 r/g$, where w and m are the weight and mass of the whole body, r is the distance from the axis about which the body is rotating to the center of gravity of the body, ω the angular velocity of the body about the axis in radians, and v the linear velocity of the center of gravity of the body.

Balancing

A rotating body is said to be in **standing balance** when its center of gravity coincides with the axis upon which it revolves. Standing balance may be obtained by resting the axis carrying the body upon two horizontal plane surfaces, as in Fig. 3.1.63. If the center of gravity of the wheel A coincides with the center of the shaft B, there will be no movement, but if the center of gravity does not coincide with the center of the shaft, the shaft will roll until the center of gravity of the wheel comes

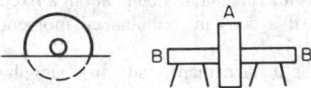

Fig. 3.1.63

directly under the center of the shaft. The center of gravity may be brought to the center of the shaft by adding or taking away weight at proper points on the diameter passing through the center of gravity and the center of the shaft. Weights may be added to or subtracted from any part of the wheel so long as its center of gravity is brought to the center of the shaft.

A rotating body may be in standing balance and not in **dynamic balance.** In Fig. 3.1.64, AA and BB are two disks whose centers of gravity are at o and p, respectively. The shaft and the disks are in standing balance if the disks are of the same weight and the distances of o and p from the center of the shaft are equal, and o and p lie in the same axial plane but on opposite sides of the shaft. Let the weight of each disk be w and the distances of o and p from the center of the shaft each be equal to

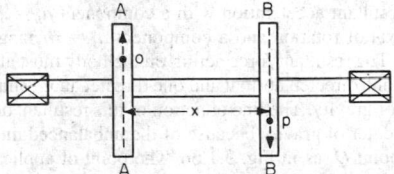

Fig. 3.1.64

r. The force exerted on the shaft by AA is equal to $w\omega^2 r/g$, where ω is the angular velocity of the shaft. Also, the force exerted on the shaft by $BB = w\omega^2 r/g$. These two equal and opposite parallel forces act at a distance x apart and constitute a couple with a moment tending to rotate the shaft, as shown by the arrows, of $(w\omega^2 r/g)x$. A couple cannot be balanced by a single force; so two forces at least must be added to or subtracted from the system to get dynamic balance.

Systems of Particles   The principles of motion for a single particle can be extended to cover a **system of particles.** In this case, **the vector resultant of all external forces acting on the system of particles must equal the total mass of the system times the acceleration of the mass center, and the direction of the resultant must be the direction of the acceleration of the mass center.** This is the **principle of motion of the mass center.**

Rotation of Solid Bodies in a Plane about Fixed Axes

For a rigid body revolving **in a plane** about a fixed axis, **the resultant moment about that axis must be equal to the product of the moment of inertia** (about that axis) **and the angular acceleration,** $\Sigma M_0 = I_0\alpha$. This is a general statement which includes the particular case of rotation about an axis that passes through the center of gravity.

Rotation about an Axis Passing through the Center of Gravity The rotation of a body about its center of gravity can only be caused or changed by a **couple.** See Fig. 3.1.65. If a single force F is applied to the wheel, the axis immediately acts on the wheel with an equal force to prevent translation, and the result is a couple (moment Fr) acting on the body and causing rotation about its center of gravity.

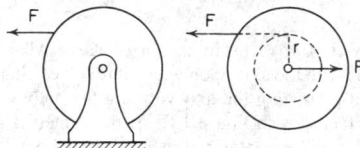

Fig. 3.1.65

General formulas for rotation of a body about a fixed axis through the center of gravity, if a constant unbalanced moment is applied (Fig. 3.1.65).

Let θ = angular displacement, rad; ω = angular velocity, rad/s; α = angular acceleration, rad/s²; M = unbalanced moment, ft · lb; I = moment of inertia (mass); g = acceleration due to gravity; t = time of application of M.

Initial angular velocity = 0	Initial angular velocity = ω_0
$M = I\alpha$	$M = I\alpha$
$\theta = \frac{1}{2}\alpha t^2$	$\theta = \omega_0 t + \frac{1}{2}\alpha t^2$
$\omega = \sqrt{2\alpha\theta}$	$\omega = \sqrt{\omega_0^2 + 2\alpha\theta}$

General Rule for Rotating Bodies Determine all the external forces acting and their moments about the axis of rotation. If these moments are balanced, there will be no change of motion. If the moments are unbalanced, this unbalanced moment, or **torque,** will cause an angular acceleration about the axis.

Rotation about an Axis Not Passing through the Center of Gravity The resultant force acting on the body must be proportional to the acceleration of the center of gravity and directed along its line of action. If the axis of rotation does not pass through the center of gravity, the center of gravity will have a resultant acceleration with a component $a_n = \omega^2 r$ directed toward the axis of rotation and a component $a_t = \alpha r$ tangential to its circular path. The resultant force acting on the body must also have two components, one directed normal and one directed tangential to the path of the center of gravity. The line of action of this resultant does not pass through the center of gravity because of the unbalanced moment $M_0 = I_0\alpha$ but at a point Q, as in Fig. 3.1.66. The point of application of this resultant is known as the **center of percussion** and may be **defined** as the point of application of the resultant of all the forces tending to cause a body to rotate about a certain axis. It is the point at which a suspended

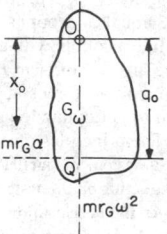

Fig. 3.1.66

body may be struck without causing any force on the axis passing through the point of suspension.

Center of Percussion The distance from the axis of suspension to the center of percussion is $q_0 = I/mx_0$, where I = moment of inertia of the body about its axis of suspension to the center of gravity of the body.

EXAMPLES. 1. Find the center of percussion of the homogeneous rod (Fig. 3.1.67) of length L and mass m, suspended at XX.

$$q_0 = \frac{I}{mx_0}$$

$$I \text{ (approx)} = \frac{m}{L}\int_0^L x^2\,dx \qquad x_0 = \frac{L}{2} \qquad \therefore q_0 = \frac{2}{L^2}\int_0^L x^2\,dx = 2L/3$$

2. Find the center of percussion of a solid cylinder, of mass m, resting on a horizontal plane. In Fig. 3.1.68, the instantaneous center of the cylinder is at A. The center of percussion will therefore be a height above the plane equal to $q_0 = I/mx_0$. Since $I = (mr^2/2) + mr^2$ and $x_0 = r$, $q_0 = 3r/2$.

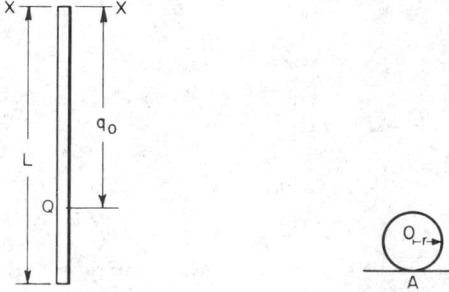

Fig. 3.1.67   **Fig. 3.1.68**

Wheel or Cylinder Rolling down a Plane In this case the component of the weight along the plane tends to make it roll down and is treated as a force causing rotation. The forces acting on the body should be resolved into components along the line of motion and perpendicular to it. If the forces are all known, their resultant is at the center of percussion. If one force is to be determined (the exact conditions as regards slipping or not slipping must be known), the center of percussion can be determined and the unknown force found.

Relation between the Center of Percussion and Radius of Gyration $q_0 = I/mx_0 = k^2/x_0 \therefore k^2 = x_0 q_0$ where k = radius of gyration. Therefore, the radius of gyration is a mean proportional between the distance from the axis of oscillation to the center of percussion and the distance from the same axis to the center of gravity.

Interchangeability of Center of Percussion and Axis of Oscillation If a body is suspended from an axis, the center of percussion for that axis can be found. If the body is suspended from this center of percussion as an axis, the original axis of suspension will then become the center of percussion. The center of percussion is sometimes known as the **center of oscillation.**

Period of Oscillation of a Compound Pendulum The length of an equivalent simple pendulum is the distance from the axis of suspension to the center of percussion of the body in question. To find the **period of oscillation** of a body about a given axis, find the distance $q_0 = I/mx_0$ from that axis to the center of percussion of the swinging body. The length of the simple pendulum that will oscillate in the same time is this distance q_0. The period of oscillation for the equivalent single pendulum is $\tau = 2\pi\sqrt{q_0/g}$.

Determination of Moment of Inertia by Experiment To find the moment of inertia of a body, suspend it from some axis not passing through the center of gravity and, by swinging it, determine the period of one complete oscillation in seconds. The known values will then be τ = time of one complete oscillation, x_0 = distance from axis to center of gravity, and m = mass of body. The length of the equivalent simple pendulum is $q_0 = I/mx_0$. Substituting this value of q_0 in $\tau = 2\pi\sqrt{q_0/g}$ gives $\tau = 2\pi\sqrt{I/mx_0 g}$, from which $\tau^2 = 4\pi^2 I/mx_0 g$, or $I = mx_0 g\tau^2/4\pi^2$.

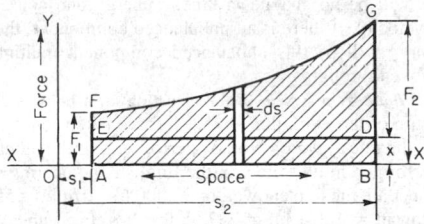

Fig. 3.1.69

Plane Motion of a Rigid Body

Plane motion may be considered to be a combination of translation and rotation (see "Kinematics"). For translation, Newton's second law of motion must always be satisfied, and the resultant of the external force system must be equal to the product of the mass times the acceleration of the center of gravity in any system of coordinates. In rotation, the body moving in plane motion will not have a fixed axis. When the methods of relative motion are being used, any point on the body may be used as a reference axis to which the motion of all other points is referred.

The sum of the moments of all external forces about the reference axis must be equal to the vector sum of the centroidal moment of inertia times the angular acceleration and the amount of the resultant force about the reference axis.

EXAMPLE. Determine the forces acting on the piston pin A and the crankpin B of the connecting rod of a reciprocating engine shown in Fig. 3.1.69 for a position of 30° from TDC. The crankshaft speed is constant at 2,000 r/min. Assume that the pressure of expanding gases on the 4-lbm (1.81-kg) piston at this point is 145 lb/in² (10⁶ N/m²). The connecting rod has a mass of 5 lbm (2.27 kg) and has a centroidal radius of gyration of 3 in (0.076 m).

The kinematics of the problem are such that the angular velocity of the crank is $\omega_{OB} = 209.4$ rad/s clockwise, the angular velocity of the connecting rod is $\omega_{AB} = 45.7$ rad/s counterclockwise, and the angular acceleration is $\alpha_{AB} = 5,263$ rad/s² clockwise. The linear acceleration of the piston is 7,274 ft/s² in the direction of the crank. From the free-body diagram of the piston, the horizontal component of the piston-pin force is $145 \times (\pi/4)(5^2) - P = (4/32.2)(7,274)$, $P = 1,943$ lbf. The acceleration of the center of gravity G is the vector sum of the component accelerations $a_G = a_B + a_{G/B}^n + a_{G/B}^t$ where $a_{G/B}^n = \omega_{GB}^2 \cdot r_{GB} = 3/12(45.7)^2 = 522$ ft/s² and $a_{G/B}^t = \alpha_{GB} \cdot r_{GB} = 3/12(5.263) = 1,316$ ft/s². The resultant acceleration of the center of gravity is 6,685 ft/s² in the x direction and 2,284 ft/s² in the negative y direction. The resultant of the external force system will have corresponding components such that $ma_{Gx} = (5/32.2)(6,685) = 1,039$ lbf and $ma_{Gy} = (5/32.2)(2,284) = 355$ lbf. The three remaining unknown forces can be found from the three equations of motion for the connecting rod.

Taking the sum of the forces in the x direction, $\epsilon F = ma_{Gx}$; $P - R_x = ma_{Gx}$, and $R_x = 905.4$ lbf. In the y direction, $\Sigma F = ma_{Gy}$; $R_y - N = ma_{gy}$; this has two unknowns, R_y, and N. Taking the sum of the movements of the external forces about the center of mass g, $\Sigma M_G = I_G \alpha AB$; $(N)(5) \cos (7.18°) - (P)(5) \sin (7.18°) + (R_y)(3) \cos (7.18°) - R_x (3) \sin (7.18°) = (5/386.4)(3)^2(5,263)$. Solving for R_y and N simultaneously, $R_y = 494.7$ lbf and $N = 140$ lbf. We could have avoided the solution of two simultaneous algebraic equations by taking the moment summation about end A, which would determine R_y independently, or about end B, which would determine N independently.

In SI units, the kinematics would be identical, the linear acceleration of the piston being 2,217 m/s² (7,274 ft/s²). From the free-body diagram of the piston, the horizontal component of the piston-pin force is $(10^6) \times (\pi/4)(0.127)^2 - P = (1.81)(2,217)$, and $P = 8,640$ N. The components of the acceleration of the center of gravity G are $a_{G/B}^N = 522$ ft/s² and $a_{G/B}^T = 1,315$ ft/s². The resultant acceleration of the center of gravity is 2,037.5 m/s² (6,685 ft/s²) in the x direction and 696.3 m/s² (2,284 ft/s²) in the negative y direction. The resultant of the external force system will have the corresponding components; $ma_{Gx} = (2.27)(2,037.5) = 4,620$ N; $ma_{Gy} = (2.27)(696.3) = 1,579$ N. $R_x = 4,027$ N, $R_y = 2,201$ N, force $N = 623$ newtons.

WORK AND ENERGY

Work When a body is displaced against resistance or accelerated, work must be done upon it. An increment of work is defined as the product of an incremental displacement and the component of the force vector in the direction of the displacement or the product of the component of the incremental displacement and the force in the direction of the force. $dU = F \cdot ds \cos \alpha$, where α is the angle between the vector displacement and the vector force. The increment of work done by a couple M acting in a body during an increment of angular rotation $d\theta$ in the plane of the couple is $dU = M \, d\theta$. In a force-displacement or moment-angle diagram, called a **work diagram** (Fig. 3.1.70), force is plotted as a function of displacement. The area under the curve represents the work done, which is equal to $\int_{s_1}^{s_2} F \, ds \cos \alpha$ or $\int_{\theta_1}^{\theta_2} M \, d\theta$.

Fig. 3.1.70

Units of Work When the force of 1 lb acts through the distance of 1 ft, 1 lb · ft of work is done. In SI units, a force of 1 newton acting through 1 metre is 1 joule of work. 1.356 N · m = 1 lb · ft.

Energy A body is said to possess energy when it can do work. A body may possess this capacity through its **position** or **condition**. When a body is so held that it can do work, if released, it is said to possess energy of position or **potential energy**. When a body is moving with some velocity, it is said to possess energy of motion or **kinetic energy**. An example of potential energy is a body held suspended by a rope; the position of the body is such that if the rope is removed work can be done by the body.

Energy is expressed in the same units as work. The kinetic energy of a particle is expressed by the formula $E = \frac{1}{2}mv^2 = \frac{1}{2}(w/g)v^2$. The kinetic energy of a rigid body in translation is also expressed as $E = \frac{1}{2}mv^2$. Since all particles of the rigid body have the same identical velocity v, the velocity v is the velocity of the center of gravity. The kinetic energy of a rigid body, rotating about a fixed axis is $E = \frac{1}{2}I_0\omega^2$, where I_0 is the mass moment of inertia about the axis of rotation. In plane motion, a rigid body has both translation and rotation. The kinetic energy is the algebraic sum of the translating kinetic energy of the center of gravity and the rotating kinetic energy about the center of gravity, $E = \frac{1}{2}mv^2 + \frac{1}{2}I\omega^2$. Here the velocity v is the velocity of the center of gravity, and the moment of inertia I is the centroidal moment of inertia.

If a force which varies acts through a space on a body of mass m, the work done is $\int_{s_1}^{s} F \, ds$, and if the work is all used in giving kinetic energy to the body it is equal to $\frac{1}{2}m(v_2^2 - v_1^2) =$ **change in kinetic energy**, where v_2 and v_1 are the velocities at distances s_2 and s_1, respectively. This is a specific statement of the law of conservation of energy. **The principle of conservation of energy requires that the mechanical energy of a system remain unchanged if it is subjected only to forces which depend on position or configuration.**

Certain problems in which the velocity of a body at any point in its straight-line path when acted upon by varying forces is required can be easily solved by the use of a **work diagram.**

In Fig. 3.1.70, let a body start from rest at A and be acted upon by a force that varies in accordance with the diagram $AFGBA$. Let the resistance to motion be a constant force $= x$. Find the velocity of the body at point B. The area $AFGBA$ represents the work done upon the body and the area $AEDBA$ ($=$ force $x \times$ distance AB) represents the work that must be done to overcome resistance. The difference of these areas, or $EFGDE$, will represent work done in excess of that required to overcome resistance, and consequently is equal to the increase in kinetic energy. Equating the work represented by the area $EFGDE$ to $\frac{1}{2}wv^2/g$ and solving for v will give the required velocity at B. If the body did not start from rest, this area would represent the change in kinetic energy, and the velocity could be obtained by the formula: Work $= \frac{1}{2}(w/g)$ $(v_1^2 - v_0^2)$, v_1 being the required velocity.

General Rule for Rectilinear Motion Resolve each force acting on the body into components, one of which acts along the line of motion of the body and the other at right angles to the line of motion. Take the sum of all the components acting in the direction of the motion and multiply this sum by the distance moved through for constant forces. (Take the average force times distance for forces that vary.) This product will be the total work done upon the body. If there is no unbalanced component, there will be no change in kinetic energy and consequently no change in velocity. If there is an unbalanced component, the change in kinetic energy will be this unbalanced component multiplied by the distance moved through.

The **work done by a system of forces acting on a body** is equal to the algebraic sum of the work done by each force taken separately.

Power is the rate at which work is performed, or the number of units of work performed in unit time. In the English engineering system, the units of power are the horsepower, or 33,000 lb · ft/min = 550 lb · ft/s, and the kilowatt = 1.341 hp = 737.55 lb · ft/s. In SI units, the unit of power is the watt, which is 1 newton-metre per second or 1 joule per second.

Friction Brake In Fig. 3.1.71 a pulley revolves under the band and in the direction of the arrow, exerting a pull of T on the spring. The friction of the band on the rim of the pulley is $(T - w)$, where w is the weight attached to one end of the band. Let the pulley make N r/min; then the work done per minute against friction by the rim of the pulley is $2\pi RN(T - w)$, and the horsepower absorbed by brake $= 2\pi RN(T - w)/33,000$.

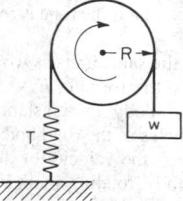

Fig. 3.1.71

IMPULSE AND MOMENTUM

The product of force and time is defined as linear impulse. The impulse of a constant force over a time interval $t_2 - t_1$ is $F(t_2 - t_1)$. If the force is not constant in magnitude but is constant in direction, the impulse is $\int_{t_1}^{t_2} F \, dt$. The dimensions of linear impulse are (force) $\times$ (time) in pound-seconds, or newton-seconds.

Impulse is a **vector** quantity which has the direction of the resultant force. Impulses may be added vectorially by means of a vector polygon, or they may be resolved into components by means of a parallelogram. The **moment** of a **linear impulse** may be found in the same manner as the moment of a force. The linear impulse is represented by a directed line segment, and the moment of the impulse is the product of the magnitude of the impulse and the perpendicular distance from the line segment to the point about which the moment is taken. **Angular impulse** over a time interval $t_2 - t_1$ is a product of the sum of applied moments on a rigid body about a reference axis and time. The dimensions for angular impulse are (force) $\times$ (time) $\times$ (displacement) in foot-pound-seconds or newton-metre-seconds. **Angular impulse and linear impulse cannot be added.**

Momentum is also a vector quantity and can be added and resolved in the same manner as force and impulse. The dimensions of linear momentum are (force) $\times$ (time) in pound-seconds or newton-seconds, and are identical to linear impulse. An alternate statement of Newton's second law of motion is that the resultant of an unbalanced force system must be equal to the time rate of change of linear momentum, $\Sigma \mathbf{F} = d(mv)/dt$.

If a variable force acts for a certain time on a body of mass m, the quantity $\int_{t_1}^{t_2} F \, dt = m(v_1 - v_2) =$ the change of momentum of the body.

The **moment of momentum** can be determined by the same methods as those used for the moment of a force or moment of an impulse. The dimensions of the moment of momentum are (force) $\times$ (time) $\times$ (displacement) in foot-pound-seconds, or newton-metre-seconds.

In **plane motion** the angular momentum of a rigid body about a reference axis perpendicular to the plane of motion is the sum of the moments of linear momenta of all particles in the body about the reference axes. Specifically, **the angular momentum of a rigid body in plane motion is the vector sum of the angular momentum about the reference axis and the moment of the linear momentum of the center of gravity about the reference axis**, $\mathbf{H}_0 = I_0\omega + \mathbf{d} \times mv$.

In three-dimensional rotation about a fixed axis, the angular momentum of a rigid body has components along three coordinate axes, which involve both the moments of inertia about the x, y, and z axes, $I_{0_{xx}}$, $I_{0_{yy}}$, and $I_{0_{zz}}$, and the products of inertia, $I_{0_{xy}}$, $I_{0_{xz}}$, and $I_{0_{yz}}$; $H_{0_x} = I_{0_{xx}} \cdot \omega_x - I_{0_{xy}} \cdot \omega_y - I_{0_{xz}} \cdot \omega_z$, $H_{0_y} = -I_{0_{xy}} \cdot \omega_x + I_{0_{yy}} \cdot \omega_y - I_{0_{yz}} \cdot \omega_z$, and $H_{0_z} = I_{0_{zx}} \cdot \omega_x - I_{0_{zy}} \cdot \omega_y + I_{0_{zz}} \cdot \omega_z$ where $\mathbf{H}_0 = \mathbf{H}_{0_x} + \mathbf{H}_{0_y} + \mathbf{H}_{0_z}$.

Impact

The collision between two bodies, where relatively large forces result over a comparatively short interval of time, is called **impact**. A straight line perpendicular to the plane of contact of two colliding bodies is called the **line of impact**. If the centers of gravity of the two colliding bodies lie on the line of contact, the impact is called **central impact**, in any other case, **eccentric impact**. If the linear momenta of the centers of gravity are also directed along the line of impact, the impact is **collinear or direct central impact**. In any other case impact is said to be **oblique**.

Collinear Impact When two masses m_1 and m_2, having respective velocities u_1 and u_2, move in the same line, they will collide if $u_2 > u_1$ (Fig. 3.1.72a). During collision (Fig. 3.1.72b), kinetic energy is ab-

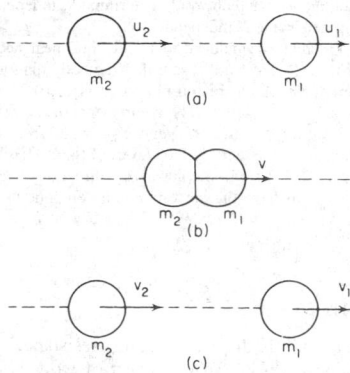

Fig. 3.1.72

sorbed in the deformation of the bodies. There follows a period of restoration which may or may not be complete. If complete restoration of the energy of deformation occurs, the impact is **elastic.** If the restoration of energy is incomplete, the impact is referred to as **inelastic.** After collision (Fig. 3.1.72c), the bodies continue to move with changed velocities of v_1 and v_2. Since the contact forces on one body are equal to and opposite the contact forces on the other, the sum of the linear momenta of the two bodies is conserved; $m_1u_1 + m_2u_2 = m_1v_1 + m_2v_2$.

The law of conservation of momentum states that the linear momentum of a system of bodies is unchanged if there is no resultant external force on the system.

Coefficient of Restitution The ratio of the velocity of separation $v_1 - v_2$ to the velocity of approach $u_2 - u_1$ is called the **coefficient of restitution** e, $e = (v_1 - v_2)/(u_2 - u_1)$.

The value of e will depend on the shape and material properties of the colliding bodies. In elastic impact, the coefficient of restitution is unity and there is no energy loss. A coefficient of restitution of zero indicates perfectly inelastic or plastic impact, where there is no separation of the bodies after collision and the energy loss is a maximum. In **oblique impact,** the coefficient of restitution applies only to those components of velocity along the line of impact or normal to the plane of impact. The coefficient of restitution between two materials can be measured by making one body many times larger than the other so that m_2 is infinitely large in comparison to m_1. The velocity of m_2 is unchanged for all practical purposes during impact and $e = v_1/u_1$. For a small ball dropped from a height H upon an extensive horizontal surface and rebounding to a height h, $e = \sqrt{h/H}$.

Impact of Jet Water on Flat Plate When a jet of water strikes a flat plate perpendicularly to its surface, the force exerted by the water on the plate is wv/g, where w is the weight of water striking the plate in a unit of time and v is the velocity. When the jet is inclined to the surface by an angle, A, the pressure is $(wv/g) \cos A$.

Variable Mass

If the mass of a body is variable such that mass is being either added or ejected, an alternate form of Newton's second law of motion must be used which accounts for changes in mass:

$$\mathbf{F} = m\frac{d\mathbf{v}}{dt} + \frac{dm}{dt}\mathbf{u}$$

The mass m is the instantaneous mass of the body, and dv/dt is the time rate of change of the absolute of velocity of mass m. The velocity u is the velocity of the mass m relative to the added or ejected mass, and dm/dt is the time rate of change of mass. In this case, care must be exercised in the choice of coordinates and expressions of sign. If mass is being added, dm/dt is plus, and if mass is ejected, dm/dt is minus.

Fields of Force—Attraction

The space within which the action of a physical force comes into play on bodies lying within its boundaries is called the **field of the force.**

The **strength** or **intensity of the field** at any given point is the relation between a force F acting on a mass m at that point and the mass. Intensity of field $= i = F/m$; $F = mi$.

The **unit of field intensity** is the same as the unit of acceleration, i.e., 1 ft/s² or 1 m/s². The intensity of a field of force may be represented by a line (or **vector**).

A field of force is said to be **homogeneous** when the intensity of all points is uniform and in the same direction.

A field of force is called a **central field of force** with a center O, if the direction of the force acting on the mass particle m in every point of the field passes through O and its magnitude is a function only of the distance r from O to m. A line so drawn through the field of force that its direction coincides at every point with that of the force prevailing at that point is called a **line of force.**

Rotation of Solid Bodies about Any Axis

The general moment equations for three-dimensional motion are usually expressed in terms of the angular momentum. For a reference axis O, which is either a fixed axis of the center of gravity, $M_{0_x} = (dH_{0_x}/dt) - H_{0_y} \cdot \omega_z + H_{0_z} \cdot \omega_y$, $M_{0_y} = (dH_{0_y}/dt) - H_{0_z} \cdot \omega_x + H_{0_x} \cdot \omega_z$, and $M_{0_z} = (dH_{0_z}/dt) - H_{0_x}\omega_y + H_{0_y}\omega_x$. If the coordinate axes are oriented to coincide with the principal axes of inertia, $I_{0_{xx}}$, $I_{0_{yy}}$, and $I_{0_{zz}}$, a similar set of three differential equations results, involving moments, angular velocity, and angular acceleration; $M_{0_x} - I_{0_{xx}}(d\omega_x/dt) + (I_{0_{zz}} - I_{0_{yy}})\omega_y \cdot \omega_z$, $M_{0_y} = I_{0_{yy}}(d\omega_y/dt) + (I_{0_{xx}} - I_{0_{zz}})\omega_z \cdot \omega_x$, and $M_{0_z} = I_{0_{zz}}(d\omega_z/dt) + (I_{0_{yy}} - I_{0_{xx}})\omega_x\omega_y$. These equations are known as **Euler's equations of motion** and may apply to any rigid body.

GYROSCOPIC MOTION AND THE GYROSCOPE

Gyroscopic motion can be explained in terms of Euler's equations. Let I_1, I_2, and I_3 represent the principal moments of inertia of a gyroscope spinning with a constant angular velocity ω, about axis 1, the subscripts 1, 2, and 3 representing a right-hand set of reference axes (Figs. 3.1.73 and 3.1.74). If the gyroscope is precessed about the third axis, a vector moment results along the second axis such that

$$M_2 = I_2(d\omega_2/dt) + (I_1 - I_3)\omega_3\omega_1$$

Where the precession and spin axes are at right angles, the term $(d\omega_2/dt)$ equals the component of $\omega_3 \times \omega_1$ along axis 2. Because of this, in the simple case of a body of symmetry, where $I_2 = I_3$, the gyroscopic

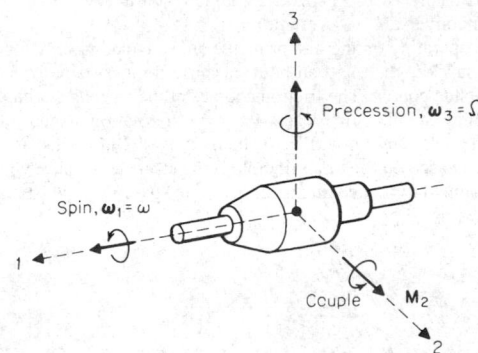

Fig. 3.1.73

moment can be reduced to the common expression $M = I\omega\Omega$, where Ω is the rate of precession, ω the rate of spin, and I the moment of inertia about the spin axis. It is important to realize that these are equations of motion and relate the applied or resulting gyroscopic moment due to forces which act *on* the rotor, as disclosed by a free-body diagram, to the resulting motion of the rotor.

Physical insight into the behavior of a steady precessing gyro with mutually perpendicular moment, spin, and precession axes is gained by recognizing from Fig. 3.1.74 that the change dH in angular momentum H is equal to the angular impulse $M\,dt$. In time dt, the angular-momen-

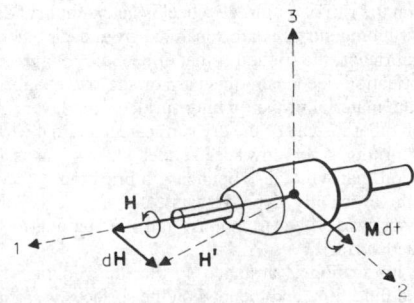

Fig. 3.1.74

tum vector swings from H to H', owing to the velocity of precession ω_3. The vector change dH in angular momentum is in the direction of the applied moment M. This fact is inherent in the basic moment-momentum equation and can always be used to establish the correct spatial relationships between the moment, precessional, and spin vectors. It is seen, therefore, from Fig. 3.1.74 that the spin axis always turns toward the moment axis. Just as the change in direction of the mass-center velocity is in the same direction as the resultant force, so does the change in angular momentum follow the direction of the applied moment.

For example, suppose an airplane is driven by a right-handed propeller (turning like a right-handed screw when moving forward). If a gust of wind or other force turns the machine to the *left,* the gyroscopic action of the propeller will make the forward end of the shaft strive to *rise;* if the wing surface is large, this motion will be practically prevented by the resistance of the air, and the gyroscopic forces become effective merely as internal stresses, whose maximum value can be computed by the formula above. Similarly, if the airplane is dipped *downward,* the gyroscopic action will make the forward end of the shaft strive to turn to the *left.*

Modern **applications** of the gyroscope are based on one of the following properties: (1) a gyroscope mounted in three gimbal rings so as to be entirely free angularly in all directions will retain its direction *in space* in the absence of outside couples; (2) if the axis of rotation of a gyroscope turns or precesses in space, a couple or torque acts on the gyroscope (and conversely on its frame).

Devices operating on the first principle are satisfactory only for short durations, say less than half an hour, because no gyroscope is entirely without outside couple. The friction couples at the various gimbal bearings, although small, will precess the axis of rotation so that after a while the axis of rotation will have changed its direction in space. The chief device based on the first principle is the **airplane compass,** which is a freely mounted gyro, keeping its direction in space during fast maneu-

vers of a fighting airplane. No magnetic compass will indicate correctly during such maneuvers. After the plane is back on an even keel in steady flight, the magnetic compass once more reads the true magnetic north, and the gyro compass has to be reset to point north again.

An example of a device operating on the second principle is the **automatic pilot** for keeping a vehicle on a given course. This device has been installed on torpedoes, ships, airplanes. When the ship or plane turns from the chosen course, a couple is exerted on the gyro axis, which makes it precess and this operates electric contacts or hydraulic or pneumatic valves. These again operate on the rudders, through relays, and bring the ship back to its course.

Another application is the ship **antirolling** gyroscope. This very large gyroscope spins about a vertical axis and is mounted in a ship so that the axis can be tipped fore and aft by means of an electric motor, the precession motor. The gyro can exert a large torque on the ship about the fore-and-aft axis, which is along the ''rolling'' axis. The sign of the torque is determined by the direction of rotation of the precession motor, which in turn is controlled by electric contacts operated by a small pilot gyroscope on the ship, which feels which way the ship rolls and gives the signals to apply a countertorque.

The **turn indicator** for airplanes is a gyro, the frame of which is held by springs. When the airplane turns, it makes the gyro axis turn with it, and the resultant couple is delivered by the springs. Thus the elongation of the springs is a measure of the rate of turn, which is suitably indicated by a pointer.

The most complicated and ingenious application of the gyroscope is the **marine compass.** This is a pendulously suspended gyroscope which is affected by gravity and also by the earth's rotation so that the gyro axis is in equilibrium only when it points north, i.e., when it lies in the plane formed by the local vertical and by the earth's north-south axis. If the compass is disturbed so that it points away from north, the action of the earth's rotation will restore it to the correct north position in a few hours.

3.2 FRICTION
by Vittorio (Rino) Castelli

REFERENCES: Bowden and Tabor, ''The Friction and Lubrication of Solids,'' Oxford. Fuller, ''Theory and Practice of Lubrication for Engineers,'' 2nd ed., Wiley. Shigley, ''Mechanical Design,'' McGraw-Hill. Rabinowicz, ''Friction and Wear of Materials,'' Wiley. Ling, Klaus, and Fein, ''Boundary Lubrication—An Appraisal of World Literature,'' ASME, 1969. Dowson, ''History of Tribology,'' Longman, 1979. Petersen and Winer, ''Wear Control Handbook,'' ASME, 1980.

Friction is the resistance that is encountered when two solid surfaces slide or tend to slide over each other. The surfaces may be either dry or lubricated. In the first case, when the surfaces are free from contaminating fluids, or films, the resistance is called **dry friction**. The friction of brake shoes on the rim of a railroad wheel is an example of dry friction.

When the rubbing surfaces are separated from each other by a very thin film of lubricant, the friction is that of **boundary** (or **greasy**) **lubrication**. The lubrication depends in this case on the strong adhesion of the lubricant to the material of the rubbing surfaces; the layers of lubricant slip over each other instead of the dry surfaces. A journal when starting, reversing, or turning at very low speed under a heavy load is an example of the condition that will cause boundary lubrication. Other examples are gear teeth (especially hypoid gears), cutting tools, wire-drawing dies, power screws, bridge trunnions, and the running-in process of most lubricated surfaces.

When the lubrication is arranged so that the rubbing surfaces are separated by a fluid film, and the load on the surfaces is carried entirely by the hydrostatic or hydrodynamic pressure in the film, the friction is

that of **complete** (or **viscous**) **lubrication**. In this case, the frictional losses are due solely to the internal fluid friction in the film. Oil ring bearings, bearings with forced feed of oil, pivoted shoe-type thrust and journal bearings, bearings operating in an oil bath, hydrostatic oil pads, oil lifts, and step bearings are instances of complete lubrication.

Incomplete lubrication or mixed lubrication takes place when the load on the rubbing surfaces is carried partly by a fluid viscous film and partly by areas of boundary lubrication. The friction is intermediate between that of fluid and boundary lubrication. Incomplete lubrication exists in bearings with drop-feed, waste-packed, or wick-fed lubrication, or on parallel-surface bearings.

STATIC AND KINETIC COEFFICIENTS OF FRICTION

In the absence of friction, the resultant of the forces between the surfaces of two bodies pressing upon each other is normal to the surface of contact. With friction, the resultant deviates from the normal.

If one body is pressed against another by a force P, as in Fig. 3.2.1, the first body will not move, provided the angle a included between the line of action of the force and a normal to the surfaces in contact does not exceed a certain value which depends upon the nature of the surfaces. The reaction force R has the same magnitude and line of action as the force P. In Fig. 3.2.1, R is resolved into two components: a force N

normal to the surfaces in contact and a force F_r parallel to the surfaces in contact. From the above statement it follows that, for motion not to occur,

$$F_r = N \tan a_0 = N f_0$$

where $f_0 = \tan a_0$ is called the **coefficient of friction of rest** (or of **static friction**) and a_0 is the **angle of friction at rest**.

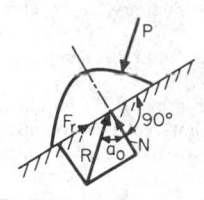

Fig. 3.2.1

If the normal force N between the surfaces is kept constant, and the tangential force F_r is gradually increased, there will be no motion while $F_r < N f_0$. A state of *impending motion* is reached when F_r nears the value of $N f_0$. If sliding motion occurs, a frictional force F resisting the motion must be overcome. The force F is commonly expressed as $F = fN$, where f is the **coefficient of sliding friction**, or **kinetic friction**. Normally, the coefficients of sliding friction are smaller than the coefficients of static friction. With small velocities of sliding and very clean surfaces, the two coefficients do not differ appreciably.

Table 3.2.4 demonstrates the typical reduction of sliding coefficients of friction below corresponding static values. Figure 3.2.2 indicates results of tests on lubricated machine tool ways showing a reduction of friction coefficient with increasing sliding velocity.

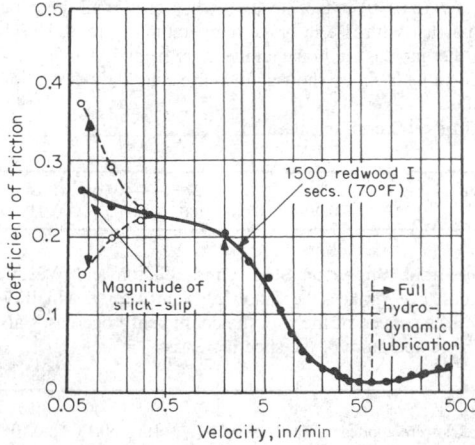

Fig. 3.2.2 Typical relationship between kinetic friction and sliding velocity for lubricated cast iron on cast iron slideways (load, 20 lb/in²; upper slider, scraped; lower slideway, scraped). *(From Birchall, Kearny, and Moss, Intl. J. Machine Tool Design Research, 1962.)*

This behavior is normal with dry friction, some conditions of boundary friction, and with the break-away friction in ball and roller bearings. This condition is depicted in Fig. 3.2.3, where the friction force decreases with relative velocity. This negative slope leads to locally unstable equilibrium and **self-excited vibrations** in systems such as the one of Fig. 3.2.4. This phenomenon takes place because, for small amplitudes, the oscillatory system displays damping in which the damping

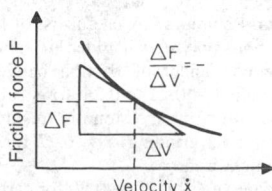

Fig. 3.2.3   Friction force decreases as velocity increases.

factor is equal to the slope of the friction curve and thus is termed **negative damping**. When the slope of the friction force versus sliding velocity is positive (**positive damping**) this type of instability is not possible. This is typical of fluid damping, squeeze films, dash pots, and fluid film bearings in general.

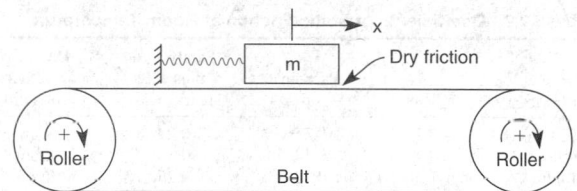

Fig. 3.2.4   Belt friction apparatus with possible self-excited vibrations.

It is interesting to note that these self-excited systems vibrate at close to their natural frequency over a large range of frictional levels and speeds. This symptom is a helpful means of identification. Another characteristic is that the moving body comes periodically to momentary relative rest, that is, zero sliding velocity. For this reason, this phenomenon is also called **stick-slip vibration**. Common examples are violin strings, chalk on blackboard, water-lubricated rubber stern tube ship bearings at low speed, squeaky hinges, and oscillating rolling element bearings, especially if they are supporting large flexible structures such as radar antennas. Control requires the introduction of fluid film bearings, viscous seals, or viscous dampers into the system with sufficient positive damping to override the effects of negative damping.

Under moderate pressures, the frictional force is proportional to the normal load on the rubbing surfaces. It is independent of the pressure per unit area of the surfaces. The direction of the friction force opposing the sliding motion is locally exactly opposite to the local relative velocity. Therefore, it takes very little effort to displace transversally two bodies which have a major direction of relative sliding. This behavior, **compound sliding,** is exploited when easing the extraction of a nail by simultaneously rotating it about its axis, and accounts for the ease with which an automobile may skid on the road or with which a plug gage can be inserted into a hole if it is rotated while being pushed in.

The coefficients of friction for dry surfaces (dry friction) depend on the materials sliding over each other and on the finished condition of the surfaces. With greasy (boundary) lubrication, the coefficients depend both on the materials and conditions of the surfaces and on the lubricants employed.

Coefficients of friction are sensitive to atmospheric dust and humidity, oxide films, surface finish, velocity of sliding, temperature, vibration, and the extent of contamination. In many instances the degree of contamination is perhaps the most important single variable. For example, in Table 3.2.1, values for the static coefficient of friction of steel on steel are listed, and, depending upon the degree of contamination of the specimens, the coefficient of friction varies effectively from ∞ (infinity) to 0.013.

The most effective boundary lubricants are generally those which react chemically with the solid surface and form an adhering film that is attached to the surface with a chemical bond. This action depends upon

Table 3.2.1   Coefficients of Static Friction for Steel on Steel

Test condition	f_0	Ref.
Degassed at elevated temp in high vacuum	∞ (weld on contact)	1
Grease-free in vacuum	0.78	2
Grease-free in air	0.39	3
Clean and coated with oleic acid	0.11	2
Clean and coated with solution of stearic acid	0.013	4

SOURCES: (1) Bowden and Young, *Proc. Roy. Soc.*, 1951. (2) Campbell, *Trans. ASME*, 1939. (3) Tomlinson, *Phil. Mag.*, 1929. (4) Hardy and Doubleday, *Proc. Roy. Soc.*, 1923.

the nature of the lubricant and upon the reactivity of the solid surface. Table 3.2.2 indicates that a fatty acid, such as found in animal, vegetable, and marine oils, reduces the coefficient of friction markedly only if it can react effectively with the solid surface. Paraffin oil is almost completely nonreactive.

Table 3.2.2 Coefficients of Static Friction at Room Temperature

Surfaces	Clean	Paraffin oil	Paraffin oil plus 1% lauric acid	Degree of reactivity of solid
Nickel	0.7	0.3	0.28	Low
Chromium	0.4	0.3	0.3	Low
Platinum	1.2	0.28	0.25	Low
Silver	1.4	0.8	0.7	Low
Glass	0.9		0.4	Low
Copper	1.4	0.3	0.08	High
Cadmium	0.5	0.45	0.05	High
Zinc	0.6	0.2	0.04	High
Magnesium	0.6	0.5	0.08	High
Iron	1.0	0.3	0.2	Mild
Aluminum	1.4	0.7	0.3	Mild

SOURCE: From Bowden and Tabor, ''The Friction and Lubrication of Solids,'' Oxford.

Values in Table 3.2.4 of sliding and static coefficients have been selected largely from investigations where these variables have been very carefully controlled. They are representative values for smooth surfaces. It has been generally observed that sliding friction between hard materials is smaller than that between softer surfaces.

Effect of Surface Films Campbell observed a lowering of the coefficient of friction when oxide or sulfide films were present on metal surfaces (*Trans. ASME,* 1939; footnotes to Table 3.2.4). The reductions listed in Table 3.2.3 were obtained with oxide films formed by heating in air at temperatures from 100 to 500° C, and sulfide films produced by immersion in a 0.02 percent sodium sulfide solution.

Table 3.2.3 Static Coefficient of Friction f_0

	Clean and dry	Oxide film	Sulfide film
Steel-steel	0.78	0.27	0.39
Brass-brass	0.88		0.57
Copper-copper	1.21	0.76	0.74

Effect of Sliding Velocity It has generally been observed that coefficients of friction reduce on dry surfaces as sliding velocity increases. (See results of railway brake-shoe tests below.) Dokos measured this reduction in friction for mild steel on medium steel. Values are for the average of four tests with high contact pressures (*Trans. ASME,* 1946; see footnotes to Table 3.2.4).

Sliding velocity, in/s	0.0001	0.001	0.01	0.1	1	10	100
f	0.53	0.48	0.39	0.31	0.23	0.19	0.18

Effect of Surface Finish The degree of surface roughness has been found to influence the coefficient of friction. Burwell evaluated this effect for conditions of boundary or greasy friction (*Jour. SAE,* 1942; see footnotes to Table 3.2.4). The values listed in Table 3.2.5 are for sliding coefficients of friction, hard steel on hard steel. The friction coefficient and wear rates of **polymers against metals** are often lowered by decreasing the surface roughness. This is particularly true of composites such as those with polytetrafluoroethylene (PTFE) which function through transfer to the counterface.

Solid Lubricants In certain applications solid lubricants are used successfully. Boyd and Robertson with pressures ranging from 50,000 to 400,000 lb/in² (344,700 to 2,757,000 kN/m²) found sliding coefficients of friction f for hard steel on hard steel as follows: powdered mica, 0.305; powdered soapstone, 0.306; lead iodide, 0.071; silver sulfate, 0.054; graphite, 0.058; molybdenum disulfide, 0.033; tungsten disulfide, 0.037; stearic acid, 0.029 (*Trans. ASME,* 1945; see footnotes to Table 3.2.4).

Coefficients of Static Friction for Special Cases

Masonry and Earth Dry masonry on brickwork, 0.6–0.7; timber on polished stone, 0.40; iron on stone, 0.3 to 0.7; masonry on dry clay, 0.51; masonry on moist clay, 0.33.

Earth on Earth Dry sand, clay, mixed earth, 0.4 to 0.7; damp clay, 1.0; wet clay, 0.31; shingle and gravel, 0.8 to 1.1.

Natural Cork On cork, 0.59; on pine with grain, 0.49; on glass, 0.52; on dry steel, 0.45; on wet steel, 0.69; on hot steel, 0.64; on oiled steel, 0.45; water-soaked cork on steel, 0.56; oil-soaked cork on steel, 0.42.

Coefficients of Sliding Friction for Special Cases

Soapy Wood Lesley gives for wood on wood, copiously lubricated with tallow, stearine, and soft soap (as used in launching practice), a starting coefficient of friction equal to 0.036, diminishing to an average value of 0.019 for the first 50 ft of motion of the ship. Rennie gives 0.0385 for wood on wood, lubricated with soft soap, under a load of 56 lb/in².

Asbestos-Fabric Brake Material The coefficient of sliding friction f of asbestos fabric against a cast-iron brake drum, according to Taylor and Holt (*NBS,* 1940) is 0.35 to 0.40 when at normal temperature. It drops somewhat with rise in brake temperature up to 300°F (149°C). With a further increase in brake temperature from 300 to 500°F (149 to 260°C) the value of f may show an increase caused by disruption of the brake surface.

Steel Tires on Steel Rails (Galton)

Speed, mi/h	Start	6.8	13.5	27.3	40.9	54.4	60
Values of f	0.242	0.088	0.072	0.07	0.057	0.038	0.027

Railway Brake Shoes on Steel Tires Galton and Westinghouse give, for cast-iron brakes, the following values for f, which decrease rapidly with the speed of the rim; the coefficient f decreases also with time, as the temperature of the shoe increases.

Speed, mi/h	10	20	30	40	50	60
f, when brakes were applied	0.32	0.21	0.18	0.13	0.10	0.06
f, after 5 s	0.21	0.17	0.11	0.10	0.07	0.05
f, after 12 s		0.13	0.10	0.08	0.06	0.05

Schmidt and Schrader confirm the marked decrease in the coefficient of friction with the increase of rim speed. They also show an irregular slight decrease in the value of f with higher shoe pressure on the wheel, but they did not find the drop in friction after a prolonged application of the brakes. Their observations are as follows:

Speed, mi/h	20	30	40	50	60
Coefficient of friction	0.25	0.23	0.19	0.17	0.16

Friction of Steel on Polymers A useful list of friction coefficients between steel and various polymers is given in Table 3.2.6.

Grindstones The coefficient of friction between coarse-grained sandstone and cast iron is $f = 0.21$ to 0.24; for steel, 0.29; for wrought iron, 0.41 to 0.46, according as the stone is freshly trued or dull; for fine-grained sandstone (wet grinding) $f = 0.72$ for cast iron, 0.94 for steel, and 1.0 for wrought iron.

Honda and Yamada give $f = 0.28$ to 0.50 for carbon steel on emery, depending on the roughness of the wheel.

Table 3.2.4 Coefficients of Static and Sliding Friction

(Reference letters indicate the lubricant used; numbers in parentheses give the sources. See footnote.)

Materials	Static Dry	Static Greasy	Sliding Dry	Sliding Greasy
Hard steel on hard steel	0.78 (1)	0.11 (1, a)	0.42 (2)	0.029 (5, h)
		0.23 (1, b)		0.081 (5, c)
		0.15 (1, c)		0.080 (5, i)
		0.11 (1, d)		0.058 (5, j)
		0.0075 (18, p)		0.084 (5, d)
		0.0052 (18, h)		0.105 (5, k)
				0.096 (5, l)
				0.108 (5, m)
				0.12 (5, a)
Mild steel on mild steel	0.74 (19)		0.57 (3)	0.09 (3, a)
				0.19 (3, u)
Hard steel on graphite	0.21 (1)	0.09 (1, a)		
Hard steel on babbitt (ASTM No. 1)	0.70 (11)	0.23 (1, b)	0.33 (6)	0.16 (1, b)
		0.15 (1, c)		0.06 (1, c)
		0.08 (1, d)		0.11 (1, d)
		0.085 (1, e)		
Hard steel on babbitt (ASTM No. 8)	0.42 (11)	0.17 (1, b)	0.35 (11)	0.14 (1, b)
		0.11 (1, c)		0.065 (1, c)
		0.09 (1, d)		0.07 (1, d)
		0.08 (1, e)		0.08 (11, h)
Hard steel on babbitt (ASTM No. 10)		0.25 (1, b)		0.13 (1, b)
		0.12 (1, c)		0.06 (1, c)
		0.10 (1, d)		0.055 (1, d)
		0.11 (1, e)		
Mild steel on cadmium silver				0.097 (2, f)
Mild steel on phosphor bronze			0.34 (3)	0.173 (2, f)
Mild steel on copper lead				0.145 (2, f)
Mild steel on cast iron		0.183 (15, c)	0.23 (6)	0.133 (2, f)
Mild steel on lead	0.95 (11)	0.5 (1, f)	0.95 (11)	0.3 (11, f)
Nickel on mild steel			0.64 (3)	0.178 (3, x)
Aluminum on mild steel	0.61 (8)		0.47 93)	
Magnesium on mild steel			0.42 (3)	
Magnesium on magnesium	0.6 (22)	0.08 (22, y)		
Teflon on Teflon	0.04 (22)			0.04 (22, f)
Teflon on steel	0.04 (22)			0.04 (22, f)
Tungsten carbide on tungsten carbide	0.2 (22)	0.12 (22, a)		
Tungsten carbide on steel	0.5 (22)	0.08 (22, a)		
Tungsten carbide on copper	0.35 (23)			
Tungsten carbide on iron	0.8 (23)			
Bonded carbide on copper	0.35 (23)			
Bonded carbide on iron	0.8 (23)			
Cadmium on mild steel			0.46 (3)	
Copper on mild steel	0.53 (8)		0.36 (3)	0.18 (17, a)
Nickel on nickel	1.10 (16)		0.53 (3)	0.12 (3, w)
Brass on mild steel	0.51 (8)		0.44 (6)	
Brass on cast iron			0.30 (6)	
Zinc on cast iron	0.85 (16)		0.21 (7)	
Magnesium on cast iron			0.25 (7)	
Copper on cast iron	1.05 (16)		0.29 (7)	
Tin on cast iron			0.32 (7)	
Lead on cast iron			0.43 (7)	
Aluminum on aluminum	1.05 (16)		1.4 (3)	
Glass on glass	0.94 (8)	0.01 (10, p)	0.40 (3)	0.09 (3, a)
		0.005 (10, q)		0.116 (3, v)
Carbon on glass			0.18 (3)	
Garnet on mild steel			0.39 (3)	
Glass on nickel	0.78 (8)		0.56 (3)	

(a) Oleic acid; (b) Atlantic spindle oil (light mineral); (c) castor oil; (d) lard oil; (e) Atlantic spindle oil plus 2 percent oleic acid; (f) medium mineral oil; (g) medium mineral oil plus ½ percent oleic acid; (h) stearic acid; (i) grease (zinc oxide base); (j) graphite; (k) turbine oil plus 1 percent graphite; (l) turbine oil plus 1 percent stearic acid; (m) turbine oil (medium mineral); (n) olive oil; (p) palmitic acid; (q) ricinoleic acid; (r) dry soap; (s) lard; (t) water; (u) rape oil; (v) 3-in-1 oil; (w) octyl alcohol; (x) triolein; (y) 1 percent lauric acid in paraffin oil.

SOURCES: (1) Campbell, *Trans. ASME*, 1939; (2) Clarke, Lincoln, and Sterrett, *Proc. API*, 1935; (3) Beare and Bowden, *Phil. Trans. Roy. Soc.*, 1935; (4) Dokos, *Trans. ASME*, 1946; (5) Boyd and Robertson, *Trans. ASME*, 1945; (6) Sachs, *Zeit f. angew. Math. und Mech.*, 1924; (7) Honda and Yamaha, *Jour. I of M*, 1925; (8) Tomlinson, *Phil. Mag.*, 1929; (9) Morin, *Acad. Roy. des Sciences*, 1838; (10) Claypoole, *Trans. ASME*, 1943; (11) Tabor, *Jour. Applied Phys.*, 1945; (12) Eyssen, General Discussion on Lubrication, *ASME*, 1937; (13) Brazier and Holland-Bowyer, General Discussion on Lubrication, *ASME*, 1937; (14) Burwell, *Jour. SAE.*, 1942; (15) Stanton, "Friction," Longmans; (16) Ernst and Merchant, Conference on Friction and Surface Finish, M.I.T., 1940; (17) Gongwer, Conference on Friction and Surface Finish, M.I.T., 1940; (18) Hardy and Bircumshaw, *Proc. Roy. Soc.*, 1925; (19) Hardy and Hardy, *Phil. Mag.*, 1919; (20) Bowden and Young, *Proc. Roy. Soc.*, 1951; (21) Hardy and Doubleday, *Proc. Roy. Soc.*, 1923; (22) Bowden and Tabor, "The Friction and Lubrication of Solids," Oxford; (23) Shooter, *Research*, **4**, 1951.

Table 3.2.4 Coefficients of Static and Sliding Friction *(Continued)*
(Reference letters indicate the lubricant used; numbers in parentheses give the sources. See footnote.)

Materials	Static Dry	Static Greasy	Sliding Dry	Sliding Greasy
Copper on glass	0.68 (8)		0.53 (3)	
Cast iron on cast iron	1.10 (16)		0.15 (9)	0.070 (9, d)
				0.064 (9, n)
Bronze on cast iron			0.22 (9)	0.077 (9, n)
Oak on oak (parallel to grain)	0.62 (9)		0.48 (9)	0.164 (9, r)
				0.067 (9, s)
Oak on oak (perpendicular)	0.54 (9)		0.32 (9)	0.072 (9, s)
Leather on oak (parallel)	0.61 (9)		0.52 (9)	
Cast iron on oak			0.49 (9)	0.075 (9, n)
Leather on cast iron			0.56 (9)	0.36 (9, t)
				0.13 (9, n)
Laminated plastic on steel			0.35 (12)	0.05 (12, t)
Fluted rubber bearing on steel				0.05 (13, t)

(*a*) Oleic acid; (*b*) Atlantic spindle oil (light mineral); (*c*) castor oil; (*d*) lard oil; (*e*) Atlantic spindle oil plus 2 percent oleic acid; (*f*) medium mineral oil; (*g*) medium mineral oil plus ½ percent oleic acid; (*h*) stearic acid; (*i*) grease (zinc oxide base); (*j*) graphite; (*k*) turbine oil plus 1 percent graphite; (*l*) turbine oil plus 1 percent stearic acid; (*m*) turbine oil (medium mineral); (*n*) olive oil; (*p*) palmitic acid; (*q*) ricinoleic acid; (*r*) dry soap; (*s*) lard; (*t*) water; (*u*) rape oil; (*v*) 3-in-1 oil; (*w*) octyl alcohol; (*x*) triolein; (*y*) 1 percent lauric acid in paraffin oil.

SOURCES: (1) Campbell, *Trans. ASME*, 1939; (2) Clarke, Lincoln, and Sterrett, *Proc. API*, 1935; (3) Beare and Bowden, *Phil. Trans. Roy. Soc.*, 1935; (4) Dokos, *Trans. ASME*, 1946; (5) Boyd and Robertson, *Trans. ASME*, 1945; (6) Sachs, *Zeit f. angew. Math. und Mech.*, 1924; (7) Honda and Yamaha, *Jour. I of M*, 1925; (8) Tomlinson, *Phil. Mag.*, 1929; (9) Morin, *Acad. Roy. des Sciences*, 1838; (10) Claypoole, *Trans. ASME*, 1943; (11) Tabor, *Jour. Applied Phys.*, 1945; (12) Eyssen, General Discussion on Lubrication, *ASME*, 1937; (13) Brazier and Holland-Bowyer, General Discussion on Lubrication, *ASME*, 1937; (14) Burwell, *Jour. SAE.*, 1942; (15) Stanton, "Friction," Longmans; (16) Ernst and Merchant, Conference on Friction and Surface Finish, M.I.T., 1940; (17) Gongwer, Conference on Friction and Surface Finish, M.I.T., 1940; (18) Hardy and Bircumshaw, *Proc. Roy. Soc.*, 1925; (19) Hardy and Hardy, *Phil. Mag.*, 1919; (20) Bowden and Young, *Proc. Roy. Soc.*, 1951; (21) Hardy and Doubleday, *Proc. Roy. Soc.*, 1923; (22) Bowden and Tabor, "The Friction and Lubrication of Solids," Oxford; (23) Shooter, *Research*, **4**, 1951.

Table 3.2.5 Coefficient of Friction of Hard Steel on Hard Steel

	Surface					
	Superfinished	Ground	Ground	Ground	Ground	Grit-blasted
Roughness, microinches	2	7	20	50	65	55
Mineral oil	0.128	0.189	0.360	0.372	0.378	0.212
Mineral oil + 2% oleic acid	0.116	0.170	0.249	0.261	0.230	0.164
Oleic acid	0.099	0.163	0.195	0.222	0.238	0.195
Mineral oil + 2% sulfonated sperm oil	0.095	0.137	0.175	0.251	0.197	0.165

Table 3.2.6 Coefficient of Friction of Steel on Polymers
Room temperature, low speeds.

Material	Condition	f
Nylon	Dry	0.4
Nylon	Wet with water	0.15
Plexiglas	Dry	0.5
Polyvinyl chloride (PVC)	Dry	0.5
Polystyrene	Dry	0.5
Low-density (LD) polyethylene, no plasticizer	Dry	0.4
LD polyethylene, no plasticizer	Wet	0.1
High-density (HD) polyethylene, no plasticizer	Dry or wet	0.15
Soft wood	Natural	0.25
Lignum vitae	Natural	0.1
PTFE, low speed	Dry or wet	0.06
PTFE, high speed	Dry or wet	0.3
Filled PTFE (15% glass fiber)	Dry	0.12
Filled PTFE (15% graphite)	Dry	0.09
Filled PTFE (60% bronze)	Dry	0.09
Polyurethane rubber	Dry	1.6
Isoprene rubber	Dry	3–10
Isoprene rubber	Wet (water and alcohol)	2–4

Inflation pressure, lb/in²	Dry pavement Static f_0	Dry pavement Sliding f	Wet pavement Static f_0	Wet pavement Sliding f
40	0.90	0.85	0.74	0.69
50	0.88	0.84	0.64	0.58
60	0.80	0.76	0.63	0.56

Tests of the Goodrich Company on wet brick pavement with tires of different treads gave the following values of f:

	Coefficients of friction			
	Static (before slipping)		Sliding (after slipping)	
Speed, mi/h	5	30	5	30
Smooth tire	0.49	0.28	0.43	0.26
Circumferential grooves	0.58	0.42	0.52	0.36
Angular grooves at 60°	0.75	0.55	0.70	0.39
Angular grooves at 45°	0.77	0.55	0.68	0.44

Rubber Tires on Pavement Arnoux gives $f = 0.67$ for dry macadam, 0.71 for dry asphalt, and 0.17 to 0.06 for soft, slippery roads. For a cord tire on a sand-filled brick surface in fair condition. Agg (*Bull.* 88, *Iowa State College Engineering Experiment Station*, 1928) gives the following values of f depending on the inflation of the tire:

Development continues using various manufacturing techniques (bias ply, belted, radial, studs), tread patterns, and rubber compounds, so that it is not possible to present average values applicable to present conditions.

Sleds For **unshod wooden runners** on smooth wood or stone surfaces, $f = 0.07$ (0.15) when tallow (dry soap) is used as a lubricant (= 0.38 when not lubricated); on snow and ice, $f = 0.035$. For **runners with metal**

shoes on snow and ice, $f = 0.02$. Rennie found for steel on ice, $f = 0.014$. However, as the temperature falls, the coefficient of friction will get larger. Bowden cites the following data for brass on ice:

Temperature, °C	f
0	0.025
−20	0.085
40	0.115
−60	0.14

ROLLING FRICTION

Rolling is substituted frequently for sliding friction, as in the case of wheels under vehicles, balls or rollers in bearings, rollers under skids when moving loads; frictional resistance to the rolling motion is substantially smaller than to sliding motion. The fact that a resistance arises to rolling motion is due to several factors: (1) the contacting surfaces are elastically deflected, so that, on the finite size of the contact, relative sliding occurs, (2) the deflected surfaces dissipate energy due to internal friction (hysteresis), (3) the surfaces are imperfect so that contact takes place on asperities ahead of the line of centers, and (4) surface adhesion phenomena. The coefficient of rolling friction $f_r = P/L$ where L is the load and P is the frictional resistance.

The frictional resistance P to the rolling of a cylinder under a load L applied at the center of the roller (Fig. 3.2.5) is inversely proportional to the radius r of the roller; $P = (k/r)L$. Note that k has the dimensions of length. Quite often k increases with load, particularly for cases involv-

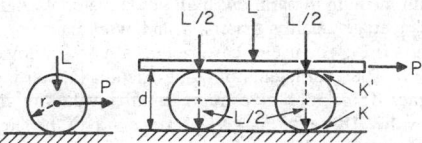

Fig. 3.2.5

ing plastic deformations. Values of k, in inches, are as follows: hardwood on hardwood, 0.02; iron on iron, steel on steel, 0.002; hard polished steel on hard polished steel, 0.0002 to 0.0004.

Data on rolling friction are scarce. Noonan and Strange give, for steel rollers on steel plates and for loads varying from light to those causing a permanent set of the material, the following values of k, in inches:

surfaces well finished and clean, 0.0005 to 0.001; surfaces well oiled, 0.001 to 0.002; surfaces covered with silt, 0.003 to 0.005; surfaces rusty, 0.005 to 0.01.

If a load L is moved on rollers (Fig. 3.2.5) and if k and k' are the respective coefficients of friction for the lower and upper surfaces, the frictional force $P = (k + k')L/d$.

McKibben and Davidson (*Agri. Eng.*, 1939) give the data in Table 3.2.7 on the rolling resistance of various types of wheels for typical road and field conditions. Note that the coefficient f_r is the ratio of resistance force to load.

Moyer found the following average values of f_r for pneumatic rubber tires properly inflated and loaded: hard road, 0.008; dry, firm, and well-packed gravel, 0.012; wet loose gravel, 0.06.

FRICTION OF MACHINE ELEMENTS

Work of Friction—Efficiency   In a simple machine or assemblage of two elements, the work done by an applied force P acting through the distance s is measured by the product Ps. The useful work done is less and is measured by the product Ll of the resistance L by the distance l through which it acts. The **efficiency** e of the machine is the ratio of the useful work performed to the total work received, or $e = Ll/Ps$. The **work expended in friction** W_f is the difference between the total work received and the useful work, or $W_f = Ps - Ll$. The lost-work ratio $= V = W_f/Ll$, and $e = 1/(1 + V)$.

If a machine consists of a train of mechanisms having the respective efficiencies $e_1, e_2, e_3 \ldots e_n$, the combined efficiency of the machine is equal to the product of these efficiencies.

Efficiencies of Machines and Machine Elements   The values for machine elements in Table 3.2.8 are from "Elements of Machine Design," by Kimball and Barr. Those for machines are from Goodman's "Mechanics Applied to Engineering." The quantities given are percentage efficiencies.

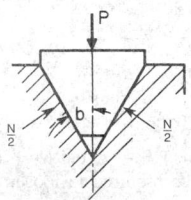

Fig. 3.2.6

Table 3.2.7   Coefficients of Rolling Friction f_r for Wheels with Steel and Pneumatic Tires

Wheel	Inflation press, lb/in²	Load, lb	Concrete	Bluegrass sod	Tilled loam	Loose sand	Loose snow 10–14 in deep
2.5 × 36 steel		1,000	0.010	0.087	0.384	0.431	0.106
4 × 24 steel		500	0.034	0.082	0.468	0.504	0.282
4.00–18 4-ply	20	500	0.034	0.058	0.366	0.392	0.210
4 × 36 steel		1,000	0.019	0.074	0.367	0.413	
4.00–30 4-ply	36	1,000	0.018	0.057	0.322	0.319	
4.00–36 4-ply	36	1,000	0.017	0.050	0.294	0.277	
5.00–16 4-ply	32	1,000	0.031	0.062	0.388	0.460	
6 × 28 steel		1,000	0.023	0.094	0.368	0.477	0.156
6.00–16 4-ply	20	1,000	0.027	0.060	0.319	0.338	0.146
6.00–16 4-ply*	30	1,000	0.031	0.070	0.401	0.387	
7.50–10 4-ply†	20	1,000	0.029	0.061	0.379	0.429	
7.50–16 4-ply	20	1,500	0.023	0.055	0.280	0.322	
7.50–28 4-ply	16	1,500	0.026	0.052	0.197	0.205	
8 × 48 steel		1,500	0.013	0.065	0.236	0.264	0.118
7.50–36 4-ply	16	1,500	0.018	0.046	0.185	0.177	0.0753
9.00–10 4-ply†	20	1,000	0.031	0.060	0.331	0.388	
9.00–16 6-ply	16	1,500	0.042	0.054	0.249	0.272	0.099

* Skid-ring tractor tire.
† Ribbed tread tractor tire.
All other pneumatic tires with implement-type tread.

Wedges

Sliding in V Guides If a wedge-shaped slide having an angle $2b$ is pressed into a V guide by a force P (Fig. 3.2.6), the total force normal to the wedge faces will be $N = P/\sin b$. A friction force F, opposing motion along the longitudinal axis of the wedge, arises by virtue of the coefficient of friction f between the contacting surface of the wedge and guides: $F = fN = fP/\sin b$. In these formulas, the fact that the elasticity of the materials permits an advance of the wedge into the guide under the load P has been neglected. The common efficiency for V guides is $e = 0.88$ to 0.90.

Taper Keys In Fig. 3.2.7 if the key is moved in the direction of the force P, the force H must be overcome. The supporting reactions K_1, K_2, and K_3 together with the required force P may be obtained by drawing the force polygon (Fig. 3.2.8). The friction angles of these faces are a_1, a_2, and a_3, respectively. In Fig. 3.2.8, draw AB parallel to H in Fig. 3.2.7, and lay it off to scale to represent H. From the point A, draw AC

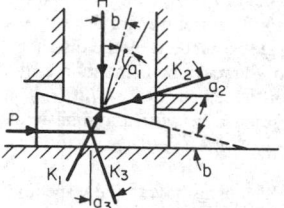

Fig. 3.2.7

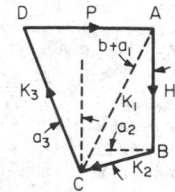

Fig. 3.2.8

parallel to K_1, i.e., making the angle $b + a_1$ with AB; from the other extremity of AB, draw BC parallel to K_2 in Fig. 3.2.7. AC and CB then give the magnitudes of K_1 and K_2, respectively. Now through C draw CD parallel to K_3 to its intersection with AD which has been drawn through A parallel to P. The magnitudes of K_3 and P are then given by the lengths of CD and DA.

By calculation,

$$K_1/H = \cos a_2/\cos (b + a_1 + a_2)$$
$$P/K_1 = \sin (b + a_1 + a_3)/\cos a_3$$
$$P/H = \cos a_2 \sin (b + a_1 + a_3)/\cos a_3 \cos (b + a_1 + a_2)$$

If $a_1 = a_2 = a_3 = a$, then $P = H \tan (b + 2a)$, and efficiency $e = \tan b/\tan (b + 2a)$. Force required to loosen the key $= P_1 = H \tan (2a - b)$. In order for the key not to slide out when force P is removed, it is necessary that $b < (a_1 + a_3)$, or $b < 2a$.

The forces acting upon the taper key of Fig. 3.2.9 may be found in a similar way (see Fig. 3.2.10).

$$P = 2H \cos a \sin (b + a)/\cos (b + 2a)$$
$$= 2H \tan (b + a)/[1 - \tan a \tan (b + a)]$$
$$= 2H \tan (b + a) \text{ approx}$$

The force to loosen the key is $P_1 = 2H \tan (a - b)$ approx, and the efficiency $e = \tan b/\tan (b + a)$. The key will be self-locking when $b < a$, or, more generally, when $2b < (a_1 + a_3)$.

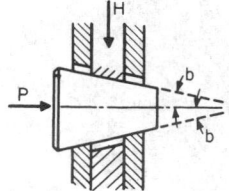

Fig. 3.2.9

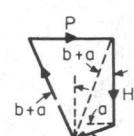

Fig. 3.2.10

Screws

Screws with Square Threads (Fig. 3.2.11) Let $r =$ mean radius of the thread $= \frac{1}{2}$ (radius at root $+$ outside radius), and $l =$ pitch (or lead

of a single-threaded screw), both in inches; $b =$ angle of inclination of thread to a plane at right angles to the axis of screw $(\tan b = l/2\pi r)$; and $f =$ coefficient of sliding friction $= \tan a$. Then for a screw in uniform motion (friction of the root and outside surfaces being neglected) there is required a force P acting at right angles to the axis at the distance r. $P = L \tan (b \pm a) = L(l \pm 2\pi rf)/(2\pi r \pm fl)$, where the upper signs are for motion in a direction opposed to that of L and the lower for motion in the same direction as that of L. When $b \leq a$, the screw will not "overhaul" (or move under the action of the load L).

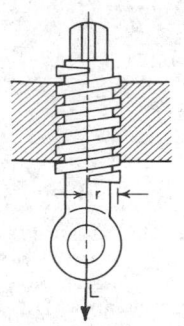

Fig. 3.2.11

The **efficiency** for motion opposed to direction in which L acts $= e = \tan b/\tan (b + a)$; for motion in the same direction in which L acts, $e = \tan (b - a)/\tan b$.

The value of e is a maximum when $b = 45° - \frac{1}{2}a$; e.g., $e_{max} = 0.81$ for $b = 42°$ and $f = 0.1$. Since e increases rapidly for values of b up to $20°$, this angle is generally not exceeded; for $b = 20°$, and $f_1 = 0.10$, $e = 0.74$. In presses, where the mechanical advantage is required to be great, b is taken down to $3°$, for which value $e = 0.34$ with $f = 0.10$.

Kingsbury found for square-threaded screws running in loose-fitting nuts, the following coefficients of friction: lard oil, 0.09 to 0.25; heavy mineral oil, 0.11 to 0.19; heavy oil with graphite, 0.03 to 0.15.

Ham and Ryan give for screws the following values of coefficients of friction, with medium mineral oil: high-grade materials and workmanship, 0.10; average quality materials and workmanship, 0.12; poor workmanship, 0.15. The use of castor oil as a lubricant lowered f from 0.10 to 0.066. The coefficients of static friction (at starting) were 30 percent higher. Table 3.2.8 gives representative values of efficiency.

Screws with V Threads (Fig. 3.2.12) Let $c =$ half the angle between the faces of a thread. Then, using the same notation as for square-threaded screws, for a screw in motion (neglecting friction of root and outside surfaces),

$$P = L(l \pm 2\pi rf \sec d)/(2\pi r \pm lf \sec d)$$

d is the angle between a plane normal to the axis of the screw through the point of the resultant thread friction, and a plane which is tangent to

Table 3.2.8 Efficiencies of Machines and Machine Elements

Common bearing (singly)	96–98
Common bearing, long lines of shafting	95
Roller bearings	98
Ball bearings	99
Spur gear, including bearings	
Cast teeth	93
Cut teeth	96
Bevel gear, including bearings	
Cast teeth	92
Cut teeth	95
Worm gear	
Thread angle, 30°	85–95
Thread angle, 15°	75–90
Belting	96–98
Pin-connected chains (bicycle)	95–97
High-grade transmission chains	97–99
Weston pulley block (½ ton)	30–47
Epicycloidal pulley block	40–45
1-ton steam hoist or windlass	50–70
Hydraulic windlass	60–80
Hydraulic jack	80–90
Cranes (steam)	60–70
Overhead traveling cranes	30–50
Locomotives (drawbar hp/ihp)	65–75
Hydraulic couplings, max	98

the surface of the thread at the same point (see Groat, *Proc. Engs. Soc. West. Penn,* **34**). Sec $d = \sec c \sqrt{1 - (\sin b \sin c)^2}$. For small values of b this reduces practically to sec $d = \sec c$, and, for all cases the approximation, $P = L(l \pm 2\pi rf \sec c)/(2\pi r \pm lf \sec c)$ is within the limits of probable error in estimating values to be used for f.

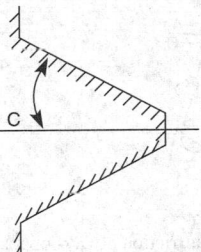

Fig. 3.2.12

The **efficiencies** are: $e = \tan b(1 - f \tan b \sec d)/(\tan b + f \sec d)$ for motion opposed to L, and $e = (\tan b - f \sec d)/\tan b(1 + f \tan b \sec d)$ for motion with L. If we let $\tan d' = f \sec d$, these equations reduce, respectively, to $e = \tan b/\tan (b + d')$ and $e = \tan (b - d')/\tan b$. Negative values in the latter case merely mean that the thread will not overhaul. Subtract the values from unity for actual efficiency, considering the external moment and not the load L as being the driver. The efficiency of a V thread is lower than that of a square thread of the same helix angle, since $d' > a$.

For a **V-threaded screw and nut,** let r_1 = outside radius of thread, r_2 = radius at root of thread, $r = (r_1 + r_2)/2$, $\tan d' = f \sec d$, r_0 = mean radius of nut seat = $1.5r$ (approx) and f' = coefficient of friction between nut and seat.

To tighten up the nut the turning moment required is $M = Pr + Lr_0 f = Lr[\tan (d' + b) + 1.5f']$. **To loosen** $M = Lr[\tan (d' - b) + 1.5f']$.

The **total tension in a bolt** due to tightening up with a moment M is $T = 2\pi M/(l + fl \sec b \sec d \operatorname{cosec} b + f'3\pi r)$. $T \div$ area at root gives unit pure tensile stress induced, S_t. There is also a unit torsional stress: $S_s = 2(M - 1.5rf'T)/\pi r_2^3$. The equivalent combined stress is $S = 0.35S_t + 0.65 \sqrt{S_t^2 + 4S_s^2}$.

Kingsbury, from tests on U.S. standard bolts, finds efficiencies for tightening up nuts from 0.06 to 0.12, depending upon the roughness of the contact surfaces and the character of the lubrication.

Toothed and Worm Gearing

The efficiency of **spur and bevel gearing** depends on the material and the workmanship of the gears and on the lubricant employed. For high-speed gears of good quality the efficiency of the gear transmission is 99 percent; with slow-speed gears of average workmanship the efficiency of 96 percent is common. On the average, efficiencies of 97 to 98 percent can be considered normal.

In **helical gears,** where considerable transverse sliding of the meshing teeth on each other takes place, the friction is much greater. If b and c are, respectively, the spiral angles of the teeth of the driving and driven helical gears (i.e., the angle between the teeth and the axis of rotation), $b + c$ is the shaft angle of the two gears, and $f = \tan a$ is the coefficient of sliding friction of the teeth, the efficiency of the gear transmission is $e = [\cos b \cos (c + a)]/[\cos c \cos (b - a)]$.

In the case of **worm gearing** when the shafts are normal to each other ($b + c = 90$), the efficiency is $e = \tan c/\tan (c + a) = (1 - pf/2\pi r)/(1 + 2\pi rf/p)$, where c is the spiral angle of the worm wheel, or the lead angle of the worm; p the lead, or pitch, of the worm thread; and r the mean radius of the worm. Typical values of f are shown in Table 3.2.9.

Journals and Bearings

Friction of Journal Bearings If P = total load on journal, l = journal length, and $2r$ = journal diameter, then $p = P/2rl$ = mean normal pressure on the projected area of the journal. Also, if f_1 is the **coefficient of journal friction,** the **moment of journal friction** for a cylindrical journal is $M = f_1Pr$. The **work expended in friction** at angular velocity ω is

$$W_f = \omega M = f_1 Pr\omega$$

For the **conical bearing** (Fig. 3.2.13) the mean radius $r_m = (r + R)/2$ is to be used.

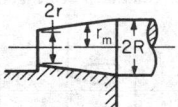

Fig. 3.2.13

Values of Coefficient of Friction For very low velocities of rotation (e.g., below 10 r/min), high loads, and with good lubrication, the coefficient of friction approaches the value of greasy friction, 0.07 to 0.15 (see Table 3.2.4). This is also the "pullout" coefficient of friction *on starting* the journal. With higher velocities, a fluid film is established between the journal and bearing, and the values of the coefficient of friction depend on the speed of rotation, the pressure on the bearing, and the viscosity of the oil. For journals running in complete bearing bushings, with a small clearance, i.e., with the diameter of the bushing slightly larger than the diameter of the journal, the experimental data of McKee give approximate values of the coefficient of friction as in Fig. 3.2.14.

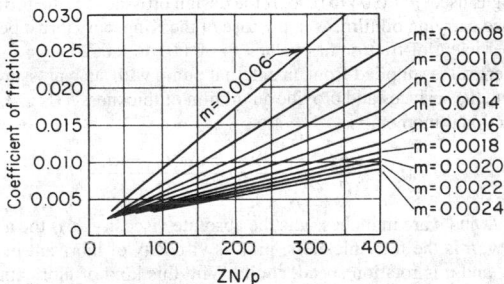

Fig. 3.2.14 Coefficient of friction of journal.

If d_1 is the diameter of the bushing in inches, d the diameter of the journal in inches, then $(d_1 - d)$ is the diametral clearance and $m = (d_1 - d)/d$ is the clearance ratio. The diagram of McKee (Fig. 3.2.14) gives the coefficient of friction as a function of the characteristic num-

Table 3.2.9 Coefficients of Friction for Worm Gears

Rubbing speed of worm, ft/min	100	200	300	500	800	1200
(m/min)	(30.5)	(61)	(91.5)	(152)	(244)	(366)
Phosphor-bronze wheel, polished-steel worm	0.054	0.045	0.039	0.030	0.024	0.020
Single-threaded cast-iron worm and gear	0.060	0.051	0.047	0.034	0.025	

ber ZN/p, where N is the speed of rotation in revolutions per minute, $p = P/(dl)$ is the average pressure in lb/in^2 on the projected area of the bearing, P is the load, l is the axial length of the bearing, and Z is the absolute viscosity of the oil in centipoises. Approximate values of Z at 100 (130)°F are as follows: light machine oil, 30 (16); medium machine oil, 60 (25); medium-heavy machine oil, 120 (40); heavy machine oil, 160 (60).

For purposes of design of ordinary machinery with bearing pressures from 50 to 300 lb/in^2 (344.7 to 2,068 kN/m^2) and speeds of 100 to 3,000 rpm, values for the coefficient of journal friction can be taken from 0.008 to 0.020.

Thrust Bearings

Frictional Resistance for Flat Ring Bearing Step bearings or pivots may be used to resist the end thrust of shafts. Let L = total load in the direction of the shaft axis and f = coefficient of sliding friction.

For a **ring-shaped flat step bearing** such as that shown in Fig. 3.2.15 (or a **collar bearing**), the **moment of thrust friction** $M = \frac{1}{3} fL(D^3 - d^3)/(D^2 - d^2)$. For a **flat circular step bearing**, $d = 0$, and $M = \frac{1}{3} fLD$.

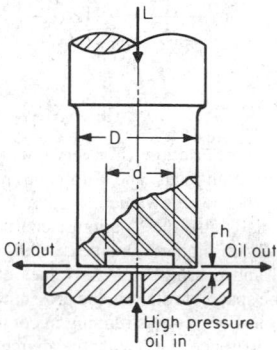

Fig. 3.2.15

The value of the coefficient of sliding friction is 0.08 to 0.15 when the speed of rotation is very slow. At higher velocities when a collar or step bearing is used, $f = 0.04$ to 0.06. If the design provides for the formation of a load carrying oil film, as in the case of the Kingsbury thrust bearing, the coefficient of friction has values $f = 0.001$ to 0.0025.

Where oil is supplied from an external pump with such pressure as to separate the surfaces and provide an oil film of thickness h (Fig. 3.2.15), the frictional moment is

$$M = \frac{Zn(D^4 - d^4)}{67 \times 10^7 h} = \frac{\pi\mu\omega(D^4 - d^4)}{32 h}$$

where D and d are in inches, μ is the absolute viscosity, ω is the angular velocity, h is the film thickness, in, Z is viscosity of lubricant in centipoises, and n is rotation speed, r/min. With this kind of lubrication the frictional moment depends upon the speed of rotation of the shaft and actually approaches zero for zero shaft speeds. The thrust load will be carried on a film of oil regardless of shaft rotation for as long as the pump continues to supply the required volume and pressure (see also Secs. 8 and 14).

EXAMPLE. A hydrostatic thrust bearing carries 101,000 lb, D is 16 in, d is 10 in, oil-film thickness h is 0.006 in, oil viscosity Z, 30 centipoises at operating temperature, and n is 750 r/min. Substituting these values, the frictional torque M is 310 in · lb (358 cm · kg). The oil supply pressure was 82.5 lb/in^2 (569 kN/m^2); the oil flow, 12.2 gal/min (46.2 l/min).

Frictional Forces in Pin Joints of Mechanisms

In the absence of friction, or when the effect of friction is negligible, the force transmitted by the link b from the driver a to the driven link c (Figs. 3.2.16 and 3.2.17) acts through the centerline OO of the pins connecting the link b with links a and c. With friction, this line of action shifts to the line AA, tangent to small circles of diameter d. The diameter

d of the circle, called the **friction circle**, for each individual joint, is equal to fD, where D is the diameter of the pin and f is the coefficient of friction between the pin and the link. The choice of the proper disposition of the tangent AA with respect to the two friction circles is dictated

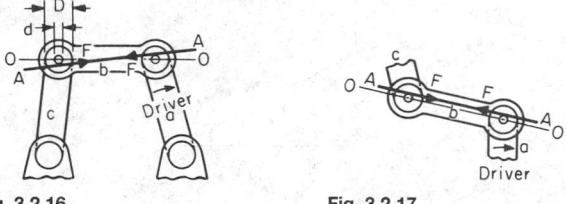

Fig. 3.2.16   **Fig. 3.2.17**

by the consideration that friction always opposes the action of the linkage. The force f opposes the motion of a; therefore, with friction it acts on a longer lever than without friction (Figs. 3.2.16 and 3.2.17). On the other hand, the force F drives the link c; friction hinders its action, and the equivalent lever is shorter with friction than without friction; the friction throws the line of action toward the center of rotation of link c.

EXAMPLE. An engine eccentric (Fig. 3.2.18) is a joint where the friction loss may be large. For the dimensions shown and with a torque of 250 in · lb applied to the rotating shaft, the resultant horizontal force, with no friction, will act through the center of the eccentric and be 250/(2.5 sin 60) or 115.5 lb. With friction coefficient 0.1, the resultant force (which for a long rod remains approximately horizontal) will be tangent to the friction circle of radius 0.1 × 5, or 0.5 in, and have a magnitude of 250/(2.5 sin 60 + 0.5), or 93.8 lb (42.6 kg).

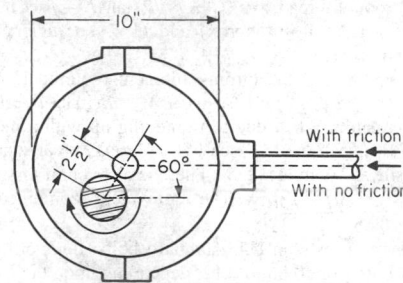

Fig. 3.2.18

Tension Elements

Frictional Resistance In Fig. 3.2.19, let T_1 and T_2 be the tensions with which a rope, belt, chain, or brake band is strained over a drum, pulley, or sheave, and let the rope or belt be on the point of slipping from T_2 toward T_1 by reason of the difference of tension $T_1 - T_2$. Then $T_1 - T_2$ = circumferential force P transferred by friction must be equal

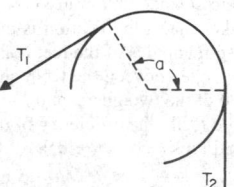

Fig. 3.2.19

to the frictional resistance W of the belt, rope, or band on the drum or pulley. Also, let a = angle subtending the arc of contact between the drum and tension element. Then, disregarding centrifugal forces,

$$T_1 = T_2 e^{fa} \quad \text{and} \quad P = (e^{fa} - 1)T_1/e^{fa} = (e^{fa} - 1)T_2 = W$$

where e = base of the napierian system of logarithms = 2.178+.

Table 3.2.10   Values of e^{fa}

$\dfrac{a°}{360°}$	0.1	0.15	0.2	0.25	0.3	0.35	0.4	0.45	0.5
0.1	1.06	1.1	1.13	1.17	1.21	1.25	1.29	1.33	1.37
0.2	1.13	1.21	1.29	1.37	1.46	1.55	1.65	1.76	1.87
0.3	1.21	1.32	1.45	1.60	1.76	1.93	2.13	2.34	2.57
0.4	1.29	1.46	1.65	1.87	2.12	2.41	2.73	3.10	3.51
0.425	1.31	1.49	1.70	1.95	2.23	2.55	2.91	3.33	3.80
0.45	1.33	1.53	1.76	2.03	2.34	2.69	3.10	3.57	4.11
0.475	1.35	1.56	1.82	2.11	2.45	2.84	3.30	3.83	4.45
0.5	1.37	1.60	1.87	2.19	2.57	3.00	3.51	4.11	4.81
0.525	1.39	1.64	1.93	2.28	2.69	3.17	3.74	4.41	5.20
0.55	1.41	1.68	2.00	2.37	2.82	3.35	3.98	4.74	5.63
0.6	1.46	1.76	2.13	2.57	3.10	3.74	4.52	5.45	6.59
0.7	1.55	1.93	2.41	3.00	3.74	4.66	5.81	7.24	9.02
0.8	1.65	2.13	2.73	3.51	4.52	5.81	7.47	9.60	12.35
0.9	1.76	2.34	3.10	4.11	5.45	7.24	9.60	12.74	16.90
1.0	1.87	2.57	3.51	4.81	6.59	9.02	12.35	16.90	23.14
1.5	2.57	4.11	6.59	10.55	16.90	27.08	43.38	69.49	111.32
2.0	3.51	6.59	12.35	23.14	43.38	81.31	152.40	285.68	535.49
2.5	4.81	10.55	23.14	50.75	111.32	244.15	535.49	1,174.5	2,575.9
3.0	6.59	16.90	43.38	111.32	285.68	733.14	1,881.5	4,828.5	12,391
3.5	9.02	27.08	81.31	244.15	733.14	2,199.90	6,610.7	19,851	59,608
4.0	12.35	43.38	152.40	535.49	1,881.5	6,610.7	23,227	81,610	286,744

NOTE: $e^{\pi} = 23.1407$, $\log e^{\pi} = 1.3643764$.

f is the static coefficient of friction (f_0) when there is no slip of the belt or band on the drum and the coefficient of kinetic friction (f) when slip takes place. For ease of computation, the values of the quantity e^{fa} are tabulated on Table 3.2.10.

Average values of f_0 for belts, ropes, and brake bands are as follows: for leather belt on cast-iron pulley, very greasy, 0.12; slightly greasy, 0.28; moist, 0.38. For hemp rope on cast-iron drum, 0.25; on wooden drum, 0.40; on rough wood, 0.50; on polished wood, 0.33. For iron brake bands on cast-iron pulleys, 0.18. For wire ropes, Tichvinsky reports coefficients of static friction, f_0, for a ⅝ rope (8 × 19) on a worn-in cast-iron groove: 0.113 (dry); for mylar on aluminum, 0.4 to 0.7.

Belt Transmissions; Effects of Belt Compliance

In the configuration of Fig. 3.2.20, pulley A drives a belt at angular velocity ω_A. Pulley B, here assumed to be of the same radius R as A, is driven at angular velocity ω_B. If the belt is extensible and the resistive torque $M = (T_1 - T_2)R$ is applied at B, ω_B will be smaller than ω_A and power will be dissipated at a rate $W = M(\omega_A - \omega_B)$. Likewise, the surface velocity V_1 of the more stretched belt will be larger than V_2. No slip will take place over the wraps A_T-A_S and B_T-B_S. The slip angles a_A

and a_B can be calculated from

$$a_A = [\ln(T_1/T_2)]/f_A \qquad a_B = [\ln(T_1/T_2)]/f_B$$

where f_A and f_B are the coefficients of friction on pulleys A and B, respectively. To calculate the above values, it is necessary to know the mean tension of the belt, $T = (T_1 + T_2)/2$. Then, $T_1/T_2 = [T + M/(2R)]/[T - M/(2R)]$. In this configuration, when the slip angles become equal to π (180°), complete slip occurs.

It is interesting to note that torque is transmitted only over the slip arcs a_A and a_B since there is no tension variation in the arcs A_T-A_S and B_T-B_S where the belt is in a uniform state of stretch.

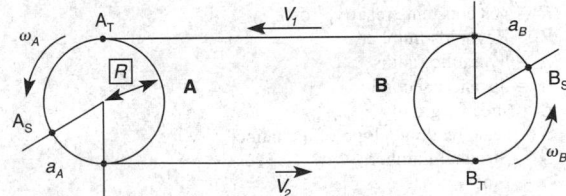

Fig. 3.2.20   Pulley transmission with extensible belt.

3.3 MECHANICS OF FLUIDS
by J. W. Murdock

REFERENCES: *Specific.* "Handbook of Chemistry and Physics," Chemical Rubber Company. "Smithsonian Physical Tables," Smithsonian Institution. "Petroleum Measurement Tables," ASTM. "Steam Tables," ASME. "American Institute of Physics Handbook," McGraw-Hill. "International Critical Tables," McGraw-Hill. "Tables of Thermal Properties of Gases," *NBS Circular 564.* Murdock, "Fluid Mechanics and its Applications," Houghton Mifflin, 1976. "Pipe Friction Manual," Hydraulic Institute. "Flow of Fluids," ASME, 1971. "Fluid Meters," 6th ed., ASME, 1971. "Measurement of Fluid Flow in Pipes Using Orifice, Nozzle, and Venturi," ASME Standard MFC-3M-1984. Murdock, ASME 64-WA/FM-6. Horton, *Engineering News,* 75, 373, 1916. Belvins, ASME 72/WA/FE-39. Staley and Graven, ASME 72PET/30. "Temperature Measurement," PTC 19.3, ASME. Moody, *Trans. ASME,* 1944, pp. 671–684. *General.* Binder, "Fluid Mechanics," Prentice-Hall. Langhaar, "Dimensional Analysis and Theory of Models," Wiley. Murdock, "Fluid Mechanics," Drexel University Press. Rouse, "Elementary Mechanics of Fluids," Wiley. Shames, "Mechanics of Fluids," McGraw-Hill. Streeter, "Fluid Mechanics," McGraw-Hill.

Notation

a = acceleration, area, exponent
A = area
c = velocity of sound
C = coefficient
$\mathbf{C}$ = Cauchy number
$\mathbf{C}_p$ = pressure coefficient
d = diameter, distance
E = bulk modulus of elasticity, modulus of elasticity (Young's modulus), velocity of approach factor, specific energy
$\mathbf{E}$ = Euler number
f = frequency, friction factor
F = dimension of force, force
$\mathbf{F}$ = Froude number
g = acceleration due to gravity
g_c = proportionality constant = 32.1740 lmb/(lbf)(ft/s^2)
G = mass velocity
h = head, vertical distance below a liquid surface
H = geopotential altitude
i = ideal
I = moment of inertia
J = mechanical equivalent of heat, 778.169 ft·lbf
k = isentropic exponent, ratio of specific heats
K = constant, resistant coefficient, weir coefficient
$\mathbf{K}$ = flow coefficient
L = dimension of length, length
m = mass, lbm
$\dot{m}$ = mass rate of flow, lbm/s
M = dimension of mass, mass (slugs)
$\dot{M}$ = mass rate of flow, slugs/s
$\mathbf{M}$ = Mach number
n = exponent for a polytropic process, roughness factor
N = dimensionless number
p = pressure
P = perimeter, power
q = heat added
$\mathbf{q}$ = flow rate per unit width
Q = volumetric flow rate
r = pressure ratio, radius
R = gas constant, reactive force
$\mathbf{R}$ = Reynolds number
R_h = hydraulic radius
s = distance, second
sp. gr. = specific gravity
S = scale reading, slope of a channel
$\mathbf{S}$ = Strouhal number
t = time
T = dimension of time, absolute temperature
u = internal energy
U = stream-tube velocity
v = specific volume
V = one-dimensional velocity, volume
$\mathbf{V}$ = velocity ratio
W = work done by fluid
$\mathbf{W}$ = Weber number
x = abscissa
y = ordinate
Y = expansion factor
z = height above a datum
Z = compressibility factor, crest height
α = angle, kinetic energy correction factor
β = ratio of primary element diameter to pipe diameter
γ = specific weight
δ = boundary-layer thickness
ε = absolute surface roughness
θ = angle
μ = dynamic viscosity
ν = kinematic viscosity
π = 3.14159 . . . , dimensionless ratio
ρ = density
σ = surface tension
τ = unit shear stress
ω = rotational speed

FLUIDS AND OTHER SUBSTANCES

Substances may be classified by their response when at rest to the imposition of a shear force. Consider the two very large plates, one moving, the other stationary, separated by a small distance y as shown in Fig. 3.3.1. The space between these plates is filled with a substance whose surfaces adhere to these plates in such a manner that its upper surface moves at the same velocity as the upper plate and the lower surface is stationary. The upper surface of the substance attains a velocity of U as the result of the application of shear force F_s. As y approaches dy, U approaches dU, and the rate of deformation of the substance becomes dU/dy. The **unit shear stress** is defined by $\tau = F_s/A_s$, where A_s is the shear or surface area. The deformation characteristics of various substances are shown in Fig. 3.3.2.

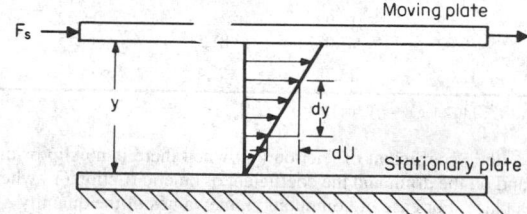

Fig. 3.3.1 Flow of a substance between parallel plates.

An ideal or **elastic solid** will resist the shear force, and its rate of deformation will be zero regardless of loading and hence is coincident with the ordinate of Fig. 3.3.2. A **plastic** will resist the shear until its yield stress is attained, and the application of additional loading will cause it to deform continuously, or flow. If the deformation rate is directly proportional to the flow, it is called an **ideal plastic**.

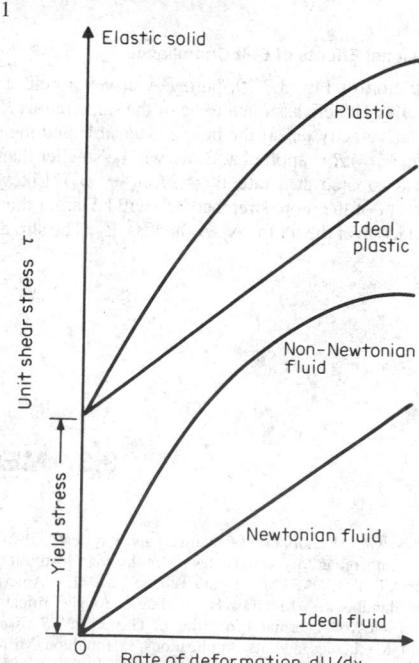

Fig. 3.3.2 Deformation characteristics of substances.

If the **substance** is unable to resist even the slightest amount of shear without flowing, it is a **fluid**. An **ideal fluid** has no internal friction, and hence its deformation rate coincides with the abscissa of Fig. 3.3.2. All real fluids have internal friction so that their rate of deformation is proportional to the applied shear stress. If it is directly proportional, it is called a **Newtonian fluid**; if not, a **non-Newtonian fluid**.

Two kinds of **fluids** are considered in this section, incompressible and compressible. A **liquid** except at very high pressures and/or temperatures may be considered incompressible. **Gases** and **vapors** are compressible fluids, but only **ideal gases** (those that follow the ideal-gas laws) are considered in this section. All others are covered in Secs. 4.1 and 4.2.

FLUID PROPERTIES

The **density** ρ of a fluid is its mass per unit volume. Its dimensions are M/L^3. In **fluid mechanics,** the units are slugs/ft^3 and lbf · s^2/ft^4 (515.3788 kg/m^3), but in **thermodynamics** (Sec. 4.1), the units are lbm/ft^3 (16.01846 kg/m^3). Numerical values of densities for selected liquids are shown in Table 3.3.1. The temperature change at 68°F (20°C) required to produce a 1 percent change in density varies from 12°F (6.7°C) for kerosene to 99°F (55°C) for mercury.

The **specific volume** v of a fluid is its volume per unit mass. Its dimensions are L^3/M. The units are ft^3/lbm. Specific volume is related to density by $v = 1/\rho g_c$, where g_c is the proportionality constant [32.1740 (lbm/lbf)(ft/s^2)]. **Specific volumes** of ideal gases may be computed from the equation of state: $v = RT/p$, where R is the gas constant in ft · lbf/(lbm)(°R) (see Sec. 4.1), T is the temperature in degrees Rankine (°F + 459.67), and p is the pressure in lbf/ft^2 abs.

The **specific weight** γ of a fluid is its weight per unit volume and has dimensions of F/L^3 or $M/(L^2)(T^2)$. The units are lbf/ft^3 or slugs/(ft^2)(s^2) (157.087 N/m^3). Specific weight is related to density by $\gamma = \rho g$, where g is the acceleration of gravity.

The **specific gravity** (sp. gr.) of a substance is a dimensionless ratio of the density of a fluid to that of a reference fluid. Water is used as the reference fluid for solids and liquids, and air is used for gases. Since the density of liquids changes with temperature for a precise definition of **specific gravity,** the temperature of the fluid and the reference fluid should be stated, for example, 60/60°F, where the upper temperature pertains to the liquid and the lower to water. If no temperatures are stated, reference is made to water at its maximum density, which occurs at 3.98°C and atmospheric pressure. The maximum density of water is 1.9403 slugs/ft^3 (999.973 kg/m^3). See Sec. 1.2 for conversion factors for API and Baumé hydrometers. For gases, it is common practice to use the ratio of the molecular weight of the gas to that of air (28.9644), thus eliminating the necessity of stating the pressure and temperature for ideal gases.

The **bulk modulus of elasticity** E of a fluid is the ratio of the pressure stress to the volumetric strain. Its dimensions are F/L^2. The units are lbf/in^2 or lbf/ft^2. E depends upon the thermodynamic process causing the change of state so that $E_x = -v(\partial p/\partial v)_x$, where x is the process. For ideal gases, $E_T = p$ for an isothermal process and $E_s = kp$ for an isentropic process where k is the ratio of specific heats. Values of E_T and E_S for liquids are given in Table 3.3.2. For liquids, a mean value is used by integrating the equation over a finite interval, or $E_{xm} = -v_1(\Delta p/\Delta v)_x - v_1(p_2 - p_1)/(v_1 - v_2)_x$.

EXAMPLE. What pressure must be applied to ethyl alcohol at 68°F (20°C) to produce a 1 percent decrease in volume at constant temperature?

$$\Delta p = -E_T(\Delta v/v) = -(130,000)(-0.01)$$
$$= 1,300 \text{ lbf/in}^2 \ (9 \times 10^6 \text{ N/m}^2)$$

In a like manner, the pressure required to produce a 1 percent decrease in the volume of mercury is found to be 35,900 lbf/in^2 (248 × 10^6 N/m^2). For most engineering purposes, liquids may be considered as incompressible fluids.

The **acoustic velocity,** or velocity of sound in a fluid, is given by $c = \sqrt{E_s/\rho}$. For an ideal gas $c = \sqrt{kp/\rho} = \sqrt{kg_cpv} = \sqrt{kg_cRT}$. Values of the speed of sound in liquids are given in Table 3.3.2.

EXAMPLE. Check the value of the velocity of sound in benzene at 68°F (20°C) given in Table 3.3.2 using the isentropic bulk modulus. $c = \sqrt{E_s/\rho} = \sqrt{144 \times 223,000/1.705} = 4,340$ ft/s (1,320 m/s). Additional information on the velocity of sound is given in Secs. 4, 11, and 12.

Application of shear stress to a fluid results in the continual and permanent distortion known as flow. **Viscosity** is the resistance of a fluid to shear motion—its internal friction. This resistance is due to two phenomena: (1) cohesion of the molecules and (2) molecular transfer from one layer to another, setting up a tangential or shear stress. In liquids, cohesion predominates, and since cohesion decreases with increasing temperature, the viscosity of liquids does likewise. Cohesion is relatively weak in gases; hence increased molecular activity with increasing temperature causes an increase in molecular transfer with corresponding increase in viscosity.

The **dynamic viscosity** μ of a fluid is the ratio of the shearing stress to the rate of deformation. From Fig. 3.3.1, $\mu = \tau/(dU/dy)$. Its dimensions are $(F)(T)/L^2$ or $M/(L)(T)$. The units are lbf · s/ft^2 or slugs/(ft)(s) [47.88026(N · s)/m^2].

In the cgs system, the unit of dynamic viscosity is the **poise,** 2,089 × 10^{-6} (lbf · s)/ft^2 [0.1 (N · s)/m^2], but for convenience the **centipoise** (1/100 poise) is widely used. The dynamic viscosity of water at 68°F (20°C) is approximately 1 centipoise.

Table 3.3.3 gives values of dynamic viscosity for selected liquids at atmospheric pressure. Values of viscosity for fuels and lubricants are given in Sec. 6. The effect of pressure on liquid viscosity is generally

Table 3.3.1   Density of Liquids at Atmospheric Pressure

Temp: °C °F	0 32	20 68	40 104	60 140	80 176	100 212
Liquid	ρ, slugs/ft^3 (515.4 kg/m^3)					
Alcohol, ethyl[f]	1.564	1.532	1.498	1.463		
Benzene[a,b]	1.746	1.705	1.663	1.621	1.579	
Carbon tetrachloride[a,b]	3.168	3.093	3.017	2.940	2.857	
Gasoline,[c] sp. gr. 0.68	1.345	1.310	1.275	1.239		
Glycerin[a,b]	2.472	2.447	2.423	2.398	2.372	2.346
Kerosene,[c] sp. gr. 0.81	1.630	1.564	1.536	1.508	1.480	
Mercury[b]	26.379	26.283	26.188	26.094	26.000	25.906
Oil, machine,[c] sp. gr. 0.907	1.778	1.752	1.727	1.702	1.677	1.651
Water, fresh[d]	1.940	1.937	1.925	1.908	1.885	1.859
Water, salt[e]	1.995	1.988	1.975			

SOURCES: Computed from data given in:
[a] "Handbook of Chemistry and Physics," 52d ed., Chemical Rubber Company, 1971–1972.
[b] "Smithsonian Physical Tables," 9th rev. ed., 1954.
[c] ASTM-IP, "Petroleum Measurement Tables."
[d] "Steam Tables," ASME, 1967.
[e] "American Institute of Physics Handbook," 3d ed., McGraw-Hill, 1972.
[f] "International Critical Tables," McGraw-Hill.

Table 3.3.2 Bulk Modulus of Elasticity, Ratio of Specific Heats of Liquids and Velocity of Sound at One Atmosphere and 68°F (20°C)

Liquid	E in lbf/in² (6,895 N/m²)		$k = c_p/c_v$	c in ft/s (0.3048 m/s)
	Isothermal E_T	Isentropic E_s		
Alcohol, ethyl[a,e]	130,000	155,000	1.19	3,810
Benzene[a,f]	154,000	223,000	1.45	4,340
Carbon tetrachloride[a,b]	139,000	204,000	1.47	3,080
Glycerin[f]	654,000	719,000	1.10	6,510
Kerosene,[a,e] sp. gr. 0.81	188,000	209,000	1.11	4,390
Mercury[e]	3,590,000	4,150,000	1.16	4,770
Oil, machine,[f] sp. gr. 0.907	189,000	219,000	1.13	4,240
Water, fresh[a]	316,000	319,000	1.01	4,860
Water, salt[a,e]	339,000	344,000	1.01	4,990

SOURCES: Computed from data given in:
[a] "Handbook of Chemistry and Physics," 52d ed., Chemical Rubber Company, 1971–1972.
[b] "Smithsonian Physical Tables," 9th rev. ed., 1954.
[c] ASTM-IP, "Petroleum Measurement Tables."
[d] "Steam Tables," ASME, 1967.
[e] "American Institute of Physics Handbook," 3d ed., McGraw-Hill, 1972.
[f] "International Critical Tables," McGraw-Hill.

Table 3.3.3 Dynamic Viscosity of Liquids at Atmospheric Pressure

Temp: °C	0	20	40	60	80	100
°F	32	68	104	140	176	212
Liquid	μ, (lbf · s)/(ft²) [47.88 (N · s)/(m²)] × 10⁶					
Alcohol, ethyl[a,e]	37.02	25.06	17.42	12.36	9.028	
Benzene[a]	19.05	13.62	10.51	8.187	6.871	
Carbon tetrachloride[e]	28.12	20.28	15.41	12.17	9.884	
Gasoline,[b] sp. gr. 0.68	7.28	5.98	4.93	4.28		
Glycerin[d]	252,000	29,500	5,931	1,695	666.2	309.1
Kerosene,[b] sp. gr. 0.81	61.8	38.1	26.8	20.3	16.3	
Mercury[a]	35.19	32.46	30.28	28.55	27.11	25.90
Oil, machine,[a] sp. gr. 0.907						
"Light"	7,380	1.810	647	299	164	102
"Heavy"	66,100	9,470	2,320	812	371	200
Water, fresh[c]	36.61	20.92	13.61	9.672	7.331	5.827
Water, salt[d]	39.40	22.61	18.20			

SOURCES: Computed from data given in:
[a] "Handbook of Chemistry and Physics," 52d ed., Chemical Rubber Company, 1971–1972.
[b] "Smithsonian Physical Tables," 9th rev. ed., 1954.
[c] "Steam Tables," ASME, 1967.
[d] "American Institute of Physics Handbook," 3d ed., McGraw-Hill, 1972.
[e] "International Critical Tables," McGraw-Hill.

Table 3.3.4 Viscosity of Gases at One Atmosphere

Temp: °C	0	20	60	100	200	400	600	800	1000
°F	32	68	140	212	392	752	1112	1472	1832
Gas	μ, (lbf · s)/(ft²) [47.88(N · s)/(m²)] × 10⁸								
Air*	35.67	39.16	41.79	45.95	53.15	70.42	80.72	91.75	100.8
Carbon dioxide*	29.03	30.91	35.00	38.99	47.77	62.92	74.96	87.56	97.71
Carbon monoxide†	34.60	36.97	41.57	45.96	52.39	66.92	79.68	91.49	102.2
Helium*	38.85	40.54	44.23	47.64	55.80	71.27	84.97	97.43	
Hydrogen*·†	17.43	18.27	20.95	21.57	25.29	32.02	38.17	43.92	49.20
Methane*	21.42	22.70	26.50	27.80	33.49	43.21			
Nitrogen*·†	34.67	36.51	40.14	43.55	51.47	65.02	76.47	86.38	95.40
Oxygen†	40.08	42.33	46.66	50.74	60.16	76.60	90.87	104.3	116.7
Steam‡		18.49	21.89	25.29	33.79	50.79	67.79	84.79	

SOURCES: Computed from data given in:
* "Handbook of Chemistry and Physics," 52d ed., Chemical Rubber Company, 1971–1972.
† "Tables of Thermal Properties of Gases," *NBS Circular* 564, 1955.
‡ "Steam Tables," ASME, 1967.

unimportant in fluid mechanics except in lubricants (Sec. 6). The viscosity of water changes little at pressures up to 15,000 lbf/in², but for animal and vegetable oils it increases about 350 percent and for mineral oils about 1,600 percent at 15,000 lbf/in² pressure.

The dynamic viscosity of gases is primarily a temperature function and essentially independent of pressure. Table 3.3.4 gives values of dynamic viscosity of selected gases.

The **kinematic viscosity** ν of a fluid is its dynamic viscosity divided by its density, or $\nu = \mu/\rho$. Its dimensions are L^2/T. The units are ft²/s (9.290304 × 10⁻² m²/s).

In the cgs system, the unit of kinematic viscosity is the **stoke** (1 × 10⁻⁴ m²/s²), but for convenience, the **centistoke** (1/100 stoke) is widely used. The kinematic viscosity of water at 68°F (20°C) is approximately 1 centistoke.

The standard device for **experimental determination of kinematic viscosity** in the United States is the **Saybolt Universal** viscometer. It consists essentially of a metal tube and an orifice built to rigid specifications and calibrated. The time required for a gravity flow of 60 cubic centimeters is called the SSU (Saybolt seconds Universal). Approximate conversions of SSU to stokes may be made as follows:

$$32 < \text{SSU} < 100 \text{ seconds, stokes} = 0.00226 \, (\text{SSU}) - 1.95/(\text{SSU})$$
$$\text{SSU} > 100 \text{ seconds, stokes} = 0.00220 \, (\text{SSU}) - 1.35/(\text{SSU})$$

For viscous oils, the **Saybolt Furol** viscometer is used. Approximate conversions of SSF (saybolt seconds Furol) may be made as follows:

$$25 < \text{SSF} < 40 \text{ seconds, stokes} = 0.0224 \, (\text{SSF}) - 1.84/(\text{SSF})$$
$$\text{SSF} > 40 \text{ seconds, stokes} = 0.0216 \, (\text{SSF}) - 0.60/(\text{SSF})$$

For exact conversions of Saybolt viscosities, see ASTM D445-71 and Sec. 6.11.

The **surface tension** σ of a fluid is the work done in extending the surface of a liquid one unit of area or work per unit area. Its dimensions are F/L. The units are lbf/ft (14.5930 N/m).

Values of σ for various interfaces are given in Table 3.3.5. Surface tension decreases with increasing temperature. Surface tension is of importance in the formation of bubbles and in problems involving atomization.

Table 3.3.5   Surface Tension of Liquids at One Atmosphere and 68°F (20°C)

Liquid	δ, lbf/ft (14.59 N/m) × 10³		
	In vapor	In air	In water
Alcohol, ethyl*	1.56	1.53	
Benzene*	2.00	1.98	2.40
Carbon tetrachloride*	1.85	1.83	3.08
Gasoline,* sp. gr. 0.68		1.3–1.6	2.7–3.6
Glycerin*	4.30	4.35	
Kerosene,* sp. gr. 0.81		1.6–2.2	
Mercury*	32.6§	32.8	25.7
Oil, machine,‡ sp. gr. 0.907	2.5	2.6	2.3–3.7
Water, fresh‡		4.99	
Water, salt ‡		5.04	

SOURCES: Computed from data given in:
* "International Critical Tables," McGraw-Hill.
† ASTM-IP, "Petroleum Measurement Tables."
‡ "American Institute of Physics Handbook," 3d ed., McGraw-Hill, 1972.
§ In vacuum.

Capillary action is due to surface tension, **cohesion** of the liquid molecules, and the **adhesion** of the molecules on the surface of a solid. This action is of importance in fluid mechanics because of the formation of a meniscus (curved section) in a tube. When the adhesion is greater than the cohesion, a liquid "wets" the solid surface, and the liquid will rise in the tube and conversely will fall if the reverse. Figure 3.3.3 illustrates this effect on manometer tubes. In the reading of a manometer, all data should be taken at the center of the meniscus.

The **vapor pressure** pv of a fluid is the pressure at which its liquid and vapor are in equilibrium at a given temperature. See Secs. 4.1 and 4.2 for further definitions and values.

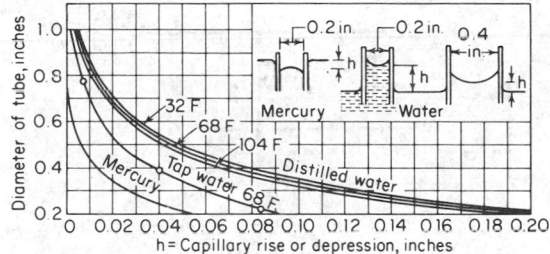

Fig. 3.3.3   Capillarity in circular glass tubes.

FLUID STATICS

Pressure p is the force per unit area exerted on or by a fluid and has dimensions of F/L^2. In fluid mechanics and in thermodynamic equations, the units are lbf/ft² (47.88026 N/m²), but engineering practice is to use units of lbf/in² (6,894.757 N/m²).

The relationship between **absolute pressure, gage pressure,** and **vacuum** is shown in Fig. 3.3.4. Most fluid-mechanics equations and **all** thermodynamic equations require the use of **absolute pressure**, and unless otherwise designated, a pressure should be understood to be **absolute pressure**. Common practice is to denote absolute pressure as lbf/ft² abs, or psfa, lbf/in² abs or psia; and in a like manner for **gage pressure** lbf/ft² g, lbf/in² g, and psig.

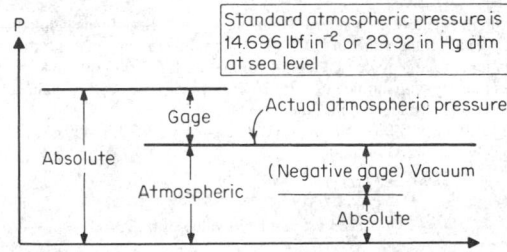

Fig. 3.3.4   Pressure relations.

According to **Pascal's** principle, the pressure in a static fluid is the same in all directions.

The **basic equation of fluid statics** is obtained by consideration of a fluid particle at rest with respect to other fluid particles, all being subjected to body-force accelerations of a_x, a_y, and a_z opposite the directions of x, y, and z, respectively, and the acceleration of gravity in the z direction, resulting in the following:

$$dp = -\rho[a_x \, dx + a_y \, dy + (a_z + g) \, dz]$$

Pressure-Height Relations   For a fluid at rest and subject only to the gravitational force, a_x, a_y, and a_z are zero and the **basic equation for fluid statics** reduces to $dp = -\rho g \, dz = \gamma \, dz$.

Liquids (Incompressible Fluids)   The pressure-height equation integrates to $(p_1 - p_2) = \rho g(z_2 - z_1) = \gamma(z_2 - z_1) = \Delta p = \gamma h$, where h is measured from the liquid surface (Fig. 3.3.5).

EXAMPLE.   A large closed tank is partly filled with 68°F (20°C) benzene. If the pressure on the surface is 10 lb/in², what is the pressure in the benzene at a depth of 11 ft below the liquid surface?

$$p_1 = \rho g h + p_2 = \frac{1.705 \times 32.17 \times 11}{144} + 10$$

$$= 14.19 \text{ lbf/in}^2 \, (9.784 \times 10^4 \text{ N/m}^2)$$

Ideal Gases (Compressible Fluids) For problems involving the upper atmosphere, it is necessary to take into account the variation of gravity with altitude. For this purpose, the **geopotential altitude** H is used, defined by $H = Z/(1 + z/r)$, where r is the radius of the earth ($\approx 21 \times$

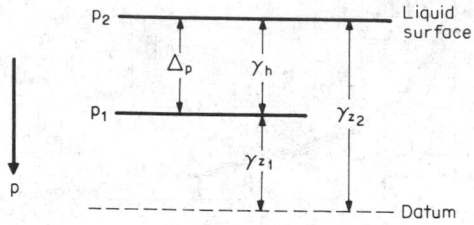

Fig. 3.3.5 Pressure equivalence.

10^6 ft $\approx 6.4 \times 10^6$ m) and z is the height above sea level. The integration of the pressure-height equation depends upon the thermodynamic process. For an **isothermal process** $p_2/p_1 = e^{-(H_2 - H_1)/RT}$ and for a **polytropic process** ($n \neq 1$)

$$\frac{p_2}{p_1} = \left[1 - \frac{(n-1)(H_2 - H_1)}{nRT_1} \right]^{n/(n-1)}$$

Temperature-height relations for a polytropic process ($n \neq 1$) are given by

$$\frac{n}{1-n} = \frac{H_2 - H_1}{R(T_2 - T_1)}$$

Substituting in the **pressure-altitude** equation,

$$p_2/p_1 = (T_2/T_1)^{(H_2 - H_1) \div R(T_1 - T_2)}$$

EXAMPLE. The U.S. Standard Atmosphere 1962 (Sec. 11) is defined as having a sea-level temperature of 59°F (15°C) and a pressure of 2,116.22 lbf/ft². From sea level to a geopotential altitude of 36,089 ft (11,000 m) the temperature decreases linearly with altitude to -69.70°F (-56.5°C). Check the value of pressure ratio at this altitude given in the standard table.

Noting that $T_1 = 59 + 459.67 = 518.67$, $T_2 = -69.70 + 459.67 = 389.97$, and $R = 53.34$ ft·lbf/(lbm)(°R),

$$p_2/p_1 = (T_2/T_1)^{(H_2 - H_1)/R(T_1 - T_2)}$$
$$= (389.97/518.67)^{(36,089-0)/53.34(518.67 - 389.97)}$$
$$= 0.2233 \text{ vs. tabulated value of } 0.2234$$

Pressure-Sensing Devices The two principal devices using liquids are the **barometer** and the **manometer**. The barometer senses absolute pressure and the manometer senses pressure differential. For discussion of the barometer and other pressure-sensing devices, refer to Sec. 16.

Manometers are a direct application of the basic equation of fluid statics and serve as a pressure standard in the range of $\frac{1}{10}$ in of water to 100 lbf/in². The most familiar type of manometer is the **U tube** shown in Fig. 3.3.6a. Because of the necessity of observing both legs simultaneously, the **well** or **cistern** type (Fig. 3.3.6b) is sometimes used. The **inclined manometer** (Fig. 3.3.6c) is a special form of the well-type manometer designed to enhance the readability of small pressure differentials. Application of the basic equation of fluid statics to each of the types results in the following equations. For the U tube, $p_1 - p_2 = (\gamma_m - \gamma_f)h$, where γ_m and γ_f are the specific weights of the manometer and sensed fluids, respectively, and h is the vertical distance between the liquid interfaces. For the **well type**, $p_1 - p_2 = (\gamma_m - \gamma_f)(z_2) \times (1 + A_2/A_1)$, where A_1 and A_2 are as shown in Fig. 3.3.6b and z_2 is the vertical distance from the fill line to the upper interface. Commercial manufacturers of well-type manometers correct for the area ratios so that $p_1 - p_2 = (\gamma_m - \gamma_f)S$, where S is the scale reading and is equal to $z_1(1 + A_2/A_1)$. For this reason, scales should not be interchanged between U type or well type or between well types without consulting the manufacturer. For inclined manometers,

$$p_1 - p_2 = (\gamma_m - \gamma_f)(A_2/A_1 + \sin \theta)R$$

where R is the distance along the inclined tube. Commercial inclined manometers also have special scales so that $p_1 - p_2 = (\gamma_m - \gamma_f)S$, where $S = (A_2/A_1 + \sin \theta)R$.

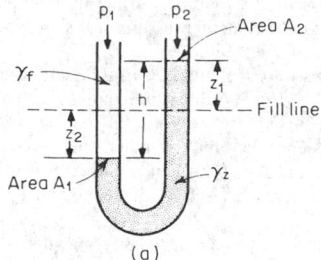

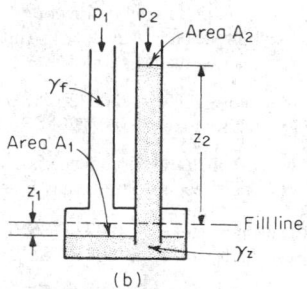

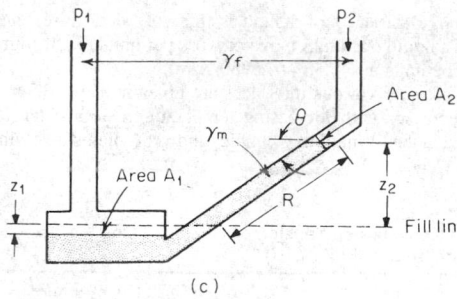

Fig. 3.3.6 (a) U-tube manometer; (b) well or cistern-type manometer; (c) inclined manometer.

EXAMPLE. A U-tube manometer containing mercury is used to sense the difference in water pressure. If the height between the interfaces is 10 in and the temperature is 68°F (20°C), what is the pressure differential?

$$p_1 - p_2 = (\gamma_m - \gamma_f)h = g(\rho_m - \rho_f)h$$
$$= 32.17(26.283 - 1.937)(10/12)$$
$$= 652.7 \text{ lbf/ft}^2 \ (3.152 \times 10^4 \text{ N/m}^2)$$

Liquid Forces The force exerted by a liquid on a **plane submerged surface** (Fig. 3.3.7) is given by $F = \int p \, dA = \gamma \int h \cdot dA = \gamma h_c A$, where h_c is the distance from the liquid surface to the center of gravity of the surface, and A is the area of the surface. The location of the center of this force is given by

$$s_F = s_c + I_G/s_c A$$

where s_F is the inclined distance from the liquid surface to the center of force, s_c the inclined distance to the center of gravity of the surface, and I_G the moment of inertia around its center of gravity. Values of I_G are given in Sec. 5.2. See also Sec. 3.1. From Fig. 3.3.7, $h = R \sin \theta$, so that the vertical center of force becomes

$$h_F = h_c + I_G(\sin \theta)^2/h_c A$$

EXAMPLE.   Determine the force and its location acting on a rectangular gate 3 ft wide and 5 ft high at the bottom of a tank containing 68°F (20°C) water, 12 ft deep, (1) if the gate is vertical, and (2) if it is inclined 30° from horizontal.

1. *Vertical gate*

$F = \gamma g h_c A = \rho g h_c A$
   $= 1.937 \times 32.17(12 - 5/2)(5 \times 3)$
   $= 8,800 \ \text{lb}_f (3.914 \times 10^4 \ \text{N})$

$h_F = h_c + I_G(\sin \theta)^2/h_c A$, from Sec. 5.2, I_G for a rectangle = (width)(height)^3/12
$h_F = (12 - 5/2) + (3 \times 5^3/12)(\sin 90°)^2/(12 - 5/2)(3 \times 5)$
$h_F = 9.719 \ \text{ft} \ (2.962 \ \text{m})$

2. *Inclined gate*

$F = \gamma h_c A = \rho g h_c A$
   $= 1.937 \times 32.17(12 - 5/2 \sin 30°)(5 \times 3)$
   $= 10,048 \ \text{lbf} \ (4.470 \times 10^4 \ \text{N})$

$h_F = h_c + I_G(\sin \theta)^2/h_c A$
   $= (12 - 5/2 \sin 30°) + (3 \times 5^3/12)(\sin 30°)^2/(12 - 5/2 \sin 30°)(3 \times 5)$
   $= 10.80 \ \text{ft} \ (3.291 \ \text{m})$

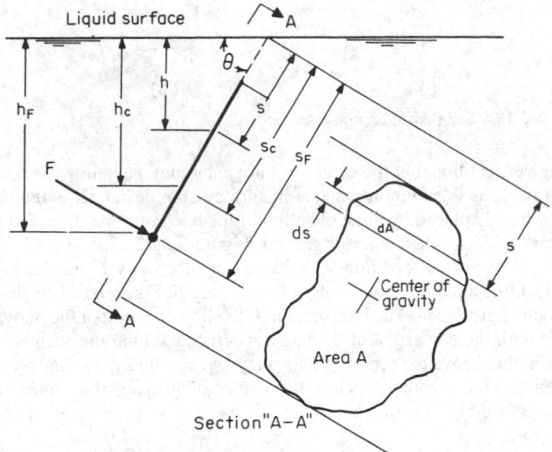

Fig. 3.3.7   Notation for liquid force on submerged surfaces.

Forces on irregular surfaces may be obtained by considering their horizontal and vertical components. The vertical component F_z equals the weight of liquid above the surface and acts through the centroid of the volume of the liquid above the surface. The horizontal component F_x equals the force on a vertical projection of the irregular surface. This force may be calculated by $F_x = \gamma h_{cx} A_x$, where h_{cx} is the distance from the surface center of gravity of the horizontal projection, and A_x is the projected area. The forces may be combined by $F = \sqrt{F_z^2 + F_z^2}$.

When fluid masses are **accelerated** without relative motion between fluid particles, the basic equation of fluid statics may be applied. For **translation** of a **liquid** mass due to uniform acceleration, the basic equation integrates to

$$p_2 - p_1 = -\rho[(x_2 - x_1)a_x + (y_2 - y_1)a_y + (z_2 - z_1)(a_z + g)]$$

EXAMPLE.   An open tank partly filled with a liquid is being accelerated up an inclined plane as shown in Fig. 3.3.8. The uniform acceleration is 20 ft/s² and the angle of the incline is 30°. What is the angle of the free surface of the liquid? Noting that on the free surface $p_2 = p_1$ and that the acceleration in the y direction is zero, the basic equation reduces to

$$(x_2 - x_1)a_x + (z_2 - z_1)(a_z + g) = 0$$

Solving for tan θ,

$$\tan \theta = \frac{z_1 - z_2}{x_2 - x_1} = \frac{a_x}{a_z + g} = \frac{a \cos \alpha}{a \sin \alpha + g}$$
$$= (20 \cos 30°)/(20 \sin 30° + 32.17) = 0.4107$$
$$\theta = 22°20'$$

For **rotation** of liquid masses with uniform rotational acceleration, the basic equation integrates to

$$p_2 - p_1 = \rho \left[\frac{\omega^2}{2}(x_2^2 - x_1^2) - g(z_2 - z_1) \right]$$

where ω is the rotational speed in rad/s and x is the radial distance from the axis of rotation.

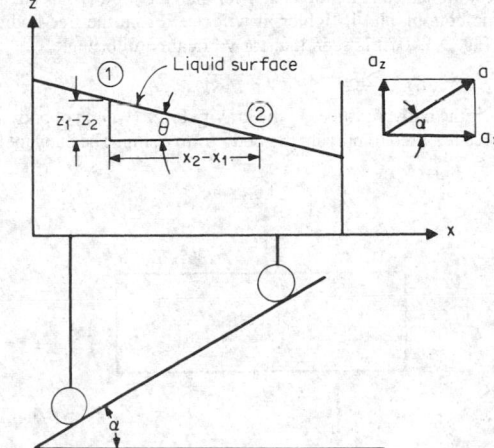

Fig. 3.3.8   Notation for translation example.

EXAMPLE.   The closed cylindrical tank shown in Fig. 3.3.9 is 4 ft in diameter and 10 ft high and is filled with 104°F (40°C) benzene. The tank is rotated at 250 r/min about an axis 3 ft from its centerline. Compute the maximum pressure differential in the tank. Analysis of the rotation equation indicates that the maxi-

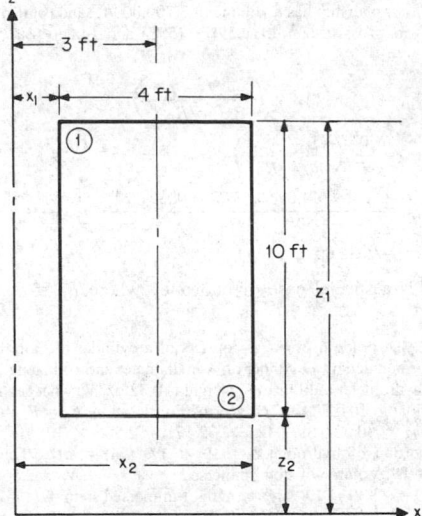

Fig. 3.3.9   Notation for rotation example.

mum pressure will occur at the maximum rotational radius and the minimum elevation and, conversely, the minimum at the minimum rotational radius and maximum elevation. From Fig. 3.3.9, $x_1 = 3 - 4/2 = 1$ ft, $x_2 = 1 + 4 = 5$ ft, $z_2 - z_1 = -10$ ft, and the rotational speed $\omega = 2\pi N/60 = 2\pi(250)/60 =$

26.18 rad/s. Substituting into the rotational equation,

$$p_2 - p_1 = \rho \left[\frac{\omega^2}{2}(x_2^2 - x_1^2) - g(z_2 - z_1) \right]$$

$$= \frac{1.663}{144}\left[\frac{(26.18)^2}{2}(5^2 - 1^2) - 32.17(-10) \right]$$

$$= 98.70 \text{ lbf/in}^2 \ (6.805 \times 10^5 \text{ N/m}^2)$$

Buoyancy Archimedes' principle states that a body immersed in a fluid is buoyed up by a force equal to the weight of the fluid displaced. If an object immersed in a fluid is heavier than the fluid displaced, it will sink to the bottom, and if lighter, it will rise. From the free-body diagram of Fig. 3.3.10, it is seen that for vertical equilibrium,

$$\Sigma F_z = 0 = F_B - F_g - F_D$$

where F_B is the buoyant force, F_g the gravity force (weight of body), and F_D the force required to prevent the body from rising. The buoyant force

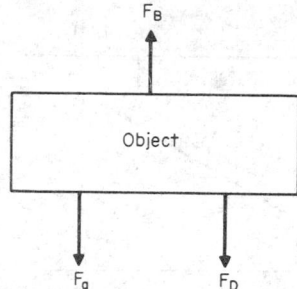

Fig. 3.3.10 Free body diagram of an immersed object.

being the weight of the displaced liquid, the equilibrium equation may be written as

$$F_D = F_B - F_g = \gamma_f V - \gamma_0 V = (\gamma_f - \gamma_0)V$$

where γ_f is the specific weight of the fluid, γ_0 is the specific weight of the object, and V is the volume of the object.

EXAMPLE. An airship has a volume of 3,700,000 ft³ and is filled with hydrogen. What is its gross lift in air at 59°F (15°C) and 14.696 psia? Noting that $\gamma = p/RT$,

$$F_D = (\gamma_f - \gamma_0)V = \left(\frac{p}{R_a T} - \frac{p}{R_{H_2} T} \right)V$$

$$= \frac{pV}{T}\left(\frac{1}{R_a} - \frac{1}{R_{H_2}} \right)$$

$$= \frac{144 \times 14.696 \times 3,700,000}{59 + 459.7}\left(\frac{1}{53.34} - \frac{1}{766.8} \right)$$

$$= 263,300 \text{ lbf} \ (1.171 \times 10^6 \text{ N})$$

Flotation is a special case of buoyancy where $F_D = 0$, and hence $F_B = F_g$.

EXAMPLE. A crude **hydrometer** consists of a cylinder of ½ in diameter and 2 in length surmounted by a cylinder of ⅛ in diameter and 10 in long. Lead shot is added to the hydrometer until its total weight is 0.32 oz. To what depth would this hydrometer float in 104°F (40°C) glycerin? For flotation, $F_B = F_g = \gamma_f V = \rho_f g V$ or $V = F_B/\rho_f g = (0.32/16)/(2.423 \times 32.17) = 2.566 \times 10^{-4}$ ft³. Volume of cylindrical portion of hydrometer $= V_c = \pi D^2 L/4 = \pi(0.5/12)^2(2/12)/4 = 2.273 \times 10^{-4}$ ft³. Volume of stem immersed $= V_S = V - V_c = 2.566 \times 10^{-4} - 2.273 \times 10^{-4} = 2.930 \times 10^{-5}$ ft³. Length of immersed stem $= L_S = 4 V_S/\pi D^2 = (4 \times 2.930 \times 10^{-5})/\pi(0.125/12)^2 = 0.3438$ ft $= 0.3438 \times 12 = 4.126$ in. Total immersion $= L + L_S = 2 + 4.126 = 6.126$ in (0.156 m).

Static Stability A body is in static equilibrium when the imposition of a small displacement brings into action forces that tend to restore the body to its original position. For **completely submerged bodies,** the center of buoyancy and the center of gravity must lie on the same vertical line and the center of buoyancy must be located above the center of gravity. Figure 3.3.11a shows a balloon and its basket in its normal position with

the center of buoyancy B above and on the same vertical line as the center of gravity G. Figure 3.3.11b shows the balloon displaced from its normal position. In this position, there is a couple $F_g x$ which tends to restore the balloon and its basket to its original position. For **floating bodies,** the center of gravity and the center of buoyancy must lie on the

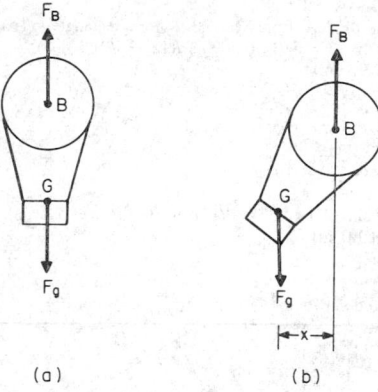

Fig. 3.3.11 Stability of an immersed body.

same vertical line, but the center of buoyancy may be below the center of gravity, as is common practice in surface-ship design. It is required that when displaced, the line of action of the buoyant force intersect the centerline above the center of gravity. Figure 3.3.12a shows a floating body in its normal position with its center of gravity G on the same vertical line and above the center of buoyancy B. Figure 3.3.12b shows the object displaced. The intersection of the line of action of the buoyant force with the centerline of the body at M is called the metacenter. As shown, this above the center of buoyancy and sets up a restoring couple. When the metacenter is below the center of gravity, the object will capsize (see Sec. 11.3).

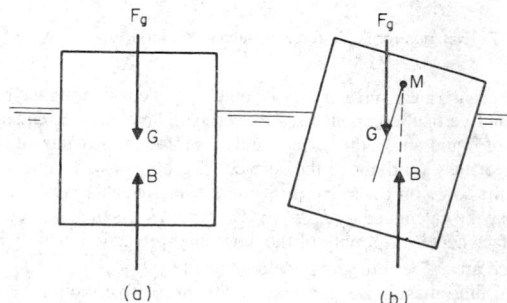

Fig. 3.3.12 Stability of a floating body.

FLUID KINEMATICS

Steady and Unsteady Flow If at every point in the fluid stream, none of the local fluid properties changes with time, the flow is said to be steady. The mathematical conditions for steady flow are met when $\partial(\text{fluid properties})/\partial t = 0$. While flow is generally unsteady by nature, many real cases of unsteady flow may be treated as steady flow by using average properties or by changing the space reference. The amount of error produced by the averaging technique depends upon the nature of the unsteady flow, but the latter technique is error-free when it can be applied.

Streamlines and Stream Tubes Velocity has both magnitude and direction and hence is a vector. A **streamline** is a line which gives the *direction* of the velocity of a fluid particle at each point in the flow stream. When streamlines are connected by a closed curve in steady flow, they will form a boundary through which the fluid particles cannot pass. The space between the streamlines becomes a **stream tube**. The stream-tube concept broadens the application of fluid-flow principles; for example, it allows treating the flow inside a pipe and the flow around an object with the same laws. A stream tube of decreasing size approaches its own axis, a central streamline; thus equations developed for a stream tube may also be applied to a streamline.

Velocity and Acceleration   In the most general case of fluid motion, the resultant **velocity** U along a streamline is a function of both distance s and time t, or $U = f(s, t)$. In differential form,

$$dU = \frac{\partial U}{\partial s}\, ds + \frac{\partial U}{\partial t}\, dt$$

An expression for **acceleration** may be obtained by dividing the velocity equation by dt, resulting in

$$\frac{dU}{dt} = \frac{\partial U}{\partial s}\frac{ds}{dt} + \frac{\partial U}{\partial t}$$

for steady flow $\partial U/\partial t = 0$.

Velocity Profile   In the flow of real fluids, the individual streamlines will have different velocities past a section. Figure 3.3.13 shows the steady flow of a fluid past a section (A-A) of a circular pipe. The **velocity profile** is obtained by plotting the velocity U of each streamline as it passes A-A. The stream tube that is formed by the space between the streamlines is the annulus whose area is dA, as shown in Fig. 3.3.13 for

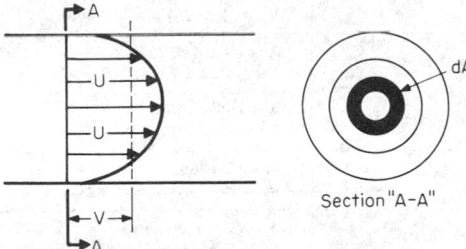

Section "A-A"

Fig. 3.3.13   Velocity profile.

the stream tube whose velocity is U. The volumetric rate of flow Q for the flow past section A-A is $Q = \int U\, dA$. All flows take place between boundaries that are **three-dimensional**. The terms **one-dimensional, two-dimensional,** and **three-dimensional** flow refer to the number of dimensions required to describe the velocity profile. For **three-dimensional** flow, a **volume** (L^3) is required; for example, the flow of a fluid in a circular pipe. For **two-dimensional** flow, an **area** (L^2) is necessary; for example, the flow between two parallel plates. For **one-dimensional** flow, a **line** (L) describes the profile. In cases of two- or three-dimensional flow, $\int U\, dA$ can be integrated either mathematically if the equations are known or graphically if velocity-measurement data are available. In many engineering applications, the **average velocity** V may be used where $V = Q/A = (1/A)\int U\, dA$.

The **continuity equation** is a special case of the general physical law of the conservation of mass. It may be simply stated for a control volume:

Mass rate entering = mass rate of storage + mass rate leaving

This may be expressed mathematically as

$$\rho U\, dA = \left[\frac{\partial}{\partial t}\left(\rho\, dA\, ds\right)\right] + \left[\rho U\, dA + \frac{\partial}{\partial s}\left(\rho U\, dA\right)ds\right]$$

where ds is an incremental distance along the control volume. For steady flow, $\partial/\partial t\,(\rho\, dA\, ds) = 0$, the general equation reduces to $d(\rho U\, dA) = 0$. Integrating the steady-flow **continuity equation** for the average velocity along a flow passage:

$$\rho VA = \text{a constant} = \rho_1 V_1 A_1 = \rho_2 V_2 A_2 = \ldots = \rho_n V_n A_n = \dot{M}$$

where $\dot{M}$ is the mass flow rate in slugs/s (14.5939 kg/s). In many engineering applications, the flow rate in pounds mass per second is desired,

so that

$$\dot{m} = \frac{V_1 A_1}{v_1} = \frac{V_2 A_2}{v_2} = \ldots = \frac{V_n A_n}{v_n}$$

where $\dot{m}$ is the flow rate in lbm/s (0.4535924 kg/s).

EXAMPLE.   Air discharges from a 12-in-diameter duct through a 4-in-diameter nozzle into the atmosphere. The pressure in the duct is 20 lbf/in², and atmospheric pressure is 14.7 lbf/in². The temperature of the air in the duct just upstream of the nozzle is 150°F, and the temperature in the jet is 147°F. If the velocity in the duct is 18 ft/s, compute (1) the mass flow rate in lbm/s and (2) the velocity in the nozzle jet. From the equation of state

$$v = RT/p$$
$$v_D = RT_D/p_D = 53.34\,(150 + 459.7)/(144 \times 20)$$
$$\quad = 11.29\ \text{ft}^3/\text{lbm}$$
$$v_J = RT_J/p_J = 53.34\,(147 + 459.7)/(144 \times 14.7)$$
$$\quad = 15.29\ \text{ft}^3/\text{lbm}$$
$$(1)\ \dot{m} = V_D A_D/v_D = 18\,[(\pi/4)(12/12)^2]/11.29$$
$$\dot{m}_J = 1.252\ \text{lbm/s}\ (0.5680\ \text{kg/s})$$
$$(2)\ VJ = m v_J/A_j = (1.252)(15.29)/[(\pi/4)(4/12)^2]$$
$$v_J = 219.2\ \text{ft/s}\ (66.82\ \text{m/s})$$

FLUID DYNAMICS

Equation of Motion   For steady one-dimensional flow, consideration of forces acting on a fluid element of length dL, flow area dA, boundary perimeter in fluid contact dP, and change in elevation dz with a unit shear stress τ moving at a velocity of V results in

$$v\, dp + \frac{V\, dV}{g_c} + \frac{g}{g_c}\, dz + v\tau\left(\frac{dP}{dA}\right)dL = 0$$

Substituting $v = g/g_c \gamma$ and simplifying,

$$\frac{dp}{\gamma} + \frac{V\, dV}{g} + dz + dh_f = 0$$

where $dh_f = (\tau/\gamma)(dP/dA)\, dL = \tau\, dL/\gamma R_h$.

The expression $1/(dP/dA)$ is the **hydraulic radius** R_h and equals the flow area divided by the perimeter of the solid boundary in contact with the fluid. This perimeter is usually called the "wetted" perimeter. The hydraulic radius of a pipe flowing full is $(\pi D^2/4)/\pi D = D/4$. Values for other configurations are given in Table 3.3.6. Integration of the equation of motion for an incompressible fluid results in

$$\frac{p_1}{\gamma} + \frac{V_1^2}{2g} + z_1 = \frac{p_2}{\gamma} + \frac{V_2^2}{2g} + z_2 + h_{1f2}$$

Each term of the equation is in feet and is equivalent to the height the fluid would rise in a tube if its energy were converted into potential energy. For this reason, in hydraulic practice, each type of energy is referred to as a **head**. The **static pressure head** is p/γ. The **velocity head** is $V^2/2g$, and the **potential head** is z. The energy loss between sections h_{1f2} is called the **lost head** or **friction head**. The **energy grade line** at any point is $\Sigma(p/\gamma + V^2/2g + z)$, and the **hydraulic grade line** is $\Sigma(p/\gamma + z)$ as shown in Fig. 3.3.14.

EXAMPLE.   A 12-in pipe (11.938 in inside diameter) reduces to a 6-in pipe (6.065 in inside diameter). Benzene at 68°F (20°C) flows steadily through this system. At section 1, the 12-in pipe centerline is 10 ft above the datum, and at section 2, the 6-in pipe centerline is 15 ft above the datum. The pressure at section 1 is 20 lbf/in² and the velocity is 4 ft/s. If the head loss due to friction is 0.05 $V_2^2/2g$, compute the pressure at section 2. Assume $g = g_c$, $\gamma = \rho g = 1.705 \times 32.17 = 54.85$ lbf/ft³. From the continuity equation,

$$\dot{M} = \rho_1 A_1 V_1 = \rho_2 A_2 V_2 \quad (p_1 = p_2)$$
$$V_2 = V_1(A_1/A_2) = V_1(\pi D_1^2/4)/(\pi D_2^2/4) = V_1(D_1/D_2)^2$$
$$V_2 = 4(11.938/6.065)^2 = 15.50\ \text{ft/s}$$

From the equation of motion,

$$\frac{p_2}{\gamma} = \frac{p_1}{\gamma} + \frac{V_1^2}{2g} + z_1 - \left(\frac{V_2^2}{2g} + z_2 + h_{1f2}\right)$$
$$\frac{p_2}{\gamma} = \frac{p_1}{\gamma} + \frac{V_1^2 - V_2^2 - 0.05V_2^2}{2g} + z_1 - z_2$$

Table 3.3.6 Values of Flow Area A and Hydraulic Radius R_h for Various Cross Sections

Cross section	Condition		Equations
	Flowing full	$h/D = 1$	$A = \pi D^2/4 \quad R_h = D/4$
	Upper half partly full	$0.5 < h/D < 1$	$\cos^{-1}(\theta/2) = (2h/D - 1)$ $A = [\pi(360 - \theta) + 180 \sin \theta](D^2/1{,}440)$ $R_h = [1 + (180 \sin \theta)/(\pi\theta)](D/4)$
		$h/D = 0.8128$	$A = 0.6839\, D^2 \quad R_h \max = 0.3043D$
	Lower half partly full	$h/D = 0.5$	$A = \pi D^2/8 \quad R_h \max = h/2$
		$0 < h/D < 0.5$	$\cos^{-1}(\theta/2) = (1 - 2h/D)$ $A = (\pi\theta - 180 \sin \theta)(D^2/1{,}440)$ $R_h = [1 - (180 \sin \theta)/(\pi\theta)](D/4)$
	Flowing full	$h/D = 1$	$A = bD \quad R_h = bD/2(b + D)$
		Square $b = D$	$A = D^2 \quad R_h = D/4$
	Partly full	$h/D < 1$ $h/b = 0.5$ $b \to \infty, h \to 0$	$A = bh \quad R_h = bh/(2h + b)$ $A = b^2/2 \quad R_h \max = h/2$ $R_h \to h$ (wide shallow stream)
	$\alpha \neq \beta$		$R_h \max = h/2$ $A = [b + 1/2h(\cot \alpha + \cot \beta)]h$ $R_h = A/[b + h(\csc \alpha + \csc \beta)]$
	$\alpha = \beta$	$\dfrac{h}{a} = \dfrac{1}{2}$ $\quad \alpha = 26°34'$	$A = (b + 2h)h$ $R_h = (b + 2h)h/(b + 4.472h)$
		$\dfrac{h}{a} = \dfrac{\sqrt{3}}{3}$ $\quad \alpha = 30°$	$A = (b + 1.732h)h$ $R_h = (b + 1.732h)h/(b + 4h)$
		$\dfrac{h}{a} = \dfrac{2}{3}$ $\quad \alpha = 33°41'$	$A = (b + 1.5h)h$ $R_h = (b + 1.5h)h/(b + 3.606h)$
		$\dfrac{h}{a} = 1$ $\quad \alpha = 45°$	$A = (b + h)h$ $R_h = (b + h)h/(b + 2.828h)$
		$\dfrac{h}{a} = \dfrac{3}{2}$ $\quad \alpha = 56°19'$	$A = (b + 0.6667h)h$ $R_h = (b + 0.6667h)h/(b + 2.404h)$
		$\dfrac{h}{a} = \sqrt{3}$ $\quad \alpha = 60°$	$A = (b + 0.5774h)h$ $R_h = (b + 0.5774h)h/(b + 2.309h)$
	$\theta = $ any angle		$A = \tan(\theta/2)h^2 \quad R_h = \sin(\theta/2)h/2$
	$\theta = 30$		$A = 0.2679h^2 \quad R_h = 0.1294h$
	$\theta = 45$		$A = 0.4142h^2 \quad R_h = 0.1913h$
	$\theta = 60$		$A = 0.5774h^2 \quad R_h = 0.2500h$
	$\theta = 90$		$A = h^2 \quad R_h = 0.3536h$

$$\frac{p_2}{\gamma} = \frac{144 \times 20}{54.85} + \frac{4^2 - 1.05(15.50)^2}{2 \times 32.17} + 10 - 15$$

$$\frac{p_2}{\gamma} = 43.83 \text{ ft}$$

$$p_2 = \frac{54.88 \times 43.83}{144} = 16.70 \text{ lbf/in}^2 \ (1.151 \times 10^5 \text{ N/m}^2)$$

Energy Equation Application of the principles of conservation of energy to a control volume for one-dimensional flow results in the following for steady flow:

$$J\, dq = dW + \frac{V\, dV}{g_c} + \frac{g}{g_c}\, dz + J\, du + d(pv)$$

where J is the mechanical equivalent of heat, 778.169 ft·lbf/Btu; q is the heat added, Btu/lbm (2,326 J/kg); W is the steady-flow shaft work

done *by* the fluid; and u is the internal energy. Btu/lbm (2,326 J/kg). If the energy equation is integrated for an incompressible fluid,

$$J_1 q_2 = \,_1 W_2 + \frac{V_2^2 - V_1^2}{2g_c} + \frac{g}{g_c}(z_2 - z_1) + J(u_2 - u_1) + v(p_2 - p_1)$$

The equation of motion does not consider thermal energy or steady-flow work; the energy equation has no terms for friction. Subtracting the differential equation of motion from the energy equation and solving for friction results in

$$dh_f = (dW + J\, du + p\, dv - J\, dq)(g_c/g)$$

Integrating for an incompressible fluid ($dv = 0$),

$$h_{1f2} = [\,_1 W_2 + J(u_2 - u_1) - J_1 q_2](g_c/g)$$

In the absence of steady-flow work in the system, the effect of friction is to increase the internal energy and/or to transfer heat from the system.

For steady frictionless, incompressible flow, both the equation of motion and the energy equation reduce to

$$\frac{p_1}{\gamma} + \frac{V_1^2}{2g} + z_1 = \frac{p_2}{\gamma} + \frac{V_2^2}{2g} + z_2$$

which is known as the **Bernoulli equation**.

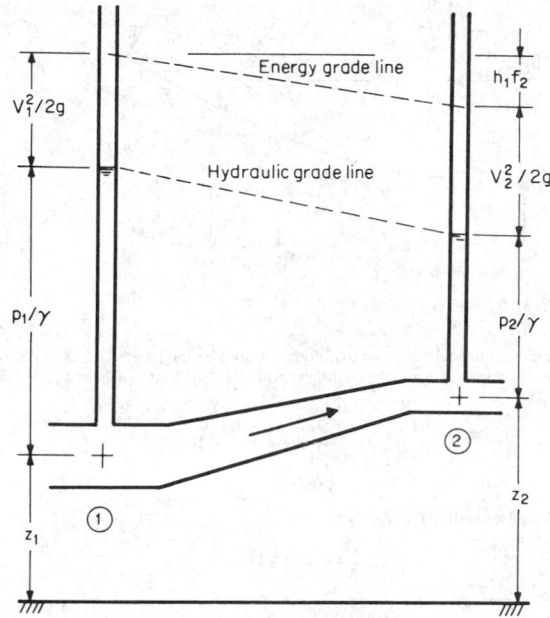

Fig. 3.3.14   Energy relations.

Area-Velocity Relations   The continuity equation may be written as $\log_e \dot{M} = \log_e V + \log_e A + \log_e \rho$, which when differentiated becomes

$$\frac{dA}{A} = -\frac{dV}{V} - \frac{d\rho}{\rho}$$

For incompressible fluids, $d\rho = 0$, so

$$\frac{dA}{A} = -\frac{dV}{V}$$

Examination of this equation indicates
1. If the area increases, the velocity decreases.
2. If the area is constant, the velocity is constant.
3. There are no critical values.

For the frictionless flow of compressible fluids, it can be demonstrated that

$$\frac{dA}{A} = -\frac{dV}{V}\left[1 - \left(\frac{V}{c}\right)^2\right]$$

Analysis of the above equation indicates:
1. *Subsonic velocity* $V < c$. If the area increases, the velocity decreases. Same as for incompressible flow.
2. *Sonic velocity* $V = c$. Sonic velocity can exist only where the change in area is zero, i.e., at the *end* of a convergent passage or at the *exit* of a constant-area duct.
3. *Supersonic velocity* $V > c$. If area increases, the velocity increases, the reverse of incompressible flow. Also, supersonic velocity can exist only in the **expanding** portion of a passage **after** a constriction where **sonic** velocity existed.

Frictionless adiabatic compressible flow of an ideal gas in a horizontal passage must satisfy the following requirements:

1. *Conservation of mass*. As expressed by the continuity equation $M = \rho_1 A_1 V_1 = \rho_2 A_2 V_2$.
2. *Conservation of energy*. As expressed by the energy equation

$$\frac{V^2}{2g_c} + Ju + pv = \frac{V_1^2}{2g_c} + Ju_1 + p_1 v_1 = \frac{V_2^2}{2g_c} + Ju_2 + p_2 v_2$$

3. *Process relationship*. For an ideal gas undergoing a frictionless adiabatic (isentropic) process,

$$pv^k = p_1 v_1^k = p_2 v_2^k$$

4. *Ideal-gas law*. The equation of state for an ideal gas

$$pv = RT$$

In an expanding supersonic flow, a **compression shock wave** will be formed if the requirements for the conservation of mass and energy are not satisfied. This type of wave is associated with large and sudden rises in pressure, density, temperature, and entropy. The shock wave is so thin that for computation purposes it may be considered as a single line. For compressible flow of gases and vapors in passages, refer to Sec. 4.1; for steam-turbine passages, Sec. 9.4; for compressible flow around immersed objects, see Sec. 11.4.

The **impulse-momentum equation** is an application of the principle of conservation of momentum and is derived from Newton's second law. It is used to calculate the forces exerted on a solid boundary by a moving stream. Because velocity and force have both magnitude and direction, they are vectors. The impulse-momentum equation may be written for all three directions:

$$\Sigma F_x = \dot{M}(V_{x2} - V_{x1})$$
$$\Sigma F_y = \dot{M}(V_{y2} - V_{y1})$$
$$\Sigma F_z = \dot{M}(V_{z2} - V_{z1})$$

Figure 3.3.15 shows a free-body diagram of a control volume. The pressure forces shown are those imposed by the boundaries on the fluid and on the atmosphere. The reactive force R is that imposed by the downstream boundary on the fluid for equilibrium. Application of the impulse-momentum equation yields

$$\Sigma F = (F_{p1} + F_{a2}) - (F_{a1} + F_{p2} + R) = \dot{M}(V_2 - V_1)$$

Solving for R,

$$R = (p_1 - p_a)A_1 - (p_2 - p_a)A_2 - \dot{M}(V_2 - V_1)$$

The impulse-momentum equation is often used in conjunction with the continuity and energy equations to solve engineering problems. Because of the wide variety of possible applications, some examples are given to illustrate the methods of attack.

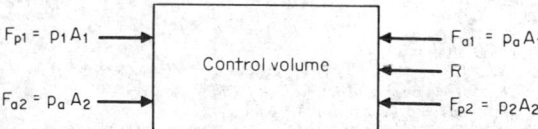

Fig. 3.3.15   Notation for impulse momentum.

EXAMPLE.   **Compressible Fluid in a Duct.** Nitrogen flows steadily through a 6-in (5.761 in inside diameter) straight, horizontal pipe at a mass rate of 25 lbm/s. At section 1, the pressure is 120 lbf/in² and the temperature is 100°F. At section 2, the pressure is 80 lbf/in² and the temperature is 110°F. Find the friction force opposing the motion. From the equation of state,

$v = RT/p$
$v_1 = 55.16\,(459.7 + 100)/(144 \times 120) = 1.787$ ft³/lbm
$v_2 = 55.16\,(459.7 + 110)/(144 \times 80) = 2.728$ ft³/lbm

Flow area of pipe $\times\ \pi D^2/4 = \pi(5.761/12)^2/4 = 0.1810$ ft²

From the continuity equation,

$v = \dot{m}V/A$
$V_1 = (25 \times 1.787)/0.1810 = 246.8$ ft/s
$V_2 = (25 \times 2.728)/0.1810 = 376.8$ ft/s

Applying the free-body equation for impulse momentum $(A = A_1 = A_2)$,

$$R = (p_1 - p_a) A_1 - (p_2 - p_a) A_2 - \dot{M}(V_2 - V_1)$$
$$= (p_1 - p_2) A - M(V_2 - V_1) = 144 \,(120 - 80) \,0.1810$$
$$- (25/32.17)(376.8 - 246.8) = 941.5 \text{ lbf } (4.188 \times 10^3 \text{ N})$$

EXAMPLE.   **Water Flow through a Nozzle.** Water at 68°F (20°C) flows through a horizontal 12- by 6-in-diameter nozzle discharging into the atmosphere. The pressure at the nozzle inlet is 65 lbf/in² and barometric pressure is 14.7 lbf/in². Determine the force exerted by the water on the nozzle.

$$A = \pi D^2/4$$
$$A_1 = \pi (12/12)^2/4 = 0.7854 \text{ ft}^2$$
$$A_2 = \pi (6/12)^2/4 = 0.1963 \text{ ft}^2$$
$$\gamma = \rho g = 1.937 \times 32.17 = 62.31 \text{ lbf/ft}^3$$

From the continuity equation $\rho_1 A_1 V_1 = \rho_2 A_2 V_2$ for $\rho_1 = \rho_2$, $V_2 = V_1 A_1/A_2 = (0.7854/0.1963)V_1 = 4V_1$. From Bernoulli's equation $(z_1 = z_2)$.

$$p_1/\gamma + V_1^2/2g = p_2/\gamma + V_2^2/2g = p_2/\gamma + (4V_1)^2/2g$$

or

$$V_1 = \sqrt{2g(p_1 - p_2)/15\gamma}$$
$$= \sqrt{2 \times 32.17 \times 144 \,(65 - 14.7)/15 \times 62.31} = 22.33 \text{ ft/s}$$
$$V_2 = 4 \times 22.33 = 89.32 \text{ ft/s}$$

Again from the equation of continuity

$$\dot{M} = \rho_1 A_1 V_1 = 1.937 \times 0.7854 \times 22.33 = 33.97 \text{ slugs/s}$$

Applying the free-body equation for impulse momentum,

$$R = (p_1 - p_a) A_1 - (p_2 - p_a) A_2 - \dot{M}(V_2 - V_1)$$
$$= 144 \,(65 - 14.7) \,0.7854 - 144 \,(14.7 - 14.7) \,0.1963$$
$$= (33.97) \,(89.32 - 22.33)$$
$$= 3,413 \text{ lbf } (1.518 \times 10^4 \text{ N})$$

EXAMPLE.   **Incompressible Flow through a Reducing Bend.** Carbon tetrachloride flows steadily without friction at 68°F (20°C) through a 90° horizontal reducing bend. The mass flow rate is 4 slugs/s, the inlet diameter is 6 in, and the outlet is 3 in. The inlet pressure is 50 lbf/in² and the barometric pressure is 14.7 lbf/in². Compute the magnitude and direction of the force required to "anchor" this bend.

$$A = \pi D^2/4$$
$$A_1 = (\pi/4)(6/12)^2 = 0.1963 \text{ ft}^2$$
$$A_2 = (\pi/4)(3/12)^2 = 0.04909 \text{ ft}^2$$

From continuity,

$$V = M/\rho A$$
$$V_1 = 4/(3.093)(0.1963) = 6.588 \text{ ft/s}$$
$$V_2 = 4/(3.093)(0.04909) = 26.35 \text{ ft/s}$$

From the Bernoulli equation $(z_1 = z_2)$,

$$\frac{p_2}{\gamma} = \frac{p_1}{\gamma} + \frac{V_1^2}{2g} - \frac{V_2^2}{2g} = \frac{144 \times 50}{3.093 \times 32.17} + \frac{(6.588)^2 - (26.35)^2}{2 \times 32.17} = 62.24 \text{ ft}$$
$$p_2 = \frac{(3.093 \times 32.17)(62.24)}{144} = 43.01 \text{ lbf/in}^2$$

From Fig. 3.3.16,

$$\Sigma F_x = (p_1 - p_a)A_1 - (p_2 - p_a)A_2 \cos \alpha - R_x$$
$$= M(V_2 \cos \alpha - V_1)$$

or

$$R_x = (p_1 - p_a)A_1 - (p_2 - p_a)A_2 \cos \alpha - \dot{M}(V_2 \cos \alpha - V_1)$$
$$\Sigma F_y = 0 - (p_2 - p_a)A_2 \sin \alpha + R_y = \dot{M}(V_2 \sin \alpha - 0)$$

or

$$R_y = (p_2 - p_a)A_2 \sin \alpha + MV_2 \sin \alpha$$
$$R_x = 144 \,(50 - 14.7) \,0.1963 - 144 \,(43.01 - 14.7)(\cos 90°)$$
$$- 4 \,(26.35 \cos 90° - 6.588)$$
$$= 1,024 \text{ lbf}$$
$$R_y = 144 \,(43.01 - 14.7)(0.04909) \sin 90° + 4(26.35)(\sin 90°)$$
$$= 305.5 \text{ lbf}$$
$$R = \sqrt{R_x^2 + R_y^2} = \sqrt{(1,024)^2 + (305.5)^2}$$
$$= 1,068 \text{ lbf } (4.753 \times 10^3 \text{ N})$$
$$\theta = \tan^{-1}(F_y/F_x) = \tan^{-1}(305.5/1,024)$$
$$= 16°37'$$

Forces on Blades and Deflectors   The forces imposed on a fluid jet whose velocity is V_J by a blade moving at a speed of V_b away from the jet are shown in Fig. 3.3.17. The following equations were developed from the application of the impulse-momentum equation for an open jet $(p_2 = p_1)$ and for **frictionless** flow:

$$F_x = \rho A_J(V_J - V_b)^2(1 - \cos \alpha)$$
$$F_y = \rho A_J(V_J - V_b)^2 \sin \alpha$$
$$F = 2\rho A_J(V_J - V_b)^2 \sin (\alpha/2)$$

EXAMPLE.   In the nozzle-blade system of Fig. 3.3.17, water at 68°F (20°C) enters a 3- by 1½-in-diameter horizontal nozzle with a pressure 23 lbf/in² and discharges at 14.7 lbf/in² (atmospheric pressure). The blade moves away from the nozzle at a velocity of 10 ft/s and deflects the stream through an angle of 80°. For

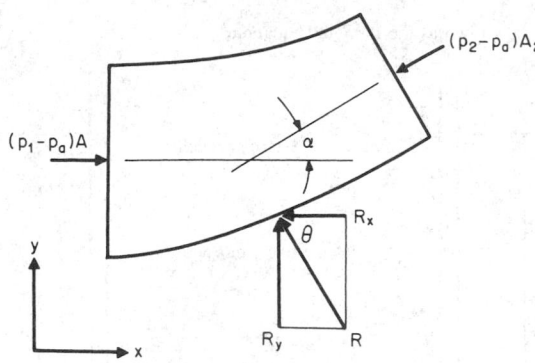

Fig. 3.3.16   Forces on a bend.

frictionless flow, calculate the total force exerted by the jet on the blade. Assume $g = g_c$; then $\gamma = \rho g$. From the continuity equation $(\rho_I = \rho_J)$, $\rho_I A_I V_I = \rho_J A_J V_J$, $V_I = (A_J/A_I)V_J$,

$$V_J = \frac{\pi D_J^2/4}{\pi D_I^2/4} V_J = \left(\frac{D_J}{D_I}\right)^2 V_J$$
$$V_I = (1.5/3)^2 V_J = V_J/4$$

From the Bernoulli equation $(z_2 = z_1)$,

$$\frac{V_J^2}{2g} = \frac{V_I^2}{2g} + \frac{p_I - p_J}{\rho g}$$
$$\frac{V_J^2 - (V_J/4)^2}{2g} = \frac{(p_I - p_J)}{\rho g}$$
$$V_J = \sqrt{\frac{2(16/15)(p_I - p_J)}{\rho}}$$
$$= \sqrt{\frac{2 \times (16/15) \,144 \,(23 - 14.7)}{1.937}} = 36.28 \text{ ft/s}$$

The total force $F = 2\rho A_J(V_J - V_b)^2 \sin (\alpha/2)$

$$F = 2 \times 1.937 \,(\pi/4)(1.5/12)^2(36.28 - 10)^2 \sin (80/2)$$
$$= 21.11 \text{ lbf } (93.90 \text{ N})$$

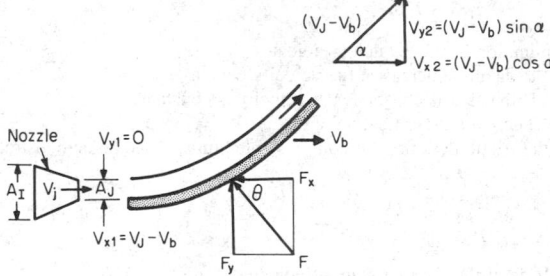

Fig. 3.3.17   Notation for blade study.

Impulse Turbine   In a turbine, the total of the separate forces acting simultaneously on each blade equals that caused by the combined mass flow rate M discharged by the nozzle or

$$\Sigma P = \Sigma F_x V_b = \dot{M}(V_J - V_b)(1 - \cos \alpha)V_b$$

The maximum value of power P is found by differentiating P with respect to V_b and setting the result equal to zero. Solving for V_b yields $V_b = V_J/2$, so that maximum power occurs when the velocity of the jet is equal to twice the velocity of the blade. Examination of the power equation also indicates that the angle α for a maximum power results when $\cos \theta = -1$ or $\alpha = 180°$. For theoretical maximum power of a blade, $2V_b = V_J$ and $\alpha = 180°$. It should be noted that in any practical impulse-turbine application, α cannot be $180°$ because the discharge interferes with the next set of blades. Substituting $V_b = V_J/2$, $\alpha = 180°$ in the power equation,

$$\Sigma P_{max} = \dot{M}(V_J - V_J/2)[1 - (-1)]V_J/2 = MV_J^2/2 = \dot{m}V_J^2/2g_c$$

or the maximum power per unit mass is equal to the total power of the jet. Application of the Bernoulli equation between the surface of a reservoir and the discharge of the turbine shows that $\Sigma P_{max} = \dot{M}\sqrt{2g(z_2 - z_1)}$. For design details, see Sec. 9.9.

Flow in a Curved Path   When a fluid flows through a bend, it is also rotated around an axis and the energy required to produce rotation must be supplied from the energy already in the fluid mass. This fluid rotation is called a **free vortex** because it is free of outside energy. Consider the fluid mass $\rho(r_o - r_i)\, dA$ of Fig. 3.3.18 being rotated as it flows through a bend of outer radius r_o, inner radius r_i, with a velocity of V. Application of Newton's second law to this mass results in

$$dF = p_o\, dA - p_i\, dA = [\rho(r_o - r_i)\, dA][V^2/(r_o + r_i)/2]$$

which reduces to

$$p_o - p_i = 2(r_o - r_i)\rho V^2/(r_o + r_i)$$

Because of the difference in fluid pressure between the inner and outer walls of the bend, secondary flows are set up, and this is the primary cause of friction loss of bends. These secondary flows set up turbulence that require 50 or more straight pipe diameters downstream to dissipate.

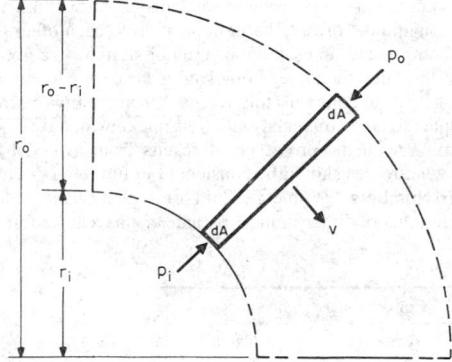

Fig. 3.3.18   Notation for flow in a curved path.

Thus this loss does not take place in the bend, but in the downstream system. These losses may be reduced by the use of splitter plates which help minimize the secondary flows by reducing $r_o - r_i$ and hence $p_o - p_i$.

EXAMPLE.   104°F (40°C) benzene flows at a rate of 8 ft³/s in a square horizontal duct. This duct makes a 90° turn with an inner radius of 1 ft and an outer radius of 2 ft. Calculate the difference between the walls of this bend. The area of this duct is $(r_o - r_i)^2 = (2 - 1)^2 = 1$ ft². From the continuity equation $V = Q/A = 8/1 = 8$ ft/s. The pressure difference

$$p_o - p_i = 2(r_o - r_i)\rho V^2/(r_o + r_i)$$
$$= 2(2 - 1)\, 1.663\, (8)^2/(2 + 1) = 70.95\ \text{lbf/ft}^2$$
$$= 70.95/144 = 0.4927\ \text{lbf/in}^2\ (3.397 \times 10^3\ \text{N/m}^2)$$

DIMENSIONLESS PARAMETERS

Modern engineering practice is based on a combination of theoretical analysis and experimental data. Often the engineer is faced with the necessity of obtaining practical results in situations where for various reasons, physical phenomena cannot be described mathematically and experimental data must be considered. The generation and use of **dimensionless parameters** provides a powerful and useful tool in (1) reducing the number of variables required for an experimental program, (2) es-

tablishing the principles of model design and testing, (3) developing equations, and (4) converting data from one system of units to another. Dimensionless parameters may be generated from (1) **physical equations**, (2) the principles of **similarity**, and (3) **dimensional analysis**. All physical equations must be dimensionally correct so that a dimensionless parameter may be generated by simply dividing one side of the equation by the other. A minimum of two dimensionless parameters will be formed, one being the inverse of the other.

EXAMPLE.   It is desired to generate a series of dimensionless parameters to describe the ratios of static pressure head, velocity head, and potential head to total head for frictionless incompressible flow. From the Bernoulli equation,

$$\frac{p}{\gamma} + \frac{V^2}{2g} + z = \Sigma h = \text{total head}$$

$$N_1 = \frac{p/\gamma + V^2/2g + z}{\Sigma h} = \frac{p/\gamma}{\Sigma h} + \frac{V^2/2g}{\Sigma h} + \frac{z}{\Sigma h} = N_p + N_v + N_z$$

or

$$N_2 = \frac{\Sigma h}{p_2/\gamma + V^2/2g + z} = N_1^{-1}$$

N_1 and N_2 are total energy ratios and N_p, N_v, and N_z are the ratios of the static pressure head, velocity head, and potential head, respectively, to the total head.

Models versus Prototypes   There are times when for economic or other reasons it is desirable to determine the performance of a structure or machine by testing another structure or machine. This type of testing is called model testing. The equipment being tested is called a **model**, and the equipment whose performance is to be predicted is called a **prototype**. A **model** may be smaller than, the same size as, or larger than the **prototype**. Model experiments on aircraft, rockets, missiles, pipes, ships, canals, turbines, pumps, and other structures and machines have resulted in savings that more than justified the expenditure of funds for the design, construction, and testing of the model. In some situations, the model and the prototype may be the same piece of equipment, for example, the laboratory calibration of a flowmeter with water to predict its performance with other fluids. Many manufacturers of fluid machinery have test facilities that are limited to one or two fluids and are forced to test with what they have available in order to predict performance with nonavailable fluids. For towing-tank testing of ship models and for wind-tunnel testing of aircraft and aircraft-component models, see Secs. 11.4 and 11.5.

Similarity Requirements   For complete similarity between a model and its prototype, it is necessary to have **geometric, kinematic,** and **dynamic** similarity. **Geometric similarity** exists between model and prototype when the **ratios** of all corresponding dimensions of the model and prototype are equal. These ratios may be written as follows:

Length:	$L_{model}/L_{prototype} = L_{ratio}$	
	$= L_m/L_p = L_r$	
Area:	$L^2_{model}/L^2_{prototype} = L^2_{ratio}$	
	$= L^2_m/L^2_p = L^2_r$	
Volume:	$L^3_{model}/L^3_{prototype} = L^3_{ratio}$	
	$= L^3_m/L^3_p = L^3_r$	

Kinematic similarity exists between model and prototype when their streamlines are geometrically similar. The kinematic ratios resulting from this condition are

Acceleration:	$a_r = a_m/a_p = L_m T_m^{-2}/L_p T_p^{-2}$
	$= L_r/T_r^{-2}$
Velocity:	$V_r = V_m/V_p = L_m T_m^{-1}/L_p T_p^{-1}$
	$= L_r/T_r^{-1}$
Volume flow rate:	$Q_r = Q_m/Q_p = L^3_m T_m^{-1}/L^3_p T_p^{-1}$
	$= L^3_r/T_r^{-1}$

Dynamic similarity exists between model and prototype having geometric and kinematic similarity when the ratios of all forces are the same. Consider the model/prototype relations for the flow around the object shown in Fig. 3.3.19. For **geometric similarity** $D_m/D_p = L_m/L_p = L_r$ and for **kinematic similarity** $U_{Am}/U_{Ap} = U_{Bm}/U_{Bp} = V_r =$

$L_r T_r^{-1}$. Next consider the three forces acting on point C of Fig. 3.3.19 without specifying their nature. From the geometric similarity of their vector polygons and Newton's law, for **dynamic similarity** $F_{1m}/F_{1p} = F_{2m}/F_{2p} = F_{3m}/F_{3p} = M_m a_{Cm}/M_p a_{Cp} = F_r$. For dynamic similarity, these force ratios must be maintained on all corresponding fluid parti-

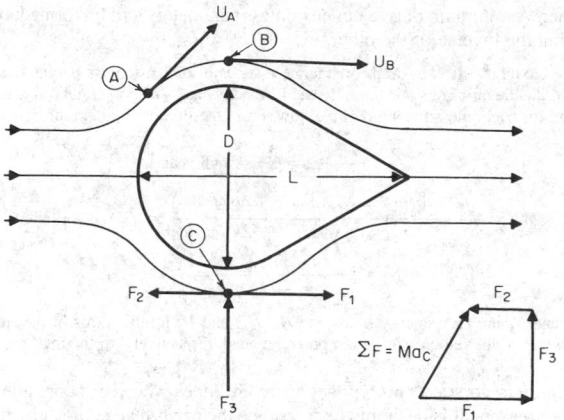

Fig. 3.3.19 Notation for dynamic similarity.

cles throughout the flow pattern. From the force polygon of Fig. 3.3.19, it is evident that $F_1 + \rightarrow F_2 + \rightarrow F_3 = M a_C$. For total model/prototype force ratio, comparisons of force polygons yield

$$F_r = \frac{F_{1m} + \rightarrow F_{2m} + \rightarrow F_{3m}}{F_{1p} + \rightarrow F_{2p} + \rightarrow F_{3p}} = \frac{M_m a_{Cm}}{M_p a_{Cp}}$$

Fluid Forces The fluid forces that are considered here are those acting on a fluid element whose mass $= \rho L^3$, area $= L^2$, length $= L$, and velocity $= L/T$.

Inertia force
$$\begin{aligned} F_i &= (\text{mass})(\text{acceleration}) \\ &= (\rho L^3)(L/T^2) = \rho L(L^2/T^2) \\ &= \rho L^2 V^2 \end{aligned}$$

Viscous force
$$\begin{aligned} F_\mu &= (\text{viscous shear stress})(\text{shear area}) \\ &= \tau L^2 = \mu(dU/dy)L^2 = \mu(V/L)L^2 \\ &= \mu L V \end{aligned}$$

Gravity force
$$\begin{aligned} F_g &= (\text{mass})(\text{acceleration due to gravity}) \\ &= (\rho L^3)(g) = \rho L^3 g \end{aligned}$$

Pressure force $\quad F_p = (\text{pressure})(\text{area}) = pL^2$

Centrifugal force
$$\begin{aligned} F_\omega &= (\text{mass})(\text{acceleration}) \\ &= (\rho L^3)(L/T^2) \\ &= (\rho L^3)(L\omega^2) = \rho L^4 \omega^2 \end{aligned}$$

Elastic force
$$\begin{aligned} F_E &= (\text{modulus of elasticity})(\text{area}) \\ &= EL^2 \end{aligned}$$

Surface-tension force $\quad F_\sigma = (\text{surface tension})(\text{length}) = \sigma L$

Vibratory force
$$\begin{aligned} F_f &= (\text{mass})(\text{acceleration}) \\ &= (\rho L^3)(L/T^2) \\ &= (\rho L^4)(T^{-2}) = \rho L^4 f^2 \end{aligned}$$

If all fluid forces were acting on a fluid element,

$$\begin{aligned} F_r &= \frac{F_{\mu m} + \rightarrow F_{gm} + \rightarrow F_{pm} + \rightarrow F_{\omega m} + \rightarrow F_{Em} + \rightarrow F_{\sigma m} + \rightarrow F_{fm}}{F_{\mu p} + \rightarrow F_{gp} + \rightarrow F_{pp} + \rightarrow F_{\omega p} + \rightarrow F_{Ep} + \rightarrow F_{\sigma p} + \rightarrow F_{fp}} \\ &= \frac{F_{im}}{F_{ip}} \end{aligned}$$

Examination of the above equation and the force polygon of Fig. 3.3.19 lead to the conclusion that dynamic similarity can be characterized by an equality of force ratios one less than the total number involved. Any force ratio may be eliminated, depending upon the quantities which are desired. Fortunately, in most practical engineering problems, not all of the eight forces are involved because some may not be acting, may be of negligible magnitude, or may be in opposition to each other in such a way as to compensate. In each application of similarity, a good understanding of the fluid phenomena involved is necessary to eliminate the irrelevant, trivial, or compensating forces. When the flow phenomenon is too complex to be readily analyzed, or is not known, only experimental verification with the prototype of results from a model test will determine what forces should be considered in future model testing.

Standard Numbers With eight fluid forces that can act in flow situations, the number of dimensionless parameters that can be formed from

Table 3.3.7 Standard Numbers

Force ratio	Equations	Result	Conventional practice		
			Form	Symbol	Name
$\dfrac{\text{Inertia}}{\text{Viscous}}$	$\dfrac{F_i}{F_\mu} = \dfrac{\rho L^2 V^2}{\mu L V}$	$\dfrac{\rho L V}{\mu}$	$\dfrac{\rho L V}{\mu}$	**R**	Reynolds
$\dfrac{\text{Inertia}}{\text{Gravity}}$	$\dfrac{F_i}{F_g} = \dfrac{\rho L^2 V^2}{\rho L^3 g}$	$\dfrac{V^2}{Lg}$	$\dfrac{V}{\sqrt{Lg}}$	**F**	Froude
$\dfrac{\text{Inertia}}{\text{Pressure}}$	$\dfrac{F_i}{F_p} = \dfrac{\rho L^2 V^2}{\rho L^2}$	$\dfrac{\rho V^2}{p}$	$\dfrac{\rho V^2}{p}$	**E**	Euler
			$\dfrac{2\Delta p}{\rho V^2}$	**C$_p$**	Pressure coefficient
$\dfrac{\text{Inertia}}{\text{Centrifugal}}$	$\dfrac{F_i}{F_\omega} = \dfrac{\rho L^2 V^2}{\rho L^4 \omega^2}$	$\dfrac{V^2}{L^2 \omega^2}$	$\dfrac{V}{DN}$	**V**	Velocity ratio
$\dfrac{\text{Inertia}}{\text{Elastic}}$	$\dfrac{F_i}{F_E} = \dfrac{\rho L^2 V^2}{EL^2}$	$\dfrac{\rho V^2}{E}$	$\dfrac{\rho V^2}{E}$	**C**	Cauchy
			$\dfrac{V}{\sqrt{E/\rho}}$	**M**	Mach
$\dfrac{\text{Inertia}}{\text{Surface tension}}$	$\dfrac{F_i}{F_\sigma} = \dfrac{\rho L^2 V^2}{\sigma L}$	$\dfrac{\rho L V^2}{\sigma}$	$\dfrac{\rho L V^2}{\sigma}$	**W**	Weber
$\dfrac{\text{Inertia}}{\text{Vibration}}$	$\dfrac{F_i}{F_f} = \dfrac{\rho L^2 V^2}{\rho L^4 f^2}$	$\dfrac{V^2}{L^2 f^2}$	$\dfrac{Lf}{V}$	**S**	Strouhal

SOURCE: Computed from data given in Murdock, "Fluid Mechanics and Its Applications," Houghton Mifflin, 1976.

their ratios is 56. However, conventional practice is to ratio the inertia force to the other fluid forces, usually by division because the inertia force is the vector sum of all the other forces involved in a given flow situation. Results obtained by dividing the inertia force by each of the other forces are shown in Table 3.3.7 compared with the standard numbers that are used in conventional practice.

DYNAMIC SIMILARITY

Vibration In the flow of fluids around objects and in the motion of bodies immersed in fluids, **vibration** may occur because of the formation of a wake caused by alternate shedding of eddies in a periodic fashion or by the vibration of the object or the body. The **Strouhal number S** is the ratio of the velocity of vibration Lf to the velocity of the fluid V. Since the vibration may be fluid-induced or structure-induced, two frequencies must be considered, the wake frequency $f\omega$ and the natural frequency of the structure f_n. Fluid-induced forces are usually of small magnitude, but as the wake frequency approaches the natural frequency of the structure, the vibratory forces increase very rapidly. When $f\omega = f_n$, the structure will go into resonance and fail. This imposes on the model designer the requirement of matching to scale the natural-frequency characteristics of the prototype. This subject is treated later under Wake Frequency. All further discussions of model/prototype relations are made under the assumption that either vibratory forces are absent or they are taken care of in the design of the model or in the test program.

Incompressible Flow Considered in this category are the flow of fluids around an object, motion of bodies immersed in incompressible fluids, and the flow of incompressible fluids in conduits. It includes, for example, a submarine traveling under water but not partly submerged, and liquids flowing in pipes and passages when the liquid completely fills them, but not when partly full as in open-channel flow. It also includes aircraft moving in atmospheres that may be considered incompressible. Incompressible flow in rotating machinery is considered separately.

In these situations the gravity force, although acting on all fluid particles, does not affect the flow pattern. Excluding rotating machinery, centrifugal forces are absent. By definition of an incompressible fluid, elastic forces are zero, and since there is no liquid-gas interface, surface-tension forces are absent.

The only forces now remaining for consideration are the inertia, viscous, and pressure. Using standard numbers, the parameters are **Reynolds number** and **pressure coefficient.** The Reynolds number may be converted into a kinematic ratio by noting that by definition $v = \mu/\rho$ and substituting in $R = \rho LV/\mu = LV/v$. In this form, Reynolds number is the ratio of the fluid velocity V and the "shear velocity" v/L. For this reason, Reynolds number is used to characterize the velocity profile. Forces and pressure losses are then determined by the pressure coefficient.

EXAMPLE. A submarine is to move submerged through 32°F (0°C) seawater at a speed of 10 knots. (1) At what speed should a 1 : 20 model be towed in 68°F (20°C) fresh water? (2) If the thrust on the model is found to be 42,500 lbf, what horsepower will be required to propel the submarine?

1. Speed of model for Reynolds-number similarity

$$R_m = R_p = \left(\frac{\rho VL}{\mu}\right)_m = \left(\frac{\rho VL}{\mu}\right)_p$$

$V_m = V_p(\rho_p/\rho_m)(L_p/L_m)(\mu_m/\mu_p)$
$V_m = (10)(1.995/1.937)(20/1)(20.92 \times 10^{-6}/39.40 \times 10^{-6})$
   $= 109.4$ knots (56.27 m/s)

2. Prototype horsepower

$$C_{pp} = C_{pm} = \left(\frac{2\Delta p}{\rho V^2}\right)_p = \left(\frac{2\Delta p}{\rho V^2}\right)_m$$

$F = \Delta p L^2, \Delta p = \dfrac{F}{L^2},$ so that $\left(\dfrac{2F}{\rho V^2 L^2}\right)_p = \left(\dfrac{2F}{\rho V^2 L^2}\right)_m$

$F_p = F_m(\rho_p/\rho_m)(V_p/V_m)^2(L_p/L_m)^2$
   $= 42,500 \ (1.995/1.937)(10/109.4)^2(20/1)^2 = 146,300$ lbf

$$P = FV = \left(\frac{146,300}{550}\right)\left(\frac{10 \times 6,076}{3,600}\right)$$
   $= 4,490$ hp $(3.35 \times 10^6$ W$)$

Compressible Flow Considered in this category are the flow of compressible fluids under the conditions specified for incompressible flow in the preceding paragraphs. In addition to the forces involved in incompressible flow, the elastic force must be added. Conventional practice is to use the square root of the inertia/elastic force ratio or **Mach number.**

Mach number is the ratio of the fluid velocity to its speed of sound and may be written $\mathbf{M} = V/c = V\sqrt{E_s/\rho}$. For an ideal gas, $\mathbf{M} = V/\sqrt{kg_cRT}$. In compressible-flow problems, practice is to use the Mach number to characterize the velocity or kinematic similarity, the Reynolds number for dynamic similarity, and the pressure coefficient for force or pressure-loss determination.

EXAMPLE. An airplane is to fly at 500 mi/h in an atmosphere whose temperature is 32°F (0°C) and pressure is 12 lbf/in². A 1 : 20 model is tested in a wind tunnel where a supply of air at 392°F (200°C) and variable pressure is available. At (1) what speed and (2) what pressure should the model be tested for dynamic similarity?

1. Speed for Mach-number similarity

$$\mathbf{M}_m = \mathbf{M}_p = \left(\frac{V}{\sqrt{E/\rho}}\right)_m = \left(\frac{V}{\sqrt{E/\rho}}\right)_p = \left(\frac{V}{\sqrt{kg_cRT}}\right)_m = \left(\frac{V}{\sqrt{kg_cRT}}\right)_p$$

$V_m = V_p(k_m/k_p)^{1/2}(R_m/R_p)^{1/2}(T_m/T_p)^{1/2}$

For the same gas $k_m = k_p, R_m = R_p,$ and

$$V_m = V_p \sqrt{T_m/T_p} = 500 \sqrt{851.7/491.7} = 658.1 \text{ mi/h}$$

2. Pressure for Reynolds-number similarity

$$\mathbf{R}_m = \mathbf{R}_p = \left(\frac{\rho VL}{\mu}\right)_m = \left(\frac{\rho VL}{\mu}\right)_p$$

$$\rho_m = \rho_p(V_p/V_m)(L_p/L_m)(\mu_m/\mu_p)$$

Since $\rho = p/g_cRT$

$$\left(\frac{p}{g_cRT}\right)_m = \left(\frac{p}{g_cRT}\right)_p (V_p/V_m)(L_p/L_m)(\mu_m/\mu_p)$$

$p_m = p_p(T_m/T_p)(V_p/V_m)(L_p/L_m)(\mu_m/\mu_p)$
$p_m = 12(851.7/491.7)(500/658.1)(20/1)(53.15 \times 10^{-6}/35.67 \times 10^{-6})$
$p_m = 470.6$ lbf/in² $(3.245 \times 10^6$ N/m²$)$

For information about wind-tunnel testing and its limitations, refer to Sec. 11.4.

Centrifugal Machinery This category includes the flow of fluids in such centrifugal machinery as compressors, fans, and pumps. In addition to the inertia, pressure, viscous, and elastic forces, centrifugal forces must now be considered. Since centrifugal force is really a special case of the inertia force, their ratio as shown in Table 3.3.7 is **velocity ratio** and is the ratio of the fluid velocity to the machine tangential velocity. In model/prototype relations for centrifugal machinery, DN (D = diameter, ft, N = rotational speed) is substituted for the velocity V, and D for L, which results in the following:

$$\mathbf{M} = DN/\sqrt{kg_cRT} \qquad \mathbf{R} = \rho D^2N/\mu \qquad C_p = 2\Delta p/\rho D^2N^2$$

EXAMPLE. A centrifugal compressor operating at 100 r/min is to compress methane delivered to it at 50 lbf/in² and 68°F (20°C). It is proposed to test this compressor with air from a source at 140°F (60°C) and 100 lbf/in². Determine compressor speed and inlet-air pressure required for dynamic similarity. Find speed for Mach-number similarity:

$$\mathbf{M}_m = \mathbf{M}_p = (DN/\sqrt{kg_cRT})_m = (DN/\sqrt{kg_cRT})_p$$
$$N_m = N_p(D_p/D_m)\sqrt{(k_m/k_p)(R_m/R_p)(T_m/T_p)}$$
   $= 100 \ (1) \sqrt{(1.40/1.32)(53.34/96.33)(599.7/527.7)}$
   $= 81.70$ r/min

Find pressure for Reynolds-number similarity:

$$\mathbf{R}_m = \mathbf{R}_p = \left(\frac{\rho D^2N}{\mu}\right)_m = \left(\frac{\rho D^2N}{\mu}\right)_p$$

For an ideal gas $\rho = p/g_c RT$, so that

$$(pD^2N/g_c RT\mu)_m = (pD^2N/g_c RT\mu)_p$$
$$p_m = p_p(D_p/D_m)^2(N_p/N_m)(R_m/R_p)(T_m/T_p)(\mu_m/\mu_p)$$
$$= 50(1)^2(100/81.70)(53.34/96.33)(599.7/527.7) \times (41.79$$
$$\times 10^{-8}/22.70 \times 10^{-8}) = 70.90 \text{ lbf/in}^2 \ (4.888 \times 10^5 \text{ N/m}^2)$$

See Sec. 14 for specific information on pump and compressor similarity.

Liquid Surfaces Considered in this category are ships, seaplanes during takeoff, submarines partly submerged, piers, dams, rivers, open-channel flow, spillways, harbors, etc. Resistance at liquid surfaces is due to surface tension and wave action. Since wave action is due to gravity, the gravity force and surface-tension force are now added to the forces that were considered in the last paragraph. These are expressed as the square root of the inertia/gravity force ratio or **Froude number** $\mathbf{F} = V/\sqrt{Lg}$ and as the inertia/surface tension force ratio or **Weber number** $\mathbf{W} = \rho L V^2/\sigma$. On the other hand, elastic and pressure forces are now absent. Surface tension is a minor property in fluid mechanics and it normally exerts a negligible effect on wave formation *except* when the waves are small, say less than 1 in. Thus the effects of surface tension on the model might be considerable, but negligible on the prototype. This type of ''scale effect'' must be avoided. For accurate results, the inertia/surface tension force ratio or Weber number should be considered. It is never possible to have complete dynamic similarity of liquid surfaces unless the model and prototype are the same size, as shown in the following example.

EXAMPLE. An ocean vessel 500 ft long is to travel at a speed of 15 knots. A 1 : 25 model of this ship is to be tested in a towing tank using seawater at design temperature. Determine the model speed required for (1) wave-resistance similarity, (2) viscous or skin-friction similarity, (3) surface-tension similarity, and (4) the model size required for complete dynamic similarity.

1. *Speed for Froude-number similarity*

$$\mathbf{F}_m = \mathbf{F}_p = (V/\sqrt{Lg})_m = (V/\sqrt{Lg})_p$$

or

$$V_m = V_p \sqrt{L_m/L_p} = 15\sqrt{1/25} = 3 \text{ knots}$$

2. *Speed for Reynolds-number similarity*

$$\mathbf{R}_m = \mathbf{R}_p = (\rho LV/\mu)_m = (\rho LV/\mu)_p$$
$$V_m = V_p(\rho_p/\rho_m)(L_p/L_m)(\mu_m/\mu_p)$$
$$V_m = 15(1)(25/1)(1) = 375 \text{ knots}$$

3. *Speed for Weber-number similarity*

$$\mathbf{W}_m = \mathbf{W}_p = (\rho LV^2/\sigma)_m = (\rho LV^2/\sigma)_p$$
$$V_m = V_p \sqrt{(\rho_p/\rho_m)(L_p/L_m)(\sigma_m/\sigma_p)}$$
$$V_m = 15\sqrt{(1)(25)(1)} = 75 \text{ knots}$$

4. *Model size for complete similarity.* First try Reynolds and Froude similarity; let

$$V_m = V_p(\rho_p/\rho_m)(L_p/L_m)(\mu_m/\mu_p) = V_p\sqrt{L_m/L_p}$$

which reduces to

$$L_m/L_p = (\rho_p/\rho_m)^{2/3}(\mu_m/\mu_p)^{2/3}$$

Next try Weber and Froude similarity; let

$$V_m = V_p\sqrt{(\rho_p/\rho_m)(L_p/L_m)(\sigma_m/\sigma_p)} = V_p\sqrt{L_m/L_p}$$

which reduces to

$$L_m/L_p = (\rho_p/\rho_m)^{1/2}(\sigma_m/\sigma_p)^{1/2}$$

For the same fluid at the same temperature, either of the above solves for $L_m = L_p$, or the model must be the same size as the prototype. For use of different fluids and/or the same fluid at different temperatures.

$$L_m/L_p = (\rho_p/\rho_m)^{2/3}(\mu_m/\mu_p)^{2/3} = (\rho_p/\rho_m)^{1/2}(\sigma_m/\sigma_p)^{1/2}$$

which reduces to

$$(\mu^4/\rho\sigma^3)_m = (\mu^4/\rho\sigma^3)_p$$

No practical way has been found to model for complete similarity. Marine engineering practice is to model for wave resistance and correct for skin-friction resistance. See Sec. 11.3.

DIMENSIONAL ANALYSIS

Dimensional analysis is the mathematics of dimensions and quantities and provides procedural techniques whereby the variables that are assumed to be significant in a problem can be formed into dimensionless parameters, the number of parameters being less than the number of variables. This is a great advantage, because fewer experimental runs are then required to establish a relationship between the parameters than between the variables. While the user is not presumed to have any knowledge of the fundamental physical equations, the more knowledgeable the user, the better the results. If any significant variable or variables are omitted, the relationship obtained from dimensional analysis will not apply to the physical problem. On the other hand, inclusion of all possible variables will result in losing the principal advantage of dimensional analysis, i.e., the reduction of the amount of experimental data required to establish relationships. Two formal methods of dimensional analysis are used, the **method of Lord Rayleigh** and **Buckingham's II theorem.**

Dimensions used in mechanics are mass M, length L, time T, and force F. Corresponding **units** for these dimensions are the slug (kilogram), the foot (metre), the second (second), and the pound force (newton). Any system in mechanics can be defined by three fundamental dimensions. Two systems are used, the force (FLT) and the mass (MLT). In the force system, mass is a derived quantity and in the mass system, force is a derived quantity. Force and mass are related by Newton's law: $F = MLT^{-2}$ and $M = FL^{-1}T^2$. Table 3.3.8 shows common variables and their dimensions and units.

Lord Rayleigh's method uses algebra to determine interrelationships among variables. While this method may be used for any number of variables, it becomes relatively complex and is not generally used for more than four. This method is most easily described by example.

EXAMPLE. In laminar flow, the unit shear stress τ is some function of the fluid dynamic viscosity μ, the velocity difference dU between adjacent laminae separated by the distance dy. Develop a relationship.

1. Write a functional relationship of the variables:

$$\tau = f(\mu, dU, \mathbf{dy})$$

Assume $\tau = K(\mu^a dU^b dy^c)$.

2. Write a dimensional equation in either FLT or MLT system:

$$(FL^{-2}) = K(FL^{-2}T)^a(LT^{-1})^b(L)^c$$

3. Solve the dimensional equation for exponents:

		τ	μ	dU	dy
Force	F	$1 =$	a	$+ 0$	$+ 0$
Length	L	$-2 =$	$-2a$	$+ b$	$+ c$
Time	T	$0 =$	a	$- b$	$+ 0$

Solution: $a = 1, b = 1, c = -1$

4. Insert exponents in the functional equation: $\tau = K(\mu^a dU^b dy^c) = K(\mu^1 du^1 dy^{-1})$, or $K = (\mu dU/\tau dy)$. This was based on the assumption of $\tau = K(\mu^a dU^b dy^c)$. The general relationship is $K = f(\mu dU/\tau dy)$. The functional relationship cannot be obtained from dimensional analysis. Only physical analysis and/or experiments can determine this. From both physical analysis and experimental data,

$$\tau = \mu \, dU/dy$$

The Buckingham II theorem serves the same purpose as the method of Lord Rayleigh for deriving equations expressing one variable in terms of its dependent variables. The II theorem is preferred when the number of variables exceeds four. Application of the II theorem results in the formation of dimensionless parameters called π ratios. These π ratios have no relation to 3.14159. . . . The II theorem will be illustrated in the following example.

EXAMPLE. Experiments are to be conducted with gas bubbles rising in a still liquid. Consider a gas bubble of diameter D rising in a liquid whose density is ρ, surface tension σ, viscosity μ, rising with a velocity of V in a gravitational field of g. Find a set of parameters for organizing experimental results.

1. List all the physical variables considered according to type: geometric, kinematic, or dynamic.

Table 3.3.8   Dimensions and Units of Common Variables

Symbol	Variable	Dimensions MLT	Dimensions FLT	Units USCS*	Units SI
		Geometric			
L	Length	L		ft	m
A	Area	L^2		ft^2	m^2
V	Volume	L^3		ft^3	m^3
		Kinematic			
t	Time	T		s	s
ω	Angular velocity				
f	Frequency	T^{-1}		s^{-1}	s^{-1}
V	Velocity	LT^{-1}		ft/s	m/s
v	Kinematic viscosity	L^2T^{-1}		ft^2/s	m^2/s
Q	Volume flow rate	L^3T^{-1}		ft^3/s	m^3/s
α	Angular acceleration	T^{-2}		s^{-2}	s^{-2}
a	Acceleration	LT^{-2}		ft/s^{-2}	m/s^2
		Dynamic			
ρ	Density	ML^{-3}	$FL^{-4}T^2$	slug/ft^3	kg/m^3
M	Mass	M	$FL^{-1}T^2$	slugs	kg
I	Moment of inertia	ML^2	FLT^2	slug · ft^2	kg · m^2
μ	Dynamic viscosity	$ML^{-1}T^{-1}$	$FL^{-2}T$	slug/ft · s	kg/m · s
M	Mass flow rate	MT^{-1}	$FL^{-1}T^{-1}$	slug/s	kg/s
MV	Momentum	MLT^{-1}	FT	lbf · s	N · s
Ft	Impulse				
$M\omega$	Angular momentum	ML^2T^{-1}	FLT	slug · ft^2/s	kg · m^2/s
γ	Specific weight	$ML^{-2}T^{-2}$	FL^{-3}	lbf/ft^3	N/m^3
p	Pressure				
τ	Unit shear stress	$ML^{-1}T^{-2}$	FL^{-2}	lbf/ft^2	N/m^2
E	Modulus of elasticity				
σ	Surface tension	MT^{-2}	FL^{-1}	lbf/ft	N/m
F	Force	MLT^{-2}	F	lbf	N
E	Energy				
W	Work	ML^2T^{-2}	FL	lbf · ft	J
FL	Torque				
P	Power	ML^2T^{-3}	FLT^{-1}	lbf · ft/s	W
v	Specific volume	$M^{-1}L^3$	$F^{-1}L^4T^{-2}$	ft^3/lbm	m^3/kg

*United States Customary System.

2. Choose either the *FLT* or *MLT* system of dimensions.
3. Select a "basic group" of variables characteristic of the flow as follows:

    a. B_G, a geometric variable
    b. B_K, a kinematic variable
    c. B_D, a dynamic variable (if three dimensions are used)

4. Assign *A* numbers to the remaining variables starting with A_1.

Type	Symbol	Description	Dimensions	Number
Geometric	D	Bubble diameter	L	B_G
Kinematic	V	Bubble velocity	LT^{-1}	B_K
	g	Acceleration of gravity	LT^{-2}	A_1
Dynamic	ρ	Liquid density	ML^{-3}	B_D
	σ	Surface tension	MT^{-2}	A_2
	μ	Liquid viscosity	$ML^{-1}T^{-1}$	A_3

5. Write the basic equation for each π ratio as follows:

$$\pi_1 = (B_G)^{x1}(B_K)^{y1}(B_D)^{z1}(A_1)$$
$$\pi_2 = (B_G)^{x2}(B_K)^{y2}(B_D)^{z2}(A_2) \ldots \pi_n = (B_G)^{xn}(B_K)^{yn}(B_D)^{zn}(A_n)$$

Note that the number of π ratios is equal to the number of *A* numbers and thus equal to the number of variables less the number of fundamental dimensions in a problem.

6. Write the dimensional equations and use the algebraic method to determine the value of exponents *x*, *y*, and *z* for each π ratio. Note that for all π ratios, the sum of the exponents of a given dimension is zero.

$$\pi_1 = (B_G)^{x1}(B_K)^{y1}(B_D)^{z1}(A_1) = (D)^{x1}(V)^{y1}(\rho)^{z1}(g)$$
$$(M^0L^0T^0) = (L^{x1})(L^{y1}T^{-y1})(M^{z1}L^{-3z1})(LT^{-2})$$

Solution: $x_1 = 1, y_1 = -2, z_1 = 0$

$$\pi_1 = D^1V^{-2}\rho^0 g = Dg/V^2$$
$$\pi_2 = (B_G)^{x2}(B_K)^{y2}(B_D)^{z2}(A_2) = (D)^{x2}(V)^{y2}(\rho)^{z2}(\sigma)$$
$$(M^0L^0T^0) = (L^{x2})(L^{y2}T^{-y2})(M^{z2}L^{-3z2})(MT^{-2})$$

Solution: $x_2 = -1, y_2 = -2, z_2 = -1$

$$\pi_2 = D^{-1}V^{-2}\rho^{-1}\sigma = \sigma/DV^2\rho$$
$$\pi_3 = (B_G)^{x3}(B_K)^{y3}(B_D)^{z3}(A_3) = (D)^{x3}(V)^{y3}(\rho)^{z3}(\mu)$$
$$(M^0L^0T^0) = (L^{x3})(L^{y3}T^{-y3})(M^{z3}L^{-3z3})(ML^{-1}T^{-1})$$

Solution: $x_3 = -1, y_3 = -1, z_3 = -1$

$$\pi_3 = D^{-1}V^{-1}\rho^{-1}\mu = \mu/DV\rho$$

7. Convert π ratios to conventional practice. One statement of the Buckingham II theorem is that any π ratio may be taken as a function of all the others, or $f(\pi_1, \pi_2, \pi_3, \ldots, \pi_n) = 0$. This equation is mathematical shorthand for a functional statement. It could be written, for example, as $\pi_2 = f(\pi_1, \pi_3, \ldots, \pi_n)$. This equation states that π_2 is some function of π_1 and π_3 through π_n but is not a statement of *what* function π_2 is of the other π ratios. This can be determined only by physical and/or experimental analysis. Thus we are free to substitute *any* function in the equation; for example, π_1 may be replaced with $2\pi_1^{-1}$ or π_n with $a\pi_n^b$.

The procedures set forth in this example are designed to produce π ratios containing the same terms as those resulting from the application of the principles of similarity so that the physical significance may be understood. However, any other combinations might have been used. The only real requirement for a "basic group" is that it contain the same number of terms as there are dimensions in a problem and that each of these dimensions be represented in it.

The π ratios derived for this example may be converted into conventional practice as follows:

$$\pi_1 = Dg/V^2$$

is recognized as the inverse of the square root of the Froude number **F**

$$\pi_2 = \sigma/DV^2\rho$$

is the inverse of the Weber number **W**

$$\pi = \mu/DV\rho$$

is the inverse of the Reynolds number **R**

Let $\qquad\qquad \pi_1 = f(\pi_2, \pi_3)$
Then $\qquad\qquad V = K(Dg)^{1\div2}$
where $\qquad\qquad K = f(\mathbf{W}, \mathbf{R})$

This agrees with the results of the dynamic-similarity analysis of liquid surfaces. This also permits a reduction in the experimental program from variations of six variables to three dimensionless parameters.

FORCES OF IMMERSED OBJECTS

Drag and Lift When a fluid impinges on an object as shown in Fig. 3.3.20, the undisturbed fluid pressure p and the velocity V change. Writing Bernoulli's equation for two points on the surface of the object, the point S being the most forward point and point A being any other point, we have, for horizontal flow,

$$p + \rho V^2/2 = p_S + \rho V_S^2/2 = p_A + \rho V_A^2/2$$

At point S, $V_S = 0$, so that $p_S = p + \rho V^2/2$. This is called the **stagnation** point, and p_S is the **stagnation pressure**. Since point A is any other point, the result of the fluid impingement is to create a pressure $p_A = p + \rho(V^2 - V_A^2)/2$ acting normal to every point on the surface of the object.

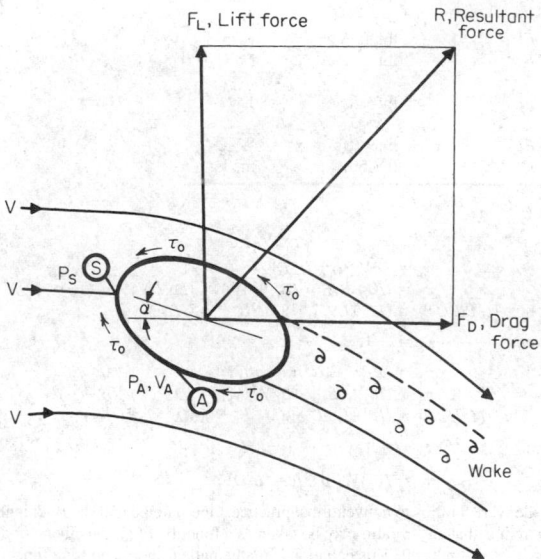

Fig. 3.3.20 Notation for drag and lift.

In addition, a frictional force $F_f = \tau_0 A_s$ tangential to the surface area A_s opposes the motion. The sum of these forces gives the resultant force R acting on the body. The resultant force R is resolved into the **drag** component F_D parallel to the flow and **lift** component F_L perpendicular to the fluid motion. Depending upon the shape of the object, a wake may be formed which sheds eddies with a frequency of f. The angle α is called the angle of attack. (See Secs. 11.4 and 11.5.)

From dimensional analysis or dynamic similarity,

$$f(\mathbf{C}_p, \mathbf{R}, \mathbf{M}, \mathbf{S}) = 0$$

The formation of a wake depends upon the Reynolds number, or $\mathbf{S} = f(\mathbf{R})$. This reduces the functional relation to $f(\mathbf{C}_p, \mathbf{R}, \mathbf{M}) = 0$.

Since the drag and lift forces may be considered independently,

$$F_D = C_D \rho V^2(A)/2$$

where $C_D = f(\mathbf{R}, \mathbf{M})$, and A = characteristic area.

$$F_L = C_L \rho V^2(A)/2$$

where $C_L = f(\mathbf{R}, \mathbf{M})$.

It is evident from Fig. 3.3.20 that C_D and C_L are also functions of the angle of attack. Since the drag force arises from two sources, the pressure or shape drag F_p and the skin-friction drag F_f due to wall shear stress τ_0, the drag coefficient is made up of two parts:

$$F_D = F_p + F_f = C_D \rho A V^2/2 = C_p \rho A V^2/2 + C_f \rho A_s V^2/2$$
or $\qquad C_D = C_p + C_f A_s/A$

where C_p is the coefficient of pressure, C_f the skin-friction coefficient, and A_s the characteristic area for shear.

Skin-Friction Drag Figure 3.3.21 shows a fluid approaching a smooth flat plate with a uniform velocity profile of V. As the fluid passes over the plate, the velocity at the plate surface is zero and increases to V at some distance δ from the surface. The region in which the velocity varies from 0 to V is called the **boundary layer**. For some

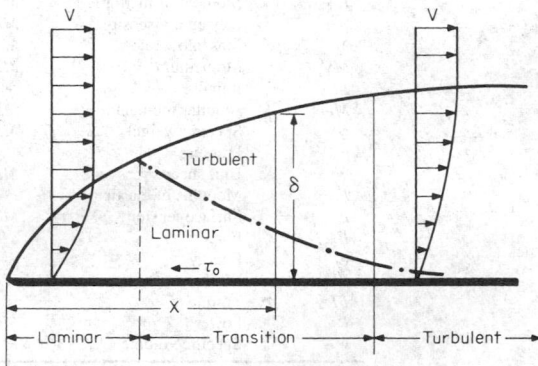

Fig. 3.3.21 Boundary layer along a smooth flat plate.

distance along the plate, the flow within the boundary layer is laminar, with viscous forces predominating, but in the transition zone as the inertia forces become larger, a turbulent layer begins to form and increases as the laminar layer decreases.

Boundary-layer thickness and skin-friction drag for incompressible flow over smooth flat plates may be calculated from the following equations, where $\mathbf{R}_X = \rho V X/\mu$:

Laminar

$\delta/X = 5.20\,\mathbf{R}_X^{-1/2} \qquad 0 < \mathbf{R}_X < 5 \times 10^5$
$C_f = 1.328\,\mathbf{R}_X^{-1/2} \qquad 0 < \mathbf{R}_X < 5 \times 10^5$

Turbulent

$\delta/X = 0.377\,\mathbf{R}_X^{-1/5} \qquad\qquad 5 \times 10^4 < \mathbf{R}_X < 10^6$
$\delta/X = 0.220\,\mathbf{R}_X^{-1/6} \qquad\qquad 10^6 < \mathbf{R}_X < 5 \times 10^8$
$C_f = 0.0735\,\mathbf{R}_X^{-1/5} \qquad\qquad 2 \times 10^5 < \mathbf{R}_X < 10^7$
$C_f = 0.455\,(\log_{10}\mathbf{R}_X)^{-2.58} \qquad 10^7 < \mathbf{R}_X < 10^8$
$C_f = 0.05863\,(\log_{10}C_f\mathbf{R}_X)^{-2} \qquad 10^8 < \mathbf{R}_X < 10^9$

Transition The Reynolds number at which the boundary layer changes depends upon the roughness of the plate and degree of turbulence. The generally accepted number is 500,000, but the transition can take place at Reynolds numbers higher or lower. (Refer to Secs. 11.4 and 11.5.) For transition at any Reynolds number $\mathbf{R}_X$,

$$C_f = 0.455\,(\log_{10}\mathbf{R}_X)^{-2.58} - (0.0735\,\mathbf{R}_t^{4/5} - 1.328\,8\,\mathbf{R}_t^{1/2})\,\mathbf{R}_X^{-1}$$

For $\mathbf{R}_t = 5 \times 10^5$, $C_f = 0.455\,(\log_{10}\mathbf{R}_X)^{-2.58} - 1{,}725\,\mathbf{R}_X^{-1}$.

Pressure Drag Experiments with sharp-edged objects placed perpendicular to the flow stream indicate that their drag coefficients are essentially constant at Reynolds numbers over 1,000. This means that the drag for $\mathbf{R}_x > 10^3$ is pressure drag. Values of C_D for various shapes are given in Sec. 11 along with the effects of Mach number.

Wake Frequency An object in a fluid stream may be subject to the downstream periodic shedding of vortices from first one side and then the other. The frequency of the resulting transverse (lift) force is a function of the stream Strouhal number. As the wake frequency approaches the natural frequency of the structure, the periodic lift force increases asymptotically in magnitude, and when resonance occurs, the structure fails. Neglecting to take this phenomenon into account in design has been responsible for failures of electric transmission lines, submarine periscopes, smokestacks, bridges, and thermometer wells. The wake-frequency characteristics of **cylinders** are shown in Fig. 3.3.22. At a Reynolds number of about 20, vortices begin to shed alternately. Behind the cylinder is a staggered stable arrangement of vortices known as the "Kármán vortex trail." At a Reynolds number of about 10^5, the flow changes from laminar to turbulent. At the end of the transition zone ($\mathbf{R} \approx 3.5 \times 10^5$), the flow becomes turbulent, the alter-

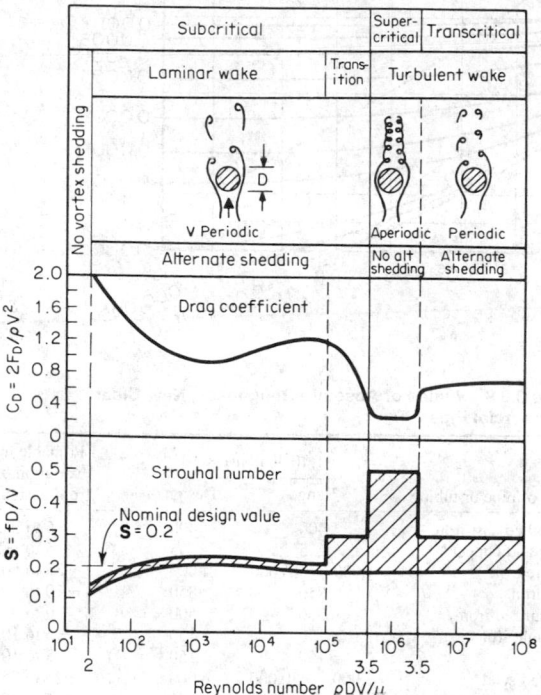

Fig. 3.3.22 Flow around a cylinder. (*From Murdock, "Fluid Mechanics and Its Applications," Houghton Mifflin, 1976.*)

nate shedding stops, and the wake is aperiodic. At the end of the supercritical zone ($\mathbf{R} \approx 3.5 \times 10^6$), the wake continues to be turbulent, but the shedding again becomes alternate and periodic.

The alternating **lift force** is given by

$$F_L(t) = C_L \rho V^2 A \sin (2\pi ft)/2$$

where t is the time. For an analysis of this force in the subcritical zone, see Belvins (Murdock, "Fluid Mechanics and Its Applications," Houghton Mifflin, 1976). For design of steel stacks, Staley and Graven (ASME 72PET/30) recommend $C_L = 0.8$ for $10^4 < \mathbf{R} < 10^5$, $C_L = 2.8 - 0.4 \log_{10} \mathbf{R}$ for $\mathbf{R} = 10^5$ to 10^6, and $C_L = 0.4$ for $10^6 < \mathbf{R} < 10^7$.

The **Strouhal number** is nearly constant to $\mathbf{R} = 10^5$, and a nominal design value of 0.2 is generally used. Above $\mathbf{R} = 10^5$, data from different experimenters vary widely, as indicated by the crosshatched zone of

Fig. 3.3.22. This wide zone is due to experimental and/or measurement difficulties and the dependence on surface roughness to "trigger" the boundary layer. Examination of Fig. 3.3.22 indicates an inverse relation of Strouhal number to drag coefficient.

Observation of actual structures shows that they vibrate at their natural frequency and with a mode shape associated with their fundamental (first) mode during vortex excitation. Based on observations of actual stacks and wind-tunnel tests, Staley and Graven recommend a constant Strouhal number of 0.2 for all ranges of Reynolds number. The ASME recommends $\mathbf{S} = 0.22$ for thermowell design ("Temperature Measurement," PTC 19.3). Until such time as the value of the Strouhal number above $\mathbf{R} = 10^5$ has been firmly established, designers of structures in this area should proceed with caution.

FLOW IN PIPES

Parameters for Pipe Flow The forces acting on a fluid flowing through and completely filling a horizontal pipe are inertia, viscous, pressure, and elastic. If the surface roughness of the pipe is ε, either similarity or dimensional analysis leads to $\mathbf{C}_p = f(\mathbf{R}, \mathbf{M}, L/D, \varepsilon/D)$, which may be written for incompressible fluids as $\Delta p = \mathbf{C}_p V^2/2 = K\rho V^2/2$, where K is the **resistance coefficient** and ε/D the **relative roughness** of the pipe surface, and the resistance coefficient $K = f(\mathbf{R}, L/D, \varepsilon/D)$. The pressure loss may be converted to the terms of **lost head**: $h_f = \Delta p/\gamma = KV^2/2g$. Conventional practice is to use the **friction factor** f, defined as $f = KD/L$ or $h_f = KV^2/2g = (fL/D)V^2/2g$, where $f = f(\mathbf{R}, \varepsilon/D)$. When a fluid flows into a pipe, the boundary layer starts at the entrance, as shown in Fig. 3.3.23, and grows continuously until it fills the pipe. From the equation of motion $dh_f = \tau \, dL/\gamma R_h$ and for circular ducts $R_h = D/4$. Comparing wall shear stress τ_0 with friction factor results in the following: $\tau_0 = f\rho V^2/8$.

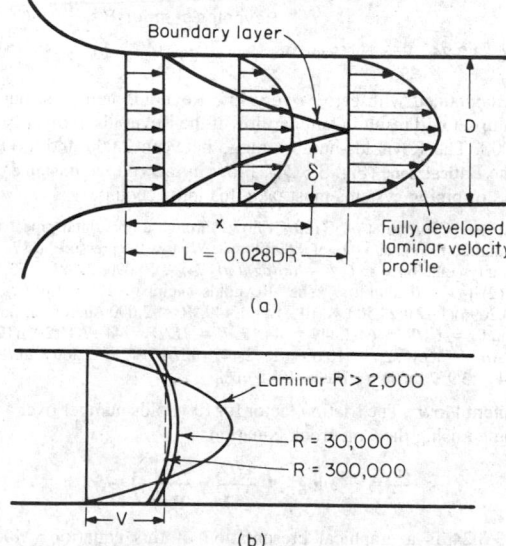

Fig. 3.3.23 Velocity profiles in pipes.

Laminar Flow In this type of flow, the resistance is due to viscous forces only so that it is independent of the pipe surface roughness, or $\tau_0 = \mu \, dU/dy$. Application of this equation to the equation of motion and the friction factor yields $f = 64/\mathbf{R}$. Experiments show that it is possible to maintain laminar flow to very high Reynolds numbers if care is taken to increase the flow gradually, but normally the slightest disturbance will destroy the laminar boundary layer if the value of Reynolds number is greater than 4,000. In a like manner, flow initially turbulent

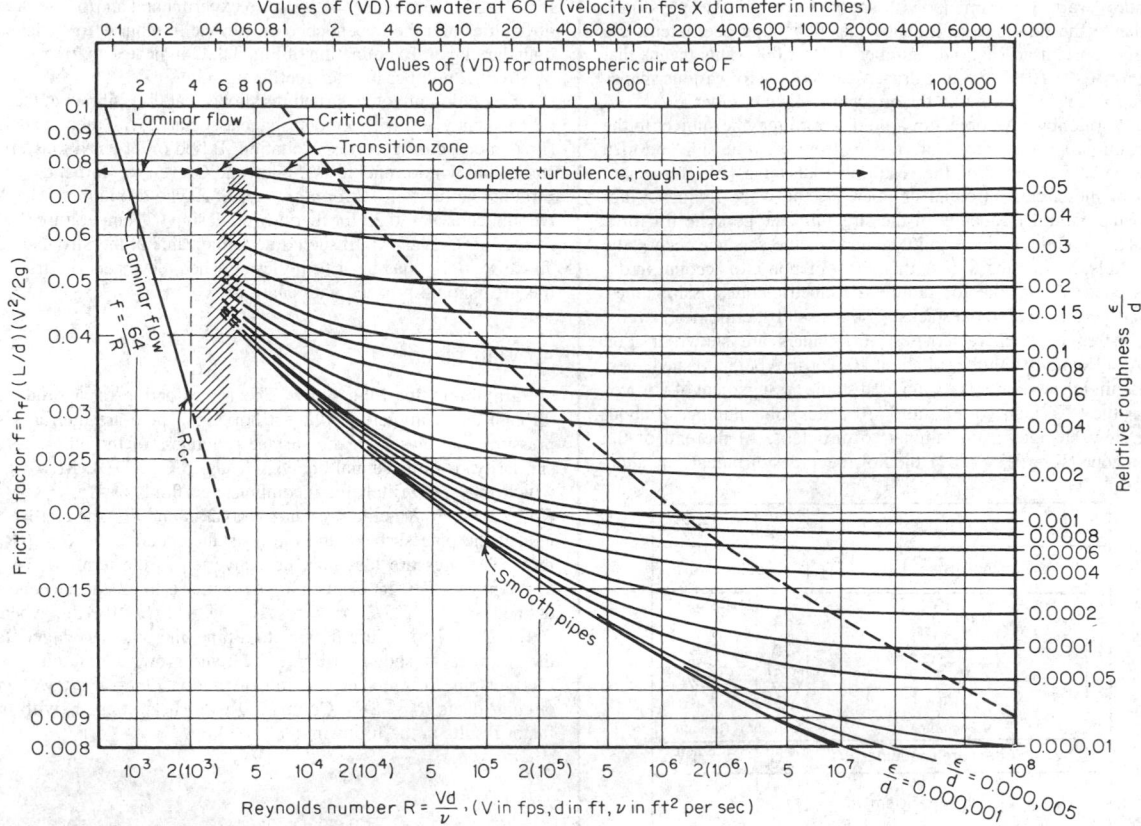

Values of (VD) for water at 60 F (velocity in fps X diameter in inches)

Values of (VD) for atmospheric air at 60 F

Fig. 3.3.24   Friction factors for flow in pipes.

can be maintained with care to very low Reynolds numbers, but the slightest upset will result in laminar flow if the Reynolds number is less than 2,000. The Reynolds-number range between 2,000 and 4,000 is called the **critical zone** (Fig. 3.3.24). Flow in the zone is **unstable**, and designers of piping systems must take this into account.

EXAMPLE.   Glycerin at 68°F (20°C) flows through a horizontal pipe 1 in in diameter and 20 ft long at a rate of 0.090 lbm/s. What is the pressure loss? From the continuity equation $V = Q/A = (m/\rho g)/(\pi D^2/4) = [0.090/(2.447 \times 32.17)]/[(\pi/4)(1/12)^2] = 0.2096$ ft/s. The Reynolds number $\mathbf{R} = \rho VD/\mu = (2.447)(0.2096)(1/12)/(29{,}500 \times 10^{-6}) = 1.449$. $\mathbf{R} < 2{,}000$; therefore, flow is laminar and $f = 64/\mathbf{R} = 64/1.449 = 44.17$. $K = fL/D = 44.17 \times 20(1/12) = 10{,}600$. $\Delta p = K\rho V^2/2 = 10{,}600 \times 2.447 \ (0.2096)^2/2 = 569.8$ lbf/ft^2 = 569.8/144 = 3.957 lbf/in^2 (2.728×10^4 N/m^2).

Turbulent Flow   The friction factor for Reynolds number over 4,000 is computed using the Colebrook equation:

$$\frac{1}{\sqrt{f}} = -2 \log_{10}\left(\frac{\varepsilon/D}{3.7} + \frac{2.51}{\mathbf{R}\sqrt{f}}\right)$$

Figure 3.3.24 is a graphical presentation of this equation (Moody, *Trans. ASME,* 1944, pp. 671–684). Examination of the Colebrook equation indicates that if the value of surface roughness ε is small compared with the pipe diameter ($\varepsilon/D \rightarrow 0$), the friction factor is a function of Reynolds number only. A **smooth pipe** is one in which the ratio $(\varepsilon/D)/3.7$ is small compared with $2.51/\mathbf{R}\sqrt{f}$. On the other hand, as the Reynolds number increases so that $2.51/\mathbf{R}\sqrt{f} \rightarrow 0$, the friction factor becomes a function of relative roughness only and the pipe is called a **rough pipe**. Thus the same pipe may be smooth under one flow condition, and rough under another. The reason for this is that as the Reynolds number increases, the thickness of the laminar sublayer decreases as shown in Fig. 3.3.21, exposing the surface roughness to flow. Values of absolute roughness ε are given in Table 3.3.9. The variation

Table 3.3.9   Values of Absolute Roughness, New Clean Commercial Pipes

| Type of pipe or tubing | ε ft (0.3048 m) $\times 10^{-6}$ | | Probable max variation of f from design, % |
	Range	Design	
Asphalted cast iron	400	400	-5 to $+5$
Brass and copper	5	5	-5 to $+5$
Concrete	1,000 10,000	4,000	-35 to 50
Cast iron	850	850	-10 to $+15$
Galvanized iron	500	500	0 to $+10$
Wrought iron	150	150	-5 to 10
Steel	150	150	-5 to 10
Riveted steel	3,000 30,000	6,000	-25 to 75
Wood stave	600 3,000	2,000	-35 to 20

SOURCE: Compiled from data given in "Pipe Friction Manual," Hydraulic Institute, 3d ed., 1961.

of friction factor shown in Fig. 3.3.9 is for new, clean pipes. The change of friction factor with **age** depends upon the chemical properties of the fluid and the piping material. Published data for flow of water through wrought-iron or cast-iron pipes show as much as 20 percent increase after a few months to 500 percent after 20 years. When necessary to allow for service life, a study of specific conditions is recommended. The calculation of friction factor to four significant figures in the examples to follow is only for numerical comparison and should not be construed to mean accuracy.

Engineering Calculations   Engineering pipe computations usually fall into one of the following classes:
1. Determine pressure loss Δp when Q, L, and D are known.
2. Determine flow rate Q when L, D, and Δp are known.
3. Determine pipe diameter D when Q, L, and Δp are known.

Pressure-loss computations may be made to engineering accuracy using an expanded version of Fig. 3.3.24. Greater precision may be obtained by using a combination of Table 3.3.9 and the Colebrook equation, as will be shown in the example to follow. Flow rate may be determined by direct solution of the Colebrook equation. Computation of pipe diameter necessitates the trial-and-error method of solution.

EXAMPLE. *Case 1:* 2,000 gal/min of 68°F (20°C) water flow through 500 ft of cast-iron pipe having an internal diameter of 10 in. At point 1 the pressure is 10 lbf/in² and the elevation 150 ft, and at point 2 the elevation is 100 ft. Find p_2.

From continuity $V = Q/A = [2{,}000 \times (231/1{,}728)/60]/[(\pi/4)(10/12)^2] = 8.170$ ft/s. Reynolds number $\mathbf{R} = \rho VD/\mu = (1.937)(8.170)(10/12)/(20.92 \times 10^{-6}) = 6.304 \times 10^5$. $\mathbf{R} > 4{,}000$ ∴ flow is turbulent. $\varepsilon/D = (850 \times 10^{-6})/(10/12) = 1.020 \times 10^{-3}$.

Determine f: from Fig. 3.3.24 by interpolation $f = 0.02$. Substituting this value on the right-hand side of the Colebrook equation,

$$\frac{1}{\sqrt{f}} = -2 \log_{10}\left(\frac{\varepsilon/D}{3.7} + \frac{2.51}{\mathbf{R}\sqrt{f}}\right)$$

$$= -2 \log_{10}\left[\frac{1.020 \times 10^{-3}}{3.7} + \frac{2.51}{(6.305 \times 10^5)\sqrt{0.02}}\right]$$

$$\frac{1}{\sqrt{f}} = 7.035 \qquad f = 0.02021$$

Resistance coefficient $K = \dfrac{fL}{D} = \dfrac{0.02021 \times 500}{10/12}$

$$K = 12.13$$
$$h_{1f2} = KV^2/2g = 12.13 \times (8.170)^2/2 \times 32.17$$
$$h_{1f2} = 12.58 \text{ ft}$$

Equation of motion: $p_1/\gamma + V_1^2/2g + z_1 = p_2/\gamma + V_2^2/2g + z_2 + h_{1f2}$. Noting that $V_1 = V_2 = V$ and solving for p_2,

$$p_2 = p_1 + \gamma(z_1 - z_2 - h_{1f2})$$
$$= 144 \times 10 + (1.937 \times 32.17)(150 - 100 - 12.58)$$
$$p_2 = 3{,}772 \text{ lbf/ft}^2 = 3{,}772/144 = 26.20 \text{ lbf/in}^2 \ (1.806 \times 10^5 \text{ N/m}^2)$$

EXAMPLE. *Case 2:* Gasoline (sp. gr. 0.68) at 68°F (20°C) flows through a 6-in schedule 40 (ID = 0.5054 ft) welded steel pipe with a head loss of 10 ft in 500 ft. Determine the flow. This problem may be solved directly by deriving equations that do not contain the flow rate Q.

From $h_f = \left(\dfrac{fL}{D}\right)\dfrac{V^2}{2g}$, $\qquad V = (2gh_fD)^{1/2}(fL)^{1/2}$

From $\mathbf{R} = \rho VD/\mu$, $\qquad V = \mathbf{R}\mu/\rho D$

Equating the above and solving,

$$\mathbf{R}\sqrt{f} = (\rho D/\mu)(2gh_fD/L)^{1/2}$$
$$= (1.310 \times 0.5054/5.98 \times 10^{-6})$$
$$\times (2 \times 32.17 \times 10 \times 0.5054/500)^{1/2} = 89{,}285$$

which is now in a form that may be used directly in the Colebrook equation:

$$\varepsilon/D = 150 \times 10^{-6}/0.5054 = 2.968 \times 10^{-4}$$

From the Colebrook equation,

$$\frac{1}{\sqrt{f}} = -2 \log_{10}\left(\frac{\varepsilon/D}{3.7} + \frac{2.51}{\mathbf{R}\sqrt{f}}\right)$$

$$= -2 \log_{10}\left(\frac{2.968 \times 10^{-4}}{3.7} + \frac{2.51}{89{,}285}\right)$$

$$\frac{1}{\sqrt{f}} = 7.931 \qquad f = 0.01590$$

$$\mathbf{R} = 89{,}285/\sqrt{f} = 89{,}285 \times 7.93 = 7.08 \times 10^5$$
$$\mathbf{R} > 4{,}000 \ ∴ \text{ flow is turbulent}$$
$$V = \mathbf{R}\mu/\rho D = (7.08 \times 10^5 \times 5.98$$
$$\times 10^{-6})/(1.310 \times 0.5054) = 6.396 \text{ ft/s}$$
$$Q = AV = (\pi/4)(0.5054)^2(6.396)$$
$$Q = 1.283 \text{ ft}^3/\text{s} \ (3.633 \times 10^{-2} \text{ m}^3/\text{s})$$

EXAMPLE. *Case 3:* Water at 68°F (20°C) is to flow at a rate of 500 ft³/s through a concrete pipe 5,000 ft long with a head loss not to exceed 50 ft. Determine the diameter of the pipe. This problem may be solved by trial and error using methods of the preceding example. First trial: Assume any diameter (say 1 ft).

$$\mathbf{R}\sqrt{f} = (\rho D/\mu)(2gh_fD/L)^{1/2}$$
$$= (1.937D/20.92 \times 10^{-6}) \times (2 \times 32.17 \times 50D/5{,}000)^{1/2}$$
$$= 74{,}269D^{3/2} = 74{,}269(1)^{3/2} = 74{,}269$$
$$\varepsilon/D_1 = 4{,}000 \times 10^{-6}/D = 4{,}000 \times 10^{-6}/(1) = 4{,}000 \times 10^{-6}$$

$$\frac{1}{\sqrt{f_1}} = -2 \log_{10}\left(\frac{\varepsilon/D_1}{3.7} + \frac{2.51}{\mathbf{R}\sqrt{f_1}}\right)$$

$$= -2 \log_{10}\left(\frac{4{,}000 \times 10^{-6}}{3.7} + \frac{2.51}{74{,}269}\right)$$

$$\frac{1}{\sqrt{f_1}} = 5.906 \qquad f_1 = 0.02867$$

$$\mathbf{R}_1 = 74{,}269/\sqrt{f_1} = 74{,}269 \times 5.906 = 438{,}600$$
$$V_1 = \mathbf{R}\mu/\rho D_1 = (438{,}600 \times 20.92 \times 10^{-6})/(1.937 \times 1)$$
$$V_1 = 4.737 \text{ ft/s}$$
$$Q_1 = A_1V_1 = [\pi(1)^2/4]4.737 = 3.720 \text{ ft}^3/\text{s}$$

For the same loss and friction factor,

$$D_2 = D_1(Q/Q_1)^{2/5} = (1)(500/3.720)^{2/5} = 7.102 \text{ ft}$$

For the second trial use $D_2 = 7.102$, which results in $Q = 502.2$ ft³/s. Since the nearest standard size would be used, additional trials are unnecessary.

Velocity Profile Figure 3.3.23*a* shows the formation of a laminar velocity profile. As the fluid enters the pipe, the boundary layer starts at the entrance and grows continuously until it fills the pipe. The flow while the boundary is growing is called **generating flow**. When the boundary layer completely fills the pipe, the flow is called **established flow**. The distance required for establishing **laminar flow** is $L/D \approx 0.028 \, \mathbf{R}$. For **turbulent flow**, the distance is much shorter because of the turbulence and not dependent upon Reynolds number, L/D being from 25 to 50.

Examination of Fig. 3.3.23*b* indicates that as the Reynolds number increases, the velocity distribution becomes "flatter" and the flow approaches **one-dimensional**. The velocity profile for laminar flow is parabolic, $U/V = 2[1 - (r/r_o)^2]$ and for **turbulent flow**, **logarithmic** (except for the very thin laminar boundary layer), $U/V = 1 + 1.43\sqrt{f} + 2.15\sqrt{f} \log_{10}(1 - r/r_o)$. The use of the average velocity produces an error in the computation of kinetic energy. If α is the **kinetic-energy correction factor**, the true kinetic-energy change per unit mass between two points on a flow system $\Delta KE = \alpha_1 V_1^2/2g_c - \alpha_2 V_2^2/2g_c$, where $\alpha = (1/AV^3)\int U^3 dA$. For **laminar flow**, $\alpha = 2$ and for **turbulent flow**, $\alpha \approx 1 + 2.7f$. Of interest is the **pipe factor** V/U_{max}; for **laminar flow**, $V/U_{max} = 1/2$ and for **turbulent flow**, $V/U_{max} = 1 + 1.43\sqrt{f}$. The location at which the local velocity equals the average velocity for **laminar flow** is $U = V$ at $r/r_o = 0.7071$ and for **turbulent flow** is $U = V$ at $r/r_o = 0.7838$.

Compressible Flow At the present time, there are no true analytical solutions for the computation of actual characteristics of compressible fluids flowing in pipes. In the real flow of a compressible fluid in a pipe, the amount of heat transferred and its direction are dependent upon the amount of insulation, the temperature gradient between the fluid and ambient temperatures, and the heat-transfer coefficient. Each condition requires an individual application of the principles of thermodynamics and heat transfer for its solution.

Conventional engineering practice is to use one of the following methods for flow computation.

1. Assume **adiabatic** flow. This approximates the flow of compressible fluids in short, insulated pipelines.

2. Assume **isothermal** flow. This approximates the flow of gases in long, uninsulated pipelines where the fluid and ambient temperatures are nearly equal.

Adiabatic Flow If the Mach number is less than ¼, results within normal engineering-accuracy requirements may be obtained by considering the fluid to be incompressible. A detailed discussion of and methods for the solution of compressible adiabatic flow are beyond the scope of this section, and any standard gas-dynamics text should be consulted.

Isothermal Flow The equation of motion for a horizontal piping system may be written as follows:

$$dp + \rho V\, dV + \gamma\, dh_f = 0$$

noting, from the continuity equation, that $\rho V = \dot{M}/A = G$, where G is the **mass velocity** in slugs/(ft^2)(s), and that $\gamma\, dh_f = [(f/D)\rho V^2/2]dL = [(f/D)GV/2]dL$. Substituting in the above equation of motion and dividing by $GV/2$ results in

$$\frac{2\rho\, dp}{G^2} + \frac{2dV}{V} + \left(\frac{f}{D}\right) dL = 0$$

Integrating for an isothermal process ($p/\rho = C$) and assuming f is a constant,

$$\frac{\rho_1 p_1}{G^2}\left[\left(\frac{p_2}{p_1}\right)^2 - 1\right] + 2\log_e\left(\frac{V_2}{V_1}\right) + \frac{fL}{D} = 0$$

Noting that $A_1 = A_2$, $V_2/V_1 = \rho_1/\rho_2 = p_1/p_2$, and solving for G,

$$G = \left\{\frac{\rho_1 p_1[1 - (p_2/p_1)^2]}{2\log_e(p_1/p_2) + fL/D}\right\}^{1/2}$$

The Reynolds number may be written as

$$\mathbf{R} = \frac{\rho VD}{\mu} = \frac{GD}{\mu} \quad \text{and} \quad G = \mathbf{R}\frac{\mu}{D}$$

The value of $\mathbf{R}\sqrt{f}$ may be obtained from the simultaneous solution of the two equations for G, assuming that $2\log_e p_1/p_2$ is small compared with fL/D.

$$\mathbf{R}\sqrt{f} \approx \left\{\frac{D^3\rho_1 p_1}{\mu^2 L}\left[1 - \left(\frac{p_2}{p_1}\right)\right]\right\}^{1/2}$$

EXAMPLE. Air at 68°F (20°C) is flowing isothermally through a horizontal straight standard 1-in steel pipe (inside diameter = 1.049 in). The pipe is 200 ft long, the pressure at the pipe inlet is 74.7 lbf/in^2, and the pressure drop through the pipe is 5 lbf/in^2. Find the flow rate in lbm/s. From the equation of state $\rho_1 = p/g_c RT = (144 \times 74.7)/(32.17 \times 53.34 \times 527.7) = 0.01188$ slugs/ft^3.

$$\mathbf{R}\sqrt{f} = \{[(D^3\rho_1 p_1/\mu^2 L)][1 - (p_2/p_1)^2]\}^{1/2} = \{[(1.049/12)^3(0.01188)$$
$$\times (144 \times 74.7)/(39.16 \times 10^{-8})^2(200)[1 - (69.7/74.7)^2]\}^{1/2}$$
$$= 18,977$$

For steel pipe $\varepsilon = 150 \times 10^{-6}$ ft, $\varepsilon/D = (150 \times 10^{-6})/(1.049/12) = 1.716 \times 10^{-3}$. From the Colebrook equation,

$$\frac{1}{\sqrt{f}} = -2\log_{10}\left(\frac{\varepsilon/D}{3.7} + \frac{2.51}{\mathbf{R}\sqrt{f}}\right)$$
$$= 2\log_{10}[(1.716 \times 10^{-3}/3.7) + (2.51)/(18,977)] = 6.449$$
$$f = 0.02404$$

$\mathbf{R} = (\mathbf{R}\sqrt{f})(1/\sqrt{f}) = (18,953)(6.449) = 122,200$
$\mathbf{R} > 4,000$ ∴ flow is turbulent

$$G = \left\{\frac{\rho_1 p_1[1 - (p_2/p_1)^2]}{2\log_e(p_1/p_2) + fL/D}\right\}^{1/2}$$
$$= \left\{\frac{(0.01188)(144 \times 74.7)[1 - (69.7/74.7)^2]}{2\log_e(74.7/69.7) + (0.02404)(200)/(1.049/12)}\right\}^{1/2}$$
$$= 0.5476 \text{ slug/(ft}^2)(s)$$
$\dot{m} = g_c AG = (32.17)(\pi/4)(1.049/12)^2(0.5476)$
$\dot{m} = 0.1057$ lbm/s (47.94 $\times 10^{-3}$ kg/s)

Noncircular Pipes For the flow of fluids in noncircular pipes, the **hydraulic diameter** D_h is used. From the definition of hydraulic radius, the diameter of a circular pipe was shown to be four times its **hydraulic radius**; thus $D_h = 4R_h$. The Reynolds number thus may be written as $\mathbf{R} = \rho VD_h/\mu = GD_h/\mu$, the relative roughness as ε/D_h, and the resistance coefficient $K = fL/D_h$. With the above modifications, flows through noncircular pipes may be computed in the same manner as for circular pipes.

EXAMPLE. Air at 68°F (20°C) and 100 lbf/in^2 enters a rectangular duct 1 by 3 ft at a rate of 720 lbm/s. The duct is horizontal, 100 ft long, and made of galvanized iron. Assuming isothermal flow, estimate the pressure loss due to

friction in this line. From the equation of state, $\rho_1 = p_1/g_c RT_1 = (144 \times 100)/(32.17)(53.34)(527.7) = 0.01590$ slug/ft^3. From Table 3.3.6, $R_h = bD/2(b + D) = 3 \times 1/2(3 + 1) = 0.375$ ft, and $D_h = 4R_h = 4 \times 0.375 = 1.5$ ft. For galvanized iron, $\varepsilon/D_h = 500 \times 10^{-6}/1.5 = 3.333 \times 10^{-4}$

$G = (\dot{m}/g_c)/A = (720/32.17)/(1 \times 3) = 7.460$ slugs/(ft^2)(s)
$\mathbf{R} = GD_h/\mu = (7.460)(1.5)/(39.16 \times 10^{-8})$
$\quad = 28,580,000 > 4,000$ ∴ flow is turbulent

From Fig. 3.3.24, $f \approx 0.015$

$$\frac{1}{\sqrt{f}} = -2\log_{10}\left(\frac{\varepsilon/D_h}{3.7} + \frac{2.51}{\mathbf{R}\sqrt{f}}\right)$$
$$= -2\log_{10}\left(\frac{3.333 \times 10^{-4}}{3.7} + \frac{2.51}{28,580,000\sqrt{0.015}}\right)$$
$$f = 0.01530$$

Solving the isothermal equation for p_2/p_1,

$$\frac{p_2}{p_1} = \left\{1 - \left(\frac{G^2}{\rho_1 p_1}\right)\left[2\log_e\left(\frac{p_1}{p_2}\right) + \frac{fL}{D_h}\right]\right\}^{1/2}$$

For first trial, assume $2\log_e(p_1/p_2)$ is small compared with fL/D:

$p_2/p_1 = \{1 - [(7.460)^2/(0.01590)(144 \times 100)][0 + (0.01530)(100)/1.5]\}^{1/2}$
$\quad = 0.8672$

Second trial using first-trial values results in 0.8263. Subsequent trials result in a balance at $p_2/p_1 = 0.8036$, $p_2 = 100 \times 0.8036 = 80.36$ lbf/in^2 (5.541 $\times 10^5$ N/m^2).

PIPING SYSTEMS

Resistance Parameters The resistance to flow of a piping system is similar to the resistance of an object immersed in a flow stream and is made up of pressure (inertia) or shape drag and skin-friction (viscous) drag. For long, straight pipes the pressure drag is characterized by the **relative roughness** ε/D and the skin friction by the **Reynolds number R.** For other piping components, two parameters are used to describe the resistance to flow, the **resistance coefficient** $K = fL/D$ and the **equivalent length** $L/D = K/f$. The resistance-coefficient method assumes that the component loss is all due to pressure drag and that the flow through the component is completely turbulent and independent of Reynold's number. The equivalent-length method assumes that resistance of the component varies in the same manner as does a straight pipe. The basic assumption then is that its pressure drag is the same as that for the relative roughness ε/D of the pipe and that the friction drag varies with the Reynolds number $\mathbf{R}$ in the same manner as the straight pipe. Both methods have the inherent advantage of simplicity in application, but neither is correct except in the fully developed turbulent region. Two excellent sources of information on the resistance of piping-system components are the Hydraulic Institute "Pipe Friction Manual," which uses the resistance-coefficient method, and the Crane Company Technical Paper 410 ("Fluid Meters," 6th ed. ASME, 1971), which uses the equivalent-length concept.

For valves, branch flow through tees, and the type of components listed in Table 3.3.10, the pressure drag is predominant, is "rougher" than the pipe to which it is attached, and will extend the completely turbulent region to lower values of Reynolds number. For bends and elbows, the loss is made up of pressure drag due to the change of direction and the consequent secondary flows which are dissipated in 50 diameters or more downstream piping. For this reason, loss through adjacent bends will not be twice that of a single bend.

In long pipelines, the effect of bends, valves, and fittings is usually negligible, but in systems where there is little straight pipe, they are the controlling factor. Under-design will result in the failure of the system to deliver the required capacity. Over-design will result in inefficient operation because it will be necessary to "throttle" one or more of the valves. For estimating purposes, Tables 3.3.10 and 3.3.11 may be used as shown in the examples. When available, the manufacturers' data should be used, particularly for valves, because of the wide variety of designs for the same type. (See also Sec. 12.4.)

Table 3.3.10 Representative Values of Resistance Coefficient K

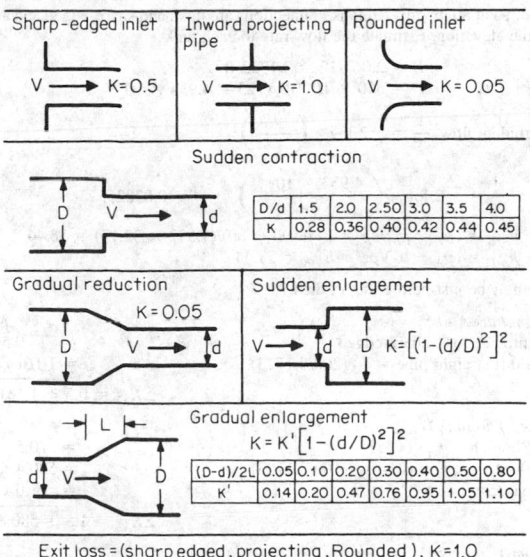

D/d	1.5	2.0	2.50	3.0	3.5	4.0
K	0.28	0.36	0.40	0.42	0.44	0.45

Gradual enlargement $K = K'\left[1-(d/D)^2\right]^2$

(D-d)/2L	0.05	0.10	0.20	0.30	0.40	0.50	0.80
K'	0.14	0.20	0.47	0.76	0.95	1.05	1.10

Exit loss = (sharp edged, projecting, Rounded), K = 1.0

Source: Compiled from data given in "Pipe Friction Manual," 3d ed., Hydraulic Institute, 1961.

Table 3.3.11 Representative Equivalent Length in Pipe Diameters (L/D) of Various Valves and Fittings

Globe valves, fully open	450
Angle valves, fully open	200
Gate valves, fully open	13
¾ open	35
½ open	160
¼ open	900
Swing check valves, fully open	135
In line, ball check valves, fully open	150
Butterfly valves, 6 in and larger, fully open	20
90° standard elbow	30
45° standard elbow	16
90° long-radius elbow	20
90° street elbow	50
45° street elbow	26
Standard tee:	
Flow through run	20
Flow through branch	60

Source: Compiled from data given in "Flow of Fluids," Crane Company Technical Paper 410, ASME, 1971.

Series Systems In a single piping system made of various sizes, the practice is to group all of one size together and apply the continuity equation, as shown in the following example.

EXAMPLE. Water at 68°F (20°C) leaves an open tank whose surface elevation is 180 ft and enters a 2-in schedule 40 steel pipe via a sharp-edged entrance. After 50 ft of straight 2-in pipe that contains a 2-in globe valve, the line enlarges suddenly to an 8-in schedule 40 steel pipe which consists of 100 ft of straight 8-in pipe, two standard 90° elbows and one 8-in angle valve. The 8-in line discharges below the surface of another open tank whose surface elevation is 100 ft. Determine the volumetric flow rate.

$D_1 = 2.067/12 = 0.1723$ ft and $D_2 = 7.981/12 = 0.6651$ ft
$\varepsilon/D_1 = 150 \times 10^{-6}/0.1723 = 8.706 \times 10^{-4}$
$\varepsilon/D_2 = 150 \times 10^{-6}/0.6651 = 2.255 \times 10^{-4}$

For turbulent flow,

$$\frac{1}{\sqrt{f}} = -2\log_{10}\left(\frac{\varepsilon/D}{3.7}\right)$$

$$\frac{1}{\sqrt{f_1}} = -2\log_{10}\left(\frac{8.706 \times 10^{-4}}{3.7}\right) \qquad f_1 = 0.01899$$

$$\frac{1}{\sqrt{f_1}} = -2\log_{10}\left(\frac{2.255 \times 10^{-4}}{3.7}\right) \qquad f_2 = 0.01407$$

1. *2-in components* $\qquad\qquad\qquad\qquad\qquad K$
   Entrance loss, sharp-edged $\qquad\qquad = 0.5$
   50 ft straight pipe $= f_1 (50/0.1723) \qquad = 290.2f_1$
   Globe valve $= f_1 (L/D) \qquad\qquad = 450.0f_1$
   Sudden enlargement $k = [-(D/D_2)^2]^2$
   $= [1 - (2.067/7.981)^2]^2 \qquad\qquad = 0.87$
   $\qquad\qquad\qquad\qquad \Sigma K_1 = \overline{1.37 + 740.2f_1}$

2. *8-in components* $\qquad\qquad\qquad\qquad\qquad K$
   100 ft of straight pipe $f_2 (100/0.6651) \qquad = 150.4f_2$
   2 standard 90° elbows $2 \times 30 f_2 \qquad\quad = 60 \quad f_2$
   1 angle valve $200 f_2 \qquad\qquad\qquad = 200 \quad f_2$
   Exit loss $\qquad\qquad\qquad\qquad\qquad = 1$
   $\qquad\qquad\qquad\qquad \Sigma K_2 = \overline{1 + 410.4f_2}$

3. *Apply equation of motion*

$$h_{1f2} = z_1 - z_2 = (\Sigma K_1)\frac{V_1^2}{2g} + (\Sigma K_2)\frac{V_2^2}{2g}$$

From continuity, $\rho_1 A_1 V_1 = \rho_2 A_2 V_2$ for $\rho_1 = \rho_2$

$V_2 = V_1(A_1/A_2) = V_1(D_1/D_2)^2$
$h_{1f2} = z_1 - z_2 = [\Sigma K_1 + \Sigma K_2(D_1/D_2)^4]V_1^2/2g$
$V_1 = \{[2g(z_1 - z_2)]/[\Sigma K_1 + \Sigma K_2(D_1/D_2)^4]\}^{1/2}$

$$V_1 = \left[\frac{2 \times 32.17 \times (180 - 100)}{(1.37 + 740.2f_1) + (1 + 410.4f_2)(2.067/7.981)^4}\right]^{1/2}$$

$$V_1 = \frac{71.74}{(1.374 + 740.2f_1 + 1.846f_2)^{1/2}}$$

4. For first trial assume f_1 and f_2 for complete turbulence

$$V_1 = \frac{71.74}{(1.374 + 740.2 \times 0.01899 + 1.846 \times 0.01407)^{1/2}}$$

$V_1 = 18.25$ ft/s
$V_2 = 18.25 (2.067/7.981)^2 = 1.224$ ft/s
$\mathbf{R}_1 = \rho_1 V_1 D_1/\mu = (1.937)(18.25)(0.1723)/(20.92 \times 10^{-6})$
$\mathbf{R}_1 = 291,100 > 4,000 \therefore$ flow is turbulent
$\mathbf{R}_2 = \rho_2 V_2 D_2/\mu_2 = (1.937)(1.224)(0.6651)/(20.92 \times 10^{-6})$
$\mathbf{R}_2 = 75,420 > 4,000 \therefore$ flow is turbulent

5. For second trial use first trial V_1 and V_2. From Fig. 3.3.24 and the Colebrook equation,

$$\frac{1}{\sqrt{f_1}} = -2\log_{10}\left(\frac{8.706 \times 10^{-4}}{3.7} + \frac{2.51}{291,100\sqrt{0.020}}\right)$$

$f_1 = 0.02008$

$$\frac{1}{\sqrt{f_2}} = -2\log_{10}\left(\frac{2.255 \times 10^{-4}}{3.7} + \frac{2.51}{75,420\sqrt{0.020}}\right)$$

$f_2 = 0.02008$

$$V_1 = \frac{71.74}{(1.374 + 740.2 \times 0.02008 + 1.864 \times 0.02008)^{1/2}}$$

$V_1 = 17.78$

A third trial results in $V = 17.77$ ft/s or $Q = A_1 V_1 = (\pi/4)(0.1723)^2(17.77) = 0.4143$ ft³/s (1.173×10^{-2} m³/s).

Parallel Systems In solution of problems involving two or more parallel pipes, the head loss for all of the pipes is the same as shown in the following example.

EXAMPLE. Benzene at 68°F (20°C) flows at a rate of 0.5 ft³/s through two parallel straight, horizontal pipes connecting two pressurized tanks. The pipes are both schedule 40 steel, one being 1 in, the other 2 in. They both are 100 ft long and have connections that project inwardly in the supply tank. If the pressure in the supply tank is maintained at 100 lbf/in², what pressure should be maintained on the receiving tank?

$D_1 = 1.049/12 = 0.08742$ ft and $D_2 = 2.067/12 = 0.1723$ ft
$\varepsilon/D_1 = 150 \times 10^{-6}/0.08742 = 1.716 \times 10^{-3}$
$\varepsilon/D_2 = 150 \times 10^{-6}/0.1723 = 8.706 \times 10^{-4}$

For turbulent flow,

$$\frac{1}{\sqrt{f}} = -2 \log_{10}\left(\frac{\varepsilon/D}{3.7}\right)$$

$$\frac{1}{\sqrt{f_1}} = -2 \log_{10}\left(\frac{1.716 \times 10^{-3}}{3.7}\right) \qquad f_1 = 0.02249$$

$$\frac{1}{\sqrt{f_2}} = -2 \log_{10}\left(\frac{8.706 \times 10^{-4}}{3.7}\right) \qquad f_2 = 0.01899$$

1. *1-in. components*

		K
Entrance loss, inward projection		$= 1.0$
100 ft straight pipe f_1 (100/0.08742)		$= 1{,}144\,f_1$
Exit loss		$= 1.0$
	$\Sigma K_1 =$	$2.0 + 1{,}144\,f_1$

2. *2-in components*

Entrance loss, inward projection		$= 1.0$
100 ft straight pipe f_2 (100/0.1723)		$= 580.4\,f_2$
Exit loss		$= 1.0$
	$\Sigma K_2 =$	$2.0 + 580.4\,f_2$

$h_f = \Sigma K_1\, V_1^2/2g = \Sigma K_2\, V_2^2/2g$
From the continuity equation, $Q = AV$

$$\Sigma K_1 \frac{Q_1^2}{2gA_1^2} = \Sigma K_2 \frac{Q_2^2}{2gA_2^2}$$

Solving for Q_1/Q_2,

$$\frac{Q_1}{Q_2} = \frac{A_1}{A_2}\sqrt{\frac{\Sigma K_2}{\Sigma K_1}} = \left(\frac{D_1}{D_2}\right)^2 \sqrt{\frac{\Sigma K_2}{\Sigma K_1}} = \left(\frac{D_1}{D_2}\right)^2 \sqrt{\frac{2.0 + 580.4\,f_2}{2.0 + 1{,}144\,f_1}}$$

For first trial assume flow is completely turbulent,

$$\frac{Q_1}{Q_2} = \left(\frac{0.08742}{0.1723}\right)^2 \sqrt{\frac{2.0 + 580.4 \times 0.01899}{2.0 + 1{,}144 \times 0.02249}}$$

$$\frac{Q_1}{Q_2} = 0.1764 \qquad Q = Q_1 + Q_2 = 0.1764\,Q_2 + Q_2$$

$$0.5000 = 1.1764\,Q_2 \qquad Q_2 = 0.4250$$
$$Q_1 = 0.5000 - 0.4250 = 0.0750$$

for the second trial use first-trial values,

$V_1 = Q_1/A_1 = 0.0750/(\pi/4)(0.08742)^2 = 12.50$
$V_2 = Q_2/A_2 = 0.4250/(\pi/4)(0.1723)^2 = 18.23$
$\mathbf{R}_1 = \rho_1 V_1 D_1/\mu_1 = (1.705)(12.50)(0.08742)/(13.62 \times 10^{-6})$
$\mathbf{R}_1 = 136{,}800 > 4{,}000 \therefore$ flow is turbulent
$\mathbf{R}_2 = \rho_2 V_2 D_2/\mu_2 = (1.705)(18.23)(0.1723)/(13.62 \times 10^{-6})$
$\mathbf{R}_2 = 393{,}200 > 4{,}000 \therefore$ flow is turbulent

Using the Colebrook equation and Fig. 3.3.24,

$$\frac{1}{\sqrt{f_1}} = -2 \log_{10}\left(\frac{1.716 \times 10^{-3}}{3.7} + \frac{2.51}{136{,}800\,\sqrt{0.024}}\right)$$

$$f_1 = 0.02389$$

$$\frac{1}{\sqrt{f_2}} = -2 \log_{10}\left(\frac{8.706 \times 10^{-4}}{3.7} + \frac{2.51}{393{,}200\,\sqrt{0.020}}\right)$$

$$f_2 = 0.01981$$

$$h_f = \Sigma K_1 \frac{V_1^2}{2g} = \Sigma K_2 \frac{V_2^2}{2g}$$

$\Sigma K_1 V_1^2/2g = (2.0 + 1{,}144 \times 0.02389)(12.50)^2/(2 \times 32.17) = 71.23$
$\Sigma K_2 V_2^2/2g = (2.0 + 580.4 \times 0.01981)(18.23)^2/(2 \times 32.17) = 69.80$

$71.23 = 69.80$; further trials not justifiable because of accuracy of f, K, L/D. Use average or 70.52, so that $\Delta p = \rho g h_f = (1.705 \times 32.17 \times 70.52)/144 = 26.86$ lbf/in^2 $= p_1 - p_2 = 100 - p_2$, $p_2 = 100 - 26.86 = 73.40$ lbf/in^2 $(5.061 \times 10^5$ N/m$^2)$.

Branch Flow Problems of a single line feeding several points may be solved as shown in the following example.

EXAMPLE. Ethyl alcohol at 68°F (20°C) flows from tank A, which is maintained at a constant pressure of 100 lb/in^2 through 200 ft of 2-in cast-iron schedule

40 pipe to a Y branch connection ($K = 0.5$) where 100 ft of 2-in pipe goes to tank B, which is maintained at 80 lbf/in^2 and 50 ft of 2-in pipe to tank C, which is also maintained at 80 lbf/in^2. All tank connections are flush and sharp-edged and are at the same elevation. Estimate the flow rate to each tank.

$$D = 2.067/12 = 0.1723 \text{ ft}$$
$$\varepsilon/D = 850 \times 10^{-6}/0.1723 = 4.933 \times 10^{-3}$$

For turbulent flow, $\dfrac{1}{\sqrt{f}} = -2 \log_{10}\left(\dfrac{\varepsilon/D}{3.7}\right)$

$$\frac{1}{\sqrt{f}} = -2 \log_{10}\left(\frac{4.933 \times 10^{-3}}{3.7}\right) \qquad f = 0.03025$$

$h_{AfB} = (p_A - p_B)/\rho g = 144(100 - 80)/(1.532 \times 32.17) = 58.44$
$h_{AfC} = (p_A - p_C)/\rho g = h_{AfB} = 58.44$

Let point X be just before the Y; then

1. *From tank A to Y*

		K
Entrance loss, sharp-edged		$= 0.5$
200 ft straight pipe $= f_{AX}$(200/0.1723)		$= 1{,}161\,f_{AX}$
	$\Sigma K_{AX} =$	$0.5 + 1{,}161\,f_{AX}$

2. *From Y to tank B*

Y branch		$= 0.5$
100 ft straight pipe $= f_{XB}$(100/0.1723)		$= 580.4\,f_{XB}$
Exit loss		$= 1.0$
	$\Sigma K_{XB} =$	$1.5 + 580.4\,f_{XB}$

3. *From Y to tank C*

Y branch		$= 0.5$
50 ft straight pipe $= f_{XC}$(50/0.1723)		$= 290.2\,f_{XC}$
Exit loss		$= 1.0$
	$\Sigma K_{XC} =$	$1.5 + 290.2\,f_{XC}$

Balance of flows:

$$Q_{AX} = Q_{XB} + Q_{XC}$$

and from continuity, $(A_{AX} = A_{XB} = A_{XC})$, $V_{AX} = V_{XB} + V_{XC}$; then

$$h_{AfB} = \Sigma K_{AX} \frac{V_{AX}^2}{2g} + \Sigma K_{XB} \frac{V_{XB}^2}{2g}$$

$$h_{AfC} = \Sigma K_{AX} \frac{V_{AX}^2}{2g} + \Sigma K_{XC} \frac{V_{XC}^2}{2g}$$

For first trial assume completely turbulent flow

$$h_{AfB} = \frac{(0.5 + 1{,}161\,f_{AX})V_{AX}^2}{2g} + \frac{(1.5 + 580.4\,f_{XB})V_{XB}^2}{2g}$$

$$58.44 = \frac{(0.5 + 1{,}161 \times 0.03025)V_{AX}^2}{2 \times 32.17} + \frac{(1.5 + 580.4 \times 0.03025)V_{XB}^2}{2 \times 32.17}$$

$$58.44 = 0.5536\,V_{AX}^2 + 0.2962\,V_{XB}^2$$

and in a like manner

$$h_{AfC} = 58.44 = 0.5536\,V_{XC}^2 + 0.1598\,V_{XC}^2$$

Equating $h_{AfB} = h_{AfC}$,

$$0.5536\,V_{AX}^2 + 0.2962\,V_{XB}^2 = 0.5536\,V_{AX}^2 + 0.1598\,V_{XC}^2$$

or $\qquad V_{XC} = 1.3615\,V_{XB}$ and since $V_{AX} = V_{XB} + V_{XC}$
$\qquad V_{AX} = V_{XB} + 1.3615\,V_{XB} = 2.3615\,V_{XB}$

so that $\qquad h_{AfB} = 58.44 = 0.5536(2.3615\,V_{XB}^2) + 0.2962\,V_{XB}^2$
$\qquad V_{XB} = 4.156$
$\qquad V_{XC} = 1.3615(4.156) = 5.658$
$\qquad V_{AX} = 4.156 + 5.658 = 9.814$

Second trial,

$$\mathbf{R}_{AX} = \frac{\rho V_{AX} D}{\mu} = \frac{1.532 \times 9.814 \times 0.1723}{25.06 \times 10^{-6}}$$

$$\mathbf{R}_{AX} = 103{,}400 > 4{,}000 \therefore \text{flow is turbulent}$$

In a like manner,

$$\mathbf{R}_{XB} = 43{,}780 \qquad \mathbf{R}_{XC} = 59{,}600$$

Using the Colebrook equation and Fig. 3.3.24,

$$\frac{1}{\sqrt{f_{AX}}} = -2 \log_{10}\left(\frac{4.933 \times 10^{-3}}{3.7} + \frac{2.51}{103{,}400\,\sqrt{0.031}}\right)$$

$$f_{AX} = 0.03116$$

In a like manner,

$$f_{XB} = 0.03231 \qquad f_{XC} = 0.03179$$

$$h_{AfB} = \frac{(0.5 + 1,161 \times 0.03116)V_{AX}^2}{2 \times 32.17} + \frac{(1.5 + 580.4 \times 0.03231)V_{XB}^2}{2 \times 32.17}$$

$$h_{AfB} = 0.5700\, V_{AX}^2 + 0.3148\, V_{XB}^2$$

$$h_{AfC} = 0.5700\, V_{AX}^2 + \frac{(1.5 + 290.2 \times 0.03179)V_{XC}^2}{2 \times 32.17}$$

$$h_{AfC} + 0.5700\, V_{AX}^2 + 0.1667\, V_{XC}^2$$
$$0.3148\, V_{XB}^2 = 0.1667\, V_{XC}^2$$
$$V_{XC} = 1.374\, V_{XB}$$
$$V_{AX} = V_{XB} + 1.374\, V_{XB} = 2.374\, V_{XB}$$

so that

$$h_{AfB} - 58.44 = 0.5700\,(2.374\, V_{XB})^2 + 0.3148\, V_{XB}^2$$
$$V_{XB} = 4.070 \qquad V_{XC} = 5.592 \qquad V_{AX} = 9.663$$

Further trials are not justified.

$$A = \pi D^2/4 = (\pi/4)(0.1723)^2 = 0.02332 \text{ ft}^2$$
$$Q_{AB} = V_{AB}A = 4.070 \times 0.02332 = 0.09491 \text{ ft}^3/\text{s } (2.686 \times 10^{-1} \text{ m}^3/\text{s})$$
$$Q_{XC} = V_{XC}A = 5.592 \times 0.02332 = 0.1304 \text{ ft}^3/\text{s } (3.693 \times 10^{-3} \text{ m}^3/\text{s})$$

Siphons are arrangements of hose or pipe which cause liquids to flow from one level A in Fig. 3.3.25 to a lower level C over an intermediate summit B. Performance of siphons may be evaluated from the equation of motion between points A and B:

$$\frac{p_A}{\gamma} + \frac{V_A^2}{2g} + z_A = \frac{p_B}{\gamma} + \frac{V_B^2}{2g} + z_B + h_{AfB}$$

Noting that on the surface $V_A = 0$ and the minimum pressure that can exist at point B is the vapor pressure pv, the maximum elevation of point B is

$$z_B - z_A = \frac{p_A}{\gamma} - \left(\frac{p_v}{\gamma} + \frac{V_B^2}{2g} + h_{AfB} \right)$$

The friction loss $h_f = \Sigma K_{AB} V_B^2/2g$, and let $V_B = V$; then

$$z_B - z_A = \frac{p_A - p_v}{\rho g} - (1 - \Sigma K_{AB}) \frac{V^2}{2g}$$

Flow under this maximum condition will be uncertain. The air pump or ejector used for priming the pipe (flow will not take place unless the siphon is full of water) might have to be operated occasionally to remove accumulated air and vapor. Values of $z_B - z_A$ less than those calculated by the above equation should be used.

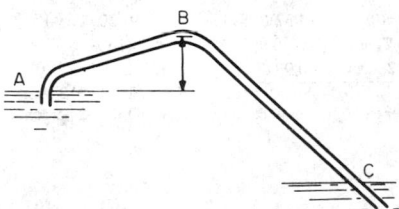

Fig. 3.3.25 Siphon.

EXAMPLE. The siphon shown in Fig. 3.3.25 is composed of 2,000 ft of 6-in schedule 40 cast-iron pipe. Reservoir A is at elevation 800 ft and C at 600 ft. Estimate the maximum height for $z_B - z_A$ if the water temperature may reach 104°F (40°C), and the amount of straight pipe from A to B is 100 ft. For the first bend $L/D = 25$ and the second (at B) $L/D = 50$. Atmospheric pressure is 14.70 lbf/in². For 6-in schedule 40 pipe $D = 6.065/12 = 0.5054$ ft, $\varepsilon/D = 850 \times 10^{-6}/0.5054 = 1.682 \times 10^{-3}$. Turbulent friction factor

$$1/\sqrt{f} = -2 \log_{10}\left(\frac{\varepsilon/D}{3.7} \right) = -2 \log_{10}(1.682 \times 10^{-3}/3.7) = 0.02238$$

1. *Components from* A *to* B. (Note loss in second bend takes place in downstream piping.)

	K
Entrance (inward projection)	$= 1.0$
100 ft straight pipe f (100/0.5054)	$= 197.9\,f$
First bend	$= 25\quad f$
ΣK_{AB}	$= 1.0 + 227.9\,f$

2. *Components from* A *to* C

ΣK_{AB}	$= 1.0 + 2,229\,f$
1,900 ft of straight pipe f (1,900/0.5054)	$= 3,759.4\,f$
Second bend	$= 50\,f$
Exit loss	$= 1$
ΣK_{AC}	$= 2.0 + 4,032\,f$

First trial assume complete turbulence. Writing the equation of motion between A and C.

$$\frac{p_A}{\gamma} + \frac{V_A^2}{2g} + z_A = \frac{p_C}{\gamma} + \frac{V_C^2}{2g} + z_C + \Sigma K_{AC}\frac{V^2}{2g}$$

Noting $V_A = V_C = 0$, and $p_A = p_C = 14.7$ lbf/in²,

$$V = \sqrt{\frac{2g(z_A - z_C)}{\Sigma K_{AC}}} = \sqrt{\frac{2g(z_A - z_C)}{2.0 + 4,032\,f}}$$
$$= \sqrt{\frac{2 \times 32.17\,(800 - 600)}{2.0 + 4,032\,f}}$$
$$= \frac{113.44}{\sqrt{2.0 + 4,032 \times 0.02238}}$$
$$= 11.81$$

Second trial, use first-trial values,

$$\mathbf{R} = \frac{\rho VD}{\mu} = (1.925)(11.81)(0.5054)/(13.61 \times 10^{-6})$$

$$\mathbf{R} = 846,200 > 4,000 \therefore \text{ flow is turbulent}$$

From Fig. 3.3.24 and the Colebrook equation,

$$\frac{1}{\sqrt{f}} = -2 \log_{10}\left(\frac{1.682 \times 10^{-3}}{3.7} + \frac{2.51}{844,200\,\sqrt{0.023}} \right)$$

$$f = 0.02263$$

$$V = \frac{113.44}{\sqrt{2.0 + 4,032 \times 0.02263}} = 11.75 \quad \text{(close check)}$$

From Sec. 4.2 steam tables at 104°F, $p_v = 1.070$ lbf/in², the maximum height

$$z_B - z_A = \frac{p_A - p_v}{\rho g} - (1 + \Sigma K_{AB})\frac{V^2}{2g}$$
$$= \frac{144(14.70 - 1.070)}{1.925 \times 32.17} - (1 + 1 + 227.9$$
$$\times 0.02262)\frac{(11.75)^2}{2 \times 32.17} = 16.58 \text{ ft } (5.053 \text{ m})$$

Note that if a $\pm$ 10 percent error exists in calculation of pressure loss, maximum height should be limited to $\sim$ 15 ft (5 m).

ASME PIPELINE FLOWMETERS

Parameters Dimensional analysis of the flow of an incompressible fluid flowing in a pipe of diameter D, surface roughness ε, through a primary element (venturi, nozzle or orifice) whose diameter is d with a velocity of V, producing a pressure drop of Δp sensed by pressure taps located a distance L apart results in $f(\mathbf{C}_p, \mathbf{R}_d, \varepsilon/D, d/D) = 0$, which may be written as $\Delta p = \mathbf{C}_p \rho V^2/2$. Conventional practice is to express the relations as $V = \mathbf{K}\sqrt{2\Delta p/\rho}$, where $\mathbf{K}$ is the **flow coefficient**, $\mathbf{K} = 1/\sqrt{\mathbf{C}_p}$, and $\mathbf{K} = f(\mathbf{R}_d, L/D, \varepsilon/d, d/D)$. The ratio of the diameter of the primary element to meter tube (pipe) diameter D is known as the **beta ratio**, where $\beta = d/D$. Application of the continuity equation leads to $Q = \mathbf{K}A_2\sqrt{2\Delta p/\rho}$, where A_2 is the area of the primary element.

Conventional practice is to base flowmeter computations on the assumption of one-dimensional frictionless flow of an incompressible fluid in a horizontal meter tube and to correct for actual conditions by the use of a coefficient for viscous effects and a factor for elastic effects. Application of the Bernoulli equation for horizontal flow from section 1 (inlet tap) to section 2 (outlet tap) results in $p_1/\rho g + V_1^2/2g = p_2/\rho g + V_2^2/2g$ or $(p_1 - p_2)/\rho = V_2^2 - V_1^2 = \Delta p/\rho$. From the equation of continuity, $Q_i = A_1 V_1 = A_2 V_2$, where Q_i is the **ideal flow rate.** Substituting, $2\Delta p/\rho = Q_i^2/A_1^2 - Q_i^2/A_2^2$, and solving for Q_i, $Q_i = A_2\sqrt{2\Delta p/\rho}/\sqrt{1 - (A_2/A_1)^2}$, noting that $A_2/A_1 = (d/D)^2 = \beta^2$, $Q_i = A_2\sqrt{2\Delta p/\rho}/\sqrt{1 - \beta^4}$. The **discharge coefficient** C is defined as the ratio of the actual flow Q to the ideal flow Q_i, or $C = Q/Q_i$, so that $Q = CQ_i = CA_2\sqrt{2\Delta p/\rho}/\sqrt{1 - \beta^4}$. It is customary to write the volumetric-flow equation as $Q = CEA_2\sqrt{2\Delta p/\rho}$, where $E = 1/\sqrt{1 - \beta^4}$. E is called the **velocity-of-approach factor** because it accounts for the one-dimensional kinetic energy at the upstream tap. Comparing the equation from dimensional analysis with the modified Bernoulli equation, $Q = KA_2\sqrt{2\Delta p/\rho} = CEA_2\sqrt{2\Delta p/\rho}$, or $K = CE$ and $C = f(\mathbf{R}_d, L/D, \beta)$.

For compressible fluids, the incompressible equation is modified by the **expansion factor** Y, where Y is defined as the ratio of the flow of a compressible fluid to that of an incompressible fluid at the same value of Reynolds number. Calculations are then based on inlet-tap-fluid properties, and the compressible equation becomes

$$Q_1 = KYA_2\sqrt{2\Delta p/\rho_1} = CEYA_2\sqrt{2\Delta p/\rho_1}$$

where $Y = f(L/D, \varepsilon/D, \beta, \mathbf{M})$. Reynolds number $\mathbf{R}_d$ is also based on inlet-fluid properties, but on the primary-element diameter or

$$\mathbf{R}_d = \rho_1 V_2 d/\mu_1 = \rho_1(Q_1/A_2)d/\mu_1 = 4\rho_1 Q_1/\pi\,d\mu_1$$

Caution The numerical values of coefficients for flowmeters given in the paragraphs to follow are based on experimental data obtained with long, straight pipes where the velocity profile approaching the primary element was fully developed. The presence of valves, bends, and fittings upstream of the primary element can cause serious errors. For approach and discharge, straight-pipe requirements, "Fluid Meters," (6th ed., ASME, 1971) should be consulted.

Venturi Tubes Figure 3.3.26 shows a typical venturi tube consisting of a cylindrical inlet, convergent cone, throat, and divergent cone. The convergent entrance has an included angle of about 21° and the divergent cone 7 to 8°. The purpose of the divergent cone is to reduce the

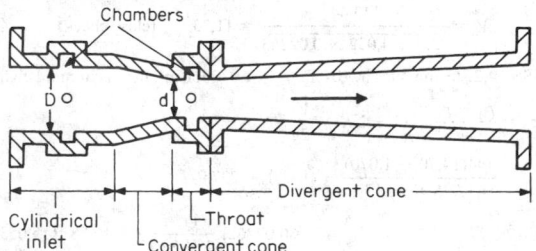

Fig. 3.3.26 Venturi tube.

overall pressure loss of the meter; its removal will have no effect on the coefficient of discharge. Pressure is sensed through a series of holes in the inlet and throat. These holes lead to an annular chamber, and the two chambers are connected to a pressure-differential sensor. Discharge coefficients for venturi tubes as established by the American Society of Mechanical Engineers are given in Table 3.3.12. Coefficients of discharge outside the tabulated limits must be determined by individual calibrations.

EXAMPLE. Benzene at 68°F (20°C) flows through a machined-inlet venturi tube whose inlet diameter is 8 in and whose throat diameter is 3.5 in. The differential pressure is sensed by a U-tube manometer. The manometer contains mercury under the benzene, and the level of the mercury in the throat leg is 4 in. Compute the volumetric flow rate. Noting that $D = 8$ in (0.6667 ft) and $\beta = 3.5/8 = 0.4375$ are within the limits of Table 3.3.12, assume $C = 0.995$, and then check $\mathbf{R}_d$ to verify if it is within limits. For a U-tube manometer (Fig. 3.3.6a), $p_2 - p_1 = (\gamma_m - \gamma_f)h = \Delta p$ and $\Delta p/\rho_1 = (\rho_m g - \rho_f g)h/\rho_f = g(\rho_m/\rho_f - 1)h = 32.17(26.283/1.705 - 1)(4/12) = 154.6$. For a liquid, $Y = 1$ (incompressible fluid), $E = 1/\sqrt{1 - \beta^4} = 1/\sqrt{1 - (0.4375)^4} = 1.019$.

$$\begin{aligned} Q_1 &= CEY A_d\sqrt{2\Delta p/\rho_1} \\ &= (0.995)(1.019)(\pi/4)(3.5/12)^2\sqrt{2 \times 154.6} \\ &= 1.192 \text{ ft}^3/\text{s}\ (3.373 \times 10^{-3} \text{ m}^3/\text{s}) \end{aligned}$$

$\mathbf{R}_d = 4\rho_1 Q_1/\pi\,d\mu_1 = 4(1.705)(1.192)/\pi(3.5/12)(13.62 \times 10^{-6})$
$\mathbf{R}_d = 651,400$, which lies between 200,000 and 1,000,000 of Table 3.3.12 $\therefore$ solution is valid.

Flow Nozzles Figure 3.3.27 shows an ASME flow nozzle. This nozzle is built to rigid specifications, and pressure differential may be sensed by either throat taps or pipe-wall taps. Taps are located one pipe diameter upstream and one-half diameter downstream from the nozzle inlet. Discharge coefficients for ASME flow nozzles may be computed from $C = 0.9975 - 0.00653\,(10^6/\mathbf{R}_d)^a$, where $a = 1/2$ for $\mathbf{R}_d < 10^6$ and $a = 1/5$ for $\mathbf{R}_d > 10^6$. Most of the data were obtained for D between 2 and 15.75 in, $\mathbf{R}_d$ between 10^4 and 10^6, and beta between 0.15 and 0.75. For values of C within these ranges, a tolerance of 2 percent may be anticipated, and outside these limits, the tolerance may be greater than 2 percent. Because slight variations in form or dimension of either pipe or nozzle may affect the observed pressures, and thus cause the exponent a and the slope term (-0.00653) to vary considerably, nozzles should be individually calibrated.

EXAMPLE. An ASME flow nozzle is to be designed to measure the flow of 400 gal/min of 68°F (20°C) water in a 6-in schedule 40 (inside diameter = 6.065 in) steel pipe. The pressure differential across the nozzle is not to exceed 75 in of water. What should be the throat diameter of the nozzle? $\Delta p = h\rho_1 g$, $\Delta p/\rho_1 = hg = (75/12)(32.17) = 201.1$, $Q = (400/60)(231/1,728) = 0.8912$ ft³/s. A trial-and-error solution is necessary to establish the values of C and E because they are dependent upon β and $\mathbf{R}_d$, both of which require that d be known. Since $\mathbf{K} = CE \approx 1$, assume for first trial that $CE = 1$. Since a liquid is involved, $Y = 1$, $A_2 = Q_1/(CE)(Y)\sqrt{2\Delta p/\rho_1} = (0.8912)/(1)(1)\sqrt{2 \times 201.1} = 0.04444$ ft², $d = \sqrt{4A_2/\pi} = \sqrt{4(0.04444)/\pi} = 0.2379$ ft or $d = 0.2379 \times 12 = 2.854$ in, $\beta = d/D = 2.854/6.065 = 0.4706$.
For second trial use first-trial value:

$$E = 1/\sqrt{1 - \beta^4} = 1/\sqrt{1 - (0.4706)^4} = 1.025$$

$\mathbf{R}_d = 4\rho_1 Q_1/\pi d\mu_1 = 4(1.937)(0.8912)/\pi(0.2379)(20.92 \times 10^{-6}) = 442,600 < 10^6$ $\therefore$ $a = 1/2$ and $C = 0.9975 - 0.00653(10^6/\mathbf{R}_d)^{1/2}$. $C = 0.9975 - 0.00653(10^6/442,600)^{1/2} = 0.9877$. $A_2 = (0.8912)/(0.9877 \times 1.025)\sqrt{2 \times 201.1} = 0.04389$, $d_2 = \sqrt{4 \times (0.04389/\pi)} = 0.2364$, $d_2 = 0.2364 \times 12 = 2.837$ in (7.205 × 10^{-2} m). Further trials are not necessary in view of the ± 2 percent tolerance of C.

Table 3.3.12   ASME Coefficients for Venturi Tubes

Type of inlet cone	Reynolds number $\mathbf{R}_d$		Inlet diam D in (2.54 × 10^{-2} m)		β		C	Tolerance, %
	Min	Max	Min	Max	Min	Max		
Machined		1×10^6	2	10		0.75	0.995	±1.0
Rough welded sheet metal	5×10^5	2×10^6	8	48	0.4	0.70	0.985	±1.5
Rough cast			4	32	0.3	0.75	0.984	±0.7

SOURCE: Compiled from data given in "Fluid Meters," 6th ed., ASME, 1971.

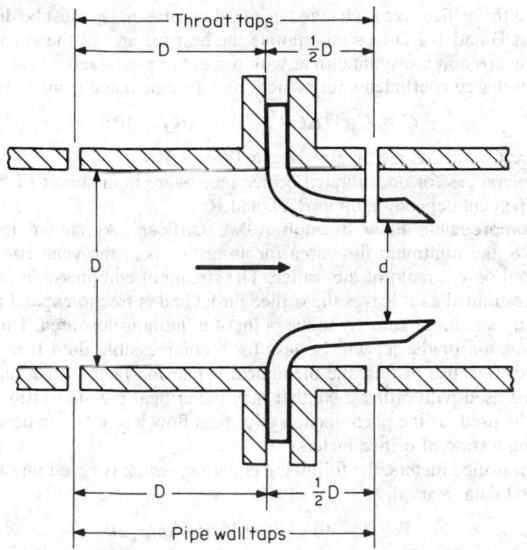

Throat taps

Fig. 3.3.27   ASME flow nozzle.

Compressible Flow—Venturi Tubes and Flow Nozzles   The expansion factor Y is computed based on the assumption of a frictionless adiabatic (isentropic) expansion of an ideal gas from the inlet to the throat of the primary element, resulting in (see Sec. 4.1)

$$Y = \left[\frac{kr^{2/k}(1 - r^{(k-1)/k})(1 - \beta^4)}{(1 - r)(k - 1)(1 - \beta^4 r^{2/k})} \right]^{1/2}$$

where $r = p_2/p_1$.

Maximum flow is obtained when the critical pressure ratio is reached. The critical pressure ratio r_c may be calculated from

$$r^{(1-k)/k} + \frac{k - 1}{2} \beta^4 r^{2/k} = \frac{k + 1}{2}$$

Table 3.3.13 gives selected values of Y_c and r_c.

EXAMPLE.   A piping system consists of a compressor, a horizontal straight length of 2-in-inside-diameter pipe, and a 1-in-throat-diameter ASME flow nozzle attached to the end of the pipe, discharging into the atmosphere. The compressor is operated to maintain a flow of air with 115 lbf/in² and 140°F (60°C) conditions in the pipe just one pipe diameter before the nozzle inlet. Barometric pressure is 14.7 lbf/in². Estimate the flow rate of the air in lbm/s.

From the equation of state, $\rho_1 = p_1/g_c RT_1 = (144 \times 115)/(32.17)(53.34)$ $(140 + 459.7) = 0.01609$ slug/ft³, $\beta = d/D = 1/2 = 0.5$, $E = 1/\sqrt{1 - \beta^4} = 1/\sqrt{1 - (0.5)^4} = 1.033$, $r = p_2/p_1 = 14.7/115 = 0.1278$, but from Table 3.3.13 at $\beta = 0.5$, $k = 1.4$, $r_c = 0.5362$, and $Y_c = 0.6973$, so that because of critical flow the throat pressure $p_c = 115 \times 0.5362 = 61.66$ lbf/in². $\Delta p_c/\rho_1 = 144(115 - 61.66)/0.01609 = 477,375$. A trial-and-error solution is necessary to obtain C. For the first trial assume $10^6/\mathbf{R}_d = 0$ or $C = 0.9975$. Then $Q_1 = CEY_c A_2 \sqrt{2\Delta p_c/\rho_1} = (0.9975)(1.033)(0.6973)(\pi/4)(1.12)^2 \sqrt{2 \times 477,375} = 3.829$ ft³/s, $\mathbf{R}_d = 4\rho_1 Q/\pi d\mu_1 = (4)(0.01609)(3.828)/\pi (1/12)(41.79 \times 10^{-8}) = 2,252,000$.

Second trial, use first-trial values:

$$\mathbf{R} > 10^6, \quad a = 115, \quad C = 0.9975 - (0.00653)(10^6/2,252,000)^{1/5}$$
$$C = 0.9919, \quad Q_1 = 3.828(0.9919/0.9975) = 3.806 \text{ ft}^3/\text{s}$$

Further trials are not necessary in view of ± 2 percent tolerance on C.

$$\dot{m} = Q_1 \rho_1 g = 3.806 \times 0.01609 \times 32.17 = 1.970 \text{ lbm/s } (0.8935 \text{ kg/s})$$

Orifice Meters   When a fluid flows through a square-edged thin-plate orifice, the minimum-flow area is found to occur downstream from the orifice plate. This minimum area is called the vena contracta, and its location is a function of beta ratio. Figure 3.3.28 shows the relative pressure difference due to the presence of the orifice plate. Because the location of the pressure taps is vital, it is necessary to specify the exact position of the downstream pressure tap. The jet con-

Table 3.3.13   Expansion Factors and Critical Pressure Ratios for Venturi Tubes and Flow Nozzles

β	k	Critical values r_c	Critical values Y_c	Expansion factor Y $r = 0.60$	$r = 0.70$	$r = 0.80$	$r = 0.90$
0	1.10	0.5846	0.6894	0.7021	0.7820	0.8579	0.9304
	1.20	0.5644	0.6948	0.7228	0.7981	0.8689	0.9360
	1.30	0.5457	0.7000	0.7409	0.8119	0.8783	0.9408
	1.40	0.5282	0.7049	0.7568	0.8240	0.8864	0.9449
0.20	1.10	0.5848	0.6892	0.7017	0.7817	0.8577	0.9303
	1.20	0.5546	0.6946	0.7225	0.7978	0.8687	0.9359
	1.30	0.5459	0.6998	0.7406	0.8117	0.8781	0.9407
	1.40	0.5284	0.7047	0.7576	0.8237	0.8862	0.9448
0.50	1.10	0.5921	0.6817	0.6883	0.7699	0.8485	0.9250
	1.20	0.5721	0.6872	0.7094	0.7864	0.8600	0.9310
	1.30	0.5535	0.6923	0.7248	0.8007	0.8699	0.9361
	1.40	0.5362	0.6973	0.7440	0.8133	0.8785	0.9405
0.60	1.10	0.6006	0.6729		0.7556	0.8374	0.9186
	1.20	0.5808	0.6784	0.6939	0.7727	0.8495	0.9250
	1.30	0.5625	0.6836	0.7126	0.7875	0.8599	0.9305
	1.40	0.5454	0.6885	0.7292	0.8006	0.8689	0.9352
0.70	1.10	0.6160	0.6570		0.7290	0.8160	0.9058
	1.20	0.5967	0.6624	0.6651	0.7469	0.8292	0.9131
	1.30	0.5788	0.6676	0.6844	0.7626	0.8405	0.9193
	1.40	0.5621	0.6726	0.7015	0.7765	0.8505	0.9247
0.80	1.10	0.6441	0.6277		0.6778	0.7731	0.8788
	1.20	0.6238	0.6331		0.6970	0.7881	0.8877
	1.30	0.6087	0.6383		0.7140	0.8012	0.8954
	1.40	0.5926	0.6433	0.6491	0.7292	0.8182	0.9021

SOURCE: Murdock, "Fluid Mechanics and Its Applications," Houghton Mifflin, 1976.

traction amounts to about 60 percent of the orifice area; so orifice coefficients are in the order of 0.6 compared with the nearly unity obtained with venturi tubes and flow nozzles.

Three pressure-differential-measuring tap locations are specified by the ASME. These are the flange, vena contracta, and the 1 D and 1/2 D. In the flange tap, the location is always 1 in from either face of the

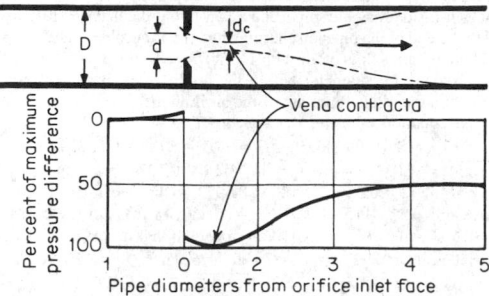

Fig. 3.3.28 Relative-pressure changes due to flow through an orifice.

orifice plate regardless of the size of the pipe. In the vena contracta tap, the upstream tap is located one pipe diameter from the inlet face of the orifice plate and the downstream tap at the location of the vena contracta. In the 1 D and 1/2 D tap, the upstream tap is located one pipe diameter from the inlet face of the orifice plate and downstream one-half pipe diameter from the inlet face of the orifice plate.

Flange taps are used because they can be prefabricated, and flanges with holes drilled at the correct locations may be purchased as off-the-shelf items, thus saving the cost of field fabrication. The disadvantage of flange taps is that they are not symmetrical with respect to pipe size. Because of this, coefficients of discharge for flange taps vary greatly with pipe size.

Vena contracta taps are used because they give the maximum differential for any given flow. The disadvantage of the vena contracta tap is

that if the orifice size is changed, a new downstream tap must be drilled. The 1 D and 1/2 D taps incorporate the best features of the vena contracta taps and are symmetrical with respect to pipe size.

Discharge coefficients for orifices may be calculated from

$$C = C_o + \Delta C \mathbf{R}_d^{-0.75} \qquad (\mathbf{R}_d > 10^4)$$

where C_o and ΔC are obtained from Table 3.3.14.

Tolerances for uncalibrated orifice meters are in the order of ± 1 to ± 2 percent depending upon β, D, and $\mathbf{R}_d$.

Compressible Flow through ASME Orifices As shown in Fig. 3.3.28, the minimum flow area for an orifice is at the vena contracta located downstream of the orifice. The stream of compressible fluid is not restrained as it leaves the orifice throat and is free to expand transversely and longitudinally to the point of minimum-flow area. Thus the contraction of the jet will be less for a compressible fluid than for a liquid. Because of this, the theoretical-expansion-factor equation may not be used with orifices. Neither may the critical-pressure-ratio equation be used, as the phenomenon of critical flow has not been observed during testing of orifice meters.

For orifice meters, the following equation, which is based on experimental data, is used:

$$Y = 1 - (0.41 + 0.35\beta^4)(\Delta p/p_1)/k$$

EXAMPLE. Air at 68°F (20°C) and 150 lbf/in² flows in a 2-in schedule 40 pipe (inside diameter = 2.067 in) at a volumetric rate of 15 ft³/min. A 0.5500-in ASME orifice equipped with flange taps is used to meter this flow. What deflection in inches could be expected on a U-tube manometer filled with 60°F water? From the equation of state, $\rho_1 = p_1/g_c RT_1 = (144 \times 150)/(32.17)(53.34)$ $(68 + 459.7) = 0.02385$ slug/ft³, $\beta = 0.5500/2.067 = 0.2661$. $Q_1 = 15/60 = 0.25$ ft³/s, $A_2 = (\pi/4)(0.5500/12)^2 = 1.650 \times 10^{-3}$ ft². $E = 1/\sqrt{1 - \beta^4} = 1/\sqrt{1 - (0.2661)^4} = 1.003$. $\mathbf{R}_d = 4\rho_1 Q_1/\pi\, d\mu_1 = 4(0.02385)(0.25)/\pi(0.5500/12)(39.16 \times 10^{-8})$. $\mathbf{R}_d = 423,000$.

From Table 3.3.14 at $\beta = 0.2661$, $D = 2.067$-in flange taps, by interpolation, $C_o = 0.5977$, $\Delta C = 9.087$, from orifice-coefficient equation $C = C_o + \Delta C \mathbf{R}_d^{-0.75}$. $C = 0.5977 + (9.087)(423,000)^{-0.75} = 0.5982$. A trial-and-error solution is required because the pressure loss is needed in order to compute Y. For the first trial, assume $Y = 1$, $\Delta p = (Q_1/CEYA_2)^2(\rho_1/2) = [(0.25)/(0.5982)(1.003)(Y)(1.650 \times 10^{-3})]^2(0.02385/2) = 760.5/Y^2 = 760.5/(1)^2 = 760.5$ lbf/ft².

Table 3.3.14 Values of C_o and ΔC for Use in Orifice Coefficient Equation

Pipe ID, in	0.20	0.25	0.30	0.35	0.40	0.45	0.50	0.55	0.60	0.65	0.70
					β						
					ΔC, all taps						
All	5.486	8.106	11.153	14.606	18.451	22.675	27.266	32.215	37.513	45.153	49.129
					C_o, vena contracta and 1D and ½D taps						
All	0.5969	0.5975	0.5983	0.5992	0.6003	0.6016	0.6031	0.6045	0.6059	0.6068	0.6069
					C_o, flange taps						
2.0	0.5969	0.5975	0.5982	0.5992	0.6003	0.6016	0.6030	0.6044	0.6056	0.6065	0.6066
2.5	0.5969	0.5975	0.5983	0.5993	0.6004	0.6017	0.6032	0.6046	0.6059	0.6068	0.6068
3.0	0.5969	0.5975	0.5983	0.5993	0.6004	0.6017	0.6031	0.6044	0.6055	0.6061	0.6057
3.5	0.5969	0.5975	0.5983	0.5993	0.6004	0.6016	0.6030	0.6042	0.6052	0.6056	0.6049
4.0	0.5969	0.5976	0.5983	0.5993	0.6004	0.6016	0.6029	0.6041	0.6050	0.6052	0.6043
5.0	0.5969	0.5976	0.5983	0.5993	0.6004	0.6016	0.6028	0.6039	0.6047	0.6047	0.6034
6.0	0.5969	0.5976	0.5983	0.5993	0.6004	0.6016	0.6028	0.6038	0.6045	0.6044	0.6029
8.0	0.5969	0.5976	0.5984	0.5993	0.6004	0.6015	0.6027	0.6037	0.6042	0.6040	0.6022
10.0	0.5969	0.5976	0.5984	0.5993	0.6004	0.6015	0.6026	0.6036	0.6041	0.6037	0.6017
12.0	0.5970	0.5976	0.5984	0.5993	0.6004	0.6015	0.6026	0.6035	0.6040	0.6035	0.6015
16.0	0.5970	0.5976	0.5984	0.5993	0.6003	0.6015	0.6026	0.6035	0.6039	0.6033	0.6011
24.0	0.5970	0.5976	0.5984	0.5993	0.6003	0.6015	0.6025	0.6034	0.6037	0.6031	0.6007
48.0	0.5970	0.5976	0.5984	0.5993	0.6003	0.6014	0.6025	0.6033	0.6036	0.6029	0.6004
∞	0.5970	0.5976	0.5984	0.5993	0.6003	0.6014	0.6025	0.6032	0.6035	0.6027	0.6000

SOURCE: Compiled from data given in ASME Standard MFC-3M-1984 "Measurement of Fluid Flow in Pipes Using Orifice, Nozzle and Venturi."

For the second trial we use first-trial values.

$$Y = 1 - (0.41 + 0.35\beta^4)\frac{\Delta p/p_1}{k}$$

$$= 1 - [0.41 + 0.35(0.2661)^4]\frac{760.5/144 \times 150}{1.4} = 0.9896$$

$$\Delta p = \frac{760.5}{Y^2} = \frac{760.5}{(0.9896)^2} = 776.1 \text{ lbf/ft}^2$$

For the third trial we use second-trial values.

$$Y = 1 - [0.41 + 0.35(0.2661)^4]\frac{776.1/144 \times 150}{1.4} = 0.9894$$

$$\Delta p = \frac{776.1}{(0.9894)^2} = 793.3 \text{ lbf/ft}^2$$

Resubstitution does not produce any further change in Y. From the U-tube-manometer equation:

$$h = \frac{\Delta p}{\gamma_m = \gamma_f} = \frac{\Delta p}{g_c(\rho_m - \rho_f)}$$

$$= \frac{793.3}{32.17}(1.937 - 0.02385) = 12.89 \text{ ft}$$

$$= 12.89 \times 12 = 154.7 \text{ in } (3.929 \text{ m})$$

PITOT TUBES

Definition A Pitot tube is a device that is shaped in such a manner that it senses stagnation pressure. The name "Pitot tube" has been applied to two general classifications of instruments, the first being a tube that measures the impact or stagnation pressures only, and the second a combined tube that measures both impact and static pressures with a single primary instrument. The combined sensor is called a Pitot-static tube.

Tube Coefficient From Fig. 3.3.29, it is evident that the Pitot tube can sense only the stagnation pressure resulting from the local stream-tube velocity U. The local ideal velocity U_i for an incompressible fluid is obtained by the application of the Bernoulli equation ($z_S = z$), $U_i^2/2g + p/\rho g = U_S^2/2g + p_S/\rho g$. Solving for U_i and noting that by definition

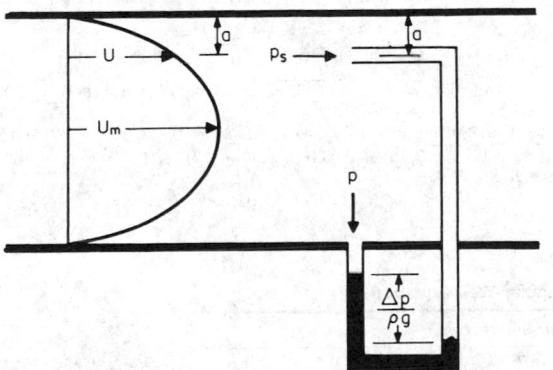

Fig. 3.3.29 Notation for Pitot tube study.

$U_S = 0$, $U_i = \sqrt{2(p_S - p)/\rho}$. Conventional practice is to define the **tube coefficient** C_T as the ratio of the actual stream-tube velocity to the ideal stream-tube velocity, or $C_T = U/U_i$ and $U = C_T U_i = C_T\sqrt{2\Delta p/\rho}$. The numerical value of C_T depends primarily upon its geometry. The value of C_T may be established (1) by calibration with a uniform velocity, (2) from published data for similar geometry, or (3) in the absence of other information, may be assumed to be unity.

Pipe Coefficient For the calculation of volumetric flow rate, it is necessary to integrate the continuity equation, $Q = \int U \, da = AV$. The

pipe coefficient C_P is defined as the ratio of the average velocity to the stream-tube velocity, or $C_P = V/U$, and $Q = C_P A_1 V = C_P C_T A_1 \sqrt{2\Delta p/\rho}$. The numerical value of C_P is dependent upon the location of the tube and the **velocity profile**. The values of C_P may be established by (1) making a "traverse" by taking data at various points in the flow stream and determining the velocity profile experimentally (see "Fluid Meters," 6th ed., ASME, 1971, for locations of traverse points), (2) using standard velocity profiles, (3) locating the Pitot tube at a point where $U = V$, and (4) assuming one-dimensional flow of $C_P = 1$ **only** in the absence of other data.

Compressible Flow For compressible flow, the **compression factor** Z is based on the assumption of a frictionless adiabatic (isentropic) compression of an ideal gas from the moving stream tube to the stagnation point (see Sec. 4.1), which results in

$$Z = \left[\frac{k}{k-1}\frac{(p_S/p)^{(k-1)/k} - 1}{(p_S/p) - 1}\right]^{1/2}$$

and the volumetric flow rate becomes

$$Q = C_P C_T Z A_1 \sqrt{2\Delta p/\rho}$$

EXAMPLE. Carbon dioxide flows at 68°F (20°C) and 20 lbf/in^2 in an 8-in schedule 40 galvanized-iron pipe. A Pitot tube located on the pipe centerline indicates a pressure differential of 6.986 lbf/in^2. Estimate the mass flow rate. For 8-in schedule 40 pipe $D = 7.981/12 = 0.6651$, $\varepsilon/D = 500 \times 10^{-6}/0.6651 = 7.518 \times 10^{-4}$, $A_1 = \pi D^2/4 = (\pi/4)(0.6651)^2 = 0.3474$ ft^2, $p_S = p + \Delta p = 20 + 6.986 = 26.986$ lbf/in^2. From the equation of state, $\rho = p/g_c RT_o = (20 \times 144)/(32.17)(35.11)(68 + 459.7) = 0.004832$,

$$Z = \left[\frac{k}{k-1}\frac{(p_S/p)^{(k-1)/k} - 1}{(p_S/p) - 1}\right]^{1/2}$$

$$= \left\{[1.3/(1.3 - 1)] \times \frac{(26.986/20)^{(1.3-1)/1.3} - 1}{(26.986/20) - 1}\right\}^{1/2} = 0.9423$$

In the absence of other data, C_T may be assumed to be unity. A trial-and-error solution is necessary to determine C_P, since f requires flow rate. For the first trial assume complete turbulence.

$$1/\sqrt{f} = -2\log_{10}(7.518 \times 10^{-4}/3.7) \qquad \sqrt{f} = 0.1354$$
$$C_P = V/U = V/U_{max} = 1/(1 + 1.43\sqrt{f}) = 1/(1 + 1.43 \times 0.1354) = 0.8378$$
$$V = C_P C_T Z\sqrt{2\Delta p/\rho} = (0.8378)(1)(0.9423)\sqrt{2 \times 144(6.987)/(0.004832)}$$
$$V = 509.4 \text{ ft/s}$$
$$\mathbf{R} = \rho VD/\mu = (0.004832)(509.4)(0.6651)/(30.91 \times 10^{-18})$$
$$\mathbf{R} = 5,296,000 > 4,000 \therefore \text{ flow is turbulent}$$

From the Colebrook equation and Fig. 3.3.24,

$$\frac{1}{\sqrt{f}} = -2\log_{10}\left(\frac{7.518 \times 10^{-4}}{3.7} + \frac{2.51}{5,296,000\sqrt{0.018}}\right)$$
$$\sqrt{f} = 0.1357$$
$$C_P = 1/(1 + 1.43 \times 0.1357) = 0.8375 \qquad \text{(close check)}$$
$$V = 509.4(0.8375/8378) = 509.2 \text{ ft/s}$$

From the continuity equation, $m = \rho A_1 V g_c = (0.004832)(0.3474)(509.2)$ $(32.17) = 27.50$ lbm/s (12.47 kg/s).

ASME WEIRS

Definitions A weir is a dam over which liquids are forced to flow. Weirs are used to measure the flow of liquids in open channels or in conduits which do not flow full; i.e., there is a free liquid surface. Weirs are almost exclusively used for measuring water flow, although small ones have been used for metering other liquids. Weirs are classified according to their notch or opening as follows: (1) **rectangular notch** (original form); (2) **V** or **triangular notch**; (3) **trapezoidal notch**, which when designed with end slopes one horizontal to four vertical is called the **Cipolletti weir**; (4) the **hyperbolic weir** designed to give a constant coefficient of discharge; and (5) the **parabolic weir** designed to give a linear relationship of head to flow. As shown in Fig. 3.3.30, the top of the weir is the **crest** and the distance from the liquid surface to the crest h is called the **head**.

The sheet of liquid flowing over the weir crest is called the **nappe**. When the nappe falls downstream of the weir plate, it is said to be free, or **aerated.** When the width of the approach channel L_c is greater than the crest length L_w, the nappe will contract so that it will have a minimum width less than the crest length. For this reason, the weir is known as a **contracted weir.** For the special case where $L_w = L_c$, the contractions do not take place, and such weirs are known as **suppressed weirs.**

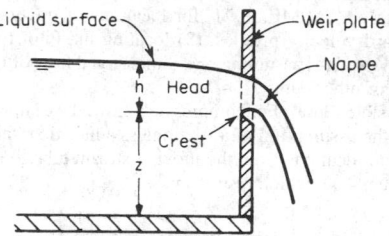

Fig. 3.3.30 Notation for weir study.

Parameters The forces acting on a liquid flowing over a weir are inertia, viscous, surface tension, and gravity. If the weir head produced by the flow is h, the characteristic length of the weir is L_w, and the channel width is L_c, either similarity or dimensional analysis leads to $f(\mathbf{F}, \mathbf{W}, \mathbf{R}, L_w/L_c) = 0$, which may be written as $V = K\sqrt{2gh}$, where K is the **weir coefficient** and $K = f(\mathbf{W}, \mathbf{R}, L_w/L_c)$. Since the weir has been almost exclusively used for metering water flow over limited temperature ranges, the effects of surface tension and viscosity have not been adequately established by experiment.

Caution The numerical values of coefficients for weirs are based on experimental data obtained from calibration of weirs with long approaches of straight channels. Head measurement should be made at a distance at least three or four times the expected maximum head h. Screens and baffles should be used as necessary to ensure steady uniform flow without waves or local eddy currents. The approach channel should be relatively wide and deep.

Rectangular Weirs Figure 3.3.31 shows a rectangular weir whose crest width is L_w. The volumetric flow rate may be computed from the continuity equation: $Q = AV = (L_w h)(K\sqrt{2gh}) = KL_w\sqrt{2g}\,h^{3/2}$. The ASME "Fluid Meters" report recommends the following equation for rectangular weirs: $Q = (\frac{2}{3})CL_a\sqrt{2g}\,h_a^{3/2}$, where C is the **coefficient of discharge** $C = f(L_w/L_c, h/Z)$, L_a is the adjusted crest length $L_a = L_w + \Delta L$, and h_a is the adjusted weir head $h_a = h + 0.003$ ft. Values of C and ΔL may be obtained from Table 3.3.15. To avoid the possibility that the liquid drag along the sides of the channel will affect side contractions, $L_c - L_w$ should be at least $4h$. The minimum crest length should be 0.5 ft to prevent mutual interference of the end contractions. The minimum head for free flow of the nappe should be 0.1 ft.

EXAMPLE. Water flows in a channel whose width is 40 ft. At the end of the channel is a rectangular weir whose crest width is 10 ft and whose crest height is 4 ft. The water flows over the weir at a height of 3 ft above the crest of the weir. Estimate the volumetric flow rate. $L_w/L_c = 10/40 = 0.25$, $h/Z = 3/4 = 0.75$, from Table 3.3.15 (interpolated), $C = 0.589$, $\Delta L = 0.008$, $L_a = L_w + \Delta L = 10 + 0.008 = 10.008$ ft, $h_a = h + 0.003 = 3 + 0.003 = 3.003$ ft, $Q = (2/3) CL_a\sqrt{2g}\,h^{3/2}$, $Q = (2/3)(0.589)(10.008)(2 \times 32.17)^{1/2}(3.003)^{3/2}$ $Q = 164.0$ ft³/s (4.644 m³/s).

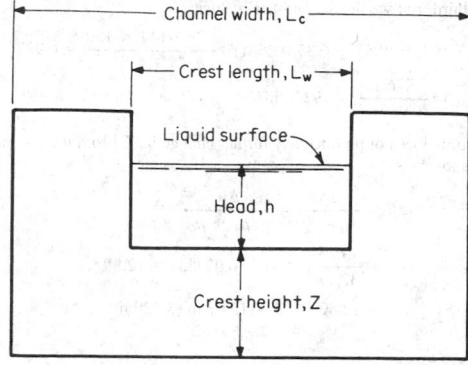

Fig. 3.3.31 Rectangular weir.

Triangular Weirs Figure 3.3.32 shows a triangular weir whose notch angle is θ. The volumetric flow rate may be computed from the continuity equation $Q = AV = (h^2 \tan \theta/2)(K\sqrt{2gh}) = K \tan(\theta/2)\sqrt{2g}\,h^{5/2}$. The ASME "Fluid Meters" report recommends the following for triangular weirs: $Q = (8/15)\,C \tan(\theta/2)\sqrt{2g}\,(h + \Delta h)^{5/2}$, where C is the coefficient of discharge $C = f(\theta)$ and Δh is the correction for head/crest ratio $\Delta h = f(\theta)$. Values of C and Δh may be obtained from Table 3.3.16.

EXAMPLE. It is desired to maintain a flow of 167 ft³/s in an open channel whose width is 20 ft at a height of 7 ft by locating a triangular weir at the end of the channel. The weir has a crest height of 2 ft. What notch angle is required to maintain these conditions? A trial-and-error solution is required. For the first trial assume $\theta = 60°$ (mean value 20 to 100°); then $C = 0.576$ and $\Delta h = 0.004$.

$$h + Z = 7 = h + 2 \therefore h = 5$$
$$Q = (8/15)\,C \tan(\theta/2)\sqrt{2g}\,(h + \Delta h)^{5/2}$$

$167 = (8/15)(0.576) \tan(\theta/2)\sqrt{2 \times 32.17}\,(5 + 0.004)^{5/2}$, $\tan^{-1}(\theta/2) = 1.20993$, $\theta = 100°51'$.

Second trial, using $\theta = 100$, $C = 0.581$, $\Delta h = 0.003$, $167 = (8/15)(0.581) \tan(\theta/2)\sqrt{2 \times 32.17}\,(5 + 0.003)^{5/2}$, $\tan^{-1}(\theta/2) = 1.20012$, $\theta = 100°39'$ (close check).

Table 3.3.15   Values of C and ΔL for Use in Rectangular-Weir Equation

h/Z	Crest length/channel width = L_w/L_c							
	0	0.2	0.4	0.6	0.7	0.8	0.9	1.0
	Coefficient of discharge C							
0	0.587	0.589	0.591	0.593	0.595	0.597	0.599	0.603
0.5	0.586	0.588	0.594	0.602	0.610	0.620	0.631	0.640
1.0	0.586	0.587	0.597	0.611	0.625	0.642	0.663	0.676
1.5	0.584	0.586	0.600	0.620	0.640	0.664	0.695	0.715
2.0	0.583	0.586	0.603	0.629	0.655	0.687	0.726	0.753
2.5	0.582	0.585	0.608	0.637	0.671	0.710	0.760	0.790
3.0	0.580	0.584	0.610	0.647	0.687	0.733	0.793	0.827
	Adjustment for crest length ΔL, ft							
Any	0.007	0.008	0.009	0.012	0.013	0.014	0.013	−0.005

SOURCE: Compiled from data given in "Fluid Meters," ASME, 1971.

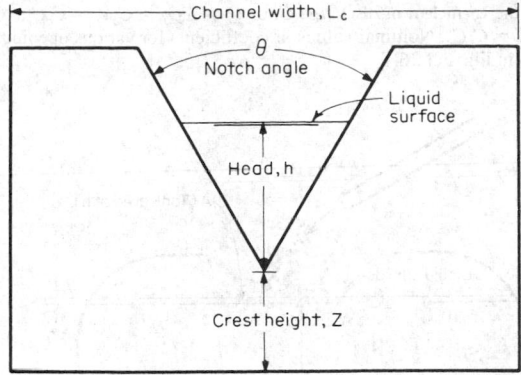

Fig. 3.3.32 Triangular weir.

Table 3.3.16 Values of C and Δh for Use in Triangular-Weir Equation

Item	Weir notch angle θ, deg						
	20	30	45	60	75	90	100
C	0.592	0.586	0.580	0.576	0.576	0.579	0.581
Δh, ft	0.010	0.007	0.005	0.004	0.003	0.003	0.003

SOURCE: Compiled from data given in "Fluid Meters," ASME, 1971.

OPEN-CHANNEL FLOW

Definitions An **open channel** is a conduit in which a liquid flows with a free surface subjected to a constant pressure. Flows of water in natural streams, artificial canals, irrigation ditches, sewers, and flumes are examples where the water surface is subjected to atmospheric pressure. The flow of any liquid in a pipe where there is a free liquid surface is an example of open-channel flow where the liquid surface will be subjected to the pressure existing in the pipe. The **slope** S of a channel is the change in elevation per unit of horizontal distance. For small slopes, this is equivalent to dividing the change in elevation by the distance L measured along the channel bottom between two sections. For steady uniform flow, the velocity distribution is the same at all sections of the channel, so that the energy grade line has the same angle as the bottom of the channel; thus:

$$S = h_f/L$$

The distance between the liquid surface and the bottom of the channel is sometimes called the **stage** and is denoted by the symbol y in Fig. 3.3.33. When the stages between the sections are not uniform, that is, $y_1 \neq y_2$ or the cross section of the channel changes, or both, the flow is said to be **varied**. When a liquid flows in a channel of uniform cross section and the slope of the surface is the same as the slope of the bottom of the channel ($y_1 = y = y_2$), the flow is said to be **uniform**.

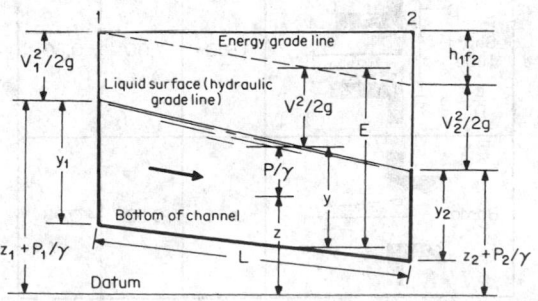

Fig. 3.3.33 Notation for open channel flow.

Parameters The forces acting on a liquid flowing in an open channel are inertia, viscous, surface tension, and gravity. If the channel has a surface roughness of ε, a hydraulic radius of R_h, and a slope of S, either similarity or dimensional analysis leads to $f(\mathbf{F}, \mathbf{W}, \mathbf{R}, \varepsilon/4R_h) = 0$, which may be written as $V = C\sqrt{R_h S}$, where $C = f(\mathbf{W}, \mathbf{R}, \varepsilon/4R_h)$ and is known as the **Chézy coefficient**. The relationship between the Chézy coefficient C and the friction factor may be determined by equating

$$V = \sqrt{8R_h h_f g/fL} = C\sqrt{R_h S} = C\sqrt{(R_h h_f)/l}$$

or $C = (8g/f)^{1/2}$. Although this establishes a relationship between the Chézy coefficient and the friction factor, it should be noted that $f = f(\mathbf{R}, \varepsilon/4R_h)$ and $C = f(\mathbf{W},\mathbf{R},\varepsilon/4R_h)$, because in open-channel flow, pressure forces are absent and in pipe flow, surface-tension and gravity forces are absent. For these reasons, data obtained in pipe flow should not be applied to open-channel flow.

Roughness Factors For open-channel flow, the Chézy coefficient is calculated by the Manning equation, which was developed from examination of experimental results of water tests. The **Manning relation** is stated as

$$C = \frac{1.486}{n} R_h^{1/6}$$

where n is a roughness factor and should be a function of Reynolds number, Weber number, and relative roughness. Since only water-test data obtained at ordinary temperatures support these values, it must be assumed that n is the value for turbulent flow only. Since surface tension is a weak property, the effects of Weber number variation are negligible, leaving n to be some function of surface roughness. Design values of n are given in Table 3.3.17. Maximum flow for a given slope will take place when R_h is a maximum, and values of R_{hmax} are given in Table 3.3.6.

Table 3.3.17 Values of Roughness Factor n for Use in Manning Equation

Surface	n	Surface	n
Brick	0.015	Earth, with stones and weeds	0.035
Cast iron	0.015		
Concrete, finished	0.012	Gravel	0.029
Concrete, unfinished	0.015	Riveted steel	0.017
Brass pipe	0.010	Rubble	0.025
Earth	0.025	Wood, planed	0.012
		Wood, unplaned	0.013

SOURCE: Compiled from data given in R. Horton, *Engineering News*, 75, 373, 1916.

EXAMPLE. It is necessary to carry 150 ft³/s of water in a rectangular unplaned timber flume whose width is to be twice the depth of water. What are the required dimensions for various slopes of the flume? From Table 3.3.6, $A = b^2/2$ and $R_h = h/2 = b/4$. From Table 3.3.17, $n = 0.013$ for unplaned wood. From Manning's equation, $C = 1.486/n$, $R_h^{1/6} = (1.486/0.013)(b^{1/6}/(4)^{1/6}) = 90.73\ b^{1/6}$. From the continuity equation, $V = Q/A = 150/(b^2/2)$, $V = 300/b^2$. From the Chézy equation, $V = C\sqrt{R_h S} = 300/b^2 = 90.73b^{1/6}\sqrt{(b/4)S}$; solving for b, $b = 2.0308/S^{3/16}$.

Assumed S: 1×10^{-1} 1×10^{-2} 1×10^{-3} 1×10^{-4} 1×10^{-5} 1×10^{-6} ft/ft
Required b: 3.127   4.816   7.416   11.42   17.59   27.08 ft

EXAMPLE. A rubble-lined trapezoidal canal with 45° sides is to carry 360 ft³/s of water at a depth of 4 ft. If the slope is 9×10^{-4} ft/ft, what should be the dimensions of the canal? From Table 3.3.17, $n = 0.025$ for rubble. From Table 3.3.6 for $\alpha = 45°$, $A = (b + h)h = 4(b + 4)$, and $R_h = (b + h)h/(b + 2.828h) = 4(b + 4)/(b + 11.312)$. From the Manning relation, $C = (1.486/n)$ $(R_h^{1/6}) = (1.486/0.025)R_h^{1/6} = 59.44\ R_h^{1/6}$. For the first trial, assume $R_h = R_{hmax} = h/2 = 4/2 = 2$; then $C = 59.44(2)^{1/6} = 66.72$ and $V = C\sqrt{R_h S} = 66.72\sqrt{2 \times 9 \times 10^{-4}} = 2.831$. From the continuity equation, $A = Q/V = 360/2.831 = 127.2 = 4(b + 4)$; $b = 27.79$ ft. Second trial, use the first trial, $R_h = 4(27.79 + 4)/(27.79 + 11.312)$, $R_h = 3.252$, $V = 59.44(3.252)^{1/6}$ $\sqrt{3.252 \times 9 \times 10^{-4}} = 3.914$. From the equation of continuity, $Q/V = 360/3.914 = 91.97 = 4(b + 4)$, $b = 18.99$. Subsequent trial-and-error solutions result in a balance at $b = 19.93$ ft (6.075 m).

Specific Energy   Specific energy is defined as the energy of the fluid referred to the bottom of the channel as the datum. Thus the specific energy E at any section is given by $E = y + V^2/2g$; from the continuity equation $V = Q/A$ or $E = y + (Q/A)^2/2g$. For a rectangular channel whose width is b, $A = by$; and if $\mathbf{q}$ is defined as the flow rate per unit width, $\mathbf{q} = Q/b$ and $E = y + (\mathbf{q}b/by)^2/2g = y + (\mathbf{q}/y)^2/2g$.

Critical Values   For rectangular channels, if the specific-energy equation is differentiated and set equal to zero, critical values are obtained; thus $dE/dy = d/dy\,[y + (\mathbf{q}/y)^2/2g] = 0 = 1 - \mathbf{q}^2/y^3 g$ or $\mathbf{q}_c^2 = y_c^3 g$. Substituting in the specific-energy equation, $E = y_c + y_c^3 g/2gy_c^2 = 3/2y_c$. Figure 3.3.34 shows the relation between depth and specific energy for a constant flow rate. If the depth is greater than critical, the flow is *subcritical;* at critical depth it is *critical* and at depths below critical the flow is *supercritical*. For a given specific energy, there is a maximum unit flow rate that can exist.

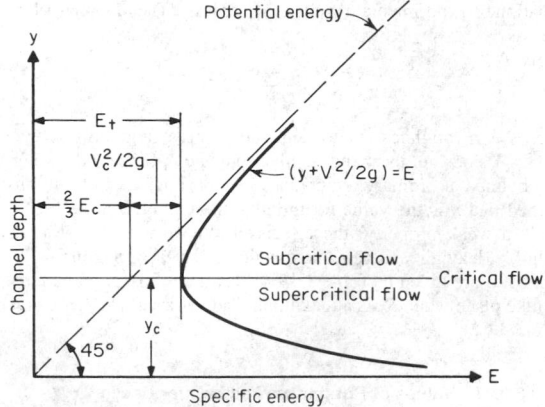

Fig. 3.3.34   Specific energy diagram, constant flow rate.

The **Froude number** $\mathbf{F} = V/\sqrt{gy}$, when substituted in the specific-energy equation, yields $E = y + (\mathbf{F}^2 gy)/2g = y(1 + \mathbf{F}^2/2)$ or $E/y = 1 + \mathbf{F}^2/2$. For critical flow, $E_c/y_c = 3/2$. Substituting $E_c/y_c = 3/2 = 1 + \mathbf{F}_c^2/2$, or $\mathbf{F} = 1$,

$$\mathbf{F} < 1 \qquad \text{Flow is subcritical}$$
$$\mathbf{F} = 1 \qquad \text{Flow is critical}$$
$$\mathbf{F} > 1 \qquad \text{Flow is supercritical}$$

It is seen that for open-channel flow the Froude number determines the type of flow in the same manner as Mach number for compressible flow.

EXAMPLE.   Water flows at a ate of 600 ft³/s in a rectangular channel 10 ft wide at a depth of 4 ft. Determine (1) specific energy and (2) type of flow.
1. from the continuity equation,

$$V = Q/A = 600/(10 \times 4) = 15 \text{ ft/s}$$
$$E = y + V^2/2g = 4 + (15)^2/2(2 \times 32.17) = 7.497 \text{ ft}$$

2. $\mathbf{F} = V/\sqrt{gy} = 15/\sqrt{32.17 \times 4} = 1.322; \mathbf{F} > 1 \therefore$ flow is supercritical.

FLOW OF LIQUIDS FROM TANK OPENINGS

Steady State   Consider the jet whose velocity is V discharging from an open tank through an opening whose area is a, as shown in Fig. 3.3.35. The liquid height above the centerline is h, and the cross-sectional area of the tank at h is A. The ideal velocity of the jet is $V_i = \sqrt{2gh}$. The ratio of the actual velocity V to the ideal velocity V_i is the **coefficient of velocity** Cv, or $V = C_v V_i = C_v \sqrt{2gh}$. The ratio of the actual opening a to the minimum area of the jet a_c is the **coefficient of contraction** C_c, or $a = C_c a_c$. The ratio of the actual discharge Q to the ideal discharge

Q_i is the **coefficient of discharge** C, or $Q = CQ_i = C_a V_i = C_c C_v a \sqrt{2gh}$, and $C = C_c C_v$. Nominal values of coefficients for various openings are given in Fig. 3.3.36.

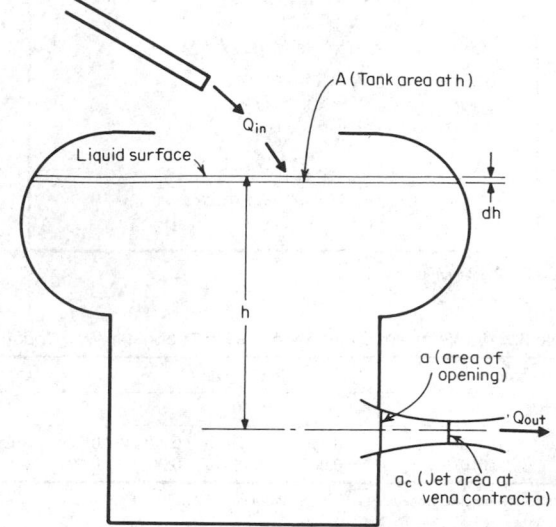

Fig. 3.3.35   Notation for tank flow.

Unsteady State   If the rate of liquid entering the tank Q_{in} is different from that leaving, the level h in the tank will change because of the change in storage. For liquids, the conservation-of-mass equation may be written as $Q_{in} - Q_{out} = Q_{stored}$; for a time interval dt, $(Q_{in} - Q_{out})dt =$

Type	Coefficient		
	c	c_c	c_v
Sharp-edged orifice	0.61	0.62	0.98
Rounded-edged orifice	0.98	1.00	0.98
Short tube $\frac{L}{D} \sim 1$	0.80	1.00	0.80
Borda	0.51	0.52	0.98

Fig. 3.3.36   Nominal coefficients of orifices.

$A\ dh$, neglecting fluid acceleration,

$$Q_{out}\ dt = Ca\sqrt{2gh}\ dt,\ \text{or}\ (Q_{in} - Ca\sqrt{2gh})\ dt$$

$$= A\ dh,\ \text{or}\ \int_{t_1}^{t_2} dt = \int_{h_1}^{h_2} \frac{A\ dh}{Q_{in} - Q_{out}}$$

$$= \int_{h_1}^{h_2} \frac{A\ dh}{Q_{in} - Ca\sqrt{2gh}}$$

EXAMPLE. An open cylindrical tank is 6 ft in diameter and is filled with water to a depth of 10 ft. A 4-in-diameter sharp-edged orifice is installed on the bottom of the tank. A pipe on the top of the tank supplies water at the rate of 1 ft³/s. Estimate (1) the steady-state level of this tank, (2) the time required to reduce the tank level by 2 ft.

1. *Steady-state level.* From Fig. 3.3.36, $C = 0.61$ for a sharp-edged orifice, $a = (\pi/4)d^2 = (\pi/4)(4/12)^2 = 0.08727$ ft². For steady state, $Q_{in} = Q_{out} = Ca\sqrt{2gh} = 1 = (0.61)(0.08727)(2 \times 32.17h)^{1/2}$; $h = 5.484$ ft.

2. Time required to lower level 2 ft, $A = (\pi/4)D^2 = (\pi/4)(6)^2 = 28.27$ ft²

$$t_2 - t_1 = \int_{h_1}^{h_2} \frac{A\ dh}{Q_{in} - Ca\sqrt{2gh}}$$

This equation may be integrated by letting $Q = Ca\sqrt{2g}\ h^{1/2}$; then $dh = 2Q\ dQ/(Ca\sqrt{2g})^2$; then

$$t_2 - t_1 = \frac{2A}{(Ca\sqrt{2g})^2}\left[Q_{in}\log_e\left(\frac{Q_{in} - Q_1}{Q_{in} - Q_2}\right) + Q_1 - Q_2\right]$$

At t_1: $Q_1 = 0.61 \times 0.08727\sqrt{2 \times 32.17 \times 10} = 1.350$ ft³/s
At t_2: $Q_2 = 0.61 \times 0.08727\sqrt{2 \times 32.17 \times 8} = 1.208$ ft³/s

$$t_2 - t_1 = \frac{2 \times 28.27}{(0.61 \times 0.08727\sqrt{2 \times 32.17})^2}$$

$$\times \left[(1)\log_e\left(\frac{1 - 1.350}{1 - 1.208}\right) + 1.350 - 1.208\right]$$

$$t_2 - t_1 = 205.4\ \text{s}$$

WATER HAMMER

Equations Water hammer is the series of shocks, sounding like hammer blows, produced **by suddenly reducing the flow** of a fluid in a pipe. Consider a fluid flowing frictionlessly in a rigid pipe of uniform area A with a velocity V. The pipe has a length L, and inlet pressure p_1 and a pressure p_2 at L. At length L, there is a valve which can suddenly reduce the velocity at L to $V - \Delta V$. The equivalent mass rate of flow of a

pressure wave traveling at sonic velocity c, $\dot{M} = \rho Ac$. From the impulse-momentum equation, $\dot{M}(V_2 - V_1) = p_2 A_2 - p_1 A_1$; for this application, $(\rho Ac)(V - \Delta V - V) = p_2 A - p_1 A$, or the increase in pressure $\Delta p = -\rho c\Delta V$. When the liquid is flowing in an elastic pipe, the equation for pressure rise must be modified to account for the expansion of the pipe; thus

$$c = \sqrt{\frac{E_s}{\rho[1 + (E_s/E_p)(D_o + D_i)/(D_o - D_i)]}}$$

where ρ = mass density of the fluid, E_s = bulk modulus of elasticity of the fluid, E_p = modulus of elasticity of the pipe material, D_o = outside diameter of pipe, and D_i = inside diameter of pipe.

Time of Closure The time for a pressure wave to travel the length of pipe L and return is $t = 2L/c$. If the time of closure $t_c \leq t$, the approximate pressure rise $\Delta p \approx -2\ \rho V(L/t_c)$. When it is not feasible to close the valve slowly, **air chambers** or **surge tanks** may be used to absorb all or most of the pressure rise. Water hammer can be very dangerous. See Sec. 9.9.

EXAMPLE. Water flows at 68°F (20°C) in a 3-in steel schedule 40 pipe at a velocity of 10 ft/s. A valve located 200 ft downstream is suddenly closed. Determine (1) the increase in pressure considering pipe to be rigid, (2) the increase considering pipe to be elastic, and (3) the maximum time of valve closure to be considered "sudden."

For water, $\rho = -1.937$ slugs/ft³ = 1.937 lb · sec²/ft⁴; $E_s = 319,000$ lb/in²; $E_p = 28.5 \times 10^6$ lb/in² (Secs. 5.1 and 6); $c = 4,860$ ft/s; from Sec. 8.7, $D_o = 3.5$ in, $D_i = 3.068$ in.

1. *Inelastic pipe*

$$\Delta p = -\rho c\Delta V = -(1.937)(4,860)(-10) = 94,138\ \text{lbf/ft}^2$$
$$= 94,138/144 = 653.8\ \text{lbf/in}^2\ (4.507 \times 10^6\ \text{N/m}^2)$$

2. *Elastic pipe*

$$c = \sqrt{\frac{E_s}{\rho[1 + (E_s/E_p)(D_o + D_i)/(D_o - D_i)]}}$$

$$= \sqrt{\frac{319,000 \times 144}{1.937\left[1 + \frac{(319,000/28.5 \times 10^6)(3.500 + 3.067)}{(3.500 - 3.067)}\right]}}$$

$$= 4,504$$

$$\Delta p = -(1.937)(4,504)(-10)$$
$$= 87,242\ \text{lbf/ft}^2 = 605.9\ \text{lbf/in}^2\ (4.177 \times 10^6\ \text{N/m}^2)$$

3. *Maximum time for closure*

$$t = 2L/c = 2 \times 200/4,860 = 0.08230\ \text{s or less than 1/10 s}$$

3.4 Vibration
by Leonard Meirovitch

REFERENCES: Harris, "Shock and Vibration Handbook," 3d ed., McGraw-Hill. Thomson, "Theory of Vibration with Applications," 4th ed., Prentice Hall. Meirovitch, "Elements of Vibration Analysis," 2d ed., McGraw-Hill. Meirovitch, "Principles and Techniques of Vibrations," Prentice-Hall.

SINGLE-DEGREE-OF-FREEDOM SYSTEMS

Discrete System Components A **system** is defined as an aggregation of components acting together as one entity. The components of a vibratory mechanical system are of three different types, and they relate forces to displacements, velocities, and accelerations. The component relating forces to displacements is known as a **spring** (Fig. 3.4.1a). For a **linear spring** the force F_s is proportional to the elongation $\delta = x_2 - x_1$, or

$$F_s = k\delta = k(x_2 - x_1) \tag{3.4.1}$$

where k represents the **spring constant**, or the **spring stiffness**, and x_1 and x_2 are the displacements of the end points. The component relating

forces to velocities is called a **viscous damper** or a **dashpot** (Fig. 3.4.1b). It consists of a piston fitting loosely in a cylinder filled with liquid so that the liquid can flow around the piston when it moves relative to the cylinder. The relation between the damper force and the velocity of the piston relative to the cylinder is

$$F_d = c(\dot{x}_2 - \dot{x}_1) \tag{3.4.2}$$

in which c is the **coefficient of viscous damping**; note that dots denote derivatives with respect to time. Finally, the relation between forces and accelerations is given by Newton's second law of motion:

$$F_m = m\ddot{x} \tag{3.4.3}$$

where m is the **mass** (Fig. 3.4.1c).

The spring constant k, coefficient of viscous damping c, and mass m represent physical properties of the components and are the **system parameters**. By implication, these properties are concentrated at points, thus they are **lumped,** or **discrete, parameters.** Note that springs and dampers are assumed to be massless and masses are assumed to be rigid.

Springs can be arranged in parallel and in series. Then, the proportionality constant between the forces and the end points is known as an

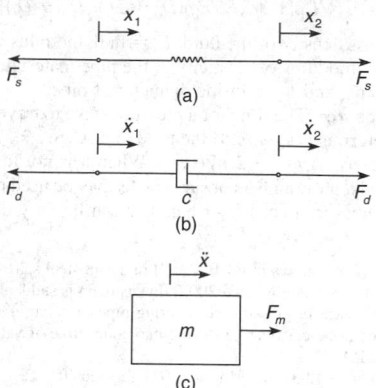

Fig. 3.4.1

Table 3.4.1 Equivalent Spring Constants

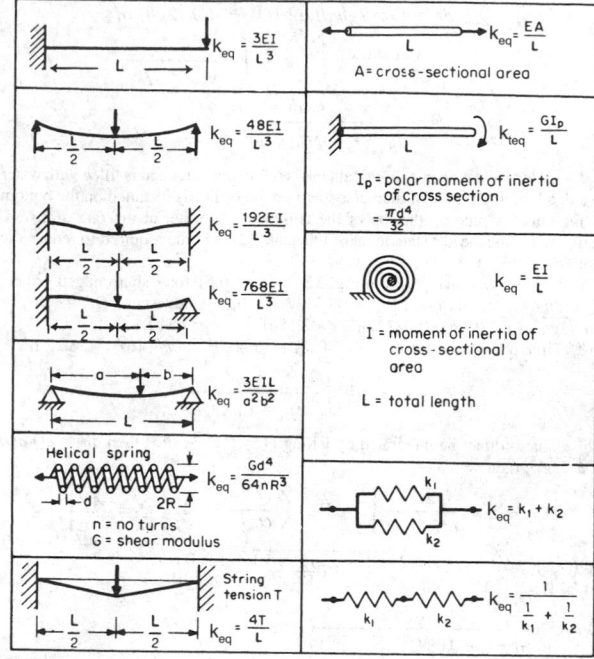

equivalent spring constant and is denoted by k_{eq}, as shown in Table 3.4.1. Certain elastic components, although distributed over a given line segment, can be regarded as lumped with an equivalent spring constant given by $k_{eq} = F/\delta$, where δ is the deflection at the point of application of the force F. A similar relation can be given for springs in torsion. Table 3.4.1 lists the equivalent spring constants for a variety of components.

Equation of Motion The dynamic behavior of many engineering systems can be approximated with good accuracy by the **mass-damper-spring model** shown in Fig. 3.4.2. Using Newton's second law in conjunction with Eqs. (3.4.1) to (3.4.3) and measuring the displacement $x(t)$ from the static equilibrium position, we obtain the differential equation of motion

$$m\ddot{x}(t) + c\dot{x}(t) + kx(t) = F(t) \qquad (3.4.4)$$

which is subject to the **initial conditions** $x(0) = x_0$, $\dot{x}(0) = v_0$, where x_0 and v_0 are the **initial displacement** and **initial velocity,** respectively. Equation (3.4.4) is in terms of a single coordinate, namely $x(t)$; the system of Fig. 3.4.2 is therefore said to be a **single-degree-of-freedom system**.

Free Vibration of Undamped Systems Assuming zero damping and external forces and dividing Eq. (3.4.4) through by m, we obtain

$$\ddot{x} + \omega_n^2 x = 0 \qquad \omega_n = \sqrt{k/m} \qquad (3.4.5)$$

In this case, the vibration is caused by the initial excitations alone. The solution of Eq. (3.4.5) is

$$x(t) = A \cos(\omega_n t - \phi) \qquad (3.4.6)$$

which represents **simple sinusoidal,** or **simple harmonic oscillation** with amplitude A, phase angle ϕ, and **frequency**

$$\omega_n = \sqrt{k/m} \qquad \text{rad/s} \qquad (3.4.7)$$

Systems described by equations of the type (3.4.5) are called **harmonic oscillators.** Because the frequency of oscillation represents an inherent property of the system, independent of the initial excitation, ω_n is called the **natural frequency.** On the other hand, the amplitude and phase angle do depend on the initial displacement and velocity, as follows:

$$A = \sqrt{x_0^2 + (v_0/\omega_n)^2} \qquad \phi = \tan^{-1} v_0/x_0\omega_n \qquad (3.4.8)$$

The time necessary to complete one cycle of motion defines the **period**

$$T = 2\pi/\omega_n \qquad \text{seconds} \qquad (3.4.9)$$

The reciprocal of the period provides another definition of the **natural frequency,** namely,

$$f_n = \frac{1}{T} = \frac{\omega_n}{2\pi} \qquad \text{Hz} \qquad (3.4.10)$$

where Hz denotes *hertz* [1 Hz = 1 cycle per second (cps)].

A large variety of vibratory systems behave like harmonic oscillators, many of them when restricted to small amplitudes. Table 3.4.2 shows a variety of harmonic oscillators together with their respective natural frequency.

Free Vibration of Damped Systems Let $F(t) = 0$ and divide through by m. Then, Eq. (3.4.4) reduces to

$$\ddot{x}(t) + 2\zeta\omega_n\dot{x}(t) + \omega_n^2 x(t) = 0 \qquad (3.4.11)$$

where

$$\zeta = c/2m\omega_n \qquad (3.4.12)$$

is the **damping factor,** a nondimensional quantity. The nature of the motion depends on ζ. The most important case is that in which $0 < \zeta < 1$.

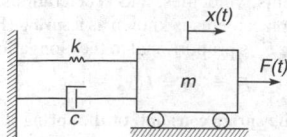

Fig. 3.4.2

Table 3.4.2 Harmonic Oscillators and Natural Frequencies

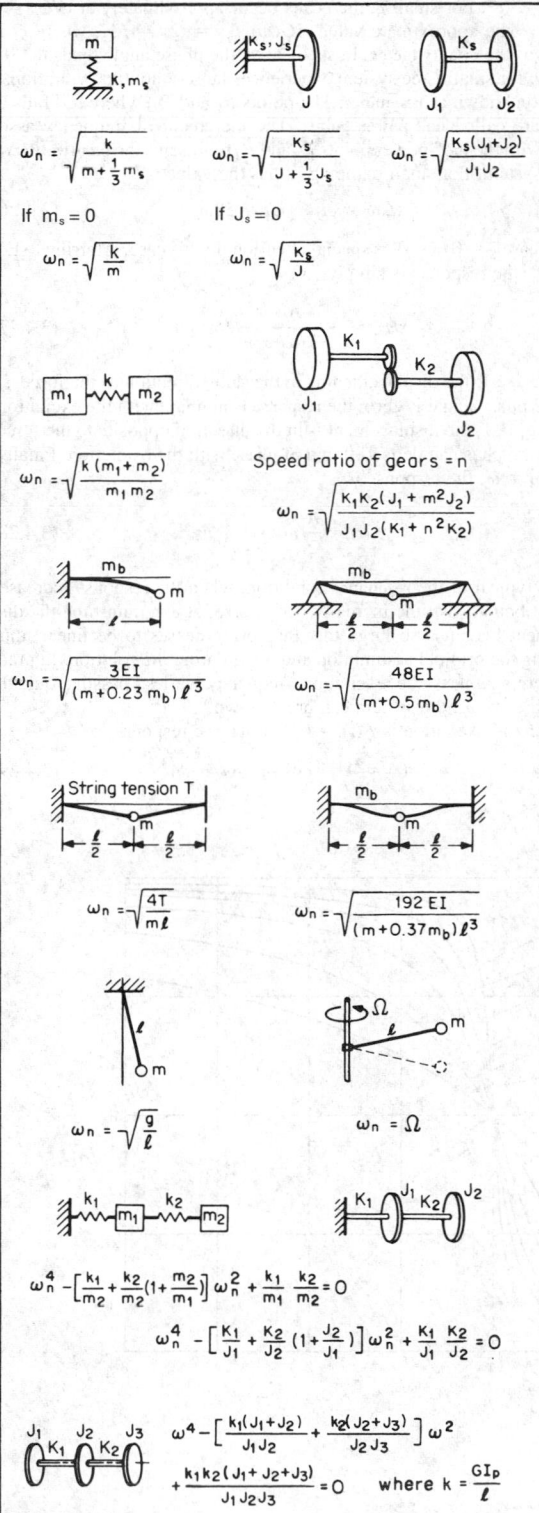

In this case, the system is said to be **underdamped** and the solution of Eq. (3.4.11) is

$$x(t) = Ae^{-\zeta\omega_n t} \cos(\omega_d t - \phi) \qquad (3.4.13)$$

where

$$\omega_d = (1 - \zeta^2)^{1/2}\omega_n \qquad (3.4.14)$$

is the **frequency of damped free vibration** and

$$T = 2\pi/\omega_d \qquad (3.4.15)$$

is the **period of damped oscillation.** The amplitude and phase angle depend on the initial displacement and velocity, as follows:

$$A = \sqrt{x_0^2 + (\zeta\omega_n x_0 + v_0)^2/\omega_d^2} \quad \phi = \tan^{-1}(\zeta\omega_n x_0 + v_0)/x_0\omega_d \quad (3.4.16)$$

The motion described by Eq. (3.4.13) represents **decaying oscillation,** where the term $Ae^{-\zeta\omega_n t}$ can be regarded as a time-dependent amplitude, providing an envelope bounding the harmonic oscillation.

When $\zeta \geq 1$, the solution represents **aperiodic decay.** The case $\zeta = 1$ represents **critical damping,** and

$$c_c = 2m\omega_n \qquad (3.4.17)$$

is the **critical damping coefficient,** although there is nothing critical about it. It merely represents the borderline between oscillatory decay and aperiodic decay. In fact, c_c is the smallest damping coefficient for which the motion is aperiodic. When $\zeta > 1$, the system is said to be **overdamped.**

Logarithmic Decrement Quite often the damping factor is not known and must be determined experimentally. In the case in which the system is underdamped, this can be done conveniently by plotting $x(t)$ versus t (Fig. 3.4.3) and measuring the response at two different times

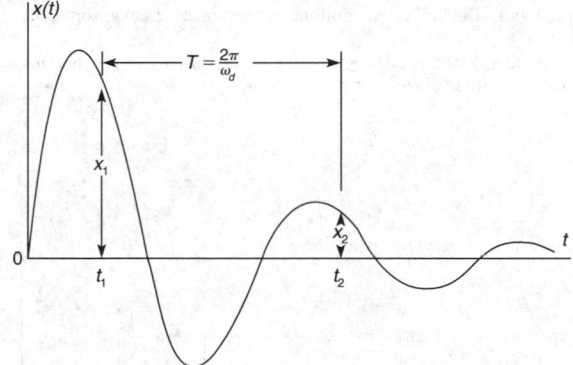

Fig. 3.4.3

separated by a complete period. Let the times be t_1 and $t_1 + T$, introduce the notation $x(t_1) = x_1$, $x(t_1 + T) = x_2$, and use Eq. (3.4.13) to obtain

$$\frac{x_1}{x_2} = \frac{Ae^{-\zeta\omega_n t_1} \cos(\omega_d t_1 - \phi)}{Ae^{-\zeta\omega_n(t_1 + T)} \cos[\omega_d(t_1 + T) - \phi]} = e^{\zeta\omega_n T} \quad (3.4.18)$$

where $\cos[\omega_d(t_1 + T) - \phi] = \cos(\omega_d t_1 - \phi + 2\pi) = \cos(\omega_d t_1 - \phi)$. Equation (3.4.18) yields the **logarithmic decrement**

$$\delta = \ln\frac{x_1}{x_2} = \zeta\omega_n T = \frac{2\pi\zeta}{\sqrt{1 - \zeta^2}} \qquad (3.4.19)$$

which can be used to obtain the damping factor

$$\zeta = \frac{\delta}{\sqrt{(2\pi)^2 + \delta^2}} \qquad (3.4.20)$$

For small damping, the logarithmic decrement is also small, and the damping factor can be approximated by

$$\zeta \approx \frac{\delta}{2\pi} \qquad (3.4.21)$$

Response to Harmonic Excitations Consider the case in which the excitation force $F(t)$ in Eq. (3.4.4) is harmonic. For convenience, express $F(t)$ in the form $kA \cos \omega t$, where k is the spring constant, A is an amplitude with units of displacement and ω is the **excitation frequency**. When divided through by m, Eq. (3.4.4) has the form

$$\ddot{x} + 2\zeta\omega_n\dot{x} + \omega_n^2 x = \omega_n^2 A \cos \omega t \qquad (3.4.22)$$

The solution of Eq. (3.4.22) can be expressed as

$$x(t) = A|G(\omega)| \cos (\omega t - \phi) \qquad (3.4.23)$$

where

$$|G(\omega)| = \frac{1}{\sqrt{[1 - (\omega/\omega_n)^2]^2 + (2\zeta\omega/\omega_n)^2}} \qquad (3.4.24)$$

is a **nondimensional magnitude factor*** and

$$\phi(\omega) = \tan^{-1} \frac{2\zeta\omega/\omega_n}{1 - (\omega/\omega_n)^2} \qquad (3.4.25)$$

is the phase angle; note that both the magnitude factor and phase angle depend on the excitation frequency ω.

Equation (3.4.23) shows that the response to harmonic excitation is also harmonic and has the same frequency as the excitation, but different amplitude $A|G(\omega)|$ and phase angle $\phi(\omega)$. Not much can be learned by plotting the response as a function of time, but a great deal of information can be gained by plotting $|G|$ versus ω/ω_n and ϕ versus ω/ω_n. They are shown in Fig. 3.4.4 for various values of the damping factor ζ.

In Fig. 3.4.4, for low values of ω/ω_n, the nondimensional magnitude factor $|G(\omega)|$ approaches unity and the phase angle $\phi(\omega)$ approaches zero. For large values of ω/ω_n, the magnitude approaches zero (see accompanying footnote about magnification factor) and the phase angle approaches 180°. The magnitude experiences peaks for $\omega/\omega_n =$

---

* The term $|G(\omega)|$ is often referred to as *magnification factor*, but this is a misnomer, as we shall see shortly.

$\sqrt{1 - 2\zeta^2}$, provided $\zeta < 1/\sqrt{2}$. The peak values are $|G(\omega)|_{max} = 1/2\zeta\sqrt{1 - \zeta^2}$. For small ζ, the peaks occur approximately at $\omega/\omega_n = 1$ and have the approximate values $|G(\omega)|_{max} = Q \approx 1/2\zeta$, where Q is known as the **quality factor**. In such cases, the phase angle tends to 90°. Clearly, for small ζ the system experiences large-amplitude vibration, a condition known as **resonance**. The points P_1 and P_2, where $|G|$ falls to $Q/\sqrt{2}$, are called **half-power points**. The increment of frequency associated with the half-power points P_1 and P_2 represents the **bandwidth** $\Delta\omega$ of the system. For small damping, it has the value

$$\Delta\omega = \omega_2 - \omega_1 \approx 2\zeta\omega_n \qquad (3.4.26)$$

The case $\zeta = 0$ deserves special attention. In this case, referring to Eq. (3.4.22), the response is simply

$$x(t) = \frac{A}{1 - (\omega/\omega_n)^2} \cos \omega t \qquad (3.4.27)$$

For $\omega/\omega_n < 1$, the displacement is in the same direction as the force, so that the phase angle is zero; the **response is in phase** with the excitation. For $\omega/\omega_n > 1$, the displacement is in the direction opposite to the force, so that the phase angle is 180° **out of phase** with the excitation. Finally, when $\omega = \omega_n$ the response is

$$x(t) = \frac{A}{2} \omega_n t \sin \omega_n t \qquad (3.4.28)$$

This is typical of the **resonance** condition, when the response increases without bounds as time increases. Of course, at a certain time the displacement becomes so large that the spring ceases to be linear, thus violating the original assumption and invalidating the solution. In practical terms, unless the excitation frequency varies, passing quickly through $\omega = \omega_n$, the system can break down.

When the excitation is $F(t) = kA \sin \omega t$, the response is

$$x(t) = A|G(\omega)| \sin (\omega t - \phi) \qquad (3.4.29)$$

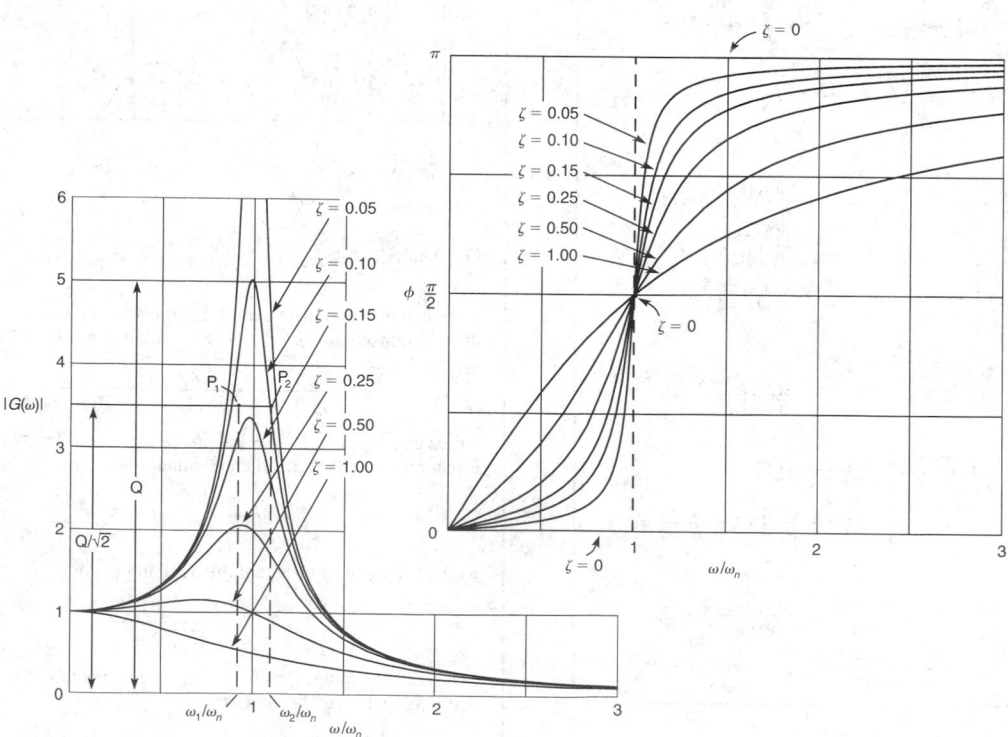

Fig. 3.4.4 Frequency response plots.

One concludes that in **harmonic response,** time plays a secondary role to the frequency. In fact, the only significant information is extracted from the magnitude and phase angle plots of Fig. 3.4.4. They are referred to as **frequency-response plots.**

Since time plays no particular role, the harmonic response is called **steady-state response.** In general, for linear systems with constant parameters, such as the mass-damper-spring system under consideration, the response to the initial excitations is added to the response to the excitation forces. The response to initial excitations, however, represents **transient response.** This is due to the fact that every system possesses some amount of damping, so that the response to initial excitations disappears with time. In contrast, steady-state response persists with time. Hence, in the case of harmonic excitations, it is meaningless to add the response to initial excitations to the harmonic response.

Vibration Isolation A problem of great interest is the magnitude of the force transmitted to the base by a system of the type shown in Fig. 3.4.2 subjected to harmonic excitation. This force is a combination of the spring force kx and the dashpot force $c\dot{x}$. Recalling Eq. (3.4.23), write

$$kx = kA|G|\cos(\omega t - \phi)$$
$$c\dot{x} = -c\omega A|G|\sin(\omega t - \phi)$$
$$= c\omega A|G|\cos\left(\omega t - \phi + \frac{\pi}{2}\right) \qquad (3.4.30)$$

so that the dashpot force is 90° out of phase with the spring force. Hence, the magnitude of the force is

$$F_{tr} = \sqrt{(kA|G|)^2 + (c\omega A|G|)^2} = kA|G|\sqrt{1 + (c\omega/k)^2}$$
$$= kA|G|\sqrt{1 + (2\zeta\omega/\omega_n)^2} \qquad (3.4.31)$$

Let the magnitude of the harmonic excitation be $F_0 = kA$; the force transmitted to the base is then

$$T = \frac{F_{tr}}{F_0} = |G|\sqrt{1 + (2\zeta\omega/\omega_n)^2}$$
$$= \sqrt{\frac{1 + (2\zeta\omega/\omega n)^2}{[1 - (\omega/\omega_n)^2]^2 + (2\zeta\omega/\omega_n)^2}} \qquad (3.4.32)$$

which represents a nondimensional ratio called **transmissibility.** Figure 3.4.5 plots F_{tr}/F_0 versus ω/ω_n for various values of ζ.

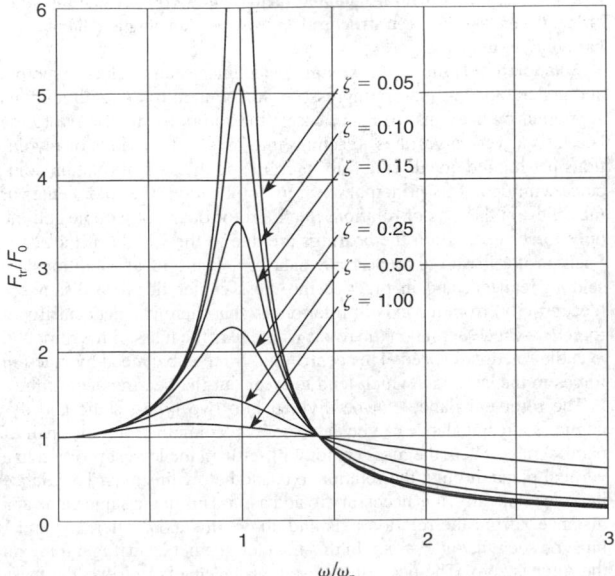

Fig. 3.4.5

The transmissibility is less than 1 for $\omega/\omega_n > \sqrt{2}$, and decreases as ω/ω_n increases. Hence, for **an isolator to perform well,** its natural frequency must be much smaller than the excitation frequency. However, for very low natural frequencies, difficulties can be encountered in isolator design. Indeed, the natural frequency is related to the static deflection δ_{st} by $\omega_n = \sqrt{k/m} = \sqrt{g/\delta_{st}}$, where g is the gravitational constant. For the natural frequency to be sufficiently small, the static deflection may have to be impractically large. The relation between the excitation frequency f measured in rotations per minute and the static deflection δ_{st} measured in inches is

$$f = 187.7\sqrt{\frac{2 - R}{\delta_{st}(1 - R)}} \qquad \text{rpm} \qquad (3.4.33)$$

where $R = 1 - T$ **represents the percent reduction** in vibration. Figure 3.4.6 shows a logarithmic plot of f versus δ_{st} with R as a parameter.

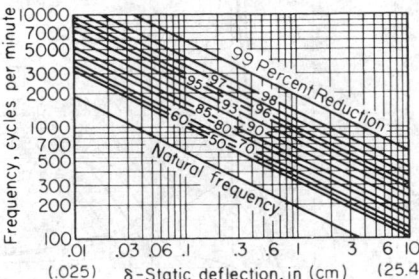

Fig. 3.4.6

Figure 3.4.7 depicts two types of isolators. In Fig. 3.4.7a, isolation is accomplished by means of springs and in Fig. 3.4.7b by rubber rings supporting the bearings. Isolators of all shapes and sizes are available commercially.

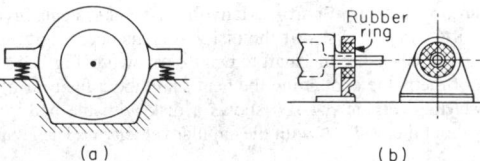

Fig. 3.4.7

Rotating Unbalanced Masses Many appliances, machines, etc., involve components spinning relative to a main body. A typical example is the clothes dryer. Under certain circumstances, the mass of the spinning component is not symmetric relative to the center of rotation, as when the clothes are not spread uniformly in the spinning drum, giving rise to harmonic excitation. The behavior of such systems can be simulated adequately by the single-degree-of-freedom model shown in Fig. 3.4.8, which consists of a main mass $M - m$, supported by two springs of combined stiffness k and a dashpot with coefficient of viscous damping c, and two eccentric masses $m/2$ rotating in opposite sense with the constant angular velocity ω. Although there are three masses, the motion of the eccentric masses relative to the main mass is prescribed, so that there is only one degree of freedom. The equation of motion for the system is

$$M\ddot{x} + c\dot{x} + kx = m\omega^2\sin\omega t \qquad (3.4.34)$$

Fig. 3.4.8

Using the analogy with Eq. (3.4.29), the solution of Eq. (3.4.34) is

$$x(t) = \frac{m}{M} l \left(\frac{\omega}{\omega_n}\right)^2 |G(\omega)| \sin(\omega t - \phi) \qquad \omega_n^2 = \frac{k}{M} \qquad (3.4.35)$$

The magnitude factor in this case is $(\omega/\omega_n)^2|G(\omega)|$, where $|G(\omega)|$ is given by Eq. (3.4.24); it is plotted in Fig. 3.4.9. On the other hand, the phase angle remains as in Fig. 3.4.4.

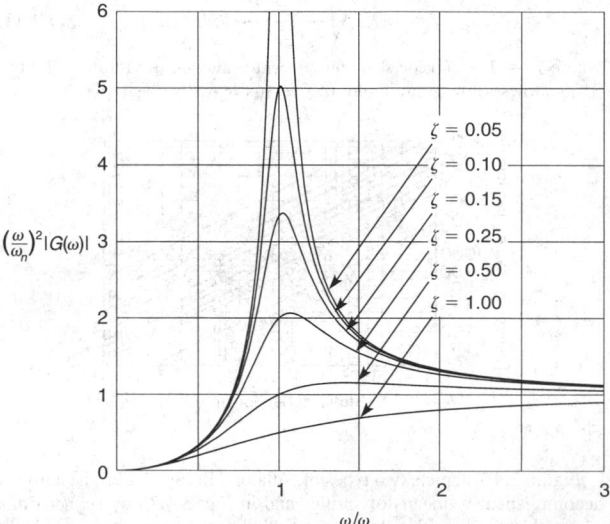

Fig. 3.4.9

Whirling of Rotating Shafts Many mechanical systems involve rotating shafts carrying disks. If the disk has some eccentricity, then the centrifugal forces cause the shaft to bend, as shown in Fig. 3.4.10a. The rotation of the plane containing the bent shaft about the bearing axis is called **whirling**. Figure 3.4.10b shows a disk with the body axes x,y rotating about the origin O with the angular velocity ω. The geometrical

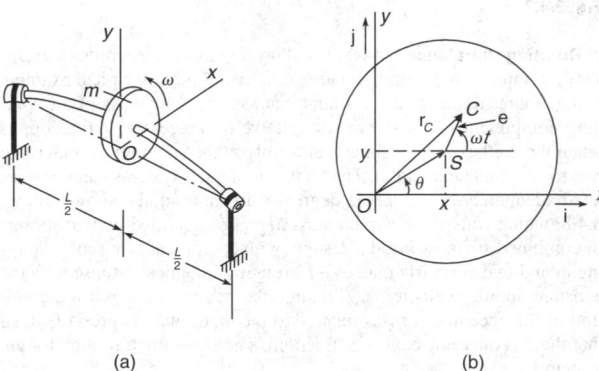

(a)   (b)

Fig. 3.4.10

center of the disk is denoted by S and the mass center by C. The distance between the two points is the eccentricity e. The shaft is massless and of stiffness k_{eq} and the disk is rigid and of mass m. The x and y components of the displacement of S relative to O are independent from one another and, for no damping, satisfy the equations of motion

$$\ddot{x} + \omega_n^2 x = e\omega^2 \cos \omega t \qquad \ddot{y} + \omega_n^2 y = e\omega^2 \sin \omega t \qquad \omega_n^2 = k_{eq}/m$$
$$(3.4.36)$$

where, assuming that the shaft is simply supported (see Table 3.4.1), $k_{eq} = 48EI/L^3$, in which E is the modulus of elasticity, I the cross-sectional area moment of inertia, and L the length of the shaft. By analogy with Eq. (3.4.27), Eqs. (3.4.36) have the solution

$$x(t) = \frac{e(\omega/\omega_n)^2}{1 - (\omega/\omega_n)^2} \cos \omega t \qquad y(t) = \frac{e(\omega/\omega_n)^2}{1 - (\omega/\omega_n)^2} \sin \omega t \qquad (3.4.37)$$

Clearly, resonance occurs when the whirling angular velocity coincides with the natural frequency. In terms of rotations per minute, it has the value

$$f_c = \frac{60}{2\pi} \omega_n = \frac{60}{2\pi} \sqrt{\frac{48EI}{mL^3}} \qquad \text{rpm} \qquad (3.4.38)$$

where f_c is called the **critical speed**.

Structural Damping Experience shows that energy is dissipated in all real systems, including those assumed to be undamped. For example, because of internal friction, energy is dissipated in real springs undergoing cyclic stress. This type of damping is called **structural damping** or **hysteretic damping** because the energy dissipated in one cycle of stress is equal to the area inside the hysteresis loop. Systems possessing structural damping and subjected to harmonic excitation with the frequency ω can be treated as if they possess viscous damping with the equivalent coefficient

$$c_{eq} = \alpha/\pi\omega \qquad (3.4.39)$$

where α is a material constant. In this case, the equation of motion is

$$m\ddot{x} + \frac{\alpha}{\pi\omega}\dot{x} + kx = kA \cos \omega t \qquad (3.4.40)$$

The solution of Eq. (3.4.40) is

$$x(t) = A|G| \cos(\omega t - \phi) \qquad (3.4.41)$$

where this time the magnitude factor and phase angle have the values

$$G = \frac{1}{\sqrt{[1 - (\omega/\omega_n)^2]^2 + \gamma^2}} \qquad \phi = \tan^{-1}\frac{\gamma\omega_n^2}{\omega[1 - (\omega/\omega_n)^2]} \qquad (3.4.42)$$

in which

$$\gamma = \frac{\alpha}{\pi k} \qquad (3.4.43)$$

is known as the **structural damping factor**. One word of caution is in order: **the analogy between structural and viscous damping is valid only for harmonic excitation**.

Balancing of Rotating Machines Machines such as electric motors and generators, turbines, compressors, etc. contain rotors with journals supported by bearings. In many cases, the rotors rotate relative to the bearings at very high rates, reaching into tens of thousands of revolutions per minute. Ideally the rotor is rigid and the axis of rotation coincides with one of its principal axes; by implication, the rotor center of mass lies on the axis of rotation. Such a rotor does not wobble and the only forces exerted on the bearings are due to the weight of the rotor. Such a rotor is said to be **perfectly balanced**. These ideal conditions are seldom realized, and in practice the mass center lies at a distance e (eccentricity) from the axis of rotation, so that there is a net centrifugal force $F = me\omega^2$ acting on the rotor, where m is the mass of the rotor and ω is the rotational speed. This centrifugal force is balanced by reaction forces in the bearings, which tend to wear out the bearings with time.

The rotor unbalance can be divided into two types, static and dynamic. Static unbalance can be detected by placing the rotor on a pair of parallel rails. Then, the mass center will settle in the lowest position in a vertical plane through the rotation axis and below this axis. To balance the rotor statically, it is necessary to add a mass m' in the same plane at a distance r from the rotation axis and above this axis, where m' and r must be such that $m'r = me$. In this manner, the net centrifugal force on the rotor is zero. The net result of **static balancing** is to cause the mass center to coincide with the rotation axis, so that the rotor will remain in any position placed on the rails. However, unless the mass m' is placed on a line containing m and at right angles with the bearings axis, the centrifugal forces on m and m' will form a couple (Fig. 3.4.11). Static balancing is suitable when the rotor is in the form of a thin disk, in which case the couple tends to be small. Automobile tires are at times balanced statically (seems), although strictly speaking they are neither thin nor rigid.

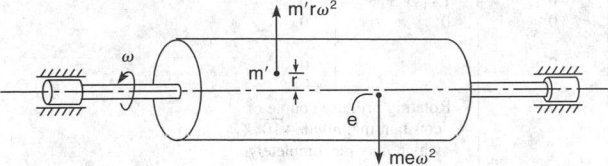

Fig. 3.4.11

In general, for practical reasons, the mass m' cannot be placed on an axis containing m and perpendicular to the bearing axis. Hence, although in static balancing the mass center lies on the rotation axis, the rotor principal axis does not coincide with the bearing axis, as shown in Fig. 3.4.12, causing the rotor to wobble during rotation. In this case, the rotor is said to be **dynamically unbalanced**. Clearly, it is highly desirable to place the mass m' so that the rotor is both statically and dynamically balanced. In this regard, note that the end planes of the rotor are conve-

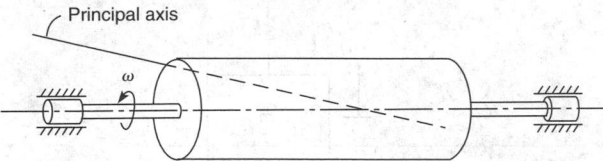

Fig. 3.4.12

nient locations to place correcting masses. In Fig. 3.4.13, if the mass center is at a distance a from the right end, then **dynamic balance** can be achieved by placing masses $m'a/L$ and $m'(L-a)/L$ on the intersection of the plane of unbalance and the rotor left end plane and right end plane, respectively. In this manner, the resultant centrifugal force is zero

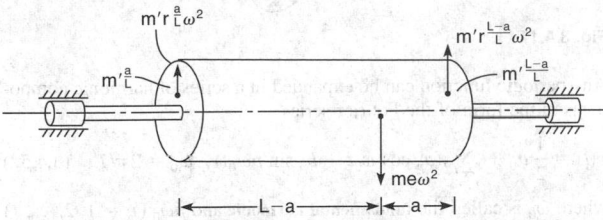

Fig. 3.4.13

and the two couples thus created are equal in value to $m'a(L-a)\omega^2/L$ and opposite in sense, so that they cancel each other. This results in a rotor completely balanced, i.e., balanced statically and dynamically.

The task of determining the magnitude and position of the unbalance is carried out by means of a balancing machine provided with elastically supported bearings permitting the rotor to spin (Fig. 3.4.14). The unbalance causes the bearings to oscillate laterally so that electrical pickups and stroboflash light can measure the amplitude and phase of the rotor with respect to an arbitrary rotor.

In cases in which the rotor is very long and flexible, the position of the unbalance depends on the elastic configuration of the rotor, which in turn depends on the speed of rotation, temperature, etc. In such cases, it is necessary to balance the rotor under normal operating conditions by means of a portable balancing instrument.

Inertial Unbalance of Reciprocating Engines   The crank-piston mechanism of a reciprocating engine produces dynamic forces capable of causing undesirable vibrations. Rotating parts, such as the crank-

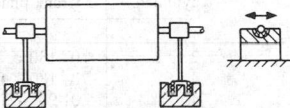

Fig. 3.4.14

shaft, can be balanced. However, translating parts, such as the piston, cannot be easily balanced, and the same can be said about the connecting rod, which executes a more complex motion of combined rotation and translation.

In the calculation of the unbalanced forces in a single-cylinder engine, the mass of the moving parts is divided into a reciprocating mass and a rotating mass. This is done by apportioning some of the mass of the connecting rod to the piston and some to the crank end. In general, this division of the connecting rod into two lumped masses tends to cause errors in the moment of inertia, and hence in the torque equation. On the other hand, the force equation can be regarded as being accurate. (See also Sec. 8.2.)

Assuming that the rotating mass is counterbalanced, only the reciprocating mass is of concern, and the inertia force for a single-cylinder engine is

$$F = m_{\text{rec}}r\omega^2 \cos \omega t + m_{\text{rec}}\frac{r^2}{L} \omega^2 \cos 2\omega t \qquad (3.4.44)$$

where m_{rec} is the reciprocating mass, r the radius of the crank, ω the angular velocity of the crank, and L the length of the connecting rod. The first component on the right side, which alternates once per revolution, is denoted by X_1 and referred to as the **primary force**, and the second component, which is smaller and alternates twice per revolution, is denoted by X_2 and is called the **secondary force**.

In addition to the inertia force, there is an unbalanced torque about the crankshaft axis due to the reciprocating mass. However, this torque is considered together with the torque created by the power stroke, and the torsional oscillations resulting from these excitations can be mitigated by means of a pendulum-type absorber (see "Centrifugal Pendulum Vibration Absorbers" below) or a torsional damper.

The analysis for the single-cylinder engine can be extended to multi-cylinder in-line and V-block engines by superposition. For the in-line engine or one block of the V engine, the inertia force becomes

$$F = m_{\text{rec}}r\omega^2 \sum_{j=1}^{n} \cos (\omega t + \phi_j)$$
$$+ m_{\text{rec}}\frac{r^2}{L} \omega^2 \sum_{j=1}^{n} \cos 2(\omega t + \phi_j) \quad (3.4.45)$$

where ϕ_j is a phase angle corresponding to the crank position associated with cylinder j and n is the number of cylinders. The vibration's force can be eliminated by proper spacing of the angular positions ϕ_j ($j = 1, 2, \ldots, n$).

Even if $F = 0$, there can be pitching and yawing moments due to the spacing of the cylinders. Table 3.4.3 gives the inertial unbalance and pitching of the primary and secondary forces for various crank-angle arrangements of n-cylinder engines.

Centrifugal Pendulum Vibration Absorbers   For a rotating system, such as the crank mechanism just discussed, the exciting torques are proportional to the rotational speed ω, which varies over a wide range. Hence, for a vibration absorber to be effective, its natural frequency must be proportional to ω. The centrifugal pendulum shown in Fig. 3.4.15 is ideally suited to this task. Strictly speaking, the system of Fig. 3.4.15 represents a two-degree-of-freedom nonlinear system. However, assuming that the motion of the wheel consists of a steady rotation ω and a small harmonic oscillation at the frequency Ω, or

$$\theta(t) = \omega t + \theta_0 \sin \Omega t \qquad (3.4.46)$$

Table 3.4.3 Inertial Unbalance of Four-Stroke-per-Cycle Engines

No. n of cylinders	Crank phase angle ϕ_j	Unbalanced forces		Unbalanced pitching moments about 1st cylinder	
		Primary	Secondary	Primary	Secondary
1		X_1	X_2	—	—
2	$0-180°$	0	$2X_2$	ℓX_1	$2\ell X_2$
4	$0-180°-180°-0$	0	$4X_2$	0	$6\ell X_2$
4	$0-90°-270°-180°$	0	0	$\ell X_1\sqrt{1+3^2}$	0
6	$0-120°-240°-240°-120°-0$	0	0	0	0
8	$0-180°-90°-270°-270°-90°-180°-0$	0	0	0	0
90° V-8	$0-90°-270°-180°$	0	0	Rotating primary couple of constant magnitude $\sqrt{10}\ell X_1$ which may be completely counterbalanced	

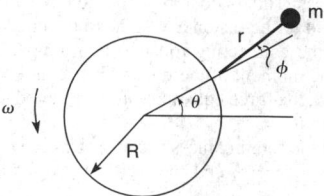

Fig. 3.4.15

and that the pendulum angle ϕ is relatively small, then the equation of motion of the pendulum reduces to the linear single-degree-of-freedom system

$$\ddot{\phi} + \omega_n^2 \phi = \frac{R+r}{r}\Omega^2\theta_0 \sin \Omega t \qquad (3.4.47)$$

where

$$\omega_n = \omega\sqrt{R/r} \qquad (3.4.48)$$

is the natural frequency of the pendulum. The torque exerted by the pendulum on the wheel is

$$T = -\frac{m(R+r)^2}{1 - r\Omega^2/R\omega^2}\ddot{\theta} \qquad (3.4.49)$$

so that the system behaves like a wheel with the effective mass moment of inertia

$$J_{\text{eff}} = -\frac{m(R+r)^2}{1 - r\Omega^2/R\omega^2} \qquad (3.4.50)$$

which becomes infinite when Ω is equal to the natural frequency ω_n. To suppress disturbing torques of frequency Ω several times larger than the rotational speed ω, the ratio r/R must be very small, which requires a short pendulum. The **bifilar pendulum** depicted in Fig. 3.4.16, which consists of a U-shaped counterweight that fits loosely and rolls on two pins of radius r_2 within two larger holes of equal radius r_1, represents a suitable design whereby the effective pendulum length is $r = r_1 - r_2$.

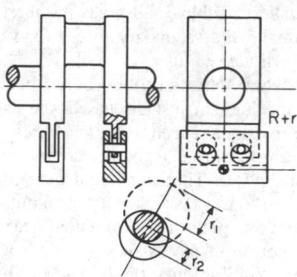

Fig. 3.4.16

Response to Periodic Excitations A problem of interest in mechanical vibrations concerns the response $x(t)$ of the cam and follower system shown in Fig. 3.4.17. As the cam rotates at a constant angular rate, the follower undergoes the periodic displacement $y(t)$, where $y(t)$ has the period T. The equation of motion is

$$m\ddot{x} + (k_1 + k_2)x = k_2 y \qquad (3.4.51)$$

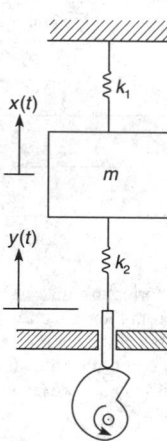

Fig. 3.4.17

Any periodic function can be expanded in a series of harmonic components in the form of the Fourier series

$$y(t) = \frac{1}{2}a_0 + \sum_{p=1}^{\infty}(a_p \cos \omega_0 t + b_p \sin p\omega_0 t) \quad \omega_0 = 2\pi/T \quad (3.4.52)$$

where ω_0 is called the **fundamental harmonic** and $p\omega_0$ ($p = 1, 2, \ldots$) are called **higher harmonics**, in which p is an integer. The coefficients have the expressions

$$a_p = \frac{2}{T}\int_0^T y(t)\cos p\omega_0 t \qquad p = 0, 1, 2, \ldots$$

$$b_p = \frac{2}{T}\int_0^T y(t)\sin p\omega_0 t \qquad p = 1, 2, \ldots \qquad (3.4.53)$$

Note that the limits of integration can be changed, as long as the integration covers one complete period. From Eq. (3.4.27), and a companion equation for the sine counterpart, the response is

$$x(t) = \frac{k_2}{k_1 + k_2}\left[\frac{1}{2}a_0 + \sum_{p=1}^{\infty}\frac{1}{1 - (p\omega_0/\omega_n)^2}\right.$$
$$\left. \times (a_p \cos p\omega_0 t + b_p \sin p\omega_0 t)\right] \qquad (3.4.54)$$

where

$$\omega_n = \sqrt{(k_1 + k_2)/m} \qquad (3.4.55)$$

is the natural frequency of the system. Equation (3.5.54) describes a steady-state response, so that a description in terms of time is not very informative. More significant information can be extracted by plotting the amplitudes of the harmonic components versus the harmonic number. Such plots are called **frequency spectra,** and there is one for the excitation and one for the response. Equation (3.4.54) leads to the conclusion that **resonance occurs for** $p\omega_0 = \omega_n$.

As an example, consider the periodic excitation shown in Fig. 3.4.18 and use Eqs. (3.4.53) to obtain the coefficients

$$a_0 = 2A, \quad a_p = 0, \quad b_p = \begin{cases} 4B/p\pi & p \text{ odd} \\ 0 & p \text{ even} \end{cases} \quad (3.4.56)$$

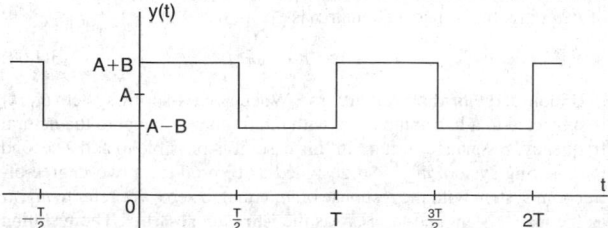

Fig. 3.4.18 Example of periodic excitation.

The excitation and response frequency spectra are displayed in Figs. 3.4.19a and b, the latter for the case in which $\omega_n = 4\omega_0$.

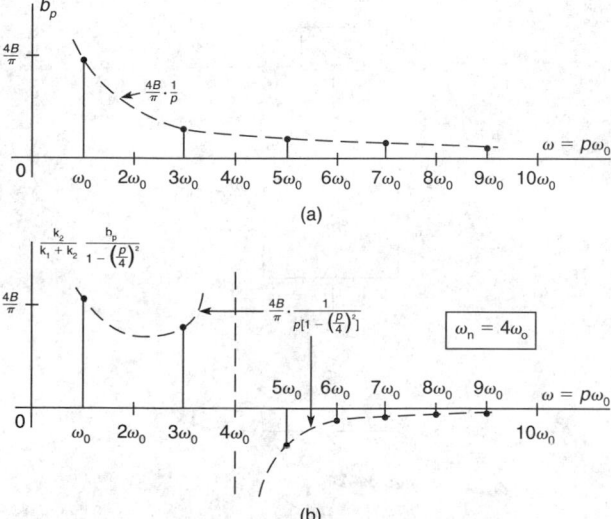

Fig. 3.4.19 (a) Excitation frequency spectrum; (b) response frequency spectrum for the periodic excitation of Fig. 3.4.18.

Unit Impulse and Impulse Response Harmonic and periodic forces represent **steady-state excitations** and **persist indefinitely.** The response to such forces is also steady state. An entirely different class of forces consists of **arbitrary,** or **transient, forces.** The term *transient* is not entirely appropriate, as some of these forces can also persist indefinitely. Concepts pivotal to the response to arbitrary forces are the unit impulse and the impulse response. The **unit impulse,** denoted by $\delta(t - a)$, represents a function of very high amplitude and defined over a very small time interval at $t = a$ such that the area enclosed is equal to 1 (Fig. 3.4.20). The **impulse response,** denoted by $g(t)$, is defined as the response of a system to a unit impulse applied at $t = 0$, with the initial conditions

being equal to zero. For the mass-damper-spring system of Fig. 3.4.2, the impulse response is

$$g(t) = \frac{1}{m\omega_d} e^{-\zeta\omega_n t} \sin \omega_d t \qquad t > 0 \qquad (3.4.57)$$

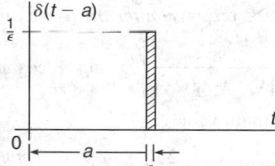

Fig. 3.4.20

Convolution Integral An arbitrary force $F(t)$ as shown in Fig. 3.4.21 can be regarded as a superposition of impulses of magnitude $F(\tau)\,d\tau$ and applied at $t = \tau$. Hence, the response to an arbitrary force can be regarded as a superposition of impulse responses $g(t - \tau)$ of magnitude $F(\tau)\,d\tau$, or

$$\begin{aligned} x(t) &= \int_0^t F(\tau)g(t - \tau)\,d\tau \\ &= \frac{1}{m\omega_d} \int_0^t F(\tau)e^{-\zeta\omega_n(t-\tau)} \sin \omega_d(t - \tau)\,d\tau \quad (3.4.58a) \end{aligned}$$

which is called the **convolution integral** or the **superposition integral;** it can also be written in the form

$$\begin{aligned} x(t) &= \int_0^t F(t - \tau)g(\tau)\,d\tau \\ &= \frac{1}{m\omega_d} \int_0^t F(t - \tau)e^{-\zeta\omega_n\tau} \sin \omega_d\tau\,d\tau \quad (3.4.58b) \end{aligned}$$

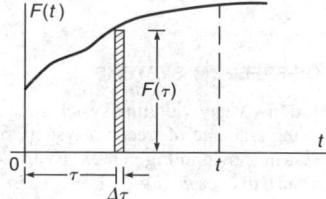

Fig. 3.4.21

Shock Spectrum Many systems are subjected on occasions to large forces applied suddenly and over periods of time that are short compared to the natural period. Such forces are capable of inflicting serious damage on a system and are referred to as **shocks.** The severity of a shock is commonly measured in terms of the maximum value of the response of a mass-spring system. The plot of the peak response versus the natural frequency is called the **shock spectrum** or **response spectrum.**

A shock $F(t)$ is characterized by its maximum value F_0, its duration T, and its shape. It is common to approximate the force by the half-sine pulse

$$F(t) = \begin{cases} F_0 \sin \omega t & \text{for } 0 < t < T = \pi/\omega \\ 0 & \text{for } t < 0 \text{ and } t > T \end{cases} \quad (3.4.59)$$

Using the convolution integral, Eq. (3.4.58b) with $\zeta = 0$, the response of a mass-spring system *during the duration of the pulse* is

$$x(t) = \frac{F_0}{k[1 - (\omega/\omega_n)^2]} \left(\sin \omega t - \frac{\omega}{\omega_n} \sin \omega_n t \right)$$
$$0 < t < \pi/\omega \quad (3.4.60)$$

The maximum response is obtained when $\dot{x} = 0$ and has the value

$$x_{\max} = \frac{F_0}{k(1 - \omega/\omega_n)} \sin \frac{2i\pi}{1 + \omega_n/\omega}$$

$$i = 1, 2, \ldots ; i < \frac{1}{2}\left(1 + \frac{\omega_n}{\omega}\right) \quad (3.4.61)$$

On the other hand, the response *after the termination of the pulse* is

$$x(t) = \frac{F_0\omega_n^2/\omega}{k[1 - (\omega_n/\omega)^2]} [\cos \omega_n t + \cos \omega_n(t - T)] \quad (3.4.62)$$

which has the maximum value

$$x_{\max} = \frac{2 F_0 \omega_n/\omega}{k[1 - (\omega_n/\omega)^2]} \cos \frac{\pi\omega_n}{2\omega} \quad (3.4.63)$$

The shock spectrum is the plot $x_{\max}$ versus ω_n/ω. For $\omega_n < \omega$, the maximum response is given by Eq. (3.4.63) and for $\omega_n > \omega$ by Eq. (3.4.61). The shock spectrum is shown in Fig. 3.4.22 in the form of the nondimensional plot $x_{\max}k/F_0$ versus ω_n/ω.

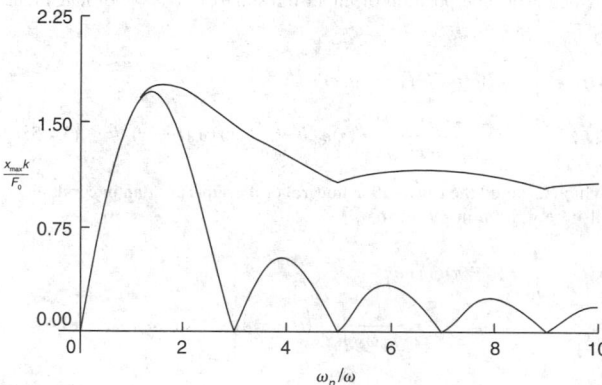

Fig. 3.4.22

MULTI-DEGREE-OF-FREEDOM SYSTEMS

Equations of Motion Many vibrating systems require more elaborate models than a single-degree-of-freedom system, such as the **multi-degree-of-freedom system** shown in Fig. 3.4.23. By using Newton's second law for each of the n masses m_i ($i = 1, 2, \ldots, n$), the equations of motion can be written in the form

$$m_i\ddot{x}_i(t) + \sum_{j=1}^{n} c_{ij}\dot{x}_j(t) + \sum_{j=1}^{n} k_{ij}x_j(t) = F_i(t)$$

$$i = 1, 2, \ldots, n \quad (3.4.64)$$

where $x_i(t)$ is the displacement of mass m_i, $F_i(t)$ is the force acting on m_i, and c_{ij} and k_{ij} are **damping** and **stiffness coefficients**, respectively. The matrix form of Eqs. (3.4.64) is

$$M\ddot{\mathbf{x}}(t) + C\dot{\mathbf{x}}(t) + K\mathbf{x}(t) = \mathbf{F}(t) \quad (3.4.65)$$

in which $\mathbf{x}(t)$ is the n-dimensional displacement vector, $\mathbf{F}(t)$ the corresponding force vector, M the **mass matrix**, C the **damping matrix**, and K

the **stiffness matrix**, all three symmetric matrices. (In the present case the mass matrix is diagonal, but in general it is not, although it is symmetric.)

Response of Undamped Systems to Harmonic Excitations Let the harmonic excitation have the form

$$\mathbf{F}(t) = \mathbf{F}_0 \sin \omega t \quad (3.4.66)$$

where $\mathbf{F}_0$ is a constant vector and ω is the excitation, or driving frequency. The response to the harmonic excitation is a **steady-state** response and can be expressed as

$$\mathbf{x}(t) = Z^{-1}(\omega)\mathbf{F}_0 \sin \omega t \quad (3.4.67)$$

where $Z^{-1}(\omega)$ is the inverse of the **impedance matrix** $Z(\omega)$. In the absence of damping, the impedance matrix is

$$Z(\omega) = K - \omega^2 M \quad (3.4.68)$$

Undamped Vibration Absorbers When a mass-spring system m_1, k_1 is subjected to a harmonic force with the frequency equal to the natural frequency, resonance occurs. In this case, it is possible to add a second mass-spring system m_2, k_2 so designed as to produce a two-degree-of-freedom system with the response of m_1 equal to zero. We refer to m_1, k_1 as the **main system** and to m_2, k_2 as the **vibration absorber**. The resulting two-degree-of-freedom system is shown in Fig. 3.4.24 and has the impedance matrix

$$Z(\omega) = \begin{bmatrix} k_1 + k_2 - \omega^2 m_1 & -k_2 \\ -k_2 & k_2 - \omega^2 m_2 \end{bmatrix} \quad (3.4.69)$$

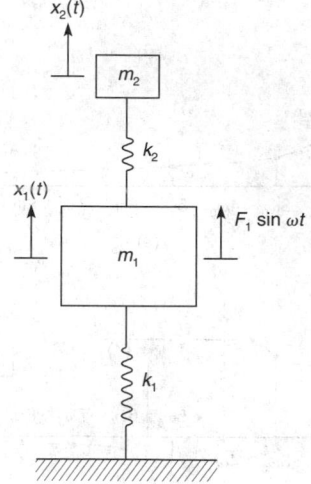

Fig. 3.4.24

Inserting Eq. (3.4.69) into Eq. (3.4.67), together with $F_1(t) = F_1 \sin \omega t$, $F_2(t) = 0$, write the steady-state response in the form

$$x_1(t) = X_1(\omega) \sin \omega t \quad (3.4.70a)$$
$$x_2(t) = X_2(\omega) \sin \omega t \quad (3.4.70b)$$

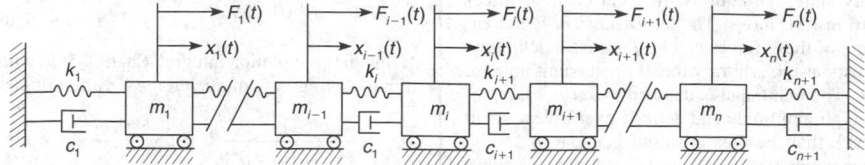

Fig. 3.4.23

where the amplitudes are given by

$$X_1(\omega) = \frac{[1 - (\omega/\omega_a)^2]x_{st}}{[1 + \mu(\omega_a/\omega_n)^2 - (\omega/\omega_n)^2][1 - (\omega/\omega_a)^2] - \mu(\omega_a/\omega_n)^2}$$

$$(3.4.71a)$$

$$X_2(\omega) = \frac{x_{st}}{[1 + \mu(\omega_a/\omega_n)^2 - (\omega/\omega_n)^2][1 - (\omega/\omega_a)^2] - \mu(\omega_a/\omega_n)^2}$$

$$(3.4.71b)$$

in which

$\omega_n = \sqrt{k_1/m_1}$ = the natural frequency of the main system alone

$\omega_a = \sqrt{k_2/m_2}$ = the natural frequency of the absorber alone

$x_{st} = F_1/k_1$ = the static deflection of the main system

$\mu = m_2/m_1$ = the ratio of the absorber mass to the main mass

From Eqs. (3.4.70a) and (3.4.71a), we conclude that if we choose m_2 and k_2 such that $\omega_a = \omega$, the response $x_1(t)$ of the main mass is zero. Moreover, from Eqs. (3.4.70b) and (3.4.71b),

$$x_2(t) = -\left(\frac{\omega_n}{\omega_a}\right)^2 \frac{x_{st}}{\mu} \sin \omega t = -\frac{F_1}{k_2} \sin \omega t \qquad (3.4.72)$$

so that the force in the absorber spring is

$$k_2 x_2(t) = -F_1 \sin \omega t \qquad (3.4.73)$$

Hence, *the absorber exerts a force on the main mass balancing exactly the applied force* $F_1 \sin \omega t$.

A vibration absorber designed for a given operating frequency ω can perform satisfactorily for operating frequencies that vary slightly from ω. In this case, the motion of m_1 is not zero, but its amplitude tends to be very small, as can be verified from a frequency response plot $X_1(\omega)/x_{st}$ versus ω/ω_n; Fig. 3.4.25 shows such a plot for $\mu = 0.2$ and $\omega_n = \omega_a$. The shaded area indicates the range in which the performance can be regarded as satisfactory. Note that the thin line in Fig. 3.4.25 represents the frequency response of the main system alone. Also note that the system resulting from the combination of the main system and the absorber has two resonance frequencies, but they are removed from the operating frequency $\omega = \omega_n = \omega_a$.

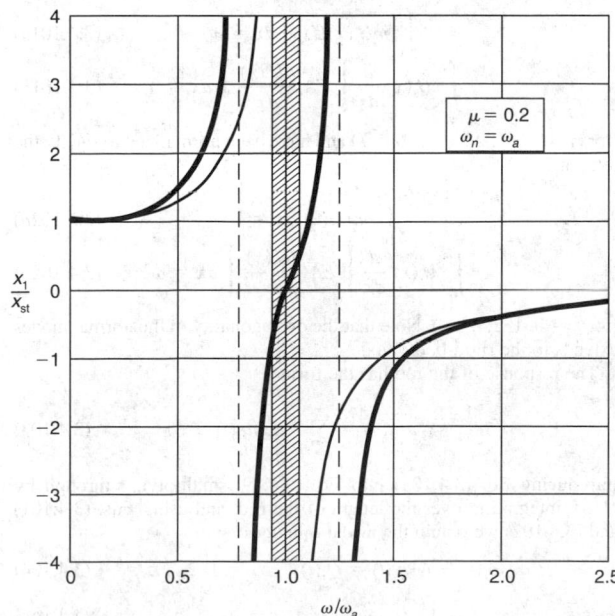

Fig. 3.4.25

Natural Modes of Vibration   In the absence of damping and external forces, Eq. (3.4.65) reduces to the free-vibration equation

$$M\ddot{x}(t) + Kx(t) = \mathbf{0} \qquad (3.4.74)$$

which has the harmonic solution

$$\mathbf{x}(t) = \mathbf{u} \cos (\omega t - \phi) \qquad (3.4.75)$$

where $\mathbf{u}$ is a constant vector, ω a frequency of oscillation, and ϕ a phase angle. Introduction of Eq. (3.4.75) into Eq. (3.4.74) and division through by $\cos (\omega t - \phi)$ results in

$$Ku = \omega^2 Mu \qquad (3.4.76)$$

which represents a set of n simultaneous algebraic equations known as the **eigenvalue problem**. It has n solutions consisting of the **eigenvalues** ω_r^2; the square roots represent the **natural frequencies** ω_r ($r = 1, 2, \ldots, n$). Moreover, to each natural frequency ω_r there corresponds a vector $\mathbf{u}_r$ ($r = 1, 2, \ldots, n$) called **eigenvector**, or **modal vector**, or **natural mode**. The modal vectors possess the **orthogonality property**, or

$$\mathbf{u}_s^T M \mathbf{u}_r = 0 \qquad (3.4.77a)$$
$$\mathbf{u}_s^T K \mathbf{u}_r = 0 \qquad (3.4.77b)$$

(for $r, s = 1, 2, \ldots, n; r \neq s$), in which $\mathbf{u}_s^T$ is the **transpose** of $\mathbf{u}_s$, a row vector. It is convenient to adjust the magnitude of the modal vectors so as to satisfy

$$\mathbf{u}_r^T M \mathbf{u}_r = 1 \qquad (3.4.78a)$$
$$\mathbf{u}_r^T K \mathbf{u}_r = \omega_r^2 \qquad (3.4.78b)$$

(for $r = 1, 2, \ldots, n$), a process known as normalization, in which case $\mathbf{u}_r$ are called **normal modes**. Note that the normalization process involves Eq. (3.4.78a) alone, as Eq. (3.4.78b) follows automatically. The solution of the eigenvalue problem can be obtained by a large variety of computational algorithms (Meirovitch, ''Principles and Techniques of Vibrations,'' Prentice-Hall). Commercially, they are available in software packages for numerical computations, such as MATLAB.

The actual solution of Eq. (3.4.74) is obtained below in the context of the transient response.

Transient Response of Undamped Systems   From Eq. (3.4.65), the vibration of undamped systems satisfies the equation

$$M\ddot{x}(t) + Kx(t) = \mathbf{F}(t) \qquad (3.4.79)$$

where $\mathbf{F}(t)$ is an arbitrary force vector. In addition, the displacement and velocity vectors must satisfy the initial conditions $\mathbf{x}(0) = \mathbf{x}_0$, $\dot{\mathbf{x}}(0) = \mathbf{v}_0$. The solution of Eq. (3.4.79) has the form

$$\mathbf{x}(t) = \sum_{r=1}^{n} \mathbf{u}_r q_r(t) \qquad (3.4.80)$$

in which $\mathbf{u}_r$ are the modal vectors and $q_r(t)$ are associated **modal coordinates**. Inserting Eq. (3.4.80) into Eq. (3.4.79), premultiplying the result by $\mathbf{u}_s^T$, and using Eqs. (3.4.77) and (3.4.78) we obtain the modal equations

$$\ddot{q}_r(t) + \omega_r^2 q_r(t) = Q_r(t) \qquad r = 1, 2, \ldots, n \qquad (3.4.81)$$

where

$$Q_r(t) = \mathbf{u}_r^T \mathbf{F}(t) \qquad r = 1, 2, \ldots, n \qquad (3.4.82)$$

are **modal forces**. Equations (3.4.81) resemble the equation of single-degree-of-freedom system and have the solution

$$q_r(t) = \frac{1}{\omega_r} \int_0^t Q_r(t - \tau) \sin \omega_r \tau \, d\tau + q_r(0) \cos \omega_r t + \frac{\dot{q}_r(0)}{\omega_r} \sin \omega_r t$$

$$r = 1, 2, \ldots, n \quad (3.4.83)$$

where

$$q_r(0) = \mathbf{u}_r^T M \mathbf{x}_0 \qquad (3.4.84a)$$
$$\dot{q}_r(0) = \mathbf{u}_r^T M \mathbf{v}_0 \qquad (3.4.84b)$$

(for $r = 1, 2, \ldots, n$) are **initial modal displacements and velocities**, respectively. The solution to both external forces and initial excitations is obtained by inserting Eqs. (3.4.83) into Eq. (3.4.80).

Systems with Proportional Damping When the system is damped, the response does not in general have the form of Eq. (3.4.80), and a more involved approach is necessary (Meirovitch, "Elements of Vibration Analysis," 2d ed., McGraw-Hill). In the special case in which the damping matrix C is proportional to the mass matrix M or the stiffness matrix K, or is a linear combination of M and K, the preceding approach yields the modal equations

$$\ddot{q}_r(t) + 2\zeta_r\omega_r\dot{q}_r(t) + \omega_r^2 q_r(t) = Q_r(t) \qquad r = 1, 2, \ldots, n \quad (3.4.85)$$

where ζ_r are **modal damping factors**. Equations (3.4.85) have the solution

$$q_r(t) = \frac{1}{\omega_{dr}} \int_0^t Q_r(t-\tau) e^{-\zeta_r\omega_r\tau} \sin \omega_{dr}\tau \, d\tau$$

$$+ \frac{q_r(0)}{\sqrt{1-\zeta_r^2}} e^{-\zeta_r\omega_r t} \cos(\omega_{dr}t - \psi_r) + \frac{\dot{q}_r(0)}{\omega_{dr}} \sin \omega_{dr}t$$

$$r = 1, 2, \ldots, n \quad (3.4.86)$$

in which

$$\omega_{dr} = \omega_r\sqrt{1-\zeta_r^2} \qquad r = 1, 2, \ldots, n \quad (3.4.87)$$

is the damped frequency in the rth mode and

$$\psi_r = \tan^{-1}\frac{\zeta_r}{\sqrt{1-\zeta_r^2}} \qquad r = 1, 2, \ldots, n \quad (3.4.88)$$

is a phase angle associated with the rth mode. The quantities $Q_r(t)$, $q_r(0)$, and $\dot{q}_r(0)$ remain as defined by Eqs. (3.4.82), (3.4.84a), and (3.4.84b), respectively.

DISTRIBUTED-PARAMETER SYSTEMS

Vibration of Rods, Shafts, and Strings The axial vibration of rods is described by the equation

$$-\frac{\partial}{\partial x}\left\{EA(x)\frac{\partial u(x,t)}{\partial x}\right\} + m(x)\frac{\partial^2 u(x,t)}{\partial t^2} = f(x,t)$$

$$0 < x < L \quad (3.4.89)$$

where $u(x, t)$ is the axial displacement, $f(x, t)$ the axial force per unit length, E the modulus of elasticity, $A(x)$ the cross-sectional area, and $m(x)$ the mass per unit length. The solution $u(x, t)$ is subject to one boundary condition at each end.

Before attempting to solve Eq. (3.4.89), consider the free vibration problem, $f(x, t) = 0$. The solution of the latter problem is harmonic and can be expressed as

$$u(x, t) = U(x) \cos(\omega t - \phi) \quad (3.4.90)$$

where $U(x)$ is the amplitude, ω the frequency, and ϕ an inconsequential phase angle. Inserting Eq. (3.4.90) into Eq. (3.4.89) with $f(x, t) = 0$ and dividing through by $\cos(\omega t - \phi)$, we conclude that $U(x)$ and ω must satisfy the *eigenvalue problem*

$$-\frac{d}{dx}\left\{EA(x)\frac{dU(x)}{dx}\right\} = \omega^2 m(x)U(x) \qquad 0 < x < L \quad (3.4.91)$$

where $U(x)$ must satisfy one boundary condition at each end. At a fixed end the displacement U must be zero and at a free end the axial force $EA\,dU/dx$ is zero.

Exact solutions of the eigenvalue problem are possible in only a few cases, mostly for uniform rods, in which case Eq. (3.4.91) reduces to

$$\frac{d^2U(x)}{dx^2} + \beta^2 U(x) = 0 \qquad \beta^2 = \frac{\omega^2 m}{EA} \qquad 0 < x < L \quad (3.4.92)$$

whose solution is

$$U(x) = A \sin \beta x + B \cos \beta x \quad (3.4.93)$$

where A and B are constants of integration, determined from specified boundary conditions. In the case of a **fixed-free** rod, the boundary conditions are

$$U(0) = 0 \quad (3.4.94a)$$

$$EA\frac{dU}{dx}\bigg|_{x=L} = 0 \quad (3.4.94b)$$

Condition (3.4.94a) gives $B = 0$ and condition (3.4.94b) yields the **characteristic equation**

$$\cos \beta L = 0 \quad (3.4.95)$$

which has the infinity of solutions

$$\beta_r L = \frac{(2r-1)\pi}{2} \qquad r = 1, 2, \ldots \quad (3.4.96)$$

where β_r represent the *eigenvalues;* they are related to the *natural frequencies* ω_r by

$$\omega_r = \beta_r\sqrt{\frac{EA}{m}} = \frac{(2r-1)\pi}{2}\sqrt{\frac{EA}{mL^2}} \qquad r = 1, 2, \ldots \quad (3.4.97)$$

From Eq. (3.4.93), the **normal modes** are

$$U_r(x) = \sqrt{\frac{2}{mL}} \sin\frac{(2r-1)\pi x}{2L} \qquad r = 1, 2, \ldots \quad (3.4.98)$$

For a **fixed-fixed** rod, the natural frequencies and normal modes are

$$\omega_r = r\pi\sqrt{\frac{EA}{mL^2}} \qquad U_r(x) = \sqrt{\frac{2}{mL}} \sin\frac{r\pi x}{L}$$

$$r = 1, 2, \ldots \quad (3.4.99)$$

and for a **free-free** rod they are

$$\omega_0 = 0 \qquad U_0 = \sqrt{\frac{1}{mL}} \quad (3.4.100a)$$

$$\omega_r = r\pi\sqrt{\frac{EA}{mL^2}} \qquad U_r(x) = \sqrt{\frac{2}{mL}} \cos\frac{r\pi x}{L}$$

$$r = 1, 2, \ldots \quad (3.4.100b)$$

Note that U_0 represents a **rigid-body mode**, with zero natural frequency.

In every case the modes are orthogonal, satisfying the conditions

$$\int_0^L mU_s(x)U_r(x) \, dx = 0 \quad (3.4.101a)$$

$$-\int_0^L U_s(x)\frac{d}{dx}\left[EA\frac{dU_r(x)}{dx}\right] dx = 0 \quad (3.4.101b)$$

(for $r, s = 0, 1, 2, \ldots, r \neq s$) and have been normalized to satisfy the relations

$$\int_0^L mU_r^2(x) \, dx = 1 \quad (3.4.102a)$$

$$-\int_0^L U_r(x)\frac{d}{dx}\left[EA\frac{dU_r(x)}{dx}\right] dx = \omega_r^2 \quad (3.4.102b)$$

(for $r = 0, 1, 2, \ldots$). Note that the orthogonality of the normal modes extends to the rigid-body mode.

The response of the rod has the form

$$u(x, t) = \sum_{r=1}^{\infty} U_r(x)q_r(t) \quad (3.4.103)$$

Introducing Eq. (3.4.103) into Eq. (3.4.89), multiplying through by $U_s(x)$, integrating over the length of the rod, and using Eqs. (3.4.101) and (3.4.102) we obtain the **modal equations**

$$\ddot{q}_r(t) + \omega_r^2 q_r(t) = Q_r(t) \qquad r = 1, 2, \ldots \quad (3.4.104)$$

where

$$Q_r(t) = \int_0^L U_r(x)f(x, t) \, dx \qquad r = 1, 2, \ldots \quad (3.4.105)$$

Table 3.4.4   Analogous Quantities for Rods, Shafts, and Strings

	Rods	Shafts	Strings
Displacement	Axial—$u(x, t)$	Torsional—$\theta(x, t)$	Transverse—$w(x, t)$
Inertia (per unit length)	Mass—$m(x)$	Mass polar moment of inertia—$I(x)$	Mass—$\rho(x)$
Stiffness	Axial—$EA(x)$ E = Young's modulus $A(x)$ = cross-sectional area	Torsional—$GJ(x)$ G = shear modulus $J(x)$ = area polar moment of inertia	Tension—$T(x)$
Load (per unit length)	Force—$f(x, t)$	Moment—$m(x, t)$	Force—$f(x, t)$

are the **modal forces.** Equations (3.4.104) resemble Eqs. (3.4.81) entirely; their solution is given by Eqs. (3.4.83). The displacement of the rod is obtained by inserting Eqs. (3.4.83) into Eq. (3.4.103).

As an example, consider the response of a uniform fixed-free rod to the uniformly distributed impulsive force

$$f(x, t) = \hat{f}_0 \delta(t) \qquad (3.4.106)$$

Inserting Eqs. (3.4.98) and (3.4.106) into Eq. (3.4.105), we obtain the modal forces

$$Q_r(t) = \sqrt{\frac{2}{mL}} \int_0^L \sin \frac{(2r - 1)\pi x}{2L} \hat{f}_0 \delta(t) \, dx$$

$$= \frac{2}{(2r - 1)\pi} \sqrt{\frac{2L}{m}} \hat{f}_0 \delta(t) \qquad r = 1, 2, \ldots \quad (3.4.107)$$

so that, from Eqs. (3.4.83), the modal displacements are

$$q_r(t) = \frac{1}{\omega_r} \frac{2}{(2r - 1)\pi} \sqrt{\frac{2L}{m}} \hat{f}_0 \int_0^t \delta(t - \tau) \sin \omega_r \tau d\tau$$

$$= \left[\frac{2}{(2r - 1)\pi} \right]^2 \sqrt{\frac{2L^3}{EA}} \hat{f}_0 \sin \frac{(2r - 1)\pi}{2} \sqrt{\frac{EA}{mL^2}} t$$

$$r = 1, 2, \ldots \quad (3.4.108)$$

Finally, from Eq. (3.4.103), the response is

$$u(x, t) = \frac{8\hat{f}_0 L}{\pi^2} \sqrt{\frac{1}{mEA}} \sum_{r=1}^{\infty} \frac{1}{(2r - 1)^2} \sin \frac{(2r - 1)\pi x}{2L}$$

$$\times \sin \frac{(2r - 1)\pi}{2} \sqrt{\frac{EA}{mL^2}} t \quad (3.4.109)$$

The torsional vibration of shafts and the transverse vibration of strings are described by the same differential equation and boundary conditions as the axial vibration of rods, except that the nature of the displacement, inertia and stiffness parameters, and external excitations differs, as indicated in Table 3.4.4.

Bending Vibration of Beams   The procedure for evaluating the response of beams in transverse vibration is similar to that for rods, the main difference arising in the stiffness term. The differential equation for beams in bending is

$$\frac{\partial^2}{\partial x^2} \left[EI(x) \frac{\partial^2 w(x, t)}{\partial x^2} \right] + m(x) \frac{\partial^2 w(x, t)}{\partial t^2}$$

$$= f(x, t) \qquad 0 < x < L \quad (3.4.110)$$

in which $w(x, t)$ is the transverse displacement, $f(x, t)$ the force per unit length, $I(x)$ the cross-sectional area moment of inertia, and $m(x)$ the mass per unit length. The solution $w(x, t)$ must satisfy two boundary conditions at each end.

The eigenvalue problem is described by the differential equation

$$\frac{d^2}{dx^2} \left[EI(x) \frac{d^2 W(x)}{dx^2} \right] = \omega^2 m(x) W(x) \qquad 0 < x < L \quad (3.4.111)$$

and two boundary conditions at each end, depending on the type of support. Some possible boundary conditions are given in Table 3.4.5. The solution of the eigenvalue problem consists of the natural frequencies ω_r and natural modes $W_r(x)$ ($r = 1, 2, \ldots$). The first five normalized natural frequencies of uniform beams with six different boundary conditions are listed in Table 3.4.6. The normal modes for the hinged-hinged beam are

$$W_r(x) = \sqrt{\frac{2}{mL}} \sin \frac{r\pi x}{L} \qquad r = 1, 2, \ldots \quad (3.4.112)$$

The normal modes for the remaining beam types are more involved and they involve both trigonometric and hyperbolic functions (Meirovitch, "Elements of Vibration Analysis," 2d ed.) The modes for every beam type are orthogonal and can be used to obtain the response $w(x, t)$ in the form of a series similar to Eq. (3.4.103).

Table 3.4.5   Quantities Equal to Zero at Boundary

Boundary type	Displacement W	Slope dW/dx	Bending moment $EI\,d^2W/dx^2$	Shearing force $d(EI\,d^2W/dx^2)/dx$
Hinged	✔		✔	
Clamped	✔	✔		
Free			✔	✔

Vibration of Membranes   A **membrane** is a very thin sheet of material stretched over a two-dimensional domain enclosed by one or two nonintersecting boundaries. It can be regarded as the two-dimensional counterpart of the string. Like a string, it derives the ability to resist transverse displacements from tension, which acts in all directions in the plane of the membrane and at all its points. It is commonly assumed that the tension is uniform and does not change as the membrane de-

Table 3.4.6   Normalized Natural Frequencies for Various Beams

Beam type	$\omega_1\sqrt{mL^4/EI}$	$\omega_2\sqrt{mL^4/EI}$	$\omega_3\sqrt{mL^4/EI}$	$\omega_4\sqrt{mL^4/EI}$	$\omega_5\sqrt{mL^2/EI}$
Hinged–hinged	π^2	$4\pi^2$	$9\pi^2$	$16\pi^2$	$25\pi^2$
Clamped–free	1.875^2	4.694^2	7.855^2	10.996^2	14.137^2
Free–free	0	0	$(1.506\pi)^2$	$(2.500\pi)^2$	$(3.500\pi)^2$
Clamped–clamped	$(1.506\pi)^2$	$(2.500\pi)^2$	$(3.500\pi)^2$	$(4.500\pi)^2$	$(5.500\pi)^2$
Clamped–hinged	3.927^2	7.069^2	10.210^2	13.352^2	16.493^2
Hinged–free	0	3.927^2	7.069^2	10.210^2	13.352^2

flects. The general procedure for calculating the response of membranes remains the same as for rods and beams, but there is one significant new factor, namely, the shape of the boundary, which dictates the type of coordinates to be used. For rectangular membranes cartesian coordinates must be used, and for circular membranes polar coordinates are indicated.

The differential equation for the transverse vibration of membranes is

$$-T\nabla^2 w + \rho \frac{\partial^2 w}{\partial t^2} = f \qquad (3.4.113)$$

which must be satisfied at every interior point of the membrane, where w is the transverse displacement, f the transverse force per unit area, T the tension, and ρ the mass per unit area. Moreover, ∇^2 is the Laplacian operator, whose expression depends on the coordinates used. The solution w must satisfy one boundary condition at every boundary point. Using the established procedure, the eigenvalue problem is described by the differential equation

$$-T\nabla^2 W = \omega^2 \rho W \qquad (3.4.114)$$

where W is the displacement amplitude; it must satisfy one boundary condition at every point of the boundary.

Consider a **rectangular membrane fixed** at $x = 0$, a and $y = 0$, b, in which case the Laplacian operator in terms of the cartesian coordinates x and y has the form

$$\nabla^2 = \frac{\partial^2}{\partial x^2} + \frac{\partial^2}{\partial y^2} \qquad (3.4.115)$$

The boundary conditions are $W(0, y) = W(a, y) = W(x, 0) = W(x, b) = 0$. The natural frequencies are

$$\omega_{mn} = \pi \sqrt{\left[\left(\frac{m}{a} \right)^2 + \left(\frac{n}{b} \right)^2 \right] \frac{T}{\rho}}$$
$$m, n = 1, 2, \ldots \quad (3.4.116)$$

and the normal modes are

$$W_{mn}(x, y) = \frac{2}{\sqrt{\rho ab}} \sin \frac{m\pi x}{a} \sin \frac{n\pi y}{b} \quad m, n = 1, 2, \ldots \quad (3.4.117)$$

The modes satisfy the orthogonality conditions

$$\int_0^a \int_0^b \rho W_{mn}(x, y) W_{rs}(x, y) \, dx \, dy = 0,$$
$$m \neq r \text{ and/or } n \neq s \quad (3.4.118a)$$

$$-\int_0^a \int_0^b W_{mn}(x, y) T\nabla^2 W_{rs}(x, y) \, dx \, dy = 0,$$
$$m \neq r \text{ and/or } n \neq s \quad (3.4.118b)$$

and have been normalized so that $\int_0^a \int_0^b \rho W_{mn}^2(x, y) \, dx \, dy = 1 (m, n = 1, 2, \ldots)$. Note that, because the problem is two-dimensional, it is necessary to identify the natural frequencies and modes by two subscripts. With this exception, the procedure for obtaining the response is the same as for rods and beams.

Next, consider a **uniform circular membrane fixed at** $r = a$. In this case, the Laplacian operator in terms of the polar coordinates r and θ is

$$\nabla^2 = \frac{\partial^2}{\partial r^2} + \frac{1}{r} \frac{\partial}{\partial r} + \frac{1}{r^2} \frac{\partial^2}{\partial \theta^2} \qquad (3.4.119)$$

The natural modes for circular membranes are appreciably more involved than for rectangular membranes. They are products of Bessel functions of $\omega_{mn} r$ and trigonometric functions of $m\theta$, where $m = 0, 1, 2, \ldots$ and $n = 1, 2, \ldots$. The modes are given in Meirovitch, "Principles and Techniques of Vibrations," Prentice-Hall. Table 3.4.7 gives the normalized natural frequencies $\omega_{mn}^* = (\omega_{mn}/2\pi)\sqrt{\rho a^2/T}$ corresponding to $m = 0, 1, 2$ and $n = 1, 2, 3$. The modes satisfy the orthogonality relations

$$\int_0^a \int_0^{2\pi} \rho W_{mn}(r, \theta) W_{rs}(r, \theta) r \, dr \, d\theta = 0$$
$$m \neq r \text{ and/or } n \neq s \quad (3.4.120a)$$

Table 3.4.7 Circular Membrane Normalized Natural Frequencies
$\omega_{mn}^* = (\omega_{mn}/2\pi)\sqrt{\rho a^2/T}$

	n		
m	1	2	3
0	0.3827	0.8786	1.3773
1	0.6099	1.1165	1.6192
2	0.8174	1.3397	1.8494

$$-\int_0^a \int_0^{2\pi} W_{mn}(r,\theta) T\nabla^2 W_{rs}(r, \theta) r \, dr \, d\theta = 0$$
$$m \neq r \text{ and/or } n \neq s \quad (3.4.120b)$$

The response of circular membranes is obtained in the usual manner.

Bending Vibration of Plates Consider **plates** whose behavior is governed by the elementary plate theory, which is based on the following assumptions: (1) deflections are small compared to the plate thickness; (2) the normal stresses in the direction transverse to the plate are negligible; (3) there is no force resultant on the cross-sectional area of a plate differential element: the middle plane of the plate does not undergo deformations and represents a neutral plane, and (4) any straight line normal to the middle plane remains so during bending. Under these assumptions, the differential equation for the bending vibration of plates is

$$D\nabla^4 w + m \frac{\partial^2 w}{\partial t^2} = f \qquad (3.4.121)$$

and is to be satisfied at every interior point of the plate, where w is the transverse displacement, f the transverse force per unit area, m the mass per unit area, $D = Eh^3/12(1 - v^2)$ the plate flexural rigidity, E Young's modulus, h the plate thickness, and v Poisson's ratio. Moreover, ∇^4 is the biharmonic operator. The solution w must satisfy two boundary conditions at every point of the boundary. The eigenvalue problem is defined by the differential equation

$$D\nabla^4 W = \omega^2 m W \qquad (3.4.122)$$

and corresponding boundary conditions.

Consider a **rectangular plate simply supported** at $x = 0$, a and $y = 0$, b. Because of the shape of the plate, we must use cartesian coordinates, in which case the biharmonic operator has the expression

$$\nabla^4 = \nabla^2 \nabla^2 = \left(\frac{\partial^2}{\partial x^2} + \frac{\partial^2}{\partial y^2} \right) \left(\frac{\partial^2}{\partial x^2} + \frac{\partial^2}{\partial y^2} \right)$$
$$= \frac{\partial^4}{\partial x^4} + 2 \frac{\partial^4}{\partial x^2 \partial y^2} + \frac{\partial^4}{\partial y^4} \qquad (3.4.123)$$

Moreover, the boundary conditions are $W = 0$ and $\partial^2 W/\partial x^2 = 0$ for $x = 0$, a and $W = 0$ and $\partial^2 W/\partial y^2 = 0$ for $y = 0$, b. The natural frequencies are

$$\omega_{mn} = \pi^2 \left[\left(\frac{m}{a} \right)^2 + \left(\frac{n}{b} \right)^2 \right] \sqrt{\frac{D}{m}}$$
$$m, n = 1, 2, \ldots \quad (3.4.124)$$

and no confusion should arise because the same symbol is used for one of the subscripts and for the mass per unit area. The corresponding normal modes are

$$W_{mn}(x, y) = \frac{2}{\sqrt{mab}} \sin \frac{m\pi x}{n} \sin \frac{n\pi y}{b} \quad m, n = 1, 2, \ldots \quad (3.4.125)$$

and they are recognized as being the same as for rectangular membranes fixed at all boundaries.

A **circular plate** requires use of polar coordinates, so that the biharmonic operator has the form

$$\nabla^4 = \nabla^2 \nabla^2 = \left(\frac{\partial^2}{\partial r^2} + \frac{1}{r} \frac{\partial}{\partial r} + \frac{1}{r^2} \frac{\partial^2}{\partial \theta^2} \right) \left(\frac{\partial^2}{\partial r^2} + \frac{1}{r} \frac{\partial}{\partial r} + \frac{1}{r^2} \frac{\partial^2}{\partial \theta^2} \right)$$
$$(3.4.126)$$

Table 3.4.8   Circular Plate Normalized Natural Frequencies
$\omega_{mn}^* = (\omega_{mn}(a/\pi)^2\sqrt{m/D}$

| | n | | |
m	1	2	3
0	1.015^2	2.007^2	3.000^2
1	1.468^2	2.483^2	3.490^2
2	1.879^2	2.992^2	4.000^2

Consider a **plate clamped** at $r = a$, in which case the boundary conditions are $W(r, \theta) = 0$ and $\partial W(r, \theta)/\partial r = 0$ at $r = a$. In addition, the solution must be finite at every interior point in the plate, and in particular at $r = 0$. The natural modes have involved expressions; they are given in Meirovitch, "Principles and Techniques of Vibrations," Prentice-Hall. Table 3.4.8 lists the normalized natural frequencies $\omega_{mn}^* = \omega_{mn}(a/\pi)^2 \sqrt{m/D}$ corresponding to $m = 0, 1, 2$ and $n = 1, 2, 3$.

The natural modes of the plates are orthogonal and can be used to obtain the response to both initial and external excitations.

APPROXIMATE METHODS FOR DISTRIBUTED SYSTEMS

Rayleigh's Energy Method   The eigenvalue problem contains vital information concerning vibrating systems, namely, the natural frequencies and modes. In the majority of practical cases, exact solutions to the eigenvalue problem for distributed systems are not possible, so that the interest lies in **approximate solutions**. This is often the case when the mass and stiffness are distributed nonuniformly and/or the boundary conditions cannot be satisfied, the latter in particular for two-dimensional systems with irregularly shaped boundaries.

When the objective is to estimate the lowest natural frequency, Rayleigh's energy method has few equals. As discussed earlier, free vibration of undamped systems is harmonic and can be expressed as

$$w(x, t) = W(x) \cos (\omega t - \phi) \qquad (3.4.127)$$

where $W(x)$ is the displacement amplitude, ω the free vibration frequency, and ϕ an inconsequential phase angle. The kinetic energy represents an integral involving the velocity squared. Hence, using Eq. (3.4.127), the kinetic energy can be written in the form

$$T(t) = \frac{1}{2} \int_0^L m(x) \left[\frac{\partial w(x, t)}{\partial t} \right]^2 dx = \omega^2 T_{\text{ref}} \sin^2(\omega t - \phi) \quad (3.4.128)$$

where

$$T_{\text{ref}} = \frac{1}{2} \int_0^L m(x) W^2(x) \, dx \qquad (3.4.129)$$

is called the **reference kinetic energy**. The form of the potential energy is system-dependent, but in general is an integral involving the square of the displacement and of its derivatives with respect to the spatial coordinates (see Table 3.4.9). It can be expressed as

$$V(t) = V_{\max} \cos^2(\omega t - \phi) \qquad (3.4.130)$$

where $V_{\max}$ is the maximum potential energy, which can be obtained by simply replacing $w(x, t)$ by $W(x)$ in $V(t)$. Using the principle of conservation of energy in conjunction with Eqs. (3.4.128) and (3.4.130), we can write

$$E = T + V = T_{\max} + 0 = 0 + V_{\max} \qquad (3.4.131)$$

in which

$$T_{\max} = \omega^2 T_{\text{ref}} \qquad (3.4.132)$$

It follows that

$$\omega^2 = \frac{V_{\max}}{T_{\text{ref}}} \qquad (3.4.133)$$

Equation (3.4.133) represents **Rayleigh's quotient,** which has the remarkable property that it has a minimum value for $W(x) = W_1(x)$, the minimum value being ω_1^2. Rayleigh's energy method amounts to selecting a trial function $W(x)$ reasonably close to the lowest natural mode $W_1(x)$, inserting this function into Rayleigh's quotient, and carrying out the indicated integrations. Then, ω^2 will be one order of magnitude closer to

the lowest eigenvalue ω_1^2 than $W(x)$ is to $W_1(x)$, thus providing a good estimate ω of the lowest natural frequency ω_1. Quite often, the static deformation of the system acted on by loads proportional to the mass distribution is a good choice. In some cases, the lowest mode of a related simpler system can yield good results.

As an example, estimate the lowest natural frequency of a uniform bar in axial vibration with a mass M attached at $x = L$ (Fig. 3.4.26) for the three trial functions (1) $U(x) = x/L$; (2) $U(x) = (1 + M/mL)(x/L) - (x/L)^2/2$, representing the static deformation; and (3) $U(x) = \sin \pi x/2L$, representing the lowest mode of the bar without the mass M. The Rayleigh quotient for this bar is

$$\omega^2 = \frac{\displaystyle\int_0^L EA(x)[dU(x)/dx]^2 \, dx}{\displaystyle\int_0^L m(x)U^2(x) \, dx + MU^2(L)} \qquad (3.4.134)$$

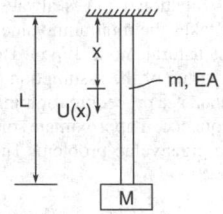

Fig. 3.4.26

The results are:

1. $\omega^2 = \dfrac{\displaystyle\int_0^L EA(1/L)^2 \, dx}{\displaystyle\int_0^L m(x/L)^2 \, dx + M} = \dfrac{EA}{(M + mL/3)L}$

2. $\dfrac{\displaystyle\int_0^L EA(1 + M/mL - x/L)^2(1/L)^2 \, dx}{\displaystyle\int_0^L m[(1 + M/mL)(x/L) - (x/L)^2/2]^2 \, dx + M(1 + 2M/mL)^2/4}$

$$= \frac{\left(\dfrac{M}{mL}\right)^2 + \dfrac{M}{mL} + \dfrac{1}{3}}{\dfrac{1}{3}\left(\dfrac{M}{mL}\right)^2 + \dfrac{5}{12}\dfrac{M}{mL} + \dfrac{2}{15} + \dfrac{M}{mL}\left(\dfrac{1}{2} + \dfrac{M}{mL}\right)^2} \frac{EA}{mL^2}$$
$$\qquad (3.4.135)$$

3. $\omega^2 = \dfrac{\displaystyle\int_0^L EA\left(\dfrac{\pi}{2L}\right)^2 \cos^2 \dfrac{\pi x}{2L} \, dx}{\displaystyle\int_0^L m \sin^2 \dfrac{\pi x}{2L} \, dx + M} = \dfrac{\pi^2}{8\left(\dfrac{1}{2} + \dfrac{M}{mL}\right)} \dfrac{EA}{mL^2}$

Table 3.4.9   Potential Energy for Various Systems

System	Potential energy* $V(t)$
Rods (also shafts and strings)	$\dfrac{1}{2}\displaystyle\int_0^L EA(x)[\partial u(x, t)/\partial x]^2 dx$
Beams	$\dfrac{1}{2}\displaystyle\int_0^L EI(x)[\partial^2 w(x, t)/\partial x^2]^2 dx$
Beams with axial force	$\dfrac{1}{2}\displaystyle\int_0^L \{EI(x)[\partial^2 w(x, t)/\partial x^2]^2 + P(x)[\partial \omega(x, t)/\partial x]^2\} dx$
Membranes	$\dfrac{1}{2}\displaystyle\int_{\text{Area}} T\{[\partial w(x, y, t)/\partial x]^2 + [\partial w(x, y, t)/\partial y]^2\} dx \, dy$
Plates	$\dfrac{1}{2}\displaystyle\int_{\text{Area}} D\{\nabla^2 w(x, y, t))^2 + 2(1 - \nu)[\partial^2 w(x, y, t)/\partial x \, \partial y\}^2$ $\qquad - (\partial^2 w(x, y, t)/\partial x^2)(\partial^2 w(x, y, t)/\partial y^2)]\} dx \, dy$

* If the distributed system has a spring at the boundary point a, then add a term $kw^2(a, t)/2$.

For comparison purposes, let $M = mL$, which yields the following estimates for the lowest natural frequency:

$$1. \quad \omega = 0.8660 \sqrt{\frac{EA}{mL^2}}$$

$$2. \quad \omega = 0.8629 \sqrt{\frac{EA}{mL^2}} \qquad (3.4.136)$$

$$2. \quad \omega = 0.9069 \sqrt{\frac{EA}{mL^2}}$$

The best estimate is the lowest one, which corresponds to case 2, with the trial function in the form of the static displacement. Note that the estimate obtained in case 1 is also quite good. It corresponds to the first case in Table 3.4.2, representing a mass-spring system in which the mass of the spring is included.

Rayleigh-Ritz Method **Rayleigh's quotient**, Eq. (3.4.133), corresponding to any trial function $W(x)$ is always larger than the lowest eigenvalue ω_1^2, and it takes the minimum value of ω_1^2 when $W(x)$ coincides with the lowest natural mode $W_1(x)$. However, this possibility must be ruled out by virtue of the assumption that W_1 is not available. The **Rayleigh-Ritz method** is a procedure for minimizing Rayleigh's quotient by means of a sequence of approximate solutions converging to the actual solution of the eigenvalue problem. The minimizing sequence has the form

$$W(x) = a_1 \phi_i(x)$$

$$W(x) = a_1 \phi_1(x) + a_2 \phi_2(x) = \sum_{j=1}^{2} a_j \phi_j(x)$$
$$\cdots \qquad (3.4.137)$$
$$W(x) = a_1 \phi_1(x) + a_2 \phi_2(x) + \cdots + a_n \phi_n(x) = \sum_{j=1}^{n} a_j \phi_j(x)$$

where a_j are undetermined coefficients and $\phi_j(x)$ are suitable trial functions satisfying all, or at least the geometric boundary conditions. The coefficients $a_j (j = 1, 2, \ldots, n)$ are determined so that Rayleigh's quotient has a minimum. With Eqs. (3.4.137) inserted into Eq. (3.4.133), Rayleigh's quotient becomes

$$\Omega^2 = \frac{\displaystyle\sum_{i=1}^{n} \sum_{j=1}^{n} k_{ij} a_i a_j}{\displaystyle\sum_{i=1}^{n} \sum_{j=1}^{n} m_{ij} a_i a_j} \qquad n = 1, 2, \ldots \qquad (3.4.138)$$

where $k_{ij} = k_{ji}$ and $m_{ij} = m_{ji}$ $(i, j = 1, 2, \ldots, n)$ are symmetric stiffness and mass coefficients whose nature depends on the potential energy and kinetic energy, respectively. The special case in which $n = 1$ represents Rayleigh's energy method. For $n \geq 2$, minimization of Rayleigh's quotient leads to the solution of the eigenvalue problem

$$\sum_{j=1}^{n} k_{ij} a_j = \Omega^2 \sum_{j=1}^{n} m_{ij} a_j$$
$$i = 1, 2, \ldots, n; n = 2, 3, \ldots \qquad (3.4.139)$$

Equations (3.4.139) can be written in the matrix form

$$K\mathbf{a} = \Omega^2 M\mathbf{a} \qquad (3.4.140)$$

in which $K = [k_{ij}]$ is the symmetric stiffness matrix and $M = [m_{ij}]$ is the symmetric mass matrix. Equation (3.4.140) resembles the eigenvalue problem for multi-degree-of-freedom systems, Eq. (3.4.76), and its solutions possess the same properties. The eigenvalues Ω_r^2 provide approximations to the actual eigenvalues ω_r^2, and approach them from above as n increases. Moreover, the eigenvectors $\mathbf{a}_r = [a_{r1} \ a_{r2} \ \ldots \ a_{rm}]^T$ can be used to obtain the approximate natural modes by writing

$$W_r(x) = a_{r1} \phi_1(x) + a_{r2} \phi_2(x) + \cdots + a_{rm} \phi_n(x) = \sum_{j=1}^{n} a_{rj} \phi_j(x)$$
$$r = 1, 2, \ldots, n; n = 2, 3, \ldots \qquad (3.4.141)$$

As an illustration, consider the same rod in axial vibration used to demonstrate Rayleigh's energy method. Insert Eqs. (3.4.137) with $W(x)$ replaced by $U(x)$ into the numerator and denominator of Eq. (3.4.134) to obtain

$$\int_0^L EA(x) \left[\frac{dU(x)}{dx}\right]^2 dx$$
$$= \sum_{i=1}^{n} \sum_{j=1}^{n} \left[\int_0^L EA(x) \frac{d\phi_i(x)}{dx} \frac{d\phi_j(x)}{dx} dx\right) a_i a_j \qquad (3.4.142a)$$

$$\int_0^L m(x) U^2(x) dx + M U^2(L)$$
$$= \sum_{i=1}^{n} \sum_{j=1}^{n} \left[\int_0^L m(x) \phi_i(x) \phi_j(x) dx + M \phi_i(L) \phi_j(L)\right] a_i a_j \qquad (3.4.142b)$$

so that the stiffness and mass coefficients are

$$k_{ij} = \int_0^L EA(x) \frac{d\phi_i(x)}{dx} \frac{d\phi_j(x)}{dx} dx \qquad i, j = 1, 2, \ldots, n \qquad (3.4.143a)$$

$$m_{ij} = \int_0^L m(x) \phi_i(x) \phi_j(x) dx + M \phi_i(L) \phi_j(L)$$
$$i, j = 1, 2, \ldots, n \qquad (3.4.143b)$$

respectively. As trial functions, use

$$\phi_j(x) = (x/L)^j \qquad j = 1, 2, \ldots, n \qquad (3.4.144)$$

which are zero at $x = 0$, thus satisfying the geometric boundary condition. Hence, the stiffness and mass coefficients are

$$k_{ij} = \frac{EAij}{L^{i+j}} \int_0^L x^{i-1} x^{j-1} dx = \frac{ij}{i+j-1} \frac{EA}{L}$$
$$i, j = 1, 2, \ldots, n \qquad (3.4.145a)$$

$$m_{ij} = \frac{m}{L^{i+j}} \int_0^L x^i x^j dx + M = \frac{mL}{i+j+1} + M$$
$$i, j = 1, 2, \ldots, n \qquad (3.4.145b)$$

so that the stiffness and mass matrices are

$$K = \frac{EA}{L} \begin{bmatrix} 1 & 1 & 1 & \cdots & 1 \\ 1 & 4/3 & 3/2 & \cdots & 2n/(n+1) \\ 1 & 3/2 & 9/5 & \cdots & 3n/(n+2) \\ \cdots\cdots\cdots\cdots\cdots\cdots\cdots\cdots\cdots \\ 1 & 2n/(n+1) & 3n/(n+2) & \cdots & n^2/(2n-1) \end{bmatrix}$$
$$(3.4.146a)$$

$$M = mL \begin{bmatrix} 1/3 & 1/4 & 1/5 & \cdots & 1/(n+2) \\ 1/4 & 1/5 & 1/6 & \cdots & 1/(n+3) \\ 1/5 & 1/6 & 1/7 & \cdots & 1/(n+4) \\ \cdots\cdots\cdots\cdots\cdots\cdots\cdots\cdots\cdots \\ 1/(n+2) & 1/(n+3) & 1/(n+4) & \cdots & 1/(2n+1) \end{bmatrix}$$
$$+ M \begin{bmatrix} 1 & 1 & 1 & \cdots & 1 \\ 1 & 1 & 1 & \cdots & 1 \\ 1 & 1 & 1 & \cdots & 1 \\ \cdots\cdots\cdots\cdots\cdots\cdots \\ 1 & 1 & 1 & \cdots & 1 \end{bmatrix} \qquad (3.4.146b)$$

For comparison purposes, consider the case in which $M = mL$. Then, for $n = 2$, the eigenvalue problem is

$$\begin{bmatrix} 1 & 1 \\ 1 & 4/3 \end{bmatrix} \begin{bmatrix} a_1 \\ a_2 \end{bmatrix} = \lambda \begin{bmatrix} 4/3 & 5/4 \\ 5/4 & 6/5 \end{bmatrix} \begin{bmatrix} a_1 \\ a_2 \end{bmatrix}$$
$$\lambda = \Omega^2 \frac{mL^2}{EA} \qquad (3.4.147)$$

which has the solutions

$$\lambda_1 = 0.7407 \qquad \mathbf{a}_1 = [1 \quad -0.1667]^T$$
$$\lambda_2 = 12.0000 \qquad \mathbf{a}_2 = [1 \quad -1.0976]^T \qquad (3.4.148)$$

Hence, the computed natural frequencies and modes are

$$\Omega_1 = 0.8607 \sqrt{\frac{EA}{mL^2}} \qquad U_1(x) = \frac{x}{L} - 0.1667 \left(\frac{x}{L}\right)^2$$

$$\Omega_2 = 3.4641 \sqrt{\frac{EA}{mL^2}} \qquad U_2(x) = \frac{x}{L} - 1.0976 \left(\frac{x}{L}\right)^2 \qquad (3.4.149)$$

Comparing Eqs. (3.4.149) with the estimates obtained by Rayleigh's energy method, Eqs. (3.4.136), note that the Rayleigh-Ritz method has produced a more accurate approximation for the lowest natural frequency. In addition, it has produced a first approximation for the second lowest natural frequency, as well as approximations for the two lowest modes, which Rayleigh's energy method is unable to produce. The approximate solutions can be improved by letting $n = 3, 4, \ldots$.

Finite Element Method In the Rayleigh-Ritz method, the trial functions extend over the entire domain of the system and tend to be complicated and difficult to work with. More importantly, they often cannot be produced, particularly for two-dimensional problems. Another version of the Rayleigh-Ritz method, the **finite element method,** does not suffer from these drawbacks. Indeed, the trial functions extending only over small subdomains, referred to as **finite elements,** are known low-degree polynomials and permit easy computer coding. As in the Rayleigh-Ritz method, a solution is assumed in the form of a linear combination of trial functions, known as **interpolation functions,** multiplied by undetermined coefficients. In the finite element method the coefficients have physical meaning, as they represent "nodal" displacements, where "nodes" are boundary points between finite elements. The computation of the stiffness and mass matrices is carried out for each of the elements separately and then the element stiffness and mass matrices are assembled into global stiffness and mass matrices. One disadvantage of the finite element method is that it requires a large number of degrees of freedom.

To illustrate the method, and for easy visualization, consider the transverse vibration of a string fixed at $x = 0$ and with a spring of stiffness K attached at $x = L$ (Fig. 3.4.27) and divide the length L into n elements of width h, so that $nh = L$. Denote the displacements of the nodal points x_e by a_e and assume that the string displacement is linear between any two nodal points. Figure 3.4.28 shows a typical element e.

Introducing Eq. (3.4.151) into Eqs. (3.4.152) and considering the boundary conditions, we obtain the element stiffness and mass matrices

$$K_1 = \frac{EA}{h} \qquad K_e = \frac{EA}{h}\begin{bmatrix} 1 & -1 \\ -1 & 1 \end{bmatrix} \qquad e = 2, 3, \ldots, n-1$$

$$K_n = \frac{EA}{h}\begin{bmatrix} 1 & -1 \\ -1 & Kh/EA \end{bmatrix} \qquad M_1 = \frac{hm}{3}$$

$$M_e = \frac{hm}{6}\begin{bmatrix} 2 & 1 \\ 1 & 2 \end{bmatrix} \qquad e = 2, 3, \ldots, n \qquad (3.4.153)$$

where K_1 and M_1 are really scalars, because the left end of the first element is fixed, so that the displacement is zero. Then, since the nodal

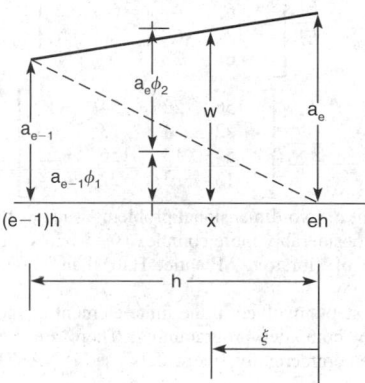

Fig. 3.4.28

displacement a_e is shared by elements e and $e + 1$ ($e = 1, 2, \ldots, n - 2$), the element stiffness and mass matrices can be assembled into the global stiffness and mass matrices

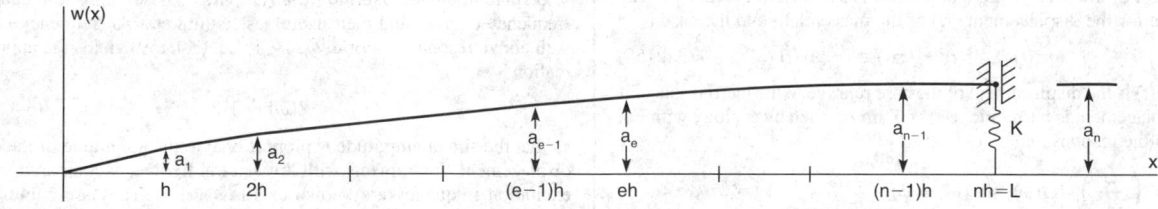

Fig. 3.4.27

The process can be simplified greatly by introducing the nondimensional local coordinate $\xi = j - x/h$. Then, considering the two linear interpolation functions

$$\phi_1(\xi) = \xi \qquad \phi_2(\xi) = 1 - \xi \qquad (3.4.150)$$

the displacement at point ξ can be expressed as

$$\omega(\xi) = a_{e-1}\phi_1(\xi) + a_e\phi_2(\xi) \qquad (3.4.151)$$

where a_{e-1} and a_e are the nodal displacements for element e. Using Eqs. (3.4.143) and changing variables from x to ξ, we can write the element stiffness and mass coefficients

$$k_{eij} = \frac{1}{h}\int_0^1 EA \frac{d\phi_i}{d\xi}\frac{d\phi_j}{d\xi}\,d\xi \qquad m_{eij} = h\int_0^1 m\phi_i\phi_j\,d\xi,$$

$$i, j = 1, 2 \quad (3.4.152)$$

$$K = \frac{EA}{h}\begin{bmatrix} 2 & -1 & 0 & \cdots & 0 & 0 \\ -1 & 2 & -1 & \cdots & 0 & 0 \\ 0 & -1 & 2 & \cdots & 0 & 0 \\ \cdots & \cdots & \cdots & \cdots & \cdots & \cdots \\ 0 & 0 & 0 & \cdots & 2 & -1 \\ 0 & 0 & 0 & \cdots & -1 & Kh/EA \end{bmatrix}$$

$$(3.4.154)$$

$$M = \frac{hm}{6}\begin{bmatrix} 4 & 1 & 0 & \cdots & 0 & 0 \\ 1 & 4 & 1 & \cdots & 0 & 0 \\ 0 & 1 & 4 & \cdots & 0 & 0 \\ \cdots & \cdots & \cdots & \cdots & \cdots & \cdots \\ 0 & 0 & 0 & \cdots & 4 & 1 \\ 0 & 0 & 0 & \cdots & 1 & 2 \end{bmatrix}$$

For beams in bending, the displacements consist of one translation and one rotation per node; the interpolation functions are the Hermite cubics

$$\phi_1(\xi) = 3\xi^2 - 2\xi^3, \; \phi_2(\xi) = \xi^2 - \xi^3$$
$$\phi_3(\xi) = 1 - 3\xi^2 + 2\xi^3, \; \phi_4(\xi) = -\xi + 2\xi^2 - \xi^3 \quad (3.4.155)$$

and the element stiffness and mass coefficients are

$$k_{eij} = \frac{1}{h^3} \int_0^1 EI \frac{d^2\phi_i}{d\xi^2} \frac{d^2\phi_j}{d\xi^2} d\xi \qquad m_{eij} = h \int_0^1 m\phi_i\phi_j d\xi$$
$$i, j = 1, 2, 3, 4 \quad (3.4.156)$$

yielding typical element stiffness and mass matrices

$$K_e = \frac{EI}{h^3} \begin{bmatrix} 12 & 6 & -12 & 6 \\ 6 & 4 & -6 & 2 \\ -12 & -6 & 12 & -6 \\ 6 & 2 & -6 & 4 \end{bmatrix}$$

$$M_e = \frac{hm}{420} \begin{bmatrix} 156 & 22 & 54 & -13 \\ 22 & 4 & 13 & -3 \\ 54 & 13 & 156 & -22 \\ -13 & -3 & -22 & 4 \end{bmatrix} \quad (3.4.157)$$

The treatment of two-dimensional problems, such as for membranes and plates, is considerably more complex (see Meirovitch, "Principles and Techniques of Vibration," Prentice-Hall) than for one-dimensional problems.

The various steps involved in the finite element method lend themselves to ready computer programming. There are many computer codes available commercially; one widely used is NASTRAN.

VIBRATION-MEASURING INSTRUMENTS

Typical quantities to be measured include acceleration, velocity, displacement, frequency, damping, and stress. Vibration implies motion, so that there is a great deal of interest in transducers capable of measuring motion relative to the inertial space. The basic transducer of many vibration-measuring instruments is a mass-damper-spring enclosed in a case together with a device, generally electrical, for measuring the displacement of the mass relative to the case, as shown in Fig. 3.4.29. The equation for the displacement $z(t)$ of the mass relative to the case is

$$m\ddot{z}(t) + c\dot{z}(t) + kz(t) = -m\ddot{y}(t) \quad (3.4.158)$$

where $y(t)$ is the displacement of the case relative to the inertial space. If this displacement is harmonic, $y(t) = Y \sin \omega t$, then by analogy with Eq. (3.4.35) the response is

$$z(t) = Y \left(\frac{\omega}{\omega_n}\right)^2 |G(\omega)| \sin(\omega t - \phi)$$
$$= Z(\omega) \sin(\omega t - \phi) \quad (3.4.159)$$

so that the magnitude factor $Z(\omega)/Y = (\omega/\omega_n)^2 |G(\omega)|$ is as plotted in Fig. 3.4.9 and the phase angle ϕ is as in Fig. 3.4.4. The plot $Z(\omega)/Y$

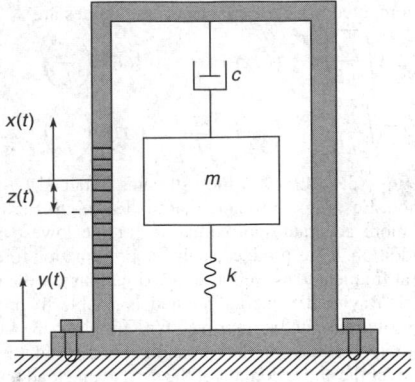

Fig. 3.4.29

versus ω/ω_n is shown again in Fig. 3.4.30 on a scale more suited to our purposes.

Accelerometers are high-natural-frequency instruments. Their usefulness is limited to a frequency range well below resonance. Indeed, for small values of ω/ω_n, Eq. (3.4.159) yields the approximation

$$Z(\omega) \approx \frac{1}{\omega_n^2} \omega^2 Y \quad (3.4.160)$$

so that the signal amplitude is proportional to the amplitude of the acceleration of the case relative to the inertial space. For $\zeta = 0.7$, the accelerometer can be used in the range $0 \le \omega/\omega_n \le 0.4$ with less than 1 percent error, and the range can be extended to $\omega/\omega_n \le 0.7$ if proper corrections, based on instrument calibration, are made.

Commonly used accelerometers are the compression-type piezoelectric accelerometers. They consist of a mass resting on a piezoelectric ceramic crystal, such as quartz, tourmaline, or ferroelectric ceramic, with the crystal acting both as spring and sensor. Piezoelectric actuators have negligible damping, so that their range must be smaller, such as $0 < \omega/\omega_n < 0.2$. In view of the fact, however, that the natural frequency is very high, about 30,000 Hz, this is a respectable range.

Displacement-Measuring Instruments These are low-natural-frequency devices and their usefulness is limited to a frequency range well above resonance. For $\omega/\omega_n \gg 1$, Eq. (3.4.159) yields the approximation

$$Z(\omega) \approx Y \quad (3.4.161)$$

so that the signal amplitude is proportional to the amplitude of the case displacement. Instruments with low natural frequency compared to the excitation frequency are known as **seismometers**. They are commonly used to measure ground motions, such as those caused by earthquakes or underground nuclear explosions. The requirement of low natural frequency dictates that the mass, referred to as **seismic mass,** be very large and the spring very soft, so that essentially the mass remains

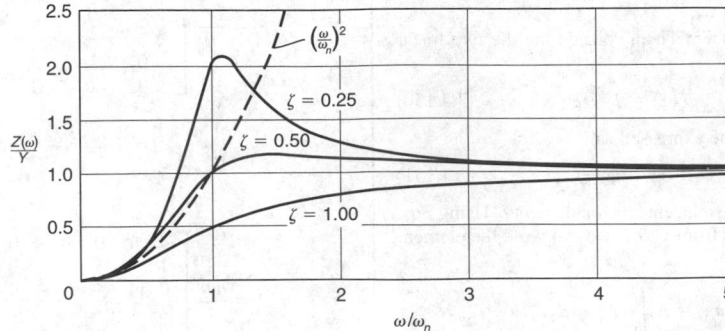

Fig. 3.4.30

stationary in an inertial space while the case attached to the ground moves relative to the mass.

Seismometers tend to be considerably larger in size than accelerometers. If a large-size instrument is undesirable, or even if size is not an issue, displacements in harmonic motion, as well as velocities, can be obtained from accelerometer signals by means of electronic integrators.

Some other transducers, not mass-damper-spring transducers, are as follows (Harris, "Shock and Vibration Handbook," 3d ed., McGraw-Hill):

Differential-transformer pickups: They consist of a core of magnetic material attached to the vibrating structure, a primary coil, and two secondary coils. As the core moves, both the inductance and induced voltage of one secondary coil increase while those of the other decrease. The output voltage is proportional to the displacement over a wide range. Such pickups are used for very low frequencies, up to 60 Hz.

Strain gages: They consist of a grid of fine wires which exhibit a change in electrical resistance proportional to the strain experienced. Their flimsiness requires that strain gages be either mounted on or bonded to some carrier material.

Heat

BY

PETER E. LILEY *Professor, School of Mechanical Engineering, Purdue University.*
HOYT C. HOTTEL *Professor Emeritus, Massachusetts Institute of Technology.*
ADEL F. SAROFIM *Lammot duPont Professor of Chemical Engineering, Massachusetts Institute of Technology.*
KENNETH A. SMITH *Edward R. Gilliland Professor of Chemical Engineering, Massachusetts Institute of Technology.*

4.1 THERMODYNAMICS
by Peter E. Liley

4.2 THERMODYNAMIC PROPERTIES OF SUBSTANCES
by Peter E. Liley

4.3 RADIANT HEAT TRANSFER
by Hoyt C. Hottel and Adel F. Sarofim

4.4 TRANSMISSION OF HEAT BY CONDUCTION AND CONVECTION
by Kenneth A. Smith

4.1 THERMODYNAMICS
by Peter E. Liley

NOTE: References are placed throughout the text for the reader's convenience. (No material is presented relating to the calibration of thermometers at fixed points, etc. Specific details of the measurement of temperature, pressure, etc. are found in Benedict, "Fundamentals of Temperature, Pressure and Flow Measurements," 3d ed. Measurement of other properties is reviewed in Maglic et al., "Compendium of Thermophysical Property Measurement Methods," vol. 1, Plenum Press. The periodical *Metrologia* presents latest developments, particularly for work of a definitive caliber.)

Thermodynamic properties of a variety of other specific materials are listed also in Secs. 4.2, 6.1, and 9.8.

THERMOMETER SCALES

Let F and C denote the readings on the Fahrenheit and Celsius (or centigrade) scales, respectively, for the same temperature. Then

$$C = \frac{5}{9}(F - 32) \qquad F = \frac{9}{5}C + 32$$

If the pressure readings of a constant-volume hydrogen thermometer are extrapolated to zero pressure, it is found that the corresponding temperature is $-273.15°C$, or $-459.67°F$. An **absolute temperature scale** was formerly used on which zero corresponding with zero pressure on the hydrogen thermometer. The basis now used is to define and give a numerical value to the temperature at a single point, the triple point of water, defined as $0.01°C$. The scales are:

Kelvins (K) = degrees Celsius + 273.15
Degrees Rankine (°R) = degrees Fahrenheit + 459.67

EXPANSION OF BODIES BY HEAT

Coefficients of Expansion The **coefficient of linear expansion** a' of a solid is defined as the increment of length in a unit of length for a rise in temperature of 1 deg. Likewise, the **coefficient of cubical expansion** a''' of a solid, liquid, or gas is the increment of volume of a unit volume for a rise of temperature of 1 deg. Denoting these coefficients by a' and a''', respectively, we have

$$a' = \frac{1}{l}\frac{dl}{dt} \qquad a''' = \frac{1}{V}\frac{dV}{dt}$$

in which l denotes length, V volume, and t temperature. For homogeneous solids $a''' = 3a'$, and the **coefficient of area expansion** $a'' = 2a'$.

The coefficients of expansion are, in general, dependent upon the temperature, but for ordinary ranges of temperature, constant mean values may be taken. If lengths, areas, and volumes at $32°F$ ($0°C$) are taken as standard, then these magnitudes at other temperatures t_1 and t_2 are related as follows:

$$\frac{l_1}{l_2} = \frac{1 + a't_1}{1 + a't_2} \qquad \frac{A_1}{A_2} = \frac{1 + a''t_1}{1 + a''t_2} \qquad \frac{V_1}{V_2} = \frac{1 + a'''t_1}{1 + a'''t_2}$$

Since for solids and liquids the expansion is small, the preceding formulas for these bodies become approximately

$$l_2 - l_1 = a'l_1(t_2 - t_1)$$
$$A_2 - A_1 = a''A_1(t_2 - t_1)$$
$$V_2 - V_1 = a'''V_1(t_2 - t_1)$$

The coefficients of cubical expansion for different gases at ordinary temperatures are about the same. From 0 to $212°F$ and at atmospheric pressure, the values multiplied by 1,000 are as follows: for NH_3, 2.11; CO, 2.04; CO_2, 2.07; H_2, 2.03; NO, 2.07. For an ideal gas, the coeffi-

cient at any temperature is the reciprocal of the (absolute) temperature. (See also Table 6.1.10.)

UNITS OF FORCE AND MASS

Force mass, length, and time are related by Newton's second law of motion, which may be expressed as

$$F \sim ma$$

In order to write this as an equality, a constant must be introduced which has magnitude and dimensions. For convenience, in the fps system, the constant may be designated as $1/g_c$. Thus,

$$F = \frac{ma}{g_c}$$

Since this equation must be homogeneous insofar as the dimensions are concerned, the units for g_c are $mL/(t^2F)$. Consider a 1-lb mass, lbm, in the earth's gravitational field, where the acceleration is 32.1740 ft/s². The force exerted on the pound mass will be defined as the pound force, lbf. This system of units gives for g_c the following magnitude and dimensions:

$$1 \text{ lbf} = \frac{(1 \text{ lbm})(32.174 \text{ ft/s}^2)}{g_c}$$

hence

$$g_c = 32.174 \text{ lbm} \cdot \text{ft}/(\text{lbf} \cdot \text{s}^2)$$

Note that g_c may be used with other units, in which case the numerical value changes. The numerical value of g_c for four systems of units is

$$g_c = 32.174 \frac{\text{lbm} \cdot \text{ft}}{\text{lbf} \cdot \text{s}^2} = 1 \frac{\text{slug} \cdot \text{ft}}{\text{lbf} \cdot \text{s}^2} = 1 \frac{\text{lbm} \cdot \text{ft}}{\text{pdl} \cdot \text{s}^2} = 1 \frac{\text{g} \cdot \text{cm}}{\text{dyn} \cdot \text{s}^2}$$

In SI, the constant is chosen to be unity and $F(N) = m(\text{kg})a(\text{m/s}^2)$. There are four possible constants, and all have been used. (See Blackman, "SI Units in Engineering," Macmillan.)

Consider now the relationship which involves weight, a gravitational force, and mass by applying the basic equation for a body of fixed mass acted upon by a gravitational force g and no other forces. The acceleration of the mass caused by the gravitational force is the acceleration due to gravity g.

Substituting gives the relationship between weight and mass

$$w = \frac{mg}{g_c}$$

If the gravitational acceleration is constant, the weight and mass are in a fixed proportion to each other; hence for accounting purposes in mass balances they can be used interchangeably. This is not possible if g is a variable.

We may now write the relation between mass m and weight w as

$$w = m\frac{g}{g_c}$$

The constant g_c is used throughout the following paragraphs. (An extensive table of conversion factors from customary units to SI units is found in Sec. 1.)

The SI unit of pressure is the newton per square metre. It is a very small pressure, as normal atmospheric pressure is 1.01325×10^5 N/m². While some use has been made of the pressure expressed in kN/m² or kPa (1 Pa = 1 N/m²) and in MN/m² or MPa, the general technical usage now seems to favor the bar = 10^5 N/m^2 = 10^5 Pa so that 1 atm = 1.01325 bar. For many approximate calculations the atmosphere and the bar can be equated.

Many representative accounts of the measurement of low and high pressure have appeared. (See, for example, Lawrance, *Chem. Eng. Progr.,* **50,** 1954, p. 155; Leck, "Pressure Measurement in Vacuum Systems," Inst. Phys., London, 1957; Peggs, "High Pressure Measurement Techniques," Appl. Sci. Publishers, Barking, Essex.)

MEASUREMENT OF HEAT

Units of Heat Many units of heat have been dependent on the experimentally determined properties of some substance. To eliminate experimental variations, the unit of heat may be defined in terms of fundamental units. The International Steam Table Conference (London, 1929) defines the Steam Table (IT) calorie as $\frac{1}{860}$ of a watthour. One British thermal unit (Btu) is defined as 251.996 IT cal, 778.26 ft·lb.

Previously, the Btu was defined as the heat necessary to raise one pound of water one degree Fahrenheit at some arbitrarily chosen temperature level. Similarly, the calorie was defined as the heat required to heat one gram of water one degree Celsius at 15°C (or at 17.5°C). These units are roughly the same in value as those mentioned above. In SI, the **joule** is the heat unit, and the newton-metre the work unit of energy. The two are equal, so that 1 J = 1 N·m; that is, in SI, the mechanical equivalent of heat is unity.

Heat Capacity and Specific Heat The **heat capacity** of a material is the amount of heat transferred to raise a unit mass of a material 1 deg in temperature. The ratio of the amount of heat transferred to raise unit mass of a material 1 deg to that required to raise unit mass of water 1 deg at some specified temperature is the **specific heat** of the material. For most engineering purposes, heat capacities may be assumed numerically equal to specific heats. Two heat capacities are generally used, that at constant pressure c_p and that at constant volume c_v. For unit mass, the instantaneous heat capacities are defined as

$$\left(\frac{\partial h}{\partial t}\right)_p = c_p \qquad \left(\frac{\partial u}{\partial t}\right)_v = c_v$$

Over a range in temperature, the mean heat capacities are given by

$$c_{pm} = \frac{1}{t_2 - t_1}\int_{t1}^{t2} c_p\, dt \qquad c_{vm} = \frac{1}{t_2 - t_1}\int_{t1}^{t2} c_v\, dt$$

Denoting by c the heat capacity, the heat required to raise the temperature of w lb of a substance from t_1 to t_2 is $Q = mc(t_2 - t_1)$, provided c is a constant.

In general, c varies with the temperature, though for moderate temperature ranges a constant mean value may be taken. If, however, c is taken as variable, $Q = m \int_{t1}^{t2} c\, dt$. The *mean* heat capacity from 0 to t deg is given by $c_m = \frac{1}{t}\int_0^t c\, dt$. If $c = a_1 + a_2 t + a_3 t^2 + \cdots$

$$c_m = a_1 + \tfrac{1}{2}a_2 t + \tfrac{1}{3}a_3 t^2 + \cdots$$

Data for the specific heat of some solids, liquids, and gases are found in Tables 4.2.22 and 4.2.27.

Specific Heat of Solids For elements near room temperature, the specific heat may be approximated by the rule of Dulong and Petit, that the specific heat at constant volume approaches $3R$. At lower temperatures, Debye's theory leads to the equation

$$\frac{C_v}{R} = 3\left(\frac{T}{\Theta}\right)\int_0^{\Theta_{max}T}\left(\frac{\Theta}{T}\right)^4\frac{\exp(\Theta/T)}{[\exp(\Theta/T)-1]^2}\,d\left(\frac{\Theta}{T}\right)$$

commonly known as Debye's function. Figure 4.1.1 shows the variation of c_v/R with T/Θ, as predicted from the equation above. The principal difficulties in using the Debye equation arise from (1) the difficulty in finding unique values of Θ for any given material and (2) the need to consider other contributions to the specific heat. For further references on the Debye equation, see Harrison and Neighbours, *U.S. Naval Post-*

graduate Lab. Rept. 49, Dec. 1964; Overton and Hancock, *Naval Research Lab. Rept. 5502,* 1960; Hilsenrath and Zeigler, *NBS Monograph 49,* 1962. For a thorough discussion of electronic, lattice, and magnetic contributions to specific heat, see Gopal, "Specific Heats at Low Temperatures," Plenum Press, New York. See also Table 6.1.11.

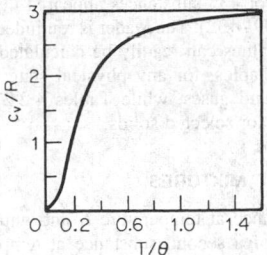

Fig. 4.1.1

For solid compounds at about room temperature, Kopp's approximation is often useful. This states that the specific heat of a solid compound at room temperature is equal to the sum of the specific heats of the atoms forming the compound.

SPECIFIC HEAT OF LIQUIDS

No general theory of any simple practical utility seems to exist for the specific heat of liquids. In "Thermophysical Properties of Refrigerants," ASHRAE, Atlanta, 1976, the interpolation device was a polynomial in temperature, usually up to T^3.

SPECIFIC HEAT OF GASES

The following table summarizes results of kinetic theory for specific heats of gases:

Gas type	c_p/R	c_v/R	c_p/c_v
Monatomic	$\frac{5}{2}$	$\frac{3}{2}$	$\frac{5}{3}$
Diatomic	$\frac{7}{2}$	$\frac{5}{2}$	$\frac{7}{5}$
n degrees of freedom	$(n + 2)/2$	$n/2$	$1 + 2/n$

Determination of the effective number of degrees of freedom limits extending this method to more comlex gases.

Properties of gases are, usually, most readily correlated on the mol basis. A **pound mol** is the mass in pounds equal to the molecular weight. Thus 1 pound mol of oxygen is 32 lb. At the same pressure and temperature, the volume of 1 mol is the same for all perfect gases, i.e., following the gas laws. Experimental findings led Avogadro (1776–1856) to formulate the microscopic hypothesis now known as **Avogadro's principle,** which states that 1 mol of any perfect gas contains the same number of molecules. The number is known as the **Avogadro number** and is equal to

$$N = 6.02214 \times 10^{26} \text{ molecules/(kg·mol)}$$
$$= 2.73160 \times 10^{26} \text{ molecules/(lb·mol)}$$

For perfect gases, $Mc_p - Mc_v = MR = 1.987$.

$$c_i = R/(k - 1) \qquad c_p = Rk(k - 1)$$

Passut and Danner [*Ind. Eng. Chem. Process Design & Dev.* (**11,** 1972, p. 543)] developed a set of thermodynamically consistent polynomials for estimating ideal gas enthalpy, entropy, and heat capacity, fittings being given for 89 compounds. The same journal reported 2 years later an extension of the work to another 57 compounds, by Huang and Daubert. Fittings of a cubic-in-temperature polynomial for 408 hydrocarbons and related compounds in the ideal gas state were reported by Thinh et al. in *Hydrocarbon Processing,* Jan. 1971, pp. 98–104. On p. 153 of a later issue (Aug. 1976), they claimed that the function $C_p = A + B\exp(-c/T^n)$ fitted for 221 hydrocarbons: graphite and hydrogen gave a more accurate fit. A cubic polynomial in temperature was also fitted for more than 700 compounds from 273 to 1,000 K by Seres et al. in *Acta Phys. Chem.*, Univ. Szegediensis (Hungary), **23**, 1977, pp. 433–468. A 1975 formula of Wilhoit was fitted for 62 substances by A. Harmens in *Proc. Conf. Chemical Thermodynamic Data on Fluids,* IPC Sci. Tech. Press, Guildford, U.K., pp. 112–120. A cubic polynomial fitting for 435 substances appeared in *J. Chem. Eng., Peking* (**2**, 1979, pp. 109–132). The reader is reminded that specific heat at constant pressure values can readily be calculated from tabulated enthalpy-temperature tables, for any physical state. Table 4.2.22 gives values for liquids and gases, while Tables 4.2.27 to 4.2.29 provide similar information for selected solids.

SPECIFIC HEAT OF MIXTURES

If w_1 lb of a substance at temperature t_1 and with specific heat c_1 is mixed with w_2 lb of a second substance at temperature t_2 and with specific heat c_2, provided chemical reaction, heat evolution, or heat absorption does not occur, the specific heat of the mixture is

$$c_m = (w_1c_1 + w_2c_2)/(w_1 + w_2)$$

and the temperature of the mixture is

$$t_m = (w_1c_1t_1 + w_2c_2t_2)/(w_1c_1 + w_2c_2)$$

In general, $t_m = \Sigma wct/\Sigma wc$.

To raise the temperature of w_1 lb of a substance having a specific heat c_1 from t_1 to t_m, the weight w_2 of a second substance required is

$$w_2 = w_1c_1(t_m - t_1)/c_2(t_2 - t_m)$$

SPECIFIC HEAT OF SOLUTIONS

For aqueous solutions of salts, the specific heat may be estimated by assuming the specific heat of the solution equal to that of the water alone. Thus, for a 20 percent by weight solution of sodium chloride in water, the specific heat would be approximately 0.8.

Although approximate calculations of mixture properties often consist simply of multiplying the mole fraction of each constituent by the property of each constituent, more accurate calculations are possible. (See "Technical Data Book—Petroleum Refining" API, Washington, DC, 1984; Daubert, "Chemical Engineering Thermodynamics," McGraw-Hill; "The Properties of Gases and Liquids," 3d ed., McGraw-Hill; Perry, "Chemical Engineers Handbook," McGraw-Hill; Walas, "Phase Equilibria in Chemical Engineering," Butterworth.)

LATENT HEAT

For pure substances, the heat effects accompanying changes in state at constant pressure are known as latent effects, because no temperature change is evident. Heat of fusion, vaporization, sublimation, and change in crystal form are examples.

The values for the heat of fusion and latent heat of vaporization are presented in Tables 4.2.21 and 4.2.28.

GENERAL PRINCIPLES OF THERMODYNAMICS

Notation

B = availability (by definition, $B = H - T_oS$)
c_p = specific heat at constant pressure
c_v = specific heat at constant volume
E, e = total energy associated with system
g = local acceleration of gravity, ft/s^2
g_c = a dimensional constant
H, h = enthalpy, Btu (by definition $h = u + p_v$)
J = mechanical equivalent of heat = 778.26 ft·lb/Btu = 4.1861 J/cal
$k = c_p/c_v$
m = mass of substance under consideration, lbm

M = molecular weight
p = absolute pressure, lb/ft^2
Q, q = quantity of heat absorbed by system from surroundings, Btu
R = ideal gas constant
R_u = universal gas constant
S, s = entropy
t = temperature, °F
$T = t + 459.69$ = absolute temperature = °R
T_0 = sink or discard temperature
U, u = internal energy
$\bar{v}$ = linear velocity
v = volume
V = total volume
w = weight of substance under consideration, lb
W = external work performed on surroundings during change of state, ft·lb
$$Y = \left(\frac{p_1}{p_2}\right)^{(k-1)k} - 1$$
z = distance above or below chosen datum
g = free energy (by definition, $g = h - Ts$)
f = Helmholtz free energy (by definition, $f = u - Ts$)

In thermodynamics, unless otherwise noted, the convention followed is that the change in any property $\psi = \Delta\psi$ = final value − initial value = $\psi_2 - \psi_1$.

In this notation, small letters usually denote magnitudes referred to a unit mass of the substance, capital letters corresponding magnitudes referred to m units of mass. Thus, v denotes the volume of 1 lb, and $V = mv$, the volume of m lb. Similarly, $U = mu$, $S = ms$, etc. Subscripts are used to indicate different states; thus, p_1, v_1, T_1, u_1, s_1 refer to state 1; p_2, v_2, T_2, u_2, s_2 refer to state 2; Q_{12} is used to denote the heat transferred during the change from state 1 to state 2, and W_{12} denotes the external work done during the same change.

Thermodynamics is the study which deals with energy, the various concepts and laws describing the conversion of one form of energy to another, and the various systems employed to effect the conversions. Thermodynamics deals in general with systems in equilibrium. By means of its fundamental concepts and basic laws, the behavior of an engineering system may be described when the various variables are altered. Thermodynamics covers a very broad field and includes many systems, for example, those dealing with chemical, thermal, mechanical, and electrical force fields and potentials. The quantity of matter within a prescribed boundary under consideration is called the **system,** and everything external to the system is spoken of as the **surroundings.** With a **closed system** there is no interchange of matter between system and surroundings; with an **open system** there is such an interchange. Any change that the system may undergo is known as a **process.** Any process or series of processes in which the system returns to its original condition or state is called a **cycle.**

Heat is energy **in transit** from one mass to another because of a temperature difference between the two. Whenever a force of any kind acts through a distance, **work** is done. Like heat, work is also energy in transit. Work is to be differentiated from the capacity of a quantity of energy to do work.

The two fundamental and general laws of thermodynamics are: (1) energy may be neither created nor destroyed, (2) it is impossible to bring about any change or series of changes the sole net result of which is transfer of energy as heat from a low to a high temperature; in other words, heat will not of itself flow from low to high temperatures.

The **first law of thermodynamics,** one of the very important laws of nature, is the law of conservation of energy. Although the law has been stated in a variety of ways, all have essentially the same meaning. The following are examples of typical statements: Whenever energy is transformed from one form to another, energy is always conserved; energy can neither be created nor destroyed; the sum total of all energy remains constant. The energy conservation hypothesis was stated by a number of investigators; however, experimental evidence was not available until the famous work of J. P. Joule. Transformation of matter into energy ($E = mc^2$), as in nuclear reactions, is ignored; within the realm of thermodynamics discussed here, mass is conserved.

It has long been the custom to designate the law of conservation of energy, the first law of thermodynamics, when it is used in the analysis of engineering systems involving heat transfer and work. Statements of the first law may be written as follows: Heat and work are mutually convertible; or, since energy can neither be created nor destroyed, the total energy associated with an energy conversion remains constant.

Before the first law may be applied to the analysis of engineering systems, it is necessary to express it in some form of expression. Thus it may be stated for an **open system** as

$$\begin{bmatrix} \text{Net amount of} \\ \text{energy added to} \\ \text{system as heat} \\ \text{and all forms} \\ \text{of work} \end{bmatrix} + \begin{bmatrix} \text{stored} \\ \text{energy} \\ \text{of mass} \\ \text{entering} \\ \text{system} \end{bmatrix} - \begin{bmatrix} \text{stored} \\ \text{energy} \\ \text{of mass} \\ \text{leaving} \\ \text{system} \end{bmatrix} = \begin{bmatrix} \text{net in-} \\ \text{crease} \\ \text{in stored} \\ \text{energy of} \\ \text{system} \end{bmatrix}$$

For an open system with fluid entering only at section 1 and leaving only at section 2 and with no electrical, magnetic, or surface-tension effects, this equation may be written as

$$Q + W + \int \left(h_1 + \frac{v_1^2}{2g_c} + \frac{gz_1}{g_c} \right) \delta m_1$$
$$- \int \left(h_2 + \frac{v_2^2}{2g_c} + \frac{gz_2}{g_c} \right) \delta m_2 = U_f - U_i$$
$$+ \frac{m_f v_f^2 - m_i v_i^2}{2g_c} + \frac{g}{g_c}(m_f z_f - m_i z_i)$$

Note that the same sign is given to both heat and work transfers. Heat and work added to the system are given a positive sign; heat lost and work output are given a negative sign. The subscripts i and f refer to entire systems before and after the process occurs, and δm refers to a differential quantity of matter.

It must be remembered that all terms in the first-law equation must be expressed in the same units.

For a **closed stationary system,** the first-law expression reduces to

$$Q + W = U_2 - U_1$$

For an **open system** fixed in position but undergoing **steady flow,** e.g., a turbine or reciprocating steam engine, for a mass flow rate of m is

$$Q + W = m \left[(h_2 - h_1) + \frac{v_2^2 - v_1^2}{2g_c} + \frac{g}{g_c}(z_2 - z_1) \right]$$

In a **steady-flow** process, the mass rate of flow into the apparatus is equal to the mass rate of flow out; in addition, at any point in the apparatus, the conditions are unchanging with time.

This condition is usually called the **continuity equation** and is written as

$$\dot{m} = \frac{dm}{dt} = \frac{A\bar{v}}{v} = \frac{A_1 \bar{v}}{v_1} = \frac{A_2 \bar{v}_2}{v_2}$$

where mass flow rate $\dot{m}$ is related to volume flow rate $\dot{V}$ by $\dot{V} = \dot{m}v$ and A is the cross-sectional area.

Since for many processes the last two terms are often negligible, they will be omitted for simplicity except when such omission would introduce appreciable error.

Work done in overcoming a fluid pressure is measured by $W = -\int p \, dv$, where p is the pressure *effectively* applied to the surroundings for doing work and dv represents the change in volume of the system.

Reversible and Irreversible Processes   A **reversible** process is one in which both the system and the surroundings may be returned to their original states. After an **irreversible** process, this is not possible. No process involving friction or an unbalanced potential can be reversible. No loss in ability to do work is suffered because of a reversible process,

but there is always a loss in ability to do work because of an irreversible process. All actual processes are irreversible. Any series of reversible processes that starts and finishes with the system in the same state is called a **reversible cycle.**

Steady-Flow Processes   With **steady flow,** the conditions at any point in an apparatus through which a fluid is flowing do not change progressively with time. Steady-flow processes involving only mechanical effects are equivalent to similar nonflow processes occurring between two weightless frictionless diaphragms or pistons moving at constant pressure with the system as a whole in motion. Under these circumstances, the total work done by or on a unit amount of fluid is made up of that done on the two diaphragms $p_2 v_2 - p_1 v_1$ and that done on the rest of the surroundings $-\int p \, dv - p_2 v_2 + p_1 v_1$. Differentiating, $-p \, dv - d(p)v = v \, dp$. The net, useful flow work done on the surroundings is $\int v \, dp$. This is often called the shaft work. The net, useful or shaft work differs from the total work by $p_2 v_2 - p_1 v_1$. The first-law equation may be written to indicate this result for a unit mass flow rate as

$$q + W_{\text{net}} = u_2 - u_1 + p_2 v_2 - p_1 v_1 + \frac{1}{2g_c}(\bar{v}_2^2 - \bar{v}_1^2) + \frac{g}{g_c}(z_2 - z_1)$$

or, since by definition

$$u + pv = h$$
$$q + W_{\text{net}} = h_2 - h_1 + \frac{1}{2g_c}(\bar{v}_2^2 - \bar{v}_1^2) + \frac{g}{g_c}(z_2 - z_1)$$

If all net work effects are mechanical,

$$q + \int v \, dp = h_2 - h_1 + \frac{1}{2g_c}(\bar{v}_2^2 - \bar{v}_1^2) + \frac{g}{g_c}(z_2 - z_1)$$

Since in evaluating $\int v \, dp$ the pressure is that *effectively* applied to the surroundings, the integration cannot usually be performed except for reversible processes.

If a fluid is passed adiabatically through a conduit (i.e., without heat exchange with the conduit), without doing any net or useful work, and if velocity and potential effects are negligible, $h_2 = h_1$. A process of the kind indicated is the Joule-Thomson flow, and the ratio $(\partial T/\partial p)_h$ for such a flow is the Joule-Thomson coefficient.

If a fluid is passed through a nonadiabatic conduit without doing any net or useful work and if velocity and potential effects are negligible, $q = h_2 - h_1$. This equation is important in the calculation of heat balances on flow apparatus, e.g., condensers, heat exchangers, and coolers.

In many engineering processes the movement of materials is not independent of time; hence the steady-flow equations do not apply. For example, the process of oxygen discharging from a storage bottle represents a transient condition. The pressure within the bottle changes as the amount of oxygen in the tank decreases. The analysis of some transient processes is very complex; however, in order to show the general approach, a simple case will be considered.

The quantity of material flowing into and out of the engineering system in Fig. 4.1.2 varies with time. The amount of work and the heat transfer crossing the system boundary are likewise dependent upon time. According to the law of conservation of mass, the rate of change of mass within the system is equal to the rate of mass flow into and out

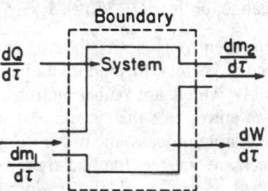

Fig. 4.1.2   Variable-flow system.

of the system. Hence, in terms of mass flow rates,

$$\frac{dm_s}{d\tau} = \frac{dm_1}{d\tau} - \frac{dm_2}{d\tau}$$

For a finite period of time, this relation may be expressed as

$$\Delta m_s = \Delta m_1 - \Delta m_2$$

The first law may be written as follows:

$$\frac{dU_s}{d\tau} = \frac{dQ}{d\tau} + \frac{dW}{d\tau} + \left(h_1 + \frac{\bar{v}_1^2}{2g_c} + \frac{g}{g_c} z_1 \right) \frac{dm_1}{d\tau} - \left(h_2 + \frac{\bar{v}_2^2}{2g_c} + \frac{g}{g_c} z_2 \right) \frac{dm_2}{d\tau}$$

Under non-steady-flow conditions the variables h, $\bar{v}$, z may change with time as well as flow rate, in which case the solution is very involved.

If steady-flow conditions prevail, then ΔU_s is equal to 0 and the integrands are independent of time, in which case the above equation reduces to the familiar steady-flow relation.

The **second law of thermodynamics** is a statement that conversion of heat to work is limited by the temperature at which conversion occurs. It may be shown that:

1. No cycle can be more efficient than a reversible cycle operating between given temperature limits.

2. The efficiency of all reversible cycles absorbing heat only at a single constant higher temperature T_1 and rejecting heat only at a single constant lower temperature T_2 must be the same.

3. For all such cycles, the efficiency is

$$\eta = \frac{W}{Q_1} = \frac{T_1 - T_2}{T_1}$$

This is usually called the **Carnot cycle efficiency**. By the first law $W = Q_1 - Q_2$,

$$(Q_1 - Q_2)/Q_1 = (T_1 - T_2)/T_1$$

By algebraic rearrangement,

$$Q_1/T_1 = Q_2/T_2$$

Clapeyron Equation

$$\frac{dp}{dT} = \frac{Q}{TV_{12}}$$

This important relation is useful in calculations relating to constant-pressure evaporation of pure substances. In that case the equation may be written

$$v_{fg} = \frac{h_{fg}}{T\,(dp/dT)}$$

ENTROPY

For reversible cyclical processes in which the temperature varies during heat absorption and rejection, i.e., for any *reversible cycle*, $\int (dQ/T) = 0$. Consequently, for any *reversible process*, $\int (dQ/T)$ is not a function of the particular *reversible path* followed. This integral is called the entropy change, or $\int_1^2 (dQ_{rev}/T) = S_2 - S_1 = S_{12}$. The entropy of a substance is dependent only on its state or condition. Mathematically, dS is a complete or perfect differential and S is a point function in contrast with Q and W which are path functions. For any reversible process, the change in entropy of the system and surroundings is zero, whereas for any irreversible process, the net entropy change is positive.

All actual processes are irreversible and therefore occur with a decrease in the amount of energy available for doing work, i.e., with an increase in unavailable energy. The increase in unavailable energy is the product of two factors, T_0 the lowest available temperature for heat discard (practically always the temperature of the atmosphere) and the net change in entropy. The increase in unavailable energy is $T_0\,\Delta S_{net}$. Any process that occurs of itself (any spontaneous process) will proceed in such a direction as to result in a net increase in entropy. This is an important concept in the application of thermodynamics to chemical processes.

Three important potentials used in the Maxwell relations are:

1. The familiar potential, known as enthalpy,

$$h = u + pv$$

2. The free energy or the Helmholtz function is defined by the following relation:

$$f = u - Ts$$

3. The free enthalpy or the Gibbs function is defined by

$$g = h - Ts = f + pv = u + pv - Ts$$

The names used for these potentials have not gained universal acceptance. In particular, the name **free energy** is used for g in many textbooks on chemical thermodynamics. One should be very cautious when referring to different books or technical papers and should verify by definition, rather than rely on the name of the potential.

Availability of a system or quantity of energy is defined as $g = h - T_0s$. In this equation, all quantities except T_0 refer to the system irrespective of the state of the surroundings. T_0 is the lowest temperature available for heat discard. The preceding definition assumes the absence of velocity, potential, and similar effects. When these are not negligible, proper allowance must be made, for example, $g = h - T_0s + \bar{v}^2/(2g_c) + (g/g_c)z$. By substitution of $Q = T_0(S_2 - S_1)$ in the appropriate first-law expressions, it may be shown that for any steady-flow process, or for any constant-pressure nonflow process, decrease in availability is equal to the maximum possible (reversible) net work effect with sink for heat discard at T_0.

The availability function g is of particular value in the thermodynamic analysis of changes occurring in the stages of a turbine and is of general utility in determining thermodynamic efficiencies, i.e., the ratio of actual work performed during a process to that which theoretically should have been performed.

Limitations of space preclude a discussion of **availability** or **exergy analysis** which, while basically simple, requires careful evaluation in some processes such as combustion. Refer to the following sources, typical of the many publications of relatively recent date: Krakow, *ASHRAE Trans. Res.*, **97**, no. 1, 1991, pp. 328–336 (dead state analysis); Szargut et al., "Exergy Analysis of Thermal, Chemical and Metallurgical Processes," Hemisphere (262 references); Kotas, *Chem. Eng. Res. Des.*, **64**, May 1986, pp. 212–230, and "The Exergy Method of Plant Analysis," Butterworth; O'Toole, *Proc. Inst. Mech. Eng.*, **204C**, 1990, pp. 329–340; Gallo and Milanez, *Energy*, **15**, no. 2, 1990, pp. 113–121; Horlock and Haywood (*Proc. Inst. Mech. Engrs.*, **199C**, 1985, pp. 11–17) analyze availability in a combined heat and power plant.

The Gibbs function is of particular importance in processes where chemical changes occur. For reversible isothermal steady-flow processes, or for reversible constant-pressure isothermal nonflow processes, change in free energy is equal to net work.

Helmholtz free energy, $f = u - Ts$, is equal to the work during a constant-volume isothermal reversible nonflow process.

All these functions g and f are point functions, and like E, h, and s their differentials are complete or perfect.

PERFECT DIFFERENTIALS. MAXWELL RELATIONS

If z is some function of x and y, in general

$$dz = \left(\frac{\partial z}{\partial x} \right)_y dx + \left(\frac{\partial z}{\partial y} \right)_x dy$$

Substituting M for $(\partial z/\partial x)_y$ and N for $(\partial z/\partial y)_x$,

$$dz = M \, dx + N \, dy$$

But $\dfrac{\partial}{\partial y}\left(\dfrac{\partial z}{\partial x}\right) = \dfrac{\partial}{\partial x}\left(\dfrac{\partial z}{\partial y}\right)$ or $\dfrac{\partial M}{\partial y} = \dfrac{\partial N}{\partial x}$. This is Euler's criterion for integrability. A perfect differential has the characteristics of dz stated above. Many important thermodynamic relations may be derived from the appropriate point function by the use of this relation; see Table 4.1.1.

From the third column of the bottom half of the table, by equating various of the terms which are equal, one may obtain

$$\left(\frac{\partial u}{\partial s}\right)_v = \left(\frac{\partial h}{\partial s}\right)_p \qquad \left(\frac{\partial u}{\partial v}\right)_s = \left(\frac{\partial f}{\partial v}\right)_T$$

$$\left(\frac{\partial h}{\partial p}\right)_s = \left(\frac{\partial g}{\partial p}\right)_T \qquad \left(\frac{\partial g}{\partial T}\right)_p = \left(\frac{\partial f}{\partial T}\right)_v$$

By mathematical manipulation of equations previously given, the following important relations may be formulated:

$$c_v = \left(\frac{\partial q}{\partial T}\right)_v = T\left(\frac{\partial s}{\partial T}\right)_v = \left(\frac{\partial u}{\partial T}\right)_v$$

$$c_p = \left(\frac{\partial q}{\partial T}\right)_p = T\left(\frac{\partial s}{\partial T}\right)_p = \left(\frac{\partial h}{\partial T}\right)_p$$

$$c_p - c_v = T\left(\frac{\partial v}{\partial T}\right)_p \left(\frac{\partial p}{\partial T}\right)_v$$

$$\left(\frac{\partial c_v}{\partial v}\right)_T = T\left(\frac{\partial^2 p}{\partial T^2}\right)_v \qquad \left(\frac{\partial c_p}{\partial T^2}\right)_p = -T\left(\frac{\partial^2 v}{\partial T^2}\right)_p$$

Relations involving q, u, h, and s:

$$dq = c_v \, dT + T\left(\frac{\partial p}{\partial T}\right)_v dv = c_p \, dT - T\left(\frac{\partial v}{\partial T}\right)_p dp$$

$$du = c_v \, dT + \left[T\left(\frac{\partial p}{\partial T}\right)_v - p\right] dv$$

$$dh = c_p \, dT - \left[T\left(\frac{\partial v}{\partial T}\right)_p - v\right] dp$$

$$ds = c_v \frac{dT}{T} + \left(\frac{\partial p}{\partial T}\right)_v dv = c_p \frac{dT}{T} - \left(\frac{v}{\partial T}\right)_p dp$$

Since $q + W = du$ and $h = u + pv$, for reversible processes,

$$du = T \, ds - p \, dv \qquad \text{and} \qquad dh = du + p \, dv + v \, dp$$

it follows that

$$v = T\left(\frac{\partial s}{\partial p}\right)_T + \left(\frac{\partial h}{\partial p}\right)_T$$

But from the Maxwell relations,

$$\left(\frac{\partial v}{\partial T}\right)_p = -\left(\frac{\partial s}{\partial p}\right)_T$$

Therefore,

$$\left(\frac{\partial h}{\partial p}\right)_T = v - T\left(\frac{\partial v}{\partial T}\right)_p$$

Similarly,

$$\left(\frac{\partial u}{\partial v}\right)_T = -\left[p - T\left(\frac{\partial p}{\partial T}\right)_v\right]$$

These last two equations give in terms of p, v, and T the necessary relations that must hold for any system, however complex. An equation in p, v, and T for the properties of a substance is called an equation of state. These two equations applicable to any substance or system are known as **thermodynamic equations of state**.

Table 4.1.1 Maxwell Relations

Function	Differential	Maxwell relation
$\Delta u = q + W$	$du = T \, ds - p \, dv$	$\left(\dfrac{\partial T}{\partial v}\right)_s = -\left(\dfrac{\partial p}{\partial s}\right)_v$
$h = u + pv$	$dh = T \, ds + v \, dp$	$\left(\dfrac{\partial T}{\partial p}\right)_s = \left(\dfrac{\partial v}{\partial s}\right)_p$
$f = u - Ts$	$df = -s \, dT - p \, dv$	$\left(\dfrac{\partial s}{\partial v}\right)_T = \left(\dfrac{\partial p}{\partial T}\right)_v$
$g = h - Ts$	$dg = -s \, dT + v \, dp$	$\left(\dfrac{\partial s}{\partial p}\right)_T = -\left(\dfrac{\partial v}{\partial T}\right)_p$

By holding certain variables constant, a second set of relations is obtained:

Differential	Independent variable held constant	Relation
$du = T \, ds - p \, dv$	s	$\left(\dfrac{\partial u}{\partial v}\right)_s = -p$
	v	$\left(\dfrac{\partial u}{\partial s}\right)_v = T$
$dh = T \, ds + v \, dp$	s	$\left(\dfrac{\partial h}{\partial p}\right)_s = v$
	p	$\left(\dfrac{\partial h}{\partial s}\right)_p = T$
$df = -s \, dT - p \, dv$	T	$\left(\dfrac{\partial f}{\partial v}\right)_T = -p$
	v	$\left(\dfrac{\partial f}{\partial T}\right)_v = -s$
$dg = -s \, dT + v \, dp$	T	$\left(\dfrac{\partial g}{\partial p}\right)_T = v$
	p	$\left(\dfrac{\partial g}{\partial T}\right)_p = -s$

Presentation of Thermal Properties Before the laws of thermodynamics can be applied and quantitative results obtained in the analysis of an engineering system, it is necessary to have available the properties of the system, some of which are temperature, pressure, internal energy entropy, and enthalpy. In general, the property of a pure substance under equilibrium conditions may be expressed as a function of two other properties. This is based on the assumption that certain effects, such as gravitational and magnetic, are not important for the condition under investigation. The various properties of a pure substance under equilibrium conditions may be expressed by an equation of state, which in general form follows:

$$p = f(T, v)$$

In this relation the pressure is shown to be a function of both the temperature and the specific volume. Many special forms of equations of state are used in the analysis of engineering systems. Plots of the properties of various pure substances are very useful in studies dealing with thermodynamics. Two-dimensional plots, such as $p - v$, $p - h$, $p - T$, $T - s$, etc., show phase relations and are important in the analysis of cycles.

The constants in the equations of state are usually based on experimental data. The properties may be presented in many different ways, some of which are:

1. As equations of state, e.g., the perfect gas laws and the van der Waals equation.

2. As charts or graphs.

3. As tables.

4. As approximations which may be useful when more reliable data are not available.

IDEAL GAS LAWS

At low pressures and high enough temperatures, in the absence of chemical reaction, all gases approach a condition such that their P-V-T properties may be expressed by the simple relation

$$pv = RT$$

If v is expressed as volume per unit weight, the value of the constant R will be different for different gases. If v is expressed as the volume of one molecular weight of gas, then R_u is the same for all gases in any chosen system of units. Hence $R = R_u/M$.

In general, for any amount of gas, the ideal gas equation becomes

$$pV = nMRT = nR_uT = \frac{m}{M} R_uT$$

where V is now the total gas volume, n is the number of moles of gas in the volume V, M is the molecular weight, and $R_u = MR$ the universal gas constant. An alternative ideal gas equation of state is $pv = R_uT/M$. It is different from the preceding in the use of specific volume v rather than total volume V.

For all ideal gases, $R_u = MR$ in $lb \cdot ft$ is 1,546. One pound mol of any perfect gas occupies a volume of 359 ft^3 at 32°F and 1 atm.

For many engineering purposes, use of the gas laws is permissible up to pressures of 100 to 200 lb/in^2 if the absolute temperatures are at least twice the critical temperatures. Below the critical temperature, errors introduced by use of the gas laws may usually be neglected up to 15 lb/in^2 pressure although errors of 5 percent are often met when dealing with saturated vapors.

The **van der Waals equation of state**, $p = BT/(v - b) - a/v^2$, is a modification of the ideal gas law which is sometimes useful at high pressures. The quantities B, a, and b are constants.

Many empirical or semiempirical equations of state have been proposed to represent the real variation of pressure with volume and temperature. The Benedict-Webb-Rubin equation is among them; see Perry, "Chemical Engineers Handbook," 6th ed., McGraw-Hill. Computer programs have been devised for the purpose; see Deutsch, "Microcomputer Programs for Chemical Engineers," McGraw-Hill. For computer programs and output for steam and other fluids, including air tables, see Irvine and Liley, "Steam and Gas Tables with Computer Programs," Academic Press.

Approximate P-V-T Relations For many gases, P-V-T data are not available. An approximation useful under such circumstances is based on the observation of van der Waals that in terms of reduced properties most gases approximate a common **reduced equation of state**. The reduced quantities are the actual ones divided by the corresponding criti-

cal quantities, e.g., the reduced temperature $T_R = T_{actual}/T_{critical}$, the reduced volume $v_R = v_{actual}/v_{critical}$, the reduced pressure $p_R = p_{actual}/p_{critical}$. The gas laws may be made to apply to any nonperfect gas by the introduction of a correction factor

$$pV = ZNR_uT$$

When the gas laws apply, $Z = 1$ and on a molal basis $Z = pV/(R_uT)$. If on a plot of Z versus p_R lines of constant T_R are drawn, for different substances these are found to fall in narrow bands. Single T_R lines may be drawn to represent approximately the various bands. This has been done in Fig. 4.1.3. To use the chart, only the critical pressure and temperature of the gas need be known.

> EXAMPLE. Find the volume of 1 lb of steam at 5,500 psia and 1200°F (by steam tables, $v = 0.1516$ ft^3/lb).
> For water, critical temperature = 705.4°F; critical pressure = 3,206.4 psia; reduced temp = 1660/1165 = 1.43; reduced pressure = 5,500/3,206.4 = 1.72; μ (see Fig. 4.1.3) = 0.83, $v = 0.83 (1,546)(1,660)/(18)(5,500)(144) = 0.149$ ft^3. Error = 100(0.152 − 0.149)/0.152 = 1.7 percent. If the gas laws had been used, the error would have been 17 percent.

No entirely satisfactory method for calculation for gaseous mixtures has been developed, but the use of average critical constants as proposed by Kay (*Ind. Eng. Chem.*, **28,** 1936, p. 1014) is easy and gives satisfactory results under conditions considerably removed from the critical. He assumes the gaseous mixture can be treated as if it were a single pure gas with a pseudocritical pressure and temperature estimated by a method of molar averaging.

$$(T_c)_{mixture} = (T_c)_a y_a + (T_c)_b y_b + (T_c)_c y_c + \cdots$$
$$(p_c)_{mixture} = (p_c)_a y_a + (p_c)_b y_b + (p_c)_c y_c + \cdots$$

where $(T_c)_a$ is the critical temperature of pure a, etc.; $(p_c)_a$ is the critical pressure of pure a, etc.; and y_a is the mole fraction of a, etc. For a gaseous mixture made up of gases a, b, c, etc., the pseudocritical constants having been determined, the gaseous mixture is handled on the μ charts as if it were a single pure gas.

IDEAL GAS MIXTURES

Many of the fluids involved in engineering systems are physical mixtures of the permanent gases or one or more of these with superheated or saturated vapors. For example, normal atmospheric air is a mixture of oxygen and nitrogen with traces of other gases, plus superheated or saturated water vapor, or at times saturated vapor and liquid. If the properties of each constituent of a mixture would have to be considered individually during an analysis of a system, the procedures would be

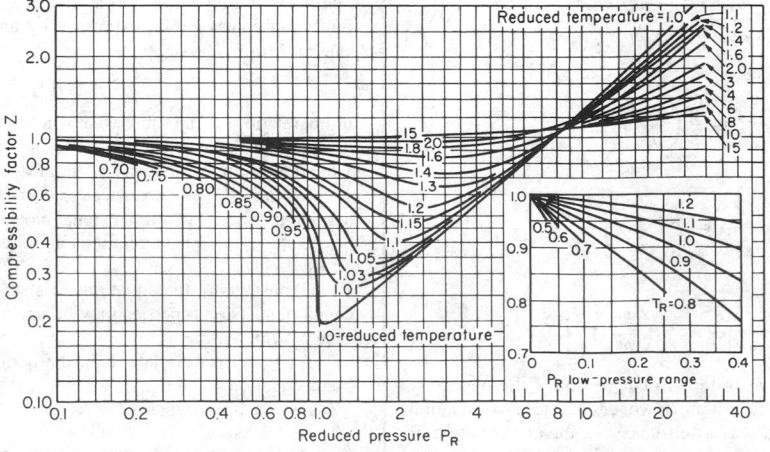

Fig. 4.1.3 Compressibility factors for gases and vapors. (*From Hougen and Watson, "Chemical Process Principles,"* Wiley.)

very complex. Experience has demonstrated that a mixture of gases may be regarded as an equivalent gas, the properties of which depend upon the kind and proportion of each of the constituents. The general relations applicable to a mixture of perfect gases will be presented. Let V denote the total volume of the mixture, $m_1, m_2, m_3, \ldots$ the masses of the constituent gases, $R_1, R_2, R_3, \ldots$ the corresponding gas constants, and R_m the constant for the mixture. The **partial pressures** of the constituents, i.e., the pressures that the constituents would have if occupying the total volume V, are $p_1 = m_1 R_1 T/V$, $p_2 = m_2 R_2 T/V$, etc.

According to Dalton's law, the **total pressure p of the mixture** is the sum of the partial pressures; i.e., $p = p_1 + p_2 + p_3 + \cdots$. Let $m = m_1 + m_2 + m_3 + \cdots$ denote the total mass of the mixture; then $pV = mR_m T$ and $R_m = \Sigma(m_i R_i)/m$. Also $p_1/p = m_1 R_1/(mR_m)$, $p_2/p = m_2 R_2/(mR_m)$, etc.

Let $V_1, V_2, V_3 + \ldots$ denote the volumes that would be occupied by the constituents at pressure p and temperature T (these are given by the volume composition of the gas). Then $V = V_1 + V_2 + V_3 + \cdots$ and the apparent molecular weight m_m of the mixture is $m_m = \Sigma(m_i V_i)/V$. Then $R_m = 1{,}546/m_m$. The subscript i denotes an individual constituent.

Volume of 1 lb at 32°F and atm pressure = $359/m_m$.

Mass of 1 ft³ at 32°F and atm pressure = $0.002788 m_m$.

The specific **heats of the mixture** are, respectively,

$$c_p = \Sigma(m_i c_{pi})/m \qquad c_v = \Sigma(m_i c_{vi})/m$$

Internal Energy, Enthalpy, and Entropy of an Ideal Gas If an ideal gas with constant specific heats changes from an initial state p_1, V_1, T_1 to a final state p_2, V_2, T_2, the following equations hold:

$$u_2 - u_1 = mc_v(T_2 - T_1) = (p_2 v_2 - p_1 v_1)(k - 1)$$

$$h_2 - h_1 = mc_p(T_2 - T_1) = k\,\frac{(p_2 v_2 - p_1 v_1)}{k - 1}$$

$$s_2 - s_1 = m\left(c_v \ln \frac{T_2}{T_1} + R \ln \frac{v_2}{v_1}\right)$$
$$= m\left(c_p \ln \frac{T_2}{T_1} - R \ln \frac{p_2}{p_1}\right) = m\left(c_p \ln \frac{v_2}{v_1} + c_v \ln \frac{p_2}{p_1}\right)$$

In general, the energy per unit mass is $u = c_v T + u_0$, the enthalpy is $h = c_p T + h_0$, and the entropy is $s = c_v \ln T + R \ln v + s_0 = c_p \ln T - R \ln p + s_0' = c_p \ln v + c_p \ln p = s_0''$.

The two fundamental equations for ideal gases are

$$dq = c_v\, dT + p\, dv \qquad dq = c_p\, dT - v\, dp$$

SPECIAL CHANGES OF STATE FOR IDEAL GASES

(Specific heats assumed constant)

In the following formulas, the subscripts 1 and 2 refer to the initial and final states, respectively.

1. *Constant volume:* $p_2/p_1 = T_2/T_1$.

$$Q_{12} = U_2 - U_1 = mc_v(t_2 - t_1) = V(p_2 - p_1)/(k - 1)$$
$$W_{12} = 0 \qquad s_2 - s_1 = mc_v \ln (T_2/T_1)$$

2. *Constant pressure:* $V_2/V_1 = T_2/T_1$.

$$W_{12} = -p(V_2 - V_1) = -mR(t_2 - t_1)$$
$$Q_{12} = mc_p(t_2 - t_1) = kW_{12}/(k - 1)$$
$$s_2 - s_1 = mc_p \ln(T_2/T_1)$$

3. *Isothermal (constant temperature):* $p_2/p_1 = V_1/V_2$.

$$U_2 - U_1 = 0 \qquad W_{12} = -mRT \ln (V_2/V_1) = -p_1 V_1 \ln (V_2/V_1)$$
$$Q_{12} = -W_{12} \qquad s_2 - s_1 = Q_{12}/T = mR \ln (V_2/V_1)$$

4. *Reversible adiabatic, isentropic:* $p_1 V_1^k = p_2 V_2^k$.

$$T_2/T_1 = (V_1/V_2)^{k-1} = (p_2/p_1)^{(k-1)/k}$$
$$W_{12} = U_1 - U_2 = mc_v(t_1 - t_2)$$
$$Q_{12} = 0 \qquad s_2 - s_1 = 0$$
$$W_{12} = (p_2 V_2 - p_1 V_1)/(k - 1)$$
$$= -p_1 V_1 [(p_2/p_1)^{(k-1)/k} - 1]/(k - 1)$$

5. *Polytropic:* This name is given to the change of state which is represented by the equation $pV^n = $ const. A polytropic curve usually represents actual expansion and compression curves in motors and air compressors for pressures up to a few hundred pounds. By giving n different values and assuming specific heats constant, the preceding changes may be made special cases of the polytropic change, thus,

For $n = 1$,	pv	$=$ const	isothermal
$n = k$,	pv^k	$=$ const	isentropic
$n = 0$,	p	$=$ const	constant pressure
$n = \infty$,	v	$=$ const	constant volume

For a polytropic change of an ideal gas (for which c_v is constant), the specific heat is given by the relation $c_n = c_v(n - k)(n - 1)$; hence for $1 < n < k$, c_n is negative. This is approximately the case in air compression up to a few hundred pounds pressure. The following are the principal formulas:

$$p_1 V_1^n = p_2 V_2^n$$
$$T_2/T_1 = (V_1/V_2)^{n-1} = (p_2/p_1)^{(n-1)/n}$$
$$W_{12} = (p_2 V_2 - p_1 V_1)/(n - 1)$$
$$= -p_1 V_1 [(p_2/p_1)^{(n-1)/n} - 1]/(n - 1)$$
$$Q_{12} = mc_n(t_2 - t_1)$$
$$W_{12} : U_2 - U_1 : Q_{12} = k - 1 : 1 - n : k - n$$

The quantity $(p_2/p_1)^{(k-1)/k} - 1$ occurs frequently in calculations for perfect gases.

Determination of Exponent n If two representative points $(p_1, V_1$ and $p_2, V_2)$ be chosen, then

$$n = (\log p_1 - \log p_2)/(\log V_2 - \log V_1)$$

Several pairs of points should be used to test the constancy of n.

Changes of State with Variable Specific Heat In case of a considerable range of temperature, the assumption of constant specific heat is not permissible, and the equations referring to changes of state must be suitably modified. (This statement does not apply to inert or monatomic gases.) Experiments on the specific heat of various gases show that the specific heat may sometimes be taken as a linear function of the temperature: thus, $c_v = a + bT$; $c_p = a' + b'T$. In that case, the following expressions apply for the change of internal energy and entropy, respectively:

$$U_2 - U_1 = m[a(T_2 - T_1) + 0.5b(T_2^2 - T_1^2)]$$
$$S_2 - S_1 = m[a \ln (T_2/T_1) + b(T_2 - T_1) + R \ln (V_2/V_1)]$$

and for an isentropic change,

$$W_{12} = U_2 - U_1$$
$$R \ln (V_1/V_2) = a \ln (T_2/T_1) + b(T_2 - T_1)$$

GRAPHICAL REPRESENTATION

The change of state of a substance may be shown graphically by taking any two of the six variables p, V, T, S, U, H as independent coordinates and drawing a curve to represent the successive values of these two variables as the change proceeds. While any pair may be chosen, there are three systems of graphical representation that are specially useful.

1. p and V. The curve (Fig. 4.1.4) represents the simultaneous values of p and V during the change (reversible) from state 1 to state

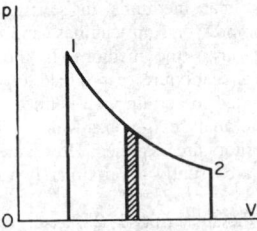

Fig. 4.1.4

2. The area between the curve and the axis OV is given by the integral $\int_{v_1}^{v_2} p \, dV$ and therefore represents the external work W_{12} done by the gas during the change. The area included by a closed cycle represents the work of the cycle (as in the indicator diagram of the steam engine).

2. T and S (Fig. 4.1.5). The absolute temperature T is taken as the ordinate, the entropy S as the abscissa. The area between the curve of change of state and the S axis is given by the integral $\int_{S_1}^{S_2} T \, dS$, and it therefore represents the heat Q_{12} absorbed by the substance from external sources provided there are no irreversible effects. On the T-S diagram, an isothermal is a straight line, as AB, parallel to the S axis; a *reversible* adiabatic is a straight line, as CD, parallel to the T axis.

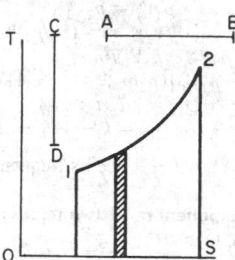

Fig. 4.1.5

In the case of internal generation of heat through friction, as in steam turbines, the increase of entropy is given by $\int_{T_1}^{T_2} (dQ'/T)$ and the area under the curve represents the heat Q' thus generated. In this case, an adiabatic is *not* a straight line parallel to the T axis.

3. H and S. In the system of representation devised by Dr. Mollier, the enthalpy H is taken as the ordinate and the entropy S as the abscissa. If on this diagram (Fig. 4.1.6) a line of constant pressure, as 12, be drawn, the heat absorbed during the change at constant pressure is given by $Q_{12} = H_2 - H_1$, and this is represented by the line segment 23. The **Mollier diagram** is specially useful in problems that involve the flow of fluids, throttling, and the action of steam in turbines.

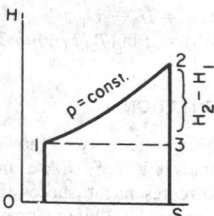

Fig. 4.1.6

IDEAL CYCLES WITH PERFECT GASES

Gases are used as heat mediums in several important types of machines. In air compressors, air engines, and air refrigerating machines, atmospheric air is the medium. In the internal-combustion engine, the medium is a mixture of products of combustion. Engines using gases are operated in certain well-defined cycles, which are described below. In the analyses given, ideal conditions that cannot be attained by actual motors are assumed. However, conclusions derived from such analyses are usually approximately valid for the modified actual cycle.

In the following, the subscripts 1, 2, 3, etc., refer to corresponding points shown in the figures. The work of the cycle is denoted by W and the net heat absorbed by Q.

Carnot Cycle The Carnot cycle (Fig. 4.1.7) is of historic interest. It consists of two isothermals and two isentropics. The heat absorbed

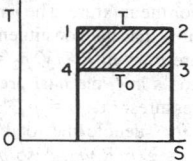

Fig. 4.1.7 Carnot cycle.

along the upper isothermal 12 is $Q_{12} = mRT \ln (V_2/V_1)$, and the heat transformed into work, represented by the cycle area, is $W = Q_{12}(1 - T_0/T)$.

$$W = -mR(T - T_0) \ln \left(\frac{V_2}{V_1} \right)$$

If the cycle is traversed in the reverse sense, $Q_{43} = mRT_0 \ln (V_3/V_4)$ is the heat absorbed from the cold body (brine), and the ratio $Q_{43} : (W) = T_0 : (T - T_0)$ is the **coefficient of performance** of the refrigerating machine.

Leff (*Amer. J. Phys.*, **55,** no. 7, 1987, pp. 602–610) showed that the **thermal efficiency** of a heat engine producing the maximum possible work per cycle consistent with its operating temperature range resulted in efficiencies equal to or well approximated by $\eta = 1 - \sqrt{T_c/T_h}$, where c = cold and h = hot, as found by Curzon and Ahlborn (*Amer. J. Phys.*, **43,** no. 1, 1975, pp. 22–24) for maximum power output. If the work output per cycle is kept fixed, the thermal efficiency can be increased by operating the heat engine at less than maximum work output per cycle, the limit being an engine of infinite size having a Carnot efficiency. Leff's paper considers Otto, Brayton-Joule, Diesel, and Atkinson cycles. Figure 4.1.8 illustrates the difference between maximum power and maximum efficiency. Detailed discussion of **finite time thermodynamics** appears in Sieniutcyz and Salamon, "Finite Time Thermodynamics and Thermoeconomics," *Advan. Thermo.*, **4,** 306 pp., 1990, Taylor & Francis, London.

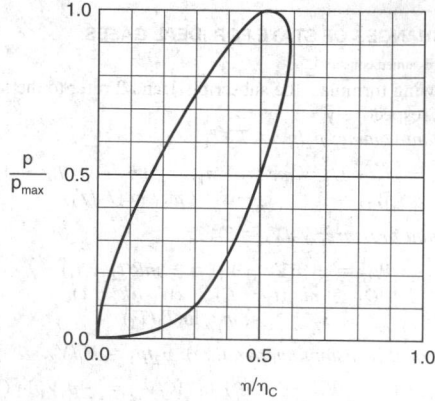

Fig. 4.1.8 Ratio of actual power to maximum power as a function of ratio of actual thermal efficiency to Carnot efficiency.

Finite time thermodynamics is a term applied to the consideration that, for any finite energy transfer, a finite time must occur. A common statement in the literature is that the analysis started from the work of

Curzon and Ahlborn (*Amer. J. Phys.*, **43**, 1975, pp. 22–24). According to Bejan (*Amer. J. Phys.*, **62**, no. 1, Jan. 1994, pp. 11–12), this statement is not true, and the original analysis was by Novikov (*At. Energy*, **3**, 1957, p. 409, and *Nucl. Energy*, pt. II, **7**, 1958, pp. 125–148). Wu (*Energy Convsn. Mgmt.*, **34**, no. 12, 1993, pp. 1239–1247) discusses the *endoreversible Carnot heat engine* being one in which all the losses are associated with the transfer of heat to and from the engine, there being no internal losses within the engine itself and refers to Wu and Kiang (*Trans. J. Eng. Gas Turbines & Power*, **113**, 1991, p. 501) for a detailed literature survey.

Otto and Diesel Cycles   The ideal cycles usually employed for internal-combustion engines may be classified in two groups: (1) explosive—Otto (the fluid is introduced in gaseous form), (2) nonexplosive—Diesel, Joule (the fluid is introduced in liquid form).

Otto Cycle   (Fig. 4.1.9 for pressure-volume plane, Fig. 4.1.10 for temperature-entropy plane)   Isentropic compression 12 is followed by ignition and rapid heating at constant volume 23. This is followed by isentropic expansion, 34. Assuming constant specific heats the following relations hold:

$$\frac{T_2}{T_1} = \frac{T_3}{T_4} = \left(\frac{p_2}{p_1}\right)^{k-1/k} = \left(\frac{p_3}{p_4}\right)^{k-1/k} = \left(\frac{V_1}{V_2}\right)^{k-1}$$

$$Q_{23} = mc_v(T_1 - T_2)$$

$$W = Q_{23}[1 - (T_1/T_2)] = mc_v(T_3 - T_4 - T_2 + T_1)$$

$$\text{Efficiency} = 1 - \frac{T_1}{T_2} = 1 - \left(\frac{V_2}{V_1}\right)^{k-1} = 1 - \left(\frac{p_1}{p_2}\right)^{(k-1)/k}$$

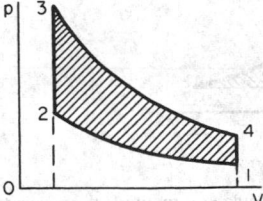

Fig. 4.1.9   Otto cycle.

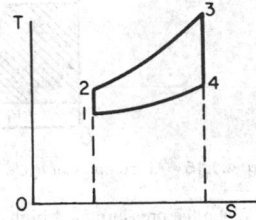

Fig. 4.1.10   Otto cycle.

If the compression and expansion curves are polytropics with the same value of n, replace k by n in the first relation above. In this case,

$$W = [(p_3V_3 - p_4V_4) - (p_2V_2 - p_1V_1)]/(n - 1)$$
$$= mR(T_3 - T_4 - T_2 + T_1)/(n - 1)$$

The **mean effective pressure** of the diagram is given by

$$p_m = ap_1(p_3/p_2 - 1)$$

where a has the values given in the following table.

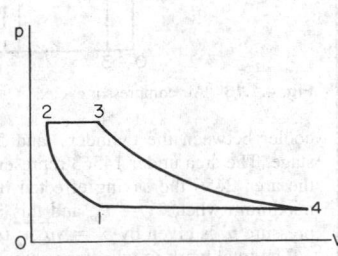

Fig. 4.1.12   Joule or Brayton cycle.

$p_2/p_1 =$	3	4	5	6	8	10	12	14	16
($n = 1.4$)	$a = 1.70$	1.94	2.13	2.31	2.62	2.88	3.10	3.31	3.50
($n = 1.3$)	$a = 1.69$	1.92	2.11	2.28	2.57	2.81	3.03	3.22	3.39
($n = 1.2$)	$a = 1.68$	1.90	2.08	2.25	2.51	2.74	2.94	3.12	3.27

A later paper by Wu and Blank (*Energy Convsn. Mgmt.*, **34**, no. 12, 1993, pp. 1255–1269) considered optimization of the endoreversible Otto cycle with respect to both power and mean effective pressure.

Diesel Cycle   In the diesel oil engine, air is compressed to a high pressure. Fuel is then injected into the air, which is at a temperature above the ignition point, and it burns at nearly constant pressure (23, in Fig. 4.1.11). Isentropic expansion of the products of combustion is followed by exhaust and suction of fresh air, as in the Otto cycle.

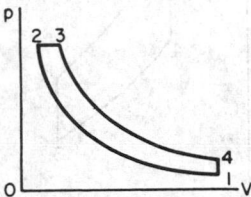

Fig. 4.1.11   Diesel cycle.

The work obtained is

$$W = m[c_p(T_3 - T_2) - c_v(T_4 - T_1)]$$

and the efficiency of the ideal cycle is

$$1 - [(T_4 - T_1)/k(T_3 - T_2)]$$

The **Joule cycle**, also called the **Brayton cycle** (Fig. 4.1.12), consists of two isentropics and two constant-pressure lines. The following relations hold:

$$V_3/V_2 = V_4/V_1 = T_3/T_2 = T_4/T_1$$

$$\frac{T_2}{T_1} = \frac{T_3}{T_4} = \left(\frac{V_1}{V_2}\right)^{k-1} = \left(\frac{V_4}{V_3}\right)^{k-1} = \left(\frac{p_2}{p_1}\right)^{k-1/k}$$

$$W = mc_p(T_3 - T_2 - T_4 + T_1)$$
$$\text{Efficiency} = W/Q_{23} = 1 - T_1/T_2$$

The Joule cycle has assumed renewed importance as a basis for analysis of gas turbine operation.

For additional information on internal combustion engines, see Campbell, "Thermodynamic Analysis of Internal Combustion Engines," Wiley; Taylor, "The Internal Combustion Engine in Theory and Practice," MIT Press. New designs for internal-combustion engines were reviewed by Wallace (*Sci. Progr., Oxford*, **75**, 1991, pp. 15–32).

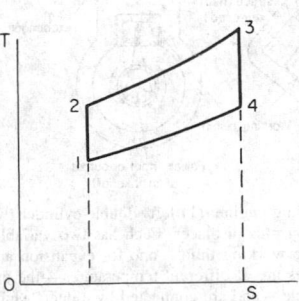

Stirling Cycle The Stirling engine may be visualized as a cylinder with a piston at each end. Between the pistons is a regenerator. The cylinder is assumed to be insulated except for a contact with a hot reservoir at one end and a contact with a cold reservoir at the other end.

Starting with state 1, Fig. 4.1.13, heat from the hot reservoir is added to the gas at T_H (or $T_H - dT$). During the reversible isothermal process, the left piston moves outward, doing work as the system volume increases and the pressure falls. Both pistons are then moved to the right at the same rate to keep the system volume constant (process 2–3). No

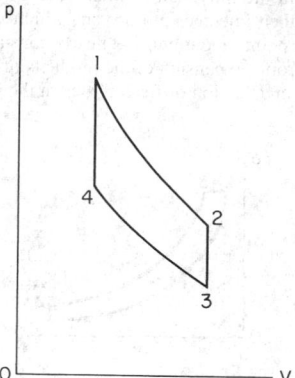

Fig. 4.1.13 Stirling cycle.

heat transfer occurs with either reservoir. As the gas passes through the regenerator, heat is transferred from the gas to the regenerator, causing the gas temperature to fall to T_L by the time the gas leaves the right end of the regenerator. For this heat-transfer process to be reversible, the temperature of the regenerator at each point must equal the gas temperature at that point. Hence there is a temperature gradient through the regenerator from T_H at the left end to T_L at the right end. No work is accomplished during this process. During the path 3–4, heat is removed from the gas at T_L (or $T_L + dT$) to the reservoir at T_L. To hold the gas temperature constant, the right piston is moved inward—doing work on the gas with a resulting increase in pressure. During process 4–1, both pistons are moved to the left at the same rate to keep the system volume constant. The pistons are closer together during this process than they were during process 2–3, since $V_4 = V_1 < V_2 = V_3$. No heat is transferred to either reservoir. As the gas passes back through the regenerator, the energy stored in the regenerator during 2–3 is returned to the gas. The gas emerges from the left end of the regenerator at temperature T_H. No work is performed during this process since the

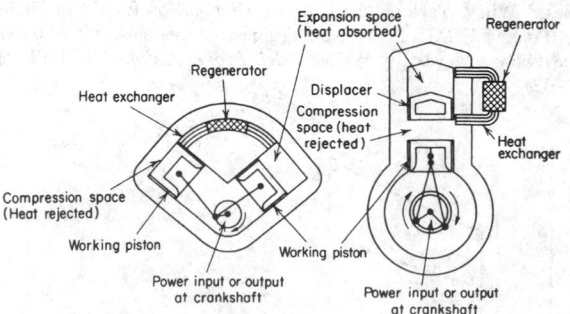

Fig. 4.1.14 Two main types of Stirling engine: (1) left, double-cylinder two piston; (2) right, single-cylinder, piston plus displacer. Each has two variable-volume working spaces filled with the working fluid—one for expansion and one for compression of the gas. Spaces are at different temperatures—the extreme temperatures of the working cycle—and are connected by a duct, which holds the regenerators and heat exchangers. (*Intl. Science and Technology,* May 1962.)

volume remains constant. Thus the cycle is completed and is externally reversible. The system exchanges a net amount of heat with only the two energy reservoirs T_H and T_L. Two types of Stirling engines are shown in Fig. 4.1.14. Extensive research-and-development effort has been devoted to the Stirling engines for future use as prime movers in space power systems operating on solar energy. (See also Sec. 9.6.)

More information can be found in Meijer, *De Ingenieur,* **81,** nos. 18 and 19, 1969; Reader and Hooper, "Stirling Engines," Spon, London; Sternlicht, *Chem. Tech.,* **13,** 1983, pp. 28–36; Walker, "Stirling Engines," Oxford Univ. Press.

AIR COMPRESSION

It is assumed that the compressor works under ideal reversible conditions without clearance and without friction losses and that the changes are over ranges where the gas laws are applicable. Where the gas laws cannot be used, analysis in terms of Z charts is convenient. If the compression from p_1 to p_2 (Fig. 4.1.15) follows the law $pV^n = $ const, the work represented by the indicator diagram is

$$W = n(p_2 V_2 - p_1 V_1)/(n - 1)$$
$$= n p_1 V_1 [(p_2/p_1)^{(n-1)/n} - 1]/(n - 1)$$

The temperature at the end of compression is given by $T_2/T_1 = (p_2/p_1)^{(n-1)/n}$. The work W is smaller the smaller the value on n, and the purpose of the water jackets is to reduce n from the isentropic value 1.4. Under usual working conditions, n is about 1.3.

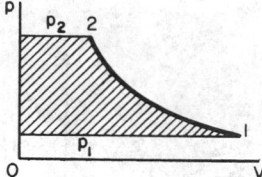

Fig. 4.1.15 Air compressor cycle.

When the pressure p_2 is high, it is advantageous to divide the process into two or more stages and cool the air between the cylinders. The saving effected is best shown on the T-S plane (Fig. 4.1.16). With single-stage compression, 12 represents the compression from p_1 to p_2, and if the constant-pressure line 23 is drawn cutting the isothermal through point 1 in point 3, the area $1'1233'$ represents the work W. When two stages are used, 14 represents the compression from p_1 to an intermediate pressure p', 45 cooling at constant pressure in the inter-

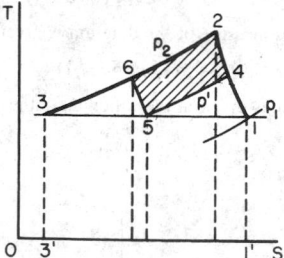

Fig. 4.1.16 Air compressor cycle.

cooler between the cylinders, and 56 the compression in the second stage. The area under 14563 represents the work of the two stages and the area 2456 the saving effected by compounding. This saving is a maximum when $T_4 = T_6$, and this is the case when the intermediate pressure p' is given by $p' = \sqrt{p_1 p_2}$ (see Sec. 14).

The total work in two-stage compression is

$$n p_1 V_1 [(p'/p_1)^{(n-1)/n} + (p_2/p')^{(n-1)/n} - 2]/(n - 1)$$

Gas Turbine The Brayton cycle, also called the Joule or constant pressure cycle, employs an air engine, a compressor, and a combustion chamber. Air enters the compressor wherein the pressure is increased. Fuel burning in the combustion chamber raises the temperature of the compressed air under constant-pressure conditions. The resulting high-temperature gases are then introduced to the engine where they expand and perform work. The excess work of the engine over that required to compress the air is available for operating other devices, such as a generator.

Basically, the simple gas-turbine cycle is the same as the Brayton cycle, except that the air compressor and engine are replaced by an axial flow compressor and gas turbine. Air is compressed in the compressor, after which it enters a combustion chamber where the temperature is increased while the pressure remain constant. The resulting high-temperature air then enters the turbine, thereby performing work.

Boyce (''Gas Turbine Engineering Handbook,'' Gulf) gives numerous examples of ideal and actual gas-turbine cycles. The graphs in this source as well as in a review by Dharmadhikari (*Chemical Engineer (London)*, Feb. 1989, pp. 16–20) show the same relation between work output and thermal efficiency as the general graph of Gordon (*Amer. J. Phys.*, **59**, no. 6, 1991, pp. 551–555). MacDonald (ASME Paper 89-GT-103, Toronto Gas Turbine Exposition, 1989) describes the increasing use of heat exchangers in gas-turbine plants and reviews the use of recuperators (i.e., regenerators) (in *Heat Recovery Systs. & CHP.* **10**, no. 1, 1990, pp. 1–30). Gas turbines as the topping cycle with steam in the bottoming cycle were described by Huang (ASME Paper 91-GT-186 and *J. Eng. Gas Turbines & Power*, **112**, Jan. 1990, pp. 117–121) and by Cerri (*Trans. ASME*, **109**, Jan. 1987, pp. 46–54). The use of steam injection in gas-turbine cycles has received renewed attention; see, e.g., Consonni (45th Congr. Nat. Assoc. Termotecnica Ital., **IIID**, 1990, pp. 49–60), Ediss (City Univ. London Conf. Paper, Nov. 1991), Lundberg (*ASME Cogen-Turbo IGTI*, **6**, 1991, pp. 9–18). Fraize and Kinney (*J. Eng. Power*, **101**, 1979, pp. 217–227), and Larson and Williams (*J. Eng. Gas Turbines & Power*, **109**, Jan. 1987, pp. 55–63).

Analysis of closed-cycle gas-turbine plant for maximum and zero power output and for maximum efficiency was made by Woods et al. (*Proc. Inst. Mech. Eng.*, **205A**, 1991, pp. 59–66). See also pp. 287–291 and ibid., **206A**, 1992, pp. 283–288. A series of papers by Najjar appeared in *Int. J. M. E. Educ.* (**15**, no. 4, 1987, pp. 267–286); *High Temp. Technol.* (**8**, no. 4, 1990, pp. 283–289); *Heat Recovery Systems & CHP* (**6**, no. 4, 1986, pp. 323–334). For more detailed information regarding the actual gas-turbine cycles see Sec. 9.

VAPORS

General Characteristics of Vapors Let a gas be compressed at constant temperature; then, provided this temperature does not exceed a certain critical value, the gas begins to liquefy at a definite pressure, which depends upon the temperature. At the beginning of liquefaction, a unit mass of gas will also have a definite volume v_g, depending on the temperature. In Fig. 4.1.17, *AB* represents the compression and the point *B* gives the **saturation** pressure and volume. If the compression is continued, the pressure remains constant with the temperature, as in-

dicated by *BC*, until at *C* the substance is in the liquid state with the volume v_f.

The curves v_f and v_g giving the volumes for various temperatures at the end and beginning of liquefaction, respectively, may be called the **limit curves.** A point *B* on curve v_g represents the state of **saturated vapor;** a point *C* on the curve v_f represents the saturated liquid state; and a point *M* between *B* and *C* represents a mixture of vapor and liquid of which the part $x = MC/BC$ is vapor and the part $1 - x = BM/BC$ is liquid. The ratio x is called the **quality of the mixture.** The region between the curves v_f and v_g is thus the region of liquid and vapor mixtures. The region to the right of curve v_g is the region of **superheated vapor.** The curve v_g dividing these regions represents the so-called **saturated vapor**.

For saturated vapor, saturated liquid, or a mixture of vapor and liquid, the pressure is a function of the temperature only, and the volume of the mixture depends upon the temperature and quality x. That is, $p = f(t)$, $v = F(t, x)$.

For the vapor in the superheated state, the volume depends on pressure and temperature [$v = F(p, t)$], and these may be varied independently.

Critical State If the temperature of the gas lies above a definite temperature t_c called the **critical temperature,** the gas cannot be liquefied by compression alone. The saturation pressure corresponding to t_c is the **critical pressure** and is denoted by p_c. At the critical states, the limit curves v_f and v_g merge; hence for temperatures above t_c, it is impossible to have a mixture of vapor and liquid. Table 4.2.21 gives the critical data for various gases; also the boiling temperature t_b, corresponding to atmospheric pressure. Study of the critical region is becoming a specialized topic. NBS Misc. Publ. 273, 1966, contains 33 papers on phenomena near critical points. The ASME symposia on thermophysical properties proceedings contain numerous papers on the subject.

Vapor Pressures At a specified temperature, a pure liquid can exist in equilibrium contact with its vapor at but one pressure, its vapor pressure. A plot of these pressures against the corresponding temperatures is known as a vapor pressure curve. As noted by Martin, ''Thermodynamic and Transport Properties of Gases, Liquids, and Solids,'' ASME, New York, p. 112, the true shape of a log vapor pressure versus reciprocal absolute temperature curve is an S shape. But if the curvature (often slight) is neglected, the equation of the curve becomes $\ln P = A + B/T$. In terms of any two pairs of values (P_1, T_1), (P_2, T_2), $A = (T_2 \ln P_2 - T_1 \ln P_1)/(T_2 - T_1)$ and $B = T_1 T_2/(T_2 - T_1) \ln (P_1/P_2)$. (Note that B is always negative.) Once the values of A and B have been determined, the equation can be used to determine P_3 at $T = T_3$ or T_3 at $P = P_3$. Algebraically, A and B can be eliminated to yield $\ln (P_3/P_1) = [T_2(T_3 - T_1)/T_3(T_2 - T_1)] \ln (P_2/P_1)$, and at any temperature T the slope of the vapor pressure curve is $dP/dT = (1/T^2)[T_1 T_2/(T_1 - T_2)] \ln (P_2/P_1)$.

A classic survey of equations for estimating vapor pressures was given by Miller in *Ind. Eng. Chem.*, **56**, 1964, pp. 46–57. The comprehensive tables of Stull in *Ind. Eng. Chem.*, **39**, 1947 pp. 517–550, are useful though slightly dated. Table 4.2.24 gives $T(K)$ for various $P(bar)$ for 50 substances; $P(bar)$ tables for various $T(K)$ for 16 elements are given in Table 4.2.29. Boublik et al., ''The Vapor Pressure of Pure Substances,'' Elsevier, presents an extensive collection of data.

THERMAL PROPERTIES OF SATURATED VAPORS AND OF VAPOR AND LIQUID MIXTURES

Notation

v_f, v_g = specific volume of 1 lb of saturated liquid and vapor, respectively

c_f, c_g = specific heat of saturated liquid and vapor, respectively

h_f, h_g = specific enthalpy of saturated liquid and vapor, respectively

u_f, u_g = specific internal energy of saturated liquid and vapor, respectively

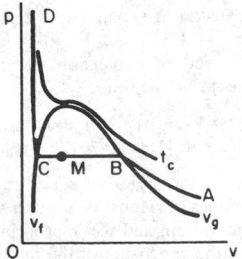

Fig. 4.1.17

s_f, s_g = specific entropy of saturated liquid and vapor, respectively

$v_{fg} = vg - v_f$ = increase of volume during vaporization

$h_{fg} = h_g - h_f$ = heat of vaporization, or heat required to vaporize a unit mass of liquid at constant pressure and temperature

Note that r may be used for h_{fg} when several heats of vaporization (as r_1, r_2, r_3, etc.) are under consideration.

$u_{fh} = u_g - u_f$ = increase of internal energy during vaporization

$s_{fg} = s_g - s_f = h_{fg}/T$ = increase of entropy during vaporization

pv_{fg} = work performed during vaporization

The energy equation applied to the vaporization process is

$$h_{fg} = u_{fg} + pv_{fg}$$

The properties of a unit mass of a mixture of liquid and vapor of quality x are given by the following expressions:

$$v = v_f + xv_{fg}$$
$$h = h_f + xh_{fg}$$
$$u = u_f + xu_{fg}$$
$$s = s_f + xs_{fg}$$

Any property ψ can be expressed as a function of the property of the saturated liquid, ψ_f, that of the saturated vapor, ψ_g, and the quality x by three entirely equivalent equations:

$$\psi = (1 - x)\psi_f + x\psi_g$$
$$= \psi_f + x\psi_{fg}$$
$$= \psi_g - (1 - x)\psi_{fg}$$

where $\psi_{fg} = \psi_g - \psi_f$. Tables of superheated vapor usually give values of v, h, and s per unit mass. If not tabulated, the internal energy u per unit mass can be found from the equation

$$u = h - pv$$

CHARTS FOR SATURATED AND SUPERHEATED VAPORS

Certain properties of vapor mixtures and superheated vapors may be shown graphically by means of charts. Such charts show the behavior of vapors and have a practical application in the solution of certain problems.

Temperature-Entropy Chart Figures 4.2.10 and 4.2.11 show the temperature-entropy chart for water vapor. The liquid curve is obtained by plotting corresponding values of T and s_f, and the saturation curve by plotting values of T and s_g. The values are taken from Tables 4.2.17 to 4.2.20. The two curves merge into each other at the critical temperature $T = 1,165.4°R$ (647 K). Between these two curves, constant pressure lines are also lines of constant temperature; but at the saturation curve the constant pressure lines show a sharp break with rising temperature. The constant quality lines $x = 0.2$, 0.4, etc., are equally spaced between the liquid and saturation curves.

Figure 4.2.1 is a temperature-entropy chart for air.

Enthalpy-Entropy Chart (Mollier Chart) In this chart, the enthalpy h is taken as the ordinate and the entropy s as the abscissa.

Enthalpy–Log Pressure Chart Previously, a chart with coordinates of enthalpy and pressure was termed a pressure-enthalpy chart. In this edition these charts are called enthalpy–log pressure charts, to more correctly identify the scale plotted for pressure. This follows modern usage. Charts with pressure per se as a coordinate have a greatly different scale and appearance.

For examples of the enthalpy–log-pressure chart, see Sec. 4.2. For enthalpy–log-pressure charts for various fluids, see ''Engineering Data Book,'' 9th ed., Gas Processors Suppliers Assoc., Tulsa, OK; Reynolds, ''Thermodynamic Properties in SI,'' Mech. Eng. Stanford Univ. Publication.

The energy-temperature diagram reported by Bucher (*Amer. J. Phys.,* **54,** 1986, pp. 850–851) for reversible cycles and by Wallingford (*Amer. J. Phys.,* **57,** 1989, pp. 379–381) for irreversible cycles was claimed by Bejan (*Amer. J. Phys.,* **62,** no. 1, Jan. 1994, pp. 11–12) to have been first reported at an earlier date (Bejan, *Mech. Eng. News,* May 1977, pp. 26–28).

CHANGES OF STATE. SUPERHEATED VAPORS AND MIXTURES OF LIQUID AND VAPOR

Isothermal In the only important cases, the fluid is a mixture of liquid and vapor in both initial and final states.

$$t = \text{const} \qquad p = \text{const}$$
$$x_1, x_2 = \text{initial and final qualities}$$
$$Q_{12} = mh_{fg}(x_2 - x_1)$$
$$W_{12} = mpv_{fg}(x_2 - x_1)$$
$$U_2 - U_1 = mu_{fg}(x_2 - x_1)$$
$$S_2 - S_1 = Q_{12}/T$$

Constant Pressure If the fluid is a mixture at the beginning and end of the change, the constant pressure change is also isothermal. If the initial state is in the mixture region and the final state is that of a superheated vapor, the following are the equations for Q_{12}, etc. Let h_2, u_2, v_2, and s_2 be the properties of 1 lb of superheated vapor in the final state 2; then

$$Q_{12} = m(h_2 - h_1)$$
$$U_2 - U_1 = m(u_2 - u_1)$$
$$S_2 - S_1 = m(s_2 - s_1)$$
$$W_{12} = -mp(v_2 - v_1)$$
$$h_1 = h_{f1} + x_1 h_{fg1}$$
$$u_1 = u_{f1} + x_1 u_{fg1}$$
$$s = s_{f1} + x_1 s_{fg1}$$
$$v_1 = v_{f1} + x_1 v_{fg1}$$

Constant Volume Since v_f the liquid volume is nearly constant,

$$x_1 v_{fg1} = x_2 v_{fg2}$$
$$x_2 = x_1 v_{fg1}/v_{fg2} \quad \text{or} \quad x_2 = x_1 v_{g1}/v_{g2} \text{ approx}$$
$$Q_{12} = U_2 - U_1 = m(u_2 - u_1) \qquad W_{12} = 0$$

Isentropic $s = \text{const}$. If the fluid is a mixture in the initial and final states,

$$s_{f1} + x_1 s_{fg1} = s_{f2} + x_2 s_{fg2}$$

If the initial state is that of superheated vapor,

$$s_1 = s_{f2} + x_2 s_{fg2}$$

in which s_1 is read from the table of superheated vapor. The final value x_2 is determined from one of these equations, and the final internal energy u_2 is then

$$u_{f2} + x_2 u_{fg2} \qquad Q_{12} = 0 \qquad W_{12} = U_2 - U_1 = m(u_2 - u_1)$$

For water vapor, the relation between p and v during an isentropic change may be represented approximately by the equation $pv^n = \text{constant}$. The exponent n is not constant, but varies with the initial quality and initial pressure, as shown in Table 4.1.2.

The isentropic expansion of superheated steam is fairly represented by $pv^n = \text{const}$, with $n = 1.315$.

The volume at the end of expansion (or compression) is $V_2 = V_1(p_1/p_2)^{1/n}$, and the external work is

$$W_{12} = (p_2 V_2 - p_1 V_1)/(n - 1)$$
$$= -p_1 V_1[1 - (p_2/p_1)^{(n-1)/n}]/(n - 1)$$

If the initial state is in the region of superheat and final state in the mixture region, two values of n must be used: $n = 1.315$ for the expansion to the state of saturation, and the appropriate value from the first row of Table 4.1.2 for the expansion of the mixture.

Table 4.1.2   Values of n (Water Vapor)

Initial quality	Initial pressure, psia											
	20	40	60	80	100	120	140	160	180	200	220	240
1.00	1.131	1.132	1.133	1.134	1.136	1.137	1.138	1.139	1.141	1.142	1.143	1.145
0.95	1.127	1.128	1.129	1.130	1.131	1.131	1.132	1.133	1.134	1.135	1.136	1.137
0.90	1.123	1.123	1.124	1.124	1.125	1.125	1.126	1.126	1.127	1.127	1.128	1.129
0.85	1.119	1.119	1.119	1.119	1.120	1.120	1.120	1.120	1.120	1.120	1.120	1.121
0.80	1.115	1.115	1.114	1.114	1.114	1.114	1.113	1.113	1.113	1.113	1.112	1.112
0.75	1.111	1.110	1.110	1.109	1.109	1.108	1.107	1.106	1.106	1.105	1.104	1.104

MIXTURES OF AIR AND WATER VAPOR

Atmospheric Humidity   The atmosphere is a mixture of air and water vapor. Dalton's law of partial pressures (for the mixture) and the ideal gas law (for each constituent) may safely be assumed to apply. The **total pressure** p_t (barometric pressure) is the sum of the **vapor pressure** p_v and the **air pressure** p_a.

The temperature of the atmosphere, as indicated by an ordinary thermometer, is the **dry-bulb temperature** t_d. If the atmosphere is cooled under constant total pressure, the partial pressures remain constant until a temperature is reached at which condensation of vapor begins. This temperature is the **dew point** t_c (condensation temperature) and is the saturation temperature, or boiling point, corresponding to the actual vapor pressure p_v. If a thermometer bulb is covered with absorbent material, e.g., linen, wet with distilled water and exposed to the atmosphere, evaporation will cool the water and the thermometer bulb to the **wet-bulb temperature** t_w. This is the temperature given by a psychrometer. The wet-bulb temperature lies between the dry-bulb temperature and the dew point. These three temperatures are distinct except for a saturated atmosphere, for which they are identical. For each of these temperatures, there is a corresponding vapor pressure. The actual vapor pressure p_v corresponds with the dew point t_c. The vapor pressures p_d and p_w, corresponding with t_d and t_w, do not represent pressures actually appearing in the atmosphere but are used in computations.

Relative humidity r is the ratio of the actual vapor pressure to the pressure of saturated vapor at the prevailing dry-bulb temperature $r = p_v/p_d$. Within the limits of usual accuracy, this equals the ratio of actual vapor density to the density of saturated vapor at dry-bulb temperature, $r = p_v/p_d$. It is to be noted that relative humidity is a property of the vapor alone; it has nothing to do with the fact that the vapor is mixed with air. It is a method of expressing the departure of the vapor from saturation. (See "ASHRAE Handbook" for information on industrial applications of relative humidity.)

Molal humidity f is the mass of water vapor in mols per 1 mol of air. The laws of Dalton and Avogadro state that the molal composition of a mixture is proportional to the distribution of partial pressures, or $f = p_v/p_a = p_v/(p_t - p_v)$.

Specific humidity (humidity ratio) W is the mass of water vapor (pounds or grains) per pound of dry air. Mass in pounds equals mass in moles multiplied by the molecular weight. The molecular weight of water is 18, and the equivalent molecular weight of air is 28.97. The ratio 28.97/18 = 1.608, or 1.61 with ample accuracy. Thus $W = f/1.61$.

Air density ρ_a is the pounds of air in one cubic foot. **Vapor density** ρ_v is the pounds of vapor in one cubic foot. **Mixture density** ρ_m is the sum of these, i.e., the pounds of air plus vapor in one cubic foot.

Notation   The subscripts a, v, m, and f apply to air, vapor, mixture, and liquid water, respectively. The subscripts d and w apply to conditions pertaining to the dry- and wet-bulb temperature, respectively.

HUMIDITY MEASUREMENTS

Many methods are in use: (1) the **dew point** method measures the temperature at which condensation begins; water-vapor pressure can then be found from steam tables. Dew point apparatus can either cool a surface or compress and expand moist air. (2) **Hygrometers** measure relative humidity, often by using the change in dimensions of a hygroscopic material such as human hair, wood, or paper; these instruments are simple and inexpensive but require frequent calibration. The electrical resistance of an electrolytic film can also be used as an indication of relative humidity. (3) The wet- and dry-bulb **psychrometer** is widely used. Humidity measurements of air flowing in ducts can be made with psychrometers that use mercury-in-glass thermometers, thermocouples, or resistance thermometers. Humidity measurements of still air can be made with sling psychrometers as aspiration psychrometers. Psychrometric wet-bulb temperatures must be corrected to obtain thermodynamic wet-bulb temperatures, or there must be adequate air motion past the wet-bulb thermometer, 800 to 900 ft/min (with duct walls at air temperature), to ensure a proper balance between radiation and convection. (4) **Chemical analysis** by the use of desiccants such as sulfuric acid, phosphorus pentoxide, lithium chloride, or silica gel can be used as primary standards of humidity measurement.

The following equations give various properties in terms of pressure in inches Hg and temperature in degrees Fahrenheit.

Relative humidity: $r = p_v/p_d$

Specific humidity: $W = p_v/1.61(p_t - p_v)$ lb/lb dry air

Volume of mixture per pound of dry air:

$$v_a = \frac{1}{\rho_a} = 0.754(t_d + 460)/(p_t - rp_d) \quad \text{ft}^3$$

Volume of mixture per pound of mixture:

$$v_m = \frac{1}{\rho m} = v_a/(1 + W) \quad \text{ft}^3$$

The **enthalpy** of a mixture of dry air and steam, when each constituent is assumed to be an ideal gas, in Btu per pound of dry air, is the sum of the enthalpy of 1 lb of dry air and the enthalpy of the W lb of steam mixed with that air. The specific enthalpy of dry air (above 0°F) is $h_a = 0.240t_d$ (up to 130°F, the specific heat of dry air is 0.240; at higher temperatures, it is larger). The specific enthalpy of low-pressure steam (saturated or superheated) is nearly independent of the vapor pressure and depends only on t_d. An empirical equation for the specific enthalpy of low-pressure steam for the range of temperatures from −40 to 250°F is

$$h_v = 1,062 + 0.44t_d \quad \text{Btu/lb}$$

The enthalpy of a mixture of air and steam is

$$h_m = 0.240t_d + W(1,062 + 0.44t_d)$$

The specific heat of a mixture of dry air and steam per pound of dry air may be called **humid specific heat** and is $0.240 + 0.44W$ Btu/lb dry air. For a steady-flow process without change of specific humidity, heat transfers per pound of dry air may be computed as the product of humid specific heat and change in dry-bulb temperature.

Thermodynamic Wet-Bulb Temperature (Temperature of Adiabatic Saturation)   The thermodynamic wet-bulb temperature $t*$ is an important property of state of mixtures of dry air and superheated steam; it is the temperature at which water (or ice), by evaporating into a mixture of air and steam, will bring the mixture to saturation at the same temperature in a steady-flow process in the absence of external heat transfer. For a mixture of dry air and saturated steam only, $t^* = t_d$; where $r < 1$, $t^* < t_d$. By writing energy and mass balances for the process of adiabatic saturation with water supplied at t^*, the following equation may be derived:

$$W = W^* - \frac{(0.240 + 0.44W^*)(t_d - t^*)}{1,094 + 0.44t_d - t^*}$$

where W^* = specific humidity for saturation at the total pressure of p_t.

The enthalpy of a mixture of dry air and **saturated** steam at the total pressure p_t and thermodynamic wet-bulb temperature t^* exceeds the enthalpy of a mixture of dry air and **superheated** steam at the same p_t and t^* for

$$h_m^* = h_m + (W^* - W)h_f^*$$

A property of the mixture that remains constant for constant p_t and t^* has been called the Σ **function,** for

$$\Sigma^* = h_m^* - W^*h_f^* = \Sigma = h_m - Wh_f^*$$

EXAMPLE.   A mixture of dry air and **saturated** steam; $p_t = 24$ inHg; $t_d = 76°F$. Partial pressure of water vapor from tables:

$$p_v = p_d = 0.905 \text{ inHg}$$

Partial pressure of dry air: $p_a = p_t - p_v = 23.095$ inHg.
Specific humidity:

$$W = 0.905/1.61(23.095) = 0.0243 \text{ lb/lb dry air}$$

Volume of mixture per pound of dry air:

$$v_a = 0.754(536)/23.095 = 17.5 \text{ ft}^3$$

Volume of mixture per pound of mixture:

$$v_m = 17.5/1.0243 = 17.1 \text{ ft}^3$$

Enthalpy of mixture:

$$h_m = 0.240(76) + 0.0243(1,095) = 44.85 \text{ Btu/lb dry air}$$

EXAMPLE.   A mixture of dry air and **superheated** steam; $p_t = 24$ inHg; $t_d = 76°F$; $t_w = t^* = 62°F$. Pressure of saturated steam at $t^* = 0.560$ inHg (from tables):

$$W^* = 0.560/1.61(23.44) = 0.01484 \text{ lb/lb dry air}$$

Specific humidity:

$$W = 0.01484 - \frac{0.2465(14)}{1,065.4} = 0.0116 \text{ lb/lb dry air}$$

Partial pressure of water vapor:

$$0.0116 = p_v/[1.61(24 - p_v)] \quad \text{and} \quad p_v = 0.44 \text{ inHg}$$

Relative humidity: $r = 0.44/0.905 = 0.486$.
Volume of mixture per pound of dry air:

$$v_a = 0.754(536)/23.56 = 17.2 \text{ ft}^3$$

Volume of mixture per pound of mixture:

$$v_m = 17.2/1.0116 = 17.0 \text{ ft}^3$$

Enthalpy of mixture:

$$h_m = 0.240(76) + 0.0116(1,095) = 30.95 \text{ Btu/lb dry air}$$

PSYCHROMETRIC CHARTS

For occasional use, algebraic equations are less confusing and more reliable; for frequent use, a **psychrometric chart** may be preferable. A disadvantage of charts is that each applies for only one value of barometric pressure, usually 760 mm or 30 inHg. Correction to other barometric readings is not simple. The equations have the advantage that the actual barometric pressure is taken into account. The equations are often more convenient for equal accuracy or more accurate for equal convenience.

Psychrometric charts are usually plotted, as indicated by Fig. 4.1.18,

with dry-bulb temperature as abscissa and specific humidity as ordinate. Since the specific humidity is determined by the vapor pressure and the barometric pressure (which is constant for a given chart), and is nearly proportional to the vapor pressure, a second ordinate scale, departing slightly from uniform graduations, will give the vapor pressure. The

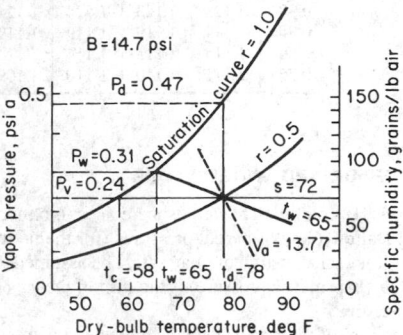

Fig. 4.1.18 Skeleton humidity chart.

saturation curve ($r = 1.0$) gives the specific humidity and vapor pressure for a mixture of air and saturated vapor. Similar curves below it give results for various values of relative humidity. Inclined lines of one set carry fixed values of the wet-bulb temperature, and those of another set carry fixed values of v_a, cubic feet per pound of air. Many charts carry additional scales of enthalpy or Σ function.

Any two values will locate the point representing the state of the atmosphere, and the desired values can be read directly.

Psychrometric charts at different temperatures and barometric pressures are useful in solving problems that fall outside the normal range indicated in Fig. 4.1.18. A collection ("trial set") of 17 different psychrometric charts in both USCS and SI units, for low, normal, and high temperatures, at sea level and at four elevations above sea level, is available from the Carrier Corp., Syracuse, NY.

AIR CONDITIONING

Air-conditioning processes alter the temperature and specific humidity of the atmosphere. The weight of dry air remains constant and consequently computations are best based upon 1 lb of dry air.

Liquid water may enter or leave the apparatus. Its weight m_f lb of air is often merely the difference between the specific humidities of the entering and leaving atmospheres. Its specific enthalpy at the observed or assumed temperature of supply or removal t_f is

$$h_f = t_f - 32 \text{ Btu/lb of liquid}$$

Because most air conditioning involves steady-flow processes, thermal results are computed by the steady flow equation, written for 1 lb of air. Using subscript 1 for entering atmosphere and liquid water, and for heat supplied; and 2 for departing atmosphere and water, and for heat abstracted; the equation becomes (in the absence of work)

$$h_{m1} + m_{f1}h_{f1} + q_1 = h_{m2} + m_{f2}h_{f2} + q_2 \qquad \text{Btu/lb air}$$

Either or both values of m_f or q may be zero.

In terms of the sigma function, the steady-flow equation becomes

$$\Sigma_1 + W_1(t_{w1} + 32) + m_{f1}h_{f1} + q_1 = \Sigma_2 + W_2(t_{w2} - 32) \\ + m_{f2}h_{f2} + q_2 \qquad \text{Btu/lb air}$$

Unit processes involved in air conditioning include heating and cooling an atmosphere above its dew point, cooling below the dew point, adiabatic saturation, and mixing of two atmospheres. These, in various sequences, make it possible to start with any given atmosphere and produce an atmosphere of any required characteristics.

Heating and cooling above the dew point entail no condensation of vapor. Barometric pressure and composition being unaltered, partial pressures remain constant. The process is represented in Fig. 4.1.19.

EXAMPLE.   Initial conditions: $p_t = 28$ inHg; $t_a = 60°F$; $t_w = 50°F$; $p_v = 0.26$ inHg; $V = 1,200$ ft^3.
Final conditions: $t_d = 82°F$.
Initial computed values. $r = 0.50$; $W = 0.0058$ lb vapor/lb air; $\rho_a = 0.0707$ lb air/ft^3; $m_a = V \times \rho_a = 1,200 \times 0.0707 = 84.9$ lb air; $h_m = 20.7$ Btu/lb air.
Final computed values: p_v, W, and m_a unaltered; $r = 0.24$; $\rho_a = 0.0679$ lb air/ft^3; $V = m_a/\rho_a = 84.9/0.0679 = 1,250$ ft^3; $h_m = 26.1$ Btu/lb air.
Heat added: $q = h_{m2} - h_{m1} = 26.1 - 20.7 = 5.4$ Btu/lb air; $Q = q \times m_a = 5.4 \times 84.9 = 458$ Btu.

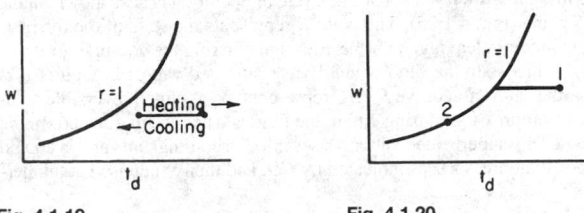

Fig. 4.1.19   **Fig. 4.1.20**

Cooling below the dew point, and **dehumidification,** entails condensation of vapor; the final atmosphere will be saturated, liquid will appear (see Fig. 4.1.20).

EXAMPLE.   Initial conditions: $p_t = 29$ inHg; $t_d = 75°F$; $t_w = 65°F$; $V = 1,500$ ft^3.
Final condition: $t_d = 45°F$.
Initial computed values: $W = 0.0113$ lb vapor/lb air; $\rho_a = 0.0706$ lb air/ft^3; $m_a = 1,500 \times 0.0706 = 106.0$ lb air; $h_m = 30.4$ Btu/lb air; $t_c = 60°F$.
Final computed values: $t_d = 45°F$; $p_v = 0.30$ inHg; $r = 1.0$; $W = 0.0065$ lb vapor/lb air; $\rho_a = 0.0754$ lb air/ft^3; $V = 106.0/0.0754 = 1,406$ ft^3; $h_m = 17.8$ Btu/lb air.
Liquid formed: $m_f = W_1 - W_2 = 0.0113 - 0.0065 = 0.0048$ lb liqud/lb air; $h_f = 50 - 32 = 18$ Btu/lb liquid (assuming that the liquid is drained out at an average temperature $t_f = 50°F$).
Heat abstracted: $q = h_{m1} - h_{m2} - m_1 h_f = 30.4 - 17.8 - 0.0048 \times 18 = 12.5$ Btu/lb air; $Q = q \times m_a = 12.5 \times 106.0 = 1,325$ Btu.

Dehumidification may be accomplished in a **surface cooler,** in which the air passes over tubes cooled by brine or refrigerant flowing through them. The solution of this type of problem is most easily handled on the chart (see Fig. 4.1.21). Locate the point representing the state of the entering atmosphere, and draw a straight line to a point on the saturation curve ($r = 1.0$) at the temperature of the cooling surface. The final state of the issuing atmosphere is approximated by a point on this line whose position on the line is determined by the heat abstracted by the cooling medium. This depends upon the extent of surface and the coefficient of heat transfer.

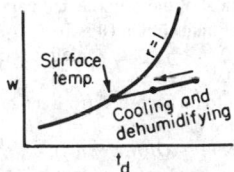

Fig. 4.1.21

Adiabatic saturation (humidification) may be conducted in a spray chamber through which atmosphere flows. A large excess of water is recirculated through spray nozzles, and evaporation is made up by a suitable water supply. After the process has been operating for some time, the water in the spray chamber will have been cooled to the temperature of adiabatic saturation, which differs from the wet-bulb temperature only because of radiation and velocity errors that affect the wet-bulb thermometer. No heat is added or abstracted; the process is adiabatic. The heat of vaporization for the water that is evaporated is supplied by the cooling of the air passing through the chamber. The

wet-bulb temperature of the atmosphere is constant throughout the chamber (Fig. 4.1.22). If the chamber is sufficiently large, the issuing atmosphere will be saturated at the wet-bulb temperature of the entering atmosphere; i.e., as the atmosphere passes through the chamber, t_w remains constant, t_d is reduced from its initial value to t_w. In a chamber of

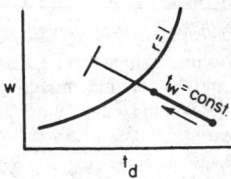

Fig. 4.1.22

commercial size, the action may terminate somewhat short of this, the precise end point being determined by the duration and effectiveness of contact between air and spray water. In any case, the weight of water evaporated equals the increase in the specific humidity of the atmosphere.

EXAMPLE.   Initial conditions $p_t = 30$ inHg; $t_d = 78°F$; $t_w = 55°F$; $r = 0.20$; $W = 28$ grains vapor/lb air.
Final conditions: $t_d = t_w = 55°F$; $r = 1.0$; $W = 64$ grains vapor/lb air.
Water evaporated: $W_2 - W_1 = 64 - 28 = 36$ grains water/lb air

The design of the spray chamber to produce this result is necessarily based upon experience with like apparatus previously built.

In practice, the spray chamber is preceded and followed by heating coils, the first to warm the entering atmosphere to the desired value of t_w, determined by the prescribed final specific humidity, the second to warm the issuing atmosphere to the desired temperature, and simultaneously to reduce its relative humidity to the desired value.

The **spray chamber** that is used for adiabatic saturation (humidification) in winter may be used for dehumidification in summer by supplying the spray nozzles with refrigerated water instead of recirculated water. In this case, the issuing atmosphere will be saturated at the temperature of the spray water, which will be held at the desired dew point. Subsequent heating of the atmosphere to an acceptable temperature will simultaneously reduce the relative humidity to the desired value.

Mixing Two Atmospheres   In recirculating ventilation systems, two atmospheres (1 and 2) are mixed to form a third (3). The state of the final atmosphere is readily found graphically on the psychrometric chart (see Fig. 4.1.23). Locate the points 1 and 2 representing the states of the initial atmospheres. Connect these points by a straight line. Locate a point that divides this line into segments inversely proportional to the weights of air in the respective atmospheres. The division point represents the state of the final mixture, so long as it falls below the saturation curve ($r = 1$). If the final point falls above the saturation curve, as in Fig. 4.1.24, condensation will ensue, and the true final point 4 is found

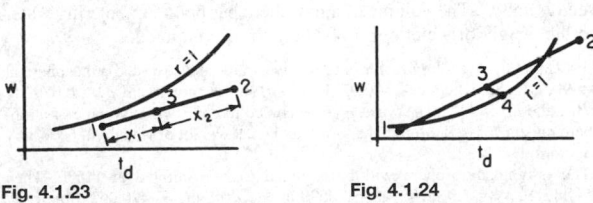

Fig. 4.1.23   **Fig. 4.1.24**

by drawing a line from the apparent point 3, parallel to the lines of constant wet-bulb temperature, to its intersection with the saturation curve. From all the points involved, readings of specific humidity may be taken, including point 3 when it falls above the saturation curve, and in this case the difference between W_3 and W_4 will be the weight of condensate, pounds per pounds air.

If the chart is sectional and the two points do not fall in the same section, or in any case in which it is preferred, the same method may be carried out arithmetically.

For adiabatic mixing in a steady-flow process of two masses of "moist" air, each at the total pressure of p_t,

$$m_{a3} = m_{a1} + m_{a2}$$

In the absence of condensation,

and
$$m_{a3}W_3 = m_{a1}W_1 + m_{a2}W_2$$
$$m_{a3}h_{m3} = m_{a1}h_{m1} + m_{a2}h_{m2}$$

When condensation occurs, assume that the condensate is removed at the final temperature t_4 and that the final mixture consists of dry air and saturated water vapor at this same temperature. The weight of condensate is

$$m_c = m_{a1}W_1 + m_{a2}W_2 - m_{a3}W_4$$

where W_4 is the specific humidity for saturation at temperature t_4 and total pressure p_t. Also

$$m_{a1}h_{m1} + m_{a2}h_{m2} = m_{a2}h_{m4} + m_c h_{f4}$$

In the case of condensation, a trial solution is necessary to find the temperature t_4 that will satisfy these relations.

EXAMPLE. Two thousand ft³ of air per min at $t_{d1} = 80°F$ and $t_{w1} = 65°F$ are mixed in an adiabatic, steady-flow process with 1,000 ft³ of air per min at $t_{d2} = 95°F$ and $t_{w2} = 75°F$; the total pressure of each mixture is 29 inHg.
By computation, $m_{a1} = 140$ lb dry air/min; $W_1 = 0.010$ lb/lb dry air; $m_{a2} = 67.6$ lb dry air/min; $W_2 = 0.0146$ lb/lb dry air.
$m_{a3} = 207.6$ lb dry air/min.
$W_3 = 0.0116$ lb/lb dry air.
$h_{m1} = 30.3$ Btu/lb dry air and $h_{m2} = 38.9$ Btu/lb dry air.
$h_{m3} = 33.1$ Btu/lb dry air.
$t_{d3} = 84.9°F$.

EXAMPLE. Fifteen hundred ft³ of air per min at $t_{d1} = 0°F$ and $r_1 = 0.8$ are mixed in an adiabatic, steady-flow process with 1,000 ft³ of air per min at $t_{d2} = 100°F$ and $r_2 = 0.9$; the total pressure of each mixture is 30 inHg.
By computation, $m_{a1} = 129.6$ lb of dry air/min; $W_1 = 0.000626$ lb/lb dry air; $m_{a2} = 66.9$ lb of dry air/min; $W_2 = 0.03824$ lb/lb dry air; $h_{m1} = 0.09$ Btu/lb dry air; $h_{m2} = 66.29$ Btu/lb dry air.
The three equations that must be satisfied by a choice of the terminal temperature, $t_4 = t_{d4} = t_{w4}$, are

$$m_c = 2.64 - 196.5W_4$$
$$4551 = 196.5h_{m4} + m_c h_{f4}$$
$$W_4 = p_{v4}/1.61(30 - p_{v4}) \quad \text{for } r_4 = 1$$

The value of t_4 that satisfies these equations is 55°F; condensation amounts to 0.84 lb/min.

The **cooling tower** is a chamber in which outdoor atmosphere flows through a spray of entering hot water, which is to be cooled. The temperature of the water is reduced in part by the warming of the air, and in greater part by the evaporation of a portion of the water. The atmosphere enters at given conditions and emerges at a higher temperature and usually saturated ($r = 1$). It is commonly possible to cool the water below the temperature of the entering air, often to about halfway between t_d and t_w. The volume of atmosphere per pound of entering water and the weight of water evaporated are to be computed.

EXAMPLE. A cooling tower is to receive water at 120°F and atmosphere at $t_d = 90$, $t_w = 80$, whence $p_v = 0.92$, $W = 0.0196$ lb vapor/lb air, $\rho_a = 0.0702$ lb air/ft³, and $h_m = 43.2$. The water is to be cooled to 85°F. What volume of atmosphere must be passed through the tower, and what weight of water will be lost by evaporation?
The issuing atmosphere will be assumed to be saturated at 115°F. Then $t_d = 115°F$, $p_v = 3.0$ inHg, $W = 0.0690$ lb vapor/lb air, $\rho_a = 0.0623$ lb air/ft³, and $h_m = 104.4$ Btu/lb air.
The two unknowns are the weight of air to be passed through the tower and the weight of water to be evaporated. The two equations are the water-weight balance and the enthalpy balance (the steady-flow equation for zero heat transfer to or from outside). Assume that 1 lb water enters, of which x lb are evaporated. The water-weight balance $1 + m_aW_1 = 1 - x + m_aW_2$ becomes $x = m_a(W_2 - W_1) = m_a(0.0690 - 0.0196) = 0.0494m_a$. The enthalpy balance $1 \times (120 - 32) + m_a h_{m1} = (1 - x)(85 - 32) + m_a h_{m2}$ becomes $88 + 43.2m_a = 53(1 - x) + 104.4m_a$; $53(1 - x) + 104.4m_a = 53 - 53x + 104.4m_a$; whence $53x = 53 - 88 + m_a(104.4 - 43.2) = -35 + 61.2m_a$.
Solving these simultaneous equations, $x = 0.0295$ lb water evaporated per pound of water entering and $m_a = 0.597$ lb air per pound water entering.

In an **evaporative condenser**, vapor is condensed within tubes that are cooled by the evaporation of water flowing over the outside of the tubes; the water evaporates into the atmosphere. The computation of results is similar to that for the cooling tower.

REFRIGERATION

Vapor Compression Machines The essential parts of a vapor-compression system are the same as in the system using air, except that the expansion cylinder is replaced by an expansion value through which the liquefied medium flows from the high-pressure condensing coils to the low-pressure brine coils. The cycle of operation is best shown on the T-S plane (Fig. 4.1.25). The point B represents the state of the refrigerating medium leaving the brine coils and entering the compressor. Usually in this state the fluid is nearly dry saturated vapor; i.e., point B is near the saturation curve S_g. BC represents the assumed reversible adiabatic compression, during which the fluid is usually superheated. In the state C, the superheated vapor passes into the cooling coils and is cooled at constant pressure, as indicated by CD, and then condensed at temper-

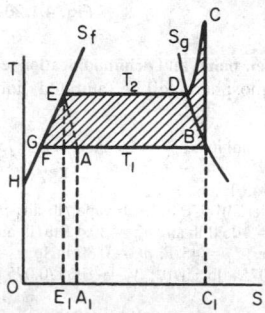

Fig. 4.1.25 Vapor compression refrigeration cycle.

ature T_2, as shown by DE. The liquid now flows through the expansion valve into the brine coils. This is a throttling process, and the final-state point A is located on the T_1-line in such a position as to make the enthalpy for state A (= area $OHGAA_1$) equal to the enthalpy at E (= area $OHEE_1$). The mixture of liquid and vapor now absorbs heat from the brine and vaporizes, as indicated by AB.

The **heat absorbed from the brine**, represented by area A_1ABC_1, is

$$Q_1 = h_b - h_a = h_b - h_e$$

The **heat rejected to the cooling water**, represented by area C_1CDEE_1, is

$$Q_2 = h_c - h_e = c_p(T_c - T_d) + r_2 \quad \text{approx}$$

where r_2 denotes the enthalpy of vaporization at the upper temperature T_2, and c_p the specific heat of the superheated vapor. The **work** that must be **supplied per pound of fluid circulated** is $W = Q_2 - Q_1 = h_c - h_b$. The ratio $Q_1/W = (h_b - h_e)/(h_c - h_b)$ is sometimes called the **coefficient of performance**.

If Q denotes the heat to be absorbed from the brine per hour, then the **quantity of fluid circulated per hour** is $m = Q/(h_b - h_a)$; or, if B is taken on the saturation curve, $m = Q/(h_{g1} - h_{f2})$.
The work per hour is $W = m(h_c - h_{g1}) = Q(h_c - h_{g1})/(h_{g1} - h_{f2})$ ft·lb, and the **horsepower required** is $H = Q(h_c - h_{g1})/2544(h_{g1} - h_{f2})$. The (U.S.) ton of refrigeration represents a cooling rate of 200 Btu/min, which is closely equivalent to that of 50 kcal/min, 210 kJ/min, or 3.5 kW. (Extensive tables of refrigerant properties are found in the "ASHRAE Handbook.")
If v_{g1} is the volume of the saturated vapor at the temperature T_1 in the brine coils, and n the number of working strokes per minute, the **displacement volume** of the **compressor cylinder** is $V = mv_{g1}/(60n)$.
The work necessary for operating a refrigerator, although usually supplied through the compressor, may be supplied in other ways. Thus in **absorption refrigerators** (see Secs. 12 and 19) an absorbent, usually water, absorbs the refrigerant, usually ammonia. The water, by its affinity for the ammonia, has, in a thermodynamic sense, ability to do work.

Having absorbed the ammonia and thereby lost its ability to do work, the water may have its work capacity restored by passing the ammonia-water solution through a rectifying column from which water and ammonia emerge. With operation under a suitable pressure, the ammonia is condensed to a liquid. This, in turn, may be evaporated, yielding refrigeration, the ammonia vapors being once again absorbed in the water. Thermodynamically the analysis for these absorption cycles is similar to that for compression cycles. See Wood, "Applications of Thermodynamics," 2d ed., Addison-Wesley, for an in-depth discussion of the absorption refrigeration cycle.

Other papers on absorption refrigeration or **heat pump systems** include Chen, *Heat Recovery Systs. & CHP,* **30,** 1988, pp. 37–51; Narodos-lawsky et al., *Heat Recovery Systs. & CHP,* **8,** no. 5, 1988, pp. 459–468, and **8,** no. 3, 1988, pp. 221–233; Egrican, **8,** no. 6, 1988, pp. 549–558; and for the aqua ammonia system, Ataer and Goguy, *Int. J. Refrig.,* **14,** Mar. 1991, pp. 86–92. Kalina (*J. Eng. Gas Turbines & Power,* **106,** 1984, pp. 737–742) proposed a new cycle using an ammonia-water solution as a bottoming cycle system. The proper selection of the composition and parameters of the working fluid was stated to be critical in the cycle design. Absorption replaces condensation of the working fluid after expansion in the turbine. Special care is also needed to regulate pressure drops between turbine stages. Chuang et al. [*AES (A.S.M.E.),* **10,** no. 3, 1989, pp. 73–77] evaluated exergy changes in the cycle while Kouremenos and Rogdaikis [*AES (A.S.M.E.),* **19,** 1990, pp. 13–19] developed a computer code for use with *h-s* and *T-s* diagrams. For heat pumps using **binary mixtures** see Hihara and Sato, *ASME/JSME Thermal Eng. Proc.,* **3,** 1991, p. 297. For **supercritical heat pump** cycles see, e.g., Angelino and Invernizzi, *Int. J. Refrig.,* **17,** no. 8, 1991; pp. 543–554.

An analysis of industrial gas separation to yield minimum overall cost, i.e., taking into account both energy and capital cost, for the processes of distillation, absorption, adsorption, and membranes was given by Haselden, *Gas Separation & Purification,* **3,** Dec. 1989, pp. 209–216. Analysis of the thermodynamic regenerator cycle with compressed-gas throttling, called the **Linde cycle,** was made by Lavrenchenko, *Cryogenics,* **33,** no. 11, 1993, pp. 1040–1045. **Orifice pulse tube** refrigerators are receiving more attention. Kittel (*Cryogenics,* **32,** no. 9, 1992, pp. 843–844) examines their thermodynamic efficiency and refers to Radebaugh (*Adv. Cryog. Eng.,* **35B,** 1990, pp. 1192–1205) for a review of these devices.

de Rossi et al. (*Proc Mtg. IIR–IIF Comm.* B1, Tel Aviv, 1990) gave an interactive computer code for refrigerant thermodynamic properties which was evaluated for 20 different refrigerants in vapor compression and a reversed Rankine cycle (*Appl. Energy,* **38,** 1991, pp. 163–180). The evaluation included an exergy analysis. A similar publication was *AES (A.S.M.E.),* **3,** no. 2, 1987, pp. 23–31. Other papers include Kumar et al. (*Heat Recovery Systems & CHP,* **9,** no. 2, 1989, pp. 151–157), Alefeld (*Int. J. Refrig.,* **10,** Nov. 1987, pp. 331–341), and Nikolaidis and Probert (*Appl. Energy,* **43,** 1992, pp. 201–220).

The problem of deciding which refrigerants will be used in the future is complex. Although recommendations exist as to the phasing out of existing substances, one cannot predict the extent to which the recommendations will be followed. The blends proposed by several manufacturers have yet to receive extensive testing. One must bear in mind that in addition to the thermodynamic suitability considerations of material compatibility, ozone depletion potential, etc. have to be taken into account. According to Dr. McLinden of the National Institute of Standards and Technology, Boulder, CO (private communication, March 1995), R 11, R 12, R 22, R 123, R 134a as well as ammonia (R 717) and propane/isobutane (R 290/R 600a) blends are likely to be important for many years. The presentation of the tables in Sec. 4.2 has been revised to include some of these plus a few other compounds used in ternary blends. The best single source for further information is the "ASHRAE Handbook—Fundamentals," 1993 or latest edition.

STEAM CYCLES

Rankine Cycle   The ideal Rankine cycle is generally employed by engineers as a standard of reference for comparing the performance of actual steam engines and steam turbines. Figure 4.1.26 shows this cycle on the *T-S* and *p-V* planes. *AB* represents the heating of the water in the boiler, *BC* represents evaporation (and superheating if there is any), *CD* the assumed isentropic expansion in the engine cylinder, and *DA* condensation in the condenser.

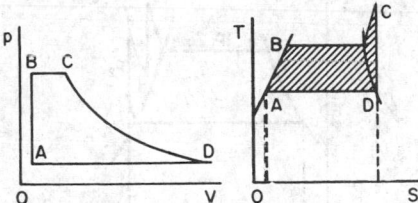

Fig. 4.1.26   Rankine cycle.

Let h_a, h_b, h_c, h_d represent the enthalpy per unit mass of steam in the four states *A, B, C,* and *D,* respectively. Then the energy transformed into work, represented by the area *ABCD,* is $h_c - h_d$ (enthalpies in Btu/lbm).

The energy expended on the fluid is $h_c - h_a$, hence the **Rankine cycle efficiency** is $e_t = (h_c - h_d)/(h_c - h_a)$.

The **steam consumption** of the ideal Rankine engine in pounds per horsepower-hour is $N_r = 2{,}544/(h_c - h_d)$. Expressed in pounds per kilowatthour, the steam consumption of the ideal Rankine cycle is $3{,}412.7/(h_c - h_d)$.

The performance of an engine is frequently stated in terms of the heat used per horsepower-hour. For the ideal Rankine engine, this is

$$Q_r = 2{,}544/e_t = 2{,}544(h_c - h_a)/(h_c - h_d)$$

Efficiency of the Actual Engine   Let Q denote the heat transformed into work per pound of steam by the actual engine; then if Q_1 is the heat furnished by the boiler per pound of steam, the **thermal efficiency of the engine** is $e_t = Q/Q_1$.

The efficiency thus defined is misleading, as it takes no account of the conditions of boiler and condenser pressure, superheat, or quality of steam. It is customary therefore to define the efficiency as the ratio Q/Q_a, where Q_a is the **available heat,** or the heat that could be transformed under ideal conditions. For steam engines and turbines, the Rankine cycle is usually taken as the ideal, and the quantity $Q/Q_a = Q/(h_c - h_d)$ is called the **engine efficiency.** For engines and turbines, this efficiency ranges from 0.50 to 0.85. The engine efficiency *e* may also be expressed in terms of steam consumed; thus, if N_a is the steam consumption of the actual engine and N_r is the steam consumption of the ideal Rankine engine under similar conditions, then $e = N_r/N_a$.

EXAMPLE.   Suppose the boiler pressure to be 180 psia, superheat 150°F, and the condenser pressure 3 in of mercury. From the steam tables or diagram, the following values are found: $h_c = 1{,}283.3$, $h_d = 942$, $h_a = 82.99$. The available heat is $Q_a = 1{,}283.3 - 942 = 341.3$ Btu, and the thermodynamic efficiency of the cycle is $341.3/(1{,}283.3 - 82.99) = 0.284$. The steam consumption per horsepower-hour is $2{,}544/341.3 = 7.46$ lb, and the heat used per horsepower-hour is $2{,}544/0.284 = 8{,}960$ Btu. If an actual engine working under the same conditions has a steam consumption of 11.4 lb/(hp · h), its efficiency is $7.46/11.4 = 0.655$, and its heat consumption per horsepower-hour is $8{,}960/0.655 = 13{,}680$ Btu.

Reheating Cycle   Let the steam after expansion from p_1 to an intermediate pressure p_2 (*cd,* Fig. 4.1.27) be reheated at constant pressure p_2, as indicated by *de.* Then follows the isentropic expansion to pressure p_3, represented by *ef.*

The energy absorbed by 1 lb of steam is $h_c - h_a$ from the boiler, and $h_e - h_d$ from the reheating. The work done, neglecting the energy required to operate the boiler feed pump, etc., is $h_c - h_d + h_e - h_f$. Hence the efficiency of the cycle is

$$e_t = \frac{h_c - h_d + h_e - h_f}{h_c - h_a + h_e - h_d}$$

Bleeding Cycle   In the **regenerative** or bleeding cycle, steam is drawn from the turbine at one or more stages and used to heat the feed water. Figure 4.1.28 shows a diagrammatic arrangement for bleeding at one stage. Entering the turbine is $1 + w$ lb of steam at p_1, t_1, and enthalpy

h_1. At the bleeding point w lb at p_2, t_2, h_2 enters the feedwater heater. The remaining 1 lb passes through the turbine and condenser and enters the feedwater heater as water at temperature t_3. Let t' denote the temperature of the water leaving the heater, and h' the corresponding en-

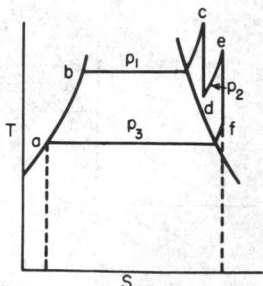

Fig. 4.1.27 Reheating cycle.

thalpy of the liquid. Then the equation for the interchange of heat in the heater is

$$w(h_2 - h') = h' - h_{f3}$$

The work done by the bled steam is $w(h_1 - h_2)$ and that by the 1 lb of steam going completely through the turbine is $h_1 - h_3$. Total work $= w(h_1 - h_2) + (h_1 - h_3)$ if work to the pumps is neglected. The heat supplied between feedwater heater and turbine is $(1 + w)(h_1 - h')$. Hence the ideal efficiency of the cycle is

$$e_t = \frac{w(h_1 - h_2) + h_1 - h_3}{(1 + w)(h - h')}$$

A **computer program** which is claimed to model the thermodynamic performance of any steam power system has been described by Thelen and Somerton [*AES, (A.S.M.E.)*, **33**, 1994, pp. 167–175], extending an earlier analysis.

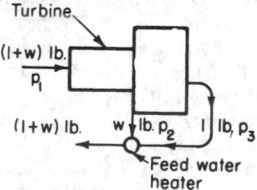

Fig. 4.1.28 Regenerative feedwater heating.

The use of selected fluid mixtures in Rankine cycles was proposed by Radermacher (*Int. J. Ht. Fluid Flow*, **10**, no. 2, June 1989, p. 90). Lee and Kim (*Energy Convsn. Mgmt.*, **33**, no. 1, 1992, pp. 59–67) describe the finite time optimization of a modified Rankine heat engine. For a steam Wankel engine see Badr et al. (*Appl. Energy*, **40**, 1991, pp. 157–190).

Heat from **nuclear reactors** can be used for heating services or, through thermodynamic cycles, for power purposes. The reactor coolant transfers the heat generated by fission so as to be used directly, or through an intermediate heat-exchange system, avoiding radioactive contamination. Steam is the preferred thermodynamic fluid in practice so that the Rankine-cycle performance standards with regenerative and reheat variations prevail. Adaptation of gas-turbine cycles, using various gases, can be expected as allowable reactor temperatures are raised.

Many engineers and scientists are actively engaged in research dealing with the location, production, utilization, transmission, storage, and distribution of new forms of energy. Examples are the study of the energy released in the fusion of hydrogen nuclei and research in solar energy. Considerable effort is being expended in studying the feasibility of combining the gas turbine with a steam-generating plant. The possi-

bility of efficiency improvement over the steam cycle is due to higher inlet temperatures associated with the gas turbine. Significant advances are predicted in the near future in expanding our energy sources and reserves.

THERMODYNAMICS OF FLOW OF COMPRESSIBLE FLUIDS

Important examples of the flow of compressible fluids are the following: (1) the flow of air and steam through orifices and short tubes or nozzles, as in the steam turbine, (2) the flow of compressed air, steam, and illuminating gas in long mains, (3) the flow of low-pressure gases, as furnace gases in ducts and chimneys or air in ventilating ducts, and (4) the flow of gases in moving channels, as in the centrifugal fan.

Notation

Let $A =$ area of section, ft^2
 $C =$ empirically determined coefficient of discharge
 $D =$ inside diameter of pipe, ft
 $d = 12D =$ inside diameter of pipe, in
 $F_{12} =$ energy expended in overcoming internal and external friction between sections A_1 and A_2
 $F' =$ energy used in overcoming friction, ft·lb/lb of fluid flowing
 $f =$ friction factor $= 4f'$
 $g = 32.2 =$ local acceleration of gravity, ft/s^2
 $g_c =$ a dimensional constant
 $h =$ enthalpy, Btu/lb
 $J = 778.3$ ft·lb/Btu
 $k = c_p/c_v$
 $L =$ equivalent length of pipe, ft
 $m =$ mass of fluid flowing past a given section per s, lb
 $\mu =$ viscosity, cP
 $P =$ pressure, lb/in^2 abs
 $\Delta P =$ differential pressure across nozzle, lb/in^2
 $p =$ pressure of fluid at given section, lb/ft^2 abs
 $p_m =$ critical flow pressure
 $Q_{12} =$ heat entering the flowing fluid between sections A_1 and A_2
 $q =$ volume of fluid flowing past section, ft^3/min
 $R =$ ideal gas constant
 $\rho =$ density, lb/ft^3
 $T =$ temperature, °R
 $\bar{v} =$ mean velocity at the given section, ft/s
 $v =$ specific volume
 $w =$ weight of fluid power flowing past a given section per s, lb
 $z =$ height from center of gravity of flow to fixed base level, ft

The cross sections of the tube or channel are denoted by A_1, A_2, etc. (Fig. 4.1.29), and the various magnitudes pertaining to these sections are denoted by corresponding subscripts. Thus, at section A_1, the velocity, specific volume, and pressure are, respectively, $\bar{v}_1$, v_1, p_1; at section A_2, they are $\bar{v}_2$, v_2, p_2.

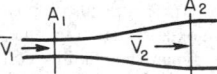

Fig. 4.1.29

Fundamental Equations In the interpretation of fluid-flow phenomena, three fundamental equations are of importance.

1. The continuity equation, or material balance,

$$\frac{A_1 \bar{v}_1}{v_1} = \frac{A_2 \bar{v}_2}{v_2} \quad \text{or} \quad \frac{dv}{v} = \frac{dA}{A} + \frac{d\bar{v}}{\bar{v}}$$

2. The first law of energy balance for steady flow,

$$q = (h_2 - h_1) + \frac{\bar{v}_2^2 - \bar{v}_1^2}{2g_c} + \frac{g}{g_c}(z_2 - z_1)$$

3. The available energy balance for a steady-flow process, based on unit weight, is

$$v\,dp + \frac{\bar{v}\,d\bar{v}}{g} + dF + dz = 0$$

In the process here discussed, no net external or shaft work is performed.

For most actual processes, the third equation cannot be integrated because the actual path is not known. Usually, adiabatic flow is assumed, but occasionally the assumption of isothermal conditions may be more nearly correct.

For **adiabatic flow of incompressible fluids,** the last equation above can be written in the more familiar form known as Bernoulli's equation:

$$(p_2 - p_1)/\rho + (\bar{v}_2^2 - \bar{v}_1^2)/2 + g/g_c(z_2 - z_1) = 0$$

Flow through Orifices and Nozzles   As a compressible fluid passes through a nozzle, drop in pressure and simultaneous increase in velocity result. By assuming the type of flow, e.g., adiabatic, it is possible to calculate from the properties of the fluid the required area for the cross section of the nozzle at any point in order that the flowing fluid may just fill the provided space. From this calculation, it is found that for all compressible fluids the nozzle form must first be converging but eventually, if the pressure drops sufficiently, a place is reached where to accommodate the increased volume due to the expansion the nozzle must become diverging in form. The smallest cross section of the nozzle is called the throat, and the pressure at the throat is the **critical flow pressure** (not to be confused with the critical pressure). If the nozzle is cut off at the throat with no diverging section and the pressure at the discharge end is progressively decreased, with fixed inlet pressure, the amount of fluid passing increases until the discharge pressure equals the critical, but further decrease in discharge pressure does not result in increased flow. This is not true for thin plate orifices. For any particular gas, the ratio of critical to inlet pressure is approximately constant. For gases, $p_m/p_1 = 0.53$ approx; for saturated steam the ratio is about 0.575; and for moderately superheated steam it is about 0.55.

Formulas for Orifice Computations   The general fundamental relation is given by the energy balance $(\bar{v}_2^2 - \bar{v}_1^2)/(2g_c) = -h_{12}$. Referring to Fig. 4.1.30, let section 2 be taken at the orifice, section 3 is somewhat

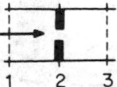

Fig. 4.1.30

beyond the orifice on the downstream side, and section 1 is before the orifice on the upstream side. Then

$$\bar{v}_2 = C\sqrt{2g_c(h_1 - h_2)}\bigg/\sqrt{1 - \left(\frac{A_2}{A_1}\right)^2\left(\frac{v_1}{v_2}\right)^2}$$

The coefficient of discharge C is discussed in Secs. 3 and 16. The volume of gas passing is $\bar{v}_2 A_2$ ft³/s, and the quantity is $\bar{v}_2 A_2 \rho$. For ideal gases, assuming reversible adiabatic expansion through the orifice,

$$\bar{v}_2 = C\frac{\sqrt{2g_c p_1 \bar{v}_1 \dfrac{k}{k-1}\left[1 - \left(\dfrac{p_2}{p_1}\right)^{(k-1)/k}\right]}}{\sqrt{1 - \left(\dfrac{A_2}{A_1}\right)^2\left(\dfrac{p_2}{p_1}\right)^{2/k}}}$$

$$m = CA_2 p_2 \frac{\sqrt{\dfrac{2g_c}{RT_1}\dfrac{k}{k-1}\left(\dfrac{p_1}{p_2}\right)^{(k-1)/k}\left[\left(\dfrac{p_1}{p_2}\right)^{(k-1)/k} - 1\right]}}{\sqrt{1 - \left(\dfrac{A_2}{A_1}\right)^2\left(\dfrac{p_2}{p_1}\right)^{2/k}}}$$

Often $\bar{v}_1$ is small compared with $\bar{v}_2$, and under these conditions the denominators in the preceding equations become approximately equal

to unity. For air, assuming $R = 53.3$, $k = 1.3937$, and $\bar{v}_1$ negligible,

$$m = 2.05CA_2 p_2 \sqrt{(1/T_1)(p_1/p_2)^{0.283}[(p_1/p_2)^{0.283} - 1]}$$

Although the preceding formulas are generally applicable under the assumed conditions, it must be remembered that irrespective of the value of p_3, p_2 cannot become less than p_m. When p_3 is less than p_m, the flow rate becomes independent of the downstream pressure; for ideal gases,

$$m = CA_2 p_1 \sqrt{\frac{g_c}{RT_1}\,k\left(\frac{2}{k+1}\right)^{(k+1)(k-1)}}$$

or for air

$$m = 0.53Cp_1 \frac{A_2}{\sqrt{T_1}}$$

The following formula is useful for calculating the flow rate, in cubic feet per minute, of any gas (provided no condensation occurs) through a nozzle for pressure drops less than the critical range:

$$q_1 = \frac{31.5Cd_n^2 Y'}{\rho_1}\sqrt{\rho_1\,\Delta P}$$

In this equation,

$$Y' = \left(\frac{k}{k-1}\right)^{1/2}\left(\frac{P_2}{P_1}\right)^{1/k}$$

$$\times \sqrt{\left[1 - \left(\frac{P_2}{P_1}\right)^{(k-1)/k}\right]\bigg/\left(1 - \frac{P_2}{P_1}\right)\left[1 - \left(\frac{d_n}{d_1}\right)^4\left(\frac{P_2}{P_1}\right)^{2/k}\right]}$$

where P_1 = static pressure on upstream side of nozzle, psia; P_2 = static pressure on downstream side of nozzle, psia; d_1 = diameter of pipe upstream of nozzle, in; d_n = nozzle throat diameter, in; ρ_1 = specific weight of gas at upstream side of nozzle, lb/ft³. Values of Y' are given in Table 4.1.3.

Where the pressure drop through the orifice is small, the hydraulic formulas applicable to incompressible fluids may be employed for gases and other compressible fluids.

In general, the formulas of the preceding section are applicable to nozzles. When so used, however, the proper value of the discharge coefficient must be employed. For steam nozzles, this may be as high as 0.94 to 0.96, although for many orifice installations it is as low as 0.50 to 0.60. Steam nozzles constitute a most important type, and calculations for these are best carried out with the aid of a Mollier or similar chart.

Formulas for Discharge of Steam   When the back pressure p_3 is less than the critical pressure p_m, the discharge depends upon the area of orifice A_2 and reservoir pressure p_1. There are three formulas widely used to express, approximately, the discharge m of **saturated** steam in terms of A_2 and p_1 as follows:

1. Napier's equation, $m = A_2 p_1/70$.
2. Grashof's formula, $m = 0.0165\,A_2 p_1^{0.97}$.
3. Rateau's formula, $m = A_2 p_1(16.367 - 0.96\log p_1)/1,000$.

In these formulas, A_2 is to be taken in square inches, p_1 in pounds per square inch. Napier's formula is merely convenient as a rough check. Formulas 2 and 3 are applicable to well-rounded convergent orifices, in which case the coefficient of discharge may be taken as 1; i.e., no correction is required.

When the back pressure p_2 is greater than the critical flow pressure p_m, the velocity and discharge are found most conveniently from the general formulas of flow. From the steam tables or from the **Mollier chart,** find the initial enthalpy h_1 and the enthalpy h_2 after isentropic expansion; also the specific volume v_2 (see Fig. 4.1.31). Then

$$\bar{v}_2 = 223.7\sqrt{h_1 - h_2} \quad \text{and} \quad m = A_2\bar{v}_2/v_2$$

The same method is used in the case of steam initially superheated.

EXAMPLE.   Required the discharge through an orifice ½ in diam of steam at 140 psi superheated 110°F, back pressure, 90 psia.

Table 4.1.3 Values for Y'

P_2/P_1	k = 1.40					k = 1.35					k = 1.30				
	d_n/d_1					d_n/d_1					d_n/d_1				
	0	0.2	0.3	0.4	0.5	0	0.2	0.3	0.4	0.5	0	0.2	0.3	0.4	0.5
1.00	1.000	1.001	1.004	1.013	1.033	1.00	1.001	1.004	1.013	1.033	1.000	1.001	1.004	1.013	1.033
0.99	0.995	0.995	0.999	1.007	1.027	0.994	0.995	0.999	1.007	1.027	0.994	0.995	0.998	1.007	1.026
0.98	0.989	0.990	0.993	1.002	1.021	0.989	0.990	0.993	1.001	1.020	0.988	0.989	0.992	1.001	1.020
0.97	0.984	0.985	0.988	0.996	1.015	0.983	0.984	0.987	0.995	1.014	0.983	0.983	0.986	0.995	1.013
0.96	0.978	0.979	0.982	0.990	1.009	0.978	0.978	0.981	0.990	1.008	0.977	0.977	0.980	0.989	1.007
0.95	0.973	0.974	0.977	0.985	1.002	0.972	0.973	0.976	0.984	1.001	0.971	0.972	0.974	0.982	1.000
0.94	0.967	0.968	0.971	0.979	0.996	0.966	0.967	0.970	0.978	0.995	0.965	0.966	0.968	0.976	0.993
0.93	0.962	0.963	0.965	0.973	0.990	0.961	0.961	0.964	0.972	0.989	0.959	0.960	0.962	0.970	0.987
0.92	0.956	0.957	0.960	0.967	0.984	0.955	0.955	0.958	0.966	0.982	0.953	0.954	0.956	0.964	0.980
0.91	0.951	0.951	0.954	0.961	0.978	0.949	0.950	0.952	0.960	0.976	0.947	0.948	0.950	0.957	0.973
0.90	0.945	0.946	0.948	0.956	0.971	0.943	0.944	0.946	0.953	0.969	0.941	0.942	0.944	0.951	0.966
0.89	0.939	0.940	0.943	0.950	0.965	0.937	0.938	0.940	0.947	0.963	0.935	0.935	0.938	0.945	0.959
0.88	0.934	0.934	0.937	0.944	0.959	0.931	0.932	0.934	0.941	0.956	0.929	0.929	0.932	0.938	0.953
0.87	0.928	0.928	0.931	0.938	0.953	0.925	0.926	0.926	0.935	0.950	0.922	0.923	0.926	0.932	0.946
0.86	0.922	0.923	0.925	0.932	0.946	0.919	0.920	0.922	0.929	0.943	0.916	0.917	0.919	0.926	0.939
0.85	0.916	0.917	0.919	0.926	0.940	0.913	0.914	0.916	0.923	0.936	0.910	0.911	0.913	0.923	0.932
0.84	0.910	0.911	0.913	0.920	0.933	0.907	0.908	0.910	0.916	0.930	0.904	0.904	0.907	0.919	0.925
0.83	0.904	0.905	0.907	0.913	0.927	0.901	0.902	0.904	0.910	0.923	0.897	0.898	0.900	0.916	0.918
0.82	0.898	0.899	0.901	0.907	0.920	0.895	0.895	0.898	0.904	0.917	0.891	0.891	0.894	0.900	0.911
0.81	0.892	0.893	0.895	0.901	0.914	0.889	0.889	0.891	0.897	0.910	0.885	0.885	0.887	0.893	0.904
0.80	0.886	0.887	0.889	0.895	0.907	0.883	0.883	0.885	0.891	0.903	0.878	0.879	0.880	0.886	0.897
0.79	0.880	0.881	0.883	0.889	0.901	0.876	0.877	0.879	0.834	0.896	0.872	0.872	0.874	0.880	0.890
0.78	0.874	0.875	0.877	0.882	0.894	0.870	0.870	0.872	0.878	0.889	0.865	0.865	0.868	0.873	0.883
0.77	0.868	0.869	0.871	0.876	0.887	0.864	0.864	0.866	0.871	0.882	0.859	0.859	0.861	0.866	0.876
0.76	0.862	0.862	0.864	0.869	0.881	0.857	0.858	0.859	0.865	0.876	0.852	0.852	0.854	0.859	0.869
0.75	0.856	0.856	0.858	0.863	0.874	0.851	0.851	0.853	0.858	0.869	0.845	0.846	0.848	0.852	0.862
0.74	0.849	0.850	0.852	0.857	0.867	0.844	0.845	0.846	0.851	0.862	0.839	0.839	0.841	0.845	0.855
0.73	0.843	0.844	0.845	0.850	0.860	0.838	0.838	0.840	0.845	0.855	0.832	0.832	0.834	0.838	0.848
0.72	0.837	0.837	0.839	0.844	0.854	0.831	0.831	0.833	0.838	0.848	0.825	0.825	0.827	0.831	0.841
0.71	0.820	0.831	0.832	0.837	0.847	0.825	0.825	0.827	0.831	0.840	0.818	0.819	0.820	0.824	0.834
0.70	0.824	0.824	0.826	0.830	0.840	0.818	0.818	0.820	0.824	0.833	0.811	0.812	0.813	0.817	0.826

If the velocity of approach is zero (as with a nozzle taking in air from the outside), d_1 is infinite and d_n/d_1 is zero.

From the Mollier chart and the steam tables, $h_1 = 1,255.7$, $h_2 = 1,214$, $v_2 = 5.30$ ft³.

$$\bar{v}_2 = 233.7\sqrt{1,255.7 - 1,214} = 1,455$$
$$A_2 = 0.1964 \text{ in}^2 = (0.1964/144)\text{ft}^2$$
$$m = A_2\bar{v}_2/v_2 = (0.1964/144) \times (1,455/5.30) = 0.372 \text{ lb/s}$$

This calculation assumes ideal conditions, and the results must be multiplied by the correct coefficient of discharge to get actual results.

Flow through Converging-Diverging Nozzles At the throat, or smallest cross section of the nozzle (Fig. 4.1.32), the pressure of saturated steam takes the value $p_m = 0.57p_1$. The quantity discharged is fixed by the area A_2 of the throat and the initial pressure p_1. For saturated steam, Grashof's or Rateau's formula (see above) may be used. The diverging part of the nozzle permits further expansion to the break pressure p_3, the velocity of the jet meanwhile increasing from $\bar{v}_m(= \bar{v}_2)$, the critical velocity at the throat, to $\bar{v}_3$ given by the fundamental equation $\bar{v}_3 = 223.7\sqrt{h_1 - h_3}$.

The frictional resistances in the nozzle have the effect of decreasing the jet energy $\bar{v}_3^2/(2g_c)$ and correspondingly increasing the enthalpy of the flowing fluid. Thus, if h_3 is the enthalpy in the final state with

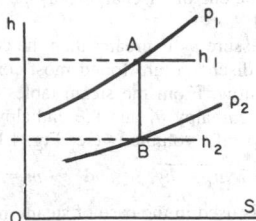

Fig. 4.1.31

frictionless expansion, h_3' ($> h_3$) is the enthalpy when friction is taken into account; hence $(\bar{v}_3')^2/(2g_c) = (h_1 - h_3')$ is less than $\bar{v}_3^2/g_c = h_1 - h_3$. The loss of kinetic energy, in Btu, is $h_3' - h_3$, and the ratio of this loss to the available kinetic energy, i.e., $(h_3' - h_3)/(h_1 - h_3)$, is denoted by y.

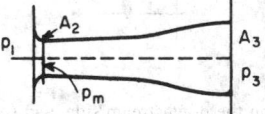

Fig. 4.1.32

The **design of a nozzle for a given discharge** m with pressures p_1 and p_3 is most conveniently effected with the aid of the Mollier chart. Determine p_m, the critical pressure, and h_1, h_m, h_3, assuming frictionless flow. Then

$$\bar{v}_m = 223.7\sqrt{h_1 - h_m}$$

and

$$\bar{v}_3' = 223.7\sqrt{(1 - y)(h_1 - h_3)}$$

Next find v_m and v_3'. Then, from the equation of continuity,

$$A_m = mv_m/\bar{v}_m$$

and

$$A_3' = mv_3'/\bar{v}_3'$$

The following example illustrates the method.

EXAMPLE. Required the throat and end sections of a nozzle to deliver 0.7 lb of steam per second. The initial pressure is 160 psia, the back pressure 15 psia, and the steam is initially superheated 100°F; $y = 0.15$.

The critical pressure is $160 \times 0.55 = 88$ lb. On the Mollier chart (Fig. 4.1.33), the point A representing the initial state is located, and line of constant entropy (frictionless adiabatic) is drawn from A. This cuts the curves $p = 88$ and $p = 15$ in the points B and C, respectively. The three values of h are found to be $h_1 = 1,253$, $h_m = 1,199$, $h_3 = 1,067$. Of the available drop in enthalpy, $h_1 - h_3 = 185.5$ Btu,

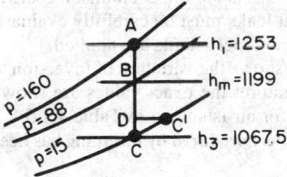

Fig. 4.1.33

15 percent or 27.9 Btu is lost through friction. Hence, $CD = 27.9$ is laid off and D is projected horizontally to point C' on the curve $p = 15$. Then C' represents the final state of the steam, and the quality is found to be $x = 0.943$. The specific volume in the state C' is $26.29 \times 0.943 = 24.8$ ft³. Likewise, the specific volume for the state B is found to be 5.29 ft³/lb.

For the velocities at throat and end sections,

$$\bar{v}_m = 223.7\sqrt{1,253 - 1,199} = 1,643 \text{ ft/s}$$
$$\bar{v}_3 = 223.7\sqrt{185.5 - 27.9} = 2,813 \text{ ft/s}$$
$$A_m = (0.7 \times 5.29)/1,643 = 0.00225 \text{ ft}^2 = 0.324 \text{ in}^2$$
$$A_3 = (0.7 \times 24.8)/2,813 = 0.00617 \text{ ft}^2 = 0.89 \text{ in}^2$$

The diameters are $d_m = 0.643$ in and $d_3 = 1.064$ in.

Divergence of Nozzles   Figure 4.1.34 gives, for various ratios of expansion, the required "divergence" of nozzle, i.e., the ratio of the area of any section to the throat area. Thus in the case of saturated steam, if the final pressure is $\frac{1}{15}$ of the initial pressure the ratio of the areas is 3.25. The curves apply to frictionless flow; the effect of friction is to increase the divergence.

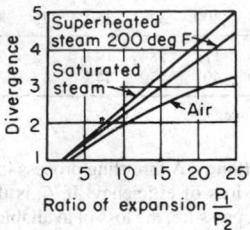

Fig. 4.1.34

Theory of Supersaturation   Certain discrepancies between the discharge of saturated steam through an orifice as calculated from the preceding theory and the discharge actually observed are explained by a hypothesis first advanced by Martin, viz., that steam when expanded rapidly, as in turbine nozzles, becomes supersaturated; in other words, the condensation required by the ordinary theory of adiabatic expansion does not occur on account of the rapidity of the expansion.

The effect of supersaturation in turbines is a loss of energy, the amount of which may be 1.5 to 3 percent of the available energy of the steam.

Flow of Wet Steam   When the steam entering a nozzle is wet, the speed of the water particles at exit is not the same as the speed of the steam. Denoting by $\bar{v}$ the speed of the steam, the speed of the water drop is $f\bar{v}$, and f may vary perhaps 0.20 to 0.05 or less, depending on the pressure. The actual velocity $\bar{v}$ of the steam is greater than the velocity $\bar{v}_0$ calculated on the usual assumption that steam and water have the same velocity. If x is the quality of the steam, the ratio of these velocities is

$$\bar{v}/\bar{v}_0 = 1/\sqrt{x + f^2(1 - x)}$$

Thus with $x = 0.92$, $f = 0.15$, $\bar{v}/\bar{v}_0 = 1.036$. Since the discharge is

practically proportional to the steam velocity, the actual discharge in this case is 3.6 percent greater than the discharge computed on the usual assumptions.

Velocity Coefficients, Loss of Energy y   On account of friction losses, the actual velocity $\bar{v}$ attained by the jet is less than the velocity $\bar{v}_0$ calculated under ideal conditions. That is, $\bar{v} = x\bar{v}_0$, where x (< 1) is a velocity coefficient. The coefficient x is connected with the coefficient y, giving the loss of energy, by the relation, $y = 1 - x^2$.

The elaborate and accurate experiments of the General Electric Co. on turbine nozzles give convergent nozzles values of x in excess of 0.98, with a corresponding loss of energy $y = 0.025$ to 0.04. For similar nozzles, the experiments of the Steam Nozzles Research Committee (of England) by a different method give values of x around 0.96, or $y = 0.08$. In the case of divergent nozzles, the velocity coefficient may be somewhat lower.

FLOW OF FLUIDS IN CIRCULAR PIPES

The fundamental equation as previously given on a unit weight bases, assuming the pipe horizontal, is

$$(\bar{v}\, d\bar{v}/g) + v\, dp + dF = 0$$

The friction term dF includes not only losses due to frictional flow along the pipe but also those due to fittings, valves, etc., as well as losses occasioned by any enlargement or contraction of the pipe as, for instance, the loss occurring when a fluid passes from a pipe into a tank. For long straight pipes of uniform diameter, dF is approximately equal to $2f'\,[\bar{v}^2\, dL/(gD)]$. It is usual to express friction due to fittings, etc., in terms of additional length of pipe, adding this to the actual pipe length to get the equivalent pipe length.

Integration of the fundamental equation leads to two sets of formulas.

1. For pressure drops, small relative to the initial pressure, the specific volume v and the velocity $\bar{v}$ may be assumed constant. Then approximately

$$p_1 - p_2 = 2f'\bar{v}^2 L/(vgD)$$

Expressing pressure in pounds per square inch, p', the diameter in inches, and $\bar{v}$ as a function of wv/d^2, this equation becomes

$$p_1' - p_2' = 174.2f'w^2vL/d^5$$

2. For considerable pressure drops, when dealing with approximately isothermal flow of gases and vapors to which the gas laws are applicable, the fundamental equation on a weight basis may be integrated to give

$$p_1^2 - p_2^2 = \frac{2w^2RT}{gA^2}\ln\frac{v_2}{v_1} + \frac{4f'RTw^2L}{gA^2D}$$

Coefficients of Friction   The coefficient of friction f is not a constant but is a function of the dimensionless expression $\mu/(\rho\bar{v}a)$ or $\mu v/(\bar{v}d)$, which is the reciprocal of the Reynolds number. McAdams and Sherwood formulate the expression

$$f' = 0.0054 + 0.375[\mu v/(\bar{v}d)]$$

This formula is applicable to water and other fluids. For high-pressure steam, the second term in the expression is small and f' is approximately equal to 0.0054. Babcock has suggested the approximation $f' = 0.0027(1 + 3.6/d)$ for steam.

Values of $f = 4f'$ as a function of pipe surface are given in Sec. 3.3.

For predicting the capacity of a given pipe operating on a chosen fluid with fixed pressure drop, the use of Fig. 4.1.35 eliminates the trial-and-error methods usually involved.

Resistances due to fittings, expressed in terms of L/D, are as follows: 90° elbows, $1-2\frac{1}{2}$ (3-6) [7-10] in, 30 (40) [50]; 90° curves, radius of centerline of curve 2-8 pipe diameters, 10; globe valves, $1-2\frac{1}{2}$ (3-6) [7-10] in, 45 (60) [75]; tees, 1-4 in, 60. The resistance in energy units, due to sudden enlargement in a pipe, is approximately $(\bar{v}_1 - \bar{v}_2)^2/(2g)$. For sudden contraction it is $1.5(1 - r)\bar{v}_2^2/[2g(3 - r)]$, where $r = A_2/A_1$.

(See "Camerons Hydraulic Data," latest edition, Ingersoll Rand Co., Woodcliff Lake, NJ; Warring, "Hydraulic Handbook," 8th ed., Gulf, Houston and Trade & Tech. Press, Morden, Surrey; Houghton and Brock, "Tables for the Compressible Flow of Dry Air," 3d ed., E. Arnold, London, with a review of basic equations and tabular data for the isentropic flow of dry air with Prandtl-Meyer expansion angles, Rayleigh flow, Fanno flow, and plane normal and oblique shock wave tables; Shapiro, "The Dynamics and Thermodynamics of Compressible Fluid Flow," Ronald Press, New York; Blevins, "Applied Fluid Dynamics Handbook," Van Nostrand Reinhold.)

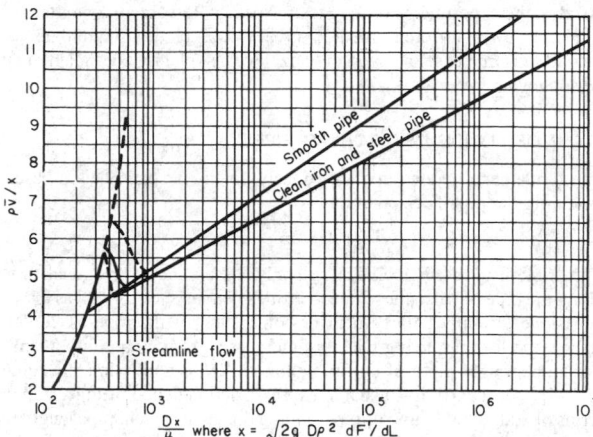

Fig. 4.1.35 Chart for estimating rate of flow from the pressure gradient.

THROTTLING

Throttling or Wire Drawing When a fluid flows from a region of higher pressure into a region of lower pressure through a valve or constricted passage, it is said to be throttled or wire-drawn. Examples are seen in the passage of steam through pressure-reducing valves, in the flow through ports and passages in the steam engine, and in the expansion valve of the refrigerating machine.

The **general equation applicable to throttling processes** is

$$(\bar{v}_2^2 - \bar{v}_1^2)/(2g_c) = h_1 - h_2$$

The velocities $\bar{v}_2$ and $\bar{v}_1$ are practically equal, and it follows that $h_1 = h_2$; i.e., in a throttling process there is no change in enthalpy.

For a mixture of liquid and vapor, $h = h_f + x h_{fg}$; hence the equation of throttling is $h_{f1} + x_1 h_{fg1} = h_{f2} + x_2 h_{fg2}$. In the case of a perfect gas, $h = c_p T + h_0$; hence the equation of throttling is $c_p T_1 + h_0 = c_p T_2 + h_0$, or $T_1 = T_2$.

Joule-Thomson Effect The investigations of Joule and Lord Kelvin showed that a gas drops in temperature when throttled. This is not universally true. For some gases, notably hydrogen, the temperature rises for throttling processes over ordinary ranges of temperature and pressure. Whether there is a rise or fall in temperature depends on the particular range of pressure and temperature over which the change

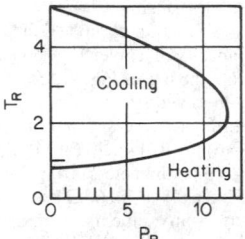

Fig. 4.1.36 Inversion curve.

occurs. For each gas, there are different values of pressure and temperature at which no temperature change occurs during a Joule-Thomson expansion. That temperature is called the inversion temperature. Below this temperature, a gas cools on throttling; above it, its temperature rises. The ratio of the observed drop in temperature to the drop in pressure, i.e., dT/dP, is the Joule-Thomson coefficient. In actual design, the effect of heat leaks must be carefully evaluated before the theoretical Joule-Thomson coefficients are applied.

Figure 4.1.36 shows the variation of inversion temperature with pressure and temperature; the exact values are shown in Table 4.1.4. The inversion locus for air is shown in Table 4.1.5.

The cooling **effect produced by throttling** has been applied to the liquefaction of gases.

Table 4.1.4 Approximate Inversion-Curve Locus in Reduced Coordinates ($T_R = T/T_c$; $P_R = P/P_c$)*

P_r	0	0.5	1	1.5	2	2.5	3	4
T_{RL}	0.782	0.800	0.818	0.838	0.859	0.880	0.903	0.953
T_{RU}	4.984	4.916	4.847	4.777	4.706	4.633	4.550	4.401
P_r	5	6	7	8	9	10	11	11.79
T_{RL}	1.01	1.08	1.16	1.25	1.35	1.50	1.73	2.24
T_{RU}	4.23	4.06	3.88	3.68	3.45	3.18	2.86	2.24

* Calculated from the best three-constant equation recommended by Miller, *Ind. Eng. Chem. Fundam.* **9**, 1970, p. 585. T_{RL} refers to the lower curve and T_{RU} to the upper curve.

Table 4.1.5 Approximate Inversion-Curve Locus for Air

P_r bar	0	25	50	75	100	125	150	175	200	225
T_L, K	(112)*	114	117	120	124	128	132	137	143	149
T_U, K	653	641	629	617	606	594	582	568	555	541

P, bar	250	275	300	325	350	375	400	425	432
T_L, K	156	164	173	184	197	212	230	265	300
T_U, K	526	509	491	470	445	417	386	345	300

* Hypothetical low-pressure limit.

Loss due to Throttling A throttling process in a cycle of operations always introduces a loss of efficiency. If T_0 is the temperature corresponding to the back pressure, the loss of available energy is the product of T_0 and the increase of entropy during the throttling process. The following example illustrates the calculation in the case of ammonia passing through the expansion valve of a refrigerating machine.

EXAMPLE. The liquid ammonia at a temperature of 70°F passes through the valve into the brine coil in which the temperature is 20 deg and the pressure is 48.21 psia. The initial enthalpy of the liquid ammonia is $h_{f1} = 120.5$, and therefore the final enthalpy is $h_{f2} + x_2 h_{fg2} = 64.7 + 553.1x_2 = 120.5$, whence $x_2 = 0.101$. The initial entropy is $s_{f1} = 0.254$. The final entropy is $s_{f2} + (x_2 h_{fg2}/T_2) = 0.144 + 0.101 \times 1.153 = 0.260$. $T_0 = 20 + 460 = 480$; hence the loss of refrigerating effect is $480 \times (0.260 - 0.254) = 2.9$ Btu.

COMBUSTION

REFERENCES: Chigier, "Energy, Combustion and Environment," McGraw-Hill. Campbell, "Thermodynamic Analysis of Combustion Engines," Wiley. Glassman, "Combustion," Academic Press. Lefebvre, "Gas Turbine Combustion," McGraw-Hill. Strehlow, "Combustion Fundamentals," McGraw-Hill. Williams et al., "Fundamental Aspects of Solid Propellant Rockets," *Agardograph,* **116,** Oct. 1969. Basic thermodynamic table type information needed in this area is found in Glushko et al., "Thermodynamic and Thermophysical Properties of Combustion Products," Moscow, and IPST translation; Gordon, NASA Technical Paper 1906, 1982; "JANAF Thermochemical Tables," NSRDS-NBS-37, 1971.

Fuels For special properties of various fuels, see Sec. 7. In general, fuels may be classed under three headings: (1) gaseous fuels, (2) liquid fuels, and (3) solid fuels.

The combustible elements that characterize fuels are carbon, hydrogen, and, in some cases, sulfur. The complete combustion of carbon gives, as a product, carbon dioxide, CO_2; the combustion of hydrogen gives water, H_2O.

Combustion of Gaseous and Liquid Fuels

Combustion Equations   The approximate molecular weights of the important elements and compounds entering into combustion calculations are:

For the elements C and H, the equations of complete combustion are

$$C + O_2 = CO_2 \qquad H_2 + \tfrac{1}{2}O_2 = H_2O$$
$$12\ lb + 32\ lb = 44\ lb \qquad 2\ lb + 16\ lb = 18\ lb$$

For a combustible compound, as CH_4, the equation may be written

$$CH_4 + x \cdot O_2 = y \cdot CO_2 + z \cdot H_2O$$

Taking, as a basis, 1 molecule of CH_4 and making a balance of the atoms on the two sides of the equation, it is seen that

$$y = 1 \qquad z = 2 \qquad 2x = 2y + z \qquad \text{or} \qquad x = 2$$

Hence,

$$CH_4 + 2O_2 = CO_2 + 2H_2O$$
$$16\ lb + 64\ lb = 44\ lb + 36\ lb$$

The coefficients in the combustion equation give the combining volumes of the gaseous components. Thus, in the last equation 1 ft³ of CH_4 requires for combustion 2 ft³ of oxygen and the resulting gaseous products of combustion are 1 ft³ of CO_2 and 2 ft³ of H_2O. The coefficients multiplied by the corresponding molecular weights give the combining weights. These are conveniently referred to 1 lb of the fuel. In the combustion of CH_4, for example, 1 lb of CH_4 requires 64/16 = 4 lb of oxygen for complete combustion and the products are 44/16 = 2.75 lb of CO_2 and 36/16 = 2.25 lb of H_2O.

Air Required for Combustion   The composition of air is approximately 0.232 O_2 and 0.768 N_2 on a pound basis, or 0.21 O_2 and 0.79 N_2 by volume. For exact analyses, it may be necessary sometimes to take account of the water vapor mixed with the air, but ordinarily this may be neglected.

The minimum amount of air required for the combustion of 1 lb of a fuel is the quantity of oxygen required, as found from the combustion equation, divided by 0.232. Likewise, the minimum volume of air required for the combustion of 1 ft³ of a fuel gas is the volume of oxygen divided by 0.21. For example, in the combustion of CH_4 the air required per pound of CH_4 is 4/0.232 = 17.24 lb and the volume of air per cubic foot of CH_4 is 2/0.21 = 9.52 ft³. Ordinarily, more air is provided than is required for complete combustion. Let a denote the minimum amount required and xa the quantity of air admitted; then $x - 1$ is the **excess coefficient**.

Products of Combustion   The products arising from the complete combustion of a fuel are CO_2, H_2O, and if sulfur is present, SO_2. Accompanying these are the nitrogen brought in with the air and the oxygen in the excess of air. Hence the products of complete combustion are principally CO_2, H_2O, N_2, and O_2. The **presence of CO indicates incomplete combustion**. In simple calculations the reaction of nitrogen with oxygen to form noxious oxides, often termed NO_x, such as nitric oxide (NO), nitrogen peroxide (NO_2), etc., is neglected. In practice, an automobile engine is run at a lower compression ratio to reduce NO_x formation. The reduced pollution is bought at the expense of reduced operating efficiency. The composition of the products of combustion is readily calculated from the combustion equations, as shown by the following illustrative example. (See also Table 4.1.7.)

EXAMPLE.   A producer gas having the volume composition given is burned with 20 percent excess of air; required the volume composition of the exhaust gases.

	V
H_2	0.08
CO	0.22
CH_4	0.024
CO_2	0.066
N_2	0.61
	1.0

Material	C	H_2	O_2	N_2	CO	CO_2	H_2O	CH_4	C_2H_4	C_2H_6O	S	NO	NO_2	SO_2
Molecular weight	12	2	32	28	28	44	18	16	28	46	32	30	46	64

Coefficients in reaction equations			Coefficients multiplied by V		
O_2	CO_2	H_2O	O_2	CO_2	H_2O
0.5	0	1	0.04	0	0.08
0.5	1	0	0.11	0.22	0
2	1	2	0.048	0.024	0.048
0	1	0	0	0.066	0
0	0	0	0	0	0
			0.198	0.31	0.128

For 1 ft³ of the producer gas, 0.198 ft³ of O_2 is required for complete combustion. The minimum volume of air required is 0.198/0.21 = 0.943 ft³ and with 20 percent excess the air supplied is 0.943 × 1.2 = 1.132 ft³. Of this, 0.238 ft³ is oxygen and 0.894 ft³ is N_2. Consequently, for 1 ft³ of the fuel gas, the exhaust gas contains

CO_2		0.31 ft³
H_2O		0.128 ft³
N_2	0.61 + 0.894 =	1.504 ft³
O_2 (excess)	0.238 − 0.198 =	0.040 ft³
		1.982 ft³

or

CO_2	15.7 percent
H_2O	6.5 percent
N_2	75.8 percent
O_2	2.0 percent
	100.0 percent

Volume Contraction   As a result of chemical action, there is often a change of volume; for example, in the reaction $2H_2 + O_2 = 2H_2O$, three volumes (two of H_2 and one of O_2) contract to two volumes of water vapor. In the example just given, the volume of producer gas and air supplied is 1 ft³ gas + 1.132 ft³ air = 2.132 ft³, and the corresponding volume of the exhaust gas is 1.982 ft³, showing a contraction of about 7 percent. For a hydrocarbon having the composition C_mH_n, the relative volume contraction is $1 - n/4$; thus for CH_4 and C_2H_4 there is no change of volume, for C_2H_2 the contraction is half the volume, and for C_2H_6 there is an increase of one-half in volume.

The change of volume accompanying a chemical reaction, such as a combustion, causes a corresponding change in the gas constant R. Let R' denote the constant for the mixture of gas and air (1 lb of gas and xa lb of air) before combustion, and R'' the constant of the mixture of resulting products of combustion. Then, if y is the resulting contraction of volume, $R''/R' = (1 + xa - y)/(1 + xa)$.

Heat of Combustion   Usually, a chemical change is accompanied by the generation or absorption of heat. The union of a combustible with oxygen produces heat, and the heat thus generated when 1 lb of combustible is completely burned is called the **heat of combustion** or the **heat value** of the combustible. Heat values are determined experimentally by calorimeters in which the products of combustion are cooled to the initial temperature and the heat absorbed by the cooling medium is measured. This is called the **high heat value**.

The heat transferred (heat of combustion) during a combustion reaction is computed on either a constant-pressure or a constant-volume basis. The first law is used in the the analysis of either process.

1. The **heat value at constant volume** (Q_v). Consider a constant-volume combustion process where several reactants combine under proper conditions to form one or more products. The heat of combustion under constant-volume conditions (Q_v) according to the first law may be expressed as follows:

$$Q_v = \Sigma(Nu)_P - \Sigma(Nu)_R$$

The term N refers to the amount of material, and the symbol u signifies the internal energy per unit quantity of material. The subscripts P and R refer to the products and reactants, respectively. Hence, it may be concluded that Q_v is equal to the change in internal energy. The heat of combustion under constant-volume conditions may also be described as the quantity of heat transferred from a calorimeter to the external surroundings when the termperature and volume of the combustion products are brought to the temperature and volume, respectively, of the gaseous mixture before burning.

2. The **heat value at constant pressure** (Q_p). For a constant-pressure process the first law may be expressed as

$$Q_p = \Sigma(Nu)_P - \Sigma(Nu)_R + pV_P - pV_R$$

Here the symbols p and V refer to the pressure and total volume, respectively. Usually in combustion reactions that part of the change in internal energy resulting from a volume change is small in comparison to the total change; hence it may usually be neglected. Assuming therefore that the internal energy change for a constant-volume reaction is approximately equal to that for a constant-pressure change, the following equation results:

$$Q_p = Q_v + p(V_P - V_R)$$

Since Q_v is equal to the change in internal energy, this relation may be changed to enthalpy values, from which it may be concluded that Q_v is equal to the change in enthalpy. The heat of combustion under constant-pressure conditions may also be described as the heat transferred from a calorimeter when the pressure and temperature of the products are brought back to the pressure and temperature, respectively, of the gaseous mixture before burning.

If the reactants and products are assumed to be ideal gases, then the relation for (Q_p) may be expressed as follows, where ΔN represents the change in number of moles and R the universal gas constant:

$$Q_p = Q_v + \Delta N\,RT$$

From this relation the heat transferred (heat of combustion) at constant pressure may be found from the heat of reaction at constant volume, or vice versa, if the temperature and molar-volume change are known.

If there is no change of volume due to the combustion, the heat values Q_p and Q_v are the same. When there is a contraction of volume, Q_p exceeds Q_v by the heat equivalent of the work done on the gas during the contraction. For example, in the burning of CO according to the equation $CO + \frac{1}{2}O_2 = CO_2$, there is a contraction of $\frac{1}{2}$ volume. Taking 62°F as the temperature, the volume of 1 lb CO at atmospheric pressure is 13.6 ft³; hence the equivalent of the work done at atmospheric pressure is $\frac{1}{2} \times 13.6 \times 2,116/778 = 18.5$ Btu, which is about 0.4 percent of the heat value of CO. Since the difference between Q_p and Q_v is small in most fuels, it is usually neglected.

It is also to be noted that heat values vary with the initial temperature (which is also the final temperature), but the variation is usually negligible.

Heat Value per Unit Volume Since the consumption of a fuel gas is more easily measured by volume than by mass, it is convenient to express heat values in terms of volumes. For this purpose, a standard temperature and pressure must be assumed. It is customary to take atmospheric pressure (14.70 psi) as standard, but there is diversity of practice in the matter of a **standard temperature**. The temperature of 68°F (20°C) is generally accepted in metric countries and has been recommended by the American delegates to the meeting of the International Committee of Weights and Measures and also by the ASME Power Test Codes Committee. The American Gas Assoc. uses 60°F as the standard temperature of reference. Conversion of density and heat values from 68 to 60°F of dry (saturated) gas is obtained by multiplying by the factor 1.0154 (1.0212). Conversion of specific volumes of dry (saturated) gas is obtained by multiplying by the factor 0.9848 (0.9792).

If the gas is at some other pressure and temperature, say p_1 psia and T_1 °R, the heat value per cubic foot is found by multiplying the heat value per cubic foot under standard conditions by $35.9 p_1/T_1$.

The heat values of a few of the more common fuels per pound and per cubic foot are given in Table 4.1.6.

Heat Value per Unit Volume of Mixture Let a denote the volume of

Table 4.1.6   Heats of Combustion

Fuel	Chemical symbol	High heat value		Low heat value	
		Btu/lb	*Btu/ft³	Btu/lb	*Btu/ft³
Carbon to CO_2	C	14,096			
Carbon to CO	C	3,960			
CO to CO_2	CO	4,346	316.0		
Sulfur to SO_2	S	3,984			
Hydrogen	H_2	61,031	319.4	51,593	270.0
Methane	CH_4	23,890	994.7	21,518	896.0
Ethane	C_2H_6	22,329	1,742.6	20,431	1,594.5
Propane	C_3H_8	21,670	2,480.1	19,944	2,282.6
Butane	C_4H_{10}	21,316	3,215.6	19,679	2,968.7
Pentane	C_5H_{12}	21,095	3,950.2	19,513	3,654.0
Hexane (liquid)	C_6H_{14}	20,675	. . . .	19,130	
Octane (liquid)	C_8H_{18}	20,529		19,029	
n-Decane (liquid)	$C_{10}H_{22}$	20,371	. . . .	19,175	
Ethylene	C_2H_4	21,646	1,576.1	20,276	1,477.4
Propene (propylene)	C_3H_6	21,053	2,299.4	19,683	2,151.3
Acetylene (ethyne)	C_2H_2	21,477	1,451.4	20,734	1,402.0
Benzene	C_6H_6	18,188	3,687.5	17,446	3,539.3
Toluene (methyl benzene)	C_7H_8	18,441	4,410.1	17,601	4,212.6
Methanol (methyl alcohol, liquid)	CH_4O	9,758	. . . .	8,570	
Ethanol (ethyl alcohol, liquid)	C_2H_6O	12,770		11,531	
Naphthalene (solid)	$C_{10}H_8$	17,310	. . . .	13,110	

* Measured as a gas at 68°F and 14.70 psia. Multiply by 1.0154 for 60°F and 14.70 psia.

air required for the combustion of 1 ft³ of fuel gas and xa the value of air actually admitted, $(x - 1)a$ being therefore the excess. Then the volume of the mixture of fuel gas and air is $1 + xa$, and the quotient $Q/(1 + xa)$ may be called the heat value per cubic foot of mixture. This magnitude is useful in comparing the relative volumes of mixture required with different fuel gases. Thus a lean gas, as blast-furnace gas or producer gas, has a low heat value Q, but the value of a is correspondingly low. On the other hand, a rich gas, like natural gas, has a high heat value but requires a large volume of air for combustion.

Low and High Heat Values   Any fuel containing hydrogen yields water as one product of combustion. At atmospheric pressure, the partial pressure of the water vapor in the resulting combustion gas mixture will usually be sufficiently high to cause water to condense out if the temperature is allowed to fall below 120 to 140°F. This causes liberation of the heat of vaporization of any water condensed. The low heat value is evaluated assuming no water vapor condensed, whereas the high heat value is calculated assuming all water vapor condensed.

To facilitate calculations of the temperature attained by combustion, it is desirable to make use of the **low heat value**. The necessity of taking into account the heat of vaporization of the water vapor and the difference between the specific heats of liquid water and of water vapor is thus avoided. The high heat of combustion exceeds the low heat of combustion by the difference between the heat actually given up on cooling the products to the initial temperature and that which would have been given up if the products had remained in the gaseous state. A

bomb calorimeter (constant volume) gives practically correct values of the high heat value; a gas calorimeter (constant pressure) gives values which, for the usual fuels, may be incorrect by a fraction of 1 percent. The quantity to be subtracted from the high heat value to obtain the low heat value will vary with the composition of the fuel; an approximate value is $1,050m$, where m is the number of pounds of H_2O formed per pound of fuel burned.

In Germany, the low heat value of the fuel is used in calculating efficiencies of internal-combustion engines. In the United States, the high value is specified by the ASME Power Test Codes.

Heat of Formation   The change in enthalpy resulting when a compound is formed from its elements isothermally and at constant pressure is numerically equal to, but of opposite sign to, the heat of formation, $\Delta H_f = -Q_f$. It is equal to the difference between the heats of combustion of the constituents forming the compound and the heat of combustion of the compound itself. The following values for heats of formation are in Btu per pound of the compound. The elements before the change and the compounds formed are assumed in their ordinary stable states at 65°F and 1 atm. A plus sign indicates heat evolved on forming the compound, a minus sign heat absorbed from the surroundings.

Fuels   Methane, CH_4 (gas), 2,001.4; ethane, C_2H_6 (gas), 1,206.1; propane, C_3H_8 (vapor), 1,008.5; acetylene, C_2H_2 (gas), -3747; ethylene, C_2H_4 (gas), -805.3; benzene, C_6H_6 (vapor), -459; toluene, C_7H_8 (vapor), -234.9; methyl alcohol, CH_3OH (liquid), 3,227.3; ethyl alcohol, C_2H_5OH (liquid), 2623.3.

Table 4.1.7   Products of Combustion

Fuel	Chemical formula	Molecular weight $O_2 = 32$	Specific weight, lb/ft³ at 68°F and 14.70 lb/in²	Volume of air necessary for combustion of unit volume of fuel at same temperature and pressure	Products of combustion of 1 ft³ of fuel in theoretical amount of air, ft³			Weight of air necessary for combustion of unit weight of fuel	Products of combustion of 1 lb of fuel in theoretical amount of air, lb		
					CO_2	H_2O	N_2		CO_2	H_2O	N_2
Oxygen	O_2	32	0.0831								
Nitrogen	N_2	28.08	0.0727								
Air			0.0753								
Hydrogen	H_2	2.016	0.0052	2.39	0	1	1.89	34.2	0.0	8.94	26.28
Steam	H_2O	18.016									
Carbon monoxide	CO	28.00	0.0727	2.39							
Carbon dioxide	CO_2	44.00	0.1142								
Methane	CH_4	16.03	0.0416	9.55	1	2	7.55	17.21	2.75	2.248	13.22
Ethane	C_2H_6	30.05	0.0779	16.71	2	3	13.21	16.07	2.93	1.799	12.34
Propane	C_3H_8	44.06	0.1142	23.87	3	4	18.87	15.65	3.00	1.635	12.02
Butane	C_4H_{10}	58.1	0.1506	30.94	4	5	24.53	15.44	3.03	1.551	11.86
Pentane	C_5H_{12}	72.1	0.1869	38.08	5	6	30.2	15.31	3.05	1.499	11.76
Hexane	C_6H_{14}	86.1	0.2232	45.3	6	7	35.8	15.22	3.07	1.465	11.69
Heptane	C_7H_{16}	100.1	0.2596	52.5	7	8	41.5	15.15	3.08	1.439	11.64
Octane	C_8H_{18}	114.1	0.2959	59.7	8	9	47.2	15.11	3.08	1.421	11.60
Nonane	C_9H_{20}	128.2	0.3323	66.8	9	10	52.8	15.07	3.09	1.406	11.57
Benzene	C_6H_6	78.0	0.2025	35.8	6	3	28.3	13.26	3.38	0.693	10.18
Toluene	C_7H_8	92.1	0.2388	42.9	7	4	34.0	13.50	3.35	0.783	10.36
Xylene	C_8H_{10}	106.2	0.2752	50.1	8	5	39.6	13.57	3.31	0.845	10.42
Cyclohexane	C_6H_{12}	84.0	0.2180	43.0	6	6	34.0	14.76	3.14	1.285	11.34
Ethylene	C_2H_4	28.03	0.0728	14.32	2	2	11.32	14.76	3.14	1.285	11.34
Propylene	C_3H_6	42.0	0.1090	21.48	3	3	16.98	14.76	3.14	1.285	11.34
Butylene	C_4H_8	64.1	0.1454	28.64	4	4	22.64	14.76	3.14	1.285	11.34
Acetylene	C_2H_2	26.02	0.0675	11.93	2	1	9.43	13.26	3.38	0.693	10.18
Allylene	C_3H_4	40.0	0.1038	19.09	3	2	15.09	13.78	3.30	0.900	10.59
Naphthalene	$C_{10}H_8$	128.1	0.3322	57.3	10	4	45.28	12.93	3.44	0.563	9.93
Methyl alcohol	CH_4O	32.0	0.0830	7.16	1	2	5.66	6.46	1.37	1.125	4.96
Ethyl alcohol	C_2H_6O	46.0	0.1194	14.32	2	3	11.32	8.99	1.91	1.174	6.90

SOURCE: Marks, "The Airplane Engine."

Table 4.1.8   Internal Energy of Gases
Btu/(lb·mol) above 520°R

Temp. °R	O_2	N_2	Air	CO_2	H_2O	H_2	CO	Apv
520	0	0	0	0	0	0	0	1,033
540	100	97	97	139	122	96	97	1,072
560	200	196	196	280	244	193	196	1,112
580	301	295	295	424	357	291	295	1,152
600	402	395	395	570	490	390	396	1,192
700	920	896	897	1,320	1,110	887	896	1,390
800	1,449	1,399	1,403	2,120	1,734	1,386	1,402	1,589
900	1,989	1,905	1,915	2,965	2,366	1,886	1,913	1,787
1,000	2,539	2,416	2,431	3,852	3,009	2,387	2,430	1,986
1,100	3,101	2,934	2,957	4,778	3,666	2,889	2,954	2,185
1,200	3,675	3,461	3,492	5,736	4,399	3,393	3,485	2,383
1,300	4,262	3,996	4,036	6,721	5,030	3,899	4,026	2,582
1,400	4,861	4,539	4,587	7,731	5,740	4,406	4,580	2,780
1,500	5,472	5,091	5,149	8,764	6,468	4,916	5,145	2,979
1,600	6,092	5,652	5,720	9,819	7,212	5,429	5,720	3,178
1,700	6,718	6,224	6,301	10,896	7,970	5,945	6,305	3,376
1,800	7,349	6,805	6,889	11,993	8,741	6,464	6,899	3,575
1,900	7,985	7,393	7,485	13,105	9,526	6,988	7,501	3,773
2,000	8,629	7,989	8,087	14,230	10,327	7,517	8,109	3,972
2,100	9,279	8,592	8,698	15,368	11,146	8,053	8,722	4,171
2,200	9,934	9,203	9,314	16,518	11,983	8,597	9,339	4,369
2,300	10,592	9,817	9,934	17,680	12,835	9,147	9,961	4,568
2,400	11,252	10,435	10,558	18,852	13,700	9,703	10,588	4,766
2,500	11,916	11,056	11,185	20,033	14,578	10,263	11,220	4,965
2,600	12,584	11,682	11,817	21,222	15,469	10,827	11,857	5,164
2,700	13,257	12,313	12,453	22,419	16,372	11,396	12,499	5,362
2,800	13,937	12,949	13,095	23,624	17,288	11,970	13,144	5,561
2,900	14,622	13,590	13,742	24,836	18,217	12,549	13,792	5,759
3,000	15,309	14,236	14,394	26,055	19,160	13,133	14,443	5,958
3,100	16,001	14,888	15,051	27,281	20,117	13,723	15,097	6,157
3,200	16,693	15,543	15,710	28,513	21,086	14,319	15,754	6,355
3,300	17,386	16,199	16,369	29,750	22,066	14,921	16,414	6,554
3,400	18,080	16,855	17,030	30,991	23,057	15,529	17,078	6,752
3,500	18,776	17,512	17,692	32,237	24,057	16,143	17,744	6,951
3,600	19,475	18,171	18,356	33,487	25,067	16,762	18,412	7,150
3,700	20,179	18,833	19,022	34,741	26,085	17,385	19,082	7,348
3,800	20,887	19,496	19,691	35,998	27,110	18,011	19,755	7,547
3,900	21,598	20,162	20,363	37,258	28,141	18,641	20,430	7,745
4,000	22,314	20,830	21,037	38,522	29,178	19,274	21,107	7,944
4,100	23,034	21,500	21,714	39,791	30,221	19,911	21,784	8,143
4,200	23,757	22,172	22,393	41,064	31,270	20,552	22,462	8,341
4,300	24,482	22,845	23,073	42,341	32,326	21,197	23,140	8,540
4,400	25,209	23,519	23,755	43,622	33,389	21,845	23,819	8,738
4,500	25,938	24,194	24,437	44,906	34,459	22,497	24,499	8,937
4,600	26,668	24,869	25,120	46,193	35,535	23,154	25,179	9,136
4,700	27,401	25,546	25,805	47,483	36,616	23,816	25,860	9,334
4,800	28,136	26,224	26,491	48,775	37,701	24,480	26,542	9,533
4,900	28,874	26,905	27,180	50,069	38,791	25,418	27,226	9,731
5,000	29,616	27,589	27,872	51,365	39,885	25,819	27,912	9,930
5,100	30,361	28,275	28,566	52,663	40,983	26,492	28,600	10,129
5,200	31,108	28,961	29,262	53,963	42,084	27,166	29,289	10,327
5,300	31,857	29,648	29,958	55,265	43,187	27,842	29,980	10,526
5,400	32,607	30,337	30,655	56,569	44,293	28,519	30,674	10,724

SOURCE: L. C. Lichty, "Internal Combustion Engines," p. 582, derived from data given by Hershey, Eberhardt, and Hottel, *Trans. SAE,* **31,** 1936, p. 409.

Inorganic Compounds Al_2O_3, 6,710; CaO, 4,869; $CaCO_3$, 5,206; FeO, 1,611; Fe_2O_3, 2,238; Fe_3O_4, 2,075; FeS_2, 532.7; HCl (gas), 1,089, HNO_3 (liquid), 1,190; H_2O (liquid), 6,827; H_2S (gas), 279.9; H_2SO_4 (liquid), 3,555.8; K_2O, 164.7; MgO, 6,522; MnO, 2,449; NO, $-1,296$; N_2O, -803.5; Na_2O, 2,888; NH_3, 1,163; NH_4CL, 1,480; NiO, 1,407; P_2O_5, 5,394; PbO (red), 423.0; PbO_2, 489.1; SO_2, 1,933; SO_3, 2,112; SnO, 904.7; ZnO, 1,847.

INTERNAL ENERGY AND ENTHALPY OF GASES

Table 4.1.8 gives the internal energy of various common gases in Btu/(lb·mol) measured above 520°R (60°F). The corresponding values of the enthalpy are obtained by adding the value of Apv from the last column.

TEMPERATURE ATTAINED BY COMBUSTION

Excluding the effect of dissociation, the temperature attained at the end of combustion may be calculated by a simple energy balance. The heat of combustion less the heat lost by conduction and radiation during the process is equal to the increase in internal energy of the products mixture if the combustion is at constant volume; or, if the combustion is at constant pressure, the difference is equal to the increase in enthalpy of the products mixture.

EXAMPLE.   To calculate the temperature of combustion of a fuel gas having the composition $H_2 = 0.50$, $CO = 0.46$, $CO_2 = 0.04$. The gas is burned with 15 percent excess air at constant volume, and the initial temperature is 62°F; i.e., $T = 522$°R.

The volume compositions of the initial mixture of fuel gas and air and of the mixture of products are, respectively,

Initial: H_2, 0.50; CO, 0.46; CO_2, 0.04; O_2, 0.552; N_2, 2.098
Products: H_2O, 0.50; CO_2, 0.50; O_2, 0.072; N_2, 2.098

Since a volume composition is also a mol composition, the products mixture may be regarded as made up of 0.5 mol each of H_2O and CO_2, 0.072 mol of O_2, and 2.098 mols of N_2. If values are taken from Tables 4.1.6 and 4.1.7, the heat generated by combustion of the fuel mixture is $0.50 \times 2 \times 51,593 + 0.46 \times 28 \times 4,346 = 107,569$ Btu. The internal energy u of the products mixture at $T = 522$ is now calculated (Table 4.1.8). For 0.5 mol H_2O + 0.5 mol CO_2 +

0.072 mol O_2 + 2.098 mol N_2 this is $6.1 + 6.95 + 0.72 + 20.34 = 34.11$ Btu.

The energy u of the mixture is next calculated for various assumed temperatures, the proper values being taken from Table 4.1.8.

If the heat of combustion, 107,569 Btu, is entirely used in the increase of energy, the temperature attained lies somewhere between 5,000 and 5,100; by interpolation, the value 5,073° is obtained.

Loss of heat during combustion may readily be taken into account; thus if 10 percent of the heat of combustion is lost, the amount available for increasing the energy of the products is $107,569 \times 0.90 = 96,812$ Btu, and this increase gives $T_2 = 4,671$°. If the fuel is burned at constant pressure, Q_p is used instead of Q_v and values of h are determined from Table 4.1.8 instead of values of u.

T_2 assumed	4,700	4,800	4,900	5,000	5,100
Energy 0.5 mol H_2O	18,308	18,851	19,396	19,943	20,429
Energy 0.5 mol CO_2	23,742	24,388	25,035	25,683	26,331
Energy 0.072 mol O_2	1,973	2,026	2,079	2,132	2,185
Energy 2.098 mol N_2	54,596	55,018	56,447	57,882	59,321
$u_2 =$	97,619	100,283	102,957	105,640	108,329
$u_1 =$	34	34	34	34	34
	97,585	100,249	102,923	105,606	108,295

EFFECT OF DISSOCIATION

The maximum temperature that can be obtained by the combustion of any fuel is limited by the dissociation of the products formed. The dissociation and equilibriums involved in high-temperature combustion are exceedingly complex, involving such chemical species as CO_2, CO, H_2O, H_2, H, OH, N_2, NO, N, O_2, and O. The equilibrium reached is a direct consequence of the second law of thermodynamics. However, the calculation of the equilibrium constant even for simple reactions is tedious. For all possible reactions $aA + bB \rightarrow cC + dD$ ($a, b, c, d \leq 2$), the excellent tables of the equilibrium constant k_p contained in the "American Institute of Physics Handbook," 3d edition, McGraw-Hill, pp. 4-31 and 4-32, are recommended to save time. Papers describing the calculation of k_p for multicomponent reacting gases are contained in the first *ASME Symposium on Thermophysical Properties Proceedings*.

Calculated flame temperatures, allowing for dissociation, for gaseous fuels with stated amounts of air present are given in Table 4.1.9. The combustion is assumed to be adiabatic and at 14.7 lb/in² absolute.

Table 4.1.9   Flame Temperatures, Deg R, at 14.7 psia, Allowing for Dissociation

	Percent of theoretical air				
Fuel	80	90	100	120	140
Hydrogen	4,210	4,330	4,390	4,000	3,670
Carbon monoxide	4,280	4,370	4,320	4,140	3,850
Methane	4,050	.....	4,010	3,660	3,330
Carbureted water gas	3,940	.....	4,150	3,820	3,510
Coal gas	3,920	.....	4,050	3,780	3,440
Natural gas	4,010	.....	4,180	3,840	3,520
Producer gas	3,040	.....	3,330	3,130	2,970
Blast furnace gas	2,810	.....	3,060	2,920	2,750

SOURCE: Satterfield, "Generalized Thermodynamics of High-Temperature Combustion," Sc.D. thesis, M.I.T., 1946.

The volumetric compositions of the fuels of Table 4.1.9 are given below:

Fuels	CO	H_2	CH_4	C_2H_4	Illuminants (assumed C_2H_4)	CO_2	O_2	N_2
H_2	....	100.0						
CO	100.0							
CH_4	....	....	100.0					
Carbureted water gas	24.1	32.5	9.0	2.2	10.3	4.6	0.6	16.7
Coal gas	5.9	53.2	29.6	....	2.7	1.4	0.7	6.5
Natural gas	....	....	78.8	14.0	......	0.4	.....	6.8
Producer gas	26.0	3.0	0.5	....	......	2.50	.....	56.0
Blast furnace gas	26.5	3.5	0.2	....	......	12.8	0.1	56.9

In the case of **explosion** in the **internal-combustion engine,** the figures in Table 4.1.9 will be somewhat changed. The effect of compression is to increase both the initial temperature and the initial pressure. The resulting increase in the explosion temperature will tend to increase the dissociation; the increase of pressure will tend to reduce it. The net effect will be a small reduction.

COMBUSTION OF LIQUID FUELS

For properties of fuel oils, heat values, etc., see Sec. 7. Calculations for the burning of liquid fuels are fundamentally the same as for gaseous fuels. Liquid fuels are almost always gasified before or during actual combustion.

COMBUSTION OF SOLID FUELS

For **properties of solid fuels,** heat values, etc., see Sec. 7.

Air Required for Combustion Let c, h, and o, denote, respectively, the parts of carbon, hydrogen, and oxygen in 1 lb of the fuel. Then the **minimum amount of oxygen** required for complete combustion is $2.67c + 8h - o$ lb, and the **minimum quantity of air** required is $a = (2.67c + 8h - o)/0.23 = 11.6[c + 3(h - o/8)]$ lb.

With air at 62°F and at atmospheric pressure, the minimum volume of air required is $v_m = 147[c + 3(h - o/8)]$ ft^3. In practice, an excess of air over that required for combustion is admitted to the furnace. The actual quantity admitted per pound of fuel may be denoted by xa. Then x = amount admitted ÷ minimum amount.

Combustion Products If v_m is the minimum volume of air required for complete combustion and xv_m the actual volume supplied, then the products will contain per pound of fuel, $O_2 = 0.21v_m(x - 1)$ ft^3, $N_2 = 0.79xv_m$ ft^3.

From the reaction equation $C + O_2 = CO_2$, the volume of CO_2 formed is equal to the volume of oxygen required for the carbon constituent alone; hence volume of $CO_2 = 0.21v_mc/[c + 3(h - 0.125o)]$.

Of the *dry* gaseous products (i.e., without water), the CO_2 content by volume is therefore given by the expression

$$CO_2 = 0.21c/[xc + (x - 0.21)3(h - 0.125o)]$$

The combined CO_2 and O_2 content is

$$CO_2 + O_2 = 0.21\left\{1 - 0.79 \middle/ \left[\frac{x + cx}{3(h - 0.125o) - 0.21}\right]\right\}$$

If the fuel is all carbon, the combined CO_2 and O_2 is by volume 21 percent of the gaseous products. The more hydrogen contained in the fuel, the smaller is the $CO_2 + O_2$ content. The CO_2 content depends in the first instance on the excess of air. Thus, for pure carbon, it is $CO_2 = 0.21/x$.

The excess of air may be calculated from the composition of the gases and that of the fuel. Thus

$$x = 0.21\left[\frac{c}{[CO_2]} + 3(h - 0.125o)\right] \middle/ [c + 3(h - 0.125o)]$$

in which [CO_2] denotes the percent by volume of the CO_2 in the dry gas.

The temperature of combustion is calculated by the same method as for gaseous fuels.

Loss due to Incomplete Combustion The loss due to incomplete combustion of the carbon in the fuel, in Btu/lb of fuel, is

$$L = 10,136C \times CO/(CO + CO_2)$$

where $10,136$ = difference in heat evolved in burning 1 lb of carbon to CO_2 and to CO; CO and CO_2 = percentages by volume of carbon monoxide and carbon dioxide as found by analysis; and C = fraction of quantity of carbon in the fuel which is actually burned and passes up the stack, either as CO or CO_2. The presence of 1 percent of CO in the flue gases will represent a decrease in the boiler efficiency of 4.5 percent. An additional loss is caused by passage through the grate to the ashpit of any unburned or partly burned fuel.

It is generally assumed that high CO_2 readings are indicative of good combustion and, hence, of high efficiencies. Such readings are not satisfactory when considered apart from the CO determination. The best percentage of CO_2 to maintain varies with different fuels and is lower for those with a high hydrogen content than for fuel mainly composed of carbon.

Hydrogen in a fuel increases the nitrogen content of the flue gases. This is due to the fact that the water vapor formed by the combustion of hydrogen will condense at the temperature at which the analysis is made, while the nitrogen which accompanied the oxygen maintains its gaseous form and passes in that form into the sampling apparatus. For this reason, where highly volatile coals containing considerable hydrogen are burned, the flue gas contains an apparently increased amount of nitrogen. The effect is even more pronounced when burning gaseous or liquid hydrocarbon fuels.

The amount of flue gases per pound of fuel, including moisture formed by the hydrogen component, is approximately $3.02[N/(CO_2 + CO)]C + (1 - A)$, where A = percent of ash found in test. The quantity of dry flue gases per pound of fuel may be approximated from the formula $W_2 = C[11CO_2 + 8O + 7(CO + N)]/3(CO_2 + CO)$. In these formulas, the amount of gas is per pound of dry or moist fuel as the percentage of C is referred to a dry or moist basis.

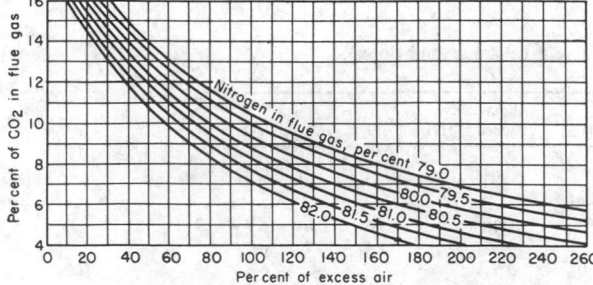

Fig. 4.1.37 Ratio of air supplied per pound of combustible to that theoretically required.

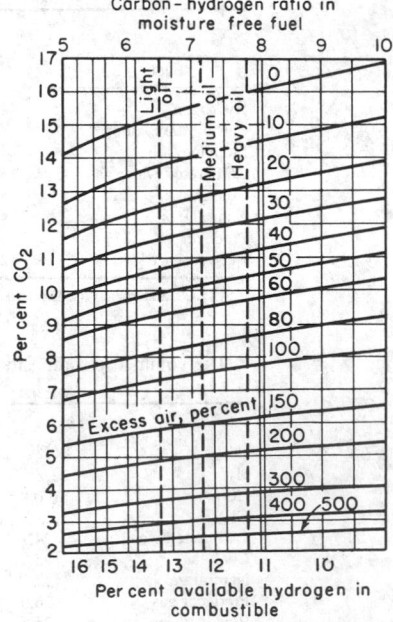

Fig. 4.1.38 Relation of CO_2 to excess air for fuels.

The ratio of air suppplied per pound of fuel to the air theoretically required is

$$\frac{W_1}{W} = \frac{3.02C[N/(CO_2 + CO)]}{34.56(C/3 + H - O/8)}$$

The ratio of air supplied per pound of combustible to that theoretically required is $N/[N - 3.782(O - \frac{1}{2}CO)]$, on the assumption that all the nitrogen in the flue gas comes from the air supplied. Figure 4.1.37 gives the value of this ratio for varying flue-gas analyses where there is no CO present.

For petroleum fuels with hydrogen content from 9 to 16 percent, the excess air can be determined from the CO_2 content of the flue gases (with no CO present) by the use of Fig. 4.1.38. The curves are based on the assumption of 0.4 percent sulfur in the oil.

4.2 THERMODYNAMIC PROPERTIES OF SUBSTANCES
by Peter E. Liley

NOTE: Thermodynamic properties of a variety of other specific materials are listed also in Secs. 4.1, 6.1, and 9.8.

Table 4.2.1 Enthalpy and Psi Functions for Ideal-Gas Air*

T, K	h, kJ/kg	Ψ	T, K	h, kJ/kg	Ψ	T, K	h, kJ/kg	Ψ
200	200.0	−0.473	800	821.9	1.679	1,400	1,515	2.653
220	220.0	−0.329	820	844.0	1.720	1,420	1,539	2.679
240	240.1	−0.197	840	866.1	1.760	1,440	1,563	2.705
260	260.1	−0.076	860	888.3	1.800	1,460	1,587	2.730
280	280.1	0.037	880	910.6	1.838	1,480	1,612	2.755
300	300.2	0.142	900	933.0	1.876	1,500	1,636	2.779
320	320.3	0.240	920	955.4	1.914	1,520	1,660	2.803
340	340.4	0.332	940	978.0	1.950	1,540	1,684	2.827
360	360.6	0.419	960	1,000.6	1.987	1,560	1,709	2.851
380	380.8	0.502	980	1,023.3	2.022	1,580	1,738	2.875
400	401.0	0.580	1,000	1,046.1	2.057	1,600	1,758	2.898
420	421.3	0.655	1,020	1,068.9	2.091	1,620	1,782	2.921
440	441.7	0.727	1,040	1,091.9	2.125	1,640	1,806	2.944
460	462.1	0.795	1,060	1,114.9	2.158	1,660	1,831	2.966
480	482.5	0.861	1,080	1,138.0	2.190	1,680	1,855	2.988
500	503.1	0.925	1,100	1,161.1	2.223	1,700	1,880	3.010
520	523.7	0.986	1,120	1,184.3	2.254	1,720	1,905	3.032
540	544.4	1.045	1,140	1,207.6	2.285	1,740	1,929	3.054
560	565.2	1.102	1,160	1,230.9	2.316	1,760	1,954	3.075
580	586.1	1.158	1,180	1,254.3	2.346	1,780	1,979	3.096
600	607.0	1.211	1,200	1,278	2.376	1,800	2,003	3.117
620	628.1	1.264	1,220	1,301	2.406	1,840	2,053	3.158
640	649.2	1.314	1,240	1,325	2.435	1,880	2,102	3.198
660	670.5	1.364	1,260	1,349	2.463	1,920	2,152	3.238
680	691.8	1.412	1,280	1,372	2.491	1,960	2,202	3.277
700	713.3	1.459	1,300	1,396	2.519	2,000	2,252	3.215
720	734.8	1.505	1,320	1,420	2.547	2,050	2,315	3.262
740	756.4	1.550	1,340	1,444	2.574	2,100	2,377	3.408
760	778.2	1.594	1,360	1,467	2.601	2,150	2,440	3.453
780	800.0	1.637	1,380	1,491	2.627	2,200	2,504	3.496

* Values rounded off from Chappell and Cockshutt, Nat. Res. Counc. Can. Rep. NRC LR 759 (NRC No. 14300), 1974. This source tabulates values of seven thermodynamic functions at 1-K increments from 200 to 2,200 K in SI units and at other increments for two other unit systems. An earlier report (NRC LR 381, 1963) gives a more detailed description of an earlier fitting from 200 to 1,400 K. In the above table h = specific enthalpy, kJ/kg, and $\Psi_2 - \Psi_1 = \log (P_2/P_1)$, for an isentrope. In terms of the Keenan and Kaye function ϕ, $\Psi = [\log (e/R)]\phi$.

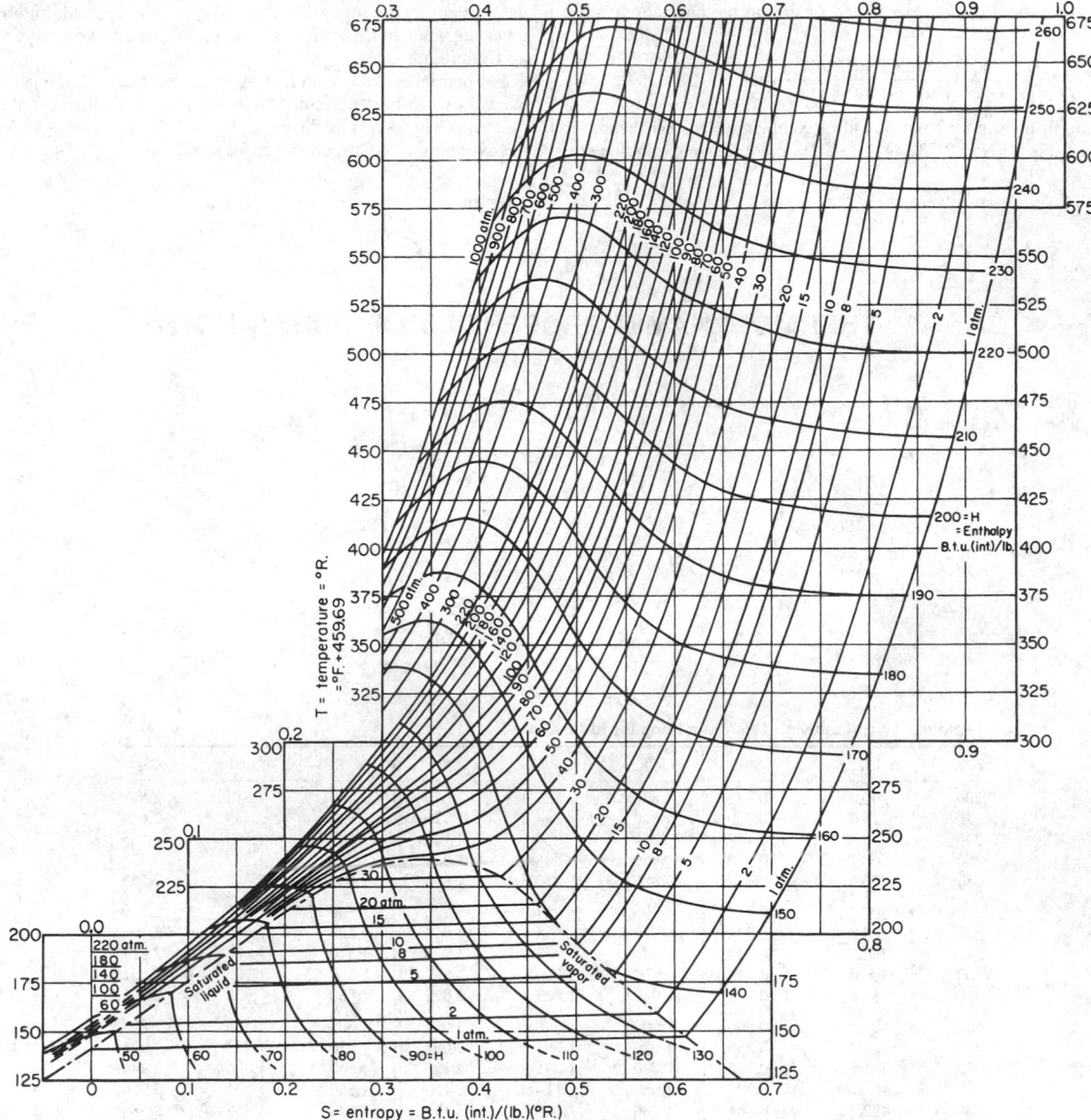

Fig. 4.2.1 Temperature-entropy diagram for air. (*Landsbaum et al., AIChE J.,* **1,** no. 3, 1955, p. 303.) *(Reproduced by permission of the authors and editor, AIChE.)*

Table 4.2.2   International ρ, Standard Atmosphere*

Z, m	T, K	P, bar	ρ, kg/m^3	g, m/s^2	M	a, m/s	λ, m	H, m
0	288.15	1.01325	1.2250	9.80665	28.964	340.29	6.63. − 8	0
1,000	281.65	0.89876	1.1117	9.8036	28.964	336.43	7.31. − 8	1,000
2,000	275.15	0.79501	1.0066	9.8005	28.964	332.53	8.07. − 8	1,999
3,000	268.66	0.70121	0.90925	9.7974	28.964	328.58	8.94. − 8	2,999
4,000	262.17	0.61660	0.81935	9.7943	28.964	324.59	9.92. − 8	3,997
5,000	255.68	0.54048	0.73643	9.7912	28.964	320.55	1.10. − 7	4,996
6,000	249.19	0.47217	0.66011	9.7882	28.964	316.45	1.23. − 7	5,994
7,000	242.70	0.41105	0.59002	9.7851	28.964	312.31	1.38. − 7	6,992
8,000	236.22	0.35651	0.52579	9.7820	28.964	308.11	1.55. − 7	7,990
9,000	229.73	0.30800	0.46706	9.7789	28.964	303.85	1.74. − 7	8,987
10,000	223.25	0.26499	0.41351	9.7759	28.964	299.53	1.97. − 7	9,984
15,000	216.65	0.12111	0.19476	9.7605	28.864	295.07	4.17. − 7	14,965
20,000	216.65	0.05529	0.08891	9.7452	28.964	295.07	9.14. − 7	19,937
25,000	221.55	0.02549	0.04008	9.7300	28.964	298.39	2.03. − 6	24,902
30,000	226.51	0.01197	0.01841	9.7147	28.964	301.71	4.42. − 6	29,859
40,000	250.35	2.87. − 3	4.00. − 3	9.6844	28.964	317.19	2.03. − 5	39,750
50,000	270.65	8.00. − 4	1.03. − 3	9.6542	28.964	329.80	7.91. − 5	49,610
60,000	247.02	2.20. − 4	3.10. − 4	9.6241	28.964	315.07	2.62. − 4	59,439
70,000	219.59	5.22. − 5	8.28. − 5	9.5942	28.964	297.06	9.81. − 4	69,238
80,000	198.64	1.05. − 5	1.85. − 5	9.5644	28.964	282.54	4.40. − 3	79,006
90,000	186.87	1.84. − 6	3.43. − 6	9.5348	28.95		2.37. − 2	88,744
100,000	195.08	3.20. − 7	5.60. − 7	9.5052	28.40		0.142	98,451
150,000	634.39	4.54. − 9	2.08. − 9	9.3597	24.10		33	146,542
200,000	854.56	8.47. − 10	2.54. − 10	9.2175	21.30		240	193,899
250,000	941.33	2.48. − 10	6.07. − 11	9.0785	19.19		890	240,540
300,000	976.01	8.77. − 11	1.92. − 11	8.9427	17.73		2,600	286,480
400,000	995.83	1.45. − 11	2.80. − 12	8.6799	15.98		1.6. + 4	376,320
500,000	999.24	3.02. − 12	5.22. − 13	8.4286	14.33		7.7. + 4	463,540
600,000	999.85	8.21. − 13	1.14. − 13	8.1880	11.51		2.8. + 5	548,252
800,000	999.99	1.70. − 13	1.14. − 14	7.7368	5.54		1.4. + 6	710,574
1,000,000	1,000.00	7.51. − 14	3.56. − 15	7.3218	3.94		3.1. + 6	864,071

* Extracted from *U.S. Standard Atmosphere, 1976, National Oceanic and Atmospheric Administration,* National Aeronautics and Space Administration and the U.S. Air Force, Washington, 1976. Z = geometric altitude, T = temperature, P = pressure, g = acceleration of gravity, M = molecular weight, a = velocity of sound, λ = mean free path, and H = geopotential altitude. The notation 1.79. − 5 signifies 1.79×10^{-5}.

Table 4.2.3   Saturated Ammonia (R 717)*

P, bar	T, °C	v_f	v_g	h_f	h_g	s_f	s_g	c_{pf}	c_{pg}	μ_f	μ_g	k_f	k_g	Pr_f	Pr_g
		m^3/kg		kJ/kg		kJ/(kg · K)		kJ/(kg · K)		μPa · s		W/(m · K)			
0.5	− 46.5	0.001438	2.175	− 9.0	1,397.9	0.1643	6.3723	4.366	2.126		7.71	0.615	0.0161		
1	− 33.6	0.001466	1.138	47.9	1,418.3	0.4080	6.1286	4.429	2.233	262.9	8.09	0.588	0.0175	1.98	1.032
1.5	− 25.2	0.001488	0.779	86.1	1,430.5	0.5610	5.9867	4.447	2.266	236.1	8.33	0.572	0.0184	1.84	1.046
2	− 18.9	0.001507	0.595	113.8	1,439.2	0.6745	5.8863	4.507	2.393	218.4	8.52	0.554	0.0191	1.78	1.060
2.5	− 13.7	0.001523	0.482	137.4	1,445.9	0.7658	5.8085	4.535	2.460	205.4	8.69	0.548	0.0199	1.70	1.074
3	− 9.2	0.001536	0.406	157.5	1,451.3	0.8426	5.7449	4.561	2.521	195.0	8.81	0.539	0.0204	1.65	1.089
4	− 1.9	0.001560	0.309	191.3	1,459.8	0.9680	5.6443	4.605	2.630	179.5	9.03	0.524	0.0215	1.59	1.104
5	4.1	0.001580	0.250	219.2	1,466.1	1.0692	5.5660	4.463	2.728	168.0	9.21	0.512	0.0225	1.54	1.118
6	9.3	0.001598	0.210	243.2	1,471.0	1.1546	5.5017	4.678	2.818	158.8	9.38	0.501	0.0234	1.48	1.129
8	17.9	0.001630	0.160	283.7	1,478.2	1.2946	5.3994	4.741	2.983	147.4	9.59	0.487	0.0247	1.43	1.160
10	24.9	0.001658	0.1285	317.4	1,483.0	1.4080	5.3189	4.798	3.133	134.6	9.86	0.469	0.0263	1.38	1.174
15	38.7	0.001719	0.0862	384.7	1,489.5	1.6258	5.1683	4.929	3.479	115.6	10.34	0.438	0.0292	1.34	1.233
20	49.4	0.001773	0.0644	437.9	1,491.1	1.7909	5.0564	5.057	3.809	104.9	10.68	0.417	0.0315	1.30	1.292
25	58.2	0.001823	0.0512	483.0	1,489.9	1.9259	4.9651	5.192	4.142	96.2	11.01	0.398	0.0335	1.26	1.360
30	65.8	0.001871	0.0421	522.6	1,486.7	2.0415	4.8864	5.340	4.488	89.3	11.31	0.381	0.0356	1.25	1.426
35	72.4	0.001918	0.03564	558.4	1,482.0	2.1434	4.8161	5.505	4.856	83.7	11.61	0.366	0.0375	1.26	1.50
40	78.4	0.001965	0.03069	591.5	1,476.1	2.2354	4.7516	5.692	5.255	78.8	11.90	0.352	0.0397	1.28	1.57
45	83.9	0.002012	0.02680	622.4	1,469.0	2.3198	4.6912	5.904	5.692	74.6	12.20	0.338	0.0419	1.30	1.66
50	88.9	0.002060	0.02364	651.7	1,461.0	2.3985	4.6338	6.148	6.181	70.8	12.49	0.326	0.0441	1.33	1.75
60	97.9	0.002161	0.01883	706.8	1,442.0	2.5431	4.5244	6.764	7.375	64.3	13.14	0.302	0.0489	1.36	1.98
80	112.9	0.002406	0.01253	810.6	1,390.7	2.8052	4.3076	9.005	11.548	53.4	14.78				
100	125.2	0.002793	0.00826	920.3	1,309.8	3.0751	4.4131	17.08	26.04	42.9	17.76				
113.4†	132.3	0.004260	0.00426	1,105.5	1,105.5	3.5006	3.5006								

* The T, P, v, h, and s values interpolated, rounded, and converted from "ASHRAE Handbook—Fundamentals," 1993. The c_p, μ, and k values from Liley and Desai, CINDAS Rep. 106, 1992. Similar values can be found in "ASHRAE Handbook—Fundamentals," 1993.
† Critical point.

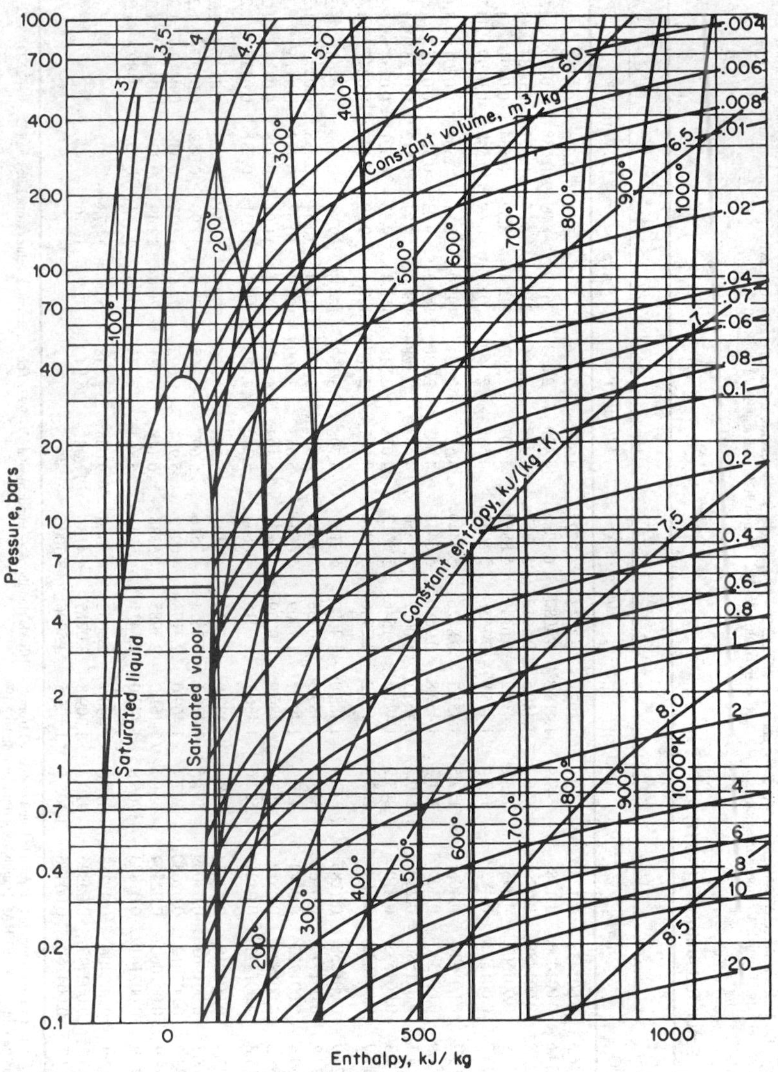

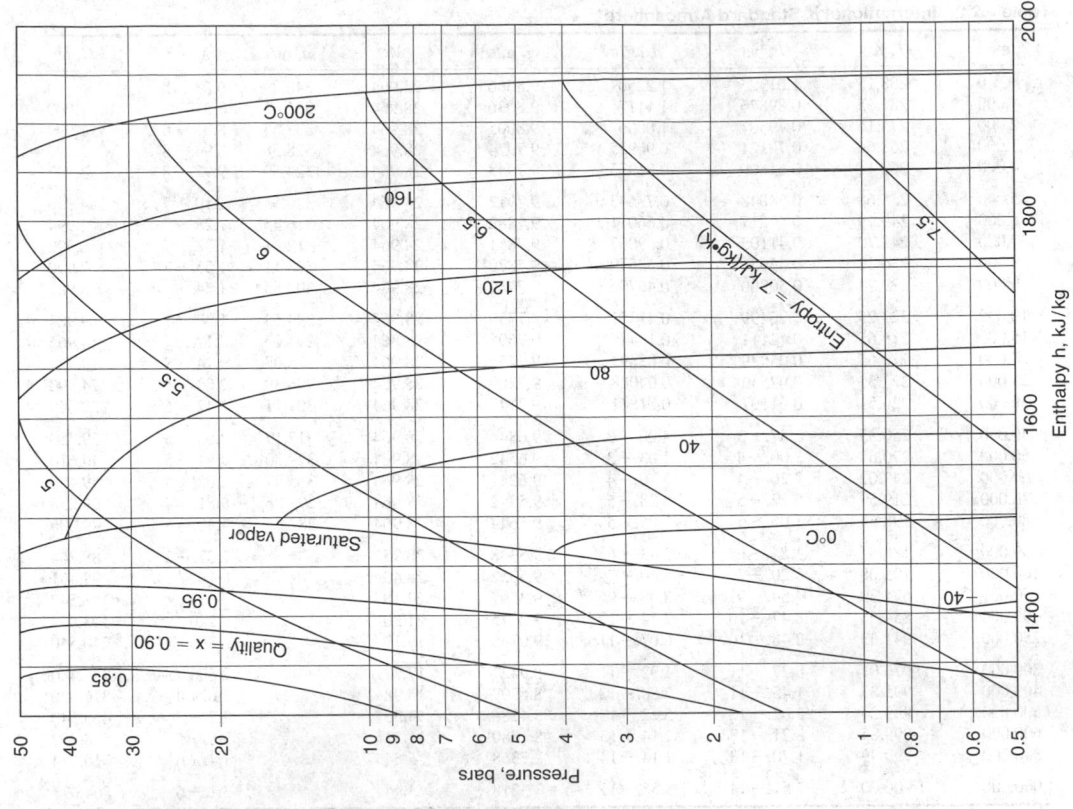

Fig. 4.2.3 Enthalpy–log pressure diagram for ammonia (R717).

Fig. 4.2.2 Enthalpy–log pressure diagram for air.

Table 4.2.4 Saturated Carbon Dioxide*

T, K	P, bar	v_f, m³/kg	v_g, m³/kg	h_f, kJ/kg	h_g, kJ/kg	s_f, kJ/(kg · K)	s_g, kJ/(kg · K)	c_{pf}, kJ/(kg · K)
216.6	5.180	8.484. − 4	0.0712	386.3	731.5	2.656	4.250	1.707
220	5.996	8.574. − 4	0.0624	392.6	733.1	2.684	4.232	1.761
225	7.357	8.710. − 4	0.0515	401.8	735.1	2.723	4.204	
230	8.935	8.856. − 4	0.0428	411.1	736.7	2.763	4.178	1.879
235	10.75	9.011. − 4	0.0357	402.5	737.9	2.802	4.152	
240	12.83	9.178. − 4	0.0300	430.2	738.9	2.842	4.128	1.933
245	15.19	9.358. − 4	0.0253	440.1	739.4	2.882	4.103	
250	17.86	9.554. − 4	0.0214	450.3	739.6	2.923	4.079	1.992
255	20.85	9.768. − 4	0.0182	460.8	739.4	2.964	4.056	
260	24.19	1.000. − 3	0.0155	471.6	738.7	3.005	4.032	2.125
270	32.03	1.056. − 3	0.0113	494.4	735.6	3.089	3.981	2.410
275	36.59	1.091. − 3	0.0097	506.5	732.8	3.132	3.954	
280	41.60	1.130. − 3	0.0082	519.2	729.1	3.176	3.925	2.887
290	53.15	1.241. − 3	0.0058	547.6	716.9	3.271	3.854	3.724
300	67.10	1.470. − 3	0.0037	585.4	690.2	3.393	3.742	
304.2†	73.83	2.145. − 3	0.0021	636.6	636.6	3.558	3.558	∞

* The notation 8.484. − 4 signifies 8.484 × 10⁻⁴.
† Critical point.

Table 4.2.5 Superheated Carbon Dioxide*

P, bar		Temperature , K								
	300	350	400	450	500	600	700	800	900	1,000
1 v	0.5639	0.6595	0.7543	0.8494	0.9439	1.1333	1.3324	1.5115	1.7005	1.8894
1 h	809.3	853.1	899.1	947.1	997.0	1,102	1,212	1,327	1,445	1,567
1 s	4.860	4.996	5.118	5.231	5.337	5.527	5.697	5.850	5.990	6.120
5 v	0.1106	0.1304	0.1498	0.1691	0.1882	0.2264	0.2645	0.3024	0.3403	0.3782
5 h	805.5	850.3	897.0	945.5	995.8	1,101	1,211	1,326	1,445	1,567
5 s	4.548	4.686	4.810	4.925	5.031	5.222	5.392	5.546	5.685	5.814
10 v	0.0539	0.0642	0.0742	0.0841	0.0938	0.1131	0.1322	0.1513	0.1703	0.1893
10 h	800.7	846.9	894.4	943.5	994.1	1,100	1,211	1,326	1,445	1,567
10 s	4.405	4.548	4.674	4.790	4.897	5.089	5.260	5.414	5.555	5.683
20 v	0.0255	0.0311	0.0364	0.0416	0.0466	0.0564	0.0661	0.0757	0.0853	0.0948
20 h	790.2	839.8	889.3	939.4	990.8	1,098	1,209	1,325	1,444	1,567
20 s	4.249	4.402	4.534	4.653	4.762	4.955	5.127	5.282	5.423	5.551
30 v	0.0159	0.0201	0.0238	0.0274	0.0309	0.0375	0.0441	0.0505	0.0570	0.0633
30 h	778.5	832.4	883.8	935.2	987.3	1,096	1,208	1,324	1,444	1,566
30 s	4.144	4.341	4.447	4.569	4.679	4.876	5.049	5.204	5.346	5.474
40 v	0.0110	0.0146	0.0175	0.0203	0.0230	0.0281	0.0331	0.0379	0.0428	0.0476
40 h	764.9	824.6	878.3	931.1	984.3	1,094	1,205	1,323	1,443	1,566
40 s	4.055	4.239	4.380	4.507	4.619	4.818	4.993	5.148	5.291	5.419
50 v	0.0080	0.0112	0.0138	0.0161	0.0183	0.0224	0.0265	0.0304	0.0343	0.0382
50 h	748.2	816.3	872.6	926.9	981.1	1,091	1,205	1,322	1,443	1,566
50 s	3.968	4.179	4.330	4.457	4.572	4.773	4.948	5.104	5.247	5.377
60 v	0.0058	0.0090	0.0113	0.0133	0.0151	0.0187	0.0221	0.0254	0.0286	0.0318
60 h	726.9	807.7	866.9	922.7	977.8	1,089	1,204	1,321	1,442	1,565
60 s	3.878	4.126	4.314	4.416	4.532	4.736	4.912	5.069	5.212	5.341
80 v		0.0062	0.0081	0.0097	0.0112	0.0140	0.0166	0.0191	0.0216	0.0240
80 h		788.4	855.1	914.2	971.3	1,085	1,201	1,320	1,441	1,565
80 s		4.029	4.208	4.347	4.468	4.675	4.854	5.011	5.155	5.286

Table 4.2.5 Superheated Carbon Dioxide* *(Continued)*

P, bar		300	350	400	450	500	600	700	800	900	1,000
							Temperature, K				
100	v		0.0045	0.0062	0.0076	0.0089	0.0111	0.0133	0.0153	0.0173	0.0193
	h		766.2	843.0	905.7	964.9	1,081	1,198	1,318	1,440	1,564
	s		3.936	4.144	4.290	4.417	4.627	4.808	4.967	5.111	5.241
150	v		0.0023	0.0038	0.0049	0.0058	0.0074	0.0089	0.0103	0.0117	0.0130
	h		704.5	811.9	884.8	949.4	1,072	1,192	1,314	1,437	1,562
	s		3.716	4.005	4.177	4.313	4.536	4.722	4.884	5.030	5.162
200	v		0.0017	0.0027	0.0035	0.0043	0.0056	0.0067	0.0078	0.0088	0.0099
	h		670.0	783.2	865.2	934.9	1,063	1,186	1,310	1,435	1,561
	s		3.591	3.894	4.088	4.234	4.468	4.668	4.824	4.970	5.104
300	v			0.0018	0.0023	0.0029	0.0038	0.0046	0.0053	0.0060	0.0067
	h			745.3	834.0	910.6	1,047	1,176	1,303	1,431	1,559
	s			3.747	3.956	4.118	4.367	4.573	4.743	4.886	5.021
400	v			0.0015	0.0018	0.0022	0.0029	0.0035	0.0041	0.0047	0.0052
	h			728.1	814.6	893.3	1,035	1,168	1,298	1,428	1,558
	s			3.663	3.867	4.033	4.292	4.497	4.671	4.824	4.960
500	v				0.0016	0.0018	0.0024	0.0029	0.0034	0.0038	0.0043
	h				803.5	881.9	1,027	1,162	1,294	1,426	1,557
	s				3.805	3.970	4.234	4.443	4.620	4.774	4.913

* Interpolated and rounded from Vukalovich and Altunin, ''Thermophysical Properties of Carbon Dioxide,'' Atomizdat, Moscow, 1965; and Collett, England, 1968. Note: v, h, and s units are the same as in Table 4.2.4.

Table 4.2.6 Saturated ISo-Butane (R 600a)*

P, bar	T, °C	v_f	v_g	h_f	h_g	s_f	s_g	c_{pf}	c_{pg}	μ_f	μ_g	k_f	k_g	Pr_f	Pr_g
		m³/kg		kJ/kg		kJ/(kg · K)		kJ/(kg · K)		μPa · s		W/(m · K)			
1	−12.13	0.001683	0.3601	288.2	655.5	3.4552	4.8626	2.24	1.56	229	6.63	0.112	0.0125	4.58	0.827
1.5	−1.42	0.001720	0.2468	312.8	668.2	3.5470	4.8615	2.30	1.63	203	6.93	0.108	0.0136	4.31	0.831
2	6.82	0.001746	0.1886	332.3	681.2	3.6166	4.8631	2.35	1.68	184	7.18	0.104	0.0145	4.16	0.832
2.5	13.60	0.001771	0.1528	348.5	690.5	3.6734	4.8658	2.38	1.73	171	7.38	0.101	0.0153	4.03	0.834
3	19.38	0.001793	0.1290	362.6	698.3	3.7214	4.8689	2.42	1.78	160	7.57	0.098	0.0159	3.95	0.847
4	29.17	0.001834	0.0978	382.3	711.5	3.8020	4.8757	2.49	1.86	144	7.89	0.094	0.0170	3.81	0.863
5	37.32	0.001870	0.0785	407.6	722.3	3.8688	4.8824	2.54	1.93	132	8.18	0.090	0.0181	3.73	0.872
6	44.28	0.001904	0.0657	425.7	731.5	3.9254	4.8889	2.60	1.99	122	8.44	0.087	0.0190	3.65	0.884
8	56.08	0.001966	0.0490	456.6	746.7	4.0213	4.9008	2.70	2.11	108	8.91	0.083	0.0206	3.51	0.913
10	65.88	0.002026	0.0389	484.2	758.8	4.1010	4.9112	2.76	2.23	97	9.34	0.079	0.0221	3.39	0.942
15	85.29	0.002220	0.0222	556.3	786.1	4.3020	4.9345	3.04	2.56	78	10.4	0.072	0.0252	3.29	1.057
20	100.38	0.002332	0.0175	588.7	795.1	4.3878	4.9405	3.38	3.01	64	11.4	0.067	0.0284	2.99	1.208
25	112.83	0.002522	0.0135	631.9	802.6	4.4980	4.9403	3.92	3.79	54	12.7	0.063	0.0326	3.36	1.476
30	123.33	0.002786	0.0095	673.7	802.2	4.6008	4.9251	6.3	7.4	44	14.3	0.061	0.0414	4.54	2.556
35	132.33	0.003312	0.0064	720.8	782.0	4.7155	4.8663			33	17.6	0.075	0.0723		
35.5†	134.85	0.004464	0.0045	752.5	752.4	4.791	4.791								

* P, T, v, h, and s are interpolated and rounded from ''ASHRAE Handbook—Fundamentals,'' 1993. c_p, μ, and k from Liley and Desai, CINDAS Rep. 106, 1992. Substantially similar values appear in the ''ASHRAE Thermophysical Properties of Refrigerants,'' 1993.
† Critical point.

Table 4.2.7   Saturated Normal Hydrogen*

T	P	v_f	v_g	h_f	h_g	s_f	s_g	c_{pf}	c_{pg}
13.95	0.072	0.0130	7.974	218.3	565.4	14.08	46.64	6.36	10.52
14	0.074	0.0130	7.205	219.6	669.3	14.17	46.30	6.47	10.54
15	0.127	0.0132	4.488	226.4	678.2	14.64	44.76	6.91	10.67
16	0.204	0.0133	2.954	233.8	686.7	15.10	43.42	7.36	10.85
17	0.314	0.0135	2.032	241.6	694.7	15.57	42.23	7.88	11.07
18	0.461	0.0137	1.449	249.9	702.1	16.03	41.16	8.42	11.34
19	0.654	0.0139	1.064	258.8	708.8	16.50	40.19	8.93	11.66
20	0.901	0.0141	0.802	268.3	714.8	16.97	39.30	9.45	12.04
21	1.208	0.0143	0.618	278.4	720.2	17.44	38.49	10.13	12.49
22	1.585	0.0146	0.483	289.2	724.4	17.92	37.71	10.82	13.03
23	2.039	0.0148	0.383	300.8	727.6	18.41	36.97	11.69	13.69
24	2.579	0.0151	0.307	313.3	729.8	18.90	36.27	12.52	14.49
25	3.213	0.0155	0.243	326.7	730.7	19.41	35.58	13.44	15.52
26	3.950	0.0159	0.203	341.2	730.2	19.93	34.90	14.80	16.85
27	4.800	0.0164	0.167	357.0	728.0	20.47	34.22	16.17	18.66
28	5.770	0.0170	0.137	374.3	723.7	21.04	33.52	18.48	21.24
29	6.872	0.0177	0.113	393.6	716.6	21.65	32.80	22.05	25.19
30	8.116	0.0185	0.092	415.4	705.9	22.31	32.00	26.59	31.99
31	9.510	0.0198	0.074	441.3	689.7	23.08	31.09	36.55	46.56
32	11.068	0.0217	0.057	474.7	663.2	24.03	29.93	65.37	87.02
33.18^c	13.130	0.0318	0.032	565.4	565.4	26.68	26.68		

* T = temperature, K; P = pressure, bar; c = critical point; v = specific volume, m³/kg; h = specific enthalpy, kJ/kg; s = specific entropy, kJ/(kg · K); c_p = specific heat at constant pressure, kJ/(kg · K); subscript f represents saturated liquid and subscript g represents saturated vapor.

Table 4.2.8   Saturated Propane (R 290)*

P, bar	T, °C	v_f (m³/kg)	v_g (m³/kg)	h_f (kJ/kg)	h_g (kJ/kg)	s_f (kJ/(kg · K))	s_g (kJ/(kg · K))	c_{pf} (kJ/(kg · K))	c_{pg} (kJ/(kg · K))	μ_f (μPa · s)	μ_g (μPa · s)	k_f (W/(m · K))	k_g (W/(m · K))	Pr_f	Pr_g
0.5	−56.95	0.001674	0.8045	388.5	831.5	3.7263	5.7747	2.181	1.374	233	6.05	0.139	0.0101	3.66	0.823
1	−42.38	0.001721	0.4186	420.9	849.0	3.8705	5.7258	2.246	1.457	198	6.46	0.130	0.0113	3.42	0.833
1.5	−32.83	0.001755	0.2871	442.8	860.5	3.9636	5.7015	2.294	1.517	178	6.74	0.124	0.0122	3.29	0.838
2	−25.48	0.001783	0.2194	460.1	869.2	4.0341	5.6861	2.336	1.568	164	6.97	0.119	0.0130	3.22	0.841
2.5	−19.43	0.001807	0.1778	474.5	876.3	4.0916	5.6753	2.371	1.610	154	7.16	0.116	0.0136	3.15	0.848
3	−14.23	0.001828	0.1498	487.1	882.4	4.1404	5.6672	2.406	1.652	146	7.33	0.113	0.0141	3.11	0.859
4	−5.53	0.001867	0.1138	508.5	892.3	4.2213	5.6556	2.467	1.723	134	7.61	0.108	0.0151	3.06	0.868
5	1.66	0.001901	0.0918	526.6	900.4	4.2875	5.6477	2.522	1.787	124	7.86	0.104	0.0160	3.01	0.878
6	7.82	0.001932	0.0771	542.4	907.1	4.3437	5.6418	2.574	1.847	116	8.08	0.101	0.0168	2.96	0.888
8	18.20	0.001990	0.0580	569.7	917.9	4.4379	5.6333	2.674	1.961	104	8.48	0.096	0.0182	2.90	0.914
10	26.86	0.002044	0.04609	593.1	926.4	4.5161	5.6270	2.769	2.072	95	9.04	0.092	0.0195	2.86	0.961
15	43.84	0.002173	0.03009	641.6	941.1	4.6697	5.6148	3.013	2.363	79	9.63	0.084	0.0225	2.83	1.011
20	57.14	0.002304	0.02165	682.3	949.9	4.7923	5.6026	3.290	2.717	67	10.4	0.078	0.0256	2.83	1.104
25	68.15	0.002450	0.01642	719.0	954.1	4.8979	5.5872	3.665	3.216	58	11.6	0.073	0.0295	2.91	1.265
30	77.67	0.002627	0.01269	753.8	953.8	4.9950	5.5654	4.270	4.041	50	12.4	0.071	0.0355	3.01	1.412
35	85.99	0.002866	0.00978	788.7	947.5	5.0895	5.5318	5.594	5.848	44	13.6	0.073	0.0412	3.37	1.93
40	93.38	0.00336	0.00685	830.0	928.9	5.200	5.470	12.12	14.25	33	17.8	0.084	0.0728	4.76	3.48
42.4†	96.65	0.00457	0.00457	879.2	879.2	5.330	5.330								

* The T, P, v, h, and s values are interpolated, rounded, and converted from "ASHRAE Handbook—Fundamentals," 1993. The c_p, μ, and k values are from Liley and Desai, CINDAS Rep. 106, 1992. Similar values can be found in "ASHRAE Handbook—Fundamentals," 1993.
† Critical point.

Table 4.2.9 Saturated Refrigerant 11*

P, bar	T, °C	v_f	v_g	h_f	h_g	s_f	s_g	c_{pf}	c_{pg}	μ_f	μ_g	k_f	k_g	Pr_f	Pr_g
		m³/kg		kJ/kg		kJ/(kg · K)		kJ/(kg · K)		μPa · s		W/(m · K)			
0.5	5.18	0.000657	0.3298	204.5	392.4	1.0162	1.6916	0.873	0.560	537	10.3	0.094	0.0083	4.35	0.695
1	23.55	0.000676	0.1731	220.3	401.8	1.0711	1.6833	0.887	0.580	440	11.0	0.090	0.0088	4.01	0.725
1.5	35.26	0.000689	0.1186	230.9	407.9	1.1060	1.6800	0.896	0.594	385	11.4	0.089	0.0091	3.80	0.736
2	44.42	0.000700	0.0908	239.1	412.5	1.1322	1.6783	0.905	0.604	341	11.8	0.085	0.0096	3.63	0.740
2.5	51.90	0.000709	0.0737	245.9	416.3	1.1532	1.6774	0.913	0.614	315	12.0	0.083	0.0099	3.47	0.745
3	58.37	0.000718	0.0620	251.8	419.4	1.1711	1.6768	0.920	0.623	288	12.3	0.081	0.0102	3.27	0.751
4	69.18	0.000733	0.0470	261.9	424.7	1.2008	1.6764	0.934	0.637	252	12.8	0.079	0.0107	2.98	0.762
5	78.07	0.000746	0.0380	270.2	428.9	1.2247	1.6763	0.947	0.652	228	13.1	0.077	0.0111	2.80	0.770
6	85.76	0.000759	0.0317	277.6	432.4	1.2482	1.6764	0.959	0.669	211	13.5	0.076	0.0115	2.66	0.779
8	98.59	0.000781	0.0239	290.1	438.0	1.2790	1.6768	0.983	0.689	187	14.1	0.073	0.0122	2.52	0.792
10	109.3	0.000802	0.0190	300.8	442.3	1.3069	1.6771	1.008	0.713	170	14.6	0.070	0.0129	2.45	0.807
15	130.3	0.000853	0.0124	322.6	449.9	1.3614	1.6770	1.076	0.783	138	15.7	0.065	0.0143	2.28	0.860
20	146.6	0.000903	0.0090	340.5	454.5	1.4038	1.6754	1.153	0.876	118	16.9	0.062	0.0158	2.19	0.937
25	160.2	0.000959	0.0068	356.4	457.0	1.4399	1.6721	1.256	1.021	101	18.1	0.059	0.0174	2.15	1.062
30	171.9	0.001024	0.0053	371.1	457.6	1.4722	1.6670	1.384	1.317	86	19.3	0.058	0.0193	2.13	1.317
35	182.2	0.001105	0.0042	385.5	456.1	1.5032	1.6583	1.82	1.84						
40	191.3	0.001246	0.0031	401.1	451.3	1.5352	1.6432	2.95	2.31						
44.1†	198.0	0.00181	0.0018	428.6	428.6	1.5933	1.5933								

* The T, P, v, h, and s values are interpolated, converted, and rounded from "ASHRAE Handbook—Fundamentals," 1993. The c_p, μ, and k are from Liley and Desai, CINDAS Rep. 106, 1992. Similar values appear in "ASHRAE Handbook—Fundamentals," 1993.
   † Critical point.

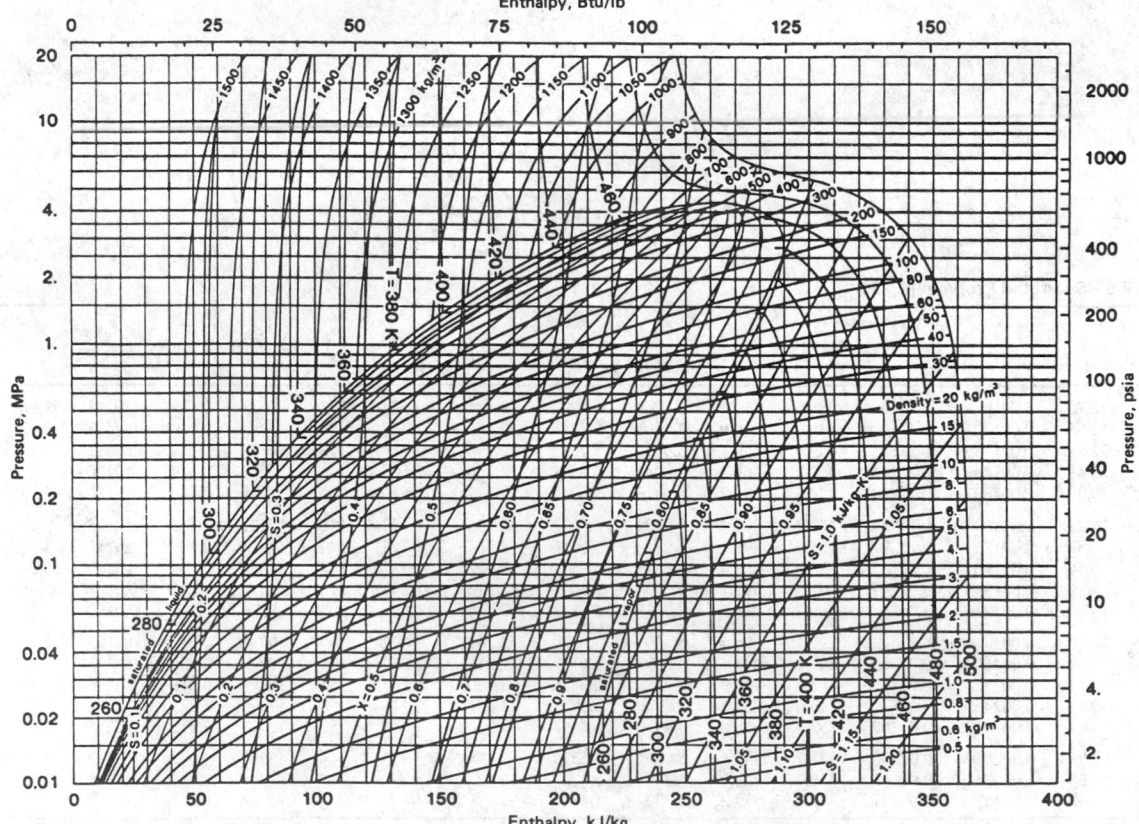

Fig. 4.2.4 Enthalpy–log pressure diagram for refrigerant 11. 1 MPa = 10 bar. *(Copyright 1981 by ASHRAE and reproduced by permission.)*

Table 4.2.10   Saturated Refrigerant 12*

P, bar	T, °C	v_f	v_g	h_f	h_g	s_f	s_g	c_{pf}	c_{pg}	μ_f	μ_g	k_f	k_g	Pr_f	Pr_g
		m³/kg		kJ/kg		kJ/(kg · K)		kJ/(kg · K)		μPa · s		W/(m · K)			
0.5	−45.24	0.000653	0.3072	159.5	331.5	0.8386	1.5936	0.883	0.545	420	9.8	0.0940	0.0062	4.08	0.861
1	−30.11	0.000672	0.1611	172.7	338.8	0.8947	1.5780	0.895	0.575	358	10.4	0.0883	0.0070	3.63	0.854
1.5	−20.15	0.000686	0.1103	181.6	345.3	0.9304	1.5702	0.906	0.595	322	11.0	0.0846	0.0075	3.45	0.873
2	−12.52	0.000697	0.0843	188.5	347.0	0.9571	1.5654	0.913	0.612	297	11.4	0.0817	0.0079	3.32	0.883
2.5	−6.22	0.000707	0.0682	194.2	349.9	0.9788	1.5619	0.921	0.627	277	11.7	0.0793	0.0083	3.22	0.834
3	0.84	0.000715	0.0574	199.2	352.3	0.9972	1.5594	0.929	0.640	262	12.0	0.0774	0.0086	3.13	0.893
4	8.19	0.000731	0.0436	207.7	356.3	1.0275	1.5556	0.945	0.663	238	12.4	0.0739	0.0091	3.04	0.903
5	15.64	0.000744	0.0351	214.8	359.4	1.0522	1.5530	0.959	0.683	221	12.8	0.0714	0.0095	2.97	0.920
6	22.01	0.000757	0.0294	220.9	362.0	1.0731	1.5510	0.969	0.702	207	13.2	0.0692	0.0098	2.90	0.946
8	32.79	0.000780	0.0221	231.7	366.2	1.1082	1.5479	0.995	0.738	186	13.9	0.0653	0.0105	2.83	0.977
10	41.70	0.000802	0.0176	240.8	369.4	1.1370	1.5455	1.021	0.773	170	14.5	0.0621	0.0111	2.80	1.01
15	59.30	0.000854	0.0141	259.6	374.7	1.1938	1.5400	1.107	0.868	143	15.9	0.0558	0.0125	2.84	1.10
20	72.99	0.000907	0.0082	275.2	377.5	1.2386	1.5341	1.225	0.993	124	17.3	0.0512	0.0137	2.97	1.25
25	84.33	0.000967	0.0062	289.2	378.4	1.2770	1.5265	1.36	1.029	108	18.9	0.0469	0.0151	3.13	1.40
30	94.05	0.001040	0.0048	302.4	377.3	1.3120	1.5162	1.51	1.55	92	20.7	0.0429	0.0167	3.24	1.92
35	102.6	0.001141	0.0036	315.7	373.5	1.3437	1.4975		2.50	75	23.2	0.0389	0.0191	3.04	3.04
40	110.1	0.001360	0.0025	332.3	362.5	1.3871	1.4659		10.9	55	28.2	0.0346	0.0222		
41.2†	111.8	0.001771	0.0018	347.4	347.4	1.4272	1.4272								

*The T, P, v, h, and s values are interpolated, converted, and rounded from "ASHRAE Handbook—Fundamentals," 1993. The c_p, μ, and k values are from Liley and Desai, CINDAS Rep. 106, 1992. Similar values appear in "ASHRAE Handbook—Fundamentals," 1993.

† Critical point.

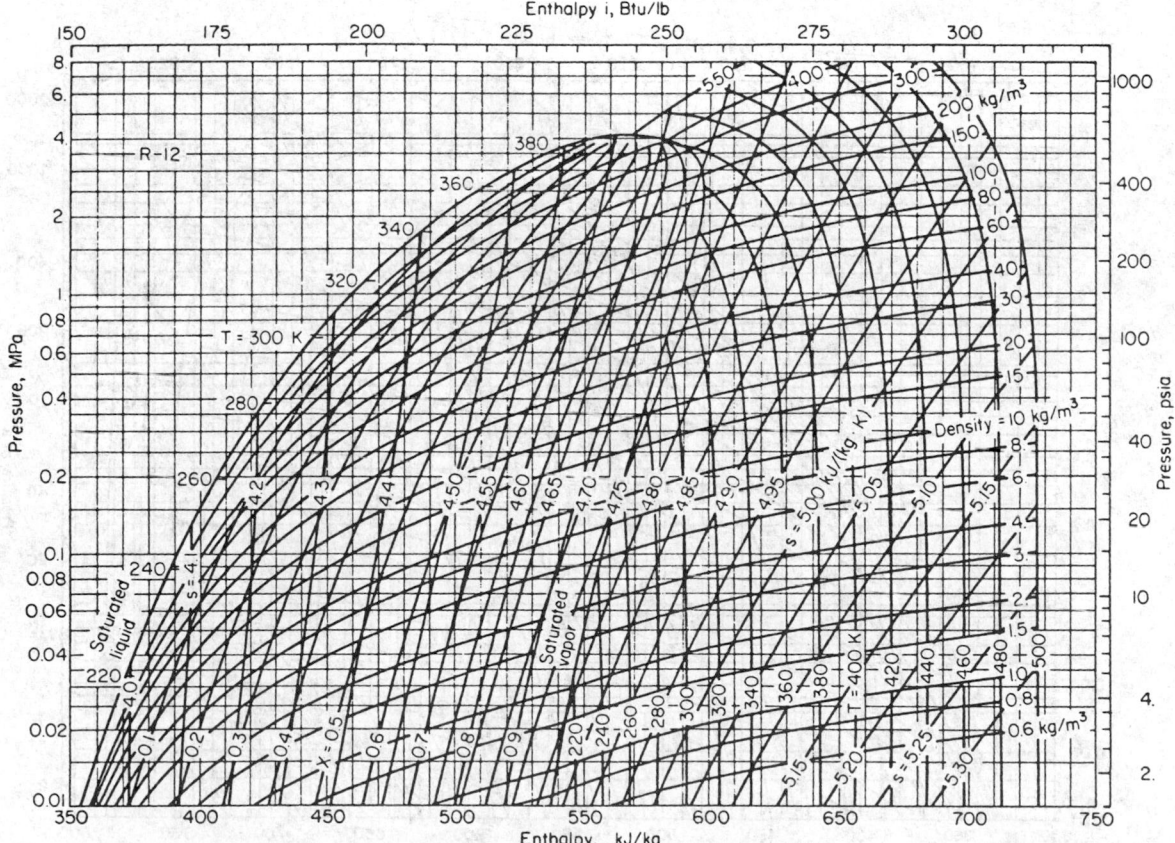

Fig. 4.2.5 Enthalpy–log pressure diagram for refrigerant 12. Prepared at the Center for Applied Thermodynamic Studies, University of Idaho, Moscow. *(Copyright by ASHRAE and reproduced by permission.)*

Table 4.2.11 Saturated Refrigerant 22*

P, bar	T, °C	v_f	v_g	h_f	h_g	s_f	s_g	c_{pf}	c_{pg}	μ_f	μ_g	k_f	k_g	Pr_f	Pr_g
		m³/kg		kJ/kg		kJ/(kg · K)		kJ/(kg · K)		μPa · s		W/(m · K)			
0.5	−54.80	0.000690	0.4264	138.7	381.1	0.7510	1.8619	1.080	0.574			0.121	0.0060		
1	−41.39	0.000709	0.2153	153.6	387.6	0.8173	1.8256	1.092	0.605			0.114	0.0069		
1.5	−32.07	0.000723	0.1472	163.6	393.5	0.8591	1.8056	1.104	0.630			0.110	0.0075		
2	−25.19	0.000734	0.1125	171.2	394.7	0.8902	1.7919	1.115	0.650			0.107	0.0079		
2.5	−19.52	0.000743	0.0910	177.5	397.2	0.9155	1.7814	1.126	0.669	258.6	11.02	0.105	0.0083	2.78	0.888
3	−14.66	0.000752	0.0766	183.1	399.2	0.9368	1.7730	1.136	0.686	245.9	11.21	0.103	0.0086	2.71	0.894
4	−6.57	0.000767	0.0582	192.3	402.4	0.9718	1.7599	1.155	0.716	225.6	11.54	0.099	0.0091	2.63	0.907
5	0.11	0.000780	0.0469	200.1	404.9	1.0005	1.7498	1.171	0.745	209.9	11.80	0.096	0.0095	2.56	0.924
6	5.85	0.000789	0.0392	206.9	405.7	1.0327	1.7417	1.189	0.771	197.2	11.97	0.094	0.0099	2.50	0.936
8	15.44	0.000815	0.0295	218.5	410.0	1.0650	1.7287	1.221	0.819	177.7	12.42	0.090	0.0104	2.42	0.974
10	23.39	0.000835	0.0236	228.3	412.3	1.0981	1.7185	1.252	0.871	163.0	12.82	0.086	0.0109	2.36	1.026
15	39.07	0.000883	0.0155	248.5	415.7	1.1628	1.6985	1.332	1.000	137.7	13.9	0.080	0.0118	2.29	1.12
20	51.23	0.000929	0.0113	265.0	417.1	1.2132	1.6822	1.426	1.149						
25	61.33	0.000978	0.0087	279.6	417.0	1.2560	1.6670	1.550	1.341						
30	70.05	0.001030	0.0068	301.3	415.7	1.2942	1.6517	1.613	2.070						
35	77.70	0.001087	0.0056	305.0	413.7	1.3275	1.6371	2.03	2.05						
40	84.53	0.001174	0.0044	318.9	408.3	1.3648	1.6150	2.67	2.96						
45	90.67	0.001326	0.0033	335.8	397.9	1.4100	1.5810	4.47	5.19						
49.9†	96.14	0.001909	0.0019	366.6	366.6	1.4918	1.4918								

* Values are interpolated and rounded from "ASHRAE Handbook—Fundamentals," 1993.
† Critical point.

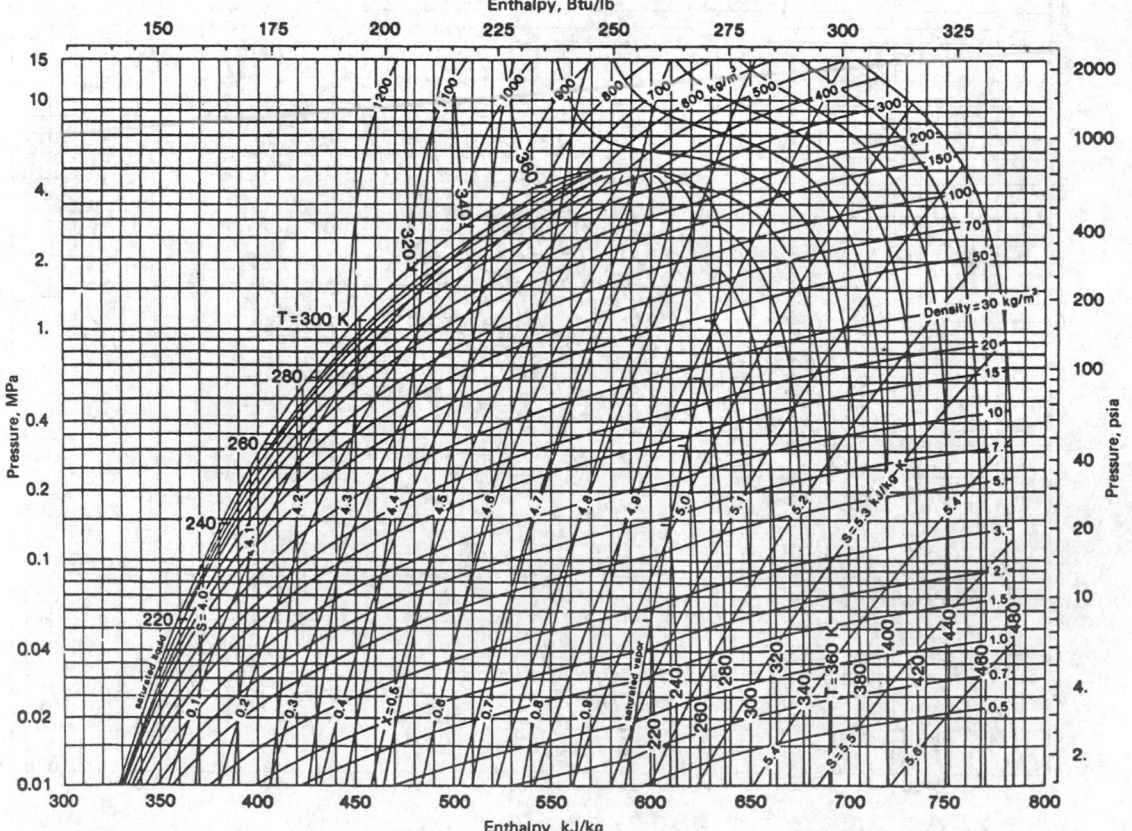

Fig. 4.2.6 Enthalpy–log pressure diagram for refrigerant 22. 1 MPa = 10 bar. *(Copyright 1981 by ASHRAE and reproduced by permission.)*

Table 4.2.12   Saturated Refrigerant 32*

P, bar	T, °C	v_f	v_g	h_f	h_g	s_f	s_g	c_{pf}	c_{pg}	μ_f	μ_g	k_f	k_g	Pr_f	Pr_g
		m³/kg		kJ/kg		kJ/(kg · K)		kJ/(kg · K)		μPa · s		W/(m · K)			
1	−51.68	0.000832	0.3361	114.6	497.5	0.6565	2.3855	1.559	0.873	278.5	10.50	0.189	0.0082	2.30	1.12
1.5	−43.66	0.000847	0.2394	127.2	501.5	0.7123	2.3435	1.576	0.911	251.6	10.61	0.181	0.0085	2.19	1.14
2	−37.35	0.000860	0.1773	137.3	504.5	0.7555	2.3127	1.601	0.955	226.6	10.70	0.175	0.0088	2.10	1.16
2.5	−32.16	0.000870	0.1433	145.7	506.8	0.7906	2.2888	1.615	1.003	218.8	10.77	0.171	0.0091	2.06	1.18
3	−27.74	0.000880	0.1205	152.7	508.6	0.8202	2.2693	1.627	1.020	207.7	10.83	0.167	0.0094	2.02	1.18
4	−20.39	0.000897	0.0914	165.2	511.3	0.8689	2.2383	1.653	1.074	192.7	10.93	0.161	0.0100	1.98	1.18
5	−14.34	0.000912	0.0736	175.3	513.3	0.9084	2.2140	1.678	1.123	180.8	11.02	0.156	0.0104	1.94	1.19
6	−9.16	0.000925	0.0616	184.2	514.7	0.9418	2.1940	1.701	1.169	170.1	11.10	0.152	0.0109	1.90	1.19
8	−0.51	0.000950	0.0463	199.1	516.7	0.9968	2.1616	1.743	1.255	154.8	11.32	0.145	0.0117	1.86	1.21
10	6.63	0.000972	0.0369	211.7	517.8	1.0415	2.1356	1.784	1.337	142.4	11.53	0.139	0.0124	1.83	1.24
15	20.64	0.001023	0.0242	237.0	518.3	1.1282	2.0855	1.895	1.541	120.7	12.08	0.128	0.0141	1.79	1.32
20	31.45	0.001072	0.0176	257.5	516.7	1.1949	2.0460	2.009	1.761	105.7	12.67	0.119	0.0156	1.78	1.43
25	40.36	0.001122	0.0136	275.3	513.7	1.2506	2.0112	2.151	2.026	94.0	13.29	0.112	0.0171	1.81	1.57
30	48.00	0.001175	0.0102	291.4	509.4	1.2997	1.9786	2.314	2.352	84.9	13.98	0.107	0.0186	1.84	1.77
35	54.69	0.001232	0.0088	306.6	503.9	1.3447	1.9463	2.524	2.791	77.7	14.74	0.101	0.0200	1.94	2.06
40	60.66	0.001299	0.0072	322.1	496.7	1.3876	1.9128	2.744	3.367	71.4	15.60	0.095	0.0215	2.06	2.44
45	66.05	0.001380	0.0060	336.5	487.8	1.4304	1.8763		4.49	66.0	16.61	0.089	0.0191		3.90
50	70.95	0.001490	0.0048	352.8	475.6	1.4759	1.8328			61.8	17.85	0.082	0.0167		
58.6†	78.41	0.002383	0.0024	413.8	413.8	1.6465	1.6465								

* The *P, T, v, h,* and *s* values are interpolated and rounded from "ASHRAE Handbook—Fundamentals," 1993; c_p values are interpolated and converted from Defbaugh et al., *J. Chem. Eng. Data,* **39,** 1994, pp. 333–340; μ_f and μ_g are interpolated from Oliveira and Wakeham, *Int. J. Thermophys.,* **14,** no. 6, 1993, pp. 1131–1143.
† Critical point.

Table 4.2.13   Saturated Refrigerant 123*

P, bar	T, °C	v_f	v_g	h_f	h_g	s_f	s_g	c_{pf}	c_{pg}	μ_f	μ_g	k_f	k_g	Pr_f	Pr_g
		m³/kg		kJ/kg		kJ/(kg · K)		kJ/(kg · K)		μPa · s		W/(m · K)			
0.5	9.72	0.000666	0.2995	209.9	387.3	1.0348	1.6626	1.002	0.668	503		0.0811		6.21	
1	27.46	0.000686	0.1564	227.7	398.0	1.0963	1.6629	1.023	0.700	410		0.0759		5.53	
1.5	39.10	0.000701	0.1068	239.7	405.0	1.1353	1.6649	1.037	0.722	362	11.36	0.0726	0.0111	5.16	0.738
2	48.05	0.000713	0.0813	249.0	410.3	1.1647	1.6670	1.049	0.741	330	11.65	0.0699	0.0119	4.78	0.728
2.5	55.38	0.000723	0.0657	256.8	414.7	1.1884	1.6692	1.059	0.756	306	11.90	0.0678	0.0122	4.78	0.736
3	61.68	0.000732	0.0552	263.5	418.4	1.2086	1.6713	1.069	0.770	287	12.11	0.0660	0.0127	4.65	0.739
4	73.20	0.000749	0.0417	274.8	424.5	1.2418	1.6751	1.086	0.796	259	12.47	0.0629	0.0134	4.47	0.742
5	80.87	0.000764	0.0335	284.3	429.4	1.2687	1.6784	1.101	0.818	238	12.75	0.0604	0.0140	4.33	0.746
6	88.34	0.000778	0.0280	292.6	433.5	1.2916	1.6815	1.116	0.840	221	13.00	0.0582	0.0145	4.23	0.754
8	100.81	0.000804	0.0217	306.7	439.3	1.3246	1.6865	1.145	0.881	195	13.41	0.0547		4.08	
10	111.15	0.000828	0.0165	318.7	445.5	1.3607	1.6906	1.175	0.922	176	13.74	0.0504		4.10	
15	131.50	0.000887	0.0106	343.2	454.7	1.4218	1.6974	1.262	1.037						
20	147.25	0.000951	0.0075	363.4	460.3	1.4697	1.7001	1.383	1.198						
25	160.24	0.001027	0.0055	381.6	463.0	1.5109	1.6990	1.590	1.481						
30	171.30	0.001131	0.0041	398.8	462.6	1.5491	1.6926	2.08	2.18						
35	180.88	0.001361	0.0027	418.4	455.4	1.5915	1.6730	5.71	7.22						
36.6†	183.68	0.001818	0.0018	437.4	437.4	1.6290	1.6290								

* The *P, T, v, h, s,* and c_p values are interpolated and rounded from Younglove and McLinden, *J. Phys. Chem. Ref. Data,* **23,** no. 5, 1994, pp. 731–779; μ and k interpolated from "ASHRAE Handbook—Fundamentals," 1993.
† Critical point.

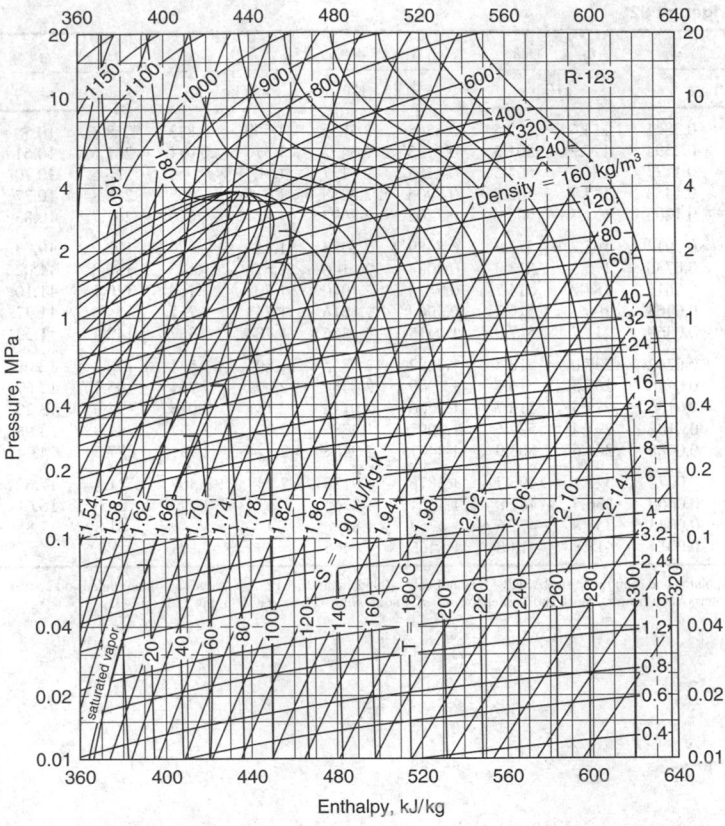

Fig. 4.2.7 Enthalpy–log pressure diagram for refrigerant 123. 1 MPa = 10 bar. *(Reprinted by permission of the ASHRAE from the 1993 "ASHRAE Handbook—Fundamentals.")*

Table 4.2.14 Saturated Refrigerant 134a*

P, bar	T, K	Spec. vol., m³/kg		Enthalpy, kJ/kg		Entropy, kJ/(kg · K)		Spec. ht. const., P, kJ/(kg · K)	
		Liquid	Vapor	Liquid	Vapor	Liquid	Vapor	Liquid	Vapor
0.5	232.7	0.000706	0.3572	148.4	374.1	0.7966	1.7640	1.254	0.748
1	246.8	0.000726	0.1926	165.4	382.6	0.8675	1.7474	1.279	0.793
1.5	256.0	0.000741	0.1313	177.4	388.3	0.9148	1.7388	1.299	0.826
2	263.1	0.000753	0.0999	186.6	392.6	0.9502	1.7334	1.315	0.854
2.5	268.9	0.000764	0.0807	194.3	396.1	0.9789	1.7295	1.330	0.878
3	273.8	0.000774	0.0677	200.9	399.0	1.0032	1.7267	1.343	0.900
4	282.1	0.000791	0.0512	212.1	403.7	1.0433	1.7225	1.367	0.940
5	288.9	0.000806	0.0411	221.5	407.5	1.0759	1.7196	1.389	0.976
6	294.7	0.000820	0.0343	229.7	410.6	1.1037	1.7184	1.411	1.010
8	304.5	0.000846	0.0256	243.6	415.5	1.1497	1.7140	1.453	1.075
10	312.5	0.000870	0.0203	255.5	419.2	1.1876	1.7112	1.495	1.139
15	328.4	0.000928	0.0131	279.8	425.2	1.2621	1.7048	1.611	1.313
20	340.6	0.000989	0.00929	300.0	428.3	1.3208	1.6975	1.761	1.539
25	350.7	0.001057	0.00694	317.8	429.0	1.3711	1.6880	1.983	1.647
30	359.4	0.001141	0.00528	334.7	427.3	1.4170	1.6748	2.388	2.527
35	366.9	0.001263	0.00237	351.9	422.2	1.4626	1.6549	3.484	4.292
40	373.5	0.001580	0.00256	375.6	405.4	1.5247	1.6045	26.33	37.63
40.6†	374.3	0.001953	0.00195	389.6	389.6	1.5620	1.5620		

* Values are rounded, converted, and interpolated from Tillner-Roth and Baehr, *J. Phys. Chem. Ref. Data*, **23**, no. 5, 1994, pp. 657–730. Liquid enthalpy and entropy at 0°C = 273.15 K are 200 kJ/kg and 1.0000 kJ/kg.K, respectively.
† Critical point.

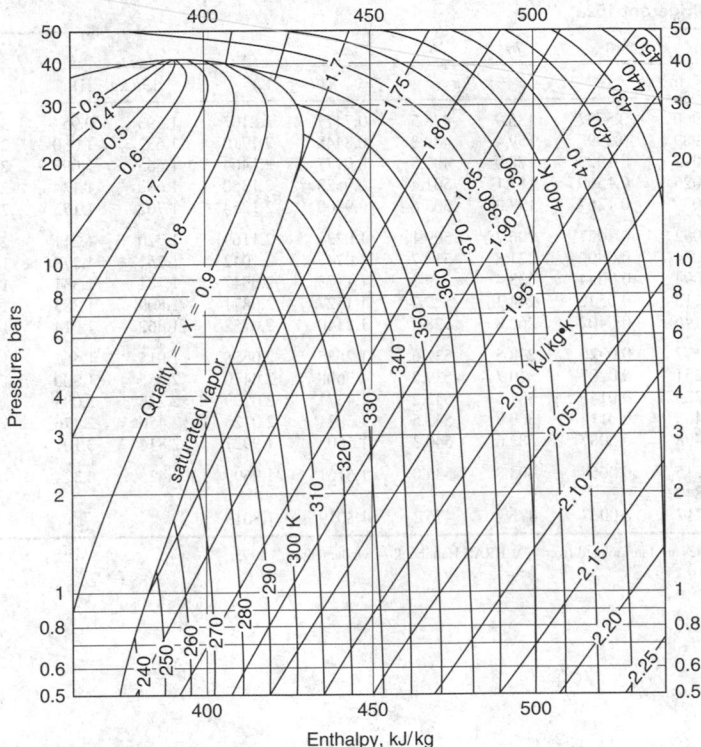

Fig. 4.2.8 Enthalpy–log pressure diagram for refrigerant 134a.

Table 4.2.15 Saturated Refrigerant 143a*

P, bar	T, °C	v_f	v_g	h_f	h_g	s_f	s_g	c_{pf}, kJ/(kg·K)	μ_f, µPa·s	k_f, W/(m·K)	Pr_f
		m³/kg		kJ/kg		kJ/(kg·K)					
0.5	−61.06	0.000840	0.4160	115.8	352.1	0.6541	1.769	1.291	314.6	0.1214	3.34
1	−47.49	0.000861	0.1977	136.4	361.9	0.7474	1.738	1.327	262.4	0.1121	3.11
1.5	−38.59	0.000875	0.1486	145.5	366.1	0.7865	1.728	1.346	243.4	0.1080	3.03
2	−31.77	0.000889	0.1113	154.7	370.2	0.8253	1.718	1.366	225.6	0.1039	2.97
2.5	−26.16	0.000903	0.0908	162.4		0.8567	1.710	1.384	212.5	0.1005	2.93
3	−21.35	0.000915	0.0755	169.2	376.3		1.706	1.401	201.8	0.0976	2.90
4	−13.34	0.000936	0.0566	180.5	380.8	0.9277	1.698	1.430	185.6	0.0928	2.86
5	−6.72	0.000955	0.0461	190.1	384.3	0.9636	1.693	1.459	173.7	0.0889	2.85
6	−1.04	0.000973	0.0382	198.5	387.2	0.9944	1.688	1.486	163.6	0.0856	2.84
8	8.47	0.001005	0.0284	212.8	391.8	1.0457	1.681	1.538	147.0	0.0800	2.83
10	16.34	0.001036	0.0224	225.1	395.2	1.088	1.675	1.590	134.2	0.0755	2.83
15	31.80	0.001112	0.0144	250.6	400.3	1.171	1.664	1.741	113.0	0.0667	2.95
20	43.75	0.001191	0.0102	271.8	402.4	1.238	1.651	1.943	99.1	0.0598	3.22
25	53.83	0.001288	0.0074	291.4	401.5	1.297	1.634	2.369	87.0	0.0538	3.83
30	61.96	0.001405	0.0056	308.9	398.0	1.350	1.616	2.93			
35	69.26	0.001616	0.0040	329.8	338.9	1.403	1.576				
38.3†	73.60	0.002311	0.0023	360.6	360.6	1.471	1.471				

* The P, T, v, h, and s values are interpolated from a tabulation as a function of temperature supplied by Dr. Friend, NIST, Boulder, CO, based on REFROP 5.
† Critical point.

Table 4.2.16 Saturated Refrigerant 152a*

P, bar	T, °C	v_f	v_g	h_f	h_g	s_f	s_g	c_{pf}	c_{pg}	μ_f	μ_g	k_f,	Pr_f
		m³/kg		kJ/kg		kJ/(kg · K)		kJ/(kg · K)		μPa · s		W/(m · K)	
0.5	−38.88	0.000960	0.5737	135.9	478.5	0.7477	2.2103	1.593	0.951	346.9			
1	−24.29	0.000989	0.2997	159.4	489.3	0.8448	2.1710	1.632	1.030	286.1	8.65	0.1280	3.65
1.5	−13.05	0.001012	0.1923	178.0	497.5	0.9177	2.1464	1.665	1.098	249.2	8.98	0.1231	3.37
2	−7.68	0.001025	0.1569	187.1	501.4	0.9523	2.1360	1.682	1.134	233.8	9.14	0.1199	3.28
2.5	−1.60	0.001039	0.1267	197.3	505.6	0.9900	2.1254	1.702	1.175	218.3	9.32	0.1164	3.18
3	3.59	0.001052	0.1057	206.2	509.1	1.0222	2.1169	1.720	1.211	205.9	9.47	0.1134	3.10
4	12.12	0.001074	0.0800	221.8	514.7	1.0747	2.1042	1.751	1.275	187.5	9.67	0.1084	3.03
5	19.14	0.001094	0.0644	233.4	519.1	1.1173	2.0947	1.781	1.331	174.2	9.91	0.1046	2.97
6	25.17	0.001113	0.0546	244.3	522.7	1.1527	2.0871	1.809	1.383	163.6	10.12	0.1010	2.93
8	35.25	0.001146	0.0402	262.8	528.4	1.2140	2.0752	1.862	1.314	147.4	10.48	0.0953	2.88
10	43.57	0.001177	0.0320	278.5	532.6	1.2635	2.0658	1.915	1.567	135.5	10.81	0.0907	2.86
15	59.95	0.001251	0.0207	310.7	539.2	1.3608	2.0470	2.055	1.800	114.6	11.56	0.0820	2.87
20	72.61	0.001325	0.0148	337.3	542.2	1.4371	2.0299	2.224	2.083		12.32	0.0758	
25	83.07	0.001405	0.0112	360.7	542.5	1.5021	2.0125	2.456	2.476		13.18		
30	92.03	0.001498	0.0087	382.6	540.3	1.5605	1.9927	2.814	3.101				
35	99.87	0.001615	0.0068	403.8	535.0	1.6161	1.9699	3.42	4.32				
40	106.76	0.001840											
45.2†	113.26	0.002717	0.0027	476.7	476.7	1.8019	1.8019						

† The P, T, v, h, s, c_p, μ, and k values are interpolated from ''ASHRAE Handbook—Fundamentals,'' 1993.
† Critical point.

Table 4.2.17 Saturated Water Substance

T, K	P, bar	v_c, m³/kg	v_g, m³/kg	h_c, kJ/kg	h_g, kJ/(kg · K)	s_c, kJ/(kg · K)	s_g, kJ/(kg · K)
250	0.00076	1.087. −3	1520	−381.5	2,459	−1.400	9.954
260	0.00196	1.088. −3	612.2	−360.5	2,477	−1.323	9.590
270	0.00469	1.090. −3	265.4	−339.6	2,496	−1.296	9.255
273.15	0.00611	1.091. −3	206.3	−333.5	2,502	−1.221	9.158
273.15	0.00611	1.000. −3	206.3	0.0	2,502	0.000	9.158
280	0.00990	1.000. −3	130.4	28.8	2,514	0.104	8.980
290	0.01917	1.001. −3	69.7	70.7	2,532	0.251	8.740
300	0.03531	1.003. −3	39.13	112.5	2,550	0.393	8.520
310	0.06221	1.007. −3	22.93	154.3	2,568	0.530	8.318
320	0.1053	1.011. −3	13.98	196.1	2,586	0.649	8.151
330	0.1719	1.016. −3	8.82	237.9	2,604	0.791	7.962
340	0.2713	1.021. −3	5.74	279.8	2,622	0.916	7.804
350	0.4163	1.027. −3	3.846	321.7	2,639	1.038	7.657
360	0.6209	1.034. −3	2.645	363.7	2,655	1.156	7.521
370	0.9040	1.041. −3	1.861	405.8	2,671	1.271	7.394
373.15	1.0133	1.044. −3	1.679	419.1	2,676	1.307	7.356
380	1.2869	1.049. −3	1.337	448.0	2,687	1.384	7.275
390	1.794	1.058. −3	0.980	490.4	2,702	1.494	7.163
400	2.455	1.067. −3	0.731	532.9	2,716	1.605	7.058
420	4.370	1.088. −3	0.425	618.6	2,742	1.810	6.865
440	7.333	1.110. −3	0.261	705.3	2,764	2.011	6.689
460	11.71	1.137. −3	0.167	793.5	2,782	2.205	6.528
480	17.90	1.167. −3	0.111	883.4	2,795	2.395	6.377
500	26.40	1.203. −3	0.0766	975.6	2,801	2.581	6.233
520	37.70	1.244. −3	0.0525	1,071	2,801	2.765	6.093
540	52.38	1.294. −3	0.0375	1,170	2,792	2.948	5.953
560	71.08	1.355. −3	0.0269	1,273	2,772	3.132	5.808
580	94.51	1.433. −3	0.0193	1,384	2,737	3.321	5.654
600	123.5	1.541. −3	0.0137	1,506	2,682	3.520	5.480
620	159.1	1.705. −3	0.0094	1,647	2,588	3.741	5.259
647.3*	221.2	3.170. −3	0.0032	2,107	2,107	4.443	4.443

Above the solid line the condensed phase is solid; below it is liquid. The notation 1.087. −3 signifies 1.087×10^{-3}. * Critical temperature.

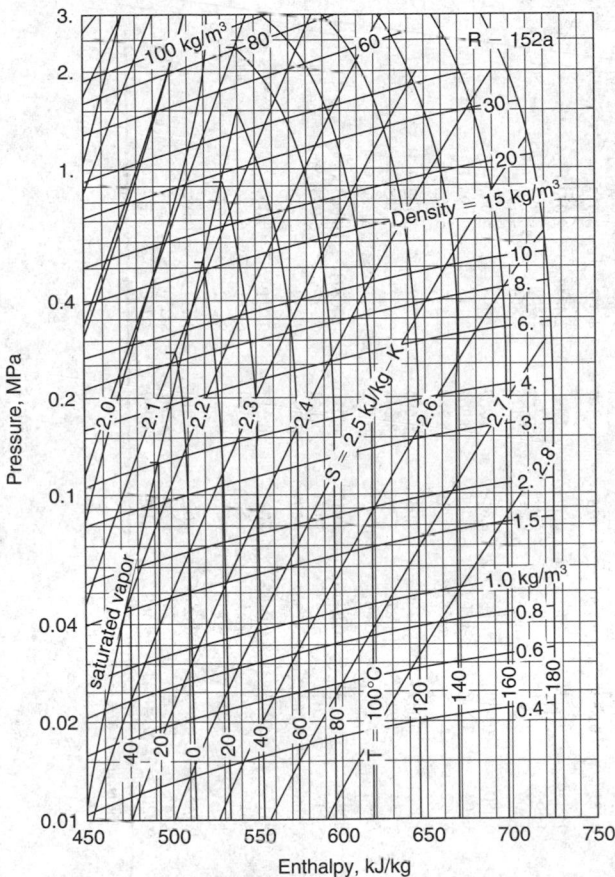

Fig. 4.2.9 Enthalpy–log pressure diagram for refrigerant 152a. *(Reprinted by permission of the ASHRAE from the 1993 "ASHRAE Handbook—Fundamentals.")*

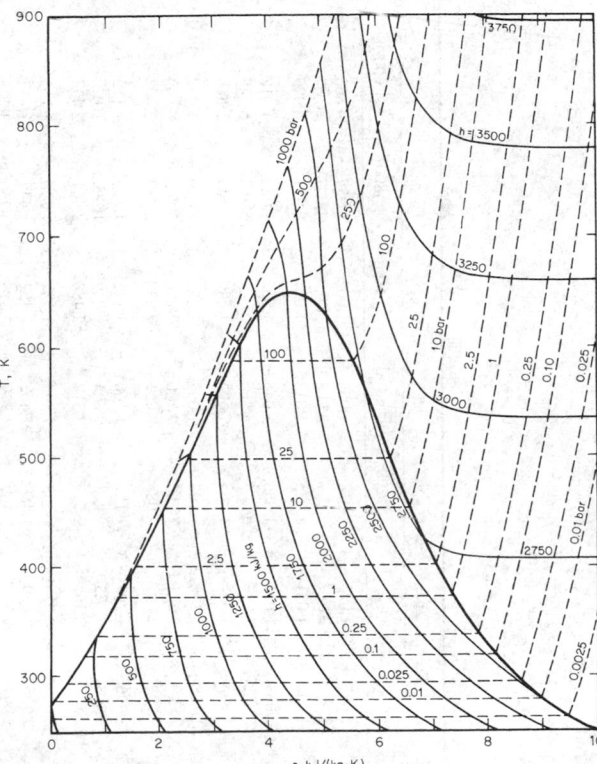

Fig. 4.2.10 Temperature-entropy diagram for water substance, SI units.

Table 4.2.18 Compressed Steam*

Temperature, K		Pressure, bar												Temperature, K
		1	10	20	40	60	80	100	200	400	600	800	1,000	
350	v	1.027.−3	1.027.−3	1.026.−3	1.025.−3	1.024.−3	1.023.−3	1.023.−3	1.018.−3	1.009.−3	1.002.−3	9.937.−4	9.865.−4	350
	h	321.8	322.5	323.3	324.9	326.4	328.1	329.7	337.7	353.8	369.7	385.7	401.7	
	s	1.037	1.037	1.036	1.035	1.034	1.032	1.031	1.025	1.013	1.001	0.991	0.979	
400	v	1.827	1.067.−3	1.066.−3	1.065.−3	1.064.−3	1.063.−3	1.061.−3	1.056.−3	1.045.−3	1.035.−3	1.027.−3	1.018.−3	400
	h	2,730	533.4	534.1	535.4	536.8	538.2	539.6	546.5	560.6	574.9	589.3	603.8	
	s	7.502	1.600	1.599	1.597	1.595	1.593	1.592	1.583	1.565	1.549	1.533	1.518	
450	v	2.063	1.124.−3	1.123.−3	1.121.−3	1.119.−3	1.118.−3	1.116.−3	1.108.−3	1.094.−3	1.082.−3	1.070.−3	1.059.−3	450
	h	2,830	749.0	749.8	750.8	751.9	753.0	754.1	759.5	771.0	783.0	795.3	807.9	
	s	7.736	2.110	2.107	2.105	2.102	2.099	2.097	2.085	2.061	2.039	2.019	2.002	
500	v	2.298	0.221	0.104	1.201.−3	1.198.−3	1.196.−3	1.193.−3	1.181.−3	1.160.−3	1.142.−3	1.126.−3	1.112.−3	500
	h	2,929	2,891.2	2,839.4	975.9	976.3	976.8	977.3	980.3	987.4	995.9	1005.3	1015.4	
	s	7.944	6.823	6.422	2.578	2.575	2.571	2.567	2.549	2.517	2.488	2.461	2.437	
600	v	2.76	0.271	0.133	0.0630	0.0396	0.0276	0.0201	1.483.−3	1.392.−3	1.337.−3	1.296.−3	1.265.−3	600
	h	3,129	3,109	3,087	3,036	2,976	2,906	2,820	1,489	1,452	1,452	1,447	1,447	
	s	8.309	7.223	6.875	6.590	6.224	5.997	5.775	3.469	3.379	3.316	3.266	3.223	
700	v	3.23	0.319	0.158	0.0769	0.0500	0.0346	0.0283	1.157.−2	2.630.−3	1.831.−3	1.639.−3	1.536.−3	700
	h	3,334	3,322	3,307	3,278	3,247	3,214	3,179	2,965	2,233	2,021	1,962	1,931	
	s	8.625	7.550	7.215	6.864	6.644	6.431	6.334	5.770	4.554	4.192	4.058	3.972	
800	v	3.69	0.367	0.182	0.0889	0.0589	0.0436	0.0343	1.575.−2	6.391.−3	3.496.−3	2.484.−3	2.072.−3	800
	h	3,546	3,537	3,526	3,506	3,485	3,464	3,442	3,325	3,047	2,734	2,567	2,465	
	s	8.908	7.837	7.507	7.151	6.965	6.809	6.685	6.252	5.654	5.175	4.864	4.701	
900	v	4.15	0.414	0.206	0.102	0.0674	0.0501	0.0398	1.899.−2	8.619.−3	5.257.−3	3.704.−3	2.907.−3	900
	h	3,764	3,757	3,750	3,737	3,719	3,704	3,688	3,609	3,440	3,269	3,113	2,995	
	s	9.165	8.097	7.770	7.462	7.237	7.092	6.975	6.587	6.119	5.780	5.510	5.305	
1,000	v	4.15	0.414	0.206	0.102	0.0674	0.0501	0.0398	2.186.−2	1.038.−2	6.605.−3	4.792.−3	3.763.−3	1,000
	h	3,990	3,984	3,978	3,967	3,955	3,944	3,935	3,874	3,756	3,640	3,532	3,435	
	s	9.402	8.336	8.011	7.682	7.486	7.345	7.233	6.867	6.453	6.172	5.951	5.727	
1,500	v	6.92	0.692	0.341	0.1730	0.1153	0.0865	0.0692	0.0346	0.0173	0.0116	0.00871	0.00700	1,500
	h	5,227	5,224	5,221	5,217	5,212	5,207	5,203	5,198	5,171	5,144	5,120	5,095	
	s	10.40	9.34	9.015	8.693	8.503	8.368	8.262	7.936	7.597	7.391	7.239	7.118	
2,000	v	9.26	0.925	0.462	0.231	0.1543	0.1157	0.0926	0.0465	0.0234	0.0157	0.0119	0.0096	2,000
	h	6,706	6,649	6,639	6,629	6,623	6,619	6,616	6,610	6,599	6,590	6,581	6,574	
	s	11.25	10.15	9.828	9.503	9.313	9.178	9.073	8.748	8.418	8.222	8.082	7.971	
2,500	v	11.90	1.171	0.583	0.291	0.1942	0.1457	0.1166	0.0584	0.0294	0.0197	0.0149	0.0120	2,500
	h	9,046	8,504	8,413	8,342	8,307	8,285	8,269	8,269	8,267	8,261	8,250	8,240	
	s	12.28	10.80	10.62	10.26	10.06	9.920	9.810	9.468	9.129	8.930	8.788	8.677	

* v = specific volume, m³/kg; h = specific enthalpy, kJ/kg; s = specific entropy, kJ/(kg · K). The notation 1.027. − 3 signifies 1.027 × 10⁻³.

Table 4.2.19 Saturated Water Substance, fps Units

Abs press, lb/in²	Temp., °F	Specific volume, ft³/lb		Enthalpy, Btu/lb			Entropy, Btu/(lb · R)			Internal energy, evap., Btu/lb
		Liquid	Vapor	Liquid	Evap.	Vapor	Liquid	Evap.	Vapor	
0.08866	32.02	0.016022	3,302	0.01	1,075.4	1,075.4	0.00000	2.1869	2.1869	1,021.2
1.0	101.70	0.016136	333.6	69.74	1,036.0	1,105.8	0.13266	1.8453	1.9779	974.3
1.5	115.65	0.016187	227.7	83.65	1,028.0	1,111.7	0.15714	1.7867	1.9438	964.8
2	126.04	0.016230	173.75	94.02	1,022.1	1,116.1	0.17499	1.7448	1.9198	957.8
3	141.43	0.016300	118.72	109.39	1,013.1	1,122.5	0.20089	1.6852	1.8861	947.2
4	152.93	0.016358	90.64	120.89	1,006.4	1,127.3	0.21983	1.6426	1.8624	939.3
5	162.21	0.016407	73.53	130.17	1,000.9	1,131.0	0.23486	1.6093	1.8441	932.9
10	193.19	0.016590	38.42	161.23	982.1	1,143.3	0.28358	1.5041	1.7877	911.0
14.696	211.99	0.016715	26.80	180.15	970.4	1,150.5	0.31212	1.4446	1.7567	897.5
15	213.03	0.016723	26.29	181.19	969.7	1,150.9	0.31367	1.4414	1.7551	896.8
20	227.96	0.016830	20.09	196.26	960.1	1,156.4	0.33580	1.3962	1.7320	885.8
25	240.08	0.016922	16.306	208.52	952.2	1,160.7	0.35345	1.3607	1.7142	876.9
30	250.34	0.017004	13.748	218.93	945.4	1,164.3	0.36821	1.3314	1.6996	869.2
35	259.30	0.017078	11.900	228.04	939.3	1,167.4	0.38093	1.3064	1.6873	862.4
40	267.26	0.017146	10.501	236.16	933.8	1,170.0	0.39214	1.2845	1.6767	856.2
45	274.46	0.017209	9.403	243.51	928.8	1,172.3	0.40218	1.2651	1.6673	850.7
50	281 03	0.017269	8.518	250.24	924.2	1,174.4	0.41129	1.2476	1.6589	845.5
55	287.10	0.017325	7.789	256.46	919.9	1,176.3	0.41963	1.2317	1.6513	840.8
60	292.73	0.017378	7.177	262.25	915.8	1,178.0	0.42733	1.2170	1.6444	836.3
65	298.00	0.017429	6.657	267.67	911.9	1,179.6	0.43450	1.2035	1.6380	832.1
70	302.96	0.017478	6.209	272.79	908.3	1,181.0	0.44120	1.1909	1.6321	828.1
75	307.63	0.017524	5.818	277.61	904.8	1,182.4	0 44749	1.1790	1.6265	824.3
80	312.07	0.017570	5.474	282.21	901.4	1,183.6	0.45344	1.1679	1.6214	820.6
85	316.29	0.017613	5.170	286.58	898.2	1,184.8	0.45907	1.1574	1.6165	817.1
90	320.31	0.017655	4.898	290.76	895.1	1,185.9	0.46442	1.1475	1.6119	813.8
95	324.16	0.017696	4.654	294.76	892.1	1,186.9	0.46952	1.1380	1.6076	810.6
100	327.86	0.017736	4.434	298.61	889.2	1,187.8	0.47439	1.1290	1.6034	807.5
150	358.48	0.018089	3.016	330.75	864.2	1,194.9	0.51422	1.0562	1.5704	781.0
200	381.86	0.018387	2.289	355.6	843.7	1,199.3	0.5440	1.0025	1.5464	759.6
250	401.04	0.018653	1.8448	376.2	825.8	1,202.1	0.5680	0.9594	1.5274	741.4
300	417.13	0.018896	1.5442	394.1	809.8	1,203.9	0.5883	0.9232	1.5115	725.1
350	431.82	0.019124	1.3267	409.9	795.0	1,204.9	0.6060	0.8917	1.4978	710.3
400	444.70	0.019340	1.1620	424.2	781.2	1,205.5	0.6218	0.8638	1.4856	696.7
450	456.39	0.019547	1.0326	437.4	768.2	1,205.6	0.6360	0.8385	1.4746	683.9
500	467.13	0.019748	0.9283	449.5	755.8	1,205.3	0.6490	0.8154	1.4645	671.7
550	477.07	0.019943	0.8423	460.9	743.9	1,204.8	0.6611	0.7941	1.4551	660.2
600	486.33	0.02013	0.7702	471.7	732.4	1,204.1	0.6723	0.7742	1.4464	649.1
700	503.23	0.02051	0.6558	491.5	710.5	1,202.0	0.6927	0.7378	1.4305	628.2
800	518.36	0.02087	0.5691	509.7	689.6	1,199.3	0.7110	0.7050	1.4160	608.4
900	532.12	0.02123	0.5009	526.6	669.5	1,196.0	0.7277	0.6750	1.4027	589.6
1,000	544.75	0.02159	0.4459	542.4	650.0	1,192.4	0.7432	0.6471	1.3903	571.5
1,500	596.39	0.02346	0.2769	611.5	557.2	1,168.7	0.8082	0.5276	1.3359	486.9
2,000	636.00	0.02565	0.18813	671.9	464.4	1,136.3	0.8623	0.4238	1.2861	404.2
2,500	668.31	0.02860	0.13059	730.9	360.5	1,091.4	0.9131	0.3196	1.2327	313.4
3,000	695.52	0.03431	0.08404	802.5	213.0	1,015.5	0.9732	0.1843	1.1575	185.4
3,203.6	705.44	0.05053	0.05053	902.5	0	902.5	1.0580	0	1.0580	0

SOURCE: Abstracted from Keenan, Keyes, Hill, and Moore, "Steam Tables," 1969.

Table 4.2.20 Compressed Water Substance, fps Units

Pressure, psia (saturation temp., °F)		Temperature of steam, °F							
		200	300	400	500	600	800	1,000	1,200
10 (193.19)	v	38.85	44.99	51.03	57.04	63.03	74.98	86.91	98.84
	h	1,146.6	1,193.7	1,240.5	1,287.7	1,335.5	1,433.3	1,534.6	1,639.4
	s	1.7927	1.8592	1.9171	1.9690	2.0164	2.1009	2.1755	2.2428
50 (281.03)	v	0.01663		8.772	10.061	11.305	14.949	17.352	19.747
	h	168.1	1,332.8	1,184.4	1,235.0	1,284.0	1,431.7	1,533.5	1,638.7
	s	0.2940	1.8371	1.6722	1.7348	1.7887	1.9225	1.9975	2.0650
100 (327.86)	v	0.01663	0.01745	4.934	5.587	6.216	7.445	8.657	9.861
	h	168.2	269.7	1,227.5	1,279.1	1,329.3	1,429.6	1,532.1	1,637.7
	s	0.2939	0.4372	1.6517	1.7085	1.7582	1.8449	1.9204	1.9882
150 (358.48)	v	0.01663	0.01744	3.221	3.679	4.111	4.944	5.759	6.566
	h	168.4	269.8	1,219.5	1,274.1	1,325.7	1,427.5	1,530.7	1,454.5
	s	0.2938	0.4371	1.5997	1.6598	1.7110	1.7989	1.8750	1636.7
200 (381.86)	v	0.01662	0.01744	2.361	2.724	3.058	3.693	4.310	4.918
	h	168.5	269.9	1,210.8	1,268.8	1,322.1	1,425.3	1,529.3	1,635.7
	s	0.2938	0.4370	1.5600	1.6239	1.6767	1.7660	1.8425	1.9109
300 (417.43)	v	0.01662	0.01743	0.01863	1.7662	2.004	2.442	2.860	3.270
	h	168.8	270.1	375.2	1,257.5	1,314.5	1,421.0	1,526.5	1,633.8
	s	0.2936	0.4368	0.5665	1.5701	1.6266	1.7187	1.7964	1.8653
400 (444.70)	v	0.01661	0.01742	0.01862	1.2843	1.4760	1.8163	2.136	2.446
	h	169.0	270.3	375.3	1,245.2	1,306.6	1,416.6	1,523.6	1,631.8
	s	0.2935	0.4366	0.5662	1.5282	1.5892	1.6844	1.7632	1.8327
500 (467.13)	v	0.01661	0.01741	0.01861	.9924	1.1583	1.4407	1.7008	1.9518
	h	169.2	270.5	375.4	1,231.5	1,298.3	1,412.1	1,520.7	1,629.8
	s	0.2934	0.4364	0.5660	1.4923	1.5585	1.6571	1.7371	1.8072
600 (486.33)	v	0.01660	0.01740	0.01860	.7947	.9456	1.1900	1.4108	1.6222
	h	169.4	270.7	375.5	1,216.2	1,289.5	1,407.6	1,517.8	1,627.8
	s	0.2933	0.4362	0.5657	1.4592	1.5320	1.6343	1.7155	1.7861
700 (503.23)	v	0.01660	0.01740	0.01859	0.02042	.7929	1.0109	1.2036	1.3868
	h	169.6	270.9	375.6	487.6	1,280.2	1,402.9	1,514.9	1,625.8
	s	0.2932	0.4360	0.5655	0.6887	1.5081	1.6145	1.6970	1.7682
800 (518.36)	v	0.01659	0.01739	0.01857	0.02040	.6776	.8764	1.0482	1.2102
	h	169.9	271.1	375.8	487.6	1,270.4	1,398.2	1,511.9	1,623.8
	s	0.2930	0.4359	0.5652	0.6883	1.4861	1.5969	1.6807	1.7526
900 (532.14)	v	0.01658	0.01739	0.01856	0.02038	.5871	.7717	.9273	1.0729
	h	170.1	271.3	375.9	487.5	1,260.0	1,393.4	1,508.9	1,621.7
	s	0.2929	0.4355	0.5650	0.6878	1.4652	1.5810	1.6662	1.7386
1,000 (544.75)	v	0.01658	0.01738	0.01855	0.02036	0.5140	0.6878	0.8305	0.9630
	h	170.3	271.5	376.0	487.5	1,248.8	1,388.5	1,505.9	1,619.7
	s	0.2928	0.4355	0.5647	0.6874	1.4450	1.5664	1.6530	1.7261
1,500 (596.39)	v	0.01655	0.01734	0.01849	0.02024	0.2816	0.4350	0.5400	0.6334
	h	171.5	272.4	376.6	487.4	1,174.8	1,362.5	1,490.3	1,609.3
	s	0.2922	0.4346	0.5634	0.6853	1.3416	1.5058	1.6001	1.6765
2,000 (636.00)	v	0.01653	0.01731	0.01844	0.02014	0.02330	0.3071	0.3945	0.4685
	h	172.6	273.3	377.2	487.3	614.0	1,333.8	1,474.1	1,598.6
	s	0.2916	0.4338	0.5622	0.6832	0.8046	1.4562	1.5598	1.6398
2,500 (668.31)	v	0.01650	0.01727	0.01839	0.02004	0.02300	0.2291	0.3069	0.3696
	h	173.8	274.3	377.8	487.3	611.6	1,301.7	1,457.2	1,587.7
	s	0.2910	0.4329	0.5609	0.6813	0.8043	1.4112	1.5262	1.6101
3,000 (695.52)	v	0.01648	0.01724	0.01833	0.01994	0.02274	0.17572	0.2485	0.3036
	h	174.9	275.2	378.5	487.3	609.6	1,265.2	1,439.6	1,576.6
	s	0.2905	0.4321	0.5597	0.6794	0.8004	1.3675	1.4967	1.5848
4,000	v	0.01643	0.01717	0.01824	0.01977	0.02229	0.1052	0.1752	0.2213
	h	177.2	277.2	379.9	487.5	606.5	1,172.9	1,402.6	1,553.9
	s	0.2931	0.4304	0.5573	0.6758	0.7936	1.2740	1.4449	1.5423
5,000	v	0.01638	0.01711	0.01814	0.01960	0.02191	0.05932	0.1312	0.1720
	h	179.5	279.1	381.3	487.9	604.2	1,042.1	1,363.4	1,530.8
	s	0.2882	0.4288	0.5551	0.6724	0.7876	1.1583	1.3988	1.5066

v = specific volume, ft³/lb; h = enthalpy, Btu/lb; s = entropy, Btu/(lb · R).

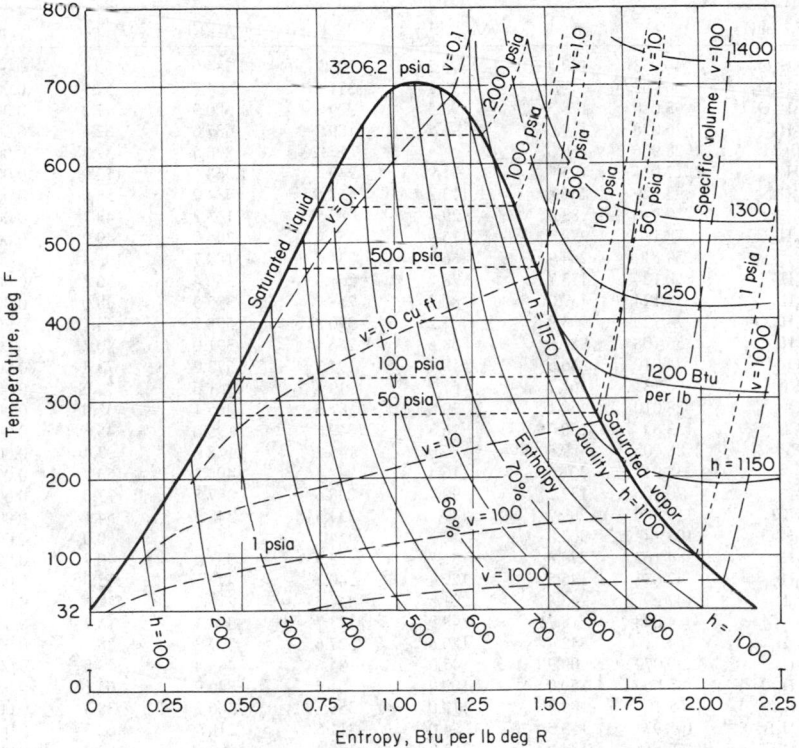

Fig. 4.2.11 Temperature-entropy diagram for water substance, fps units. *(Data from Keenan and Keyes, "Thermodynamic Properties of Steam," Wiley.)*

Table 4.2.21 Phase Transition and Other Data for 100 Fluids*

Name	Formula	M	T_m, K	Δh_{fus}, kJ/kg	T_b, K	Δh_{vap}, kJ/kg	P_c, bar	v_c, m³/kg	T_c, K	Z_c
Acetaldehyde	C_2H_4O	44.053	149.7	73.2	293.7	584.0	55.4	0.00382	461	0.243
Acetic acid	$C_2H_4O_2$	60.053		195.2	391.2	404.7	57.9	0.00285	594.5	0.200
Acetone	C_3H_6O	58.080	178.5	98.5	329.3	500.9	47.2	0.00360	508.2	0.232
Acetylene	C_2H_2	26.038	179.0	96.5	189.2	687.0	62.4	0.00435	308.3	0.276
Air	Mixed	28.966	60.0		79, 82	206.5	37.7	0.00320	132.6	0.263
Ammonia	NH_3	17.031	195.4	331.9	239.7	1,368	112.9	0.00427	405.7	0.244
Aniline	C_6H_7N	93.129	266.8	113.2	457.5	484.9	53.0	0.00340	698.8	0.289
Argon	A	39.948	83.8	29.4	87.5	163.2	48.7	0.00187	150.8	0.290
Benzene	C_6H_6	78.114	278.7	125.9	353.3	394.0	49.2	0.00332	562.1	0.273
Bromine	Br_2	159.81	264.9	66.2	331.6	187.7	103.4	0.00079	584.2	0.270
Butane, n	C_4H_{10}	58.124	113.7	78.2	261.5	366.4	36.5	0.00452	408.1	0.283
Butane, iso	C_4H_{10}	58.124	137.0	80.2	272.7	385.5	38.0	0.00439	425.2	0.274
Butanol	$C_4H_{10}O$	74.123	183.9	121.2	390.8	593.2	44.1	0.00370	562.9	0.259
Butylene	C_4H_8	56.108	87.8	68.6	266.9	391.0	40.2	0.00428	419.6	0.277
Carbon dioxide	CO_2	44.010	216.6	18.4	194.7	573.2	73.8	0.00214	304.1	0.274
Carbon disulfide	CS_2	76.131	161.1	57.7	319.4	351.6	79.0	0.00223	552.0	0.293
Carbon monoxide	CO	28.010	68.1	29.8	81.6	215.1	35.0	0.00332	132.9	0.295
Carbon tetrachloride	CCl_4	153.82	250.3	16.3	349.8	195.0	45.6	0.00170	556.4	0.258
Carbon tetrafluoride	CF_4	88.005	89.5	8.0	145.2	138	37.4	0.00156	228.0	0.272
Cesium	Cs	132.91	301.6	16.4	942.4	494.3	153.7	0.00230	2,043	0.240
Chlorine	Cl_2	70.906	172.2	90.4	238.6	287.5	77.0	0.00175	417.2	0.278
Chloroform	$CHCl_3$	119.38	209.7	77.1	334.5	248.5	54.5	0.00202	536.0	0.294
o-Cresol	C_7H_8O	108.14	303.8		464.1		50.0	0.00291	697.6	0.271
Cyclohexane	C_6H_{12}	84.162	279.8	31.7	353.9	357.4	40.7	0.00366	553.4	0.273
Cyclopropane	C_3H_6	42.081	145.5	129.4	240.3	477	55.0	0.00387	397.8	0.271
Decane	$C_{10}H_{22}$	142.29	243.4	201.8	447.3	276.2	21.0	0.00424	617.5	0.247
Deuterium	D_2	4.028	18.71	48.9	23.7	304.4	16.7	0.00143	38.34	0.301
Diphenyl	$C_{12}H_{10}$	154.21	342.4	120.6	527.6	317.3	38.5	0.00326	789.0	0.295
Ethane	C_2H_6	30.070	89.9	45.0	184.6	488.4	48.8	0.00486	305.4	0.281
Ethanol	C_2H_6O	46.069	159.0	109.0	351.5	840.9	63.8	0.00362	516.2	0.248
Ethyl acetate	$C_4H_8O_2$	88.107	189.4	119.0	350.3	366.2	38.3	0.00325	523.2	0.252
Ethyl bromide	C_2H_5Br	108.97	153.5	54.0	311.5	218	62.3	0.00197	503.9	0.320
Ethyl chloride	C_2H_5Cl	64.515	134.9	68.9	285.4	382.5	52.0	0.00309	460.4	0.270
Ethyl ether	$C_4H_{10}O$	74.123	150.0	93.1	307.8	358.9	36.8	0.00377	466.8	0.262
Ethyl formate	$C_3H_6O_2$	74.080	193.8	124.3	327.4	405.8	47.4	0.00310	508.5	0.257
Ethylene	C_2H_4	28.054	104.0	119.5	169.5	480.0	50.5	0.00455	283.1	0.274
Ethylene oxide	C_2H_4O	44.054	160.6	117.5	283.6	580.0	71.9	0.00318	468.9	0.258
Fluorine	F_2	37.997	53.5	13.4	85.1	172.1	52.2	0.00174	144.3	0.288
Helium 4	He	4.003			4.3	20.6	2.3	0.0144	5.189	0.303
Heptane	C_7H_{16}	100.20	182.5	139.9	371.6	316.3	27.4	0.0043	540.2	0.263
Hexane	C_6H_{14}	86.178	177.8	151.2	341.9	334.8	30.3	0.0043	507.6	0.265
Hydrazine	N_2H_4	32.045	274.7	395.0	386.7	1,207	147.0		653.2	0.284
Hydrogen	H_2	2.016	14.0	58.0	20.4	454.0	13.0	0.0323	33.3	0.305
Hydrogen bromide	HBr	80.912	186.3	37.4	206.4	217.5	85.5	0.00124	363.2	0.284
Hydrogen chloride	HCl	36.461	160.0	54.7	188.1	443.0	83.1	0.00022	324.7	0.249
Hydrogen fluoride	HF	20.006	181.8	196.3	272.7	374.3	64.9	0.00345	461.2	0.120
Hydrogen iodide	HI	127.91	222.4	22.4	237.8	154.0	83.1	0.00106	423.9	0.318
Hydrogen sulfide	H_2S	34.076	187.5	69.8	213.0	248.0	89.4	0.00289	373.1	0.283

Table 4.2.21 Phase Transition and Other Data for 100 Fluids* (Continued)

Name	Formula	M	T_m, K	Δh_{fus}, kJ/kg	T_b, K	Δh_{vap}, kJ/kg	P_c, bar	v_c, m³/kg	T_c, K	Z_c
Iodine	I_2	253.81	387.0	62.1	457.5	164.3	117.5	0.00054	785.0	0.248
Krypton	Kr	83.80	116.0	19.5	121.4	107.9	55.0	0.00109	209.4	0.288
Lithium	Li	6.940	453.8		1,615	1,945			3,750	
Mercury	Hg	200.59	234.3	11.4	630.1	295.6	1,510	0.00018	1,763	
Methane	CH_4	16.043	90.7	58.4	111.5	511.8	46.0	0.00617	190.5	0.287
Methanol	CH_4O	32.042	175.5	98.9	337.7	1,104	79.5	0.00368	512.6	0.220
Methyl acetate	$C_3H_6O_2$	74.080	175		330.3	410.0	46.9	0.00308	506.9	0.254
Methyl bromide	CH_3Br	94.939	179.5	62.8	276.7	252.0		0.00173	467.2	
Methyl chloride	CH_3Cl	50.49	175.4	127.4	249.4	428.5	66.8	0.00270	416.3	0.277
Methyl formate	$C_2H_4O_2$	60.053	173.4	125.5	304.7	481.2	60.0	0.00287	487.2	0.255
Methylene chloride	CH_2Cl_2	84.922	176.5	54.4	312.9	328	61.3	0.00197	510.	0.255
Naphthalene	$C_{10}H_8$	128.17	353.4	148.1	491.1	341	40.5	0.00321	748.4	0.270
Neon	Ne	20.179	24.5	16.6	27.3	91.3	27.6	0.00207	44.4	0.311
Nitric oxide	NO	30.006	111	76.6	121.4	460	64.9	0.00192	180	0.249
Nitrogen	N_2	28.013	63.1	25.7	77.3	197.6	34.0	0.00318	126.2	0.287
Nitrogen peroxide	NO_2	46.006	263	159.5	294.5	414.4	101.3	0.00180	431.4	0.233
Nitrous oxide	N_2O	44.013	176	148.6	184.7	376.0	72.4	0.00221	309.6	0.274
Octane	C_8H_{18}	114.23	216.4	161.6	398.9	302.7	25.0	0.00426	508.9	0.258
Oxygen	O_2	31.999	54.4	13.9	90.0	212.5	50.4	0.00229	154.6	0.288
Pentane, iso	C_5H_{12}	72.151	113.7	71.1	301.0	341.0	33.5	0.00427	460.4	0.270
Pentane, n	C_5H_{12}	72.151	143.7	116.6	309.2	357.2	33.7	0.00431	469.6	0.268
Potassium	K	39.098	336.4	59.8	1,030		167		2,265	
Propane	C_3H_8	44.097	86	80.0	231.1	425.7	42.6	0.00453	369.8	0.277
Propanol	C_3H_8O	60.096	147.0	86.5	370.4	695.8	51.7	0.00364	536.7	0.253
Propylene	C_3H_6	42.081	87.9	71.4	225.5	437.5	46.0	0.00429	365.1	0.275
Radon	Rn	224	201	12.3	211	82.8	65.5	0.00063	377.0	0.293
Refrigerant 11	$CFCl_3$	137.37	162.2	50.2	296.9	180.2	44.1	0.00181	471.2	0.280
Refrigerant 12	CF_2Cl_2	120.91	115.4	34.3	243.4	165.1	41.2	0.00179	385.2	0.278
Refrigerant 13	CF_3Cl	104.46	92.1		191.7	148.4	38.7	0.00173	302.0	0.279
Refrigerant 13B1	CF_3Br	148.91	105.4		215.4	118.7	39.6	0.00134	340.2	0.280
Refrigerant 21	$CHFCl_2$	102.91	138.2		282.1	242.1	51.7	0.00192	451.4	0.271
Refrigerant 22	CHF_2Cl	86.469	113.2	47.6	232.4	233.6	49.8	0.00191	369.2	0.267
Refrigerant 23	CHF_3	70.014	118.0	58.0	191.2	239	48.4	0.00190	299.1	0.259
Refrigerant 113	$C_2F_3Cl_3$	187.38	238.2		320.8	146.8	34.1	0.00174	487.3	0.274
Refrigerant 114	$C_2F_4Cl_2$	170.92	179.2		276.7	136.0	32.6	0.00172	418.9	0.275
Refrigerant 115	C_2F_5Cl	154.47	171	12.2	234.0	124.1	31.6	0.00163	353.1	0.271
Refrigerant 142b	$C_2F_2H_3Cl$	100.50	142.4	26.7	263.9	223	41.5	0.00232	410.0	0.279
Refrigerant 152a	$C_2F_2H_4$	66.051	156		248	326	45.0	0.00274	386.7	0.253
Refrigerant 216	$C_3F_6Cl_2$	220.93			308	117.3	27.5	0.00174	453.2	0.281
Refrigerant C318	C_4F_8	200.03	233.0		267	116	27.8	0.00161	388.5	0.272
Refrigerant 500	Mix	99.303	114.3		239.7	201.1	44.3	0.00201	378.7	0.281
Refrigerant 502	Mix	111.63			237	172.2	40.7	0.00178	355.3	0.275
Refrigerant 503	Mix	87.267			184	172.9			293	
Refrigerant 504	Mix	79.240			216	242.9			339	
Refrigerant 505	Mix	103.43			243.6		47.3		391	
Refrigerant 506	Mix	93.69			260.7		51.6		416	
Rubidium	Rb	85.468	312.6		959.4	811.3		0.00288	2,083	
Sodium	Na	22.990	371.0		1,155	3,880			2,730	
Sulfur dioxide	SO_2	64.059	197.8		268.4	368.3	78.8	0.00190	430.7	0.269
Sulfur hexafluoride	SF_6	146.051					37.8	0.00137	318.7	0.285
Toluene	C_7H_8	92.141	178.2		383.8	339.0	40.5	0.00345	594.0	0.260
Water	H_2O	18.015	273.2		373.2	2,256	221.2	0.00315	647.3	0.234
Xenon	Xe	131.36	161.5		165	99.2	58.7	0.00091	290	0.290

* M = molecular weight, T_m = normal melting temperature, Δh_{fus} = enthalpy (or latent heat) of fusion, T_b = normal boiling point temperature, Δh_{vap} = enthalpy (or latent heat) of vaporization, T_c = critical temperature, P_c = critical pressure, v_c = critical volume, Z_c = critical compressibility factor.
   SOURCE: Prepared by the author and abstracted from Rohsenow et al., "Handbook of Heat Transfer Fundamentals," McGraw-Hill.

Table 4.2.22 Specific Heat at Constant Pressure [kJ/(kg · K)] of Liquids and Gases

Substance	Temperature, K												
	200	225	250	275	300	325	350	375	400	425	450	475	500
Acetylene	1.457	1.517	1.575	1.635	1.695	1.747							
Air	1.048	1.036	1.029	1.025	1.021	1.021	1.020	1.021	1.022	1.024	1.027	1.030	1.035
Ammonia	4.605	4.360	2.210	2.176	2.169	2.183	2.210	2.247	2.289	2.331	2.381	2.429	2.477
Argon	0.524	0.523	0.522	0.522	0.522	0.521	0.521	0.521	0.521	0.521	0.521	0.521	0.521
Butane, i	1.997	2.099	2.207	1.590	1.694	1.810	1.921	2.035	2.149	2.258	2.367	2.436	2.571
Butane, n	2.066	2.134	2.214	1.642	1.731	1.835	1.941	2.047	2.154	2.253	2.361	2.461	2.558
Carbon dioxide		0.763	0.791	0.818	0.845	0.870	0.894	0.917	0.938	0.958	0.978	0.995	1.014
Ethane	1.419	1.492	1.574	1.658	1.762	1.871	1.968	2.081	2.184	2.287	2.393	2.492	2.590
Ethylene	1.296	1.340	1.397	1.475	1.560	1.656	1.748	1.839	1.928	2.014	2.097	2.179	2.258
Fluorine	0.785	0.793	0.803	0.815	0.827								
Helium	5.193	5.193	5.193	5.193	5.193	5.193	5.193	5.193	5.193	5.193	5.193	5.193	5.193
Hydrogen, n	13.53	13.83	14.05	14.20	14.31	14.38	14.43	14.46	14.48	14.49	14.50	14.50	14.51
Krypton	0.252	0.251	0.250	0.249	0.249	0.249	0.249	0.249	0.249	0.249	0.249	0.249	0.248
Methane	2.105	2.122	2.145	2.184	2.235	2.297	2.375	2.454	2.534	2.617	2.709	2.797	2.892
Neon	1.030	1.030	1.030	1.030	1.030	1.030	1.030	1.030	1.030	1.030	1.030	1.030	1.030
Nitrogen	1.043	1.042	1.042	1.041	1.041	1.041	1.042	1.043	1.045	1.047	1.050	1.053	1.056
Oxygen	0.915	0.914	0.915	0.917	0.920	0.924	0.929	0.935	0.942	0.949	0.956	0.964	0.972
Propane	2.124	2.220	1.500	1.596	1.695	1.805	1.910	2.030	2.140	2.249	2.355	2.458	2.558
Propylene	2.094	2.132	1.436	1.481	1.536	1.625	1.709	1.793	1.890	1.981	2.070	2.153	2.245
Refrigerant 12			0.549	0.578	0.602	0.625	0.647	0.666	0.685	0.701	0.716	0.728	0.741
Refrigerant 21	0.973	0.982	0.991	1.015	0.594	0.617	0.641	0.661	0.683	0.701	0.720	0.735	0.752
Refrigerant 22	1.065	1.085	0.588	0.617	0.647	0.676	0.704	0.729	0.757	0.782	0.806	0.826	0.848
Water substance	1.545	1.738	1.935	4.211	4.179	4.182	4.195	2.035	1.996	1.985	1.981	1.980	1.983
Xenon	0.165	0.163	0.162	0.161	0.160	0.160	0.160	0.159	0.159	0.159	0.159	0.159	0.159

Values for the saturated liquid are tabulated up to the normal boiling point. Higher temperature values are for the dilute gas. Values for water substance below 275 K are for ice.

Table 4.2.23 Specific Heat Ratio c_p/c_v for Liquids and Gases at Atmospheric Pressure

Substance	Temperature, K												
	200	225	250	275	300	325	350	375	400	425	450	475	500
Acetylene	1.313	1.289	1.269	1.250	1.234	1.219	1.205						
Air	1.399	1.399	1.399	1.399	1.399	1.399	1.398	1.397	1.395	1.393	1.391	1.388	1.386
Ammonia					1.327	1.302	1.295	1.285	1.278	1.269	1.262	1.256	1.249
Argon	1.663	1.665	1.666	1.666	1.666	1.666	1.666	1.666	1.666	1.666	1.666	1.666	1.666
Butane, i	1.357	1.356	1.359	1.116	1.103	1.094	1.086	1.080	1.075	1.070	1.066	1.063	1.060
Butane, n	1.418	1.412	1.407	1.114	1.103	1.094	1.086	1.080	1.075	1.071	1.067	1.063	1.061
Carbon dioxide		1.344	1.323	1.302	1.290	1.279	1.269	1.260	1.252	1.245	1.239	1.233	1.229
Carbon monoxide	1.405	1.404	1.403	1.402	1.401	1.401	1.400	1.398	1.396	1.394	1.392	1.390	1.387
Ethane		1.246	1.226	1.210	1.193	1.180	1.167	1.157	1.148	1.139	1.132	1.126	1.120
Ethylene			1.275	1.254	1.236	1.220	1.206	1.194	1.183	1.173	1.165	1.158	1.151
Fluorine	1.393	1.386	1.377	1.370	1.362								
Helium	1.667	1.667	1.667	1.667	1.667	1.667	1.667	1.667	1.667	1.667	1.667	1.667	1.667
Hydrogen, n	1.439	1.426	1.415	1.410	1.406	1.403	1.401	1.400	1.399	1.398	1.398	1.397	1.397
Krypton	1.649	1.655	1.658	1.660	1.662	1.662	1.662	1.662	1.662	1.663	1.664	1.665	1.667
Methane	1.337	1.332	1.325	1.316	1.306	1.295	1.282	1.271	1.258	1.247	1.237	1.228	1.219
Neon	1.667	1.667	1.667	1.667	1.667	1.667	1.667	1.667	1.667	1.667	1.667	1.667	1.667
Nitrogen	1.399	1.399	1.399	1.399	1.399	1.399	1.399	1.398	1.397	1.396	1.394	1.393	1.391
Oxygen	1.398	1.397	1.396	1.395	1.394	1.391	1.388	1.385	1.381	1.377	1.373	1.369	1.365
Propane	1.513	1.504	1.164	1.148	1.135	1.124	1.114	1.107	1.100	1.094	1.089	1.085	1.081
Propylene			1.171	1.160	1.150	1.140	1.131	1.123	1.116	1.110	1.105	1.100	1.096
Refrigerant 12			1.165	1.114	1.101	1.088	1.077	1.065	1.055	1.044	1.034	1.025	1.071
Refrigerant 21					1.179	1.164	1.152	1.144	1.137	1.132	1.127	1.124	1.120
Refrigerant 22				1.207	1.190	1.172	1.164	1.155	1.148	1.143	1.138	1.133	1.129
Steam								1.322	1.319	1.317	1.314	1.312	1.309
Xenon	1.623	1.634	1.642	1.650	1.655	1.659	1.662	1.662	1.662	1.662	1.662	1.662	1.662

Table 4.2.24   Saturation Temperature, in Kelvins, of Selected Substances

Pressure, bar	Acetone C₃H₆O	Acetylene, C₂H₂	Air, sat. liquid	Air, sat. vapor	Ammonia, NH₃	Aniline C₆H₇N	Argon, Ar	Benzene, C₆H₆	Butanol, C₄H₁₀O	n-Butane, C₄H₁₀	Carbon dioxide, CO₂	Carbon tetrachloride, CCl₄	Carbon tetrafluoride, CF₄
0.010	237.8					337.2			299.1				
0.015	243.8					344.2			305.4	196.5		255.1	
0.020	247.1					350.0			310.0	200.3		259.7	106.4
0.025	251.6					353.6			313.8	203.1		263.2	107.9
0.030	254.3					357.1			316.9	205.7		266.0	109.3
0.04	259.0					363.1		276.5	321.8	209.8		270.9	111.5
0.05	262.7					368.1		280.0	325.6	212.9		275.0	113.2
0.06	266.8					372.3		283.2	328.8	215.8		278.5	114.7
0.08	270.9				198.9	379.0		288.6	333.9	220.3		284.6	117.2
0.10	275.1		62.3	66.3	201.9	385.2		293.0	337.8	224.1		289.1	119.1
0.15	283.2		64.6	68.5	207.6	397.1		301.7	345.8	231.0		298.1	122.9
0.20	289.6		66.4	70.2	211.8	404.5		308.3	351.8	236.3		304.8	125.8
0.25	294.1		67.9	71.5	215.2	410.6		313.7	356.8	240.7		310.2	128.1
0.30	298.0		69.1	72.7	218.1	415.8		318.2	360.8	244.4		314.9	130.1
0.40	304.7		71.2	74.6	222.8	424.8		325.8	367.4	250.5		322.2	133.3
0.5	310.1		72.8	76.2	226.6	432.1		331.8	372.8	255.4		327.8	136.0
0.6	315.0		74.3	77.5	229.9	438.2		337.1	377.2	259.6		333.1	138.2
0.8	322.8		76.6	79.7	235.2	448.8	85.1	345.7	384.3	266.6		342.0	142.0
1.0	329.0		78.6	81.6	239.6	456.5	87.2	352.7	390.0	272.2		349.2	145.0
1.5	341.0	195.1	82.3	85.3	247.9	472.1	91.2	366.5	402.6	283.4		363.8	151.0
2.0	350.1	200.1	85.2	88.1	254.3	484.2	94.3	377.0	412.0	291.9		374.7	155.6
2.5	358.9	204.8	87.6	90.4	259.5	494.0	96.8	385.7	419.8	299.0		383.0	159.3
3.0	365.9	208.8	89.7	92.4	263.9	502.0	99.0	393.1	426.1	305.0		390.2	162.1
4.0	377.1	215.3	93.1	95.7	271.3	516.0	102.6	405.6	436.8	315.1		402.0	168.0
5.0	386.1	220.9	96.0	98.4	277.3	527.0	105.8	415.8	445.1	323.4		412.8	172.5
6	393.9	225.8	98.4	100.8	282.4	536.2	108.4	424.7	452.9	330.6	220.0	422.0	175.5
8	406.4	233.9	102.7	104.8	291.0	551.5	110.8	439.9	465.4	342.6	227.1	433.0	182.9
10	417.3	240.3	106.1	108.1	298.1	564.7	116.6	451.6	475.8	352.6	233.0	449.5	187.7
15	435.9	252.5	113.1	114.7	311.9	593.0	124.0	475.8	496.0	372.2	244.6	474.3	199.0
20	452.0	262.3	118.5	119.9	322.5	613.6	129.8	494.6	510.1	387.6	253.6	495.0	207.4
25	465.8	270.3	123.0	124.1	331.3	631.2	134.6	510.1		400.2	260.6	511.2	214.3
30	476.9	277.6	127.0	127.8	338.9	645.8	138.7	523.5		410.9	269.6	524.0	220.2
40	496.0	289.5			351.6	672.0	145.7	545.8			278.5	547.6	
50		299.0			362.1	693.0					287.4		
60		306.9			371.1						295.1		
80					386.1								
100					398.4								

Table 4.2.24   Saturation Temperature, in Kelvins, of Selected Substances   (Continued)

Pressure, bar	Chlorine, Cl_2	Cyclohexane, C_6H_{12}	Cyclopropane, C_3H_6	Diphenyl, $C_{12}H_{10}$	Ethane, C_2H_6	Ethanol C_2H_6O	Ethyl chloride C_2H_5Cl	Ethylene, C_2H_4	Fluorine, F_2	Helium, He	Heptane, C_7H_{16}	Hexane, C_6H_{14}	Hydrazine, N_2H_4
0.010			172.9	384.3	127.6	266.7		117.4	58.2		266.9	244.2	285.9
0.015	173.4		176.9	393.5	131.1	272.7	207.9	120.4	59.8		273.2	250.0	293.2
0.020	176.6		179.9	400.1	133.6	277.1	212.0	122.8	61.0		277.9	254.2	298.3
0.025	179.1		182.3	405.9	135.7	280.6	215.1	124.7	62.0		281.8	257.7	302.4
0.030	181.2		184.4	410.2	137.3	283.1	217.6	126.1	62.7		285.0	260.7	306.0
0.04	184.7		187.7	418.0	140.4	287.7	221.9	128.8	64.1		290.1	265.6	311.7
0.05	186.4		190.1	423.8	142.5	291.2	225.0	130.8	65.1		294.4	269.7	315.9
0.06	189.8	282.2	192.5	428.9	144.6	294.3	227.8	132.6	66.1	2.25	298.1	273.0	319.7
0.08	193.8	288.0	196.2	436.8	147.6	299.2	232.4	136.9	67.6	2.38	304.2	278.4	325.5
0.10	197.0	292.6	199.2	443.6	150.4	303.1	236.1	138.0	68.8	2.49	309.0	282.6	330.1
0.15	203.0	301.4	205.0	456.3	155.3	310.4	243.4	143.0	71.1	2.71	318.3	291.5	339.1
0.20	207.7	308.0	209.5	466.1	158.9	315.7	248.8	146.6	72.9	2.82	325.1	298.3	345.9
0.25	211.4	313.6	213.2	473.9	161.9	320.1	253.1	149.3	74.3	3.03	330.5	304.0	350.8
0.30	214.8	318.1	216.1	480.4	164.6	323.9	256.9	151.7	75.5	3.15	335.0	308.6	355.2
0.40	219.9	325.7	221.2	491.5	168.8	330.1	263.1	155.3	77.6	3.37	343.0	316.6	362.3
0.5	224.0	331.8	225.1	499.6	172.4	335.2	267.9	158.3	79.2	3.55	349.2	321.4	368.1
0.6	227.7	337.3	228.7	506.6	175.3	339.2	272.1	161.0	80.6	3.71	354.8	326.4	372.8
0.8	233.8	346.9	234.6	518.2	180.2	345.9	279.5	165.6	82.9	3.98	363.9	334.6	380.2
1.0	238.8	353.7	239.3	528.7	184.3	351.4	285.1	169.2	84.8	4.21	371.0	341.3	386.9
1.5	248.3	368.3	249.0	548.0	192.2	362.3	297.6	176.5	88.6	4.67	385.3	355.2	399.1
2.0	255.9	378.9	260	562.2	198.2	370.6	306.9	181.9	91.4	5.03	396.7	361.0	408.1
2.5	262.1	387.7	269	573.8	203.1	377.5	314.6	186.3	93.7		406.0	373.7	415.6
3.0	267.2	395.0	275	583.8	207.4	382.9	319.6	190.6	95.8		413.5	381.7	422.0
4.0	275.9	407.9	282	600.3	214.5	392.0	329.4	197.7	99.2		426.0	393.3	432.4
5.0	283.0	418.6	287	614.3	220.4	399.1	337.0	202.9	102.0		431.9	404.0	441.2
6	289.2	427.8	292	626.6	225.5	405.0	343.5	207.2	104.4		441.5	412.2	449.0
8	290.6	443.0	301	646.6	234.0	415.6	354.7	215.0	108.4		462.1	427.0	462.5
10	307.8	455.2	309	664.2	241.1	424.1	364.1	221.2	111.8		474.9	438.9	473.7
15	325.0	481.1	327	697.8	255.1	442.4	385.9	234.1	118.5		498.9	462.0	494.3
20	338.0	499.9	342	722.5	266.0	456.0	400.4	243.9	123.7		518.9	479.1	510.2
25	349.4	516.3	355	744.2	275.1	467.0	412.8	252.3	128.1		535.0	495.0	522.2
30	359.2	528.8	365	763.4	282.9	476.0	422.2	259.9	131.8			506.3	531.7
40	376.0	551.1	379	794.0	296.0	490.4	442.4	272.7	138.1				547.8
50	389.0			818.2		502.8	456.5	282.7	143.3				561.8
60	400.7					513.4							574.0
80													598.0
100													616.2

Table 4.2.24   Saturation Temperature, in Kelvins, of Selected Substances   (*Continued*)

Pressure, bar	r-Hydrogen, H_2	Hydrochloric acid, HCl	Hydrogen sulfide, H_2S	Mercury, Hg	Methane, CH_4	Methanol	Methyl chloride, CH_3Cl	Naphthalene	Nitrogen, N_2	Octane, C_8H_{18}	Oxygen, O_2	Pentane, C_5H_{12}	Potassium, K
0.010				448.6		252.6	173.2	353.0		286.4	62.2	219.1	699
0.015				460.5		258.2	178.4	361.2		293.8	63.0	225.0	719
0.020				468.9		262.3	182.6	367.6		299.4	64.3	228.8	732
0.025				475.3		265.6	185.9	372.8		303.7	65.2	232.0	743
0.030				481.6		268.4	188.6	376.9		306.8	66.2	234.7	752
0.04				491.2		272.9	192.7	384.3		312.4	67.6	239.2	764
0.05				498.2		276.5	196.0	389.9		316.9	68.8	242.8	776
0.06				504.7		278.8	198.6	394.8		320.8	69.8	245.9	788
0.08	14.0			515.2		284.4	202.8	402.7		327.1	71.4	250.8	811
0.10	14.4			523.6		288.4	206.0	408.9		332.0	72.7	254.9	829
0.15	15.2	161.0		538.9	92.6	295.9	211.9	420.9	64.1	341.6	75.2	263.0	861
0.20	15.8	164.5		551.0	95.0	301.4	216.5	429.5	65.8	348.7	77.1	269.1	883
0.25	16.3	167.2	189.8	561.0	97.0	305.9	220.5	437.1	67.2	354.6	78.7	274.0	899
0.30	16.7	169.3	192.5	569.0	98.6	309.7	223.8	443.3	68.3	359.7	80.0	277.9	916
0.40	17.5	172.7	197.0	582.2	101.4	315.9	229.1	454.0	70.2	368.3	82.2	284.9	940
0.5	18.1	175.2	200.4	592.7	103.7	320.8	233.4	462.1	71.8	374.9	83.9	290.0	960
0.6	18.6	177.5	203.1	601.8	105.6	325.0	237.0	469.4	73.2	380.6	85.5	294.7	977
0.8	19.5	181.0	208.0	616.8	108.8	331.9	243.1	481.1	75.4	390.2	88.0	302.1	1,006
1.0	20.2	187.8	212.5	628.9	111.5	337.5	248.0	490.2	77.2	398.0	90.1	308.6	1,029
1.5	21.7	195.6	222.0	652.1	116.6	348.1	257.9	509.0	80.8	413.3	94.1	321.6	1,074
2.0	22.8	201.5	227.9	670.0	120.6	356.1	265.4	523.3	83.6	424.6	97.2	331.1	1,108
2.5	23.8	206.1	234.3	684.5	123.9	362.6	271.5	535.0	85.9	434.4	99.8	339.2	1,137
3.0	24.6	210.0	237.9	696.7	126.7	368.1	276.6	544.8	87.9	442.1	102.0	345.7	1,176
4.0	26.0	216.7	245.1	717.1	131.4	377.2	285.5	560.3	91.2	456.3	105.7	356.7	1,203
5.0	27.1	222.0	250.9	733.5	135.3	384.6	292.6	573.2	94.0	467.4	108.8	365.4	1,238
6	28.1	226.2	256.1	747.8	138.7	390.9	298.9	584.7	96.4	477.0	111.4	373.6	1,268
8	29.8	233.6	264.9	770.8	144.4	401.4	309.4	605.0	100.4	494.0	115.9	387.6	1,318
10	31.2	241.6	272.1	790.2	149.1	410.0	318.1	620.7	103.8	507.0	119.6	398.2	1,359
15		254.2	283.3	827.5	158.5	426.5	332.7	652.6	110.4	532.4	127.0	420.7	1,440
20		263.7	299.8	857.2	165.8	439.1	348.6	677.9	115.6	553.6	132.8	436.8	1,505
25		272.3	309.4	879.2	172.0	449.4	359.4	699.0	119.9		137.6	450.4	1,558
30		279.9	316.7	901.0	177.2	458.2	368.6	718.4	123.6		141.7	462.1	1,606
40		291.9	330.0	934.4	186.1	473.0	384.3				148.6		1,689
50		301.1	340.4	962.8		484.9	397.2				154.4		1,759
60		309.5	350.3	987.6		495.3	408.5						1,819
80		322.9	366.9	1,028		511.9							1,920
100				1,065									2,000

Table 4.2.24 Saturation Temperature, in Kelvins, of Selected Substances (Continued)

Pressure, bar	Propane, C_3H_8	Propanol, C_3H_8O	Propylene, C_3H_6	Refrigerant 11, $CFCl_3$	Refrigerant 12, CF_2Cl_2	Refrigerant 13, CF_3Cl	Refrigerant 21, $CHFCl_2$	Refrigerant 22, CHF_2Cl	Sodium, Na	Sulfur dioxide, SO_2	Toluene, C_7H_8	Water, H_2O
0.010	162.0	284.0	157.8	209.9	171.5	134.1	201.8	165.7	804	190.6	275.1	280.1
0.015	166.2	289.8	161.9	215.0	175.8	137.6	206.6	170.0	825	195.9	282.0	286.1
0.020	169.4	294.0	165.1	219.1	179.1	140.3	210.5	173.1	841	199.6	286.3	290.6
0.025	171.9	297.3	167.4	222.2	181.8	142.4	212.9	175.6	854	202.6	290.0	294.2
0.030	174.1	299.9	169.6	224.7	184.1	144.2	215.5	177.8	865	204.8	293.4	297.2
0.04	177.4	304.2	172.8	229.0	187.8	147.1	219.6	180.9	883	208.7	298.9	302.1
0.05	180.1	307.8	175.7	232.6	190.3	149.4	222.9	183.5	898	211.8	303.3	306.0
0.06	182.4	310.7	177.9	235.5	193.0	151.3	225.7	185.8	910	214.3	307.1	309.3
0.08	186.2	315.7	181.7	240.7	197.0	154.6	230.2	189.6	930	218.6	313.7	314.7
0.10	189.4	319.7	184.9	244.7	200.1	157.1	233.9	192.6	946	222.0	318.6	319.0
0.15	195.6	327.4	190.7	252.1	206.3	162.1	241.0	198.5	977	228.2	328.5	327.3
0.20	200.1	333.2	195.2	257.8	211.1	165.9	246.3	202.9	1,000	232.9	335.1	333.2
0.25	203.9	337.8	198.7	262.4	214.9	169.0	250.6	206.6	1,019	236.2	341.6	338.1
0.30	207.0	341.7	201.7	266.4	218.2	171.7	254.2	209.6	1,034	239.8	346.3	342.3
0.40	212.1	348.1	206.8	273.1	223.5	176.0	260.2	214.7	1,061	244.8	354.1	349.0
0.5	216.3	352.9	211.0	278.2	227.9	179.5	265.1	218.6	1,082	248.7	360.3	354.5
0.6	219.9	357.3	214.5	282.8	231.7	182.5	269.2	220.2	1,099	252.2	366.1	359.1
0.8	225.9	364.6	221.4	290.3	237.9	187.1	276.1	227.7	1,129	258.0	375.6	366.7
1.0	230.7	370.1	225.2	296.6	243.0	191.4	281.7	232.2	1,153	262.8	383.2	372.8
1.5	240.3	381.4	234.5	308.6	253.0	199.3	292.6	241.2	1,199	272.6	398.2	384.5
2.0	247.7	389.9	241.6	317.9	260.6	205.4	299.4	248.1	1,235	279.6	409.3	393.4
2.5	253.2	396.9	247.3	325.5	266.9	210.4	307.8	253.8	1,264	285.4	419.0	400.6
3.0	258.9	402.3	252.6	331.8	272.3	214.6	313.6	258.6	1,289	290.0	427.0	406.7
4.0	267.6	411.9	262.6	342.7	281.3	221.8	323.5	266.5	1,330	298.6	441.7	416.8
5.0	274.8	419.8	267.2	351.1	288.8	227.1	331.6	273.0	1,364	305.3	451.4	425.0
6	281.0	426.8	273.9	358.8	295.2	232.8	338.5	279.0	1,393	311.1	460.9	432.0
8	291.4	438.5	284.0	372.2	306.0	241.4	350.2	289.1	1,430	320.7	476.8	445.6
10	300.0	447.8	292.3	382.7	314.9	248.5	361.0	297.1	1,480	328.6	490.0	453.0
15	317.0	466.6	308.9	403.7	332.6	262.7	379.1	312.3	1,556	345.2	515.8	472.0
20	330.3	481.4	321.8	419.2	346.3	273.7	394.0	324.9	1,623	357.8	535.5	485.5
25		493.6	332.5	432.1	357.5	282.8	406.4	335.1	1,676	368.2	552.3	497.1
30		503.8	341.7	444.8	367.2	290.7	417.1	343.4	1,720	377.0	566.4	507.0
40		511.3	357.0	463.9	383.3		434.7	358.3	1,795	391.6	589.6	523.5
50		536.1					449.2		1,859	403.8		537.1
60									1,913	414.5		548.7
80									2,010			568.1
100									2,085			584.1

Table 4.2.25 Color Scale of Temperature for Iron or Steel

	Temperature	
Color	°F	K
Dark blood red, black red	1,000	810
Dark red, blood red, low red	1,050	840
Dark cherry red	1,175	910
Medium cherry red	1,250	950
Cherry, full red	1,375	1,020
Light cherry, light red	1,550	1,120
Orange, free scaling heat	1,650	1,170
Light orange	1,725	1,210
Yellow	1,825	1,270
Light yellow	1,975	1,350
White	2,200	1,475

Table 4.2.26 Melting Points of Refractory Materials

	Temperature	
Material	°F	K
Aluminum nitride, AlN	4,060	2,500
Aluminum oxide, Al_2O_3	3,720	2,320
Aluminum oxide–beryllium oxide, Al_2O_3-BeO	3,400	2,140
Beryllium carbide, Be_2C	3,810	2,370
Beryllium nitride, Be_3N_4	4,000	2,480
Beryllium oxide, BeO	4,570	2,790
Beryllium silicide, $2BeO \cdot SiO_2$	3,630	2,270
Borazon, BN	5,430	3,270
Calcia (lime), CaO	4,660	2,840
Graphite, C	6,700	3,980
Hafnia, HfO_2	5,090	3,090
Magnesia, MgO	5,070	3,070
Niobium carbide, NbC	6,330	3,770
Silica, SiO_2	3,110	1,980
Silicon carbide, SiC	3,990	2,470
Thoria, ThO_2	5,830	3,490
Titanium carbide, TiC	5,680	3,410
Zirconium aluminide, $ZrAl_2$	3,000	1,920
Zirconium beryllide, $ZrBe_{13}$	3,180	2,020
Zirconium carbide, ZrC	6,400	3,810
Zirconium disilicide, $ZrSi_2$	3,090	1,970
Zirconium nitride, ZrN	5,400	3,260
Zirconium oxide, ZrO_2	4,900	2,980
Zirconium silicides, $Zr_3Si_2, Zr_4Si_3, Zr_6Si_5$	≈4,050	≈2,500

Table 4.2.27 Mean Specific Heats of Various Solids (32–212°F, 273–373 K)

Solid	c, Btu/(lb · °F)	c, kJ/(kg · K)
Alumina	0.183	0.77
Asbestos	0.20	0.84
Ashes	0.20	0.84
Bakelite	≈0.35	1.50
Basalt (lava)	0.20	0.84
Bell metal	0.086	0.36
Bismuth-tin	0.043	0.18
Borax	0.229	0.96
Brass, yellow	0.088	0.37
Brass, red	0.090	0.38
Brick	0.22	0.92
Bronze	0.104	0.44
Carbon-coke	0.203	0.85
Chalk	0.215	0.90
Charcoal	0.20	0.84
Cinders	0.18	0.75
Coal	≈0.30	≈1.25
Concrete	0.156	0.65
Constantan	0.098	0.41
Cork	0.485	2.03
Corundum	0.198	0.83
D'Arcet's metal	0.050	0.21
Dolomite	0.222	0.93
Ebonite	0.33	1.38
German silver	0.095	0.40
Glass, crown	0.16	0.70
Glass, flint	0.12	0.50
Glass, normal	0.20	0.84
Gneiss	0.18	0.75
Granite	0.20	0.84
Graphite	0.20	0.84
Gypsum	0.26	1.10
Hornblende	0.20	0.84
Humus (soil)	0.44	1.80
India rubber (para)	≈0.37	1.50
Kaolin	0.224	0.94
Lead oxide (PbO)	0.055	0.23
Limestone	0.217	0.91
Lipowitz's metal	0.040	0.17
Magnesia	0.222	0.93
Magnesite (Fe_3O_4)	0.168	0.70
Marble	0.210	0.88
Nickel steel	0.109	0.46
Paraffin wax	0.69	2.90
Porcelain	0.22	0.92
Quartz	≈0.23	0.96
Quicklime	0.217	0.91
Rose's metal	0.050	0.21
Salt, rock	0.21	0.88
Sand	0.195	0.82
Sandstone	0.22	0.92
Serpentine	0.25	1.05
Silica	0.191	0.80
Soda	0.231	0.97
Solders (Pb + Sn)	0.043	0.18
Sulfur	0.180	0.75
Talc	0.209	0.87
Tufa	0.33	1.40
Type metal	0.039	0.16
Vulcanite	0.331	1.38
Wood, fir	0.65	2.70
Wood, oak	0.57	2.40
Wood, pine	0.67	2.80
Wood's metal	0.040	0.17

Table 4.2.28   Phase Transition and Other Data for the Elements

Name	Symbol	Formula weight	T_m, K	Δh_{fus}, kJ/kg	T_b, K	Δh_{vap}, kJ/kg	T_c, K	P_c, bar	T_{tr}, K
Actinium	Ac	227.028	1,323	63	3,475	1,750			
Aluminum	Al	26.9815	933.5	398	2,750	10,875	7,850	4,800	
Antimony	Sb	121.75	903.9	163	1,905		5,700	3,200	368, 686
Argon	Ar	39.948	83	30	87.2	163	151	50	
Arsenic	As	74.9216					1,703	2,100	
Barium	Ba	137.33	1,002	55.8		1,099	4,450	720	643
Beryllium	Be	9.01218	1,560	1,355	2,750	32,450	6,200	4,600	1,530
Bismuth	Bi	208.980	544.6	54.0	1,838	725	4,450	1,400	
Boron	B	10.81	2,320	1,933	4,000		3,300		1,473
Bromine	Br	159.808	266	66.0	332	188	584		
Cadmium	Cd	112.41	594	55.1	1,040	886	2,690	1,680	
Calcium	Ca	40.08	1,112	213.1	1,763	3.833	4,300	1,000	720
Carbon	C	12.011	3,810		4,275		7,200	11,500	
Cerium	Ce	140.12	1,072	390		2,955	9,750	3,350	103, 263, 1,003
Cesium	Cs	132.905	301.8	16.4	951	496	2,015	125	
Chlorine	Cl$_2$	70.906	172	180.7	239	576	417		
Chromium	Cr	51.996	2,133	325.6	2,950	6.622	5,500		2,113
Cobalt	Co	58.9332	1,766	274.7	3,185	6,390	6,300		700, 1,400
Copper	Cu	63.546	1,357	206.8	2,845	4,726	8,280	7,400	
Dysprosium	Dy	162.50	1,670	68.1	2,855	1,416	6,925	2,500	1,659
Erbium	Er	167.26	1,795	119.1	3,135	1,563	7,250		1,643
Europium	Eu	151.96	1,092	60.6	1,850	944	4,350	690	
Fluorine	F$_2$	37.997	53.5	13.4	85.0	172	144		46
Gadolinium	Gd	157.25	1,585	63.8	3,540	2,285	8,670		1,537
Gallium	Ga	69.72	303	80.1	2,500	3,688	7,125	4,150	276
Germanium	Ge	72.59	1,211	508.9	3,110	4,558	8,900	5,300	
Gold	Au	196.967	1,337	62.8	3,130	1,701	7,250	5,450	
Hafnium	Hf	178.49	2,485	134.8	4,885	3,211	10,400		2,000
Helium	He	4.00260	3.5	2.1	4.22	21	5.2	2.3	2.2
Holmium	Ho	164.930	1,744	73.8	2,968	1,461	7,575		1,703
Hydrogen	H$_2$	2.0159	14.0		20.4				
Indium	In	114.82	430	28.5	2,346	2,019	6,150	2,550	
Iodine	I$_2$	253.809	387	125.0	457		785		
Iridium	Ir	192.22	2,718	13.7	4,740	3,185	7,800		
Iron	Fe	55.847	1,811	247.3	3,136	6,259	8,500	10,000	1,183, 1,671
Krypton	Kr	83.80	115.8	19.6	119.8	108	209.4	55	
Lanthanum	La	138.906	1,194	44.6	3,715	2,978	10,500		550, 1,134
Lead	Pb	207.2	601	23.2	2,025	858	5,500	1,650	
Lithium	Li	6.941	454		1,607	21,340	3,700	1,000	80
Lutetium	Lu	174.967	1,937	106.6	3.668	2,034			
Magnesium	Mg	24.305	922	368.4	1,364	5,242	3,850	1,750	
Manganese	Mn	54.9380	1,518	219.3	2,334	4,112	4,325	560	1,374, 1,447
Mercury	Hg	200.59	234.6	11.4	630	293	1,720	1,500	194
Molybdenum	Mo	95.94	2,892	290.0	4,900		1,450	12,000	
Neodymium	Nd	144.24	1,290	49.6	3,341	1,891	7,900		1,132, 1,297
Neon	Ne	20.179	24.5	16.4	27.1	89	44.5	26.6	
Neptunium	Np	237.048	910		4,160		12,000		551, 847
Nickel	Ni	58.70	1,728	297.6	3,190	6,308	8,000	11,100	631
Niobium	Nb	92.9064	2,740	283.7	5,020	7,341	12,500		
Nitrogen	N$_2$	28.013	63.2	25.7	77.3	198	126.2	34.0	35.6
Osmium	Os	190.2	3,310	150.0	5,300	3,310	12,700		
Oxygen	O$_2$	31.9988	54.4	13.8	90.2	213	154.8		23.8, 43.8
Palladium	Pd	106.4	1,826	165.0	3,240	3,358	7,700	7,100	
Phosphorus	P	30.9738	317		553		995	81	196, 298
Platinum	Pt	195.09	2,045	101	4,100	2,612	10,700	11,000	
Plutonium	Pu	244	913	11.7	3,505	1,409	10,500	3,250	395, 480, 588, 730
Potassium	K	39.0983	336.4	60.1	1,032	2,052	2,210	170	
Praseodymium	Pr	140.908	1,205	49	3,785	2,105	8,900		1,066
Promethium	Pm	145	1,353		2,730				
Protactinium	Pa	231	1,500	64.8	4,300	2,036			
Radium	Ra	226.025	973		1,900				
Radon	Rn	222	202	12.3	211	83	377	66	
Rhenium	Re	186.207	3,453	177.8	5,920	3,842	18,900	14,900	
Rhodium	Rh	102.906	2,236	209.4	3,980	4,798	7,000		
Rubidium	Rb	85.4678	312.6	26.4	964	810	2,070	168	

Table 4.2.28   Phase Transition and Other Data for the Elements   (Continued)

Name	Symbol	Formula weight	T_m, K	Δh_{fus}, kJ/kg	T_b, K	Δh_{vap}, kJ/kg	T_c, K	P_c, bar	T_{tr}, K
Ruthenium	Ru	101.07	2,525	256.3	4,430	5,837	9,600		1,300, 1,475, 1,775
Samarium	Sm	150.4	1,345	57.3	2,064	1,107	5,050	1,780	1,190
Scandium	Sc	44.9559	1,813	313.6	3,550	6,989	6,410	3,750	1,608
Selenium	Se	78.96	494	66.2	958	1,210	1,810	320	398, 425
Silicon	Si	28.0855	1,684	1,802	3,540	14,050	5,160	540	
Silver	Ag	107.868	1,234	104.8	2,435	2,323	6,400	4,450	
Sodium	Na	22.9898	371	113.1	1,155	4,263	2,500	370	
Strontium	Sr	87.62	1,043	1,042	1,650	1,585	4,275	375	505, 893
Sulfur	S	32.06	388	53.4	718		1,210	130	369, 374
Tantalum	Ta	180.948	3,252	173.5	5,640	4,211	16,500	12,000	
Technetium	Tc	98	2,447	232	4,550	5,830	11,500		
Tellurium	Te	127.60	723	137.1	1,261	895	2,330		
Terbium	Tb	158.925	1,631	67.9	3,500	2,083	8,470		228, 1,575
Thallium	Tl	204.37	577	20.1	1,745	806	4,550	1,700	507
Thorium	Th	232.038	2,028	69.4	5,067	2,218	14,400	6,165	1,670
Thulium	Tm	168.934	1,819	99.6	2,220	1,129	6,450		
Tin	Sn	118.69	505	58.9	2,890	2,496	7,700	2,250	286, 476
Titanium	Ti	47.90	1,943	323.6	3,565	8,787	5,850		1,162, 1,353
Tungsten	W	183.85	3,660	192.5	5,890	4,483	15,500	15,000	
Uranium	U	238.029	1,406	35.8	4,422	1,949	12,500	5,000	938, 1,046
Vanadium	V	50.9415	2,191	410.7	3,680	8,870	11,300	10,300	
Xenon	Xe	131.30	161.3	17.5	164.9	96	290	58	
Ytterbium	Yb	173.04	1,098	44.2	1,467	745	4,080	1,150	1,050
Yttrium	Y	88.9059	1,775	128.2	3,610	4,485	8,950		1,758
Zinc	Zn	65.38	692.7	113.0	1,182	1,768			
Zirconium	Zr	91.22	2,125	185.3	4,681	6,376	10,500		1,135

Formula weight = molecular weight, T_m = normal melting temperature, Δh_{fus} = enthalpy (or latent heat) of fusion, T_b = normal boiling-point temperature, Δh_{vap} = enthalpy (or latent heat) of vaporization, T_c = critical temperature, P_c = critical pressure, T_{tr} = transition temperature.
SOURCE: Prepared by the author and abstracted from Rohsenow et al., "Handbook of Heat Transfer Fundamentals," McGraw-Hill.

Table 4.2.29 Thermophysical Properties of Selected Solid Elements

Element	Property	Temperature, K							
		100	200	300	400	500	600	800	1,000
Al	P, bar			2.1. − 43	4.9. − 31	1.1. − 23	1.0. − 18	1.4. − 12	6.6. − 9
	ρ, kg/m³	2,732	2,719	2,701	2,681	2,661	2,639	2,591	2,365
	h, kJ/kg			4,623	4,716	4,812	4,913	5,131	5,768
	s, kJ/(kg · K)			1.056	1.323	1.539	1.723	2.035	2.728
	c_p, kJ/(kg · K)	0.481	0.797	0.902	0.949	0.997	1.042	1.134	0.921
	λ, W/(m · K)	300	237	237	240	236	231	218	
	α, m²/s	2.3. − 4	1.1. − 4	9.7. − 5	9.4. − 5	8.9. − 5	8.4. − 5	7.4. − 5	6.6. − 5
Cr	P, bar			4.6. − 62	8.9. − 45	2.1. − 34	1.6. − 27	6.3. − 19	9.1. − 14
	ρ, kg/m³	7,155	7,145	7,135	7,120	7,110	7,080	7,040	7,000
	h, kJ/kg			4,113	4,160	4,210	4,263	4,375	4,495
	s, kJ/(kg · K)			0.457	0.591	0.703	0.800	0.962	1.094
	c_p, kJ/(kg · K)	0.190	0.382	0.450	0.501	0.537	0.565	0.611	0.653
	λ, W/(m · K)	160	110	94	91	86	81	71	65
	α, m²/s	1.2. − 4	4.1. − 5	2.9. − 5	2.6. − 5	2.3. − 5	2.0. − 5	1.7. − 5	1.4. − 5
Cu	P, bar			1.1. − 52	9.0. − 38	5.5. − 29	3.8. − 23	7.6. − 16	1.7. − 11
	ρ, kg/m³	9,009	8,973	8,930	8,884	8,837	8,787	8,686	8,568
	h, kJ/kg			5,067	5,106	5,146	5,188	5,273	5,361
	s, kJ/(kg · K)			0.524	0.637	0.726	0.802	0.924	1.022
	c_p, kJ/(kg · K)	0.254	0.357	0.386	0.396	0.406	0.431	0.448	0.466
	λ, W/(m · K)	480	413	401	393	386	379	366	352
	α, m²/s	2.2. − 4	1.3. − 4	1.2. − 4	1.1. − 4	1.1. − 4	1.0. − 4	9.0. − 5	8.0. − 5
Au	P, bar			6.7. − 58	6.3. − 42	2.8. − 32	3.6. − 25	6.5. − 18	3.7. − 13
	ρ, kg/m³	19,460	19,380	19,300	19,210	19,130	19,040	18,860	18,660
	h, kJ/kg			6,046	6,059	6,072	6,086	6,113	6,142
	s, kJ/(kg · K)			0.241	0.279	0.309	0.333	0.373	0.404
	c_p, kJ/(kg · K)	0.109	0.124	0.129	0.131	0.133	0.136	0.141	0.147
	λ, W/(m · K)	327	323	317	311	304	298	284	270
	α, m²/s	1.5. − 4	1.34. − 4	1.27. − 4	1.23. − 4	1.19. − 4	1.15. − 4	1.07. − 4	9.8. − 5
Fe	P, bar			3.1. − 65	6.3. − 54	3.9. − 47	1.5. − 36	6.6. − 20	1.5. − 14
	ρ, kg/m³	7,900	7,880	7,860	7,830	7,800	7,760	7,690	7,650
	h, kJ/kg			4,523	4,570	4,621	4,676	4,801	4,958
	s, kJ/(kg · K)			0.491	0.626	0.740	0.840	1,018	1.193
	c_p, kJ/(kg · K)	0.216	0.384	0.450	0.491	0.524	0.555	0.692	1.034
	λ, W/(m · K)	134	94	80	70	61	55	43	32
	α, m²/s	8.2. − 5	3.1. − 5	2.2. − 5	1.8. − 5	1.5. − 5	1.3. − 5	1.1. − 5	1.0. − 5
Pb	P, bar			6.3. − 29	1.8. − 20	2.2 − 15	5.4. − 12	6.2. − 8	1.6. − 5
	ρ, kg/m³	11,520	11,430	11,330	11,230	11,130	11,010	10,430	10,190
	h, kJ/kg			6,929	6,942	6,955	6,969	7,022	7,050
	s, kJ/(kg · K)			0.314	0.351	0.381	0.406	0.487	0.519
	c_p, kJ/(kg · K)	0.118	0.125	0.129	0.132	0.137	0.142	0.145	0.142
	λ, W/(m · K)	39.7	36.7	35.3	34.0	32.8	31.4		
	α, m²/s	2.9. − 5	2.6. − 5	2.4. − 5	2.3. − 5	2.2. − 5	2.0. − 5	1.3. − 5	1.5. − 5
Li	P, bar								
	ρ, kg/m³	546	541	533	526	492	482	462	442
	h, kJ/kg			4,615	4,919	5,844	6,274	7,113	7,945
	s, kJ/(kg · K)			4.214	5.289	7.182	7.967	9.173	10.102
	c_p, kJ/(kg · K)	1.923	3.105	3.54	3.76	4.34	4.26	4.17	4.15
	λ, W/(m · K)	105	90	85	80				
	α, m²/s	1.0. − 4	5.4. − 5	4.5. − 5	3.2. − 5	2.1. − 5	2.3. − 5	2.8. − 5	3.3. − 5
Ni	P, bar			1.1. − 67	5.8. − 49	9.8. − 38	2.7. − 30	5.5. − 21	2.1. − 15
	ρ, kg/m³	8,960	8,930	8,900	8,860	8,820	8,780	8,690	8,610
	h, kJ/kg			4,837	4,883	4,934	4,990	5,099	5,207
	s, kJ/(kg · K)			0.512	0.645	0.758	0.859	1.017	1.137
	c_p, kJ/(kg · K)	0.232	0.383	0.444	0.490	0.540	0.590	0.530	0.556
	λ, W/(m · K)	165	105	91	80	72	66	68	72
	α, m²/s	8.0. − 5	3.1. − 5	2.3. − 5	1.9. − 5	1.5. − 5	1.3. − 5	1.4. − 5	1.5. − 5
Pt	P, bar			3.2. − 91	1.3. − 66	7.3. − 52	5.0. − 42	9.7. − 30	2.3. − 22
	ρ, kg/m³	21,550	21,500	21,450	21,380	21,330	21,270	21,140	21,010
	h, kJ/kg			5,837	5,850	5,864	5,878	5,907	5,937
	s, kJ/(kg · K)			0.214	0.253	0.283	0.309	0.350	0.383
	c_p, kJ/(kg · K)	0.101	0.127	0.134	0.136	0.138	0.140	0.146	0.152
	λ, W/(m · K)	78	73	72	72	72	73	76	79
	α, m²/s	3.6. − 5	2.7. − 5	2.5. − 5	2.5. − 5	2.5. − 5	2.5. − 5	2.5. − 5	2.5. − 5

Table 4.2.29   Thermophysical Properties of Selected Solid Elements   (Continued)

Element	Property	Temperature, K							
		100	200	300	400	500	600	800	1,000
Rh	P, bar			5.5. − 89	6.5. − 65	1.8. − 50	7.1. − 41	7.0. − 29	1.1. − 21
	ρ, kg/m^3	12,480	12,460	12,430	12,400	12,360	12,330	12,250	12,170
	h, kJ/kg			4,921	4,946	4,972	4,999	5,055	5,115
	s, kJ/(kg · K)			0.308	0.380	0.437	0.487	0.568	0.635
	c_p, kJ/(kg · K)	0.147	0.220	0.246	0.257	0.265	0.274	0.290	0.307
	λ, W/(m · K)	190	154	150	146	141	136	127	121
	α, m^2/s	1.0. − 4	5.6. − 5	4.9. − 5	4.6. − 5	4.3. − 5	4.0. − 5	3.6. − 5	3.2. − 5
Ag	P, bar			2.1. − 43	4.9. − 31	1.1. − 23	1.0. − 18	1.4. − 12	6.6. − 9
	ρ, kg/m^3	10,600	10,550	10,490	10,430	10.360	10,300	10,160	10,010
	h, kJ/kg			5,791	5,815	5,839	5,864	5,915	5,968
	s, kJ/(kg · K)			0.3959	0.4641	0.5180	0.5630	0.6365	0.6964
	c_p, kJ/(kg · K)	0.187	0.225	0.236	0.240	0.245	0.251	0.264	0.276
	λ, W/(m · K)	450	430	429	425	419	412	396	379
	α, m^2/s	2.3. − 4	1.8. − 4	1.7. − 4	1.7. − 4	1.7. − 4	1.6. − 4	1.5 − 4	1.3. − 4
Ti	P, bar			1.0. − 74	4.6. − 54	5.2. − 42	7.2. − 35	1.2. − 23	1.4. − 17
	ρ, kg/m^3	4,530	4,520	4,510	4,490	4,480	4,470	4,440	4,410
	h, kJ/kg			4,857	4,911	4,967	5,025	5,147	5,278
	s, kJ/(kg · K)			0.643	0.797	0.922	1.028	1.205	1.350
	c_p, kJ/(kg · K)	0.295	0.464	0.525	0.555	0.578	0.597	0.627	0.670
	λ, W/(m · K)	31	25	21	20	20	19	19	21
	α, m^2/s								
W	P, bar			3.2. − 141	2.9. − 104	4.3. − 82	2.7. − 67	8.7. − 49	1.1. − 37
	ρ, kg/m^3	19,310	19,290	19,270	19,240	19,220	19,190	19,130	19,080
	h, kJ/kg			6,255	6,268	6,282	6,296	6,325	6,354
	s, kJ/(kg · K)			0.178	0.217	0.248	0.273	0.315	0.347
	c_p, kJ/(kg · K)	0.089	0.125	0.135	0.137	0.139	0.140	0.144	0.148
	λ, W/(m · K)	208	185	174	159	146	137	125	118
	α, m^2/s								
V	P, bar			3.0. − 82	7.1. − 60	1.9. − 46	1.6. − 37	2.4. − 26	1.2. − 19
	ρ, kg/m^3	6,074	6,062	6,050	6,030	6,010	6,000	5,960	5,920
	h, kJ/kg			4,740	4,790	4,843	4,896	5,006	5,121
	s, kJ/(kg · K)			0.571	0.716	0.832	0.930	1.088	1.216
	c_p, kJ/(kg · K)	0.257	0.434	0.483	0.512	0.528	0.540	0.563	0.598
	λ, W/(m · K)	36	31	31	31	32	33	36	38
	α, m^2/s	2.3. − 5	1.2. − 5	1.1. − 5	1.0. − 5	1.0. − 5	1.0. − 5	1.1. − 5	1.1. − 5
Zn	P, bar			3.7. − 17	1.6. − 11	3.7. − 8	6.7. − 6	3.4. − 3	1.2. − 1
	ρ, kg/m^3	7,260	7,200	7,135	7,070	7,000	6,935	6,430	6,260
	h, kJ/kg			5,690	5,730	5,771	5,813	5,970	6,114
	s, kJ/(kg · K)			0.639	0.753	0.844	0.922	1.216	1.323
	c_p, kJ/(kg · K)	0.295	0.366	0.389	0.404	0.419	0.435	0.479	
	λ, W/(m · K)	117	118	116	111	107	103		
	α, m^2/s	5.5. − 5	4.7. − 5	4.1. − 5	3.9. − 5	3.7. − 5	3.4. − 5	1.8. − 5	2.2. − 5
Zr	P, bar			2.8. − 99	8.6. − 73	6.6. − 57	2.7. − 46	4.6. − 33	4.1. − 25
	ρ, kg/m^3	6,535	6,525	6,515	6,510	6,490	6,480	6,450	6,420
	h, kJ/kg			5,540	5,569	5,600	5,632	5,698	5,768
	s, kJ/(kg · K)			0.429	0.513	0.590	0.640	0.735	0.813
	c_p, kJ/(kg · K)	0.120	0.126	0.130	0.136	0.143	0.153	0.153	0.153
	λ, W/(m · K)	33	25	23	22	21	21	21	23
	α, m^2/s								

P = saturation vapor pressure; ρ = density; h = enthalpy; s = entropy; c_p = specific heat at constant pressure; λ = thermal conductivity; α = thermal diffusivity.

4.3 RADIANT HEAT TRANSFER
by Hoyt C. Hottel and Adel F. Sarofim

REFERENCES: Hottel and Sarofim, "Radiative Transfer," McGraw-Hill. Siegel and Howell, "Thermal Radiation Heat Transfer," McGraw-Hill, 3d ed. Modest, "Radiative Heat Transfer," McGraw-Hill.

A heated body loses energy continuously by radiation, at a rate dependent on the shape, the size, and, particularly, the temperature of the body. In contrast to conductive energy transport, such emitted radiation is capable of passage to a distant body, where it may be absorbed, reflected, scattered, or transmitted.

Consider a pencil of radiation, defined as all the rays passing through each of two small, widely separated areas dA_1 and dA_2. The rays at dA_1 will have a solid angle of divergence $d\Omega_1$, equal to the apparent area of dA_2 viewed from dA_1, divided by the square of the separating distance. Let the normal to dA_1 make the angle θ_1 with the pencil. The flux density q [energy/(time)(area normal to beam)] per unit solid angle of divergence is called the **intensity I**, and the flux dQ_1 (energy/time) through area dA_1 (of apparent area $dA_1 \cos \theta_1$ normal to the beam) is therefore given by

$$\dot{Q}_1 = dA_1 \cos \theta_1 q_1 = I \, dA_1 \cos \theta_1 \, d\Omega_1 \quad (4.3.1)$$

The intensity I along a pencil, in the absence of absorption or scatter is constant (unless the beam passes into a medium of different refractive index n; $I_1/n_1^2 = I_2/n_2^2$). The **emissive power** of a surface is the flux density [energy/(time)(surface area)] due to emission from it throughout a hemisphere. If the intensity I of emission from a surface is independent of the angle of emission, Eq. (4.3.1) may be used to show that the surface emissive power is πI, though the emission is throughout 2π steradians.

BLACKBODY RADIATION

Engineering calculations of thermal radiation from surfaces are best keyed to the radiation characteristics of the **blackbody**, or **ideal radiator**. The characteristic properties of a blackbody are that it absorbs all the radiation incident on its surface and that the quality and intensity of the radiation it emits are completely determined by its temperature. The total radiative flux throughout a hemisphere from a black surface of area A and absolute temperature T is given by the **Stefan-Boltzmann law:** $Q = A\sigma T^4$ or $q = \sigma T^4$. The Stefan-Boltzmann constant σ has the value 5.67×10^{-8} W/m²(K)⁴, 0.1713×10^{-8} Btu/(ft)²(h)(°R)⁴ or 1.356×10^{-12} cal/(cm)²(s)(K)⁴. From the above definition of emissive power, σT^4 is the total emissive power of a blackbody, called E; and the intensity I_B of emission from a blackbody is E/π, or $\sigma T^4/\pi$.

The spectral distribution of energy flux from a blackbody is expressed by Planck's law

$$E_\lambda \, d\lambda = \frac{2\pi h c^2 n^2 \lambda^{-5}}{e^{hc/(k\lambda T)} - 1} \, d\lambda \equiv \frac{n^2 c_1 \lambda^{-5}}{e^{c_2/(\lambda T)} - 1} \quad (4.3.2)$$

wherein $E_\lambda \, d\lambda$ is the hemispherical flux density in W/m² lying in the wavelength range λ to $\lambda + d\lambda$; h is Planck's constant, 6.6262×10^{-34} J·s; c is the velocity of light in vacuo, 2.9979×10^8 m/s; k is the Boltzmann constant, 1.3807×10^{-23} J/K; λ is the wavelength measured in vacuo, m; n is the refractive index of the emitter; c_1 and c_2, the first and second Planck's law constants, are 3.7418×10^{-16} W·m² and 1.4388×10^{-2} m·K. To show how E_λ varies with wavelength or temperature, Planck's law may be cast in the form

$$\frac{E_\lambda}{n^2 T^5} = \frac{c_1 (\lambda T)^{-5}}{e^{c_2/(\lambda T)} - 1} \quad (4.3.3)$$

* Variously called, in the literature, **emittance, total hemispherical intensity, radiant flux density** or **exitance.**

i.e., when $n \cong 1$ (e.g., in a gas), E_λ/T^5 is a unique function of λT. And E_λ is a maximum at $\lambda T = 2,898 \ \mu m \cdot K$ (Wien's displacement law). A more useful displacement law: Half of blackbody radiation lies on either side of $\lambda T = 4,107 \ \mu m \cdot K$. Another: The maximum intensity per unit fractional change in wavelength *or* frequency is at $\lambda T = 3,670 \ \mu m \cdot K$. Integration of E_λ over λ shows that the fraction f of blackbody radiation lying at wavelengths below λ depends only on λT. Values of f versus λT appear in Table 4.3.1. A twofold range of λT geometrically centered on $\lambda T = 3,670 \ \mu m \cdot K$ spans about half the energy.

A limiting form of the Planck equation as $\lambda T \to 0$ is $E_\lambda = n^2 c_1 \lambda^{-5} e^{-c_2/(\lambda T)}$, the Wien equation, less than 1 percent in error when λT is less than $3,000 \ \mu m \cdot K$. This is useful for optical pyrometry (red screen $\lambda = 0.65 \ \mu m$) when $T < 4,800$ K.

Table 4.3.1 Fraction f of Blackbody Radiation below λ
$\lambda T = \mu m \cdot K$

λT	1,200	1,600	1,800	2,000	2,200	2,400	2,600	2,800
f	0.002	0.020	0.039	0.067	0.101	0.140	0.183	0.228
λT	3,000	3,200	3,400	3,600	3,800	4,000	4,200	4,500
f	0.273	0.318	0.362	0.404	0.443	0.480	0.516	0.564
λT	4,800	5,100	5,500	6,000	6,500	7,000	7,600	8,400
f	0.608	0.646	0.691	0.738	0.776	0.808	0.839	0.871
λT	10,000	12,000	14,000	20,000	50,000			
f	0.914	0.945	0.963	0.986	0.999			

RADIATIVE EXCHANGE BETWEEN SURFACES OF SOLIDS

The ratio of the total radiating power of a real surface to that of a black surface at the same temperature is called the **emittance** of the surface (for a perfectly plane surface, the **emissivity**), designated by ε. Subscripts λ, θ, and n may be assigned to differentiate monochromatic, directional, and surface-normal values, respectively, from the total hemispherical value. If radiation is incident on a surface, the fraction absorbed is called the **absorptance (absorptivity)**, a term in which two subscripts may be appended, the first to identify the temperature of the surface and the second to identify the quality of the incident radiation. According to **Kirchhoff's law,** the emissivity and absorptivity of a surface *in surroundings at its own temperature* are the same, for both monochromatic and total radiation. When the temperatures of the surface and its surroundings differ, the total emissivity and absorptivity of the surface are found often to be different, but because absorptivity is substantially independent of irradiation density, the monochromatic emissivity and absorptivity of surfaces are for all practical purposes the same. The difference between total emissivity and absorptivity depends on the variation, with wavelength, of ε_λ and on the difference between the emitter temperature and the effective source temperature.

Consider radiative exchange between a body of area A_1 and temperature T_1 and its black surroundings at T_2. The net interchange is given by

$$\dot{Q}_{1 \rightleftharpoons 2} = A_1 \int_0^\alpha [\varepsilon_\lambda E_\lambda(T_1) - \alpha_\lambda E_\lambda(T_2)] \, d\lambda$$
$$= A_1(\varepsilon_1 \sigma T_1^4 - \alpha_{12} \sigma T_2^4) \quad (4.3.4)$$

where

$$\varepsilon_1 = \int_0^1 \varepsilon_\lambda \, df_{\lambda T_1} \quad \text{and} \quad \alpha_{12} = \int_0^1 \varepsilon_\lambda \, df_{\lambda T_2} \quad (4.3.5)$$

i.e., ε_1 (or α_{12}) is the area under a curve of ε_λ versus f, read as a function of λT at T_1 (or T_2) from Table 4.3.1. If ε_λ does not change with wavelength, the surface is called **gray**, and $\varepsilon_1 = \alpha_{12} = \varepsilon_\lambda$. A selective surface is one whose ε_λ changes dramatically with wavelength. If this change is monotonic, ε_1 and α_{12} are, according to Eqs. (4.3.4) and (4.3.5), markedly different when the absolute temperature ratio is far from 1; e.g., when $T_1 = 293$ K (ambient temperature) and $T_2 = 5,800$ K (effective solar temperature), $\varepsilon_1 = 0.9$ and $\alpha_{12} = 0.1-0.2$ for a white paint, but ε_1 can be as low as 0.12 and α_{12} above 0.9 for a thin layer of copper oxide on bright aluminum, or of chromic oxide on bright nickel.

Although values of emittances and absorptances depend in very complex ways on the real and imaginary components of the refractive index and on the geometric structure of the surface layer, some generalizations are possible.

Polished Metals (1) ε_λ is quite low in the infrared and, for $\lambda > 8$ μm, can be adequately approximated by $0.00365\sqrt{r/\lambda}$, where r is the resistivity in ohm·cm and λ is in micrometres; at shorter wavelengths, ε_λ increases and, for many metals, has values of 0.4 to 0.8 in the visible (0.4–0.7 μm). ε_λ is approximately proportional to the square root of the absolute temperature ($\varepsilon_\lambda \propto \sqrt{r}$ and $r \propto T$) in the far infrared ($\lambda > 8$ μm), is temperature insensitive in the near infrared (0.7–1.5 μm) and, in the visible, decreases slightly as temperature increases. (2) Total emittance is substantially proportional to absolute tempera-

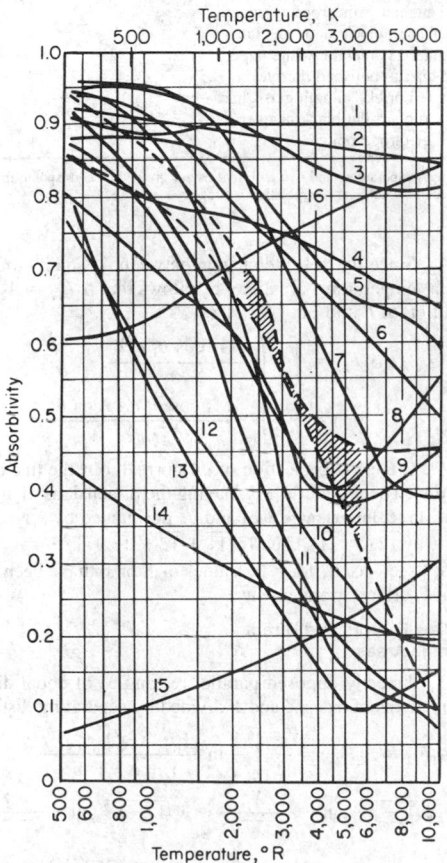

Fig. 4.3.1 Variation of absorptivity with temperature of radiation source. (1) Slate composition roofing; (2) linoleum, red-brown; (3) asbestos slate (asbestos use is obsolete, but may be encountered in existing construction); (4) soft rubber, gray; (5) concrete; (6) porcelain; (7) vitreous enamel, white; (8) red brick; (9) cork; (10) white Dutch tile; (11) white chamotte; (12) MgO, evaporated; (13) anodized aluminum; (14) aluminum paint; (15) polished aluminum; (16) graphite. The two dashed lines bound the limits of data for gray paving brick, asbestos paper (asbestos use is obsolete, but may be encountered in existing construction), wood, various cloths, plaster of paris, lithopone, and paper.

ture; at moderate temperature, $\varepsilon_n = 0.58T\sqrt{r_0/T_0}$, where T is in kelvins. (3) Total absorptance of a metal at T_1 for radiation from a black or gray source at T_2 is equal to the emissivity evaluated at the geometric mean of T_1 and T_2. (4) The ratio of hemispherical to normal emittance (absorptance) varies from 1.33 at very lower ε's (α's) to about 1.03 at an ε (α) of 0.4.

Unless extraordinary pains are taken to prevent oxidation, however, a metallic surface may exhibit several times the emittance or absorptance of a polished specimen. The emittance of iron and steel, for example, varies widely with degree of oxidation and roughness—clean metallic surfaces have an emittance of from 0.05–0.45 at ambient temperatures to 0.4–0.7 at high temperatures; oxidized and/or rough surfaces range from 0.6–0.95 at low temperatures to 0.9–0.95 at high temperatures.

Refractory Materials Grain size and concentration of trace impurities are important. (1) Most refractory materials have an ε_λ of 0.8 to 1.0 at wavelengths beyond 2 to 4 μm; ε_λ decreases rapidly toward shorter wavelengths for materials that are white in the visible but retains its high value for black materials such as FeO and Cr_2O_3. Small concentrations of FeO and Cr_2O_3 or other colored oxides can cause marked increases in the emittance of materials that are normally white. ε_λ for refractory materials varies little with temperature. (2) Refractory materials generally have a total emittance which is high (0.7 to 1.0) at ambient temperatures and decreases with increase in temperature; a change from 1,000 to 1,600°C may cause a decrease in ε of one-fourth to one-third. (3) The emittance and absorptance increase with increase in grain size over a grain-size range of 1–200 μm. (4) The ratio $\varepsilon/\varepsilon_n$ of hemispherical to normal emissivity of polished surfaces varies with refractive index from 1 at $n = 1$ to 0.95 at $n = 1.5$ (common glass) and back to 0.98 at $n = 3.5$. (5) The ratio $\varepsilon/\varepsilon_n$ for a surface composed of particulate matter which scatters isotropically varies with ε from 1 when $\varepsilon = 1$ to 0.8 when $\varepsilon = 0.07$. (6) The total absorptance shows a decrease with increase in temperature of the radiation source similar to the decrease in emittance with increase in the specimen temperature. Figure 4.3.1 shows the effect of the temperature of the radiation source on the absorptance of surfaces of various materials at room temperature. It will be noted that polished aluminum (line 15) and anodized aluminum (line 13), representative of metals and nonmetals, respectively, respond oppositely to a change in the temperature of the radiation source. The absorptance of surfaces for sunlight may be read from the right of Fig. 4.3.1, assuming sunlight to consist of blackbody radiation from a source at 10,440°R (5,800 K).

When T_2 is not too different from T_1, α_{12} may be expressed as $\varepsilon_1(T_2/T_1)^n$, with n determined from Fig. 4.3.1. For this case, Eq. (4.3.4) becomes

$$\dot{Q}_{1,\text{net}} = \sigma A_1 \varepsilon_{AV}(1 + n/4)(T_1^4 - T_2^4) \qquad (4.3.6)$$

where ε_{AV} is evaluated at the arithmetic mean of T_1 and T_2.

Table 4.3.2 gives the emittance of various surfaces and emphasizes the variation possible in a single material. The values in the table apply, with a few exceptions, to normal radiation from the surface.

For opaque materials, the **reflectance** ρ is the complement of the absorptance. The directional distribution of the reflected radiation depends on the material, its degree of roughness or grain size, and if a metal, its state of oxidation. Polished surfaces of homogeneous materials reflect specularly. In contrast, the intensity of the radiation reflected from a **perfectly diffuse**, or **Lambert**, surface is independent of direction. The directional distribution of reflectance of many oxidized metals, refractory materials, and natural products approximates that of a perfectly diffuse reflector. A better model, adequate for many calculational purposes, is achieved by assuming that the total reflectance ρ is the sum of diffuse and specular components ρ_D and ρ_S (Hottel and Sarofim, p. 180).

Black Surface Enclosures When several surfaces are present, the need arises for evaluating a geometric factor F, called the **direct-view factor**. Restriction is temporarily to black surfaces, the intensity from which is independent of angle of emission. Define F_{12} as the fraction of the radiation leaving surface A_1 in all directions which is intercepted by

Table 4.3.2   Emissivity of Surfaces

Surface	Temp.,* °C	Emissivity*
Metals and their oxides		
Aluminum:		
Highly polished	230–580	0.039–0.057
Polished	23	0.040
Rough plate	26	0.055–0.07
Oxidized at 600°C	200–600	0.11–0.19
Oxide	280–830	0.63–0.26
Alloy 75ST	24	0.10
75ST, repeated heating	230–480	0.22–0.16
Brass:		
Highly polished	260–380	0.03–0.04
Rolled plate, natural	22	0.06
Rolled, coarse-emeried	22	0.20
Oxidized at 600°C	200–600	0.61–0.59
Chromium	40–540	0.08–0.26
Copper:		
Electrolytic, polished	80	0.02
Comm'l plate, polished	20	0.030
Heated at 600°C	200–600	0.57–0.57
Thick oxide coating	25	0.78
Cuprous oxide	800–1,100	0.66–0.54
Molten copper	1,080–1,280	0.16–0.13
Dow metal, cleaned, heated	230–400	0.24–0.20
Gold, highly polished	230–630	0.02–0.04
Iron and steel:		
Pure Fe, polished	180–980	0.05–0.37
Wrought iron, polished	40–250	0.28
Smooth sheet iron	700–1,040	0.55–0.60
Rusted plate	20	0.69
Smooth oxidized iron	130–530	0.78–0.82
Strongly oxidized	40–250	0.95
Molten iron and steel	1,500–1,770	0.40–0.45
Lead:		
99.96%, unoxidized	130–230	0.06–0.08
Gray, oxidized	24	0.28
Oxidized at 190°C	190	0.63
Mercury, pure clean	0–100	0.09–0.12
Molybdenum filament	730–2590	0.10–0.29
Monel metal, K5700		
Washed, abrasive soap	24	0.17
Repeated heating	230–875	0.46–0.65
Nickel and alloys:		
Electrolytic, polished	23	0.05
Electroplated, not polished	20	0.11
Wire	190–1,010	0.10–0.19
Plate, oxid. at 600°C	200–600	0.37–0.48
Nickel oxide	650–1,250	0.59–0.86
Copper-nickel, polished	100	0.06
Nickel-silver, polished	100	0.14
Nickelin, gray oxide	21	0.26
Nichrome wire, bright	50–1,000	0.65–0.79
Nichrome wire, oxide	50–500	0.95–0.98
ACI-HW (60Ni, 12Cr); firm black ox, coat	270–560	0.89–0.82
Platinum, polished plate	230–1,630	0.05–0.17
Silver, pure polished	230–630	0.02–0.03
Stainless steels:		
Type 316, cleaned	24	0.28
316, repeated heating	230–870	0.57–0.66
304, 42 h at 520°C	220–530	0.62–0.73
310, furnace service	220–530	0.90–0.97
Allegheny #4, polished	100	0.13
Tantalum filament	1,330–3,000	0.194–0.33
Thorium oxide	280–830	0.58–0.21
Tin, bright	24	0.04–0.06
Tungsten, aged filament	25–3,320	0.03–0.35
Zinc, 99.1%, comm'l, polished	230–330	0.05
Galv., iron, bright	28	0.23
Galv. gray oxide	24	0.28

Table 4.3.2   Emissivity of Surfaces   (*Continued*)

Surface	Temp.,* °C	Emissivity*
Refractories, building materials, paints, misc.		
Alumina	260–680	0.6–0.33
Alumina, 50-μm grain size	1,010–1,570	0.39–0.28
Alumina-silica, cont'g	1,010–1,570	
0.4% Fe_2O_3		0.61–0.43
1.7% Fe_2O_3		0.73–0.62
2.9% Fe_2O_3		0.78–0.68
Al paints (vary with amount of lacquer body, age)	100	0.27–0.67
Asbestos	40–370	0.93–0.95
Calcium oxide	750–1,100	0.29–0.28
Candle soot; lampblack-waterglass	20–370	0.95 ± 0.01
Carbon plate, heated	130–630	0.81–0.79
Ferric oxide (Fe_2O_3)	500–900	0.8–0.43
Magnesium oxide, 1 μm	260–760	0.67–0.41
Oil layers		
Lube oil, 0.01 in on pol. Ni	20	0.82
Linseed, 1–2 coats on Al	20	0.56–0.57
Rubber, soft gray reclaimed	24	0.86
Silica, 3 μm	260–740	0.7–0.5
Misc. I: shiny black lacquer, planed oak, white enamel, serpentine, gypsum, white enamel paint, roofing paper, lime plaster, black matte shellac	21	0.87–0.91
Misc. II: glazed porcelain, white paper, fused quartz, polished marble, rough red brick, smooth glass, hard glossy rubber, flat black lacquer, water, electrographite	21	0.92–0.96

*When two temperatures and two emissivities are given they correspond, first to first and second to second, and linear interpolation is suggested.

surface A_2. Since the net interchange between A_1 and A_2 must be zero when their temperatures are alike, it follows that $A_1F_{12} = A_2F_{21}$. From the definition of F and Eq. (4.3.1),

$$A_1F_{12} = \int_{A1} \int_{\Omega} \frac{dA_1 \cos\theta_1 \, d\Omega_1}{\pi}$$

$$= \int_{A1} \int_{A2} \frac{dA_1 \cos\theta_1 \, dA_2 \cos\theta_2}{\pi r^2} \qquad (4.3.7)$$

where $dA \cos\theta$ is the projection of dA normal to r, the line connecting dA_1 and dA_2. The product A_1F_{12}, having the dimensions of area, will be called the **direct-interchange area** and be designated by $\overline{s_1 s_2}$, sometimes for brevity by $\overline{12}$ ($\equiv \overline{21}$). Clearly, $\overline{11} + \overline{12} + \overline{13} + \cdots = A_1$; and when A_1 cannot "see" itself, $\overline{11} = 0$. Values of F or $\overline{ss}$ have been calculated for various surface arrangements.

Direct-View Factors and Direct Interchange Areas

CASE 1.   Directly opposed parallel rectangles of equal dimensions, and with lengths of sides X and Y divided by separating distance z:

$$\overline{s_1 s_2} (\equiv A_1F_{12} \equiv A_2F_{21}) = \frac{z^2}{2}\left[\frac{1}{2}\ln\frac{(1 + X^2)(1 + Y^2)}{1 + X^2 + Y^2}\right.$$
$$+ X\sqrt{1 + Y^2}\tan^{-1}\frac{X}{\sqrt{1 + Y^2}} + Y\sqrt{1 + X^2}\tan^{-1}\frac{Y}{\sqrt{1 + X^2}}$$
$$\left. - X\tan^{-1}X - Y\tan^{-1}Y\right]$$

See also Fig. 4.3.2.

CASE 2.   Parallel circular disks with centers on a common normal and with radii R_1 and R_2 divided by separating distance z:

$$\overline{s_1 s_2} (\equiv A_1F_{12} \equiv A_2F_{21}) = \frac{\pi z^2}{2}\left[1 + R_1^2 + R_2^2\right.$$
$$\left. - \sqrt{(1 + R_1^2 + R_2^2)^2 - 4R_1^2 R_2^2}\right]$$

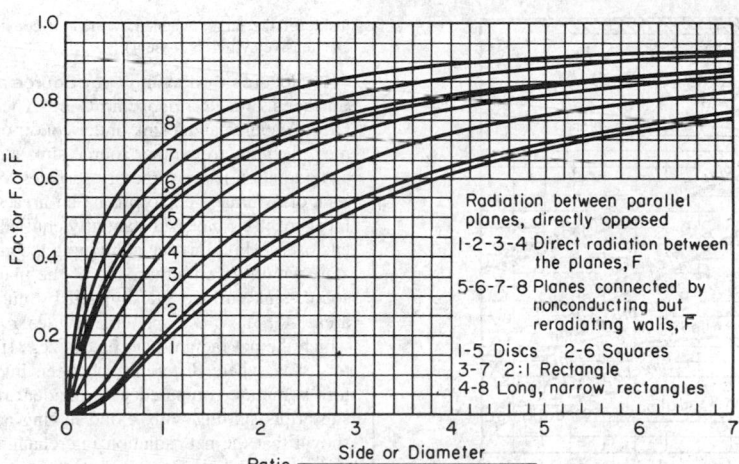

Fig. 4.3.2 Variation of the factor F or $\overline{F}$ for parallel planes directly opposed.

CASE 3. Rectangles in perpendicular planes of area A_1 and A_2, with a common edge l and with other dimension divided by $l = W_1$ and W_2:

$$\overline{s_1 s_2} \ (\equiv A_1 F_{12} \equiv A_2 F_{21})$$

$$= \frac{l^2}{\pi} \left(\frac{1}{4} \ln \frac{(1 + W_1^2)(1 + W_2^2)}{1 + W_1^2 + W_2^2} \left[\frac{W_1^2(1 + W_1^2 + W_2^2)}{(1 + W_1^2)(W_1^2 + W_2^2)} \right]^{W_1^2} \right.$$

$$\times \left[\frac{W_2^2(1 + W_1^2 + W_2^2)}{(1 + W_2^2)(W_1^2 + W_2^2)} \right]^{W_2^2} \right]$$

$$\left. + W_1 \tan^{-1} \frac{1}{W_1} + W_2 \tan^{-1} \frac{1}{W_2} - \sqrt{W_1^2 + W_2^2} \tan^{-1} \frac{1}{\sqrt{W_1^2 + W_2^2}} \right)$$

CASE 4. Circular cylinder of radius r_1 surrounded by cylinder of radius r_2, both of equal length l and on a common axis:

$$A_1 = 2\pi r_1 l \qquad A_2 = 2\pi r_2 l$$

Let $r_1/l = R_1$; $r_2/l = R_2$; $[1/R_1^2 - (R_2/R_1)^2 + 1] = B$; $[1/R_1^2 + (R_2/R_1)^2 + 1] = D$; and $r_2/r_1 = R$.

$$\overline{s_1 s_2} \ (\equiv A_1 F_{12} \equiv A_2 F_{21})$$

$$= l^2 \left\{ R_1^2 \left[\sqrt{(D+2)^2 - 4R^2} \cos^{-1} \frac{B}{DR} + B \sin^{-1} \frac{1}{R} - \frac{\pi}{2} D \right] \right.$$

$$\left. + 2R_1 \left(\pi - \cos^{-1} \frac{B}{D} \right) \right\}$$

$$\overline{s_2 s_2} \ (\equiv A_2 F_{22}) = l^2 R_1 \left\{ 2\pi(R - 1) + 4 \tan^{-1} (2R_1 \sqrt{R^2 - 1}) \right.$$

$$- \sqrt{4R^2 + \frac{1}{R_1^2}} \left(\frac{\pi}{2} + \sin^{-1} \frac{4(R^2 - 1) + 1/[R_1^2(1 - 2/R^2)]}{4(R^2 - 1) + 1/R_1^2} \right)$$

$$\left. + \frac{1}{R_1} \left[\sin^{-1} \left(1 - \frac{2}{R^2} \right) + \frac{\pi}{2} \right] \right\}$$

CASE 5. Two closed surfaces, one enclosing the other and neither having any negative curvature; A_1 is inside. Since $F_{12} = 1$,

$$\overline{s_1 s_2} \ (\equiv A_1 F_{12} \equiv A_2 F_{21}) = A_1$$

$$F_{21} = \frac{A_1}{A_2} \qquad F_{22} = 1 - \frac{A_1}{A_2}$$

CASE 6. Sphere of total inside area A_T; radiative exchange between sphere segments of areas A_1 and A_2. Application of Eq. (4.3.7) shows that, independent of relative position,

$$\overline{s_1 s_2} \ (\equiv A_1 F_{12} \equiv A_2 F_{21}) = \frac{A_1 A_2}{A_T} \qquad F_{12} = \frac{A_2}{A_T}$$

CASE 7. Two dimensional surfaces A_1 and A_2 per unit length normal to cross section, with each area defined by the length of stretched string, on inside face, between ends (i.e., elimination of negative curvature). Graphical exact solution: $\overline{s_1 s_2}$ (per unit normal length) = sum of lengths of crossed stretched strings between ends of A_1 and A_2 minus sum of uncrossed strings, all divided by 2. If an obstruction lies between A_1 and A_2, there may be two sets of strings to represent views on both sides of the obstruction, with results added. The relations for cases 8, 9, and 10 are the results of three among many applications of this principle.

CASE 8. Exchange among inside surfaces of hollow triangular shape of infinite length and areas A_1, A_2, and A_3:

$$\overline{s_1 s_2} \ (\equiv A_1 F_{12} \equiv A_2 F_{21}) = \frac{A_1 + A_2 - A_3}{2} \qquad F_{12} = \frac{A_1 + A_2 - A_3}{2A_1}$$

CASE 9. Exchange between two long parallel circular tubes of diameter D and center-to-center distance C, having areas A_{1a} and A_{1b} per unit length:

$$\overline{s_{1a} s_{1b}} \ (\text{per unit length}) = D \left[\sin^{-1} \frac{D}{C} + \sqrt{\left(\frac{C}{D} \right)^2 - 1} - \frac{C}{D} \right]$$

$$F_{1a \to 1b} = \frac{1}{\pi} \left[\sin^{-1} \frac{D}{C} + \sqrt{\left(\frac{C}{D} \right)^2 - 1} - \frac{C}{D} \right]$$

CASE 10. Exchange between a row of tubes and a plane parallel to it. Consider a unit length along tube axes, with single tube area $A_{1a} = \pi D$ and associated plane area $A_p = C$. A tube sees two tubes and two plane areas:

$$A_{1a} = 2\overline{s_{1a} s_{1b}} + 2\overline{s_{1a} s_p}$$

$$\overline{s_{1a} s_p} \ (\equiv \overline{s_p s_{1a}} \equiv A_p F_{p1}) = \frac{A_{1a}}{2} - \overline{s_{1a} s_{1b}}$$

Substituting from previous example (case 9) yields

$$F_{p1} = 1 - \frac{D}{C} \left[\sin^{-1} \frac{D}{C} + \sqrt{\left(\frac{C}{D} \right)^2 - 1} - \frac{\pi}{2} \right]$$

The value from case 10 appears as line 1 of Fig. 4.3.3. The same figure gives the fraction going to the second row. Additional curves in Fig. 4.3.3 can be obtained by considering the refractory backing as radiatively adiabatic, i.e., by assuming that the radiation that is not absorbed directly is reflected or reradiated, undergoing the same fractional absorption as the incoming beam. In a furnace chamber one zone of which is one or two rows of tubes backed by a refractory, one may visualize the zone as a continuous plane of area A_p at a temperature T_T,

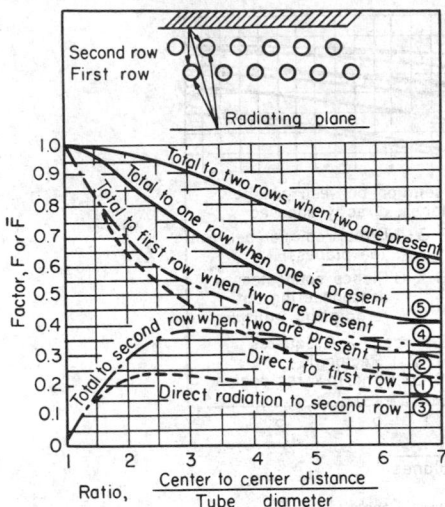

Fig. 4.3.3 Values of F or $\overline{F}$ for a plane parallel to rows of tubes.

the tube surface-temperature, and having an effective absorptivity or emissivity $\varepsilon(=\mathscr{F}_{pT})$ that is equal to the value read from Fig. 4.3.3, line 5 or 6; in total exchange area nomenclature, it is $(\overline{S_pS_T})_R/A_p$. Its complement is headed back toward the emitter, which is whatever faces the replaced tube zone—radiating gas or surfaces or a mixture of them.

When the tubes are gray,

$$\frac{A_p}{(\overline{S_pS_T})_R} = \frac{A_p}{(\overline{S_pS_T})_R}\bigg)_{\substack{black \\ tubes}} + \frac{\rho_T}{\varepsilon_T} \qquad (4.3.8)$$

When $C/D = 2$, the treatment of a single tube row system with the tubes divided into two zones, front and rear half, reduces $(\overline{S_pS_T})_R$ or $\mathscr{F}_{pT}$ below the value given by Eq. (4.3.8) by only 1.7 percent (3 percent) when ε_T is 0.8 (0.6).

For other cases, see References.

The view factor F may often be evaluated from that for simpler configurations by the application of three principles: that of reciprocity, $A_iF_{ij} = A_jF_{ji}$; that of conservation, $\Sigma F_{ij} = 1$; and that due to Yamauti, showing that the exchange areas AF between two pairs of surfaces are equal when there is a one-to-one correspondence for all sets of symmetrically placed pairs of elements in the two surface combinations (Hottel and Sarofim, p. 60).

EXAMPLE. The exchange area between the two squares 1 and 4 of Fig. 4.3.4 is to be evaluated. The following exchange areas may be obtained from the values of F for common-side rectangles (case 3, direct-view factors): $\overline{13} = 0.24$, $\overline{24} = 2 \times 0.29 = 0.58$, $\overline{(1+2)(3+4)} = 3 \times 0.32 = 0.96$. Expression of $\overline{(1+2)(3+4)}$ in terms of its components yields $\overline{(1+2)(3+4)} = \overline{13} + \overline{14} + \overline{23} + \overline{24}$. And by the Yamauti principle $\overline{14} = \overline{23}$, since for every pair of elements in 1 and 4, there is a corresponding pair in 2 and 3. Therefore,

$$\overline{14} = [\overline{(1+2)(3+4)} - \overline{13} - \overline{24}]/2 = 0.07$$

Case 1 may be modified in the same way. Another example is the evaluation of AF for exchange between the outside of the smaller of two coaxial cylinders and the

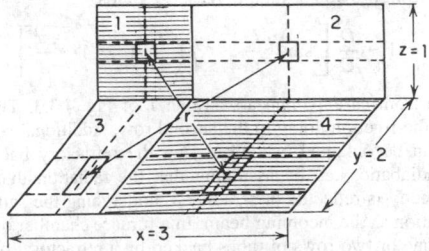

Fig. 4.3.4 Illustration of the Yamauti principle.

inside of the larger when they are *not* coextensive, given the view factor for coextensive cylinders (case 4).

Enclosures Containing Gray Source and Sink Surfaces, Refractory Surfaces, and No Absorbing Gas The calculation of interchange between a source and a sink under conditions involving successive multiple reflections from other source-sink surfaces in the enclosure, as well as reradiation from refractory surfaces, can become complicated. Let a **zone** of a furnace enclosure be an area small enough to make all elements of itself have substantially equivalent "views" of the rest of the enclosure. (In a furnace containing a symmetry plane, parts of a single zone would lie on either side of the plane.) Zones are of two classes, source-sink surfaces, designated by numerical subscripts and having areas $A_1, A_2, \ldots$ and emissivities $\varepsilon_1, \varepsilon_2, \ldots$; and surfaces at which the net radiant-heat flux is zero (fulfilled by the average refractory wall where difference between internal convection and external loss is minute compared with incident radiation), designated by letter subscripts starting with r, and having areas $A_r, A_s, \ldots$. It may be shown that the net radiation interchange between source-sink zones i and j is given by

$$\dot{Q}_{i=j} = \overline{S_rS_j}\sigma T_i^4 - \overline{S_jS_i}\sigma T_j^4 \qquad (4.3.9)$$

The term $\overline{S_rS_j}$ is called the **total interchange area** shared by areas A_i and A_j and depends on the shape of the enclosure and the emissivity and absorptivity of the source and sink zones. It is sometimes called $A_i\mathscr{F}_{ij}$. Restriction here is to gray source-sink zones, for which $\overline{S_iS_j} = \overline{S_jS_i}$; the more general case is treated elsewhere (Hottel and Sarofim, Chaps. 3 and 5).

Evaluation of the $\overline{SS}$'s that characterize an enclosure involves solution of a system of radiation balances on the surfaces. If at a surface the total **leaving flux density**, emitted plus reflected, is denoted by W, radiation balances take the form for source-sink surface j:

$$A_j\varepsilon_jE_j + \rho_j \sum_i \overline{(ij)}W_i = A_jW_j \qquad (4.3.10)$$

and for adiabatic surface r:

$$\sum_i \overline{(ir)}W_i = A_rW_r \qquad (4.3.11)$$

where ρ is reflectance and the summation is over all surfaces in the enclosure. These equations apply to surfaces which emit and reflect diffusely (i.e., their leaving intensity W_i/π is independent of its direction. Most nonmetallic, tarnished, or rough metal surfaces correspond reasonably well to this restriction (but see p. 4-72). In matrix notation, Eqs. (4.3.10) and (4.3.11) become

$$\begin{bmatrix} \overline{11}-\dfrac{A_1}{\rho_1} & \overline{12} & \cdots & \overline{1r} & \overline{1s} & \cdots \\ \overline{12} & \overline{22}-\dfrac{A_2}{\rho_2} & \cdots & \overline{2r} & \overline{2s} & \cdots \\ \cdots & \cdots & \cdots & \cdots & \cdots & \cdots \\ \overline{1r} & \overline{2r} & \cdots & \overline{rr}-A_r & \overline{rs} & \cdots \\ \overline{1s} & \overline{2s} & \cdots & \overline{rs} & \overline{ss}-A_s & \cdots \\ \cdots & \cdots & \cdots & \cdots & \cdots & \cdots \end{bmatrix} \begin{bmatrix} W_1 \\ W_2 \\ \cdots \\ W_r \\ W_s \\ \cdots \end{bmatrix} = \begin{bmatrix} -\dfrac{A_1\varepsilon_1}{\rho_1}E_1 \\ -\dfrac{A_2\varepsilon_2}{\rho_2}E_2 \\ \cdots \\ 0 \\ 0 \\ \cdots \end{bmatrix}$$

$$\qquad (4.3.12)$$

This represents a system of simultaneous equations equal in number to the number of rows of the square matrix. Each equation consists, on the left, of the sum of the products of the members of a row of the square matrix and the corresponding members of the W-column matrix, and, on the right, of the member of that row in the third matrix. With the above set of equations solved for W_i, the net flux at any surface A_i is given by

$$\dot{Q}_{i,net} = \frac{A_i\varepsilon_i}{\rho_i}(E_i - W_i) \qquad (4.3.13)$$

Refractory temperature is obtained from $W_r = E_r = \sigma T_r^4$.

The more general use of Eq. (4.3.12) is to obtain the set of total-interchange areas $\overline{SS}$ which constitute a complete description of the effect of shape, size, and emissivity on radiative flux, independent of the presence or absence of other transfer mechanisms. It may be shown that

$$\overline{S_iS_j} \equiv \overline{S_jS_i} \equiv A_i\mathscr{F}_{ij} = \frac{A_i\varepsilon_i}{\rho_i}\frac{A_j\varepsilon_i}{\rho_j}\left(-\frac{D_{ij}{}'}{D}\right) \quad (4.3.14)$$

where D is the determinant of the square coefficient matrix in Eq. (4.3.12) and $D_{ij}{}'$ is the cofactor of its ith row and jth column, or -1^{i+j} times the minor of D formed by crossing out the ith row and jth column.

As an example, consider radiation between two surfaces A_1 and A_2 which together form a complete enclosure. Equation (4.3.12) takes the form

$$A_1\mathscr{F}_{12} = \frac{A_1\varepsilon_1}{\rho_1}\frac{A_2\varepsilon_2}{\rho_2}\frac{\overline{12}}{\begin{vmatrix} \overline{11} - \dfrac{A_1}{\rho_1} & \overline{12} \\ \overline{12} & \overline{22} - \dfrac{A_2}{\rho_2} \end{vmatrix}} \quad (4.3.15)$$

Only one direct-view factor F_{12} or direct exchange area $\overline{12}$ is needed because F_{11} equals $1 - F_{12}$ and F_{22} equals $1 - F_{21}$ or $1 - F_{12}A_1/A_2$. Then $\overline{11}$ equals $A_1 - \overline{12}$, and $\overline{22}$ equals $A_2 - \overline{21}$. With the above substitutions, Eq. (4.3.15) becomes

$$A_1\mathscr{F}_{12} = \frac{A_1}{1/F_{12} + 1/\varepsilon_1 - 1 + (A_1/A_2)(1/\varepsilon_2 - 1)} \quad (4.3.16)$$

Special cases include:
1. Parallel plates, large compared to clearance. Substitution of $F_{12} = 1$ and $A_1 = A_2$ gives

$$A_1\mathscr{F}_{12} = \frac{A_1}{1/\varepsilon_1 + 1/\varepsilon_2 - 1} \quad (4.3.17)$$

2. Sphere of area A_1 concentric with surrounding sphere of area A_2. $F_{12} = 1$. Then

$$A_1\mathscr{F}_{12} = \frac{A_1}{1/\varepsilon_1 + (A_1/A_2)(1/\varepsilon_2 - 1)} \quad (4.3.18)$$

3. Body of surface A_1 having no negative curvature, surrounded by very much larger surface A_2. $F_{12} = 1$ and $A_1/A_2 \rightarrow 0$. Then

$$\mathscr{F}_{12} = \varepsilon_1 \quad (4.3.19)$$

Many furnace problems are adequately handled by dividing the enclosure into but two source-sink zones A_1 and A_s, and any number of no-flux zones, A_r, A_s, ... For this case Eq. (4.3.14) yields

$$\frac{1}{\overline{S_1S_2}}\left(\equiv \frac{1}{\overline{S_2S_1}}\right) = \frac{1}{A_1}\left(\frac{1}{\varepsilon_1} - 1\right)$$
$$+ \frac{1}{A_2}\left(\frac{1}{\varepsilon_2} - 1\right) = \frac{1}{(\overline{S_1S_2})_B} \quad (4.3.20)$$

Here the expression $(\overline{S_1S_2})_B \; [\equiv (\overline{S_2S_1})_B]$ represents the total interchange area for the limiting case of a black source and black sink (the refractory emissivity is of no moment). The factor $(\overline{S_1S_2})_B/A_1$, called $\overline{F}_{12}$, is known exactly for a few geometrically simple cases and may be approximated for others. If A_1 and A_2 are equal parallel disks, squares, or rectangles, connected by nonconducting but reradiating refractory walls, then $\overline{F}$ is given by Fig. 4.3.2, lines 5 to 8. If A_1 represents an infinite plane and A_2 is one or two rows of infinite parallel tubes in a parallel plane, and if the only other surface is a refractory surface behind the tubes, $\overline{F}_{12}$ is given by line 5 or 6 of Fig. 4.3.3. If an enclosure may be divided into several radiant-heat sources or sinks A_1, A_2, etc., and the rest of the enclosure (reradiating refractory surface) may be lumped together as A_r at a uniform temperature T_r, then the total interchange area for zone pairs in the black system is given by

$$(\overline{S_1S_2})_B(\equiv A_1\overline{F}_{12}) = \overline{12} + \frac{(\overline{1r})\,(\overline{r2})}{A_r - \overline{rr}} \quad (4.3.21)$$

For the two-source-sink-zone system to which Eq. (4.3.20) applies, Eq. (4.3.21) simplifies to $(\overline{S_1S_2})_B = \overline{12} + 1/[1/\overline{1r} + 1/(\overline{2r})]$; and if A_1 and A_2 each can see none of itself, there is further simplification to

$$(\overline{S_1S_2})_B = \overline{12} + \frac{1}{1/(A_1 - \overline{12}) + 1/(A_2 - \overline{12})}$$
$$= \frac{A_1A_2 - (\overline{12})^2}{A_1 + A_2 - 2(\overline{12})} \quad (4.3.22)$$

which necessitates the evaluation of but one direct-view factor F.

Equation (4.3.20) covers many problems of radiant heat interchange between source and sink in furnace enclosures involving no radiating gas. The error due to single zoning of source and sink is small even if the "views" of the enclosure from different parts of each zone are quite different, provided the emissivity is fairly high; the error in $\overline{F}$ is zero if it is obtainable from Fig. 4.3.2 or 4.3.3, small if Eq. (4.3.21) is used and the variation in temperature over the refractory is small. Approach to any desired accuracy can be made by use of Eq. (4.3.14) with division of the surfaces into more zones.

From the definitions of F, $\overline{F}$, and $\mathscr{F}$ or of $\overline{ss}$, $(\overline{SS})_B$, and $\overline{SS}$ it is to be noted that

$$\left.\begin{array}{l} F_{11} + F_{12} + \cdots + F_{1r} + F_{1s} + \cdots = 1 \\ \overline{F}_{11} + \overline{F}_{12} + \cdots \qquad\qquad\quad = 1 \\ \mathscr{F}_{11} + \mathscr{F}_{12} + \cdots \qquad\qquad\quad = \varepsilon_1 \end{array}\right\}$$

or

$$\begin{array}{l} \overline{s_1s_2} + \overline{s_1s_2} + \cdots + \overline{s_1s_r} + \overline{s_1s_s} + \cdots = A_1 \\ (\overline{S_1S_1})_B + (\overline{S_1S_2})_B + \cdots = A_1 \\ \overline{S_1S_1} + \overline{S_1S_2} + \cdots = A_1\varepsilon_1 \end{array}$$

EXAMPLE. A furnace chamber of rectangular parallelepipedal form is heated by the combustion of gas inside vertical radiant tubes lining the side walls. The tubes are on centers 2.4 diameters apart. The stock forms a continuous plane on the hearth. Roof and end walls are refractory. Dimensions are shown in Fig. 4.3.5. The radiant tubes and stock are gray bodies having emissivities 0.8 and 0.9, respectively. What is the net rate of heat transmission to the stock by radiation when the mean temperature of the tube surface is 1,500°F (1,089 K) and that of the stock is 1,200°F (922 K)?

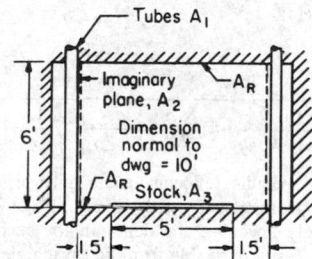

Fig. 4.3.5 Dimensions of a furnace chamber.

This problem must be broken up into two parts, first considering the walls with their refractory-backed tubes. To imaginary planes A_2 of area 6×10 ft and located parallel to and inside the rows of radiant tubes, the tubes emit radiation $\sigma T_1^4 A_1 \mathscr{F}_{12}$, which equals $\sigma T_1^4 A_2 \mathscr{F}_{21}$. To find $\mathscr{F}_{21}$ use Fig. 4.3.3, line 5, from which $\overline{F}_{21} = 0.81$. Then from Eq. (4.3.20).

$$\mathscr{F}_{21} = 1/[(1/0.81) + (1/1 - 1) + (2.4/\pi)\,(1/0.8 - 1)] = 0.702$$

This amounts to saying that the system of refractory-backed tubes is equal in radiating power to a continuous plane A_2 replacing the tubes and refractory back of them, having a temperature equal to that of the tubes and an equivalent or effective emissivity of 0.702.

The new simplified furnace now consists of an enclosure formed by two 6×10 ft radiating side walls (area A_2, of emissivity 0.702), a 5×10 ft receiving plane on the floor (A_3), and refractory surfaces (A_R) to complete the enclosure (ends, roof, and floor side strips); the desired heat transfer is

$$q_{2=3} = \sigma(T_1^4 - T_3^4)A_2\mathscr{F}_{23}$$

To evaluate $\mathscr{F}_{23}$, start with the direct interchange factor F_{23}. $F_{23} = F$ from A_2 to (A_3 + a strip of A_R alongside A_3 which has a common edge with A_2) minus F from

A_2 to the strip only. These two F's may be evaluated from case 3 for direct-view factors. For the first F, $Y/X = 6/10$, $Z/X = 6.5/10$, $F = 0.239$; for the second F, $Y/X = 6/10$, $Z/X = 1.5/10$, $F = 0.100$. Then $F_{23} = 0.239 - 0.10 = 0.139$. Now $\overline{F}$ may be evaluated. From Eq. (4.3.21) et seq.,

$$A_2 \overline{F}_{23} = \overline{23} + \frac{1}{1/\overline{2r} + 1/\overline{3r}}$$

$$\overline{F}_{23} = F_{23} + \frac{1}{1/F_{2r} + (A_2/A_3)(1/F_{3r})}$$

Since A_2 "sees" A_r, A_3, and some of itself (the plane opposite), $F_{2r} = 1 - F_{22} - F_{23}$. F_{22}, the direct interchange factor between parallel 6×10 ft rectangles separated by 8 ft, may be taken as the geometric mean of the factors for 6-ft squares separated by 8 ft, and 10-ft squares separated by 8 ft. These come from Fig. 4.3.2, line 2, according to which $F_{22} = \sqrt{0.13 \times 0.255} = 0.182$. Alternatively, the first of the 10 cases listed above under "Direct-View Factors" may be used. Then $F_{2r} = 1 - 0.182 - 0.139 = 0.679$. The other required direct factor is $F_{3r} = 1 - F_{32} = 1 - F_{23}A_2/A_3 = 1 - 0.139 \times {}^{120}/_{50} = 0.666$. Then $\overline{F}_{23} = 0.139 \{1/[(1/0.679) + (120/50)(1/0.666)]\} = 0.336$. Having $\overline{F}_{23}$, we may now evaluate the factor $\mathscr{F}_{23}$ using Eq. (4.3.20) with $A_1 \rightarrow A_2$, $A_2 \rightarrow A_3$, and $[\overline{S_1 S_2}]_{13} \rightarrow A_2 \overline{F}_{23}$.

$$\mathscr{F}_{23} = \frac{1}{1/0.336 + 1/0.702 - 1 + (120/50)(1/0.9 - 1)}$$

$$= 0.273$$

$$\dot{Q}_{net} = \sigma(T_1^4 - T_3^4)A_2\mathscr{F}_{23} = 0.171(19.6^4 - 16.6^4)(120)(0.273)$$
$$= 402,000 \text{ Btu/h}$$

In SI units

$$\dot{Q}_{net} = 5.67(10.89^4 - 9.22^4)(120 \times .3048^2)(0.273) = 118,000 \text{ W}$$

A result of interest is obtained by dividing the term $A_2\mathscr{F}_{23}(120 \times 0.273$, or 32.7 ft^2) by the actual area A_1 of the radiating tubes $[(\pi/2.4) \times 60 \times 2 = 157 \text{ ft}^2]$. This is $32.7/157 = 0.208$; i.e., the net radiation from a tube to the stock is 20.8 percent as much as if the tube were black and completely surrounded by black stock.

Enclosures of Surfaces That Are Not Diffuse Reflectors   The total interchange-area concept has been generalized to include surfaces the reflectance ρ of which can be divided into a diffuse, or Lambert-reflecting, component ρ_D and a specular component ρ_S independent of angle of incidence, with $\varepsilon + \rho_S + \rho_D = 1$. In application to concentric spheres or infinite cylinders, with A_1 the inner surface, the method yields (Hottel and Sarofim, p. 181)

$$A_1 \mathscr{F}_{12} \equiv \overline{S_1 S_2}$$

$$= \frac{1}{\dfrac{1}{A_1 \varepsilon_1} + \dfrac{1}{A_2}\left(\dfrac{1}{\varepsilon_2} - 1\right) + \dfrac{\rho_{S2}}{1 - \rho_{S2}}\left(\dfrac{1}{A_1} - \dfrac{1}{A_2}\right)} \quad (4.3.23)$$

When there is no specular reflectance, the third term in the denominator drops out, in agreement with Eq. (4.3.18). When the reflectance is exclusively specular, the denominator becomes $1/(A_1\varepsilon_1) + \rho_{S2}/[A_1(1 - \rho_{S2})]$, easily derivable from first principles.

RADIATION FROM FLAMES, COMBUSTION PRODUCTS, AND PARTICLE CLOUDS

The radiation from a flame consists of (1) radiation throughout the spectrum from burning soot particles of microscopic and submicroscopic dimensions, from suspended larger particles of coal, coke, or ash, all contributing to what is spoken of as flame luminosity, (2) infrared radiation, mostly from the water vapor and carbon dioxide in the hot gaseous combustion products, and (3) nonequilibrium radiation associated with the combustion process itself, called chemiluminescence and not a significant contributor to the total radiation. A major problem is the effect of the shape of the emitting volume on the radiative flux; this will be considered first.

Mean Beam Lengths   Evaluation of radiation from a nonisothermal volume is beyond the scope of this section (see Hottel and Sarofim, Chap. 11). If a volume emitter is isothermal and at a temperature T, the ratio of the emission from an element of its volume subtending the solid angle $d\Omega$ at a receiver element dA, and making the angle θ with the

normal thereto, to blackbody radiation arriving from within the same solid angle is called the **gas emissivity**. Clearly, ε depends on the path length L through the volume to dA. A hemispherical volume radiating to a spot on the center of its base represents the case in which L is independent of direction. Flux at that spot relative to hemispherical blackbody flux is thus an alternative way to visualize emissivity. The flux density to an area of interest on the envelope of an emitter volume of any shape can be matched by that at the base of a hemispherical volume of some radius L, which will be called the **mean beam length**. It is found that, although the ratio of L to a characteristic dimension D of the shape varies with opacity, the variation is small enough for most engineering purposes to permit use of a constant ratio, L_M/D, where L_M is the **average mean beam length**. L_M can be defined to apply either to a spot on the envelope or to any finite portion of its area. An important limiting case is that of opacity approaching zero ($pD \rightarrow 0$, where p = partial pressure of the emitter constituent). For this case, L (called L_0) equals $4 \times$ ratio of gas volume to bounding area when interest is in radiation to the entire envelope. For the range of pD encountered in practice, L (now L_M) is always less. For various shapes, 0.8 to 0.95 times L_0 has been found optimum (see Table 4.3.3); for shapes not reported in Table 4.3.3, a factor of 0.88 (or $L_M = 0.88L_0 = 3.5V/A$) is recommended.

Soot luminosity is important where combustion occurs under such conditions that the hydrocarbons in the flame are subject to heat in the absence of sufficient air well mixed on a molecular scale. Because soot particles are small relative to the wavelength of radiation of interest (diameters 20 to 140 nm), the monochromatic emissivity ε_λ depends on the total particle volume per unit volume of space f_v, regardless of particle size. It is given by

$$\varepsilon_\lambda = 1 - e^{-Kf_v L/\lambda}$$

where L is the path length.

Use of the perfect gas law and a material balance enables the restatement of the above as

$$\varepsilon_\lambda = 1 - e^{-KPSL/(\lambda T)} \quad (4.3.24)$$

where P is the total pressure, atm, and S is the mole fraction of soot in the gas. Here S depends on the fractional conversion of f_c of the fuel carbon to soot, and it is the mole fraction, wet basis, of carbon in gaseous form (CO_2, CO, CH_4, etc.) times $f_c/(1 - f_c)$ or, with negligible error, times f_c, which is a very small number (more later on this). Evaluation of K is complex, and its numerical value depends somewhat on the age of the soot, the temperature at which it is formed, and its hydrogen content. It is recommended that $K = 0.526$ [K/atm] be used in the absence of specific information on the soot in question.

The total emissivity of soot ε_s is obtained by integration over the wavelength spectrum (Felske and Tien, *Comb. Sci. & Tech.*, **7**, no. 2, 1973), giving

$$\varepsilon_s = 1 - \frac{15}{4}[\psi^{(3)}(1 + KPSL/c_2)] \quad (4.3.25)$$

where $\psi^{(3)}(x)$ is the pentagamma function of x. It may be shown that an excellent approximation to Eq. (4.3.25) is

$$\varepsilon_s = 1 - (1 + 34.9SPL)^{-4} \quad (4.3.26)$$

where PL is in atm $\cdot$ m. The error is less the lower ε_s and is only 0.5 percent at $\varepsilon_s = 0.5$; 0.8 percent at 0.67. Expression of ε_s in e-power form is feasible but of lower accuracy than Eq. (4.3.25) or (4.3.26). In that form, with L in metres,

$$\varepsilon_s = 1 - e^{-143SPL} \pm 8\% \quad (4.3.27)$$

There is at present no method of predicting soot concentration of a luminous flame analytically; reliance must be placed on experimental measurement on flames similar to that of interest. Visual observation is misleading; a flame so bright as to hide the wall behind it may be far from a "black" radiator. The International Flame Foundation at Ijmuiden has recorded data on many luminous flames from gas, oil, and coal (see *Jour. Inst. Fuel*, 1956–present).

Table 4.3.3   Mean Beam Lengths for Volume Radiation

Shape	Characteristic dimension D	L_0/D	L_M/D
Sphere	Diameter	0.67	0.63
Infinite cylinder	Diameter	1	0.94
Semi-infinite cylinder, radiating to:			
Center of base	Diameter	1	0.90
Entire base	Diameter	0.81	0.65
Right-circle cylinder, ht = diam, radiating to:			
Center of base	Diameter	0.76	0.71
Whole surface	Diameter	0.67	0.60
Right-circle cylinder, ht = 0.5 diam, radiating to:			
End	Diameter	0.47	0.43
Side	Diameter	0.52	0.46
Total surface	Diameter	0.50	0.45
Right-circle cylinder, ht = 2 × diam, radiating to:			
End	Diameter	0.73	0.60
Side	Diameter	0.82	0.76
Total surface	Diameter	0.80	0.73
Infinite cylinder, half-circle cross section, radiating to spot on middle of flat side	Radius		1.26
Rectangular parallelepipeds			
1 : 1 : 1 (cube)	Edge	0.67	0.60
1 : 1 : 4, radiating to:			
1 × 4 face	Shortest edge	0.90	0.82
1 × 1 face	Shortest edge	0.86	0.71
Whole surface	Shortest edge	0.89	0.81
1 : 2 : 6, radiating to:			
2 × 6 face	Shortest edge	1.18	
1 × 6 face	Shortest edge	1.24	
1 × 2 face	Shortest edge	1.18	
Whole surface	Shortest edge	1.2	
Infinite parallel planes	Clearance	2.00	1.76
Space outside infinite bank of tubes, centers on equilateral triangles; tube diam = clearance	Clearance	3.4	2.8
Same, except tube diam = 0.5 clearance	Clearance	4.45	3.8
Same, except tube centers on squares, diam = clearance	Clearance	4.1	3.5

The chemical kinetics and fluid mechanics of soot burnout have not progressed far enough to evaluate the soot fraction f_c for relatively complex systems. Additionally, the soot in a combustion chamber is highly localized, and a mean value is needed for calculation of the radiative heat transfer performance of the chamber. On the basis of limited experience with fitting data to a model, the following procedure is recommended when total combustion chamber performance is being estimated: (1) When pitch or a highly aromatic fuel is burned, 1 percent of the fuel carbon appears as soot. This produces values of ε_s of 0.4 to 0.5 and ε_{G+s} of 0.6 to 0.7. These values are lower than some measurements on pitch flames, but the measurements are usually taken through the flame at points of high luminosity. (2) When no. 2 fuel oil is burned, ⅓ percent of the fuel carbon appears as soots, but that number varies greatly with burner design. (3) When natural gas is burned, any soot contribution to emissivity may be ignored. Admittedly the numbers given should be functions of burner design and excess air, and they should be considered tentative, subject to change when good data show they are off target.

Clouds of Large Black Particles   The emissivity of a cloud of particles depends on their area projected along the line of sight. The projected area per unit volume of space is the projected area A of a particle times the particle number concentration c, or the volume fraction f_v of space occupied by particles times b/d, the projected-surface/volume ratio, where d is the characteristic dimension. [For any randomly oriented particles without dimples, A/(total area) is ¼; for spheres, $b = \frac{3}{2}$.] The emissivity of a particle cloud is then given by the alternative formulations

$$\varepsilon = 1 - e^{-bf_v L/d} = 1 - e^{-cAL} \qquad (4.3.28)$$

As an example, consider heavy fuel oil ($CH_{1.5}$, s.g. 0.95) atomized to a surface mean particle diameter of d μm, burned with 20 percent excess

air to produce coke residue particles having the original drop diameter, and suspended in combustion products at 1,500 K. From stoichiometry, $f_v = 1.27 \times 10^{-5}$. For spherical particles $b = \frac{3}{2}$, and the flame emissivity due to the particles along a path L will be $1 - e^{-1.9 \times 10 - 2L/d}$. With 200-μm particles and an L of 3 m, the particle contribution to emissivity will be 0.25. Soot luminosity will increase this; particle burnout will decrease it. The combined emissivity due to several kinds of emitters will be treated later. The correction for nonblackness of the particles is complicated by multiple scatter of the radiation reflected by each particle. The emissivity ε_M of a cloud of gray particles of individual surface emissivity ε_1 can be estimated by the use of Eq. (4.3.28) with its exponent multiplied by ε_1 if the optical thickness cAL does not exceed about 2.

Gaseous Combustion Products   Radiation from water vapor and carbon dioxide occurs in spectral bands in the infrared. Its magnitude is 3 to 10 times that of convection at furnace temperatures. It depends on gas temperature T_G, on the partial pressure-beam length products $p_w L$ and $p_c L$ (subscripts w and c refer to water vapor and carbon dioxide), and to a much lesser extent on total and partial pressure. The gas emissivity ε_G is the sum of the separate contributions due to H_2O and CO_2, corrected for pressure broadening of the spectral bands and for band overlap (Hottel and Sarofim, Chap. 6). The elaborate calculations can be combined for a restricted set of conditions, here taken to be the practically important cases of 1-atm total pressure and partial pressures representative of fossil fuel combustion in air. In the range of furnace operating conditions the product $\varepsilon_G T_G$ varies much less than ε_G with T_G, and $\varepsilon_G T_G$ depends primarily on $(p_w + p_c)L$, much less on $p_w/(p_w + p_c)$, and so little on T_G as to permit linear interpolation between widely separated T_G's. An equation of the form

$$\log \overline{\varepsilon_G T_G} = a_0 + a_1 \log pL + a_2 \log^2 pL + a_3 \log^3 pL \qquad (4.3.29)$$

where p is the sum of partial pressures $p_w + p_c$ atm and L is the mean beam length, has been found capable of fitting emissivity data over a 1000-fold range of pL, from 0.01 to 10 m $\cdot$ atm (0.03 to 30 ft $\cdot$ atm).

Table 4.3.4, section 2, gives values of the constants representing the results of an averaging of all the available total and integrated spectral data on CO_2 and H_2O, together with corrections for spectral band broadening and overlap. Equation (4.3.29) represents the original data with a precision greater than their accuracy. The constants are given for computation in either metres and kelvins or feet and degrees Rankine for mixtures, in nonradiating gases, of water vapor alone, CO_2 alone, and four p_w/p_c mixtures. Four suffice, since a change halfway from one mixture ratio to the adjacent one changes the emissivity by a maximum of only 5 percent; linear interpolation may be used if necessary. The

constants are given for only three temperatures, which is adequate for linear interpolation since $\overline{\varepsilon_G}T$ changes a maximum of only one-sixth due to a change from one temperature base halfway to the adjacent one. Based on metre atmospheres and kelvins, the interpolation relation, with T_H and T_L representing the higher and lower base temperatures bracketing T, and with the brackets in the term $[A(x)]$ indicating that the parentheses refer not to a multiplier but to an argument, is

$$\overline{\varepsilon_G T_G} = \frac{[\overline{\varepsilon_G T_H}(pL)](T_G - T_L) + [\overline{\varepsilon_G T_L}(pL)](T_H - T_G)}{500} \quad (4.3.30)$$

Extrapolation to a temperature which is above the highest or below the lowest of the three base temperatures in Table 4.3.4 uses the same

Table 4.3.4   Emissivity of ε_G of H_2O-CO_2 Mixtures

Section 1: Limited range for furnaces, valid over 25-fold range of $p_{w+c}L$, 0.046–1.15 m $\cdot$ atm (0.15–3.75 ft $\cdot$ atm)

p_w/p_c	0	0.5		1		2		3		∞
$\dfrac{p_w}{p_w + p_c}$	0	⅓(0.2–0.42)		½(0.42–0.6)		⅔(0.6–0.7)		¾(0.7–0.8)		1
	CO_2 only	Corresponding to $(CH)_x$, covering coal, heavy oils, pitch		Corresponding to $(CH_2)_x$, covering distillate oils, paraffins, olefines		Corresponding to CH_4, covering natural gas and refinery gas		Corresponding to $(CH_6)_x$, covering future high-H_2 fuels		H_2O only

Constants b and n of equation $\overline{\varepsilon_G T} = b(pL - 0.015)^n$, pL in m $\cdot$ atm, T in K

T, K	b	n	b	n	b	n	b	n	b	n	b	n
1,000	188	0.209	384	0.33	416	0.34	444	0.34	455	0.35	416	0.400
1,500	252	0.256	448	0.38	495	0.40	540	0.42	548	0.42	548	0.523
2,000	267	0.316	451	0.45	509	0.48	572	0.51	594	0.52	632	0.640

Constants b and n of equation $\overline{\varepsilon_G T} = b(pL - 0.05)^n$, pL in ft $\cdot$ atm, T in °R

T, °R	b	n	b	n	b	n	b	n	b	n	b	n
1,800	264	0.209	467	0.33	501	0.34	534	0.34	541	0.35	466	0.400
2,700	335	0.256	514	0.38	555	0.40	591	0.42	600	0.42	530	0.523
3,600	330	0.316	476	0.45	519	0.48	563	0.51	577	0.52	532	0.640

Section 2: Full range, valid over 2000-fold range of $p_{w+c}L$, 0.005–10.0 m $\cdot$ atm (0.016–32.0 ft $\cdot$ atm)
Constants of equation, $\log \overline{\varepsilon_G T} = a_0 + a_1 \log pL + a_2 \log^2 pL + a_3 \log^3 pL$

$\dfrac{p_w}{p_c}$	$\dfrac{p_w}{p_w + p_c}$	pL in m $\cdot$ atm, T in K					pL in ft $\cdot$ atm, T in °R				
		T, K	a_0	a_1	a_2	a_3	T, °R	a_0	a_1	a_2	a_3
0	0	1,000	2.2661	0.1742	−0.0390	0.0040	1,800	2.4206	0.2176	−0.0452	0.0040
		1,500	2.3954	0.2203	−0.0433	0.00562	2,700	2.5248	0.2695	−0.0521	0.00562
		2,000	2.4104	0.2602	−0.0651	−0.00155	3,600	2.5143	0.3621	−0.0627	−0.00155
$\dfrac{1}{2}$	$\dfrac{1}{3}$	1,000	2.5754	0.2792	−0.0648	0.0017	1,800	2.6691	0.3474	−0.0674	0.0017
		1,500	2.6461	0.3418	−0.0685	−0.0043	2,700	2.7074	0.4091	−0.0618	−0.0043
		2,000	2.6504	0.4279	−0.0674	−0.0120	3,600	2.6686	0.4879	−0.0489	−0.0120
1	$\dfrac{1}{2}$	1,000	2.6090	0.2799	−0.0745	−0.0006	1,800	2.7001	0.3563	−0.0736	−0.0006
		1,500	2.6862	0.3450	−0.0816	−0.0039	2,700	2.7423	0.4261	−0.0756	−0.0039
		2,000	2.7029	0.4440	−0.0859	−0.0135	3,600	2.7081	0.5210	−0.0650	−0.0135
2	$\dfrac{2}{3}$	1,000	2.6367	0.2723	−0.0804	0.0030	1,800	2.7296	0.3577	−0.0850	0.0030
		1,500	2.7178	0.3386	−0.0990	−0.0030	2,700	2.7724	0.4384	−0.0944	−0.0030
		2,000	2.7482	0.4464	−0.1086	−0.0139	3,600	2.7461	0.5474	−0.0871	−0.0139
3	$\dfrac{3}{4}$	1,000	2.6432	0.2715	−0.0816	0.0052	1,800	2.7359	0.3599	−0.0896	0.0052
		1,500	2.7257	0.3355	−0.0981	0.0045	2,700	2.7811	0.4403	−0.1051	0.0045
		2,000	2.7592	0.4372	−0.1122	−0.0065	3,600	2.7599	0.5478	−0.1021	−0.0065
∞	1	1,000	2.5995	0.3015	−0.0961	0.0119	1,800	2.6720	0.4102	−0.1145	0.0119
		1,500	2.7083	0.3969	−0.1309	0.00123	2,700	2.7238	0.5330	−0.1328	0.00123
		2,000	2.7709	0.5099	−0.1646	−0.0165	3,600	2.7215	0.6666	−0.1391	−0.0165

NOTE: Values of $p_w/(p_w + p_c)$ of ⅓, ½, ⅔, ¾ may be used to cover the ranges 0.2–0.42, 0.42–0.6, 0.6–0.7, and 0.7–0.8, respectively, with a maximum error in ε_G of 5 percent at $pL = 6.5$ m $\cdot$ atm, less at lower pL's. Linear interpolation reduces the error generally to less than 1 percent. Linear interpolation or extrapolation on T introduces an error generally below 2 percent, less than the accuracy of the original data.

formulation, but one of the terms becomes negative. Linearization on the constants a_0 to a_3 rather than on $\overline{\varepsilon_G T}$ may be preferable if fuel quality is unchanging.

When pL lies in the 25-fold range of 0.046 to 1.15 m · atm (0.15 to 3.75 ft · atm), adequate for furnaces, a much simpler two-constant relation is adequate.

$$\overline{\varepsilon_G T} = \begin{cases} b(pL - 0.015)^n & \text{with } T = \text{K}, \ pL = \text{m} \cdot \text{atm} \\ b(pL - 0.05)^n & \text{with } T = {}^\circ\text{R}, \ pL = \text{ft} \cdot \text{atm} \end{cases}$$

The constants are given in Table 4.3.4, section 1.

Combined Radiation from Gases and Suspended Solids

The total emissivity of gases and suspended solids is less than the sum of the separate contributions because of interference between overlapping spectral emissions. The spectral overlap of H_2O and CO_2 radiation has been taken into account by the constants of Table 4.3.4 used for obtaining ε_G. Additional overlap occurs when soot emissivity ε_s is added. If the emission bands of water vapor and CO_2 were randomly placed in the spectrum and soot radiation were gray, the combined emissivity would be $\varepsilon_G + \varepsilon_s$ minus an overlap correction $\varepsilon_G \varepsilon_s$. Monochromatic soot emissivity is higher as the wavelength gets shorter, and in a highly sooted flame at 1,500 K half the soot emission lies below 2.5 μm where H_2O and CO_2 emission is negligible. Then the correction $\varepsilon_G \varepsilon_s$ must be reduced, and the following is recommended:

$$\varepsilon_{G+s} = \varepsilon_G + \varepsilon_s - M\varepsilon_G \varepsilon_s \qquad (4.3.31)$$

where M depends mostly on T_G and to a much lesser extent on the optical density SPL. Values that have been calculated from this simple model can be represented with acceptable error by

$$M = 1.07 + 18SPL - 0.27(T/1,000)$$

If, in addition to gas and soot, massive particles such as fly ash, coal char, or carbonaceous cenospheres from heavy fuel oil of emissivity ε_M are present, it is recommended that the total emissivity be approximated by

$$\varepsilon_{\text{total}} = \varepsilon_{G+s} + \varepsilon_M - \varepsilon_{G+s}\varepsilon_M \qquad (4.3.32)$$

Radiant interchange between a gas and a *completely bounding black surface* at T_1 produces a surface flux density q given by

$$q = \sigma(\varepsilon_G T_G^4 - \alpha_{G1} T_1^4) \qquad (4.3.33)$$

where α_{G1} is the **absorptivity** of the gas at T_G for radiation from a surface at T_1. The absorptivity of water vapor–CO_2 mixtures may also be obtained from the constants for emissivities. The product $\overline{\alpha_{G1} T_1}$—the absorptivity of gas at T_G for black radiation at T_1 times the surface temperature—is the product $\varepsilon_G T_1$ with ε_G evaluated at surface temperature T_1 instead of T_G and at pLT_1/T_G instead of pL, then multiplied by $(T_G/T_1)^{0.5}$, or

$$\overline{\alpha_{G1} T_1} = [\overline{\varepsilon_G T_1}(pLT_1/T_G)](T_G/T_1)^{0.5} \qquad (4.3.34)$$

The exponent 0.5 is an adequate average of the exponents for the pure components. The interpolation relation for absorptivity is

$$\overline{\alpha_{G1} T_1} = \left[\overline{\varepsilon_G T_H}\left(\frac{pLT_1}{T_G}\right) \right]\left(\frac{T_G}{T_H}\right)^{0.5}\frac{T_1 - T_L}{500}$$
$$+ \left[\overline{\varepsilon_G T_L}\left(\frac{pLT_L}{T_G}\right) \right]\left(\frac{T_G}{T_L}\right)^{0.5}\frac{T_H - T_1}{500} \qquad (4.3.35)$$

The base temperature pair T_H and T_L can be different for the evaluation of ε_G and α_{G1} if T_G and T_1 are far enough apart. Extrapolation from the lowest T_G in Eq. (4.3.35) to a much lower T_1 to obtain α_{G1} may yield too high a value for it. That occurs, however, only when $T_1 \ll T_G$, and the fourth-power temperature relation makes the error in q negligible.

If the surface is not black, the right-hand side of Eq. (4.3.33) must be modified. If the surface is gray, multiplication by $\alpha_1 (\equiv \varepsilon_1)$ allows for reduction in the primary beam from gas to surface and surface to gas, but some of the gas radiation initially reflected from the surface has further opportunity for absorption at the surface because the gas is but incompletely opaque to the reflected beam. Consequently, the factor to allow for surface lies between absorptance α_1 and unity, nearer the latter the more transparent the gas (low pL) and the more convoluted the surface. In the absorptance range of most industrial surfaces, 0.7 to 1.0, an adequate approximation consists in use of an effective absorptance α_1' halfway between the actual value and unity. If the surface is not gray, q depends much more on surface absorptance, which modifies $\varepsilon_G T_G^4$, than on emittance, which modifies $\alpha_{G1} T_1^4$. Absorption is treated more rigorously later in the section.

EXAMPLE. Flue gas containing 9.5 percent CO_2 and 7.1 percent H_2O, wet basis, flows through a bank of tubes of 1.5-in OD on equilateral triangular centers 4.5 in apart. In a section in which the gas and tube surface temperatures are 1,700 and 1,000°F, what is the heat transfer rate per square foot of tube area, due to gas radiation only? Tube surface absorptance = 0.8.

$T_G = 2,160°\text{R} \ (1,200 \text{ K}); \ T_S = 1,460°\text{R} \ (811 \text{ K})$
$p_w/(p_w + p_c) = 7.1/16.6 = 0.428;$ use 0.5
$pL = 0.166 \ [3.8(4.5 - 1.5)/12] = 0.158 \text{ ft} \cdot \text{atm} \ (0.0480 \text{ m} \cdot \text{atm})$
$pL(T_s/T_G) = 0.1580(1,460/2,160) = 0.1066 \text{ ft} \cdot \text{atm} \ (0.0325 \text{ m} \cdot \text{atm})$
From Table 4.3.4, for $T_G = 1,500$ K and $pL = 0.0480$ and 0.0325, $\varepsilon T = 125$ and 101 K, and for $T_G = 1,000$ K and $pL = 0.0480$ and 0.0325, $\varepsilon T = 129$ and 107 K. Then

$$\varepsilon_G = \frac{1}{1,200}\ \frac{125(1,200 - 1,000) + 129(1,500 - 1,200)}{1,500 - 1,000} = 0.106$$

$$\alpha_{G1} = \frac{1}{811}\ \frac{101(811 - 1,000) + 107(1,500 - 1,000)}{1,500 - 1,000} = 0.135$$

The effective surface absorptance factor $\alpha_1 = (0.8 + 1)/2 = 0.9$. From Eq. (4.3.33), modified,

$$q = 0.9 \times 0.1713(0.106 \times 21.6^4 - 0.135 \times 14.6^4)$$
$$= 2,612 \text{ Btu}/(\text{ft}^2 \cdot \text{h})$$

or

$$q = 0.9 \times 5.67(0.106 \times 12^4 - 0.135 \times 8.111^4) = 8,235 \text{ W/m}^2$$

This is equivalent to a convection coefficient of $2,612/700$ or 3.73 Btu/(ft²·F · h) or 21.2 W/(m²·K). The emissivity of an equivalent gray flame is $(0.106 \times 21.6^4 - 0.135 \times 14.6^4)/(21.6^4 - 14.6^4) = 0.098$.

RADIATIVE EXCHANGE IN ENCLOSURES OF RADIATING GAS

The so-called *radiant section* of a furnace presents a heat-transfer problem in which there enters the combined action of direct radiation from the flame to the stock or heat sink and radiation from the flame to refractory surfaces and thence back through the flame (with partial absorption) to the sink, convection, and external losses. Solutions of the problem based on varying degrees of simplification are available, including allowance for temperature variation in both gas and refractory walls (Hottel and Sarofim, Chap. 14). A less rigorous treatment suffices, however, for handling many problems. There are two limiting cases: the long chamber with gas temperature varying only in the direction of gas flow and the compact chamber containing a gas or flame at a uniform temperature. The latter, with variations, will be considered first.

Total Exchange Areas $\overline{SS}$ and $\overline{GS}$ The arguments leading to the development of the interchange factor $A_i \mathcal{F}_{ij} (= \overline{S_i S_j})$ between surface zones [Eq. (4.3.14) et seq.] apply to the case of absorption within the gas volume if, in the evaluation of the direct exchange area, allowance is made for attenuation of the radiant beam through the gas. This necessitates nothing more than the redefinition, in Eqs. (4.3.7) to (4.3.22), of every term $\overline{ij} (\equiv \overline{s_i s_j} \equiv A_i F_{ij})$ to represent, per unit of black emissive power, flux from A_i through an absorbing gas to A_j; that is, the prior F_{ij} must be multiplied by a mean transmittance τ_{ij} of the gas ($= 1 - \overline{\alpha_{ij}} = 1 - \varepsilon_G$ for a gray gas). In a system containing an isothermal gas and source-sink boundaries of areas $A_1, A_2, \ldots, A_n$, the total emission from A_1 per unit of its black emissive power is $A_1\varepsilon_1$, of which $\overline{S_1S_1} + \overline{S_1S_2} + \cdots + \overline{S_1S_n}$ is absorbed in the surfaces by all mechanisms, direct and indirect. The difference has been absorbed in the gas; it is called the **gas surface total exchange area** $\overline{GS_1}$:

$$\overline{GS_1} = A_1\varepsilon_1 - \sum_i \overline{S_1S_i} \qquad (4.3.36)$$

The letters identifying total exchange areas are, of course, commutative; $\overline{GS_1} \equiv \overline{S_1G}$. Note that although $\overline{S_1S_1}$ is never used in calculating radiative interchange, its value is needed for use of Eq. (4.3.36) in calculating $\overline{GS_1}$. $\overline{GS_1}$ embraces the full effect of radiation complexities on radiative exchange between gas and A_1, including multiple reflection at all surfaces, and it is capable of including the effects of gas nongrayness and of assistance given by refractory surfaces to gas-A_1 interchange. It is but mildly temperature-sensitive and is independent of any changes in conduction, convection, mass flow, and energy balances except for their effect on the temperature used in evaluating it.

If the gas volume is not isothermal, the principles used here can be extended to setting up balances on a zoned gas volume (see, e.g., Hottel and Sarofim, "Radiative Transfer," McGraw-Hill, Chap. 11).

Systems with a Single Gas Zone and Two Surface Zones

An enclosure consisting of but one isothermal gas zone and two gray surface zones, when properly specified, can model so many industrially important radiation problems as to merit detailed presentation. One can evaluate the total radiation flux between any two of the three zones, including multiple reflection at all surfaces.

$$\dot{Q}_{G \leftrightarrow 1} = \overline{GS_1} \sigma (T_G^4 - T_1^4)$$
$$\dot{Q}_{1 \leftrightarrow 2} = \overline{S_1S_2} \sigma (T_1^4 - T_2^4) \qquad (4.3.37)$$

The total exchange area takes a relatively simple closed form, even when important allowance is made for gas radiation not being gray and when a reduction of the number of system parameters is introduced by assuming that one of the surface zones, if refractory, is radiatively adiabatic (see later). Before allowance is made for these factors, the case of a gray gas enclosed by two source-sink surface zones will be presented. Modification of Eqs. (4.3.7) to (4.3.22), discussed previously, combined with the assumption that a single mean beam length applies to all transfers, i.e., that there is but one gas transmittance $\tau (= 1 - \varepsilon_G)$, gives

$$\overline{S_1S_2} = \frac{A_1 \varepsilon_1 \varepsilon_2 F_{12}}{1/\tau + \tau \rho_1 \rho_2 (1 - F_{12}/C_2) - \rho_1 (1 - F_{12}) - \rho_2 (1 - F_{21})}$$
$$(4.3.38)$$

$$\overline{S_1S_1} = \frac{A_1 \varepsilon_1^2 [F_{11} + \rho_2 \tau (F_{12}/C_2 - 1)]}{\text{same denominator}} \qquad (4.3.39)$$

$$\overline{GS_1} = \frac{A_1 \varepsilon_1 \varepsilon_G [1/\tau + \rho_2 (F_{12}/C_2 - 1)]}{\text{same denominator}} \qquad (4.3.40)$$

(Here C is the area expressed as a ratio to the total enclosure area A_T; $C_1 = A_1/A_T$, $C_2 = A_2/A_T$; $C_1 + C_2 = 1$.) The three equations above suffice to formulate total exchange areas for gas-enclosing arrangements which include, e.g., the four geometric cases illustrated in Table 4.3.5, to be discussed later.

An additional surface arrangement of importance is a single zone surface fully enclosing gas. With the gas assumed gray, the simplest derivation of $\overline{GS_1}$ is to note that the emission from surface A_1 per unit of its blackbody emissive power is $A_1 \varepsilon_1$, of which the fractions ε_G and $(1 - \varepsilon_G) \varepsilon_1$ are absorbed by the gas and the surface, respectively, and the surface reflected residue always repeats this distribution. Therefore,

$$\overline{GS}_{\substack{\text{single surface} \\ \text{zone surrounding} \\ \text{gray gas}}} \equiv \overline{GS_1} = A_1 \varepsilon_1 \frac{\varepsilon_G}{\varepsilon_G + (1 - \varepsilon_G) \varepsilon_1} = \frac{A_1}{1/\varepsilon_G + 1/\varepsilon_1 - 1}$$
$$(4.3.41)$$

Alternatively, $\overline{GS_1}$ could be obtained from case 1 of Table 4.3.5 by letting plane area A_1 approach 0, leaving A_2 as the sole surface zone.

Although departure of gas from grayness has a marked effect on radiative transfer, the subject is complex and will be presented in stages, as the cases shown in Table 4.3.5 are discussed.

Partial Allowance for the Effect of Gas Nongrayness on Total Exchange Areas

A radiating gas departs from grayness in two ways: (1) Gas emissivity ε_G and absorptivity α_{G1} are not the same unless T_1 equals T_G. (2) The

fractional transmittance τ of radiation through successive path lengths L_m due to surface reflection, instead of being constant, keeps increasing because at the wavelengths of high absorption the incremental absorption decreases with increasing path length. The first of these effects is sufficiently straightforward to be introduced at this point, coupled with allowance for refractory surfaces being substantially radiatively adiabatic. The second, much more complicated effect will be introduced later; it sometimes changes the computed flux significantly.

In the simplest case of gas-surface radiative exchange—a gas at T_G completely enclosed by a black surface at T_1—the net flux $\dot{Q}_{G \leftrightarrow 1}$ is given by

$$\dot{Q}_{G \leftrightarrow 1} = \sigma (\varepsilon_G T_G^4 - \alpha_{G1} T_1^4) \equiv \sigma \varepsilon_{G,e} (T_G^4 - T_1^4)$$

The evaluation of the absorptivity α_G was covered in Eqs. (4.3.34) and (4.3.35). The second form of the above equation defines $\varepsilon_{G,e}$, the equivalent gray-gas emissivity

$$\varepsilon_{G,e} = \frac{\varepsilon_G - \alpha_{G1}(T_1^4/T_G^4)}{1 - (T_1/T_G)^4} \qquad (4.3.42)$$

Although this introduction of $\varepsilon_{G,e}$ has added no information, the evaluation of $\dot{Q}_{G \leftrightarrow 1}$ in terms of $\varepsilon_{G,e}$ rather than ε_G and α_{G1} gives a better structure for trial-and-error solutions of problems in which either T_G or T_1 is not known and a second energy relation is available.

With partial allowance for gas nongrayness having been made, the evaluation of radiative flux $\dot{Q}_{G \leftrightarrow 1}$ or $\dot{Q}_{1 \leftrightarrow 2}$ [Eq. (4.3.37)] for cases falling in one of the categories of Table 4.3.5 is straightforward if both A_1 and A_2 are source-sink surfaces. Wherever ε_G or τ appears in the table, or in Eqs. (4.3.38) to (4.3.41), use $\varepsilon_{G,e}$ or $1 - \varepsilon_{G,e}$ instead.

EXAMPLE (FIRST APPROXIMATION TO NONGRAYNESS). Methane is burned to completion with 20 percent excess air (air half saturated with water vapor at 298 K (60°F), 0.0088 mol H_2O/mol dry air) in a furnace chamber with floor dimensions of 3×10 m and 5 m high. The whole surface is a gray energy sink of emissivity 0.8 at 1,000 K, surrounding gas at 1,500 K, well stirred. Find the effective gas emissivity $\varepsilon_{G,e}$ and the surface radiative flux density, assuming that the only correction necessary for gas nongrayness is use of $\varepsilon_{G,e}$ rather than ε_G.

SOLUTION. Combustion is 1 CH_4 + 2 × 1.2 O_2 + 1.2 × (79/21)N_2 + 2 × 1.2 × 100/21 × 0.0088 H_2O going to 1 CO_2 + [2 + 2 × 1.2 × (100/21) × 0.0088] H_2O + 0.4 O_2 + 9.03 N_2 = 12.53 mol/mol of CH_4. And $P_C + P_W$ = (1 + 2.1)/12.53 = 0.2474 atm. The mean beam length $L_m = 0.88 × 4V/A_T = 0.88 ×$ $4(10 × 3 × 5)/\{2[2 × (10 × 3 + 10 × 5 + 3 × 5)]\} = 2.779$ m. And $pL_m = 0.2474 × 2.779 = 0.6875$ m · atm. From emissivity Table 4.3.4, $b(1,500) = 540$; $n(1,500) = 0.42$; $b(1,000) = 444$; $n(1,000) = 0.34$. Also $\varepsilon_G(pL) = 540(0.6875 - 0.015)^{0.42}/1,500 = 0.3047$, and $\alpha_{G1}(pL) = 444(0.6875 × 1,000/1,500 - 0.015)^{0.34}(1,500/1,000)^{0.5}/1,000 = 0.4124$. Then $\varepsilon_{G,e}(pL) = [0.3047 - 0.4124(1,000/1,500)^4]/[1 - (1,000/1,500)^4] = 0.2782$. From Eq. (4.3.41), with ε_G replaced by $\varepsilon_{G,e}$, $(\overline{GS_1}/A_1) = 1/(1/0.2782 + 1/0.8 - 1) = 0.2601$. Then $\dot{Q}_{G \leftrightarrow 1}/A_1 = 56.7 × 0.2601[(1,500/1,000)^4 - (1,000/1,000)^4] = 59.91$ kW/m² [18,990 Btu/(ft² · h)].

Refractory Surfaces If one of the surfaces A_r of an enclosure of gas is refractory, an extra temperature T_{refr} and an extra heat transfer equation are needed to determine the fluxes *unless* A_r can be assumed to be radiatively adiabatic. Consider the facts that irradiation of A_r plus convection from gas to it must equal back radiation plus conduction through it if steady state exists, and irradiation is enormous compared to convection. It then follows that the difference between convection and conduction is so minute compared to irradiation or back radiation as to make A_r substantially radiatively adiabatic; assume that A_1 is a source-sink zone and A_2 a radiatively adiabatic zone, and call it A_r. The condition for adiabaticity of A_r is

$$\overline{GS_r}(T_G^4 - T_r^4) = \overline{S_rS_1}(T_r^4 - T_1^4)$$

or, to eliminate T_r,

$$\frac{T_G^4 - T_r^4}{1/\overline{GS_r}} = \frac{T_r^4 - T_1^4}{1/\overline{S_rS_1}} = \frac{T_G^4 - T_1^4}{1/\overline{GS_r} + 1/\overline{S_rS_1}} \qquad (4.3.43)$$

The net flux from gas G is $\overline{GS_1}\sigma(T_G^4 - T_1^4) + \overline{GS_r}\sigma(T_G^4 - T_r^4)$ which, with replacement of the last term, using Eq. (4.3.43), gives the single

Table 4.3.5 Total Exchange Areas for Four Arrangements of Two-Zone-Surface Enclosures of a Gray Gas*

Case 1	Case 2	Case 3	Case 4

Case 1

A plane surface A_1, and a surface A_2 complete the enclosure

$$F_{12} = 1$$

$$\frac{\overline{S_1 S_2}}{A_1} = \frac{\varepsilon_1 \varepsilon_2}{D_1}$$

$$\frac{\overline{GS_1}}{A_1} = \frac{\varepsilon_1 \varepsilon_G (1/\tau + \rho_2 C_1/C_2)}{D_1}$$

$$\frac{\overline{GS_2}}{A_2} = \frac{\varepsilon_2 \varepsilon_G (1/\tau + \rho_1 C_1/C_2)}{D_1}$$

$$\frac{\overline{S_1 S_1}}{A_1} = \frac{\varepsilon_1^2 \tau \rho_2 C_1/C_2}{D_1} \left[1 - \frac{C_1}{C_2}(1 - \tau \rho_1) \right]$$

$$D_1 \equiv \frac{1}{\tau} - \rho_2 \quad \tau = 1 - \varepsilon_G$$

Case 2

Infinite parallel planes

$$F_{12} = F_{21} = 1$$

$$\frac{\overline{S_1 S_2}}{A_1} = \frac{\varepsilon_1 \varepsilon_2}{D_2}$$

$$\frac{\overline{GS_1}}{A_1} = \frac{\varepsilon_1 \varepsilon_G (1/\tau + \rho_2)}{D_2}$$

$$\frac{\overline{S_1 S_1}}{A_1} = \frac{\varepsilon_1^2 \rho_2 \tau}{D_2}$$

$$D_2 \equiv \frac{1}{\tau} - \tau \rho_1 \rho_2$$

Case 3

Concentric spherical or infinite cylindrical surface zones, A_1 inside

$$F_{12} = 1 \qquad F_{21} = \frac{A_1}{A_2} = \frac{C_1}{C_2}$$

$$\frac{\overline{S_1 S_2}}{A_1} = \frac{\varepsilon_1 \varepsilon_2}{D_3}$$

$$\frac{\overline{GS_1}}{A_1} = \frac{\varepsilon_1 \varepsilon_G (1/\tau + \rho_2 C_1/C_2)}{D_3}$$

$$\frac{\overline{GS_2}}{A_2} = \frac{\varepsilon_2 \varepsilon_G (1/\tau + \rho_1 C_1/C_2)}{D_3}$$

$$\frac{\overline{S_1 S_1}}{A_1} = \frac{\varepsilon_1^2 \rho_2 \tau C_1/C_2}{D_3}$$

$$D_3 \equiv \frac{1}{\tau} - \rho_2 \left[1 - \frac{C_1}{C_2}(1 - \tau \rho_1) \right]$$

Case 4

Two-surface-zone enclosure, each zone in one or more parts, any shape

Case A
Rigorous evaluation of F's, with i and j representing parts of A_1 and A_2:

$$A_1 F_{12} = \sum_i \left(A_i \sum_j F_{ij} \right)$$

$$\left. \begin{array}{c} \dfrac{\overline{S_1 S_2}}{\overline{GS_1}} \\ \overline{S_1 S_1} \end{array} \right\} = \left\{ \begin{array}{l} \text{same as in base} \\ \text{case, Eqs. (4.3.38)} \\ \text{to (4.3.40)} \end{array} \right.$$

Case B
Assume that enclosing surface is a speckled enclosure, or spherical

$$F_{12} = F_{22} = C_2$$
$$F_{21} = F_{11} = C_1$$

$$\frac{\overline{S_1 S_2}}{A_1} = \frac{\varepsilon_1 \varepsilon_2 C_2}{D_4}$$

$$\frac{\overline{GS_1}}{A_1} = \frac{\varepsilon_1 \varepsilon_G/\tau}{D_4}$$

$$\frac{\overline{S_1 S_1}}{A_1} = \frac{\varepsilon_1^2 C_1}{D_4}$$

$$D_4 \equiv \frac{1}{\tau} - \rho_1 C_1 - \rho_2 C_2$$

* All equations above come from Eqs. (4.3.38) to (4.3.40), with substitutions for view factor F given before equations.

term multiplying a fourth-power temperature difference.

$$\dot{Q}_{G\leftrightarrow 1} = \sigma(T_G^4 - T_1^4)\left[\overline{GS}_1 + \frac{1}{1/\overline{GS}_r + 1/\overline{S_rS}_1}\right]$$
$$\equiv (\overline{GS}_1)_R \sigma(T_G^4 - T_1^4) \quad (4.3.44)$$

The bracketed term is called $(\overline{GS}_1)_R$, the total exchange area from G to A_1 with assistance from a refractory surface. Table 4.3.5 supplies the forms for the three total exchange area terms needed to formulate $(\overline{GS}_1)_R$, with A_r substituted for A_2 and with ε_G or τ replaced by $\varepsilon_{G,e}$ or $1 - \varepsilon_{G,e}$.

A general expression for a gray gas enclosure of two surfaces, one of which is radiatively adiabatic, comes from Eq. (4.3.44), in which $(\overline{GS}_1)_R$ becomes $\overline{GS}_1$ because $\overline{S_rS}_1$ and $\overline{GS}_r$ are zero, and then from Eq. (4.3.36), which becomes $\overline{GS}_1 [= (\overline{GS}_1)_R] = A_1\varepsilon_1 - \overline{S_1S}_1$. With $\overline{S_1S}_1$ coming from Eq. (4.3.39), one finally obtains

$$\frac{(\overline{GS}_1)_R}{A_1} = \frac{1}{\rho_1/\varepsilon_1 + 1/\{\varepsilon_G[1 + 1/(C_1/C_2 + \varepsilon_G/\tau\, F_{1r})]\}} \quad (4.3.45)$$

Note that since the first denominator term is zero when A_1 is black, the denominator of the second term is $(\overline{GS}_1)_R/A_1$ for a black surface.

The above equation is perfectly general when the **gas is gray** and the two enclosing surfaces are a sink and a radiatively adiabatic surface. When A_1 is a plane (simulation of slab or billet heating furnaces and glass tanks), $(\overline{GS}_1)_R/A_1$ is Eq. (4.3.45) with $F_{1r} = 1$. A different but completely equivalent form is

$$\left[\frac{(\overline{GS}_1)_R}{A_1}\right]_{\substack{A_1\text{ is}\\ \text{plane}}} = \frac{1}{1/\varepsilon_1 + (1/\varepsilon_G - 1)^2/[1/(C_1\varepsilon_G) - 1]} \quad (4.3.46)$$

For most refinery **processing furnaces**, with sink and refractory assumed to form a **speckled** enclosure, $(\overline{GS}_1)_R/A_1$ is Eq. (4.3.45) with $F_{1r} = C_r$. A better but completely equivalent form is

$$\left[\frac{(\overline{GS}_1)_R}{A_1}\right]_{\substack{\text{surface is}\\ \text{speckled}}} = \frac{1}{1/\varepsilon_1 + C_1(1/\varepsilon_G - 1)} \quad (4.3.47)$$

As previously stated, partial allowance for the gas not being gray is made by evaluating $\overline{GS}$ or $(\overline{GS})_R$ with use of $\varepsilon_{G,e}$ rather than ε_G, and $1 - \varepsilon_{G,e}$ rather than τ, in Eqs. (4.3.45) to (4.3.47). A slightly better but more tedious allowance for partial **nongrayness** in evaluating $\dot{Q}_{rad}$ is to replace $\overline{GS}_1\sigma(T_G^4 - T_1^4)$, with $\overline{GS}_1$ evaluated by using $\varepsilon_{G,e}$, by $\overline{GS}_s T_G^4 - \overline{GS}_1 T_1^4$, where $\overline{GS}_s$ and GS_s are evaluated by using ε_G and $\alpha_{G,1}$, respectively.

This completes the presentation of procedures for evaluating the total exchange area between gas and sink surface when the gas is gray or by making approximate allowance for the nongrayness by using ε_G and $\alpha_{G,1}$. These exchange areas can be used in the formulations below on furnace chamber performance. Methods will be presented first for a more rigorous treatment of gas nongrayness.

Full Allowance for Gas Nongrayness The above paragraphs failed to allow for the previously discussed change in gas transmittance on successive passages of reflected radiation through the gas. In many radiative transfer problems, interchange between many different pairs of radiators creates a system of simultaneous equations to be solved for the energy fluxes, and a shift from use of the Stefan-Boltzmann to the Planck equation would enormously increase the difficulty of solution. Use will be made of the fact that total emissivity of a real gas, the spectral emissivity and absorptivity ε_λ of which vary in any way with λ, can be expressed *rigorously* as the a-weighted mean of a suitable number of gray gas emissivity or absorptivity terms $\varepsilon_{G,i}$ or $\alpha_{G,i}$ representing the gray gas emissivity or absorptivity in the energy fractions a_i of the blackbody spectrum. Then

$$\varepsilon_G = \sum_0^n a_i\varepsilon_{G,i} = \sum_0^n a_i(1 - e^{-k_i pL}) \quad (4.3.48)$$

The linearity between flux $\dot{Q}$ and blackbody emissive power E_B allows the above relations to be used for converting $\overline{GS}$ or $\overline{SS}$ for a gray gas

to a form allowing for nongrayness. For a real gas $\overline{GS}$ or $\overline{SS}$ is the a_i-weighted sum of its values based on each of the $\varepsilon_{G,i}$ values of Eq. (4.3.48). Into an expression for $\overline{GS}_{\text{gray}}$ replace ε_G by $\varepsilon_{G,i}$, or τ by $1 - \varepsilon_{G,i}$, multiply the result by a_i and sum the resultant $\overline{GS}$ values to obtain $\overline{GS}_{\text{real gas}}$. Obviously, the number of terms n should be as small as possible while consistent with small error. Consider an n of 2, with the gas modeled as the sum of one gray gas plus a clear gas, with the gray gas of absorption coefficient k occupying the energy fraction a of the blackbody spectrum and the clear gas ($k = 0$) the fraction $1 - a$. Then, at path lengths L and $2L$,

$$[\varepsilon_G(pL)] = a(1 - e^{-kpL}) + (1 - a)(0)$$
$$[\varepsilon_G(2pL)] = a(1 - e^{-2kpL}) + (1 - a)(0) \quad (4.3.49)$$

Solution of these gives

$$a = \frac{\varepsilon_G(pL)}{2 - \varepsilon_G(2pL)/\varepsilon_G(pL)}$$
$$kpL = -\ln\left[1 - \frac{\varepsilon_G(pL)}{a}\right] \quad (4.3.50)$$

The equivalent gray gas emissivity in the spectral range a is $1 - e^{-kpL}$, from Eq. (4.3.48), and from Eq. (4.3.49) that is $\varepsilon_G(pL)/a$; in the spectral range $1 - a$, the equivalent gray gas emissivity is zero. This simple model will be correct for the contribution of the direct gas emission from path length L and for that of the once-reflected emission (path length $2L$); and the added contributions due to increasing numbers of reflections will be attenuated sufficiently by surface reflections to make errors in them unimportant. Note that a and k are not general constants; they are specific to the **subject mean beam length** L_m and come from basic data, such as Table 4.3.4. Note also that when full allowance for nongrayness is to be made by replacing ε_G by $\varepsilon_{G,e}$, then a is also changed to a_e, which comes from Eq. (4.3.50), with $\varepsilon_{G,e}$ replacing ε_G.

Conversion of gray gas total exchange areas $\overline{GS}$ and $\overline{SS}$ to their nongray forms is carried out as follows when the nongray model is gray-plus-clear gas: From Eq. (4.3.49) the equivalent gray gas emissivity in the spectral energy fraction a_e is $\varepsilon_{G,e}(pL)/a_e$, which replaces ε_G wherever it or its complement τ occurs in $\overline{GS}$; the result is then multiplied by a_e. There is no contribution from the clear gas energy fraction. Conversion of $\overline{SS}$ from gray to gray plus clear involves making the same substitution, but for $\overline{SS}$ another term must be added. For the clear gas contribution, 0 and 1 are substituted for ε_G and τ, and the result is multiplied by the weighting factor $1 - a_e$; this $\overline{SS}$ is added to the preceding one to give $\overline{SS}_{g+c}$.

The simplest application of this gray-plus-clear model of gas radiation is the case of a single gas zone surrounded by a single surface zone, the case covered for a gray gas by Eq. (4.3.51) and illustrated in the last numerical example, where radiation from methane combustion products is surrounded by a single-zone sink surface. That example will be repeated using the gray-plus-clear gas model, for which the total exchange area is

$$\frac{\overline{GS}_1}{A_1} = \frac{a_e}{a_e/\varepsilon_{G,e} + 1/\varepsilon_1 - 1} \quad (4.3.51)$$

EXAMPLE (GRAY-PLUS-CLEAR GAS RADIATION FROM METHANE COMBUSTION PRODUCTS). The computations of the previous example, radiative flux from methane combustion products, will be repeated with the more rigorous treatment of nongrayness, and the results will be compared with the more approximate calculations for the case of a wall emissivity ε_1 of 0.4 and 0.8.

SOLUTION. Repeat the calculations given, in the earlier example, of $\varepsilon_G(pL)$, $\alpha_{G1}(pL)$, and $\varepsilon_{G,e}(pL)$ for $pL = 2 \times 0.6875$, to give $\varepsilon_G(2pL) = 0.4096$, $\alpha_{G1}(2pL) = 0.5250$, and $\varepsilon_{G,e}(2pL) = 0.3812$. Then $a_e = 0.2782/(2 - 0.3812/0.2782) = 0.4418$, and the emissivity substitute for the gray gas portion of the gray-plus-clear gas model is $0.2782/0.4418 = 0.6297$. For a single enveloping surface zone, the total exchange area comes from Eq. (4.3.51): $\overline{GS}_1/A_1 = a_e/(a_e/\varepsilon_{G,e} + 1/\varepsilon_1 - 1) = 0.4418/(0.4418/0.2782 + 1/0.8 - 1) = 0.2404$. The flux density is $Q/A = q = (\overline{GS}_1/A)\sigma(T_G^4 - T_1^4) = 0.2404 \times 56.7 \times [(1,500/1,000)^4 - (1,000/1,000)^4] = 55.37$ kW/m² [17,550 Btu/(ft² · h)]. This is 7.6 percent lower than it is when only the difference between ε_G and $\alpha_{G,1}$ is allowed for in finding the effect of gas nongrayness. In some problems the difference is as high as 20 percent. (Note that allowing for average humidity in air adds 5 percent to H_2O and about 2 percent to the gas emissivity.) Changing ε_1 from 0.8 to 0.4 changes the approximate solution for $\overline{GS_1}/A_1$ from 0.2501 to 0.1963 and the gray-plus-clear treatment from 0.2404 to 0.1431.

The procedures for introducing the nongray gas model can be used to convert the total exchange areas for the basic one-gas two-surface model, Eqs. (4.3.38) to (4.3.40), as used to evaluate the cases in Table 4.3.5, to the following gray-plus-clear-gas model forms:

$$\frac{\overline{S_1S_2}}{A_1} = F_{12}\varepsilon_1\varepsilon_2\left(\frac{a_e}{D_a} + \frac{1 - a_e}{D_b}\right) \qquad (4.3.52)$$

$$\frac{\overline{S_1S_2}}{A_1} = F_{12}\varepsilon_1^2\left\{\frac{a_e[1 - F_{12} + \rho_2(1 - \varepsilon_{G,e}/a_e)(F_{12}/C_2 - 1)]}{D_a}\right.$$
$$\left. + \frac{(1 - a_e)[1 - F_{12} + \rho_2(F_{12}/C_2 - 1)]}{D_b}\right\} \qquad (4.3.53)$$

$$\frac{\overline{GS_1}}{A_1} = \frac{\varepsilon_1\varepsilon_{G,e}[1/(1 - \varepsilon_{G,e}/a_e) + \rho_2(F_{12}/C_2 - 1)]}{D_a} \qquad (4.3.54)$$

$$D_a = \frac{1}{1 - \varepsilon_{G,e}/a_e} + \left(1 - \frac{\varepsilon_{G,e}}{a_e}\right)\rho_1\rho_2(1 - F_{12}C_2)$$
$$- \rho_1(1 - F_{12}) - \rho_2(1 - F_{21})$$

$$D_b = 1 + \rho_1\rho_2\left(1 - \frac{F_{12}}{C_2}\right) - \rho_1(1 - F_{12}) - \rho_2(1 - F_{21})$$

$$\text{or} \quad = \varepsilon_1\varepsilon_2 + F_{12}\left[\varepsilon_2 + \frac{\varepsilon_1(C_1 - \varepsilon_2)}{C_2}\right]$$

The above relations, with the view factor F_{12} specified, may be used to convert the geometric cases of Table 4.3.5 to their more nearly correct forms with gray gas replaced by the gray-plus-clear gas model. That has been done in Table 4.3.6, which covers a moderate idealization of many practical industrial systems.

Effect of Gas Nongrayness on Refractory Zones Full allowance for the effect of gas nongrayness on enclosures in which part of the enclosing surface is radiatively adiabatic is straightforward but sometimes tedious. The term of Eq. (4.3.44) must be evaluated. It is tempting to use Eq. (4.3.45), but that is invalid because, although total radiative interchange at zone A_r is 0, the gas nongrayness makes A_r a net absorber in the spectral energy fraction a (or a_e) and a net emitter in the clear gas fraction $1 - a$. It is necessary, then, to use the basic equation

$$(\overline{GS_1})_R = \left(\overline{GS_1} + \frac{1}{1/\overline{GS_r} + 1/\overline{S_rS_1}}\right) \qquad (4.3.55)$$

evaluating each of the right-hand members of a geometric system of interest, such as found in Table 4.3.5 (where A_r is A_2). As previously discussed, $\varepsilon_{G,e}/a_e$ is substituted for ε_G (or its complement for τ), and the result is weighted by the factor a_e; and for $\overline{S_rS_1}$ an additional term based on ε_G being replaced by 0 or τ by 1, with weighting $1 - a_e$, is added.

Of the cases covered in Table 4.3.5, only two will be evaluated to make A_2 represent the radiatively adiabatic zone A_r. The first is for the case of heat sink A_1 in a plane—the simulation of a slab-heating furnace. Insertion into Eq. (4.3.44) of the gray-plus-clear terms $\overline{GS_1}$, $\overline{GS_r}$, and $\overline{S_rS_1}$ from Table 4.3.6 (with subscript r replacing subscript 2) and rearrangement gives.

$$\left[\frac{(\overline{GS_1})_R}{A_1}\right]_{\substack{A_1 \text{ in a} \\ \text{plane}}} = \frac{\varepsilon_G}{D_1}\left[\varepsilon_1\left(\rho_r\frac{C_1}{C_r} + \frac{1}{1 - \varepsilon_G/a}\right)\right.$$
$$\left. + \frac{\varepsilon_r}{\dfrac{1}{\rho_1 + \dfrac{C_r}{C_1}\left(1 - \dfrac{\varepsilon_G}{a}\right)} + \dfrac{\varepsilon_G/\varepsilon_1}{a + \dfrac{(1 - a)D_1}{\varepsilon_r + \rho_r\varepsilon_1 C_1/C_r}}}\right] \qquad (4.3.56)$$

where $D_1 = 1/(1 - \varepsilon_G/a) - \rho_r[1 - (C_1/C_r)(\varepsilon_1 + \rho_1\varepsilon_G/a)]$. Although ε_G and a are used here, $\varepsilon_{G,e}$ and a_e should be used if allowance is to be made for the difference between gas emissivity and absorptivity. Com-

parison of Eq. (4.3.56) with it gray gas equivalent, Eq. (4.3.46), shows the complexity introduced by allowance for gas nongrayness. $[(\overline{GS_1})_R$ for the gray-plus-clear gas model is about 15 percent higher than for gray gas when $\varepsilon_1 = 0.8$, $\varepsilon_G = 0.3$, $C_1 = \frac{1}{3}$, $a = 0.4$, and $\varepsilon_r = 0.6$, but only 1 percent higher when $\varepsilon_r = 1$.]

The second conversion of $\overline{GS}$ to $(\overline{GS_1})_R$ will be case 4B of Table 4.3.5, the two-surface-zone enclosure with the computation simplified by assuming that the direct-view factor from any spot to a surface equals the fraction of the whole enclosure which the surface occupies (the speckled furnace model). This case can be considered an idealization of many processing furnaces such as distilling and cracking coil furnaces, with parts of the enclosure tube-covered and part left refractory. (The refractory under the tubes is not to be classified as part of the refractory zone.) Again, one starts with substitution, into Eq. (4.3.44), of the terms $\overline{GS_1}$, $\overline{GS_r}$, and $\overline{S_rS_1}$ from Table 4.3.5, case 4B, with all terms first converted to their gray-plus-clear form. To indicate the procedure, one of the components, $\overline{S_rS_1}$, will be formulated.

$$\frac{\overline{S_rS_1}}{A_1} = a\frac{C_r\varepsilon_1\varepsilon_r}{D_4'} + (1 - a)\frac{C_r\varepsilon_1\varepsilon_r}{1 - \rho_1C_1 - \rho_rC_r}$$
$$= \frac{C_r\varepsilon_1\varepsilon_r}{D_4'}\left[1 + \frac{\varepsilon_G(1 - a)(a - \varepsilon_G)}{1 - \rho_1C_1 - \rho_rC_r}\right]$$

With $D_4' = 1/(1 - \varepsilon_G/a) - \rho_1C_1 - \rho_rC_r$, the result of the full substitution simplifies to

$$\frac{(\overline{GS_1})_R}{A_1} = \frac{1}{C_1\left(\dfrac{1}{\varepsilon_G} - \dfrac{1}{a}\right) + \dfrac{1}{\varepsilon_1} + \dfrac{1/a - 1}{\varepsilon_1 + \varepsilon_r(C_r/C_1)}} \qquad (4.3.57)$$

For a gray gas ($a = 1$) the above becomes

$$\frac{(\overline{GS_1})_R}{A_1} = \frac{1}{C_1(1/\varepsilon_G - 1) + 1/\varepsilon_1} \qquad (4.3.58)$$

Equation (4.3.57) has wide applicability.

The beginning of this subsection mentions compact chambers (just treated) and long chambers as limiting cases. The latter will now be treated.

The Long Combustion Chamber If a chamber is long enough in the x direction compared to its mean hydraulic radius, the local flux from gas to wall sink comes substantially from gas at its local temperature, with $\overline{GS_1}$ [or $(\overline{GS_1})_R$] calculated by methods just described but based on a two-dimensional structure; i.e., the opposed upstream and downstream fluxes through the flow cross section will substantially cancel. That limiting case will be considered, with $(\overline{GS_1})_R/A_1$ evaluated by using local mean values of T_G and T_1. The local $(\overline{GS_1})_R$ applicable to a surface element of length dx and perimeter P is then $[(\overline{GS_1})_R/A_1]P\,dx$. Let $T_{G,\text{in}}$, $T_{G,\text{out}}$, $T_{1,\text{in}}$, and $T_{1,\text{out}}$ be specified; furnace length L is to be determined. Assume a constant sink temperature T_1 equal to the arithmetic mean gas temperature minus the logarithmic mean of the temperature difference, gas to sink, at the ends. The equation of heat transfer in the furnace length element $P\,dx$ is then

$$-\dot{m}C_p\,dT_G = P\,dx\left[\frac{(\overline{GS_1})_R}{A_1}\sigma(T_G^4 - T_1^4) + h(T_G - T_1)\right] \qquad (4.3.59)$$

The second of the heat-transfer terms is an order of magnitude smaller than the first, and to permit ready integration, $h(T_G - T_1)$ will be set equal to $b\sigma(T_G^4 - T_1^4)$, from which

$$b\sigma = \frac{h}{T_G^3 + T_G^2T_1 + T_GT_1^2 + T_1^3} = \frac{h}{4T_{G1}^3}$$

T_{G1} is the mean value of T_G and T_1, and a 10 percent error in T_1 will make but a 1 percent error in the calculated heat transfer. Then Equation (4.3.59) becomes

$$-\dot{m}C_p\,dT_G = P\,dx\left[\frac{(\overline{GS_1})_R}{A_1} + \frac{h}{4\sigma T_{G1}^3}\right]\sigma(T_G^4 - T_1^4) \qquad (4.3.60)$$

Table 4.3.6 Conversion of Total Exchange Areas for Cases of Table 4.3.5 to Their Gray-plus-Clear Values

Case 1: Plane slab A_1 and surface A_2 completing an enclosure of gas; $F_{12} = 1$

$$\frac{\overline{S_1S_2}}{A_1} = \frac{a\varepsilon_1\varepsilon_2}{D_1} + \frac{(1-a)\varepsilon_1\varepsilon_2}{1 - \rho_2(1 - \varepsilon_1 C_1/C_2)}$$

where

$$D_1 = \frac{1}{(1 - \varepsilon_G/a)} - \rho_2 \left[1 - \frac{C_1}{C_2}\left(\varepsilon_1 + \frac{\rho_1\varepsilon_G}{a} \right) \right]$$

$$\frac{\overline{GS_1}}{A_1} = \varepsilon_1\varepsilon_G \left[\frac{1/(1 - \varepsilon_G/a) - \rho_2 C_1/C_2}{D_1} \right]$$

$$\frac{\overline{GS_2}}{A_2} = \varepsilon_2\varepsilon_G \left[\frac{1/(1 - \varepsilon_G/a) + \rho_1 C_1/C_2}{D_1} \right]$$

Case 2: Infinite parallel planes, gas between; $F_{12} = F_{21} = 1$

$$\frac{\overline{S_1S_2}}{A_1} = \frac{a\varepsilon_1\varepsilon_2}{D_2} + \frac{(1-a)\varepsilon_1\varepsilon_2}{1 - \rho_1\rho_2}$$

where

$$D_2 = \frac{1}{1 - \varepsilon_G/a} - \left(1 - \frac{\varepsilon_G}{a} \right)\rho_1\rho_2$$

$$\frac{\overline{GS_1}}{A_1} = \varepsilon_1\varepsilon_G \left[\frac{1/(1 - \varepsilon_G/a) + \rho_2}{D_2} \right]$$

Case 3: Concentric spherical or infinite cylindrical surface zones, A_1 inside; $F_{12} = 1$; $F_{21} = A_1/A_2 \equiv C_1/C_2$

$$\frac{\overline{S_1S_2}}{A_1} = \frac{a\varepsilon_1\varepsilon_2}{D_3} + \frac{(1-a)\varepsilon_1\varepsilon_2}{1 - \rho_2(1 - \varepsilon_1 C_1/C_2)}$$

where

$$D_3 = \frac{1}{1 - \varepsilon_G/a} - \rho_2 \left[1 - \left(\frac{C_1}{C_2} \right)\left(\varepsilon_1 + \frac{\rho_1\varepsilon_G}{a} \right) \right]$$

$$\frac{\overline{GS_1}}{A_1} = \varepsilon_1\varepsilon_G \left[\frac{1/(1 - \varepsilon_G/a) + \rho_2 C_1/C_2}{D_3} \right]$$

$$\frac{\overline{GS_2}}{A_2} = \varepsilon_2\varepsilon_G \left[\frac{1/(1 - \varepsilon_G/a) + \rho_1 C_1/C_2}{D_3} \right]$$

Case 4A: Two-surface-zone enclosure, with F values exact

$$\frac{\overline{S_1S_1}}{A_T} = \frac{aC_1\varepsilon_1^2[1 - F_{12} + \rho_2(1 - \varepsilon_G/a)(F_{12}/C_2 - 1)]}{D_a} + \frac{(1-a)C_1\varepsilon_1^2[1 + F_{12} + \rho_2(F_{12}/C_2 - 1)]}{(F_{12}/C_2)(C_1\varepsilon_1\rho_2 + C_2\varepsilon_2\rho_1) + \varepsilon_1\varepsilon_2}$$

$$\frac{\overline{S_1S_2}}{A_T} = \frac{aC_1\varepsilon_1\varepsilon_2 F_{12}}{D_a} + \frac{(1-a)C_1\varepsilon_1\varepsilon_2 F_{12}}{(F_{12}/C_2)(C_1\varepsilon_1\rho_2 + C_2\varepsilon_2\rho_1) + \varepsilon_1\varepsilon_2}$$

$$\frac{\overline{GS_1}}{A_T} = \frac{C_1\varepsilon_1\varepsilon_G[1/(1 - \varepsilon_G/a) + \rho_2(F_{12}/C_2 - 1)]}{D_a}$$

where

$$D_a \equiv \frac{\varepsilon_G}{a} \left[\frac{1}{1 - \varepsilon_G/a} + \rho_1\rho_2 \left(\frac{F_{12}}{C_2} - 1 \right) \right] + \frac{F_{12}}{C_2}(C_1\varepsilon_1\rho_2 + C_2\varepsilon_2\rho_1) + \varepsilon_1\varepsilon_2$$

Case 4B: Spherical enclosure of two surface zones *or* speckled $A_1 : A_2$ enclosure; $F_{12} = F_{22} = C_2$; $F_{21} = F_{11} = C_1$

$$\frac{\overline{S_1S_2}}{A_1} = \frac{a\varepsilon_1\varepsilon_2 C_2}{D_4} + \frac{(1-a)\varepsilon_1\varepsilon_2 C_2}{1 - \rho_1 C_1 - \rho_2 C_2}$$

where

$$D_4 = \frac{1}{1 - \varepsilon_G/a} - \rho_1 C_1 - \rho_2 C_2$$

$$C_1 = \frac{A_1}{A_1 + A_2}$$

$$\frac{\overline{GS_1}}{A_1} = \frac{\varepsilon_1\varepsilon_G/(1 - \varepsilon_G/a)}{D_4}$$

Integration of T_G from $T_{G,\text{in}}$ to $T_{G,\text{out}}$ and of x from 0 to L, and solution for L give

$$L = \frac{\dot{m}\overline{C}_p \left(\tan^{-1}\frac{T_{G,\text{out}}}{T_1} - \tan^{-1}\frac{T_{G,\text{in}}}{T_1} - \frac{1}{2}\ln\frac{T_{G,\text{out}} - T_1 T_{G,\text{in}} + T_1}{T_{G,\text{out}} + T_1 T_{G,\text{in}} - T_1} \right)}{2PT_1^3\sigma\left[\frac{(\overline{GS_1})_R}{A_1} + \frac{h}{4\sigma T_{G1}^3} \right]}$$

(4.3.61)

Trial and error are necessary if L is specified and $T_{G,\text{out}}$ is to be found. If

L is not long, axial radiative flux becomes important and a much more complex treatment is necessary. Use of a multigas zone system is one possibility.

Partially Stirred Model of Furnace Chamber Performance An equation representing an energy balance on a combustion chamber of two surface zones—a heat sink A_1 at temperature T_1 and a refractory surface A_r assumed radiatively adiabatic at T_r—is most simply solved if the total enthalpy input H is expressed as $\dot{m}\overline{C}_p(T_F - T_0)$; $\dot{m}$ is the mass rate of fuel plus air, and T_F is a pseudo-adiabatic flame temperature based on a mean specific heat from base temperature T_0 *up to the gas exit temperature T_E* rather than up to T_F. Assume that enough stirring occurs in the chamber to produce two temperatures—the heat-transfer temperature T_G and the leaving gas enthalpy temperature T_E—the two differing by an empirical amount, zero if the stirring were perfect.

Of the many ways tried to introduce this empiricism, the best is to assume that $T_G - T_E$, expressed as a ratio to T_F, is a constant Δ. Although Δ will vary with burner type, the effects of excess air and firing rate are small, except that for very small chambers or abnormally low firing rates the predicted radiative transfer is excessive. For such an abnormal situation, wall cooling reduces the effective size of the chamber. These conditions excepted and in the absence of performance data on the subject furnace type, assume $\Delta = 0.08$, or

$$\frac{T_G - T_E}{T_F} = \Delta = 0.08$$

This assumption bypasses complex allowance for temperature variations in the chamber gas and for the effects of fluid mechanics and combustion kinetics, but at the cost of not permitting evaluation of **flux distribution** over the surface. The heat-transfer rate Q out of the gas is then $\dot{H} - \dot{m}\overline{C}_p(T_E - T_0)$ or $\dot{m}\overline{C}_p(T_F - T_E)$. A combination of energy balance and heat transfer, with the ambient temperature taken as the enthalpy base temperature T_0, gives

$$(\dot{Q} =) \ \dot{H} - \dot{m}\overline{C}_p(T_E - T_0) = \overline{(GS_1)}_R \, \sigma(T_G^4 - T_1^4) \\ + h_1 A_1(T_G - T_1) + A_0 F_0 \, \sigma(T_G^4 - T_0^4) + UA_r(T_G - T_0) \quad (4.3.62)$$

where U is the overall convection coefficient, gas through refractory to ambient.

To make the relations dimensionless, divide through by $\overline{(GS_1)}_R \sigma T_F^4$, and let all temperatures, expressed as ratios to T_F, be called T^*. For clarity the terms are tabulated:

$$\dot{m}\overline{C}_p/\overline{(GS_1)}_R\sigma T_F^3 \ [\equiv \dot{H}/\overline{(GS_1)}_R \, \sigma T_F^4(1 - T_0^*)]$$
$$= \text{dimensionless firing density } D$$

After the division, the left-hand side term of Eq. (4.3.62) = $D(1 - T_E^*)$ and the first right-hand side term = $T_G^{*4} - T_1^{*4}$.

$$\frac{h_1 A_1}{\overline{(GS_1)}_R\sigma T_F^3} = N_c, \text{ convection number (dimensionless)}$$

$$\frac{A_0 F_0}{\overline{(GS_1)}_R} = L_o, \text{ wall openings loss number (dimensionless)}$$

$$\frac{UA_r}{\overline{(GS_1)}_R\sigma T_F^3} = L_r, \text{ refractory wall loss number (dimensionless)}$$

The equation then becomes

$$D(1 - T_E^*) = T_G^{*4} - T_1^{*4} + N_c(T_G^* - T_1^*) \\ + L_o(T_G^{*4} - T_o^{*4}) + L_r(T_G^* - T_o^*) \quad (4.3.63)$$

The two unknowns T_G^* and T_E^* are reduced to one by expressing T_E^* in terms of T_G^* and Δ. Equation (4.3.63), with coefficients of T_G^{*4} and T_G^* collected, then becomes

$$T_G^{*4} + \frac{D + N_c + L_r}{1 + L_o} T_G^* \\ - \frac{T_1^{*4} + N_c T_1^* + L_o T_o^{*4} + L_r T_o^* + D(1 + \Delta)}{1 + L_o} = 0 \quad (4.3.64)$$

Although Eq. (4.3.64) is a quartic equation, it is capable of explicit solution because of the absence of second- and third-degree terms (see end of subsection). Trial and error enter, however, because $\overline{(GS_1)}_R$ and $\overline{C}_p$ are mild functions of T_G and related T_E, respectively, and a preliminary guess of T_G is necessary. Ambiguity can exist in the interpretation of terms. If part of the enclosure surface consists of screen tubes over the chamber gas exit to a convection section, radiative transfer to those tubes is included in the chamber energy balance but convection is not, because it has no effect on the chamber gas temperature. Although the results must be considered approximations, depending as they do on the empirical Δ, the equation may be used to find the effect of firing rate, excess air, and air preheat on efficiency. With some performance data available, the small effect of various factors on Δ may be found.

For the commonly encountered case of the sink consisting of a row of tubes mounted on a refractory wall, A_1 is the area of the whole plane in which the tubes lie, T_1 is tube surface temperature, and ε_1 is the effective emissivity of the tube-row-refractory-wall combination, as in the earlier numerical example associated with Fig. 4.3.6, where $\varepsilon_1 = 0.702$. A

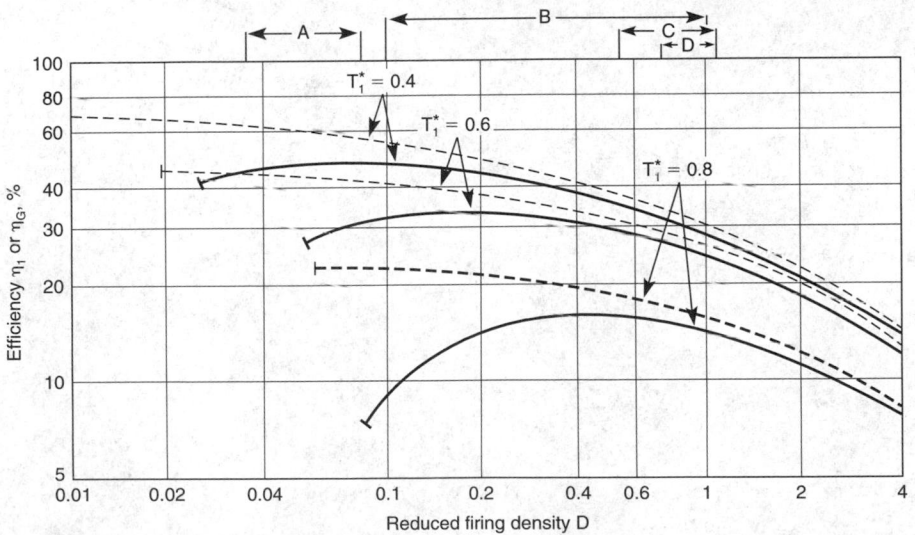

Fig. 4.3.6 The thermal performance of well-stirred furnace chambers. Conditions: $L_R = UA_R/(\overline{GS_1}\sigma T_F^3) = 0.016$; $N_c = hA_1/(\overline{GS_1}\sigma T_F^3) = 0.04$; $L_O = A_O F_O/\overline{GS_1} = 0$; $D = \dot{m}\overline{C}_p/(\overline{GS_1}\sigma T_F^3)$. Dotted lines: $\eta_G =$ (heat flux from gas)/(entering enthalpy in fuel and oxidant). Solid lines: $\eta_1 =$ (heat flux to sink)/(entering enthalpy in fuel and oxidant). Approximate range of D for various furnace classes: A, open hearths, $T_1^* = 0.7$ to 0.8; B, oil processing furnaces, $T_1^* \approx 0.4$; C, domestic boiler combustion chambers, $T_1^* \approx 0.2$; D, soaking pits, $T_1^* \approx 0.6$; gas-turbine combustors, off scale at right.

further simplification is to replace A_1/A_T by C, the "cold" fraction of the wall ($A_T \equiv A_1 + A_r$).

With T_G^* known, the chamber efficiency η_G based on heat transfer from the gas is given by

$$\eta_G = \frac{(1 - T_G^* + \Delta)}{1 - T_O^*} \tag{4.3.65}$$

The efficiency based on energy to the sink is

$$\eta_1 = \eta_G - \frac{L_O(T_G^{*4} - T_O^{*4}) + L_r(T_G^* - T_O^*)}{D(1 - T_O^*)} \tag{4.3.66}$$

or

$$\eta_1 = \frac{(GS_1)_R \sigma(T_G^4 - T_1^4) + h_1 A_1(T_G - T_1)}{H}$$

All heat transferred to the sink is included in η_1, and losses from its backside to the ambient must be subtracted.

Furnace Chamber Performance—General Although the chamber efficiency η depends on D, N_c, L_r, L_O, T_1^*, and T_O^*, the reduced firing density D is the dominant factor; it makes allowance for such operating variables as fuel type, excess air or air preheat—which affect flame temperature or gas emissivity, for fractional occupancy of the walls by sink surfaces, and for sink emissivity. Variation in the normalized sink temperature T_1^* has little effect until it exceeds 0.3; T_O^* is generally about ⅛; L_O is often negligible; N_c and L_r, though significant, are secondary.

Solution of Eq. (4.3.63) gives the relation between D and T_G^*, and Eqs. (4.3.65) and (4.3.66) give the relation between η_1 and D. As an example, Fig. 4.3.6 gives D versus η_1 (solid lines) and versus η_G (dotted lines), for values of N_c, L_r, L_O, and T_O^* of 0.04, 0.016, 0, and ⅛. Approximate operating regimes of various classes of furnaces are shown at the top of Fig. 4.3.6. Note the significant properties of the functions presented: (1) As firing rate D goes down, η_G rises, and so does η_1 until the losses due to L_O and L_r cause it to decrease. T_G^* approaches T_1^* in the limit as D decreases to $[L_O(T_1^{*4} - T_O^{*4}) + L_r(T_1^* - T_O^*)]/(1 - T_1^*)$, where $\eta_1 = 0$. (2) Changing T_1^* has a large effect only when it exceeds about 0.4. (3) As the furnace walls approach complete coverage by a black sink ($C = \varepsilon_1 = 1$), $\overline{GS_1}$ becomes $\varepsilon_G A_T$ and $D \propto 1/\varepsilon_G$. Thus, at very high firing rates where η_G approaches inverse proportionality to D, the efficiency of heat transfer varies directly as ε_G (gas-turbine chambers), but at low firing rates ε_G has relatively little effect. (4)

When $C\varepsilon_1 \ll 1$ because of a nonblack sink or much refractory surface, the effect of changing flame emissivity is to produce a much less than proportional effect on heat flux.

Equations (4.3.63) to (4.3.66) predict the effects of excess air, air preheat, and fuel quality on performance, through the effect on T_F; through $\overline{GS_1}$ they show the effect of gas and sink emissivity and the fraction of the chamber walls occupied by heat sink; through L_O and L_R they allow for external losses. They serve as a framework for correlating the performance of furnaces with flow patterns—plug flow, parabolic profile, and recirculatory flow—differing from the well-stirred model (Hottel and Sarofim, Chap. 14). As expected, plug-flow furnaces show somewhat higher efficiency, mild recirculation types somewhat lower efficiency, and strong recirculation furnaces an efficiency closely similar to that of the well-stirred model.

Explicit Solution of Limited Quartics Equations like (4.3.64) have the general form $ax^4 + bx = c$, which can be converted to $y^4 + y = B$, with y and B defined by $y \equiv (a/b)^{1/3}x$ and $B \equiv (c/a)(a/b)^{4/3}$.

An explicit solution comes from

$$k = \{[\sqrt{(B/3)^3 + 1/256} + 1/16]^{1/3} - [\sqrt{(B/3)^3 + 1/256} - 1/16]^{1/3}\}/2$$
$$y = \sqrt{\sqrt{4k^2 + B} - k} - \sqrt{k}$$

Refractory Temperature T_r Though the average value of T_r in a combustion chamber is not involved in the evaluation of T_G or η, it is sometimes of interest. It assumes a mean value between T_G and T_1, given by

$$\left(\frac{T_r}{T_G}\right)^4 = \frac{1 + E(T_1/T_G)^4}{1 + E}$$

For the speckled furnace gray gas model

$$E = C\varepsilon_1 \left(\frac{1}{\varepsilon_G} - 1\right)$$

Allowance is made for the difference between emissivity and absorptivity by changing ε_G to $\varepsilon_{G,e}$ in the above equation. For the gray-plus-clear-gas model,

$$E = C\varepsilon_1 \left[\frac{1}{\varepsilon_G} - \frac{1}{a} + \frac{1/a - 1}{C\varepsilon_1 + (1 + C)\varepsilon_R}\right]$$

4.4 TRANSMISSION OF HEAT BY CONDUCTION AND CONVECTION

by Kenneth A. Smith

REFERENCES: McAdams, "Heat Transmission," McGraw-Hill. Eckert and Drake, "Analysis of Heat and Mass Transfer," McGraw-Hill. Carslaw and Jaeger, "Conduction of Heat in Solids," Oxford. Jakob, "Heat Transfer," vols. I and II, Wiley. Kays and Crawford, "Convective Heat and Mass Transfer," 3d ed., McGraw-Hill. Wilkes, "Heat Insulation," Wiley. Kays and London, "Compact Heat Exchangers," McGraw-Hill. "Thermophysical Properties Data Book," Purdue University.

Notation and Units

The units are based on feet, pounds, hours, degrees Fahrenheit, and Btu. Any other consistent set may be used in the dimensionless relations given, but for the dimensional equations the units of this table must be used.

A = area of heat-transfer surface, ft^2
A_i = inside area
A_o = outside area
A_m = average value of A, ft^2
a = empirical constant
C_p = specific heat at constant pressure, $Btu/lb \cdot °F$
D = diameter, ft
D_o = outside diameter, ft
D_i = inside diameter, ft
D' = diameter, in
D'_o = outside diameter, in
D'_i = inside diameter, in
G = mass velocity, equals w/S, $lb/h \cdot ft^2$ of cross section occupied by fluid
G_{max} = mass velocity through minimum free area in a row of pipes normal to fluid stream, $lb/h \cdot ft^2$
g_c = conversion factor, equal to 4.18×10^8 (mass lb)(ft)/ (force lb)(h)^2
g_L = local acceleration due to gravity, $4.18 \times 10^8 ft/h^2$ at sea level
h = local individual coefficient of heat transfer, equals $dq/dA\ \Delta t$, $Btu/h(ft^2)(°F)$diff
$h_c + h_r$ = combined coefficient by conduction, convection, and radiation between surface and surroundings
h_m = mean value of h for entire surface, based on $(\Delta t)_m$
$h_{a.m.}$ = average h, arbitrarily based on arithmetic-mean temperature difference
h_s = heat-transfer coefficient through scale deposits
J = mechanical equivalent of heat, 778 ft · lb/Btu
K = empirical constant
k = thermal conductivity, $Btu/h \cdot ft \cdot °F$
$$k_m = -\frac{1}{t_1 - t_2}\int_1^2 k\,dt$$
k_f = k at the "film" temperature, $t_f = (t + t_w)/2$
l = thickness of material normal to heat flow, ft
L = length of heat-transfer surface, heated length, ft
N = number of rows of tubes
N_{Gr} = Grashof number, $L_c^3\rho_f^2 g_L\beta_f(\Delta t)_s/\mu_f^2$
q = total rate of heat flow, Btu/h
$\dot{q}$ = heat-flux vector, $Btu/h \cdot ft^2$
$\dot{q}_x$ = x component of heat flux vector
R = thermal resistances, $1/(UA)$, $1/(hA)$, $1/(h_c + h_r)A_0$
$\mathcal{R}$ = recovery factor
r = radius, ft
S = cross section, filled by fluid, in plane normal to direction of fluid flow, ft^2
T = temperature, $°R = t + 460$

T_1, T_2 = inlet and outlet bulk temperatures, respectively, of warmer fluid, °F
t = bulk temperature (based on heat balance), °F
$t_{a.w.}$ = temperature of adiabatic wall
t_∞ = temperature at infinity
t_w = wall temperature, °F
t_1, t_2 = inlet and outlet bulk temperatures of colder fluid, °F
t_i, t_o = temperatures of fluid inside and outside, °F
t_f = $(t + t_w)/2$
t_{sat} = saturation temperature, °F
U = overall coefficient of heat transfer, $Btu/h \cdot ft^2 \cdot °F$; U_i, U_o based on inside and outside surface, respectively
V = mean velocity, ft/h
V_s = average velocity, volumetric rate divided by cross section filled by fluid, ft/s
V_{sm} = maximum velocity, through minimum cross section, ft/s
x = one of the axes of a Cartesian reference frame, ft
X = $(t_2 - t_1)/(T_1 - t_1)$
w = mass rate of flow per tube, lb/h/tube
Z = $(T_1 - T_2)/(t_2 - t_1)$
β = volumetric coefficient of thermal expansion, $°F^{-1}$
Γ = mass rate of flow, $lb/(h)$ (ft of wetted periphery measured on a plane normal to direction of fluid flow); = $w/\pi D$ for a vertical and $w/2L$ for a horizontal tube
γ = ratio of specific heats, c_p/c_v; 1.4 for air
∇ = gradient operator
Δt = temperature difference, °F
$(\Delta t)_{ave}, (\Delta t)_{l.m.}$ = arithmetic and logarithmic means of terminal temperature differences, respective, °F
$(\Delta t)_m$ = true mean value of the terminal temperature differences, °F
$(\Delta t)_o$ = overall temperature difference, °F
$(\Delta t)_s$ = temperature difference between surface and surroundings, °F
λ = latent heat (enthalpy) of vaporization, Btu/lb
μ = viscosity at bulk temperature, $lbm/h \cdot ft$; equals 2.42 times centipoises; equals 116,000 times viscosity in (lb force)(s)/ft^2
μ_f = viscosity, $lbm/h \cdot ft$, at arithmetic mean of wall and fluid temperatures
μ_w = viscosity at wall temperature, $lbm/h \cdot ft$
ρ = density, lbm/ft^3
σ = surface tension, lb force/ft

Subscripts:

l = liquid
v = vapor

Preliminary Statements   The transfer of heat is usually considered to occur by three processes:

1. **Conduction** is the transfer of heat from one part of a body to another part or to another body by short-range interaction of molecules and/or electrons.

2. **Convection** is the transfer of heat by the combined mechanisms of fluid mixing and conduction.

3. **Radiation** is the emission of energy in the form of electromagnetic waves. All bodies above absolute zero temperature radiate. Radiation incident on a body may be absorbed, reflected, and transmitted. (See Sec. 4.3.)

Table 4.4.1 Thermal Conductivities of Metals*

$k = Btu/h \cdot ft \cdot °F$

$k_t = k_{t_0} - a(t - t_0)$

Substance	Temp range, °F	k_{t_0}	a	Substance	Temp range, °F	k_{t_0}	a
Metals				Tin	60–212	36	0.0135
Aluminum	70–700	130	0.03	Titanium	70–570	9	0.001
Antimony	70–212	10.6	0.006	Tungsten	70–570	92	0.02
Beryllium	70–700	80	0.027	Uranium	70–770	14	−0.007
Cadmium	60–212	53.7	0.01	Vanadium	70	20	—
Cobalt	70	28	—	Zinc	60–212	65	0.007
Copper	70–700	232	0.032	Zirconium	32	11	—
Germanium	70	34	—	Alloys:			
Gold	60–212	196	—	Admiralty metal	68–460	58.1	−0.054
Iron, pure	70–700	41.5	0.025	Brass	−265–360	61.0	−0.066
Iron, wrought	60–212	34.9	0.002	(70% Cu, 30% Zn)	360–810	84.6	0
Steel (1% C)	60–212	26.2	0.002	Bronze, 7.5% Sn	130–460	34.4	−0.042
Lead	32–500	20.3	0.006	7.7% Al	68–392	39.1	−0.038
Magnesium	32–370	99	0.015	Constantan	−350–212	12.7	−0.0076
Mercury	32	4.8	—	(60% Cu, 40% Ni)	212–950	10.1	−0.019
Molybdenum	32–800	79	0.016	Dural 24S (93.6% Al, 4.4% Cu,	−321–550	63.8	−0.083
Nickel	70–560	36	0.0175	1.5% Mg, 0.5% Mn)	550–800	130.	+0.038
Palladium	70	39	—	Inconel X (73% Ni, 15% Cr, 7%	27–1,070	7.62	−0.0068
Platinum	70–800	41	0.0014	Fe, 2.5% Ti)			
Plutonium	70	5	—	Manganin (84% Cu, 12% Mn,	1,070–1,650	3.35	−0.0111
Rhodium	70	88	—	4% Ni)	−256–212	11.5	−0.015
Silver	70–600	242	0.058	Monel (67.1% Ni, 29.2% Cu,	−415–1,470	12.0	−0.008
Tantalum	212	32	—	1.7% Fe, 1.0% Mn)			
Thallium	32	29	—	Nickel silver (64% Cu, 17% Zn,	68–390	18.1	−0.0156
Thorium	70–570	17	−0.0045	18% Ni)			

* For refractories see Sec. 6; for pipe coverings, Sec. 8; for building materials, Sec. 4. Conversion factors for various units are given in Sec. 1. Tables 4.4.3–4.4.7 were revised by G. B. Wilkes.

CONDUCTION

See Tables 4.4.1 to 4.4.7 and 4.4.10

The **basic Fourier conduction law** for an isotropic material is

$$\dot{q} = -k\nabla t \qquad (4.4.1)$$

In cartesian coordinates, the x component of this equation is $\dot{q}_x = -k(\partial t/\partial x)$, and if the heat flow is unidimensional, $q = \dot{q}A(x) = -kA(x)(dt/dx)$. This states that the steady-state rate of heat conduction q is proportional to the cross-sectional area $A(x)$ normal to the direction of flow and to the temperature gradient $\partial t/\partial x$ along the conduction path. The proportionality constant k is called the **"true" thermal conductivity** of the material.

The thermal conductivity of a given material varies with temperature, and the mean thermal conductivity is defined by

$$k_m = \frac{1}{t_0' - t_i'}\int_{t_0'}^{t_i'} k\, dt$$

Over moderate range, k varies linearly with t, and hence k_m is the value of k at the arithmetic mean of t_i' and t_0'.

Thermal Conductivity of Nickel-Chromium Alloys with Iron

$k_t = k_{t0} - a(t - t_0)$

ANSI number	Temp, range, °F	k_{t0}	a
301, 302, 303, 304 (303 Se, 304 L)	95–1,650	8.08	−0.0052
310 (3105)	32–1,650	6.85	−0.0072
314	80–572	10.01	−0.00124
	572–1,650	8.20	−0.0045
316 (316 L)	−60–1,750	7.50	−0.0042
321, 347 (348)	−100–1,650	8.22	−0.0050
403, 410 (416, 416 Se, 420)	−100–1,850	15.0	0
430 [430 F, 430 F (Se)]	122–1,650	12.60	−0.0012
440 C	212–932	12.77	−0.0043
446	32–1,850	12.96	−0.0050
501, 502	80–1,520	21.4	+0.0037

For unidimensional heat flow through a material of thickness l

$$q\int_0^l \frac{dx}{A(x)} = \int_{t_i'}^{t_0'} k\, dt = k_m(t_i' - t_0') \qquad (4.4.2)$$

with an obvious definition for the mean area:

$$\frac{q}{A_m} = k_m(t_i' - t_0')$$

For flat plates, $A_m = A_i = A_0$; for hollow cylinders, $A_m = (A_0 - A_i)/\ln(A_0/A_i)$; for hollow spheres, $A_m = \sqrt{A'A_0}$. For more complex shapes, Eq. (4.4.1) must be employed. For other configurations, mean areas may often be found elsewhere, e.g., Kutateladze and Borishanskei, "A Concise Encyclopedia of Heat Transfer," Pergamon Press, pp. 36–44.

CONDUCTION AND CONVECTION

Phenomena of Heat Transmission In many practical cases of heat transmission—e.g., boilers, condensers, the cooling of engine cylinders—heat is transmitted from one fluid to another through a wall separating the two. The processes occurring in the fluids may be extremely complex. However, to facilitate discussion, it is convenient to imagine that most of the fluid offers no resistance to heat transmission but that a thin film of fluid adjacent to the wall offers considerable resistance. This situation is depicted in Fig. 4.4.1. Then, by definition,

$$q = h_i A_i(t_i - t_i') = \frac{k}{l}A_m(t_i' - t_0') = h_0 A_0(t_0' - t_0)$$

The terms h_i and h_0 are the **film coefficients,** or **unit conductances,** of the films f_1 and f_2, respectively, and k is the thermal conductivity of the wall. Since q, A, $t_i - t_i'$, and $t_0' - t_0$ are susceptible to direct measurement, h_i and h_0 are simply defined quantities and the propriety of the above equation does not rest upon the heuristic film concept. Indeed, for laminar flow, the film concept is a gross misrepresentation, and yet the definition of a film coefficient (or heat-transfer coefficient) remains convenient and valid.

Properties of Molten Metals*

Metal (melting point)	Temperature, °F	k, Btu / (h)(ft)(°F)	ρ, lb / cu ft	c_p, Btu / (lb)(°F)	μ, lb / (ft)(h)
Bismuth (520°F)	600	9.5	625	0.0345	3.92
	1,000	9.0	608	0.0369	2.66
	1,400	9.0	591	0.0393	1.91
Lead (621°F)	700	10.5	658	0.038	5.80
	900	11.4	650	0.037	4.65
	1,300	—	633	—	3.31
Mercury (−38°F)	50	4.7	847	0.033	3.85
	300	6.7	826	0.033	2.66
	600	8.1	802	0.032	2.09
Potassium (147°F)	300	26.0	50.4	0.19	0.90
	800	22.8	46.3	0.18	0.43
	1,300	19.1	42.1	0.18	0.31
Sodium (208°F)	200	49.8	58.0	0.33	1.69
	700	41.8	53.7	0.31	0.68
	1,300	34.5	48.6	0.30	0.43
Na, 56 wt % K, 44 wt % (66.2°F)	200	14.8	55.4	0.270	1.40
	700	15.9	51.3	0.252	0.570
	1,300	16.7	46.2	0.249	0.389
Na, 22 wt % K, 78 wt % (12°F)	200	14.1	53.0	0.226	1.19
	750	15.4	48.4	0.210	0.500
	1,400	—	43.1	0.211	0.353
Pb, 44.5 wt % Bi, 55.5 wt % (257°F)	300	5.23	657	0.035	
	700	6.85	639	0.035	3.71
	1,200	—	614	—	2.78

* Based largely on "Liquid-Metals Handbook," 2d ed., Government Printing Office, Washington.

If t_i' and t_0' are eliminated from the above equation, a relation is obtained for steady flow through several resistances in series:

$$q = \frac{t_i - t_0}{1/(h_i A_i) + 1/(kA_m) + 1/(h_0 A_0)} \quad (4.4.3)$$

Each of the terms in the denominator represents a resistance to heat transfer. There may also be a resistance, $1/(h_s A_s)$, due to the presence of a scale deposit on the surface. Thus, if the overall heat transfer is given by $q = UA(t_i - t_0)$, then the total thermal resistance is given by

$$1/(UA) = 1/(h_i A_i) + 1/(kA_m) + 1/(h_o A_o) + 1/(h_s A_s) \quad (4.4.4)$$

Coefficients for scale deposits are given in Table 4.4.9.

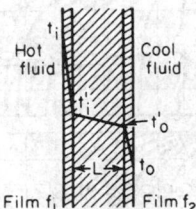

Fig. 4.4.1 Temperature gradients in heat flow through a wall.

Mean Temperature Difference The basic equation for any steadily operated heat exchanger is $dq = U(\Delta t)_o \, dA$, in which U is the overall coefficient [Eq. (4.4.4)], $(\Delta t)_o$ is the overall temperature difference between hot and cold fluids, and dq/dA is the local rate of flow per unit surface. In order to apply this relation to a finite exchanger, it is necessary to integrate it. The assumptions usually made are constant U, constant mass rates of flow, no changes in phase, constant specific heats,

and negligible heat losses. The resulting equation for parallel or countercurrent flow of fluids is

$$q = UA(\Delta t)_m = UA[(\Delta t)_{01} - (\Delta t)_{02}]/\ln [(\Delta t)_{01}/(\Delta t)_{02}] \quad (4.4.5a)$$

in which $(\Delta t)_m$ is the logarithmic mean of the terminal temperature differences, $(\Delta t)_{01}$ and $(\Delta t)_{02}$, between hot and cold fluid. The value of UA is evaluated from the resistance concept of Eq. (4.4.4) and the values of h are obtained from the following pages. For more complicated flow geometries, the logarithmic mean is not appropriate, and the true mean temperature difference may be obtained from Fig. 4.4.2, where

$$Y = \text{ordinate}$$
$$= \frac{\text{true mean temp difference}}{\text{logarithmic mean temp difference for counterflow}}$$

For the other symbols see. p. 4-79. (From *Trans. ASME,* **62,** 1940, pp. 283–294.)

The above discussion focuses on the concepts of an overall coefficient and a mean-temperature difference. An alternative approach focuses on the concepts of effectiveness and the number of transfer units. The alternatives are basically equivalent, but one or the other may enjoy a computational advantage. The latter method is presented in detail by Kays and London and by Mickley and Korchak (*Chem. Eng.,* **69,** 1962, pp. 181–188 and 239–242).

EXAMPLE. Assume an exchanger in which the hot fluid enters at 400°F and leaves at 327°F; the cold fluid enters at 100°F and leaves at 283°F. Assuming U independent of temperature, what will be the true mean temperature difference from hot to cold fluid, (1) for counterflow and (2) for a reversed current apparatus with one well-baffled pass in the shell and two equal passes in the tubes?

1. With counterflow, the terminal differences are $400 - 283 = 117°F$ and $327 - 100 = 227°F$; the logarithmic mean difference is $110/0.662 = 166°F$.

2. $Z = (400 - 327)/(283 - 100) = 0.4$; $X = (283 - 100)/(400 - 100) = 0.61$; from section A of Fig. 4.4.2, $Y = 0.9 = (\Delta t)_m/166$; $(\Delta t)_m = 149°F$.

Table 4.4.2 Thermal Conductivities of Liquids and Gases

Substance	Temp, °F	k	Substance	Temp, °F	k
Liquids:			Gases:		
Acetone	68	0.103	Air (see below)	32	0.0140
Ammonia	45	0.29	Ammonia, vapor	32	0.0126
Aniline	32	0.104	Ammonia	212	0.0192
Benzol	86	0.089	Argon	32	0.00915
Carbon bisulfide	68	0.0931	Carbon dioxide	32	0.0084
Ethyl alcohol	68	0.105		212	0.0128
Ether	68	0.0798	Carbon monoxide	32	0.0135
Glycerin, USP, 95%	68	0.165	Chlorine	32	0.0043
Kerosene	68	0.086	Ethane	32	0.0106
Methyl alcohol	68	0.124	Ethylene	32	0.0101
n-Pentane	68	0.0787	Helium	32	0.0818
Petroleum ether	68	0.0758	n-Hexane	32	0.0072
Toluene	86	0.086	Hydrogen	32	0.0966
Water	32	0.343		212	0.124
	140	0.377	Methane	32	0.0175
Oil, castor	39	0.104	Neon	32	0.0267
Oil, olive	39	0.101	Nitrogen	32	0.0140
Oil, turpentine	54	0.0734	Nitrous oxide	32	0.0088
Vaseline	59	0.106		212	0.0090
			Nitric oxide	32	0.0138
			Oxygen	32	0.0142
			n-Pentane	32	0.0074
			Sulphur dioxide	32	0.005

Thermal Conductivities of Air and Steam

Temperature, °F	32	200	400	600	800	1,000
Air, 1 atm	0.0140	0.0181	0.0225	0.0266	0.0303	0.0337
Steam, 1 lb/in² absolute	—	0.0132	0.0184	0.0238	0.0292	0.0345

SOURCE: F. G. Keyes, *Tech. Rept. 37,* Project Squid (Apr. 1, 1952).

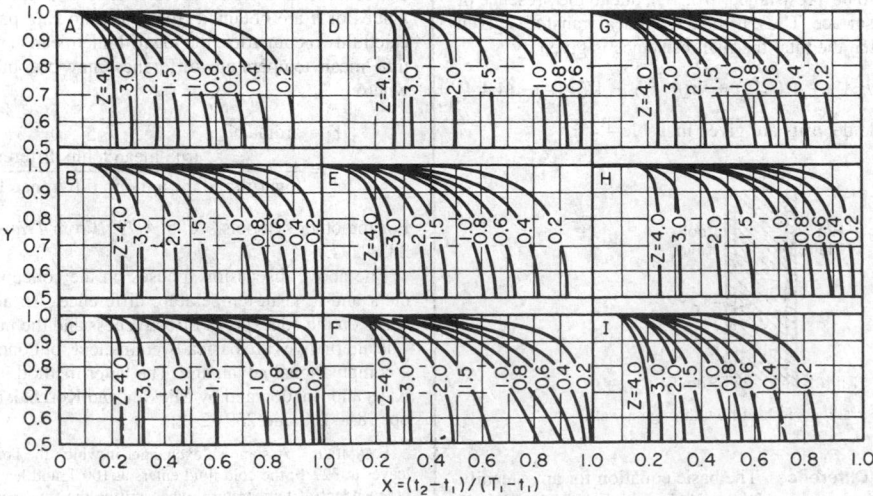

Fig. 4.4.2 (*A*) One shell pass and two tube passes; (*B*) two shell passes and four tube passes; (*C*) three shell passes and six tube passes; (*D*) four shell passes and eight tube passes; (*E*) cross flow, one shell pass and one tube pass, both fluids mixed; (*F*) single-pass cross-flow exchanger, both fluids unmixed; (*G*) single-pass cross-flow exchanger, one fluid mixed, the other unmixed; (*H*) two-pass cross-flow exchanger, shell fluid mixed, tube fluid unmixed, shell fluid first crossing the second tube pass; (*I*) same as (*H*), but shell fluid first crosses the first tube pass.

Table 4.4.3   Thermal Conductivities of Miscellaneous Solid Substances*
Values of k are to be regarded as rough average values for the temperature range indicated

Material	Bulk density, lb/ft^3	Temp, °F	k	Material	Bulk density, lb/ft^3	Temp, °F	k
Asbestos board, compressed asbestos and cement	123.	86.	0.225	Quartz, crystal, parallel to C axis	...	−300.	25.0
						0.	8.3
						300.	4.2
Asbestos millboard	60.5	86.	0.070	Rubber, hard	74.3	100.	0.092
Asbestos wool	25.	212.	0.058	Rubber, soft, vulcanized	68.6	86.	0.08
Ashes, soft wood	12.5	68.	0.018	Sand, dry	94.8	68.	0.188
Ashes, volcanic	51.	300.	0.123	Sawdust, dry	13.4	68.	0.042
Carbon black	12.	133.	0.012	Silica, fused	...	200.	0.83
Cardboard, corrugated	...	...	0.037	Silica gel, powder	32.5	131.	0.049
Celluloid	87.3	86.	0.12	Soil, dry	...	68.	0.075
Cellulose sponge, du Pont	3.4	82.	0.033	Soil, dry, including stones	127.	68.	0.30
Concrete, sand, and gravel	142.	75.	1.05	Snow	7–31	32.	0.34–1.3
Concrete, cinder	97.	75.	0.41	Titanium oxide, finely ground	52.	1000.	0.041
Charcoal, powder	11.5	63.	0.029	Wool, pure	5.6	86.	0.021
Cork, granulated	5.4	23.	0.028	Zirconia grain	113.	600.	0.11
Cotton wool	5.0	100.	0.035	Woods, oven dry, across grain†:			
Diamond	151.	70.	320.	Aspen	26.	85.	0.069
Earth plus 42% water	108.	0.	0.62	Bald cypress	24.	85.	0.063
Fiber, red	80.5	68.	0.27	Balsa	10.	85.	0.034
Flotofoam (U.S. Rubber Co.)	1.6	92.	0.017	Basswood	24.	85.	0.058
Glass, pyrex	139	200.	0.59	Douglas Fir	29.	85.	0.063
Glass, soda lime	...	200.	0.59	Elm, rock	48.	85.	0.097
Graphite, solid	93.5	122.	87.	Fir, white	26.	85.	0.069
Gravel	116.	68.	0.22	Hemlock	29.	85.	0.066
Gypsum board	51.	99.	0.062	Larch, western	36.	85.	0.078
Ice	57.5	...	1.26	Maple, sugar	43.	85.	0.094
Kaolin wool	10.6	800.	0.059	Oak, red	42.	85.	0.099
Leather, sole	62.4	...	0.092	Pine, southern yellow	35.	85.	0.078
Mica	122.	...	0.25	Pine, white	25.	85.	0.060
Pearlite, Arizona, spherical shell of siliceous material	9.1	112.	0.035	Red cedar, western	21.	85.	0.053
Polystyrene, expanded "Styrofoam"	1.7	...	0.021	Redwood	25.	85.	0.062
				Spruce	21.	85.	0.052
Pumice, powdered	49.	300.	0.11				
Quartz, crystal, perpendicular to C axis	...	−300.	12.5				
		0.	4.3				
		300.	2.3				

* The thermal conductivity of different materials varies greatly. For metals and alloys k is high, while for certain insulating materials, such as glass wool, cork, and kapok, it is very low. In general, k varies with the temperature, but in the case of metals, the variation is relatively small. With most other substances, k increases with rising temperatures, but in the case of many crystalline materials, the reverse is true.

† With heat flow parallel to the grain, k may be 2 to 3 times that with heat flow perpendicular to the grain, the values for wool are taken chiefly from J. D. MacLean, *Trans. ASHRAE*, **47**, 1941, p. 323.

If one of the temperatures remains constant, as in a condenser or in an evaporative cooler, Eq. (4.4.5a) applies for parallel flow, counterflow, reversed current, and cross flow.

If U varies considerably with temperature, the apparatus should be considered to be divided into stages, in each of which the variation of U with temperature or temperature difference is linear. Then for parallel or counterflow operation, the following relation may be applied to each stage:

$$q = \frac{A[U_2(\Delta t)_{01} - U_1(\Delta t)_{02}]}{\ln [U_2(\Delta t)_{01}/U_1(\Delta t)_{02}]} \qquad (4.4.5b)$$

FILM COEFFICIENTS

The important physical properties which affect film coefficients (see Sec. 4.1) are thermal conductivity, viscosity, density, and specific heat. Factors within the control of the designer include fluid velocity and shape and arrangement of the heating surface. With forced flow of gases or water, under the conditions usually met in practice, the flow is turbulent (see Sec. 3) and under these conditions the film coefficient can be greatly increased by increasing the velocity of the fluid at the expense of a greater power requirement. For a given velocity and fluid, the film coefficient depends upon the direction of flow of fluid relative to the heating surface. With free or natural convection, for a given arrangement of surface, the film coefficient depends on an additional fluid property, the coefficient of thermal expansion, on the temperature difference between surface and fluid, and on the local gravitational acceleration. With forced convection at low rates of flow, particularly with viscous fluids such as oils, laminar motion may prevail and the film coefficient depends on thermal conductivity, specific heat, mass rate of flow per tube, and length and diameter of the tube. In any event, the film coefficients h are correlated in terms of dimensionless groups of the controlling factors.

Turbulent Flow inside Clean Tubes (No Change in Phase), $DG/\mu_f > 7,000$

$$\frac{h_m}{C_p G}\left(\frac{C_p \mu_f}{k_f}\right)^{2/3} = \frac{0.023}{(DG/\mu_f)^{0.2}} \qquad (4.4.6a)$$

Table 4.4.4   Thermal Conductivities for Building Insulation

Material	Bulk density, lb/ft^3	Temp, °F	k
Balsam wool, blanket	3.6	70.	0.021
Cabot's Quilt, eelgrass	15.6	86.	0.027
Glass wool, blanket	3.25	100.	0.022
Hairfelt, blanket	11.0	86.	0.022
Insulating boards, Insulite, Celotex, etc.	12–19	100.	0.027–0.031
Kapok, DryZero, blanket	1.6	75.	0.019
Redwood bark, loose, shredded, Palco Bark	4.0	100.	0.025
Rock wool, loose	7.	117.	0.024
Sil-O-Cel powder	10.6	86.	0.026
Vermiculite, loose, Zonolite	8.2	60.	0.038

Table 4.4.5   Thermal Conductivities of Material for Refrigeration and Extreme Low Temperatures

Material	Bulk density, lb/ft^3	Temp, °F	k
Corkboard	6.9	100	0.022
		− 100	0.018
		− 300	0.010
Fiberglas with asphalt coating (board)	11.0	100	0.023
		− 100	0.014
		− 300	0.007
Glass blocks, expanded cellular glass	8.5	100	0.033
		− 100	0.024
		− 300	0.016
Mineral wool board, Rockcork	14.3	100	0.024
		− 100	0.017
		− 300	0.008
Silica aerogel, powder, Santocel	5.3	100	0.013
		0	0.012
		− 100	0.010
Vegetable fiberboard, asphalt coating	14.4	100	0.028
		− 100	0.021
		− 300	0.013
Foams:			
Polystyrene*	2.9	− 100	0.015
Polyurethane†	5.0	− 100	0.019

* Test space pressure, 1.0 atm; $k = 0.0047$ at 10^{-5} mmHg.
† Test space pressure, 1.0 atm; $k = 0.007$ at 10^{-3} mmHg.

For L/D less than 60, multiply the right-hand side of Eqs. (4.4.6a), (4.4.6b), and (4.4.6c) by $1 + (D/L)^{0.7}$.

Turbulent Flow of Gases inside Clean Tubes, $DG/\mu_f > 7,000$

$$h_m = 0.024 C_p G^{0.8}/(D_i')^{0.2} \qquad (4.4.6b)$$

Turbulent Flow of Water inside Clean Tubes, $DG/\mu_f > 7,000$

$$h_m = 160(1 + 0.012 t_f)V_s^{0.8}/(D_i')^{0.2} \qquad (4.4.6c)$$

Turbulent Flow of Liquid Metals inside Clean Tubes, $C_p \mu/k < 0.05$ The equation of Sleicher and Tribus ("Recent Advances in Heat Transfer," p. 281, McGraw-Hill, 1961) is recommended for isothermal tube walls:

$$\frac{h_m D}{k} = 6.3 + 0.016 \left(\frac{DGC_p}{k}\right)^{0.91} \left(\frac{C_p \mu}{k}\right)^{0.3} \qquad (4.4.6d)$$

Turbulent Flow of Gases or Water in Annuli   Use Eq. (4.4.6b) or (4.4.6c), with D' taken as the clearance, inches. If the clearance is comparable to the diameter of the inner tube, see Kays and Crawford.

Water in Coiled Pipes   Multiply h_m for the staight pipe by the term $1 + 3.5 D_i/D_c$, where D_i is the inside diameter of the pipe and D_c is that of the coil.

Turbulent Boundary Layer on a Flat Plate, $V_\infty \rho_f x/\mu_f > 4 \times 10^5$, no pressure gradient

$$\frac{h}{\rho_f C_p V_\infty} \left(\frac{C_p \mu}{k}\right)^{2/3} = \frac{0.0148}{(\rho_f V_\infty x/\mu_f)^{0.2}} \qquad (4.4.6e)$$

$$\frac{h_m}{\rho_f C_p V_\infty} \left(\frac{C_p \mu}{k}\right)_f^{2/3} = \frac{0.0185}{(\rho_f V_\infty L/\mu_f)^{0.2}} \qquad (4.4.6f)$$

Fluid Flow Normal to a Single Tube, $D_o G/\mu_f$ from 1,000 to 50,000

$$\frac{h_m D_o}{k_f} = 0.26 \left(\frac{D_o G}{\mu_f}\right)^{0.6} \left(\frac{C_p \mu}{k}\right)_f^{0.3} \qquad (4.4.7)$$

Gas Flow Normal to a Single Tube, $D_o G/u_f$ from 1,000 to 50,000

$$h_m = 0.30 C_p G^{0.6}/(D_o')^{0.4} \qquad (4.4.7a)$$

Fluid Flow Normal to a Bank of Staggered Tubes, $D_o G_{max}/\mu_f$ from 2,000 to 40,000

$$\frac{h_m D_o}{k_f} = K \left(\frac{C_p \mu}{k}\right)_f^{1/3} \left(\frac{D_o G_{max}}{\mu_f}\right)^{0.6} \qquad (4.4.8)$$

Values of K are given in Table 4.4.8.

Water Flow Normal to a Bank of Staggered Tubes, $D_o G_{max}/\mu_f$ from 2,000 to 40,000

$$h_m = 370(1 + 0.0067 t_f)V_{sm}^{0.6}/(D_o')^{0.4} \qquad (4.4.8a)$$

Table 4.4.6   Thermal Conductivities of Insulating Materials for High Temperatures

Material	Bulk density, lb/ft^3	Max temp, °F	k 100°F	300°F	500°F	1,000°F	1,500°F	2,000°F
Asbestos paper, laminated	22.	400	0.038	0.042				
Asbestos paper, corrugated	16.	300	0.031	0.042				
Diatomaceous earth, silica, powder	18.7	1,500	0.037	0.045	0.053	0.074		
Diatomaceous earth, asbestos and bonding material	18.	1,600	0.045	0.049	0.053	0.065		
Fiberglas block, PF612	2.5	500	0.023	0.039				
Fiberglas block, PF614	4.25	500	0.021	0.033				
Fiberglas block, PF617	9.	500	0.020	0.033				
Fiberglas, metal mesh blanket, #900	......	1,000	0.020	0.030	0.040			
Cellular glass blocks, ave. value	8.5	900	0.033	0.045	0.062			
Hydrous calcium silicate, "Kaylo"	11.	1,200	0.032	0.038	0.045			
85% magnesia	12.	600	0.029	0.035				
Micro-quartz fiber, blanket	3.	3,000	0.021	0.028	0.042	0.075	0.108	0.142
Potassium titanate, fibers	71.5	.....	.....	0.022	0.024	0.030		
Rock wool, loose	8–12	.....	0.027	0.038	0.049	0.078		
Zirconia grain	113.	3,000	.....	.....	0.108	0.129	0.163	0.217

Table 4.4.7 Thermal Conductance across Airspaces
Btu/(h)(ft²)—Reflective insulation

Airspace, in	Direction of heat flow	Temp diff, °F	Mean temp, °F	Aluminum surfaces, $\varepsilon = 0.05$	Ordinary surfaces, nonmetallic, $\varepsilon = 0.90$
Horizontal, ¾–4 across	Upward	20.	80.	0.60	1.35
Vertical, ¾–4 across	Across	20.	80.	0.49	1.19
Horizontal, ¾ across	Downward	20.	75.	0.30	1.08
Horizontal, 4 across	Downward	20.	80.	0.19	0.93

Table 4.4.8 Values of K for N Rows Deep

N	1	2	3	4	5	6	7	10
K	0.24	0.25	0.27	0.29	0.30	0.31	0.32	0.33

Table 4.4.9 Heat-Transfer Coefficients h_s for Scale Deposits from Water*
For use in Eq. (4.4.4)

	Temp of heating medium			
	Up to 240°F		240 to 400°F	
	Temp of water			
	125°F or less		Above 125°F	
	Water velocity, ft/s			
	3 and less	Over 3	3 and less	Over 3
Distilled	2,000	2,000	2,000	2,000
Sea water	2,000	2,000	1,000	1,000
Treated boiler feedwater	1,000	2,000	500	1,000
Treated makeup for cooling tower	1,000	1,000	500	500
City, well, Great Lakes	1,000	1,000	500	500
Brackish, clean river water	500	1,000	330	500
River water, muddy, silty†	330	500	250	330
Hard (over 15 grains per gal)	330	330	200	200
Chicago Sanitary Canal	130	170	100	130

Miscellaneous cases: Refrigerating liquids, brine clean petroleum distillates, organic vapors, 1,000; refrigerant vapor, 500; vegetable oils, 330; fuel oil (topped crude), 200.
* From standards of Tubular Exchanger Manufacturers Assoc., 1952.
† Delaware, East River (NY), Mississippi, Schuylkill, and New York Bay.

For baffled exchangers, to allow for leakage of fluids around the baffles, use 60 percent of the values of h_m from Eq. (4.4.8); for tubes in line, deduct 25 percent from the values of h_m given by Eq. (4.4.8).

Water Flow in Layer Form over Horizontal Tubes, $4\Gamma/\mu < 2,100$

$$h_{a.m.} = 150(\Gamma/D_o')^{1/3} \tag{4.4.9}$$

for Γ ranging from 100 to 1,000 lb of water per h per ft (each side).

Water Flow in Layer down Vertical Tubes, $w/\pi D > 500$

$$h_m = 120\Gamma^{1/3} \tag{4.4.9a}$$

Heat Transfer to Gases Flowing at Very High Velocities If a nonreactive gas stream is brought to rest adiabatically, as at the true stagnation point of a blunt body, the temperature rise will be

$$t_s - t_\infty = V^2/(2g_c JC_p) \tag{4.4.9b}$$

where t_s is the stagnation temperature and t_∞ is the temperature of the free stream moving at velocity V. At every other point on the body, the gas is brought to rest partly by pressure changes and partly by viscous effects in the boundary layer. In general, this process is not adiabatic, even though the body transfers no heat. The thermal conductivity of the gas will transfer heat from one layer of gas to another. At an insulated surface, the gas temperature will therefore be neither the free-stream temperature nor the stagnation temperature. In general, the rise in gas temperature will be given by the equation

$$t_{aw} - t_\infty = \mathscr{R}(t_s - t_\infty) = \mathscr{R}V^2/(2g_c JC_p) \tag{4.4.9c}$$

where t_{aw} is the gas temperature at the adiabatic wall and $\mathscr{R}$ is the recovery factor.

If a given point on the surface of a body is not at the temperature t_{aw} given by Eq. (4.4.9c) with the proper local value of $\mathscr{R}$ inserted, there will be a transfer of heat to or from the body. This suggests defining the coefficient of heat transfer in the usual way, except that the difference $t_w - t_{aw}$ should be used:

$$q/A = h(t_\infty - t_{aw}) = h\{t_w - [t + \mathscr{R}V^2/(2g_c JC_p)]\} \tag{4.4.9d}$$

where t_w is the surface temperature of the heated wall. With this modification, it is found that the correlations for h are nearly independent of Mach number; e.g., Eq. (4.4.6a) may be used for turbulent, compressible flow in a pipe. Obviously, $\mathscr{R} = 1.0$ at a forward stagnation point. For flows parallel to surfaces which have little or no curvature in the direction of flow, the following are recommended:

Laminar flow $\mathscr{R} = \left(\dfrac{C_p\mu}{k}\right)^{1/2}$

Turbulent flow $\mathscr{R} = \left(\dfrac{C_p\mu}{k}\right)^{1/3}$

Very little is presently known about point values of the recovery factor for flow over more complex shapes. Thus, special thermocouples should be used to measure the temperature of high-velocity gas streams (Hottel and Kalitinsky, *Jour. Applied Mechanics,* 1945, pp. A25–A32;

Table 4.4.10 Typical Values of h_m for Heating and Cooling, Forced Convention
$D_0' = 1.31$ ln, $D_i' = 1.05$ in

Fluid and arrangement	t_f, °F	Velocity*		Btu/(h · ft² · °F)	Eq. no.
		ft/s	lb/(h · ft²)		
Air inside tubes	.....	$V_S = 31.8$, $G = 8,600$		8.0	4.4.6b
Air normal to staggered tubes	170	$V_S = 8.92$, $G_m = 2,000$		7.5	4.4.8
Water inside tubes	100	$V_S = 5.0$, $G = 1.12 \times 10^6$		1260	4.4.6c
Water normal to staggered tubes	100	$V_S = 2.0$, $G_m = 0.448 \times 10^6$		800	4.4.8a
Trickle cooler, water	.....	$\Gamma = 100$ lb/(h · ft)		640	4.4.9
Falling water film vertical tube	.....	$\Gamma = 1,000$ lb/(h · ft)		1200	4.4.9a

* Velocity in ft/s at 70°F and 1 atm = $G/3,600\rho$.

and Franz, *Jahrb* 1938 *deut. Luftfahrt-Forsch* II, pp. 215–218). Eckert (*Trans. ASME,* **78,** 1956, pp. 1273–1283) recommends that all property values be evaluated at a film temperature defined by

$$t_f = (t_\infty + t_w)/2 + 0.22(t_{aw} - t_\infty) \qquad (4.4.9e)$$

Nielsen (*NACA Wartime Rep.* L-179) gives graphs for predicting the heat transfer and pressure drop for airflow at Mach numbers up to 1.0, in tubes having a uniform wall temperature.

Heat transfer from a reacting gas to a surface is treated by Lees ("Recent Advances in Heat and Mass Transfer," p. 161, McGraw-Hill).

LAMINAR FLOW

PIPE FLOW, $DG/\mu < 2{,}100$. Use the Sieder-tate modification of the Graetz equation for isothermal tube walls and $wC_p/kL > 10$:

$$h_{a.m.}D/k = 2.0(wC_p/kL)^{1/3}(\mu/\mu_w)^{0.14} \qquad (4.4.10)$$

or

$$(h_{a.m.}/C_p G)(C_p\mu/k)^{2/3}(\mu_w/\mu)^{0.14} = 1.85(D/L)^{1/3}(DG/\mu)^{-2/3} \qquad (4.4.10a)$$

As shown in Fig. 4.4.3, as DG/μ increases from 2,100 to 7,000, the effect of L/D diminishes and finally becomes negligible for $L/D > 60$.

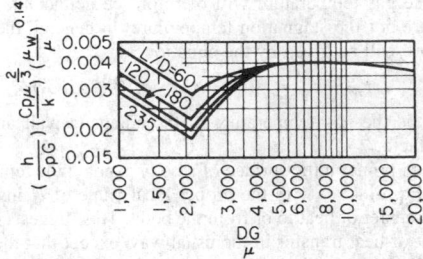

Fig. 4.4.3 Heating and cooling of viscous oils flowing inside tubes. [The curves for DG/μ below 2,100 are based on Eq. (4.4.10).]

LAMINAR BOUNDARY LAYER ON A FLAT PLATE. $\rho V_\infty x/\mu < 4 \times 10^3$, isothermal plate, no pressure gradient

$$\frac{h}{\rho_f C_p V_\infty}\left(\frac{C_p\mu}{k}\right)_f^{2/3} = \frac{0.332}{(\rho_f V_\infty x/\mu_f)^{1/2}} \qquad (4.4.10b)$$

$$\frac{h_m}{\rho_f C_p V_\infty}\left(\frac{C_p\mu}{k}\right)_f^{2/3} = \frac{0.664}{(\rho_f V_\infty L/\mu_f)^{1/2}}$$

Extended Surfaces Fin efficiency is defined as the ratio of the mean temperature difference from surface to fluid divided by the temperature difference from fin to fluid at the base or root of the fin. Graphs of fin efficiency for extended surfaces of various types are given by Gardner (*Trans. ASME,* **67,** pp. 621–628, 1945) and in numerous texts, e.g., by Eckert and Drake, pp. 92–93. Heat-transfer coefficients for various extended surfaces are given by Kays and London.

Natural Convection Heat transfer by natural convection is governed by relations of the form

$$\frac{h_m L_c}{k_f} = f[L_c^3 \rho_f^2 g_L \beta_f (\Delta t)_s/\mu_f^2, \ (C_p\mu/k)_f] \qquad (4.4.11)$$

where β_f is defined by the equation $\rho_f = \rho_\infty[1 - \beta_f(\Delta t)_s]$. For perfect gases, $\beta_f = 1/T_\infty$. The dimensionless group $L_c^3\rho_f^2 g_L\beta_f(\Delta t)_s/\mu_f^2 \equiv N_{Gr}$ represents the ratio of the product (inertial force times buoyant force) to (viscous force squared).

If the flow is of the laminar boundary layer type and if $(C_p\mu/k)_f > 1$, an effective correlation is

$$\frac{h_m L_c}{k_f} = B_1[N_{Gr}(C_p\mu/k)_f]^{0.25} \qquad (4.4.11a)$$

where B_1 is a weak function of $(C_p\mu/k)_f$. Similarly, for $(C_p\mu/k)_f < 1$,

$$\frac{h_m L_c}{k_f} = B_2[N_{Gr}(C_p\mu/k)_f^2]^{0.25} \qquad (4.4.11b)$$

VERTICAL FLAT PLATES. For this case, $L = L_c$ and the flow of the laminar boundary layer type will be laminar if

$$(C_p\mu/k)_f > 1 \qquad 10^9 > N_{Gr}(C_p\mu/k)_f > 10^4$$
$$(C_p\mu/k)_f < 1 \qquad ? > N_{Gr}(C_p\mu/k)_f^2 > 10^4$$

Lefevre (*Rept. Heat* 113, National Engineering Laboratory, Great Britain, Aug. 1956) gives an interpolation formula which contains the proper limiting forms and is in complete agreement with existing numerical results:

$$\frac{h_m L_c}{k_f} = \left[\frac{N_{Gr}(C_p\mu/k)^2}{2.435 + 4.884(C_p\mu/k)^{1/2} + 4.953(C_p\mu/k)_f}\right]^{0.25} \qquad (4.4.11c)$$

If $(C_p\mu/k)_f$ is in the vicinity of unity and if $N_{Gr}(C_p\mu/k)_f > 10^9$, the boundary layer will be turbulent and

$$\frac{hL}{k_f} = 0.13[N_{Gr}(C_p\mu/k)_f]^{1/3} \qquad (4.4.11d)$$

HORIZONTAL CYLINDERS. Replace L in the vertical flat plate formulas by $\pi D_o/2$.

HEATED HORIZONTAL PLATES FACING UPWARD OR COOLED HORIZONTAL PLATES FACING DOWNWARD.

$$2 \times 10^7 > N_{Gr}(C_p\mu/k)_f > 10^5$$
$$\frac{h_m L}{k_f} = 0.54[N_{Gr}(C_p\mu/k)_f]^{0.25} \qquad (4.4.11e)$$
$$N_{Gr}(C_p\mu/k)_f > 2 \times 10^7$$
$$\frac{h_m L}{k_f} = 0.14[N_{Gr}(C_p\mu/k)_f]^{1/3} \qquad (4.4.11f)$$

HEATED HORIZONTAL PLATES FACING DOWNWARD OR COOLED HORIZONTAL PLATES FACING UPWARD.

$$3 \times 10^{10} > N_{Gr}(C_p\mu/k)_f > 3 \times 10^5$$
$$\frac{hL}{k_f} = 0.27[N_{Gr}(C_p\mu/k)_f]^{0.25} \qquad (4.4.11g)$$

Equations (4.4.11e) to (4.4.11g) should not be considered reliable if $(C_p\mu/k)_f$ differs greatly from unity.

For more complex systems, it is best to consult plots of experimental data (McAdams).

For any particular fluid, the above equations may be greatly simplified. For air which is at room temperature and atmospheric pressure and is subjected to the gravitational attraction at sea level:

VERTICAL PLATES.

$$10^3 > L^3(\Delta t)_s > 10^{-2}$$
$$h_m = 0.28[(\Delta t)_s/L]^{0.25} \qquad (4.4.12a)$$
$$L^3(\Delta t)_s > 10^3$$
$$h_m = 0.19(\Delta t)_s^{1/3} \qquad (4.4.12b)$$

HORIZONTAL CYLINDERS.

$$10^2 > D^3(\Delta t)_s > 10^{-3}$$
$$h_m = 0.25[(\Delta t)_s/D]^{0.25} \qquad (4.4.12c)$$
$$D^3(\Delta t)_s > 10^2$$
$$h_m = 0.19(\Delta t)_s^{1/3} \qquad (4.4.12d)$$

HEATED HORIZONTAL PLATES FACING UPWARD OR COOLED HORIZONTAL PLATES FACING DOWNWARD.

$$10 > L^3(\Delta t)_s > 0.1$$
$$h_m = 0.27[(\Delta t)_s/L]^{0.25} \qquad (4.4.12e)$$
$$10^4 > L^3(\Delta t)_s > 10$$
$$h_m = 0.22(\Delta t)_s^{1/3} \qquad (4.4.12f)$$

Heated Horizontal Plates Facing downward or Cooled Horizontal Plates Facing upward.

$$10^4 > L^3(\Delta t)_s > 0.1 \qquad (4.4.12g)$$
$$h_m = 0.12[(\Delta t)_s/L]^{0.25}$$

Condensing Vapors If the condensate of a single pure vapor, saturated or supersaturated, wets the surface, film-type condensation is obtained. The rate of heat transfer equals $h_m(\Delta t)_m$, where $(\Delta t)_m$ is the mean difference between the saturation temperature and the temperature of the surface. As long as the condensate flow is laminar ($4\Gamma/\mu_f < 2{,}100$), the following dimensionless equations may be used:

For **horizontal tubes,**

$$h_m D/k = 0.73[D^3\rho^2\lambda g_L/k\mu_f N(\Delta t)_m]^{0.25}$$
$$= 0.76(D^3\rho^2 g_L/\mu_f\Gamma)^{1/3} \qquad (4.4.13)$$

For **vertical tubes,**

$$h_m L/k = 0.94[L^3\rho^2\lambda g_L/k\mu_f(\Delta t)_m]^{0.25}$$
$$= 0.93(L^3\rho^2 g_L/\mu_f\Gamma)^{1/3} \qquad (4.4.13a)$$

The equations show that a tube of given dimensions, for the usual case where $L/(ND)$ is greater than 2.76, is more effective in a horizontal than in a vertical position. Thus for $L/(ND) = 100$, a horizontal tube gives an average h which is 2.5 times that for a vertical tube. Since there is but little variation in the thermal conductivity or viscosity of the condensate at the condensing temperature at 1 atm, there is little variation in h_m. With horizontal tubes, use h_m from 200 to 400 Btu/h · ft² · °F for the following vapors condensing at atmospheric pressure: benzene, carbon tetrachloride, dichlormethane, dichlordifluoromethane, diphenyl ethyl alcohol, heptane, hexane, methyl alcohol, octane, toluene, and xylene. Ammonia gives h_m of 1,000, and mixtures of steam and organic vapors, forming immiscible condensates, give h_m ranging from 250 to 750, increasing with increasing proportion of steam. With film-type condensation of clean steam on horizontal tubes, h_m ranges from 1,000 to 3,000; see Eq. (4.4.13). With vertical tubes 10 to 20 ft long, ripples form in the film; values of h_m from Eq. (4.4.13a) should be increased 20 percent.

For long vertical tubes, $4\Gamma/\mu_f$ may exceed 2,100; in that case:

$$h_m(\mu_f^2/k_f^3\rho_f^2 g_L)^{1/3} = 0.0077(4\Gamma/\mu_f)^{0.4} \qquad (4.4.13b)$$

The presence of **noncondensible gas,** such as air, seriously reduces h, and consequently all vapor-heated apparatus should be well vented.

With steam, small traces of certain promoters (Nagle, U.S. Patent 1,995,361) such as oleic acid and benzyl mercaptan become adsorbed in a very thin layer on the surface of the tubes, preventing the condensate from wetting the metal and inducing dropwise condensation, which gives much higher values of h_m (7,000 to 70,000) than film-type condensation. However, with dirty or corroded surfaces, it is difficult to maintain dropwise condensation. Figure 4.4.4 shows overall coefficients U_o for condensing steam at 1 atm on a vertical 10-ft length of copper tube, ⅝ in OD, 0.049-in wall, at various water velocities.

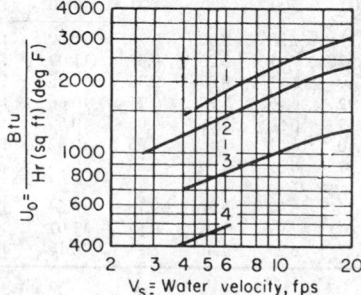

Fig. 4.4.4 Overall coefficients between condensing steam and water. Curve 1, chromium-plated copper, oleic acid; curve 2, copper, benzyl mercaptan; curve 3, copper, oleic acid; curve 4, admiralty metal, no promoter.

Boiling Liquids The nature of the heat transfer from a submerged heater to a pool of boiling water is shown in Fig. 4.4.5. Other liquids exhibit the same qualitative features. In the range AB, heat transfer to the liquid occurs solely by natural convection, and evaporation occurs at the free surface of the pool. In the range BC, **nucleate boiling** occurs. Bubbles form at active nuclei on the heating surface, detach, and rise to the pool surface. At point C, the heat flux passes through a maximum at

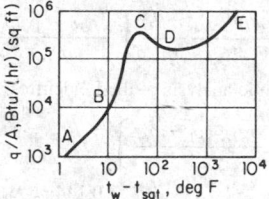

Fig. 4.4.5 Boiling of water at 212°F on a platinum surface.

a temperature difference called the **critical** Δt. In the range CD, transitional boiling occurs. At point D, the transition is complete and the heating surface is completely blanketed by a vapor film. This is the point of minimum heat flux, or the **Leidenfrost point.** In the range DE, the heating surface continues to be blanketed by a vapor film.

The range AB is adequately correlated by the usual natural-convection equations. No truly adequate correlation is available for the range BC because the complex processes of nucleation and interfacial interaction are only partially understood. However, the relation due to Rohsenow (*Trans. ASME,* **74,** 1952, pp. 969–976) is one of the best and can be reliably used for modest extrapolations of existing data.

$$\frac{C_{p,l}(t_w - t_{\text{sat}})}{\lambda} = C_{fs}\left[\frac{q/A}{\mu_l\lambda}\sqrt{\frac{g_c\sigma}{g_L(\rho_l - \rho_v)}}\right]^{1/3}\left(\frac{C_{p,l}\mu_l}{k_l}\right)^{1.7} \qquad (4.4.14a)$$

The value of the constant C_{fs} is intimately dependent on the nature of the particular fluid-solid pair and must be determined by experiment. It usually assumes values in the range $0.003 < C_{fs} < 0.05$ and is not affected by moderate subcooling or the shape of the heating surface.

Zuber (*USAEC Rep.* AECU-4439, June, 1959) has presented a theoretical equation for the maximum heat flux from a flat, horizontal surface. The analysis is based on considerations of hydrodynamic stability. For saturated liquids,

$$(q/A)_{\text{max}} = K_1\rho_v\lambda\left[\frac{\sigma g_L g_c(\rho_l - \rho_v)}{\rho_v^2}\right]^{1/4}\left(\frac{\rho_l}{\rho_l + \rho_v}\right)^{1/2} \qquad (4.4.14b)$$
$$0.12 < K_1 < 0.157 \qquad \text{(theoretical)}$$

Berenson (Sc.D. thesis, Mechanical Engineering Department, MIT, 1960) used a similar analysis and obtained a relation which is identical for $\rho_l \gg \rho_v$, but he found that $K_1 = 0.18$ gives better agreement with the data. The theoretical basis of this equation has been subject to attack,

Table 4.4.11 Maximum Flux and Corresponding Overall Temperature Difference for Liquids Boiled at 1 atm with a Submerged Horizontal Steam-Heated Tube

Liquid	Aluminum		Copper		Chromium-plated copper		Steel	
	$\frac{q/A}{1{,}000}$	$(\Delta t)_o$	$\frac{q/A}{1{,}000}$	$(\Delta t)_o$	$\frac{q/A}{1{,}000}$	$(\Delta t)_o$	$\frac{q/A}{1{,}000}$	$(\Delta t)_o$
Ethyl acetate	41	70	61	55	77	55		
Benzene	51	80	58	70	73	100	82	100
Ethyl alcohol	55	80	85	65	124	65		
Methyl alcohol	.....	....	100	95	110	110	155	110
Distilled water	.....	....	230	85	350	75	410	150

but the correlation appears to be the best available. Zuber also performed an analysis for subcooled liquids and proposed a modification which is also in excellent agreement with experiment:

$$\left(\frac{q}{A}\right)_{\text{max}} = K_1\rho_v[\lambda + C_{p,l}(t_{\text{sat}} - t_l)]\left[\frac{\sigma g_L g_c(\rho_l - \rho_v)}{\rho_v^2}\right]^{0.25}$$

$$\times \left(\frac{\rho_l}{\rho_l + \rho_v}\right)^{1/2}$$

$$\times \left\{1 + \frac{5.33(\rho_l C_{p,l} k_l)^{1/2}(t_{\text{sat}} - t_l)}{\rho_v[\lambda + C_{p,l}(t_{\text{sat}} - t_l)]}\left[\frac{g_L(\rho_l - \rho_v)\rho_v^2}{\sigma^3 g_c^3}\right]^{1/8}\right\} \quad (4.4.14c)$$

Zuber's hydrodynamic analysis of the Leidenfrost point yields

$$(q/A)_{\text{min}} = K_2\lambda\rho_v\left[\frac{\sigma g_L g_c(\rho_l - \rho_v)}{\rho_l^2}\right]^{1/4}$$

$$0.144 < K_2 < 0.177 \quad (4.4.14d)$$

Berenson finds better agreement with the data if $K_2 = 0.09$. For very small wires, the heat flux will exceed that predicted by this flat-plate formula. A reliable prediction of the critical temperature is not available.

For nucleate boiling accompanied by forced convection, the heat flux may be approximated by the sum of the heat flux for pool boiling alone and the heat flux for forced convection alone. This procedure will not be satisfactory at high qualities, and no satisfactory correlation exists for the maximum heat flux.

For a given liquid and system pressure, the nature of the surface may substantially influence the flux at a given (Δt), Table 4.4.11. These data may be used as rough approximations for a bank of submerged tubes. Film coefficients for scale deposits are given in Table 4.4.9.

For **forced-circulation evaporators**, vapor binding is also encountered. Thus with liquid benzene entering a 4-pass steam-jacketed pipe at 0.9 ft/s, up to the point where 60 percent by weight was vaporized, the maximum flux of 60,000 Btu/h · ft² was obtained at an overall temperature difference of 60°F; beyond this point, the coefficient and flux decreased rapidly, approaching the values obtained in superheating vapor, see Eq. (4.4.6b). For comparison, in a natural convection evaporator, a maximum flux of 73,000 Btu/h · ft² was obtained at $(\Delta t)_v$ of 100°F.

Combined Convection and Radiation Coefficients In some cases of heat loss, such as that from bare and insulated pipes, where loss is by convection to the air and radiation to the walls of the enclosing space, it is convenient to use a combined convection and radiation coefficient $h_c + h_r$. The rate of heat loss thus becomes

$$q = (h_c + h_r)A(\Delta t)_s \quad (4.4.15)$$

where $(\Delta t)_s$ is the temperature difference, °F, between the surface of the hot body and the walls of the space. In evaluating $(h_c + h_r)$, h_c should be calculated by the appropriate convection formula [see Eqs. (4.4.11c) to (4.4.11g)] and h_r from the equation

$$h_r = 0.00685\varepsilon(T_{\text{av}}/100)^3$$

where ε is the emissivity of the radiating surface (see Sec. 4.3). T_{av} is the average temperature of the surface and the enclosing walls, °R. For oxidized bare steel pipe, the sum $h_c + h_r$ may be taken directly from Table 4.4.12.

Heat Transmission through Pipe Insulation (McMillan, *Trans. ASME*, 1915) For any number of layers of insulation on any size of pipe, Eqs. (4.4.2), (4.4.4), and (4.4.15) combine to give

$$\frac{q_o}{A_o} = \frac{(\Delta t)_o}{\dfrac{r_o}{k_1}\ln\dfrac{r_2}{r_1} + \dfrac{r_o}{k_2}\ln\dfrac{r_3}{r_2} + \cdots + \dfrac{1}{h_c + h_r}} \quad (4.4.16)$$

where q_o/A_o is the Btu/(h · ft²) of outer surface of the last layer; $(\Delta t)_o$ is the overall temperature difference (°F) between pipe and air; r_o is the radius, feet, of the outer surface; r_1 is the outside radius (ft) of the pipe, $r_2 = r_1 + $ thickness of first layer of insulation, foot; $r_3 = r_2$ plus thickness of second layer, etc.; and k_1, k_2, k_3, etc., are the conductivities of the respective layers. For average indoor conditions, $h_c + h_r$ is often taken as 2 as an approximation, since a substantial error in $h_c + h_r$ will have but little effect on the overall loss of heat. Figure 4.4.6 shows the variation in U_o with pipe size and thickness of insulation (for $k = 0.042$) for pipe and air temperatures of 375 and 75°F, respectively.

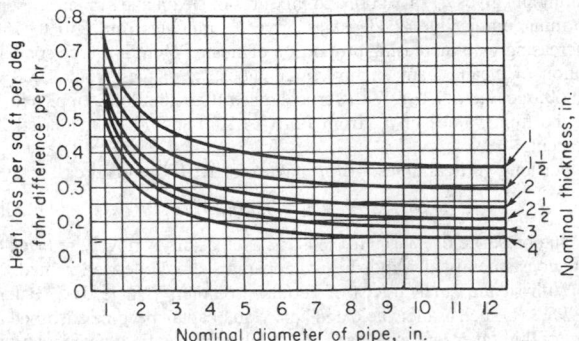

Fig. 4.4.6 Variation with pipe size of overall coefficient U_o for a given thickness of insulation, for $k = 0.042$.

Table 4.4.12 Values of $h_c + h_r$
For horizontal bare or insulated standard steel pipe of various sizes and for flat plates in a room at 80°F

Nominal pipe diam, in	$(\Delta t)_s$, temperature difference, °F, from surface to room														
	50	100	150	200	250	300	400	500	600	700	800	900	1,000	1,100	1,200
½	2.12	2.48	2.76	3.10	3.41	3.75	4.47	5.30	6.21	7.25	8.40	9.73	11.20	12.81	14.65
1	2.03	2.38	2.65	2.98	3.29	3.62	4.33	5.16	6.07	7.11	8.25	9.57	11.04	12.65	14.48
2	1.93	2.27	2.52	2.85	3.14	3.47	4.18	4.99	5.89	6.92	8.07	9.38	10.85	12.46	14.28
4	1.84	2.16	2.41	2.72	3.01	3.33	4.02	4.83	5.72	6.75	7.89	9.21	10.66	12.27	14.09
8	1.76	2.06	2.29	2.60	2.89	3.20	3.88	4.68	5.57	6.60	7.73	9.05	10.50	12.10	13.93
12	1.71	2.01	2.24	2.54	2.82	3.13	3.83	4.61	5.50	6.52	7.65	8.96	10.42	12.03	13.84
24	1.64	1.93	2.15	2.45	2.72	3.03	3.70	4.48	5.37	6.39	7.52	8.83	10.28	11.90	13.70
Flat plates															
Vertical	1.82	2.13	2.40	2.70	2.99	3.30	4.00	4.79	5.70	6.72	7.86	9.18	10.64	12.25	14.06
HFU*	2.00	2.35	2.65	2.97	3.26	3.59	4.31	5.12	6.04	7.07	8.21	9.54	11.01	12.63	14.45
HFD*	1.58	1.85	2.09	2.36	2.63	2.93	3.61	4.38	5.27	6.27	7.40	8.71	10.16	11.76	13.57

* HFU = horizontal facing upward; HFD = horizontal facing downward.

Strength of Materials

BY

JOHN SYMONDS *Fellow Engineer (Retired), Oceanic Division, Westinghouse Electric Corporation.*

J. P. VIDOSIC *Regents' Professor Emeritus of Mechanical Engineering, Georgia Institute of Technology.*

HAROLD V. HAWKINS *Late Manager, Product Standards and Services, Columbus McKinnon Corporation, Tonawanda, N.Y.*

DONALD D. DODGE *Supervisor (Retired), Product Quality and Inspection Technology, Manufacturing Development, Ford Motor Company.*

5.1 MECHANICAL PROPERTIES OF MATERIALS
by John Symonds, Expanded by Staff

REFERENCES: Davis et al., "Testing and Inspection of Engineering Materials," McGraw-Hill, Timoshenko, "Strength of Materials," pt. II, Van Nostrand. Richards, "Engineering Materials Science," Wadsworth. Nadai, "Plasticity," McGraw-Hill. Tetelman and McEvily, "Fracture of Structural Materials," Wiley. "Fracture Mechanics," ASTM STP-833. McClintock and Argon (eds.), "Mechanical Behavior of Materials," Addison-Wesley. Dieter, "Mechanical Metallurgy," McGraw-Hill. "Creep Data," ASME. ASTM Standards, ASTM. Blaznynski (ed.), "Plasticity and Modern Metal Forming Technology," Elsevier Science.

STRESS-STRAIN DIAGRAMS

The Stress-Strain Curve The engineering tensile stress-strain curve is obtained by **static loading** of a standard specimen, that is, by applying the load slowly enough that all parts of the specimen are in equilibrium at any instant. The curve is usually obtained by controlling the loading rate in the tensile machine. ASTM Standards require a loading rate not exceeding 100,000 lb/in² (70 kgf/mm²)/min. An alternate method of obtaining the curve is to specify the strain rate as the independent variable, in which case the loading rate is continuously adjusted to maintain the required strain rate. A strain rate of 0.05 in/in/(min) is commonly used. It is measured usually by an extensometer attached to the gage length of the specimen. Figure 5.1.1 shows several stress-strain curves.

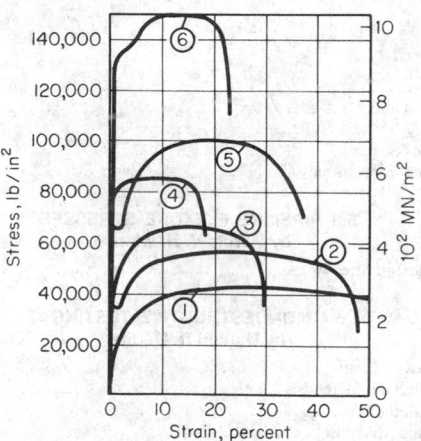

Fig. 5.1.1. Comparative stress-strain diagrams. (1) Soft brass; (2) low carbon steel; (3) hard bronze; (4) cold rolled steel; (5) medium carbon steel, annealed; (6) medium carbon steel, heat treated.

For most engineering materials, the curve will have an initial linear elastic region (Fig. 5.1.2) in which deformation is reversible and time-independent. The slope in this region is **Young's modulus** E. The **proportional elastic limit** (PEL) is the point where the curve starts to deviate from a straight line. The **elastic limit** (frequently indistinguishable from PEL) is the point on the curve beyond which plastic deformation is present after release of the load. If the stress is increased further, the stress-strain curve departs more and more from the straight line. Unloading the specimen at point X (Fig. 5.1.2), the portion XX'' is linear and is essentially parallel to the original line OX''. The horizontal distance OX' is called the **permanent set** corresponding to the stress at X. This is the basis for the construction of the arbitrary **yield strength.** To determine the yield strength, a straight line XX' is drawn parallel to the initial elastic line OX'' but displaced from it by an arbitrary value of

permanent strain. The permanent strain commonly used is 0.20 percent of the original gage length. The intersection of this line with the curve determines the stress value called the yield strength. In reporting the yield strength, the amount of permanent set should be specified. The arbitrary yield strength is used especially for those materials not exhibiting a natural yield point such as nonferrous metals; but it is not limited to these. Plastic behavior is somewhat time-dependent, particularly at high temperatures. Also at high temperatures, a small amount of time-dependent reversible strain may be detectable, indicative of **anelastic** behavior.

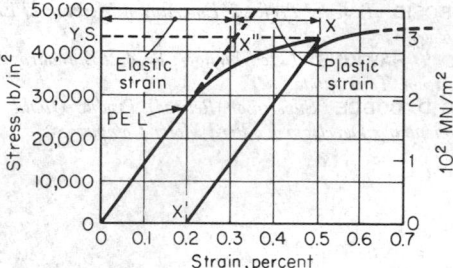

Fig. 5.1.2. General stress-strain diagram.

The **ultimate tensile strength** (UTS) is the maximum load sustained by the specimen divided by the original specimen cross-sectional area. The percent **elongation at failure** is the plastic extension of the specimen at failure expressed as (the change in original gage length × 100) divided by the original gage length. This extension is the sum of the **uniform** and **nonuniform** elongations. The uniform elongation is that which occurs prior to the UTS. It has an unequivocal significance, being associated with uniaxial stress, whereas the nonuniform elongation which occurs during localized extension (necking) is associated with triaxial stress. The nonuniform elongation will depend on geometry, particularly the ratio of specimen gage length L_0 to diameter D or square root of cross-sectional area A. ASTM Standards specify test-specimen geometry for a number of specimen sizes. The ratio $L_0/\sqrt{A}$ is maintained at 4.5 for flat- and round-cross-section specimens. The original gage length should always be stated in reporting elongation values.

The specimen percent **reduction in area** (RA) is the contraction in cross-sectional area at the fracture expressed as a percentage of the original area. It is obtained by measurement of the cross section of the broken specimen at the fracture location. The RA along with the load at fracture can be used to obtain the **fracture stress,** that is, fracture load divided by cross-sectional area at the fracture. See Table 5.1.1.

The type of fracture in tension gives some indications of the quality of the material, but this is considerably affected by the testing temperature, speed of testing, the shape and size of the test piece, and other conditions. Contraction is greatest in tough and ductile materials and least in brittle materials. In general, fractures are either of the **shear** or of the **separation** (loss of cohesion) type. Flat tensile specimens of ductile metals often show shear failures if the ratio of width to thickness is greater than 6:1. A completely shear-type failure may terminate in a chisel edge, for a flat specimen, or a point rupture, for a round specimen. Separation failures occur in brittle materials, such as certain cast irons. Combinations of both shear and separation failures are common on round specimens of ductile metal. Failure often starts at the axis in a necked region and produces a relatively flat area which grows until the material shears along a cone-shaped surface at the outside of the speci-

Table 5.1.1   Typical Mechanical Properties at Room Temperature
(Based on ordinary stress-strain values)

Metal	Tensile strength, 1,000 lb/in^2	Yield strength, 1,000 lb/in^2	Ultimate elongation, %	Reduction of area, %	Brinell no.
Cast iron	18–60	8–40	0	0	100–300
Wrought iron	45–55	25–35	35–25	55–30	100
Commercially pure iron, annealed	42	19	48	85	70
Hot-rolled	48	30	30	75	90
Cold-rolled	100	95			200
Structural steel, ordinary	50–65	30–40	40–30		120
Low-alloy, high-strength	65–90	40–80	30–15	70–40	150
Steel, SAE 1300, annealed	70	40	26	70	150
Quenched, drawn 1,300°F	100	80	24	65	200
Drawn 1,000°F	130	110	20	60	260
Drawn 700°F	200	180	14	45	400
Drawn 400°F	240	210	10	30	480
Steel, SAE 4340, annealed	80	45	25	70	170
Quenched, drawn 1,300°F	130	110	20	60	270
Drawn 1,000°F	190	170	14	50	395
Drawn 700°F	240	215	12	48	480
Drawn 400°F	290	260	10	44	580
Cold-rolled steel, SAE 1112	84	76	18	45	160
Stainless steel, 18-S	85–95	30–35	60–55	75–65	145–160
Steel castings, heat-treated	60–125	30–90	33–14	65–20	120–250
Aluminum, pure, rolled	13–24	5–21	35–5		23–44
Aluminum-copper alloys, cast	19–23	12–16	4–0		50–80
Wrought, heat-treated	30–60	10–50	33–15		50–120
Aluminum die castings	30		2		
Aluminum alloy 17ST	56	34	26	39	100
Aluminum alloy 51ST	48	40	20	35	105
Copper, annealed	32	5	58	73	45
Copper, hard-drawn	68	60	4	55	100
Brasses, various	40–120	8–80	60–3		50–170
Phosphor bronze	40–130		55–5		50–200
Tobin bronze, rolled	63	41	40	52	120
Magnesium alloys, various	21–45	11–30	17–0.5		47–78
Monel 400, Ni-Cu alloy	79	30	48	75	125
Molybdenum, rolled	100	75	30		250
Silver, cast, annealed	18	8	54		27
Titanium 6–4 alloy, annealed	130	120	10	25	352
Ductile iron, grade 80-55-06	80	55	6		225–255

NOTE: Compressive strength of cast iron, 80,000 to 150,000 lb/in^2.
Compressive yield strength of all metals, except those cold-worked = tensile yield strength.
Stress 1,000 lb/in^2 × 6.894 = stress, MN/m^2.

men, resulting in what is known as the cup-and-cone fracture. Double cup-and-cone and rosette fractures sometimes occur. Several types of tensile fractures are shown in Fig. 5.1.3.

Annealed or hot-rolled mild steels generally exhibit a **yield point** (see Fig. 5.1.4). Here, in a constant strain-rate test, a large increment of extension occurs under constant load at the elastic limit or at a stress just below the elastic limit. In the latter event the stress drops suddenly from the **upper yield point** to the **lower yield point**. Subsequent to the drop, the yield-point extension occurs at constant stress, followed by a rise to the UTS. Plastic flow during the yield-point extension is discontinuous;

Fig. 5.1.3.   Typical metal fractures in tension.

successive zones of plastic deformation, known as **Luder's bands** or **stretcher strains,** appear until the entire specimen gage length has been uniformly deformed at the end of the yield-point extension. This behavior causes a banded or stepped appearance on the metal surface. The exact form of the stress-strain curve for this class of material is sensitive

to test temperature, test strain rate, and the characteristics of the tensile machine employed.

The plastic behavior in a uniaxial tensile test can be represented as the **true stress-strain curve.** The **true stress** σ is based on the instantaneous

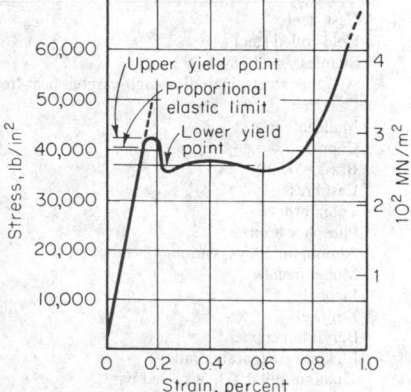

Fig. 5.1.4.   Yielding of annealed steel.

cross section A, so that $\sigma =$ load$/A$. The instantaneous **true strain** increment is $-dA/A$, or dL/L prior to necking. Total true strain ε is

$$\int_{A_0}^{A} -\frac{dA}{A} = \ln\left(\frac{A_0}{A}\right)$$

or $\ln (L/L_0)$ prior to necking. The true stress-strain curve or flow curve obtained has the typical form shown in Fig. 5.1.5. In the part of the test subsequent to the maximum load point (UTS), when necking occurs, the true strain of interest is that which occurs in an infinitesimal length at the region of minimum cross section. True strain for this element can still be expressed as $\ln (A_0/A)$, where A refers to the minimum cross

section. Methods of constructing the true stress-strain curve are described in the technical literature. In the range between initial yielding and the neighborhood of the maximum load point the relationship between plastic strain ε_p and true stress often approximates

$$\sigma = k\varepsilon_p^n$$

where k is the **strength coefficient** and n is the **work-hardening exponent.** For a material which shows a yield point the relationship applies only to the rising part of the curve beyond the lower yield. It can be shown that at the maximum load point the slope of the true stress-strain curve equals the true stress, from which it can be deduced that for a material obeying the above exponential relationship between ε_p and n, $\varepsilon_p = n$ at the maximum load point. The exponent strongly influences the spread between YS and UTS on the engineering stress-strain curve. Values of n and k for some materials are shown in Table 5.1.2. A point on the flow curve indentifies the **flow stress** corresponding to a certain strain, that is, the stress required to bring about this amount of plastic deformation. The concept of true strain is useful for accurately describing large amounts of plastic deformation. The linear strain definition $(L - L_0)/L_0$ fails to correct for the continuously changing gage length, which leads to an increasing error as deformation proceeds.

During extension of a specimen under tension, the change in the specimen cross-sectional area is related to the elongation by **Poisson's ratio** μ, which is the ratio of strain in a transverse direction to that in the longitudinal direction. Values of μ for the elastic region are shown in Table 5.1.3. For plastic strain it is approximately 0.5.

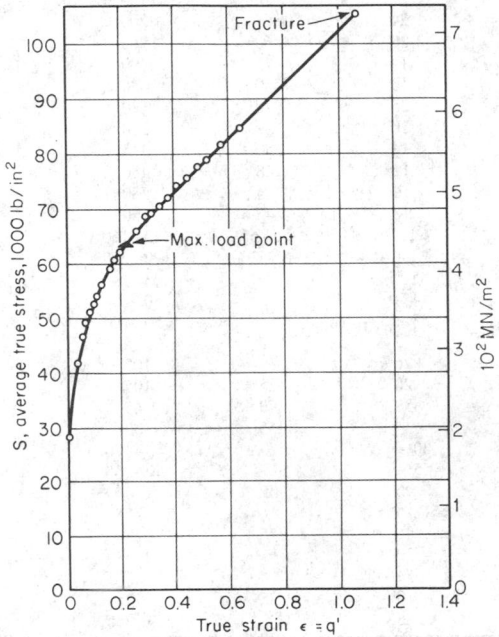

Fig. 5.1.5. True stress-strain curve for 20°C annealed mild steel.

Table 5.1.2 Room-Temperature Plastic-Flow Constants for a Number of Metals

Material	Condition	k, 1,000 in^2 (MN/m^2)	n
0.40% C steel	Quenched and tempered at 400°F (478K)	416 (2,860)	0.088
0.05% C steel	Annealed and temper-rolled	72 (49.6)	0.235
2024 aluminum	Precipitation-hardened	100 (689)	0.16
2024 aluminum	Annealed	49 (338)	0.21
Copper	Annealed	46.4 (319)	0.54
70–30 brass	Annealed	130 (895)	0.49

SOURCE: Reproduced by permission from "Properties of Metals in Materials Engineering," ASM, 1949.

Table 5.1.3 Elastic Constants of Metals
(Mostly from tests of R. W. Vose)

Metal	E Modulus of elasticity (Young's modulus). 1,000,000 lb/in^2	G Modulus of rigidity (shearing modulus). 1,000,000 lb/in^2	K Bulk modulus. 1,000,000 lb/in^2	μ Poisson's ratio
Cast steel	28.5	11.3	20.2	0.265
Cold-rolled steel	29.5	11.5	23.1	0.287
Stainless steel 18–8	27.6	10.6	23.6	0.305
All other steels, including high-carbon, heat-treated	28.6–30.0	11.0–11.9	22.6–24.0	0.283–0.292
Cast iron	13.5–21.0	5.2–8.2	8.4–15.5	0.211–0.299
Malleable iron	23.6	9.3	17.2	0.271
Copper	15.6	5.8	17.9	0.355
Brass, 70–30	15.9	6.0	15.7	0.331
Cast brass	14.5	5.3	16.8	0.357
Tobin bronze	13.8	5.1	16.3	0.359
Phosphor bronze	15.9	5.9	17.8	0.350
Aluminum alloys, various	9.9–10.3	3.7–3.9	9.9–10.2	0.330–0.334
Monel metal	25.0	9.5	22.5	0.315
Inconel	31	11		0.27–0.38
Z-nickel	30	11		±0.36
Beryllium copper	17	7		±0.21
Elektron (magnesium alloy)	6.3	2.5	4.8	0.281
Titanium (99.0 Ti), annealed bar	15–16	6.5		0.34
Zirconium, crystal bar	11–14			
Molybdenum, arc-cast	48–52			

The general effect of increased strain rate is to increase the resistance to plastic deformation and thus to raise the flow curve. Decreasing test temperature also raises the flow curve. The effect of strain rate is expressed as **strain-rate sensitivity** m. Its value can be measured in the tension test if the strain rate is suddenly increased by a small increment during the plastic extension. The flow stress will then jump to a higher value. The strain-rate sensitivity is the ratio of incremental changes of $\log \sigma$ and $\log \dot{\varepsilon}$

$$m = \left(\frac{\delta \log \sigma}{\delta \log \dot{\varepsilon}} \right)_{\varepsilon}$$

For most engineering materials at room temperature the strain rate sensitivity is of the order of 0.01. The effect becomes more significant at elevated temperatures, with values ranging to 0.2 and sometimes higher.

Compression Testing   The compressive stress-strain curve is similar to the tensile stress-strain curve up to the yield strength. Thereafter, the progressively increasing specimen cross section causes the compressive stress-strain curve to diverge from the tensile curve. Some ductile metals will not fail in the compression test. Complex behavior occurs when the direction of stressing is changed, because of the **Bauschinger effect,** which can be described as follows: If a specimen is first plastically strained in tension, its yield stress in compression is reduced and vice versa.

Combined Stresses   This refers to the situation in which stresses are present on each of the faces of a cubic element of the material. For a given cube orientation the applied stresses may include shear stresses over the cube faces as well as stresses normal to them. By a suitable rotation of axes the problem can be simplified: applied stresses on the new cubic element are equivalent to three mutually orthogonal **principal stresses** σ_1, σ_2, σ_3 alone, each acting normal to a cube face. Combined stress behavior in the elastic range is described in Sec. 5.2, Mechanics of Materials.

Prediction of the conditions under which plastic yielding will occur under combined stresses can be made with the help of several empirical theories. In the **maximum-shear-stress theory** the criterion for yielding is that yielding will occur when

$$\sigma_1 - \sigma_3 = \sigma_{ys}$$

in which σ_1 and σ_3 are the largest and smallest principal stresses, respectively, and σ_{ys} is the uniaxial tensile yield strength. This is the simplest theory for predicting yielding under combined stresses. A more accurate prediction can be made by the **distortion-energy theory,** according to which the criterion is

$$(\sigma_1 - \sigma_2)^2 + (\sigma_2 - \sigma_3)^2 + (\sigma_2 - \sigma_1)^2 = 2(\sigma_{ys})^2$$

Stress-strain curves in the plastic region for combined stress loading can be constructed. However, a particular stress state does not determine a unique strain value. The latter will depend on the stress-state path which is followed.

Plane strain is a condition where strain is confined to two dimensions. There is generally stress in the third direction, but because of mechanical constraints, strain in this dimension is prevented. Plane strain occurs in certain metalworking operations. It can also occur in the neighborhood of a crack tip in a tensile loaded member if the member is sufficiently thick. The material at the crack tip is then in triaxial tension, which condition promotes brittle fracture. On the other hand, ductility is enhanced and fracture is suppressed by triaxial compression.

Stress Concentration   In a structure or machine part having a notch or any abrupt change in cross section, the maximum stress will occur at this location and will be greater than the stress calculated by elementary formulas based upon simplified assumptions as to the stress distribution. The ratio of this maximum stress to the nominal stress (calculated by the elementary formulas) is the stress-concentration factor K_t. This is a constant for the particular geometry and is independent of the material, provided it is isotropic. The stress-concentration factor may be determined experimentally or, in some cases, theoretically from the mathematical theory of elasticity. The factors shown in Figs. 5.1.6 to 5.1.13 were determined from both photoelastic tests and the theory of elasticity. Stress concentration will cause failure of brittle materials if

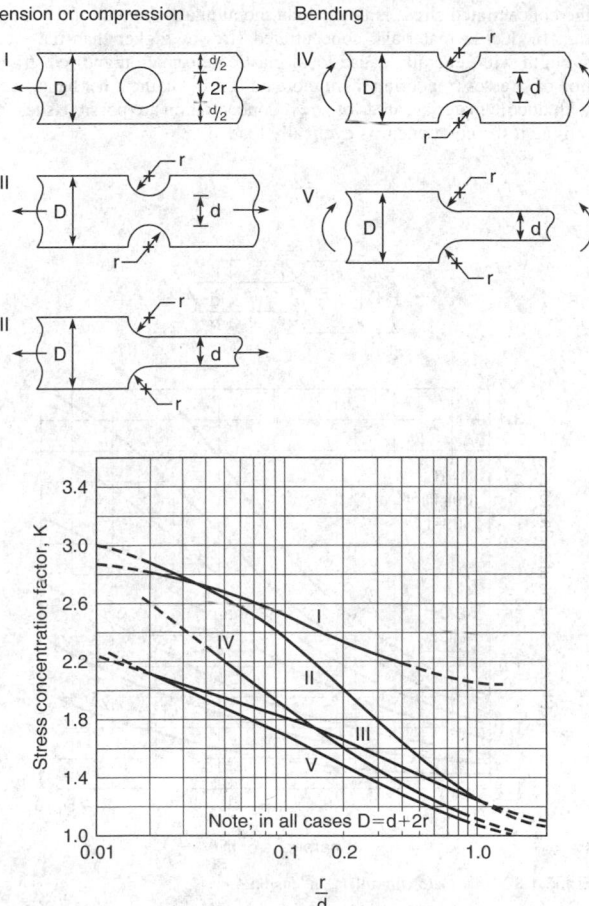

Fig. 5.1.6.   Flat plate with semicircular fillets and grooves or with holes. I, II, and III are in tension or compression; IV and V are in bending.

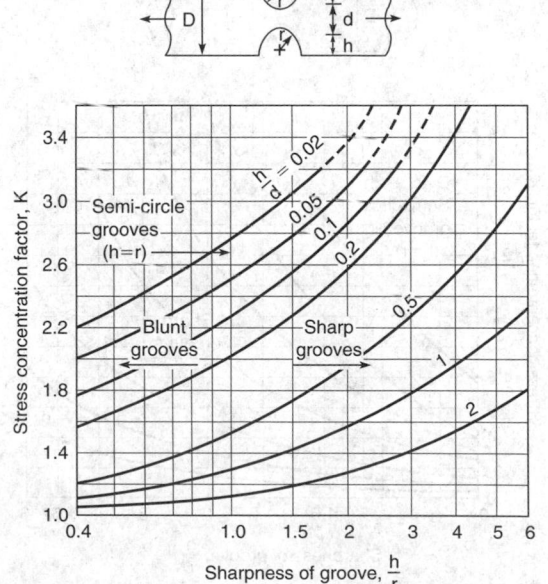

Fig. 5.1.7.   Flat plate with grooves, in tension.

the concentrated stress is larger than the ultimate strength of the material. In ductile materials, concentrated stresses higher than the yield strength will generally cause local plastic deformation and redistribution of stresses (rendering them more uniform). On the other hand, even with ductile materials areas of stress concentration are possible sites for fatigue if the component is cyclically loaded.

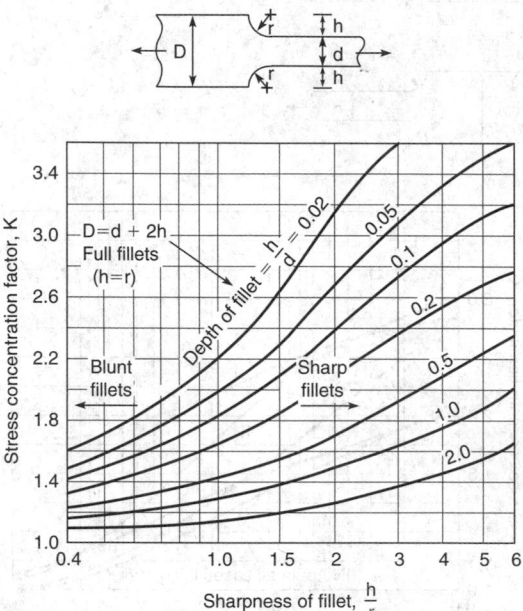

Fig. 5.1.8. Flat plate with fillets, in tension.

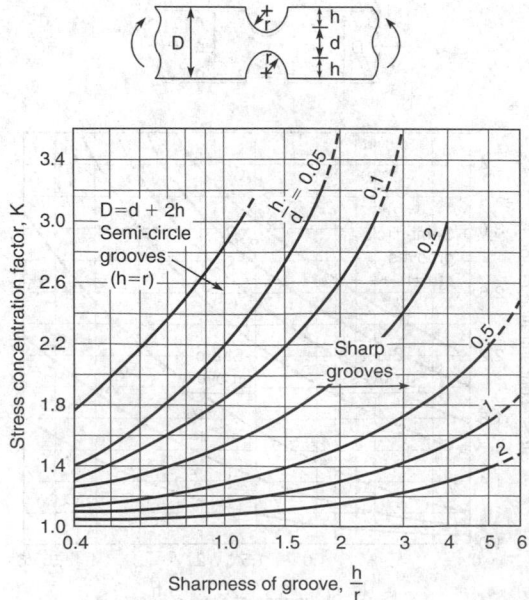

Fig. 5.1.9. Flat plate with grooves, in bending.

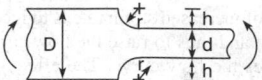

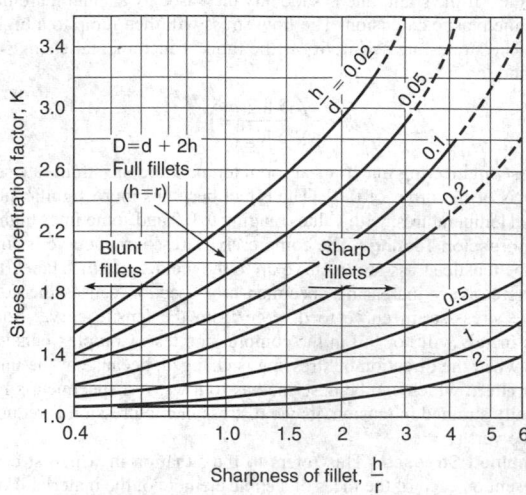

Fig. 5.1.10. Flat plate with fillets, in bending.

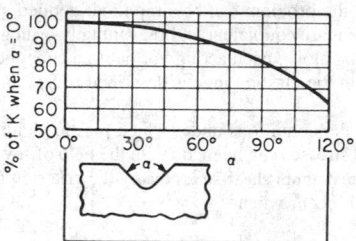

Fig. 5.1.11. Flat plate with angular notch, in tension or bending.

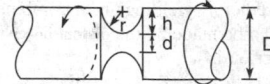

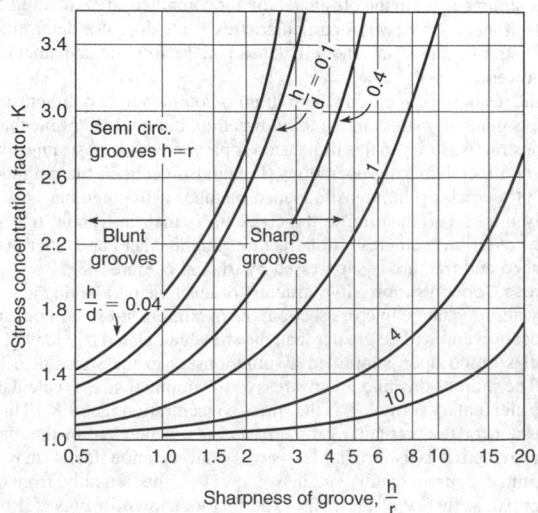

Fig. 5.1.12. Grooved shaft in torsion.

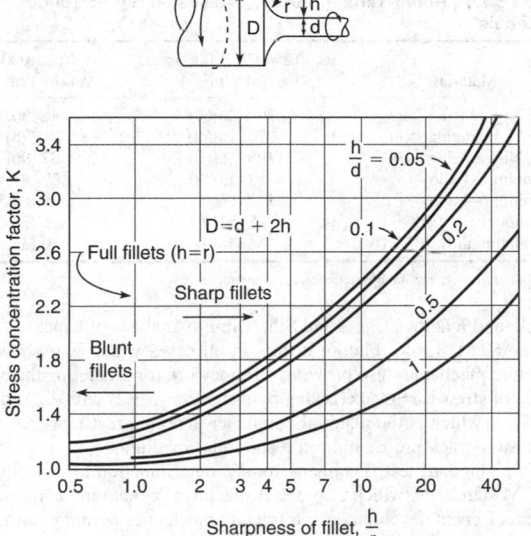

Fig. 5.1.13.   Filleted shaft in torsion.

FRACTURE AT LOW STRESSES

Materials under tension sometimes fail by rapid fracture at stresses much below their strength level as determined in tests on carefully prepared specimens. These **brittle, unstable,** or **catastrophic failures** originate at preexisting stress-concentrating flaws which may be inherent in a material.

The **transition-temperature approach** is often used to ensure fracture-safe design in structural-grade steels. These materials exhibit a characteristic temperature, known as the **ductile brittle transition** (DBT) temperature, below which they are susceptible to brittle fracture. The transition-temperature approach to fracture-safe design ensures that the transition temperature of a material selected for a particular application is suitably matched to its intended use temperature. The DBT can be detected by plotting certain measurements from tensile or impact tests against temperature. Usually the transition to brittle behavior is complex, being neither fully ductile nor fully brittle. The range may extend over 200°F (110 K) interval. The **nil-ductility temperature** (NDT), determined by the **drop weight test** (see ASTM Standards), is an important reference point in the transition range. When NDT for a particular steel is known, temperature-stress combinations can be specified which define the limiting conditions under which catastrophic fracture can occur.

In the **Charpy V-notch** (CVN) impact test, a notched-bar specimen (Fig. 5.1.26) is used which is loaded in bending (see ASTM Standards). The energy absorbed from a swinging pendulum in fracturing the specimen is measured. The pendulum strikes the specimen at 16 to 19 ft (4.88 to 5.80 m)/s so that the specimen deformation associated with fracture occurs at a rapid strain rate. This ensures a conservative measure of toughness, since in some materials, toughness is reduced by high strain rates. A CVN impact energy vs. temperature curve is shown in Fig. 5.1.14, which also shows the transitions as given by percent brittle fracture and by percent lateral expansion. The CVN energy has no analytical significance. The test is useful mainly as a guide to the fracture behavior of a material for which an empirical correlation has been established between impact energy and some rigorous fracture criterion. For a particular grade of steel the CVN curve can be correlated with NDT. (See ASME Boiler and Pressure Vessel Code.)

Fracture Mechanics   This analytical method is used for ultra-high-strength alloys, transition-temperature materials below the DBT temperature, and some low-strength materials in heavy section thickness.

Fracture mechanics theory deals with crack extension where plastic effects are negligible or confined to a small region around the crack tip. The present discussion is concerned with a through-thickness crack in a tension-loaded plate (Fig. 5.1.15) which is large enough so that the crack-tip stress field is not affected by the plate edges. Fracture mechanics theory states that unstable crack extension occurs when the work required for an increment of crack extension, namely, surface energy and energy consumed in local plastic deformation, is exceeded by the elastic-strain energy released at the crack tip. The elastic-stress

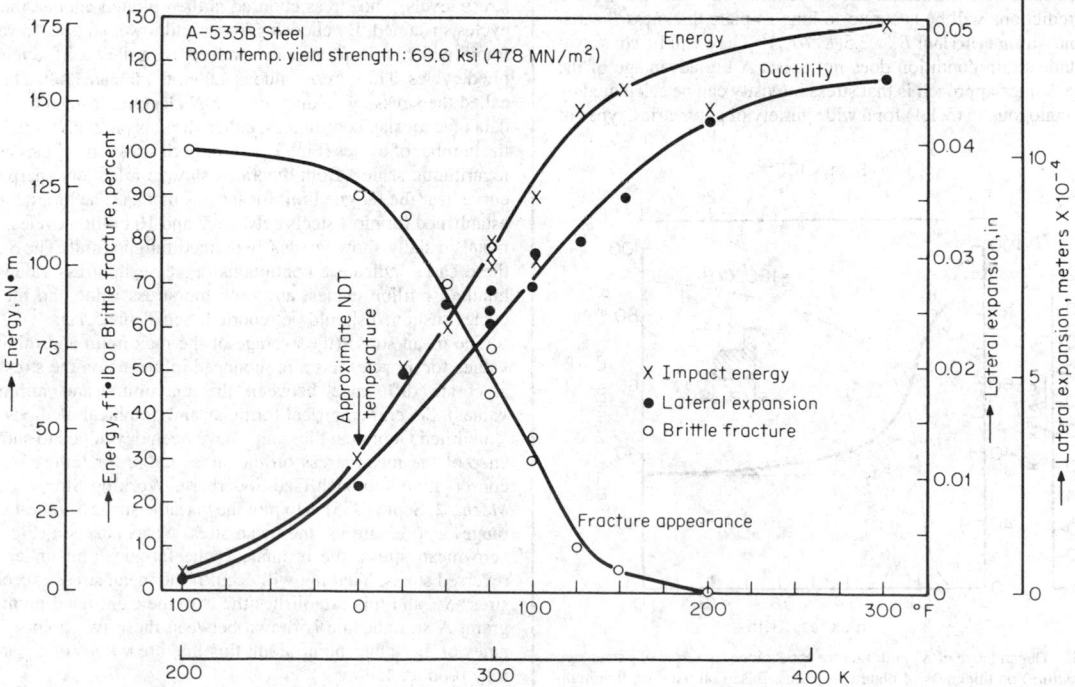

Fig. 5.1.14.   CVN transition curves. *(Data from Westinghouse R & D Lab.)*

field surrounding one of the crack tips in Fig. 5.1.15 is characterized by the **stress intensity** K_I, which has units of (lb $\sqrt{in}$) /in^2 or (N$\sqrt{m}$) /m^2. It is a function of applied nominal stress σ, crack half-length a, and a geometry factor Q:

$$K_I^2 = Q\sigma^2\pi a \qquad (5.1.1)$$

for the situation of Fig. 5.1.15. For a particular material it is found that as K_I is increased, a value K_c is reached at which unstable crack propa-

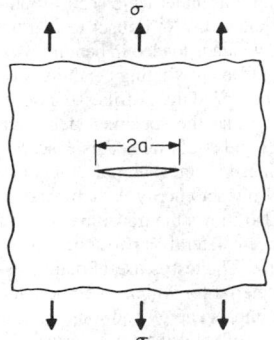

Fig. 5.1.15. Through-thickness crack geometry.

gation occurs. K_c depends on plate thickness B, as shown in Fig. 5.1.16. It attains a constant value when B is great enough to provide plane-strain conditions at the crack tip. The low plateau value of K_c is an important material property known as the **plane-strain critical stress intensity** or **fracture toughness** K_{Ic}. Values for a number of materials are shown in Table 5.1.4. They are influenced strongly by processing and small changes in composition, so that the values shown are not necessarily typical. K_{Ic} can be used in the critical form of Eq. (5.1.1):

$$(K_{Ic})^2 = Q\sigma^2\pi a_{cr} \qquad (5.1.2)$$

to predict failure stress when a maximum flaw size in the material is known or to determine maximum allowable flaw size when the stress is set. The predictions will be accurate so long as plate thickness B satisfies the **plane-strain criterion:** $B \geq 2.5(K_{Ic}/\sigma_{ys})^2$. They will be conservative if a plane-strain condition does not exist. A big advantage of the fracture mechanics approach is that stress intensity can be calculated by equations analogous to (5.1.1) for a wide variety of geometries, types of

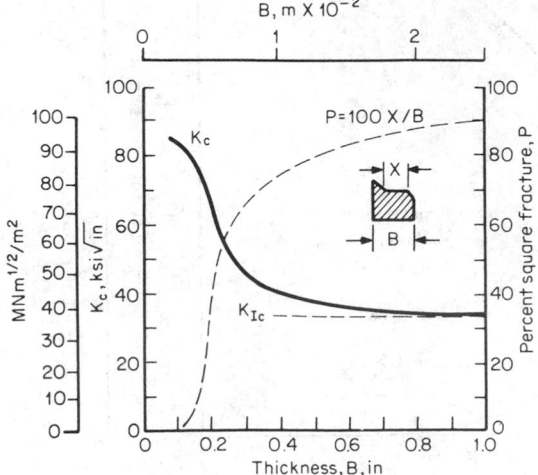

Fig. 5.1.16. Dependence of K_c and fracture appearance (in terms of percentage of square fracture) on thickness of plate specimens. Based on data for aluminum 7075-T6. (*From Scrawly and Brown, STP-381, ASTM.*)

Table 5.1.4 Room-Temperature K_{Ic} Values on High-Strength Materials*

Material	0.2% YS, 1,000 in^2 (MN/m^2)	K_{Ic}, 1,000 in^2 $\sqrt{in}$ (MN m$^{1/2}$/m^2)
18% Ni maraging steel	300 (2,060)	46 (50.7)
18% Ni maraging steel	270 (1,850)	71 (78)
18% Ni maraging steel	198 (1,360)	87 (96)
Titanium 6-4 alloy	152 (1,022)	39 (43)
Titanium 6-4 alloy	140 (960)	75 (82.5)
Aluminum alloy 7075-T6	75 (516)	26 (28.6)
Aluminum alloy 7075-T6	64 (440)	30 (33)

* Determined at Westinghouse Research Laboratories.

crack, and loadings (Paris and Sih, "Stress Analysis of Cracks," STP-381, ASTM, 1965). Failure occurs in all cases when K_I reaches K_{Ic}. Fracture mechanics also provides a framework for predicting the occurrence of **stress-corrosion cracking** by using Eq. (5.1.2) with K_{Ic} replaced by K_{Iscc}, which is the material parameter denoting resistance to stress-corrosion-crack propagation in a particular medium.

Two standard test specimens for K_{Ic} determination are specified in ASTM standards, which also detail specimen preparation and test procedure. Recent developments in fracture mechanics permit treatment of crack propagation in the ductile regime. (See "Elastic-Plastic Fracture," ASTM.)

FATIGUE

Fatigue is generally understood as the gradual deterioration of a material which is subjected to repeated loads. In fatigue testing, a specimen is subjected to periodically varying constant-amplitude stresses by means of mechanical or magnetic devices. The applied stresses may alternate between equal positive and negative values, from zero to maximum positive or negative values, or between unequal positive and negative values. The most common loading is alternate tension and compression of equal numerical values obtained by rotating a smooth cylindrical specimen while under a bending load. A series of fatigue tests are made on a number of specimens of the material at different stress levels. The stress endured is then plotted against the number of cycles sustained. By choosing lower and lower stresses, a value may be found which will not produce failure, regardless of the number of applied cycles. This stress value is called the **fatigue limit**. The diagram is called the stress-cycle diagram or *S-N* diagram. Instead of recording the data on cartesian coordinates, either stress is plotted vs. the logarithm of the number of cycles (Fig. 5.1.17) or both stress and cycles are plotted to logarithmic scales. Both diagrams show a relatively sharp bend in the curve near the fatigue limit for ferrous metals. The fatigue limit may be established for most steels between 2 and 10 million cycles. Nonferrous metals usually show no clearly defined fatigue limit. The *S-N* curves in these cases indicate a continuous decrease in stress values to several hundred million cycles, and both the stress value and the number of cycles sustained should be reported. See Table 5.1.5.

The mean stress (the average of the maximum and minimum stress values for a cycle) has a pronounced influence on the stress range (the algebraic difference between the maximum and minimum stress values). Several empirical formulas and graphical methods such as the "modified Goodman diagram" have been developed to show the influence of the mean stress on the stress range for failure. A simple but conservative approach (see Soderberg, Working Stresses, *Jour. Appl. Mech.*, **2**, Sept. 1935) is to plot the variable stress S_v (one-half the stress range) as ordinate vs. the mean stress S_m as abscissa (Fig. 5.1.18). At zero mean stress, the ordinate is the fatigue limit under completely reversed stress. Yielding will occur if the mean stress exceeds the yield stress S_o, and this establishes the extreme right-hand point of the diagram. A straight line is drawn between these two points. The coordinates of any other point along this line are values of S_m and S_v which may produce failure.

Surface defects, such as roughness or scratches, and notches or

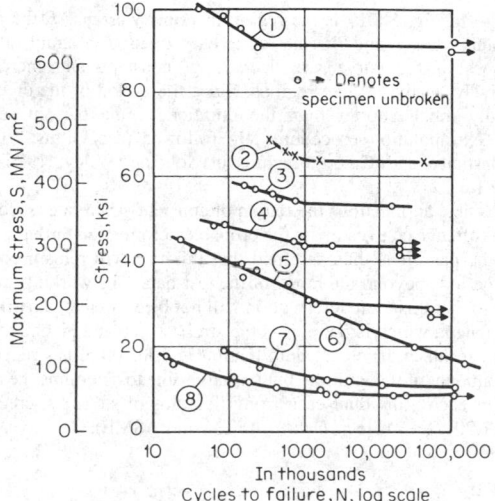

Fig. 5.1.17. The *S-N* diagrams from fatigue tests. (1) 1.20% C steel, quenched and drawn at 860°F (460°C); (2) alloy structural steel; (3) SAE 1050, quenched and drawn at 1,200°F (649°C); (4) SAE 4130, normalized and annealed; (5) ordinary structural steel; (6) Duralumin; (7) copper, annealed; (8) cast iron (reversed bending).

shoulders all reduce the fatigue strength of a part. With a notch of prescribed geometric form and known concentration factor, the reduction in strength is appreciably less than would be called for by the concentration factor itself, but the various metals differ widely in their susceptibility to the effect of roughness and concentrations, or **notch sensitivity**.

For a given material subjected to a prescribed state of stress and type of loading, notch sensitivity can be viewed as the ability of that material to resist the concentration of stress incidental to the presence of a notch. Alternately, notch sensitivity can be taken as a measure of the degree to which the geometric stress concentration factor is reduced. An attempt is made to rationalize notch sensitivity through the equation $q = (K_f - 1)/(K - 1)$, where q is the notch sensitivity, K is the geometric stress concentration factor (from data similar to those in Figs. 5.1.5 to 5.1.13 and the like), and K_f is defined as the ratio of the strength of unnotched material to the strength of notched material. Ratio K_f is obtained from laboratory tests, and K is deduced either theoretically or from laboratory tests, but both must reflect the same state of stress and type of loading. The value of q lies between 0 and 1, so that (1) if $q = 0$, $K_f = 1$ and the material is not notch-sensitive (soft metals such as copper, aluminum, and annealed low-strength steel); (2) if $q = 1$, $K_f = K$, the material is fully notch-sensitive and the full value of the geometric stress concentration factor is not diminished (hard, high-strength steel). In practice, q will lie somewhere between 0 and 1, but it may be hard to quantify.

Accordingly, the pragmatic approach to arrive at a solution to a design problem often takes a conservative route and sets $q = 1$. The exact material properties at play which are responsible for notch sensitivity are not clear.

Further, notch sensitivity seems to be higher, and ordinary fatigue strength lower in large specimens, necessitating full-scale tests in many cases (see Peterson, Stress Concentration Phenomena in Fatigue of

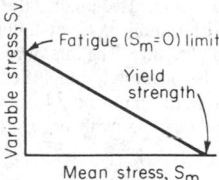

Fig. 5.1.18. Effect of mean stress on the variable stress for failure.

Metals, *Trans. ASME,* **55,** 1933, p. 157, and Buckwalter and Horger, Investigation of Fatigue Strength of Axles, Press Fits, Surface Rolling and Effect of Size, *Trans. ASM,* **25,** Mar. 1937, p. 229). **Corrosion** and **galling** (due to rubbing of mating surfaces) cause great reduction of fatigue strengths, sometimes amounting to as much as 90 percent of the original endurance limit. Although any corroding agent will promote severe corrosion fatigue, there is so much difference between the effects of "sea water" or "tap water" from different localities that numerical values are not quoted here.

Overstressing specimens above the fatigue limit for periods shorter than necessary to produce failure at that stress reduces the fatigue limit in a subsequent test. Similarly, **understressing** below the fatigue limit may increase it. Shot peening, nitriding, and cold work usually improve fatigue properties.

No very good overall correlation exists between fatigue properties and any other mechanical property of a material. The best correlation is between the fatigue limit under completely reversed bending stress and the ordinary tensile strength. For many ferrous metals, the fatigue limit is approximately 0.40 to 0.60 times the tensile strength if the latter is below 200,000 lb/in². Low-alloy high-yield-strength steels often show higher values than this. The fatigue limit for nonferrous metals is approximately to 0.20 to 0.50 times the tensile strength. The fatigue limit in reversed shear is approximately 0.57 times that in reversed bending.

In some very important engineering situations components are cyclically stressed into the plastic range. Examples are thermal strains resulting from temperature oscillations and notched regions subjected to secondary stresses. Fatigue life in the plastic or "**low-cycle**" fatigue range has been found to be a function of plastic strain, and low-cycle fatigue testing is done with strain as the controlled variable rather than stress. Fatigue life N and cyclic plastic strain ε_p tend to follow the relationship

$$N\varepsilon_p^2 = C$$

where C is a constant for a material when $N < 10^5$. (See Coffin, A Study

Table 5.1.5   Typical Approximate Fatigue Limits for Reversed Bending

Metal	Tensile strength, 1,000 lb/in²	Fatigue limit, 1,000 lb/in²	Metal	Tensile strength, 1,000 lb/in²	Fatigue limit, 1,000 lb/in²
Cast iron	20–50	6–18	Copper	32–50	12–17
Malleable iron	50	24	Monel	70–120	20–50
Cast steel	60–80	24–32	Phosphor bronze	55	12
Armco iron	44	24	Tobin bronze, hard	65	21
Plain carbon steels	60–150	25–75	Cast aluminum alloys	18–40	6–11
SAE 6150, heat-treated	200	80	Wrought aluminum alloys	25–70	8–18
Nitralloy	125	80	Magnesium alloys	20–45	7–17
Brasses, various	25–75	7–20	Molybdenum, as cast	98	45
Zirconium crystal bar	52	16–18	Titanium (Ti-75A)	91	45

NOTE: Stress, 1,000 lb/in² × 6.894 = stress, MN/m².

of Cyclic-Thermal Stresses in a Ductile Material, *Trans. ASME,* **76,** 1954, p. 947.)

The type of physical change occurring inside a material as it is repeatedly loaded to failure varies as the life is consumed, and a number of stages in fatigue can be distinguished on this basis. The early stages comprise the events causing nucleation of a crack or flaw. This is most likely to appear on the surface of the material; fatigue failures generally originate at a surface. Following nucleation of the crack, it grows during the crack-propagation stage. Eventually the crack becomes large enough for some rapid terminal mode of failure to take over such as ductile rupture or brittle fracture. The rate of crack growth in the crack-propagation stage can be accurately quantified by fracture mechanics methods. Assuming an initial flaw and a loading situation as shown in Fig. 5.1.15, the rate of crack growth per cycle can generally be expressed as

$$da/dN = C_0 (\Delta K_I)^n \qquad (5.1.3)$$

where C_0 and n are constants for a particular material and ΔK_I is the range of stress intensity per cycle. K_I is given by (5.1.1). Using (5.1.3), it is possible to predict the number of cycles for the crack to grow to a size at which some other mode of failure can take over. Values of the constants C_0 and n are determined from specimens of the same type as those used for determination of K_{Ic} but are instrumented for accurate measurement of slow crack growth.

Constant-amplitude fatigue-test data are relevant to many rotary-machinery situations where constant cyclic loads are encountered. There are important situations where the component undergoes variable loads and where it may be advisable to use **random-load testing**. In this method, the load spectrum which the component will experience in service is determined and is applied to the test specimen artificially.

CREEP

Experience has shown that, for the design of equipment subjected to sustained loading at elevated temperatures, little reliance can be placed on the usual short-time tensile properties of metals at those temperatures. Under the application of a constant load it has been found that materials, both metallic and nonmetallic, show a gradual flow or **creep** even for stresses below the proportional limit at elevated temperatures. Similar effects are present in low-melting metals such as lead at room temperature. The deformation which can be permitted in the satisfactory operation of most high-temperature equipment is limited.

In metals, creep is a plastic deformation caused by slip occurring along crystallographic directions in the individual crystals, together with some flow of the grain-boundary material. After complete release of load, a small fraction of this plastic deformation is recovered with time. Most of the flow is nonrecoverable for metals.

Since the early creep experiments, many different types of tests have come into use. The most common are the **long-time creep test** under constant tensile load and the **stress-rupture** test. Other special forms are the **stress-relaxation test** and the **constant-strain-rate test**.

The **long-time creep test** is conducted by applying a dead weight to one end of a lever system, the other end being attached to the specimen surrounded by a furnace and held at constant temperature. The axial deformation is read periodically throughout the test and a curve is plotted of the strain ε_0 as a function of time t (Fig. 5.1.19). This is repeated for various loads at the same testing temperature. The portion of the

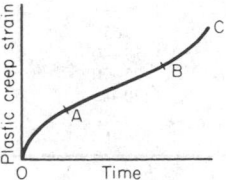

Fig. 5.1.19. Typical creep curve.

curve OA in Fig. 5.1.19 is the region of **primary creep**, AB the region of **secondary creep**, and BC that of **tertiary creep**. The strain rates, or the slopes of the curve, are decreasing, constant, and increasing, respectively, in these three regions. Since the period of the creep test is usually much shorter than the duration of the part in service, various extrapolation procedures are followed (see Gittus, "Creep, Viscoelasticity and Creep Fracture in Solids," Wiley, 1975). See Table 5.1.6.

In practical applications the region of constant-strain rate (secondary creep) is often used to estimate the probable deformation throughout the life of the part. It is thus assumed that this rate will remain constant during periods beyond the range of the test-data. The working stress is chosen so that this total deformation will not be excessive. An **arbitrary creep strength**, which is defined as the stress which at a given temperature will result in 1 percent deformation in 100,000 h, has received a certain amount of recognition, but it is advisable to determine the proper stress for each individual case from diagrams of stress vs. creep rate (Fig. 5.1.20) (see "Creep Data," ASTM and ASME).

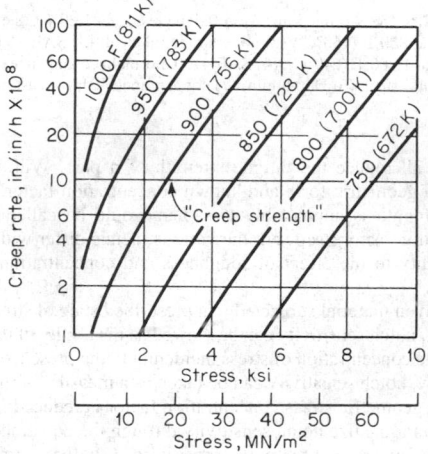

Fig. 5.1.20. Creep rates for 0.35% C steel.

Additional temperatures (°F) and stresses (in 1,000 lb/in²) for stated creep rates (percent per 1,000 h) for wrought nonferrous metals are as follows:

60-40 Brass: Rate 0.1, temp. 350 (400), stress 8 (2); rate 0.01, temp 300 (350) [400], stress 10 (3) [1].

Phosphor bronze: Rate 0.1, temp 400 (550) [700] [800], stress 15 (6) [4] [4]; rate 0.01, temp 400 (550) [700], stress 8 (4) [2].

Nickel: Rate 0.1, temp 800 (1000), stress 20 (10).

70 Cu, 30 Ni. Rate 0.1, temp 600 (750), stress 28 (13–18); rate 0.01, temp 600 (750), stress 14 (8–9).

Aluminum alloy 17 *S* (Duralumin): Rate 0.1, temp 300 (500) [600], stress 22 (5) [1.5].

Lead pure (commercial) (0.03 percent Ca): At 110°F, for rate 0.1 percent the stress range, lb/in², is 150–180 (60–140) [200–220]; for rate of 0.01 percent, 50–90 (10–50) [110–150].

Stress, 1,000 lb/in² × 6.894 = stress, MN/m², $t_k = \frac{5}{9}(t_F + 459.67)$.

Structural changes may occur during a creep test, thus altering the metallurgical condition of the metal. In some cases, premature rupture appears at a low fracture strain in a normally ductile metal, indicating that the material has become embrittled. This is a very insidious condition and difficult to predict. The **stress-rupture test** is well adapted to study this effect. It is conducted by applying a constant load to the specimen in the same manner as for the long-time creep test. The nominal stress is then plotted vs. the time for fracture at constant temperature on a log-log scale (Fig. 5.1.21).

Table 5.1.6   Stresses for Given Creep Rates and Temperatures*

Material Temp, °F	Creep rate 0.1% per 1,000 h					Creep rate 0.01% per 1,000 h				
	800	900	1,000	1,100	1,200	800	900	1,000	1,100	1,200
Wrought steels:										
SAE 1015	17–27	11–18	3–12	2–7	1	10–18	6–14	3–8	1	
0.20 C, 0.50 Mo	26–33	18–25	9–16	2–6	1–2	16–24	11–22	4–12	2	1
0.10–0.25 C, 4–6 Cr + Mo	22	15–18	9–11	3–6	2–3	14–17	11–15	4–7	2–3	1–2
SAE 4140	27–33	20–25	7–15	4–7	1–2	19–28	12–19	3–8	2–4	1
SAE 1030–1045	8–25	5–15	5	2	1	5–15	3–7	2–4	1	
Commercially pure iron	7		4		3	5		2		
0.15 C, 1–2.5 Cr, 0.50 Mo	25–35	18–28	8–20	6–8	3–4	20–30	12–18	3–12	2–5	1–2
SAE 4340	20–40	15–30	2–12	1–3		8–20		1–6		
SAE X3140	7–10		5–4			3–8		1–2		
0.20 C, 4–6 Cr	30	10–20	7–10	1				3–5		
0.25 C, 4–6 Cr + W	30	10–15	4–10	2–8			6–11	2–7		
0.16 C, 1.2 Cu		18	10–15	3	1		10–18	7–12		
0.20 C, 1 Mo	35	27	12			25	12	6		
0.10–0.40 C, 0.2–0.5 Mo, 1–2 Mn	30–40	12–20	4–14			25–28	8–15	2–8		0.5
SAE 2340	7–12	5	2							
SAE 6140	30	12	4			7	6	1		
SAE 7240	30	21	6–15	2		30	11	3–9	1	
Cr + Va + W, various	20–70	14–30	5–15			18–50	8–18	2–13		

Temp, °F	1,100	1,200	1,300	1,400	1,500	1,000	1,100	1,200	1,300	1,400
Wrought chrome-nickel steels:										
18-8†	10–18	5–11	3–10	2–5	2.5	11–16	5–12	2–10		1–2
10–25 Cr, 10–30 Ni‡	10–20	5–15	3–10	2–5			6–15	3–10	2–8	1–3

Temp, °F	800	900	1,000	1,100	1,200	800	900	1,000	1,100	1,200
Cast steels:										
0.20–0.40 C	10–20	5–10	3			8–15		1		
0.10–0.30 C, 0.5–1 Mo	28	20–30	6–12	2		20	10–15	2–5		
0.15–0.30 C, 4–6 Cr + Mo	25–30	15–25	8–15	8		20–25	9–15	2–7	2	
18-8§			20–25	15	10			20	15	8
Cast iron	20	8	4			10		2		
Cr Ni cast iron			9					3		

* Based on 1,000-h tests. Stresses in 1,000 lb/in².
† Additional data. At creep rate 0.1 percent and 1,000 (1,600)°F the stress is 18–25 (1); at creep rate 0.01 percent at 1,500°F, the stress is 0.5.
‡ Additional data. At creep rate 0.1 percent and 1,000 (1,600)°F the stress is 10–30 (1).
§ Additional data. At creep rate 0.1 percent and 1,600°F the stress is 3; at creep rate 0.01 and 1,500°F, the stress is 2–3.

The stress reaction is measured in the **constant-strain-rate test** while the specimen is deformed at a constant strain rate. In the **relaxation test,** the decrease of stress with time is measured while the total strain (elastic + plastic) is maintained constant. The latter test has direct application to the loosening of turbine bolts and to similar problems. Although some correlation has been indicated between the results of these various types of tests, no general correlation is yet available, and it has been found necessary to make tests under each of these special conditions to obtain satisfactory results.

The interrelationship between strain rate and temperature in the form of a velocity-modified temperature (see MacGregor and Fisher, A Velocity-modified Temperature for the Plastic Flow of Metals, *Jour. Appl. Mech.,* Mar. 1945) simplifies the creep problem in reducing the number of variables.

Superplasticity Superplasticity is the property of some metals and alloys which permits extremely large, uniform deformation at elevated temperature, in contrast to conventional metals which neck down and subsequently fracture after relatively small amounts of plastic deformation. Superplastic behavior requires a metal with small equiaxed grains, a slow and steady rate of deformation (strain

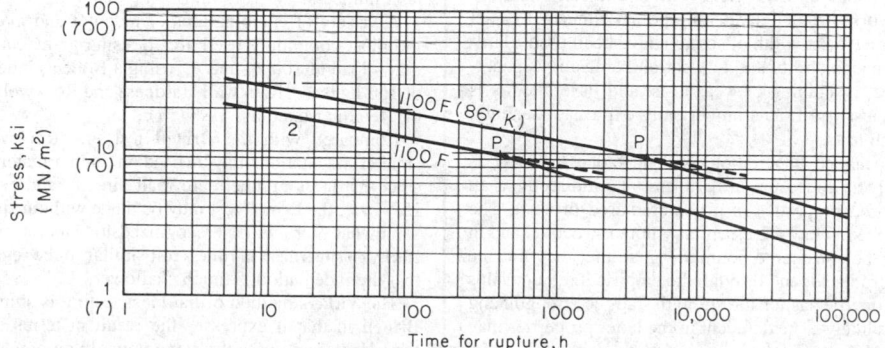

Fig. 5.1.21   Relation between time to failure and stress for a 3% chromium steel. (1) Heat treated 2 h at 1,740°F (950°C) and furnace cooled; (2) hot rolled and annealed 1,580°F (860°C).

rate), and a temperature elevated to somewhat more than half the melting point. With such metals, large plastic deformation can be brought about with lower external loads; ultimately, that allows the use of lighter fabricating equipment and facilitates production of finished parts to near-net shape.

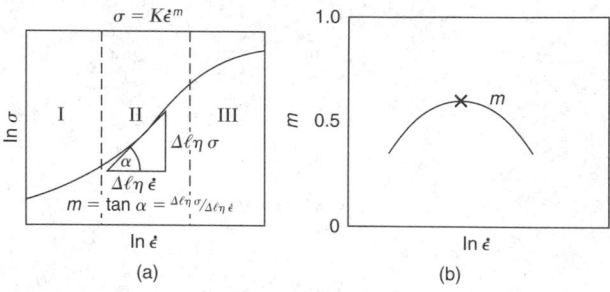

Fig. 5.1.22. Stress and strain rate relations for superplastic alloys. (*a*) Log-log plot of $\sigma = K\dot{\varepsilon}^m$; (*b*) *m* as a function of strain rate.

Stress and strain rates are related for a metal exhibiting superplasticity. A factor in this behavior stems from the relationship between the applied stress and strain rates. This factor *m*—the strain rate sensitivity index—is evaluated from the equation $\sigma = K\dot{\varepsilon}^m$, where σ is the applied stress, *K* is a constant, and $\dot{\varepsilon}$ is the strain rate. Figure 5.1.22*a* plots a stress/strain rate curve for a superplastic alloy on log-log coordinates. The slope of the curve defines *m*, which is maximum at the point of inflection. Figure 5.1.22*b* shows the variation of *m* versus $\ln \dot{\varepsilon}$. Ordinary metals exhibit low values of *m*—0.2 or less; for those behaving superplastically, $m = 0.6$ to $0.8 +$. As *m* approaches 1, the behavior of the metal will be quite similar to that of a newtonian viscous solid, which elongates plastically without necking down.

In Fig. 5.1.22*a*, in region I, the stress and strain rates are low and creep is predominantly a result of diffusion. In region III, the stress and strain rates are highest and creep is mainly the result of dislocation and slip mechanisms. In region II, where superplasticity is observed, creep is governed predominantly by grain boundary sliding.

HARDNESS

Hardness has been variously defined as resistance to local penetration, to scratching, to machining, to wear or abrasion, and to yielding. The multiplicity of definitions, and corresponding multiplicity of hardness-measuring instruments, together with the lack of a fundamental definition, indicates that hardness may not be a fundamental property of a material but rather a composite one including yield strength, work hardening, true tensile strength, modulus of elasticity, and others.

Scratch hardness is measured by **Mohs scale** of minerals (Sec. 1.2) which is so arranged that each mineral will scratch the mineral of the next lower number. In recent mineralogical work and in certain microscopic metallurgical work, jeweled scratching points either with a set load or else loaded to give a set width of scratch have been used. Hardness in its relation to machinability and to wear and abrasion is generally dealt with in direct machining or wear tests, and little attempt is made to separate hardness itself, as a numerically expressed quantity, from the results of such tests.

The resistance to localized penetration, or **indentation hardness,** is widely used industrially as a measure of hardness, and indirectly as an indicator of other desired properties in a manufactured product. The indentation tests described below are essentially nondestructive, and in most applications may be considered nonmarring, so that they may be applied to each piece produced; and through the empirical relationships of hardness to such properties as tensile strength, fatigue strength, and impact strength, pieces likely to be deficient in the latter properties may be detected and rejected.

Brinell hardness is determined by forcing a hardened sphere under a known load into the surface of a material and measuring the diameter of the indentation left after the test. The **Brinell hardness number,** or simply the **Brinell number,** is obtained by dividing the load used, in kilograms, by the actual surface area of the indentation, in square millimeters. The result is a pressure, but the units are rarely stated.

$$\text{BHN} = P \left/ \left[\frac{\pi D}{2} (D - \sqrt{D^2 - d^2}) \right] \right.$$

where BHN is the Brinell hardness number; *P* the imposed load, kg; *D* the diameter of the spherical indenter, mm; and *d* the diameter of the resulting impression, mm.

Hardened-steel bearing balls may be used for hardness up to 450, but beyond this hardness specially treated steel balls or jewels should be used to avoid flattening the indenter. The standard-size ball is 10 mm and the standard loads 3,000, 1,500, and 500 kg, with 100, 125, and 250 kg sometimes used for softer materials. If for special reasons any other size of ball is used, the load should be adjusted approximately as follows: for iron and steel, $P = 30D^2$; for brass, bronze, and other soft metals, $P = 5D^2$; for extremely soft metals, $P = D^2$ (see "Methods of Brinell Hardness Testing," ASTM). Readings obtained with other than the standard ball and loadings should have the load and ball size appended, as such readings are only approximately equal to those obtained under standard conditions.

The size of the specimen should be sufficient to ensure that no part of the plastic flow around the impression reaches a free surface, and in no case should the thickness be less than 10 times the depth of the impression. The load should be applied steadily and should remain on for at least 15 s in the case of ferrous materials and 30 s in the case of most nonferrous materials. Longer periods may be necessary on certain soft materials that exhibit creep at room temperature. In testing thin materials, it is not permissible to pile up several thicknesses of material under the indenter, as the readings so obtained will invariably be lower than the true readings. With such materials, smaller indenters and loads, or different methods of hardness testing, are necessary.

In the standard Brinell test, the diameter of the impression is measured with a low-power hand microscope, but for production work several testing machines are available which automatically measure the depth of the impression and from this give readings of hardness. Such machines should be calibrated frequently on test blocks of known hardness.

In the **Rockwell method** of hardness testing, the depth of penetration of an indenter under certain arbitrary conditions of test is determined. The indenter may be either a steel ball of some specified diameter or a spherical-tipped conical diamond of 120° angle and 0.2-mm tip radius, called a "Brale." A *minor load* of 10 kg is first applied which causes an initial penetration and holds the indenter in place. Under this condition, the dial is set to zero and the major load applied. The values of the latter are 60, 100, or 150 kg. Upon removal of the major load, the reading is taken while the minor load is still on. The hardness number may then be read directly from the scale which measures penetration, and this scale is so arranged that soft materials with deep penetration give low hardness numbers.

A variety of combinations of indenter and major load are possible; the most commonly used are R_B using as indenter a $\frac{1}{16}$-in ball and a major load of 100 kg and R_C using a Brale as indenter and a major load of 150 kg (see "Rockwell Hardness and Rockwell Superficial Hardness of Metallic Materials," ASTM).

Compared with the Brinell test, the Rockwell method makes a smaller indentation, may be used on thinner material, and is more rapid, since hardness numbers are read directly and need not be calculated. However, the Brinell test may be made without special apparatus and is somewhat more widely recognized for laboratory use. There is also a **Rockwell superficial hardness test** similar to the regular Rockwell, except that the indentation is much shallower.

The **Vickers** method of hardness testing is similar in principle to the Brinell in that it expresses the result in terms of the pressure under the indenter and uses the same units, kilograms per square millimeter. The indenter is a diamond in the form of a square pyramid with an apical angle of 136°, the loads are much lighter, varying between 1 and 120 kg, and the impression is measured by means of a medium-power compound microscope.

$$V = P/(0.5393d^2)$$

where V is the Vickers hardness number, sometimes called the **diamond-pyramid hardness** (DPH); P the imposed load, kg; and d the diagonal of indentation, mm. The Vickers method is more flexible and is considered to be more accurate than either the Brinell or the Rockwell, but the equipment is more expensive than either of the others and the Rockwell is somewhat faster in production work.

Among the other hardness methods may be mentioned the **Scleroscope,** in which a diamond-tipped "hammer" is dropped on the surface and the rebound taken as an index of hardness. This type of apparatus is seriously affected by the resilience as well as the hardness of the material and has largely been superseded by other methods. In the **Monotron** method, a penetrator is forced into the material to a predetermined depth and the load required is taken as the indirect measure of the hardness. This is the reverse of the Rockwell method in principle, but the loads and indentations are smaller than those of the latter. In the **Herbert pendulum,** a 1-mm steel or jewel ball resting on the surface to be tested acts as the fulcrum for a 4-kg compound pendulum of 10-s period. The swinging of the pendulum causes a rolling indentation in the material, and from the behavior of the pendulum several factors in hardness, such as **work hardenability,** may be determined which are not revealed by other methods. Although the Herbert results are of considerable significance, the instrument is suitable for laboratory use only (see Herbert, The Pendulum Hardness Tester, and Some Recent Developments in Hardness Testing, *Engineer,* **135,** 1923, pp. 390, 686). In the **Herbert cloudburst** test, a shower of steel balls, dropped from a predetermined height, dulls the surface of a hardened part in proportion to its softness and thus reveals defective areas. A variety of **mutual indentation methods,** in which crossed cylinders or prisms of the material to be tested are forced together, give results comparable with the Brinell test. These are particularly useful on wires and on materials at high temperatures.

The relation among the scales of the various hardness methods is not exact, since no two measure exactly the same sort of hardness, and a relationship determined on steels of different hardnesses will be found only approximately true with other materials. The **Vickers-Brinell relation** is nearly linear up to at least 400, with the Vickers approximately 5 percent higher than the Brinell (actual values run from + 2 to + 11 percent) and nearly independent of the material. Beyond 500, the values become more widely divergent owing to the flattening of the Brinell ball. The **Brinell-Rockwell relation** is fairly satisfactory and is shown in Fig. 5.1.23. Approximate relations for the **Shore Scleroscope** are also given on the same plot.

The **hardness of wood** is defined by the ASTM as the load in pounds required to force a ball 0.444 in in diameter into the wood to a depth of 0.222 in, the speed of penetration being ¼ in/min. For a summary

of the work in hardness see Williams, "Hardness and Hardness Measurements," ASM.

TESTING OF MATERIALS

Testing Machines   Machines for the mechanical testing of materials usually contain elements (1) for gripping the specimen, (2) for deforming it, and (3) for measuring the load required in performing the deformation. Some machines (ductility testers) omit the measurement of load and substitute a measurement of deformation, whereas other machines include the measurement of both load and deformation through apparatus either integral with the testing machine (stress-strain recorders) or auxiliary to it (strain gages). In most general-purpose testing machines, the deformation is controlled as the independent variable and the resulting load measured, and in many special-purpose machines, particularly those for light loads, the load is controlled and the resulting deformation is measured. Special features may include those for constant rate of loading (pacing disks), for constant rate of straining, for constant load maintenance, and for cyclical load variation (fatigue).

In modern testing systems, the load and deformation measurements are made with load-and-deformation-sensitive transducers which generate electrical outputs. These outputs are converted to load and deformation readings by means of appropriate electronic circuitry. The readings are commonly displayed automatically on a recorder chart or digital meter, or they are read into a computer. The transducer outputs are typically used also as feedback signals to control the test mode (constant loading, constant extension, or constant strain rate). The load transducer is usually a load cell attached to the test machine frame, with electrical output to a bridge circuit and amplifier. The load cell operation depends on change of electrical resistivity with deformation (and load) in the transducer element. The deformation transducer is generally an extensometer clipped on to the test specimen gage length, and operates on the same principle as the load cell transducer: the change in electrical resistance in the specimen gage length is sensed as the specimen deforms. Optical extensometers are also available which do not make physical contact with the specimen. Verification and classification of extensometers is controlled by ASTM Standards. The application of load and deformation to the specimen is usually by means of a screw-driven mechanism, but it may also be applied by means of hydraulic and servohydraulic systems. In each case the load application system responds to control inputs from the load and deformation transducers. Important features in test machine design are the methods used for reducing friction, wear, and backlash. In older testing machines, test loads were determined from the machine itself (e.g., a pressure reading from the machine hydraulic pressure) so that machine friction made an important contribution to inaccuracy. The use of machine-independent transducers in modern testing has eliminated much of this source of error.

Grips should not only hold the test specimen against slippage but should also apply the load in the desired manner. Centering of the load is of great importance in compression testing, and should not be neglected in tension testing if the material is brittle. Figure 5.1.24 shows the theoretical errors due to off-center loading; the results are directly applicable to compression tests using swivel loading blocks. Swivel (ball-and-socket) holders or compression blocks should be used with all except the most ductile materials, and in compression testing of brittle materials (concrete, stone, brick), any rough faces should be smoothly capped with plaster of paris and one-third portland cement. Serrated grips may be used to hold ductile materials or the shanks of other holders in tension; a taper of 1 in 6 on the wedge faces gives a self-tightening action without excessive jamming. Ropes are ordinarily held by wet eye splices, but braided ropes or small cords may be given several turns over a fixed pin and then clamped. Wire ropes should be zinced into forged sockets (solder and lead have insufficient strength). Grip selection for tensile testing is described in ASTM standards.

Accuracy and Calibration   ASTM standards require that commercial machines have errors of less than 1 percent within the "loading range" when checked against acceptable standards of comparison at at least five suitably spaced loads. The "loading range" may be any range

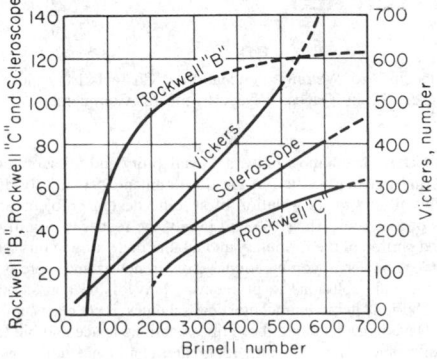

Fig. 5.1.23.   Hardness scales.

through which the preceding requirements for accuracy are satisfied, except that it shall not extend below 100 times the least load to which the machine will respond or which can be read on the indicator. The use of calibration plots or tables to correct the results of an otherwise inaccurate machine is not permitted under any circumstances. Machines with errors less than 0.1 percent are commercially available (Tate-Emery and others), and somewhat greater accuracy is possible in the most refined research apparatus.

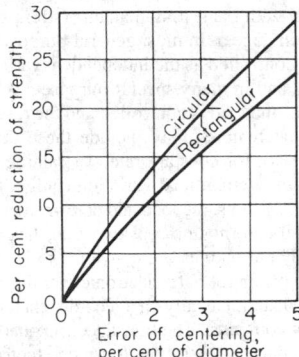

Fig. 5.1.24. Effect of centering errors on brittle test specimens.

Dead loads may be used to check machines of low capacity; accurately calibrated proving levers may be used to extend the range of available weights. Various elastic devices (such as the Morehouse proving ring) made of specially treated steel, with sensitive disortion-measuring devices, and calibrated by dead weights at the NIST (formerly Bureau of Standards) are mong the most satisfactory means of checking the higher loads.

Two **standard forms of test specimens** (ASTM) are shown in Figs. 5.1.25 and 5.1.26. In wrought materials, and particularly in those which

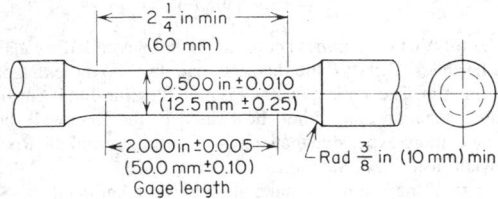

Fig. 5.1.25. Test specimen, 2-in (50-mm) gage length, ½-in (12.5-mm) diameter. Others available for 0.35-in (8.75-mm) and 0.25-in (6.25-mm) diameters. *(ASTM).*

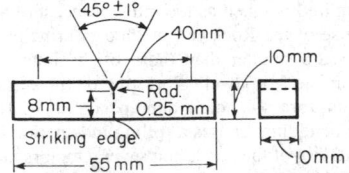

Fig. 5.1.26. Charpy V-notch impact specimens. *(ASTM.)*

have been cold-worked, different properties may be expected in different directions with respect to the direction of the applied work, and the test specimen should be cut out from the parent material in such a way as to give the strength in the desired direction. With the exception of fatigue specimens and specimens of extremely brittle materials, surface finish is of little practical importance, although extreme roughness tends to decrease the ultimate elongation.

5.2 MECHANICS OF MATERIALS
by J. P. Vidosic

REFERENCES: Timoshenko and MacCullough, "Elements of Strength of Materials," Van Nostrand. Seeley, "Advanced Mechanics of Materials," Wiley. Timoshenko and Goodier, "Theory of Elasticity," McGraw-Hill. Phillips, "Introduction to Plasticity," Ronald. Van Den Broek, "Theory of Limit Design," Wiley. Hetényi, "Handbook of Experimental Stress Analysis," Wiley. Dean and Douglas, "Semi-Conductor and Conventional Strain Gages," Academic. Robertson and Harvey, "The Engineering Uses of Holography," University Printing House, London. Sellers, "Basic Training Guide to the New Metrics and SI Units," National Tool, Die and Precision Machining Association. Roark and Young, "Formulas for Stress and Strain," McGraw-Hill. Perry and Lissner, "The Strain Gage Primer," McGraw-Hill. Donnell, "Beams, Plates, and Sheets," Engineering Societies Monographs, McGraw-Hill. Griffel, "Beam Formulas" and "Plate Formulas," Ungar. Durelli et al., "Introduction to the Theoretical and Experimental Analysis of Stress and Strain," McGraw-Hill. "Stress Analysis Manual," Department of Commerce, Pub. no. AD 759 199, 1969. Blodgett, "Welded Structures," Lincoln Arc Welding Foundation. "Characteristics and Applications of Resistance Strain Gages," Department of Commerce, *NBS Circ.* 528, 1954.

EDITOR'S NOTE: The almost universal availability and utilization of computers in engineering practice has led to the development of many forms of software individually tailored to the solution of specific design problems in the area of mechanics of materials. Their use will permit the reader to amplify and supplement a good portion of the formulary and tabular collection in this section, as well as utilize those powerful computational tools in newer and more powerful techniques to facilitate solutions to problems. Many of the approximate methods, involving laborious iterative mathematical schemes, have been supplanted by the computer. Developments along those lines continue apace and bid fair to expand the types of problems handled, all with greater confidence in the results obtained thereby.

Main Symbols

Unit Stress

S = apparent stress
S_v or S_s = pure shearing
T = true (ideal) stress
S_p = proportional elastic limit
S_y = yield point
S_M = ultimate strength, tension
S_c = ultimate compression
S_v = vertical shear in beams
S_R = modulus of rupture

Moment

M = bending
M_t = torsion

External Action

P = force
G = weight of body
W = weight of load
V = external shear

Modulus of Elasticity

E = longitudinal
G = shearing
K = bulk
U_p = modulus of resilience
U_R = ultimate resilience

Geometric

l = length
A = area
V = volume
v = velocity
r = radius of gyration
I = rectangular moment of inertia
I_P or J = polar moment of inertia

Deformation

e, e' = gross deformation
$\varepsilon, \varepsilon'$ = unit deformation; strain
d or α = unit, angular
s' = unit, lateral
μ = Poisson's ratio
n = reciprocal of Poisson's ratio
r = radius
f = deflection

SIMPLE STRESSES AND STRAINS

Deformations are changes in form produced by external forces or loads that act on nonrigid bodies. Deformations are **longitudinal**, e, a lengthening ($+$) or shortening ($-$) of the body; and **angular**, α, a change of angle between the faces.

Unit deformation (dimensionless number) is the deformation in unit distance. Unit longitudinal deformation (longitudinal strain), $\varepsilon = e/l$ (Fig. 5.2.1). Unit angular-deformation tan α equals α approx (Fig. 5.2.2).

The accompanying lateral deformation results in unit lateral deformation (lateral strain) $\varepsilon' = e'/l'$ (Fig. 5.2.1). For homogeneous, isotropic material operating in the elastic region, the ratio ε'/ε is a constant and is a definite property of the material; this ratio is called **Poisson's ratio** μ.

A fundamental relation among the **three interdependent constants** E, G, and μ for a given material is $E = 2G(1 + \mu)$. Note that μ cannot be larger than 0.5; thus the shearing modulus G is always smaller than the elastic modulus E. At the extremes, for example, $\mu \approx 0.5$ for rubber and

paraffin; $\mu \approx 0$ for cork. For concrete, μ varies from 0.10 to 0.20 at working stresses and can reach 0.25 at higher stresses; μ for ordinary glass is about 0.25. In the absence of definitive data, μ for most structural metals can be taken to lie between 0.25 and 0.35. Extensive listings of Poisson's ratio are found in other sections; see Tables 5.1.3 and 6.1.9.

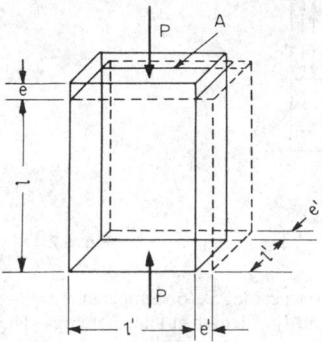

Fig. 5.2.1

Stress is an internal distributed force, or, force per unit area; it is the internal mechanical reaction of the material accompanying deformation. Stresses always occur in pairs. Stresses are **normal** [tensile stress ($+$) and compressive stress ($-$)]; and **tangential**, or **shearing**.

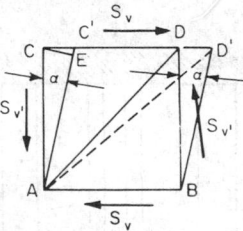

Fig. 5.2.2

Intensity of stress, or **unit stress**, S, lb/in² (kgf/cm²), is the amount of force per unit of area (Fig. 5.2.3). P is the load acting through the center of gravity of the area. The uniformly distributed normal stress is

$$S = P/A$$

When the stress is not uniformly distributed, $S = dP/dA$.

A **long rod** will stretch under its own weight G and a terminal load P (see Fig. 5.2.4). The total elongation e is that due to the terminal load plus that due to one-half the weight of the rod considered as acting at the end.

$$e = (Pl + Gl/2)/(AE)$$

The maximum stress is at the upper end.

When a **load** is **carried by several paths** to a support, the different paths take portions of the load in proportion to their stiffness, which is controlled by material (E) and by design.

EXAMPLE. Two pairs of bars rigidly connected (with the same elongation) carry a load P_0 (Fig. 5.2.5). A_1, A_2 and E_1, E_2 and P_1, P_2 and S_1, S_2 are cross sections, moduli of elasticity, loads, and stresses of the bars, respectively; e = elongation.

$$e = P_1 l/(E_1 A_1) = P_2 l/(E_2 A_2)$$
$$P_0 = 2P_1 + 2P_2$$
$$S_2 = P_2/A_2 = \tfrac{1}{2}[P_0 E_2/(E_1 A_1 + E_2 A_2)]$$
$$S_1 = \tfrac{1}{2}[P_0 E_1/(E_1 A_1 + E_2 A_2)]$$

Temperature Stresses   When the deformation arising from change of temperature is prevented, temperature stresses arise that are proportional to the amount of deformation that is prevented. Let a = coeffi-

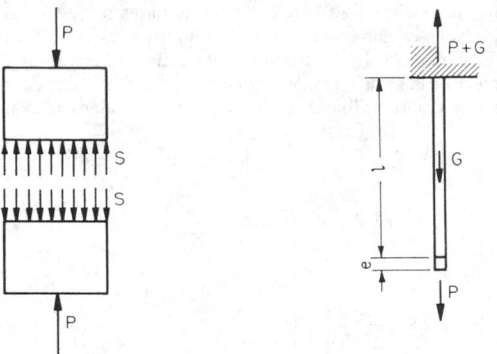

Fig. 5.2.3   **Fig. 5.2.4**

cient of expansion per degree of temperature, l_1 = length of bar at temperature t_1, and l_2 = length at temperature t_2. Then

$$l_2 = l_1[1 + a(t_2 - t_1)]$$

If, subsequently, the bar is cooled to a temperature t_1, the proportionate deformation is $s = a(t_2 - t_1)$ and the corresponding unit stress $S = Ea(t_2 - t_1)$. For **coefficients of expansion,** see Sec. 4. In the case of steel, a change of temperature of 12°F (6.7 K, 6.7°C) will cause in general a unit stress of 2,340 lb/in² (164 kgf/cm²).

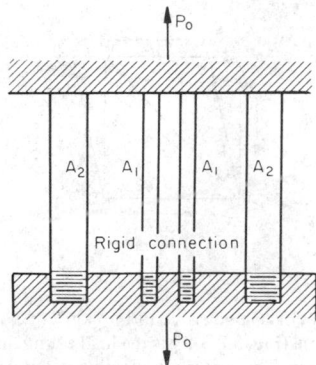

Fig. 5.2.5

Shearing stresses (Fig. 5.2.2) act tangentially to surface of contact and do not change length of sides of elementary volume; they **change the angle between faces** and the **length of diagonal.** Two pairs of shearing stresses must act together. **Shearing stress intensities are of equal magnitude on all four faces of an element.** $S_v = S_v'$ (Fig. 5.2.6).

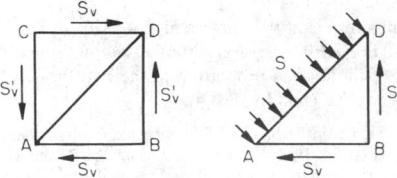

Fig. 5.2.6

In the presence of **pure shear** on external faces (Fig. 5.2.6), the **resultant stress** S on one diagonal plane at 45° is pure tension and on the other diagonal plane pure compression; $S = S_v = S_v'$. S on diagonal plane is called ''diagonal tension'' by writers on reinforced concrete. Failure under pure shear is difficult to produce experimentally, except under torsion and in certain special cases. Figure 5.2.7 shows an ideal case,

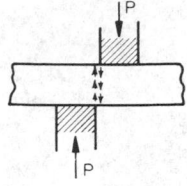

Fig. 5.2.7

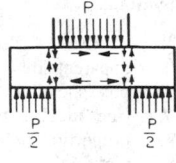

Fig. 5.2.8

and Fig. 5.2.8 a common form of test piece that introduces bending stresses.

Let Fig. 5.2.9 represent the symmetric section of area A with a shearing force V acting through its centroid. If pure shear exists, $S_v = V/A$, and this shear would be uniformly distributed over the area A. When this **shear** is **accompanied by bending (transverse shear in beams),** the unit shear

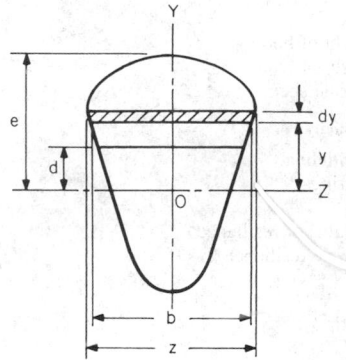

Fig. 5.2.9

S_v increases from the extreme fiber to its maximum, which may or may not be at the neutral axis OZ. The unit shear parallel to OZ at a point d distant from the neutral axis (Fig. 5.2.9) is

$$S_v = \frac{V}{Ib} \int_d^e yz\, dy$$

where z = the section width at distance y; and I is the moment of inertia of the **entire** section about the neutral axis OZ. Note that $\int_d^e yz\, dy$ is the first moment of the area above d with respect to axis OZ. For a **rectangular cross section** (Fig. 5.2.10a),

$$S_v = \frac{3}{2}\frac{V}{bh}\left[1 - \left(\frac{2y}{h}\right)^2\right]$$

$$S_v \text{ (max)} = \frac{3}{2}\frac{V}{bh} = \frac{3}{2}\frac{V}{A} \qquad \text{for } y = 0$$

For a **circular cross section** (Fig. 5.2.10b),

$$S_v = \frac{4}{3}\frac{V}{\pi r^2}\left[1 - \left(\frac{y}{r}\right)^2\right]$$

$$S_v \text{ (max)} = \frac{4}{3}\frac{V}{\pi r^2} = \frac{4}{3}\frac{V}{A} \qquad \text{for } y = 0$$

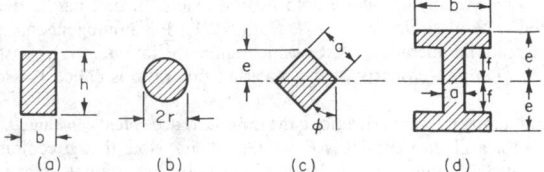

Fig. 5.2.10

Table 5.2.1   Resilience per Unit of Volume U_p
(S = longitudinal stress; S_v = shearing stress; E = tension modulus of elasticity; G = shearing modulus of elasticity)

Tension or compression	$\frac{1}{2}S^2/E$	Torsion	
Shear	$\frac{1}{2}S_v^2/G$	Solid circular	$\frac{1}{4}S_v^2/G$
Beams (free ends)		Hollow, radii R_1 and R_2	$\dfrac{R_1^2 + R_2^2}{R_1^2}\dfrac{1}{4}\dfrac{S_v^2}{G}$
Rectangular section, bent in arc of circle; no shear	$\frac{1}{6}S^2/E$		
Ditto, circular section	$\frac{1}{8}S^2/E$	Springs	
Concentrated center load; rectangular cross section	$\frac{1}{18}S^2/E$	Carriage	$\frac{1}{6}S^2/E$
		Flat spiral, rectangular section	$\frac{1}{24}S^2/E$
Ditto, circular cross section	$\frac{1}{24}S^2/E$	Helical: axial load, circular wire	$\frac{1}{4}S_v^2/G$
Uniform load, rectangular cross section	$\frac{5}{36}S^2/E$	Helical: axial twist	$\frac{1}{8}S^2/E$
I-beam section, concentrated center load	$\frac{3}{32}S^2/E$	Helical: axial twist, rectangular section	$\frac{1}{6}S^2/E$

For a **circular ring** (thickness small in comparison with the major diameter), $S_v(\text{max}) = 2V/A$, for $y = 0$.

For a **square cross section** (diagonal vertical, Fig. 5.2.10c),

$$S_v = \frac{V\sqrt{2}}{a^2}\left[1 + \frac{y\sqrt{2}}{a} - 4\left(\frac{y}{a}\right)^2\right]$$

$$S_v(\text{max}) = 1.591\,\frac{V}{A} \quad \text{for } y = \frac{e}{4}$$

For an **I-shaped cross section** (Fig. 5.2.10d),

$$S_v(\text{max}) = \frac{3}{4}\frac{V}{a}\left[\frac{be^2 - (b-a)f^2}{be^3 - (b-a)f^3}\right] \quad \text{for } y = 0$$

Elasticity is the ability of a material to return to its original dimensions after the removal of stresses. The **elastic limit** S_p is the limit of stress within which the deformation completely disappears after the removal of stress; i.e., no set remains.

Hooke's law states that, within the elastic limit, deformation produced is proportional to the stress. Unless modified, the deduced formulas of mechanics apply only within the elastic limit. Beyond this, they are modified by experimental coefficients, as, for instance, the modulus of rupture.

The **modulus of elasticity**, lb/in^2 (kgf/cm^2), is the ratio of the increment of unit stress to increment of unit deformation within the elastic limit.

The **modulus of elasticity in tension,** or **Young's modulus,**

$$E = \text{unit stress/unit deformation} = Pl/(Ae)$$

The modulus of elasticity in compression is similarly measured.

The **modulus of elasticity in shear** or **coefficient of rigidity,** $G = S_v/\alpha$ where α is expressed in radians (see Fig. 5.2.2).

The **bulk modulus of elasticity** K is the ratio of normal stress, applied to all six faces of a cube, to the change of volume.

Change of volume under normal stress is so small that it is rarely of significance. For example, given a body with length l, width b, thickness d, Poisson's ratio μ, and longitudinal strain ε, $V = lbd$ = original volume. The deformed volume = $(1 + \varepsilon)l\,(1 - \mu\varepsilon)b(1 - \mu\varepsilon)d$. Neglecting powers of ε, the deformed volume = $(1 + \varepsilon - 2\mu\varepsilon)V$. The change in volume is $\varepsilon(1 - 2\mu)V$; the **unit volumetric strain** is $\varepsilon(1 - 2\mu)$. Thus, a steel rod ($\mu = 0.3$, $E = 30 \times 10^6$ lb/in^2) compressed to a stress of 30,000 lb/in^2 will experience $\varepsilon = 0.001$ and a unit volumetric strain of 0.0004, or 1 part in 2,500.

The following relationships exist between the modulus of elasticity in tension or compression E, modulus of elasticity in shear G, bulk modulus of elasticity K, and Poisson's ratio μ:

$$E = 2G(1 + \mu)$$
$$G = E/[2(1 + \mu)]$$
$$\mu = (E - 2G)/(2G)$$
$$K = E/[3(1 - 2\mu)]$$
$$\mu = (3K - E)/(6K)$$

Resilience U (in·lb)[(cm·kgf)] is the potential energy stored up in a deformed body. The amount of resilience is equal to the work required to deform the body from zero stress to stress S. When S does not exceed the elastic limit. For normal stress, resilience = work of deformation = average force times deformation = $\frac{1}{2}Pe = \frac{1}{2}AS \times Sl/E = \frac{1}{2}S^2V/E$.

Modulus of resilience U_p (in·lb/in^3) [(cm·kgf/cm^3)], or **unit resilience,** is the elastic energy stored up in a cubic inch of material at the elastic limit. For normal stress,

$$U_p = \frac{1}{2}S_p^2/E$$

The unit resilience for any other kind of stress, as shearing, bending, torsion, is a constant times one-half the square of the stress divided by the appropriate modulus of elasticity. For values, see Table 5.2.1.

Unit rupture work U_R, sometimes called **ultimate resilience,** is measured by the area of the stress-deformation diagram to rupture.

$$U_R = \frac{1}{3}e_u(S_y + 2S_M) \quad \text{approx}$$

where e_u is the total deformation at rupture.

For structural steel, $U_R = \frac{1}{3} \times \frac{27}{100} \times [35,000 + (2 \times 60,000)] = 13,950$ in·lb/in^3 (982 cm·kgf/cm^3).

EXAMPLE 1.   A load $P = 40,000$ lb compresses a wooden block of cross-sectional area $A = 10$ in^2 and length = 10 in, an amount $e = \frac{4}{100}$ in. Stress $S = \frac{1}{10} \times 40,000 = 4,000$ lb/in^2. Unit elongation $s = \frac{4}{100} \div 10 = \frac{1}{250}$. Modulus of elasticity $E = 4,000 \div \frac{1}{250} = 1,000,000$ lb/in^2. Unit resilience $U_p = \frac{1}{2} \times 4,000 \times 4,000/1,000,000 = 8$ in·lb/in^3 (0.563 cm·kgf/cm^3).

EXAMPLE 2.   A weight $G = 5,000$ lb falls through a height $h = 2$ ft; $V =$ number of cubic inches required to absorb the shock without exceeding a stress of 4,000 lb/in^2. Neglect compression of block. Work done by falling weight = $Gh = 5,000 \times 2 \times 12$ in·lb ($2,271 \times 61$ cm·kgf) Resilience of block = $V \times 8$ in·lb = $5,000 \times 2 \times 12$. Therefore, $V = 15,000$ in^3 (245,850 cm^3).

Thermal Stresses   A bar will change its length when its temperature is raised (or lowered) by the amount $\Delta l_0 = \alpha l_0(t_2 - 32)$. The linear coefficient of thermal expansion α is assumed constant at normal temperatures and l_0 is the length at 32°F (273.2 K, 0°C). If this expansion (or contraction) is prevented, a **thermal-time stress** is developed, equal to $S = E\alpha(t_2 - t_1)$, as the temperature goes from t_1 to t_2. In thin flat plates the stress becomes $S = E\alpha(t_2 - t_1)/(1 - \mu)$; μ is Poisson's ratio. Such stresses can occur in castings containing large and small sections. Similar stresses also occur when heat flows through members because of the difference in temperature between one point and another. The heat flowing across a length b as a result of a linear drop in temperature Δt equals $Q = k\,A\Delta t/b$ Btu/h (cal/h). The thermal conductivity k is in Btu/(h)(ft^2)(°F)/(in of thickness) [cal/(h)(m^2)(k)/(m)]. The **thermal-flow stress** is then $S = E\alpha Qb/(kA)$. Note, when Q is substituted the stress becomes $S = E\alpha\,\Delta t$ as above, only t is now a function of distance rather than time.

EXAMPLE.   A cast-iron plate 3 ft square and 2 in thick is used as a fire wall. The temperature is 330°F on the hot side and 160°F on the other. What is the thermal-flow stress developed across the plate?

$$S = E\alpha\,\Delta t = 13 \times 10^6 \times 6.5 \times 10^{-6} \times 170$$
$$= 14,360 \text{ lb/in}^2 \text{ (1,010 kgf/cm}^2)$$

or
$$Q = 2.3 \times 9 \times 170/2 = 1,760 \text{ Btu/h}$$

and
$$S = 13 \times 10^6 \times 6.5 \times 10^{-6} \times 1,760 \times 2/2.3 \times 9$$
$$= 14,360 \text{ lb/in}^2 \text{ (1,010 kgf/cm}^2)$$

COMBINED STRESSES

In the discussion that follows, the element is subjected to stresses lying in one plane; this is the case of **plane stress, or two-dimensional stress**.

Simple stresses, defined as such by the flexure and torsion theories, lie in planes normal or parallel to the line of action of the forces. Normal, as well as shearing, stresses may, however, exist in other directions. A particle out of a loaded member will contain normal and shearing stresses as shown in Fig. 5.2.11. Note that the four shearing stresses must be of the same magnitude, if equilibrium is to be satisfied.

If the particle is "cut" along the plane AA, equilibrium will reveal that, in general, normal as well as shearing stresses act upon the plane AC (Fig. 5.2.12). The normal stress on plane AC is labeled S_n, and shearing S_s. The application of equilibrium yields

$$S_n = \frac{S_x + S_y}{2} + \frac{S_x - S_y}{2} \cos 2\theta + S_{xy} \sin 2\theta$$

and

$$S_s = \frac{S_x - S_y}{2} \sin 2\theta - S_{xy} \cos 2\theta$$

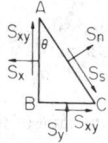

Fig. 5.2.11   Fig. 5.2.12

A *sign convention* must be used. A tensile stress is positive while compression is negative. A shearing stress is positive when directed as on plane AB of Fig. 5.2.12; i.e., when the shearing stresses on the vertical planes form a clockwise couple, the stress is positive.

The planes defined by $\tan 2\theta = 2S_{xy}/S_x - S_y$, the *principal planes,* contain the **principal stresses**—the maximum and minimum normal stresses. These stresses are

$$S_M, S_m = \frac{S_x + S_y}{2} \pm \sqrt{\left(\frac{S_x - S_y}{2}\right)^2 + S_{xy}^2}$$

The maximum and minimum shearing stresses are represented by the quantity

$$S_{s\,M,m} = \pm \sqrt{\left(\frac{S_x - S_y}{2}\right)^2 + S_{xy}^2}$$

and they act on the planes defined by

$$\tan 2\theta = -\frac{S_x - S_y}{2S_{xy}}$$

EXAMPLE. The steam in a boiler subjects a paticular particle on the outer surface of the boiler shell to a circumferential stress of 8,000 lb/in² and a longitudinal stress of 4,000 lb/in² as shown in Fig. 5.2.13. Find the stresses acting on the plane XX, making an angle of 60° with the direction of the 8,000 lb/in² stress. Find the principal stresses and locate the principal planes. Also find the maximum and minimum shearing stresses.

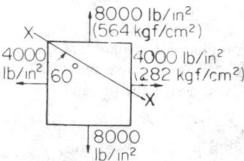

Fig. 5.2.13

$$60° \; S_n = \frac{4,000 + 8,000}{2} + \frac{4,000 - 8,000}{2}(-0.5000) + 0$$
$$= 7,000 \text{ lb/in}^2$$

$$60° \; S_s = \frac{4,000 - 8,000}{2}(0.8660) - 0 = -1,732 \text{ lb/in}^2$$

$$S_{M,m} = \frac{4,000 + 8,000}{2} \pm \sqrt{\left(\frac{4,000 - 8,000}{2}\right)^2 + 0}$$
$$= 6,000 \pm 2,000$$
$$= 8,000 \text{ and } 4,000 \text{ lb/in}^2 \; (564 \text{ and } 282 \text{ kgf/cm}^2)$$

$$\text{at } \tan 2\theta = \frac{2 \times 0}{4,000 - 8,000} = 0 \quad \text{or} \quad \theta = 90° \text{ and } 0°$$

$$S_{s\,M,m} = \pm \sqrt{\left(\frac{4,000 - 8,000}{2}\right)^2 + 0}$$
$$= \pm 2,000 \text{ lb/in}^2 \; (\pm 141 \text{ kgf/cm}^2)$$

Mohr's Stress Circle The biaxial stress field with its combined stresses can be represented graphically by the Mohr stress circle. For instance, for the particle given in Fig. 5.2.11, Mohr's circle is as shown in Fig. 5.2.14. The stress *sign convention* previously defined must be adhered to. Furthermore, in order to locate the point (on Mohr's circle) that yields the stresses on a plane $\theta°$ from the vertical side of the particle (such as plane AA in Fig. 5.2.11), $2\theta°$ must be laid off in the same

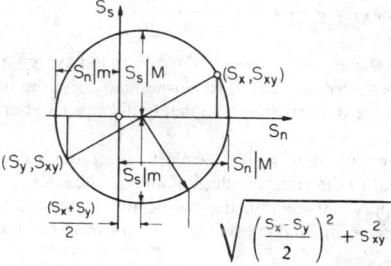

Fig. 5.2.14

direction from the radius to (S_x, S_{xy}). For the previous example, Mohr's circle becomes Fig. 5.2.15.

Eight special stress fields are shown in Figs. 5.2.16 to 5.2.23, along with Mohr's circle for each.

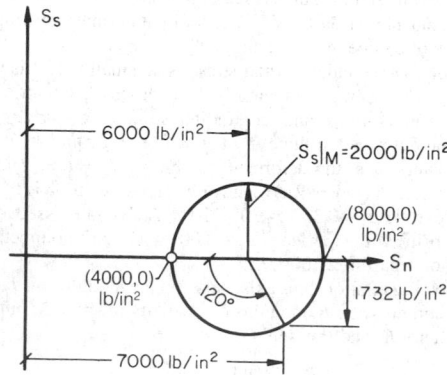

Fig. 5.2.15

Combined Loading **Combined flexure and torsion** arise, for instance, when a shaft twisted by a torque M_t is bent by forces produced by belts or gears. An element on the surface, such as $ABCD$ on the shaft of Fig. 5.2.24, is subjected to a flexure stress $S_x = Mc/I = 8Fl(\pi d^3)$ and a

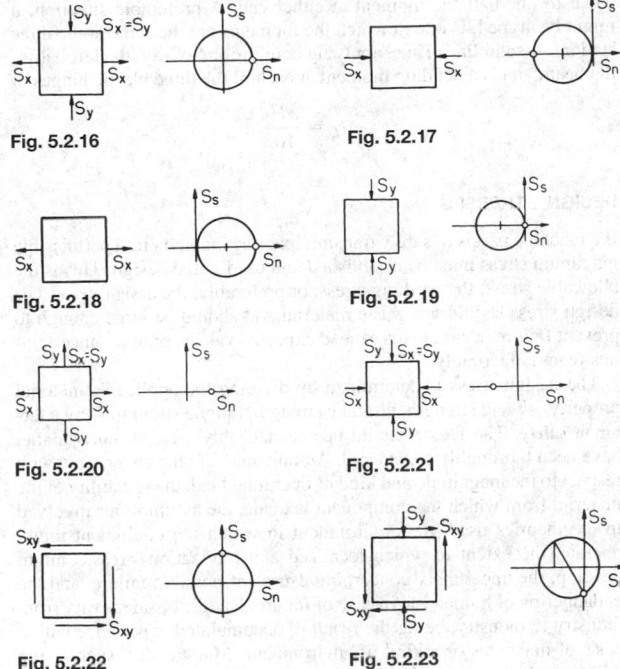

Fig. 5.2.16

Fig. 5.2.17

Fig. 5.2.18

Fig. 5.2.19

Fig. 5.2.20

Fig. 5.2.21

Fig. 5.2.22

Fig. 5.2.23

torsional shearing stress $S_{xy} = M_t c/J = 16 M_t/(\pi d^3)$. These stresses will induce combined stresses. The maximum combined stresses will be

$$S_n = \tfrac{1}{2}(S_x \pm \sqrt{S_x^2 + 4S_{xy}^2})$$

and

$$S_s = \pm \tfrac{1}{2}\sqrt{S_x^2 + 4S_{xy}^2}$$

The above situation applies to any case of normal stress with shear, as when a bolt is under both tension and shear. A beam particle subjected to both flexure and transverse shear is another case.

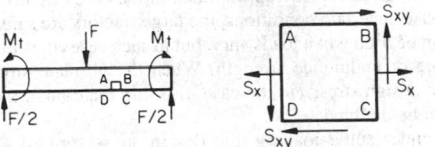

Fig. 5.2.24

Combined torsion and longitudinal loads exist on a propeller shaft. A particle on this shaft will contain a tensile stress computed using $S = F/A$ and a torsion shearing stress equal to $S_s = M_t c/J$. The free body of a particle on the surface of a vertical turbine shaft is subjected to direct compression and torsion.

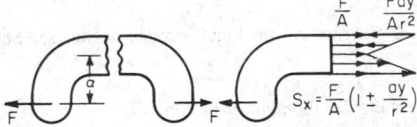

Fig. 5.2.25

When combined loading results in stresses of the same type and direction, the addition is algebraic. Such a situation exists on an offset link like that of Fig. 5.2.25.

Mohr's Strain Circle   Strain equations can also be derived for plane-strain fields. Strains e_x and e_y are the extensional strains (tension or compression) occurring at a point in two right-angle directions, and the change of the angle between them is γ_{xy}. The strain e at the point in any direction a at an angle θ with the x direction derives as

$$e_a = \frac{e_x + e_y}{2} + \frac{e_x - e_y}{2}\cos 2\theta + \frac{\gamma_{xy}}{2}\sin 2\theta$$

Similarly, the shearing strain γ_{ab} (change in the original right angle between directions a and b) is defined by

$$\gamma_{ab} = (e_x - e_y)\sin 2\theta + \gamma_{xy}\cos 2\theta$$

Inspection easily reveals that the above equations for e_a and γ_{ab} are mathematically identical to those for S_n and S_s. Thus, once a sign convention is established, a Mohr circle for strain can be constructed and used as the stress circle is used. The strain e is positive when an extension and negative when a contraction. If the direction associated with the first subscript a rotates counterclockwise during straining with respect to the direction indicated by the second subscript b, the shearing strain is positive; if clockwise, it is negative. In constructing the circle, positive extensional strains will be plotted to the right as abscissas and positive **half-shearing** strains will be plotted upward as ordinates.

For the strains shown in Fig. 5.2.26a, Mohr's strain circle becomes that shown in Fig. 5.2.26b. The extensional strain in the direction a, making an angle of θ_a with the x direction, is e_a, and the shearing strain is γ_{ab} counterclockwise. The strain 90° away is e_b. The maximum principal strain is e_M at an angle θ_M clockwise from the x direction. The other principal or minimum strain is e_m 90° away.

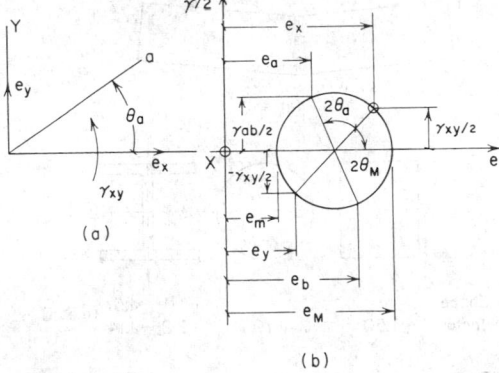

Fig. 5.2.26

PLASTIC DESIGN

Early efforts in stress analysis were based on limit loads, that is, loads which stress a member "wholly" to the yield strength. Euler's famous paper on column action ("Sur la Force des Colonnes," Academie des Sciences de Berlin, 1757) deals with the column problem this way. More recently, the concept of limit loads, referred to as **limit,** or **plastic, design,** has found strong application in the design of certain structures. The theory presupposes a ductile material, absence of stress raisers, and fabrication free of embrittlement. Local load overstress is allowed, provided the structure does not deform appreciably.

To visualize the limit-load approach, consider a simple beam of uniform section subjected to a concentrated load of midspan, as depicted in Fig. 5.2.27a. According to elastic theory, the outermost fiber on each side and at midspan—the section of maximum bending moment—will first reach the yield-strength value. Across the depth of the beam, the stress distribution will, of course, follow the triangular pattern, becoming zero at the neutral axis. If the material is ductile, the stress in the outermost fibers will remain at the yield value until every other fiber reaches the same value as the load increases. Thus the stress distribution assumes the rectangular pattern before the *plastic hinge* forms and failure ensues.

The problem is that of finding the final limit load. Elastic-flexure theory gives the maximum load—triangular distribution—as

$$F_y = \frac{2S_y bh^2}{3l}$$

For the rectangular stress distribution, the limit load becomes

$$F_L = \frac{S_y bh^2}{l}$$

The ratio $F_L/F_y = 1.50$—an increase of 50 percent in load capability. The ratio F_L/F_y has been named **shape factor** (Jenssen, Plastic Design in Welded Structures Promises New Economy and Safety, *Welding Jour.*, Mar. 1959). See Fig. 5.2.27b for shape factors for some other sections. The shape factor may also be determined by dividing the first moment of area about the neutral axis by the section modulus.

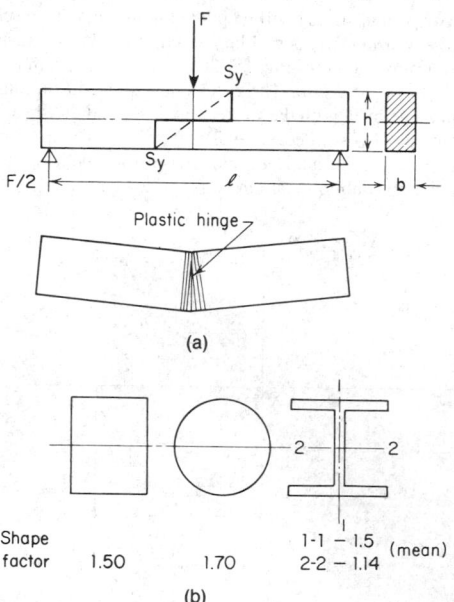

(a)

Shape		1-1 — 1.5 (mean)
factor	1.50	1.70	2-2 — 1.14

(b)

Fig. 5.2.27

A constant-section beam with both ends fixed, supporting a uniformly distributed load, illustrates another application of the plastic-load approach. The bending-moment diagram based on the elastic theory drawn in Fig. 5.2.28 (broken line) shows a moment at the center

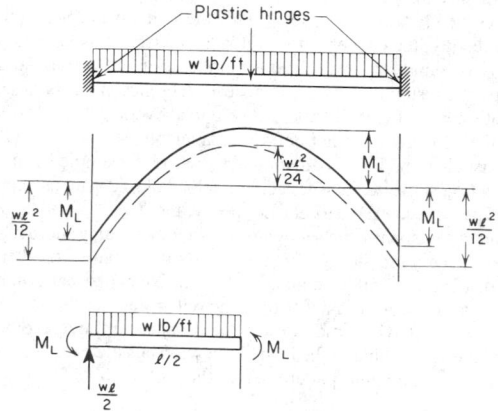

Fig. 5.2.28

equal to one-half the moment at either end. A preferable situation, it might be argued, is one in which the moments are the same at the three stations—solid line. Thus, applying equilibrium to, say, the left half of the beam yields a bending moment at each of the three plastic hinges of

$$M_L = \frac{wl^2}{16}$$

DESIGN STRESSES

If a machine part is to safely transmit loads acting upon it, a permissible maximum stress must be established and used in the design. This is the allowable stress, the working stress, or preferably, the **design stress**. The design stress should not waste material, yet should be large enough to prevent failure in case loads exceed expected values, or other uncertainties react unfavorably.

The design stress is determined by dividing the applicable material property—yield strength, ultimate strength, fatigue strength—by a **factor of safety**. The factor should be selected only after all **uncertainties** have been thoroughly considered. Among these are the uncertainty with respect to the magnitude and kind of operating load, the reliability of the material from which the component is made, the assumptions involved in the theories used, the environment in which the equipment might operate, the extent to which localized and fabrication stresses might develop, the uncertainty concerning causes of possible failure, and the endangering of human life in case of failure. Factors of safety vary from industry to industry, being the result of accumulated experience with a class of machines or a kind of environment. Many codes, such as the ASME code for power shafting, recommend design stresses found safe in practice.

In general, the **ductility** of the material determines the property upon which the factor should be based. Materials having an elongation of over 5 percent are considered ductile. In such cases, the factor of safety is based upon the yield strength or the endurance limit. For materials with an elongation under 5 percent, the ultimate strength must be used because these materials are **brittle** and so fracture without yielding.

Factors of safety based on yield are often taken **between 1.5 and 4.0**. For more reliable materials or well-defined design and operating conditions, the lower factors are appropriate. In the case of untried materials or otherwise uncertain conditions, the larger factors are safer. The same values can be used when loads vary, but in such cases they are applied to the fatigue or endurance strength. When the ultimate strength determines the design stress (in the case of brittle materials), the factors of safety can be doubled.

Thus, under static loading, the design stress for, say, SAE 1020, which has a yield strength of 45,000 lb/in² (3,170 kgf/cm²) may be taken at 45,000/2, or 22,500 lb/in² (1,585 kgf/cm²), if a reasonably certain design condition exists. A Class 30 cast-iron part might be designed at 30,000/5 or 6,000 lb/in² (423 kgf/cm²). A 2017S-0 aluminum-alloy component (13,000 lb/in² endurance strength) could be computed at a design stress of 13,000/2.5 or 5,200 lb/in² (366 kgf/cm²) in the usual fatigue-load application.

BEAMS

For properties of structural steel and wooden beams, see Sec. 12.2.

Notation

I = rectangular moment of inertia
I_p = polar moment of inertia
I/c = section modulus
M = bending moment
P, W' = concentrated load
Q or V = total vertical shear
R = reaction
S = unit normal stress
S_s or S_v = transverse shearing stress
W = total distributed load

f = deflection
i = slope
l = distance between supports
r = radius of gyration
r_c = radius of curvature
w = distributed load per longitudinal unit

A **simple beam** rests on supports at its ends which permit rotation. A **cantilever beam** is fixed (no rotation) at one end. When computing reactions and moments, distributed loads may be replaced by their resultants acting at the center of gravity of the distributed-load area.

Reactions are the forces and/or couples acting at the supports and holding the beam in place. In general, the weight of the beam should be accounted for.

The **bending moment** (pound-feet or pound-inches) (kgf · m) at any section is the algebraic sum of the external forces and moments acting on the beam on one side of the section. It is also equal to the moment of the internal-stress forces at the section, $M = \int s\, dA/y$. A bending moment that bends a beam convex downward (tensile stress on bottom fiber) is considered **positive**, while convex upward (compression on bottom) is **negative**.

The **vertical shear** V (lb) (kgf) effective on a section is the algebraic sum of all the forces acting parallel to and on one side of the section, $V = \Sigma F$. It is also equal to the sum of the transverse shear stresses acting on the section, $V = \int S_s\, dA$.

Moment and **shear diagram** may be constructed by plotting to scale the particular entity as the ordinate for each section of the beam. Such diagrams show in continuous form the variation along the length of the beam.

Moment-Shear Relation The shear V is the first derivative of moment with respect to distance along the beam, $V = dM/dx$. This relationship does not, however, account for any sudden changes in moment.

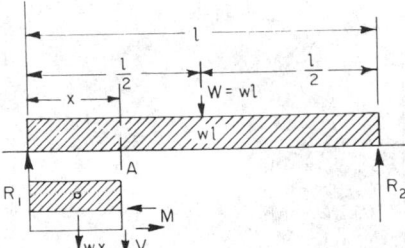

Fig. 5.2.29

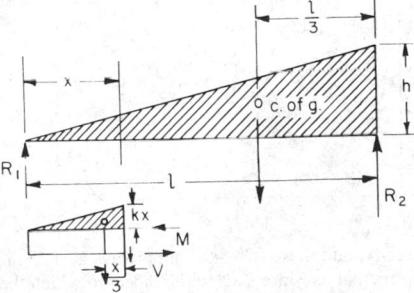

Fig. 5.2.30

EXAMPLES. Figure 5.2.29 illustrates a simple beam subjected to a uniform load. $M = R_1 x - wx \times \dfrac{x}{2} = \dfrac{w/x}{2} - \dfrac{wx^2}{2}$ and $V = R_1 - wx = \dfrac{wl}{2} - wx$. Note also that $V = \dfrac{d}{dx}\left(\dfrac{wlx}{2} - \dfrac{wx^2}{2}\right) = \dfrac{wl}{2} - wx$.

Figure 5.2.30 is a simple beam carrying a uniformly varying load; $M = R_1 x - $

$h\dfrac{x}{1} \times \dfrac{x}{2} \times \dfrac{x}{3} = \dfrac{hlx}{6} - \dfrac{hx^3}{6l}$, if h is in pounds per foot and weight of beam is neglected. The vertical shear $V = R_1 - \dfrac{hx}{l} \times \dfrac{x}{2} = \dfrac{hl}{6} - \dfrac{hx^2}{2l}$. Note again that $V = \dfrac{d}{dx}\left(\dfrac{hlx}{6} - \dfrac{hx^3}{6l}\right) = \dfrac{hl}{6} - \dfrac{hx^2}{2l}$.

Table 5.2.2 gives the reactions, bending-moment equations, vertical shear equations, and the deflection of some of the more common types of beams.

Maximum Safe Load on Steel Beams See Table 5.2.3   To obtain maximum safe load (or maximum deflection under maximum safe load) for any of the conditions of loading given in Table 5.2.5, multiply the corresponding coefficient in that table by the greatest safe load (or deflection) for distributed load for the particular section under consideration as given in Table 5.2.4.

The following approximate factors for reducing the load should be used when beams are long in comparison with their breadth:

Ratio of unsupported (lateral) length to flange width or breadth	20	30	40	50	60	70
Ratio of greatest safe load to calculated load	1	0.9	0.8	0.7	0.6	0.5

Theory of Flexure   A bent beam is shown in Fig. 5.2.31. The concave side is in compression and the convex side in tension. These are divided by the **neutral plane** of zero stress $A'B'BA$. The intersection of the neutral plane with the face of the beam is in the **neutral line or elastic curve** AB. The intersection of the neutral plane with the cross section is the **neutral axis** NN'.

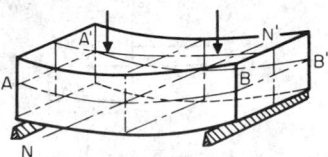

Fig. 5.2.31

It is assumed that a beam is prismatic, of a length at least 10 times its depth, and that the external forces are all at right angles to the axis of the beam and in a plane of symmetry, and that flexure is slight. Other assumptions are: (1) That the material is homogeneous, and obeys Hooke's law. (2) That **stresses are within the elastic limit**. (3) That every layer of material is free to expand and contract longitudinally and laterally under stress as if separate from other layers. (4) That the tensile and compressive moduli of elasticity are equal. (5) That the cross section remains a plane surface. (The assumption of plane cross sections is strictly true only when the shear is constant or zero over the cross section, and when the shear is constant throughout the length of the beam.)

It follows then that: (1) The internal forces are in horizontal balance. (2) The **neutral axis contains the center of gravity** of the cross section, where there is no resultant axial stress. (3) The stress intensity varies directly with the distance from the neutral axis.

The moment of the elastic forces about the neutral axis, i.e., the **stress moment** or **moment of resistance**, is $M = SI/c$, where S is an elastic unit stress at outer fiber whose distance from the neutral axis is c; and I is the rectangular moment of inertia about the neutral axis. I/c is the **section modulus**.

This formula is for the **strength of beams**. For rectangular beams, $M = \frac{1}{6}Sbh^2$, where b = breadth and h = depth; i.e., the elastic **strength of beam sections** varies as follows: (1) for equal width, as the square of the depth; (2) for equal depth, directly as the width; (3) for equal depth and width, directly as the strength of the material; (4) if span varies, then for equal depth, width, and material, inversely as the span.

Table 5.2.2   Beams of Uniform Cross Section, Loaded Transversely

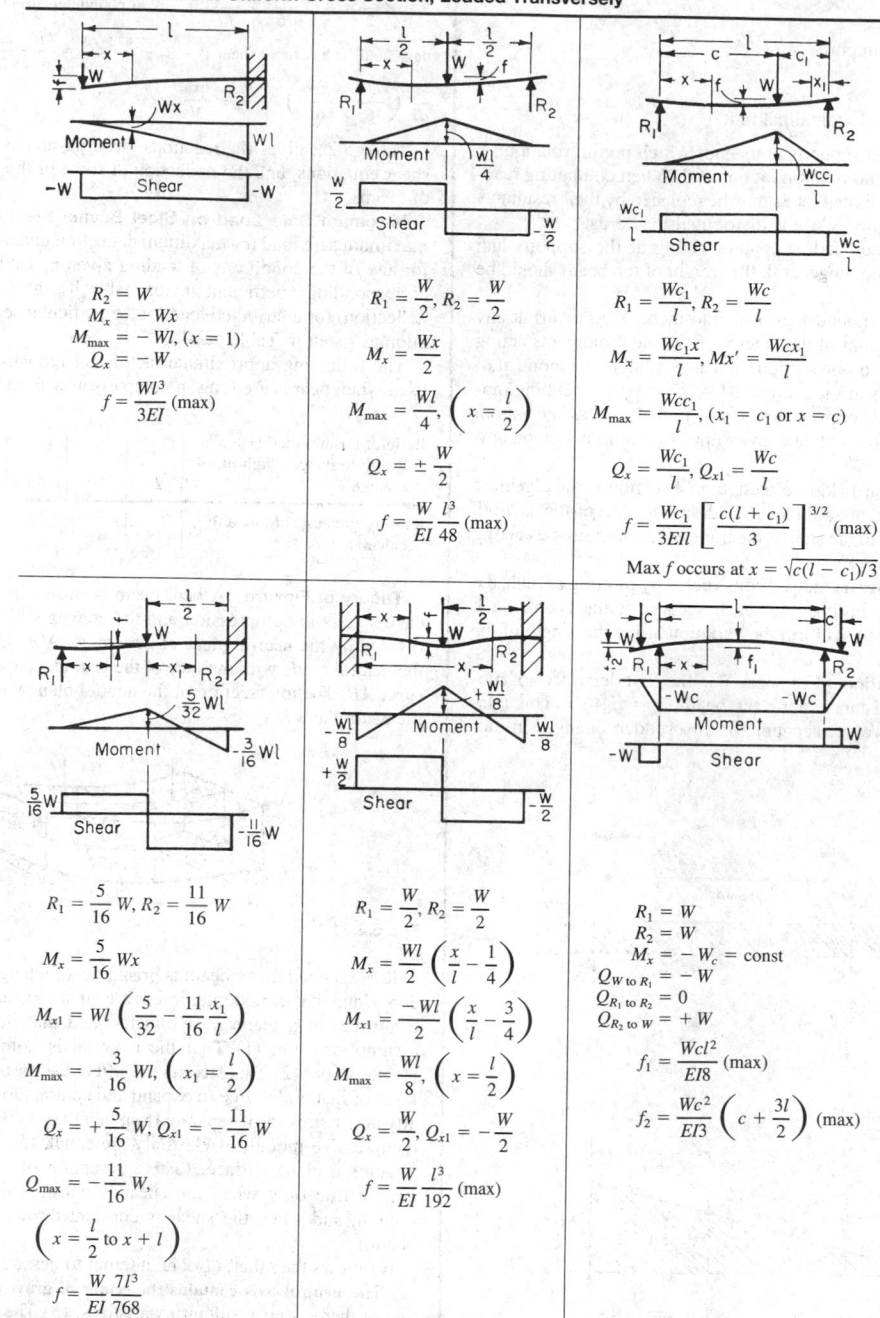

$$R_2 = W$$
$$M_x = -Wx$$
$$M_{max} = -Wl, \ (x = 1)$$
$$Q_x = -W$$

$$f = \frac{Wl^3}{3EI} \ (\text{max})$$

$$R_1 = \frac{W}{2}, R_2 = \frac{W}{2}$$

$$M_x = \frac{Wx}{2}$$

$$M_{max} = \frac{Wl}{4}, \ \left(x = \frac{l}{2} \right)$$

$$Q_x = \pm \frac{W}{2}$$

$$f = \frac{W}{EI} \frac{l^3}{48} \ (\text{max})$$

$$R_1 = \frac{Wc_1}{l}, R_2 = \frac{Wc}{l}$$

$$M_x = \frac{Wc_1 x}{l}, Mx' = \frac{Wcx_1}{l}$$

$$M_{max} = \frac{Wcc_1}{l}, (x_1 = c_1 \text{ or } x = c)$$

$$Q_x = \frac{Wc_1}{l}, Q_{x1} = \frac{Wc}{l}$$

$$f = \frac{Wc_1}{3EIl} \left[\frac{c(l + c_1)}{3} \right]^{3/2} (\text{max})$$

Max f occurs at $x = \sqrt{c(l - c_1)/3}$

$$R_1 = \frac{5}{16} W, R_2 = \frac{11}{16} W$$

$$M_x = \frac{5}{16} Wx$$

$$M_{x1} = Wl \left(\frac{5}{32} - \frac{11}{16} \frac{x_1}{l} \right)$$

$$M_{max} = -\frac{3}{16} Wl, \ \left(x_1 = \frac{l}{2} \right)$$

$$Q_x = +\frac{5}{16} W, Q_{x1} = -\frac{11}{16} W$$

$$Q_{max} = -\frac{11}{16} W,$$

$$\left(x = \frac{l}{2} \text{ to } x + l \right)$$

$$f = \frac{W}{EI} \frac{7l^3}{768}$$

$$R_1 = \frac{W}{2}, R_2 = \frac{W}{2}$$

$$M_x = \frac{Wl}{2} \left(\frac{x}{l} - \frac{1}{4} \right)$$

$$M_{x1} = \frac{-Wl}{2} \left(\frac{x}{l} - \frac{3}{4} \right)$$

$$M_{max} = \frac{Wl}{8}, \ \left(x = \frac{l}{2} \right)$$

$$Q_x = \frac{W}{2}, Q_{x1} = -\frac{W}{2}$$

$$f = \frac{W}{EI} \frac{l^3}{192} \ (\text{max})$$

$$R_1 = W$$
$$R_2 = W$$
$$M_x = -W_c = \text{const}$$
$$Q_{W \text{ to } R_1} = -W$$
$$Q_{R_1 \text{ to } R_2} = 0$$
$$Q_{R_2 \text{ to } W} = +W$$

$$f_1 = \frac{Wcl^2}{EI8} \ (\text{max})$$

$$f_2 = \frac{Wc^2}{EI3} \left(c + \frac{3l}{2} \right) \ (\text{max})$$

If a beam is cut in halves vertically, the two halves laid side by side each will carry only one-half as much as the original beam.

Tables 5.2.6 to 5.2.8 give the properties of various beam cross sections. For properties of structural-steel shapes, see Sec. 12.2.

Oblique Loading   It should be noted that Table 5.2.6 includes certain cases for which the horizontal axis is not a neutral axis, assuming the common case of vertical loading. The rectangular section with the diagonal as a horizontal axis (Table 5.2.6) is such a case. These cases must be handled by the principles of oblique loading.

Every section of a beam has two principal axes passing through the center of gravity, and these two axes are always at right angles to each other. The principal axes are axes with respect to which the moment of inertia is, respectively, a maximum and a minimum, and for which the product of inertia is zero. For symmetrical sections, axes of symmetry are always principal axes. For unsymmetrical sections, like a **rolled angle** section (Fig. 5.2.32), the inclination of the principal axis with the X axis may be found from the formula $\tan 2\theta = 2I_{xy}/(I_y - I_x)$, in which θ = angle of inclination of the principal axis to the X axis, I_{xy} = the product of inertia of the section with respect to the X and Y axes, I_y = moment of inertia of the section with respect to the Y axis, I_x = moment of inertia of

Table 5.2.2 Beams of Uniform Cross Section, Loaded Transversely (Continued)

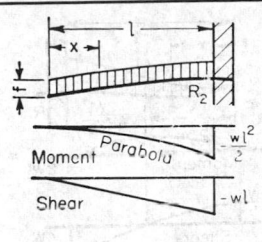

$$R_2 = W = wl$$

$$M_x = -\frac{wx^2}{2}$$

$$M_{max} = -\frac{wl^2}{2}, (x = l)$$

$$Q_x = -wx$$
$$Q_{max} = -wl, (x = l)$$

$$f = \frac{W}{EI}\frac{l^3}{8} \text{ (max)}$$

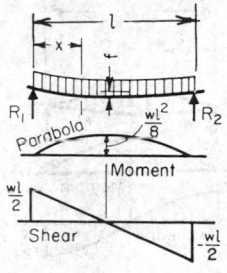

$$R_1 = \frac{W}{2} = \frac{wl}{2}$$

$$R_2 = \frac{W}{2} = \frac{wl}{2}$$

$$M_x = \frac{wx}{2}(l - x)$$

$$M_{max} = \frac{wl^2}{8}, (x = \tfrac{1}{2}l)$$

$$Q_x = \frac{wl}{2} - wx$$

$$Q_{max} = \frac{wl}{2}, (x = 0)$$

$$f = \frac{W}{EI}\frac{5l^3}{384} \text{ (max)}$$

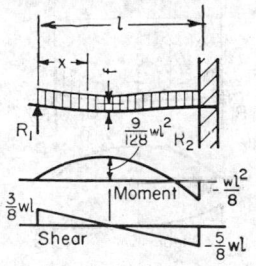

$$R_1 = \frac{3}{8}W = \frac{3}{8}wl$$

$$R_2 = \frac{5}{8}W = \frac{5}{8}wl$$

$$M_x = \frac{wx}{2}\left(\frac{3}{4}l - x\right)$$

$$M_{max} = \frac{9}{128}wl^2, \left(x = \frac{3}{8}l\right)$$

$$M_{max} = -\frac{wl^2}{8}, (x = l)$$

$$Q_x = \frac{3}{8}wl - wx$$

$$Q_{max} = -\frac{5}{8}wl$$

$$f = \frac{W}{EI}\frac{l^3}{185} \text{ (max)}$$

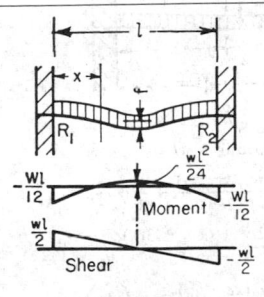

$$R_1 = \frac{W}{2} = \frac{wl}{2}, R_2 = \frac{W}{2} = \frac{wl}{2}$$

$$M_x = -\frac{wl^2}{2}\left(\frac{1}{6} - \frac{x}{l} + \frac{x^2}{l^2}\right)$$

$$M_{max} = -\frac{1}{12}wl^2, (x = 0, \text{ or } x = l)$$

$$Q_x = \frac{wl}{2} - wx$$

$$Q_{max} = \pm\frac{wl}{2}$$

$$f = \frac{W}{EI}\frac{l^3}{384} \text{ (max)}$$

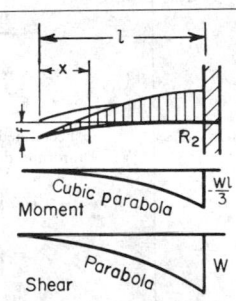

$$R_2 = W = \text{total load}$$

$$M_x = -\frac{W}{3}\frac{x^3}{l^2}$$

$$M_{max} = -\frac{Wl}{3}$$

$$Q_x = -\frac{Wx^2}{l^2}$$

$$Q_{max} = -W$$

$$f = \frac{W}{EI}\frac{l^3}{15} \text{ (max)}$$

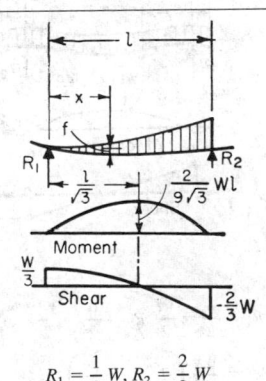

$$R_1 = \frac{1}{3}W, R_2 = \frac{2}{3}W$$

$$M_x = \frac{Wx}{3}\left(1 - \frac{x^2}{l^2}\right)$$

$$M_{max} = \frac{2}{9\sqrt{3}}Wl, \left(x = \frac{1}{\sqrt{3}}\right)$$

$$Q_x = W\left(\frac{1}{3} - \frac{x^2}{l^2}\right)$$

$$Q_{max} = -\frac{2}{3}W, (x = l)$$

$$f = 0.01304\frac{Wl^3}{EI} \text{ (max)}$$

Table 5.2.2 Beams of Uniform Cross Section, Loaded Transversely (*Continued*)

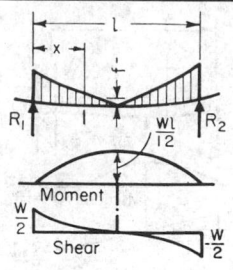

$$R_1 = \frac{W}{2},\ R_2 = \frac{W}{2}$$

$$M_x = Wx\left(\frac{1}{2} - \frac{x}{l} + \frac{2x^2}{3l^2}\right)$$

$$M_{max} = \frac{Wl}{12},\ \left(x = \frac{1}{2}l\right)$$

$$Q_x = W\left(\frac{1}{2} - \frac{2x}{l} + \frac{2x^2}{l^2}\right)$$

$$Q_{max} = \pm\frac{W}{2},\ (x = 0)$$

$$f = \frac{W}{EI}\frac{3l^3}{320}\ (\text{max})$$

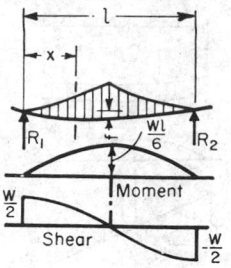

$$R_1 = \frac{W}{2},\ R_2 = \frac{W}{2}$$

$$M_x = Wx\left(\frac{1}{2} - \frac{2}{3}\frac{x^2}{l^2}\right)$$

$$M_{max} = \frac{Wl}{6},\ \left(x = \frac{1}{2}l\right)$$

$$Q_x = W\left(\frac{1}{2} - \frac{2x^2}{l^2}\right)$$

$$Q_{max} = \pm\frac{W}{2},\ (x = 0)$$

$$f = \frac{W}{EI}\frac{l^3}{60}\ (\text{max})$$

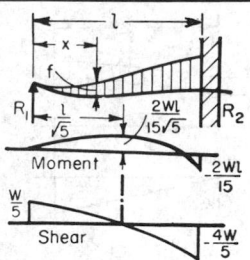

$$R_1 = \frac{W}{5},\ R_2 = \frac{4W}{5}$$

$$M_x = Wx\left(\frac{1}{5} - \frac{x^2}{3l^2}\right)$$

$$M_{max} = -\frac{2}{15}Wl \text{ at support 2}$$

$$Q_x = W\left(\frac{1}{5} - \frac{x^2}{l^2}\right)$$

$$Q_{max} = -\frac{4W}{5}$$

$$f = \frac{16Wl^3}{1,500\sqrt{5}EI}$$

$$= \frac{0.00477Wl^3}{EI}\ (\text{max})$$

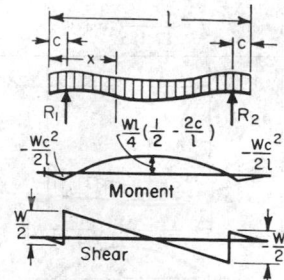

$$R_1 = \frac{W}{2} = \frac{wl}{2},\ R_2 = \frac{W}{2} = \frac{wl}{2}$$

$$M_x = \frac{Wx}{2}\left(1 - \frac{c}{x} - \frac{x}{l}\right),\ (x > c)$$

$$M_x = -\frac{Wx^2}{2l},\ (x \le c)$$

$$M_{max} = \frac{Wl}{4}\left(\frac{1}{2} - \frac{2c}{l}\right),\ c \le \left(\frac{\sqrt{2}-1}{2}\right)l$$

$$Q_x = \frac{W}{2} - wx\ (x > c)$$

$$Q_x = -wx\ (x \le c)$$

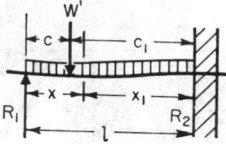

Concentrated load W'
Uniformly dist. load $W = wl$

$$R_1 = W'\frac{c_1^2(3c + 2c_1)}{2l^3} + \frac{3}{8}W$$

$$R_2 = W'\frac{(2c^2 + 6cc_1 + 3c_1^2)c}{2l^3} + \frac{5}{8}W$$

$$M_2 = W'\frac{cc_1(2c + c_1)}{2l^2} + W\left(\frac{l}{8}\right)$$

$$M_{W'} = W'\frac{cc_1^2(3c + 2c_1)}{2l^3} + W\frac{(3c_1 - c)c}{8l}$$

$$(a)\quad \frac{W'}{W} < \frac{l^2}{4c_1^2}\frac{5c - 3c_1}{3c + 2c_1}$$

$$M_{c\,max} = \frac{R_1^2}{2W}l,\ \left(x = \frac{R_1 l}{W}\right)$$

$$(b)\quad \frac{W'}{W} < \frac{l^2(3c_1 - 5c)}{4c(2c^2 + 6cc_1 + 3c_1^2)}$$

$$M_{c_1\,max} = W'c + \frac{(R_1 - W')^2}{2W}l,\ \left(x = \frac{R_1 - W'}{W}l\right)$$

Deflection under W'

$$f = \frac{W'}{EI}\frac{c^2c_1^3(4c + 3c_1)}{12l^3} + \frac{W}{EI}\frac{cc_1^2(3c + c_1)}{48l}$$

Table 5.2.2   Beams of Uniform Cross Section, Loaded Transversely   (Continued)

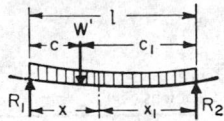
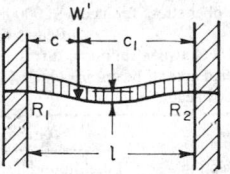

Concentrated load W'
Uniformly dist. load $W = wl$; $c < c_1$

$$R_1 = W'\frac{c_1}{l} + \frac{W}{2}$$

$$R_2 = W'\frac{c}{l} + \frac{W}{2}$$

(a) $\dfrac{W'}{W} < \dfrac{c_1 - c}{2c}$

$$M_{max} = R_2\frac{x_1}{2} = \frac{R_2^2 l}{2W}, \left(x_1 = \frac{R_2 l}{W}\right)$$

(b) $\dfrac{W'}{W} > \dfrac{c_1 - c}{2c}$

$$M_{max} = \left(W' + \frac{W}{2}\right)\frac{cc_1}{t}, (x_1 = c_1)$$

Deflection of beam under W':

$$f = \left(W' + \frac{l^2 + cc_1}{8cc_1}W\right)\frac{c^2 c_1^2}{3EIl}$$

$c < c_1$

$$R_1 = W'\frac{(3c + c_1)c_1^2}{l^3} + \frac{W}{2}$$

$$R_2 = W'\frac{(c + 3c_1)c^2}{l^3} + \frac{W}{2}$$

$$M_{max} = M_1 = W'\frac{cc_1^2}{l^2} + \frac{Wl}{12}$$

Deflection under W'

$$f = \frac{1}{EI}\left(W'\frac{c^3 c_1^3}{3l^3} + W\frac{c^2 c_1^2}{24l}\right)$$

Table 5.2.3   Uniformly Distributed Loads on Simply Supported Rectangular Beams 1-in Wide*
(Laterally Supported Sufficiently to Prevent Buckling)
[Calculated for unit fiber stress at 1,000 lb/in² (70 kgf/cm²): nominal size]
Total load in pounds (kgf)† including the weight of beam

Span, ft	Depth of beam, in (cm)§										
(m)‡	6	7	8	9	10	11	12	13	14	15	16
5	800	1,090	1,420	1,800	2,220	2,690	3,200	3,750	4,350	5,000	5,690
6	670	910	1,180	1,500	1,850	2,240	2,670	3,130	3,630	4,170	4,740
7	570	780	1,010	1,290	1,590	1,920	2,280	2,680	3,110	3,570	4,060
8	500	680	890	1,120	1,390	1,680	2,000	2,350	2,720	3,130	3,560
9	440	600	790	1,000	1,230	1,490	1,780	2,090	2,420	2,780	3,160
10	400	540	710	900	1,110	1,340	1,600	1,880	2,180	2,500	2,840
11	360	490	650	820	1,010	1,220	1,450	1,710	1,980	2,270	2,590
12	330	450	590	750	930	1,120	1,330	1,560	1,810	2,080	2,370
13	310	420	550	690	850	1,030	1,230	1,440	1,680	1,920	2,190
14	290	390	510	640	790	960	1,140	1,340	1,560	1,790	2,030
15	270	360	470	600	740	900	1,070	1,250	1,450	1,670	1,900
16	250	340	440	560	690	840	1,000	1,170	1,360	1,560	1,780
17	230	320	420	530	650	790	940	1,100	1,280	1,470	1,670
18	220	300	400	500	620	750	890	1,040	1,210	1,390	1,580
19	210	290	380	470	590	710	840	990	1,150	1,320	1,500
20	200	270	360	450	560	670	800	940	1,090	1,250	1,420
22	180	250	320	410	500	610	730	850	990	1,140	1,290
24	160	230	290	370	460	560	670	780	910	1,040	1,180
26	150	210	270	340	420	520	610	720	840	960	1,090
28	140	190	250	320	390	480	570	670	780	890	1,010
30	130	180	240	300	370	450	530	630	730	830	950

* This table is convenient for wooden beams. For any other fiber stress S', multiply the values in table by $S'/1,000$. See Sec. 12.2 for properties of wooden beams of commercial sizes.
† To change to kgf, multiply by 0.454.
‡ To change to m, multiply by 0.305.
§ To change to cm, multiply by 2.54.

the section with respect to the X axis. When this principal axis has been found, the other principal axis is at right angles to it.

Calling the moments of inertia with respect to the principal axes I'_x and I'_y, the unit stress existing anywhere in the section at a point whose coordinates are x and y (Fig. 5.2.33) is $S = My \cos \alpha/I'_x + Mx \sin \alpha/I'_y$, in which M = bending moment with respect to the section in question, α = the angle which the plane of bending moment or the plane of the loads makes with the y axis, $M \cos \alpha$ = the component of bending moment causing bending about the principal axis which has been designated as the X axis, $M \sin \alpha$ = the component of bending moment causing bending about the principal axis which has been designated as the Y axis. The sign of the two terms for unit stress may be determined by inspection in the usual way, and the result will be tension or compression as determined by the algebraic sum of the two terms.

Table 5.2.4 Approximate Safe Loads in Pounds (kgf) on Steel Beams,* Simply Supported, Single Span

Allowable fiber stress for steel, 16,000 lb/in² (1,127 kgf/cm²) (basis of table)

Beams simply supported at both ends.

L = distance between supports, ft (m)
A = sectional area of beam, in² (cm²)
D = depth of beam, in (cm)
a = interior area, in² (cm²)
d = interior depth, in (cm)
w = total working load, net tons (kgf)

Shape of section	Greatest safe load, lb		Deflection, in	
	Load in middle	Load distributed	Load in middle	Load distributed
Solid rectangle	$\dfrac{890AD}{L}$	$\dfrac{1,780AD}{L}$	$\dfrac{wL^3}{32AD^2}$	$\dfrac{wL^3}{52AD^2}$
Hollow rectangle	$\dfrac{890(AD-ad)}{L}$	$\dfrac{1,780(AD-ad)}{L}$	$\dfrac{wL^3}{32(AD^2-ad^2)}$	$\dfrac{wL^3}{52(AD^2-ad^2)}$
Solid cylinder	$\dfrac{667AD}{L}$	$\dfrac{1,333AD}{L}$	$\dfrac{wL^3}{24AD^2}$	$\dfrac{wL^3}{38AD^2}$
Hollow cylinder	$\dfrac{667(AD-ad)}{L}$	$\dfrac{1,333(AD-ad)}{L}$	$\dfrac{wL^3}{24(AD^2-ad^2)}$	$\dfrac{wL^3}{38(AD^2-ad^2)}$
I beam	$\dfrac{1,795AD}{L}$	$\dfrac{3,390AD}{L}$	$\dfrac{wL^3}{58AD^2}$	$\dfrac{wL^3}{93AD^2}$

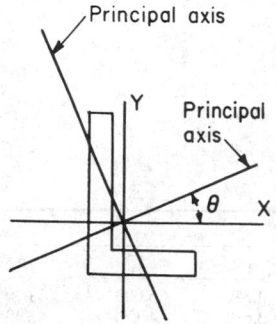

Fig. 5.2.32

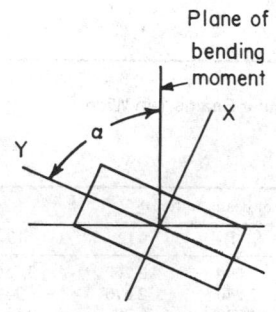

Fig. 5.2.33

In general, it may be stated that when the plane of the bending moment coincides with one of the principal axes, the other principal axis is the neutral axis. This is the ordinary case, in which the ordinary formula for unit stress may be applied. When the plane of the bending moment does not coincide with one of the principal axes, the above formula for oblique loading may be applied.

Internal Moment Beyond the Elastic Limit

Ordinarily, the expression $M = SI/c$ is used for stresses above the elastic limit, in which case S becomes an experimental coefficient S_R, the **modulus of rupture**, and the formula is empirical. The true relation is obtained by applying to the cross section a stress-strain diagram from a tension and compression test, as in Fig. 5.2.34. Figure 5.2.34 shows the side of a beam of depth d under flexure beyond its elastic limit; line 1–1 shows the distorted cross section; line 3–3, the usual rectilinear

Table 5.2.5 Coefficients for Correcting Values in Table 5.2.4 for Various Methods of Support and of Loading, Single Span

Conditions of loading	Max relative safe load	Max relative deflection under max relative safe load
Beam supported at ends:		
Load uniformly distributed over span	1.0	1.0
Load concentrated at center of span	½	0.80
Two equal loads symmetrically concentrated	$l/4c$	
Load increasing uniformly to one end	0.974	0.976
Load increasing uniformly to center	¾	0.96
Load decreasing uniformly to center	3/2	1.08
Beam fixed at one end, cantilever:		
Load uniformly distributed over span	¼	2.40
Load concentrated at end	⅛	3.20
Load increasing uniformly to fixed end	⅜	1.92
Beam continuous over two supports equidistant from ends:		
Load uniformly distributed over span		
1. If distance $a > 0.2071l$	$l^2/(4a^2)$	
2. If distance $a < 0.2071l$	$\dfrac{l}{l-4a}$	
3. If distance $a = 0.2071l$	5.83	
Two equal loads concentrated at ends	$l/(4a)$	

NOTE: l = length of beam; c = distance from support to nearest concentrated load; a = distance from support to end of beam.

Table 5.2.6 Properties of Various Cross Sections*
(I = moment of inertia; I/c = section modulus; $r = \sqrt{I/A}$ = radius of gyration)

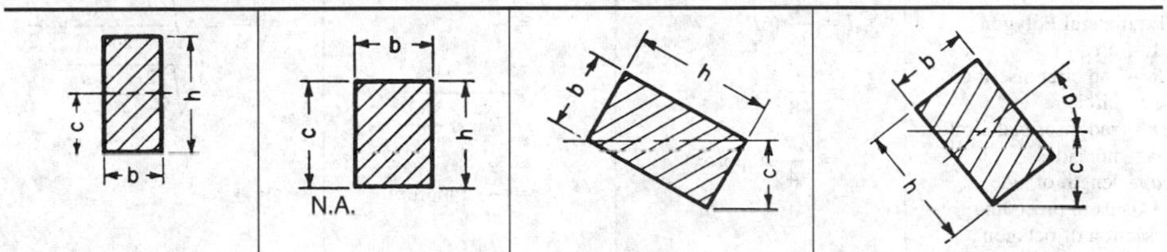

$I = \dfrac{bh^3}{12}$	$\dfrac{bh^3}{3}$	$\dfrac{b^3h^3}{6(b^2 + h^2)}$	$\dfrac{bh}{12}(h^2 \cos^2 a + b^2 \sin^2 a)$
$\dfrac{I}{c} = \dfrac{bh^2}{6}$	$\dfrac{bh^2}{3}$	$\dfrac{b^2h^2}{6\sqrt{b^2 + h^2}}$	$\dfrac{bh}{6}\left(\dfrac{h^2 \cos^2 a + b^2 \sin^2 a}{h \cos a + b \sin a}\right)$
$r = \dfrac{h}{\sqrt{12}} = 0.289h$	$\dfrac{h}{\sqrt{3}} = 0.577h$	$\dfrac{bh}{\sqrt{6(b^2 + h^2)}}$	$\sqrt{\dfrac{h^2 \cos^2 a + b^2 \sin^2 a}{12}}$

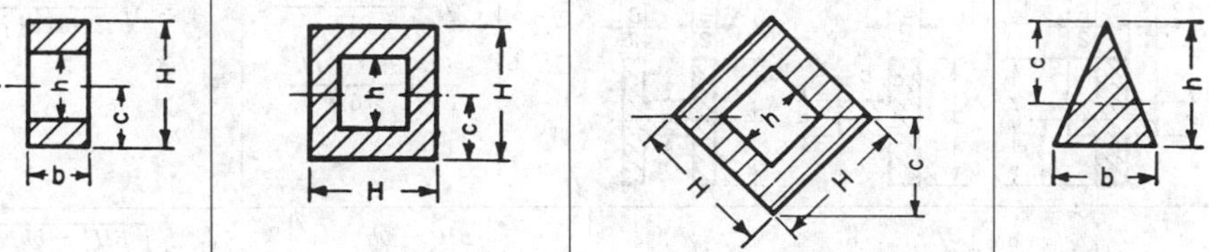

$I = \dfrac{b}{12}(H^3 - h^3)$	$\dfrac{H^4 - h^4}{12}$	$\dfrac{H^4 - h^4}{12}$	$\dfrac{bh^3}{36}; c = \dfrac{2}{3}h$
$\dfrac{I}{c} = \dfrac{b}{6}\dfrac{H^3 - h^3}{H}$	$\dfrac{1}{6}\dfrac{H^4 - h^4}{H}$	$\dfrac{\sqrt{2}}{12}\dfrac{H^4 - h^4}{H}$	$\dfrac{bh^2}{24}$
$r = \sqrt{\dfrac{H^3 - h^3}{12(H - h)}}$	$\sqrt{\dfrac{H^2 + h^2}{12}}$	$\sqrt{\dfrac{H^2 + h^2}{12}}$	$\dfrac{h}{\sqrt{18}}$

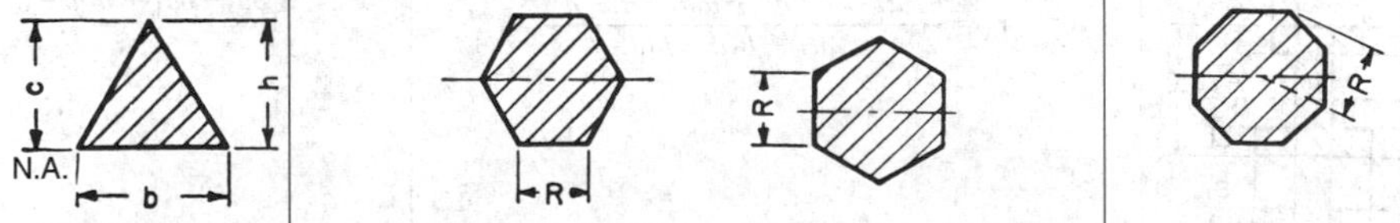

$I = \dfrac{bh^3}{12}$	$\dfrac{5\sqrt{3}}{16}R^4$	$\dfrac{5\sqrt{3}}{16}R^4$	$\dfrac{1 + 2\sqrt{2}}{6}R^4$
$\dfrac{I}{c} = \dfrac{bh^2}{12}$	$\tfrac{5}{8}R^3$	$\dfrac{5\sqrt{3}}{16}R^3$	$0.6906R^3$
$r = \dfrac{h}{\sqrt{6}}$	$\sqrt{\dfrac{5}{24}}R$	$\sqrt{\dfrac{5}{24}}R$	$0.475R$

NOTE: Square, axis same as first rectangle, side = h; $I = h^4/12$; $I/c = h^3/6$; $r = 0.289h$.
Square, diagonal taken as axis: $I = h^4/12$; $I/c = 0.1179h^3$; $r = 0.289h$.

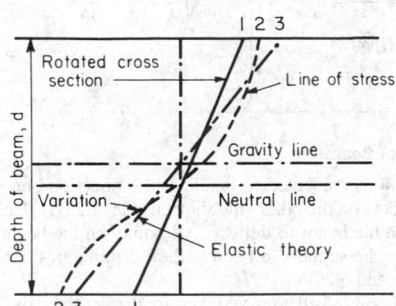

Fig. 5.2.34

relation of stress to strain; and line 2–2, an actual stress-strain diagram, applied to the cross section of the beam, compression above and tension below. The neutral axis is then below the gravity axis. The **outer material** may be expected to develop **greater ultimate strength** than in simple stress, because of the reinforcing action of material nearer the neutral axis that is not yet overstrained. This leads to an **equalization of stress** over the cross section. S_R exceeds the ultimate strength S_M in tension as follows: for cast iron, $S_R = 2S_M$; for sandstone, $S_R = 3S_M$; for concrete, $S_R = 2.2S_M$; for wood (green), $S_R = 2.3S_M$.

In the case of steel I beams, failure begins practically when the elastic limit in the compression flange is reached.

Because of the support of adjoining material, the **elastic limit in flexure** S_p is also greater than in tension, depending upon the relation of breadth to depth of section. For the same breadth, the difference decreases with

Table 5.2.6   Properties of Various Cross Sections*   (Continued)

Equilateral Polygon
A = area
R = rad circumscribed circle
r = rad inscribed circle
n = no. sides
a = length of side
Axis as in preceding section of octagon

$$I = \frac{A}{24}(6R^2 - a^2)$$
$$= \frac{A}{48}(12r^2 + a^2)$$
$$= \frac{AR^2}{4} \text{ (approx)}$$

$$\frac{I}{c} = \frac{I}{r}$$
$$= \frac{I}{R\cos\dfrac{180°}{n}}$$
$$= \frac{AR}{4} \text{ (approx)}$$

$$\sqrt{\frac{I}{A}} = \sqrt{\frac{6R^2 - a^2}{24}} \approx \frac{R}{2}$$
$$= \sqrt{\frac{12r^2 + a^2}{48}}$$

$$I = \frac{6b^2 + 6bb_1 + b_1^2}{36(2b + b_1)}h^3$$
$$c = \frac{1}{3}\frac{3b + 2b_1}{2b + b_1}h$$

$$\frac{I}{c} = \frac{6b^2 + 6bb_1 + b_1^2}{12(3b + 2b_1)}h^2$$

$$r = \frac{h\sqrt{12b^2 + 12bb_1 + 2b_1^2}}{6(2b + b_1)}$$

$$I = \frac{BH^3 + bh^3}{12}$$
$$\frac{I}{c} = \frac{BH^3 + bh^3}{6H}$$

$$r = \sqrt{\frac{BH^3 + bh^3}{12(BH + bh)}}$$

$$I = \frac{BH^3 - bh^3}{12}$$
$$\frac{I}{c} = \frac{BH^3 - bh^3}{6H}$$

$$r = \sqrt{\frac{BH^3 - bh^3}{12(BH - bh)}}$$

$$I = \tfrac{1}{3}(Bc_1^3 - B_1h^3 + bc_3^3 - b_1h_1^3)$$
$$c_1 = \frac{1}{2}\frac{aH^2 + B_1d^2 + b_1d_1(2H - d_1)}{aH + B_1d + b_1d_1}$$

$$r = \sqrt{\frac{I}{Bd + bd_1 + a(h + h_1)}}$$

$$I = \tfrac{1}{3}(Bc_1^3 - bh^3 + ac_2^3)$$
$$c_1 = \frac{1}{2}\frac{aH^2 + bd^2}{aH + bd}$$
$$c_2 = H - c_1$$
$$r = \sqrt{\frac{I}{Bd + a(H - d)}}$$

$$I = \frac{\pi d^4}{64} = \frac{\pi r^4}{4} = \frac{A}{4}r^2$$
$$= 0.05d^4 \text{ (approx)}$$

$$\frac{I}{c} = \frac{\pi d^3}{32} = \frac{\pi r^3}{4} = \frac{A}{4}r$$
$$= 0.1d^3 \text{ (approx)}$$

$$\sqrt{\frac{I}{A}} = \frac{r}{2} = \frac{d}{4}$$

increase of height. No difference will occur in the case of an I beam, or with hard materials.

Wide plates will not expand and contract freely, and the value of E will be increased on account of side constraint. As a consequence of lateral contraction of the fibers of the tension side of a beam and lateral swelling of fibers at the compression side, the cross section becomes distorted to a trapezoidal shape, and the neutral axis is at the center of gravity of the trapezoid. Strictly, this shape is one with a curved perimeter, the radius being r_c/μ, where r_c is the radius of curvature of the neutral line of the beam, and μ is Poisson's ratio.

Deflection of Beams

When a beam is subjected to bending, the fibers on one side elongate, while the fibers on the other side shorten (Fig. 5.2.35). These changes in length cause the beam to deflect. All points on the beam except those directly over the support fall below their original position, as shown in Figs. 5.2.31 and 5.2.35.

The **elastic curve** is the curve taken by the neutral axis. The radius of curvature at any point is

$$r_c = EI/M$$

Table 5.2.6 Properties of Various Cross Sections* (Continued)

$d_m = \frac{1}{2}(D + d)$ $s = \frac{1}{2}(D - d)$	$I = \frac{\pi}{64}(D^4 - d^4)$ $= \frac{\pi}{4}(R^4 - r^4)$ $= \frac{1}{4}A(R^2 + r^2)$ $= 0.05(D^4 - d^4)$ $\qquad$ (approx)	$\frac{I}{c} = \frac{\pi}{32}\frac{D^4 - d^4}{D}$ $= \frac{\pi}{4}\frac{R^4 - r^4}{R}$ $= 0.8d_m^2 s$ (approx) when $\frac{s}{d_m}$ is very small	$\sqrt{\frac{I}{A}} = \frac{\sqrt{R^2 + r^2}}{2} = \frac{\sqrt{D^2 + d^2}}{4}$
	$I = r^4\left(\frac{\pi}{8} - \frac{8}{9\pi}\right)$ $= 0.1098r^4$	$\frac{I}{c_2} = 0.1908r^3$ $\frac{I}{c_1} = 0.2587r^2$ $c_1 = 0.4244r$	$\sqrt{\frac{I}{A}} = \frac{\sqrt{9\pi^2 - 64}}{6\pi}r = 0.264r$
	$I = 0.1098(R^4 - r^4)$ $\quad - \frac{0.283R^2 r^2(R - r)}{R + r}$ $= 0.3tr_1^3$ (approx) when $\frac{t}{r_1}$ is very small	$c_1 = \frac{4}{3\pi}\frac{R^2 + Rr + r^2}{R + r}$ $c_2 = R - c_1$	$\sqrt{\frac{I}{A}} = \sqrt{\frac{2I}{\pi(R^2 - r^2)}}$ $= 0.31r_1$ (approx)
	$I = \frac{\pi a^3 b}{4} = 0.7854a^3 b$	$\frac{I}{c} = \frac{\pi a^2 b}{4} = 0.7854a^2 b$	$r = \frac{a}{2}$
	$I = \frac{\pi}{4}(a^3 b - a_1^3 b_1)$ $= \frac{\pi}{4}a^2(a + 3b)t$ $\qquad$ (approx)	$\frac{I}{c} = \frac{\pi}{4}a(a + 3b)t$ $\qquad$ (approx)	$r = \sqrt{\frac{I}{(\pi ab - a_1 b_1)}}$ $= \frac{a}{2}\sqrt{\frac{a + 3b}{a + b}}$ $\qquad$ (approx)
	$I = \frac{1}{12}\left[\frac{3\pi}{16}d^4 + b(h^3 - d^3) + b^3(h - d)\right]$ $\frac{I}{c} = \frac{1}{6h}\left[\frac{3\pi}{16}d^4 + b(h^3 + d^3) + b^3(h - d)\right]$		$r = \sqrt{\frac{I}{\pi\frac{d^2}{4} + 2b(h - d)}}$ $\qquad$ (approx)
	$I = \frac{t}{4}\left(\frac{\pi B^3}{16} + B^2 h + \frac{\pi B h^2}{2} + \frac{2}{3}h^3\right)$ $h = H - \frac{1}{2}B$ $\frac{I}{c} = \frac{2I}{H + t}$		$r = \sqrt{\frac{I}{2\left(\frac{\pi B}{4} + h\right)t}}$

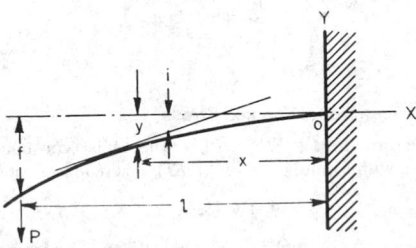

Fig. 5.2.35

A beam bent to a **circular curve** of constant radius has a constant bending moment.

Replacing r_c in the equation by its approximate geometrical value, $1/r_c = d^2 y/dx^2$, the fundamental equation from which the elastic curve of a bent beam can be developed and the deflection of any beam obtained is,

$$M = EI\,d^2 y/dx^2 \quad \text{(approx)}$$

Substituting the value of M, in terms of x, and integrating once, gives the slope of the tangent to the elastic curve of the beam at point x;

$\tan i = dy/dx = \int_0^x M\,dx/(EI)$. Since i is usually small, $\tan i = i$,

Table 5.2.6   Properties of Various Cross Sections*   (Continued)

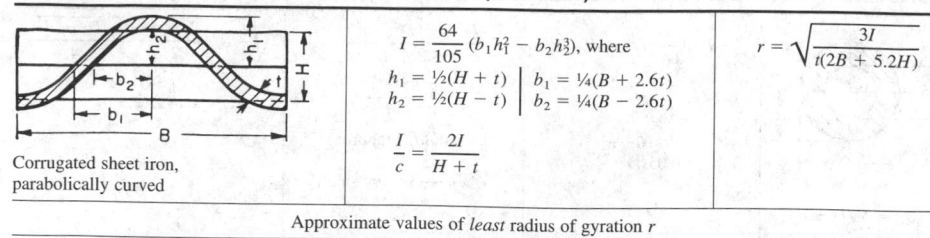

Corrugated sheet iron, parabolically curved	$I = \dfrac{64}{105}(b_1 h_1^2 - b_2 h_2^3)$, where $h_1 = \frac{1}{2}(H + t)$ $b_1 = \frac{1}{4}(B + 2.6t)$ $h_2 = \frac{1}{2}(H - t)$ $b_2 = \frac{1}{4}(B - 2.6t)$ $\dfrac{I}{c} = \dfrac{2I}{H + t}$	$r = \sqrt{\dfrac{3I}{t(2B + 5.2H)}}$

Approximate values of *least* radius of gyration r

I-beam	Channel	Deck beam

$r = $ 0.36336D 0.295D $D/4.58$ $D/3.54$ $D/6$

T-beam	Angle Equal legs	Angle Unequal legs	Cross

$r = $ $D/4.74$ $D/5$ $BD/[2.6(B + D)]$ $D/4.74$

* Some of the cross sections depicted in this table will be encountered most often in machinery as castings, forgings, or individual sections assembled and joined mechanically (or welded). A number of the sections shown are obsolete and will be encountered mainly in older equipment and/or building structures.

expressed in radians. A second integration gives the vertical deflection of any point of the elastic curve from its original position.

EXAMPLE.   In the cantilever beam shown in Fig. 5.2.35, the bending moment at any section $= -P(l - x) = EI\, d^2y/(dx)^2$. Integrate and determine constant by the condition that when $x = 0$, $dy/dx = 0$. Then $EI\, dy/dx = -Plx + \frac{1}{2}Px^2$. Integrate again; and determine constant by the condition that when $x = 0$, $y = 0$. Then $EIy = -\frac{1}{2}Plx^2 + Px^3/6$. This is the equation of the elastic curve. When $x = l$, $y = f = -Pl^3/(3EI)$. In general, the two constants of integration must be determined simultaneously.

Deflection in general, f, may be expressed by the equation $f = Pl^3/(mEI)$, where m is a coefficient. See Tables 5.2.2 and 5.2.4 for values of f for beams of various sections and loadings. For coefficients of deflection of wooden beams and structural steel shapes, see Sec. 12.2.

Since I varies as the cube of the depth, the **stiffness,** or inverse deflection, of various **beams** varies, other factors remaining constant, inversely as the load, inversely as the cube of the span, and directly as the cube of the depth. This deflection is due to bending moment only. In general, however, the bending of beams involves transverse shearing stresses which cause **shearing strains** and thus **add to the total deflection.** The contribution of shearing strain to overall deflection becomes significant only when the beam span is very short. These strains may affect substantially the strength as well as the deflection of beams. When deflection due to transverse shear is to be accounted for, the differential equation of the elastic curve takes the form

$$EI\frac{d^2y}{dx^2} = EI\left(\frac{d^2y_b}{dx^2} + \frac{d^2y_s}{dx^2}\right) = M - \frac{kEI}{AG}\times\frac{d^2M}{dx^2}$$

where k is a factor dependent upon the beam cross section. Sergius Sergev, in ''The Effect of Shearing Forces on the Deflection and Strength of Beams'' (*Univ. Wash. Eng. Exp. Stn. Bull.* 114) gives $k = 1.2$ for rectangular sections, 10/9 for circular sections, and 2.4 for I beams. He also points out that in the case of a deep, rectangular-section cantilever, carrying a concentrated load at the free end, the deflection due to shear may be up to 3.1 percent of that due to bending moment; if this beam supports a uniformly distributed load, it may be up to 4.1 percent. A deep, simple beam deflection may increase up to 15.6 percent when carrying a uniformly distributed load and up to 12.5 percent when the load is concentrated at midspan.

Design of beams may be based on **strength** (stress) or on **stiffness** if deflection must be limited. Deflection rather than stress becomes the criterion for design, e.g., of machine tools, for which the relative positions of tool and workpiece must be maintained while the cutting loads are applied during operation. Similarly, large steam-turbine shafts supported on two end bearings must maintain alignment and tight critical clearances between the rotating blade assemblies and the stationary stator blades during operation. When more than one beam shares a load, each beam will assume a portion of the load that is proportional to its stiffness. **Superposition** may be used in connection with both stresses and deflections.

EXAMPLE.   (Fig. 5.2.36). Two wooden stringers—one (A) 8×16 in in cross section and 20 ft in span, the other (B) 8 in $\times$ 8 in $\times$ 16 ft—carrying the center load $P_0 = 22,000$ lb are required, the load carried by each stringer. The deflections f of the two stringers must be equal. Load on $A = P_1$, and on $B = P_2$. $f = P_1 l_1^3/(48EI_1) = P_2 l_2^3/(48EI_2)$. Then $P_1/P_2 = l_2^3 I_1/(l_1^3 I_2) = 4$. $P_0 = P_1 + P_2 = 4P_2 + P_2$, whence $P_2 = 22,000/5 = 4,400$ lb (1,998 kgf) and $P_1 = 4 \times 4,400 = 17,600$ lb (7,990 kgf).

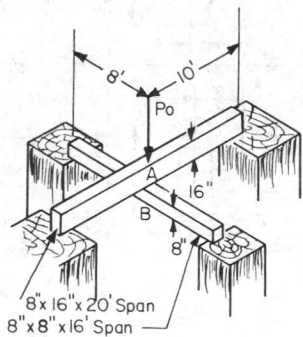

8"x 16"x 20' Span
8"x 8"x 16' Span

Fig. 5.2.36

Relation between Deflection and Stress

Combine the formula $M = SI/c = Pl/n$, where n is a constant, $P =$ load, and $l =$ span, with formula $f = Pl^3/(mEI)$, where m is a constant. Then

$$f = C'' Sl^2/(Ec)$$

where C'' is a new constant $= n/m$. Other factors remaining the same, the **deflection varies directly as the stress and inversely as** E. If the span is

Table 5.2.7

Beam	Load	n	m	C''
Cantilever	Concentrated at end	1	3	1/3
Cantilever	Uniform	2	8	1/4
Simple	Concentrated at center	4	48	1/12
Simple	Uniform	8	384/5	5/48
Fixed ends	Concentrated at center	8	192	1/24
Fixed ends	Uniform	12	384	1/32
One end fixed, one end supported	Concentrated at center	16/3	768/7	7/144
One end fixed, one end supported	Uniform	128/9	185	1/13
Simple	Uniformly varying, maximum at center	6	60	1/10

constant, a shallow beam will submit to greater deformations than a deeper beam without exceeding a safe stress. If depth is constant, a beam of double span will attain a given deflection with only one-quarter the stress. Values of n, m, and C'' are given in Table 5.2.7 (for other values, see Table 5.2.2).

Graphical Relations

Referring to Fig. 5.2.37, the shear V acting at any section is equal to the total load on the right of the section, or

$$V = \int w \, dx$$

Since $w \, dx$ is the product of w, a loading intensity (which is expressed as a vertical height in the load diagram), and dx, an elementary length along the horizontal, evidently $w \, dx$ is the area of a small vertical strip of the **load diagram**. Then $\int w \, dx$ is the summation of all such vertical strips between two indefinite points. Thus, to obtain the shear in any

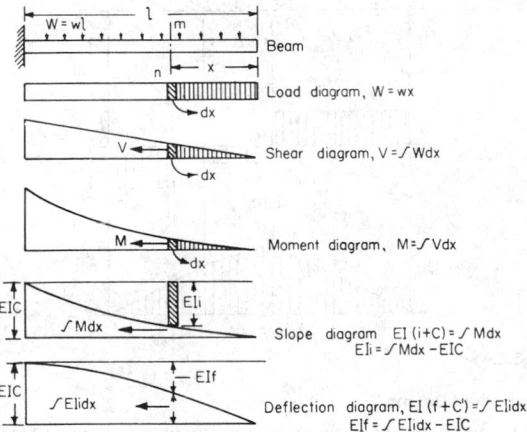

Fig. 5.2.37

section mn, find the area of the load diagram up to that section, and draw a second diagram called the **shear diagram**, any ordinate of which is proportional to the shear, or to the area in the load diagram to the right of mn. Since $V = dM/dx$,

$$\int V \, dx = M$$

By similar reasoning, a **moment diagram** may be drawn, such that the ordinate at any point is proportional to the area of the shear diagram to the right of that point. Since $M = EI \, d^2f/dx^2$,

$$\int M \, dx = EI \, (df/dx + C) = EI(i + C)$$

if I is constant. Here C is a constant of integration. Thus i, the slope or grade of the elastic curve at any point, is proportional to the area of the

moment diagram $\int M \, dx$ up to that point; and a **slope diagram** may be derived from the moment diagram in the same manner as the moment diagram was derived from the shear diagram.

If I is not constant, draw a new curve whose ordinates are M/I and use these M/I ordinates just as the M ordinates were used in the case where I was constant; that is, $\int (M/I)dx = E(i + C)$. The ordinate at any point of the slope curve is thus proportional to the area of the M/I curve to the right of that point. Again, since $iE = E \, df/dx$.

$$\int iE \, dx = \int E \, df = E(f + C)$$

and thus the ordinate f to the elastic curve at any point is proportional to the area of the slope diagram $\int i \, dx$ up to that point. The equilibrium polygon may be used in drawing the **deflection curve** directly from the M/I diagram.

Thus, the five curves of load, shear, moment, slope, and deflection are so related that each curve is derived from the previous one by a process of graphical integration, and with proper regard to scales the deflection is thereby obtained.

The vertical distance from any point A (Fig. 5.2.38) on the elastic curve of a beam to the tangent at any other point B equals the moment of the area of the $M/(EI)$ diagram from A to B about A. This distance, the **tangential deviation** t_{AB}, may be used with the slope-area relation and the geometry of the elastic curve to obtain deflections. These theorems, together with the equilibrium equations, can be used to compute reactions in the case of statically **indeterminate beams**.

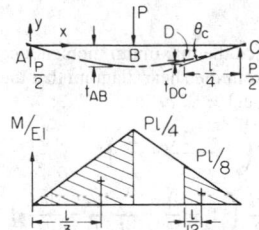

Fig. 5.2.38

EXAMPLE. The deflections of points B and D (Fig. 5.2.38) are

$$y_B = -t_{AB} = \text{moment area} \left. \frac{M}{EI} \right|_A^B$$

$$= -\frac{1}{EI} \times \frac{Pl}{4} \times \frac{l}{4} \times \frac{l}{3} = -\frac{Pl^3}{48EI}$$

$$\theta_C = \nabla \theta \left.\right|_B^C = \text{area} \left. \frac{M}{EI} \right|_B^C = \frac{1}{EI} \times \frac{Pl}{4} \times \frac{l}{4} = \frac{Pl^2}{16EI}$$

$$y_D = -\left(\theta_C \times \frac{l}{4} - t_{DC} \right)$$

$$= -\frac{Pl^2}{16EI} \times \frac{l}{4} + \frac{l}{EI} \times \frac{Pl}{8} \times \frac{l}{8} \times \frac{l}{12} = -\frac{11Pl^3}{768EI}$$

Resilience of Beams

The external work of a load gradually applied to a beam, and which increases from zero to P, is $\frac{1}{2}Pf$ and equals the **resilience** U. But, from the formulas $P = nSI/(cl)$ and $f = nSl^2/(mcE)$, where n and m are constants that depend upon loading and supports, S = fiber stress, c = distance from neutral axis to outer fiber, and l = length of span. Substitute for P and f, and

$$U = \frac{n^2}{m}\left(\frac{k}{c}\right)^2 \frac{S^2 V}{2E}$$

where k is the radius of gyration and V the volume of the beam. For values of U, see Table 5.2.1.

The resilience of beams of similar cross section at a given stress is proportional to their volumes. The **internal resilience,** or the elastic deformation energy in the material of a beam in a length x is dU, and

$$U = \frac{1}{2}\int M^2\, dx/(EI) = \frac{1}{2}\int M\, di$$

where M is the moment at any point x, and di is the angle between the tangents to the elastic curve at the ends of dx. The values of resilience and deflection in special cases are easily developed from this equation.

Rolling Loads

Rolling or **moving loads** are those loads which may change their position on a beam. Figure 5.2.39 represents a beam with two equal concentrated moving loads, such as two wheels on a crane girder, or the wheels of a

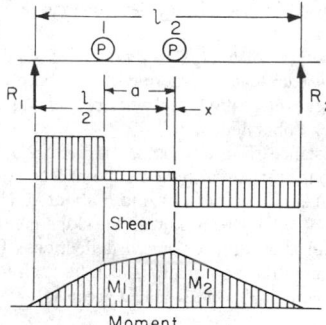

Fig. 5.2.39

truck on a bridge. Since the maximum moment occurs where the shear is zero, it is evident from the shear diagram that the maximum moment will occur under a wheel. $x < a/2$:

$$R_1 = P\left(1 - \frac{2x}{l} + \frac{a}{l}\right)$$

$$M_2 = \frac{Pl}{2}\left(1 - \frac{a}{l} + \frac{2x}{l}\frac{a}{l} - \frac{4x^2}{l^2}\right)$$

$$R_2 = P\left(1 + \frac{2x}{l} - \frac{a}{l}\right)$$

$$M_1 = \frac{Pl}{2}\left(1 - \frac{a}{l} - \frac{2a^2}{l^2} + \frac{2x}{l}\frac{3a}{l} - \frac{4x^2}{l^2}\right)$$

$$M_2 \text{ max when } x = \tfrac{1}{4}a$$
$$M_1 \text{ max when } x = \tfrac{3}{4}a$$

$$M_{\text{max}} = \frac{Pl}{2}\left(1 - \frac{a}{2l}\right)^2 = \frac{P}{2l}\left(l - \frac{a}{2}\right)^2$$

EXAMPLE. Two wheel loads of 3,000 lb each, spaced on 5-ft centers, move on a span of l = 15 ft, x = 1.25 ft, and R_2 = 2,500 lb. $\therefore M_{\text{max}} = M_2 = 2,500 \times 6.25$ (1,135 × 1.90) = 15,600 lb · ft (2,159 kgf · m).

Figure 5.2.40 shows the condition when two equal loads are equally distant on opposite sides of the center. The moment is equal under the two loads.

If the **two moving loads** are **of unequal weight,** the condition for **maximum moment** is that the maximum moment will occur under the heavy wheel when the center of the beam bisects the distance between the resultant of the loads and the heavy wheel. Figure 5.2.41 shows this position and the shear and moment diagrams.

When **several wheel loads** constituting a system occur, the several suspected wheels must be examined in turn to determine which will cause the greatest moment. The **position for** the **greatest moment** that can occur under a given wheel is, as stated, when the center of the span bisects the distance between the wheel in question and the resultant of all the loads then on the span. The **position for maximum shear** at the support will be when one wheel is passing off the span.

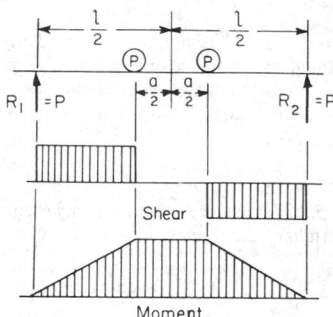

Fig. 5.2.40

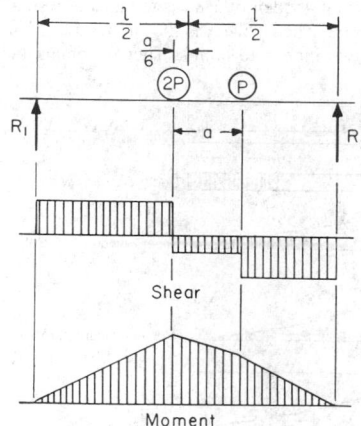

Fig. 5.2.41

Constrained Beams

Constrained beams are those so held or "built in" at one or both ends that the tangent to the elastic curve remains fixed in direction. These beams are held at the ends in such a manner as to allow free horizontal motion, as illustrated by Fig. 5.2.42. A constrained beam is stiffer than a simple beam of the same material, because of the modification of the moment by an end resisting moment. Figure 5.2.43 shows the two most common cases of constrained beams. See also Table 5.2.2.

Fig. 5.2.42   **Fig. 5.2.43**

Continuous Beams

A continuous beam is one resting upon several supports which may or may not be in the same horizontal plane. The general discussion for beams holds for continuous beams. $S_v A = V$, $SI/c = M$, and $d^2f/dx^2 = M/(EI)$. The **shear** at any section is equal to the algebraic sum of the components parallel to the section of all external forces on either side of the section. The bending moment at any section is equal to the moment of all external forces on either side of the section. The relations stated above between shear and moment diagrams hold true for continuous beams. The bending moment at any section is equal to the bending moment at any other section, plus the shear at that section times its arm, plus the product of all the intervening external forces times their respective arms. To illustrate (Fig. 5.2.44):

$$V_x = R_1 + R_2 + R_3 - P_1 - P_2 - P_3$$
$$M_x = R_1(l_1 + l_2 + x) + R_2(l_2 + x) + R_3 x$$
$$\qquad - P_1(l_2 + c + x) - P_2(b + x) - P_3 a$$
$$M_x = M_3 + V_3 x - P_3 a$$

Table 5.2.8 gives the value of the moment at the various supports of a uniformly loaded continuous beam over equal spans, and it also gives the values of the shears on each side of the supports. Note that the shear is of opposite sign on either side of the supports and that the sum of the two shears is equal to the reaction.

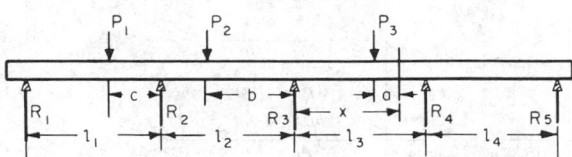

Fig. 5.2.44

Figure 5.2.45 shows the relation between the moment and shear diagrams for a uniformly loaded continuous beam of four equal spans (see Table 5.2.8). Table 5.2.8 also gives the **maximum bending moment** which will occur **between supports,** and in addition the position of this moment and the points of inflection (see Fig. 5.2.46).

Figure 5.2.46 shows the values of the functions for a uniformly loaded continuous beam resting on three equal spans with four supports.

Continuous beams are stronger and much stiffer than simple beams. However, a small, unequal subsidence of piers will cause serious

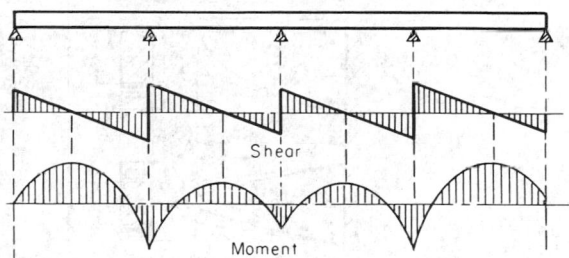

Fig. 5.2.45

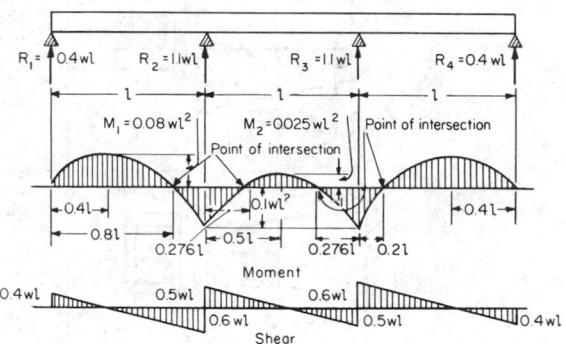

Fig. 5.2.46

Table 5.2.8   Uniformly Loaded Continuous Beams over Equal Spans
(Uniform load per unit length = w; length of equal span = l)

Number of supports	Notation of support of span	Shear on each side of support. L = left, R = right. Reaction at any support is $L + R$ — L	R	Moment over each support	Max moment in each span	Distance to point of max moment, measured to right from support	Distance to point of inflection, measured to right from support
2	1 or 2	0	½	0	0.125	0.500	None
3	1	0	³⁄₈	0	0.0703	0.375	0.750
	2	⁵⁄₈	⁵⁄₈	⅛	0.0703	0.625	0.250
4	1	0	⁴⁄₁₀	0	0.080	0.400	0.800
	2	⁶⁄₁₀	⁵⁄₁₀	¹⁄₁₀	0.025	0.500	0.276, 0.724
5	1	0	¹¹⁄₂₈	0	0.0772	0.393	0.786
	2	¹⁷⁄₂₈	¹⁵⁄₂₈	³⁄₂₈	0.0364	0.536	0.266, 0.806
	3	¹³⁄₂₈	¹³⁄₂₈	²⁄₂₈	0.0364	0.464	0.194, 0.734
6	1	0	¹⁵⁄₃₈	0	0.0779	0.395	0.789
	2	²³⁄₃₈	²⁰⁄₃₈	⁴⁄₃₈	0.0332	0.526	0.268, 0.783
	3	¹⁸⁄₃₈	¹⁹⁄₃₈	³⁄₃₈	0.0461	0.500	0.196, 0.804
7	1	0	⁴¹⁄₁₀₄	0	0.0777	0.394	0.788
	2	⁶³⁄₁₀₄	⁵⁵⁄₁₀₄	¹¹⁄₁₀₄	0.0340	0.533	0.268, 0.790
	3	⁴⁹⁄₁₀₄	⁵¹⁄₁₀₄	⁸⁄₁₀₄	0.0433	0.490	0.196, 0.785
	4	⁵³⁄₁₀₄	⁵³⁄₁₀₄	⁹⁄₁₀₄	0.0433	0.510	0.215, 0.804
8	1	0	⁵⁶⁄₁₄₂	0	0.0778	0.394	0.789
	2	⁸⁶⁄₁₄₂	⁷⁵⁄₁₄₂	¹⁵⁄₁₄₂	0.0338	0.528	0.268, 0.788
	3	⁶⁷⁄₁₄₂	⁷⁰⁄₁₄₂	¹¹⁄₁₄₂	0.0440	0.493	0.196, 0.790
	4	⁷²⁄₁₄₂	⁷¹⁄₁₄₂	¹²⁄₁₄₂	0.0405	0.500	0.215, 0.785
Values apply to:		wl	wl	wl^2	wl^2	l	l

NOTE: The numerical values given are coefficients of the expressions at the foot of each column.

Table 5.2.9 Beams of Uniform Strength (in Bending)

Beam	Cross section	Elevation and plan	Formulas
1. Fixed at one end, load P concentrated at other end			
	Rectangle: width (b) constant, depth (g) variable	Elevation: 1, top, straight line; bottom, parabola. 2, complete parabola. Plan: rectangle	$y^2 = \dfrac{6P}{bS_s} x$ $h = \sqrt{\dfrac{6Pl}{bS_s}}$ Deflection at A: $f = \dfrac{8P}{bE}\left(\dfrac{l}{h}\right)^3$
	Rectangle: width (y) variable, depth (h) constant	Elevation: rectangle. Plan: triangle	$y = \dfrac{6P}{h^2 S_s} x$ $b = \dfrac{6Pl}{h^2 S_s}$ Deflection at A: $f = \dfrac{6P}{bE}\left(\dfrac{l}{h}\right)^3$
	Rectangle: width (z) variable, depth (y) variable. $\dfrac{z}{y} = k$(const.)	Elevation: cubic parabola. Plan: cubic parabola	$y^3 = \dfrac{6P}{kS_s} x$ $z = ky$ $h = \sqrt[3]{\dfrac{6Pl}{kS_s}}$ $b = kh$
	Circle: diam (y) variable	Elevation: cubic parabola. Plan: cubic parabola	$y^3 = \dfrac{32P}{\pi S_s} x$ $d = \sqrt[3]{\dfrac{32Pl}{\pi S_s}}$
2. Fixed at one end, load of total magnitude P uniformly distributed			
	Rectangle: width (b) constant, depth (y) variable	Elevation: triangle. Plan: rectangle	$y = x\sqrt{\dfrac{3P}{blS}}$ $h = \sqrt{\dfrac{3Pl}{bS_s}}$ $f = 6\dfrac{P}{bE}\left(\dfrac{l}{h}\right)^3$
	Rectangle: width (y) variable, depth (h) constant	Elevation: rectangle. Plan: two parabolic curves with vertices at free end	$y = \dfrac{3Px^2}{lS_s h^2}$ $b = \dfrac{3Pl}{S_s h^2}$ Deflection at A: $f = \dfrac{3P}{bE}\left(\dfrac{l}{h}\right)^3$

Table 5.2.9   Beams of Uniform Strength (in Bending)   (Continued)

Beam	Cross section	Elevation and plan	Formulas
2. Fixed at one end, load of total magnitude P uniformly distributed			
	Rectangle: width (z) variable, depth (y) variable, $\dfrac{z}{y} = k$	Elevation: semi-cubic parabola Plan: semicubic parabola	$y^3 = \dfrac{3Px^2}{kS_s l}$ $z = ky$ $h = \sqrt[3]{\dfrac{3Pl}{kS_s}}$ $b = kh$
	Circle: diam (y) variable	Elevation: semi-cubic parabola Plan: semicubic parabola	$y^3 = \dfrac{16P}{\pi l S_s} x^2$ $d = \sqrt[3]{\dfrac{16Pl}{\pi S_s}}$
3. Supported at both ends, load P concentrated at point C			
	Rectangle: width (b) constant, depth (y) variable	Elevation: two parabolas, vertices at points of support Plan: rectangle	$y = \sqrt{\dfrac{3P}{S_s b}}\, x$ $h = \sqrt{\dfrac{3Pl}{2bS_s}}$ $f = \dfrac{P}{2Eb}\left(\dfrac{l}{h}\right)^3$
	Rectangle: width (y) variable, depth (h) constant	Elevation: rectangle Plan: two triangles, vertices at points of support	$y = \dfrac{3P}{S_s h^2}\, x$ $b = \dfrac{3Pl}{2S_s h^2}$ $f = \dfrac{3Pl^3}{8Ebh^3}$
	Rectangle: width (b) constant, depth (y or y_1) variable	Elevation: two parabolas, vertices at points of support Plan: rectangle	$y^2 = \dfrac{6P(l-p)}{blS_s}\, x$ $y_1{}^2 = \dfrac{6P_p}{blS_s}\, x_1$ $h = \sqrt{\dfrac{6P(l-p)p}{blS_s}}$
Load P moving across span			
	Rectangle: width (b) constant, depth (y) variable	Elevation: ellipse Major axis = l Minor axis = $2h$ Plan: rectangle	$\dfrac{x^2}{\left(\dfrac{l}{2}\right)^2} + \dfrac{y^2}{\dfrac{3Pl}{2bS_s}} = 1$ $h = \sqrt{\dfrac{3Pl}{2bS_s}}$

Table 5.2.9 Beams of Uniform Strength (in Bending) (Continued)

Beam	Cross section	Elevation and plan	Formulas
4. Supported at both ends, load of total magnitude P uniformly distributed			
	Rectangle: width (b) constant, depth (y) variable	Elevation: ellipse Plan: rectangle	$\dfrac{x^2}{\left(\dfrac{l}{2}\right)^2} + \dfrac{y^2}{\dfrac{3Pl}{4bS_s}} = 1$ $h = \sqrt{\dfrac{3Pl}{4bS_s}}$
		Deflection at O:	$f = \dfrac{1}{64}\dfrac{Pl^3}{EI}$ $= \dfrac{3}{16}\dfrac{P}{bE}\left(\dfrac{l}{h}\right)^3$
	Rectangle: width (y) variable, depth (h) constant	Elevation: rectangle Plan: two parabolas with vertices at center of span	$y = \dfrac{3P}{S_s h^2}\left(x - \dfrac{x^2}{l}\right)$ $b = \dfrac{3Pl}{4S_s h^2}$

changes in sign and magnitude of the bending stresses, reactions, and shears.

Maxwell's Theorem When a number of loads rest upon a beam, the deflection at any point is equal to the sum of the deflections at this point due to each of the loads taken separately. Maxwell's theorem states that if unit loads rest upon a beam at two points A and B, the deflection at A due to the unit load at B equals the deflection at B due to the unit load at A.

Castigliano's theorem states that the deflection of the point of application of an external force acting on a beam is equal to the partial derivative of the work of deformation with respect to this force. Thus, if P is the force, f the deflection, and U the work of deformation, which equals the resilience,

$$dU/dP = f$$

According to the **principle of least work,** the deformation of any structure takes place in such a manner that the work of deformation is a minimum.

Beams of Uniform Strength

Beams of uniform strength vary in section so that the unit stress S remains constant, the I/c varies as M. For **rectangular beams,** of breadth b and depth d, $I/c = bd^2/6$; and $M = Sbd^2/6$. Thus, for a cantilever beam of rectangular cross section, under a load P, $Px = Sbd^2/6$. If b is constant, d^2 varies with x, and the profile of the shape of the beam will be a parabola, as Fig. 5.2.47. If d is constant, b will vary as x and the beam will be triangular in plan, as shown in Fig. 5.2.48.

Shear at the end of a beam necessitates a modification of the forms determined above. The area required to resist shear will be P/S_v in a cantilever and R/S_v in a simple beam. The dotted extensions in Figs. 5.2.47 and 5.2.48 show the changes necessary to enable these cantilevers to resist shear. The extra material and cost of fabrication, however, make many of the forms impractical.

Table 5.2.9 shows some of the simple **sections of uniform strength.** In none of these, however, is shear taken into account.

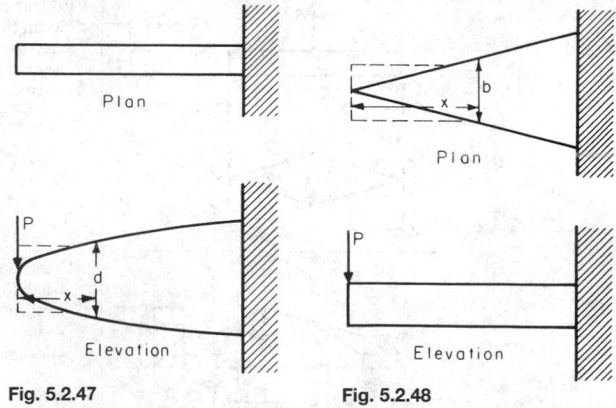

Fig. 5.2.47   **Fig. 5.2.48**

TORSION

Under torsion, a bar (Fig. 5.2.49) is twisted by a couple of magnitude Pp. Elements of the surface becomes helices of angle d, and a radius rotates through an angle θ in a length l, both d and θ being expressed in radians. S_v = shearing unit stress at distance r from center; I_p = *polar moment of inertia;* G = shearing modulus of elasticity. It is assumed that the cross sections remain plane surfaces. The strain on the cross section is wholly tangential, and is zero at the center of the section. Note that $ld = r\theta$.

In the case of a **circular cross section,** the stress S_v increases directly as the distance of the strained element from the center.

The polar moment of inertia I_p for any section may be obtained from $I_p = I_1 + I_2$, where I_1 and I_2 are the rectangular moments of inertia of the section about any two lines at right angles to each other, through the center of gravity.

The **external twisting** ~~...~~ **nt** M_t is balanced by the internal resisting moment.

For strength, $M_t = S_v l$ ...

For stiffness, $M_t = a l$ ...

The torsional resilience l.

$$U = \frac{1}{2}Ppa = S_v^2 I_p l/(2r^2 G) = a^2 G I_p/(2l).$$

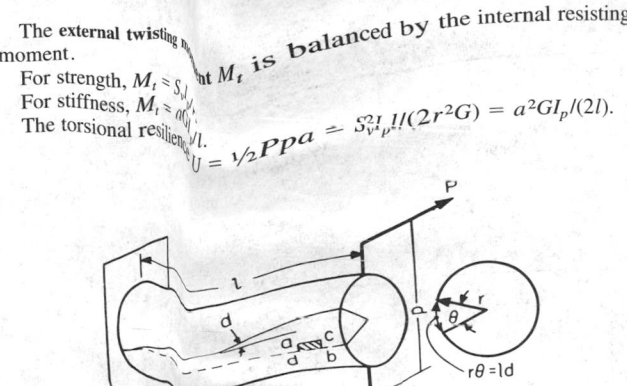

Fig. 5.2.49

The state of stress on an element taken from the surface of the shaft, as in Fig. 5.2.50, is pure shear. Pure tension exists at right angles to one 45° helix and pure compression at right angles to the opposite helix.

Reduced **formulas for shafts of various sections** are given in Table 5.2.11. When the ratio of shaft length to the largest lateral dimension in the cross section is less than approximately 2, the **end effects** may drastically affect the torsional stresses calculated.

Failure under torsion in brittle materials is a tensile failure at right angles to a helical element on the surface. Plastic materials twist off squarely. Fibrous materials separate in long strips.

Torsion of Noncircular Sections When a section is not circular, the unit stress no longer varies directly as the distance from the center. Cross sections become warped, and the greatest unit stress usually occurs at a point on the perimeter of the cross section *nearest* the axis of twist; thus, there is no stress at the corners of square and rectangular sections. The analyses become complex for noncircular sections, and the methods for solution of design problems using them most often admit only of approximations.

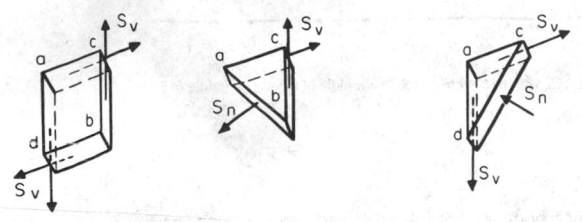

Fig. 5.2.50

Torsion problems have been solved for many different noncircular cross sections by utilizing the **membrane analogy,** due to Prandtl, which makes use of the fact that the mathematical treatment of a twisted bar is governed by the same equations as for a membrane stretched over a hole

which has the same shape as the bar and then inflated. The resulting three-dimensional surface provides the following: (1) The torque transmitted is proportional to twice the volume under the inflated membrane, and (2) the shear stress at any point is proportional to the slope of the curve measured perpendicular to that slope. In recent years, several other mathematical techniques have become widely used, especially with the aid of faster computational methods available from electronic computers.

By using **finite-difference methods,** the differential operators of the governing equations are replaced with difference operators which are related to the desired unknown values at a gridwork of points in the outline of the cross section being investigated.

The **finite-element method,** commonly referred to as **FEM,** deals with a spatial approximation of a complex shape which is then analyzed to determine deformations, stresses, etc. By using FEM, the exact structure is replaced with a set of simple structural elements interconnected at a finite number of nodes. The governing equations for the approximate structure can be solved exactly. Note that inasmuch as there is an exact solution for an approximate structure, the end result must be viewed, and the results thereof used, as approximate solutions to the real structure.

Using a finite-difference approach to Poisson's partial differential equation, which defines the stress functions for solid and hollow shafts with generalized contours, along with Prandtl's membrane analogy, Isakower has developed a series of practical design charts (ARRADCOM-MISD Manual UN 80-5, January 1981, Department of the Army). Dimensionless charts and tables for transmitted torque and maximum shearing stress have been generated. Information for circular shafts with rectangular and circular keyways, external splines and milled flats, as well as rectangular and X-shaped torsion bars, is presented.

Assuming the stress distribution from the point of maximum stress to the corner to be parabolic, Bach derived the approximate expression, $S_s M = 9M_t/(2b^2 h)$ for a rectangular section, b by h, where $h > b$. For closer results, the shearing stresses for a **rectangular section** (Fig. 5.2.51) may be expressed $S_A = M_t/(\alpha_A b^2 h)$ and $S_B = M_t/(\alpha_B b^2 h)$. The angle of twist for these shafts is $\theta = M_t l/(\beta G b^3 h)$. The factors α_A, α_B, and β are functions of the ratio h/b and are given in Table 5.2.10.

In the case of **composite sections,** such as a tee or angle, the torque that can be resisted is $M_t = G\theta \Sigma \beta h b^3$; the summation applies to each of the rectangles into which the section can be divided. The maximum stress occurs on the component rectangle having the largest b value. It is computed from

$$S_A = M_t \beta_A b_A/(\alpha_A \Sigma \beta h b^3)$$

Torque, deflection, and work relations for some additional sections are given in Table 5.2.11.

Fig. 5.2.51

Table 5.2.10 Factors for Torsion of Rectangular Shafts (Fig. 5.2.51)

b/b	1.00	1.50	1.75	2.00	2.50	3.00	4.00	5.00	6.00	8.00	10.0	∞
α_A	0.208	0.231	0.239	0.246	0.258	0.267	0.282	0.291	0.299	0.307	0.312	0.333
α_B	0.208	0.269	0.291	0.309	0.336	0.355	0.378	0.392	0.402	0.414	0.421	
β	0.141	0.196	0.214	0.229	0.249	0.263	0.281	0.291	0.299	0.307	0.312	0.333

Table 5.2.11 Torsion of Shafts of Various Cross Sections
(For strength and stiffness of shafts, see Sec. 8.2)

Cross section	Torsional resisting moment M_t	Angular twist, θ_1 (length = 1 in, radius = 1 in) In terms of torsional moment	In terms of max shear	Work of torsion (V = volume)
Solid circle, dia d	$\dfrac{\pi}{16} d^3 S_v$	$\dfrac{M_t}{GI_P} = \dfrac{32}{\pi d^4}\dfrac{M_t}{G}$	$2\dfrac{S_{v_{max}}}{G}\dfrac{1}{d}$	$\dfrac{1}{4}\dfrac{S_{v_{max}}^2}{G}V$ (Note 1)
Hollow circle D, d	$\dfrac{\pi}{16}\dfrac{D^4 - d^4}{D} S_v$	$\dfrac{32}{\pi(D^4 - d^4)}\dfrac{M_t}{G}$	$2\dfrac{S_{v_{max}}}{G}\dfrac{1}{D}$	$\dfrac{1}{4}\dfrac{S_{v_{max}}^2}{G}\dfrac{D^2 + d^2}{D^2}V$ (Note 2)
Ellipse B, A, h, b	$\dfrac{\pi}{16} b^2 h S_v$ $(h > b)$	$\dfrac{16}{\pi}\dfrac{b^2 + h^2}{b^3 h^3}\dfrac{M_t}{G}$	$\dfrac{S_{v_{max}}}{G}\dfrac{b^2 + h^2}{bh^2}$	$\dfrac{1}{8}\dfrac{S_{v_{max}}^2}{G}\dfrac{b^2 + h^2}{h^2}V$ (Note 3)
Rectangle h, b	$\dfrac{2}{9} b^2 h S_v$ $(h > b)$	$3.6*\dfrac{b^2 + h^2}{b^3 h^3}\dfrac{M_t}{G}$	$0.8*\dfrac{S_{v_{max}}}{G}\dfrac{b^2 + h^2}{bh^2}$	$\dfrac{4}{45}\dfrac{S_{v_{max}}^2}{G}\dfrac{b^2 + h^2}{h^2}V$ (Note 4)
Square h	$\dfrac{2}{9} h^3 S_v$	$7.2\dfrac{1}{h^4}\dfrac{M_t}{G}$	$1.6\dfrac{S_{v_{max}}}{G}\dfrac{1}{h}$	$\dfrac{8}{45}\dfrac{S_{v_{max}}^2}{G}V$ (Note 5)
Equilateral triangle b	$\dfrac{b^3}{20} S_v$	$46.2\dfrac{1}{b^4}\dfrac{M_t}{G}$	$2.31\dfrac{S_{v_{max}}}{G}\dfrac{1}{b}$	
Hexagon b	$\dfrac{b^3}{1.09} S_v$	$0.967\dfrac{1}{b^4}\dfrac{M_t}{G}$	$0.9\dfrac{S_{v_{max}}}{G}\dfrac{1}{b}$	

*When h/b =	1	2	4	8
Coefficient 3.6 becomes =	3.56	3.50	3.35	3.21
Coefficient 0.8 becomes =	0.79	0.78	0.74	0.71

NOTES: (1) $S_{v_{max}}$ at circumference. (2) $S_{v_{max}}$ at outer circumference. (3) $S_{v_{max}}$ at A; $S_{vB} = 16M_t/\pi bh^2$. (4) $S_{v_{max}}$ at middle of side h; in middle of b, $S_v = 9M_t/2bh^2$. (5) $S_{v_{max}}$ at middle of side.

COLUMNS

Members subjected to direct compression can be grouped into three classes. **Compression blocks** are so short (slenderness ratios below 30) that bending of member is unlikely. At the other limit, columns so slender that bending is primary, are the **long columns** defined by Euler's theory. The intermediate columns, quite common in practice, are called **short columns**.

Long columns and the more slender short columns usually fail by **buckling** when the **critical load** is reached. This is a matter of **instability**; that is, the column may continue to yield and deflect even though the load is not increased above critical. The **slenderness ratio** is the unsupported length divided by the least radius of gyration, parallel to which it can bend.

Long columns are handled by **Euler's column formula**,

$$P_{cr} = n\pi^2 EI/l^2 = n\pi^2 EA/(l/r)^2$$

The **coefficient** n accounts for **end conditions**. When the column is pivoted at both ends, $n = 1$; when one end is fixed and other rounded, $n = 2$; when both are fixed, $n = 4$; and when one end is fixed with the other free, $n = \frac{1}{4}$. The slenderness ratio that separates long columns from short ones depends upon the modulus of elasticity and the yield strength of the column material. When Euler's formula results in $(P_{cr}/A) > S_y$, strength rather than buckling causes failure, and the column ceases to be long. In round numbers, this **critical slenderness ratio** falls between 120 and 150. Table 5.2.12 gives additional facts concerning long columns.

Short Columns The stress in a short column may be considered partly due to compression and partly due to bending. A theoretical equation has not been derived. Empirical, though rational, expressions are, in general, based on the assumption that the permissible stress must be reduced below that which could be permitted were it due to compression only. The manner in which this reduction is made determines the type of equation as well as the slenderness ratio beyond which the equation does not apply. Figure 5.2.52 illustrates the situation. Some typical formulas are given in Table 5.2.13.

EXAMPLE. A machine member unsupported for a length of 15 in has a square cross section 0.5 in on a side. It is to be subjected to compression. What maximum safe load can be applied centrally, according to the AISC formula? At the com-

Table 5.2.12 Strength of Round-ended Columns according to Euler's Formula*

Material	Cast iron	Wrought iron[†]	Low-carbon steel	Medium-carbon steel
Ultimate compressive strength, lb/in^2	107,000	53,400	62,600	89,000
Allowable compressive stress, lb/in^2 (maximum)	7,100	15,400	17,000	20,000
Modulus of elasticity	14,200,000	28,400,000	30,600,000	31,300,000
Factor of safety	8	5	5	5
Smallest I allowable at worst section, in^4	$\dfrac{Pl^2}{17,500,000}$	$\dfrac{Pl^2}{56,000,000}$	$\dfrac{Pl^2}{60,300,000}$	$\dfrac{Pl^2}{61,700,000}$
Limit of ratio, l/r	50.0	60.6	59.4	55.6
Rectangle ($r = b\sqrt{1/12}$), $l/b >$	14.4	17.5	17.2	16.0
Circle ($r = \frac{1}{4}d$), $l/d >$	12.5	15.2	14.9	13.9
Circular ring of small thickness ($r = d\sqrt{1/8}$), $l/d >$	17.6	21.4	21.1	19.7

* P = allowable load, lb; l = length of column, in; b = smallest dimension of a rectangular section, in; d = diameter of a circular section, in; r = least radius of gyration of section.
† This material is no longer manufactured but may be encountered in existing structures and machinery.

Table 5.2.13 Typical Short-Column Formulas

Formula	Material	Code	Slenderness ratio
$S_w = 17,000 - 0.485 \left(\dfrac{l}{r}\right)^2$	Carbon steels	AISC	$l/r < 20$
$S_w = 16,000 - 70\,(l/r)$	Carbon steels	Chicago	$l/r < 120$
$S_w = 15,000 - 50 \left(\dfrac{l}{r}\right)$	Carbon steels	AREA	$l/r < 150$
$S_w = 19,000 - 100\,(l/r)$	Carbon steels	Am. Br. Co.	$60 < \dfrac{l}{r} < 120$
$S_{cr}{}^* = 135,000 - \dfrac{15.9}{c}\left(\dfrac{l}{r}\right)^2$	Alloy-steel tubing	ANC	$\dfrac{1}{\sqrt{cr}} < 65$
$S_w = 9,000 - 40\left(\dfrac{l}{r}\right)$	Cast iron	NYC	$\dfrac{l}{r} < 70$
$S_{cr}{}^* = 34,500 - \dfrac{245}{\sqrt{c}}\left(\dfrac{l}{r}\right)$	2017ST Aluminum	ANC	$\dfrac{1}{\sqrt{cr}} < 94$
$S_{cr}{}^* = 5,000 - \dfrac{0.5}{c}\left(\dfrac{l}{r}\right)^2$	Spruce	ANC	$\dfrac{l}{\sqrt{cr}} < 72$
$S_{cr}{}^* = S_y\left[1 - \dfrac{S_y}{4n\pi^2 E}\left(\dfrac{l}{r}\right)^2\right]$	Steels	Johnson	$\dfrac{l}{r} < \sqrt{\dfrac{2n\pi^2 E}{S_y}}$
$S_{cr}{}^*† = \dfrac{S_y}{1 + \dfrac{ec}{r^2}\sec\left(\dfrac{l}{r}\sqrt{\dfrac{P}{4AE}}\right)}$	Steels	Secant	$\dfrac{l}{r} <$ critical

* S_{cr} = theoretical maximum, c = end fixity coefficient,
$c = 2$, both ends pivoted; $c = 2.86$, one pivoted, other fixed;
$c = 4$, both ends fixed; $c = 1$, one end fixed, one end free.
† e is initial eccentricity at which load is applied to center of column cross section.

puted load, what size section (also square) would be needed, if it were to be designed according to the AREA formula?

$$l/r = 15/0.5/\sqrt{12} = 104 \quad \therefore \text{ short column}$$

$$P/A = 17,000 - 0.485\,(104)^2 = 11,730$$

or

$$P = 0.25 \times 11,730 = 2,940 \text{ lb } (1,335 \text{ kgf})$$

and

$$\frac{2,940}{a^2} \le 15,000 - 50\left(\frac{15la}{\sqrt{12}}\right) = 15,000 - \frac{2,600}{a}$$

thus

$$a^2 - 0.173a - 0.196 = 0 \quad \text{or} \quad a = 0.536 \text{ in } (1.36 \text{ cm})$$

Combined Flexure and Longitudinal Force Figure 5.2.53 shows a bar under flexure due to transverse and longitudinal loads. The maximum fiber stress S is made up of S_0, due to the direct action of load P, and S_b, due to the entire bending moment M. M is the algebraic sum of two bending moments, M_1 due to longitudinal load (+ for compression and − for tension), and M_2 due to transverse load. $M = M_2 \pm M_1$. Here $M_1 = Pf$ and $f = CS_b l^2/(Ec)$.

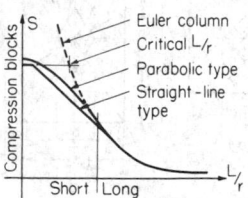

Fig. 5.2.52

For the Case of Longitudinal Compression. $S_b l/c = M_2 + CPS_b l^2/(Eo)$, or $S_b = M_2 c(I - CPl^2/E)$. The maximum stress is $S = S_b + S_0$ compression. The constant C for the case of Fig. 5.2.53 is derived from the equations $P'l/4 = S_b l/c$ and $f = P'l^3/(48EI)$. Solving for f; $f = \frac{1}{12} S_b l^2/(Ec)$, or $C = \frac{1}{12}$. For a beam supported at the ends and uniformly loaded, $C = \frac{5}{48}$. Other cases can be similarly calculated.

For the Case of Longitudinal Tension. $M = M_2 - Pf$, and $S_b = M_2 c/(I + CPl^2/E)$. The maximum stress is $S = S_b + S_0$, tension.

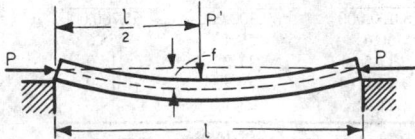

Fig. 5.2.53

ECCENTRIC LOADS

When **short blocks** are **loaded eccentrically** in compression or in tension, i.e., not through the center of gravity (cg), a combination of axial and bending stress results. The maximum unit stress S_M is the algebraic sum of these two unit stresses.

In Fig. 5.2.54 a load P acts in a line of symmetry at the distance e from cg; $r =$ radius of gyration. The unit stresses are (1) S_c, due to P, as if it acted through cg, and (2) S_b, due to the bending moment of P acting with a leverage of e about cg. Thus unit stress S at any point y is

$$S = S_c \pm S_b$$
$$= P/A \pm Pey/I$$
$$= S_c(1 \pm ey/r^2)$$

y is positive for points on the same side of cg as P, and negative on the opposite side. For a **rectangular cross section** of width b, the maximum stress $S_M = S_c (1 + 6e/b)$. When P is outside the middle third of width b and is a compressive load, tensile stresses occur.

For a **circular cross section** of diameter d, $S_M = S_c(1 + 8e/d)$. The stress due to the weight of the solid will modify these relations.

Note. In these formulas e is measured from the gravity axis, and gives tension when e is greater than one-sixth the width (measured in the same direction as e), for rectangular sections; and when greater than one-eighth the diameter for solid circular sections.

If, as in certain classes of masonry construction, the **material cannot withstand tensile stress** and thus no tension can occur, the center of moments (Fig. 5.2.55) is taken at the center of stress. For a **rectangular section,** P acts at distance k from the nearest edge. Length under compression $= 3k$, and $S_M = \frac{2}{3}P/(hk)$. For a **circular section,** $S_M = [0.372 + 0.056(k/r)]P/k\sqrt{rk}$, where $r =$ radius and $k =$ distance of P from circumference. For a **circular ring,** $S =$ average compressive stress on cross section produced by P; $e =$ eccentricity of P; $z =$ length of diameter under compression (Fig. 5.2.56). Values of z/r and of the ratio of S_{max} to average S are given in Tables 5.2.14 and 5.2.15.

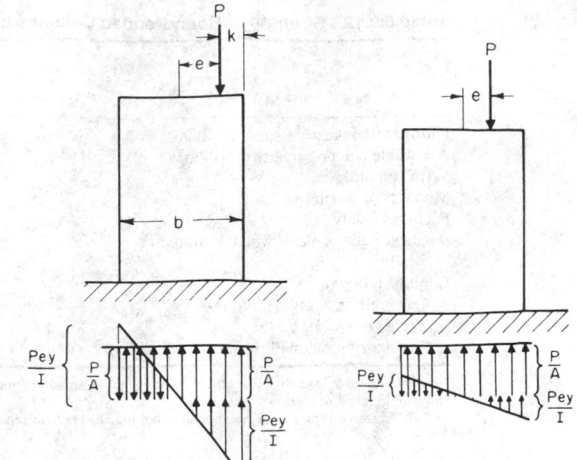

Fig. 5.2.54

Chimney Problem. Weight of chimney = 563,000 lb; $e = 1.56$ ft; OD of chimney = 10 ft 8 in; ID = 6 ft 6½ in. Overturning moment = $Pe = 878,000$ ft · lb, $r_1/r = 0.6$. $e/r = 0.29$. This gives $z/r > 2$. Therefore, the entire area of the base is under compression. Area under compression = 55.8 ft²; $I = 546$; $S = 563,000/55.8 \pm (878,000 \times 5.33)/546 = 18,700$ (max) and 1,500 (min) lb compression per ft². From Table 5.2.15, by interpolation, $S_{max}/S_{av} = 1.85$. $\therefore S_{max} = (563,000/55.8) \times 1.85 = 18,685$ lb/ft² (91,313 kgf/m²).

The **kern** is the area around the center of gravity of a cross section within which any load applied will produce stress of only one sign throughout the entire cross section. Outside the kern, a load produces stresses of different sign. Figure 5.2.57 shows kerns (shaded) for various sections.

For a **circular ring,** the radius of the kern $r = D[1 + (d/D)^2]/8$.

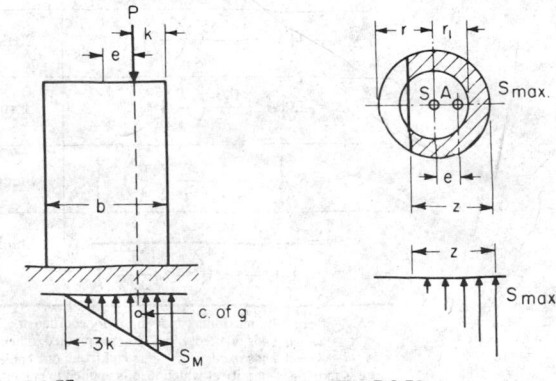

Fig. 5.2.55 | **Fig. 5.2.56**

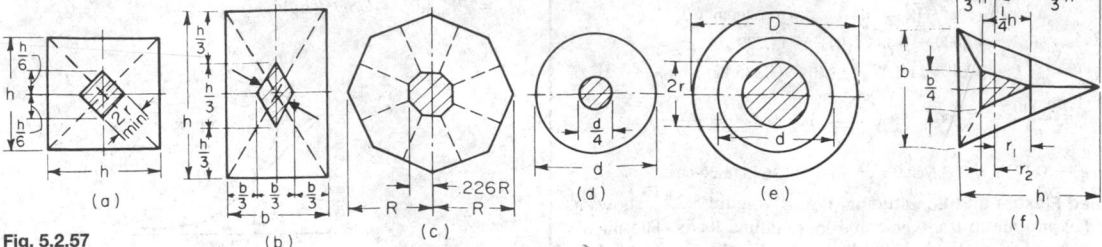

Fig. 5.2.57

Table 5.2.14 Values of the Ratio z/r (Fig. 5.2.56)

$\dfrac{e}{r}$	$\dfrac{r_1}{r}$ 0.0	0.5	0.6	0.7	0.8	0.9	1.0	$\dfrac{e}{r}$
0.25	2.00							0.25
0.30	1.82							0.30
0.35	1.66	1.89	1.98					0.35
0.40	1.51	1.75	1.84	1.93				0.40
0.45	1.37	1.61	1.71	1.81	1.90			0.45
0.50	1.23	1.46	1.56	1.66	1.78	1.89	2.00	0.50
0.55	1.10	1.29	1.39	1.50	1.62	1.74	1.87	0.55
0.60	0.97	1.12	1.21	1.32	1.45	1.58	1.71	0.60
0.65	0.84	0.94	1.02	1.13	1.25	1.40	1.54	0.65
0.70	0.72	0.75	0.82	0.93	1.05	1.20	1.35	0.70
0.75	0.59	0.60	0.64	0.72	0.85	0.99	1.15	0.75
0.80	0.47	0.47	0.48	0.52	0.61	0.77	0.94	0.80
0.85	0.35	0.35	0.35	0.36	0.42	0.55	0.72	0.85
0.90	0.24	0.24	0.24	0.24	0.24	0.32	0.49	0.90
0.95	0.12	0.12	0.12	0.12	0.12	0.12	0.25	0.95

Table 5.2.15 Values of the Ratio S_{max}/S_{avg}
(In determining S average, use load P divided by total area of cross section)

$\dfrac{r_1}{r}$	$\dfrac{r_1}{r}$ 0.0	0.5	0.6	0.7	0.8	0.9	1.0	$\dfrac{e}{r}$
0.00	1.00	1.00	1.00	1.00	1.00	1.00	1.00	0.00
0.05	1.20	1.16	1.15	1.13	1.12	1.11	1.10	0.05
0.10	1.40	1.32	1.29	1.27	1.24	1.22	1.20	0.10
0.15	1.60	1.48	1.44	1.40	1.37	1.33	1.30	0.15
0.20	1.80	1.64	1.59	1.54	1.49	1.44	1.40	0.20
0.25	2.00	1.80	1.73	1.67	1.61	1.55	1.50	0.25
0.30	2.23	1.96	1.88	1.81	1.73	1.66	1.60	0.30
0.35	2.48	2.12	2.04	1.94	1.85	1.77	1.70	0.35
0.40	2.76	2.29	2.20	2.07	1.98	1.88	1.80	0.40
0.45	3.11	2.51	2.39	2.23	2.10	1.99	1.90	0.45
0.50	3.55	2.80	2.61	2.42	2.26	2.10	2.00	0.50
0.55	4.15	3.14	2.89	2.67	2.42	2.26	2.17	0.55
0.60	4.96	3.58	3.24	2.92	2.64	2.42	2.26	0.60
0.65	6.00	4.34	3.80	3.30	2.92	2.64	2.42	0.65
0.70	7.48	5.40	4.65	3.86	3.33	2.95	2.64	0.70
0.75	9.93	7.26	5.97	4.81	3.93	3.33	2.89	0.75
0.80	13.87	10.05	8.80	6.53	4.93	3.96	3.27	0.80
0.85	21.08	15.55	13.32	10.43	7.16	4.50	3.77	0.85
0.90	38.25	30.80	25.80	19.85	14.60	7.13	4.71	0.90
0.95	96.10	72.20	62.20	50.20	34.60	19.80	6.72	0.95
1.00	∞	∞	∞	∞	∞	∞	∞	1.00

For a **hollow square** (H and h = lengths of outer and inner sides), the kern is a square similar to Fig. 5.2.57a, where

$$r_{min} = \frac{H}{6}\frac{1}{\sqrt{2}}\left[1 + \left(\frac{h}{H}\right)^2\right] = 0.1179H\left[1 + \left(\frac{h}{H}\right)^2\right]$$

For a **hollow octagon** R_a and R_i = radii of circles circumscribing the outer and inner sides; thickness of wall = $0.9239(R_a - R_i)$, the kern is an octagon similar to Fig. 5.2.57c, where $0.2256R$ becomes $0.2256R_a[1 + (R_i/R_a)^2]$.

CURVED BEAMS

The application of the flexure formula for a straight beam to the case of a curved beam results in error. When all "fibers" of a member have the same center of curvature, the **concentric** or common type of curved beam exists (see Fig. 5.2.58). Such a beam is defined by the Winkler-Bach theory. The stress at a point y units from the centroidal axis is

$$S = \frac{M}{AR}\left[1 + \frac{y}{Z(R + y)}\right]$$

M is the bending moment, positive when it increases curvature; Y is positive when measured toward the convex side; A is the cross-sectional area; R is the radius of the centroidal axis; Z **is a cross-section property** defined by

$$Z = -\frac{1}{A} \int \frac{y}{R + y} \, dA$$

Analytical expressions for Z of certain sections are given in Table 5.2.16. Also Z can be found by **graphical** integration methods (see any advanced

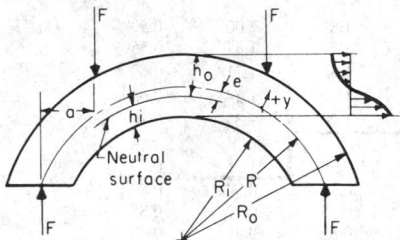

Fig. 5.2.58

strength book). The **neutral surface** shifts toward the center of curvature, or inside fiber, an amount equal to $e = ZR/(Z + 1)$. The Winkler-Bach theory, though practically satisfactory, disregards radial stresses as well as lateral deformations and assumes pure bending. The **maximum stress** occurring on the inside fiber is $S = Mh_i/(AeR_i)$, while that on the outside fiber is $S = Mh_0/(AeR_0)$.

EXAMPLE. A split steel ring of rectangular cross section is subjected to a diametral force of 1,000 lb as shown in Fig. 5.2.59a. Compute the stress at the point 0.5 in from the outside fiber on plane mm. Also compute the maximum stress.

$$Z = -1 + \frac{R}{h} \ln \frac{R + C}{R - C}$$

$$= -1 + \frac{10}{4} \ln \frac{10 + 2}{10 - 2} = 0.0133$$

$$S_{1.5} = \frac{M}{AR} \left[1 + \frac{y}{Z(R + y)} \right] + \frac{F}{A}$$

$$= \frac{-1,000 \times 10}{8 \times 10} \left[1 + \frac{1.5}{0.0133(10 + 1.5)} \right] + \frac{1,000}{8}$$

$$= -1,250 + 125 = -1,125 \text{ lb/in}^2 \text{ (compr.)(79 kgf/cm}^2)$$

$$S_M = \frac{-1,000}{8} \left[1 + \frac{-2}{0.0133(10 - 2)} \right] + \frac{1,000}{8}$$

$$= 2,230 + 125 = 2,355 \text{ lb/in}^2 \text{ (166 kgf/cm}^2)$$

or

$$e = \frac{ZR}{Z + 1} = \frac{0.0133 \times 10}{0.0133 + 1} = 0.131$$

and

$$S_M = \frac{Mh_i}{AeR_i} + \frac{F}{A} = \frac{1,000 \times 10 \times 1.87}{8 \times 0.131 \times 8} + \frac{1,000}{8}$$

$$= 2,355 \text{ lb/in}^2 \text{ (166 kgf/cm}^2)$$

The **deflection** in curved beams can be computed by means of the moment-area theory. If the origin of axes is taken at the point whose deflection is wanted, it can be shown that the component displacements in the x and y directions are

$$\Delta_x = \int_0^s \frac{My \, ds}{EI} \quad \text{and} \quad \Delta_y = \int_0^s \frac{Mx \, ds}{EI}$$

The resultant deflection is then equal to $\Delta_0 = \sqrt{\Delta_x^2 + \Delta_y^2}$ in the direction defined by $\tan \theta = \Delta_y/\Delta_x$. Deflections can also be found conveniently by use of **Castigliano's theorem**. It states that in an elastic system the displacement in the direction of a force (or couple) and due to that force (or couple) is the partial derivative of the strain energy with respect to the force (or couple). Stated mathematically, $\Delta_z = \partial U/\partial F_z$. If a force does not exist at the point and/or in the direction desired, a dummy force

may be applied. This force must then be eliminated by equating it to zero at the end.

EXAMPLE. A quadrant of radius R is fixed at one end as shown in Fig. 5.2.59b. The force F is applied in the radial direction at the free end B. Find the deflection of B.

By moment area:

$$y = R \sin \theta \qquad x = R(1 - \cos \theta)$$
$$ds = R \, d\theta \qquad M = FR \sin \theta$$

$$_B\Delta_x = \frac{FR^3}{EI} \int_0^{\pi/2} \sin^2 \theta \, d\theta = \frac{\pi FR^3}{4EI}$$

$$_B\Delta_y = \frac{FR^3}{EI} \int_0^{\pi/2} \sin \theta (1 - \cos \theta) \, d\theta = -\frac{FR^3}{2EI}$$

and

$$\Delta_B = \frac{FR^3}{2EI} \sqrt{1 + \frac{\pi^2}{4}}$$

at

$$\theta_x = \tan^{-1} \left(-\frac{FR^3}{2EI} \times \frac{4EI}{\pi FR^3} \right) = \tan^{-1} \frac{2}{\pi} = 32.5°$$

By Castigliano:

$$_B\Delta_x = \frac{\partial U}{\partial F} = \frac{\partial}{\partial F_x} \int_0^{\pi/2} \frac{F^2 R^3}{2EI} \sin^2 \theta \, d\theta = \frac{\pi FR^3}{4EI}$$

$$_B\Delta_y = \frac{\partial U}{\partial F_y} = \frac{\partial}{\partial F_y} \int_0^{\pi/2} \frac{[FR \sin \theta - F_y R (1 - \cos \theta)]^2 \, R \, d\theta}{2EI}$$

$$= -\frac{FR^3}{2EI}$$

The F_y, assumed downward, is equated to zero, after the integration and differentiation are performed to find $_B\Delta_y$. The remainder of the computation is exactly as in the moment-area method.

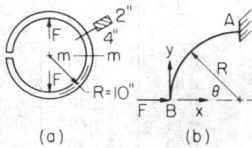

Fig. 5.2.59

Eccentrically Curved Beams These beams (Fig. 5.2.60) are bounded by arcs having different centers of curvature. In addition, it is possible for either radius to be the larger one. The one in which the section depth shortens as the central section is approached may be called the **arch beam**. When the central section is the largest, the beam is of the crescent type. **Crescent I** denotes the beam of larger outside radius and **crescent II** of larger inside radius. The stress at the **central section** of such beams may be found from $S = KMC/I$. In the case of rectangular cross section, the equation becomes $S = 6KM/(bh^2)$ where M is the bending moment, b is the width of the beam section, and h its height. The **stress factors** K for the **inner boundary**, established from photoelastic data, are given in Table 5.2.17. The outside radius is denoted by R_o and the inside by R_i. The geometry of crescent beams is such that the stress can be larger in **off-center sections**. The stress at the central section determined above must then be multiplied by the **position factor** k, given in Table 5.2.18. As in the concentric beam, the **neutral surface** shifts slightly toward the inner boundary (see Vidosic, Curved Beams with Eccentric Boundaries, *Trans. ASME*, 79, pp., 1317–1321).

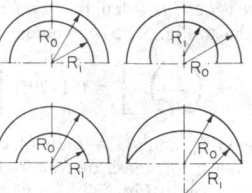

Fig. 5.2.60

Table 5.2.16   Analytical Expressions for Z

Section	Expression

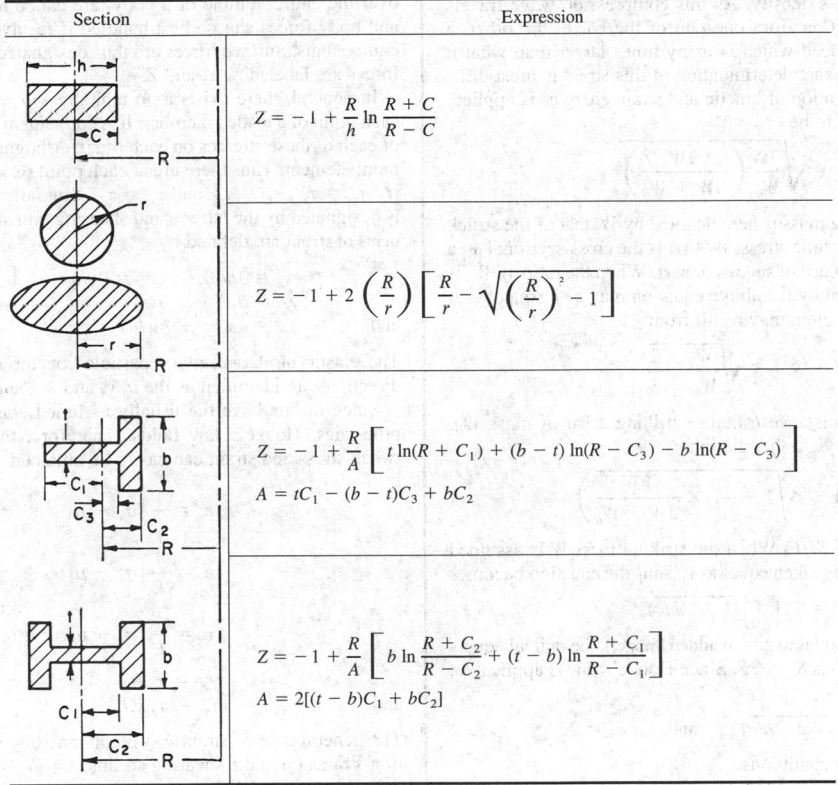

$$Z = -1 + \frac{R}{h} \ln \frac{R + C}{R - C}$$

$$Z = -1 + 2\left(\frac{R}{r}\right)\left[\frac{R}{r} - \sqrt{\left(\frac{R}{r}\right)^2 - 1}\right]$$

$$Z = -1 + \frac{R}{A}\left[t\ln(R + C_1) + (b - t)\ln(R - C_3) - b\ln(R - C_3)\right]$$
$$A = tC_1 - (b - t)C_3 + bC_2$$

$$Z = -1 + \frac{R}{A}\left[b\ln\frac{R + C_2}{R - C_2} + (t - b)\ln\frac{R + C_1}{R - C_1}\right]$$
$$A = 2[(t - b)C_1 + bC_2]$$

Table 5.2.17   Stress Factors for Inner Boundary at Central Section (See Fig. 5.2.60)

1. For the arch-type beams

(a) $K = 0.834 + 1.504\,\dfrac{h}{R_o + R_i}$ if $\dfrac{R_o + R_i}{h} < 5$.

(b) $K = 0.899 + 1.181\,\dfrac{h}{R_o + R_i}$ if $5 < \dfrac{R_o + R_i}{h} < 10$.

(c) In the case of larger section ratios use the equivalent beam solution.

2. For the crescent I-type beams

(a) $K = 0.570 + 1.536\,\dfrac{h}{R_o + R_i}$ if $\dfrac{R_o + R_i}{h} < 2$.

(b) $K = 0.959 + 0.769\,\dfrac{h}{R_o + R_i}$ if $2 < \dfrac{R_o + R_i}{h} < 20$.

(c) $K = 1.092\left(\dfrac{h}{R_o + R_i}\right)^{0.0298}$ if $\dfrac{R_o + R_i}{h} > 20$.

3. For the crescent II-type beams

(a) $K = 0.897 + 1.098\,\dfrac{h}{R_o + R_i}$ if $\dfrac{R_o + R_i}{h} < 8$.

(b) $K = 1.119\left(\dfrac{h}{R_o + R_i}\right)^{0.0378}$ if $8 < \dfrac{R_o + R_i}{h} < 20$.

(c) $K = 1.081\left(\dfrac{h}{R_o + R_i}\right)^{0.0270}$ if $\dfrac{R_o + R_i}{h} > 20$.

Table 5.2.18   Crescent-Beam Position Stress Factors (See Fig. 5.2.60)

Angle θ, deg	k Inner	k Outer
10	$1 + 0.055H/h$	$1 + 0.03H/h$
20	$1 + 0.164H/h$	$1 + 0.10H/h$
30	$1 + 0.365H/h$	$1 + 0.25H/h$
40	$1 + 0.567H/h$	$1 + 0.467H/h$
50	$1.521 - \dfrac{(0.5171 - 1.382H/h)^{1/2}}{1.382}$	$1 + 0.733H/h$
60	$1.756 - \dfrac{(0.2416 - 0.6506H/h)^{1/2}}{0.6506}$	$1 + 1.123H/h$
70	$2.070 - \dfrac{(0.4817 - 1.298H/h)^{1/2}}{0.6492}$	$1 + 1.70H/h$
80	$2.531 - \dfrac{(0.2939 - 0.7084H/h)^{1/2}}{0.3542}$	$1 + 2.383H/h$
90		$1 + 3.933H/h$

NOTE: All formulas are valid for $0 < H/h \leq 0.325$. Formulas for the inner boundary, except for 40 deg, may be used to $H/h \leq 0.36$. H = distance between centers.

IMPACT

A force or stress is considered **suddenly applied** when the duration of load application is less than one-half the **fundamental natural period** of vibration of the member upon which the force acts. Under impact, a compression wave propagates through the member at a velocity $c = \sqrt{E/\rho}$, where ρ is the mass density. As this compression wave travels back and forth by reflection from one end of the bar to the other, a maximum stress is produced which is many times larger than what it would be statically. An exact determination of this stress is most difficult. However, if conservation of kinetic and strain energies is applied, the **impact stress** is found to be

$$S' = S \sqrt{\frac{W}{W_b} \left(\frac{3W}{3W + W_b} \right)}$$

The weight of the striking mass is here denoted by W, that of the struck bar by W_b, while S is the static stress, W/A (A is the cross-sectional area of the bar). Above is the case of **sudden impact**. When the ratio W/W_b is small, the stress computed by the above equation may be erroneous. A better solution of this problem may result from

$$S' = S + S \sqrt{\frac{W}{W_b} + \frac{2}{3}}$$

If a weight W falls a distance h before striking a bar of mass W_b, energy conservation will yield the relation

$$S' = S \left(1 + \sqrt{1 + \frac{2h}{e} \times \frac{3W}{3W + W_b}} \right)$$

The elongation $e = \varepsilon l = Sl/E$. When the striking mass W is assumed rigid, the elasticity factor is taken equal to 1. Thus the equation becomes

$$S' = S (1 + \sqrt{1 + 2h/e})$$

If, in addition, h is taken equal to zero (**sudden impact**), the radical equals 1, and so the stress becomes $S' = 2S$. Since Hooke's law is applicable, the relations

$$e' = e (1 + \sqrt{1 + 2h/e}) \quad \text{and} \quad e' = 2e$$

are also true for the same conditions.

The expression may be converted, by using $v^2 = 2gh$, to

$$S' = S [1 + \sqrt{1 + v^2/(eg)}]$$

This might be called the **energy impact** form. If the **natural frequency** f_n of the bar is used, the stress equation is

$$S' = S (1 + \sqrt{1 + 0.204 h f_n^2})$$

In general, the **maximum impact stress** in a **beam** and a **shaft** can be approximated from the simplified falling-weight equation. It is necessary, though, to substitute the maximum **deflection** y for e, in the case of beams, and for the **angle of twist** θ in the case of shafts. Of course $S = Mc/I$ and $M_t c/J$, respectively. Thus

$$S' = S \left[1 + \sqrt{1 + \frac{2h}{y \text{ (or } \theta)}} \right]$$

For a more exact solution, elastic yield in each member must be considered. The theory then yields

$$S' = S \left[1 + \sqrt{1 + \frac{2h}{y} \left(\frac{35W}{35W + 17W_b} \right)} \right]$$

for a **simply supported beam** struck in the middle by a weight W.

THEORY OF ELASTICITY

Loaded members in which the stress distribution cannot be estimated fail of solution by elementary strength-of-material methods. To such cases, the more advanced mathematical principles of the **theory of elasticity** must be applied. When this is not possible, experimental stress analysis has to be used. Because of the complexity of solution, only some of the more practical problems have been solved by the theory of elasticity. The more general concepts and methods are presented.

Two kinds of forces may act on a body. **Surface forces** are distributed over the surface as the result of, for instance, the pressure of one body

on another. Forces due to gravity, inertia, magnetism, etc., which act over the entire volume of a body, are called **body forces**. Both surface and body forces can be best handled if resolved into three orthogonal components. Surface forces are thus designated $\overline{X}$, $\overline{Y}$, and $\overline{Z}$, while body forces are labeled X, Y, and Z.

In general, there exists a normal stress σ and a shearing stress τ at each point of a loaded member. It is convenient to deal with components of each of these stresses on each of six orthogonal planes that bound the point element. Thus there are at each point **six stress components**, σ_x, σ_y, σ_z, $\tau_{yx} = \tau_{xy}$, $\tau_{xz} = \tau_{zx}$, and $\tau_{yz} = \tau_{zy}$. Similarly, if the normal unit strain is designated by the letter ε and shearing unit strain by γ, the **six components of strain** are defined by

$$\varepsilon_x = \partial u/\partial x \quad \varepsilon_y = \partial v/\partial y \quad \varepsilon_z = \partial w/\partial z$$
$$\gamma_{xy} = \partial u/\partial y + \partial v/\partial x \quad \gamma_{yz} = \partial v/\partial z + \partial w/\partial y$$
and
$$\gamma_{xz} = \partial u/\partial z + \partial w/\partial x$$

The elastic **displacements** of particles on the body in the x, y, and z directions are identified as the u, v, and w **components**, respectively.

Since metals have the usually assumed elastic as well as isotropic properties, Hooke's law holds. Therefore, the interrelationships between stress and strain can easily be obtained.

$$\varepsilon_x = \frac{1}{E} [\sigma_x - \mu(\sigma_y + \sigma_z)]$$

$$\varepsilon_y = \frac{1}{E} [\sigma_y - \mu(\sigma_x + \sigma_z)]$$

$$\varepsilon_z = \frac{1}{E} [\sigma_z - \mu(\sigma_x + \sigma_y)]$$

$$\gamma_{xy} = \tau_{xy}/G \quad \gamma_{xz} = \tau_{xz}/G$$
and
$$\gamma_{yz} = \tau_{yz}/G$$

The general case of strain can be obtained by superposing the elongation strains upon the shearing strains.

Problems depending upon theories of elasticity are considerably simplified if the stresses are all parallel to one plane or if all deformations occur in planes perpendicular to the length of the member. The first case is one of **plane stress**, as when a thin plate of uniform thickness is subjected to central, boundary forces parallel to the plane of the plate. The second is a case of **plane strain**, such as a gate subjected to hydrostatic pressure, the intensity of which does not vary along the gate's length. All particles therefore displace at right angles to the length, and so cross sections remain plane. In plane-stress problems, three of the six stress components vanish, thus leaving only σ_x, σ_y, and τ_{xy}. Similarly, in plane strain, only ε_x, ε_y, and γ_{xy} will not equal zero; thus the same three stresses σ_x, σ_y, and τ_{xy} remain to be considered. Plane problems can thus be represented by the element shown in Fig. 5.2.61. Equilibrium considerations applied to this particle result in the **differential equations of equilibrium** which reduce to

$$\frac{\partial \sigma_x}{\partial x} + \frac{\partial \tau_{xy}}{\partial y} + X = 0$$
and
$$\frac{\partial \sigma_y}{\partial y} + \frac{\partial \tau_{xy}}{\partial x} + Y = 0$$

Since the two differential equations of equilibrium are insufficient to find the three stresses, a third equation must be used. This is the **compatibility equation** relating the three strain components. It is

$$\frac{\partial^2 \varepsilon_x}{\partial y^2} + \frac{\partial^2 \varepsilon_y}{\partial x^2} = \frac{\partial^2 \gamma_{xy}}{\partial x \, \partial y}$$

If strains are expressed in terms of the stresses, the compatibility equation becomes

$$\left(\frac{\partial^2}{\partial x^2} + \frac{\partial^2}{\partial y^2} \right) (\sigma_x + \sigma_y) = 0$$

Now, in any **two-dimensional** problem, the compatibility equation along with the differential equilibrium equations must be simultaneously solved for the three unknown stresses. This is accomplished using **stress functions,** which permit the integration and satisfy boundary conditions in each particular situation.

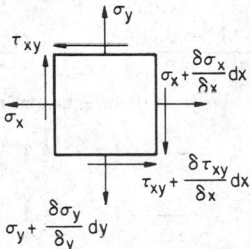

Fig. 5.2.61

In **three-dimensional** problems, the third dimension must be considered. This results in three differential equations of equilibrium, as well as three compatibility equations. The six stress components can thus be found. The complexity involved in the solution of these equations is such, however, that only a few special cases have been solved.

In certain problems, such as rotating circular disks, **polar coordinates** become more convenient. In such cases, the stress components in a two-dimensional field are the **radial stress** σ_r, the **tangential stress** σ_θ, and the **shearing stress** $\tau_{r\theta}$. In terms of these stresses the **polar differential equations** become

$$\frac{\partial \sigma_r}{\partial r} + \frac{1}{r}\frac{\partial \tau_{r\theta}}{\partial \theta} + \frac{\sigma_r - \sigma_\theta}{r} + R = 0$$

and

$$\frac{1}{r}\frac{\partial \sigma_\theta}{\partial \theta} + \frac{\partial \tau_{r\theta}}{\partial r} + \frac{2\tau_{r\theta}}{r} = 0$$

The body force per unit volume is represented by R. The **compatibility equation** in polar coordinates is

$$\left(\frac{\partial^2}{\partial r^2} + \frac{1}{r}\frac{\partial}{\partial r} + \frac{1}{r^2}\frac{\partial^2}{\partial \theta^2}\right)\left(\frac{\partial^2 \phi}{\partial r^2} + \frac{1}{r}\frac{\partial \phi}{\partial r} + \frac{1}{r^2}\frac{\partial^2 \phi}{\partial \theta^2}\right) = 0$$

ϕ is again a stress function of r and θ that will provide a solution of the differential equations and satisfy boundary conditions. As an example, the exact solution of a **simply supported beam** carrying a uniformly distributed load w yields

$$\sigma_x = \frac{w}{2I}(l^2 - x^2)\,y + \frac{w}{2I}\left(\frac{2y^3}{3} - \frac{2c^2 y}{5}\right)$$

The origin of coordinates is at the center of the beam, $2c$ is the beam depth, and $2l$ is the span length. Thus the maximum stress at $x = 0$ and $y = c$ is $\sigma_x = \dfrac{wl^2 c}{2I} + \dfrac{2}{15}\dfrac{wc^3}{I}$. The first term represents the stress as obtained by the elementary flexure theory; the second is a correction. The second term becomes negligible when c is small compared to l.

The important case of a **flat plate** of unit width with a **circular hole** of diameter $2a$ at its center, subjected to a uniform tensile load, has been solved using polar coordinates. If S is the uniform stress at some distance from the hole, r is measured from the center of the hole, and θ is the angle of r with respect to the longitudinal axis of the member, the stresses are

$$\sigma_r = \frac{S}{2}\left(1 - \frac{a^2}{r^2}\right) + \frac{S}{2}\left(1 + \frac{3a^4}{r^4} - \frac{4a^2}{r^2}\right)\cos 2\theta$$

$$\sigma_\theta = \frac{S}{2}\left(1 + \frac{a^2}{r^2}\right) - \frac{S}{2}\left(1 + \frac{3a^4}{r^4}\right)\cos 2\theta$$

$$\tau_r\theta = -\frac{S}{2}\left(1 - \frac{3a^4}{r^4} + \frac{2a^2}{r^2}\right)\sin 2\theta$$

CYLINDERS AND SPHERES

A thin-wall cylinder has a wall thickness such that the assumption of constant stress across the wall results in negligible error. Cylinders having internal-diameter-to-thickness (D/t) ratios greater than 10 are usually considered thin-walled. Boilers, drums, tanks, and pipes are often treated as such. Equilibrium equations reveal the circumferential, or hoop, stress to be $S = pr/t$ under an internal pressure p (see Fig. 5.2.62). If the cylinder is closed at the ends, a longitudinal stress of $pr/(2t)$ is developed. The tensile stress developed in a thin hollow sphere subjected to internal pressure is also $pr/(2t)$.

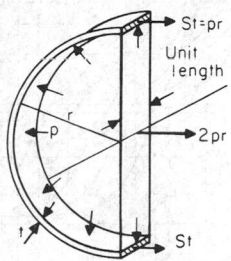

Fig. 5.2.62

When thin-walled cylinders, such as vacuum tanks and submarines, are subjected to **external pressure, collapse** becomes the mode of failure. The shell is assumed perfectly round and of uniform thickness, the material obeys Hooke's law, the radial stress is negligible, and the normal stress distribution is linear. Other, lesser assumptions are also made. Using the theory of elasticity, R. G. Sturm (*Univ. Ill. Eng. Exp. Stn. Bull.,* no 12, Nov. 11, 1941) derived the **collapsing pressure** as

$$W_c = KE\left(\frac{t}{D}\right)^3 \quad \text{lb/in}^2$$

The factor K, a numerical coefficient, depends upon the L/R and D/t ratios (D is outside-shell diameter), the kind of end support, and whether pressure is applied radially only, or at the ends as well. Figures 5.2.63 to 5.2.66, reproduced from the bulletin, supply the K values. N on these charts indicates the number of lobes into which the shell collapses. These values are for materials having Poisson's ratio $\mu = 0.3$. It may also be pointed out that in the case of long cylinders (infinitely long, theoretically) the value of K approaches $2/(1 - \mu^2)$.

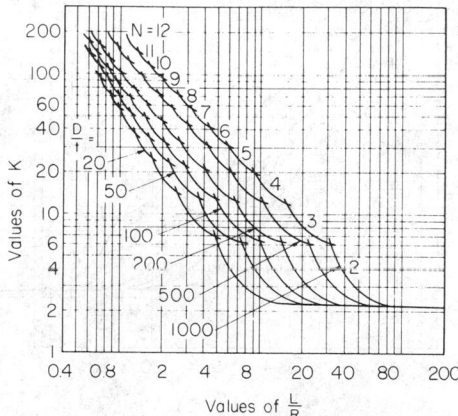

Fig. 5.2.63 Radial external pressure with simply supported edges.

When the **cylinder** is **stiffened with rings,** the shell may be assumed to be divided into a series of shorter shells, equal in length to the ring spacing. The previous equation can then be applied to a ring-to-ring length of cylinder. However, the flexural rigidity of the combined stiffener and shell EI_c necessary to withstand the pressure is $EI_c = W_s D^3 L_s / 24$. W_s is the pressure, L_s the length between rings, and I_c the combined moment of inertia of the ring and that portion of the shell assumed acting with the ring.

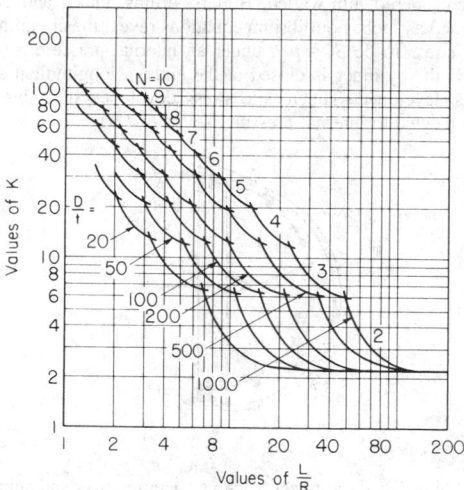

Fig. 5.2.64 Radial external pressure with fixed edges.

In some instances, cylinders collapse only after a stress in excess of the elastic limit has been reached; that is, **plastic range** stresses are present. In such cases the same equation applies, but the modulus of elasticity must be modified.

When the average stress S_a is less than the proportional limit S_p, and the maximum stress (direct, plus bending) is S, the modified modulus

$$E' = E \left[1 - \frac{1}{4} \left(\frac{S - S_l}{S_u - S_a} \right)^2 \right]$$

S_u is the modulus of rupture. When the average stress is larger than the proportional limit, the modified modulus is taken as the tangent at the average stress.

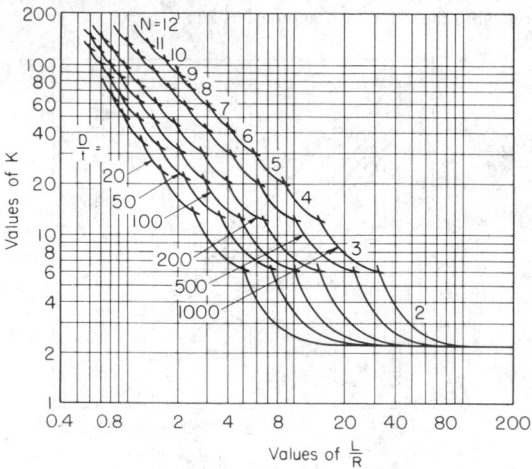

Fig. 5.2.65 Radial and end external pressure with simply supported edges.

In **thick-walled cylinders** (Fig. 5.2.67a) the circumferential, hoop, or tangential stress S_t is not uniform. In addition a radial stress S_r is present. When equilibrium is applied to the annulus taken out of Fig. 5.2.67a and

shown in Fig. 5.2.67b and the equation is integrated, the general tangential and radial stress relations, called the **Lamé** equations, are derived.

$$S_t = \frac{r_1^2 p_1 - r_2^2 p_2 + (p_1 - p_2) r_1^2 r_2^2 / r^2}{r_2^2 - r_1^2}$$

and

$$S_r = \frac{r_1^2 p_1 - r_2^2 p_2 - (p_1 - p_2) r_1^2 r_2^2 / r^2}{r_2^2 - r_1^2}$$

When the external pressure $p_2 = 0$, the equations reduce to

$$S_t = \frac{r_1^2 p_1}{r_2^2 - r_1^2} \left(1 + \frac{r_2^2}{r^2} \right)$$

and

$$S_r = \frac{r_1^2 p_1}{r_2^2 - r_1^2} \left(1 - \frac{r_2^2}{r^2} \right)$$

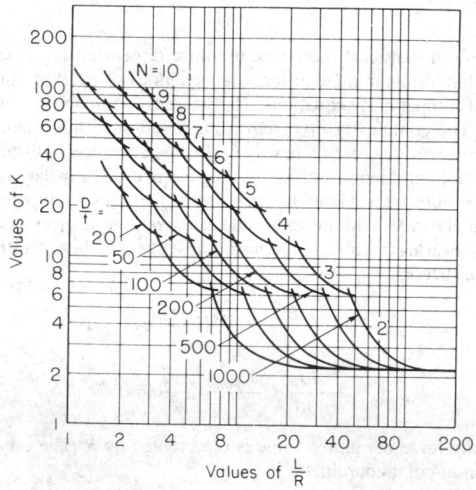

Fig. 5.2.66 Radial and end external pressure with fixed edges.

At the inner boundary the tangential elongation ε_t is equal to

$$\varepsilon_t = (S_t - \mu S_r)/E$$

The increase in the bore radius Δr_1 resulting therefrom is

$$\Delta r_1 = \frac{r_1 p_1}{E_h} \left(\frac{1 + r_1^2/r_2^2}{1 - r_1^2/r_2^2} + \mu \right)$$

Similarly a solid shaft of r radius under external pressure p_2 will have its radius decreased by the amount

$$\Delta r = -\frac{r p_2}{E_s} (1 - \mu)$$

In the case of a **press or shrink fit**, $p_1 = p_2 = p$. The sum of Δr_1 and Δr_1 absolute is the radial interference; twice this sum is the **diametral interference** Δ or

$$\Delta = 2 r_1 p \left[\frac{1}{E_h} \left(\frac{1 + r_1^2/r_2^2}{1 - r_1^2/r_2^2} + \mu \right) + \frac{1 - \mu}{E_s} \right]$$

If the hub and shaft materials are the same, $E_h = E_s = E$, and

$$\Delta = \frac{4 r_1 r_2^2 p}{E (r_2^2 - r_1^2)}$$

If the equation is solved for p and this value is substituted in Lamé's equation, the maximum tangential stress on the inner surface of the hub is found to be

$$S_t = \frac{E \Delta}{4 r_1 r_2^2} (r_2^2 + r_1^2)$$

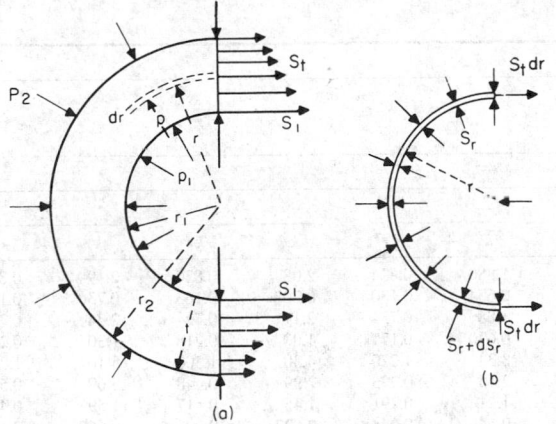

Fig. 5.2.67

EXAMPLE. The barrel of a field gun has an outside diameter of 9 in and a bore of 4.7 in. An internal pressure of 16,000 lb/in² is developed during firing. What maximum stress occurs in the barrel? An investigation of Lamé's equations for internal pressure reveals the maximum stress to be the tangential one on the inner surface. Thus,

$$S_t = p_1 \frac{r_1^2 + r_2^2}{r_2^2 - r_1^2} = \frac{16{,}000\,(2.35^2 + 4.5^2)}{4.5^2 - 2.35^2}$$

$$= 28{,}000 \text{ lb/in}^2 \ (1{,}972 \text{ kgf/cm}^2)$$

Oval Hollow Cylinders In Fig. 5.2.68, let a and b be the semiminor and semimajor axes. The bending moments at A and C will then be

$$M_0 = pa^2/2 - pI_x/(2S) - pI_y/(2S)$$
$$M_1 = M_0 - p(a^2 - b^2)/2$$

where I_x and I_y are the moments of inertia of the arc AC about the x and y axes, respectively. The **bending moment at any point** will be

$$M = M_0 - pa^2/2 + px^2/2 + py^2/2$$

Thick Hollow Spheres With an **internal pressure** p, where $p <$ $T/0.65$.

$$r_2 = r_1[(T + 0.4p)/(T - 0.65p)]^{1/3}$$

The maximum tensile stress is on the inner surface, in the direction of the circumference. With an **external pressure** p, where $p < T/1.05$,

$$r_2 = r_1[T/(T - 1.05p)]^{1/3}$$

In both cases T is the true stress.

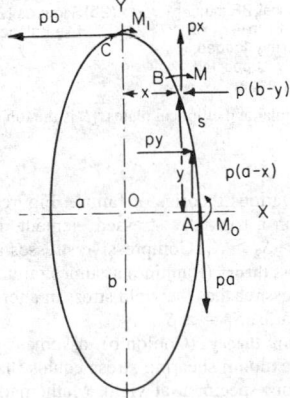

Fig. 5.2.68

PRESSURE BETWEEN BODIES WITH CURVED SURFACES

(See Hertz, "Gesammelte Werke," vol. 1, pp. 159 *et seq.*, Barth.)

Two Spheres The radius A of the compressed area is obtained from the formula $A^3 = 0.68P(c_1 + c_2)/(1/r_1 + 1/r_2)$, in which P is the compressing force, c_1 and c_2 (= $1/E_1$ and $1/E_2$) are reciprocals of the respective moduli of elasticity, and r_1 and r_2 are the radii. (Reciprocal of Poisson's ratio is assumed to be $n = 10/3$.) The **greatest contact pressure** in the middle of the compressed surface will be $S_{max} = 1.5(P/\pi A^2)$, and

$$S_{max}^3 = 0.235P(1/r_1 + 1/r_2)^2/(c_1 + c_2)^2$$

The **total deformation** of the two spheres will be Y, which is obtained from

$$Y^3 = 0.46P^2(c_1 + c_2)^2(1/r_1 + 1/r_2)$$

For $c_1 = c_2 = 1/E$, i.e., **two spheres with the same modulus of elasticity,** it follows that $A^3 = 1.36\ P/E(1/r_1 + 1/r_2)$, $S_{max}^3 = 0.059PE^2(1/r_1 + 1/r_2)^2$, and $Y^3 = 1.84P^2(1/r_1 + 1/r_2)/E^2$. If the radii of these spheres are also equal, $A^3 = 0.68Pr/E = 0.34Pd/E$; $S_{max}^3 = 0.235PE^2/r^2 = 0.94PE^2/d^2$; and $Y^3 = 3.68P^2/(E^2r) = 7.36P^2/(E^2d)$.

Sphere and Flat Plate In this case $r_1 = r$ and $r_2 = \infty$, and the above formulas become $A^3 = 0.68Pr(c_1 + c_2) = 1.36Pr/E$, and

$$S_{max}^3 = 0.235P/[r^2(c_1 + c_2)^2] = 0.059PE^2/r^2$$
$$Y^3 = 0.46P^2(c_1 + c_2)^2/r = 1.84P^2/(E^2r)$$

Two Cylinders The width b of the rectangular pressure surface is obtained from $(b/4)^2 = 0.29P(c_1 + c_2)/l[(1/r_1) + (1/r_2)]$, where r_1 and r_2 are the radii, and l the length

$$S_{max}^2 = [4P/(\pi bl)]^2 = 0.35P(1/r_1 + 1/r_2)/l(c_1 + c_2)]$$

For cylinders with the same moduli of elasticity, $c_1 = c_2 = 1/E$, and $(b/4)^2 = 0.58P/El[(1/r_1) + (1/r_2)]$; and $S_{max}^2 = 0.175PE(1/r_1 + 1/r_2)/l$. When $r_1 = r_2 = r$, $(b/4)^2 = 0.29Pr/(El)$, and $S_{max}^2 = 0.35PE/(lr)$.

Cylinder and Flat Plate Here $r_1 = r$, $r_2 = \infty$, and the above formulas reduce to $(b/4)^2 = 0.29Pr(c_1 + c_2)/l = 0.58Pr/(El)$, and

$$S_{max}^2 = 0.35P/[lr(c_1 + c_2)] = 0.175PE/(lr)$$

For application to ball and roller bearings and to gear teeth, see Sec. 8.

FLAT PLATES

The analysis of flat plates subjected to **lateral loads** is very involved because plates bend in all vertical planes. Strict mathematical derivations have therefore been accomplished only in some special cases. Most of the available formulas contain some amount of rational empiricism. Plates may be classified as (1) **thick plates,** in which transverse shear is important; (2) **average-thickness plates,** in which flexure stress predominates; (3) **thin plates,** which depend in part upon direct tension; and (4) **membranes,** which are subject to direct tension only. However, exact lines of demarcation do not exist.

The flat-plate formulas given apply primarily to symmetrically loaded **average-thickness plates** of constant thickness. They are valid only if the maximum deflection is small relative to the plate thickness; usually, $y \leq 0.4t$. In the mathematical analyses, allowance for stress redistribution, because of slight local yielding, is usually not made. Since this yielding, especially in ductile materials, is beneficial, the formulas generally err on the side of safety. Certain cases of symmetrically loaded **circular and rectangular plates** are presented in Figs. 5.2.69 and 5.2.70. The maximum stresses are calculated from

$$S_M = k\frac{wR^2}{t^2} \qquad S_M = k\frac{P}{t^2} \qquad \text{or} \qquad S_M = k\frac{C}{t^2}$$

The first equation is for a uniformly distributed load w, lb/in²; the second supports a concentrated load P, lb; and the third a couple C, per unit length, uniformly distributed along the edge. Combinations of these loadings may be treated by superposition. The factors k and k_1 are given

Table 5.2.19 Coefficients k and k_1 for Circular Plates
($\mu = 0.3$)

Case	k	k_1
1	1.24	0.696
2	0.75	0.171
3	6.0	4.2

	R/r											
	1.25		1.5		2		3		4		5	
Case	k	k_1	k	k_1	k	k_1	k	k_1	k	k_1	k	k_1
4	0.592	0.184	0.976	0.414	1.440	0.664	1.880	0.824	2.08	0.830	2.19	0.813
5	0.105	0.0025	0.259	0.0129	0.481	0.057	0.654	0.130	0.708	0.163	0.730	0.176
6	1.10	0.341	1.26	0.519	1.48	0.672	1.88	0.734	2.17	0.724	2.34	0.704
7	0.195	0.0036	0.320	0.024	0.455	0.081	0.670	0.171	1.00	0.218	1.30	0.238
8	0.660	0.202	1.19	0.491	2.04	0.902	3.34	1.220	4.30	1.300	5.10	1.310
9	0.135	0.0023	0.410	0.0183	1.04	0.0938	2.15	0.293	2.99	0.448	3.69	0.564
10	0.122	0.00343	0.336	0.0313	0.740	0.1250	1.21	0.291	1.45	0.417	1.59	0.492
11	0.072	0.00068	0.1825	0.005	0.361	0.023	0.546	0.064	0.627	0.092	0.668	0.112
12	6.865	0.2323	7.448	0.6613	8.136	1.493	8.71	2.555	8.930	3.105	9.036	3.418
13	6.0	0.196	6.0	0.485	6.0	0.847	6.0	0.940	6.0	0.801	6.0	0.658
14	0.115	0.00129	0.220	0.0064	0.405	0.0237	0.703	0.062	0.933	0.092	1.13	0.114
15	0.090	0.00077	0.273	0.0062	0.710	0.0329	1.54	0.110	2.23	0.179	2.80	0.234

in Tables 5.2.19 and 5.2.20; R is the radius of circular plates or one side of rectangular plates, and t is the plate thickness.
[In Figs. 5.2.69 and 5.2.70, $r = R$ for circular plates and $r =$ smaller side rectangular plates.]

The **maximum deflection** for the same cases is given by

$$y_M = k_1 \frac{wR^4}{Et^3} \qquad y_M = k_1 \frac{PR^2}{Et^3} \quad \text{and} \quad y_M = k_1 \frac{CR^2}{Et^3}$$

The factors k_1 are also given in the tables. For additional information, including shells, refer to ASME Handbook, ''Metals Engineering: Design,'' McGraw-Hill.

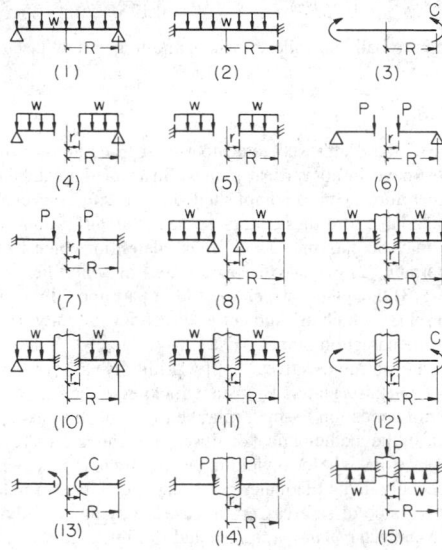

Fig. 5.2.69 Circular plates. Cases (4), (5), (6), (7), (8), and (13) have central hole of radius r; cases (9), (10), (11), (12), (14), and (15) have a central piston of radius r to which the plate is fixed.

THEORIES OF FAILURE

Material properties are usually determined from tests in which specimens are subjected to **simple stresses** under static or fluctuating loads. The attempt to apply these data to **bi- or triaxial stress fields** has resulted

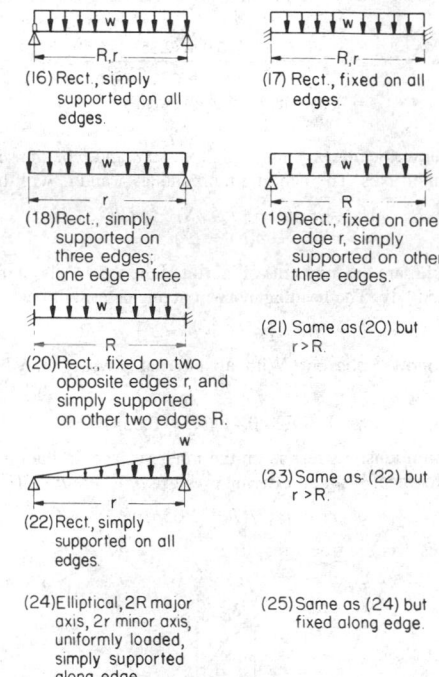

Fig. 5.2.70 Rectangular and elliptical plates. [R is the longer dimension except in cases (21) and (23).]

in the proposal of various theories of failure. Figure 5.2.71 shows the principal stresses on a triaxially stressed element. It is assumed, for simplicity, that $S_1 > S_2 > S_3$. Compressive stresses are negative.

1. **Maximum-stress theory** (Rankine) assumes failure occurs when the largest principal stress reaches the yield stress in a tension (or compression) specimen. That is, $S_1 = \pm S_y$.

2. **Maximum-shear theory** (Coulomb) assumes yielding (failure) occurs when the maximum shearing stress equals that in a simple tension (or compression) specimen at yield. Mathematically, $S_1 - S_3 = \pm S_y$.

3. **Maximum-strain-energy theory** (Beltrami) assumes failure occurs when the energy absorbed per unit volume equals the strain energy per

Table 5.2.20   Coefficients k and k_1 for Rectangular and Elliptical Plates
($\mu = 0.3$)

Case	R/r									
	1.0		1.5		2.0		3.0		4.0	
	k	k_1	k	k_1	k	k_1	k	k_1	k	k_1
16	0.287	0.0443	0.487	0.0843	0.610	0.1106	0.713	0.1336	0.741	0.1400
17	0.308	0.0138	0.454	0.0240	0.497	0.0277	0.500	0.028	0.500	0.028
18	0.672	0.140	0.768	0.160	0.792	0.165	0.798	0.166	0.800	0.166
19	0.500	0.030	0.670	0.070	0.730	0.101	0.750	0.132	0.750	0.139
20	0.418	0.0209	0.626	0.0582	0.715	0.0987	0.750	0.1276	0.750	
21*	0.418	0.0216	0.490	0.0270	0.497	0.0284	0.500	0.0284	0.500	0.0284
22	0.160	0.0221	0.260	0.0421	0.320	0.0553	0.370	0.0668	0.380	0.0700
23*	0.160	0.0220	0.260	0.0436	0.340	0.0592	0.430	0.0772	0.490	0.0908
24	1.24	0.70	1.92	1.26	2.26	1.58	2.60	1.88	2.78	2.02
25	0.75	0.171	1.34	0.304	1.63	0.379	1.84	0.419	1.90	0.431

* Length ratio is r/R in cases 21 and 23.

unit volume in a tension (or compression) specimen at yield. Mathematically, $S_1^2 + S_2^2 + S_3^2 - 2\mu(S_1S_2 + S_2S_3 + S_3S_1) = S_y^2$.

4. **Maximum-distortion-energy theory** (Huber, von Mises, Hencky) assumes yielding occurs when the distortion energy equals that in simple tension at yield. The distortion energy, that portion of the total energy which causes distortion rather than volume change, is

$$U_d = \frac{1 + \mu}{3E}(S_1^2 + S_2^2 + S_3^2 - S_1S_2 - S_2S_3 - S_3S_1)$$

Thus failure is defined by

$$S_1^2 + S_2^2 + S_3^2 - (S_1S_2 + S_2S_3 + S_3S_1) = S_y^2$$

5. **Maximum-strain theory** (Saint-Venant) claims failure occurs when the maximum strain equals the strain in simple tension at yield or $S_1 - \mu(S_2 + S_3) = S_y$.

6. **Internal-friction theory** (Mohr). When the ultimate strengths in tension and compression are the same, this theory reduces to that of maximum shear. For principal stresses of opposite sign, failure is defined by $S_1 - (S_{uc}/S_u) S_2 = -S_{uc}$; if the signs are the same $S_1 = S_u$ or $-S_{uc}$, where S_{uc} is the ultimate strength in compression. If the principal stresses are both either tension or compression, then the larger one, say S_1, must equal S_u when S_1 is tension or S_{uc} when S_1 is compression.

A **graphical representation** of the first four theories applied to a biaxial stress field is presented in Fig. 5.2.72. Stresses outside the bounding lines in the case of each theory mean failure (yield or fracture). A comparison with experimental data proves the distortion-energy theory (4) best for ductile materials of equal tension-compression properties. When these properties are unequal, the internal friction theory (6) appears best. In practice, judging by some accepted codes, the maximum-

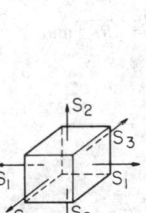

Fig. 5.2.71

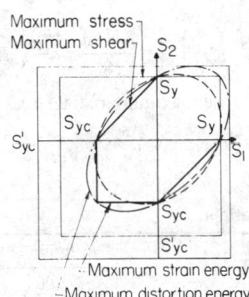

Maximum stress
Maximum shear
Maximum strain energy
Maximum distortion energy

Fig. 5.2.72

shear theory (2) is generally used for ductile materials, and the maximum-stress theory (1) for brittle materials.

Fatigue failures cannot be related, theoretically, to elastic strength and thus to the theories described. However, experimental results justify this, at least to a limited extent. Therefore, the theory evaluation given

above holds for **fluctuating stresses**, provided that principal stresses at the maximum load are used and the **endurance strength** in simple bending is substituted for the yield strength.

EXAMPLE.   A steel shaft, 4 in in diameter, is subjected to a bending moment of 120,000 in · lb, as well as a torque. If the yield strength in tension is 40,000 lb/in², what maximum torque can be applied under the (1) maximum-shear theory and (2) the distortion-energy theory?

$$S_x = \frac{M_c}{I} = \frac{120,000 \times 2}{12.55}$$

$$= 19,100 \text{ lb/in}^2 \qquad S_{xy} = \frac{TC}{J} = \frac{T \times 2}{25.1} = 0.0798T$$

and

$$S_{M,m} = \frac{S_x}{2} \pm \sqrt{\left(\frac{S_x}{2}\right)^2 + S_{xy}^2}$$

$$S_M - S_m = S_y \quad \text{or} \quad 2\sqrt{\left(\frac{19,100}{2}\right)^2 + (0.0798T)^2} \qquad (1)$$

$$= (40,000)^2$$

or

$$T = 221,000 \text{ in} \cdot \text{lb} \ (254,150 \text{ cm} \cdot \text{kgf})$$

$$S_M^2 + S_m^2 - S_M S_m = S_y^2 \qquad (2)$$

substituting and simplifying,

$$(9,550)^2 + 3\sqrt{\left(\frac{19,100}{2}\right)^2 + (0.0798T)^2} = (40,000)^2$$

or

$$T = 255,000 \text{ in} \cdot \text{lb} \ (293,250 \text{ cm} \cdot \text{kgf})$$

PLASTICITY

The reaction of materials to stress and strain in the plastic range is not fully defined. However, some concepts and theories have been proposed.

Ideally, a **purely elastic material** is one complying explicitly with Hooke's law. In a **viscous material,** the shearing stress is proportional to the shearing strain. The **purely plastic material** yields indefinitely, but only after reaching a certain stress. Combinations of these are the **elasto-viscous** and the **elastoplastic** materials.

Engineering materials are not ideal, but usually contain some of the elastoplastic characteristics. The **total strain** ε_t is the sum of the **elastic strain** ε_o plus the **plastic strain** ε_p, as shown in Fig. 5.2.73, where the stress-strain curve is approximated by two straight lines. The **natural strain**, which is at the same time the total strain, is $\varepsilon = \int_{l_o}^{l} dl/l = \ln (l/l_o)$. In this equation, l is the instantaneous length, while l_o is the original length. In terms of the normal strain, the natural strain becomes $\bar{\varepsilon} = \ln(1 + \varepsilon_o)$. Since it is assumed that the volume remains constant, $l/l_o = A_o/A$, and so the natural stress becomes $\bar{S} = P/A = (P/A_o)(1 + \varepsilon_o)$. A_o is the original cross-sectional area. If the natural stress is plotted against strain on log-log paper, the graph is very nearly a straight line. The plastic-range relation is thus approximated by $\bar{S} = K\bar{\varepsilon}^n$, where the proportionality factor K and the **strain-hardening coefficient** n are determined from best fits to experimental data. Values of K and n determined by Low and Garofalo (*Proc. Soc. Exp. Stress Anal.,* vol. IV, no. 2, 1947) are given in Table 5.2.21.

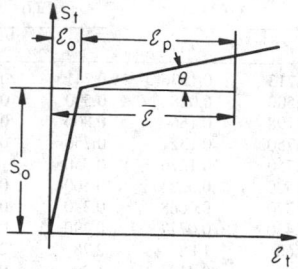

Fig. 5.2.73

The geometry of Fig. 5.2.73 can be used to arrive at a second approximate relation

$$\overline{S} = S_o + (\varepsilon_p - \varepsilon_o)\tan\theta = S_o\left(1 - \frac{H}{E}\right) + \varepsilon_p H$$

where $H = \tan\theta$ is a kind of **plastic modulus**.

The **deformation theory of plastic flow** for the general case of combined stress is developed using the above concepts. Certain additional assumptions involved include: principal plastic-strain directions are the same as principal stress directions; the elastic strain is negligible compared to plastic strain; and the ratios of the three principal shearing strains—$(\overline{\varepsilon}_1 - \overline{\varepsilon}_2)$, $(\overline{\varepsilon}_2 - \overline{\varepsilon}_3)$, $(\overline{\varepsilon}_3 - \overline{\varepsilon}_1)$—to the principal shearing stresses—$(\overline{S}_1 - \overline{S}_2)/2$, $(\overline{S}_2 - \overline{S}_3)/2$, $(\overline{S}_3 - \overline{S}_1)/2$—are equal. The relations between the principal strains and stresses in terms of the simple tension quantities become

$$\overline{\varepsilon}_1 = \overline{\varepsilon}/\overline{S}[\overline{S}_1 - (\overline{S}_2 + \overline{S}_3)/2]$$
$$\overline{\varepsilon}_2 = \overline{\varepsilon}/\overline{S}[\overline{S}_2 - (\overline{S}_3 + \overline{S}_1)/2]$$
$$\overline{\varepsilon}_3 = \overline{\varepsilon}/\overline{S}[\overline{S}_3 - (\overline{S}_1 + \overline{S}_2)/2]$$

If these equations are added, the plastic-flow theory is expressed:

$$\frac{\overline{S}}{\overline{\varepsilon}} = \sqrt{\frac{[(\overline{S}_1 - \overline{S}_2)^2 + (\overline{S}_2 - \overline{S}_3)^2 + (\overline{S}_3 - \overline{S}_1)^2]/2}{2(\overline{\varepsilon}_1^2 + \overline{\varepsilon}_2^2 + \overline{\varepsilon}_3^2)/3}}$$

In the above equation

$$\sqrt{[(\overline{S}_1 - \overline{S}_2)^2 + (\overline{S}_2 - \overline{S}_3)^2 + (\overline{S}_3 - \overline{S}_1)^2]/2} = \overline{S}_e$$

and

$$\sqrt{2(\overline{\varepsilon}_1^2 + \overline{\varepsilon}_2^2 + \overline{\varepsilon}_3^2)/3} = \varepsilon_e$$

are the effective, or significant, stress and strain, respectively.

EXAMPLE. An annealed, stainless-steel type 430 tank has a 41-in inside diameter and has a wall 0.375 in thick. The ultimate strength of the stainless steel is 85,000 lb/in². Compute the maximum strain as well as the pressure at fracture.

The tank constitutes a biaxial stress field where $S_1 = pd/(2t)$, $S_2 = pd/(4t)$, and $S_3 = 0$. Taking the power stress-strain relation

$$\overline{S}_e = K\overline{\varepsilon}_e^n \qquad \text{or} \qquad \overline{\varepsilon}/\overline{S} = \overline{S}_e^{(1-n)/n}/K^{1/n}$$

thus

$$\overline{\varepsilon}_1 = \frac{\overline{S}_e^{(1-n)/n}}{K^{1/n}}\left(\frac{3}{4}\overline{S}_1\right) \qquad \overline{\varepsilon} = 0,$$

and

$$\overline{\varepsilon}_3 = \frac{\overline{S}_e^{(1-n)/n}}{K^{1/n}}\left(-\frac{3}{4}\overline{S}_1\right) = -\overline{\varepsilon}_1$$

The maximum-shear theory, which is applicable to a ductile material under combined stress, is acceptable here. Thus rupture will occur at $\overline{S}_1 - \overline{S}_3 = \overline{S}_u$, and

$$\overline{S}_e = \sqrt{\frac{1}{2}\left[\left(\frac{\overline{S}_1}{2}\right)^2 + \left(\frac{\overline{S}_1}{2}\right)^2 + \overline{S}_1^2\right]} = \sqrt{\frac{3}{4}\overline{S}_1^2} = \left(\frac{3}{4}\right)^{1/2}\overline{S}_u$$

$$\overline{\varepsilon}_1 = \frac{[(3/4)^{1/2}\overline{S}_u]^{(1-n)/n}}{K^{1/n}}\left(\frac{3}{4}\overline{S}_u\right) = \left(\frac{3}{4}\right)^{1+0.229/0.458}\left(\frac{85,000}{143,000}\right)^{1/0.229}$$

$$= 0.0475 \text{ in/in } (0.0475 \text{ cm/cm})$$

Since

$$\overline{S}_u = S_1 = \frac{pd}{2t}, \text{ then } p = \frac{2t\overline{S}_u}{d}$$

or

$$p = \frac{2 \times 0.375 \times 85,000}{41} = 1,550 \text{ lb/in}^2 \text{ (109 kgf/cm}^2\text{)}$$

ROTATING DISKS

Rotating circular disks may be of various profiles, of constant or variable thickness, with or without centrally and noncentrally located holes, and with radial, tangential, and shearing stresses.

Solution starts with the differential equations of equilibrium and compatibility and the subsequent application of appropriate boundary conditions for the derivation of working-stress equations.

If the disk thickness is small compared with the diameter, the variation of stress with thickness can be assumed to be negligible, and symmetry eliminates the shearing stress. In the rotating case, the disk weight is neglected, but its inertia force becomes the body-force term in the equilibrium equations.

Thus solved, the stress components in a solid disk become

$$\sigma_r = \frac{3+\mu}{8}\rho\omega^2(R^2 - r^2)$$

$$\sigma_\theta = \frac{3+\mu}{8}\rho\omega^2 R^2 - \frac{1+3\mu}{8}\rho\omega^2 r^2$$

where μ = Poisson's ratio; ρ = mass density, lb · s²/in⁴; ω = angular speed, rad/s; R = outside disk radius; and r = radius to point in question.

The largest stresses occur at the center of the solid disk and are

$$\sigma_r = \sigma_\theta = \frac{3+\mu}{8}\rho\omega^2 R^2$$

A disk with a central hole of radius r_h (no external forces) is subjected to the following stresses:

$$\sigma_r = \frac{3+\mu}{8}\rho\omega^2\left(R^2 + r_h^2 - \frac{R^2 r_h^2}{r^2} - r^2\right)$$

$$\sigma_\theta = \frac{3+\mu}{8}\rho\omega^2\left(R^2 + r_h^2 + \frac{R^2 r_h^2}{r^2} - \frac{1+3\mu}{3+\mu}r^2\right)$$

The maximum radial stress $\sigma_r|_M$ occurs at $r = \sqrt{Rr_h}$, and

$$\sigma_r|_M = \frac{3+\mu}{8}\rho\omega^2(R - r_h)^2$$

Table 5.2.21 Constants K and n for Sheet Materials

Material	Treatment	K, lb/in²	n
0.05%C rimmed steel	Annealed	77,100	0.261
0.05%C killed steel	Annealed and tempered	73,100	0.234
Decarburized 0.05%C steel	Annealed in wet H_2	75,500	0.284
0.05/0.07% phos. low C	Annealed	93,330	0.156
SAE 4130	Annealed	169,400	0.118
SAE 4130	Normalized and tempered	154,500	0.156
Type 430 stainless	Annealed	143,000	0.229
Alcoa 24-S	Annealed	55,900	0.211
Reynolds R-301	Annealed	48,450	0.211

The largest tangential stress $\sigma_\theta|_M$ exists at the inner boundary, and

$$\sigma_\theta|_M = \frac{3 + \mu}{4}\,\rho\omega^2\left(R^2 + \frac{1 - \mu}{3 + \mu}\,r_h^2\right)$$

As the hole radius r_h approaches zero, the tangential stress assumes a value twice that at the center of a rotating solid disk, given above.

Stresses in Turbine Disks   Explicit solutions for cases other than those cited are not available; so approximate solutions, such as those proposed by Stodola, Thomson, Hetényi, and Robinson, are necessary. Manson uses the calculus of finite differences. See commentary under previous discussion of torsion for alternate methods of approximate solution. The problem illustrated below is a prime example of the elegance of the combination of approximate methods and electronic computers, which allow a rapid solution to be obtained. The speed with which the repetitive calculations are done allows equally rapid solutions with changes in design variables.

The customary, simplifying assumptions of axial symmetry—no variation of stress in the thickness direction and a completely elastic stress situation—are made. The differential equations of equilibrium and comparibility are rewritten in finite-difference form.

Solution of the finite-difference equations, appreciation of their linear nature, and successive application of them yield the stresses at any station in terms of those at a boundary station such as r_0. The equations thus derived are

$$\sigma_{r,n} = A_{r,n}\sigma_{t,r_0} + B_{r,n}$$
$$\sigma_{t,n} = A_{t,n}\sigma_{t,r_0} + B_{t,n} \tag{5.2.1}$$

The finite-difference expressions yield Eqs. (5.2.2), which permit the coefficients at station n to be computed from those at station $n - 1$.

$$\begin{aligned}
A_{r,n} &= K_n A_{r,n-1} + L_n A_{t,n-1} \\
A_{t,n} &= K'_n A_{r,n-1} + L'_n A_{t,n-1} \\
B_{r,n} &= K_n B_{r,n-1} + L_n B_{t,n-1} + M_n \\
B_{t,n} &= K'_n B_{r,n-1} + L'_n B_{t,n-1} + M'_n
\end{aligned} \tag{5.2.2}$$

The coefficients at the first station can be established by inspection. For a solid disk, for instance, where both stresses are equal to the tangential stress at the center, the coefficients in Eqs. (5.2.1) are $A_{r,n} = A_{t,n} = 1$ and $B_{r,n} = B_{t,n} = 0$. In the case of the disk with a central hole, where $\sigma_{r,rh} = 0$, $A_{r,rh} = B_{r,rh} = B_{t,rh} = 0$ and $A_{t,rl} = 1$. Knowing these, all others can be found from Eqs. (5.2.2).

At the outer boundary, $\sigma_{r,R} = A_{r,R}\sigma_{t,r_0} + B_{r,R}$ and $\sigma_{t,r_0} = (\sigma_{r,R} - B_{r,R})/A_{r,R}$. The radial and tangential stresses at each station are successively obtained, knowing σ_{t,r_0} and all the coefficients, using Eqs. (5.2.1).

The remaining coefficients in Eqs. (5.2.2), extracted from the finite-difference equations, are defined below, where E is Young's modulus at the temperature of the point in question, h is the profile thickness, α is the thermal coefficient of expansion, ΔT is the temperature increment above that at which the thermal stress is zero, μ is Poisson's ratio, ω is angular velocity of disk, and ρ is the mass density of disk material.

$$\begin{aligned}
C_n &= r_n/h_n \\
C'_n &= \mu_n/E_n + (1 + \mu_n)(r_n - r_{n-1})/(2E_n r_n) \\
D_n &= \tfrac{1}{2}(r_n - r_{n-1})h_n \\
D'_n &= 1/E_n + (1 + \mu_n)(r_n - r_{n-1})/(2E_n r_n) \\
F_n &= r_{n-1}h_{n-1} \\
F'_n &= (\mu_{n-1}/E_{n-1}) - (1 + \mu_{n-1})(r_n - r_{n-1})/(2E_{n-1}r_{n-1}) \\
G_n &= \tfrac{1}{2}(r_n - r_{n-1})h_{n-1} \\
G'_n &= (1/E_{n-1}) - (1 + \mu_{n-1})(r_n - r_{n-1})/(2E_{n-1}r_{n-1}) \\
H_n &= \tfrac{1}{2}\omega^2(r_n - r_{n-1})(\rho_n h_n r_n^2 + \rho_{n-1}h_{n-1}r_{n-1}^2) \\
H'_n &= \alpha_n\Delta T_n - \alpha_{n-1}\Delta T_{n-1} \\
K_n &= (F'_n D_n - F_n D'_n)/(C'_n D_n - C_n D'_n) \\
K'_n &= (C_n F'_n - C'_n F_n)/(C'_n D_n - C_n D'_n) \\
L_n &= -(G'_n D_n + G_n D'_n)/(C'_n D_n - C_n D'_n) \\
L'_n &= -(C'_n G_n + C_n G'_n)/(C'_n D_n - C_n D'_n) \\
M_n &= (H'_n D_n + H_n D'_n)/(C'_n D_n - C_n D'_n) \\
M'_n &= (C'_n H_n + C_n H'_n)/(C'_n D_n - C_n D'_n)
\end{aligned}$$

Situations need not be equally spaced between the two boundaries. It is best to space them more closely where the profile, temperature, or other property is changing rapidly. In cases of sudden or abrupt section changes, it is best to fair in across the change; the material density should, however, be adjusted to give a total mass equal to the actual. Six to ten stations are often sufficient.

The modulus of elasticity has a significant effect, and its exact value at the temperature of each station should be used. The coefficients of thermal expansion are usually averaged for the temperature between the station and at which no thermal stress occurs.

The first two Eqs. (5.2.2) and the last two must be worked simultaneously.

At the outer boundary, loads external to the disk may be imposed, e.g., the radial stress $\sigma_{r,R}$ from the centrifuged pull of a bucket. At the center, the disk may be shrunk on a shaft with the fit pressures causing a radial external push at this boundary.

Numerical solutions are most expeditiously accomplished by use of a table with column-to-column procedures. This technique lends itself readily to programmable computers or calculators.

Disks with Noncentral Holes   This case has not been solved explicitly, but approximations are useful (e.g., Armstrong, Stresses in Rotating Tapered Disks with Noncentral Holes, Ph.D. dissertation, Iowa State University, 1960). The area between the holes is considered removed and replaced by uniform spokes, each one with a cross-sectional area equal to the original minimum spoke area and with a length equal to the diameter of the noncentral holes. The higher stress in such a spoke results in an additional extension, which is then applied to the outer annulus according to thin-ring theory and based on the average radius of the ring. The additional stress is considered constant and is added to the tangential stress which would be present in a disk of the same dimensions but filled (that is, no noncentral holes).

The stress in the substitute spoke is computed by adjusting the stress at the hole-center radius in the solid or filled disk in proportion to the areas, or $S_{sp} = \sigma_{r,h}(A_g/A_{sp})$, where $\sigma_{r,h}$ is the radial stress in the filled disk at the radius of the hole circle, A_g is the gross circumferential area at the same radius of the filled disk, and A_{sp} is the area of the substitute spoke. The increase in total strain is $\delta = \sigma_{r,h}/[E(A_g/A_{sp} - 1)l_{sp}]$, where l_{sp} is the length of the substitute spoke.

The spoke-effect correction to be applied to the tangential stress is therefore $\sigma\theta_c = \delta E/r'$, where r' is the average outer-rim or annulus radius. This is added to the tangential stress found at the corresponding radius in the filled disk. The final step is to adjust the tangential and radial stresses as determined for stress concentrations caused by the holes in the actual disk. The factors for this adjustment are those in an infinite plate of uniform thickness having the same size hole. The method is claimed to yield stresses within 5 percent of those measured photoelastically at points of highest stress.

EXPERIMENTAL STRESS ANALYSIS

Analytical methods of stress analysis can reach limits of applicability. Many experimental techniques have been suggested and tried; several have been developed to a state of great usefulness, e.g., photoelasticity, strain-gage measurement, brittle coating, birefringent coating, and holography.

Photoelasticity

Most transparent materials exhibit temporary double refraction, or **birefringence,** when stressed. Light is resolved into components along the two principal plane directions. The effect is temporary as long as the elastic stress is not exceeded and is in direct proportion to the applied load. The stress magnitude can be established by the amount of component wave retardation, as given in the white and black band field (fringe pattern) obtained when a monochromatic light source is used. The polariscope, consisting of the light source, the polarizer, the model in a loading frame, an analyzer (same as polarizer), and a screen or camera, is used to produce and evaluate the fringe effect. Quarter-wave plates may be placed on either side of the model, making the light components through the model independent of the absolute orientation of polarizer and analyzer. The polarizer is a plane polariscope and yields the directions of principal stresses (the isoclinics); the analyzer is a circular polariscope yielding the fringes (isochromatics) as well.

Figure 5.2.74 shows the fringe pattern and the 20° isoclinics of a disk loaded radially at four places.

The isochromatics in the fringe pattern depict the **difference** between principal stresses. At free boundaries where the normal stress is zero, the difference automatically becomes the tangential stress. Starting at such a boundary and proceeding into the interior, the stresses can be separated by numerical calculation.

(a) Fringe pattern

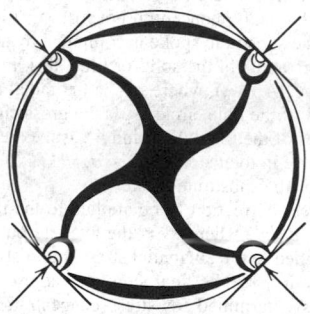

(b) 20° isoclinics

Fig. 5.2.74

The Stress-Optic Law In a transparent, isotropic plate subjected to a biaxial stress field within the elastic limit, the relative retardation R_t between the two components produced by temporary double refraction is $R_t = Ct(p - q) = n\lambda$, where C is the stress-optic coefficient, t is the plate thickness, p and q are the principal stresses, n is the fringe order (the number of fringes which have passed the point during application of load), and λ is the wavelength of monochromatic light used. Thus,

$$(p - q)/2 = \tau|_M = n\lambda/(2Ct) = nf/t$$

If the **material-fringe value** f is determined with the same light source (generally a mercury-vapor lamp emitting light having a wavelength of 5,461 Å) as used in the model study, the maximum shearing stress, or one-half the difference between the principal stresses, is directly determined. The calibration is a matter of obtaining the material-fringe value in lb/in² per fringe per inch (kgf/cm² per fringe per cm).

Isoclinics, or the direction of the principal planes, can be obtained with a plane polariscope. A new isoclinic parameter is observed each time the polarizer and analyzer are rotated simultaneously into a new position. A white-light source reveals a more distinct isoclinic, as the black curve is more distinguishable against a colored background.

Isostatics, or stress trajectories, are curves the tangents to which represent the progressive change in principal-plane directions. They are constructed graphically using the isoclinics. Since there are two princi-

pal planes at each point, two families of orthogonal curves are drawn. Care must be exercised in the drawing of trajectories for practical accuracy.

Stress Separation If knowledge of each principal stress is required, the photoelastic data must be treated to separate the stresses from the difference given by the data. If the sum of the two stresses is also obtained somehow, a simultaneous solution of the sum and difference values will yield each principal stress. One can also start at a boundary where the normal stress value is zero. There, the photoelastic reading gives the principal stress parallel to the boundary. Starting with the single value, methods have been developed which can be used to proceed with the separation. Typical of the former are lateral-extensometer, iteration, and membrane-analogy techniques; typical of the latter are the slope-equilibrium, shear-difference, graphical-integration, alternating-summation methods, and oblique incidence. Often, however, the surface stresses are the maximum valued ones. (See Frocht, ''Photoelasticity,'' McGraw-Hill.)

EXAMPLE. The fringe pattern of a Homalite disk 1.31 in in diam, 0.282 in thick, and carrying four radial loads of 155 lb each is shown in Fig. 5.2.74.

A closed solution is not known. However, by counting the fringe order at any point, the stress can be determined photoelastically. For instance, the dark spot at the center marks a fringe of zero order, as do the disk edges except in the immediate vicinity of the concentrated loads. The point at the center, which remained dark throughout the loading, is an isotropic point (zero stress difference and normal stresses are equal in all directions). Counting out from the center toward the load, the first ''circular'' fringe is of order 3. Therefore, anywhere along it $(p - q)/2 = \tau|_M = nf/t = 3 \times 65/0.282 = 692$ lb/in² (49 kgf/cm²). Carefully inspected, fringe 12 can be counted at the point of load application. Therefore, $\tau|_M = 12 \times 65/0.282 = 2{,}770$ lb/in² (195 kgf/cm²).

THREE-DIMENSIONAL PHOTOELASTICITY

Stress ''freezing'' and slicing, wherein a plastic model is brought up to its critical temperature, loaded as desired, and while loaded, slowly brought back to room temperature, are techniques which freeze the fringe pattern into the model. The model can be cut into slices without disturbing the ''frozen'' strains. Two-dimensional models are usually machined from plate stock, and three-dimensional models are cast. The frozen stress model is sliced so that the desired information can be obtained by normal incidence using the previous formulations.

When normal incidence is not possible, **oblique incidence** becomes necessary. Oblique-incidence patterns are usable in two-dimensional as well as three-dimensional stress separation. The measurement of fractional fringes is often required when using oblique incidence. With a crossed, circular, monochromatic polariscope, oriented to the principal stresses at a point, the analyzer is rotated through some angle ϕ until extinction occurs. The fringe value n is $n = n_n \pm \phi/180$, where n_n is the order of the last visible fringe. Whether the fractional term is added or subtracted depends upon the direction in which the analyzer is rotated (established by inspection).

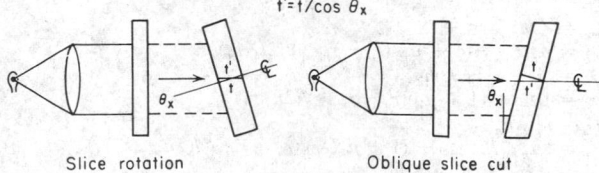

Slice rotation   Oblique slice cut

Fig. 5.2.75

Oblique-incidence calculations are based on the stress-optic law: $n_n = R_1 = t(p - q)/f = tp/f = tq/f = n_p - n_q$. Also, when polarized light is directed through the slice at an angle θ_x to a principal plane, either by rotating the slice away from normal to the light ray or by cutting it at the angle θ_x (see Fig. 5.2.75), the fringe order becomes

$$n_{\theta x} = \frac{t'}{f}(p' - q') = \frac{t}{f\cos\theta_x}(p - q\cos^2\theta_x)$$
$$= (n_p - n_q\cos^2\theta_x)/\cos\theta_x$$

Solving algebraically,

$$n_p = (n_{\theta_x} \cos \theta_x - n_n \cos^2 \theta_x)/\sin^2 \theta_x$$

and

$$n_q = (n_{\theta x} \cos \theta_x - n_n)/\sin^2 \theta_x$$

If orders n_n and $n_{\theta x}$ are thus measured at a point, n_p and n_q can be computed. The principal stresses are then determined from $p = f n_p/t$ and $q = f n_q/t$.

The material-fringe value f in these equations is at the "freezing" temperature (critical temperature). The angle of incidence, as well as the fringe orders, must be accurately measured if errors are to be minimized.

Bonded Metallic Gages

Strain measurements down to one-millionth inch per inch (one-millionth cm/cm) are possible with electrical-resistance wire gages. Such gages can be used to measure surface strains (stress by Hooke's law) on any shape or size of object. Figure 5.2.76 illustrates schematically the gage construction with a grid of fine alloy wire or thin foil, bonded to paper and covered for protection with a felt pad. In use, the gage is cemented rigidly to the surface of the member to be analyzed. The strain relation is $\varepsilon = (\Delta R/R)(1/G_f)$ in/in (cm/cm). Thus, if the resistance R and gage factor G_f (given by the gage manufacturer) are known and the change in resistance ΔR is measured, the strain which caused the resistance change can be determined and Hooke's law can be applied to determine the stress.

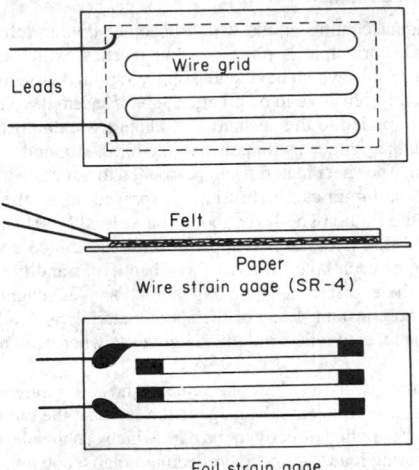

Fig. 5.2.76

Gages must be properly selected in accordance with manufacturer's recommendations. The surface to which the gage is applied must be clean, the proper cement must be used, and the gage assembly must be coated for protection against environmental conditions (e.g., moisture).

A gaging unit, usually a Wheatstone bridge or a ballast circuit (see Fig. 5.2.77 and Sec. 15), is needed to detect the signal resulting from the change in resistance of the strain gage. The strain and, therefore, the signal are often too small for direct handling, so that amplification is needed, with a metering discriminator for magnitude evaluation.

The signal is read or recorded by a galvanometer, oscilloscope, or other device. Equipment specifically constructed for strain measurement is available to indicate or record the signal directly in strain units.

Static strains are best gaged on a **Wheatstone bridge,** with strain gages wired to it as indicated in Fig. 5.2.77a. With the bridge set so that the only unbalance is the change of resistance in the active-strain gage, the potential difference between the output terminals becomes a measurement of strain. Since the gage is sensitive to temperature as well as strain, it will measure the combined effect. However, if a "dummy" gage, cemented to an unstressed piece of the same metal subjected to the same climatic conditions, is wired into the bridge leg adjacent to the one containing the "active" gage, the electric-resistance temperature effect is canceled out. Thus the active gage reports only that which is taking place in the stressed plate. The power supply can be either ac or dc.

It is sometimes useful to make both gages active—e.g., mounted on opposite sides of a beam, with one gage subjected to tension and the other to compression. Temperature effects are still compensated, but the bridge output is doubled. In other instances, it may be desirable to make all four bridge arms active gages. The experimenter must determine the most practical arrangement for the problem at hand and must bear in mind that the bridge unbalances in proportion to the difference in the strains of gages located in adjacent legs and to the sum of strain in gages located in opposite legs.

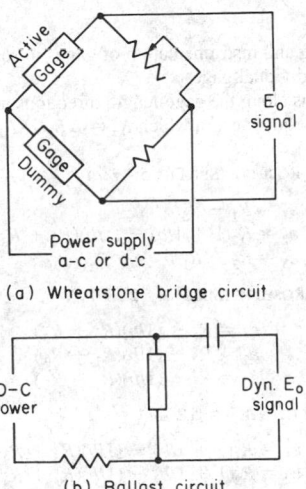

(a) Wheatstone bridge circuit

(b) Ballast circuit

Fig. 5.2.77

Dynamic strains can be detected using circuits such as the **ballast** type shown in Fig. 5.2.77b. The capacitor coupling passes only rapidly varying or dynamic strains. The capacitor's infinite impedance to a steady voltage filters out any static effects or strains. The circuit is dc powered.

Transverse Sensitivity Grid-type gages possess some strain sensitivity in the direction perpendicular to the gage axis. In a uniaxial stress field, this transverse sensitivity is of no concern because the gage factor was obtained in such a field. However, in a biaxial stress field, neglect of transverse sensitivity will give slightly erroneous strains. When accounted for, the true strains in the axial direction of gage, ε_1, and at right angles to it, ε_2, are $\varepsilon_1 = (1 - \mu k)(\varepsilon_{a1} - k\varepsilon_{a2})/(1 - k^2)$ and $\varepsilon_2 = (1 - \mu k)(\varepsilon_{a2} - k\varepsilon_{a1})/(1 - k^2)$, where the apparent strains are $\varepsilon_{a1} = \Delta R_1/(RG_f)$ and $\varepsilon_{a2} = \Delta R_2/(RG_f)$, measured by cementing a gage in each direction 1 and 2. The factor μ is Poisson's ratio of the material to which gages are cemented, and k (usually provided by the gage manufacturer) is the coefficient of transverse sensitivity of the gage. The gage is cemented to the test piece, a uniaxial stress is applied in its axial direction, and the resistance change and strain are measured. The gage factor $G_1 = \Delta R_1/(R\varepsilon_1)$ is computed. A uniaxial stress is next applied transversely to the gage. Again the resistance change and strain are measured and G_2 computed. Then $k = (G_2 + \mu G_1)/(G_1 + \mu G_2)$.

Strain Rosettes In a general biaxial stress field, the principal plane directions, as well as the stresses, are unknown. Thus, three gages mounted in three differing directions are needed if the three unknowns are to be determined. Three standard gage combinations, called strain rosettes, are commercially available and are best for the purpose. These are the *rectangular strain rosette* (Fig. 5.2.78a), which covers a minimum of area and is therefore best where the strain gradient is high; the *equiangular strain rosette* (Fig. 5.2.78b), where the gages do not overlap and which can be used where the strain gradient is low; the *T-delta strain rosette* (Fig. 5.2.78c), which occupies no more area than the equiangular rosette and which provides an extra check, or "insurance"

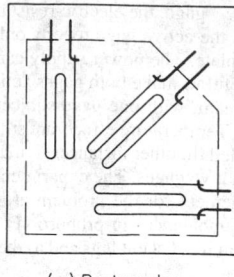
(a) Rectangular

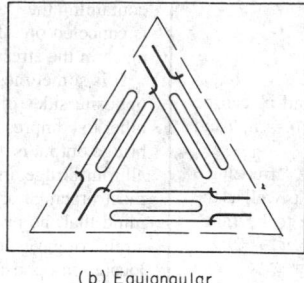
(b) Equiangular

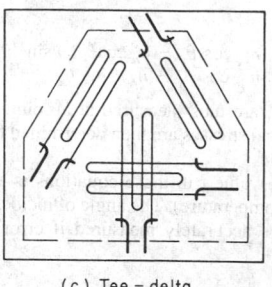
(c) Tee — delta

Fig. 5.2.78

gage. The wiring and instrumentation of gages in rosettes do not differ from those of individual gages.

The true strains along the gage-length directions are found according to the following equations, in which $R_n = \Delta R_n/[RF_1(1 - k^2)]$ and $b = 1/k$.

RECTANGULAR ROSETTE (SEE FIG. 5.2.78a)

$$\varepsilon_1 = R_1 - R_3/b$$
$$\varepsilon_2 = R_2(1 + 1/b) - (1/b)(R_1 + R_3)$$
$$\varepsilon_3 = R_3 - R_1/b$$

EQUIANGULAR ROSETTE (SEE FIG. 5.2.78b)

$$\varepsilon_1 = R_1 - (1/b)(R_2 + R_3)$$
$$\varepsilon_2 = R_2 - (1/b)(R_1 + R_3)$$
$$\varepsilon_3 = R_3 - (1/b)(R_1 + R_3)$$

T-DELTA ROSETTE (SEE FIG. 5.2.78c)

$$\varepsilon_1 = R_1(1 + 1/b) - (1/b)(R_3 + R_4)$$
$$\varepsilon_2 = R_2(1 + 1/b) - (1/b)(R_3 + R_4)$$
$$\varepsilon_3 = R_3 - (1/b)R_4$$
$$\varepsilon_4 = R_4 - (1/b)R_3$$

Foil Gages Foil gages are produced from thin foil by photoetching techniques and are applied, instrumented, read, and evaluated just like the wire-grid type. Foil gages, being much thinner, may be applied easily to curved surfaces, have lower transverse sensitivity, exhibit negligible hysteresis under cycling loads, creep little under sustained loads, and can be stacked on top of each other.

Brittle-Coating Analysis

Brittle coatings which adhere to the surface well can reveal the strain in the underlying material. Probably the first such coating used was mill scale, a thin iron oxide which forms on hot-rolled steel stock. Many coatings such as whitewash, portland cement, and shellac have been tried.

The most popular of presently available strain-indicating brittle coatings are the wood-rosin lacquers supplied by the Magnaflux Corporation under the trade name **Stresscoat**. Several Stresscoat compositions are available; the suitability of a particular lacquer depends upon the prevailing temperature and humidity. The lacquer is usually sprayed to a thickness of 0.004 to 0.008 in (0.01 to 0.02 cm) upon the surface, which must be clean and free of grease and loose particles. Calibration bars are sprayed at the same time. Both must be dried at an even temperature for up to 24 h. To facilitate observation of cracks, an undercoating of bright aluminum is often applied.

When the cured test piece is subjected to loads, the lacquer will first begin to crack at its threshold sensitivity in the area of the largest principal stress, with the parallel cracks perpendicular to the principal stress. This information is often sufficient, as it reveals the critical area and the direction of normal stress.

The threshold sensitivity of Stresscoat lacquers is 600 to 800 microinches per inch (600 to 800 microcentimeters per centimeter) in a uniaxial stress field. Exact control of lacquer selection, thickness, curing, and testing temperatures may reduce the threshold to 400 mi-

croinches per inch (400 microcentimeters per centimeter). If desired, the approximate strain (probably within 10 percent) may be established using the calibration strip sprayed with the test part. The strip is placed in a loading device and bent as a cantilever beam by means of a cam at the free end, causing the coating to crack on the tension surface. Crack spacing varies with the strain, being close at the fixed end and diminishing toward the free end down to threshold sensitivity values. The strip is placed in a holder containing strain graduations. A visual comparison of cracks on the testpart surface with those on the strip reveals the strain magnitude which caused the cracks.

Birefringent Coatings

A birefringent coating is one which becomes double refractive when strained. The principle is quite old, but plastics, which adhere to all kinds of materials, which have stable optical-strain constants, and which are sufficiently sensitive to be practical, are of recent development. The trade name applied to this technique is **Photostress**. Photostress plastics can be obtained either as thin sheets (0.040, 0.080, and 0.20 in) or in liquid form. The sheet material can be bonded to a surface with a special adhesive. The liquid can be brushed or sprayed on, or the part can be dipped in the liquid. The layer should be at least 0.004 in (0.010 cm) thick. It is often necessary to apply several successive coatings, with heat curing of each layer in turn. Two sheet types and two liquids are available; these differ in stretching ability and in magnitude of the strain-optical constant. Each of the sheet materials is available metallized on one face, to reflect polarized light even when cemented to a dull surface.

The principles involved are the same as those for conventional photoelasticity. One frequent advantage is the fact that the plastic (sheet or liquid) can be applied directly to the part, which can then be subjected to actual operating loads. A special reflecting polariscope must be used. It contains only one polarizer and quarter-wave disk because the light passes back through the same pair after reflection by the stressed surface-plastic interface. The only limitation rests in the geometry of the structural component to be examined; not only must it be possible to apply the plastic to the surface, but the surface must be accessible to light.

The strain-optic law, since the light passes the plastic thickness twice, becomes

$$p - q = \frac{n}{2t}\frac{E}{K(1 + \mu)}$$

where n is fringe order, E is modulus, μ is Poisson's ratio of workpiece material, and K (supplied by the manufacturer) is the strain-optic coefficient of the plastic. As in conventional photoelasticity, isoclinics are present as well.

Holography

A more recently developed technique applicable to stress, or rather, strain analysis as well as to many other purposes is that of holography. It is made possible by the laser, an instrument which produces a highly concentrated, thin beam of light of single wavelength. The helium-neon (He-Ne) laser, emitting at the red end of the visible spectrum at a wavelength of 633 nanometers, has found much favor. The output of a helium-cadmium (He-Cd) laser is at half the wavelength of the He-Ne laser; accordingly, the He-Ne laser is twice as sensitive to displacements.

The laser beam is split into two components, one of which is directed upon the object (or specimen) and then onto the photographic plate. It is identified as the object beam. The other component, referred to as the reference beam, propagates directly to the plate. Interference between the beams resulting from retardations caused by displacements or strains forms fringes which in turn provide a measure of the disturbance. Spacing of such fringes depends upon Bragg's law:

$$d = \frac{\lambda}{2 \sin (\theta/2)}$$

where d is the distance between fringes, λ is the wavelength of the light source, and θ is the angle between the object and reference ray at the plate.

A simple holographic setup consists of the laser source, beam splitter, reflecting surfaces, filters, and the recording plate. A possible arrangement is depicted in Fig. 5.2.79. Some arrangement for loading the specimen must also be provided. Additional auxiliary and refining hardware becomes necessary as the analysis assumes greater complexity. Thus the system layout is limited only by test requirements and the experimenter's imagination. However, only a thorough understanding of the laws of optics and interferometry will make possible a reliable investigation and interpretation of results.

Stability of setup must be assured via a rigid optical bench and supporting brackets. Component instruments must be spaced upon the bench so that beam coherence is assured, required coherence depth satisfied, and the object/reference angle θ consistent with the fringe spacing desired. The film must also possess adequate sensitivity in the spectral range of the laser beam used. It is important to recognize the inherent hazards of the high-intensity radiation in laser beams and to practice every precaution in the use of lasers.

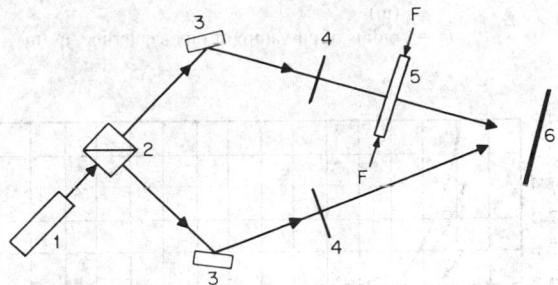

Fig. 5.2.79 Simple holographic setup. (1) Laser source; (2) beam splitter; (3) reflecting surfaces; (4) circular polarizers; (5) loaded specimen (birefringent); (6) photographic plate.

Holography, using pulsed lasers, can be used to measure transient disturbances. Thus vibration studies are possible. Fatigue detection using holographic techniques has also been undertaken. Holography has been used in acoustical studies and in automatic gaging as well. It is a versatile engineering tool.

5.3 PIPELINE FLEXURE STRESSES
by Harold V. Hawkins

EDITOR'S NOTE: The almost universal availability and utilization of personal computers in engineering practice has led to the development of many competing and complementary forms of piping stress analysis software. Their use is widespread, and individual packaged software allows analysis and design to take into account static and dynamic conditions, restraint conditions, aboveground and buried configurations, etc. The reader is referred to the technical literature for the most suitable and current software available for use in solving the immediate problems at hand.

The brief discussion in this section addresses the fundamental concepts entailed and sets forth the solution of simple systems as an exercise in application of the principles.

REFERENCES: Shipman, Design of Steam Piping to Care for Expansion, *Trans. ASME*, 1929, Wahl, Stresses and Reactions in Expansion Pipe Bends, *Trans. ASME*, 1927. Hovgaard, The Elastic Deformation of Pipe Bends, *Jour. Math. Phys.*, Nov. 1926, Oct. 1928, and Dec. 1929. M. W. Kellog Co., "The Design of Piping Systems," Wiley.

For details of pipe and pipe fittings see Sec. 8.7.

Nomenclature (see Figs. 5.3.1 and 5.3.2)

M_0 = end moment at origin, in·lb (N·m)
M = max moment, in·lb (N·m)
F_x = end reaction at origin in x direction, lb (N)
F_y = end reaction at origin in y direction, lb (N)
S_l = $(Mr/I)\alpha$ = max unit longitudinal flexure stress, lb/in² (N/m²)
S_t = $(Mr/I)\beta$ = max unit transverse flexure stress, lb/in² (N/m²)
S_s = $(Mr/I)\gamma$ = max unit shearing stress, lb/in² (N/m²)
Δx = relative deflection of ends of pipe parallel to x direction caused by either temperature change or support movement, or both, in (m)

Δy = same as Δx but parallel to y direction, in. Note that Δx and Δy are positive if under the change in temperature the end opposite the origin tends to move in a positive x or y direction, respectively.
t = wall thickness of pipe, in (m)
r = mean radius of pipe cross section, in (m)
λ = constant = tR/r^2
I = moment of inertia of pipe cross section about pipe centerline, in⁴ (m⁴)
E = modulus of elasticity of pipe at *actual working* temperature, lb/in² (N/m²)
K = flexibility index of pipe. $K = 1$ for all straight pipe sections, $K = (10 + 12\lambda^2)/(1 + 12\lambda^2)$ for all curved pipe sections where $\lambda > 0.335$ (see Fig. 5.3.3)
α, β, γ = ratios of actual max longitudinal flexure, transverse flexure, and shearing stresses to Mr/I for curved sections of pipe (see Fig. 5.3.3)

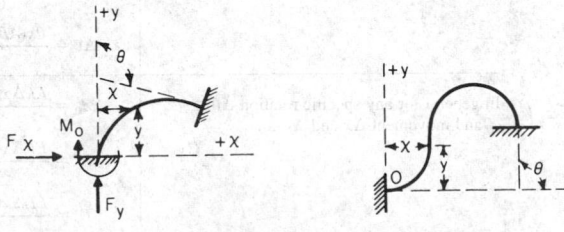

Fig. 5.3.1   **Fig. 5.3.2**

A, B, C, F, G, H = constants given by Table 5.3.2

θ = angle of intersection between tangents to direction of pipe at reactions

$\Delta\theta$ = change in θ caused by movements of supports, or by temperature change, or both, rad

ds = an infinitesimal element of length of pipe

s = length of a particular curved section of pipe, in (m)

R = radius of curvature of pipe centerline, in (m)

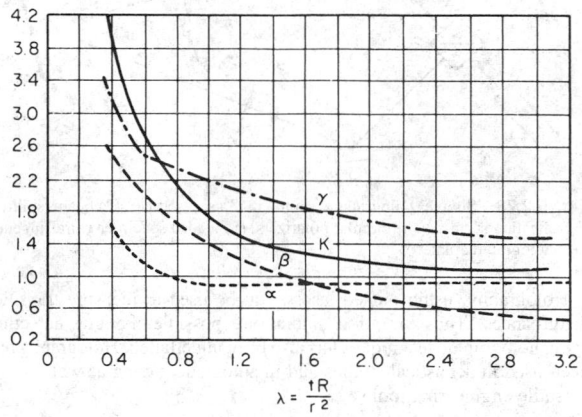

Fig. 5.3.3 Flexure constants of initially curved pipes.

General Discussion

Under the effect of changes in temperature of the pipeline, or of movement of support reactions (either translation or rotation), or both, the determination of stress distribution in a pipe becomes a statically indeterminate problem. In general the problem may be solved by a slight modification of the standard arch theory: $\Delta x = -K\int My\,ds/(EI)$, $\Delta y = K\int Mx\,ds/(EI)$, and $\Delta\theta = K\int M\,ds/(EI)$ where the constant K is introduced to correct for the increased flexibility of a curved pipe, and where the integration is over the entire length of pipe between supports. In Table 5.3.1 are given equations derived by this method for moment and thrust at one reaction point for pipes in one plane that are fully fixed, hinged at both ends, hinged at one end and fixed at the other, or partly fixed. If the reactions at one end of the pipe are known, the moment distribution in the entire pipe then can be obtained by simple statics.

Since an initially curved pipe is more flexible than indicated by its moment of inertia, the constant K is introduced. Its value may be taken from Fig. 5.3.3, or computed from the equation given below. $K = 1$ for all straight pipe sections, since they act according to the simple flexure theory.

In Fig. 5.3.3 are given the flexure constants K, α, β, and γ for initially curved pipes as functions of the quantity $\lambda = tR/r^2$. The flexure constants are derived from the equations.

$$K = (10 + 12\lambda^2)/(1 + 12\lambda^2) \qquad \text{when } \lambda > 0.335$$
$$\alpha = \tfrac{2}{3}K\sqrt{(5 + 6\lambda^2)}/18 \qquad \lambda \leq 1.472$$
$$\alpha = K(6\lambda^2 - 1)/(6\lambda^2 + 5) \qquad \lambda > 1.472$$
$$\beta = 18\lambda/(1 + 12\lambda^2)$$
$$\gamma = [8\lambda - 36\lambda^3 + (32\lambda^2 + 20/3)$$
$$\times \sqrt{(4/3)\lambda^2 + 5/18}] \div (1 + 12\lambda^2) \qquad \text{when } \lambda < 0.58$$
$$= (12\lambda^2 + 18\lambda - 2)/(1 + 12\lambda^2) \qquad \text{when } \lambda > 0.58$$

Table 5.3.1 General Equations for Pipelines in One Plane (See Figs. 5.3.1 and 5.3.2)

Type of supports	Unsymmetric	Symmetric about y-axis
Both ends fully fixed	$M_0 = \dfrac{EI\,\Delta x(CF - AB) + EI\,\Delta y(BF - AG)}{2ABF + CGH - B^2H - A^2G - CF^2}$	$M_0 = \dfrac{EI\,\Delta x\,F}{GH - F^2}$
	$F_x = \dfrac{EI\,\Delta x(CH - A^2) + EI\,\Delta y(BH - AF)}{2ABF + CGH - B^2H - A^2G - CF^2}$	$F_x = \dfrac{EI\,\Delta x\,H}{GH - F^2}$
	$F_y = \dfrac{EI\,\Delta x(BH - AF) + EI\,\Delta y(GH - F^2)}{2ABF + CGH - B^2H - A^2G - CF^2}$	$F_y = 0$ $\Delta\theta = 0$
	$\Delta\theta = 0$	
Both ends hinged	$M_0 = 0$	$M_0 = 0$
	$F_x = \dfrac{EI\,\Delta x\,C + EI\,\Delta y\,B}{CG - B^2}$	$F_x = \dfrac{EI\,\Delta x}{G}$
	$F_y = \dfrac{EI\,\Delta x\,B + EI\Delta y\,G}{CG - B^2}$	$F_y = 0$
	$\Delta\theta = \dfrac{\Delta x(AB - CF) + \Delta y(AG - BF)}{CG - B^2}$	$\Delta\theta = \dfrac{-\Delta x F}{G}$
Origin end only hinged, other end fully fixed	$M_0 = 0$	
	$F_x = \dfrac{EI\,\Delta x\,C + EI\,\Delta y\,B}{CG - B^2}$	
	$F_y = \dfrac{EI\,\Delta x\,B + EI\,\Delta y\,G}{CG + B^2}$	
	$\Delta\theta = \dfrac{\theta x(AB - CF) + \Delta y(AG - BF)}{CG - B^2}$	
In general for any specific rotation $\Delta\theta$ and movement Δx and $\Delta y \ldots$	$M_0 = \dfrac{EI\,\Delta x(CF - AB) + EI\,\Delta y(BF - AG) + EI\,\Delta\theta(CG - B^2)}{2ABF + CGH - A^3G - CF^2 - B^2H}$	
	$F_x = \dfrac{EI\,\Delta x(CH - A^2) + EI\,\Delta y(BH - AF) + EI\,\Delta\theta(CF - AB)}{2ABF + CGH - A^2G - CF^2 - B^2H}$	
	$F_y = \dfrac{EI\,\Delta x(BH - AF) + EI\,\Delta y(GH - F^2) + EI\,\Delta\theta(BF - AG)}{2ABF + CGH - A^2G - CF^2 - B^2H}$	

The increased flexibility of the curved pipe is brought about by the tendency of its cross section to flatten. This flattening causes a transverse flexure stress whose maximum is S_r. Because the maximum longitudinal and maximum transverse stresses do not occur at the same point in the pipe's cross section, the resulting maximum shear is not one-half the difference of S_l and S_t; it is S_s. In the straight sections of the pipe, $\alpha = 1$, the transverse stress disappears, and $\lambda = \frac{1}{2}$. This discussion of S_s does not include the uniform transverse or longitudinal tension stresses induced by the internal pressure in the pipe; their effects should be added if appreciable.

Table 5.3.2 gives values of the constants A, B, C, F, G, and H for use in equations listed in Table 5.3.1. The values may be used (1) for the solution of any pipeline or (2) for the derivation of equations for standard shapes composed of straight sections and arcs of circles as of Fig. 5.3.5. Equations for shapes not given may be obtained by algebraic addition of those given. All measurements are from the left-hand end of the pipeline. Reactions and stresses are greatly influenced by end conditions. Formulas are given to cover the extreme conditions. The following suggestions and comments should be considered when laying out a pipeline:

Avoid expansion bends, and design the entire pipeline to take care of its own expansion.

The movement of the equipment to which the ends of the pipeline are attached must be included in the Δx and Δy of the equations.

Maximum flexibility is obtained by placing **supports** and **anchors** so that they will not interfere with the natural movement of the pipe.

That shape is most efficient in which the **maximum length of pipe is working** at the maximum safe stress.

Excessive **bending moment at joints** is more likely to cause trouble than excessive stresses in pipe walls. Hence, keep pipe joints away from points of high moment.

Reactions and stresses are greatly influenced by **flattening of the cross section** of the curved portions of the pipeline.

It is recommended that **cold springing** allowances be discounted in stress calculations.

Application to Two- and Three-Plane Pipelines Pipelines in more than one plane may be solved by the successive application of the preceding data, dividing the pipeline into two or more one-plane lines.

EXAMPLE 1. The unsymmetric pipeline of Fig. 5.3.4 has fully fixed ends. From Table 5.3.2 use $K = 1$ for all sections, since only straight segments are involved.

Upon introduction of $a = 120$ in (3.05 m), $b = 60$ in (1.52 m), and $c = 180$ in (4.57 m), into the preceding relations (Table 5.3.3) for A, B, C, F, G, H, the equations for the reactions at 0 from Table 5.3.1 become

$$M_0 = EI\,\Delta x\,(-7.1608 \times 10^{-5}) + EI\,\Delta y\,(-8.3681 \times 10^{-5})$$
$$F_x = EI\,\Delta x\,(+1.0993 \times 10^{-5}) + EI\,\Delta y\,(+3.1488 \times 10^{-6})$$
$$F_y = EI\,\Delta x\,(+3.1488 \times 10^{-6}) + EI\,\Delta y\,(+1.33717 \times 10^{-6})$$

Also it follows that

$$M_1 = M_0 + F_y a = EI\,\Delta x\,(+3.0625 \times 10^{-4}) + EI\,\Delta y\,(+7.6779 \times 10^{-5})$$
$$M_2 = M_1 - F_x b = EI\,\Delta x\,(-3.5333 \times 10^{-4}) + EI\,\Delta y\,(-1.1215 \times 10^{-4})$$
$$M_3 = M_2 + F_y c = EI\,\Delta x\,(+2.1345 \times 10^{-4}) + EI\,\Delta y\,(+1.2854 \times 10^{-4})$$

Thus the maximum moment M occurs at 3.

The total maximum longitudinal fiber stress ($a = 1$ for straight pipe)

$$S_l = \frac{F_x}{2\pi rt} \pm \frac{M_3 r}{I}$$

There is no transverse flexure stress since all sections are straight. The maximum shearing stress is either (1) one-half of the maximum longitudinal fiber stress as given above, (2) one-half of the hoop-tension stress caused by an internal radial pressure that might exist in the pipe, or (3) one-half the difference of the maximum longitudinal fiber stress and hoop-tension stress, whichever of these three possibilities is numerically greatest.

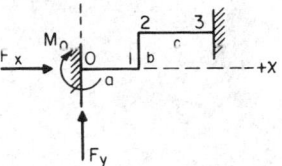

Fig. 5.3.4

EXAMPLE 2. The equations of Table 5.3.1 may be employed to develop the solution of generalized types of pipe configurations for which Fig. 5.3.5 is a typical example. If only temperature changes are considered, the reactions for the right-angle pipeline (Fig. 5.3.5) may be determined from the following equations:

$$M_0 = C_1 EI\,\Delta x/R^2$$
$$F_x = C_2 EI\,\Delta x/R^3$$
$$F_y = C_3 EI\,\Delta x/R^3$$

In these equations, Δx is the x component of the deflection between reaction points caused by temperature change only. The values of C_1, C_2, and C_3 are given in Fig. 5.3.6 for $K = 1$ and $K = 2$. For other values of K, interpolation may be employed.

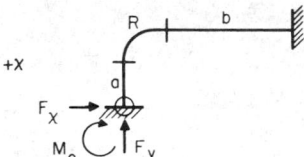

Fig. 5.3.5   Right angle pipeline.

EXAMPLE 3. With $a/R = 20$ and $b/R = 3$, the value of C_1 is 0.185 for $K = 1$ and 0.165 for $K = 2$. If $K = 1.75$, the interpolated value of C_1 is 0.175.

Elimination of Flexure Stresses Pipeline flexure stresses that normally would result from movement of supports or from the tendency of the pipes to expand under temperature change often may be avoided entirely through the use of expansion joints (Sec. 8.7). Their use may simplify both the design of the pipeline and the support structure. When using expansion joints, the following suggestions should be considered: (1) select expansion joint carefully for maximum temperature range (and deflection) expected so as to prevent damage to expansion fitting; (2) provide guides to limit movement at expansion joint to direction permitted by joint; (3) provide adequate anchors at one end of each straight section or along their midlength, forcing movement to occur at expansion joint yet providing adequate support for pipeline; (4) mount expansion joints adjacent to an anchor point to prevent sagging of the pipeline under its own weight and do not depend upon the expansion joint for stiffness—it is intended to be flexible; (5) give consideration to effects of corrosion, since corrugated character of expansion joints makes cleaning difficult.

Table 5.3.2 Values of A, B C, F, G, and H for Various Piping Elements

	$A = K\int x\,ds$	$B = K\int xy\,ds$	$C = K\int x^2\,ds$	$F = K\int y\,ds$	$G = K\int y^2\,ds$	$H = K\int ds$
segment s, (x_1,y_1) (x_2,y_2)	$\dfrac{s}{2}(x_1 + x_2)$	Ay	$s\left(\dfrac{s^2}{3} + x_1 x_2\right)$	sy	Fy	s
segment (x,y_2) (x,y_1), s	sx	$\dfrac{A}{2}(y_1 + y_2)$	Ax	$\dfrac{s}{2}(y_1 + y_2)$	$s\left(\dfrac{s^2}{3} + y_1 y_2\right)$	s
diagonal (x_2,y_2) to (x_1,y_1), s	$\dfrac{s}{2}(x_1 + x_2)$	$\dfrac{A}{3}(y_1 + y_2) + \dfrac{s}{6}(x_1 y_1 + x_2 y_2)$	$\dfrac{s}{3}(x_1^2 + x_1 x_2 + x_2^2)$	$\dfrac{s}{2}(y_1 + y_2)$	$\dfrac{s}{3}(y_1^2 + y_1 y_2 + y_2^2)$	s
semicircle (x,y)	$\pi K R x$	$A\left(y + \dfrac{2R}{\pi}\right)$	$A\left(x + \dfrac{R^2}{2x}\right)$	$(\pi y + 2R)KR$	$Fy + \left(2y + \dfrac{\pi}{2}R\right)KR^2$	πKR
semicircle (x,y)		$A\left(y - \dfrac{2R}{\pi}\right)$		$(\pi y - 2R)KR$	$Fy - \left(2y - \dfrac{\pi}{2}R\right)KR^2$	
quarter circle (x,y)	$\left(\dfrac{\pi x}{2} - R\right)KR$	$Ay + \left(x - \dfrac{R}{2}\right)KR^2$	$Ax + \left(\dfrac{\pi R}{4} - x\right)KR^2$	$\left(\dfrac{\pi y}{2} + R\right)KR$	$Fy + \left(\dfrac{\pi R}{4} + y\right)KR^2$	$\dfrac{\pi KR}{2}$

(x,y)	$\left(\dfrac{\pi x}{2}+R\right)KR$	$Ay+\left(x+\dfrac{R}{2}\right)KR^2$	$Ax+\left(\dfrac{\pi R}{4}+x\right)KR^2$	$\left(\dfrac{\pi y}{2}+R\right)KR$	$Fy+\left(\dfrac{\pi R}{4}+y\right)KR^2$
(x,y)	$\left(\dfrac{\pi x}{2}-R\right)KR$	$Ay-\left(x+\dfrac{R}{2}\right)KR^2$	$Ax+\left(\dfrac{\pi R}{4}-x\right)KR^2$	$\left(\dfrac{\pi y}{2}-R\right)KR$	$Fy+\left(\dfrac{\pi R}{4}-y\right)KR^2$
(x,y)		$Ay-\left(x-\dfrac{R}{2}\right)KR^2$	$Ax+\left(\dfrac{\pi R}{4}-x\right)KR^2$	$\left(\dfrac{\pi y}{2}-R\right)KR$	$Fy+\left(\dfrac{\pi R}{4}-y\right)KR^2$
(x,y) 45°	$\left(\dfrac{\pi x}{4}-\dfrac{R}{\sqrt2}\right)KR$	$Ay+\left[\left(1-\dfrac{\sqrt2}{2}\right)x-\dfrac{R}{4}\right]KR^2$	$Ax+\left[\left(\dfrac{\pi}{8}+\dfrac14\right)R-\dfrac{x\sqrt2}{2}\right]KR^2$	$\left[\dfrac{\pi y}{4}+\left(1-\dfrac{\sqrt2}{2}\right)R\right]KR$	$Fy+\left[\left(1-\dfrac{\sqrt2}{2}\right)y+\left(\dfrac{\pi}{8}-\dfrac14\right)R\right]KR^2$
(x,y) 45°		$Ay-\left[\left(1-\dfrac{\sqrt2}{2}\right)x-\dfrac{R}{4}\right]KR^2$		$\left[\dfrac{\pi y}{4}-\left(1-\dfrac{\sqrt2}{2}\right)R\right]KR$	$Fy-\left[\left(1-\dfrac{\sqrt2}{2}\right)y-\left(\dfrac{\pi}{8}-\dfrac14\right)R\right]KR^2$
(x,y) 45°	$\left(\dfrac{\pi x}{4}+\dfrac{R}{\sqrt2}\right)KR$	$Ay+\left[\left(1-\dfrac{\sqrt2}{2}\right)x+\dfrac{R}{4}\right]KR^2$	$Ax+\left[\left(\dfrac{\pi}{8}+\dfrac14\right)R+\dfrac{x\sqrt2}{2}\right]KR^2$	$\left[\dfrac{\pi y}{4}+\left(1-\dfrac{\sqrt2}{2}\right)R\right]KR$	$Fy+\left[\left(1-\dfrac{\sqrt2}{2}\right)y+\left(\dfrac{\pi}{8}-\dfrac14\right)R\right]KR^2$
(x,y) 45°		$Ay-\left[\left(1-\dfrac{\sqrt2}{2}\right)x+\dfrac{R}{4}\right]KR^2$		$\left[\dfrac{\pi y}{4}-\left(1-\dfrac{\sqrt2}{2}\right)R\right]KR$	$Fy-\left[\left(1-\dfrac{\sqrt2}{2}\right)y-\left(\dfrac{\pi}{8}-\dfrac14\right)R\right]KR^2$
θ_1,θ_2 (x,y)	$[r(\theta_2-\theta_1)-R(\sin\theta_2-\sin\theta_1)]KR$	$Ay-\left[x(\cos\theta_2-\cos\theta_1)+\dfrac{R}{2}(\sin^2\theta_2-\sin^2\theta_1)\right]KR^2$	$Ax-\left[x(\sin\theta_2-\sin\theta_1)-\dfrac{R}{4}(\sin2\theta_2-\sin2\theta_1)-\dfrac{R}{2}(\theta_2-\theta_1)\right]KR^2$	$[y(\theta_2-\theta_1)-R(\cos\theta_2-\cos\theta_1)]KR$	$Fy-\left[y(\cos\theta_2-\cos\theta_1)-\dfrac{R}{4}(\sin2\theta_2-\sin2\theta_1)-\dfrac{R}{2}(\theta_2-\theta_1)\right]KR^2$

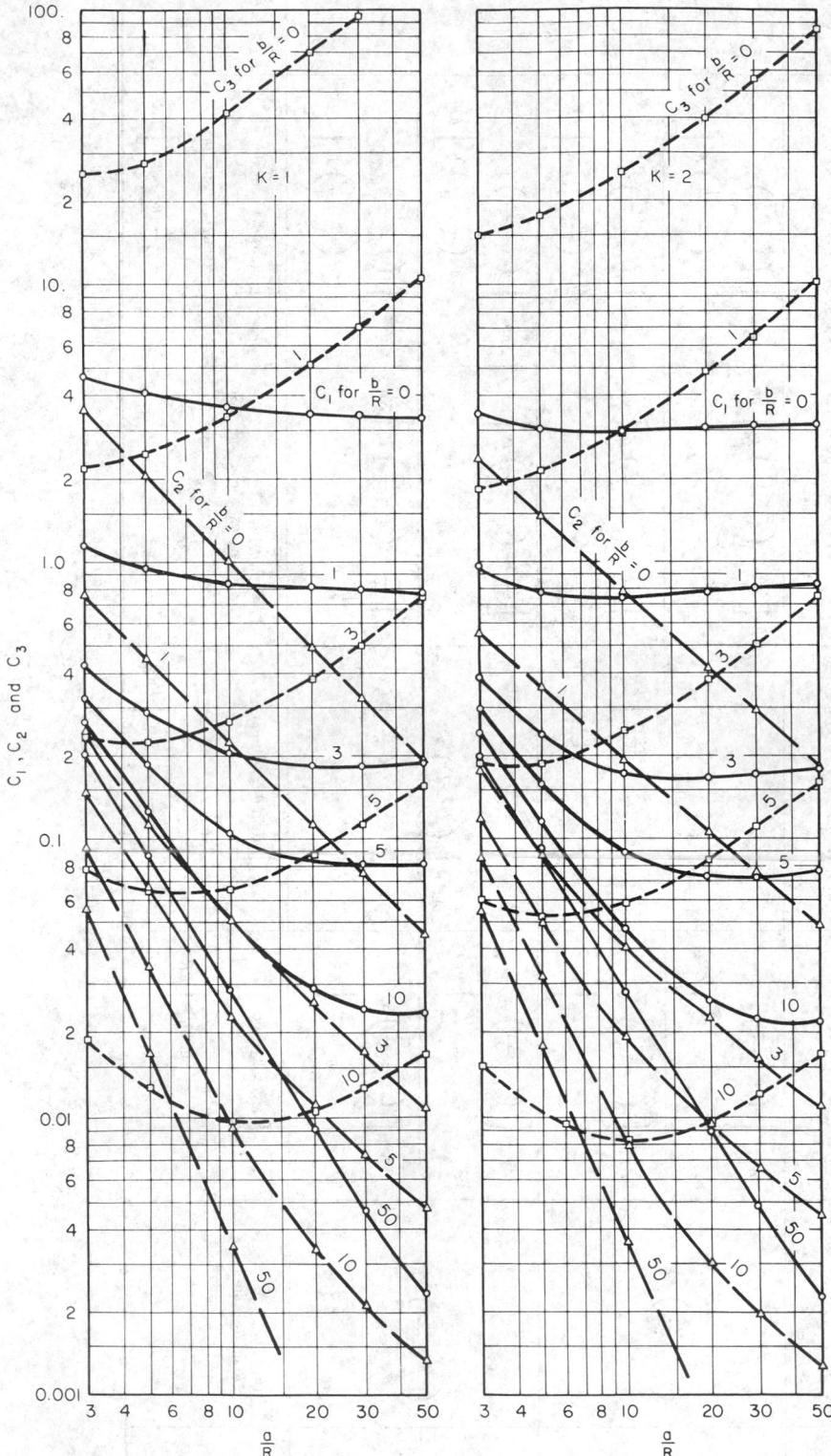

Fig. 5.3.6 Reactions for right-angle pipelines.

Table 5.3.3 Example 1 Showing Determination of Integrals

Part of pipe	Values of integrals					
	A	B	C	F	G	H
0–1	$\dfrac{a^2}{2}$	0	$\dfrac{a^3}{3}$	0	0	a
1–2	ab	$\dfrac{ab^2}{2}$	a^2b	$\dfrac{b^2}{2}$	$\dfrac{b^3}{3}$	b
2–3	$\dfrac{c}{2}(2a+c)$	$\dfrac{bc}{2}(2a+c)$	$\dfrac{c^3}{3}+ac(a+c)$	bc	b^2c	c
Total 0–3	$\dfrac{a^2}{2}+ab$ $+\dfrac{c}{2}(2a+c)$	$\dfrac{ab^2}{2}+\dfrac{bc}{2}(2a+c)$	$\dfrac{a^3}{3}+a^2b+\dfrac{c^3}{3}$ $+ac(a+c)$	$\dfrac{b^2}{2}+bc$	$\dfrac{b^3}{3}+b^2c$	$a+b+c$

5.4 NONDESTRUCTIVE TESTING
by Donald D. Dodge

REFERENCES: Various authors, "Nondestructive Testing Handbook," 8 vols., American Society for Nondestructive Testing. Boyer, "Metals Handbook," vol. 11, American Society for Metals. Heuter and Bolt, "Sonics," Wiley. Krautkramer, "Ultrasonic Testing of Materials," Springer-Verlag. Spanner, "Acoustic Emission: Techniques and Applications," Intex, American Society for Nondestructive Testing. Crowther, "Handbook of Industrial Radiography," Arnold. Wiltshire, "A Further Handbook of Industrial Radiography," Arnold. "Standards," vol. 03.03, ASTM, Boiler and Pressure Vessel Code, Secs. III, V, XI, ASME. ASME Handbook, "Metals Engineering—Design," McGraw-Hill. SAE Handbook, Secs. J358, J359, J420, J425–J428, J1242, J1267, SAE. Materials Evaluation, *Jour. Am. Soc. Nondestructive Testing*.

Nondestructive tests are those tests that determine the usefulness, serviceability, or quality of a part or material without limiting its usefulness. Nondestructive tests are used in machinery maintenance to avoid costly unscheduled loss of service due to fatigue or wear; they are used in manufacturing to ensure product quality and minimize costs. Consideration of test requirements early in the design of a product may facilitate testing and minimize testing cost. Nearly every form of energy is used in nondestructive tests, including all wavelengths of the electromagnetic spectrum as well as vibrational mechanical energy. Physical properties, composition, and structure are determined; flaws are detected; and thickness is measured. These tests are here divided into the following basic methods: **magnetic particle, penetrant, radiographic, ultrasonic, eddy current, acoustic emission, microwave, and infrared.** Numerous techniques are utilized in the application of each test method. Table 5.4.1 gives a summary of many nondestructive test methods.

MAGNETIC PARTICLE METHODS

Magnetic particle testing is a nondestructive method for detecting discontinuities at or near the surface in **ferromagnetic materials.** After the test object is properly magnetized, finely divided magnetic particles are applied to its surface. When the object is properly oriented to the induced magnetic field, a discontinuity creates a leakage flux which attracts and holds the particles, forming a visible indication. Magnetic-field direction and character are dependent upon how the magnetizing force is applied and upon the type of current used. For best sensitivity, the **magnetizing current** must flow in a direction parallel to the principal direction of the expected defect. Circular fields, produced by passing current through the object, are almost completely contained within the test object. Longitudinal fields, produced by coils or yokes, create external poles and a general-leakage field. Alternating, direct, or half-wave direct current may be used for the location of surface defects. Half-wave direct current is most effective for locating subsurface defects. Magnetic particles may be applied dry or as a wet suspension in a liquid such as kerosene or water. Colored **dry powders** are advantageous when testing for subsurface defects and when testing objects that have rough surfaces, such as castings, forgings, and weldments. **Wet particles** are preferred for detection of very fine cracks, such as fatigue, stress corrosion, or grinding cracks. Fluorescent wet particles are used to inspect objects with the aid of ultraviolet light. Fluorescent inspection is widely used because of its greater sensitivity. Application of particles while magnetizing current is on (continuous method) produces stronger indications than those obtained if the particles are applied after the current is shut off (residual method). Interpretation of subsurface-defect indications requires experience. Demagnetization of the test object after inspection is advisable.

Magnetic flux leakage is a variation whereby leakage flux due to flaws is detected electronically via a Hall-effect sensor. Computerized signal interpretation and data imaging techniques are employed.

Electrified particle testing indicates minute cracks in nonconducting materials. Particles of calcium carbonate are positively charged as they are blown through a spray gun at the test object. If the object is metal-backed, such as porcelain enamel, no preparation other than cleaning is necessary. When it is not metal-backed, the object must be dipped in an aqueous penetrant solution and dried. The penetrant remaining in cracks provides a mobile electron supply for the test. A readily visible powder indication forms at a crack owing to the attraction of the positively charged particles.

PENETRANT METHODS

Liquid penetrant testing is used to locate flaws open to the surface of **nonporous materials.** The test object must be thoroughly cleaned before testing. Penetrating liquid is applied to the surface of a test object by a brush, spray, flow, or dip method. A time allowance (1 to 30 min) is required for liquid penetration of surface flaws. Excess penetrant is then carefully removed from the surface, and an absorptive coating, known as developer, is applied to the object to draw penetrant out of flaws, thus showing their location, shape, and approximate size. The developer is typically a fine powder, such as talc usually in suspension in a liquid. Penetrating-liquid types are (1) for test in visible light, and (2) for test under ultraviolet light (3,650 Å). Sensitivity of penetrant testing is greatest when a fluorescent penetrant is used and the object is observed

Table 5.4.1   Nondestructive Test Methods*

Method	Measures or defects	Applications	Advantages	Limitations
Acoustic emission	Crack initiation and growth rate Internal cracking in welds during cooling Boiling or cavitation Friction or wear Plastic deformation Phase transformations	Pressure vessels Stressed structures Turbine or gearboxes Fracture mechanics research Weldments Sonic-signature analysis	Remote and continuous surveillance Permanent record Dynamic (rather than static) detection of cracks Portable Triangulation techniques to locate flaws	Transducers must be placed on part surface Highly ductile materials yield low-amplitude emissions Part must be stressed or operating Interfering noise needs to be filtered out
Acoustic-impact (tapping)	Debonded areas or delaminations in metal or nonmetal composites or laminates Cracks under bolt or fastener heads Cracks in turbine wheels or turbine blades Loose rivets or fastener heads Crushed core	Brazed or adhesive-bonded structures Bolted or riveted assemblies Turbine blades Turbine wheels Composite structures Honeycomb assemblies	Portable Easy to operate May be automated Permanent record or positive meter readout No couplant required	Part geometry and mass influences test results Impactor and probe must be repositioned to fit geometry of part Reference standards required Pulser impact rate is critical for repeatability
D-Sight (Diffracto)	Enhances visual inspection for surface abnormalities such as dents protrusions, or waviness' Crushed core Lap joint corrosion Cold-worked holes Cracks	Detect impact damage to composites or honeycomb corrosion in aircraft lap joints Automotive bodies for waviness	Portable Fast, flexible Noncontact Easy to use Documentable	Part surface must reflect light or be wetted with a fluid
Eddy current	Surface and subsurface cracks and seams Alloy content Heat-treatment variations Wall thickness, coating thickness Crack depth Conductivity Permeability	Tubing Wire Ball bearings "Spot checks" on all types of surfaces Proximity gage Metal detector Metal sorting Measure conductivity in % IACS	No special operator skills required High speed, low cost Automation possible for symmetric parts Permanent-record capability for symmetric parts No couplant or probe contact required	Conductive materials Shallow depth of penetration (thin walls only) Masked or false indications caused by sensitivity to variations such as part geometry Reference standards required Permeability variations
Magneto-optic eddy-current imager	Cracks Corrosion thinning in aluminum	Aluminum aircraft Structure	Real-time imaging Approximately 4-in area coverage	Frequency range of 1.6 to 100 kHz Surface contour Temperature range of 32 to 90°F Directional sensitivity to cracks
Eddy-sonic	Debonded areas in metal-core or metal-faced honeycomb structures Delaminations in metal laminates or composites Crushed core	Metal-core honeycomb Metal-faced honeycomb Conductive laminates such as boron or graphite-fiber composites Bonded-metal panels	Portable Simple to operate No couplant required Locates far-side debonded areas Access to only one surface required May be automated	Specimen or part must contain conductive materials to establish eddy-current field Reference standards required Part geometry
Electric current	Cracks Crack depth Resistivity Wall thickness Corrosion-induced wall thinning	Metallic materials Electrically conductive materials Train rails Nuclear fuel elements Bars, plates other shapes	Access to only one surface required Battery or dc source Portable	Edge effect Surface contamination Good surface contact required Difficult to automate Electrode spacing Reference standards required
Electrified particle	Surface flaws in nonconducting material Through-to-metal pinholes on metal-backed material Tension, compression, cyclic cracks Brittle-coating stress cracks	Glass Porcelain enamel Nonhomogeneous materials such as plastic or asphalt coatings Glass-to-metal seals	Portable Useful on materials not practical for penetrant inspection	Poor resolution on thin coatings False indications from moisture streaks or lint Atmospheric conditions High-voltage discharge

Table 5.4.1 Nondestructive Test Methods* (*Continued*)

Method	Measures or defects	Applications	Advantages	Limitations
Filtered particle	Cracks Porosity Differential absorption	Porous materials such as clay, carbon, powdered metals, concrete Grinding wheels High-tension insulators Sanitary ware	Colored or fluorescent particles Leaves no residue after baking part over 400°F Quickly and easily applied Portable	Size and shape of particles must be selected before use Penetrating power of suspension medium is critical Particle concentration must be controlled Skin irritation
Infrared (radiometers) (thermography)	Hot spots Lack of bond Heat transfer Isotherms Temperature ranges	Brazed joints Adhesive-bonded joints Metallic platings or coatings; debonded areas or thickness Electrical assemblies Temperature monitoring	Sensitive to 0.1°F temperature variation Permanent record or thermal picture Quantitative Remote sensing; need not contact part Portable	Emissivity Liquid-nitrogen-cooled detector Critical time-temperature relationship Poor resolution for thick specimens Reference standards required
Leak testing	Leaks: Helium Ammonia Smoke Water Air bubbles Radioactive gas Halogens	Joints: Welded Brazed Adhesive-bonded Sealed assemblies Pressure or vacuum chambers Fuel or gas tanks	High sensitivity to extremely small, light separations not detectable by other NDT methods Sensitivity related to method selected	Accessibility to both surfaces of part required Smeared metal or contaminants may prevent detection Cost related to sensitivity
Magnetic particle	Surface and slightly subsurface flaws; cracks, seams, porosity, inclusions Permeability variations Extremely sensitive for locating small tight cracks	Ferromagnetic materials; bar, plate, forgings, weldments, extrusions, etc.	Advantage over penetrant is that it indicates subsurface flaws, particularly inclusions Relatively fast and low-cost May be portable	Alignment of magnetic field is critical Demagnetization of parts required after tests Parts must be cleaned before and after inspection Masking by surface coatings
Magnetic field (also magnetic flux leakage)	Cracks Wall thickness Hardness Coercive force Magnetic anisotropy Magnetic field Nonmagnetic coating thickness on steel	Ferromagnetic materials Ship degaussing Liquid-level control Treasure hunting Wall thickness of nonmetallic materials Material sorting	Measurement of magnetic material properties May be automated Easily detects magnetic objects in nonmagnetic material Portable	Permeability Reference standards required Edge effect Probe lift-off
Microwave (300 MHz– 300 GHz)	Cracks, holes, debonded areas, etc., in nonmetallic parts Changes in composition, degree of cure, moisture content Thickness measurement Dielectric constant Loss tangent	Reinforced plastics Chemical products Ceramics Resins Rubber Wood Liquids Polyurethane foam Radomes	Between radio waves and infrared in electromagnetic spectrum Portable Contact with part surface not normally required Can be automated	Will not penetrate metals Reference standards required Horn-to-part spacing critical Part geometry Wave interference Vibration
Liquid penetrants (dye or fluorescent)	Flaws open to surface of parts; cracks, porosity, seams, laps, etc. Through-wall leaks	All parts with nonabsorbing surfaces (forgings, weldments, castings, etc.). Note: Bleed-out from porous surfaces can mask indications of flaws	Low cost Portable Indications may be further examined visually Results easily interpreted	Surface films such as coatings, scale, and smeared metal may prevent detection of flaws Parts must be cleaned before and after inspection Flaws must be open to surface
Fluoroscopy (cinefluorography) (kinefluorography)	Level of fill in containers Foreign objects Internal components Density variations Voids, thickness Spacing or position	Flow of liquids Presence of cavitation Operation of valves and switches Burning in small solid-propellant rocket motors	High-brightness images Real-time viewing Image magnification Permanent record Moving subject can be observed	Costly equipment Geometric unsharpness Thick specimens Speed of event to be studied Viewing area Radiation hazard

Table 5.4.1   Nondestructive Test Methods*   (_Continued_)

Method	Measures or defects	Applications	Advantages	Limitations
Neutron radiology (thermal neutrons from reactor, accelerator, or Californium 252)	Hydrogen contamination of titanium or zirconium alloys Defective or improperly loaded pyrotechnic devices Improper assembly of metal, nonmetal parts Corrosion products	Pyrotechnic devices Metallic, nonmetallic assemblies Biological specimens Nuclear reactor fuel elements and control rods Adhesive-bonded structures	High neutron absorption by hydrogen, boron, lithium, cadmium, uranium, plutonium Low neutron absorption by most metals Complement to X-ray or gamma-ray radiography	Very costly equipment Nuclear reactor or accelerator required Trained physicists required Radiation hazard Nonportable Indium or gadolinium screens required
Gamma radiology (cobalt 60, iridium 192)	Internal flaws and variations, porosity, inclusions, cracks, lack of fusion, geometry variations, corrosion thinning Density variations Thickness, gap, and position	Usually where X-ray machines are not suitable because source cannot be placed in part with small openings and/or power source not available Panoramic imaging	Low initial cost Permanent records; film Small sources can be placed in parts with small openings Portable Low contrast	One energy level per source Source decay Radiation hazard Trained operators needed Lower image resolution Cost related to source size
X-ray radiology	Internal flaws and variations; porosity, inclusions, cracks, lack of fusion, geometry variations, corrosion Density variations Thickness, gap, and position Misassembly Misalignment	Castings Electrical assemblies Weldments Small, thin, complex wrought products Nonmetallics Solid-propellant rocket motors Composites Container contents	Permanent records; film Adjustable energy levels (5 kV–25 meV) High sensitivity to density changes No couplant required Geometry variations do not affect direction of X-ray beam	High initial costs Orientation of linear flaws in part may not be favorable Radiation hazard Depth of flaw not indicated Sensitivity decreases with increase in scattered radiation
Radiometry X-ray, gamma ray, beta ray) (transmission or backscatter)	Wall thickness Plating thickness Variations in density or composition Fill level in cans or containers Inclusions or voids	Sheet, plate, strip, tubing Nuclear reactor fuel rods Cans or containers Plated parts Composites	Fully automatic Fast Extremely accurate In-line process control Portable	Radiation hazard Beta ray useful for ultrathin coatings only Source decay Reference standards required
Reverse-geometry digital X-ray	Cracks Corrosion Water in honeycomb Carbon epoxy honeycomb Foreign objects	Aircraft structure	High-resolution 10^6 pixel image with high contrast	Access to both sides of object Radiation hazard
X-ray computed tomography (CT)	Small density changes Cracks Voids Foreign objects	Solid-propellant rocket motors Rocket nozzles Jet-engine parts Turbine blades	Measures X-ray opacity of object along many paths	Very expensive Trained operator Radiation hazard
Shearography electronic	Lack of bond Delaminations Plastic deformation Strain Crushed core Impact damage Corrosion in Al honeycomb	Composite-metal honeycomb Bonded structures Composite structures	Large area coverage Rapid setup and operation Noncontacting Video image easy to store	Requires vacuum thermal, ultrasonic, or microwave stressing of structure to cause surface strain
Thermal (thermochromic paint, liquid crystals)	Lack of bond Hot spots Heat transfer Isotherms Temperature ranges Blockage in coolant passages	Brazed joints Adhesive-bonded joints Metallic platings or coatings Electrical assemblies Temperature monitoring	Very low initial cost Can be readily applied to surfaces which may be difficult to inspect by other methods No special operator skills	Thin-walled surfaces only Critical time-temperature relationship Image retentivity affected by humidity Reference standards required
Sonic (less than 0.1 MHz)	Debonded areas or delaminations in metal or nonmetal composites or laminates Cohesive bond strength under controlled conditions Crushed or fractured core Bond integrity of metal insert fasteners	Metal or nonmetal composite or laminates brazed or adhesive-bonded Plywood Rocket-motor nozzles Honeycomb	Portable Easy to operate Locates far-side debonded areas May be automated Access to only one surface required	Surface geometry influences test results Reference standards required Adhesive or core-thickness variations influence results

Table 5.4.1   Nondestructive Test Methods*   *(Continued)*

Method	Measures or defects	Applications	Advantages	Limitations
Ultrasonic (0.1–25 MHz)	Internal flaws and variations; cracks, lack of fusion, porosity, inclusions, delaminations, lack of bond, texturing Thickness or velocity Poisson's ratio, elastic modulus	Metals Welds Brazed joints Adhesive-bonded joints Nonmetallics In-service parts	Most sensitive to cracks Test results known immediately Automating and permanent-record capability Portable High penetration capability	Couplant required Small, thin, or complex parts may be difficult to inspect Reference standards required Trained operators for manual inspection Special probes
Thermoelectric probe	Thermoelectric potential Coating thickness Physical properties Thompson effect *P-N* junctions in semiconductors	Metal sorting Ceramic coating thickness on metals Semiconductors	Portable Simple to operate Access to only one surface required	Hot probe Difficult to automate Reference standards required Surface contaminants Conductive coatings

* From Donald J. Hagemaier, "Metal Progress Databook," Douglas Aircraft Co., McDonnell-Douglas Corp., Long Beach, CA.

in a semidarkened location. After testing, the penetrant and developer are removed by washing with water, sometimes aided by an emulsifier, or with a solvent.

In **filtered particle** testing, cracks in **porous objects** (100 mesh or smaller) are indicated by the difference in absorption between a cracked and a flaw-free surface. A liquid containing suspended particles is sprayed on a test object. If a crack exists, particles are filtered out and concentrate at the surface as liquid flows into the additional absorbent area created by the crack. Fluorescent or colored particles are used to locate flaws in unfired dried clay, certain fired ceramics, concrete, some powdered metals, carbon, and partially sintered tungsten and titanium carbides.

RADIOGRAPHIC METHODS

Radiographic test methods employ X-rays, gamma rays, or similar penetrating radiation to reveal flaws, voids, inclusions, thickness, or structure of objects. Electromagnetic energy wavelengths in the range of 0.01 to 10 Å (1 Å = 10^{-8} cm) are used to examine the interior of opaque materials. Penetrating radiation proceeds from its source in straight lines to the test object. Rays are differentially absorbed by the object, depending upon the energy of the radiation and the nature and thickness of the material.

X-rays of a variety of wavelengths result when high-speed electrons in a vacuum tube are suddenly stopped. An X-ray tube contains a heated filament (cathode) and a target (anode); radiation intensity is almost directly proportional to filament current (mA); tube voltage (kV) determines the penetration capability of the rays. As tube voltage increases, shorter wavelengths and more intense X-rays are produced. When the energy of penetrating radiation increases, shorter wavelengths and more intense X-rays are produced. Also, when the energy of penetrating radiation increases, the difference in attenuation between materials decreases. Consequently, more film-image contrast is obtained at lower voltage, and a greater range of thickness can be radiographed at one time at higher voltage.

Gamma rays of a specific wavelength are emitted from the disintegrating nuclei of natural radioactive elements, such as radium, and from a variety of artificial radioactive isotopes produced in nuclear reactors. Cobalt 60 and iridium 192 are commonly used for industrial radiography. The half-life of an isotope is the time required for half of the radioactive material to decay. This time ranges from a few hours to many years.

Radiographs are photographic records produced by the passage of penetrating radiation onto a film. A void or reduced mass appears as a darker image on the film because of the lesser absorption of energy and the resulting additional exposure of the film. The quantity of X-rays absorbed by a material generally increases as the atomic number increases.

A radiograph is a shadow picture, since X-rays and gamma rays follow the laws of light in shadow formation. Four factors determine the best geometric sharpness of a picture: (1) The effective focal-spot size of the radiation source should be as small as possible. (2) The source-to-object distance should be adequate for proper definition of the area of the object farthest from the film. (3) The film should be as close as possible to the object. (4) The area of interest should be in the center of and perpendicular to the X-ray beams and parallel to the X-ray film.

Radiographic films vary in speed, contrast, and grain size. Slow films generally have smaller grain size and produce more contrast. Slow films are used where optimum sharpness and maximum contrast are desired. Fast films are used where objects with large differences in thickness are to be radiographed or where sharpness and contrast can be sacrificed to shorten exposure time. Exposure of a radiographic film comes from direct radiation and scattered radiation. **Direct radiation** is desirable, image-forming radiation; **scattered radiation,** which occurs in the object being X-rayed or in neighboring objects, produces undesirable images on the film and loss of contrast. **Intensifying screens** made of 0.005- or 0.010-in- (0.13-mm or 0.25-mm) thick lead are often used for radiography at voltages above 100 kV. The lead filters out much of the low-energy scatter radiation. Under action of X-rays or gamma rays above 88 kV, a lead screen also emits electrons which, when in intimate contact with the film, produce additional coherent darkening of the film. Exposure time can be materially reduced by use of intensifying screens above and below the film.

Penetrameters are used to indicate the contrast and definition which exist in a radiograph. The type generally used in the United States is a small rectangular plate of the same material as the object being X-rayed. It is uniform in thickness (usually 2 percent of the object thickness) and has holes drilled through it. ASTM specifies hole diameters 1, 2, and 4 times the thickness of the penetrameter. Step, wire, and bead penetrameters are also used. (See ASTM Materials Specification E94.)

Because of the variety of factors that affect the production and measurements of an X-ray image, operating factors are generally selected from reference tables or graphs which have been prepared from test data obtained for a range of operating conditions.

All materials may be inspected by radiographic means, but there are limitations to the configurations of materials. With optimum techniques, wires 0.0001 in (0.003 mm) in diameter can be resolved in small electrical components. At the other extreme, welded steel pressure vessels with 20-in (500-mm) wall thickness can be routinely inspected by use of high-energy accelerators as a source of radiation. **Neutron radiation** penetrates extremely dense materials such as lead more readily than X-rays or gamma rays but is attenuated by lighter-atomic-weight materials such as plastics, usually because of their hydrogen content.

Radiographic standards are published by ASTM, ASME, AWS, and API, primarily for detecting lack of penetration or lack of fusion in welded objects. Cast-metal objects are radiographed to detect conditions such as shrink, porosity, hot tears, cold shuts, inclusions, coarse structure, and cracks.

The usual method of utilizing penetrating radiation employs film. However, Geiger counters, semiconductors, phosphors (fluoroscopy), photoconductors (xeroradiography), scintillation crystals, and vidicon tubes (image intensifiers) are also used. Computerized digital radiography is an expanding technology.

The **dangers** connected with exposure of the human body to X-rays and gamma rays should be fully understood by any person responsible for the use of radiation equipment. NIST is a prime source of information concerning radiation safety. NRC specifies maximum permissible exposure to be a 1.25 R/¼ year.

ULTRASONIC METHODS

Ultrasonic nondestructive test methods employ high-frequency mechanical vibrational energy to detect and locate structural discontinuities or differences and to measure thickness of a variety of materials. An electric pulse is generated in a test instrument and transmitted to a transducer, which converts the electric pulse into mechanical vibrations. These low-energy-level vibrations are transmitted through a coupling liquid into the test object, where the ultrasonic energy is attenuated, scattered, reflected, or resonated to indicate conditions within material. Reflected, transmitted, or resonant sound energy is reconverted to electrical energy by a transducer and returned to the test instrument, where it is amplified. The received energy is then usually displayed on a cathode-ray tube. The presence, position, and amplitude of echoes indicate conditions of the test-object material.

Materials capable of being tested by ultrasonic energy are those which transmit vibrational energy. Metals are tested in dimensions of up to 30 ft (9.14 m). Noncellular plastics, ceramics, glass, new concrete, organic materials, and rubber can be tested. Each material has a characteristic sound velocity, which is a function of its density and modulus (elastic or shear).

Material characteristics determinable through ultrasonics include structural discontinuities, such as flaws and unbonds, physical constants and metallurgical differences, and thickness (measured from one side). A common application of ultrasonics is the inspection of welds for inclusions, porosity, lack of penetration, and lack of fusion. Other applications include location of unbond in laminated materials, location of fatigue cracks in machinery, and medical applications. Automatic testing is frequently performed in manufacturing applications.

Ultrasonic systems are classified as either **pulse-echo,** in which a single transducer is used, or **through-transmission,** in which separate sending and receiving transducers are used. Pulse-echo systems are more common. In either system, ultrasonic energy must be transmitted into, and received from, the test object through a **coupling medium,** since air will not efficiently transmit ultrasound of these frequencies. Water, oil, grease, and glycerin are commonly used couplants. Two types of testing are used: contact and immersion. In contact testing, the transducer is placed directly on the test object. In immersion testing, the transducer and test object are separated from one another in a tank filled with water or by a column of water or by a liquid-filled wheel. Immersion testing eliminates transducer wear and facilitates scanning of the test object. Scanning systems have paper-printing or computerized video equipment for readout of test information.

Ultrasonic transducers are piezoelectric units which convert electric energy into acoustic energy and convert acoustic energy into electric energy of the same frequency. Quartz, barium titanate, lithium sulfate, lead metaniobate, and lead zirconate titanate are commonly used transducer crystals, which are generally mounted with a damping backing in a housing. Transducers range in size from ¹⁄₁₆ to 5 in (0.15 to 12.7 cm) and are circular or rectangular. Ultrasonic beams can be focused to improve resolution and definition. Transducer characteristics and beam patterns are dependent upon frequency, size, crystal material, and construction.

Test frequencies used range from 40 kHz to 200 MHz. Flaw-detection and thickness-measurement applications use frequencies between 500 kHz and 25 MHz, with 2.25 and 5 MHz being most commonly employed for flaw detection. Low frequencies (40 kHz to 1.0 MHz) are used on materials of low elastic modulus or large grain size. High frequencies (2.25 to 25 MHz) provide better resolution of smaller defects and are used on fine-grain materials and thin sections. Frequencies above 25 MHz are employed for investigation and measurement of physical properties related to acoustic attenuation.

Wave-vibrational modes other than longitudinal are effective in detecting flaws that do not present a reflecting surface to the ultrasonic beam, or other characteristics not detectable by the longitudinal mode. They are useful also when large areas of plates must be examined. Wedges of plastic, water, or other material are inserted between the transducer face and the test object to convert, by refraction, to shear, transverse, surface, or Lamb vibrational modes. As in optics, Snell's law expresses the relationship between incident and refracted beam angles; i.e., the ratio of the sines of the angle from the normal, of the incident and refracted beams in two mediums, is equal to the ratio of the mode acoustic velocities in the two mediums.

Limiting conditions for ultrasonic testing may be the test-object shape, surface roughness, grain size, material structure, flaw orientation, selectivity of discontinuities, and the skill of the operator. Test sensitivity is less for cast metals than for wrought metals because of grain size and surface differences.

Standards for acceptance are published in many government, national society, and company specifications (see references above). Evaluation is made by comparing (visually or by automated electronic means) received signals with signals obtained from reference blocks containing flat bottom holes between ¹⁄₆₄ and ⁵⁄₆₄ in (0.40 and 0.325 cm) in diameter, or from parts containing known flaws, drilled holes, or machined notches.

EDDY CURRENT METHODS

Eddy current nondestructive tests are based upon correlation between electromagnetic properties and physical or structural properties of a test object. Eddy currents are induced in metals whenever they are brought into an ac magnetic field. These eddy currents create a secondary magnetic field, which opposes the inducing magnetic field. The presence of discontinuities or material variations alters eddy currents, thus changing the apparent impedance of the inducing coil or of a detection coil. Coil impedance indicates the magnitude and phase relationship of the eddy currents to their inducing magnetic-field current. This relationship is dependent upon the mass, conductivity, permeability, and structure of the metal and upon the frequency, intensity, and distribution of the alternating magnetic field. Conditions such as heat treatment, composition, hardness, phase transformation, case depth, cold working, strength, size, thickness, cracks, seams, and inhomogeneities are indicated by eddy current tests. Correlation data must usually be obtained to determine whether test conditions for desired characteristics of a particular test object can be established. Because of the many factors which cause variation in electromagnetic properties of metals, care must be taken that the instrument response to the condition of interest is not nullified or duplicated by variations due to other conditions.

Alternating-current **frequencies** between 1 and 5,000,000 Hz are used for eddy current testing. Test frequency determines the depth of current penetration into the test object, owing to the ac phenomenon of "skin effect." One "standard depth of penetration" is the depth at which the eddy currents are equal to 37 percent of their value at the surface. In a plane conductor, depth of penetration varies inversely as the square root of the product of conductivity, permeability, and frequency. High-frequency eddy currents are more sensitive to surface flaws or conditions while low-frequency eddy currents are sensitive also to deeper internal flaws or conditions.

Test coils are of three general types: the **circular coil,** which surrounds an object; the **bobbin coil,** which is inserted within an object; and the **probe coil,** which is placed on the surface of an object. Coils are further classified as **absolute,** when testing is conducted without direct comparison with a reference object in another coil; or **differential,** when comparison is made through use of two coils connected in series opposition. Many variations of these coil types are utilized. Axial length of a circular test coil should not be more than 4 in (10.2 cm), and its shape should correspond closely to the shape of the test object for best results. Coil diameter should be only slightly larger than the test-object diameter for consistent and useful results. Coils may be of the air-core or magnetic-core type.

Instrumentation for the analysis and presentation of electric signals resulting from eddy current testing includes a variety of means, ranging from meters to oscilloscopes to computers. Instrument meter or alarm circuits are adjusted to be sensitive only to signals of a certain electrical phase or amplitude, so that selected conditions are indicated while others are ignored. Automatic and automated testing is one of the principal advantages of the method.

Thickness measurement of metallic and nonmetallic coatings on metals is performed using eddy current principles. Coating thicknesses measured typically range from 0.0001 to 0.100 in (0.00025 to 0.25 cm). For measurement to be possible, coating conductivity must differ from that of the base metal.

MICROWAVE METHODS

Microwave test methods utilize electromagnetic energy to determine characteristics of nonmetallic substances, either solid or liquid. Frequencies used range from 1 to 3,000 GHz. Microwaves generated in a test instrument are transmitted by a waveguide through air to the test object. Analysis of reflected or transmitted energy indicates certain material characteristics, such as moisture content, composition, structure, density, degree of cure, aging, and presence of flaws. Other applications include thickness and displacement measurement in the range of 0.001 in (0.0025 cm) to more than 12 in (30.4 cm). Materials that can be tested include most solid and liquid nonmetals, such as chemicals, minerals, plastics, wood, ceramics, glass, and rubber.

INFRARED METHODS

Infrared nondestructive tests involve the detection of infrared electromagnetic energy emitted by a test object. Infrared radiation is produced naturally by all matter at all temperatures above absolute zero. Materials emit radiation at varying intensities, depending upon their temperature and surface characteristics. A **passive** infrared system detects the natural radiation of an unheated test object, while an **active** system employs a source to heat the test object, which then radiates infrared energy to a detector. Sensitive indication of temperature or temperature distribution through infrared detection is useful in locating irregularities in materials, in processing, or in the functioning of parts. Emission in the infrared range of 0.8 to 15 μm is collected optically, filtered, detected, and amplified by a test instrument which is designed around the characteristics of the detector material. Temperature variations on the order of 0.1°F can be indicated by meter or graphic means. Infrared theory and instrumentation are based upon radiation from a blackbody; therefore, **emissivity correction** must be made electrically in the test instrument or arithmetically from instrument readings.

ACOUSTIC SIGNATURE ANALYSIS

Acoustic signature analysis involves the analysis of sound energy emitted from an object to determine characteristics of the object. The object may be a simple casting or a complex manufacturing system. A passive test is one in which **sonic** energy is transmitted into the object. In this case, a mode of resonance is usually detected to correlate with cracks or structure variations, which cause a change in effective modulus of the object, such as a nodular iron casting. An active test is one in which the object emits sound as a result of being struck or as a result of being in operation. In this case, characteristics of the object may be correlated to damping time of the sound energy or to the presence or absence of a certain frequency of sound energy. Bearing wear in rotating machinery can often be detected prior to actual failure, for example. More complex analytical systems can monitor and control manufacturing processes, based upon analysis of emitted sound energy.

Acoustic emission is a technology distinctly separate from acoustic signature analysis and is one in which strain produces bursts of energy in an object. These are detected by **ultrasonic** transducers coupled to the object. Growth of microcracks, and other flaws, as well as incipient failure, is monitored by counting the pulses of energy from the object or recording the time rate of the pulses of energy in the ultrasonic range (usually a discrete frequency between 1 kHz and 1 MHz).

Section **6**

Materials of Engineering

BY

HOWARD S. BEAN   *Late Physicist, National Bureau of Standards*
HAROLD W. PAXTON   *United States Steel Professor Emeritus, Carnegie Mellon University*
JAMES D. REDMOND   *President, Technical Marketing Resources, Inc.*
MALCOLM BLAIR   *Technical & Research Director, Steel Founders Society of America*
ROBERT E. EPPICH   *Vice President, Technology, American Foundrymen's Society*
L. D. KUNSMAN   *Late Fellow Engineer, Research Labs, Westinghouse Electric Corp.*
C. L. CARLSON   *Late Fellow Engineer, Research Labs, Westinghouse Electric Corp.*
J. RANDOLPH KISSELL   *Partner, The TGB Partnership*
FRANK E. GOODWIN   *Vice President, Materials Science, II.ZRO, Inc.*
DON GRAHAM   *Manager, Turning Programs, Carboloy, Inc.*
ARTHUR COHEN   *Manager, Standards and Safety Engineering, Copper Development Assn.*
JOHN H. TUNDERMANN   *Vice President, Research & Technology, INCO Alloys International, Inc.*
JAMES D. SHEAROUSE, III   *Senior Development Engineer, The Dow Chemical Co.*
PETER K. JOHNSON   *Director, Marketing & Public Relations, Metal Powder Industries Federation*
JOHN R. SCHLEY   *Manager, Technical Marketing, RMI Titanium Co.*
ROBERT D. BARTHOLOMEW   *Engineer, Powell Labs, Ltd.*
DAVID A. SHIFLER   *Metallurgist, Naval Surface Warfare Center*
HAROLD M. WERNER   *Consultant*
RODNEY C. DEGROOT   *Research Plant Pathologist, Forest Products Lab, USDA*
DAVID W. GREEN   *Supervisory Research General Engineer, Forest Products Lab, USDA*
ROLAND HERNANDEZ   *Research General Engineer, Forest Products Lab, USDA*
RUSSELL C. MOODY   *Supervisory Research General Engineer, Forest Products Lab, USDA*
JOSEPH F. MURPHY   *Supervisory General Engineer, Forest Products Lab, USDA*
ROBERT J. ROSS   *Supervisory Research General Engineer, Forest Products Lab, USDA*
WILLIAM T. SIMPSON   *Research Forest Products Technologist, Forest Products Lab, USDA*
ANTON TENWOLDE   *Research Physicist, Forest Products Lab, USDA*
ROBERT H. WHITE   *Supervisory Wood Scientist, Forest Products Lab, USDA*
ANTONIO F. BALDO   *Professor of Mechanical Engineering, Emeritus, The City College, The City University of New York*
WILLIAM L. GAMBLE   *Professor of Civil Engineering, University of Illinois at Urbana-Champaign*
ARNOLD S. VERNICK   *Associate, Geraghty & Miller, Inc.*
JULIAN H. DANCY   *Consulting Engineer. Formerly Senior Technologist, Technology Division, Fuels and Lubricants Research Department, Texaco, Inc.*
STEPHEN R. SWANSON   *Professor of Mechanical Engineering, University of Utah*

6.1 GENERAL PROPERTIES OF MATERIALS
by Howard S. Bean

REFERENCES: "International Critical Tables," McGraw-Hill. "Smithsonian Physical Tables," Smithsonian Institution. Landolt, "Landolt-Börnstein, Zahlenwerte und Funktionen aus Physik, Chemie, Astronomie, Geophysik und Technik," Springer. "Handbook of Chemistry and Physics," Chemical Rubber Co. "Book of ASTM Standards," ASTM. "ASHRAE Refrigeration Data Book," ASHRAE. Brady, "Materials Handbook," McGraw-Hill. Mantell, "Engineering Materials Handbook," McGraw-Hill. International Union of Pure and Applied Chemistry, Butterworth Scientific Publications. "U.S. Standard Atmosphere," Government Printing Office. Tables of Thermodynamic Properties of Gases, *NIST Circ.* 564, ASME Steam Tables.

Thermodynamic properties of a variety of other specific materials are listed also in Secs. 4.1, 4.2, and 9.8. Sonic properties of several materials are listed in Sec. 12.6.

CHEMISTRY

Every **elementary substance** is made up of exceedingly small particles called **atoms** which are all alike and which cannot be further subdivided or broken up by **chemical processes**. It will be noted that this statement is virtually a definition of the term elementary substance and a limitation of the term chemical process. There are as many different classes or families of atoms as there are **chemical elements**. See Table 6.1.1.

Two or more atoms, either of the same kind or of different kinds, are, in the case of most elements, capable of uniting with one another to form a higher order of distinct particles called **molecules**. If the molecules or atoms of which any given material is composed are all exactly alike, the material is a **pure substance**. If they are not all alike, the material is a **mixture**.

If the atoms which compose the molecules of any pure substances are all of the same kind, the substance is, as already stated, an **elementary substance**. If the atoms which compose the molecules of a pure chemical substance are not all of the same kind, the substance is a **compound substance**. The atoms are to be considered as the smallest particles which occur separately in the structure of molecules of either compound or elementary substances, so far as can be determined by ordinary chemical analysis. The molecule of an element consists of a definite (usually small) number of its atoms. The molecule of a compound consists of one or more atoms of each of its several elements, the numbers of the

Table 6.1.1 Chemical Elements[a]

Element	Symbol	Atomic no.	Atomic weight[b]	Valence
Actinium	Ac	89		
Aluminum	Al	13	26.9815	3
Americium	Am	95		
Antimony	Sb	51	121.75	3, 5
Argon[c]	Ar	18	39.948	0
Arsenic[d]	As	33	74.9216	3, 5
Astatine	At	85		
Barium	Ba	56	137.34	2
Berkelium	Bk	97		
Beryllium	Be	4	9.0122	2
Bismuth	Bi	83	208.980	3, 5
Boron[d]	B	5	10.811[l]	3
Bromine[e]	Br	35	79.904[m]	1, 3, 5
Cadmium	Cd	48	112.40	2
Calcium	Ca	20	40.08	2
Californium	Cf	98		
Carbon[d]	C	6	12.01115[l]	2, 4
Cerium	Ce	58	140.12	3, 4
Cesium[k]	Cs	55	132.905	1
Chlorine[f]	Cl	17	35.453[m]	1, 3, 5, 7
Chromium	Cr	24	51.996[m]	2, 3, 6
Cobalt	Co	27	58.9332	2, 3
Columbium (see Niobium)				
Copper	Cu	29	63.546[m]	1, 2
Curium	Cm	96		
Dysprosium	Dy	66	162.50	3
Einsteinium	Es	99		
Erbium	Er	68	167.26	3
Europium	Eu	63	151.96	2, 3
Fermium	Fm	100		
Fluorine[g]	F	9	18.9984	1
Francium	Fr	87		
Gadolinium	Gd	64	157.25	3
Gallium[k]	Ga	31	69.72	2, 3
Germanium	Ge	32	72.59	2, 4
Gold	Au	79	196.967	1, 3
Hafnium	Hf	72	178.49	4
Helium[c]	He	2	4.0026	0
Holmium	Ho	67	164.930	3
Hydrogen[h]	H	1	1.00797[i]	1
Indium	In	49	114.82	1, 2, 3
Iodine[d]	I	53	126.9044	1, 3, 5, 7

Table 6.1.1   Chemical Elements[a]   (*Continued*)

Element	Symbol	Atomic no.	Atomic weight[b]	Valence
Iridium	Ir	77	192.2	2, 3, 4, 6
Iron	Fe	26	55.847[m]	2, 3
Krypton[c]	Kr	36	83.80	0
Lanthanum	La	57	138.91	3
Lead	Pb	82	207.19	2, 4
Lithium[i]	Li	3	6.939	1
Lutetium	Lu	71	174.97	3
Magnesium	Mg	12	24.312	2
Manganese	Mn	25	54.9380	2, 3, 4, 6, 7
Mendelevium	Md	101		
Mercury[e]	Hg	80	200.59	1, 2
Molybdenum	Mo	42	95.94	3, 4, 5, 6
Neodymium	Nd	60	144.24	3
Neon[c]	Ne	10	20.183	0
Neptunium	Np	93		
Nickel	Ni	28	58.71	2, 3, 4
Niobium	Nb	41	92.906	2, 3, 4, 5
Nitrogen[f]	N	7	14.0067	3, 5
Nobelium	No	102		
Osmium	Os	76	190.2	2, 3, 4, 6, 8
Oxygen[f]	O	8	15.9994[l]	2
Palladium	Pd	46	106.4	2, 4
Phosphorus[d]	P	15	30.9738	3, 5
Platinum	Pt	78	195.09	2, 4
Plutonium	Pu	94		
Polonium	Po	84		2, 4
Potassium	K	19	39.102	1
Praseodymium	Pr	59	140.907	3
Promethium	Pm	61		5
Protactinium	Pa	91		
Radium	Ra	88		2
Radon[j]	Rn	86		0
Rhenium	Re	75	186.2	1, 4, 7
Rhodium	Rh	45	102.905	3, 4
Rubidium	Rb	37	85.47	1
Ruthenium	Ru	44	101.07	3, 4, 6, 8
Samarium	Sm	62	150.35	3
Scandium	Sc	21	44.956	3
Selenium[d]	Se	34	78.96	2, 4, 6
Silicon[d]	Si	14	28.086[l]	4
Silver	Ag	47	107.868[m]	1
Sodium	Na	11	22.9898	1
Strontium	Sr	38	87.62	2
Sulfur[d]	S	16	32.064[l]	2, 4, 6
Tantalum	Ta	73	180.948	4, 5
Technetium	Tc	43		
Tellurium[d]	Te	52	127.60	2, 4, 6
Terbium	Tb	65	158.924	3
Thallium	Tl	81	204.37	1, 3
Thorium	Th	90	232.038	3
Thulium	Tm	69	168.934	3
Tin	Sn	50	118.69	2, 4
Titanium	Ti	22	47.90	3, 4
Tungsten	W	74	183.85	3, 4, 5, 6
Uranium	U	92	238.03	4, 6
Vanadium	V	23	50.942	1, 2, 3, 4, 5
Xenon[c]	Xe	54	131.30	0
Ytterbium	Yb	70	173.04	2, 3
Yttrium	Y	39	88.905	3
Zinc	Zn	30	65.37	2
Zirconium	Zr	40	91.22	4

[a] All the elements for which atomic weights listed are metals, except as otherwise indicated. No atomic weights are listed for most radioactive elements, as these elements have no fixed value.
[b] The atomic weights are based upon nuclidic mass of $C^{12} = 12$.
[c] Inert gas. [d] Metalloid. [e] Liquid. [f] Gas. [g] Most active gas. [h] Lightest gas. [i] Lightest metal. [j] Not placed. [k] Liquid at 25°C.
[l] The atomic weight varies because of natural variations in the isotopic composition of the element. The observed ranges are boron, ± 0.003; carbon, ± 0.00005; hydrogen, ± 0.00001; oxygen, ± 0.0001; silicon, ± 0.001; sulfur, ± 0.003.
[m] The atomic weight is believed to have an experimental uncertainty of the following magnitude: bromine, ± 0.001; chlorine, ± 0.001; chromium, ± 0.001; copper, ± 0.001; iron, ± 0.003; silver, ± 0.001. For other elements, the last digit given is believed to be reliable to ± 0.5.
SOURCE: Table courtesy IUPAC and Butterworth Scientific Publications.

various kinds of atoms and their arrangement being definite and fixed and determining the character of the compound. This notion of molecules and their constituent atoms is useful for interpreting the observed fact that chemical reactions—e.g., the analysis of a compound into its elements, the synthesis of a compound from the elements, or the changing of one or more compounds into one or more different compounds —take place so that the masses of the various substances concerned in a given reaction stand in definite and fixed ratios.

It appears from recent researches that some substances which cannot by any available means be decomposed into simpler substances and which must, therefore, be defined as elements, are continually undergoing spontaneous changes or **radioactive transformation** into other substances which can be recognized as physically and chemically different from the original substance. Radium is an element by the definition given and may be considered as made up of atoms. But it is assumed that these atoms, so called because they resist all efforts to break them up and are, therefore, apparently indivisible, nevertheless split up spontaneously, at a rate which scientists have not been able to influence in any way, into other atoms, thus forming other elementary substances of totally different properties. See Table 6.1.3.

The view generally accepted at present is that the atoms of all the chemical elements, including those not yet known to be radioactive, consist of several kinds of still smaller particles, three of which are known as **protons, neutrons,** and **electrons**. The protons are bound together in the atomic nucleus with other particles, including neutrons, and are positively charged. The neutrons are particles having approximately the mass of a proton but are uncharged. The electrons are negatively charged particles, all alike, external to the nucleus, and sufficient in number to neutralize the nuclear charge in an atom. The differences between the atoms of different chemical elements are due to the different numbers of these smaller particles composing them. According to the original Bohr theory, an ordinary atom is conceived as a stable system of such electrons revolving in closed orbits about the nucleus like the planets of the solar system around the sun. In a hydrogen atom, there is 1 proton and 1 electron; in a radium atom, there are 88 electrons surrounding a nucleus 226 times as massive as the hydrogen nucleus. Only a few, in general the outermost or valence electrons of such an atom, are subject to rearrangement within, or ejection from, the atom, thereby enabling it, because of its increased energy, to combine with other atoms to form molecules of either elementary substances or compounds. The **atomic number** of an element is the number of excess positive charges on the nucleus of the atom. The essential feature that dis-

Table 6.1.2   Solubility of Inorganic Substances in Water
(Number of grams of the anhydrous substance soluble in 1,000 g of water. The common name of the substance is given in parentheses)

Substance	Composition	Temperature, °F (°C)		
		32 (0)	122 (50)	212 (100)
Aluminum sulfate	$Al_2(SO_4)_3$	313	521	891
Aluminum potassium sulfate (potassium alum)	$Al_2K_2(SO_4)_4 \cdot 24H_2O$	30	170	1,540
Ammonium bicarbonate	NH_4HCO_3	119		
Ammonium chloride (sal ammoniac)	NH_4Cl	297	504	760
Ammonium nitrate	NH_4NO_3	1,183	3,440	8,710
Ammonium sulfate	$(NH_4)_2SO_4$	706	847	1,033
Barium chloride	$BaCl_2 \cdot 2H_2O$	317	436	587
Barium nitrate	$Ba(NO_3)_2$	50	172	345
Calcium carbonate (calcite)	$CaCO_3$	0.018*		0.88
Calcium chloride	$CaCl_2$	594		1,576
Calcium hydroxide (hydrated lime)	$Ca(OH)_2$	1.77		0.67
Calcium nitrate	$Ca(NO_3)_2 \cdot 4H_2O$	931	3,561	3,626
Calcium sulfate (gypsum)	$CaSO_4 \cdot 2H_2O$	1.76	2.06	1.69
Copper sulfate (blue vitriol)	$CuSO_4 \cdot 5H_2O$	140	334	753
Ferrous chloride	$FeCl_2 \cdot 4H_2O$	644§	820	1,060
Ferrous hydroxide	$Fe(OH)_2$	0.0067‡		
Ferrous sulfate (green vitriol or copperas)	$FeSO_4 \cdot 7H_2O$	156	482	
Ferric chloride	$FeCl_3$	730	3,160	5,369
Lead chloride	$PbCl_2$	6.73	16.7	33.3
Lead nitrate	$Pb(NO_3)_2$	403		1,255
Lead sulfate	$PbSO_4$	0.042†		
Magnesium carbonate	$MgCO_3$	0.13‡		
Magnesium chloride	$MgCl_2 \cdot 6H_2O$	524		723
Magnesium hydroxide (milk of magnesia)	$Mg(OH)_2$	0.009‡		
Magnesium nitrate	$Mg(NO_3)_2 \cdot 6H_2O$	665	903	
Magnesium sulfate (Epsom salts)	$MgSO_4 \cdot 7H_2O$	269	500	710
Potassium carbonate (potash)	K_2CO_3	893	1,216	1,562
Potassium chloride	KCl	284	435	566
Potassium hydroxide (caustic potash)	KOH	971	1,414	1,773
Potassium nitrate (saltpeter or niter)	KNO_3	131	851	2,477
Potassium sulfate	K_2SO_4	74	165	241
Sodium bicarbonate (baking soda)	$NaHCO_3$	69	145	
Sodium carbonate (sal soda or soda ash)	$NaCO_3 \cdot 10H_2O$	204	475	452
Sodium chloride (common salt)	$NaCl$	357	366	392
Sodium hydroxide (caustic soda)	$NaOH$	420	1,448	3,388
Sodium nitrate (Chile saltpeter)	$NaNO_3$	733	1,148	1,755
Sodium sulfate (Glauber salts)	$Na_2SO_4 \cdot 10H_2O$	49	466	422
Zinc chloride	$ZnCl_2$	2,044	4,702	6,147
Zinc nitrate	$Zn(NO_3)_2 \cdot 6H_2O$	947		
Zinc sulfate	$ZnSO_4 \cdot 7H_2O$	419	768	807

* 59°F.
† 68°F.
‡ In cold water.
§ 50°F.

Table 6.1.3 Periodic Table of the Elements

Key:
- Atomic Number (top)
- Element (name)
- Symbol
- Atomic weight based on $C^{12} = 12.00$
- ( ) denotes mass number of most stable known isotope
- Valence (bottom)

Category labels: Light metals · Brittle metals · Ductile metals · Low melting · Nonmetallic elements · Inert gases · Rare earth elements · LANTHANIDE SERIES · ACTINIDE SERIES · Transuranium elements

Light metals		Brittle metals						Ductile metals			Low melting		Nonmetallic elements				Inert gases
1 Hydrogen H 1.00797 1																	2 Helium He 4.0026 0
3 Lithium Li 6.939 1	4 Beryllium Be 9.0122 2											5 Boron B 10.811 3	6 Carbon C 12.01115 2,4	7 Nitrogen N 14.0067 3,5	8 Oxygen O 15.9994 2	9 Fluorine F 18.9984 1	10 Neon Ne 20.183 0
11 Sodium Na 22.9898 1	12 Magnesium Mg 24.312 2											13 Aluminum Al 26.9815 3	14 Silicon Si 28.086 4	15 Phosphorus P 30.9738 3,5	16 Sulphur S 32.064 2,4,6	17 Chlorine Cl 35.453 1,3,5,7	18 Argon Ar 39.948 0
19 Potassium K 39.102 1	20 Calcium Ca 40.08 2	21 Scandium Sc 44.956 3	22 Titanium Ti 47.90 3,4	23 Vanadium V 50.942 1,2,3,4,5	24 Chromium Cr 51.996 2,3,6	25 Manganese Mn 54.938 2,3,4,6,7	26 Iron Fe 55.847 3,4,6,8	27 Cobalt Co 58.9332 3,4	28 Nickel Ni 58.71 2,3,4	29 Copper Cu 63.546 1,2	30 Zinc Zn 65.37 2	31 Gallium Ga 69.72 2,3	32 Germanium Ge 72.59 2,4	33 Arsenic As 74.9216 3,5	34 Selenium Se 78.96 2,4,6	35 Bromine Br 79.904 1,3,5	36 Krypton Kr 83.80 0
37 Rubidium Rd 85.47 1	38 Strontium Sr 87.62 2	39 Yttrium Y 88.905 3	40 Zirconium Zr 91.22 4	41 Niobium Nb 92.906 2,3,4,5	42 Molybdenum Mo 95.94 3,4,5,6	43 Technetium Tc (99)	44 Ruthenium Ru 101.07 3,4,6,8	45 Rhodium Rh 103.905 3,4	46 Palladium Pd 106.4 2,4	47 Silver Ag 107.868 1	48 Cadmium Cd 112.40 2	49 Indium In 114.82 1,2,3	50 Tin Sn 118.69 2,4	51 Antimony Sb 121.75 3,5	52 Tellurium Te 127.60 2,4,6	53 Iodine I 126.9044 1,3,5,7	54 Xenon Xe 131.30 0
55 Cesium Cs 132.905 1	56 Barium Ba 137.34 2	57 Lanthanum La 137.91 3	72 Hafnium Hf 178.49 4	73 Tantalum Ta 180.948 4,5	74 Tungsten W 183.85 3,4,5,6	75 Rhenium Re 186.2 1,4,7	76 Osmium Os 190.2 2,3,4,6,8	77 Iridium Ir 192.2 2,3,4,6	78 Platinum Pt 195.09 2,4	79 Gold Au 196.967 1,3	80 Mercury Hg 200.59 1,2	81 Thallium Tl 204.37 1,3	82 Lead Pb 207.19 2,4	83 Bismuth Bi 208.98 3,5	84 Polonium Po (210) 2,4	85 Astatine At (210)	86 Radon Rn (212) 0
87 Francium Fr (223) 1	88 Radium Ra (227) 2	89 Actinium Ac (227)															103 Lawrencium Lw (257)

LANTHANIDE SERIES (Rare earth elements)

58 Cerium Ce 140.12 3,4	59 Praseodymium Pr 140.907 3	60 Neodymium Nd 144.24 3	61 Promethium Pm (147)	62 Samarium Sm 150.35 3	63 Europium Eu 151.96 2,3	64 Godalinium Gd 157.25 3	65 Terbium Tb 158.924 3	66 Dysprosium Dy 162.50 3	67 Holmium Ho 164.93 3	68 Erbium Er 167.26 3	69 Thulium Tm 168.934 3	70 Ytterbium Yb 173.04 2,3	71 Lutetium Lu 174.97 3

ACTINIDE SERIES (Transuranium elements)

90 Thorium Th 232.038	91 Protactinium Pa (231)	92 Uranium U 238.03	93 Neptunium Np (237)	94 Plutonium Pu (242)	95 Americium Am (243)	96 Curium Cm (245)	97 Berkelium Bk (249)	98 Californium Cf (249)	99 Einsteinium Es (254)	100 Fermium Fm (252)	101 Mendelevium Md (256)	102 Nobelium No (254)	

Table 6.1.4   Solubility of Gases in Water
(By volume at atmospheric pressure)

	t, °F (°C)				t, °F (°C)		
	32 (0)	68 (20)	212 (100)		32 (0)	68 (20)	212 (100)
Air	0.032	0.020	0.012	Hydrogen	0.023	0.020	0.018
Acetylene	1.89	1.12		Hydrogen sulfide	50	2.8	0.87
Ammonia	1,250	700		Hydrochloric acid	560	480	
Carbon dioxide	1.87	0.96	0.26	Nitrogen	0.026	0.017	0.0105
Carbon monoxide	0.039	0.025		Oxygen	0.053	0.034	0.185
Chlorine	5.0	2.5	0.00	Sulfuric acid	87	43	

tinguishes one element from another is this charge of the nucleus. It also determines the position of the element in the periodic table. Modern researches have shown the existence of **isotopes,** that is, two or more species of atoms having the same atomic number and thus occupying the same place in the periodic system, but differing somewhat in atomic weight. These isotopes are chemically identical and are merely different species of the same chemical element. Most of the ordinary inactive elements have been shown to consist of a mixture of isotopes. This convenient atomic model should be regarded as only a working hypothesis for coordinating a number of phenomena about which much yet remains to be known.

Calculation of the Percentage Composition of Substances   Add the atomic weights of the elements in the compound to obtain its molecular weight. Multiply the atomic weight of the element to be calculated by the number of atoms present (indicated in the formula by a subscript number) and by 100, and divide by the molecular weight of the compound. For example, hematite iron ore (Fe_2O_3) contains 69.94 percent of iron by weight, determined as follows: Molecular weight of Fe_2O_3 = $(55.84 \times 2) + (16 \times 3) = 159.68$. Percentage of iron in compound = $(55.84 \times 2) \times 100/159.68 = 69.94$.

SPECIFIC GRAVITIES AND DENSITIES

Table 6.1.5   Approximate Specific Gravities and Densities
(Water at 39°F and normal atmospheric pressure taken as unity) For more detailed data on any material, see the section dealing with the properties of that material. Data given are for usual room temperatures.

Substance	Specific gravity	Avg density	
		lb/ft³	kg/m³
Metals, Alloys, Ores*			
Aluminum, cast-hammered	2.55–2.80	165	2,643
Brass, cast-rolled	8.4–8.7	534	8,553
Bronze, aluminum	7.7	481	7,702
Bronze, 7.9–14% Sn	7.4–8.9	509	8,153
Bronze, phosphor	8.88	554	8,874
Copper, cast-rolled	8.8–8.95	556	8,906
Copper ore, pyrites	4.1–4.3	262	4,197
German silver	8.58	536	8,586
Gold, cast-hammered	19.25–19.35	1,205	19,300
Gold coin (U.S.)	17.18–17.2	1,073	17,190
Iridium	21.78–22.42	1,383	22,160
Iron, gray cast	7.03–7.13	442	7,079
Iron, cast, pig	7.2	450	7,207
Iron, wrought	7.6–7.9	485	7,658
Iron, spiegeleisen	7.5	468	7,496
Iron, ferrosilicon	6.7–7.3	437	6,984
Iron ore, hematite	5.2	325	5,206
Iron ore, limonite	3.6–4.0	237	3,796
Iron ore, magnetite	4.9–5.2	315	5,046
Iron slag	2.5–3.0	172	2,755
Lead	11.34	710	11,370
Lead ore, galena	7.3–7.6	465	7,449
Manganese	7.42	475	7,608
Manganese ore, pyrolusite	3.7–4.6	259	4,149
Mercury	13.546	847	13,570
Monel metal, rolled	8.97	555	8,688
Nickel	8.9	537	8,602

Table 6.1.5   Approximate Specific Gravities and Densities   (Continued)

Substance	Specific gravity	Avg density	
		lb/ft³	kg/m³
Platinum, cast-hammered	21.5	1,330	21,300
Silver, cast-hammered	10.4–10.6	656	10,510
Steel, cold-drawn	7.83	489	7,832
Steel, machine	7.80	487	7,800
Steel, tool	7.70–7.73	481	7,703
Tin, cast-hammered	7.2–7.5	459	7,352
Tin ore, cassiterite	6.4–7.0	418	6,695
Tungsten	19.22	1,200	18,820
Uranium	18.7	1,170	18,740
Zinc, cast-rolled	6.9–7.2	440	7,049
Zinc, ore, blende	3.9–4.2	253	4,052
Various Solids			
Cereals, oats, bulk	0.41	26	417
Cereals, barley, bulk	0.62	39	625
Cereals, corn, rye, bulk	0.73	45	721
Cereals, wheat, bulk	0.77	48	769
Cork	0.22–0.26	15	240
Cotton, flax, hemp	1.47–1.50	93	1,491
Fats	0.90–0.97	58	925
Flour, loose	0.40–0.50	28	448
Flour, pressed	0.70–0.80	47	753
Glass, common	2.40–2.80	162	2,595
Glass, plate or crown	2.45–2.72	161	2,580
Glass, crystal	2.90–3.00	184	1,950
Glass, flint	3.2–4.7	247	3,960
Hay and straw, bales	0.32	20	320
Leather	0.86–1.02	59	945
Paper	0.70–1.15	58	929
Plastics (see Sec. 6.12)			
Potatoes, piled	0.67	44	705
Rubber, caoutchouc	0.92–0.96	59	946
Rubber goods	1.0–2.0	94	1,506
Salt, granulated, piled	0.77	48	769
Saltpeter	2.11	132	2,115
Starch	1.53	96	1,539
Sulfur	1.93–2.07	125	2,001
Wool	1.32	82	1,315
Timber, Air-Dry			
Apple	0.66–0.74	44	705
Ash, black	0.55	34	545
Ash, white	0.64–0.71	42	973
Birch, sweet, yellow	0.71–0.72	44	705
Cedar, white, red	0.35	22	352
Cherry, wild red	0.43	27	433
Chestnut	0.48	30	481
Cypress	0.45–0.48	29	465
Fir, Douglas	0.48–0.55	32	513
Fir, balsam	0.40	25	401
Elm, white	0.56	35	561
Hemlock	0.45–0.50	29	465
Hickory	0.74–0.80	48	769
Locust	0.67–0.77	45	722
Mahogany	0.56–0.85	44	705
Maple, sugar	0.68	43	689
Maple, white	0.53	33	529
Oak, chestnut	0.74	46	737
Oak, live	0.87	54	866

Table 6.1.5   Approximate Specific Gravities and Densities *(Continued)*

Substance	Specific gravity	Avg density lb/ft³	Avg density kg/m³
Oak, red, black	0.64–0.71	42	673
Oak, white	0.77	48	770
Pine, Oregon	0.51	32	513
Pine, red	0.48	30	481
Pine, white	0.43	27	433
Pine, Southern	0.61–0.67	38–42	610–673
Pine, Norway	0.55	34	541
Poplar	0.43	27	433
Redwood, California	0.42	26	417
Spruce, white, red	0.45	28	449
Teak, African	0.99	62	994
Teak, Indian	0.66–0.88	48	769
Walnut, black	0.59	37	593
Willow	0.42–0.50	28	449
Various Liquids			
Alcohol, ethyl (100%)	0.789	49	802
Alcohol, methyl (100%)	0.796	50	809
Acid, muriatic, 40%	1.20	75	1,201
Acid, nitric, 91%	1.50	94	1,506
Acid, sulfuric, 87%	1.80	112	1,795
Chloroform	1.500	95	1,532
Ether	0.736	46	738
Lye, soda, 66%	1.70	106	1,699
Oils, vegetable	0.91–0.94	58	930
Oils, mineral, lubricants	0.88–0.94	57	914
Turpentine	0.861–0.867	54	866
Water			
Water, 4°C, max density	1.0	62.426	999.97
Water, 100°C	0.9584	59.812	958.10
Water, ice	0.88–0.92	56	897
Water, snow, fresh fallen	0.125	8	128
Water, seawater	1.02–1.03	64	1,025
Gases (see Sec. 4)			
Ashlar Masonry			
Granite, syenite, gneiss	2.4–2.7	159	2,549
Limestone	2.1–2.8	153	2,450
Marble	2.4–2.8	162	2,597
Sandstone	2.0–2.6	143	2,290
Bluestone	2.3–2.6	153	2,451
Rubble Masonry			
Granite, syenite, gneiss	2.3–2.6	153	2,451
Limestone	2.0–2.7	147	2,355
Sandstone	1.9–2.5	137	2,194
Bluestone	2.2–2.5	147	2,355
Marble	2.3–2.7	156	2,500
Dry Rubble Masonry			
Granite, syenite, gneiss	1.9–2.3	130	2,082
Limestone, marble	1.9–2.1	125	2,001
Sandstone, bluestone	1.8–1.9	110	1,762
Brick Masonry			
Hard brick	1.8–2.3	128	2,051
Medium brick	1.6–2.0	112	1,794
Soft brick	1.4–1.9	103	1,650
Sand-lime brick	1.4–2.2	112	1,794
Concrete Masonry			
Cement, stone, sand	2.2–2.4	144	2,309
Cement, slag, etc.	1.9–2.3	130	2,082
Cement, cinder, etc.	1.5–1.7	100	1,602
Various Building Materials			
Ashes, cinders	0.64–0.72	40–45	640–721
Cement, portland, loose	1.5	94	1,505
Portland cement	3.1–3.2	196	3,140
Lime, gypsum, loose	0.85–1.00	53–64	849–1,025
Mortar, lime, set	1.4–1.9	103	1,650
		94	1,505
Mortar, portland cement	2.08–2.25	135	2,163
Slags, bank slag	1.1–1.2	67–72	1,074–1,153
Slags, bank screenings	1.5–1.9	98–117	1,570–1,874
Slags, machine slag	1.5	96	1,538
Slags, slag sand	0.8–0.9	49–55	785–849

Table 6.1.5   Approximate Specific Gravities and Densities *(Continued)*

Substance	Specific gravity	Avg density lb/ft³	Avg density kg/m³
Earth, etc., Excavated			
Clay, dry	1.0	63	1,009
Clay, damp, plastic	1.76	110	1,761
Clay and gravel, dry	1.6	100	1,602
Earth, dry, loose	1.2	76	1,217
Earth, dry, packed	1.5	95	1,521
Earth, moist, loose	1.3	78	1,250
Earth, moist, packed	1.6	96	1,538
Earth, mud, flowing	1.7	108	1,730
Earth, mud, packed	1.8	115	1,841
Riprap, limestone	1.3–1.4	80–85	1,282–1,361
Riprap, sandstone	1.4	90	1,441
Riprap, shale	1.7	105	1,681
Sand, gravel, dry, loose	1.4–1.7	90–105	1,441–1,681
Sand, gravel, dry, packed	1.6–1.9	100–120	1,602–1,922
Sand, gravel, wet	1.89–2.16	126	2,019
Excavations in Water			
Sand or gravel	0.96	60	951
Sand or gravel and clay	1.00	65	1,041
Clay	1.28	80	1,281
River mud	1.44	90	1,432
Soil	1.12	70	1,122
Stone riprap	1.00	65	1,041
Minerals			
Asbestos	2.1–2.8	153	2,451
Barytes	4.50	281	4,504
Basalt	2.7–3.2	184	2,950
Bauxite	2.55	159	2,549
Bluestone	2.5–2.6	159	2,549
Borax	1.7–1.8	109	1,746
Chalk	1.8–2.8	143	2,291
Clay, marl	1.8–2.6	137	2,196
Dolomite	2.9	181	2,901
Feldspar, orthoclase	2.5–2.7	162	2,596
Gneiss	2.7–2.9	175	2,805
Granite	2.6–2.7	165	2,644
Greenstone, trap	2.8–3.2	187	2,998
Gypsum, alabaster	2.3–2.8	159	2,549
Hornblende	3.0	187	2,998
Limestone	2.1–2.86	155	2,484
Marble	2.6–2.86	170	2,725
Magnesite	3.0	187	2,998
Phosphate rock, apatite	3.2	200	3,204
Porphyry	2.6–2.9	172	2,758
Pumice, natural	0.37–0.90	40	641
Quartz, flint	2.5–2.8	165	2,645
Sandstone	2.0–2.6	143	2,291
Serpentine	2.7–2.8	171	2,740
Shale, slate	2.6–2.9	172	2,758
Soapstone, talc	2.6–2.8	169	2,709
Syenite	2.6–2.7	165	2,645
Stone, Quarried, Piled			
Basalt, granite, gneiss	1.5	96	1,579
Limestone, marble, quartz	1.5	95	1,572
Sandstone	1.3	82	1,314
Shale	1.5	92	1,474
Greenstone, hornblend	1.7	107	1,715
Bituminous Substances			
Asphaltum	1.1–1.5	81	1,298
Coal, anthracite	1.4–1.8	97	1,554
Coal, bituminous	1.2–1.5	84	1,346
Coal, lignite	1.1–1.4	78	1,250
Coal, peat, turf, dry	0.65–0.85	47	753
Coal, charcoal, pine	0.28–0.44	23	369
Coal, charcoal, oak	0.47–0.57	33	481
Coal, coke	1.0–1.4	75	1,201
Graphite	1.64–2.7	135	2,163
Paraffin	0.87–0.91	56	898
Petroleum	0.87	54	856

Table 6.1.5   Approximate Specific Gravities and Densities   (Continued)

Substance	Specific gravity	Avg density lb/ft³	kg/m³
Petroleum, refined (kerosene)	0.78–0.82	50	801
Petroleum, benzine	0.73–0.75	46	737
Petroleum, gasoline	0.70–0.75	45	721
Pitch	1.07–1.15	69	1,105
Tar, bituminous	1.20	75	1,201
Coal and Coke, Piled			
Coal, anthracite	0.75–0.93	47–58	753–930
Coal, bituminous, lignite	0.64–0.87	40–54	641–866
Coal, peat, turf	0.32–0.42	20–26	320–417
Coal, charcoal	0.16–0.23	10–14	160–224
Coal, coke	0.37–0.51	23–32	369–513

* See also Sec. 6.4.

Compressibility of Liquids

If v_1 and v_2 are the volumes of the liquids at pressures of p_1 and p_2 atm, respectively, at any temperature, the coefficient of compressibility b is given by the equation

$$b = \frac{1}{v_1}\frac{v_1 - v_2}{p_2 - p_1}$$

The value of $b \times 10^6$ for oils at low pressures at about 70°F varies from about 55 to 80; for mercury at 32°F, it is 3.9; for chloroform at 32°F, it is 100 and increases with the temperature to 200 at 140°F; for ethyl alcohol, it increases from about 100 at 32°F and low pressures to 125 at 104°F; for glycerin, it is about 24 at room temperature and low pressure.

Table 6.1.6   Specific Gravity and Density of Water at Atmospheric Pressure*
(Weights are in vacuo)

Temp, °C	Specific gravity	Density lb/ft³	kg/m³	Temp, °C	Specific gravity	Density lb/ft³	kg/m³
0	0.99987	62.4183	999.845	40	0.99224	61.9428	992.228
2	0.99997	62.4246	999.946	42	0.99147	61.894	991.447
4	1.00000	62.4266	999.955	44	0.99066	61.844	990.647
6	0.99997	62.4246	999.946	46	0.98982	61.791	989.797
8	0.99988	62.4189	999.854	48	0.98896	61.737	988.931
10	0.99973	62.4096	999.706	50	0.98807	61.682	988.050
12	0.99952	62.3969	999.502	52	0.98715	61.624	987.121
14	0.99927	62.3811	999.272	54	0.98621	61.566	986.192
16	0.99897	62.3623	998.948	56	0.98524	61.505	985.215
18	0.99862	62.3407	998.602	58	0.98425	61.443	984.222
20	0.99823	62.3164	998.213	60	0.98324	61.380	983.213
22	0.99780	62.2894	997.780	62	0.98220	61.315	982.172
24	0.99732	62.2598	997.304	64	0.98113	61.249	981.113
26	0.99681	62.2278	996.793	66	0.98005	61.181	980.025
28	0.99626	62.1934	996.242	68	0.97894	61.112	978.920
30	0.99567	62.1568	995.656	70	0.97781	61.041	977.783
32	0.99505	62.1179	995.033	72	0.97666	60.970	976.645
34	0.99440	62.0770	994.378	74	0.97548	60.896	975.460
36	0.99371	62.0341	993.691	76	0.97428	60.821	974.259
38	0.99299	61.9893	992.973	78	0.97307	60.745	973.041

* See also Secs. 4.2 and 6.10.

OTHER PHYSICAL DATA

Table 6.1.7   Average Composition of Dry Air between Sea
Level and 90-km (295,000-ft) Altitude

Element	Formula	% by Vol.	% by Mass	Molecular weight
Nitrogen	N_2	78.084	75.55	28.0134
Oxygen	O_2	20.948	23.15	31.9988
Argon	Ar	0.934	1.325	39.948
Carbon Dioxide	CO_2	0.0314	0.0477	44.00995
Neon	Ne	0.00182	0.00127	20.183
Helium	He	0.00052	0.000072	4.0026
Krypton	Kr	0.000114	0.000409	83.80
Methane	CH_4	0.0002	0.000111	16.043

From 0.0 to 0.00005 percent by volume of nine other gases.
Average composite molecular weight of air 28.9644.
SOURCE: "U.S. Standard Atmosphere," Government Printing Office.

Table 6.1.8   Volume of Water as a Function of Pressure and Temperature

Temp, °F (°C)	Pressure, atm								
	0	500	1,000	2,000	3,000	4,000	5,000	6,500	8,000
32 (0)	1.0000	0.9769	0.9566	0.9223	0.8954	0.8739	0.8565	0.8361	
68 (20)	1.0016	0.9804	0.9619	0.9312	0.9065	0.8855	0.8675	0.8444	0.8244
122 (50)	1.0128	0.9915	0.9732	0.9428	0.9183	0.8974	0.8792	0.8562	0.8369
176 (80)	1.0287	1.0071	0.9884	0.9568	0.9315	0.9097	0.8913	0.8679	0.8481

SOURCE: "International Critical Tables."

Table 6.1.9 Basic Properties of Several Metals
(Staff contribution)*

Material	Density,[†] g/cm³	Coefficient of linear thermal expansion,[‡] in/(in · °F) × 10⁻⁶	Thermal conductivity, Btu/(h · ft · °F)	Specific heat,[‡] Btu/(lb · °F)	Approx melting temp, °F	Modulus of elasticity, lb/in² × 10⁶	Poisson's ratio	Yield stress, lb/in² × 10³	Ultimate stress, lb/in² × 10³	Elongation, %
Aluminum 2024-T3	2.77	12.6	110	0.23	940	10.6	0.33	50	70	18
Aluminum 6061-T6	2.70	13.5	90	0.23	1,080	10.6	0.33	40	45	17
Aluminum 7079-T6	2.74	13.7	70	0.23	900	10.4	0.33	68	78	14
Beryllium, QMV	1.85	6.4–10.2	85	0.45	2,340	40–44	0.024–0.030	27–38	33–51	1–3.5
Copper, pure	8.90	9.2	227	0.092	1,980	17.0	0.32		See "Metals Handbook"	30
Gold, pure	19.32		172	0.031	1,950	10.8	0.42		18	
Lead, pure	11.34	29.3	21.4	0.031	620	2.0	0.40–0.45	1.3	2.6	20–50
Magnesium AZ31B-H24 (sheet)	1.77	14.5	55	0.25	1,100	6.5	0.35	22	37	15
Magnesium HK31A-H24	1.79	14.0	66	0.13	1,100	6.4	0.35	29	37	8
Molybdenum, wrought	10.3	3.0	83	0.07	4,730	40.0	0.32	80	120–200	Small
Nickel, pure	8.9	7.2	53	0.11	2,650	32.0	0.31§		See "Metals Handbook"	35–40
Platinum	21.45	5.0	40	0.031	3,217	21.3	0.39		20–24	Small
Plutonium, alpha phase	19.0–19.7	30.0	4.8	0.034	1,184	14.0	0.15–0.21	40	60	48
Silver, pure	10.5	11.0	241	0.056	1,760	10–11	0.37	8	18	36
Steel, AISI C1020 (hot-worked)	7.85	6.3	27	0.10	2,750	29–30	0.29	48	65	65
Steel, AISI 304 (sheet)	8.03	9.9	9.4	0.12	2,600	28	0.29	39	87	1–40
Tantalum	16.6	3.6	31	0.03	5,425	27.0	0.35		50–145	34
Thorium, induction melt	11.6	6.95	21.7	0.03	3,200	7–10	0.27	21	32	9
Titanium, B 120VCA (aged)	4.85	5.2	4.3	0.13	3,100	14.8	0.3	190	200	1–3
Tungsten	19.3	2.5	95	0.033	6,200	50	0.28	28	18–600	4
Uranium D-38	13.97	4.0–8.0	17	0.028	2,100	24	0.21	28	56	

Room-temperature properties are given. For further information, consult the "Metals Handbook" or a manufacturer's publication.
* Compiled by Anders Lundberg, University of California, and reproduced by permission.
† To obtain the preferred density units, kg/m³, multiply these values by 1,000.
‡ See also Tables 6.1.10 and 6.1.11.
§ At 25°C.

Table 6.1.10   Coefficient of Linear Thermal Expansion for Various Materials
[Mean values between 32 and 212°F except as noted; in/(in · °F) × 10^{-4}]

Metals		Other Materials		Paraffin:	
Aluminum bronze	0.094	Bakelite, bleached	0.122	32–61°F	0.592
Brass, cast	0.104	Brick	0.053	61–100°F	0.724
Brass, wire	0.107	Carbon—coke	0.030	100–120°F	2.612
Bronze	0.100	Cement, neat	0.060	Porcelain	0.02
Constantan (60 Cu, 40 Ni)	0.095	Concrete	0.060	Quarts:	
German silver	0.102	Ebonite	0.468	Parallel to axis	0.044
Iron:		Glass:		Perpend. to axis	0.074
Cast	0.059	Thermometer	0.045	Quarts, fused	0.0028
Soft forged	0.063	Hard	0.033	Rubber	0.428
Wire	0.080	Plate and crown	0.050	Vulcanite	0.400
Magnalium (85 Al, 15 Mg)	0.133	Flint	0.044	Wood (‖ to fiber):	
Phosphor bronze	0.094	Pyrex	0.018	Ash	0.053
Solder	0.134	Granite	0.04–0.05	Chestnut and maple	0.036
Speculum metal	0.107	Graphite	0.044	Oak	0.027
Steel:		Gutta percha	0.875	Pine	0.030
Bessemer, rolled hard	0.056	Ice	0.283	Across the fiber:	
Bessemer, rolled soft	0.063	Limestone	0.023–0.05	Chestnut and pine	0.019
Nickel (10% Ni)	0.073	Marble	0.02–0.09	Maple	0.027
Type metal	0.108	Masonry	0.025–0.050	Oak	0.030

Table 6.1.11   Specific Heat of Various Materials
[Mean values between 32 and 212°F; Btu/(lb · °F)]

Solids				Wood:	
Alloys:		Granite	0.195	Fir	0.65
Bismuth-tin	0.040–0.045	Graphite	0.201	Oak	0.57
Bell metal	0.086	Gypsum	0.259	Pine	0.67
Brass, yellow	0.0883	Hornblende	0.195	**Liquids**	
Brass, red	0.090	Humus (soil)	0.44	Acetic acid	0.51
Bronze	0.104	Ice:		Acetone	0.544
Constantan	0.098	−4°F	0.465	Alcohol (absolute)	0.58
German silver	0.095	32°F	0.487	Aniline	0.49
Lipowits's metal	0.040	India rubber (Para)	0.27–0.48	Bensol	0.40
Nickel steel	0.109	Kaolin	0.224	Chloroform	0.23
Rose's metal	0.050	Limestone	0.217	Ether	0.54
Solders (Pb and Sn)	0.040–0.045	Marble	0.210	Ethyl acetate	0.478
Type metal	0.0388	Oxides:		Ethylene glycol	0.602
Wood's metal	0.040	Alumina (Al_3O_2)	0.183	Fusel oil	0.56
40 Pb + 60 Bi	0.0317	Cu_2O	0.111	Gasoline	0.50
25 Pb + 75 Bi	0.030	Lead oxide (PbO)	0.055	Glycerin	0.58
Asbestos	0.20	Lodestone	0.156	Hydrochloric acid	0.60
Ashes	0.20	Magnesia	0.222	Kerosene	0.50
Bakelite	0.3–0.4	Magnetite (Fe_3O_4)	0.168	Naphthalene	0.31
Basalt (lava)	0.20	Silica	0.191	Machine oil	0.40
Borax	0.229	Soda	0.231	Mercury	0.033
Brick	0.22	Zinc oxide (ZnO)	0.125	Olive oil	0.40
Carbon-coke	0.203	Paraffin wax	0.69	Paraffin oil	0.52
Chalk	0.215	Porcelain	0.22	Petroleum	0.50
Charcoal	0.20	Quarts	0.17–0.28	Sulfuric acid	0.336
Cinders	0.18	Quicklime	0.21	Sea water	0.94
Coal	0.3	Malt, rock	0.21	Toluene	0.40
Concrete	0.156	Sand	0.195	Turpentine	0.42
Cork	0.485	Sandstone	0.22	Molten metals:	
Corundum	0.198	Serpentine	0.25	Bismuth (535–725°F)	0.036
Dolomite	0.222	Sulfur	0.180	Lead (590–680°F)	0.041
Ebonite	0.33	Talc	0.209	Sulfur (246–297°F)	0.235
Glass:		Tufa	0.33	Tin (460–660°F)	0.058
Normal	0.199	Vulcanite	0.331		
Crown	0.16				
Flint	0.12				

6.2 IRON AND STEEL
by Harold W. Paxton

REFERENCES: "Metals Handbook," ASM International, 10th ed., ASTM Standards, pt. 1. SAE Handbook. "Steel Products Manual," AISI. "Making, Shaping and Treating of Steel," AISE, 10th ed.

CLASSIFICATION OF IRON AND STEEL

Iron (Fe) is not a high-purity metal commercially but contains other chemical elements which have a large effect on its physical and mechanical properties. The amount and distribution of these elements are dependent upon the method of manufacture. The most important commercial forms of iron are listed below.

Pig iron is the product of the blast furnace and is made by the reduction of iron ore.

Cast iron is an alloy of iron containing enough carbon to have a low melting temperature and which can be cast to close to final shape. It is not generally capable of being deformed before entering service.

Gray cast iron is an iron which, as cast, has combined carbon (in the form of cementite, Fe_3C) not in excess of a eutectoid percentage—the balance of the carbon occurring as graphite flakes. The term "gray iron" is derived from the characteristic gray fracture of this metal.

White cast iron contains carbon in the combined form. The presence of cementite or iron carbide (Fe_3C) makes this metal hard and brittle, and the absence of graphite gives the fracture a white color.

Malleable cast iron is an alloy in which all the combined carbon in a special white cast iron has been changed to free or temper carbon by suitable heat treatment.

Nodular (ductile) cast iron is produced by adding alloys of magnesium or cerium to molten iron. These additions cause the graphite to form into small nodules, resulting in a higher-strength, ductile iron.

Ingot iron, electrolytic iron (an iron-hydrogen alloy), and **wrought iron** are terms for low-carbon materials which are no longer serious items of commerce but do have considerable historical interest.

Steel is an alloy predominantly of iron and carbon, usually containing measurable amounts of manganese, and often readily formable.

Carbon steel is steel that owes its distinctive properties chiefly to the carbon it contains.

Alloy steel is steel that owes its distinctive properties chiefly to some element or elements other than carbon, or jointly to such other elements and carbon. Some alloy steels necessarily contain an important percentage of carbon, even as much as 1.25 percent. There is no complete agreement about where to draw the line between the alloy steels and the carbon steels.

Basic oxygen steel and **electric-furnace steel** are steels made by the basic oxygen furnace and electric furnace processes, irrespective of carbon content; the effective individual alloy content in engineering steels can range from 0.05 percent up to 3 percent, with a total usually less than 5 percent. Open-hearth and Bessemer steelmaking are no longer practiced in the United States.

Iron ore is reduced in a blast furnace to form **pig iron,** which is the raw material for practically all iron and steel products. Formerly, nearly 90 percent of the iron ore used in the United States came from the Lake Superior district; the ore had the advantages of high quality and the cheapness with which it could be mined and transported by way of the Great Lakes. With the rise of global steelmaking and the availability of high-grade ores and pellets (made on a large scale from low-grade ores) from many sources, the choice of feedstock becomes an economic decision.

The modern **blast furnace** consists of a vertical shaft up to 10 m or 40 ft in diameter and over 30 m (100 ft) high containing a descending column of iron ore, coke, and limestone and a large volume of ascending hot gas. The gas is produced by the burning of coke in the hearth of the furnace and contains about 34 percent carbon monoxide. This gas reduces the iron ore to metallic iron, which melts and picks up considerable quantities of carbon, manganese, phosphorus, sulfur, and silicon. The **gangue** (mostly silica) of the iron ore and the ash in the coke combine with the limestone to form the blast-furnace slag. The pig iron and slag are drawn off at intervals from the hearth through the iron notch and cinder notch, respectively. Some of the larger blast furnaces produce around 10,000 tons of pig iron per day. The blast furnace produces a liquid product for one of three applications: (1) the huge majority passes to the steelmaking process for refining; (2) pig iron is used in foundries for making castings; and (3) ferroalloys, which contain a considerable percentage of another metallic element, are used as addition agents in steelmaking. Compositions of commercial pig irons and two ferroalloys (ferromanganese and ferrosilicon) are listed in Table 6.2.1.

Physical Constants of Unalloyed Iron Some physical properties of iron and even its dilute alloys are sensitive to small changes in composition, grain size, or degree of cold work. The following are reasonably accurate for "pure" iron at or near room temperature; those with an asterisk are sensitive to these variables perhaps by 10 percent or more. Those with a dagger (†) depend measurably on temperature; more extended tables should be consulted.

Specific gravity, 7.866; melting point, 1,536°C (2,797°F); heat of fusion 277 kJ/kg (119 Btu/lbm); thermal conductivity 80.2 W/(m · C) [557 Btu/(h · ft² · in · °F)*†; thermal coefficient of expansion 12 × 10^{-6}/°C (6.7 × 10^{-6}/°F)†; electrical resistivity 9.7 $\mu\Omega$ · cm*†; and temperature coefficient of electrical resistance 0.0065/°C (0.0036/°F).†

Mechanical Properties Representative mechanical properties of annealed low-carbon steel (often similar to the former ingot iron) are as follows: yield strength 130 to 150 MPa (20 to 25 ksi); tensile strength 260 to 300 MPa (40 to 50 ksi); elongation 20 to 45 percent in 2 in; reduction in area of 60 to 75 percent; Brinell hardness 65 to 100. These figures are at best approximate and depend on composition (especially trace additives) and processing variables. For more precise data, suppliers or broader databases should be consulted.

Young's modulus for ingot iron is 202,000 MPa (29,300,000 lb/in²) in both tension and compression, and the shear modulus is 81,400 MPa (11,800,000 lb/in²). Poisson's ratio is 0.28. The **effect of cold rolling** on the tensile strength, yield strength, elongation, and shape of the stress-strain curve is shown in Fig. 6.2.1, which is for Armco ingot iron but would not be substantially different for other low-carbon steels.

Uses Low-carbon materials weld evenly and easily in all processes, can be tailored to be readily paintable and to be enameled, and with other treatments make an excellent low-cost soft magnetic material with high permeability and low coercive force for mass-produced motors and transformers. Other uses, usually after galvanizing, include culverts, flumes, roofing, siding, and housing frames; thin plates can be used in oil and water tanks, boilers, gas holders, and various nondemanding pipes; enameled sheet retains a strong market in ranges, refrigerators, and other household goods, in spite of challenges from plastics.

STEEL

Steel Manufacturing

Steel is produced by the removal of impurities from pig iron in a basic oxygen furnace or an electric furnace.

Basic Oxygen Steel This steel is produced by blowing pure (99 percent) oxygen either vertically under high pressure (1.2 MPa or

Table 6.2.1 Types of Pig Iron for Steelmaking and Foundry Use

Designation	Chemical composition, %*				Principal use
	Si	P	Mn	C†	
Basic pig, northern	1.50 max	0.400 max	1.01–2.00	3.5–4.4	Basic oxygen steel
In steps of	0.25		0.50		
Foundry, northern	3.50 max	0.301–0.700	0.50–1.25	3.0–4.5	A wide variety of castings
In steps of	0.25		0.25		
Foundry, southern	3.50 max	0.700–0.900	0.40–0.75	3.0–4.5	Cast-iron pipe
In steps of	0.25		0.25		
Ferromanganese (3 grades)	1.2 max	0.35 max	74–82	7.4 max	Addition of manganese to steel or cast iron
Ferrosilicon (silvery pig)	5.00–17.00	0.300 max	1.00–2.00	1.5 max	Addition of silicon to steel or cast iron

* Excerpted from ''The Making, Shaping and Treating of Steel,'' AISE, 1984; further information in ''Steel Products Manual,'' AISI, and ASTM Standards, Pt. 1.
† Carbon content not specified—for information only. Usually S is 0.05 max (0.06 for ferrosilicon) but S and P for basic oxygen steel are typically much lower today.

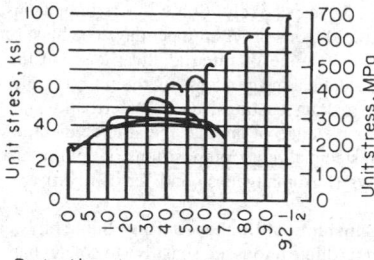

Fig. 6.2.1 Effect of cold rolling on the stress-strain relationship of Armco ingot iron. *(Kenyon and Burns.)*

175 lb/in²) onto the surface of molten pig iron (BOP) or through tuyeres in the base of the vessel (the Q-BOP process). Some facilities use a combination depending on local circumstances and product mix. This is an autogenous process that requires no external heat to be supplied. The furnaces are similar in shape to the former Bessemer converters but range in capacity to 275 metric tons (t) (300 net tons) or more. The barrel-shaped furnace or vessel may or may not be closed on the bottom, is open at the top, and can rotate in a vertical plane about a horizontal axis for charging and for pouring the finished steel. Selected scrap is charged into the vessel first, up to 30 percent by weight of the total charge. Molten pig iron (often purified from the raw blast-furnace hot metal to give lower sulfur, phosphorus, and sometimes silicon) is poured into the vessel. In the Q-BOP, oxygen must be flowing through the bottom tuyeres at this time to prevent clogging; further flow serves to refine the charge and carries in fluxes as powders. In the BOP process, oxygen is introduced through a water-cooled lance introduced through the top of the vessel. Within seconds after the oxygen is turned on, some iron in the charge is converted to ferrous oxide, which reacts rapidly with the impurities of the charge to remove them from the metal. As soon as reaction starts, limestone is added as a flux. Blowing is continued until the desired degree of purification is attained. The reactions take place very rapidly, and blowing of a heat is completed in about 20 min in a 200-net-ton furnace. Because of the speed of the process, a computer is used to calculate the charge required for making a given heat of steel, the rate and duration of oxygen blowing, and to regulate the quantity and timing of additions during the blow and for finishing the steel. Production rates of well over 270 t per furnace hour (300 net tons) can be attained. The comparatively low investment cost and low cost of operation have already made the basic oxygen process the largest producer of steel in the world, and along with electric furnaces, it almost completely replaces the basic open hearth as the major steelmaking process. No open hearths operate in the United States today.

Electric Steel The biggest change in steelmaking over the last 20 years is the fraction of steel made by remelting scrap in an electric furnace (EF), originally to serve a relatively nondemanding local market, but increasingly moving up in quality and products to compete with mills using the blast-furnace/oxygen steelmaking route. The eco-

nomic competition is fierce and has served to improve choices for customers.

Early processes used three-phase alternating current, but increasingly the movement is to a single dc electrode with a conducting hearth. The high-power densities necessitate water cooling and improved basic refractory linings. Scrap is charged into the furnace, which usually contains some of the last heat to improve efficiency. Older practices often had a second slag made after the first meltdown and refining by oxygen blowing, but today, final refining takes place outside the melting unit in a ladle furnace, which allows refining, temperature control, and alloying additions to be made without interfering with the next heat. The materials are continuously cast with various degrees of sophistication including slabs only 50 mm (2 in) thick.

The degree to which electric melting can replace more conventional methods is of great interest and depends in large part on the availability of sufficiently pure scrap at an attractive price and some improvements in surface quality to be able to make the highest-value products. Advances in EF technology are countered aggressively by new developments and cost control in traditional steelmaking; it may well be a decade or more before the pattern clarifies.

The **induction furnace** is simply a fairly small melting furnace to which the various metals are added to make the desired alloy, usually quite specialized. When steel scrap is used as a charge, it will be a high-grade scrap the composition of which is well known (see also Sec. 7).

Ladle Metallurgy One of the biggest contributors to quality in steel products is the concept of refining liquid steel outside the first melting unit—BOP, Q-BOP, or EF, none of which is well designed to perform the refining function. In this separate unit, gases in solution (oxygen, hydrogen, and, to a lesser extent, nitrogen) can be reduced by vacuum treatment, carbon can be adjusted to desirable very low levels by reaction with oxygen in solution, alloy elements can be added, the temperature can be adjusted, and the liquid steel can be stirred by inert gases to float out inclusions and provide a homogeneous charge to the continuous casters which are now virtually ubiquitous. Reducing oxygen in solution means a ''cleaner'' steel (fewer nonmetallic inclusions) and a more efficient recovery of alloying elements added with a purpose and which otherwise might end up as oxides.

Steel Ingots With the advent of continuous casters, ingot casting is now generally reserved for the production of relatively small volumes of material such as heavy plates and forgings which are too big for current casters. Ingot casting, apart from being inefficient in that the large volume change from liquid to solid must be handled by discarding the large void space usually at the top of the ingot (the **pipe**), also has several other undesirable features caused by the solidification pattern in a large volume, most notably significant differences in composition throughout the piece (**segregation**) leading to different properties, **inclusions** formed during solidification, and surface flaws from poor mold surfaces, splashing and other practices, which if not properly removed lead to defects in finished products (**seams, scabs, scale,** etc.). Some defects can be removed or attenuated, but others cannot; in general, with the exception of some very specialized tool and bearing steels, products from ingots are no longer state-of-the-art unless they are needed for size.

Continuous Casting   This concept, which began with Bessemer in the 1850s, began to be a reliable production tool around 1970 and since then has replaced basically all ingot casting. Industrialized countries all continuously cast well above 90 percent of their production. Sizes cast range from 2-m (80-in)—or more—by 0.3-m (12-in) **slabs** down to 0.1-m (4-in) square or round **billets**. Multiple strands are common where production volume is important. Many heats of steel can be cast in a continuous string with changes of width possible during operation. Changes of composition are possible in succeeding ladles with a discard of the short length of mixed composition. There is great interest in casting much thinner slabs or even casting directly to sheet; this is currently possible with some quality loss, but major efforts around the world to reduce differences are underway.

By intensive process control, it is often possible to avoid cooling the cast slabs to room temperature for inspection, enabling energy savings since the slabs require less reheating before hot rolling. If for some reason the slabs are cooled to room temperature, any surface defects which might lead to quality problems can be removed—usually by **scarfing** with an oxyacetylene torch or by grinding. Since this represents a yield loss, there is a real economic incentive to avoid the formation of such defects by paying attention to casting practices.

Mechanical Treatment of Steel

Cast steel, in the form of slabs, billets, or bars (these latter two differ somewhat arbitrarily in size) is treated further by various combinations of hot and cold deformation to produce a finished product for sale from the mill. Further treatments by fabricators usually occur before delivery to the final customer. These treatments have three purposes: (1) to change the shape by deformation or metal removal to desired tolerances; (2) to break up—at least partially—the segregation and large grain sizes inevitably formed during the solidification process and to redistribute the nonmetallic inclusions which are present; and (3) to change the properties. For example, these may be functional—strength or toughness—or largely aesthetic, such as reflectivity.

These purposes may be separable or in many cases may be acting simultaneously. An example is hot-rolled sheet or plate in which often the rolling schedule (reductions and temperature of each pass, and the cooling rate after the last reduction) is a critical path to obtain the properties and sizes desired and is often known as "heat treatment on the mill."

Most steels are reduced after appropriate heating (to above 1,000°C) in various multistand hot rolling mills to produce **sheet, strip, plate, tubes, shaped sections,** or **bars**. More specialized deformation, e.g., by **hammer forging**, can result in working in more than one direction, with a distribution of inclusions which is not extended in one direction. Rolling, e.g., more readily imparts anisotropic properties. **Press forging** at slow strain rates changes the worked structure to greater depths and is preferred for high-quality products. The degree of reduction required to eliminate the cast structure varies from 4:1 to 10:1; clearly smaller reductions would be desirable but are currently not usual.

The slabs, blooms, and billets from the caster must be reheated in an atmosphere-controlled furnace to the working temperature, often from room temperature, but if practices permit, they may be charged hot to save energy. Coupling the hot deformation process directly to slabs at the continuous caster exit is potentially more efficient, but practical difficulties currently limit this to a small fraction of total production.

The steel is oxidized during heating to some degree, and this oxidation is removed by a combination of light deformation and high-pressure water sprays before the principal deformation is applied. There are differences in detail between processes, but as a representative example, the conventional production of wide "hot-rolled sheet" [> 1.5 m (60 in)] will be discussed.

The slab, about 0.3 m (12 in) thick at about 1200°C is passed through a *scale breaker* and high-pressure water sprays to remove the oxide film. It then passes through a set of *roughing passes* (possibly with some modest width reduction) to reduce the thickness to just over 25 mm (1 in), the ends are sheared perpendicular to the length to remove irregularities, and finally they are fed into a series of up to seven roll stands each of which creates a reduction of 50 to 10 percent passing

along the train. Process controls allow each mill stand to run sufficiently faster than the previous one to maintain tension and avoid pileups between stands. The temperature of the sheet is a balance between heat added by deformation and that lost by heat transfer, sometimes with interstand water sprays. Ideally the temperature should not vary between head and tail of the sheet, but this is hard to accomplish.

The deformation encourages recrystallization and even some grain growth between stands; even though the time is short, temperatures are high. Emerging from the last stand between 815 and 950°C, the austenite may or may not recrystallize, depending on the temperature. At higher temperatures, when austenite does recrystallize, the grain size is usually small (often in the 10- to 20-μm range). At lower exit temperatures austenite grains are rolled into "pancakes" with the short dimension often less than 10 μm. Since several ferrite grains nucleate from each austenite grain during subsequent cooling, the ferrite grain size can be as low as 3 to 6 μm (ASTM 14 to 12). We shall see later that small ferrite grain sizes are a major contributor to the superior properties of today's carbon steels, which provide good strength and superior toughness simultaneously and economically.

Some of these steels also incorporate strong carbide and nitride formers in small amounts to provide extra strength from precipitation hardening; the degree to which these are undissolved in austenite during hot rolling affects recrystallization significantly. The subject is too complex to treat briefly here; the interested reader is referred to the ASM "Metals Handbook," 10th ed., vol. 1, pp. 389–423.

After the last pass, the strip may be cooled by programmed water sprays to between 510 and 730°C so that during coiling, any desired precipitation processes may take place in the coiler. The finished coil, usually 2 to 3 mm (0.080 to 0.120 in) thick and sometimes 1.3 to 1.5 mm (0.052 to 0.060 in) thick, which by now has a light oxide coating, is taken off line and either shippped directly or retained for further processing to make higher value-added products. Depending on composition, typical values of yield strength are from 210 up to 380 MPa (30 to 55 ksi), UTS in the range of 400 to 550 MPa (58 to 80 ksi), with an elongation in 200 mm (8 in) of about 20 percent. The higher strengths correspond to low-alloy steels.

About half the sheet produced is sold directly as hot-rolled sheet. The remainder is further cold-worked after scale removal by **pickling** and either is sold as cold-worked to various **tempers** or is recrystallized to form a very formable product known as **cold-rolled and annealed,** or more usually as **cold-rolled,** sheet. Strengthening by cold work is common in sheet, strip, wire, or bars. It provides an inexpensive addition to strength but at the cost of a serious loss of ductility, often a better surface finish, and finished product held to tighter tolerances. It improves springiness by increasing the yield strength, but does not change the elastic moduli. Examples of the effect of cold working on carbon-steel drawn wires are shown in Figs. 6.2.2 and 6.2.3.

To make the highest class of formable sheet is a very sophisticated operation. After pickling, the sheet is again reduced in a multistand (three, four, or five) mill with great attention paid to tolerances and surface finish. Reductions per pass range from 25 to 45 percent in early passes to 10 to 30 percent in the last pass. The considerable heat generated necessitates an oil-water mixture to cool and to provide the necessary lubrication. The finished coil is degreased prior to annealing.

The purpose of **annealing** is to provide, for the most demanding applications, pancake-shaped grains after recrystallization of the cold-worked ferrite, in a matrix with a very sharp crystal texture containing little or no carbon or nitrogen in solution. The exact metallurgy is complex but well understood. Two types of annealing are possible: slow heating, holding, and cooling of coils in a hydrogen atmosphere (box annealing) lasting several days, or continuous feeding through a furnace with a computer-controlled time-temperature cycle. The latter is much quicker but very capital-intensive and requires careful and complex process control.

As requirements for formability are reduced, production controls can be relaxed. In order of increasing cost, the series is commercial quality (CQ), drawing quality (DQ), deep drawing quality (DDQ), and extra deep drawing quality (EDDQ). Even more formable steels are possible, but they are not often commercially interesting.

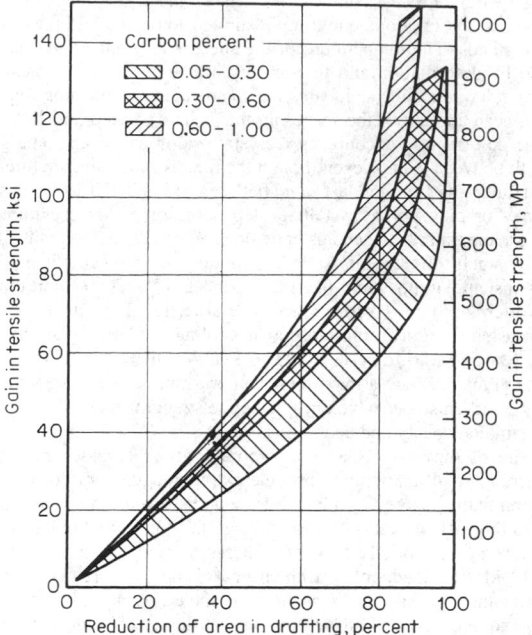

Fig. 6.2.2 Increase of tensile strength of plain carbon steel with increasing amounts of cold working by drawing through a wire-drawing die.

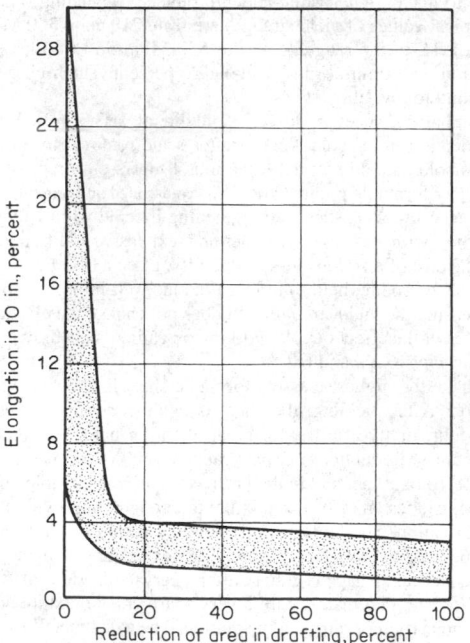

Fig. 6.2.3 Reduction in ductility of plain carbon steel with increasing amounts of cold working by drawing through a wire-drawing die.

Some other deformation processes are occasionally of interest, such as wire drawing, usually done cold, and extrusion, either hot or cold. **Hot extrusion** for materials that are difficult to work became practical through the employment of a glass lubricant. This method allows the hot extrusion of highly alloyed steels and other exotic alloys subjected to service at high loads and/or high temperatures.

Constitution and Structure of Steel

As a result of the methods of production, the following elements are always present in steel: carbon, manganese, phosphorus, sulfur, silicon, and traces of oxygen, nitrogen, and aluminum. Various alloying elements are frequently added, such as nickel, chromium, copper, molybdenum, niobium (columbium), and vanadium. The most important of the above elements in steel is carbon, and it is necessary to understand the effect of carbon on the internal structure of steel to understand the heat treatment of carbon and low-alloy steels.

The iron–iron carbide equilibrium diagram in Fig. 6.2.4 shows the phases that are present in steels of various carbon contents over a range of temperatures under equilibrium conditions. Pure iron when heated to 910°C (1,670°F) changes its internal crystalline structure from a body-centered cubic arrangement of atoms, **alpha iron,** to a face-centered cubic structure, **gamma iron.** At 1,390°C (2,535°F), it changes back to the body-centered cubic structure, **delta iron,** and at 1,539°C (2,802°F) the iron melts. When carbon is added to iron, it is found that it has only slight solid solubility in alpha iron (much less than 0.001 percent at room temperature at equilibrium). These small amounts of carbon, however, are critically important in many high-tonnage applications where formability is required. On the other hand, gamma iron will hold up to 2.0 percent carbon in solution at 1,130°C (2,066°F). The alpha iron containing carbon or any other element in solid solution is called **ferrite,** and the gamma iron containing elements in solid solution is called **austenite.** Usually when not in solution in the iron, the carbon forms a compound Fe_3C (iron carbide) which is extremely hard and brittle and is known as **cementite.**

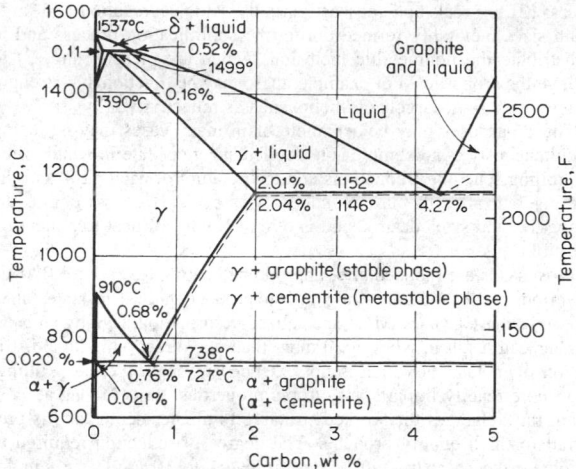

Fig. 6.2.4 Iron–iron carbide equilibrium diagram, for carbon content up to 5 percent. (Dashed lines represent equilibrium with cementite, or iron carbide; adjacent solid lines indicate equilibrium with graphite.)

The temperatures at which the phase changes occur are called **critical points** (or temperatures) and, in the diagram, represent equilibrium conditions. In practice there is a lag in the attainment of equilibrium, and the critical points are found at lower temperatures on cooling and at higher temperatures on heating than those given, the difference increasing with the rate of cooling or heating.

The various critical points have been designated by the letter A; when obtained on cooling, they are referred to as Ar, on the heating as Ac. The subscripts r and c refer to *refroidissement* and *chauffage,* respectively, and reflect the early French contributions to heat treatment. The various critical points are distinguished from each other by numbers after the letters, being numbered in the order in which they occur as the temperature increases. Ac_1 represents the beginning of transformation of ferrite to austenite on heating; Ac_3 the end of transformation of ferrite to austenite on heating, and Ac_4 the change from austenite to delta iron on heating. On cooling, the critical points would be referred to as Ar_4, Ar_3, and Ar_1, respectively. The subscript 2, not mentioned here, refers to a magnetic transformation. It must be remembered that the diagram represents the pure iron-iron carbide system at equilibrium. The varying amounts of impurities in commercial steels affect to a considerable extent the position of the curves and especially the lateral position of the eutectoid point.

Carbon steel in equilibrium at room temperature will have present both ferrite and cementite. The physical properties of ferrite are approximately those of pure iron and are characteristic of the metal. Cementite is itself hard and brittle; its shape, amount, and distribution control many of the mechanical properties of steel, as discussed later. The fact that the carbides can be dissolved in austenite is the basis of the heat treatment of steel, since the steel can be heated above the A_1 critical temperature to dissolve all the carbides, and then suitable cooling through the appropriate range will produce a wide and predictable range of the desired size and distribution of carbides in the ferrite.

If austenite with the **eutectoid** composition at 0.76 percent carbon (Fig. 6.2.4) is cooled slowly through the critical temperature, ferrite and cementite are rejected simultaneously, forming alternate plates or lamellae. This microstructure is called **pearlite**, since when polished and etched it has a pearly luster. When examined under a high-power optical microscope, however, the individual plates of cementite often can be distinguished easily. If the austenite contains less carbon than the eutectoid composition (i.e., **hypoeutectoid** compositions), free ferrite will first be rejected on slow cooling through the critical temperature until the remaining austenite reaches eutectoid composition, when the simultaneous rejection of both ferrite and carbide will again occur, producing pearlite. A hypoeutectoid steel at room temperature will be composed of areas of free ferrite and areas of pearlite; the higher the carbon percentage, the greater the amount of pearlite present in the steel. If the austenite contains more carbon than the eutectoid composition (i.e., **hypereutectoid** composition) and is cooled slowly through the critical temperature, then cementite is rejected and appears at the austenitic grain boundaries, forming a continuous cementite network until the remaining austenite reaches eutectoid composition, at which time pearlite is formed. A hypereutectoid steel, when slowly cooled, will exhibit areas of pearlite surrounded by a thin network of cementite, or iron carbide.

As the cooling rate is increased, the spacing between the pearlite lamellae becomes smaller; with the resulting greater dispersion of carbide preventing slip in the iron crystals, the steel becomes harder. Also, with an increase in the rate of cooling, there is less time for the separation of excess ferrite or cementite, and the equilibrium amount of these constituents will not be precipitated before the austenite transforms to pearlite. Thus with a fast rate of cooling, pearlite may contain less or more carbon than given by the eutectoid composition. When the cooling rate becomes very rapid (as obtained by quenching), the carbon does not have sufficient time to separate out in the form of carbide, and the austenite transforms to a highly elastically stressed structure supersaturated with carbon called **martensite**. This structure is exceedingly hard but brittle and requires **tempering** to increase the ductility. Tempering consists of heating martensite to some temperature below the critical temperature, causing the carbide to precipitate in the form of small spheroids, or especially in alloy steels, as needles or platelets. The higher the tempering temperature, the larger the carbide particle size, the greater the ductility of the steel, and the lower the hardness.

In a carbon steel, it is possible to have a structure consisting either of parallel plates of carbide in a ferrite matrix, the distance between the plates depending upon the rate of cooling, or of carbide spheroids in a ferrite matrix, the size of the spheroids depending upon the temperature to which the hardened steel was heated. (Some spheroidization occurs when pearlite is heated, but only at high temperatures close to the critical temperature range.)

Heat-Treating Operations

The following definitions of terms have been adopted by the ASTM, SAE, and ASM in substantially identical form.

Heat Treatment An operation, or combination of operations, involving the heating and cooling of a metal or an alloy in the solid state, for the purpose of obtaining certain desirable conditions or properties.

Quenching Rapid cooling by immersion in liquids or gases or by contact with metal.

Hardening Heating and quenching certain iron-base alloys from a temperature either within or above the critical range for the purpose of producing a hardness superior to that obtained when the alloy is not quenched. Usually restricted to the formation of martensite.

Annealing A heating and cooling operation implying usually a relatively slow cooling. The purpose of such a heat treatment may be (1) to remove stresses; (2) to induce softness; (3) to alter ductility, toughness, electrical, magnetic, or other physical properties; (4) to refine the crystalline structure; (5) to remove gases; or (6) to produce a definite microstructure. The temperature of the operation and the rate of cooling depend upon the material being heat-treated and the purpose of the treatment. Certain specific heat treatments coming under the comprehensive term annealing are as follows:

Full Annealing Heating iron base alloys above the critical temperature range, holding above that range for a proper period of time, followed by slow cooling to below that range. The annealing temperature is usually about 50°C ($\approx$ 100°F) above the upper limit of the critical temperature range, and the time of holding is usually not less than 1 h for each 1-in section of the heaviest objects being treated. The objects being treated are ordinarily allowed to cool slowly in the furnace. They may, however, be removed from the furnace and cooled in some medium that will prolong the time of cooling as compared with unrestricted cooling in the air.

Process Annealing Heating iron-base alloys to a temperature below or close to the lower limit of the critical temperature range followed by cooling as desired. This heat treatment is commonly applied in the sheet and wire industries, and the temperatures generally used are from 540 to 705°C (about 1,000 to 1,300°F).

Normalizing Heating iron base alloys to approximately 40°C (about 100°F) above the critical temperature range followed by cooling to below that range in still air at ordinary temperature.

Patenting Heating iron base alloys above the critical temperature range followed by cooling below that range in air or a molten mixture of nitrates or nitrites maintained at a temperature usually between 425 and 565°C (about 800 to 1,050°F), depending on the carbon content of the steel and the properties required of the finished product. This treatment is applied in the wire industry to medium- or high-carbon steel as a treatment to precede further wire drawing.

Spheroidizing Any process of heating and cooling steel that produces a rounded or globular form of carbide. The following spheroidizing methods are used: (1) Prolonged heating at a temperature just below the lower critical temperature, usually followed by relatively slow cooling. (2) In the case of small objects of high-carbon steels, the spheroidizing result is achieved more rapidly by prolonged heating to temperatures alternately within and slightly below the critical temperature range. (3) Tool steel is generally spheroidized by heating to a temperature of 750 to 805°C (about 1,380 to 1,480°F) for carbon steels and higher for many alloy tool steels, holding at heat from 1 to 4 h, and cooling slowly in the furnace.

Tempering (also termed **Drawing**) Reheating hardened steel to some temperature below the lower critical temperature, followed by any desired rate of cooling. Although the terms *tempering* and *drawing* are practically synonymous as used in commercial practice, the term *tempering* is preferred.

Transformation Reactions in Carbon Steels Much of, but not all, the heat treatment of steel involves heating into the region above Ac_3 to form austenite, followed by cooling at a preselected rate. If the parts are large, heat flow may limit the available cooling rates. As an example selected for simplicity rather than volume of products, we may follow the possible transformations in a eutectoid steel over a range of temperature. (The reactions to produce ferrite in hypoeutectoid steels, which are by far the most common, do not differ in principle; the products are, of course, softer.) The curve is a derivative of the TTT (time-temperature-transformation) curves produced by a systematic study of austenite transformation rates isothermally on specimens thin enough to avoid heat flow complications. The data collected for many steels are found in the literature.

Figure 6.2.5 summarizes the rates of decomposition of a eutectoid carbon steel over a range of temperatures. Various cooling rates are shown diagrammatically, and it will be seen that the faster the rate of cooling, the lower the temperature of transformation, and the harder the product formed. At around 540°C (1,000°F), the austenite transforms rapidly to fine pearlite; to form martensite it is necessary to cool very rapidly through this temperature range to avoid the formation of pearlite before the specimen reaches the temperature at which the formation of martensite begins (M_s). The minimum rate of cooling that is required to form a fully martensite structure is called the **critical cooling rate**. No matter at what rate the steel is cooled, the only products of transformation of this steel will be pearlite or martensite. However, if the steel can be given an interrupted quench in a molten bath at some temperature between 205 and 540°C (about 400 and 1,000°F), an acicular structure, called **bainite**, of considerable toughness, combining high strength with

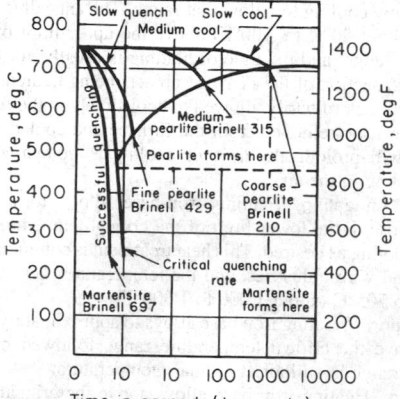

Fig. 6.2.5 Influence of cooling rate on the product of transformation in a eutectoid carbon steel.

high ductility, is obtained, and this heat treatment is known as austempering. A somewhat similar heat treatment called martempering can be utilized to produce a fully martensitic structure of high hardness, but free of the cracking, distortion, and residual stresses often associated with such a structure. Instead of quenching to room temperature, the steel is quenched to just above the martensitic transformation temperature and held for a short time to permit equalization of the temperature gradient throughout the piece. Then the steel may be cooled relatively slowly through the martensitic transformation range without superimposing thermal stresses on those introduced during transformation. The limitation of austempering and martempering for carbon steels is that these two heat treatments can be applied only to articles of small cross section, since the rate of cooling in salt baths is not sufficient to prevent the formation of pearlite in samples with diameter of more than ½ in.

The maximum hardness obtainable in a high-carbon steel with a fine pearlite structure is approximately 400 Brinell, although a martensitic structure would have a hardness of approximately 700 Brinell. Besides being able to obtain structures of greater hardness by forming martensite, a spheroidal structure will have considerably higher proof stress (i.e., stress to cause a permanent deformation of 0.01 percent) and ductility than a lamellar structure of the same tensile strength and hardness. It is essential, therefore, to form martensite when optimum properties are desired in the steel. This can be done with a piece of steel having a small cross section by heating the steel above the critical and quenching in water; but when the cross section is large, the cooling rate at the center of the section will not be sufficiently rapid to prevent the formation of pearlite. The characteristic of steel that determines its capacity to harden throughout the section when quenched is called **hardenability**. In

the discussion that follows, it is pointed out that hardenability is significantly affected by most alloying elements. This term should not be confused with the ability of a steel to attain a certain hardness. The intensity of hardening, i.e., the maximum hardness of the martensite formed, is very largely dependent upon the carbon content of the steel.

Determination of Hardenability A long-established test for hardenability is the **Jominy test** which performs controlled water cooling on one end of a standard bar. Since the thermal conductivity of steel does not vary significantly, each distance from the quenched end (DQE) corresponds to a substantially unique cooling rate, and the structure obtained is a surrogate for the TTT curve on cooling. For a detailed account of the procedure, see the SAE Handbook. The figures extracted from the Jominy test can be extended to many shapes and quenching media. In today's practices, the factors discussed below can often be used to calculate hardenability from chemical composition with considerable confidence, leaving the actual test as a referee or for circumstances where a practice is being developed.

Three main factors affect the hardenability of steel: (1) austenite composition; (2) austenite grain size; and (3) amount, nature, and distribution of undissolved or insoluble particles in the austenite. All three determine the rate of decomposition, in the range of 540°C (about 1,000°F). The slower the rate of decomposition, the larger the section that can be hardened throughout, and therefore the greater the hardenability of the steel. Everything else being equal, the higher the carbon content, the greater the hardenability; this approach, however, is often counterproductive in that other strength properties may be affected in undesirable ways. The question of **austenitic grain size** is of considerable importance in any steel that is to be heat-treated, since it affects the properties of the steel to a considerable extent. When a steel is heated to just above the critical temperature, small polyhedral grains of austenite are formed. With increase in temperature, there is an increase in the size of grains, until at temperatures close to the melting point the grains are very large. Since the transformation of austenite to ferrite and pearlite usually starts at grain boundaries, a fine-grained steel will transform more rapidly than a coarse-grained steel because the latter has much less surface area bounding the grains than a steel with a fine grain size.

The grain size of austenite at a particular temperature depends primarily on the "pinning" of the boundaries by undissolved particles. These particles, which can be aluminum nitride from the deoxidation or various carbides and/or nitrides (added for their effect on final properties), dissolve as temperature increases, allowing grain growth. While hardenability is increased by large austenite grain size, this is not usually favored since some properties of the finished product can be seriously downgraded. Small particles in the austenite will act as nuclei for the beginning of transformation in a manner similar to grain boundaries, and therefore the presence of a large number of small particles (sometimes submicroscopic in size) will also result in low hardenability.

Determination of Austenitic Grain Size The subject of austenite grain size is of considerable interest because of the fact noted above that the grain size developed during heat treatment has a large effect on the physical properties of the steel. In steels of similar chemical analysis, the steel developing the finer austenitic grain size will have a lower hardenability but will, in general, have greater toughness, show less tendency to crack or warp on quenching, be less susceptible to grinding cracks, have lower internal stresses, and retain less austenite than coarse-grained steel. There are several methods of determining the grain-size characteristics of a steel.

The **McQuaid-Ehn test** (ASTM E112), which involves the outlining of austenite grains by cementite after a specific carburizing treatment, is still valid. It has been largely replaced, however, by quenching from the austenitizing treatment under investigation and observing the grain size of the resulting martensite after light tempering and etching with Vilella's reagent. There are several ways to report the grain size observed under the microscope, the one used most extensively being the ASTM index numbers. In fps or English units, the numbers are based on the formula: number of grains per square inch at $100x = 2^{N-1}$, in which N is the grain-size index. The usual range in steels will be from 1 to 128 grains/in^2 at $100x$, and the corresponding ASTM numbers will be 1 to 8, although today grain sizes up to 12 are common. Whereas at one time "coarse" grain sizes were 1 to 4, and "fine" grain sizes 5 to 8, these would not serve modern requirements. Most materials in service would be no coarser than 7 or 8, and the ferritic low-alloy high-strength steels routinely approach 12 or smaller. Grain-size relationships in SI units are covered in detail in Designation E112 of ASTM Standards. For further information on grain size, refer to the ASM "Metals Handbook."

EFFECT OF ALLOYING ELEMENTS ON THE PROPERTIES OF STEEL

When relatively large amounts of alloying elements are added to steel, the characteristic behavior of carbon steels is not lost. Most **alloy steel** is medium- or high-carbon steel to which various elements have been added to modify its properties to an appreciable extent; the alloys as a minimum allow the properties characteristic of the carbon content to be fully realized even in larger sections, and in some cases may provide additional benefits. The percentage of alloy element required for a given purpose ranges from a few hundredths of 1 percent to possibly as high as 5 percent.

When ready for service, these steels will usually contain only two constituents, ferrite and carbide. The only way that an alloying element can affect the properties of the steel is to change the dispersion of carbide in the ferrite, change the properties of the ferrite, or change the characteristics of the carbide. The effect on the distribution of carbide is the most important factor, since in sections amenable to close control of structure, carbon steel is only moderately inferior to alloy steel. However, in large sections where carbon steels will fail to harden throughout the section even on a water quench, the hardenability of the steel can be increased by the addition of any alloying element (except possibly cobalt). The increase in hardenability permits the hardening of a larger section of alloy steel than of plain carbon steel. The quenching operation does not have to be so drastic. Consequently, there is a smaller difference in temperature between the surface and center during quenching, and cracking and warping resulting from sharp temperature gradients in a steel during hardening can be avoided. The elements most effective in increasing the hardenability of steel are manganese, silicon, and chromium, or combinations of small amounts of several elements such as chromium, nickel, and molybdenum in SAE 4340, where the joint effects are greater than alloys acting singly.

Elements such as molybdenum, tungsten, and vanadium are effective in increasing the hardenability when dissolved in the austenite, but not when present in the austenite in the form of carbides. When dissolved in austenite, and thus contained in solution in the resulting martensite, they can modify considerably the rate of coarsening of carbides in tempered martensite. Tempering relieves the internal stresses in the hardened steel in part by precipitating various carbides of iron at fairly low temperature, which coarsen as the tempering temperature is increased. The increasing particle separation results in a loss of hardness and strength accompanied by increased ductility. See Fig. 6.2.6. Alloying elements can cause slower coarsening rates or, in some cases at temperatures from 500 to 600°C, can cause dissolution of cementite and the precipitation of a new set of small, and thus closely spaced, alloy carbides which in some cases can cause the hardness to actually rise again with no loss in toughness or ductility. This is especially important in tool steels. The presence of these stable carbide-forming elements enables higher tempering temperatures to be used without sacrificing strength. This permits these alloy steels to have a greater ductility for a given strength, or, conversely, greater strength for a given ductility, than plain carbon steels.

The third factor which contributes to the strength of alloy steel is the presence of the alloying element in the ferrite. Any element in solid solution in a metal will increase the strength of the metal, so that these elements will materially contribute to the strength of hardened and tempered steels. The elements most effective in strengthening the ferrite are phosphorus, silicon, manganese, nickel, molybdenum, and chromium. Carbon and nitrogen are very strong ferrite strengtheners but generally are not present in interstitial solution in significant amounts, and there are other processing reasons to actively keep the amount in solution small by adding strong carbide and/or nitride formers to give interstitial-free (IF) steels.

A final important effect of alloying elements discussed above is their influence on the austenitic grain size. Martensite formed from a fine-grained austenite has considerably greater resistance to shock than when formed from a coarse-grained austenite.

In Table 6.2.2, a summary of the effects of various alloying elements is given. Remember that this table indicates only the trends of the various elements, and the fact that one element has an important influence on one factor does not prevent it from having a completely different influence or another one.

PRINCIPLES OF HEAT TREATMENT OF IRON AND STEEL

When heat-treating a steel for a given part, certain precautions have to be taken to develop optimum mechanical properties in the steel. Some of the major factors that have to be taken into consideration are outlined below.

Heating The first step in the heat treatment of steel is the heating of the material to above the critical temperature to make it fully austenitic. The **heating rate** should be sufficiently slow to avoid injury to the material through excessive thermal and transformational stresses. In general, hardened steel should be heated more slowly and uniformly than is necessary for soft stress-free materials. Large sections should not be placed in a hot furnace, the allowable size depending upon the carbon and alloy content. For high-carbon steels, care should be taken in heating sections as small as 50-mm (2-in) diameter, and in medium-carbon steels precautions are required for sizes over 150-mm (6-in) diameter. The **maximum temperature** selected will be determined by the chemical composition of the steel and its grain-size characteristics. In hypoeutectoid steel, a temperature about 25 to 50°C above the upper critical range is used, and in hypereutectoid steels, a temperature between the lower and the upper critical temperature is generally used to retain enough carbides to keep the austenite grain size small and preserve what is often limited toughness. Quenching temperatures are usually a little closer to the critical temperature for hypoeutectoid steels than to those for normalizing; annealing for softening is carried out just below Ac_1 for steels up to 0.3 percent C and just above for higher-carbon steels. Tables of suggested temperatures can be found in the ASM "Metals Handbook," or a professional heat treater may be consulted.

The **time** at maximum temperature should be such that a uniform temperature is obtained throughout the cross section of the steel. Care should be taken to avoid undue length of time at temperature, since this will result in undesirable grain growth, scaling, or decarburization of the surface. A practical figure often given for the total time in the hot furnace is 12 min/cm (about ½ h/in) of cross-sectional thickness. When the steel has attained a uniform temperature, the **cooling rate** must be such as to develop the desired structure; slow cooling rates (furnace or air cooling) to develop the softer pearlitic structures and high cooling rates (quenching) to form the hard martensitic structures. In selecting a **quenching medium** (see ASM "Metals Handbook"), it is important to select the quenching medium for a particular job on the basis of size, shape, and allowable distortion before choosing the steel composition. It is convenient to classify steels in two groups on the basis of depth of hardening: shallow hardening and deep hardening. **Shallow-hardening steels** may be defined as those which, in the form of 25-mm- (1-in-) diameter rounds, have, after brine quenching, a completely martensitic shell not deeper than 6.4 mm (¼ in). The shallow-hardening steels are those of low or no alloy content, whereas the deep-hardening steels have a substantial content of those alloying elements that increase penetration of hardening, notably chromium, manganese, and nickel. The high cooling rates required to harden shallow-hardening steel produce severe distortion and sometimes quench cracking in all but simple, symmetric shapes having a low ratio of length to diameter or thickness. Plain carbon steels cannot be used for complicated shapes where distortion must be avoided. In this case, water quenching must be abandoned and a

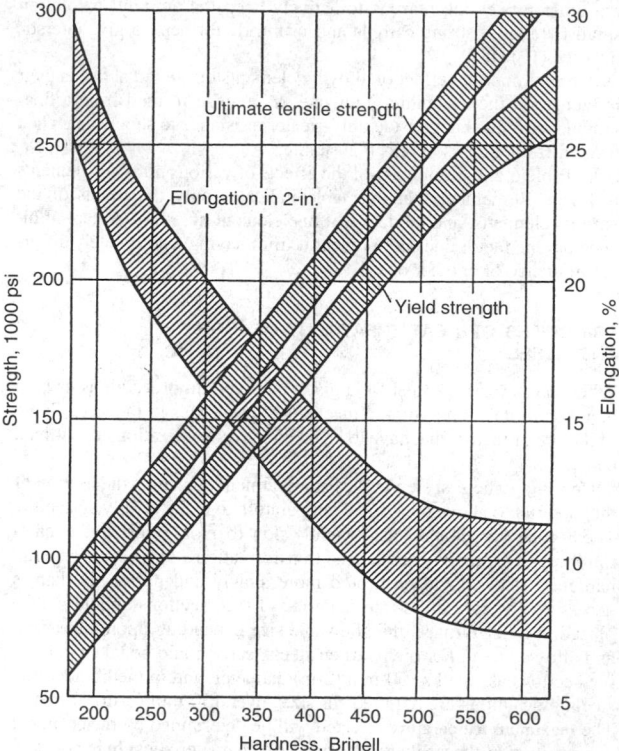

Fig. 6.2.6   Range of tensile properties in several quenched and tempered steels at the same hardness values. *(Janitzky and Baeyertz. Source: ASM; reproduced by permission.)*

less active quench used which materially reduces the temperature gradient during quenching. Certain oils are satisfactory but are incapable of hardening shallow-hardening steels of substantial size. A change in steel composition is required with a change from water to an oil quench. Quenching in oil does not entirely prevent distortion. When the degree of distortion produced by oil quenching is objectionable, recourse is taken to air hardening. The cooling rate in air is very much slower than in oil or water; so an exceptionally high alloy content is required. This means that a high price is paid for the advantage gained, in terms of both

metal cost and loss in machinability, though it may be well justified when applied to expensive tools or dies. In this case, danger of cracking is negligible.

Liquids for Quenching Shallow-Hardening Steels   Shallow-hardening steels require extremely rapid surface cooling in the quench, particularly in the temperature range around 550°C (1,020°F). A submerged water spray will give the fastest and most reproducible quench practicable. Such a quench is limited in application to simple short objects which are not likely to warp. Because of difficulty in obtaining symmetric flow of the water relative to the work, the spray quench is conducive to warping. The ideal practical quench is one that will give the required surface cooling without agitation of the bath. The addition of ordinary salt, sodium chloride, greatly improves the performance of water in this respect, the best concentration being around 10 percent. Most inorganic salts are effective in suppressing the formation of vapor at the surface of the steel and thus aid in cooling steel uniformly and eliminating the formation of soft spots. To minimize the formation of vapor, water-base quenching liquids must be kept cold, preferably under 20°C (about 70°F). The addition of some other soluble materials to water such as soap is extremely detrimental because of increased formation of vapor.

Liquids for Quenching Deeper-Hardening Steels   When oil quenching is necessary, use a steel of sufficient alloy content to produce a completely martensitic structure at the surface over the heaviest section of the work. To minimize the possibility of cracking, especially when hardening tool steels, keep the quenching oil warm, preferably between 40 and 65°C (about 100 and 150°F). If this expedient is insufficient to prevent cracking, the work may be removed just before the start of the hardening transformation and cooled in air. Whether or not transformation has started can be determined with a permanent magnet, the work being completely nonmagnetic before transformation.

The cooling characteristics of quenching oils are difficult to evaluate and have not been satisfactorily correlated with the physical properties of the oils as determined by the usual tests. The standard tests are important with regard to secondary requirements of quenching oils. Low viscosity assures free draining of oil from the work and therefore low oil loss. A high flash and fire point assures a high boiling point and reduces the fire hazard which is increased by keeping the oil warm. A low carbon residue indicates stability of properties with continued use and little sludging. The steam-emulsion number should be low to ensure low water content, water being objectionable because of its vapor-film-forming tendency and high cooling power. A low saponification number assures that the oil is of mineral base and not subject to organic deterioration of fatty oils which give rise to offensive odors. Viscosity index is a valuable property for maintenance of composition.

Table 6.2.2   Trends of Influence of Some Alloying Elements

Element	Strengthening as dissolved in ferrite	Hardenability effects if dissolved in austenite	Effect on grain coarsening in austenite if undissolved as compound	Effects on tempered hardness, strength, and toughness
Al	*	*	†	None
Cr	*	†	†	*
Co	†	Negative	None	None
Cu	†	*	None	None
Mn	†	*	*	*
Mo	*	†	†	None
Nb (Cb)	None	†‡	†	†
Ni	*	*	None	None
P	†	*	None	None
Si	*	*	None	None
Ta	‡	†‡	†	†
Ti	†	†	†	‡
W	*	†	†	†
V	*	†	†	†

* Effects are moderate at best.
† Effects are strong to very strong.
‡ Effects not clear, or not used significantly.

In recent years, polymer-water mixtures have found application because of their combination of range of heat abstraction rates and relative freedom from fire hazards and environmental pollution. In all cases, the balance among productivity, danger of distortion and cracking, and minimum cost to give adequate hardenability is not simple; even though some general guides are available (e.g., "Metals Handbook," vol. 1, 10th ed.), consultation with experienced professionals is recommended.

Effect of the Condition of Surface   The factors that affect the depth of hardening are the hardenability of the steel, the size of specimen, the quenching medium, and finally the condition of the surface of the steel before quenching. Steel that carries a heavy coating of scale will not cool so rapidly as a steel that is comparatively scale-free, and soft spots may be produced; or, in extreme cases, complete lack of hardening may result. It is therefore essential to minimize scaling as much as possible. Decarburization can also produce undesirable results such as nonuniform hardening and thus lowers the resistance of the material to alternating stresses (i.e., fatigue).

Tempering, as noted above, relieves quenching stresses and offers the ability to obtain useful combinations of properties through selection of tempering temperature. The ability of alloying elements to slow tempering compared to carbon steel allows higher temperatures to be used to reach a particular strength. This is accompanied by some usually modest increases in ductility and toughness. Certain high-hardenability steels are subject to delayed cracking after quenching and should be tempered without delay. Data on tempering behavior are available from many sources, such as ASM "Metals Handbook" or Bain and Paxton, "Functions of the Alloying Elements in Steel," 3d ed., ASM International.

Relation of Design to Heat Treatment

Care must be taken in the design of a machine part to prevent cracking or distortion during heat treatment. With proper design the entire piece may be heated and cooled at approximately the same rate during the heat-treating operation. A light section should never be joined to a heavy section. Sharp reentrant angles should be avoided. Sharp corners and inadequate fillets produce serious stress concentration, causing the actual service stresses to build up to a point where they amount to two to five times the normal working stress calculated by the engineer in the original layout. The use of generous fillets is especially desirable with all high-strength alloy steels.

The modulus of elasticity of all commercial steels, either carbon or alloy, is the same so far as practical designing is concerned. The deflection under load of a given part is, therefore, entirely a function of the section of the part and is not affected by the composition or heat treatment of the steel. Consequently if a part deflects excessively, a change in design is necessary; either a heavier section must be used or the points of support must be increased.

COMPOSITE MATERIALS

For some applications, it is not necessary or even desirable that the part have the same composition throughout. The oldest method of utilizing this concept is to produce a high-carbon surface on low-carbon steel by **carburizing,** a high-temperature diffusion treatment, which after quenching gives a wear-resistant case 1 or 2 mm (0.04 or 0.08 in) thick on a fairly shock-resistant core. Clearly, this is an attractive process for gears and other complex machined parts. Other processes which produce a similar product are **nitriding** (which can be carried out at lower temperatures) and **carbonitriding** (a hybrid), or where corrosion resistance is important, by **chromizing**. Hard surfaces can also be obtained on a softer core by using selective heating to produce surface austenite (induction, flames, etc.) before quenching, or by depositing various hard materials on the surface by welding.

Many other types of surface treatment which provide corrosion protection and sometimes aesthetic values are common, beginning with paint or other polymeric films and ranging through enamels (a glass film); films such as zinc and its alloys which can be applied by dipping in molten baths or can be deposited electrolytically on one or both sides (galvanizing); tinplate for cans and other containers; chromium plating; or a light coherent scale of iron oxide. These processes are continually being improved, and they may be used in combinations, e.g., paint on a galvanized surface for exposed areas of automobiles. (See Sec. 6.6.)

Carburizing   Various methods are available depending on the production volume. Pack carburizing can handle diverse feedstock by enclosing the parts in a sealed heat-resistant alloy box with carbonates and carbonaceous material, and by heating for several hours at about 925°C. The carbon dioxide evolved reacts with carbon to form carbon monoxide as the carrier gas, which does the actual carburizing. While the process can be continuous, much of it is done as a batch process, with consequent high labor costs and uncertain quality to be balanced against the flexibility of custom carburizing.

For higher production rates, furnaces with controlled atmospheres involving hydrocarbons and carbon monoxide provide better controls, low labor costs, shorter times to produce a given case depth [4 h versus 9 h for a 1-mm (0.04-in) case], and automatic quenching. Liquid carburizing using cyanide mixtures is even quicker for thin cases, but often it is not as economical for thicker ones; its great advantage is flexibility and control in small lots.

To provide the inherent value in this material of variable composition, it must be treated to optimize the properties of case and core by a double heat treatment. In many cases, to avoid distortion of precision parts, the material is first annealed at a temperature above the carburizing temperature and is cooled at a rate to provide good machinability. The part is machined and then carburized; through careful steel selection, the austenite grain size is not large at this point, and the material can be quenched directly without danger of cracking or distortion. Next the part is tempered; any retained austenite from the low M_s of the high-carbon case has a chance to precipitate carbides, raise its M_s, and transform during cooling after tempering. Care is necessary to avoid internal stresses if the part is to be ground afterward. An alternative approach is to cool the part reasonably slowly after carburizing and then to heat-treat the case primarily by suitable quenching from a temperature typical for hypereutectoid steels between A_1 and A_{cm}. This avoids some retained austenite and is helpful in high-production operations. The core is often carbon steel, but if alloys are needed for hardenability, this must be recognized in the heat treatment.

Nitriding   A very hard, thin case can be produced by exposing an already quenched and tempered steel to an ammonia atmosphere at about 510 to 540°C, but unfortunately for periods of 50 to 90 h. The nitrogen diffuses into the steel and combines with strong nitride formers such as aluminum and chromium, which are characteristically present in steels where this process is to be used. The nitrides are small and finely dispersed; since quenching is not necessary after nitriding, dimensional control is excellent and cracking is not an issue. The core properties do not change since tempering at 550°C or higher temperature has already taken place. A typical steel composition is as follows: C, 0.2 to 0.3 percent; Mn, 0.04 to 0.6 percent; Al, 0.9 to 1.4 percent; Cr, 0.9 to 1.4 percent; and Mo, 0.15 to 0.25 percent.

Carbonitriding (Cyaniding)   An interesting intermediate which rapidly adds both carbon and nitrogen to steels can be obtained by immersing parts in a cyanide bath just above the critical temperature of the core followed by direct quenching. A layer of about 0.25 mm (0.010 in) can be obtained in 1 h. The nitrides add to the wear resistance.

Local Surface Hardening   For some parts which do not readily fit in a furnace, the surface can be hardened preferentially by local heating using flames, induction coils, electron beams, or lasers. The operation requires skill and experience, but in proper hands it can result in very good local control of structure, including the development of favorable surface compressive stresses to improve fatigue resistance.

Clad steels can be produced by one of several methods, including simple cladding by rolling a sandwich out of contact with air at a temperature high enough to bond (1,200°C); by explosive cladding where the geometry is such that the energy of the explosive causes a narrow molten zone to traverse along the interface and provide a good fusion bond; and by various casting and welding processes which can deposit a wide variety of materials (ranging from economical, tough, corrosion-resistant, or high-thermal-conductivity materials to hard and stable carbides in a suitable matrix). Many different product shapes lend themselves to these practices.

Chromizing Chromizing of low-carbon steel is effective in improving corrosion resistance by developing a surface containing up to 40 percent chromium. Some forming operations can be carried out on chromized material. Most chromizing is accomplished by packing the steel to be treated in a powdered mixture of chromium and alumina and then heating to above 1,260°C (2,300°F) for 3 or 4 h in a reducing atmosphere. Another method is to expose the parts to be treated to gaseous chromium compounds at temperatures above 845°C (about 1,550°F). Flat rolled sheets for corrosive applications such as auto mufflers can thus be chromized in open-coil annealing facilities.

THERMOMECHANICAL TREATMENT

The effects of mechanical treatment and heat treatment on the mechanical properties of steel have been discussed earlier in this section. Thermomechanical treatment consists of combining controlled (sometimes large) amounts of plastic deformation with the heat-treatment cycle to achieve improvements in yield strength beyond those attainable by the usual rolling practices alone or rolling following by a separate heat treatment. The tensile strength, of course, is increased at the same time as the yield strength (not necessarily to the same degree), and other properties such as ductility, toughness, creep resistance, and fatigue life can be improved. However, the high strength and hardness of thermomechanically treated steels limit their usefulness to the fabrication of components that require very little cold forming or machining, or very simple shapes such as strip and wire that can be used as part of a composite structure. Although the same yield strength may be achieved in a given steel by different thermomechanical treatments, the other mechanical properties (particularly the toughness) are not necessarily the same.

There are many possible combinations of deformation schedules and time-temperature relationships in heat treatment that can be used for thermomechanical treatment, and individual treatments cannot be discussed here. Table 6.2.3 classifies broadly thermomechanical treatments into three principal groups related to the time-temperature dependence of the transformation of austenite discussed earlier under heat treatment. The names in parentheses following the subclasses in the table are those of some types of thermomechanical treatments that have been used commercially or have been discussed in the literature. At one time it was thought that these procedures would grow in importance, but in fact they are still used in a very minor way with the important exception of class 1a and class 1c, which are critically important in large tonnages.

Table 6.2.3 Classification of Thermomechanical Treatments

Class I. Deformation before austenite transformation
  a. Normal hot-working processes
     (hot/cold working)
  b. Deformation before transformation to martensite
     (ausforming, austforming, austenrolling, hot-cold working, marworking, warm working)
  c. Deformation before transformation to ferrite-carbide aggregates (austentempering)
Class II. Deformation during austenite transformation
  a. Deformation during transformation to martensite (Zerolling and Ardeform processes)
  b. Deformation during transformation to ferrite-carbide aggregates (flow tempering of bainite and isoforming)
Class III. Deformation after austenite transformation
  a. Deformation of martensite followed by tempering
  b. Deformation of tempered martensite followed by aging (flow tempering, marstraining, strain tempering, tempforming, warm working)
  c. Deformation of isothermal transformation products (patenting, flow tempering, warm working)

SOURCE: Radcliffe Kula, Syracuse University Press, 1964.

COMMERCIAL STEELS

The wide variety of applications of steel for engineering purposes is due to the range of mechanical properties obtainable by changes in carbon content and heat treatment. Some typical applications of carbon steels are given in Table 6.2.4. Carbon steels can be subdivided roughly into three groups: (1) low-carbon steel, 0.01 to 0.25 percent carbon, for use where only moderate strength is required together with considerable plasticity; (2) machinery steels, 0.30 to 0.55 percent carbon, which can be heat-treated to develop high strength; and (3) tool steels, containing from 0.60 to 1.30 percent carbon (this range also includes rail and spring steels).

Table 6.2.4 Some Typical Applications of Carbon Steels

Percent C	Uses
0.01–0.10	Sheet, strip, tubing, wire nails
0.10–0.20	Rivets, screws, parts to be case-hardened
0.20–0.35	Structural steel, plate, forgings such as camshafts
0.35–0.45	Machinery steel—shafts, axles, connecting rods, etc.
0.45–0.55	Large forgings—crankshafts, heavy-duty gears, etc.
0.60–0.70	Bolt-heading and drop-forging dies, rails, setscrews
0.70–0.80	Shear blades, cold chisels, hammers, pickaxes, band saws
0.80–0.90	Cutting and blanking punches and dies, rock drills, hand chisels
0.90–1.00	Springs, reamers, broaches, small punches, dies
1.00–1.10	Small springs and lathe, planer, shaper, and slotter tools
1.10–1.20	Twist drills, small taps, threading dies, cutlery, small lathe tools
1.20–1.30	Files, ball races, mandrels, drawing dies, razors

The importance of various types of steel products, as measured by consumption over the last 20 years, is shown in Fig. 6.2.7. Of approximately 100,000,000 tons used annually, some 60 percent is now low-carbon sheet and strip, roughly equally divided between hot-rolled and cold-rolled. When bars (of all compositions), plates, and structurals are added (note the log scale), it is seen that the tonnages of all others are minor. However, some of the specialized forms of steel products are very important, and their capabilities are sometimes unique. Challenges frequently arise from newer nonmetallic materials; e.g., the use of oxide cutting tools to compete with high-service-temperature tool steels. The combination of property, performance, and price must be evaluated for each case.

The chemical compositions and mechanical and physical properties of many of the steels whose uses are listed in Table 6.2.4 are covered by specifications adopted by the American Society for Testing and Materials (ASTM), the Society of Automotive Engineers (SAE), and the American Society of Mechanical Engineers (ASME); other sources include government specifications for military procurement, national specifications in other industrial countries, and smaller specialized groups with their own interests in mind. The situation is more complex than necessary because of mixtures of chemical composition ranges and property and geometric ranges. Simplification will take a great deal of time to develop, and understanding among users, when and if it is reached, will require a major educational endeavor. For the time being, most of the important specifications and equivalents are shown in the ASM "Metals Handbook," 10th ed., vol. 1.

Understanding Some Mechanical Properties of Steels

Before we give brief outlines of some of the more important classes of steels, a discussion of the important factors influencing selection of steels and an appreciation of those for relevent competitive materials may be helpful. The easy things to measure for a material, steel or otherwise, are chemical composition and a stress-strain curve, from which one can extract such familiar quantities as yield strength, tensile strength, and ductility expressed as elongation and reduction of area. Where the application is familiar and the requirements are not particularly crucial, this is still appropriate and, in fact, may be overkill; if the necessary properties are provided, the chemical composition can vary outside the specification with few or no ill effects. However, today's

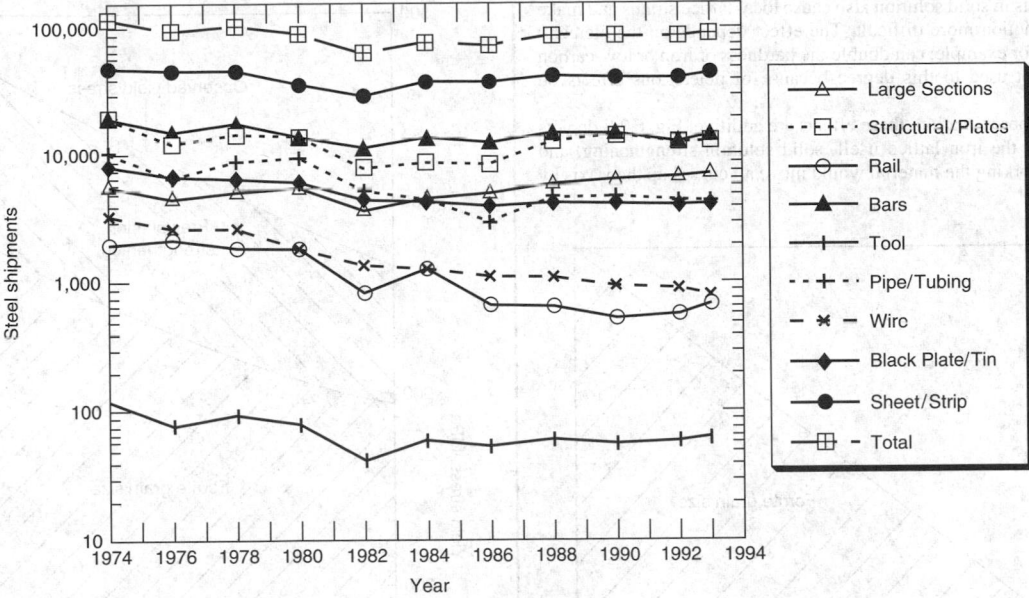

Fig. 6.2.7   Annual U.S. steel shipments, thousands of tons from 1974 to 1994.

designs emphasize total life-cycle cost and performance and often will require consideration of several other factors such as manufacturing and assembly methods; costs involving weldability (or more broadly join-ability), formability, and machinability; corrosion resistance in various environments; fatigue, creep, and fracture; dimensional tolerances; and acceptable disposability at the end of the life cycle. These factors inter-act with each other in such complex ways that merely using tabulated data can invite disaster in extreme cases. It is not possible to cover all these issues in detail, but a survey of important factors for some of the most common ones may be helpful.

Yield Strength   In technical terms the yield stress, which measures the onset of plastic deformation (for example, 0.1 or 0.2 percent perma-nent extension), occurs when significant numbers of "dislocations" move in the crystal lattice of the major phase present, usually some form of ferrite. Dislocations are line discontinuities in the lattice which allow crystal planes to slide over each other; they are always present at levels of 10^4 to 10^6 per square centimeter in undeformed metals, and this dislocation density can increase steadily to 10^{10} to 10^{12} per square cen-timeter after large deformations. Any feature which interferes with dis-location movement increases the yield stress. Such features include the following:

1. The lattice itself provides a lower limit (the **Peierls' stress**) by exhibiting an equivalent viscosity for movement. As a practical matter, this may be up to about 20 ksi (135 MPa) and is the same for all iron-based materials.

2. Dislocations need energy to cut through each other; thus, the stress necessary to continue deformation rises continuously as the dislocation density increases (work hardening). This is an important, inexpensive source of strength as various tempers are produced by rolling or draw-ing (see Table 6.2.5); it is accompanied by a significant loss of ductility.

3. Precipitates interfere with dislocation movement. The magnitude of interference depends sensitively on precipitate spacing, rising to a maximum at a spacing from 10 to 30 nm for practical additions. If a given heat treatment cannot develop these fine spacings (e.g., because the piece is so large that transformations take place at high tempera-tures), the strengthening is more limited. Precipitates are an important source of strength; martensite is used as an intermediate structure for carbon and alloy steels precisely because the precipitate spacing can be accurately controlled by tempering.

4. Small grain sizes, by interfering with the passage of dislocations across the boundaries, result in important increases in yield strength. To a very good approximation, the increase in yield is proportional to (grain diameter)$^{-1/2}$. A change from ASTM 8 to 12, for example, in-creases the yield strength by about 100 MPa.

Table 6.2.5   Approximate Mechanical Properties for Various Tempers of Cold-Rolled Carbon Strip Steel

	Tensile strength		Elongation in 50 mm or 2 in for 1.27-mm (0.050-in) thickness of strip, %	
Temper	MPa	1,000 lb/in²		Remarks
No. 1 (hard)	621 ± 69	90 ± 10		A very stiff cold-rolled strip intended for flat blanking only, and not requiring ability to withstand cold forming
No. 2 (half-hard)	448 ± 69	65 ± 10	10 ± 6	A moderately stiff cold-rolled strip intended for limited bending
No. 3 (quarter-hard)	379 ± 69	55 ± 10	20 ± 7	A medium-soft cold-rolled strip intended for limited bending, shallow drawing, and stamping
No. 4 (skin-rolled)	331 ± 41	48 ± 6	32 ± 8	A soft ductile cold-rolled strip intended for deep drawing where no stretcher strains or fluting are permissible
No. 5 (dead-soft)	303 ± 41	44 ± 6	39 ± 6	A soft ductile cold-rolled strip intended for deep drawing where stretcher strains or fluting are permissible. Also for extrusions

SOURCE: ASTM A109. Complete specification should be consulted in ASTM Standards (latest edition).

5. Elements in solid solution also cause local lattice strains and make dislocation motion more difficult. The effect depends on the element; 1 percent P, for example, can double the hardness of iron or low-carbon steel. It is not used to this degree because of deleterious effects on toughness.

To a first approximation, these effects are additive; Fig. 6.2.8 depicts the effects of the iron lattice itself, solid solution strengthening, and grain size. Working the material would move all curves up the y axis by

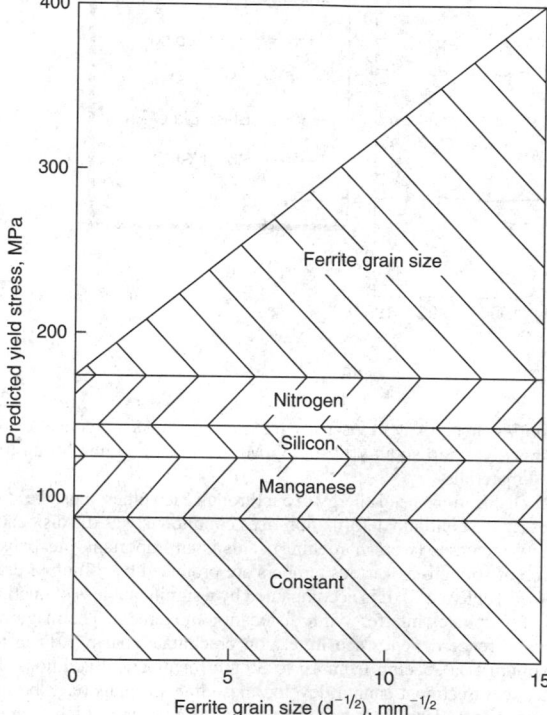

Fig. 6.2.8   The components of yield stress predicted for an air-cooled carbon-manganese steel containing 1.0 percent manganese, 0.25 percent silicon, and 0.01 percent nitrogen. (*Source: Union Carbide Corp.; reproduced by permission.*)

adding a work-hardening term. Figure 6.2.9 shows typical effects in an LAHS steel (see later) on the separated effects of manganese on yield strength. Manganese provides a little solid solution hardening and by its effect on ferrite grain size provides nonlinear strengthening. Figure 6.2.10 is a schematic of some combinations to obtain desired yield strengths and toughness simultaneously.

Tensile Properties   Materials fail in tension when the increment of nominal stress from work hardening can no longer support the applied load on a decreasing diameter. The load passes through a maximum [the **ultimate tensile stress** (UTS)], and an instability (**necking**) sets in at that point. The triaxial stresses thus induced encourage the formation of internal voids nucleated at inclusions or, less commonly, at other particles such as precipitates. With increasing strain, these voids grow until they join and ultimately lead to a ductile failure.

Some plastic behavior of steels is sensitive to the number and type of inclusions. In a tensile test the UTS and **elongation to failure** are not affected much by increasing the number of inclusions (although the **uniform ductility** prior to necking is), but the **reduction in area** is; of greater importance, the energy absorbed to propagate a crack (related to the **fracture toughness**) is very sensitive to inclusion content. This relates directly to steelmaking practices, and to ladle metallurgy in particular, which has proved extremely effective in control of deleterious inclusions (not all inclusions are equally bad).

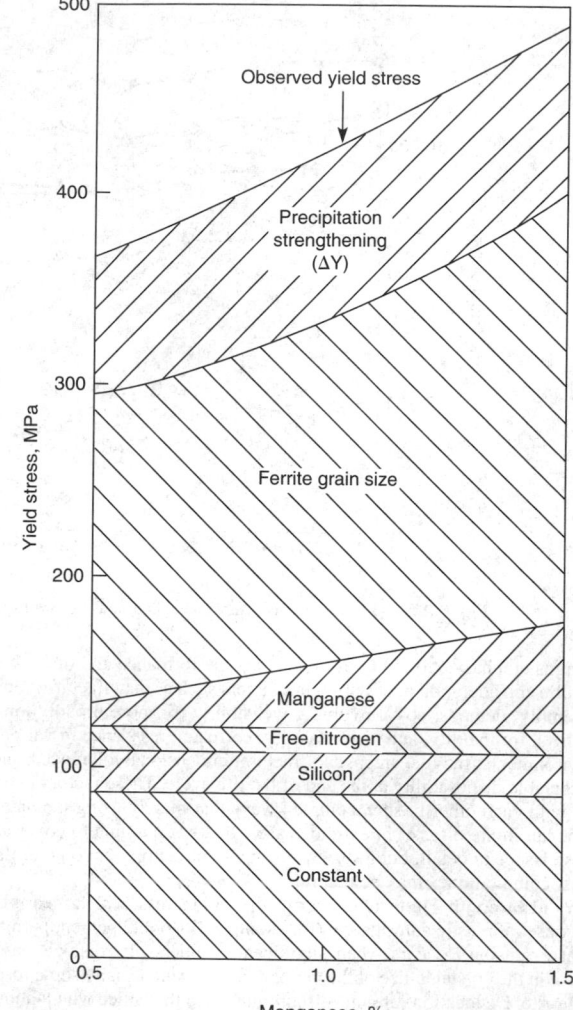

Fig. 6.2.9   The effect of increasing manganese content on the components of the yield stress of steels containing 0.2 percent carbon, 0.2 percent silicon, 0.15 percent vanadium, and 0.015 percent nitrogen, normalized from 900°C (1650°F). (*Source: Union Carbide Corp., reproduced by permission.*)

Toughness   A full treatment of this topic is not possible here, but we note the following:

1. As strength increases, toughness falls in all cases except where strengthening arises from grain-size reduction. Fine grain size thus is a double blessing; it will increase strength and toughness simultaneously.

2. The energy involved in crack growth depends on carbon content (Fig. 6.2.11). High-carbon materials have not only much lower propagation energies (e.g., the "shelf" in a Charpy test, or the value of K_c), but also higher **impact transition temperatures** (ITTs). Below the ITT the energy for crack propagation becomes very small, and we may loosely describe the steel as "brittle." There is, then, a real incentive to keep carbon in steel as low as possible, especially because pearlite does not increase the yield strength, but does increase the ITT, often to well above room temperature.

3. **Welding** is an important assembly technique. Since the hardenability of the steel can lead to generally undesirable martensite in the **heat-affected zone** (HAZ) and possibly to cracks in this region, the toughness of welds necessitates using steels with the lowest possible carbon

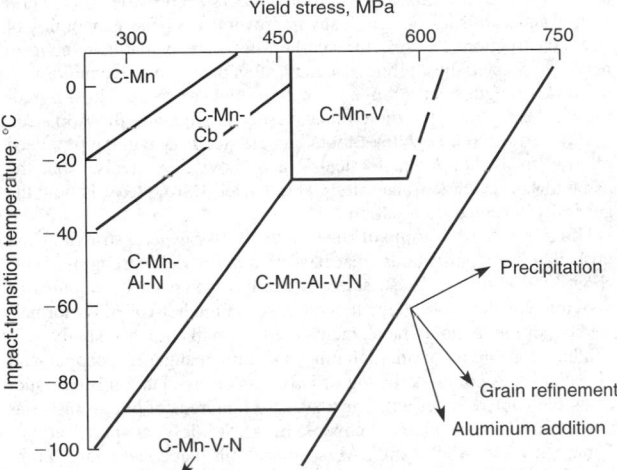

Fig. 6.2.10 Combinations of yield strength and impact transition temperature available in normalized high strength, low alloy (HSLA) steels. *(Source: Union Carbide Corp.; reproduced by permission.)*

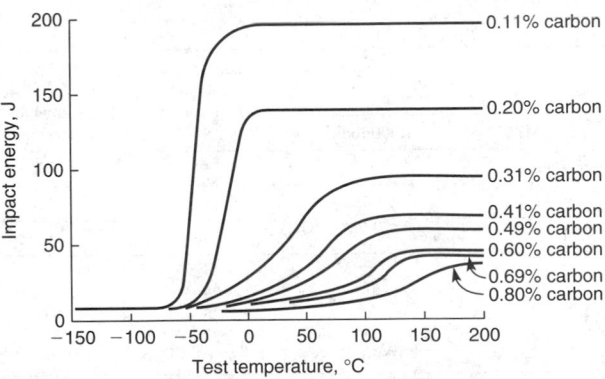

Fig. 6.2.11 The effect of carbon, and hence the pearlite content on impact transition temperature curves of ferrite-pearlite steels. *(Source: Union Carbide Corp.; reproduced by permission.)*

and whatever is placed on it without buckling. This involves consideration of the minimum weight and maximum aspect ratio of the legs. Examination of the mechanics shows that for minimum weight, we need a maximum value of the quantity $M_1 = E^{1/2}/\rho$, where E is Young's modulus and ρ is the density. For resistance to buckling, we need the maximum value of $M_2 = E$. If we plot E versus ρ of several materials, we can evaluate regions where both M_1 and M_2 are large. In doing so, we find attractive candidates are carbon-fiber-reinforced polymers (CFRPs) and certain engineering ceramics. They win out over wood, steel, and aluminum by their combination of high modulus and light weight. If we had other constraints such as cost, the CFRPs would be eliminated; as for toughness, the engineering ceramics would be unsuitable. Whenever many constraints must be satisfied simultaneously, their priority must be established. The net result is a compromise—but one which is better than an off-the-cuff guess!

This simple example can be generalized and used to develop a short list of very different alternatives in a reasonably quantitative and nonjudgmental manner for a wide variety of necessary properties. Ashby (op. cit.) provides a large collection of data.

Low-Carbon Steels Of the many low-carbon-steel products, sheet and strip steels are becoming increasingly important. The consumption of steel in the sheet and tinplate industry has accounted for approximately 60 percent of the total steel production in the United States (Fig. 6.2.7). This large production has been made possible by refinement of continuous sheet and strip rolling mills. Applications in which large quantities of sheet are employed are tinplate for food containers; black, galvanized, and terne-coated sheets for building purposes; and high-quality sheets for automobiles, furniture, refrigerators, and countless other stamped, formed, and welded products. The difference between **sheet** and **strip** is based on width and is arbitrary. Cold working produces a better surface finish, improves the mechanical properties, and permits the rolling of thinner-gage material than hot rolling. Some hot-rolled products are sold as sheet or strip directly after coiling from the hot mill. Coils may be pickled and oiled for surface protection. Thicker coiled material up to 2 m (80 in) or more wide may be cut into plates, while thicker plates are produced directly from mills which may involve rolling a slab in two dimensions to produce widths approaching 5 m (200 in). Some plates are subject to heat treatment, occasionally elaborate, to produce particular combinations of properties for demanding applications. Structurals (beams, angles, etc.) are also produced directly by hot rolling, normally without further heat treating. Bars with various cross sections are also hot-rolled; some may undergo further heat treatment depending on service requirements. Wire rods for further cold drawing are also hot-rolled.

Roughly one-half of the total hot-rolled sheet produced is cold-rolled further; the most demanding forming applications (automobiles and trucks, appliances, containers) require careful annealing, as discussed earlier. See Table 6.2.5.

Steels for deep-drawing applications must have a low yield strength and sufficient ductility for the intended purpose. This ductility arises from having a very small amount of interstitial atoms (carbon and nitrogen) in solution in the ferrite (**IF** or **interstitial-free** steels) and a sharp preferred orientation of the pancake-shaped ferrite grains. The procedures to obtain these are complex and go all the way back to steelmaking practices; they will not be discussed further here. They must also have a relatively fine grain size, since a large grain size will cause a rough finish, an "orange-peel" effect, on the deep drawn article. The sharp yield point characteristic of conventional low-carbon steel must be eliminated to prevent sudden local elongations in the sheet during forming, which result in strain marks called **stretcher strains** or **Lüders lines**. This can be done by cold working (Fig. 6.2.1), a reduction of only 1 percent in thickness usually being sufficient. This cold reduction is usually done by cold rolling, known as **temper rolling,** followed by alternate bending and reverse bending in a roller leveler. Temper rolling must always precede roller leveling because soft annealed sheets will "break" (yield locally) in the roller leveler. An important phenomenon in these temper-rolled low-carbon sheets is the return, partial or complete, of the sharp yield point after a period of

equivalent. **Carbon equivalent** is based on a formula which includes C and strong hardenability agents such as Mn, Ni, and Cr (see Sec. 6.3). Furthermore, the natural tendency of austenite to grow to a very coarse grain size near the weld metal may be desirably restrained by adding elements such as Ti, which as undissolved carbides, interfere with grain growth. In critical applications, finer grain sizes after transformation during cooling increase toughness in the HAZ.

Other mechanical properties such as fatigue strength, corrosion resistance, and formability are frequently important enough to warrant consideration, but too short a description here may be misleading; standard texts and/or professional journals and literature should be consulted when necessary.

Steel is not necessarily the only material of choice in today's competitive world. The issue revolves on the method to make a legitimate and rational choice between two or more materials with many different properties and characteristics. Consider a very simple example (see Ashby, "Materials Selection in Mechanical Design," Pergamon Press, Oxford).

A furniture designer conceives of a lightweight table—a flat sheet of toughened glass supported on slender unbraced cylindrical legs. The legs must be solid, as light as possible, and must support the tabletop

time. This is known as **aging** in steel. The return of the yield point is accompanied by an increase in hardness and a loss in ductility. While a little more expensive, IF steels do not display a yield point and would not really require temper rolling, although they normally receive about 0.25 percent to improve surface finish. (The 1 percent typically used for non-IF steels would cause harmful changes in the stress-strain curve.) With these options available, aging is no longer the problem it used to be.

High-strength hot-rolled, cold-rolled, and galvanized sheets are now available with specified yield strengths. By the use of small alloy additions of niobium, vanadium, and sometimes copper, it is possible to meet the requirements of ASTM A607-70 for hot-rolled and cold-rolled sheets—345 MPa (50,000 lb/in²) yield point, 483 MPa (70,000 lb/in²) tensile strength, and 22 percent elongation in 2 in. By further alloy additions, sheets are produced to a minimum of 448 MPa (65,000 lb/in²) yield point, 552 MPa (80,000 lb/in²) tensile strength, and 16 percent elongation in 2 in. The sheets are available in coil form and are used extensively for metallic buildings and for welding into tubes for construction of furniture, etc.

Structural Carbon Steels Bridges and buildings frequently are constructed with structural carbon steel meeting the requirements of ASTM A36 (Table 6.2.6). This steel has a minimum yield point of 248 MPa (36,000 lb/in²) and was developed to fill the need for a higher-strength structural carbon steel than the steels formerly covered by ASTM A7 and A373. The controlled composition of A36 steel provides good weldability and furnishes a significant improvement in the economics of steel construction. The structural carbon steels are available in the form of plates, shapes, sheet piling, and bars, all in the hot-rolled condition. A uniform strength over a range of section thickness is provided by adjusting the amount of carbon, manganese, and silicon in the A36 steel.

High-Strength Low-Alloy Steels These steels have in the past been referred to as "high-tensile steels" and "low-alloy steels," but the name **high-strength low-alloy steels,** abbreviated **HSLA steels,** is now the generally accepted designation.

HSLA steels are a group of steels, intended for general structural and miscellaneous applications, that have minimum yield strengths above about 40,000 lb/in². These steels typically contain small amounts of alloying elements to achieve their strength in the hot-rolled or normalized condition. Among the elements used in small amounts, singly or in combination, are niobium, titanium, vanadium, manganese, copper, and phosphorus. A complete listing of HSLA steels available from producers in the United States and Canada shows hundreds of brands or variations, many of which are not covered by ASTM or other specifications. Table 6.2.7 lists ASTM and SAE specifications that cover a large number of some common HSLA steels. These steels generally are available as sheet, strip, plates, bars, and shapes and often are sold as proprietary grades.

HSLA steels have characteristics and properties that result in econo-

Table 6.2.6 Mechanical Properties of Some Constructional Steels*

ASTM designation	Thickness range, mm (in)	Yield point, min		Tensile strength		Elongation in 200 mm (8 in) min, %	Suitable for welding?
		MPa	1,000 lb/in²	MPa	1,000 lb/in²		
Structural carbon-steel plates							
ASTM A36	To 100 mm (4 in), incl.	248	36	400–552	58–80	20	Yes
Low- and intermediate-tensile-strength carbon-steel plates							
ASTM A283	(structural quality)						
Grade A	All thicknesses	165	24	310	45	28	Yes
Grade B	All thicknesses	186	27	345	50	25	Yes
Grade C	All thicknesses	207	30	379	55	22	Yes
Grade D	All thicknesses	228	33	414	60	20	Yes
Carbon-silicon steel plates for machine parts and general construction							
ASTM A284							
Grade A	To 305 mm (12 in)	172	25	345	50	25	Yes
Grade B	To 305 mm (12 in)	159	23	379	55	23	Yes
Grade C	To 305 mm (12 in)	145	21	414	60	21	Yes
Grade D	To 200 mm (8 in)	145	21	414	60	21	Yes
Carbon-steel pressure-vessel plates							
ASTM A285							
Grade A	To 50 mm (2 in)	165	24	303–379	44–55	27	Yes
Grade B	To 50 mm (2 in)	186	27	345–414	50–60	25	Yes
Grade C	To 50 mm (2 in)	207	30	379–448	55–65	23	Yes
Structural steel for locomotives and railcars							
ASTM A113							
Grade A	All thicknesses	228	33	414–496	60–72	21	No
Grade B	All thicknesses	186	27	345–427	50–62	24	No
Grade C	All thicknesses	179	26	331–400	48–58	26	No
Structural steel for ships							
ASTM A131 (all grades)		221	32	400–490	58–71	21	No
Heat-treated constructional alloy-steel plates							
ASTM A514	To 64 mm (2½ in), incl.	700	100	800–950	115–135	18†	Yes
	Over 64 to 102 mm (2½ to 4 in), incl.	650	90	750–950	105–135	17†	Yes

* See appropriate ASTM documents for properties of other plate steels, shapes, bars, wire, tubing, etc.
† Elongation in 50 mm (2 in), min.

Table 6.2.7   Specifications of ASTM and SAE for Some High-Strength Low-Alloy (HSLA) Steels

Society	Designation	Min yield point[a]		Min tensile strength		Min thickness[b]	
		MPa	1,000 lb/in^2	MPa	1,000 lb/in^2	mm	in
SAE	J410b grade 42X	290	42	414	60	9.5	⅜
ASTM	A572 grade 42	290	42	414	60	101.6	4
SAE	J410b grade 945X	310	45	414	60	9.5	⅜
ASTM	A572 grade 45	310	45	414	60	38.1	1½
ASTM	A607 grade 45	310	45	414	60	c	c
ASTM	A606	310	45	448	65	c	c
SAE	J410b grades 945A, C[d]	310	45	448	65[e]	12.7	½[f]
SAE	J410b grade 950X	345	50	448	65	9.5	⅜
ASTM	A572 grade 50	345	50	448	65	38.1	1½
ASTM	A607 grade 50	345	50	448	65	c	c
SAE	J410b grades 950A, B, C, D[d]	345	50	483	70	38.1	1½[f]
ASTM	A242	345	50	483	70	19.1	¾[f]
ASTM	A440[d]	345	50	483	70	19.1	¾[f]
ASTM	A441	345	50	483	70	19.1	¾[f]
ASTM	A588	345	50	483	70	101.6	4[f]
SAE	J410b grade 955X	379	55	483	70	9.5	⅜
ASTM	A572 grade 55	379	55	483	70	38.1	1½
ASTM	A607 grade 55	378	55	483	70	c	c
SAE	J410b grade 960X	414	60	517	75	9.5	⅜
ASTM	A572 grade 60	414	60	517	75	25.4	1
ASTM	A607 grade 60	414	60	517	75	c	c
SAE	J410b grade 965X	448	65	552	80	9.5	⅜
ASTM	A572 grade 65	448	65	552	80	12.7	½
ASTM	A607 grade 65	448	65	552	80	c	c
SAE	J410b grade 970X	483	70	586	85	9.5	⅜
ASTM	A607 grade 70	483	70	586	85	c	c
SAE	J410b grade 980X	552	80	655	95	9.5	⅜

[a] SAE steels specify minimum yield strength.
[b] Applies to plates and bars, approximate web thickness for structurals.
[c] ASTM A606 and A607 apply to sheet and strip only.
[d] SAE J410b grades 945C and 950C and ASTM A440 steels are high-strength carbon-manganese steels rather than HSLA steels.
[e] Reduced 34.5 MPa (5,000 ibf/in^2) for sheet and strip.
[f] Available in heavier thickness at reduced strength levels.

mies to the user when the steels are properly applied. They are considerably stronger, and in many instances tougher, than structural carbon steel, yet have sufficient ductility, formability, and weldability to be fabricated successfully by customary shop methods. In addition, many of the steels have improved resistance to corrosion, so that the necessary equal service life in a thinner section or longer life in the same section is obtained in comparison with that of a structural carbon steel member. Good resistance to repeated loading and good abrasion resistance in service may be other characteristics of some of the steels. While high strength is a common characteristic of all HSLA steels, the other properties mentioned above may or may not be, singly or in combination, exhibited by any particular steel.

HSLA steels have found wide acceptance in many fields, among which are the construction of railroad cars, trucks, automobiles, trailers, and buses; welded steel bridges; television and power-transmission towers and lighting standards; columns in highrise buildings; portable liquefied petroleum gas containers; ship construction; oil storage tanks; air conditioning equipment; agricultural and earthmoving equipment.

Dual-Phase Steels   The good properties of HSLA steels do not provide sufficient cold formability for automotive components which involve stretch forming. **Dual-phase steels** generate a microstructure of ferrite plus islands of austenite-martensite by quenching from a temperature between A_1 and A_3. This leads to continuous yielding (rather than a sharp yield point) and a high work-hardening rate with a larger elongation to fracture. Modest forming strains can give a yield strength in the deformed product of 350 MPa (about 50 ksi), comparable to HSLA steels.

Quenched and Tempered Low-Carbon Constructional Alloy Steels   These steels, having yield strengths at the 689-MPa (100,000-lb/in^2) level, are covered by ASTM A514, by military specifications,

and for pressure-vessel applications, by ASME Code Case 1204. They are available in plates, shapes, and bars and are readily welded. Since they are heat-treated to a tempered martensitic structure, they retain excellent toughness at temperatures as low as $-45°C$ ($-50°F$). Major cost savings have been effected by using these steels in the construction of pressure vessels, in mining and earthmoving equipment, and for major members of large steel structures.

Ultraservice Low-Carbon Alloy Steels   (Quenched and Tempered) The need for high performance materials with higher strength-to-weight ratios for critical military needs, for hydrospace explorations, and for aerospace applications has led to the development of quenched and tempered ultraservice alloy steels. Although these steels are similar in many respects to the quenched and tempered low-carbon constructional alloy steels described above, their significantly higher notch toughness at yield strengths up to 965 MPa (140,000 lb/in^2) distinguishes the ultraservice alloy steels from the constructional alloy steels. These steels are not included in the AISI-SAE classification of alloy steels. There are numerous proprietary grades of ultraservice steels in addition to those covered by ASTM designations A543 and A579. Ultraservice steels may be used in large welded structures subjected to unusually high loads, and must exhibit excellent weldability and toughness. In some applications, such as hydrospace operations, the steels must have high resistance to fatigue and corrosion (especially stress corrosion) as well.

Maraging Steels   For performance requiring high strength and toughness and where cost is secondary, age-hardening low-carbon martensites have been developed based on the essentially carbon-free iron-nickel system. The as-quenched martensite is soft and can be shaped before an aging treatment. Two classes exist: (1) a nominal 18 percent nickel steel containing cobalt, molybdenum, and titanium with yield strengths of 1,380 to 2,070 MPa (200 to 300 ksi) and an outstanding resistance to stress corrosion cracking (normally a major problem at these strengths) and (2) a nominal 12 percent nickel steel containing chromium, molybdenum, titanium, and aluminum adjusted to yield strengths of 1,034 to 1,379 MPa (150 to 200 ksi). The toughness of this series is the best available of any steel at these yield strengths.

Cryogenic-Service Steels   For the economical construction of cryogenic vessels operating from room temperature down to the temperature of liquid nitrogen ($-195°C$ or $-320°F$) a 9 percent nickel alloy steel has been developed. The mechanical properties as specified by ASTM A353 are 517 MPa (75,000 lb/in^2) minimum yield strength and 689 to 827 MPa (100,000 to 120,000 lb/in^2) minimum tensile strength. The minimum Charpy impact requirement is 20.3 J (15 ft·lbf) at $-195°C$ ($-320°F$). For lower temperatures, it is necessary to use austenitic stainless steel.

Machinery Steels   A large variety of carbon and alloy steels is used in the automotive and allied industries. Specifications are published by AISI and SAE on all types of steel, and these specifications should be referred to for detailed information.

A numerical index is used to identify the compositions of AISI (and SAE) steels. Most AISI and SAE alloy steels are made by the basic oxygen or basic electric furnace processes; a few steels that at one time were made in the electric furnace carry the prefix E before their number, i.e., E52100. However, with the almost complete use of ladle furnaces,

this distinction is rarely necessary today. Steels are "melted" in the BOP or EF and "made" in the ladle. A series of four numerals designates the composition of the AISI steels; the first two indicate the steel type, and the last two indicate, as far as feasible, the average carbon content in "points" or hundredths of 1 percent. Thus 1020 is a carbon steel with a carbon range of 0.18 to 0.23 percent, probably made in the basic oxygen furnace, and E4340 is a nickel-chromium molybdenum steel with 0.38 to 0.43 percent carbon made in the electric-arc furnace. The compositions for the standard steels are listed in Tables 6.2.8 and 6.2.9. A group of steels known as **H steels,** which are similar to the standard AISI steels, are being produced with a specified Jominy hardenability; these steels are identified by a suffix H added to the conventional series number. In general, these steels have a somewhat greater allowable variation in chemical composition but a smaller variation in hardenability than would be normal for a given grade of steel. This smaller variation in hardenability results in greater reproducibility of the mechanical properties of the steels on heat treatment; therefore, H steels have become increasingly important in machinery steels.

Boron steels are designated by the letter B inserted between the second and third digits, e.g., 50B44. The effectiveness of boron in increasing hardenability was a discovery of the late thirties, when it was noticed that heats treated with complex deoxidizers (containing boron) showed exceptionally good hardenability, high strength, and ductility after heat treatment. It was found that as little as 0.0005 percent of boron in-

Table 6.2.8   Chemical Composition of AISI Carbon Steels

AISI grade designation	C	Mn	P	S	AISI grade designation	C	Mn	P	S
	Chemical composition limits (ladle analyses), %					Chemical composition limits (ladle analyses), %			
1006	0.08 max	0.25–0.40	0.04 max	0.05 max		Resulfurized (free-machining) steels*			
1008	0.10 max	0.30–0.50			1108	0.08–0.13	0.50–0.80	0.04 max	0.08–0.13
1010	0.08–0.13	0.30–0.60			1109	0.08–0.13	0.60–0.90		0.08–0.13
1012	0.10–0.15	0.30–0.60			1117	0.14–0.20	1.00–1.30		0.08–0.13
1015	0.13–0.18	0.30–0.60			1118	0.14–0.20	1.30–1.60		0.08–0.13
1016	0.13–0.18	0.60–0.90			1119	0.14–0.20	1.00–1.30		0.24–0.33
1017	0.15–0.20	0.30–0.60			1132	0.27–0.34	1.35–1.65		0.08–0.13
1018	0.15–0.20	0.60–0.90			1137	0.32–0.39	1.35–1.65		0.08–0.13
1019	0.15–0.20	0.70–1.00			1139	0.35–0.43	1.35–1.65		0.13–0.20
1020	0.18–0.23	0.30–0.60			1140	0.37–0.44	0.70–1.00		0.08–0.13
1021	0.18–0.23	0.60–0.90			1141	0.37–0.45	1.35–1.65		0.08–0.13
1022	0.18–0.23	0.70–1.00			1144	0.40–0.48	1.35–1.65		0.24–0.33
1023	0.20–0.25	0.30–0.60			1145	0.42–0.49	0.70–1.00		0.04–0.07
1025	0.22–0.28	0.30–0.60			1146	0.42–0.49	0.70–1.00		0.08–0.13
1026	0.22–0.28	0.60–0.90			1151	0.48–0.55	0.70–1.00		0.08–0.13
1030	0.28–0.34	0.60–0.90			Rephosphorized and resulfurized (free-machining) steels*			
1035	0.32–0.38	0.60–0.90			1110	0.08–0.13	0.30–0.60	0.04 max	0.08–0.13
1037	0.32–0.38	0.70–1.00			1211	0.13 max	0.60–0.90	0.07–0.12	0.10–0.15
1038	0.35–0.42	0.60–0.90			1212	0.13 max	0.70–1.00	0.07–0.12	0.16–0.23
1039	0.37–0.44	0.70–1.00			1213	0.13 max	0.70–1.00	0.07–0.12	0.24–0.33
1040	0.37–0.44	0.60–0.90			1116	0.14–0.20	1.10–1.40	0.04 max	0.16–0.23
1042	0.40–0.47	0.60–0.90			1215	0.09 max	0.75–1.05	0.04–0.09	0.26–0.35
1043	0.40–0.47	0.70–1.00			12L14	0.15 max	0.85–1.15	0.04–0.09	0.26–0.35
1045	0.43–0.50	0.60–0.90			High-manganese carbon steels			
1046	0.43–0.50	0.70–0.90			1513	0.10–0.16	1.10–1.40	0.04 max	0.05 max
1049	0.46–0.53	0.60–0.90			1518	0.15–0.21	1.10–1.40		
1050	0.48–0.55	0.60–0.90			1522	0.18–0.24	1.10–1.40		
1055	0.50–0.60	0.60–0.90			1524	0.19–0.25	1.35–1.65		
1060	0.55–0.65	0.60–0.90			1525	0.23–0.29	0.80–1.10		
1064	0.60–0.70	0.50–0.80			1526	0.22–0.29	1.10–1.40		
1065	0.60–0.70	0.60–0.90			1527	0.22–0.29	1.20–1.50		
1070	0.65–0.75	0.60–0.90			1536	0.30–0.37	1.20–1.50		
1078	0.72–0.85	0.30–0.60			1541	0.36–0.44	1.35–1.65		
1080	0.75–0.88	0.60–0.90			1547	0.43–0.51	1.35–1.65		
1084	0.80–0.93	0.60–0.90			1548	0.44–0.52	1.10–1.40		
1086	0.80–0.93	0.30–0.50			1551	0.45–0.56	0.85–1.15		
1090	0.85–0.98	0.60–0.90			1552	0.47–0.55	1.20–1.50		
1095	0.90–1.03	0.30–0.50			1561	0.55–0.65	0.75–1.05		
					1566	0.60–0.71	0.85–1.15		
					1572	0.65–0.76	1.00–1.30		

*Except for free-machining steels, the nominal 0.04 percent P and 0.05 percent S are much lower in modern steels; S contents of less than 0.01 percent are readily achieved.

Table 6.2.9 Alloy-Steel Compositions[a,b,c,d]

AISI no.	Chemical composition limits (ladle analyses), %[c,d]								
	C	Mn	P, max[f]	S, max[f]	Si	Ni	Cr	Mo	V
1330	0.28–0.33	1.60–1.90	0.035	0.040	0.20–0.35				
1335	0.33–0.38	1.60–1.90	0.035	0.040	0.20–0.35				
1340	0.38–0.43	1.60–1.90	0.035	0.040	0.20–0.35				
1345	0.43–0.48	1.60–1.90	0.035	0.040	0.20–0.35				
4012	0.09–0.14	0.75–1.00	0.035	0.040	0.20–0.35			0.15–0.25	
4023	0.20–0.25	0.70–0.90	0.035	0.040	0.20–0.35			0.20–0.30	
4024	0.20–0.25	0.70–0.90	0.035	0.035–0.050	0.20–0.35			0.20–0.30	
4027	0.25–0.30	0.70–0.90	0.035	0.040	0.20–0.35			0.20–0.30	
4028	0.25–0.30	0.70–0.90	0.035	0.035–0.050	0.20–0.35			0.20–0.30	
4037	0.35–0.40	0.70–0.90	0.035	0.040	0.20–0.35			0.20–0.30	
4047	0.45–0.50	0.70–0.90	0.035	0.040	0.20–0.35			0.20–0.30	
4118	0.18–0.23	0.70–0.90	0.035	0.040	0.20–0.35		0.40–0.60	0.08–0.15	
4130	0.28–0.33	0.40–0.60	0.035	0.040	0.20–0.35		0.80–1.10	0.15–0.25	
4137	0.35–0.40	0.70–0.90	0.035	0.040	0.20–0.35		0.80–1.10	0.15–0.25	
4140	0.38–0.43	0.75–1.00	0.035	0.040	0.20–0.35		0.80–1.10	0.15–0.25	
4142	0.40–0.45	0.75–1.00	0.035	0.040	0.20–0.35		0.80–1.10	0.15–0.25	
4145	0.43–0.48	0.75–1.00	0.035	0.040	0.20–0.35		0.80–1.10	0.15–0.25	
4147	0.45–0.50	0.75–1.00	0.035	0.040	0.20–0.35		0.80–1.10	0.15–0.25	
4150	0.48–0.53	0.75–1.00	0.035	0.040	0.20–0.35		0.80–1.10	0.15–0.25	
4320	0.17–0.22	0.45–0.65	0.035	0.040	0.20–0.35	1.65–2.00	0.40–0.60	0.20–0.30	
4340	0.38–0.43	0.60–0.80	0.035	0.040	0.20–0.35	1.65–2.00	0.70–0.90	0.20–0.30	
4419	0.18–0.23	0.45–0.65	0.035	0.040	0.20–0.35			0.45–0.60	
4615	0.13–0.18	0.45–0.65	0.035	0.040	0.20–0.35	1.65–2.00		0.20–0.30	
4620	0.17–0.22	0.45–0.65	0.035	0.040	0.20–0.35	1.65–2.00		0.20–0.30	
4621	0.18–0.23	0.70–0.90	0.035	0.040	0.20–0.35	1.65–2.00		0.20–0.30	
4626	0.24–0.29	0.45–0.65	0.035	0.040	0.20–0.35	0.70–1.00		0.15–0.25	
4718	0.16–0.21	0.70–0.90				0.90–1.20	0.35–0.55	0.30–0.40	
4720	0.17–0.22	0.50–0.70	0.035	0.040	0.20–0.35	0.90–1.20	0.35–0.55	0.15–0.25	
4815	0.13–0.18	0.40–0.60	0.035	0.040	0.20–0.35	3.25–3.75		0.20–0.30	
4817	0.15–0.20	0.40–0.60	0.035	0.040	0.20–0.35	3.25–3.75		0.20–0.30	
4820	0.18–0.23	0.50–0.70	0.035	0.040	0.20–0.35	3.25–3.75		0.20–0.30	
5015	0.12–0.17	0.30–0.50	0.035	0.040	0.20–0.35		0.30–0.50		
50B44[e]	0.43–0.48	0.75–1.00	0.035	0.040	0.20–0.35		0.40–0.60		
50B46[e]	0.44–0.49	0.75–1.00	0.035	0.040	0.20–0.35		0.20–0.35		
50B50[e]	0.48–0.53	0.75–1.00	0.035	0.040	0.20–0.35		0.40–0.60		
50B60[e]	0.56–0.64	0.75–1.00	0.035	0.040	0.20–0.35		0.40–0.60		
5120	0.17–0.22	0.70–0.90	0.035	0.040	0.20–0.35		0.70–0.90		
5130	0.28–0.33	0.70–0.90	0.035	0.040	0.20–0.35		0.80–1.10		
5132	0.30–0.35	0.60–0.80	0.035	0.040	0.20–0.35		0.75–1.00		
5135	0.33–0.38	0.60–0.80	0.035	0.040	0.20–0.35		0.80–1.05		
5145	0.43–0.48	0.70–0.90	0.035	0.040	0.20–0.35		0.70–0.90		
5147	0.46–0.51	0.70–0.95	0.035	0.040	0.20–0.35		0.85–1.15		
5150	0.48–0.53	0.70–0.90	0.035	0.040	0.20–0.35		0.70–0.90		
5155	0.51–0.59	0.70–0.90	0.035	0.040	0.20–0.35		0.70–0.90		
5160	0.56–0.64	0.75–1.00	0.035	0.040	0.20–0.35		0.70–0.90		
51B60[e]	0.56–0.64	0.75–1.00	0.035	0.040	0.20–0.35		0.70–0.90		
51100[e]	0.98–1.10	0.25–0.45	0.025	0.025	0.20–0.35		0.90–1.15		
52100[e]	0.98–1.10	0.25–0.45	0.025	0.025	0.20–0.35		1.30–1.60		
6118	0.16–0.21	0.50–0.70	0.035	0.040	0.20–0.35		0.50–0.70		0.10–0.15
6150	0.48–0.53	0.70–0.90	0.035	0.040	0.20–0.35		0.80–1.10		0.15
81B45[e]	0.43–0.48	0.75–1.00	0.035	0.040	0.20–0.35	0.20–0.40	0.35–0.55	0.08–0.15	
8615	0.13–0.18	0.70–0.90	0.035	0.040	0.20–0.35	0.40–0.70	0.40–0.60	0.15–0.25	
8617	0.15–0.20	0.70–0.90	0.035	0.040	0.20–0.35	0.40–0.70	0.40–0.60	0.15–0.25	
8620	0.18–0.23	0.70–0.90	0.035	0.040	0.20–0.35	0.40–0.70	0.40–0.60	0.15–0.25	
8622	0.20–0.25	0.70–0.90	0.035	0.040	0.20–0.35	0.40–0.70	0.40–0.60	0.15–0.25	
8625	0.23–0.28	0.70–0.90	0.035	0.040	0.20–0.35	0.40–0.70	0.40–0.60	0.15–0.25	
8627	0.25–0.30	0.70–0.90	0.035	0.040	0.20–0.35	0.40–0.70	0.40–0.60	0.15–0.25	
8630	0.28–0.33	0.70–0.90	0.035	0.040	0.20–0.35	0.40–0.70	0.40–0.60	0.15–0.25	
8637	0.35–0.40	0.75–1.00	0.035	0.040	0.20–0.35	0.40–0.70	0.40–0.60	0.15–0.25	
8640	0.38–0.43	0.75–1.00	0.035	0.040	0.20–0.35	0.40–0.70	0.40–0.60	0.15–0.25	
8642	0.40–0.45	0.75–1.00	0.035	0.040	0.20–0.35	0.40–0.70	0.40–0.60	0.15–0.25	
8645	0.43–0.48	0.75–1.00	0.035	0.040	0.20–0.35	0.40–0.70	0.40–0.60	0.15–0.25	
8655	0.51–0.59	0.75–1.00	0.035	0.040	0.20–0.35	0.40–0.70	0.40–0.60	0.15–0.25	
8720	0.18–0.23	0.70–0.90	0.035	0.040	0.20–0.35	0.40–0.70	0.40–0.60	0.20–0.30	
8740	0.38–0.43	0.75–1.00	0.035	0.040	0.20–0.35	0.40–0.70	0.40–0.60	0.20–0.30	

Table 6.2.9 Alloy-Steel Compositions[a,b,c,d] *(Continued)*

AISI no.	Chemical composition limits (ladle analyses), %[c,d]								
	C	Mn	P, max[f]	S, max[f]	Si	Ni	Cr	Mo	V
8822	0.20–0.25	0.75–1.00	0.035	0.040	0.20–0.35	0.40–0.70	0.40–0.60	0.30–0.40	
9255	0.51–0.59	0.70–0.95	0.035	0.040	1.80–2.20				
9260	0.56–0.64	0.75–1.00	0.035	0.040	1.80–2.20				
94B17[e]	0.15–0.20	0.75–1.00	0.035	0.040	0.20–0.35	0.30–0.60	0.30–0.50	0.08–0.15	
94B30[e]	0.28–0.33	0.75–1.00	0.035	0.040	0.20–0.35	0.30–0.60	0.30–0.50	0.08–0.15	

[a] These tables are subject to change from time to time, with new steels sometimes added, other steels eliminated, and compositions of retained steels occasionally altered. Current publications of AISI and SAE should be consulted for latest information.

[b] Applicable to blooms, billets, slabs, and hot-rolled and cold-rolled bars.

[c] These steels may be produced by the basic oxygen or basic electric steelmaking process.

[d] Small quantities of certain elements which are not specified or required may be found in alloy steels. These elements are considered to be incidental and are acceptable up to the following maximum amounts: copper to 0.35 percent, nickel to 0.25 percent, chromium to 0.20 percent, and molybdenum to 0.06 percent.

[e] Boron content is 0.0005 percent minimum.

[f] P max and S max can easily be much lower.

creased the hardenability of steels with 0.15 to 0.60 carbon, whereas boron contents of over 0.005 percent had an adverse effect on hot workability. Boron steels achieve special importance in times of alloy shortages, for they can replace such critical alloying elements as nickel, molybdenum, chromium, and manganese and, when properly heat-treated, possess physical properties comparable to the alloy grades they replace. Additional advantages for the use of boron in steels are a decrease in susceptibility to flaking, formation of less adherent scale, greater softness in the unhardened condition, and better machinability. It is also useful in low-carbon bainite and acicular ferrite steels.

Specific applications of these steels cannot be given, since the selection of a steel for a given part must depend upon an intimate knowledge of factors such as the availability and cost of the material, the detailed design of the part, and the severity of the service to be imposed. However, the mechanical properties desired in the part to be heat-treated will determine to a large extent the carbon and alloy content of the steel. Table 6.2.10 gives a résumé of mechanical properties that can be expected on heat-treating AISI steels, and Table 6.2.11 gives an indication of the effect of size on the mechanical properties of heat-treated steels.

The low-carbon AISI steels are used for carburized parts, cold-headed bolts and rivets, and for similar applications where high quality is required. The AISI 1100 series are low-carbon **free-cutting** steels for high-speed screw-machine stock and other machining purposes. These steels have high sulfur present in the steel in the form of manganese sulfide inclusions causing the chips to break short on machining. Manganese and phosphorus harden and embrittle the steel, which also contributes toward free machining. The high manganese contents are intended to ensure that all the sulfur is present as manganese sulfide. Lead was a common additive, but there are environmental problems; experiments have been conducted with selenium, tellurium, and bismuth as replacements.

Cold-finished carbon-steel bars are used for bolts, nuts, typewriter and cash register parts, motor and transmission power shafting, piston pins, bushings, oil-pump shafts and gears, etc. Representative mechanical properties of cold-drawn steel are given in Table 6.2.12. Besides improved mechanical properties, cold-finished steel has better machining properties than hot-rolled products. The surface finish and dimensional accuracy are also greatly improved by cold finishing.

Forging steels, at one time between 0.30 and 0.40 percent carbon and used for axles, bolts, pins, connecting rods, and similar applications, can now contain up to 0.7 percent C with microalloying additions to refine the structure (for better toughness) and to deliver precipitation strengthening. In some cases, air cooling can be used, thereby saving the cost of alloy steels. These steels are readily forged and, after heat treatment, develop considerably better mechanical properties than low-carbon steels. For heavy sections where high strength is required, such as in crankshafts and heavy-duty gears, the carbon may be increased and sufficient alloy content may be necessary to obtain the desired hardenability.

TOOL STEELS

The application of tool steels can generally be fitted into one of the following categories or types of operations: cutting, shearing, forming, drawing, extruding, rolling, and battering. Each of these operations requires in the tool steel a particular physical property or a combination of such metallurgical characteristics as hardness, strength, toughness, wear resistance, and resistance to heat softening, before optimum performance can be realized. These considerations are of prime importance in tool selection; but hardenability, permissible distortion, surface decarburization during heat treatment, and machinability of the tool steel are a few of the additional factors to be weighed in reaching a final decision. In actual practice, the final selection of a tool steel represents a compromise of the most desirable physical properties with the best overall economic performance. Tool steels have been identified and classified by the SAE and the AISI into six major groups, based upon quenching methods, applications, special characteristics, and use in specific industries. These six classes are water-hardening, shock-resisting, cold-work, hot-work, high-speed, and special-purpose tool steels. A simplified classification of these six basic types and their subdivisions is given in Table 6.2.13.

Water-hardening tool steels, containing 0.60 to 1.40 percent carbon, are widely used because of their low cost, good toughness, and excellent machinability. They are shallow-hardening steels, unsuitable for nondeforming applications because of high warpage, and possess poor resistance to softening at elevated temperatures. Water-hardening tool steels have the widest applications of all major groups and are used for files, twist drills, shear knives, chisels, hammers, and forging dies.

Shock-resisting tool steels, with chromium-tungsten, silicon-molybdenum, or silicon-manganese as the dominant alloys, combine good hardenability with outstanding toughness. A tendency to distort easily is their greatest disadvantage. However, oil quenching can minimize this characteristic.

Cold-work tool steels are divided into three groups: oil-hardening, medium-alloy air-hardening, and high-carbon, high-chromium. In general, this class possesses high wear resistance and hardenability, develops little distortion, but at best is only average in toughness and in resistance to heat softening. Machinability ranges from good in the oil-hardening grade to poor in the high-carbon, high-chromium steels.

Hot-work tool steels are either chromium- or tungsten-based alloys possessing good nondeforming, hardenability, toughness, and resistance to heat-softening characteristics, with fair machinability and wear resistance. Either air or oil hardening can be employed. Applications are blanking, forming, extrusion, and casting dies where temperatures may rise to 540°C (1,000°F).

High-speed tool steels, the best-known tool steels, possess the best combination of all properties except toughness, which is not critical for high-speed cutting operations, and are either tungsten or molybdenum-base types. Cobalt is added in some cases to improve the cutting qualities in roughing operations. They retain considerable hardness at a

Table 6.2.10   Mechanical Properties of Certain AISI Steels with Various Heat Treatments
Sections up to 40 mm (or 1½ diam or thickness)

Tempering temp		Tensile strength		Yield point			Elongation	
°C	°F	MPa	1,000 lb/in²	MPa	1,000 lb/in²	Reduction of area, %	in 50 mm (2 in), %	Brinell hardness
AISI 1040 quenched in water from 815°C (1,500°F)								
315	600	862	125	717	104	46	11	260
425	800	821	119	627	91	53	13	250
540	1,000	758	110	538	78	58	15	220
595	1,100	745	108	490	71	60	17	216
650	1,200	717	104	455	66	62	20	210
705	1,300	676	98	414	60	64	22	205
AISI 1340 normalized at 865°C (1,585°F), quenched in oil from 845°C (1,550°F)								
315	600	1,565	227	1,420	206	43	11	448
425	800	1,248	181	1,145	166	51	13	372
540	1,000	966	140	834	121	58	17.5	297
595	1,100	862	125	710	103	62	20	270
650	1,200	793	115	607	88	65	23	250
705	1,300	758	110	538	78	68	25.5	234
AISI 4042 normalized at 870°C (1,600°F), quenched in oil from 815°C (1,500°F)								
315	600	1,593	231	1,448	210	41	12	448
425	800	1,207	175	1,089	158	50	14	372
540	1,000	966	140	862	125	58	19	297
595	1,100	862	125	758	110	62	23	260
650	1,200	779	113	683	99	65	26	234
705	1,800	724	105	634	92	68	30	210

Table 6.2.11   Effect of Size of Specimen on the Mechanical Properties of Some AISI Steels

Diam of section		Tensile strength		Yield point			Elongation	
in	mm	MPa	1,000 lb/in²	MPa	1,000 lb/in²	Reduction of area, %	in 50 mm (2 in), %	Brinell hardness
AISI 1040, water-quenched, tempered at 540°C (1,000°F)								
1	25	758	110	538	78	58	15	230
2	50	676	98	448	65	49	20	194
3	75	641	93	407	59	48	23	185
4	100	621	90	393	57	47	24.5	180
5	125	614	89	372	54	46	25	180
AISI 4140, oil-quenched, tempered at 540°C (1,000°F)								
1	25	1,000	145	883	128	56	18	297
2	50	986	143	802	125	58	19	297
3	75	945	137	814	118	59	20	283
4	100	869	126	758	110	60	18	270
5	125	841	122	724	105	59	17	260
AISI 8640, oil-quenched, tempered at 540°C (1,000°F)								
1	25	1,158	168	1,000	145	44	16	332
2	50	1,055	153	910	132	45	20	313
3	75	951	138	807	117	46	22	283
4	100	889	129	745	108	46	23	270
5	125	869	126	724	105	45	23	260

red heat. Very high heating temperatures are required for the heat treatment of high-speed steel and, in general, the tungsten-cobalt high-speed steels require higher quenching temperatures than the molybdenum steels. High-speed steel should be tempered at about 595°C (1,100°F) to increase the toughness; owing to a secondary hardening effect, the hardness of the tempered steels may be higher than as quenched.

Special-purpose tool steels are composed of the low-carbon, low-alloy, carbon-tungsten, mold, and other miscellaneous types.

SPRING STEEL

For small springs, steel is often supplied to spring manufacturers in a form that requires no heat treatment except perhaps a low-temperature anneal to relieve forming strains. Types of previously treated steel wire for small helical springs are **music wire** which has been given a special heat treatment called patenting and then cold-drawn to develop a high yield strength, **hard-drawn wire** which is of lower quality than music wire since it is usually made of lower-grade material and is seldom patented, and **oil-tempered wire** which has been quenched and tempered.

Table 6.2.12 Representative Average Mechanical Properties of Cold-Drawn Steel

AISI no.	Tensile strength		Yield strength		Elongation in 50 mm (2 in), %	Reduction of area, %	Brinell hardness
	MPa	1,000 lb/in²	MPa	1,000 lb/in²			
1010	462	67	379	55.0	25.0	57	137
1015	490	71	416	60.3	22.0	55	149
1020	517	75	439	63.7	20.0	52	156
1025	552	80	469	68.0	18.5	50	163
1030	600	87	509	73.9	17.5	48	179
1035	634	92	539	78.2	17.0	45	187
1040	669	97	568	82.4	16.0	40	197
1045	703	102	598	86.7	15.0	35	207
1117	552	80	469	68.0	19.0	51	163
1118	569	82.5	483	70.1	18.5	50	167
1137	724	105	615	89.2	16.0	35	217
1141	772	112	656	95.2	14.0	30	223

Sizes 16 to 50 mm (⅝ to 2 in) diam, test specimens 50 × 13 mm (2 × 0.505 in).
SOURCE: ASM "Metals Handbook."

Table 6.2.13 Simplified Tool-Steel Classification*

Major grouping	Symbol	Types
Water-hardening tool steels	W	
Shock-resisting tool steels	S	
Cold-work tool steels	O	Oil hardening
	A	Medium-alloy air hardening
	D	High-carbon, high-chromium
Hot-work tool steels	H	H10–H19 chromium base
		H20–H39 tungsten base
		H40–H59 molybdenum base
High-speed tool steels	T	Tungsten base
	M	Molybdenum base
Special-purpose tool steels	F	Carbon-tungsten
	L	Low-alloy
	P	Mold steels
		P1–P19 low carbon
		P20–P39 other types

* Each subdivision is further identified as to type by a suffix number which follows the letter symbol.

The wire usually has a Brinell hardness between 352 and 415, although this will depend on the application of the spring and the severity of the forming operation. Steel for small flat springs has either been cold-rolled or quenched and tempered to a similar hardness.

Steel for both helical and flat springs which is hardened and tempered after forming is usually supplied in an annealed condition. Plain carbon steel is satisfactory for small springs; for large springs it is necessary to use alloy steels such as chrome-vanadium or silicon-manganese steel in order to obtain a uniform structure throughout the cross section. Table 6.2.14 gives the chemical composition and heat treatment of several spring steels. It is especially important for springs that the surface of the steel be free from all defects and decarburization, which lowers fatigue strength.

SPECIAL ALLOY STEELS

Many steel alloys with compositions tailored to specific requirements are reported periodically. Usually, the compositions and/or treatments are patented, and they are most likely to have registered trademarks and trade names. They are too numerous to be dealt with here in any great detail, but a few of the useful properties exhibited by some of those special alloys are mentioned.

Iron-silicon alloys with minimum amounts of both carbon and other alloying elements have been used by the electrical equipment industry for a long time; often these alloys are known as **electrical sheet steel.** Iron-nickel alloys with high proportions of nickel, and often with other alloying elements, elicit properties such as nonmagnetic behavior, high permeability, low hysteresis loss, low coefficient of expansion, etc.; iron-cobalt alloys combined with other alloying elements can result in materials with low resistivity and high hysteresis loss. The reader interested in the use of materials with some of these or other desired properties is directed to the extensive literature and data available.

STAINLESS STEELS
by James D. Redmond

REFERENCES: "Design Guidelines for the Selection and Use of Stainless Steel," Specialty Steel Institute of North America (SSINA), Washington, DC. Publications of the Nickel Development Institute, Toronto, Ontario, Canada. "Metals Handbook," 10th ed., ASM International. ASTM Standards.

When the chromium content is increased to about 11 percent in an iron-chromium alloy, the resulting material is generally classified as a stainless steel. With that minimum quantity of chromium, a thin, protective, passive film forms spontaneously on the steel. This passive film acts as a barrier to prevent corrosion. Further increases in chromium content strengthen the passive film and enable it to repair itself if it is damaged in a corrosive environment. Stainless steels are also heat-resistant

Table 6.2.14 Type of Steel and Heat Treatment for Large Hot-Formed Flat, Leaf, and Helical Springs

AISI steel no.	Normalizing temp*		Quenching temp†		Tempering temp	
	°C	°F	°C	°F	°C	°F
1095	860–885	1,575–1,625	800–830	1,475–1,525	455–565	850–1,050
6150	870–900	1,600–1,650	870–900	1,600–1,650	455–565	850–1,050
9260	870–900	1,600–1,650	870–900	1,600–1,650	455–565	850–1,050
5150	870–900	1,600–1,650	800–830	1,475–1,525	455–565	850–1,050
8650	870–900	1,600–1,650	870–900	1,600–1,650	455–565	850–1,050

* These normalizing temperatures should be used as the forming temperature whenever feasible.
† Quench in oil at 45 to 60°C (110 to 140°F).

because exposure to high temperatures (red heat and above) causes the formation of a tough oxide layer which retards further oxidation.

Other alloying elements are added to stainless steels to improve corrosion resistance in specific environments or to modify or optimize mechanical properties or characteristics. Nickel changes the crystal structure to improve ductility, toughness, and weldability. Nickel improves corrosion resistance in reducing environments such as sulfuric acid. Molybdenum increases pitting and crevice corrosion resistance in chloride environments. Carbon and nitrogen increase strength. Aluminum and silicon improve oxidation resistance. Sulfur and selenium are added to improve machinability. Titanium and niobium (columbium) are added to prevent **sensitization** by preferentially combining with carbon and nitrogen, thereby preventing intergranular corrosion.

Stainless Steel Grades   There are more than 200 different **grades** of stainless steel. Each is alloyed to provide a combination of corrosion resistance, heat resistance, mechanical properties, or other specific characteristics. There are five **families** of stainless steels: austenitic, ferritic, duplex, martensitic, and precipitation hardening. Table 6.2.15 is a representative list of stainless steel grades by family and chemical composition as typically specified in ASTM Standards. The mechanical properties of many of these grades are summarized in Table 6.2.16. Note that the number designation of a stainless steel does not describe its performance when utilized to resist corrosion.

Austenitic stainless steels are the most widely used, and while most are designated in the 300 series, some of the highly alloyed grades, though austenitic, have other identifying grade designations, e.g., alloy 20, 904L, and the 6 percent molybdenum grades. These latter are often known by their proprietary designation. These stainless steel grades are available in virtually all wrought product forms, and many are employed as castings. Some austenitic grades in which manganese and nitrogen are substituted partially for nickel are also designated in the 200 series. Types 304 (18 Cr, 8 Ni) and 316 (17 Cr, 10 Ni, 2 Mo) are the workhorse grades, and they are utilized for a broad range of equipment and structures. Austenitic grades can be hardened by cold work but not by heat treatment; cold work increases strength with an accompanying decrease in ductility. They provide excellent corrosion resistance, respond very well to forming operations, and are readily welded. When fully annealed, they are not magnetic, but may become slightly magnetic when cold-worked. Compared to carbon steel, they have higher coefficients of thermal expansion and lower thermal conductivities. When austenitic grades are employed to resist corrosion, their performance may vary over a wide range, depending on the particular corrosive media encountered. As a general rule, increased levels of chromium, molybdenum, and nitrogen result in increased resistance to pitting and crevice corrosion in chloride environments. Type 304 is routinely used for atmospheric corrosion resistance and to handle low-chloride potable water. At the other end of the spectrum are the 6% Mo austenitic stainless steels, which have accumulated many years of service experience handling seawater in utility steam condensers and in piping systems on offshore oil and gas platforms. Highly alloyed austenitic stainless steels are subject to precipitation reactions in the range of 1,500 to 1,800°F (815 to 980°C), resulting in a reduction of their corrosion resistance and ambient-temperature impact toughness.

Low-carbon grades, or **L grades,** are restricted to very low levels of carbon (usually 0.030 wt % maximum) to reduce the possibility of sensitization due to chromium carbide formation either during welding or when exposed to a high-temperature thermal cycle. When a sensitized stainless steel is exposed subsequently to a corrosive environment, intergranular corrosion may occur. Other than improved resistance to intergranular corrosion, the low-carbon grades have the same resistance to pitting, crevice corrosion, and chloride stress corrosion cracking as the corresponding grade with the higher level of carbon (usually 0.080 wt % maximum).

There is a trend to dual-certify some pairs of stainless-steel grades. For example, an 18 Cr – 8 Ni stainless steel low enough in carbon to meet the requirements of an L grade and high enough in strength to qualify as the standard carbon version may be **dual-certified.** ASTM specifications allow such a material to be certified and marked, for example, 304/304L and S30400/S30403.

Ferritic stainless steels are iron-chromium alloys with 11 to 30 percent chromium. They can be strengthened slightly by cold working. They are magnetic. They are difficult to produce in plate thicknesses, but are readily available in sheet and bar form. Some contain molybdenum for improved corrosion resistance in chloride environments. They exhibit excellent resistance to chloride stress corrosion cracking. Resistance to pitting and crevice corrosion is a function of the total chromium and molybdenum content. The coefficient of thermal expansion is similar to that of carbon steel. The largest quantity produced is Type 409 (11 Cr), used extensively for automobile catalytic converters, mufflers, and exhaust system components. The most highly alloyed ferritic stainless steels have a long service history in seawater-cooled utility condensers.

Duplex stainless steels combine some of the best features of the austenitic and the ferritic grades. Duplex stainless steels have excellent chloride stress corrosion cracking resistance and can be produced in the full range of product forms typical of the austenitic grades. Duplex stainless steels are comprised of approximately 50 percent ferrite and 50 percent austenite. They are magnetic. Their yield strength is about double that of the 300-series austenitic grades, so economies may be achieved from reduced piping and vessel wall thicknesses. The coefficient of thermal expansion of duplex stainless steels is similar to that of carbon steel and about 30 to 40 percent less than that of the austenitic grades. The general-purpose duplex grade is 2205 (22 Cr, 5 Ni, 3 Mo, 0.15 N). Because they suffer embrittlement with prolonged high-temperature exposure, ASME constructions using the duplex grades are limited to a maximum 600°F (315°C) service temperature. Duplex stainless steel names often reflect their chemical composition or are proprietary designations.

Martensitic stainless steels are straight chromium grades with relatively high levels of carbon. The martensitic grades can be strengthened by heat treatment. To achieve the best combination of strength, corrosion resistance, ductility, and impact toughness, they are tempered in the range of 300 to 700°F (150 to 370°C). They have a 400-series designation; 410 is the general-purpose grade. They are magnetic. The martensitic grades containing up to about 0.15 wt % carbon—e.g., grades 403, 410, and 416 (a free-machining version) can be hardened to about 45 RC. The high-carbon martensitics, such as Types 440A, B, and C, can be hardened to about 60 RC. The martensitics typically exhibit excellent wear or abrasion resistance but limited corrosion resistance. In most environments, the martensitic grades have less corrosion resistance than Type 304.

Precipitation-hardening stainless steels can be strengthened by a relatively low-temperature heat treatment. The low-temperature heat treatment minimizes distortion and oxidation associated with higher-temperature heat treatments. They can be heat-treated to strengths greater than can the martensitic grades. Most exhibit corrosion resistance superior to the martensitics and approach that of Type 304. While some precipitation-hardening stainless steels have a 600-series designation, they are most frequently known by names which suggest their chemical composition, for example, 17-4PH, or by proprietary names.

The commonly used stainless steels have been approved for use in ASME boiler and pressure vessel construction. Increasing the carbon content increases the maximum **allowable stress values.** Nitrogen additions are an even more powerful strengthening agent. This may be seen by comparing the allowable stresses for 304L (0.030 C maximum), 304H (0.040 C minimum), and 304N (0.10 N minimum) in Table 6.2.17. In general, increasing the total alloy content—especially chromium, molybdenum, and nitrogen—increases the allowable design stresses. However, precipitation reactions which can occur in the most highly alloyed grades limit their use to a maximum of about 750°F (400°C). The duplex grade 2205 has higher allowable stress values than even the most highly alloyed of the austenitic grades, up to a maximum service temperature of 600°F (315°C). The precipitation-hardening grades exhibit some of the highest allowable stress values but also are limited to about 650°F (345°C).

Copious amounts of information and other technical data regarding physical and mechanical properties, examples of applications and histories of service lives in specific corrosive environments, current costs, etc. are available in the references and elsewhere in the professional and trade literature.

Table 6.2.15 ASTM/AWS/AMS Chemical Composition (Wt. Pct.)[a]

UNS number	Grade	C	Cr	Ni	Mo	Cu	N	Other
				Austenitic stainless steels				
S20100	201	0.15	16.00–18.00	3.50–5.50	b	—	0.25	Mn: 5.50–7.50
S20200	202	0.15	17.00–19.00	4.00–6.00	—	—	0.25	Mn: 7.50–10.00
S30100	301	0.15	16.00–18.00	6.00–8.00	—	—	0.10	—
S30200	302	0.15	17.00–19.00	8.00–10.00	—	—	0.10	—
S30300	303	0.15	17.00–19.00	8.00–10.00	—	—	—	S: 0.15 min
S30323	303Se	0.15	17.00–19.00	8.00–10.00	—	—	—	Se: 0.15 min
S30400	304	0.08	18.00–20.00	8.00–10.50	—	—	0.10	—
S30403	304L	0.030	18.00–20.00	8.00–12.00	—	—	0.10	—
S30409	304H	0.04–0.10	18.00–20.00	8.00–10.50	—	—	—	—
S30451	304N	0.08	18.00–20.00	8.00–10.50	—	—	0.10–0.16	—
S30500	305	0.12	17.00–19.00	10.50–13.00	—	—	—	—
S30800	308	0.08	19.00–21.00	10.00–12.00	—	—	—	—
S30883	308L	0.03	19.50–22.00	9.00–11.00	0.05	0.75	—	Si: 0.39–0.65
S30908	309S	0.08	22.00–24.00	12.00–15.00	—	—	—	—
S31008	310S	0.08	24.00–26.00	19.00–22.00	—	—	—	—
S31600	316	0.08	16.00–18.00	10.00–14.00	2.00–3.00	—	0.10	—
S31603	316L	0.030	16.00–18.00	10.00–14.00	2.00–3.00	—	0.10	—
S31609	316H	0.04–0.10	16.00–18.00	10.00–14.00	2.00–3.00	—	0.10	—
S31703	317L	0.030	18.00–20.00	11.00–15.00	3.00–4.00	—	—	—
S31726	317LMN	0.030	17.00–20.00	13.50–17.50	4.0–5.0	—	0.10–0.20	—
S32100	321	0.080	17.00–19.00	9.00–12.00	—	—	0.10	Ti: 5(C + N) min; 0.70 max
S34700	347	0.080	17.00–19.00	9.00–13.00	—	—	—	Nb: 10C min; 1.00 max
N08020	Alloy 20	0.07	19.00–21.00	32.00–38.00	2.00–3.00	3.00–4.00	—	Nb + Ta: 8C min; 1.00 max
N08904	904L	0.020	19.00–23.00	23.00–28.00	4.0–5.0	1.0–2.0	0.10	—
S31254	254 SMO[c]	0.020	19.50–20.50	17.50–18.50	6.00–6.50	0.50–1.00	0.18–0.22	—
N08367	AL-6XN[d]	0.030	20.00–22.00	23.50–25.50	6.00–7.00	0.75	0.18–0.25	—
N08926	25-6MO (3)/1925hMo[e]	0.020	19.00–21.00	24.00–26.00	6.0–7.0	0.5–1.0	0.15–0.25	—
S32654	654 SMO[e]	0.020	24.00–25.00	21.00–23.00	7.00–8.00	0.30–0.60	0.45–0.55	Mn: 2.00–4.00
				Duplex stainless steels				
S31260	DP-3[g]	0.030	24.0–26.0	5.50–7.50	2.50–3.50	0.20–0.80	0.10–0.30	W 0.10–0.50
S31803	2205	0.030	21.0–23.0	4.50–6.50	2.50–3.50	—	0.08–0.20	—
S32304	2304	0.030	21.5–24.5	3.00–5.00	0.05–0.60	0.05–0.60	0.05–0.20	—
S32550	Ferralium 255[h]	0.040	24.0–27.0	4.50–6.50	2.90–3.90	1.5–2.5	0.10–0.25	—
S32750	2507	0.030	24.0–26.0	6.00–8.00	3.00–5.00	0.50	0.24–0.32	—
S32900	329	0.08	23.00–28.00	2.50–5.00	1.00–2.00	—	—	—

Ferritic stainless steels

UNS No.	Designation	C	Cr	Ni	Mo	Cu	N	Other elements
S40500	405	0.08	11.50–14.50	0.60	—	—	—	Al: 0.10–0.30
S40900	409	0.08	10.50–11.75	0.50	—	—	—	Ti: 6xC min; 0.75 max
S43000	430	0.12	16.00–18.00	—	—	—	—	—
S43020	430F	0.12	16.00–18.00	—	—	—	—	S: 0.15 min
S43035	439	0.07	17.00–19.00	0.5	—	—	0.04	Ti: 0.20 + 4(C + N) min; 1.10 max Al: 0.15 max
S43400	434	0.12	16.00–18.00	—	0.75–1.25	—	—	—
S44400	444 (18 Cr–2 Mo)	0.025	17.5–19.5	1.00	1.75–2.50	—	0.035	Ti + Nb: 0.20 + 4(C + N) min; 0.80 min
S44600	446	0.20	23.00–27.50	—	—	—	0.25	—
S44627	E-BRITE 26-1[d]	0.01	25.00–27.00	—	0.75–1.50	0.20	0.015	Nb: 0.05–0.20
S44660	SEA-CURE[i]	0.030	25.00–28.00	1.0–3.50	3.00–4.00	—	0.040	Ti + Nb: 0.20–1.00 and 6(C + N) min
S44735	AL 29-4C[c]	0.030	28.00–30.00	1.00	3.60–4.20	—	0.045	Ti + Nb: 0.20–1.00 and 6(C + N) min

Martensitic stainless steels

UNS No.	Designation	C	Cr	Ni	Mo	Cu	N	Other elements
S40300	403	0.15	11.50–13.50	—	—	—	—	—
S41000	410	0.15	11.50–13.50	0.75	—	—	—	—
S41008	410S	0.08	11.50–13.50	0.60	—	—	—	—
S41600	416	0.15	12.00–14.00	—	—	—	—	—
S41623	416Se	0.15	12.00–14.00	—	—	—	—	Se: 0.15 min
S42000	420	0.15 min	12.00–14.00	—	—	—	—	—
S42020	420F	0.08	12.00–14.00	0.50	0.50	—	—	S: 0.15 min
S43100	431	0.20	15.00–17.00	1.25–2.50	—	—	—	—
S44002	440A	0.60–0.75	16.00–18.00	—	0.75	—	—	—
S44003	440B	0.75–0.95	16.00–18.00	—	0.75	—	—	—
S44004	440C	0.95–1.20	16.00–18.00	—	0.75	—	—	—
S44020	440F	0.95–1.20	16.00–18.00	0.50	0.6	—	—	S: 0.15 min

Precipitation-hardening stainless steels

UNS No.	Designation	C	Cr	Ni	Mo	Cu	N	Other elements
S13800	XM-13/13-8Mo PH	0.05	12.25–13.25	7.50–8.50	2.00–2.50	—	0.01	Al: 0.90–1.35
S15700	632/15-7PH	0.09	14.00–16.00	6.50–7.75	2.00–3.00	—	—	Al: 0.75–1.00
S17400	630/17-4PH	0.07	15.00–17.50	3.00–5.00	—	3.00–5.00	—	Nb + Ta: 0.15–0.45
S17700	631/17-7PH	0.09	16.00–18.00	6.50–7.75	—	—	—	Al: 0.75–1.00
S35000	AMS 350	0.07–0.11	16.00–17.00	4.00–5.00	2.50–3.25	—	0.07–0.13	—
S35500	634/AMS 355	0.10–0.15	15.00–16.00	4.00–5.00	2.50–3.25	—	0.07–0.13	Mn: 0.50–1.25
S45000	XM-25/Custom 450[j]	0.05	14.00–16.00	5.00–7.00	0.50–1.00	—	—	Nb: 8C min
S45500	XM-16/Custom 455[j]	0.03	11.00–12.50	7.50–9.50	0.5	—	—	Ti: 0.90–1.40

[a] Maximum unless range or minimum is indicated.
[b] None required in the specification.
[c] Trademark of Avesta Sheffield AB.
[d] Trademark of Allegheny Ludlum Corp.
[e] Trademark of the INCO family of companies.
[f] Trademark of Krupp-VDM.
[g] Trademark of Sumitomo Metals.
[h] Trademark of Langley Alloys Ltd.
[i] Trademark of Crucible Materials Corp.
[j] Trademark of Carpenter Technology Corp.

Table 6.2.16 ASTM Mechanical Properties (A 240, A 276, A 479, A 564, A 582)

UNS no.	Grade	Condition*	Tensile strength, min ksi	Tensile strength, min MPa	Yield strength, min ksi	Yield strength, min MPa	Elongation in 2 in or 50 mm min, %	Hardness, max Brinell	Hardness, max Rockwell B†
				Austenitic stainless steels					
S20100	201	Annealed	95	655	38	260	40.0	—‡	95
S20200	202	Annealed	90	620	38	260	40.0	241	—
S30100	301	Annealed	75	515	30	205	40.0	217	95
S30200	302	Annealed	75	515	30	205	40.0	201	92
S30300	303	Annealed	—	—	—	—	—	262	—
S30323	303Se	Annealed	—	—	—	—	—	262	—
S30400	304	Annealed	75	515	30	205	40.0	201	92
S30403	304L	Annealed	70	485	25	170	40.0	201	92
S30409	304H	Annealed	75	515	30	205	40.0	201	92
S30451	304N	Annealed	80	550	35	240	30.0	201	92
S30500	305	Annealed	75	515	30	205	40.0	183	88
S30908	309S	Annealed	75	515	30	205	40.0	217	95
S31008	310S	Annealed	75	515	30	205	40.0	217	95
S31600	316	Annealed	75	515	30	205	40.0	217	95
S31603	316L	Annealed	70	485	30	170	40.0	217	95
S31609	316H	Annealed	75	515	30	205	40.0	217	95
S31703	317L	Annealed	75	515	30	205	40.0	217	95
S31726	317LMN	Annealed	80	550	35	240	40.0	223	96
S32100	321	Annealed	75	515	30	205	40.0	217	95
S34700	347	Annealed	75	515	30	205	40.0	201	92
N08020	Alloy 20	Annealed	80	551	35	241	30.0	217	95
N08904	904L	Annealed	71	490	31	215	35.0	—	—
S31254	254 SMO	Annealed	94	650	44	300	40.0	223	96
N08367	AL-6XN	Annealed	95	655	45	310	30.0	233	—
N08926	25-6MO/1925hMo	Annealed	94	650	43	295	35.0	—	—
S32654	654 SMO	Annealed	109	750	62	430	40.0	250	—
				Duplex stainless steels					
S31260	DP-3	Annealed	100	690	70	485	20.0	290	—
S31803	2205	Annealed	90	620	65	450	25.0	293	31HRC
S32304	2304	Annealed	87	600	58	400	25.0	290	32HRC
S32550	Ferralium 255	Annealed	110	760	80	550	15.0	302	32HRC
S32750	2507	Annealed	116	795	80	550	15.0	310	32HRC
S32900	329	Annealed	90	620	70	485	15.0	269	28HRC
				Ferritic stainless steels					
S40500	405	Annealed	60	415	25	170	20.0	179	88
S40900	409	Annealed	55	380	25	205	20.0	179	88
S43000	430	Annealed	65	450	30	205	22.0	183	89
S43020	430F	Annealed	—	—	—	—	—	262	—
S43035	439	Annealed	60	415	30	205	22.0	183	89
S43400	434	Annealed	65	450	35	240	22.0	—	—
S44400	444 (18 Cr–2 Mo)	Annealed	60	415	40	275	20.0	217	96
S44600	446	Annealed	65	515	40	275	20.0	217	96
S44627	E-BRITE 26-1	Annealed	65	450	40	275	22.0	187	90
S44660	SEA-CURE	Annealed	85	585	65	450	18.0	241	100
S44735	AL 29-4C	Annealed	80	550	60	415	18.0	255	25HRC
				Martensitic stainless steels					
S40300	403	Annealed	70	485	40	275	20.0	223	—
S41000	410	Annealed	65	450	30	205	20.0	217	96
S41008	410S	Annealed	60	415	30	205	22.0	183	89
S41600	416	Annealed	—	—	—	—	—	262	—
S41623	416Se	Annealed	—	—	—	—	—	262	—
S42000	420	Annealed	—	—	—	—	—	241	—
S42020	420F	Annealed	—	—	—	—	—	262	—
S43100	431	Annealed	—	—	—	—	—	285	—
S44002	440A	Annealed	—	—	—	—	—	269	—
S44003	440B	Annealed	—	—	—	—	—	269	—
S44004	440C	Annealed	—	—	—	—	—	269	—
S44020	440F	Annealed	—	—	—	—	—	285	—

Table 6.2.16   ASTM Mechanical Properties (A 240, A 276, A 479, A 564, A 582) (Continued)

UNS no.	Grade	Condition*	Tensile strength, min ksi	MPa	Yield strength, min ksi	MPa	Elongation in 2 in or 50 mm min, %	Hardness, max Brinell	Rockwell B†
					Precipitation-hardening stainless steels				
S13800	XM-13/13-8Mo PH	Annealed	—	—	—	—	—	363	38HRC
		H950	220	1,520	205	1,410	10.0	430	45HRC
		H1000	205	1,410	190	1,310	10.0	400	43HRC
		H1050	175	1,210	165	1,140	12.0	372	40HRC
		H1150	135	930	90	620	14.0	283	30HRC
S15700	632/15-7PH	Annealed	—	—	—	—	—	269	100
		RH950	200	1,380	175	1,210	7.0	415	—
S17400	630/17-4PH	Annealed	—	—	—	—	—	363	38HRC
		H900	190	1,310	170	1,170	10.0	388	40HRC
		H925	170	1,170	155	1,070	10.0	375	38HRC
		H1025	155	1,070	145	1,000	12.0	331	35HRC
		H1100	140	965	115	795	14.0	302	31HRC
S17700	631/17-7PH	Annealed	—	—	—	—	—	229	98
		RH950	185	1,275	150	1,035	6.0	388	41HRC
S35500	634/AMS 355	Annealed	—	—	—	—	—	363	—
		H1000	170	1,170	155	1,070	12.0	341	37HRC
S45000	XM-25/Custom 450	Annealed	130	895	95	655	10.0	321	32HRC
		H900	180	1,240	170	1,170	10.0	363	39HRC
		H950	170	1,170	160	1,100	10.0	341	37HRC
		H1000	160	1,100	150	1,030	12.0	331	36HRC
		H1100	130	895	105	725	16.0	285	30HRC
S45500	XM-16/Custom 455	Annealed	—	—	—	—	—	331	36HRC
		H900	235	1,620	220	1,520	8.0	444	47HRC
		H950	220	1,520	205	1,410	10.0	415	44HRC
		H1000	205	1,410	185	1,280	10.0	363	40HRC

* Condition defined in applicable ASTM specification.
† Rockwell B unless hardness Rockwell C (HRC) is indicated.
‡ None required in ASTM specifications.

Table 6.2.17   Maximum Allowable Stress Values (ksi), Plate: ASME Section I; Section III, Classes 2 and 3; Section VIII, Division 1

UNS no.	Grade	−20 to 100°F	300°F	400°F	500°F	600°F	650°F	700°F	750°F	800°F	850°F	900°F
					Austenitic stainless steels							
S30403	304L	16.7	12.8	11.7	10.9	10.3	10.5	10.0	9.8	9.7	—*	—
S30409	304H	18.8	14.1	12.9	12.1	11.4	11.2	11.1	10.8	10.6	10.4	10.2
S30451	304N	20.0	16.7	15.0	13.9	13.2	13.0	12.7	12.5	12.3	12.1	11.8
S31603	316L	16.7	12.7	11.7	10.9	10.4	10.2	10.0	9.8	9.6	9.4	—
N08904	904L	17.8	15.1	13.8	12.7	12.0	11.7	11.4	—	—	—	—
S31254	254 SMO	23.5	21.4	19.9	18.5	17.9	17.7	17.5	17.3	—	—	—
					Duplex stainless steel							
S31803	2205	22.5	21.7	20.9	20.4	20.2	—	—	—	—	—	—
					Ferritic stainless steels							
S40900	409	13.8	12.7	12.2	11.8	11.4	11.3	11.1	10.7	10.2	—	—
S43000	430	16.3	15.0	14.4	13.9	13.5	13.3	13.1	12.7	12.0	11.3	10.5
					Precipitation-hardening stainless steel							
S17400	630/17-4	35.0	35.0	34.1	33.3	32.8	32.6	—	—	—	—	—

* No allowable stress values at this temperature; material not recommended for service at this temperature.

6.3 IRON AND STEEL CASTINGS
by Malcolm Blair and Robert E. Eppich

REFERENCES: "Metals Handbook," ASM. "Iron Casting Handbook," Iron Casting Society. "Malleable Iron Castings," Malleable Founders' Society. ASTM Specifications. "Ductile Iron Handbook," American Foundrymen's Society. "Modern Casting," American Foundrymen's Society. "Ductile Iron Data for Design Engineers," Quebec Iron and Titanium (QIT). "History Cast in Metal," American Foundrymen's Society. "Steel Castings Handbook," 5th ed., Steel Founders' Society of America.

CLASSIFICATION OF CASTINGS

Cast-Iron Castings The term **cast iron** covers a wide range of iron-carbon-silicon alloys containing from 2.0 to 4.0 percent carbon and from 0.5 to 3.0 percent silicon. The alloy typically also contains varying percentages of manganese, sulfur, and phosphorus along with other alloying elements such as chromium, molybdenum, copper, and titanium. For a type and grade of cast iron, these elements are specifically and closely controlled. Cast iron is classified into five basic types; each type is generally based on graphite morphology as follows:

Gray iron
Ductile iron
Malleable iron
Compacted graphite iron
White iron

Even though chemistry is important to achieve the type of cast iron, the molten metal processing and cooling rates play major roles in developing each type. Different mechanical properties are generally associated with each of the five basic types of cast iron.

Figure 6.3.1 illustrates the range of carbon and silicon for each type.

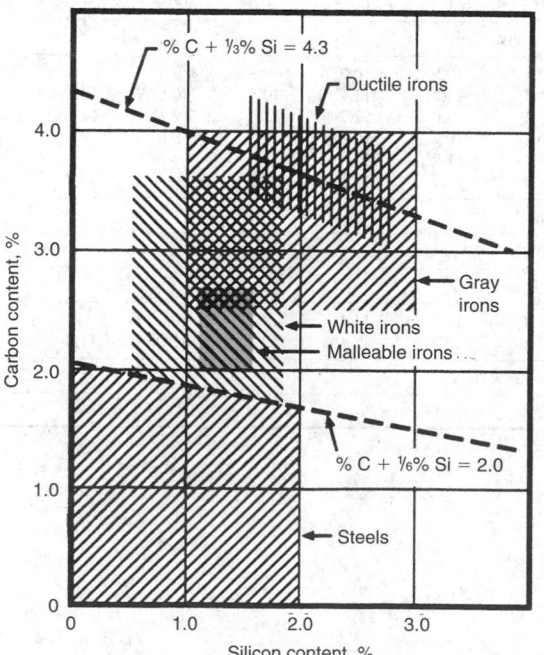

Fig. 6.3.1 Approximate ranges of carbon and silicon for steels and various cast irons. (*QIT-Fer Titane, Inc.*)

The chemistry for a specific grade within each type is usually up to the foundry producing the casting. There are situations where the chemistry for particular elements is specified by ASTM, SAE, and others; this should always be fully understood prior to procuring the casting.

Steel Castings There are two main classes of steel castings: carbon and low-alloy steel castings, and high-alloy steel castings. These classes may be broken down into the following groups: (1) **low-carbon steels** (C < 0.20 percent), (2) **medium-carbon steels** (C between 0.20 and 0.50 percent), (3) **high-carbon steels** (C > 0.50 percent), (4) **low-alloy steels** (total alloy ≤ 8 percent), and (5) **high-alloy steels** (total alloy > 8 percent). The tensile strength of cast steel varies from 60,000 to 250,000 lb/in² (400 to 1,700 MPa) depending on the composition and heat treatment. Steel castings are produced weighing from ounces to over 200 tons. They find universal application where strength, toughness, and reliability are essential.

CAST IRON

The engineering and physical properties of cast iron vary with the type of iron; the designer must match the engineering requirements with all the properties of the specific type of cast iron being considered. Machinability, e.g., is significantly affected by the type of cast iron specified. Gray cast iron is the most machinable; the white cast irons are the least machinable.

Composition The properties of cast iron are controlled primarily by the graphite morphology and the quantity of graphite. Also to be considered is the matrix microstructure, which may be established either during cooling from the molten state (as-cast condition) or as a result of heat treatment. Except where previous engineering evaluations have established the need for specific chemistry or the cast iron is to be produced to a specific ASTM specification, such that chemistry is a part of that specification, the foundry normally chooses the composition that will meet the specified mechanical properties and/or graphite morphology.

Types of Cast Iron

Gray Iron Gray iron castings have been produced since the sixth century B.C. During solidification, when the composition and cooling rate are appropriate, carbon will precipitate in the form of graphite flakes that are interconnected within each eutectic cell. Figure 6.3.2 is a photomicrograph of a typical gray iron structure. The flakes appear black in the photograph. The sample is unetched; therefore the matrix microstructure does not show. Gray iron fractures primarily along the graphite. The fracture appears gray; hence the name *gray iron*. This graphite morphology establishes the mechanical properties. The cooling rate of the casting while it solidifies plays a major role in the size and shape of the graphite flakes. Thus the mechanical properties in a single casting can vary by as much as 10,000 lb/in² with thinner sections having higher strength (Fig. 6.3.3).

The flake graphite imparts unique characteristics to gray iron, such as excellent machinability, ability to resist galling, excellent wear resistance (provided the matrix microstructure is primarily pearlite), and excellent damping capacity. This same flake graphite exhibits essentially zero ductility and very low impact strength.

Mechanical Properties Gray iron **tensile strength** normally ranges from 20,000 to 60,000 lb/in² and is a function of chemistry, graphite morphology, and matrix microstructure. ASTM A48 recognizes nine grades based on the tensile strength of a separately cast test bar whose dimensions are selected to be compatible with the critical controlling section size of the casting. These dimensions are summarized in Table

6.3.1. Test bars are machined to the specified size prior to testing. Table 6.3.2 summarizes the various specifications under ASTM A48. Most engineering applications utilize gray iron with tensile strengths of 30,000 to 35,000 lb/in² (typically a B bar). An alternate method of specifying gray iron is to utilize SAE J431. This specification, shown in Table 6.3.3, is based on hardness (BHN) rather than tensile test bars.

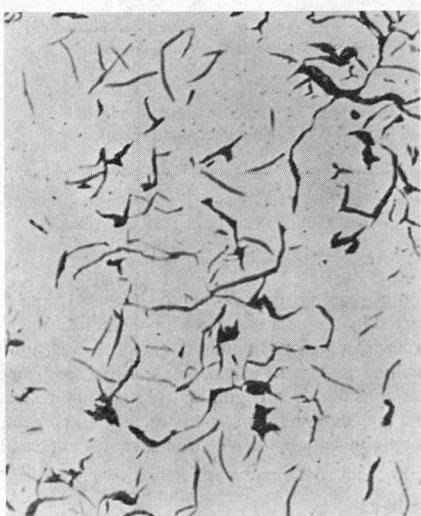

Fig. 6.3.2 Gray iron with flake graphite. Unetched, 100×.

Yield strength is not normally shown and is not specified because it is not a well-defined property of the material. There is no abrupt change in the stress-strain relation as is normally observed in other material. A proof stress or load resulting in 0.1 percent permanent strain will range form 65 to 80 percent of the tensile strength.

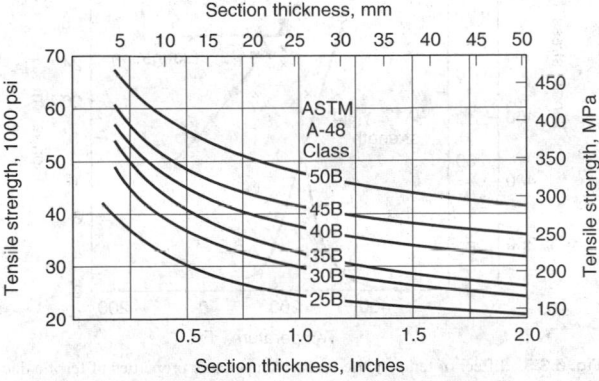

Fig. 6.3.3 The influence of casting section thickness on the tensile strength and hardness for a series of gray irons classified by their strength as cast in 1.2-in-(30-mm-) diameter, B bars. *(Steel Founders Society of America.)*

Table 6.3.4 summarizes the mechanical properties of gray iron as they are affected by a term called the **carbon equivalent**. Carbon equivalent is often represented by the equation

$$CE = \text{carbon (C)} + \tfrac{1}{3} \text{ silicon (Si)} + \tfrac{1}{3} \text{ phosphorus (P)}$$

Typical applications for gray iron are engine blocks, compressor housings, transmission cases, and valve bodies.

Ductile Iron Whereas gray iron has been used as an engineering material for hundreds of years, ductile iron was developed recently. It was invented in the late 1940s and patented by the International Nickel Company. By adding magnesium to the molten metal under controlled

Table 6.3.1 ASTM A48: Separately Cast Test Bars for Use when a Specific Correlation Has Not Been Established between Test Bar and Casting

Thickness of wall of controlling section of casting, in (mm)	Test bar diameter (See Table 6.3.2)
Under 0.25 (6)	S
0.25 to 0.50 (6 to 12)	A
0.51 to 1.00 (13 to 25)	B
1.01 to 2 (26 to 50)	C
Over 2 (50)	S

SOURCE: Abstracted from ASTM data, by permission.

conditions, the graphite structure becomes spherical or nodular. The material resulting has been variously named *spheroidal graphite iron, sg iron, nodular iron,* and *ductile iron.* **Ductile iron** is preferred since it also describes one of its characteristics. The structure is shown in Fig. 6.3.4. As would be expected, this nodular graphite structure, when compared to the graphite flakes in gray iron, shows significant improvement in mechanical properties. However, the machinability and thermal conductivity of ductile iron are lower than those for gray iron.

Table 6.3.2 ASTM A48: Requirements for Tensile Strength of Gray Cast Irons in Separately Cast Test Bars

Class no.	Tensile strength, min, ksi (MPa)	Nominal test bar diameter, in (mm)
20 A	20 (138)	0.88 (22.4)
20 B		1.2 (30.5)
20 C		2.0 (50.8)
20 S		Bars S*
25 A	25 (172)	0.88 (22.4)
25 B		1.2 (30.5)
25 C		2.0 (50.8)
25 S		Bars S*
30 A	30 (207)	0.88 (22.4)
30 B		1.2 (30.5)
30 C		2.0 (50.8)
30 S		Bars S*
35 A	35 (241)	0.88 (22.4)
35 B		1.2 (30.5)
35 C		2.0 (50.8)
35 S		Bars S*
40 A	40 (276)	0.88 (22.4)
40 B		1.2 (30.5)
40 C		2.0 (50.8)
40 S		Bars S*
45 A	45 (310)	0.88 (22.4)
45 B		1.2 (30.5)
45 C		2.0 (50.8)
45 S		Bars S*
50 A	50 (345)	0.88 (22.4)
50 B		1.2 (30,5)
50 C		2.0 (50.8)
50 S		Bars S*
55 A	55 (379)	0.88 (22.4)
55 B		1.2 (30.5)
55 C		2.0 (50.8)
55 S		Bars S*
60 A	60 (414)	0.88 (22.4)
60 B		1.2 (30.5)
60 C		2.0 (50.8)
60 S		Bars S*

* All dimensions of test bar S shall be as agreed upon between the manufacturer and the purchaser.

SOURCE: Abstracted from ASTM data, by permission.

Table 6.3.3 SAE J431: Grades of Automotive Gray Iron Castings Designed by Brinell Hardness

SAE grade	Specified hardness BHN*	Minimum tensile strength (for design purposes)		Other requirements
		lb/in²	MPa	
G1800	187 max	18,000	124	
G2500	170–229	25,000	173	
G2500a†	170–229	25,000	173	3.4% min C and microstructure specified
G3000	187–241	30,000	207	
G3500	207–255	35,000	241	
G3500b†	207–255	35,000	241	3.4% min C and microstructure specified
G3500c†	207–255	35,000	241	3.5% min C and microstructure specified
G4000	217–269	40,000	276	

* Hardness at a designated location on the castings.
† For applications such as brake drums, disks, and clutch plates to resist thermal shock.
SOURCE: Abstracted from SAE data, by permission.

The generally accepted grades of ductile iron are shown in Table 6.3.5, which summarizes ASTM A-536. In all grades the nodularity of the graphite remains the same. Thus the properties are controlled by the casting microstructure, which is influenced by both the chemistry and

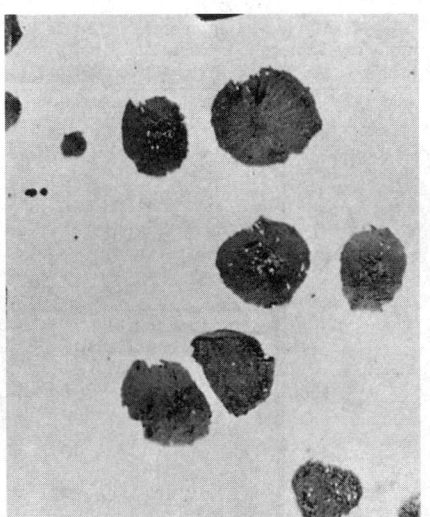

Fig. 6.3.4 Ductile iron with spherulitic graphite. Unetched, 100 ×.

the cooling rate after solidification. The grades exhibiting high elongation with associated lower tensile and yield properties can be produced as cast, but annealing is sometimes used to enhance these properties. Grades 60-40-18 and 60-45-12 have a ferrite microstructure. Both major elements (C and S) and minor elements (Mn, P, Cr, and others) are controlled at the foundry in order to achieve the specified properties.

Figure 6.3.5 illustrates the mechanical properties and impact strength as they are affected by temperature and test conditions.

An important new material within the nodular iron family has been developed within the last 20 years. This material, **austempered ductile iron (ADI)**, exhibits tensile strengths up to 230,000 lb/in², but of greater significance is higher elongation, approximately 10 percent, compared to 2 percent for conventional ductile iron at equivalent tensile strength.

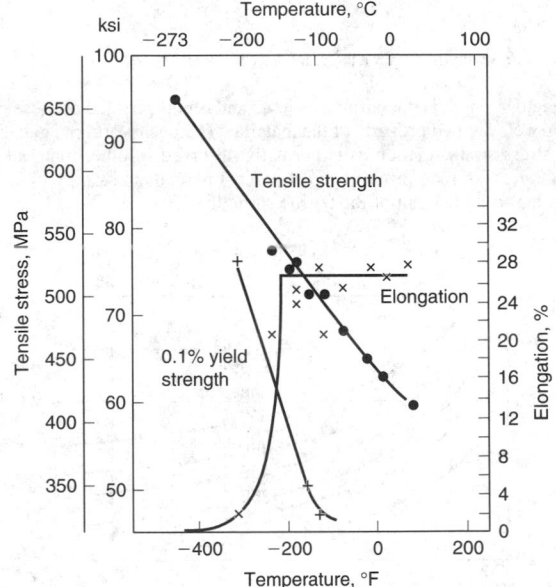

Fig. 6.3.5 Effect of temperature on low-temperature properties of ferritic ductile iron. (*QIT-Fer Titane, Inc.*)

Table 6.3.4 Effect of Carbon Equivalent on Mechanical Properties of Gray Cast Irons

Carbon equivalent	Tensile strength, lb/in²*	Modulus of elasticity in tension at ½ load, lb/in² × 10⁶†	Modulus of rupture, lb/in²*	Deflection in 18-in span, in‡	Shear strength, lb/in²*	Endurance limit, lb/in²*	Compressive strength, lb/in²*	Brinell hardness, 3,000-kg load	Izod impact, unnotched, ft·lb
4.8	20,400	8.0	48,300	0.370	29,600	10,000	72,800	146	3.6
4.6	22,400	8.7	49,200	0.251	33,000	11,400	91,000	163	3.6
4.5	25,000	9.7	58,700	0.341	35,500	11,800	95,000	163	4.9
4.3	29,300	10.5	63,300	0.141	37,000	12,300	90,900	179	2.2
4.1	32,500	13.6	73,200	0.301	44,600	16,500	128,800	192	4.2
4.0	35,100	13.3	77,000	0.326	47,600	17,400	120,800	196	4.4
3.7	40,900	14.8	84,200	0.308	47,300	19,600	119,100	215	3.9
3.3	47,700	20.0	92,000	0.230	60,800	25,200	159,000	266	4.4

* × 6.89 = kPa.   † × 0.00689 = MPa.   ‡ × 25.4 = mm.

Table 6.3.5   ASTM A536: Ductile Irons. Tensile Requirements

	Grade 60-40-18	Grade 65-45-12	Grade 80-55-06	Grade 100-70-03	Grade 120-90-02
Tensile strength, min, lb/in²	60,000	65,000	80,000	100,000	120,000
Tensile strength, min, MPa	414	448	552	689	827
Yield strength, min, lb/in²	40,000	45,000	55,000	70,000	90,000
Yield strength, min, MPa	276	310	379	483	621
Elongation in 2 in or 50 mm, min, %	18	12	6.0	3.0	2.0

SOURCE: Abstracted from ASTM data, by permission.

The material retains the spherical graphite nodules characteristic of ductile iron, but the matrix microstructure of the heat-treated material now consists of acicular ferrite in a high-carbon austenite. Since 1990, ADI has been specified per ASTM 897. Properties associated with this specification are shown in Table 6.3.6 and are summarized in Fig. 6.3.6. Significant research has resulted in excellent documentation of other important properties such as fatigue resistance (Fig. 6.3.7), abrasion resistance (Fig. 6.3.8), and heat-treating response to specific compositions.

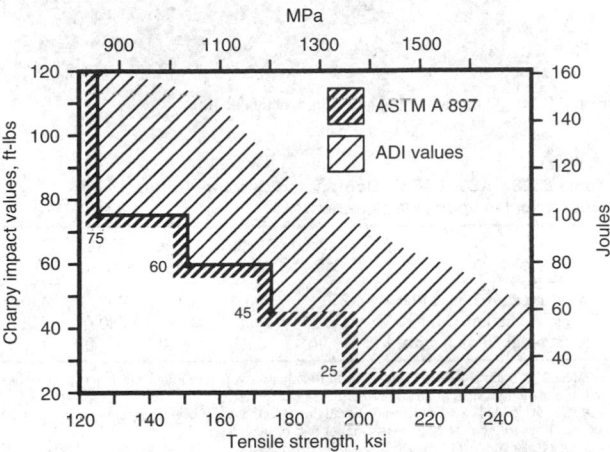

Fig. 6.3.6   Austempered ductile iron mechanical properties. (*QIT-Fer Titane, Inc.*)

Malleable Iron   Whiteheart malleable iron was invented in Europe in 1804. It was cast as white iron and heat-treated for many days, resulting in a material with many properties of steel. Blackheart malleable iron was invented in the United States by Seth Boyden. This material contains no free graphite in the as-cast condition, but when subjected to an extended annealing cycle (6 to 10 days), the graphite is present as irregularly shaped nodules called **temper carbon** (Fig. 6.3.9). All malleable iron castings produced currently are of blackheart malleable iron; the terms blackheart and whiteheart are obsolete.

The properties of several grades of malleable iron are shown in Table 6.3.7. The properties are similar to those of ductile iron, but the annealing cycles still range from 16 to 30 h. Production costs are higher than for ductile iron, so that virtually all new engineered products are designed to be of ductile iron. Malleable iron accounts for only 3 percent of total cast-iron production and is used mainly for pipe and electrical fittings and replacements for existing malleable-iron castings.

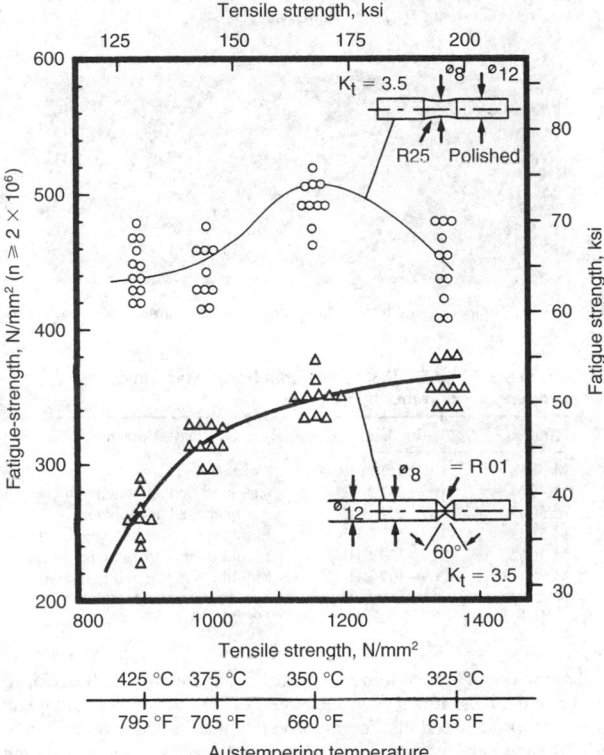

Fig. 6.3.7   Fatigue properties of notched and unnotched ADI. (*QIT-Fer Titane, Inc.*)

Table 6.3.6   ASTM A897: Austempered Ductile Irons. Mechanical Property Requirements

	Grade 125/80/10	Grade 150/100/7	Grade 175/125/4	Grade 200/155/1	Grade 230/185/—
Tensile strength, min, ksi	125	150	175	200	230
Yield strength, min, ksi	80	100	125	155	185
Elongation in 2 in, min %	10	7	4	1	*
Impact energy, ft · lb†	75	60	45	25	*
Typical hardness, BHN, kg/mm²‡	269–321	302–363	341–444	388–477	444–555

* Elongation and impact requirements are not specified. Although grades 200/155/1 and 230/185/— are both primarily used for gear and wear resistance applications, Grade 200/155/1 has applications where some sacrifice in wear resistance is acceptable in order to provide a limited amount of ductility and toughness.
† Unnotched charpy bars tested at 72 ± 7°F. The values in the table are a minimum for the average of the highest three test values of the four tested samples.
‡ Hardness is not mandatory and is shown for information only.
SOURCE: Abstracted from ASTM data, by permission.

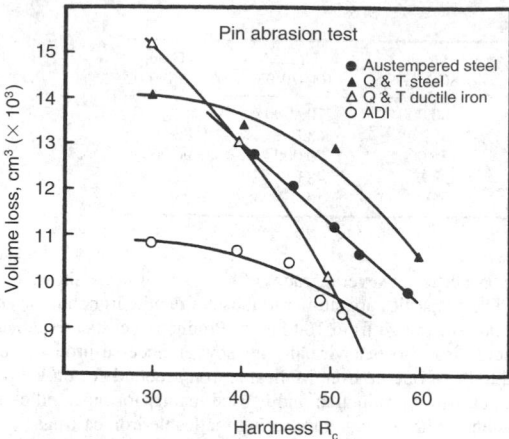

Fig. 6.3.8 Comparison of pin abrasion test results of ADI, ductile iron, and two abrasion-resistant steels. *(QIT-Fer Titane, Inc.)*

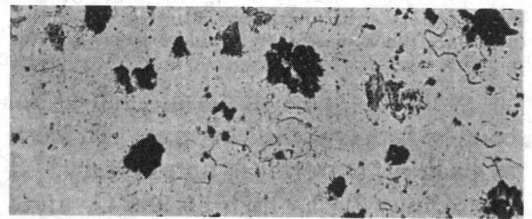

Fig. 6.3.9 Temper carbon form of graphite in malleable iron. Unetched, $100 \times$.

Table 6.3.7   SAE J158a: Malleable Irons. Mechanical Property Requirements

Grade	Casting hardness	Heat treatment
M 3210	156 BHN max.	Annealed
M 4504	163–217 BHN	Air-quenched and tempered
M 5003	187–241 BHN	Air-quenched and tempered
M 5503	187–241 BHN	Liquid-quenched and tempered
M 7002	229–269 BHN	Liquid-quenched and tempered
M 8501	269–302 BHN	Liquid-quenched and tempered

SOURCE: Abstracted from SAE data, by permission.

Compacted Graphite Iron   Compacted graphite iron was developed in the mid-1970s. It is a controlled class of iron in which the graphite takes a form between that in gray iron and that in ductile iron. Figure 6.3.10 shows the typical graphite structure that is developed during casting. It was desired to produce a material that, without extensive alloying, would be stronger than the highest-strength grades of gray iron but would be more machinable than ductile iron. These objectives were met. The properties of the various grades of compacted iron are summarized in Table 6.3.8, which summarizes ASTM A842. Several other properties of this material should be considered. For example, its galling resistance is superior to that of ductile iron, and it is attractive for hydraulic applications; its thermal conductivity is superior to that of ductile iron but inferior to that of gray iron. Compacted graphite iron properties have resulted in its applications to satisfy a number of unique requirements.

White Iron   When white iron solidifies, virtually all the carbon appears in the form of carbides. White irons are hard and brittle, and they break with a white fracture. These irons are usually alloyed with chromium and nickel; Table 6.3.9 summarizes ASTM A532. Their hardness is in the range of 500 to 600 BHN. The specific alloying that is required is a function of section size and application; there must be coordination

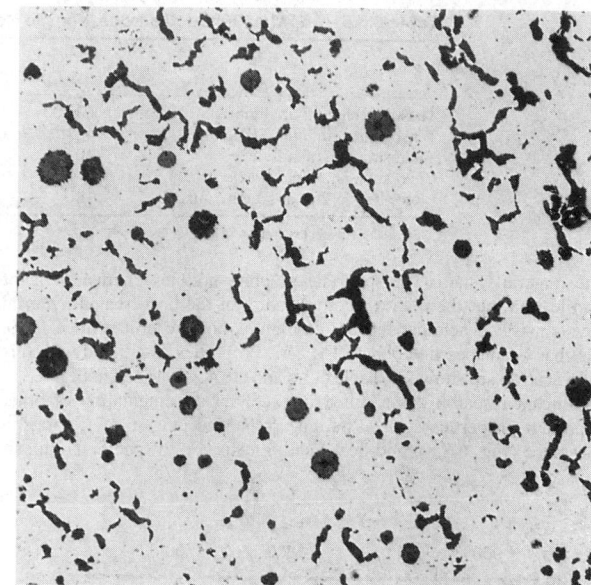

Fig. 6.3.10   Compacted graphite iron. Unetched, $100 \times$.

Table 6.3.8   ASTM A842: Compacted Graphite Irons. Mechanical Property Requirements

	Grade* 250	Grade 300	Grade 350	Grade 400	Grade† 450
Tensile strength, min, MPa	250	300	350	400	450
Yield strength, min, MPa	175	210	245	280	315
Elongation in 50 mm, min, %	3.0	1.5	1.0	1.0	1.0

* The 250 grade is a ferritic grade. Heat treatment to attain required mechanical properties and microstructure shall be the option of the manufacturer.

† The 450 grade is a pearlitic grade usually produced without heat treatment with addition of certain alloys to promote pearlite as a major part of the matrix.

SOURCE: Abstracted from ASTM data, by permission.

between designer and foundry. These irons exhibit outstanding wear resistance and are used extensively in the mining industry for ballmill shell liners, balls, impellers, and slurry pumps.

Specialty Irons   There are a number of highly alloyed gray and ductile iron grades of cast iron. These grades typically contain between 20 and 30 percent nickel along with other elements such as chromium and molybdenum. They have the graphite morphology characteristic of conventional gray or ductile iron, but alloying imparts significantly enhanced performance when wear resistance or corrosion resistance is required. Tables 6.3.10 and 6.3.11 summarize ASTM A436 for austenitic gray iron and ASTM A439 for austenitic ductile iron. This family of alloys is best known by its trade name *Ni-Resist*. Corrosion resistance of several of *Ni-Resist* alloys is summarized in Tables 6.3.12 and 6.3.13. Castings made from these alloys exhibit performance comparable to many stainless steels.

Several other specialty irons utilize silicon or aluminum as the major alloying elements. These also find specific application in corrosion or elevated-temperature environments.

Allowances for Iron Castings

Dimension allowances that must be applied in the production of castings arise from the fact that different metals contract at different rates during solidification and cooling. The shape of the part has a major influence on the as-cast dimensions of the casting. Some of the principal allowances are discussed briefly here.

Table 6.3.9   ASTM A532: White Irons. Chemical Analysis Requirements

Class	Type	Designation	Carbon	Manganese	Silicon	Nickel	Chromium	Molybdenum	Copper	Phosphorus	Sulfur
I	A	Ni-Cr-Hc	2.8–3.6	2.0 max	0.8 max	3.3–5.0	1.4–4.0	1.0 max	—	0.3 max	0.15 max
I	B	Ni-Cr-Lc	2.4–3.0	2.0 max	0.8 max	3.3–5.0	1.4–4.0	1.0 max	—	0.3 max	0.15 max
I	C	Ni-Cr-GB	2.5–3.7	2.0 max	0.8 max	4.0 max	1.0–2.5	1.0 max	—	0.3 max	0.15 max
I	D	Ni-HiCr	2.5–3.6	2.0 max	2.0 max	4.5–7.0	7.0–11.0	1.5 max	—	0.10 max	0.15 max
II	A	12% Cr	2.0–3.3	2.0 max	1.5 max	2.5 max	11.0–14.0	3.0 max	1.2 max	0.10 max	0.06 max
II	B	15% Cr-Mo	2.0–3.3	2.0 max	1.5 max	2.5 max	14.0–18.0	3.0 max	1.2 max	0.10 max	0.06 max
II	D	20% Cr-Mo	2.0–3.3	2.0 max	1.0–2.2	2.5 max	18.0–23.0	3.0 max	1.2 max	0.10 max	0.06 max
III	A	25% Cr	2.0–3.3	2.0 max	1.5 max	2.5 max	23.0–30.0	3.0 max	1.2 max	0.10 max	0.06 max

SOURCE: Abstracted from ASTM data, by permission.

Table 6.3.10   ASTM A436: Austenitic Gray Irons. Chemical Analysis and Mechanical Property Requirements

Element	Composition, %							
	Type 1	Type 1b	Type 2	Type 2b	Type 3	Type 4	Type 5	Type 6
Carbon, total, max	3.00	3.00	3.00	3.00	2.60	2.60	2.40	3.00
Silicon	1.00–2.80	1.00–2.80	1.00–2.80	1.00–2.80	1.00–2.00	5.00–6.00	1.00–2.00	1.50–2.50
Manganese	0.5–1.5	0.5–1.5	0.5–1.5	0.5–1.5	0.5–1.5	0.5–1.5	0.5–1.5	0.5–1.5
Nickel	13.50–17.50	13.50–17.50	18.00–22.00	18.00–22.00	28.00–32.00	29.00–32.00	34.00–36.00	18.00–22.00
Copper	5.50–7.50	5.50–7.50	0.50 max	0.50 max	0.50 max	0.50 max	0.50 max	3.50–5.50
Chromium	1.5–2.5	2.50–3.50	1.5–2.5	3.00–6.00*	2.50–3.50	4.50–5.50	0.10 max	1.00–2.00
Sulfur, max	0.12	0.12	0.12	0.12	0.12	0.12	0.12	0.12
Molybdenum, max	—	—	—	—	—	—	—	1.00
	Type 1	Type 1b	Type 2	Type 2b	Type 3	Type 4	Type 5	Type 6
Tensile strength, min, ksi (MPa)	25 (172)	30 (207)	25 (172)	30 (207)	25 (172)	25 (172)	20 (138)	25 (172)
Brinell hardness (3,000 kg)	131–183	149–212	118–174	171–248	118–159	149–212	99–124	124–174

* Where some matching is required, the 3.00 to 4.00 percent chromium range is recommended.
SOURCE: Abstracted from ASTM data, by permission.

Table 6.3.11   ASTM A439: Austenitic Ductile Irons. Chemical Analysis and Mechanical Property Requirements

Element	Type								
	D-2*	D-2B	D-2C	D-3*	D-3A	D-4	D-5	D-5B	D-5S
	Composition, %								
Total carbon, max	3.00	3.00	2.90	2.60	2.60	2.60	2.40	2.40	2.30
Silicon	1.50–3.00	1.50–3.00	1.00–3.00	1.00–2.80	1.00–2.80	5.00–6.00	1.00–2.80	1.00–2.80	4.90–5.50
Manganese	0.70–1.25	0.70–1.25	1.80–2.40	1.00 max†	1.00 max†	1.00 max†	1.00 max†	1.00 max†	1.00 max
Phosphorus, max	0.08	0.08	0.08	0.08	0.08	0.08	0.08	0.08	0.08
Nickel	18.00–22.00	18.00–22.00	21.00–24.00	28.00–32.00	28.00–32.00	28.00–32.00	34.00–36.00	34.00–36.00	34.00–37.00
Chromium	1.75–2.75	2.75–4.00	0.50 max†	2.50–3.50	1.00–1.50	4.50–5.50	0.10 max	2.00–3.00	1.75–2.25
Element	Type								
	D-2	D-2B	D-2C	D-3	D-3A	D-4	D-5	D-5B	D-5S
	Properties								
Tensile strength, min, ksi (MPa)	58 (400)	58 (400)	58 (400)	55 (379)	55 (379)	60 (414)	55 (379)	55 (379)	65 (449)
Yield strength (0.2% offset), min, ksi (MPa)	30 (207)	30 (207)	28 (193)	30 (207)	30 (207)	—	30 (207)	30 (207)	30 (207)
Elongation in 2 in or 50 mm, min, %	8.0	7.0	20.0	6.0	10.0	—	20.0	6.0	10
Brinell hardness (3,000 kg)	139–202	148–211	121–171	139–202	131–193	202–273	131–185	139–193	131–193

* Additions of 0.7 to 1.0% of molybdenum will increase the mechanical properties above 800°F (425°C).
† Not intentionally added.
SOURCE: Abstracted from ASTM data, by permission.

Shrinkage allowances for iron castings are of the order of ⅛ in/ft (0.1 mm/cm). The indiscriminate adoption of this shrinkage allowance can lead to problems in iron casting dimensions. It is wise to discuss shrinkage allowances with the foundry that makes the castings before patterns are made.

Casting finish allowances (unmachined) and **machine finish allowances** are based on the longest dimension of the casting, although the weight of the casting may also influence these allowances. Other factors which affect allowances include pattern materials and the method of pattern construction. A useful guide is the international standard. ISO 8062 (Castings—System of Dimensional Tolerances and Machining Allowances). It is strongly recommended that designers discuss dimensional

and machine finish allowance requirements with the foundry to avoid unnecessary costs.

STEEL CASTINGS

Composition   There are five classes of commercial steel castings: low-carbon steels (C < 0.20 percent), medium-carbon steels (C between 0.20 and 0.50 percent), high-carbon steels (C > 0.50 percent), low-alloy steels (total alloy ≤ 8 percent), and high-alloy steels (total alloy > 8 percent).

Carbon Steel Castings   Carbon steel castings contain less than 1.00 percent C, along with other elements which may include Mn (0.50 to

Table 6.3.12   Corrosion Resistance of Ductile Ni-Resist Irons D-2 and D-2C

Corrosive media	Penetration, in/yr	
	Type D-2C	Type D-2
Ammonium chloride solution: 10% NH_4Cl, pH 5.15, 13 days at 30°C (86°F), 6.25 ft/min	0.0280	0.0168
Ammonium sulfate solution: 10% $(NH_4)2SO_4$, pH 5.7, 15 days at 30°C (86°F), 6.25 ft/min	0.0128	0.0111
Ethylene vapors & splash: 38% ethylene glycol, 50% diethylene glycol, 4.5% H_2O, 4% Na_2SO_4, 2.7% NaCl, 0.8% Na_2CO_3 + trace NaOH, pH 8 to 9, 85 days at 275–300°F	0.0023	0.0019*
Fertilizer: commercial "5-10-5", damp, 290 days at atmospheric temp	—	0.0012
Nickel chloride solution: 15% $NiCl_2$, pH 5.3, 7 days at 30°C (86°F), 6.25 ft/min	0.0062	0.0040
Phosphoric acid, 86%, aerated at 30°C (86°F), velocity 16 ft/min, 12 days	0.213	0.235
Raw sodium chloride brine, 300 gpl of chlorides, 2.7 gpl CaO, 0.06 gpl NaOH, traces of NH_3 & H_2S, pH 6–6.5, 61 days at 50°F, 0.1 to 0.2 fps	0.0023	0.0020*
Seawater at 26.6°C (80°F), velocity 27 ft/s, 60-day test	0.039	0.018
Soda & brine: 15.5% NaCl, 9.0% NaOH, 1.0% Na_2SO_4, 32 days at 180°F	0.0028	0.0015
Sodium bisulfate solution: 10% $NaHSO_4$, pH 1.3, 13 days at 30°C (86°F), 6.25 ft/min	0.0431	0.0444
Sodium chloride solution: 5% NaCl, pH 5.6, 7 days at 30°C (86°F), 6.25 ft/min	0.0028	0.0019
Sodium hydroxide:		
50% NaOH + heavy conc. of suspended NaCl, 173 days at 55°C (131°F), 40 gal/min.	0.0002	0.0002
50% NaOH saturated with salt, 67 days at 95°C (203°F), 40 gal/min	0.0009	0.0006
50% NaOH, 10 days at 260°F, 4 days at 70°F	0.0048	0.0049
30% NaOH + heavy conc. of suspended NaCl, 82 days at 85°C (185°F)	0.0004	0.0005
74% NaOH, 19¾ days, at 260°F	0.005	0.0056
Sodium sulfate solution: 10% Na_2SO_4, pH 4.0, 7 days at 30°C (86°F), 6.25 ft/min	0.0136	0.0130
Sulfuric acid: 5%, at 30°C (86°F) aerated, velocity 14 ft/min, 4 days	0.120	0.104
Synthesis of sodium bicarbonate by Solvay process: 44% solid $NaHCO_3$ slurry plus 200 gpl NH_4Cl, 100 gpl NH_4HCO_3, 80 gpl NaCl, 8 gpl $NaHCO_3$, 40 gpl CO_2, 64 days at 30°C (86°F)	0.0009	0.0003
Tap water aerated at 30°F, velocity 16 ft/min, 28 days	0.0015	0.0023
Vapor above ammonia liquor: 40% NH_3, 9% CO_2, 51% H_2O, 109 days at 85°C (185°F), low velocity	0.011	0.025
Zinc chloride solution: 20% $ZnCl_2$, pH 5.25, 13 days at 30°C (86°F), 6.25 ft/min	0.0125	0.0064

*Contains 1% chromium.
SOURCE: QIT-Fer Titane, Inc.

Table 6.3.13   Oxidation Resistance of Ductile Ni-Resist, Ni-Resist, Conventional and High-Silicon Ductile Irons, and Type 309 Stainless Steel

Oxidation resistance		
	Penetration (in/yr)	
	Test 1	Test 2
Ductile iron (2.5 Si)	0.042	0.50
Ductile iron (5.5 Si)	0.004	0.051
Ductile Ni-Resist type D-2	0.042	0.175
Ductile Ni-Resist type D-2C	0.07	—
Ductile Ni-Resist type D-4	0.004	0.0
Conventional Ni-Resist type 2	0.098	0.30
Type 309 stainless steel	0.0	0.0

Test 1—Furnace atmosphere—air, 4,000 h at 1,300°F.
Test 2—Furnace atmosphere—air, 600 h at 1,600–1,700°F, 600 h between 1,600–1,700°F, and 800–900°F, 600 h at 800–900°F.
SOURCE: QIT-Fer Titane, Inc.

1.00 percent), Si (0.20 to 0.70 percent), max P (0.05 percent), and max S (0.06 percent). In addition, carbon steels, regardless of whether they are cast or wrought, may contain small percentages of other elements as residuals from raw materials or additives incorporated in the steelmaking process.

Low-Alloy Steel Castings   In a low-alloy steel casting, the alloying elements are present in percentages greater than the following: Mn, 1.00; Si, 0.70; Cu, 0.50; Cr, 0.25; Mo, 0.10; V, 0.05; W, 0.05; and Ti, 0.05. Limitations on phosphorus and sulfur contents apply to low-alloy steels as well as carbon steels unless they are specified to be different for the purpose of producing some desired effect, e.g., free machining. Carbon and low-alloy steels account for approximately 85 percent of the steel castings produced in the United States.

High-Alloy Steel Castings   Steel castings with total alloy content greater than 8 percent are generally considered to be high-alloy, and usually the steel castings industry requires that the composition contain greater than 11 percent Cr. This Cr content requirement eliminates the potential confusion arising from including austenitic manganese steels (Hadfield) in this group. High-alloy steels include corrosion-resistant, heat-resistant, and duplex stainless steels. Many of the cast corrosion-resistant and duplex stainless steels are similar to the wrought stainless steels, but their performance may not be the same. Table 6.3.14 lists cast high-alloy steels and similar wrought grades.

Weight Range   Steel castings weigh from a few ounces to over 200 tons. Steel castings may be made in any thickness down to 0.25 in (6 mm), and in special processes a thickness of 0.080 in (2 mm) has been achieved. In investment castings, wall thicknesses of 0.060 in (1.5 mm) with sections tapering down to 0.030 in (0.75 mm) are common.

Mechanical Properties   The outstanding mechanical properties of cast steel are high strength, ductility, resistance to impact, stiffness, endurance strength, and resistance to both high and low temperatures. It is also weldable. The mechanical properties of carbon steel castings are shown in Figs. 6.3.11 and 6.3.12; those of low-alloy cast steels are shown in Figs. 6.3.13 and 6.3.14.

Tensile and Yield Strength   Ferritic steels of a given hardness or hardenability have the same tensile strength whether cast, wrought, or forged regardless of alloy content. For design purposes involving tensile and yield properties, cast, wrought, and forged properties are considered the same.

Ductility   If the ductility values of steels are compared with the hardness values, the cast, wrought, and forged values are almost identical. The longitudinal properties of wrought and forged steels are slightly higher than those for castings. The transverse properties of the wrought and forged steels are lower, by an amount that depends on the degree of working. Since most service conditions involve several directions of loading, the isotropic behavior of cast steels is particularly advantageous (Fig. 6.3.15).

Impact   The notched-bar impact test is used often as a measure of the toughness of materials. Cast steels have excellent impact properties at normal and low temperatures. Generally wrought steels are tested in the direction of rolling, show higher impact values than cast steels of similar composition. Transverse impact values will be 50 to 70 percent lower than these values. Cast steels do not show directional properties. If the directional properties are averaged for wrought steels, the values obtained are comparable to the values obtained for cast steels of similar

Table 6.3.14   ACI Designations for Heat- and Corrosion-Resistant High-Alloy Castings
(See also Secs. 6.2 and 6.4.)

Cast desig	UNS	Wrought alloy type*	Cast desig	UNS	Wrought alloy type*
CA15	J91150	410†	CA15M	J91151	—
CA28MWV	J91422	422†	CA40	J91153	420†
CA6NM	J91540	F6NM¶	CA6N	J91650	—
CB30	J91803	431†	CB7Cu-1	J92180	17-4§
CB7Cu-2	J92110	15-5§	CC50	J92615	446†
CD3MWCuN	J93380	—			
CD3MN	J92205	2205§	CD4MCu	J93370	255§
CD6MN	J93371	—	CE3MN	J93404	—
CE30	J93423	—	CE8MN	J93345	—
CF3	J92500	304L†	CF8	J92600	304†
CF10SMnN	J92972	Nitronic60‡*			
CF20	J92602	302†	CF3M	J92800	316L†
CF3MN	J92804	316LN†	CF8M	J92900	D319(316)†
CF8C	J92710	347†	CF16F	J92701	303†
CG6MMN	J93790	Nitronic50‡			
CG12	J93001	—	CG8M	J93000	317†
CH20	J93402	309†	CK20	J94202	310†
CK3MCuN	J93254	254SMO‡²	CN3MN	—	AL6XN‡³
CN7M	N08007	—	CN7MS	N02100	—
CU5MCuC	—	825§	CW2M	N26455	C4§
CW6M	N30107	—	CW6MC	N26625	625§
CW12MW	N30002	C§	CX2MW	N26022	C22§
CY5SnBiM	N26055	—	HA	—	—
HC	—	446†	HD	—	327†
HE	—	—	HF	—	302B†
HH	—	309†	HI	—	—
HK	—	310†	HL	—	—
HN	—	—	HP	—	—
HT	—	330†	HU	—	—
HW	—	—	HX	—	—
M25S	N24025	—	M30C	N24130	—
M30H	N24030	—	M35-1	N24135	400§
M35-2	N04020	400§	N7M	N30007	B2§
N12MV	N30012	B§			

* Wrought alloy type references are listed only for the convenience of those who want corresponding wrought and cast grades. This table does not imply that the performance of the corresponding cast and wrought grades will be the same. Because the cast alloy chemical composition ranges are not the same as the wrought composition ranges, buyers should use cast alloy designations for proper identification of castings.
  † Common description, formerly used by AISI.
  ‡ Proprietary trademark: ‡¹ Armco, Inc, ‡² Avesta Sheffield, ‡³ Allegheny Ludlum Corp.
  § Common name used by two or more producers; not a trademark.
  ¶ ASTM designation.
  NOTE: Most of the standard grades listed are covered for general applications by ASTM specifications.
  SOURCE: *Steel Founders Society of America Handbook*, 5th ed.

composition. The **hardenability** of cast steels is influenced by composition and other variables in the same manner as the hardenability of wrought steels. The ratio of the **endurance limit** to the tensile strength for cast steel varies from 0.42 to 0.50, depending somewhat on the composition and heat treatment of the steel. The notch-fatigue ratio varies from 0.28 to 0.32 for cast steels and is the same for wrought steels (Fig. 6.3.16).

In **wear testing**, cast steels react similarly to wrought steels and give corresponding values depending on composition, structure, and hardness. Carbon cast steels of approximately 0.50 percent C and low-alloy cast steels of the chromium, chromium-molybdenum, nickel-chromium, chromium-vanadium, and medium-manganese types, all of which contain more than 0.40 percent C, exhibit excellent resistance to wear in service.

Corrosion Resistance   Cast and wrought steels of similar composition and heat treatment appear to be equally resistant to corrosion in the same environments. Small amounts of copper in cast steels increase the resistance of the steel to atmospheric corrosion. High-alloy steels of chromium, chromium-nickel, and chromium-nickel-molybdenum types are normally used for corrosion service.

Heat Resistance   Although not comparable to the high-alloy steels of the nickel-chromium types developed especially for heat resistance, the 4.0 to 6.5 percent Cr cast steels, particularly with additions of 0.75 to 1.25 percent W, or 0.40 to 0.70 percent Mo, and 0.75 to 1.00 percent Ti, show good strength and considerable resistance to scaling up to 1,000°F (550°C).

Many cast high-alloy nickel-chromium steels have **creep** and **rupture** properties superior to those of wrought materials. In many cases wrought versions of the cast grades are not available due to hot-working difficulties. These cast materials may be used in service conditions up to 2,000°F (1,100°C), with some grades capable of withstanding temperatures of 2,200°F (1,200°C).

Machinability of carbon and alloy cast steels is comparable to that of wrought steels having equivalent strength, ductility, hardness, and similar microstructure. Factors influencing the machinability of cast steels are as follows: (1) Microstructure has a definite effect on machinability of cast steels. In some cases it is possible to improve machinability by as much as 100 to 200 percent through heat treatments which alter the microstructure. (2) Generally speaking, hardness alone cannot be taken as the criterion for predicting tool life in cutting cast steels. (3) In general, for a given structure, plain carbon steels machine better than alloy steels. (4) When machining cast carbon steel (1040) with carbides, tool life varies with the ratio of ferrite to pearlite in the microstructure of the casting, the 60:40 ratio resulting in optimum tool life.

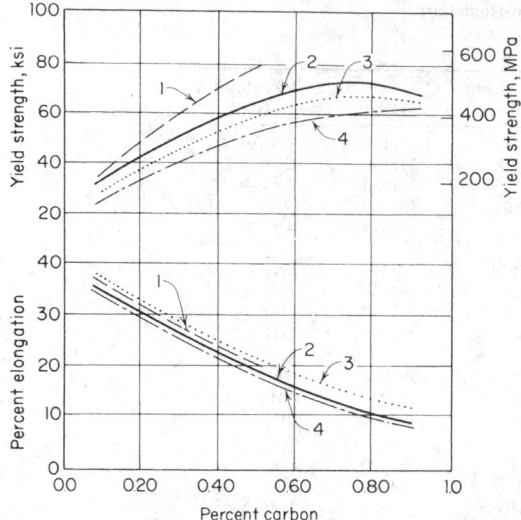

Fig. 6.3.11 Yield strength and elongation versus carbon content for cast carbon steels. (1) Water quenched and tempered at 1,200°F; (2) normalized; (3) normalized and tempered at 1,200°F; (4) annealed.

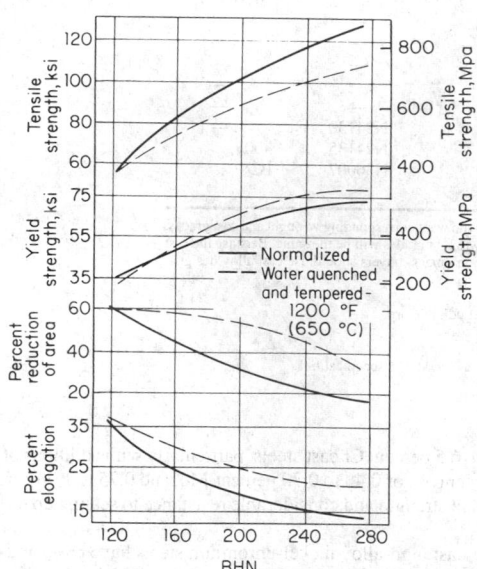

Fig. 6.3.12 Tensile properties of carbon steel as a function of hardness.

For equal tool life, the skin of a steel casting should be machined at approximately one-half the cutting speed recommended for the base metal. The machinability of various carbon and low-alloy cast steels is given in Table 6.3.15.

Welding steel castings presents the same problems as welding wrought steels. It is interesting to note that cast low-alloy steels show greater resistance to underbead cracking than their wrought counterparts.

Purchase Specifications for Steel Castings Many steel castings are purchased according to mechanical property specifications, although more frequent use of specifications which indicate the use of steel castings may be more helpful. Note that it is now possible to order steel castings as cast equivalents of the SAE/AISI chemical composition

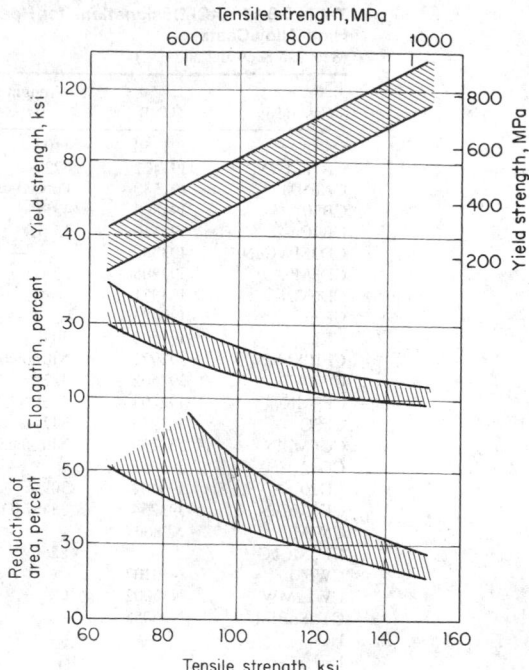

Fig. 6.3.13 Properties of normalized and tempered low-alloy cast steels.

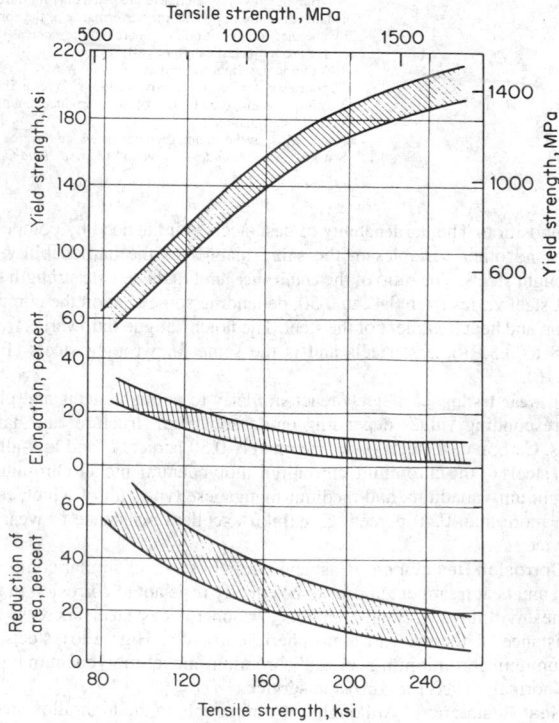

Fig. 6.3.14 Properties of quenched and tempered low-alloy cast steels.

Table 6.3.15 Machinability Index for Cast Steels

Steel	BHN	Conventional		Metcut*	
		Carbide	HSS	HSS	Carbide
B1112 Free machining steel (wrought)	179	. . .	100		
1020 Annealed	122	10	90	160	400
1020 Normalized	134	6	75	135	230
1040 Double normalized	185	11	70	130	400
1040 Normalized and annealed	175	10	75	135	380
1040 Normalized	190	6	65	120	325
1040 Normalized and oil-quenched	225	6	45	80	310
1330 Normalized	187	2	40	75	140
1330 Normalized and tempered	160	3	65	120	230
4130 Annealed	175	4	55	95	260
4130 Normalized and spheroidized	175	3	50	90	200
4340 Normalized and annealed	200	3	35	60	210
4340 Normalized and spheroidized	210	6	55	95	290
4340 Quenched and tempered	300	2	25	45	200
4340 Quenched and tempered	400	½	20	35	180
8430 Normalized and tempered at 1,200°F (660°C)	200	3	50	90	200
8430 Normalized and tempered at 1,275°F (702°C)	180	4	60	110	240
8630 Normalized	240	2	40	75	180
8630 Annealed	175	5	65	120	290

* The Metcut speed index number is the actual cutting speed (surface ft/min) which will give 1 h tool life in turning.

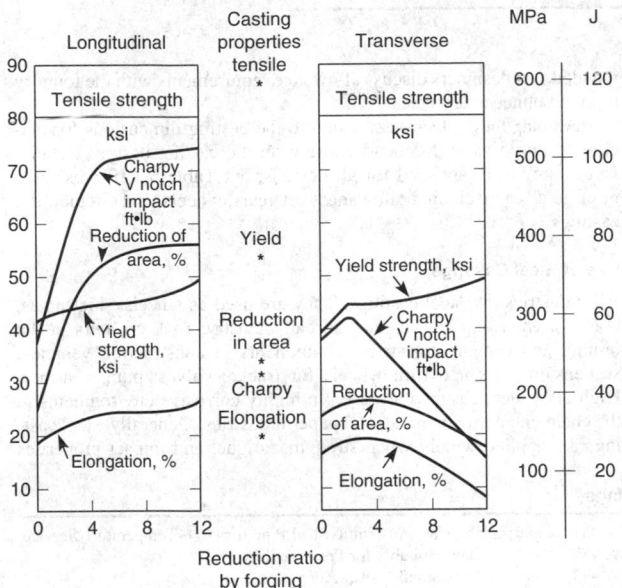

Fig. 6.3.15 Influence of forging reduction on anisotropy for a 0.35 percent carbon wrought steel. Properties for a cast of 0.35 percent carbon steel are shown with an asterisk for comparison purposes. (*Steel Founders Society of America Handbook, 5th ed.*)

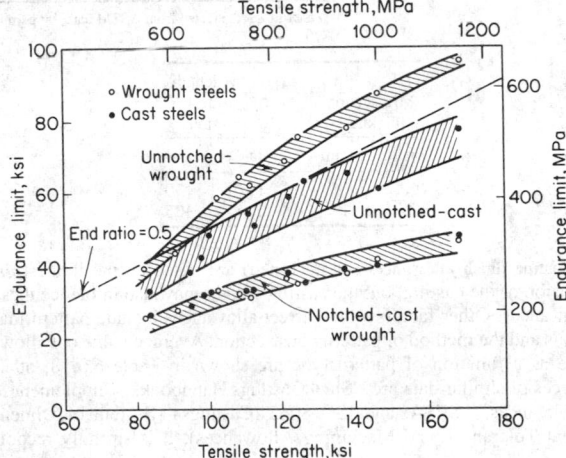

Fig. 6.3.16 Variations of endurance limit with tensile strength for comparable carbon and low-alloy cast and wrought steels.

ranges using ASTM A915. It is preferable to order steel castings to conform with national standards such as ASTM. ASTM standards are prepared by purchasers and suppliers and are well understood by steel foundries. The property levels specified in ASTM specifications A27 and A148 are shown in Table 6.3.16. Table 6.3.17 lists current applicable ASTM steel castings standards.

Steel Melting Practice Steel for steel castings is produced commercially by processes similar to those used for wrought steel production, but electric arc and electric induction melting furnaces predominate. Some steel foundries also use secondary refining processes such as **argon oxygen decarburization** (AOD) and/or **vacuum oxygen degassing** (VOD). Wire injection is also used selectively depending on service and process requirements for the steel casting. In addition to the materials listed in the ASTM standards, the steel casting industry will produce compositions tailored especially for particular service applications, and even in small quantity.

Allowances for Steel Castings

Dimensional allowances that must be applied in the production of castings are due to the fact that different metals contract at different rates during solidification and cooling. The shape of the part has a major influence on the as-cast dimensions of the casting. Some of the principal allowances are discussed briefly here.

Shrinkage allowances for steel castings vary from ³⁄₃₂ to ¹⁄₁₆ in/ft (0.03 to 0.005 cm/cm). The amount most often used is ¼ in/ft (0.021 cm/cm), but if it is applied indiscriminately, it can lead to problems in steel casting dimensions. The best policy is to discuss shrinkage allowances with the foundry that makes the castings before patterns are made.

Minimum Section Thickness The fluidity of steel when compared to other metals is low. In order that all sections may be completely filled, a minimum value of section thickness must be adopted that is a function of the largest dimension of the casting. These values are suggested for design use:

Table 6.3.16   ASTM Requirements for Steel Castings in A27 and A148

Grade	Tensile strength, min, ksi (MPa)	Yield point min, ksi (MPa)	Elongation, min, %	Reduction of area, min, %
		ASTM A27		
U-60-30	60 (415)	30 (205)	22	30
60-30	60 (415)	30 (205)	24	35
65-35	65 (450)	35 (240)	24	35
70-36	70 (485)	36 (250)	22	30
70-40	70 (485)	40 (275)	22	30
		ASTM A148		
80-40	80 (550)	40 (275)	18	30
80-50	80 (550)	50 (345)	22	35
90-60	90 (620)	60 (415)	20	40
105-85	105 (725)	85 (585)	17	35
115-95	115 (795)	95 (655)	14	30
130-115	130 (895)	115 (795)	11	25
135-125	135 (930)	125 (860)	9	22
150-135	150 (1,035)	135 (930)	7	18
160-145	160 (1,105)	145 (1,000)	6	12
165-150	165 (1,140)	150 (1,035)	5	20
165-150L	165 (1,140)	150 (1,035)	5	20
210-180	210 (1,450)	180 (1,240)	4	15
210-180L	210 (1,450)	180 (1,240)	4	15
260-210	260 (1,795)	210 (1,450)	3	6
260-210L	260 (1,795)	210 (1,450)	3	6

SOURCE: Abstracted from ASTM data, by permission.

Min section thickness		Max length of section	
in	mm	in	cm
¼	6	12	30
½	13	50	125

Casting finish allowances (unmachined) are based on the longest dimension of the casting, but the casting weight may also influence these allowances. Other factors which affect allowances include pattern materials and the method of pattern construction. Average values of allowance as a function of pattern type are shown in Table 6.3.18; other sources of similar data are "Steel Castings Handbook," Supplement 3, "Tolerances," SFSA, and ISO 8062 (Castings—System of Dimensional Tolerances and Machining Allowances). It is strongly recom-

mended that designers discuss allowance requirements with the foundry to avoid unnecessary costs.

Machining finish allowances added to the casting dimensions for machining purposes will depend entirely on the casting design. Definite values are not established for all designs, but Table 6.3.19 presents a brief guide to machining allowances on gears, wheels, and circular, flat castings.

Use of Steel Castings

All industries use steel castings. They are used as wheels, sideframes, bolsters, and couplings in the railroad industry; rock crushers in the mining and cement industries; components in construction vehicles; suspension parts and fifth wheels for trucks; valves; pumps; armor. High-alloy steel castings are used in highly corrosive environments in the chemical, petrochemical, and paper industries. Generally, steel castings are applied widely when strength, fatigue, and impact properties

Table 6.3.17   A Summary of ASTM Specifications for Steel and Alloy Castings

A 27	Steel Castings, Carbon, for General Application
A 128	Steel Castings, Austenitic Manganese
A 148	Steel Castings, High Strength, for Structural Purposes
A 216	Steel Castings, Carbon, Suitable for Fusion Welding, for High-Temperature Service
A 217	Steel Castings, Martensitic Stainless and Alloy, for Pressure-Containing Parts, Suitable for High-Temperature Service
A 297	Steel Castings, Iron-Chromium and Iron-Chromium-Nickel, Heat-Resistant, for General Application
A 351	Steel Castings, Austenitic, Austenitic-Ferritic (Duplex), for Pressure-Containing Parts
A 352	Steel Castings, Ferritic and Martensitic, for Pressure-Containing Parts, Suitable for Low-Temperature Service
A 356	Steel Castings, Carbon, Low-Alloy, and Stainless Steel, Heavy-Walled for Steam Turbines
A 389	Steel Castings, Alloy, Specially Heat-Treated, for Pressure-Containing Parts, Suitable for High-Temperature Service
A 426	Centrifugally Cast Ferritic Alloy Steel Pipe for High-Temperature Service
A 447	Steel Castings, Chromium-Nickel-Iron Alloy (25-12 Class), for High-Temperature Service
A 451	Centrifugally Cast Austenitic Steel Pipe for High-Temperature Service
A 487	Steel Castings, Suitable for Pressure Service
A 494	Castings, Nickel and Nickel Alloy
A 560	Castings, Chromium-Nickel Alloy
A 660	Centrifugally Cast Carbon Steel Pipe for High-Temperature Service
A 703	Steel Castings, General Requirements, for Pressure-Containing Parts
A 732	Castings, Investment, Carbon and Low-Alloy Steel for General Application, and Cobalt Alloy for High Strength at Elevated Temperatures
A 743	Castings, Iron-Chromium, Iron-Chromium-Nickel, Corrosion-Resistant, for General Application
A 744	Castings, Iron-Chromium, Iron-Chromium-Nickel, Corrosion-Resistant, for Severe Service
A 747	Steel Castings, Stainless, Precipitation Hardening
A 757	Steel Castings, Ferritic and Martensitic, for Pressure-Containing and Other Applications, for Low-Temperature Service
A 781	Castings, Steel and Alloy, Common Requirements, for General Industrial Use
A 890	Castings, Iron-Chromium-Nickel-Molybdenum, Corrosion-Resistant, Duplex (Austenitic/Ferritic) for General Application
A 915	Steel Castings, Carbon and Alloy, Chemical Requirements Similar to Standard Wrought Grades

SOURCE: Abstracted from ASTM data, by permission.

Table 6.3.18   Dimensional Tolerances for Steel Castings*
(Deviation from the design dimension)

Pattern type	Drawing dimension, in†			
	0–3.0	3.1–7.0	7.1–20.0	20.1–100.0
Metal match plate	+ 1/32, − 1/16	+ 3/32, − 1/16	+ 1/8, − 1/16	+ 1/8, − 1/8
Metal pattern mounted on cope and drag boards	+ 1/16, − 1/16	+ 3/32, − 3/32	+ 1/8, − 3/32	+ 7/32, − 1/8
Hardwood pattern mounted on cope and drag boards	+ 3/32, − 1/16	+ 1/8, − 3/32	+ 1/8, − 3/32	+ 1/4, − 3/32

* Surfaces that are not to be machined.   † × 2.54 = cm.

Table 6.3.19   Machining Allowances for Steel Castings

Specific dimension or reference line (plane) distance, in		Greatest dimension of casting							
		10 in		10–20 in		20–100 in		Over 100 in*	
Greater than	But not exceeding	0.xx in*	X/32 in*	0.xx in*	X/32 in*	0.xx in*	X/32 in*	0.xx in*	X/32 in*
		+	+	+	+	+	+	+	+
0	2	0.187	6	0.218	7	0.25	8	0.312	10
2	5	0.25	8	0.281	9	0.312	10	0.437	14
5	10	0.312	10	0.344	11	0.375	12	0.531	17
10	20			0.406	13	0.468	15	0.625	20
20	50					0.562	18	0.75	24
50	75					0.656	21	0.875	28
75	100					0.75	24	1.00	32
100	500							1.25	40

		Machine tolerances on section thickness	
		0.XXX in*	X/32 in*
		+	+
0	1/2	0.062	2
1/2	1/2	0.092	3
1 1/2	4	0.187	6
4	7	0.25	8
7	10	0.344	11
10		0.50	16

* × 2.54 = cm.

are required, as well as weldability, corrosion resistance, and high-temperature strength. Often, overall economy of fabrication will dictate a steel casting as opposed to a competing product.

Casting Design   Maximum service and properties can be obtained from castings only when they are properly designed. Design rules for castings have been prepared in detail to aid design engineers in preparing efficient designs. Engineers are referred to the following organizations for the latest publications available:

Steel Founders' Society of America, Des Plaines, IL 60016.
American Foundrymen's Society, Des Plaines, IL 60016.
Investment Casting Institute, Dallas, TX 75206.

6.4   NONFERROUS METALS AND ALLOYS; METALLIC SPECIALTIES

REFERENCES: Current edition of "Metals Handbook," American Society for Materials. Current listing of applicable ASTM, AISI, and SAE Standards. Publications of the various metal-producing companies. Publications of the professional and trade associations which contain educational material and physical property data for the materials promoted.

INTRODUCTION

by L. D. Kunsman and C. L. Carlson; Amended by Staff

Seven **nonferrous metals** are of primary commercial importance: copper, zinc, lead, tin, aluminum, nickel, and magnesium. Some 40 other elements are frequently alloyed with these to make the commercially important alloys. There are also about 15 minor metals that have important specific uses. The properties of these elements are given in Table 6.4.1. (See also Sec. 4, 5, and 6.1.)

Metallic Properties   Metals are substances that characteristically are opaque crystalline solids of high reflectivity having good electrical and thermal conductivities, a positive chemical valence, and, usually, the important combination of considerable strength and the ability to flow before fracture. These characteristics are exhibited by the metallic elements (e.g., iron, copper, aluminum, etc.), or **pure metals,** and by combinations of elements (e.g., steel, brass, dural), or **alloys.** Metals are composed of many small **crystals,** which grow individually until they fill the intervening spaces by abutting neighboring crystals. Although the external shape of these crystals and their orientation with respect to each other are usually random, within each such **grain,** or crystal, the atoms are arranged on a regular three-dimensional lattice. Most metals are arranged according to one of the three common types of **lattice,** or **crystal structure:** face-centered cubic, body-centered cubic, or hexagonal close-packed. (See Table 6.4.1.)

Table 6.4.1   Physical Constants of the Principal Alloy-Forming Elements

	Atomic no.	Atomic weight	Density, lb/in³	Melting point, °F	Boiling point, °F	Specific heat*	Latent heat of fusion, Btu/lb	Linear coef of thermal exp, per °F × 10⁻⁶	Thermal conductivity (near 68°F), Btu/(ft²·h·°F/in)	Electrical resistivity, μΩ·cm	Modulus of elasticity (tension), lb/in² × 10⁶	Crystal structure†	Transition temp, °F	Symbol
Aluminum	13	26.97	0.09751	1,220.4	4,520	0.215	170	13.3	1,540	2.655	10	FCC		Al
Antimony	51	121.76	0.239	1,166.9	2,620	0.049	68.9	4.7–7.0	131	39.0	11.3	Rhom		Sb
Arsenic	33	74.91	0.207	1,497	1,130	0.082	159	2.6		35	11	Rhom		As
Barium	56	137.36	0.13	1,300	2,980	0.068		(10)		50	1.8	BCC		Ba
Beryllium	4	9.02	0.0658	2,340	5,020	0.52	470	6.9	1,100	5.9	37	HCP		Be
Bismuth	83	209.00	0.354	520.3	2,590	0.029	22.5	7.4	58	106.8	4.6	Rhom		Bi
Boron	5	10.82	0.083	3,812	4,620	0.309		4.6		1.8×10^{22}		O		B
Cadmium	48	112.41	0.313	609.6	1,409	0.055	23.8	16.6	639	6.83	8	HCP		Cd
Calcium	20	40.08	0.056	1,560	2,625	0.149	100	12	871	3.43	3	FCC/BCC	867	Ca
Carbon	6	12.010	0.0802	6,700	8,730	0.165	27.2	0.3–2.4	165	1,375	0.7	Hex/D		C
Cerium	58	140.13	0.25	1,460	4,380	0.042		3.4		78		HCP/FCC	572/1328	Ce
Chromium	24	52.01	0.260	3,350	4,500	0.11	146	3.4	464	13	36	BCC/FCC	3344	Cr
Cobalt	27	58.94	0.32	2,723	6,420	0.099	112	6.8	479	6.24	30	HCP/FCC/HCP	783/2048	Co
Columbium (niobium)	41	92.91	0.310	4,380	5,970	0.065		4.0		13.1	15	BCC		Cb
Copper	29	63.54	0.324	1,981.4	4,700	0.092	91.1	9.2	2,730	1.673	16	FCC		Cu
Gadolinium	64	156.9	0.287									HCP		Gd
Gallium	31	69.72	0.216	85.5	3,600	0.0977	34.5	10.1	232	56.8	1	O		Ga
Germanium	32	72.60	0.192	1,756	4,890	0.086	205.7	(3.3)		60×10^{6}	11.4	D		Ge
Gold	79	197.2	0.698	1,945.4	5,380	0.031	29.0	7.9	2,060	2.19	12	FCC		Au
Hafnium	72	178.6	0.473	3,865	9,700	0.0351		(3.3)		32.4	20	HCP/BCC	3540	Hf
Hydrogen	1	1.0080	3.026×10^{-6}	-434.6	-422.9	3.45	27.0		1.18			Hex		H
Indium	49	114.76	0.264	313.5	3,630	0.057	12.2	18	175	8.37	1.57	FCT		In
Iridium	77	193.1	0.813	4,449	9,600	0.031	47	3.8	406	5.3	75	FCC		Ir
Iron	26	55.85	0.284	2,802	4,960	0.108	117	6.50	523	9.71	28.5	BCC/FCC/BCC	1663/2554	Fe
Lanthanum	57	138.92	0.223	1,535	8,000	0.0448				59	5	HCP/FCC/?	662/1427	La
Lead	82	207.21	0.4097	621.3	3,160	0.031	11.3	16.3	241	20.65	2.6	FCC		Pb
Lithium	3	6.940	0.019	367	2,500	0.79	286	31	494	11.7	1.7	BCC		Li
Magnesium	12	24.32	0.0628	1,202	2,030	0.25	160	14	1,100	4.46	6.5	HCP		Mg
Manganese	25	54.93	0.268	2,273	3,900	0.115	115	12		185	23	CCX/CCX/FCT	1340/2010/2080	Mn

Mercury	80	200.61	0.4896	−37.97	675	0.033	4.9	33.8	58	94.1			Rhom	Hg
Molybdenum	42	95.95	0.369	4,750	8,670	0.061	126	3.0	1,020	5.17	50		BCC	Mo
Nickel	28	58.69	0.322	2,651	4,950	0.105	133	7.4	639	6.84	30		FCC	Ni
Niobium (see Columbium)														Nb
Nitrogen	7	14.008	0.042×10^{-3}	−346	−320.4	0.247	11.2		0.147		80		Hex	N
Osmium	76	190.2	0.813	4,900	9,900	0.031		2.6	607	9.5			HCP	Os
Oxygen	8	16.000	0.048×10^{-3}	−361.8	−297.4	0.218	5.9	6.6	0.171				C	O
Palladium	46	106.7	0.434	2,829	7,200	0.058	69.5	70	494	10.8	17		FCC	Pd
Phosphorus	15	30.98	0.0658	111.4	536	0.177	9.0	9.0		10^{17}			C	P
Platinum	78	195.23	0.7750	3,224.3	7,970	0.032	49	4.9	494	9.83	21		FCC	Pt
Plutonium	94	239	0.686	1,229			50–65	46		150		6 forms	MC	Pu
Potassium	19	39.096	0.031	145	1,420	0.177	26.1	46	697	6.15	0.5		BCC	K
Radium	88	226.05	0.18	1,300		0.0326	76						?	Ra
Rhenium	75	186.31	0.765	5,733	10,700	0.059	21	4.6	377	21	75		HCP	Re
Rhodium	45	102.91	0.4495	3,571	8,100	0.084		4.5	610	4.5	54		FCC	Rh
Selenium	34	78.96	0.174	428	1,260	0.056	29.6	21	3	12	8.4	248	MC/Hex	Se
Silicon	14	28.06	0.084	2,605	4,200	0.162	607	1.6–4.1	581	10^{5}	16		D	Si
Silver	47	107.88	0.379	1,760.9	4,010	0.056	45.0	10.9	2,900	1.59	11		FCC	Ag
Sodium	11	22.997	0.035	207.9	1,638	0.295	49.5	39	929	4.2	1.3		BCC	Na
Sulfur	16	32.066	0.0748	246.2	832.3	0.175	16.7	36	1.83	2×10^{23}		204	Rhom/FCC	S
Tantalum	73	180.88	0.600	5,420	9,570	0.036		3.6	377	12.4	27		BCC	Ta
Tellurium	52	127.61	0.225	840	2,530	0.047	13.1	9.3	41	2×10^{5}	6		Hex	Te
Thallium	81	204.39	0.428	577	2,655	0.031	9.1	16.6	18	18	1.2	446	HCP/BCC	Tl
Thorium	90	232.12	0.422	3,348	8,100	0.0355	35.6	6.2	18.6	18.6	11.4		FCC	Th
Tin	50	118.70	0.264	449.4	4,120	0.054	26.1	13	464	11.5	6	55	D/BCT	Sn
Titanium	22	47.90	0.164	3,074	6,395	0.139	187	4.7	119	47.8	16.8	1650	HCP/BCC	Ti
Tungsten	74	183.92	0.697	6,150	10,700	0.032	79	2.4	900	5.5	50		BCC	W
Uranium	92	238.07	0.687	2,065	7,100	0.028	19.8	11.4	186	29	29.7	1229/1427	O/Tet/BCC	U
Vanadium	23	50.95	0.217	3,452	5,430	0.120	43.3	9.4–22	215	26	18.4	2822	BCC	V
Zinc	30	65.38	0.258	787	1,663	0.092	43.3	16.6	784	5.92	12		HCP	Zn
Zirconium	40	91.22	0.23	3,326	9,030	0.066	3.1	3.1	116	41.0	11	1585	HCP/BCC	Zr

NOTE: See Sec. 1 for conversion factors to SI units.

*Cal/(g·°C) at room temperature equals Btu/(lb·°F) at room temperature.

†FCC = face-centered cubic; BCC = body-centered cubic; C = cubic; HCP = hexagonal closest packing; Rhomb = rhombohedral; Hex = hexagonal; FCT = face-centered tetragonal; O = orthorhombic; FCO = face-centered orthorhombic; CCX = cubic complex; D = diamond cubic; MC = monoclinic; BCT = body-centered tetragonal.

SOURCE: ASM "Metals Handbook," revised and supplemented where necessary from Hampel's "Rare Metals Handbook" and elsewhere.

The several properties of metals are influenced to varying degrees by the testing or service **environment** (temperature, surrounding medium) and the **internal structure** of the metal, which is a result of its chemical composition and previous history such as casting, hot rolling, cold extrusion, annealing, and heat treatment. These relationships, and the discussions below, are best understood in the framework of the several phenomena that may occur in metals processing and service and their general effect on metallic structure and properties.

Ores, Extractions, Refining All metals begin with the mining of ores and are successively brought through suitable physical and chemical processes to arrive at commercially useful degrees of purity. At the higher-purity end of this process sequence, scrap metal is frequently combined with that derived from ore. Availability of suitable ores and the extracting and refining processes used are specific for each metal and largely determine the price of the metal and the impurities that are usually present in commercial metals.

Melting Once a metal arrives at a useful degree of purity, it is brought to the desired combination of shape and properties by a series of physical processes, each of which influences the internal structure of the metal. The first of these is usually melting, during which several elements can be combined to produce an alloy of the desired composition. Depending upon the metal, its container, the surrounding atmosphere, the addition of alloy-forming materials, or exposure to vacuum, various chemical reactions may be utilized in the melting stage to achieve optimum results.

Casting The molten metal of desired composition is poured into some type of mold in which the heat of fusion is dissipated and the melt becomes a solid of suitable shape for the next stage of manufacture. Such castings either are used directly (e.g., sand casting, die casting, permanent-mold casting) or are transferred to subsequent operations (e.g., ingots to rolling, billets to forging or extrusion). The principal advantage of castings is their design flexibility and low cost. The typical casting has, to a greater or lesser degree, a large crystal size (**grain size**), extraneous **inclusions** from slag or mold, and **porosity** caused by gas evolution and/or shrinking during solidification. The foundryman's art lies in minimizing the possible defects in castings while maximizing the economies of the process.

Metalworking (See also Sec. 13.) The major portion of all metals is subjected to additional shape and size changes in the solid state. These metalworking operations substantially alter the internal structure and eliminate many of the defects typical of castings. The usual sequence involves first **hot working** and then, quite often, **cold working**. Some metals are only hot-worked, some only cold-worked; most are both hot- and cold-worked. These terms have a special meaning in metallurgical usage.

A cast metal consists of an aggregate of variously oriented grains, each one a single crystal. Upon deformation, the grains flow by a process involving the slip of blocks of atoms over each other, along definite crystallographic planes. The metal is hardened, strengthened, and rendered less ductile, and further deformation becomes more difficult. This is an important method of increasing the strength of nonferrous metals. The effect of progressive **cold rolling** on brass (Fig. 6.4.1) is typical. Terms such as "soft," "quarter hard," "half hard," and "full hard" are frequently used to indicate the degree of hardening produced by such working. For most metals, the hardness resulting from cold working is stable at room temperature, but with lead, zinc, or tin, it will decrease with time.

Upon **annealing, or heating the cold-worked metal,** the first effect is to relieve macrostresses in the object without loss of strength; indeed, the strength is often increased slightly. Above a certain temperature, softening commences and proceeds rapidly with increase in temperature (Fig. 6.4.2). The cold-worked distorted metal undergoes a change called **recrystallization.** New grains form and grow until they have consumed the old, distorted ones. The temperature at which this occurs in a given time, called the **recrystallization temperature,** is lower and the resulting grain size finer, the more severe the working of the original piece. If the original working is carried out above this temperature, it does not harden the piece, and the operation is termed **hot working.** If the annealing temperature is increased beyond the recrystallization temperature or

if the piece is held at that temperature for a long time, the average grain size increases. This generally softens and decreases the strength of the piece still further.

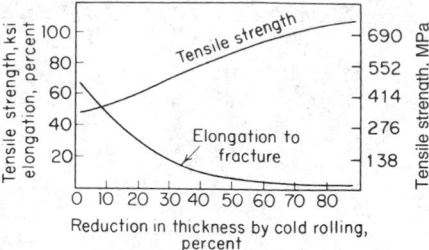

Fig. 6.4.1 Effect of cold rolling on annealed brass (Cu 72, Zn 28).

Hot working occurs when deformation and annealing proceed simultaneously, so that the resulting piece of new size and shape emerges in a soft condition roughly similar to the annealed condition.

Alloys The above remarks concerning structure and property control through casting, hot working, cold working, and annealing gener-

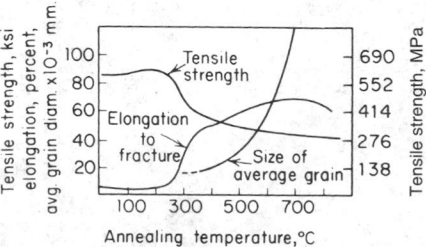

Fig. 6.4.2 Effect of annealing on cold-rolled brass (Cu 72, Zn 28).

ally apply both to pure metals and to alloys. The addition of alloying elements makes possible other means for controlling properties. In some cases, the addition to a metal A of a second element B simply results in the appearance of some new crystals of B as a **mixture** with crystals of A; the resulting properties tend to be an average of A and B. In other cases, an entirely new substance will form—the **intermetallic compound AB,** having its own set of distinctive properties (usually hard and brittle). In still other cases, element B will simply dissolve in element A to form the **solid solution A(B).** Such solid solutions have the characteristics of the **solvent A** modified by the presence of the **solute B,** usually causing increased hardness, strength, electrical resistance, and recrystallization temperature. The most interesting case involves the combination of solid solution A(B) and the precipitation of a second constituent, either B or AB, brought about by the precipitation-hardening heat treatment, which is particularly important in the major nonferrous alloys.

Precipitation Hardening Many alloys, especially those of aluminum but also some alloys of copper, nickel, magnesium, and other metals, can be hardened and strengthened by heat treatment. The heat treatment is usually a two-step process which involves (1) a solution heat treatment followed by rapid quenching and (2) a precipitation or aging treatment to cause separation of a second phase from solid solution and thereby cause hardening. After a solution treatment, these alloys are comparatively soft and consist of homogeneous grains of solid solution generally indistinguishable microscopically from a pure metal. If very slowly cooled from the solution treatment temperature, the alloy will deposit crystals of a second constituent, the amount of which increases as the temperature decreases. Rapid cooling after a solution treatment will retain the supersaturated solution at room temperature, but if the alloy is subsequently reheated to a suitable temperature, fine particles of a new phase will form and in time will grow to a microscopically resolvable size. At some stage in this precipitation process the hardness, the tensile strength, and particularly the yield strength of the alloy, will be increased considerably. If the reheating treatment is carried out too long, the alloy will overage and soften. The temperature and time for both solution and precipitation heat treatments must be closely controlled to obtain the best results. To some extent, precipitation hardening may be superimposed upon hardening resulting from cold working. Precipitation-hardened alloys have an unusually high ratio of proportional limit to tensile strength, but the endurance limit in fatigue is not increased to nearly the same extent.

Effect of Environment The properties of metals under "normal" conditions of 70°F and 50 percent relative humidity in clean air can be markedly changed under other conditions, with the various metals differing greatly in their degree of response to such changing conditions. The topic of **corrosion** is both important and highly specific (see Sec. 6.5). **Oxidation** is a chemical reaction specific to each metal and, wherever important, is so treated. The **effect of temperature**, however, can be profitably considered in general terms. The primary consequence of increase in temperature is increased atom movement, or **diffusion**. In the discussion above, the effects of recrystallization, hot working, solution treatment, and precipitation were all made possible by increased diffusion at elevated temperatures. The temperature at which such atom movements become appreciable is roughly proportional to the melting point of the metal. If their melting points are expressed on an absolute-temperature scale, then various metals can be expected to exhibit comparable effects at about the same fraction of their melting points. Thus the recrystallization temperatures of lead, zinc, aluminum, copper, nickel, molybdenum, and tungsten are successively higher. The consequent rule of thumb is that alloys do not have useful structural strength at temperatures above about 0.5 of their absolute melting temperatures.

ALUMINUM AND ITS ALLOYS

by J. Randolph Kissell

REFERENCES: Kissell and Ferry, "Aluminum Structures," Wiley. Aluminum Association Publications: "Aluminum Standards and Data, 1993," "Aluminum Design Manual," "Aluminum with Food and Chemicals," and "Standards for Aluminum Sand and Permanent Mold Castings, 1992."

Aluminum owes most of its applications to its low density [about 0.1 lb/in³ (0.16 kg/m³)] and to the relatively high strength of its alloys. Other uses depend upon its comparatively good corrosion resistance, good working properties, high electrical and thermal conductivity, reflectivity, and toughness at low temperatures. Designs utilizing aluminum should take into account its relatively low modulus of elasticity (10×10^3 ksi) (69×10^3 MPa) and high coefficient of thermal expansion [13×10^{-6}/°F (2.3×10^{-5}/°C)]. Commercially **pure aluminum** contains a minimum of 99 percent aluminum and is a soft and ductile metal used for many applications where high strength is not necessary. **Aluminum alloys,** on the other hand, possess better casting and machining characteristics and better mechanical properties than the pure metal and, therefore, are used more extensively.

Aluminum alloys are divided into two general classes: wrought alloys and cast alloys, each with its own alloy designation system as specified in ANSI H35.1, Alloy and Temper Designation Systems for Aluminum, maintained by the Aluminum Association.

Aluminum Alloy Designation System **Wrought** aluminum and aluminum alloys are designated by a four-digit number. The first digit indicates the alloy group according to the main alloying element as follows:

Aluminum, 99.00 percent or more	1xxx
Copper	2xxx
Manganese	3xxx
Silicon	4xxx
Magnesium	5xxx
Magnesium and silicon	6xxx
Zinc	7xxx
Other element	8xxx
Unused series	9xxx

The last two digits identify the aluminum alloy or indicate the aluminum purity in the case of the 1xxx series. The second digit indicates

modifications of the original alloy or impurity limits. Experimental alloys are indicated by the prefix X.

Cast aluminum and aluminum alloys are also designated by a four-digit number, but with a decimal point between the last two digits. The first digit indicates the alloy group according to the main alloying element as follows:

Aluminum, 99.00 percent or more	1xx.x
Copper	2xx.x
Silicon, with added copper and/or magnesium	3xx.x
Silicon	4xx.x
Magnesium	5xx.x
Zinc	7xx.x
Tin	8xx.x
Other element	9xx.x
Unused series	6xx.x

The second two digits identify the aluminum alloy or indicate the aluminum purity in the case of the 1xx.x series. The last digit indicates whether the product form is a casting (designated by 0) or ingot (designated by 1 or 2). A modification of the original alloy or impurity limits is indicated by a serial letter prefix before the numerical designation. The serial letters are assigned in alphabetical order but omitting I, O, Q, and X. The prefix X is used for experimental alloys.

The **temper** designation system applies to all aluminum and aluminum alloys, wrought and cast, except ingot. The temper designation follows the alloy designation, separated by a hyphen. The basic tempers are designated by letters as follows:

F	as fabricated (no control over thermal conditions or strain hardening)
O	annealed (the lowest-strength temper of an alloy)
H	strain-hardened (applies to wrought products only)
T	thermally treated

The H and T tempers are always followed by one or more numbers, say, T6 or H14, indicating specific sequences of treatments. The properties of alloys of H and T tempers are discussed further below.

Wrought Aluminum Alloys The alloys listed in Table 6.4.2 are divided into two classes: those which may be strengthened by strain hardening only (*non-heat-treatable,* designated by the H temper) and those which may be strengthened by thermal treatment (*heat-treatable,* designated by the T temper).

The heat-treatable alloys (see Table 6.4.3) first undergo **solution heat treatment** at elevated temperature to put the soluble alloying elements into solid solution. For example, alloy 6061 is heated to 990°F (530°C). This is followed by rapidly dropping the temperature, usually by quenching in water. At this point, the alloy is very workable, but if it is held at room temperature or above, strength gradually increases and ductility decreases as precipitation of constituents of the supersaturated solution begins. This gradual change in properties is called **natural aging.** Alloys that have received solution heat treatment only are designated T4 temper and may be more readily cold-worked in this condition. Some materials that are to receive severe forming operations, such as cold-driven rivets, are held in this workable condition by storing at freezing temperatures.

By applying a second heat treatment, **precipitation heat treatment** or **artificial aging,** at slightly elevated temperatures for a controlled period, further strengthening is achieved and properties are stabilized. For example, 6061 sheet is artificially aged when held at 320°F (160°C) for 18 h. Material so treated is designated T6 temper. Artificial aging changes the characteristic shape of the stress-strain curve above yielding, thus affecting the tangent modulus of elasticity. For this reason, inelastic buckling strengths are determined differently for artificially aged alloys and for alloys that have not received such treatment.

Non-heat-treatable alloys cannot be strengthened by heating, but may be given a heat treatment to stabilize mechanical properties. Strengthening is achieved by cold-working, also called *strain hardening.*

All wrought alloys may be annealed by heating to a prescribed temperature. For example, 6061 is annealed when held at 775°F (415°C) for approximately 2 to 3 h. **Annealing** reduces the alloy to its lowest strength but increases ductility. Strengths are degraded whenever strain-hard-

Table 6.4.2 Composition and Typical Room-Temperature Properties of Wrought Aluminum Alloys

Aluminum Assoc. alloy designation*	Nominal composition, % (balance aluminum)					Temper†	Density, lb/in^2	Electrical conductivity, % IACS	Brinell hardness, 500-kg load, 10-mm ball	Tensile yield strength,‡ 1,000 lb/in^2	Tensile strength, 1,000 lb/in^2	Elong, %, in 2 in§
	Cu	Si	Mn	Mg	Other elements							
Work-hardenable alloys												
1100	0.12					0	0.098	59	23	5	13	35
						H14			32	17	18	9
						H18		57	44	22	24	5
3003	0.12		1.2			0	0.099	50	28	6	16	30
						H14		41	40	21	22	8
						H18		40	55	27	29	4
5052				2.5	0.25 Cr	0	0.097	35	47	13	28	25
						H34			68	31	38	10
						H38		35	77	37	42	7
5056			0.12	5.0	0.12 Cr	0	0.095	29	65	22	42	35
						H18		27	105	59	63	10
Heat-treatable alloys												
2011	5.5				0.4 Pb	T3	0.102	39	95	43	55	15
					0.4 Bi	T8		45	100	45	59	12
2014	4.4	0.8	0.8	0.5		0	0.101	50	45	14	27	18
						T4, 451		34	105	42	62	20
						T6, 651		40	135	60	70	13
Alclad 2014						0	0.101	40		10	25	21
						T4, 451				37	61	22
						T6, 651				60	68	10
2017	4.0	0.5	0.7	0.6		0	0.101	50	45	10	26	22
						T4, 451		34	105	40	62	22
2024	4.4		0.6	1.5		0	0.100	50	47	11	27	20
						T3		30	120	50	70	18
						T4, 351		30	120	47	68	20
Alclad 2024						0	0.100			11	26	20
						T3				45	65	18
						T4, 351				42	64	19
2025	4.4	0.8	0.8			T6	0.101	40	110	37	58	19
2219	6.3		0.3		0.18 Zr	0		44		11	25	18
					0.10 V	T351				36	52	17
					0.06 Ti	T851				51	66	10
6053		0.7		1.2	0.25 Cr	0	0.097	45	26	8	16	35
						T6		42	80	32	37	13
6061	0.28	0.6		1.0	0.20 Cr	0	0.098	47	30	8	18	25
						T6, 651		43	95	40	45	12
6063		0.4		0.7		0	0.097	58	25	7	13	25
						T6		53	73	31	35	12
7075	1.6			2.5	5.6 Zn	0	0.101		60	15	33	17
					0.23 Cr	T6, 651		33	150	73	83	11
Alclad 7075						0	0.101			14	32	17
						T6, 651				67	76	11
7178	2.0			2.8	0.23 Cr, 6.8 Zn	T6, 651	0.102	31		78	88	10

NOTE: See Sec. 1 for conversion factors for SI units.
* Aluminum Association Standardized System of Alloy Designation adopted October 1954.
† Standard temper designations: 0 = fully annealed. H14 and H34 correspond to half-hard and H18 to H38 to hard strain-hardened tempers. T3 = solution treated and then cold-worked; T4 = solution heat-treated; T6 = solution treated and then artificially aged; T-51 = stretcher stress-relieved; T8 = solution treated, cold-worked, and artificially aged.
‡ At 0.2% offset.
§ Percent in 2 in, 1/16-in thick specimen.

Table 6.4.3 Conditions for Heat Treatment of Aluminum Alloys

Alloy	Product	Solution heat treatment		Precipitation heat treatment		
		Temp, °F	Temper designation	Temp, °F	Time of aging	Temper designation
2014	Extrusions	925–945	T4	315–325	18 h	T6
2017	Rolled or cold finished wire, rod and bar	925–945	T4	Room	4 days	
2024	Coiled Sheet	910–930	T4	Room	4 days	
6063	Extrusions	940–960	T4	340–360	8 h	T6
6061	Sheet	980–1,000	T4	310–330	18 h	T6
7075	Forgings	860–880	W	215–235 } * 340–360 }	6–8 h 8–10 h	T73

°C = (°F − 32)/1.8.
* This is a two-stage treatment.

ened or heat-treated material is reheated. The loss in strength is proportional to the time at elevated temperature. For example, 6061-T6 heated to 400°F (200°C) for no longer than 30 min or to 300°F (150°C) for 100 to 200 h suffers no appreciable loss in strength. Table 6.4.4 shows the effect of elevated temperatures on the strength of aluminum alloys.

Some common wrought alloys (by major alloying element) and applications are as follows:

1xxx Series: Aluminum of 99 percent purity is used in the electrical and chemical industries for its high conductivity, corrosion resistance, and workability. It is available in extruded or rolled forms and is hardened by cold working but not by heat treatment.

2xxx Series: These alloys are heat-treatable and may attain strengths comparable to those of steel alloys. They are less corrosion-resistant than other aluminum alloys and thus are often clad with pure aluminum or an alloy of the 6xxx series (see alclad aluminum, below). Alloys 2014 and 2024 are popular, 2024 being perhaps the most widely used aircraft alloy. Many of the alloys in this group, including 2014, are not usually welded.

3xxx Series: Alloys in this series are non-heat-treatable. Alloys 3003, 3004, and 3105 are popular general-purpose alloys with moderate strengths and good workability, and they are often used for sheet metal work.

4xxx Series: Silicon added to alloys in this group lowers the melting point, making these alloys suitable for use as weld filler wire (such as 4043) and brazing alloys.

5xxx Series: These alloys attain moderate-to-high strengths by strain hardening. They usually have the highest welded strengths among aluminum alloys and good corrosion resistance. Alloys 5083, 5086, 5154, 5454, and 5456 are used in welded structures, including pressure vessels. Alloy 5052 is a popular sheet metal alloy.

6xxx Series: Although these alloys usually are not as strong as those in the other heat-treatable series, 2xxx and 7xxx, they offer a good combination of strength and corrosion resistance. Alloys 6061 and 6063 are used widely in construction, with alloy 6061 providing better strength at a slightly greater cost.

7xxx Series: Heat-treating alloys in this group produces some of the highest-strength alloys, frequently used in aircraft, such as 7050, 7075, 7178, and 7475. And 7178-T651 plate has a minimum ultimate tensile strength of 84 ksi (580 MPa). Corrosion resistance is fair. Many alloys in this group (such as 7050, 7075, and 7178) are not arc-welded.

Aluminum-lithium alloys have been produced with higher strength and modulus of elasticity than those of any of the alloys previously available, but they are not as ductile.

Wrought alloys are available in a number of product forms. **Extrusions,** produced by pushing the heated metal through a die opening, are among the most useful. A great variety of custom shapes, as well as standard shapes such as I beams, angles, channels, pipe, rectangular tube, and many others, are extruded. Extrusion cross-sectional sizes may be as large as those fitting within a 31-in (790-mm) circle, but more commonly are limited to about a 12-in (305-mm) circle. Alloys extruded include 1100, 1350, 2014, 2024, 3003, 5083, 5086, 5454, 5456, 6005, 6061, 6063, 6101, 6105, 6351, 7005, 7075, and 7178. The most common are 6061 and 6063. Rod, bar, and wire are also produced rolled or cold-finished as well as extruded. Tubes may be drawn or extruded.

Flat rolled products include foil, sheet, and plate. **Foil** is defined as rolled product less than 0.006 in (0.15 mm) thick. **Sheet** thickness is less than 0.25 in (6.4 mm) but not less than 0.006 in; aluminum sheet gauge thicknesses are different from those used for steel, and decimal thicknesses are preferred when ordering. Sheet is available flat and coiled. **Plate** thickness is 0.25 in and greater, and it ranges up to about 6 in (150 mm). Minimum bend radii for sheet and plate depend on alloy and temper, but are generally greater than bend radii for mild carbon steel. Commercial roofing and siding sheet is available in a number of profiles, including corrugated, ribbed, and V-beam.

Aluminum **forgings** are produced by open-die and closed-die methods. Minimum mechanical strengths are not published for open-die forgings, so structural applications usually require closed-die forgings. Like castings, forgings may be produced in complex shapes, but have more uniform properties and better ductility than castings and are used for products such as wheels and aircraft frames. For some forging alloys, minimum mechanical properties are slightly lower in directions other than parallel to the grain flow. Alloy 6061-T6 is popular for forgings.

Minimum **mechanical properties** (typically tensile ultimate and yield strengths and elongation) are specified for most wrought alloys and tempers by the Aluminum Association in "Aluminum Standards and Data." These minimum properties are also listed in ASTM specifica-

Table 6.4.4 Typical Tensile Properties of Aluminum Alloys at Elevated Temperatures

Alloy and temper	Property*	Temp, °F*				
		75	300	400	500	700
			Wrought			
1100-H18	T.S.	24	18	6	4	2.1
	Y.S.	22	14	3.5	2.6	1.6
	El.	15	20	65	75	85
2017-T4	T.S.	62	40	16	9	4.3
	Y.S.	40	30	13	7.5	3.5
	El.	22	15	35	45	70
2024-T4	T.S.	68	45	26	11	5
	Y.S.	47	36	19	9	4
	El.	19	17	27	55	100
3003-H18	T.S.	29	23	14	7.5	2.8
	Y.S.	27	16	9	4	1.8
	El.	10	11	18	60	70
3004-H38	T.S.	41	31	22	12	5
	Y.S.	36	27	15	7.5	3
	El.	6	15	30	50	90
5052-H34	T.S.	38	30	24	12	5
	Y.S.	31	27	15	7.5	3.1
	El.	16	27	45	80	130
6061-T6	T.S.	45	34	19	7.5	3
	Y.S.	40	31	15	5	1.8
	El.	17	20	28	60	95
6063-T6	T.S.	35	21	9	4.5	2.3
	Y.S.	31	20	6.5	3.5	2
	El.	18	20	40	75	105
7075-T6	T.S.	83	31	16	11	6
	Y.S.	73	27	13	9	4.6
	El.	11	30	55	65	70
		Temp, °F*				
		75	300	400	500	600
			Sand castings			
355-T51	T.S.	28	24	14	10	6
	Y.S.	23	19	10	5	3
	El.	2	3	8	16	36
356-T6	T.S.	33	23	12	8	4
	Y.S.	24	20	9	5	3
	El.	4	6	18	35	60
		Permanent-mold castings				
355-T51	T.S.	30	23	15	10	6
	Y.S.	24	20	10	5	3
	El.	2	4	9	33	98
356-T6	T.S.	38	21	12	8	4
	Y.S.	27	17	9	5	3
	El.	5	10	30	55	70

*T.S., tensile strength ksi. Y.S., yield strength, ksi, 0.2 percent offset. El., elongation in 2 in, percent.
ksi × 6.895 = MPa °C = (°F − 32)/1.8.
Tensile tests made on pieces maintained at elevated temperatures for 10,000 h under no load. Stress applied at 0.05 in/(in/min) strain rate.

tions, but are grouped by ASTM by product (such as sheet and plate) rather than by alloy. The minimum properties are established at levels at which 99 percent of the material is expected to conform at a confidence level of 0.95. The strengths of aluminum members subjected to axial force, bending, and shear under static and fatigue loads, listed in the Aluminum Association "Specifications for Aluminum Structures," are calculated using these minimum properties. Aluminum and some of its alloys are also used in structural applications such as storage tanks at temperatures up to 400°F (200°C), but strengths are reduced due to creep and the annealing effect of heat.

Cast Aluminum Alloys These are used for parts of complex shapes by sand casting, permanent mold casting, and die casting. The compositions and minimum mechanical properties of some cast aluminum alloys are given in Tables 6.4.5*a* to 6.4.5*d*. Castings generally exhibit more variation in strength and less ductility than wrought products. Tolerances, draft requirements, heat treatments, and quality standards are given in the Aluminum Association "Standards for Aluminum Sand and Permanent Mold Castings." Tolerances and the level of quality and frequency of inspection must be specified by the user if desired.

Sand castings (see ASTM B26) produce larger parts—up to 7,000 lb (3,200 kg)—in relatively small quantities at slow solidification rates. The sand mold is used only once. Tolerances and minimum thicknesses for sand castings are greater than those for other casting types.

Permanent mold castings (see ASTM B108) are produced by pouring the molten metal into a reusable mold, sometimes with a vacuum to assist the flow. While more expensive than sand castings, permanent mold castings can be used for parts with wall thicknesses as thin as about 0.09 in (2.3 mm).

In **die casting** (see ASTM B85), aluminum is injected into a reusable steel mold, or die, at high velocity; fast solidification rates are achieved.

The lowest-cost general-purpose casting alloy is 356-T6, while A356-T6 is common in aerospace applications. Alloy A444-T4 provides excellent ductility, exceeding that of many wrought alloys. These three rank among the most weldable of the casting alloys. The cast alloys utilizing copper (2xx.x) generally offer the highest strengths at elevated temperatures. In addition to end use, alloy selection should take into account fluidity, resistance to hot cracking, and pressure tightness.

Machining (see Sec. 13) Many aluminum alloys are easily machined without special technique at cutting speeds generally much higher than those for other metals. Pure aluminum and alloys of aluminum-manganese (3xxx) and aluminum-magnesium (5xxx) are harder to machine than alloys of aluminum-copper (2xxx) and aluminum-zinc (7xxx). The most machinable wrought alloy is 2011 in the T3, T4, T6, and T8 tempers, producing small broken chips and excellent finish; they are used where physical properties are subordinate to high machinability, such as for screw machine products. Castings are also machined; aluminum-copper (such as 201, 204, and 222), aluminum-magnesium (5xx.x), aluminum-zinc (7xx.x), and aluminum-tin (8xx.x) alloys are among the best choices.

Joining Mechanical fasteners (including rivets, bolts, and screws) are the most common methods of joining, because the application of heat during welding decreases the strength of aluminum alloys. Aluminum **bolts** (usually 2024-T4, 6061-T6, or 7075-T73) are available in diameters from ¼ in (6.4 mm) to 1 in (25 mm), with properties conforming to ASTM F468. Mechanical properties are given in Table 6.4.6. Aluminum nuts (usually 2024-T4, 6061-T6, and 6262-T9) are also available. Galvanized steel and 300-series stainless steel bolts are also used to join aluminum. Hole size usually exceeds bolt diameter by ¹⁄₁₆ in (1.6 mm) or less. **Rivets** are used to resist shear loads only; they do not develop sufficient clamping force between the parts joined and thus cannot reliably resist tensile loads. In general, rivets of composition similar to the base metal are used. Table 6.4.7 lists common aluminum rivet alloys and their minimum ultimate shear strengths. Hole diameter for cold-driven rivets may be no larger than 4 percent greater than the nominal rivet diameter; hole diameter for hot-driven rivets may be no larger than 7 percent greater than the nominal rivet diameter. **Screws** of 2024-T4, 7075-T73 aluminum, or 300-series stainless steels are often used to fasten aluminum sheet. Holes for fasteners may be punched,

drilled, or reamed, but punching is not used if the metal thickness is greater than the diameter of the hole. Applications of adhesive joining are increasing.

Welding (see Sec. 13) Most wrought aluminum alloys are weldable by experienced operators using either the fusion or resistance method. **Fusion welding** is typically by **gas tungsten arc welding** (GTAW), commonly called **TIG** (for tungsten inert gas) welding, or **gas metal arc welding** (GMAW), referred to as **MIG** (for metal inert gas) welding. TIG welding is usually used to join parts from about ¹⁄₃₂ to ⅛ in 0.8 to 3.2 mm) thick; MIG welding is usually used to weld thicker parts. The American Welding Society Standard D1.2, Structural Welding Code—Aluminum, provides specifications for structural applications of aluminum fusion-welding methods. Filler rod alloys must be chosen carefully for strength, corrosion resistance, and compatibility with the parent alloys to be welded. Cast alloys may also be welded, but are more susceptible to cracking than wrought alloy weldments. All aluminum alloys suffer a reduction in strength in the heat-affected weld zone, although this reduction is less in some of the aluminum-magnesium (5xxx series) alloys. Postweld heat treatment may be used to counter the reduction in strength caused by welding, but extreme care must be taken to avoid embrittling or warping the weldment. **Resistance welding** includes **spot welding,** often used for lap joints, and seam welding. Methods of **nondestructive testing of aluminum welds** include dye-penetrant methods to detect flaws accessible to the surface and ultrasonic and radiographic inspection.

Brazing is also used to join aluminum alloys with relatively high melting points, such as 1050, 1100, 3003, and 6063, using aluminum-silicon alloys such as 4047 and 4145. Aluminum may also be **soldered,** but corrosion resistance of soldered joints is inferior to that of welded, brazed, or mechanically fastened joints.

Corrosion Resistance Although aluminum is chemically active, the presence of a rapidly forming and firmly adherent self-healing oxide surface coating inhibits corrosive action except under conditions that tend to remove this surface film. Concentrated nitric and acetic acids are handled in aluminum not only because of its resistance to attack but also because any resulting corrosion products are colorless. For the same reason, aluminum is employed in the preparation and storage of foods and beverages. Hydrochloric acid and most alkalies dissolve the protective surface film and result in fairly rapid attack. Moderately alkaline soaps and the like can be used with aluminum if a small amount of sodium silicate is added. Aluminum is very resistant to sulfur and most of its gaseous compounds.

Galvanic corrosion may occur when aluminum is electrically connected by an electrolyte to another metal. Aluminum is more anodic than most metals and will be sacrificed for the benefit of the other metal, which is thereby cathodically protected from attack. Consequently, aluminum is usually isolated from other metals such as steel (but not stainless steel) where moisture is present.

Another form of corrosion is **exfoliation,** a delamination or peeling of layers of metal in planes approximately parallel to the metal surface, caused by the formation of corrosion product. Alloys with more than 3 percent magnesium (such as 5083, 5086, 5154, and 5456) and held at temperatures above 150°F (65°C) for extended periods are susceptible to this form of attack.

Care should be taken to store aluminum in a manner to avoid trapping water between adjacent flat surfaces, which causes water stains. These stains, which vary in color from dark gray to white, do not compromise strength but are difficult to remove and may be cosmetically unacceptable.

Ordinary atmospheric corrosion is resisted by aluminum and most of its alloys, and they may be used outdoors without any protective coating. (An exception is 2014-T6, which is usually painted when exposed to the elements.) The pure metal is most resistant to attack, and additions of alloying elements usually decrease corrosion resistance, particularly after heat treatment. Under severe conditions of exposure such as may prevail in marine environments or where the metal is continually in contact with wood or other absorbent material in the presence of moisture, a protective coat of paint will provide added protection.

Alclad Aluminum The corrosion resistance of aluminum alloys may

Table 6.4.5a   Composition Limits of Aluminum Casting Alloys (Percent)

Alloy	Product*	Silicon	Iron	Copper	Manganese	Magnesium	Chromium	Nickel	Zinc	Titanium	Tin	Others Each	Others Total
201.0	S	0.10	0.15	4.0–5.2	0.20–0.50	0.15–0.55	—	—	—	0.15–0.35	—	0.05	0.10
204.0	S&P	0.20	0.35	4.2–5.0	0.10	0.15–0.35	—	0.05	0.10	0.15–0.30	—	0.05	0.15
208.0	S&P	2.5–3.5	1.2	3.5–4.5	0.50	0.10	—	0.35	1.0	0.25	—	—	0.50
222.0	S&P	2.0	1.5	9.2–10.7	0.50	0.15–0.35	—	0.50	0.8	0.25	—	—	0.35
242.0	S&P	0.7	1.0	3.5–4.5	0.35	1.2–1.8	0.25	1.7–2.3	0.35	0.25	—	0.05	0.15
295.0	S	0.7–1.5	1.0	4.0–5.0	0.35	0.03	—	—	0.35	0.25	—	0.05	0.15
296.0	P	2.0–3.0	1.2	4.0–5.0	0.35	0.05	—	0.35	0.50	0.25	—	—	0.35
308.0	P	5.0–6.0	1.0	4.0–5.0	0.50	0.10	—	—	1.0	0.25	—	—	0.50
319.0	S&P	5.5–6.5	1.0	3.0–4.0	0.50	0.10	—	0.35	1.0	0.25	—	—	0.50
328.0	S	7.5–8.5	1.0	1.0–2.0	0.20–0.6	0.20–0.6	0.35	0.25	1.5	0.25	—	—	0.50
332.0	P	8.5–10.5	1.2	2.0–4.0	0.50	0.50–1.5	—	0.50	1.0	0.25	—	—	0.50
333.0	P	8.0–10.0	1.0	3.0–4.0	0.50	0.05–0.50	—	0.50	1.0	0.25	—	—	0.50
336.0	P	11.0–13.0	1.2	0.50–1.5	0.35	0.7–1.3	—	2.0–3.0	0.35	0.25	—	0.05	—
354.0	S&P	8.6–9.4	0.20	1.6–2.0	0.10	0.40–0.6	—	—	0.10	0.20	—	0.05	0.15
355.0	S&P	4.5–5.5	0.6	1.0–1.5	0.50	0.40–0.6	0.25	—	0.35	0.25	—	0.05	0.15
C355.0	S&P	4.5–5.5	0.20	1.0–1.5	0.10	0.40–0.6	—	—	0.10	0.20	—	0.05	0.15
356.0	S&P	6.5–7.5	0.6	0.25	0.35	0.20–0.45	—	—	0.35	0.25	—	0.05	0.15
A356.0	S&P	6.5–7.5	0.20	0.20	0.10	0.25–0.45	—	—	0.10	0.20	—	0.05	0.15
357.0	S&P	6.5–7.5	0.15	0.05	0.03	0.45–0.6	—	—	0.05	0.20	—	0.05	0.15
A357.0	S&P	6.5–7.5	0.20	0.20	0.10	0.40–0.7	—	—	0.10	0.04–0.20	—	0.05	0.15
359.0	S&P	8.5–9.5	0.20	0.20	0.10	0.50–0.7	—	—	0.10	0.20	—	0.05	0.15
360.0	D	9.0–10.0	2.0	0.6	0.35	0.40–0.6	—	0.50	0.50	—	0.15	—	0.25
A360.0	D	9.0–10.0	1.3	0.6	0.35	0.40–0.6	—	0.50	0.50	—	0.15	—	0.25
380.0	D	7.5–9.5	2.0	3.0–4.0	0.50	0.10	—	0.50	3.0	—	0.35	—	0.50
A380.0	D	7.5–9.5	1.3	3.0–4.0	0.50	0.10	—	0.50	3.0	—	0.35	—	0.50
383.0	D	9.5–11.5	1.3	2.0–3.0	0.50	0.10	—	0.30	3.0	—	0.15	—	0.50
384.0	D	10.5–12.0	1.3	3.0–4.5	0.50	0.10	—	0.50	3.0	—	0.35	—	0.50
390.0	D	16.0–18.0	1.3	4.0–5.0	0.10	0.45–0.65	—	—	0.10	0.20	—	0.05	0.20
B390.0	D	16.0–18.0	1.3	4.0–5.0	0.50	0.45–0.65	—	0.10	1.5	0.10	—	0.05	0.20
392.0	D	18.0–20.0	1.5	0.40–0.80	0.20–0.60	0.80–1.20	—	0.50	0.50	0.20	0.30	—	0.50
413.0	D	11.0–13.0	2.0	1.0	0.35	0.10	—	0.50	0.50	—	0.15	—	0.25
A413.0	D	11.0–13.0	1.3	1.0	0.35	0.10	—	0.50	0.50	—	0.15	—	0.25
C433.0	D	4.5–6.0	2.0	0.6	0.35	0.10	—	0.50	0.50	—	0.15	—	0.25
443.0	S&P	4.5–6.0	0.8	0.6	0.50	0.05	0.25	—	0.50	0.25	—	0.05	0.35
B443.0	S&P	4.5–6.0	0.8	0.15	0.35	0.05	—	—	0.35	0.25	—	0.05	0.15
A444.0	P	6.5–7.5	0.20	0.20	0.10	0.05	—	—	0.10	0.20	—	0.05	0.15
512.0	S	1.4–2.2	0.6	0.35	0.8	3.5–4.5	0.25	—	0.35	0.25	—	0.05	0.15
513.0	P	0.30	0.40	0.10	0.30	3.5–4.5	—	—	1.4–2.2	0.20	—	0.05	0.15
514.0	S	0.35	0.50	0.15	0.35	3.5–4.5	—	—	0.15	0.25	—	0.05	0.15
518.0	D	0.35	1.8	0.25	0.35	7.5–8.5	—	0.15	0.15	—	0.15	0.05	0.25
520.0	S	0.25	0.30	0.25	0.15	9.5–10.6	—	—	0.15	0.25	—	0.05	0.15
535.0	S&P	0.15	0.15	0.05	0.10–0.25	6.2–7.5	—	—	—	0.10–0.25	—	0.05	0.15
705.0	S&P	0.20	0.8	0.20	0.40–0.6	1.4–1.8	0.20–0.40	—	2.7–3.3	0.25	—	0.05	0.15
707.0	S&P	0.20	0.8	0.20	0.40–0.6	1.8–2.4	0.20–0.40	—	4.0–4.5	0.25	—	0.05	0.15
710.0	S	0.15	0.50	0.35–0.65	0.05	0.6–0.8	—	—	6.0–7.0	0.25	—	0.05	0.15
711.0	P	0.30	0.7–1.4	0.35–0.65	0.05	0.25–0.45	—	—	6.0–7.0	0.20	—	0.05	0.15
712.0	S	0.30	0.50	0.25	0.10	0.50–0.65	0.40–0.6	—	5.0–6.5	0.15–0.25	—	0.05	0.20
713.0	S&P	0.25	1.1	0.40–1.0	0.6	0.20–0.50	0.35	0.15	7.0–8.0	0.25	—	0.10	0.25
771.0	S	0.15	0.15	0.10	0.10	0.8–1.0	0.06–0.20	—	6.5–7.5	0.10–0.20	—	0.05	0.15
850.0	S&P	0.7	0.7	0.7–1.3	0.10	0.10	—	0.7–1.3	—	0.20	—	—	0.30
851.0	S&P	2.0–3.0	0.7	0.7–1.3	0.10	0.10	—	0.30–0.7	—	0.20	—	—	0.30
852.0	S&P	0.40	0.7	1.7–2.3	0.10	0.6–0.9	—	0.9–1.5	—	0.20	—	—	0.30

* S = sand casting, P = permanent mold casting, D = die casting.

Table 6.4.5b Mechanical Properties of Aluminum Sand Castings

Alloy	Temper	Ultimate ksi	Ultimate (MPa)	Yield (0.2% offset) ksi	Yield (0.2% offset) (MPa)	% Elongation in 2 in, or 4 times diameter
201.0	T7	60.0	(414)	50.0	(345)	3.0
204.0	T4	45.0	(310)	28.0	(193)	6.0
208.0	F	19.0	(131)	12.0	(83)	1.5
222.0	O	23.0	(159)	—	—	—
222.0	T61	30.0	(207)	—	—	—
242.0	O	23.0	(159)	—	—	—
242.0	T571	29.0	(200)	—	—	—
242.0	T61	32.0	(221)	20.0	(138)	—
242.0	T77	24.0	(165)	13.0	(90)	1.0
295.0	T4	29.0	(200)	13.0	(90)	6.0
295.0	T6	32.0	(221)	20.0	(138)	3.0
295.0	T62	36.0	(248)	28.0	(193)	—
295.0	T7	29.0	(200)	16.0	(110)	3.0
319.0	F	23.0	(159)	13.0	(90)	1.5
319.0	T5	25.0	(172)	—	—	—
319.0	T6	31.0	(214)	20.0	(138)	1.5
328.0	F	25.0	(172)	14.0	(97)	1.0
328.0	T6	34.0	(234)	21.0	(145)	1.0
354.0	*	—	—	—	—	—
355.0	T51	25.0	(172)	18.0	(124)	—
355.0	T6	32.0	(221)	20.0	(138)	2.0
355.0	T7	35.0	(241)	—	—	—
355.0	T71	30.0	(207)	22.0	(152)	—
C355.0	T6	36.0	(248)	25.0	(172)	2.5
356.0	F	19.0	(131)	—	—	2.0
356.0	T51	23.0	(159)	16.0	(110)	—
356.0	T6	30.0	(207)	20.0	(138)	3.0
356.0	T7	31.0	(214)	29.0	(200)	—
356.0	T71	25.0	(172)	18.0	(124)	3.0
A356.0	T6	34.0	(234)	24.0	(165)	3.5
357.0	*	—	—	—	—	—
A357.0	*	—	—	—	—	—
359.0	*	—	—	—	—	—
443.0	F	17.0	(117)	7.0	(49)	3.0
B433.0	F	17.0	(117)	6.0	(41)	3.0
512.0	F	17.0	(117)	10.0	(69)	—
514.0	F	22.0	(152)	9.0	(62)	6.0
520.0	T4	42.0	(290)	22.0	(152)	12.0
535.0	F or T5	35.0	(241)	18.0	(124)	9.0
705.0	F or T5	30.0	(207)	17.0	(117)	5.0
707.0	T5	33.0	(228)	22.0	(152)	2.0
707.0	T7	37.0	(255)	30.0	(207)	1.0
710.0	F or T5	32.0	(221)	20.0	(138)	2.0
712.0	F or T5	34.0	(234)	25.0	(172)	4.0
713.0	F or T5	32.0	(221)	22.0	(152)	3.0
771.0	T5	42.0	(290)	38.0	(262)	1.5
771.0	T51	32.0	(221)	27.0	(186)	3.0
771.0	T52	36.0	(248)	30.0	(207)	1.5
771.0	T53	36.0	(248)	27.0	(186)	1.5
771.0	T6	42.0	(290)	35.0	(241)	5.0
771.0	T71	48.0	(331)	45.0	(310)	2.0
850.0	T5	16.0	(110)	—	—	5.0
851.0	T5	17.0	(117)	—	—	3.0
852.0	T5	24.0	(165)	18.0	(124)	—

Values represent properties obtained from separately cast test bars. Average properties of specimens cut from castings shall not be less than 75% of tensile and yield strength values and shall not be less than 25% of elongation values given above.
  * Mechanical properties for these alloys depend on casting process. Consult individual foundries.

be augmented by coating the material with a surface layer of high-purity aluminum or, in some cases, a more corrosion-resistant alloy of aluminum. Such products are referred to as **alclad**. This cladding becomes an integral part of the material, is metallurgically bonded, and provides cathodic protection in a manner similar to zinc galvanizing on steel. Because the cladding usually has lower strength than the base metal, alclad products have slightly lower strengths than uncoated material. Cladding thickness varies from 1.5 to 10 percent. Products available clad are 3003 tube, 5056 wire, and 2014, 2024, 2219, 3003, 3004, 6061, 7075, 7178, and 7475 sheet and plate.

Anodizing The corrosion resistance of any of the alloys may also be improved by anodizing, done by making the parts to be treated the anode in an electrolytic bath such as sulfuric acid. This process produces a tough, adherent coating of aluminum oxide, usually 0.4 mil (0.01 mm) thick or greater. Any welding should be performed before anodizing, and filler alloys should be chosen judiciously for good ano-

Table 6.4.5c Mechanical Properties of Aluminum Permanent Mold Castings

| Alloy | Temper | Tensile strength | | | | % Elongation in 2 in, or 4 times diameter |
| | | Ultimate | | Yield (0.2% offset) | | |
		ksi	(MPa)	ksi	(MPa)	
204.0	T4	48.0	(331)	29.0	(200)	8.0
208.0	T4	33.0	(228)	15.0	(103)	4.5
208.0	T6	35.0	(241)	22.0	(152)	2.0
208.0	T7	33.0	(228)	16.0	(110)	3.0
222.0	T551	30.0	(207)	—	—	—
222.0	T65	40.0	(276)	—	—	—
242.0	T571	34.0	(234)	—	—	—
242.0	T61	40.0	(276)	—	—	—
296.0	T6	35.0	(241)	—	—	2.0
308.0	F	24.0	(165)	—	—	—
319.0	F	28.0	(193)	14.0	(97)	1.5
319.0	T6	34.0	(234)	—	—	2.0
332.0	T5	31.0	(214)	—	—	—
333.0	F	28.0	(193)	—	—	—
333.0	T5	30.0	(207)	—	—	—
333.0	T6	35.0	(241)	—	—	—
333.0	T7	31.0	(214)	—	—	—
336.0	T551	31.0	(214)	—	—	—
336.0	T65	40.0	(276)	—	—	—
354.0	T61	48.0	(331)	37.0	(255)	3.0
354.0	T62	52.0	(359)	42.0	(290)	2.0
355.0	T51	27.0	(186)	—	—	—
355.0	T6	37.0	(255)	—	—	1.5
355.0	T62	42.0	(290)	—	—	—
355.0	T7	36.0	(248)	—	—	—
355.0	T71	34.0	(234)	27.0	(186)	—
C355.0	T61	40.0	(276)	30.0	(207)	3.0
356.0	F	21.0	(145)	—	—	3.0
356.0	T51	25.0	(172)	—	—	—
356.0	T6	33.0	(228)	22.0	(152)	3.0
356.0	T7	25.0	(172)	—	—	3.0
356.0	T71	25.0	(172)	—	—	3.0
A356.0	T61	37.0	(255)	26.0	(179)	5.0
357.0	T6	45.0	(310)	—	—	3.0
A357.0	T61	45.0	(310)	36.0	(248)	3.0
359.0	T61	45.0	(310)	34.0	(234)	4.0
359.0	T62	47.0	(324)	38.0	(262)	3.0
443.0	F	21.0	(145)	7.0	(49)	2.0
B443.0	F	21.0	(145)	6.0	(41)	2.5
A444.0	T4	20.0	(138)	—	—	20.0
513.0	F	22.0	(152)	12.0	(83)	2.5
535.0	F	35.0	(241)	18.0	(124)	8.0
705.0	T5	37.0	(255)	17.0	(117)	10.0
707.0	T7	45.0	(310)	35.0	(241)	3.0
711.0	T1	28.0	(193)	18.0	(124)	7.0
713.0	T5	32.0	(221)	22.0	(152)	4.0
850.0	T5	18.0	(124)	—	—	8.0
851.0	T5	17.0	(117)	—	—	3.0
851.0	T6	18.0	(124)	—	—	8.0
852.0	T5	27.0	(186)	—	—	3.0

Values represent properties obtained from separately cast test bars. Average properties of specimens cut from castings shall not be less than 75% of tensile and yield strength values and shall not be less than 25% of elongation values given above.

dized color match with the base alloy. The film is colorless on pure aluminum and tends to be gray or colored on alloys containing silicon, copper, or other constituents. To provide a consistent color appearance after anodizing, **AQ (anodizing quality) grade** may be specified in certain alloys. Where appearance is the overriding concern, 5005 sheet and 6063 extrusions are preferred for anodizing. If a colored finish is desired, the electrolytically oxidized article may be treated with a dye solution. Care must be taken in selecting the dye when the part will be exposed to the weather, for not all have proved to be colorfast. Two-step electrolytic coloring, produced by first clear anodizing and then electrolytically depositing another metal oxide, can produce shades of bronze, burgundy, and blue.

Painting When one is painting or lacquering aluminum, it is impor-

tant that the surface be properly prepared prior to the application of paint. A thin anodic film makes an excellent paint base. Alternately, the aluminum surface may be chemically treated with a dilute phosphoric acid solution. Abrasion blasting may be used on parts thicker than $\frac{1}{8}$ in (3.2 mm). Zinc chromate is frequently used as a primer, especially for corrosive environments. Most paints are baked on. That affects the strength of the metal, for baking tends to anneal it, and must be taken into account where strength is a factor. Sometimes the paint baking process is used as the artificial aging heat treatment. Aluminum sheet is available with factory-baked paint finish; minimum mechanical properties must be obtained from the supplier. High-, medium-, and low-gloss paint finishes are available and are determined in accordance with ASTM D523.

Table 6.4.5d Mechanical Properties of Aluminum Die Castings

Alloy		Typical tensile strength, ksi	Typical yield strength (0.2% offset), ksi	Typical elongation in 2 in, %
ANSI	ASTM			
360.0	SG100B	44	25	2.5
A360.0	SG100A	46	24	3.5
380.0	SC84B	46	23	2.5
A380.0	SC84A	47	23	3.5
383.0	SC102A	45	22	3.5
384.0	SC114A	48	24	2.5
390.0	SC174A	40.5	35	<1
B390.0	SC174B	46	36	<1
392.0	S19	42	39	<1
413.0	S12B	43	21	2.5
A413.0	S12A	42	19	3.5
C443.0	S5C	33	14	9.0
518.0	G8A	45	28	5

ksi × 6.895 = MPa

Table 6.4.6 Mechanical Properties of Aluminum Fasteners

Alloy and temper	Minimum tensile strength, ksi	Minimum shear strength, ksi
2024-T4	62	37
6061-T6	42	25
7075-T73	68	41

ksi × 6.895 = MPa

Table 6.4.7 Minimum Shear Strength of Aluminum Rivets

Alloy and temper before driving	Minimum shear strength, ksi
1100-H14	9.5
2017-T4	33
2117-T4	26
5056-H32	25
6053-T61	20
6061-T6	25
7050-T7	39

ksi × 6.895 = MPa

Aluminum Conductors On a weight basis, aluminum has twice the electrical conductance of copper; on a volume basis, the conductivity of aluminum is about 62 percent that of copper. For electrical applications, a special, high-purity grade of aluminum is used (designated 1350, also referred to as EC) or alloyed to improve strength with minimum sacrifice in conductivity. Table 6.4.8 lists common conductor alloys and gives their strengths and conductivities in various forms and treatments. In power transmission lines, the necessary strength for long spans is

Table 6.4.8 Aluminum Electrical Conductors

Product, alloy, treatment* (size)	Ultimate tensile strength, ksi	Min. electrical conductivity, % IACS
Drawing stock (rod)		
1350-O (0.375- to 1.000-in diam)	8.5–14.0	61.8
1350-H12 and H22 (0.375- to 1.000-in diam)	12.0–17.0	61.5
1350-H14 and H24 (0.375- to 1.000-in diam)	15.0–20.0	61.4
1350-H16 and H26 (0.375- to 1.000-in diam)	17.0–22.0	61.3
5005-O (0.375-in diam)	14.0–20.0	54.3
5005-H12 and H22 (0.375-in diam)	17.0–23.0	54.0
5005-H14 and H24 (0.375-in diam)	20.0–26.0	53.9
5005-H16 and H26 (0.375-in diam)	24.0–30.0	53.8
8017-H12 and H22 (0.375-in diam)	16.0–22.0	58.0
8030-H12 (0.375-in diam)	16.0–20.5	60.0
8176-H14 (0.375-in diam)	16.0–20.0	59.0
8177-H13 and H23 (0.375-in diam)	16.0–22.0	58.0
Wire		
1350-H19 (0.0801- to 0.0900-in diam)	26.0 min	61.0
5005-H19 (0.0801- to 0.0900-in diam)	37.0 min	53.5
6201-T81 (0.0612- to 0.1327-in diam)	46.0 min	52.5
8176-H24 (0.0500- to 0.2040-in diam)	15.0 min	61.0
Extrusions		
1350-H111 (all)	8.5 min	61.0
6101-H111 (0.250 to 2.000 in thick)	12.0 min	59.0
6101-T6 (0.125 to 0.500 in thick)	29.0 min	55.0
6101-T61 (0.125 to 0.749 in thick)	20.0 min	57.0
6101-T61 (0.750 to 1.499 in thick)	18.0 min	57.0
6101-T61 (1.500 to 2.000 in thick)	15.0 min	57.0
Rolled bar		
1350-H12 (0.125 to 1.000 in thick)	12.0 min	61.0
Sawed plate bar		
1350-H112 (0.125 to 0.499 in thick)	11.0 min	61.0
1350-H112 (0.500 to 1.000 in thick)	10.0 min	61.0
1350-H112 (1.001 to 1.500 in thick)	9.0 min	61.0
Sheet		
1350-O (0.006 to 0.125 in thick)	8.0–14.0	61.8

ksi × 6.895 = MPa, in × 25.4 = mm
IACS = International Annealed Copper Standard.
* Treatments: O = annealed; H = cold-worked; T = heat-treated

obtained by stranding aluminum wires about a core wire of steel (ACSR) or a higher-strength aluminum alloy.

BEARING METALS
by Frank E. Goodwin

REFERENCES: Current edition of ASM "Metals Handbook." Publications of the various metal producers. Trade association literature containing material properties. Applicable current ASTM and SAE Standards.

Babbitt metal is a general term used for soft tin and lead-base alloys which are cast as bearing surfaces on steel, bronze, or cast-iron shells. Babbits have excellent **embedability** (ability to embed foreign particles in itself) and **conformability** (ability to deform plastically to compensate for irregularities in bearing assembly) characteristics. These alloys may be run satisfactorily against a soft-steel shaft. The limitations of Babbit alloys are the tendency to spread under high, steady loads and to fatigue under high, fluctuating loads. These limitations apply more particularly at higher temperatures, for increase in temperature between 68 and 212°F (20 and 100°C) reduces the metal's strength by 50 percent. These limitations can be overcome by properly designing the thickness and rigidity of the backing material, properly choosing the Babbitt alloy for good mechanical characteristics, and ensuring a good bond between backing and bearing materials.

The important tin- and lead-base (Babbitt) bearing alloys are listed in Table 6.4.9. Alloys 1 and 15 are used in internal-combustion engines. Alloy 1 performs satisfactorily at low temperatures, but alloy 15, an arsenic alloy, provides superior performance at elevated temperatures by virtue of its better high-temperature hardness, ability to support higher loads, and longer fatigue life. Alloys 2 and 3 contain more antimony, are harder, and are less likely to pound out. Alloys 7 and 8 are lead-base Babbitts which will function satisfactorily under moderate conditions of load and speed. Alloy B is used for diesel engine bearings. In general, increasing the lead content in tin-based Babbitt provides higher hardness, greater ease of casting, but lower strength values.

Silver lined bearings have an excellent record in heavy-duty applications in aircraft engines and diesels. For reciprocating engines, silver bearings normally consist of electrodeposited silver on a steel backing with an overlay of 0.001 to 0.005 in of lead. An indium flash on top of the lead overlay is used to increase corrosion resistance of the material.

Aluminum and **zinc alloys** are used for high-load, low-speed applications but have not replaced Babbits for equipment operating under a steady high-speed load. The Al 20 to 30, Sn 3 copper alloy is bonded to a steel bearing shell. The Al 6.25, Sn 1, Ni 1, copper alloy can be either used as-cast or bonded to a bearing shell. The Al 3 cadmium alloy with varying amounts of Si, Cu, and Ni can also be used in either of these two ways. The Zn 11, Al 1, Cu 0.02, magnesium (ZA-12) and Zn 27, Al 2, Cu 0.01 magnesium (ZA-27) alloys are used in cast form, especially in continuous-cast hollow-bar form.

Mention should be made of **cast iron** as a bearing material. The flake graphite in cast iron develops a glazed surface which is useful at surface speeds up to 130 ft/min and at loads up to 150 lb/in² approx. Because of the poor conformability of cast iron, good alignment and freedom from dirt are essential.

Copper-base bearing alloys have a wide range of bearing properties that fit them for many applications. Used alone or in combination with steel, Babbitt (white metal), and graphite, the bronzes and copper-leads meet the conditions of load and speed given in Table 6.4.10. **Copper-lead alloys** are cast onto steel backing strips in very thin layers (0.02 in) to provide bearing surfaces.

Three families of copper alloys are used for bearing and wear-resistant alloys in cast form: **phosphor bronzes** (Cu-Sn), Cu-Sn-Pb alloys, and **manganese bronze, aluminum bronze,** and **silicon bronze**. Typical compositions and applications are listed in Table 6.4.10. Phosphor bronzes have residual phosphorus ranging from 0.1 to 1 percent. Hardness increases with phosphorus content. Cu-Sn-Pb alloys have high resistance to wear, high hardness, and moderate strength. The high lead compositions are well suited for applications where lubricant may be deficient. The Mn, Al, and Si bronzes have high tensile strength, high hardness, and good resistance to shock. They are suitable for a wide range of bearing applications.

Porous bearing materials are used in light- and medium-duty applications as small-sized bearings and bushings. Since they can operate for long periods without an additional supply of lubricant, such bearings are useful in inaccessible or inconvenient places where lubrication would be difficult. Porous bearings are made by pressing mixtures of copper and tin (bronze), and often graphite, Teflon, or iron and graphite, and sintering these in a reducing atmosphere without melting. Iron-based, oil-impregnated sintered bearings are often used. By controlling the conditions under which the bearings are made, porosity may be adjusted so that interconnecting voids of up to 35 percent of the total volume may be available for impregnation by lubricants. Applicable specifications for these bearings are given in Tables 6.4.11 and 6.4.12.

Miscellaneous A great variety of materials, e.g., rubber, wood, phenolic, carbon-graphite, ceramets, ceramics, and plastics, are in use for special applications. Carbon-graphite is used where contamination by oil or grease lubricants is undesirable (e.g., textile machinery, pharmaceutical equipment, milk and food processing) and for elevated-temperature applications. Notable among plastic materials are Teflon and nylon, the polycarbonate Lexan, and the acetal Delrin. Since Lexan and Delrin can be injection-molded easily, bearings can be formed quite economically from these materials.

CEMENTED CARBIDES
by Don Graham

REFERENCE: Schwartzkopf and Kieffer, "Refractory Hard Metals," Macmillan. German, "Powder Metallurgy Science," 2d ed., Metal Powder Industries Federation. "Powder Metallurgy Design Manual," 2d ed., Metal Powder Industries Federation, Princeton, NJ. Goetzal, "Treatise on Powder Metallurgy," Interscience. Schwartzkopf, "Powder Metallurgy," Macmillan.

Table 6.4.9   Compositions and Properties of Some Babbitt Alloys

ASTM B23 grade	Composition, %					Compressive ultimate strength		Brinell hardness	
	Sn	Sb	Pb	Cu	As	68°F, lb/in²	20°C, MPa	68°F (20°C)	212°F (100°C)
1	91.0	4.5	0.35	4.5	—	12.9	88.6	17	8
2	89.0	7.5	0.35	3.5	—	14.9	103	24.5	12
3	84.0	8.0	0.35	8.0	—	17.6	121	27.0	14.5
7	10.0	15.0	74.5	0.5	0.45	15.7	108	22.5	10.5
8	5.0	15.0	79.5	0.5	0.45	15.6	108	20.0	9.5
15	1.0	16.0	82.5	0.5	1.10	—	—	21.0	13.0
Other alloys									
B	0.8	12.5	83.3	0.1	3.05	—	—		
ASTM B102, alloy Py 1815A	65	15	18	2	0.15	15	103	23	10

SOURCE: ASTM, reprinted with permission.

Table 6.4.10   Compositions, Properties, and Applications of Some Copper-Base Bearing Metals

| | Composition, % | | | | | | Minimum tensile strength | | |
Specifications	Cu	Sn	Pb	Zn	P	Other	ksi	MPa	Applications
C86100, SAE J462	64	—	—	24	—	3 Fe, 5 Al, 4 Mn	119	820	Extra-heavy-duty bearings
C87610, ASTM B30	90	—	0.2	5	—	4 Si	66	455	High-strength bearings
C90700, ASTM B505	89	11	0.3	—	0.2	—	44	305	Wormgears and wheels; high-speed low-load bearings
C91100	84	16	—	—	—	—	35	240	Heavy load, low-speed bearings
C91300	81	19	—	—	—	—	35	240	Heavy load, low-speed bearings
C93200, SAE J462	83	7	7	3	—	—	35	240	General utility bearings, automobile fittings
C93700, ASTM B22	80	10	10	—	—	—	35	240	High-speed, high-pressure bearings, good corrosion resistance
C93800, ASTM B584	78	7	15	—	—	—	30	205	General utility bearings, backing for Babbit bearings
C94300, ASTM B584	70	5	25	—	—	—	27	185	Low-load, high-speed bearings

Source: ASTM, reprinted with permisstion.

Cemented carbides are a commercially and technically important class of composite material. Comprised primarily of hard tungsten carbide (WC) grains cemented together with a cobalt (Co) binder, they can contain major alloying additions of titanium carbide (TiC), titanium carbonitride (TiCN), tantalum carbide (TaC), niobium carbide (NbC), chromium carbide (CrC), etc., and minor additions of other elements. Because of the extremely high hardness, stiffness, and wear resistance of these materials, their primary use is in cutting tools for the material (usually metal, wood, composite, or rock) removal industry. Secondary applications are as varied as pen balls, food processing equipment, fuel pumps, wear surfaces, fishing line guide rings, and large high-pressure components. Since their introduction in the mid-1920s, a very large variety of grades of carbides have been developed and put into use in a diverse number of applications. They are ubiquitous throughout industry, particularly so in the high-production metalworking sector.

Most cemented carbide parts are produced by **powder metallurgical techniques**. Powders of WC, Co, and sometimes TiC and TaC are blended by either ball or attritor milling. The mixed powder is then **compacted** under very high pressure in appropriately shaped molds. Pressed parts are **sintered** at temperatures between 1,250 and 1,500°C, depending on the composition of the powder. During the sintering process, cobalt melts and wets the carbide particles. On cooling, cobalt "cements" the carbide grains together. Essentially complete densification takes place during sintering. While binder metals such as iron or nickel are sometimes used, cobalt is preferred because of its ability to wet the tungsten carbide. A cemented carbide workpiece should be as close as possible to its final shape before sintering, for the final product is extremely hard and can be shaped further only by grinding with silicon carbide or diamond wheels.

The metal removal industry consumes most of the cemented carbide that is produced. Tips having unique geometries, called **inserts**, are pressed and sintered. Many of these pressed and sintered parts are ready for use directly after sintering. Others are finish-ground and/or honed to close tolerances, often with a slight radius imparted to the cutting edge. Both procedures induce longer tool life between sharpenings.

The first cemented carbides developed consisted only of WC and Co. As cutting speeds increased, temperatures at the tool/workpiece interface increased. In the presence of iron or steel at high temperatures, carbide tools interact chemically with the work material and result in "cratering," or removal of tool material at the cutting edge. Additions of secondary carbides like TiC and TaC minimize cratering by virtue of their greater chemical stability. Because of the desire for even higher metal removal rates (and accompanying higher temperatures), **chemical vapor deposition** (CVD) overlay coatings were developed. Titanium carbide coatings became available in 1969; titanium nitride coatings followed shortly after that. Aluminum oxide coatings for very high-speed operations first appeared in 1973. Coatings usually result in tool life increases of 50 percent; in some extreme cases, phenomenal increases of 10,000 percent have been reported. The CVD coatings have the synergistic effect not only to increase wear resistance (i.e., reduce cratering), but also to allow much higher cutting speeds (i.e., operating cutting temperatures).

In the late 1980s, **physical vapor deposition** (PVD) coatings became popular for specific applications. Coatings such as titanium nitride (TiN), titanium aluminum nitride (TiAlN), titanium carbonitride (TiCN), zirconium nitride (ZrN), chromium-based coatings, and amorphous coatings are used for special-purpose applications such as machining of high-temperature alloys, machining of ductile irons at low speeds, and as wear surfaces. About 70 percent of all cemented carbide tools sold today are coated, strong testimony to the effectiveness of coatings.

Design Considerations

Cemented carbide, in common with all **brittle materials,** may fragment in service, particularly under conditions of impact or upon release from high compressive loading. Precautionary measures must be taken to

Table 6.4.11   Oil-Impregnated Iron-Base Sintered Bearings (ASTM Standard B439-83)

| | Composition, % | | | |
Element	Grade 1	Grade 2	Grade 3	Grade 4
Copper			7.0–11.0	18.0–22.0
Iron	96.25 min	95.9 min	Remainder*	Remainder*
Total other elements by difference, max	3.0	3.0	3.0	3.0
Combined carbon† (on basis of iron only)	0.25 max	0.25–0.60	—	—
Silicon, max	0.3	0.3	—	—
Aluminum, max	0.2	0.2	—	—

* Total of iron plus copper shall be 97% min.
† The combined carbon may be a metallographic estimate of the carbon in the iron.
Source: ASTM, reprinted with permission.

Table 6.4.12   Oil-Impregnated Sintered Bronze Bearings (ASTM Standard B438-83)

| | Composition, % | | | |
| | Grade 1 | | Grade 2 | |
Element	Class A	Class B	Class A	Class B
Copper	87.5–90.5	87.5–90.5	82.6–88.5	82.6–88.5
Tin	9.5–10.5	9.5–10.5	9.5–10.5	9.5–10.5
Graphite, max	0.1	1.75	0.1	1.75
Lead	*	*	2.0–4.0	2.0–4.0
Iron, max	1.0	1.0	1.0	1.0
Total other elements by difference, max	0.5	0.5	1.0	1.0

* Included in other elements.
SOURCE: ASTM, reprinted with permission.

ensure that personnel and equipment are protected from flying fragments and sharp edges when working with carbides.

Failure of brittle materials frequently occurs as a result of tensile stress at or near the surface; thus the strength of brittle materials is very dependent on surface conditions. Chips, scratches, thermal cracks, and grinding marks, which decrease the strength and breakage resistance of cemented carbide parts, should be avoided. Even electric-discharge machined (EDM) surfaces should be ground or lapped to a depth of 0.05 mm to remove damaged material.

Special care must be taken when one is grinding cemented carbides. Adequate ventilation of grinding spaces must comply with existing government regulations applicable to the health and safety of the workplace environment. During the fabrication of cemented carbide parts, particularly during grinding of tools and components, adequate provisions must be made for the removal and collection of generated dusts and cutting fluid mists that contain microscopic metal particles, even though those concentrations may be very low. Although the elements contained in these alloys are not radioactive, note that they may become radioactive when exposed to a sufficiently strong radiation source. The cobalt in the cemented carbide alloys, in particular, could be made radioactive.

Physical and Mechanical Properties

Typical property data are shown in Table 6.4.13. In addition to being an effective binder, cobalt is the primary component that determines mechanical properties. For example, as cobalt content increases, toughness increases but hardness and wear resistance decrease. This tradeoff is illustrated in Table 6.4.13. A secondary factor that affects mechanical properties is WC grain size. Finer grain size leads to increased hardness and wear resistance but lower toughness.

Mechanical Properties   The outstanding feature of these materials is their hardness, high compressive strength, stiffness, and wear resistance. Unfortunately, but as might be expected of hard materials, toughness and ductility of carbides are low. Table 6.4.13 lists some **mechanical** and **physical property** data for typical and popular cemented carbide compositions.

Hardness and Wear Resistance   Most applications of cemented carbide alloys involve wear and abrasive conditions. In general, the wear resistance of these materials can be estimated based on hardness. Note that wear is a complex process and can occur by means other than normal abrasion. Should wear occur by microchipping, fracture, chemical attack, or metallurgical interactions, it will be obvious that performance does not correlate directly with hardness.

The data presented in Table 6.4.13 under the heading Abrasion Resistance were determined by using an abrasive wheel. Results from this simplified test should be used only as a rough guide in the selection of those alloys, since abrasive wear that occurs under actual service conditions is usually complex and varies greatly with particular circumstances. The performance characteristics of a given grade are, of course, best determined under actual service conditions.

In many applications, cemented carbides are subjected to relatively high operating temperatures. For example, the temperatures of the interface between the chip and the cutting edge of a carbide cutting tool may reach between 500 and 1,100°C. Wear parts may have point contact temperatures in this range. The extent to which the cemented carbide maintains its hardness at the elevated temperatures encountered in use is an important consideration. The cemented carbides not only have high room-temperature hardness, but also maintain hardness at elevated temperatures better than do steels and cast alloys.

Corrosion Resistance   The corrosion resistance of the various cemented carbide alloys is fairly good when compared with other materials, and they may be employed very advantageously in some corrosive environments where outstanding wear resistance is required. Generally they are not employed where corrosion resistance alone is the requirement because other materials usually can be found which are either cheaper or more corrosion-resistant.

Corrosion may cause strength deterioration due to the preferential attack on the binder matrix phase. This results in the creation of microscopic surface defects to which cemented carbides, in common with other brittle materials, are sensitive.

Special corrosion-resistant grades have been developed in which the cobalt binder has been modified or replaced by other metals or alloys, resulting in improved resistance to acid attack. Other factors such as temperature, pressure, and surface condition may significantly influence corrosive behavior, so that specific tests under actual operating conditions should be performed when possible to select a cemented carbide grade for service under corrosive conditions.

Micrograin Carbides

As **carbide grain size decreases,** hardness increases but toughness decreases. However, once the WC grain size diameter drops below approximately 1 μm, this tradeoff becomes much more favorable. Further decreases in grain size result in further increase in hardness but are accompanied by a much smaller drop in toughness. As a result, **micrograin cemented carbides** have outstanding hardness, wear resistance, and compressive strength combined with surprising toughness. This combination of properties makes these grades particularly useful in machining nickel-, cobalt-, and iron-based superalloys and other difficult-to-machine alloys, such as refractories and titanium alloys.

Cermets

Another class of cemented carbides is called **cermets**. These alloys usually are composed of TiC or TiCN with a binder of nickel, nickel-iron, or nickel-molybdenum with small amounts of other elements such as cobalt. These alloys usually find application in the high-speed finish machining of steels, stainless steels, and occasionally, cast irons and high-temperature alloys. As a rule, cermets are wear-resistant but less tough than traditional cemented tungsten carbides. Recent processing improvements, such as sinter HIP (hot isostatic pressing) and a better understanding of the alloying characteristics of cermets, have helped improve their toughness immensely, so that very tough cermets are currently available.

Table 6.4.13 Properties of Cemented Carbides

Composition, wt. %	Grain size, μm	Hardness, Ra	Abrasion resistance, 1/vol. loss cm³	Density, g/cm³	Transverse rupture strength, 1,000 lb/in²	Ultimate compressive strength, 1,000 lb/in²	Ultimate tensile strength, 1,000 lb/in²	Modulus of elasticity, 10^6 lb/in²	Proportional limit, 1,000 lb/in²	Ductility, % elong	Fracture toughness, (lb/in²)($\sqrt{\text{in}}$)	Thermal expansion 75 to 400°F, in/(in · °F) $\times 10^{-6}$	Thermal conductivity, cal/(s · °C · cm)	Electrical resistivity, μΩ · cm
WC-3% Co	1.7	93	60	15.3	290	850		98	350			2.2	0.3	17
WC-6% Co	0.8	93.5	62	15	335	860		92	370			2.9		
WC-6% Co	1.1	92.8	60	15	380	790		92	280		9,200	2.5	0.25	17
WC-6% Co	2.1	92	35	15	400	750	160	92	210	0.2	11,500	2.4	0.25	17
WC-6% Co	3	91	15	14.7	425	660		87	140		13,000	2.7	0.25	
WC-9% Co	4	89.5	10	14.6	440	750	220	85	230		11,500		0.25	
WC-10% Co	1.8	91	13	14.5	460	630		85	130			2.5		17
WC-10% Co	5.2	89	7		500	600		81	140					17
WC-13% Co	4	88.2	4	14.2	500	560		77	100	0.3	14,500	3	0.02	
WC-16% Co	4	86.8	3	13.9	440	450	270	67	60	0.4	15,800	3.2	0.2	
WC-25% Co	3.7	84	2	13							21,000	3.5		18
Other materials*														
Tool steel (T-8)		85 (66 R_c)	2	8.4	575	600	300	34				6.5	0.12	20
Carbon steel (1095)		79 (55 R_c)		7.8	105			30						
Cast iron				7.3				15–30						
Copper				8.9								9.2	0.94	1.67

*Data for other materials are included to aid in comparing carbide properties with those of the referenced materials.

COPPER AND COPPER ALLOYS
by Arthur Cohen

REFERENCES: ASM "Metals Handbook," Applicable current ASTM and SAE Standards. Publications of the Copper Development Association Standards promulgated by industrial associations related to specific types of products.

Copper and copper alloys constitute one of the major groups of commercial metals. They are widely used because of their excellent electrical and thermal conductivities, outstanding corrosion resistance, and ease of fabrication. They are generally nonmagnetic and can be readily joined by conventional soldering, brazing, and welding processes (gas and resistance).

Primary Copper

Copper used in the manufacture of fabricated brass mill or wire and cable products may originate in the ore body of open-pit or underground mines or from rigidly controlled recycled high-grade copper scrap.

Traditional sulfide ores require conventional crushing, milling, concentration by flotation, smelting to form a matte, conversion to blister copper, and final refining by either the electrolytic or the electrowinning processes to produce copper cathode, which is fully described in ASTM B115 Electrolytic Cathode Copper. This electrochemical refining step results in the deposition of virtually pure copper, with the major impurity being oxygen.

Probably the single most important innovation in the copper industry in the past 30 years has been the introduction of continuous cast wire rod technology. Within this period, the traditional 250-lb wire bar has been almost completely replaced domestically (and largely internationally) with continuous cast wire rod product produced to meet the requirements of ASTM B49 Copper Redraw Rod for Electrical Purposes.

Generic Classification of Copper Alloys

The most common way to categorize wrought and cast copper and copper alloys is to divide them into six families: coppers, high-copper alloys, brasses, bronzes, copper nickels, and nickel silvers. They are further subdivided and identified via the Unified Numbering System (UNS) for metals and alloys.

The designation system is an orderly method of defining and identifying the alloys. The copper section is administered by the Copper Development Association, but it is not a specification. It eliminates the limitations and conflicts of alloy designations previously used and at the same time provides a workable method for the identification marking of mill and foundry products.

In the designation system, numbers from C10000 through C79999 denote wrought alloys. Cast alloys are numbered from C80000 through C999999. Within these two categories, compositions are further grouped into families of coppers and copper alloys shown in Table 6.4.14.

Coppers These metals have a designated minimum copper content of 99.3 percent or higher.

High-Copper Alloys In wrought form, these are alloys with designated copper contents less than 99.3 percent but more than 96 percent which do not fall into any other copper alloy group. The cast high-copper alloys have designated copper contents in excess of 94 percent, to which silver may be added for special properties.

Brasses These alloys contain zinc as the principal alloying element with or without other designated alloying elements such as iron, aluminum, nickel, and silicon. The wrought alloys comprise three main families of brasses: copper-zinc alloys; copper-zinc-lead alloys (leaded brasses); and copper-zinc-tin alloys (tin brasses). The cast alloys comprise four main families of brasses: copper-tin-zinc alloys (red, semired, and yellow brasses); manganese bronze alloys (high-strength yellow brasses); leaded manganese bronze alloys (leaded high-strength yellow brasses); and copper-zinc-silicon alloys (silicon brasses and bronzes). Ingot for remelting for the manufacture of castings may vary slightly from the ranges shown.

Bronzes Broadly speaking, bronzes are copper alloys in which the major element is not zinc or nickel. Originally *bronze* described alloys with tin as the only or principal alloying element. Today, the term generally is used not by itself, but with a modifying adjective. For wrought alloys, there are four main families of bronzes: copper-tin-phosphorus alloys (phosphor bronzes); copper-tin-lead-phosphorus alloys (leaded phosphor bronzes); copper-aluminum alloys (aluminum bronzes); and copper-silicon alloys (silicon bronzes).

The cast alloys comprise four main families of bronzes: copper-tin alloys (tin bronzes); copper-tin-lead alloys (leaded and high leaded tin

Table 6.4.14 Generic Classification of Copper Alloys

Generic name	UNS Nos.	Composition
Wrought alloys		
Coppers	C10100–C15815	>99% Cu
High-copper alloys	C16200–C19900	>96% Cu
Brasses	C21000–C28000	Cu-Zn
Leaded brasses	C31200–C38500	Cu-Zn-Pb
Tin brasses	C40400–C48600	Cu-Zn-Sn-Pb
Phosphor bronzes	C50100–C52400	Cu-Sn-P
Leaded phosphor bronzes	C53200–C54400	Cu-Sn-Pb-P
Copper-phosphorus and copper-silver-phosphorus alloys	C55180–C55284	Cu-P-Ag
Aluminum bronzes	C60800–C64210	Cu-Al-Ni-Fe-Si-Sn
Silicon bronzes	C64700–C66100	Cu-Si-Sn
Other copper-zinc alloys	C66400–C69710	—
Copper nickels	C70100–C72950	Cu-Ni-Fe
Nickel silvers	C73500–C79800	Cu-Ni-Zn
Cast alloys		
Coppers	C80100–C81200	>99% Cu
High-copper alloys	C81400–C82800	>94% Cu
Red and leaded red brasses	C83300–C84800	Cu-Zn-Sn-Pb (75–89% Cu)
Yellow and leaded yellow brasses	C85200–C85800	Cu-Zn-Sn-Pb (57–74% Cu)
Manganese bronzes and leaded manganese bronzes	C86100–C86800	Cu-Zn-Mn-Fe-Pb
Silicon bronzes, silicon brasses	C87300–C87800	Cu-Zn-Si
Tin bronzes and leaded tin bronzes	C90200–C94500	Cu-Sn-Zn-Pb
Nickel-tin bronzes	C94700–C94900	Cu-Ni-Sn-Zn-Pb
Aluminum bronzes	C95200–C95900	Cu-Al-Fe-Ni
Copper nickels	C96200–C96900	Cu-Ni-Fe
Nickel silvers	C97300–C97800	Cu-Ni-Zn-Pb-Sn
Leaded coppers	C98200–C98840	Cu-Pb
Miscellaneous alloys	C99300–C99750	—

Table 6.4.15 ASTM B601 Temper Designation Codes for Copper and Copper Alloys

Temper designation	Former temper name or material conditioned
Cold-worked tempers[a]	
H00	1/8 hard
H01	1/4 hard
H02	1/2 hard
H03	3/4 hard
H04	Hard
H06	Extra hard
H08	Spring
H10	Extra spring
H12	Special spring
H13	Ultra spring
H14	Super spring
Cold-worked tempers[b]	
H50	Extruded and drawn
H52	Pierced and drawn
H55	Light drawn
H58	Drawn general-purpose
H60	Cold heading; forming
H63	Rivet
H64	Screw
H66	Bolt
H70	Bending
H80	Hard-drawn
H85	Medium hard-drawn electrical wire
H86	Hard-drawn electrical wire
H90	As-finned
Cold-worked and stress-relieved tempers	
HR01	H01 and stress-relieved
HR02	H02 and stress-relieved
HR04	H04 and stress-relieved
HR08	H08 and stress-relieved
HR10	H10 and stress-relieved
HR20	As-finned
HR50	Drawn and stress-relieved
Cold-rolled and order-strengthened tempers[c]	
HT04	H04 and treated
HT08	H08 and treated
As-manufactured tempers	
M01	As sand cast
M02	As centrifugal cast
M03	As plaster cast
M04	As pressure die cast
M05	As permanent mold cast
M06	As investment cast
M07	As continuous cast
M10	As hot-forged and air-cooled
M11	As forged and quenched
M20	As hot-rolled
M30	As hot-extruded
M40	As hot-pierced
M45	As hot-pierced and rerolled

Temper designation	Former temper name or material condition
Annealed tempers[d]	
O10	Cast and annealed (homogenized)
O11	As cast and precipitation heat-treated
O20	Hot-forged and annealed
O25	Hot-rolled and annealed
O30	Hot-extruded and annealed
O31	Extruded and precipitation heat-treated
O40	Hot-pierced and annealed
O50	Light anneal
O60	Soft anneal
O61	Annealed
O65	Drawing anneal
O68	Deep-drawing anneal
O70	Dead soft anneal
O80	Annealed to temper-1/8 hard
O81	Annealed to temper-1/4 hard
O82	Annealed to temper-1/2 hard
Annealed tempers[e]	
OS005	Nominal Avg. grain size, 0.005 mm
OS010	Nominal Avg. grain size, 0.010 mm
OS015	Nominal Avg. grain size, 0.015 mm
OS025	Nominal Avg. grain size, 0.025 mm
OS035	Nominal Avg. grain size, 0.035 mm
OS050	Nominal Avg. grain size, 0.050 mm
OS060	Nominal Avg. grain size, 0.060 mm
OS070	Nominal Avg. grain size, 0.070 mm
OS100	Nominal Avg. grain size, 0.100 mm
OS120	Nominal Avg. grain size, 0.120 mm
OS150	Nominal Avg. grain size, 0.150 mm
OS200	Nominal Avg. grain size, 0.200 mm
Solution-treated temper	
TB00	Solution heat-treated (A)
Solution-treated and cold-worked tempers	
TD00	TB00 cold-worked: 1/8 hard
TD01	TB00 cold-worked: 1/4 hard
TD02	TB00 cold-worked: 1/2 hard
TD03	TB00 cold-worked: 3/4 hard
TD04	TB00 cold-worked: hard
Solution treated and precipitation-hardened temper	
TF00	TB00 and precipitation-hardened temper
Cold-worked and precipitation-hardened tempers	
TH01	TD01 and precipitation-hardened
TH02	TD02 and precipitation-hardened
TH03	TD03 and precipitation-hardened
TH04	TD04 and precipitation-hardened
Precipitation-hardened or spinodal heat treated and cold-worked tempers	
TL00	TF00 cold-worked: 1/8 hard
TL01	TF00 cold-worked: 1/4 hard
TL02	TF00 cold-worked: 1/2 hard
TL04	TF00 cold-worked: hard
TL08	TF00 cold-worked: spring
TL10	TF00 cold-worked: extra spring

Temper designation	Former temper name or material condition
Mill-hardened tempers	
TM00	AM
TM01	1/4 HM
TM02	1/2 HM
TM04	HM
TM06	XHM
TM08	XHMS
Quench-hardened tempers	
TQ00	Quench-hardened
TQ30	Quench-hardened and tempered
TQ50	Quench-hardened and temper-annealed
TQ55	Quench-hardened and temper-annealed, cold-drawn and stress-relieved
TQ75	Interrupted quench-hardened
Precipitation-hardened, cold-worked, and thermal-stress-relieved tempers	
TR01	TL01 and stress-relieved
TR02	TL02 and stress-relieved
TR04	TL04 and stress-relieved
Tempers of welded tubing[f]	
WH00	Welded and drawn: 1/8 hard
WH01	Welded and drawn: 1/4 hard
WH02	Welded and drawn: 1/2 hard
WH03	Welded and drawn: 3/4 hard
WH04	Welded and drawn: full hard
WH06	Welded and drawn: extra hard
WM00	As welded from H00 (1/8-hard) strip
WM01	As welded from H01 (1/4-hard) strip
WM02	As welded from H02 (1/2-hard) strip
WM03	As welded from H03 (3/4-hard) strip
WM04	As welded from H04 (hard) strip
WM06	As welded from H06 (extra hard) strip
WM08	As welded from H08 (spring) strip
WM10	As welded from H10 (extra spring) strip
WM15	WM50 and stress-relieved
WM20	WM00 and stress-relieved
WM21	WM01 and stress-relieved
WM22	WM02 and stress-relieved
WM50	Welded and light annealed
WO50	Welded from annealed strip
WR00	WM00; drawn and stress-relieved
WR01	WM01; drawn and stress-relieved
WR02	WM02; drawn and stress-relieved
WR03	WM03; drawn and stress-relieved
WR04	WM04; drawn and stress-relieved
WR06	WM06; drawn and stress-relieved

[a] Cold-worked tempers to meet standard requirements based on cold rolling or cold drawing.
[b] Cold-worked tempers to meet standard requirements based on temper names applicable to specific products.
[c] Tempers produced by controlled amounts of cold work followed by a thermal treatment to produce order strengthening.
[d] Annealed to meet mechanical properties.
[e] Annealed to meet nominal average grain size.
[f] Tempers of fully finished tubing that has been drawn or annealed to produce specified mechanical properties or that has been annealed to produce a prescribed nominal average grain size are commonly identified by the property H, O, or OS temper designation.
SOURCE: ASTM, reprinted with permission.

bronzes); copper-tin-nickel alloys (nickel-tin bronzes); and copper-aluminum alloys (aluminum bronzes).

The family of alloys known as **manganese bronzes,** in which zinc is the major alloying element, is included in the brasses.

Copper-Nickels These are alloys with nickel as the principal alloying element, with or without other designated alloying elements.

Copper-Nickel-Zinc Alloys Known commonly as **nickel silvers,** these are alloys which contain zinc and nickel as the principal and secondary alloying elements, with or without other designated elements.

Leaded Coppers These comprise a series of cast alloys of copper with 20 percent or more lead, sometimes with a small amount of silver, but without tin or zinc.

Miscellaneous Alloys Alloys whose chemical compositions do not fall into any of previously described categories are combined under "miscellaneous alloys."

Temper Designations

Temper designations for wrought copper and copper alloys were originally specified on the basis of cold reduction imparted by the rolling of sheet or drawing of wire. Designations for rod, seamless tube, welded tube, extrusions, castings and heat-treated products were not covered. In 1974, ASTM B601 Standard Practice for Temper Designations for Copper and Copper Alloys—Wrought and Cast, based on an alphanumeric code, was created to accommodate this deficiency. The general temper designation codes listed in ASTM B601 by process and product are shown in Table 6.4.15.

Coppers

Coppers include the **oxygen-free coppers** (C10100 and C10200), made by melting prime-quality cathode copper under nonoxidizing conditions. These coppers are particularly suitable for applications requiring high electrical and thermal conductivities coupled with exceptional ductility and freedom from hydrogen embrittlement.

The most common copper is C11000—**electrolytic tough pitch.** Its electrical conductivity exceeds 100 percent IACS and is invariably selected for most wire and cable applications. Selective properties of wire produced to ASTM B1 Hard Drawn Copper Wire, B2 Medium Hard Drawn Copper Wire, and B3 Soft or Annealed Copper Wire are shown in Table 6.4.16.

Where resistance to softening along with improved fatigue strength is required, small amounts of silver are added. This permits the silver-containing coppers to retain the effects of cold working to a higher temperature than pure copper (about 600°F versus about 400°F), a property particularly useful where comparatively high temperatures are to be withstood, as in soldering operations or for stressed conductors designed to operate at moderately elevated temperatures.

If superior **machinability** is required, C14500 (tellurium copper), C14700 (sulfur copper), or C18700 (leaded copper) can be selected. With these coppers, superior machinability is gained at a modest sacrifice in electrical conductivity.

Similarly, chromium and zirconium are added to increase elevated-temperature strength with little decrease in conductivity.

Copper-beryllium alloys are **precipitation-hardening alloys** that combine moderate conductivity with very high strengths. To achieve these properties, a typical heat treatment would involve a solution heat treatment for 1 h at 1,450°F (788°C) followed by water quenching, then a precipitation heat treatment at 600°F (316°C) for 3 h.

Wrought Copper Alloys (Brasses and Bronzes)

There are approximately 230 wrought brass and bronze compositions. The most widely used is alloy C26000, which corresponds to a 70 : 30 copper-zinc composition and is most frequently specified unless high corrosion resistance or special properties of other alloys are required. For example, alloy C36000—free-cutting brass—is selected when extensive machining must be done, particularly on automatic screw machines. Other alloys containing aluminum, silicon, or nickel are specified for their outstanding corrosion resistance. The other properties of greatest importance include mechanical strength, fatigue resistance, ability to take a good finish, corrosion resistance, electrical and thermal conductivities, color, ease of fabrication, and machinability.

The bronzes are divided into five alloy families: phosphor bronzes, aluminum bronzes, silicon bronzes, copper nickels, and nickel silvers.

Phosphor Bronzes Three tin bronzes, commonly referred to as **phosphor bronzes,** are the dominant alloys in this family. They contain 5, 8, and 10 percent tin and are identified, respectively, as alloys C51000, C52100, and C52400. Containing up to 0.4 percent phosphorus, which improves the casting qualities and hardens them somewhat, these alloys have excellent elastic properties.

Aluminum Bronzes These alloys with 5 and 8 percent aluminum find application because of their high strength and corrosion resistance, and sometimes because of their golden color. Those with 10 percent aluminum content or higher are very plastic when hot and have exceptionally high strength, particularly after heat treatment.

Silicon Bronzes There are three dominant alloys in this family in which silicon is the primary alloying agent but which also contain appreciable amounts of zinc, iron, tin, or manganese. These alloys are as corrosion-resistant as copper (slightly more so in some solutions) and possess excellent hot workability with high strengths. Their outstanding characteristic is that of ready weldability by all methods. The alloys are extensively fabricated by arc or acetylene welding into tanks and vessels for hot-water storage and chemical processing.

Copper Nickels These alloys are extremely malleable and may be worked extensively without annealing. Because of their excellent corrosion resistance, they are used for condenser tubes for the most severe service. Alloys containing nickel have the best high-temperature properties of any copper alloy.

Nickel Silvers Nickel silvers are white and are often applied because of this property. They are tarnish resistant under atmospheric conditions. Nickel silver is the base for most silver-plated ware.

Overall, the primary selection criteria can be met satisfactorily by one or more of the alloys listed in Table 6.4.17.

Table 6.4.16 Mechanical Properties of Copper Wire

| Diameter | | Hard-drawn | | | Medium hard | | | Soft or annealed |
in	mm	Tensile strength (nominal) ksi	Tensile strength (nominal) MPa	Elongation (nominal) in 10 in (250 mm), % min	Tensile strength ksi	Tensile strength MPa	Elongation in 10 in (250 mm), % min	Elongation (nominal) in 10 in (250 mm), % min
0.460	11.7	49.0	340	3.8	42.0–49.0	290–340	3.8	35
0.325	8.3	54.5	375	2.4	45.0–52.0	310–360	3.0	35
0.229	5.8	59.0	405	1.7	48.0–55.0	330–380	2.2	30
0.162	4.1	62.1	430	1.4	49.0–56.0	340–385	1.5	30
0.114	2.9	64.3	445	1.2	50.0–57.0	345–395	1.3	30
0.081	2.05	65.7	455	1.1	51.0–58.0	350–400	1.1	25
0.057	1.45	66.4	460	1.0	52.0–59.0	360–405	1.0	25
0.040	1.02	67.0	460	1.0	53.0–60.0	365–415	1.0	25

SOURCE: ASTM, abstracted with permission.

Table 6.4.17 Composition and Properties of Selected Wrought Copper and Copper Alloys

Alloy no. (and name)	Nominal composition, %	Commercial forms[a]	Tensile strength ksi	Tensile strength MPa	Yield strength ksi	Yield strength MPa	Elongation in 2 in (50 mm), %[b]
C10200 (oxygen-free copper)	99.95 Cu	F, R, W, T, P, S	32–66	221–455	10–53	69–365	55–4
C11000 (electrolytic tough pitch copper)	99.90 Cu, 0.04 O	F, R, W, T, P, S	32–66	221–455	10–53	69–365	55–4
C12200 (phosphorus-deoxidized copper, high residual phosphorus)	99.90 Cu, 0.02 P	F, R, T, P	32–55	221–379	10–50	69–345	45–8
C14500 (phosphorus-deoxidized tellurium-bearing copper)	99.5 Cu, 0.50 Te, 0.008 P	F, R, W, T	32–56	221–386	10–51	69–352	50–3
C14700 (sulfur-bearing copper)	99.6 Cu, 0.40 S	R, W	32–57	221–393	10–55	69–379	52–8
C15000 (zirconium copper)	99.8 Cu, 0.15 Zr	R, W	29–76	200–524	6–72	41–496	54–1.5
C17000 (beryllium copper)	99.5 Cu, 1.7 Be, 0.20 Co	F, R	70–190	483–1,310	32–170	221–1,172	45–3
C17200 (beryllium copper)	99.5 Cu, 1.9 Be, 0.20 Co	F, R, W, T, P, S	68–212	469–1,462	25–195	172–1,344	48–1
C18200 (chromium copper)	99.0 Cu[c], 1.0 Cr	F, W, R, S, T	34–86	234–593	14–77	97–531	40–5
C18700 (leaded copper)	99.0 Cu, 1.0 Pb	R	32–55	221–379	10–50	69–345	45–8
C19400	97.5 Cu, 2.4 Fe, 0.13 Zn, 0.03 P	F	45–76	310–524	24–73	165–503	32–2
C21000 (gilding, 95%)	95.0 Cu, 5.0 Zn	F, W	34–64	234–441	10–58	69–400	45–4
C22000 (commercial bronze, 90%)	90.0 Cu, 10.0 Zn	F, R, W, T	37–72	255–496	10–62	69–427	50–3
C23000 (red brass, 85%)	85.0 Cu, 15.0 Zn	F, W, T, P	39–105	269–724	10–63	69–434	55–3
C24000 (low brass, 80%)	80.0 Cu, 20.0 Zn	F, W	42–125	290–862	12–65	83–448	55–3
C26000 (cartridge brass, 70%)	70.0 Cu, 30.0 Zn	F, R, W, T	44–130	303–896	11–65	76–448	66–3
C26800, C27000 (yellow brass)	65.0 Cu, 35.0 Zn	F, R, W	46–128	317–883	14–62	97–427	65–3
C28000 (Muntz metal)	60.0 Cu, 40.0 Zn	F, R, T	54–74	372–510	21–55	145–379	52–10
C31400 (leaded commercial bronze)	89.0 Cu, 1.8 Pb, 9.2 Zn	F, R	37–60	255–414	12–55	83–379	45–10
C33500 (low-leaded brass)	65.0 Cu, 0.5 Pb, 34.5 Zn	F	46–74	317–510	14–60	97–414	65–8
C34000 (medium-leaded brass)	65.0 Cu, 1.0 Pb, 34.0 Zn	F, R, W, S	47–88	324–607	15–60	103–414	60–7
C34200 (high-leaded brass)	64.5 Cu, 2.0 Pb, 33.5 Zn	F, R	49–85	338–586	17–62	117–427	52–5
C35000 (medium-leaded brass)	62.5 Cu, 1.1 Pb, 36.4 Zn	F, R	45–95	310–655	13–70	90–483	66–1
C35300 (high-leaded brass)	62.0 Cu, 1.8 Pb, 36.2 Zn	F, R	49–85	338–586	17–62	117–427	52–5
C35600 (extra-high-leaded brass)	63.0 Cu, 2.5 Pb, 34.5 Zn	F	49–74	338–510	17–60	117–414	50–7
C36000 (free-cutting brass)	61.5 Cu, 3.0 Pb, 35.5 Zn	F, R, S	49–68	338–469	18–45	124–310	53–18
C36500 to C36800 (leaded Muntz metal)[c]	60.0 Cu[e], 0.6 Pb, 39.4 Zn	F	54	372	20	138	45
C37000 (free-cutting Muntz metal)	60.0 Cu, 1.0 Pb, 39.0 Zn	T	54–80	372–552	20–60	138–414	40–6
C37700 (forging brass)[d]	59.0 Cu, 2.0 Pb, 39.0 Zn	R, S	52	359	20	138	45
C38500 (architectural bronze)[d]	57.0 Cu, 3.0 Pb, 40.0 Zn	R, S	60	414	20	138	30
C40500	95.0 Cu, 1.0 Sn, 4.0 Zn	F	39–78	269–538	12–70	83–483	49–3
C41300	90.0 Cu, 1.0 Sn, 9.0 Zn	F, R, W	41–105	283–724	12–82	83–565	45–2
C43500	81.0 Cu, 0.9 Sn, 18.1 Zn	F, T	46–80	317–552	16–68	110–469	46–7
C44300, C44400, C44500 (inhibited admiralty)	71.0 Cu, 28.0 Zn, 1.0 Sn	F, W, T	48–55	331–379	18–22	124–152	65–60
C46400 to C46700 (naval brass)	60.0 Cu, 39.3 Zn, 0.7 Sn	F, R, T, S	55–88	379–607	25–66	172–455	50–17
C48200 (naval brass, medium-leaded)	60.5 Cu, 0.7 Pb, 0.8 Sn, 38.0 Zn	F, R, S	56–75	386–517	25–53	172–365	43–15
C48500 (leaded naval brass)	60.0 Cu, 1.8 Pb, 37.5 Zn, 0.7 Sn	F, R, S	55–77	379–531	25–53	172–365	40–15
C51000 (phosphor bronze, 5% A)	95.0 Cu, 5.0 Sn, trace P	F, R, W, T	47–140	324–965	19–80	131–552	64–2
C51100	95.6 Cu, 4.2 Sn, 0.2 P	F	46–103	317–710	50–80	345–552	48–2
C52100 (phosphor bronze, 8% C)	92.0 Cu, 8.0 Sn, trace P	F, R, W	55–140	379–965	24–80	165–552	70–2
C52400 (phosphor bronze, 10% D)	99.0 Cu, 10.0 Sn, trace P	F, R, W	66–147	455–1,014	28	193	70–3
C54400 (free-cutting phosphor bronze)	88.0 Cu, 4.0 Pb, 4.0 Zn, 4.0 Sn	F, R	44–75	303–517	19–63	131–434	50–16
C60800 (aluminum bronze, 5%)	95.0 Cu, 5.0 Al	T	60	414	27	186	55
C61000	92.0 Cu, 8.0 Al	R, W	70–80	483–552	30–55	207–379	65–25
C61300	92.7 Cu, 0.3 Sn, 7.0 Al	F, R, T, P, S	70–85	483–586	30–58	207–400	42–35
C61400 (aluminum bronze, D)	91.0 Cu, 7.0 Al, 2.0 Fe	F, R, W, T, P, S	76–89	524–614	33–60	228–414	45–32
C63000	82.0 Cu, 3.0 Fe, 10.0 Al, 5.0 Ni	F, R	90–118	621–814	50–75	345–517	20–15
C63200	82.0 Cu, 4.0 Fe, 9.0 Al, 5.0 Ni	F, R	90–105	621–724	45–53	310–365	25–20
C64200	91.2 Cu, 7.0 Al	F, R	75–102	517–703	35–68	241–469	32–22
C65100 (low-silicon bronze, B)	98.5 Cu, 1.5 Si	R, W, T	40–95	276–655	15–69	103–476	55–11
C65500 (high-silicon bronze, A)	97.0 Cu, 3.0 Si	F, R, W, T	56–145	386–1,000	21–70	145–483	63–3
C67500 (manganese bronze, A)	58.5 Cu, 1.4 Fe, 39.0 Zn, 1.0 Sn, 0.1 Mn	R, S	65–84	448–579	30–60	207–414	33–19
C68700 (aluminum brass, arsenical)	77.5 Cu, 20.5 Zn, 2.0 Al, 0.1 As	T	60	414	27	186	55
C70600 (copper nickel, 10%)	88.7 Cu, 1.3 Fe, 10.0 Ni	F, T	44–60	303–414	16–57	110–393	42–10
C71500 (copper nickel, 30%)	70.0 Cu, 30.0 Ni	F, R, T	54–75	372–517	20–70	138–483	45–15
C72500	88.2 Cu, 9.5 Ni, 2.3 Sn	F, R, W, T	55–120	379–827	22–108	152–745	35–1
C74500 (nickel silver, 65–10)	65.0 Cu, 25.0 Zn, 10.0 Ni	F, W	49–130	338–896	18–76	124–524	50–1
C75200 (nickel silver, 65–18)	65.0 Cu, 17.0 Zn, 18.0 Ni	F, R, W	56–103	386–710	25–90	172–621	45–3
C77000 (nickel silver, 55–18)	55.0 Cu, 27.0 Zn, 18.0 Ni	F, R, W	60–145	414–1,000	27–90	186–621	40–2
C78200 (leaded nickel silver, 65–8–2)	65.0 Cu, 2.0 Pb, 25.0 Zn, 8.0 Ni	F	53–91	365–627	23–76	159–524	40–3

[a] F, flat products; R, rod; W, wire; T, tube; P, pipe; S, shapes.

[b] Ranges are from softest to hardest commercial forms. The strength of the standard copper alloys depends on the temper (annealed grain size or degree of cold work) and the section thickness of the mill product. Ranges cover standard tempers for each alloy.

[c] Values are for as-hot-rolled material.

[d] Values are for as-extruded material.

[e] Rod, 61.0 Cu min.

SOURCE: Copper Development Association Inc.

Machinability rating, %[c]	Melting point		Density, lb/in³	Specific gravity	Coefficient of thermal expansion × 10⁻⁶ at 77–572°F (25–300°C)	Electrical conductivity (annealed) % IACS	Thermal conductivity Btu·ft/(h·ft²·°F) [cal·cm/(s·cm²·°C)]
	Solidus, °F (°C)	Liquidus, °F (°C)					
20	1,981 (1,083)	1,981 (1,083)	0.323	8.94	9.8 (17.7)	101	226 (0.934)
20	1,949 (1,065)	1,981 (1,083)	0.321–0.323	8.89–8.94	9.8 (17.7)	101	226 (0.934)
20	1,981 (1,083)	1,981 (1,083)	0.323	8.94	9.8 (17.7)	85	196 (0.81)
85	1,924 (1,051)	1,967 (1,075)	0.323	8.94	9.9 (17.8)	93	205 (0.85)
85	1,953 (1,067)	1,969 (1,076)	0.323	8.94	9.8 (17.7)	95	216 (0.89)
20	1,796 (980)	1,976 (1,080)	0.321	8.89	9.8 (17.7)	93	212 (0.876)
20	1,590 (865)	1,800 (980)	0.304	8.41	9.9 (17.8)	22	62–75 (0.26–0.31)
20	1,590 (865)	1,800 (980)	0.298	8.26	9.9 (17.8)	22	62–75 (0.26–0.31)
20	1,958 (1,070)	1,967 (1,075)	0.321	8.89	9.8 (17.6)	80	187 (0.77)
85	1,747 (953)	1,976 (1,080)	0.323	8.94	9.8 (17.6)	96	218 (0.93)
20	1,980 (1,080)	1,990 (1,090)	0.322	8.91	9.8 (17.9)	65	150 (0.625)
20	1,920 (1,050)	1,950 (1,065)	0.320	8.86	10.0 (18.1)	56	135 (0.56)
20	1,870 (1,020)	1,910 (1,045)	0.318	8.80	10.2 (18.4)	44	109 (0.45)
30	1,810 (990)	1,880 (1,025)	0.316	8.75	10.4 (18.7)	37	92 (0.38)
30	1,770 (965)	1,830 (1,000)	0.313	8.67	10.6 (19.1)	32	81 (0.33)
30	1,680 (915)	1,750 (955)	0.308	8.53	11.1 (19.9)	28	70 (0.29)
30	1,660 (905)	1,710 (930)	0.306	8.47	11.3 (20.3)	27	67 (0.28)
40	1,650 (900)	1,660 (905)	0.303	8.39	11.6 (20.8)	28	71 (0.29)
80	1,850 (1,010)	1,900 (1,040)	0.319	8.83	10.2 (18.4)	42	104 (0.43)
60	1,650 (900)	1,700 (925)	0.306	8.47	11.3 (20.3)	26	67 (0.28)
70	1,630 (885)	1,700 (925)	0.306	8.47	11.3 (20.3)	26	67 (0.28)
90	1,630 (885)	1,670 (910)	0.306	8.47	11.3 (20.3)	26	67 (0.28)
70	1,640 (895)	1,680 (915)	0.305	8.44	11.3 (20.3)	26	67 (0.28)
90	1,630 (885)	1,670 (910)	0.306	8.47	11.3 (20.3)	26	67 (0.28)
100	1,630 (885)	1,660 (905)	0.307	8.50	11.4 (20.5)	26	67 (0.28)
100	1,630 (885)	1,650 (900)	0.307	8.50	11.4 (20.5)	26	67 (0.28)
60	1,630 (885)	1,650 (900)	0.304	8.41	11.6 (20.8)	28	71 (0.29)
70	1,630 (885)	1,650 (900)	0.304	8.41	11.6 (20.8)	27	69 (0.28)
80	1,620 (880)	1,640 (895)	0.305	8.44	11.5 (20.7)	27	69 (0.28)
90	1,610 (875)	1,630 (890)	0.306	8.47	11.6 (20.8)	28	71 (0.29)
20	1,875 (1,025)	1,940 (1,060)	0.319	8.83	— (—)	41	95 (0.43)
20	1,850 (1,010)	1,900 (1,038)	0.318	8.80	10.3 (18.6)	30	77 (0.32)
30	1,770 (965)	1,840 (1,005)	0.313	8.66	10.8 (19.4)	28	— (—)
30	1,650 (900)	1,720 (935)	0.308	8.53	11.2 (20.2)	25	64 (0.26)
30	1,630 (885)	1,650 (900)	0.304	8.41	11.8 (21.2)	26	67 (0.28)
50	1,630 (885)	1,650 (900)	0.305	8.44	11.8 (21.2)	26	67 (0.28)
70	1,630 (885)	1,650 (900)	0.305	8.44	11.8 (21.2)	26	67 (0.28)
20	1,750 (950)	1,920 (1,050)	0.320	8.86	9.9 (17.8)	15	40 (0.17)
20	1,785 (975)	1,945 (1,060)	0.320	8.86	9.9 (17.8)	20	48.4 (0.20)
20	1,620 (880)	1,880 (1,020)	0.318	8.80	10.1 (18.2)	13	36 (0.15)
20	1,550 (845)	1,830 (1,000)	0.317	8.78	10.2 (18.4)	11	29 (0.12)
80	1,700 (930)	1,830 (1,000)	0.321	8.89	9.6 (17.3)	19	50 (0.21)
20	1,920 (1,050)	1,945 (1,063)	0.295	8.17	10.0 (18.1)	17	46 (0.19)
20	— —	1,905 (1,040)	0.281	7.78	9.9 (17.9)	15	40 (0.17)
30	1,905 (1,040)	1,915 (1,045)	0.287	7.95	9.0 (16.2)	12	32 (0.13)
20	1,905 (1,040)	1,915 (1,045)	0.285	7.89	9.0 (16.2)	14	39 (0.16)
30	1,895 (1,035)	1,930 (1,054)	0.274	7.58	9.0 (16.2)	7	22.6 (0.09)
30	1,905 (1,040)	1,940 (1,060)	0.276	7.64	9.0 (16.2)	7	20 (0.086)
60	1,800 (985)	1,840 (1,005)	0.278	7.69	10.0 (18.1)	8	26 (0.108)
30	1,890 (1,030)	1,940 (1,060)	0.316	8.75	9.9 (17.9)	12	33 (0.14)
30	1,780 (970)	1,880 (1,025)	0.308	8.53	10.0 (18.0)	7	21 (0.09)
30	1,590 (865)	1,630 (890)	0.302	8.36	11.8 (21.2)	24	61 (0.26)
30	1,740 (950)	1,765 (965)	0.296	8.33	10.3 (18.5)	23	58 (0.24)
20	2,010 (1,100)	2,100 (1,150)	0.323	8.94	9.5 (17.1)	9.0	26 (0.11)
20	2,140 (1,170)	2,260 (1,240)	0.323	8.94	9.0 (16.2)	4.6	17 (0.07)
20	1,940 (1,060)	2,065 (1,130)	0.321	8.89	9.2 (16.5)	11	31 (0.13)
20	— (—)	1,870 (1,020)	0.314	8.69	9.1 (16.4)	9.0	26 (0.11)
20	1,960 (1,070)	2,030 (1,110)	0.316	8.73	9.0 (16.2)	6.0	19 (0.08)
30	— (—)	1,930 (1,055)	0.314	8.70	9.3 (16.7)	5.5	17 (0.07)
60	1,780 (970)	1,830 (1,000)	0.314	8.69	10.3 (18.5)	10.9	28 (0.11)

Cast Copper-Base Alloys

Casting makes it possible to produce parts with shapes that cannot be achieved easily by fabrication methods such as forming or machining. Often it is more economical to produce a part as a casting than to fabricate it by other means.

Copper alloy castings serve in applications that require superior corrosion resistance, high thermal or electrical conductivity, good bearing surface qualities, or other special properties.

All copper alloys can be successfully **sand-cast,** for this process allows the greatest flexibility in casting size and shape and is the most economical and widely used casting method, especially for limited production quantities. Permanent mold casting is best suited for tin, silicon, aluminum, and manganese bronzes, as well as yellow brasses. Most copper alloys can be cast by the centrifugal process.

Brass die castings are made when great dimensional accuracy and/or a better surface finish is desired. While inferior in properties to hot-pressed parts, die castings are adaptable to a wider range of designs, for they can be made with intricate coring and with considerable variation in section thickness.

The estimated market distribution of the dominant copper alloys by end-use application is shown in Table 6.4.18. Chemical compositions and selective mechanical and physical properties of the dominant alloys used to produce sand castings are listed in Table 6.4.19.

Test results obtained on **standard test bars** (either attached to the casting or separately poured) indicate the quality of the metal used but not the specific properties of the casting itself because of variations of thickness, soundness, and other factors. The ideal casting is one with a fairly uniform metal section with ample fillets and a gradual transition from thin to thick parts.

Properties and Processing of Copper Alloys

Mechanical Properties **Cold working** copper and copper alloys increases both tensile and yield strengths, with the more pronounced increase imparted to yield strength. For most alloys, tensile strength of the hardest cold-worked temper is approximately twice the tensile strength of the annealed temper, whereas the yield strength of the hardest cold-worked temper can be up to 5 to 6 times that of the annealed temper. While hardness is a measure of temper, it is not an accurate one, because the determination of hardness is dependent upon the alloy, its strength level, and the method used to test for hardness.

All the brasses may be **hot-worked** irrespective of their lead content. Even for alloys having less than 60 percent copper and containing the beta phase in the microstructure, the process permits more extensive changes in shape than cold working because of the plastic (ductile) nature of the beta phase at elevated temperatures, even in the presence of lead. Hence a single hot-working operation can often replace a sequence of forming and annealing operations. Alloys for **extrusion, forging,** or **hot pressing** contain the beta phase in varying amounts.

Extruded sections of many copper alloys are made in a wide variety of shapes. In addition to architectural applications of extrusions, extrusion is an important production process since many objects such as hinges, pinions, brackets, and lock barrels can be extruded directly from bars.

While the copper-zinc alloys (brasses) may contain up to 40 percent zinc, those with 30 to 35 percent zinc find the greatest application for they exhibit high ductility and can be readily cold-worked. With decreasing zinc content, the mechanical properties and corrosion resistance of the alloys approach those of copper. The properties of these alloys are listed in Table 6.4.20.

Heat Treating Figures 6.4.3 and 6.4.4 show the progressive effects of cold rolling and annealing of alloy C26000 flat products. **Cold rolling** clearly increases the hardness and the tensile and yield strengths while concurrently decreasing the ductility.

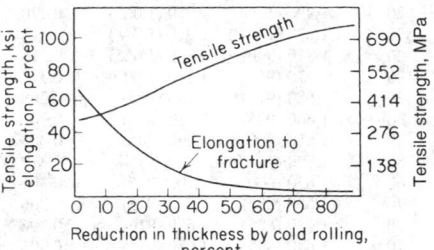

Fig. 6.4.3 Effect of cold working on annealed brass (alloy C26000).

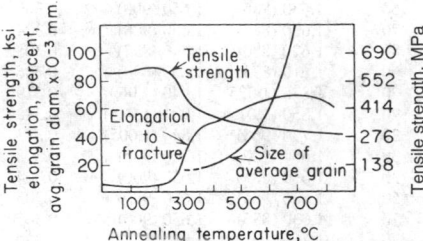

Fig. 6.4.4 Effect of annealing on cold-rolled brass (alloy C26000).

Annealing below a certain minimum temperature has practically no effect, but when the temperature is in the recrystallization range, a rapid decrease in strength and an increase in ductility occur. With a proper anneal, the effects of cold working are almost entirely removed. Heating beyond this point results in grain growth and comparatively little further increase in ductility. Figures 6.4.5 and 6.4.6 show the variation of properties of various brasses after annealing at the temperatures indicated.

Table 6.4.18 Estimated Casting Shipment Distribution by Alloy for End-Use Application

End use	Dominant copper alloys—Estimated
Plumbing and heating	C84400 (75%), C83800/C84800 (25%)
Industrial valves and fittings	C83600 (45%), C83800/C84400 (15%)
	C90300 (5%), C92200 (15%), Misc. (20%)
Bearings and bushings	C93200 (80%), C93700 (20%)
Water meters, hydrants, and water system components	C83600 (25%), C84400 (75%)
Electrical connectors and components	C83300 (50%), C83600 (20%), C94600 (15%)
Pumps, impellers, and related components	C92200 (20%), C95400/C95500 (50%), Misc. (30%)
Outdoor sprinkler systems	C84400 (100%)
Heavy equipment components	C86300 (80%), C95500 (20%)
Hardware, plaques, and giftware	C83600 (15%), C85200/C85400/C85700 (50%), C92200 (35%)
Marine	C86500 (30%), C95800 (50%), C96200 (5%), C96400 (15%)
Locksets	C85200 (90%), Misc. (10%)
Railroad journal bearings and related parts	C93200 (90%), Misc. (10%)
Other	C87300 (35%), C87610 (65%)

Table 6.4.19 Composition and Properties of Selected Cast Copper Alloys

Alloy Number (and name)	Nominal composition, %	Mechanical properties					Melting point		Density, lb/in³	Specific gravity	Coefficient of thermal expansion × 10⁻⁶ at 77–572°F (25–300°C)	Electrical conductivity (annealed), % IACS	Thermal conductivity, Btu·ft/(h·ft²·°F) [cal·cm/(s·cm²·°C)]
		Tensile strength		Yield strength		Elongation in 50 mm (2 in), %[b]	Solidus °F (°C)	Liquidus °F (°C)					
		ksi	MPa	ksi	MPa								
C83600	85 Cu, 5 Sn, 5 Pb, 5 Zn	37	255	17	117	30	1,570 (855)	1,850 (1,010)	0.318	8.83	10.0 (18.0)	15	41.6 (0.172)
C83800	83 Cu, 4 Sn, 6 Pb, 7 Zn	35	240	16	110	25	1,550 (845)	1,840 (1,005)	0.312	8.64	10.0 (18.0)	15	41.9 (0.173)
C84400	81 Cu, 3 Sn, 7 Pb, 9 Zn	34	235	15	105	26	1,540 (840)	1,840 (1,005)	0.314	8.70	10.0 (18.0)	16.4	41.9 (0.173)
C84800	76 Cu, 3 Sn, 6 Pb, 15 Zn	37	255	14	97	35	1,530 (832)	1,750 (954)	0.310	8.58	10.4 (18.7)	16.4	41.6 (0.172)
C85200	72 Cu, 1 Sn, 3 Pb, 24 Zn	38	260	13	90	35	1,700 (925)	1,725 (940)	0.307	8.50	11.5 (21)	18.6	48.5 (0.20)
C85400	67 Cu, 1 Sn, 3 Pb, 29 Zn	34	235	12	83	35	1,700 (925)	1,725 (940)	0.305	8.45	11.2 (20.2)	19.6	51 (0.21)
C86300	62 Cu, 26 Zn, 3 Fe, 6 Al, 3 Mn	119	820	67	460	18	1,625 (885)	1,693 (923)	0.278	7.69	12 (22)	9	21 (0.085)
C87300	95 Cu, 1 Mn, 4 Si	55	380	25	170	30	1,680 (916)	1,510 (821)	0.302	8.36	10.9 (19.6)	6.7	28 (0.113)
C87610	92 Cu, 4 Zn, 4 Si	55	380	25	170	30	— (—)	— (—)	0.302	8.36	— (—)	6.0	— (—)
C90300	88 Cu, 8 Sn, 4 Zn	45	310	21	145	30	1,570 (854)	1,830 (1,000)	0.318	8.80	10.0 (18.0)	12	43 (0.18)
C92200	88 Cu, 6 Sn, 1.5 Pb, 3.5 Zn	40	275	20	140	30	1,520 (825)	1,810 (990)	0.312	8.64	10.0 (18.0)	14.3	40 (0.17)
C93200	83 Cu, 7 Sn, 7 Pb, 3 Zn	35	240	18	125	20	1,570 (855)	1,790 (975)	0.322	8.93	10.0 (18.0)	12	34 (0.14)
C93700	80 Cu, 10 Sn, 10 Sn	35	240	18	125	20	1,403 (762)	1,705 (930)	0.323	8.95	10.3 (18.5)	10	27 (0.11)
C95400	88.5 Cu, 4 Fe, 10.5 Al	75	515	30	205	12	1,880 (1,025)	1,900 (1,040)	0.269	7.45	9.0 (16.2)	13	34 (0.14)
C95400 (TQ50)	88.5 Cu, 4 Fe, 10.5 Al	90	620	45	310	6	1,880 (1,025)	1,900 (1,040)	0.269	7.45	9.0 (16.2)	13	34 (0.14)
C95500	81 Cu, 4 Fe, 11 Al, 4 Ni	90	620	40	275	6	1,930 (1,055)	1,900 (1,040)	0.272	7.53	9.0 (16.2)	8.5	24 (0.10)
C95500 (TQ50)	81 Cu, 4 Fe, 11 Al, 4 Ni	110	760	60	415	5	1,930 (1,055)	1,900 (1,040)	0.272	7.53	9.0 (16.2)	8.5	24 (0.10)
C95800	81.5 Cu, 4 Fe, 9 Al, 4 Ni, 1.5 Mn	85	585	35	240	15	1,910 (1,045)	1,940 (1,060)	0.276	7.64	9.0 (16.2)	7.1	21 (0.085)
C96200	88.5 Cu, 1.5 Fe, 10 Ni	45	310	25	172	20	2,010 (1,100)	2,100 (1,150)	0.323	8.94	9.5 (17.3)	11	26 (0.11)
C96400	69 Cu, 1 Fe, 30 Ni	68	470	27	255	28	2,140 (1,170)	2,260 (1,240)	0.323	8.94	9.0 (16.2)	5	17 (0.068)
C97600	64 Cu, 4 Pb, 20 Ni, 4 Sn, 8 Zn	45	310	24	165	20	2,027 (1,108)	2,089 (1,143)	0.321	8.90	9.3 (17)	5	13 (0.075)

Table 6.4.20   Mechanical Properties of Rolled Yellow Brass (Alloy C26800)

Standard temper designation (ASTM B 601)*	Nominal grain size, mm	Tensile strength		Approximate Rockwell hardness†	
		ksi	MPa	F	30T
OS120	0.120	—	—	50–62	21 max
OS070	0.070	—	—	52–67	3–27
OS050	0.050	—	—	61–73	20–35
OS035	0.035	—	—	65–76	25–38
OS025	0.025	—	—	67–79	27–42
OS015	0.015	—	—	72–85	33–50

Standard temper designation (ASTM B 601)*	Former temper designation	Tensile strength		Approximate Rockwell hardness‡			
				B scale		Superficial 30-T	
		ksi	MPa	0.020 (0.058) to 0.036 in (0.914 mm) incl	Over 0.036 in (0.914 mm)	0.012 (0.305) to 0.028 in (0.711 mm) incl	Over 0.028 in (0.711 mm)
M20	As hot-rolled	40–50	275–345	—	—	—	—
H01	Quarter-hard	49–59	340–405	40–61	44–65	43–57	46–60
H02	Half-hard	55–65	380–450	57–71	60–74	54–64	56–66
H03	Three-quarters hard	62–72	425–495	70–77	73–80	65–69	67–71
H04	Hard	68–78	470–540	76–82	78–84	68–72	69–73
H06	Extra hard	79–89	545–615	83–87	85–89	73–75	74–76
H08	Spring	86–95	595–655	87–90	89–92	75–77	76–78
H10	Extra spring	90–99	620–685	88–91	90–93	76–78	77–79

* Refer to Table 6.4.15 for definition of temper designations.
† Rockwell hardness values apply as follows: The F scale applies to metal 0.020 in (0.508 mm) in thickness and over; the 30-T scale applies to metal 0.015 in (0.381 mm) in thickness and over.
‡ Rockwell hardness values apply as follows: The B scale values apply to metal 0.020 in (0.508 mm) and over in thickness, and the 30-T scale values apply to metal 0.012 in (0.305 mm) and over in thickness.
SOURCE: ASTM, abstracted with permission.

Work-hardened metals are restored to a soft state by annealing. Single-phase alloys are transferred into unstressed crystals by **recovery, recrystallization,** and **grain growth**. In severely deformed metal, recrystallization occurs at lower temperatures than in lightly deformed metal. Also, the grains are smaller and more uniform in size when severely deformed metal is recrystallized.

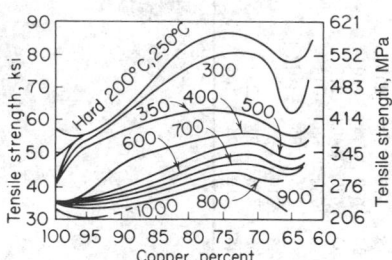

Fig. 6.4.5   Tensile strengths of copper-zinc alloys.

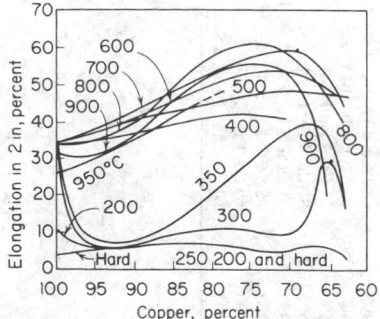

Fig. 6.4.6   Percent of elongation in 2 in of copper-zinc alloys.

Grain size can be controlled by proper selection of cold-working and annealing practices. Large amounts of prior cold work, rapid heating to annealing temperature, and short annealing times foster formation of fine grains. Larger grains are normally produced by a combination of limited deformation and long annealing times.

Practically all wrought copper alloys are used in the cold-worked condition to gain additional strength. Articles are often made from annealed stock and depend on the cold work of the forming operation to shape and harden them. When the cold work involved is too small to do this, brass rolled to a degree of temper consistent with final requirements should be used.

Brass for springs should be rolled as hard as is consistent with the subsequent forming operations. For articles requiring sharp bends, or for deep-drawing operations, annealed brass must be used. In general, smaller grain sizes are preferred.

Table 6.4.20 summarizes the ASTM B36 specification requirements for alloy C26800 (65 : 35 copper-zinc) for various rolled and annealed tempers.

Effect of Temperature   Copper and all its alloys increase in strength slightly and uniformly as temperature decreases from room temperature. No low-temperature brittleness is encountered. Copper is useless for prolonged-stress service much above 400°F (204°C), but some of its alloys may be used up to 550°F (287°C). For restricted service above this temperature, only copper-nickel and copper-aluminum alloys have satisfactory properties.

For specific data, see Elevated-Temperature Properties of Copper Base Alloys, ASTM STP 181; Low-Temperature Properties of Copper and Selected Copper Alloys, *NBS Mono.* 101, 1967; and Wilkens and Bunn, "Copper and Copper Base Alloys," McGraw-Hill, 1943.

Electrical and Thermal Conductivities   The brasses (essentially copper-zinc alloys) have relatively good **electrical and thermal conductivities,** although the levels are somewhat lower than those of the coppers because of the effect of solute atoms on the copper lattice. For this reason, copper and high-copper alloys are preferred over the brasses containing more than a few percent total alloy content when high electrical or thermal conductivity is required for the application.

Machinability Machinability of copper alloys is governed by their metallurgical structure and related properties. In that regard, they are divided into three subgroups. All leaded brasses are yellow, have moderate electrical conductivity and fair hot-working qualities, but are relatively poor with respect to fabrication when compared with C26000.

While they all have excellent to satisfactory machinability ratings, the most important alloy for machined products is C36000, **free-cutting brass**, which is the standard material for automatic screw machine work where the very highest machinability is necessary.

ASTM B16 Free-Cutting Brass Rod, Bar, and Shapes for Use in Screw Machines is the dominant specification. Use of this material will often result in considerable savings over the use of steel. Most copper alloys are readily machined by usual methods using standard tools designed for steel, but machining is done at higher cutting speeds. Consideration of the wide range of characteristics presented by various types of copper alloys and the adaptation of machining practice to the particular material concerned will improve the end results to a marked degree.

Group A is composed of alloys of homogeneous structure: copper, wrought bronzes up to 10 percent tin, brasses and nickel silvers up to 37 percent zinc, aluminum bronzes up to 8 percent aluminum, silicon bronzes, and copper nickel. These alloys are all tough and ductile and form long, continuous chips. When they are severely cold-worked, they approach the group B classification in their characteristics.

Group B includes lead-free alloys of duplex structure, some cast bronzes, and most of the high-strength copper alloys. They form continuous but brittle chips by a process of intermittent shearing against the tool edge. Chatter will result unless work and tool are rigid.

Many of the basic **Group C** brasses and bronzes are rendered particularly adaptable for machining operations by the addition of 0.5 to 3.0 percent lead, which resides in the structure as minute, uniformly distributed droplets. They serve to break the chip and lubricate the tool cutting edge. Chips are fine, almost needlelike, and are readily removed. Very little heat is evolved, but the tendency to chatter is greater than for Group A alloys. Lead additions may be made to most copper alloys, but its low melting point makes hot working impossible. Tellurium and sulfur additions are used in place of lead when the combination of hot workability and good machinability is desired.

For **turning operations**, the tough alloys of Group A need a sharp top rake angle (20° to 30° for copper and copper nickel; 12° to 16° for the brasses, bronzes, and silicon bronzes with high-speed tools; 8° to 12° with carbide tools, except for copper, for which 16° is recommended). Type C (leaded) materials require a much smaller rake angle to minimize chatter, a maximum of 8° with high-speed steels and 3° to 6° with carbide. Type B materials are intermediate, working best with 6° to 12° rake angle with high-speed tools and 3° to 8° with carbide, the higher angle being used for the tougher materials. Side clearance angle should be 5° to 7° except for tough, "sticky" materials like copper and copper nickel, where a side rake of 10° to 15° is better.

Many copper alloys **drill** satisfactorily with standard helix angle drills. Straight fluted tools (helix angle 0°) are preferable for the Group C leaded alloys. A helix angle of 10° is preferred for Group B and 40° for copper and copper nickel. Feed rates generally are 2 to 3 times faster than those used for steel. With Type B alloy, a fairly coarse feed helps to break the chip, and with Type A, a fine feed and high speed give best results, provided that sufficient feed is used to prevent rubbing and work hardening.

Corrosion Resistance For over half a century, copper has been the preferred material for tubing used, to convey potable water in domestic plumbing systems because of its outstanding corrosion resistance. In outdoor exposure, however, copper alloys containing less than 20 percent zinc are preferred because of their resistance to stress corrosion cracking, a form of corrosion originally called **season cracking**. For this type of corrosion to occur, a combination of tensile stress (either residual or applied) and the presence of a specific chemical reagent such as ammonia, mercury, mercury compounds, and cyanides is necessary. Stress relief annealing after forming alleviates the tendency for stress corrosion cracking. Alloys that are either zinc-free or contain less than 15 percent zinc generally are not susceptible to stress corrosion cracking.

Arsenical copper (C14200), arsenical admiralty metal (C44300), and arsenical aluminum brass (C68700) were once popular **condenser tube alloys** for freshwater power plant applications. They have been successfully replaced largely with alloy 706 (90 : 10 copper-nickel). In seawater or brackish water, copper, admiralty metal, and aluminum brass are even less suitable because of their inability to form protective films.

Caution must be exercised, for copper and its alloys are not suitable for use in oxidizing acidic solutions or in the presence of moist ammonia and its compounds. Specific alloys are also limited to maximum flow velocities.

Fabrication Practically any of the copper alloys listed in Table 6.4.17 can be obtained in sheet, rod, and wire form, and many can be obtained as tubes. Most, in the annealed condition, will withstand extensive amounts of cold work and may be shaped to the desired form by deep drawing, flanging, forming, bending, and similar operations. If extensive cold work is planned, the material should be purchased in the annealed condition, and it may need intermediate annealing either to avoid metal failure or to minimize power consumption. Annealing is done at 900 to 1,300°F (482 to 704°C), depending on the alloy, and is usually followed by air cooling. Because of the ready workability of brass, it is often less costly to use than steel. Brass may be drawn at higher speeds than ferrous metals and with less wear on the tools. In cupping operations, a take-in of 45 percent is usual and on some jobs, it may be larger. Brass hardened by cold working is softened by annealing at about 1,100 F (593°C).

Joining by Welding, Soldering, and Brazing

Welding Deoxidized copper will **weld** satisfactorily by the oxyacetylene method. Sufficient heat input to overcome its high thermal conductivity must be maintained by the use of torches considerably more powerful than those customary for steel, and preferably by additional preheating. The filler rod must be deoxidized. Gas-shielded arc welding is preferred. Tough-pitch copper will not result in high-strength welds because of embrittlement due to the oxygen content. Copper may be arc-welded, using shielded metal arc, gas metal arc (MIG), or gas tungsten arc (TIG) welding procedures using experienced operators. Filler rods of phosphor bronze or silicon bronze will give strong welds more consistently and are used where the presence of a weld of different composition and corrosion-resistance characteristics is not harmful. Brass may be welded by the oxyacetylene process but not by arc welding. A filler rod of about the same composition is used, although silicon is frequently added to prevent zinc fumes.

Copper-silicon alloys respond remarkably well to welding by all methods. The conductivity is not too high, and the alloy is, to a large extent, self-fluxing.

Applicable specifications for joining copper and copper alloys by welding include

ANSI/AWS A5.6   Covered Copper and Copper Alloy Arc Welding Electrodes
ANSI/AWS A5.7   Copper and Copper Alloy Bare Welding Rods and Electrodes
ANSI/AWS A5.27   Copper and Copper Alloy Rods for Oxyfuel Gas Welding

The **fluxes** used for brazing copper joints are water-based and function by dissolving and removing residual oxides from the metal surface; they protect the metal from reoxidation during heating and promote wetting of the joined surfaces by the brazing filler metal.

Soldering Soldered joints with **capillary fittings** are used in plumbing for water and sanitary drainage lines. Such joints depend on capillary action drawing free-flowing molten solder into the gap between the fitting and the tube.

Flux acts as a cleaning and wetting agent which permits uniform spreading of the molten solder over the surfaces to be joined.

Selection of a solder depends primarily on the operating pressure and temperature of the system. Lead-free solders required for joining copper tube and fittings in potable water systems are covered by ASTM B32 Solder Metal.

As in brazing, the functions of soldering flux are to remove residual oxides, promote wetting, and protect the surfaces being soldered from oxidation during heating.

Fluxes best suited for soldering copper and copper alloy tube should meet the requirements of ASTM B813 Liquid and Paste Fluxes for Soldering Applications of Copper and Copper Alloy Tube with the joining accomplished per ASTM B828 Making Capillary Joints by Soldering of Copper and Copper Alloy Tube and Fittings.

JEWELRY METALS
Staff Contribution

REFERENCES: ASM "Metals Handbook." Publications of the metal-producing companies. Standards applicable to classes of metals available from manufacturers' associations. Trade literature pertinent to each specific metal.

Gold is used primarily as a monetary standard; the small amount put to metallurgical use is for jewelry or decorative purposes, in dental work, for fountain-pen nibs, and as an electrodeposited protecting coating. Electroplated gold is widely used for electronic junction points (transistors and the like) and at the mating ends of telephone junction wires. An alloy with palladium has been used as a platinum substitute for laboratory vessels, but its present price is so high that it does not compete with platinum.

Silver has the highest electrical conductivity of any metal, and it found some use in bus bars during World War II, when copper was in short supply. Since its density is higher than that of copper and since government silver frequently has deleterious impurities, it offers no advantage over copper as an electrical conductor. Heavy-duty electrical contacts are usually made of silver. It is used in aircraft bearings and solders. Its largest commercial use is in tableware as sterling silver, which contains 92.5 percent silver (the remainder is usually copper). United States coinage used to contain 90 percent silver, 10 percent copper, but coinage is now manufactured from a debased alloy consisting of copper sandwiched between thin sheets of nickel.

Platinum has many uses because of its high melting point, chemical inertness, and catalytic activity. It is the standard catalyst for the oxidation of sulfur dioxide in the manufacture of sulfuric acid. Because it is inert toward most chemicals, even at elevated temperatures, it can be used for laboratory apparatus. It is the only metal that can be used for an electric heating element about 2,300°F without a protective atmosphere. Thermocouples of platinum with platinum-rhodium alloy are standard for high temperatures. Platinum and platinum alloys are used in large amounts in feeding mechanisms of glass-working equipment to ensure constancy of the orifice dimensions that fix the size of glass products. They are also used for electrical contacts, in dental work, in aircraft spark-plug electrodes, and as jewelry. It has been used for a long time as a catalyst in refining of gasoline and is in widespread use in automotive catalytic converters as a pollution control device.

Palladium follows platinum in importance and abundance among the platinum metals and resembles platinum in most of its properties. Its density and melting point are the lowest of the platinum metals, and it forms an oxide coating at a dull-red heat so that it cannot be heated in air above 800°F (426°C) approx. In the finely divided form, it is an excellent hydrogenation catalyst. It is as ductile as gold and is beaten into leaf as thin as gold leaf. Its hardened alloys find some use in dentistry, jewelry, and electrical contacts.

Iridium is one of the platinum metals. Its chief use are as a hardener for platinum jewelry alloys and as platinum contacts. Its alloys with osmium are used for tipping fountain-pen nibs. Isotope Ir 192 is one of the basic materials used in radiation therapy and is widely employed as a radioactive implant in oncological surgical procedures. It is the most corrosion-resistant element known.

Rhodium is used mainly as an alloying addition to platinum. It is a component of many of the pen-tipping alloys. Because of its high reflectivity and freedom from oxidation films, it is frequently used as an electroplate for jewelry and for reflectors for motion-picture projectors, aircraft searchlights, and the like.

LOW-MELTING-POINT METALS AND ALLOYS
by Frank E. Goodwin

REFERENCES: Current edition of ASM "Metals Handbook." Current listing of applicable ASTM and AWS standards. Publications of the various metal-producing companies.

Metals with low-melting temperatures offer a diversity of industrial applications. In this field, much use is made of the **eutectic**-type alloy, in which two or more elements are combined in proper proportion so as to have a minimum melting temperature. Such alloys melt at a single, fixed temperature, as does a pure metal, rather than over a range of temperatures, as with most alloys.

Liquid Metals A few metals are used in their liquid state. **Mercury** [mp, $-39.37°F$ $(-40°C)$] is the only metal that is liquid below room temperature. In addition to its use in thermometers, scientific instruments, and electrical contacts, it is a constituent of some very low-melting alloys. Its application in dental amalgams is unique and familiar to all. It has been used as a heat-exchange fluid, as have **sodium** and the sodium-potassium alloy **NaK** (see Sec. 9.8).

Moving up the temperature scale, **tin** melts at 449.4°F (232°C) and finds its largest single use in coating steel to make tinplate. Tin may be applied to steel, copper, or cast iron by hot-dipping, although for steel this method has been replaced largely by electrodeposition onto continuous strips of rolled steel. Typical tin coating thicknesses are 40 μin (1 μm). **Solders** constitute an important use of tin. Sn-Pb solders are widely used, although Sn-Sb and Sn-Ag solders can be used in applications requiring lead-free compositions. Tin is alloyed in bronzes and battery grid material. Alloys of 12 to 25 percent Sn, with the balance lead, are applied to steel by hot-dipping and are known as **terneplate**.

Modern **pewter** is a tarnish-resistant alloy used only for ornamental ware and is composed of 91 to 93 percent Sn, 1 to 8 percent Sb, and 0.25 to 3 percent Cu. Alloys used for casting are lower in copper than those used for spinning hollowware. Pewter does not require intermediate annealing during fabrication. Lead was a traditional constituent of pewter but has been eliminated in modern compositions because of the propensity for lead to leach out into fluids contained in pewter vessels.

Pure **lead** melts at 620°F (327°C), and four grades of purities are recognized as standard; see Table 6.4.21. Corroding lead and common lead are 99.94 percent pure, while chemical lead and copper-bearing lead are 99.9 percent pure. Corroding lead is used primarily in the chemical industry, with most used to manufacture white lead or litharge (lead monoxide, PbO). Chemical lead contains residual copper that improves both its corrosion resistance and stiffness, allowing its extensive employment in chemical plants to withstand corrosion, particularly from sulfuric acid. Copper-bearing lead also has high corrosion resistance because of its copper content. Common lead is used for alloying and for battery oxides. Lead is strengthened by alloying with antimony at levels between 0 and 15 percent. Lead-tin alloys with tin levels over the entire composition range are widely used.

The largest single use of lead is in **lead-acid batteries**. Deep cycling batteries typically use Pb 6 7 antimony, while Pb 0.1 Ca 0.3 tin alloys are used for maintenance-free batteries found in automobiles. Battery alloys usually have a minimum tensile strength of 6,000 lb/in² (41 MPa) and 20 percent elongation. Pb 0.85 percent antimony alloy is also used for sheathing of high-voltage power cables. Other cable sheathing alloys are based on Sn-Cd, Sn-Sb, or Te-Cu combined with lead. The lead content in cable sheathing alloys always exceeds 99 percent.

Lead sheet for construction applications, primarily in the chemical industries, is based on either pure lead or Pb 6 antimony. When cold-rolled, this alloy has a tensile strength of 4,100 lb/in² (28.3 MPa) and an elongation of 47 percent. Lead sheet is usually 3/64 in (1.2 mm) thick and weighs 3 lb/ft² (15 kg/m²). **Architectural uses for lead** are numerous,

Table 6.4.21   Composition Specifications for Lead, According to ASTM B-29-79

Element	Composition, wt %			
	Corroding lead	Common lead	Chemical lead	Copper-bearing lead
Silver, max	0.015	0.005	0.020	0.020
Silver, min	—	—	0.002	—
Copper, max	0.0015	0.0015	0.080	0.080
Copper, min	—	—	0.040	0.040
Silver and copper together, max	0.025	—	—	—
Arsenic, antimony, and tin together, max	0.002	0.002	0.002	0.002
Zinc, max	0.001	0.001	0.001	0.001
Iron, max	0.002	0.002	0.002	0.002
Bismuth, max	0.050	0.050	0.005	0.025
Lead (by difference), min	99.94	99.94	99.90	99.90

SOURCE: ASTM, reprinted with permission.

Table 6.4.22   Compositions and Properties of Type Metal

Service	Composition, %			Liquidus		Solidus		Brinell hardness*
	Sn	Sb	Pb	°F	°C	°F	°C	
Electrotype	4	3	93	561	294	473	245	12.5
Linotype	5	11	84	475	246	462	239	22
Stereotype	6.5	13	80.5	485	252	462	239	22
Monotype	8	17	75	520	271	462	239	27

* 0.39-in (10-mm) ball, 550-lb (250-kg) load
SOURCE: "Metals Handbook," ASM International, 10th ed.

primarily directed to applications where water is to be contained (shower and tiled bathing areas overlaid by ceramic tile). Often, copper is lead-coated to serve as roof flashing or valleys; in those applications, the otherwise offending verdigris coating which would develop on pure copper sheet is inhibited.

Ammunition for both sports and military purposes uses alloys containing percentages up to 8 Sb and 2 As. **Bullets** have somewhat higher alloy content than shot.

Type metals use the hardest and strongest lead alloys. Typical compositions are shown in Table 6.4.22. A small amount of copper may be added to increase hardness. Electrotype metal contains the lowest alloy content because it serves only as a backing to the shell and need not be hard. Linotype, or slug casting metal, must be fluid and capable of rapid solidification for its use in composition of newspaper type; thus metal of nearly eutectic composition is used. It is rarely used as the actual printing surface and therefore need not be as hard as stereotype and monotype materials. Repeated remelting of type metal gradually results in changes in the original composition due to oxidation and loss of original ingredients. Likewise, the alloy is often contaminated by shop dirt and residue from fluxes. In the ordinary course of its use, type metal composition is checked regularly and adjusted as required to restore its original properties. (Note that printing using type metal is termed *hot metal,* in contrast to printing methods which eliminate type metal altogether, or *cold type.* Hot-metal techniques are obsolescent, although they are still employed where the existing machinery exists and is in use.)

Fusible Alloys (See Table 6.4.23.)   These alloys are used typically as fusible links in sprinkler heads, as electric cutouts, as fire-door links, for making castings, for patterns in making match plates, for making electroforming molds, for setting punches in multiple dies, and for dyeing cloth. Some fusible alloys can be cast or sprayed on wood, paper, and other materials without damaging the base materials, and many of these alloys can be used for making hermetic seals. Since some of these alloys melt below the boiling point of water, they can be used in bending tubing. The properly prepared tubing is filled with the molten alloy and allowed to solidify, and after bending, the alloy is melted out by immer-

Table 6.4.23   Compositions, Properties, and Applications of Fusible Alloys

Composition, %					Solidus		Liquidus		Typical application
Sn	Bi	Pb	Cd	In	°F	°C	°F	°C	
8.3	44.7	22.6	5.3	19.10	117	47	117	47	Dental models, part anchoring, lens chucking
13.3	50	26.7	10.0	—	158	70	158	70	Bushings and locators in jigs and fixtures, lens chucking, reentrant tooling, founding cores and patterns, light sheet-metal embossing dies, tube bending
—	55.5	44.5	—	—	255	124	255	124	Inserts in wood, plastics, bolt anchors, founding cores and patterns, embossing dies, press-form blocks, duplicating plaster patterns, tube bending, hobbyist parts
12.4	50.5	27.8	9.3	—	163	73	158	70	Wood's metal-sprinkler heads
14.5	48.0	28.5	(9% Sb)	—	440	227	217	103	Punch and die assemblies, small bearings, anchoring for machinery, tooling, forming blocks, stripper plates in stamping dies
60	40	—	—	—	338	170	281	138	Locator members in tools and fixtures, electroforming cores, dies for lost-wax patterns, plastic casting molds, prosthetic development work, encapsulating avionic components, spray metallizing, pantograph tracer molds

SOURCE: "Metals Handbook," ASM International, 10th ed.

sion of the tube in boiling water. The volume changes during the solidification of a fusible alloy are, to a large extent, governed by the bismuth content of the alloy. As a general rule, alloys containing more than about 55 percent bismuth expand and those containing less than about 48 percent bismuth contract during solidification; those containing 48 to 55 percent bismuth exhibit little change in volume. The change in volume due to cooling of the solid metal is a simple linear shrinkage, but some of the fusible alloys owe much of their industrial importance to other volume changes, caused by change in structure of the solid alloy, which permit the production of castings having dimensions equal to, or greater than, those of the mold in which the metal was cast.

For **fire-sprinkler heads,** with a rating of 160°F (71°C) **Wood's metal** is used for the fusible-solder-alloy link. Wood's metal gives the most suitable degree of sensitivity at this temperature, but in tropical countries and in situations where industrial processes create a hot atmosphere (e.g., baking ovens, foundries), solders having a higher melting point must be used. Alloys of eutectic compositions are used since they melt sharply at a specific temperature.

Fusible alloys are also used as molds for thermoplastics, for the production of artificial jewelry in pastes and plastic materials, in foundry patterns, chucking glass lenses, as hold-down bolts, and inserts in plastics and wood.

Solders are filler metals that produce coalescence of metal parts by melting the solder while heating the base metals below their melting point. To be termed *soldering* (versus brazing), the operating temperature must not exceed 840°F (450°C). The filler metal is distributed between the closely fitting faying surfaces of the joint by capillary action. Compositions of solder alloys are shown in Table 6.4.24. The Pb 63 tin eutectic composition has the lowest melting point—361°F (183°C)—of the binary tin-lead solders. For joints in copper pipe and cables, a wide melting range is needed. Solders containing less than 5 percent Sn are used to seal precoated containers and in applications where the surface temperatures exceed 250°F (120°C). Solders with 10 to 20 percent Sn are used for repairing automobile radiators and bodies, while compositions containing 40 to 50 percent Sn are used for manufacture of automobile radiators, electrical and electronic connections, roofing seams, and heating units. Silver is added to tin-lead solders for electronics applications to reduce the dissolution of silver from the substrate. Sb-containing solders should not be used to join base metals containing zinc, including galvanized iron and brass. A 95.5 percent Sn 4 Cu 0.5 Ag solder has supplanted lead-containing solders in potable water plumbing.

Table 6.4.25 Brazing Filler Metals

AWS-ASTM filler-metal classification	Base metals joined
BAlSi (aluminum-silicon)	Aluminum and aluminum alloys
BCuP (copper-phosphorus)	Copper and copper alloys; limited use on tungsten and molybdenum; should not be used on ferrous or nickel-base metals
BAg (silver)	Ferrous and non-ferrous metals except aluminum and magnesium; iron, nickel, cobalt-base alloys; thin-base metals
BCu (copper)	Ferrous and non-ferrous metals except aluminum and magnesium
RBCuZn (copper-zinc)	Ferrous and non-ferrous metals except aluminum and magnesium; corrosion resistance generally inadequate for joining copper, silicon, bronze, copper, nickel, or stainless steel
BMg (magnesium)	Magnesium-base metals
BNi (nickel)	AISI 300 and 400 stainless steels; nickel- and cobalt-base alloys; also carbon steel, low-alloy steels, and copper where specific properties are desired

Table 6.4.26 Compositions and Melting Ranges of Brazing Alloys

AWS designation	Nominal composition, %	Melting temp, °F
BAg-1	45 Ag, 15 Cu, 16 Zn, 24 Cd	1,125–1,145
BNI-7	13 Cr, 10 P, bal. Ni	1,630
BAu-4	81.5 Au, bal. Ni	1,740
BAlSi-2	7.5 Si, bal. Al	1,070–1,135
BAlSi-5	10 Si, 4 Cu, 10 Zn, bal. Al	960–1,040
BCuP-5	80 Cu, 15 Ag, 5 P	1,190–1,475
BCuP-2	93 Cu, 7 P	1,310–1,460
BAg-1a	50 Ag, 15.5 Cu, 16.5 Zn, 18 Cd	1,160–1,175
BAg-7	56 Ag, 22 Cu, 17 Zn, 5 Sn	1,145–1,205
RBCuZn-A	59 Cu, 40 Zn, 0.6 Sn	1,630–1,650
RBCuZn-D	48 Cu, 41 Zn, 10 Ni, 0.15 Si, 0.25 P	1,690–1,715
BCu-1	99.90 Cu, min	1,980
BCu-1a	99.9 Cu, min	1,980
BCu-2	86.5 Cu, min	1,980

°C = (°F − 32)/1.8.
SOURCE: "Metals Handbook," ASM.

Table 6.4.24 Compositions and Properties of Selected Solder Alloys

Composition, %		Solidus temperature		Liquidus temperature		Uses
Tin	Lead	°C	°F	°C	°F	
2	98	316	601	322	611	Side seams for can manufacturing
5	95	305	581	312	594	Coating and joining metals
10	90	268	514	302	576	Sealing cellular automobile radiators, filling seams or dents
15	85	227	440	288	550	Sealing cellular automobile radiators, filling seams or dents
20	80	183	361	277	531	Coating and joining metals, or filling dents or seams in automobile bodies
25	75	183	361	266	511	Machine and torch soldering
30	70	183	361	255	491	
35	65	183	361	247	477	General-purpose and wiping solder
40	60	183	361	238	460	Wiping solder for joining lead pipes and cable sheaths; also for automobile radiator cores and heating units
45	55	183	361	227	441	Automobile radiator cores and roofing seams
50	50	183	361	216	421	Most popular general-purpose solder
60	40	183	361	190	374	Primarily for electronic soldering applications where low soldering temperatures are required
63	37	183	361	183	361	Lowest-melting (eutectic) solder for electronic applications
40	58 (2% Sb)	185	365	231	448	General-purpose, not recommended on zinc-containing materials
95	0 (5% Sb)	232	450	240	464	Joints on copper, electrical plumbing, heating, not recommended on zinc-containing metals
0	97.5 (2.5% Ag)	304	580	308	580	Copper, brass, not recommended in humid environments
95.5	0 (4% Cu, 0.5% Ag)	226	440	260	500	Potable water plumbing

SOURCE: "Metals Handbook," vol. 2, 10th ed., p. 553.

Brazing is similar to soldering but is defined as using filler metals which melt at or above 840°F (450°C). As in soldering, the base metal is not melted. Typical applications of brazing filler metals classified by AWS and ASTM are listed in Table 6.4.25; their compositions and melting points are given in Table 6.4.26.

METALS AND ALLOYS FOR USE AT ELEVATED TEMPERATURES

by John H. Tundermann

REFERENCES: "High Temperature High Strength Alloys," AISI. Simmons and Krivobok, Compilation of Chemical Compositions and Rupture Strength of Super-strength Alloys, *ASTM Tech. Pub.* 170-A. "Metals Handbook," ASM. Smith, "Properties of Metals at Elevated Temperatures," McGraw-Hill. Clark, "High-Temperature Alloys," Pittman. Cross, Materials for Gas Turbine Engines, *Metal Progress,* March 1965. "Heat Resistant Materials," ASM Handbook, vol. 3, 9th ed., pp. 187–350. Lambert, "Refractory Metals and Alloys," ASM Handbook, vol. 2, 10th ed., pp. 556–585. Watson et al., "Electrical Resistance Alloys," ASM Handbook, vol. 2, 10th ed., pp. 822–839.

Some of data presented here refer to materials with names which are proprietary and registered trademarks. They include Inconel, Incoloy, and Nimonic (Inco Alloys International); Hastelloy (Haynes International); MAR M (Martin Marietta Corp.); and René (General Electric Co.).

Metals are used for an increasing variety of applications at elevated temperatures, "elevated" being a relative term that depends upon the specific metal and the specific service environment. Elevated-temperature properties of the common metals and alloys are cited in the several subsections contained within this section. This subsection deals with metals and alloys whose prime use is in **high-temperature applications.** [Typical operating temperatures are compressors, 750°F (399°C); steam turbines, 1,100°F (593°C); gas turbines, 2,000°F (1,093°C); resistance-heating elements, 2,400°F (1,316°C); electronic vacuum tubes, 3,500°F (1,926°C), and lamps, 4,500°F (2,482°C).] In general, alloys for high-temperature service must have melting points above the operating temperature, low vapor pressures at that temperature, resistance to attack (oxidation, sulfidation, corrosion) in the operating environment, and sufficient strength to withstand the applied load for the service life without deforming beyond permissible limits. At high temperatures, atomic diffusion becomes appreciable, so that time is an important factor with respect to surface chemical reactions, to **creep** (slow deformation under constant load), and to internal changes within the alloy during service. The effects of time and temperature are conveniently combined by the empirical **Larson-Miller parameter** $P = T(C + \log t) \times 10^{-3}$,

where T = test temp in °R (°F + 460) and t = test time, h. The constant C depends upon the material but is frequently taken to be 20.

Many alloys have been developed specifically for such applications. The selection of an alloy for a specific high-temperature application is strongly influenced by service conditions (stress, stress fluctuations, temperature, heat shock, atmosphere, service life), and there are hundreds of alloys from which to choose. The following illustrations, data, and discussion should be regarded as examples. Vendor literature and more extensive references should be consulted.

Figure 6.4.7 indicates the general stress-temperature range in which various alloy types find application in elevated-temperature service. Figure 6.4.8 indicates the important effect of time on the strength of alloys at high temperatures, comparing a familiar stainless steel (type 304) with **superalloy** (M252).

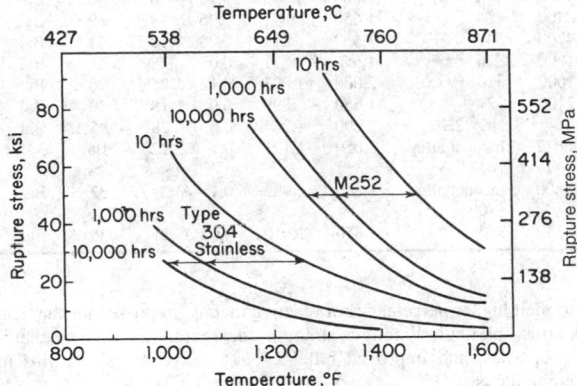

Fig. 6.4.8 Effect of time on rupture strength of type 304 stainless steel and alloy M252.

Common Heat-Resisting Alloys A number of alloys containing large amounts of **chromium and nickel** are available. These have excellent oxidation resistance at elevated temperatures. Several of them have been developed as electrical-resistance heating elements; others are modifications of stainless steels, developed for general corrosion resistance. Selected data on these alloys are summarized in Table 6.4.27. The maximum temperature value given is for resistance to oxidation with a reasonable life. At higher temperatures, failure will be rapid because of scaling. At lower temperatures, much longer life will be obtained. At the maximum useful temperature, the metal may be very weak and frequently must be supported to prevent sagging. Under load, these alloys are generally useful only at considerably lower temperatures, say up to 1,200°F (649°C) max, depending upon permissible creep rate and the load.

Superalloys were developed largely to meet the needs of aircraft gas turbines, but they have also been used in other applications demanding high strength at high temperatures. These alloys are based on nickel and/or cobalt, to which are added (typically) chromium for oxidation resistance and a complex of other elements which contribute to hot strength, both by solid solution hardening and by forming relatively stable dispersions of fine particles. Hardening by cold work, hardening by precipitation-hardening heat treatments, and hardening by deliberately arranging for slow precipitation during service are all methods used to enhance the properties of these alloys. **Fabrication** of these alloys is difficult since they are designed to resist distortion even at elevated temperatures. Forging temperatures of about 2,300°F (1,260°C) are used with small reductions and slow rates of working. Many of these alloys are fabricated by precision casting. Cast alloys that are given a strengthening heat treatment often have better properties than wrought alloys, but the shapes that can be made are limited.

Vacuum melting is an important factor in the production of superalloys. The advantages which result from its application include the abil-

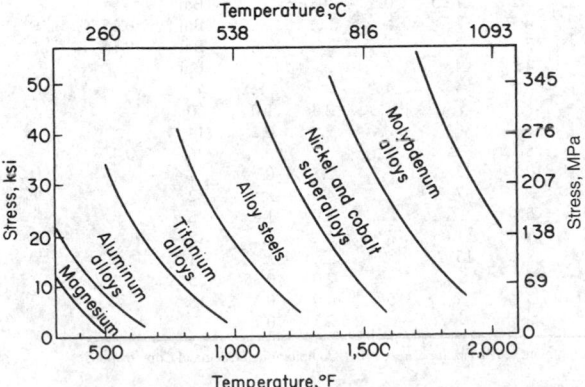

Fig. 6.4.7 Stress-temperature application ranges for several alloy types; stress to produce rupture in 1,000 h.

Table 6.4.27 Properties of Common Heat-Resisting Alloys

Alloy UNS no.	Name	Max temp for oxidation resistance °F	°C	C	Cr	Ni	Fe	Other	Specific gravity	Coef of thermal expansion per °F × 10⁻⁶ (0–1,200°F)	Stress to rupture in 1,000 h, ksi 1,200°F (649°C)	1,500°F (816°C)	1,800°F (982°C)	0.2 percent offset yield strength, ksi Room temp	1,000°F (538°C)	1,200°F (649°C)
N 06004	62 Ni, 15 Cr	1,700	926		15	62	Bal		8.19	9.35						
N 06003	80 Ni, 20 Cr	2,100	1,148		20	80			8.4	9.8						
	Kanthal	2,450	1,371		25		Bal	3 Co, 5 Al	7.15					63		
N 06600	Inconel alloy 600	2,050	1,121		13	79	Bal		8.4		14.5	3.7		36	22	22
G 10150	1,015	1,000	537	0.15			Bal		7.8	8.36	2.7			42	20	10
S 50200	502	1,150	621	0.12	5		Bal	0.5 Mo	7.8	7.31	6	1.5		27	18	12
S 44600	446	2,000	1,093	0.12	26	0.3	Bal		7.6	6.67	4	1.2				
S 30400	304	1,650	871	0.06	18	9	Bal		7.9	10.4	11	3.5		32	14	11
S 34700	347	1,650	871	0.08	18.5	11.5	Bal	0.8 Nb	8.0	10.7	20			41	31	26
S 31600	316	1,650	871	0.07	18	13	Bal	2.5 Mo	8.0	10.3	25	7		41	22	21
S 31000	310	2,000	1,093	0.12	25	20	Bal		7.9	9.8	13.2	3.0	2.7	34	28	25
S 32100	321	1,650	899	0.06	18	10	Bal	0.5 Ti	8.0	10.7	17.5	3.7		39	27	25
	NA 22H	2,200	1,204	0.5	28	48	Bal	5W		8.6	30	18	3.6			
N 06617	Inconel alloy 617	2,050	1,121	0.1	22	Bal	3	9 Mo, 12.5 Co, 1.2 Al, 0.6 Ti	8.36	6.4	52	14	4	43	29	25
N 08800	Inconel alloy 800	1,800	982	0.1	21	32	Bal		7.94	7.9	24	7	2	36	26	26
N 08330	330	1,850	1,010	0.05	19	36	Bal		8.08	8.3		5	1.3	40	26	21

ity to melt higher percentages of reactive metals, improved mechanical properties (particularly fatigue strength), decreased scatter in mechanical properties, and improved billet-to-bar stock-conversion ratios in wrought alloys.

Table 6.4.28 to 6.4.31 list compositions and properties of some superalloys, and Fig. 6.4.9 indicates the temperature dependence of the rupture strength of a number of such alloys.

Cast tool alloys are another important group of materials having high-temperature strength and wear resistance. They are principally alloys of cobalt, chromium, and tungsten. They are hard and brittle, and they must be cast and ground to shape. Their most important application is for hard facing, but they compete with high-speed steels and cemented carbides for many applications, in certain instances being superior to both. (See Cemented Carbides, Sec. 6.4.) Typical compositions are given in Table 6.4.32; typical properties are: elastic modulus = 35 ×

10⁶ lb/in² (242 GPa); specific gravity = 8.8; hardness, BHN at room temp = 660; 1,500°F (815°C), 435, and 2,000°F (1,093°C) 340.

For still higher-temperature applications, the possible choices are limited to a few metals with high melting points, all characterized by limited availability and by difficulty of extraction and fabrication. Advanced alloys of metals such as chromium, niobium, molybdenum, tantalum, and tungsten are still being developed. Table 6.4.33 cites properties of some of these alloys. Wrought tool steel materials are described in "Tool Steels," ASM Handbook, vol. 3, 9th ed., pp. 421–447.

Chromium with room-temperature ductility has been prepared experimentally as a pure metal and as strong high-temperature alloys. It is not generally available on a commercial basis. Large quantities of chromium are used as melt additions to steels, stainless steels, and nickel-base alloys.

Molybdenum is similar to tungsten in most of its properties. It can be

Table 6.4.28 Wrought Superalloys, Compositions*

Common designation	C	Mn	Si	Cr	Ni	Co	Mo	W	Nb	Ti	Al	Fe	Other
19-9D L	0.2	0.5	0.6	19	9	....	1.2	1.25	0.3	0.3	....	Bal	
Timken	0.1	0.5	0.5	16	25	....	6	.....	...	...	....	Bal	0.15 N
L.C. N-155	0.1	0.5	0.5	20	20	20	3	2	1	...	....	Bal	0.15 N
S-590	0.4	0.5	0.5	20	20	20	4	4	4	...	....	Bal	
S-816	0.4	0.5	0.5	20	20	Bal	4	4	4	...	....	4	
Nimonic alloy 80A	0.05	0.7	0.5	20	76	....	.....	.....	...	2.3	1.0	0.5	
K-42-B	0.05	0.7	0.7	18	42	22	....	.....	...	2.0	0.2	14	
Hastelloy alloy B	0.1	0.5	0.5	..	65	....	28	.....	...	...	....	6	0.4 V
M 252†	0.10	0.7	0.7	19	53.5	10	10	.....	...	2.5	0.75	2	
J1570	0.20	0.1‡	0.2‡	20	29	37.5		7	...	4.1	....	2	
HS-R235	0.10	....	....	16	Bal	1.5	6		...	3.0	1.8	8	
Hastelloy alloy X	0.10	....	....	21	48	2.0	9	1.0	...	...	....	18	
HS 25 (L605)	0.05	1.5	1.0‡	20	10	53	....	15	...	...	....	1.0	
A-286	0.05	....	....	15	26	....	1.3	.....	...	2.0	0.35	55	0.3 V
Inconel alloy 718	0.05	0.2	0.2	19	52.5	....	3	.....	5	0.9	0.5	18.5	
Inconel alloy X-750	0.1	1.0	0.5	15	72	1.0	....	.....	...	2.5	0.7	7	

* For a complete list of superalloys, both wrought and cast, experimental and under development, see *ASTM Spec. Pub.* 170, "Compilation of Chemical Compositions and Rupture Strengths of Superstrength Alloys."
† Waspaloy has a similar composition (except for lower Mo content), and similar properties.
‡ Max.

Table 6.4.29   Wrought Superalloys, Properties

Alloy (UNS)	Alloy (Name)	Max temp under load °F	Max temp under load °C	Coef of thermal expansion, in/(in·°F) × 10⁻⁶ (70–1,200°F)	Specific gravity	Stress to rupture, ksi — At 1,200°F 1,000 h	Stress to rupture, ksi — At 1,500°F 100 h	Stress to rupture, ksi — At 1,500°F 1,000 h	Yield strength 0.2% offset — Room	Yield strength 0.2% offset — 1,200°F	Yield strength 0.2% offset — 1,500°F	Yield strength 0.2% offset — 1,800°F	Tensile strength — Room	Tensile strength — 1,200°F	Tensile strength — 1,500°F	Tensile strength — 1,800°F
K63198	19-9 DL	1,200	649	9.7	7.75	38	17	10	115	39	30		140	75	33	13
	Timken	1,350	732	9.25	8.06	36	13.5	9	96*	70*	16*		134	90	40	18
R30155	L.C. N-155	1,400	760	9.4	8.2	48	20.0	15	53	40	33	12	115	53	35	19
	S-590	1,450	788	8.0	8.34	40	19.0	15	78*	70	47	20	140	82	60	22
R30816	S-816	1,500	816	8.3	8.66	53	29.0	18	63*	45	41		140	112	73	25
N07080	Nimonic alloy 80A	1,450	788	7.56		56	24	15	80*	73	47		150	101	69	
—	K-42-B	1,400	760	8.5	8.23	38	22	15	105	84	52		158	117	54	
N10001	Hastelloy alloy B	1,450	788	6.9	9.24	36	17	10	58	42			135	94	66	24
N07252	M 252	1,500	816	7.55	8.25	70	26	18	90	75	65		160	140	80	20
—	J 1570	1,600	871	8.42	8.66	84	34	24	81	70	71	17	152	135	82	20
—	HS-R235	1,600	871	8.34	7.88	70	35	23	100	90	80	22	170	145	83	25
N06002	Hastelloy alloy X	1,350	732			30	14	10	56	41	37		113	83	52	15
R30605	HS 25 (L605)	1,500	816			54	22	18	70	35				75	50	23
K66286	A-286	1,350	732			45	13.8	7.7								
N07718	Inconel alloy 718	1,400	760	7.2	8.19	54	19	13	152	126			186	149		
N07750	Inconel alloy X-750	1,300	704	7.0	8.25				92	82	45		162	120	52	

* 0.02 percent offset.

prepared in the massive form by powder metallurgy techniques, by inert-atmosphere or vacuum-arc melting. Its most serious limitation is its ready formation of a volatile oxide at temperatures of 1,400°F approx. In the worked form, it is inferior to tungsten in melting point, tensile strength, vapor pressure, and hardness, but in the recrystallized condition, the ultimate strength and elongation are higher. Tensile strengths up to 350,000 lb/in² (2,413 MPa) have been reported for hard-drawn wire, and to 170,000 lb/in² (1,172 MPa) for soft wire. In the hard-drawn condition, molybdenum has an elongation of 2 to 5 percent, but after recrystallizing, this increases to 10 to 25 percent. Young's modulus is 50 × 10⁶ lb/in² (345 GPa). It costs about the same as tungsten, per pound, but its density is much less. It has considerably better forming properties than tungsten and is extensively used for anodes, grids, and supports in vacuum tubes, lamps, and X-ray tubes. Molybde-

num is generally used for winding electric furnaces for temperatures up to 3,000°F (1,649°C). As it must be protected against oxidation, such furnaces are usually operated in hydrogen. It is the common material for cathodes for radar devices, heat radiation shields, rocket nozzles and other missile components, and hot-working tools and die-casting cores in the metalworking industry. Its principal use is still as an alloying addition to steels, especially tool steels and high-temperature steels.

Molybdenum alloys have found increased use in aerospace and commercial structural applications, where their high stiffness, microstructural stability, and creep strength at high temperatures are required. They are generally stronger than niobium alloys but are not ductile in the welded condition. Molybdenum alloys are resistant to alkali metal corrosion.

Niobium (also known as **columbium**) is available as a pure metal and

Table 6.4.30   Cast Superalloys, Nominal Composition

Common designation	Chemical composition, % C	Mn	Si	Cr	Ni	Co	Mo	W	Nb	Fe	Other*
Vitallium (Haynes 21)	0.25	0.6	0.6	27	2	Bal	6	—	—	1	—
61 (Haynes 23)	0.4	0.6	0.6	26	1.5	Bal	—	5	—	1	—
422-19 (Haynes 30)	0.4	0.6	0.6	26	16	Bal	6	—	—	1	—
X-40 (Haynes 31)	0.4	0.6	0.6	25	10	Bal	—	7	—	1	—
S-816	0.4	0.6	0.6	20	20	Bal	4	4	4	5	—
HE 1049	0.45	0.7	0.7	25	10	45	—	15	—	1.5	0.4B
Hastelloy alloy C	0.10	0.8	0.7	16	56	1	17	4	—	5.0	0.3V
B 1900 + Hf	0.1	—	—	8	Bal	10	6	—	—	—	6 Al 4 Ta / 1 Ti 1.5 Hf
MAR-M 247	0.15	—	—	8.5	Bal	10	0.5	10	—	—	5.5 Al 3Ta / 1 Ti 1.5 Hf
René 80	0.15	—	—	14	Bal	9.5	4	4	—	—	3 Al 5 Ti
MO-RE 2	0.2	0.3	0.3	32	48	—	—	15	—	Bal	1 Al

* Trace elements not shown.

Table 6.4.31 Cast Superalloys, Properties

Common designation	Max temp under load °F	Max temp under load °C	Coef of thermal expansion, in/(in · °F) × 10⁻⁶ (70–1,200°F)	Specific gravity	Stress to rupture in 1,000 h, ksi 1,200°F	Stress to rupture in 1,000 h, ksi 1,500°F	Stress to rupture in 1,000 h, ksi 1,800°F	Short-time tensile properties, ksi Yield strength, 0.2% offset Room 68°F (20°C)	1,200°F (650°C)	1,500°F (816°C)	1,800°F (982°C)	Tensile strength Room 68°F (20°C)	1,200°F (650°C)	1,500°F (816°C)	1,800°F (982°C)
Vitallium (Haynes 21)	1,500	816	8.35	8.3	44	14	7	82	71	49		101	89	59	33
61 (Haynes 23)	1,500	816	8.5	8.53	47	22	5.5	58	74	40	33	105	97	58	45
422-19 (Haynes 30)	1,500	816	8.07	8.31		21	7	55	37*	48		98	59*	64	37
N-40 (Haynes 31)	1,500	816	8.18	8.60	46	23	9.8	74	37*	44		101	77*	59	29
S-816	1,500	816	8.27	8.66	46	21	9.8					112			
HE 1049	1,650	899		8.9	75	35	7	80	72	62	36	90	82	81	52
Hastelloy alloy C				8.91	42.5	14.5	1.4	54	50			130	87	51	19
B 1900 + Hf	1,800	982	7.73	8.22		55	15	120	134	109	60	140	146	126	80
MAR-M 247	1,700	927	7.7	8.53			18	117	117	108	48	140	152	130	
René 80	1,700	927		8.16			15	124	105	90		149	149	123	76

* Specimen not aged.

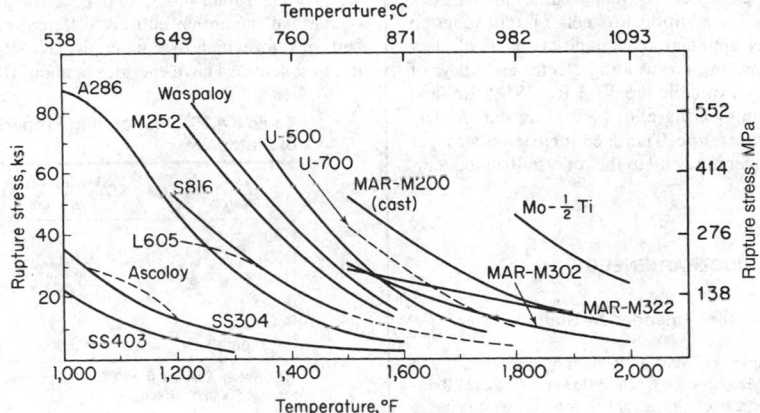

Fig. 6.4.9 Temperature dependence of strengths of some high-temperature alloys; stress to produce rupture in 1,000 h.

Table 6.4.32   Typical Compositions of Cast Tool Alloys

Name	C	Cr	W	Co	Fe	V	Ta	B	Other
Rexalloy	3	32	20	45					
Stellite 98 M2	3	28	18	35	9	4	0.1	0.1	3 Ni
Tantung G-2	3	15	21	40	...	...	19	0.2	
Borcoloy no. 6	0.7	5	18	20	Bal	1.3	....	0.7	6 Mo
Colmonoy WCR 100	...	10	15	...	Bal	...	....	3	

as the base of several niobium alloys. The alloys are used in nuclear applications at temperatures of 1,800 to 2,200°F (980 to 1,205°C) because of their low thermal neutron cross section, high strength, and good corrosion resistance in liquid or gaseous alkali metal atmospheres. Niobium alloys are used in leading edges, rocket nozzles, and guidance structures for hypersonic flight vehicles. A niobium-46.5 percent titanium alloy is used as a low-temperature superconductor material for magnets in magnetic resonance imaging machines.

Tantalum has a melting point that is surpassed only by tungsten. Its early use was as an electric-lamp filament material. It is more ductile than molybdenum or tungsten; the elongation for annealed material may be as high as 40 percent. The tensile strength of annealed sheet is about 50,000 lb/in² (345 MPa). In this form, it is used primarily in electronic capacitors. It is also used in chemical-processing equipment, where its high rate of heat transfer (compared with glass or ceramics) is particularly important, although it is equivalent to glass in corrosion resistance. Its corrosion resistance also makes it attractive for surgical implants. Like tungsten and molybdenum, it is prepared by powder metallurgy techniques; thus, the size of the piece that can be fabricated is limited by

the size of the original pressed compact. Tantalum carbide is used in cemented-carbide tools, where it decreases the tendency to seize and crater. The stability of the anodic oxide film on tantalum leads to rectifier and capacitor applications. Tantalum and its **alloys** become competitive with niobium at temperatures above 2,700°F (1,482°C). As in the case of niobium, these materials are resistant to liquid or vapor metal corrosion and have excellent ductility even in the welded condition.

Tungsten has a high melting point, which makes it useful for high-temperature structural applications. The massive metal is usually prepared by powder metallurgy from hydrogen-reduced powder. As a metal, its chief use is as filaments in incandescent lamps and electronic tubes, since its vapor pressure is low at high temperatures. Tensile strengths over 600,000 lb/in² (4,136 MPa) have been reported for fine tungsten wires; in larger sizes (0.040 in), tensile strength is only 200,000 lb/in² (1,379 MPa). The hard-drawn wire has an elongation of 2 to 4 percent, but the recrystallized wire is brittle. Young's modulus is 60 million lb/in² (415 GPa). It can be sealed directly to hard glass and so is used for lead-in wires. A considerable proportion of tungsten rod and sheet is used for electrical contacts in the form of disks cut from rod or

Table 6.4.33   Properties of Selected Refractory Metals

Alloy	Melting temp, °F	Density, lb/in³ at 75°F	Elastic modulus at 75°F	Ductile-to-brittle transition temp, °F	Tensile strength, ksi (recrystallized condition)				
					75°F	1,800°F	2,200°F	3,000°F	3,500°F
Chromium	3,450	0.760	42	625		12	8		
Columbium	4,474	0.310	16	−185	45	12	10		
F48 (15 W, 5 Mo, 1 Zr, 0.05 C)			25		121	75	50		
Cb74 or Cb752 (10 W, 2 Zr)		0.326			84	60	36		
Molybdenum	4,730	0.369	47	85	75	34	22	6	3
Mo (½ Ti)					110	70	48	8	4
TZM (0.5 Ti, 0.08 Zr)					130	85	70	14	5
Tantalum	5,425	0.600	27	< −320	30	22	15	8	4
Ta (10 W)						65	35	12	8
Rhenium	5,460	0.759	68	<75	170	85	60	25	11
Tungsten	6,170	0.697	58	645	85	36	32	19	10
W (10 Mo)								28	10
W (2 ThO₂)							40	30	25

sheet and brazed to supporting elements. The major part of the tungsten used is made into ferroalloys for addition to steels or into tungsten carbide for cutting tools. Other applications include elements of electronic tubes, X-ray tube anodes, and arc-welding electrodes. **Alloys** of tungsten that are commercially available are W-3 Re, W-25 Re, and thoriated tungsten. The rhenium-bearing alloys are more ductile than unalloyed tungsten at room temperature. Thoriated tungsten is stronger than unalloyed tungsten at temperatures up to the recrystallization temperature.

METALS AND ALLOYS FOR NUCLEAR ENERGY APPLICATIONS

by L. D. Kunsman and C. L. Carlson; Amended by Staff

REFERENCES: Glasstone, ''Principles of Nuclear Reactor Engineering,'' Van Nostrand. Hausner and Roboff, ''Materials for Nuclear Power Reactors,'' Reinhold. Publications of the builders of reactors (General Electric Co., Westinghouse Corp., General Atomic Co.) and suppliers of metal used in reactor components. See also Sec. 9.8 commentary and other portions of Sec. 6.4.

The advent of atomic energy not only created a demand for new metals and alloys but also focused attention on certain properties and combinations of properties which theretofore had been of little consequence. Reactor technology requires special materials for fuels, fuel cladding, moderators, reflectors, controls, heat-transfer mediums, operating mechanisms, and auxiliary structures. Some of the properties pertinent to such applications are given in a general way in Table 6.4.34. In general, outside the reflector, only normal engineering requirements need be considered.

Nuclear Properties A most important consideration in the design of a nuclear reactor is the control of the number and speed of the neutrons resulting from fission of the fuel. The designer must have knowledge of the effectiveness of various materials in slowing down neutrons or in capturing them. The slowing-down power depends not only on the relative energy loss per atomic collision but also on the number of collisions per second per unit volume. The former will be larger, the lower the atomic weight, and the latter larger, the greater the atomic density and the higher the probability of a scattering collision. The effectiveness of a moderator is frequently expressed in terms of the moderating ratio, the ratio of the slowing-down power to the capture cross section. The capturing and scattering tendencies are measured in terms of nuclear cross section in barns (10^{-24} cm^2). Data for some of the materials of interest

are given in Tables 6.4.35 to 6.4.37. The absorption data apply to slow, or thermal, neutrons; entirely different cross sections obtain for fast neutrons, for which few materials have significantly high capture affinity. Fast neutrons have energies of about 10^6 eV, while slow, or thermal,

Table 6.4.35 Moderating Properties of Materials

Moderator	Slowing-down power, cm^{-1}	Moderating ratio
H_2O	1.53	70
D_2O	0.177	21,000
He	1.6×10^{-5}	83
Be	0.16	150
BeO	0.11	180
C (graphite)	0.063	170

SOURCE: Adapted from Glass, ''Principles of Nuclear Reactor Engineering,'' Van Nostrand.

neutrons have energies of about 2.5×10^{-2} eV. Special consideration must frequently be given to the presence of small amounts of high cross-section elements such as Co or W which are either normal incidental impurities in nickel alloys and steels or important components of high-temperature alloys.

Table 6.4.36 Slow-Neutron Absorption by Structural Materials

Material	Relative neutron absorption per cm^3 $\times 10^3$	Relative neutron absorpton for pipes of equal strength, 68°F (20°C)	Melting point °F	Melting point °C
Magnesium	3.5	10	1,200	649
Aluminum	13	102	1,230	666
Stainless steel	226	234	2,730	1,499
Zirconium	12.6	16	3,330	1,816

SOURCE:: Leeser, *Materials & Methods*, **41**, 1955, p. 98.

Effects of Radiation Irradiation affects the properties of solids in a number of ways: dimensional changes; decrease in density; increase in hardness, yield, and tensile strengths; decrease in ductility; decrease in electrical conductivity; change in magnetic susceptibility. Another consideration is the activation of certain alloying elements by irradiation. Tantalum[181] and Co[60] have moderate radioactivity but long half-

Table 6.4.34 Requirements of Materials for Nuclear Reactor Components

Component	Neutron absorption cross section	Effect in slowing neutrons	Strength	Resistance to radiation damage	Thermal conductivity	Corrosion resistance	Cost	Other	Typical material
Moderator and reflector	Low	High	Adequate	—	—	High	Low	Low atomic wt	H_2O graphite, Be
Fuel*	Low	—	Adequate	High	High	High	Low	—	U, Th, Pu
Control rod	High	—	Adequate	Adequate	High	—	—	—	Cd, B_4C
Shield	High	High	High	—	—	—	—	High γ radiation absorption	Concrete
Cladding	Low	—	Adequate	—	High	High	Low	—	Al, Zr, stainless steel
Structural	Low	—	High	Adequate	—	High	—	—	Zr, stainless steel
Coolant	Low	—	—	—	High	—	—	Low corrosion rate, high heat capacity	H_2O, Na, NaK, CO_2, He

U, Th, and Pu are used as fuels in the forms of metals, oxides, and carbides.

Table 6.4.37 Slow-Neutron Absorption Cross Sections

Low		Intermediate		High	
Element	Cross section, barns	Element	Cross section, barns	Element	Cross section, barns
Oxygen	0.0016	Zinc	1.0	Manganese	12
Carbon	0.0045	Columbium	1.2	Tungsten	18
Beryllium	0.009	Barium	1.2	Tantalum	21
Fluorine	0.01	Strontium	1.3	Chlorine	32
Bismuth	0.015	Nitrogen	1.7	Cobalt	35
Magnesium	0.07	Potassium	2.0	Silver	60
Silicon	0.1	Germanium	2.3	Lithium	67
Phosphorus	0.15	Iron	2.4	Gold	95
Zirconium	0.18	Molybdenum	2.4	Hafnium	100
Lead	0.18	Gallium	2.8	Mercury	340
Aluminum	0.22	Chromium	2.9	Iridium	470
Hydrogen	0.32	Thallium	3.3	Boron	715
Calcium	0.42	Copper	3.6	Cadmium	3,000
Sodium	0.48	Nickel	4.5	Samarium	8,000
Sulphur	0.49	Tellurium	4.5	Gadolinium	36,000
Tin	0.6	Vanadium	4.8		
		Antimony	5.3		
		Titanium	5.8		

SOURCE: Leeser, *Materials & Methods*, **41**, 1955, p. 98.

lives; isotopes having short lives but high-activity levels include Cr^{51}, Mn^{56}, and Fe^{59}.

Metallic Coolants The need for the efficient transfer of large quantities of heat in a reactor has led to use of several metallic coolants. These have raised new problems of pumping, valving, and corrosion. In addition to their thermal and flow properties, consideration must also be given the nuclear properties of prospective coolants. Extensive thermal, flow, and corrosion data on metallic coolants are given in the "Liquid Metals Handbook," published by USNRC. The resistances of common materials to liquid sodium and NaK are given qualitatively in Table 6.4.38. Water, however, remains the most-used coolant; it is used under pressure as a single-phase liquid, as a boiling two-phase coolant, or as steam in a superheat reactor.

Fuels There are, at present, only three fissionable materials, U^{233}, U^{235}, and Pu^{239}. Of these, only U^{235} occurs naturally as an isotopic "impurity" with natural uranium U^{238}. Uranium^{233} may be prepared from natural thorium, and Pu^{239} from U^{238} by neutron bombardment. Both uranium and thorium are prepared by conversion of the oxide to the tetrafluoride and subsequent reduction to the metal.

The properties of **uranium** are considerably affected by the three allotropic changes which it undergoes, the low-temperature forms being highly anisotropic. The strength of the metal is low (see Table 6.4.39) and decreases rapidly with increasing temperature. The corrosion resistance is also poor. Aluminum-base alloys containing uranium-aluminum intermetallic compounds have been used to achieve improved properties. **Thorium** is even softer than uranium but is very ductile. Like

uranium it corrodes readily, particularly at elevated temperatures. Owing to its crystal structure, its properties are isotropic. **Plutonium** is unusual in possessing six allotropic forms, many of which have anomalous physical properties. The electrical resistivity of the alpha phase is greater than that for any other metallic element. Because of the poor corrosion resistance of nuclear-fuel metals, most fuels in power reactors today are in the form of oxide UO_2, ThO_2, or PuO_2, or mixtures of these. The carbides UC and UC_2 are also of interest in reactors using liquid-metal coolants.

Control-Rod Materials Considerations of neutron absorption cross sections and melting points limit the possible control-rod materials to a very small group of elements, of which only four—boron, cadmium, hafnium, and gadolinium—have thus far been prominent. **Boron** is a very light metal of high hardness ($\sim$ 3,000 Knoop) prepared by thermal or hydrogen reduction of BCl_3. Because of its high melting point, solid shapes of boron are prepared by powder-metallurgy techniques. Boron has a very high electrical resistivity at room temperature but becomes conductive at high temperatures. The metal in bulk form is oxidation-resistant below 1,800°F (982°C) but reacts readily with most halogens at only moderate temperatures. Rather than the elemental form, boron is generally used as boron steel or as the carbide, oxide, or nitride. **Cadmium** is a highly ductile metal of moderate hardness which is recovered as a by-product in zinc smelting. Its properties greatly resemble those of zinc. The relatively low melting point renders it least attractive as a control rod of the four metals cited. **Hafnium** metal is reduced from hafnium tetrachloride by sodium and subsequently purified by the io-

Table 6.4.38 Resistance of Materials to Liquid Sodium and NaK*

	Good	Limited	Poor
< 1,000°F	Carbon steels, low-alloy steels, alloy steels, stainless steels, nickel alloys, cobalt alloys, refractory metals, beryllium, aluminum oxide, magnesium oxide, aluminum bronze	Gray cast iron, copper, aluminum alloys, magnesium alloys, glasses	Sb, Bi, Cd, Ca, Au, Pb, Se, Ag, S, Sn, Teflon
1,000–1,600°F	Armco iron, stainless steels, nickel alloys,† cobalt alloys, refractory metals‡	Carbon steels, alloy steels, Monel, titanium, zirconium, beryllium, aluminum oxide, magnesium oxide	Gray cast iron, copper alloys, Teflon, Sb, Bi, Cd, Ca, Au, Pb, Se, Ag, S, Sn, Pt, Si, Magnesium alloys

* For more complete details, see "Liquid Metals Handbook."
† Except Monel.
‡ Except titanium and zirconium.

Table 6.4.39 Mechanical Properties of Metals for Nuclear Reactors

Material	Room temp					Charpy impact ft · lb	Elevated temp*		
	Longitudinal			Transverse					
	Yield strength, ksi	Ult strength, ksi	Elong,† %	Ult str, ksi	Elong,† %		°F	Ult str, ksi	Elong, %
Beryllium:									
Cast, extruded and annealed		40	1.82	16.6	0.18		392	62	23.5
Flake, extruded and annealed		63.7	5.0	25.5	0.30		752	43	29
Powder, hot-extruded	39.5	81.8	15.8	45.2	2.3	4.1	1,112	23	8.5
Powder, vacuum hot-pressed	32.1	45.2	2.3	45.2	2.3	0.8	1,472	5.2	10.5
Zirconium:									
Kroll-50% CW		82.6				14.8	250	32	
Kroll-annealed		49.0					500	23	
Iodide	15.9	35.9	31			2.5–6.0	700	17	
							900	12	
							1,500	3	
Uranium	25	53	<10			15	302	27	
							1,112	12	
Thorium	27	37.5	40				570	22	
							930	17.5	

* All elevated-temperature data on beryllium for hot-extruded powder; on zirconium for iodide material.

† Beryllium is extremely notch-sensitive. The tabulated data have been obtained under very carefully controlled conditions, but ductility values in practice will be found in general to be much lower and essentially zero in the transverse direction.

dide hot-wire process. It is harder and less readily worked than zirconium, to which it is otherwise very similar in both chemical and physical properties. Hafnium reacts easily with oxygen, and its properties are sensitive to traces of most gases. The very high absorption cross section of **gadolinium** renders it advantageous for fast-acting control rods. This metal is one of the rare earths and as yet is of very limited availability. It is most frequently employed as the oxide.

Beryllium Great interest attaches to beryllium because it is unique among the metals with respect to its very low neutron-absorption cross section and high neutron-slowing power. It may also serve as a source of neutrons when subjected to alpha-particle bombardment. Beryllium is currently prepared almost entirely by magnesium reduction of the fluoride, although fused-salt electrolysis is also practicable and has been used. The high affinity of the metal for oxygen and nitrogen renders its processing and fabrication especially difficult. At one time, beryllium was regarded as almost hopelessly brittle. Special techniques, such as vacuum hot pressing, or vacuum casting followed by hot extrusion of clad slugs, led to material having marginally acceptable, though highly directional, mechanical properties. The extreme toxicity of beryllium powder necessitates special precautions in all operations.

Zirconium Zirconium's importance in nuclear technology derives from its low neutron-absorption cross section, excellent corrosion resistance, and high strength at moderate temperatures. The metal is produced by magnesium reduction of the tetrachloride (Kroll process). The Kroll product is usually converted to ingot form by consumable-electrode-arc melting. An important step in the processing for many applications is the difficult chemical separation of the 1½ to 3 percent hafnium with which zirconium is contaminated. Unless removed, this small hafnium content results in a prohibitive increase in absorption cross section from 0.18 to 3.5 barns. The mechanical properties of zirconium are particularly sensitive to impurity content and fabrication technique. In spite of the high melting point, the mechanical properties are poor at high temperatures, principally because of the allotropic transformation at 1,585°F (863°C). A satisfactory annealing temperature is 1,100°F (593°C). The low-temperature corrosion behavior is excellent but is seriously affected by impurities. The oxidation resistance at high temperatures is poor. Special alloys (Zircoloys) have been developed having greatly improved oxidation resistance in the intermediate-temperature range. These alloys have a nominal content of 1.5 percent Sn and minor additions of iron-group elements.

Stainless Steel Although stainless steel has a higher thermal neutron-absorption cross section than do the zirconium alloys, its good corrosion resistance, high strength, low cost, and ease of fabrication make it a strong competitor with zirconium alloys as a fuel-cladding material for water-cooled power-reactor applications. Types 348 and 304 stainless are used in major reactors.

MAGNESIUM AND MAGNESIUM ALLOYS
by James D. Shearouse, III

REFERENCES: "Metals Handbook," ASM. Beck, "Technology of Magnesium and Its Alloys." Annual Book of Standards, ASTM. Publications of the Dow Chemical Co.

Magnesium is the lightest metal of structural importance [108 lb/ft³ (1.740 g/cm³)]. The principal uses for **pure magnesium** are in aluminum alloys, steel desulfurization, and production of nodular iron. Principal uses for **magnesium alloys** are die castings (for automotive and computer applications), wrought products, and, to a lesser extent, gravity castings (usually for aircraft and aerospace components). Because of its chemical reactivity, magnesium can be used in pyrotechnic material and for sacrificial galvanic protection of other metals. Since magnesium in its molten state reacts with the oxygen in air, a protective atmosphere containing sulfur hexafluoride is employed as a controlled atmosphere.

Commercially pure magnesium contains 99.8 percent magnesium and is produced by extraction from seawater or by reduction from magnesite and dolomite ores. Chief impurities are iron, silicon, manganese, and aluminum. The major use of magnesium, consuming about 50 percent of total magnesium production, is as a component of aluminum alloys for beverage can stock. Another use for primary magnesium is in steel desulfurization. Magnesium, usually mixed with lime or calcium carbide, is injected beneath the surface of molten steel to reduce the sulfur content.

A growing application for **magnesium** alloys is die cast automotive components. Their light weight can be used to help achieve reduced fuel consumption, while high-ductility alloys are used for interior components which must absorb impact energy during collisions. These alloys can be cast by hot or cold chamber die-casting methods, producing a diverse group of components ranging from gear cases to steering wheel frames. Some vehicles use magnesium die castings as structural members to serve as instrument panel support beams and steering columns.

Designs employing magnesium must account for the relatively low value of the modulus of elasticity (6.5×10^6 lb/in^2) and high thermal coefficient of expansion [14×10^{-6} per °F at 32°F (0°C) and 16×10^{-6} for 68 to 752°F (20 to 400°C)]. (See Tables 6.4.40 and 6.4.41.)

The development of jet engines and high-velocity aircraft, missiles, and spacecraft led to the development of magnesium alloys with improved **elevated-temperature properties**. These alloys employ the addition of some combination of rare earth elements, yttrium, manganese, zirconium, or, in the past, thorium. Such alloys have extended the useful temperatures at which magnesium can serve in structural applications to as high as 700 to 800°F.

Magnesium alloy forgings are used for applications requiring properties superior to those obtainable in castings. They are generally **press-forged**. Alloy AZ61 is a general-purpose alloy, while alloys AZ80A-T5 and ZK60A-T5 are used for the highest-strength press forgings of simple design. These alloys are aged to increase strength.

A wide range of extruded shapes are available in a number of compositions. Alloys AZ31B, AZ61A, AZ80A-T5, and ZK60A-T5 increase in cost and strength in the order named. **Impact extrusion** produces smaller, symmetric tubular parts.

Sheet is available in several alloys (see Table 6.4.41), in both the soft (annealed) and the hard (cold-rolled) tempers. Magnesium alloy sheet usually is hot-formed at temperatures between 400 and 650°F, although simple bends of large radius are made cold.

Joining Magnesium alloys are joined by **riveting** or **welding**. Riveting is most common. Aluminum alloy rivets are used; alloy 5052 is preferred, to minimize contact corrosion. Other aluminum alloy rivets can be used, but they are not as effective in inhibiting contact corrosion. Rivets should be anodized to prevent contact corrosion. Adhesive bonding is another accepted method for joining magnesium. It saves weight and improves fatigue strength and corrosion resistance.

Arc welding with inert-gas (helium or argon) shielding of the molten metal produces satisfactory joints. Butt joints are preferred, but any type of welded joint permissible for mild steel can be used. After welding, a stress relief anneal is necessary. Typical stress relief anneal times are 15 min at 500°F (260°C) for annealed alloys and 1 h at 400°F (204°C) for cold-rolled alloys.

Machining Magnesium in all forms is a free-machining metal. Standard tools such as those used for brass and steel can be used with slight modification. Relief angles should be from 7° to 12°; rake angles from 0° to 15°. High-speed steel is satisfactory and is used for most drills, taps, and reamers. Suitable grades of cemented carbides are better for production work and should be used where the tool design permits it. Finely divided magnesium constitutes a fire hazard, and good housekeeping in production areas is essential. (see also Sec. 13.4.)

Corrosion Resistance and Surface Protection All magnesium alloys display good resistance to ordinary inland atmospheric exposure, to most alkalies, and to many organic chemicals. High-purity alloys introduced in the mid-1980s demonstrate excellent corrosion resistance and are used in under-vehicle applications without coatings. However, galvanic couples formed by contact with most other metals, or by contamination of the surface with other metals during fabrication, can cause rapid attack of the magnesium when exposed to salt water. Protective treatments or coated fasteners can be used to isolate galvanic couples and prevent this type of corrosion.

Table 6.4.40   Typical Mechanical Properties of Magnesium Casting Alloys

Alloy	Condition or temper*	Nominal composition, % (balance Mg plus trace elements)					Tensile strength, ksi	Tensile yield strength, ksi	Elongation, %, in 2 in	Shear strength, ksi	Strength, ksi		Hardness BHN	Electrical conductivity, % IACS‡
		Al	Zn	Mn, min	Zr	Other					Tensile	Yield		
Permanent mold and sand casting alloys:														
AM100A	−T6	10.0		0.10			40	22	1	22			70	14
AZ63A	−F	6.0	3.0	0.15			29	14	6	18	60	40	50	14
	−T4						40	13	12	17	60	44	55	12
	−T5						30	14	4	17	60	40	55	
	−T6						40	19	5	20	75	52	73	15
AZ81A	−T4	7.5	0.7	0.13			40	12	15	17	60	44	55	12
AZ91E	−F	8.7	0.7	0.17			24	14	2	18	60	40	52	13
	−T4						40	12	14	17	60	44	53	11
	−T5						26	17	3					
	−T6						40	19	5	20	75	52	66	13
AZ92A	−F	9.0	2.0	0.10			24	14	2	18	50	46	65	12
	−T4						40	14	9	20	68	46	63	10
	−T5						26	16	2	19	50	46	70	
	−T6						40	21	2	22	80	65	84	14
EZ33A	−T5		2.5		0.8	3.5 RE†	23	15	3	22	57	40	50	25
K1A	−F				0.7		25	7	19	8				31
QE22A	−T6				0.7	2.0 RE,† 2.5 Ag	40	30	4	23				25
ZE41A	−T5		4.0		0.7	1.3 RE†	30	20	3.5	22	70	51	62	31
ZK51A	−T5		4.5		0.8		40	24	8	22	72	47	65	27
ZK61A	−T6		6.0		0.8		45	28	10	26			70	27
Die casting alloys:														
AZ91D	−F	9.0	0.7	0.15			34	23	3	20			75	
AM60B	−F	6.0	0.22 max	0.24			32	19	6–8				62	
AM50A	−F	5.0	0.22 max	0.26			32	18	8–10				57	
AS41B	−F	4.2	0.12 max	0.35		1.0 Si	31	18	6				75	
AE42X1	−F	4.0	0.22 max	0.25		2.4 RE†	33	20	8–10				57	

ksi $\times$ 6.895 = MPa
* −F = as cast; −T4 = artificially aged; −T5 = solution heat-treated; −T6 = solution heat-treated.
† RE = rare-earth mixture.
‡ Percent electrical conductivity/100 approximately equals thermal conductivity in cgs units.

Table 6.4.41 Properties of Wrought Magnesium Alloys

| Alloy and temper* | Nominal composition, % (balance Mg plus trace elements) | | | | Physical properties | | | Tensile strength, ksi | Room-temperature mechanical properties (typical) | | | | Bearing strength, ksi | | Hardness BHN |
	Al	Mn	Zn	Zr	Density, lb/in³	Thermal conductivity, cgs units, 68°F (20°C)	Electrical resistivity, μΩ·cm, 68°F (20°C)		Tensile yield strength, ksi	Elongation in 2 in, %	Compressive yield strength, ksi	Shear strength, ksi	Tensile	Yield	
Extruded bars, rods, shapes															
AZ31B-F	3.0		1.0		0.0639	0.18	9.2	38	29	15	14	19	56	33	49
AZ61A-F	6.5		1.0		0.0647	0.14	12.5	45	33	16	19	22	68	40	60
AZ80A-T5	8.5		0.5		0.0649	0.12	14.5	55	40	7	35	24	60	58	82
ZK60A-F			5.7	0.5	0.0659	0.28	6.0	49	38	14	33	24	76	56	75
-T5			5.7	0.5	0.0659	0.29	5.7	53	44	11	36	26	79	59	82
Extruded tube															
AZ31B-F	3.0		1.0		0.0639	0.18	9.2	36	24	16	12				46
AZ61A-F	6.5		1.0		0.0647	0.14	12.5	41	24	14	16				50
ZK60A-F			5.7	0.5	0.0659	0.28	6.0	47	35	13	25				75
-T5			5.7	0.5	0.0659	0.29	5.7	50	40	11	30				82
M1A		1.2			0.0635	0.31	5.4	37	26	11	12				
Sheet and plate															
AZ31B-H24	3.0		1.0		0.0639	0.18	9.2	42	32	15	26	29	77	47	73
								40	29	17	23	28	72	45	
AZ31B-O	3.0		1.0		0.0639	0.18	9.2	39	27	19	19	27	70	40	
								37	22	21	16	26	66	37	
M1A		1.2			0.0635	0.31	5.4	37	26	11	12				56
Tooling plate															
AZ31B	3.0		1.0		0.0639	0.18	9.2	35	19	12	10				
Tread plate															
AZ31B	3.0		1.0		0.0639	0.18	9.2	35	19	14	11				52

See Sec. 1 for conversion factors to SI units.

NOTE: For all above alloys: coefficient of thermal expansion = 0.0000145; modulus of elasticity = 6,500,000 lb/in²; modulus of rigidity = 2,400,000 lb/in²; Poisson's ratio = 0.35.

* Temper: −F = as fabricated; H24 = strain-hardened, then partially annealed; −O = fully annealed; −T5 = artificially aged.

POWDERED METALS

by Peter K. Johnson

REFERENCES: German, ''Powder Metallurgy Science,'' 2d ed., Metal Powder Industries Federation, Princeton, NJ. ''Powder Metallurgy Design Manual,'' 2d ed., Metal Powder Industries Federation.

Powder metallurgy (PM) is an automated manufacturing process to make precision metal parts from metal powders or particulate materials. Basically a net or near-net shape metalworking process, a PM operation usually results in a finished part containing more than 97 percent of the starting raw material. Most PM parts weigh less than 5 lb (2.26 kg), although parts weighing as much as 35 lb (15.89 kg) can be fabricated in conventional PM equipment.

In contrast to other metal-forming techniques, PM parts are shaped directly from powders, whereas castings originate from molten metal, and wrought parts are shaped by deformation of hot or cold metal, or by machining (Fig. 6.4.10). The PM process is cost-effective in manufacturing simple or complex shapes at, or very close to, final dimensions in production rates which can range from a few hundred to several thousand parts per hour. Normally, only a minimum amount of machining is required.

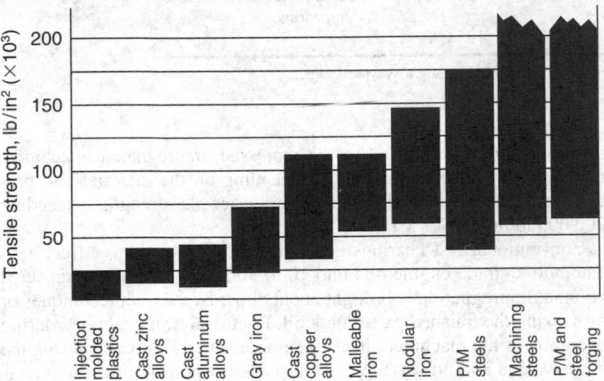

Fig. 6.4.10 Comparison of material strengths.

Powder metallurgy predates melting and casting of iron and other metals. Egyptians made iron tools, using PM techniques from at least 3000 B.C. Ancient Inca Indians made jewelry and artifacts from precious metal powders. The first modern PM product was the tungsten filament for electric light bulbs, developed in the early 1900s. Oil-impregnated PM bearings were introduced for automotive use in the 1920s. This was followed by tungsten carbide cutting tool materials in the 1930s, automobile parts in the 1960s and 1970s, aircraft gas-turbine engine parts in the 1980s, and **powder forged** (PF) and **metal injection molding** (MIM) parts in the 1990s.

PM parts are used in a variety of end products such as lock hardware, garden tractors, snowmobiles, automobile automatic transmissions and engines, auto antilock brake systems (ABS), washing machines, power tools, hardware, firearms, copiers, and postage meters, off-road equipment, hunting knives, hydraulic assemblies, X-ray shielding, oil and gas drilling well-head components, and medical equipment. The typical five- or six-passenger car contains about 30 lb of PM parts, a figure that could increase within the next several years. Iron powder is used as a carrier for toner in electrostatic copying machines. People in the United States consume about 2 million lb of iron powder annually in iron-enriched cereals and breads. Copper powder is used in antifouling paints for boat hulls and in metallic pigmented inks for packaging and printing. Aluminum powder is used in solid-fuel booster rockets for the space shuttle program. The spectrum of applications of powdered metals is very wide indeed.

There are five major processes that consolidate metal powders into precision shapes: conventional powder metallurgy (PM) (the dominant sector of the industry), metal injection molding (MIM), powder forging (PF), hot isostatic pressing (HIP), and cold isostatic pressing (CIP).

PM processes use a variety of alloys, giving designers a wide range of material properties (Tables 6.4.42 to 6.4.44). Major alloy groups include powdered iron and alloy steels, stainless steel, bronze, and brass. Powders of aluminum, copper, tungsten carbide, tungsten, and heavy-metal alloys, tantalum, molybdenum, superalloys, titanium, high-speed tool steels, and precious metals are successfully fabricated by PM techniques.

A PM microstructure can be designed to have controlled microporosity. This inherent advantage of the PM process can often provide special useful product properties. Sound and vibration damping can be enhanced. Components are impregnated with oil to function as self-lubricating bearings, resin impregnated to seal interconnecting microporosity, infiltrated with a lower-melting-point metal to increase strength and impact resistance, and steam-treated to increase corrosion resistance and seal microporosity. The amount and characteristics of the microporosity can be controlled within limits through powder characteristics, powder composition, and the compaction and sintering processes.

Powder metallurgy offers the following advantages to the designer:
1. PM eliminates or minimizes machining.
2. It eliminates or minimizes material losses.
3. It maintains close dimensional tolerances.
4. It offers the possibility of utilizing a wide variety of alloyed materials.
5. PM produces good surface finishes.
6. It provides components that may be heat-treated for increased strength or wear resistance.
7. It provides part-to-part reproducibility.
8. It provides controlled microporosity for self-lubrication or filtration.
9. PM facilitates the manufacture of complex or unique shapes that would be impractical or impossible with other metalworking processes.
10. It is suitable for moderate- to high-volume production.
11. It offers long-term performance reliability for parts in critical applications.

Metal powders are materials precisely engineered to meet a wide range of performance requirements. Major metal powder production processes include atomization (water, gas, centrifugal); chemical (reduction of oxides, precipitation from a liquid, precipitation from a gas) and thermal; and electrolytic. **Particle shape,** size, and size distribution strongly influence the characteristics of powders, particularly their behavior during die filling, compaction, and sintering. The range of shapes covers highly spherical, highly irregular, flake, dendritic (needlelike), and sponge (porous). Metal powders are classified as elemental, partially alloyed, or prealloyed.

The three basic steps for producing conventional-density PM parts are mixing, compacting, and sintering.

Elemental or prealloyed metal powders are mixed with lubricants and/or other alloy additions to produce a homogeneous mixture.

In compacting, a controlled amount of mixed powder is automatically gravity-fed into a precision die and is compacted, usually at room temperatures at pressures as low as 10 or as high as 60 or more tons/in^2 (138 to 827 MPa), depending on the density requirements of the part. Normally compacting pressures in the range of 30 to 50 tons/in^2 (414 to 690 MPa) are used. Special mechanical or hydraulic presses are equipped with very rigid dies designed to withstand the extremely high loads experienced during compaction. Compacting loose powder produces a *green compact* which, with conventional pressing techniques, has the size and shape of the finished part when ejected from the die and is sufficiently rigid to permit in-process handling and transport to a sintering furnace. Other specialized compacting and alternate forming methods can be used, such as powder forging, isostatic pressing, extrusion, injection molding, and spray forming.

During sintering, the green part is placed on a wide mesh belt and moves slowly through a controlled-atmosphere furnace. The parts are

Table 6.4.42 Properties of Powdered Ferrous Metals

Material	Density, g/cm^3	Tensile strength* MPa	Tensile strength* lb/in^2	Comments
Carbon steel				
MPIF F-0008	6.9	370†	54,000†	Often cost-effective
0.8% combined carbon (c.c.)	6.9	585‡	85,000‡	
Copper steel				
MPIF FC-0208	6.7	415†	60,000†	Good sintered strength
2% Cu, 0.8% c.c.	6.7	585‡	85,000‡	
Nickel steel				
MPIF FN-0205	6.9	345†	50,000†	Good heat-treated strength,
2% Ni, 0.5% c.c.	6.9	825‡	120,000‡	impact energy
MPIF FN-0405	6.9	385†	56,000†	
4% Ni, 0.5% c.c.	6.9	845‡	123,000‡	
Infiltrated steel				
MPIF FX-1005	7.3	530†	77,000†	Good strength, closed-off
10% Cu, 0.5% c.c.	7.3	825‡	120,000‡	internal porosity
MPIF FX-2008	7.3	550†	80,000†	
20% Cu, 0.8% c.c.	7.3	690‡	100,000‡	
Low-alloy steel, prealloyed Ni, Mo, Mn				
MPIF FL-4605 HT	6.95	895‡	130,000‡	Good hardenability, consistency in heat treatment
Stainless steels				
MPIF SS-316 N2 (316 stainless)	6.5	415†	60,000†	Good corrosion resistance, appearance
MPIF SS-410 HT (410 stainless)	6.5	725‡	105,000‡	

* Reference: MPIF Standard 35, Materials Standards for P/M Structural Parts. Strength and density given as typical values.
† As sintered.
‡ Heat-treated.

Table 6.4.43 Properties of Powdered Nonferrous Metals

Material	Use and characteristics
Copper	Electrical components
Bronze	Self-lubricating bearings: @ 6.6 g/cm^3 oiled
MPIF CTG-1001	density, 160 MPa (23,000 lb/in^2) K strength,*
10% tin, 1% graphite	17% oil content by volume
Brass	Electrical components, applications requiring good corrosion resistance, appearance, and ductility @ 7.9 g/cm^3, yield strength 75 MPa
MPIF CZ-1000	(11,000 lb/in^2)
10% zinc	
MPIF CZ-3000	@7.9 g/cm^3, yield strength 125 MPa (18,300
30% zinc	lb/in^2)
Nickel silver	Improved corrosion resistance, toughness @ 7.9
MPIF CNZ-1818	g/cm^3, yield strength 140 MPa (20,000 lb/in^2)
18% nickel, 18% zinc	
Aluminum alloy	Good corrosion resistance, lightweight, good electrical and thermal conductivity
Titanium	Good strength/mass ratio, corrosion resistance

* Radial crushing constant. See MPIF Standard 35, Materials Standards for P/M Self-Lubricating Bearings.

Table 6.4.44 Properties of Powdered Soft Magnetic Metals

Material	Density, g/cm^3	Maximum magnetic induction, kG*	Coercive field, Oe
Iron, low-density	6.6	10.0	2.0
Iron, high-density	7.2	12.5	1.7
Phosphorous iron, 0.45% P	7.0	12.0	1.5
Silicon iron, 3% Si	7.0	11.0	1.2
Nickel iron, 50% Ni	7.0	11.0	0.3
410 Stainless steel	7.1	10.0	2.0
430 Stainless steel	7.1	10.0	2.0

* Magnetic field—15 oersteds (Oe).

heated to below the melting point of the base metal, held at the sintering temperature, and then cooled. Basically a solid-state diffusion process, sintering transforms compacted mechanical bonds between the powder particles to metallurgical bonds. Sintering gives the PM part its primary functional properties. PM parts generally are ready for use after sinter-

ing. If required, parts can also be repressed, impregnated, machined, tumbled, plated, or heat-treated. Depending on the material and processing technique, PM parts demonstrate tensile strengths exceeding 200,000 lb/in^2 (1,379 MPa).

Conventional PM bearings generally can absorb additive-free, nonautomotive-grade engine oils into 10 to 30 percent of their compacted volume. Impregnation is brought about either by vacuum techniques or by soaking the finished parts in hot oil. Frictional heat generated during operation of the machinery heats the impregnated PM part, causing the oil to expand and flow to the bearing surface. Upon cooling, the flow is reversed, and oil is drawn into the pores by capillary action.

Development continues in the five major processes cited previously that function with metal powders as the raw material. The basic advantages of the processes make it attractive to seek wider application in a host of new products, to increase the size and complexity of parts that can be handled, and to expand further the types of powders that can be processed. Included among the newer materials are advanced particulates such as intermetallics, cermets, composites, nanoscale materials, and aluminides.

NICKEL AND NICKEL ALLOYS

by John H. Tundermann

REFERENCES: Mankins and Lamb, "Nickel and Nickel Alloys," ASM Handbook, vol. 2, 10th ed., pp. 428–445. Tundermann et al., "Nickel and Nickel Alloys," Kirk Othmer Encyclopedia of Chemical Technology, vol. 17. Frantz, "Low Expansion Alloys," ASM Handbook, vol. 2, 10th ed., pp. 889–896.

Nickel is refined from sulfide and lateritic (oxide) ores using pyrometallurgical, hydrometallurgical, and other specialized processes. Sulfide ores are used to produce just over 50 percent of the nickel presently used in the world. High-purity nickel (99.7$^+$ percent) products are produced via electrolytic, carbonyl, and powder processes.

Electroplating accounts for about 10 percent of the total annual consumption of nickel. Normal nickel electroplate has properties approximating those of wrought nickel, but special baths and techniques can give much harder plates. Bonds between nickel and the base metal are usually strong. The largest use of nickel in plating is for corrosion

Table 6.4.45 Nominal Compositions, %, of Selected Nickel Alloys

Alloy	Ni	Cu	Fe	Cr	Mo	Al	Si	Mn	W	C	Co	Nb	Ti	Y_2O_3
Nickel 200	99.5	0.05	0.1				0.05	0.25		0.05				
Duranickel alloy 301	93.5	0.2	0.2			4.4	0.5	0.3		0.2			0.5	
Monel alloy 400	65.5	31.5	1.5				0.25	1.0		0.15				
Monel alloy K-500	65.5	29.5	1.0			3.0	0.15	0.5		0.15				
Hastelloy alloy C-276	56		5	15.5	16		0.05	1.0	4	0.02	2.5			
Inconel alloy 600	75.5	0.5	8	15.5			0.2	0.5		0.08				
Inconel alloy 625	62		2.5	22	9	0.2	0.2	0.2		0.05		3.5	0.2	
Inconel alloy 718	52.5		18.5	19	3	0.5	0.2	0.2		0.04		5	0.9	
Inconel alloy X-750	71	0.5	7	15.5		0.7	0.5	1.0		0.08		1.0	2.5	
Inconel alloy MA 754	78.5			20		0.3				0.05			0.5	0.6
Incoloy alloy 825	42	2.2	30	21	3	0.1	0.25	0.5		0.03			0.9	
Incoloy alloy 909	38		42				0.4			0.1	13.0	4.7	1.5	
Hastelloy alloy X	45		19.5	22	9		1.0	1.0	0.6	0.1	1.5			

protection of iron and steel parts and zinc-base die castings for automotive use. A 0.04- to 0.08-mm (1.5- to 3-mil) nickel plate is covered with a chromium plate only about one hundredth as thick to give a bright, tarnish-resistant, hard surface. Nickel electrodeposits are also used to facilitate brazing of chromium-containing alloys, to reclaim worn parts, and for electroformed parts.

Wrought nickel and nickel alloys are made by several melting techniques followed by hot and cold working to produce a wide range of product forms such as plate, tube, sheet, strip, rod, and wire.

The nominal composition and typical properties of selected commercial nickel and various nickel alloys are summarized in Tables 6.4.45 to 6.4.48.

Commercially pure nickels, known as Nickel 200 and 201, are available as sheet, rod, wire, tubing, and other fabricated forms. They are used where the thermal or electrical properties of nickel are required and where corrosion resistance is needed in parts that have to be worked extensively.

Commercial nickel may be forged or rolled at 871 to 1,260°C (1,600 to 2300°F). It becomes increasingly harder below 871°C (1,600°F) but has no brittle range. The recrystallization temperature of cold-worked pure nickel is about 349°C (660°F), but commercial nickel recrystallizes at about 538°C (1,100° F) and is usually annealed at temperatures between 538 and 954°C (1,100 and 1,750°F).

The addition of certain elements to nickel renders it responsive to precipitation or aging treatments to increase its strength and hardness. In the unhardened or quenched state, Duranickel alloy 301 fabricates almost as easily as pure nickel, and when finished, it may be hardened by heating for about 8 h at 538 to 593°C (1,000 to 1,100°F). Intermediate anneals during fabrication are at 900 to 954°C (1,650 to 1,750°F). The increase in strength due to aging may be superimposed on that due to cold work.

Nickel castings are usually made in sand molds by using special techniques because of the high temperatures involved. The addition of

1.5 percent silicon and lesser amounts of carbon and manganese is necessary to obtain good casting properties. Data on nickel and nickel alloy castings can be found in Spear, "Corrosion-Resistant Nickel Alloy Castings," ASM Handbook, vol. 3, 9th ed, pp. 175–178.

Nickel has good to excellent resistance to corrosion in caustics and nonoxidizing acids such as hydrochloric, sulfuric, and organic acids, but poor resistance to strongly oxidizing acids such as nitric acid. Nickel is resistant to corrosion by chlorine, fluorine, and molten salts.

Nickel is used as a catalyst for the hydrogenation or organic fats and for several industrial processes. Porous nickel electrodes are used for battery and fuel cell applications.

Nickel-Copper Alloys Alloys containing less than 50 percent nickel are discussed under "Copper and Copper Alloys." **Monel** alloy 400 (see Tables 6.4.45 to 6.4.47) is a nickel-rich alloy that combines high strength, high ductility, and excellent resistance to corrosion. It is a homogeneous solid-solution alloy; hence its strength can be increased by cold working alone. In the annealed state, its tensile strength is about 480 MPa (70 ksi), and this may be increased to 1,170 MPa (170 ksi) in hard-drawn wire. It is available in practically all fabricated forms. Alloy 400 is hot-worked in the range 871 to 1,177°C (1,600 to 2,150°F) after rapid heating in a reducing, sulfur-free atmosphere. It can be cold-worked in the same manner as mild steel, but requires more power. Very heavily cold-worked alloy 400 may commence to recrystallize at 427°C (800°F), but in normal practice no softening will occur below 649°C (1,200°F). Annealing can be done for 2 to 5 h at about 760°C (1,400°F) or for 2 to 5 min at about 940°C (1,725°F). Nonscaling, sulfur-free atmospheres are required.

Because of its toughness, alloy 400 must be machined with high-speed tools with slower cutting speeds and lighter cuts than mild steel. A special grade containing sulfur, Monel alloy 405, should be used where high cutting speeds must be maintained. This alloy is essentially the same as the sulfur-free alloy in mechanical properties and corrosion resistance, and it can be hot-forged.

Table 6.4.46 Typical Physical Properties of Selected Alloys

Alloy	UNS no.	Density, g/cm³	Elastic modulus, GPa	Specific heat (20°C), J/(Kg·K)	Thermal expansion (20°–93°C), μm/(m·K)	Thermal conductivity (20°C), W/(m·K)	Electrical resistivity (annealed), $\mu\Omega\cdot$m
Nickel 200	N02200	8.89	204	456	13.3	70	0.096
Duranickel alloy 301	N03301	8.25	207	435	13.0	23.8	0.424
Monel alloy 400	N04400	8.80	180	427	13.9	21.8	0.547
Monel alloy K-500	N05500	8.44	180	419	13.9	17.5	0.615
Hastelloy alloy C-276	N10276	8.89	205	427	11.2	9.8	1.29
Inconel alloy 600	N06600	8.47	207	444	13.3	14.9	1.19
Inconel alloy 625	N06625	8.44	207	410	12.8	9.8	1.29
Inconel alloy 718	N07718	8.19	211	450	13.0	11.4	1.25
Inconel alloy X-750	N07750	8.25	207	431	12.6	12.0	1.22
Inconel alloy MA 754	N07754	8.30	160	440	12.2	14.3	1.075
Incoloy alloy 825	N08825	8.14	206	440	14.0	11.1	1.13
Incoloy alloy 909	N19909	8.30	159	427	7.7	14.8	0.728
Hastelloy alloy X	N06002	8.23	205	461	13.3	11.6	1.16

Table 6.4.47 Typical Mechanical Properties of Selected Nickel Alloys

Alloy	Yield strength, MPa	Tensile strength, MPa	Elongation, % in 2 in	Hardness
Nickel 201	150	462	47	110 HB
Duranickel alloy 301	862	1,170	25	35 HRC
Monel alloy 400	240	550	40	130 HB
Monel alloy K-500	790	1,100	20	300 HB
Hastelloy alloy C-276	355	790	60	90 HRB
Inconel alloy 600	310	655	40	36 HRC
Inconel alloy 625	520	930	42	190 HB
Inconel alloy 718	1,036	1,240	12	45 HRC
Inconel alloy X-750	690	1,137	20	330 HB
Inconel alloy MA 754	585	965	22	25 HRC
Incoloy alloy 825	310	690	45	75 HRB
Incoloy alloy 909	1,035	1,275	15	38 HRC
Hasteloy alloy X	355	793	45	90 HRB

The short-time tensile strengths of alloy 400 at elevated temperatures are summarized in Table 6.4.48.

The fatigue endurance limit of alloy 400 is about 240 MPa (34 ksi) when annealed and 325 MPa (47 ksi) when hard-drawn. The action of corrosion during fatigue is much less drastic on alloy 400 than on steels of equal or higher endurance limit.

Alloy 400 is highly resistant to atmospheric action, seawater, steam, foodstuffs, and many industrial chemicals. It deteriorates rapidly in the presence of moist chlorine and ferric, stannic, or mercuric salts in acid solutions. It must not be exposed when hot to molten metals, sulfur, or gaseous products of combustion containing sulfur.

Small additions of aluminum and titanium to the alloy 400 base nickel-copper alloy produces precipitation-hardenable Monel alloy K-500. This alloy is sufficiently ductile in the annealed state to permit drawing, forming, bending, or other cold-working operations but work-hardens rapidly and requires more power than mild steel. It is hot-worked at 927 to 1,177°C (1,700 to 2,150°F) and should be quenched from 871°C (1,600°F) if the metal is to be further worked or hardened. Heat treatment consists of quenching from 871°C (1,600°F), cold working if desired, and reheating for 10 to 16 h at 538 to 593°C (1,000 to 1,100°F). If no cold working is intended, the quench may be omitted on sections less than 50 mm (2 in) thick and the alloy hardened at 593°C (1,100°F). The properties of the heat-treated alloy remain quite stable, at least up to 538°C (1,000°F). It is nonmagnetic down to −101°C (−150°F).

Heat-Resistant Nickel-Chromium and Nickel-Chromium-Iron Alloys See "Metals and Alloys for Use at Elevated Temperatures" for further information on high-temperature properties of nickel alloys.

The addition of chromium to nickel improves strength and corrosion resistance at elevated temperature. Nickel chromium alloys such as 80 Ni 20 Cr are extensively used in electrical resistance heating applications. Typically, additions of up to 4 percent aluminum and 1 percent yttrium are made to these alloys to increase hot oxidation and corrosion resistance. Incorporating fine dispersions of inert oxides to this system significantly enhances high-temperature properties and microstructural stability. **Inconel** alloy MA 754, which contains 0.6 percent Y_2O_3, exhibits good fatigue strength and corrosion resistance at 1,100°C (2,010°F) and is used in gas-turbine engines and other extreme service applications.

Inconel alloy 600 is a nickel-chromium-iron alloy. Alloy 600 is a high-strength nonmagnetic [at −40°C (−40°F)] alloy which is used widely for corrosion- and heat-resisting applications at temperatures up to 1,204°C (2,200°F) in sulfidizing atmospheres. In sulfidizing atmospheres, the maximum recommended temperature is 816°C (1,500°F). Inconel alloy X-750 is an age-hardenable modification suitable for stressed applications at temperatures up to 649 to 816°C (1,200 to 1,500°F). The addition of molybdenum to this system provides alloys, e.g., Hastelloy alloy X, with additional solid-solution strengthening and good oxidation resistance up to 1,200°C (2,200°F).

The combination of useful strength and oxidation resistance makes nickel alloys frequent choices for high-temperature service. Table 6.4.48 indicates the changes in mechanical properties with temperature.

Corrosion-Resistant Nickel-Molybdenum and Nickel-Iron-Chromium Alloys A series of solid-solution nickel alloys containing molybdenum exhibit superior resistance to corrosion by hot concentrated acids such as hydrochloric, sulfuric, and nitric acids. Hastelloy alloy C-276, which also contains chromium, tungsten, and cobalt, is resistant to a wide range of chemical process environments including strong oxidizing acids, chloride solutions, and other acids and salts. Incoloy alloy 825, a nickel-iron-chromium alloy with additions of molybdenum, is especially resistant to sulfuric and phosphoric acids. These alloys are used in pollution control, chemical processing, acid production, pulp and paper production, waste treatment, and oil and gas applications. Although these types of alloys are used extensively in aqueous environments, they also have good elevated-temperature properties.

Table 6.4.48 Short-Time High-Temperature Properties of Hot-Rolled Nickel and Its Alloys*

	Temperature, °C							
	21	316	427	540	650	816	982	1,093
Nickel 200†								
Tensile strength, MPa	505	575	525	315	235	170	55	
Yield strength 0.2% offset, MPa	165	150	145	115	105			
Elongation, % in 2 in	40	50	52	55	57	65	91	
Monel alloy 400								
Tensile strength, MPa	560	540	490	350	205	110	55	
Yield strength 0.2% offset, MPa	220	195	200	160	125	60		
Elongation, % in 2 in	46	51	52	29	34	58	45	
Inconel alloy 600								
Tensile strength, MPa	585	545	570	545	490	220	105	75
Yield strength 0.2% offset, MPa	385	185	195	150	150			
Elongation, % in 2 in	50	51	50	21	5	23	51	67
Inconel alloy 718								
Tensile strength, MPa	1,280			1,140	1,030			
Yield strength 0.2% offset, MPa	1,050			945	870			
Elongation, % in 2 in	22			26	15			

* See also data on metals and alloys for use at elevated temperature elsewhere in Sec. 6.4.
† Nickel 200 is not recommended for use above 316°C; low-carbon nickel 201 is the preferred substitute.

Nickel-Iron Alloys   Nickel is slightly ferromagnetic but loses its magnetism at a temperature of 368°C (695°F) when pure. For commercial nickel, this temperature is about 343°C (650°F). Monel alloy 400 is lightly magnetic and loses all ferromagnetism above 93°C (200°F). The degree of ferromagnetism and the temperature at which ferromagnetism is lost are very sensitive to variations in composition and mechanical and thermal treatments.

An important group of soft magnetic alloys is the nickel irons and their modifications, which exhibit high initial permeability, high maximum magnetization, and low residual magnetization. Nickel-iron alloys also exhibit low coefficients of thermal expansion that closely match those of many glasses and are therefore often used for glass-sealing applications. Nickel-iron alloys such as Incoloy alloy 909, which contains cobalt, niobium, and titanium, have found wide applications in gas-turbine and rocket engines, springs and instruments, and as controlled-expansion alloys designed to provide high strength and low coefficients of thermal expansion up to 650°C (1,200°F).

Low-Temperature Properties of Nickel Alloys   Several nickel-base alloys have very good properties at low temperatures, in contrast to ferrous alloys whose impact strength (the index of brittleness) falls off very rapidly with decreasing temperature. The impact strength remains nearly constant with most nickel-rich alloys, while the tensile and yield strengths increase as they do in the other alloys. Specific data on low-temperature properties may be found in White and Siebert, ''Literature Survey of Low-Temperature Properties of Metals,'' Edwards, and ''Mechanical Properties of Metals at Low Temperatures,'' *NBS Circ* 520, 1952.

Welding Alloys   Nickel alloys can be welded with similar-composition welding materials and basic welding processes such as gas tungsten arc welding (GTAW), gas metal arc welding (GMAW), shielded metal arc welding (SMAW), brazing, and soldering. The procedures used are similar to those for stainless steels. Nickel-base welding products are also used to weld dissimilar materials. Nickel-base filler metals, especially with high molybdenum contents, are typically used to weld other alloys to ensure adequate pitting and crevice corrosion resistance in final weld deposits. Nickel and nickel iron welding electrodes are used to weld cast irons.

(*Note:* Inconel, Incoloy, and Monel are trademarks of the Inco family of companies. Hastelloy is a trademark of Haynes International.)

TITANIUM AND ZIRCONIUM

by John R. Schley

REFERENCES: J. Donachie, ''Titanium, A Technical Guide,'' ASM International, Metals Park, OH. ''Titnanium, The Choice,'' rev. 1990, Titanium Development Association, Boulder, CO. ASM Metals Handbook, ''Corrosion,'' vol. 13, 1987 edition, ASM International, Metals Park, OH. Schemel, ''ASTM Manual on Zirconium and Hafnium,'' ASTM STP-639, Philadelphia, PA. ''Corrosion of Zirconium Alloys,'' ASTM STP 368, Philadelphia. ''Zircadyne Properties and Applications,'' Teledyne Wah Chang Albany, Albany, OR. ''Basic Design Facts about Titanium,'' RMI Titanium Company, Niles, OH.

Titanium

The metal titanium is the ninth most abundant element in the earth's surface and the fourth most abundant structural metal after aluminum, iron, and magnesium. It is a soft, ductile, silvery-gray metal with a density about 60 percent of that of steel and a melting point of 1,675°C (3,047°F). A combination of low density and high strength makes titanium attractive for structural applications, particularly in the aerospace industry. These characteristics, coupled with excellent corrosion resistance, have led to the widespread use of titanium in the chemical processing industries and in oil production and refining.

Like iron, titanium can exist in two crystalline forms: hexagonal close-packed (hcp) below 883°C (1,621°F) and body-centered cubic (bcc) at higher temperatures up to the melting point. Titanium combines readily with many common metals such as aluminum, vanadium, zirconium, tin, and iron and is available in a variety of alloy compositions offering a broad range of mechanical and corrosion properties. Certain of these compositions are heat-treatable in the same manner as steels.

Production Processes

The commercial production of titanium begins with an ore, either ilmenite ($FeTiO_3$) or rutile (TiO_2), more often the latter, which passes through a series of chemical reduction steps leading to elemental titanium particles, referred to as **titanium sponge**. The sponge raw material then follows a sequence of processing operations resembling those employed in steelmaking operations. Beginning with the melting of titanium or titanium alloy ingots, the process follows a hot-working sequence of ingot breakdown by forging, then rolling to sheet, plate, or bar product. Titanium is commercially available in all common mill product forms such as billet and bar, sheet and plate, as well as tubing and pipe. It also is rendered into castings.

Mechanical Properties

Commercial titanium products fall into three structural categories corresponding to the allotropic forms of the metal. The alpha structure is present in **commercially pure** (CP) products and in certain alloys containing alloying elements that promote the alpha structure. The alpha-beta form occurs in a series of alloys that have a mixed structure at room temperature, and the beta class of alloys contains elements that stabilize the beta structure down to room temperature. Strength characteristics of the alpha-beta and beta alloys can be varied by heat treatment. Typical alloys in each class are shown in Table 6.4.49, together with their nominal compositions and tensile properties in the annealed condition. Certain of these alloys can be solution-heat-treated and aged to higher strengths.

Corrosion Properties

Titanium possesses outstanding **corrosion resistance** to a wide variety of environments. This characteristic is due to the presence of a protective, strongly adherent oxide film on the metal surface. This film is normally transparent, and titanium is capable of healing the film almost instantly in any environment where a trace of moisture or oxygen is present. It is very stable and is only attacked by a few substances, most notably hydrofluoric acid. Titanium is unique in its ability to handle certain chemicals such as chlorine, chlorine chemicals, and chlorides. It is essentially inert in seawater, for example. The unalloyed CP grades typically are used for corrosion applications in industrial service, although the alloyed varieties are used increasingly for service where higher strength is required.

Fabrication

Titanium is cast and forged by conventional practices. Both means of fabrication are extensively used, particularly for the aerospace industry. Wrought titanium is readily fabricated in the same manner as and on the same equipment used to fabricate stainless steels and nickel alloys. For example, cold forming and hot forming of sheet and plate are done on press brakes and hydraulic presses with practices modified to accommodate the forming characteristics of titanium. Titanium is machinable by all customary methods, again using practices recommended for titanium. Welding is the most common method used for joining titanium, for the metal is readily weldable. The standard welding process is **gas tungsten arc welding** (GTAW), or **TIG**. Two other welding processes, **plasma arc welding** (PAW) and **gas metal arc welding** (GMAW), or MIG, are used to a lesser extent. Since titanium reacts readily with the atmospheric gases oxygen and nitrogen, precautions must be taken to use an inert-gas shield to keep air away from the hot weld metal during welding. Usual standard practices apply to accomplish this. Titanium can be torch-cut by **oxyacetylene flame** or by **plasma torch,** but care must be taken to remove contaminated surface metal. For all titanium fabricating practices cited above, instructional handbooks and other reference materials are available. Inexperienced individuals are well advised to consult a titanium supplier or fabricator.

Applications

Titanium as a material of construction offers the unique combination of wide-ranging corrosion resistance, low density, and high strength. For this reason, titanium is applied broadly for applications that generally

Table 6.4.49   Typical Commercial Titanium Alloys

Designation	Tensile strength MPa	Tensile strength ksi	0.2% Yield strength MPa	0.2% Yield strength ksi	Elonga-tion, %	Al	V	Sn	Zr	Mo	Other	N (max)	C (max)	H (max)	Fe (max)	O (max)
Unalloyed (CP grades:																
ASTM grade 1	240	35	170	25	24	—	—	—	—	—	—	0.03	0.10	0.015	0.20	0.18
ASTM grade 2	340	50	280	40	20	—	—	—	—	—	—	0.03	0.10	0.015	0.30	0.25
ASTM grade 7	340	50	280	40	20	—	—	—	—	—	0.2 Pd	0.03	0.10	0.015	0.30	0.25
Alpha and near-alpha alloys:																
Ti-6 Al-2 Sn-4 Zr-2 Mo	900	130	830	120	10	6.0	—	2.0	4.0	2.0	—	0.05	0.05	0.0125	0.25	0.15
Ti-8 Al-1 Mo-1 V	930	135	830	120	10	8.0	1.0	—	—	1.0	—	0.05	0.08	0.012	0.30	0.12
Ti-5 Al-2 5 Sn	830	120	790	115	10	5.0	—	2.5	—	—	—	0.05	0.08	0.02	0.50	0.20
Alpha-beta alloys:																
Ti-3 Al-2 5 V	620	90	520	75	15	3.0	2.5	—	—	—	—	0.015	0.05	0.015	0.30	0.12
Ti-6 Al-2 V*	900	130	830	120	10	6.0	4.0	—	—	—	—	0.05	0.10	0.0125	0.30	0.20
Ti-6 Al-6 V-2 Sn*	1,030	150	970	140	8	6.0	6.0	2.0	—	—	0.75 Cu	0.04	0.05	0.015	1.0	0.20
Ti-6 Al-2 Sn-4 Zr-6 Mo†	1,170	170	1,100	160	8	6.0	—	2.0	4.0	6.0	—	0.04	0.04	0.0125	0.15	0.15
Beta alloys:																
Ti-10 V-2 Fe-3 Al†	1,170	180	1,100	170	8	3.0	10.0	—	—	—	—	0.05	0.05	0.015	2.2	0.13
Ti-15 V-3 Al-3 Sn-3 Cr†	1,240	180	1,170	170	10	3.0	15.0	3.0	—	—	3.0 Cr	0.03	0.03	0.015	0.30	0.13
Ti-3 Al-8 V-6 Cr-4 Mo-4 Zr*	900	130	860	125	16	3.0	8.0	—	4.0	4.0	6.0 Cr	0.03	0.05	0.020	0.25	0.12

* Mechanical properties given for annealed condition; may be solution-treated and aged to increase strength.
† Mechanical properties given for solution-treated and aged condition.

are categorized as aerospace and nonaerospace, the latter termed *industrial*. The **aerospace applications** consume about 70 percent of the annual production of titanium, and the titanium alloys rather than the CP grades predominate since these applications depend primarily on titanium's high strength/weight ratio. This category is best represented by **aircraft gas-turbine engines,** the largest single consumer of titanium, followed by **airframe structural members** ranging from fuselage frames, wing beams, and landing gear to springs and fasteners. **Spacecraft** also take a growing share of titanium. Nonaerospace or **industrial applications** more often use titanium for its corrosion resistance, sometimes coupled with a requirement for high strength. The unalloyed, commercially pure grades predominate in most corrosion applications and are typically represented by equipment such as heat exchangers, vessels, and piping.

A recent and growing trend is the application of titanium in the offshore oil industry for service on and around offshore platforms. As applied there, titanium uses are similar to those onshore, but with a growing trend toward large-diameter pipe for subsea service. Strength is a major consideration in these applications, and titanium alloys are utilized increasingly.

In the automotive field, the irreversible mandate for lightweight, fuel-efficient vehicles appears to portend a large use of titanium and its alloys.

Zirconium

Zirconium was isolated as a metal in 1824 but was only developed commercially in the late 1940s, concurrently with titanium. Its first use was in nuclear reactor cores, and this is still a major application. Zirconium bears a close relationship to titanium in physical and mechanical properties, but has a higher density, closer to that of steel. Its melting point is 1,852°C (3,365°F) and, like titanium, it exhibits two allotropic crystalline forms. The pure metal is body-centered cubic (bcc) above 865°C (1,590°F) and hexagonal close-packed (hcp) below this temperature. Zirconium is available commercially in unalloyed form and in several alloyed versions, combined most often with small percentages of niobium or tin. It is corrosion-resistant in a wide range of environments.

Zirconium is derived from naturally occurring zircon sand processed in a manner similar to titanium ores and yielding a **zirconium sponge**. The sponge is converted to mill products in the same manner as titanium sponge. The principal commercial compositions intended for corrosion applications together with corresponding mechanical properties are listed in Tables 6.4.50 and 6.4.51.

Table 6.4.50   Chemical Compositions of Zirconium Alloys (Percent)

Grade (ASTM designation)	R60702	R60704	R60705	R60706
Zr + Hf, min	99.2	97.5	95.5	95.5
Hafnium, max	4.5	4.5	4.5	4.5
Fe + Cr	max 0.20	0.2–0.4	max 0.2	max 0.2
Tin	—	1.0–2.0	—	—
Hydrogen, max	0.005	0.005	0.005	0.005
Nitrogen, max	0.025	0.025	0.025	0.025
Carbon, max	0.05	0.05	0.05	0.05
Niobium	—	—	2.0–3.0	2.0–3.0
Oxygen, max	0.16	0.18	0.18	0.16

Table 6.4.51   Minimum ASTM Requirements for the Mechanical Properties of Zirconium at Room Temperature (Cold-Worked and Annealed)

Grade (ASTM designation)	R60702	R60704	R60705	R60706
Tensile strength, min, ksi (MPa)	55 (379)	60 (413)	80 (552)	74 (510)
Yield strength, min, ksi (MPa)	30 (207)	35 (241)	55 (379)	50 (345)
Elongation (0.2% offset), min, percent	16	14	16	20
Bend test radius*	5T	5T	3T	2.5T

* Bend tests are not applicable to material over 0.187 in (4.75 mm) thick, and T equals the thickness of the bend test sample.

Zirconium is a reactive metal that owes its **corrosion resistance** to the formation of a dense, stable, and highly adherent oxide on the metal surface. It is resistant to most organic and mineral acids, strong alkalies, and some molten salts. A notable corrosion characteristic is its resistance to both strongly reducing hydrochloric acid and highly oxidizing nitric acid, relatively unique among metallic corrosion-resistant materials.

Zirconium and its alloys can be machined, formed, and welded by conventional means and practices much like those for stainless steels and identical to those for titanium. Cast shapes are available for equipment such as pumps and valves, and the wrought metal is produced in all standard mill product forms. Persons undertaking the fabrication of zirconium for the first time are well advised to consult a zirconium supplier.

The traditional application of zirconium in the nuclear power industry takes the form of fuel element cladding for reactor fuel bundles. The zirconium material used here is an alloy designated **Zircaloy,** containing about 1.5 percent tin and small additions of nickel, chromium, and iron. It is selected because of its low neutron cross section coupled with the necessary strength and resistance to water corrosion at reactor operating temperatures. Most zirconium is employed here. An interesting noncorrosion zirconium application in photography had been its use as a foil in flash bulbs, but the advent of the electronic flash has curtailed this. An active and growing corrosion application centers on fabricated chemical processing equipment, notably heat exchangers, but also a variety of other standard equipment.

ZINC AND ZINC ALLOYS

by Frank E. Goodwin

REFERENCES: Porter, "Zinc Handbook," Marcel Dekker, New York. ASM "Metals Handbook," 10th ed., vol. 2. ASTM, "Annual Book of Standards." "NADCA Product Specification Standards for Die Casting," Die Casting Development Council.

Zinc, one of the least expensive nonferrous metals, is produced from sulfide, silicate, or carbonate ores by a process involving concentration and roasting followed either by thermal reduction in a zinc-lead blast furnace or by leaching out the oxide with sulfuric acid and electrolyzing the solution after purification. Zinc distilled from blast-furnace production typically contains Pb, Cd, and Fe impurities that may be eliminated by fractional redistillation to produce zinc of 99.99 + percent purity. Metal of equal purity can be directly produced by the electrolytic process. Zinc ingot range from cast balls weighing a fraction of a pound to a 1.1-ton [1 metric ton (t)] "jumbo" blocks. Over 20 percent of zinc metal produced each year is from recycled scrap.

The three standard grades of zinc available in the United States are described in ASTM specification B-6 (see Table 6.4.52). **Special high grade** is overwhelmingly the most commercially important and is used in all applications except as a coating for steel articles galvanized after fabrication. In other applications, notably die casting, the higher levels of impurities present in grades other than special high grade can have harmful effects on corrosion resistance, dimensional stability, and formability. Special high grade is also used as the starting point for brasses containing zinc.

Galvanized Coatings

Protective coatings for steel constitute the largest use of zinc and rely upon the galvanic or sacrificial property of zinc relative to steel. Lead can be added to produce the solidified surface pattern called **spangle** preferred on unpainted articles. The addition of aluminum in amounts of 5 and 55 percent results in coatings with improved corrosion resistance termed **Galfan** and **Galvalume,** respectively.

Zinc Die Castings

Zinc alloys are particularly well suited for making die castings since the melting point is reasonably low, resulting in long die life even with ordinary steels. High accuracies and good surface finish are possible.

Table 6.4.52   ASTM Specification B-6-77 for Slab Zinc

Grade	Composition, %			
	Lead, max	Iron, max	Cadmium, max	Zinc, min, by difference
Special high grade*	0.003	0.003	0.003	99.990
High grade	0.03	0.02	0.02	99.90
Prime western†	1.4	0.05	0.20	98.0

* Tin in special high grade shall not exceed 0.001%.
† Aluminum in prime western zinc shall not exceed 0.05%.
SOURCE: ASTM, reprinted with permission.

Alloys currently used for die castings in the United States are covered by ASTM Specifications B-86 and B-669. Nominal compositions and typical properties of these compositions are given in Table 6.4.53. The low limits of impurities are necessary to avoid disintegration of the casting by intergranular corrosion under moist atmospheric conditions. The presence of magnesium or nickel prevents this if the impurities are not higher than the specification values. The mechanical properties shown in Table 6.4.53 are from die-cast test pieces of 0.25-in (6.4-mm) section thickness. Zinc die castings can be produced with section thicknesses as low as 0.03 in (0.75 mm) so that considerable variations in properties from those listed must be expected. These die-casting alloys generally increase in strength with increasing aluminum content. The 8, 12, and 27 percent alloys can also be gravity-cast by other means.

Increasing copper content results in growth of dimensions at elevated temperatures. A measurement of the expansion of the die casting after exposure to water vapor at 203°F (95°C) for 10 days is a suitable index of not only stability but also freedom from susceptibility to intergranular corrosion. When held at room temperature, the copper-containing alloys begin to shrink immediately after removal from the dies; total **shrinkage** will be approximately two-thirds complete in 5 weeks. The maximum extent of this **shrinkage** is about 0.001 in/in (10 μm/cm). The copper-free alloys do not exhibit this effect. Stabilization may be effected by heating the alloys for 3 to 6 h at 203°F (95°C), followed by air cooling to room temperature.

Wrought Zinc

Zinc rolled in the form of sheet, strip, or plate of various thicknesses is used extensively for automobile trim, dry-cell battery cans, fuses, and plumbing applications. Compositions of standard alloys are shown in Table 6.4.54. In addition, a comparatively new series of alloys containing titanium, exhibiting increased strength and creep resistance together with low thermal expansion, has become popular in Europe. A typical analysis is 0.5 to 0.8 percent Cu, 0.08 to 0.16 percent Ti, and maximum values of 0.2 percent Pb, 0.015 percent Fe, 0.01 percent Cd, 0.01 percent Mn, and 0.02 percent Cr. All alloys are produced by hot rolling followed by cold rolling when some stiffness and temper are required. Deep-drawing or forming operations are carried out on the softer grades, while limited forming to produce architectural items, plumbing, and automotive trim can be carried out using the harder grades.

Alloys containing 0.65 to 1.025 percent Cu are significantly stronger than unalloyed zinc and can be work-hardened. The addition of 0.01 percent Mg allows design stresses up to 10,000 lb/in^2 (69 MPa). The Zn-Cu-Ti alloy is much stronger, with a typical tensile strength of 25,000 lb/in^2 (172 MPa).

Rolled zinc may be easily formed by all standard techniques. The deformation behavior of rolled zinc and its alloys varies with direction; its crystal structure renders it nonisotropic. In spite of this, it can be formed into parts similar to those made of copper, aluminum, and brass, and usually with the same tools, provided the temperature is not below 70°F (21°C). More severe operations can best be performed at temperatures up to 125°F (52°C). When a cupping operation is performed, a take-in of 40 percent on the first draw is usual. Warm, soapy water is widely used as a lubricant. The soft grades are self-annealing at room temperature, but harder grades respond to deformation better if they are annealed between operations. The copper-free zincs are annealed at 212°F (100°C) and the zinc-copper alloys at 440°C (220°C). Welding is

Table 6.4.53 Composition and Typical Properties of Zinc-Base Die Casting Alloys

Element	Composition, %					
	UNS Z33521 (AG40A) Alloy 3	UNS Z33522 (AG40B) Alloy 7	UNS Z35530 (AC41A) Alloy 5	UNS Z25630 ZA-8	UNS Z35630 ZA-12	UNS Z35840 ZA-27
Copper	0.25 max	0.25 max	0.75–1.25	0.8–1.3	0.5–1.25	2.0–2.5
Aluminum	3.5–4.3	3.5–4.3	3.5–4.3	8.0–8.8	10.5–11.5	25.0–28.0
Magnesium	0.020–0.05	0.005–0.020	0.03–0.08	0.015–0.030	0.015–0.030	0.010–0.020
Iron, max	0.100	0.075	0.100	0.10	0.075	0.10
Lead, max	0.005	0.0030	0.005	0.004	0.004	0.004
Cadmium, max	0.004	0.0020	0.004	0.003	0.003	0.003
Tin, max	0.003	0.0010	0.003	0.002	0.002	0.002
Nickel	—	0.005–0.020	—	—	—	—
Zinc	Rest	Rest	Rest	Rest	Rest	Rest
Yield strength						
lb/in^2	32	32	39	41–43	45–48	52–55
MPa	221	221	269	283–296	310–331	359–379
Tensile strength						
lb/in^2	41.0	41.0	47.7	54	59	62
MPa	283	283	329	372	400	426
Elongation in 2 in (5 cm), %	10	14	7	6–10	4–7	2.0–3.5
Charpy impact on square specimens						
ft/lb	43	43	48	24–35	15–27	7–12
J	58	58	65	32–48	20–37	9–16
Brinell Hardness number on square specimens, 500-kg load, 10-mm ball, 30 s	82	80	91	100–106	95–105	116–122

SOURCE: ASTM, reprinted with permission.
Compositions from ASTM Specifications B-86-83 and B-669-84. Properties from NADCA Product Specification Standards for Die Castings.

Table 6.4.54 Typical Composition of Rolled Zinc ASTM Specification B69-66, %

Lead	Iron, max	Cadmium	Copper	Magnesium	Zinc
0.05 max	0.010	0.005 max	0.001 max	—	Remainder
0.05–0.12	0.012	0.005 max	0.001 max	—	Remainder
0.30–0.65	0.020	0.20–0.35	0.005 max	—	Remainder
0.05–0.12	0.012	0.005 max	0.65–1.25	—	Remainder
0.05–0.12	0.015	0.005 max	0.75–1.25	0.007–0.02	Remainder

SOURCE: ASTM, reprinted with permission.

possible with a wire of composition similar to the base metals. Soldering with typical tin-lead alloys is exceptionally easy. The impact extrusion process is widely used for producing battery cups and similar articles.

Effect of Temperature

Properties of zinc and zinc alloys are very sensitive to temperature. **Creep resistance** decreases with increasing temperature, and this must be considered in designing articles to withstand continuous load. When steel screws are used to fasten zinc die castings, maximum long-term clamping load will be reached if an engagement length 4 times the diameter of the screw is used along with cut (rather than rolled) threads on the fasteners.

6.5 CORROSION
by Robert D. Bartholomew and David A. Shifler

REFERENCES: Uhlig, "The Corrosion Handbook," Wiley, New York. Evans, "An Introduction to Metallic Corrosion," 3d ed., Edward Arnold & ASM International, Metals Park, OH. van Delinder (ed.), "Corrosion Basics—An Introduction," NACE International, Houston. Wranglen, "An Introduction to Corrosion and Protection of Metals," Chapman and Hall, London. Fontana, "Corrosion Engineering," 3d ed., McGraw-Hill, New York. Jones, "Principles and Prevention of Corrosion," Macmillan, New York. Scully, "The Fundamentals of Corrosion," 3d ed., Pergamon, New York. Kaesche, "Metallic Corrosion," 2d ed., NACE International, Houston. Sheir (ed.), "Corrosion," 2d ed., Newnes-Butterworths, London. "Corrosion," vol. 13, "Metals Handbook," 9th ed., ASM International, Metals Park, OH. Gellings, "Introduction to Corrosion Prevention and Control," Delft University Press, Delft, The Netherlands. Pourbaix, "Atlas of Electrochemical Equilibria in Aqueous Solutions," NACE International, Houston. Prentice, "Electrochemical Engineering Principles," Prentice-Hall, Englewoods Cliffs, NJ. Bockris and Reddy, "Modern Electrochemistry," Plenum, New York. Bockris and Khan, "Surface Electrochemistry—A Molecular Level Approach," Plenum, New York.

INTRODUCTION

Corrosion is the deterioration of a material or its properties due to its reaction with the environment. Materials may include metals and alloys, nonmetals, woods, ceramics, plastics, composites, etc. Although corrosion as a science is barely 150 years old, its effects have affected people for thousands of years. The importance of understanding the causes, initiation, and propagation of corrosion and the methods for controlling its degradation is threefold. First, corrosion generates an **economic impact** through materials losses and failures of various structures and components. An SSINA-Battelle study estimated that the combined economic loss to the United States in 1995 due to metal and alloy corrosion (nonmetallics not included) was $300 billion [Payer, *Mater. Perf.* v. 34 (1995)]. Studies of corrosion costs in other countries have determined that 3 to 4 percent of GNP is related to economic losses from corrosion. These estimates consider only the direct economic costs. Indirect costs are difficult to assess but can include plant downtime, loss of products or services, lowered efficiency, contamination, and overdesign of structures and components. Application of current corrosion control technology (discussed later) could recover or avoid 15 to 20 percent of the costs due to corrosion.

Second, an important consideration of corrosion prevention or control is improved **safety**. Catastrophic degradation and failures of pressure vessels, petrochemical plants, boilers, airplane sections, tanks containing toxic materials, and automotive parts have led to thousands of personal injuries and deaths, which often result in subsequent litigation. If the safety of individual workers or the public is endangered in some manner by selection of a material or use of a corrosion control method, that approach should be abandoned. Once assurance is given that this criterion is met through proper materials selection and improved design and corrosion control methods, other factors can be evaluated to provide the optimum solution.

Third, understanding factors leading to corrosion can **conserve resources**. The world's supply of easily extractable raw materials is limited. Corrosion also imparts wastage of energy water resources, and raw or processed materials.

Corrosion can occur through chemical or electrochemical reactions and is usually an interfacial process between a material and its environment. Deterioration solely by physical means such as erosion, galling, or fretting is not generally considered corrosion. **Chemical corrosion** may be a heterogeneous reaction which occurs at the metal/environment interface and involves the material (metal) itself as one of the reactants. Such occurrences may include metals in strong acidic or alkaline solutions, high-temperature oxidation metal/gas reactions where the oxide or compound is volatile, breakdown of chemical bonds in polymers, dissolution of solid metals or alloys in a liquid metal (e.g., aluminum in mercury), or dissolution of metals in fused halides or nonaqueous solutions.

However, even in acidic or alkaline solutions and in most high-temperature environments, **electrochemical corrosion** consisting of two or more partial reactions involves the transfer of electrons or charges. An electrochemical reaction requires (1) an anode, (2) a cathode, (3) an electrolyte, and (4) a complete electric circuit. In the reaction of a metal (M) in hydrochloric acid, the metal is oxidized and electrons are generated at the anode [see Eq. (6.5.1)]. At the same time, hydrogen cations are reduced and electrons are consumed, to evolve hydrogen gas at the cathode—the **hydrogen evolution reaction** (HER) described in Eq. (6.5.2).

Anodic (oxidation) reaction: $\qquad$ $M \rightarrow M^{+n} + ne^-$ $\qquad$ (6.5.1)

Cathodic (reduction) reaction: $\qquad$ $2H^+ + 2e^- \rightarrow H_2$ $\qquad$ (6.5.2)

Generally, oxidation occurs at the anode, while reduction occurs at the cathode. Corrosion is usually involved at the anode where the metal or alloy is oxidized; this causes dissolution, wastage, and penetration. Cathodic reactions significant to corrosion are few. These may include, in addition to HER, the following:

Oxygen reduction

(acidic solutions): $\qquad$ $O_2 + 4H^+ + 4e^- \rightarrow 2H_2O$ $\qquad$ (6.5.3)

Oxygen reduction

(neutral or basic): $\qquad$ $O_2 + 2H_2O \rightarrow 4e^- \rightarrow 4OH^-$ $\qquad$ (6.5.4)

Metal-iron reduction: $\qquad$ $Fe^{+3} + e^- \rightarrow Fe^{2+}$ $\qquad$ (6.5.5)

Metal deposition: $\qquad$ $Cu^{2+} + 2e^- \rightarrow Cu$ $\qquad$ (6.5.6)

THERMODYNAMICS OF CORROSION

For an anodic oxidation reaction of metal to occur, a simultaneous reduction must take place. In corroding metal systems, the anodic and cathodic half-reactions are mutually dependent and form a galvanic or spontaneous cell. A cell in which electrons are driven by an external energy source in the direction counter to the spontaneous half-reactions is termed an **electrolytic cell**. A metal establishes a potential or emf with respect to its environment and is dependent on the ionic strength and composition of the electrolyte, the temperature, the metal or alloy itself, and other factors. The potential of a metal at the anode in solution arises from the release of positively charged metal cations together with the creation of a negatively charged metal. The standard potential of a metal is defined by fixing the equilibrium concentration of its ions at unit activity and under reversible (zero net current) and standard conditions (1 atm, 101 kPa; 25°C). At equilibrium, the net current density ($\mu A/cm^2$) of an electrochemical reaction is zero. The nonzero anodic and cathodic currents are equal and opposite, and the absolute magnitude of either current at equilibrium is equal to the exchange current density (that is, $i_{an} = i_{cath} = i_o$).

The potential, a measure of the driving influence of an electrochemical reaction, cannot be evaluated in absolute terms, but is determined by the difference between it and another reference electrode. Common **standard reference electrodes** used are saturated calomel electrode (SCE, 0.2416 V) and saturated Cu/CuSO$_4$ (0.337 V) whose potentials are measured relative to **standard hydrogen electrode** (SHE), which by definition is 0.000 V under standard conditions. Positive electrochemical potentials of half-cell reactions versus the SHE (written as a reduction reaction) are more easily reduced and noble, while negative values signify half-cell reactions that are more difficult to reduce and, conversely, more easily oxidized or active than the SHE. The signs of the potentials are reversed if the SHE and half-cell reactions of interest are written as oxidation reactions. The potential of a galvanic cell is the sum of the potentials of the anodic and cathodic half-cells in the environment surrounding them.

From thermodynamics, the potential of an electrochemical reaction is a measure of the **Gibbs free energy** $\Delta G = -nFE$, where n is the number of electrons participating in the reaction, F is Faraday's constant (96,480 C/mol), and E is the electrode potential. The potential of the galvanic cell will depend on the concentrations of the reactants and products of the respective partial reactions, and on the pH in aqueous solutions. The potential can be related to the Gibbs free energy by the **Nernst equation**

$$\Delta E = \Delta E° + \frac{2.3RT}{nF} \log \frac{(ox)^x}{(red)^r}$$

where $\Delta E°$ is the standard electrode potential, (ox) is the activity of an oxidized species, (red) is the activity of the reduced species, and x and r are stoichiometric coefficients involved in the respective half-cell reactions. Corrosion will not occur unless the spontaneous direction of the reaction (that is, $\Delta G < 0$) indicates metal oxidation. The application of thermodynamics to corrosion phenomena has been generalized by use of **potential-pH plots** or **Pourbaix diagrams**. Such diagrams are constructed from calculations based on the Nernst equation and solubility data for various metal compounds. As shown in Fig. 6.5.1, it is possible to differentiate regions of potential versus pH in which iron either is immune or will potentially passivate from regions in which corrosion will thermodynamically occur. The main uses of these diagrams are to: (1) predict the spontaneous directions of reactions, (2) estimate the composition of corrosion products, and (3) predict environmental changes that will prevent or reduce corrosion. The major limitations of Pourbaix diagrams are that (1) only pure metals, and not alloys, are usually considered; (2) pH is assumed to remain constant, whereas HER may alter the pH; (3) they do not provide information on metastable films; (4) possible aggressive solutions containing Cl^-, Br^-, I^-, or NH_4^+ are not usually considered; and (5) they do not predict the kinetics or corrosion rates of the different electrochemical reactions. The predictions possible are based on the metal, solution, and temperature designated in the diagrams. Higher-temperature potential-pH diagrams

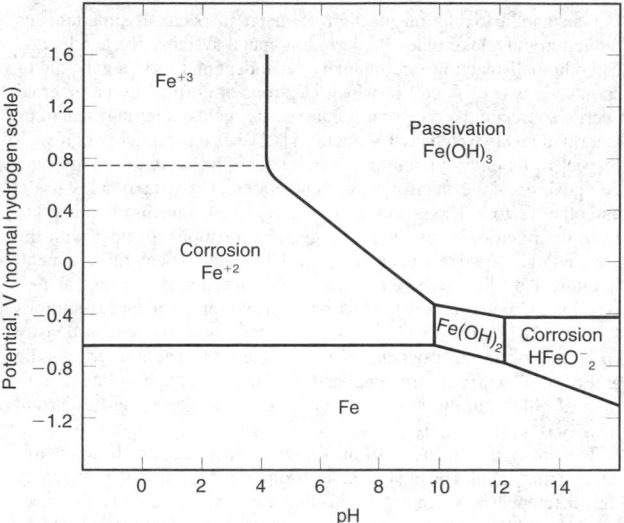

Fig. 6.5.1 Simplified potential-pH diagram for the Fe-H_2O system. (*Pourbaix, "Atlas of Electrochemical Equilibria in Aqueous Solutions."*)

have been developed (*Computer-Calcuated Potential-pH Diagrams to 300°C*, NP-3137, vol. 2. Electric Power Research Institute, Palo Alto, CA). **Ellingham diagrams** provide thermodynamically derived data for pure metals in gaseous environments to predict stable phases, although they also do not predict the kinetics of these reactions.

CORROSION KINETICS

Corrosion is thermodynamically possible for most environmental conditions. It is of primary importance to assess the kinetics of corrosion. While free energy and electrode potential are thermodynamic parameters of an electrochemical reaction, the equilibrium exchange current density i_0 is a fundamental kinetic property of the reaction. It is dependent on the material, surface properties, and temperature.

Anodic or cathodic electrochemical reactions such as established in Eqs. (6.5.1) to (6.5.6) often proceed at finite rates. When a cell is short-circuited, net oxidation and reduction occur at the anode and cathode, respectively. The potentials of these electrodes will no longer be at equilibrium. This deviation from **equilibrium potential** is termed **polarization** and results from the flow of a net current. Overvoltage or overpotential is a measure of polarization. Deviations from the equilibrium potential may be caused by (1) concentration polarization, (2) activation polarization, (3) resistance polarization, or (4) mixed polarization. **Concentration polarization** η_c is caused by the limiting diffusion of the electrolyte and is generally important only for cathodic reactions. For example, at high reduction rates of HER, the cathode surface will become depleted of hydrogen ions. If the reduction is increased further, the diffusion rate of hydrogen ions to the cathode will approach a limiting diffusion current density i_L. This limiting current density can be increased by agitation, higher temperatures, and higher reactant concentrations.

Activation polarization η_a refers to the electrochemical process that is controlled by the slow rate-determining step at the metal/electrolyte interface. Activation polarization is characteristic of cathodic reactions such as HER and metal-ion deposition or dissolution. The relationship between activation polarization and the rate of reaction i_A at the anode is $\eta_{a(A)} = \beta_A \log i_A/i_o$, where β_A represents the Tafel slope (~ 0.60 to

0.120 V per decade of current) of the anodic half-reaction. A similar expression can be written for the cathodic half-reaction.

Resistance polarization η_R includes the ohmic potential drop through a portion of the electrolyte surrounding the electrode, or the ohmic resistance in a metal-reaction product film on the electrode surface, or both. High-resistivity solutions and insulating films deposited at either the cathode or anode restrict or completely block contact between the metal and the solution and will promote a high-resistance polarization. **Mixed polarization** occurs in most systems and is the sum of η_c, η_a, and η_R. Given β, i_L, and i_o, the kinetics of almost any corrosion reaction can be described.

When polarization occurs mostly at the anodes of a cell, the corrosion reaction is anodically controlled. When polarization develops mostly at the cathode, the corrosion rate is cathodically controlled. Resistance control occurs when the electrolyte resistance is so high that the resultant current is insufficient to polarize either the anode or the cathode. Mixed control is common in most systems in which both anodes and cathodes are polarized to some degree.

Hydrogen overpotential is a dominant factor in controlling the corrosion rate of many metals either in deaerated water or in nonoxidizing acids at cathodic areas; the overpotential is dependent on the metal, temperature, and surface roughness. The result of hydrogen overpotential often will be a surface film of hydrogen atoms. Decreasing pH usually will promote the HER reaction and corrosion rate; acids and acidic salts create a corrosive environment for many metals. Conversely, high-pH conditions may increase hydrogen overpotential and decrease corrosion and may provide excellent protection even without film formation, such as when caustic agents, ammonia, or amines are added to condensate or to demineralized or completely softened water.

Oxygen dissolved in water reacts with the atomic hydrogen film on cathodic regions of the metal surface by depolarization and enables corrosion to continue. The rate of corrosion is approximately limited by the rate at which oxygen diffuses toward the cathode (**oxygen polarization**); thus, extensive corrosion occurs near or at the waterline of a partially immersed or filled steel specimen. In systems where both hydrogen ions and oxygen are present, the initial predominant cathodic reaction will be the one that is most thermodynamically and kinetically favorable.

Corrosion rates have been expressed in a variety of ways. Mils per year (mils/yr) and μm/yr (SI) are desirable forms in which to measure corrosion or penetration rates and to predict the life of a component from the weight loss data of a corrosion test, as shown by Eqs. (6.5.7a) and (6.5.7b).

$$\text{mils/yr} = \frac{534W}{DAT} \qquad (6.5.7a)$$

$$\mu\text{m/yr} = \frac{87,600W}{DAT} \qquad (6.5.7b)$$

where W = weight loss, mg; D = density of specimen, g/cm³; A = area of specimen (mpy; sq. in.) (μm/yr., sq. cm); T = exposure time, h. The corrosion rate considered detrimental for a metal or an alloy will depend on its initial cost, design life, cost of materials replacement, environment, and temperature.

During metallic corrosion, the sum of the anodic currents must equal the sum of the cathodic currents. The corrosion potential of the metal surface is defined by the intersection of the anodic and cathodic polarization curves, where anodic and cathodic currents are equal. Figure 6.5.2 illustrates an idealized polarization curve common for most metals, while Fig. 6.5.3 displays typical anodic dissolution behavior of **active/passive metals** such as iron, chromium, cobalt, molybdenum, nickel, titanium, and alloys containing major amounts of these elements. Anodic currents increase with higher anodic potentials until a critical anodic current density i_{pp} is reached at the primary passive potential E_{pp}. Higher anodic potentials will result in formation of a **passive film**.

Generally, metals and alloys (with the possible exception of gold) are not in their most stable thermodynamic state. Corrosion tends to convert these pure materials back to their stable thermodynamic form. The major factor controlling corrosion of metals and alloys is the nature of the **protective, passive film** in the environment. **Passivity** may be defined as the loss of chemical reactivity under certain environmental conditions. A second definition is illustrated by Fig. 6.5.3; a metal becomes passive if, on increasing its potential to more anodic or positive values, the rate of anodic dissolution decreases (even though the rate should increase as the potential increases). A metal can passivate by (1) chemisorption of the solvent, (2) salt film formation, (3) oxide/oxyhydroxide

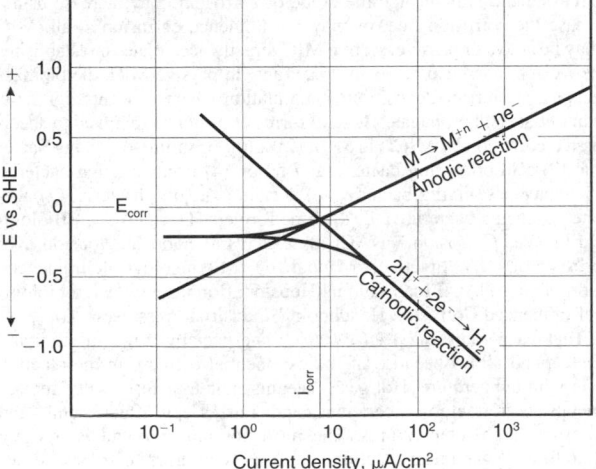

Fig. 6.5.2 Idealized polarization common for most metals at E_{corr}, where anodic ($i_{corr} = i_a$) and cathodic currents are equal. *(Adapted from "Fontana's Book on Corrosive Engineering," McGraw-Hill.)*

formation (Hoar et al., *Corr. Sci.,* **5,** 1965, p. 279; Agladze et al., *Prot. of Metals,* **22,** 1987, p. 404), and (4) electropolymerization (Shifler et al., *Electrochim. Acta,* **38,** 1993, p. 881). The stability of the passive film will depend on its chemical, ionic, and electronic properties; its degree of crystallinity; the film's flexibility; and the film mechanical properties in a given environment [Frankenthal and Kruger (eds.), "Passivity of Metals— Proc. of the Fourth International Symposium on Passivity," Electrochemical Society, Pennington, NJ]. The **oxide/oxyhydroxide** film is the most common passive film at ambient temperatures in most environments. The passive layer thickness, whether considered as an absorbed oxygen structure or an oxide film, is generally 50 nm or less.

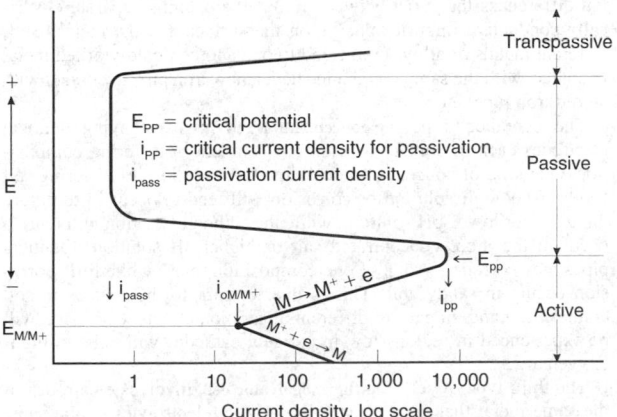

Fig. 6.5.3 Idealized anodic polarization curve for active/passive metals that exhibit passivity. Three different potential regions are identified. *(Adapted from "Fontana's Book on Corrosive Engineering," McGraw-Hill.)*

FACTORS INFLUENCING CORROSION

A number of factors influence the stability and breakdown of the **passive film,** and include (1) surface finish, (2) metallurgy, (3) stress, (4) heat treatment, and (5) environment. Polished surfaces generally resist corrosion initiation better than rough surfaces. Surface roughness tends to increase the kinetics of the corrosion reaction by increasing the exchange current density of the HER or the oxygen reduction at the cathodes.

Metallurgical structures and properties often have major effects on corrosion. Regions of varying composition exist along the surface of most metals or alloys. These local compositional changes have different potentials that may initiate local-action cells. Nonmetallic inclusions, particularly sulfide inclusions, are known to initiate corrosion on carbon and stainless steels. The size, shape, and distribution of sulfide inclusions in 304 stainless steel may have a large impact on dissolution kinetics and pitting susceptibility and growth. Elements such as chromium and nickel improve the corrosion resistance of carbon steel by improving the stability of the passive oxide film. Copper (0.2 percent) added to carbon steel improves the atmospheric corrosion resistance of weathering steels.

Stresses, particularly tensile stresses, affect corrosion behavior. These stresses may be either applied or residual. **Residual stresses** may either arise from dislocation pileups or stacking faults due to deforming or cold-working the metal; these may arise from forming, heat-treating, machining, welding, and fabrication operations. Cold working increases the stresses applied to the individual grains by distorting the crystals. Corrosion at cold-worked sites is not increased in natural waters, but increases severalfold in acidic solutions. Possible segregation of carbon and nitrogen occurs during cold working. Welding can induce residual stresses and provide sites that are subject to preferential corrosion and cracking. The welding of two dissimilar metals with different thermal expansion coefficients may restrict expansion of one member and induce applied stresses during service. Thermal fluctuations of fabrication bends or attachments welded to components may cause applied stresses to develop that may lead to cracking.

Improper **heat treatment** or welding can influence the microstructure of different alloys by causing either precipitation of deleterious phases at alloy grain boundaries or (as is the case of austenitic stainless steels) depletion of chromium in grain boundary zones, which decreases the local corrosion resistance. **Welding** also can cause phase transformations, formation of secondary precipitates, and induce stresses in and around the weld. Rapid quenching of steels from austenizing temperatures may form martensite, a distorted tetragonal structure, that often suffers from preferential corrosion.

The nature of the **environment** can affect the rate and form of corrosion. Environments include (1) natural and treated waters, (2) the atmosphere, (3) soil, (4) microbiological organisms, and (5) high temperature. Corrosivity in freshwater varies with oxygen content, hardness, chloride and sulfur content, temperature, and velocity. Water contains colloidal or suspended matter and dissolved solids and gases. All these constituents may stimulate or suppress corrosion either by affecting the cathodic or anodic reaction or by forming a protective barrier. Oxygen is probably the most significant constituent affecting corrosion in neutral and alkaline solutions; hydrogen ions are more significant in acidic solutions. **Freshwater** can be hard or soft. In **hard waters,** calcium carbonate often deposits on the metal surface and protects it; pitting may occur if the calcareous coating is not complete. **Soft waters** are usually more corrosive because protective deposits do not form. Several saturation indices are used to provide scaling tendencies. Nitrates and chlorides increase aqueous conductivity and reduce the effectiveness of natural protective films. The chloride/bicarbonate ratio has been observed to predict the probability of dezincification. Sulfides in polluted waters tend to cause pitting. Deposits on metal surfaces may lead to local stagnant conditions which may lead to pitting. High velocities usually increase corrosion rates by removing corrosion products which otherwise might suppress the anodic reaction and by stimulating the cathodic reaction by providing more oxygen. **High-purity water,** used in nuclear and high-pressure power units, decreases corrosion by increasing the electrical resistance of the fluid. Temperature effects in aqueous systems are complex, depending on the nature of the cathodic and anodic reactions.

Seawater is roughly equivalent to 3½ percent sodium chloride, but also contains a number of other major constituents and traces of almost all naturally occurring elements. Seawater has a higher conductivity than freshwater, which alone can increase the corrosion of many metals. The high conductivity permits larger areas to participate in corrosion reactions. The high chloride content in seawater can increase localized breakdown of oxide films. The pH of seawater is usually 8.1 to 8.3. Plant photosynthesis and decomposition of marine organisms can raise or lower the pH, respectively. Because of the relatively high pH, the most important cathodic reaction in seawater corrosion processes is oxygen reduction. Highly aerated waters, such as tidal splash zones, are usually regions of very high corrosion rates. Barnacles attached to metal surfaces can cause localized attack if present in discontinuous barriers. Copper and copper alloys have the natural ability to suppress barnacles and other **microfouling organisms**. Sand, salt, and abrasive particles suspended in seawater can aggravate erosion corrosion. Increased seawater temperatures generally increase corrosivity (Laque, "Marine Corrosion—Causes and Prevention," Wiley-Interscience, New York).

In general, **atmospheric corrosion** is the result of the conjoint action of oxygen and water, although contaminants such as sodium chloride and sulfur dioxide accelerate corrosion rates. In the absence of moisture (below 60 to 70 percent relative humidity), most metals corrode very slowly at ambient temperatures. Water is required to provide an electrolyte for charge transfer. Damp corrosion requires moisture from the atmosphere; wet corrosion occurs when water pockets or visible water layers are formed on the metal surface due to salt spray, rain, or dew formation. The solubility of the corrosion products can affect the corrosion rate during wet corrosion; soluble corrosion products usually increase corrosion rates.

Time of wetness is a critical variable that determines the duration of the electrochemical corrosion processes in atmospheric corrosion. Temperature, climatic conditions, relative humidity, and surface shape and conditions that affect time of wetness also influence the corrosion rate. Metal surfaces that retain moisture generally corrode faster than surfaces exposed to rain. Atmospheres can be classified as rural, marine, or industrial. Rural atmospheres tend to have the lowest corrosion rates. The presence of NaCl near coastal shores increases the aggressiveness of the atmosphere. Industrial atmospheres are more corrosive than rural atmospheres because of sulfur compounds. Under humid conditions SO_2 can promote the formation of sulfurous or sulfuric acid. Other contaminants include nitrogen compounds, H_2S, and dust particles. Dust particles adhere to the metal surface and absorb water, prolonging the time of wetness; these particles may include chlorides that tend to break down passive films [Ailor (ed.), "Atmospheric Corrosion," Wiley-Interscience, New York].

Soil is a complex, dynamic environment that changes continuously, both chemically and physically, with the seasons of the year. Characterizing the **corrosivity of soil** is difficult at best. **Soil resistivity** is a measure of the concentration and mobility of ions required to migrate through the soil electrolyte to a metal surface for a corrosion reaction to continue. Soil resistivity is an important parameter in underground corrosion; high resistivity values often suggest low corrosion rates. The mineralogical composition and earth type affect the grain size, effective surface area, and pore size which, in turn, affect soil corrosivity. Further, soil corrosivity can be strongly influenced by certain chemical species, microorganisms, and soil acidity or alkalinity. A certain water content in soil is required for corrosion to occur. Oxygen also is generally required for corrosion processes, although steel corrosion can occur under oxygen-free, anaerobic conditions in the presence of **sulfate-reducing bacteria** (SRB). Soil water can regulate the oxygen supply and its transport. [Romanoff, Underground Corrosion, *NBS Circ.,* **579,** 1957, available from NACE International, Houston; Escalante (ed.), "Underground Corrosion," STP 741, ASTM, Philadelphia].

Almost all commercial alloys are affected by **microbiological in-**fluenced corrosion (MIC). Most MIC involves localized corrosion. Biological organisms are present in virtually all natural aqueous environments (freshwater, brackish water, seawater, or industrial water) and in some soils. In these environments, the tendency is for the microorganisms to attach to and grow as a biofilm on the surface of structural materials. Environmental variables (pH, velocity, oxidizing power, temperature, electrode polarization, and concentration) under a biofilm can be vastly different from those in the bulk environment. MIC can occur under aerobic or anaerobic conditions. Biofilms may cause corrosion under conditions that otherwise would not cause dissolution in the environment, may change the mode of corrosion, may increase or decrease the corrosion rate, or may not influence corrosion at all. MIC may be active or passive. Active MIC directly accelerates or establishes new electrochemical corrosion reactions. In passive MIC, the biomass acts as any dirt or deposition accumulation where concentration cells can initiate and propagate. Several forms of bacteria are linked to accelerated corrosion by MIC: (1) SRB, (2) sulfur or sulfide-oxidizing bacteria, (3) acid-producing bacteria and fungi, (4) iron-oxidizing bacteria, (5) manganese-fixing bacteria, (6) acetate-oxidizing bacteria, (7) acetate-producing bacteria, and (8) slime formers [Dexter (ed.), "Biological Induced Corrosion," NACE-8, NACE International, Houston; Kobrin (ed.), "A Practical Manual on Microbiological Influenced Corrosion," NACE International Houston; Borenstein, "Microbiological Influenced Corrosion Handbook," Industrial Press, New York].

High-temperature service ($\geq 100°C$) is especially damaging to many metals and alloys because of the exponential increase in the reaction rate with temperature. Hot gases, steam, molten or fused salts, molten metals, or refractories, ceramics, and glasses can affect metals and alloys at high temperatures. The most common reactant is oxygen; therefore all gas-metal reactions are usually referred to as oxidations, regardless of whether the reaction involves oxygen, steam, hydrogen sulfide, or combustion gases. High-temperature oxidation reactions are generally not electrochemical; diffusion is a fundamental property involved in oxidation reactions. The corrosion rate can be influenced considerably by the presence of contaminants, particularly if they are adsorbed on the metal surface. Pressure generally has little effect on the corrosion rate unless the dissociation pressure of the oxide or scale constituent lies within the pressure range involved. Stress may be important when attack is intergranular. Differential mechanical properties between a metal and its scale may cause periodic scale cracking which leads to accelerated oxidation. **Thermal cycling** can lead to cracking and flaking of scale layers. A variety of molten salt baths are used in industrial processes. Salt mixtures of nitrates, carbonates, or halides of alkaline or alkaline-earth metals may adsorb on a metal surface and cause beneficial or deleterious effects.

There are three major types of **cells** that transpire in corrosion reactions. The first is the **dissimilar electrode cell**. This is derived from potential differences that exist between a metal containing separate electrically conductive impurity phases on the surface and the metal itself, different metals or alloys connected to one another, cold-worked metal in contact with the same metal annealed, a new iron pipe in contact with an old iron pipe, etc.

The second cell type, the **concentration cell,** involves having identical electrodes each in contact with an environment of differing composition. One kind of concentration cell involves solutions of differing salt levels. Anodic dissolution or corrosion will tend to occur in the more dilute salt or lower-pH solution, while the cathodic reaction will tend to occur in the more concentrated salt or higher-pH solution. Identical pipes may corrode in soils of one composition (clay) while little corrosion occurs in sandy soil. This is due, in part, to differences in soil composition and, in part, to differential aeration effects. Corrosion will be experienced in regions low in oxygen; cathodes will exist in high-oxygen areas.

The third type of cell, the **thermogalvanic cell,** involves electrodes of the same metal that are exposed, initially, in electrolytes of the same composition but at different temperatures. Electrode potentials change with temperature, but temperature also may affect the kinetics of dissolution. Depending on other aspects of the environment, thermogalvanic cells may accelerate or slow the rate of corrosion. Denickelification of copper-nickel alloys may occur in hot areas of a heat exchanger in brackish water, but are unaffected in colder regions.

FORMS OF CORROSION

It is convenient to classify corrosion by the various forms in which it exists, preferably by visual examination, although in many cases analysis may require more sophisticated methods. An arbitrary list of **corrosion types** includes (1) uniform corrosion, (2) galvanic corrosion, (3) crevice corrosion, (4) pitting, (5) intergranular corrosion, (6) dealloying or selective leaching, (7) erosion corrosion, (8) environmentally induced cracking, and (9) high-temperature corrosion.

Uniform attack is the most common form of corrosion. It is typified by an electrochemical or chemical attack that affects the entire exposed surface. The metal becomes thinner and eventually fails. Unlike many of the other forms of corrosion, uniform corrosion can be easily measured by weight-loss tests and by using equations such as (6.5.7a) and (6.5.7b). The life of structures and components suffering from uniform corrosion then can be accurately estimated.

Stray-current corrosion or **stray-current electrolysis** is caused by externally induced electric currents and is usually independent of environmental factors such as oxygen concentration of pH. Stray currents are escaped currents that follow paths other than their intended circuit via a low-resistance path through soil, water, or any suitable electrolyte. Direct currents from electric mass-transit systems, welding machines, and implied cathodic protection systems (discussed later) are major sources of stray-current corrosion. At the point where stray currents enter the unintended structure, the sites will become cathodic because of changes in potential. Areas where the stray currents leave the metal structure become anodic and become sites where serious corrosion can occur. Alternating currents generally cause less severe damage than direct currents. Stray-current corrosion may vary over short periods, and it sometimes looks similar to galvanic corrosion.

Galvanic corrosion occurs when a metal or an alloy is electrically coupled to another metal, alloy, or conductive nonmetal in a common, conductive medium. A potential difference usually exists between dissimilar metals, which causes a flow of electrons between them. The corrosion rate of the less corrosion-resistant (active) anodic metal is increased, while that of the more corrosion-resistant (noble) cathodic metal or alloy is decreased. Galvanic corrosion may be affected by many factors, including (1) electrode potential, (2) reaction kinetics, (3) alloy composition, (4) protective-film characteristics, (5) mass transport (migration, diffusion, and/or convection), (6) bulk solution properties and environment, (7) cathode/anode ratio and total geometry, (8) polarization, (9) pH, (10) oxygen content and temperature, and (11) type of joint [Hack (ed.), "Galvanic Corrosion," STP 978, ASTM, Philadelphia].

The driving force for corrosion or current flow is the potential difference formed between dissimilar metals. The extent of accelerated corrosion resulting from galvanic corrosion is related to potential differences between dissimilar metals or alloys, the nature of the environment, the polarization behavior of metals involved in the galvanic couple, and the geometric relationship of the component metals. A **galvanic series of metals and alloys** is useful for predicting the possibility of galvanic corrosion (see Fig. 6.5.4) (Baboian, "Galvanic and Pitting Corrosion—Field and Laboratory Studies," STP 576, ASTM, Philadelphia). A galvanic series must be arranged according to the potentials of the metals or alloys in a specific electrolyte at a given temperature and conditions. Separation of two metals in the galvanic series may indicate the possible magnitude of the galvanic corrosion. Since corrosion-product films and other changes in the surface composition may occur in different environments, no singular value may be assigned for a particular metal or alloy. This is important since polarity reversal may occur. For example, at ambient temperatures, zinc is more active than steel and is used to cathodically protect steel; however, above 60°C (e.g., in hot-water heaters), corrosion products formed on zinc may cause the steel to become anodic and corrode preferentially to zinc.

Corrosion from galvanic effects is usually worse near the couple junction and decreases with distance from the junction. The distance influenced by galvanic corrosion will be affected by solution conductivity. In low-conductivity solutions, galvanic attack may be a sharp, localized groove at the junction, while in high-conductivity solutions, galvanic corrosion may be shallow and may spread over a relatively long distance. Another important factor in galvanic corrosion is the cathodic-to-anodic area ratio. Since anodic and cathodic currents (μA) must be equal in a corrosion reaction, an unfavorable area ratio is composed of a large cathode and a small anode. The current density ($\mu A/cm^2$) and corrosion rate are greater for the small anode than for the large cathode. Corrosion rates depending on the area ratio may be 100 to 1,000 times greater than if the cathode and anode areas were equal. The method by which dissimilar metals or alloys are joined may affect the degree of galvanic corrosion. Welding, where a gradual transition from one material to another often exists, could react differently in a system where two materials are insulated by a gasket but electrically connected elsewhere through the solution, or differently still in a system connected by fasteners.

Galvanic corrosion may be minimized by (1) selecting combinations of metals or alloys near each other in the galvanic series; (2) avoiding unfavorably large cathode/anode area ratios; (3) completely insulating dissimilar metals whenever possible to disrupt the electric circuit; (4) adding inhibitors to decrease the aggressiveness of the environment; (5) avoiding the use of threaded joints to connect materials far apart in the galvanic series; (6) designing for readily replaceable anodic parts; and (7) adding a third metal that is anodic to both metals or alloys in the galvanic couple.

Crevice corrosion is the intensive localized corrosion that may occur because of the presence of narrow openings or gaps between metal-to-metal (under bolts, lap joints, or rivet heads) or nonmetal-to-metal (under gaskets, surface deposits, or dirt) components and small volumes of stagnant solution. Crevice corrosion sometimes is referred to as **underdeposit corrosion** or **gasket corrosion**. Resistance to crevice corrosion can vary from one alloy/environmental system to another. Crevice corrosion may range from near uniform attack to severe localized dissolution at the metal surface within the crevice and may occur in a variety of metals and alloys ranging from relatively noble metals such as silver and copper to active metals such as aluminum and titanium. Stainless steels are susceptible to crevice corrosion. Fibrous gasket materials that draw solution into a crevice by capillary action are particularly effective in promoting crevice corrosion. The environment can be any neutral or acidic aggressive solution, but solutions containing chlorides are particularly conducive to crevice corrosion. In cases where the bulk environment is particularly aggressive, general corrosion may deter localized corrosion at a crevice site.

A condition for crevice corrosion is a crevice or gap that is wide enough to permit solution entry but is sufficiently narrow to maintain a stagnant zone within the crevice to restrict entry of cathodic reactants and removal of corrosion products through diffusion and migration. Crevice gaps are usually 0.025 to 0.1 mm (0.001 to 0.004 in) wide. The critical depth for crevice corrosion is dependent on the gap width and may be only 15 to 40 μm deep. Most cases of crevice corrosion occur in near-neutral solutions in which dissolved oxygen is the cathodic reactant. In seawater, localized corrosion of copper and its alloys (due to differences in Cu^{2+} concentration) is different from that of the stainless steel group because the attack occurs outside the crevice rather than within it. In acidic solutions where hydrogen ions are the cathodic reactant, crevice corrosion actually occurs at the exposed surface near the crevice. Decreasing crevice width or increasing crevice depth, bulk chloride concentration, and/or acidity will increase the potential for crevice corrosion. In some cases, deep crevices may restrict propagation because of the voltage drop through the crevice solution (Lee et al., Corrosion/83, paper 69, NACE International, Houston).

A condition common to crevice corrosion is the development of an occluded, localized environment that differs considerably from the bulk solution. The basic overall mechanism of crevice corrosion is the dissolution of a metal [Eq. (6.5.1)] and the reduction of oxygen to hydroxide

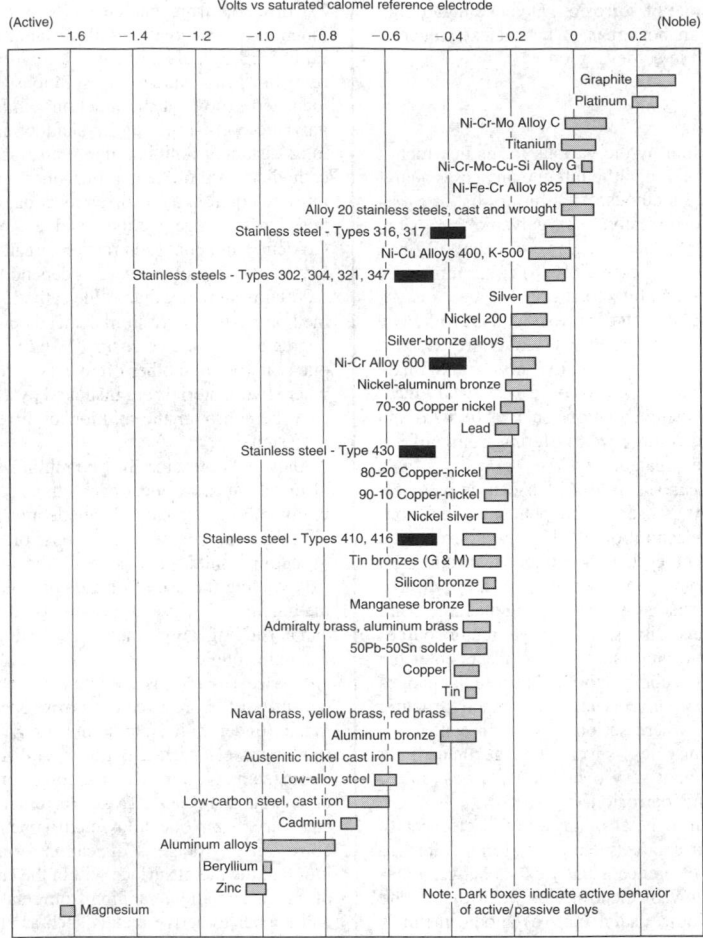

Volts vs saturated calomel reference electrode

Fig. 6.5.4   Galvanic series for metals and alloys in seawater. Flowing seawater at 2.4 to 4.0 m/s; immersion for 5 to 15 days at 5 to 30°C. *(Source: ASTM.)*

ions [Eq. (6.5.4)]. Initially, these reactions take place uniformly, both inside and outside the crevice. However, after a short time, oxygen within the crevice is depleted because of restricted convection and oxygen reduction ceases in this area. Because the area within the crevice is much smaller than the external area, oxygen reduction remains virtually unchanged. Once oxygen reduction ceases within the crevice, the continuation of metal dissolution tends to produce an excess of positive charge within the crevice. To maintain charge neutrality, chloride or possibly hydroxide ions migrate into the crevice. This results in an increased concentration of metal chlorides within the crevice, which hydrolyzes in water according to Eq. (6.5.8) to form an insoluble hydroxide and a free acid.

$$M^+Cl^- + H_2O \rightarrow MOH \downarrow + H^+Cl^- \qquad (6.5.8)$$

The increased acidity boosts the dissolution rates of most metals and alloys, which increases migration and results in a rapidly accelerating and autocatalytic process. The chloride concentration within crevices exposed to neutral chloride solutions is typically 3 to 10 times higher than that in the bulk solution. The pH within the crevice is 2 to 3. The pH drop with time and critical crevice solution that will cause breakdown in stainless steels can be calculated (Oldfield and Sutton, *Brit. Corrosion J.*, **13**, 1978, p. 13).

Optimum crevice corrosion resistance can be achieved with an active/passive metal that possesses (1) a narrow active-passive transition, (2) a small critical current density, and (3) an extended passive region. Crevice corrosion may be minimized by avoiding riveted, lap, or bolted

joints in favor of properly welded joints, designing vessels for complete drainage and removal of sharp corners and stagnant areas, removing deposits frequently, and using solid, **nonabsorbent gaskets** (such as Teflon) wherever possible.

Pitting is a form of extremely localized attack forming a cavity or hole in the metal or alloy. Deterioration by pitting is one of the most dangerous types of localized corrosion, but its unanticipated occurrences and propagation rates are difficult to consider in practical engineering designs. Pits are often covered by corrosion products. Depth of pitting is sometimes expressed by the **pitting factor,** which is the ratio of deepest metal penetration to average metal penetration, as determined by weight-loss measurements. A pitting factor of 1 denotes uniform corrosion.

Pitting is usually associated with the breakdown of a passive metal. Breakdown involves the existence of a critical potential E_b, induction time at a potential $>E_b$, presence of aggressive species (Cl$^-$, Br$^-$, ClO$_4^-$, etc.), and discrete sites of attack. Pitting is associated with local imperfections in the passive layer. Nonmetallic inclusions, second-phase precipitates, grain boundaries, scratch lines, dislocations, and other surface inhomogeneities can become initiation sites for pitting (Szklarska-Smialowska, "Pitting of Metals," NACE International, Houston). The induction time before pits initiate may range from days to years. The induction time depends on the metal, the aggressiveness of the environment, and the potential. The induction time tends to decrease if either the potential or the concentration of aggressive species increases.

Once pits are initiated, they continue to grow through an autocatalytic process (i.e., the corrosion conditions within the pit are both stimulating and necessary for continuing pit growth). Pit growth is controlled by the rate of depolarization at the cathodic areas. Oxygen in aqueous environments and ferric chloride in various industrial environments are effective depolarizers. The propagation of pits is virtually identical to crevice corrosion. Factors contributing to localized corrosion such as crevice corrosion and pitting of various materials are found in other publications ["Localized Corrosion—Cause of Metal Failure," STP 516, ASTM, Philadelphia; Brown et al. (eds.), "Localized Corrosion," NACE-3, NACE International, Houston; Isaacs et al. (eds.), "Advances in Localized Corrosion," NACE-9, NACE International, Houston; Frankel and Newman (eds.), "Critical Factors in Localized Corrosion," vol. 92-9, The Electrochemical Society, Pennington, NJ].

Copper and its alloys suffer localized corrosion by nodular pitting in which the attacked areas are covered with small mounds or nodules composed of corrosion product and $CaCO_3$ precipitated from water. The pit interior is covered with CuCl which prevents formation of a protective Cu_2O layer. Pitting occurs most often in cold, moderately hard to hard waters, but it also happens in soft waters above 60°C. Pitting is associated with the presence of very thin carbon (cathodic) films from lubricants used during tube manufacture. Copper pitting is favored by a high sulfate/chloride ratio. Copper-nickel tubing suffers from accelerated corrosion in seawater due to the presence of both sulfides and oxygen. Sulfides prevent the formation of a protective oxide layer, which allows the anodic reaction to proceed unabated and supported by the oxygen reduction reaction.

Caustic corrosion, frequently referred to as **caustic gouging** or **ductile gouging,** is a form of localized corrosion that occurs as a result of fouled heat-transfer surfaces of water-cooled tubes and the presence of an active corrodent in the water of high-pressure boilers. Porous deposits that form and grow in these high-heat-input areas produce conditions that allow sodium hydroxide to permeate the deposits by a process called **wick boiling** and that concentrate to extremely high levels (for example, 100 ppm bulk water concentrated to 220,000 ppm NaOH in water film under these deposits) which dissolves the protective magnetite and then reacts directly with the steel to cause rapid corrosion. A smooth, irregular thinning occurs, usually downstream of flow disruptions or in horizontal or inclined tubing.

Intergranular corrosion is localized attack that follows a narrow path along the grain boundaries of a metal or an alloy. Intergranular corrosion is caused by the presence of impurities, precipitation of one of the alloying elements, or depletion of one of these elements at or along the grain boundaries. The driving force of intergranular corrosion is the difference in corrosion potential that exists between a thin grain boundary zone and the bulk of immediately adjacent grains. Intergranular corrosion most commonly occurs in aluminum or copper alloys and austenitic stainless steels.

When austenitic stainless steels such as 304 are heated or placed in service at temperatures of 950 to 1450°C, they become sensitized and susceptible to intergranular corrosion, because chromium carbide ($Cr_{26}C_6$) is formed. This effectively removes chromium from the vicinity of the grain boundaries, thereby approaching the corrosion resistance of carbon steel in these chromium-depleted areas. Methods to avoid **sensitization** and to control intergranular corrosion of austenitic stainless steels (1) employ high-temperature heat treatment (solution quenching) at 1,950 to 2,050°C; (2) add elements that are stronger carbide formers (stabilizers such as columbium, columbium plus tantalum, or titanium) than chromium; and (3) lower the carbon content below 0.03 percent. Stabilized austenitic stainless steels can experience intergranular corrosion if heated above 2,250°C, where columbium carbides dissolve, as may occur during welding. Intergranular corrosion of stabilized stainless steels is called **knife-line attack.** This attack transpires in a narrow band in the parent metal adjacent to a weld. Reheating the steel to around 1,950 to 2,250°C reforms the stabilized carbides.

Surface carburization also can cause intergranular corrosion in cast austenitic stainless steels. Overaging of aluminum alloys can form precipitates such as $CuAl_2$, $FeAl_3$, $MgAl_3$, $MgZn_2$, and $MnAl_6$ along grain boundaries and can promote intergranular corrosion. Some nickel alloys can form intergranular precipitates of carbides and intermetallic phases during heat treatment or welding. Alloys with these intergranular precipitates are subject to intergranular corrosion.

Dealloying involves the selective removal of the most electrochemically active component metal in the alloy. Selective removal of one metal can result in either localized attack, leading to possible perforation (plug dealloying), or a more uniform attack (layer dealloying), resulting in loss of component strength. Although selective removal of metals such as Al, Fe, Si, Co, Ni, and Cr from their alloys has occurred, the most common dealloying is the selective removal of zinc from brasses, called **dezincification.** Dezincification of brasses (containing 15 percent or more zinc) takes place in soft waters, especially if the carbon dioxide content is high. Other factors that increase the susceptibility of brasses to dealloying are high temperatures, waters with high chloride content, low water velocity, crevices, and deposits on the metal surface. Dezincification is experienced over a wide pH range. Layer dezincification is favored when the environment is acidic and the brass has a high zinc content. Plug dezincification tends to develop when the environment is slightly acidic, neutral, or alkaline and the zinc content is relatively low. It has been proposed that either zinc is selectively dissolved from the alloy, leaving a porous structure of metallic copper in situ within the brass lattice, or both copper and zinc are dissolved initially, but copper immediately redeposits at sites close to where the zinc was dissolved.

Dezincification can be minimized by reducing oxygen in the environment or by cathodic protection. Dezincification of α-brass (70 percent Cu, 30 percent Zn) can be minimized by alloy additions of 1 percent tin. Small amounts (~ 0.05 percent) of arsenic, antimony, or phosphorus promote effective inhibition of dezincification for 70/30 α-brass (inhibited admiralty brass). Brasses containing less than 15 percent zinc have not been reported to suffer dezincification.

Gray cast iron selectively leaches iron in mild environments, particularly in underground piping systems. Graphite is cathodic to iron, which sets up a galvanic cell. The dealloying of iron from gray cast iron leaves a porous network of rust, voids, and graphite termed **graphitization** or **graphite corrosion.** The cast iron loses both its strength and its metallic properties, but it experiences no apparent dimensional changes. Graphitization does not occur in nodular or malleable cast irons because a continuous graphite network is not present. White cast iron also is immune to graphitization since it has essentially no available free carbon.

Dealloying also can occur at high temperatures. Exposure of stainless steels to low-oxygen atmospheres at 1,800°F (980°C) results in the selective oxidation of chromium, the creation of a more protective scale, and the depletion of chromium under the scale in the substrate metal. Decarburization can occur in carbon steel tubing if it is heated above 1,600°F (870°C). Denickelification of austenitic stainless steels occurs in liquid sodium.

Erosion corrosion is the acceleration or increase of attack due to the relative movement between a corrosive fluid and the metal surface. Mechanically, the conjoint action of erosion and corrosion damages not only the protective film but also the metal or alloy surface itself by the abrasive action of a fluid (gas or liquid) at high velocity. The inability to maintain a protective passive film on the metal or alloy surface raises the corrosion rate. Erosion corrosion is characterized by grooves, gullies, waves, rounded holes, and valleys—usually exhibited in a directional manner. Components exposed to moving fluids and subject to erosion corrosion include bends, elbows, tees, valves, pumps, blowers, propellers, impellers, agitators, condenser tubes, orifices, turbine blades, nozzles, wear plates, baffles, boiler tubes, and ducts.

The hardness of the alloy or metal and/or the mechanical or protective nature of the passive film will affect the resistance to erosion corrosion in a particular environment. Copper and brasses are particularly susceptible to erosion corrosion. Often, an escalation in the fluid velocity increases the attack by erosion corrosion. The effect may be nil until a critical velocity is reached, above which attack may increase very rapidly. The critical velocity is dependent on the alloy, passive layer, temperature, and nature of the fluid environment. The increased attack may be attributed, in part, to increasing cathodic reactant concentration to the metal surface. Increased velocity may decrease attack by increasing the effectiveness of inhibitors or by preventing silt or dirt from depositing and causing crevice corrosion.

Erosion corrosion includes impingement, cavitation, and fretting corrosion. **Impingement attack** occurs when a fluid strikes the metal surface at high velocity at directional-change sections such as bends, tees, turbine blades, and inlets of pipes or tubes. In the majority of cases involving impingement attack, a geometric feature of the system results in turbulence. Air bubbles or solids present in turbulent flow will accentuate attack. **Cavitation damage** is caused by the formation and collapse of vapor bubbles in a liquid near a metal surface. Cavitation occurs where high-velocity water and hydrodynamic pressure changes are encountered, such as with ship propellers or pump impellers. If pressure on a liquid is lowered drastically by flow divergence, water vaporizes and forms bubbles. These bubbles generally collapse rapidly, producing shock waves with pressures as high as 60,000 lb/in² (410 MPa), which destroys the passive film. The newly exposed metal surface corrodes, re-forms a passive film which is destroyed again by another cavitation bubble, and so forth. As many as 2 million bubbles may collapse over a small area in 1 s. Cavitation damage causes both corrosion and mechanical effects. **Fretting corrosion** is a form of damage caused at the interface of two closely fitting surfaces under load when subjected to vibration and slip. Fretting is explained as (1) the surface oxide is ruptured at localized sites prompting reoxidation or (2) material wear and friction cause local oxidation. The degree of plastic deformation is greater for softer materials than for hard; seizing and galling may often occur. Lubrication, increased hardness, increased friction, and the use of gaskets can reduce fretting corrosion.

Environmentally induced cracking is a brittle fracture of an otherwise ductile material due to the conjoint action of tensile stresses in a specific environment over a period of time. Environmentally induced cracking includes (1) stress corrosion cracking, (2) hydrogen damage, (3) corrosion fatigue, (4) liquid-metal embrittlement, and (5) solid-metal-induced embrittlement.

Stress corrosion cracking (SCC) refers to service failures due to the joint interaction of tensile stress with a specific corrodent and a susceptible, normally ductile material. The observed crack propagation is the result of mechanical stress and corrosion reactions. Stress can be externally applied, but most often it is the result of residual tensile stresses. SCC can initiate and propagate with little outside evidence of corrosion or macroscopic deformation and usually provides no warning of impending failure. Cracks often initiate at surface flaws that were preexisting or were formed during service. Branching is frequently associated with SCC; the cracks tend to be very fine and tight and are visible only with special techniques or microscopic instrumentation. Cracking can be intergranular or transgranular. Intergranular SCC is often associated with grain boundary precipitation or grain boundary segregation of alloy constituents such as found in sensitized austenitic stainless steels. Transgranular SCC is related to crystal structure, anisotropy, grain size and shape, dislocation density and geometry, phase composition, yield strength, ordering, and stacking fault energies. The specific corrodents of SCC include (1) carbon steels, such as sodium hydroxide; (2) stainless steels, such as NaOH or chlorides; (3) α-brass, such as ammoniacal solutions; (4) aluminum alloys, such as aqueous Cl^-, Br^-, I^- solutions; (5) titanium alloys, such as aqueous Cl^-, Br^-, I^- solutions and organic liquids; and (6) high-nickel alloys, high-purity steam [Gangloff and Ives (eds.), "Environment-Induced Cracking of Metals," NACE-10, NACE International, Houston; Bruemmer et al. (eds.), "Parkins Symposium on Fundamental Aspects of Stress Corrosion Cracking," The Minerals, Metals, and Materials Society, Warrendale, PA].

Hydrogen interactions can affect most engineering metals and alloys. Terms to describe these interactions have not been universally agreed upon. Hydrogen damage or hydrogen attack is caused by the diffusion of hydrogen through carbon and low-alloy steels. Hydrogen reacts with carbon, either in elemental form or as carbides, to form methane gas. Methane accumulates at grain boundaries and can cause local pressures to exceed the yield strength of the material. Hydrogen damage causes intergranular, discontinuous cracks and decarburization of the steel. Once cracking occurs, the damage is irreversible. In utility boilers, hydrogen damage has been associated with fouled heat-transfer surfaces.

Hydrogen damage is temperature dependent with a threshold temperature of about 400°F (204°C). Hydrogen damage has been observed in boilers operating at pressures of 450 to 2,700 lb/in² (3.1 to 18.6 MPa) and tube metal temperatures of 600 to 950°F (316 to 510°C). Hydrogen damage may occur in either acidic or alkaline conditions, although this contention is not universally accepted.

Hydrogen embrittlement results from penetration and adsorption of atomic hydrogen into an alloy matrix. This process results in a decrease of toughness and ductility of alloys due to hydrogen ingress. Hydrogen may be removed by baking at 200 to 300°F (93 to 149°C), which returns the alloy mechanical properties very nearly to those existing before hydrogen entry. Hydrogen embrittlement can affect most metals and alloys. Hydrogen blistering results from atomic hydrogen penetration at internal defects near the surface such as laminations or nonmetallic inclusions where molecular hydrogen forms. Pressure from H_2 can cause local plastic deformation or surface exfoliation [Moody and Thompson (eds.), "Hydrogen Effects on Material Behavior," The Minerals, Metals, and Materials Society, Warrendale, PA].

When a metal or an alloy is subjected to cyclic, changing stresses, then cracking and fracture can occur at stresses lower than the yield stress through fatigue. A clear relationship exists between stress amplitude and number of cyclic loads; an endurance limit is the stress level below which fatigue will not occur indefinitely. Fracture solely by fatigue is not viewed as an example of environmental cracking. However, the conjoint action of a corrosive medium and cyclic stresses on a material is termed **corrosion fatigue**. Corrosion fatigue failures are usually "thick-walled," brittle ruptures showing little local deformation. An endurance limit often does not exist in corrosion fatigue failures. Cracking often begins at pits, notches, surface irregularities, welding defects, or sites of intergranular corrosion. Surface features of corrosion fatigue cracking vary with alloys and specific environmental conditions. Crack paths can be transgranular (carbon steels, aluminum alloys, etc.) or intergranular (copper, copper alloys, etc.); exceptions to the norm are found in each alloy system. "Oyster shell" markings may denote corrosion fatigue failures, but corrosion products and corrosion often cover and obscure this feature. Cracks usually propagate in the direction perpendicular to the principal tensile stress; multiple, parallel cracks are usually present near the principal crack which led to failure. Applied cyclic stresses are caused by mechanical restraints or thermal fluctuations; susceptible areas may also include structures that possess residual stresses from fabrication or heat treatment. Low pH and high stresses, stress cycles, and oxygen content or corrosive species concentrations will increase the probability of corrosion fatigue cracking [Crooker and Leis (eds.), "Corrosion Fatigue-Mechanics, Metallurgy, Electrochemistry, and Engineering," STP-801, ASTM, Philadelphia; McEvily and Staehle (eds.), "Corrosion Fatigue-Chemistry, Mechanics, and Microstructure," NACE International, Houston].

Liquid-metal embrittlement (LME) or liquid-metal-induced embrittlement is the catastrophic brittle fracture of a normally ductile metal when coated by a thin film of liquid metal and subsequently stressed in tension. Cracking may be either intergranular or transgranular. Usually, a solid that has little or no solubility in the liquid and forms no intermetallic compounds with the liquid constitutes a couple with the liquid. Fracture can occur well below the yield stress of the metal or alloy (Jones, "Engineering Failure Analysis," **1**, 1994, p. 51). A partial list of examples of LME includes (1) aluminum (by mercury, gallium, indium, tin-zinc, lead-tin, sodium, lithium, and inclusions of lead, cadmium, or bismuth in aluminum alloys); (2) copper and select copper alloys (mercury, antimony, cadmium, lead, sodium, lithium, and bismuth); (3) zinc (mercury, gallium, indium, Pb-Sn solder); (4) titanium or titanium alloys (mercury, zinc, and cadmium); (5) iron and carbon steels (aluminum, antimony, cadmium, copper, gallium, indium, lead, lithium, zinc, and mercury); and (6) austenitic and nickel-chromium steels (tin and zinc).

Solid-metal-induced embrittlement (SMIE) occurs below the melting temperature T_m of the solid in various LME couples. Many instances of loss of ductility or strength and brittle fracture have taken place with electroplated metals or coatings and inclusions of low-melting-point alloys below T_m. Cadmium-plated steel, leaded steels, and titanium embrittled by cadmium, silver, or gold are materials affected by SMIE. The prerequisites for SMIE and LME are similar, although multiple cracks are formed in SMIE and SMIE crack propagation is 100 to 1,000 times slower than LME cracking. More detailed descriptions of LME and SMIE are given elsewhere [Kambar (ed.), "Embrittlement of Metals," American Institute of Mining, Metallurgical and Petroleum Engineers, St. Louis].

High-temperature corrosion reactions are generally not electrochemical; diffusion is a fundamental property involved in the reactions. Oxide growth advances with time at parabolic, linear, cubic, or logarithmic rates. The corrosion rate can be influenced considerably by the presence of contaminants, particularly if they are adsorbed on the metal surface. Each alloy has an upper temperature limit where the oxide scale, based on its chemical composition and mechanical properties, cannot protect the underlying alloy. Catastrophic oxidation refers to metal-oxygen reactions that occur at continuously increasing rates or break away from protective behavior very rapidly. Molybdenum, tungsten, vanadium, osmium, rhenium, and alloys containing Mo or V may oxidize catastrophically. Internal oxidation may take place in copper- or silver-based alloys containing small levels of Al, Zn, Cd, or Be when more stable oxides are possible below the surface of the metal-scale interface. Sulfidation forms sulfide phases with less stable base metals, rather than oxide phases, when H_2S, S_2, SO_2, and other gaseous sulfur species have a sufficiently high concentration. In corrosion-resistant alloys, sulfidation usually occurs when Al or Cr is tied up with sulfides, which interferes with the process of developing a protective oxide. Carburization of high-temperature alloys is possible in reducing carbon-containing environments. Hot corrosion involves the high-temperature, liquid-phase attack of alloys by eutectic alkali-metal sulfates that solubilize a protective alloy oxide, and exposes the bare metal, usually iron, to oxygen which forms more oxide and promotes subsequent metal loss at temperatures of 1,000 to 1,300°F (538 to 704°C). Vanadium pentoxide and sodium oxide can also form low-melting-point [~ 1,000°F (538°C)] liquids capable of causing alloy corrosion. These are usually contaminants in fuels, either coal or oil.

Alloy strength decreases rapidly when metal temperatures approach 900 to 1,000°F and above. Creep entails a time-dependent deformation involving grain boundary sliding and atom movements at high temperatures. When sufficient strain has developed at the grain boundaries, voids and microcracks develop, grow, and coalesce to form larger cracks until failure occurs. The creep rate will increase and the projected time to failure decrease when stress and/or the tube metal temperature is increased [Rapp (ed.), "High Temperature Corrosion," NACE-10, NACE International, Houston; Lai, "High Temperature Corrosion of Engineering Alloys," ASM International, Materials Park, OH].

CORROSION TESTING

Corrosion testing can reduce cost, improve safety, and conserve resources in industrial, commercial, and personal components, processes, and applications. Corrosion evaluation includes laboratory, pilot-plant, and field monitoring and testing. The selection of materials for corrosion resistance in specific industrial services is best made by field testing and actual service performance. However, uncontrollable environmental factors may skew precise comparison of potential alloys or metals, and test times may be unrealistically long. Accelerated corrosion tests in the laboratory can provide practical information to assess potential material candidates in a given environment, predict the service life of a product or component, evaluate new alloys and processes, assess the effects of environmental variations or conditions on corrosion and corrosion control methods, provide quality control, and study corrosion mechanisms.

Careful planning is essential to obtain meaningful, representative, and reliable test results. Metallurgical factors, environmental variables, statistical treatment, and proper interpretation and correlation of accelerated test results to actual field conditions are considerations in the planning and conducting of corrosion tests. The chemical composition, fabrication and metallurgical history, identification, and consistency of materials should be known. Even though, ideally, the surface condition of the test specimen should mirror the plant conditions, a standard test condition should be selected and maintained. This often requires prescribed procedures for cutting, surface-finishing, polishing, cleaning, chemical treatment or passivation, measuring and weighing, and handling. The exposure technique should provide easy assess of the environment to the sample and should not directly cause conditions for test corrosion (e.g., crevice corrosion) unrelated to field conditions.

The type, interval, and duration of the tests should be seriously considered. Corrosion rates of the material may decrease with time because of the formation of protective films or the removal of a less resistant surface metal. Corrosion rates may increase with time if corrosion-inducing salts or scales are produced or if a more resistant metal is removed. The environment may change during testing because of the decrease or increase in concentration of a corrosive species or inhibitor, or the formation of autocatalytic products or other metal-catalyzed variations in the solution. These test solution variations should be recognized and related to field-condition changes over time, if possible. Testing conditions should prescribe the presence or absence of oxygen in the system. Aeration, or the presence of oxygen, can influence the corrosion rate of an alloy in solution. Aeration may increase or decrease the corrosion rate according to the influence of oxygen in the corrosion reaction or the nature of the passive films which develop on different metals and alloys. Temperature is an extremely important parameter in corrosion reactions. The temperature at the specimen surface should be known; a rough rule of thumb is that the corrosion rate doubles for each 10°C rise in temperature.

Laboratory accelerated corrosion tests are used to predict corrosion behavior when service history is unavailable and cost and time prohibit simulated field testing. Laboratory tests also can provide screening evaluations prior to field testing. Such tests include nonelectrochemical and sophisticated electrochemical tests. Nonelectrochemical laboratory techniques include immersion and various salt spray tests that are used to evaluate the corrosion of ferrous and nonferrous metals and alloys, as well as the degree of protection afforded by both organic and inorganic coatings. Since these are accelerated tests by design, the results must be interpreted cautiously. **Standardized test methods** are very effective for both specifications and routine tests used to evaluate experimental or candidate alloys, inhibitors, coatings, and other materials. Many such tests are described in the *Annual Book of ASTM Standards* (vol. 3.02, "Metal Corrosion, Erosion, and Wear") and in standard test methods from the NACE International Book of Standards.

Electrochemical techniques are attractive because (1) they allow a direct method of accelerating corrosion processes without changing the environment, (2) they can be used as a nondestructive tool to evaluate corrosion rates, and (3) they offer in situ (field) or ex situ (laboratory) investigations. Most typical forms of corrosion can be investigated by electrochemical techniques. Electrochemical tests may include linear polarization resistance, ac electrochemical impedance, electrochemical noise, cyclic voltammetry, cyclic potentiodynamic polarization, potentiodynamic and potentiostatic polarization, and scratch repassivation. AC impedance has been used to monitor coating integrity and to evaluate the effectiveness of cathodic protection and coating systems in underground piping. There are complications involved in various electrochemical test methods which must be resolved or eliminated before interpretation is possible. The methods used above employ a potentiostat, an instrument that regulates the electrode potential and measures current as a function of potential. A galvanostat varies potential as a function of current.

Corrosion coupons are generally easy to install, reflect actual environmental conditions, and can be used for long-exposure tests. Coupons can evaluate inhibitor programs and are designed to measure specific forms of corrosion. A simple embrittlement detector covered by ASTM D807 permits concentration of boiler water on a stressed steel specimen to detect caustic cracking. Field coupon testing has several limitations: (1) It cannot detect brief process upsets. (2) It cannot guarantee initiation of localized corrosion rates before the coupons are removed. (3) It cannot directly correlate the calculated coupon corrosion rate to equipment or component corrosion. (4) It cannot detect certain forms of corrosion. Corrosion rate sensors created to monitor the corrosion of large infrastructural components in real time have been developed for aqueous, soil, and concrete environments using linear polarization resistance (Ansuini et al., Corrosion/95, paper 14, NACE International, Houston). Ultrasonic thickness, radiography, or eddy current measurements can evaluate corrosion rates in situ. The component surfaces must be clean (free of dirt, paint, and corrosion products) to make measurements. Ultrasonic attenuation techniques have been applied to evaluate the extent of hydrogen damage. A noncontact, nondestructive inspection technique for pitting, SCC, and crevice corrosion has been developed using digital speckle correlation (Jin and Chiang, Corrosion/95, paper 535, NACE International, Houston). An ASTM publication [Baboian (ed.), "Manual 20—Corrosion Tests and Standards: Application and Interpretation," ASTM, Philadelphia] provides a detailed, comprehensive reference for field and laboratory testing in different environments for metals and alloys, coatings, and composites. It includes testing for different corrosion forms and for different industrial applications.

CORROSION PREVENTION OR REDUCTION METHODS

The basis of **corrosion prevention** or reduction methods involves restricting or controlling anodic and/or cathodic portions of corrosion reactions, changing the environmental variables, or breaking the electrical contact between anodes and cathodes. Corrosion prevention or reduction methods include (1) proper materials selection, (2) design, (3) coatings, (4) use of inhibitors, (5) anodic protection, and (6) cathodic protection [Treseder et al. (eds.), "NACE Corrosion Engineer's Reference Book," 2d ed., NACE International, Houston].

The most common method to reduce corrosion is *selecting the right material* for the environmental conditions and applications. Ferrous and nonferrous metals and alloys, thermoplastics, nonmetallic linings, resin coatings, composites, glass, concrete, and nonmetal elements are a few of the materials available for selection. Experience and data, either in-house or from outside vendors and fabricators, may assist in the materials selection process. Many materials can be eliminated by service conditions (temperature, pressure, strength, chemical compatibility). Corrosion data for a particular chemical or environment may be obtained through a literature survey ("Corrosion Abstracts," NACE International; "Corrosion Data Survey—Metals Section," 6th ed., "Corrosion Data Survey—Nonmetals Section," NACE International, Houston) or from expert systems or databases. Different organizations (ASTM, ASM International, ASME, NACE International) may offer references to help solve problems and make predictions about the corrosion behavior of candidate materials. The material should be cost-effective and resistant to the various forms of corrosion (discussed earlier) that may be encountered in its application. General rules may be applied to determine the resistance of metals and alloys. For reducing or nonoxidizing environments (such as air-free acids and aqueous solutions), nickel, copper, and their alloys are usually satisfactory. For oxidizing conditions, chromium alloys are used, while in extremely powerful oxidizing environments, titanium and its alloys have shown superior resistance. The corrosion resistance of chromium and titanium can be enhanced in hot, concentrated oxidizer-free acid by small additions of platinum, palladium, or rhodium.

Many costs related to corrosion could be eliminated by **proper design.** The designer should have a credible knowledge of corrosion or should work in cooperation with a materials and corrosion engineer. The design should avoid gaps or structures where dirt or deposits could easily form crevices; horizontal faces should slope for easy drainage. Tanks should be properly supported and should employ the use of drip skirts and dished, fatigue-resistant bottoms ("Guidelines for the Welded Fabrication of Nickel Alloys for Corrosion-Resistant Service," part 3, Nickel Development Institute, Toronto, Canada). Connections should not be made of dissimilar metals or alloys widely separated in the galvanic series for lap joints or fasteners, if possible. Welded joints should avoid crevices, microstructural segregation, and high residual or applied stresses. Positioning of parts should take into account prevailing environmental conditions. The design should avoid local differences in concentration, temperature, and velocity or turbulence. All parts requiring maintenance or replacement must be easily accessible. All parts to be coated must be accessible for the coating application.

Protective coatings are used to provide an effective barrier to corrosion and include metallic, chemical conversion, inorganic nonmetallic, and organic coatings. Before the application of an effective coating on metals and alloys, it is necessary to clean the surface carefully to remove, dirt, grease, salts, and oxides such mill scale and rust. Metallic coatings may be applied by cladding, electrodeposition (copper, cadmium, nickel, tin, chromium, silver, zinc, gold), hot dipping (tin, zinc-galvanizing, lead, and aluminum), diffusion (chromium-chromizing, zinc-sherardizing, boron, silicon, aluminum-calorizing or aluminizing, tin, titanium, molybdenum), metallizing or flame spraying (zinc, aluminum, lead, tin, stainless steels), vacuum evaporation (aluminum), ion implantation, and other methods. All commercially prepared coatings are porous to some degree. Corrosion or galvanic action may influence the performance of the metal coating. Noble (determined by the galvanic series) coatings must be thicker and have a minimum number of pores and small pore size to delay entry of any deleterious fluid. Some pores of noble metal coatings are filled with an organic lacquer or a second lower-melting-point metal is diffused into the initial coating. Sacrificial or cathodic coatings such as cadmium zinc, and in some environments, tin and aluminum, cathodically protect the base metal. Porosity of sacrificial coatings is not critical as long as cathodic protection of the base metal continues. Higher solution conductivity allows larger defects in the sacrificial coating; thicker coatings provide longer times of effective cathodic protection.

Conversion coatings are produced by electrochemical reaction of the metal surface to form adherent, protective corrosion products. Anodizing aluminum forms a protective film of Al_2O_3. Phosphate and chromate treatments provide temporary corrosion resistance and usually a basis for painting. Inorganic nonmetallic coatings include vitreous enamels, glass linings, cement, or porcelain enamels bonded on metals. Susceptibility to mechanical damage and cracking by thermal shock are major disadvantages of these coatings. An inorganic coating must be perfect and defect-free unless cathodic protection is applied.

Paints, lacquers, coal-tar or asphalt enamels, waxes, and varnishes are typical organic coatings. Converted epoxy, epoxy polyester, moisture-cured polyurethane, vinyl, chlorinated rubber, epoxy ester, oil-modified phenolics, and zinc-rich organic coatings are other barriers used to control corrosion. Paints, which are a mixture of insoluble particles of pigments (for example, TiO_2, Pb_3O_4, Fe_2O_3, and $ZnCrO_4$) in a continuous organic or aqueous vehicle such as linseed or tung oil, are the most common organic coatings. All paints are permeable to water and oxygen to some degree and are subject to mechanical damage and eventual breakdown. Inhibitor and antifouling agents are added to improve protection against corrosion. Underground pipelines and tanks should be covered with thicker coatings of asphalt or bituminous paints, in conjunction with cloth or plastic wrapping. Polyester and polyethylene have been used for tank linings and as coatings for tank bottoms (Munger, "Corrosion Prevention by Protective Coatings," NACE International, Houston).

Inhibitors, when added above a threshold concentration, decrease the corrosion rate of a material in an environment. If the level of an inhibitor is below the threshold concentration, corrosion could occur more quickly than if the inhibitor were completely absent. Inhibitors may be organic or inorganic and fall into several different classes: (1) passivators or oxidizers, (2) precipitators, (3) cathodic or anodic, (4) organic adsorbents, (5) vapor phase, and (6) slushing compounds. Any one inhibitor may be found in one or more classifications. Factors such as temperature, fluid velocity, pH, salinity, cost, solubility, interfering species, and metal or alloy may determine both the effectiveness and the choice of inhibitor. Federal EPA regulations may limit inhibitor choices (such as chromate use) based on its toxicity and disposal requirements.

Passivators in contact with the metal surface act as depolarizers, initiating high anodic current densities and shifting the potential into the passivation range of active/passive metal and alloys. However, if present in concentrations below 10^{-3} to 10^{-4} M, these inhibitors can cause pitting. Passivating inhibitors are anodic inhibitors and may include oxidizing anions such as chromate, nitrite, and nitrate. Phosphate, molybdate, and tungstate require oxygen to passivate steel and are considered nonoxidizing. Nonoxidizing sodium benzoate, cinnamate, and polyphosphate compounds effectively passivate iron in the near-neutral range by facilitating oxygen adsorption. Alkaline compounds (NaOH, Na_3PO_4, $Na_2B_4O_7$, and Na_2O-$nSiO_4$) indirectly assist iron passivation by enhancing oxygen adsorption.

Precipitators are film-forming compounds which create a general action over the metal surface and subsequently indirectly interfere with both anodes and cathodes. Silicates and phosphates in conjunction with oxygen provide effective inhibition by forming deposits. Calcium and magnesium interfere with inhibition by silicates; 2 to 3 ppm of polyphosphates is added to overcome this. The levels of calcium and phosphate must be balanced for effective calcium phosphate inhibition. Addition of a zinc salt often improves polyphosphate inhibition.

Cathodic inhibitors (cathodic poisons, cathodic precipitates, oxygen scavengers) are generally cations which migrate toward cathode surfaces where they are selectively precipitated either chemically or electrochemically to increase circuit resistance and restrict diffusion of reducible species to the cathodes. Cathodic poisons interfere with and slow the HER, thereby slowing the corrosion process. Depending on the pH level, arsenic, bismuth, antimony, sulfides, and selenides are useful cathodic poisons. Sulfides and arsenic can cause hydrogen blistering and hydrogen embrittlement. Cathodic precipitation-type inhibitors such as calcium and magnesium carbonates and zinc sulfate in natural waters require adjustment of the pH to be effective. Oxygen scavengers help inhibit corrosion, either alone or with another inhibitor, to prevent cathodic depolarization. Sodium sulfite, hydrazine, carbohydrazide, hydroquinone, methylethylketoxime, and diethylhydroxylamine are oxygen scavengers.

Organic inhibitors, in general, affect the entire surface of a corroding metal and affect both the cathode and anode to differing degrees, depending on the potential and the structure or size of the molecule. Cationic, positively charged inhibitors such as amines and anionic, negatively charged inhibitors such as sulfonates will be adsorbed preferentially depending on whether the metal surface is negatively or positively charged. Soluble organic inhibitors form a protective layer only a few molecules thick; if an insoluble organic inhibitor is added, the film may become about 0.003 in (76 μm) thick. Thick films show good persistence by continuing to inhibit even when the inhibitor is no longer being injected into the system.

Vapor-phase inhibitors consist of volatile aliphatic and cyclic amines and nitrites that possess high vapor pressures. Such inhibitors are placed in the vicinity of the metal to be protected (e.g., inhibitor-impregnated paper), which transfers the inhibiting species to the metal by sublimation or condensation where they adsorb on the surface and retard the cathodic, anodic, or both corrosion processes. This inhibitor protects against water and/or oxygen. Vapor-phase inhibitors are usually effective only if used in closed spaces and are used primarily to retard atmospheric corrosion. Slushing compounds are polar inorganic or organic additives in oil, grease, or wax that adsorb on the metal surface to form a continuous protective film. Suitable additives include organic amines, alkali and alkaline-earth metal salts of sulfonated oils, sodium nitrite, and organic chromates [Nathan (ed.), "Corrosion Inhibitors," NACE International, Houston].

Anodic protection is based on the formation of a protective film on metals by externally applied anodic currents. Figure 6.5.5 illustrates the effect. The applied anodic current density is equal to the difference between the total oxidation and reduction rates of the system, or $i_{app} = i_{oxid} - i_{red}$. The potential range in which anodic protection is achieved is the protection range or passive region. At the optimum potential E_A, the applied current is approximately 1 μA/cm^2. Anodic protection is limited to active/passive metals and alloys and can be utilized in environments ranging from weak to very aggressive. The applied current is usually equal to the corrosion of the protected system which can be used to monitor the instantaneous corrosion rate. Operating conditions in the

field usually can be accurately and quickly determined by electrochemical laboratory tests (Riggs and Locke, "Anodic Protection—Theory and Practice in the Prevention of Corrosion," Plenum Press, New York).

Cathodic protection (CP) controls corrosion by supplying electrons to a metal structure, thereby suppressing metal dissolution and increasing hydrogen evolution. Figure 6.5.5 shows that when an applied cathodic current density ($i_{app} = i_{red} - i_{oxid}$) of 10,000 μA/cm^2 on a bare metal surface shifts the potential in the negative or active direction to E_c, the corrosion rate has been reduced to 1 μA/cm^2. CP is applicable to all

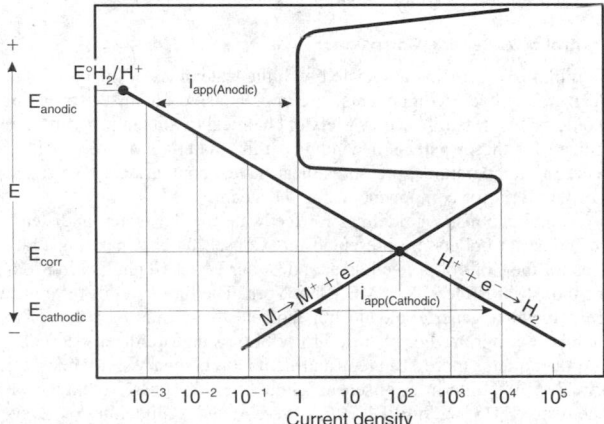

Fig. 6.5.5   Effect of applied anodic and cathodic currents on behavior of an active/passive alloy necessary for anodic and cathodic protection, respectively. *(Adapted from "Fontana's Book on Corrosive Engineering," McGraw-Hill.)*

metals and alloys and is common for use in aqueous and soil environments. CP can be accomplished by (1) impressed current from an external power source through an inert anode or (2) the use of galvanic couplings by sacrificial anodes. Sacrificial anodes include zinc, aluminum, magnesium, and alloys containing these metals. The total protective current is directly proportional to the surface area of the metal being protected; hence, CP is combined with surface coatings where only the coating pores or holidays and damaged spots need be cathodically protected [e.g., for 10 miles of bare pipe, a current of 500 A would be required, while for 10 miles of a superior coated pipe (5×10^6 Ω/ft^2) 0.03 A would be required]. The polarization potential can be measured versus a reference, commonly saturated $Cu/CuSO_4$. The criterion for proper CP is -0.85 V versus this reference electrode. Overprotection by CP (denoted by potentials less than -0.85 V) can lead to blistering and debonding of pipe coatings and hydrogen embrittlement of steel from hydrogen gas evolution. Another problem with CP involves stray currents to unintended structures. The proper CP for a system is determined empirically and must resolve a number of factors to be effective (Peabody, "Control of Pipeline Corrosion," NACE International, Houston; Morgan, "Cathodic Protection," 2d ed., NACE International, Houston).

CORROSION IN INDUSTRIAL AND UTILITY STEAM-GENERATING SYSTEMS

Boiler Corrosion

Corrosion can be initiated from the fireside or the waterside of the surfaces in the boiler. In addition to corrosion, there are a number of other types of failure mechanisms for boiler components, including fatigue, erosion, overheating, manufacturing defects, and maintenance problems. Some of the actual failure mechanisms are combinations of mechanical and chemical mechanisms, such as corrosion fatigue and stress corrosion cracking (SCC). The identification, cause(s), and corrective action(s) for each type of failure mechanism are presented in

"The Boiler Tube Failure Metallurgical Guide," vol. 1, Tech. Rep. TR-102433-V1, EPRI, Palo Alto, CA.

Corrosion in Other Steam/Water Cycle Components

Corrosion in turbines, condensers, heat exchangers, tanks, and piping in the steam/water cycle can occur by general corrosion, pitting, crevice corrosion, intergranular corrosion, SCC, corrosion fatigue, erosion corrosion, dealloying (e.g., dezincification), or galvanic corrosion. A detailed (370-page) discussion of corrosion of steam/water cycle materials is presented in the ASME "Handbook on Water Technology for Thermal Power Systems," chap. 9, ASME, New York.

Control of Waterside Corrosion

To minimize corrosion associated with the watersides of the boiler and steam/water cycle components, a comprehensive chemistry program should be instituted. A wide variety of chemical treatment programs are utilized for these systems depending on the materials of construction, operating temperatures, pressures, heat fluxes, contaminant levels, and purity criteria for components using the steam.

General chemistry guidelines for feedwater, boiler water, and steam for industrial boilers are presented in "Consensus on Operating Practices for the Control of Feedwater and Boiler Water Chemistry in Modern Industrial Boilers," ASME, New York. The limits in this reference are somewhat generic and do not indicate the details of a complete chemistry program. For electric utility boilers and fluidized-bed combustion boilers, more stringent and detailed chemistry guidelines are provided in "Interim Consensus Guidelines on Fossil Plant Cycle Chemistry," CS-4629, EPRI, Palo Alto, CA, and "Guidelines on Cycle Chemistry for Fluidized-Bed Combustion Plants," TR-102976, EPRI, Palo Alto. As indicated later in this section, subsequent publications detail updated versions of the treatment programs.

A good chemistry control program requires a high-purity makeup water to the steam/water cycle. For high-pressure electric utility boilers, two-stage demineralization (e.g., cation/anion/mixed-bed ion exchange) is generally utilized. Consult "Guidelines for Makeup Water Treatment," GS-6699, EPRI, Palo Alto, for information on the design of makeup treatment systems. For industrial boilers, one- or two-stage demineralization or softening (sometimes with dealkalization) are utilized for makeup water treatment, depending on the boiler pressure. Unsoftened makeup water has been utilized in small, low-pressure [e.g., 15 lb/in^2 (gage) (100 kPa)] boiler systems which receive large amounts of condensate return, but this practice is not optimal.

In addition to providing a high-purity makeup water, condensate returns often require **purification** or **polishing**. Condensate polishing can be achieved with specially designed filters, softeners, or mixed-bed ion exchangers. Industrial boilers typically use softeners and/or filters (e.g., electromagnetic). Once-through boilers, full-flow mixed-bed polishers (preferred), or precoat (with crushed ion-exchange resin) filters are standard practice. Also, for high-pressure, drum-type boilers, mixed-bed condensate polishers are often used.

Minimizing the transport of corrosion products to the boiler is essential for corrosion control in the boiler. Corrosion products from the feedwater deposit on boiler tubes, inhibit cooling of tube surfaces, and can provide an evaporative-type concentration mechanism of dissolved salts under boiler tube deposits.

Except for oxygenated treatment (discussed later), feedwater treatment relies on raising the feedwater pH to 8.5 to 10 and eliminating dissolved oxygen. In this environment, the predominant oxide formed on steel is a mixed metal oxide composed primarily of magnetite (Fe_3O_4). The mixed oxide layer formed in deoxygenated solutions resists dissolution or erosion and thus minimizes iron transport to the boiler and corrosion of the underlying metal.

Oxygen removal is normally achieved through thermal deaeration followed by the addition of chemical oxygen scavengers. **Thermal deaerators** are direct-contact steam heaters which spray feedwater in fine droplets and strip the dissolved oxygen from the water with steam. Deaerators normally operate at pressures of 5 to 150 lb/in^2 (gage) (35 to 1,035 kPa), although a few deaerator designs maintain a vacuum.

A wide variety of **oxygen scavengers** are utilized. For boilers below 700 to 900 lb/in^2 (4.8 to 6.2 MPa), sodium sulfite (often catalyzed with cobalt) is commonly used although other nonvolatile (e.g., erythorbate) or volatile oxygen scavengers can be effective. A slight excess of sodium sulfite is applied to the feedwater, resulting in a residual of sulfite in the boiler water. The exact level of sulfite maintained depends primarily on the boiler pressure and oxygen concentrations in the feedwater, although other factors such as boiler water pH, steam and condensate purity, and feedwater temperatures are involved. Sodium sulfite should not be used in boilers operating above 900 lb/in^2 (gage) (6.2 MPa), because a significant amount of the sulfite decomposes to volatile sulfur species at elevated temperatures. These species are acidic and lower steam and condensate pH values, contributing to increased corrosion in steam/water cycle components. Sulfur species in steam may also contribute to SCC.

Commonly used volatile oxygen scavengers include hydrazine, carbohydrazide, hydroquinone, diethylhydroxylamine, and methylethylketoxime. These are normally applied to obtain a free residual of the scavenger in the feedwater at the economizer inlet to the boiler. Due to the slower reaction rates, routine dissolved-oxygen monitoring is more crucial for boilers using volatile oxygen scavengers than for boilers using sulfite. One benefit of some of the volatile oxygen scavengers listed is their ability to enhance metal passivation. Sulfite is an effective oxygen scavenger, but has little effect on passivation beyond the effect of oxygen removal.

Aqueous solutions of ammonia or volatile amines are usually applied to the feedwater to control feedwater, steam, and condensate pH. In all steel cycles, these pH levels are typically controlled at 9.2 to 9.6. In boiler systems containing both steel and copper alloys, pH levels in the feedwater, steam, and condensate are typically controlled from 8.5 to 9.2.

Amines utilized in steam/water cycles include cyclohexylamine, morpholine, ethanolamine, diethanolamine, diethylaminoethanol, dimethyl isopropanolamine, and methoxypropylamine, and the list is expanding. For large steam distribution systems, a blend of two to four amines with different distribution coefficients (ratio of amine in steam to condensate) is often applied to provide pH control throughout the steam and condensate system.

In addition to the **feedwater treatment** program, chemistry must be controlled in the boiler to minimize corrosion. Currently, the basic types of boiler treatment programs used in the United States are coordinated phosphate treatment, congruent phosphate treatment (CPT), equilibrium phosphate treatment (EPT), phosphate treatment with low-level caustic, caustic treatment, all-volatile treatment (AVT), and oxygenated treatment (OT). With the exception of oxygenated treatment, all these treatment programs are based on oxygen-free feedwater and the formation of a mixed metal oxide of predominantly magnetite (Fe_3O_4).

A wide variety of phosphate and caustic treatments are currently utilized in industrial and electric utility boilers. The treatments are designed to provide a reserve of alkalinity to minimize pH fluctuations. Caustic and acid concentrating underneath deposits in boiler tubing contributes to most types of waterside corrosion. It is generally desired to control the pH in a range which keeps the level of free caustic alkalinity and acidic constituents within acceptable values.

Research studies indicate the optimal boiler water treatment program varies considerably with boiler pressure (Tremaine et al., Phosphate Interactions with Metal Oxides under High-Performance Boiler Hideout Conditions, paper 35, International Water Conference, Pittsburgh, PA, 1993). Currently, there is not a consensus regarding optimal treatment programs. Phosphate treatment programs used by electric utility boilers are presented in the literature "Cycle Chemistry Guidelines for Fossil Plants: Phosphate Treatment for Drum Units," TR-103665, EPRI, Palo Alto; "Sodium Hydroxide for Conditioning the Boiler Water of Drum-Type Boilers," TR-104007, EPRI, Palo Alto).

AVT is an extension of the feedwater treatment program. Boiler water pH control is maintained by the portion of amines or ammonia from the feedwater which remains in the boiler water. Due to the low concentrations and poor buffering capacity of ammonia or amines in boiler water, strict purity limits are maintained on the feedwater and boiler water. AVT is generally only utilized in high-pressure boilers ($\sim$ 1,000 lb/in^2 (gage) (12.4 kPa)] with condensate polishers, although some lower-pressure boilers [850 lb/in^2 (gage) (5.9 kPa)] also utilize this treatment.

Although it was used for many years in Germany and Russia, OT was not utilized in the United States until the 1990s. It involves a completely different approach from the other treatment programs since it relies on the formation of a gamma ferric oxide (Fe_2O_3) or gamma ferric oxide hydrate (FeOOH) layer on the steel surfaces in the preboiler cycle.

Originally, oxygenated treatment was based on neutral pH values (7 to 8) and oxygen levels of 50 to 250 ppb in pure feedwater (cation conductivities < 0.15 μS/cm). Later, a program based on slightly alkaline pH values (8.0 to 8.5) and 30 to 150 ppb of dissolved oxygen was developed. This treatment program was originally referred to as **combined water treatment,** but it is now commonly referred to as **oxygenated treatment** (OT) in the United States. Its application is primarily limited to once-through boilers with all-steel steam/water cycles (if full-flow condensate polishing is provided, as recommended, copper alloys are permitted in the condensers). The primary benefit of OT appears to be minimization of deposit-induced pressure drop through the boiler during service. The reduction in iron oxide transport into and deposition in the boiler can also significantly reduce the frequency of chemical cleanings. The details of oxygenated treatment programs for once-through boilers are presented in "Cycle Chemistry Guidelines for Fossil Plants: Oxygenated Treatment," TR-102285, EPRI, Palo Alto. Although this document also presents guidance for utilizing OT for drum-type boilers, currently there is insufficient information to indicate whether this is an appropriate application of this treatment philosophy.

Control of Fireside Corrosion

Corrosive fuel constituents in coal or oil at appropriate metal temperatures may promote fireside corrosion in boiler tube steel. The corrosive ingredients can form liquids that solubilize the oxide film on tubing and react with the underlying metal to reduce the tube wall thickness. Coal ash or oil ash corrosion and waterwall fireside corrosion occur at different areas of the boiler and at different temperatures. Dew point corrosion may occur on the fireside surfaces of the economizer or on other low-temperature surfaces when condensation can form acidic products (Reid, "External Corrosion and Deposits—Boiler and Gas Turbine," American Elsevier, New York; Barna et al., "Fireside Corrosion Inspections of Black Liquor Recovery Boilers," 1993 Kraft Recovery Short Course Notes, TAPPI, Atlanta).

Fireside corrosion often can be mitigated by (1) material selection, (2) purchase of fuels with low impurity levels (limit levels of sulfur, chloride, sodium, vanadium, etc.), (3) purification of fuels (e.g., coal washing), (4) application of fuel additives (e.g., magnesium oxide), (5) adjustment of operating conditions (e.g., percent excess air, percent solids in recovery boilers), and (6) modification of lay-up practices (e.g., dehumidification systems).

CORROSION IN HEATING AND COOLING WATER SYSTEMS AND COOLING TOWERS

Control of corrosion in the waterside of cooling and heating water systems begins with proper preconditioning of the system, as covered in "Standard Recommended Practice: Initial Conditioning of Cooling Water Equipment," RP0182-85, NACE, Houston, 1985, and "Guidelines for Treatment of Galvanized Cooling Towers to Prevent White Rust," Cooling Tower Institute, Houston, 1994. Objectives of the subsequent cooling water treatment program are as follows:

Prevent corrosion of metals in the system.
Prevent the formation of scale (e.g., calcium carbonate, calcium sulfate, calcium phosphate).
Minimize the amount and deposition of suspended solids.
Minimize biological growths.

Substantial emphasis in a cooling tower treatment program is placed on preventing deposits of scale, suspended solids, or biological matter

from forming on system components ("Design and Operating Guidelines Manual for Cooling-Water Treatment," CS-2276, EPRI, Palo Alto, 1982; "Special Report: Cooling-Water Treatment for Control of Scaling, Fouling, Corrosion," *Power,* McGraw-Hill, New York, June 1984). The primary reason for this emphasis is to maintain the performance of the cooling system. However, minimizing deposition is integral to corrosion control to avoid problems with underdeposit corrosion or MIC attack. Also, suspended particles can lead to erosion of piping, particularly copper tubing.

Depending on the inherent corrosivity of the cooling water, corrosion control in open recirculating cooling water systems may consist merely of controlling the pH, conductivity, and alkalinity levels. However, in systems with extensive piping networks, applications of corrosion inhibitors are often utilized to protect materials in the system. These inhibitors can include the following compounds: zinc salts, molybdate salts, triazoles, orthophosphates, polyphosphates, phosphonates, and polysilicates. Often, two or more of these inhibitors are used together. Oxidizing or nonoxidizing biocides also are required in cooling tower systems to inhibit biological activity.

For closed cooling and heating water systems, corrosion inhibitors are usually applied and pH levels are controlled at 8 to 11. Corrosion inhibitors utilized in closed cooling and heating water systems include the inhibitors listed for open recirculating cooling water systems except for zinc salts. Also, nitrite salts are often used in closed water systems.

Corrosion control in closed cooling systems is greatly facilitated by minimizing water loss from the system. Makeup water introduced to replace leakage introduces oxygen, organic matter to support biological activity, and scale constituents. Utilization of pumps with mechanical seals (with filtered seal water) rather than packing seals can greatly reduce water losses from closed systems. Control of water leakage is probably the single most important operational factor in reducing corrosion in these systems.

In heating and cooling water systems, concerns regarding corrosion control generally focus on the waterside or interior of piping and components. However, substantial corrosion can occur on the exterior of these piping systems underneath insulation. External corrosion of piping generally requires a source of external moisture. In chill water and secondary water systems, moisture can be introduced through exposure to the air and subsequent condensation of moisture on the pipe surface.

While chill water piping systems generally are installed with vapor barriers, breaks in the vapor barrier can exist due to inadequate installation or subsequent maintenance activities. Secondary water system piping is sometimes not installed with vapor barriers because the cooling water temperature is designed to be above the expected dew point. However, humidity levels in the building can sometimes be elevated and can lead to condensation at higher temperatures than initially expected.

Leaks from piping at fittings or joints can saturate the insulation with moisture and can lead to corrosion. This type of moisture source can be particularly troublesome in hot water piping because the moisture can be evaporated to dryness and wet/dry corrosion can result. Corrosion rates during this transition from wetness to dryness can be many times higher than that of a fully wetted surface [Pollock and Steely (eds.), "Corrosion under Wet Thermal Insulation," NACE International, Houston]. If the moisture can be eliminated, corrosion underneath insulation should be negligible. However, if the moisture cannot be eliminated, coatings may need to be applied to the metal surface. Inhibitors are also available to mitigate corrosion underneath insulation.

CORROSION IN THE CHEMICAL PROCESS INDUSTRY

Corrosion control in the chemical process industry is important to provide continuous operation in order to meet production schedules and to maintain purity of manufactured products. Also, protection of personnel and the environment is of primary importance to the chemical process industry. Many of the chemicals utilized have high toxicity, and their release due to unexpected corrosion failures of piping and vessels is not permitted. Corrosion prevention and control in the chemical process industry are largely focused on material selection. Nondestructive evaluation methods are routinely used to evaluate material integrity. (See Sec. 5.4.)

Due to the wide range of chemicals and operating conditions in the chemical process industry, the types of corrosion cannot be adequately addressed here. The following references address the corrosivity of the various environments and appropriate material selection and operational and maintenance practices: Moniz and Pollock, "Process Industries Corrosion," NACE, Houston. Degnan, "Corrosion in the Chemical Processing Industry," "Metals Handbook." Aller et al., "First International Symposium on Process Industry Piping," Materials Technology Institute, Houston. Dillon, "Corrosion Control in the Chemical Process Industries," McGraw-Hill, New York.

6.6 PAINTS AND PROTECTIVE COATINGS
by Harold M. Werner and Expanded by Staff

REFERENCES: Keane (ed.). "Good Painting Practice," Steel Structures Painting Council. Levinson and Spindel, "Recent Developments in Architectural and Maintenance Painting." Levinson, "Electrocoat, Powder Coat, Radiate." Reprints, *Journal of Coatings Technology*, 1315 Walnut St., Philadelphia. Roberts, "Organic Coatings—Their Properties, Selection and Use," Government Printing Office. "Paint/Coatings Dictionary," Federation of Societies for Coatings Technology. Weismantel (ed.), "Paint Handbook," McGraw-Hill. Publications of the National Paint and Coatings Association.

Paint is a mixture of filmogen (film-forming material, binder) and pigment. The pigment imparts color, and the filmogen, continuity; together, they create opacity. Most paints require volatile thinner to reduce their consistencies to a level suitable for application. An important exception is the powder paints made with fusible resin and pigment.

In **conventional oil-base paint,** the filmogen is a vegetable oil. Driers are added to shorten drying time. The thinner is usually petroleum spirits.

In **water-thinned paints,** the filmogen may be a material dispersible in water, such as solubilized linseed oil or casein, an emulsified polymer, such as butadiene-styrene or acrylate, a cementitious material, such as portland cement, or a soluble silicate.

Varnish is a blend of resin and drying oil, or other combination of filmogens, in volatile thinner. A solution of resin alone is a **spirit varnish,** e.g., shellac varnish. A blend of resin and drying oil is an **oleoresinous varnish,** e.g., spar varnish. A blend of nonresinous, nonoleaginous filmogens requiring a catalyst to promote the chemical reaction necessary to produce a solid film is a **catalytic coating.**

Enamel is paint that dries relatively harder, smoother, and glossier than the ordinary type. These changes come from the use of varnish or synthetic resins instead of oil as the liquid portion. The varnish may be oleoresinous, spirit, or catalytic.

Lacquer is a term that has been used to designate several types of painting materials; it now generally means a spirit varnish or enamel, based usually on cellulose nitrate or cellulose acetate butyrate, acrylate or vinyl.

PAINT INGREDIENTS

The ingredients of paints are drying and semidrying vegetable oils, resins, plasticizers, thinners (solvents), driers or other catalysts, and pigments.

Drying oils dry (become solid) when exposed in thin films to air. The drying starts with a chemical reaction of the oil with oxygen. Subsequent or simultaneous polymerization completes the change. The most important drying oil is linseed oil. Addition of small percentages of driers shortens the drying time.

Soybean oil, with poorer drying properties than linseed oil, is classed as a **semidrying oil. Tung oil** dries much faster than linseed oil, but the film wrinkles so much that it is reminiscent of frost; heat treatment eliminates this tendency. Tung oil gels after a few minutes cooking at high temperatures. It is more resistant to alkali than linseed oil. Its chief use is in oleoresinous varnishes. **Oiticica oil** resembles tung oil in many of its properties. **Castor oil** is nondrying, but heating chemically "dehydrates" it and converts it to a drying oil. The dehydrated oil resembles tung oil but is slower drying and less alkali-resistant. **Nondrying oils,** like coconut oil, are used in some baking enamels. An oxidizing hydrocarbon oil, made from petroleum, contains no esters or fatty acids.

The drying properties of oils are linked to the amount and nature of unsaturated compounds. By molecular distillation or solvent extraction the better drying portions of an oil can be separated. Drying is also improved by changing the structure of the unsaturated compounds (isomerization) and by modifying the oils with small amounts of such compounds as phthalic acid and maleic acid.

Driers are oil-soluble compounds of certain metals, mainly lead, manganese, and cobalt. They accelerate the drying of coatings made with oil. The metals are introduced into the coatings by addition of separately prepared compounds. Certain types of synthetic nonoil coatings (catalytic coatings) dry by baking or by the action of a **catalyst** other than conventional drier. Drier action begins only after the coating has been applied. Catalyst action usually begins immediately upon its addition to the coating; hence addition is delayed until the coating is about to be used.

Resins Both natural and synthetic resins are used. **Natural resins** include fossil types from trees now extinct, recent types (rosin, Manila, and dammar), lac (secretion of an insect), and asphalts (gilsonite). **Synthetic resins** include ester gum, phenolic, alkyd, urea, melamine, amide, epoxy, urethane, vinyl, styrene, rubber, petroleum, terpene, cellulose nitrate, cellulose acetate, and ethyl cellulose.

Latex filmogens contain a stable aqueous dispersion of synthetic resin produced by emulsion polymerization as the principal constituent.

Plasticizers In an oleoresinous varnish the oil acts as a softening agent or plasticizer for the resin. There are other natural compounds and substances, and many synthetic ones, for plasticizing film formers such as the cellulosics. Plasticizers include dibutyl phthalate, tricresyl phosphate, dibutyl sebacate, dibutyltartrate, tributyl citrate, methyl abietate, and chlorinated biphenyl.

Thinners As the consistency of the film-forming portion of most paints and varnishes is too high for easy application, thinners (**solvents**) are needed to reduce it. If the nonvolatile portion is already liquid, the term thinner is used; if solid, the term solvent is used. **Petroleum** or **mineral spirits** has replaced turpentine for thinning paints in the factory. The coal-tar products—**toluene, xylene,** and **solvent naphtha**—are used where solvency better than that given by petroleum spirits is needed. Esters, alcohols, and ketones are standard for cellulosic or vinyl lacquers. Finally, water is used to thin latex paints, to be thinned at point of use.

Pigments may be natural or synthetic, organic or inorganic, opaque or nonopaque, white or colored, chemically active or inert. Factors entering into the selection of a pigment are color, opacity, particle size, compatibility with other ingredients, resistance to light, heat, alkali, and acid, and cost. The most important pigments are: **white**—zinc oxide, leaded zinc oxide, titanium dioxide; **red**—iron oxides, red lead* (for rust prevention rather than color), chrome orange, molybdate orange, toluidine red, para red; **yellow**—iron oxide, chrome yellow, zinc yellow (for rust prevention rather than color), Hansa; **green**—chrome green, chrome oxide; **blue**—iron, ultramarine, phthalocyanine; **extenders** (non-opaque), i.e., magnesium silicate, calcium silicate, calcium carbonate, barium sulfate, aluminum silicate.

Miscellany   Paint contains many other ingredients, minor in amount but important in function, such as emulsifiers, antifoamers, leveling agents, and thickeners.

Ecological Considerations   Many jurisdictions restrict the amounts of volatile organic compounds that may be discharged into the atmosphere during coating operations. Users should examine container labels for applicable conditions.

PAINTS

Aluminum paint is a mixture of aluminum pigment and varnish, from 1 to 2 lb of pigment per gal of varnish. The aluminum is in the form of thin flakes. In the paint film, the flakes "leaf"; i.e., they overlap like leaves fallen from trees. Leafing gives aluminum paint its metallic appearance and its impermeability to moisture. Aluminum paint ranks high as a reflector of the sun's radiation and as a retainer of heat in hot-air or hot-water pipes or tanks.

Bituminous Paint   Hard asphalts, like gilsonite, cooked with drying oils, and soft asphalts and coal tar, cut back with thinner, may be used to protect metal and masonry wherever their black color is not objectionable.

Chemical-Resistant Paint   Chemical resistance is obtained by use of resins such as vinyls, epoxies and urethanes. Vinyls are air-drying types. Epoxies are catalytically cured with acids or amino resins. The latter are two-component paints to be mixed just prior to use. Urethanes can be two-component also, or single-component paints that react with moisture in the air to cure.

Emulsion paints are emulsions in water of oil or resin base mixed with pigment. The **latex paints** (acrylic, butadiene-styrene, polyvinylacetate, etc.) belong to this class. Emulsion or latex paints are designed for both exterior and interior use.

Strippable coatings are intended to give temporary protection to articles during storage or shipment.

Powder coatings are 100 percent solids coatings applied as a dry powder mix of resin and pigment and subsequently formed into a film with heat. The solid resin binder melts upon heating, binds the pigment, and results in a pigment coating upon cooling. The powder is applied either by a electrostatic spray or by passing the heated object over a fluidized bed of powder with subsequent oven heating to provide a smooth continuous film.

Marine Paint   Ocean-going steel ships require antifouling paint over the anticorrosive primer. Red lead or zinc yellow is used in the primer. **Antifouling** paints contain ingredients, such as cuprous oxide and mercuric oxide, that are toxic to barnacles and other marine organisms. During World War II, the U.S. Navy developed **hot plastic** ship-bottom paint. It is applied by spray gun in a relatively thick film and dries or sets as it cools. **Copper powder paint** is suitable for small wooden craft.

Fire-Retardant Paint   Most paint films contain less combustible matter than does wood. To this extent, they are fire-retardant. In addition they cover splinters and fill in cracks. However, flames and intense heat will eventually ignite the wood, painted or not. The most that ordinary paint can do is to delay ignition. Special compositions fuse and give off flame-smothering fumes, or convert to spongy heat-insulating masses, when heated. The relatively thick film and low resistance to abrasion and cleaning make these coatings unsuited for general use as paint, unless conventional decorative paints are applied over them. However, steady improvement in paint properties is being made.

Traffic paint is quick-drying, so that traffic is inconvenienced as little as possible. Rough texture and absence of gloss increase visibility. Tiny

* The use of lead pigment (red lead) is prohibited by law. (See discussion below under Painting Steel.)

glass beads may be added for greater night visibility. Catalytically cured resinous coating are also a recent development for fast-drying traffic paints.

Heat-Reflecting Paint   Light colors reflect more of the sun's radiation than do dark colors. White is somewhat better than aluminum, but under some conditions, aluminum may retain its reflecting power longer.

Heat-Resistant Paint   Silicone paint is the most heat-resistant paint yet developed. Its first use was as an insulation varnish for electric motors, but types for other uses, such as on stoves and heaters, have been developed. Since it must be cured by baking, it is primarily a product finish.

Fungicides should be added to paint that is to be used in bakeries, breweries, sugar refineries, dairies, and other places where fungi flourish. Until ecological considerations forbade their use, mercury compounds were important fungicides. Their place has been filled by chlorinated phenols and other effective chemical compounds. Fungi-infected surfaces should be sterilized before they are painted. Scrub them with mild alkali; if infection is severe, add a disinfectant.

Wood preservatives of the paintable type are used extensively to treat wood windows, doors, and cabinets. They reduce rotting that may be promoted by water that enters at joints. They comprise solutions of compounds, such as chlorinated phenols, and copper and zinc naphthenates in petroleum thinner. Water repellants, such as paraffin wax, are sometimes added but may interfere with satisfactory painting.

Clear **water repellents** are treatments for masonry to prevent wetting by water. Older types are solutions of waxes, drying oils, or metallic soaps. Silicone solutions, recently developed, are superior, both in repelling water and in effect on appearance.

Zinc-Rich Primer   Anticorrosive primers for iron and steel, incorporating zinc dust in a concentration sufficient to give electrical conductivity in the dried film, enable the zinc metal to corrode preferentially to the substrate, i.e., to give cathodic protection. The binders can be organic resins or inorganic silicates.

Painting

Exterior Architectural Painting   Surfaces must be clean and dry, except that wood to be painted with emulsion paint will tolerate a small amount of moisture and masonry to be painted with portland cement—base paint should be damp. Scrape or melt the resin from the knots in wood, or scrub it off with paint thinner or alcohol. As an extra precaution, seal the knots with shellac varnish, aluminum paint, or proprietary knot sealer. When painting wood for the first time, fill nail holes and cracks with putty, after the priming coat has dried. If the wood has been painted before, remove loose paint with a scraper, wire brush, or sandpaper. Prime all bare wood. In bad cases of cracking and scaling, remove all the paint by dry scraping or with paint and varnish remover or with a torch.

Paint on new wood should be from 4 to 5 mils (0.10 to 0.13 mm) thick. This usually requires three coats. A system consisting of nonpenetrating primer and special finish coat may permit the minimum to be reached with two coats. Repainting should be frequent enough to preserve a satisfactory thickness of the finish coat. Nonpenetrating primers stick on some types of wood better than regular house paint does.

Interior Architectural Painting   Interior paints are used primarily for decoration, illumination, and sanitation. Enamels usually are used for kitchen and bathroom walls, where water resistance and easy cleaning are needed; semiglossy and flat types are used on walls and ceilings where it is desired to avoid glare. Wet plaster will eventually destroy oil-base paints. Fresh plaster must be allowed to dry out. Water from leaks and condensation must be kept away from aged plaster. Several coats of paint with low permeability to water vapor make an effective vapor barrier.

Painting Concrete and Masonry   Moisture in concrete and masonry brings alkali to the surface where it can destroy oil paint. Allow 2 to 6 months for these materials to dry out before painting; a year or more, if they are massive. Emulsion or latex paints may be used on damp concrete or masonry if there is reason to expect the drying to continue.

Portland-cement paint, sometimes preferred for masonry, is especially suitable for first painting of porous surfaces such as cinder block. Before applying this paint, wet the surface so that capillarity will not extract the water from the paint. Scrubbing the paint into the surface with a stiff brush of vegetable fiber gives the best job, but good jobs are also obtained with usual brushes or by spraying. Keep the paint damp for 2 or 3 days so that it will set properly. Latex and vinyl paints are much used on concrete, because of their high resistance to alkali.

Painting Steel Preparation of steel for painting, type of paint, and condition of exposure are closely related. Methods of preparation, arranged in order of increasing thoroughness, include (1) removal of oil with solvent; (2) removal of dirt, loose rust, and loose mill scale with scraper or wire brush; (3) flame cleaning; (4) sandblasting; (5) pickling; (6) phosphating. Exposure environments, arranged in order of increasing severity, include (a) dry interiors, or arid regions; (b) rural or light industrial areas, normally dry; (c) frequently wet; (d) continuously wet; (e) corrosive chemical. Paint systems for condition (a) often consist of a single coat of low-cost paint. For other conditions, the systems comprise one or two coats of rust-inhibitive primer, such as a zinc-rich primer, and one or more finish coats, selected according to severity of conditions.

The primer contains one or more rust-inhibitive pigments, selected mainly from red lead (found mostly in existing painted surfaces), zinc yellow, and zinc dust. It may also contain zinc oxide, iron oxide, and extender pigments. Of equal importance is the binder, especially for the top coats. For above-water surfaces, linseed oil and alkyd varnish give good service; for underwater surfaces, other binders, like phenolic and vinyl resin, are better. Chemical-resistant binders include epoxy, synthetic rubber, chlorinated rubber, vinyl, urethanes, and neoprene.

Paint for structural steel is normally air-drying. Large percentages for factory-finished steel products are catalytic-cured, or are baked.

By virtue of the health problems associated with lead-based paint, this product has effectively been proscribed by law from further use. A problem arises in the matter of removal and repainting areas which have previously been coated with lead-based paints; this is particularly true in the case of steel structures such as bridges, towers, and the like.

Removal of leaded paints from such structures requires that extreme care be exercised in the removal of the paint, due diligence being paid to the protection of not only the personnel involved therein but also the general public and that, further, the collection and disposal of the hazardous waste be effectuated in accordance with prescribed practice. Reference to current applicable OSHA documents will provide guidance in this regard.

The removal of leaded paints can be accomplished by chipping, sand blasting, use of solvents (strippers), and so forth. In all cases, stringent precautions must be emplaced to collect all the residue of old paint as well as the removal medium. Alternatively, recourse may be had to topcoating the existing leaded paint, effectively sealing the surface of the old paint and relying on the topcoating medium to keep the leaded undercoat from flaking away. In this type of application, urethane coatings, among others, have proved suitable. While the topcoating procedure will alleviate current problems in containment of existing lead-painted surfaces, it is not likely that this procedure will prove to be the final solution. Ultimately, as with any painted surface, there will come a time when the leaded-paint subcoat can no longer be contained in situ and must be removed. It is expected that better and more economical procedures will evolve in the near future, so that the extant problems associated with removal of leaded paint will become more tractable. The AISI and AASHTO have collaborated in this regard and have issued a guide addressing these problems.

Galvanized Iron Allow new galvanized iron to weather for 6 months before painting. If there is not enough time, treat it with a proprietary etching solution. For the priming coat, without pretreatment, a paint containing a substantial amount of zinc dust or portland cement should be used. If the galvanizing has weathered, the usual primers for steel are also good.

Painting Copper The only preparation needed is washing off any grease and roughening the surface, if it is a polished one. Special primers are not needed. Paint or varnish all copper to prevent corrosion products from staining the adjoining paint.

Painting Aluminum The surface must be clean and free from grease. Highly polished sheet should be etched with phosphoric acid or chromic acid. Zinc-yellow primers give the best protection against corrosion. The only preparation needed for interior aluminum is to have the surface clean; anticorrosive primers are not needed.

Magnesium and its alloys corrode readily, especially in marine atmospheres. Red-lead primers must not be used. For factory finishing, it is customary to chromatize the metal and then apply a zinc-yellow primer.

Water-Tank Interiors Among the best paints for these are asphaltic compositions. For drinking-water tanks, select one that imparts no taste. Zinc-dust paints made with phenolic varnish are also good.

Wood Products Finishes may be lacquer or varnish, or their corresponding enamels. High-quality clear finishes may require many operations, such as sanding, staining, filling, sealing, and finishing. Furniture finishing is done mostly by spray. Small articles are often finished by tumbling; shapes like broom handles, by squeegeeing.

Plastics Carefully balanced formulations are necessary for satisfactory adhesion, to avoid crazing and to prevent migration of plasticizer.

Paint Deterioration Heavy dew, hot sun, and marine atmospheres shorten the life of paint. Industrial zones where the atmosphere is contaminated with hygroscopic and acidic substances make special attention to painting programs necessary. Dampness within masonry and plaster walls brings alkali to the surface where it can destroy oil-base paints. Interior paints on dry surfaces endure indefinitely; they need renewal to give new color schemes or when it becomes impractical to wash them. Dry temperate climates are favorable to long life of exterior paints.

Application Most industrial finishing is done with spray guns. In electrostatic spraying, the spray is charged and attracted to the grounded target. Overspray is largely eliminated. Other methods of application include dipping, electrocoating, flowing, tumbling, doctor blading, rolling, fluid bed, and screen stenciling.

An increasing proportion of maintenance painting is being done with spray and hand roller. The spray requires up to 25 percent more paint than the brush, but the advantage of speed is offsetting. The roller requires about the same amount of paint as the brush.

Dry finishing is done by flame-spraying powdered pigment-filmogen compositions or by immersing heated articles in a **fluid bed** of the powdered composition.

Spreading Rates When applied by brush, approximate spreading rates for paints on various surfaces are as shown in Table 6.6.1.

OTHER PROTECTIVE AND DECORATIVE COATINGS

Porcelain enamel as applied to metal or clay products is a vitreous, inorganic coating which is set by firing. Enameling is done in several different ways, but the resulting surface qualities and properties are largely similar. The coating finds utility in imparting to the product a surface which is hard and scratch-resistant, possesses a remarkable resistance to general corrosion, and perhaps even has the constituents of the coating material tailored for the enamel to be resistant to specific working environments.

Frit is the basic unfired raw coating material, composed of silica, suitable fluxes, and admixtures to impart either opacity or color.

The surface to be enameled must be rendered extremely clean, after which a tightly adhering undercoat is applied. The surface is covered with powdered frit, and the product finally is fired and subsequently cooled. Despite its demonstrated advantageous properties, porcelain enamel has effectively zero tolerance for deformation, and it will craze or spall easily when mechanically distressed.

Fused dry resin coatings employing polymers as the active agent can be bonded to most metals, with the resin applied in a fluidized bed or by means of electrostatic deposition. The surface to be treated must be extremely clean. Resinous powders based on vinyl, epoxy, nylon, poly-

Table 6.6.1   Spreading Rates for Brushed Paints

Surface	Type	First coat ft²/gal	First coat m²/L	Second and third coats ft²/gal	Second and third coats m²/L
Wood	Oil emulsion	400–500	10–12	500–600	12–15
Wood, primed	Emulsion	500–600	12–15	500–600	12–15
Structural steel	Oil	450–600	11–15	650–900	16–22
Sheet metal	Oil	500–600	12–15	550–650	13–16
Brick, concrete	Oil	200–300	5–7	400 500	10–12
Brick, concrete	Cement	100–125	2.5–3	125–175	3–4.5
Brick, concrete	Emulsion	200–300	5–7	400–500	10–12
Smooth plaster	Oil	550–650	13–16	550–650	13–16
Smooth plaster	Emulsion	400–500	10–12	500–600	12–15
Concrete floor	Enamel	400–500	10–12	550–650	13–16
Wood floor	Varnish	550–650	13–16	550–650	13–16

ethylene, and the like are used. Intermediate bonding agents may be required to foster a secure bond between the metal surface and the resin. After the resin is applied, the metal piece is heated to fuse the resin into a continuous coating. The tightly bonded coating imparts corrosion resistance and may serve decorative functions.

Bonded phosphates provide a corrosion-resistant surface coat to metals, principally steel. Usually, the phosphated surface layer exhibits a reduced coefficient of friction and for this reason can be applied to sheet steel to be deep-drawn, resulting in deeper draws per die pass. Passing the raw sheet metal through an acid phosphate bath results in the deposition of an insoluble crystalline phosphate on the surface.

Hot dipping permits the deposition of metal on a compatible substrate and serves a number of purposes, the primary ones being to impart corrosion resistance and to provide a base for further surface treatment, e.g., paint. In **hot dip galvanizing,** a ferrous base metal, usually in the form of sheet, casting, or fabricated assembly, is carefully cleaned and then passed through a bath of molten zinc. The surface coating is metallographically very complex, but the presence of zinc in the uppermost surface layer provides excellent corrosion resistance. In the event that the zinc coating is damaged, leaving the base metal exposed to corrosive media, galvanic action ensues and zinc will corrode preferentially to the base metal. This sacrificial behavior thereby protects the base metal. (See Secs. 6.4 and 6.5.)

Automotive requirements in recent years have fostered the use of galvanized sheet steel in the fabrication of many body components, often with the metal being galvanized on one side only.

Hardware and fittings exposed to salt water historically have been hot dip galvanized to attain maximum service life.

Terne, a lead coating applied to steel (hence, **terne sheet**), is applied by hot dipping. The sheet is cleaned and then passed through a bath of molten lead alloyed with a small percentage of tin. The product is widely applied for a variety of fabricated pieces and is employed architecturally when its dull, though lasting, finish is desired. It exhibits excellent corrosion resistance, is readily formed, and lends itself easily to being soldered.

Lead-coated copper is used largely as a substitute for pure copper when applied as an architectural specialty for roof flashing, valleys, and the like, especially when the design requires the suppression of green copper corrosion product (verdigris). Its corrosion resistance is effectively equivalent to that of pure lead. It is produced by dipping copper sheet into a bath of molten lead.

Electroplating, or **electrodeposition,** employs an electric current to coat one metal with another. The method can be extended to plastics, but requires that the plastic surface be rendered electrically conductive by some means. In the basic electroplating circuit, the workpiece is made the cathode (negative), the metal to be plated out is made the anode (positive), and both are immersed in an electrolytic bath containing salts of the metal to be plated out. The circuit is completed with the passage of a direct current, whence ions migrate from anode to cathode, where they lose their charge and deposit as metal on the cathode. (In essence, the anode becomes sacrificial to the benefit of the cathode, which accretes the metal to be plated out.) Most commercial metals and some of the more exotic ones can be electroplated; they include, among others, cadmium, zinc, nickel, chromium, copper, gold, silver, tin, lead, rhodium, osmium, selenium, platinum, and germanium. This list is not exclusive; other metals may be pressed into use as design requirements arise from time to time.

Plating most often is used to render the workpiece more corrosion-resistant, but decorative features also are provided or enhanced thereby. On occasion, a soft base metal is plated with a harder metal to impart wear resistance; likewise, the resultant composite material may be designed to exhibit different and/or superior mechanical or other physical properties.

Electroplating implies the addition of metal to a substrate. Accordingly, the process lends itself naturally to buildup of worn surfaces on cutting and forming tools and the like. This buildup must be controlled because of its implications for the finished dimensions of the plated part.

The most familiar example we observe is the universal use of chromium plating on consumer goods and pieces of machinery. Gold plating and silver plating abound. Germanium is electroplated in the production of many solid-state electronic components. Brass, while not a pure metal, can be electroplated by means of a special-composition electrolyte.

Of particular concern in the current working environment is the proper handling and disposition of spent plating solutions. There are restrictions in this regard, and those people engaged in plating operations must exercise stringent precautions in the matter of their ultimate disposal.

Aluminum normally develops a natural thin surface film of aluminum oxide which acts to inhibit corrosion. In **anodizing,** this surface film is thickened by immersion in an electrolytic solution. The protection provided by the thicker layer of oxide is increased, and for many products, the oxide layer permits embellishment with the addition of a dye to impart a desired color. (See Sec. 6.4.)

Vacuum deposition of metal onto metallic or nonmetallic substrates provides special functional or decorative properties. Both the workpiece to be coated and the metal to be deposited are placed in an evacuated chamber, wherein the deposition metal is heated sufficiently to enable atom-size particles to depart from its surface and impinge upon the surface to be coated. The atomic scale of the operation permits deposition of layers less than 1 μm thick; accordingly, it becomes economical to deposit expensive jewelry metals, e.g., gold and silver, on inexpensive substrates. Some major applications include production of highly reflective surfaces, extremely thin conductive surface layers for electronic components, and controlled buildup of deposited material on a workpiece.

A companion process, **sputtering,** operates in a slightly different fashion, but also finds major application in deposition of 1-μm thicknesses on substrates. Other than for deposition of 1-μm thicknesses on solid-state components and circuitry, it enables deposition of hard metals on edges of cutting tools.

Chemical vapor deposition (CVD) is employed widely, but is noted here for its use to produce diamond vapors at ambient temperatures, which are deposited as continuous films from 1 to 1,000 μm thick. The CVD diamond has a dense polycrystalline structure with discrete diamond grains, and in this form it is used to substitute for natural and artificial diamonds in abrasive tools.

VARNISH

Oleoresinous varnishes are classed according to oil length, i.e., the number of gallons of oil per 100 lb of resin. Short oil varnishes contain up to 10 gal of oil per 100 lb of resin; medium, from 15 to 25 gal; long, over 30 gal.

Floor varnish is of medium length and is often made with modified phenolic resins, tung, and linseed oils. It should dry overnight to a tough hard film. Some floor varnishes are rather thin and penetrating so that they leave no surface film. They show scratches less than the orthodox type. Moisture-cured urethane or epoxy ester varnishes are more durable types.

Spar varnish is of long oil length, made usually with phenolic or modified-phenolic resins, tung or dehydrated castor oil, and linseed oil. Other spar varnishes are of the alkyd and urethane types. Spar varnishes dry to a medium-hard glossy film that is resistant to water, actinic rays of the sun, and moderate concentrations of chemicals.

Chemical-resistant varnishes are designed to withstand acid, alkali, and other chemicals. They are usually made of synthetic resins, such as chlorinated rubber, cyclized rubber, phenolic resin, melamine, urea-aldehyde, vinyl, urethane, and epoxy resin. Some of these must be dried by baking. Coal-tar epoxy and coal-tar urethane combinations are also used. **Catalytic varnish** is made with a nonoxidizing film former, and is cured with a catalyst, such as hydrochloric acid or certain amines.

Flat varnish is made by adding materials, such as finely divided silica or metallic soap, to glossy varnish. Synthetic latex or other emulsion, or aqueous dispersion, such as glue, is sometimes called **water varnish.**

LACQUER

The word *lacquer* has been used for (1) spirit varnishes used especially for coating brass and other metals, (2) Japanese or Chinese lacquer, (3) coatings in which cellulose nitrate (**pyroxylin**), cellulose acetate, or cellulose acetate butyrate is the dominant ingredient, (4) oleoresinous baking varnishes for interior of food cans.

Present-day lacquer primarily refers to cellulosic coatings, clear or pigmented. These lacquers dry by evaporation. By proper choice of solvents, they are made to dry rapidly. Besides cellulosic compounds, they contain resins, plasticizers, and solvent. Cellulose acetate, cellulose acetate butyrate, and cellulose acetate propionate lacquers are nonflammable, and the clear forms have better exterior durability than the cellulose nitrate type. Although cellulosic and acrylic derivatives dominate the lacquer field, compositions containing vinyls, chlorinated hydrocarbons, or other synthetic thermoplastic polymers are of growing importance.

Lac is a resinous material secreted by an insect that lives on the sap of certain trees. Most of it comes from India. After removal of dirt, it is marketed in the form of grains, called seed-lac; cakes, called button lac; or flakes, called shellac. Lac contains up to 7 percent of wax, which is removed to make the refined grade. A bleached, or white, grade is also available.

Shellac varnish is made by "cutting" the resin in alcohol; the cut is designated by the pounds of lac per gallon of alcohol, generally 4 lb. Shellac varnish should always be used within 6 to 12 months of manufacture, as some of the lac combines with the alcohol to form a soft, sticky material.

6.7 WOOD

by Staff, Forest Products Laboratory, USDA Forest Service. Prepared under the direction of David W. Green.

(*Note:* This section was written and compiled by U.S. government employees on official time. It is, therefore, in the public domain and not subject to copyright.)

REFERENCES Freas and Selbo, Fabrication and Design of Glued Laminated Wood Structural Members, *U.S. Dept. Agr. Tech. Bull.* no. 1069, 1954. Hunt and Garratt, "Wood Preservation," McGraw-Hill. "Wood Handbook," *U.S. Dept. Agr. Handbook.* "Design Properties of Round, Sawn and Laminated Preservatively Treated Construction Poles as Posts," ASAE Eng. Practice EP 388.2, American Society of Agricultural Engineering Standards. AF&PA, "National Design Specification for Wood Construction," American Forest & Paper Association, Washington, 1994. AITC, "Recommended Practice for Protection of Structural Glued Laminated Timber during Transit, Storage and Erection," AITC III, American Institute of Timber Construction, Englewood, CO, 1994. AITC, "Laminated Timber Design Guide," American Institute of Timber Construction, Englewood, CO, 1994. AITC, "Timber Construction Manual," Wiley, New York, 1994. ANSI, "American National Standards for Wood Products—Structural Glued Laminated Timber," ANSI/AITC A190.1-1992, American National Standards Institute, New York, 1992. ASTM, "Annual Book of Standards," vol. 04.09, "Wood," American Society for Testing and Materials, Philadelphia, PA, 1993: "Standard Terminology Relating to Wood," ASTM D9; "Standard Practice for Establishing Structural Grades and Related Allowable Properties for Visually Graded Dimension Lumber," ASTM D245; "Standard Definitions of Terms Relating to Veneer and Plywood," ASTM D1038; "Standard Nomenclature of Domestic Hardwoods and Softwoods," ASTM D1165; "Standard Practice for Establishing Allowable Properties for Visually Graded Dimension Lumber from In-Grade Tests of Full Size Specimens," ASTM D1990; "Standard Methods for Establishing Clear Wood Strength Values," ASTM D2555; "Standard Practice for Evaluating Allowable Properties of Grades of Structural Lumber," ASTM D2915; "Standard Test Methods for Mechanical Properties of Lumber and Wood Based Structural Material," ASTM D4761; "Standard Test Method for Surface Burning Characteristics of Building Materials," ASTM E84; "Standard Test Methods for Fire Tests of Building Construction and Materials," ASTM E119. ASTM, "Annual Book of Standards," vol. 04.10. American Society for Testing and Materials, Philadelphia, PA, 1995: "Standard Method for Establishing Design Stresses for Round Timber Piles," ASTM D2899; "Standard Practice for Establishing Stress Grades for Structural Members Used in Log Buildings," ASTM D3957; "Standard Practice for Establishing Stresses for Structural Glued Laminated Timber," ASTM D3737; "Standard Specification for Evaluation of Structural Composite Lumber Products," ASTM D5456. AWPA, "Book of Standards," American Wood Preserver's Association, Stubensville, MD, 1994. American Wood Systems, "Span tables for glulam timber." APA-EWS, Tacoma, WA, 1994. Cassens and Feist, Exterior Wood in the South—Selection, Applications, and Finishes, *Gen. Tech. Rep.* FPL-GTR-69, USDA Forest Service, Forest Products Laboratory, Madison, WI, 1991. Chudnoff, Tropical Timbers of the World, *Agric. Handb.* no. 607, 1984. Forest Products Laboratory, Wood Handbook: Wood as an Engineering Material, *Agric. Handb.,* no. 72. rev., 1987. Green, Moisture Content and the Shrinkage of Lumber, *Res. Pap.* FPL-RP-489, USDA Forest Service, Forest Products Laboratory, Madison, WI, 1989. Green and Kretschmann, Moisture Content and the Properties of Clear Southern Pine, *Res. Pap.* FPL-RP-531, USDA Forest Service, Forest Products Laboratory, Madison, WI, 1995. James, Dielectric Properties of Wood and Hardboard: Variation with Temperature, Frequency, Moisture Content, and Grain Orientation, *Res. Pap.* FPL-RP-245, USDA Forest Service, Forest Products Laboratory, Madison, WI, 1975. James, Electric Moisture Meters for Wood, *Gen. Tech. Rep.* FPL-GTR-6, USDA Forest Service, Forest Products Laboratory, Madison, WI, 1988. MacLean, Thermal Conductivity of Wood, *Heat. Piping Air Cond.,* **13,** no. 6, 1941, pp. 380–391. NFPA, "Design Values for Wood Construction: A Supplement to the

National Design Specifications,'' National Forest Products Association, Washington, 1991. Ross and Pellerin, Nondestructive Testing for Assessing Wood Members in Structures: A Review, *Gen. Tech. Rep.* FPL-GTR-70, USDA Forest Service, Forest Products Laboratory, Madison, WI, 1994. TenWolde et al., "Thermal Properties of Wood and Wood Panel Products for Use in Buildings," ORNL/Sub/87-21697/1, Oak Ridge National Laboratory, Oak Ridge, TN, 1988. Weatherwax and Stamm, The Coefficients of Thermal Expansion of Wood and Wood Products, *Trans. ASME,* **69,** no. 44, 1947, pp. 421–432. Wilkes, Thermo-Physical Properties Data Base Activities at Owens-Corning Fiberglas, in "Thermal Performance of the Exterior Envelopes of Buildings," ASHRAE SP 28, American Society of Heating, Refrigerating, and Air-Conditioning Engineers, Atlanta, 1981.

COMPOSITION, STRUCTURE, AND NOMENCLATURE
by David W. Green

Wood is a naturally formed organic material consisting essentially of elongated tubular elements called **cells** arranged in a parallel manner for the most part. These cells vary in dimensions and wall thickness with position in the tree, age, conditions of growth, and kind of tree. The walls of the cells are formed principally of chain molecules of cellulose, polymerized from glucose residues and oriented as a partly crystalline material. These chains are aggregated in the cell wall at a variable angle, roughly parallel to the axis of the cell. The cells are cemented by an amorphous material called lignin. The complex structure of the gross wood approximates a rhombic system. The direction parallel to the grain and the axis of the stem is longitudinal (L), the two axes across the grain are radial (R) and tangential (T) with respect to the cylinder of the tree stem. This anisotropy and the molecular orientation account for the major differences in physical and mechanical properties with respect to direction which are present in wood.

Natural variability of any given physical measurement in wood approximates the normal probability curve. It is traceable to the differences in the growth of individual samples and at present cannot be controlled. For engineering purposes, statistical evaluation is employed for determination of safe working limits.

Lumber is classified as **hardwood,** which is produced by the broad-level trees (*angiosperms*), such as oak, maple, ash; and **softwood,** the product of coniferous trees (*gymnosperms*), such as pines, larch, spruce, hemlock. The terms "hard" and "soft" have no relation to actual hardness of the wood. **Sapwood** is the living wood of pale color on the outside of the stem. **Heartwood** is the inner core of physiologically inactive wood in a tree and is usually darker in color, somewhat heavier, due to infiltrated material, and more decay-resistant than the sapwood. Other terms relating to wood, veneer, and plywood are defined in ASTM D9, D1038 and the "Wood Handbook."

Standard nomenclature of lumber is based on commercial practice which groups woods of similar technical qualities but separate botanical identities under a single name. For listings of domestic hardwoods and softwoods see ASTM D1165 and the "Wood Handbook."

The **chemical composition** of woody cell walls is generally about 40 to 50 percent cellulose, 15 to 35 percent lignin, less than 1 percent mineral, 20 to 35 percent hemicellulose, and the remainder extractable matter of a variety of sorts. Softwoods and hardwoods have about the same cellulose content.

PHYSICAL AND MECHANICAL PROPERTIES OF CLEAR WOOD
by David Green, Robert White, Anton TenWolde, William Simpson, Joseph Murphy, and Robert Ross

Moisture Relations

Wood is a hygroscopic material which contains water in varying amounts, depending upon the relative humidity and temperature of the surrounding atmosphere. **Equilibrium conditions** are established as shown in Table 6.7.1. The standard reference condition for wood is **oven-dry weight,** which is determined by drying at 100 to 105°C until there is no significant change in weight.

Moisture content is the amount of water contained in the wood, usually expressed as a percentage of the mass of the oven-dry wood. Mois-

ture can exist in wood as **free water** in the cell cavities, as well as water **bound** chemically within the intermolecular regions of the cell wall. The moisture content at which cell walls are completely saturated but at which no water exists in the cell cavities is called the **fiber saturation point.** Below the fiber saturation point, the cell wall shrinks as moisture is removed, and the physical and mechanical properties begin to change as a function of moisture content. **Air-dry wood** has a moisture content of 12 to 15 percent. **Green wood** is wood with a moisture content above the fiber saturation point. The moisture content of green wood typically ranges from 40 to 250 percent.

Dimensional Changes

Shrinkage or **swelling** is a result of change in water content within the cell wall. Wood is dimensionally stable when the moisture content is above the fiber saturation point (about 28 percent for shrinkage estimates). Shrinkage is expressed as a percentage of the dimensional change based on the green wood size. Wood is an anisotropic material with respect to shrinkage. **Longitudinal shrinkage** (along the grain) ranges from 0.1 to 0.3 percent as the wood dries from green to oven-dry and is usually neglected. Wood shrinks most in the direction of the annual growth rings **(tangential shrinkage)** and about half as much across the rings **(radial shrinkage).** Average shrinkage values for a number of commercially important species are shown in Table 6.7.2. Shrinkage to any moisture condition can be estimated by assuming that the change is linear from green to oven-dry and that about half occurs in drying to 12 percent.

Swelling in polar liquids other than water is inversely related to the size of the molecule of the liquid. It has been shown that the tendency to hydrogen bonding on the dielectric constant is a close, direct indicator of the swelling power of water-free organic liquids. In general, the strength values for wood swollen in any polar liquid are similar when there is equal swelling of the wood.

Swelling in aqueous solutions of sulfuric and phosphoric acids, zinc chloride, and sodium hydroxide above pH 8 may be as much as 25 percent greater in the transverse direction than in water. The transverse swelling may be accompanied by longitudinal shrinkage up to 5 percent. The swelling reflects a chemical change in the cell walls, and the accompanying strength changes are related to the degradation of the cellulose.

Dimensional stabilization of wood cannot be completely attained. Two or three coats of varnish, enamel, or synthetic lacquer may be 50 to 85 percent efficient in preventing short-term dimensional changes. Metal foil embedded in multiple coats of varnish may be 90 to 95 percent efficient in short-term cycling. The best long-term stabilization results from internal bulking of the cell wall by the use of materials such as phenolic resins polymerized in situ or water solutions of polyethylene glycol (PEG) on green wood. The presence of the bulking agents alters the properties of the treated wood. Phenol increases electrical resistance, hardness, compression strength, weight, and decay resistance but lowers the impact strength. Polyethylene glycol maintains strength values at the green-wood level, reduces electric resistance, and can be finished only with polyurethane resins.

Mechanical Properties

Average **mechanical properties** determined from tests on clear, straight-grained wood at 12 percent moisture content are given in Table 6.7.2. Approximate standard deviation(s) can be estimated from the following equation.

$$s = CX$$

where X = average value for species

$$C = \begin{cases} 0.10 & \text{for specific gravity} \\ 0.22 & \text{for modulus of elasticity} \\ 0.16 & \text{for modulus of rupture} \\ 0.18 & \text{for maximum crushing strength parallel to grain} \\ 0.14 & \text{for compression strength perpendicular to grain} \\ 0.25 & \text{for tensile strength perpendicular to grain} \\ 0.25 & \text{for impact bending strength} \\ 0.10 & \text{for shear strength parallel to grain} \end{cases}$$

Table 6.7.1 Moisture Content of Wood in Equilibrium with Stated Dry-Bulb Temperature and Relative Humidity

Temperature (dry-bulb)		Moisture content, % at various relative-humidity levels																			
°F	(°C)	5	10	15	20	25	30	35	40	45	50	55	60	65	70	75	80	85	90	95	98
30	−1.3	1.4	2.6	3.7	4.6	5.5	6.3	7.1	7.9	8.7	9.5	10.4	11.3	12.4	13.5	14.9	16.5	18.5	21.0	24.3	26.9
40	4.2	1.4	2.6	3.7	4.6	5.5	6.3	7.1	7.9	8.7	9.5	10.4	11.3	12.3	13.5	14.9	16.5	18.5	21.0	24.3	26.9
50	9.8	1.4	2.6	3.6	4.6	5.5	6.3	7.1	7.9	8.7	9.5	10.3	11.2	12.3	13.4	14.8	16.4	18.4	20.9	24.3	26.9
60	15	1.3	2.5	3.6	4.6	5.4	6.2	7.0	7.8	8.6	9.4	10.2	11.1	12.1	13.3	14.6	16.2	18.2	20.7	24.1	26.8
70	21	1.3	2.5	3.5	4.5	5.4	6.2	6.9	7.7	8.5	9.2	10.1	11.0	12.0	13.1	14.4	16.0	17.9	20.5	23.9	26.6
80	26	1.3	2.4	3.5	4.4	5.3	6.1	6.8	7.6	8.3	9.1	9.9	10.8	11.7	12.9	14.2	15.7	17.7	20.2	23.6	26.3
90	32	1.2	2.3	3.4	4.3	5.1	5.9	6.7	7.4	8.1	8.9	9.7	10.5	11.5	12.6	13.9	15.4	17.3	19.8	23.3	26.0
100	38	1.2	2.3	3.3	4.2	5.0	5.8	6.5	7.2	7.9	8.7	9.5	10.3	11.2	12.3	13.6	15.1	17.0	19.5	22.9	25.6
110	43	1.1	2.2	3.2	4.0	4.9	5.6	6.3	7.0	7.7	8.4	9.2	10.0	11.0	12.0	13.2	14.7	16.6	19.1	22.4	25.2
120	49	1.1	2.1	3.0	3.9	4.7	5.4	6.1	6.8	7.5	8.2	8.9	9.7	10.6	11.7	12.9	14.4	16.2	18.6	22.0	24.7
130	54	1.0	2.0	2.9	3.7	4.5	5.2	5.9	6.6	7.2	7.9	8.7	9.4	10.3	11.3	12.5	14.0	15.8	18.2	21.5	24.2
140	60	0.9	1.9	2.8	3.6	4.3	5.0	5.7	6.3	7.0	7.7	8.4	9.1	10.0	11.0	12.1	13.6	15.3	17.7	21.0	23.7
150	65	0.9	1.8	2.6	3.4	4.1	4.8	5.5	6.1	6.7	7.4	8.1	8.8	9.7	10.6	11.8	13.1	14.9	17.2	20.4	23.1
160	71	0.8	1.6	2.4	3.2	3.9	4.6	5.2	5.8	6.4	7.1	7.8	8.5	9.3	10.3	11.4	12.7	14.4	16.7	19.9	22.5
170	76	0.7	1.5	2.3	3.0	3.7	4.3	4.9	5.6	6.2	6.8	7.4	8.2	9.0	9.9	11.0	12.3	14.0	16.2	19.3	21.9
180	81	0.7	1.4	2.1	2.8	3.5	4.1	4.7	5.3	5.9	6.5	7.1	7.8	8.6	9.5	10.5	11.8	13.5	15.7	18.7	21.3
190	88	0.6	1.3	1.9	2.6	3.2	3.8	4.4	5.0	5.5	6.1	6.8	7.5	8.2	9.1	10.1	11.4	13.0	15.1	18.1	20.7
200	93	0.5	1.1	1.7	2.4	3.0	3.5	4.1	4.6	5.2	5.8	6.4	7.1	7.8	8.7	9.7	10.9	12.5	14.6	17.5	20.0
210	99	0.5	1.0	1.6	2.1	2.7	3.2	3.8	4.3	4.9	5.4	6.0	6.7	7.4	8.3	9.2	10.4	12.0	14.0	16.9	19.3
220	104	0.4	0.9	1.4	1.9	2.4	2.9	3.4	3.9	4.5	5.0	5.6	6.3	7.0	7.8	8.8	9.9	*	*	*	*
230	110	0.3	0.8	1.2	1.6	2.1	2.6	3.1	3.6	4.2	4.7	5.3	6.0	6.7	*	*	*	*	*	*	*
240	115	0.3	0.6	0.9	1.3	1.7	2.1	2.6	3.1	3.5	4.1	4.6	*	*	*	*	*	*	*	*	*
250	121	0.2	0.4	0.7	1.0	1.3	1.7	2.1	2.5	2.9	*	*	*	*	*	*	*	*	*	*	*
260	126	0.2	0.3	0.5	0.7	0.9	1.1	1.4	*	*	*	*	*	*	*	*	*	*	*	*	*
270	132	0.1	0.1	0.2	0.3	0.4	0.4	*	*	*	*	*	*	*	*	*	*	*	*	*	*

* Conditions not possible at atmospheric pressure.
SOURCE: "Wood Handbook," Forest Products Laboratory, 1987.

Table 6.7.2 Strength and Related Properties of Wood at 12% Moisture Content (Average Values from Tests on Clear Pieces 2 × 2 inches in Cross Section per ASTM D143)

Kind of wood	Specific gravity, oven-dry volume	Density at 12% m.c., lb/ft³	Shrinkage, % from green to oven-dry condition based on dimension when green — Rad.	Tan.	Static bending — Modulus of rupture, lb/in²	Static bending — Modulus of elasticity, ksi	Max crushing strength parallel to grain, lb/in²	Compression perpendicular to grain at proportional limit, lb/in²	Tensile strength perpendicular to grain, lb/in²	Impact bending, height of drop in inches for failure with 50-lb hammer	Shear strength parallel to grain, lb/in²	Hardness perpendicular to grain, avg of R and T
Hardwoods												
Ash, white	0.60	42	4.9	7.8	15,400	1,740	7,410	1,160	940	43	1,910	1,320
Basswood	0.37	26	6.6	9.3	8,700	1,460	4,730	370	350	16	990	410
Beech	0.64	45	5.5	11.9	14,900	1,720	7,300	1,010	1,010	41	2,010	1,300
Birch, yellow	0.62	43	7.3	9.5	16,600	2,010	8,170	970	920	55	1,880	1,260
Cherry, black	0.50	35	3.7	7.1	12,300	1,490	7,110	690	560	29	1,700	950
Cottonwood, eastern	0.40	28	3.9	9.2	8,500	1,370	4,910	380	580	20	930	430
Elm, American	0.50	35	4.2	9.5	11,800	1,340	5,520	690	660	39	1,510	830
Elm, rock	0.63	44	4.8	8.1	14,800	1,540	7,050	1,230		56	1,920	1,320
Sweetgum	0.52	36	5.4	10.2	12,500	1,640	6,320	620	760	32	1,600	850
Hickory, shagbark	0.72	50	7.0	10.5	20,200	2,160	9,210	1,760		67	2,430	
Maple, sugar	0.63	44	4.8	9.9	15,800	1,830	7,830	1,470		39	2,330	1,450
Oak, red, northern	0.63	44	4.0	8.6	14,300	1,820	6,760	1,010	800	43	1,780	1,290
Oak, white	0.60	48	5.6	10.5	15,200	1,780	7,440	1,070	800	37	2,000	1,360
Poplar, yellow	0.42	29	4.6	8.2	10,100	1,580	5,540	500	540	24	1,190	540
Tupelo, black	0.50	35	4.2	7.6	9,600	1,200	5,520	930	500	22	1,340	810
Walnut, black	0.35	38	5.5	7.8	14,600	1,680	7,580	1,010	690	34	1,370	1,010
Softwoods												
Cedar, western red	0.32	23	2.4	5.0	7,500	1,110	4,560	460	220	17	990	350
Cypress, bald	0.46	32	3.8	6.2	10,600	1,440	6,360	730	300	24	1,900	510
Douglas-fir, coast	0.48	34	4.8	7.6	12,400	1,950	7,230	800	340	31	1,130	710
Hemlock, eastern	0.40	28	3.0	6.8	8,900	1,200	5,410	650		21	1,060	500
Hemlock, western	0.45	29	4.2	7.8	11,300	1,630	7,200	550	340	26	1,250	540
Larch, western	0.52	38	4.5	9.1	13,000	1,870	7,620	930	430	35	1,360	830
Pine, red	0.46	31	3.8	7.2	11,000	1,630	6,070	600	460	26	1,210	560
Pine, ponderosa	0.40	28	3.9	6.2	9,400	1,290	5,320	580	420	19	1,130	460
Pine, eastern white	0.35	24	2.1	6.1	8,600	1,240	4,800	440	310	18	900	380
Pine, western white	0.38	27	4.1	7.4	9,700	1,460	5,040	470		23	1,040	420
Pine, shortleaf	0.51	36	4.6	7.7	13,100	1,750	7,270	820	470	33	1,390	690
Redwood	0.40	28	2.6	4.4	10,000	1,340	6,150	700	240	19	940	480
Spruce, sitka	0.40	28	4.3	7.5	10,200	1,570	5,610	580	370	25	1,150	510
Spruce, black	0.42	29	4.1	6.8	10,800	1,610	5,960	550		20	1,230	520

SOURCE: Tabulated from "Wood Handbook," "Tropical Woods" no. 95, and unpublished data from the USDA Forest Service, Forest Products Laboratory.

Relatively few data are available on tensile strength parallel to the grain. The modulus of rupture is considered to be a conservative estimate for tensile strength.

Mechanical properties remain constant as long as the moisture content is above the fiber saturation point. Below the fiber saturation point, properties generally increase with decreasing moisture content down to about 8 percent. Below about 8 percent moisture content, some properties, principally tensile strength parallel to the grain and shear strength, may decrease with further drying. An approximate adjustment for clear wood properties between about 8 percent moisture and green can be obtained by using an annual compound-interest type of formula:

$$P_2 = P_1 \left(1 + \frac{C}{100} \right)^{-(M_2 - M_1)}$$

where P_1 is the known property at moisture content M_1, P_2 is the property to be calculated at moisture content M_2, and C is the assumed percentage change in property per percentage change in moisture content. Values of P_1 at 12 percent moisture content are given in Table 6.7.2, and values of C are given in Table 6.7.3. For the purposes of property adjustment, green is assumed to be 23 percent moisture content. The formula should not be used with redwood and cedars. A more accurate adjustment formula is given in "Wood Handbook." Additional data and tests on green wood can be found in the "Wood Handbook." Data on foreign species are given in "Tropical Timbers of the World."

Specific Gravity and Density

Specific gravity G_m of wood at a given moisture condition, m, is the ratio of the weight of the oven-dry wood W_o to the weight of water displaced by the sample at the given moisture condition w_m.

$$G_m = W_o / w_m$$

This definition is required because volume and weight are constant only under special conditions. The **weight density of wood** D (unit weight) at any given moisture content is the weight of oven-dry wood and the contained water divided by the volume of the piece at that same moisture content. Average values for specific gravity oven-dry and weight density at 12 percent moisture content are given in Table 6.7.2. Specific gravity of solid, dry wood substance based on helium displacement is 1.46, or about 91 lb/ft³.

Conversion of weight density from one moisture condition to another can be accomplished by the following equation ("Standard Handbook for Mechanical Engineers," 9th ed., McGraw-Hill).

$$D_2 = D_1 \frac{100 + M_2}{100 + M_1 + 0.0135 D_1 (M_2 - M_1)}$$

D_1 is the weight density, lb/ft³, which is known for some moisture condition M_1. D_2 is desired weight density at a moisture content M_2. Moisture contents M_1 and M_2 are expressed in percent.

Specific gravity and strength properties vary directly in an exponential relationship $S = KG^n$. Table 6.7.3 gives values of K and the exponent n for various strength properties. The equation is based on more than 160 kinds of wood and yields estimated average values for wood in general. This relationship is the best general index to the quality of defect-free wood.

Load Direction and Relation to Grain of Wood

All strength properties vary with the orthotropic axes of the wood in a manner which is approximated by the **Hankinson's formula** ("Wood Handbook"):

$$N = \frac{PQ}{P \sin^2 \theta + Q \cos^2 \theta}$$

where N = allowable stress induced by a load acting at an angle to the grain direction, lb/in²; P = allowable stress parallel to the grain, lb/in²; Q = allowable stress perpendicular to the grain, lb/in²; and θ = angle between direction of load and direction of grain.

The deviation of the grain from the long axis of the member to which the load is applied is known as the **slope of grain** and is determined by measuring the length of run in inches along the axis for a 1-in deviation of the grain from the axis. The effect of grain slope on the important strength properties is shown by Table 6.7.4.

Rheological Properties

Wood exhibits viscoelastic characteristics. When first loaded, a wood member deforms elastically. If the load is maintained, additional time-dependent deformation occurs. Because of this time-dependent relation, the rate of loading is an important factor to consider in the testing and use of wood. For example, the load required to produce failure in 1 s is approximately 10 percent higher than that obtained in a standard 5-min strength test. Impact and dynamic measures of elasticity of small specimens are about 10 percent higher than those for static measures. Impact strengths are also affected by this relationship. In the impact bending test, a 50-lb (23-kg) hammer is dropped upon a beam from increasing heights until complete rupture occurs. The maximum height, as shown in Table 6.7.2, is for comparative purposes only.

When solid material is strained, some mechanical energy is dissipated as heat. **Internal friction** is the term used to denote the mechanism that causes this energy dissipation. The internal friction of wood is a complex function of temperature and moisture content. The value of internal friction, expressed by logarithmic decrement, ranges from 0.1 for hot, moist wood to less than 0.02 for hot, dry wood. Cool wood, regardless of moisture content, has an intermediate value.

The term **fatigue** in engineering is defined as progressive damage that occurs in a material subjected to cyclic loading. **Fatigue life** is a term used to define the number of cycles sustained before failure. Researchers at the Forest Products Laboratory of the USDA Forest Service have found that small cantilever bending specimens subjected to fully reversed stresses, at 30 Hz with maximum stress equal to 30 percent of

Table 6.7.3   Functions Relating Mechanical Properties to Specific Gravity and Moisture Content of Clear, Straight-Grained Wood

Property	Specific gravity–strength relation*				Change for 1% change in moisture content, %
	Green wood		Wood at 12% moisture content		
	Softwood	Hardwood	Softwood	Hardwood	
Static bending					
Modulus of elasticity (10^6 lb/in²)	$2.331G^{0.76}$	$2.02G^{0.72}$	$2.966G^{0.84}$	$2.39G^{0.70}$	2.0
Modulus of rupture (lb/in²)	$15,889G^{1.01}$	$17,209G^{1.16}$	$24,763G^{1.01}$	$24,850G^{1.13}$	5
Maximum crushing strength parallel to grain (lb/in²)	$7,207G^{0.94}$	$7,111G^{1.11}$	$13,592G^{0.97}$	$11,033G^{0.89}$	6.5
Shear parallel to grain (lb/in²)	$1,585G^{0.73}$	$2,576G^{1.24}$	$2,414G^{0.85}$	$3,174G^{1.13}$	4.0
Compression perpendicular to grain at proportional limit (lb/in²)	$1,360G^{1.60}$	$2,678G^{2.48}$	$2,393G^{1.57}$	$3,128G^{2.09}$	6.5
Hardness perpendicular to grain (lb)	$1,399G^{1.41}$	$3,721G^{2.31}$	$1,931G^{1.50}$	$3,438G^{2.10}$	3.0

* The properties and values should be read as equations; e.g., modulus of rupture for green wood of softwoods = $15,889G^{1.01}$, where G represents the specific gravity of wood, based on the oven-dry weight and the volume at the moisture condition indicated.

Table 6.7.4   Strength of Wood Members with Various Grain Slopes as Percentages of Straight-Grained Members

Maximum slope of grain in member	Static bending		Impact bending; drop height to failure (50-lb hammer), %	Maximum crushing strength parallel to grain, %
	Modulus of rupture, %	Modulus of elasticity, %		
Straight-grained	100	100	100	100
1 in 25	96	97	95	100
1 in 20	93	96	90	100
1 in 15	89	94	81	100
1 in 10	81	89	62	99
1 in 5	55	67	36	93

SOURCE: "Wood Handbook."

estimated static strength and at 12 percent moisture content and 75°F (24°C), have a fatigue life of approximately 30 million cycles.

Thermal Properties

The **coefficients of thermal expansion** in wood vary with the structural axes. According to Weatherwax and Stamm (*Trans. ASME*, **69**, 1947, p. 421), the longitudinal coefficient for the temperature range +50°C to −50°C averages 3.39×10^{-6}/°C and is independent of specific gravity. Across the grain, for an average specific gravity oven-dry of 0.46, the radial coefficient α_r is 25.7×10^{-6}/°C and the tangential α_t is 34.8×10^{-6}/°C. Both α_r and α_t vary with specific gravity approximately to the first power. Thermal expansions are usually overshadowed by the larger dimensional changes due to moisture.

Thermal conductivity of wood varies principally with the direction of heat with respect to the grain. **Approximate transverse conductivity** can be calculated with a linear equation of the form

$$k = G(B + CM) + A$$

where G is specific gravity, based on oven-dry weight and volume at a given moisture content M percent. For specific gravities above 0.3, temperatures around 75°F (24°C), and moisture contents below 25 percent, the values of constants, A, B, and C are $A = 0.129$, $B = 1.34$, and $C = 0.028$ in English units, with k in Btu · in/(h · ft² · °F) (TenWolde et al., 1988). Conductivity in watts per meter per kelvin is obtained by multiplying the result by 0.144. The effect of temperature on thermal conductivity is relatively minor and increases about 1 to 2 percent per 10°F (2 to 3 percent per 10°C). **Longitudinal conductivity** is considerably greater than transverse conductivity, but reported values vary widely. It has been reported as 1.5 to 2.8 times larger than transverse conductivity, with an average of about 1.8.

Specific heat of wood is virtually independent of specific gravity and varies principally with temperature and moisture content. Wilkes found that the approximate specific heat of dry wood can be calculated with

$$c_{p0} = a_0 + a_1 T$$

where $a_0 = 0.26$ and $a_1 = 0.000513$ for English units (specific heat in Btu per pound per degree Fahrenheit and temperature in degrees Fahrenheit) or $a_0 = 0.103$ and $a_1 = 0.00387$ for SI units [specific heat in kJ/(kg · K) and temperature in kelvins]. The specific heat of moist wood can be derived from

$$c_p = \frac{c_{p0} + 0.01 M c_{p,w}}{1 + 0.01 M} + A$$

where $c_{p,w}$ is the specific heat of water [1 Btu/lb · °F), or 4.186 kJ/ (kg · K), M is the moisture content (percent), and A is a correction factor, given by

$$A = M(b_1 + b_2 T + b_3 M)$$

with $b_1 = -4.23 \times 10^{-4}$, $b_2 = 3.12 \times 10^{-5}$, and $b_3 = -3.17 \times 10^{-5}$ in English units, and $b_1 = -0.06191$, $b_2 = 2.36 \times 10^{-4}$, and $b_3 =$

-1.33×10^{-4} in SI units. These formulas are valid for wood below fiber saturation at temperatures between 45°F (7°C) and 297°F (147°C). T is the temperature at which c_{p0} is desired.

The **fuel value** of wood depends primarily upon its dry density, moisture content, and chemical composition. Moisture in wood decreases the fuel value as a result of latent heat absorption of water vaporization. An approximate relation for the fuel value of moist wood (Btu per pound on wet weight basis)(2,326 Btu/lb = 1 J/kg) is

$$H_{w} = H_D \left(\frac{100 - u/7}{100 + u} \right)$$

where H_D is higher fuel value of dry wood, averaging 8,500 Btu/lb for hardwoods and 9,000 Btu/lb for conifers, and u is the moisture content in percent. The actual fuel value of moist wood in a furnace will be less since water vapor interferes with the combustion process and prevents the combustion of pyrolytic gases. (See Sec. 7 for fuel values and Sec. 4 for combustion.)

Wood undergoes thermal degradation to volatile gases and char when it is exposed to elevated temperature. When wood is directly exposed to the standard fire exposure of ASTM E 199, the **char rate** is generally considered to be 1½ in/h (38 mm/h). The temperature at the base of the char layer is approximately 550°F (300°C). Among other factors, the ignition of wood depends on the intensity and duration of exposure to elevated temperatures. Typical values for rapid ignition are 570 to 750°F (300 to 400°C). In terms of heat flux, a surface exposure to 1.1 Btu/ft² (13 kW/m²) per second is considered sufficient to obtain piloted ignition. Recommended "maximum safe working temperatures" for wood exposed for prolonged periods range from 150 to 212°F (65 to 100°C). **Flame spread** values as determined by ASTM E 84 generally range from 65 to 200 for nominal 1-in- (25-mm-) thick lumber. Flame spread can be reduced by impregnating the wood with fire-retardant chemicals or applying a fire-retardant coating.

The **reversible effect of temperature on the properties of wood** is a function of the change in temperature, moisture content of the wood, duration of heating, and property being considered. In general, the mechanical properties of wood decrease when the wood is heated above normal temperatures and increase when it is cooled. The magnitude of the change is greater for green wood than for dry. When wood is frozen, the change in property is reversible; i.e., the property will return to the value at the initial temperature. At a constant moisture content and below about 150°F (65°C), mechanical properties are approximately linearly related to temperature. The change in property is also reversible if the wood is heated for a short time at temperatures below about 150°F. Table 6.7.5 lists the changes in properties at −58°F (−50°C) and 122°F (50°C) relative to those at 68°F (20°C).

Permanent loss in properties occurs when wood is exposed to higher temperatures for prolonged periods and then is cooled and tested at normal temperatures. If the wood is tested at a higher temperature after prolonged exposure, the actual strength loss is the sum of the reversible and permanent losses in properties. Permanent losses are higher for

Table 6.7.5   Approximate Middle-Trend Effects of Temperature on Mechanical Properties of Clear Wood at Various Moisture Conditions

Property	Moisture condition	Relative change in mechanical property from 68°F, %	
		At − 58°F	At + 122°F
Modulus of elasticity parallel to grain	0	+ 11	− 6
	12	+ 17	− 7
	>FSP*	+ 50	—
Modulus of rupture	≤ 4	+ 18	− 10
	11–15	+ 35	− 20
	18–20	+ 60	− 25
	>FSP*	+ 110	− 25
Tensile strength parallel to grain	0–12	—	− 4
Compressive strength parallel to grain	0	+ 20	− 10
	12–45	+ 50	− 25
Shear strength parallel to grain	>FSP*	—	− 25
Compressive strength perpendicular to grain at proportional limit	0–6	—	− 20
	≥ 10	—	− 35

* Moisture content higher than the fiber saturation point (FSP).
$T_C = (T_F - 32) \, 0.55.$

heating in steam than in water, and higher when heated in water than when heated in air. Repeated exposure to elevated temperatures is assumed to have a cumulative effect on wood properties. For example, at a given temperature the property loss will be about the same after six exposures of 1-year duration as it would be after a single exposure of 6 years. Figure 6.7.1 illustrates the effect of heating at 150°F (65°C) at 12 percent moisture content on the modulus of rupture relative to the strength at normal temperatures for two grades of spruce-pine-fir and clear southern pine. Over the 3-year period, there was little or no change in the modulus of elasticity.

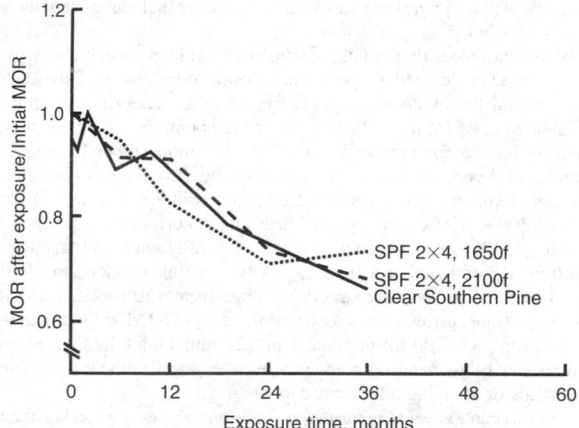

Fig. 6.7.1   Permanent loss in bending strength at 12 percent moisture content. Specimens exposed at 150°F (65°C) and tested at 68°F (20°C). SPF = spruce/pine/fir.

Electrical Properties

The important electrical properties of wood are **conductivity** (or **resistivity**), **dielectric constant,** and **dielectric power factor** (see James, 1988).

Resistivity approximately doubles for each 10°C decrease in temperature. As moisture content increases from zero to the fiber saturation point (FSP), the resistivity decreases by 10^{10} to 10^{13} times in an approximately linear relationship between the logarithm of each. The resistivity is about 10^{14} to 10^{16} $\Omega \cdot$ m for oven-dry wood and 10^3 to 10^4 $\Omega \cdot$ m for wood at FSP. As the moisture content increases up to complete saturation, the decrease in resistivity is a factor of only about 50. Wood species also affect resistivity (see James), and the resistivity perpendicular to the grain is about twice that parallel to the grain. Water-soluble salts (some preservatives and fire-retardants) reduce resistivity by only a minor amount when the wood has 8 percent moisture content or less, but they have a much larger effect when moisture content exceeds 10 to 12 percent.

The **dielectric constant** of oven-dry wood ranges from about 2 to 5 at room temperature, and it decreases slowly with increasing frequency. The dielectric constant increases as either temperature or moisture content increases. There is a negative interaction between moisture and frequency: At 20 Hz, the dielectric constant may range from about 4 for dry wood to 10^6 for wet wood; at 1 kHz, from 4 dry to 5,000 wet; and at 1 MHz, from about 3 dry to 100 wet. The dielectric constant is about 30 percent greater parallel to the grain than perpendicular to it.

The **power factor** of wood varies from about 0.01 for dry, low-density woods to as great as 0.95 for wet, high-density woods. It is usually greater parallel to the grain than perpendicular. The power factor is affected by complex interactions of frequency, moisture content, and temperature (James, 1975).

The change in electrical properties of wood with moisture content has led to the development of moisture meters for nondestructive estimation of moisture content. Resistance-type meters measure resistance between two pins driven into the wood. Dielectric-type meters depend on the correlation between moisture content and either dielectric constant or power factor, and they require only contact with the wood surface, not penetration.

Wood in Relation to Sound

The **transmission of sound** and vibrational properties in wood are functions of the grain angle. The speed of sound transmission is described by the expression $v = E/\rho$, in which v is the speed of sound in wood, in/s, E is the dynamic Young's modulus, lb/in^2, and ρ is the density of the wood, slugs/in^3 (Ross and Pellerin, 1994). Various factors influence the speed of sound transmission; two of the most important factors are grain angle and the presence of degradation from decay. Hankinson's formula, cited previously, adequately describes the relationship between speed of sound transmission and grain angle. The dynamic modulus is about 10 percent higher than the static value and varies inversely with moisture changes by approximately 1.3 percent for each percentage change in moisture content.

Degradation from biological agents can significantly alter the speed at which sound travels in wood. Speed of sound transmission values are greatly reduced in severely degraded wood members. Sound transmission characteristics of wood products are used in one form of **nondestructive testing** to assess the performance characteristics of wood products. Because speed of sound transmission is a function of the extent of degradation from decay, this technique is used to estimate the extent of severe degradation in large timbers.

PROPERTIES OF LUMBER PRODUCTS
by Russell Moody and David Green

Visually Graded Structural Lumber

Stress-graded structural lumber is produced under two systems: visual grading and machine grading. **Visual structural grading** is the oldest stress grading system. It is based on the premise that the mechanical properties of lumber differ from those of clear wood because many growth characteristics of lumber affect its properties; these characteristics can be seen and judged by eye (ASTM D245). The principal growth features affecting lumber properties are the size and location of knots, sloping grain, and density

Grading rules for lumber nominally 2 × 4 in (standard 38 × 89 mm) thick (**dimension lumber**) are published by grading agencies (listing and addresses are given in "National Design Specification," American Forest & Paper Association, 1992 and later). For most species, allowable properties are based on test results from full-size specimens graded by agency rules, sampled according to ASTM D2915, and tested according to ASTM D4761. Procedures for deriving allowable properties from these tests are given in ASTM D1990. Allowable properties for visually graded hardwoods and a few softwoods are derived from clearwood data following principles given in ASTM D2555. Derivation of the allowable strength properties accounts for within-species variability by starting with a nonparametric estimate of the 5th percentile of the data. Thus, 95 of 100 pieces would be expected to be stronger than the assigned property. The allowable strength properties are based on an assumed normal duration of load of 10 years. Tables 6.7.6 and 6.7.7 show the grades and allowable properties for the four most commonly used species groupings sold in the United States. The allowable strength values in bending, tension, shear, and compression parallel to the grain can be multiplied by factors for other load durations. Some commonly used factors are 0.90 for permanent (50-year) loading, 1 15 for snow loads (2 months), and 1.6 for wind/earthquake loading (10 min). The most recent edition of "National Design Specification" should be consulted for updated property values and for property values for other species and size classifications.

Allowable properties are assigned to visually graded dimension lumber at two moisture content levels: green and 19 percent maximum moisture content (assumed 15 percent average moisture content). Because of the influence of knots and other growth characteristics on lumber properties, the effect of moisture content on lumber properties is generally less than its effect on clear wood. The C_M factors of Table 6.7.8 are for adjusting the properties in Tables 6.7.6 and 6.7.7 from 15 percent moisture content to green. The Annex of ASTM D1990 provides formulas that can be used to adjust lumber properties to any moisture content between green and 10 percent. Below about 8 percent moisture content, some properties may decrease with decreasing values, and care should be exercised in these situations (Green and Kretschmann, 1995).

Shrinkage in commercial lumber differs from that in clear wood pri-

Table 6.7.6   Base Design Values for Visually Graded Dimension Lumber*
(Tabulated design values are for normal load duration and dry service conditions.)

Species and commercial grade	Size classification, in	Design values, (lb/in²)						Grading rules agency
		Bending F_b	Tension parallel to grain F_t	Shear parallel to grain F_v	Compression perpendicular to grain $F_{c\perp}$	Compression parallel to grain F_c	Modulus of elasticity E	
Douglas-Fir-Larch								
Select structural		1,000	1,000	95	625	1,700	1,900,000	
No. 1 and better	2–4 thick	1,150	775	95	625	1,500	1,800,000	
No. 1		1,000	675	95	625	1,450	1,700,000	
No. 2	2 and wider	875	575	95	625	1,300	1,600,000	
No. 3		500	325	95	625	750	1,400,000	WCLIB
Stud		675	450	95	625	825	1,400,000	WWPA
Construction	2–4 thick	1,000	650	95	625	1,600	1,500,000	
Standard		550	375	95	625	1,350	1,400,000	
Utility	2–4 wide	275	175	95	625	875	1,300,000	
Hem-Fir								
Select structural		1,400	900	75	405	1,500	1,600,000	
No. 1 and better	2–4 thick	1,050	700	75	405	1,350	1,500,000	
No. 1		950	600	75	405	1,300	1,500,000	
No. 2	2 and wider	850	500	75	405	1,250	1,300,000	
No. 3		500	300	75	405	725	1,200,000	WCLIB
Stud		675	400	75	405	800	1,200,000	WWPA
Construction	2–4 thick	975	575	75	405	1,500	1,300,000	
Standard		550	325	75	405	1,300	1,200,000	
Utility	2–4 wide	250	150	75	405	850	1,100,000	
Spruce-Pine-Fir								
Select structural	2–4 thick	1,250	675	70	425	1,400	1,500,000	
No. 1–No. 2		875	425	70	425	110	1,400,000	
No. 3	2 and wider	500	250	70	425	625	1,200,000	
Stud		675	325	70	425	675	1,200,000	NLGA
Construction	2–4 thick	975	475	70	425	1,350	1,300,000	
Standard		550	275	70	425	1,100	1,200,000	
Utility	2–4 wide	250	125	70	425	725	1,100,000	

* Lumber dimensions—Tabulated design values are applicable to lumber that will be used under dry conditions such as in most covered structures. For 2- to 4-in-thick lumber, the *DRY* dressed sizes shall be used regardless of the moisture content at the time of manufacture or use. In calculating design values, the natural gain in strength and stiffness that occurs as lumber dries has been taken into consideration as well as the reduction in size that occurs when unseasoned lumber shrinks. The gain in load-carrying capacity due to increased strength and stiffness resulting from drying more than offsets the design effect of size reductions due to shrinkage. Size factor C_F, repetitive-member factor C_r, flat-use factor C_{fu}, and wet-use factor C_M are given in Table 6.7.8.

SOURCE: Table used by permission of the American Forest & Paper Association.

Table 6.7.7   Design Values for Visually Graded Southern Pine Dimension Lumber*
(Tabulated design values are for normal load duration and dry service conditions.)

Species and commercial grade	Size classification, in	Design values, lb/in²						Grading rules agency
		Bending F_b	Tension parallel to grain F_t	Shear parallel to grain F_v	Compression perpendicular to grain $F_{c\perp}$	Compression parallel to grain F_c	Modulus of elasticity E	
Dense select structural	2–4 thick	3,050	1,650	100	660	2,250	1,900,000	SPIB
Select structural		2,850	1,600	100	565	2,100	1,800,000	
Nondense select structural	2–4 wide	2,650	1,350	100	480	1,950	1,700,000	
No. 1 dense		2,000	1,100	100	660	2,000	1,800,000	
No. 1		1,850	1,050	100	565	1,850	1,700,000	
No. 1 nondense		1,700	900	100	480	1,700	1,600,000	
No. 2 dense		1,700	875	90	660	1,850	1,700,000	
No. 2		1,500	825	90	656	1,650	1,600,000	
No. 2 nondense		1,350	775	90	480	1,600	1,400,000	
No. 3		850	475	90	565	975	1,400,000	
Stud		875	500	90	565	975	1,400,000	
Construction	2–4 thick	1,100	625	100	565	1,800	1,500,000	
Standard		625	350	90	565	1,500	1,300,000	
Utility	4 wide	300	175	90	565	975	1,300,000	
Dense select structural	2–4 thick	2,700	1,500	90	660	2,150	1,900,000	
Select structural		2,550	1,400	90	565	2,000	1,800,000	
Nondense select structural	5–6 wide	2,350	1,200	90	480	1,850	1,700,000	
No. 1 dense		1,750	950	90	660	1,900	1,800,000	
No. 1		1,650	900	90	565	1,750	1,700,000	
No. 1 nondense		1,500	800	90	480	1,600	1,600,000	
No. 2 dense		1,450	775	90	660	1,750	1,700,000	
No. 2		1,250	725	90	565	1,600	1,600,000	
No. 2 nondense		1,150	675	90	480	1,500	1,400,000	
No. 3		750	425	90	565	925	1,400,000	
Stud		775	425	90	565	925	1,400,000	
Dense select structural	2–4 thick	2,450	1,350	90	660	2,050	1,900,000	
Select structural		2,300	1,300	90	565	1,900	1,800,000	
Nondense select structural	8 wide	2,100	1,100	90	480	1,750	1,700,000	
No. 1 dense		1,650	875	90	660	1,800	1,800,000	
No. 1		1,500	825	90	565	1,650	1,700,000	
No. 1 nondense		1,350	725	90	480	1,550	1,600,000	
No. 2 dense		1,400	675	90	660	1,700	1,700,000	
No. 2		1,200	650	90	565	1,550	1,600,000	
No. 2 nondense		1,100	600	90	480	1,450	1,400,000	
No. 3		700	400	90	565	875	1,400,000	
Dense select structural	2–4 thick	2,150	1,200	90	660	2,000	1,900,000	
Select structural		2,050	1,100	90	565	1,850	1,800,000	
Nondense select structural	10 wide	1,850	950	90	480	1,750	1,700,000	
No. 1 dense		1,450	775	90	660	1,750	1,800,000	
No. 1		1,300	725	90	565	1,600	1,700,000	
No. 1 nondense		1,200	650	90	480	1,500	1,600,000	
No. 2 dense		1,200	625	90	660	1,650	1,700,000	
No. 2		1,050	575	90	565	1,500	1,600,000	
No. 2 nondense		950	550	90	480	1,400	1,400,000	
No. 3		600	325	90	565	850	1,400,000	
Dense select structural	2–4 thick	2,050	1,100	90	660	1,950	1,900,000	
Select structural		1,900	1,050	90	565	1,800	1,800,000	
Nondense select structural	12 wide	1,750	900	90	480	1,700	1,700,000	
No. 1 dense		1,350	725	90	660	1,700	1,800,000	
No. 1		1,250	675	90	565	1,600	1,700,000	
No. 1 nondense		1,150	600	90	480	1,500	1,600,000	
No. 2 dense		1,150	575	90	660	1,600	1,700,000	
No. 2		975	550	90	565	1,450	1,600,000	
No. 2 nondense		900	525	90	480	1,350	1,400,000	
No. 3		575	325	90	565	825	1,400,000	

* For size factor C_F, appropriate size adjustment factors have already been incorporated in the tabulated design values for most thicknesses of southern pine dimension lumber. For dimension lumber 4 in thick, 8 in and wider, tabulated bending design values F_b shall be permitted to be multiplied by the size factor $C_F = 1.1$. For dimension lumber wider than 12 in, tabulated bending, tension, and compression parallel-to-grain design values for 12-in-wide lumber shall be multiplied by the size factor $C_F = 0.9$. Repetitive-member factor C_r, flat-use factor C_{fu}, and wet-service factor C_M are given in Table 6.7.8.

SOURCE: Table used by permission of the American Forest & Paper Association.

Table 6.7.8   Adjustment Factors

Size factor C_F for Table 6.7.6 (Douglas–Fir–Larch, Hem–Fir, Spruce–Pine–Fir)

Tabulated bending, tension, and compression parallel to grain design values for dimension lumber 2 to 4 inches thick shall be multiplied by the following size factors:

		F_b Thickness, in			
Grades	Width, in	2 & 3	4	F_t	F_c
Select structural no. 1 and	2, 3, and 4	1.5	1.5	1.5	1.15
better no. 1, no. 2, no. 3	5	1.4	1.4	1.4	1.1
	6	1.3	1.3	1.3	1.1
	8	1.2	1.3	1.2	1.05
	10	1.1	1.2	1.1	1.0
	12	1.0	1.1	1.0	1.0
	14 and wider	0.9	1.0	0.9	0.9
Stud	2, 3, and 4	1.1	1.1	1.1	1.05
	5 and 6	1.0	1.0	1.0	1.0
Construction and standard	2, 3, and 4	1.0	1.0	1.0	1.0
Utility	4	1.0	1.0	1.0	1.0
	2 and 3	0.4	—	0.4	0.6

Repetitive-member factor C_r for Tables 6.7.6 and 6.7.7

Bending design values F_b for dimension lumber 2 to 4 in thick shall be multiplied by the repetitive factor $C_r = 1.15$, when such members are used as joists, truss chords, rafters, studs, planks, decking, or similar members which are in contact or spaced not more than 24 in on centers, are not less than 3 in number and are joined by floor, roof, or other load-distributing elements adequate to support the design load.

Flat-use factor C_{fu} for Tables 6.7.6 and 6.7.7

Bending design values adjusted by size factors are based on edgewise use (load applied to narrow face). When dimension lumber is used flatwise (load applied to wide face), the bending design value F_b shall also be multiplied by the following flat-use factors:

	Thickness, in	
Width, in	2 and 3	4
2 and 3	1.0	—
4	1.1	1.0
5	1.1	1.05
6	1.15	1.05
8	1.15	1.05
10 and wider	1.2	1.1

Wet-use factor C_M for Tables 6.7.6 and 6.7.7

When dimension lumber is used where moisture content will exceed 19 percent for an extended period, design values shall be multiplied by the appropriate wet service factors from the following table:

F_b	F_t	F_v	$F_{c\perp}$	F_c	E
0.85*	1.0	0.97	0.67	0.8†	0.9

* When $F_b C_F \leq 1150$ lb/in^2, $C_M = 1.0$.
† When $F_c C_F \leq 750$ lb/in^2, $C_M = 1.0$.
SOURCE: Used by permission of the American Forest & Paper Association.

marily because the grain in lumber is seldom oriented in purely radial and tangential directions. Approximate formulas used to estimate shrinkage of lumber for most species are

$$S_w = 6.031 - 0.215M$$
$$S_t = 5.062 - 0.181M$$

where S_w is the shrinkage across the wide [8-in (203-mm)] face of the lumber in a 2 × 8 (standard 38 × 184 mm), S_t is the shrinkage across the narrow [2-in (51-mm)] face of the lumber, and M = moisture con-

tent (percent). As with clear wood, shrinkage is assumed to occur below a moisture content of 28 percent. Because extractives make wood less hygroscopic, less shrinkage is expected in redwood, western redcedar, and northern white cedar (Green, 1989).

The effect of **temperature** on lumber properties appears to be similar to that on clear wood. For simplicity, "National Design Specification" uses conservative factors to account for reversible reductions in properties as a result of heating to 150°F (65°C) or less (Table 6.7.9). No increase in properties is taken for temperatures colder than normal be-

Table 6.7.9   Temperature Factors C_t for Short-Term Exposure

		C_t		
Design values	In-service moisture conditions	$T \leq 100°F$	$100°F < T \leq 125°F$	$125°F < T \leq 150°F$
F_t, E	Green or dry	1.0	0.9	0.9
F_b, F_v, F_c, and $F_{c\perp}$	≤ 19% green	1.0	0.8	0.7
		1.0	0.7	0.5

SOURCE: Table used by permission of the American Forest & Paper Association.

Table 6.7.10   Design Values for Mechanically Graded Dimension Lumber
(Tabulated design values are for normal load duration and dry service conditions.)

Species and commercial grade	Size classification, in	Design values, lb/in²				Grading rules agency
		Bending	Tension parallel	Compression parallel	MOE	
Machine-stress-rated lumber						
900f-1.0E		900	350	1,050	1,000,000	WCLIB
1200f-1.2E		1,200	600	1,400	1,200,000	NLGA, SPIB, WCLIB, WWPA
1250f-1.4E		1,250	800	1,450	1,400,000	WCLIB
1350f-1.3E		1,350	750	1,600	1,300,000	SPIB, WCLIB, WWPA
1400f-1.2E		1,400	800	1,600	1,200,000	SPIB
1450f-1.3E		1,450	800	1,625	1,300,000	NLGA, WCLIB, WWPA
1500f-1.3E		1,500	900	1,650	1,300,000	SPIB
1500f-1.4E		1,500	900	1,650	1,400,000	NLGA, SPIB, WCLIB, WWPA
1600f-1.4E		1,600	950	1,675	1,400,000	SPIB
1650f-1.4E		1,650	1,020	1,700	1,400,000	SPIB
1650f-1.5E		1,650	1,020	1,700	1,500,000	NLGA, SPIB, WCLIB, WWPA
1650f-1.6E		1,650	1,075	1,700	1,600,000	WCLIB
1800f-1.5E		1,800	1,300	1,750	1,500,000	SPIB
1800f-1.6E	2 and less in thickness	1,800	1,175	1,750	1,600,000	NLGA, SPIB, WCLIB, WWPA
1950f-1.5E		1,950	1,375	1,800	1,500,000	SPIB
1950f-1.7E		1,950	1,375	1,800	1,700,000	NLGA, SPIB, WCLIB, WWPA
2000f-1.6E	2 and wider	2,000	1,300	1,825	1,600,000	SPIB
2100f-1.8E		2,100	1,575	1,875	1,800,000	NLGA, SPIB, WCLIB, WWPA
2250f-1.6E		2,250	1,750	1,925	1,600,000	SPIB
2250f-1.9E		2,250	1,750	1,925	1,900,000	NLGA, SPIB, WCLIB, WWPA
2400f-1.7E		2,400	1,925	1,975	1,700,000	SPIB
2400f-1.8E		2,400	1,925	1,975	1,800,000	SPIB
2400f-2.0E		2,400	1,925	1,975	2,000,000	NLGA, SPIB, WCLIB, WWPA
2500f-2.2E		2,500	1,750	2,000	2,200,000	WCLIB
2550f-2.1E		2,550	2,050	2,025	2,100,000	NLGA, SPIB, WWPA
2700f-2.0E		2,700	1,800	2,100	2,000,000	WCLIB
2700f-2.2E		2,700	2,150	2,100	2,200,000	NLGA, SPIB, WCLIB, WWPA
2850f-2.3E		2,850	2,300	2,150	2,300,000	SPIB, WWPA
3000f-2.4E		3,000	2,400	2,200	2,400,000	NLGA, SPIB
3150f-2.5E		3,150	2,500	2,250	2,500,000	SPIB
3300f-2.6E		3,300	2,650	2,325	2,600,000	SPIB
900f-1.2E		900	350	1,050	1,200,000	NLGA, WCLIB
1200f-1.5E	2 and less in thickness	1,200	600	1,400	1,500,000	NLGA, WCLIB
1350f-1.8E		1,350	750	1,600	1,800,000	NLGA
1500f-1.8E	6 and wider	1,500	900	1,650	1,800,000	WCLIB
1800f-2.1E		1,800	1,175	1,750	2,100,000	NLGA, WCLIB
Machine-evaluated lumber						
M-10		1,400	800	1,600	1,200,000	
M-11		1,550	850	1,650	1,500,000	
M-12		1,600	850	1,700	1,600,000	
M-13		1,600	950	1,700	1,400,000	
M-14		1,800	1,000	1,750	1,700,000	
M-15		1,800	1,100	1,750	1,500,000	
M-16		1,800	1,300	1,750	1,500,000	
M-17	2 and less in thickness	1,950	1,300	2,050	1,700,000	
M-18		2,000	1,200	1,850	1,800,000	SPIB
M-19	2 and wider	2,000	1,300	1,850	1,600,000	
M-20		2,000	1,600	2,100	1,900,000	
M-21		2,300	1,400	1,950	1,900,000	
M-22		2,350	1,500	1,950	1,700,000	
M-23		2,400	1,900	2,000	1,800,000	
M-24		2,700	1,800	2,100	1,900,000	
M-25		2,750	2,000	2,100	2,200,000	
M-26		2,800	1,800	2,150	2,000,000	
M-27		3,000	2,000	2,400	2,100,000	

Lumber dimensions: Tabulated design values are applicable to lumber that will be used under dry conditions such as in most covered structures. For 2- to 4-in-thick lumber, the *dry* dressed sizes shall be used regardless of the moisture content at the time of manufacture or use. In calculating design values, natural gain in strength and stiffness that occurs as lumber dries had been taken into consideration as well as reduction in size that occurs when unseasoned lumber shrinks. The gain in load-carrying capacity due to increased strength and stiffness resulting from drying more than offsets the design effect of size reductions due to shrinkage. **Shear parallel to grain F_v and compression perpendicular to grain $F_{c\perp}$:** Design values for shear parallel to grain F_{v1} and compression perpendicular to grain $F_{c\perp}$ are identical to the design values given in Tables 6.7.6 and 6.7.7 for No. 2 visually graded lumber of the appropriate species. When the F_v or $F_{c\perp}$ values shown on the grade stamp differ from the values shown in the tables, the values shown on the grade stamp shall be used for design. **Modulus of elasticity E and tension parallel to grain F_t:** For any given bending design value F_b, the average modulus of elasticity E and tension parallel to grain F_t design value may vary depending upon species, timber source, or other variables. The E and F_t values included in the F_b and E grade designations are those usually associated with each F_b level. Grade stamps may show higher or lower values if machine rating indicates the assignment is appropriate. When the E or F_t values shown on a grade stamp differ from the values in Table 6.7.10 the values shown on the grade stamp shall be used for design. The tabulated F_b and F_c values associated with the designated F_b value shall be used for design.
   SOURCE: Table used by permission of the American Forest & Paper Association.

cause in practice it is difficult to ensure that the wood temperature remains consistently low.

Mechanically Graded Structural Lumber

Machine-stress-rated (MSR) lumber and **machine-evaluated lumber (MEL)** are two types of mechanically graded lumber. The three basic components of both mechanical grading systems are (1) sorting and prediction of strength through machine-measured nondestructive determination of properties coupled with visual assessment of growth characteristics, (2) assignment of allowable properties based upon strength prediction, and (3) quality control to ensure that assigned properties are being obtained. Grade names for MEL lumber start with an M designation. Grade "names" for MSR lumber are a combination of the allowable bending stress and the average modulus of elasticity [e.g., 1650f-1.4E means an allowable bending stress of 1,650 lb/in^2 (11.4 MPa) and modulus of elasticity of 1.4×10^6 lb/in^2 (9.7 GPa)]. Grades of mechanically graded lumber and their allowable properties are given in Table 6.7.10.

Structural Composite Lumber

Types of Structural Composite Lumber Structural composite lumber refers to several types of reconstituted products that have been developed to meet the demand for high-quality material for the manufacture of engineered wood products and structures. Two distinct types are commercially available: laminated veneer lumber (LVL) and parallel-strand lumber (PSL).

Laminated veneer lumber is manufactured from layers of veneer with the grain of all the layers parallel. This contrasts with plywood, which consists of adjacent layers with the grain perpendicular. Most manufacturers use sheets of 1/10- to 1/6-in- (2.5- to 4.2-mm-) thick veneer. These veneers are stacked up to the required thickness and may be laid end to end to the desired length with staggered end joints in the veneer. Waterproof adhesives are generally used to bond the veneer under pressure. The resulting product is a billet of lumber that may be up to 1¾ in (44 mm) thick, 4 ft (1.2 m) wide, and 80 ft (24.4 m) long. The billets are then ripped to the desired width and cut to the desired length. The common sizes of LVL closely resemble those of sawn dimension lumber.

Parallel-strand lumber is manufactured from strands or elongated flakes of wood. One North American product is made from veneer clipped to ½ in (13 mm) wide and up to 8 ft (2.4 m) long. Another product is made from elongated flakes and technology similar to that used to produce oriented strandboard. A third product is made from mats of interconnected strands crushed from small logs that are assembled into the desired configuration. All the products use waterproof adhesive that is cured under pressure. The size of the product is controlled during manufacture through adjustments in the amount of material and pressure applied. Parallel-strand lumber is commonly available in the same sizes as structural timbers or lumber.

Properties of Structural Composite Lumber Standard design values have not been established for either LVL or PSL. Rather, standard procedures are available for developing these design values (ASTM D5456). Commonly, each manufacturer follows these procedures and submits supporting data to the appropriate regulatory authority to establish design properties for the product. Thus, design information for LVL and PSL varies among manufacturers and is given in their product literature. Generally, the engineering design properties compare favorably with or exceed those of high-quality solid dimension lumber. Example design values accepted by U.S. building codes are given in Table 6.7.11.

Glulam Timber

Structural **glued-laminated (glulam)** timber is an engineered, stress-rated product of a timber laminating plant, consisting of two or more layers of wood glued together with the grain of all layers (or laminations) approximately parallel. Laminations are typically made of specially selected and prepared sawn lumber. Nominal 2-in (standard 38-mm) lumber is used for straight or slightly curved members, and nominal 1-in (standard 19-mm) lumber is used for other curved mem-

Table 6.7.11 Example Design Values for Structural Composite Lumber

Product	Bending stress		Modulus of elasticity		Horizontal shear	
	lb/in^2	MPa	$\times 10^3$ lb/in^2	GPa	lb/in^2	MPa
LVL	2,800	19.2	2,000	13.8	190	1.31
PSL, type A	2,900	20.0	2,000	13.8	210	1.45
PSL, type B	1,500	10.3	1,200	8.3	150	1.03

bers. A national standard, ANSI A190.1, contains requirements for production, testing, and certification of the product in the United States.

Manufacture Straight members up to 140 ft (42 m) long and more than 7 ft (2.1 m) deep have been manufactured with size limitations generally resulting from transportation constraints. Curved members have been used in domed structures spanning over 500 ft (152 m), such as the Tacoma Dome. Manufacturing and design standards cover many softwoods and hardwoods; Douglas-fir and southern pine are the most commonly used softwood species.

Design standards for glulam timber are based on either dry or wet use. Manufacturing standards for dry use, which is defined as use conditions resulting in a moisture content of 16 percent or less, permits manufacturing with nonwaterproof adhesives; however, nearly all manufacturers in North America use waterproof adhesives exclusively. For wet-use conditions, these waterproof adhesives are required. For wet-use conditions in which the moisture content is expected to exceed 20 percent, pressure preservative treatment is recommended (AWPA C28). Lumber can be pressure-treated with water-based preservatives prior to gluing, provided that special procedures are followed in the manufacture. For treatment after gluing, oil-based preservatives are generally recommended. Additional information on manufacture is provided in the "Wood Handbook."

Glulam timber is generally manufactured at a moisture content below 16 percent. For most dry-use applications, it is important to protect the glulam timber from increases in moisture content. End sealers, surface sealers, primer coats, and wrappings may be applied at the manufacturing plant to provide protection from changes in moisture content. Protection will depend upon the final use and finish of the timber.

Special precautions are necessary during handling, storage, and erection to prevent structural damage to glulam members. Padded or nonmarring slings are recommended; cable slings or chokers should be avoided unless proper blocking protects the members. AITC 111 provides additional details on protection during transit, storage, and erection.

Design Glulam timber beams are available in standard sizes with standardized design properties. The following standard widths are established to match the width of standard sizes of lumber, less an allowable amount for finishing the edges of the manufactured beams:

3 or 3⅛ in (76 or 79 mm)
5 or 5⅛ in (127 or 130 mm)
6¾ in (171 mm)
8½ or 8¾ in (216 or 222 mm)
10½ or 10¾ in (267 or 273 mm)

Standard beam depths are common multiples of lamination thickness of either 1⅜ or 1½ in (35 or 38 mm). There are no standard beam lengths, although most uses will be on spans where the length is from 10 to 20 times the depth. Allowable spans for various loadings of the standard sizes of beams are available from either the American Institute of Timber Construction or American Wood Systems.

The design stresses for beams in bending for dry-use applications are standardized in multiples of 200 lb/in^2 (1.4 MPa) within the range of 2,000 to 3,000 lb/in^2 (13.8 to 20.7 MPa). Modulus of elasticity values associated with these design stresses in bending vary from 1.6 to 2.0 $\times$ 10^6 lb/in^2. A bending stress of 2,400 lb/in^2 (16.5 MPa) and a modulus of elasticity of 1.8 $\times$ 10^6 lb/in^2 (12.4 GPa) are most commonly specified, and the designer needs to verify the availability of beams with higher

Table 6.7.12   Design Stresses for Selected Species of Round Timbers for Building Construction

Type of timber and species	Design stress					
	Bending		Compression		Modulus of elasticity	
	lb/in²	MPa	lb/in²	MPa	× 10⁶ lb/in²	GPa
Poles*						
Southern pine and Douglas-fir	2,100	14.5	1,000	6.9	1.5	10.3
Western redcedar	1,400	9.6	800	5.5	0.9	6.2
Piles†						
Southern pine	2,400	16.5	1,200	8.3	1.5	10.3
Douglas-fir	2,450	16.9	1,250	8.6	1.5	10.3
Red pine	1,900	13.1	900	6.2	1.3	8.8

* From ''Timber Construction Manual'' (AITC, 1994).
† From ''National Design Specification'' (AF&PA, 1991).

values. Design properties must be adjusted for wet-use applications. Detailed information on other design properties for beams as well as design properties and procedures for arches and other uses are given in ''National Design Specification'' (AF&PA).

Round Timbers

Round timbers in the form of poles, piles, or construction logs represent some of the most efficient uses of forest products because of the minimum of processing required. Poles and piles are generally debarked or peeled, seasoned, graded, and treated with a preservative prior to use. Construction logs are often shaped to facilitate their use. See Table 6.7.12.

Poles   The primary use of wood poles is to support utility and transmission lines. An additional use is for building construction. Each of these uses requires that the poles be pressure-treated with preservatives following the applicable AWPA standard (C1). For utility structures, pole length may vary from 30 to 125 ft (9.1 to 38.1 m). Poles for building construction rarely exceed 30 ft (9.1 m). Southern pines account for the highest percentage of poles used in the United States because of their favorable strength properties, excellent form, ease of treatment, and availability. Douglas-fir and western redcedar are used for longer lengths; other species are also included in the ANSI O5.1 standard (ANSI 1992) that forms the basis for most pole purchases in the United States.

Design procedures for the use of ANSI O5.1 poles in utility structures are described in the ''National Electric Safety Code'' (NESC). For building construction, design properties developed based on ASTM D2899 (see ASTM, 1995) are provided in ''Timber Construction Manual'' (AITC, 1994) or ASAE EP 388.

Piles   Most piles used for foundations in the United States utilize either southern pine or Douglas-fir. Material requirements for timber piles are given in ASTM D25, and preservative treatment should follow the applicable AWPA standard (C1 or C3). Design stress and procedures are provided in ''National Design Specifications.''

Construction Logs   Log buildings continue to be a popular form of construction because nearly any available species of wood can be used. Logs are commonly peeled prior to fabrication into a variety of shapes. There are no standardized design properties for construction logs, and when they are required, log home suppliers may develop design properties by following an ASTM standard ASTM D3957).

PROPERTIES OF STRUCTURAL PANEL PRODUCTS

by Roland Hernandez

Structural panel products are a family of wood products made by bonding veneer, strands, particles, or fibers of wood into flat sheets. The members of this family are (1) plywood, which consists of products made completely or in part from wood veneer; (2) flakeboard, made from strands, wafers, or flakes; (3) particleboard, made from particles; and (4) fiberboard and hardboard, made from wood fibers. Plywood and flakeboard make up a large percentage of the panels used in structural applications such as roof, wall, and floor sheathing; thus, only those two types will be described here.

Plywood

Plywood is the name given to a wood panel composed of relatively thin layers or **plies** of **veneer** with the wood grain of adjacent layers at right angles. The outside plies are called **faces** or **face and back plies**, the inner plies with grain parallel to that of the face and back are called **cores** or **centers,** and the plies with grain perpendicular to that of the face and back are called **crossbands.** In four-ply plywood, the two center plies are glued with the grain direction parallel to each ply, making one center layer. Total panel thickness is typically not less than $\frac{1}{16}$ in (1.6 mm) nor more than 3 in (76 mm). Veneer plies may vary as to number, thickness, species, and grade. Stock plywood sheets usually measure 4 by 8 ft (1.2 by 2.4 m), with the 8-ft (2.4-m) dimension parallel to the grain of the face veneers.

The alternation of grain direction in adjacent plies provides plywood panels with dimensional stability across their width. It also results in fairly similar axial strength and stiffness properties in perpendicular directions within the panel plane. The laminated construction results in a distribution of defects and markedly reduces splitting (compared to solid wood) when the plywood is penetrated by fasteners.

Two general classes of plywood, covered by separate standards, are available: **construction and industrial plywood** and **hardwood and decorative plywood.** Construction and industrial plywood are covered by Product Standard PS 1-83, and hardwood and decorative plywood are covered by ANSI/HPVA HP-1-1994. Each standard recognizes different exposure durability classifications, which are primarily based on the moisture resistance of the glue used, but sometimes also address the grade of veneer used.

The exposure durability classifications for construction and industrial plywood specified in PS-1 are **exterior, exposure 1, intermediate glue (exposure 2),** and **interior.** Exterior plywood is bonded with exterior (waterproof) glue and is composed of C-grade or better veneers throughout. Exposure 1 plywood is bonded with exterior glue, but it may include D-grade veneers. Exposure 2 plywood is made with glue of intermediate resistance to moisture. Interior-type plywood may be bonded with interior, intermediate, or exterior (waterproof) glue. D-grade veneer is allowed on inner and back plies of certain interior-type grades.

The exposure durability classifications for hardwood and decorative plywood specified in ANSI/HPVA HP-1-1994 are, in decreasing order of moisture resistance, as follows: **technical (exterior), type I (exterior), type II (interior),** and **type III (interior).** Hardwood and decorative plywood are not typically used in applications where structural performance is a prominent concern. Therefore, most of the remaining discussion of plywood performance will concern construction and industrial plywood.

A very significant portion of the market for construction and industrial plywood is in residential construction. This market reality has resulted in the development of performance standards for sheathing and single-layer subfloor or underlayment for residential construction by the American Plywood Association (APA). Plywood panels conforming to these performance standards for sheathing are marked with grade stamps such as those shown in Fig. 6.7.2 (example grade stamps are shown for different agencies). As seen in this figure, the grade stamps must show (1) conformance to the plywood product standards; (2) recognition as a quality assurance agency by the National Evaluation Service (NES), which is affiliated with the Council of American Building Officials; (3) exposure durability classification; (4) thickness of panel; (5) span rating, 32/16, which refers to the maximum allowable roof support spacing of 32 in (813 mm) and maximum floor joist spacing of 16 in (406 mm); (6) conformance to the performance-rated standard of the agency; (7) manufacturer's name or mill number; and (8) grades of face and core veneers.

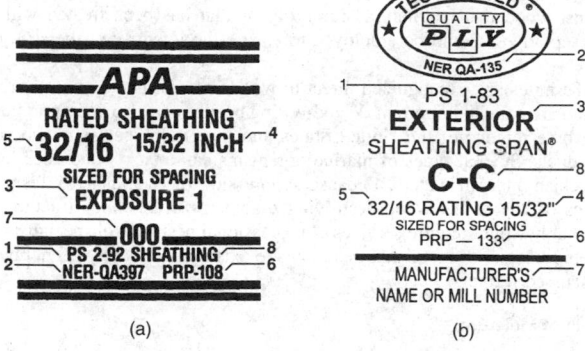

Fig. 6.7.2   Typical grade marks for *(a)* sheathing-grade plywood conforming to Product Standard PS 1-83 and *(b)* sheathing-grade structural-use panel conforming to Product Standard PS 2-92. (1) Conformance to indicated product standard, (2) recognition as a quality assurance agency, (3) exposure durability classification, (4) thickness, (5) span rating, (6) conformance to performance-rated product, (7) manufacturer's name or mill number, and (8) grade of face and core veneers.

All hardwood plywood represented as conforming to American National Standard ANSI/HPVA-HP-1-1994 is identified by one of two methods—either marking each panel with the HPVA plywood grade stamp (Fig. 6.7.3) or including a written statement with this information with the order or shipment. The HPVA grade stamp shows (1) HPVA trademark, (2) standard that governs manufacture, (3) HPVA mill num-

ber, (4) plywood bond-line type, (5) flame spread index class, (6) description of layup, (7) formaldehyde emission characteristics, (8) face species, and (9) veneer grade of face.

The span-rating system for plywood was established to simplify specification of plywood without resorting to specific structural engineering design. This system indicates performance without the need to refer to species group or panel thickness. It gives the allowable span when the face grain is placed across supports.

If design calculations are desired, a design guide is provided by APA-EWS in "Plywood Design Specifications" (PDS). The design guide contains tables of grade stamp references, section properties, and allowable stresses for plywood used in construction of buildings and

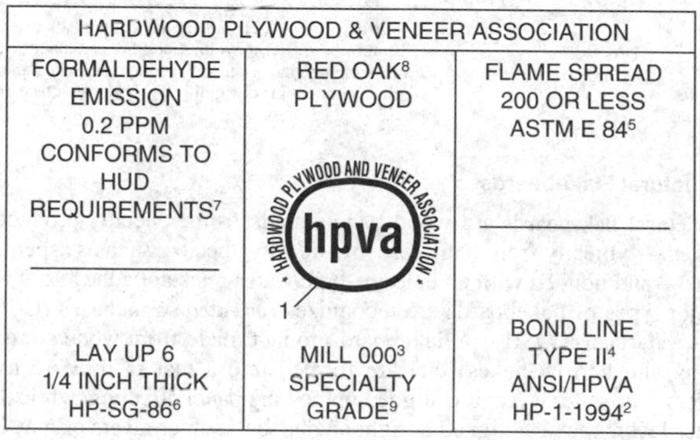

HARDWOOD PLYWOOD & VENEER ASSOCIATION		
FORMALDEHYDE EMISSION 0.2 PPM CONFORMS TO HUD REQUIREMENTS[7]	RED OAK[8] PLYWOOD hpva 1	FLAME SPREAD 200 OR LESS ASTM E 84[5]
LAY UP 6 1/4 INCH THICK HP-SG-86[6]	MILL 000[3] SPECIALTY GRADE[9]	BOND LINE TYPE II[4] ANSI/HPVA HP-1-1994[2]

Fig. 6.7.3   Grade stamp for hardwood plywood conforming to ANSI/HPVA HP-1-1994. (1) Trademark of Hardwood Plywood and Veneer Association, (2) standard that governs manufacture, (3) HPVA mill number, (4) plywood bond-line type, (5) flame spread index class, (6) layup description, (7) formaldehyde emission characteristics, (8) face species, and (9) veneer grade of face.

similar related structures. For example, given the grade stamp shown in Fig. 6.7.2, the grade stamp reference table in the PDS specifies that this particular plywood is made with veneer from species group 1, has section property information based on unsanded panels (Table 6.7.13), and is assigned allowable design stresses from the S-3 grade level (Table 6.7.14). Design information for grade stamps other than that shown in Fig. 6.7.2 is available in the PDS.

If calculations for the actual physical and mechanical properties of plywood are desired, formulas relating the properties of the particular wood species in the component plies to the laminated panel are provided in "Wood Handbook" (Forest Products Laboratory, 1987). These formulas could be applied to plywood of any species, provided the basic mechanical properties of the species were known. Note, however, that the formulas yield predicted actual properties (not design values) of plywood made of defect-free veneers.

Table 6.7.13   Effective Section Properties for Plywood—Unsanded Panels*

Nominal thickness, in	Approximate weight, lb/ft²	t_s Effective thickness for shear, in	Stress applied parallel to face grain				Stress applied perpendicular to face grain			
			A Area, in²/ft	I Moment of inertia, in⁴/ft	KS Effective section modulus, in³/ft	Ib/Q Rolling shear constant, in²/ft	A Area, in²/ft	I Moment of inertia, in⁴/ft	KS Effective section modulus, in³/ft	Ib/Q Rolling shear constant, in²/ft
5/16-U	1.0	0.268	1.491	0.022	0.112	2.569	0.660	0.001	0.023	4.497
3/8-U	1.1	0.278	1.868	0.039	0.152	3.110	0.799	0.002	0.033	5.444
15/32 & 1/2-U	1.5	0.298	2.292	0.067	0.213	3.921	1.007	0.004	0.056	2.450
19/32 & 5/8-U	1.8	0.319	2.330	0.121	0.379	5.004	1.285	0.010	0.091	3.106
23/32 & 3/4-U	2.2	0.445	3.247	0.234	0.496	6.455	1.563	0.036	0.232	3.613
7/8-U	2.6	0.607	3.509	0.340	0.678	7.175	1.950	0.112	0.397	4.791
1-U	3.0	0.842	3.916	0.493	0.859	9.244	3.145	0.210	0.660	6.533
1 1/8-U	3.3	0.859	4.725	0.676	1.047	9.960	3.079	0.288	0.768	7.931

* 1 in = 25.4 mm; 1 ft = 0.3048 m; 1 lb/ft² = 4.882 kg/m².

Table 6.7.14   Allowable Stresses for Construction and Industrial Plywood (Species Group 1)*

Type of stress, lb/in²	Grade stress level				
	S-1		S-2		S-3
	Wet	Dry	Wet	Dry	Dry only
F_b and F_t	1,430	2,000	1,190	1,650	1,650
F_c		970	1,640	900	1,540
F_v	155	190	155	190	160
F_s	63	75	63	75	—
G	70,000	90,000	70,000	90,000	82,000
F_c	210	340	210	340	340
E	1,500,000	1,800,000	1,500,000	1,800,000	1,800,000

* Stresses are based on normal duration of load and on common structural applications where panels are 24 in (610 mm) or greater in width. For other use conditions, see PDS for modifications. F_b is extreme fiber stress in bending; F_t, tension in plane of plies; F_c, compression in plane of plies; F_v, shear through the thickness; F_s, rolling shear in plane of plies; G, modulus of rigidity; F_c, bearing on face; and E, modulus of elasticity in plane of plies. 1 lb/in² = 6.894 kPa.

Structural Flakeboards

Structural flakeboards are wood panels made from specially produced flakes—typically from relatively low-density species, such as aspen or pine—and bonded with an exterior-type water-resistant adhesive. Two major types of flakeboards are recognized, **oriented strandboard** (OSB) and **waferboard**. OSB is a flakeboard product made from wood strands (long and narrow flakes) that are formed into a mat of three to five layers. The outer layers are aligned in the long panel direction, while the inner layers may be aligned at right angles to the outer layers or may be randomly aligned. In waferboard, a product made almost exclusively from aspen wafers (wide flakes), the flakes are not usually oriented in any direction, and they are bonded with an exterior-type resin. Because flakes are aligned in OSB, the bending properties (in the aligned direction) of this type of flakeboard are generally superior to those of waferboard. For this reason, OSB is the predominant form of structural flakeboard. Panels commonly range from 0.25 to 0.75 in (6 to 19 mm) thick and 4 by 8 ft (1 by 2 m) in surface dimension. However, thicknesses up to 1.125 in (28.58 mm) and surface dimensions up to 8 by 24 ft (2 by 7 m) are available by special order.

A substantial portion of the market for structural flakeboard is in residential construction. For this reason, structural flakeboards are usually marketed as conforming to a product standard for sheathing or single-layer subfloor or underlayment and are graded as a performance-rated product (PRP-108) similar to that for construction plywood. The Voluntary Product Standard PS 2-92 is the performance standard for wood-based structural-use panels, which includes such products as plywood, composites, OSB, and waferboard. The PS 2-92 is not a replacement for PS 1-83, which contains necessary veneer grade and glue bond requirements as well as prescriptive layup provisions and includes many plywood grades not covered under PS 2-92.

Design capacities of the APA performance-rated products, which include OSB and waferboard, can be determined by using procedures outlined in the APA-EWS *Technical Note* N375A. In this reference, allowable design strength and stiffness properties, as well as nominal thicknesses and section properties, are specified based on the span rating of the panel. Additional adjustment factors based on panel grade and construction are also provided.

Because of the complex nature of structural flakeboards, formulas for determining actual strength and stiffness properties, as a function of the component material, are not available.

DURABILITY OF WOOD IN CONSTRUCTION

by Rodney De Groot and Robert White

Biological Challenge

In the natural ecosystem, wood residues are recycled into the nutrient web through the action of wood-degrading fungi, insects, and other organisms. These same natural recyclers may pose a practical biological challenge to wood used in construction under conditions where one or more of these microorganisms or insects can thrive. Under those conditions, wood that has natural durability or that has been treated with preservatives should be employed to ensure the integrity of the structure.

Termites are a recognized threat to wood in construction, but decay fungi are equally important. Wood-boring beetles can also be important in some regions of the United States and in certain species of wood products. Several types of marine organisms can attack wood used in brackish and salt waters. Because suppression of established infestations of any of the wood-destroying organisms in existing structures probably would require services of professional pest control specialists, methodologies for remedial treatments to existing structures will not be discussed further.

Role of Moisture

Three environmental components govern the development of wood-degrading organisms within terrestrial wood construction: moisture, oxygen supply, and temperature. Of these, moisture content of wood seems most directly influenced by design and construction practices. The oxygen supply is usually adequate except for materials submerged below water or deep within the soil. Temperature is largely a climatic function; within the global ecosystem, a variety of wood-degrading organisms have evolved to survive within the range of climates where trees grow. For these reasons, most of the following discussion will focus on relationships between wood moisture and potential for wood deterioration.

Soil provides a continuing source of moisture. Nondurable wood in contact with the ground will decay most rapidly at the groundline where the moisture from soil and the supply of oxygen within the wood support growth of decay fungi. Deep within the soil, as well as in wood submerged under water, a limited supply of oxygen prevents growth of decay fungi. As the distance from groundline increases above ground, wood dries out and the moisture content becomes limiting for fungal growth, unless wood is wetted through exposure to rain or from water entrapped as a consequence of design and/or construction practices that expose wood to condensate from air conditioners, plumbing failures, etc. Wood absorbs water through exposed, cut ends about 11 times faster than through lateral surfaces. Consequently decay fungi, which require free water within the wood cells (above 20 to 25 percent moisture content) to survive, develop first at the joints in aboveground construction.

Naturally Durable Woods

The heartwood of old-growth trees of certain species, such as bald cypress, redwood, cedars, and several white oaks, is naturally resistant or very resistant to decay fungi. Heartwood of several other species, such as Douglas-fir, longleaf pine, eastern white pine, and western larch, is moderately resistant to wood decay fungi. Similarly, these species are not a wood of choice for subterranean termites. A more complete listing of naturally durable woods is given in ''Wood Handbook.'' These woods historically have been used to construct durable buildings, but some of these species are becoming less available as building materials. Consequently, other forms of protection are more frequently used in current construction. (See section on protection from decay).

Methods for Protecting Wood

Protection with Good Design   The most important aspect to consider when one is protecting structural wood products is their **design.** Many wood structures are several hundred years old, and we can learn from the principles used in their design and construction. For example, in nearly all those old buildings, the wood has been kept dry by a barrier over the structure (roof plus overhang), by maintaining a separation between the ground and the wood elements (foundation), and by preventing accumulation of moisture in the structure (ventilation). Today's engineered wood products will last for centuries if good design practices are used.

Protection from Weathering   The combination of sunlight and other **weathering agents** will slowly remove the surface fibers of wood products. This removal of fibers can be greatly reduced by providing a wood finish; if the finish is properly maintained, the removal of fibers can be nearly eliminated. Information on wood finishes is available in Cassens and Feist, ''Exterior Wood in the South.''

Protection from Decay   As naturally durable woods become less available in the marketplace, greater reliance is being placed on preservative-treated wood. Wood that is treated with a pressure-impregnated chemical is used for most load-bearing applications. In non-load-bearing applications such as exterior millwork around windows and doors, wood is usually protected with water-repellent preservative treatments that are applied by nonpressure processes. Standards for preservative treatment are published by the American Wood-Preservers' Association, and detailed information on wood preservation is given in ''Wood Handbook.''

An extremely low oxygen content in wood submerged below water will prevent growth of decay fungi, but other microorganisms can slowly colonize submerged wood over decades or centuries of exposure. Thus, properties of such woods need to be reconfirmed when old, submerged structures are retrofitted.

Protection from Insects   Subterranean **termites,** native to the U.S. mainland, establish a continuous connection with soil to maintain adequate moisture in wood that is being attacked above ground. One fundamental approach most often utilized in the protection of buildings from subterranean termites is to establish a physical or chemical barrier between soil and building. The function of that barrier is to bar access of termites to the building. The use of preservative-treated wood is relied upon to protect wood products such as poles, piling, and bridges that cannot be protected from exposure to termites via other mechanisms.

Formosan termites have the capacity to use sources of aboveground moisture without establishing a direct connection with the soil. Thus, where these termites occur, good designs and construction practices that eliminate sources of aboveground moisture are particularly important.

Dry-wood termites survive only in tropical or neotropical areas with sufficiently high relative humidity to elevate the wood moisture content to levels high enough to provide moisture for insect metabolism. Where these termites occur, use of naturally durable wood or preservative-treated wood warrants consideration.

Wood-destroying beetles may occur in wood members where the ambient moisture content is high enough for them to complete certain phases of their life cycle. Many beetles attack only certain species or groups of woods that may be used in specialty items such as joinery. Consequently, the first step in good construction practice is to use wood that is not preinfested at the time of construction. The next step is to

utilize designs and construction practices that will keep wood at low moisture content in use. Finally, chemical treatments may be needed for certain specific applications.

Protection from Fire   In general, proper design for fire safety allows the use of untreated wood. When there is a need to reduce the potential for heat contribution or flame spread, **fire-retardant treatments** are available. Although fire-retardant coatings or dip treatments are available, effective treatment often requires that the wood be pressure-impregnated with the fire-retardant chemicals. These chemicals include inorganic salts such as monoammonium and diammonium phosphate, ammonium sulfate, zinc chloride, sodium tetraborate, and boric acid. Resin polymerized after impregnation into wood is used to obtain a leach-resistant treatment. Such amino resin systems are based on urea, melamine, dicyandiamide, and related compounds. An effective treatment can reduce the ASTM E84 flame spread to less than 25. When the external source of heat is removed, the flames from fire-retardant-treated wood will generally self-extinguish. Many fire-retardant treatments reduce the generation of combustible gases by lowering the thermal degradation temperature. Fire-retardant treatments may increase the hygroscopic properties of wood.

Protection from Marine Organisms   **Pressure treatment** of native wood species with wood preservatives is required to protect wood used in marine or brackish waters from attack by marine borers.

Effect of Long-Term Exposure

Long exposure of wood to the atmosphere also causes changes in the cellulose. A study by Kohara and Okamoto of sound old timbers of a softwood and hardwood of known ages from temple roof beams shows that the percentage of cellulose decreases steadily over a period up to 1,400 years while the lignin remains almost constant. These changes are reflected in strength losses (Fig. 6.7.4). Impact properties approximate a loss that is nearly linear with the logarithm of time. Allowable working stresses for preservative-treated lumber usually need not be reduced to account for the effect of the treating process. Tests made by the USDA Forest Service, Forest Products Laboratory, of preservative-treated

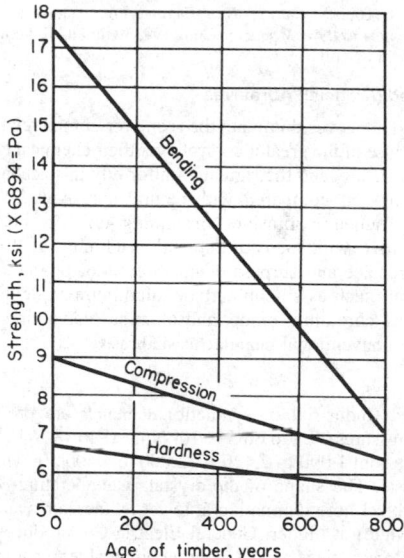

Fig. 6.7.4   Strength loss with age in a hardwood *(Zelkowa serrata). (From Sci. Rpts. Saikyo Univ., no. 7, 1955.)*

lumber when undergoing bending, tension, and compression perpendicular to grain show reductions in mean extreme fiber stress from a few percent up to 25 percent, but few reductions in working stresses. Compression parallel to grain is affected less and modulus of elasticity very little. The effect on horizontal shear can be estimated by inspection for an increase in shakes and checks after treatment. AWPA Standards keep temperatures, heating periods, and pressures to a minimum for required penetration and retention, which precludes the need for adjustment in working stresses.

COMMERCIAL LUMBER STANDARDS

Standard abbreviations for lumber description and **size standards** for yard lumber are given in "Wood Handbook."

Cross-sectional **dimensions** and **section properties** for beams, stringers, joists, and planks are given in the National Design Specification.

Standard patterns for finish lumber are shown in publications of the grading rules for the various lumber associations.

Information and **specifications for construction and industrial plywood** are given in Product Standard PS 1-83 and in ANSI/HPVA HP-1-1994 for hardwood and decorative plywood. Information and specifications for structural flakeboard are given in PS 2-92.

6.8 NONMETALLIC MATERIALS
by Antonio F. Baldo

ABRASIVES

REFERENCES: "Abrasives: Their History and Development," The Norton Co. Searle, "Manufacture and Use of Abrasive Materials," Pitman. Heywood, "Grinding Wheels and Their Uses," Penton. "Boron Carbide," The Norton Co. "Abrasive Materials," annual review in "Minerals Yearbook," U.S. Bureau of Mines. "Abrasive Engineering," Hitchcock Publishing Co. Coated Abrasives Manufacturer's Institute, "Coated Abrasives—Modern Tool of Industry," McGraw-Hill. Wick, Abrasives; Where They Stand Today, *Manufacturing Engineering and Management,* **69,** no. 4, Oct. 1972. Burls, "Diamond Grinding; Recent Research and Development," Mills and Boon, Ltd., London. Coes, Jr., "Abrasives," Springer-Verlag, New York. *Proceedings of the American Society for Abrasives Methods.* 1971, ANSI B74.1 to B74.3 1977 to 1993. Washington Mills Abrasive Co., Technical Data, *Mechanical Engineering,* Mar. 1984. "Modern Abrasive Recipes," *Cutting Tool Engineering,* April 1994. "Diamond Wheels Fashion Carbide Tools," *Cutting Tool Engineering,* August 1994. Synthetic Gemstones, *Compressed Air Magazine,* June 1993. Krar and Ratterman, "Superabrasives," McGraw-Hill.

Manufactured (Artificial) Abrasives

Manufactured abrasives dominate the scene for commercial and industrial use, because of the greater control over their chemical composition and crystal structure, and their greater uniformity in size, hardness, and cutting qualities, as compared with natural abrasives. Technical advances have resulted in abrasive "machining tools" that are economically competitive with many traditional machining methods, and in some cases replace and surpass them in terms of productivity. Fused electrominerals such as silicon carbide, aluminum oxide, alumina zirconia, alumina chromium oxide, and alumina titania zirconia are the most popular conventional **manufactured abrasives.**

Grains

Manufactured (industrial or synthetic) **diamonds** are produced from graphite at pressures 29.5 to 66.5×10^6 N/m² (8 to 18×10^5 lb/in²) and temperatures from 1,090 to 2,420°C (2,000 to 4,400°F), with the aid of metal catalysts. The shape of the crystal is temperature-controllable, with cubes (black) predominating at lower temperatures and octahedra (yellow to white) at higher. General Electric Co. produces synthetics reaching 0.01 carat sizes and of quality comparable to natural diamond powders. Grade MBG-11 (a blocky powder) is harder and tougher and is used for cutting wheels. Diamond hardness (placed at 10 on the Mohs scale) ranges from 5,500 to 7,000 on the Knoop scale. Specific gravity is 3.521.

Recently, chemical vapor deposition (CVD) has produced diamond vapors at ambient temperatures, which are then deposited on a substrate as a continuous film which may be from 1 to as much as 1,000 μm thick. Microscopically, CVD diamond is a dense polycrystalline structure with discrete diamond grains. Deposition can be directly on tools such as wheels or other surfaces. CVD diamond can be polished to any required surface texture and smoothness.

Crystalline alumina, as chemically purified Al_2O_3, is very adaptable in operations such as precision grinding of sensitive steels. However, a tougher grain is produced by the addition of TiO_2, Fe_2O_3, SiO_2, ZrO_2. The percentage of such additions and the method of cooling the pig greatly influence grain properties. Advantage is taken of this phenomenon to "custom" make the most desirable grain for particular machining needs. Alumina crystals have conchoidal fracture, and the grains when crushed or broken reveal sharp cutting edges and points. Average properties are: density ~3.8, coefficient of expansion ~0.81 × 10^{-5}/°C (0.45 × 10^{-5}/°F), hardness ~9 (Mohs scale), and melting temperature ~2,040°C (3,700°F). Applications cover the grinding of high-tensile-strength materials such as soft and hard steels, and annealed malleable iron.

Silicon carbide (SiC) corresponds to the mineral moissanite, and has a hardness of about 9.5 (Mohs scale) or 2,500 (Knoop), and specific gravity 3.2. It is insoluble in acid and is infusible but decomposes above 2,230°C (4,060°F). It is manufactured by fusing together coke and sand in an electric furnace of the resistance type. Sawdust is used also in the batch and burns away, leaving passages for the carbon monoxide to escape. The grains are characterized by great brittleness. Abrasives of silicon carbide are best adapted to the grinding of low-tensile-strength materials such as cast iron, brass, bronze, marble, concrete, stone, and glass. It is available under several trade names such as **Carborundum, Carbolon, Crystolon.**

Boron carbide (B_4C), a black crystal, has a hardness of about 9.32 (Mohs scale) or about 2,800 (Knoop), melts at about 2,460°C (4,478°F) but reacts with oxygen above 983°C (1,800°F), and is not resistant to fused alkalies. Boron carbide powder for grinding and lapping is obtainable in standard mesh sizes to 240, and to 800 in special finer sizes. It is being used in loose grain form for the lapping of cemented-carbide tools. In the form of molded shapes, it is used for pressure blast nozzles, wire-drawing dies, bearing surfaces for gages, etc.

Boron nitride (BN) when produced at extremely high pressures and temperatures forms tiny reddish to black grains of cubic crystal structure having hardness equal to diamond, and moreover is stable to 1,930°C (3,500°F). Abrasive powders, such as Borazon, are used extensively for coated-abrasive applications as for grinding tool and die steels and high-alloy steels, particularly where chemical reactivity of diamonds is a problem. Ease of penetration and free-cutting action minimize heat generation, producing superior surface integrity.

Crushed steel is made by heating high-grade crucible steel to white heat and quenching in a bath of cold water. The fragments are then crushed to sizes ranging from fine powder to $\frac{1}{16}$ in diam. They are classified as diamond crushed steel, diamond steel, emery, and steelite, used chiefly in the stone, brick, glass, and metal trades.

Rouge and crocus are finely powdered oxide of iron used for buffing and polishing. Rouge is the red oxide; crocus is purple.

Natural Abrasives

Diamond of the **bort** variety, crushed and graded into usable sizes and bonded with synthetic resin, metal powder, or vitrified-type bond, is used extensively for grinding tungsten- and tantalum-carbide cutting tools, and glass, stone, and ceramics.

Corundum is a mineral composed chiefly of crystallized alumina (93 to 97 percent Al_2O_3). It has been largely replaced by the manufactured variety.

Emery, a cheap and impure form of natural corundum which has been used for centuries as an abrasive, has been largely superseded by manufactured aluminum oxide for grinding. It is still used to some extent in the metal- and glass-polishing trades.

Garnet Certain deposits of garnet having a hardness between quartz and corundum are used in the manufacture of abrasive paper. **Quartz** is also used for this purpose. Garnet costs about twice as much as quartz and generally lasts proportionately longer.

Buhrstones and millstones are generally made from cellular quartz. Chasers (or stones running on edge) are also made from the same mineral.

Natural oilstones, the majority being those quarried in Arkansas, are of either the hard or the soft variety. The hard variety is used for tools requiring an extremely fine edge like those of surgeons, engravers, and dentists. The soft variety, more porous and coarser, is used for less exacting applications.

Pumice, of volcanic origin, is extensively used in leather, felt, and woolen industries and in the manufacture of polish for wood, metal, and stone. An artificial pumice is made from sand and clay in five grades of hardness, grain, and fineness.

Infusorial earth or tripoli resembles chalk or clay in physical properties. It can be distinguished by absence of effervescence with acid, is generally white or gray in color, but may be brown or even black. Owing to its porosity, it is very absorptive. It is used extensively in polishing powders, scouring soaps, etc., and, on account of its porous structure, in the manufacture of dynamite as a holder of nitroglycerin, also as a nonconductor for steam pipes and as a filtering medium. It is also known as **diatomaceous silica.**

Grinding Wheels For complete coverage and details see current ANSI/ASTM Standards D896 to D3808.

Vitrified Process In wheels, segments, and other abrasive shapes of this type, the abrasive grains are bonded with a glass or porcelain obtained by mixing the grains with such materials as clays and feldspars in various proportions, molding the wheel, drying, and firing at a temperature of 1,370°C (2,500°F) approx. It is possible to manufacture wheels as large as 60 in diam by this process, and even larger wheels may be obtained by building up with segments. Most of the grinding wheels and shapes (segments, cylinders, bricks, etc.) now manufactured are of the vitrified type and are very satisfactory for general grinding operations.

Silicate Process Wheels and shapes of the silicate type are manufactured by mixing the abrasive grain with sodium silicate (water glass) and fillers that are more or less inert, molding the wheel by tamping, and baking at a moderate temperature. Silicate bonded wheels are considered relatively "mild acting" and, in the form of large wheels, are still used to some extent for grinding-edge tools in place of the old-fashioned sandstone wheels.

Organic Bonded Wheels Organic bonds are used for high-speed wheels, and are equally well adapted to the manufacture of very thin wheels because of their flexibility compared with vitrified wheels. There are three distinct types in the group. The **shellac process** consists of mixing abrasive grains with shellac, heating the mass until the shellac is viscous, stirring, cooling, crushing, forming in molds, and reheating sufficiently to permit the shellac to set firmly upon cooling. Wheels made by this process are used for saw gumming, roll grinding, ball-race

and cam grinding, and in the cutlery trade. In the **rubber process** the bond is either natural or synthetic rubber. The initial mixture of grain, rubber, and sulfur (and such special ingredients as accelerators, fillers, and softeners) may be obtained by rolling or other methods. Having formed the wheel, the desired hardness is then developed by vulcanization. Wheels can be made in a wide variety of grain combinations and grades and have a high factor of safety as regards resistance to breakage in service. Wheels made by the rubber process are used for cutoff service on wet-style machines, ball-race punchings, feed wheels and centerless grinders, and for grinding stainless-steel billets and welds, which usually require a high-quality finish. With the **resinoid process,** the practice is to form the wheel by the cold-press process using a synthetic resin. After heating, the resultant bond is an insoluble, infusible product of notable strength and resiliency. Resinoid bond is used for the majority of high-speed wheels in foundries, welding shops, and billet shops, and also for cutoff wheels. The rate of stock removal is generally in direct proportion to the peripheral speed. Resinoid-bonded wheels are capable of being operated at speeds as high as 2,900 m/min (9,500 surface ft/min), as contrasted with 1,980 m/min (6,500 surface ft/min) for most vitrified bonds.

Silicone-coated abrasives, such as Silkote are reported to resist deleterious coolant effects on the bond between resin and grain, because of the silicone's ability to repel entry of coolant.

The **grain size** or **grit** of a wheel is determined by the size or combination of sizes of abrasive grain used. The Grinding Wheel Institute has standardized sizes 8, 10, 12, 14, 16, 20, 24, 30, 36, 46, 54, 60, 70, 80, 90, 100, 120, 150, 180, 200, 220, and 240. The finer sizes, known as flours, are designated as 280, 320, 400, 500, 600, 800, 900, or as F, FF, FFF, and XF.

The **grade** is the hardness or relative strength of bonding of a grinding wheel. The wheel from which grain particles are easily broken away, causing it to wear rapidly, is called soft, and one that is able to retain its particles longer is called hard. The complete range of grade letters used for the order of increasing hardness is

Soft		Hard
A, B, C, D, E, F, G, H, I, J, K, L, M, N, O, P, Q, R, S, T, U, V, W, X, Y, Z		

Table 6.8.1 provides a guide to surface finishes attainable with diamond grit.

Table 6.8.1   Guide to Surface Finish Attainable with Diamond Grit

Grit size	Surface finish		Recommended max DOC* per pass	
	μ in (AA)	μm R_a	in	μm
80	26–36			
100	24–32			
105	18–30			
100S	16–26	0.4–0.8	0.001–0.002	25–50
110				
120	16–18			
150	14–16			
180	12–14	0.25–0.50	0.0007–0.0010	17–25
220	10–12			
240	8–10			
320	8	0.12–0.25	0.0004–0.0006	10–15
400	7–8	0.08–0.15	0.0003–0.0005	8–12
30–40 μm	7			
20–30	6–7	0.05–0.10	0.0002–0.0004	5–10
	6			
10–20	5–6			
8–16				
6–12	3–5			
4–8	2–4	0.013–0.025	0.00005–0.00007	1–2
3–6	2			
0–2				

* DOC = depth of cut.
SOURCE: *Cutting Tool Engineering*, Aug. 1994.

Coated Abrasives

Coated abrasives are "tools" consisting of an abrasive grit, a backing, and an adhesive bond. Grits are generally one of the manufactured variety listed above, and are available in mesh sizes ranging from the coarsest at 12 to the finest at 600. Backings can be cloth, paper, fiber, or combinations and are made in the form of belts and disks for power-operated tools and in cut sheets for both manual and power usage. Adhesive bonds consist of two layers, a "make coat" and a "size coat." Both natural and synthetic adhesives serve for bond materials. Abrasive coatings can be closed-coat, in which the abrasive grains are adjacent to one another without voids, or open-coat, in which the grains are set at a predetermined distance from one another. The flex of the backing is obtained by a controlled, directional, spaced backing of the adhesive bond.

Abrasive Waterjets

Such tough-to-cut materials as titanium, ceramics, metallic honeycomb structures, glass, graphite, and bonding compounds can be successfully cut with abrasive waterjets. Abrasive waterjets use pressurized water, up to 60 lb/in², to capture solid abrasives by flow momentum, after which the mixture is expelled through a sapphire nozzle to form a highly focused, high-velocity cutting "tool." Advantages of this cutting method include: minimal dust, high cutting rates, multidirectional cutting capability, no tool dulling, no deformation or thermal stresses, no fire hazards, ability to cut any material, small power requirements, avoidance of delamination, and reduction of striation. (See also Sec. 13.)

Abrasive waterjet cutting, however, has some limitations including high noise level (80 to 100 dB), safety problems, low material removal rates, inability to machine blind holes or pockets, damage to accidentally exposed machine elements by the particles and/or high-pressure water, and the size of the overall system.

Industrial Sharpening Stones

Sharpening stones come in several forms such as bench stones, files, rubbing bricks, slip stones, and specialties, and are made chiefly from aluminum oxide, such as Alundum, or silicon carbide, such as Crystolon. Grit sizes of fine, medium, and coarse are available.

ADHESIVES

REFERENCES: Cagle (ed.), "Handbook of Adhesive Bonding," McGraw-Hill. Bikerman, "The Science of Adhesive Joints," Academic. Shields, "Adhesives Handbook," CRC Press (Division of The Chemical Rubber Co.). Cook, "Construction Sealants and Adhesives," Wiley. Patrick, "Treatise on Adhesives," Marcel Dekker. NASA SP-5961 (01) Technology Utilization, "Chemistry Technology: Adhesives and Plastics," National Technical Information Services, Virginia. Simonds and Church, "A Concise Guide to Plastics," Reinhold. Lerner, Kotshar, and Sheckman, "Adhesives Red Book," Palmerton Publishing Co., New York. *Machine Design,* June 1976. ISO 6354-1982. "1994 Annual Book of ASTM Standards," vol. 15.06 (Adhesives). ANSI/ASTM Standards D896–D3808.

Adhesives are substances capable of holding materials together in a useful manner by surface attachment. Some of the advantages and disadvantages of adhesive bonding are as follows:

Advantages Ability to bond similar or dissimilar materials of different thicknesses; fabrication of complex shapes not feasible by other fastening means; smooth external joint surface; economic and rapid assembly; uniform distribution of stresses; weight reduction; vibration damping; prevention or reduction of galvanic corrosion; insulating properties.

Disadvantages Surface preparation; long cure times; optimum bond strength not realized instantaneously, service-temperature limitations; service deterioration; assembly fire or toxicity; tendency to creep under sustained load.

A broad scheme of classification is given in Table 6.8.2a.

For the vocabulary of adhesives the reader should refer to ISO 6354-1982, and for standard definitions refer to ASTM D907-82. Procedures for the testing of adhesive strength viscosity, storage life, fatigue properties, etc. can be found in ANSI/ASTM D950–D3808.

Thermoplastic adhesives are a general class of adhesives based upon

Table 6.8.2a Classification of Adhesives

Origin and basic type		Adhesive material
Natural	Animal	Albumen, animal glue (including fish), casein, shellac, beeswax
	Vegetable	Natural resins (gum arabic, tragacanth, colophony, Canada balsam, etc.); oils and waxes (carnauba wax, linseed oils); proteins (soybean); carbohydrates (starch, dextrins)
	Mineral	Inorganic materials (silicates, magnesia, phosphates, litharge, sulfur, etc.); mineral waxes (paraffin); mineral resins (copal, amber); bitumen (including asphalt)
Synthetic	Elastomers	Natural rubber (and derivatives, chlorinated rubber, cyclized rubber, rubber hydrochloride)
		Synthetic rubbers and derivatives (butyl, polyisobutylene, polybutadiene blends (including styrene and acylonitrile), polyisoprenes, polychloroprene, polyurethane, silicone, polysulfide, polyolefins (ethylene vinyl chloride, ethylene polypropylene)
		Reclaim rubbers
	Thermoplastic	Cellulose derivatives (acetate, acetate-butyrate, caprate, nitrate, methyl cellulose, hydroxy ethyl cellulose, ethyl cellulose, carboxy methyl cellulose)
		Vinyl polymers and copolymers (polyvinyl acetate, alcohol, acetal, chloride, polyvinylidene chloride, polyvinyl alkyl ethers)
		Polyesters (saturated) [polystyrene, polyamides (nylons and modifications)]
		Polyacrylates (methacrylate and acrylate polymers, cyanoacrylates, acrylamide)
		Polyethers (polyhydroxy ether, polyphenolic ethers)
		Polysulfones
	Thermosetting	Amino plastics (urea and melamine formaldehydes and modifications)
		Epoxides and modifications (epoxy polyamide, epoxy bitumen, epoxy polysulfide, epoxy nylon)
		Phenolic resins and modifications (phenol and resorcinol formaldehydes, phenolic-nitrile, phenolic-neoprene, phenolic-epoxy)
		Polyesters (unsaturated)
		Polyaromatics (polyimide, polybenzimidazole, polybenzothiazole, polyphenylene)
		Furanes (phenol furfural)

SOURCE: Shields, "Adhesives Handbook," CRC Press (Division of the Chemical Rubber Co.).

long-chained polymeric structure, and are capable of being softened by the application of heat.

Thermosetting adhesives are a general class of adhesives based upon *cross-linked* polymeric structure, and are incapable of being softened once solidified. A recent development (1994) in cross-linked aromatic polyesters has yielded a very durable adhesive capable of withstanding temperatures of 700°F before failing. It retains full strength through 400°F. It is likely that widespread applications for it will be found first in the automotive and aircraft industries.

Thermoplastic and thermosetting adhesives are cured (set, polymerized, solidified) by heat, catalysis, chemical reaction, free-radical activity, radiation, loss of solvent, etc., as governed by the particular adhesive's chemical nature.

Elastomers are a special class of thermoplastic adhesive possessing the common quality of substantial flexibility or elasticity.

Anaerobic adhesives are a special class of thermoplastic adhesive (polyacrylates) that set *only* in the *absence* of air (oxygen). The two basic types are: (1) *machinery*—possessing shear strength only, and (2) *structural*—possessing both tensile and shear strength.

Pressure-sensitive adhesives are permanently (and aggressively) tacky (sticky) solids which form immediate bonds when two parts are brought together under pressure. They are available as films and tapes as well as hot-melt solids.

The relative performance of a number of adhesives is given in Table 6.8.2b.

For high-performance adhesive applications (engineering or machine parts) the following grouping is convenient.

Thread locking	Anaerobic acrylic
Hub mounting	Anaerobic acrylic—compatible materials or flow migration unimportant.
	Modified acrylic—large gaps or migration must be avoided.
	Epoxy—maximum strength at high temperatures
Bearing mounting	Anaerobic acrylic—compatible materials necessary and flow into bearing area to be prevented.
	Modified acrylic—for lowest cost
Structural joining	Epoxies and modified epoxies—for maximum strength (highest cost)
	Acrylics—anaerobic or modified cyanacrylates
Gasketing	Silicones—primarily anaerobic

Table 6.8.3 presents a sample of a number of adhesives (with practical information) that are available from various sources. The table is adapted from the rather extensive one found in J. Shields, ''Adhesives Handbook,'' CRC Press (Division of The Chemical Rubber Co.), 1970, by permission of the publisher. Domestic and foreign trade sources are listed there (pages 332–340) and appear coded (in parentheses) in the second column. For other extensive lists of trade sources, the reader is referred to Charles V. Cagle (ed.), ''Handbook of Adhesive Bonding,'' McGraw-Hill, and ''Adhesives Red Book,'' Palmerton Publishing Co., New York.

BRICK, BLOCK, AND TILE

REFERENCES: Plummer and Reardon, ''Principles of Brick Engineering, Handbook of Design,'' Structural Clay Products Institute. Stang, Parsons, and McBurney, Compressive Strength of Clay Brick Walls, *B. of S. Research Paper* 108. Hunting, ''Building Construction,'' Wiley. Amrhein, ''Reinforced Masonry Engineering Handbook,'' Masonry Institute of America. Simpson and Horrbin (eds.), ''The Weathering and Performance of Building Materials,'' Wiley. SVCE and Jeffers (eds.), ''Modern Masonry Panel Construction Systems,'' Cahners Books (Division of Cahners Publishing Co.). ASTM Standards C67-C902. ANSI/ASTM Standards C62-C455.

Brick

In the case of structural and road building material, a small unit, solid or practically so, commonly in the form of a rectangular prism, formed

Table 6.8.2b   Performance of Adhesive Resins
(Rating 1 = poorest or lowest, 10 = best or highest)

Adhesive resin	Adherence to					Resistance			
	Paper	Wood	Metal	Ceramics	Rubbers	Water	Solvents	Alkali	Acids
Alkyd	6	7	5	6	7	7	2	2	5
Cellulose acetate	4	3	1	3	5	2	3	1	3
Cellulose acetate butyrate	3	3	1	4	5	2	3	1	3
Cellulose nitrate	5	5	1	5	5	3	2	2	4
Ethyl cellulose	3	3	1	3	5	2	3	3	3
Methyl cellulose	5	1	1	3	3	1	6	3	3
Carboxy methyl cellulose	6	1	2	3	2	1	6	1	4
Epoxy resin	10	10	8	8	8	8	9	9	8
Furane resin	8	7	1	8	7	8	9	10	8
Melamine resin	10	10	2	2	2	7	9	5	5
Phenolic resins	9	8	2	6	7	8	10	7	8
Polyester, unsaturated	6	8	2	5	7	7	6	1	6
Polyethylacrylate	3	4	3	5	6	8	2	6	7
Polymethylmethacrylate	2	3	2	3	6	8	3	8	7
Polystyrene	1	3	2	2	5	8	1	10	8
Polyvinylacetate	8	7	7	7	3	3	3	4	6
Polyvinyl alcohol	6	2	2	4	6	1	7	1	3
Polyvinyl acetal	5	7	8	7	7	8	5	3	5
Polyvinyl chloride	5	7	6	7	6	8	6	10	9
Polyvinyl acetate chloride	6	8	6	7	5	8	5	9	9
Polyvinylidene copolymer	4	7	6	7	7	8	7	10	9
Silicone T.S.	4	6	7	7	8	10	7	6	6
Urethane T.S.	8	10	10	9	10	7	8	4	4
Acrylonitrile rubber	3	6	8	6	9	7	5	8	8
Polybutene rubber	3	3	6	2	8	8	3	10	9
Chlorinated rubber	3	5	7	4	7	6	3	10	9
Styrene rubber	5	7	6	5	8	7	3	10	9

SOURCE: Adapted from Herbert R. Simonds and James M. Church, ''A Concise Guide to Plastics,'' 2d ed., Reinhold, 1963, with permission of the publisher.

Table 6.8.3 Properties and Uses of Various Adhesives

Basic type	Curing cycle, time at temp	Service temp range, °C	Adherends	Main uses	Remarks
			Animal		
Animal (hide)	Melted at 70–75°C. Sets on cooling		Paper, wood, textiles	Woodworking, carpet materials, paper, bookbinding	May be thinned with water
Animal (hide) + plasticizers	Applied as a melt at 60°C	<60	Paper, cellulosic materials	Bookbinding, stationery applications	Cures to permanent flexible film
Fish glue	1 h at 20°C	60	Wood, chipboard, paper	General-purpose for porous materials	Rapid setting. Good flexibility Moderate resistance to water. High tack.
Casein	Cold setting after 20-min standing period on mixing		Timber with moisture content	Laminated timber arches and beams, plybox beams, and engineering timber work	Full bond strength developed after seasoning period of 48 h
Casein + 60% latex	Cold setting after 20-min standing period on mixing		Aluminum, wood, phenolic formaldehyde (rigid), leather, rubber	Bonding of dissimilar materials to give flexible, water-resistant bond	Flexible
			Vegetable		
Dextrine	Air drying		Paper, cardboard, leather, wood, pottery	General-purpose glue for absorbent materials	Medium drying period of 2–3 h
Dextrine-starch blend	Applied above 15°C air drying	48	Cellulosic materials, cardboard, paper	Labeling, carton sealing, spiral-tube winding	Fast setting. May be diluted with water
Gum arabic	Cold setting		Paper, cardboard	Stationery uses	Fast drying
			Mineral		
Silicate	8 h at 20°C	10–430	Asbestos, magnesia	Lagging asbestos cloth on high-temperature insulation	Unsuitable where moisture: not recommended for glass or painted surfaces
Silicate with china-clay filler	Dried at 80°C before exposure to heat	−180–1,500	Asbestos, ceramics, brickwork, glass, silver, aluminum, steel (mild)-steel	General-purpose cement for bonding refractory materials and metals. Furnace repairs and gastight jointing of pipe work. Heat-insulating materials	Resistant to oil, gasoline, and weak acids
Sodium silicate	Dried at 20–80°C before exposure to heat	0–850	Aluminum (foil), paper, wood-wood	Fabrication of corrugated fiberboard. Wood bonding, metal foil to paper lamination	Suitable for glass-to-stone bonding
Aluminum phosphate + silica filler	Dried ½ h at 20°C, then ½ h at 70°C + ½ h at 100°C + 1 h at 200°C + 1 h at 250°C. Repeat for 2 overcoatings and finally cure 1 h at 350°C	750	Steels (low-alloy), iron, brass, titanium, copper, aluminum	Strain-gage attachment to heat-resistant metals. Heater-element bonding	Particularly suited to heat-resistant steels where surface oxidation of metal at high temperatures is less detrimental to adhesion
Bitumen/latex emulsion	Dried in air to a tacky state	0–66	Cork, polystyrene (foam), polyvinyl chloride, concrete, asbestos	Lightweight thermal-insulation boards, and preformed sections to porous and nonporous surfaces. Building applications	Not recommended for constructions operated below 0°C
			Elastomers		
Natural rubber	Air-dried 20 min at 20°C and heat-cured 5 min at 140°C		Rubber (styrene butadiene), rubber (latex), aluminum, cardboard, leather, cotton	Vulcanizing cement for rubber bonding to textiles and rubbers	May be thinned with toluene
Natural rubber in hydrocarbon solvent	Air-dried 10 min at 20°C and heat-cured for 20 min at 150°C	100	Hair (keratin), bristle, polyamide fiber	Brush-setting cement for natural- and synthetic-fiber materials	Resistant to solvents employed in oil, paint and varnish industries. Can be nailed without splitting
Rubber latex	Air drying within 15 min		Canvas, paper, fabrics, cellulosic materials	Bonding textiles, papers, packaging materials. Carpet bonding	Resistant to heat. Should be protected from frosts, oils
Chlorinated rubber in hydrocarbon solvents	Air-dried 10 min at 20°C and contact bonded	−20–60	Polyvinyl chloride acrylonitrile butadiene styrene, polystyrene, rubber, wood	General-purpose contact adhesive	Resistant to aging, water, oils, petroleum

Table 6.8.3 Properties and Uses of Various Adhesives (*Continued*)

Basic type	Curing cycle, time at temp	Service temp range, °C	Adherends	Main uses	Remarks
			Elastomers (*Continued*)		
Styrene-butadiene rubber lattices	Air drying		Polystyrene (foam), wood, hardboard, asbestos, brickwork	Bonding polystyrene foams to porous surface	
Neoprene/nitrile rubbers in	Dried 30 min in air and bonded under pressure while tacky		Wood, linoleum, leather, paper, metals, nitrile rubbers, glass, fabrics	Cement for bonding synthetic rubbers to metals, woods, fabrics	May be thinned with ketones
Acrylonitrile rubber + phenolic resin	Primer air-dried 60 min at 20°C film cured 60 min at 175°C under pressure. Pressure released on cooling at 50°C	$-10-130$	Aluminum (alloy)-aluminum to DTD 746	Metal bonding for structural applications at elevated temperatures	Subject to creep at 150°C for sustained loading
Polysulfide rubber in ketone solvent and catalyst	3 days at 25°C	$-50-130$, withstands higher temps. for short periods	Metals	Sealant for fuel tanks and pressurized cabins in aircraft, where good weatherproof and waterproof properties are required	Resistant to gasoline, oil, hydraulic fluids, ester lubricants. Moderate resistance to acids and alkalies
Silicone rubber	24 h at 20°C (20% R.H.). Full cure in 5 days	$-65-260$	Aluminum, titanium, steel (stainless), glass, cork, silicone rubber, cured rubber-aluminum, cured rubber-titanium, cured rubber-steel (stainless), aluminum-aluminum (2024 Alclad), cork-cork (phenolic bonded)	General-purpose bonding and sealing applications. Adhesive/sealant for situations where material is expected to support considerable suspended weight. High pressure exposure conditions	Resistant to weathering and moisture
Reclaim rubber	Contact bonded when tacky		Fabric, leather, wood, glass, metals (primed)	General industrial adhesive for rubber, fabric, leather, porous materials	May be thinned with toluene
Polychloroprene	Air-dried 10–20 min at 20°C		Rubber, steel, wood, concrete	Bonding all types of rubber flooring to metals, woods, and masonry	Good heat resistance
Modified polyurethane	3 h at 18°C to 16 h at -15°C	$-80-110$	Concrete, plaster, ceramics, glass, hardboards, wood, polyurethane (foam), phenol formaldehyde (foam), polystyrene (foam), copper, lead, steel, aluminum	Bonding rigid and semirigid panels to irregular wall surfaces, wall cladding and floor laying. Building industry applications	Foam remains flexible on aging even at elevated temperatures. Will withstand a 12% movement
			Thermoplastic		
Nitrocellulose in ester solvent	Heat set 1 h at 60°C after wet bonding	60	Paper, leather, textiles, silicon carbide, metals	Labeling, general bonding of inorganic materials including metals	Good resistance to mineral oils
Modified methyl cellulose	Dries in air		Vinyl-coated paper, polystyrene foam	Heavy-duty adhesive. Decorating paper and plastics	Contains fungicide to prevent biodeterioration
Ethylene vinyl acetate copolymer + resins	Film transfer at 70–80°C followed by bonding at 150–160°C	60 or 1 h at 90	Cotton (duck)-cotton, resin rubber-leather, melamine laminate—plywood, steel (mild)-steel, acrylic (sheet)-acrylic	Metals, laminated plastics, and textiles. Fabrication of leather goods. Lamination work	Good electrical insulation
Polyvinyl acetate	Rapid setting		Paper, cardboard	Carton sealing in packaging industry	Resistant to water
Synthetic polymer blend	Applied as a melt at 177°C	71	Paper, cardboard, polythene (coated materials)	Carton and paper-bag sealing. Packaging	Resistant to water
Polychloroprene/resin blend in solvent	Air-dried 10 min at 20°C and cured 4 days at 20°C to 7 h at 75°C		Chlorosulfonated polythene, polychloroprene fabrics, polyamide fabrics, leather, wood, textiles	Bonding synthetic rubbers and porous materials. Primer for polyamide-coated fabrics such as nylon, terylene	
Polychloroprene	Air-dried 10–20 min at 20°C		Rubber, steel, wood, concrete	Bonding all types of rubber flooring to metals, woods, and masonry	Good heat resistance
Saturated polyester + isocyanate catalyst in ethyl acetate	Solvent evaporation and press cured at 40–80°C when tacky		Cellulose, cellulose acetate, polyolefins (treated film), polyvinyl chloride (rigid), paper, aluminum (foil), copper (foil)	Lamination of plastic films to themselves and metal foils for packaging industry, printed circuits	Resistant to heat, moisture, and many solvents

Table 6.8.3 Properties and Uses of Various Adhesives *(Continued)*

Basic type	Curing cycle, time at temp	Service temp range, °C	Adherends	Main uses	Remarks
		Thermoplastic *(Continued)*			
Cyanoacrylate *(anaerobic)*	15 s to 10 min at 20°C substrate-dependent	Melts at 165	Steel-steel, steel-aluminum, aluminum-aluminum, butyl rubber-phenolic	Rapid assembly of metal, glass, plastics, rubber components	*Anaerobic* adhesive. Curing action is based on the rapid polymerization of the monomer under the influence of basic catalysts. Absorbed water layer on most surfaces suffices to initiate polymerization and brings about bonding
Polyacrylate resin *(anaerobic)*	3 min at 120°C to 45 min at 65°C or 7 days at 20°C	−55–95	Aluminum-aluminum	Assembly requirements requiring high resistance to impact or shock loading. Metals, glass, and thermosetting plastics	*Anaerobic* adhesive
		Thermosetting			
Urea formaldehyde	9 h at 10°C to 1 h at 21°C after mixing powder with water (22%)		Wood, phenolic laminate	Wood gluing and bonding on plastic laminates to wood. Plywood, chipboard manufacture. Boat building and timber engineering	Excess glue may be removed with soapy water
Phenolic formaldehyde + catalyst PX-2Z	Cold setting		Wood	Timber and similar porous materials for outdoor-exposure conditions. Shop fascia panels	Good resistance to weathering and biodeterioration
Resorcinol formaldehyde + catalyst RXS-8	Cured at 16°C to 80°C under pressure		Wood, asbestos, aluminum, phenolic laminate, polystyrene (foam), polyvinyl chloride, polyamide (rigid)	Constructional laminates for marine craft. Building and timber applications. Aluminum-plywood bonding. Laminated plastics	Recommended for severe outdoor-exposure conditions
Epoxy resin + catalyst	24–48 h at 20°C to 20 min at 120°C	100	Steel, glass, polyester-glass fiber composite, aluminum-aluminum	General-purpose structural adhesive	
Epoxy resin + catalyst	8 h at 24°C to 2 h at 66°C to 45 min at 121°C	65	Steel, copper, zinc, silicon carbide, wood, masonry, polyester-glass fiber composite, aluminum-aluminum	Bonding of metals, glass, ceramics, and plastic composites	Cures to strong, durable bond
Epoxy + steel filler (80% w/w)	1–2 h at 21°C	120	Iron, steel, aluminum, wood, concrete, ceramics, aluminum-aluminum	Industrial maintenance repairs. Metallic tanks, pipes, valves, engine casings, castings	Good resistance to chemicals, oils, water
Epoxy + amine catalyst (ancamine LT)	2–7 days at 20°C for 33% w/w catalyst content	−5–60	Concrete, stonework	Repair of concrete roads and stone surfaces	Excellent pigment-wetting properties. Effective under water and suited to applications under adverse wet or cold conditions
Epoxy resin (modified)	4–5 h at 149°C to 20 min at 230°C to 7 min at 280°C	150	Aluminum, steel, ceramics	One-part structural adhesive for high-temperature applications	Good gap-filling properties for poorly fitting joints. Resistant to weather, galvanic action
Epoxy	45 s at 20°C		Gem stones, glass, steel, aluminum-aluminum	Rapid assembly of electronic components, instrument parts, printed circuits. Stone setting in jewelry, and as an alternative to soldering	
Epoxy resin in solvent + catalyst	8 h at 52°C to ½ h at 121°C	−270–371	Aluminum and magnesium alloys for elevated-temperature service	Strain gages for cryogenic and elevated-temperature use. Micro measurement strain gages	Cured material resists outgassing in high vacuum
Epoxy polyamide	8 h at 20°C to 15 min at 100°C	100	Copper, lead, concrete, glass, wood, fiberglass, steel-steel, aluminum-aluminum	Metals, ceramics, and plastics bonding. Building and civil engineering applications	Resists water, acids, oils, greases

Table 6.8.3 Properties and Uses of Various Adhesives (Continued)

Basic type	Curing cycle, time at temp	Service temp range, °C	Adherends	Main uses	Remarks
			Thermosetting (Continued)		
Epoxy/polysulfide	24 h at 20°C to 3 h at 60°C to 20 min at 100°C		Asbestos (rigid), ceramics, glass-fiber composites, carbon, polytetrafluoroethylene (treated), polyester (film), polystyrene (treated), rubber (treated), copper (treated), tungsten carbide, magnesium alloys, aluminum-aluminum, steel (stainless)-steel	Cold-setting adhesive especially suitable for bonding materials with differing expansion properties	Cures to flexible material. Resistant to water, petroleum, alkalies, and mild acids
Phenol furfural + acid catalyst	2 days at 21°C		Alumina, carbon (graphite)	Formulation of chemically resistant cements. Bedding and joining chemically resistant ceramic tiles	Extremely resistant to abrasion and heat
			Pressure-sensitive		
	Heated by air drying for several hours or 15–30 min at 210°F		Teflon-Teflon, Teflon-metal		Good resistance to acids and alkalies. Excellent electrical properties
			Miscellaneous		
Ceramic-based	Dried for ½ h at 77°C and cured ½ h at 100°C + 1 h at 200°C + 1 h at 250°C. Postcured, 1 h at 350°C	816	Metals	Strain gages, temperature sensors for elevated-temperature work	

SOURCE: Adapted from J. Shields, "Adhesives Handbook," CRC Press (Division of The Chemical Rubber Co., 1970), with the permission of the publisher.

from inorganic, nonmetallic substances and hardened in its finished shape by heat or chemical action. Note that the term is also used collectively for a number of such units, as "a carload of brick." In the present state of the art, the term brick, when used without a qualifying adjective, should be understood to mean such a unit, or a collection of such units, made from clay or shale hardened by heat. When other substances are used, the term brick should be suitably qualified unless specifically indicated by the context.

It is recognized that unless suitably qualified, a brick is a unit of burned clay or shale.

Brick (Common) Any brick made primarily for building purposes and not especially treated for texture or color, but including clinker and oven-burn brick.

Brick (Facing) A brick made especially for facing purposes, usually treated to produce surface texture or made of selected clays or otherwise treated to produce the desired color.

Brick are manufactured by the dry-press, the stiff-mud, or the soft-mud process. The **dry-press brick** are made in molds under high pressure and from relatively dry clay mixes. Usually all six surfaces are smooth and even, with geometrical uniformity. The **stiff-mud brick** are made from mixes of clay or shale with more moisture than in the dry-press process, but less moisture than used in the soft-mud process. The clay is extruded from an auger machine in a ribbon and cut by wires into the required lengths. These brick may be side-cut or end-cut, depending on the cross section of the ribbon and the length of the section cut off. The two faces cut by wires are rough in texture, the other faces may be smooth or artificially textured. The **soft-mud process** uses a wet mix of clay which is placed in molds under slight pressure.

Brick are highly resistant to freezing and thawing, to attacks of acids and alkalies, and to fire. They furnish good thermal insulation and good insulation against sound transference.

Paving brick are made of clay or shale, usually by the stiff-mud or dry-press process. Brick for use as paving brick are burned to vitrification. The common requirements are as follows: size, $8\frac{1}{2} \times 2\frac{1}{2} \times 4$, $8\frac{1}{2} \times 3 \times 3\frac{1}{2}$, $8\frac{1}{2} \times 3 \times 4$, with permissible variations of $\frac{1}{8}$ in in either transverse dimension, and $\frac{1}{4}$ in in length.

Although brick have always been used in construction, their use has been limited, until recently, to resisting compressive-type loadings. By adding steel in the mortar joints to take care of tensile stresses, **reinforced brick masonry** extends the use of brick masonry to additional types of building construction such as floor slabs.

Sand-lime brick are made from a mixture of sand and lime, molded under pressure and cured under steam at 200°F. They are usually a light gray in color and are used primarily for backing brick and for interior facing.

Cement brick are made from a mixture of cement and sand, manufactured in the same manner as sand-lime brick. In addition to their use as backing brick, they are used where there is no danger of attack from acid or alkaline conditions.

Firebrick (see this section, Refractories).

Specialty Brick A number of types of specialty brick are available for important uses, particularly where refractory characteristics are needed. These include **alumina** brick, **silicon carbide** brick, and **boron carbide** brick.

Other Structural Blocks Important structural units which fall outside the classic definition of brick are as follows: **Concrete blocks** are made with portland cement as the basic binder. Often, these are referred to as **concrete masonry units** (CMUs). The choice of aggregate ranges from relatively dense aggregate such as small crushed stone, small gravel, or coal cinders (the latter, if available) to light aggregate such as sand, limestone tailings, or the like. **Gypsum blocks** are generally used for fire protection in non-load-bearing situations. **Glass blocks** are available where transparency is desired. **Structural clay tile** is widely used in both load-bearing and non-load-bearing situations; a variety of complex shapes are available for hollow-wall construction. For classification of all these products with respect to dimensions and specifications for usage, the references should be consulted.

Brick and block can come with porcelain glazed finish for appearance, ease of cleaning, resistance to weathering, corrosion, etc.

Brick panels (prefabricated brick walls) are economic and find extensive use in modern high-rise construction. Such panels can be constructed on site or in factories.

New tile units requiring no mortar joint, only a thin epoxy line, form walls which resemble brick and can be installed three times as fast as conventional masonry.

For further definitions relating to structural clay products, the reader is referred to ANSI/ASTM C43-70(75).

See Table 6.8.4 for brick properties.

CERAMICS

REFERENCES: Kingery, "Introduction to Ceramics," Wiley. Norton, "Elements of Ceramics," Addison-Wesley. "Carbon Encyclopedia of Chemical Technology," Interscience. Humenik, "High-Temperature Inorganic Coatings," Reinhold. Kingery, "Ceramic Fabrication Processes," MIT. McCreight, Rauch, Sr., Sutton, "Ceramic and Graphite Fibers and Whiskers; A Survey of the Technology," Academic. Rauch, Sr., Sutton, McCreight, "Ceramic Fibers and Fibrous Composite Materials," Academic. Hague, Lynch, Rudnick, Holden, Duckworth (compilers and editors), "Refractory Ceramics for Aerospace; A Material Selection Handbook," The American Ceramic Society. Waye, "Introduction to Technical Ceramics," Maclaren and Sons, Ltd., London. Hove and Riley, "Ceramics for Advanced Technologies," Wiley. McMillan, "Glass-Ceramics," Academic. *Machine Design,* May 1983. "1986 Annual Book of ASTM Standards," vols. 15.02, 15.04 (Ceramics).

Ceramic materials are a diverse group of nonmetallic, inorganic solids with a wide range of compositions and properties. Their structure may be either crystalline or glassy. The desired properties are often achieved by high-temperature treatment (firing or burning).

Traditional ceramics are products based on the silicate industries, where the chief raw materials are naturally occurring minerals such as the clays, silica, feldspar, and talc. While silicate ceramics dominate the industry, newer ceramics, sometimes referred to as **electronic** or **technical** ceramics, are playing a major role in many applications. **Glass ceramics** are important for electrical, electronic, and laboratory-ware uses. Glass ceramics are melted and formed as glasses, then converted, by controlled nucleation and crystal growth, to polycrystalline ceramic materials.

Manufacture Typically, the manufacture of traditional-ceramic products involves blending of the finely divided starting materials with water to form a plastic mass which can be formed into the desired shape. The plasticity of clay constituents in water leads to excellent forming properties. Formation processes include extrusion, pressing, and ramming. Unsymmetrical articles can be formed by "slip-casting" techniques, where much of the water is taken up by a porous mold. After the water content of formed articles has been reduced by drying, the ware is **fired** at high temperature for fusion and/or reaction of the components and for attainment of the desired properties. Firing temperatures can usually be considerably below the fusion point of the pure components through the use of a **flux,** often the mineral feldspar. Following burning, a vitreous ceramic coating, or **glaze,** may be applied to render the surface smooth and impermeable.

Quite different is the **fusion casting** of some refractories and refractory blocks and most glasses, where formation of the shaped article is carried out after fusion of the starting materials.

Properties The physical properties of ceramic materials are strongly dependent on composition, microstructure (phases present and their distribution), and the history of manufacture. Volume pore concentration can vary widely (0 to 30 percent) and can influence shock resistance, strength, and permeability. Most traditional ceramics have a glassy phase, a crystalline phase, and some porosity. The last can be eliminated at the surface by glazing. Most ceramic materials are resistant to large compressive stresses but fail readily in tension. Resistance to abrasion, heat, and stains, chemical stability, rigidity, good weatherability, and brittleness characterize many common ceramic materials.

Products Many traditional-ceramic products are referred to as **whiteware** and include pottery, semivitreous wares, electrical porcelains, sanitary ware, and dental porcelains. Building products which are ceramic include brick and structural tile and conduit, while refractory blocks, as well as many abrasives, are also ceramic in nature. Porcelain **enamels** for metals are opacified, complex glasses which are designed to match the thermal-expansion properties of the substrate. Important **technical ceramics** include magnetic ceramics, with magnetic properties but relatively high electrical resistance; nuclear ceramics, including uranium dioxide fuel elements; barium titanate as a material with very high dielectric constant. Several of the pure oxide ceramics with superior physical properties are being used in electrical and missile applications where high melting and deformation temperatures and stability in oxygen are important. Fibrous ceramic composed of ziconium oxide fibers, such as Zircar, provide optimum combination of strength, low thermal conductivity, and high temperature resistance to about 2,490°C (4,500°F). Partially stabilized zirconia, because of its steel-strong and crack-resisting properties, is finding high-temperature applications, for example, in diesel engines.

Various forms of ceramics are available from manufacturers such as paper, "board," blankets, tapes, and gaskets.

A fair amount of development is under way dealing with **reinforcing ceramics** to yield **ceramic matrix composite** materials. The driving interest here is potential applications in high-temperature, high-stress environments, such as those encountered in aircraft and aerospace structural components. None of the end products are available yet "off the shelf," and applications are likely to be quite circumscribed and relegated to solution of very specialized design problems.

Ceramic inserts suitably shaped and mechanically clamped in tool holders are applied as cutting tools, and they find favor especially for cutting abrasive materials of the type encountered in sand castings or forgings with abrasive oxide crusts. They offer the advantage of resisting abrasion at high tool/chip interface temperatures, and consequently they exhibit relatively long tool life between sharpenings. They tend to be somewhat brittle and see limited service in operations requiring interrupted cuts; when they are so applied, tool life between sharpenings is very short.

Table 6.8.4 Building Brick Made from Clay or Shale
(Standard specifications, ANSI/ASTM C62-81)

Designation*	Min compressive strength (brick flatwise), lb/in² gross area		Max water absorption by 5-h boil, %		Max saturation coefficient†	
	Avg of 5	Indiv.	Avg of 5	Indiv.	Avg of 5	Indiv.
Grade SW	3,000	2,500	17.0	20.0	0.78	0.80
Grade MW	2,500	2,200	22.0	25.0	0.88	0.90
Grade NW	1,500	1,250	No limit		No limit	

* Grade SW includes brick intended for use where a high degree of resistance to frost action is desired and the exposure is such that the brick may be frozen when permeated with water.
 Grade MW includes brick intended for use where exposed to temperatures below freezing but unlikely to be permeated with water or where a moderate and somewhat nonuniform degree of resistance to frost action is permissible.
 Grade NW includes brick intended for use as backup or interior masonry, or if exposed, for use where no frost action occurs; or if frost action occurs, where the average annual precipitation is less than 20 in.
 † The saturation coefficient is the ratio of absorption by 24 h submersion in cold water to that after 5 h submersion in boiling water.

CLEANSING MATERIALS

REFERENCES: McCutcheon, "Detergents and Emulsifiers," McCutcheon, Inc. Schwartz, Perry, and Berch, "Surface Active Agents and Detergents," Interscience. *ASTM Special Tech. Pub.* 197. Niven, "Industrial Detergency," Reinhold. McLaughlin, "The Cleaning, Hygiene, and Maintenance Handbook," Prentice-Hall. Hackett, "Maintenance Chemical Specialties," Chemical Publishing Co. Bennett (ed.), "Cold Cleaning with Halogenated Solvents," *ASTM Special Technical Publication* 403, 1966. ANSI/ASTM D459, D534-1979 (definitions and specifications).

Cleansing is the removal of dirt, soil, and impurities from surfaces of all kinds. Means of soil attachment to the surface include simple entrapment in interstices, electrostatically held dirt, wetting of the surface with liquid soils, and soil-surface chemical reaction. A variety of cleansing systems has been developed which are difficult to classify since several soil-removal mechanisms are often involved. A liquid suspending medium, an active cleansing agent, and mechanical action are usually combined. The last may involve mechanical scrubbing, bath agitation, spray impingement, or ultrasonic energy. Cleansing involves detachment from the surface, suspension of solids or emulsification of liquids, or dissolution, either physical or by chemical reaction.

Organic Solvents

Both **petroleum solvents** (mineral spirits or naphtha) and **chlorinated hydrocarbons** (trichloroethylene and perchloroethylene) are used to remove solvent-soluble oils, fats, waxes, and greases as well as to flush away insoluble particles. Petroleum solvents are used in both soak-tank and spray equipment. Chlorinated hydrocarbons are widely used in vapor degreasing (where the metal stock or part is bathed in the condensing vapor), their high vapor density and nonflammability being advantageous. Solvent recovery by distillation can be used if the operation is of sufficient size. Both solvent types are used in garment **dry cleaning** with solvent-soluble detergents.

Fluorocarbons such as UCON solvent 113-LRI (trichlorotrifluoroethane) are nonflammable and are ideal for critical cleaning of mechanical electrical and electronic equipment, especially for white-room conditions.

Fluorocarbon solvents are highly proscribed and regulated and are to be used only with utmost care, under close supervision and accountable control. The reader should become familiar with the several references pertaining to health hazards of industrial materials listed under Chlorinated Solvents in this section.

Emulsifiable solvent cleaners contain a penetrating solvent and dissolved emulsifying agent. Following soaking, the surface is flushed with hot water, the resulting emulsion carrying away both soil and solvent. These cleaners are usually extended with kerosinelike solvents.

Alkali Cleansers

Alkali cleansers are water-soluble inorganic compounds, often strong cleansers. Carbonates, phosphates, pyrophosphates, and caustic soda are common, with numerous applications in plant maintenance, material processing, and process-water treatment. Cleansing mechanisms vary from beneficial water softening and suspension of solids to chemical reaction in the solubilization of fats and oils. Certain of these compounds are combined with **detergents** to improve efficiency; these are referred to as **builders**. Boiler and process equipment scales can sometimes be controlled or removed with selected alkali cleansers.

Synthetic Detergents

Detergents concentrate strongly at a solid-liquid or liquid-liquid interface and are thus characterized as **surface active**. In contrast to soaps, they can be tailored to perform over a wide range of conditions of temperature, acidity, and presence of dissolved impurities with little or no foaming. Detergents promote **wetting** of the surface by the suspending medium (usually water), **emulsification** of oils and greases, and **suspension** of solids without redeposition, the last function being the prime criterion of a good detergent. Detergents are classified as **anionic** (nega-

tively charged in solution), **cationic** (positively charged), and **nonionic**. Germicidal properties may influence detergent choice.

For specific applications, the supplier should be consulted since there is a large variety of available formulations and since many detergent systems contain auxiliary compounds which may be diluents, foam promotors, or alkali chemicals. Additives are designed for pH control, water softening, and enhanced suspending power. Formulations have been developed for cleaning food and dairy process equipment, metals processing, metal cleaning prior to electroplating, textile fiber and fabric processing, and industrial building maintenance. Strong, improperly selected detergent systems can cause deterioration of masonry or marble floors, aluminum window frames, water-based paints, and floor tiles, whereas detergents matched to the job at hand can result in increased plant efficiency.

Soaps, the oldest surface-active cleansers, lack the versatility of synthetic detergents but are widely used in the home and in the laundry industry. Properties depend on the fat or oil and alkali used in their preparation and include solvent-soluble soaps for dry cleaning.

Chemical cleaners which attack specific soils include dilute acid for metal oxide removal, for the cleaning of soldered or brazed joints, and for the removal of carbonate scale in process equipment. Oxalic acid (usually with a detergent) is effective on rust.

Chelating agents are organic compounds which complex with several metal ions and can aid in removal of common boiler scales and metal oxides from metal surfaces.

Steam cleaning in conjunction with a detergent is effective on grease-laden machinery.

There are numerous cleansers for the **hands,** including Boraxo, soap jelly and sawdust, lard for loosening oil grime, and linseed oil for paints; repeated use of solvents can be hazardous.

Alcohol ethoxylates are nonionic surfactants suitable for use in the production of maintenance and institutional cleaners. Products such as Tergitol and Neodol 23-6.5 serve in the formulation of liquid detergents for household and industrial uses.

Alcohol ethoxysulfates are anionic surfactants which are suited to the formulation of high-foaming liquid detergents, as for manual dishwashing. They offer the advantages of excellent solubility and biodegradability. Neodol 25-3S40 also exhibits low sensitivity to water hardness.

Enzyme cleaners containing a combination of bacteria culture, enzymes, and nutrients are used to dissolve grease, human waste, and protein stains.

CORDAGE

REFERENCE: Himmelfarb, "The Technology of Cordage Fibers and Rope," Textile Book Publishers, Inc. (division of Interscience Publishers, Inc.).

The term **cordage** denotes any flexible string or line. Usage includes wrapping, baling, hauling, and power transmission in portable equipment. **Twine** and **cord** generally imply lines of ⅜ in diam or less, with larger sizes referred to as **rope**.

Natural fibers used in cordage are abaca, sisal, hemp, cotton, and jute. For heavy cordage abaca and manila predominate. Hemp is used for small, tarred lines, and henequen for agricultural binder twine.

Rope is made by twisting yarns into strands, with the strands (usually three) **twisted** (laid) into a line. The line twist may be S or Z (see Sec. 10), generally opposite to the twist of the strands, which, in turn, is opposite to the twist of the yarns. The term **lay** designates the number of turns of the strands per unit length of rope but may also characterize the rope properties, a function of the degree of twist of each component. Grades range from **soft lay** (high ultimate strength) to **hard lay** (high abrasion resistance). Cable-laid rope results from twisting together conventional, three-strand rope.

Synthetic fibers are used in cordage because of resistance to rot, high strength, and other special properties. These fibers include nylon for strength, polyester (Dacron) for strength and dimensional stability, vinyls for chemical-plant use, fiberglass for electrical stability, vinyls for chemical-plant use, fiberglass for electrical and chemical properties, and polypropylene for strength and flotation (see Sec. 8).

Braided cordage has been used largely for small diameter lines such as sash cord and clothesline. However, braided lines are now available in larger diameters between 2 and 3 in. The strength of braided rope is slightly superior, and the line has less tendency to elongate in tension and cannot rotate or unlay under load. These properties are balanced against a somewhat higher cost than that of twisted rope.

A no. 1 common-lay rope will conform to the strength and weight table of the Federal Specification TR601A listed below.

For comparison, a **nylon rope** of a given diameter will have about 3 times the breaking strength given above and a **polyester rope** about 2½ times the strength of manila. Both have substantially greater flex and abrasion resistance.

See Table 6.8.5 for manila rope properties.

ELECTRICAL INSULATING MATERIALS

REFERENCES: Plastics Compositions for Dielectrics, *Ind. Eng. Chem.*, **38**, 1946, p. 1090. High Dielectric Ceramics, *Ind. Eng. Chem.*, **38**, 1946, p. 1097. Polystyrene Plastics as High Frequency Dielectrics, *Ind. Eng. Chem.*, **38**, 1946, p. 1121. Paper Capacitors Containing Chlorinated Impregnants, *Ind. Eng. Chem.*, **38**, 1946, p. 1110. "Contributions of the Chemist to Dielectrics," National Research Council, 1947. National Research Council, Conference on Electrical Insulation, (1) Annual Report, (2) Annual Digest of Literature on Dielectrics. Von Hippel, "Dielectric Materials and Applications," Wiley. Birks (ed.), "Modern Dielectric Materials," Academic. Saums and Pendelton, "Materials for Electrical Insulating and Dielectric Functions," Hayden Book, Inc. Licari, "Plastic Coatings for Electronics," McGraw-Hill. Clark, "Insulating Materials for Design and Engineering Practice," Wiley. Mayofis, "Plastic Insulating Materials," Illiffe, Ltd., London. Bruins (ed.), "Plastic for Electrical Insulation," Interscience Publishers. Swiss Electrochemical Committee, "Encyclopedia of Electrical Insulating Materials," *Bulletin of the Swiss Association of Electrical Engineers,* vol. 48, 1958. ISO 455-3-1 1981. "1986 Annual Book of ASTM Standards," vols. 10.01–10.03 (Electrical Insulating Materials).

The insulating properties of any material are dependent upon **dielectric strength,** or the ability to withstand high voltages without breakdown; **ohmic resistance,** or the ability to prevent leakage of small currents; and **power loss,** or the absorption of electrical energy that is transformed into heat. Power loss depends upon a number of influences, particularly the molecular symmetry of the insulation and frequency of the voltage, and is the basis of power factor, an important consideration whenever efficient handling of alternating currents is concerned, and a dominating consideration when high frequencies are used, as in radio circuits. Materials may have one of these qualities to a far greater extent than the other; e.g., air has a very high specific resistance but very little dielectric strength and no power loss at any frequency; glass has great dielectric strength yet much lower resistance than air. The **ideal insulator** is one having the maximum dielectric strength and resistance, minimum power loss, and also mechanical strength and chemical stability. Moisture is by far the greatest enemy of insulation; consequently the absence of hygroscopic quality is desirable.

The common insulating materials are described below. For their electrical properties, see Sec. 15.

Rubber See Rubber and Rubberlike Materials.

Mica and Mica Compounds Mica is a natural mineral varying widely in color and composition, and occurs in sheets that can be subdivided down to a thickness of 0.00025 in. White mica is best for electrical purposes. The green shades are the softest varieties, and the white amber from Canada is the most flexible. Mica has high insulating qualities, the best grades having a dielectric strength of 12,000 V per 0.1 mm. Its lack of flexibility, its nonuniformity, and its surface leakage are disadvantages. To offset these, several mica products have been developed, in which small pieces of mica are built up into finished shapes by means of binders such as shellac, gum, and phenolic resins.

Micanite consists of thin sheets of mica built into finished forms with insulating cement. It can be bent when hot and machined when cold, and is obtainable in thicknesses of 0.01 to 0.12 in. Flexible micanite plates, cloth, and paper are also obtainable in various thicknesses. **Megohmit** is similar to micanite except that it is claimed not to contain adhesive matter. It can be obtained in plates, paper, linen, and finished shapes. **Megotalc,** built up from mica and shellac, is similar to the above-named products and is obtainable in similar forms.

Insulating Varnishes Two general types of insulating varnish are used: (1) asphalt, bitumen, or wax, in petroleum solvent, and (2) drying-oil varnishes based on natural oils compounded with resins from natural or synthetic sources. Varnishes have changed greatly in the last few years, since new oils have become available, and particularly since phenolic and alkyd resins have been employed in their manufacture.

Silicone varnishes harden by baking and have electrical properties similar to those of the phenol-aldehyde resins. They are stable at temperatures up to 300°F (138°C). They can be used as wire coatings, and such wire is used in the manufacture of motors that can operate at high temperatures.

Impregnating Compounds Bitumens and waxes are used to impregnate motor and transformer coils, the melted mix being forced into the coil in a vacuum tank, forming a solid insulation when cooled. Brittle compounds, which gradually pulverize owing to vibration in service, and soft compounds, which melt and run out under service temperatures, should be avoided as far as possible.

Oil Refined grades of petroleum oils are extensively used for the insulation of transformers, switches, and lightning arresters. The following specification covers the essential points:

Table 6.8.5 Weight and Strength of Different Sizes of Manila Rope Specification Values

Approx diam, in*	Circumference, in*	Max net weight, lb/ft	Min breaking strength, lb†	Approx diam, in*	Circumference, in*	Max net weight, lb/ft	Min breaking strength, lb	Approx diam, in	Circumference, in	Max net weight, lb/ft	Min breaking strength, lb
3/16	5/8	0.015	450	13/16	2½	0.195	6,500	1 13/16	5½	0.895	26,500
¼	¾	0.020	600	15/16	2¾	2.225	7,700	2	6	1.08	31,000
5/16	1	0.029	1,000	1	3	0.270	9,000	2¼	7	1.46	41,000
3/8	1⅛	0.041	1,350	1 1/16	3¼	0.313	10,500	2⅝	8	1.91	52,000
7/16	1¼	0.053	1,750	1 3/16	3½	0.360	12,000	3	9	2.42	64,000
½	1½	0.075	2,650	1¼	3¾	0.418	13,500	3¼	10	2.99	77,000
9/16	1¾	0.104	3,450	1 5/16	4	0.480	15,000	3⅝	11	3.67	91,000
11/16	2	0.133	4,400	1½	4½	0.600	18,500	4	12	4.36	105,000
¾	2¼	0.167	5,400	1⅝	5	0.744	22,500				

The approximate length of coil is 1,200 ft for diam 7/16 in and larger. For smaller sizes it is longer, up to 3,000 ft for 3/16 in diam.
* 1 in = 0.0254 m; 1 ft = 0.3048 m.
† 1 lbf = 4.448 N.
SOURCE: U.S. government specification TR601A, dated Nov. 26, 1935, formulated jointly by cordage manufacturers and government representatives.

Specific gravity, 0.860; flash test, not less than 335°F; cold test, not more than −10°C (14°F); viscosity (Saybolt) at 37.8°C (100°F), not more than 120 s; loss on evaporation (8 h at 200°F), not more than 0.5 percent; dielectric strength, not less than 35,000 V; freedom from water, acids, alkalies, saponifiable matter, mineral matter or free sulfur. Moisture is particularly dangerous in oil.

Petroleum oils are used for the impregnation of kraft or manila paper, after wrapping on copper conductors, to form high-voltage power cables for services up to 300,000 V. Oil-impregnated paper insulation is sensitive to moisture, and such cables must be lead-sheathed. The transmission of power at high voltages in underground systems is universally accomplished by such cables.

Chlorinated Hydrocarbons Chlorinated hydrocarbons have the advantage of being nonflammable and are used as filling compounds for transformers and condensers where this property is important. Chlorinated naphthalene and chlorinated diphenyl are typical of this class of material. They vary from viscous oils to solids, with a wide range of melting points. The reader is cautioned to investigate the use of such insulation carefully because of alleged carcinogenic propensities.

Impregnated Fabrics Fabrics serve as a framework to hold a film of insulating material and must therefore be of proper thickness, texture, and mechanical strength, and free from nap and acidity. A wide variety of drying varnishes is used for the impregnation, and the dipping is followed by baking in high-temperature towers. Varnished cambric is used for the wrapping of coils and for the insulation of conductors. These cables have high power factor and must be kept free of moisture, but they are desirable for resisting electrical surges.

Thermosetting substances of the phenol-aldehyde type and of the urea-formaldehyde type first soften and then undergo a chemical reaction which converts them quickly to a strong infusible product. Good properties are available in the phenolics, while numerous special types have been developed for high heat resistance, low-power-factor arc resistance, and other specialized properties. A wide variety of resins is available. The urea plastics are lacking in heat resistance but are suitable for general-purpose molding and have fair arc resistance.

Thermoplastic resins (see Plastics, below) are used for molding and extruding electrical insulations. They differ from the thermosetting resins in that they do not become infusible. **Polyethylene** softens between 99 and 116°C (210 and 240°F). Its dielectric strength and resistivity are high, its power factor is only 0.0003, and its dielectric constant is 2.28. It is used extensively in high-frequency and radar applications and as insulation on some power and communication cables. **Teflon** is a fluorocarbon resin which has electrical properties similar to those of polyethylene. Its softening point is 750°F approx, and it extrudes and molds with difficulty; however, more tractable grades of Teflon are now available. It is resistant to nearly all chemicals and solvents.

Nylon is a synthetic plastic with interesting mechanical and electrical properties. It has only fair water resistance. Its melting point is 198 to 249°C (390 to 480°F), and is used for the molding of coil forms. It can be extruded onto wire in thin layers. Such wires are used in place of the conventional varnish-coated magnet wires in coils and motors. Operation may be at temperatures up to 127°C (260°F).

Paper Except in lead-sheathed telephone cables, the present tendency is to use paper only as a backing or framework for an insulating film or compound, owing to its hygroscopic qualities. **Manila** and **kraft** papers possess the best dielectric and mechanical strength and, when coated with good insulating varnish, are excellent insulators. Various types of paraffined paper are used in condensers. (See also Paper, below.)

Silicone rubber (see Silicones) is a rubberlike material of good physical properties [tensile strength, 400 to 700 lb/in² (2.78 to 4.85 × 10⁶ N/m²) elongation, 200 percent]. It can be operated for long periods of time at temperatures up to 138°C (300°F) or intermittently up to 249°C (480°F).

Ceramics and **glasses** find wide usage as insulating materials where brittleness and lack of flexibility can be tolerated.

Polymeric (plastic) films, particularly the polyester and fluorocarbon types, are being used increasingly where fabrication of the electrical component permits either wrapping or insertion of film chips.

Heat-shrinkable tubing of polyvinylchloride, polyolefin, or polytetrafluoroethylene compositions allow for very tight-fitting insulation around a member, thus affording efficient protection of electrical wires or cables.

FIBERS AND FABRICS
(See also Cordage.)

REFERENCES: Matthews-Mauersberger, "The Textile Fibers," Wiley. Von Bergen and Krauss, "Textile Fiber Atlas," Textile Book Publishers, Inc. Hess, "Textile Fibers and Their Use," Lippincott. Sherman and Sherman, "The New Fibers," Van Nostrand. Kaswell, "Textile Fibers, Yarns and Fabrics," Reinhold. "Harris' Handbook of Textile Fibers," Waverly House. Kaswell, "Wellington Sears Handbook of Industrial Textiles," Wellington Sears Co. Fiber Charts, *Textile World*, McGraw-Hill. ASTM Standards. C. Z. Carroll-Porcynski, "Advanced Materials; Refractory Fibers, Fibrous Metals, Composites," Chemical Publishing Co. Marks, Atlas, and Cernia, "Man-Made Fibers, Fibrous Metals, Composites," Chemical Publishing Co. Marks, Atlas, and Cernia, "Man-Made Fibers, Science and Technology," Interscience Publishers. Frazer, High Temperature Resistant Fibers, *Jour. Polymer Sci.*, Part C, Polymer Symposia, no. 19, 1966. Moncrieff, "Man-Made Fibers," Wiley. Preston and Economy, "High Temperature and Flame Resistant Fibers," Wiley. ANSI/ASTM Standards D1175-D 3940. ANSI/AATCC Standards 79, 128. "1986–1994 Annual Book of ASTM Standards," vols. 07.01, 07.02 (Textiles).

Fibers are threadlike structural materials, adaptable for spinning, weaving, felting, and similar applications. They may be of natural or synthetic origin and of inorganic or organic composition. **Fabrics** are defined by the ASTM as "planar structures produced by interlacing yarns, fibers, or filaments." A **bonded fabric** (or nonwoven fabric) consists of a web of fibers held together with a cementing medium which does not form a continuous sheet of adhesive material. A **braided fabric** is produced by interlacing several ends of yarns such that the paths of the yarns are not parallel to the fabric axis. A **knitted fabric** is produced by interlooping one or more ends of yarn. A **woven fabric** is produced by interlacing two or more sets of yarns, fibers or filaments such that the elements pass each other essentially at right angles and one set of elements is parallel to the fabric axis. A woven narrow fabric is 12 in or less in width and has a selvage on either side.

Inorganic Fibers

Asbestos is the only mineral fiber of natural origin. Its demonstrated carcinogenic properties, when its particles lodge in the lung, have led to its removal from most consumer products. Formerly ubiquitously applied as a thermal insulator, e.g., it has now been proscribed from that use and has been supplanted largely by fiberglass. One of the few remaining products in which it is still used is a component in automotive and aircraft brake shoe material. In that service, its superior heat-resistant property is paramount in the design of the brake shoes, for which application there is no readily available peer, as to both performance and economy. The quest for an equivalent substitute continues; eventually, asbestos may well be relegated to history or to some limited experimental and/or laboratory use. In this environmental climate regarding asbestos, for further information see Natalis, "The Asbestos Product Guide," A&E Insul-Consult Inc., Box 276, Montvale, NJ.

Synthetic mineral fibers are **spun glass, rock wool,** and **slag wool**. These fibers can endure high temperatures without substantial loss of strength. Glass fibers possess a higher strength-to-weight ratio at the elastic limit than do other common engineering materials.

Metal filaments (wires) are used as textile fibers where their particular material properties are important.

Carbon (graphite) fibers, such as Thornel, are high-modulus, highly oriented structures characterized by the presence of carbon crystallites (polycrystalline graphite) preferentially aligned parallel to the fiber axis. Depending on the particular grade, tensile strengths can range from 10.3 to 241 × 10⁷ N/m² (15,000 to 350,000 lb/in²). Hybrid composites of carbon and glass fibers find aerospace and industrial applications.

Zirconia fibers (ZrO_2 + HFO_2 + stabilizers), such as Zircar, have excellent temperature resistance to about 2,200°C (4,000°F), and fabrics woven from the fibers serve as thermal insulators.

Natural Organic Fibers

The **animal fibers** include **wool** from sheep, **mohair** from goats, **camel's hair**, and **silk**. Wool is the most important of these, and it may be processed to reduce its susceptibility to moth damage and shrinkage. Silk is no longer of particular economic importance, having been supplanted by one or another of the synthetic fibers for most applications.

The **vegetable fibers** of greatest utility consist mainly of cellulose and may be classified as follows: seed hairs, such as cotton; bast fibers, such as flax, hemp, jute, and ramie; and vascular fibers. Those containing the most cellulose are the most flexible and elastic and may be bleached white most easily. Those which are more lignified tend to be stiff, brittle, and hard to bleach. Vegetable fibers are much less hygroscopic than wool or silk.

Mercerized cotton is cotton fiber that has been treated with strong caustic soda while under tension. The fiber becomes more lustrous, stronger, and more readily dyeable.

Synthetic Organic Fibers

In recent years, a large and increasing portion of commercial fiber production has been of synthetic or semisynthetic origin. These fibers provide generally superior mechanical properties and greater resistance to degradation than do natural organic fibers and are available in a variety of forms and compositions for particular end-use applications. Generic categories of man-made fibers have been established by the Textile Fiber Products Identification Act, 15 USC 70, 72 Stat. 1717.

Among the important fibers are **rayons**, made from regenerated cellulose, plain or acetylated, which has been put into a viscous solution and extruded through the holes of a spinneret into a setting bath. The types most common at present are **viscose** and **acetate**. Cuprammonium rayon, saponified acetate rayon, and high-wet-modulus rayon are also manufactured and have properties which make them suitable for particular applications. Rayon is generally less expensive than other synthetic fibers.

In contrast to these regenerated fibers are a variety of polymer fibers which are chemically synthesized. The most important is **nylon**, including nylon 6.6, a condensation polymer of hexamethylene diamine-adipic acid, and nylon 6, a polymer of caprolactam. Nylon possesses outstanding mechanical properties and is widely used in industrial fabrics. Nomex, a high-temperature-resistant nylon retains its most important properties at continuous operating temperatures up to 260°C (500°F). Other polymer fibers possess mechanical or chemical properties which make them a specific material of choice for specialized applications. These include **polyester** fibers, made from polyethylene terephthalate; **acrylic** and **modacrylic** fibers, made from copolymers of acrylonitrile and other chemicals; **Saran**, a polymer composed essentially of vinylidene chloride; and the **olefins**, including polyethylenes and polypropylene.

Teflon, a fluorocarbon resin, has excellent chemical resistance to acids, bases, or solvents. Its useful physical properties range from cryogenic temperatures to 260°C (500°F), high dielectric strength, low dissipation factor, and high resistivity, along with the lowest coefficient of friction of any solid. It is nonflammable and is inert to weather and sunlight.

As part of their manufacture, substantially all synthetic fibers are **drawn** (hot- or cold-stretched) after extrusion to achieve desirable changes in properties. Generally, increases in draw increase breaking strength and modulus and decrease ultimate elongation.

Proximate Identification of Fibers Fibers are most accurately distinguished under the microscope, with the aid of chemical reagents and stains. A useful rough test is burning, in which the odor of burned meat distinguishes animal fibers from vegetable and synthetic fibers. Animal fibers, cellulose acetate, and nylon melt before burning and fuse to hard rounded beads. Cellulose fibers burn off sharply. Cellulose acetate dissolves in either acetone or chloroform containing some alcohol.

Heat Endurance Fibers of organic origin lose strength when heated over long periods of time above certain temperatures: cellulose, 149°C (300°F); cellulose acetate, 93.5°C (200°F); nylon, 224°C (435°F); casein, 100°C (212°F); and glass, 316°C (600°F).

Creep Textile fibers exhibit the phenomenon of creep at relatively low loads. When a textile fiber is subjected to load, it suffers three kinds of distortion: (1) an elastic deformation, closely proportioned to load and fully and instantly recoverable upon load removal; (2) a primary creep, which increases at a decreasing rate with time and which is fully, but not instantaneously, recoverable upon load removal; and (3) a secondary creep, which varies obscurely with time and load and is completely nonrecoverable upon load removal. The relative amounts of these three components, acting to produce the total deformation, vary with the different fibers. The two inelastic components give rise to mechanical hysteresis on loading and unloading.

Felts and Fabrics

A **felt** is a compacted formation of randomly entangled fibers. Wool felt is cohesive because the scaly structure of the wool fibers promotes mechanical interlocking of the tangled fibers. Felts can be made with blends of natural or synthetic fibers, and they may be impregnated with resins, waxes or lubricants for specific mechanical uses. Felts are available in sheets or cut into washers or shaped gaskets in a wide range of thicknesses and densities for packing, for vibration absorption, for heat insulation, or as holders of lubricant for bearings.

Fabrics are woven or knitted from **yarns**. Continuous-filament synthetic fibers can be made into monofilament or multifilament yarns with little or no twist. Natural fibers, of relatively short length, and synthetic **staple** fibers, which are purposely cut into short lengths, must be twisted together to form yarns. The amount of twist in yarns and the tension and arrangement of the weaving determine the appearance and mechanical properties of fabrics. Staple-fiber fabrics retain less than 50 percent of the intrinsic fiber strength, but values approaching 100 percent are retained with continuous-filament yarns. Industrial fabrics can be modified by mechanical and chemical treatments, as well as by coatings and impregnations, to meet special demands for strength and other mechanical, chemical, and electrical properties or to resist insect, fungus, and bacterial action and flammability.

Nomenclature There are literally dozens of different numbering systems for expressing the relationship between yarn weight and length, all differing and each used in connection with particular fiber types or in different countries. The most common currently used unit is the **denier**, which is the weight in grams of 9,000 m of yarn. A universal system, based on the **tex** (the weight in grams of 1,000 m of yarn), has been approved by the International Standards Organization and by the ASTM. Relative strengths of fibers and yarns are expressed as the **tenacity**, which reflects the specific gravity and the average cross section of the yarn. Units are grams per denier (gpd) or grams per tex (gpt).

Many textiles can sustain high-energy impact loads because of their considerable elongation before rupture. The total work done per unit length on a fiber or yarn which is extended to the point of rupture can be approximated by multiplying the specific strength by one-half the final extension of that length.

Yarn **twist** direction is expressed as **S** or **Z** twist, with the near-side helical paths of a twisted yarn held in a vertical position comparable in direction of slope to the center portion of one of these letters. Amount of twist is expressed in turns per inch (tpi).

Fabrics are characterized by the composition of the fiber material, the type of weave or construction, the count (the number of yarns per inch in the warp and the filling directions), and the weight of the fabric, usually expressed in ounces per running yard. **Cover factor** is the ratio of fabric surface covered by yarn to the total fabric surface. **Packing factor** is the ratio of fiber volume to total fabric volume.

Tables 6.8.6 to 6.8.8 give physical data and other information about commercial fibers and yarns. The tabulated quantities involving the denier should be regarded as approximate; they are not absolute values such as are used in engineering calculations.

With the continuing phasing out of asbestos products, one can substitute, under appropriate conditions, fiberglass woven fabrics. See Fiberglass in this section.

Table 6.8.6 Fiber Properties*

Kind	Source	Length of fiber, in	Width or diam of cells, μm	Specific gravity	Moisture regain,‡ %	Chemical description	Principal uses
Cotton	Plant seed hair	5/8–2	8–27	1.52	8.5	Cellulose	Industrial, household, apparel
Jute†	Plant bast	50–80	15–20	1.48	13.7	Lignocellulose	Bagging, twine, carpet backing
Wool	Animal	2–16	10–50	1.32	17	Protein	Apparel, household, industrial
Viscose	Manufactured	Any	8–43	1.52	11	Regenerated cellulose	Apparel, industrial, household
Cellulose acetate	Manufactured	Any	12–46	1.33	6	Cellulose ester	Apparel, industrial, household
Nylon	Manufactured	Any	8	1.14	4.2	Polyamide	Apparel, industrial, household
Casein	Manufactured	Any	11–28	1.3	4.1	Protein	Apparel
Flax†	Plant bast	12–36	15–17	1.5	12	Cellulose	Household, apparel, industrial
Hemp†	Plant bast	—	18–23	1.48	12	Cellulose	Twine, halyards, rigging
Sisal†	Plant leaf	30–48	10–30	—	—	Lignocellulose	Twine, cordage
Manila†	Plant leaf	60–140	10–30	—	—	Lignocellulose	Rope, twine, cordage
Ramie†	Plant bast	3–10	24–70	1.52	—	Cellulose	Household, apparel, seines
Silk	Silkworm	Any	5–23	1.35	11	Protein	Apparel, household, industrial
Glass	Manufactured	Any	3	2.5	0	Fused metal oxides	Industrial, household
Dacron	Manufactured	Any	8	1.38	0.4	Polyester	Apparel, industrial, household

1 in $= 0.025 + $ m; 1 $\mu = 10^{-6}$ m. The more up-to-date term for the micron (μ) is the micrometer (μm).
* Adapted from Smith, Textile Fibers, *proc. ASTM*, 1944; Appel, A Survey of the Synthetic Fibers, *Am. Dyestuff Reporter*, **34**, 1945, pp. 21–26; and other sources.
† These fibers are commercially used as bundles of cells. They vary greatly in width. Width figures given are for the individual cells.
‡ In air at 70°F and 65 percent relative humidity.

Table 6.8.7 Tensile Properties of Single Fibers

Fiber	Breaking tenacity, gpd	Extension at break, %	Elastic recovery at corresponding strain, %	Elastic modulus,* gpd
Glass	6.0–7.3	3.0–4.0	100 at 2.9	200–300
Fortisan (rayon)	6.0–7.0		100 at 1.2 60 at 2.4	150–200
Flax	2.6–7.7	2.7–3.3	65 at 2	
Nylon 6,6	4.6–9.2	16–32	100 at 8	25–50
Nylon 6	4.5–8.6	16–40	100 at 8	25–50
Silk	2.4–5.1	10–25	92 at 2	75–125
Saran	1.1–2.3	15–25	95 at 10	
Cotton	3.0–4.9	3–7	74 at 2	50–100
Steel (90,000 lb/in² T.S.)	0.9	28	—	300
Steel (music wire)	3.5	8	—	300
Viscose rayon	1.5–5.0	15–30	82 at 2	50–150
Wool	1.0–1.7	25–35	99 at 2	25–40
Acetate rayon	1.3–1.5	23–34	100 at 1	25–40
Polyester	4.4–7.8	10–25	100 at 2	50–80
Polypropylene	4.0–7.0	15–25	95 at 7	15–50
Polytetrafluoroethylene	1.7	13		—

* From Kaswell, "Textile Fibers, Yarns, and Fabrics," Reinhold.
SOURCE: Kaswell "Wellington Sears Handbook of Industrial Textiles," Wellington Sears Co., Inc.

FREEZING PREVENTIVES

Common salt is sometimes used to prevent the freezing of water; it does not, however, lower the freezing point sufficiently to be of use in very cold weather, and in concentrated solution tends to "creep" and to crystallize all over the receptacle. It also actively corrodes metals and has deleterious effects on many other materials, e.g., concrete. For freezing temperatures, see Sec. 18.

Calcium chloride ($CaCl_2$) is a white solid substance widely used for preventing freezing of solutions and (owing to its great hygroscopic power) for keeping sizing materials and other similar substances moist. It does not "creep" as in the case of salt. It does not rust metal but attacks solder.

Calcium chloride solutions are much less corrosive on metal if made alkaline by the use of lime, and also if a trace of sodium chromate is present. They are not suitable for use in automobile radiators, because of corrosive action while hot, and because of tendency of any spray therefrom to ruin the insulation of spark plugs and high-tension cables. For freezing temperatures, see Sec. 18.

Glycerol is a colorless, viscid liquid without odor and miscible with water in all proportions. It should have a specific gravity of approximately 1.25. It has no effect on metals but disintegrates rubber and loosens up iron rust.

Denatured alcohol is free from the disadvantages of calcium chloride, salt, and glycerin solutions, but is volatile from water mixtures which

Table 6.8.8 Temperature and Chemical Effects on Textiles

Fiber	Temperature limit, °F	Resistance to chemicals	Resistance to mildew
Cotton	Yellows 250; decomposes 300	Poor resistance to acids	Attacked
Flax	275	Poor resistance to acids	Attacked
Silk	Decomposes 300	Attacked	Attacked
Glass	Softens 1,350	Resists	Resists
Nylon 6	Sticky 400; melts 420–430	Generally good	Resists
Nylon 6,6	Sticky 455; melts 482	Generally good	Resists
Viscose rayon	Decomposes 350–400	Poor resistance to acids	Attacked
Acetate	Sticky 350–400; melts 500	Poor resistance to acids	Resists
Wool	Decomposes 275	Poor resistance to alkalies	Attacked
Asbestos	1,490	Resists	Resists
Polyester	Sticky 455; melts 480	Generally good	Resists
Polypropylene	Softens 300–310; melts 325–335	Generally good	Resists
Polyethylene	Softens 225–235; melts 230–250	Generally good	Resists
Jute	275	Poor resistance to acids	Attacked

Temperature conversion: $t_c = (5/9)(t_F - 32)$.
SOURCE: Fiber Chart, *Textile World*, McGraw-Hill, 1962.

Table 6.8.9 Nonfreezing Percentages by Volume in Solution

	Temperature, °C (°F)			
	−6.7 (20)	−12.2 (10)	−17.8 (0)	−23.4 (−10)
Methyl alcohol	13	20	25	30
Prestone	17	25	32	38
Denatured alcohol	17	26	34	42
Glycerol	22	33	40	47

run hot. A solution containing 50 percent alcohol becomes flammable, but it is rarely necessary to use more than 30 percent.

Methyl alcohol solutions were sold widely in the past for use as automotive antifreeze. It is an effective antifreeze, but its fumes are poisonous and it must be used with extreme care. It has been largely supplanted for this use by ethylene glycol solutions, which are applied in automotive practice and for stationary applications where, e.g., it is circulated in subgrade embedded pipes and thus inhibits formation of ice on walking surfaces.

Ethylene glycol (Prestone) is used as a freezing preventive and also permits the use of high jacket temperatures in aircraft and other engines. Sp gr 1.125 (1.098) at 32 (77)°F; boiling point, 387°F; specific heat, 0.575 (0.675) at 68 (212)°F. Miscible with water in all proportions.

See Table 6.8.9 for nonfreezing percentages by volume.

Winter-summer concentrates, such as Prestone 11, an ethylene glycol base combined with patented inhibitor ingredients, give maximum freezing protection to about −92°F (−62°C) at 68 percent mixture, and to about −34°F (−37°C) at 50 percent mixture.

Anti-icing of fuels for aircraft is accomplished by adding ethylene glycol monomethyl ether mixtures as Prist.

GLASS

REFERENCES: *Journal of American Ceramic Society,* Columbus, Ohio. *Journal of the Society of Glass Technology,* Sheffield, England. Morey, "Properties of Glass," Reinhold. Scholes, "Handbook of the Glass Industry," Ogden-Watney. "Non-Silica Glasses," *Chem. & Met. Eng.,* Mar. 1946. Phillips, "Glass the Miracle-Maker," Pitman. Long, "Propriétés physiques et fusion du verre," Dunod. Eitel-Pirani-Scheel, "Glastechnische Tabellen," Springer. Jebsen-Marwedel, "Glastechnische Fabrikationsfehler," Springer. Shand, "Glass Engineering Handbook," McGraw-Hill. Phillips, "Glass: Its Industrial Applications," Reinhold. Persson, "Flat Glass Technology," Plenum Press, 1969. McMillan, "Glass-Ceramics," Academic. Pye, Stevens, and La Course (eds.), "Introduction to Glass Science," Plenum Press. Jones, "Glass," Chapman & Hall. Doremus, "Glass Science," Wiley. Technical Staffs Division, "Glass," Corning Glass Works. ANSI/ASTM C "1986 Annual Book of ASTM Standards," vol. 15.02 (Glass).

Glass is an inorganic product of fusion which has cooled to a rigid condition without crystallizing. It is obtained by melting together silica, alkali, and stabilizing ingredients, such as lime, alumina, lead, and barium. Bottle, plate, and window glass usually contain SiO_2, Al_2O_3, CaO, and Na_2O. Small amounts of the oxides of manganese and selenium are added to obtain colorless glass.

Special glasses, such as fiberglass, laboratory ware, thermometer glass, and optical glass, require different manufacturing methods and different compositions. The following oxides are either substituted for or added to the above base glass: B_2O_3, ZnO, K_2O, As_2O_3, PbO, etc., to secure the requisite properties. Colored glasses are obtained by adding the oxides of iron, manganese, copper, selenium, cobalt, chromium, etc., or colloidal gold.

Molten glass possesses the ability to be fabricated in a variety of ways and to be cooled down to room temperature rapidly enough to prevent crystallization of the constituents. It is a rigid material at ordinary temperatures but may be remelted and molded any number of times by the application of heat. Ordinary glass is melted at about 1,430°C (2,600°F) and will soften enough to lose its shape at about 594°C (1,100°F).

Window glass is a soda-lime-silica glass, fabricated in continuous sheets up to a width of 6 ft. The sheets are made in two thicknesses, SS and DS, which are, respectively, $\frac{1}{16}$ and $\frac{1}{8}$ in. Both thicknesses are made in A, B, C, and D grades.

Reflective-coated glass, such as Vari-Tran, insulates by reflecting hot sun in summer. Various colors are available, with heat transmissions ranging from 8 to 50 percent.

Borosilicate glass is the oldest type of glass to have significant heat resistance. It withstands higher operating temperatures than either limed or lead glasses and is also more resistant to chemical attack. It is used for piping, sight glasses, boiler-gage glasses, sealed-beam lamps, laboratory beakers, and oven cooking ware. This type of glass was the first to carry the Pyrex trademark.

Aluminosilicate glass is similar to borosilicate glass in behavior but is able to withstand higher operating temperatures. It is used for top-of-stove cooking ware, lamp parts, and when coated, as resistors for electronic circuitry.

Plate glass and float glass are similar in composition to window glass. They are fabricated in continuous sheets up to a width of 15 ft and are polished on both sides. They may be obtained in various thicknesses and grades, under names like Parallel-O-Plate and Parallel-O-Float.

Before the introduction of float glass, plate glass sheets were polished mechanically on both sides, which turned out to be the most expensive part of the entire production process. Pilkington, in England, invented the concept of pouring molten glass onto a bed of molten tin and to float it thereon—hence the name *float glass*. In that process, the underside of the glass conforms to the smooth surface of the molten tin surface, and the top surface flows out to a smooth surface. The entire grinding and polishing operations are eliminated, with attendant impact on reducing the cost of the final product. With rare exceptions, all plate glass is currently produced as float glass.

Safety glass consists of two layers of plate glass firmly held together by an intermediate layer of organic material, such as polyvinyl butyral. Safety glass is ordinarily ¼ in thick but can be obtained in various thicknesses. This plate is shatter proof and is used for windshields, bank cashier's windows, etc.

Tempered glass is made from sheet glass in thicknesses up to 1 in. It possesses great mechanical strength, which is obtained by rapidly chilling the surfaces while the glass is still hot. This process sets up a high compression on the glass surfaces, which have the capacity of withstanding very high tensile forces.

Wire glass is a glass having an iron wire screen thoroughly embedded in it. It offers about 1½ times the resistance to bending that plain glass does; even thin sheets may be walked on. If properly made, it does not fall apart when cracked by shocks or heat, and is consequently fire-resistant. It is used for flooring, skylights, fireproof doors, fire walls, etc.

Pressed glass is made by forming heat-softened glass in molds under pressure. Such articles as tableware, lenses, insulators, and glass blocks are made by this process.

Glass blocks find wide application for building purposes, where they form easy-to-clean, attractive, airtight, light-transmitting panels. Glass blocks, such as Vistabrik, are solid throughout, and are exceptionally rugged and virtually indestructible. Glass blocks manufactured to have an internal, partially depressurized air gap have energy-conserving, insulating qualities, with "thermal resistance" or R values ranging from 1.1 to almost 2.3 (h·ft²·°F)/Btu. Solar heat transmission can be varied within wide limits by using different colored glasses and by changing the reflection characteristics by means of surface sculpting. Standard blocks are 3⅞ in thick, and can be had in 6 by 6, 8 by 8, 12 by 12, 3 by 6, 4 by 8, and 4 by 12-in sizes. Thinner blocks are also available.

Fiberglass is a term used to designate articles that consist of a multitude of tiny glass filaments ranging in size from 0.0001 to 0.01 in in diam. The larger fibers are used in air filters; those 0.0005 in diam, for thermal insulation; and the 0.0001- to 0.0002-in-diam fibers, for glass fabrics, which are stronger than ordinary textiles of the same size. Insulating tapes made from glass fabric have found wide application in electrical equipment, such as motors and generators and for mechanical uses. Woven fiberglass has supplanted woven asbestos, which is no longer produced.

Glass is used for many structural purposes, such as store fronts and tabletops, and is available in thicknesses upward of $\frac{5}{16}$ in and in plates up to 72×130 in. Plates with customized dimensions to suit specific architectural requirements are available on special order.

Cellular glass is a puffed variety with about 14–15 million cells per cubic foot. It is a good heat insulator and makes a durable marine float. See also Sec. 4.

Vycor is a 96 percent silica glass having extreme heat-shock resistance to temperatures up to 900°C (1,652°F), and is used as furnace sight glasses, drying trays, and space-vehicle outer windows.

Fused silica (silicon dioxide) in the noncrystalline or amorphous state shows the maximum resistance to heat shock and the highest permissible operating temperature, 900°C (1,652°F) for extended periods, or 1,200°C (2,192°F) for short periods. It has maximum ultraviolet transmission and the highest chemical-attack resistance. It is used for astronomical telescopes, ultrasonic delay lines, and crucibles for growing single metal crystals.

Vitreous silica, also called **fused quartz** is made by melting rock crystal or purest quartz sand in the electric furnace. It is unaffected by changes of temperature, is fireproof and acid-resistant, does not conduct electricity, and has practically no expansion under heat. It is used considerably for high-temperature laboratory apparatus. See Plastics, below, for glass substitutes.

Glass ceramics, such as Pyroceram, are melted and formed as glasses, then converted by controlled nucleation and crystal growth to polycrystalline ceramic materials. Most are opaque and stronger than glass, and may also be chemically strengthened.

Leaded glass used to fabricate fine crystal (glasses, decanters, vases, etc.) has a lead content which has been shown to leach out into the liquid contents if the two are in contact for a long time, certainly in the range of months or longer. Often the optical properties of this glass are enhanced by elaborate and delicate designs cut into the surface. The addition of lead to the base glass results in a softer product which lends itself more readily to such detailed artisanry.

Glass beads are used extensively for reflective paints.

Properties of Glass Glass is a brittle material and can be considered perfectly elastic to the fracture point. The range of Young's modulus is 4 to 14×10^3 kg/mm² (6 to 20×10^6 lb/in²), with most commercial glasses falling between 5.5 and 9.0×10^3 kg/mm² (8 and 13×10^6 lb/in²). Theory predicts glass strength as high as 3,500 kg/mm² (5×10^6 lb/in²); fine glass fibers have shown strengths around 700 kg/mm² (10^6 lb/in²); however, glass products realize but a fraction of such properties owing to surface-imperfection stress-concentration effects. Often design strengths run from 500 to 1,000 times lower than theoretical. Glass strength also deteriorates when held under stress in atmospheric air (static fatigue), an apparent result of reaction of water with glass; high-strength glasses suffer the greater penalty. Glass also exhibits a time-load effect, or creep, and may even break when subjected to sustained loads, albeit the stresses induced may be extremely low. There exist samples of very old glass (centuries old) which, while not broken, show evidence of plastic flow albeit at ambient temperature all the while. A cylindrical shape, e.g., will slowly degenerate into a teardrop shape, with the thickening at the bottom having been caused by the deadweight alone of the piece of glass. At room temperatures the thermal conductivity of glasses ranges from 1.6×10^{-3} to 2.9×10^{-3} cal/(s·cm²·°C/cm) [4.65 to 8.43 Btu/(h·ft²·°F/in)].

NATURAL STONES

REFERENCES: Kessler, Insley, and Sligh, *Jour. Res. NBS,* **25,** pp. 161–206. Birch, Schairer, and Spicer, Handbook of Physical Constants, *Geol. Soc. Am. Special Paper* 36. Currier, Geological Appraisal of Dimension Stone Deposits, *USGS Bull.* 1109. ANSI/ASTM Standards C99–C880. ''1986 Annual Book of ASTM Standards,'' vol. 04.08 (Building Stones).

A **stone** or a **rock** is a naturally occurring composite of minerals. Stone has been used for thousands of years as a major construction material because it possesses qualities of strength, durability, architectural adaptability, and aesthetic satisfaction. There are two principal branches of the natural-stone industry—**dimension** stone and **crushed** or broken stone. The uses of the latter vary from aggregate to riprap, in which stones in a broad range of sizes are used as structural support in a matrix or to provide weathering resistance. Dimension stones are blocks or slabs of stone processed to specifications of size, shape, and surface finish. The largest volume today lies in the use of slabs varying from 1 to 4 in in thickness that are mounted on a structure as a protective and aesthetic veneer.

There are two major types of natural stone: **igneous** and **metamorphic** stones, composed of tightly interlocking crystals of one or more minerals, and **sedimentary** rocks, composed of cemented mineral grains in which the cement may or may not be of the same composition as the grains. The major groups of natural stone used commercially are:

Granite, a visibly crystalline rock made of silicate minerals, primarily feldspar and quartz. Commercially, ''granite'' refers to all stones geologically defined as plutonic, igneous, and gneissic.

Marble, generally a visibly carbonate rock; however, microcrystalline rocks, such as onyx, travertine, and serpentine, are usually included by the trade as long as they can take a polish.

Limestone, a sedimentary rock composed of calcium or magnesium carbonate grains in a carbonate matrix.

Sandstone, a sedimentary rock composed chiefly of cemented, sand-sized quartz grains. In the trade, quartzites are usually grouped with sandstones, although these rocks tend to fracture through, rather than around, the grains. **Conglomerate** is a term used for a sandstone containing aggregate in sizes from the gravel range up.

The above stones can be used almost interchangeably as dimension stone for architectural or structural purposes.

Slate, a fine-grained rock, is characterized by marked cleavages by which the rock can be split easily into relatively thin slabs. Because of this characteristic, slate was at one time widely used for roofing tiles. See Roofing Materials, later in this section.

It is still widely applied for other building uses, such as steps, risers, spandrels, flagstones, and in some outdoor sculptured work.

Formerly, it was used almost universally for blackboards and electrical instrument panels, but it has been supplanted in these applications by plastic materials. Plastic sheets used as a writing surface for chalk are properly called **chalkboards;** the surface on which writing is done with crayons or fluid tip markers is termed **marker board.**

Miscellaneous stones, such as traprock (fine-grained black volcanic rock), greenstone, or argillite, are commonly used as crushed or broken stone but rarely as dimension stone.

Table 6.8.10 lists physical properties and contains the range of values that can be obtained from stones in various orientations relative to their textural and structural anisotropy. For a particular application, where one property must be exactly determined, the value must be obtained along a specified axis.

Selected Terms Applying to the Use of Dimension Stone

Anchor, a metal tie or rod used to fasten stone to backup units.

Arris, the meeting of two surfaces producing an angle, corner, or edge.

Ashlar, a facing of square or rectangular stones having sawed, dressed, or squared beds.

Bond stones, stones projecting a minimum of 4 in laterally into the backup wall; used to tie the wall together.

Cut stone, finished dimension stone—ready to set in place. The finish may be polished, honed, grooved (for foot traffic), or broken face.

Bearing wall, a wall supporting a vertical load in addition to its own weight.

Cavity wall, a wall in which the inner and outer parts are separated by an air space but are tied together with cross members.

Composite wall, a wall in which the facing and backing are of different materials and are united with bond stones to exert a common reaction under load. It is considered preferable, however, not to require the facing to support a load; thus the bond stones merely tie the facing to the supporting wall, as in the case of a veneer.

Table 6.8.10 Physical and Thermal Properties of Common Stones

Type of stone	Density, lb/ft^3	Compressive strength $\times 10^{-3}$, lb/in^2	Rupture modulus $\times 10^{-3}$, lb/in^2 (ASTM C99-52)	Shearing strength $\times 10^{-3}$, lb/in^2	Young's modulus $\times 10^{-6}$, lb/in^2	Modulus of rigidity $\times 10^{-6}$, lb/in^2	Poisson's ratio	Abrasion-hardness index (ASTM C241-51)	Porosity, vol %	48-h water absorption (ASTM C97-47)	Thermal conductivity, Btu/(ft · h · °F)	Coefficient of thermal expansion $\times 10^{-6}$, per °F
Granite	160–190	13–55	1.4–5.5	3.5–6.5	4–16	2–6	0.05–0.2	37–88	0.6–3.8	0.02–0.58	20–35	3.6–4.6
Marble	165–179	8–27	0.6–4.0	1.3–6.5	5–11.5	2–4.5	0.1–0.2	8–42	0.4–2.1	0.02–0.45	8–36	3.0–8.5
Slate	168–180	9–10	6–15	2.0–3.6	6–16	2.5–6	0.1–0.3	6–12	0.1–1.7	0.01–0.6	12–26	3.3–5.6
Sandstone	119–168	5–20	0.7–2.3	0.3–3.0	0.7–10	0.3–4	0.1–0.3	2–26	1.9–27.3	2.0–12.0	4–40	3.9–6.7
Limestone	117–175	2.5–28	0.5–2.0	0.8–3.6	3–9	1–4	0.1–0.3	1–24	1.1–31.0	1.0–10.0	20–32	2.8–4.5

Conversion factors: 1 lbm/ft^3 = 16,018 kg/m^3. 1 lb/in^2 = 6,894.8 N/m^2. 1 Btu/(ft · h · °F) = 623 W/(m · °C).

Veneer or **faced wall,** a wall in which a thin facing and the backing are of different materials but are not so bonded as to exert a common reaction under load.

Definitions may be found in ANSI/ASTM, C119-74 (1980).

PAPER

REFERENCES: Calkin, "Modern Pulp and Papermaking," Reinhold. Griffin and Little, "Manufacture of Pulp and Paper," McGraw-Hill. Guthrie, "The Economics of Pulp and Paper," State College of Washington Press. "Dictionary of Paper," American Pulp and Paper Assoc. Sutermeister, "Chemistry of Pulp and Papermaking," Wiley. Casey, "Pulp and Paper—Chemistry and Technology," Interscience, TAPPI Technical Information Sheets. TAPPI Monograph Series. "Index of Federal Specifications, Standards and Handbooks," GSA. Lockwoods Directory of the Paper and Allied Trades. Casey, "Pulp and Paper; Chemistry and Chemical Technology," 3 vols., Interscience. Libby (ed.), "Pulp and Paper Science and Technology," 2 vols., McGraw-Hill. ASTM Committee D6 on Paper and Paper Products, "Paper and Paperboard Characteristics, Nomenclature, and Significance of Tests," ASTM Special Technical Publication 60-B, 3d ed., 1963. Britt, "Handbook of Pulp and Paper Technology," Van Nostrand Reinhold. Johnson, "Synthetic Paper from Synthetic Fibers," Noyes Data Corp., New Jersey. "1986 Annual Book of ASTM Standards," vol. 15.09 (Paper).

Paper Grades

Specific paper **qualities** are achieved in a number of ways: (1) By selecting the composition of the furnish for the paper machine. Usually more than one pulp (prepared by different pulping conditions or processes) is required. The ratio of long-fibered pulp (softwood) to short-fibered pulp (hardwood or mechanical type), the reused-fiber content, and the use of nonfibrous fillers and chemical additives are important factors. (2) By varying the paper-machine operation. Fourdrinier wire machines are most common, although multicylinder machines and high-speed tissue machines with Yankee driers are also used. (3) By using various finishing operations (e.g., calendering, supercalendering, coating, and laminating).

There is a tremendous number of paper grades, which are, in turn, used in a wide variety of converted products. The following broad classifications are included as a useful guide:

Sanitary papers are tissue products characterized by bulk, opacity, softness, and water absorbency.

Glassine, greaseproof, and **waxing papers**—in glassine, high transparency and density, low grease penetration, and uniformity of formation are important requirements.

Food-board products require a good brightness and should be odorless and have good tear, tensile, and fold-endurance properties, opacity, and printability. More stringent sanitary requirements in current practice have caused this product to be supplanted largely by foamed plastics (polyethylene and the like).

Boxboard, misnomered **cardboard,** and a variety of other board products are made on multicylinder machines, where layers of fiber are built up to the desired thickness. Interior plies are often made from wastepaper furnishes, while surface plies are from bleached, virgin fiber. Boards are often coated for high brightness and good printing qualities.

Printing papers include publication papers (magazines), book papers, bond and ledger papers, newsprint, and catalog papers. In all cases, printability, opacity, and dimensional stability are important.

Linerboard and **bag paper** are principally unbleached, long-fiber, kraft products of various weights. Their principal property requirements are high tensile and bursting strengths.

Corrugating medium, in combined fiberboard, serves to hold the two linerboards apart in rigid, parallel separation. Stiffness is the most important property, together with good water absorbency for ease of corrugation.

Corrugated paper, often misnomered **corrugated cardboard,** is an end product of kraft paper, in which a corrugated layer is sandwiched and glued between two layers of cover paper. The whole assembly, when the glue has set, results in a material having a high flexural strength (although it may be directional, depending on whether the material is flexed parallel or crosswise to the corrugations), and high penetration strength. For heavier-duty applications, several sandwiches are themselves glued together and result in a truly superior load-carrying material.

Pulps

The most important source of fiber for paper pulps is wood, although numerous other vegetable substances are used. Reused fiber (wastepaper) constitutes about 30 percent of the total furnish used in paper, principally in boxboard and other packaging or printing papers.

There are three basic processes used to convert wood to papermaking fibers: mechanical, chemical, and "chemimechanical."

Mechanical Pulping (Groundwood) Here, the entire log is reduced to fibers by grinding against a stone cylinder, the simplest route to papermaking fiber. Wood fibers are mingled with extraneous materials which can cause weakening and discoloration in the finished paper. Nonetheless, the resulting pulp imparts good bulk and opacity to a printing sheet. Some long-fibered chemical pulp is usually added. Groundwood is used extensively in low-cost, short-service, and throwaway papers, e.g., newsprint and catalog paper.

Chemical Pulping Fiber separation can also be accomplished by chemical treatment of wood chips to dissolve the lignin that cements the fibers together. From 40 to 50 percent of the log is extracted, resulting in relatively pure cellulosic fiber. There are two major chemical-pulping processes, which differ both in chemical treatment and in the nature of the pulp produced. In **sulfate pulping,** also referred to as the **kraft** or **alkaline** process, the pulping chemicals must be recovered and reused for economic reasons. Sulfate pulps result in papers of high physical strength, bulk, and opacity; low unbleached brightness; slow beating rates; and relatively poor sheet-formation properties. Both bleached and unbleached sulfate pulps are used in packaging papers, container board, and a variety of printing and bond papers. In **sulfite pulping,** the delignifying agents are sulfurous acid and an alkali, with several variations of the exact chemical conditions in commercial use. In general, sulfate pulps have lower physical strength properties and lower bulk and opacity than kraft pulps, with higher unbleached brightness and better sheet-formation properties. The pulps are blended with groundwood for **newsprint** and are used in printing and bond papers, tissues, and glassine.

"**Chemimechanical**" **processes** combine both chemical and mechanical methods of defibration, the most important commercial process being the NSSC (Neutral Sulfite Semi Chemical). A wide range of yields and properties can be obtained.

Bleaching

Chemical pulps may be bleached to varying degrees of brightness, depending on the end use. During some bleaching operations, the remaining lignin is removed and residual coloring matter destroyed. Alternatively, a nondelignifying bleach lightens the color of high-lignin pulps.

Refining

The final character of the pulp is developed in the **refining** (beating) operation. Pulp fibers are fibrillated, hydrated, and cut. The fibers are roughened and frayed, and a gelatinous substance is produced. This results in greater fiber coherence in the finished paper.

Sizing, Loading, and Coating

Sizing is used, principally in book papers, to make the paper water-repellant and to enhance interfiber bonding and surface characteristics. Sizing materials may be premixed with the pulp or applied after sheet formation.

Loading materials, or **fillers,** are used by the papermaker to smooth the surface, to provide ink affinity, to brighten color, and to increase opacity. The most widely used fillers are clay and kaolin.

Coatings consisting of pigment and binder are often applied to the base stock to create better printing surfaces. A variety of particulate, inorganic materials are combined with binders such as starch or casein in paper coatings. Coating is generally followed by calendering.

Converting and Packaging

A host of products are made from paper; the converting industry represents a substantial portion of the total paper industry. Principal products are in the fine-paper and book-paper fields. Slush pulps are used to make molded pulp products such as egg cartons and paper plates.

Combinations of paper laminated with other materials such as plastic film and metal foil have found wide use in the packaging market.

Plastic-Fiber Paper

A tough, durable product, such as Tyvek, can be made from 100 percent high-density polyethylene fibers by spinning very fine fibers and then bonding them together with heat and pressure. Binders, sizes, or fillers are not required. Such sheets combine some of the best properties of the fabrics, films, and papers with excellent puncture resistance. They can readily be printed by conventional processes, dyed to pastel colors, embossed for decorative effects, coated with a range of materials, and can be folded, sheeted, die-cut, sewn, hot-melt-sealed, glued, and pasted. Nylon paper, such as Nomex type 410, is produced from short fibers (floc) and smaller binder particles (fibrids) of a high-temperature-resistant polyamide polymer, formed into a sheet product without additional binders, fillers, or sizes, and calendered with heat and pressure to form a nonporous structure. It possesses excellent electrical, thermal, and mechanical properties, and finds use in the electrical industry. Fiber impregnation of ordinary paper, using glass or plastic fibers, produces a highly tear-resistant product.

Recycling

In the current environmentally conscious climate, almost 50 percent of newsprint itself contains a portion of recycled old newsprint. The remainder, together with almost all other paper products, finds a ready market for recycling into other end products. The small portion that ends up discarded is increasingly used as feedstock for incineration plants which produce power and/or process steam by converting "scrap paper" and other refuse to refuse-derived fuel (RDF). See Sec. 7.4. Only in those geographic areas where it is impractical to collect and recycle paper does the residue finally find its way to landfill dumps.

ROOFING MATERIALS

REFERENCES: Abraham, "Asphalts and Allied Substances," Van Nostrand. Grondal, "Certigrade Handbook of Red Cedar Shingles," Red Cedar Shingle Bureau, Seattle, Wash. ASTM *Special Technical Publication* 409, "Engineering Properties of Roofing Systems," ASTM. Griffin, Jr., "Manual of Built-up Roof Systems," McGraw-Hill. "1986 Annual Book of ASTM Standards," vol. 04.04 (Roofing, Waterproofing and Bituminous Materials).

Asphalt Asphalts are bitumens, and the one most commonly seen in roofing and paving is obtained from petroleum residuals. These are obtained by the refining of petroleum. The qualities of asphalt are affected by the nature of the crude and the process of refining. When the flux asphalts obtained from the oil refineries are treated by blowing air through them while the asphalt is maintained at a high temperature, a material is produced which is very suitable and has good weathering properties.

Coal Tar Coal tar is more susceptible to temperature change than asphalt; therefore, for roofing purposes its use is usually confined to flat decks.

Asphalt prepared roofing is manufactured by impregnating a dry roofing felt with a hot asphaltic saturant. A coating consisting of a harder asphalt compounded with a fine mineral **filler** is applied to the weather side of the saturated felt. Into this coating is embedded mineral **surfacing** such as mineral granules, powdered talc, mica, or soapstone. The reverse side of the roofing has a very thin coating of the same asphalt which is usually covered with powdered talc or mica to prevent the roofing from sticking in the package. The surfacing used on **smooth-surfaced roll roofing** is usually powdered talc or mica. The surfacing used on **mineral-** or **slate-surfaced roll roofing** is roofing granules either in natural colors prepared from slate or artificial colors usually made by applying a coating to a rock granule base. **Asphalt shingles** usually have a granular surfacing. They are made in strips and as individual shingles. The different shapes and sizes of these shingles provide single, double, and triple coverage of the roof deck.

Materials Used in Asphalt Prepared Roofing The **felt** is usually composed of a continuous sheet of felted fibers of selected rag, specially prepared wood, and high-quality waste papers. The constituents may be varied to give a felt with the desired qualities of strength, absorbency, and flexibility. (See above, Fibers and Fabrics.)

The most satisfactory roofing **asphalts** are obtained by air-blowing a steam- or vacuum-refined petroleum residual. Saturating asphalts must possess a low viscosity in order for the felt to become thoroughly impregnated. Coating asphalts must have good weather-resisting qualities and possess a high fusion temperature in order that there will be no flowing of the asphalt after the application to the roof.

Asphalt built-up roof coverings usually consist of several layers of asphalt-saturated felt with a continuous layer of hot-mopped asphalt between the layers of felt. The top layer of such a roof covering may consist of a hot mopping of asphalt only, a top pouring of hot asphalt with slag or gravel embedded therein, or a mineral-surfaced cap sheet embedded in a hot mopping of asphalt.

Wood shingles are usually manufactured in three different lengths: 16, 18, and 24 in. There are three grades in each length: no. 1 is the best, and no. 3 is intended for purposes where the presence of defects is not objectionable. Red-cedar shingles of good quality are obtainable from the Pacific Coast; in the South, red cypress from the Gulf states is preferable. Redwood shingles come 5½ butts to 2 in; lesser thicknesses are more likely to crack and have shorter life. Shingles 8 in wide or over should be split before laying. Dimension shingles of uniform width are obtainable. Various stains are available for improved weathering resistance and altered appearance.

Asbestos shingles and (siding) are no longer made, but will be found from applications made in the past. The following is presented for information in the event the reader has occasion to deal with this product. They are composed of portland cement reinforced with asbestos fiber and are formed under pressure. They resist the destructive effects of time, weather, and fire. Asbestos shingles (American method) weigh about 500 lb per square (roofing to cover 100 ft^2) and carry Underwriter's class A label. Asbestos shingles are made in a variety of colors and shapes. Asbestos roofing shingles have either a smooth surface or a textured surface which represents wood graining.

Slate should be hard and tough and should have a well-defined vein that is not too coarse. Roofing slates in place will weigh from 650 to over 1,000 lb per square, depending on the thickness and the degree of overlap. Prudent roofing practice in installing slate roofs requires use of copper (or stainless steel) nails and, as appropriate, the installation of snow guards to prevent snow slides. When applied thus, slate proved long-lasting and able to maintain weathertight integrity in all climates. The quarrying is labor-intensive, however, as is the splitting required to render it into suitably thin pieces; installation labor is extremely expensive, and the substructure must be of heavier than usual construction to withstand the higher dead load of the roofing slates. For these reasons, among others, slate is used rarely for roofing in the current market. Slate roofing is available in a variety of sizes and colors and has good fire resistance. (See above, Natural Stones.)

Metallic roofings are usually laid in large panels, often strengthened by corrugating, but they are sometimes cut into small sizes bent into interlocking shapes and laid to interlock with adjacent sheets or shingles. Metallic roofing panels of both aluminum and steel are available with a variety of prefinished surface treatments to enhance weatherability. Metal tile and metal shingles are usually made of copper, copper-bearing galvanized steel, tinplate, zinc, or aluminum. The lightest metal shingle is the one made from aluminum, which weighs approximately 40 lb per square. The metal radiates solar heat, resulting in lower temperatures beneath than with most other types of uninsulated roofs. Terne-coated stainless steel is unusually resistant to weathering and corrosion.

Roofing cements and **coatings** are usually made from asphalt, a fibrous filler, and solvents to make the cement workable. Asbestos fibers traditionally were used as filler, but asbestos is no longer used; instead, fiberglass largely has replaced it. The cements are used for flashings and repairs and contain slow-drying oils so that they will remain plastic on long exposure. Roof coatings are used to renew old asphalt roofings.

Asphalt-base aluminum roof coatings are used to renew old asphalt roofs, and to prolong the service life of smooth-surfaced roofs, new or old.

Tile Hard-burned clay tiles with overlapping or interlocking edges cost about the same as slate. They should have a durable glaze and be well made. Unvitrified tiles with slip glaze are satisfactory in warm climates, but only vitrified tiles should be used in the colder regions. Tile roofs weigh from 750 to 1,200 lb per square. Properly made, tile does not deteriorate, is a poor conductor of heat and cold, and is not as brittle as slate. Fiberglass, saturated with asphalt and embedded with ceramic granules, is made into roof shingles having an exceptionally long life. Such shingles possess class A fire-resistant qualities, and tend also to be dimensionally very stable. Glass-fiber mats coated with weathering-grade asphalt can be had in rolls.

Elastomers

The trend toward irregular roof surfaces—folded plates, hyperbolic paraboloids, domes, barrel shells—has brought the increased use of plastics or synthetic rubber elastomers (applied as fluid or sheets) as roofing membranes. Such membranes offer light weight, adaptability to any roof slope, good heat reflectivity, and high elasticity at moderate temperatures. Negatively, elastomeric membranes have a more limited range of satisfactory substrate materials than conventional ones. Table 6.8.11 presents several such membranes, and the method of use.

RUBBER AND RUBBERLIKE MATERIALS (ELASTOMERS)

REFERENCES: Dawson and Porritt, "Rubber: Physical and Chemical Properties," Research Association of British Rubber Manufacturers. Davis and Blake, "The Chemistry and Technology of Rubber," Reinhold. ASTM Standards on Rubber Products. "Rubber Red Book Directory of the Rubber Industry," *The Rubber Age.* Flint, "The Chemistry and Technology of Rubber Latex," Van Nostrand. "The Vanderbilt Handbook," R. T. Vanderbilt Co. Whitby, Davis, and Dunbrook, "Synthetic Rubber," Wiley. "Rubber Bibliography," Rubber Division, American Chemical Society. Noble, "Latex in Industry," *The Rubber Age.* "Annual Report on the Progress of Rubber Technology," Institution of the Rubber Industry. Morton (ed.). "Rubber Technology," Van Nostrand Reinhold. ISO 188-1982. "1986 Annual Book of ASTM Standards," vols. 09.01, 09.02 (Rubber).

To avoid confusion by the use of the word **rubber** for a variety of natural and synthetic products, the term **elastomer** has come into use, particularly in scientific and technical literature, as a name for both natural and

Table 6.8.11 Elastomeric Membranes

	Material	Method of application	Number of coats or sheets
Fluid-applied	Neoprene-Hypalon*	Roller, brush, or spray	2 + 2
	Silicone	Roller, brush, or spray	2
	Polyurethane foam, Hypalon* coating	Spray	2
	Clay-type asphalt emulsion reinforced with chopped glass fibers†	Spray	1
Sheet-applied	Chlorinated polyethylene on foam	Adhesive	1
	Hypalon* on asbestos felt	Adhesive	1
	Neoprene-Hypalon*	Adhesive	1 + surface paint
	Tedlar‡ on asbestos felt	Adhesive	1
	Butyl rubber	Adhesive	1
Traffic decks	Silicone plus sand	Trowel	1 + surface coat
	Neoprene with aggregate§	Trowel	1 + surface coat

* Registered trademark of E. I. du Pont de Nemours & Co., for chlorosulfonated polyethylene.
† Frequently used with coated base sheet.
‡ Registered trademark of E. I. du Pont de Nemours & Co. for polyvinyl fluoride.
§ Aggregate may be flint, sand, or crushed walnut shells.
SOURCE: C. W. Griffin, Jr., "Manual of Built-up Roof Systems," McGraw-Hill, 1970, with permission of the publisher.

synthetic materials which are elastic or resilient and in general resemble natural rubber in feeling and appearance.

The utility of rubber and synthetic elastomers is increased by compounding. In the raw state, elastomers are soft and sticky when hot, and hard or brittle when cold. **Vulcanization** extends the temperature range within which they are flexible and elastic. In addition to vulcanizing agents, ingredients are added to make elastomers stronger, tougher, and harder, to make them age better, to color them, and in general to modify them to meet the needs of service conditions. Few rubber products today are made from rubber or other elastomers alone.

The elastomers of greatest commercial and technical importance today are natural rubber, GR-S, Neoprene, nitrile rubbers, and butyl.

Natural rubber of the best quality is prepared by coagulating the latex of the *Hevea brasilensis* tree, cultivated chiefly in the Far East. This represents nearly all of the natural rubber on the market today.

Unloaded vulcanized rubber will stretch to approximately 10 times its length and at this point will bear a **load** of 13.8×10^6 N/m² (10 tons/in²). It can be compressed to one-third its thickness thousands of times without injury. When most types of vulcanized rubber are stretched, their resistance increases in greater proportion than the extension. Even when stretched almost to the point of rupture, they recover very nearly their original dimensions on being released, and then gradually recover a part of the residual distortion.

Freshly cut or torn raw rubber possesses the power of **self-adhesion** which is practically absent in vulcanized rubber. Cold water preserves rubber, but if exposed to the air, particularly to the sun, rubber goods tend to become hard and brittle. Dry heat up to 49°C (120°F) has little deteriorating effect; at temperatures of 181 to 204°C (360 to 400°F) rubber begins to melt and becomes sticky; at higher temperatures, it becomes entirely carbonized. Unvulcanized rubber is **soluble** in gasoline, naphtha, carbon bisulfide, benzene, petroleum ether, turpentine, and other liquids.

Most rubber is **vulcanized**, i.e., made to combine with sulfur or sulfur-bearing organic compounds or with other chemical cross-linking agents. Vulcanization, if properly carried out, improves mechanical properties, eliminates tackiness, renders the rubber less susceptible to temperature changes, and makes it insoluble in all known solvents. It is impossible to dissolve vulcanized rubber unless it is first decomposed. Other ingredients are added for general effects as follows:

To increase tensile strength and resistance to abrasion: carbon black, precipitated pigments, as well as organic vulcanization accelerators.

To cheapen and stiffen: whiting, barytes, talc, silica, silicates, clays, fibrous materials.

To soften (for purposes of processing or for final properties): bituminous substances, coal tar and its products, vegetable and mineral oils, paraffin, petrolatum, petroleum oils, asphalt.

Vulcanization accessories, dispersion and wetting mediums, etc.: magnesium oxide, zinc oxide, litharge, lime, stearic and other organic acids, degras, pine tar.

Protective agents (natural aging, sunlight, heat, flexing): condensation amines, waxes.

Coloring pigments: iron oxides, especially the red grades, lithopone, titanium oxide, chromium oxide, ultramarine blue, carbon and lampblacks, and organic pigments of various shades.

Specifications should state suitable physical tests. Tensile strength and extensibility tests are of importance and differ widely with different compounds.

GR-S is an outgrowth and improvement of German Buna S. The quantity now produced far exceeds all other synthetic elastomers. It is made from butadiene and styrene, which are produced from petroleum. These two materials are copolymerized directly to GR-S, which is known as a butadiene-styrene copolymer. GR-S has *recently* been improved, and now gives excellent results in tires.

Neoprene is made from acetylene, which is converted to vinylacetylene, which in turn combines with hydrogen chloride to form chloroprene. The latter is then polymerized to Neoprene.

Nitrile rubbers, an outgrowth of German Buna N or Perbunan, are made by a process similar to that for GR-S, except that acrylonitrile is used instead of styrene. This type of elastomer is a butadiene-acrylonitrile copolymer.

Butyl, one of the most important of the synthetic elastomers, is made from petroleum raw materials, the final process being the copolymerization of isobutylene with a very small proportion of butadiene or isoprene.

Polysulfide rubbers having unique resistance to oxidation and to softening by solvents are commercially available and are sold under the trademark Thiokol.

Ethylene-propylene rubbers are notable in their oxidation resistance. **Polyurethane** elastomers can have a tensile strength up to twice that of conventional rubber, and solid articles as well as foamed shapes can be cast into the desired form using prepolymer shapes as starting materials. **Silicone** rubbers have the advantages of a wide range of service temperatures and room-temperature curing. **Fluorocarbon** elastomers are available for high-temperature service. Polyester elastomers have excellent impact and abrasion resistance.

No one of these elastomers is satisfactory for all kinds of service conditions, but rubber products can be made to meet a large variety of service conditions.

The following examples show some of the important properties required of rubber products and some typical services where these properties are of major importance:

Resistance to abrasive wear: auto-tire treads, conveyor-belt covers, soles and heels, cables, hose covers, V belts.

Resistance to tearing: auto inner tubes, tire treads, footwear, hot-water bags, hose covers, belt covers, V belts.

Resistance to flexing: auto tires, transmission belts, V belts, (see Sec. 8.2 for information concerning types, sizes, strengths, etc., of V belts), mountings, footwear.

Resistance to high temperatures: auto tires, auto inner tubes, belts conveying hot materials, steam hose, steam packing.

Resistance to cold: airplane parts, automotive parts, auto tires, refrigeration hose.

Minimum heat buildup: auto tires, transmission belts, V belts, mountings.

High resilience: auto inner tubes, sponge rubber, mountings, elastic bands, thread, sandblast hose, jar rings, V belts.

High rigidity: packing, soles and heels, valve cups, suction hose, battery boxes.

Long life: fire hose, transmission belts, tubing, V belts.

Electrical resistivity: electricians' tape, switchboard mats, electricians' gloves.

Electrical conductivity: hospital flooring, nonstatic hose, matting.

Impermeability to gases: balloons, life rafts, gasoline hose, special diaphragms.

Resistance to ozone: ignition distributor gaskets, ignition cables, windshield wipers.

Resistance to sunlight: wearing apparel, hose covers, bathing caps.

Resistance to chemicals: tank linings, hose for chemicals.

Resistance to oils: gasoline hose, oil-suction hose, paint hose, creamery hose, packinghouse hose, special belts, tank linings, special footwear.

Stickiness: cements, electricians' tapes, adhesive tapes, pressure-sensitive tapes.

Low specific gravity: airplane parts, forestry hose, balloons.

No odor or taste: milk tubing, brewery and wine hose, nipples, jar rings.

Special colors: ponchos, life rafts, welding hose.

Table 6.8.12 gives a comparison of some important characteristics of the most important elastomers when vulcanized. The lower part of the table indicates, for a few representative rubber products, preferences in the use of different elastomers for different service conditions without consideration of cost.

Specifications for rubber goods may cover the chemical, physical, and mechanical properties, such as elongation, tensile strength, permanent set, and oven tests, minimum rubber content, exclusion of reclaimed rubber, maximum free and combined sulfur contents, maxi-

Table 6.8.12 Comparative Properties of Elastomers

	Natural rubber	GR-S	Neoprene	Nitrile rubbers	Butyl	Thiokol
Tensile properties	Excellent	Good	Very good	Good	Good	Fair
Resistance to abrasive wear	Excellent	Good	Very good	Good	Good	Poor
Resistance to tearing	Very good	Poor	Good	Fair	Very good	Poor
Resilience	Excellent	Good	Good	Fair	Poor	Poor
Resistance to heat	Good	Fair	Good	Excellent	Good	Poor
Resistance to cold	Excellent	Good	Good	Good	Excellent	Poor
Resistance to flexing	Excellent	Good	Very good	Good	Excellent	Poor
Aging properties	Excellent	Excellent	Good	Good	Excellent	Good
Cold flow (creep)	Very low	Low	Low	Very low	Fairly low	High
Resistance to sunlight	Fair	Fair	Excellent	Good	Excellent	Excellent
Resistance to oils and solvents	Poor	Poor	Good	Excellent	Fair	Excellent
Permeability to gases	Fairly low	Fairly low	Low	Fairly low	Very low	Very low
Electrical insulation	Fair	Excellent	Fair	Poor	Excellent	Good
Flame resistance	Poor	Poor	Good	Poor	Poor	Poor
Auto tire tread	Preferred	Alternate				
Inner tube	Alternate				Preferred	
Conveyor-belt cover	Preferred		Alternate			
Tire sidewall	Alternate	Preferred				
Transmission belting	Preferred		Alternate			
Druggist sundries	Preferred					
Gasoline and oil hose				Preferred		
Lacquer and paint hose						Preferred
Oil-resistant footwear			Preferred			
Balloons	Alternate		Preferred			
Jar rings	Alternate	Preferred				
Wire and cable insulation	Alternate	Preferred				

mum acetone and chloroform extracts, ash content, and many construction requirements. It is preferable, however, to specify properties such as resilience, hysteresis, static or dynamic shear and compression modulus, flex fatigue and cracking, creep, electrical properties, stiffening, heat generation, compression set, resistance to oils and chemicals, permeability, brittle point, etc., in the temperature range prevailing in service, and to leave the selection of the elastomer to a competent manufacturer.

Latex, imported in stable form from the Far East, is used for various rubber products. In the manufacture of such products, the latex must be compounded for vulcanizing and otherwise modifying properties of the rubber itself. Important products made directly from compounded latex include surgeons' and household gloves, thread, bathing caps, rubberized textiles, balloons, and sponge. A recent important use of latex is for "foam sponge," which may be several inches thick and used for cushions, mattresses, etc.

Gutta-percha and **balata,** also natural products, are akin to rubber chemically but more leathery and thermoplastic, and are used for some special purposes, principally for submarine cables, golf balls, and various minor products.

Rubber Derivatives

Rubber derivatives are chemical compounds and modifications of rubber, some of which have become of commercial importance.

Chlorinated rubber, produced by the action of chlorine on rubber in solution, is nonrubbery, incombustible, and extremely resistant to many chemicals. As commercial **Parlon,** it finds use in corrosion-resistant paints and varnishes, in inks, and in adhesives.

Rubber hydrochloride, produced by the action of hydrogen chloride on rubber in solution, is a strong, extensible, tear-resistant, moisture-resistant, oil-resistant material, marketed as **Pliofilm** in the form of tough transparent films for wrappers, packaging material, etc.

Cyclized rubber is formed by the action of certain agents, e.g., sulfonic acids and chlorostannic acid, on rubber, and is a thermoplastic, nonrubbery, tough or hard product. One form, **Thermoprene,** is used in the **Vulcalock process** for adhering rubber to metal, wood, and concrete, and in chemical-resistant paints. **Pliolite,** which has high resistance to many chemicals and has low permeability, is used in special paints, paper, and fabric coatings. **Marbon-B** has exceptional electrical properties and is

valuable for insulation. **Hypalon** (chlorosulphonated polyethylene) is highly resistant to many important chemicals, notably ozone and concentrated sulfuric acid, for which other rubbers are unsuitable.

SOLVENTS

REFERENCES: Sax, "Dangerous Properties of Industrial Materials," Reinhold. Perry, "Chemical Engineers' Handbook," McGraw-Hill. Doolittle, "The Technology of Solvents and Plasticizers," Wiley. Riddick and Bunger, "Organic Solvents," Wiley. Mellan, "Industrial Solvents Handbook," Noges Data Corp., New Jersey. "1986 Annual Book of ASTM Standards," vol: 15.05 (Halogenated Organic Solvents).

The use of solvents has become widespread throughout industry. The health of personnel and the fire hazards involved should always be considered. Generally, solvents are **organic liquids** which vary greatly in solvent power, flammability, volatility, and toxicity.

Solvents for Polymeric Materials

A wide choice of solvents and solvent combinations is available for use with organic polymers in the manufacture of polymer-coated products and unsupported films. For a given polymer, the choice of solvent system is often critical in terms of solvent power, cost, safety, and evaporation rate. In such instances, the supplier of the base polymer should be consulted.

Alcohols

Methyl alcohol (methanol) is now made synthetically. It is completely miscible with water and most organic liquids. It evaporates rapidly and is a good solvent for dyes, gums, shellac, nitrocellulose, and some vegetable waxes. It is widely used as an antifreeze for automobiles, in shellac solutions, spirit varnishes, stain and paint removers. It is toxic; imbibition or prolonged breathing of the vapors can cause blindness. It should be used only in well-ventilated spaces. Flash point 11°C (52°F).

Ethyl alcohol (ethanol) is produced by fermentation and synthetically. For industrial use it is generally denatured and sold under various trade names. There are numerous formulations of specially denatured alcohols which can legally be used for specified purposes.

This compound is miscible with water and most organic solvents. It evaporates rapidly and, because of its solvent power, low cost, and agreeable odor, finds a wide range of uses. The common uses are antifreeze (see Freezing Preventives, above), shellac solvent, in mixed solvents, spirit varnishes, and solvent for dyes, oils, and animal greases.

Denatured alcohols are toxic when taken orally. Ethyl alcohol vapors when breathed in high concentration can produce the physiological effects of alcoholic liquors. It should be used in well-ventilated areas. Flash point 15.3 to 16.7°C (960 to 62°F).

Isopropyl alcohol (isopropanol) is derived mainly from petroleum gases. It is not as good a solvent as denatured alcohol, although it can be used as a substitute for ethyl alcohol in some instances. It is used as a rubbing alcohol and in lacquer thinners. Flash point is 11°C (52°F).

Butyl alcohol (normal butanol) is used extensively in lacquer and synthetic resin compositions and also in penetrating oils, metal cleaners, insect sprays, and paints for application over asphalt. It is an excellent blending agent for otherwise incompatible materials. Flash point is 29°C (84°F).

Esters

Ethyl acetate dissolves a large variety of materials, such as nitrocellulose oils, fats, gums, and resins. It is used extensively in nitrocellulose lacquers, candy coatings, food flavorings, and in chemical synthesis. On account of its high rate of evaporation, it finds a use in paper, leather, and cloth coatings and cements. Flash point −4.5°C (24°F).

Butyl acetate is the acetic-acid ester of normal butanol. This ester is used extensively for dissolving various cellulose esters, mineral and vegetable oils, and many synthetic resins, such as the vinylites, polystyrene, methyl methacrylate, and chlorinated rubber. It is also a good solvent for natural resins. It is the most important solvent used in lacquer manufacture. It is useful in the preparation of perfumes and synthetic flavors. Flash point 22.3°C (72°F).

Amyl acetate, sometimes known as banana oil, is used mainly in lacquers. Its properties are somewhat like those of butyl acetate. Flash point 21.1°C (70°F).

Hydrocarbons

Some industrial materials can be dangerous, toxic, and carcinogenic. The reader should become familiar with the references listed under Chlorinated Solvents in this section, which discuss the health hazards of industrial materials.

The **aromatic hydrocarbons** are derived from coal-tar distillates, the most common of which are benzene, toluene, xylene (also known as benzol, toluol, and xylol), and high-flash solvent naphtha.

Benzene is an excellent solvent for fats, vegetable and mineral oils, rubber, chlorinated rubber; it is also used as a solvent in paints, lacquers, inks, paint removers, asphalt, coal tar. This substance should be used with caution. Flash point 12°F.

Toluene General uses are about the same as benzol, in paints, lacquers, rubber solutions, and solvent extractions. Flash point 40°F.

Xylene is used in the manufacture of dyestuffs and other synthetic chemicals and as a solvent for paints, rubber, lacquer, and varnishes. Flash point 63°F.

Hi-flash naphtha or coal-tar naphtha is used mainly as a diluent in lacquers, synthetic enamels, paints, and asphaltic coatings. Flash point 100°F.

Petroleum

These hydrocarbons, derived from petroleum, are, next to water, the cheapest and most universally used solvents.

V. M. & P. naphtha, sometimes called **benzine,** is used by paint and varnish makers as a solvent or diluent. It finds wide use as a solvent for fats, oils, greases and is used as a diluent in paints and lacquers. It is also used as an extractive agent as well as in some specialized fields of cleansing (fat removal). It is used to compound rubber cement, inks, varnish removers. It is relatively nontoxic. Flash point 20 to 45°F.

Mineral spirits, also called **Stoddard solvent,** is extensively used in dry cleaning because of its high flash point and clean evaporation. It is also

widely used in turpentine substitutes for oil paints. Flash point 100 to 110°F.

Kerosine, a No. 1 fuel oil, is a good solvent for petroleum greases, oils, and fats. Flash point 100 to 165°F.

Chlorinated Solvents

Solvents can pose serious health and environmental problems, therefore the reader's attention is called to the following publications: (1) "Industrial Health and Safety; Mutagens and Carcinogens; Solution; Toxicology," vols. 1 to 3, Wiley. (2) "Patty's Industrial Hygiene and Toxicology," vol. 1 (General Principles), vol. 2 (Toxicology), vol. 3 (Industrial Hygiene Practices). (3) Peterson, "Industrial," Prentice-Hall. (4) Sax and Lewis, "Dangerous Properties of Industrial Materials," 7th ed., Van Nostrand Reinhold. Section 18 of this handbook.

The chlorinated hydrocarbons and fluorocarbon solvents are highly proscribed for use in industry and in society in general because of the inherent health and/or environmental problems ascribed to them. These materials are highly regulated and when used should be handled with utmost care and under close supervision and accountable control.

Carbon tetrachloride is a colorless nonflammable liquid with a chloroformlike odor. It is an excellent solvent for fats, oils, greases, waxes, and resins and was used in dry cleaning and in degreasing of wool, cotton waste, and glue. It is also used in rubber cements and adhesives, as an extracting agent, and in fire extinguishers. It should be used only in well-ventilated spaces; prolonged inhalation is extremely dangerous.

Trichlorethylene is somewhat similar to carbon tetrachloride but is slower in evaporation rate. It is the solvent most commonly used for vapor degreasing of metal parts. It is also used in the manufacture of dyestuffs and other chemicals. It is an excellent solvent and is used in some types of paints, varnishes, and leather coatings.

Tetrachlorethylene, also called **perchlorethylene,** is nonflammable and has uses similar to those of carbon tetrachloride and trichlorethylene. Its chief use is in dry cleaning; it is also used in metal degreasing.

Ketones

Acetone is an exceptionally active solvent for a wide variety of organic materials, gases, liquids, and solids. It is completely miscible with water and also with most of the organic liquids. It can also be used as a blending agent for otherwise immiscible liquids. It is used in the manufacture of pharmaceuticals, dyestuffs, lubricating compounds, and pyroxylin compositions. It is a good solvent for cellulose acetate, ethyl cellulose, vinyl and methacrylate resins, chlorinated rubber, asphalt, camphor, and various esters of cellulose, including smokeless powder, cordite, etc. Some of its more common uses are in paint and varnish removers, the storing of acetylene, and the dewaxing of lubricating oils. It is the basic material for the manufacture of iodoform and chloroform. It is also used as a denature for ethyl alcohol. Flash point 0°F.

Methylethylketone (MEK) can be used in many cases where acetone is used as a general solvent, e.g., in the formulation of pyroxylin cements and in compositions containing the various esters of cellulose. Flash point 30°F.

Glycol ethers are useful as solvents for cellulose esters, lacquers, varnishes, enamels, wood stains, dyestuffs, and pharmaceuticals.

THERMAL INSULATION

REFERENCES: "Guide and Data Book," ASHRAE. Glaser, "Aerodynamically Heated Structures," Prentice-Hall. Timmerhaus, "Advances in Cryogenic Engineering," Plenum. Wilkes, "Heat Insulation," Wiley. Wilson, "Industrial Thermal Insulation," McGraw-Hill. Technical Documentary Report ML-TDR-64-5, "Thermophysical Properties of Thermal Insulating Materials," Air Force Materials Laboratory, Research and Technology Division, Air Force Systems Command; Prepared by Midwest Research Institute, Kansas City, MO. Probert and Hub (eds.), "Thermal Insulation," Elsevier. Malloy, "Thermal Insulation," Van Nostrand Reinhold. "1986 Annual Book of ASTM Standards," vol. 04.06 (Thermal Insulation). ANSI/ASTM Standards.

Thermal insulation consisting of a single material, a mixture of materials, or a composite structure is chosen to reduce heat flow. **Insulating effectiveness** is judged on the basis of thermal conductivity and depends on the physical and chemical structure of the material. The heat transferred through an insulation results from **solid conduction, gas conduction,** and **radiation.** Solid conduction is reduced by small particles or fibers in loose-fill insulation and by thin cell walls in foams. Gas conduction is reduced by providing large numbers of small pores (either interconnected or closed off from each other) of the order of the mean free paths of the gas molecules, by substituting gases of low thermal conductivity, or by evacuating the pores to a low pressure. Radiation is reduced by adding materials which absorb, reflect, or scatter radiant energy. (See also Sec. 4, Transmission of Heat by Conduction and Convection.)

The **performance** of insulations depends on the temperature of the bounding surfaces and their emittance, the insulation density, the type and pressure of gas within the pores, the moisture content, the thermal shock resistance, and the action of mechanical loads and vibrations. In **transient** applications, the heat capacity of the insulation (affecting the rate of heating or cooling) has to be considered.

The **form** of the **insulations** can be loose fill (bubbles, fibers, flakes, granules, powders), flexible (batting, blanket, felt, multilayer sheets, and tubular), rigid (block, board, brick, custom-molded, sheet and pipe covering), cemented, foamed-in-place, or sprayed.

The choice of **insulations** is dictated by the service-temperature range as well as by design criteria and economic considerations.

Cryogenic Temperatures [below −102°C (−150°F)]

(See also Sec. 19.)

At the low temperatures experienced with cryogenic liquids, **evacuated multilayer insulations,** consisting of a series of radiation shields of high reflectivity separated by low-conductivity spacers, are effective materials. Radiation shield materials are aluminum foils or aluminized polyester films used in combination with spacers of thin polyester fiber or glass-fiber papers; radiation shields of crinkled, aluminized polyester film without spacers are also used. To be effective, multilayer insulations require a vacuum of at least 10^{-4} mmHg. **Evacuated powder** and **fiber insulations** can be effective at gas pressures up to 0.1 mmHg over a wide temperature range. **Powders** include colloidal silica (8×10^{-7} in particle diam), silica aerogel (1×10^{-6} in), synthetic calcium silicate (0.001 in), and perlite (an expanded form of glassy volcanic lava particles, 0.05 in diam). Powder insulations can be **opacified** with copper, aluminum, or carbon particles to reduce radiant energy transmission. **Fiber insulations** consist of mats of fibers arranged in ordered parallel layers either without binders or with a minimum of binders. Glass fibers (10^{-5} in diam) are used most frequently. For large process installations and cold boxes, unevacuated perlite powder or mineral fibers are useful.

Cellular glass (see Glass, this section) can be used for temperatures as low as −450°F (−268°C) and has found use on liquefied natural gas tank bases.

Foamed organic plastics, using either fluorinated hydrocarbons or other gases as expanding agents, are partially evacuated when gases within the closed cells condense when exposed to low temperatures. Polystyrene and polyurethane foams are used frequently. **Gastight barriers** are required to prevent a rise in thermal conductivity with aging due to diffusion of air and moisture into the foam insulation. Gas barriers are made of aluminum foil, polyester film, and polyester film laminated with aluminum foil.

Refrigeration, Heating, and Air Conditioning up to 120°C (250°F)

At temperatures associated with commercial refrigeration practice and building insulation, **vapor barriers** resistant to the diffusion of water vapor should be installed on the warm side of most types of insulations if the temperature within the insulation is expected to fall below the dew point (this condition would lead to condensation of water vapor within the insulation and result in a substantial decrease in insulating effectiveness). Vapor barriers include oil- or tar-impregnated paper, paper laminated with aluminum foil, and polyester films. Insulations which have an impervious outer skin or structure require a **vaportight sealant** at exposed joints to prevent collection of moisture or ice underneath the insulation. (See also Secs. 12 and 19.)

Loose-fill insulations include powders and granules such as perlite, vermiculite (an expanded form of mica), silica aerogel, calcium silicate, expanded organic plastic beads, granulated cork (bark of the cork tree), granulated charcoal, redwood wool (fiberized bark of the redwood tree), and synthetic fibers. The most widely used fibers are those made of glass, rock, or slag produced by centrifugal attenuation or attenuation by hot gases.

Flexible or **blanket insulations** include those made from organically bonded glass fibers; rock wool, slag wool, macerated paper, or hair felt placed between or bonded to paper laminate (including vapor-barrier material) or burlap; foamed organic plastics in sheet and pad form (polyurethane, polyethylene); and elastomeric closed-cell foam in sheet, pad, or tube form.

Rigid or **board insulations** (obtainable in a wide range of densities and structural properties) include foamed organic plastics such as polystyrene (extended or molded beads); polyurethane, polyvinyl chloride, phenolics, and ureas; balsa wood, cellular glass, and corkboard (compressed mass of baked-cork particles).

Moderate Temperatures [up to 650°C (1200°F)]

The widest use of a large variety of insulations is in the temperature range associated with power plants and industrial equipment. Inorganic insulations are available for this temperature range, with several capable of operating over a wider temperature range.

Loose-fill insulations include diatomaceous silica (fossilized skeletons of microscopic organisms), perlite, vermiculite, and fibers of glass, rock, or slag. Board and blanket insulations of various shapes and degrees of flexibility and density include glass and mineral fibers, asbestos paper, and millboard [asbestos is a heat-resistant fibrous mineral obtained from Canadian (chrysotile) or South African (amosite) deposits]; asbestos fibers bonded with sodium silicate, 85 percent basic magnesium carbonate, expanded perlite bonded with calcium silicate, calcium silicate reinforced with asbestos fibers, expanded perlite bonded with cellulose fiber and asphalt, organic bonded mineral fibers, and cellulose fiberboard.

Sprayed insulation (macerated paper or fibers and adhesive or frothed plastic foam), **insulating concrete** (concrete mixed with expanded perlite or vermiculite), and **foamed-in-place plastic insulation** (prepared by mixing polyurethane components, pouring the liquid mix into the void, and relying on action of generated gas or vaporization of a low boiling fluorocarbon to foam the liquid and fill the space to be insulated) are useful in special applications.

Reflective insulations form air spaces bounded by surfaces of high reflectivity to reduce the flow of radiant energy. Surfaces need not be mirror bright to reflect long-wavelength radiation emitted by objects below 500°F. Materials for reflective insulation include aluminum foil cemented to one or both sides of kraft paper and aluminum particles applied to the paper with adhesive. Where several reflective surfaces are used, they have to be separated during the installation to form airspaces.

Cellular glass can be used up to 900°F (482°C) alone and to 1,200°F (649°C) or higher when combined with other insulating materials and jackets.

High Temperatures [above 820°C (1,500°F)]

At the high temperatures associated with furnaces and process applications, physical and chemical stability of the insulation in an oxidizing, reducing, or neutral atmosphere or vacuum may be required. **Loose-fill insulations** include glass fibers 538°C (1,000°F) useful temperature limit, asbestos fibers 650°C (1,200°F), fibrous potassium titanate 1,040°C (1,900°F), alumina-silica fibers 1,260°C (2,300°F), microquartz fibers 1,370°C (2,500°F), opacified colloidal alumina 1,310°C (2,400°F in vacuum), zirconia fibers 1,640°C (3,000°F), alumina bubbles 1,810°C (3,300°F), zirconia bubbles 2,360°C (4,300°F), and carbon and graphite fibers 2,480°C (4,500°F) in vacuum or an inert atmosphere.

Rigid insulations include reinforced and bonded colloidal silica 1,090°C (2,000°F), bonded diatomaceous earth brick 1,370°C (2,500°F), insulating firebrick (see Refractories, below), and anisotropic pyrolitic graphite (100 : 1 ratio of thermal conductivity parallel to surface and across thickness).

Reflective insulations, forming either an airspace or an evacuated chamber between spaced surfaces, include stainless steel, molybdenum, tantalum, or tungsten foils and sheets.

Insulating cements are based on asbestos, mineral, or refractory fibers bonded with mixtures of clay or sodium silicate. Lightweight, castable insulating materials consisting of mineral or refractory fibers in a calcium aluminate cement are useful up to 2,500°F.

Ablators are composite materials capable of withstanding high temperatures and high gas velocities for limited periods with minimum erosion by subliming and charring at controlled rates. Materials include asbestos, carbon, graphite, silica, nylon or glass fibers in a high-temperature resin matrix (epoxy or phenolic resin), and cork compositions.

SILICONES

(See also discussion in Sealants later.)

Silicones are organosilicon oxide polymers characterized by remarkable temperature stability, chemical inertness, waterproofness, and excellent dielectric properties. The investigations of Prof. Kipping in England for over forty years established a basis for the development of the numerous industrially important products now being made by the Dow-Corning Corp. of Medland, Mich., the General Electric Co., and by others. Among these products are the following:

Water repellents in the form of extremely thin films which can be formed on paper, cloth, ceramics, glass, plastics, leather, powders, or other surfaces. These have great value for the protection of delicate electrical equipment in moist atmospheres.

Oils with high flash points [above 315°C (600°F)], low pour points, −84.3°C (−120°F), and with a constancy of viscosity notably superior to petroleum products in the range from 260 to −73.3°C (500 to −100°F). These oil products are practically incombustible. They are in use for hydraulic servomotor fluids, damping fluids, dielectric liquids for transformers, heat-transfer mediums, etc., and are of special value in aircraft because of the rapid and extreme temperature variations to which aircraft are exposed. At present they are not available as lubricants except under light loads.

Greases and compounds for plug cocks, spark plugs, and ball bearings which must operate at extreme temperatures and speeds.

Varnishes and resins for use in electrical insulation where temperatures are high. Layers of glass cloth impregnated with or bounded by silicone resins withstand prolonged exposure to temperatures up to 260°C (500°F). They form paint finishes of great resistance to chemical agents and to moisture. They have many other industrial uses.

Silicone rubbers which retain their resiliency for −45 to 270°C (−50 to 520°F) but with much lower strength than some of the synthetic rubbers. They are used for shaft seals, oven gaskets, refrigerator gaskets, and vacuum gaskets.

Dimethyl silicone fluids and emulsions; organomodified silicone fluids and emulsions; organomodified reactive silicone fluids and emulsions; and silicone antifoam fluids, compounds, and emulsions find applications in coatings, paints, inks, construction, electrical and electronic industries, the foundry industry, etc.

REFRACTORIES

REFERENCES: Buell, "The Open-Hearth Furnace," 3 vols., Penton. *Bull.* R-2-E, The Babcock & Wilcox Co. "Refractories," General Refractories Co. Chesters, "Steel Plant Refractories," United Steel Companies, Ltd. "Modern Refractory Practice," Harbison Walker Refractories Co. Green and Stewart, "Ceramics: A Symposium," British Ceramic Soc. ASTM Standards on Refractory Materials (Committee C-8). Trinks, "Industrial Furnaces," vol. I, Wiley. Campbell, "High Temperature Technology," Wiley. Budnikov, "The Technology of Ceramics and Refractories," MIT Press. Campbell and Sherwood (eds.), "High-Temperature Materials and Technology," Wiley. Norton, "Refractories," 4th ed., McGraw-Hill. Clauss, "Engineer's Guide to High-Temperature Materials," Addison-Wesley. Shaw, "Refractories and Their Uses," Wiley. Chesters, "Refractories; Production and Properties," The Iron and Steel Institute, London. "1986 Annual Book of ASTM Standards," vol. 15.01 (Refractories). ANSI Standards B74.10 - B74.8.

Types of Refractories

Fire-Clay Refractories Fire-clay brick is made from fire clays, which comprise all refractory clays that are not white burning. Fire clays can be divided into plastic clays and hard flint clays; they may also be classified as to alumina content.

Firebricks are usually made of a blended mixture of flint clays and plastic clays which is then formed, after mixing with water, to the required shape. Some or all of the flint clay may be replaced by highly burned or calcined clay, called **grog.** A large proportion of modern bricks are molded by the dry press or power press process where the forming is carried out under high pressure and with a low water content. Some extruded and hand-molded bricks are still made.

The dried bricks are burned in either periodic or tunnel kilns at temperatures varying between 1,200 and 1,480°C (2,200 and 2,700°F). Tunnel kilns give continuous production and a uniform temperature of burning.

Fire-clay bricks are used for boiler settings, kilns, malleable-iron furnaces, incinerators, and many portions of steel and nonferrous metal furnaces. They are resistant to spalling and stand up well under many slag conditions, but are not generally suitable for use with high lime slags, fluid-coal ash slags, or under severe load conditions.

High-alumina bricks are manufactured from raw materials rich in alumina, such as diaspore and bauxite. They are graded into groups with 50, 60, 70, 80, and 90 percent alumina content. When well fired, these bricks contain a large amount of mullite and less of the glassy phase than is present in firebricks. Corundum is also present in many of these bricks. High-alumina bricks are generally used for unusually severe temperature or load conditions. They are employed extensively in lime kilns and rotary cement kilns, the ports and regenerators of glass tanks, and for slag resistance in some metallurgical furnaces; their price is higher than that for firebrick.

Silica bricks are manufactured from crushed ganister rock containing about 97 to 98 percent silica. A bond consisting of 2 percent lime is used, and the bricks are fired in periodic kilns at temperatures of 1,480 to 1,540°C (2,700 to 2,800°F) for several days until a stable volume is obtained. They are especially valuable where good strength is required at high temperatures. Recently, superduty silica bricks are finding some use in the steel industry. They have a lowered alumina content and often a lowered porosity.

Silica brick are used extensively in coke ovens, the roofs and walls of open-hearth furnaces, in the roofs and sidewalls of glass tanks, and as linings of acid electric steel furnaces. Although silica brick is readily **spalled** (cracked by a temperature change) below red heat, it is very stable if the temperature is kept above this range, and for this reason stands up well in regenerative furnaces. Any structure of silica brick should be heated up slowly to the working temperature; a large structure often requires 2 weeks or more to bring up.

Magnesite bricks are made from crushed magnesium oxide which is produced by calcining raw magnesite rock to high temperatures. A rock containing several percent of iron oxide is preferable, as this permits the rock to be fired at a lower temperature than if pure materials were used. Magnesite bricks are generally fired at a comparatively high temperature in periodic or tunnel kilns, though large tonnages of unburned bricks are now produced. The latter are made with special grain sizing and a bond such as an oxychloride. A large proportion of magnesite brick made in this country uses raw material extracted from seawater.

Magnesite bricks are basic and are used whenever it is necessary to resist high lime slags, e.g., formerly in the basic open-hearth furnace. They also find use in furnaces for the lead and copper refining industry. The highly pressed unburned bricks find extensive use as linings for cement kilns. Magnesite bricks are not so resistant to spalling as fireclay bricks.

Dolomite This rock contains a mixture of $Mg(OH)_2$ and $Ca(OH_2)$, is calcined, and is used in granulated form for furnace bottoms.

Chrome bricks are manufactured in much the same way as magnesite bricks but are made from natural chromite ore. Commercial ores always contain magnesia and alumina. Unburned hydraulically pressed chrome bricks are also made.

Chrome bricks are very resistant to all types of slag. They are used as separators between acid and basic refractories, also in soaking pits and floors of forging furnaces. The unburned hydraulically pressed bricks now find extensive use in the walls of the open-hearth furnace and are often enclosed in a metal case. Chrome bricks are used in sulfite recovery furnaces and to some extent in the refining of nonferrous metals. Basic bricks combining various proportions of magnesite and chromite are now made in large quantities and have advantages over either material alone for some purposes.

The **insulating firebrick** is a class of brick which consists of a highly porous fire clay or kaolin. They are lightweight (about one-half to one-sixth that of fireclay), low in thermal conductivity, and yet sufficiently resistant to temperature to be used successfully on the hot side of the furnace wall, thus permitting thin walls of low thermal conductivity and low heat content. The low heat content is particularly valuable in saving fuel and time on heating up, allows rapid changes in temperature to be made, and permits rapid cooling. These bricks are made in a variety of ways, such as mixing organic matter with the clay and later burning it out to form pores; or a bubble structure can be incorporated in the clay-water mixture which is later preserved in the fired brick. The insulating firebricks are classified into several groups according to the maximum use limit; the ranges are up to 872, 1,090, 1,260, 1,420, and above 1,540°C (1,600, 2,000, 2,300, 2,600, and above 2,800°F).

Insulating refractories are used mainly in the heat-treating industry for furnaces of the periodic type; the low heat content permits noteworthy fuel savings as compared with firebrick. They are also extensively in stress-relieving furnaces, chemical-process furnaces, oil stills or heaters, and in the combustion chambers of domestic-oilburner furnaces. They usually have a life equal to the heavy bricks that they replace.

They are particularly suitable for constructing experimental or laboratory furnaces because they can be cut or machined readily to any shape. However, they are not resistant to fluid slag.

There are a number of types of **special brick**, obtainable from individual manufactories. High burned kaolin refractories are particularly valuable under conditions of severe temperature and heavy load, or severe spalling conditions, as in the case of high temperature oil-fired boiler settings, or piers under enameling furnaces. Another brick for the same uses is a high-fired brick of Missouri aluminous clay.

There are a number of bricks on the market made from electrically fused materials, such as fused mullite, fused alumina, and fused zircon. These bricks, although high in cost, are particularly suitable for certain severe conditions, such as bottoms and walls of glass-melting furnaces.

Bricks of **silicon carbon,** either nitride or clay bonded, have a high thermal conductivity and find use in muffle walls and as a slag-resisting material.

Other types of refractory that find certain limited use are **forsterite** and **zirconia.** Acid-resisting bricks consisting of a dense body like stoneware are used for lining tanks and conduits in the chemical industry. Carbon blocks are used as linings for the crucibles of blast furnaces.

The **chemical composition** of some of the refractories is given in Table 6.8.13. The **physical properties** are given in Table 6.8.14. Reference should be made to ASTM Standards for details of standard tests, and to ANSI Standards for further specifications and properties.

Standard and Special Shapes

There are a large number of **standard refractory shapes** carried in stock by most manufacturers. Their catalogs should be consulted in selecting these shapes, but the common ones are shown in Table 6.8.15. These shapes have been standardized by the American Refractories Institute and by the Bureau of Simplification of the U.S. Department of Commerce.

Regenerator tile sizes $a \times b \times c$ are: 18×6 or 9×3; 18×9 or 12×4; $22\frac{1}{2} \times 6$ or 9×3; $22\frac{1}{2} \times 9$ or 12×4; $27 \times 9 \times 3$; 27×9 or 12×4; $31\frac{1}{2} \times 12 \times 4$; $36 \times 12 \times 4$.

Table 6.8.13 Chemical Composition of Typical Refractories*

No.	Refractory type	SiO_2	Al_2O_3	Fe_2O_3	TiO_2	CaO	MgO	Cr_2O_3	SiC	Alkalies	Siliceous steel-slag	High-lime steel-slag	Fused mill-scale	Coal-ash slag
											Resistance to:			
1	Alumina (fused)	8–10	85–90	1–1.5	1.5–2.2	...	....	..	......	0.8–1.3‡	E	G	F	G
2	Chrome	6	23	15†	....	...	17	38	......	....	G	E	E	G
3	Chrome (unburned)	5	18	12†	....	...	32	30	......	....	G	E	E	G
4	Fireclay (high-heat duty)	50–57	36–42	1.5–2.5	1.5–2.5	...	....	..	......	1–3.5‡	F	P	P	F
5	Fireclay (super-duty)	52	43	1	2	...	....	..	......	2‡	F	P	F	F
6	Forsterite	34.6	0.9	7.0	....	1.3	55.4							
7	High-alumina	22–26	68–72	1–1.5	3.5	...	....	..	......	1–1.5‡	G	F	F	F
8	Kaolin	52	45.4	0.6	1.7	0.1	0.2	..	......	....	F	P	G§	F
9	Magnesite	3	2	6	....	3	86	..	......	....	P	E	E	E
10	Magnesite (unburned)	5	7.5	8.5	....	2	64	10	......	....	P	E	E	E
11	Magnesite (fused)	......	......	....	....	...	....	..	......	....	F	E	E	E
12	Refractory porcelain	25–70	25–60	....	....	...	....	..	......	1–5	G	F	F	F
13	Silica	96	1	1	....	2	....	..	......	....	E	P	F	P
14	Silicon carbide (clay-bonded)	7–9	2–4	0.3–1	1	...	....	..	85–90	....	E	G	F	E
15	Sillimanite (mullite)	35	62	0.5	1.5	...	....	..	......	0.5§	G	F	F	F
16	Insulating firebrick (2,600°F)	57.7	36.8	2.4	1.5	0.6	0.5	..	......	....	P	P	G¶	F

E = excellent. G = good. F = fair. P = poor.

* Many of these data have been taken from a table prepared by Trostel, *Chem. Met. Eng.,* Nov. 1938.
† As FeO.
‡ Includes lime and magnesia.
§ Excellent if left above 1200°F.
¶ Oxidizing atmosphere.

Table 6.8.14 Physical Properties of Typical Refractories*
(Refractory numbers refer to Table 6.8.13)

Refractory no.	Fusion point °F	Pyrometric cone	Deformation under load, % at °F and lb/in²	Spalling resistance	Reheat shrinkage after 5 h, % at (°F)	Wt. of straight 9-in brick, lb
1	3,390+	39+	1 at 2,730 and 50	Good	0.5 (2,910)	9–10.6
2	3,580	41+	Shears 2,740 and 28	Poor	−0.5 to 1.0 (3,000)	11.0
3	3,580+	41+	Shears 2,955 and 28	Fair	−0.5 to 1.0 (3,000)	11.3
4	3,060–3,170	31–33	2.5–10 at 2,460 and 25	Good	±0 to 1.5 (2,550)	7.5
5	3,170–3,200	33–34	2–4 at 2,640 and 25	Excellent	±0 to 1.5 (2,910)	8.5
6	3,430	40	10 at 2,950	Fair	....	9.0
7	3,290	36	1–4 at 2,640 and 25	Excellent	−2 to 4 (2,910)	7.5
8	3,200	34	0.5 at 2,640 and 25	Excellent	−0.7 to 1.0 (2,910)	7.7
9	3,580+	41+	Shears 2,765 and 28	Poor	−1 to 2 (3,000)	10.0
10	3,580+	41+	Shears 2,940 and 28	Fair	−0.5 to 1.5 (3,000)	10.7
11	3,580+	41+		Fair	....	10.5
12	2,640–3,000	16–30		Good		
13	3,060–3,090	31–32	Shears 2,900 and 25	Poor†	+0.5 to 0.8 (2,640)	6.5
14	3,390	39	0–1 at 2,730 and 50	Excellent	+2‡ (2,910)	8–9.3
15	3,310–3,340	37–38	0–0.5 at 2,640 and 25	Excellent	−0 to 0.8 (2,910)	8.5
16	2,980–3,000	29–30	0.3 at 2,200 and 10	Good	−0.2 (2,600)	2.25

Refractory no.	Porosity	Specific heat at 60–1,200°F	Mean coefficient of thermal expansion from 60°F shrinkage point × 10⁵	Mean thermal conductivity, Btu/(ft²·h·°F·in) Mean temperatures between hot and cold face, °F 200	400	800	1,200	1,600	2,000	2,400
1	20–26	0.20	0.43	...	20	22	24	27	30	32
2	20–26	0.20	0.56	...	8	9	10	11	12	12
3	10–12	0.21								
4	15–25	0.23	0.25–0.30	5	6	7	8	10	11	12
5	12–15	0.23	0.25–0.30	6	7	8	9	10	12	13
6	23–26	0.25								
7	28–36	0.23	0.24	6	7	8	9	10	12	13
8	18	0.22	0.23	...	....	11	12	13	13	14
9	20–26	0.27	0.56–0.83	...	40	35	30	27	26	25
10	10–12	0.26								
11	20–30	0.27	0.56–0.80							
12	....	0.23	0.30	...	14	15	17	18	19	20
13	20–30	0.23	0.46§	...	8	10	12	13	14	15
14	13–28	0.20	0.24	...	....	100	80	65	55	50
15	20–25	0.23	0.30	...	10	11	12	13	14	15
16	75	0.22	0.25	...	1.6	2.0	2.6	3.2	3.8	

* Many of these data have been taken from a table prepared by Trostel, *Chem. Met. Eng.*, Nov. 1938.
† Excellent if left above 1,200°F
‡ Oxidizing atmosphere.
§ Up to 0.56 at red heat.

The following **arch, wedge,** and **key bricks** have maximum dimensions, $a \times b \times c$ of $9 \times 4\frac{1}{2} \times 2\frac{1}{2}$ in. The minimum dimensions a', b', c', are as noted: no. 1 arch, $c' = 2\frac{1}{8}$; no. 2 arch, $c' = 1\frac{3}{4}$; no. 3 arch, $c' = 1$; no. 1 wedge, $c' = 1\frac{7}{8}$; no. 2 wedge, $c' = 1\frac{1}{2}$; no. 3 wedge, $c' = 2$; no. 1 key, $b' = 4$; no. 2 key, $b' = 3\frac{1}{2}$; no. 3 key, $b' = 3$; no. 4 key, $b' = 2\frac{1}{4}$; edge skew, $b' = 1\frac{1}{2}$; feather edge, $c' = \frac{1}{8}$; no. 1 neck, $a' = 3\frac{1}{2}$, $c' = \frac{5}{8}$;

no. 2 neck, $a' = 12\frac{1}{2}$; $c = \frac{5}{8}$; no. 3 neck, $a' = 0$, $c' = \frac{5}{8}$; end skew, $a' = 6\frac{3}{4}$; side skew, $b' = 2\frac{1}{4}$; jamb brick, $9 \times 2\frac{1}{2}$; bung arch, $c' = 2\frac{3}{8}$.

Special shapes are more expensive than the standard refractories, and, as they are usually hand-molded, will not be so dense or uniform in structure as the regular brick. When special shapes are necessary, they should be laid out as simply as possible and the maximum size should be kept down below 30 in if possible. It is also desirable to make all special shapes with the vertical dimension as an even multiple of 2½ in plus one joint so that they will bond in with the rest of the brickwork.

Refractory Mortars, Coatings, Plastics, Castables, and Ramming Mixtures

Practically all brickwork is laid up with some type of jointing material to give a more stable structure and to seal the joints. This material may be ground fire clay or a specially prepared mortar containing grog to reduce the shrinkage. The bonding mortars may be divided into three general classes. The first are **air-setting mortars** which often contain chemical or organic binder to give a strong bond when dried or fired at comparatively low temperatures. Many of the air-setting mortars should not be used at extremely high temperatures because the fluxing action of the air-setting ingredient reduces the fusion point. The second class is called **heat-setting mortar** and requires temperatures of over 1,090°C (2,000°F) to produce a good bond. These mortars vary in vitrifying point, some producing a strong bond in the lower temperature ranges,

Table 6.8.15 Shapes of Firebricks

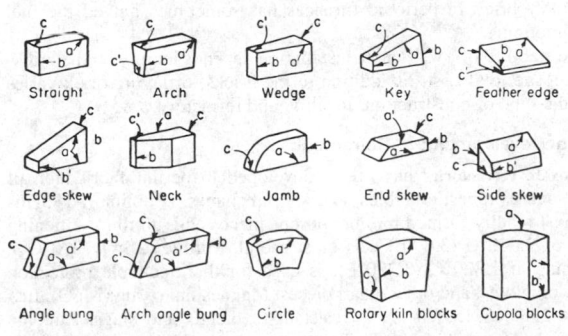

Straight Arch Wedge Key Feather edge

Edge skew Neck Jamb End skew Side skew

Angle bung Arch angle bung Circle Rotary kiln blocks Cupola blocks

and the others requiring very high temperatures to give good strength. The third classification comprises **special base mortars** such as silica, magnesite, silicon carbide, or chrome, which are specially blended for use with their respective bricks. The chrome-base mortar may be satisfactorily used with fire-clay bricks in many cases.

The refractory bonding mortars should preferably be selected on the advice of the manufacturer of the refractory to obtain good service, although there are a considerable number of independent manufacturers of mortars who supply an excellent product. From 1,330 to 1,778 N (300 to 400 lb) of dry mortar per 1,000 bricks is required for thin joints, which are desirable in most furnace construction. For thicker trowel joints, up to 2,220 N (500 lb) per 1,000 bricks is required. In the case of chrome base mortars, 2,660 N (600 lb) per 1,000 bricks should be allowed, and for magnesite cement 3,550 N (800 lb) per 1,000 bricks.

The working properties of the bonding mortar are important. Mortars for insulating refractories should be carefully selected, as many of the commercial products do not retain water sufficiently long to enable a good joint to be made. There are special mortars for this purpose which are entirely satisfactory.

Coatings are used to protect the hot surface of the refractories, especially when they are exposed to dust-laden gases or slags. These coatings usually consist of ground grog and fireclay of a somewhat coarser texture than the mortar. There are also chrome-base coatings which are quite resistant to slags, and in a few cases natural clays containing silica and feldspar are satisfactory.

The coatings can be applied to the surface of the brickwork with a brush in thin layers about 1/16 in thick, or they may be sprayed on with a cement gun, the latter method generally giving the best results. Some types of coating can be put on in much thicker layers, but care should be taken to assure that the coating selected will fit the particular brick used; otherwise it is apt to peel off in service. The coating seals the pores and openings in the brickwork and presents a more continuous and impervious service to the action of the furnace gases and slag. It is not a cure-all for refractory troubles.

Plastics and ramming mixtures are generally a mixture of fire clay and coarse grog of somewhat the same composition as the original fire-clay brick. They are used in repairing furnace walls which have been damaged by spalling or slag erosion, and also for making complete furnace walls in certain installations such as small boiler furnaces. They are also used to form special or irregular shapes, in temporary wooden forms, in the actual furnace construction.

Some of the plastics and ramming mixtures contain silicate of soda and are air-setting, so that a strong structure is produced as soon as the material is dry. Others have as a base chrome ore or silicon carbide, which make a mixture having a high thermal conductivity and a good resistance to slag erosion. These mixtures are often used in the water walls of large boiler furnaces; they are rammed around the tubes and held in place by small studs welded to the tube walls. The chrome plastic has been used with good success for heating-furnace floors and subhearths of open-hearth furnaces.

Castable mixes are a refractory concrete usually containing high-alumina cement to give the setting properties. These find considerable use in forming intricate furnace parts in wooden molds; large structures have been satisfactorily cast by this method. This type of mixture is much used for baffles in boilers where it can be cast in place around the tubes. Lightweight castables with good insulating properties are used to line furnace doors.

Furnace Walls

The modern tendency in furnace construction is to make a comparatively thin wall, anchored and supported at frequent intervals by castings or heat-resisting alloys which, in turn, are held by a structural framework and does not rest on the base. The wall may be made of heavy refractories backed up with insulating material, or of insulating refractory. Table 6.8.16 gives heat losses and heat contents of a number of wall combinations and may enable designers to pick out a wall section to suit their purpose. Solid walls built with standard 9-in brick are made up in various ways, but the hot face has usually four header courses and one stretcher course alternating.

Many modern furnaces are constructed with air-cooled walls, with refractory blocks held in place against a casing by alloy steel holders. Sectional walls made up with steel panels having lightweight insulating refractories attached to the inner surface are also used and are especially valuable for use in the upper parts of large boiler furnaces, oil stills, and similar types of construction. The sections can be made up at the plant and shipped as a unit. They have the advantage of low cost because of the light ironwork required to support them.

Many failures in furnace construction result from improper expansion joints. **Expansion joints** should usually be installed at least every 10 ft, although in some low-temperature structures the spacing may be greater. For high-temperature construction, the expansion joint allowance per foot in inches should be as follows: fire clay, 1/16 to 3/32; high alumina, 3/32 to 1/8; silica, 1/8 to 3/16; magnesite, 1/4; chrome, 5/32; forsterite, 1/4. Corrugated cardboard is often used in the joints.

The **roof** of the furnace is usually either a sprung arch or a suspended arch. A **sprung arch** is generally made of standard shapes using an inside radius equal to the total span. In most cases, it is necessary to build a form on which the arch is sprung.

In the case of arches with a considerable rise, it has been found that an inverted catenary shape is better than a circular shape for stability, and it is possible to run the sidewalls of the furnace right down to the floor in one continuous arch with almost complete elimination of the ironwork. The catenary can be readily laid out by hanging a flexible chain from two points of a vertical wall.

The **suspended arch** is used when it is desirable to have a flat roof (curved suspended arches are also made); it presents certain advantages in construction and repair but is more difficult to insulate than the sprung arch. Special suspended arch shapes are commercially available. The insulating refractory is suited to this type of construction because the steel supports are light and the heat loss is low.

Selection of Refractories

The selection of the most suitable refractory for a given purpose demands experience in furnace construction. A brick that costs twice as much as another brand and gives twice the life is preferable since the total cost includes the laying cost. Furthermore, a brick that gives longer service reduces the shutdown period of the furnace. Where slag or abrasion is severe, brick with a dense structure is desirable. If spalling conditions are important, a brick with a more flexible structure is better, although there are cases where a very dense structure gives better spalling resistance than a more open one.

High-lime slag can be taken care of with magnesite, chrome, or high-alumina brick, but if severe temperature fluctuations are encountered also, no brick will give long life. For coal-ash slag, dense fire-clay bricks give fairly good service if the temperature is not high. At the higher temperatures, a chrome-plastic or silicon carbide refractory often proves successful. When the conditions are unusually severe, air- or water-cooled walls must be resorted to; the water-cooled stud-tube wall has been very successful in boiler furnaces.

With a general freedom from slag, it is often most economical to use an insulating refractory. Although this brick may cost more per unit, it allows thinner walls, so that the total construction cost may be no greater than the regular brick. The substitution of insulating refractory for heavy brick in periodic furnaces has sometimes halved the fuel consumption.

The stability of a refractory installation depends largely on the bricklaying. The total cost, in addition to the bricks, of laying brick varies with the type of construction, locality, and refractory.

Recent Developments in Refractories

Pure-oxide refractories have been developed to permit fabrication of parts such as tubes, crucibles, and special shapes. Alumina (Al_2O_3) is the most readily formed into nonporous pieces and, up to its softening point of 2,040°C (3,690°F), is most useful. Mullite ($2SiO_2 : 3Al_2O_3$), softening at 1,820°C (3,290°F), is used for thermocouple protection tubes, crucibles, and other small pieces. Magnesium oxide (MgO), fusing at 2,800°C (5,070°F), is resistant to metals and slags. Zirconia (ZrO_2), softening at 4,600°F, is very sensitive to temperature changes

Table 6.8.16 Transmitted Heat Losses and Heat-Storage Capacities of Wall Structures under Equilibrium Conditions
(Based on still air at 80°F)

Thickness, in		Hot face temperature, °F									
		1,200		1,600		2,000		2,400		2,800	
Wall	Of insulating refractory and firebrick	HL	HS	HL	HS	HL	HS	HL	HS	HL	HS
	4½ 20	355	1,600	537	2,300	755	2,900				
4½	4½ 28	441	2,200	658	3,100	932	4,000	1,241	4,900	1,589	5,900
	4½ FB	1,180	8,400	1,870	11,700	2,660	14,800	3,600	18,100	4,640	21,600
7	4½ 28 + 2½ 20	265	3,500	408	4,900	567	6,500	751	8,100	970	9,800
	4½ FB	423	12,500	660	17,700	917	23,000	1,248	28,200		
9	4½ 28 + 4½ 20	203	4,100	311	5,900	432	7,900	573	9,900	738	12,200
	4½ FB + 4½ 20	285	13,700	437	19,200	615	24,800				
	9 20	181	3,100	280	4,300	395	5,500				
	9 28	233	4,100	349	5,800	480	7,500	642	9,300	818	11,100
	9 FB	658	15,800	1,015	21,600	1,430	27,600	1,900	34,000	2,480	40,300
11½	9 28 + 2½ 20	169	5,700	260	8,000	364	10,500	484	13,100	623	15,800
	9 FB + 2½ 20	335	22,300	514	31,400	718	40,600	962	50,400	1,233	60,300
	9 28 + 4½ 20	143	6,500	217	9,300	305	12,300	404	15,300	514	18,700
	9 FB + 4½ 20	241	24,100	367	34,500	516	44,800	690	55,100		
13½	9 20 + 4½ FB	165	5,300	255	7,300	348	9,900				
	9 28 + 4½ FB	200	6,900	302	9,700	415	12,600	556	15,700	710	19,100
	13½ FB	452	22,300	700	31,000	980	39,900	1,310	49,100	1,683	58,300
16	13½ FB + 2½ 20	275	31,200	423	43,300	588	56,300	780	70,000	994	84,200
18	9 20 + 9 FB	147	8,500	225	11,900	319	15,700				
	9 28 + 9 FB	175	10,700	266	15,100	375	19,700	493	24,600	635	29,800
	13½ FB + 4½ 20	210	34,100	318	48,400	440	62,600	587	77,500	753	92,600
	18 FB	355	28,800	532	40,300	745	52,200	1,000	64,200	1,283	76,500
20½	18 FB + 2½ 20	234	39,000	356	55,400	500	72,000	665	89,200	847	107,000
22½	18 FB + 4½ 20	182	43,200	281	61,000	392	79,200	519	97,700	667	117,600
	22½ FB	287	36,000	435	49,500	612	64,100	814	78,800	1,040	93,400

Conversion factors: $t_c = \frac{5}{9}(t_F - 32)$; 1 in = 0.0254 m.
HL = heat loss in Btu/(ft^2)/(h). HS = heat storage capacity in Btu/ft^2. 20 = 2,000°F insulating refractory. 28 = 2,800°F insulating refractory. FB = fireclay brick.
SOURCE: Condensed from "B & W Insulating Firebrick" Bulletin of The Babcock & Wilcox Co.

but can be stabilized with a few percent of lime. Beryllium oxide (BeO), softening at 2,570°C (4,660°F), has a very high thermal conductivity but must be fabricated with great care because of health hazards. Thoria (ThO_2) softens at 3,040°C (5,520°F) and has been used in crucibles for melting active metals and as a potential nuclear fuel.

Refractory carbides, sulfides, borides, silicides, and nitrides have been developed for special uses. Many have high softening points, but all have limited stability in an oxidizing atmosphere. Silicon carbide (SiC) is the most used because of its high thermal and electrical conductivity, its resistance against certain slags, and its relatively good stability in air. Molybdenum silicide ($MoSi_2$) also has considerable resistance to oxidation and, like SiC, can be used for metal-melting crucibles. Cerium sulfide (CeS_2) is a metallic-appearing material of high softening point but no resistance to oxidation. Zirconium nitride (ZrN) and titanium nitride (TiN) are also metalliclike but are not stable when heated in air. Graphite has valuable and well-known refractory properties but is not resistant to oxidation.

Refractory fibers are coming into use quite extensively. Fibers of silica-alumina glass have a use limit of about 1,090°C (2,000°F). They are used for insulating blankets, expansion joints, and other high temperature insulation. Development of higher-temperature fibers is being carried out on a small scale for use as high temperature insulation or mechanical reenforcement.

Nuclear fuels of uranium, thorium, and plutonium oxides or carbides are now extensively used in high temperature reactors.

Space vehicles are using nozzles of refractories of various kinds to withstand the high temperatures and erosion. Nose cones of sintered alumina are now used extensively because of their excellent refractory and electrical properties. Heat shields to protect space vehicles upon reentry are an important use of special refractories.

Physical Properties of High-Purity Refractories

In Table 6.8.17 are shown the properties of some of the more important pure refractory materials. It should be realized that, as purer materials become available and testing methods become more refined, some of these values will be changed.

SEALANTS

REFERENCES: Damusis (ed.), "Sealants," Reinhold. Flick, "Adhesives and Sealant Compound Formulations," 2d ed., Noyes Publ. *Mech. Eng.*, April 1991. *Engr. News Record,* Nov. 12, 1987; 1972–1993 "Annual Book of ASTM Standards." Baldwin, Selecting High Performance Building Sealants, *Plant Eng.*, Jan. 1976, pp. 58–62. Panek and Cook, "Construction Sealants and Adhesives," 2d ed, Wiley-Interscience. Panek (ed.), Building Seals and Sealants, *ASTM Spec. Tech. Publ.* 606, American Society for Testing Materials, Philadelphia. Klosowski, "Sealants in Construction," Marcel Dekker.

Klosowski's book is an excellent source of information on sealants and serves as an excellent reference. The reader should treat the following material as a guide and should consult the references and manufacturers' catalogs for expanded coverage.

Classifying sealants is somewhat arbitrary, depending as such on a multitude of properties. One of these properties tends to be more important than the rest, and that is *the ability to take cyclic movement.* Low movement ability together with short useful lifetimes tends to be found for the lower-cost sealants, and ability to take cyclic movement along with long useful lifetimes is found for the higher-cost sealants. The remaining key properties are adhesion/cohesion, hardness, modulus, stress relaxation, compression set, and weather resistance. Table 6.8.18 provides a rough comparison of sealant classes.

Table 6.8.17   Physical Properties of Some Dense,* Pure Refractories

Material	Modulus of rupture, 10^3 lb/in^2 at 70°F	Modulus of rupture, 10^3 lb/in^2 at 1,800°F	Modulus of elasticity, 10^6 lb/in^2 at 70°F	Fusion point, °F	Linear coef of expansion, 10^{-6} in/(in · °F) between 65 and 1,800°F	Thermal conductivity, Btu/(ft^2 · h · °F · in), at 212°F	Thermal conductivity, Btu/(ft^2 · h · °F · in), at 1,800°F	Thermal stress resistance
Al_2O_3	100	60	53	3,690	5.0	210	55	Good
BeO	20	10	45	4,660	4.9	1,450	130	Very good
MgO	14	12	31	5,070	7.5	240	47	Poor
ThO_2	12	7	21	5,520	5.0	62	20	Fair
ZrO_2	20	15	22	4,600	5.5	15	15	Fair
UO_2	12		25	5,070	5.6	58	20	Fair
SiC	24	24	68	5,000†	2.2	390	145	Excel.
BC	50	40	42	4,440	2.5	200	145	Good
BN	7	1	12	5,000	2.6	150	130	Good
$MoSi_2$	100	40	50	3,890	5.1	220	100	Good
C	3	4	2	7,000	2.2	870	290	Good

Conversion factors: 1 lb/in^2 = 6,894.8 N/m^2 · t_c = ⁵⁄₉(t_F − 32). 1 Btu/(ft^2 · h · °F · in) = 225 W/(m^2 · °C · m). 1 Btu/(lbm · °F) = 4,190 kg · °C.
* Porosity, 0 to 5 percent.
† Stabilized.
SOURCES: Norton, "Refractories," 3d ed.; Green and Stewart, "Ceramics: A Symposium"; Ryschkewitsch, "Oxydekramik, der Finstuffsystemme," Springer; Campbell, "High Temperature Technology"; Kingery, "Property Measurements at High Temperatures," Wiley.

Key Properties

Adhesion/Cohesion   A sealant must stick or adhere to the joint materials to prevent fluid penetration. The sealant must also stick or cohere to itself internally so that it does not split apart and thus allow leakage to occur. The ASTM C-719 test method is recommended for crucial designs and applications.

Hardness   The resistance of a sealant to a blunt probe penetration on a durometer is a measure of its hardness. Any change in this hardness over time will alert the user to check the sealant's useful performance span.

Modulus   The springiness or elastic quality of a sealant is defined by the ratio of force (stress) needed to unit elongation (strain). A high-modulus sealant can exert a rather high force on a joint, so that for weak or marginal substrates it becomes a decided disadvantage. For instance, concrete joints, having rather low tensile strength, will perform poorly with high-modulus sealants.

Stress-Relaxation   Under stress some sealants relax internally over time and stretch; an extreme example is bubble gum. Certain sealants have internal polymer chains that exhibit stress relaxation over time; mastics behave likewise, and low modulus sealants are more likely to do so than are high-modulus sealants.

A small amount of relaxation is useful in continuously tensed joints since it lowers the bond line force. In moving joints, however, a high degree of stress relaxation is problematical because the sealant tends to recover its original shape slowly and incompletely. Such joints tend to pump out their sealants.

Table 6.8.18   Sealant Classes by Movement Capabilities*

Range or movement capability	Sealant types	Comments

Low—near or at 0% of joint movement	1. Oil-based
	2. Resin-based
	3. Resinous caulks
	4. Bituminous-based mastics
	5. Polybutene-based
	6. PVA (vinyl) latex
		Short service life
		Low cost
		Major component usually low-cost mineral fillers.
		  For example, putty is mostly finely ground chalk
		  mixed with oil to form a doughlike entity.
		Some recent ones may contain about 0.2–2% silicone

Medium—0–12½% of joint width	1. Butyls 2. Latex acrylics 3. Neoprenes 4. Solvent-release acrylics	Longer service life Medium cost For protected environments and low movement, service life is perhaps 10–15 years. Typical is 3–10 years Common disadvantage is shrinkage, which may be near 30% in some products. Plasticized types tend to discolor walls
High—greater than 12½% of joint width	1. Polysulfides 2. Urethanes 3. Silicones 4. Proprietary modifications of above	Long service life Higher cost Some advertised as allowing movements of ± 25 to ± 50%, or + 100 to − 50% Generally used in commercial jobs Expanding market into do-it-yourself and over-the-counter trade

SOURCE: Adapted from Klosowski, "Sealants in Construction," Dekker.

Compression Set If a sealant is unable to reexpand to its original shape after it is compressed, it has experienced **compression set**; i.e., the compressed shape becomes permanent. When the joint reopens, the sealant must sustain high internal and bonding stresses which can cause the sealant to tear off the joint and/or cause internal tearing in the sealant itself as well as possible surface failure of the joint substrate. Polysulfide sealants are prone to such failure. Urethane sealants under combined conditions of joint movement and weathering will exhibit such failure.

Resistance to Weathering Construction sealants are designed to resist weathering. The nature of the polymer system in silicones provides that resistance, while urethane and polysulfides achieve equivalent resistance only by the use of heavy filler loads. Screening from the sun is important. Heavy filler loads or chemical sunscreens help stop radiation penetration to the internal polymer. Deep joints and opaque joint materials help sealants weather successfully.

Sealant Types

Oil-Based Caulks Combining a drying oil (such as linseed) with mineral fillers will result in a doughlike material (putty) that can be tooled into a joint. The oils dry and/or oxidize and in about 24 h develop sufficient skin to receive paint. Movement ability is about $\pm 2\frac{1}{2}$ percent of the joint. To prevent porous substrates from wicking in the caulk's oil, a primer should be applied.

Butyl Sealants (Almost Totally Cured Systems Dispersed in a Solvent) Chemically combining isobutylene and isoprene results in a butyl rubber. End-product variations are gotten by varying both the proportion of starting ingredients and the polymer chain length. Carbon black serves as a reinforcer and stabilizer in the final product. Chlorbutyl rubber is a similar sealant.

Butyl sealants are available as solvent-release caulks, soft deformable types, more rigid gaskets, and hot melts. The better butyls have movement capabilities of $\pm 12\frac{1}{2}$ percent. For applications needing 20 to 30 percent compression, butyl tapes serve well and are used extensively in glazing applications.

Advantages of butyl sealants include moderate cost, good water resistance (not for immersion), and good adhesion without primers.

Disadvantages include poor extension recovery, limited joint movement capability, and odor plus stickiness during application. They also pick up dirt and cause staining. They tend to soften under hot sun conditions such as found in cars with closed windows.

Acrylic Latex (Almost Fully Reacted in a Cartridge) Acrylic polymers are made by using a surfactant, water, and a catalyst plus appropriate monomers, which results in a high molecular weight polymer, coated with surfactant and dispersed in water. The addition of fillers, plasticizers, and other additives such as silanes for adhesion and ethyl glycol for freeze/thaw stability completes the product. Once extruded, it dries rapidly and can be painted in 30 to 50 min. Movement capability lies between $\pm 7\frac{1}{2}$ and $\pm 12\frac{1}{2}$ percent of joint width.

Advantages include excellent weathering, ease of application, ease of cleanup, low odor, low toxicity, no flammability during cure, and ability to be applied to damp substrates.

Disadvantages include the need for a primer for concrete, most wood, and plastics where joints will move and cause sealant stress. Since they harden when cold and soften when hot, they are not recommended for extremes of temperature cycling. They are not recommended for below-grade, underwater, or chronically damp applications.

Solvent Acrylic Sealants (Solvent Release Sealants) (Almost Totally Cured Systems Dispersed in a Solvent) This sealant is an acrylic polymer of short chain length, and it is based on an alkyl ester of acrylic and methacrylic acids with various modifications. The system does not quite cure to be a true elastomer, winding up somewhat more like a dry-feeling mastic, and so it exhibits stress relaxation. The best can tolerate $\pm 12\frac{1}{2}$ percent of joint movement, with the more typical between ± 8 percent and ± 10 percent.

Advantages include excellent adhesion without priming to most substrates (and, by some claims, through dirt and light oils and perhaps even damp surfaces), good weathering, and moderate cost.

Disadvantages include the need to heat the cartridge or dispensing container for easy application in cool weather, offensive odor during cure (thus requiring ventilation in restricted quarters), and slow recovery from either extension or contraction. The cured seal gets very hard in cold weather and loses flexibility, thus compromising movement capability.

Two-Part Polysulfide Sealants These sealants are the pioneers of high-range products with life expectancies (in certain environments) of up to 20 years and movement capability of up to ± 25 percent of joint width. It is widely used in curtain wall construction and in concrete/masonry applications. In glazing applications shielded from direct sunlight, polysulfide sealant is suitable. Since polysulfides also can be produced with excellent solvent resistance, they are used as fuel tank sealants and in other fuel contact applications. Fillers are generally carbon blacks, clays, and mineral materials. Curing agents may be lead peroxide, cyclic amides, diisocyanates, etc. Silanes act as adhesion promoters and stabilizers. Most polysulfides are mixed and dispensed on the job at the time of use.

Advantages include good extensibility, recovery, cure with two-package systems (tack-free in about 36 to 48 h and about 1 week for full cure), adhesion, and resistance to weathering and aging.

Disadvantages include the need for site mixing, requirements for primers on porous surfaces, sun exposure cracking, and compression set effects which emerge in about 5 to 10 years.

One-Part Polysulfide Sealants These sealants are almost on a par in cured performance to two-part polysulfide sealants. They need no mixing at the time of use and are thus ready to apply. Curing depends upon moisture or oxygen from the air and is relatively slow, requiring days, weeks, or more, after which they become tack-free. Dry and cold weather can lengthen the curing process.

Two-Part Polyurethane Sealants These sealants are second-generation premium products, and they outperform the polysulfides. They weather well, perform well in expansion joints by tolerating large movement (± 25 percent of joint movement), and are tough and resilient. The top of the product line can achieve 85 to 90 percent recovery from compression set. Adhesion is very good. They stay clean for almost the life of the sealant, or about 10 to 20 years. Urethanes tend to be water-sensitive and to bubble if they contact water at the curing surface, and they tend to stiffen with age. Urethane sealants are widely used in construction and for nonglazing joints in walkways and pedestrian traffic areas.

One-Part Urethane Sealants These sealants are comparable to two-part urethanes, but come in a single package ready to use. Package stability is problematical because of moisture sensitivity. One-part systems take a long time to cure, especially at low temperatures and low humidities.

Silicone Sealants In some aspects of performance, these sealants represent third-generation products, and others are at least second-generation in quality. Only silicone polymers comprise a true silicone sealant. They generally contain mineral or other inorganic fillers along with functional silane or siloxane cross-crosslinker, with special additives included for specific purposes.

In terms of temperature exposure, silicone polymers are about one-tenth as sensitive as typical hydrocarbons or polyether chains. Thus they extrude easily at both low and high temperatures. Temperature stability ranges from about -40 to $250°F$ and for some to more than $400°F$. Once cured, they maintain elasticity very well at both temperature extremes and weather quite well (lasting 12 to 20 times longer than typical organic sealants, as indicated by weatherometer tests). Warranties often extend from 20 to 50 years. Joint movement performance embraces a wide range from ± 12 to $+100/-50$ percent of joint width. Adhesion qualities are excellent, allowing some common silicones unprimed application to most substrates (including concrete), and some are applicable for service under conditions of total submersion in water.

These performance levels have made silicone sealants the industry standard to which others are compared. Competitive products marketed using terms such as *siliconized, siliconelike, modified with silicone, modified silicone,* and so forth contain a minimal amount of true sili-

Table 6.8.19 Summary of Sealant Properties*

Type of sealant	Low performance: Oil-based, one-part	Medium performance: Latex (acrylic), one-part	Medium performance: Butyl skimming, one-part	Acrylic: Solvent-release	Polysulfide: One-part	Polysulfide: Two-part	Urethane: One-part	Urethane: Two-part	Silicone: One-part	Silicone: Two-part
Movement ability in percent of joint width (recommended maximum joint movement)	±3	±5, ±12½	±7.5	±10–12½	±25	±25	±25	±25	±25 to +100/−50	±12½–±50
Life expectancy, years†	2–10	2–10	5–15	5–20	10–20	10–20	10–20	10–20	10–50	10–50
Service temperature range, °F (°C)	−20 to +150 (−29 to +66)	−20 to +180 (−20 to +82)	−40 to +80 (−40 to +82)	−20 to +180 (−29 to +82)	−40 to +180 (−40 to +82)	−60 to +180 (−51 to +82)	−40 to +180 (−40 to +82)	−25 to +180 (−32 to +82)	−65 to +400 (−54 to +200)	−65 to +400¶ (−54 to +200)
Recommended application temperature range, °F‡	40–120	40–120	40–120	40–180	40–120	40–120	40–120	40–180	−20 to +160	−20 to +160
Cure time§ to a tack-free condition, h	6	½–1	24	36	24§	36–48§	12–36	24	1–3	½–2
Cure time§ to specified performance, days	Continues	5	Continues	14	30–45	7	8–21	3–5	5–14	¼–3
Shrinkage, %	5	20	20	10–15	8–12	0–10	0–5	0–5	0–5	0–5
Hardness, new (1–6 mo), A scale at 75°F		15–40	10–30	10–25	20–40	20–45	20–45	10–45	15–40	15–40
Hardness, old (5 yr), A scale at 75°F		30–45	30–50	30–55	30–55	20–55	30–55	20–60	15–40	15–50
Resistance to extension at low temperature	Low to moderate	Moderate to high	Moderate to high	High	Low to high	Low to moderate	Low to high	Low to high	Low	Low
Primer required for sealant bond to: Masonry	No	Yes	No	No	Yes	Yes	Yes	Yes	No	No
Metal	No	Sometimes	No	No	Yes	Yes	No	No	?	?
Glass	No	No	No	No	No	No	No	No	No	No
Applicable specifications: United States	TT-C-00593b	ASTM TTS-00230	ASTM TT-S-001657	ASTM C-920 ITS-00230	ASTM C-920 TT-C-00230C	ASTM C-920 TTS-00227E	ASTM C-920 TTS-00230C	ASTM C-920 TTS-00227C	ASTM C-920 TTS-00230C TTS-001543A	ASTM C-920 TTS-001543A TTS-00230C
Canada	CAN 2-19.2M		CGSB 19-GP-14M	CGSB 19-GP-SM	CAN 2-19.13M		CAN 2-19.13M		CAN 2-19.18M	

* Data from manufacturer's data sheets; U.S.-made sealants are generally considered.
† Affected by conditions of exposure.
‡ Some sealants may require heating in low temperatures.
§ Affected by temperature and humidity.
¶ The wide range in performance of the various types of silicone sealants is discussed briefly in the text preceding this table.
SOURCE: Adapted from J. M. Klosowski, "Sealants in Construction," Marcel Dekker, 1989. By permission.

cone, usually from 0.1 to 10 percent. Most silicones are fully elastic and exhibit the smallest compression or tension set, ranging from 85 to 99 percent recovery.

Some silicones pick up dirt to varying degrees, but owing to their relatively fast curing times (tack-free in 15 min to 3 h), they are less prone to do so when compared with polyurethane and polysulfides, which have longer cure times. After curing, however, because of their relatively soft surface, silicones tend to accumulate dirt faster. Unfortunately, the dirt cannot be completely washed off. To obtain the advantage of silicone's durability and to avoid dirt pickup, one resorts to overcoating the silicone sealant with a hard silicone resin. Alternately, one can dust the surface of the uncured, tacky silicone with powdered chalk or some similar material; this can be used as a base upon which to apply paint. Some silicone sealants are formulated to be paintable, but, in general, silicones simply will not accept paint. Silicones abrade easily and are not used in heavily trafficked areas. Silicones which release acetic acid should not be used on marble, galvanized metal, copper, cementaceous substances, or other corrodable materials. Silicones which release amines and the like must not be used on copper and for electrical applications. Neutral-cure systems are available which can be safely used on almost all substrates. The reader is directed to consult the supplier when in doubt about compatibility.

Water-Based Silicones Water-based silicones combine the durability and longevity of silicones with ease of cleanup, ease of application, and paintability. They are true silicone systems, being dispersions of polymers, crosslinkers, and catalysts in water.

Table 6.8.19 gives a summary of sealant properties.

Miscellaneous

Hot Pours Several categories of materials are softened by heating and so can be injected or poured into joints (almost exclusively horizontal). The majority of these materials are filled out with bituminous substances such as asphalt tars, coal tars, etc. Such bituminous bases can be blended with urethanes, polysulfides, polyvinyl chlorides, etc. Highway joint sealing constitutes the largest use of hot-pour sealants. They are inexpensive and easy to use, but deteriorate rapidly in typical U.S. climate and suffer surface crazing and lose elasticity (stiffen) in cold weather.

Forever Tacky Nonhardening sealants, such as Hylomar, remain tacky for years and retain their original sealing qualities very well. They were originally developed for jet-engine joints in the 1950s, but have been applied to resist extreme vibrations and to make emergency repairs. Continued improvements in this product are directed toward broadening its compatibility with a wider range of chemicals and oils. Its stable, tacky condition for long periods enables easy disassembly of parts it has bonded.

Formed-in-Place Gasket Replacing die-cut gaskets can reduce costs, and for this reason, formed-in-place gaskets have been developed. One such product, Dynafoam, a curable thermoplastic elastomer, forms in-place gaskets by pouring the sealant into the desired shape. Automotive applications include taillight assemblies and sunroof gaskets to seal out moisture, wind, and dust. This type of sealant is designed to fill the niche between traditional hot-melt adhesives and gaskets, such as butyl rubbers or ethyl vinyl acetates, and curable silicone and urethane sealants. Curing time is within a few minutes, and they resist softening up to about 280°F (continuous exposure) and up to 400°F (for periods of 1 h or less).

Foamed-in-Place Sealants In commercial operations, a sealant such as Dynafoam is heated to 180°F, pumped through a heated hose into a pressurized, dry gas chamber (usually nitrogen), and pumped further through a heated hose and dispensing nozzle. Upon exiting the nozzle, the sealant expands to bead size as the entrained nitrogen expands. The wormlike bead is then applied directly to a part and forms a gasket. The sealant begins curing upon contact with air.

Similar materials in small pressurized canisters contain a room-temperature curing sealant. Upon release, the sealant expands into a foam, fills the joint, and cures in place. It is applied easily to provide local insulation, to fill crevices (e.g., to block wind), and so forth. For extensive operations of this kind, this material is supplied in much larger, yet portable, containers.

6.9 CEMENT, MORTAR, AND CONCRETE
by William L. Gamble

REFERENCES: Neville, "Properties of Concrete," Wiley. Sahlin, "Structural Masonry," Prentice-Hall. Mindless and Young, "Concrete," Prentice-Hall. "BOCA Basic Building Code," Building Officials and Code Administrators International. "Concrete Manual," U.S. Bureau of Reclamation. ACI 318, "Building Code Requirements for Reinforced Concrete"; ACI 211.1, "Standard Practice for Selecting Proportions for Normal, Heavyweight, and Mass Concrete"; ACI 304, "Recommended Practice for Measuring, Mixing, Transporting, and Placing Concrete"; American Concrete Institute, Detroit. "Building Code Requirements for Masonry Structures (ACI 530/ TMS 402/ASCE 5)," American Concrete Institute, The Masonry Society, and American Society of Civil Engineers.

CEMENT

Normal portland cement is used for concrete, for reinforced concrete, and either with or without lime, for mortar and stucco. It is made from a mixture of about 80 percent carbonate of lime (limestone, chalk, or marl) and about 20 percent clay (in the form of clay, shale, or slag). After being intimately mixed, the materials are finely ground by a wet or dry process and then calcined in kilns to a clinker. When cool, this clinker is ground to a fine powder. During the grinding, a small amount of gypsum is usually added to regulate the setting of the cement. The chemical analysis of 32 American type I cements gives the following average percentage composition: silica (SiO_2), 21.92; alumina (Al_2O_3), 6.91; iron oxide (Fe_2O_3), 2.91; calcium oxide (CaO), 62.92; magnesium oxide (MgO), 2.54; sulfuric oxide (SO_3), 1.72; alkalies (R_2O_3), 0.82; loss on ignition, 1.50, insoluble residue 0.20.

Types and Kinds of Cements Five types of portland cements are covered by ASTM specification C150.

Normal portland cement, type I, is used for purposes for which another type having special properties is not required. Most structures, pavements, and reservoirs are built with type I cement.

Modified portland cement, type II, generates less heat from its hydration and is more resistant to sulfate attacks than type I. This cement is used in structures having large cross sections, such as large abutments and heavy retaining walls. It may also be used in drainage where a moderate sulfate concentration exists.

High-early-strength portland cement, type III, is used when high strengths are required in a few days. Use of high-early-strengths will allow earlier removal of forms and shorter periods of curing.

Low-heat portland cement, type IV, generates less heat during hydration than type II and is used for mass concrete construction such as large dams where large temperature increases would create special problems. Type IV cement gains strength more slowly than type I. The tricalcium aluminate content is limited to 7 percent.

Sulfate-resisting portland cement, type V, is a special cement, not readily available, to be used when concrete is exposed to severe sulfate attack. Type V cements gain strength more slowly than type I cement. The tricalcium aluminate content is limited to a maximum of 5 percent.

Air-entraining portland cements purposely cause air, in minute, closely spaced bubbles, to occur in concrete. Entrained air makes the concrete more resistant to the effects of repeated freezing and thawing and of the deicing agents used on pavements. To obtain such cements, air-entraining agents are interground with the cement clinker during manufacture. Types I to III can be obtained as air-entraining cements and are then designated as types IA, IIA, and IIIA, under ASTM C150.

Portland blast-furnace slag cements are made by grinding granulated high-quality slag with portland-cement clinker. Portland blast-furnace slag cement type IS and air-entraining portland blast-furnace slag cement type IS-A are covered by ASTM specification C595. Provisions are also made for moderate-heat-of-hydration cements (MH) and moderate-sulfate-resistance cements (MS), or both (MH-MS). Type IS cements initially gain strength more slowly but have about the same 28-day strength as type I cements.

White portland cement is used for architectural and ornamental work because of its white color. It is high in alumina and contains less than 0.5 percent iron. The best brands are true portlands in composition.

Portland-pozzolan cement is a blended cement made by intergrinding portland cement and pozzolanic materials. Two types, type IP (portland-pozzolan cement) and type IP-A (air-entraining portland-pozzolan cement), are covered in ASTM specification C595.

Masonry cement, ASTM specification C91, is a blended cement used in place of job cement-lime mixtures to reduce the number of materials handled and to improve the uniformity of the mortar. These cements are made by combining either natural or portland cements with fattening materials such as hydrated lime and, sometimes, with air-entraining admixtures.

Waterproofed cement is sometimes used where a waterproof or water-repellent concrete or mortar is particularly desirable. It is cement ground with certain soaps and oils. The effectiveness is limited to 3 or 4 ft of water pressure.

Shrinkage-compensated cements are special portland cements which expand slightly during the moist curing period, compensating for the shrinkage accompanying later drying. They are used primarily to aid in producing crack-free concrete floor slabs and other members. ASTM C845 covers these cements.

Regulated-set cements are special portland cements formulated to set in very short times, producing usable concrete strengths in regulated times of as little as 1 h or less. Such concretes are obviously well suited to repair work done when it is important to minimize downtime.

Portland Cement Tests Cement should be tested for all but unimportant work. Tests should be made in accordance with the standard specifications of the ASTM or with the federal specifications where they apply. Samples should be taken at the mill, and tests completed before shipments are made. When this is not possible, samples should be taken at random from sound packages, one from every 10 bbl or 40 bags, and mixed. The total sample should weigh about 6 lb. ASTM requirements for standard portland cements are given in ASTM specification C150.

The **autoclave soundness test** consists of determining the expansion of a 1 in sq neat cement bar 10 in long which, after 24-h storage in 90 percent or greater humidity, is placed in an autoclave, where the pressure is raised to 295 lb/in^2 in about 1 h, maintained for 3 h, and then brought back to normal in 1½ h. Cements that show over 1 percent expansion may show unsoundness after some years of service; the ASTM allows a maximum of 0.80 percent.

Time of Setting Initial set should not be less than 45 min when Vicat needle is used or 60 min when Gilmore needle is used. Final set should be within 10 h. Cement paste must remain plastic long enough to be properly placed and yet submit to finishing operations in a reasonable time.

Compressive Strength Minimum requirements for average compressive strength of not less than three 2-in cubes composed of 1 part (by weight) cement and 2.75 parts standard graded mortar sand tested in accordance with ASTM method C109, are shown in Table 6.9.1.

LIME

Common lime, or **quicklime,** when slaked or hydrated, is used for interior plastering and for lime mortar. Mixed with cement, it is used for lime and cement mortar and for stucco. Mortars made with lime alone are not satisfactory for thick walls because of slow-setting qualities. They must never be used under water. Quicklime slakes rapidly with water with much heat evolution, forming calcium hydrate (CaH_2O_2). With proper addition of water, it becomes plastic, and the volume of putty obtained is 2 or 3 times the loose volume of the lime before slaking, and its weight is about 2½ times the weight of the lime. Plastic lime sets by drying, by crystallization of calcic hydrate, and by absorbing carbonic acid from the air. The process of hardening is very slow. Popping is likely to occur in plaster unless the lime is sound, as indicated by an autoclave test at 120 lb/in^2 pressure for 2 h.

Magnesium lime, used for the same purposes as common or high-calcium lime, contains more than 20 percent magnesium oxide. It slakes more slowly, evolves less heat, expands less, sets more rapidly, and produces higher-strength mortars than does high-calcium quicklime.

Pulverized and granulated limes slake completely much more quickly than ordinary lump lime. They are sometimes waterproofed by the addition of stearates and other compounds similar to those used in cement for the same purpose. The waterproofing treatment retards the slaking.

Hydrated lime is a finely divided white powder manufactured by slaking quicklime with the requisite amount of water. It has the advantage over lime slaked on the job of giving a more uniform product, free from unslaked lime. It does not have plasticity or water retention equal to freshly slaked quicklime.

Hydraulic hydrated lime is used for blending with portland cement and as a masonry cement. It is the hydrated product of calcined impure limestone which contains enough silica and alumina to permit the formation of calcium silicates.

Quicklime is covered by ASTM C5 and hydrated limes by ASTM C6, C206, and C821. The testing methods are covered by other ASTM specifications referenced in the above specifications. ASTM C5 contains instructions and cautions for slaking quicklime, as the chemical reactions following the addition of water to quicklime are exothermic and potentially dangerous.

Table 6.9.1 Minimum Requirements for Average Compressive Strength

Age of test, days	Storage of test pieces	Compressive strength, lb/in^2				
		Normal	Moderate heat	High early strength	Low heat	Sulfate-resistant
1	1 day moist air	....	....	1,800	....	....
3	1 day moist air, 2 days water	1,800	1,500	3,500	....	1,200
7	1 day moist air, 6 days water	2,800	2,500	....	1,000	2,200
28	1 day moist air, 27 days water	....	....	....	2,500	3,000

1,000 lb/in^2 = 6.895 MPa.
SOURCE: Reprinted from ASTM C150, with permission from ASTM.

Table 6.9.2

Sieve no.	100	50	30	16	8	4
Sieve opening, in	0.0059	0.0117	0.0234	0.0469	0.0937	0.187
Wire diam, in	0.0043	0.0085	0.0154	0.0256	0.0394	0.0606
Sieve size, in	⅜	¾	1	1½	2	3
Sieve opening, in	0.375	0.750	1.00	1.50	2.00	3.00
Wire diam, in	0.0894	0.1299	0.1496	0.1807	0.1988	0.2283

AGGREGATES

Sand

Sand to be used for mortar, plaster, and concrete should consist of clean, hard, uncoated grains free from organic matter, vegetable loam, alkali, or other deleterious substances. Vegetable or organic impurities are particularly harmful. A quantity of vegetable matter so small that it cannot be detected by the eye may render a sand absolutely unfit for use with cement. Stone screenings, slag, or other hard inert material may be substituted for or mixed with sand. Sand for concrete should range in size from fine to coarse, with not less than 95 percent passing a no. 4 sieve, not less than 10 percent retained on a no. 50 sieve, and not more than 5 percent (or 8 percent, if screenings) passing a no. 100 sieve. A straight-line gradation on a graph, with percentages passing plotted as ordinates to normal scale and sieve openings as abscissas to logarithmic scale, gives excellent results.

The grading of sand for mortar depends upon the width of joint, but normally not less than 95 percent should pass a no. 8 sieve, and it should grade uniformly from coarse to fine without more than 8 percent passing a no. 100 sieve.

Sand for plaster should have at least 90 percent passing a no. 8 sieve and not more than 5 percent passing the no. 100 sieve.

Silt or clayey material passing a no. 200 sieve in excess of 2 percent is objectionable.

Test of Sand   Sand for use in important concrete structures should always be tested. The strength of concrete and mortar depends to a large degree upon the quality of the sand and the coarseness and relative coarseness of the grains. Sand or other fine aggregate when made into a mortar of 1 part portland cement to 3 parts fine aggregate by weight should show a tensile strength at least equal to the strength of 1:3 mortar of the same consistency made with the same cement and standard sand. If the aggregate is of poor quality, the proportion of cement in the mortar or concrete should be increased to secure the desired strength. If the strength is less than 90 percent that of Ottawa sand mortar, the aggregate should be rejected unless compression tests of concrete made with selected aggregates pass the requirements. The standard Ottawa sand gradation is described in ASTM specification C109. This sand is supplied by the Ottawa Silica Co., Ottawa, Ill. The compressive strength of 2-in cubes made from a cement and sand mixture with a 0.9 water-cement ratio and a flow of 100 percent should equal 90 percent of the strength of similar cubes made with graded Ottawa sand.

The ASTM standard test (C40) for the presence of injurious organic compounds in natural sands for cement mortar or concrete is as follows: A 12-oz graduated glass prescription bottle is filled to the 4½-oz mark with the sand to be tested. A 3 percent solution of sodium hydroxide (NaOH) in water is then added until the volume of sand and liquid, after shaking, gives a total volume of 7 liquid oz. The bottle is stoppered, shaken thoroughly, and then allowed to stand for 24 h. A standard-reference-color solution of potassium dichromate in sulfuric acid is prepared as directed in ASTM D154. The color of the clear liquid above the sand is then compared with the standard-color solution; if the liquid is darker than the standard color, further tests of the sand should be made before it is used in mortar or concrete. The standard color is similar to light amber.

Coarse Aggregate

Broken Stone and Gravel   Coarse aggregate for concrete may consist of broken stone, gravel, slag, or other hard inert material with similar characteristics. The particles should be clean, hard, durable, and free from vegetable or organic matter, alkali, or other deleterious matter and should range in size from material retained on the no. 4 sieve to the coarsest size permissible for the structure. For reinforced concrete and small masses of unreinforced concrete, the maximum size should be that which will readily pass around the reinforcement and fill all parts of the forms. Either 1- or 1½-in diam is apt to be the maximum. For heavy mass work, the maximum size may run up to 3 in or larger.

The coarse aggregate selected should have a good performance record, and especially should not have a record of alkali-aggregate reaction which may affect opaline and chert rocks.

Lightweight aggregates are usually pumice, lava, slag, burned clay or shale, or cinders from coal and coke. It is recommended that lightweight fine aggregate not be used in conjunction with lightweight coarse aggregate unless it can be demonstrated, from either previous performance or suitable tests, that the particular combination of aggregates results in concrete that is free from soundness and durability problems. In case of doubt, the concrete mix should be designed using sand fine aggregate, and lightweight coarse aggregate. Their application is largely for concrete units and floor slabs where saving in weight is important and where special thermal insulation or acoustical properties are desired.

Heavyweight aggregates are generally iron or other metal punchings, ferrophosphate, hematite, magnetite, barite, limenite, and similar heavy stones and rocks. They are used in concrete for counterweights, dry docks, and shielding against rays from nuclear reactions.

Fineness Modulus   The fineness modulus, which is used in the Abrams method as an index of the characteristics of the aggregates, is the sum of the cumulative percentages (divided by 100) which would be retained by all the sieves in a special sieve analysis. The sieves used in this method are nos. 100, 50, 30, 16, 8, and 4 for fine aggregates and these plus the ⅜-, ¾-, 1½-, and 3-in sizes for coarse aggregates. A high fineness modulus indicates a relatively low surface area because the particles are relatively large, which means less water required and, therefore, a higher concrete strength. Aggregates of widely different gradation may have the same fineness modulus.

ASTM standard sieves for analysis of aggregates for concrete have the sizes of opening and wire shown in Table 6.9.2.

WATER

Water for concrete or mortar should be clean and free from oil, acid, alkali, organic matter, or other deleterious substance. Cubes or briquettes made with it should show strength equal to those made with distilled water. Water fit for drinking is normally satisfactory for use with cement. However, many waters not suitable for drinking may be suitable for concrete. Water with less than 2,000 ppm of total dissolved solids can usually be used safely for making concrete. (See Sec. 6.10, Water.)

Seawater can be used as mixing water for plain concrete, although 28-day strength may be lower than for normal concrete. If seawater is used in reinforced concrete, care must be taken to provide adequate cover with a dense air-entrained concrete to minimize risks of corrosion. Seawater should not be used with prestressed concrete.

ADMIXTURES

Admixtures are substances, other than the normal ingredients, added to mortars or concrete for altering the normal properties so as to improve them for a particular purpose. Admixtures are frequently used to entrain air, increase workability, accelerate or retard setting, provide a pozzolanic reaction with lime, reduce shrinkage, and reduce bleeding. However, before using an admixture, consideration must be given to its effect on properties other than the one which is being improved. Most important is consideration of possible changes in the basic mix which might make the admixture unnecessary. Particular care must be used when using two or more admixtures in the same concrete, such as a retarding agent plus an air-entraining agent, to ensure that the materials are compatible with each other when mixed in concrete. The properties of chemical admixtures should meet ASTM C494 and air-entraining agents should meet ASTM C260.

Air-entraining agents constitute one of the most important groups of admixtures. They entrain air in small, closely spaced, separated bubbles in the concrete, greatly improving resistance to freezing and thawing and to deicing agents.

Accelerators are used to decrease the setting time and increase early strength. They permit shorter curing periods, earlier form removal, and placing at lower temperatures. Calcium chloride is the most frequently used accelerator and can be used in amounts up to 2 percent of the weight of the cement, but must never be used with prestressed concrete.

Retarders increase the setting time. They are particularly useful in hot weather and in grouting operations.

Water reducers are used to increase the workability of concrete without an accompanying increase in the water content. The most recent development is the use of high-range water reducers, or superplasticizers, which are polymer liquid materials added to the concrete in the mixer. These materials lead to spectacular temporary increases in slump for a given water content, and may be used to obtain several different results. A normal mix can be transformed into a "flowing" concrete which will practically level itself. Or the addition of the superplasticizer can allow the mix to be redesigned for the same consistency at the time of mixing, but considerably less water will be required and consequently a higher concrete strength can be reached without adding cement, or the same strength can be reached with a lower cement content.

Fly ash from coal-burning power plants can be added to concrete mixes to achieve several effects. The very fine material tends to act as a lubricant in the wet concrete, and thus may be added primarily to increase the workability. Fly ash also enters the chemical reactions involved in the setting of concrete and leads to higher strengths. Fly ash is often used in mass concrete such as dams as a replacement for part of the cement, in order to save some material costs, to reduce the rate at which the hydration produces heat, and to increase the long-term strength gain potential of the concrete. Fly ash is a pozzolan, and should meet requirements of ASTM C618. Free carbon from incomplete combustion must be strictly limited.

Silica Fume   Silica fume, also called **condensed silica fume** or **microsilica**, is another pozzolan which may be added to concrete as a supplement or partial replacement to the cement. This material reacts with the lime in the cement and helps lead to very high-strength, low-permeability concrete. The silica-lime reaction can occur very quickly, and retarders are often necessary in concretes containing silica fume. Concretes containing silica fume tend to be quite dark gray compared with ordinary concrete mixes, and concretes containing fly ash tend to be lighter than the usual concrete gray.

Ground Granulated Blast-Furnace Slag   Ground granulated blast-furnace slag has been used in Europe as a supplement to or partial replacement for portland cement for several decades but has become available in North America relatively recently. Concretes made with this material tend to develop very low permeabilities, which improves durability by retarding oxygen and chloride penetration, which in turn

protects the reinforcement from corrosion for longer periods. Chloride sources include both deicing chemicals and seawater, and these concretes have been successfully used to resist both these severe environments. They are also sulfate-resistant, which is important in those geographic areas which have sulfate-bearing groundwater.

Other admixtures may be classed as gas-forming agents, pozzolanic materials, curing aids, water-repelling agents, and coloring agents.

MORTARS

Properties desirable in a mortar include (1) good plasticity or workability, (2) low volume change or volume change of the same character as the units bonded, (3) low absorption, (4) low solubility and thus freedom from efflorescence, (5) good strength in bond and ample strength to withstand applied loads, (6) high resistance to weathering.

Mortar Types   There are several different types of mortar which are suitable for masonry construction of different kinds, uses, and exposure conditions. The BOCA Basic Building Code lists several mortar types, and their permitted uses are given in Table 6.9.3. The makeup of these mortars is specified in terms of volumes of materials, as listed in Table 6.9.4, also from BOCA. There is considerable leeway given for the proportions, and the materials are mixed with water to the consistency desired by the mason. The portland cement, masonry cement, and lime may be purchased separately or blended, and for small jobs a dry mix mortar containing everything except the water may be purchased in bags. The hydrated lime tends to give the fresh mortar plasticity, or stickiness, which is necessary for the proper bedding of the masonry or concrete units being laid.

Additional mortar types may be used for reinforced masonry, in which steel bars are used to increase the strength in the same way that concrete is reinforced. Other mortars may be used for filling the cavities in concrete block construction, or concrete masonry unit construction, and this mortar may be referred to as grout.

Design procedures and requirements for concrete and clay masonry are contained in ACI 530/TMS 402/ASCE 5. **Unreinforced masonry** cannot be relied on to resist tension stresses. Compressive stresses permitted under the empirical design rules vary widely, depending on the strength and type of masonry unit. The highest stress permitted is 350 lb/in^2 for high-strength solid bricks, while the lowest is 60 lb/in^2 for the lowest-strength hollow-concrete blocks, where the stresses are computed on the gross area and the units are laid with type M or S mortar. Type N mortar leads to stresses which are 10 to 15 percent lower.

Mortars for Plastering and Stucco   Interior plastering is much less common than it was in the past because of the use of gypsum board products for walls, but is done in various instances. Common lime plaster consists of hydrated lime (usually purchased in prepared, bagged

Table 6.9.3   Masonry and Mortar Types

Type of masonry	Type of mortar permitted
Masonry in contact with earth	M or S
Grouted and filled cell masonry	M or S
Masonry above grade or interior masonry	
Piers of solid units	M, S, or N
Piers of hollow units	M or S
Walls of solid units	M, S, N, or O
Walls of hollow units	M, S, or N
Cavity walls and masonry bonded hollow walls	
Design wind pressure exceeds 20 lb/ft^2	M or S
Design wind pressure 20 lb/ft^2 or less	M, S, or N
Glass block masonry	S or N
Non-load-bearing partitions and fireproofing	M, S, N, O, or gypsum
Firebrick	Refractory air-setting mortar
Linings of existing masonry, above or below grade	M or S
Masonry other than above	M, S, or N

1 lb/ft^2 = 47.9 Pa.
SOURCE: BOCA Basic Building Code, 1984.

Table 6.9.4   Mortar Proportions Specification Requirements (Parts by Volume)

Mortar type	Portland cement	Masonry cement	Hydrated lime or lime putty Min	Max	Damp loose aggregate
M	1			¼	
	1	1			
S	1		¼	½	Not less than 2¼ and not more than 3 times the sum of the volumes of the cements and lime used
	½	1			
N	1		½	1¼	
		1			
O		1			
	1		1¼	2½	

SOURCE: BOCA Basic Building Code, 1984.

form), clean coarse sand, and hair or fiber. The plastering is normally done in two or three layers, with the base coats containing about equal volumes of lime and sand, plus hair or fiber, and the final layer containing less sand. In three-layer work, the first layer is the scratch coat, the second the brown coat, and the last the white or skim coat. The scratch coat is applied directly to either a masonry wall or the lath in a frame wall.

Skim coat is a finish coat composed of lime putty and fine white sand. It is placed in two layers and troweled to a hard finish. **Gaged skim coat** is skimming mixed with a certain amount of plaster of paris, which makes it a hard finish. **Hard finish** consists of 1 part lime putty to 1 or 2 parts plaster of paris.

Keene's cement, which is an anhydrous calcined gypsum with an accelerator, is much used as a hard-finish plaster.

Gypsum plaster (ASTM C28) lacks the plasticity and sand-carrying capacity of lime plaster but is widely used because of its more rapid hardening and drying and because of the uniformity obtainable as the result of its being put up in bags ready-mixed for use.

Gypsum ready-mixed plaster should contain not more than 3 ft³ of mineral aggregate per 100 lb of calcined gypsum plaster, to which may be added fiber and material to control setting time and workability. Gypsum neat plaster used in place of sanded plaster for second coat should contain at least 66 percent $CaSO_4 : \frac{1}{2}H_2O$; the remainder may be fiber and retarders. Calcined gypsum for finishing coat may be white or gray. If it contains no retarder, it should set between 20 and 40 min; if retarded, it should set between 40 min and 6 h.

Cement plaster is used where a very hard or strong plaster is required, e.g., for thin metal-lath partitions or as a fire protection. It should contain not more than 2 parts sand, by dry and loose volume, to 1 part portland cement. Lime putty or hydrated lime is added up to 15 percent by volume of the cement.

Under moisture conditions where the plaster will not dry rapidly, **curing** at temperatures above 60 and below 90°F is absolutely necessary if cracking is to be avoided.

Stucco   Stucco is used for exterior plastering and is applied to brick or stone or is plastered onto wood or metal lath. For covering wooden buildings, the stucco is plastered either on wood lath or on metal lath in three coats, using mortar similar to that for brick or stone. Concrete in northern climates exposed to frost should never be plastered but should be finished by rubbing down with carborundum brick or similar tool when the surface is comparatively green. It may also be tooled in various ways. Whenever stucco is used, extreme care must be taken to get a good bond to the supporting surface. While stucco has traditionally been applied with a trowel, it can also be applied pneumatically. Similar material used for lining metal pipes can be applied with a centrifugal device which "slings" the material onto the pipe wall.

For three-coat work on masonry or wood lath, the first or scratch coat should average ¼ in thick outside the lath or surface of the brick. The thickness of the second coat should be ⅜ to ½ in, while the finish coat should be thin, i.e., ⅛ in or not more than ¼ in. The second coat should generally be applied 24 h after the first or scratch coat. The finish coat should not be applied in less than 1 week after the second coat. Proportions of mix for all coats may be ⅕ part hydrated lime, 1 part portland cement, 3 parts fairly coarse sand, measured by volume. Stucco work should not be put on in freezing weather and must be kept moist for at least 7 days after application of the mortar.

Mineral colors for stucco, if used, should be of such composition that they will not be affected by cement, lime, or the weather. The best method is to use colored sands when possible. The most satisfactory results with colored mortar are obtained by using white portland cement. Prepared patented stuccos which are combinations of cement, sand, plasticizers, waterproofing agents, and pigment are widely used.

CONCRETE

Concrete is made by mixing cement and an aggregate composed of hard inert particles of varying size, such as a combination of sand or broken-stone screenings, with gravel, broken stone, lightweight aggregate, or other material. Portland cement should always be used for reinforced concrete, for mass concrete subjected to stress, and for all concrete laid under water.

Proportioning Concrete   Compressive strength is generally accepted as the principal measure of the quality of concrete, and although this is not entirely true, there is an approximate relation between compressive strength and the other mechanical properties. Methods of proportioning generally aim to give concrete of a predetermined compressive strength.

The concrete mixture is proportioned or designed for a particular condition in various ways: (1) arbitrary selection based on experience and common practice, such as 1 part cement, 2 parts sand, 4 parts stone (written **1 : 2 : 4**); (2) proportioning on the basis of the water/cement ratio, either assumed from experience or determined by trial mixtures with the given materials and conditions; (3) combining materials on the basis of either the voids in the aggregates or mechanical-analysis curves so as to obtain the least voids and thus concrete of the maximum density for a given cement content.

Concrete mixes for small jobs can generally be determined by consultation with a ready-mixed concrete supplier, on the basis of the supplier's experience. If this information is not available, the mixes recommended by ACI Committee 211 and reproduced in Table 6.9.5 can be used as long as the required compressive strength is not higher than perhaps 3,500 lb/in². It is important that the mix be kept as dry as can be satisfactorily compacted, by vibration, into the form work since excess water reduces the eventual compressive strength. And the air entrainment is important for concretes which will be exposed to freeze-thaw cycles after curing.

For larger jobs and in all cases where high compressive strength is necessary, the mix must be designed on a more thorough basis. Again,

Table 6.9.5   Concrete Mixes for Small Jobs

Procedure: Select the proper maximum size of aggregate. Use mix B, adding just enough water to produce a workable consistency. If the concrete appears to be undersanded, change to mix A, and, if it appears oversanded, change to mix C.

| | | | Approximate weights of solid ingredients per ft³ of concrete, lb | | | |
| | | | Sand* | | Coarse aggregate | |
Maximum size of aggregate, in	Mix designation	Cement	Air-entrained concrete†	Concrete without air	Gravel or crushed stone	Iron blast-furnace slag
⅛	A	25	48	51	54	47
	B	25	46	49	56	49
	C	25	44	47	58	51
¾	A	23	45	49	62	54
	B	23	43	47	64	56
	C	23	41	45	66	58
1	A	22	41	45	70	61
	B	22	39	43	72	63
	C	22	37	41	74	65
1½	A	20	41	45	75	65
	B	20	39	43	77	67
	C	20	37	41	79	69
2	A	19	40	45	79	69
	B	19	38	43	81	71
	C	19	36	41	83	72

* Weights are for dry sand. If damp sand is used, increase tabulated weight of sand 2 lb and, if very wet sand is used, 4 lb.
† Air-entrained concrete should be used in all structures which will be exposed to alternate cycles of freezing and thawing. Air-entrainment can be obtained by the use of an air-entraining cement or by adding an air-entraining admixture. If an admixture is used, the amount recommended by the manufacturer will, in most cases, produce the desired air content.
SOURCE: ACI 211.

local suppliers of ready-mixed concrete may have records which will be of considerable help. The concrete mix has several contradictory requirements placed on it, so the final product represents a compromise. The hardened concrete must be strong enough for its intended use, it must be durable enough for its expected exposure conditions, and the freshly mixed concrete must be workable enough to be placed and compacted in the forms.

For a given aggregate type and size, the strength is controlled primarily by the water/cement ratio, with a decrease in water content, relative to cement, leading to an increase in strength, as long as the concrete remains workable enough to be placed. The durability is controlled by the water/cement ratio and the air content, assuming a suitable aggregate for the exposure condition. Table 6.9.6, from ACI Committee 211, gives the maximum water/cement ratios which should be used in severe exposure conditions, and Table 6.9.7 gives the water/cement ratios required for various concrete strengths.

For the severe-exposure cases, the air content should be between 4.5 and 7.5 percent, with the smaller value being for larger maximum sized aggregate. These tables establish two of the constraints on the mix design. A third constraint is the consistency of the concrete, as determined by the slump test which is described later. Table 6.9.8 gives the same committee's recommendations about the maximum and minimum slump values for various kinds of members. Stiffer mixes, with slumps less than 1 in, can be placed only with very heavy vibration and with great care to achieve the necessary compaction, but the dry mixes can produce very high concrete strengths with only moderate cement contents. Greater slumps are sometimes necessary when the reinforcement is very congested or the members small, such as thin walls or cast-in-place piling. The higher slumps may be produced with the high-range water reducers, or superplasticizers, without the penalty of requiring excessive water content.

The next constraint on the mix is the maximum size of aggregate. Larger aggregate sizes tend to lead to lower cement contents, but the maximum size used is limited by what is available, and in addition usually should be limited to not more than one-fifth the narrowest width between forms, one-third the thickness of slabs, three-fourths the mini-

Table 6.9.6   Maximum Permissible Water/Cement Ratios for Concrete in Severe Exposures

Type of structure	Structure wet continuously or frequently and exposed to freezing and thawing*	Structure exposed to seawater or sulfates
Thin sections (railings, curbs, sills, ledges, ornamental work) and sections with less than 1-in cover over steel	0.45	0.40†
All other structures	0.50	0.45†

* Concrete should also be air-entrained.
† If sulfate-resisting cement (type II or type V of ASTM C150) is used, permissible water/cement ratio may be increased by 0.05.
SOURCE: ACI 211.

Table 6.9.7   Relationships between Water/Cement Ratio and Compressive Strength of Concrete

| | Water/cement ratio, by weight | |
Compressive strength at 28 days, lb/in²*	Non-air-entrained concrete	Air-entrained concrete
6,000	0.41	—
5,000	0.48	0.40
4,000	0.57	0.48
3,000	0.68	0.59
2,000	0.82	0.74

* Values are estimated average strengths for concrete containing not more than the percentage of air shown in Table 6.9.9. For a constant water/cement ratio, the strength of concrete is reduced as the air content is increased.
Strength is based on 6 × 12 in cylinders moist-cured 28 days at 73.4 ± 3°F (23 ± 1.7°C) in accordance with Section 9(b) of ASTM C31 for Making and Curing Concrete Compression and Flexure Test Specimens in the Field.
Relationship assumes maximum size of aggregate about ¾ to 1 in; for a given source, strength produced for a given water/cement ratio will increase as maximum size of aggregate decreases.
SOURCE: ACI 211.

Table 6.9.8   Recommended Slumps for Various Types of Construction

Types of construction	Slump, in	
	Maximum*	Minimum
Reinforced foundation walls and footings	3	1
Plain footings, caissons, and substructure walls	3	1
Beams and reinforced walls	4	1
Building columns	4	1
Pavements and slabs	3	1
Mass concrete	2	1

* May be increased 1 in for methods of consolidation other than vibration.
SOURCE: ACI 211.

mum clear spacing between reinforcing steel. Extremely strong concretes will require smaller rather than larger maximum aggregate sizes. Once the maximum-size stone has been selected, the water content to produce the desired slump can be estimated from Table 6.9.9, and once the water content has been determined, the cement content is determined from the required water/cement ratio determined earlier. The volume of coarse aggregate is then determined from Table 6.9.10, and when the fraction is multiplied by the dry rodded unit weight, the weight of coarse aggregate per unit volume of concrete can be found.

The weight of sand needed to make a cubic yard of concrete can then be estimated by adding up the weights of materials determined so far,

Table 6.9.10   Volume of Coarse Aggregate per Unit Volume of Concrete

Maximum size of aggregate, in	Volume of dry-rodded coarse aggregate* per unit volume of concrete for different fineness moduli of sand			
	2.40	2.60	2.80	3.00
⅜	0.50	0.48	0.46	0.44
½	0.59	0.57	0.55	0.53
¾	0.66	0.64	0.62	0.60
1	0.71	0.69	0.67	0.65
1½	0.75	0.73	0.71	0.69
2	0.78	0.76	0.74	0.72
3	0.82	0.80	0.78	0.76
6	0.87	0.85	0.83	0.81

[a] Volumes are based on aggregates in dry-rodded condition as described in ASTM C 29 for Unit Weight of Aggregate.

These volumes are selected from empirical relationships to produce concrete with a degree of workability suitable for usual reinforced construction. For less workable concrete such as required for concrete pavement construction they may be increased about 10 percent.
SOURCE: ACI 211.

and subtracting from the expected weight of a cubic yard of concrete, which might be estimated at 3,900 lb for an air-entrained concrete and 4,000 lb for a mix without air. This weight should be the surface-dry saturated weight, with a correction made for the moisture content of the sand by increasing the weight of sand and decreasing the amount of mix

Table 6.9.9   Approximate Mixing Water and Air Content Requirements for Different Slumps and Nominal Maximum Sizes of Aggregates

Slump, in	Water, lb/yd³ of concrete for indicated nominal maximum sizes of aggregate							
	⅜ in[a]	½ in[a]	¾ in[a]	1 in[a]	1½ in[a]	2 in[a,b]	3 in[b,c]	6 in[b,c]
Non-air-entrained concrete								
1–2	350	335	315	300	275	260	220	190
3–4	385	365	340	325	300	285	245	210
6–7	410	385	360	340	315	300	270	—
Approximate amount of entrapped air in non-air-entrained concrete, %	3	2.5	2	1.5	1	0.5	0.3	0.2
Air-entrained concrete								
1–2	305	295	280	270	250	240	205	180
3–4	340	325	305	295	275	265	225	200
6–7	365	345	325	310	290	280	260	—
Recommended average[d] total air content, percent for level of exposure:								
Mild exposure	4.5	4.0	3.5	3.0	2.5	2.0	1.5[e,g]	1.0[e,g]
Moderate exposure	6.0	5.5	5.0	4.5	4.5	4.0	3.5[e,g]	3.0[e,g]
Extreme exposure[g]	7.5	7.0	6.0	6.0	5.5	5.0	4.5[e,g]	4.0[e,g]

[a] These quantities of mixing water are for use in computing cement factors for trial batches. They are maxima for reasonably well-shaped angular coarse aggregates graded within limits of accepted specifications.

[b] The slump values for concrete containing aggregate larger than 1½ in are based on slump tests made after removal of particles larger than 1½ in by wet-screening.

[c] These quantities of mixing water are for use in computing cement factors for trial batches when 3 in or 6 in nominal maximum size aggregate is used. They are average for reasonably well-shaped coarse aggregates, well-graded from coarse to fine.

[d] Additional recommendations for air content and necessary tolerances on air content for control in the field are given in a number of ACI documents, including ACI 201, 345, 318, 301, and 302. ASTM C 94 for ready-mixed concrete also gives air-content limits. The requirements in other documents may not always agree exactly so in proportioning concrete consideration must be given to selecting an air content that will meet the needs of the job and also meet the applicable specifications.

[e] For concrete containing large aggregates which will be wet-screened over the 1½ in sieve prior to testing for air content, the percentage of air expected in the 1½ in minus material should be as tabulated in the 1½ in column. However, initial proportioning calculations should include the air content as a percent of the whole.

[f] When using large aggregate in low cement factor concrete, air entrainment need not be detrimental to strength. In most cases mixing water requirement is reduced sufficiently to improve the water/cement ratio and to thus compensate for the strength reducing effect of entrained air concrete. Generally, therefore, for these large maximum sizes of aggregate, air contents recommended for extreme exposure should be considered even though there may be little or no exposure to moisture and freezing.

[g] These values are based on the criteria that 9 percent air is needed in the mortar phase of the concrete. If the mortar volume will be substantially different from that determined in this recommended practice, it may be desirable to calculate the needed air content by taking 9 percent of the actual mortar volume.
SOURCE: ACI 211.

water. Table 6.9.11 gives the expected ranges of water content of fine and coarse aggregates. Concretes with small-maximum-size aggregates will be lighter than the values just cited, and concretes with aggregates larger than 1 in will probably be heavier, and the kind of stone making up the coarse aggregate will also make some difference.

Table 6.9.11 Free Moisture Content

	Dry	Damp	Wet
Gravel	0.2	1.0	2.0
Sand	2.0	4.0	7.0

The mix properties just determined represent a starting point, and the final mix will usually be adjusted somewhat after trials have been conducted. Ideally, the trials should be done in a laboratory before the job-site concrete work starts, and the trial mixes should be evaluated for slump, tendency for segregation, ease with which the surface can be finished (especially important for slabs), air content, actual unit weight, actual water requirements for the desired slump, and the compressive strength of the concrete. It must be recognized that the various tables are based on average conditions, and that a particular combination of cement and aggregate may produce a concrete considerably stronger or weaker than expected from these values, and this can be true even when all of the materials meet the appropriate limitations in the ASTM specifications. Admixtures will also change the required mix proportions, and water reducers have the potential of allowing significant reductions in cement content. The tables include values for very large aggregate, up to 6-in maximum size, but aggregate larger than 1½ in will seldom be used in ordinary structures, and such concretes should be left to specialists.

It must also be recognized that consistent quality will be achieved only when care is taken to ensure that the various materials are of consistently good quality. Variations in the size of the sand and coarse aggregate, dirty aggregate, improperly stored cement, very high or low temperatures, and carelessness in any of the batching and mixing operations can reduce the uniformity and quality of the resulting concrete. The addition of extra water "to make it easier to place" is a too-common field error leading to poor concrete, since the extra water increases the water/cement ratio, which reduces the strength and durability. Some of the tests used are described in the following paragraphs.

The **dry rodded weight** is determined by filling a container, of a diameter approximately equal to the depth, with the aggregate in three equal layers, each layer being rodded 25 times using a bullet-pointed rod ⅝ in diam and 24 in long.

Measurement of materials is usually done by weight. The bulking effect of moisture, particularly on the fine aggregate, makes it difficult to keep proportions uniform when volume measurement is used. Water is batched by volume or weight.

Mixing In order to get good concrete, the cement and aggregates must be thoroughly mixed so as to obtain a homogeneous mass and coat all particles with the cement paste. Mixing may be done either by hand or by machine, although hand mixing is rare today.

Quality Control of Concrete Control methods include measuring materials by weight, allowing for the water content of the aggregates; careful limitation of the total water quantity to that designed; frequent tests of the aggregates and changing the proportions as found necessary to maintain yield and workability; constant checks on the consistency by the slump test, careful attention to the placing of steel and the filling of forms; layout of the concrete distribution system so as to eliminate segregation; check on the quality of the concrete as placed by means of specimens made from it as it is placed in the forms; and careful attention to proper curing of the concrete. The field specimens of concrete are usually 6 in diam by 12 in high for aggregates up to 1½ in max size and 8 by 16 in for 3-in aggregate. They are made by rodding the concrete in three layers, each layer being rodded 25 times using a ⅝-in diam bullet-pointed rod 24 in long.

Machine-mixed concrete is employed almost universally. The mixing time for the usual batch mixer of 1 yd³ or less capacity should not be less than 1 min from the time all materials are in the mixer until the time of discharge. Larger mixers require 25 percent increase in mixing time per ½-yd³ increase in capacity. Increased mixing times up to 5 min increase the workability and the strength of the concrete. Mixing should always continue until the mass is homogeneous.

Concrete mixers of the batch type give more uniform results; few continuous mixers are used. Batch mixers are either (1) rotating mixers, consisting of a revolving drum or a square box revolving about its diagonal axis and usually provided with deflectors and blades to improve the mixing; or (2) paddle mixers, consisting of a stationary box with movable paddles which perform the mixing. Paddle mixers work better with relatively dry, high-sand, small-size aggregate mixtures and mortars and are less widely used than rotating mixers.

Ready-mixed concrete is (1) proportioned and mixed at a central plant (central-mixed concrete) and transported to the job in plain trucks or agitator trucks, or (2) proportioned at a central plant and mixed in a mixer truck (truck-mixed concrete) equipped with water tanks, during transportation. Ready-mixed concrete is largely displacing job-mixed concrete in metropolitan areas. The truck agitators and mixers are essentially rotary mixers mounted on trucks. There is no deleterious effect on the concrete if it is used within 1 h after the cement has been added to the aggregates. Specifications often state the minimum number of revolutions of the mixer drum following the last addition of materials.

Materials for concrete are also **centrally batched,** particularly for road construction, and transported to the site in batcher trucks with compartments to keep aggregates separated from the cement. The truck discharges into the charging hopper of the job mixer. Road mixers sometimes are arranged in series—the first mixer partly mixes the materials and discharges into a second mixer, which completes the mixing.

Consistency of Concrete The consistency to be used depends upon the character of the structure. The proportion of water in the mix is of vital importance. A very wet mixture of the same cement content is much weaker than a dry or mushy mixture. Dry concrete can be employed in dry locations for mass foundations provided that it is carefully spread in layers not over 6 in thick and is thoroughly rammed. Medium, or quaking, concrete is adapted for ordinary mass-concrete uses, such as foundations, heavy walls, large arches, piers, and abutments. Mushy concrete is suitable as rubble concrete and reinforced concrete, for such applications as thin building walls, columns, floors, conduits, and tanks. A medium, or quaking, mixture has a tenacious, jellylike consistency which shakes on ramming; a mushy mixture will settle to a level surface when dumped in a pile and will flow very sluggishly into the forms or around the reinforcing bars; a dry mixture has the consistency of damp earth.

The two methods in common use for measuring the consistency or workability of concrete are the slump test and the flow test. In the **slump test,** which is the more widely used of the two, a form shaped as a frustum of a cone is filled with the concrete and immediately removed. The slump is the subsidence of the mass below its height when in the cone. The form has a base of 8-in diam, a top of 4-in diam, and a height of 12 in. It is filled in three 4-in layers of concrete, each layer being rodded by 25 strokes of a ⅝-in rod, 24 in long and bullet-pointed at the lower end.

A test using the penetration of a half sphere, called the **Kelly ball test,** is sometimes used for field control purposes. A 1-in penetration by the Kelly ball corresponds to about 2 in of slump.

Usual limitations on the consistency of concrete as measured by the slump test are given in Table 6.9.8.

The **consistency** of concrete has some relation to its **workability,** but a lean mix may be unworkable with a given slump and a rich mix may be very workable. Certain admixtures tend to lubricate the mix and, therefore, increase the workability at certain slumps. The slump at all times should be as small as possible consistent with the requirements of handling and placing. A slump over 7 in is usually accompanied by segregation and low strength of concrete, unless the high slump was produced with the aid of a superplasticizer.

Forms for concrete should maintain the lines required and prevent leakage of mortar. The pressure on forms is equivalent to that of a liquid with the same density as the concrete, and of the depth placed within 2 h. Dressed lumber or plywood is used for exposed surfaces, and rough lumber for unexposed areas. Wood or steel forms should be oiled before placing concrete.

Placement of concrete for most structures is by chutes or buggies. Chutes should have a slope not less than one vertical to two horizontal; the use of flatter slopes encourages the use of excess water, leading to segregation and low strengths. Buggies are preferable to chutes because they handle drier concrete and allow better placement control. Drop-bottom buckets are desirable for large projects and dry concrete. Concrete pumped through pipelines by mechanically applied pressure is sometimes economical for construction spread over large areas. When concrete is to be pumped, it should be a fairly rich mix, with slump of 4 in or more and aggregate not over about ¾ in, unless it has been determined that the pump can handle harsher concretes. Concrete for tunnels is placed by pneumatic pumps. Underwater concrete is deposited by drop-bottom bucket or by a tremie or pipe. Such concrete should have a cement-content increase of 15 percent to allow for loss of cement in placement.

Compaction of concrete (working it into place) is accomplished by tampers or vibrators. Vibrators are applied to the outside of the forms and to the surface or interior of the concrete; they should be used with care to avoid producing segregation. The frequency of vibrators is usually between 3,000 and 10,000 pulsations per minute.

Curing of concrete is necessary to ensure proper hydration. Concrete should be kept moist for a period of at least 7 days, and the temperature should not be allowed to fall below 50°F for at least 3 days. Sprayed-on membrane curing compounds may be used to retain moisture. Special precautions must be taken in cold weather and in hot weather.

Weight of Concrete   The following are average weights, lb/ft^3, of portland cement concrete:

Sand-cinder concrete	112	Limestone concrete	148
Burned-clay or shale concrete	105	Sandstone concrete	143
Gravel concrete	148	Traprock concrete	155

Watertightness   Concrete can be made practically impervious to water by proper proportioning, mixing, and placing. Leakage through concrete walls is usually due to poor workmanship or occurs at the joints between 2 days' work or through cracks formed by contraction. New concrete can be bonded to old by wetting the old surface, plastering it with neat cement, and then placing the concrete before the neat cement has set. It is almost impossible to prevent contraction cracks entirely, although a sufficient amount of reinforcement may reduce their width so as to permit only seepage of water. For best results, a low-volume-change cement should be used with a concrete of a quaking consistency; the concrete should be placed carefully so as to leave no visible stone pockets, and the entire structure should be made without joints and preferably in one continuous operation. The best waterproofing agent is an additional proportion of cement in the mix. The concrete should contain not less than 6 bags of cement per yd^3.

For maximum watertightness, mortar and concrete may require more fine material than would be used for maximum strength. Gravel produces a more watertight concrete than broken stone under similar conditions. Patented compounds are on the market for producing watertight concrete, but under most conditions, equally good results can be obtained for less cost by increasing the percentage of cement in the mix. Membrane waterproofing, consisting of asphalt or tar with layers of felt or tarred paper, or plastic or rubber sheeting in extreme cases, is advisable where it is expected that cracks will occur. Mortar troweled on very hard may produce watertight work.

Concrete to be placed through water should contain at least 7 bags of cement per yd^3 and should be of a quaking consistency.

According to Fuller and Thompson (*Trans. ASCE*, **59,** p. 67), watertightness increases (1) as the percentage of cement is increased and in a very much larger ratio; (2) as the maximum size of stone is increased,

provided the mixture is homogeneous; (3) materially with age; and (4) with thickness of the concrete, but in a much larger ratio. It decreases uniformly with increase in pressure and rapidly with increase in the water/cement ratio.

Air-Entrained Concrete   The entrainment of from 3 to 6 percent by volume of air in concrete by means of vinsol resin or other air-bubble-forming compounds has, under certain conditions, improved the resistance of concrete in roads to frost and salt attack. The air entrainment increases the workability and reduces the compressive strength and the weight of the concrete. As the amount of air entrainment produced by a given percentage of vinsol resin varies with the cement, mixture, aggregates, slump, and mixing time, good results are dependent on very careful control.

Concrete for Masonry Units   Mixtures for concrete masonry units, which are widely used for walls and partitions, employ aggregates of a maximum size of ½ in and are proportioned either for casting or for machine manufacture. Cast units are made in steel or wooden forms and employ concrete slumps of from 2 to 4 in. The proportions used are 1 part cement and 3 to 6 parts aggregate by dry and loose volumes. The forms are stripped after 24 h, and the blocks piled for curing. Most **blocks** are made by machine, using very dry mixtures with only enough water present to enable the concrete to hold together when formed into a ball. The proportions used are 1 part cement and 4 to 8 parts aggregate by dry and loose volumes. The utilization of such a lean and dry mixture is possible because the blocks are automatically tamped and vibrated in steel molds. The blocks are stripped from the molds at once, placed on racks on trucks, and cured either in air or in steam. High-pressure-steam (50 to 125 lb/in^2) curing develops the needed strength of the blocks in less than 24 h. Most concrete blocks are made with lightweight aggregates such as cinders and burned clay or shale in order to reduce the weight and improve their acoustical and thermal insulating properties. These aggregates not only must satisfy the usual requirements for gradation and soundness but must be limited in the amount of coal, iron, sulfur, and phosphorus present because of their effect on durability, discoloration and staining, fire resistance, and the formation of "pops." Pops form on the surface of concrete blocks using cinders or burned clay and shale as a result of the increase in volume of particles of iron, sulfur, and phosphorus when acted on by water, oxygen, and the alkalies of the cement.

Strength of Concrete

The strength of concrete increases (1) with the quantity of cement in a unit volume, (2) with the decrease in the quantity of mixing water relative to the cement content, and (3) with the density of the concrete. Strength is decreased by an excess of sand over that required to fill the voids in the stone and give sufficient workability. The volume of fine aggregate should not exceed 60 percent of that of coarse aggregate 1½-in max size or larger.

Compressive Strength   Table 6.9.7 gives the results obtainable with first-class materials and under first-class conditions. Growth in strength with age depends in a large measure upon the consistency characteristics of the cement and upon the curing conditions. Table 6.9.12 gives the change in relative strength with age for several water/cement ratios and a wide range of consistencies for a cement with a good age-strength gain relation. Many normal portland cements today show very little gain in strength after 28 days.

Tensile Strength   The tensile strength of concrete is of less importance than the crushing strength, as it is seldom relied upon. The true

Table 6.9.12   Variation of Compressive Strength with Age
(Strength at 28 days taken as 100)

Water-cement ratio by weight	3 days	7 days	28 days	3 months	1 year
0.44	40	75	100	125	145
0.62	30	65	100	135	155
0.80	25	50	100	145	165

tensile strength is about 8 percent of the compression strength and must not be confused with the tensile fiber stress in a concrete beam, which is greater.

The tensile strength is not easily measured, and an ''indirect tension'' test is used, in which a cylinder is loaded along a diametral line. Details of the test are given in ASTM C496. Nearly all the concrete in the plane of the two load strips is in tension. The tensile strength may be computed as $T = 2P/(\pi l d)$, where T = splitting tensile strength, lb/in^2; P = maximum load, lb; l = length of cylinder, in; d = diameter of cylinder.

Transverse Strength There is an approximate relationship between the tensile fiber stress of plain concrete beams and their compressive strength. The modulus of rupture is greatly affected by the size of the coarse aggregate and its bond and transverse strength. Quartzite generally gives low-modulus-of-rupture concrete.

The transverse or beam test is generally used for checking the quality of concrete used for roads. The standard beam is 6 by 6 by 20 in, tested on an 18-in span and loaded at the one-third points.

The tensile strength of the concrete as determined by the split cylinder (indirect tension) test is likely to be in the range of 4 to 5 times $\sqrt{f'_c}$, where both f'_c and $\sqrt{f'_c}$ have units of lb/in^2. The modulus of rupture values are likely to be somewhat higher, in the range of 6 to 7.5 times $\sqrt{f'_c}$.

The strength of concrete in direct **shear** is about 20 percent of the compressive strength.

Deformation Properties Young's modulus for concrete varies with the aggregates used and the concrete strength but will usually be 3 to 5 $\times 10^6$ lb/in^2 in short-term tests. According to the ACI Code, the modulus may be expressed as

$$E_c = 33w^{3/2} \sqrt{f'_c}$$

where w = unit weight of concrete in lb/ft^3 and both f'_c and $\sqrt{f'_c}$ have lb/in^2 units. For normal-weight concretes, $E_c = 57,000 \sqrt{f'_c}$ lb/in^2.

In addition to the instantaneous deformations, concrete shrinks with drying and is subjected to creep deformations developing with time. Shrinkage strains, which are independent of external stress, may reach 0.05 percent at 50 percent RH, and when restrained may lead to cracking. Creep strains typically reach twice the elastic strains. Both creep and shrinkage depend strongly on the particular aggregates used in the concrete, with limestone usually resulting in the lowest values and river gravel or sandstone in the largest. In all cases, the longer the period of wet curing of the concrete, the lower the final creep and shrinkage values.

Deleterious Actions and Materials

Freezing retards the setting and hardening of portland-cement concrete and is likely to lower its strength permanently. On exposed surfaces such as walls and sidewalks placed in freezing weather, a thin scale may crack from the surface. Natural cement is completely ruined by freezing.

Concrete laid in freezing weather or when the temperature is likely to drop to freezing should have the materials heated and should be protected from the frost, after laying, by suitable covering or artificial heat. The use of calcium chloride, salt, or other ingredients in sufficient quantities to lower the temperature of freezing significantly is not permitted, since the concrete would be adversely affected.

Mica and Clay in Sand Mica in sand, if over 2 percent, reduces the density of mortar and consequently its strength, sometimes to a very large extent. In crushed-stone screenings, the effect of the same percentage of mica in the natural state is less marked. Black mica, which has a different crystalline form, is not injurious to mortar. Clay in sand may be injurious because it may introduce too much fine material or form balls in the concrete. When not excessive in quantity, it may increase the strength and watertightness of a mortar of proportions 1 : 3 or leaner.

Mineral oils which have not been disintegrated by use do not injure concrete when applied externally. Animal fats and vegetable oils tend to disintegrate concrete unless it has thoroughly hardened. Concrete after it has thoroughly hardened resists the attack of diluted organic acids but is disintegrated by even dilute inorganic acids; protective treatments are magnesium fluorosilicate, sodium silicate, or linseed oil. Green concrete is injured by manure but is not affected after it has thoroughly hardened.

Electrolysis injures concrete under certain conditions, and electric current should be prevented from passing through it. (See also Sec. 6.5, Corrosion.)

Seawater attacks cement and may disintegrate concrete. Deleterious action is greatly accelerated by frost. To prevent serious damage, the concrete must be made with a sulfate-resisting cement, a rich mix (not leaner than 1 : 2 : 4), and exceptionally good aggregates, including a coarse sand, and must be allowed to harden thoroughly, at least 7 days, before it is touched by the seawater. Although tests indicate that there is no essential difference in the strengths of mortars gaged with fresh water and with seawater, the latter tends to retard the setting and may increase the tendency of the reinforcement to rust.

6.10 WATER
by Arnold S. Vernick

(See also Secs. 4.1, 4.2, and 6.1.)

REFERENCES: ''Statistical Abstract of the United States, 1994,'' 114th ed., U.S. Dept. of Commerce, Economics and Statistics Administration, Bureau of the Census, Bernan Press, Lanham, MD. van der Leeden, Troise, and Todd, ''The Water Encyclopedia,'' Lewis Publishers, Inc., Chelsea, MI. ''Hold the Salt,'' *Compressed Air Magazine*, March 1995. ''Water Treatment Plant Design,'' 2d ed., ASCE and AWWA, McGraw-Hill. Corbitt, ''Standard Handbook of Environmental Engineering,'' McGraw-Hill. AID Desalination Manual, Aug. 1980, U.S. Department of State. Saline Water Conversion Report, *OSW Annual Reports*, U.S. Department of the Interior. ''Water Quality and Treatment,'' AWWA, 1971. ''Safe Drinking Water Act,'' PL-93-523, Dec. 1974 as amended through 1993. ''National Primary and Secondary Drinking Water Regulations,'' Title 40 CFR Parts 141–143, U.S. Environmental Protection Agency. ''The A-B-C of Desalting,'' OSW, U.S. Department of the Interior, 1977. ''New Water for You,'' OSW, U.S. Department of the Interior, 1970. Howe, ''Fundamentals of Water Desalination,'' Marcel Dekker, New York. Ammerlaan, ''Seawater Desalting Energy Requirements,'' *Desalination*, **40**, pp. 317–326, 1982. ''National Water Summary 1985,'' U.S. Geological Survey. ''Second National Water Assessment, The Nation's Water Resources 1975–2000,'' U.S. Water Resources Council, December 1978. Vernick and Walker, ''Handbook of Wastewater Treatment Processes,'' Marcel Dekker, New York.

WATER RESOURCES

Oceans cover 70 percent of the earth's surface and are the basic source of all water. Ocean waters contain about 3½ percent by weight of dissolved materials, generally varying from 32,000 to 36,000 ppm and as high as 42,000 ppm in the Persian Gulf.

About 50 percent of the sun's energy falling on the ocean causes evaporation. The vapors form clouds which precipitate pure water as rain. While most rain falls on the sea, land rainfall returns to the sea in rivers or percolates into the ground and back to the sea or is reevaporated. This is known as the **hydrological cycle.** It is a closed distillation cycle, without additions or losses from outer space or from the interior of the earth.

Water supply to the United States depends on an annual rainfall averaging 30 in (76 cm) and equal to 1,664,800 billion gal (6,301 Gm^3) per year, or 4,560 billion gal (17.3 Gm^3) per day. About 72 percent returns to the atmosphere by direct evaporation and transpiration from trees and plants. The remaining 28 percent, or 1,277 billion gal (4.8 Gm^3) daily, is the maximum supply available. This is commonly called runoff and properly includes both surface and underground flows. Between 33 and 40 percent of runoff appears as groundwater.

Two-thirds of the runoff passes into the ocean as flood flow in one-third of the year. By increased capture, it would be possible to retain about one-half of the 1,277 billion gal (4.8 Gm³), or 638 billion gal (2.4 Gm³), per day as **maximum usable water.**

Withdrawal use is the quantity of water removed from the ground or diverted from a body of surface water. **Consumptive use** is the portion of such water that is discharged to the atmosphere or incorporated in growing vegetation or in industrial or food products.

The estimated withdrawal use of water in the United States in 1990 was 408 billion gal per day, including some saline waters (see Table 6.10.1). This figure increased historically from 140 billion gal per day to a maximum of 440 billion gal per day in 1980. The reduction of 7.3 percent in the decade since then reflects an increased emphasis on water conservation and reuse.

Fresh-water withdrawal is about 30 percent of runoff, and consumptive use is about 8 percent of runoff. The consumptive use of water for irrigation is about 55 percent, but another 15 percent is allowed for transmission and distribution losses.

Dividing total water withdrawal by the population of the United States shows a 1990 water use, the **water index,** of 1,620 gal (6,132 L) per capita day (gpcd). It reflects the great industrial and agricultural uses of water, because the withdrawal and consumption of water for domestic purposes is relatively small. The average domestic consumption of water in urban areas of the United States in 1980 was 38 gpcd (144 Lpcd). However, since public water utilities also supply industrial and commercial customers, the average withdrawal by U.S. municipal water systems in 1990 was 195 gpcd (738 Lpcd). The reduction in water withdrawal since 1980 cited above is also reflected in water consumption. Consumption per capita per day peaked at 451 gpcd (1,707 Lpcd) in 1975 and was 370 gpcd (1,400 Lpcd) in 1990, a reduction of 18 percent.

Additions to water resources for the future can be made by (1) increase in storage reservoirs, (2) injection of used water or flood water into underground strata called **aquifers,** (3) covering reservoirs with films to reduce evaporation, (4) rainmaking, (5) saline-water conversion, or (6) wastewater renovation and reuse.

It is equally important to improve the efficient use of water supplies by (1) multiple use of cooling water, (2) use of air cooling instead of water cooling, (3) use of cooling towers, (4) reclamation of waste waters, both industrial and domestic, (5) abatement of pollution by treatment rather than by dilution, which requires additional fresh water.

MEASUREMENTS AND DEFINITIONS

Water **quantities** in this country are measured by U.S. gallons, the larger unit being 1,000 U.S. gal. For agriculture and irrigation, water use is measured in acre-feet, i.e., the amount of water covering 1 acre of surface to a depth of 1 ft. In the British-standard area, the imperial gallon is used. In SI or metric-system areas, water is measured in litres

(L) and the larger unit is the cubic metre, expressed as m³, which equals a metric ton of water.

Conversion Table

1 acre · ft =	325,850 U.S. gal or 326,000 approx
1 acre · ft =	1,233 t or m³
1 acre · ft =	43,560 ft³
1 imperial gal =	1.20 U.S. gal
1 m³ =	1000 L
1 L =	0.2642 gal
1 metric ton (t) =	1,000 kg
=	2,204 lb
=	(264.2 U.S. gal)
=	220 imperial gal
1 U.S. ton =	240 U.S. gal
1,000,000 U.S. gal =	3.07 acre · ft

For stream flow and hydraulic purposes, water is measured in cubic feet per second.

$$1,000,000 \text{ U.S. gal per day} = 1.55 \text{ ft}^3/\text{s } (0.044 \text{ m}^3/\text{s})$$
$$= 1,120 \text{ acre} \cdot \text{ft } (1,380,000 \text{ m}^3)$$
$$\text{per year}$$

Water **costs** are expressed in terms of price per 1,000 gal, per acre · foot, per cubic foot, or per cubic metre.

$$10 \text{¢ per } 1,000 \text{ gal} = \$32.59 \text{ per acre} \cdot \text{ft}$$
$$= 0.075 \text{¢ per ft}^3$$
$$= 2.64 \text{¢ per m}^3$$

Water **quality** is measured in terms of solids, of any character, which are suspended or dissolved in water. The concentration of solids is usually expressed in parts per million or in grains per gallon. One grain equals 1/7,000 lb (64.8 mg). Therefore, 17.1 ppm = 1 grain per U.S. gal. In the metric system, 1 ppm = 1 g/m³ = 1 mg/L.

Standards for drinking water, or **potable water,** have been established by the U.S. Environmental Protection Agency. The recommended limit for a specific containment is shown in Table 6.10.2. The limitations indicated in the table apply to public water systems, with the primary regulations being established to protect public health, while the secondary regulations control aesthetic qualities relating to the public acceptance of drinking water.

For **agricultural water,** mineral content of up to 700 mg/L is considered excellent to good. However, certain elements are undesirable, particularly sodium and boron. The California State Water Resources Board limits class I irrigation water to:

	Max mg/L
Sodium, as % of total sodium, potassium, magnesium, and calcium equivalents	60
Boron	0.5
Chloride	177
Sulfate	960

Class II irrigation water may run as high as 2,100 mg/L total dissolved solids, with higher limits on the specific elements, but whether

Table 6.10.1 U.S. Water Withdrawals and Consumption per Day by End Use, 1990

| | Total, 10⁹ gal[a] | Per capita,[b] gal | Irrigation, 10⁹ gal | Public water utilities[c] | | Rural domestic,[d] 10⁹ gal | Industrial and misc.,[e] 10⁹ gal | Steam electric utilities, 10⁹ gal |
				Total, 10⁹ gal	Per capita,[b] gal			
Withdrawal	408	1,620	137	41	195	7.9	30	195
Consumption	96	370	76	7.1[f]	38[f]	8.9	6.7	4.0

[a] Conversion factor: gal × 3.785 = L.
[b] Based on Bureau of the Census resident population as of July 1.
[c] Includes domestic and commercial water withdrawals.
[d] Rural farm and nonfarm household and garden use, and water for farm stock and dairies.
[e] Includes manufacturing, mining, and mineral processing, ordnance, construction, and miscellaneous.
[f] 1980 data—latest available.
SOURCE: "Statistical Abstract of the United States, 1994," U.S. Dept. of Commerce, Economics and Statistics Administration, Bureau of the Census, 114th ed., Bernan Press, Lanham, MD.

Table 6.10.2 Potable Water Contaminant Limitations

Primary standards			
Organic contaminants		**Inorganic contaminants**	
Contaminant	MCL, mg/L	Contaminant	MCL, mg/L
Vinyl chloride	0.002	Fluoride	4.0
Benzene	0.005	Asbestos	7 million fibers/L (longer than 10 μm)
Carbon tetrachloride	0.005		
1,2-Dichloroethane	0.005	Barium	2
Trichloroethylene	0.005	Cadmium	0.005
para-Dichlorobenzene	0.075	Chromium	0.1
1,1-Dichloroethylene	0.007	Mercury	0.002
1,1,1-Trichloroethane	0.2	Nitrate	10 (as nitrogen)
cis-1,2-Dichloroethylene	0.07	Nitrite	1 (as nitrogen)
1,2-Dichloropropane	0.005	Total nitrate and nitrite	10 (as nitrogen)
Ethylbenzene	0.7	Selenium	0.05
Monochlorobenzene	0.1	Antimony	0.006
o-Dichlorobenzene	0.6	Beryllium	0.004
Styrene	0.1	Cyanide (as free cyanide)	0.2
Tetrachloroethylene	0.005	Nickel	0.1
Toluene	1	Thallium	0.002
trans-1,2-Dichloroethylene	0.1	Arsenic	0.05
Xylenes (total)	10	Lead	0.015
Dichloromethane	0.005	Copper	1.3
1,2,4-Trichlorobenzene	0.07		
1,1,2-Trichloroethane	0.005	**Radionuclides**	
Alachlor	0.002		
Aldicarb	0.003	Contaminant	MCL, pCi/L*
Aldicarb sulfoxide	0.004	Gross alpha emitters	15
Aldicarb sulfone	0.003	Radium 226 plus 228	5
Atrazine	0.003	Radon	300
Carbofuran	0.04	Gross beta particle and photon	4m Rem
Chlordane	0.002	Emitters	
Dibromochloropropane	0.0002	Uranium	20 μL (equivalent to
2,4-D	0.07		30 pCi/L)
Ethylene dibromide	0.00005		
Heptachlor	0.0004	Bacteria, Viruses, Turbidity, Disinfectants and By-products—	
Heptachlor epoxide	0.0002	Refer to 40CFR141	
Lindane	0.0002		
Methoxychlor	0.04	**Secondary standards†**	
Polychlorinated biphenyls	0.0005	Contaminant	Level
Pentachlorophenol	0.001	Aluminum	0.05 to 0.2 mg/L
Toxaphene	0.003	Chloride	250 mg/L
2,4,5-TP	0.05	Color	15 color units
Benzo(a)pyrene	0.0002	Copper	1.0 mg/L
Dalapon	0.2	Corrosivity	Noncorrosive
Di(2-ethylhexyl)adipate	0.4	Fluoride	2.0 mg/L
Di(2-ethylhexyl)phthalate	0.006	Foaming agents	0.5 mg/L
Dinoseb	0.007	Iron	0.3 mg/L
Diquat	0.02	Manganese	0.05 mg/L
Endothall	0.1	Odor	3 threshold odor number
Endrin	0.002	pH	6.5–8.5
Glyphosate	0.7	Silver	0.1 mg/L
Hexachlorobenzene	0.001	Sulfate	250 mg/L
Hexachlorocyclopentadiene	0.05	Total dissolved solids (TDS)	500 mg/L
Oxamyl (Vydate)	0.2	Zinc	5 mg/L
Picloram	0.5		
Simazine	0.004		
2,3,7,8-TCDD (Dioxin)	$3 \times 10\text{--}10^{-8}$		

* Except as noted.

† Not federally enforceable; intended as guidelines for the states.

SOURCE: EPA Primary Drinking Water Regulations (40CFR141). EPA Secondary Drinking Water Regulations (40CFR143).

such water is satisfactory or injurious depends on the character of soil, climate, agricultural practice, and type of crop.

Waters containing dissolved salts are called **saline waters,** and the lower concentrations are commonly called **brackish;** these waters are defined, in mg/L, as follows:

Saline	All concentrations up to 42,000
Slightly brackish	1,000–3,000
Brackish	3,000–10,000
Seawaters, average	32,000–36,000
Brine	Over 42,000

Hardness of water refers to the content of calcium and magnesium salts, which may be bicarbonates, carbonates, sulfates, chlorides, or nitrates. Bicarbonate content is called **temporary hardness,** as it may be removed by boiling. The salts in "hard water" increase the amount of soap needed to form a lather and also form deposits or "scale" as water is heated or evaporated.

Hardness is a measure of calcium and magnesium salts expressed as equivalent calcium carbonate content and is usually stated in mg/L (or in grains per gal) as follows: very soft water, less than 15 mg/L; soft water, 15 to 50 mg/L, slightly hard water, 50 to 100 mg/L; hard water, 100 to 200 mg/L; very hard water, over 220 mg/L.

Table 6.10.3 Water Use in U.S. Manufacturing by Industry Group, 1983

Industry	Establishments reporting*	Gross water used — Total Quantity, 10^9 gal	Gross water used — Total Average per establishment, 10^6 gal	Water intake, 10^9 gal	Water recycled,† 10^9 gal	Water discharged Quantity, 10^9 gal	Water discharged Percent untreated
Food and kindred products	2,656	1,406	529	648	759	552	64.5
Tobacco products	20	34	1,700	5	29	4	N/A
Textile mill products	761	333	438	133	200	116	52.6
Lumber and wood products	223	218	978	86	132	71	63.4
Furniture and fixtures	66	7	106	3	3	3	100.0
Paper and allied products	600	7,436	12,393	1,899	5,537	1,768	27.1
Chemicals and allied products	1,315	9,630	7,323	3,401	6,229	2,980	67.0
Petroleum and coal products	260	6,177	23,758	818	5,359	699	46.2
Rubber, miscellaneous plastic products	375	328	875	76	252	63	63.5
Leather and leather products	69	7	101	6	1	6	N/A
Stone, clay, and glass products	602	337	560	155	182	133	75.2
Primary metal products	776	5,885	7,584	2,363	3,523	2,112	58.1
Fabricated metal products	724	258	356	65	193	61	49.2
Machinery, excluding electrical	523	307	587	120	186	105	67.6
Electrical and electronic equipment	678	335	494	74	261	70	61.4
Transportation equipment	380	1,011	2,661	153	859	139	67.6
Instruments and related products	154	112	727	30	82	28	50.0
Miscellaneous manufacturing	80	15	188	4	11	4	N/A
Total	10,262	33,835	3,297	10,039	23,796	8,914	54.9

* Establishments reporting water intake of 20 million gal or more, representing 96 percent of the total water use in manufacturing industries.
† Refers to water recirculated and water reused.
N/A = not available.
SOURCE: "The Water Encyclopedia," 1990.

INDUSTRIAL WATER

The use of water within a given industry varies widely because of conditions of price, availability, and process technology (see Table 6.10.3).

When a sufficient water supply of suitable quality is available at low cost, plants tend to use maximum volumes. When water is scarce and costly at an otherwise desirable plant site, improved processes and careful water management can reduce water usage to the minimum. Industrial water may be purchased from local public utilities or self-supplied. Approximately 79 percent of the water used by the chemical industry in the United States comes from company systems, and 66 percent of those self-supplied plants utilize surface water as their source of supply as shown in Table 6.10.4.

The cost of usable water is highly sensitive to a variety of factors including geographic location, proximity to an abundant water source, quality of the water, regulatory limitations, and quantity of water used. Raw surface water not requiring much treatment prior to use can cost as little as 7¢ to 15¢ per 1,000 gal (1.8¢ to 4¢ per m³) in areas of the United States where water is plentiful, or as much as $1.75 to $2.00 per 1,000 gal (46¢ to 53¢ per m³) in water-scarce areas. These costs include collection, pumping, storage tanks, distribution and fire-protection facilities, but treatment, if needed, would add materially to these figures.

The cost of water obtained from public supply systems for industrial use also varies considerably depending upon geographic location and water usage. An EPA survey of over 58,000 public water systems in 1984 indicated water costs as high as $2.30 per 1,000 gal (61¢ per m³) in New England for water use of 3,750 gal, to a low of 88¢ per 1,000 gal (23¢ per m³) in the northwest for water use of 750,000 gal. Water cost in 1983 in Houston, Texas, for the highest volume users, which includes refineries and chemical plants, was $1.14 per 1,000 gal (30¢ per m³).

The use of water by the chemical industry is reflected in Table 6.10.5.

About 80 percent of industrial water is used for cooling, mostly on a once-through system. An open recirculation system with cooling tower or spray pond (Fig. 6.10.1) reduces withdrawal use of water by over 90 percent but increases consumptive use by 3 to 8 percent because of evaporation loss. Even more effective reduction in water demand can be achieved by multiple reuse. An example is shown in Fig. 6.10.2.

Quality requirements for general plant use (nonprocess) are that the water be low in suspended solids to prevent clogging, low in total dis-

Table 6.10.4 Sources of Water for Chemical Plants

	Quantity, billion gal/year ($\times 3.785 \times 10^{-3}$ = billion m³/year)
By source	
Public water systems	727
Company systems	3,399
Surface water	2,251
Groundwater	366
Tidewater	782
Other	200
By type	
Fresh	3,425
Brackish	182
Salt	719

Table 6.10.5 How Water Is Used in Chemical Plants

Use	Quantity, billion gal/year ($\times 3.785 \times 10^{-3}$ = billion m³/year)
Process	754
Cooling and condensing	
Electric power generation	480
Air conditioning	71
Other	2,760
Sanitary service	38
Boiler feed	112
Other	81

solved solids to prevent depositions, free of organic growth and color, and free of iron and manganese salts. Where the water is also used for drinking, quality must meet the Environmental Protection Agency regulations.

Cooling service requires that water be nonclogging. Reduction in suspended solids is made by settling or by using a coagulating agent such as alum and then settling. For recirculating-type cooling systems, corrosion inhibitors such as polyphosphates and chromates are added; algicides and biocides may be needed to control microorganism growth; for cooling jackets on equipment, hardness may cause scaling and should be reduced by softening.

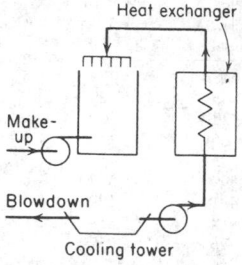

Fig. 6.10.1 Open recirculating system.

Process-water quality requirements are often more exacting than potable-water standards; e.g., boiler feedwater must have less than 1 mg/L of dissolved solids (see Sec. 6, Corrosion, and Sec. 9, Steam Boilers). The required quality may be met by the general plant water as available or must be provided by treatment.

Table 6.10.6 shows methods and objectives for industrial-water treatment.

WATER POLLUTION CONTROL

A pollutant is defined in the Federal Clean Water Act as "dredge spoil, solid waste, incinerator residue, sewage, garbage, sewage sludge, munitions, chemical wastes, biological materials, heat, wrecked or discarded equipment, rock, sand, cellar dirt, and industrial, municipal, and agricultural waste discharged into water."

Control of water pollution in the United States is a joint responsibility of the U.S. Environmental Protection Agency and the state environmental protection departments. Promulgation of regulations controlling both industrial and municipal wastewater discharges has been accomplished by the EPA as mandated by the Clean Water Act. Enforcement has been largely delegated to the States, with the EPA serving in an oversight role. Water quality standards for specific streams and lakes have been established by the States, or interstate agencies where appropriate, which frequently mandate tighter control of discharges where local conditions warrant such action.

For the support of fish and aquatic life, water must contain a supply of **dissolved oxygen** (DO). Organic wastes consume oxygen by microbiological action, and this effect is measured by the **biochemical oxygen-demand** (BOD) test.

The discharge of industrial wastewater in the United States is controlled by permits issued by the States and/or the EPA to each facility. The permits set specific limits for applicable pollutant parameters such as suspended solids, BOD, COD, color, pH, oil and grease, metals, ammonia, and phenol. In order to achieve these limitations, most industrial facilities must treat their wastewater effluent by either physical, chemical or biological methods.

The investment by industry in water pollution abatement facilities and the annual cost of operating these facilities have increased dramatically over the last two decades. In 1992, manufacturing industries in the United States spent a total of $2.5 billion in capital expenditures and $6.6 billion in operating costs on water pollution abatement facilities (see Table 6.10.7). These amounts represented 32 and 38 percent, respectively, of the total amount spent by industry for all pollution abatement and environmental protection activities. The chemical industry was the major contributor to these totals, spending approximately 40 percent of its capital expenditures and 30 percent of its operating costs for water pollution abatement.

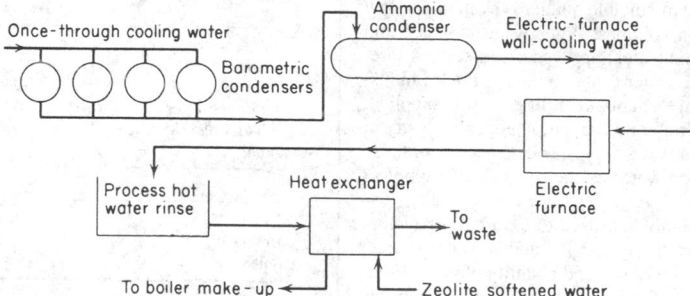

Fig. 6.10.2 Stepwise or cascade cooling system.

Table 6.10.6 Water Treatment

Treatment	Method	Objective
Clarification	Presedimentation Coagulation Settling Filtering	Remove suspended solids and reduce turbidity, color, organic matter
Disinfection	Add chlorine, 5 to 6 ppm, or continuous-feed to maintain 0.2 to 0.3 ppm residual-free Cl_2	Prevent algae and slime growth
Softening	Cold-lime process	Reduce temporary hardness to 85 ppm; also reduce iron and manganese
	Hot-lime-soda process	Reduce total hardness to 25 ppm
	Zeolite	Reduce total hardness to 5 ppm
Membrane processes	Reverse osmosis	Partial removal of ions; can reduce seawater and brackish water to 550 ppm or less
	Electrodialysis	Partial removal of ions; can reduce 10,000 ppm brackish water to 500 ppm or less
Demineralization	Ion exchange: two-stage or mixed-bed	Remove both positive and negative ions (cations and anions) to provide very pure water
Distillation	Evaporation using steam heat	Produce very pure water—10 ppm or less total solids

Table 6.10.7 Water Pollution Abatement Capital Expenditures and Operating Costs of Manufacturing Industry Groups, 1992

Industry group	Pollution abatement capital expenditures, million $		Pollution abatement gross operating costs, million $*	
	Total†	Water	Total†	Water
Food and kindred products	316.8	202.6	1,312.0	835.7
Lumber and wood products	94.5	18.9	243.0	49.5
Paper and allied products	1,004.6	373.4	1,860.7	822.7
Chemicals and allied products	2,120.9	1,017.3	4,425.1	1,946.8
Petroleum and coal products	2,685.0	492.6	2,585.4	742.8
Stone, clay, and glass products	138.8	20.2	491.2	94.6
Primary metal industries	525.7	123.5	1,993.4	575.0
Fabricated metal products	103.3	42.4	761.2	284.3
Machinery, excluding electrical	150.3	31.7	463.7	160.5
Electrical, electronic equipment	126.6	45.6	657.1	288.8
Transportation equipment	281.0	69.2	1,171.7	347.0
Instruments, related products	89.1	18.8	331.9	89.4
All industries‡	7,866.9	2,509.8	17,466.4	6,576.9

* Includes payments to governmental units.
† Includes air, water, hazardous waste, and solid waste.
‡ Includes industries not shown separately; excludes apparel and other textile products, and establishments with less than 20 employees.
SOURCE: "Statistical Abstract of the United States," 1994.

Physical treatment includes screening, settling, flotation, equalization, centrifuging, filtration, and carbon adsorption. **Chemical treatment** includes coagulation or neutralization of acids with soda ash, caustic soda, or lime. Alkali wastes are treated with sulfuric acid or inexpensive waste acids for neutralization. Chemical oxidation is effective for certain wastes. **Biological treatment** is accomplished by the action of two types of microorganisms: aerobic, which act in the presence of oxygen, and anaerobic, which act in the absence of oxygen. Most organic wastes can be treated by biological processes.

The principal wastewater treatment techniques include the activated sludge process, trickling filters and aerated lagoons, which all employ aerobic microorganisms to degrade the organic waste material. Anaerobic processes are mainly employed for digestion of the sludge produced by biological wastewater treatment.

WATER DESALINATION

An *average seawater* contains 35,000 ppm of dissolved solids, equal to 3½ percent by weight of such solids, or 3.5 lb per 100 lb; in 1,000 gal, there are 300 lb of dissolved chemicals in 8,271 lb of pure water. The principal ingredient is sodium chloride (common salt), which accounts for about 80 percent of the total. Other salts are calcium sulfate (gypsum), calcium bicarbonate, magnesium sulfate, magnesium chloride, potassium chloride, and more complex salts. Because these dissolved chemicals are dissociated in solution, the composition of seawater is best expressed by the concentration of the major ions (see Table 6.10.8). Even when the content of total solids in seawater varies because of dilution or concentration, for instance Arabian Gulf water with total dissolved solids concentrations of 45,000 to 50,000 ppm, the proportion of the ions remains almost constant.

The **composition of brackish waters** varies so widely that no average analysis can be given. In addition to sodium chloride and sulfate, brackish well waters often contain substantial amounts of calcium, magnesium, bicarbonate, iron, and silica.

Desalination is the process by which seawater and brackish water are purified. It has been practiced for centuries, primarily aboard sailing vessels, but recently its application has grown significantly in the form of land-based facilities. The International Desalination Association reports that there are more than 9,000 commercial desalination plants in operation worldwide in 1995, whose total capacity exceeds approximately 4 billion gal per day.

Purification of seawater normally requires reduction of 35,000 ppm of total dissolved solids to less than 500 ppm, or a reduction of 70 to 1.

For potable water, it is desirable to have a chloride content below 250 mg/L. The oldest method of purification of seawater is **distillation.** This technique has been practiced for over a century on oceangoing steamships. Distillation is used in over 95 percent of all seawater conversion plants and is principally accomplished by multiple-effect distillation, flash distillation, and vapor-compression distillation.

Purification of brackish water requires dissolved solids reductions ranging from a low of 3 to 1, to a high of 30 to 1. In this range, **membrane** desalting processes are more economical. Over 95 percent of all brackish water plants use either the reverse-osmosis process or the electrodialysis process.

In **flash distillation,** the salt water is heated in a tubular brine heater and then passed to a separate chamber where a pressure lower than that in the heating tubes prevails. This causes some of the hot salt water to vaporize, or "flash," such vapor then being condensed by cooler incoming salt water to produce pure distilled water. Flash distillation can be carried on in a number of successive stages (Fig. 6.10.3) wherein the heated salt water flashes to vapor in a series of chambers, each at a lower pressure than the preceding one. The higher the number of stages in such a **multistage flash** (MSF) system, the better the overall yield per heat unit. Today's large-scale MSF plants are designed with 15 to 30 stages and operate at performance ratios of 6 to 10, which means that for every 1 lb (kg) of steam, 6 to 10 lb (kg) of product water is produced.

An advantageous feature of flash distillation is that the separate heating of salt water without boiling causes fewer scale deposits. No scale occurs in the flash chambers as they contain no heated surfaces, the increased concentration of salts remaining in the seawater. Designs of MSF evaporators provide a great number of horizontal stages in one

Table 6.10.8 Ions in Seawater

Ions	ppm	lb/1,000 gal	mg/L
Chloride	19,350	165.6	19,830
Sodium	10,600	91.2	10,863
Sulfate	2,710	23.2	2,777
Magnesium	1,300	11.1	1,332
Calcium	405	3.48	415
Potassium	385	3.30	395
Carbonate and bicarbonate	122	1.05	125
Total principal ingredients	34,872	298.9	35,737
Others	128	1.1	131
Total dissolved solids	35,000	300.0	35,868

SOURCE: Compiled from Spiegler, "Salt Water Purification," and Ellis, "Fresh Water from the Ocean."

vessel by putting vertical partition walls inside the vessel (Fig. 6.10.3). All large MSF plants are designed this way.

If a saline water is made to boil, by heating it with steam through a heat-transfer surface, and the vapors from the saline water are subsequently condensed, a distillate of practically pure water (< 10 ppm salt) is obtained. This is known as **single-effect** distillation.

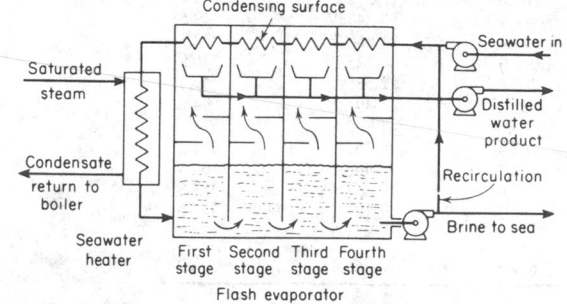

Fig. 6.10.3 Multiple-stage (four-stage) flash distillation process.

Multiple-effect distillation is achieved by using the condensation heat of the saline water vapor for boiling additional seawater in a following vessel or **effect** and repeating this step several more times. Each subsequent effect is operating at a somewhat lower temperature and vapor pressure than the previous one. In a single effect, 1 lb (kg) of steam will produce nearly 0.9 lb (kg) of distilled water, a double effect will yield 1.75 lb (kg) of product, and so on. In commercial seawater desalting plants, up to 17 effects have been employed.

Older multieffect plants used tubes submerged in seawater (**submerged-tube evaporators**), but scaling was always a serious problem and this type of design has been phased out in favor of thin-film designs. Only for small, low-temperature waste heat evaporators is a submerged-tube design sometimes applied. In a **thin-film** evaporator, saline water flows as a thin film of water through the inside of a vertical tube (VTE), or over the outside of a bundle of horizontal tubes (HT) after being sprayed on the bundle. Heating of the thin film of seawater is accomplished by condensing steam on the opposite side of the tube wall. See Figs. 6.10.4 and 6.10.5. Horizontal tube designs with vertically stacked effects have been developed.

In **vapor-compression distillation** (Fig. 6.10.6), the energy is supplied by a compressor which takes the vapor from boiling salt water and compresses it to a higher pressure and temperature to furnish the heat for vaporization of more seawater. In so doing, the vapor is condensed to yield distilled water. Vapor compression is theoretically a more efficient method of desalination than other distillation methods. The principle was widely applied during World War II in the form of small, portable units, and still is often used for relatively small plants of up to 200,000 gpd (32 m³/h). The disadvantages of this process lie in the cost, mechanical operation, and maintenance of the compressors.

Reverse osmosis (Figs. 6.10.7 and 6.10.8), in which a membrane permits fresh water to be forced through it but holds back dissolved solids, has now emerged as the most commonly used process for desalting brackish waters containing up to 10,000 ppm of total dissolved solids.

Two commercially available membrane configurations exist: (1) hollow thin-fiber membrane cartridges and (2) spiral-wound sheet membrane modules. Brackish water plants operate at pressures ranging from 250 to 400 lb/in² (17 to 27 bar), which is sufficient to reverse the osmotic flow between brackish water and fresh water.

In the last few years, high-pressure membranes capable of operating at 800 to 1,000 lb/in² (55 to 70 bar) have been commercialized for one-stage desalting of seawater. Several reverse-osmosis seawater plants exceeding 1 mgd (160 m³/h) are presently in operation in Saudi Arabia, Malta, Bahrain, and Florida. Recovery of power from the high-pressure brine discharged from the reverse-osmosis plant is economically attractive for seawater plants because of the high operating pressure and the relatively large quantities of brine (65 to 80 percent of feed) discharged.

Electrodialysis (Fig. 6.10.9) is a proven method of desalination, but where used on ocean water, it is not competitive. The purification of brackish or low-saline waters lends itself most advantageously to the process. The ions of dissolved salts are pulled out of saline water by electric forces and pass from the salt-water compartment into adjacent compartments through membranes which are alternately permeable to positively charged ions or to negatively charged ions.

Electrodialysis equipment typically contains 100 or more compartments between one set of electrodes. The number of cells and the amount of electric current required increase with the amount of purification to be done. A brackish water of 5,000 ppm max dissolved salts can be reduced to potable water with less than 500 ppm. Over 600 electrodialysis plants have been installed since the commercialization of the process 40 years ago. Most of these plants are relatively small: 50,000 gpd (8 m³/h) or less.

Other methods of purifying seawater include freezing and solar evaporation. The freezing process yields ice crystals of pure water, but a certain amount of salt water is trapped in the crystals which then has to be removed by washing, centrifugation, or other means. This process is promising because theoretically it takes less energy to freeze water than to distill it; but it is not yet in major commercial application because in several pilot plants the energy consumption was much higher, and the yield of ice much lower, than anticipated.

Solar evaporation produces about 0.1 gpd of water per ft² of area [0.17 L/(m²·h)], depending on climatic conditions. Capital and maintenance costs for the large areas of glass that are required so far have not made solar evaporation competitive.

A major problem in desalting processes, especially distillation, is the concentration of dissolved salts. Heating seawater above 150°F (65°C) causes **scale**, a deposition of insoluble salts on the heating surfaces, which rapidly reduces the heat transfer. For initial temperatures of 160 to 195°F (71 to 90°C), the addition of polyphosphates results in the formation of a sludge rather than a hard scale. This practice is widely used in ships and present land-based plants but is limited to 195°F (90°C). Some plants use acid treatment that permits boiling of seawater at 240 to 250°F (115 to 121°C). This improves the performance of the process and reduces costs.

In recent years, polymer additives have become available that allow operating temperatures of 230 to 250°F (110 to 121°C) and eliminate the hazardous handling of strong acids and the possibility of severe corrosion of equipment. In reverse osmosis and electrodialysis, scale formation on the membranes is also controlled by adding acid and other additives or by the prior removal of the hardness from the feed by lime softening. Reversing the polarity of the applied dc power can greatly minimize or entirely eliminate the need for pretreatment and additives for the electrodialysis process.

Energy requirements for the various commercial desalting processes are shown in Fig. 6.10.10 in terms of primary energy sources, such as oil, coal, or gas. The method and arrangement by which primary energy is converted to usable energy, e.g., mechanical shaft power, steam, or electricity, has a profound impact on the energy requirements and cost of every desalting process.

Cost of desalting equipment at the 1-mgd level (160 m³/h), complete and installed, generally ranges from $6 to $8 per gpd ($37,500 to $50,000 per m³/h) installed capacity for both distillation and reverse-osmosis seawater plants.

Electrodialysis and reverse-osmosis brackish water plants of similar capacity usually range from $1.50 to $2.00 per gpd ($9,400 to $12,500 per m³/h). **Total operating costs**, including capital depreciation, power and chemical consumption, maintenance and operating personnel, for a 1-mgd (160-m³/h) facility fall into three categories: (1) seawater single-purpose distillation and reverse osmosis without power recovery = $6 to $8 per 1,000 gal ($1.60 to $2.10 per m³), (2) seawater dual-purpose distillation and reverse osmosis with power recovery = $4 to $6 per 1,000 gal ($1.00 to $1.50 per m³), and (3) brackish water electrodialysis and reverse osmosis = $1 to $2 per 1,000 gal ($0.25 to $0.50 per m³).

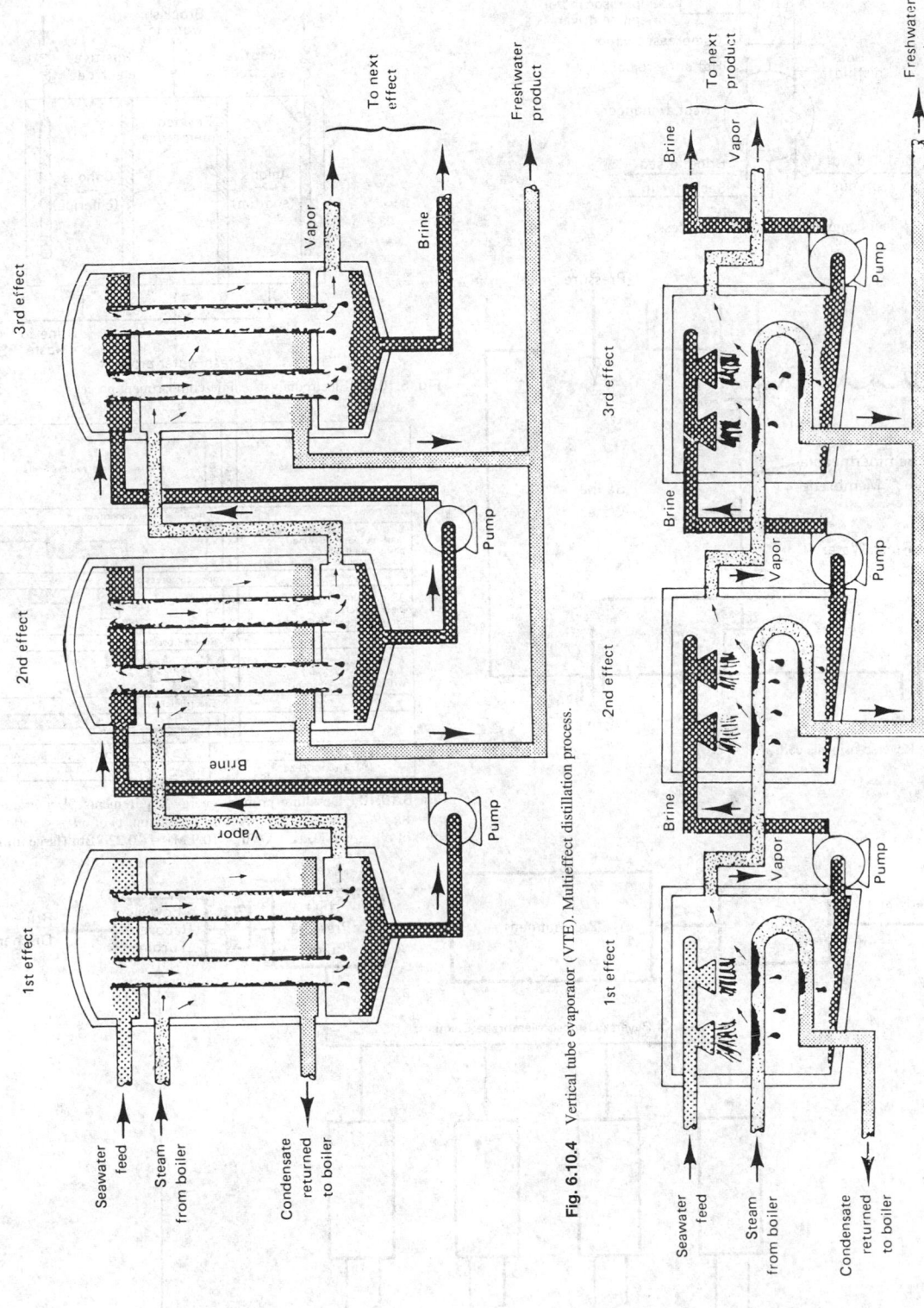

Fig. 6.10.4 Vertical tube evaporator (VTE). Multieffect distillation process.

Fig. 6.10.5 Horizontal tube multieffect evaporator (HTME). Multieffect distillation process.

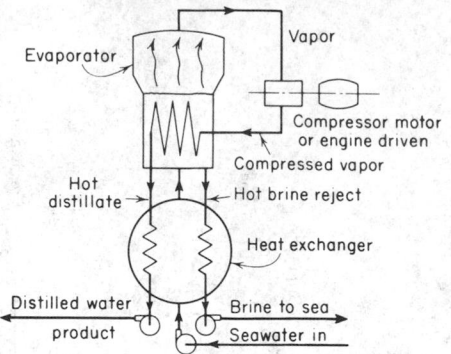

Fig. 6.10.6 Vapor compression distillation process.

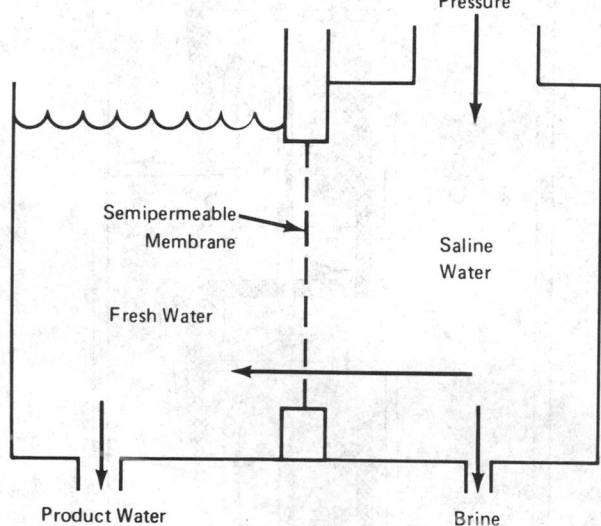

Fig. 6.10.7 Reverse-osmosis cell.

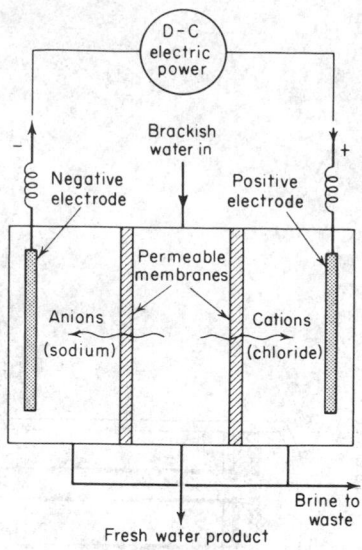

Fig. 6.10.9 Electrodialysis single-compartment process.

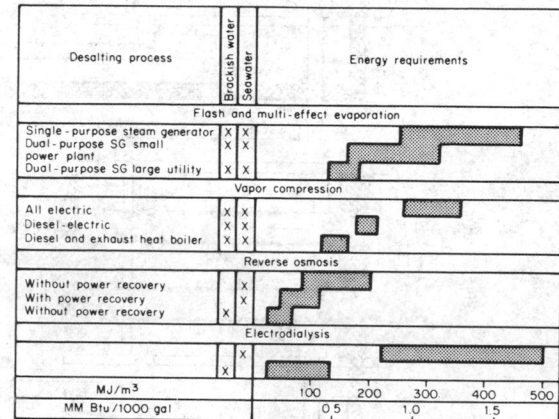

Fig. 6.10.10 Desalting primary energy requirements *(Ammerlaan, Desalination,* **40**, *1982, pp. 317–326.)* Note: 1 metric ton (t) of oil = 43,800 MJ; 1 barrel (bbl) of oil = 6.2 TBtu; 1 kWh = 10.9 MJ = 10,220 Btu (large utility).

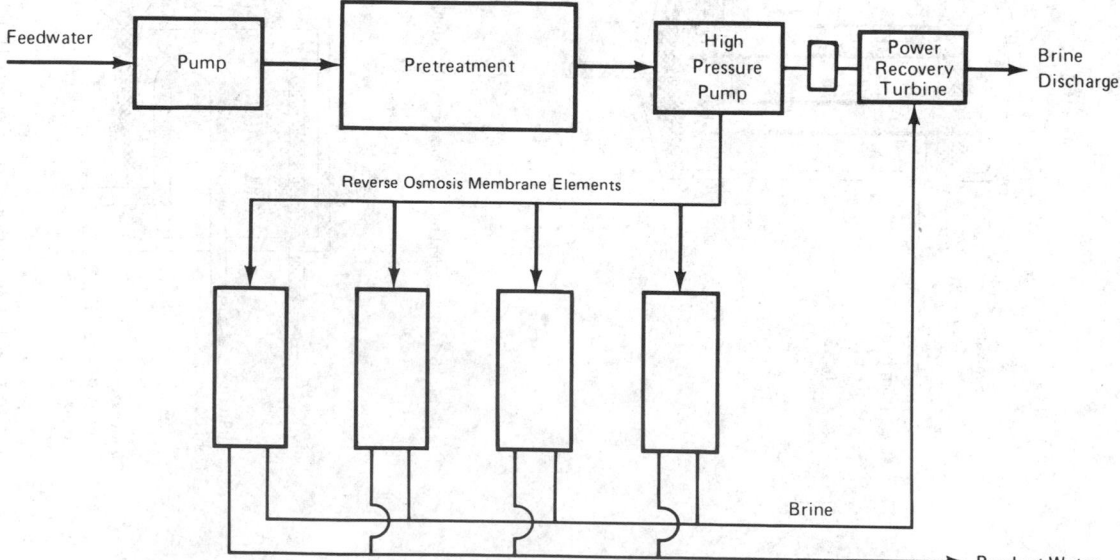

Fig. 6.10.8 Reverse-osmosis plant block flow diagram.

6.11 LUBRICANTS AND LUBRICATION

by Julian H. Dancy

(For general discussion of friction, viscosity, bearings, and coefficients of friction, see Secs. 3 and 8.)

REFERENCES: STLE, "Handbook of Lubrication," 3 vols., CRC Press. ASME, "Wear Control Handbook." ASM, "ASM Handbook," vol. 18, "Friction, Lubrication, and Wear Technology." ASTM, "Annual Book of ASTM Standards," 3 vols., "Petroleum Products and Lubricants." NLGI, "Lubricating Grease Guide." Fuller, "Theory and Practice of Lubrication for Engineers," Wiley. Shigley and Mischke, "Bearings and Lubrication: A Mechanical Designers' Workbook," McGraw-Hill. Stipina and Vesely, "Lubricants and Special Fluids," Elsevier. Miller, "Lubricants and Their Applications," McGraw-Hill. Shubkin (ed.), "Synthetic Lubricants and High Performance Functional Fluids," Marcel Dekker. Bartz (ed.), "Engine Oils and Automotive Lubrication," Dekker. Hamrock, "Fundamentals of Fluid Film Lubrication," McGraw-Hill. Neale, "Lubrication: A Tribology Handbook," Butterworth-Heinemann. Ramsey, "Elastohydrodynamics," Wiley. Arnell, "Tribology, Principles and Design Applications," Springer-Verlag. Czichos, "Tribology, A Systems Approach to . . . Friction, Lubrication and Wear," Elsevier. Bhushan and Gupta, "Handbook of Tribology: Materials, Coatings, and Surface Treatments," McGraw-Hill. Yamaguchi, "Tribology of Plastic Materials," Elsevier.

Tribology, a term introduced in the 1960s, is the science of rubbing surfaces and their interactions, including friction, wear, and lubrication phenomena. **Lubrication** primarily concerns modifying friction and reducing wear and damage at the surface contacts of solids rubbing against one another. Anything introduced between the surfaces to accomplish this is a **lubricant.**

LUBRICANTS

Lubricants can be liquids, solids, or even gases, and they are most often oils or greases. Liquid lubricants often provide many functions in addition to controlling friction and wear, such as scavenging heat, dirt, and wear debris; preventing rust and corrosion; transferring force; and acting as a sealing medium.

Engineers are called upon to select and evaluate lubricants, to follow their performance in service, and to use them to best advantage in the design of equipment. Lubricant manufacturers and distributors may have hundreds of lubricants in their product line, each described separately in the product literature as to intended applications, properties, and benefits, as well as performance in selected standard tests. Lubricants are selected according to the needs of the particular application. Careful lubricant selection helps obtain improved performance, lower operating cost, and longer service life, for both the lubricant and the equipment involved. Industry's demands for efficient, competitive equipment and operations, which meet the latest environmental regulations, create continued demand for new and improved lubricants. Equipment manufacturers and suppliers specify lubricants that suit their particular equipment and its intended operating conditions, and their recommendations should be followed.

LIQUID LUBRICANTS

A liquid lubricant consists of (1) a mixture of selected base oils and additives, (2) blended to a specific viscosity, with (3) the blend designed to meet the performance needs of a particular type of service. A lubricant may contain several base oils of different viscosities and types, blended to meet viscosity requirements and to help meet performance requirements, or to minimize cost. Additives may be used to help meet viscosity as well as various performance needs.

Petroleum Oils Lubricants made with petroleum base oils are widely used because of their general suitability to much of existing equipment and availability at moderate cost. Most petroleum base oils, often called **mineral oils,** are prepared by conventional refining processes from naturally occurring hydrocarbons in crude oils. The main crude oil types are paraffinic- and naphthenic-based, terms referring to the type of chemical structure of most molecules. Paraffinic oils are most often preferred as lubricants, although naphthenic oils are important in certain applications. A few new types of petroleum base oils have become available, produced by more severe processing. These oils (hydrocracked oils and hydroisomerized wax oils) cost more, and availability is rather limited, but they have many improved properties, approaching those of more expensive synthetic oils. The improvements include better oxidation and thermal stability, lower volatility, higher viscosity index, and lower levels of sulfur, aromatics, and nitrogen compounds. Use of these oils is destined to grow as the trend toward higher-performance lubricants continues.

Synthetic oils are artificially made, as opposed to naturally occurring petroleum fluids. Synthetic oil types include polyalphaolefins (PAOs), diesters, polyol esters, alkylbenzenes, polyalkylene glycols, phosphate esters, silicones, and halogenated hydrocarbons. Synthetic oils include diverse types of chemical compounds, so few generalities apply to all synthetic oils. Generally, synthetic oils are organic chemicals. Synthetic oils cost much more than petroleum oils, and usually they have some lubricant-related properties that are good or excellent, but they are not necessarily strong in all properties. Their stronger properties often include greater high-temperature stability and oxidation resistance, wider temperature operating range, and improved energy efficiency. As with petroleum base oils, some of their weaknesses are correctable by additives. Some synthetic esters, such as polyol esters, have good biodegradability, a property of increasing need for environmental protection.

Synthetic lubricants, because of their high cost, often are used only where some particular property is essential. There are instances where the higher cost of synthetics is justified, even though petroleum lubricants perform reasonably well. This may occur, e.g., where use of synthetics reduces the oxidation rate in high-temperature service and as a result extends the oil life substantially. Properties of some synthetic lubricants are shown in Table 6.11.1, the assessments based on formulated lubricants, not necessarily the base oils. Synthetic oil uses include lubricants for gears, bearings, engines, hydraulic systems, greases, compressors, and refrigeration systems, although petroleum oils tend to dominate these applications. Lubricants blended with both petroleum and synthetic base oils are called **semisynthetic lubricants.**

Vegetable oils have very limited use as base stocks for lubricants, but are used in special applications. Naturally occurring esters, such as from rapeseed or castor plants, are used as the base oils in some hydraulic fluids, and are used where there are environmental concerns of possibility of an accidental fluid spill or leakage. These oils have a shorter life and lower maximum service temperature, but have a high degree of biodegradability.

Fatty oils, extracted from vegetable, animal, and fish sources, have excellent lubricity because of their glyceride structure, but are seldom used as base oils because they oxidize rapidly. A few percent of fatty oils are sometimes added to petroleum oils to improve lubricity (as in worm-gear applications), or to repel moisture (as in steam engines), and this use is often referred to as **compounding.** They also are used widely in forming soaps which, in combination with other oils, set up grease structures.

Additives are chemical compounds added to base oils to modify or enhance certain lubricant performance characteristics. The amount of additives used varies with the type of lubricant, from a few parts per million in some types of lubricants to over 20 percent in some engine oils. The types of additives used include antioxidants, antiwear agents, extreme-pressure agents, viscosity index improvers, dispersants, detergents, pour point depressants, friction modifiers, corrosion inhibitors,

Table 6.11.1 Relative Properties of Synthetic Lubricants*

	Viscosity index	High-temperature stability	Lubricity	Low-temperature properties	Hydrolytic stability	Fire resistance	Volatility
Polyalphaolefins	Good	Good	Good	Good	Excellent	Poor	Good
Diesters	Varies	Excellent	Good	Excellent	Fair	Fair	Average
Polyol esters	Good	Excellent	Good	Good	Good	Poor	Average
Alkylbenzenes	Poor	Fair	Good	Good	Excellent	Poor	Average
Polyalkylene glycols	Excellent	Good	Good	Good	Good	Poor	Good
Phosphate esters	Poor	Excellent	Good	Varies	Fair	Excellent	Average
Silicones	Excellent	Excellent	Poor	Excellent	Fair	—	Good
Fluorinated lubes	Excellent	Excellent	Varies	Fair	Excellent	Excellent	Average

* Comparisons made for typical formulated products with additive packages, not for the base oils alone.
SOURCE: *Courtesy of Texaco's magazine Lubrication,* 78, no. 4, 1992.

rust inhibitors, metal deactivators, tackiness agents, antifoamants, air-release agents, demulsifiers, emulsifiers, odor control agents, and biocides.

LUBRICATION REGIMES

The viscosity of a lubricant and its additive content are to a large extent related to the lubrication regimes expected in its intended application. It is desirable, but not always possible, to design mechanisms in which the rubbing surfaces are totally separated by lubricant films. **Hydrostatic** lubrication concerns equipment designed such that the lubricant is supplied under sufficient pressure to separate the rubbing surfaces. **Hydrodynamic** lubrication is a more often used regime, and it occurs when the motion of one surface over lubricant on the other surface causes sufficient film pressure and thickness to build up to separate the surfaces. Achieving this requires certain design details, as well as operation within a specific range of speed, load, and lubricant viscosity. Increasing speed or viscosity increases the film thickness, and increasing load reduces film thickness. Plain journal bearings, widely used in automotive engines, operate in this regime. For concentrated nonconforming contacts, such as rolling-element bearings, cams, and certain type of gears, the surface shape and high local load squeeze the oil film toward surface contact. Under such conditions, **elastohydrodynamic** lubrication (EHL) may occur, where the high pressures in the film cause it to become very viscous and to elastically deform and separate the surfaces.

Boundary lubrication occurs when the load is carried almost entirely by the rubbing surfaces, separated by films of only molecular dimensions. **Mixed-film lubrication** occurs when the load is carried partly by the oil film and partly by contact between the surfaces' roughness peaks (asperities) rubbing one another. In mixed-film and boundary lubrication, lubricant viscosity becomes less important and chemical composition of the lubricant film more important. The films, although very thin, are needed to prevent surface damage. At loads up to a certain level, antiwear additives, the most common one being zinc dithiophosphate (ZDTP), are used in the oil to form films to help support the load and reduce wear. ZDTP films can have thicknesses from molecular to much larger dimensions. In very highly loaded contacts, such as hypoid gears, **extreme-pressure** (EP) additives are used to minimize wear and prevent scuffing. EP additives may contain sulfur, phosphorus, or chlorine, and they react with the high-temperature contact surfaces to form molecular-dimension films. Antiwear, EP, and friction modifier additives function only in the mixed and boundary regimes.

LUBRICANT TESTING

Lubricants are manufactured to have specific characteristics, defined by physical and chemical properties, and performance characteristics. Most of the tests used by the lubricants industry are described in ASTM (American Society for Testing and Materials) Standard Methods. **Physical tests** are frequently used to characterize petroleum oils, since lubricant performance often depends upon or is related to physical properties. Common physical tests include measurements of viscosity, density,

pour point, API gravity, flash point, fire point, odor, and color. **Chemical tests** measure such things as the amount of certain elements or compounds in the additives, or the acidity of the oil, or the carbon residue after heating the oil at high temperature. **Performance tests** evaluate particular aspects of in-service behavior, such as oxidation stability, rust protection, ease of separation from water, resistance to foaming, and antiwear and extreme-pressure properties. Many of the tests used are small-size or bench tests, for practical reasons. To more closely simulate service conditions, full-scale standardized performance tests are required for certain types of lubricants, such as oils for gasoline and diesel engines. Beyond these industry standard tests, many equipment builders require tests in specific machines or in the field. The full scope of lubricant testing is rather complex, as simple physical and chemical tests are incapable of fully defining in-service behavior.

VISCOSITY TESTS

Viscosity is perhaps the single most important property of a lubricant. Viscosity needs to be sufficient to maintain oil films thick enough to minimize friction and wear, but not so viscous as to cause excessive efficiency losses. Viscosity reduces as oil temperature increases. It increases at high pressure and, for some fluids, reduces from shear, during movement in small clearances.

Kinematic viscosity is the most common and fundamental viscosity measurement, and it is obtained by oil flow by gravity through capillary-type instruments at low-shear-rate flow conditions. It is measured by ASTM Standard Method D445. A large number of commercially available capillary designs are acceptable. Automated multiple-capillary machines also are sold. Kinematic viscosity is typically measured at 40 and 100°C and is expressed in centistokes, cSt (mm²/s).

Saybolt viscosity, with units of Saybolt universal seconds (SUS), was once widely used. There is no longer a standard method for direct measurement of SUS. Some industrial consumers continue to use SUS, and it can be determined from kinematic viscosity using ASTM D2161.

Dynamic viscosity is measured for some lubricants at low temperatures (ASTM D4684, D5293, and D2983); it is expressed in centipoise cP (mPa·s). Kinematic and dynamic viscosities are related by the equation cSt = cP/density, where density is in g/cm³.

Viscosity grades, not specific viscosities, are used to identify viscosities of lubricants in the marketplace, each viscosity grade step in a viscosity grade system having minimum and maximum viscosity limits. There are different viscosity grade systems for automotive engine oils, gear oils, and industrial oils. Grade numbers in any one system are independent of grade numbers in other systems.

ISO Viscosity Grades The International Standards Organization (ISO) has a viscosity grade system (ASTM D2422) for industrial oils. Each grade is identified by ISO VG followed by a number representing the nominal 40°C (104°F) kinematic viscosity in centistokes for that grade. This system contains 18 viscosity grades, covering the range from 2 to 1,500 cSt at 40°C, each grade being about 50 percent higher in viscosity than the previous one. The other viscosity grade systems are described later.

The **variation of viscosity with temperature** for petroleum oils can be determined from the viscosities at any two temperatures, such as 40 and 100°C kinematic viscosity, using a graphical method or calculation (ASTM D341). The method is accurate only for temperatures above the wax point and below the initial boiling point, and for oils not containing certain polymeric additives.

The **viscosity index** (VI) is an empirical system expressing an oil's rate of change in viscosity with change in temperature. The system, developed some years ago, originally had a 0 to 100 VI scale, based on two oils thought to have the maximum and minimum limits of viscosity-temperature sensitivity. Subsequent experience identified oils far outside the VI scale in both directions. The VI system is still used, and the VI of an oil can be calculated from its 40 and 100°C kinematic viscosity using ASTM D2270, or by easy-to-use tables (ASTM Data Series, DS 39B). Lubricants with high VI have less change in viscosity with temperature change, desirable for a lubricant in service where temperatures vary widely.

Base oils with more than 100 VI can be made from a wide variety of crude oil distillates by solvent refining, by hydrogenation, by selective blending of paraffinic base oils, by adding a few percent of high-molecular-weight polymeric additives called **viscosity index improvers**, or by combinations of these methods. A few highly processed petroleum base oils and many synthetic oils have high VIs.

Effect of Pressure on Viscosity   When lubricating oils are subjected to high pressures, thousands of lb/in², their viscosity increases. When oil-film pressures are in this order of magnitude, their influence on viscosity becomes significant. Empirical equations are available to estimate the effect of pressure on viscosity. In rolling-contact bearings, gears, and other machine elements, the high film pressures will influence viscosity with an accompanying increase in frictional forces and load-carrying capacity.

Effect of Shear on Viscosity   The viscosity of lubricating oils not containing polymeric additives is independent of **shear rate**, except at low temperatures where wax is present. Where polymeric additives are used, as in multigrade engine oils, high shear rates in lubricated machine elements cause **temporary viscosity loss**, as the large polymer molecules temporarily align in the direction of flow. Engine oils consider this effect by high-temperature/high-shear-rate measurements (ASTM D4683 or ASTM D4741). Such oils can suffer **permanent viscosity loss** in service, as polymer molecules are split by shear stresses into smaller molecules with less thickening power. The amount of the permanent viscosity loss depends on the type of polymeric additive used, and it is measurable by comparing the viscosity of the new oil to that of the used oil, although other factors need considering, such as any oxidative thickening of the oils.

OTHER PHYSICAL AND CHEMICAL TESTS

Cloud Point   Petroleum oils, when cooled, may become plastic solids, either from wax formation or from the fluid congealing. With some oils, the initial wax crystal formation becomes visible at temperatures slightly above the solidification point. When that temperature is reached at specific test conditions, it is known as the **cloud point** (ASTM D2500). The cloud point cannot be determined for those oils in which wax does not separate prior to solidification or in which the separation is invisible.

The cloud point indicates the temperature below which clogging of filters may be expected in service. Also, it identifies a temperature below which viscosity cannot be predicted by ASTM D2270, as the wax causes a higher viscosity than is predicted.

Pour point is the temperature at which cooled oil will just flow under specific test conditions (ASTM D97). The pour point indicates the lowest temperature at which a lubricant can readily flow from its container. It provides only a rough guide to its flow in machines. Pour point depressant additives are often used to reduce the pour point, but they do not affect the cloud point. Other tests, described later, are used to more accurately estimate oil flow properties at low temperatures in automotive service.

Density and Gravity   ASTM D1298 may be used for determining density (mass), specific gravity, and the API (American Petroleum Institute) gravity of lubricating oils. Of these three, petroleum lubricants tend to be described by using API gravity. The specific gravity of an oil is the ratio of its weight to that of an equal volume of water, both measured at 60°F (16°C), and the API gravity of an oil can be calculated by

$$\text{API gravity, deg} = (141.5/\text{sp. gr. @ }60°/60°\text{F}) - 131.5$$

The gravity of lubricating oils is of no value in predicting their performance. Low-viscosity oils have higher API gravities than higher-viscosity oils of the same crude-oil series. Paraffinic oils have the lowest densities or highest API gravities, naphthenic are intermediate, and animal and vegetable oils have the heaviest densities or lowest API gravities.

Flash and Fire Points   The flash point of an oil is the temperature to which an oil has to be heated until sufficient flammable vapor is driven off to flash when brought into momentary contact with a flame. The fire point is the temperature at which the oil vapors will continue to burn when ignited. ASTM D92, which uses the Cleveland open-cup (COC) tester, is the flash and fire point method used for lubricating oils. In general, for petroleum lubricants, the open flash point is 30°F or (17°C) higher than the closed flash (ASTM D93), and the fire point is some 50 to 70°F or (28 to 39°C) above the open flash point.

Flash and fire points may vary with the nature of the original crude oil, the narrowness of the distillation cut, the viscosity, and the method of refining. For the same viscosities and degree of refinement, paraffinic oils have higher flash and fire points than naphthenic oils. While these values give some indication of fire hazard, they should be taken as only one element in fire risk assessment.

Color   The color of a lubricating oil is obtained by reference to transmitted light; the color by reflected light is referred to as **bloom**. The color of an oil is not a measure of oil quality, but indicates the uniformity of a particular grade or brand. The color of an oil normally will darken with use. ASTM D1500 is for visual determination of the color of lubricating oils, heating oils, diesel fuels, and petroleum waxes, using a standardized colorimeter. Results are reported in terms of the ASTM color scale and can be compared with the former ASTM union color. The color scale ranges from 0.5 to 8; oils darker than color 8 may be diluted with kerosene by a prescribed method and then observed. Very often oils are simply described by visual assessment of color: brown, black, etc. For determining the color of petroleum products lighter than 0.5, the ASTM D156 Test for Saybolt Color of Petroleum Products can be used.

Carbon residue, the material left after heating an oil under specified conditions at high temperature, is useful as a quality control tool in the refining of viscous oils, particularly residual oils. It does not correlate with carbon-forming tendencies of oils in internal-combustion engines. Determination is most often made by the Conradson procedure (ASTM D189). Values obtained by the more complex Ramsbottom procedure (ASTM D524) are sometimes quoted. The correlation of the two values is given in both methods.

Ash   Although it is unlikely that well-refined oils that do not contain metallic additives will yield any appreciable ash from impurities or contaminants, measurement can be made by ASTM D482. A more useful determination is **sulfated ash** by ASTM D874, as applied to lubricants containing metallic additives.

Neutralization number and **total acid number** are interchangeable terms indicating a measure of acidic components in oils, as determined by ASTM D664 or D974. The original intent was to indicate the degree of refining in new oils, and to follow the development of oxidation in service, with its effects on deposit formation and corrosion. However, many modern oils contain additives which, in these tests, act similar to undesirable acids and are indistinguishable from them. Caution must be exercised in interpreting results without knowledge of additive behavior. Change in acid number is more significant than the absolute value in assessing the condition of an oil in service. Oxidation of an oil is usually accompanied by an increase in acid number.

Total base number (TBN) is a measure of alkaline components in oils, especially those additives used in engine oils to neutralize acids formed during fuel combustion. Some of these acids get in the crankcase with gases that blow by the rings. Today, TBN is generally measured by ASTM D2896 or D4739, as it provides a better indication of an oil's ability to protect engines from corrosive wear than ASTM D664, which was previously used for TBN measurement.

Antifoam The foaming characteristic of crankcase, turbine, or circulating oils is checked by a foaming test apparatus (ASTM D892), which blows air through a diffuser in the oil. The test measures the amount of foam generated and the rate of settling. Certain additive oils tend to foam excessively in service. Many types of lubricants contain a minute quantity of antifoam inhibitor to prevent excessive foaming in service. At this writing, a new high-temperature foam test is under development in ASTM for passenger-car engine oil use, and a Navistar engine test is used to test for diesel-engine-oil aeration.

Oxidation Tests Lubricating oils may operate at relatively high temperatures in the presence of air and catalytically active metals or metallic compounds. Oxidation becomes significant when oil is operated above 150°F (66°C). Some lubricating-oil sump temperatures may exceed 250°F (121°C). The oxidation rate doubles for about every 18°F (10°C) rise in oil temperature above 150°F (66°C). The resultant oil oxidation increases viscosity, acids, sludge, and lacquer.

To accelerate oxidation to obtain practical test length, oxidation tests generally use temperatures above normal service, large amounts of catalytic metals, and sufficient oxygen. Two standard oxidation tests, ASTM D943 and D2272, are often applied to steam turbine and similar industrial oils. There are many other standard oxidation tests for other specific lubricants. Because test conditions are somewhat removed from those encountered in service, care should be exercised in extrapolating results to expected performance in service.

Rust Prevention Lubricating oils are often expected to protect ferrous surfaces from rusting when modest amounts of water enter the lubricating system. Many oils contain additives specifically designed for that purpose. Steam turbine and similar industrial oils often are evaluated for rust prevention by ASTM D665 and D3603. Where more protection is required, as when a thin oil film must protect against a moisture-laden atmosphere, ASTM D1748 is a useful method.

Water Separation In many applications, it is desirable for the oil to have good water-separating properties, to enable water removal before excessive amounts accumulate and lead to rust and lubrication problems. The demulsibility of an oil can be measured by ASTM D1401 for light- to moderate-viscosity oils and by ASTM D2711 for heavier-viscosity oils.

Wear and EP Tests A variety of test equipment and methods have been developed for evaluating the wear protection and EP properties (**load-carrying capacity before scuffing**) of lubricant fluids and greases under different types of heavy-duty conditions. The tests are simplified, accelerated tests, and correlations to service conditions are seldom available, but they are important in comparing different manufacturers' products and are sometimes used in lubricant specifications. Each apparatus tends to emphasize a particular characteristic, and a given lubricant will not necessarily show the same EP properties when tested on different machines. ASTM methods D2783 (fluids) and D2509 (grease) use the **Timken EP tester** to characterize the EP properties of industrial-type gear lubricants, in tests with incremented load steps. ASTM D3233 (fluids) uses the **Falex pin and vee block tester** for evaluating the EP properties of oils and gear lubricants, using either continuous (method A) or incremented (method B) load increase during the test. Evaluation of wear-prevention properties also can be conducted using this apparatus, by maintaining constant load. ASTM methods D2783 (fluids) and D2596 (grease) use the **four-ball EP tester** for evaluating the EP properties of oils and greases. ASTM methods D4172 (fluids) and D2266 (grease) use the **four-ball wear tester** to evaluate the wear-prevention properties of oils and grease.

GREASES

Lubricating greases consist of lubricating liquid dispersed in a thickening agent. The lubricating liquid is about 70 to 95 wt % of the finished grease and provides the principal lubrication, while the thickener holds the oil in place. Additives are often added to the grease to impart special properties. Grease helps seal the lubricating fluid in and the contaminants out. Grease also can help reduce lubrication frequency, particularly in intermittent operation.

The lubricating liquid is usually a napthenic or paraffinic petroleum oil or a mixture of the two. Synthetic oils are used also, but because of their higher cost, they tend to be used only for specialty greases, such as greases for very low or high temperature use. Synthetic oils used in greases include polyalphaolefins, dialkylbenzenes, dibasic acid esters, polyalkylene glycols, silicones, and fluorinated hydrocarbons. The viscosity of the lubricating liquid in the grease should be the same as would be selected if the lubricant were an oil.

Soaps are the most common thickeners used in greases. Complex soaps, pigments, modified clays, chemicals (such as polyurea), and polymers are also used, alone or in combination. Soaps are formed by reaction of fatty material with strong alkalies, such as calcium hydroxide. In this saponification reaction, water, alcohol, and/or glycerin may be formed as by-products. Soaps of weak alkalies, such as aluminum, are formed indirectly through further reactions. Metallic soaps, such as sodium, calcium, and lithium, are the most widely used thickeners today.

Additives may be incorporated in the grease to provide or enhance tackiness, to provide load-carrying capability, to improve resistance to oxidation and rusting, to provide antiwear or EP properties, and to lessen sensitivity to water. The additives sometimes are solid lubricants such as graphite, molybdenum disulfide, metallic powders, or polymers, which grease can suspend better than oil. Solid fillers may be used to improve grease performance under extremely high loads or shock loads.

The **consistency** or **firmness** is the characteristic which may cause grease to be chosen over oil in some applications. Consistency is determined by the depth to which a cone penetrates a grease sample under specific conditions of weight, time, and temperature. In ASTM D217, a standardized cone is allowed to drop into the grease for 5 s at 77°F (25°C). The resulting depth of fall, or **penetration,** is measured in tenths of millimeters. If the test is made on a sample simply transferred to a standard container, the results are called **unworked penetration.** To obtain a more uniform sample, before the penetration test, the grease is worked for 60 strokes in a mechanical worker, a churnlike device. Variations on unworked and worked penetrations, such as prolonged working, undisturbed penetration, or block penetration, also are described in ASTM D217. The worked 60-stroke penetration is the most widely used indication of grease consistency. The National Lubricating Grease Institute (NLGI) used this penetration method to establish a system of **consistency numbers** to classify differences in grease consistency or firmness. The consistency numbers range from 000 to 6, each representing a range of 30 penetration units, and with a 15 penetration unit gap between each grade, as shown in Table 6.11.2.

As grease is heated, it may change gradually from a semisolid to a liquid state, or its structure may weaken until a significant amount of oil is lost. Grease does not exhibit a true melting point, which implies a sharp change in state. The weakening or softening behavior of grease when heated to a specific temperature can be defined in a carefully controlled heating program with well-defined conditions. This test gives a temperature called the **dropping point,** described in ASTM D566 or

Table 6.11.2 Grease Consistency Ranges

Consistency no.	Appearance	Worked penetration
000	Semifluid	445–475
00	Semifluid	400–430
0	Semifluid	355–385
1	Soft	310–340
2	Medium	265–295
3	Medium hard	220–250
4	Hard	175–205
5	Very hard	130–160
6	Block type	85–115

D2265. Typical dropping points are given in Table 6.11.3. At its **dropping point,** a grease already has become fluid, so the maximum application temperature must be well below the dropping point shown in the table.

The oil constituent of a grease is loosely held. This is necessary for some lubrication requirements, such as those of ball bearings. As a result, on opening a container of grease, free oil is often seen. The tendency of a grease to **bleed** oil may be measured by ASTM D1742. Test results are related to bleeding in storage, not in service or at elevated temperatures.

Texture of a grease is determined by formulation and processing. Typical textures are shown in Table 6.11.3. They are useful in identification of some products, but are of only limited assistance for predicting behavior in service. In general, smooth, buttery, short-fiber greases are preferred for rolling-contact bearings and stringy, fibrous products for sliding service. Stringiness or tackiness imparted by polymeric additives helps control leakage, but this property may be diminished or eliminated by the shearing action encountered in service.

Grease **performance** depends only moderately on physical test characteristics, such as penetration and dropping point. The likelihood of suitable performance may be indicated by these and other tests. However, the final determination of how a lubricating grease will perform in service should be based on observations made in actual service.

SOLID LUBRICANTS

Solid lubricants are materials with low coefficients of friction compared to metals, and they are used to reduce friction and wear in a variety of applications. There are a large number of such materials, and they include graphite, molybdenum disulfide, polytetrafluoroethylene, talc, graphite fluoride, polymers, and certain metal salts. The many diverse types have a variety of different properties, operating ranges, and methods of application. The need for lubricants to operate at extremes of temperature and environment beyond the range of organic fluids, such as in the space programs, helped foster development of solid lubricants.

One method of using solid lubricants is to apply them as thin films on the bearing materials. The film thicknesses used range from 0.0002 to 0.0005 in (0.005 to 0.013 mm). There are many ways to form the films. Surface preparation is very important in all of them. The simplest method is to apply them as **unbonded solid lubricants,** where granular or powdered lubricant is applied by brushing, dipping, or spraying, or in a liquid or gas carrier for ease of application. Burnishing the surfaces is beneficial. The solid lubricant adheres to the surface to some degree by mechanical or molecular action. More durable films can be made by using **bonded solid lubricants,** where the lubricant powders are mixed with binders before being applied to the surface. Organic binders, if used, can be either room-temperature air-cured type or those requiring thermal setting. The latter tend to be more durable. Air-cured films are generally limited to operating temperatures below 300°F (260°C), while some thermoset films may be satisfactory to 700°F (371°C). Inorganic ceramic adhesives combined with certain powdered metallic solid lubricants permit film use at temperatures in excess of 1,200°F (649°C). The performance of solid-lubricant films is influenced by the solid lubricant used, the method of application, the bearing-surface-material finish and its hardness, the binder-solid lubricant mix, the film and surface pretreatment, and the application's operating conditions. Solid-lubricant films can be evaluated in certain standard tests, including ASTM D2510 (adhesion), D2511 (thermal shock sensitivity), D2625 (wear endurance and load capacity), D2649 (corrosion characteristics), and D2981 (wear life by oscillating motion).

Many plastics are also solid lubricants compared to the friction coefficients typical of metals. Plastics are used without lubrication in many applications. Strong plastics can be compounded with a variety of solid lubricants to make plastics having both strength and low friction. Solid-lubricant powders and plastics can be mixed, compacted, and sintered to form a lubricating solid, such as for a bearing. Such materials also can be made by combining solid-lubricant fibers with other stronger plastic fibers, either woven together or chopped, and used in compression-molded plastics. Bushings made of low-friction plastic materials may be press-fitted into metal sleeves. The varieties of low-friction plastic materials are too numerous to mention. Bearings made with solid low-friction plastic materials are commercially available.

Solid lubricants may also be embedded in metal matrices to form solid-lubricant composites of various desirable properties. Graphite is the most common solid lubricant used, and others include molybdenum disulfide and other metal-dichalcogenides. The base metals used include copper, aluminum, magnesium, cadmium, and others. These materials are made by a variety of processes, and their uses range from electrical contacts to bearings in heavy equipment.

LUBRICATION SYSTEMS

There are many positive methods of applying lubricating oils and greases to ensure proper lubrication. Bath and circulating systems are common. There are constant-level lubricators, gravity-feed oilers, multiple-sight feeds, grease cups, forced-feed lubricators, centralized lubrication systems, and air-mist lubricators, to name a few. Selecting the method of lubricant application is as important as the lubricant itself. The choice of the device and the complexity of the system depend upon many factors, including the type and quantity of the lubricant, reliability and value of the machine elements, maintenance schedules, accessibility of the lubrication points, labor costs, and other economic considerations, as well as the operating conditions.

The importance of removing foreign particles from circulating oil has been increasingly recognized as a means to minimize wear and avoid formation of potentially harmful deposits. Adequate **filtration** must be provided to accomplish this. In critical systems involving closely fitting parts, monitoring of particles on a regular basis may be undertaken.

LUBRICATION OF SPECIFIC EQUIPMENT

Internal-combustion-engine oils are required to carry out numerous functions to provide adequate lubrication. Crankcase oils, in addition to reducing friction and wear, must keep the engine clean and free from rust and corrosion, must act as a coolant and sealant, and must serve as a hydraulic oil in engines with hydraulic valve lifters. The lubricant may

Table 6.11.3   General Characteristics of Greases

Base	Texture	Typical dropping point, °F (°C)	Max usable temperature,* °F (°C)	Water resistance	Primary uses
Sodium	Fibrous	325–350 (163–177)	200 (93)	P–F	Older, slower bearings with no water
Calcium	Smooth	260–290 (127–143)	200 (93)	E	Moderate temperature, water present
Lithium 12-hydroxystearate	Smooth	380–395 (193–202)	250 (121)	G	General bearing lubrication
Polyurea	Smooth	450+ (232+)	350 (177)	G–E	Sealed-for-life bearings
Calcium complex	Smooth	500+ (260+)	350 (177)	F–E	Used at high temperatures, with frequent relubrication
Lithium complex	Smooth	450+ (232+)	325 (163)	G–E	
Aluminum complex	Smooth	500+ (260+)	350 (177)	G–E	
Organoclay	Smooth	500+ (260+)	250 (121)	F–E	

* Continuous operation with relatively infrequent lubrication.

function over a wide temperature range and in the presence of atmospheric dirt and water, as well as with combustion products that blow by the rings into the crankcase. It must be resistant to oxidation, sludge, and varnish formation in a wide range of service conditions.

Engine oils are heavily fortified with additives, such as **detergents,** to prevent or reduce deposits and corrosion by neutralizing combustion by-product acids; **dispersants,** to help keep the engine clean by solubilizing and dispersing sludge, soot, and deposit precursors; **oxidation inhibitors,** to minimize oil oxidation, particularly at high temperatures; **corrosion inhibitors,** to prevent attack on sensitive bearing metals; **rust inhibitors,** to prevent attack on iron and steel surfaces by condensed moisture and acidic corrosion products, aggravated by low-temperature stop-and-go operation; **pour point depressants,** to prevent wax gelation and improve low-temperature flow properties; **viscosity-index improvers,** to help enable adequate low-temperature flow, along with sufficient viscosity at high temperatures; **antiwear** additives, to minimize wear under boundary lubrication conditions, such as cam and lifter, and cylinder-wall and piston-ring surfaces; **defoamants,** to allow air to break away easily from the oil; and **friction modifiers,** to improve fuel efficiency by reducing friction at rubbing surfaces.

The **SAE viscosity grade system for engine oils** (SAE J300), shown in Table 6.11.4 provides W (winter) viscosity grade steps (for example, 5W) for oils meeting certain primarily low-temperature viscosity limits, and non-W grade steps (for example, 30) for oils meeting high-temperature viscosity limits. The system covers both single-grade oils (for example, SAE 30), not designed for winter use, and multigrade oils (for example, SAE 5W-30), formulated for year-round service. This system is subject to revision as improvements are made, and the table shows the December 1995 version. Each W grade step has maximum low-temperature viscosity limits for cranking (D5293) and pumping (D4684) bench tests, and a minimum 100°C kinematic viscosity limit. The lower the W grade, the colder the potential service use. Pumping viscosity limits are based on temperatures 10°C below those used for cranking limits, to provide a safety margin so that an oil that will allow engine starting at some low temperature also will pump adequately at that temperature. Each non-W grade step has a specific 100°C kinematic viscosity range and a minimum 150°C high-shear-rate viscosity limit. High-temperature high-shear-rate viscosity limits are a recent addition to J300, added in recognition that critical lubrication in an engine occurs largely at these conditions. Two 150°C high-shear-rate viscosity limits are shown for 40 grade oils, each for a specific set of W grades, the limit being higher for viscosity grades used in heavy-duty diesel engine service.

Gasoline Engines Two systems are used at this time to define gasoline-engine-oil performance levels, not counting systems outside the United States. The API service category system has been used for many years and is the result of joint efforts of many organizations. Since 1970, gasoline engine performance categories in this system have used the letter S (spark ignition) followed by a second letter indicating the specific performance level. Performance category requirements are based primarily on performance in various standard engine tests. The categories available change periodically as higher performance categories are needed for newer engines, and as older categories become obsolete when tests on which they are based are no longer available. At this writing, SH is the highest performance level in this system.

The newer ILSAC system is based primarily on the same tests, but has a few additional requirements. This system was developed by the International Lubricant Standardization and Approval Committee (ILSAC), a joint organization of primarily U.S. and Japanese automobile and engine manufacturers. The first category is in use, ILSAC GF-1.

There are similarities as well as distinct differences between the two systems. Both are displayed on oil containers using API licensed marks. The API service category system uses the API service symbol, and ILSAC system uses the API certification mark, both shown in Fig. 6.11.1. Either or both marks can be displayed on an oil container. In the API service symbol, the performance level (for example, SH), viscosity grade (for example, SAE 5W-30), and energy-conserving level (for example, energy-conserving II) are shown. None of these are shown in the API certification mark of the ILSAC system. The API certification mark remains unchanged as new performance levels are developed and regardless of the viscosity grade of the oil in the container. ILSAC categories require the highest energy-conserving level, and only allow a few viscosity grades, those recommended for newer engines. The vis-

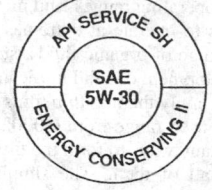

API Service Symbol   API Certification Mark

Fig. 6.11.1 API marks for engine oil containers. *(From API Publication 1509, 12th ed., 1993. Reprinted by courtesy of the American Petroleum Institute.)*

Table 6.11.4   SAE Viscosity Grades for Engine Oil[a] (SAE J300 DEC95)

SAE viscosity grade	Low temperature viscosity, cP, max, at temp, °C, max		Viscosity,[d] cSt, at 100°C		High-shear-rate viscosity,[e] cP, min, at 150°C and 10⁶ s⁻¹
	Cranking[b]	Pumping[c]	Min	Max	
0W	3,250 at −30	60,000 at −40	3.8	—	—
5W	3,500 at −25	60,000 at −35	3.8	—	—
10W	3,500 at −20	60,000 at −30	4.1	—	—
15W	3,500 at −15	60,000 at −25	5.6	—	—
20W	4,500 at −10	60,000 at −20	5.6	—	—
25W	6,000 at −5	60,000 at −15	9.3	—	—
20	—	—	5.6	< 9.3	2.6
30	—	—	9.3	<12.5	2.9
40	—	—	12.5	<16.3	2.9[f]
40	—	—	12.5	<16.3	3.7[g]
50	—	—	16.3	<21.9	3.7
60	—	—	21.9	<26.1	3.7

[a] All values are critical specifications as defined in ASTM D3244.
[b] ASTM D5293.
[c] ASTM D4684. Note that the presence of any yield stress detectable by this method constitutes a failure, regardless of viscosity.
[d] ASTM D445   Note: 1 cP = 1 mPa · s   1 cSt = 1 mm²/s
[e] ASTM D4683, CEC L-36-A-90 (ASTM D4741)
[f] 0W-40, 5W-40, and 10W-40 grade oils.
[g] 15W-40, 20W-40, 25W-40, and 40 grade oils.

cosity grades allowed are SAE 0W-20, 5W-20, 5W-30, and 10W-30. The viscosity grade may be displayed separately. In the ILSAC system, to ensure high-performance-level oils, the API certification mark is licensed annually and only for oils meeting the latest ILSAC performance level.

Diesel Engines   API performance-level categories for commercial heavy-duty diesel engine oils, using the API service category system, begin with the letter C (compression ignition), followed by additional letters and numbers indicating the performance level and engine type, the performance letter changing as category improvements are needed. At this writing, CG-4 is the highest performance category for four-cycle low-sulfur-fueled turbocharged heavy-duty diesel engines, and CF-2 is the highest for two-cycle turbocharged heavy-duty diesel engines. ILSAC intends to develop a new performance category system for heavy-duty diesel engine oils and to use a different design API mark for this system. Viscosity grades used depend on engine manufacturer's recommendations for the ambient temperature range. SAE 15W-40 oils are very popular for most four-cycle diesel engines, and single-grade oils, such as SAE 40, are used primarily in hotter climates and for two-cycle diesel engines.

The engine oil is increasingly becoming a major part in the total design, affecting performance of critical components as well as engine life. Emission control restrictions are tightening up on diesel engines, and this is affecting the engine oils. Designs of pistons are changing, and rings are being located higher on the pistons and are more subject to carbon deposits. Deposits are being controlled partly by use of proper engine oils. One new engine design with reduced emissions uses the engine oil as hydraulic fluid to operate and improve control of the fuel injectors, and this has created a need for improved oil deaeration performance. Most engine manufacturers maintain lists of oils that perform well in their engines.

Larger diesel engines found in marine, railroad, and stationary service use crankcase oils that are not standardized as to performance levels, unlike the oils discussed above. They are products developed by reputable oil suppliers working with major engine builders. In general, they are formulated along the same lines as their automotive counterparts. However, particularly in marine service where fuels often contain relatively high levels of sulfur, the detergent additive may be formulated with a high degree of alkalinity to neutralize sulfuric acid resulting from combustion.

Gas engines pose somewhat different lubrication requirements. They burn a clean fuel which gives rise to little soot, but conditions of operation are such that nitrogen oxides formed during combustion can have a detrimental effect on the oil. Suitable lubricants may contain less additive than those which must operate in a sootier environment. However, the quality of the base oil itself assumes greater importance. Special selection of crude source and a high degree of refining must be observed to obtain good performance. The most common viscosity is SAE 40, although SAE 30 is also used. Where cold starting is a factor, SAE 15W-40 oils are available.

Steam Turbines   Although they do not impose especially severe lubrication requirements, steam turbines are expected to run for very long periods, often measured in many years, on the same oil charge. Beyond that, the oil must be able to cope with ingress of substantial amounts of water. Satisfactory products consist of highly refined base oils with rust and oxidation (R&O) inhibitors, and they must show good water-separating properties (demulsifiability). There is demand for longer-life turbine oil in the power generation industry, such as 6,000 h or more of life in the ASTM D943 oxidation bench test. This demand is being met by using more severely processed base oils and with improved R&O inhibitors. Most large utility turbines operate with oil viscosity ISO VG 32. Where gearing is involved, ISO VG 100 is preferred. Smaller industrial turbines, which do not have circulating systems typical of large units, may operate with ISO VG 68. Circulating systems should be designed with removal of water in mind, whether by centrifuge, coalescer, settling tank, or a combination of these methods. Where the turbine governor hydraulic system is separate from the central lubrication system, fire-resistant phosphate ester fluids are often used in the governor hydraulic system.

Gas turbines in industrial service are lubricated with oils similar to the ISO VG 32 product recommended for steam turbines; but to obtain acceptable service life under higher-temperature conditions, they may be inhibited against oxidation to a greater extent. There is demand for longer-life oils for gas turbines, and, as for steam turbines, the demand is being met by using more severely processed base oils and improved R&O inhibitors. Gas turbines aboard aircraft operate at still higher bearing temperatures which are beyond the capability of petroleum oils. Synthetic organic esters are used, generally complying with military specifications.

Gears   API System of Lubricant Service Designations for Automotive Manual Transmissions and Axles (SAE J308B) defines service levels for automotive gear oils. The service levels include API GL-1 to GL-6 and API PL-1 to PL-2, related to gear types and service conditions. Also, there are SAE Viscosity Grades for Axle and Transmission Lubricants (SAE J306C) for gear oils in automotive service. In this system, there are winter grades from SAE 70W to 85W, based on low-temperature viscosity (ASTM D2983) limits and 100°C kinematic viscosity (ASTM D445) limits, and grades from SAE 90 to 250, based on 100°C kinematic viscosity limits. Many multigrade oils are marketed based on it, capable of year-round operation.

The American Gear Manufacturers Association (AGMA) has classification systems for industrial gear lubricants, defined in AGMA 250.04 and 251.02. These systems specify lubricant performance grades by AGMA lubricant numbers, each grade linked to a specific ISO viscosity grade. There are three systems for enclosed gear drives, and they cover (1) rust- and oxidation-inhibited, (2) extreme pressure, and (3) compounded lubricants. Three systems for open gear drives cover (1) rust- and oxidation-inhibited, (2) extreme pressure, and (3) residual lubricants.

The various gear types in use differ in severity of lubrication requirements. For spur, bevel, helical, and herringbone gears, clastohydrodynamic lubrication occurs at rated speeds and loads. Oils used for these types of gears need not contain additives that contribute to oil film strength, but may contain antioxidants, rust inhibitors, defoamants, etc., depending on the application. Worm gears need oils with film strength additives, such as compounded lubricants (containing fatty oils in a petroleum base), and use oils of relatively high viscosity, such as ISO VG 460. Other gear types, such as spur, use oils of widely varying viscosities, from ISO VG 46 to 460 or higher. The choice relates to operating conditions, as well as any need to lubricate other mechanisms with the same oil. Hypoid gears, widely used in automobiles, need oils containing effective EP additives. There is increased use of synthetic gear oils where wide operating temperature ranges and need for long life are involved. For example, polyalkylene glycol gear oils, which have high viscosity indices and excellent load capacities, are being used in industrial applications involving heavily loaded or high-temperature worm gears. Synthetic gear oils based on PAO base stocks have performed well in closed gearbox applications. With open gears, the lubricant often is applied sporadically and must adhere to the gear for some time, resisting being scraped off by meshing teeth. Greases, and viscous asphaltic-base lubricants, often containing EP additives, are among the gear lubricants used in this service. Asphaltic products may be diluted with a solvent to ease application, the solvent evaporating relatively quickly from the gear teeth.

Rolling-Contact Bearings   These include ball bearings of various configurations, as well as roller bearings of the cylindrical, spherical, and tapered varieties and needle bearings. They may range in size from a few millimetres to several metres. They are basically designed to carry load under EHL conditions, but considerable sliding motion may exist as well as rolling motion at points of contact between rolling elements and raceways. In addition, sliding occurs between rolling elements and separators and, in roller bearings, between the ends of rollers and raceway flanges. In many instances, it is desirable to incorporate antiwear additives to deal with these conditions.

A choice can frequently be made between oil and grease. Where the lubricant must carry heat away from the bearings, oil is the obvious selection, especially where large quantities can be readily circulated. Also, where contamination by water and solids is difficult to avoid, oil is preferred, provided an adequate purification system is furnished. If circulation is not practical, then grease may be the better choice. Where access to bearings is limited, grease may be indicated. Grease is also preferred in instances where leakage of lubricant might be a problem.

Suitable oils generally contain rust and oxidation inhibitors, and antiwear additives. Viscosity is selected on the basis of EHL considerations or builder recommendations for the particular equipment, and it may range anywhere from ISO VG 32 to 460, typically perhaps ISO VG 68.

Where grease is employed, the selection of type will be governed by operating conditions including temperature, loading, possibility of water contamination, and frequency and method of application. Lithium 12-hydroxystearate grease, particularly of no. 2 consistency grade, is widely used. Where equipment is lubricated for life at the time of manufacture, such as electric motors in household appliances, polyurea greases are often selected.

Air Compressors **Reciprocating** compressors require lubrication of the cylinder walls, packing, and bearings. Temperatures at the cylinder wall are fairly high, and sufficient viscosity must be provided. Oils of ISO VG from about 68 to 460 may be specified. Since water condensation may be encountered, rust inhibitor is needed in addition to oxidation inhibitor. A principal operating problem concerning the lubricant is development of carbonaceous deposits on the valves and in the piping. This can seriously interfere with valve operation and can lead to disastrous fires and explosions. It is essential to choose an oil with a minimum tendency to form such deposits. Conradson and Ramsbottom carbon residue tests are of little value in predicting this behavior, and builders and reputable oil suppliers should be consulted. Maintenance procedures need to ensure that any deposits are cleaned on a regular basis and not be allowed to accumulate. In units which call for all-loss lubrication to the cylinders, feed rates should be reduced to the minimum recommended levels to minimize deposit-forming tendencies. Increasingly, certain synthetic oils, such as fully formulated diesters, polyalkylene glycols, and PAOs, are being recommended for longer, trouble-free operation.

Screw compressors present a unique lubrication problem in that large quantities of oil are sprayed into the air during compression. Exposure of large surfaces of oil droplets to hot air is an ideal environment for oxidation to occur. This can cause lacquer deposition to interfere with oil separator operation which is essential to good performance. Although general-purpose rust- and oxidation-inhibited oils, as well as crankcase oils, are often used, they are not the best choice. Specially formulated petroleum oils, in ISO VG 32 to 68, are available for these severe conditions. For enhanced service life, synthetic organic ester fluids of comparable viscosity are often used.

Refrigeration compressors vary in their lubricant requirements depending on the refrigerant gas involved, particularly as some lubricant may be carried downstream of the compressor into the refrigeration system. If any wax from the lubricant deposits on evaporator surfaces, performance is seriously impaired. Ammonia systems can function with petroleum oils, even though ammonia is only poorly miscible with such oils. Many lubricants of ISO VG 15 to 100 are suitable, provided that the pour point is somewhat below evaporator temperature and that it does not contain additives that react with ammonia. Miscibility of refrigerant and lubricant is important to lubrication of some other types of systems. Chlorinated refrigerants (CFCs and HCFCs) are miscible with oil, and highly refined, low-pour, wax-free naphthenic oils of ISO VG 32 to 68 are often used, or certain wax-free synthetic lubricants. CFC refrigerants, which deplete the ozone layer, will cease being produced after 1995. Chlorine-free hydrofluorocarbon (HFC) refrigerants are now widely used, as they are non-ozone-depleting. HFCs are largely immiscible in petroleum-based lubricants, but partial miscibility of refrigerant and lubricant is necessary for adequate lubrication in these systems. Synthetic lubricants based on polyol ester or polyalkylene glycol base oils have miscibility with HFCs and are being used in HFC air conditioning and refrigeration systems.

Hydraulic Systems Critical lubricated parts include pumps, motors, and valves. When operated at rated load, certain pumps and motors are very sensitive to the lubricating quality of the hydraulic fluid. When the fluid is inadequate in this respect, premature wear occurs, leading to erratic operation of the hydraulic system. For indoor use, it is best to select specially formulated antiwear hydraulic oils. These contain rust and oxidation inhibitors as well as antiwear additives, and they provide good overall performance in ISO VG 32 to 68 for many applications. Newer hydraulic systems have higher operating temperatures and pressures and longer drain intervals. These systems require oils with improved oxidation stability and antiwear protection. Fluids with good filterability are being demanded to allow use of fine filters to protect the critical clearances of the system. Hydraulic fluids may contain VI improver and/or pour point depressant for use in low-temperature outside service, and antifoamants and demulsifiers for rapid release of entrained air and water. Hydraulic equipment on mobile systems has greater access to automotive crankcase oils and automatic transmission fluids and manufacturers may recommend use of these oils in their equipment. Where biodegradability is of concern, fluids made of vegetable oil (rapeseed) are available.

Systems Needing Fire-Resistant Fluids Where accidental rupture of an oil line may cause fluid to splash on a surface above about 600°F (310°C), a degree of fire resistance above that of petroleum oil is desirable. Four classes of fluids, generally used in hydraulic systems operating in such an environment, are available. **Phosphate esters** offer the advantages of a good lubricant requiring little attention in service, but they require special seals and paints and are quite expensive. **Water-glycol** fluids contain some 40 to 50 percent water, in a uniform solution of diethylene glycol, or glycol and polyglycol, to achieve acceptable fire resistance. They require monitoring in service to ensure proper content of water and rust inhibitor. **Invert emulsions** contain 40 to 50 percent water dispersed in petroleum oil. With oil as the outside phase of the emulsion, lubrication properties are fairly good, but the fluid must be monitored to maintain water content and to ensure that the water remains adequately dispersed. Finally, there are **conventional emulsions** which contain 5 to 10 percent petroleum oil dispersed in water. With the water as the outside phase, the fluid is a rather poor lubricant, and equipment requiring lubrication must be designed and selected to operate with these emulsions. Steps must also be taken to ensure that problems such as rusting, spoilage, and microbial growth are controlled.

Note that *fire-resistant* is a relative term, and it does not mean nonflammable. Approvals of the requisite degree of fire resistance are issued by Factory Mutual Insurance Company and by the U.S. Bureau of Mines where underground mining operations are involved.

Metal Forming The functions of fluids in machining operation are (1) to cool and (2) to lubricate. Fluids remove the heat generated by the chip/tool rubbing contact and/or the heat resulting from the plastic deformation of the work. Cooling aids tool life, preserves tool hardness, and helps to maintain the dimensions of the machined parts. Fluids lubricate the chip/tool interface to reduce tool wear, frictional heat, and power consumption. Lubricants aid in the reduction of metal welding and adhesion to improve surface finish. The fluids may also serve to carry away chips and debris from the work as well as to protect machined surfaces, tools, and equipment from rust and corrosion.

Many types of fluids are used. Most frequently they are (1) mineral oils, (2) soluble oil emulsions, or (3) chemicals or synthetics. These are often compounded with additives to impart specific properties. Some metalworking operations are conducted in a controlled gaseous atmosphere (air, nitrogen, carbon dioxide).

The choice of a **metalworking fluid** is very complicated. Factors to be considered are (1) the metal to be machined, (2) the tools, and (3) the type of operation. Tools are usually steel, carbide, or ceramic. In operations where chips are formed, the relative motion between the tool and chip is high-speed under high load and often at elevated temperature. In chipless metal forming—drawing, rolling, stamping, extruding, spinning—the function of the fluid is to (1) lubricate and cool the die and work material and (2) reduce adhesion and welding on dies. In addition to fluids, solids such as talc, clay, and soft metals may be used in drawing operations.

In addition to the primary function to lubricate and cool, the cutting fluids should not (1) corrode, discolor, or form deposits on the work;

(2) produce undesirable fumes, smoke, or odors; (3) have detrimental physical effects on operators. Fluids should also be stable, resist bacterial growth, and be foam-resistant.

In the machining operations, many combinations of tools, workpieces, and operating conditions may be encountered. In some instances, straight mineral oils or oils with small amounts of additives will suffice, while in more severe conditions, highly compounded oils are required. The effectiveness of the additives depends upon their chemical activity with newly formed, highly reactive surfaces; these combined with the high temperatures and pressures at the contact points are ideal for chemical reactions. Additives include fatty oils, sulfur, sulfurized fatty oils, and sulfurized, chlorinated, and phosphorus additives. These agents react with the metals to form compounds which have a lower shear strength and may possess EP properties. Oils containing additives, either dark or transparent, are most widely used in the industry.

Oils with sulfur- and chlorine-based additives raise some issues of increasing concern, as the more active sulfurized types stain copper and the chlorinated types have environmental concerns and are not suitable for titanium. New cutting oils have been developed which perform well and avoid these concerns.

Water is probably the most effective coolant available but can seldom be used as an effective cutting fluid. It has little value as a lubricant and will promote rusting. One way of combining the cooling properties of water with the lubricating properties of oil is through the use of soluble oils. These oils are compounded so they form a stable emulsion with water. The main component, water, provides effective cooling while the oil and compounds impart desirable lubricating, EP, and corrosion-resistance properties.

The ratio of water to oil will influence the relative lubrication and cooling properties of the emulsion. For these nonmiscible liquids to form a stable emulsion, an emulsifier must be added. Water hardness is an important consideration in the forming of an emulsion, and protection against bacterial growth should be provided. Care in preparing and handling the emulsion will ensure more satisfactory performance and longer life. Overheating, freezing, water evaporation, contamination, and excessive air mixing will adversely affect the emulsion. When one is discarding the used emulsion, it is often necessary to "break" the emulsion into its oil and water components for proper disposal.

Newer-technology soluble oils, using nonionic emulsifiers instead of anionic emulsifiers, have improved properties over earlier-technology oils, and have less tendency to form insoluble soaps with hard water, which results in fewer filter changes, less machine downtime, and lower maintenance costs. They also have longer emulsion life, from greater bacterial and fungal resistance.

Water-soluble chemicals, and synthetics, are essentially solutions or microdispersions of a number of ingredients in water. These materials are used for the same purposes described above for mineral and soluble oils. Water-soluble synthetic fluids are seeing increased use and have greater bioresistance than soluble oils and provide longer coolant life and lower maintenance.

Health Considerations   Increased attention is being focused on hazards associated with manufacture, handling, and use of all types of industrial and consumer materials, and lubricants are no exception. Much progress has been made in removing substances suspected of creating adverse effects on health. Suppliers can provide information on general composition and potential hazards requiring special handling of their products, in the form of Material Safety Data Sheets (MSDSs). It is seldom necessary to go further to ascertain health risks. It is necessary that personnel working with lubricants observe basic hygienic practices. They should avoid wearing oil-soaked clothing, minimize unnecessary exposure of skin, and wash exposed skin frequently with approved soaps.

Used-Lubricant Disposal   The nonpolluting disposal of used lubricants is becoming increasingly important and requires continual attention. More and more legislation and control are being enacted, at local, state, and national levels, to regulate the disposal of wastes. Alternative methods of handling specific used lubricants may be recommended, such as in situ purification and refortifying the oil and returning it to service, often in mobile units provided by the lubricant distributor. Also, waste lubricating oils may be reprocessed into base oils for re-refined lubricant manufacture. The use of waste lubricating oils as fuels is being increasingly regulated because of air pollution dangers. If wastes are dumped on the ground or directly into sewers, they may eventually be washed into streams and water supplies and become water pollutants. They may also interfere with proper operation of sewage plants. Improper burning may contribute to air pollution. Wastes must be handled in such a way as to ensure nonpolluting disposal. Reputable lubricant suppliers are helpful in suggesting general disposal methods, although they cannot be expected to be knowledgeable about all local regulations.

Lubricant management programs are being used increasingly by industry to minimize operating cost, including lubricant, labor, and waste disposal costs. These programs can be managed by lubricant user/supplier teams. Fluid management programs include the following activities: lubrication selection; lubricant monitoring during service; reclamation and refortification; and disposal. The supplier and/or lubricant service firms often have the major role in performing the last three activities. These programs can be beneficial to both lubricant user and supplier.

6.12   PLASTICS
(Staff Contribution)

REFERENCES: Billmyer, "Textbook of Polymer Science," 3d ed. Brydson, "Plastics Materials," 4th ed. Birley, Heath, and Scott, "Plastics Materials: Properties and Applications." Current publications of and sponsored by the Society of the Plastics Industry (SPI) and similar plastics industry organizations. Manufacturers' specifications, data, and testing reports. "Modern Plastics Encyclopedia," McGraw-Hill.

GENERAL OVERVIEW OF PLASTICS

Plastics are ubiquitous engineering materials which are wholly or in part composed of long, chainlike molecules called **high polymers.** While carbon is the element common to all commercial high polymers, hydrogen, oxygen, nitrogen, sulfur, halogens, and silicon can be present in varying proportions. High polymers may be divided into two classes: **thermoplastic** and **thermosetting.** The former reversibly melt to become highly viscous liquids and solidify upon cooling (Fig. 6.12.1). The resultant solids will be elastic, ductile, tough, or brittle, depending on the structure of the solid as evolved from the molten state. Thermosetting polymers are infusible without thermal or mechanical degradation. Thermosetting polymers (often termed **thermosets**) cure by a chain-linking chemical reaction usually initiated at elevated temperature and pressure, although there are types which cure at room temperature through the use of catalysts. The more highly cured the polymer, the higher its heat distortion temperature and the harder and more brittle it becomes.

As a class, plastics possess a combination of physical and mechanical properties which are attractive to the designer. There are certain proper-

Table 6.12.1 Properties of Plastic Resins and Compounds

	Properties	ASTM test method	Extrusion grade	ABS/Nylon	Heat-resistant	Medium-impact	High-impact	Platable grade	20% glass fiber-reinforced
					ABS[a] — Injection molding grades (Continued)				
Processing	1a. Melt flow, g/10 min	D1238	0.85–1.0		1.1–1.8	1.1–1.8	1.1–18	1.1	
	1. Melting temperature, °C. T_m (crystalline)								
	T_g (amorphous)		88–120		110–125	102–115	91–110	100–110	100–110
	2. Processing temperature range, °F. (C = compression; T = transfer; I = injection; E = extrusion)		E: 350–500	I: 460–520	C: 325–500 I: 475–550	C: 325–350 I: 390–525	C: 325–350 I: 380–525	C: 325–400 I: 350–500	C: 350–500 I: 350–500
	3. Molding pressure range, 10^3 lb/in²			8–25	8–25	8–25	8–25	8–25	15–30
	4. Compression ratio		2.5–2.7	1.1–2.0	1.1–2.0	1.1–2.0	1.1–2.0	1.1–2.0	
	5. Mold (linear) shrinkage, in/in	D955	0.004–0.007	0.003–0.010	0.004–0.009	0.004–0.009	0.004–0.009	0.005–0.008	0.001–0.002
Mechanical	6. Tensile strength at break, lb/in²	D638[b]	2,500–8,000	4,000–6,000	4,800–7,500	5,500–7,500	4,400–6,300	5,200–6,400	10,500–13,000
	7. Elongation at break, %	D638[b]	20–100	40–300	3–45	5–60	5–75		2–3
	8. Tensile yield strength, lb/in²	D638[b]	4,300–6,400	4,300–6,300	4,300–7,000	5,000–7,200	2,600–5,900	6,700	
	9. Compressive strength (rupture or yield), lb/in²	D695	5,200–10,000		7,200–10,000	1,800–12,500	4,500–8,000		13,000–14,000
	10. Flexural strength (rupture or yield), lb/in²	D790	4,000–14,000	8,800–10,900	9,000–13,000	7,100–13,000	5,400–11,000	10,500–11,500	14,000–17,500
	11. Tensile modulus, 10^3 lb/in²	D638[b]	130–420	260–320	285–360	300–400	150–350	320–380	740–880
	12. Compressive modulus, 10^3 lb/in²	D695	150–390		190–440	200–450	140–300		800
	13. Flexural modulus, 10^3 lb/in² 73°F	D790	130–440	250–310	300–400	310–400	179–375	340–390	650–800
	200°F	D790							
	250°F	D790							
	300°F	D790							
	14. Izod impact, ft · lb/in of notch (⅛-in-thick specimen)	D256A	1.5–12	15–20	2.0–6.5	3.0–6.0	6.0–9.3	4.0–8.3	1.1–1.4
	15. Hardness Rockwell	D785	R75–115	R93–105	R100–115	R102–115	R85–106	R103–109	M85–98, R107
	Shore/Barcol	D2240/ D2583							
Thermal	16. Coef. of linear thermal expansion, 10^6 in/(in · °C)	D696	60–130	90–110	60–93	80–100	95–110	47–53	20–21
	17. Deflection temperature under flexural load, °F 264 lb/in²	D648	170–220	130–150	220–240 annealed 181–193[g]	200–220 annealed	205–215 annealed	190–222 annealed	210–220
	66 lb/in²	D648	170–235	180–195	230–245 annealed	215–225 annealed	210–225 annealed	215–222 annealed	220–230
	18. Thermal conductivity, 10^{-4} cal · cm/(s · cm² · °C)	C177			4.5–8.0				4.8
Physical	19. Specific gravity	D792	1.02–1.08	1.06–1.07	1.05–1.08	1.03–1.06	1.01–1.05	1.04–1.07	1.18–1.22
	20. Water absorption (⅛-in-thick specimen), % 24 h	D570	0.20–0.45		0.20–0.45	0.20–0.45	0.20–0.45		0.18–0.20
	Saturation	D570							
	21. Dielectric strength (⅛-in-thick specimen), short time, V/mil	D149	350–500		350–500	350–500	350–500	420–550	450–460

SOURCE: Abstracted from "Modern Plastics Encyclopedia," 1995.
NOTE: Footnotes [a] to [f] are at end of table.

Homopolymer	Copolymer	Extrusion and blow molding grade (terpolymer)	20% Glass-reinforced homopolymer	25% Glass-coupled copolymer	2-20% PTFE-filled copolymer	Chemically lubricated homopolymer
			Acetal			
1-20	1-90	1.0	6.0			6
172-184	160-175	160-170	175-181	160-180	160-175	175
I:380-470	C:340-400 I:360-450	E:360-400	I:350-480	I:365-480	I:350-445 I:325-500 E:360-500	I:400-440
10-20	8-20		10-20	8-20	8-20	
3.0-4.5	3.0-4.5	3.0-4.0		3.0-4.5	3.0-4.5	
0.018-0.025	0.020 (Avg.)	0.02	0.009-0.012	0.004 (flow) 0.018 (trans.)	0.018-0.029	
9,700-10,000			8,500-9,000	16,000-18,500	8,300	9,500
10-75	15-75	67	6-12	2-3	30	40
9,500-12,000	8,300-10,400	8,700	7,500-8,250	16,000	8,300	9,500
15,600-18,000 @ 10%	16,000 @ 10%	16,000	18,000 @ 10%	17,000 @ 10%	11,000-12,600	
13,600-16,000	13,000	12,800	10,700-16,000	18,000-28,000	11,500	13,000
400-520	377-464		900-1,000	1,250-1,400	250-280	450
670	450					
380-490	370-450	350	600-730	1,100	310-360	400
120-135			300-360			130
75-90			250-270			80
1.1-2.3	0.8-1.5	1.7	0.5-1.0	1.0-1.8	0.5-1.0	1.4
M92-94, R120	M75-90	M84	M90	M79-90, R110	M79	M90
50-112	61-110		33-81	17-44	52-68	
253-277	185-250	205	315	320-325	198-225	257
324-342	311-330	318	345	327-331	280-325	329
5.5	5.5				4.7	
1.42	1.40	1.41	1.54-1.56	1.58-1.61	1.40	1.42
0.25-1	0.20-0.22	0.22	0.25	0.22-0.29	0.15-0.26	0.27
0.90-1	0.65-0.80	0.8	1.0	0.8-1.0	0.5	1.00
400-500 (90 mil)	500 (90 mil)		490 (125 mil)	480-580	400-410	400 (125 mils)

Table 6.12.1 Properties of Plastic Resins and Compounds (Continued)

	Properties	ASTM test method	Sheet Cast	Acrylic Molding and extrusion compounds PMMA	Impact-modified	Heat-resistant	Acrylonitrile Molding and extrusion	Injection
	1a. Melt flow, g/10 min	D1238		1.4–27	1–11	1.6–8.0		12
	1. Melting temperature, °C. T_m (crystalline)						135	
	T_g (amorphous)		90–105	85–105	80–103	100–165	95	
	2. Processing temperature range, °F. (C = compression; T = transfer; I = injection; E = extrusion)			C:300–425 I:325–500 E:360–500	C:300–400 I:400–500 E:380–480	C:350–500 I:400–625 E:360–550	C:320–345 I:410 E:350–410	380–420
	3. Molding pressure range, 10^3 lb/in^2			5–20	5–20	5–30	20	20
	4. Compression ratio			1.6–3.0		1.2–2.0	2	2–2.5
	5. Mold (linear) shrinkage, in/in	D955	1.7	0.001–0.004 (flow) 0.002–0.008 (trans.)	0.002–0.008	0.002–0.008	0.002–0.005	0.002–0.005
	6. Tensile strength at break, lb/in^2	D638[b]	66–11,000	7,000–10,500	5,000–9,000	9,300–11,500	9,000	
	7. Elongation at break, %	D638[b]	2–7	2–5.5	4.6–70	2–10	3–4	3–4
	8. Tensile yield strength, lb/in^2	D638[b]		7,800–10,600	5,500–8,470	10,000	7,500	9,500
	9. Compressive strength (rupture or yield), lb/in^2	D695	11,000–19,000	10,500–18,000	4,000–14,000	15,000–17,000	12,000	12,000
	10. Flexural strength (rupture or yield), lb/in^2	D790	12,000–17,000	10,500–19,000	7,000–14,000	12,000–18,000	14,000	14,000
	11. Tensile modulus, 10^3 lb/in^2	D638[b]	450–3,100	325–470	200–500	350–650	510–580	500–550
	12. Compressive modulus, 10^3 lb/in^2	D695	390–475	370–460	240–370	450		
	13. Flexural modulus, 10^3 lb/in^2 73°F	D790	390–3,210	325–460	200–430	450–620	500–590	480
	200°F	D790				150–440		
	250°F	D790				350–420		
	300°F	D790						
	14. Izod impact, ft · lb/in of notch (⅛-in-thick specimen)	D256A	0.3–0.4	0.2–0.4	0.40–2.5	0.2–0.4	2.5–6.5	2.5
	15. Hardness Rockwell	D785	M80–102	M68–105	M35–78	M94–100	M72–78	M60
	Shore/Barcol	D2240/ D2583						
	16. Coef. of linear thermal expansion, 10^6 in/(in · °C)	D696	50–90	50–90	48–80	40–71	66	66
	17. Deflection temperature under flexural load, °F 264 lb/in^2	D648	98–215	155–212	165–209	190–310	164	151
	66 lb/in^2	D648	165–235	165–225	180–205	200–315	172	166
	18. Thermal conductivity, 10^{-4} cal · cm/(s · cm^2 · °C)	C177	4.0–6.0	4.0–6.0	4.0–5.0	2.0–4.5	6.2	6.1
	19. Specific gravity	D792	1.17–1.20	1.17–1.20	1.11–1.18	1.16–1.22	1.15	1.15
	20. Water absorption (⅛-in-thick specimen), % 24 h	D570	0.2–0.4	0.1–0.4	0.19–0.8	0.2–0.3	0.28	
	Saturation	D570						
	21. Dielectric strength (⅛-in-thick specimen), short time, V/mil	D149	450–550	400–500	380–500	400–500	220–240	220–240

	Cellulosic				Epoxy				
Ethyl cellulose molding compound and sheet	Cellulose acetate		Cellulose acetate butyrate	Bisphenol molding compounds		Casting resins and compounds			
	Sheet	Molding compound	Molding compound	Glass fiber-reinforced	Mineral-filled	Unfilled	Aluminum-filled	Flexibilized
135	230	230	140	Thermoset	Thermoset	Thermoset	Thermoset	Thermoset
C:250–390 I:350–500		C:260–420 I:335–490	C:265–390 I:335–480	C:300–330 T:280–380	C:250–330 T:250–380			
8–32		8–32	8–32	1–5	0.1–3			
1.8–2.4		1.8–2.6	1.8–2.4	3.0–7.0	2.0–3.0			
0.005–0.009		0.003–0.010	0.003–0.009	0.001–0.008	0.002–0.010	0.001–0.010	0.001–0.005	0.001–0.010
2,000–8,000	4,500–8,000	1,900–9,000	2,600–6,900	5,000–20,000	4,000–10,800	4,000–13,000	7,000–12,000	2,000–10,000
5–40	20–50	6–70	40–88	4		3–6	0.5–3	20–85
		2,500–6,300						
		3,000–8,000	2,100–7,500	18,000–40,000	18,000–40,000	15,000–25,000	15,000–33,000	1,000–14,000
4,000–12,000	6,000–10,000	2,000–16,000	1,800–9,300	8,000–30,000	6,000–18,000	13,000–21,000	8,500–24,000	1,000–13,000
			50–200	3,000	350	350		
					650			1–350
		1,200–4,000	90–300	2,000–4,500	1,400–2,000			
0.4	2.0–8.5	1.0–7.8	1.0–10.9	0.3–10.0	0.3–0.5	0.2–1.0	0.4–1.6	2.3–5.0
R50–115	R85–120	R17–125	R31–116	M100–112	M100–M112	M80–110	M55–85	
								Shore D65–89
100–200	100–150	80–180	110–170	11–50	20–60	45–65	5.5	20–100
115–190		111–195	113–202	225–500	225–500	115–550	190–600	73–250
		120–209	130–227					
3.8–7.0	4–8	4–8	4–8	4.0–10.0	4–35	4.5	15–25	
1.09–1.17	1.28–1.32	1.22–1.34	1.15–1.22	1.6–2.0	1.6–2.1	1.11–1.40	1.4–1.8	0.96–1.35
0.8–1.8	2.0–7.0	1.7–6.5	0.9–2.2	0.04–0.20	0.03–0.20	0.08–0.15	0.1–4.0	0.27–0.5
350–500	250–600	250–600	250–400	250–400	250–420	300–500		235–400

Table 6.12.1 Properties of Plastic Resins and Compounds (*Continued*)

	Properties	ASTM test method	Polychlorotrifluoroethylene	Polytetrafluoroethylene Granular	Polytetrafluoroethylene 25% Glass fiber-reinforced	Polyvinylidene fluoride Molding and extrusion	EMI shielding (conductive); 30% PAN carbon fiber	Modified PE-TFE 25% Glass fiber-reinforced
					Fluoroplastics			
Processing	1a. Melt flow, g/10 min	D1238						
	1. Melting temperature, °C. T_m (crystalline)			327	327	141–178		270
	T_g (amorphous)		220			−60 to −20		
	2. Processing temperature range, °F. (C = compression; T = transfer; I = injection; E = extrusion)		C: 460–580 I: 500–600 E: 360–590			C: 360–550 I: 375–550 E: 375–550	I: 430–500	C: 575–625 I: 570–650
	3. Molding pressure range, 10^3 lb/in^2		1–6	2–5	3–8	2–5		2–20
	4. Compression ratio		2.6	2.5–4.5		3		
	5. Mold (linear) shrinkage, in/in	D955	0.010–0.015	0.030–0.060	0.018–0.020	0.020–0.035	0.001	0.002–0.030
Mechanical	6. Tensile strength at break, lb/in^2	D638[b]	4,500–6,000	3,000–5,000	2,000–2,700	3,500–7,250	14,000	12,000
	7. Elongation at break, %	D638[b]	80–250	200–400	200–300	12–600	0.8	8
	8. Tensile yield strength, lb/in^2	D638[b]	5,300			2,900–8,250		
	9. Compressive strength (rupture or yield), lb/in^2	D695	4,600–7,400	1,700	1,000–1,400 @ 1% strain	8,000–16,000		10,000
	10. Flexural strength (rupture or yield), lb/in^2	D790	7,400–11,000		2,000	9,700–13,650	19,800	10,700
	11. Tensile modulus, 10^3 lb/in^2	D638[b]		58–80	200–240	200–80,000	2,800	1,200
	12. Compressive modulus, 10^3 lb/in^2	D695	150–300	60		304–420		
	13. Flexural modulus, 10^3 lb/in^2 73°F	D790	170–200	80	190–235	170–120,000	2,100	950
	200°F	D790	180–260					450
	250°F	D790						310
	300°F	D790						200
	14. Izod impact, ft · lb/in of notch (⅛-in-thick specimen)	D256A	2.5–5	3	2.7	2.5–80	1.5	9.0
	15. Hardness Rockwell	D785	R75–112			R79–83, 85		R74
	Shore/Barcol	D2240/ D2583	Shore D75–80	Shore D50–65	Shore D60–70	Shore D80, 82 65–70		
Thermal	16. Coef. of linear thermal expansion, 10^6 in/(in · °C)	D696	36–70	70–120	77–100	70–142		10–32
	17. Deflection temperature under flexural load, °F 264 lb/in^2	D648		115		183–244	318	410
	66 lb/in^2	D648	258	160–250		280–284		510
	18. Thermal conductivity, 10^{-4} cal · cm/(s · cm^2 · °C)	C177	4.7–5.3	6.0	8–10	2.4–3.1		
Physical	19. Specific gravity	D792	2.08–2.2	2.14–2.20	2.2–2.3	1.77–1.78	1.74	1.8
	20. Water absorption (⅛-in-thick specimen), % 24 h	D570	0	<0.01		0.03–0.06	0.12	0.02
	Saturation	D570						
	21. Dielectric strength (⅛-in-thick specimen), short time, V/mil	D149	500–600	480	320	260–280		425

	Phenolic				
Furan	Molding compounds, phenol-formaldehyde				Casting resins
		High-strength glass fiber-reinforced	Impact-modified		
Asbestos-filled	Wood flour-filled		Fabric- and rag-filled	Cellulose-filled	Unfilled
			0.5–10		
Thermoset	Thermoset	Thermoset	Thermoset	Thermoset	Thermoset
C: 275–300	C: 290–380 I: 330–400	C: 300–380 I: 330–390 T: 300–350	C: 290–380 I: 330–400 T: 300–350	C: 290–380 I: 330–400	
0.1–0.5	2–20	1–20	2–20	2–20	
	1.0–1.5	2.0–10.0	1.0–1.5	1.0–1.5	
	0.004–0.009	0.001–0.004	0.003–0.009	0.004–0.009	
3,000–4,500	5,000–9,000	7,000–18,000	6,000–8,000	3,500–6,500	5,000–9,000
	0.4–0.8	0.2	1–4	1–2	1.5–2.0
10,000–13,000	25,000–31,000	16,000–70,000	20,000–28,000	22,000–31,000	12,000–15,000
600–9,000	7,000–14,000	12,000–60,000	10,000–14,000	5,500–11,000	11,000–17,000
1,580	800–1,700	1,900–3,300	900–1,100		400–700
		2,740–3,500			
	1,000–1,200	1,150–3,300	700–1,300	900–1,300	
	0.2–0.6	0.5–18.0	0.8–3.5	0.4–1.1	0.24–0.4
R110	M100–115	E54–101	M105–115	M95–115	M93–120
		Barcol 72			
	30–45	8–34	18–24	20–31	68
	300–370	350–600	325–400	300–350	165–175
	4–8	8–14	9–12	6–9	3.5
1.75	1.37–1.46	1.69–2.0	1.37–1.45	1.38–1.42	1.24–1.32
0.01–0.02	0.3–1.2	0.03–1.2	0.6–0.8	0.5–0.9	0.1–0.36
		0.12–1.5			
	260–400	140–400	200–370	300–380	250–400

Table 6.12.1 Properties of Plastic Resins and Compounds (Continued)

	Properties	ASTM test method	Molding and extrusion compound	30–35% Glass fiber-reinforced	Toughened 33% Glass fiber-reinforced	High-impact copolymers and rubber-modified compounds	Impact-modified; 30% glass fiber-reinforced	Cast
					Polyamide — **Nylon, Type 6**			
Processing	1a. Melt flow, g/10 min	D1238	0.5–10			1.5–5.0		
	1. Melting temperature, °C. T_m (crystalline)		210–220	210–220	210–220	210–220	220	227–238
	T_g (amorphous)							
	2. Processing temperature range, °F. (C = compression; T = transfer; I = injection; E = extrusion)		I: 440–550 E: 440–525	I: 460–550	I: 520–550	I: 450–580 E: 450–550	I: 480–550	
	3. Molding pressure range, 10^3 lb/in²		1–20	2–20		1–20	3–20	
	4. Compression ratio		3.0–4.0	3.0–4.0		3.0–4.0	3.0–4.0	
	5. Mold (linear) shrinkage, in/in	D955	0.003–0.015	0.001–0.005	0.001–0.003	0.008–0.026	0.003–0.005	
Mechanical	6. Tensile strength at break, lb/in²	D638[b]	6,000–24,000	24–27,600[c]; 18,900[d]	17,800[c]	6,300–11,000[c]	21,000[c]; 14,500[d]	12,500
	7. Elongation at break, %	D638[b]	30–100[c]; 300[d]	2.2–3.6[c]	4.0[c]	150–270[c]	5[c]–8[d]	20–30
	8. Tensile yield strength, lb/in²	D638[b]	13,100[c]; 7,400[d]			9,000[c]–9,500		
	9. Compressive strength (rupture or yield), lb/in²	D695	13,000–16,000[c]	19,000–24,000[c]		3,900[c]		17,000
	10. Flexural strength (rupture or yield), lb/in²	D790	15,700[c]; 5,800[d]	34–3,600[c]; 21,000[d]	25,800[c]	5,000–12,000[c]		16,500
	11. Tensile modulus, 10^3 lb/in²	D638[b]	380–464[c]; 100–247[d]	1,250–1,600[c]; 1,090[d]			1,220[c]–754[d]	500
	12. Compressive modulus, 10^3 lb/in²	D695	250[d]					325
	13. Flexural modulus, 10^3 lb/in² 73°F	D790	390–410[c]; 140[d]	1,250–1,400[c]; 800–950[d]	1,110[c]	110–320[c]; 130[d]	1,160[c]–600[d]	430
	200°F	D790				60–130[c]		
	250°F	D790						
	300°F	D790						
	14. Izod impact, ft · lb/in of notch (⅛-in-thick specimen)	D256A	0.6–2.2[c]; 3.0[d]	2.1–3.4[c]; 3.7–5.5[d]	3.5[c]	1.8–No break[c] 1.8–No break[d]	2.2[c]–6[d]	0.7–0.9
	15. Hardness Rockwell	D785	R119[c]; M100–105[c]	M93–96c; M78[d]		R81–113[c]; M50		R115–125
	Shore/Barcol	D2240/ D2583						D78.83
Thermal	16. Coef. of linear thermal expansion, 10^6 in/(in · °C)	D696	80–83	16–80		72–120	20–25	50
	17. Deflection temperature under flexural load, °F 264 lb/in²	D648	155–185[c]	392–420[c]	400[c]	113–140[c]	410[c]	330–400
	66 lb/in²	D648	347–375[c]	420–430[c]	430[c]	260–367[c]	428[c]	400–430
	18. Thermal conductivity, 10^{-4} cal · cm/(s · cm² · °C)	C177	5.8	5.8–11.4				
Physical	19. Specific gravity	D792	1.12–1.14	1.35–1.42	1.33	1.07–1.17	1.33	1.15–1.17
	20. Water absorption (⅛-in-thick specimen), % 24 h	D570	1.3–1.9	0.90–1.2	0.86	1.3–1.7	2.0	0.3–0.4
	Saturation	D570	8.5–10.0	6.4–7.0		8.5	6.2	5–6
	21. Dielectric strength (⅛-in-thick specimen), short time, V/mil	D149	400[c]	400–450[c]		450–470[c]		500–600

Polyamide (*Continued*)

Nylon, Type 66						
		Toughened	Lubricated			
Molding compound	30–33% Glass fiber-reinforced	15–33% Glass fiber-reinforced	Antifriction molybdenum disulfide-filled	5% Silicone	30% PTFE	5% Molybdenum disulfide and 30% PTFE
255–265	260–265	256–265	249 -265	260–265	260–265	260–265
I: 500–620	I: 510–580	I: 530–575	I: 500–600	I: 530–570	I: 530–570	I: 530–570
1–25	5–20		5–25			
3.0–4.0	3.0–4.0					
0.007–0.018	0.002–0.006	0.0025–0.0045[c]	0.007–0.018	0.015	0.007	0.01
13,700[c]; 11,000[d]	27,600[c]; 20,300[d]	10,900–20,300[c]; 14,500[d]	10,500–13,700[c]	8,500[c]	5,500[c]	7,500[c]
15–80[c]; 150–300[d]	2.0–3.4[c]; 3–7[d]	4.7[c]; 8[d]	4.4–40[c]			
8,000–12,000[c]; 6,500–8,500[d]	25,000[c]					
12,500–15,000[c] (yld.)	24,000–40,000[c]	15,000[c]–20,000	12,000–12,500[c]			
17,900–1,700[c]; 6,100[d]	40,000[c]; 29,000[d]	17,400–29,900[c]	15,000–20,300[c]	15,000[c]	8,000[c]	1,200[c]
230–550[c]; 230–500[d]	1,380[c]; 1,090[d]	1,230[c]; 943[d]	350–550[c]			
410–470[c]; 185[d]	1,200–1,450[c]; 800[d]; 900	479–1,100[c]	420–495[c]	300[c]	460[c]	400[c]
0.55–1.0[c]; 0.85–2.1[d]	1.6–4.5[c]; 2.6–3.0[d]	>3.2–5.0	0.9–4.5[c]	1.0[c]	0.5[c]	0.6[c]
R120[c]; M83[c]; M95–105[d]	R101–119[c]; M101–102[c]; M96[d]	R107[c]; R115; R116; M86[c]; M70[d]	R119[c]			
80	15–54	43	54	63.0	45.0	
158–212[c]	252–490[c]	446–470	190–260[c]	170	180	185
425–474[c]	260–500[c]	480–495	395–430			
5.8	5.1–11.7					
1.13–1.15	1.15–1.40	1.2–1.34	1.15–1.18	1.16	1.34	1.37
1.0–2.8	0.7–1.1	0.7–1.5	0.8–1.1	1.0	0.55	0.55
8.5	5.5–6.5	5	8.0			
600[c]	360–500		360[c]			

Table 6.12.1 Properties of Plastic Resins and Compounds (*Continued*)

Properties	ASTM test method	Polycarbonate			
		Unfilled molding and extrusion resins — High viscosity	Glass fiber-reinforced — 10% glass	Impact-modified polycarbonate/ polyester blends	Lubricated — 10–15% PTFE, 20% glass fiber-reinforced
Processing					
1a. Melt flow, g/10 min	D1238	3–10	7.0		
1. Melting temperature, °C. T_m (crystalline)					
T_g (amorphous)		150	150		150
2. Processing temperature range, °F. (C = compression; T = transfer; I = injection; E = extrusion)		I: 560	I: 520–650	I: 475–560	I: 590–650
3. Molding pressure range, 10^3 lb/in²		10–20	10–20	15–20	
4. Compression ratio		1.74–5.5		2–2.5	
5. Mold (linear) shrinkage, in/in	D955	0.005–0.007	0.002–0.005	0.006–0.009	0.002
Mechanical					
6. Tensile strength at break, lb/in²	D638[b]	9,100–10,500	7,000–10,000	7,600–8,500	12,000–15,000
7. Elongation at break, %	D638[b]	110–120	4–10	120–165	2
8. Tensile yield strength, lb/in²		9,000	8,500–11,600	7,400–8,300	
9. Compressive strength (rupture or yield), lb/in²	D695	10,000–12,500	12,000–14,000	7,000	11,000
10. Flexural strength (rupture or yield), lb/in²	D790	12,500–13,500	13,700–16,000	10,900–12,500	18,000–23,000
11. Tensile modulus, 10^3 lb/in²	D638[b]	345	450–600		1,200
12. Compressive modulus, 10^3 lb/in²	D695	350	520		
13. Flexural modulus, 10^3 lb/in² 73°F	D790	330–340	460–580	280–325	850–900
200°F	D790	275	440		
250°F	D790	245	420		
300°F	D790				
14. Izod impact, ft · lb/in of notch (⅛-in-thick specimen)	D256A	12–18 @ ⅛ in 2.3 @ ¼ in	2–4	2–18	1.8–3.5
15. Hardness Rockwell	D785	M70–M75	M62–75; R118–122	R114–122	
Shore/Barcol	D2240/ D2583				
Thermal					
16. Coef. of linear thermal expansion, 10^6 in/(in · °C)	D696	68	32–38	80–95	21.6–23.4
17. Deflection temperature under flexural load, °F 264 lb/in²	D648	250–270	280–288	190–250	280–290
66 lb/in²	D648	280–287	295	223–265	290
18. Thermal conductivity, 10^{-4} cal · cm/(s · cm² · °C)	C177	4.7	4.6–5.2	4.3	
Physical					
19. Specific gravity	D792	1.2	1.27–1.28	1.20–1.22	1.43–1.5
20. Water absorption (⅛-in-thick specimen), % 24 h	D570	0.15	0.12–0.15	0.12–0.16	0.11
Saturation	D570	0.32–0.35	0.25–0.32	0.35–0.60	
21. Dielectric strength (⅛-in-thick specimen), short time, V/mil	D149	380–>400	470–530	440–500	

Polyester, thermoplastic				Polyester, thermoset and alkyd		
Polybutylene terephthalate		Polyethylene terephthalate				Glass fiber-reinforced
Unfilled	30% Glass fiber-reinforced	Unfilled	30% Glass fiber-reinforced	Cast		Preformed, chopped roving
				Rigid	Flexible	
220–267	220–267	212–265	245–265	Thermoset	Thermoset	Thermoset
		68–80				
I:435–525	I:440–530	I:440–660 E:520–580	I:510–590			C:170–320
4–10	5–15	2–7	4–20			0.25–2
		3.1	2–3			1.0
0.009–0.022	0.002–0.008	0.002–0.030	0.002–0.009			0.0002–0.002
8,200–8,700	14,000–19,000	7,000–10,500	20,000–24,000	600–13,000	500–3,000	15,000–30,000
50–300	2–4	30–300	2–7	<2.6	40–310	1–5
8,200–8,700		8,600	23,000			
8,600–14,500	18,000–23,500	11,000–15,000	25,000	13,000–30,000		15,000–30,000
12,000–16,700	22,000–29,000	12,000–18,000	30,000–36,000	8,500–23,000		10,000–40,000
280–435	1,300–1,450	400–600	1,300–1,440	300–640		800–2,000
375	700					
330–400	850–1,200	350–450	1,200–1,590	490–610		1,000–3,000
			520			
			390			
0.7–1.0	0.9–2.0	0.25–0.7	1.5–2.2	0.2–0.4	>7	2–20
M68–78	M90	M94–101; R111	M90; M100			
				Barcol 35–75	Shore D84–94	Barcol 50–80
60–95	15–25	65×10^{-6}	18–30	55–100		20–50
122–185	385–437	70–150	410–440	140–400		>400
240–375	421–500	167	470–480			
4.2–6.9	7.0	3.3–3.6	6.0–7.6			
1.30–1.38	1.48–1.53	1.29–1.40	1.55–1.70	1.04–1.46	1.01–1.20	1.35–2.30
0.08–0.09	0.06–0.08	0.1–0.2	0.05	0.15–0.6	0.5–2.5	0.01–1.0
0.4–0.5	0.35	0.2–0.3				
420–550	460–560	420–550	430–650	380–500	250–400	350–500

Table 6.12.1 Properties of Plastic Resins and Compounds (Continued)

		Polyethylene and ethylene copolymers					
		Low and medium density		High density			Cross-linked
			LDPE copolymers				
Properties	ASTM test method	Linear copolymer	Ethylene-vinyl acetate	Poly-ethylene homopolymer	Ultrahigh molecular weight	30% Glass fiber-reinforced	Molding grade
1a. Melt flow, g/10 min	D1238		1.4–2.0	5–18			
1. Melting temperature, °C. T_m (crystalline)		122–124	103–110	130–137	125–138	120–140	
T_g (amorphous)							
2. Processing temperature range, °F. (C = compression; T = transfer; I = injection; E = extrusion)		I:350–500 E:450–600	C:200–300 I:350–430 E:300–380	I:350–500 E:350–525	C:400–500	I:350–600	C:240–450 I:250–300
3. Molding pressure range, 10^3 lb/in²		5–15	1–20	12–15	1–2	10–20	
4. Compression ratio		3		2			
5. Mold (linear) shrinkage, in/in	D955	0.020–0.022	0.007–0.035	0.015–0.040	0.040	0.002–0.006	0.007–0.090
6. Tensile strength at break, lb/in²	D638[b]	1,900–4,000	2,200–4,000	3,200–4,500	5,600–7,000	7,500–9,000	1,600–4,600
7. Elongation at break, %	D638[b]	100–965	200–750	10–1,200	420–525	1.5–2.5	10–440
8. Tensile yield strength, lb/in²	D638[b]	1,400–2,800	1,200–6,000	3,800–4,800	3,100–4,000		
9. Compressive strength (rupture or yield), lb/in²	D695			2,700–3,600		6,000–7,000	2,000–5,500
10. Flexural strength (rupture or yield), lb/in²	D790					11,000–12,000	2,000–6,500
11. Tensile modulus, 10^3 lb/in²	D638[b]	38–75	7–29	155–158		700–900	50–500
12. Compressive modulus, 10^3 lb/in	D695						50–150
13. Flexural modulus, 10^3 lb/in² 73°F	D790	40–105	7.7	145–225	130–140	700–800	70–350
200°F	D790						
250°F	D790						
300°F	D790						
14. Izod impact, ft · lb/in of notch (⅛-in-thick specimen)	D256A	1.0–No break	No break	0.4–4.0	No break	1.1–1.5	1–20
15. Hardness Rockwell	D785				R50	R75–90	
Shore/Barcol	D2240/ D2583	Shore D55–56	Shore D17–45	Shore D66–73	Shore D61–63		Shore D55–80
16. Coef. of linear thermal expansion, 10^6 in/(in. · °C)	D696		160–200	59–110	130–200	48	100
17. Deflection temperature under flexural load, °F 264 lb/in²	D648				110–120	250	105–145
66 lb/in²	D648			175–196	155–180	260–265	130–225
18. Thermal conductivity, 10^{-4} cal · cm/(s · cm² · °C)	C177			11–12		8.6–11	
19. Specific gravity	D792	0.918–0.940	0.922–0.943	0.952–0.965	0.94	1.18–1.28	0.95–1.45
20. Water absorption (⅛-in-thick specimen), % 24 h	D570		0.005–0.13	<0.01	<0.01	0.02–0.06	0.01–0.06
Saturation	D570						
21. Dielectric strength (⅛-in-thick specimen), short time, V/mil	D149		620–760	450–500	710	500–550	230–550

| Polyimide | | | | Polypropylene | | | | |
| Thermoplastic | | Thermoset | | Homopolymer | | | Copolymer | |
Unfilled	30% Glass fiber-reinforced	Unfilled	50% Glass fiber-reinforced	Unfilled	10–30% Glass fiber-reinforced	Impact-modified, 40% mica-filled	Unfilled	10–20% Glass fiber-reinforced
4.5–7.5				0.4–38.0	1–20		0.6–44.0	0.1–20
388	388	Thermoset	Thermoset	160–175	168	168	150–175	160–168
250–365	250			−20			−20	
C:625–690 I:734–740 E:734–740	I:734–788	460–485	C:460 I:390 T:390	I:375–550 E:400–500	I:425–475	I:350–470	I:375–550 E:400–500	I:350–480
3–20	10–30	7–29	3–10	10–20			10–20	
1.7–4	1.7–2.3	1–1.2		2.0–2.4			2–2.4	
0.0083	0.0044	0.001–0.01	0.002	0.010–0.025	0.002–0.008	0.007–0.008	0.010–0.025	0.003–0.01
10,500–17,100	24,000	4,300–22,900	6,400	4,500–6,000	6,500–13,000	4,500	4,000–5,500	5,000–8,000
7.5–90	3	1		100–600	1.8–7	4	200–500	3.0–4.0
12,500–13,000		4,300–22,900		4,500–5,400	7,000–10,000		3,000–4,300	
17,500–40,000	27,500	19,300–32,900	34,000	5,500–8,000	6,500–8,400		3,500–8,000	5,500–5,600
10,000–28,800	35,200	6,500–50,000	21,300	6,000–8,000	7,000–20,000	7,000	5,000–7,000	7,000–11,000
300–400	1,720	460–4,650		165–225	700–1,000	700	130–180	
315–350	458	421		150–300				
360–500	1,390	422–3,000	1,980	170–250	310–780	600	130–200	355–510
				50			40	
				35			30	
210	1,175	1,030–2,690						
1.5–1.7	2.2	0.65–15	5.6	0.4–1.4	1.0–2.2	0.7	1.1–14.0	0.95–2.7
E52–99, R129, M95	R128, M104	110M–120M	M118	R80–102	R92–115		R65–96	R100–103
							Shore D70–73	
45–56	17–53	15–50	13	81–100	21–62		68–95	
460–680	469	572–>575	660	120–140	253–288	205	130–140	260–280
				225–250	290–320		185–220	305
2.3–4.2	8.9	5.5–12	8.5	2.8	5.5–6.2		3.5–4.0	
1.33–1.43	1.56	1.41–1.9	1.6–1.7	0.900–0.910	0.97–1.14	1.23	0.890–0.905	0.98–1.04
0.24–0.34	0.23	0.45–1.25	0.7	0.01–0.03	0.01–0.05		0.03	0.01
1.2								
415–560	528	480–508	450	600			600	

Table 6.12.1 Properties of Plastic Resins and Compounds (Continued)

		Polystyrene and styrene copolymers					
		Polystyrene homopolymers			Rubber-modified	Styrene copolymers	
						Styrene-acrylonitrile (SAN)	High heat-resistant copolymers
Properties	ASTM test method	High and medium flow	Heat-resistant	20% Long and short glass fiber-reinforced	High-impact	Molding and extrusion	20% glass fiber-reinforced
1a. Melt flow, g/10 min	D1238				5.8		
1. Melting temperature, °C. T_m (crystalline)							
T_g (amorphous)		74–105	100–110	115	9.3–105	100–200	
2. Processing temperature range, °F. (C = compression; T = transfer; I = injection; E = extrusion)		C:300–400 I:350–500 E:350–500	C:300–400 I:350–500 E:350–500	I:400–550	I:350–525 E:375–500	C:300–400 I:360–550 E:360–450	I:425–550
3. Molding pressure range, 10^3 lb/in²		5–20	5–20	10–20	10–20	5–20	
4. Compression ratio		3	3–5		4	3	
5. Mold (linear) shrinkage, in/in	D955	0.004–0.007	0.004–0.007	0.001–0.003	0.004–0.007	0.003–0.005	0.003–0.004
6. Tensile strength at break, lb/in²	D638[b]	5,200–7,500	6,440–8,200	10,000–12,000	1,900–6,200	10,000–11,900	10,000–14,000
7. Elongation at break, %	D638[b]	1.2–2.5	2.0–3.6	1.0–1.3	20–65	2–3	1.4–3.5
8. Tensile yield strength, lb/in²	D638[b]		6,440–8,150		2,100–6,000	9,920–12,000	
9. Compressive strength (rupture or yield), lb/in²	D695	12,000–13,000	13,000–14,000	16,000–17,000		14,000–15,000	
10. Flexural strength (rupture or yield), lb/in²	D790	10,000–14,600	13,000–14,000	14,000–18,000	3,300–10,000	11,000–19,000	16,300–22,000
11. Tensile modulus, 10^3 lb/in²	D638[b]	330–475	450–485	900–1,200	160–370	475–560	850–900
12. Compressive modulus, 10^3 lb/in²	D695	480–490	495–500			530–580	
13. Flexural modulus, 10^3 lb/in² 73°F	D790	380–490	450–500	950–1,100	160–390	500–610	800–1,050
200°F	D790						
250°F	D790						
300°F	D790						
14. Izod impact, ft · lb/in of notch (⅛-in-thick specimen)	D256A	0.35–0.45	0.4–0.45	0.9–2.5	0.95–7.0	0.4–0.6	2.1–2.6
15. Hardness Rockwell	D785	M60–75	M75–84	M80–95, R119	R50–82; L–60	M80, R83	
Shore/Barcol	D2240/ D2583						
16. Coef. of linear thermal expansion, 10^6 in/(in · °C)	D696	50–83	68–85	39.6–40	44.2	65–68	20
17. Deflection temperature under flexural load, °F 264 lb/in²	D648	169–202	194–217	200–220	170–205	214–220	231–247
66 lb/in²	D648	155–204	200–224	220–230	165–200	220–224	
18. Thermal conductivity, 10^{-4} cal · cm/(s · cm² · °C)	C177	3.0	3.0	5.9		3.0	
19. Specific gravity	D792	1.04–1.05	1.04–1.05	1.20	1.03–1.06	1.06–1.08	1.20–1.22
20. Water absorption (⅛-in-thick specimen), % 24 h	D570	0.01–0.03	0.01	0.07–0.01	0.05–0.07	0.15–0.25	0.1
Saturation	D570	0.01–0.03	0.01	0.3		0.5	
21. Dielectric strength (⅛-in-thick specimen), short time, V/mil	D149	500–575	500–525	425		425	

Polyurethane					Silicone		
Thermoset			Thermoplastic		Casting resins	Liquid injection molding	Molding and encapsulating compounds
Casting resins		55–65% Mineral-filled potting and casting compounds	Unreinforced molding	10–20% Glass fiber-reinforced molding compounds	Flexible (including RTV)	Liquid silicone rubber	Mineral- and/or glass-filled
Liquid	Unsaturated						
Thermoset	Thermoset	Thermoset	75–137		Thermoset	Thermoset	Thermoset
				120–160			
C:43–250		25 (casting)	I:430–500 E:430–510	I:360–410		I:360–420	C:280–360 I:330–370 T:330–370
0.1–5				8–11		1–2	0.3–6
							2.0–8.0
0.020		0.001–0.002	0.004–0.006	0.004–0.010	0.0–0.006	0.0–0.005	0.0–0.005
175–10,000	10,000–11,000	1,000–7,000	7,200–9,000	4,800–7,500	350–1,000	725–1,305	500–1,500
100–1,000	3–6	5–55	60–180	3–70	20–700	300–1,000	80–800
			7,800–11,000				
20,000				5,000			
700–4,500	19,000		10,200–15,000	1,700–6,200			
10–100			190–300	0.6–1.40			
10–100							
10–100	610		235–310	40–90			
25 to flexible	0.4		1.5–1.8	10–14-No break			
			R > 100; M48	R45–55			
Shore A10–13, D90	Barcol 30–35	Shore A90, D52–85			Shore A10–70	Shore A20–70	Shore A10–80
100–200		71–100		34	10–19	10–20	20–50
Varies over wide range	190–200		158–260	115–130			>500
			176–275	140–145			
5		6.8–10			3.5–7.5		7.18
1.03–1.5	1.05	1.37–2.1	1.2	1.22–1.36	0.97–2.5	1.08–1.14	1.80–2.05
0.2–1.5	0.1–0.2	0.06–0.52	0.17–0.19	0.4–0.55	0.1		0.15
			0.5–0.6	1.5			0.15–0.40
300–500		500–750 @ 1/16 in.	400	600	400–550		200–550

Table 6.12.1   Properties of Plastic Resins and Compounds   (*Continued*)

Materials		ASTM test method	PVC molding compound, 20% glass fiber-reinforced	Vinyl polymers and copolymers			Molding and extrusion compounds
				PVC and PVC-acetate MC, sheets, rods, and tubes			Chlorinated polyvinyl chloride
				Rigid	Flexible, unfilled	Flexible, filled	
Processing	1a. Melt flow, g/10 min	D1238					
	1. Melting temperature, °C. T_m (crystalline)						
	T_g (amorphous)		75–105	75–105	75–105	75–105	110
	2. Processing temperature range, °F. (C = compression; T = transfer; I = injection; E = extrusion)		I: 380–400 E: 390–400	C: 285–400 I: 300–415	C: 285–350 I: 320–385	C: 285–350 I: 320–385	C: 350–400 I: 395–440 E: 360–420
	3. Molding pressure range, 10^3 lb/in^2		5–15	10–40	8–25	1–2	15–40
	4. Compression ratio		1.5–2.5	2.0–2.3	2.0–2.3	2.0–2.3	1.5–2.5
	5. Mold (linear) shrinkage, in/in	D955	0.001	0.002–0.006	0.010–0.050	0.008–0.035 0.002–0.008	0.003–0.007
Mechanical	6. Tensile strength at break, lb/in^2	D638[b]	8,600–12,800	5,900–7,500	1,500–3,500	1,000–3,500	6,800–9,000
	7. Elongation at break, %	D638[b]	2–5	40–80	200–450	200–400	4–100
	8. Tensile yield strength, lb/in^2	D638[b]		5,900–6,500			6,000–8,000
	9. Compressive strength (rupture or yield), lb/in^2	D695		8,000–13,000	900–1,700	1,000–1,800	9,000–22,000
	10. Flexural strength (rupture or yield), lb/in^2	D790	14,200–22,500	10,000–16,000			14,500–17,000
	11. Tensile modulus, 10^3 lb/in^2	D638[b]	680–970	350–600			341–475
	12. Compressive modulus, 10^3 lb/in^2	D695					335–600
	13. Flexural modulus, 10^3 lb/in^2 73°F	D790	680–970	300–500			380–450
	200°F	D790					
	250°F	D790					
	300°F	D790					
	14. Izod impact, ft · lb/in of notch (⅛-in-thick specimen)	D256A	1.0–1.9	0.4–22	Varies over wide range	Varies over wide range	1.0–5.6
	15. Hardness Rockwell	D785	R108–119				R117–112
	Shore/Barcol	D2240/ D2583	Shore D85–89	Shore D65–85	Shore A50–100	Shore A50–100	
Thermal	16. Coef. of linear thermal expansion, 10^6 in/(in · °C)	D696	24–36	50–100	70–250		62–78
	17. Deflection temperature under flexural load, °F 264 lb/in^2	D648	165–174	140–170			202–234
	66 lb/in^2	D648		135–180			215–247
	18. Thermal conductivity, 10^{-4} cal · cm/(s · cm^2 · °C)	C177		3.5–5.0	3–4	3–4	3.3
Physical	19. Specific gravity	D792	1.43–1.50	1.30–1.58	1.16–1.35	1.3–1.7	1.49–1.58
	20. Water absorption (⅛-in-thick specimen), % 24 h	D570	0.01	0.04–0.4	0.15–0.75	0.5–1.0	0.02–0.15
	Saturation	D570					
	21. Dielectric strength (⅛-in-thick specimen), short time, V/mil	D149		350–500	300–400	250–300	600–625

[a] Acrylonitrile-butadiene-styrene.   [b] Tensile test method varies with material: D638 is standard for thermoplastics; D651 for rigid thermoset plastics; D412 for elastomeric plastics; D882 for thin plastics sheeting.   [c] Dry, as molded (approximately 0.2% moisture content).   [d] As conditioned to equilibrium with 50% relative humidity.   [e] Test method is ASTM D4092.   [f] *Pseudo* indicates that the thermoset and thermoplastic components were mixed in the form of pellets or powder prior to fabrication.

ties of plastics which cannot be replicated in metals, including light weight and density (specific gravity rarely greater than 2, with the normal value in the range of 1.1 to 1.7—compare this with magnesium, the lightest structural metal of significance, whose specific gravity is 1.75); optical properties which may range from complete clarity to complete opacity, with the ability to be compounded with through colors in an

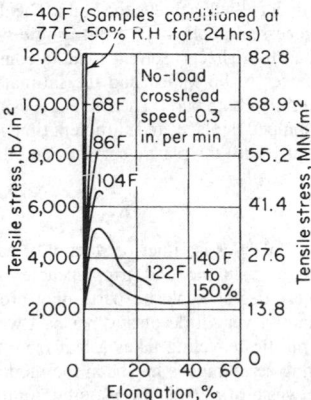

Fig. 6.12.1   Tensile stress-strain curve for thermoplastic Lucite. *(The Du Pont Co.)*

almost limitless range; low thermal conductivity and good electrical resistance; the ability to impart excellent surface finish to parts made by many of the primary production processes. Conversely, when compared to metals, plastics generally have lower elastic modulus—and thus, are inherently more flexible; have lower flexural and impact strength, and inferior toughness; and have lower dimensional stability generally than for most metals.

Some properties of plastics which are most desirable are high strength/weight ratios and the ability to process raw materials through to a finished size and shape in one of several basic operations; this last item has a large impact on the overall costs of finished products by virtue of eliminating secondary operations.

The seemingly endless variety of all types of plastics which are available in the marketplace is continually augmented by new compositions; if an end product basically lends itself to the use of plastics, there is most likely to be some existing composition available to satisfy the design requirements. See Table 6.12.1.

The properties of a given plastic can often be modified by the incorporation of additives into the basic plastic resin which include colorants, stabilizers, lubricants, and fibrous reinforcement. Likewise, where local strength requirements are beyond the capacity of the plastic, ingenious configurations of **metallic inserts** can be incorporated into the manufacture of the plastic part and become structurally integral with the part itself; female threaded inserts and male threaded studs are the most common type of inserts found in plastic part production.

Plastic resins may themselves be used as **adhesives** to join other plastics or other paired materials including wood/wood, wood/plastic, metal/plastic, metal/metal, and wood/metal. Structural **sandwich panels,** with the inner layer of sheet or foamed plastic, are an adaptation of composite construction.

Foamed plastic has found widespread use as thermal insulation, a volume filler, cushioning material, and many lightweight consumer items.

The ability to be recycled is an important attribute of thermoplastics; thermosets are deficient in this regard. In view of the massive amounts of plastic in the consumer stream, **recycling** has become more the norm than the exception; indeed, in some jurisdictions, recycling of spent plastic consumables is mandated by law, while in others, the incentive to recycle manifests itself as a built-in cost of the product—a tax, as it were, imposed at the manufacturing phase of production to spur recycling efforts. (See Recycling on next page.)

In those cases where secondary operations are required to be performed on plastic parts, the usual chip-producing material-removal processes are employed. In those regards, due diligence must be paid to the nature of the material being cut. Thermoplastics will soften from heat generated during cutting, and their low flexural modulus may require backing to prevent excessive deflection in response to cutting forces; thermosets may prove abrasive (even without fillers) and cause rapid wear of cutting edges. Cutting operations on some soft thermoplastics will experience elastic flow of the material beyond the cutting region, necessitating multiple cuts, each of smaller magnitude.

The raw-materials cost for plastics will vary. The economies of quantity production are self-evident from a study of applicable statistics of plastics production. Consider that in 1993, global production of plastic raw materials was in excess of 100 million metric tons (t). The final overall cost of a finished plastic part is impacted by the cost of raw materials, of course, but further, the use of plastic itself affects the design, manufacture, and shipping components of the final cost; these latter are not inconsequential in arriving at cost comparisons of production with plastics vis-à-vis metals. Suffice it to say that in general, when all costs are accounted for and when design requirements can be met with plastics, the end cost of a plastic item is often decidedly favorable.

RAW MATERIALS

The source of virgin raw materials, or **feedstock,** is generally petroleum or natural gas. The feedstock is converted to monomers, which then become the basis for plastic resins. As recycling efforts increase, conversion of postconsumer plastic items (i.e., scrap) may result in increasing amounts of plastics converted back to feedstock by application of hydrolytic and pyrolitic processes.

The plastic is supplied to primary processers usually in the form of powder, solid pellets, or plastisols (liquid or semisolid dispersions of finely powdered polymer in a nonmigrating liquid).

Compounding, mixing, and blending prepare plastic materials to be fed into any of the various fabrication processes.

PRIMARY FABRICATION PROCESSES

A number of processes are used to achieve finished plastic parts, including some conventional ones usually associated with metalworking and others synonymous with plastics processing. Casting, blow molding, extrusion, forming in metal molds (injection, compression, and transfer molding), expanded-bead molding to make foams, thermoforming of sheet plastic, filament winding over a form, press laminating, vacuum forming, and open molding (hand layup) are widespread plastics fabrication processes.

ADDITIVES

The inherent properties of most plastics can be tailored to impart other desired properties or to enhance existing ones by the introduction of additives. A wide variety of additives allow compounding a specific type of plastic to meet some desired end result; a few additive types are listed below.

Fillers   Wood flour, mica, silica, clay, and natural synthetic fibers reduce weight, reduce the volume of bulk resin used, impart some specific strength property (and often, directionality of strength properties), etc.

Blowing or Foaming Agents   Compressed gas or a liquid which will evolve gas when heated is the agent which causes the network of interstices in expanded foam. A popular foaming agent, CFC (chlorinated fluorocarbon) is being displaced because of environmental concerns. Non-CFC foaming agents will increasingly replace CFCs.

Mold-Release Compounds   These facilitate the removal of molded parts from mold cavities with retention of surface finish on the finished parts and elimination of pickup of molding material on mold surfaces.

Lubricants   They ease fabrication and can serve to impart lubricant to plastic parts.

Antistatic Compounds   The inherent good electrical insulation property of plastic often leads to buildup of static electric charges. The static charge enhances dust pickup and retention to the plastic surface. At the very least, this may inhibit the action of a mold-release agent; at worst, the leakage of static electric charge may be an explosion or fire hazard.

Antibacterial Agents   These act to inhibit the growth of bacteria, especially when the type of filler used can be an attractive host to bacterial action (wood flour, for example).

Colorants   Color is imparted to the resin by dyes or pigments. Dyes allow a wider range of color. Colorants enhance the utility of plastic products where cosmetic and/or aesthetic effects are required. Ultraviolet (uv) radiation from sunlight degrades the color unless a uv-inhibiting agent is incorporated into the plastic resin. (See below.)

Flame Retardants   These are introduced primarily for safety, and they work by increasing the ignition temperature and lowering the rate of combustion.

Heat Stabilizers   Plastics will degrade when subjected to high temperatures; thermoplastics will soften and eventually melt; thermosets will char and burn. Either type of plastic part is subjected to heat during processing and may see service above room temperature during its useful life. The judicious addition of an appropriate heat-stabilizing compound will prevent thermal degradation under those circumstances. Some of the more effective heat-stabilizing agents contain lead, antimony, or cadmium and are subject to increasingly stringent environmental limits.

Impact Modifiers   When added to the basic resin, these materials increase impact strength or toughness.

Plasticizers   These materials increase the softness and flexibility of the plastic resin, but often there is an accompanying loss of tensile strength, increased flammability, etc.

Ultraviolet Stabilizers   These additives retard or prevent the degradation of some strength properties as well as color, which result from exposure to sunlight.

ADHESIVES AND ASSEMBLY

Most adhesives operate to effect solvent-based **bonding,** and they are either one- or two-part materials. While these adhesives cure, the evaporation of volatile organic compounds (VOCs) constitutes an environmental hazard which is subject to increasing regulation. Water-based bonding agents, on the other hand, emit no VOCs and effectively are environmentally benign. Adhesives which are constituted of all solid material contain no solvents or water, hence no evaporants. A small quantity of bonding is performed via **radiation-curable adhesives.** The source of energy is ultraviolet radiation or electron beam radiation.

The adhesives cited above, and others, display a range of properties which must be assessed as to suitability for the application intended.

Welding of plastics most often implies bonding or sealing of plastic sheet through the agent of ultrasonic energy, which, when applied locally, transforms high-frequency ultrasonic energy to heat, which, in turn, fuses the plastic locally to achieve the bond. The design of the joint is critical to the successful application of ultrasonic welding when the pieces are other than thin sheets. **Solvent welding** consists of localized softening of the pieces to be joined and the intimate mixture of the plastic surface material of both parts. The solvent is usually applied with the parts fixed or clamped in place, and sufficient time is allowed for the solvent to react locally with the plastic surfaces and then evaporate.

RECYCLING

The feedstock for all plastics is either petroleum or natural gas. These materials constitute a valuable and irreplaceable natural resource. Plastic materials may degrade; plastic parts may break or no longer be of service; certainly, very little plastic per se "wears away." The basic idea behind plastic recycling takes a leaf from the metal industries, where recycling scrap metals has been the modus operandi from ancient times. The waste of a valuable resource implicit in discarding scrap plastic in landfills or the often environmentally hazardous incineration of plastics has led to very strong efforts toward recycling plastic scrap.

Separation of plastic by types, aided by unique identification symbols molded into consumer items, is mandated by law in many localities which operate an active recycling program. Different plastic types present a problem when comingled during recycling, but some postcycling consumer goods have found their way to the market. Efforts will continue as recycling of plastics becomes universal; the costs associated with the overall recycling effort will become more attractive as economies of large-scale operations come into play.

Ideally, plastic recycling responds best to clean scrap segregated as to type. In reality, this is not the case, and for those reasons, a number of promising scrap processing directions are being investigated. Among those is depolymerization, whereby the stream of plastic scrap is subjected to pyrolitic or hydrolytic processing which results in reversion of the plastic to feedstock components. This regenerated feedstock can then be processed once more into the monomers, then to virgin polymeric resins. Other concepts will come under consideration in the near future.

6.13   FIBER COMPOSITE MATERIALS
by Stephen R. Swanson

REFERENCES: Jones, "Mechanics of Composite Materials," Hemisphere Publishing. Halpin, "Primer on Composite Materials Analysis," 2d ed., Technomic Pub. Co. "Engineered Materials Handbook," vol. 1, "Composites," ASM. Hull, "An Introduction to Composite Materials," Cambridge Univ. Press. Schwartz (ed.), "Composite Materials Handbook," 2d ed., McGraw-Hill.

INTRODUCTION

Composite materials are composed of two or more discrete constituents. Examples of engineering use of composites date back to the use of straw in clay by the Egyptians. Modern composites using fiber-reinforced matrices of various types have created a revolution in high-performance structures in recent years. Advanced composite materials offer significant advantages in strength and stiffness coupled with light weight, relative to conventional metallic materials. Along with this structural performance comes the freedom to select the orientation of the fibers for optimum performance. Modern composites have been described as being revolutionary in the sense that the material can be designed as well as the structure.

The **stiffness** and **strength-to-weight** properties make materials such as carbon fiber composites attractive for applications in aerospace and sporting goods. In addition, composites often have superior resistance to environmental attack, and glass fiber composites are used extensively in the chemical industries and in marine applications because of this advantage. Both glass and carbon fiber composites are being considered for infrastructure applications, such as for bridges and to reinforced concrete, because of this environmental resistance.

The cost competitiveness of composites depends on how important the weight reduction or environmental resistance provided by them is to the overall function of the particular application. While glass fibers usually cost less than aluminum on a weight basis, carbon fibers are still considerably more expensive. Equally important or more important than the material cost is the cost of manufacturing. In some cases composite structures can achieve significant cost savings in manufacturing, often by reducing the number of parts involved in a complex assembly. There is a large variability in cost and labor content between the various methods of composite manufacture, and much attention is currently given to reduce manufacturing costs.

TYPICAL ADVANCED COMPOSITES

Modern composite materials usually, but not always, utilize a **reinforcement** phase and a **binder** phase, in many cases with more rigid and higher-strength fibers embedded and dispersed in a more compliant matrix. Modern applications started with **glass fibers**, although the word **advanced** fibers often means the more **high-performance fibers** that followed, such as carbon, aramid, boron, and silicon carbide. A typical example is carbon fiber that is being widely introduced into aerospace and sporting goods applications. The tensile strength of carbon fiber varies with the specific type of fiber being considered, but a typical range of values is 3.1 to 5.5 GPa (450 to 800 ksi) for fiber tensile strength and stiffness on the order of 240 GPa (35 Msi), combined with a specific gravity of 1.7. Thus the fiber itself is stronger than 7075 T6 aluminum by a factor of 5 to 10, is stiffer by a factor of 3.5, and weighs approximately 60 percent as much. The potential advantages of high-performance fiber structures in mechanical design are obvious. On the other hand, current costs for carbon fibers are on the order of several times to an order of magnitude or more higher than those for aluminum. The cost differential implies that composite materials will be utilized in demanding applications, where increases in performance justify the increased material cost. However, the material cost is only part of the story, for manufacturing costs must also be considered. In many instances it has been possible to form composite parts with significantly fewer individual components compared to metallic structures, thus leading to an overall lower-cost structure. On the other hand, composite components may involve a significant amount of hand labor, resulting in higher manufacturing costs.

FIBERS

Glass Fiber   Glass fiber in a **polymeric matrix** has been widely used in commercial products such as boats, sporting goods, piping, and so forth. It has relatively low stiffness, high elongation, moderate strength and weight, and generally lower cost relative to other composites. It has been used extensively where corrosion resistance is important, for piping for the chemical industry, and in marine applications.

Aramid Fiber   Aramid fibers (sold under the trade name **Kevlar**) offer higher strength and stiffness relative to glass, coupled with light weight, high tensile strength but lower compressive strength. Both glass fiber and aramid fiber composites exhibit good toughness in impact environments. Aramid fiber is used as a higher-performance replacement for glass fiber in industrial and sporting goods and in protective clothing.

Carbon Fiber   Carbon fibers are used widely in aerospace and for some sporting goods, because of their relatively high stiffness and high strength/weight ratios. Carbon fibers vary in strength and stiffness with processing variables, so that different grades are available (high modulus or intermediate modulus), with the tradeoff being between high modulus and high strength. The intermediate-modulus and high-strength grades are almost universally made from a **PAN (polyacrylonitrile) precursor,** which is then heated and stretched to align the structure and remove noncarbon material. Higher-modulus but lower-cost fibers with much lower strength are made from a petroleum pitch precursor.

Other Fibers   Boron fibers offer very high stiffness, but at very high cost. These fibers have been used in specialized applications in both aluminum and polymeric matrices. A new fiber being used in textile applications is oriented polyethylene, marketed under the trademark **Spectra**. This fiber combines high strength with extremely light weight. The fiber itself has a specific gravity of 0.97. It is limited to a very low range of temperature, and the difficulty of obtaining adhesion to matrix materials has limited its application in structural composites. A number of other fibers are under development for use with ceramic matrices, to enable use in very high-temperature applications such as engine components. An example is **silicon carbide fiber,** used in **whisker** form.

Table 6.13.1 lists data pertinent to fibers currently available.

MATRICES

Most current applications utilize polymeric matrices. **Thermosetting polymers** (such as **epoxies**) are widely utilized, and a large amount of characterization data are available for these materials. Epoxies provide superior performance but are more costly. Typical cure temperatures are in the range of 121 to 177°C (250 to 350°F). A recent development has been high-toughness epoxies, which offer significantly improved resistance to damage from accidental impact, but at higher cost. Polyester and vinyl ester are often used in less demanding applications, and polyurethane matrices are being considered because of short cure times in high-production-rate environments. **Thermoplastic matrices** are under development and have had limited applications to date. Wider use awaits accumulation of experience in their application and is inhibited in some instances by high material costs. Their use is likely to increase because of the increased toughness they may provide, as well as the potential for forming complex shapes. Composites using thermoplastics often are formed to final shape at temperatures of about 315°C (600°F), and they need no cure cycle.

Polymeric matrix materials have a limited temperature range for practical application, with epoxies usually limited to 150°C (300°F) or less, depending on the specific material. Higher-temperature polymers available, such as polyimides, usually display increased brittleness. They are used for cowlings and ducts for jet engines.

Metal matrix composites are utilized for higher-temperature applications than is possible with polymeric matrices. Aluminum matrix has been utilized with boron and carbon fibers. Ceramic matrix materials are being developed for service at still higher temperatures. In this case, the fiber is not necessarily higher in strength and stiffness than the matrix, but is used primarily to toughen the ceramic matrix.

Table 6.13.1   Mechanical Properties of Typical Fibers

Fiber	Fiber diameter, μm	Fiber density		Tensile strength		Tensile modulus	
		lb/in³	g/cm³	ksi	GPa	Msi	GPa
E-glass	8–14	0.092	2.54	500	3.45	10.5	72.4
S-glass	8–14	0.090	2.49	665	4.58	12.5	86.2
Polyethylene	10–12	0.035	0.97	392	2.70	12.6	87.0
Aramid (Kevlar 49)	12	0.052	1.44	525	3.62	19.0	130.0
HS carbon, T300	7	0.063	1.74	514	3.54	33.6	230
AS4 carbon	7	0.065	1.80	580	4.00	33.0	228
IM7 carbon	5	0.065	1.80	785	5.41	40.0	276
GY80 carbon	8.4	0.071	1.96	270	1.86	83.0	572
Boron	50–203	0.094	2.60	500	3.44	59.0	407

MATERIAL FORMS AND MANUFACTURING

Composite materials come in a wide variety of material forms. The fiber itself may be used in continuous form or as a **chopped fiber.** Chopped glass fibers are used typically to reinforce various polymers, with concomitant lower strength and stiffness relative to continuous fiber composites. Chopped fibers in conjunction with automated fabrication techniques have been utilized to fabricate automotive body parts at high production rates.

Continuous-fiber materials are available in a number of different forms, with the specific form utilized depending on the manufacturing process. Thus it is useful to consider both the material and the manufacturing process at the same time. The fibers themselves have a small diameter, with sizes of 5 to 7 μm (0.0002 to 0.0003 in) typical for carbon fibers. A large number of fibers, from 2,000 to 12,000, are gathered in the manufacturing process to form a **tow** (also called a **roving** or **yarn**). The filament winding process utilizes these tows directly. The tows may be further processed by **prepregging,** the process of coating the individual fibers with the matrix material. This process is widely used with thermosetting polymeric resins. The resin is partially cured, and the resulting "ply" is placed on a paper backing. The prepregged material is available in continuous rolls in widths of 75 to 1,000 mm (3 to 40 in). These rolls must be kept refrigerated until they are utilized and the assembled product is cured. Note that the ply consists of a number of fibers through its thickness, and that the fibers are aligned and continuous. Typical volume fractions of fiber are on the order of 60 percent. The material forms discussed here are used in a variety of specific manufacturing techniques. Some of the more popular techniques are described briefly below.

Filament Winding The process consists of winding the fiber around a mandrel to form the structure. Usually the mandrel rotates while fiber placement is synchronized to proceed in a longitudinal direction. The matrix may be added to the fiber by passing the fiber tow through a matrix bath at the time of placement, a process called **wet winding;** or the tows may be prepregged prior to winding. Filament winding is widely used to make glass fiber pipe, rocket motor cases, and similar products. Filament winding is a highly automated process, with typical low manufacturing costs. Obviously, it lends itself most readily to axisymmetric shapes, but a number of specialized techniques are being considered for nonaxisymmetric shapes.

Prepreg Layup This common procedure involves laying together individual plies of prepregged composite into the final laminated structure. A mold may be used to control the part geometry. The plies are laid down in the desired pattern, and then they are wrapped with several additional materials used in the curing process. The objective is to remove volatiles and excess air to facilitate consolidation of the laminate. To this end, the laminate is covered with a **peel ply,** for removal of the other curing materials, and a **breather ply,** which is often a fiberglass net. Optionally, a bleeder may be used to absorb excess resin, although the net resin process omits this step. Finally, the assembly is covered with a vacuum bag and is sealed at the edges. A vacuum is drawn, and after inspection heat is applied. If an autoclave is used, pressure on the order of 0.1 to 0.7 MPa (20 to 100 lb/in²) is applied to ensure final consolidation. Autoclave processing ensures good lamination but requires a somewhat expensive piece of machinery. Note that the individual plies are relatively thin [on the order of 0.13 mm (0.005 in)], so that a large number of plies will be required for thick parts. The lamination process is often performed by hand, although automated tape-laying machines are available. Although unidirectional plies have been described here, cloth layers can also be used. The bends in the individual fibers that occur while using cloth layers carry a performance penalty, but manufacturing considerations such as drapeability may make cloth layers desirable.

Automated Tape and Tow Placement Automated machinery is used for tape layup and fiber (tow) placement. These machines can be large enough to construct wing panels or other large structures. Thermosetting matrices have been used extensively, and developments using thermoplastic matrices are underway.

Textile Forms The individual tows may be combined in a variety of textile processes such as **braiding** and **weaving.** Preforms made in these ways then can be impregnated with resin, often called **resin transfer molding** (RTM). These textile processes can be designed to place fibers in the through-the-thickness direction, to impart higher strength in this direction and to eliminate the possibility of delamination. There is considerable development underway in using braided and/or stitched preforms with RTM because of the potential for automation and high production rates.

DESIGN AND ANALYSIS

The fundamental way in which fiber composites, and in particular continuous-fiber composites, differ from conventional engineering materials such as metals is that their properties are **highly directional.** Stiffness and strength in the fiber direction may be higher than in the direction transverse to the fibers by factors of 20 and 50, respectively. Thus a basic principle is to align the fibers in directions where stiffness and strength are needed. A well-developed theoretical basis called **classical lamination theory** is available to predict stiffness and to calculate stresses within the layers of a laminate. This theory is often applied to filament-wound structures and to textile preform structures, if allowances are made for the undulations in the fiber path. The basic assumption is that fiber composites are orthotropic materials. Thin composite layers require four independent elasticity constants to characterize the stiffness; those are the fiber direction modulus, transverse direction modulus, in-plane shear modulus, and one of the two Poisson ratios, with the other related through symmetry of the orthotropic stress-strain matrix. The material constants are routinely provided by material suppliers. Lamination theory is then used to predict the overall stiffness of the laminate and to calculate stresses within the individual layers under mechanical and thermal loads.

Transverse properties are much lower than those in the fiber direction, so that fibers must usually be oriented in more than one direction, even if only to take care of secondary loads. For example, a laminate may consist of fibers in an axial direction combined with fibers oriented at $\pm 60°$ to this direction, commonly designated as an $[0_m/\pm 60_n]s$ laminate, where m and n refer to the number of plies in the axial and $\pm 60°$ directions and s stands for symmetry with respect to the midplane of the laminate. Although not all laminates are designed to be symmetric, residual stresses will cause flat panels to curve when cooled from elevated-temperature processing unless they are symmetric. Other popular laminates have fibers oriented at $0°$, $\pm 45°$, and $90°$. If the relative amounts of fibers are equal in the $[0/\pm 60]$ or the $[0/\pm 45/90]$ directions, the laminate has in-plane stiffness properties equal in all directions and is thus termed **quasi-isotropic.**

Note that the **stiffness of composite laminates** is less than that of the fibers themselves for two reasons. First, the individual plies contain fibers and often a much less stiff matrix. With high-stiffness fibers and polymeric matrices, the contribution of the matrix can be neglected, so that the stiffness in the fiber direction is essentially the **fiber volume fraction** VF times the fiber stiffness. Fiber volume fractions are often on the order of 60 percent. Second, laminates contain fibers oriented in more than one direction, so that the stiffness in any one direction is less than it would be in the fiber direction of a unidirectional laminate. As an example, the in-plane stiffness of a quasi-isotropic carbon fiber/polymeric matrix laminate is about 45 percent of unidirectional plies, which in turn is approximately 60 percent of the fiber modulus for a 60 percent fiber volume fraction material.

Strength Properties Fiber composites typically have excellent strength-to-weight properties; see Table 6.13.1. Like stiffness properties, strength properties are reduced by dilution by the weaker matrix, and because not all the fibers can be oriented in one direction. The overall reduction in apparent strength because of these two factors is similar to that for stiffness, although this is a very rough guide. As an example, the strength of a quasi-isotropic carbon/epoxy laminate under uniaxial tensile load has been reported to be about 35 to 40 percent of that provided by the same material in a unidirectional form.

Fiber composite materials can fail in several different modes. The design goal is usually to have the failure mode be **in-plane failure** due to fiber failure, as this takes advantage of the strong fibers. Unanticipated loads, or poorly designed laminates, can lead to ultimate failure of the matrix, which is much weaker. For example, in-plane shear loads on a [0/90] layup would be resisted by the matrix, with the potential result of failure at low applied load. Adding [± 45] fibers will resist this shear loading effectively.

Fiber composites respond differently to compressive loads than they do to tensile loads. Some fibers, such as aramid and the very high-modulus carbon fibers, are inherently weaker in compression than in tension. It is believed that this is not the case with intermediate-modulus carbon fibers, but even these fibers often display lower compressive strength. The mechanism is generally believed to involve bending or buckling of the very small-diameter fibers against the support provided by the matrix.

Matrix cracking usually occurs in the transverse plies of laminates with polymeric matrices, often at an applied load less than half that for ultimate fiber failure. This is caused by the lower failure strain of the matrix and the stress concentration effect of the fibers, and it can also depend on residual thermal stresses caused by the differences in coefficients of thermal expansion in the fibers and matrix. In some cases these matrix cracks are tolerated, while in other cases they produce undesirable effects. An example of the latter is that matrix cracking increases the permeability in unlined pressure vessels, which can result in pipes weeping.

Another potential failure mode is **delamination.** Delamination is often caused by high interlaminar shear and through-the-thickness tensile stresses, perhaps combined with manufacturing deficiencies such as inadequate cure or matrix porosity. Delamination can start at the edges of laminates due to "edge stresses" which have no counterpart in isotropic materials such as metals.

Fiber composites usually have excellent fatigue properties. However, stress concentrations or accidental damage can result in localized damage areas that can lead to failure under fatigue loading.

Section **7**

Fuels and Furnaces

BY

MARTIN D. SCHLESINGER *Wallingford Group, Ltd.*
KLEMENS C. BACZEWSKI *Consulting Engineer*
GLENN W. BAGGLEY *Manager, Regenerative Systems, Bloom Engineering Co., Inc.*
CHARLES O. VELZY *Consultant*
ROGER S. HECKLINGER *Project Director, Roy F. Weston of New York, Inc.*
GEORGE J. RODDAM *Sales Engineer, Lectromelt Furnace Division, Salem Furnace Co.*

7.1 FUELS

by Martin D. Schlesinger and Associates

COAL
by Martin D. Schlesinger
Wallingford Group, Ltd.

REFERENCES: Petrography of American Coals, *U.S. BuMines Bull.* 550. Lowry, "Chemistry of Coal Utilization," Wiley. ASTM, "Standards on Gaseous Fuels, Coal and Coke." Methods of Analyzing and Testing Coal and Coke, *U.S. BuMines Bull.* 638. Karr, "Analytical Methods for Coal and Coal Products," Academic. Preprints, Division of Fuel Chemistry, American Chemical Society.

Coal is a black or brownish-black combustible solid formed by the decomposition of vegetation in the absence of air. Microscopy can identify plant tissues, resins, spores, etc. that existed in the original structure. It is composed principally of carbon, hydrogen, oxygen, and small amounts of sulfur and nitrogen. Associated with the organic matrix are water and as many as 65 other chemical elements. Many trace elements can be determined by spectrometric method D-3683. Coal is used directly as a fuel, a chemical reactant, and a source of organic chemicals. It can also be converted to liquid and gaseous fuels.

Classification and Description

Coal may be classified by rank, by variety, by size and sometimes by use. Rank classification takes into account the degree of metamorphism or progressive alteration in the natural series from lignite to anthracite. Table 7.1.1 shows the classification of coals by rank adopted as standard by the ASTM (method D-388). The basic scheme is according to fixed carbon (FC) and heating value (HV) from a proximate analysis, calculated on the mineral-matter-free (mmf) basis. The higher-rank coals are classified according to the FC on a dry basis and the lower-rank according to HV in Btu on a moist basis. **Agglomerating** character is used to differentiate between certain adjacent groups. Coals are considered agglomerating if, in the test to determine volatile matter, they produce either a coherent button that will support a 500-g weight or a button that shows swelling or cell structure.

For classifying coals according to rank, FC and HV can be calculated to a moisture-free basis by the **Parr formulas**, Eqs. (7.1.1) to (7.1.3) below:

$$FC \text{ (dry, mmf)} = \frac{FC - 0.15S}{100 - (M + 1.08A + 0.55S)} \times 100 \quad (7.1.1)$$

$$VM \text{ (dry, mmf)} = 100 - FC \quad (7.1.2)$$

$$HV \text{ (moist, mmf)} = \frac{Btu - 50S}{100 - (1.08A + 0.55S)} \quad (7.1.3)$$

where FC = percentage of fixed carbon, VM = percentage of volatile matter, M = percentage of moisture, A = percentage of ash, S = percentage of sulfur, all on a moist basis. "Moist" coal refers to the natural bed moisture, but there is no visible moisture on the surface. HV = heating value, Btu/lb (Btu/lb × 0.5556 = g·cal/g). Because of its complexity, the analysis of coal requires care in sampling, preparation, and selection of the method of analysis.

Figure 7.1.1 shows representative proximate analyses and heating values of various ranks of coal in the United States. The analyses were calculated to an ash-free basis because ash is not a function of rank. Except for anthracite, FC and HV increase from the lowest to the highest rank as the percentages of volatile matter and moisture decrease. The sources and analyses of coals representing various ranks are given in Table 7.1.2.

Meta-anthracite is a high-carbon coal that approaches graphite in structure and composition. It usually is slow to ignite and difficult to burn. It has little commercial importance.

Anthracite, sometimes called hard coal, is hard, compact, and shiny black, with a generally conchoidal fracture. It ignites with some difficulty and burns with a short, smokeless, blue flame. Anthracite is used primarily for space heating and as a source of carbon. It is also used in

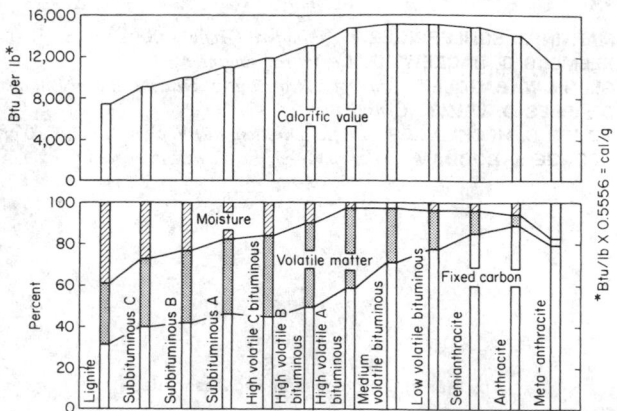

Fig. 7.1.1 Proximate analysis and heating values of various ranks of coal (ash-free basis).

electric power generating plants in or close to the anthracite-producing area. The iron and steel industry uses some anthracite in blends with bituminous coal to make coke, for sintering iron-ore fines, for lining pots and molds, for heating, and as a substitute for coke in foundries.

Semianthracite is dense, but softer than anthracite. It burns with a short, clean, bluish flame and is somewhat more easily ignited than anthracite. The uses are about the same as for anthracite.

Low-volatile bituminous coal is grayish black, granular in structure and friable on handling. It cakes in a fire and burns with a short flame that is usually considered smokeless under all burning conditions. It is used for space heating and steam raising and as a constituent of blends for improving the coke strength of higher-volatile bituminous coals. Low-volatile bituminous coals cannot be carbonized alone in slot-type ovens because they expand on coking and damage the walls of the ovens.

Medium-volatile bituminous coal is an intermediate stage between high-volatile and low-volatile bituminous coal and therefore has some of the characteristics of both. Some are fairly soft and friable, but others are hard and do not disintegrate on handling. They cake in a fuel bed and smoke when improperly fired. These coals make cokes of excellent strength and are either carbonized alone or blended with other bituminous coals. When carbonized alone, only those coals that do not expand appreciably can be used without damaging oven walls.

High-volatile A bituminous coal has distinct bands of varying luster. It is hard and handles well with little breakage. It includes some of the best steam and coking coal. On burning in a fuel bed, it cakes and gives off smoke if improperly fired. The coking property is often improved by blending with more strongly coking medium- and low-volatile bituminous coal.

High-volatile B bituminous coal is similar to high-volatile A bituminous coal but has slightly higher bed moisture and oxygen content and is less strongly coking. It is good coal for steam raising and space heating.

Table 7.1.1 Classification of Coals by Rank (ASTM D388)*

Class	Group	Fixed-carbon limits, percent (dry, mineral-matter-free basis)		Volatile-matter limits, percent (dry, mineral-matter-free basis)		Calorific value limits, Btu/lb (moist,† mineral-matter-free basis)		Agglomerating character
		Equal or greater than	Less than	Greater than	Equal or less than	Equal or greater than	Less than	
I. Anthracitic	1. Meta-anthracite	98	......	......	2	......	......	
	2. Anthracite	92	98	2	8	......	......	Nonagglomerating‡
	3. Semianthracite	86	92	8	14	......	......	
II. Bituminous	1. Low-volatile bituminous coal	78	86	14	22	......	......	
	2. Medium-volatile bituminous coal	69	78	22	31	......	......	
	3. High-volatile A bituminous coal	...	69	31	...	14,000§	......	Commonly agglomerating¶
	4. High-volatile B bituminous coal	...	...	...	...	13,000§	14,000	
	5. High-volatile C bituminous coal	...	...	...	...	11,500	13,000	
						10,500	11,500	Agglomerating
III. Subbituminous	1. Subbituminous A coal	...	...	...	...	10,500	11,500	Nonagglomerating
	2. Subbituminous B coal	...	...	...	...	9,500	10,500	
	3. Subbituminous C coal	...	...	...	...	8,300	9,500	
IV. Lignitic	1. Lignite A	...	...	...	...	6,300	8,300	
	2. Lignite B	...	...	...	...	......	6,300	

* This classification does not include a few coals, principally nonbanded varieties, which have unusual physical and chemical properties and which come within the limits of fixed-carbon or calorific value of the high-volatile bituminous and subbituminous ranks. All of these coals either contain less than 48 percent dry, mineral-matter-free fixed carbon or have more than 15,500 moist, mineral-matter-free British thermal units per pound. Btu/lb × 2.323 = kJ/kg.
† Moist refers to coal containing its natural inherent moisture but not including visible water on the surface of the coal.
‡ If agglomerating, classify in low-volatile group of the bituminous class.
§ Coals having 69 percent or more fixed carbon on the dry, mineral-matter-free basis are classified according to fixed carbon, regardless of calorific value.
¶ It is recognized that there may be nonagglomerating varieties in these groups of the bituminous class, and there are notable exceptions in the high-volatile C bituminous group.

Table 7.1.2 Sources and Analyses of Various Ranks of Coal as Received

Classification by rank	State	County	Bed	Proximate, %				Ultimate, %					Calorific value, Btu/lb*
				Moisture	Volatile matter	Fixed carbon	Ash†	Sulfur	Hydrogen	Carbon	Nitrogen	Oxygen	
Meta-anthracite	Rhode Island	Newport	Middle	13.2	2.6	65.3	18.9	0.3	1.9	64.2	0.2	14.5	9,310
Anthracite	Pennsylvania	Lackawanna	Clark	4.3	5.1	81.0	9.6	0.8	2.9	79.7	0.9	6.1	12,880
Semianthracite	Arkansas	Johnson	Lower Hartshorne	2.6	10.6	79.3	7.5	1.7	3.8	81.4	1.6	4.0	13,880
Low-volatile bituminous coal	West Virginia	Wyoming	Pocahontas no. 3	2.9	17.7	74.0	5.4	0.8	4.6	83.2	1.3	4.7	14,400
Medium-volatile bituminous coal	Pennsylvania	Clearfield	Upper Kittanning	2.1	24.4	67.4	6.1	1.0	5.0	81.6	1.4	4.9	14,310
High-volatile A bituminous coal	West Virginia	Marion	Pittsburgh	2.3	36.5	56.0	5.2	0.8	5.5	78.4	1.6	8.5	14,040
High-volatile B bituminous coal	Kentucky, western field	Muhlenburg	No. 9	8.5	36.4	44.3	10.8	2.8	5.4	65.1	1.3	14.6	11,680
High-volatile C bituminous coal	Illinois	Sangamon	No. 5	14.4	35.4	40.6	9.6	3.8	5.8	59.7	1.0	20.1	10,810
Subbituminous A coal	Wyoming	Sweetwater	No. 3	16.9	34.8	44.7	3.6	1.4	6.0	60.4	1.2	27.4	10,650
Subbituminous B coal	Wyoming	Sheridan	Monarch	22.2	33.2	40.3	4.3	0.5	6.9	53.9	1.0	33.4	9,610
Subbituminous C coal	Colorado	El Paso	Fox Hill	25.1	30.4	37.7	6.8	0.3	6.2	50.5	0.7	35.5	8,560
Lignite	North Dakota	McLean	Unnamed	36.8	27.8	29.5	5.9	0.9	6.9	40.6	0.6	45.1	7,000

* Btu/lb × 2.325 = J/g; Btu/lb × 0.5556 = g·cal/g.
† Ash is part of both the proximate and ultimate analyses.

Some of it is blended with more strongly coking coals for making metallurgical coke.

High-volatile C bituminous coal is a stage lower in rank than the B bituminous coal and therefore has a progressively higher bed moisture and oxygen content. It is used primarily for steam raising and space heating.

Subbituminous coals usually show less evidence of banding than bituminous coals. They have a high moisture content, and on exposure to air, they disintegrate or "slack" because of shrinkage from loss of moisture. They are noncaking and noncoking, and their primary use is for steam raising and space heating.

Lignites are brown to black in color and have a bed moisture content of 30 to 45 percent with a resulting lower heating value than higher-rank coals. Like subbituminous coals, they have a tendency to "slack" or disintegrate during air drying. They are noncaking and noncoking. Lignite can be burned on traveling or spreader stokers and in pulverized form.

The principal ranks of coal mined in the major coal-producing states are shown in Table 7.1.3. Their analyses depend on several factors, e.g., source, size of coal, and method of preparation. Periodic reports are issued by the U.S. Department of Energy, Energy Information Agency. They provide statistics on production, distribution, end use, and analytical data.

Composition and Characteristics

Proximate analysis, sulfur content, and calorific values are the analytical determination most commonly used for industrial characterization of coal. The **proximate analysis** is the simplest means for determining the distribution of products obtained during heating. It separates the products into four groups: (1) water or moisture, (2) volatile matter consisting of gases and vapors, (3) fixed carbon consisting of the carbonized residue less ash, and (4) ash derived from the mineral impurities in the coal. ASTM methods D3712 and D5142 are used; the latter is an instrumental method.

Moisture is the loss in weight obtained by drying the coal at a temperature between 104 and 110°C (220 and 230°F) under prescribed conditions. Further heating at higher temperatures may remove more water, but this moisture usually is considered part of the coal substance. The moisture obtained by the standard method consists of (1) surface or extraneous moisture that may come from external sources such as percolating waters in the mine, rain, condensation from the air, or water from a coal washery; (2) inherent moisture, sometimes called **bed moisture,** which is so closely held by the coal substance that it does not separate these two types of moisture. A coal may be air-dried at room temperature or somewhat above, thereby determining an "air-drying loss," but this result is not the extraneous moisture because part of the inherent moisture also vaporizes during the drying.

Mine samples taken at freshly exposed faces in the mine, which are free from visible surface moisture, give the best information as to inherent or bed moisture content. Such moisture content ranges in value from 2 to 4 percent for anthracite and for bituminous coals of the eastern Appalachian field, such as the Pocahontas, Sewell, Pittsburgh, Freeport, and Kittaning beds. In the western part of this field, especially in Ohio, the inherent moisture ranges from 4 to 10 percent. In the interior fields of Indiana, Illinois, western Kentucky, Iowa, and Missouri, the range is from 8 to 17 percent. In subbituminous coals the inherent moisture ranges from 15 to 30 percent, and in lignites from 30 to 45 percent. The total amount of moisture in commercial coal may be greater or less than that of the coal in the mine. Freshly mined subbituminous coal and lignite lose moisture rapidly when exposed to the air. The extraneous or surface moisture in coal is a function of the surface exposed, each surface being able to hold a film of moisture. Fine sizes hold more moisture than lump. Coal which in the mine does not contain more than 4 percent moisture may in finer sizes hold as much as 15 percent; the same coal in lump sizes, even after underwater storage, may contain little more moisture than originally in the mine.

In the standard method of analysis, the **volatile matter** is taken as the loss in weight, less moisture, obtained by heating the coal for 7 min in a covered crucible at about 950°C (1,742°F) under specified conditions. Volatile matter does not exist in coal as such but is produced by decomposition of the coal when heated. It consists chiefly of the combustible gases, hydrogen, carbon monoxide, methane and other hydrocarbons, tar vapors, volatile sulfur compounds, and some noncombustible gases such as carbon dioxide and water vapor. The composition of the volatile

Table 7.1.3 Principal Ranks of Coal Mined in Various States*

State	Anthracite	Semi-anthracite	Low-volatile bituminous	Medium-vol. bituminous	High-vol. A bituminous	High-vol. B bituminous	High-vol. C bituminous	Subbituminous A	Subbituminous B	Subbituminous C	Lignite
Alabama				x	x						
Alaska						x	x	x	x	x	x
Arkansas		x	x	x	x						x
Colorado					x	x	x	x	x	x	x
Illinois					x	x	x				
Indiana						x	x				
Iowa							x				
Kansas					x	x					
Kentucky											
Eastern					x	x					
Western					x	x	x				
Maryland			x	x	x						
Missouri							x				
Montana						x	x	x	x	x	x
New Mexico					x	x			x		
North Dakota											x
Ohio					x	x	x				
Oklahoma			x	x	x	x	x				
Pennsylvania	x	x	x	x	x						
South Dakota											x
Tennessee				x	x						
Texas								x	x		x
Utah					x	x	x				
Virginia		x	x	x	x						
Washington					x	x	x		x		x
West Virginia			x	x	x						
Wyoming						x	x	x	x	x	

* Compiled largely from *Typical Analyses of Coals of the United States*, *BuMines Bull.* 446, and *Coal Reserves of the United States*, *Geol. Survey Bull.* 1136.

matter varies greatly with different coals: the amount can vary with the rate of heating. The inert or noncombustible gas may range from 4 percent of the total volatile matter in low-volatile coals to 40 percent in subbituminous coals.

The standard method of determining the **fixed carbon** is to subtract from 100 the sum of the percentages of the moisture, volatile matter, and ash of the proximate analysis. It is the carbonaceous residue less ash remaining in the test crucible in the determination of the volatile matter. It does not represent the total carbon in the coal because a considerable part of the carbon is expelled as volatile matter in combination with hydrogen as hydrocarbons and with oxygen as carbon monoxide and carbon dioxide. It also is not pure carbon because it may contain several tenths percent of hydrogen and oxygen, 0.4 to 1.0 percent of nitrogen, and about half of the sulfur that was in the coal.

In the standard method, **ash** is the inorganic residue that remains after burning the coal in a muffle furnace to a final temperature of 700 to 750°C (1,292 to 1,382°F). It is composed largely of compounds of silicon, aluminum, iron, and calcium, with smaller quantities of compounds of magnesium, titanium, sodium, and potassium. The ash as determined is usually less than the inorganic mineral matter originally present in the coal. During incineration, various weight changes take place, such as loss of water of constitution of the silicate minerals, loss of carbon dioxide from carbonate minerals, oxidation of iron pyrites to iron oxide, and fixation of a part of the oxides of sulfur by bases such as calcium and magnesium.

The chemical composition of coal ash varies widely depending on the mineral constituents associated with the coal. Typical limits of ash composition of U.S. bituminous coals are as follows:

Constituent	Percent
Silica, SiO_2	20–40
Alumina, M_2O_3	10–35
Ferric oxide, Fe_2O_3	5–35
Calcium oxide, CaO	1–20
Magnesium oxide, MgO	0.3–4
Titanium dioxide, TiO_2	0.5–2.5
Alkalies, Na_2O and K_2O	1–4
Sulfur trioxide, SO_3	0.1–12

The ash of subbituminous coals may have more CaO, MgO, and SO_3 than the ash of bituminous coals; the trend may be even more pronounced for lignite ash.

Ultimate analysis expresses the composition of coal as sampled in percentages of carbon, hydrogen, nitrogen, sulfur, oxygen, and ash. The carbon includes that present in the organic coal substance as well as a minor amount that may be present as mineral carbonates. In ASTM practice, the hydrogen and oxygen values include those of the organic coal substance as well as those present in the form of moisture and the water of constitution of the silicate minerals. In certain other countries, the values for hydrogen and oxygen are corrected for the moisture in the coal and are reported separately. The ash is the same as reported in the proximate analysis; the sulfur, carbon, hydrogen, and nitrogen are determined chemically. Oxygen in coal is usually estimated by subtracting the sum of carbon, hydrogen, nitrogen, sulfur, and ash from 100. Many of the analyses discussed can be performed with modern instruments in a short time. ASTM methods are applied when referee data are required.

Sulfur occurs in three forms in coal: (1) **pyritic sulfur,** or sulfur combined with iron as pyrite or marcasite; (2) **organic sulfur,** or sulfur combined with coal substance as a heteroatom or as a bridge atom; (3) **sulfate sulfur,** or sulfur combined mainly with iron or calcium together with oxygen as iron sulfate or calcium sulfate. Pyrite and marcasite are recognized by their metallic luster and pale brass-yellow color, although some marcasite is almost white. Organic sulfur may comprise from about 20 to 85 percent of the total sulfur in the coal. Most freshly mined coal contains only very small quantities of sulfate sulfur; it increases in weathered coal. The total sulfur content of coal mined in the United States varies from about 0.4 to 5.5 percent by weight on a dry coal basis.

The **gross calorific value** of a fuel expressed in Btu/lb of fuel is the heat produced by complete combustion of a unit quantity, at constant volume, in an oxygen bomb calorimeter under standard conditions. It includes the latent heat of the water vapor in the products of combustion. Since the latent heat is not available for making steam in actual operation of boilers, a net caloric value is sometimes determined, although not in usual U.S. practice, by the following formula:

Net calorific value, Btu/lb = gross calorific value, Btu/lb
$$- (92.70 \times \text{total hydrogen, \% in coal})$$

The gross calorific value may also be approximated by Dulong's formula

$$\text{Btu/lb} = 14,544C + 62,028 \left(H - \frac{O}{8} \right) + 4,050S$$

$$(\text{Btu/lb} \times 2.328 = \text{kJ/kg})$$

where C, H, O, and S are weight fractions from the ultimate analysis. For anthracites, semianthracites, and bituminous coals, the calculated values are usually with 1½ percent of those determined by the bomb calorimeter. For subbituminous and lignitic coals, the calculated values show deviations often reaching 4 and 5 percent.

Because coal ash is a mixture of various components, it does not have a definite melting point; the gradual softening and fusion of the ash is not merely the successive melting of the various ash constituents but is a more complicated process in which reactions involving the formation of new and more fusible compounds take place.

The **fusibility of coal ash** is determined by heating a triangular pyramide (cone), ¾ in high and ¼ in wide at each side of the base, made up of the ash together with a small amount of organic binder. As the cone is heated, three temperatures are noted: (1) the initial deformation temperature (IDT), or the temperature at which the first rounding of the apex or the edges of the cone occurs; (2) the softening temperature (ST), or the temperature at which the cone has fused down to a spherical lump; and (3) the fluid temperature (FT), or the temperature at which the cone has spread out in a nearly flat layer. The softening interval is the degrees of temperature difference between (2) and (1), the flowing interval the difference between (3) and (2), and the fluidity range the difference between (3) and (1). Of the three, the softening temperature is most widely used. Table 7.1.4 shows ash-fusion data typical of some important U.S. coals.

Data on ash fusion characteristics are useful to the combustion engineer concerned with evaluation of the clinkering tendencies of coals used in combustion furnaces and with corrosion of metal surfaces in boilers due to slag deposits. The kinds of mineral matter occurring in different coals are not well related to rank or geographic location, al-

Table 7.1.4 Fusibility of Ash from Some Coals

Seam	Pocahontas no. 3	Ohio no. 9	Pittsburgh	Illinois	Utah	Wyoming	Texas
Type	Low volatile	High volatile	High volatile	High volatile	High volatile	Subbituminous	Lignite
Ash, %	12.3	14.1	10.9	17.4	6.6	6.6	12.8
Temperature, °C*							
Initial deformation	>1,600	1,325	1,240	1,260	1,160	1,200	1,190
Softening		1,430	1,305	1,330		1,215	1,200
Fluid		1,465	1,395	1,430	1,350	1,260	1,255

* In an oxidizing atmosphere.

though there is a tendency for midcontinent coals (Indiana to Oklahoma) to have low ash fusion temperatures. Significance is attached to all of the previously indicated fusion temperatures and the intervals between them. The IDT is sometimes identified with surface stickiness, the ST with plastic distortion or sluggish flow, and the FT with liquid mobility. Long fusion intervals often produce tough, dense, slags; short intervals favor porous, friable structures.

Most bituminous coals, when heated at uniformly increasing temperatures in the absence or partial absence of air, fuse and become plastic. These coals may be designated as either caking or coking in different degrees. **Caking** usually refers to the fusion process in a boiler furnace. **Coking coals** are those that make good coke, suitable for metallurgical purposes where the coke must withstand the burden of the ore and flux above it. Coals that are caking in a fuel bed do not necessarily make good coke in a coke oven. Subbituminous coal, lignite, and anthracite are noncaking.

The **free-swelling index** test measures the free-swelling properties of coal and gives an indication of the caking characteristics of the coal when burned on fuel beds. It is not intended to determine the expansion of coals in coke ovens. The test consists in heating 1 g of pulverized coal in a silica crucible over a gas flame under prescribed conditions to form a coke button, the size and shape of which are then compared with a series of standard profiles numbered 1 to 9 in increasing order of swelling.

The **specific gravity** of coal is the ratio of the weight of solid coal to the weight of an equal volume of water. It is useful in calculating the weight of solid coal as it occurs in the ground for estimating the tonnage of coal per acre of surface. An increase in ash-forming mineral matter increases the specific gravity; e.g., bituminous coals of Alabama, ranging from 2 to 15 percent ash and from 2 to 4.5 percent moisture, vary in specific gravity from 1.26 to 1.37.

Bulk density is the weight per cubic foot of broken coal. It varies according to the specific gravity of the coal, its size distribution, its moisture content, and the amount of orientation when piled. The range of weight from subbituminous coal to anthracite is from 44 to 59 lb/ft^3 when loosely piled; when piled in layers and compacted, the weight per cubic foot may increase as much as 25 percent. The weight of fuel in a pile can usually be determined to within 10 to 15 percent by measuring its volume. Typical weights of coal, as determined by shoveling it loosely into a box of 8 ft^3 capacity, are as follows: anthracite, 50 to 58 lb/ft^3; low- and medium-volatile bituminous coal, 49 to 57 lb/ft^3; high-volatile bituminous and subbituminous coal, 42 to 57 lb/ft^3.

The **grindability** of coal, or the ease with which it can be ground fine enough for use as a pulverized fuel, is a composite of several specific physical properties such as hardness, tensile strength, and fracture. A laboratory procedure adopted by ASTM (D409) for evaluating grindability, known as the **Hardgrove** machine method, uses a specially designed grinding apparatus to determine the relative grindability or ease of pulverizing coal in comparison with a standard coal, chosen as 100 grindability. Primarily, the ASTM Hardgrove grindability test is used for estimating how various coals affect the capacity of commercial pulverizers. A general relationship exists between grindability of coal and its rank. Coals that are easiest to grind (highest grindability index) are those of about 14 to 30 percent volatile matter on a dry, ash-free basis. Coals of either lower or higher volatile-matter content usually are more difficult to grind. The relationship of grindability and rank, however, is not sufficiently precise for grindability to be estimated from the chemical analysis, partly because of the variation in grindability of the various petrographic and mineral components. Grindability indexes of U.S. coals range from about 20 for an anthracite to 120 for a low-volatile bituminous coal.

Mining

Coal is mined by either underground or surface methods. In underground mining the coal beds are made accessible through shaft, drift, or slope entries (vertical, horizontal, or inclined, respectively), depending on location of the bed relative to the terrain.

The most widely used methods of coal mining in the United States are termed continuous and conventional mining. The former makes use of **continuous miners** which break the coal from the face and load it onto conveyors, shuttle cars, or railcars in one operation. Continuous miners are of ripping, boring, or milling types or hybrid combinations of these. In conventional mining the coal is usually broken from the face by means of blasting agents or by pressurized air or carbon dioxide devices. In preparation for breaking, the coal may be cut horizontally or vertically by cutting machines and holes drilled for charging explosives. The broken coal is then picked up by loaders and discharged to conveyors or cars. A method that is increasing in use is termed **long-wall mining**. It employs shearing or plowing machines to break coal from more extensive faces. Eighty long-wall mines are now in operation. Pillars to support the roof are not needed because the roof is caved under controlled conditions behind the working face. About half the coal presently mined underground is cut by machine and nearly all the mined coal is loaded mechanically.

An important requirement in all mining systems is **roof support**. When the roof rock consists of strong sandstone or limestone, relatively uncommon, little or no support may be required over large areas. Most mine roofs consist of shales and must be reinforced. Permanent supports may consist of arches, crossbars and legs, or single posts made of steel or wood. Screw or hydraulic jacks, with or without crossbars, often serve as temporary supports. Long roof bolts, driven into the roof and anchored in sound strata above, are used widely for support, permitting freedom of movement for machines. Drilling and insertion of bolts is done by continuous miners or separate drilling machines. **Ventilation** is another necessary factor in underground mining to provide a proper atmosphere for personnel and to dilute or remove dangerous concentrations of methane and coal dust. The ventilation system must be well-designed so that adequate air is supplied across the working faces without stirring up more dust.

When coal occurs near the surface, **strip or open-pit mining** is often more economical than underground mining. This is especially true in states west of the Mississippi River where coal seams are many feet thick and relatively low in sulfur. The proportion of coal production from surface mining has been increasing rapidly and now amounts to over 60 percent.

In preparation for surface mining, core drilling is conducted to survey the underlying coal seams, usually with dry-type rotary drills. The overburden must then be removed. It is first loosened by ripping or drilling and blasting. Ripping can be accomplished by bulldozers or scrapers. Overburden and coal are then removed by shovels, draglines, bulldozers, or wheel excavators. The first two may have bucket capacities of 200 yd^3 (153 m^3). Draglines are most useful for very thick cover or long dumping ranges. Hauling of stripped coal is usually done by trucks or tractor-trailers with capacities up to 240 short tons [218 metric tons (t)]. Reclamation of stripped coal land is becoming increasingly necessary. This involves returning the land to near its original contour, replanting with ground cover or trees, and sometimes providing water basins and arable land.

Preparation

About half the coal presently mined in the United States is cleaned mechanically to remove impurities and supply a marketable product. Mechanical mining has increased the proportion of fine coal and noncoal minerals in the product. At the preparation plant run-of-mine coal is usually given a preliminary size reduction with roll crushers or rotary breakers. Large or heavy impurities are then removed by hand picking or screening. Tramp iron is usually removed by magnets. Before washing, the coal may be given a preliminary size fractionation by screening. Nearly all preparation practices are based on density differences between coal and its associated impurities. **Heavy-medium** separators using magnetite or sand suspensions in water come closest to ideal gravity separation conditions. **Mechanical devices** include jigs, classifiers, washing tables, cyclones, and centrifuges. Fine coal, less than ¼ in (6.3 mm) is usually treated separately, and may be cleaned by froth flotation. Dewatering of the washed and sized coal may be accomplished by screening, centrifuging, or filtering, and finally, the fine coal may be heated to complete the drying. Before shipment the coal may be dust-proofed and freezeproofed with oil or salt.

Removal of sulfur from coal is an important aspect of preparation because of the role of sulfur dioxide in air pollution. Pyrite, the main inorganic sulfur mineral, is partly removed along with other minerals in conventional cleaning. Processes to improve pyrite removal are being developed. These include magnetic and electrostatic separation, chemical leaching, and specialized froth flotation.

Storage

Coal may heat spontaneously, with the likelihood of self-heating greatest among coals of lowest rank. The heating begins when freshly broken coal is exposed to air. The process accelerates with increase in temperature, and active burning will result if the heat from oxidation is not dissipated. The finer sizes of coal, having more surface area per unit weight than the larger sizes, are more susceptible to spontaneous heating.

The **prevention of spontaneous heating** in storage poses a problem of minimizing oxidation and of dissipating any heat produced. Air may carry away heat, but it also brings oxygen to create more heat. Spontaneous heating can be prevented or lessened by (1) storing coal underwater; (2) compressing the pile in layers, as with a road roller, to retard access of air; (3) storing large-size coal; (4) preventing any segregation of sizes in the pile; (5) storing in small piles; (6) keeping the storage pile as low as possible (6 ft is the limit for many coals); (7) keeping storage away from any external sources of heat; (8) avoiding any draft of air through the coal; (9) using older portions of the storage first and avoiding accumulations of old coal in corners. It is desirable to watch the temperature of the pile. A thermometer inserted in an iron tube driven into the coal pile will reveal the temperature. When the coal reaches a temperature of 50°C (120°F), it should be moved. Using water to put out a fire, although effective for the moment, may only delay the necessity of moving the coal. Furthermore, this may be dangerous because steam and coal can react at high temperatures to form carbon monoxide and hydrogen.

Sampling

Because coal is a heterogeneous material, collection and handling of samples that adequately represent the bulk lot of coal are required if the analytical and test data are to be meaningful. Coal is best sampled when in motion, as it is being loaded or unloaded from belt conveyors or other coal-handling equipment, by collecting increments of uniform weight evenly distributed over the entire lot. Each increment should be sufficiently large and so taken to represent properly the various sizes of the coal.

Two procedures are recognized: (1) commercial sampling and (2) special-purpose sampling, such as classification by rank or performance. The **commercial sampling** procedure is intended for an accuracy such that, if a large number of samples were taken from a large lot of coal, the test results in 95 out of 100 cases would fall within ± 10 percent of the ash content of these samples. For commercial sampling of lots up to 1,000 tons, it is recommended that one gross sample represent the lot taken. For lots over 1,000 tons, the following alternatives may be used: (1) Separate gross samples may be taken for each 1,000 tons of coal or fraction thereof, and a weighted average of the analytical determinations of these prepared samples may be used to represent the lot. (2) Separate gross samples may be taken for each 1,000 tons or fraction thereof, and the − 20 or − 60 mesh samples taken from the gross samples may be mixed together in proportion to the tonnage represented by each sample and one analysis carried out on the composite sample. (3) One gross sample may be used to represent the lot, provided that at least four times the usual minimum number of increments are taken. In **special-purpose sampling**, the increment requirements used in the commercial sampling procedure are increased according to prescribed rules.

Specifications

Specifications for the purchase of coal vary widely depending on the intended use, whether for coke or a particular type of combustion unit, and whether it meets the standards imposed by customers abroad. Attempts at international standardization have met with only limited success. Most of the previously discussed factors in this section must be taken into consideration, for example, the sulfur content and swelling properties of a coal used for metallurgical coke production, the heating value of a coal used for steam generation, and the slagging properties of its ash formed after combustion.

Statistics

Coal production in 1994 was 1.03 billion short tons, 0.5 percent anthracite, 93 percent bituminous, and 6.5 percent lignite. The industry employed about 228,000 miners, 150,000 of them underground. Over 6000 coal mines are in operation in 26 states but over half the production comes from Kentucky, West Virginia, Wyoming, and Pennsylvania. Of the total production, 62 percent is from surface mines. **Transportation** to the point of consumption is primarily by rail (67 percent) followed by barges (11 percent) and trucks (10 percent); about 1 percent moves through pipelines. About 11 percent of the coal consumed is burned at mine-mouth power plants, for it is cheaper and easier to transport electric power than bulk coal.

Several long-distance coal-slurry pipelines are proposed but only one, 273 mi (439 km) long, is in commercial use. The Black Mesa pipeline runs from a coal mine near Kayenta, AZ to the Mohave Power Plant in Nevada. Nominal capacity is 4.8 million short tons (4.35 t) per year, but it usually operates at 3 to 4 million tons per year. The coal concentration is about 47 percent and the particle size distribution is controlled carefully. Other pipelines will be constructed where feasible and when the problems of eminent domain are resolved.

Overall energy statistics for the United States show that coal accounts for about 30 percent of the total energy production, with the balance coming from petroleum and natural gas. Below is a distribution of 1 billion tons of coal produced:

Electric power	87.2 percent
Industrial	8.4 percent
Coke	3.8 percent
Commercial and residential	0.6 percent

In 1994 over 100 million tons was exported. The industrial use is primarily for power in the production of food, cement, paper, chemicals, and ceramics.

Reserves of coal in the United States are ample for several hundred years even allowing for the increased production of electric power and the synthesis of fuels and chemicals. The total reserve base is estimated to be almost 4 million tons, of which 1.7 trillion tons is identified resources.

BIOMASS FUELS
by Martin D. Schlesinger
Wallingford Group Ltd.

REFERENCES: Peat, "U.S. Bureau of Mines Mineral Commodities Summaries." Fryling, "Combustion Engineering," Combustion Engineering, Inc., New York. "Standard Classification of Peats, Mosses, Humus and Related Products," ASTM D2607. Lowry, "Chemistry of Coal Utilization," Wiley. United Nations Industrial Development Organization publications.

Biomass conversion to energy continues to be a subject of intensive study for both developed and less developed countries. In the United States, combustion of biomass contributes only a few percent of the total U.S. energy supply, and it is mostly in the form of agricultural wastes and paper. Almost any plant material can be the raw material for gasification from which a variety of products can be created catalytically from the carbon monoxide, carbon dioxide, and hydrogen. Typical commercial products are methanol, ethanol, methyl acetate, acetic anhydride, and hydrocarbons. Alcohols are of particular interest as fuels for transportation and power generation. Producer gas has been introduced into small compression ignition engines, and the diesel oil feed has been reduced. Technical and environmental problems are still not solved in the large-scale use of biomass.

Plants and vegetables are another source of biomass-derived oils. Fatty acid esters have been used in diesel engines alone and in blends with diesel oil, and although the esters are effective, some redesign and changes in operation are required. In developing countries where fossil fuels are costly, deforestation has led to land erosion and its consequences.

Useful biomass materials include most of the precoal organic vegetation such as peat, wood, and food processing wastes like bagasse from sugar production. Digestable food processing wastes can be converted to biogas.

Peat, an early stage in the metamorphosis of vegetable matter into coal, is the product of partial decomposition and disintegration of plant remains in water bogs, swamps or marshlands, and in the absence of air. Like all material of vegetable origin, peat is a complex mixture of carbon, hydrogen, and oxygen in a ratio similar to that of cellulose and lignin. Generally peat is low in essential growing elements (K, S, Na, P) and ash. Trace elements, when found in the peat bed, are usually introduced by leaching from adjacent strata. The Federal Trade Commission specifies that to be so labeled, peat must contain at least 75 percent peat with the rest composed of normally related soil materials. The water content of undrained peat in a bog is 92 to 95 percent but it is reduced to 10 to 50 percent when peat is used as a fuel. Peat is harvested by large earth-moving equipment from a drained bog dried by exposure to wind and sun. The most popular method used in Ireland and the Soviet Union involves harrowing of drum-cut peat and allowing it to field dry before being picked up mechanically or pneumatically.

The **chemical and physical properties of peat** vary considerably, depending on the source and the method of processing. Typical ranges are:

Processed peat	Air-Dried	Mulled	Briquettes
Moisture, wt %	25–50	50–55	10–12
Density, lb/ft³	15–25		30–60
Caloric value, Btu/lb	6,200	3,700–5,300	8,000

Proximate analyses of samples, calculated back to a dry basis, follow a similar broad pattern: 55 to 70 percent volatile matter, 30 to 40 percent fixed carbon and 2 to 10 percent ash. The dry, ash-free ultimate analysis ranges from 53 to 63 percent carbon, 5.5. to 7 percent hydrogen, 30 to 40 percent oxygen, 0.3 to 0.5 percent sulfur, and 1.2 to 1.5 percent nitrogen.

World production of peat in 1993 was about 150 million tons, mostly from the former Soviet states. Other significant producers are Ireland, Finland, and Germany. Annual U.S. imports are primarily from Canada, about 650,000 tons per annum. Several states produce peat, Florida and Michigan being the larger producers. Over a 10-year period, U.S. production declined from 730,000 to about 620,000 short tons (590,000 t) per year. The value of production was $16 million from 67 operations.

By type, peat was about 66 percent reed sedge with the balance distributed between humus, sphagnum, and hypnum. The main uses for peat in the United States are soil improvement, mulch, filler for fertilizers, and litter for domestic animals.

Wood, when used as a fuel, is often a by-product of the sawmill or papermaking industries. The conversion of logs to lumber results in 50 percent waste in the form of bark, shavings, and sawdust. Fresh timber contains 30 to 50 percent moisture, mostly in the cell structure of the wood, and after air drying for a year, the moisture content reduces to 18 to 25 percent. Kiln-dried wood contains about 8 percent moisture. A typical analysis range is given in Table 7.1.5. When additional fuel is required, supplemental firing of coal, oil, or gas is used.

Combustion systems for wood are generally designed specially for the material or mixture of fuels to be burned. When the moisture content is high, 70 to 80 percent, the wood must be mixed with low moisture fuel so that enough energy enters the boiler to support combustion. Dry wood may have a heating value of 8,750 Btu/lb but at 80 percent moisture a pound of wet wood has a heating value of only 1,750 Btu/lb. The heat required just to heat the fuel and evaporate the water is over 900

Table 7.1.5   Typical Analysis of Dry Wood Fuels

	Most woods, range
Proximate analysis, %:	
Volatile matter	74–82
Fixed carbon	17–23
Ash	0.5–2.2
Ultimate analysis, %:*	
Carbon	49.6–53.1
Hydrogen	5.8–6.7
Oxygen	39.8–43.8
Heating value	
Btu/lb	8,560–9,130
kJ/kg	19,900–21,250
Ash-fusion temperature, °F:	
Initial	2,650–2,760
Fluid	2,730–2,830

* Typically, wood contains no sulfur and about 0.1 percent nitrogen. Cellulose = 44.5 percent C, 6.2 percent H, 49.3 percent O.

Btu, and combustion may not occur. Table 7.1.6 shows the moisture-energy relationship. The usual practice when burning wood is to propel the wood particles into the furnace through injectors along with preheated air with the purpose of inducing high turbulence to the boiler. Furthermore, the wood is injected high enough in the combustion chamber so that it is dried, and all but the largest particles are burned before they reach the grate at the bottom of the furnace. Spreader stokers and cyclone burners work well.

Table 7.1.6   Available Energy in Wood

Moisture, %	Heating value, Btu/lb	Wt water/wt wood
0	8,750	0
20	7,000	0.25
50	4,375	1.00
80	1,750	4.00

Wood for processing or burning is usually sold by the cord, an ordered pile 8 ft long, 4 ft high, and 4 ft wide or 128 ft³ (3.625 m³). Its actual solid content is only about 70 percent, or 90 ft³. Other measures for wood are the cord run, which is measured only by the 8-ft length and 4-ft height; the width may vary. Sixteen-inch-long wood is called **stovewood** or **blockwood.**

Small wood-burning power plants and home heating became popular, but in some areas there was an adverse environmental impact under adverse weather conditions.

Wood charcoal is made by heating wood to a high temperature in the absence of air. Wood loses up to 75 percent of its weight and 50 percent of its volume owing to the elimination of moisture and volatile matter. As a result, charcoal has a higher heating value per cubic foot than the original wood, especially if the final product is compacted in the form of briquettes. Charcoal is marketed in the form of lumps, powder, or briquettes and finds some use as a fuel for curing, restaurant cooking, and a picnic fuel. Its nonfuel uses, particularly in the chemical industry, are as an adsorption medium for purifying gas and liquid streams and as a decolorizing agent.

In addition to peat and wood, several lesser-known fuels are in common use for the generation of industrial steam and power. Aside from their value as a fuel, the burning of wastes minimizes a troublesome disposal problem that could have serious environmental impact. Nearly all these waste fuels are cellulosic in character, and the heating value is a function of the carbon content. Ash content is generally low, but much moisture could be present from processing, handling, and storage. On a moisture- and ash-free basis the heating values can be estimated at 8,000 Btu/lb; more resinous materials about 9,000 Btu/lb. Table 7.1.7 is a list of some typical by-product solid fuels.

Table 7.1.7 By-product Fuels

	Heating value, Btu/lb (dry)	Moisture, % as received	Ash, % moisture-free
Black liquor (sulfate)	6,500	35	40–45
Cattle manure	7,400	50–75	17
Coffee grounds	10,000	65	1.5
Corncobs	9,300	10	1.5
Cottonseed cake	9,500	10	8
Municipal refuse	9,500	43	8
Pine bark	9,500	40–50	5–10
Rice straw or hulls	6,000	7	15
Scrap tires	16,400	0.5	6
Wheat straw	8,500	10	4

Bagasse is the fibrous material left after pressing the juice from sugarcane or harvesting the seeds from sunflowers. The chopped waste usually contains about 50 percent moisture and is burned in much the same manner as wood waste. Spreader stokers or cyclone burners are used. Supplemental fuel is added sometimes to maintain steady combustion and to provide energy for the elimination of moisture. Bagasse can usually supply all the fuel requirements of raw sugar mills. A typical analysis of dry bagasse from Puerto Rico is 44.47 percent C, 6.3 percent H, 49.7 percent O, and 1.4 percent ash. Its heating value is 8,390 Btu/lb.

Furnaces have been developed to burn particular wastes, and some preferences emerge by virtue of particular operating characteristics. Spreader stokers are preferred for wood waste and bagasse. Tangential firing seems to be used for coffee grounds, rice hulls, some wood waste, and chars from coal or lignite. Traveling-grate stokers are used for industrial wastes and coke breeze.

Biogas is readily produced by the anaerobic digestion of wastes. The process is cost-effective in areas remote from natural gas lines. In Asia, for example, there are millions of family biodigesters with a capacity of 8 to 10 m³. Larger-capacity systems of about 2,000 m³ are installed where industrial biodegradable wastes are generated as in communes, feed lots, wineries, food processors, etc. Not only is the gas useful, but also the sludge is a good fertilizer. Remaining parasite eggs and bacteria are destroyed by lime or ammonia treatment. Harvest increases of 10 to 35 percent are reported for rice and corn.

Biogas contains an average of 62 percent methane and 36 percent carbon dioxide; it also has a small amount of nitrogen and hydrogen sulfide. Raw gas heating value is about 600 Btu/ft³ (5,340 cal/m³). The raw gas will burn in an engine, but corrosion can occur unless proper materials of construction are selected. Gas from a garbage site in California is treated to remove acid gases, and the methane is sold to a pipeline system. Most sites, however, do not produce enough gas to use economically and merely flare the collected gas.

Other methods have been demonstrated for converting biomass of variable composition to an energy source of relatively consistent composition. One procedure is to process bulk volume wastes with pressurized carbon monoxide. A yield of about 2 barrels of oil per ton of dry feed is obtained, with a heating value of about 15,000 Btu/lb. Heavy fuel oil from petroleum has a heating value of 18,000 Btu/lb. Pyrolysis is also possible at temperatures up to 1,000°C in the absence of air. The gas produced has a heating value of 400 to 500 Btu/ft³ (about 4,000 kcal/m³). The oil formed has a heating value about 10,500 Btu/lb (24,400 J/g).

Refuse-derived fuels (RDFs) usually refer to waste material that has been converted to a fuel of consistent composition for commercial application. Although several sources exist, the problem lies in gathering the raw material, processing it into a usable form and composition, and delivering it to the point of combustion. Almost any carbonaceous material and a suitable binder can be converted to a form that can be fired into a pulverized fuel or a stoker boiler. Each application should be evaluated to eliminate carryover of low-density material, such as loose paper, and where in the boiler that combustion takes place.

Some installations are cofired with fossil fuels. Most of the by-product fuels in Table 7.1.7 might be used alone or with a binder. An ideal situation would be nearby waste streams from a major production facility. After mixing and extrusion, the pellets would be storable and transported only a short distance. If the pellet density is close to the normal boiler feed, the problems of separation during transportation and firing are reduced or eliminated. The problem of hazardous emissions remains if the waste streams are contaminated; if they are clean-burning, hazardous emissions might reduce the apparent cleanliness of the effluent. (See Sec. 7.4.).

PETROLEUM AND OTHER LIQUID FUELS
by James G. Speight

Western Research Institute

REFERENCES: ASTM, "Annual Book of ASTM Standards," 1993, vols. 05.01, 05.02, 05.03, and 05.04. Bland and Davidson, "Petroleum Processing Handbook," McGraw-Hill, New York. Gary and Handwerk, "Petroleum Refining: Technology and Economics," Marcel Dekker, New York. Hobson and Pohl, "Modern Petroleum Technology," Applied Science Publishers, Barking, England. Mushrush and Speight, "Petroleum Products: Instability and Incompatibility," Taylor & Francis, Washington. Speight, "The Chemistry and Technology of Petroleum," 2d ed., Marcel Dekker, New York.

Petroleum and Petroleum Products

Petroleum accumulates over geological time in porous underground rock formations called reservoirs, where it has been trapped by overlying and adjacent impermeable rock. Oil reservoirs sometimes exist with an overlying gas "cap" in communication with aquifers or with both. The oil resides together with water, and sometimes free gas, in very small holes (pore spaces) and fractures. The size, shape, and degree of interconnection of the pores vary considerably from place to place in an individual reservoir. The anatomy of a reservoir is complex, both microscopically and macroscopically. Because of the various types of accumulations and the existence of wide ranges of both rock and fluid properties, reservoirs respond differently and must be treated individually.

Petroleum occurs throughout the world, and commercial fields have been located on every continent. Reservoir depths vary, but most reservoirs are several thousand feet deep, and the oil is produced through wells that are drilled to penetrate the oil-bearing formations.

Petroleum is an extremely complex mixture and consists predominantly of hydrocarbons as well as compounds containing nitrogen, oxygen, and sulfur. Most petroleums also contain minor amounts of nickel and vanadium.

Petroleum may be qualitatively described as brownish green to black liquids of specific gravity from about 0.810 to 0.985 and having a boiling range from about 20°C (68°F) to above 350°C (660°F), above which active decomposition ensues when distillation is attempted. The oils contain from 0 to 35 percent or more of components boiling in the gasoline range, as well as varying proportions of kerosene hydrocarbons and higher-boiling-point constituents up to the viscous and nonvolatile compounds present in lubricants and the asphalts. The composition of the petroleum obtained from the well is variable and depends on both the original composition of the petroleum in situ and the manner of production and stage reached in the life of the well or reservoir.

The chemical and physical properties of petroleum vary considerably because of the variations in composition. The specific gravity of petroleum ranges from 0.8 (45.3 degrees API) for the lighter crude oils to over 1 (< 10 degrees API) for the near-solid bitumens which are found in many tar (oil) sand deposits. There is also considerable variation in viscosity; lighter crude oils range from 2 to 100 cSt, bitumens have viscosities in excess of 50,000 cSt.

The ultimate analysis (elemental composition) of petroleum is not reported to the same extent as it is for coal since there is a tendency for the ultimate composition of petroleum to vary over narrower limits— carbon: 83.0 to 87.0 percent; hydrogen: 10.0 to 14.0 percent; nitrogen: 0.1 to 1.5 percent; oxygen: 0.1 to 1.5 percent; sulfur: 0.1 to 5.0 percent; metals (nickel plus vanadium): 10 to 500 ppm. The **heat content of petro-**

Table 7.1.8 Analyses and Heat Values of Petroleum and Petroleum Products

Product	Gravity, deg API	Specific gravity at 60°F	Wt lb/gal	High-heat value, Btu/lb*	Ultimate analysis, % C	H	S	N	O
California crude	22.8	0.917	7.636	18,910	84.00	12.70	0.75	1.70	1.20
Kansas crude	22.1	0.921	7.670	19,130	84.15	13.00	1.90	0.45	
Oklahoma crude (east)	31.3	0.869	7.236	19,502	85.70	13.11	0.40	0.30	
Oklahoma crude (west)	31.0	0.871	7.253	19,486	85.00	12.90	0.76		
Pennsylvania crude	42.6	0.813	6.769	19,505	86.06	13.88	0.06	0.00	0.00
Texas crude	30.2	0.875	7.286	19,460	85.05	12.30	1.75	0.70	0.00
Wyoming crude	31.5	0.868	7.228	19,510					
Mexican crude	13.6	0.975	8.120	18,755	83.70	10.20	4.15		
Gasoline	67.0	0.713	5.935	......	84.3	15.7			
Gasoline	60.0	0.739	6.152	20,750	84.90	14.76	0.08		
Gasoline-benzene blend	46.3	0.796	6.627	......	88.3	11.7			
Kerosine	41.3	0.819	6.819	19,810					
Gas oil	32.5	0.863	7.186	19,200					
Fuel oil (Mex.)	11.9	0.987	8.220	18,510	84.02	10.06	4.93		
Fuel oil (midcontinent)	27.1	0.892	7.428	19,376	85.62	11.98	0.35	0.50	0.60
Fuel oil (Calif.)	16.7	0.9554	7.956	18,835	84.67	12.36	1.16		

* Btu/lb $\times$ 2.328 = kJ/kg.

leum generally varies from 18,000 to 20,000 Btu/lb while the heat content of petroleum products may exceed 20,000 Btu/lb (Table 7.1.8).

Refining Crude Oil Crude oils are seldom used as fuel because they are more valuable when refined to petroleum products. Distillation separates the crude oil into fractions equivalent in boiling range to gasoline, kerosene, gas oil, lubricating oil, and residual. Thermal or catalytic cracking is used to convert kerosene, gas oil, or residual to gasoline, lower-boiling fractions, and a residual coke. Catalytic reforming, isomerization, alkylation, polymerization, hydrogenation, and combinations of these catalytic processes are used to upgrade the various refinery intermediates into improved gasoline stocks or distillates. The major finished products are usually blends of a number of stocks, plus additives. Distillation curves for these products are shown in Fig. 7.1.2.

Physical Properties of Petroleum Products Petroleum products are sold in the United States by barrels of 42 gal corrected to 60°F, Table 7.1.9. Their *specific gravity* is expressed on an arbitrary scale termed degrees API.

The **high heat value** (hhv) of petroleum products is determined by combustion in a bomb with oxygen under pressure (ASTM D240). It may also be calculated, in products free from impurities, but the formula

$$Q_v = 22,320 - 3,780d^2$$

in which Q_v is the hhv at constant volume in Btu/lb and d is the specific gravity at 60/60°F.

The low heat value at constant pressure Q_p may be calculated by the relation

$$Q_p = Q_v - 90.8H$$

where H is the weight percentage of hydrogen and can be obtained from the relation

$$H = 26 - 15d$$

Typical heats of combustion of petroleum oils free from water, ash, and sulfur vary (within an estimated accuracy of 1 percent) with the API gravity (i.e., with the "heaviness" or "lightness" of the material) (Table 7.1.10). The heat value should be corrected when the oil contains sulfur by using an hhv of 4,050 Btu/lb for sulfur (ASTM D1405).

The **specific heat** c of petroleum products of specific gravity d and at temperature t (°F) is given by the equation

$$c = (0.388 + 0.00045t)/\sqrt{d}$$

The **heat of vaporation L** (Btu/lb) may be calculated from the equation

$$L = (110.9 - 0.09t)/d$$

The heat of vaporization per gallon (measured at 60°F) is

$$8.34Ld = 925 - 0.75t$$

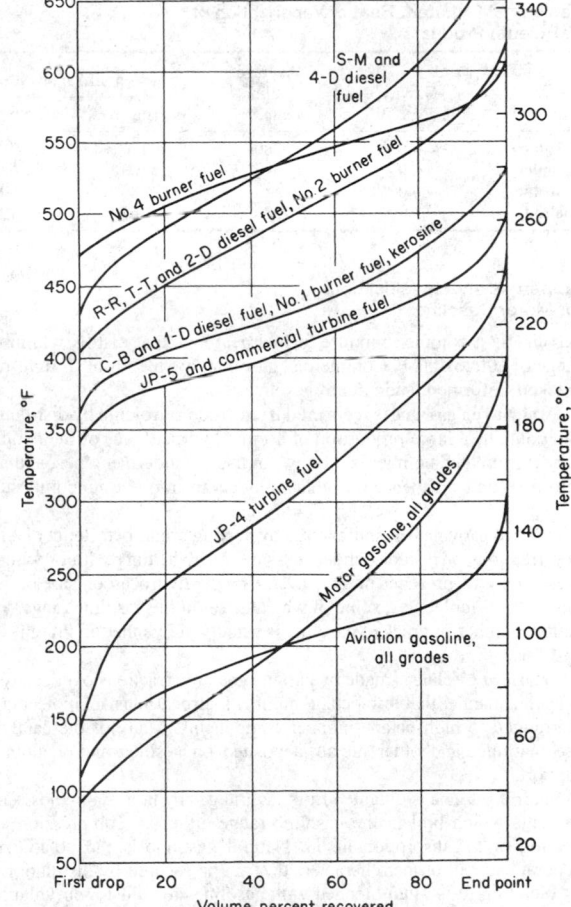

Fig. 7.1.2 Typical distillation curves.

indicating that the heat of vaporization per gallon depends only on the temperature of vaporization t and varies over the range of 450 for the heavier products to 715 for gasoline (Table 7.1.11). These data have an estimated accuracy within 10 percent when the vaporization is at constant temperature and at pressures below 50 lb/in² without chemical change.

Table 7.1.9 Coefficients of Expansion of Petroleum Products at 60°F

Deg API	< 15	15–34.9	35–50.9	51–63.9	64–78.9	79–88.9	89–93.9	94–100
Coef of expansion	0.00035	0.0004	0.0005	0.0006	0.0007	0.0008	0.00085	0.0009

Table 7.1.10 Heat Content of Different Petroleums and Petroleum Products

Deg API at 60°F	Density, lb/gal*	High heat value at constant volume Q_v, Btu		Low heat value at constant pressure Q_p, Btu	
		Per lb	Per gal	Per lb	Per gal
10	8.337	18,540	154,600	17,540	146,200
20	7.787	19,020	148,100	17,930	139,600
30	7.305	19,420	141,800	18,250	133,300
40	6.879	19,750	135,800	18,510	127,300
50	6.500	20,020	130,100	18,720	121,700
60	6.160	20,260	124,800	18,900	116,400
70	5.855	20,460	119,800	19,020	112,500
80	5.578	20,630	115,100	19,180	107,000

* Btu/lb × 2.328 = kJ/kg; Btu/gal × 279 = kJ/m³.

Table 7.1.11 Latent Heat of Vaporization of Petroleum Products

Product	Gravity, deg API	Average boiling temp, °F	Heat of vaporization	
			Btu/lb	Btu/gal
Gasoline	60	280	116	715
Naphtha	50	340	103	670
Kerosine	40	440	86	595
Fuel oil	30	580	67	490

Properties and Specifications for Motor Gasoline

Gasoline is a complex mixture of hydrocarbons that distills within the range of 100 to 400°F. Commercial gasolines are blends of straight-run, cracked, reformed, and natural gasolines.

Straight-run gasoline is recovered from crude petroleum by distillation and contains a large proportion of normal hydrocarbons of the paraffin series. Its octane number is too low for use in modern engines, and it is reformed and blended with other products to improve its combustion properties.

Cracked gasoline is manufactured by heating crude-petroleum distillation fractions or residues under pressure, or by heating with or without pressure in the presence of a catalyst. Heavier hydrocarbons are broken into smaller molecules, some of which distill in the gasoline range. The octane number of cracked gasoline is usually above that of straight-run gasoline.

Reformed gasoline is made by passing gasoline fractions over catalysts in such a manner that low-octane-number hydrocarbons are molecularly rearranged to high-octane-number components. Many of the catalysts use platinum and other metals deposited on a silica and/or alumina support.

Natural gasoline is obtained from natural gas by liquefying those constituents which boil in the gasoline range either by compression and cooling or by absorption in oil. Natural gasoline is too volatile for general use, but proper characteristics can be secured by distillation or by blending. It is often blended with gasolines to adjust their volatility characteristics to meet climatic conditions.

Catalytic hydrogenation is used extensively to upgrade gasoline and cracking stocks for blending or further refining. Hydrogenation improves octane number, removes sulfur and nitrogen, and increases storage stability.

The specifications for gasoline (ASTM D439) provide for various volatility classes, varying from low-volatility gasolines to minimize vapor lock to high-volatility gasoline that permits easier starting during cold weather.

Antiknock characteristics of gasolines are very important because engine power output and fuel economy are limited by the antiknock characteristics of the fuel. The antiknock index is currently defined as the average of the research octane number (RON) and motor octane number (MON). The **research octane number** is a measure of antiknock performance under mild operating conditions at low to medium engine speeds. The **motor octane number** is indicative of antiknock performance under more severe conditions, such as those encountered during power acceleration at relatively high engine speeds.

Reformulated motor gasoline is believed to be the answer to many environmental issues that arise from the use of automobiles. In fact, there has been a serious effort to produce reformulated gasoline components from a variety of processes (Table 7.1.12). In addition, methyl-*t*-butyl ether (MTBE), an additive to maintain the octane ratings of gasoline in the absence of added lead, is claimed to reduce (through more efficient combustion of the hydrocarbons) the emissions of unburned hydrocarbons during gasoline use. However, the ether (MTBE) is believed to have an adverse effect insofar as it appears that aldehyde emissions may be increased.

Table 7.1.12 Production of Reformulated Gasoline Constituents

Process	Objective
Catalytic reformer prefraction	Reduce benzene
Reformate fractionation	Reduce benzene
Isomerization	Increase octane
Aromatics saturation	Reduce total aromatics
Catalytic reforming	Oxygenate for octane enhancement
MTBE synthesis	Provide oxygenates
Isobutane dehydrogenation	Feedstock for oxygenate synthesis
Catalytic cracker naphtha fractionation	Increase alkylate
	Increase oxygenates
	Reduce olefins and sulfur
Feedstock hydrotreating	Reduce sulfur

Diesel Fuel

Diesel is a liquid product distilling over the range of 150 to 400°C (300 to 750°F). The carbon number ranges on average from about C_{13} to about C_{21}.

The chemical composition of a typical diesel fuel and how it applies to the individual specifications—API gravity, distillation range, freezing point, and flash point—are directly attributable to both the carbon number and the compound classes present in the finished fuel (Tables 7.1.13 and 7.1.14).

Aviation Gasolines Gasolines for aircraft piston engines have a

Table 7.1.13 General ASTM Specifications for Various Types of Diesel Fuels

Specification	Military*	No. 1-D†	No. 2-D‡	No. 1§	No. 2¶
API gravity, deg	40	34.4	40.1	35	30
Total sulfur, percent	0.5	0.5	0.5	0.5	0.5
Boiling point, °C	357	288	338	288	338
Flash point, °C	60	38	52	38	38
Pour point, °C	−6	−18	−6	−18	−6
Hydrogen, wt %	12.5	—	—		
Cetane number	43	40	40	—	—
Acid number	0.3	0.3	0.3	—	—

* MIL-F-16884J (1993) is also NATO F-76.
† High-speed, high-load engines.
‡ Low-speed, high-load engines.
§ Special-purpose burners.
¶ General-purpose heating fuel oil.

narrower boiling range than motor gasolines; i.e., they have fewer low-boiling and fewer high-boiling components. The three grades of aviation gasolines are indicated in Table 7.1.15. Specifications applicable to all three grades of military gasoline are given in Table 7.1.16.

Table 7.1.14   ASTM Methods for Determining Fuel Properties (see also Table 7.1.13)

Specification	ASTM method
API gravity	D1298
Total sulfur, percent	D129
Boiling point, °C	D86
Flash point, °C	D93
Pour point, °C	D97
Hydrogen, wt %	D3701
Cetane number	D613, D976
Acid number	D974

Jet or **aviation turbine fuels** are not limited by antiknock requirements, and they have wider boiling-point ranges to provide greater availability (ASTM D1655). Fuel JP-4 is a relatively wide-boiling-point range distillate that encompasses the boiling point range of gasoline and kerosene. The average initial boiling point is about 140°F, and the endpoint is about 455°F. Fuel JP-5 is a high-flash point kerosine type of fuel with an initial boiling point of 346°F and endpoint of 490°F.

Table 7.1.15   Grades of Aviation Gasoline (ASTM D910)

Grade	Color	Tetraethyl lead content mL/gal, max
80	Red	0.6
100	Green	4.0
100 LL	Blue	2.0

There are a variety of grades (Table 7.1.15) and specifications (Table 7.1.16) for jet fuel because of its use as a commercial and a military fuel (Table 7.1.17). The chemical composition of each jet fuel type, API gravity, distillation range, freezing point, and flash point are directly attributable to both the carbon number and the compound classes present in the finished fuel.

Kerosine is less volatile than gasoline and has a higher flash point, to provide greater safety in handling. Other quality tests are specific gravity, color, odor, distillation range, sulfur content, and burning quality. Most kerosine is used for heating, ranges, and illumination, so it is treated with sulfuric acid to reduce the content of aromatics, which burn

Table 7.1.17   Commercial and Military Specifications for Jet Fuels

Fuel types	Specification
Jet-A	ASTM D1655
JP-4	Mil-T-5624
JP-5	Mil-T-5624
JP-8	Mil-T-83133
JP-10	Mil-P-87107

with a smoky flame. Specification tests for quality control include flash point (minimum 115°F), distillation endpoint (maximum 572°F), sulfur (maximum 0.13 percent), and color (minimum + 16) (ASTM D187).

Diesel fuel for diesel engines requires variability in its performance since the engines range from small, high-speed engines used in trucks and buses to large, low-speed stationary engines for power plants. Thus several grades of diesel fuel are needed (Table 7.1.18) (ASTM D975) for different classes of service:

Grade 1-D: A volatile distillate fuel for engines in service requiring frequent speed and load changes

Grade 2-D: A distillate fuel of lower volatility for engines in industrial and heavy mobile service

Grade 4-D: A fuel for low- and medium-speed engines

An additional guide to fuel selection is the grouping of fuels according to these types of service:

Type C-B: Diesel fuel oils for city bus and similar operations

Type T-T: Fuels for diesel engines in trucks, tractors, and similar service

Type R-R: Fuels for railroad diesel engines

Type S-M: Heavy-distillate and residual fuels for large stationary and marine diesel engines

The combustion characteristics of diesel fuels are expressed in terms of the cetane number, a measure of ignition delay. A short ignition

Table 7.1.16   Specifications for Aviation Gasoline

Test	Test limit	Test method
Distillation:		
Fuel evaporated, 10% min at	167°F (75°C)	D86
Fuel evaporated, 40% max at	167°F (75°C)	
Fuel evaporated, 50% min at	221°F (105°C)	
Fuel evaporated, 90% min at	275°F (135°C)	
Endpoint, max	338°F (170°C)	
Sum of 10% and 50% evaporated temp, min	307°F	
Residue, vol, max %	1.5	
Distillation loss, vol, max %	1.5	
Existent gum, max, mg/100 mL	3.0	D381
Potential gum, 16 h aging, max, mg/100 mL	6.0	D873
Precipitate, max, mg/100 mL	2.0	D873
Sulfur, max, wt %	0.05	D1266 or D2622
Reid vapor pressure at 100°F, min, lb/in²	5.5	D323
Reid vapor pressure at 100°F, max, lb/in²	7.0	D323
Freezing point, max	−76°F (−60°C)	D2386
Copper corrosion, max	No. 1	D130
Water reaction:		
Interface rating, max	2	D1094
Vol. change, max, mL	2	D1094
Heating value:		
Net heat of combustion, min, Btu/lb	18,700	D1405
Aniline-gravity product, min	7,500	D611 and D287

Table 7.1.18   Specifications* for Diesel Fuels

| Test | ASTM method | ASTM grade of diesel fuel | | | U.S. Military spec. Mil-F-16884G |
| | | 1-D | 2-D | 4-D | |
		Limit			
Flash point, min, °F	D93	100 or legal	125 or legal	130 or legal	140
Water and sediment, vol %, max	D1796	Trace	0.10	0.50	
Viscosity, kinematic, cSt, 100°F	D445				
Min		1.3	1.9	5.5	1.8
Max		2.4	4.1	24.0	4.5
Carbon residue on 10% residuum, % max	D524	015	0.35		0.20
Ash, wt %, max	D482	0.01	0.01	0.10	0.005
Sulfur, wt %, max	D129	0.50	0.50	2.0	1.00
Ignition quality, cetane number, min	D613	40	40	30	45
Distillation temp, °F, 90% evaporated:	D86				
Min			540		
Max		550	640		

* See ASTM D975 and Mil-F-16884G specifications for full details.

delay, i.e., the time between injection and ignition, is desirable for a smooth-running engine. Some diesel fuels contain cetane improvers, which usually are alkyl nitrates. The cetane number is determined by an engine test (ASTM D613) or an approximate value, termed the **cetane index** (ASTM D976), can be calculated for fuels that do not contain cetane improvers.

The list of additives used in diesel fuels has grown in recent years because of the increased use of catalytically cracked fuels, rather than exclusively straight-run distillate. In addition to cetane improvers, the list includes antioxidants, corrosion inhibitors, and dispersants. The dispersants are added to prevent agglomeration of gum or sludge deposits so these deposits can pass through filters, injectors, and engine parts without plugging them.

Gas-Turbine Fuels

Five grades of gas-turbine fuels are specified (ASTM D2880) according to the types of service and engine:

Grade O-GT: A naphtha or other low-flash-point hydrocarbon liquid which also includes jet B fuel
Grade 1-GT: A volatile distillate for gas turbines that requires a fuel that burns cleaner than grade 2-GT
Grade 2-GT: A distillate fuel of low ash and medium volatility, suitable for turbines not requiring grade 1-GT
Grade 3-GT: A low-volatility, low-ash fuel that may contain residual components
Grade 4-GT: A low-volatility fuel containing residual components and having higher vanadium content than grade 3-GT

Grade 1-GT corresponds in physical properties to no. 1 burner fuel and grade 1-D diesel fuel. Grade 2-GT corresponds in physical properties to no. 2 burner fuel and grade 2-D diesel fuel. The viscosity ranges of grades 3-GT and 4-GT bracket the viscosity ranges of no. 4, no. 5 light, no. 5 heavy, and no. 6 burner fuels.

Fuel Oils

The characteristics of the grades of fuel oil (ASTM D396) are as follows:

No. 1: A distillate oil intended for vaporizing pot-type burners and other burners requiring this grade of fuel.
No. 2: A distillate oil for general-purpose domestic heating in burners not requiring no. 1 fuel oil.
No. 4 and no. 4 light: Preheating not usually required for handling or burning.
No. 5 light: Preheating may be required depending upon climate and equipment.
No. 5 heavy: Preheating may be required for burning and, in cold climates, may be required for handling.
No. 6: Preheating required for burning and handling.

The sulfur content of no. 1 and no. 2 fuel oil is limited to 0.5 percent (ASTM D396). The sulfur content of fuels heavier than no. 2 must meet the legal requirements of the locality in which they are to be used. The additional refinery processing needed by some residual fuels to meet low-sulfur-content regulations may lower the viscosity enough to cause the fuels to change the grade classification.

Fuel oil used for domestic purposes or for small heating installations will have lower viscosities and lower sulfur content. In large-scale industrial boilers, heavier-grade fuel oil is used with sulfur content (ASTM D129 and D1552) requirements regulated according to the environmental situation of each installation and the local environmental regulations.

The **flash point** (ASTM D93) is usually limited to 60°C (140°F) minimum because of safety considerations. Asphaltene content (ASTM D3279), carbon residue value (ASTM D189 and D524), ash (ASTM D482), water content (ASTM D95), and metal content requirements are included in some specifications.

The **pour point** (ASTM D97), indicating the lowest temperature at which the fuel will retain its fluidity, is limited in the various specifications according to local requirements and fuel-handling facilities. The upper limit is sometimes 10°C (50°F), in warm climates somewhat higher.

Another important specification requirement is the heat of combustion (ASTM D240). Usually, specified values are 10,000 cal/kg (gross) or 9,400 cal/kg (net).

Because of economic considerations residual fuel oil has been replacing diesel fuel for marine purposes. Viscosity specifications had to be adjusted to the particular operational use, and some additional quality requirements had to be allowed for. The main problem encountered during the use of residual fuel oils for marine purposes concerns the stability properties (sludge formation) and even more so the compatibility properties of the fuel.

Blending of fuel oils to obtain lower viscosity values and to mix fuel oils of different chemical characteristics is a source of deposit and sludge formation in the vessel fuel systems. This incompatibility is mainly observed when high-asphaltene fuel oils are blended with diluents or other fuel oils of a paraffinic nature.

Gas oil is a general term applied to distillates boiling between kerosine and lubricating oils. The name was derived from its initial use for making illuminating gas, but it is now used as burner fuel, diesel fuel, and catalytic-cracker charge stock.

GASEOUS FUELS
by James G. Speight
Western Research Institute

REFERENCES: Bland and Davidson, ''Petroleum Processing Handbook,'' McGraw-Hill, New York. Francis and Peters, ''Fuels and Fuel Technology,'' 2d ed., Pergamon, New York, Sec. C, Gaseous Fuels. Goodger, ''Alternative Fuels: Chemical Energy Resources,'' Wiley, New York. Kohl and Riesenfeld, ''Gas Purification,'' Gulf Publishing Co., Houston, TX. Kumar, ''Gas Production Engineering,'' Gulf Publishing Co., TX. Reid, Prausnitz, and Sherwood, ''The Proper-

Table 7.1.19 Composition of Natural Gases*

	Natural gas from oil or gas wells						Natural gas from pipelines				
Sample no.	299	318	393	522	732	1177	1214	1225	1249	1276	1358
Composition, mole percent:											
Methane	92.1	96.3	67.7	63.2	43.6	96.9	94.3	72.3	88.9	75.4	85.6
Ethane	3.8	0.1	5.6	3.1	18.3	1.7	2.1	5.9	6.3	6.4	7.8
Propane	1.0	0.0	3.1	1.7	14.2	0.3	0.4	2.7	1.8	3.6	1.4
Normal butane	0.3	0.0	1.5	0.5	8.6	0.1	0.2	0.3	0.2	1.0	0.0
Isobutane	0.3	0.0	1.2	0.4	2.3	0.0	0.0	0.2	0.1	0.6	0.1
Normal pentane	0.1	0.0	0.6	0.1	2.7	0.3	Tr	Tr	0.0	0.1	0.0
Isopentane	Tr	0.0	0.4	0.2	3.3	0.0	Tr	0.2	Tr	0.2	0.1
Cyclopentane	Tr	0.0	0.2	Tr	0.9	Tr	Tr	0.0	Tr	Tr	0.0
Hexanes plus	0.2	0.0	0.7	0.1	2.0	0.1	Tr	Tr	Tr	0.1	Tr
Nitrogen	0.9	1.0	17.4	27.9	3.0	0.6	0.0	17.8	2.2	12.0	4.7
Oxygen	0.2	0.0	Tr	0.1	0.5	Tr	Tr	Tr	Tr	Tr	Tr
Argon	Tr	Tr	0.1	0.1	Tr	0.0	0.0	Tr	0.0	Tr	Tr
Hydrogen	0.0	0.2	0.0	0.0	0.1	0.0	Tr	0.1	0.1	0.0	0.0
Carbon dioxide	1.1	2.3	0.1	0.4	0.5	0.0	2.8	0.1	0.1	0.1	0.2
Helium	Tr	Tr	1.4	2.1	Tr	Tr	Tr	0.4	0.1	0.4	0.1
Heating value†	1,062	978	1,044	788	1,899	1,041	1,010	934	1,071	1,044	1,051
Origin of sample	La.	Miss.	N. Mex.	Okla.	Tex.	W. Va.	Colo.	Kan.	Kan.	Okla.	Tex.

* Analyses from *BuMines Bull.* 617 (Tr = trace).
† Calculated total (gross) Btu/ft³, dry, at 60°F and 30 in Hg.

ties of Gases and Liquids,'' McGraw-Hill. Shekhtman, ''Gasdynamic Functions of Real Gases,'' Hemisphere Publishing Corp., Washington. Speight, ''Fuel Science and Technology Handbook,'' J. G. Speight, ed., Marcel Dekker, New York, pt. 5, pp. 1055 et seq. Speight, ''Gas Processing: Environmental Aspects and Methods,'' Butterworth-Heinemann, Oxford, England. Sychev et al., ''Thermodynamic Properties of Propane,'' Hemisphere Publishing Corp.

Natural gas, which is predominantly methane, occurs in underground reservoirs separately or in association with crude petroleum. But **manufactured gas** is a fuel-gas mixture made from other solid, liquid, or gaseous materials, such as coal, coke, oil, or natural gas. The principal types of manufactured gas are retort coal gas, coke oven gas, water gas, carbureted water gas, producer gas, oil gas, reformed natural gas, and reformed propane or liquefied petroleum gas (LPG). Several processes for making substitute natural gas (SNG) from coal have been developed. **Mixed gas** is a gas prepared by adding natural gas or liquefied petroleum gas to a manufactured gas, giving a product with better utility and higher heat content or Btu value.

Liquefied petroleum gas (LPG) is the term applied to certain specific hydrocarbons and their mixtures, which exist in the gaseous state under atmospheric ambient conditions but can be converted to the liquid state under conditions of moderate pressure at ambient temperature. Thus liquefied petroleum gas is a hydrocarbon mixture usually containing propane ($CH_3.CH_2.CH_3$), n-butane ($CH_3.CH_2.CH_2.CH_3$), isobutane [$CH_3CH(CH_3)CH_3$], and to a lesser extent propylene ($CH_3.CH:CH_2$) or butylene ($CH_3CH_2CH:CH_2$). The most common commercial products are propane, butane, or some mixture of the two, and they are generally extracted from natural gas or crude petroleum. Propylene

and butylenes result from cracking other hydrocarbons in a petroleum refinery and are two important chemical feedstocks.

Composition of Gaseous Fuels The principal constituent of natural gas is methane (CH_4) (Table 7.1.19). Other constituents are paraffinic hydrocarbons such as ethane, propane, and the butanes. Many natural gases contain nitrogen as well as carbon dioxide and hydrogen sulfide. Trace quantities of argon, hydrogen, and helium may also be present. A portion of the heavier hydrocarbons, carbon dioxide, and hydrogen sulfide are removed from natural gas prior to its use as a fuel. Typical natural-gas analyses are given in Table 7.1.19. Manufactured gases contain methane, ethane, ethylene, propylene, hydrogen, carbon monoxide, carbon dioxide, and nitrogen, with low concentrations of water vapor, oxygen, and other gases.

Specifications Since the composition of natural, manufactured, and mixed gases can vary so widely, no single set of specifications could cover all situations. The requirements are usually based on performances in burners and equipment, on minimum heat content, and on maximum sulfur content. Gas utilities in most states come under the supervision of state commissions or regulatory bodies, and the utilities must provide a gas that is acceptable to all types of consumers and that will give satisfactory performance in all kinds of consuming equipment. Specifications for liquefied petroleum gases (Table 7.1.20) (ASTM D1835) depend upon the required volatility.

Odorization Since natural gas as delivered to pipelines has practically no odor, the addition of an odorant is required by most regulations in order that the presence of the gas can be detected readily in case of accidents and leaks. This odorization is provided by the addition of trace amounts of some organic sulfur compounds to the gas before it reaches

Table 7.1.20 Specifications* for Liquefied Petroleum Gas

	Product designation			
	Propane	Butane	PB mixtures	Test method
Vapor pressure at 100°F, max, psig	210	70		D1267 or D2598
Volatile residue:				
Butane and heavier, %, max	2.5			D2163
Pentane and heavier, %, max		2.0	2.0	D2163
Residual matter:				
Residue on evaporation, + 100 mL, max mL	0.05	0.05	0.05	D2158
Oil-stain observation	Pass	Pass	Pass	D2158
Specific gravity at 60°/60°F		To be reported		D1657 or D2598
Corrosion, copper strip, max	No. 1	No. 1	No. 1	D1838
Sulfur, grains/100 ft³, max	15	15	15	D1266
Moisture content		To be reported		
Free-water content		None	None	D1657

* Refer to ASTM D1835 for full details.

the consumer. The standard requirement is that a user will be able to detect the presence of the gas by odor when the concentration reaches 1 percent of gas in air. Since the lower limit of flammability of natural gas is approximately 5 percent, this 1 percent requirement is essentially equivalent to one-fifth the lower limit of flammability. The combustion of these trace amounts of odorant does not create any serious problems of sulfur content or toxicity.

Analysis

The different methods for gas analysis include absorption, distillation, combustion, mass spectroscopy, infrared spectroscopy, and gas chromatography (ASTM D2163, D2650, and D4424). Absorption methods involve absorbing individual constituents one at a time in suitable solvents and recording of contraction in volume measured. Distillation methods depend on the separation of constituents by fractional distillation and measurement of the volumes distilled. In combustion methods, certain combustible elements are caused to burn to carbon dioxide and water, and the volume changes are used to calculate composition. Infrared spectroscopy is useful in particular applications. For the most accurate analyses, mass spectroscopy and gas chromatography are the preferred methods.

ASTM has adopted a number of methods for gas analysis, including ASTM D2650, D2163, and D1717.

Physical Constants   The specific gravity of gases, including LP gases, may be determined conveniently by a number of methods and a variety of instruments (ASTM D1070 and D4891).

The heat value of gases is generally determined at constant pressure in a flow calorimeter in which the heat released by the combustion of a definite quantity of gas is absorbed by a measured quantity of water or air. A continuous recording calorimeter is available for measuring heat values of natural gases (ASTM D1826).

Flammability   The lower and upper limits of flammability indicate the percentage of combustible gas in air below which and above which flame will not propagate. When flame is initiated in mixtures having compositions within these limits, it will propagate and therefore the mixtures are flammable. A knowledge of flammable limits and their use in establishing safe practices in handling gaseous fuels is important, e.g., when purging equipment used in gas service, in controlling factory or mine atmospheres, or in handling liquefied gases.

Many factors enter into the experimental determination of flammable limits of gas mixtures, including the diameter and length of the tube or vessel used for the test, the temperature and pressure of the gases, and the direction of flame propagation—upward or downward. For these and other reasons, great care must be used in the application of the data. In monitoring closed spaces where small amounts of gases enter the atmosphere, often the maximum concentration of the combustible gas is limited to one fifth of the concentration of the gas at the lower limit of flammability of the gas-air mixture. (See Table 7.1.21.)

The calculation of flammable limits is accomplished by Le Chatelier's modification of the mixture law, which is expressed in its simplest form as

$$L = \frac{100}{p_1/N_1 + p_2/N_2 + \cdots + p_n/N_n}$$

where L is the volume percentage of fuel gas in a limited mixture of air and gas; $p_1, p_2, \ldots, p_n$ are the volume percentages of each combustible gas present in the fuel gas, calculated on an air- and inert-free basis so that $p_1 + p_2 + \cdots + p_n = 100$; and $N_1, N_2, \ldots, N_n$ are the volume percentages of each combustible gas in a limit mixture of the individual gas and air. The foregoing relation may be applied to gases with inert content of 10 percent or less without introducing an absolute error of more than 1 or 2 percent in the calculated limits.

The **rate of flame propagation** or **burning velocity** in gas-air mixtures is of importance in utilization problems, including those dealing with burner design and rate of energy release. There are several methods that have been used for measuring such burning velocities, in both laminar and turbulent flames. Results by the various methods do not agree, but

Table 7.1.21   Flammability Limits of Gases in Air

Gas	Flammable limits in air, vol %	
	Lower	Upper
Methane	5.0	15.0
Ethane	3.0	12.4
Propane	2.1	9.5
Butane	1.8	8.4
Isobutane	1.8	8.4
Pentane	1.4	7.8
Isopentane	1.4	7.6
Hexane	1.2	7.4
Ethylene	2.7	36.0
Propylene	2.4	11.0
Butylene	1.7	9.7
Acetylene	2.5	100.0
Hydrogen	4.0	75.0
Carbon monoxide	12.5	74.2
Ammonia	15.0	28.0
Hydrogen sulfide	4.0	44.0
Natural	4.8	13.5
Producer	20.2	71.8
Blast-furnace	35.0	73.5
Water	6.9	70.5
Carbureted-water	5.3	40.7
Coal	4.8	33.5
Coke-oven	4.4	34.0
High-Btu oil	3.9	20.1

any one method does give relative values of utility. Maximum burning velocities for turbulent flames are greater than those for laminar flames. The bunsen flame method gives results that have significance in gas utilization problems. In this method, the burning velocity is determined by dividing the volume rate of gas flow from the bunsen burner by the area of the inner cone of the flame.

The use of other fuels and equipment to supplement the regular supply of gas during periods of peak demand or in emergencies is known as **peak shaving**. Most gas utilities, particularly natural-gas utilities at the end of long-distance transmission lines, maintain peak-shaving or standby equipment. Also, gas utilities in many cases have established natural-gas storage facilities close to their distribution systems. This allows gas to be stored underground near the point of consumption during periods of low demand, as in summer, and then withdrawn to meet peak or emergency demands, as may occur in winter. Propane-air mixtures are the major supplements to natural gas for peak shaving use. Gas manufactured by cracking various petroleum distillates supplies most of the remaining peak requirements.

SYNTHETIC FUELS
by Martin D. Schlesinger
Wallingford Group Ltd.

REFERENCES: U.S. Department of Energy, Fossil Energy Reports. Whitehurst et al., "Coal Liquefaction," Academic. Speight, "The Chemistry and Technology of Coal," Marcel Dekker. Preprints, Division of Fuel Chemistry, American Chemical Society. Annual International Conferences on Coal Gasification, Liquefaction, and Conversion to Electricity, Department of Chemical and Petroleum Engineering, University of Pittsburgh.

Liquid and gaseous fuels in commercial use can be produced from sources other than petroleum and natural gas. Much new technology has been developed during the twentieth century as the threat of petroleum shortages or isolation loomed periodically. The result was the improvement of known processes and the introduction of new concepts.

A generalized flow sheet Fig. 7.1.3, shows the routes that can be followed from coal to finished products. Liquid and gaseous fuels are formed from coal by increasing its hydrogen to carbon ratio. Primary

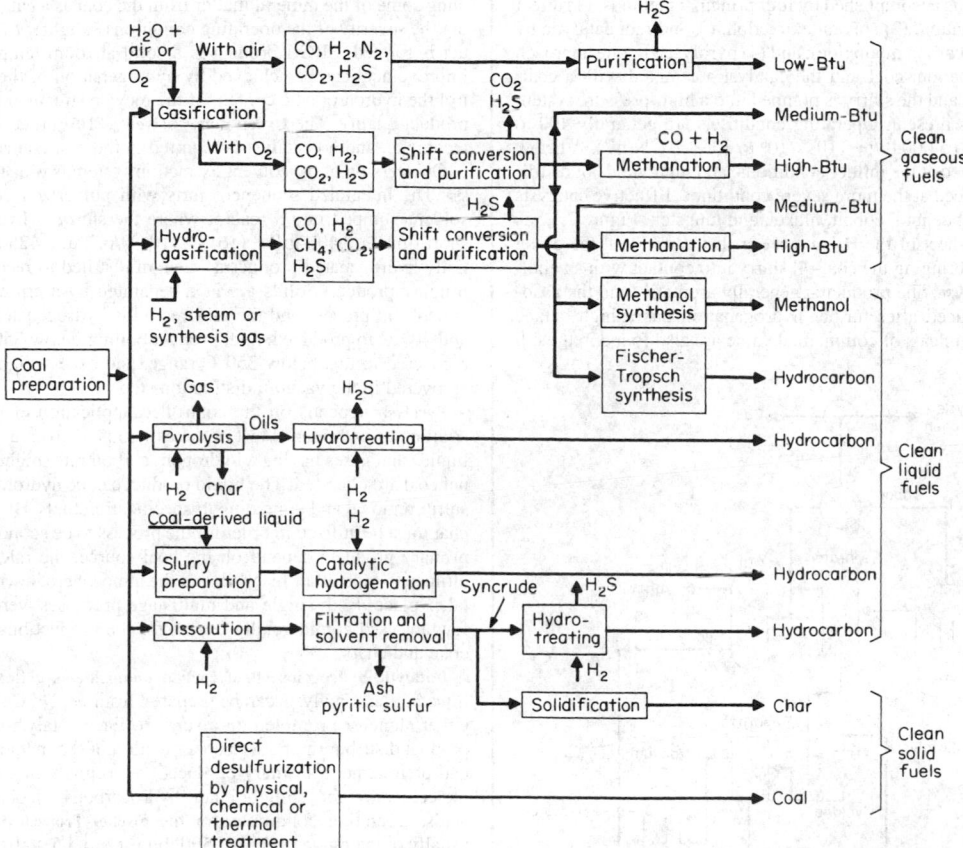

Fig. 7.1.3 Clean fuels from coal.

conversion products are deashed, desulfurized, and further upgraded to a wide range of clean fuels. Gasification can yield clean gases for combustion or synthesis gas with a controlled ratio of hydrogen to carbon monoxide. Catalytic conversion, of synthesis gas to fluids (indirect liquefaction) can be carried out in fixed and fluidized beds and in dilute phase systems. Both gases and liquids are used as the temperature control medium for the exothermic reactions.

Direct liquefaction is accomplished by pyrolysis or hydrogenation and several processes are available for each approach for adding hydrogen to the coal and removing undesirable constituents. The amount of hydrogen consumed is determined by the properties desired in the final product, whether it be a heavy fuel oil, diesel oil, jet fuel, gasoline, or gases. Clean solid fuels, i.e, with low ash and low sulfur contents, can be produced as a product of pyrolysis, from liquefaction residues, or by chemical treatment of the coal.

What must be done with some naturally occurring sources of fuels is exemplified in Fig. 7.1.4. Upgrading involves the addition of hydrogen and more hydrogen is required to upgrade coal than to process petroleum into finished products. The ranges are generalized to indicate the relative need for processing and the range of product distributions. Coal contains very little hydrogen, averaging 0.8 H:C atomic ratio and petroleum has about twice the relative amounts. Premium fuels are in the kerosene/gasoline range, which includes diesel and jet fuels. Commercial transportation fuels contain 15 to 18 wt % hydrogen. At the upper end of the scale is methane with an H:C ratio of 4:1. Not considered in the above is the elimination of mineral matter and constituents such as oxygen, sulfur, and nitrogen, where hydrogen is consumed to form water, hydrogen sulfide, and ammonia. Some oxygen is removed as carbon dioxide.

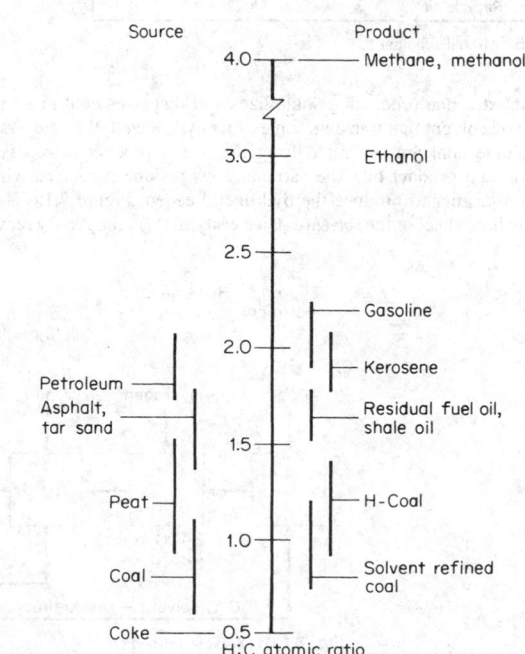

Fig. 7.1.4 Upgrading of carbonaceous sources.

Coal liquefaction is accomplished by four principal methods: (1) direct catalytic hydrogenation, (2) solvent extraction, (3) indirect catalytic by hydrogenation (of carbon monoxide), and (4) pyrolysis. In one approach to direct hydrogenation, coal and the catalyst are mixed with a coal-derived recycle oil and the slurry is pumped into a high-pressure system where hydrogen is present. Operating conditions are generally 400 to 480°C and 1,500 to 5,000 lb/in² (10×10^6 to 35×10^6 N/m²). A heavy syncrude is produced at the milder conditions, and high yields of distillable oils are produced at the more severe conditions. Effective catalysts contain iron, molybdenum, cobalt, nickel, and tungsten. Figure 7.1.5 is a schematic flow sheet of the **H-coal process** that carries out the direct hydrogenation by bringing the coal-oil slurry into contact with an ebullating bed of catalyst. The product is generally aromatic, and the gasoline fraction produced after further hydrogenation has a high octane number. Many chemicals of commercial value are also found in the oil.

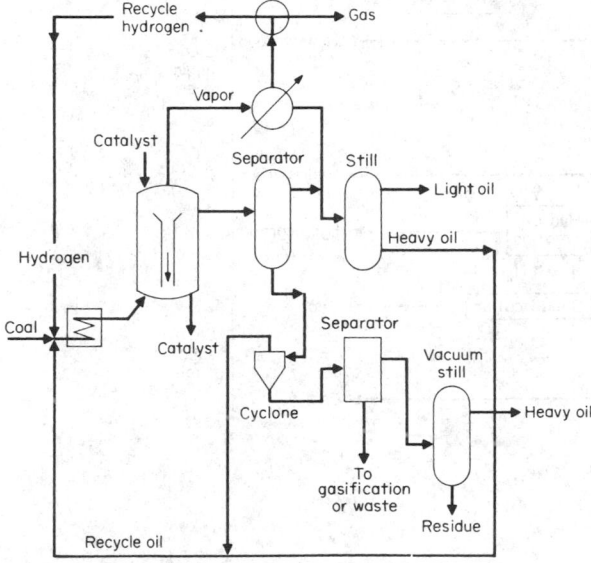

Fig. 7.1.5 H-coal schematic.

Solvent extraction processing solubilizes and disperses coal in a hydroaromatic solvent that transfers some of its hydrogen to the coal. Ash and insoluble coal are separated from the liquid product to recover recycle oil and product oil. The carbonaceous residue is reacted with steam in a gasifier to produce the hydrogen needed. Figure 7.1.6 is a schematic flow sheet of the **solvent-refined coal (SRC I)** process. By recycling some of the mineral matter from the coal as a catalyst and increasing the severity of the operating conditions, a lighter hydrocarbon product is formed. SRC I product is a solid at room temperature. Further improvement can be achieved by hydrogenation of the solvent to control the hydroaromatic content of the recycle stream and to improve the product quality. The **Exxon donor solvent (EDS)** process uses this latter technique, and no catalyst is required in the first contacting vessel.

One version of the solvent extraction system is **two-stage hydrogenolysis**. The integrated sequence starts with pulverized coal in a recycle solvent pumped into a reactor where the slurry is hydrogenated for a short time at 2,400 lb/in² (16.5×10^6 N/m³) and 425 and 450°C. Partially hydrogenated product is vacuum distilled to recover solvent and primary product. Solids are next separated by a critical solvent or an antisolvent process and the ash-free oil is hydrogenated at 2,700 lb/in² and 400°C to produce a final product boiling below 350°C. Some oil is recovered in this below 350°C range, and excess higher-boiling oil is recovered in the vacuum distillation step.

Pyrolysis depends on the controlled application of heat without the addition of hydrogen. Most of the carbon is recovered as solid product; liquids and gases having a hydrogen : carbon ratio higher than the original coal are liberated. The liquid product can be hydrotreated further for sulfur removal and upgrading to specification fuels. By-product gas and char must be utilized in order for the process to be economical. Yields of primary products depend on the coal source, the rate of heating, the ultimate temperature reached, and the atmosphere in which the reaction takes place. Both single and multistage processes were developed but few are used commercially except for the production of metallurgical coke and chars.

Catalytic hydrogenation of carbon monoxide is a flexible method of liquefaction. Catalysts can be prepared from Fe, Ni, Co, Ru, Zn, and Th either alone or promoted on a support. Each catalyst gives a different product distribution that is also a function of the method of preparation and pretreatment. Primary products are normally methane and higher-molecular-weight, straight-chain hydrocarbons, alcohols, and organic acids. Operating conditions for the **Fischer-Tropsch-type synthesis** are usually in the range of 300 to 500 lb/in² (2 to 3.5×10^6 N/m³) and 200 to 400°C. Temperature of the exothermic reaction [-39.4 kcal/(g·mol)] is controlled by carrying out the reaction in fixed beds, fluidized beds, slurry, and dilute phase systems with heat removal. Diesel oil from this type of synthesis has a high cetane number; the paraffin waxes can be of high quality. Generally, the gasoline produced by indirect synthesis has a low octane number because of its aliphatic nature. One method of producing a high-octane gasoline is to make methanol from 2 : 1 H : CO in a first stage and then process the alcohol over a zeolite catalyst. The hydrocarbon product has a narrow boiling range and contains about 30 percent aromatics.

In South Africa there are three commercial coal conversion plants that use the indirect liquefaction method. About 28 million t/yr of coal is gasified under pressure in 97 Lurgi gasifiers; and the synthesis gas, after

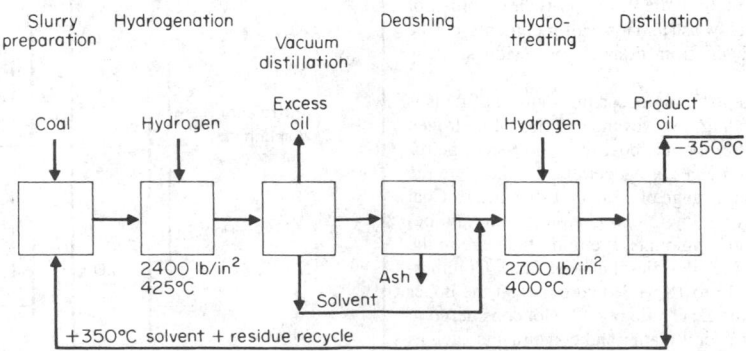

Fig. 7.1.6 Solvent-refined coal.

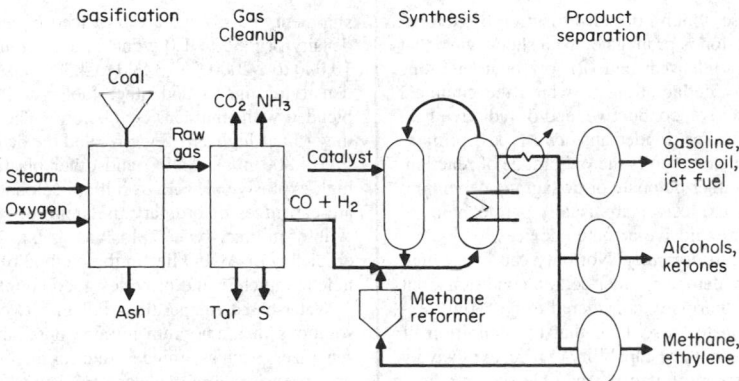

Fig. 7.1.7   Flow sheet of SASOL II and III.

cleaning and adjustment of the hydrogen/carbon monoxide ratio, is passed through catalyst beds. A total output of 150,000 bpd (23,850 m³) of automotive fuel provides about one-half of the nation's needs. There is also produced about 1,600 tpd of chemicals. The simplified flow sheet in Fig. 7.1.7 shows some of the main features of the plants.

SASOL I (SASOL Chemical Industries) uses fixed-bed reactors and dilute phase systems; the fixed bed makes higher-molecular-weight products. **SASOL II** and **III** (SASOL synthetic fuels) do not produce the fixed-bed heavy hydrocarbons but maximize gasoline and diesel oil formation by hydrotreating and reforming.

Research is reported to be continuing on the application of the slurry process and direct hydrogenation to coal conversion technology.

Shale oil is readily produced by the thermal processing of many shales. The basic technology is available and commercial plants are operated in many parts of the world. The first modern plant in the United States was put on stream in 1983 by the Union Oil Co. in Colorado. About 12,500 tons of raw shale, averaging 35 gal/ton (0.145 m³/ 1,000 kg), crushed to 5-cm particles, is pushed upward into the retort each day, and at 400 to 500°C crude shale oil is produced from the kerogen in the shale.

The refined product yield was 7,000 bpd (1,115 m³/d) of diesel oil and 3,000 bpd (475 m³/d) of jet fuel. The present low cost of petroleum has not justified the continued development of this and other systems. Estimated reserves are equivalent to about 3 billion barrels of shale oil. Figure 7.1.8 is a generalized flow sheet of the process.

Tar sand is a common term for oil-impregnated sediments that can be found in almost every continent. High-grade tar sands have a porosity of 25 to 35 percent and contain about 18 percent by weight of bitumen. The sand grains are wetted by about 2 percent of water, making them hydrophilic and thus more amenable to hot-water extraction. Solvent extraction, thermal retorting, in situ combustion, and steam injection methods have been tested.

Reserves of tar sands in the United States are equivalent to about 30 billion barrels of petroleum. Most of the deposits are too deep for surface mining and require in situ treatment before extraction. Some of the surface deposits are worked to produce asphalt for highway application and other minor uses.

Tar sand is a source of hydrocarbon fuels at Syncrude Canada in Alberta. Two commercial plants with combined capacity of 190,000 bpd of synthetic crude oil operate in this region and use the principle of ore mining, hot-water extraction, coking (delayed and fluid), and distillate hydrogenation. Methods for modification of the bitumen in situ and recovery without mining are also under investigation. Properties of conventional petroleum, tar sand bitumen, and synthetic crude oil from the bitumen are given in Table 7.1.22.

Table 7.1.22   Comparison of Tar Sand Bitumen and Synthetic Crude Oil from the Bitumen with Petroleum

	Petroleum	Tar sand bitumen	Synthetic crude oil
API gravity	25–27°	8°	
Viscosity			
cSt at 100°F	3–7	120,000	6
cSt at 210°F		2,000	
Carbon, wt %	86.0	83.1	86.3
Hydrogen, wt %	13.5	10.6	13.4
Nickel, ppm	2–10	100	0
Sulfur, wt %	1–2	4.8	0.15
Nitrogen	0.2	0.4	0.06
Vanadium, ppm	2–10	250	0
Ash, wt %	0	1.0	0
Carbon residue, wt %	1–5	14.0	0
Pentane insolubles, wt %	<5	17.0	0

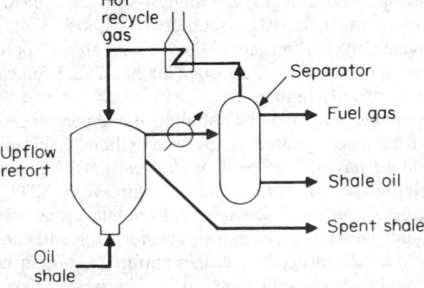

Fig. 7.1.8   Shale oil processing.

EXPLOSIVES
by J. Edmund Hay

U.S. Department of the Interior

REFERENCES: Meyer, "Explosives," Verlag Chemie. Cook, "The Science of High Explosives," Reinhold. Johansson and Persson, "Detonics of High Explosives," Academic. Davis, "The Chemistry of Powder and Explosives," Wiley. "Manual on Rock Blasting." Aktiebolaget Atlas Diesel, Stockholm. Dick, Fletcher, and D'Andrea, Explosives and Blasting Procedures Manual, *BuMines Inf. Circ.* 8925, 1983.

The term *explosives* refers to any substance or article which is able to function by explosion (i.e., the extremely rapid release of gas and heat) by chemical reaction within itself.

Explosive substances are commonly divided into two types: (1) high or **detonating explosives** are those which normally function by detonation, in which the chemical reaction is propagated by a shock wave that in turn is driven by the energy released; and (2) low or **deflagrating explosives,** normally function by deflagration, in which the chemical reaction is propagated by convective, conductive, and/or radiative heat transfer. High explosives are further divided into two types: primary explosives are those for which detonation is the only mode of reaction, and secondary explosives may either detonate or deflagrate, depending on a variety of conditions. Low explosives are usually pyrotechnics or propellants for guns, rockets, or explosive-actuated devices.

It is important to note the word *normally*. Not only can many high explosives deflagrate rather than detonate under certain conditions, but also some explosives which are normally considered to be "low" explosives can be forced to detonate. Also note that the definition of *explosive* is itself somewhat elastic—the capability to react explosively is strongly dependent on the geometry, density, particle size, confinement, and initiating stimulus, as well as the chemical composition. Historically, immense grief has resulted from ignorance of these facts. In normal use, the term *explosive* refers to those substances or articles which have been classified as explosive by the test procedures recommended by the United Nations (UN) Committee on the Transport of Dangerous Goods.

According to the UN scheme of classification, most high explosives are designated class 1.1 (formerly called *class A*), and most low explosives are designated class 1.3 (formerly called *class B*). Explosive devices of minimal hazard (formerly called *class C*) are designated class 1.4.

However, a very important subclass of secondary high explosives is designated class 1.5 (formerly called *blasting agents* or *nitrocarbonitrates*). The distinction is based primarily on **sensitivity:** In simplified terms, the initiation of detonation of a class 1.5 explosive requires a stronger stimulus than that provided by a detonator with a 0.45-g PETN base charge which exhibits a low tendency to the deflagration-to-detonation transition.

The important physical properties of high explosives include their bulk density, detonation rate, critical and "ideal" diameters (or thickness), "sensitivity," and strength.

The detonation rate is the linear speed at which the detonation propagates through the explosive, and it ranges from about 6,000 ft/s (1.8 km/s) to about 28,000 ft/s (8.4 km/s), depending on the density and other properties of the explosive.

The *critical diameter* (or thickness, in the case of a sheet explosive) is the minimum diameter or thickness at which detonation can propagate through the material. For most class 1.1 explosives, this is between 1 and 30 mm; for class 1.5 explosives, it is usually greater than 50 mm. This value depends to some extent on the confinement provided by the material adjacent to the explosive. For explosives whose diameter (thickness) is only slightly above the critical value, the detonation rate increases with increasing diameter (thickness), finally attaining a value for which no increase in rate is observed for further increases in dimension. The diameter or thickness at which this occurs is called the *ideal diameter* or *thickness*.

Sensitivity and strength are two of the most misunderstood properties of explosives, in that there is a persistent belief that each of these terms refers to a property with a unique value. Each of these properties can be quantitatively determined by a particular test or procedure, but there are far more such procedures than can even be listed in this space, and there are large deviations from correlation between the values determined by these different tests.

Descriptions of some of the more common explosives or types of high explosive are given below.

Ammonium nitrate–fuel oil (ANFO) blasting agents are the most widely used type of explosive product, accounting for about 90 percent of explosives in the United States. They usually contain 5.5 to 6 percent fuel oil, typically no. 2 diesel fuel. If used underground, the oil content must be carefully regulated to minimize the production of toxic fumes. Some ANFO compositions contain aluminum or densifying agents. Premixed ANFO is usually shipped in 50- to 100-lb bags, although bulk shipment and storage are practiced in certain operations. ANFO has a density of 0.85 to 1.0 g/cm^3 and a detonation velocity in the range of 10,000 to 14,000 ft/s (3,000 to 4,300 m/s). ANFO may also incorporate densifying agents and other fuels such as aluminum, and it may be blended with **emulsions** (see below). The density of such compositions may run as high as 1.5 g/cm^3, and the detonation rate as high as 16,000 ft/s (4,800 m/s). ANFO and other blasting agents require the use of high-explosive **primers** to initiate detonation. Commonly used primers are cartridges of ordinary dynamite or specially cast charges of ¼ to ¾ lb of military-type explosives such as **composition B or pentolite**. The efficiency of ANFO lies in the method of loading, which fills the borehole completely and provides good coupling with the burden.

Water-based compositions fall into two types. **Water gels** are gelled solutions of ammonium nitrate containing other oxidizers, fuels, and sensitizers such as amine nitrates or finely milled aluminum, in solution or suspension. **Emulsions** are emulsions of ammonium nitrate solution with oil, and they may contain additional oxidizers, fuels, and sensitizers. These compositions are classified either as explosives or as blasting agents depending on their sensitivity. Both types contain a thickening agent to prevent segregation of suspended solids. For larger operations, mobile mixing trucks capable of high-speed mixing and loading of the composition directly into the large-diameter vertical boreholes have become popular. Another advantage of this on-site mixing and loading is the ability to change composition and strength between bottom and top loads. The density of the explosive-sensitized composition is usually about 1.4 g/cm^3 but may be as high as 1.7 g/cm^3 for the aluminum-sensitized type; detonation velocities vary from 10,000 to 17,000 ft/s (3,000 to 5,200 m/s).

Dynamite is a generic term covering a multitude of nitroglycerine-sensitized mixtures of carbonaceous materials (wood, flour, starch) and oxygen-supplying salts such as ammonium nitrate and sodium nitrate. The nitroglycerin contains ethylene glycol dinitrate or other nitrated compounds to lower its freezing point, and antacids, such as chalk or zinc oxide, are divided into nongelatinous or granular and gelatinous types, the latter containing nitrocellulose. All dynamites are capsensitive.

Straight dynamites are graded by the percentage of explosive oil they contain; this may be as low as 15 percent and as high as 60 percent. A typical percentage formulation for a 40 percent straight dynamite is: nitroglycerin, 40; sodium nitrate, 44; antacid, 2; carbonaceous material, 14. The rate of detonation increases with grade from 9,000 to 19,000 ft/s (2,700 to 5,800 m/s). Straight dynamites now find common use only in ditching where propagation by influence is practiced.

Ammonia dynamites differ from straight dynamites in that some of the sodium nitrate and much of the explosive oil have been replaced by ammonium nitrate. Strength of ammonia dynamites ranges from 15 to 60 percent, each grade having the same weight strength as the corresponding straight dynamite when compared in the ballistic mortar. A typical percentage formula for a 40 percent ammonia dynamite is: explosive oil, 14; ammonium nitrate, 36; sodium nitrate, 33; antacid, 1; carbonaceous material, 16. The rate of detonation, 4,000 to 17,000 ft/s (1,200 to 5,200 m/s), again increases with grade. Low-density, high-weight-strength compositions are popular in many applications, but ANFO has displaced them in numerous operations.

Blasting gelatin is the strongest and highest-velocity explosive used in industrial operations. It consists essentially of explosive oil (nitroglycerin plus ethylene glycol dinitrate) colloided with about 7 percent nitrocellulose. It is completely water-resistant but has a poor fume rating and consequently finds only limited use.

Gelatin dynamites correspond to straight dynamites except that the explosive oil has been gelatinized by nitrocellulose; this results in a cohesive mixture having improved water resistance. Under confinement, the gelatins develop high velocity, ranging from 8,500 to 22,000 ft/s (2,600 to 6,700 m/s) and increasing between the grades of 20 and 90 percent. An approximate percentage composition for a 40 percent grade is: explosive oil, 32; nitrocellulose, 0.7; sulfur, 2; sodium nitrate, 52; antacid, 1.5; and carbonaceous material, 11. In the common grades of 40 and 60 percent, fume characteristics are good, making these types useful for underground hard-rock blasting.

Ammonia gelatin dynamites are similar to the ammonia dynamites except for their nitrocellulose content. These used to be popular in quarrying and hard-rock mining. Their excellent fume characteristics make them suitable for use underground, but again ANFO is widely used. The rates of detonation of 7,000 to 20,000 ft/s (2,000 to 6,000 m/s) are somewhat less than the straight gelatins. A typical percentage composition for the 40 percent grade is: gelatinized explosive oil, 21; ammonium nitrate, 14; sodium nitrate, 49; with antacid and combustible making up the remainder.

The **semigels** are important variants of the ammonia gels; these contain less explosive oil, sodium nitrate, and nitrocellulose and more ammonium nitrate than the corresponding grade of ammonia gel. Rates of detonation fall in the limited range of 10,000 to 13,000 ft/s (3,000 to 4,000 m/s). These powders are cohesive and have good water resistance and good fume characteristics.

Permissible explosives are powders especially designed for use in underground coal mines, which have passed a series of tests established by the Bureau of Mines. The most important of these tests concern incendivity of the explosives—their tendency to ignite methane-air or methane-coal dust-air mixtures. Permissible explosives are either granular or gelatinous; the granular type makes up the bulk of the powders used today. Typically, a granular permissible contains the following, in percent: explosive oil, 9; ammonium nitrate, 65; sodium nitrate, 5; sodium chloride, 10; carbonaceous material, 10; and antacid, 1. Gels contain nitrocellulose for improved water resistance and more explosive oil. Detonation velocities for the granular grades vary from 4,500 to 11,000 ft/s (1,400 to 3,400 m/s), and for the gels from 10,500 to 18,500 ft/s (3,200 to 5,600 m/s). Many water-based permissible formulations with comparable physical and safety properties are now marketed as well.

Liquid oxygen explosives (LOX) once saw considerable use in coal strip mines but have been completely displaced by ANFO or water-based compositions. LOX consisted of bags of pressed carbon black or specially processed char that were saturated with liquid oxygen just before loading into the borehole. The rate of detonation ranged from 12,000 to 18,000 ft/s (3,700 to 5,500 m/s).

Military explosives, originally developed for such uses as bomb, shell, and mine loads and demolition work, have been adapted to many industrial explosive applications. The more common military explosives are listed in Table 7.1.23, with their compositions, ballistic mortar strengths, and detonation velocities.

Amatol was used early in World War II, largely because of the short supply of TNT. Modifications of amatol have been used as industrial

Table 7.1.23 Physical Characteristics of Military Explosives

Explosive	Composition, %	Ballistic mortar strength (TNT = 100)	Density, g/cm³	Rate of detonation, m/s
80/20 amatol	Ammonium nitrate, 80; TNT, 20	117	Cast	4,500
50/50 amatol	Ammonium nitrate, 50; TNT, 50	122	Cast	5,600
Composition A (pressed)	RDX, 91; wax, 9	134	0.80	4,560
			1.20	6,340
			1.50	7,680
			1.60	8,130
Composition B (cast)	RDX, 59.5; TNT, 39.5; wax, 1	130	1.65	7,660
Composition C-3 (plastic)	RDX, 77; tetryl, 3; mononitrotoluene, 5; dinitrotoluene, 10; TNT, 4; nitrocellulose, 1	145	1.55	8,460
Composition C-4 (plastic)	RDX, 91; dioctyl sebacate, 5.3; polyisobutylene, 2.1; oil, 1.6	—	1.59	8,000
Explosive D	Ammonium picrate	97	0.80	4,000
			1.20	5,520
			1.50	6,660
			1.60	7,040
HBX-1	RDX, 40; TNT, 38; aluminum, 17; desensitizer, 5	130	1.70	7,310
Lead azide*	Lead azide	—	2.0	4,070
			3.0	4,630
			4.0	5,180
PETN	Pentaerythritol tetranitrate	145	0.80	4,760
			1.20	6,340
			1.50	7,520
			1.60	7,920
50/50 Pentolite	PETN, 50; TNT, 50	120	1.20	5,410
			1.50	7,020
			1.60	7,360
			Cast	7,510
Picric acid	Trinitrophenol	108	1.20	5,840
			1.50	6,800
			Cast	7,350
RDX (cyclonite)	Cyclotrimethylene trinitramine	150	0.80	5,110
			1.20	6,550
			1.50	7,650
			1.60	8,000
			1.65	8,180
Tetryl	Trinitrophenylmethylnitramine	121	0.80	4,730
			1.20	6,110
			1.50	7,160
			1.60	7,510
75/25 Tetrytol	Tetryl, 75; TNT, 25	113	1.60	7,400
TNT	Trinitrotoluene	100	0.80	4,170
			1.20	5,560
			1.50	6,620
			1.60	6,970
			Cast	6,790

* Primary compound for blasting caps.

blasting agents. **Explosive D,** or ammonium picrate, by virtue of its extreme insensitivity, was used in explosive-filled armor-piercing shells and bombs.

RDX, a very powerful explosive compound, was widely used during World War II in many compositions, of which Compositions A, B, and C were typical. **Composition A** was used as a shell loading; **B** was used as a bomb and shaped charge filling; **C,** being plastic enough to allow molding to desired shapes, was developed for demolition work. Compositions that also contained aluminum powder were developed for improved underwater performance **(Torpex, HBX).** RDX has found limited commercial application as the base charge in some detonators, the filling for special-purpose detonating fuses or cordeau detonants, and the explosive in small shaped charges used as oil well perforators and tappers for openhearth steel furnaces.

PETN has never found wide military application because of its sensitivity and relative instability. It is used extensively, however, as the core of detonating fuses and in caps and, mixed with **TNT,** in boosters.

Tetryl, once widely used by the military as a booster loading and commercially as a base charge in detonators, has been displaced by other compositions. **Tetrytol** found limited application as a demolition charge.

TNT is a very widely used military explosive. Its stability, insensitivity, convenient melting point (81°C), and relatively low cost have made it the explosive of choice either alone or in admixture with other materials for loadings which are to be cast. A free-flowing pelletized form has found application in certain types of blasting requiring high loading density, where it is used to fill the cavity formed in sprung holes or the free space around the column of other explosives in the borehole.

DUST EXPLOSIONS
by Harry C. Verakis and John Nagy (Retired)
Mine Safety and Health Administration

REFERENCES: Nagy and Verakis, "Development and Control of Dust Explosions," Dekker. "Classification of Dusts Relative to Electrical Equipment in Class II Hazardous Locations," NMAB 353-4, National Academy of Sciences, Washington. "Fire Protection Handbook," 17th ed., National Fire Protection Assoc. *BuMines Rept. Inv.* 4725, 5624, 5753, 5971, 7132, 7208, 7279, 7507. Eckhoff, "Dust Explosions in the Process Industries," Butterworth-Heinemann.

A dust explosion hazard exists where combustible dusts accumulate or are processed, handled, or stored. The possibility of a dust explosion may often be unrecognized because the material in bulk form presents little or no explosion hazard. However, if the material is dispersed in the atmosphere, the potential for a dust explosion is increased significantly. The first well-recorded dust explosion occurred in a flour mill in Italy in 1785. Dust explosions continue to plague industry and cause serious disasters with loss of life, injuries, and property damage. For example, there were about 100 reportable dust explosions (excluding grain dust) from 1970 to 1980 which caused about 25 deaths and a yearly property loss averaging about $20 million. A complete and accurate record of the number of dust explosions, deaths, injuries, and property damage is unavailable because reporting of each incident is not required unless a fatality occurs or more than five persons are seriously injured. In recent years, most dust explosions have involved wood, grain, resins and plastics, starch, and aluminum. Most of the incidents occurred during crushing or pulverizing, buffing or grinding, conveying, and at dust collectors.

Despite the well-recognized hazards inherent with explosible dusts, the vast amount of technical data accumulated and published, and standards for prevention of and protection from dust explosions, severe property damage and loss of life occur every year. As an example of the severity of dust explosions, there was a series of explosions in four grain elevators during December 1977, which caused 59 deaths, 47 injuries, and nearly $60 million in property damage.

A **dust explosion** is the rapid combustion of a cloud of particulate matter in a confined or partially confined space in which heat is generated at a higher rate than it is dissipated. In a confined space, the explosion is characterized by relatively rapid development of pressure with the evolution of large quantities of heat and reaction products. The condition necessary for a dust explosion to occur is the simultaneous presence of a dust cloud of proper concentration in air or gas that will support combustion and an ignition source.

Dust means particles of materials smaller than 0.016 in in diameter, or those particles passing a no. 40 U.S. standard sieve, 425 μm (this definition relates to the limiting size, not to the average particle size of the material); and *explosible* dust means a dust which, when dispersed, is ignited by spark, flame, heated coil, or in the Godbert-Greenwald furnace at or below 730°C, when tested in accordance with the equipment and procedures described in *BuMines Rept. Inv.* 5624.

Explosibility Factors

Empirical methods and experimental data are the chief guides in evaluating relative dust explosion hazards. A mathematical model correlating some of the numerous interrelated factors affecting dust explosion development in closed vessels has been developed. Details of the model are presented in Nagy and Verakis, "Development and Control of Dust Explosions."

Dust Composition Many industrial dusts are not pure compounds. The severity of a dust explosion varies with the chemical constitution and certain physical properties of the dust. High percentages of noncombustible material, such as mineral matter or moisture, reduce the ease of oxidation, and oxygen requirements influence the explosibility of dusts. Volatile, combustible components in such materials as coals, asphalts, and pitches increase explosibility.

Dust composition also affects the amount and type of products produced in an explosion. Organic materials evolve new gaseous products, whereas most metals form solid oxides during combustion in an air atmosphere.

Particle Size and Surface Area Explosibility of dusts increases with a decrease in particle size. Fine dust particles have greater surface area, more readily disperse into a cloud, mix better with air, remain longer in suspension, and oxidize more rapidly and completely than coarse particles. Decrease of particle size generally results in lower ignition temperature, lower igniting energy, lower minimum explosive concentration, and higher pressure and rates of pressure rise. Some metals, such as chromium, become explosive only at very fine particle sizes (average particle diameter of 3 μm), and almost all metals become pyrophoric if reduced to very fine powder.

Range of Explosibility Most combustible dusts have a well-defined lower limit, but the upper limit is usually indefinite. The upper limit has been determined for only a few dusts, but these data have only limited importance in practice. The range of dust explosibility is normally 0.015 to greater than 10 oz/ft³ (10 kg/m³). The optimum concentration producing the strongest dust explosions is about 0.5 to 1.0 oz/ft³. Table 7.1.24 gives explosion characteristics for a number of dusts at a concentration of 0.5 oz/ft³. A typical example of the effect of dust concentration on maximum pressure and maximum rate of pressure rise from explosions in closed vessels of various size and shape is shown in Figs. 7.1.9 and 7.1.10.

Ignition Source Ignition sources known to have initiated dust explosions in industry include electric sparks and arcs in fuses, faulty wiring, motors and other appliances, static electrical fuses, faulty wiring, motors and other appliances, static electrical discharges, open flames, frictional, or metallic sparks, glowing particles, overheated bearings and other machine parts; hot electric bulbs, overheated driers, and other hot surfaces; dust layers may also ignite by these sources as well as by spontaneous ignition. Ignition temperatures of many dust clouds are given in Table 7.1.24. Normally the ignition temperature of a dust layer is considerably less than for a dust cloud. The position and intensity of the ignition source affect dust-explosion development; detailed information on these factors is presented in *BuMines Rept. Inv.* 7507.

Turbulence Turbulence has a slight effect on maximum pressure, but a marked effect on the rates of pressure rise for dust explosions.

Table 7.1.24   Explosive Characteristics of Various Dusts*

Type of dust	Ignition temperature of dust cloud, °C	Minimum igniting energy, J	Minimum explosive concentration, oz/ft³	Maximum explosion pressure, lb/in² gage	Maximum rate of pressure rise, lb/(m²)(s)	Terminal oxygen concentration, %†
Agricultural:						
Alfalfa	530	0.320	0.105	92	2,200	
Cereal grass	550	0.800	0.250	52	500	
Cinnamon	440	0.030	0.060	114	3,900	
Citrus peel	730	0.045	0.065	71	2,000	
Cocoa	500	0.120	0.065	55	900	
Coffee	720	0.160	0.085	53	300	
Corn	400	0.040	0.055	95	6,000	
Corncob	480	0.080	0.040	110	3,100	
Corn dextrine	410	0.040	0.040	105	7,000	
Cornstarch	390	0.030	0.040	115	9,000	
Cotton linters	520	1.920	0.500	48	150	
Cottonseed	530	0.120	0.055	96	3,000	
Egg white	610	0.640	0.140	58	500	
Flax shive	430	0.080	0.080	81	800	
Garlic	360	0.240	0.100	80	2,600	
Grain, mixed	430	0.030	0.055	115	5,500	
Grass seed	490	0.260	0.290	34	400	
Guar seed	500	0.060	0.040	98	2,400	
Gum, Manila (copal)	360	0.030	0.030	88	5,600	
Hemp hurd	440	0.035	0.040	103	10,000	
Malt, brewers	400	0.035	0.055	92	4,400	
Milk, skim	490	0.050	0.050	83	2,100	
Pea flour	560	0.040	0.050	95	3,800	
Peanut hull	460	0.050	0.045	82	4,700	
Peat, sphagnum	460	0.050	0.045	84	2,200	
Pecan nutshell	440	0.050	0.030	106	4,400	
Pectin	410	0.035	0.075	112	8,000	
Potato starch	440	0.025	0.045	97	8,000	
Pyrethrum	460	0.080	0.100	82	1,500	
Rauwolfia vomitoria root	420	0.045	0.055	106	7,500	
Rice	440	0.050	0.050	93	2,600	
Safflower	460	0.025	0.055	84	2,900	
Soy flour	550	0.100	0.060	111	1,600	15
Sugar, powdered	370	0.030	0.045	91	1,700	
Walnut shell, black	450	0.050	0.030	97	3,300	
Wheat flour	440	0.060	0.050	104	4,400	
Wheat, untreated	500	0.060	0.065	98	4,400	
Wheat starch	430	0.025	0.045	100	6,500	
Wheat straw	470	0.050	0.055	99	6,000	
Yeast, torula	520	0.050	0.050	105	2,500	
Carbonaceous:						
Asphalt, resin, volatile content 57.5%	510	0.025	0.025	94	4,600	
Charcoal, hardwood mix, volatile content 27.1%	530	0.020	0.140	100	1,800	18
Coal, Colo., Brookside, volatile content, 38.7%	530	0.060	0.045	88	3,200	
Coal, Ill., no. 7, volatile content 48.6%	600	0.050	0.040	84	1,800	15
Coal, Ky., Breek, volatile content 40.6%	610	0.030	0.050	88	4,000	
Coal, Pa., Pittsburgh, volatile content 37.0%	610	0.060	0.055	83	2,300	17
Coal, Pa., Thick Freeport, volatile content, 35.6%	595	0.060	0.060	77	2,200	
Coal, W. Va., no. 2 Gas, volatile content 37.1%	600	0.060	0.060	82	1,600	
Coal, Wyo., Laramie no. 3, volatile content 43.3%	575	0.050	0.050	92	2,000	
Gilsonite, Utah, volatile content 86.5%	580	0.025	0.020	78	3,700	
Lignite, Calif., volatile content 60.4%	450	0.030	0.030	90	8,000	
Pitch, coal tar, volatile content 58.1%	710	0.020	0.035	88	6,000	
Chemical compounds:						
Benzoic acid, C_6H_5COOH	620	0.020	0.030	74	5,500	
Phosphorus pentasulfide, P_2S_5, slowly cooled to give single crystals	280	0.015	0.050	54	10,000+	
Phosphorus pentasulfide, P_2S_5, cooled quickly	290	0.015	0.050	58	7,500	13

Table 7.1.24 Explosive Characteristics of Various Dusts* (Continued)

Type of dust	Ignition temperature of dust cloud, °C	Minimum igniting energy, J	Minimum explosive concentration, oz/ft³	Maximum explosion pressure, lb/in² gage	Maximum rate of pressure rise, lb/(m²)(s)	Terminal oxygen concentration, %†
Chemical compounds (*Continued*):						
Phthalimide, $C_6H_4(CO)_2NH$	630	0.050	0.030	79	4,500	
Potassium bitartrate, $KHC_4H_4O_6$	520					
Salicylanilide, $o\text{-}HOC_6H_4CONHC_6H_5$	610	0.020	0.040	61	4,400	
Sodium thiosulfate, anhydrous, $Na_2S_2O_3$	510					
Sorbic acid, $CH_3(CH{:}CH)_2 COOH$	470	0.015	0.020	88	10,000+	
Sucrose, $C_{12}H_{22}O_{11}$	420	0.040	0.045	82	4,200	14
Sulfur, S_8 100% finer than 44 μm	210	0.020	0.045	56	3,100	
Sulfur, S_8, avg particle size 4 μm	190	0.015	0.035	78	4,700	12
Drugs:						
Aspirin (acetylsalicylic acid), $o\text{-}CH_3COOC_6H_4COOH$, fine	660	0.025	0.050	83	10,000+	
Mannitol (hexahydric alcohol), $CH_2OH(CHOH)_4CH_2OH$	460	0.040	0.065	82	2,800	
Secobarbital sodium, $C_{12}H_{17}N_2O_3Na$	520	0.960	0.100	54	500	
Vitamin C, ascorbic acid, $C_6H_8O_6$	460	0.060	0.070	88	4,800	15
Explosives and related compounds:						
Dinitrobenzamide	500	0.045	0.040	94	6,500	
Dinitrobenzoic acid	460	0.045	0.040	92	4,300	
Dinitro-sym-diphenyl-urea (dinitrocarbanilide)	550	0.060	0.095	87	2,500	
Dinitrotoluamide (3,5-dinitro-ortho-toluamide)	500	0.015	0.050	106	10,000−	13
Metals:						
Aluminum	650	0.015	0.045	100	10,000+	2
Antimony	420	1.920	0.420	8	100	16
Boron	470	0.060	0.100	90	2,400	
Cadmium	570	4.000				
Chromium	580	0.140	0.230	56	5,000	14
Cobalt	760					
Copper	900					
Iron	420	0.020	0.100	46	6,000	10
Lead	710					
Magnesium	520	0.020	0.020	94	10,000+	0
Molybdenum	720					
Nickel	950+					
Selenium	950+					
Silicon	780	0.080	0.100	106	10,000+	12
Tantalum	630	0.120	0.200	51	3,700	
Tellurium	550					
Thorium	270	0.005	0.075	48	3,300	0
Tin	630	0.080	0.190	37	1,300	15
Titanium	460	0.010	0.045	80	10,000+	0
Tungsten	950+					
Uranium	20	0.045	0.060	53	3,400	0
Vanadium, 86%	500	0.060	0.220	48	600	13
Zinc	600	0.640	0.480	48	1,800	9
Zirconium	20	0.005	0.045	65	9,000	0
Alloys and compounds:						
Aluminum-cobalt	950	0.100	0.180	78	8,500	
Aluminum-copper	930	1.920	0.280	27	500	
Aluminum-iron	550	0.720	0.500	21	100	
Aluminum-magnesium	430	0.020	0.020	90	10,000	0
Aluminum-nickel	940	0.080	0.190	79	10,000	14
Aluminum-silicon, 12% Si	670	0.060	0.040	74	7,500	
Calcium silicide	540	0.130	0.060	73	10,000+	8
Ferrochromium, high-carbon	790		2.000			19
Ferromanganese, medium-carbon	450	0.080	0.130	47	4,200	
Ferrosilicon, 75% Si	860	0.400	0.420	87	3,600	16
Ferrotitanium, low-carbon	370	0.080	0.140	53	9,500	13

Table 7.1.24 Explosive Characteristics of Various Dusts* (Continued)

Type of dust	Ignition temperature of dust cloud, °C	Minimum igniting energy, J	Minimum explosive concentration, oz/ft³	Maximum explosion pressure, lb/in² gage	Maximum rate of pressure rise, lb/(m²)(s)	Terminal oxygen concentration, %†
Alloys and compounds (*Continued*):						
Ferrovanadium	440	0.400	1.300			17
Thorium hydride	260	0.003	0.080	60	6,500	6
Titanium hydride	440	0.060	0.070	96	10,000+	13
Uranium hydride	20	0.005	0.060	43	6,500	0
Zirconium hydride	350	0.060	0.085	69	9,000	8
Plastics:						
Acetal resin (polyformaldehyde)	440	0.020	0.035	89	4,100	11
Acrylic polymer resin Methyl methacrylate-ethyl acrylate	480	0.010	0.030	85	6,000	11
Alkyd resin Alkyd molding compound	500	0.120	0.155	15	150	15
Allyl resin, allyl alcohol derivative, CR-39	500	0.020	0.035	106	10,000+	13
Amino resin, urea-formaldehyde molding compound	450	0.080	0.075	89	3,600	17
Cellulosic fillers, wood flour	430	0.020	0.035	110	5,500	17
Cellulosic resin, ethyl cellulose molding compound	320	0.010	0.025	102	6,000	11
Chlorinated polyether resin, chlorinated polyether alcohol	460	0.160	0.045	66	1,000	
Cold-molded resin, petroleum resin	510	0.030	0.025	94	4,600	
Coumarone-indene resin	520	0.010	0.015	93	10,000+	14
Epoxy resin	530	0.020	0.020	86	6,000	12
Fluorocarbon resin, fluorethylene polymer	600					
Furane resin, phenol furfural	520	0.010	0.025	90	8,500	14
Ingredients, hexamethylenetetramine	410	0.010	0.015	98	10,000+	14
Miscellaneous resins, petrin acrylate monomer	220	0.020	0.045	104	10,000+	
Natural resin, rosin, DK	390	0.010	0.015	87	10,000+	14
Nylon polymer resin	500	0.020	0.030	89	7,000	13
Phenolic resin, phenol-formaldehyde molding compound	500	0.020	0.030	92	10,000+	14
Polycarbonate resin	710	0.020	0.025	78	4,700	15
Polyester resin, polyethylene terephthalate	500	0.040	0.040	91	5,500	13
Polyethylene resin	410	0.010	0.020	83	5,000	12
Polymethylene resin, carboxypolymethylene	520	0.640	0.115	70	5,500	
Polypropylene resin	420	0.030	0.020	76	5,000	
Polyurethane resin, polyurethane foam	510	0.020	0.025	88	3,700	
Rayon (viscose) flock	520	0.240	0.055	88	1,700	
Rubber, synthetic	320	0.030	0.030	93	3,100	15
Styrene polymer resin, polystyrene latex	500	0.020	0.020	91	7,000	13
Vinyl polymer resin, polyvinyl butyral	390	0.010	0.020	84	2,000	14

* Data taken from the following *Bureau of Mines Reports of Investigations:* RI 5753, ''Explosibility of Agricultural Dusts''; RI 5971, ''Explosibility of Dusts Used in the Plastics Industry''; RI 6516, ''Explosibility of Metal Powders''; RI 7132, ''Dust Explosibility of Chemicals, Drugs, Dyes and Pesticides''; RI 7208, ''Explosibility of Miscellaneous Dusts.'' The data were obtained using the equipment described in RI 5624, ''Laboratory Equipment and Test Procedures for Evaluating Explosibility of Dusts.''
† The terminal oxygen concentration is the limiting oxygen concentration in air-CO_2 atmosphere required to prevent ignition of dust clouds by electric spark.

Experiments show the maximum rate of pressure rise in a highly turbulent dust-air mixture can be as much as 8 times higher than in a nonturbulent mixture (*BuMines Repts. Inv.* 5815 and 7507 and Nagy and Verakis).

Moisture and Other Inerts Moisture in a dust absorbs heat and tends to reduce the explosibility of a dust. A high concentration of moisture in the dust also tends to reduce the dispersibility of a dust. An increase in moisture content causes an increase in ignition temperature and a reduction in maximum pressure and rates of pressure rise. However, the amount of moisture required to produce a marked lowering of the explosibility parameters is higher than can ordinarily be tolerated in industrial processes. Most mineral inert dusts admixed with a combustible absorb heat during the combustion reaction and reduce explosibility similar to the action of water. Some chemical compounds, such as sodium and potassium carbonates, act as inhibitors and are more effective than mineral inerts; the limiting inert dust concentration required to prevent ignition and explosion depends on the strength of the igniting source.

Atmospheric Oxygen Concentration The pressure and rate of pressure development decrease as the oxygen concentration in the atmosphere decreases. The ignition sensitivity of dusts decreases with decrease in oxygen concentration and for most dusts, ignition and explosion can be prevented by reducing the oxygen concentration to a safe value. Carbon dioxide, nitrogen, argon, helium, and water vapor are effective diluents. For highly reactive metal powders, only argon and helium are chemically inert. Limiting oxygen concentrations using carbon dioxide as a diluent are given in Table 7.1.24 for many dusts. With carbon dioxide as a diluent, a reduction of oxygen in the atmosphere to 11 percent is sufficient to prevent ignition by sparks for all dusts tested except the metallic powders. With nitrogen as the diluent, ignition of nonmetallic dusts is prevented by diluting the atmosphere to 8 percent oxygen. Some metal dusts, such as magnesium, titanium, and zirconium, ignite by spark in a pure carbon dioxide atmosphere. Freon and halons are sometimes used as diluent gases, but if metal dusts are involved, they can intensify rather than suppress ignition. The limiting oxygen concentration decreases as the dust becomes finer in particle

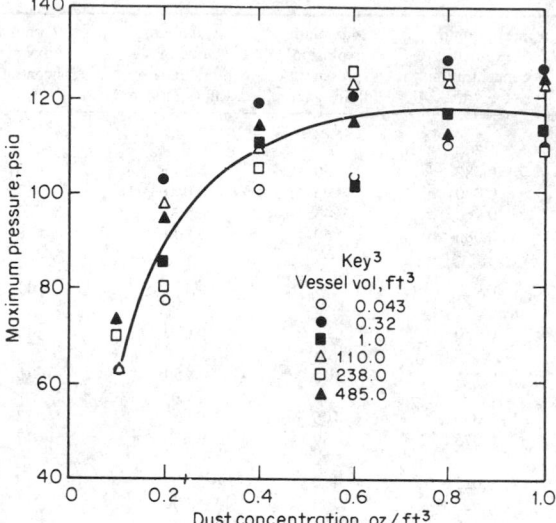

Fig. 7.1.9   Effect of dust concentration on maximum pressure produced by explosions of cellulose acetate dust in closed vessels.

size; limiting oxygen concentration varies slightly with dust concentration and is lowest at concentrations two to five times the stoichiometric mixture.

Relative Dust Explosion Hazards

Table 7.1.24 gives test results of selected dusts whose explosive characteristics have been evaluated in the laboratory by the Bureau of Mines. The data were obtained for dusts passing a no. 200 sieve and represent the most hazardous of the specific materials tested. The values are relative rather than absolute since the test apparatus and experimental procedures affect the results to some degree. The samples were dried before testing only if the moisture content exceeded 5 percent. Detailed description of the equipment and procedures for the small-scale testing are given in *BuMines Rept. Inv. 5624.*

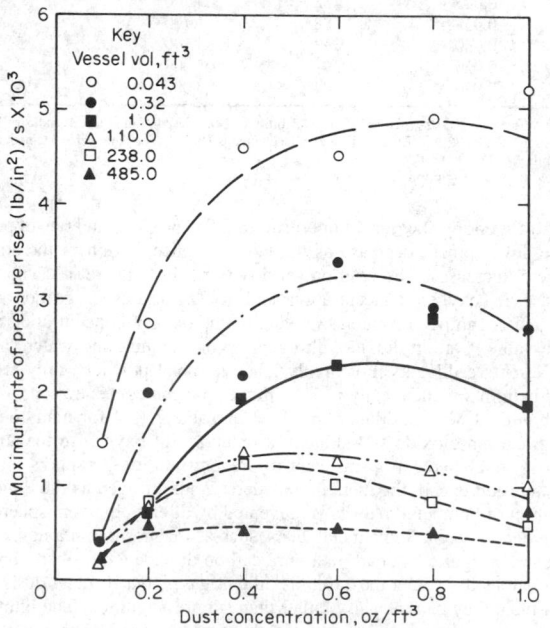

Fig. 7.1.10   Effect of dust concentration on maximum rate of pressure rise by explosions of cellulose acetate dust in closed vessels.

Ignition Temperature   The ignition temperature of a dust cloud was determined by dispersing dust through a heated cylindrical furnace. The ignition temperature is the minimum furnace temperature at which flame appears at the bottom of the furnace in one or more trials in a group of four.

Minimum Energy   The minimum electrical spark energy required to ignite a dust cloud was determined by dispersing the dust in a vertically mounted, 2¾-in-diameter, 12-in-long tube. The dust is dispersed by an air blast and then a condenser of known capacitance and voltage is discharged through a spark gap located within the dust dispersion. The top of the tube is enclosed with a paper diaphragm. The minimum energy for ignition of the dust cloud is the least amount of energy required to produce flame propagation 4 in or longer in the tube.

Minimum Concentration   The minimum explosive concentration or the lower explosive limit of a dust cloud was determined in a vertically mounted, 2¾-in diameter, 12-in-long tube using a continuous spark-igniting source. A known weight of dust was dispersed within the tube by an air blast. The lowest weight of dust at which sufficient pressure develops to burst a paper diaphragm enclosing the tube or which causes flame to fill the tube is used to calculate the minimum explosive concentration; this calculation is made utilizing the tube volume.

Maximum Pressure and Rates of Pressure Rise   The maximum pressure and rates of pressure rise developed by a dust explosion were determined by dispersing dust in a closed steel tube. A continuous spark is used for ignition. A transcribed pressure-time record is obtained during the test. The maximum rate is the steepest slope of the pressure-time curve. Normally, explosion tests are made at dust concentrations of 0.10 to 2.0 oz/ft³. Maximum pressure is primarily dependent on dust composition and independent of vessel size and shape. The maximum rate of pressure rise increases as vessel size decreases.

Explosibility Index   The overall explosion hazard of a dust is related to the ignition sensitivity and to explosion severity and is characterized by empirical indexes. The ignition sensitivity of a dust cloud depends on the ignition temperature, minimum energy, and minimum concentration. The explosion severity of a dust depends on the maximum pressure and maximum rate of pressure rise. The indexes are not derived from theoretical considerations, but provide a numerical rating consistent with research observations and practical experience. Results obtained for a sample dust are compared with values for a standard Pittsburgh-seam coal dust. The indexes are defined as follows:

Ignition sensitivity =

$$\frac{(\text{ign temp} \times \text{min energy} \times \text{min conc}) \text{ Pittsburgh coal dust}}{(\text{ign temp} \times \text{min energy} \times \text{min conc}) \text{ sample dust}}$$

Explosion severity =

$$\frac{\begin{array}{c}(\text{max explosive pressure}\\ \times \text{ max rate of pressure rise}) \text{ sample dust}\end{array}}{\begin{array}{c}(\text{max explosive pressure}\\ \times \text{ max rate of pressure rise}) \text{ Pittsburgh coal dust}\end{array}}$$

Explosibility index = ignition sensitivity × explosion severity

A dust having ignition and explosion characteristics equivalent to the standard Pittsburgh-seam coal has an explosibility index of unity. The relative hazard of dusts is further classified by the following adjective ratings: fire, weak, moderate, strong, or severe. The notation ≪ 0.1 designates a combustible dust presenting primarily a fire hazard as ignition of the dust cloud is not obtained by a spark or flame source, but only by an intense, heated surface source. These ratings are correlated with the empirical indexes as follows:

Type of explosion	Ignition sensitivity	Explosion severity	Index of explosibility
Fire			<0.1
Weak	<0.2	<0.5	<0.1
Moderate	0.2–1.0	0.50–1.0	0.1–1.0
Strong	1.0–5.0	1.0–2.0	1.0–10
Severe	>5.0	>2.0	>10

Prevention of Dust Explosions

(See also Sec. 12.1)

There are codes published by the National Fire Protection Association which contain recommendations for a number of dust-producing industries.

Additional sources of information may be found in the NFPA "Fire Protection Handbook," the Factory Mutual "Handbook of Industrial Loss Prevention," *BuMines Rept. Inv.* 6543, Nagy and Verakis, and Eckhoff.

Safeguards against explosions include, but are not limited to the following:

Good housekeeping: An excellent means to minimize the potential for and extent of an explosion is good housekeeping. Control of dust spillage or leakage and elimination of dust accumulations removes the fuel required for an explosion.

Limited personnel: Wherever a hazardous operation must be performed, the number of persons should be limited to the minimum required for safe operation.

Elimination of Ignition Sources All sources of ignition should be eliminated from equipment containing combustible dust and from adjacent areas. Open flames or lights and smoking should be prohibited. The use of electric or gas cutting and welding equipment for repairs should be avoided unless dust-producing machinery is shut down and all dust has been removed from the machines and from their vicinity. Proper control methods should be instituted for materials susceptible to spontaneous combustion. Additional safety measures to follow are ground and bonding of all equipment to prevent the accumulation of static electrical charges; strict adherence to the National Electric Code when installing electrical equipment and wiring in hazardous locations; use of magnetic separators to prevent entrance of ferrous materials into dust-grinding mills; use of nonferrous blades in fans through which dust passes; and avoidance of spark-producing tools in certain industries and of high-speed shafting and belts. Safeguards against ignition by lightning should also be considered. (See also NFPA no. 77, Static Electricity and NFPA no. 78, Lightning Protection.)

Building and Equipment Construction Buildings should be constructed to minimize the collection of dust on beams, ledges, and other surfaces, particularly overhead. Vacuum cleaning is preferable to other methods for dust removal, but soft push brooms may be used without serious hazard. Buildings, including inside partitions, where combustible dusts are handled or stored should be detached units of incombustible construction. Hazardous units within buildings should be separated by substantial fire walls.

Grinders, conveyors, elevators, collectors, and other equipment which may produce dust clouds should be as dust-tight as possible; they should have the smallest practical interior volume and should be constructed to withstand dust explosion pressures. The degree of turbulence within and around an enclosure should be kept to a minimum to prevent dust from being suspended.

Dust collectors should preferably be located outside of buildings or detached rooms and near the dust source. The choice of a suitable dust collector depends on the particle size, dryness, explosibility, dust concentration, gas velocity and temperature, efficiency and space requirements, and economic considerations. (See also Secs. 9 and 18.)

Inerted Atmosphere Equipment such as grinders, conveyors, pulverizers, mixers, dust collectors, and sacking machines can frequently be protected by using an inerted atmosphere or explosion suppression systems. The inert gas for this purpose may be obtained by dilution of air with flue gases from boilers, internal-combustion engines, or other sources, or by dilution with carbon dioxide, nitrogen, helium from high-pressure cylinders, or gas from inert-gas generators. The amount and rate of application of inert gas required depend upon the permissible oxygen concentration, leakage loss, atmospheric and operating conditions, equipment to be protected, and application method. Addition of inert dusts to the combustible dust may also prevent explosive dust-air mixtures from forming in and around equipment. (See NFPA no. 69, Explosion Prevention Systems.)

Relief Venting To reduce structural damage and to protect personnel from dust explosions, dust collectors and other equipment and the rooms in which dust-producing machinery is located should be provided with relief vents. Relief vents properly designed and located will sufficiently relieve explosion pressures in most instances and direct explosion gases away from occupied areas. The vents may be unrestricted or free openings; hinged or pivoted sash that swing outward at a low internal pressure; fixed sash with light wall anchorages; scored glass panes; light wall panels; monitors or skylights; paper, metal foil, or other diaphragms that burst at low pressures; poppet-type vent closures; pullout diaphragms; or other similar arrangements. Vents should be located near potential sources of ignition to keep explosion pressure at a minimum and to prevent a dust explosion from developing into a detonation in long ducts.

Empirical formulas and mathematical methods for calculating the vent area to limit pressure from an explosion have been developed. Unfortunately, because of the many factors involved in determining venting requirements, none of these methods can be considered entirely satisfactory to cover the complete range of situations confronting an equipment or building designer. For example, the maximum pressure that can develop in a vented enclosure during a dust explosion is affected by factors such as the chemical affinity of the combustible material with oxygen, heat of combustion, particle size distribution, degree of turbulence, uniformity of the dust cloud, size and energy of the igniting source, location of the igniting source relative to the vent, area of the vent opening, bursting strength or inertial resistance of the vent closure, initial pressure and initial temperature within an enclosure, and the oxygen concentration of the atmosphere. Because of the complexity of the phenomena during explosion in a vented vessel, information on the required vent area to limit the excess or explosion pressure to be a safe value for a given vessel or structure under specific conditions is still estimated from data obtained by physical tests usually made under severe test conditions. Extremely reactive dusts such as magnesium and aluminum are difficult or nearly impossible to vent successfully if an explosion occurs under optimum conditions. Agricultural dusts, most plastic-type dusts, and other metallic dusts can usually be vented successfully. Materials containing oxygen or a mixture with an oxidant should be subjected to tests before venting is attempted. Information and recommendations on venting are given in NFPA no. 68, "Guide for Venting of Deflagrations" (1994). A mathematical analysis showing the relationship of numerous factors affecting the venting of explosions is presented in "Development and Control of Dust Explosions" (Nagy and Verakis, Dekker).

Information published by others shows that higher values of maximum pressures and rates of pressure rise are normally obtained in vessels larger and differently shaped than the Hartmann apparatus described in *BuMines RI 5624*. Data on maximum pressures and rates obtained from the Hartmann apparatus are shown in Table 7.1.24. A comparison of test data from the Hartmann apparatus and a 1-m³ vessel is shown in "Development and Control of Dust Explosions." NFPA no. 68, "Guide for Venting of Deflagrations" (1994), recommends using the rate of pressure rise data obtained from closed vessels, 1-m³ or larger, in venting calculations.

Combating Dust Fires

The following points should be observed when one is dealing with dust fires, in addition to the usual recommendations for fire prevention and firefighting, including sprinkler protection (see also Secs. 12 and 18).

1. Attention should be directed to the potential hazard of spontaneous heating of dust products, particularly when grinding or pulverizing processes are used.

2. First-aid and firefighting equipment should be installed. Small hoses with spray nozzles or automatic sprinkler systems fitted with spray or fog nozzles are particularly satisfactory. The fine spray wets the dust and is not so likely to raise a dust cloud as with a solid stream. Portable extinguishers used to combat dust fires should be provided with similar devices for safe discharge.

3. Large hose of fire department size giving solid water streams should be used with caution; a dust cloud may be formed with consequent risk of explosion. Plant employees and the fire department should

be advised of this potential hazard in advance. Hose equipped with spray or fog nozzles should be provided and kept ready for an emergency.

4. Fires involving aluminum, magnesium, or some other metal powders are difficult to extinguish. Sand, talc, or other dry inert materials, and special proprietary powders designed for this purpose should be used. These materials should be applied gently to smother the fire. Materials such as hard pitch can completely seal the dust from oxygen and may be used. (See NFPA Code nos. 48, 65, and 651.)

ROCKET FUELS
by Randolph T. Johnson
Indian Head Division, Naval Surface Warfare Center

REFERENCES: Billig, Tactical Missile Design Concepts, Johns Hopkins Applied Physics Laboratory, *Technical Digest*, 1983, **4**, no. 3. Roth and Capener, Propellants, Solid, "Encyclopedia of Explosives and Related Items," Kave, ed., PATR 2700, vol. 8, U.S. Army Armament Research and Development Command, Dover, NJ, 1978. "Solid Propellant Selection and Characterization," NASA Design Criteria Guide, NASA SP 8064, 1971.

Until the development of the solid-fueled Polaris missile in the late 1950s and early 1960s, rocket motors tended to be divided into two distinct categories: rocket motors greater than 3 ft in diameter were liquid-fueled, while smaller rocket motors were solid-fueled. Since that time, a number of quite large rocket motors have been developed which use solid fuel; these include the Minuteman series and, most recently, the solid rocket boosters used on the space shuttle, each of which has over 1,000,000 lb of propellant. There are exceptions to the rule in the other direction, too; e.g., the air-launched Bullpup rocket motor started out as a liquid-fueled unit, changed to a solid fuel, then back to a liquid fuel.

Design Criteria

Before selecting the proper propellant system for a given rocket motor, the designer must consider the parameters by which the design is constrained. The following criteria generally need to be considered when choosing the propellant(s) for a given rocket motor:

Envelope Constraints The designer must consider the volume, mass and shape limits within which the rocket motor is constrained. For example, an air-launched rocket may be limited by the carrying capacity of the aircraft, the size of the launcher, and the size of the payload.

Performance Requirements In general, the requirements imposed on a rocket motor are expressed as the minimum required and/or the maximum allowed to complete the mission and the maximum allowed to prevent damage to the payload, launcher, etc. Such parameters include velocity, range, burn time, and acceleration.

Environmental Conditions The environments seen by various rocket motors differ dramatically, depending on the intended use. For example, a submarine-launched strategic missile lives a pampered life in near ideal conditions while a field-launched barrage rocket or an air-launched missile see near worse-case environments. Some of the parameters to consider in the choice of propellants include temperature limits of storage and operation, vibration and shock spectra to be experienced and survived, and the moisture and corrosion environments the rocket motor (and possibly the propellant) may be expected to encounter. It may be possible to reduce the effects of these environments on the propellant by rocket motor design, but the propellant chemist and the rocket motor designer must be willing to work together to come up with a viable combination.

Safety Requirements Safety considerations include toxic and explosion hazards in the manufacture of the rocket motor and in servicing, handling, and use. The use of the rocket must be considered in the safety margin built into its design. If it is to be used in a "human-rated" system, such as an ejection seat, it must be of more conservative design than if it were to be used in a barrage rocket, for instance.

Service Life The designer must consider the duration over which the rocket motor will be in service. It is, in general, less expensive to make a number of rocket motors at a time and then store them, than to make the same number in several small lots over a period of time with the accordingly increased start-up and shutdown costs. A very short service life leads to heightened costs for repetitive shipping to and from storage, as well as rework and replacement of overaged units. There is yet another hidden cost of a very short service life: the risk that an overaged unit will be inadvertently used with potentially catastrophic results.

Maintenance The availability of maintenance will affect the choice of propellants for the given unit. The unit with readily available maintenance facilities will have far less severe constraints than the unit which must function after years of storage and/or far from the reach of service facilities.

Smokiness The choice of propellants must be influenced by whether or not the mission will allow the use of a propellant which leaves a smoky trail. Propellants containing a metal fuel, as well as certain other ingredients, leave a large smoky trail which reveals the location of the launch point. This is particularly undesirable in tactical rocket systems for it leaves the user revealed to the enemy. The smoke from metal fuels is termed **primary** smoke as it is a product of the primary combustion, **secondary** smoke comes from the reaction of propellant combustion products with the atmosphere, such as the reaction of hydrogen chloride with moisture in the air to leave a hydrochloric acid cloud, or water vapor condensing in cold air to leave a contrail.

Cost The cost of a rocket motor must be divided into a number of categories including design, test, ingredients, processing, components, surveillance, maintenance, rework, and disposal. The hidden cost of nonfunction (reliability) should also be factored into the equation.

Liquid Propellants versus Solid Propellants

The first choice to be made in the selection of propellants is whether to use a liquid or a solid propellant. Each of these comes with its own set of advantages and disadvantages which must be weighed in the selection process.

Liquid propellants offer the possibility of extremely high impulse per unit weight of propellant. The thrust of a liquid-propellant rocket motor may be easily modulated by controlling the flow rate of the propellant into the combustion chamber.

Liquid propellants, however, tend to be of low density, which leads to large packages for a given total energy. Most liquid propellants have a very limited usable temperature range because of freezing or vaporization. Because they require the use of valves, pipes, pumps, and the like, liquid-propellant rocket motors are relatively complex and require a high level of maintenance.

The use of liquid rocket propellants is predominant in older strategic rocket motors, such as Titan, and in space flight, such as the space shuttle main engines and attitude-control motors.

Solid propellants offer the possibility of use over a wide range of environmental conditions. They offer a significantly higher density than liquid propellants. Solid-propellant rocket motors are much simpler than liquid rocket motors, as the propellant grain forms at least one wall of the combustion chamber. This simplicity of design leads to low maintenance and high reliability. Safety is somewhat higher than with liquid propellants, for there are no volatile and hazardous liquids to spill during handling and storage.

Solid propellants, however, do not attain the impulse levels of the more energetic liquid propellants. Furthermore, once ignited, the thrust-time profile will be as dictated by the propellant surface history and propellant burn rate for the nozzle given; this profile is not easily altered, unlike that of a liquid-propellant rocket motor suitably equipped.

Solid-propellant rocket motors are now used in every size from small thrusters on the Dragon antitank round to the boosters on the space shuttle. They are used when a preprogrammed thrust-time history is appropriate.

Liquid Propellants Once the decision has been made to use liquid propellant, the designer is confronted with the decision of whether to use a monopropellant or a bipropellant system.

A **monopropellant** is a fuel which requires no separate oxidizer, but provides its propulsive energy through its own decomposition. The advantage of a monopropellant is the inherent simplicity of having only one liquid to supply to the combustion chamber. The principal disadvantage of the commonly used monopropellants is their very low impulse compared to most bipropellant or solid-propellant systems. Two of the more commonly used monopropellants are ethylene oxide and hydrogen peroxide.

Bipropellant systems use two liquids, an oxidizer and a fuel, which are merged and burned in a combustion chamber. There is a much wider selection of fuel constituents for bipropellant systems than for monopropellant systems. Bipropellants offer much higher impulse values than do the monopropellants, and higher than those available with solid propellants. The primary disadvantage of bipropellant systems is the need for a far more complex piping and metering system than is required for monopropellants.

Bipropellants may be subdivided further into two categories, **hypergolic** (in which the two constituents ignite on contact) and **nonhypergolic**. Hypergolic systems eliminate the need for a separate igniter, thus decreasing system complexity; however, this increases the fire hazard if the fuel system should leak.

When **cryogenic liquids** such as liquid hydrogen and liquid oxygen are used, the problems of storage become of major significance. A significant penalty in weight and complexity must be paid to store these liquids and hold them at temperature. Furthermore, these liquids charge a significant penalty, because their very low density enlarges the packaging requirements even more. This penalty is in the parasitic weight of the packaging and the increased drag induced by the increased skin area. These problems have limited the use of these propellants to systems where immediate response is not required, such as space launches, as opposed to strategic or tactical systems.

Fuels for bipropellant systems include methyl alcohol, ethyl alcohol, aniline, turpentine, unsymmetrical dimethylhydrazine (UDMH), hydrazine, JP-4, kerosene, hydrogen, and ammonia. **Oxidizers** for bipropellant systems include nitric acid, hydrogen peroxide, oxygen, fluorine, and nitrogen tetroxide. In some cases, mixtures of these will yield superior properties to either ingredient used alone; e.g., 50:50 mixture of UDMH with hydrazine is sometimes used in lieu of either alone.

Solid Propellants   Once the decision has been made to use a solid propellant, one must decide between a case-bonded and cartridge-loaded propellant grain. The decision then must be made among composite, double-base, and composite modified double-base propellants.

Case-Bonded versus Cartridge-Loaded   Case bonding refers to the technique by which the propellant grain is mechanically (adhesively) linked to the motor case. Cartridge-loaded propellant grains are retained in the motor case by purely mechanical means.

Case bonding takes advantage of the strength of the motor case to support the propellant grain radially and longitudinally; this permits the use of low-modulus propellants. This technique also yields high volumetric efficiency.

Since the propellant-to-case bond effectively inhibits the outer surface of the propellant, the quality of this bond becomes critical to the proper function of the rocket motor. This outer inhibition also tends to limit the potential propellant grain surface configurations, thus limiting the interior ballistician's leeway in tailoring the rocket motor ballistics. Since the propellant grain is bonded to the motor case, the grain must be cast into the motor case or secondarily bonded to it. The former method leads to reduced flexibility in scheduling motor manufacture. The second technique can lead to quality control problems if a bare, uninhibited grain is bonded to the motor case, or to a loss in volumetric loading efficiency if a bare grain is inhibited or cast into a premolded form that is bonded to the motor case. Bonding the propellant grain to the motor case also makes it difficult to dispose of the rock motor when it reaches the end of its service life, especially in regard to making the motor case suitable for reuse.

Cartridge loading of the propellant grain offers flexibility of manufacture of the rocket motor, as the motor-case manufacture and propellant-grain manufacture may be pursued independently. This technique

also allows the interior ballistician a free hand in configuring the propellant surface to obtain optimal ballistics. Since the propellant grain is held in the motor case mechanically, it is a simple matter to remove the propellant at the end of its service life and reuse the motor case and associated hardware.

The cartridge-loaded propellant grain must be inhibited in a separate operation, as opposed to the case-bonded grain. The free-standing grain must be supported so that it is not damaged by vibration and shock; correspondingly, the propellant must have substantial strength of its own to withstand the rigors of the support system and the vibrations and shocks that filter through the system. The presence of the mechanical support system and the inhibitor, and the space required to slide the propellant grain into the motor case, prevent the cartridge-loaded propellant grain from having the volumetric loading efficiency of a case bonded propellant grain.

Composite versus Double-Base versus Composite Modified Double-Base   Composite propellants consist primarily of a binder material such as polybutadiene (artificial rubber) and finely ground solid fuels (such as aluminum) and oxidizers (such as ammonium perchlorate). Double-base propellants consist primarily of stabilized nitrocellulose and nitroglycerine. Composite modified double-base propellants use a double-base propellant for a binder, with the solid fillers commonly found in composite propellants.

Composite propellants offer moderately high impulse levels and widely tailorable physical properties. They may be designed to function over a wide range of temperatures and, depending upon the fillers, have high ignition temperatures and consequently very favorable safety features.

Composite propellants at the present time are limited to cast applications. The binders are petroleum-based, and subject to the vagaries of oil availability and price. Many of the formulations for their binders contain toxic ingredients as well as ingredients which are sensitive to moisture. The water sensitivity carries over to the filler materials which tend to dissolve or agglomerate in the presence of moisture in the air. Composite propellants also are notoriously difficult to adequately inhibit once the propellant has fully cured.

Historically, a number of binder materials have been used in the manufacture of composite propellants. Among these are asphalt, polysulfides, polystyrene-polyester, and polyurethanes. Propellants of recent development have been predominantly products of the polybutadiene family: carboxy-terminated polybutadiene (CTPB), hydroxy-terminated polybutadiene (HTPB), and carboxy-terminated polybutadiene-acrylonitrile (CTBN). It should be pointed out that the binder material also serves as a fuel, so a satisfactory propellant for many purposes may be manufactured without a separate fuel added to it. While the addition of metallic fuels to the propellant significantly increases the energy of the propellant, it also produces primary smoke in the form of metal oxides. The most commonly added metal is aluminum, but magnesium, beryllium, and other metals have been tried. The most commonly used oxidizer (which makes up the preponderance of the weight of the propellant) is ammonium perchlorate. Other oxidizers which have been used include ammonium nitrate, potassium nitrate, and potassium perchlorate.

Double-base propellants have been used in guns since Alfred Nobel's discovery of ballistite in the nineteenth century. Their first significant use in rocketry came during World War II in barrage rockets. All our modern double-base propellants are descended from JPN, of World War II vintage. These propellants may be cast or extruded, use relatively nontoxic ingredients which are relatively insensitive to water, do not use petroleum derivatives to any appreciable extent, and are easily inhibited by solvent bonding an inert material to the surface of the propellant. It is possible to attain very low temperature coefficients of burn rate with these propellants; it is also possible to attain a double-base propellant which declines in burn rate with pressure over a portion of its usable pressure range, as opposed to the more usual monotonic increase in burn rate. This "mesa" burning, as it is called, allows the interior ballistician to more easily stabilize the pressure and thrust of the rocket motor.

Double-base propellants tend to have low to moderate impulse levels and are limited in temperature range not only by a high glass transition temperature, but also by their tendency to soften, liberate nitroglycerine, and decompose at higher temperatures. These propellants tend to auto-ignite at low temperatures and tend to be explosive hazards. Their physical properties are essentially fixed by the properties of nitrocellulose and are not easily tailored to changing requirements.

Double-base propellants contain nitrocellulose of various nitration levels (usually 12.6 percent nitrogen), nitroglycerine, and a stabilizer. Various inert plasticizers are added to modify either the flame temperature or the physical properties of the propellant.

Composite modified double-base propellants offer the highest energy levels presently available with solid propellants. This energy comes at an extremely high price, however: the storage and operating temperature limits only differ from minimum to maximum by about 20°F. These propellants are also extremely shock-sensitive and have very low auto-ignition temperatures. They are limited in use to systems such as submarine-launched ballistic missiles, where the temperature and vibration conditions are closely controlled. Classically, these compositions have contained nitrocellulose, nitroglycerine, aluminum, ammonium perchlorate, and the explosive HMX, which serves both as an oxidizer and gas-producing additive.

Propellant Properties and Interior Ballistics

The thrust and pressure profiles of a rocket motor must be controlled in order to meet the design criteria noted earlier. With liquid and solid propellants, the nominal controls differ dramatically, but in the elimination of spurious pulses and the control of nozzle erosion, the two types of propellant are more similar than different.

Pressure and Thrust Control In design of a liquid-propellant rocket motor, the prime considerations are properly sizing the combustor and nozzle and metering the propellant flow to attain the desired thrust. If variations in thrust are desired, these may be carried out either through preprogramming the flow by orifice size or by pump pressure; similarly, the flow may be varied on command by the operator.

The situation with a solid rocket motor is less clear-cut. The rate of gas generation inside the rocket motor is controlled by the burn rate of the propellant and the amount of burn surface available.

The burn rate of the propellant may be varied by the addition of various chemical catalysts; iron compounds are sometimes added to composite propellants and lead compounds to double-base propellants in order to speed burning. Coolants such as oxamide are added to composite propellants and inert plasticizers are added to double-base propellants to slow burn rates. The burn rate of composite propellants may be changed by changing oxidizers or by modifying the size distribution of the oxidizer. In efforts to achieve ultra-high burn rates, silver or aluminum wires have been added to propellants to increase heat conduction and so speed burning.

The surface area of a propellant grain is controlled by its shape and by the amount and areas in which the propellant grain is inhibited. This surface area tends to change in configuration as the propellant burns. The propellant grain designer must evolve a grain configuration which yields the desired pressure-time and thrust-time history. Most often, the objective in grain design is to achieve a neutral to slightly regressive (constant to slightly decreasing) thrust-time history.

In some cases (as where a high-thrust phase is to be followed by a low- to moderate-thrust phase), it is necessary to develop a grain or grains with varying geometries and/or varying burn rates to achieve the desired thrust-time profile. These results may require such techniques as using tandem or coaxial grains with different geometries and/or different burn rates and perhaps different propellants.

The pressure within the combustion chamber depends not only on the rate of combustion, but also on the size of the nozzle throat. Throat size tends to decrease with heating, but the surface tends to wear away with the hot gases and embedded particles eroding material from the surface of the throat and nozzle exit cone. In some cases, this loss of material may serve to assist the interior ballistician in the quest for the ideal thrust-time profile.

Combustion Instability Rocket motors are sometimes given to sudden, erratic pressure excursions for a number of reasons. Liquid-propellant fuel may surge; solid propellant grains may crack; material may be ejected from the motor and temporarily block the nozzle. Sometimes these excursions cannot be explained by any of the above possibilities, but will be attributed to "unstable combustion." The causes of unstable or "resonant" combustion are still under investigation.

A number of techniques have been developed to reduce or eliminate the pressure excursions brought on by unstable combustion, but there is, as yet, no panacea.

Nozzle Erosion As the propellant gases pass out of the rocket motor through the nozzle, their heat is partially transferred to the nozzle. The heated nozzle material softens and tends to be eroded by the mechanical and chemical action of the propellant gases. The degree of erosion is heightened when a metal fuel is added to the propellant gases, particulate matter is contained in the gases, the gases are corrosive, or the propellant gases are oxidizing.

One technique widely used in reducing nozzle erosion is to add coolant to the propellant formulation. Ablatives are sometimes added to the nozzle or chamber ahead of the nozzle throat; gases generated by the ablating material form a boundary layer to protect the nozzle from the hot propellant gases in the core flow. In liquid propellant rockets, the fuel (where stability permits) may be used as a coolant fluid. Nozzle inserts are probably the most commonly used technique to limit nozzle erosion. The nozzle shell is usually made of aluminum or steel when inserts are used and the nozzle throat is made of heat-resistant material. For a small amount of permissible erosion, carbon inserts are used; when no erosion is acceptable, tungsten or molybdenum inserts are used.

7.2 CARBONIZATION OF COAL AND GAS MAKING
by Klemens C. Baczewski

REFERENCES: Porter, "Coal Carbonization," Reinhold, Morgan, "Manufactured Gas," J. J. Morgan, New York. *BuMines Monogr.* 5 and other papers. Powell, "Future Possibilities in Methods of Gas Manufacture," and Russell, "The Selection of Coals for the Manufacture of Coke," papers presented to the AGA Production and Chemical Conference, Wilson and Wells, "Coal, Coke and Coal Chemicals," McGraw-Hill. Elliott, High-Btu Gas from Coal, *Coal Utilization,* Dec. 1961. Osthaus, Town Gas Production from Coal by the Koppers-Totzek Process, *Gas and Coke,* Aug. 1962. "Clean Fuels from Coal Symposium," IGT, Sept. 1973. Kirk-Othmar, "Encyclopedia of Chemical Technology," 2d ed., Barker, Possible Alternate Methods for the Manufacture of Solid Fuel for the Blast Furnace, *Jour. Iron Steel Inst.,* Feb. 1971. Potter, Presidential Address, 1970, Formed Coke, *Jour. Inst. Fuel,* Dec. 1970, International Congress, Coke in Iron and Steel Industry, Charleroi, "1966 Gas Engineers Handbook," The Industrial Press. Quarterly *Coal Reports* 1982, 1983, U.S. Department of Energy. Elliott, "Chemistry of Coal Utilization," 2d supp. vol., Wiley, 1981. Davis, Selection of Coals for Coke Making, *U.S. BuMines Rep. Inv.* 3601, 1942. Wolfson, Birge, and Walters, Comparison of Coke Produced by BM-AGA and Industrial Methods, *U.S. BuMines Rep. Inv.* 6354, 1964. Iron and Steel Society, Inc., *Ironmaking Conf. Proc.,* 1988, 1992, 1993; "Marks' Standard Handbook for Mechanical Engineers," 9th ed., McGraw-Hill.

Carbonization of coal, or the breaking down of its constituent substances by heat in the absence of air, is carried on for the production of coke for metallurgical, gas-making, and general fuel purposes; and gas of industrial and public-utility use. Coal chemicals recovered in this country include tar from which are produced crude chemicals and materials for creosoting, road paving, roofing, and waterproofing; light oils, mostly benzene and its homologues, used for motor fuels and chemical synthesis; ammonia, usually as ammonium sulfate, used mostly for fertilizer; to a lesser extent, tar acids (phenol), tar bases (pyridine), and various other chemicals. Developments in new designs, pollution control equipment, and the production of formed coke are discussed.

Gas making, as treated here, includes gas from coal carbonization, gasification of solid carbonaceous feedstocks via fixed and fluid-bed units, gasification in suspension or entrainment, and gasification of liquid hydrocarbons.

CARBONIZATION OF COAL

Coke is the infusible, cellular, coherent, solid material obtained from the thermal processing of coal, pitch, and petroleum residues, and from some other carbonaceous materials, such as the residue from destructive distillation. This residue has a characteristic structure resulting from the decomposition and polymerization of a fused or semiliquid mass. Specific varieties of coke, other than those from coal, are distinguished by prefixing a qualifying word to indicate their source, such as "petroleum coke" and "pitch coke." A prefix may also be used to indicate the process by which coke is manufactured, e.g., "coke from coal," "slot oven coke," "beehive coke," "gashouse coke," and "formcoke." See Table 7.2.1.

High-temperature coke, for blast-furnace or foundry use, is the most common form in the United States. In 1982, almost 100 percent of the production of high-temperature coke was from slot-type ovens, with minimal quantities being produced from nonrecovery beehive and other types of ovens. Blast furnaces utilized 93.1 percent; foundries, 4.3 percent; and other industries, the remainder. Low- and medium-temperature cokes have limited production in the United States because of a limited market for low-temperature tar and virtually no market for the coke.

The following data on the properties of blast-furnace coke were obtained from a survey of plants representing 30 percent of the U.S. production; volatile matter of the cokes ranged from 0.6 to 1.4 percent; ash, from 7.5 to 10.7 percent; sulfur, from 0.6 to 1.1 percent; 2-in shatter index, from 59 to 82; 1½-in shatter index, from 83 to 91; 1-in tumbler (stability factor), from 35 to 57; and ¼-in tumbler (hardness factor), 61 to 68. Comparison of this survey with a prior survey made in 1949 indicates that the quality of blast-furnace coke has been improved by reducing the ash and sulfur contents and increasing the average ASTM tumbler stability from 39 to 52. The tumbler stability is the principal index for evaluating the physical properties of blast-furnace coke in the United States. Other tests for determining the physical properties of blast-furnace and foundry cokes that are cited in export specifications are the MICUM, IRSID, ISG, and JIS methods.

During the coking process, several additional products of commercial value are produced. If the plant is large enough to recover these products, their value can approach 35 percent of the coal cost. The more

valuable products are fuel gas with a heating value of 550 Btu/ft³ (20,500 kJ/m³); tar and light oils that contain benzene, toluene, xylene, and naphthalene; ammonia; phenols; etc.

The requirements for foundry coke are somewhat different from those for blast-furnace coke. Chemically, in the cupola the only function of the coke is to furnish heat to melt the iron, whereas in the blast furnace the function is twofold: to supply carbon monoxide for reducing the ore and to supply heat to melt the iron. Foundry coke should be of large size (more than 3 in or 75 mm) and strong enough to prevent excessive degradation by impact of the massive iron charged into the cupola shaft. The following characteristics are desired in foundry coke: volatile matter, not over 2 percent; fixed carbon, not under 86 percent; ash, not over 12 percent; and sulfur, not over 1 percent. In the coke production survey, foundry coke from two plants showed the following properties; volatile matter, 0.6 and 1.4 percent; fixed carbon, 89.6 and 91.4 percent; ash, 8.7 and 7.5 percent; and sulfur, 0.6 percent. The 1½- and 2-in shatter indexes, which are a measure of the ability of coke to withstand breakage by impact, were 98 and 97, respectively, for both cokes.

Pitch coke is made from coal tar pitch, whereas petroleum coke is made from petroleum-refining residues. Both are characterized by high-carbon and low-ash contents and are used primarily for the production of electrode carbon. Coke consumption in the United States was approximately 29.2 million tons in 1989 and 23.9 million tons in 1993. The major user is blast-furnace operations, with others using 100,000 tons. (American Iron and Steel Inst., Annual Statistical Report, Washington, 1993). This trend is expected to continue to decrease as electric-arc furnace (EAF) use increases and as direct reduction and coal injection systems are installed. (Chem. Engrg., March 1995, p. 37.)

High-temperature carbonization or coking is carried on in ovens or retorts with flue-wall temperatures of ± 1,800°F (980°C) for the production of foundry coke and up to ± 2,550°F (1,400°C) for the production of blast-furnace coke. Typical yields from carbonizing 2,204 lb [1.0 metric ton (t)] of dry coal, containing 30 to 31 percent volatile matter, in a modern oven are: coke, 1,590 lb (720 kg); gas, 12,350 ft³ (330 m³); tar, 10 gal (37.85 L); water, 10.5 gal (39.8 L); light oil, 3.3 gal (12.5 L); ammonia, 4.9 lb (2.22 kg).

Coal Characteristics   Despite the vast coal reserves in the United States, most of the coal is not coking coal. Coking coals are only those coals which, according to the ASTM classification by rank, fall into the class of bituminous coal and are in the low-volatile, medium-volatile, high-volatile A or high-volatile B groups and which, when heated in the absence of air, pass through a plastic state and resolidify into a porous mass that is termed coke. In determining if an unknown coal is a coking coal, prime importance is placed upon obtaining a freshly mined sample since all coking coals experience oxidation or weathering which can cause a loss in coking ability.

Laboratory tests are used by coal investigators to determine if particular coals have coking properties and how they can best be used to make coke. The most common of these tests is proximate analysis (ASTM D3172), which provides the coal rank and ash content. It is desirable to have low ash coals (below 8 percent), since the ash does not contribute to the blast furnace or foundry processes. Sulfur content (ASTM D3177) passes through the coking process and appears in the final coke and in the evolved gases during coking. High sulfur contents (above 1.0 per-

Table 7.2.1   Analyses of Cokes

| | "As-received" basis | | | | | | | | | |
| | Proximate, % | | | | Ultimate, % | | | | | High heat value, Btu/lb† |
Coke type	Moisture	Volatile matter	Fixed carbon	Ash*	Hydrogen	Carbon	Nitrogen	Oxygen	Sulfur	
By-product coke	0.4	1.0	89.6	9.0	0.7	87.7	1.5	0.1	1.0	13,200
Beehive coke	0.5	1.2	88.8	9.5	0.7	87.5	1.1	0.2	1.0	13,100
Low-temperature coke	0.9	9.6	80.3	9.2	3.1	81.0	1.9	2.8	1.0	12,890
Pitch coke	0.3	1.1	97.6	1.0	0.6	96.6	0.7	0.6	0.5	14,100
Petroleum coke	1.1	7.0	90.7	1.2	3.3	90.8	0.8	3.1	0.8	15,050

* Ash is part of both the proximate and ultimate analyses.
† Btu/lb × 2.328 = kJ/kg.

cent) affect the iron quality and require additional gas processing for removal. **Free-swelling index** (ASTM D720) is a fast method to determine if a coke will form a coherent mass. Some observers feel the size of the "button" produced is important while others use it only as a screening device. **Gieseler plastomer** (ASTM D2639) is the most popular of several dilatometer testers which measure the fluid properties of the coal through the plastic state and into the solidification phase. It is generally agreed that the test is useful, although some investigators feel it only indicates that a coal is coking. Others feel that the temperatures at which the coal begins to soften and then resolidify serve as guides to establish which coals are suitable for blending. Still others couple the use of these data with petrographic analysis to predict coke strengths of various blends. While Gieseler is widely used in the United States, two other methods, Audibert-Arnu and the Ruhr test, are frequently used throughout the world. **Petrographic composition** is determined by the examination of coal under a microscope. Results were first reported in 1919 but it was not until 1960 that a method for predicting coke strength was introduced by Schapiro and Gray (Petrographic Constituents of Coal, *Illinois Mining Institute Proc.*, 1960); it has become an important and popular test for the selection of coal for coke manufacturing. The method consists of determining which portion of the coal becomes plastic during heating (reactive entities) and which portion does not undergo plastic change (inert entities). These observations are then correlated with pilot oven tests and are used to predict coke strength.

Coal Blending The use of a single coal to produce strong metallurgical coke without resultant high coking pressures and oven wall damage is very rare, and coke producers rely on the blending of coals of varying coking properties to produce strong coke. It is common practice in the coking industry to mix two or more coals to make a better grade of coke or to avoid excessive expansion pressures in the oven. One coal is usually of high volatile content (31 to 40 percent) and the other of low volatile content (15 to 22 percent). The amount of low-volatile coal in the mixture is generally in the range of 15 to 25 percent although as much as 50 percent may be used in mixtures for producing foundry coke. High-volatile coals tend to shrink during coking, while low-volatile coals tend to expand. Examinations of the plastic properties of coals when heated are valuable in selecting the best types for blending. Because of the risks of oven wall damage from unknown coal mixtures, most operators will not rely solely on laboratory tests but will insist on some pilot-scale oven testing.

Pilot-Scale Tests A number of designs of pilot-scale ovens which closely approximate commercial coke ovens have been developed. A few have been used solely to produce coke for testing but most have been designed with one fixed wall and one movable wall so that data about carbonization pressures in the oven during coking can be collected while making coke for testing. These ovens usually hold between 400 and 1,000 lb (180 to 450 kg) of coal. Another test oven in use was developed by W. T. Brown (*Proc. ASTM*, **43**, 1943, pp. 314–316) and differs from the movable wall oven by applying a constant pressure on the charge while heating the coal from only one side.

The European Cokemaking Technology Center (EEZK, Essen Germany) has operated a mini-jumbo reactor, which has a chamber width of 34 in (864 mm) and length and height of 40 in (101.6 mm). A demonstration facility, the Jumbo Coking Reactor, is operational, at 100 mt/d. The chambers of this unit are 34 in (864 mm) wide, 32.8 ft (10 m) high, and 65.6 ft (20 m) long. The concept of such large chambers is supported by full-scale tests conducted on 17-, 24-, and 30-in-wide (450-, 610-, and 760-mm-wide) ovens and is based on preheated coal charging. (Bertling, Rohde, and Weissiepe, *51st Ironmaking Conf. Proc.*, Toronto, Apr. 1992.)

Coking Process Coal produces coke because the particles of coal soften and fuse together when sufficiently heated. Initial softening of the coal, as determined by plastometer tests, occurs at 570 to 820°F (300 to 440°C). At or near the softening temperature gases of decomposition begin to appear in appreciable quantities, gas evolution increasing rapidly as the temperature is raised. This evolution of gases within the plastic mass causes the phenomenon that finally results in the cellular structure which is so characteristic of coke. Further temperature

increase and decomposition cause hardening into coherent porous coke.

As a result of the low thermal conductivity of coal (less than one-sixth of that of fire clay), and also of semicoke, heat penetrates slowly into the pieces and through the plastic layer; uniform plasticity or complete coalescence does not appear until the temperature (at the heated side of a softening layer) is considerably higher than the softening point of the coal. (See Fig. 7.2.1.)

The **plastic zone** moves slowly from the hot wall of the oven toward the center, at a rate first decreasing with the distance from the wall and then increasing again at the middle of the oven. For several hours after charging a red-hot oven, the center of the charge remains cool. The plastic layer's temperature variation, from one border to the other, is from 700 to 875°F (370 to 468°C), and its thickness is ⅜ to ¾ in depending on the coal, the charge density, and the oven temperature.

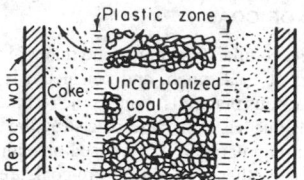

Fig. 7.2.1 Diagrammatic illustration of the progress of carbonization and of composition of the plastic zone.

In the modern coal-chemical-recovery coke oven, the average rate of travel of the plastic zone is about 0.70 in (17.8 mm) per h, and the average **coking rate**, to finished coke in the center, is 0.50 to 0.58 in (12.7 to 14.7 mm); i.e., a 17-in (432-mm) oven may be run on a net coking time of 16 to 17 h. (See Fig. 7.2.2.)

New designs of wide ovens with chamber widths of 21.6 in (550 mm) have been built. The coking time increases slightly, to about 23 h, but not in proportion to the width increase. (Beckmann and Meyer, *52d Ironmaking Conf. Proc.*, Dallas, Mar. 1993.)

The gases and tar vapors travel chiefly outward toward the wall from the plastic layer and from the intermediate partly coked material, finding exit upward through coke and semicoke. Exit through the center core of uncoked coal, except for a very small fraction of the early formed gases, is barred by the relative impermeability of the plastic layer. The final chemical products, including gas, are the result of secondary decompositions and interreactions in the course of this travel.

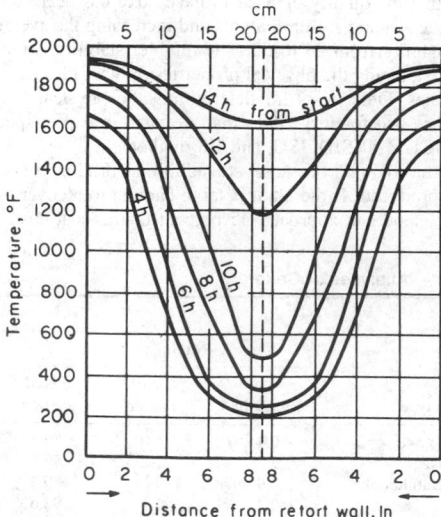

Fig. 7.2.2 Temperature gradients in a cross section of a coal-chemical-recovery coke oven, 17 in wide, at about middepth on a 17-h coking time.

Average temperatures in various parts of the carbonizing system, for modern rapid coal-chemical-recovery oven operation, are about as follows: heating flues, at bottom, 2,500 to 2,600°F (1,370 to 1,425°C); heating flues, upper part, 2,150 to 2,450°F (1,175 to 1,345°C); oven wall, inner side (average final), 1,850 to 2,100°F (1,010 to 1,150°C).

Temperature Effects during Carbonization In industrial carbonizing, higher temperatures at the oven wall and in the outer layers of the charge produce higher gas yield and less tar. The gases and tar vapors change in quality and quantity continuously during the carbonizing period. The percentage content of hydrocarbons and condensables in the oven gases decreases and that of hydrogen increases. Passage through the highly heated free space above the charge has the effect of increasing the yield of gas and light oil (reducing, however, the toluene and xylenes) and lowering tar yield, with increase of naphthalene, anthracene, and lowering tar yield, with increase of naphthalene, anthracene, and free carbon. Modern ovens tend to exercise control of the temperature in this free space. The rate of decomposition of ammonia increases above 1,450°F (788°C).

The progressive change in gas yield and composition during the carbonizing period for a good gas-making coal is about as in Table 7.2.2. Some gas yields and composition are shown in Table 7.2.3.

The **overall thermal efficiency** of industrial coal carbonization (useful recovery of heat from total input of heat) is between 86 and 92 percent approx. External sensible-heat efficiencies are between 65 and 80 percent approx. Typical heat balances on coal-chemical-recovery ovens are shown in Table 7.2.4.

Heat Used for Carbonization The total heating value of the gas burned in the flues to heat the ovens varies from 950 to 1,250 Btu/lb (528 to 695 kcal/kg) of wet coal carbonized in efficient installations depending on the heat required by the individual coals in the blend. Producer and blast-furnace gases are called **lean** gases. When underfiring with lean gas, both air and gas are regenerated (preheated) in order to get sufficient flame temperature for the flues. Natural, refinery, and I.P. gases have been used to a limited extent for underfiring. These and coke-oven gas are rich gases and are not preheated, as their flame temperatures are sufficiently high and regeneration would crack their hydrocarbons. Air is preheated in all cases.

Table 7.2.4 Heat Balance as a Function of the Volatile Matter Content*

	% Volatile matter (DAF)			
	23.8	26.5	28.7	33.2
Moisture, %	10.6	10.3	9.7	10.1
Consumption of heat, kcal/kg	492	496	522	559
Waste-gas loss, %	10.9	10.7	10.8	9.6
Surface loss, %	10.1	9.9	9.5	8.8
Total loss, %	21.0	20.6	20.3	18.4
kcal/kg	103.3	102.2	106.0	102.9
Effective heat (heat of coking), kcal/kg	388.7	393.8	416.0	456.1
Sensible heat in coke, kcal/kg	284.1	278.2	281.0	260.2
Sensible heat in gas, kcal/kg	144.1	152.3	164.9	188.3
Total sensible heat, kcal/kg	428.4	430.5	445.9	448.5
Heat of reaction (exothermic)	+ 39.7	+ 36.7	+ 29.9	− 7.6

The heat of reaction is the difference between effective heat and loss by sensible heats.
* W. Weskamp, Influence of the Properties of Coking Coal as a Raw Material on High Temperature Coking in Horizontal Slot Ovens, *Glückauf*, **103** (5), 1967, 215–225.

CARBONIZING APPARATUS

The current trend in design of **coal-chemical-recovery coke ovens** is to larger-capacity ovens and improved, oven-wall liner brick and wall design to afford increased heat transfer from the heating flues to the oven chamber. Formerly, ovens were usually about 40 ft (12 m) long and from 12 to 16 ft (3.7 to 5.0 m) high. Modern ovens are about 50 ft (15 m) long and 20 to 23 ft (6 to 7 m) high and hold a charge of 35 tons (32 mt) or more. Average oven width is 16 to 19 in (400 to 475 mm), usually 18 in (450 mm), with a taper of 3.0 to 4.5 in (75 to 115 mm) from the pushing end to the coke-discharge end of the oven. New de-

Table 7.2.2 Variation in Gas Yield during Carbonization

Period	Volume m³/Mt	kcal/m³	Volume ft³/ton	Btu/ft³	Approx composition, % Hydrocarbons	Hydrogen	Oxides of carbon
First quarter	1,130	5,800	3,630	651	41	46	7
Second quarter	1,000	3,400	3,190	610	37	53	7
Third quarter	1,010	5,050	3,250	567	32	59	6
Fourth quarter	585	3,230	1,875	363	8	82	5

Table 7.2.3 Gas Yields and Composition from Various Coal Types with High-Temperature Carbonization

Coal	Temp in inner wall, °F (°C)	Gas yield, ft³/ton (m³/Mt)	Gas composition, % Carbon dioxide	Carbon monoxide	Unsat'd hydro-carbons	Methane	Ethane, etc.	Hydrogen	Nitrogen
Pittsburgh bed, Fayette Co., Pa., V.M.* 33.6	1,950 (1,065)	11,700 (365)	1.3	6.8	3.2	31.1	0	56.5	1.1
Elkhorn bed, Letcher Co., Ky., V.M. 36.6	1,950 (1,065)	11,500 (358)	1.1	7.7	4.0	31.0	0.2	55.0	1.0
Sewell bed, W. Va., V.M. 26.5	1,950 (1,065)	12,000 (375)	0.7	5.5	2.5	26.5	0	64.8	1.0
Pocahontas no. 4, W. Va., V.M. 16.4	1,950 (1,065)	11,900 (372)	0.4	5.0	1.1	18.0	0	75.0	0.5
Illinois, Franklin Co., V.M. 32.1	1,950 (1,065)	12,000 (375)	3.8	14.5	2.8	21.0	0	56.9	1.0
Utah, Sunnyside, V.M. 38.8	1,950 (1,065)	12,600 (394)	3.0	14.5	3.7	26.0	0.5	51.3	1.0

*V.M. = percentage of volatile matter.

signs, tending toward larger ovens, have been built with oven widths from 21.6 to 24.0 in (550 to 610 mm), 24.75 ft (7.85 m) high, and 54 ft (16.5 m) long. Coke production of the 21.6-in oven is about 47.3 tons (43 mt) per charge. The larger oven yields more coke per push, reducing environmental problems while increasing productivity (Hermann and Schonmuth, *52d Ironmaking Conf. Proc.*, Detroit, Mar. 1993; Beckmann and Meyer, op. cit.).

Various oven designs are used, distinguished chiefly by their arrangements of the vertical heating flues and wasteheat ducts. Two basic designs are used for heating with rich gas (coke-oven gas): (1) the gun-flue design wherein the fuel gas is introduced via horizontal ducts atop the regenerators and thence through nozzles which meter the gas into the vertical flues, and (2) the underjet design which incorporates a basement, located underneath the regenerators, in the battery structure. Horizontal headers running parallel with the vertical heating walls convey the fuel gas via riser pipes through the regenerators to the vertical heating flues. One design recirculates waste combustion gas from the adjacent heating wall through a duct underneath the oven and regenerator by the jet action of the fuel gas through specially designed nozzles. This provides a leaner gas at the place combustion occurs and affords a more even vertical heat distribution for tall ovens. Designs for lean gas (blast-furnace gas, producer gas) employ sole flues beneath the regenerators for introducing the fuel gas. Ports control the quantity of gas fed from the sole flues into the regenerators. To prevent equalization of gas distribution after the gas leaves the sole flues, the regenerator chambers are divided into compartments. All modern ovens use regenerators for preheating the combustion air and lean gas. Average wastegas (stack) temperatures range from 450 to 700°F (230 to 370°C).

Coke ovens are built in batteries of 15 to 106 and arranged so that each row of heating flues, or wall, heats half of two adjacent ovens. Modern practice is to build batteries of the maximum number of ovens in a single battery that can be operated by a single work crew to optimize productivity. This is in the range of 79 to 85 ovens per battery. Coal is charged from a larry car through openings in the top of the oven. After coking, doors are removed from both ends of the oven and coke is pushed out of the oven horizontally by a ram operated by a pusher machine. Gas is removed continuously at constant pressure (few millimeters of water column) via oven standpipes connected to a gas collecting main.

Computerized control is being used for new facilities and is being applied to existing ones to improve efficiency. This integrates control of variables such as charge weight and moisture, excess combustion air, flue temperature, and coke temperature. The performance improvement in one case was the reduction in heating requirements from about 1,500 Btu/lb (832 Kcal/kg) dry coal to under 1,200 Btu/lb (666 Kcal/kg) dry coal and the stabilization of coking times, ranging from 18 to 40 to 24 h (Pfeiffer, *47th Ironmaking Conf. Proc.*, Toronto, Apr. 1988).

Coal Chemical Recovery The gas is first cooled in either direct or indirect coolers which condense most of the tar and water from the gas. Some ammonia is absorbed in the water, forming a weak ammonia liquor. Exhausters (usually centrifugal) follow and operate from 6- to 12-in (150- to 300-mm) water column suction at the inlet to 50- to 80-in (1,250- to 2,000-mm) pressure discharge. Electrical precipitators remove the last traces of tar fog. The gas, combined with ammonia vapor stripped from the weak ammonia liquor, then passes through dilute sulfuric acid in saturators or scrubbers which recover the ammonia as ammonium sulfate. Phosphoric acid may be used as the absorbent to recover the ammonia as mono- or diammonium phosphate. Low-cost synthetic ammonia produced via reforming of natural gas and hydrocarbons has made recovery of coke-oven ammonia uneconomical. Some recent coke plants have been designed with ammonia destruction units.

After direct cooling with water, which removes much of the naphthalene from the gas, the gas is scrubbed of light oil (benzol, toluol, xylol, and solvents) with a petroleum oil. The enriched petroleum oil is stripped of the light oil by steam distillation, and the light oil is usually sold to the local oil companies. Only a few coke plants continue to refine light oil. Phenols and tar acids are recovered from the ammonia liquor and tar. Pyridine and tar bases are recovered

from the ammonia saturator liquor and tar. Distillation of the tar produces cresols, naphthalene, and various grades of road tar and pitches. Acid gases (hydrogen sulfide, hydrogen cyanide) are removed from the gas. The hydrogen sulfide is converted to elemental sulfur. The cyanogen may be recovered as sodium cyanide.

Pollution Control Enactment of pollution-control laws has motivated many developments of control devices and new operating techniques. The continuous emissions caused by leakage from oven doors, charging hold lids, standpipe lids, and oven stacks are being controlled by closer attention to operations and maintenance. Emissions from charging are being contained by the use of a second collecting main, jumper pipes and U-tube cars.

Pushing emissions equipment includes hood duct arrangements with wet scrubbers or baghouses, mobile hoods with wet scrubbers, and coke side enclosures with baghouses. There has been much interest in Japan and the Soviet Union in dry quenching of coke with inert gases for pollution control and coke quality improvements, but so far in the United States the capital and maintenance costs have dampened any enthusiasm for dry quenching. In the coal chemical plant area, proposed laws for benzene emissions will require many plants to revise and rebuild major portions of their equipment. Sulfur emissions standards have caused most plants to install sulfur recovery equipment.

Systems for charging **preheated coal** into coke ovens by either pipeline (Marting and Auvil, Pipeline Charging Preheated Coal to Coke Ovens, UNEC Symposium, Rome, Mar. 1973), hot conveyors, or hot larry cars have been built to eliminate charging emissions, improve oven productivity, and to use larger amounts of weakly coking coals. It appears that the economics are positive only if very poor and cheap coking coals are available. No systems are operating in the United States because good coking coal is readily available. The Japanese have looked at preheating of coal and have decided that it is more effective to **briquette** a portion of the poorer-quality coals and mix them with the normal coal mixture being charged to the ovens. Briquettes usually make up about 30 percent of the blend, and many Japanese plants have adopted a form of briquette blending. Preheating is becoming important again in the development of wider ovens and jumbo reactors, which are based on using this technique.

Form coke, the production of shaped coke pieces by extrusion or briquetting of coal fines followed by carbonization, has been practiced for many years in the United Kingdom and Europe to provide a "smokeless" fuel primarily for domestic heating. Form coke for use in low shaft blast furnaces is produced commercially from brown coal (lignite) at large plants in Lauchhammer and Schwarze-Pumpe, East Germany. As much as 60 to 70 percent of form coke is used in combination with conventional slot oven coke.

A major development is **FMC coke process** which is operating commercially at the FMC plant in Kemmerer, Wyoming. The facility is producing coke for use in the elemental phosphorus plant in Pocatello, Idaho. The process produces a low-volatile char, called **calcinate,** and a pitch binder. Crushed coal is dried, carbonized, and calcined at successively higher temperatures, while solids flow continuously through a series of fluid-bed reactors. Operating temperatures range from 300 to 600°F (150 to 315°C) in the first bed to 1,500 to 2,200°F (815 to 1,200°C) in the third. The calcinate is cooled before being mixed with the pitch binder. The mix is briquetted into pillow shapes up to 2-in sizes and is sent to curing ovens to be devolatilized and hardened into the finished coke product. Off-gases from all systems are cooled and cleaned, recycling dusts into the process to be included in the product. Cleaned gases have a heating value of 100 to 140 Btu/SCF (890 to 1,240 kcal/m³) and can be used as best suit the local conditions. Typically, it is fuel for process needs, and for steam generation, while residual gas can be used in cogeneration applications.

The process is continuous, and fully contained, so that environmental controls can be met with conventional equipment. The Kemmerer plant fully meets EPA and OSHA regulations. Coal types suitable for processing include lignites and extend to anthracites. Different volatility, fluidity, and swelling characteristics can be accommodated by appropriate adjustment of operating parameters. This permits the use of non-metallurgical coals for coke production. The coke product has been tested in blast furnaces for many years. A major effort was the 20,000-ton trial at Indiana Harbor Works of Inland Steel by a consortium of steel companies. Furnace operation was normal up to 50 percent of FMC product in the coke burden, with indications that higher proportions would also be satisfactory. The process is illustrated in Fig. 7.2.3. (FMC personal communications)

The **Bergbau-Forschung process** [Peters, Status of Development of Bergbau-Forschung Process for Continuous Production of Formed Coke, *Glückauf*, **103** (25), 1967] involves devolatilization of low-rank coal to yield a hot char. The hot char is mixed with a fluid coking coal (about 70 percent char and 30 percent coking coal) which becomes plastic at the mixing temperature, and the mixture is briquetted hot in roll presses to produce "green" briquettes containing 7 to 8 percent volatile matter. If required, further devolatilization of the briquettes is accomplished in a vertical hot-sand carbonizer. The **Ancit process** (Goosens and Hermann, "The Production of Blast Furnace Fuel by the Hot Briquetting Process of Eschweiler Bergwerks-Verein," ECEC, Rome, Mar. 1973). The process is similar to the B-F process in that it uses about 70 percent noncoking coal with about 30 percent coking coal as binder. The noncoking coal component is conveyed pneumatically from bunkers and introduced at two locations into a horizontal, parallel-flight stream reactor heated by hot products of combustion. The coal is heated to 600°C in a fraction of a second and is thermally decomposed by the rapid evolution of water and volatile matter. Coal and gas are separated in a cyclone with the gas passing to a second reactor (installed in tandem arrangement) into which the coking coal is fed. Coal and gas pass to a second cyclone for separation. The two heated coals are fed by screw feeders into a vertical cylindrical mixer, and the mixture is fed to roll presses and briquetted. The **Consolidation Coal process** employs a heated rotary kiln to produce medium-temperature coke pellets from a mixture of char and coking coal. Pellets are then subjected to final high-temperature carbonization in a vertical-shaft unit. Other processes receiving attention were the Sumitomo process, Japan; Auscoke (BHP), Australia; the Sapozhnikov process, Russia; and INIEX, Belgium. After years of work on these various processes, it became apparent that production rates, costs, and product quality could not compete with conventional by-product coke ovens and the test facilities in the United States were shut down.

GASIFICATION

Producer gas and carbureted water gas were common in Europe and the United States for many years and were based on coal and coke. These units gasified the solid fuels by the reaction of oxygen (as air or enriched oxygen) and steam. Oil injection was practiced to improve the heating value. With the widespread distribution of natural gas, these plants have all been closed.

Gasification is achieved by partial oxidation of carbon to CO (exothermic reaction). To obtain a mixture of CO and H_2, water is introduced, typically as steam, which reacts endothermically with the coal. The partial oxidation supplies heat to the endotherm. These reactions are described in detail in Elliott, "Chemistry of Coal Utilization" (2d suppl. vol., Wiley-Interscience). The heating values of the producer gas were approximately 120 Btu/SCF (1,068 kcal/m³) for air-blow units, 250 Btu/SCF (2,225 kcal/m³) or more for oxygen-blown units, and as much as 500 Btu/SCF (4,450 kcal/m³) for the oil-carbureted units. The development of abundant natural-gas supplies and their distribution to most areas of the world have supplanted these processes. In 1974, an oil supply crisis combined with a distribution pinch on natural gas. Interest in converting coal to gaseous and oil fuels was rekindled in the United States. Many processes were piloted by government and industry. These included **Hi-Gas, Bi-Gas, CO_2 Acceptor, Synthane, Atgas,** and **molten salt processes**. A demonstration program was established by the U.S. government as the Synthetic Fuels Corporation. The program funded commercial-size units for **Cogas** and **Slagging Lurgi** for synthetic natural gas (SNG), and **H-coal** and **solvent-refined coal** (SRC) for liquids. As natural-gas distribution and the oil supplies improved, the urgency diminished,

and the projects were canceled during the latter part of their design. A notable exception is the commercial-scale Lurgi plant, which was built as the Great Plains Project in North Dakota and funded by the Department of Energy to produce SNG. Dakota Gasification Inc. operates this plant and is planning ammonia production due to low natural-gas pricing.

Interest in gasification continued for chemicals manufacture, and power, for several reasons. With respect to power, environmental regulations could be met more readily for sulfur emissions by treating a smaller stream than the corresponding flue gas from a fossil-fueled plant, while yielding a saleable product as opposed to landfill material. The potential efficiency improvement in **combined cycles** with gas turbines and steam turbines would further reduce the size of the treated streams, because of the need for less fuel. Corollary benefits ensue in reducing CO_2 emissions, and allow the use of high-sulfur coals. This has led to the construction of several **integrated gasification combined cycle** (IGCC) plants for power.

As part of this continuity of interest, research on hot gas desulfurization is in progress to reduce thermal losses in gas cooling. Other approaches such as the demonstration of **underground gasification** are being pursued.

Gasifiers come in three types:

1. Fixed-bed with the coal supported by a rotating grate
2. Fluidized-bed, in which the fuel is supported by gaseous reactants
3. Entrained flow gasifiers that use very fine particles suspended in a high-velocity gas stream

Fixed-Bed Gasifiers   Representative units are **Lurgi, Wellman-Galusha, Koppers-Kerpley, Heurtey,** and **Woodall Duckham.** Early units were air-blown and used coke, with a number built for anthracite. Agitators were added to permit feeding of bituminous coal. Coal feed is a sized lump, with 1½ in (37 mm) being typical. Fines are tolerated only in small amounts. Early units operated at essentially atmospheric pressure, but later units operate at elevated pressure. Temperatures are limited to avoid softening and clinkering of the ash. With air, gas having heating values of 120 to 150 Btu/SCF (1,068 to 1,335 kcal/m³) was produced. Use of oxygen allowed gas heating values of 250 to 300 Btu/SCF (2,225 to 2,673 kcal/m³). Coal is fed through lock hoppers to prevent loss of gases and to permit charging into pressurized units. The gasifiers have a distinct upper reduction zone, which dries and preheats the coal. The gasification reactions take place at 1,150 to 1,600°F (620 to 870°C), and the gases leave the unit at 700 to 1,100°F (370 to 595°C). Under these conditions, methane and other light hydrocarbons, naphtha, phenols, tars, oils, and ammonia are generated. The CH_4 is an advantage in SNG production. Tars and oils are removed from the gas stream before further processing to absorb ammonia and acid gases, including carbon dioxide and hydrogen sulfide.

The devolatilized coal passes into the lower combustion zone, reaching temperatures of 1,800 to 2,500°F (980 to 1,370°C), depending on the ash softening temperatures. It is removed by a rotating grate through lock hoppers.

The **Lurgi process** advanced this concept of a pressurized, oxygen-blown system. Gasifier pressure is 350 to 450 psig (24 to 31 bar). Typical composition of gas from the gasifier with oxygen blowing is as follows:

	Vol % dry basis
C_2H_4	0.42
C_2H_6	0.62
CH_4	11.38
CO	20.24
H_2	37.89
N_2 + A	0.33
CO_2	28.69
H_2S + COS	0.49

After removal of the acid gases (CO_2, H_2S, COS), the gas can be used as fuel gas or can be upgraded to SNG by using the CO shift reaction to adjust the H_2/CO ratio for methanation. This can also be shifted to ratios

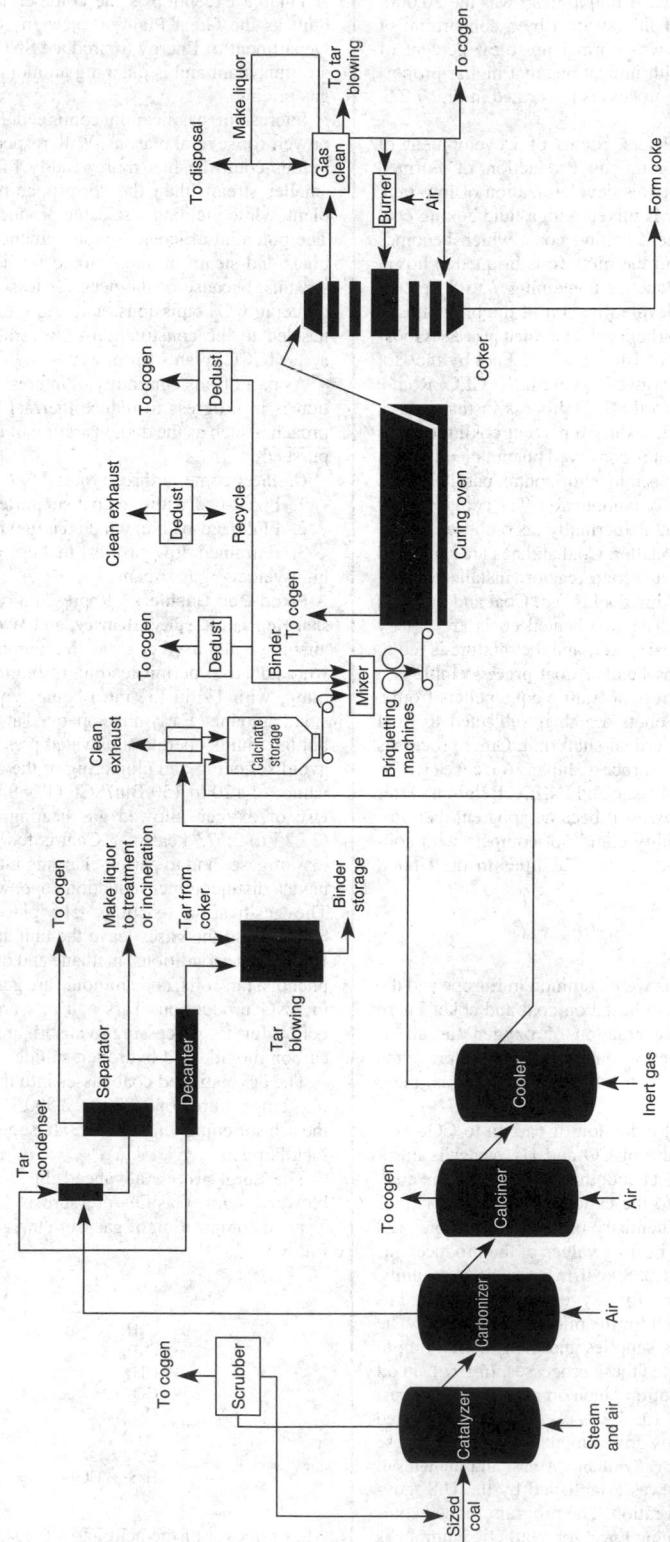

Fig. 7.2.3 FMC form coke flow diagram. (*FMC Corp.*)

suited to the synthesis of methanol, or ammonia if H_2 is suitably optimized. A flow diagram of the process is shown in Fig. 7.2.4. This process has been used in a number of commercial plants (Rudolph, *Oil & Gas Jour.* Jan. 22, 1973) including one at Sasolburg in South Africa and one in North Dakota.

A later development is the **British Gas/Lurgi slagging gasifier.** Coal is fed with a size distribution of 2 in by 0 with up to 35 percent minus ¼ in. The operation is similar to the dry bottom unit except that molten slag is removed through a slag tap, is water-quenched, and is discharged through a lock hopper. Tars, oils, and naphtha can be recycled to the gasifier. Gas composition differs from the conventional dry ash unit in that water vapor, CO_2, and CH_4 are lower, and CO is higher, resulting in cold gas efficiencies of 88 percent or more. The unit was tested extensively on a variety of coals including caking types at the British Gas Town Gas Plant in Westfield, Scotland. This was to have been one of the demonstration plants of the Synfuels Corp. (Lurgi Corp., private communication).

Fluid-Bed Gasifiers The Winkler and Kellogg KRW (formerly Westinghouse) gasifiers constitute this design. Since the bed is truly fluidized, it permits the flexibility of processing solids such as coal and coke. Particle sizes of ⅛ to ⅜ in (3 to 10 mm) are required. Lock hoppers are the method of coal feed. The **Winkler process** employs the fluidized-bed technique, and it has been commercialized in a number of plants (Banchik, "Clean Fuels from Coal," IGT symposium, Chicago, Sept. 1973). Caking coals may be preoxidized to avoid agglomeration. The mixing of the bed causes a uniform temperature, so that the distinct regions of oxidation and reduction of the fixed-bed units are absent. To avoid agglomeration by softening of the ash and loss of fluidization, temperatures are limited to 1,800 to 2,000°F (980 to 1,095°C). Because of the relatively low temperature, these units are primarily useful with reactive coals such as lignite and subbituminous. It has been further developed in a pressure mode as the **high-temperature Winkler (HTW) process** by Rheinbraun, Uhde, and Lurgi. This process is being used for a 300-MW ICGG plant at 25 bar (362 lb/in²) and is scheduled for start-up in 1995. (Adloch et al., The Development of the HTW Coal Gasification Process, Rheinbraun, Uhde, Lurgi brochure.)

The **Kellogg gasifier** is an extension of the design developed by **Westinghouse,** which used limestone or dolomite to capture sulfur in the bed, similar to fluid-bed combustion. The gasifier is shown in Fig. 7.2.5. The coal is quickly pyrolyzed in the jet, which supplies the endothermic heat for reaction. This permits a high proportion of fines to be used. The agitation of this region and the rapid approach to high temperature permit the use of highly caking coals. Operating conditions are 1,900 to 2,000°F (1,040 to 1,050°C), at pressures to 300 psig (21 bar). Further, the combination of retention time and temperature cracks tars and oils to CH_4, CO, and H_2. Product gas is removed through cyclones, where carbon dust and ash are collected and recycled to the gasifier. The gas has a residual concentration of H_2S and COS so that desulfurization may be required. Regenerable **hot-gas desulfurization (HGD) systems,** with zinc reagents, have been developed which recover the sulfur as SO_2. The gas is further cleaned by removing residual fines by ceramic filter candles. The gases then go to gas turbines in a combined cycle.

The residual solids contain carbon, sulfided sorbent, and ash. With the alkaline components, the mixture forms eutectics with melting points of 1,000 to 2,000°F (540 to 1,090°C). In the zone between the combustion jet and the fluid bed, the smaller particles tend to agglomerate, and fall, while char particles rise into the reaction zone. The solids, called **lash,** contain sulfided components which are converted to sulfates after leaving the gasifier in a sulfator/combustor.

This design has been selected by the Sierra Pacific Power Company for commercial demonstration as a 100-MW IGCC plant at the Piñon Pine Station near Reno, Nevada. In addition, further development of the process is proceeding with a transport gasifier, based on petroleum fluid catalytic cracking technology. The particle size in this unit is smaller than that in the fluid bed, improving reaction kinetics and allowing shorter residence times. Limestone is used as the sulfur sorbent. This technique is also being applied to the **hot-gas desulfurizer** (HGD) to incorporate the regeneration of zinc sorbent on a continuous basis. The **transport HGD** is incorporated into the Sierra Pacific project. The transport gasifier concept is under test operation at Southern Company Services, in Wilsonville, Alabama, under a cooperative agreement with the Department of Energy. (Campbell, "Kellogg's KRW Fluid Bed Process

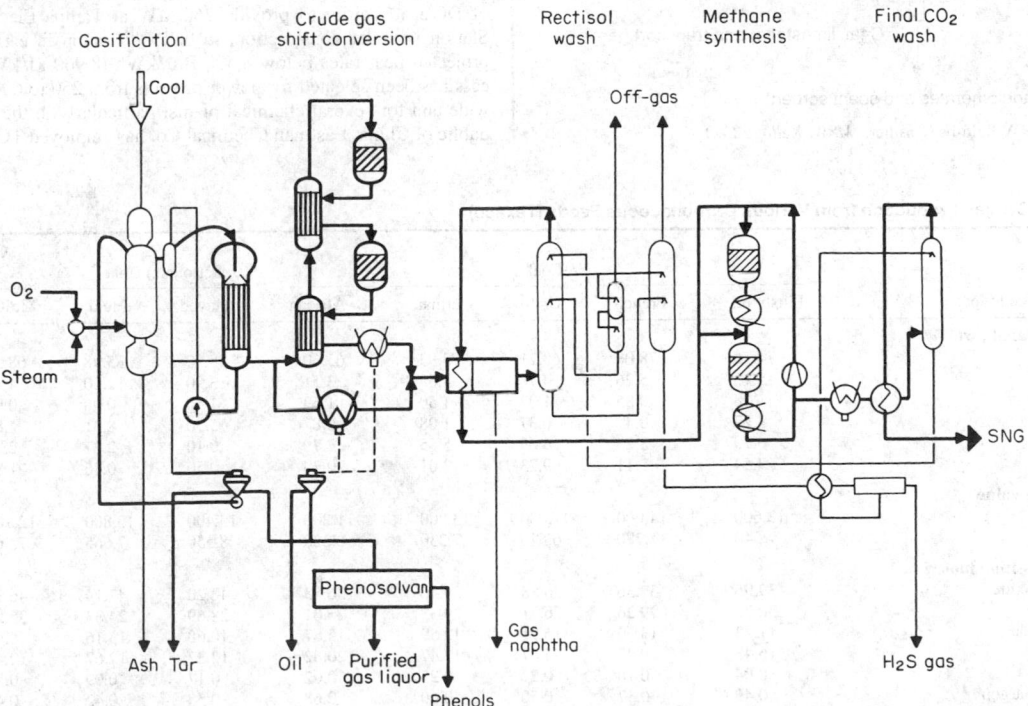

Fig. 7.2.4 Lurgi process.

for Gasification of Petroleum Coke," The M. W. Kellogg Co., December 1994.)

Entrained Flow Gasifiers These have been developed as coal gasifiers, and as partial oxidation units to produce synthesis gas from liquid hydrocarbons, petroleum residues, or coke. They are characterized by a

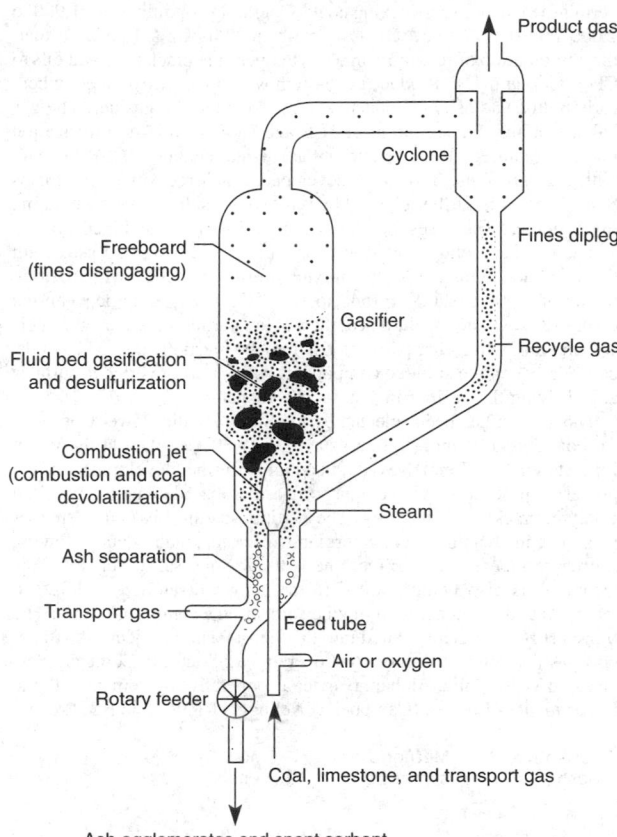

Fig. 7.2.5 KRW fluid-bed gasifier. (*M. W. Kellogg Co.*)

short residence time (< 1 s) and concurrent flow of the feed and gasifying agents. Operating temperatures are high, from 2,200 to 3,500°F (1,200 to 1,927°C) and pressures are as high as 80 bar (1,200 psig). Coal feed systems may be dry with lock hoppers and pneumatic transport, or water slurries of coal. Thus, any feedstock which can be pulverized and dispersed can be processed, including highly caking coals. Oxygen is used as the oxidant reactant to achieve the high temperatures required. This has the additional advantage of not diluting the product gas with nitrogen, particularly if the gas is for synthesis. Product gases are free of tars, condensable hydrocarbons, phenols, and ammonia. Sulfur compounds such as H_2S and COS must be scrubbed from the gas. The high temperature results in slagging operation of these units. Commercial processes include **Texaco, Shell,** and **Destec.** All operate at elevated pressure. The **Koppers-Totzek** was an atmospheric unit, but it has been developed into a pressurized system, now the **PRENFLO gasifier.**

The **Texaco gasification process** (TGP) is the application of the Texaco partial oxidation process to the use of a slurry feed of coal (60 to 70 wt %) in water. The unit is operated at temperatures of 1,200 to 1,500°C (2,200 to 2,700°F) at pressures of 27 to 80 bar (400 to 1,200 psig). The slurry is fed with oxygen through a special injection nozzle into the refractory-lined gasifier to produce syngas, while slagging the ash. The syngas goes to a water-quenched unit or a waste heat boiler. The latter is typical for an IGCC facility. Fine ash is removed by a scrubber before conventional removal of H_2S. The molten ash or slag exits the gasifier, is water-quenched, and is removed through lock hoppers for disposal. The slag forms a glassy solid which is nonhazardous. Typical gasifier products are shown in Table 7.2.5. The cleaned gas is used for gas-turbine fuel or for chemicals manufacture. The plant configuration can be modified for optimal heat recovery for a power cycle or for maximum H_2 and CO generation for **chemical production.** A configuration for **power application** is shown in Fig. 7.2.6. The process was demonstrated at a commercial-size (110-MW) combined-cycle (IGCC) unit at the Coolwater project, which tested many coals during its operation from 1984 to 1989. Emissions of SO_2 from the Coolwater demonstration plant were as low as 0.076 lb/TBtu (0.033 kgs/10^6 kJ) and of NO_x of 0.07 lb/TBtu (0.03 kg/10^6 kJ). Under the Clean Coal Program of DOE, a facility to provide 260 MW at Tampa Electric Co. Polk Station is under construction, to be in operation in early 1996, with projected heat rates below 8,500 Btu/kWh (8,960 kJ/kWh). The process has been selected for power projects from 250 to 600 MW worldwide and for several chemical plants, particularly in the People's Republic of China. Eastman Chemical Co. has employed TGP for over 12

Table 7.2.5 Syngas Production from Various Carbonaceous Feeds (Texaco)

Feed type:	Coal					Petroleum Coke		Coal Liquef. Residue	
	Pittsb. no. 8	French	Utah	German	S. African	Delayed	Fluid	Molten	Slurry
Feedstock dry anal., wt. %									
Carbon	74.16	78.08	68.21	73.93	65.60	88.50	85.98	68.39	68.39
Hydrogen	5.15	5.26	4.78	4.65	3.51	3.90	2.00	4.75	4.75
Nitrogen	1.18	0.85	1.22	1.50	1.53	1.50	0.98	0.98	0.98
Sulfur	3.27	0.47	0.37	1.08	0.87	5.50	8.31	1.87	1.87
Oxygen	6.70	8.23	15.69	5.85	7.79	0.10	2.27	2.21	2.21
Ash	9.54	7.11	9.73	13.01	20.70	0.50	0.46	21.80	21.80
Higher heating value									
Btu/lb	13,600	14,000	11,800	13,200	11,200	15,400	13,800	12,700	12,700
kcal/kg	7,540	7,780	6,570	7,330	6,220	8,550	7,665	7,060	7,060
Product composition mol %									
Carbon monoxide	39.95	37.36	30.88	39.46	36.53	46.20	47.14	46.31	33.48
Hydrogen	30.78	29.26	26.71	29.33	26.01	28.69	24.33	35.54	28.56
Carbon dioxide	11.43	13.30	15.91	12.59	15.67	10.68	13.16	6.41	13.09
Water	16.43	19.43	25.67	17.47	20.82	12.37	12.67	10.46	23.72
Methane	0.04	0.16	0.22	0.25	0.02	0.17	0.09	0.27	0.23
Nitrogen and argon	0.49	0.37	0.50	0.60	0.68	0.55	0.42	0.45	0.42
Hydrogen sulfide + carbonyl sulfide	0.88	0.12	0.11	0.30	0.27	1.34	2.19	0.56	0.50

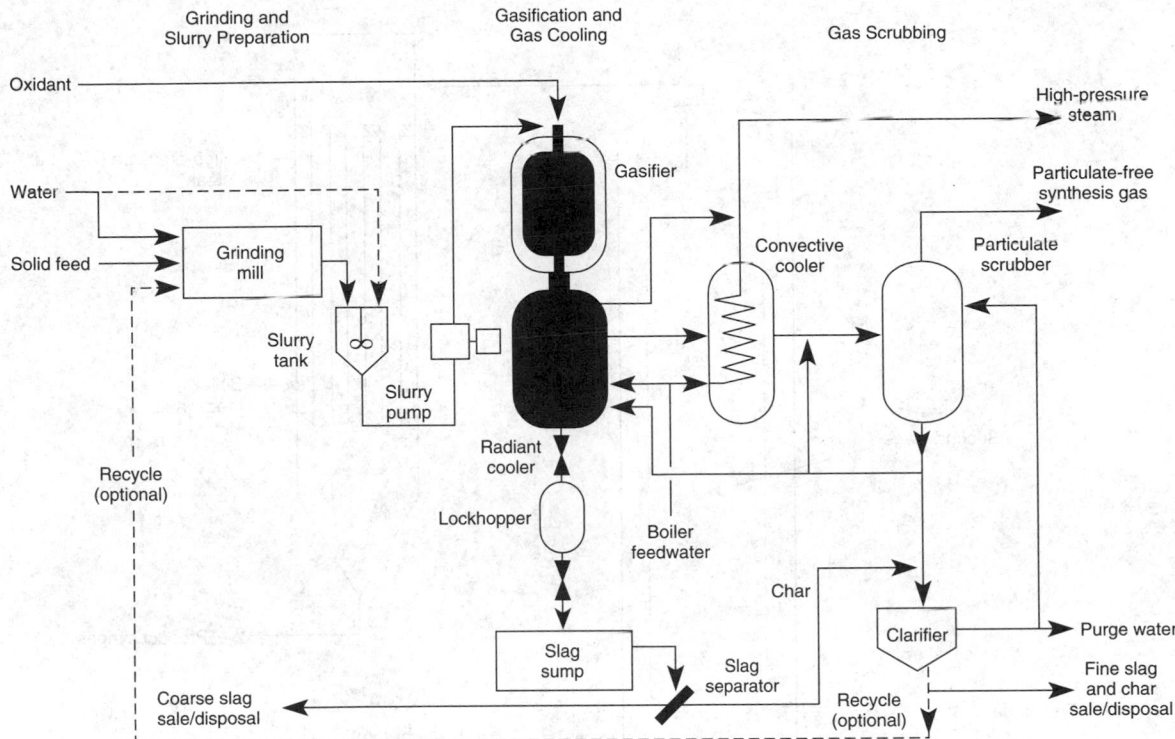

Fig. 7.2.6 TGP-Gas cooler mode. *(Texaco Development Corp.)*

years, using 1,150 tons/d of coal to make acetic anhydride for photographic films and chemicals. Ube Industries produces 1,000 Mt/d of ammonia in Japan. (Gerstbrein and Guenther, International VGB Conference, Dortmund, May 1991; Janke, American Power Conference, Chicago, Apr. 1995; Texaco Gasification Process for Solid Feedstocks, *Texaco Dev't Corp. Bull.* Z-2154, 1993; Watts, 6th Annual International Coal Conference, Pittsburgh, Sept. 1989.)

Variations in the configuration of other gasifiers are made to improve thermal efficiency. An example is the **Destec** unit, which has two stages. The coal slurry is fed to both stages, that going to the first providing the exothermic heat by partial oxidation by the oxidant. This is absorbed later, in the second stage, by the endothermic gasification of the coal, without oxidant. Operating conditions of the first stage are 1,450°C (2,600°F) at 27 bar (400 psig), the exit gas being about 1,040°C (1,900°F). The syngas from the gasifier is cooled, generating high-pressure steam used in steam turbines. Particulates are removed, and the gas is scrubbed to remove H_2S before it is being fed to the gas turbine-generator sets. The **Destec gasifier** is illustrated in Fig. 7.2.7. The system has been operating in a 160-MW facility at Dow Chemical, Plaquemine, LA, since 1987. The Destec technology is being used at the PSI Energy Inc. Wabash River plant to combine a 100-MW steam-turbine facility with a gas turbine to yield 262 MW. The net plant heat rate of this unit is 9,000 Btu/kWh (9,500 kJ/kWh), compared to typical values of 10,500 Btu/kWh (11,077 kJ/kWh) for a new coal-fired plant with SO_2 scrubbers. (Destec Energy, Inc., *Tecnotes* nos. 8 and 14, personal communication.)

New Developments   The research in coal conversion has been limited by available resources, i.e., an abundant oil and gas supply. One effort which has continued involves is **underground coal gasification** (UGC), wherein the coal seam is both reactor and reactant in place. It has been practiced in Russia for some time for local industrial and residential heating. During the oil crisis of the 1970s, this research program was supported by the U.S. Department of Energy and a consortium of industrial companies. The technology depends on the evaluation of several characteristics: coal seam characteristics such as dip, thickness, partings and rock lenses; coal chemistry; its agglomerating and free-swelling properties; water, ash, and sulfur content; boundary strata; overburden height; faults; bulking factor; and hydrology. The program was begun in Wyoming with a series of test burns, which continue under private auspices. Design is underway in New Zealand for an IGCC based on test burns of their coal seams. (Energy International Corp, private communication.)

Gasification of Liquid Hydrocarbons   In the era of manufactured gas in the United States, both base-load and peak-shaving gases were produced by gasifying oils via thermal cracking techniques. Most of the processes produced gases having calorific values compatible with coal gas (coke-oven gas). As natural gas became available, some processes were modified to produce a high-heating-value gas interchangeable with natural gas. These oil-gas units operated on a cyclic (heat-make) basis. To supply the heat for the endothermic thermal cracking of oil, a mass of checker brick was heated to 1,300 to 1,700°F (705 to 925°C) by burning oil and deposited carbon with air. During the "make" cycle, steam and oil were introduced to produce a mixture of hydrogen, methane, saturated and unsaturated hydrocarbons, aromatic oils, tar, and carbon.

The supply of natural gas throughout the world displaced the manufactured gas plants by virtue of lower cost, operational simplicity, and the reduction of emissions. Some manufactured gas and coke-oven gas were distributed into the early 1980s, but this is no longer current practice.

Increased demand for chemicals resulted in the development of processes for production of carbon monoxide, hydrogen, and carbon dioxide. These are for the chemical synthesis of ammonia and methanol, which are feedstocks for many other products. The processes include **partial oxidation, catalytic reforming,** and **hydrogasification**

Partial oxidation processes were developed to produce syngas from liquid feedstocks of any weight, particularly heavy residual oils. These processes produce principally a carbon-monoxide-rich gas which is

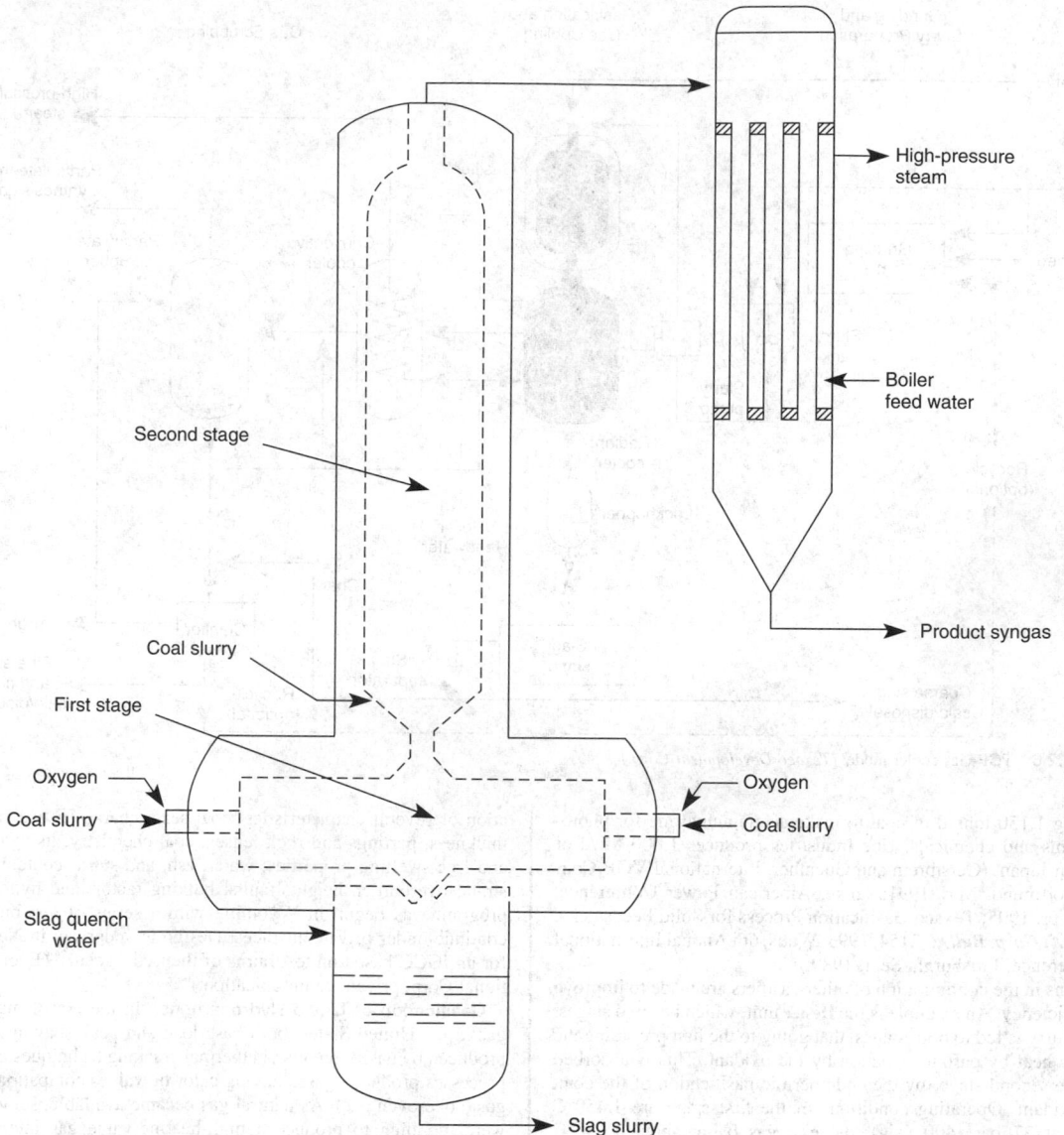

Fig. 7.2.7 Dow two-stage gasifier. *(Destec Energy, Inc.)*

reacted with water in a shift reactor to add hydrogen. Catalytic systems then produce methane or other chemicals. **Texaco** and **Shell** developed their processes originally for these reasons and later adapted them to coal gasification.

Catalytic reforming of naphtha, natural-gas liquids, and LPG is presently applied commercially for the production of synthetic natural gas. Over 30 plants with a total capacity of 6.5×10^9 SCF/d (184 million m³/d) are planned, but the actual number of installations is limited by availability of feedstock. Commercial processes available are the **CRG** (**catalytic-rich gas**), British Gas Council; **MRG** (**methane-rich gas**), Japan Gasoline Co.; and **Gasynthan**, BASF/Lurgi. Processes are similar in that each uses steam reforming of light hydrocarbons over a bed of nickel catalyst. The product gas is a mixture of methane, carbon monoxide, carbon dioxide, and hydrogen. Upgrading to synthetic natural gas requires methanation steps. The four basic steps of the process are desulfurization, gasification, methanation, and purification (CO_2 removal and drying).

Hydrogasification The British Gas Corporation has developed the **GRH** (**gas recycle hydrogenator**) for hydrogenating vaporizable oils to produce synthetic natural gas, and the **FHB** (**fluid-bed hydrogenation**) of gasifying crudes or heavy oils for synthetic natural-gas production. In the **GRH process,** naphthas, middle distillates, and gas oils that need not be desulfurized are reacted directly with hydrogen-rich gas prepared by steam reforming a rich gas sidestream. Exothermic reactions decompose paraffins and naphthenes into methane and ethane. In the **FBH process,** crude or heavy oil is preheated and atomized in the presence of coke particles fluidized by a supply of preheated hydrogen-rich gas. Paraffins and naphthenes are hydrogenated to methane and ethane, and an aromatic condensate is recovered. Desulfurization, followed by secondary hydrogenation, allows reduction of hydrogen and ethane to produce synthetic natural gas.

7.3 COMBUSTION FURNACES

by Glenn W. Baggley

REFERENCE: Trinks-Mawhinney, "Industrial Furnaces," vols. 1 and 2, Wiley.

FUELS

The selection of the best fuel should be based upon a study of the comparative prepared costs, cleanliness of operation, adaptability to temperature control, labor required, and the effects of each fuel upon the material to be heated and upon the furnace lining. Attention must be paid to the quantity to be burned in each burner, the atmosphere (fuel/air ratio) desired in the furnace, and the uniformity of temperature distribution required, which determines the number and the location of the burners. Common methods of burning furnace fuels are as follows:

Solid Fuels (Almost Entirely Bituminous Coals)

Coal was once a common fuel for industrial furnaces, either hand-fired, stoker-fired, or with powdered coal burners. With the increasing necessity for accurate control of temperature and atmosphere in industrial heating, coal has been almost entirely replaced by liquid and gaseous fuels. It can be expected that methods will be developed for the production of a synthetic gas (natural-gas equivalent) from coal.

Liquid Fuels (Fuel Oil and Tar)

To burn liquid fuels effectively, first it is necessary to atomize the oil into tiny droplets which then vaporize and burn. Atomization can be accomplished mechanically or with the aid of steam or air. With heavy oils and tar, it is important to maintain the proper viscosity of the oil at the atomizer by preheating the fuel.

For larger industrial burners, combustion air is supplied by fans of appropriate capacity and pressure. Combustion air is induced with some smaller burner designs.

Gaseous Fuels

Burners for refined gases (natural gas, synthetic gas, coke-oven gas, clean producer gas, propane, butane):

Two-pipe systems: Include blast burners (open or closed setting), nozzle mixing, luminous flame, excess air (tempered flame), baffle, and radiant-tube burners, all for low-pressure gas and air.

Premix systems: Air and gas mixed in a blower and supplied through one pipe.

Proportioning low-pressure mixers: Air and gas supplied under pressure and proportioned automatically (air aspirating gas or gas inspirating air). The resulting mixture is burned in tunnel burners, radiant-cup, baffle, radiant-tube, ribbon, and line burners.

Pilot flames are generally used to ensure ignition for gas and oil burners. Insurance frequently requires additional safety provision in two main categories: an interconnected pressure system to prevent lighting if any burner in a zone is open, and burner monitors using heat or light to permit ignition.

Burners for crude gas (raw producer-gas, blast-furnace gas, or coke-oven gas):

Simple mixing systems with large orifices and simple mechanisms which cannot become clogged by tar and dirt contained in these gases.

Separate gas and air supplies to the furnace, with all mixture taking place within the furnace.

TYPES OF INDUSTRIAL HEATING FURNACES

Heating furnaces are usually classified according to (1) the purpose for which the material is heated, (2) the nature of the transfer of heat to the material, (3) the method of firing the furnace, or (4) the method of handling material through the furnace.

Purpose Primarily a metallurgical distinction, according as the furnace is intended for tempering, annealing, carburizing, cyaniding, case hardening, forging, heating for forming or rolling, enameling, or for some other purpose.

Transfer of Heat The principal varieties are **oven furnaces,** in which the heat is transferred from the products of combustion of the fuel, in direct contact with the heated material, by convection and direct radiation from the hot gases or by reradiation from the hot walls of the furnace; **muffle furnaces,** in which the heat is conducted through a metal or refractory muffle which protects the heated material from contact with the gases, and is then transferred from the interior of the muffle by radiation to the heated material, which is sometimes surrounded by inert gases to exclude air; or **liquid-bath furnaces,** in which a metal pot is heated on the outside or by immersion. This pot contains a liquid heating or processing medium which transfers heat to the material contained in it. This type includes low-temperature tempering furnaces with oil as the heating medium, hardening furnaces using a bath of lead, hardening and cyaniding furnaces with baths of special salts, and galvanizing or tinning furnaces for coating the heating material with zinc or tin. The generally accepted form of muffle is the **radiant-tube** fired furnace, in which the fuel is burned in metal or refractory tubes which radiate heat to the charge. An important form of furnace for temperatures below 1,300°F (700°C) is the **recirculating** type, in which the atmosphere (products of combustion, air, or protective gases) is recirculated rapidly through the heating chamber. **Forced convection** heating is accomplished by a large number of jets of hot gas at high velocity. In **high-speed** heating (or **patterned combustion**), premixed burners are arranged for close application of heat, and with a high-temperature head, very rapid heating is achieved.

Method of Firing This classification applies principally to the oven type of furnace, and it indicates whether the furnace is direct-fired, overfired, underfired, or heated by radiant tubes. Figure 7.3.1 shows the principles of each of these types. The **direct-fired** method finds increased utilization from constant improvement in the design and control of gas and oil burners, especially for temperatures above 1,200°F (650°C). In

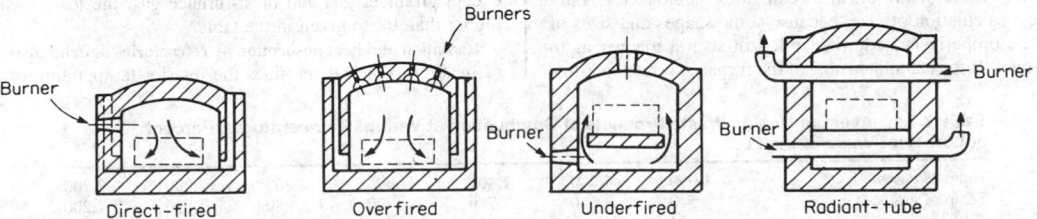

Fig. 7.3.1 Methods of firing oven furnaces.

overfired furnaces, radiant burners fire through the roof and are arranged in patterns to obtain the best temperature distribution. The **underfired** furnace is excellent for temperatures between 800 and 1,800°F (ca. 400 to 1,000°C) because the heated product is protected from the burning fuel. The temperature and atmosphere can be closely controlled, but the temperature is limited by the life of the refractories to about 1,800°F (1,000°C). Many furnaces are now designed for the use of special protective atmospheres and involve the use of **radiant tubes** to avoid any contact with the combustion gases. These fuel-fired tubes of heat-resisting alloy may be horizontal across the furnace above and below the heated material or may be vertical on the sidewalls of the furnace.

Method of Material Handling In the **batch** type, the heated material charged into the furnace remains in the same position until it is withdrawn after sufficient heating. In a **continuous furnace,** the material is moved through the furnace by mechanical means which include pushers, chain conveyors, reciprocating hearths, rotating circular hearths, cars, walking beams, and roller hearths. Continuous furnaces are principally labor-saving devices and may or may not save fuel.

SIZE AND ECONOMY OF FURNACES

The size of furnace required depends upon the amount of material to be heated per hour, the heating time required, the size of the pieces to be heated, and the amount of heat that can be liberated without excessive damage to the furnace. The efficiency and refractory life obtained depend upon the correctness of furnace size.

Heating Time For the usual relation of refractory area to stock area, time to heat steel plate from one side for each ⅛ in (3.18 mm) of thickness varies from 3 min for high-speed heating and 6 to 12 min for heating for forming by usual methods to 20 min for heat treating. Steel cylinders will be heated in one-half these times per ⅛-in (3.18-mm) diam. Below 800°F (ca. 400°C), the time may be 2 to 3 times these values. Brass requires about one-half as long as steel to heat, copper 40 percent as long, and aluminum 85 percent as long. The preceding heating times are based on a furnace temperature 50 to 100°F (ca. 25 to 50°C) higher than the final temperature of the heated material. It is assumed that the material is fully exposed to the heat of the furnace. Piling of material in a furnace lengthens the heating time by an amount that must be determined by actual trial. In addition to simple heating, there is frequently additional **time required for soaking** (holding at furnace temperature) to cause metallurgical changes in the material or for some other reason.

The **weight of material in the furnace** at any time is the product of weight of material per hour multiplied by the heating time in hours. If the weight and sizes of pieces involved are known, the **area of the furnace** can then be fixed. The width and length of the furnace to produce this hearth are fixed by the method of firing to be used and by the method of handling material.

The life of a furnace at given temperature depends upon the rate of heating, which may be expressed in pounds per square foot of hearth area per hour. The maximum allowable **rate of heating steel** is about 35 lb/(ft² · h) for heat treating, 70 lb for in-and-out rolling-mill furnaces, 100 lb for single-zone continuous furnaces, and 150 lb for multiple-zone furnaces. These are upper limits which should not be used if long life of furnace refractories is expected. These rates are for heating mild steel; they may be about twice as great when heating brass, 2½ times as great for copper, 0.7 as great for alloy steel, and 1.1 times as great when heating aluminum. These maximum allowable rates should be used only for checking the calculation of size, because some shapes and sizes of pieces cannot be properly heated when piled in such a manner as to produce these rates. If the calculated size of the furnace corresponds to a rate of heating that is too great, it should be reduced by making the furnace larger. If the rate is too small, it can sometimes be increased by piling material in a smaller furnace.

EXAMPLE. To determine furnace size. If a furnace is required to heat 20 pieces per h weighing 30 lb each and requiring a heating time of ½ h, the furnace must be large enough to hold ½ × 20 = 10 pieces. If each piece requires an area of 2 ft², the area of the hearth will be 2 × 10 = 20 ft² for a single layer of pieces in the furnace. If the furnace is of the batch type, a size of 4 ft wide × 5 ft deep would probably be about right for convenient handling. On checking, the rate of heating is 20 pieces per h × 30 lb/20 ft² = 30 lb/(ft³ · h). For this rate an underfired furnace would be satisfactory, although for other methods of firing, a smaller furnace could be used if the pieces could be more densely piled without seriously interfering with the circulation in the furnace.

The heat released by the fuel in a furnace (heat input) is equal to the sum of the heat required in the heating process (useful heat) plus the heat losses from the furnace. **Heat input** includes the heat of combustion of the fuel, sensible heat in preheated air or fuel, and heat in the material charged. Low-heat values of the fuel are used, and the sensible heat can be calculated from the specific heats of the preheated air, fuel, or material. **Useful heat** includes the heat absorbed by the material in the furnace. Figure 7.3.2 gives heat contents for different metals. In the simple heating of metals, the useful heat applied to the metal includes only the heat absorption, as given in Fig. 7.3.2; but there are many processes that include other requirements, such as drying, where moisture must be heated and evaporated, heating of chemical products where heat is utilized to cause chemical changes, and other special cases. (See also Sec. 4.3.)

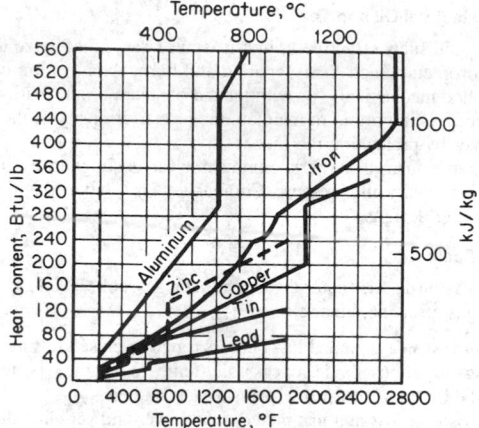

Fig. 7.3.2 Heat content of metals.

Heat losses in a heating furnace include heat lost in waste gases, radiation from and heat absorbed by refractories, heat carried out of the furnace by containers or conveyors, heat lost through openings, and heat in unburned fuel escaping with the products of combustion. The **heat contained in waste gases** depends upon the temperature of these gases as they leave the heating chamber. Table 7.3.1 gives the approximate percentage of heat contained in the flue gases from perfect combustion at different temperatures. These values are about the same for most fuels except producer gas and blast-furnace gas, the losses with which are higher than those given in the table.

Radiation and heat absorption by refractories depend also upon the rate of heating (which determines the interior temperature of the refracto-

Table 7.3.1 Average Heat in Waste Products of Combustion at Various Temperatures, Percent of Low Heat Value of the Fuel

Temp of gases, °F	1,000	1,200	1,400	1,600	1,800	2,000	2,200
Temp of gases, °C	540	650	760	870	980	1,090	1,200
% of low heat value in gases	24	28	34	38	45	50	55

Table 7.3.2   Radiation through Openings in Furnace Walls, kBtu/h

	Furnace temp, °F (°C)					
	1,400 (760)			2,200 (1,200)		
	Wall thickness, in*			Wall thickness, in*		
Size of opening, in*	4½	9	18	4½	9	18
4½ × 4½	1.4	1.1	0.8	5.1	4.1	2.8
9 × 9	7.8	6.1	4.5	28.5	22.7	16.8
18 × 18	37	30.5	24.3	137	114	90
24 × 24	71	60	48	264	225	180
36 × 36	173	150	124	650	560	465

* × 25.4 = mm.

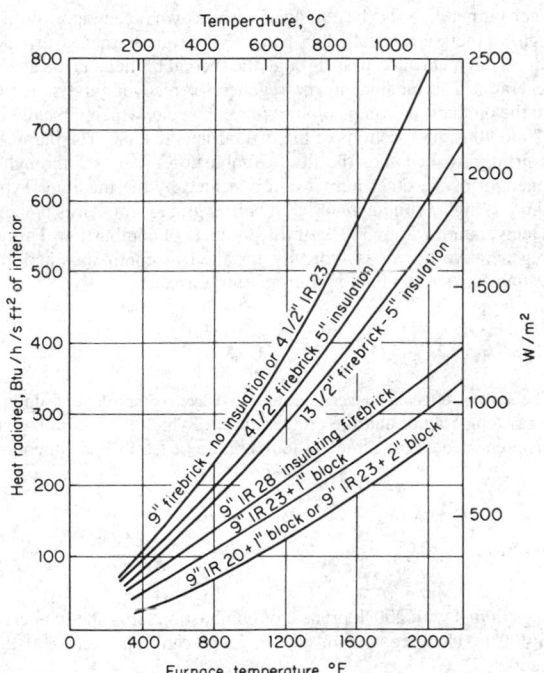

Fig. 7.3.3   Heat loss from thoroughly heated walls, based on interior area.

ries) and upon the refractory area and thickness. Figure 7.3.3 shows the heat radiated through walls of different thickness at various furnace temperatures, for equilibrium conditions, when the wall has reached steady temperatures throughout (see also references at the beginning of this section and Keller, "Flow of Heat through Furnace Hearths," ASME). The **heat carried out by containers and conveyors** is the sensible heat content of these items as they leave the heating chamber. Such losses include the heat in carburizing boxes, pans, chain conveyors, and furnace cars. **Radiation from furnace openings** depend upon the size and shape of the opening and the thickness of the walls in which they are located, as well as upon the temperature of the furnace. Some idea of the magnitude of these losses is given by the values in Table 7.3.2.

The **heat lost in unburned fuel** escaping with the flue gases is small in most furnaces because the fuel can be almost completely consumed.

The **efficiency of an industrial furnace** is the ratio of the heat absorbed by the heated material to the heat of combustion of the fuel burned. The magnitude of the various heat losses is indicated in Table 7.3.3.

Table 7.3.3   Heat Balances for Various Furnace Types, Percent of Heat of Combustion

Disposition of heat	Type I	Type II	Type III
Heat to material, or efficiency	16	49	23
Heat to refractories	20	17	22
Heat lost in flue gases	44	19	40
Heat to water cooling	—	5	—
Heat through openings	20	10	15

Column 1 is for a high-temperature batch-type billet-heating furnace, heating 4,200 lb of billets, per hour, a furnace load at a time, to 2,300°F,

Table 7.3.4   Average Net Efficiencies and Fuel Requirements of Various Furnace Types with Good Operation

Type	Temp, °F	Temp, °C	Avg efficiency, %	Avg heat required from fuel, Btu/lb,* of steel
Ingot heating, soaking pits, recuperative	2,000–2,400	1,100–1,300	20	500
Billet heating for forming:				
Batch, in-and-out	2,000–2,400	1,100–1,300	20	1,750
Continuous	2,000–2,400	1,100–1,300	32	1,100
Wire annealing of coils, hood type	1,300–1,500	700–800	16	1,350
Wire annealing of strands, in lead	1,300–1,500	700–800	19	1,100
Wire patenting, strands	1,650	900	21	1,250
Wire baking, coils, continuous	450	230	20	250
Tube annealing, continuous, bright	1,300–1,500	700–800	35	600
Skelp heating, butt weld, continuous	2,900	1,600	25	1,500
Slab heating, continuous, recuperative	2,400	1,300	42	800
Strip coil annealing, hood type	1,250–1,400	680–760	30	600
Hardening, continuous conveyor	1,650	900	21	1,250
Drawing, continuous conveyor	900–1,100	500–600	20	750
Carburizing, gas, continuous	1,750	950	19	1,500

* × 2.326 = kJ/kg.

at a rate of 25 lb/(ft^2 · h), averaged over 10 h of operation, and with a fuel consumption of 30 gal of oil per ton of steel heated. Column II is for a large continuous billet-heating furnace of the usual pusher type with a flow of gases opposite to that of the steel, and operating at a rate of 60 lb of steel heated to 2,300°F/ft^2 of hearth area per hour. Column III is for an underfired batch-type furnace, heating steel to 1,600°F for annealing, at a rate of 30 lb/(ft^3 · h).

Table 7.3.4 gives average requirements in fuel of typical industrial heating furnaces. The values are for furnaces without heat-saving appliances (recuperators, regenerators, or waste-heat boilers) except as noted and show the efficiency and the Btu required in the fuel per net pound of steel heated. To obtain the average amount of any fuel required, this latter figure is divided by the low heat value of the fuel. The values are for average rates of heating. Fuel economy is of small importance as compared with the quality of the product.

FURNACE CONSTRUCTION

When furnace refractories are made up largely of standard bricks and shapes, it is advisable to specify furnace dimensions that can be built with a minimum of cutting. Horizontal flues are made a multiple of 2½ or 3 in (63 or 76 mm) in height, and most other flue dimensions are multiples of 4½ in (114 mm) to correspond to the width and length of standard bricks. The area of furnace flues must be large enough to avoid excessive pressures at maximum fuel rates. Flues should be located so as to promote the circulation of gases in all parts of the furnace. Average allowable **velocities in flues** for furnaces without stacks are:

Furnace temp, °F (°C)	200 (93)	1,000 (538)	1,500 (816)	2,000 (1,093)
Allowable velocity (hot gases), ft/s (m/s)	9 (2.74)	13 (3.97)	15 (4.57)	17 (5.2)

The total **flue areas required** in in^2/ft^3 of fuel/h (or per gal/h for fuel oil) for furnaces without stacks as temperatures of the products of combustion of 1,000 and 2,000°F are as follows:

Temp, °F (°C)	Fuel oil	Natural gas	Artificial gas	Coke-oven gas	Raw producer gas
1,000 (538)	14.0	0.11	0.06	0.05	0.02
2,000 (1,093)	19.0	0.15	0.08	0.06	0.02

The **metal parts of a furnace,** consist of the steel and cast-iron binding, alloy parts exposed to the direct heat of the furnace, and water-cooled steel members. The alloy parts are of nickel or chromium alloys and must be made heavy enough to offset the loss of strength at high temperatures. They are resistant to oxidation at temperatures below 2,000°F (1,093°C). To reduce heat losses, water-cooled members must be insulated.

HEAT-SAVING METHODS

Methods of conserving heat include the use of recuperators or regenerators, waste-heat boilers (see Sec. 9), insulation of refractories, automatic control of temperature and atmosphere, and special attention to the construction and operation of the furnace.

Recuperators and regenerators extract some heat from the escaping flue gases and return it to the furnace by preheating the combustion air or the entering fuel. In **recuperators,** continuous flow of hot gases and cold entering air or gas is maintained through metal or refractory ducts which keep the two gas streams apart but which conduct heat from the hotter stream to the colder. Recuperators are built in the form of self-enclosed units set above the ground or in pits below floor level, and are made of fire-clay tile, silicon carbide, or heat-resisting metal. Overall coefficients of heat transfer in metallic recuperators are between 2.5 and 6.0 Btu/(ft^2 · h · °F) [14 and 34 W/(m^2 · °C)] and in silicon carbide recuperators about the same; the coefficient for fire-clay recuperators is considerably less than these values. Usual velocities of hot air in recuperators do not exceed 12 ft/s (3.6 m/s) in order to keep pressure drop to a reasonable value.

Regenerators are used where the high temperature of air preheat is required to maintain a high furnace temperature or to obtain high thermal efficiency. When one regenerator serves an entire furnace, it is usually constructed of fire brick and consists of two chambers completely filled with a checkerwork. The flow of flue gases and that of air or gas to be heated are periodically reversed so that the hot gases and cold gases alternately flow through the two sets of chambers. The checkerwork retains the heat of the hot gases and gives it up to the cold gases with each reversal. Another regenerator design employs metal plates. Regenerators are frequently used with glass-melting furnaces and are used almost exclusively where open-hearth furnaces are still employed. Overall coefficient of heat transfer in regenerators is from 1.5 to 2.5 Btu/ft^2 of checkerbrick surface per h per °F temperature difference ([8.5 to 14 W/(m^2 · C)], and the usual mass velocity of hot gas through the openings of the checker is about 0.065 lb/(ft^2 · s) [0.32 kg/(m^2 · s)].

Each burner may also be equipped with its own regenerator. With this design, burners are installed in pairs. When one burner is firing, the products of combustion pass through the second burner and the attached regenerator. The medium in the regenerator recovers and stores heat from the products of combustion. After one cycle, which typically lasts for 20 to 90 s, the functions of the two burners reverse. The burner that was firing now becomes the flue. Combustion air passes through the regenerator of the other burner and is heated by the medium. Typical medium is high-alumina material in ball or grain form. Air is preheated to a temperature within 300°F of the products of combustion. This type of regenerative burner is typically installed in continuous and batch reheating furnaces and aluminum-melting furnaces.

The **savings effected by recuperators or regenerators** depend upon the flue gas temperature and the temperature to which the incoming air or gas is preheated. With a flue gas temperature of 1,600°F, the theoretical savings in fuel with 200°F preheat of combustion air is about 4 percent, with 400°F, 11 percent; with 600°F, 15 percent; and with 800°F, 19 percent.

A recuperator or a regenerator installation, to be a good investment, must show a satisfactory net savings after all costs of investment, repairs, and associated shutdown time lost by such repairs are subtracted from the savings in fuel used or investment savings related to the heat recovery system. For example, it is often possible to reduce the length of continuous furnaces using regenerative burners compared to more conventional designs.

Automatic control prevents the waste of heat by unnecessarily high temperatures, preventable cold periods, and excessive air or unburned fuel from poor combustion. Of even greater importance is the prevention of damage to the heated product from overheating, excessive oxidation, and objectionable chemical reaction between furnace atmosphere and the product (principally decarburization and recarburization). Automatic **temperature** controllers are actuated by thermocouples in the furnace. The thermocouple must not be located in the direct path of the flames, which are not only several hundred degrees hotter than the furnace temperature but are also of extremely variable temperature and not an indication of the average temperature. Automatic control of **atmosphere** for the consistent maintenance of good combustion is accomplished by properly proportioning the fuel and combustion air as they enter the furnace. This is accomplished by the utilization of some characteristic of the flow of one fluid to regulate the flow of the other fluid. Automatic **pressure** control operates the flue dampers of a furnace to maintain a constant predetermined pressure [usually about 0.01 to 0.05 in (0.25 to 1.25 mm) water] in the heating chamber, which excludes free oxygen from the surrounding atmosphere.

Table 7.3.5 Protective Gas Atmospheres

Type	Typical analysis					Dew point, °F (°C)
	CO_2	CO	CH_4	H_2	N_2	
I. Hydrogen, purified				100.0		−60 (−51)
II. Dissociated ammonia				75.0–5.0	25.0–95.0	
III. Rich hydrocarbon gas, not conditioned	5.5	9.0	0.8	15.0	69.7	+50 (10)
IV. Lean hydrocarbon gas, not conditioned	11.5	0.7		0.7	87.1	+50 (10)
V. Rich hydrocarbon gas, completely conditioned	0.1	9.5	0.8	15.8	73.8	−60 (−51)
VI. Lean hydrocarbon gas, completely conditioned	0.1	2.8		3.9	93.2	−60 (−51)
VII. Endothermic generator gas	0.5	20.0	1.0	38.0	40.5	+50 (10)

Care in furnace construction, operation, and maintenance is the simplest but often most neglected of all methods of heat savings. A large quantity of fuel can be saved by care in the construction of furnace refractories so that they will remain tight, by attention to the sealing of doors, by taking care that the doors and other openings are closed when not in use, and by maintaining insulation on any water-cooled members in the furnace.

SPECIAL ATMOSPHERES

(See also Sec. 7.5.)

In an increasing number of heat-treating operations, the necessity for improved quality has created a demand for clean- or bright-heating furnaces, in which the heating material is surrounded by a suitable protective gas while it is heated by radiation from electric resistors, radiant tubes, or the walls of a muffle. Table 7.3.5 gives the chemical analysis of common protective gases used in the heat-treating industry.

Type I Purified hydrogen is used for annealing, brazing, and other treatment of low-carbon steel; for the sintering of low-carbon ferrous powders; for the treatment of silicon iron (electrical sheets and strip); for the bright annealing of stainless steels, and the sintering of molybdenum, tungsten, and other metals.

Type II Ammonia is dissociated by steam or electric heat, and is dried by chemical driers. By partial combustion the relative percentages of hydrogen and nitrogen may be varied as shown in Table 7.3.5. The resulting gases from this treatment are cheaper and are used for brazing and sintering copper alloys, and for annealing low-carbon steels. Dissociated ammonia without combustion is used for annealing stainless steels containing nickel, short-cycle heating of all carbon and alloy steels, treatment of silicon iron, and the treatment of cuprous products.

Type III Rich hydrocarbon gas is produced by combustion with about 60 percent of theoretical air (6:1 air/gas ratio when using natural gas) in the presence of a nickel catalyst, followed by cooling to reduce the moisture content. It is used for the annealing of low-carbon steels, for short-cycle hardening of low-carbon steels, for clean annealing of chrome-type stainless steels, for treatment of silicon iron, and for brazing of copper alloys.

Type IV This gas is similar to type III except that about 90 percent of theoretical air is used for combustion. It is used for bright annealing of copper (straight N_2 and CO_2 can also be used for this purpose) and for clean heating of brass and bronze.

Type V This gas is the same as type III but is conditioned by chemical removal of carbon dioxide by monoethanolamine and by drying in chemical driers. It is used for short-cycle treatment of all carbon, alloy, and high-speed steels; for sintering of all ferrous powders; and as a carrier gas for carburizing and carbon restoration with the addition of natural gas or propane.

Type VI This gas is similar to type V except that about 90 percent of theoretical air is used in the combustion. The resulting gas is used for long-cycle treatment of all ferrous materials except stainless steels containing nickel, and is effective in controlling decarburization in all carbon and alloy steels. It is also used for the annealing of brass and bronze.

Type VII This endothermic gas is made in an externally heated generator with only 25 percent of theoretical air and is cooled to reduce moisture. It is used for short-cycle (under 2 h) heat treating and brazing, usually with small furnace installations. It is also used for dry cyaniding and as a carrier gas for carburizing and carbon restoration.

7.4 INCINERATION

by Charles O. Velzy and Roger S. Hecklinger

REFERENCES: *Proc. Biennial ASME National Waste Processing Conf.*, 1964–1994. "Design Considerations in Heat Recovery from Refuse," International Symposium on Energy Recovery from Refuse, 1975. Velzy et al., eds., "CRC Handbook on Energy Efficiency," chap. 4, Waste-to-Energy Combustion, in press. "Steam, Its Generation and Use," 40th ed., Babcock and Wilcox Co. Kirklin et al., The Variability of Municipal Solid Waste and Its Relationship to the Determination of the Calorific Value of Refuse Derived Fuels, *Resources and Conservation,* **9,** 1982, pp. 281–300.

Incineration is a method for processing of solid wastes by the burning of the combustible portions. It reduces the volume of solid wastes and eliminates the possibility of pollution of groundwater from putrescible organic waste, and the residue may serve as a source of mineral constituents and as a fill. With the application of boilers, beneficial use of the energy generated from burning of the waste is possible.

NATURE OF THE FUEL

The refuse which is received at an incinerator today will contain a high proportion of paper; plastics; some wood; vegetable and animal waste; and varying amounts of cloth, leather, and rubber—together with metal cans, glass, and other noncombustible matter. Collections may also include metal appliances, furniture, tree limbs, other yard waste, waste building material, broken concrete, and other coarse waste matter, commonly classified as rubbish. With little or no regulation of the handling of refuse by the homeowner, there may be a wide variation in moisture content of refuse, depending on the weather. Thus, after a storm, the moisture content may be so high that it is difficult to sustain combustion. Industrial and hazardous waste should be specifically identified and combustion facilities designed for the particular waste.

TYPES OF FURNACES

The type of furnace for incinerators is dictated largely by the type of grate around which the furnace is built. Except in small plants, the modern furnace is equipped with a mechanical grate.

The "Controlled Air" Furnace In the late 1970s, a type of combustion unit, batch fed, utilizing two chambers for staged combustion and followed by a waste heat boiler, was installed in smaller communities in the United States. Such units, while less efficient than larger furnaces, can be factory assembled in large segments (or modular components) and therefore are less expensive than large, field-erected units. Thus they extended the economics of energy-from-waste plants to the smaller communities. In such plants, the refuse is normally dumped on a receiving floor and is then pushed into a ram for feeding into the combustion units. The smallest units do not have grates, while larger units are fitted with rams to move the material through the furnace as it is burned.

The Rectangular Furnace Mechanical-grate furnaces are rectangular in shape, with movement provided by travel of the grate or by a reciprocating or rocking action of the grate sections. Refuse is fed by gravity through a vertical chute, by a ram or similar arrangements.

PLANT DESIGN

Capacity The capacity to be provided is a function of (1) the area and population to be served; (2) the number of shifts (one, two, or three) the plant is to operate; and (3) the rate of refuse production for the population served. If records of collections have been kept, the capacity can be determined and forecasts made; lacking records, the quantity of refuse may be estimated as approximately 4 lb (1.8 kg) per capita per day, when there is little or no waste from industry, to 5 lb (2.3 kg) per capita per day where there is some waste from industry. A small plant (100 tons/d) [90 metric tons per day (t/d)] will probably operate one shift per day; for capacities above 400 tons (360 t) per day, economic considerations usually dictate three-shift operation.

Location An isolated site may be preferred to avoid the possible objections of neighbors to the proximity of a waste disposal plant. However, well-designed and well-operated incinerators which do not present a nuisance may also be installed in light industrial and commercial areas, thereby avoiding the economic burden of extended truck routes. Since considerable vertical distance is involved in passing refuse through an incinerator, there is an advantage in a sloping or hillside site. Collection trucks can then deliver refuse at the higher elevation while the residue trucks operate at the lower elevation with a minimum of site grading.

Refuse-Handling Facilities Scales should be provided for recording the weight of material delivered by collection trucks. Trucks should then proceed to the tipping floor at the edge of the storage pit. This area, which may be open or enclosed, must be large enough to permit more than one truck at a time to maneuver to and from the dumping position.

Since collections usually are limited to one 8-h daily shift (with partial weekend operation) while burning may be continuous over 24 h, ample **storage** must be provided. Seasonal and cyclic variations must also be considered in establishing the storage requirements.

Refuse storage in larger plants is normally in long, narrow, and deep pits extending either along the front of the furnaces, or split in two halves extending from either side of the front end of the furnace. If the pit is much over 25 ft in width, it is generally necessary to rehandle refuse dumped from trucks. In smaller plants, floor dumping and storage of refuse is common practice.

Feeding the Furnaces In a large incinerator (pit and crane type), burning continuously, refuse is transferred from storage pit to furnace hopper by a crane equipped with a grapple. (See Sec. 10.) Batch feeding or batch discharge of residue is undesirable because of the resulting variations in furnace temperatures, adverse impact on furnace side walls, and increased air emissions.

In a more modern furnace using mechanized grates, a vertical charging chute, 12 to 14 ft (3.6 to 4.2 m) long, leads from the hopper to the front end of the furnace. This chute is kept full of refuse; feeding is accomplished by the operation of the mechanical grate, or by a ram; the front of the furnace is sealed from cold air; and the fuel is spread over the grate in a relatively thin bed.

In some newer plants, conveyors, live-bottom bins, and shredding and pneumatic handling of the combustible fraction of the refuse have been utilized to produce a refuse-derived fuel (RDF). See Fig. 7.4.1.

FURNACE DESIGN

The basic design factors which determine furnace capacity are grate area and furnace volume. Both provision for and quantity of underfire air, and provision for quantity and method of applying overfire air influence capacity. The required grate area depends upon the selected burning rate, which varies between 60 and 90 lb/(ft² · h) of refuse in practice. Conservative design, with reasonable reserve capacity and reasonable refractory maintenance, calls for a burning rate between 60 and 70 lb/ft² of grate area.

Furnace volume is a function of the rate of heat release from the fuel. A commonly accepted minimum volume is that which results from a heat release of 20,000 Btu/(ft³ · h). Thus, at this rate, if the fuel has a heat content of 5,000 Btu/lb, the burning rate would be 4 lb/(h · ft³) of furnace volume. A conservative design, allowing for some overload and possible quantities of refuse of high heat content, would be from 30 to 35 ft³/ton of rated capacity.

The primary objective of a mechanical **grate** is to convey the refuse automatically from the point of feed through the burning zone to the point of residue discharge with a proper depth of fuel and in a period of time to accomplish complete combustion. The rate of movement of the grate or its parts should be adjustable to meet varying conditions.

A secondary, but important, objective is to stir gently or tumble the refuse to aid in completeness of combustion. In the United States, there are several types of mechanical grates: (1) traveling, (2) rocking, (3) reciprocating, and (4) a proprietary water-cooled rotary combustor. With the traveling grate, stirring is accomplished by building the grate in two or more sections with a drop between sections to tumble the material. The reciprocating and rocking grates tumble the material by movement of the grate elements. The rotary grate slowly rotates to tumble the material which is inside the cylinder. In Europe, variations of the U.S. designs as well as other types have been developed. The Volund incinerator (Danish) uses a slowly rotating, refractory-lined cylinder or kiln through which the fuel passes as it is burned; the so-called Duesseldorf grate uses a series of rotating cylindrical grates in an inclined arrangement.

Furnace configuration is largely dictated by the type of grate used. When built with a mechanical grate, the furnace is rectangular in plan and the height is dependent upon the volume required by the limiting rate of heat release.

The total **air capacity** provided in a refractory-walled incinerator must be more than the theoretical amount required for combustion in order to obtain complete combustion and to control temperatures—particularly with dry, high-heat-content refuse. The total combustion-air requirements may range to 10 lb of air/lb of refuse. For the modern mechanical-grate furnace chamber, two blower systems should be provided to supply combustion air to the furnace. Blower capacities can be divided, with half or more from the underfire blower and somewhat less than half from the overfire blower and with dampers on fan inlets and air distribution ducts for control. The pressure on the underfire system for most U.S. grate systems approximates 3 in of water. The pressure on the overfire air should be high enough so that the jet effect on passage through properly proportioned and distributed nozzles in the furnace roof and walls produces sufficient turbulence and retains the gases in the primary furnace chamber long enough to ensure complete combustion.

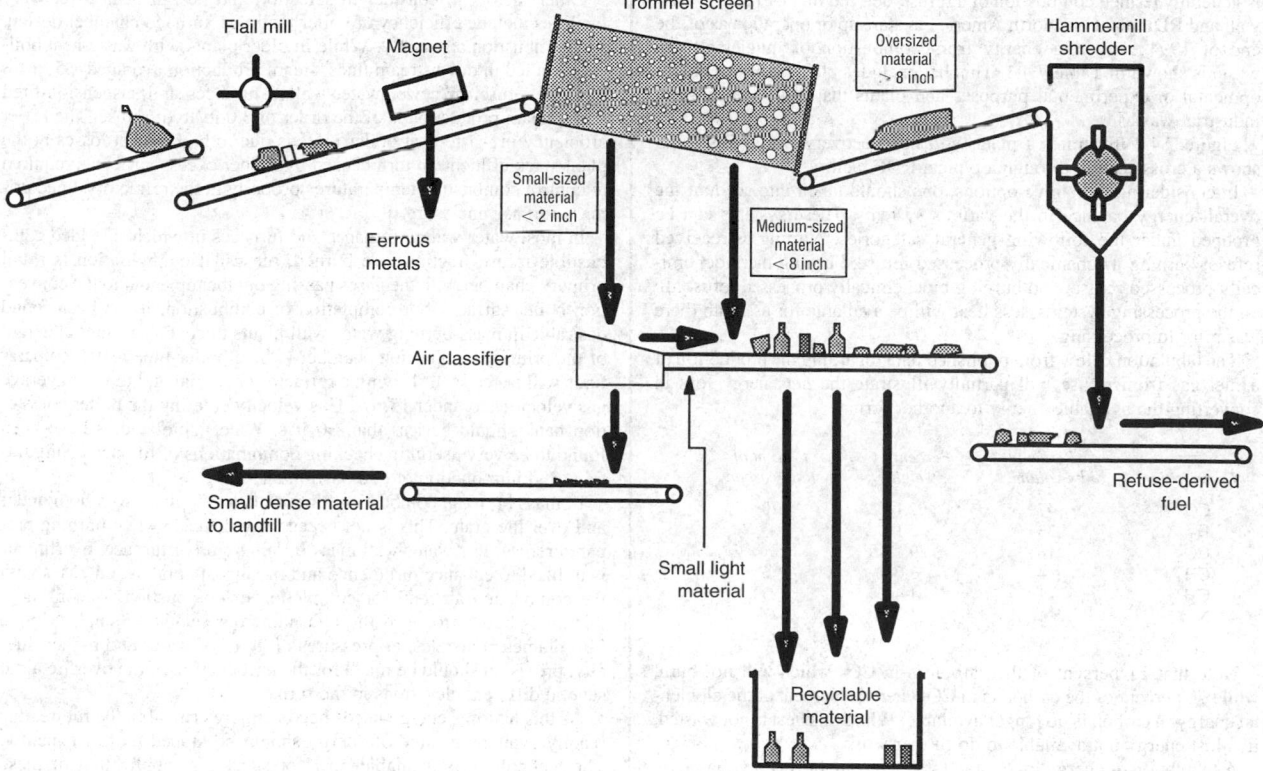

Fig. 7.4.1 Process diagram for a typical RDF system. *(Roy F. Weston, Inc.)*

Heat Recovery Perhaps the most potentially attractive form of recovery, or extraction, of resources from municipal solid wastes is recovery of energy from the incineration process.

Several options exist when one is considering recovery of energy from incineration. These options include mass burning in a refractory-walled furnace with a waste-heat boiler inserted in the flue downstream; mass burning in a water-walled furnace with the convection surface immediately downstream; and refuse preprocessing and separation of the combustible fraction with combustion taking place in a utility-type boiler partially in suspension and partially on a grate. This latter option

Table 7.4.1 Energy-from-Waste Plants Larger than 300 tons/day Commissioned 1990–1995

Plant location	Plant size (tons/d)	Start-up year	Type	Energy sold
Broward County, FL, north	2,250	1991	Mass burn water wall	Electricity
Broward County, FL, south	2,250	1991	Mass burn water wall	Electricity
Camden, NJ	1,050	1991	Mass burn water wall	Electricity
Chester, PA	2,688	1991	Rotary kiln water wall	Electricity
Essex County, NJ	2,250	1990	Mass burn water wall	Electricity
Fort Myers, FL (Lee County)	1,200	1994	Mass burn water wall	Electricity
Gloucester County, NJ	575	1990	Mass burn water wall	Electricity
Honolulu, HI	2,165	1990	RDF water wall	Electricity
Hudson Falls, NY	400	1992	Mass burn water wall	Electricity
Huntington, NY	750	1991	Mass burn water wall	Electricity
Huntsville, AL	690	1990	Mass burn water wall	Steam
Kent County, MI	625	1990	Mass burn water wall	Electricity/steam
Lake County, FL	528	1991	Mass burn water wall	Electricity
Lancaster County, PA	1,200	1991	Mass burn water wall	Electricity
Long Beach, CA	1,380	1990	Mass burn water wall	Electricity
Lorton, VA (Fairfax County)	3,000	1990	Mass burn water wall	Electricity
Montgomery County, PA	1,200	1991	Mass burn water wall	Electricity
Pasco County, FL	1,050	1991	Mass burn water wall	Electricity
Rochester, MA	2,700	1988/1993	RDF water wall	Electricity
Southeast CT (Preston, CT)	600	1992	Mass burn water wall	Electricity
Spokane, WA	800	1991	Mass burn water wall	Electricity
Union County, NJ	1,440	1994	Mass burn water wall	Electricity
Wallingford, CT	420	1990	Mass burn water wall	Electricity/steam

is generally termed combustion of a refuse-derived fuel. A list of water-wall and RDF plants in North America in start-up or operation as of the end of 1994 extracting energy from combustion of municipal-type waste is shown in Table 7.4.1. This list excludes plants built for developmental or experimental purposes and plants that utilize specialized industrial wastes.

Figure 7.4.2 illustrates a plant with heat recovery, while Fig. 7.4.3 shows a cross section through a typical RDF facility.

In considering the above options, one should taken into account the overall energy balance in the various systems. These systems can be grouped under the following general categories: burning as-received refuse; burning mechanically processed refuse; burning thermochemically processed refuse; and burning biochemically processed refuse. In all the processing systems, less heat will be available for use than there was prior to processing.

The tabulation below from published data regarding the production of a fuel gas from refuse will partially illustrate the net energy loss in converting the available energy to another form:

Composition	Percent by volume	Percent by weight	Percent of total carbon
CO	47	62.1	70
H_2	33	3.1	
CO_2	14	29.1	21
CH_4	4	3.0	6
C_2H_2	1	1.4	3
N_2	1	1.3	

Note that 21 percent of the carbon is in CO_2 which will not burn, while 70 percent of the carbon is in CO where 30 percent of the elemental energy in carbon is no longer available. While this heat is not wasted, it is lost energy not available to do further work.

A tabulation of energy losses and total net available energy, based on information published in 1974–1975 for refuse with an initial heat content of 4,400 Btu/lb, is given in Table 7.4.2.

Of the 4,400 Btu/lb in the refuse as received, the tabulated data indicate the useful energy that may be made available through combustion. While the data are not absolute, the relative magnitudes are meaningful, provided similar degrees of design efficiency and sophistication of control are used for each process.

Other factors to consider in selection and design of heat-recovery facilities include efficiency of boiler facilities, furnace-chamber design, and combustion air supply. While in older plants with waste-heat boilers installed in downstream flues, steam production averaged 1.5 to 1.8 lb/lb of refuse, in newer water-walled furnaces and suspension-fired units, steam production is of the order of 3.0 lb/lb of refuse. The lower efficiency in waste-heat boiler units is due to higher heat losses in the plant stack effluent, in turn caused by higher excess air levels required to control combustion temperatures properly in the refractory-lined primary furnace enclosure.

In most water-walled furnaces and furnaces in which shredded combustible refuse fractions are burned, the usual configuration is a tall primary chamber with the gases passing out the top and into the convection boiler surface after completion of combustion. It has been found desirable in mass-burning water-wall plants to coat a substantial height of the primary combustion chamber (where boiler-tube metal temperatures will exceed 500°F) with a refractory material and to limit average gas velocities to under 15 ft/s. Gas velocity entering the boiler convection bank should be less than 30 ft/s. Water-table studies have been found to be very useful in checking combinations of furnace configurations and introduction of combustion air.

In mass burning, combustion air is usually supplied from both under and over the grate. This is not necessarily the case when burning prepared refuse. In a water-wall mass-burning type of furnace, overfire air is utilized to enhance turbulence and mixing of combustion gases with the combustion air, and for completion of combustion. Accordingly, this air is best introduced through numerous relatively small (1½- to 3-in-diameter) nozzles, at pressures of 20 in of water and higher. Ideally, provision should be made for the introduction of the overfire air at several different elevations in the furnace.

As this nation's energy needs become more critical in the future, this readily available source of energy should be tapped more frequently. The technology is available now for successful application of these techniques if provision is made for adequate funding and properly trained operating staffs.

Flues and chambers beyond the furnace convey gases to the stack and house facilities for removal of fly ash and other pollutants. The draft for an incinerator furnace may be provided by a stack of adequate diameter and height or by an induced-draft fan. (See Secs. 4 and 14.) Most modern plants include heat recovery equipment and extensive air pollution

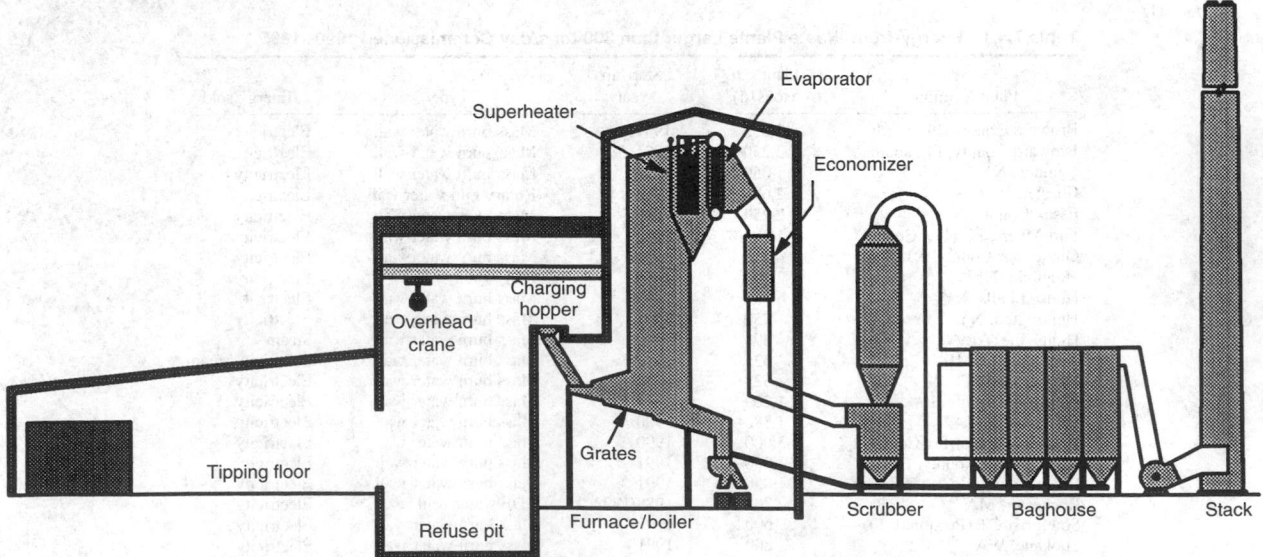

Fig. 7.4.2 Typical cross section of a mass burn facility. *(Roy F. Weston, Inc.)*

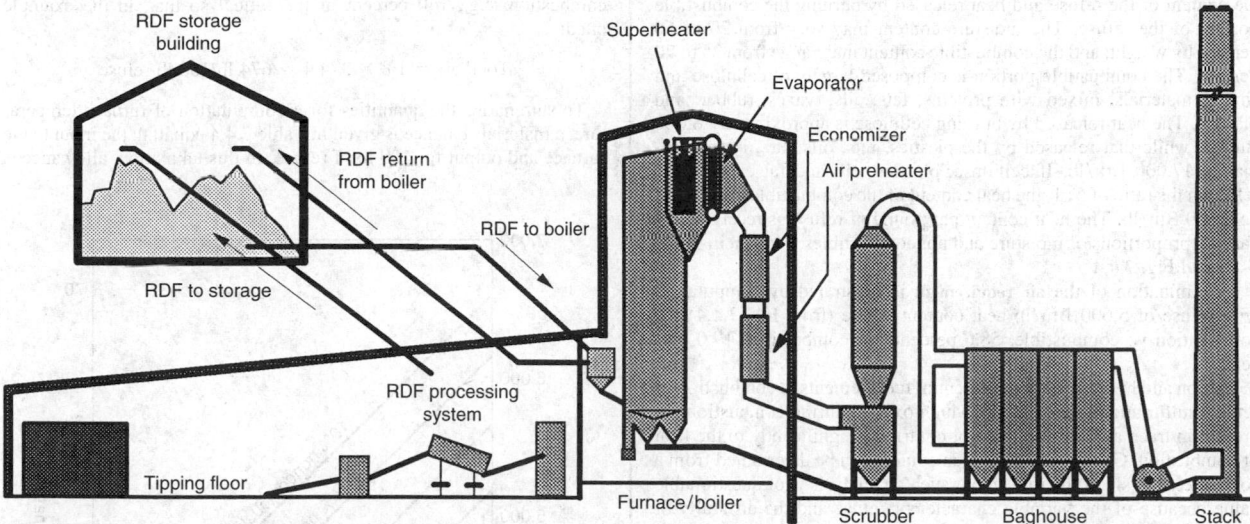

Fig. 7.4.3 Typical cross section of an RDF facility. *(Roy F. Weston, Inc.)*

control equipment. In these plants, draft losses are so high that an induced-draft fan is required.

Air Pollution Control   The federal new source emission standard for air pollution control at new or enlarged municipal sized incinerators is 0.015 grain per dry standard cubic foot (gr/DSCF) (34 mg/DSCM) corrected to 7 percent O_2. (This is approximately 0.03 lb/1,000 lb of flue gas corrected to 50 percent excess air.) Some jurisdictions have promulgated emission requirements lower than the federal standards. Designers of modern plants will also have to include facilities to reduce acid gas emissions and, at some plants, NO_x and mercury emissions.

To meet current emission requirements, two basic techniques have been utilized on incinerators: electrostatic precipitation and fabric filters preceded by lime addition in a slurry, so-called "dry scrubbing." Electrostatic precipitation has performed reliably and has given predictable emission test results. The fabric filter/dry scrubber air pollution control system is considered by many to represent the most efficient combination of pollution control systems currently available for particulates and acid gases. Nonselective catalytic reduction using ammonia has been applied at some plants for NO_x control, while limited experience has been gained in the use of activated carbon for control of mercury emissions. Good combustion control is used to limit emission of organics.

Residue Discharge and Disposal   The residue from refuse burning consists of relatively fine, light ash mixed with items such as burned tin cans, partly melted glass, and pieces of metal. Discharge from furnaces may be through manually operated dump grates or from mechanically operated grates to a hopper, where it is quenched and delivered to a

truck through a bottom gate. The residue may also be discharged through a chute into a conveyor trough filled with water for quenching and then carried by flight conveyor to an elevated storage hopper for truck delivery. Usually there are two conveyor troughs, so arranged that the residue can be discharged to either, one trough being used at a time. A European system uses a ram discharger submerged in a water-filled container.

The lower end of the discharge chute leading to the trough is submerged in a water seal to prevent entrance of cold air to the furnace. In design of the conveyor mechanism, the proportions should be large because of the nature of the material handled, and the metal used should be selected to withstand severe abrasive service. Final disposal of the residue is by dumping at a suitable location, which for modern plants usually means a monofill. Volume required for disposal is 5 to 15 percent of that required for dumping raw refuse.

Miscellaneous Facilities   Good working environment and reasonable comfort for the staff should be provided.

COMBUSTION CALCULATIONS

Among the factors directly affecting design are **moisture** and **combustible content** of refuse as received, heat released by combustion, temperature control, and water requirements. The design of furnaces, chambers, flues, and other plant elements should be based on characteristics which result in large sizes. Controls should provide satisfactory operation for loads below the maximum. The computations which follow are for relatively high heat releases.

The prime factors in **heat calculations** are the moisture and combusti-

Table 7.4.2   Energy Losses and Total Net Available Energy for Refuse with Initial Heat Content of 4,400 Btu/lb, 1974–1975

Process	Energy loss, %			Total net available energy Btu/lb
	Processing	Combustion	Total	
As received	1	39	40	2,640
Dry shredding	18	30	48	2,288
Wet shredding	35	21	56	1,936
Pyrolysis, oil	62	9	71	1,276
Pyrolysis, gas	32	25	57	1,892
Pyrolysis with oxygen	37	15	52	2,112
Anaerobic digestion	72	6	78	968

ble content of the refuse and heat released by burning the combustible portion of the refuse. The moisture content may vary from 20 to 50 percent by weight, and the combustible content may range from 25 to 70 percent. The combustible portion is composed largely of cellulose and similar materials, mixed with proteins, fats, oils, waxes, rubber, and plastics. The heat released by burning cellulose is approximately 8,000 Btu/lb, while that released by the plastics, fats, oils, etc., is approximately 17,000 Btu/lb. If cellulose, plastics, oil, and fat exist in the refuse in the ratio of 5 : 1, the heat content of the combustible matter will be 9,500 Btu/lb. The heat content per pound of refuse as received, for varying proportions of moisture and noncombustibles, is given in Table 7.4.3 and Fig. 7.4.4

Determination of the **air requirement** is illustrated by computation with refuse of 5,000 Btu/lb heat content where (from Fig. 7.4.4) the composition is: combustible, 58.6 percent; noncombustible, 19.0 percent.

Carbon and hydrogen are the essential fuel elements in combustion of refuse; sulfur and other elements which oxidize during combustion are present in trace amounts and do not contribute significantly to the heat of combustion. Carbon and hydrogen content can be determined from a complete analysis of the refuse, but such an analysis is of questionable value because of the variable character of refuse and the difficulty of obtaining representative samples. For the purpose of this computation, a typical analysis is used in which the total carbon is 28 lb and the hydrogen 1.5 lb/100 lb of refuse. It is probable that 1 to 3 lb of combustible material per 100 lb of refuse will escape unburned with the residue. For the sake of clarity in the illustrated computations, complete combustion is assumed.

Oxygen requirements and products of combustion can be determined from the reactions as follows:

Cellulose
Atomic wt:
$$C_6H_{10}O_5 + 6O_2 \rightarrow 6CO_2 + 5H_2O$$
$$72 + 10 + 80 + 192 = 264 + 90$$
$$162 + 192 = 264 + 90$$

Ratios:
Referred to carbon: $1 + 0.14 + 1.11 + 2.667 = 3.667 + 1.25$
Referred to cellulose: $1 + 1.185 = 1.63 + 0.555$

Carbon
Atomic wt: $C + O_2 \rightarrow CO_2$
$12 + 32 = 44$
Ratio: $1 + 2.667 = 3.667$

Hydrogen
Atomic wt: $2H_2 + O_2 \rightarrow 2H_2O$
$4 + 32 = 36$
Ratio: $1 + 8 = 9$

The theoretical air required per 100 lb of refuse follows from these figures where air is considered to contain 23.15 percent oxygen.

Air required $= 28 \times 2.667/0.2315 + 1.5$
$$\times 8/0.2315 = 374.4 \text{ lb}/100 \text{ lb refuse}$$

For incineration, furnace temperature must be controlled to minimize refractory maintenance. With no other provision for heat absorption, it is necessary to introduce excess air well beyond the needs for complete combustion, e.g., 140 percent of theoretical so that, in the example cited:

$$\text{Total air} = 1.8 \times 374.4 = 674 \text{ lb}/100 \text{ lb refuse}$$

To summarize the quantities for a computation of furnace temperature, a materials balance is given in Table 7.4.4 equating the input to the furnace and output for 100 lb of refuse. In this tabulation, allowance is

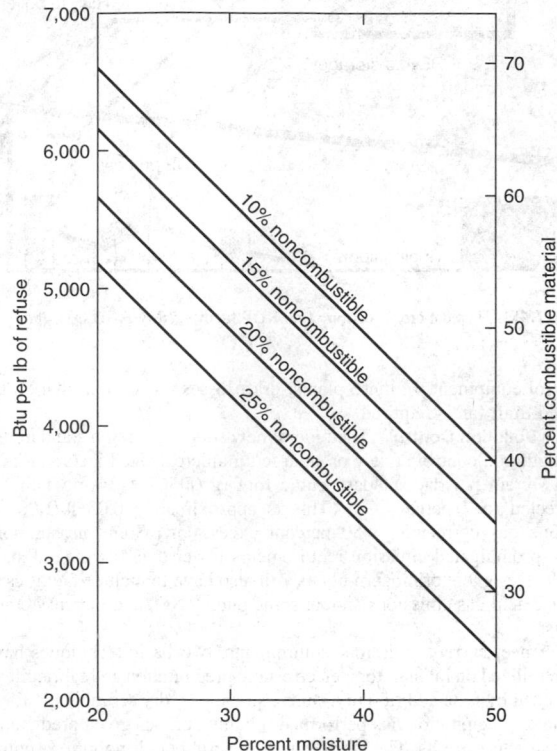

Fig. 7.4.4 Moisture–heat content relation with 9,500 Btu/lb combustible material.

made for moisture in the air at a commonly accepted rate of 0.0132 lb/lb of dry air. Some residue quench water will be evaporated, and the moisture added to the flue gases is estimated at 5 lb for each 100 lb of refuse burned. Since the assumed analyses are not precise, an exact balance is not obtained, but the indicated computations are sufficiently accurate for incinerator design.

Table 7.4.3 Heat Content of Refuse, as Received

| | Noncombustible, % | | | | | | | |
| | 10 | | 15 | | 20 | | 25 | |
Moisture, %	Comb., %	Heat content*	Comb., %	Heat content*	Comb., %	Heat content*	Comb., %	Heat content*
50	40	3,800	35	3,325	30	2,850	25	2,375
40	50	4,750	45	4,275	40	3,800	35	3,325
30	60	5,700	55	5,225	50	4,750	45	4,275
20	70	6,650	65	6,175	60	5,700	55	5,225

* Btu/lb.

Table 7.4.4   Materials Balance for Furnace lb/100 lb of Refuse

Input:			
Refuse			
Combustible material			
Cellulose	43.75		
Plastics, oils, fats, etc.	8.75	52.5	
Moisture		25.0	
Noncombustible		22.5	100.0
Total air, at 140% excess air			
Oxygen	156.0		
Nitrogen	517.9	673.9	
Moisture in air		8.9	
Residue quench water		5.0	
Total		787.8	
Output:			
CO₂ (28 × 3.667)		102.7	
Air			
Oxygen (156-87)	69.0		
Nitrogen	517.9	586.9	
Moisture			
In refuse	25.0		
From burning cellulose	24.3		
From burning hydrogen	13.5		
In air	8.9		
In residue quench water	5.0	76.7	
Furnace gas subtotal		766.3	
Noncombustible material		22.5	
Unaccounted for		− 1.0	
Total		787.8	

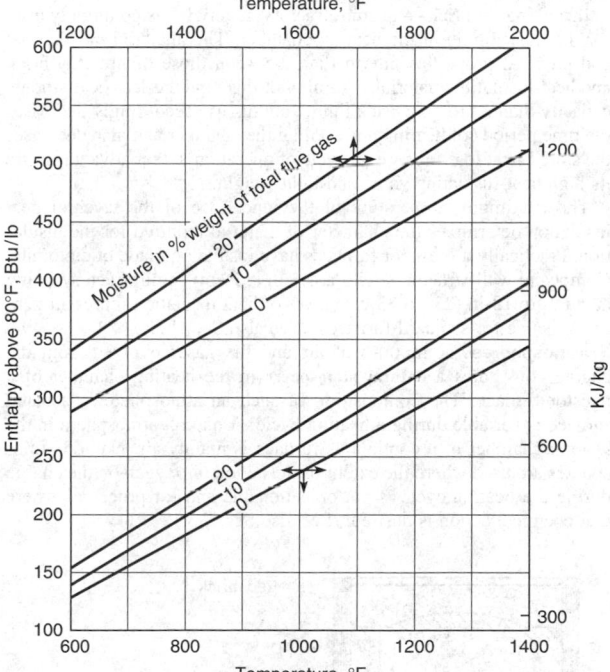

Fig. 7.4.5 Enthalpy of flue gas above 80°F.

In Table 7.4.4, total air is broken down into oxygen and nitrogen on the basis that 23.15 percent of the air is oxygen. To compute the air in the "output," or flue gas, the nitrogen is the same as the "input." Oxygen is diminished by the amount consumed in combustion. Since carbon and hydrogen unite with oxygen during combustion, the oxygen consumed per 100 lb refuse is:

$$\text{For carbon, } 28 \times 2.667 = 74.68 \text{ lb}$$
$$\text{For hydrogen, } 1.5 \times 8 = \underline{12.00 \text{ lb}}$$
$$\text{Total}\quad 86.68, \text{ say } 87 \text{ lb}$$

The moisture from burning cellulose and hydrogen is: for cellulose, 0.555 × 43.75 = 29.3 lb; for hydrogen, 9 × 1.5 = 13.5 lb.

Abiabatic flame temperature is the maximum theoretical temperature that can be reached by the products of combustion of a specific fuel-air combination. To calculate this temperature, the total heat input in the fuel is adjusted to subtract the heat input required to vaporize moisture in the fuel, moisture produced in combustion of cellulose and hydrogen, and the residue quench water that is vaporized. A loss is assumed to account for incomplete combustion and other small losses. The remaining heat energy is the sensible heat available in the furnace gas. It can be calculated per 100 lb of refuse from the data of Table 7.4.4 and the enthalpy data of Fig. 7.4.5 as follows:

Input, 100 lb × 5,000 Btu/lb		500,000 Btu
Losses:		
Heat of vaporization deduction		
Moisture		
In refuse	25.0	
From burning cellulose	24.3	
From burning hydrogen	13.5	
From reside quench	5.0	
	67.8	
67.8 × 1,050 Btu/lb	71,900	
Assumed loss due to incomplete combustion and other losses 2%	10,000	
Total deduction		− 81,190
Sensible heat available in furnace gas		418,810 Btu
Enthalpy of gas, 418,810 Btu ÷ 766.3 lb		547 Btu/lb
% Moisture in furnace gas		
Moisture vaporized	67.8	
Moisture in air	8.9	
	76.7	
76.7 ÷ 766.3 = 10.0%		
Temperature of furnace gas at 547 Btu/lb and 10.0% moisture from Fig. 7.4.5		1,950°F

RECOVERY

Many modern waste-to-energy (WTE) plants incorporate waste processing/materials recovery facilities. Such facilities are a natural adjunct at plants producing RDF. Relatively active stable markets have developed for newsprint, glass, metals, and certain specific plastics. Recent surveys of waste composition indicate that after recycling, paper and noncombustibles have decreased slightly while high-heat-value plastics have increased significantly as a percentage of municipal solid waste going to disposal. The impact of these changes in waste composition has been to increase slightly the heat content of waste available for processing in WTE plants and to remove some of the more troublesome materials from a materials handling standpoint.

Fly ash has been used to a limited extent as a concrete additive and as a road base. **Incinerator residue** has been used for land reclamation in low areas and, in some cases, as a road-base material. Generally, before such use, the fly ash and residue must be tested to ensure that they cannot be classified as hazardous waste.

7.5 ELECTRIC FURNACES AND OVENS
by George J. Roddam

REFERENCES: Robiette, "Electric Melting and Smelting Practice," Griffin. Campbell, "High-Temperature Technology," Wiley. "Electric-Furnace Steel Proceedings," Annual, AIME. Paschkis, "Industrial Electric Furnaces and Appliances," Interscience. Stansel, "Induction Heating," McGraw-Hill. Ess, The Modern Arc Furnace, *Iron Steel Eng.*, Feb. 1944.

CLASSIFICATION AND SERVICE

The furnaces and ovens addressed in this section generally are those small and medium-size units used in general foundry practice, heat treating, and associated processes. The larger units are generally used for melting large quantities of metal as part of specific production processes such as the production of high-purity alloy steels, processing batches of processed parts receiving vitreous enamel, annealing glass, and so on.

In **resistor furnaces and ovens,** heat is developed by the passage of current through distributed resistors (heating units) mounted apart from the charge. Alternating current of a standard power frequency is used. The furnace service is for heat applications to solids such as heat treatment of metals, annealing glass, and firing of vitreous enamel. Oven service is limited to drying and baking processes usually below 500°F (260°C).

In **induction heaters** heat is developed by currents induced in the charge. The service is heating metals to temperatures below the melting points.

In **induction furnaces** heat is developed by currents induced in the charge. The service is melting metals and alloys.

In **arc furnaces** heat is developed by an arc, or arcs, drawn either to the charge or above the charge. Direct-arc furnaces are those in which the arcs are drawn to the charge itself. In indirect-arc furnaces the arc is drawn between the electrodes and above the charge. A standard power frequency is used in either case. Direct-current (dc) electric power is an alternative source of energy. The general service is melting and refining metals and alloys. The ladle arc furnace is used particularly when a charge of metal is to be processed primarily to refine its chemistry.

In **resistance furnaces** of the submerged-arc type, heat is developed by the passage of current from electrode to electrode through the charge. The manufacture of basic products, such as ferroalloys, graphite, calcium carbide, and silicon carbide, is the general service. Alternating current at a standard power frequency is used. An exception is the use of direct current where the product is obtained by electrolytic action in a molten bath, e.g., in the production of aluminum.

The **characteristics of electric heat** are:

1. Precision of the control of the development of heat and of its distribution.

2. The heat development is independent of the nature of the gases surrounding the charge. This atmosphere can be selected at will with reference to the nature of the charge and the chemistry of the heat process. This freedom is often a primary reason for the use of electric heat.

3. The maximum temperature is limited only by the nature of the material of the charge.

The first two characteristics underlie the design of all electric heating apparatus. The third is utilized in thermal processes for the production of certain materials not obtainable in any other way.

RESISTOR FURNACES

Resistor furnaces may be either the batch or the continuous type. Batch furnaces include box furnaces, elevator furnaces, car-bottom furnaces, and bell furnaces. Continuous furnaces include belt-conveyor furnaces, chain-conveyor furnaces, rotary-hearth furnaces, and roller-hearth furnaces.

Standard resistor furnaces are designed to operate at temperatures within the range 1,000 to 2,000°F (550 to 1,200°C). For higher heating chamber temperatures, see Resistors, later in this section.

The **heating chamber** of a standard furnace is an enclosure with a refractory lining, a surrounding layer of heat insulation, and an outer casing of steel plate, or for large furnaces an outer layer of brick or tile, as indicated by Figs. 7.5.1 and 7.5.2. The hearth of a batch furnace often is constructed of a heat-resisting alloy, made in sections to prevent warping. In some continuous furnaces the conveyor forms the hearth; in others a separate hearth is required.

Insulating firebrick—a semirefractory material—is commonly used for the inner lining of the heating chamber. This material has thermal and physical properties intermediate between those of fire-clay brick and heat-insulating materials. A lining of this kind has less heat-storage capacity than a fire-clay brick lining, and its use accordingly decreases the time periods of heating and cooling the chamber and also decreases the stored-heat loss for a given cycle of operation. Other advantages are its high heat-insulating value and light weight.

The maximum temperature of the inner face of the layer of heat insulation determines the character of material required for the insulation. Practically all resistor furnaces have insulation made of diatomite. Composite wall structures with a 4½-in (11-cm) semirefractory lining and a 9- to 13-in (23- to 33-cm) layer of heat insulation represent general practice for standard furnaces. (See also Sec. 4.3.)

Atmospheres A mixture of air and the gases evolved from the charge constitutes a **natural atmosphere** in the heating chamber of a resistor furnace. The composition of such an atmosphere in a batch furnace is variable during a heating cycle. A natural atmosphere in the heating chamber of a continuous furnace is mainly air. Natural atmospheres are used where the extent of the action of oxygen on the charge during the heating cycle is not objectionable and for processes where that chemical action is desired. (See also Sec. 7.3.)

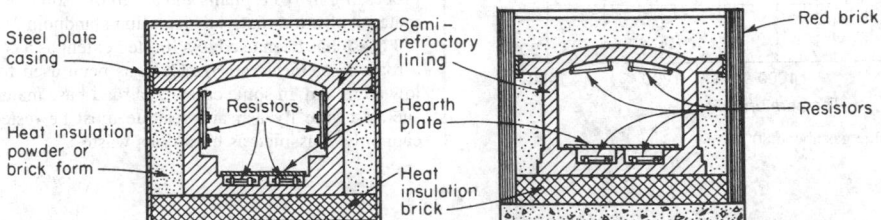

Fig. 7.5.1 Heating chamber with sidewall and hearth resistors.

Fig. 7.5.2 Heating chamber with roof and hearth resistors.

The basis of an **artificial atmosphere** is the exclusion of oxygen (air) from the heating chamber by the substitution of some other gas or mixture of gases. This gas or mixture of gases is selected with reference to the chemical activity of that atmosphere on the charge at the temperature of the heat application. A definite chemical action may be desired, for example, the reduction of any metallic oxide present on the charge, or it may be required that the artificial atmosphere to be chemically inactive. Thus artificial atmospheres are divided into (1) active or process atmospheres and (2) inactive or protective atmospheres. The term "controlled" atmosphere refers generally to a protective atmosphere, but it also includes artificial atmospheres of some degree of chemical activity. An example of a process atmosphere is the use of a hydrocarbon gas to carburize steel. Examples of controlled atmospheres are: the bright annealing of metals, the prevention of decarburization of steel during a heat application, the use of a reducing gas (hydrogen or carbon monoxide) in a copper brazing furnace, etc. In this last example the reducing gas serves to clean the faces of the joint to be made (by removal of any oxide present) and to maintain that cleanliness during the operation. The primary gases for controlled atmospheres are hydrogen and carbon monoxide and nitrogen.

The main uses of controlled atmospheres are (1) the prevention of the formation of oxides on the material of the charge, or conversely the reduction of any oxides present, and (2) the prevention of a change in the carbon content of a steel undergoing a heat treatment. Each of these uses denotes a chemical system in which the reactions are reversible.

The chemical systems relating to metallic oxides are:

A: Oxide + hydrogen ⇌ metal + water vapor
B: Oxide + carbon monoxide ⇌ metal + carbon dioxide

The chemical systems relating to carbon in steel are

E: Methane ⇌ hydrogen + carbon
F: Carbon monoxide ⇌ carbon dioxide + carbon

In artificial atmospheres the volume ratio of the two gases in the heating chamber should be so maintained as to correspond to the desired direction of the chemical activity of the system, or, if no chemical action is desired, to maintain that volume ratio at (or near) its equilibrium value for the temperature of the heat application. The equilibrium volume ratios for each of the four chemical systems A, B, E, and F for carbon steel over the usual range of temperature of heat-treatment processes and for atmospheric pressure are shown in Fig. 7.5.3. There is little tendency toward a change in the carbon content of a steel below the critical range. Oxidation is active down to about 1,100°F (650°C).

Curves E and F of Fig. 7.5.3 show the volume ratios of systems E and F for equilibriums with graphite. The equilibrium volume ratios of these two chemical systems for carbon in solid solution in steel (austenite) depend in each case on the carbon content of the steel. For the methane-hydrogen-carbon system (E) the volume ratio of the two gases at equilibrium with carbon in an unsaturated steel at a given temperature is less than the value shown by curve E. For the carbon monoxide-car-bon dioxide-carbon system (F) the volume ratio of the two gases at equilibrium with the carbon in low- and medium-carbon steel at a given temperature is somewhat greater than the value shown by curve F; for high-carbon steels the equilibrium volume-ratios approach the values of curve F.

In the case of the hydrogen-iron oxide reaction, curve A, the water vapor content of the mixture of gases at equilibrium decreases with decrease of temperature. Hence if a steel is to be cooled in a controlled atmosphere of this kind, the permissible water vapor content of the controlled atmosphere is dictated by the lowest temperature of the operation. The reverse is true of the carbon monoxide–iron oxide reaction, curve B. Thus if at a given temperature the carbon dioxide content of the mixture of carbon monoxide and carbon dioxide is less than the volume for equilibrium at that temperature it will be less than the volume for equilibrium at any lower temperatures and the steel can be cooled in that atmosphere without oxidation.

In the use of mixtures of the gases of the chemical systems noted to form controlled atmospheres for the heat treatment of steel, the interactions of the gases at elevated temperatures must be controlled by removal of all or nearly all the carbon dioxide and water vapor from the heating chamber.

The available data concerning controlled atmospheres for the protection of alloy steels during heat-treatment processes indicate that the technique for alloy steels is much the same as for carbon steels; i.e., a controlled atmosphere suitable for a carbon steel would, in general, be suitable for an alloy steel of the same carbon content.

In the heat treatment of nonferrous metals and alloys the use of either chemical system A or B requires for each oxide a knowledge of the equilibrium volume ratios of the chemical system used over the range of the operating temperature. Individual problems may arise. For example, copper can be bright-annealed in an atmosphere of dry steam—an inactive gas for this application—but the resultant staining of the copper during cooling may be objectionable. Copper usually contains a small percentage of oxide, and when annealing such copper in an atmosphere containing a reducing gas the temperature of the metal must be kept below about 750°F (420°C); otherwise the oxide will be reduced and the copper made brittle.

The foregoing discussion of atmosphere in heating chambers is intended to indicate the principles involved in the use of gases at elevated temperatures. The terms oxidation, reduction, carburization, and decarburization refer here to the chemical condition of a particular atmosphere and not to the extent of its effect on a charge. In all cases the concentration of the active gas or gases, time, temperature, in case of steel the carbon content and the gas pressure, and the catalytic action of hot surfaces within the chamber are important factors in the result obtained.

Bath Heating   Heating for local hardening of edge tools is the most general service. The lead-bath furnace has a working temperature range of 650 to 1,700°F (360 to 950°C). The salt-bath furnace can be adapted to working-temperature ranges within a total range of 300 to 2,350°F (170 to 1,300°C) by the selection of suitable mixture of salts. The two salt baths most generally used are cyanide mixtures and chloride mixtures. The rate of heating by immersion is much faster than obtained by radiation. The rate of heat transfer in a salt bath is about one-half that in the lead bath. An additional use of the salt bath is for cyaniding, in effect a process atmosphere.

Resistors   The resistor of a standard furnace is a sinuous winding mounted on the inner surfaces of the heating chamber as shown in Figs. 7.5.1 and 7.5.2. The resistor winding covers practically the entire surface of the space chosen. Resistors are applied on the basis of 2 to 3 kW/ft² (20 to 30 kW/m²) of wall surface in general practice. The basis of resistor location is radiation to all surfaces of the charge. Hence, the height and width dimensions of the heating chamber indicate the choice between sidewall and roof resistors. In some cases both locations are used. Uniform distribution of heat flow to the charge is obtained by a designed distribution of the surfaces of the resistors supplemented by reradiation from the inner surfaces of the chamber.

Resistors for the majority of standard furnaces are made of 80 Ni,

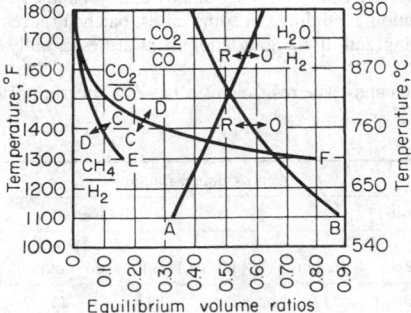

Fig. 7.5.3   Equilibrium volume ratios of chemical systems A, B, E, and F for steel. C = carburizing condition; O = oxidizing; D = decarburizing; R = reducing.

20 Cr alloy. A nickel-chromium-iron alloy is used in some furnaces for operation only over the lower portion of the furnace temperature range. Both ribbon and cast shapes are in use. The effort in each case is to obtain the maximum surface area per unit length of resistor and at the same time retain sufficient mechanical strength in the resistor winding.

The 80 Ni, 20 Cr alloy is self-protecting against oxidation, but this protection decreases with rise of temperature. The operating temperature of a resistor should be no higher than is needed in each case and should always be at a safe margin below the softening point of the alloy, which is about 2,500°F (1,390°C). This corresponds to a maximum furnace temperature of about 2,100°F (1,170°C). The life of a resistor is also affected by the frequency of heating and cooling. Barring accidents, the resistor of a standard furnace under average conditions of operation has a long life, usually measured in years of service.

The nickel-chromium resistor is used in artificial atmospheres as well as in natural atmospheres. This alloy is not resistant to compounds of sulfur and is affected to some extent by carbon monoxide.

The electric insulation of the resistor circuit is that of its refractory supports at elevated temperatures. This limits the voltage of the circuit to about 600 V. Small furnaces are usually designed for 110 V, medium sizes for 220 V, and the larger units for 440 V. Single phase up to 25 or 30 kW and three phase for higher ratings is general practice.

The resistivity-temperature coefficient of the nickel-chromium alloy permits the operation of resistors of this material on constant-voltage circuits. The rate of heat development in a resistor is proportional to the square of the applied voltage; hence maintenance of normal voltage is desirable. Voltage regulation is not as important as for other types of electrical apparatus because of the heat-storage capacity of the structure of the heating chamber. The power factor of the resistor circuit is practically unity.

High-Temperature Furnaces Silicon carbide is the basis of a type of nonmetallic resistor for heating-chamber temperatures up to about 2,800°F (1,560°C). The material is formed into rods. Resistors of this material do not require protection against oxidation and are operated on constant-voltage circuits.

Molybdenum resistors are suitable for temperatures up to 3,000°F (1,670°C). Above that temperature the metal begins to vaporize. A molybdenum resistor cannot be operated in a natural atmosphere, and also it must be protected from reactions with silica and carbon. The metal is immune from reactions with sulfur compounds, nitrogen, and water vapor. Hydrogen is the most common artificial atmosphere used with molybdenum resistors. The difference between the cold and hot resistances of the circuit makes a starting device necessary.

Other materials used to some extent for resistors are iron, tungsten, and graphite. These require protection against oxidation.

Temperature Regulation The temperature of the heating chamber of a resistor furnace is in most cases regulated by a more or less intermittent application of current—the on-and-off method—which is made automatic by instrument control. This method utilizes the heat-storage capacity of the inner lining of the heating chamber as a temperature equalizer. The variation from the normal temperature of the chamber can be kept within less than 7°F (4°C) plus or minus, without undue wear of the temperature-control equipment. Temperature regulation by voltage control is equally applicable to resistor furnaces, and the trend is toward the use of this more accurate method particularly for the more important installations.

Temperature protection for resistor furnaces is obtained by means of a temperature fuse mounted in the heating chamber and connected in the control circuit of the power supply to the furnace.

Multiple-Temperature Control The resistors of the larger furnaces are divided into two or more circuits. Each circuit can be equipped with individual temperature control. That arrangement provides temperature regulation at more than one location in the heating chamber and is an aid toward maintaining uniform temperature distribution within the chamber.

The subdivision of resistor circuits is used also for zone heating—and zone cooling where needed—in continuous furnaces.

Melting Pots Resistor heating is applied to melting pots for the soft metals and alloys and for lead baths and for salt baths. The immersion heating unit is used for temperatures up to 950°F (530°C). For higher temperatures the metal pot is heated by resistors mounted outside and around the pot. The assembly in each case includes a heat-insulating wall similar to that of a resistor furnace. Another method of heating applicable only to salt baths is the passage of alternating current (of any frequency) between electrodes immersed in the bath.

Tempering Furnaces The temperature is comparatively low—below 1,300°F (720°C). Electrically heated oil baths and salt baths are used for tempering many kinds of small parts. Another form of tempering furnace is a vertical resistor furnace with the addition of a removable metal cylinder (or basket) to contain the charge and to provide an annular passageway for the circulation of air (by a fan mounted on the furnace) over the resistors and thence through the charge—an application of forced convection heating.

Sizes The electrical rating is the general method of expressing the size of a resistor furnace. Sizes up to 100 kW predominate. 100- to 500-kW furnaces are common, and others within the range 500 to 1,000 kW are in service. The data in Table 7.5.1 refer to common sizes of so-called box furnaces for general service.

The **losses** from a resistor furnace for a given heating chamber temperature are as follows: The open-door loss is a variable depending on the area of the door (or doors) and the percentage of the time that the door is open—from a continuous furnace this loss also varies with the type and speed of the conveyor; with artificial atmospheres, the loss of heat in the gases discharged for atmosphere control; the stored-heat loss, a variable that depends on the extent and frequency of the cooling of the furnace within a given period of operation; the heat dissipated from the outer surfaces of the furnace.

The **operating efficiency** is expressed as either pounds of material treated or kWh or kWh per ton. Representative values for average service for the heat treatment of steel range from 7 to 12 lb/kWh. Corresponding values for nonferrous metals and alloys are within the range 12 to 22 lb/kWh.

The general field of the batch furnace is defined by the following conditions: (1) intermittent and varied production; (2) long periods of heating (and in some cases slow cooling); (3) heating service beyond the range of the handling capacity of furnace conveyors; and (4) supplementary heating service. A continuous furnace is indicated where the flow of material to be heated is reasonably uniform and continuous, i.e., mass-production conditions. In some cases, batch furnaces with automatic charging and discharging equipment are essentially continuous furnaces.

Resistor Ovens The resistor oven is a modification of the resistor

Table 7.5.1 Box Resistor Furnaces, 1,850°F (1,030°C) Class

| | | | | | Approx dimension, in | | | | | | |
| | | | | | Inside | | | Overall | | | |
Connected load, kW	Power supply, 200 V	Steel, lb/h at 1,500°F	Time to heat to 1,500°F (830°C) when used previous day, min	Radiation, kWh/h, at 1,500°F	Width	Depth	Height	Width	Depth	Height, door closed	Height, door open
29	1-phase	300	35	4.9	18	36	18	55	89	86	97
45	3-phase	500	35	6.9	24	54	20	61	108	90	101
60	3-phase	650	25	7.8	30	63	23	78	125	90	98
72	3-phase	750	25	9.1	36	72	23	84	135	90	98

furnace to correspond to the low temperatures of drying and baking processes. The heating chamber is an insulated metal structure with a fresh-air inlet and an exhaust fan for ventilation (the removal of vapors and gases evolved from the charge). A refractory lining is not required. Ovens may be of the batch type with conventional methods of handling the charge of the continuous type, usually with chain conveyors.

The most common type of electric oven is heated by resistors mounted in a separate compartment of the heating-chamber enclosure. The heat transfer is by forced convection which is accomplished by recirculation of the chamber atmosphere by a motor-driven fan.

A resistor oven with filament-type lamps as heating units provides what is generally known as infrared heating. The lamps, usually with self-contained reflectors, surround the charge, and the heat transfer is by radiation, mainly in the infrared portion of the spectrum. This type of oven is best adapted to the continuous heating of charges which present a large surface area in proportion to the mass and which require only surface heating, e.g., baking finishes on sheet products.

Ventilation The vapors and gases evolved from the charge during baking processes are often flammable, and the continuous discharge of these products from the oven chamber is essential for protection against explosions. For detailed recommendations, see *Pamphlet* 74 of the Assoc. Factory Mutual Fire Ins. Cos., Boston.

DIELECTRIC HEATING

The term relates to the heat developed in dielectric materials, such as rubber, glue, textiles, paper, and plastics, when exposed to an alternating electric field. The material to be heated is placed between plate-form electrodes, as indicated in Fig. 7.5.4. It is not necessary that the electrodes be in contact with the charge; hence continuous heating is often practicable.

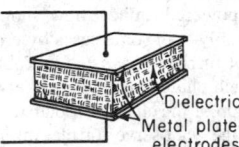

Fig. 7.5.4 Assembly for dielectric heating.

If the material of the charge is homogeneous and the electric field uniform, heat is developed uniformly and simultaneously throughout the mass of the charge. The thermal conductivity of the material is a negligible factor in the rate of heating. The temperatures and services are within the oven classification.

The frequency and voltage for this class of service depend in each case on the electrical properties of the material of the charge at the temperature specified for the heat application. The frequencies in use range from 2 to 40 MHz; the most common frequencies are from 10 to 30 MHz. It is advisable to select the frequency for heating by trial.

The upper limit of voltage across the electrodes is fixed by the spark-over value and by corona. The permissible voltage gradient depends on the nature of the material of the charge. Values within the range 2,000 to 6,000 V/in (790 to 2,400 V/cm) are found in practice; the voltage across the electrodes should not exceed 15,000 V.

Applications of dielectric heating include setting glue as in plywood manufacture, curing rubber, drying textiles, and the heat treatment of plastics.

INDUCTION HEATING

In induction heating, the lateral surface of the charge is exposed to an alternating magnetic flux. The currents thus induced in the charge flow wholly within its mass. The term "eddy-current heating" is sometimes applied to the method.

A common assembly, if the charge is to be heated to a temperature below its melting point, is to place the charge within a coil as indicated in Fig. 7.5.5. An alternating current in the coil establishes the required alternating magnetic flux around the charge.

A peculiar feature of such assemblies, termed "induction heaters," is the absence of heat insulation; the coil is water-cooled. Thus, the charge is heated in the open air, or an artificial atmosphere can be used, if the assembly is enclosed. This requires rapid heating with heat cycles measured in minutes or seconds.

The frequency required is a function of the electric and magnetic properties of the charge at the temperature specified for the heat application and of the radius, or one-half the thickness, of the charge. This frequency for a given material increases with decrease of the dimension noted. The frequency in any case is not critical. In practice, 480, 960, 3,000, and 9,600 Hz and around 450 kHz suffice for the entire range of

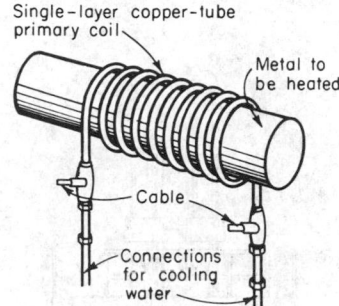
Single-layer copper-tube primary coil
Metal to be heated
Cable
Connections for cooling water

Fig. 7.5.5 Assembly for induction heating.

induction heating. The highest frequencies needed are those for heating steel charges to temperatures above the Curie point. About ½ in (1¼ cm) diam in this case is the lower limit for 9,600 Hz. This limit dimension is decreased for steel heated to temperatures below the Curie point and for all charges of nonferrous materials.

The operation can be either batch heating or continuous heating as required. Applications include heating for forging, for annealing, for hardening steel, for brazing, soldering, and strain relief. As most of the heat is developed within the annular zone of the charge, the method is particularly well adapted to heating steel parts for surface hardening. A recent application of induction heating is the raising in temperature of billet-size ingots for rolling into merchant bars.

ARC FURNACES

Two types of arc furnaces are in common use: (1) the three-phase furnace and (2) the single-phase furnace. The general field of the three-phase furnace is the melting and refining of carbon and alloy steels; that of the single-phase furnace is the melting of nonferrous alloys. There is an increasing amount of arc-furnace capacity used for melting and refining various types of iron.

Three-Phase Arc Furnaces The general design of this type of furnace is shown in Fig. 7.5.6. In operation, each heat is started by swinging the furnace roof aside and then loading the refractory-lined furnace body with scrap dropped from a crane-handled clamshell charging bucket. Arcs next are drawn between the lower ends of the graphite electrodes and the scrap; melting proceeds under automatic control until the hearth carries the molten metal. This fluidizing stage is effected at about 85 percent thermal efficiency. Several charges usually are needed to build up the bath—particularly in ingot practice. The furnace tilts forward for pouring; the back tilt serves in the removal of slag and permits the furnace hearth to be kept in proper condition. The slagging door is opposite the pouring spout. Large furnaces frequently also have a side door, known as a working door.

Refractories Furnaces that produce foundry steels operate with acid lining. This means silica brick form the walls; the hearth is of gannister or the equivalent. Silica-brick roofs are the more widely used although, for intermittent operation, clay brick may be preferred. The slags of acid-lining practice remove no phosphorus or sulfur. Essentially all ingot operations are carried on with basic linings. This means magnesite bottom and sidewalls, so that the limey slags employed will not erode them. Entry ports for the electrodes may be of extra-quality refractories to prolong roof life, particularly where the furnace is in continuous operation. In basic practice phosphorus joins the slag readily; sulfur can be removed next by a second slag, when this slag has been made highly reducing. Slag covering the molten bath serves in refining the metal and reduces the heating of wall and roof brick. In modern arc furnace operations, foamy slag practice is employed, wherein a deep, foamy slag prevents the arcs from damaging the wall and roof linings. Superrefractories find application in high-temperature, long-refining operations. Electric irons are made in acid-lined furnaces.

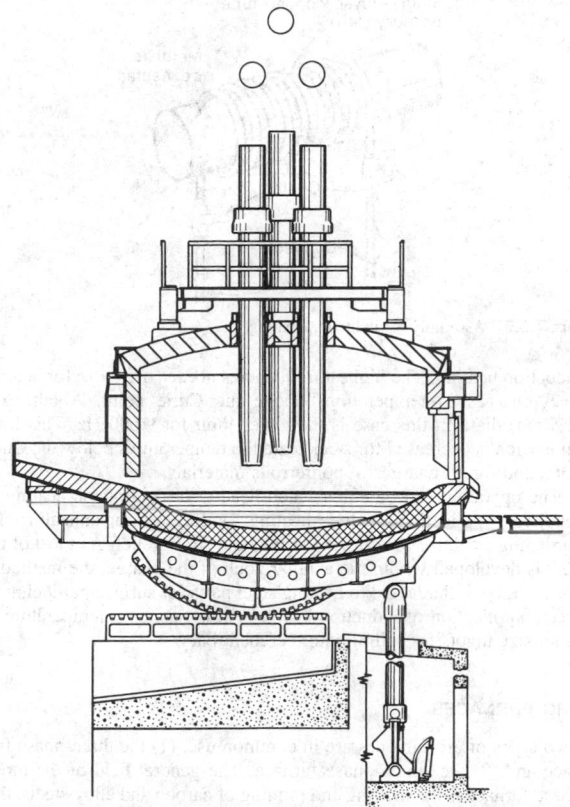

Fig. 7.5.6 Three-phase arc furnace with basic lining.

Temperature Arcs approximate 6,300°F (3,500°C); hence operation must be carried out so as to protect the refractories as much as possible. As the top-charge furnace now has supplanted the door-charge furnace in nearly all cold-melt work, the conditions for shielding the refractories during the melt-down stage of each heat are good. With the furnace filled to the top with scrap, the electrodes bore down through that scrap, and the heat of the arcs is liberated right in the metallic charge itself. When the charge, and any back charges made, approach the fluid stage it is customary to reduce both the power input and the length of arcs employed. During the finishing stages, roof and sidewalls are protected both by the slag and by the "umbrella" effect of the electrodes themselves. Deserving mention is the expanding use of oxygen to gain speed in production, which makes for increasing furnace temperatures. The higher sidewalls of modern furnaces aid in obtaining good roof life. Additionally, water-cooled sidewalls extend refractory life and thus minimize the cost of replacing refractory.

Charges The three-phase arc furnace is primarily a unit for con-

verting scrap charges into steel for pouring into ingots, castings, or a continuous caster. This type of equipment finds increasing use also in the cold melting and duplexing of gray and white irons. Hand and chute charging have practically disappeared, at least insofar as furnaces of a ton charge size upward are concerned. One of the main advantages of the top-charge furnace is that the scrap used does not need to be cut to door size, as was the case formerly.

Although first employed only for the more expensive grades of steel, the arc furnace now is used widely in making ingots for rolling into merchant bars and similar grades and supplies liquid metal fed into a continuous caster. The speed of production on this type of working—termed single-slag dephosphorizing basic practice—can be double that obtained with the same furnace used to make two slag dephosphorized and desulfurized basic alloy steels. Acid working on foundry steels generally approximates the same speed as single-slag dephosphorizing basic practice, and some alloy steels require about half again as much time. While most carbon steel for castings is made on an acid hearth, a basic bottom is regularly used for making manganese steels, for refining nickel and copper, and for the furnacing of many heat-resistant alloys. Section 13.1 discusses steel-foundry practice.

In general, approximately 320 kWh at 100 percent thermal efficiency will be needed to melt 1 ton of cold steel scrap. This means about 400 kWh will be needed to fluidize each ton. Additionally, about 100 kWh/ ton will be needed to finish the heat and superheat the bath—this in the case of ordinary plain carbon steels made on single-slag acid or basic practice. Double-slag steel heats will require no more power than others for fluidizing the scrap charge, but the additional power needed for melting new slag, refining, melting added alloys, etc., may require as much as 250 kWh/ton of bath, or even more.

Three-phase arc furnaces are usually given an hourly productive rating in terms of acid foundry steels when these equipments are supplied in sizes up to and including the 11-ft (3.4-m) diam unit. However, with many furnaces extra-powered, quite a few shops exceed the normal hourly rating considerably—in some cases by essentially 100 percent. Representative sizes of furnaces are listed in Table 7.5.2.

Arcs The arc in each phase is maintained between the lower end of the electrode and the top of the charge (or bath, after the molten state is reached). Higher voltages can serve for melting as the size of the furnace increases; thus, where a 7-ft (2.1-m) diam furnace employs 215 V as its highest melting potential, a 15-ft furnace would use 290 V or higher as the top tap. For such a furnace constructed with water-cooled sidewall and roof panels, the application of 500 V would be normal. The furnace transformer is universally of the motor-operated tap-changer type, and in the case of, say, a 10,000 kVA at 55°C rise substation, a secondary voltage variation of more than 150 V is customary. The range of lower voltages used for refining the molten metal is obtained by changing the primary of the main transformer from delta to star connection; this reduces both voltage and capacity to 58 percent of their values with delta primary connection. If, say, 12 tons of steel scrap are to be melted down to fluid in 1 h, then the electric energy needed will approximate 5,000 kWh. With 245 V used as the principal melt-down voltage, the current per phase will have to average close to

Table 7.5.2 Sizes of Three-Phases Arc Furnaces

Diam of shell, ft	Normal charge, tons	Normal powering, kVA	Normal productive rate, single-slag steels, tons/h*
5	1½	600	½
7	3½	1,500	1½
9	8	3,000	3
11	16	6,000	6
12½	27	9,000	9
15	50	12,500	13
20	115	25,000	27
24	225	36,000	40

* Many users exceed these outputs, particularly those using burners and oxygen to speed operations.

12,000 A. A 12-in (30-cm) diam graphite electrode would amply carry this current. Small furnaces operate with 600 kVa and even higher powering per ton of charge, whereas in the case of the larger equipments the electrical backing of the furnace normally does not exceed 300 kVA/ton of charge.

Reactance is required in the circuits of an arc furnace to give stability and to limit the current when an electrode makes contact with the metallic charge. The inherent reactance (impedance) in the instance of 10,000-kVa installations and above normally is sufficient. The total stabilizing reactance provided in the case of a 1,000-kVa load normally approximates 30 percent.

Regulation The characteristics of an arc furnace circuit for a given applied voltage are shown in Fig. 7.5.7. For each voltage there is a value of current that gives maximum power in the furnace. This optimum current is the basis of the regulation of the circuit.

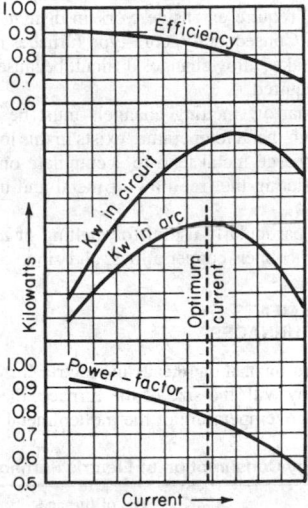

Fig. 7.5.7 Characteristics of an arc furnace circuit.

The control of the power input into direct-arc electric furnaces is effected by the adjustment of the arc length. To accomplish this, the electrode arms are positioned in the "raise" or in the "lower" direction by an automatic regulator. This regulator, which responds within a few cycles, causes the electrode arms to be lowered by extra-fast motor-driven winches when voltage is obtained by closing the circuit breaker. As soon as contact between electrode and scrap charge is established, melting current flows, and this current, whenever excessive, functions immediately through the medium of the winch motor to elevate that particular electrode arm and electrode by the distance corresponding to the diminution in power input needed just at that instant.

Formerly, the so-called contactor regulator was used universally to energize the winch motors. More recently the rotary regulator—this, in effect, being a particularly responsive motor generator set for each of the three phases—has forged to the forefront by reason of giving more precise control with minimized maintenance. Currently, even faster response and electrode-travel speed are provided by low-inertia static-regulating equipment.

Single-Phase Arc Furnaces Single-phase arc furnaces usually are manufactured in the two-electrode type. When the electrodes operate vertically, the furnace melts much as a three-phase direct-arc furnace does. However, most vertical-electrode single-phase furnaces are of laboratory size—that is, up to 150 kVa in powering.

When two electrodes are mounted horizontally in a rocking furnace an indirect-arc unit is obtained. Many rocking furnaces serve well in the melting of brasses, bronzes, and in similar work. Volatiles are reincor-

porated in the metal since the bath washes over much of the interior of a rocking furnace. The oscillation approximates 200°.

Rocking furnaces usually do not exceed 500 kW in powering. A single operating voltage can suffice. In regulating a rocking furnace, only one electrode need be movable, on a carriage under automatic control, to maintain the requisite amperage by varying the length, and therefore the resistance, of the arc gap.

INDUCTION FURNACES

There are two basic types of metal-melting induction furnaces: (1) coreless and (2) core-type. Both types utilize the principle of a transformer. The high-voltage circuit is coupled with that of the low voltage without directly connecting the two circuits. The element responsible for this coupling effect is the magnetic field. Induction heating utilizes the property of the magnetic field, which enables heat to be transferred without direct contact. By correctly disposing the high-voltage winding, which in the case of the induction furnace would be an induction coil or inductor, the magnetic field is directed so that the metal to be heated or melted is made to absorb energy. The temperature attainable is limited solely by the resistance to heat of the surrounding lining material. Induction heating enables any temperature to be achieved while providing for excellent regulation of temperature and metallurgical properties. Any metal which will conduct electric current can be melted in an induction furnace.

Coreless Induction Furnaces (See Fig. 7.5.8.) This type of furnace consists of a crucible, copper coil, and framework on supports arranged for tilting and pouring. The specially designed induction coil acts as the primary of the transformer. The crucible conforms to conventional refractory practice. A rammed crucible is used for furnaces above 50 kW, and preformed crucibles are used on smaller furnaces such as laboratory units.

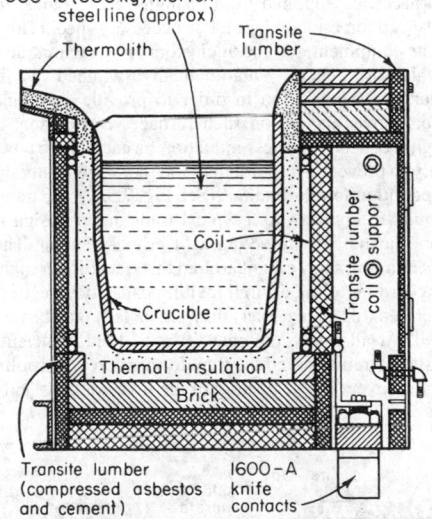

Fig. 7.5.8 Coreless induction furnace.

The principle of operation is essentially the same as that of the induction heater previously described. The initial charge in the furnace is cold scrap metal—pieces of assorted dimensions and shapes and a large percentage of voids. As the power is applied and the heat cycle progresses, the charge changes to a body of molten metal; additional cold metal is added until the molten-metal level is brought to the desired temperature and metallurgical chemistry. The furnace then is tapped.

When the metal in the furnace becomes fluid, depending on whether a line frequency or medium-frequency supply by means of convectors is used, a certain electromagnetic stirring action will occur. This stirring action is peculiar to the induction furnace and aids in the production of certain types of alloys. The stirring action increases as the frequency is reduced.

Line-frequency applications are generally reserved to furnaces having a metal-holding capacity of 800 lb (360 kg) and above. There is always an ideal relationship between the size of a coreless furnace and its operating frequency. As a general rule, a small furnace gives best results at high to medium frequencies and large furnaces work best at the lower frequencies. A frequency is suited to a given furnace when it yields good, fast melting with a gentle stirring action. Too high or too low frequencies are accompanied by undesirable side effects. The tabulation below gives the charge weights and frequencies generally to be used:

Charge weight, lb	Frequency, Hz
2–50	9,600
12–500	3,000
200–15,000	960
800–75,000	60

The coreless induction furnace is usually charged full and tapped empty, although at line frequencies, it may be necessary to retain a certain amount of metal in the furnace to continue the operation, since it is difficult to start the furnace with small metal particles, such as turnings and borings, in a cold crucible. As a result, it is general practice to retain a heel in the furnace of about one-third its molten-metal volume. This problem can be avoided in furnaces of higher frequencies, where start-up can be performed with small-size metal charges without carrying the heel.

Coreless induction furnaces are particularly attractive for melting charges and alloys of known analysis; in essence, the operation becomes one of metal melting with rapidly absorbed electric heat without disturbing the metallurgical properties of the initial charge.

These furnaces are supplied from a single-phase source. In order to obtain a balanced three-phase input, it is necessary specifically to design the electrical equipment for the inclusion of capacitors and suitable reactors, which are generally automatically switched (by inductance changes) during the operation in order to provide a reasonably high power factor. Power factors on such furnaces can be kept at or near unity. In high-frequency coreless induction furnaces, high power factors are necessary to prevent overburdening the motor-generator equipment.

Core-Type Induction Furnaces (See Fig. 7.5.9.) The transformer is actually wound to conform to a typical transformer design having an iron core and layers of wire acting as a primary circuit. The melting channel acts as a ring short circuit around this transformer in the melting chamber. According to the desired melting capacity, one, two, or three such transformers (or **inductors**, as they are called) may be added to the furnace shell. At all times, the channel must hold sufficient metal to maintain a short circuit around the transformer core. Air cooling is used as required to prevent undue heating of the inductor coils and magnetic cores.

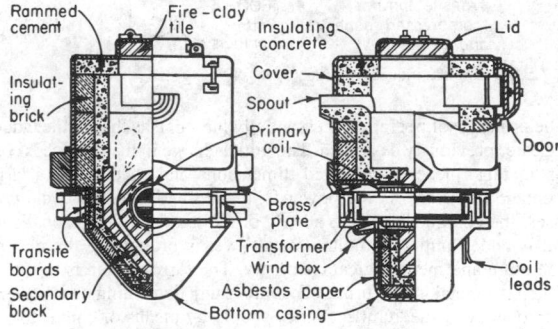

Fig. 7.5.9 Core-type induction furnace.

The melting output is controlled by varying the voltage supplied to the inductors with the aid of a variable-voltage transformer connected to the primary circuit of the supply. Core-type furnaces always use line frequencies. Voltage or power-input regulation, therefore, can be performed by adjusting the tap setting of the transformer feeding the furnace transformer attached to the furnace shell. These transformers are single-phase units, and by using three such units, a balanced three-phase input can be obtained. The current flowing through the primary inductors by transformation causes a much larger current in the metal loop, whose resistance creates heat for melting.

The core-type furnace is the most efficient type of induction furnace because its iron core concentrates magnetic flux in the area of the magnetic loop, ensuring maximum power transfer from primary to secondary. Efficiency in the use of power can be as high as 95 to 98 percent.

The essential loop of metal must always be maintained in the core-type furnace. If this loop is allowed to freeze by cooling, extreme care is necessary in remelting because the loop may rupture and disrupt the circuit. This could require extensive work in dismantling the coil and restoring the loop. Consequently, core-type furnaces rarely are permitted to cool. This makes alloy changes difficult because a heel of molten metal always is required.

The relatively narrow melting channels must be kept as clean as possible since a high metal temperature exists in this loop. Nonmetallics or tramps in the charge metal tend to accumulate on the walls in the channel area, restricting the free flow of metal and ultimately closing the passage.

This furnace is particularly useful for melting of nonferrous metals such as aluminum, copper, copper alloys, and zinc.

POWER REQUIREMENTS FOR ELECTRIC FURNACES

The **energy required for melting metals** in electric furnaces varies for a given metal or alloy with the size of the furnace, the thickness of the refractory lining, the temperature of the molten metal, the rate of melt-

Table 7.5.3 Energy Consumption of Electric Furnaces

Process	Type of furnace	lb/kWh
Baking finishes on sheet metal	Batch oven	10–18
Baking finishes on sheet metal	Continuous oven	25–30
Baking bread	Continuous oven	10–12
Annealing brass and copper	Batch furnace	10–25
Annealing steel	Batch furnace	5–15
Hardening steel	Batch furnace	7–11
Tempering steel	Batch furnace	15–25
Annealing glass	Continuous furnace	40–100
Vitreous enameling, single coat	Batch furnace	5–8
Vitreous enameling, single coat	Continuous furnace	10–15
Galvanizing	Batch furnace	12–20

Melting metals	Type of furnace	kWh/ton (2,000 lb)
Lead	Resistor	40–50
Solder 50–50	Resistor	40–50
Tin	Resistor	35–50
Zinc	Induction	80–100
Brass	Arc and induction	250–400
Steel, melting only	Arc and induction	450–700
Steel, melting and refining	Arc	600–750
Gray iron	Arc and induction	450–600

Furnace products	Type of furnace	kWh/ton (2,000 lb)
Aluminum	Electrolytic	22,000–27,000
Calcium carbide	Resistance	3,000–6,000
Ferroalloys	Resistance	4,000–8,000
Graphite	Resistance	3,000–8,000
Phosphoric acid	Resistance	5,000–6,000
Silicon carbide	Resistance	8,500–10,000
Smelting iron ore	Resistance	1,650–2,400

ing, and with the degree of the continuity of the operation of the furnace. An estimated efficiency of 50 to 60 percent is often used for preliminary purposes. As is well known, 3- to 6-ton direct-arc furnaces often are used to tap acid foundry steels with the consumption of less than 500 kWh to the ton, and large ingot furnaces of this same type, operating basic-lined on common steels for ingots, give even better results despite the call for several more charges of scrap per heat.

Average values in kWh/ton of molten metal are as follows: yellow brass, 200 to 350; red brass, 250 to 400; copper, 250 to 400; lead, 30 to 50; steel melting, when making high-quality double-slag basic heats, 650 to 800 (Table 7.5.3).

Electrode consumption varies considerably in arc furnaces because of their different constructions and operations. Average values in pounds of electrode per ton of molten metal are: steel melting, with graphite electrodes, 5 to 10; brass melting, with graphite electrodes, 3 to 5. Graphite electrodes have largely superseded carbon electrodes.

SUBMERGED-ARC AND RESISTANCE FURNACES

The resistance furnace is essentially a refractory-lined chamber with electrodes—movable or fixed—buried in the charge. This simplicity permits a wide range of designs and much latitude in dimensions. The general service is heating charges of a refractory nature to bring about chemical reactions or changes in the physical structure of the material of the charge. The energy requirement of each of such processes is a large item in the cost of production. Large units and a favorable power location are the rule. Resistance furnaces also are termed submerged-arc furnaces and/or, in quite a few instances, smelting-type furnaces.

The only limit on the temperature to which a charge can be heated by this method is the temperature at which the materials of the charge are vaporized. For temperatures beyond the limit of refractory linings, the materials of the charge are used to form a protective layer between the core of the charge (through which the current passes) and the walls of the furnace.

Resistance furnaces with movable electrodes may be either single-phase or polyphase. The materials of the charge are fed more or less continuously, and the product is discharged intermittently or continuously as required. In some cases the product is in the molten state; in others the product is a vapor. The usual method of operation is the use of a single operating voltage and a constant power input. The power is regulated by adjustment of the depths of the electrodes in the charge. The load is fairly uniform and, if polyphase, is kept reasonably well balanced.

The **resistance furnace with fixed electrodes** is designed for heating materials in batches and is usually rectangular in shape with an electrode at each end for single-phase operation. The length and cross-sectional area of the path of the current are proportioned to suit the power characteristics of the charge. Refractory materials have negative temperature-resistance coefficients, and hence to maintain constant power in the furnace circuit the applied voltage must be reduced as the temperature of the charge rises in proportion to the square root of the ratio of the initial resistance of the furnace circuit to the resistance of the furnace circuit at the end of the heat cycle. If the materials of the charge are nonconductors at room temperature, a starting circuit is provided by means of a core of carbon—usually coke—placed in the charge. The heat cycles of furnaces of this class generally extend over a period of several days.

Some of the more common **uses of the resistance furnace** are:

Calcium carbide furnaces are charged continuously with lime and coke. These equipments can be either open or closed top. This type of furnace has been built up to 70,000 kVA in electrical powering—covered and sealed for gas collection.

Ferroalloy furnaces for the production of ferrochrome, ferrosilicon, ferromanganese, etc., are usually three-phase furnaces with movable electrodes and are similar in construction to the three-phase arc furnace. The charge is a mixture of the ore (oxide) of the selected metal, scrap iron, and a reducing agent, generally carbon except for very low carbon content alloys, for which some other reducing agent such as aluminum or silicon is required. Six-electrode furnaces often are used for power inputs of 15,000 kVA and more.

The **graphitizing furnace** is of the single-phase batch type. Artificial graphite is made by heating amorphous carbon (coal or coke) while shielded from air to a high temperature—around 4,500°F (2,500°C). The presence of some metallic impurity, such as iron oxide, in the charge appears to be necessary for the conversion of amorphous carbon to graphitic carbon. The raw material for making bulk graphite constitutes both the charge and the protective layer around the core of the charge. Graphite shapes are made from the corresponding shapes of amorphous carbon which are embedded—between the electrodes—in raw material as noted for the manufacture of bulk graphite.

The **silicon carbide furnace** is similar to the graphitizing furnace. The charge is a mixture of sand (silica), coke, sawdust, and a small amount of salt. This mixture is packed around a core of granulated coke to form the initial circuit between the electrodes. The sand and coke are the reacting materials. The sawdust serves to make the charge porous so that the gases formed during the heating of the charge can escape freely. The salt vaporizes and removes impurities, such as iron, in the form of chlorides. The temperature of the process is 2,700 to 3,400°F (1,500 to 1,880°C).

Machine Elements

BY

HEARD K. BAUMEISTER *Senior Engineer, Retired, International Business Machines Corporation.*

ANTONIO F. BALDO *Professor of Mechanical Engineering, Emeritus, The City College, The City University of New York.*

GEORGE W. MICHALEC *Consulting Engineer. Formerly Professor and Dean of Engineering and Science, Stevens Institute of Technology.*

VITTORIO (RINO) CASTELLI *Senior Research Fellow, Xerox Corp.*

MICHAEL J. WASHO *Engineering Associate, Eastman Kodak Company, Kodak Park, Engineering Division.*

JOHN W. WOOD, JR. *Applications Specialist, Fluidtec Engineered Products, Coltec Industries.*

HELMUT THIELSCH *President, Thielsch Engineering Associates.*

C. H. BERRY *Late Gordon McKay Professor of Mechanical Engineering, Emeritus, Harvard University.*

8.6 PACKING AND SEALS
by John W. Wood, Jr.

8.7 PIPE, PIPE FITTINGS, AND VALVES
by Helmut Thielsch

8.8 PREFERRED NUMBERS
by C. H. Berry

8.1 MECHANISM
by Heard K. Baumeister, Amended by Staff

REFERENCES: Beggs, ''Mechanism,'' McGraw-Hill. Hrones and Nelson, ''Analysis of the Four Bag Linkage,'' Wiley. Jones, ''Ingenious Mechanisms for Designers and Inventors,'' 4 vols., Industrial Press. Moliam, ''The Design of Cam Mechanisms and Linkages,'' Elsevier. Chironis, ''Gear Design and Application,'' McGraw-Hill.

NOTE: The reader is referred to the current and near-past professional literature for extensive material on linkage mechanisms. The vast number of combinations thereof has led to the development of computer software programs to aid in the design of specific linkages.

Definition A **mechanism** is that part of a machine which contains two or more pieces so arranged that the motion of one compels the motion of the others, all in a fashion prescribed by the nature of the combination.

LINKAGES

Links may be of any form so long as they do not interfere with the desired motion. The simplest form is four bars *A, B, C,* and *D,* fastened together at their ends by cylindrical pins, and which are all movable in parallel planes. If the links are of different lengths and each is fixed in

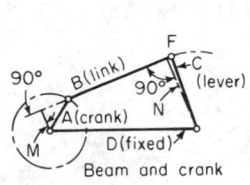

Fig. 8.1.1 Beam-and-crank mechanism.

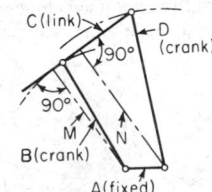

Fig. 8.1.2 Drag-link mechanism.

turn, there will be four possible combinations; but as two of these are similar there will be produced three mechanisms having distinctly different motions. Thus, in Fig. 8.1.1, if *D* is fixed *A* can rotate and *C* oscillate, giving the **beam-and-crank** mechanism, as used on side-wheel steamers. If *B* is fixed, the same motion will result; if *A* is fixed (Fig. 8.1.2), links *B* and *D* can rotate, giving the **drag-link** mechanism used to feather the floats on paddle wheels. Fixing link *C* (Fig. 8.1.3), *D* and *B* can only oscillate, and a **rocker** mechanism sometimes used in straight-line motions is produced. It is customary to call a rotating link a **crank;** an oscillating link a **lever,** or beam; and the connecting link a **connecting rod,** or **coupler.** Discrete points on the coupler, crank, or lever can be pressed into service to provide a desired motion. The fixed link is often enlarged and used as the supporting frame.

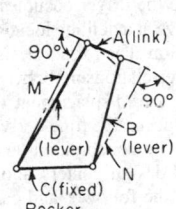

Fig. 8.1.3 Rocker mechanism.

If in the linkage (Fig. 8.1.1) the pin joint *F* is replaced by a slotted piece *E* (Fig. 8.1.4), no change will be produced in the resulting motion, and if the length of links *C* and *D* is made infinite, the slotted piece *E* will become straight and the motion of the slide will be that of pure translation, thus obtaining the engine, or **sliding-block,** linkage (Fig. 8.1.5).

If in the sliding-block linkage (Fig. 8.1.5) the long link *B* is fixed

(Fig. 8.1.6), *A* will rotate and *E* will oscillate and the infinite links *C* and *D* may be indicated as shown. This gives the **swinging-block linkage.** When used as a quick-return motion the slotted piece and slide are usually interchanged (Fig. 8.1.7) which in no way changes the resulting motion. If the short link *A* is fixed (Fig. 8.1.8), *B* and *E* can both rotate,

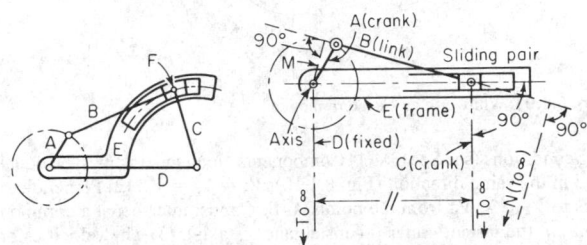

Figs. 8.1.4 and 8.1.5 Sliding-block linkage.

and the mechanism known as the **turning-block linkage** is obtained. This is better known under the name of the **Whitworth quick-return motion,** and is generally constructed as in Fig. 8.1.9. The **ratio of time of advance to time of return** *H*/*K* of the two quick-return motions (Figs. 8.1.7 and

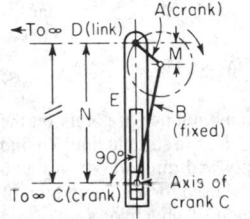

Fig. 8.1.6 Swinging-block linkage.

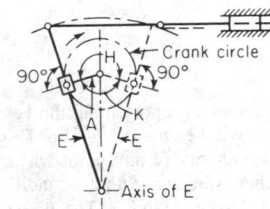

Fig. 8.1.7 Slow-advance, quick-return linkage.

8.1.9) may be found by locating, in the case of the swinging block (Fig. 8.1.7), the two tangent points (*t*) and measuring the angles *H* and *K* made by the two positions of the crank *A*. If *H* and *K* are known, the axis of *E* may be located by laying off the angles *H* and *K* on the crank circle

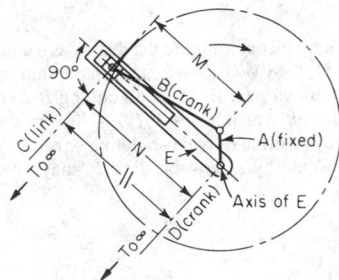

Fig. 8.1.8 Turning-block linkage.

and drawing the tangents *E*, their intersection giving the desired point. For the turning-block linkage (Fig. 8.1.9), determine the angles *H* and *K* made by the crank *B* when *E* is in the horizontal position; or, if the angles are known, the axis of *E* may be determined by drawing a horizontal line through the two crankpin positions (S) for the given angle, and the point where a line through the axis of B cuts E perpendicularly will be the axis of E.

Velocities of any two or more points on a link must fulfill the follow-

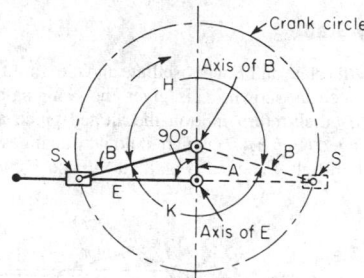

Fig. 8.1.9   Whitworth quick-return motion.

ing conditions (see Sec. 3). (1) Components along the link must be equal and in the same direction (Fig. 8.1.10): $V_a = V_b = V_c$. (2) Perpendiculars to V_A, V_B, V_C from the points A, B, C must intersect at a common point d, the **instant center** (or instantaneous axis). (3) The velocities of points A, B, and C are directly proportional to their distances from this center (Fig. 8.1.11): $V_A/a = V_B/b = V_C/c$. For a straight link the tips of

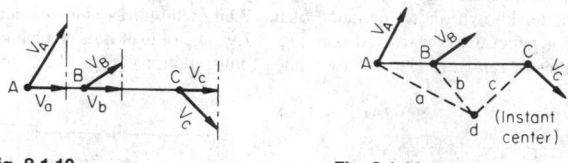

Fig. 8.1.10   **Fig. 8.1.11**

the vectors representing the velocities of any number of points on the link will be on a straight line (Fig. 8.1.12); abc = a straight line. To find the velocity of any point when the velocity and direction of any two other points are known, condition 2 may be used, or a combination of conditions 1 and 3. The **linear velocity ratio** of any two points on a

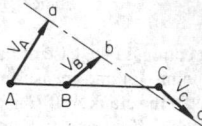

Fig. 8.1.12

linkage may be found by determining the distances e and f to the instant center (Fig. 8.1.13); then $V_c/V_b = e/f$. This may often be simplified by noting that a line drawn parallel to e and cutting B forms two similar triangles efB and sAy, which gives $V_c/V_b = e/f = s/A$. The **angular velocity ratio** for any position of two oscillating or rotating links A and C (Fig. 8.1.1), connected by a movable link B, may be determined by

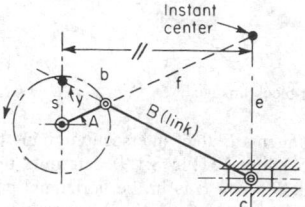

Fig. 8.1.13

scaling the length of the perpendiculars M and N from the axes of rotation to the centerline of the movable link. The angular velocity ratio is inversely proportional to these perpendiculars, or $O_C/O_A = M/N$. This method may be applied directly to a linkage having a sliding pair if the two infinite links are redrawn perpendicular to the sliding pair, as indicated in Fig. 8.1.14. M and N are shown also in Figs. 8.1.1, 8.1.2, 8.1.3, 8.1.5, 8.1.6, 8.1.8. In Fig. 8.1.5 one of the axes is at infinity; therefore, N is infinite, or the slide has pure translation.

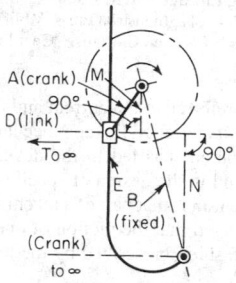

Fig. 8.1.14

Forces   A mechanism must deliver as much work as it receives, neglecting friction; therefore, the force at any point F multiplied by the velocity V_F in the direction of the force at that point must equal the force at some other point P multiplied by the velocity V_P at that point; or the forces are inversely as their velocities and $F/P = V_P/V_F$. It is at times more convenient to equate the moments of the forces acting around each axis of rotation (sometimes using the instant center) to determine the force acting at some other point. In Fig. 8.1.15, $F \times a \times c/(b \times d) = P$.

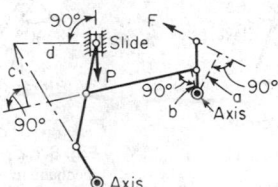

Fig. 8.1.15

CAMS

Cam Diagram   A cam is usually a plate or cylinder which communicates motion to a follower as dictated by the geometry of its edge or of a groove cut in its surface. In the practical design of cams, the follower (1) must assume a definite series of positions while the driver occupies a corresponding series of positions or (2) must arrive at a definite location by the time the driver arrives at a particular position. The former design may be severely limited in speed because the interrelationship between the follower and cam positions may yield a follower displacement vs. time function that involves large values for the successive time derivatives, indicating large accelerations and forces, with concomitant large impacts and accompanying noise. The second design centers about finding that particular interrelationship between the follower and cam positions that results in the minimum forces and impacts so that the speed may be made quite large. In either case, the desired interrelationship must be put into hardware as discussed below. In the case of high-speed machines, small irregularities in the cam surface or geometry may be severely detrimental.

A stepwise displacement in time for the follower running on a cam driven at constant speed is, of course, impossible because the follower would require infinite velocities. A step in velocity for the follower would result in infinite accelerations; these in turn would bring into being forces that approach infinite magnitudes which would tend to destroy the machine. A step in acceleration causes a large jerk and large

Table 8.1.1 Displacement, Velocity, Acceleration, and Jerk for Some Cams

	Polynomial forms				Trigonometric forms	
	Second power Reg. Reg.	Third power Reg. Reg. Reg.	Fourth power Reg. Reg.	Fifth power	Simple harmonic	Cycloidal
Displacement x						
Velocity $\dfrac{dx}{dt}$	2	2	2	1.9	1.6	2
Acceleration $\dfrac{d^2x}{dt^2}$	4 4	8 8	6 6	5.8 5.8	4.9 4.9	6.3 6.3
Jerk $\dfrac{d^3x}{dt^3}$	∞ ∞	32 32	48 48	60 30	∞ ∞ 15.5	40 40

Time ⟶

SOURCE: Adapted from Gutman, *Mach. Des.*, Mar. 1951.

shock waves to be transmitted and reflected throughout the parts that generate noise and would tend to limit the life of the machine. A step in jerk, the third derivative of the follower displacement with respect to time, seems altogether acceptable. In those designs requiring or exhibiting clearance between the follower and cam (usually at the bottom of the stroke), as gentle and slow a ramp portion as can be tolerated must be inserted on either side of the clearance region to limit the magnitude of the acceleration and jerk to a minimum. The tolerance on the clearance adjustment must be small enough to assure that the follower will be left behind and picked up gradually by the gentle ramp portions of the cam.

Table 8.1.1 shows the comparable and relative magnitudes of velocity, acceleration, and jerk for several high-speed cam, where the displacements are all taken as 1 at time 1 without any overshoot in any of the derivatives.

The three most common forms of motion used are uniform motion (Fig. 8.1.16), harmonic motion (Fig. 8.1.17), and uniformly accelerated and retarded motion (Fig. 8.1.18). In plotting the diagrams (Fig. 8.1.18) for this last motion, divide *ac* into an even number of equal parts and *bc*

into the same number of parts with lengths increasing by a constant increment to a maximum and then decreasing by the same decrement, as, for example, 1, 3, 5, 5, 3, 1, or 1, 3, 5, 7, 9, 9, 7, 5, 3, 1. In order to prevent shock when the direction of motion changes, as at *a* and *b* in the uniform motion, the harmonic motion may be used; if the cam is to be operated at high speed, the uniformly accelerated and retarded motion should preferably be employed; in either case there is a very gradual change of velocity.

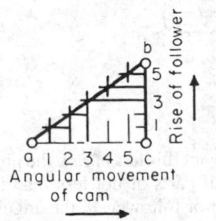

Fig. 8.1.16

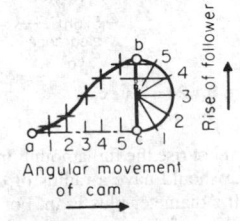

Fig. 8.1.17

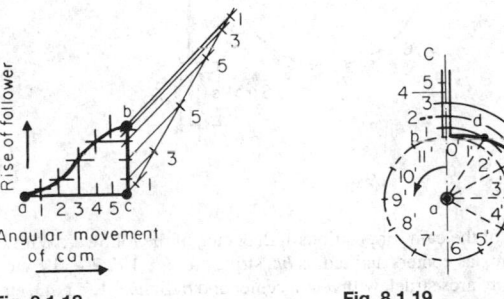

Fig. 8.1.18   **Fig. 8.1.19**

Pitch Line The actual pitch line of a cam varies with the type of motion and with the position of the follower relative to the cam's axis. Most cams as ordinarily constructed are covered by the following four cases.

FOLLOWER ON LINE OF AXIS. (Fig. 8.1.19). To draw the pitch line, subdivide the motion *bc* of the follower in the manner indicated in Figs. 8.1.16, 8.1.17, and 8.1.18. Draw a circle with a radius equal to the smallest radius of the cam *a*0 and subdivide it into angles 0*a*1′, 0*a*2′,

$0a3'$, etc., corresponding with angular displacements of the cam for positions 1, 2, 3, etc., of the follower. With a as a center and radii $a1$, $a2$, $a3$, etc., strike arcs cutting radial lines at d, e, f, etc. Draw a smooth curve through points d, e, f, etc.

OFFSET FOLLOWER (Fig. 8.1.20). Divide bc as indicated in Figs. 8.1.16, 8.1.17, and 8.1.18. Draw a circle of radius ac (highest point of rise of follower) and one tangent to cb produced. Divide the outer circle into parts $1'$, $2'$, $3'$, etc., corresponding with the angular displacement of

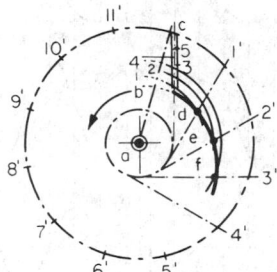

Fig. 8.1.20

the cam for positions 1, 2, 3, etc., of the follower, and draw tangents from points $1'$, $2'$, $3'$, etc., to the small circle. With a as a center and radii $a1$, $a2$, $a3$, etc., strike arcs cutting tangents at d, e, f, etc. Draw a smooth curve through d, e, f, etc.

ROCKER FOLLOWER (Fig. 8.1.21). Divide the stroke of the slide S in the manner indicated in Figs. 8.1.16, 8.1.17, and 8.1.18, and transfer these points to the arc bc as points 1, 2, 3, etc. Draw a circle of radius ak and divide it into parts $1'$, $2'$, $3'$, etc., corresponding with angular dis-

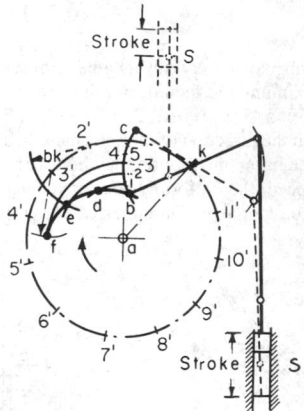

Fig. 8.1.21

placements of the cam for positions 1, 2, 3, etc., of the follower. With k, $1'$, $2'$, $3'$, etc., as centers and radius bk, strike arcs kb, $1'd$, $2'e$, $3'f$, etc., cutting at $bdef$ arcs struck with a as a center and radii ab, $a1$, $a2$, $a3$, etc. Draw a smooth curve through b, d, e, f, etc.

CYLINDRICAL CAM (Fig. 8.1.22). In this type of cam, more than one complete turn may be obtained, provided in all cases the follower returns to its starting point. Draw rectangle $wxyz$ (Fig. 8.1.22) representing the development of cylindrical surface of the cam. Subdivide the desired motion of the follower bc horizontally in the manner indicated in Figs. 8.1.16, 8.1.17, and 8.1.18, and plot the corresponding angular displacement $1'$, $2'$, $3'$, etc., of the cam vertically; then through the intersection of lines from these points draw a smooth curve. This may best be shown by an example, assuming the following data for the

diagram in Fig. 8.1.22: Total motion of follower = bc; circumference of cam = $2\pi r$. Follower moves harmonically 4 units to right in 0.6 turn, then rests (or "dwells") 0.4 turn, and finishes with uniform motion 6 units to right and 10 units to left in 2 turns.

Cam Design In the practical design of cams the following points must be noted. If only a small force is to be transmitted, sliding contact may be used, otherwise **rolling contact**. For the latter the pitch line must

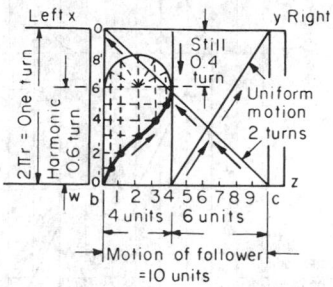

Fig. 8.1.22 Cylindrical cam.

be corrected in order to get the true slope of the cam. An approximate construction (Fig. 8.1.23) may be employed by using the pitch line as the center of a series of arcs the radii of which are equal to that of the follower roll to be used; then a smooth curve drawn tangent to the arcs will give the slope desired for a roll working on the periphery of the cam

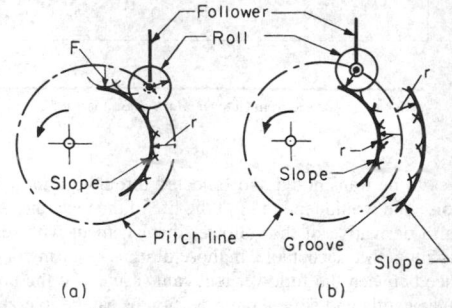

Fig. 8.1.23

(Fig. 8.1.23a) or in a groove (Fig. 8.1.23b). For plate cams the roll should be a small cylinder, as in Fig. 8.1.24a. In cylindrical cams it is usually sufficiently accurate to make the roll conical, as in Fig. 8.1.24b, in which case the taper of the roll produced should intersect the axis of the cam. If the pitch line abc is made too sharp (Fig. 8.1.25) the follower

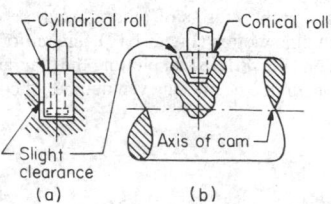

Fig. 8.1.24 Plate cam.

will not rise the full amount. In order to prevent this **loss of rise**, the pitch line should have a radius of curvature at all parts of not less than the roll's diameter plus ⅛ in. For the same rise of follower, a, the angular motion of the cam, O, the slope of the cam changes considerably, as indicated by the heavy lines A, B, and C (Fig. 8.1.26). Care should be taken to keep a moderate slope and thereby keep down the side thrust on the follower, but this should not be carried too far, as the cam would become too large and the friction increase.

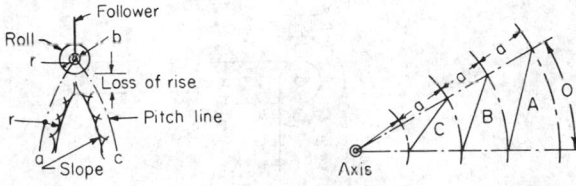

Fig. 8.1.25   **Fig. 8.1.26**

ROLLING SURFACES

In order to connect two shafts so that they shall have a definite angular velocity ratio, rolling surfaces are often used; and in order to have no slipping between the surfaces they must fulfill the following two conditions: the line of centers must pass through the point of contact, and the arcs of contact must be of equal length. The angular velocities, expressed usually in r/min, will be inversely proportional to the radii: $N/n = r/R$. The two surfaces most commonly used in practice, and the only ones having a constant angular velocity ratio, are cylinders where the shafts are parallel, and cones where the shafts (projected) intersect at an angle. In either case there are two possible directions of rotation, depending upon whether the surfaces roll in opposite directions (external contact) or in the same direction (internal contact). In Fig. 8.1.27, $R = nc/(N + n)$ and $r = Nc/(N + n)$; in Fig. 8.1.28, $R = nc/(N - n)$ and $r = Nc/(N - n)$. In Fig. 8.1.29, $\tan B = \sin A/(n/N + \cos A)$ and $\tan C = \sin A/$

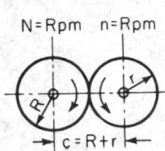

Fig. 8.1.27

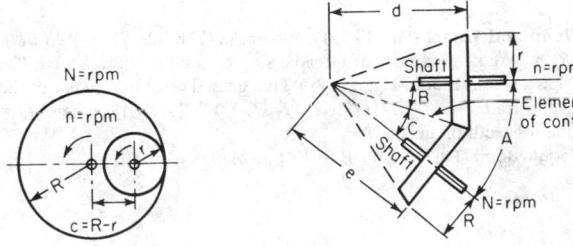

Fig. 8.1.28   **Fig. 8.1.29**

$(N/n + \cos A)$; in Fig. 8.1.30, $\tan B = \sin A/(N/n - \cos A)$, and $\tan C = \sin A/(n/N - \cos A)$. With the above values for the angles B and C, and the length d or e of one of the cones, R and r may be calculated.

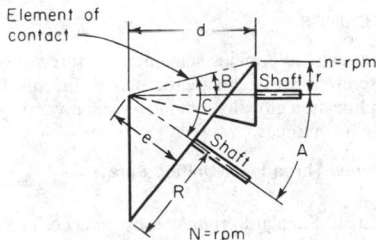

Fig. 8.1.30

The natural limitations of **rolling without slip**, with the use of pure rolling surfaces limited to the transmission of very small amounts of torque, led historically to the alteration of the geometric surfaces to include teeth and tooth spaces, i.e., **toothed wheels**, or simply **gears**.

Modern gear tooth systems are described in greater detail in Sec. 8.3. This brief discussion is limited to the kinematic considerations of some common gear combinations.

EPICYCLIC TRAINS

Epicyclic trains are combinations of gears in which some of or all the gears have a motion compounded of rotation about an axis and a translation or revolution of that axis. The gears are usually connected by a link called an arm, which often rotates about the axis of the first gear. Such trains may be calculated by first considering all gears locked and the arm turned; then the arm locked and the gears rotated. The algebraic sum of the separate motions will give the desired result. The following examples and method of tabulation will illustrate this. The figures on each gear refer to the number of teeth for that gear.

	A	B	C	D
Gear locked, Fig. 8.1.31	+1	+1	+1	+1
Arm locked, Fig. 8.1.31	0	−1	+1 × $^{50}/_{20}$	−1 × $^{50}/_{20}$ × $^{20}/_{40}$
Addition, Fig. 8.1.31	+1	0	+3½	−¼
Gears locked, Fig. 8.1.32	+1	+1	+1	+1
Arm locked, Fig. 8.1.32	0	−1	+1 × $^{30}/_{20}$	+1 × $^{30}/_{20}$ × $^{20}/_{70}$
Addition, Fig. 8.1.32	+1	0	+2½	+1³⁄₇

In Figs. 8.1.31 and 8.1.32 lock the gears and turn the arm A right-handed through 1 revolution $(+1)$; then lock the arm and turn the gear B back to where it started (-1); gears C and D will have rotated the amount indicated in the tabulation. Then the algebraic sum will give the relative turns of each gear. That is, in Fig. 8.1.31, for one turn of the

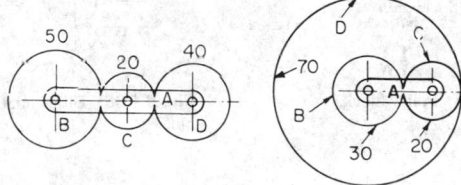

Figs. 8.1.31 and 8.1.32 Epicyclic trains.

arm, B does not move and C turns in the same direction 3½ r, and D in the opposite direction ¼ r; whereas in Fig. 8.1.32, for one turn of the arm, B does not turn, but C and D turn in the same direction as the arm, respectively, 2½ and 1³⁄₇ r. (Note: The arm in the above case was turned +1 for convenience, but any other value might be used.)

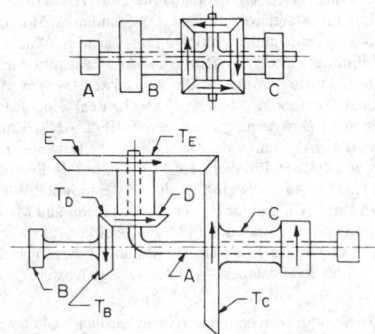

Figs. 8.1.33 and 8.1.34 Bevel epicyclic trains.

Bevel epicyclic trains are epicyclic trains containing bevel gears and may be calculated by the preceding method, but it is usually simpler to use the general formula which applies to all cases of epicyclic trains:

$$\frac{\text{Turns of } C \text{ relative to arm}}{\text{Turns of } B \text{ relative to arm}} = \frac{\text{absolute turns of } C - \text{turns of arm}}{\text{absolute turns of } B - \text{turns of arm}}$$

The left-hand term gives the value of the train and can always be expressed in terms of the number of teeth (T) on the gears. Care must be used, however, to express it as either plus ($+$) or minus ($-$), depending upon whether the gears turn in the same or opposite directions.

$$\frac{\text{Relative turns of } C}{\text{Relative turns of } B} = \frac{C - A}{B - A} = -1 \quad \text{(in Fig. 8.1.33)}$$

$$= +\frac{T_E}{T_C} \times \frac{T_B}{T_D} \quad \text{(Fig. 8.1.34)}$$

HOISTING MECHANISMS

Pulley Block (Fig. 8.1.35) Given the weight W to be raised, the force F necessary is $F = V_W W / V_F = W/n = $ load/number of ropes, V_W and V_F being the respective velocities of W and F.

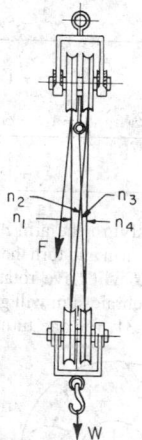

Fig. 8.1.35 Pulley block.

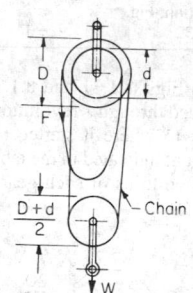

Fig. 8.1.36 Differential chain block.

Differential Chain Block (Fig. 8.1.36)

$$F = V_W W / V_F = W(D - d)/(2D)$$

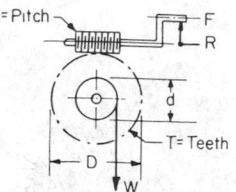

Fig. 8.1.37 Worm and worm wheel.

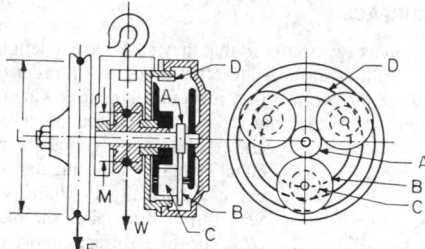

Fig. 8.1.38 Triplex chain block.

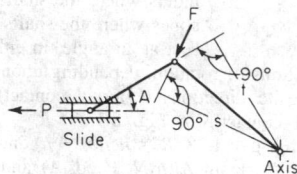

Fig. 8.1.39 Toggle joint.

Worm and Wheel (Fig. 8.1.37) $F = \pi d(n/T)W/(2\pi R) = WP(d/D)/(2\pi R)$, where $n =$ number of threads, single, double, triple, etc.

Triplex Chain Block (Fig. 8.1.38) This geared hoist makes use of the epicyclic train. $W = FL/\{M[1 + (T_D/T_C) \times (T_B/T_A)]\}$, where $T =$ number of teeth on gears.

Toggle Joint (Fig. 8.1.39) $P = Fs \ (\cos A)/t$.

8.2 MACHINE ELEMENTS
by Antonio F. Baldo

REFERENCES: American National Standards Institute (ANSI) Standards. International Organization for Standardization (ISO) Standards. Morden, "Industrial Fasteners Handbook," Trade and Technical Press. Parmley, "Standard Handbook of Fastening and Joining," McGraw-Hill. Bickford, "An Introduction to the Design and Behavior of Bolted Joints," Marcel Dekker. Maleev, "Machine Design," International Textbook. Shigley, "Mechanical Engineering Design," McGraw-Hill. *Machine Design* magazine, Penton/IPC. ANSI/Rubber Manufacturers Assn. (ANSI/RMA) Standards. "Handbook of Power Transmission Flat Belting," Goodyear Rubber Products Co. "Industrial V-Belting," Goodyear Rubber Products Co. Carlson, "Spring Designer's Handbook," Marcel Dekker. American Chain Assn., "Chains for Power Transmission and Material Handling—Design and Applications Handbook," Marcel Dekker. "Power Transmission Handbook," DAYCO. "Wire Rope User's Manual," American Iron and Steel Institute. Blake, "Threaded Fasteners—Materials and Design," Marcel Dekker.

NOTE. At this writing, conversion to metric hardware and machine elements continues. SI units are introduced as appropriate, but the bulk of the material is still presented in the form in which the designer or reader will find it available.

SCREW FASTENINGS

At present there exist two major standards for screw threads, namely Unified inch screw threads and metric screw threads. Both systems enjoy a wide application globally, but movement toward a greater use of the metric system continues.

Unified Inch Screw Threads (or Unified Screw Threads)

The Unified Thread Standard originated by an accord of screw thread standardization committees of Canada, the United Kingdom, and the United States in 1984. The Unified Screw-Thread Standard was published by ANSI as American Unified and American Screw Thread Publication B1.1-1974, revised in 1982 and then again in 1989. Revisions did not tamper with the basic 1974 thread forms. In conjunction with Technical Committee No. 1 of the ISO, the Unified Standard was adopted as an ISO Inch Screw Standard (ISO 5864-1978).

Of the numerous and different screw thread forms, those of greatest consequence are

UN—unified (no mandatory radiused root)
UNR—unified (mandatory radiused root; minimum $0.108 = p$)
UNJ—unified (mandatory larger radiused root; recommended $0.150 = p$)
M—metric (inherently designed and manufactured with radiused root; has $0.125 = p$)
MJ—metric (mandatory larger radiused root; recommended $0.150 = p$)

The basic American screw thread profile was standardized in 1974, and it now carries the UN designations (UN = unified). ANSI publishes these standards and all subsequent revisions. At intervals these standards are published with a "reaffirmation date" (that is, R1988). In 1969 an *international* basic thread profile standard was established, and it is designated as M. The ISO publishes these standards with yearly updates. The UN and M profiles are the same, but **UN screws** are manufactured to **inch** dimensions while **M screws** are manufactured to **metric** dimensions.

The **metric system** has only the two thread forms: **M**, standard for commercial uses, and **MJ**, standard for aerospace use and for aerospace-quality commercial use.

Certain groups of diameter and pitch combinations have evolved over time to become those most used commercially. Such groups are called **thread series.** Currently there are 11 UN series for inch products and 13 M series for metric products.

The Unified standard comprises the following two parts:

1. **Diameter-pitch combinations.** (See Tables 8.2.1 to 8.2.5.)
   a. UN inch series:

Coarse	UNC or UNRC
Fine	UNF or UNRF
Extra-fine	UNEF or UNREF
Constant-pitch	UN or UNR

   b. Metric series:

Coarse	M
Fine	M

   NOTE: Radiused roots apply only to external threads. The preponderance of important commercial use leans to UNC, UNF, 8UN (eight-threaded), and metric coarse M. Aerospace and aerospace-quality applications use UNJ and MJ.

2. **Tolerance classes.** The amounts of tolerance and allowance distinguish one thread class from another. Classes are designated by one of three numbers (1, 2, 3), and either letter A for external threads or letter B for internal threads. Tolerance decreases as class number increases. Allowance is specified only for classes 1A and 2A. Tolerances are based on engagement length equal to nominal diameter. 1A/1B—liberal tolerance and allowance required to permit easy assembly even with dirty or nicked threads. 2A/2B—most commonly used for general applications, including production of bolts, screws, nuts, and similar threaded fasteners. Permits external threads to be plated. 3A/3B—for closeness of fit and/or accuracy of thread applications where zero allowance is needed. 2AG—allowance for rapid assembly where high-temperature expansion prevails or where lubrication problems are important.

   Unified screw threads are designated by a set of numbers and letter symbols signifying, in sequence, the nominal size, threads per inch, thread series, tolerance class, hand (only for left hand), and in some instances in parentheses a Thread Acceptability System Requirement of ANSI B1.3.

   EXAMPLE. ¼-20 UNC-2A-LH (21), or optionally 0.250-20 UNC-2A-LH (21), where ¼ = nominal size (fractional diameter, in, or screw number, with decimal equivalent of either being optional); 20 = number of threads per inch, n; UNC = thread form and series; 2A = tolerance class; LH = left hand (no symbol required for right hand); (21) = thread gaging system per ANSI B1.3.

3. **Load considerations**
   a. Static loading. Only a slight increase in tensile strength in a screw fastener is realized with an increase in *root* rounding radius, because minor diameter (hence cross-sectional area at the root) growth is small. Thus the basic tensile stress area formula is used in stress calculations for all thread forms. See Tables 8.2.2, 8.2.3, and 8.2.4. The designer should take into account such factors as stress concentration as applicable.

   b. Dynamic loading. Few mechanical joints can remain absolutely free of some form of fluctuating stress, vibration, stress reversal, or impact. Metal-to-metal joints of very high-modulus materials or non-elastic-gasketed high-modulus joints plus preloading at assembly (preload to be greater than highest peak of the external fluctuating load) can realize absolute static conditions inside the screw fastener. For ordinary-modulus joints and elastic-gasketed joints, a fraction of the external fluctuating load will be transmitted to the interior of the screw fastener. Thus the fastener must be designed for fatigue according to a **static plus fluctuating load** model. See discussion under "Strength" later.

   Since **fatigue failures** generally occur at locations of high **stress concentration,** screw fasteners are especially vulnerable because of the abrupt change between head and body, notchlike conditions at the thread roots, surface scratches due to manufacturing, etc. The highest stress concentrations occur at the thread roots. The stress concentration factor can be very large for nonrounded roots, amounting to about 6 for sharp or flat roots, to less than 3 for UNJ and MJ threads which are generously rounded. This can effectively double the fatigue life. UNJ and MJ threads are especially well suited for dynamic loading conditions.

Screw Thread Profile

Basic Profile The basic profiles of UN and UNR are the same, and these in turn are identical to those of ISO metric threads. Basic thread shape (60° thread angle) and basic dimensions (major, pitch, and minor diameters; thread height; crest, and root flats) are defined. See Fig. 8.2.1.

Design Profile Design profiles define the maximum material (no allowance) for external and internal threads, and they are derived from the basic profile. UN threads (external) may have either flat or rounded crests and roots. UNR threads (external) must have rounded roots, but may have flat or rounded crests. UN threads (internal) *must* have rounded roots. Any rounding must clear the basic flat roots or crests.

Basic major diameter	Largest diameter of basic screw thread.
Basic minor diameter	Smallest diameter of basic screw thread.
Basic pitch diameter	Diameter to imaginary lines through thread profile and parallel to axis so that thread and groove widths are equal. These three definitions apply to both external and internal threads.
Maximum diameters (external threads)	Basic diameters minus allowance.
Minimum diameters (internal threads)	Basic diameters.
Pitch	$1/n$ (n = number of threads per inch).
Tolerance	Inward variation tolerated on maximum diameters of external threads and outward variation tolerated on minimum diameters of internal threads.

Metric Screw Threads

Metric screw thread standardization has been under the aegis of the International Organization for Standardization (ISO). The ISO basic profile is essentially the same as the Unified screw thread basic form,

Table 8.2.1 Standard Series Threads (UN/UNR)*

Nominal size, in — Primary	Secondary	Basic major diameter, in	Series with graded pitches — Coarse UNC	Fine UNF	Extra-fine UNEF	Series with constant pitches (threads per inch) — 4UN	6UN	8UN	12UN	16UN	20UN	28UN	32UN	Nominal size, in
0		0.0600	—	80	—	—	—	—	—	—	—	—	—	0
	1	0.0730	64	72	—	—	—	—	—	—	—	—	—	1
2		0.0860	56	64	—	—	—	—	—	—	—	—	—	2
	3	0.0990	48	56	—	—	—	—	—	—	—	—	—	3
4		0.1120	40	48	—	—	—	—	—	—	—	—	—	4
5		0.1250	40	44	—	—	—	—	—	—	—	—	—	5
6		0.1380	32	40	—	—	—	—	—	—	—	—	UNC	6
8		0.1640	32	36	—	—	—	—	—	—	—	—	UNC	8
10		0.1900	24	32	—	—	—	—	—	—	—	—	UNF	10
	12	0.2160	24	28	32	—	—	—	—	—	—	UNF	UNEF	12
1/4		0.2500	20	28	32	—	—	—	—	—	UNC	UNF	UNEF	1/4
5/16		0.3125	18	24	32	—	—	—	—	—	20	28	UNEF	5/16
3/8		0.3750	16	24	32	—	—	—	—	UNC	20	28	UNEF	3/8
7/16		0.4375	14	20	28	—	—	—	—	16	UNF	UNEF	32	7/16
1/2		0.5000	13	20	28	—	—	—	—	16	20	UNEF	32	1/2
9/16		0.5625	12	18	24	—	—	—	12	16	20	28	32	9/16
5/8		0.6250	11	18	24	—	—	—	12	16	20	28	32	5/8
	11/16	0.6875	—	—	24	—	—	—	12	16	20	28	32	11/16
3/4		0.7500	10	16	20	—	—	—	12	16	UNEF	28	32	3/4
	13/16	0.8125	—	—	20	—	—	—	12	16	20	28	32	13/16
7/8		0.8750	9	14	20	—	—	—	12	16	20	28	32	7/8
	15/16	0.9375	—	—	20	—	—	—	12	16	20	28	32	15/16
1		1.0000	8	12	20	—	—	UNC	UNF	16	20	28	32	1
	1 1/16	1.0625	—	—	18	—	—	8	12	16	20	28	—	1 1/16
1 1/8		1.1250	7	12	18	—	—	8	UNF	16	20	28	—	1 1/8
	1 3/16	1.1875	—	—	18	—	—	8	12	16	20	28	—	1 3/16
1 1/4		1.2500	7	12	18	—	—	8	UNF	16	20	28	—	1 1/4
	1 5/16	1.3125	—	—	18	—	—	8	12	16	20	—	—	1 5/16
1 3/8		1.3750	6	12	18	—	UNC	8	UNF	16	20	—	—	1 3/8
	1 7/16	1.4375	—	—	18	—	6	8	12	16	20	—	—	1 7/16
1 1/2		1.5000	6	12	18	—	UNC	8	UNF	16	20	—	—	1 1/2
	1 9/16	1.5625	—	—	18	—	6	8	12	16	20	—	—	1 9/16
1 5/8		1.6250	—	—	18	—	6	8	12	16	20	—	—	1 5/8
	1 11/16	1.6875	—	—	18	—	6	8	12	16	20	—	—	1 11/16
1 3/4		1.7500	5	—	—	—	6	8	12	16	20	—	—	1 3/4
	1 13/16	1.8125	—	—	—	—	6	8	12	16	20	—	—	1 13/16
1 7/8		1.8750	—	—	—	—	6	8	12	16	20	—	—	1 7/8
	1 15/16	1.9375	—	—	—	—	6	8	12	16	20	—	—	1 15/16

Threads per Inch*

Primary	Secondary	Decimal Equiv.	UNC	UNF	UNEF	4	6	8	12	16	20	28	32
2		2.0000	4½	—	—	4	6	8	12	16	20	—	—
	2⅛	2.1250	—	—	—	4	6	8	12	16	20	—	—
2¼		2.2500	4½	—	—	4	6	8	12	16	20	—	—
	2⅜	2.3750	—	—	—	4	6	8	12	16	20	—	—
2½		2.5000	4	—	—	UNC	6	8	12	16	20	—	—
	2⅝	2.6250	—	—	—	4	6	8	12	16	20	—	—
2¾		2.7500	4	—	—	UNC	6	8	12	16	20	—	—
	2⅞	2.8750	—	—	—	4	6	8	12	16	—	—	—
3		3.0000	4	—	—	UNC	6	8	12	16	—	—	—
	3⅛	3.1250	—	—	—	4	6	8	12	16	—	—	—
3¼		3.2500	4	—	—	UNC	6	8	12	16	—	—	—
	3⅜	3.3750	—	—	—	4	6	8	12	16	—	—	—
3½		3.5000	4	—	—	UNC	6	8	12	16	—	—	—
	3⅝	3.6250	—	—	—	4	6	8	12	16	—	—	—
3¾		3.7500	4	—	—	UNC	6	8	12	16	—	—	—
	3⅞	3.8750	—	—	—	4	6	8	12	16	—	—	—
4		4.0000	4	—	—	UNC	6	8	12	16	—	—	—
	4⅛	4.1250	—	—	—	4	6	8	12	16	—	—	—
4¼		4.2500	—	—	—	4	6	8	12	16	—	—	—
	4⅜	4.3750	—	—	—	4	6	8	12	16	—	—	—
4½		4.5000	—	—	—	4	6	8	12	16	—	—	—
	4⅝	4.6250	—	—	—	4	6	8	12	16	—	—	—
4¾		4.7500	—	—	—	4	6	8	12	16	—	—	—
	4⅞	4.8750	—	—	—	4	6	8	12	16	—	—	—
5		5.0000	—	—	—	4	6	8	12	16	—	—	—
	5⅛	5.1250	—	—	—	4	6	8	12	16	—	—	—
5¼		5.2500	—	—	—	4	6	8	12	16	—	—	—
	5⅜	5.3750	—	—	—	4	6	8	12	16	—	—	—
5½		5.5000	—	—	—	4	6	8	12	16	—	—	—
	5⅝	5.6250	—	—	—	4	6	8	12	16	—	—	—
5¾		5.7500	—	—	—	4	6	8	12	16	—	—	—
	5⅞	5.8750	—	—	—	4	6	8	12	16	—	—	—
6		6.0000	—	—	—	4	6	8	12	16	—	—	—

* Series designation shown indicates the UN thread form; however, the UNR thread form may be specified by substituting UNR in place of UN in all designations for external use only.

SOURCE: ANSI B1.1-1982; reaffirmed in 1989, reproduced by permission.

Table 8.2.2 Basic Dimensions for Coarse Thread Series (UNC/UNRC)

Nominal size, in	Basic major diameter D, in	Threads per inch n	Basic pitch diameter* E, in	UNR design minor diameter external† K_s, in	Basic minor diameter internal K, in	Section at minor diameter at $D - 2h_b$, in^2	Tensile stress area,‡ in^2
1 (0.073)§	0.0730	64	0.0629	0.0544	0.0561	0.00218	0.00263
2 (0.086)	0.0860	56	0.0744	0.0648	0.0667	0.00310	0.00370
3 (0.099)§	0.0990	48	0.0855	0.0741	0.0764	0.00406	0.00487
4 (0.112)	0.1120	40	0.0958	0.0822	0.0849	0.00496	0.00604
5 (0.125)	0.1250	40	0.1088	0.0952	0.0979	0.00672	0.00796
6 (0.138)	0.1380	32	0.1177	0.1008	0.1042	0.00745	0.00909
8 (0.164)	0.1640	32	0.1437	0.1268	0.1302	0.01196	0.0140
10 (0.190)	0.1900	24	0.1629	0.1404	0.1449	0.01450	0.0175
12 (0.216)§	0.2160	24	0.1889	0.1664	0.1709	0.0206	0.0242
¼	0.2500	20	0.2175	0.1905	0.1959	0.0269	0.0318
⁵⁄₁₆	0.3125	18	0.2764	0.2464	0.2524	0.0454	0.0524
⅜	0.3750	16	0.3344	0.3005	0.3073	0.0678	0.0775
⁷⁄₁₆	0.4375	14	0.3911	0.3525	0.3602	0.0933	0.1063
½	0.5000	13	0.4500	0.3334	0.4167	0.1257	0.1419
⁹⁄₁₆	0.5625	12	0.5084	0.4633	0.4723	0.162	0.182
⅝	0.6250	11	0.5660	0.5168	0.5266	0.202	0.226
¾	0.7500	10	0.6850	0.6309	0.6417	0.302	0.334
⅞	0.8750	9	0.8028	0.7427	0.7547	0.419	0.462
1	1.0000	8	0.9188	0.8512	0.8647	0.551	0.606
1⅛	1.1250	7	1.0322	0.9549	0.9704	0.693	0.763
1¼	1.2500	7	1.1572	1.0799	1.0954	0.890	0.969
1⅜	1.3750	6	1.2667	1.1766	1.1946	1.054	1.155
1½	1.5000	6	1.3917	1.3016	1.3196	1.294	1.405
1¾	1.7500	5	1.6201	1.5119	1.5335	1.74	1.90
2	2.0000	4½	1.8557	1.7353	1.7594	2.30	2.50
2¼	2.2500	4½	2.1057	1.9853	2.0094	3.02	3.25
2½	2.5000	4	2.3376	2.2023	2.2294	3.72	4.00
2¾	2.7500	4	2.5876	2.4523	2.4794	4.62	4.93
3	3.0000	4	2.8376	2.7023	2.7294	5.62	5.97
3¼	3.2500	4	3.0876	2.9523	2.9794	6.72	7.10
3½	3.5000	4	3.3376	3.2023	3.2294	7.92	8.33
3¾	3.7500	4	3.5876	3.4523	3.4794	9.21	9.66
4	4.0000	4	3.8376	3.7023	3.7294	10.61	11.08

* British: effective diameter.
† See formula under definition of tensile stress area in Appendix B of ANSI B1.1-1987.
‡ Design form. See Fig. 2B in ANSI B1.1-1982 or Fig. 1 in 1989 revision.
§ Secondary sizes.
SOURCE: ANSI B1.1-1982, revised 1989; reproduced by permission.

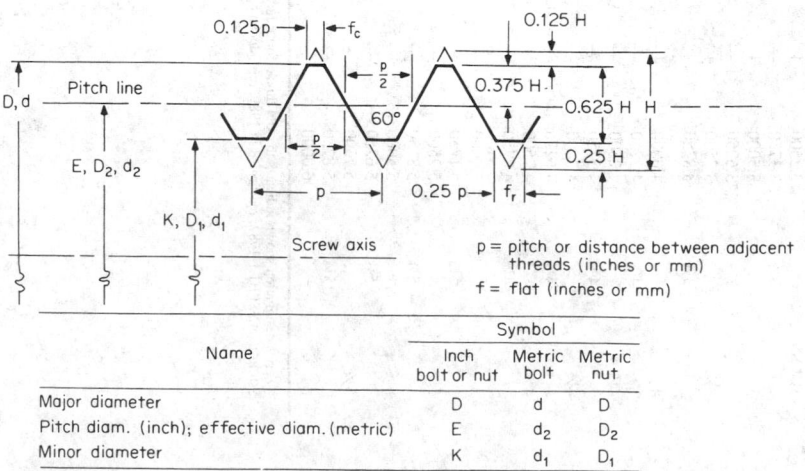

p = pitch or distance between adjacent threads (inches or mm)
f = flat (inches or mm)

Name	Inch bolt or nut	Metric bolt	Metric nut
		Symbol	
Major diameter	D	d	D
Pitch diam. (inch); effective diam. (metric)	E	d_2	D_2
Minor diameter	K	d_1	D_1

Fig. 8.2.1 Basic thread profile.

Table 8.2.3   Basic Dimensions for Fine Thread Series (UNF/UNRF)

Nominal size, in	Basic major diameter D, in	Threads per inch n	Basic pitch diameter* E, in	UNR design minor diameter external† K_s, in	Basic minor diameter internal K, in	Section at minor diameter at $D - 2h_b$, in²	Tensile stress area,‡ in²
0 (0.060)	0.0600	80	0.0519	0.0451	0.0465	0.00151	0.00180
1 (0.073)§	0.0730	72	0.0640	0.0565	0.0580	0.00237	0.00278
2 (0.086)	0.0860	64	0.0759	0.0674	0.0691	0.00339	0.00394
3 (0.099)§	0.0990	56	0.0874	0.0778	0.0797	0.00451	0.00523
4 (0.112)	0.1120	48	0.0985	0.0871	0.0894	0.00566	0.00661
5 (0.125)	0.1250	44	0.1102	0.0979	0.1004	0.00716	0.00830
6 (0.138)	0.1380	40	0.1218	0.1082	0.1109	0.00874	0.01015
8 (0.164)	0.1640	36	0.1460	0.1309	0.1339	0.01285	0.01474
10 (0.190)	0.1900	32	0.1697	0.1528	0.1562	0.0175	0.0200
12 (0.216)§	0.2160	28	0.1928	0.1734	0.1773	0.0226	0.0258
¼	0.2500	28	0.2268	0.2074	0.2113	0.0326	0.0364
⁵⁄₁₆	0.3125	24	0.2854	0.2629	0.2674	0.0524	0.0580
⅜	0.3750	24	0.3479	0.3254	0.3299	0.0809	0.0878
⁷⁄₁₆	0.4375	20	0.4050	0.3780	0.3834	0.1090	0.1187
½	0.5000	20	0.4675	0.4405	0.4459	0.1486	0.1599
⁹⁄₁₆	0.5625	18	0.5264	0.4964	0.5024	0.189	0.203
⅝	0.6250	18	0.5889	0.5589	0.5649	0.240	0.256
¾	0.7500	16	0.7094	0.6763	0.6823	0.351	0.373
⅞	0.8750	14	0.8286	0.7900	0.7977	0.480	0.509
1	1.0000	12	0.9459	0.9001	0.9098	0.625	0.663
1⅛	1.1250	12	1.0709	1.0258	1.0348	0.812	0.856
1¼	1.2500	12	1.1959	1.1508	1.1598	1.024	1.073
1⅜	1.3750	12	1.3209	1.2758	1.2848	1.260	1.315
1½	1.5000	12	1.4459	1.4008	1.4098	1.521	1.581

* British: effective diameter.
† See formula under definition of tensile stress area in Appendix B of ANSI B1.1-1982.
‡ Design form. See Fig. 2B of ANSI B1.1-1982 or Fig. 1 in 1989 revision.
§ Secondary sizes.
SOURCE: ANSI B1.1-1982, revised 1989; reproduced by permission.

Table 8.2.4   Basic Dimensions for Extra-Fine Thread Series (UNEF/UNREF)

Nominal size, in Primary	Nominal size, in Secondary	Basic major diameter D, in	Threads per inch n	Basic pitch diameter* E, in	UNR design minor diameter external† K_s, in	Basic minor diameter internal K, in	Section at minor diameter at $D - 2h_b$, in²	Tensile stress area,‡ in²
	12 (0.216)	0.2160	32	0.1957	0.1788	0.1822	0.0242	0.0270
¼		0.2500	32	0.2297	0.2128	0.2162	0.0344	0.0379
⁵⁄₁₆		0.3125	32	0.2922	0.2753	0.2787	0.0581	0.0625
⅜		0.3750	32	0.3547	0.3378	0.3412	0.0878	0.0932
⁷⁄₁₆		0.4375	28	0.4143	0.3949	0.3988	0.1201	0.1274
½		0.5000	28	0.4768	0.4573	0.4613	0.162	0.170
⁹⁄₁₆		0.5625	24	0.5354	0.5129	0.5174	0.203	0.214
⅝		0.6250	24	0.5979	0.5754	0.5799	0.256	0.268
	¹¹⁄₁₆	0.6875	24	0.6604	0.6379	0.6424	0.315	0.329
¾		0.7500	20	0.7175	0.6905	0.6959	0.369	0.386
	¹³⁄₁₆	0.8125	20	0.7800	0.7530	0.7584	0.439	0.458
⅞		0.8750	20	0.8425	0.8155	0.8209	0.515	0.536
	¹⁵⁄₁₆	0.9375	20	0.9050	0.8780	0.8834	0.598	0.620
1		1.0000	20	0.9675	0.9405	0.9459	0.687	0.711
	1¹⁄₁₆	1.0625	18	1.0264	0.9964	1.0024	0.770	0.799
1⅛		1.1250	18	1.0889	1.0589	1.0649	0.871	1.901
	1³⁄₁₆	1.1875	18	1.1514	1.1214	1.1274	0.977	1.009
1¼		1.2500	18	1.2139	1.1839	1.1899	1.090	1.123
	1⁵⁄₁₆	1.3125	18	1.2764	1.2464	1.2524	1.208	1.244
1⅜		1.3750	18	1.3389	1.3089	1.3149	1.333	1.370
	1⁷⁄₁₆	1.4375	18	1.4014	1.3714	1.3774	1.464	1.503
1½		1.5000	18	1.4639	1.4339	1.4399	1.60	1.64
	1⁹⁄₁₆	1.5625	18	1.5264	1.4964	1.5024	1.74	1.79
1⅝		1.6250	18	1.5889	1.5589	1.5649	1.89	1.94
	1¹¹⁄₁₆	1.6875	18	1.6514	1.6214	1.6274	2.05	2.10

* British: effective diameter.
† Design form. See Fig. 2B in ANSI B1.1-1982 or Fig. 1 in 1989 revision.
‡ See formula under definition of tensile stress area in Appendix B in ANSI B1.1-1982.
SOURCE: ANSI B1.1-1982 revised 1989; reproduced by permission.

Table 8.2.5 ISO Metric Screw Thread Standard Series

Nominal size diam, mm — Column* 1	2	3	Series with graded pitches — Coarse	Fine	6	4	3	2	1.5	1.25	1	0.75	0.5	0.35	0.25	0.2	Nominal size diam, mm
0.25			0.075	—	—	—	—	—	—	—	—	—	—	—	—	—	0.25
0.3			0.08	—	—	—	—	—	—	—	—	—	—	—	—	—	0.3
	0.35		0.09	—	—	—	—	—	—	—	—	—	—	—	—	—	0.35
0.4			0.1	—	—	—	—	—	—	—	—	—	—	—	—	—	0.4
	0.45		0.1	—	—	—	—	—	—	—	—	—	—	—	—	—	0.45
0.5			0.125	—	—	—	—	—	—	—	—	—	—	—	—	—	0.5
	0.55		0.125	—	—	—	—	—	—	—	—	—	—	—	—	—	0.55
0.6			0.15	—	—	—	—	—	—	—	—	—	—	—	—	—	0.6
	0.7		0.175	—	—	—	—	—	—	—	—	—	—	—	—	—	0.7
0.8			0.2	—	—	—	—	—	—	—	—	—	—	—	—	—	0.8
	0.9		0.225	—	—	—	—	—	—	—	—	—	—	—	—	—	0.9
1			0.25	—	—	—	—	—	—	—	—	—	—	—	—	0.2	1
	1.1		0.25	—	—	—	—	—	—	—	—	—	—	—	—	0.2	1.1
1.2			0.25	—	—	—	—	—	—	—	—	—	—	—	—	0.2	1.2
	1.4		0.3	—	—	—	—	—	—	—	—	—	—	—	—	0.2	1.4
1.6			0.35	—	—	—	—	—	—	—	—	—	—	—	—	0.2	1.6
	1.8		0.35	—	—	—	—	—	—	—	—	—	—	—	—	0.2	1.8
2			0.4	—	—	—	—	—	—	—	—	—	—	—	0.25	—	2
	2.2		0.45	—	—	—	—	—	—	—	—	—	—	—	0.25	—	2.2
2.5			0.45	—	—	—	—	—	—	—	—	—	—	0.35	—	—	2.5
3			0.5	—	—	—	—	—	—	—	—	—	—	0.35	—	—	3
	3.5		0.6	—	—	—	—	—	—	—	—	—	—	0.35	—	—	3.5
4			0.7	—	—	—	—	—	—	—	—	—	0.5	—	—	—	4
	4.5		0.75	—	—	—	—	—	—	—	—	—	0.5	—	—	—	4.5
5			0.8	—	—	—	—	—	—	—	—	—	0.5	—	—	—	5
		5.5	—	—	—	—	—	—	—	—	—	—	0.5	—	—	—	5.5
6			1	—	—	—	—	—	—	—	—	0.75	—	—	—	—	6
		7	1	—	—	—	—	—	—	—	—	0.75	—	—	—	—	7
8			1.25	1	—	—	—	—	—	—	1	0.75	—	—	—	—	8
		9	1.25	—	—	—	—	—	—	—	1	0.75	—	—	—	—	9
10			1.5	1.25	—	—	—	—	—	1.25	1	0.75	—	—	—	—	10
		11	1.5	—	—	—	—	—	—	—	1	0.75	—	—	—	—	11
12			1.75	1.25	—	—	—	—	1.5	1.25	1	—	—	—	—	—	12
	14		2	1.5	—	—	—	—	1.5	1.25†	1	—	—	—	—	—	14
		15	—	—	—	—	—	—	1.5	—	1	—	—	—	—	—	15
16			2	1.5	—	—	—	—	1.5	—	1	—	—	—	—	—	16
		17	—	—	—	—	—	—	1.5	—	1	—	—	—	—	—	17
	18		2.5	1.5	—	—	—	2	1.5	—	1	—	—	—	—	—	18
20			2.5	1.5	—	—	—	2	1.5	—	1	—	—	—	—	—	20
	22		2.5	1.5	—	—	—	2	1.5	—	1	—	—	—	—	—	22
24			3	2	—	—	—	2	1.5	—	1	—	—	—	—	—	24
		25	—	—	—	—	—	2	1.5	—	1	—	—	—	—	—	25
		26	—	—	—	—	—	—	1.5	—	1	—	—	—	—	—	26
	27		3	2	—	—	—	2	1.5	—	1	—	—	—	—	—	27
		28	—	—	—	—	—	2	1.5	—	1	—	—	—	—	—	28
30			3.5	2	—	—	(3)	2	1.5	—	1	—	—	—	—	—	30
		32	—	2	—	—	—	2	1.5	—	—	—	—	—	—	—	32
	33		3.5	2	—	—	(3)	2	1.5	—	—	—	—	—	—	—	33
		35‡	—	—	—	—	—	—	1.5	—	—	—	—	—	—	—	35‡
36			4	3	—	—	—	2	1.5	—	—	—	—	—	—	—	36
		38	—	—	—	—	—	—	1.5	—	—	—	—	—	—	—	38
	39		4	3	—	—	—	2	1.5	—	—	—	—	—	—	—	39
		40	—	—	—	—	3	2	1.5	—	—	—	—	—	—	—	40
42			4.5	3	—	4	3	2	1.5	—	—	—	—	—	—	—	42
	45		4.5	3	—	4	3	2	1.5	—	—	—	—	—	—	—	45

* Thread diameter should be selected from column 1, 2 or 3; with preference being given in that order.
† Pitch 1.25 mm in combination with diameter 14 mm has been included for spark plug applications.
‡ Diameter 35 mm has been included for bearing locknut applications.
NOTE: The use of pitches shown in parentheses should be avoided wherever possible. The pitches enclosed in the bold frame, together with the corresponding nominal diameters in columns 1 and 2, are those combinations which have been established by ISO Recommendations as a selected ''coarse'' and ''fine'' series for commercial fasteners. Sizes 0.25 mm through 1.4 mm are covered in ISO Recommendation R68 and, except for the 0.25-mm size, in ANSI B1.10.
SOURCE: ISO 261-1973, reproduced by permission.

Table 8.2.6 Limiting Dimensions of Standard Series Threads for Commercial Screws, Bolts, and Nuts

Nominal size diam, mm	Pitch p, mm	Basic thread designation	Ext. Tol. class	Allowance	Ext. Major diam Max	Ext. Major diam Min	Ext. Pitch diam Max	Ext. Pitch diam Min	Ext. Pitch diam Tol.	Ext. Minor diam Max*	Ext. Minor diam Min†	Int. Tol. class	Int. Minor diam Min	Int. Minor diam Max	Int. Pitch diam Max	Int. Pitch diam Min	Int. Pitch diam Tol.	Major diam, min
1.6	0.35	M1.6	6g	0.019	1.581	1.496	1.354	1.291	0.063	1.151	1.063	6H	1.221	1.321	1.458	1.373	0.085	1.600
1.8	0.35	M1.8	6g	0.019	1.781	1.696	1.554	1.491	0.063	1.351	1.263	6H	1.421	1.521	1.658	1.573	0.085	1.800
2	0.4	M2	6g	0.019	1.981	1.886	1.721	1.654	0.067	1.490	1.394	6H	1.567	1.679	1.830	1.740	0.090	2.000
2.2	0.45	M2.2	6g	0.020	2.180	2.080	1.888	1.817	0.071	1.628	1.525	6H	1.713	1.838	2.003	1.908	0.095	2.200
2.5	0.45	M2.5	6g	0.020	2.480	2.380	2.188	2.117	0.071	1.928	1.825	6H	2.013	2.138	2.303	2.208	0.095	2.500
3	0.5	M3	6g	0.020	2.980	2.874	2.655	2.580	0.075	2.367	2.256	6H	2.459	2.599	2.775	2.675	0.100	3.000
3.5	0.6	M3.5	6g	0.021	3.479	3.354	3.089	3.004	0.085	2.742	2.614	6H	2.850	3.010	3.222	3.110	0.112	3.500
4	0.7	M4	6g	0.022	3.978	3.838	3.523	3.433	0.090	3.119	2.979	6H	3.242	3.422	3.663	3.545	0.112	4.000
4.5	0.75	M4.5	6g	0.022	4.478	4.338	3.991	3.901	0.090	3.558	3.414	6H	3.688	3.878	4.131	4.013	0.118	4.500
5	0.8	M5	6g	0.024	4.976	4.826	4.456	4.361	0.095	3.994	3.841	6H	4.134	4.334	4.605	4.480	0.118	5.000
6	1	M6	6g	0.026	5.974	5.794	5.324	5.212	0.112	4.747	4.563	6H	4.917	5.153	5.500	5.350	0.125	6.000
7	1	M7	6g	0.026	6.974	6.794	6.324	6.212	0.112	5.747	5.563	6H	5.917	6.153	6.500	6.350	0.150	7.000
8	1.25	M8	6g	0.028	7.972	7.760	7.160	7.042	0.118	6.439	6.231	6H	6.647	6.912	7.348	7.188	0.160	8.000
	1	M8 × 1	6g	0.026	7.974	7.794	7.324	7.212	0.112	6.747	6.563	6H	6.918	7.154	7.500	7.350	0.150	8.000
10	1.5	M10	6g	0.032	9.968	9.732	8.994	8.862	0.132	8.127	7.879	6H	8.376	8.676	9.206	9.026	0.180	10.000
	1.25	M10 × 1.25	6g	0.028	9.972	9.760	9.160	9.042	0.118	8.439	8.231	6H	8.646	8.911	9.348	9.188	0.160	10.000
12	1.75	M12	6g	0.034	11.966	11.701	10.829	10.679	0.150	9.819	9.543	6H	10.106	10.441	11.063	10.863	0.200	12.000
	1.25	M12 × 1.25	6g	0.028	11.972	11.760	11.160	11.028	0.118	10.439	10.217	6H	10.646	10.911	11.368	11.188	0.180	12.000
14	2	M14	6g	0.038	13.962	13.682	12.663	12.503	0.160	11.508	11.204	6H	11.835	12.210	12.913	12.701	0.212	14.000
	1.5	M14 × 1.5	6g	0.032	13.968	13.732	12.994	12.854	0.140	12.127	11.879	6H	12.376	12.676	13.216	13.026	0.190	14.000
16	2	M16	6g	0.038	15.962	15.682	14.663	14.503	0.160	13.508	13.204	6H	13.385	14.210	14.913	14.701	0.212	16.000
	1.5	M16 × 1.5	6g	0.032	15.968	15.732	14.994	14.854	0.140	14.127	13.879	6H	14.376	14.676	15.216	15.026	0.190	16.000
18	2.5	M18	6g	0.038	17.958	17.623	16.334	16.164	0.170	14.891	14.541	6H	15.294	15.744	16.600	16.376	0.224	18.000
	1.5	M18 × 1.5	6g	0.032	17.968	17.732	16.994	16.854	0.140	16.127	15.879	6H	16.376	16.676	17.216	17.026	0.190	18.000
20	2.5	M20	6g	0.042	19.958	19.623	18.334	18.164	0.170	16.891	16.541	6H	17.294	17.744	18.600	18.376	0.224	20.000
	1.5	M20 × 1.5	6g	0.032	19.968	19.732	18.994	18.854	0.140	18.127	17.879	6H	18.376	18.676	19.216	19.026	0.190	20.000
22	2.5	M22	6g	0.042	21.958	21.623	20.334	20.164	0.170	18.891	18.541	6H	19.294	19.744	20.600	20.376	0.224	22.000
	1.5	M22 × 1.5	6g	0.032	21.968	21.732	20.994	20.854	0.140	20.127	19.879	6H	20.376	20.676	21.216	21.026	0.190	22.000
24	3	M24	6g	0.048	23.952	23.577	22.003	21.803	0.200	20.271	19.855	6H	20.752	21.252	22.316	22.051	0.265	24.000
	2	M24 × 2	6g	0.038	23.962	23.682	22.663	22.493	0.170	21.508	21.194	6H	21.835	22.210	22.925	22.701	0.224	24.000
27	3	M27	6g	0.048	26.952	26.577	25.003	24.803	0.200	23.271	22.855	6H	23.752	24.252	25.316	25.051	0.265	27.000
	2	M27 × 2	6g	0.038	26.962	26.682	25.663	25.493	0.170	24.508	24.194	6H	24.835	25.210	25.925	25.701	0.224	27.000
30	3.5	M30	6g	0.053	29.947	29.522	27.674	27.462	0.212	25.653	25.189	6H	26.211	26.771	28.007	27.727	0.280	30.000
	2	M30 × 2	6g	0.038	29.962	29.682	28.663	28.493	0.170	27.508	27.194	6H	27.835	28.210	28.925	28.701	0.224	30.000
33	3.5	M33	6g	0.053	32.947	32.522	30.674	30.462	0.212	28.653	28.189	6H	29.211	29.771	31.007	30.727	0.280	33.000
	2	M33 × 2	6g	0.038	32.962	32.682	31.663	31.493	0.170	30.508	30.194	6H	30.835	31.210	31.925	31.701	0.224	33.000
36	4	M36	6g	0.060	35.940	35.465	33.342	33.118	0.224	31.033	30.521	6H	31.670	32.270	33.702	33.402	0.300	36.000
	3	M36 × 3	6g	0.048	35.952	35.577	34.003	33.803	0.200	32.271	31.855	6H	32.752	33.252	34.316	34.051	0.265	36.000
39	4	M39	6g	0.060	38.940	38.465	36.342	36.118	0.224	34.033	33.521	6H	34.670	35.270	36.702	36.402	0.300	39.000
	3	M39 × 3	6g	0.048	38.952	38.577	37.003	36.803	0.200	35.271	34.855	6H	35.752	36.252	37.316	37.051	0.265	39.000

* Design form, see Figs. 2 and 5 of ANSI B1.13M-1979 (or Figs. 1 and 4 in 1983 revision).

† Required for high-strength applications where rounded root is specified.

SOURCE: [Appeared in ASME/SAE Interpretive document, Metric Screw Threads, B1.13 (Nov. 3, 1966). pp. 9, 10.] ISO 261-1973, reproduced by permission.

and it is shown in Fig. 8.2.1. The ISO thread series (see Table 8.2.5) are those published in ISO 261-1973. Increased overseas business sparked U.S. interest in metric screw threads, and the ANSI, through its Special Committee to Study Development of an Optimum Metric Fastener System, in joint action with an ISO working group (ISO/TC 1/TC 2), established compromise recommendations regarding metric screw threads. The approved results appear in ANSI B1.13-1979 (Table 8.2.6). This ANSI metric thread series is essentially a selected subset (boxed-in portion of Table 8.2.5) of the larger ISO 261-1973 set. The M profiles of tolerance class 6H/6g are intended for metric applications where inch class 2A/2B has been used.

Metric Tolerance Classes for Threads Tolerance classes are a selected combination of tolerance grades and tolerance positions applied to length-of-engagement groups.

Tolerance grades are indicated as numbers for crest diameters of nut and bolt and for pitch diameters of nut and bolt. Tolerance is the acceptable variation permitted on any such diameter.

Tolerance positions are indicated as letters, and are allowances (fundamental deviations) as dictated by field usage or conditions. Capital letters are used for internal threads (nut) and lower case for external threads (bolt).

There are three established groups of length of thread engagement, S (short), N (normal), and L (long), for various diameter-pitch combinations. Normal length of thread engagement is calculated from the formula $N = 4.5pd^{0.2}$, where p is pitch and d is the smallest nominal size within each of a series of groupings of nominal sizes.

In conformance with coating (or plating) requirements and demands of ease of assembly, the following tolerance positions have been established:

Bolt	Nut	
e		Large allowance
g	G	Small allowance
h	H	No allowance

See Table 8.2.7 for preferred tolerance classes.

ISO metric screw threads are designated by a set of number and letter symbols signifying, in sequence, metric symbol, nominal size, × (symbol), pitch, tolerance grade (on pitch diameter), tolerance position (for pitch diameter), tolerance grade (on crest diameter), and tolerance position (for crest diameter).

EXAMPLE. M6 × 0.75-5g6g, where M = metric symbol; 6 = nominal size, × = symbol; 0.75 = pitch-axial distance of adjacent threads measured between corresponding thread points (millimeters); 5 = tolerance grade (on pitch diameter); g = tolerance position (for pitch diameter); 6 = tolerance grade (on crest diameter); g = tolerance position (for crest diameter).

Power Transmission Screw Threads: Forms and Proportions

The **Acme** thread appears in four series [ANSI B1.8-1973 (revised 1988) and B1.5-1977]. Generalized dimensions for the series are given in Table 8.2.8.

The 29° general-purpose thread (Fig. 8.2.2) is used for all Acme thread applications outside of special design cases.

The 29° stub thread (Fig. 8.2.3) is used for heavy-loading designs and where space constraints or economic factors make a shallow thread advantageous.

The 60° stub thread (Fig. 8.2.4) finds special applications in the machine-tool industry.

The 10° modified square thread (Fig. 8.2.5) is, for all practical purposes, equivalent to a "square" thread.

For selected Acme diameter-pitch combinations, see Table 8.2.9.

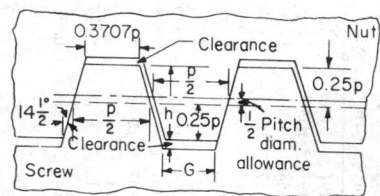

Fig. 8.2.2 29° Acme general-purpose thread.

Table 8.2.7 Preferred Tolerance Classes

	Length of engagement																	
	External threads (bolts)									Internal threads (nuts)								
	Tolerance position e (large allowance)			Tolerance position g (small allowance)			Tolerance position h (no allowance)			Tolerance position G (small allowance)			Tolerance position H (no allowance)		
Quality	Group S	Group N	Group L	Group S	Group N	Group L	Group S	Group N	Group L	Group S	Group N	Group L	Group S	Group N	Group L
Fine							3h4h	4h	4h5h					4H	5H	6H
Medium		6e	7e6e	5g6g	6g	7g6g	5h6h	6h	7h6h	5G	6G	7G	5H	6H	7H
Coarse					8g	9g8g					7G	8G		7H	8H

NOTE: Fine quality applies to precision threads where little variation in fit character is permissible. Coarse quality applies to those threads which present manufacturing difficulties, such as the threading of hot-rolled bars or tapping deep blind holes.
SOURCE: ISO 261-1973, reproduced by permission.

Table 8.2.8 Acme Thread Series
(D = outside diam, p = pitch. All dimensions in inches.)
(See Figs. 8.2.2 to 8.2.5.)

	Thread dimensions			
Symbols	29° general purpose	29° stub	60° stub	10° modified
t = thickness of thread	$0.5p$	$0.5p$	$0.5p$	$0.5p$
R = basic depth of thread	$0.5p$	$0.3p$	$0.433p$	$0.5p*$
F = basic width of flat	$0.3707p$	$0.4224p$	$0.250p$	$0.4563p†$
G = (see Figs. 8.2.2, 8.2.3, 8.2.4)	$F - (0.52 \times \text{clearance})$	$F - (0.52 \times \text{clearance})$	$0.227p$	$F - (0.17 \times \text{clearance})$
E = basic pitch diam	$D - 0.5p$	$D - 0.3p$	$D - 0.433p$	$D - 0.5p$
K = basic minor diam	$D - p$	$D - 0.6p$	$D - 0.866p$	$D - p$
Range of threads, per inch	$1-16$	$2-16$	$4-16$	

* A clearance of at least 0.010 in is added to h on threads of 10-pitch and coarser, and 0.005 in on finer pitches, to produce extra depth, thus avoiding interference with threads of mating parts of a minor or major diameters.
† Measured at crest of screw thread.

Table 8.2.9 Acme Thread Diameter-Pitch Combinations
(See Figs. 8.2.2 to 8.2.5.)

Size	Threads per inch	Size	Threads per inch	Size	Threads per inch	Size	Threads per inch	Size	Threads per inch
¼	16	⅜	8	1¼	5	2¼	3	4	2
⁵⁄₁₆	14	¾	6	1⅜	4	2½	3	4½	2
⅜	12	⅞	6	1½	4	2¾	3	5	2
⁷⁄₁₆	12	1	5	1¾	4	3	2		
½	10	1⅛	5	2	4	3½	2		

Three classes (2G, 3G, 4G) of general-purpose threads have clearances on all diameters for free movement. A fourth class (5G) of general-purpose threads has no allowance or clearance on the pitch diameter for purposes of minimum end play or backlash.

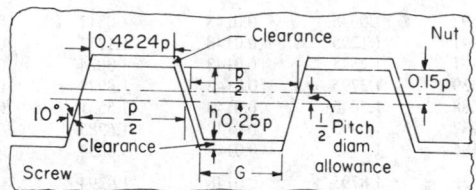

Fig. 8.2.3 20° stub Acme thread.

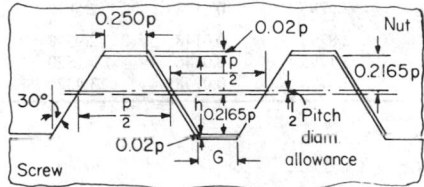

Fig. 8.2.4 60° stub Acme thread.

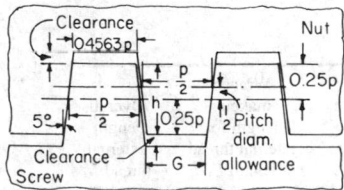

Fig. 8.2.5 10° modified square thread.

High-Strength Bolting Screw Threads

High-strength bolting applications include pressure vessels, steel pipe flanges, fittings, valves, and other services. They can be used for either hot or cold surfaces where high tensile stresses are produced when the joints are made up. For sizes 1 in and smaller, the ANSI coarse-thread series is used. For larger sizes, the ANSI 8-pitch thread series is used (see Table 8.2.10).

Screw Threads for Pipes

American National Standard Taper Pipe Thread (ANSI/ASME B1.20.1-1983) This thread is shown in Fig. 8.2.6. It is made to the following specifications: The taper is 1 in 16 or 0.75 in/ft. The basic length of the external taper thread is determined by $L_2 = p(0.8D + 6.8)$, where D is the basic outside diameter of the pipe (see Table 8.2.11). Thread designation and notation is written as: nominal size, number of threads per inch, thread series. For example: ⅜-18 NPT, ⅛-27 NPSC, ½-14 NPTR, ⅛-27 NPSM, 1-11.5 NPSL, where N = National (American) Standard, T = taper, C = coupling, S = straight, M = mechanical, L = locknut, H = hose coupling, and R = rail fittings. Where pressure-tight joints are required, it is intended that taper pipe threads be made up wrench-tight with a sealant. Descriptions of thread

series include: NPSM = free-fitting mechanical joints for fixtures, NPSL = loose-fitting mechanical joints with locknuts, NPSH = loose-fitting mechanical joints for hose coupling.

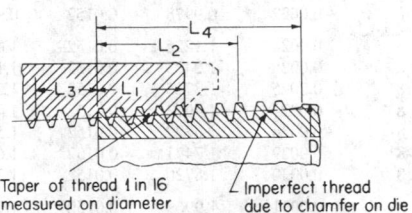

Fig. 8.2.6 American National Standard taper pipe threads.

American National Standard Straight Pipe Thread (ANSI/ASME B1.20.1-1983) This thread can be used to advantage for the following: (1) pressure-tight joints with sealer; (2) pressuretight joints without sealer for drain plugs, filler plugs, etc.; (3) free-fitting mechanical joints for fixtures; (4) loose-fitting mechanical joints with locknuts; and (5) loose-fitting mechanical joints for hose couplings. Dimensions are shown in Table 8.2.12.

American National Standard Dry-Seal Pipe Threads (ANSI B1.20.3-1976 (inch), ANSI B1.20.4-1976 (metric translation) Thread designation and notation include nominal size, number of threads per inch, thread series, class. For example, ⅛-27 NPTF-1, ⅛-27 NPTF-2, ⅛-27 PTF-SAE short, ⅛-27 NPSI, where N = National (American) standard, P = pipe, T = taper, S = straight, F = fuel and oil, I = intermediate. NPTF has two classes: class 1 = specific inspection of root and crest truncation *not* required; class 2 = specific inspection of root and crest truncation *is* required. The series includes: NPTF for all types of service; PTF-SAE short where clearance is not sufficient for full thread length as NPTF; NPSF, nontapered, economical to produce, and used with soft or ductile materials; NPSI nontapered, thick sections with little expansion.

Dry-seal pipe threads resemble tapered pipe threads except the form is truncated (see Fig. 8.2.7), and $L_4 = L_2 + 1$ (see Fig. 8.2.6). Although these threads are designed for nonlubricated joints, as in automobile work, under certain conditions a lubricant is used to prevent galling. Table 8.2.13 lists truncation values.

Tap drill sizes for tapered and straight pipe threads are listed in Table 8.2.14.

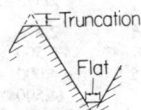

Fig. 8.2.7 American National Standard dry-seal pipe thread.

Wrench bolt heads, nuts, and wrench openings have been standardized (ANSI 18.2-1972). Wrench openings are given in Table 8.2.15; bolt head and nut dimensions are in Table 8.2.16.

Machine Screws

Machine screws are defined according to head types as follows:

Flat Head This screw has a flat surface for the top of the head with a

Table 8.2.10 Screw Threads for High-Strength Bolting
(All dimensions in inches)

Size	Threads per inch	Allowance (minus)	Major diam	Major diam tolerance	Max pitch diam*	Max pitch diam tolerance	Minor diam max	Nut max minor diam	Nut max minor diam tolerance	Nut max pitch diam*	Nut max pitch diam tolerance
¼	20	0.0010	0.2490	0.0072	0.2165	0.0026	0.1877	0.2060	0.0101	0.2211	0.0036
5⁄16	18	0.0011	0.3114	0.0082	0.2753	0.0030	0.2432	0.2630	0.0106	0.2805	0.0041
⅜	16	0.0013	0.3737	0.0090	0.3331	0.0032	0.2990	0.3184	0.0111	0.3389	0.0045
7⁄16	14	0.0013	0.4362	0.0098	0.3898	0.0036	0.3486	0.3721	0.0119	0.3960	0.0049
½	13	0.0015	0.4985	0.0104	0.4485	0.0037	0.4041	0.4290	0.0123	0.4552	0.0052
9⁄16	12	0.0016	0.5609	0.0112	0.5068	0.0040	0.4587	0.4850	0.0127	0.5140	0.0056
⅝	11	0.0017	0.6233	0.0118	0.5643	0.0042	0.5118	0.5397	0.0131	0.5719	0.0059
¾	10	0.0019	0.7481	0.0128	0.6831	0.0045	0.6254	0.6553	0.0136	0.6914	0.0064
⅞	9	0.0021	0.8729	0.0140	0.8007	0.0049	0.7366	0.7689	0.0142	0.8098	0.0070
1	8	0.0022	0.9978	0.0152	0.9166	0.0054	0.8444	0.8795	0.0148	0.9264	0.0076
1⅛	8	0.0024	1.1226	0.0152	1.0414	0.0055	0.9692	1.0045	0.0148	1.0517	0.0079
1¼	8	0.0025	1.2475	0.0152	1.1663	0.0058	1.0941	1.1295	0.0148	1.1771	0.0083
1⅜	8	0.0025	1.3725	0.0152	1.2913	0.0061	1.2191	1.2545	0.0148	1.3024	0.0086
1½	8	0.0027	1.4973	0.0152	1.4161	0.0063	1.3439	1.3795	0.0148	1.4278	0.0090
1⅝	8	0.0028	1.6222	0.0152	1.5410	0.0065	1.4688	1.5045	0.0148	1.5531	0.0093
1¾	8	0.0029	1.7471	0.0152	1.6659	0.0068	1.5937	1.6295	0.0148	1.6785	0.0097
1⅞	8	0.0030	1.8720	0.0152	1.7908	0.0070	1.7186	1.7545	0.0148	1.8038	0.0100
2	8	0.0031	1.9969	0.0152	1.9157	0.0073	1.8435	1.8795	0.0148	1.9294	0.0104
2⅛	8	0.0032	2.1218	0.0152	2.0406	0.0075	1.9682	2.0045	0.0148	2.0545	0.0107
2¼	8	0.0033	2.2467	0.0152	2.1655	0.0077	2.0933	2.1295	0.0148	2.1798	0.0110
2½	8	0.0035	2.4965	0.0152	2.4153	0.0082	2.3431	2.3795	0.0148	2.4305	0.0117
2¾	8	0.0037	2.7463	0.0152	2.6651	0.0087	2.5929	2.6295	0.0148	2.6812	0.0124
3	8	0.0038	2.9962	0.0152	2.9150	0.0092	2.8428	2.8795	0.0148	2.9318	0.0130
3¼	8	0.0039	3.2461	0.0152	3.1649	0.0093	3.0927	3.1295	0.0148	3.1820	0.0132
3½	8	0.0040	3.4960	0.0152	3.4148	0.0093	3.3426	3.3795	0.0148	3.4321	0.0133

The Unified form of thread shall be used. Pitch diameter tolerances include errors of lead and angle.
* The maximum pitch diameters of screws are smaller than the minimum pitch diameters of nuts by these amounts.

Table 8.2.11   ANSI Taper Pipe Thread
(All dimensions in inches)
(See Fig. 8.2.6.)

Nominal pipe size	OD of pipe	Threads per inch	Pitch of thread	Hand-tight engagement length L_1	Effective thread external length L_2	Wrench makeup length for internal thread length L_3	Overall length external thread L_4
1⁄16	0.3125	27	0.03704	0.160	0.2611	0.1111	0.3896
⅛	0.405	27	0.03704	0.180	0.2639	0.1111	0.3924
¼	0.540	18	0.05556	0.200	0.4018	0.1667	0.5946
⅜	0.675	18	0.05556	0.240	0.4078	0.1667	0.6006
½	0.840	14	0.07143	0.320	0.5337	0.2143	0.7815
¾	1.050	14	0.07143	0.339	0.5457	0.2143	0.7935
1	1.315	11½	0.08696	0.400	0.6828	0.2609	0.9845
1¼	1.660	11½	0.08696	0.420	0.7068	0.2609	1.0085
1½	1.900	11½	0.08696	0.420	0.7235	0.2609	1.0252
2	2.375	11½	0.08696	0.436	0.7565	0.2609	1.0582
2½	2.875	8	0.12500	0.682	1.1375	0.2500	1.5712
3	3.500	8	0.12500	0.766	1.2000	0.2500	1.6337
3½	4.000	8	0.12500	0.821	1.2500	0.2500	1.6837
4	4.500	8	0.12500	0.844	1.3000	0.2500	1.7337
5	5.563	8	0.12500	0.937	1.4063	0.2500	1.8400
6	6.625	8	0.12500	0.958	1.5125	0.2500	1.9462
8	8.625	8	0.12500	1.063	1.7125	0.2500	2.1462
10	10.750	8	0.12500	1.210	1.9250	0.2500	2.3587
12	12.750	8	0.12500	1.360	2.1250	0.2500	2.5587
14 OD	14.000	8	0.12500	1.562	2.2500	0.2500	2.6837
16 OD	16.000	8	0.12500	1.812	2.4500	0.2500	2.8837
18 OD	18.000	8	0.12500	2.000	2.6500	0.2500	3.0837
20 OD	20.000	8	0.12500	2.125	2.8500	0.2500	3.2837
24 OD	24.000	8	0.12500	2.375	3.2500	0.2500	3.6837

Table 8.2.12   ANSI Straight Pipe Threads
(All dimensions in inches)

Nominal pipe size (1)	Threads per inch (2)	Pressure-tight with seals		Pressure-tight without seals		Free-fitting (NPSM)				Loose-fitting (NPSL)			
						External		Internal		External		Internal	
		Pitch diam, max (3)	Minor diam, min (4)	Pitch diam, max (5)	Minor diam, min (6)	Pitch diam, max (7)	Major diam, max (8)	Pitch diam, max (9)	Minor diam, min (10)	Pitch diam, max (11)	Major diam, max (12)	Pitch diam, max (13)	Minor diam, min (14)
⅛	27	0.3782	0.342	0.3736	0.3415	0.3748	0.399	0.3783	0.350	0.3840	0.409	0.3989	0.362
¼	18	0.4951	0.440	0.4916	0.4435	0.4899	0.527	0.4951	0.453	0.5038	0.541	0.5125	0.470
⅜	18	0.6322	0.577	0.6270	0.5789	0.6270	0.664	0.6322	0.590	0.6409	0.678	0.6496	0.607
½	14	0.7851	0.715	0.7784	0.7150	0.7784	0.826	0.7851	0.731	0.7963	0.844	0.8075	0.753
¾	14	0.9956	0.925	0.9889	0.9255	0.9889	1.036	0.9956	0.941	1.0067	1.054	1.0179	0.964
1	11½	1.2468	1.161	1.2386	1.1621	1.2386	1.296	1.2468	1.181	1.2604	1.318	1.2739	1.208
1¼	11½	1.5915	1.506	—	—	1.5834	1.641	1.5916	1.526	1.6051	1.663	1.6187	1.553
1½	11½	1.8305	1.745	—	—	1.8223	1.880	1.8305	1.764	1.8441	1.902	1.8576	1.792
2	11½	2.3044	2.219	—	—	2.2963	2.354	2.3044	2.238	2.3180	2.376	2.3315	2.265
2½	8	2.7739	2.650	—	—	2.7622	2.846	2.7739	2.679	2.7934	2.877	2.8129	2.718
3	8	3.4002	3.277	—	—	3.3885	3.472	3.4002	3.305	3.4198	3.503	3.4393	3.344
3½	8	3.9005	3.777	—	—	3.8888	3.972	3.9005	3.806	3.9201	4.003	3.9396	3.845
4	8	4.3988	4.275	—	—	4.3871	4.470	4.3988	4.304	4.4184	4.502	4.4379	4.343
5	8					5.4493	5.533	5.4610	5.366	5.4805	5.564	5.5001	5.405
6	8	—	—			6.5060	6.589	6.5177	6.423	6.5372	6.620	6.5567	6.462
8	8	—	—			—	—	—	—	8.5313	8.615	8.5508	8.456
10	8	—	—			—	—	—	—	10.6522	10.735	10.6717	10.577
12	8	—	—			—	—	—	—	12.6491	12.732	12.6686	12.574

countersink angle of 82°. It is standard for machine screws, cap screws, and wood screws.

Round Head   This screw has a semielliptical head and is standard for machine screws, cap screws, and wood screws except that for the cap screw it is called *button head*.

Fillister Head   This screw has a rounded surface for the top of the head, the remainder being cylindrical. The head is standard for machine screws and cap screws.

Oval Head   This screw has a rounded surface for the top of the head and a countersink angle of 82°. It is standard for machine screws and wood screws.

Hexagon Head   This screw has a hexagonal head for use with external wrenches. It is standard for machine screws.

Socket Head   This screw has an internal hexagonal socket in the head for internal wrenching. It is standard for cap screws.

These screw heads are shown in Fig. 8.2.8; pertinent dimensions are in Table 8.2.17. There are many more machine screw head shapes available to the designer for special purposes, and many are found in the literature. In addition, lots of different screw head configurations have been developed to render fasteners ''tamperproof''; these, too, are found in manufacturers' catalogs or the trade literature.

Eyebolts

Eyebolts are classified as rivet, nut, or screw, and can be had on a swivel. See Fig. 8.2.9 and Table 8.2.18. The safe working load may be obtained for each application by applying an appropriate factor of safety.

Driving recesses come in many forms and types and can be found in company catalogs. Figure 8.2.10 shows a representative set.

Setscrews are used for fastening collars, sheaves, gears, etc. to shafts to prevent relative rotation or translation. They are available in a variety

of head and point styles, as shown in Fig. 8.2.11. A complete tabulation of dimensions is found in ANSI/ASME B18.3-1982 (R86), ANSI 18.6.2-1977 (R93), and ANSI 18.6.3-1977 (R91). Holding power for various sizes is given in Table 8.2.19.

Locking Fasteners

Locking fasteners are used to prevent loosening of a threaded fastener in service and are available in a wide variety differing vastly in design, performance, and function. Since each has special features which may make it of particular value in the solution of a given machine problem, it is important that great care be exercised in the selection of a particular

Table 8.2.13   ANSI Dry-Seal Pipe Threads*
(See Fig. 8.2.7.)

Threads per inch n	Truncation, in		Width of flat	
	Min	Max	Min	Max
27 Crest	0.047p	0.094p	0.054p	0.108p
Root	0.094p	0.140p	0.108p	0.162p
18 C	0.047p	0.078p	0.054p	0.090p
R	0.078p	0.109p	0.090p	0.126p
14 C	0.036p	0.060p	0.042p	0.070p
R	0.060p	0.085p	0.070p	0.098p
11½ C	0.040p	0.060p	0.046p	0.069p
R	0.060p	0.090p	0.069p	0.103p
8 C	0.042p	0.055p	0.048p	0.064p
R	0.055p	0.076p	0.064p	0.088p

* The *truncation* and *width-of-flat* proportions listed above are also valid in the metric system.

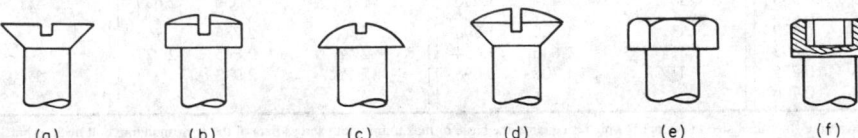

Fig. 8.2.8   Machine screw heads. (*a*) Flat; (*b*) fillister; (*c*) round; (*d*) oval; (*e*) hexagonal; (*f*) socket.

Table 8.2.14 Suggested Tap Drill Sizes for Internal Pipe Threads

	Taper pipe thread							Straight pipe thread			
		Minor diameter at distance		Drill for use without reamer		Drill for use with reamer		Minor diameter			
Size	Probable drill oversize cut (mean)	L_1 from large end	$L_1 + L_3$ from large end	Theoretical drill size*	Suggested drill size†	Theoretical drill size‡	Suggested drill size†	NPSF	NPSI	Theoretical drill size§	Suggested drill size†
	1	2	3	4	5	6	7	8	9	10	11
					Inch						
1/16–27	0.0038	0.2443	0.2374	0.2405	"C" (0.242)	0.2336	"A" (0.234)	0.2482	0.2505	0.2444	"D" (0.246)
1/32–27	0.0044	0.3367	0.3298	0.3323	"Q" (0.332)	0.3254	21/64 (0.328)	0.3406	0.3429	0.3362	"R" (0.339)
¼–18	0.0047	0.4362	0.4258	0.4315	7/17 (0.438)	0.4211	27/64 (0.422)	0.4422	0.4457	0.4375	7/16 (0.438)
⅜–18	0.0049	0.5708	0.5604	0.5659	9/16 (0.562)	0.5555	9/16 (0.563)	0.5776	0.5811	0.5727	27/64 (0.578)
½–14	0.0051	0.7034	0.6901	0.6983	45/64 (0.703)	0.6850	11/16 (0.688)	0.7133	0.7180	0.7082	45/64 (0.703)
¾–14	0.0060	0.9127	0.8993	0.9067	29/32 (0.906)	0.8933	57/64 (0.891)	0.9238	0.9283	0.9178	59/64 (0.922)
1–11½	0.0080	1.1470	1.1307	1.1390	1 9/64 (1.141)	1.1227	1⅛ (1.125)	1.1600	1.1655	1.1520	1 5/32 (1.156)
1¼–11½	0.0100	1.4905	1.4742	1.4805	1 31/64 (1.484)	1.4642	1 11/32 (1.469)				
1½–11½	0.0120	1.7295	1.7132	1.7175	1 23/32 (1.719)	1.7012	1 45/64 (1.703)				
2–11½	0.0160	2.2024	2.1861	2.1864	2 3/16 (2.188)	2.1701	2 11/64 (2.172)				
2½–8	0.0180	2.6234	2.6000	2.6054	2 39/64 (2.609)	2.5820	2 27/64 (2.578)				
3–8	0.0200	3.2445	3.2211	3.2245	3 15/64 (3.234)	3.2011	3 13/64 (3.203)				
					Metric						
1/16–27	0.097	6.206	6.029	6.109	6.1	5.932	6.0	6.304	6.363	6.207	6.2
⅛–27	0.112	8.551	8.363	8.438	8.4	8.251	8.2	8.651	8.710	8.539	8.5
¼–18	0.119	11.080	10.816	10.961	11.0	10.697	10.8	11.232	11.321	11.113	11.0
⅜–18	0.124	14.499	14.235	14.375	14.5	14.111	14.0	14.671	14.760	14.547	14.5
½–14	0.130	17.867	17.529	17.737	17.5	17.399	17.5	18.118	18.237	17.988	18.0
¾–14	0.152	23.182	22.842	23.030	23.0	22.690	23.0	23.465	23.579	23.212	23.0
1–11½	0.203	29.134	28.720	28.931	29.0	28.517	28.5	29.464	29.604	29.261	29.0
1¼–11½	0.254	37.859	37.444	37.605	37.5	37.190	37.0				
1½–11½	0.305	43.929	43.514	43.624	43.5	43.209	43.5				
2–11½	0.406	55.941	55.527	55.535	56.0	55.121	55.0				
2½–8	0.457	66.634	66.029	66.177	66.0	65.572	65.0				
3–8	0.508	82.410	81.815	81.902	82.0	81.307	81.0				

* Column 4 values equal column 2 values minus column 1 values.
† Some drill sizes listed may not be standard drills, and in some cases, standard metric drill sizes may be closer to the theoretical inch drill size and standard inch drill sizes may be closer to the theoretical metric drill size.
‡ Column 6 values equal column 3 values minus column 1 values.
§ Column 10 values equal column 8 values minus column 1 values.
SOURCE: ANSI B1.20.3-1976 and ANSI B1.20.4-1976, reproduced by permission.

Table 8.2.15 Wrench Bolt Heads, Nuts, and Wrench Openings
(All dimensions in inches)

Basic or max width across flats, bolt heads, and nuts	Wrench openings Max	Wrench openings Min	Basic or max width across flats, bolt heads, and nuts	Wrench openings Max	Wrench openings Min	Basic or max width across flats, bolt heads, and nuts	Wrench openings Max	Wrench openings Min	Basic or max width across flats, bolt heads, and nuts	Wrench openings Max	Wrench openings Min
5/32	0.163	0.158	13/16	0.826	0.818	1 13/16	1.835	1.822	3	3.035	3.016
3/16	0.195	0.190	⅞	0.888	0.880	1⅞	1.898	1.885	3⅛	3.162	3.142
¼	0.257	0.252	15/16	0.953	0.944	2	2.025	2.011	3⅜	3.414	3.393
5/16	0.322	0.316	1	1.015	1.006	2 1/16	2.088	2.074	3½	3.540	3.518
11/32	0.353	0.347	1 1/16	1.077	1.068	2 3/16	2.225	2.200	3¾	3.793	3.770
⅜	0.384	0.378	1⅛	1.142	1.132	2¼	2.277	2.262	3⅞	3.918	3.895
7/16	0.446	0.440	1¼	1.267	1.257	2⅜	2.404	2.388	4⅛	4.172	4.147
½	0.510	0.504	1 5/16	1.331	1.320	2 7/16	2.466	2.450	4¼	4.297	4.272
9/16	0.573	0.566	1⅜	1.394	1.383	2 9/16	2.593	2.576	4½	4.550	4.524
19/32	0.605	0.598	1 7/16	1.457	1.446	2⅝	2.656	2.639	4⅝	4.676	4.649
⅝	0.636	0.629	1½	1.520	1.508	2¾	2.783	2.766	5	5.055	5.026
11/16	0.699	0.692	1⅝	1.646	1.634	2 13/16	2.845	2.827	5⅜	5.434	5.403
¾	0.763	0.755	1 11/16	1.708	1.696	2 15/16	2.973	2.954	5¾	5.813	5.780
25/32	0.794	0.786							6⅛	6.192	6.157

Wrenches shall be marked with the "nominal size of wrench" which is equal to the basic or maximum width across flats of the corresponding bolt head or nut.
Allowance (min clearance) between maximum width across flats of nut or bolt head and jaws of wrench equals $1.005W + 0.001$. Tolerance on wrench opening equals plus $0.005W + 0.004$ from minimum (W equals nominal size of wrench).

Table 8.2.16   Width Across Flats of Bolt Heads and Nuts
(All dimensions in inches)

Nominal size or basic major diam of thread	Dimensions of regular bolt heads unfinished, square, and hexagon		Dimensions of heavy bolt heads unfinished, square, and hexagon		Dimensions of cap-screw heads hexagon		Dimensions of setscrew heads		Dimensions of regular nuts and regular jam nuts, unfinished, square, and hexagon (jam nuts, hexagon only)		Dimensions of machine-screw and stove-bolt nuts, square and hexagon		Dimensions of heavy nuts and heavy jam nuts, unfinished, square, and hexagon (jam nuts, hexagon only)	
	Max	Min	Max	Min	Max	Min	Max	Min	Max	Min	Max	Min	Max	Min
No. 0	—	—	—	—	—	—	—	—	—	—	0.1562	0.150		
No. 1	—	—	—	—	—	—	—	—	—	—	0.1562	0.150		
No. 2	—	—	—	—	—	—	—	—	—	—	0.1875	0.180		
No. 3	—	—	—	—	—	—	—	—	—	—	0.1875	0.180		
No. 4	—	—	—	—	—	—	—	—	—	—	0.2500	0.241		
No. 5	—	—	—	—	—	—	—	—	—	—	0.3125	0.302		
No. 6	—	—	—	—	—	—	—	—	—	—	0.3125	0.302		
No. 8	—	—	—	—	—	—	—	—	—	—	0.3438	0.332		
No. 10	—	—	—	—	—	—	—	—	—	—	0.3750	0.362		
No. 12	—	—	—	—	—	—	—	—	—	—	0.4375	0.423		
¼	0.3750	0.362	—	—	0.4375	0.428	0.2500	0.241	0.4375	0.425	0.4375	0.423	0.5000	0.488
5⁄16	0.5000	0.484	—	—	0.5000	0.489	0.3125	0.302	0.5625	0.547	0.5625	0.545	0.5938	0.578
3⁄8	0.5625	0.544	—	—	0.5625	0.551	0.3750	0.362	0.6250	0.606	0.6250	0.607	0.6875	0.669
7⁄16	0.6250	0.603	—	—	0.6250	0.612	0.4375	0.423	0.7500	0.728	—	—	0.7812	0.759
½	0.7500	0.725	0.8750	0.850	0.7500	0.736	0.5000	0.484	0.8125	0.788	—	—	0.8750	0.850
9⁄16	0.8750	0.847	0.9375	0.909	0.8125	0.798	0.5625	0.545	0.8750	0.847	—	—	0.9375	0.909
5⁄8	0.9375	0.906	1.0625	1.031	0.8750	0.860	0.6250	0.606	1.0000	0.969	—	—	1.0625	1.031
¾	1.1250	1.088	1.2500	1.212	1.0000	0.983	0.7500	0.729	1.1250	1.088	—	—	1.2500	1.212
7⁄8	1.3125	1.269	1.4375	1.394	1.1250	1.106	0.8750	0.852	1.3125	1.269	—	—	1.4375	1.394
1	1.5000	1.450	1.6250	1.575	1.3125	1.292	1.0000	0.974	1.5000	1.450	—	—	1.6250	1.575
1⅛	1.6875	1.631	1.8125	1.756	1.5000	1.477	1.1250	1.096	1.6875	1.631	—	—	1.8125	1.756
1¼	1.8750	1.812	2.0000	1.938	1.6875	1.663	1.2500	1.219	1.8750	1.812	—	—	2.0000	1.938
1⅜	2.0625	1.994	2.1857	2.119	—	—	1.3750	1.342	2.0625	1.994	—	—	2.1875	2.119
1½	2.2500	2.175	2.3750	2.300	—	—	1.5000	1.464	2.2500	2.175	—	—	2.3750	2.300
1⅝	2.4375	2.356	2.5625	2.481	—	—	—	—	2.4375	2.356	—	—	2.5625	2.481
1¾	2.6250	2.538	2.7500	2.662	—	—	—	—	2.6250	2.538	—	—	2.7500	2.662
1⅞	2.8125	2.719	2.9375	2.844	—	—	—	—	2.8125	2.719	—	—	2.9375	2.844
2	3.0000	2.900	3.1250	3.025	—	—	—	—	3.0000	2.900	—	—	3.1250	3.025
2¼	3.3750	3.262	3.5000	3.388	—	—	—	—	3.3750	3.262	—	—	3.5000	3.388
2½	3.7500	3.625	3.8750	3.750	—	—	—	—	3.7500	3.625	—	—	3.8750	3.750
2¾	4.1250	3.988	4.2500	4.112	—	—	—	—	4.1250	3.988	—	—	4.2500	4.112
3	4.5000	4.350	4.6250	4.475	—	—	—	—	4.5000	4.350	—	—	4.6250	4.475
3¼	—	—	—	—	—	—	—	—	—	—	—	—	5.0000	4.838
3½	—	—	—	—	—	—	—	—	—	—	—	—	5.3750	5.200
3¾	—	—	—	—	—	—	—	—	—	—	—	—	5.7500	5.562
4	—	—	—	—	—	—	—	—	—	—	—	—	6.1250	5.925

Regular bolt heads are for general use. Unfinished bolt heads are not finished on any surface. Semifinished bolt heads are finished under head.
Regular nuts are for general use. Semifinished nuts are finished on bearing surface and threaded. Unfinished nuts are not finished on any surface but are threaded.

design in order that its properties may be fully utilized. These fasteners may be divided into six groups, as follows: seating lock, spring stop nut, interference, wedge, blind, and quick-release. The **seating-lock type** locks only when firmly seated and is therefore free-running on the bolt. The **spring stop-nut type** of fastener functions by a spring action clamping down upon the bolt. The **prevailing torque type** locks by elastic or plastic flow of a portion of the fastener material. A recent development employs an adhesive coating applied to the threads. The **wedge type** locks by relative wedging of either elements or nut and bolt. The **blind type** usually utilizes spring action of the fastener, and the **quick-release type** utilizes a quarter-turn release device. An example of each is shown in Fig. 8.2.12.

One such specification developed for prevailing torque fasteners by the Industrial Fasteners Institute is based on locking torque and may form a precedent for other types of fasteners as well.

Coach and lag screws find application in wood, or in masonry with an expansion anchor. Figure 8.2.13a shows two types, and Table 8.2.20 lists pertinent dimensions.

Wood screws [ANSI B18.22.1-1975 (R81)] are made in lengths from ¼ to 5 in for steel and from ¼ to 3½ in for brass screws, increasing by ⅛ in up to 1 in, by ¼ in up to 3 in, and by ½ in up to 5 in. Sizes are given in Table 8.2.21. Screws are made with flat, round, or oval heads. Figure 8.2.13b shows several heads.

Washers [ANSI B18.22.1-1975 (R81)] for bolts and lag screws, either round or square, are made to the dimensions given in Table 8.2.22. For other types of washers, see Fig. 8.2.14a and b.

Self-tapping screws are available in three types. **Thread-forming** tapping screws plastically displace material adjacent to the pilot hole. **Thread-cutting** tapping screws have cutting edges and chip cavities (flutes) and form a mating thread by removing material adjacent to the pilot hole. Thread-cutting screws are generally used to join thicker and harder materials and require a lower driving torque than thread-forming screws. **Metallic drive** screws are forced into the material by pressure and are intended for making permanent fastenings. These three types are further classified on the basis of thread and point form as shown in Table 8.2.23. In addition to these body forms, a number of different

Table 8.2.17   Head Diameters (Maximum), In

			Machine screws			
Nominal size	Screw diam	Flat head	Round head	Fillister head	Oval head	Hexagonal head across flats
2	0.086	0.172	0.162	0.140	0.172	0.125
3	0.099	0.199	0.187	0.161	0.199	0.187
4	0.112	0.225	0.211	0.183	0.225	0.187
5	0.125	0.252	0.236	0.205	0.252	0.187
6	0.138	0.279	0.260	0.226	0.279	0.250
8	0.164	0.332	0.309	0.270	0.332	0.250
10	0.190	0.385	0.359	0.313	0.385	0.312
12	0.216	0.438	0.408	0.357	0.438	0.312
¼	0.250	0.507	0.472	0.414	0.507	0.375
5⁄16	0.3125	0.636	0.591	0.519	0.636	0.500
⅜	0.375	0.762	0.708	0.622	0.762	0.562

			Cap screws		
Nominal size	Screw diam	Flat head	Button head	Fillister head	Socket head
¼	0.250	½	7⁄16	⅜	⅜
5⁄16	0.3125	⅝	9⁄16	7⁄16	7⁄16
⅜	0.375	¾	⅝	9⁄16	9⁄16
7⁄16	0.4375	13⁄16	¾	⅝	⅝
½	0.500	⅞	13⁄16	¾	¾
9⁄16	0.5625	1	15⁄16	13⁄16	13⁄16
⅝	0.625	1⅛	1	⅞	⅞
¾	0.750	1⅜	1¼	1	1
⅞	—	—	—	1⅛	1⅛
1	—	—	—	1 5⁄16	1 5⁄16

head types are available. Basic dimensional data are given in Table 8.2.24.

Carriage bolts have been standardized in ANSI B18.5-1971, revised 1990. They come in styles shown in Fig. 8.2.15. The range of bolt diameters is no. 10 (= 0.19 in) to 1 in, no. 10 to ¾ in, no. 10 to ½ in, and no. 10 to ¾ in, respectively.

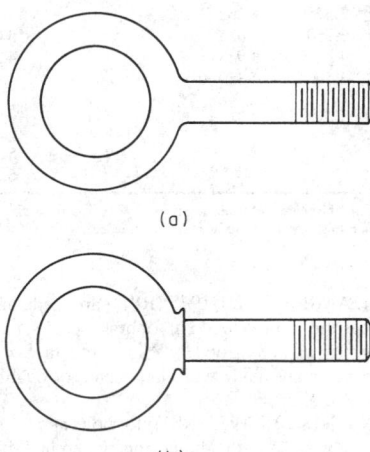

(a)

(b)

Fig. 8.2.9   Eyebolts.

Materials, Strength, and Service Adaptability of Bolts and Screws Materials

Table 8.2.25 shows the relationship between selected metric bolt classes and SAE and ASTM grades. The first number of a metric bolt class equals the minimum tensile strength (ultimate) in megapascals (MPa) divided by 100, and the second number is the approximate ratio between minimum yield and minimum ultimate strengths.

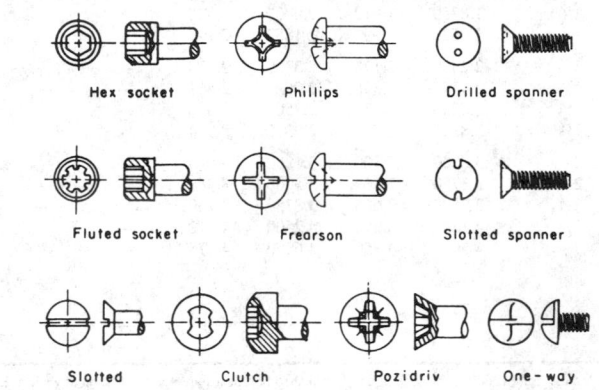

Hex socket   Phillips   Drilled spanner

Fluted socket   Frearson   Slotted spanner

Slotted   Clutch   Pozidriv   One-way

Fig. 8.2.10   Driving recesses. *(Adapted, with permission, from Machine Design.)*

EXAMPLE.   Class 5.8 has a minimum ultimate strength of approximately 500 MPa and a minimum yield strength approximately 80 percent of minimum ultimate strength.

Strength   The fillet between head and body, the thread runout point, and the first thread to engage the nut all create stress concentrations causing local stresses much greater than the average tensile stress in the bolt body. The complexity of the stress patterns renders ineffective the ordinary design calculations based on yield or ultimate stresses. Bolt strengths are therefore determined by laboratory tests on bolt-nut assemblies and published as **proof loads.** Fastener manufacturers are required to periodically repeat such tests to ensure that their products meet the original standards.

In order that a bolted joint remain firmly clamped while carrying its external load P, the bolt must be tightened first with sufficient torque to induce an initial tensile preload F_i. The total load F_B experienced by the bolt is then $F_B = F_i + \varepsilon P$. The fractional multiplier ε is given by $\varepsilon =$

Table 8.2.18 Regular Nut Eyebolts—Selected Sizes
(Thomas Laughlin Co., Portland, Me.)
(All dimensions in inches)
(See Fig. 8.2.9.)

Diam and shank length	Thread length	Eye dimension		Approx breaking strength, lb	Diam and shank length	Thread length	Eye dimension		Approx breaking strength, lb
		ID	OD				ID	OD	
¼ × 2	1½	½	1	2,200	¾ × 6	3	1½	3	23,400
5/16 × 2¼	1½	5/8	1¼	3,600	¾ × 10	3	1½	3	23,400
3/8 × 4½	2½	¾	1½	5,200	¾ × 15	5	1½	3	23,400
½ × 3¼	1½	1	2	9,800	7/8 × 8	4	1¾	3½	32,400
½ × 6	3	1	2	9,800	1 × 6	3	2	4	42,400
½ × 10	3	1	2	9,800	1 × 9	4	2	4	42,400
5/8 × 4	2	1¼	2½	15,800	1 × 18	7	2	4	42,400
5/8 × 6	3	1¼	2½	15,800	1¼ × 8	4	2½	5	67,800
5/8 × 10	3	1¼	2½	15,800	1¼ × 20	6	2½	5	67,800

Table 8.2.19 Cup-Point Setscrew Holding Power

Nominal screw size	Seating torque, lb·in	Axial holding power, lb
No. 0	0.5	50
No. 1	1.5	65
No. 2	1.5	85
No. 3	5	120
No. 4	5	160
No. 5	9	200
No. 6	9	250
No. 8	20	385
No. 10	33	540
¼ in	87	1,000
5/16 in	165	1,500
3/8 in	290	2,000
7/16 in	430	2,500
½ in	620	3,000
9/16 in	620	3,500
5/8 in	1,225	4,000
¾ in	2,125	5,000
7/8 in	5,000	6,000
1 in	7,000	7,000

NOTES: 1. Torsional holding power in inch-pounds is equal to one-half of the axial holding power times the shaft diameter in inches.

2. Experimental data were obtained by seating an alloy-steel cup-point setscrew against a steel shaft with a hardness of Rockwell C 15. Screw threads were class 3A, tapped holes were class 2B. Holding power was defined as the minimum load necessary to produce 0.01 in of relative movement between the shaft and the collar.

3. Cone points will develop a slightly greater holding power; flat, dog, and oval points, slightly less.

4. Shaft hardness should be at least 10 Rockwell C points less than the setscrew point.

5. Holding power is proportional to seating torque. Torsional holding power is increased about 6% by use of a flat on the shaft.

6. Data by F. R. Kull, Fasteners Book Issue, *Mach. Des.*, Mar. 11, 1965.

$K_B/(K_B + K_M)$, where K_B = elastic constant of the bolt and $1/K_M = 1/K_N + 1/K_W + 1/K_G + 1/K_J$. K_N = elastic constant of the nut; K_W = elastic constant of the washer; K_G = elastic constant of the gasket; K_J = elastic constant of the clamped surfaces or joint.

By manipulation, the fractional multiplier can be written $\varepsilon = 1/(1 + K_M/K_B)$. When K_M/K_B approaches 0, $\varepsilon \to 1$. When K_M/K_B approaches infinity, $\varepsilon \to 0$. Generally, K_N, K_W, and K_J are much stiffer than K_B, while K_G can vary from very soft to very stiff. In a metal-to-metal joint, K_G is effectively infinity, which causes K_M to approach infinity and ε to approach 0. On the other hand, for a very soft gasketed joint, $K_M \to 0$ and $\varepsilon \to 1$. For a metal-to-metal joint, then, $F_B = F_i + 0 \times P = F_i$; thus no fluctuating load component enters the bolt. In that case, the bolt remains at *static force* F_i at all times, and the static design will suffice. For a very soft gasketed joint, $F_B = F_i + 1 \times P = F_i + P$, which means that if P is a dynamically fluctuating load, it will be superimposed onto the static value of F_i. Accordingly, one must use the *fatigue design* for the bolt. Of course, for conditions between $\varepsilon = 0$ and $\varepsilon = 1$, the load within the bolt body is $F_B = F_i + \varepsilon P$, and again the *fatigue design* must be used.

In general, one wants as much preload as a bolt and joint will tolerate, without damaging the clamped parts, encouraging stress corrosion, or reducing fatigue life. For ungasketed, unpressurized joints under static loads using high-quality bolt materials, such as SAE 3 or better, the preload should be about 90 percent of proof load.

The **proof strength** is the stress obtained by dividing **proof load** by **stress area**. Stress area is somewhat larger than the root area and can be found in thread tables, or calculated approximately from a diameter which is the mean of the root and pitch diameters.

Initial sizing of bolts can be made by calculating area = (% × proof load)/(proof strength). See Table 8.2.26 for typical physical properties.

NOTE. In European practice, proof stress of a given grade is independent of diameter and is accomplished by varying chemical composition with diameters.

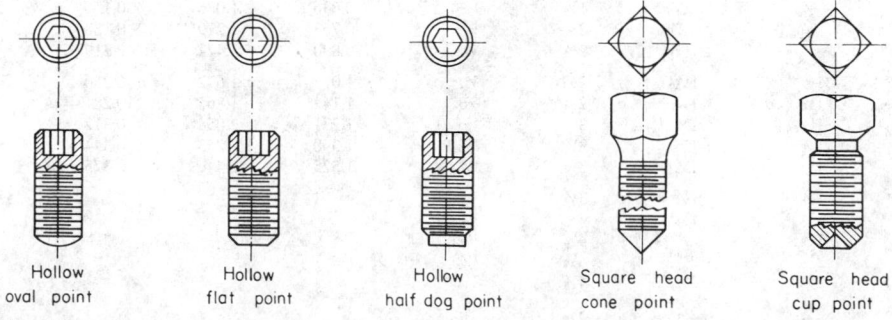

Hollow oval point   Hollow flat point   Hollow half dog point   Square head cone point   Square head cup point

Fig. 8.2.11 Setscrews.

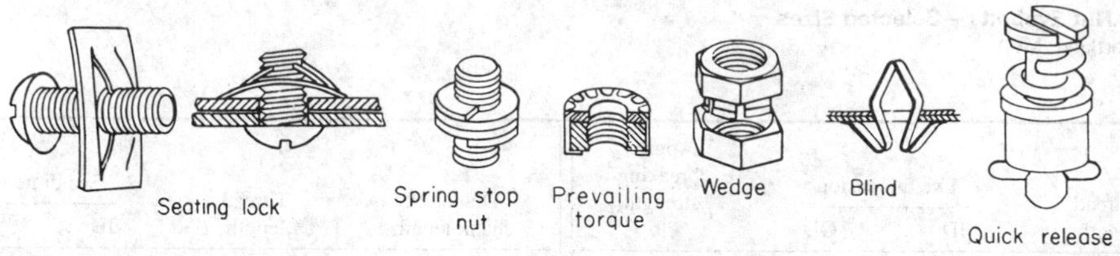
Seating lock
Spring stop nut
Prevailing torque
Wedge
Blind
Quick release

Fig. 8.2.12 Locking fasteners.

General Notes on the Design of Bolted Joints

Bolts subjected to shock and sudden change in load are found to be more serviceable when the unthreaded portion of the bolt is turned down or drilled to the area of the root of the thread. The drilled bolt is stronger in torsion than the turned-down bolt.

When a **number of bolts** are **employed in fastening** together two parts of a machine, such as a cylinder and cylinder head, the load carried by each bolt depends on its relative tightness, the tighter bolts carrying the greater loads. When the conditions of assembly result in differences in tightness, lower working stresses must be used in designing the bolts than otherwise are necessary. On the other hand, it may be desirable to have the bolts the weakest part of the machine, since their breakage from overload in the machine may result in a minimum replacement cost. In such cases, the breaking load of the bolts may well be equal to the load which causes the weakest member of the machine connected to be stressed up to the elastic limit.

Table 8.2.20 Coach and Lag Screws

Diam of screw, in	¼	5⁄16	⅜	7⁄16	½	⅝	¾	⅞	1
No. of threads per inch	10	9	7	7	6	5	4½	4	3½
Across flats of hexagon and square heads, in	⅜	15⁄32	9⁄16	21⁄32	¾	15⁄16	1⅛	1 5⁄16	1½
Thickness of hexagon and square heads, in	3⁄16	¼	5⁄16	⅜	7⁄16	17⁄32	⅝	¾	⅞

Length of threads for screws of all diameters

Length of screw, in	1½	2	2½	3	3½	4	4½
Length of thread, in	To head	1½	2	2¼	2½	3	3½
Length of screw, in	5	5½	6	7	8	9	10–12
Length of thread, in	4	4	4½	5	6	6	7

Table 8.2.21 American National Standard Wood Screws

Number	0	1	2	3	4	5	6	7	8
Threads per inch	32	28	26	24	22	20	18	16	15
Diameter, in	0.060	0.073	0.086	0.099	0.112	0.125	0.138	0.151	0.164
Number	9	10	11	12	14	16	18	20	24
Threads per inch	14	13	12	11	10	9	8	8	7
Diameter, in	0.177	0.190	0.203	0.216	0.242	0.268	0.294	0.320	0.372

Table 8.2.22 Dimensions of Steel Washers, in

	Plain washer			Lock washer		
Bolt size	Hole diam	OD	Thickness	Hole diam	OD	Thickness
3⁄16	¼	9⁄16	3⁄64	0.194	0.337	0.047
¼	5⁄16	¾	1⁄16	0.255	0.493	0.062
5⁄16	⅜	⅞	1⁄16	0.319	0.591	0.078
⅜	7⁄16	1	5⁄64	0.382	0.688	0.094
7⁄16	½	1¼	5⁄64	0.446	0.781	0.109
½	9⁄16	1⅜	3⁄32	0.509	0.879	0.125
9⁄16	⅝	1½	3⁄32	0.573	0.979	0.141
⅝	11⁄16	1¾	⅛	0.636	1.086	0.156
¾	13⁄16	2	⅛	0.763	1.279	0.188
⅞	15⁄16	2¼	5⁄32	0.890	1.474	0.219
1	1 1⁄16	2½	5⁄32	1.017	1.672	0.250
1⅛	1¼	2¾	5⁄32	1.144	1.865	0.281
1¼	1⅜	3	5⁄32	1.271	2.058	0.312
1⅜	1½	3¼	11⁄64	1.398	2.253	0.344
1½	1⅝	3½	11⁄64	1.525	2.446	0.375
1⅝	1¾	3¾	11⁄64			
1¾	1⅞	4	11⁄64			
1⅞	2	4¼	11⁄64			
2	2⅛	4½	11⁄64			
2¼	2⅜	4¾	3⁄16			
2½	2⅝	5	7⁄32			

Table 8.2.23   Tapping Screw Forms

ASA type and thread form	Description and recommendations
 AB	Spaced thread, with gimlet point, designed for use in sheet metal, resin-impregnated plywood, wood, and asbestos compositions. Used in pierced or punched holes where a sharp point for starting is needed.
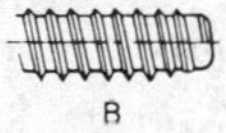 B	Type B is a blunt-point spaced-thread screw, used in heavy-gage sheet-metal and nonferrous castings.
 BP	Same as type B, but has a 45 deg included angle unthreaded cone point. Used for locating and aligning holes or piercing soft materials.
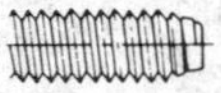 C	Blunt point with threads approximating machine-screw threads. For applications where a machine-screw thread is preferable to the spaced-thread form. Unlike thread-cutting screws, type C makes a chip-free assembly.
 U	Multiple-threaded drive screw with steep helix angle and a blunt, unthreaded starting pilot. Intended for making permanent fastenings in metals and plastics. Hammered or mechanically forced into work. Should not be used in materials less than one screw diameter thick.
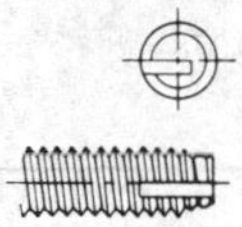 D	Blunt point with single narrow flute and threads approximating machine-screw threads. Flute is designed to produce a cutting edge which is radial to screw center. For low-strength metals and plastics; for high-strength brittle metals; and for rethreading clogged pretapped holes.
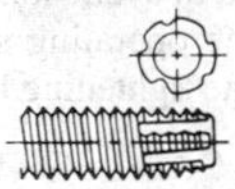 F	Approximate machine-screw thread and blunt point.
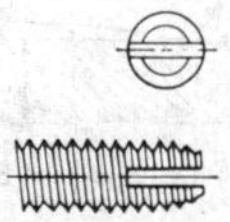 G	Approximate machine-screw thread with single through slot which forms two cutting edges. For low-strength metals and plastics.
 T	Same as type D with single wide flute for more chip clearance.
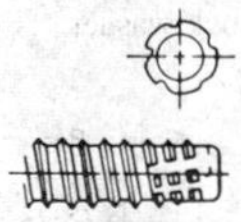 BF	Spaced thread with blunt point and five evenly spaced cutting grooves and chip cavities. Wall thickness should be 1½ times major diameter of screw. Reduces stripping in brittle plastics and die castings.
 BT	Same as type BF except for single wide flute which provides room for twisted, curly chips.
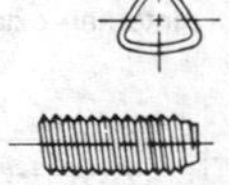 TT	Thread-rolling screws roll-form clean, screw threads. The plastic movement of the material it is driven into locks it in place. The Taptite form is shown here.

SOURCE: *Mach. Des.,* Mar. 11, 1965.

Bolts screwed up tight have an initial stress due to the tightening (preload) before any external load is applied to the machine member. The initial tensile load due to screwing up for a tight joint varies approximately as the diameter of the bolt, and may be estimated at 16,000 lb/in of diameter. The actual value depends upon the applied torque and the efficiency of the screw threads. Applying this rule to bolts of 1-in diam or less results in excessively high stresses, thus demonstrating why bolts of small diameter frequently fail during assembly. It is advisable to use as large-diameter bolts as possible in pressure-tight joints requiring high tightening loads.

In pressure-tight joints without a gasket the force on the bolt under load is essentially never greater than the initial tightening load. When a gasket is used, the total bolt force is approximately equal to the initial tightening load plus the external load. In the first case, deviations from the rule are a result of elastic behavior of the joint faces without a gasket, and inelastic behavior of the gasket in the latter case. The fol-

Table 8.2.24   Self-Tapping Screws

Screw size	Basic major diam, in	Threads per inch					Type U	
		AB	B, BP	C	D, F, G, T	BF, BT	Max outside diam, in	Number of thread starts
00	—	—	—	—	—	—	0.060	6
0	0.060	40	48	—	—	48	0.075	6
1	0.073	32	42	—	—	42		
2	0.086	32	32	56 and 64	56 and 64	32	0.100	8
3	0.099	28	28	48 and 56	48 and 56	28		
4	0.112	24	24	40 and 48	40 and 48	24	0.116	7
5	0.125	20	20	40 and 44	40 and 44	20		
6	0.138	18	20	32 and 40	32 and 40	20	0.140	7
7	0.151	16	19	—	—	19	0.154	8
8	0.164	15	18	32 and 36	32 and 36	18	0.167	8
10	0.190	12	16	24 and 32	24 and 32	16	0.182	8
12	0.216	11	14	24 and 28	24 and 28	14	0.212	8
14	0.242	10	—	—	—	—	0.242	9
¼	0.250	—	14	20 and 28	20 and 28	14		
16	0.268	10						
18	0.294	9						
⁵⁄₁₆	0.3125	—	12	18 and 24	18 and 24	12	0.315	14
20	0.320	9						
24	0.372	9						
⅜	0.375	—	12	16 and 24	16 and 24	12	0.378	12
⁷⁄₁₆	0.4375	—	10	—	—	10		
½	0.500	—	10	—	—	10		

lowing generalization will serve as a guide. If the bolt is more yielding than the connecting members, it should be designed simply to resist the initial tension or the external load, whichever is greater. If the probable yielding of the bolt is 50 to 100 percent of that of the connected members, take the resultant bolt load as the initial tension plus one-half the external load. If the yielding of the connected members is probably four to five times that of the bolt (as when certain packings are used), take the resultant bolt load as the initial tension plus three-fourths the external load.

In cases where bolts are subjected to cyclic loading, an increase in the initial tightening load decreases the operating stress range. In certain applications it is customary to fix the tightening load as a fraction of the yield-point load of the bolt.

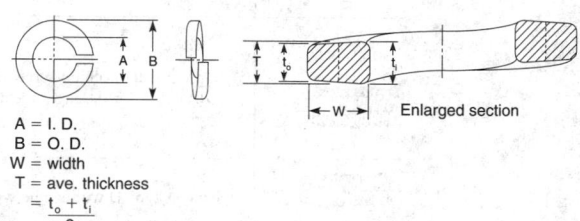

A = I. D.
B = O. D.
W = width
T = ave. thickness
$= \dfrac{t_o + t_i}{2}$

Fig. 8.2.14a   Regular helical spring lock washer.

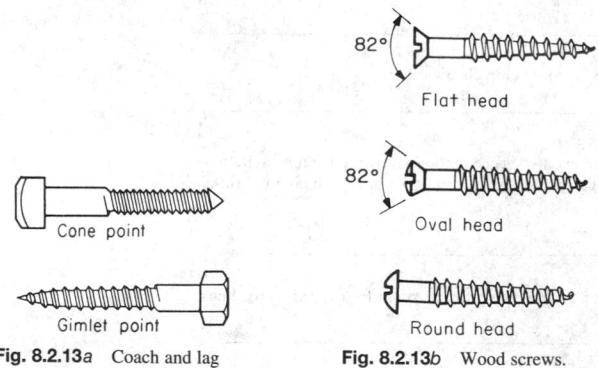

Fig. 8.2.13a   Coach and lag screws.

Fig. 8.2.13b   Wood screws.

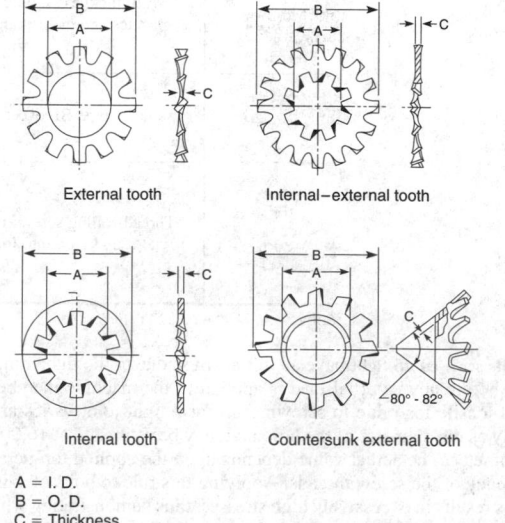

External tooth

Internal–external tooth

Internal tooth

Countersunk external tooth

A = I. D.
B = O. D.
C = Thickness

Fig. 8.2.14b   Toothed lock washers.

Table 8.2.25   ISO Metric Fastener Materials

Metric bolt	Metric nut class normally used	Roughly equivalent U.S. bolt materials	
		SAE J429 grades	ASTM grades
4.6	4 or 5	1	A193, B8; A307, grade A
4.8	4 or 5	2	
5.8	5	2	
8.8	8	5	A325, A449
9.9	9	5 +	A193, B7 and B16
10.9	10 or 12	8	A490; A354, grade 8D
12.9	10 or 12		A540; B21 through B24

SOURCE: Bickford, "An Introduction to the Design and Behavior of Bolted Joints," Marcel Dekker, 1981; reproduced by permission. See Appendix G for additional metric materials.

Table 8.2.26a   Specifications and Identification Markings for Bolts, Screws, Studs, Sems,[a] and U Bolts[b]
(Multiply the strengths in kpsi by 6.89 to get the strength in MPa.)

SAE grade	ASTM grade	Metric[c] grade	Nominal diameter, in	Proof strength, kpsi	Tensile strength, kpsi	Yield[d] strength, kpsi	Core hardness, Rockwell min/max	Products[a]
1	A307	4.6	¼ thru 1½	33	60	36	B70/B100	B, Sc, St
2		5.8	¼ thru ¾	55	74	57	B80/B100	B, Sc, St
		4.6	Over ¾ thru 1½	33	60	36	B70/B100	B, Sc, St
4		8.9	¼ thru 1½	65[f]	115	100	C22/C32	St
5	A449 or A325 type 1	8.8	¼ thru 1	85	120	92	C25/C34	B, Sc, St
		7.8	Over 1 thru 1½	74	105	81	C19/C30	B, Sc, St
		8.6	Over 1½ to 3	55	90	58		B, Sc, St
5.1		8.8	No. 6 thru ⅝	85	120		C25/C40	Se
		8.8	No. 6 thru ½	85	120		C25/C40	B, Sc, St
5.2	A325 type 2	8.8	¼ thru 1	85	120	92	C26/C36	B, Sc
7[g]		10.9	¾ thru 1½	105	133	115	C28/C34	B, Sc
8	A354 Grade BD	10.9	¼ thru 1½	120	150	130	C33/C39	B, Sc, St
8.1		10.9	¼ thru 1½	120	150	130	C32/C38	St
8.2		10.9	¼ thru 1	120	150	130	C35/C42	B, Sc
	A574	12.9	0 thru ½	140	180	160	C39/C45	SHCS
		12.9	⅝ thru 1½	135	170	160	C37/C45	SHCS

NOTE: Company catalogs should be consulted regarding *proof loads*. However, approximate values of proof loads may be calculated from: proof load = proof strength × stress area.
[a] Sems are screw and washer assemblies.
[b] Compiled from ANSI/SAE J429j; ANSI B18.3.1-1978; and ASTM A307, A325, A354, A449, and A574.
[c] Metric grade is *xx.x* where *xx* is approximately 0.01 S_{ut} in MPa and *x* is the ratio of the minimum S_y to S_{ut}.
[d] Yield strength is stress at which a permanent set of 0.2% of gage length occurs.
[e] B = bolt, Sc = Screws, St = studs, Se = sems, and SHCS = socket head cap screws.
[f] Entry appears to be in error but conforms to the standard, ANSI/SAE J429j.
[g] Grade 7 bolts and screws are roll-threaded after heat treatment.
SOURCE: Shigley and Mitchell, "Mechanical Engineering Design," 4th ed., McGraw-Hill, 1983, by permission.

In order to avoid the possibility of bolt failure in pressure-tight joints and to obtain uniformity in bolt loads, some means of determining initial bolt load (**preload**) is desirable. Calibrated torque wrenches are available for this purpose, reading directly in inch-pounds or inch-ounces. Inaccuracies in initial bolt load are possible even when using a torque wrench, owing to variations in the coefficient of friction between the nut and the bolt and, further, between the nut or bolthead and the abutting surface.

An exact method to determine the preload in a bolt requires that the bolt elongation be measured. For a through bolt in which both ends are accessible, the elongation is measured, and the preload force *P* is obtained from the relationship

$$P = AEe \div l$$

where *E* = modulus of elasticity, *l* = original length, *A* = cross-sectional area, *e* = elongation. In cases where both ends of the bolt are not accessible, strain-gage techniques may be employed to determine the strain in the bolt, and thence the preload.

High-strength bolts designated ASTM A325 and A490 are almost exclusively employed in the assembly of structural steel members, but they are applied in mechanical assemblies such as flanged joints. The direct tension indicator (DTI) (Fig. 8.2.16), for use with high-strength bolts, allows **bolt preload** to be applied rapidly and simply. The device is a hardened washer with embossed protrusions (Fig. 8.2.16a). Tightening the bolt causes the protrusions to flatten and results in a decrease in the gap between washer and bolthead. The prescribed degree of bolt-tightening load, or preload, is obtained when the gap is reduced to a predetermined amount (Fig. 8.2.16c). A feeler gage of a given thickness is used to determine when the gap has been closed to the prescribed amount (Fig. 8.2.16b and c). With a paired bolt and DTI, the degree of gap closure is proportional to bolt preload. The system is reported to provide bolt-preload force accuracy within + 15 percent of that prescribed (Fig. 8.2.16d). The devices are available in both inch and metric series and are covered under ASTM F959 and F959M.

Preload-indicating bolts and nuts provide visual assurance of preload in that tightening to the desired preload causes the wavy flange to flatten flush with the clamped assembly (Fig. 8.2.17).

In drilling and tapping cast iron for steel studs, it is necessary to tap to a

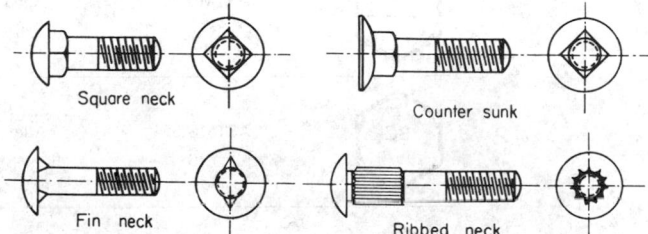

Fig. 8.2.15   Carriage bolts.

depth equal to 1½ times the stud diameter so that the strength of the cast-iron threads in shear may equal the tensile strength of the stud. Drill sizes and depths of hole and thread are given in Table 8.2.27.

It is not good practice to drill holes to be tapped through the metal into pressure spaces, for even though the bolt fits tightly, leakage will result that is difficult to eliminate.

Screw thread inserts made of high-strength material (Fig. 8.2.18) are useful in many cases to provide increased thread strength and life. Soft or ductile materials tapped to receive thread inserts exhibit improved load-carrying capacity under static and dynamic loading conditions. Holes in which threads have been stripped or otherwise damaged can be restored through the use of thread inserts.

Holes for thread inserts are drilled oversize and specially tapped to receive the insert selected to mate with the threaded fastener used. The standard material for inserts is 18-8 stainless steel, but other materials are available, such as phosphor bronze and Inconel. Recommended insert lengths are given in Table 8.2.28.

Drill Sizes   Unified thread taps are listed in Table 8.2.29.

RIVET FASTENINGS

Forms and Proportion of Rivets   The forms and proportions of small and large rivets have been standardized and conform to ANSI B18.1.1-1972 (R89) and B18.1.2-1972 (R89) (Figs. 8.2.19a and b).

Materials Specifications for Rivets and Plates   See Sec. 6 and 12.2.

Conventional signs to indicate the form of the head to be used and

Table 8.2.26b   ASTM and SAE Grade Head Markings for Steel Bolts and Screws

Grade marking	Specification	Material
No mark	SAE grade 1	Low- or medium-carbon steel
	ASTM A307	Low-carbon steel
	SAE grade 2	Low- or medium-carbon steel
	SAE grade 5	Medium-carbon steel, quenched and tempered
	ASTM A449	
	SAE grade 5.2	Low-carbon martensite steel, quenched and tempered
A 325	ASTM A325 type 1	Medium-carbon steel, quenched and tempered; radial dashes optional
A 325	ASTM A325 type 2	Low-carbon martensite steel, quenched and tempered
A 325	ASTM A325 type 3	Atmospheric corrosion (weathering) steel, quenched and tempered
BC	ASTM A354 grade BC	Alloy steel, quenched and tempered
	SAE grade 7	Medium-carbon alloy steel, quenched and tempered, roll-threaded after heat treatment
	SAE grade 8	Medium-carbon alloy steel, quenched and tempered
	ASTM A354 grade BD	Alloy steel, quenched and tempered
	SAE grade 8.2	Low-carbon martensite steel, quenched and tempered
A 490	ASTM A490 type 1	Alloy steel, quenched and tempered
A 490	ASTM A490 type 3	Atmospheric corrosion (weathering) steel, quenched and tempered

SOURCE: ANSI B18.2.1–1981 (R92), Appendix III, p. 41. By permission.

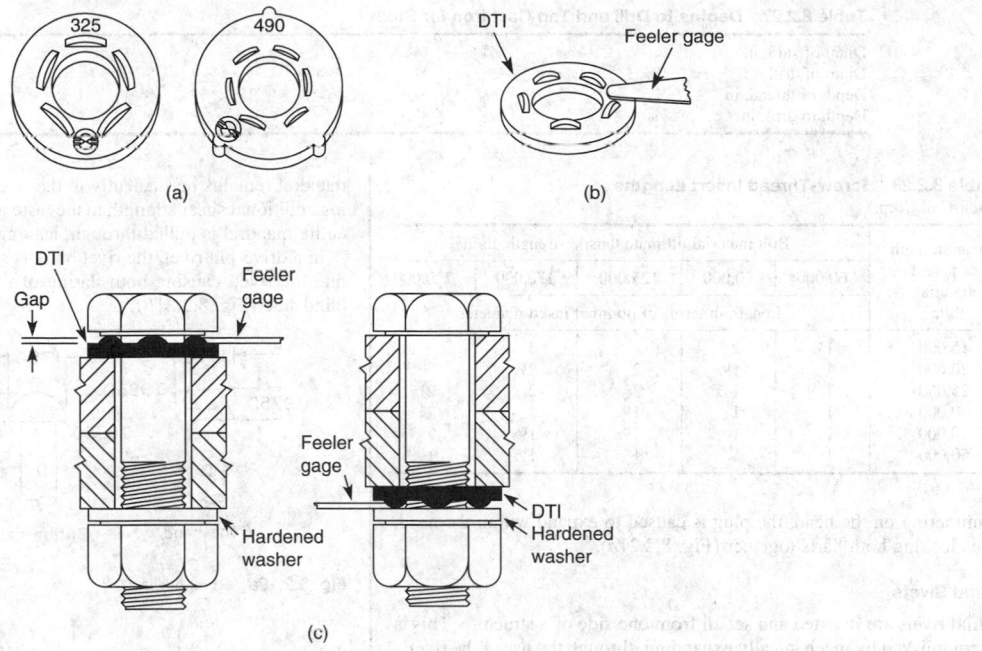

(a)

(b)

(c)

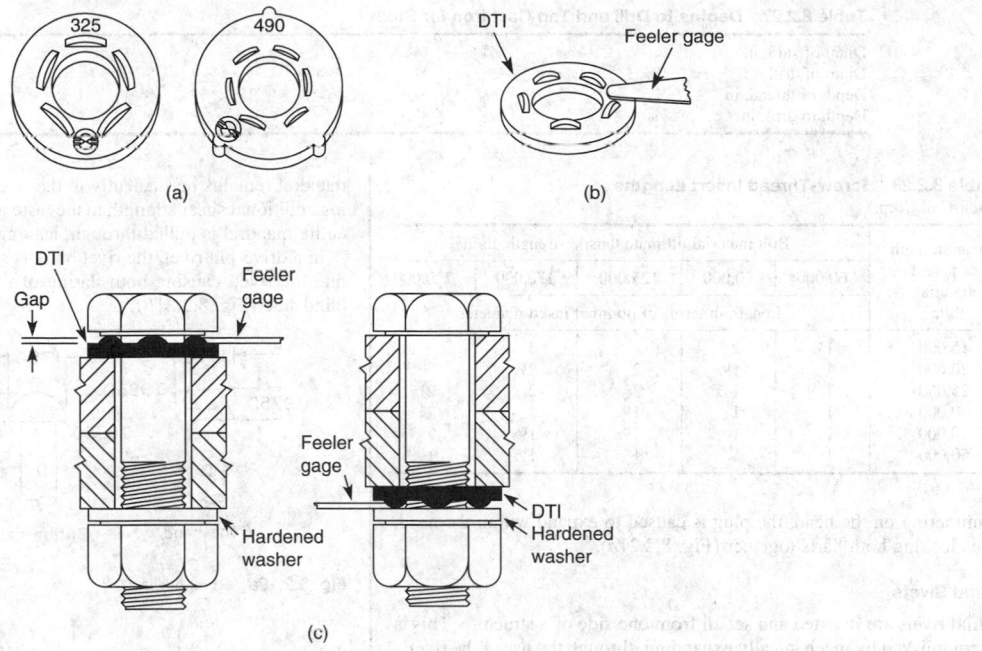

DTI Gaps To Give Required Minimum Bolt Tension

DTI Fitting	325	490
Under bolt head		
Plain finish DTIs	0.015″ (.40 mm)	0.015″
Mechanically galvanized DTIs	0.005″ (.125 mm)	—
Epoxy coated on mechanically galvanized DTIs	0.005″ (.125 mm)	—
Under turned Element		
With hardened washers (plain finish)	0.005″ (.125 mm)	0.005″

With average gaps equal or less than above, bolt tensions will be greater than in adjacent listing

Minimum Bolt Tensions
In thousands of pounds (Kips)

Bolt dia.	A325	A490
1/2″	12	—
5/8″	19	—
3/4″	28	35
7/8″	39	49
1 ″	51	64
1 1/8″	56	80
1 1/4″	71	102
1 3/8″	85	121
1 1/2″	103	—

(d)

Fig. 8.2.16 Direct tension indicators; gap and tension data. *(Source: J & M Turner Inc.)*

whether the rivet is to be driven in the shop or the field at the time of erection are given in Fig. 8.2.20. **Rivet lengths** and **grips** are shown in Fig. 8.2.19*b*.

For **structural riveting,** see Sec. 12.2.

Punched vs. Drilled Plates   Holes in plates forming parts of riveted structures are punched, punched and reamed, or drilled. Punching, while

Before tightening   Full preload

Fig. 8.2.17   Load-indicating wavy-flange bolt (or nut).

cheaper, is objectionable. The holes in different plates cannot be spaced with sufficient accuracy to register perfectly on being assembled. If the hole is punched out, say 1/16 in smaller than is required and then reamed to size, the metal injury by cold flow during punching will be removed. Drilling, while more expensive, is more accurate and does not injure the metal.

Tubular Rivets

In tubular rivets, the end opposite the head is made with an axial hole (partway) to form a thin-walled, easily upsettable end. As the material at the edge of the rivet hole is rolled over against the surface of the joint, a **clinch** is formed (Fig. 8.2.21*a*).

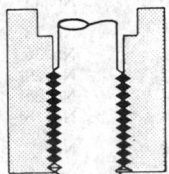

Fig. 8.2.18   Screw thread insert.

Two-part tubular rivets have a thin-walled head with attached thin-walled rivet body and a separate thin-walled expandable plug. The head-body is inserted through a hole in the joint from one side, and the plug from the other. By holding an anvil against the plug bottom and

Table 8.2.27 Depths to Drill and Tap Cast Iron for Studs

Diam of stud, in	¼	5⁄16	⅜	7⁄16	½	9⁄16	⅝	¾	⅞	1
Diam of drill, in	13⁄64	17⁄64	5⁄16	⅜	27⁄64	31⁄64	17⁄32	41⁄64	¾	55⁄64
Depth of thread, in	⅜	15⁄32	9⁄16	21⁄32	¾	27⁄32	15⁄16	1⅛	1 5⁄16	1½
Depth to drill, in	7⁄16	17⁄32	⅝	23⁄32	27⁄32	15⁄16	1 1⁄32	1¼	1 7⁄16	1⅝

Table 8.2.28 Screw-Thread Insert Lengths
(Heli-Coil Corp.)

Shear strength of parent material, lb/in²	Bolt material ultimate tensile strength, lb/in²				
	60,000	90,000	125,000	170,000	220,000
	Length in terms of nominal insert diameter				
15,000	1½	2	2½	3	
20,000	1	1½	2	2½	3
25,000	1	1½	2	2	2½
30,000	1	1	1½	2	2
40,000	1	1	1½	1½	2
50,000	1	1	1	1½	1½

hammering on the head, the plug is caused to expand within the head, thus locking both parts together (Fig. 8.2.21*a*).

Blind Rivets

Blind rivets are inserted and set all from one side of a structure. This is accomplished by mechanically expanding, through the use of the rivet's built-in mandrel, the back (blind side) of the rivet into a bulb or upset head after insertion. Blind rivets include the *pull type* and *drive-pin type*.

The pull-type rivet is available in two configurations: a self-plugging type and a pull-through type. In the self-plugging type, part of the mandrel remains permanently in the rivet body after setting, contributing additional shear strength to the fastener. In the pull-through type, the entire mandrel is pulled through, leaving the installed rivet empty.

In a drive-pin rivet, the rivet body is slotted. A pin is driven forward into the rivet, causing both flaring of the rivet body and upset of the blind side (Fig. 8.2.21*b*).

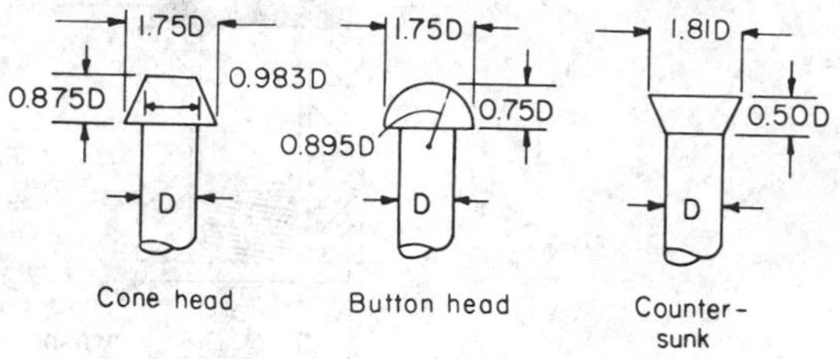

Fig. 8.2.19*a* Rivet heads.

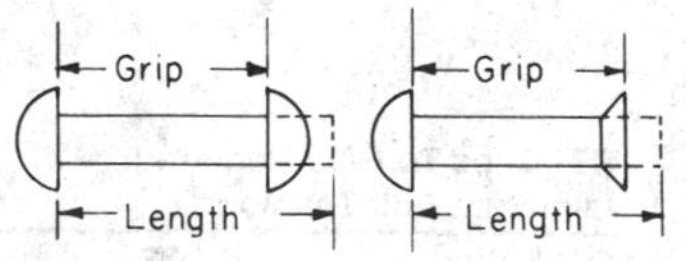

Fig. 8.2.19*b* Rivet length and grip.

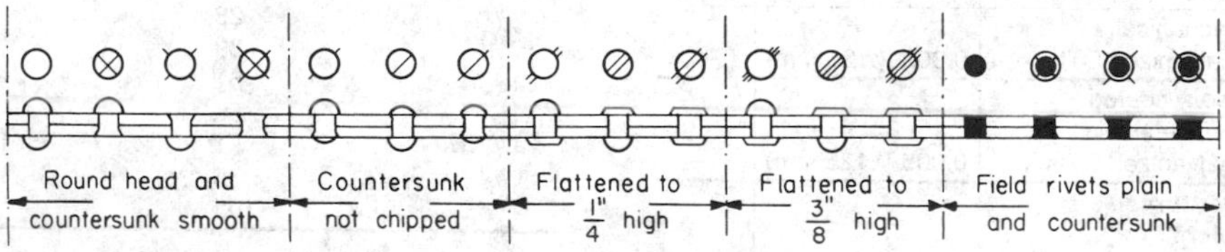

Fig. 8.2.20 Conventional signs for rivets.

Table 8.2.29 Tap-Drill Sizes for American National Standard Screw Threads
(The sizes listed are the commercial tap drills to produce approx 75% full thread)

Size	Coarse-thread series: Threads per inch	Coarse-thread series: Tap drill size	Fine-thread series: Threads per inch	Fine-thread series: Tap drill size	Size	Coarse-thread series: Threads per inch	Coarse-thread series: Tap drill size	Fine-thread series: Threads per inch	Fine-thread series: Tap drill size
No. 0	—	—	80	3⁄64	¾	10	21⁄32	16	11⁄16
No. 1	64	No. 53	72	No. 53	⅞	9	49⁄64	14	13⁄16
No. 2	56	No. 50	64	No. 50	1	8	⅞	14	15⁄16
No. 3	48	No. 47	56	No. 45	1⅛	7	63⁄64	12	1 3⁄64
No. 4	40	No. 43	48	No. 42	1¼	7	1 7⁄64	12	1 11⁄64
No. 5	40	No. 38	44	No. 37	1⅜	6	1 7⁄32	12	1 19⁄64
No. 6	32	No. 36	40	No. 33	1½	6	1 21⁄64	12	1 27⁄64
No. 8	32	No. 29	36	No. 29	1¾	5	1 35⁄64		
No. 10	24	No. 25	32	No. 21	2	4½	1 25⁄32		
No. 12	24	No. 16	28	No. 14	2¼	4½	2 1⁄32		
¼	20	No. 7	28	No. 3	2½	4	2¼		
5⁄16	18	F	24	I	2¾	4	2½		
⅜	16	5⁄16	24	Q	3	4	2¾		
7⁄16	14	U	20	25⁄64	3¼	4	3		
½	13	27⁄64	20	29⁄64	3½	4	3¼		
9⁄16	12	31⁄64	18	33⁄64	3¾	4	3½		
⅝	11	17⁄32	18	37⁄64	4	4	3¾		

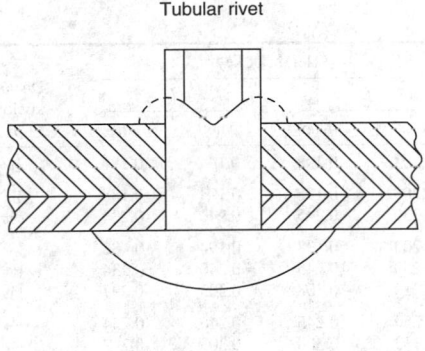

Tubular rivet

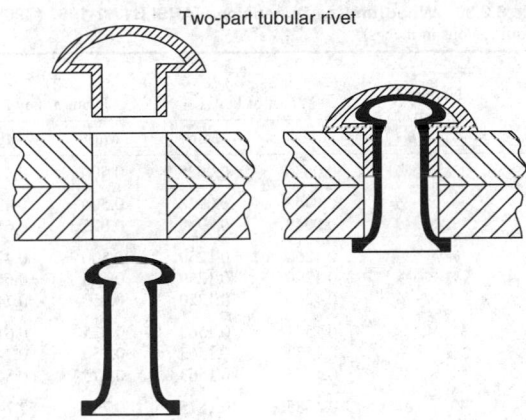

Two-part tubular rivet

Fig. 8.2.21a   Tubular rivets.

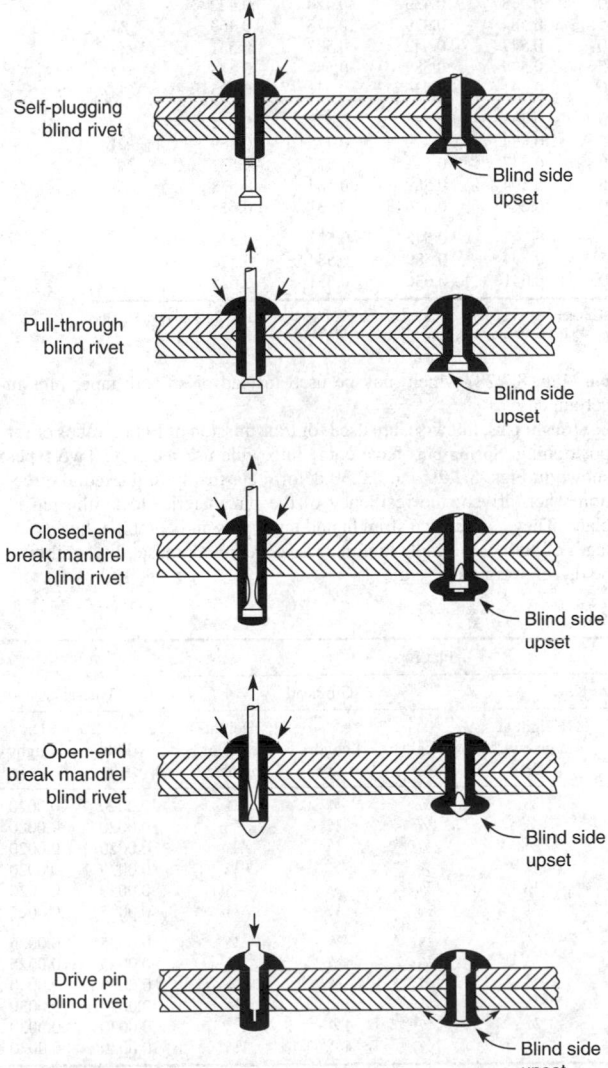

Self-plugging blind rivet — Blind side upset

Pull-through blind rivet — Blind side upset

Closed-end break mandrel blind rivet — Blind side upset

Open-end break mandrel blind rivet — Blind side upset

Drive pin blind rivet — Blind side upset

Fig. 8.2.21b   Blind rivets.

KEYS, PINS, AND COTTERS

Keys and key seats have been standardized and are listed in ANSI B17.1-1967 (R89). Descriptions of the principal key types follow.

Woodruff keys [ANSI B17.2-1967 (R90)] are made to facilitate removal of pulleys from shafts. They should not be used as sliding keys. Cutters for milling out the key seats, as well as special machines for using the cutters, are to be had from the manufacturer. Where the hub of the gear or pulley is relatively long, two keys should be used. Slightly rounding the corners or ends of these keys will obviate any difficulty met with in removing pulleys from shafts. The key is shown in Fig. 8.2.22 and the dimensions in Table 8.2.30.

Square and flat plain taper keys have the same dimensions as gib-head keys (Table 8.2.31) up to the dotted line of Fig. 8.2.23. **Gib-head keys** (Fig. 8.2.23) are necessary when the smaller end is inaccessible for drifting out and the larger end is accessible. It can be used, with care,

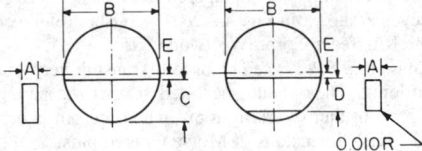

Fig. 8.2.22   Woodruff key.

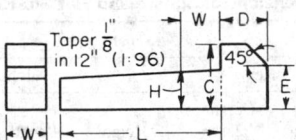

Fig. 8.2.23   Gib-head taper stock key.

with all sizes of shafts. Its use is forbidden in certain jobs and places for safety reasons. Proportions are given in Table 8.2.31.

The minimum stock length of keys is 4 times the key width, and maximum stock length of keys is 16 times the key width. The increments of increase of length are 2 times the width.

Sunk keys are made to the form and dimensions given in Fig. 8.2.24 and Table 8.2.32. These keys are adapted particularly to the case of fitting adjacent parts with neither end of the key accessible. **Feather keys** prevent parts from turning on a shaft while allowing them to move in a lengthwise direction. They are of the forms shown in Fig. 8.2.25 with dimensions as given in Table 8.2.32.

In **transmitting large torques,** it is customary to use two or more keys.

Table 8.2.30 Woodruff Key Dimensions [ANSI B17.2-1967 (R90)]
(All dimensions in inches)

Key no.	Nominal key size, $A \times B$	Width of key A Max	Width of key A Min	Diam of key B Max	Diam of key B Min	Height of key C Max	Height of key C Min	Height of key D Max	Height of key D Min	Distance below E
204	1/16 × 1/2	0.0635	0.0625	0.500	0.490	0.203	0.198	0.194	0.188	3/64
304	3/32 × 1/2	0.0948	0.0928	0.500	0.490	0.203	0.198	0.194	0.188	3/64
305	3/32 × 5/8	0.0948	0.0938	0.625	0.615	0.250	0.245	0.240	0.234	1/16
404	1/8 × 1/2	0.1260	0.1250	0.500	0.490	0.203	0.198	0.194	0.188	3/64
405	1/8 × 5/8	0.1260	0.1250	0.625	0.615	0.250	0.245	0.240	0.234	1/16
406	1/8 × 3/4	0.1260	0.1250	0.750	0.740	0.313	0.308	0.303	0.297	1/16
505	5/32 × 5/8	0.1573	0.1563	0.625	0.615	0.250	0.245	0.240	0.234	1/16
506	5/32 × 3/4	0.1573	0.1563	0.750	0.740	0.313	0.308	0.303	0.297	1/16
507	5/32 × 7/8	0.1573	0.1563	0.875	0.865	0.375	0.370	0.365	0.359	1/16
606	3/16 × 3/4	0.1885	0.1875	0.750	0.740	0.313	0.308	0.303	0.297	1/16
607	3/16 × 7/8	0.1885	0.1875	0.875	0.865	0.375	0.370	0.365	0.359	1/16
608	3/16 × 1	0.1885	0.1875	1.000	0.990	0.438	0.433	0.428	0.422	1/16
609	3/16 × 1 1/8	0.1885	0.1875	1.125	1.115	0.484	0.479	0.475	0.469	5/64
807	1/4 × 7/8	0.2510	0.2500	0.875	0.865	0.375	0.370	0.365	0.359	1/16
808	1/4 × 1	0.2510	0.2500	1.000	0.990	0.438	0.433	0.428	0.422	1/16
809	1/4 × 1 1/8	0.2510	0.2500	1.125	1.115	0.484	0.479	0.475	0.469	5/64
810	1/4 × 1 1/4	0.2510	0.2500	1.250	1.240	0.547	0.542	0.537	0.531	5/64
811	1/4 × 1 3/8	0.2510	0.2500	1.375	1.365	0.594	0.589	0.584	0.578	3/32
812	1/4 × 1 1/2	0.2510	0.2500	1.500	1.490	0.641	0.636	0.631	0.625	7/64
1008	5/16 × 1	0.3135	0.3125	1.000	0.990	0.438	0.433	0.428	0.422	1/16
1009	5/16 × 1 1/8	0.3135	0.3125	1.125	1.115	0.484	0.479	0.475	0.469	5/64
1010	5/16 × 1 1/4	0.3135	0.3125	1.250	1.240	0.547	0.542	0.537	0.531	5/64
1011	5/16 × 1 3/8	0.3135	0.3125	1.375	1.365	0.594	0.589	0.584	0.578	3/32
1012	5/16 × 1 1/2	0.3135	0.3125	1.500	1.490	0.641	0.636	0.631	0.625	7/64
1210	3/8 × 1 1/4	0.3760	0.3750	1.250	1.240	0.547	0.542	0.537	0.531	5/64
1211	3/8 × 1 3/8	0.3760	0.3750	1.375	1.365	0.594	0.589	0.584	0.578	3/32
1212	3/8 × 1 1/2	0.3760	0.3750	1.500	1.490	0.641	0.636	0.631	0.625	7/64

Numbers indicate the nominal key dimensions. The last two digits give the nominal diameter (B) in eighths of an inch, and the digits preceding the last two give the nominal width (A) in thirty-seconds of an inch. Thus, 204 indicates a key 2/32 × 4/8 or 1/16 × 1/2 in; 1210 indicates a key 12/32 × 10/8 or 3/8 × 1 1/4 in.

Another means for fastening gears, pulleys flanges, etc., to shafts is through the use of mating pairs of tapered sleeves known as **grip springs**. A set of sleeves is shown in Fig. 8.2.26. For further references see data issued by the Ringfeder Corp., Westwood, NJ.

Tapered pins (Fig. 8.2.27) can be used to transmit very small torques or for positioning. They should be fitted so that the parts are drawn together to prevent their working loose when the pin is driven home. Table 8.2.33 gives dimensions of Morse tapered pins.

The Groov-Pin Corp., New Jersey, has developed a special **grooved pin** (Fig. 8.2.28) which may be used instead of smooth taper pins in certain cases.

Straight pins, likewise, are used for transmission of light torques or for positioning. **Spring pins** have come into wide use recently. Two types shown in Figs. 8.2.29 and 8.2.30 deform elastically in the radial direction when driven; the resiliency of the pin material locks the pin in place. They can replace straight and taper pins and combine the advantages of both, i.e., simple tooling, ease of removal, reusability, ability to be driven from either side.

Table 8.2.31 Dimensions of Square and Flat Gib-Head Taper Stock Keys, in

Shaft diam	Square type — Key Max width W	Square type — Key Height at large end,† H	Square type — Gib head Height C	Square type — Gib head Length D	Square type — Height edge of chamfer E	Flat type — Key Max width W	Flat type — Key Height at large end,† H	Flat type — Gib head Height C	Flat type — Gib head Length D	Flat type — Height edge of chamfer E	Tolerance On width (−)	Tolerance On height (+)
1/2–9/16	1/8	1/8	1/4	7/32	5/32	1/8	3/32	3/16	1/8	1/8	0.0020	0.0020
5/8–7/8	3/16	3/16	5/16	9/32	7/32	3/16	1/8	1/4	3/16	5/32	0.0020	0.0020
15/16–1 1/4	1/4	1/4	7/16	11/32	11/32	1/4	3/16	5/16	1/4	3/16	0.0020	0.0020
1 5/16–1 3/8	5/16	5/16	9/16	13/32	13/32	5/16	1/4	3/8	5/16	1/4	0.0020	0.0020
1 7/16–1 3/4	3/8	3/8	11/16	15/32	15/32	3/8	1/4	7/16	3/8	5/16	0.0020	0.0020
1 13/16–2 1/4	1/2	1/2	7/8	19/32	5/8	1/2	3/8	5/8	1/2	7/16	0.0025	0.0025
2 5/16–2 3/4	5/8	5/8	1 1/16	23/32	3/4	5/8	7/16	3/4	5/8	1/2	0.0025	0.0025
2 7/8–3 1/4	3/4	3/4	1 1/4	7/8	7/8	3/4	1/2	7/8	3/4	5/8	0.0025	0.0025
3 3/8–3 3/4	7/8	7/8	1 1/2	1	1	7/8	5/8	1 1/16	7/8	3/4	0.0030	0.0030
3 7/8–4 1/2	1	1	1 3/4	1 3/16	1 3/16	1	3/4	1 1/4	1	1 3/16	0.0030	0.0030
4 3/4–5 1/2	1 1/4	1 1/4	2	1 7/16	1 7/16	1 1/4	7/8	1 1/2	1 1/4	1	0.0030	0.0030
5 3/4–6	1 1/2	1 1/2	2 1/2	1 3/4	1 3/4	1 1/2	1	1 3/4	1 1/2	1 1/4	0.0030	0.0030

* Stock keys are applicable to the general run of work and the tolerances have been set accordingly. They are not intended to cover the finer applications where a closer fit may be required.
† This height of the key is measured at the distance W equal to the width of the key, from the gib head.

Cotter pins (Fig. 8.2.31) are used to secure or lock nuts, clevises, etc. Driven into holes in the shaft, the eye prevents complete passage, and the split ends, deformed after insertion, prevent withdrawal.

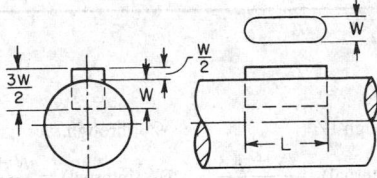

Fig. 8.2.24   Sunk key.

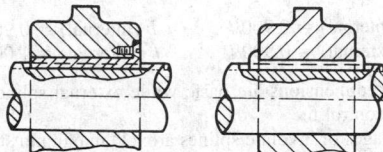

Fig. 8.2.25   Feather key.

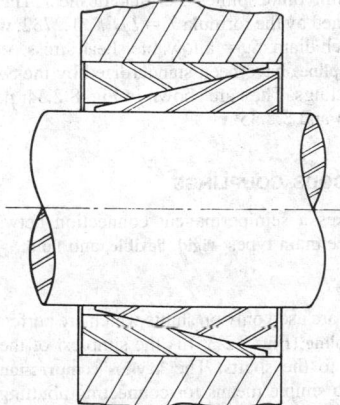

Fig. 8.2.26   Grip springs.

When two rods are to be joined so as to permit movement at the joint, a round pin is used in place of a cotter. In such cases, the proportions may be as shown in Fig. 8.2.32 (knuckle joint).

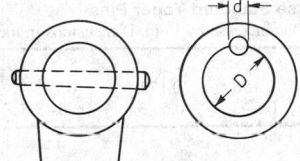

Fig. 8.2.27   Taper pins.

Fig. 8.2.28   Grooved pins.

Fig. 8.2.29   Roll pin.

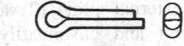

Fig. 8.2.30   Spiral pins.

Fig. 8.2.31   Cotter pin.

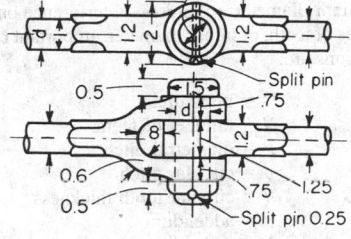

Fig. 8.2.32   Knuckle joint.

SPLINES

Involute spline proportions, dimensions, fits, and tolerances are given in detail in ANSI B92.1-1970. External and internal involute splines (Fig. 8.2.33) have the same general form as involute gear teeth, except that the teeth are one-half the depth of standard gear teeth and the pressure angle is 30°. The spline is designated by a fraction in which the numerator is the diametral pitch and the denominator is always twice the numerator.

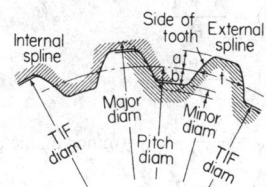

Fig. 8.2.33   Involute spline.

Table 8.2.32   Dimensions of Sunk Keys
(All dimensions in inches. Letters refer to Fig. 8.2.24)

Key no.	L	W	Key no.	L	W	Key no.	L	W	Key no.	L	W
1	½	¹⁄₁₆	13	1	³⁄₁₆	22	1⅜	¼	54	2¼	¼
2	½	³⁄₃₂	14	1	⁷⁄₃₂	23	1⅜	⁵⁄₁₆	55	2¼	⁵⁄₁₆
3	½	⅛	15	1	¼	F	1⅜	⅜	56	2¼	⅜
4	⅝	³⁄₃₂	B	1	⁵⁄₁₆	24	1½	¼	57	2¼	⁷⁄₁₆
5	⅝	⅛	16	1⅛	³⁄₁₆	25	1½	⁵⁄₁₆	58	2½	⁵⁄₁₆
6	⅝	⁵⁄₃₂	17	1⅛	⁷⁄₃₂	G	1½	⅜	59	2½	⅜
7	¾	⅛	18	1⅛	¼	51	1¾	¼	60	2½	⁷⁄₁₆
8	¾	⁵⁄₃₂	C	1⅛	⁵⁄₁₆	52	1¾	⁵⁄₁₆	61	2½	½
9	¾	³⁄₁₆	19	1¼	³⁄₁₆	53	1¾	⅜	30	3	⅜
10	⅞	⁵⁄₃₂	20	1¼	⁷⁄₃₂	26	2	³⁄₁₆	31	3	⁷⁄₁₆
11	⅞	³⁄₁₆	21	1¼	¼	27	2	¼	32	3	½
12	⅞	⁷⁄₃₂	D	1¼	⁵⁄₁₆	28	2	⁵⁄₁₆	33	3	⁹⁄₁₆
A	⅞	¼	E	1¼	⅜	29	2	⅜	34	3	⅝

Table 8.2.33 Morse Standard Taper Pins
(Taper, ⅛ in/ft. Lengths increase by ¼ in. Dimensions in inches)

Size no.	0	1	2	3	4	5	6	7	8	9	10
Diam at large end	0.156	0.172	0.193	0.219	0.250	0.289	0.341	0.409	0.492	0.591	0.706
Length	0.5–3	0.5–3	0.75–3.5	0.75–3.5	0.75–4	0.75–4	0.75–5	1–5	1.25–5	1.5–6	1.5–6

There are 17 series, as follows: 2.5/5, 3/6, 4/8, 5/10, 6/12, 8/16, 10/20, 12/24, 16/32, 20/40, 24/48, 32/64, 40/80, 48/96, 64/128, 80/160, 128/256. The number of teeth within each series varies from 6 to 50. Both a flat-root and a fillet-root type are provided. There are three **types of fits**: (1) **major diameter**—fit controlled by varying the major diameter of the external spline; (2) **sides of teeth**—fit controlled by varying tooth thickness and customarily used for fillet-root splines; (3) **minor diameter**—fit controlled by varying the minor diameter of the internal spline. Each type of fit is further divided into three classes: (*a*) **sliding**—clearance at all points; (*b*) **close**—close on either major diameter, sides of teeth, or minor diameter; (*c*) **press**—interference on either the major diameter, sides of teeth, or minor diameter. Important basic formulas for tooth proportions are:

$$D = \text{pitch diam}$$
$$N = \text{number of teeth}$$
$$P = \text{diametral pitch}$$
$$p = \text{circular pitch}$$
$$t = \text{circular tooth thickness}$$
$$a = \text{addendum}$$
$$b = \text{dedendum}$$
$$D_O = \text{major diam}$$
$$\text{TIF} = \text{true involute form diam}$$
$$D_R = \text{minor diam}$$

Flat and Fillet Roots

$$D = N/P$$
$$p = \pi/P$$
$$t = p/2$$
$$a = 0.5000/P$$
$$D_O \text{ (external)} = \frac{N + 1}{P}$$
$$\text{TIF (internal)} = \frac{N + 1}{P}$$
$$D_R = \frac{N + 1}{P} \quad \text{(minor-diameter fits only)}$$
$$\text{TIF (external)} = \frac{N - 1}{P}$$

$b = 0.600/P + 0.002$ (For major-diameter fits, the internal spline dedendum is the same as the addendum; for minor-diameter fits, the dedendum of the external spline is the same as the addendum.)

Fillet Root Only

½ through ¹²⁄₂₄

$$D_O \text{ (internal)} = \frac{N + 1.8}{P}$$
$$D_R \text{ (external)} = \frac{N - 1.8}{P}$$
$$b \text{ (internal)} = 0.900/P$$
$$b \text{ (external)} = 0.900/P$$

¹⁶⁄₃₂ through ⁴⁸⁄₉₆

$$D_O \text{ (internal)} = \frac{N + 1.8}{P}$$
$$D_R \text{ (external)} = \frac{N - 2}{P}$$
$$b \text{ (internal)} = 0.900/P$$
$$b \text{ (external)} = 1.000/P$$

Internal spline dimensions are basic while external spline dimensions are varied to control fit.

The advantages of involute splines are: (1) maximum strength at the minor diameter, (2) self-centering equalizes bearing and stresses among all teeth, and (3) ease of manufacture through the use of standard gear-cutting tools and methods.

The design of involute splines is critical in shear. The torque capacity may be determined by the formula $T = LD^2S_s/1.2732$, where L = spline length, D = pitch diam, S_s = allowable shear stress.

Parallel-side splines have been standardized by the SAE for 4, 6, 10, and 16 spline fittings. They are shown in Fig. 8.2.34; pertinent data are in Tables 8.2.34 and 8.2.35.

DRY AND VISCOUS COUPLINGS

A **coupling** makes a semipermanent connection between two shafts. They are of three main types: **rigid**, **flexible**, and **fluid**.

Rigid Couplings

Rigid couplings are used only on shafts which are perfectly aligned. The **flanged-face coupling** (Fig. 8.2.35) is the simplest of these. The flanges must be keyed to the shafts. The **keyless compression coupling** (Fig. 8.2.36) affords a simple means for connecting abutting shafts without the necessity of key seats on the shafts. When drawn over the slotted tapered sleeve the two flanges automatically center the shafts and provide sufficient contact pressure to transmit medium or light loads. **Ribbed-clamp couplings** (Fig. 8.2.37) are split longitudinally and are bored to the shaft diameter with a shim separating the two halves. It is necessary to key the shafts to the coupling.

Flexible Couplings

Flexible couplings are designed to connect shafts which are misaligned either laterally or angularly. A secondary benefit is the absorption of

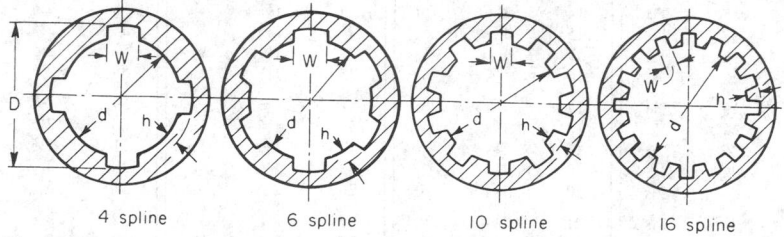

Fig. 8.2.34   Parallel-sided splines.

4 spline   6 spline   10 spline   16 spline

Table 8.2.34   Dimensions of Spline Fittings, in
(SAE Standard)

Nominal diam	4-spline for all fits		6-spline for all fits		10-spline for all fits		16-spline for all fits	
	D max*	W max†	D max*	W max†	D max*	W max†	D max*	W max†
¾	0.750	0.181	0.750	0.188	0.750	0.117		
⅞	0.875	0.211	0.875	0.219	0.875	0.137		
1	1.000	0.241	1.000	0.250	1.000	0.156		
1⅛	1.125	0.271	1.125	0.281	1.125	0.176		
1¼	1.250	0.301	1.250	0.313	1.250	0.195		
1⅜	1.375	0.331	1.375	0.344	1.375	0.215		
1½	1.500	0.361	1.500	0.375	1.500	0.234		
1⅝	1.625	0.391	1.625	0.406	1.625	0.254		
1¾	1.750	0.422	1.750	0.438	1.750	0.273		
2	2.000	0.482	2.000	0.500	2.000	0.312	2.000	0.196
2¼	2.250	0.542	2.250	0.563	2.250	0.351		
2½	2.500	0.602	2.500	0.625	2.500	0.390	2.500	0.245
3	3.000	0.723	3.000	0.750	3.000	0.468	3.000	0.294
3½	—	—	—	—	3.500	0.546	3.500	0.343
4	—	—	—	—	4.000	0.624	4.000	0.392
4½	—	—	—	—	4.500	0.702	4.500	0.441
5	—	—	—	—	5.000	0.780	5.000	0.490
5½	—	—	—	—	5.500	0.858	5.500	0.539
6	—	—	—	—	6.000	0.936	6.000	0.588

* Tolerance allowed of − 0.001 in for shafts ¾ to 1¾ in, inclusive; of − 0.002 for shafts 2 to 3 in, inclusive; − 0.003 in for shafts 3½ to 6 in, inclusive, for 4-, 6-, and 10-spline fittings; tolerance of − 0.003 in allowed for all sizes of 16-spline fittings.
† Tolerance allowed of − 0.002 in for shafts ¾ in to 1¾ in, inclusive; of − 0.003 in for shafts 2 to 6 in, inclusive, for 4-, 6-, and 10-spline fittings; tolerance of − 0.003 allowed for all sizes of 16-spline fittings.

Table 8.2.35   Spline Proportions

No. of splines	W for all fits	Permanent fit		To slide when not under load		To slide under load	
		h	d	h	d	h	d
4	0.241D	0.075D	0.850D	0.125D	0.750D		
6	0.250D	0.050D	0.900D	0.075D	0.850D	0.100D	0.800D
10	0.156D	0.045D	0.910D	0.070D	0.860D	0.095D	0.810D
16	0.098D	0.045D	0.910D	0.070D	0.860D	0.095D	0.810D

impacts due to fluctuations in shaft torque or angular speed. The **Old-ham**, or **double-slider, coupling** (Fig. 8.2.38) may be used to connect shafts which have only lateral misalignment. The **"Fast" flexible coupling** (Fig. 8.2.39) consists of two hubs each keyed to its respective shaft. Each hub has generated splines cut at the maximum possible

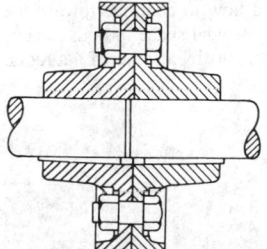

Fig. 8.2.35   Flanged face coupling.

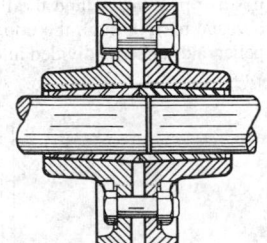

Fig. 8.2.36   Keyless compression coupling.

distance from the shaft end. Surrounding the hubs is a casing or sleeve which is split transversely and bolted by means of flanges. Each half of this sleeve has generated internal splines cut on its bore at the end opposite to the flange. These internal splines permit a definite error of alignment between the two shafts.

Another type, the Waldron coupling (Midland-Ross Corp.), is shown in Fig. 8.2.40.

The chain coupling shown in Fig. 8.2.41 uses silent chain, but stan-

dard roller chain can be used with the proper mating sprockets. Nylon links enveloping the sprockets are another variation of the chain coupling.

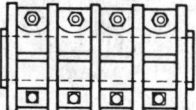

Fig. 8.2.37   Ribbed-clamp coupling.

Steelflex couplings (Fig. 8.2.42) are made with two grooved steel hubs keyed to their respective shafts. Connection between the two halves is secured by a specially tempered alloy-steel member called the "grid."

Fig. 8.2.38   Double-slider coupling.

In the rubber flexible coupling shown in Fig. 8.2.43, the torque is transmitted through a comparatively soft rubber section acting in shear. The type in Fig. 8.2.44 loads the intermediate rubber member in compression. Both types permit reasonable shaft misalignment and are recommended for light loads only.

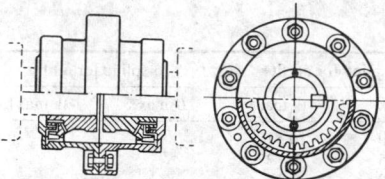

Fig. 8.2.39 "Fast" flexible coupling.

Universal joints are used to connect shafts with much larger values of misalignment than can be tolerated by the other types of flexible couplings. Shaft angles up to 30° may be used. The Hooke's-type joint (Fig. 8.2.45) suffers a loss in efficiency with increasing angle which

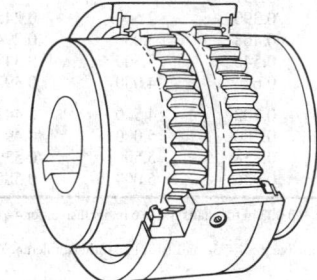

Fig. 8.2.40 Waldron coupling.

may be approximated for angles up to 15° by the following relation: efficiency = $100(1 - 0.003\theta)$, where θ is the angle between the shafts. The velocity ratio between input and output shafts with a single universal joint is equal to

$$\omega_2/\omega_1 = \cos\theta/1 - \sin^2\theta\sin^2(\alpha + 90°)$$

where ω_2 and ω_1 are the angular velocities of the driven and driving shafts respectively, θ is the angle between the shafts, and α is the angu-

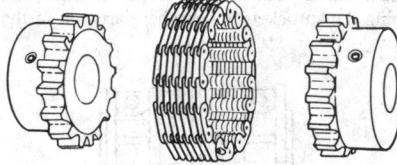

Fig. 8.2.41 Chain coupling.

lar displacement of the driving shaft from the position where the pins on the drive-shaft yoke lie in the plane of the two shafts. A velocity ratio of 1 may be obtained at any angle using two Hooke's-type joints and an intermediate shaft. The intermediate shaft must make equal angles with the main shafts, and the driving pins on the yokes attached to the intermediate shaft must be set parallel to each other.

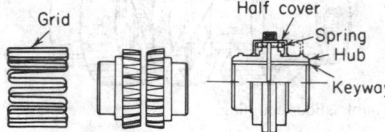

Fig. 8.2.42 Falk Steelflex coupling.

The Bendix-Weiss "rolling-ball" universal joint provides constant angular velocity. Torque is transmitted between two yokes through a set of four balls such that the centers of all four balls lie in a plane which

bisects the angle between the shafts. Other variations of constant velocity universal joints are found in the Rzeppa, Tracta, and double Cardan types.

Fig. 8.2.43 Rubber flexible coupling, shear type.

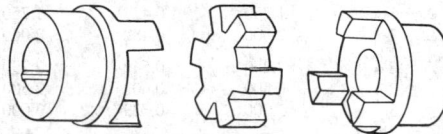

Fig. 8.2.44 Rubber flexible coupling, compression type.

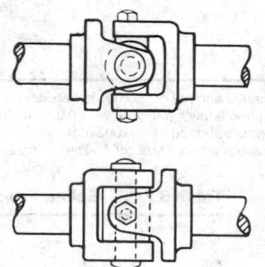

Fig. 8.2.45 Hooke's universal joint.

Fluid Couplings

(See also Sec. 11.)

Fluid couplings (Fig. 8.2.46) have two basic parts—the input member, or impeller, and the output member, or runner. There is no mechanical connection between the two shafts, power being transmitted by kinetic energy in the operating fluid. The impeller B is fastened to the flywheel A and turns at engine speed. As this speed increases, fluid within the impeller moves toward the outer periphery because of centrifugal force. The circular shape of the impeller directs the fluid toward the runner C, where its kinetic energy is absorbed as torque delivered by shaft D. The positive pressure behind the fluid causes flow to continue toward the hub and back through the impeller. The toroidal space in both the impeller and runner is divided into compartments by a series of flat radial vanes.

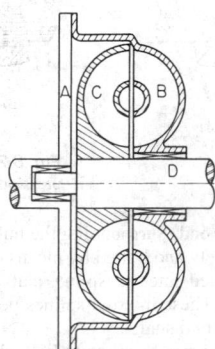

Fig. 8.2.46a Fluid coupling. (*A*) Flywheel; (*B*) impeller; (*C*) runner; (*D*) output shaft.

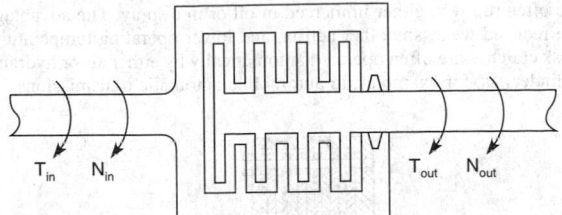

Fig. 8.2.46*b* Schematic of viscous coupling.

The torque capacity of a fluid coupling with a full-load slip of about 2.5 percent is $T = 0.09n^2D^5$, where n is the impeller speed, hundreds of r/min, and D is the outside diameter, ft. The output torque is equal to the input torque over the entire range of input-output speed ratios. Thus the prime mover can be operated at its most effective speed regardless of the speed of the output shaft. Other advantages are that the prime mover cannot be stalled by application of load and that there is no transmission of shock loads or torsional vibration between the connected shafts.

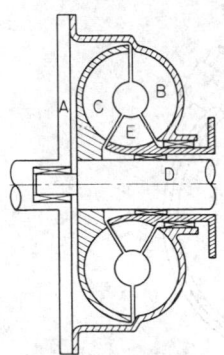

Fig. 8.2.47 Hydraulic torque converter. (*A*) Flywheel; (*B*) impeller; (*C*) runner; (*D*) output shaft; (*E*) reactor.

A **hydraulic torque converter** (Fig. 8.2.47) is similar in form to the hydraulic coupling, with the addition of a set of stationary guide vans, the reactor, interposed between the runner and the impeller. All blades in a converter have compound curvature. This curvature is designed to control the direction of fluid flow. Kinetic energy is therefore transferred as a result of both a scalar and vectorial change in fluid velocity. The blades are designed such that the fluid will be moving in a direction parallel to the blade surface at the entrance (Fig. 8.2.48) to each section. With a design having fixed blading, this can be true at only one value of runner and impeller velocity, called the design point. Several design modifications are possible to overcome this difficulty. The angle of the blades can be made adjustable, and the elements can be divided into sections operating independently of each other according to the load requirements. Other refinements include the addition of multiple stages in the runner and reactor stages as in steam reaction turbines (see Sec. 9). The advantages of a torque converter are the ability to multiply starting torque 5 to 6 times and to serve as a stepless transmission. As in the coupling, torque varies as the square of speed and the fifth power of diameter.

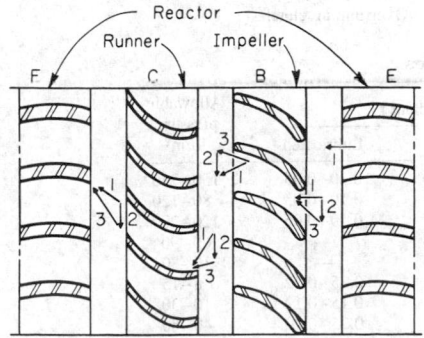

Fig. 8.2.48 Schematic of converter blading. (1) Absolute fluid velocity; (2) velocity vector—converter elements; (3) fluid velocity relative to converter elements.

Optimum efficiency (Fig. 8.2.49) over the range of input-output speed ratios is obtained by a combination converter coupling. When the output speed rises to the point where the torque multiplication factor is 1.0, the clutch point, the torque reaction on the reactor element reverses direction. If the reactor is mounted to freewheel in this opposite direction, the unit will act as a coupling over the higher speed ranges. An automatic friction clutch (see "Clutches," below) set to engage at or near the clutch point will also eliminate the poor efficiency of the converter at high output speeds.

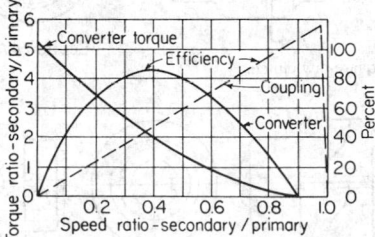

Fig. 8.2.49 Hydraulic coupling characteristic curves. (*Heldt, "Torque Converters and Transmissions," Chilton.*)

Viscous couplings are becoming major players in mainstream front-wheel-drive applications and are already used in four-wheel-drive vehicles.

Torque transmission in a viscous coupling relies on shearing forces in an entrapped fluid between axially positioned disks rotating at different angular velocities (Fig. 8.2.46*b*), all encased in a lifetime leakproof housing. A hub carries the so-called inner disks while the housing carries the so-called outer disks. Silicone is the working fluid.

Operation of the coupling is in normal (slipping) mode when torque is being generated by viscous shear. However, prolonged slipping under severe starting conditions causes heat-up, which in turn causes the fluid, which has a high coefficient of thermal expansion, to expand considerably with increasing temperature. It then fills the entire available space, causing a rapid pressure increase, which in turn forces the disks together into metal-to-metal frictional contact. Torque transmission now increases substantially. This self-induced torque amplification is known as the **hump effect.** The point at which the hump occurs can be set by the design and coupling setup. Under extreme conditions, 100 percent lockup occurs, thus providing a self-protecting relief from overheating as fluid shear vanishes. This effect is especially useful in autos using viscous couplings in their limited-slip differentials, when one wheel is on low-friction surfaces such as ice. The viscous coupling transfers torque to the other gripping wheel. This effect is also useful when one is driving up slopes with uneven surface conditions, such as rain or snow, or on very rough surfaces. Such viscous coupling differentials have allowed a weight and cost reduction of about 60 percent. A fuller account can be found in Barlage, Viscous Couplings Enter Main Stream Vehicles, *Mech. Eng.,* Oct. 1993.

CLUTCHES

Clutches are couplings which permit the disengagement of the coupled shafts during rotation.

Positive clutches are designed to transmit torque without slip. The **jaw clutch** is the most common type of positive clutch. These are made with **square jaws** (Fig. 8.2.50) for driving in both directions or **spiral jaws** (Fig. 8.2.51) for unidirectional drive. Engagement speed should be limited to 10 r/min for square jaws and 150 r/min for spiral jaws. If disengagement under load is required, the jaws should be finish-machined and lubricated.

Friction clutches are designed to reduce coupling shock by slipping during the engagement period. They also serve as safety devices by slipping when the torque exceeds their maximum rating. They may be divided into two main groups, **axial** and **rim clutches,** according to the direction of contact pressure.

The **cone clutch** (Fig. 8.2.52) and the **disk clutch** (Fig. 8.2.53) are examples of axial clutches. The disk clutch may consist of either a

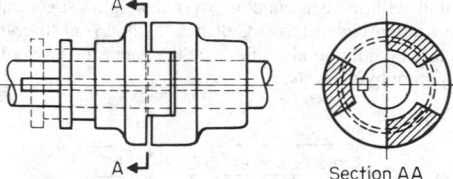

Fig. 8.2.50 Square-jaw clutch.

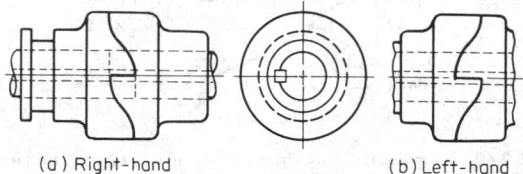

Fig. 8.2.51 Spiral-jaw clutch.

single plate or multiple disks. Table 8.2.36 lists typical friction materials and important design data. The torque capacity of a disk clutch is given by $T = 0.5 i f F_a D_m$, where T is the torque, i the number of pairs of contact surfaces, f the applicable coefficient of friction, F_a the axial

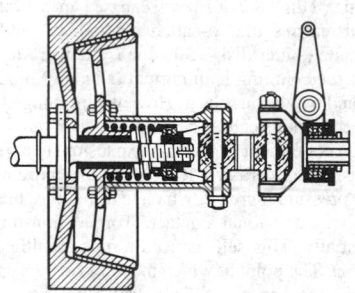

Fig. 8.2.52 Cone clutch.

engaging force, and D_m the mean diameter of the clutch facing. The spring forces holding a disk clutch in engagement are usually of relatively high value, as given by the allowable contact pressures. In order to lower the force required at the operating lever, elaborate linkages are required, usually having lever ratios in the range of 10 to 12. As these linkages must rotate with the clutch, they must be adequately balanced and the effect of centrifugal forces must be considered. Disk clutches

are often run wet, either immersed in oil or in a spray. The advantages are reduced wear, smoother action, and lower operating temperatures. Disk clutches are often operated automatically by either air or hydraulic cylinders as, for examples, in automobile automatic transmissions.

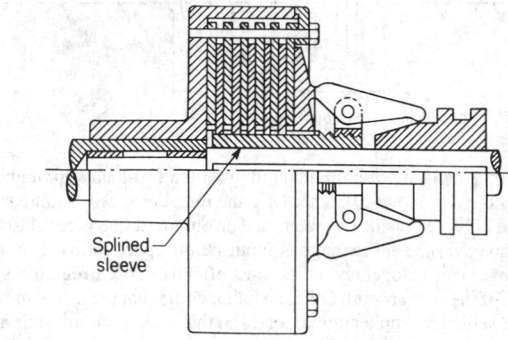

Fig. 8.2.53 Multidisk clutch.

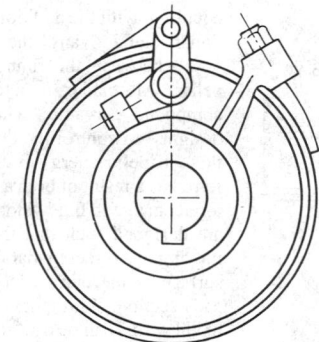

Fig. 8.2.54 Band clutch.

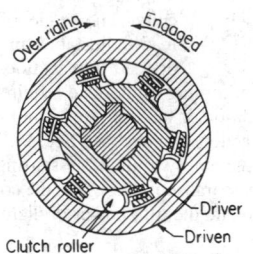

Fig. 8.2.55 Overrunning clutch.

Table 8.2.36 Friction Coefficients and Allowable Pressures

Materials in contact	Friction coefficient f			Allowable pressure, lb/in²
	Dry	Greasy	Lubricated	
Cast iron on cast iron	0.2–0.15	0.10–0.06	0.10–0.05	150–250
Bronze on cast iron	—	0.10–0.05	0.10–0.05	80–120
Steel on cast iron	0.30–0.20	0.12–0.07	0.10–0.06	120–200
Wood on cast iron	0.25–0.20	0.12–0.08	—	60–90
Fiber on metal	—	0.20–0.10	—	10–30
Cork on metal	0.35	0.30–0.25	0.25–0.22	8–15
Leather on metal	0.5–0.3	0.20–0.15	0.15–0.12	10–30
Wire asbestos on metal	0.5–0.35	0.30–0.25	0.25–0.20	40–80
Asbestos blocks on metal	0.48–0.40	0.30–0.25	—	40–160
Asbestos on metal, short action	—	—	0.25–0.20	200–300
Metal on cast iron, short action	—	—	0.10–0.05	200–300

SOURCE: Maleev, *Machine Design,* International Textbook, by permission.

Rim clutches may be subdivided into two groups: (1) those employing either a band or block (Fig. 8.2.54) in contact with the rim and (2) overrunning clutches (Fig. 8.2.55) employing the wedging action of a roller or sprag. Clutches in the second category will automatically engage in one direction and freewheel in the other.

HYDRAULIC POWER TRANSMISSION

Hydraulic power transmission systems comprise machinery and auxiliary components which function to generate, transmit, control, and utilize hydraulic power. The **working fluid**, a pressurized incompressible liquid, is usually either a petroleum base or a fire-resistant type. The latter are water and oil emulsions, glycol-water mixtures, or synthetic liquids such as silicones or phosphate esters.

Liquid is pressurized in a **pump** by virtue of its resistance to flow; the pressure difference between pump inlet and outlet results in flow. Most hydraulic applications employ positive-displacement pumps of the gear, vane, screw, or piston type; piston pumps are axial, radial, or reciprocating (see Sec. 14).

Power is transmitted from pump to controls and point of application through a combination of **conduit and fittings** appropriate to the particular application. Flow characteristics of hydraulic circuits take into account fluid properties, pressure drop, flow rate, and pressure-surging tendencies. Conduit systems must be designed to minimize changes in flow velocity, velocity distribution, and random fluid eddies, all of which dissipate energy and result in pressure drops in the circuit (see Sec. 3). Pipe, tubing, and flexible hose are used as hydraulic power conduits; suitable fittings are available for all types and for transition from one type to another.

Controls are generally interposed along the conduit between the pump and point of application (i.e., an actuator or motor), and act to control pressure, volume, or flow direction.

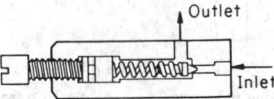

Fig. 8.2.56   Relief valve.

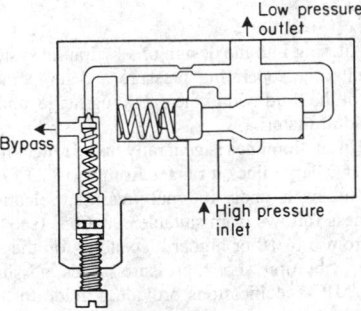

Fig. 8.2.57   Reducing valve.

Pressure control valves, of which an ordinary safety valve is a common type (normally closed), include relief and reducing valves and pressure switches (Figs. 8.2.56 and 8.2.57). Pressure valves, normally closed, can be used to control sequential operations in a hydraulic circuit. **Flow control valves** throttle flow to or bypass flow around the unit being controlled, resulting in pressure drop and temperature increase as pressure energy is dissipated. Figure 8.2.58 shows a simple needle valve with variable orifice usable as a flow control valve. **Directional control valves** serve primarily to direct hydraulic fluid to the point of application. Directional control valves with rotary and sliding spools are shown in Figs. 8.2.59 and 8.2.60.

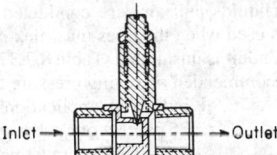

Fig. 8.2.58   Needle valve.

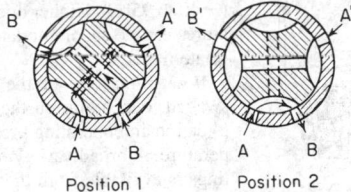

Fig. 8.2.59   Rotary-spool directional flow valve.

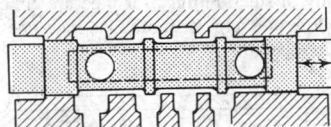

Fig. 8.2.60   Sliding-spool directional flow valve.

A poppet (valve) mechanism is shown in Fig. 8.2.61, a diaphragm valve in Fig. 8.2.62, and a shear valve in Fig. 8.2.63.

Accumulators are effectively "hydraulic flywheels" which store potential energy by accumulating a quantity of pressurized hydraulic fluid

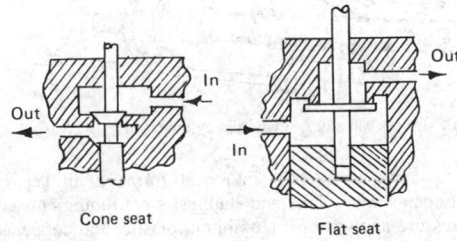

Fig. 8.2.61   Poppet valve.

in a suitable enclosed vessel. The bag type shown in Fig. 8.2.64 uses pressurized gas inside the bag working against the hydraulic fluid outside the bag. Figure 8.2.65 shows a piston accumulator.

Pressurized hydraulic fluid acting against an **actuator** or motor converts fluid pressure energy into mechanical energy. Motors providing

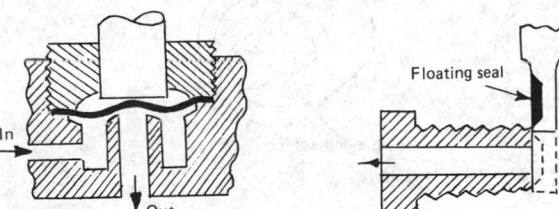

Fig. 8.2.62   Diaphragm valve.   **Fig. 8.2.63**   Shear valve.

continuous rotation have operating characteristics closely related to their pump counterparts. A linear actuator, or cylinder (Fig. 8.2.66), provides straight-line reciprocating motion; a rotary actuator (Fig. 8.2.67) provides arcuate oscillatory motion. Figure 8.2.68 shows a one-shot booster (linear motion) which can be used to deliver sprays through a nozzle.

Hydraulic fluids (liquids and air) are conducted in **pipe, tubing,** or **flexible hose.** Hose is used when the lines must flex or in applications in which fixed, rigid conduit is unsuitable. Table 8.2.37 lists SAE standard hoses. Maximum recommended operating pressure for a broad range of industrial applications is approximately 25 percent of rated bursting pressure. Due consideration must be given to the operating-temperature range; most applications fall in the range from − 40 to 200°F (− 40 to 95°C). Higher operating temperatures can be accommodated with appropriate materials.

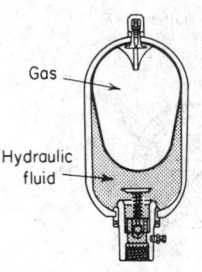

Fig. 8.2.64 Bag accumulator.

Hose fittings are of the screw-type or swaged, depending on the particular application and operating pressure and temperature. A broad variety of hose-end fittings is available from the industry.

Pipe has the advantage of being relatively cheap, is applied mainly in straight runs, and is usually of steel. Fittings for pipe are either standard pipe fittings for fairly low pressures or more elaborate ones suited to leak-proof high-pressure operation.

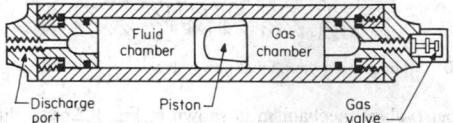

Fig. 8.2.65 Piston accumulator.

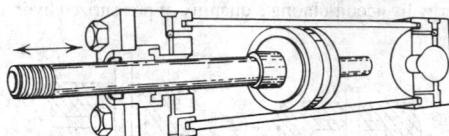

Fig. 8.2.66 Linear actuator or hydraulic cylinder.

Tubing is more easily bent into neat forms to fit between inlet and outlet connections. Steel and stainless-steel tubing is used for the highest-pressure applications; aluminum, plastic, and copper tubing is also used as appropriate for the operating conditions of pressure and temperature. Copper tubing hastens the oxidation of oil-base hydraulic fluids; accordingly, its use should be restricted either to air lines or with liquids which will not be affected by copper in the operating range.

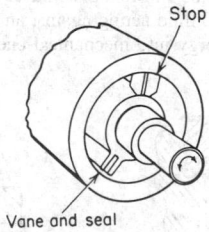

Fig. 8.2.67 Rotary actuator.

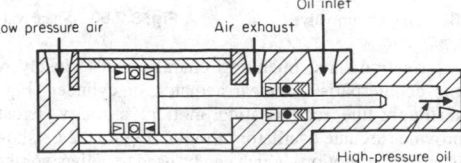

Fig. 8.2.68 One-shot booster.

Table 8.2.37 SAE Standard Hoses

100R1A	One-wire-braid reinforcement, synthetic rubber cover
100R1T	Same as R1A except with a thin, nonskive cover
100R2A	Two-wire-braid reinforcement, synthetic rubber cover
100R2B	Two spiral wire plus one wire-braid reinforcement, synthetic rubber cover
100R2AT	Same as R2A except with a thin, nonskive cover
100R2BT	Same as R2B except with a thin, nonskive cover
100R3	Two rayon-braid reinforcement, synthetic rubber cover
100R5	One textile braid plus one wire-braid reinforcement, textile braid cover
100R7	Thermoplastic tube, synthetic fiber reinforcement, thermoplastic cover (thermoplastic equivalent to SAE 100R1A)
100R8	Thermoplastic tube, synthetic fiber reinforcement, thermoplastic cover (thermoplastic equivalent to SAE 100R2A)
100R9	Four-ply, light-spiral-wire reinforcement, synthetic rubber cover
100R9T	Same as R9 except with a thin, nonskive cover
100R10	Four-ply, heavy-spiral-wire reinforcement, synthetic rubber cover
100R11	Six-ply, heavy-spiral-wire reinforcement, synthetic rubber cover

Tube fittings for permanent connections allow for brazed or welded joints. For temporary or separable applications, **flared** or **flareless fittings** are employed (Figs. 8.2.69 and 8.2.70). ANSI B116.1-1974 and B116.2-1974 pertain to tube fittings. The variety of fittings available is vast; the designer is advised to refer to manufacturers' literature for specifics.

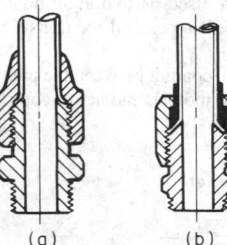

Fig. 8.2.69 Flared tube fittings. (*a*) A 45° flared fitting; (*b*) Triple-lok flared fitting. (*Parker-Hannafin Co.*)

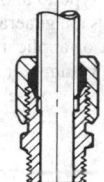

Fig. 8.2.70 Ferulok flareless tube fitting. (*Parker-Hannafin Co.*)

Parameters entering into the design of a hydraulic system are volume of flow per unit time, operating pressure and temperature, viscosity characteristics of the fluid within the operating range, and compatibility of the fluid/conduit material.

Flow velocity in suction lines is generally in the range of 1 to 5 ft/s (0.3 to 1.5 m/s); in discharge lines it ranges from 10 to 25 ft/s (3 to 8 m/s).

The pipe or tubing is under internal pressure. Selection of material and wall thickness follows from suitable equations (see Sec. 5). Safety factors range from 6 to 10 or higher, depending on the severity of the application (i.e., vibration, shock, pressure surges, possibility of physical abuse, etc.). JIC specifications provide a guide to the designer of hydraulic systems.

BRAKES

Brakes may be classified as: (1) rim type—internally expanding or externally contracting shoes, (2) band type, (3) cone type, (4) disk or axial type, (5) miscellaneous.

Rim Type—Internal Shoe(s) (Fig. 8.2.71)

$$F = \begin{cases} \dfrac{M_N - M\mu}{d} & \text{clockwise rotation} \\[2mm] \dfrac{M_N + M\mu}{d} & \text{counterclockwise rotation} \end{cases}$$

where

$$M\mu = \frac{\mu P_a Br}{\sin \theta_a} \int_{\theta_1}^{\theta_2} (\sin \theta)(r - d \cos \theta)\, d\theta$$

B = face width of frictional material; P_a = maximum pressure; θ_a = angle to point of maximum pressure (if $\theta_2 > 90°$; then $\theta_a = 90°$; $\theta_2 < 90°$, then $\theta_a = \theta_2$); μ = coefficient of friction; r = radius of drum; d = distance from drum center to brake pivot;

$$M_N = \frac{P_a Brd}{\sin \theta_a} \int_{\theta_1}^{\theta_2} \sin^2 \theta\, d\theta$$

and torque on drum is

$$T = \frac{\mu P_a Br^2(\cos \theta_1 - \cos \theta_2)}{\sin \theta_a}$$

Self-locking of the brake ($F = 0$) will occur for clockwise rotation when $M_N = M\mu$. This self-energizing phenomenon can be used to advantage without actual locking if μ is replaced by a larger value μ' so that $1.25\, \mu \le \mu' \le 1.50$, from which the pivot position a can be solved.

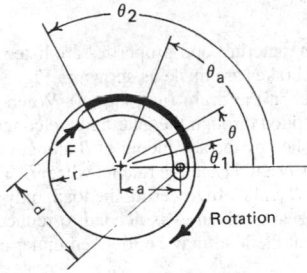

Fig. 8.2.71 Rim brake: internal friction shoe.

In automative use there are two shoes made to pivot in opposition, so that self-energization is present and can be used to great advantage (Fig. 8.2.72).

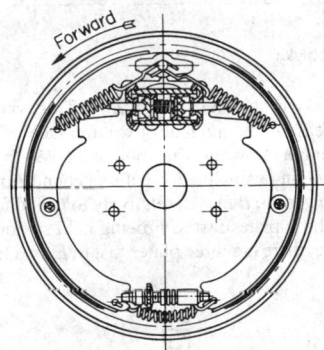

Fig. 8.2.72 Internal brake.

Rim Type—External Shoe(s) (Fig. 8.2.73)

The equations for M_N and $M\mu$ are the same as above:

$$F = \begin{cases} \dfrac{M_N + M\mu}{d} & \text{clockwise rotation} \\[2ex] \dfrac{M_N - M\mu}{d} & \text{counterclockwise rotation} \end{cases}$$

Self-locking ($F = 0$) can occur for counterclockwise rotation at $M_N = M\mu$.

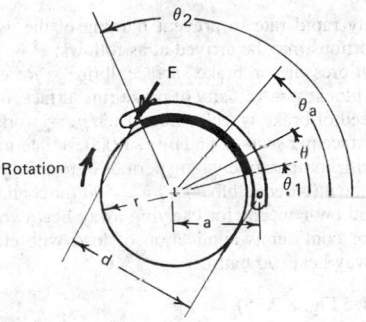

Fig. 8.2.73 Rim brake: external friction shoe.

Band Type (Fig. 8.2.74a, b, and c)

Flexible band brakes are used in power excavators and in hoisting. The bands are usually of an asbestos fabric, sometimes reinforced with copper wire and impregnated with asphalt.

In Fig. 8.2.74a, F = force at end of brake handle; P = tangential force at rim of wheel; f = coefficient of friction of materials in contact;

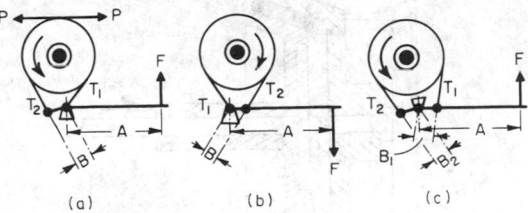

Fig. 8.2.74 Band brakes.

a = angle of wrap of band, deg; T_1 = total tension in band on tight side; T_2 = total tension in band on slack side. Then $T_1 - T_2 = P$ and $T_1/T_2 = 10^{0.0076fa} = 10^b$, where $b = 0.0076fa$. Also, $T_2 = P/(10^b - 1)$ and $T_1 = P \times 10^b/(10^b - 1)$. The values of $10^{0.0076fa}$ are given in Fig. 8.2.90 for a in radians.

For the arrangement shown in Fig. 8.2.74a,

$$FA = T_2 B = PB/(10^b - 1)$$
and
$$F = PB/[A(10^b - 1)]$$

For the construction illustrated in Fig. 8.2.74b,

$$F = PB/\{A[10^b/(10^b - 1)]\}$$

For the differential brake shown in Fig. 8.2.74c,

$$F = (P/A)[(B_2 - 10^b B_1)/(10^b - 1)]$$

In this arrangement, the quantity $10^b B_1$ must always be less than B_2, or the band will grip the wheel and the brake, or part of the mechanism to which it is attached, will rupture.

It is usual in practice to have the leverage ratio A/B for band brakes about 10:1.

If f for wood on iron is taken at 0.3 and the angle of wrap for the band is 270°, i.e., subtends three-fourths of the circumference, then $10^b = 4$ approx; the loads required for a given torque will be as follows for the cases just considered and for the leverage ratios stated above:

Band brake, Fig. 8.2.74a	$F = 0.033P$
Band brake, Fig. 8.2.74b	$F = 0.133P$
Band brake, Fig. 8.2.74c	$F = 0.016P$

In the case of Fig. 8.2.74c, the dimension B_2 must be greater than $B_1 \times 10^b$. Accordingly, B_1 is taken as ¼, A as 10, and, since $10^b = 4$, B_2 is taken as 1½.

The principal function of a brake is to absorb energy. This energy appears at the surface of the brake as heat, which must be carried away at a sufficiently rapid rate to prevent burning of the wooden blocks. Suitable proportions may be arrived at as follows:

Let p = unit pressure on brake surface, lb/in^2 = R (reaction against block)/area of block; v = velocity of brake rim surface, ft/s = $2\pi rn/60$, where n = speed of brake wheel, r/min. Then pv = work absorbed per in^2 of brake surface per second, and $pv \leq 1,000$ for intermittent applications of load with comparatively long periods of rest and poor means for carrying away heat (wooden blocks); $pv \leq 500$ for continuous application of load and poor means for carrying away heat (wooden blocks); $pv \leq 1,400$ for continuous application of load with effective means for carrying away heat (oil bath).

Cone Brake (see Fig. 8.2.75)

Uniform Wear

$$F = \frac{\pi P_a d}{2}(D - d)$$

$$T = \frac{\pi \mu P_a d}{8 \sin \alpha}(D^2 - d)$$

where P_a = maximum pressure occurring at $d/2$.

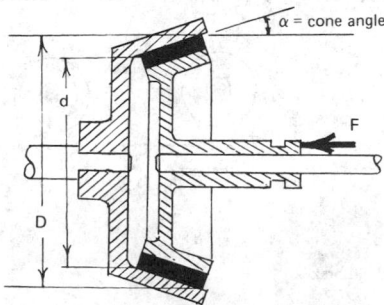

Fig. 8.2.75 Cone break.

Uniform Pressure

$$F = \frac{\pi P_a}{4}(D^2 - d^2)$$

$$T = \frac{\pi \mu P_a}{12 \sin \alpha}(D^3 - d^3) = \frac{F\mu}{3 \sin \alpha} \times \frac{D^3 - d^3}{D^2 - d^2}$$

Figure 8.2.76 shows a cone brake arrangement used for lowering heavy loads.

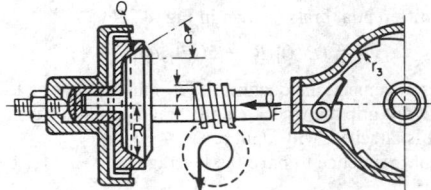

Fig. 8.2.76 Cone brake for lowering loads.

Disk Brakes (see Fig. 8.2.77)

Disk brakes are free from "centrifugal" effects, can have large frictional areas contained in small space, have better heat dissipation qualities than the rim type, and enjoy a favorable pressure distribution.

Uniform Wear

$$F = \frac{\pi P_a d}{2}(D - d)$$

$$T = \frac{F\mu}{4}(D + d)$$

Uniform Pressure

$$F = \frac{\pi P_a}{4}(D^2 - d^2)$$

$$T = \frac{F\mu}{3} \times \frac{D^3 - d^3}{D^2 - d^2}$$

These relations apply to a single surface of contact. For caliper disk, or multidisk brakes, the above relations are applied for each surface of contact.

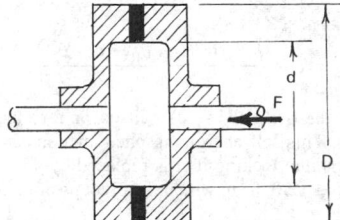

Fig. 8.2.77 Disk brake.

Selected friction materials and properties are listed in Table 8.2.38.

Frequently disk brakes are made as shown in Fig. 8.2.78. The pinion Q engages the gear in the drum (not shown). When the load is to be raised, power is applied through the gear and the connection between B and C is accomplished by the advancing of B along A and the clamping of the friction disks D and D and the ratchet wheel E. The reversal of the motor disconnects B and C. In lowering the load, only as much reversal of rotation of the gear is given as is needed to reduce the force in the friction disks so that the load may be lowered under control.

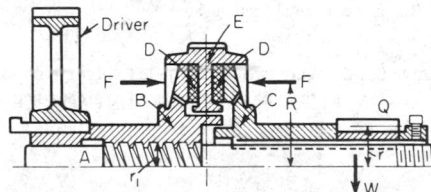

Fig. 8.2.78 Disk brake.

A **multidisk brake** is shown in Fig. 8.2.79. This type of construction results in an increase in the number of friction faces. The drum shaft is geared to the pinion A, while the motive power for driving comes through the gear G. In raising the load, direct connection is had between G, B, and A. In lowering, B moves relatively to G and forces the friction plates together, those plates fast to E being held stationary by the pawl on E. In the figure, there are three plates fast to E, one fast to G, and one fast to C.

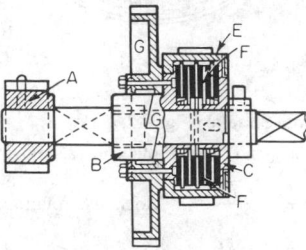

Fig. 8.2.79 Multidisk brake.

Eddy-current brakes (Fig. 8.2.80) are used with flywheels where quick braking is essential, and where large kinetic energy of the rotating

Table 8.2.38   Selected Friction Materials and Properties

Material	Opposing material	Friction coefficient			Max pressure		Max temperature	
		Dry	Wet	In oil	lb/in²	kPa	°F	°C
Sintered metal	Cast iron or steel	0.1–0.4	0.05–.01	0.05–0.08	150–250	1,000–1,720	450–1,250	232–677
Wood	Cast iron or steel	0.2–0.35	0.16	0.12–0.16	60–90	400–620	300	149
Leather	Cast iron or steel	0.3–0.5	0.12		10–40	70–280	200	93
Cork	Cast iron or steel	0.3–0.5	0.15–0.25	0.15–0.25	8–14	55–95	180	82
Felt	Cast iron or steel	0.22	0.18		5–10	35–70	280	138
Asbestos-woven	Cast iron or steel	0.3–0.6	0.1–0.2	0.08–0.10	50–100	350–700	400–500	204–260
Asbestos-molded	Cast iron or steel	0.2–0.5	0.08–0.12	0.06–0.09	50–150	350–1,000	400–500	204–260
Asbestos-impregnated	Cast iron or steel	0.32	0.12					
Cast iron	Cast iron	0.15–0.20	0.05	0.03–0.06	150–250	1,000–1,720	500	260
Cast iron	Steel			0.03–0.06	100–250	690–1,720	500	260
Graphite	Steel	0.25	0.05–0.1	0.12 (av)	300	2,100	370–540	188–282

masses precludes the use of block brakes due to excessive heating, as in reversible rolling mills. A number of poles *a* are electrically excited (north and south in turn) and create a magnetic flux which permeates the gap and the iron of the rim, causing eddy current. The flywheel energy

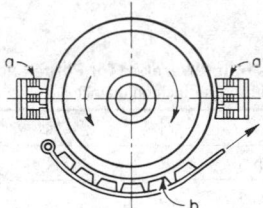

Fig. 8.2.80   Eddy-current brake.

is converted through these currents into heat. The hand brake *b* may be used for quicker stopping when the speed of the wheel is considerably decreased; i.e., when the eddy-current brake is inefficient. Two brakes are provided to avoid bending forces on the shaft.

Electric brakes are often used in cranes, bridges, turntables, and machine tools, where an automatic application of the brake is important as soon as power is cut off. The brake force is supplied by an adjustable spring which is counteracted by the force of a solenoid or a centrifugal thrustor. Interruption of current automatically applies the spring-activated brake shoes. Figures 8.2.81 and 8.2.82 show these types of electric brake.

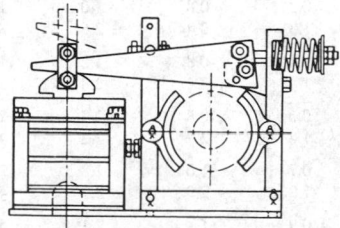

Fig. 8.2.81   Solenoid-type electric brake.

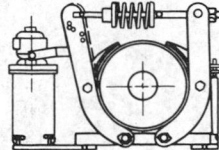

Fig. 8.2.82   Thrustor-type electric brake.

SHRINK, PRESS, DRIVE, AND RUNNING FITS

Inch Systems   ANSI B4.1-1967 (R87) recommends preferred sizes, allowances, and tolerances for fits between plain cylindrical parts. Such fits include bearing, shrink and drive fits, etc. Terms used in describing fits are defined as follows: **Allowance:** minimum clearance (positive allowance) or maximum interference (negative allowance) between mating parts. **Tolerance:** total permissible variation of size. **Limits of size:** applicable maximum and minimum sizes. **Clearance fit:** one having limits of size so prescribed that a clearance always results when mating parts are assembled. **Interference fit:** In this case, limits are so prescribed that interference always results on assembly. **Transition fit:** This may have either a clearance or an interference on assembly. **Basic size:** one from which limits of size are derived by the application of allowances and tolerances. **Unilateral tolerance:** In this case a variation in size is permitted in only one direction from the basic size.

Fits are divided into the following general classifications: (1) running and sliding fits, (2) locational clearance fits, (3) transition fits, (4) locational interference fits, and (5) force or shrink fits.

1. **Running and sliding fits.**   These are intended to provide similar running performance with suitable lubrication allowance throughout the range of sizes. These fits are further subdivided into the following classes:

Class RC1: close-sliding fits.   Intended for accurate location of parts which must assemble without perceptible play.

Class RC2: sliding fits.   Parts made with this fit move and turn easily but are not intended to run freely; also, in larger sizes they may seize under small temperature changes.

Class RC3: precision-running fits.   These are intended for precision work at slow speeds and light journal pressures but are not suitable where appreciable temperature differences are encountered.

Class RC4: close-running fits.   For running fits on accurate machinery with moderate surface speeds and journal pressures, where accurate location and minimum play is desired.

Classes RC5 and RC6: medium-running fits.   For higher running speeds or heavy journal pressures.

Class RC7: free-running fits.   For use where accuracy is not essential, or where large temperature variations are likely to be present, or both.

Classes RC8 and RC9: loose-running fits.   For use with materials such as cold-rolled shafting or tubing made to commercial tolerances.

Limits of clearance given in ANSI B4.1-1967 (R87) for each of these classes are given in Table 8.2.39. Hole and shaft tolerances are listed on a unilateral tolerance basis in this reference to give the clearance limits of Table 8.2.39, the hole size being the basic size.

2. **Locational clearance fits.**   These are intended for normally stationary parts which can, however, be freely assembled or disassembled. These are subdivided into various classes which run from snug fits for parts requiring accuracy of location, through medium clearance fits (spigots) to the looser fastener fits where freedom of assembly is of prime importance.

Table 8.2.39 Limits of Clearance for Running and Sliding Fits (Basic Hole)
(Limits are in thousandths of an inch on diameter)

Nominal size range, in	Class								
	RC1	RC2	RC3	RC4	RC5	RC6	RC7	RC8	RC9
0–0.12	0.1 0.45	0.1 0.55	0.3 0.95	0.3 1.3	0.6 1.6	0.6 2.2	1.0 2.6	2.5 5.1	4.0 8.1
0.12–0.24	1.5 0.5	0.15 0.65	0.4 1.2	0.4 1.6	0.8 2.0	0.8 2.7	1.2 3.1	2.8 5.8	4.5 9.0
0.24–0.40	0.2 0.6	0.2 0.85	0.5 1.5	0.5 2.0	1.0 2.5	1.0 3.3	1.6 3.9	3.0 6.6	5.0 10.7
0.40–0.71	0.25 0.75	0.25 0.95	0.6 1.7	0.6 2.3	1.2 2.9	1.2 3.8	2.0 4.6	3.5 7.9	6.0 12.8
0.71–1.19	0.3 0.95	0.3 1.2	0.8 2.1	0.8 2.8	1.6 3.6	1.6 4.8	2.5 5.7	4.5 10.0	7.0 15.5
1.19–1.97	0.4 1.1	0.4 1.4	1.0 2.6	1.0 3.6	2.0 4.6	2.0 6.1	3.0 7.1	5.0 11.5	8.0 18.0
1.97–3.15	0.4 1.2	0.4 1.6	1.2 3.1	1.2 4.2	2.5 5.5	2.5 7.3	4.0 8.8	6.0 13.5	9.0 20.5
3.15–4.73	0.5 1.5	0.5 2.0	1.4 3.7	1.4 5.0	3.0 6.6	3.0 8.7	5.0 10.7	7.0 15.5	10.0 24.0

3. **Transition fits.** These are for applications where accuracy of location is important, but a small amount of either clearance or interference is permissible.

4. **Locational interference fits.** Used where accuracy of location is of prime importance and for parts requiring rigidity and alignment with no special requirements for bore pressure.

Data on clearance limits, interference limits, and hole and shaft diameter tolerances for locational clearance fits, transition fits, and locational interference fits are given in ANSI B4.1-1967 (R87).

5. **Force or shrink fits.** These are characterized by approximately constant bore pressures throughout the range of sizes; interference varies almost directly as the diameter, and the differences between maximum and minimum values of interference are small. These are divided into the following classes:

Class FN1: light-drive fits. For applications requiring light assembly pressures (thin sections, long fits, cast-iron external members).

Class FN2: medium-drive fits. Suitable for ordinary steel parts or for shrink fits on light sections. These are about the tightest fits that can be used on high-grade cast-iron external members.

Class FN3: heavy-drive fits. For heavier steel parts or shrink fits in medium sections.

Classes FN4 and FN5: force fits. These are suitable for parts which can be highly stressed. Shrink fits are used instead of press fits in cases where the heavy pressing forces required for mounting are impractical.

In Table 8.2.40 are listed the limits of interference (maximum and minimum values) for the above classes of force or shrink fits for various diameters, as given in ANSI B4.1-1967 (R87). Hole and shaft tolerances to give these interference limits are also listed in this reference.

Metric System ANSI B4.2-1978 (R94) and ANSI B4.3-1978 (R94) define limits and fits for cylindrical parts, and provide tables listing preferred values.

The standard ANSI B4.2-1978 (R94) is essentially in accord with ISO R286.

ANSI B4.2 provides 22 basic deviations, each for the shaft (a to z plus js), and the hole (A to Z plus Js). International has 18 tolerance grades: IT 01, IT 0, and IT 1 through 16.

IT grades are roughly applied as follows: measuring tools, 01 to 7; fits, 5 to 11; material, 8 to 14; and large manufacturing tolerances, 12 to 16. See Table 8.2.42 for metric preferred fits.

Basic size—The basic size is the same for both members of a fit, and is the size to which limits or deviations are assigned. It is designated by 40 in 40H7.

Table 8.2.40 Limits of Interference for Force and Shrink Fits
(Limits are in thousandths of an inch on diameter)

Nominal size range, in	Class				
	FN 1	FN 2	FN 3	FN 4	FN 5
0.04–0.12	0.05 0.5	0.2 0.85		0.3 0.95	0.3 1.3
0.12–0.24	0.1 0.6	0.2 1.0		0.95 1.2	1.3 1.7
0.24–0.40	0.1 0.75	0.4 1.4		0.6 1.6	0.5 2.0
0.40–0.56	0.1 0.8	0.5 1.6		0.7 1.8	0.6 2.3
0.56–0.71	0.2 0.9	0.5 1.6		0.7 1.8	0.8 2.5
0.71–0.95	0.2 1.1	0.6 1.9		0.8 2.1	1.0 3.0
0.95–1.19	0.3 1.2	0.6 1.9	0.8 2.1	1.0 2.3	1.3 3.3
1.19–1.58	0.3 1.3	0.8 2.4	1.0 2.6	1.5 3.1	1.4 4.0
1.58–1.97	0.4 1.4	0.8 2.4	1.2 2.8	1.8 3.4	2.4 5.0
1.97–2.56	0.6 1.8	0.8 2.7	1.3 3.2	2.3 4.2	3.2 6.2
2.56–3.15	0.7 1.9	1.0 2.9	1.8 3.7	2.8 4.7	4.2 7.2
3.15–3.94	0.9 2.4	1.4 3.7	2.1 4.4	3.6 5.9	4.8 8.4
3.94–4.73	1.1 2.6	1.6 3.9	2.6 4.9	4.6 6.9	5.8 9.4
4.73–5.52	1.2 2.9	1.9 4.5	3.4 6.0	5.4 8.0	7.5 11.6
5.52–6.30	1.5 3.2	2.4 5.0	3.4 6.0	5.4 8.0	9.5 13.6
6.30–7.09	1.8 3.5	2.9 5.5	4.4 7.0	6.4 9.0	9.5 13.6

Deviation—The algebraic difference between a size and the corresponding basic size.

Upper deviation—The algebraic difference between the maximum limit of size and the corresponding basic size.

Lower deviation—The algebraic difference between the minimum limit of size and the corresponding basic size.

Fundamental deviation—That one of the two deviations closest to the basic size. It is designated by the letter H in 40H7.

Tolerance—The difference between the maximum and minimum size limits on a part.

International tolerance grade (IT)—A group of tolerances which vary depending on the basic size, but which provide the same relative level of accuracy within a grade. It is designated by 7 in 40H7 (IT 7).

Tolerance zone—A zone representing the tolerance and its position in relation to the basic size. The symbol consists of the fundamental deviation letter and the tolerance grade number (i.e., H7).

Hole basis—The system of fits where the minimum hole size is basic. The fundamental deviation for a hole basis system is H.

Shaft basis—Maximum shaft size is basic in this system. Fundamental deviation is h. NOTE: Capital letters refer to the hole and lowercase letters to the shaft.

Clearance fit—A fit in which there is clearance in the assembly for all tolerance conditions.

Interference fit—A fit in which there is interference for all tolerance conditions. Table 8.2.41 lists preferred metric sizes. Table 8.2.42 lists preferred tolerance zone combinations for clearance, transition and interference fits. Table 8.2.43 lists dimensions for the grades corresponding to preferred fits. Table 8.2.44a and b lists limits (numerical) of preferred hole-basis clearance, transitions, and interference fits.

Stresses Produced by Shrink or Press Fit

STEEL HUB ON STEEL SHAFT.   The maximum equivalent stress, pounds per square inch, set up by a given press-fit allowance (in inches per inch of shaft diameter) is equal to $3x \times 10^7$, where x is the allowance per inch of shaft diameter (Baugher, *Trans. ASME,* 1931, p. 85). The press-fit pressures set up between a steel hub and shaft, for various ratios d/D between shaft and hub outside diameters, are given in Fig. 8.2.83. These

Table 8.2.41   Preferred Sizes (Metric)

Basic size, mm		Basic size, mm		Basic size, mm	
First choice	Second choice	First choice	Second choice	First choice	Second choice
1		10		100	
	1.1		11		110
1.2		12		120	
	1.4		14			140
1.6		16		160	
	1.8		18			180
2		20		200	
	2.2		22			220
2.5		25		250	
	2.8		28			280
3		30		300	
	3.5		35			350
4		40		400	
	4.5		45			450
5		50		500	
	5.5		55			550
6		60		600	
7			70		700
8		80		800	
9			90		900
				1000	

SOURCE: ANSI B 4.2-1978 (R94), reproduced by permission.

curves are accurate to 5 percent even if the shaft is hollow, provided the inside shaft diameter is not over 25 percent of the outside. The equivalent stress given above is based on the maximum shear theory and is numerically equal to the radial-fit pressure added to the tangential tension in the hub. Where the shaft is hollow, with an inside diameter equal to more than about 25 percent of the outside diameter, the allowance in inches per inch to obtain an equivalent hub stress of 30,000 lb/in² may be determined by using Lamé's thick-cylinder formulas (*Jour. Appl.*

Table 8.2.42   Description of Preferred Fits (Metric)

ISO symbol		
Hole basis	Shaft basis	Description
		Clearance fits
H11/c11	C11/h11	*Loose-running* fit for wide commercial tolerances or allowances on external members.
H9/d9	D9/h9	*Free-running* fit not for use where accuracy is essential, but good for large temperature variations, high running speeds, or heavy journal pressures.
H8/f7	F8/h7	*Close-running* fit for running on accurate machines and for accurate location at moderate speeds and journal pressures.
H7/g6	G7/h6	*Sliding fit* not intended to run freely, but to move and turn freely and locate accurately.
H7/h6	H7/h6	*Locational clearance* fit provides snug fit for locating stationary parts; but can be freely assembled and disassembled.
		Transition fits
H7/k6	K7/h6	*Locational transition* fit for accurate location, a compromise between clearance and interference.
H7/n6	N7/h6	*Locational transition* fit for more accurate location where greater interference is permissible.
		Interference fits
H7/p6*	P7/h6	*Locational interference* fit for parts requiring rigidity and alignment with prime accuracy of location but without special bore pressure requirements.
H7/s6	S7/h6	*Medium-drive* fit for ordinary steel parts or shrink fits on light sections, the tightest fit usable with cast iron.
H7/u6	U7/h6	*Force* fit suitable for parts which can be highly stressed or for shrink fits where the heavy pressing forces required are impractical.

More clearance →

More interference →

* Transition fit for basic sizes in range from 0 through 3 mm.
SOURCE: ANSI B4.2-1978 (R94), reproduced by permission.

Table 8.2.43 International Tolerance Grades

Basic sizes		Tolerance grades, mm					
Over	Up to and including	IT6	IT7	IT8	IT9	IT10	IT11
0	3	0.006	0.010	0.014	0.025	0.040	0.060
3	6	0.008	0.012	0.018	0.030	0.048	0.075
6	10	0.009	0.015	0.022	0.036	0.058	0.090
10	18	0.011	0.018	0.027	0.043	0.070	0.110
18	30	0.013	0.021	0.033	0.052	0.084	0.130
30	50	0.016	0.025	0.039	0.062	0.100	0.160
50	80	0.019	0.030	0.046	0.074	0.120	0.190
80	120	0.022	0.035	0.054	0.087	0.140	0.220
120	180	0.025	0.040	0.063	0.100	0.160	0.250
180	250	0.029	0.046	0.072	0.115	0.185	0.290
250	315	0.032	0.052	0.081	0.130	0.210	0.320
315	400	0.036	0.057	0.089	0.140	0.230	0.360
400	500	0.040	0.063	0.097	0.155	0.250	0.400
500	630	0.044	0.070	0.110	0.175	0.280	0.440
630	800	0.050	0.080	0.125	0.200	0.320	0.500
800	1,000	0.056	0.090	0.140	0.230	0.360	0.560
1,000	1,250	0.066	0.105	0.165	0.260	0.420	0.660
1,250	1,600	0.078	0.125	0.195	0.310	0.500	0.780
1,600	2,000	0.092	0.150	0.230	0.370	0.600	0.920
2,000	2,500	0.110	0.175	0.280	0.440	0.700	1.100
2,500	3,150	0.135	0.210	0.330	0.540	0.860	1.350

SOURCE: ANSI B4.2-1978 (R94), reproduced by permission.

Mech., 1937, p. A-185). It should be noted that these curves hold only when the maximum equivalent stress is below the yield point; above the yield point, plastic flow occurs and the stresses are less than calculated.

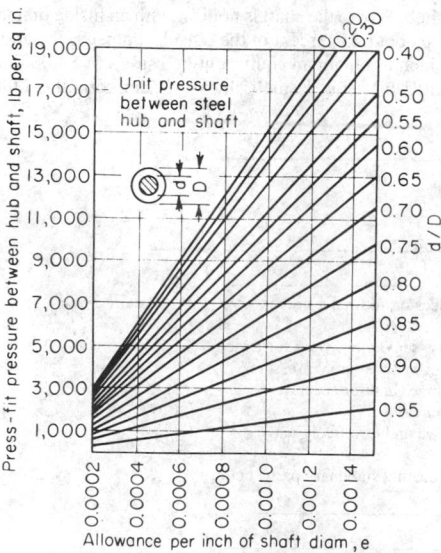

Fig. 8.2.83 Press-fit pressures between steel hub and shaft.

Cast-Iron Hub on Steel Shaft Where the shaft is solid, or hollow with an inside diameter not over 25 percent of the outside diameter, Fig. 8.2.84 may be used to determine maximum tensile stresses in the cast-iron hub, resulting from the press-fit allowance; for various ratios *d/D*, Fig. 8.2.85 gives the press-fit pressures. These curves are based on a modulus of elasticity of 30×10^6 lb/in² for steel and 15×10^6 for cast iron. For a hollow shaft with an inside diameter more than about ¼ the outside, the Lamé formulas may be used.

Pressure Required in Making Press Fits The force required to press a hub on the shaft is given by $\pi f p d l$, where *l* is length of fit, *p* the unit press-fit pressure between shaft and hub, *f* the coefficient of fric-

tion, and *d* the shaft diameter. Values of *f* varying from 0.03 to 0.33 have been reported, the lower values being due to yielding of the hub as a consequence of too high a fit allowance; the average is around 0.10 to 0.15. (For additional data see Horger and Nelson, "Design Data and Methods," ASME, 1953, pp. 87–91.)

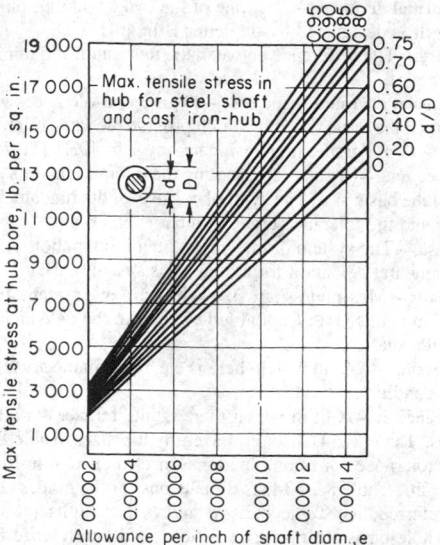

Fig. 8.2.84 Variation of tensile stress in cast-iron hub in press-fit allowance.

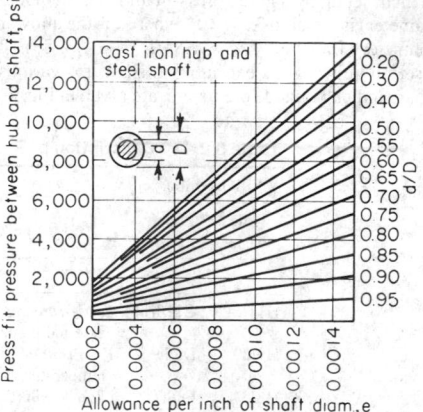

Fig. 8.2.85 Press-fit pressures between cast-iron hub and shaft.

Torsional Holding Ability The torque required to cause complete slippage of a press fit is given by $T = \frac{1}{2}\pi f p l d^2$. Local slippage will usually occur near the end of the fit at much lower torques. If the torque is alternating, stress concentration and rubbing corrosion will occur at the hub face so that, eventually, fatigue failure may occur at considerably lower torques. Only in cases of static torque application is it justifiable to use ultimate torque as a basis for design.

A designer can often improve shrink-, press-, and slip-fit cylindrical assemblies with adhesives. When applied, adhesives can achieve high frictional force with attendant greater torque transmission without extra bulk, and thus augment or even replace press fits, compensate for differential thermal expansion, make fits with leakproof seals, eliminate backlash and clearance, etc.

The adhesives used are the anaerobic (see Sec. 6.8, "Adhesives") variety, such as Loctite products. Such adhesives destabilize and tend to harden when deprived of oxygen. Design suggestions on the use of such adhesives appear in industrial catalogs.

SHAFTS, AXLES, AND CRANKS

Most shafts are subject to combined bending and torsion, either of which may be steady or variable. Impact conditions, such as sudden starting and stopping, will cause momentary peak stresses greater than those related to the steady or variable portions of operation.

Design of shafts requires a theory of failure to express a stress in terms of loads and shaft dimensions, and an allowable stress as fixed by material strength and safety factor. Maximum shear theory of failure and distortion energy theory of failure are the two most commonly used in shaft design. Material strengths can be estimated from any one of several analytic representations of combined-load fatigue test data, starting from the linear (Soderberg, modified Goodman) which tend to give conservative designs to the nonlinear (Gerber parabolic, quadratic, Kececioglu, Bagci) which tend to give less conservative designs.

When linear representations of material strengths are used, and where both bending and torsion stresses have steady and variable components, the maximum shear theory and the distortion energy theory lead to somewhat similar formulations:

$$d = \left\{ \varepsilon \, \frac{n}{\pi} \left[\left(\frac{T_a}{S_{se}} + \frac{T_m}{S_{sy}} \right)^2 + \left(\frac{M_a}{S_{se}} + \frac{M_m}{S_{sy}} \right)^2 \right]^{1/2} \right\}^{1/3}$$

where $\varepsilon = 32$ (maximum shear theory) or 48 (distortion energy theory); d = shaft diameter; n = safety factor; T_a = amplitude torque = $(T_{max} - T_{min})/2$; T_m = mean torque = $(T_{max} + T_{min})/2$; M_a = amplitude bending moment = $(M_{max} - M_{min})/2$; M_m = mean bending moment = $(M_{max} + M_{min})/2$; S_{sy} = yield point in shear; S_{se} = completely corrected shear endurance limit = $S'_{se}k_ak_bk_ck_d/K_f$; S'_{se} = statistical average endurance limit of mirror finish, standard size, laboratory test specimen at standard room temperature; k_a = surface factor, a decimal to adjust S'_e for other than mirror finish; k_b = size factor, a decimal to adjust S'_e for other than standard test size; k_c = reliability factor, a decimal to adjust S'_{se} to other than its implied statistical average of 50 percent safe, 50 percent fail rate (50 percent reliability); k_d = temperature factor, decimal, to adjust S'_{se} to other than room temperature; $K_f = 1 + q(K_t - 1)$ = actual or fatigue stress concentration factor, a number greater than unity to adjust the nominal stress implied by T_a and M_a to a peak stress as induced by stress-raising conditions such as holes, fillets, keyways, press fits, etc. (K_f for T_a need not necessarily be the same as K_f for M_a); q = notch sensitivity; K_t = theoretical or geometric stress concentration factor.

For specific values of endurance limits and various factors the reader is referred to the technical literature (e.g., ASTM, NASA technical reports, ASME technical papers) or various books [e.g., "Machinery's Handbook" (Industrial Press, New York) and machine design textbooks].

If one allows for a variation of at least 15 percent, the following approximation is useful for the endurance limit in bending: $S'_e = 0.5S_{ut}$. This becomes for maximum shear theory $S'_{se} = 0.5(0.5S_{ut})$ and for distortion energy theory $S'_{se} = 0.577(0.5S_{ut})$.

Representative or approximate values for the various factors mentioned above were abstracted from Shigley, "Mechanical Engineering Design," McGraw-Hill, and appear below with permission.

Surface Factor k_a

Surface condition	S_{ut}, kips			
	60	120	180	240
Polished	1.00	1.00	1.00	1.00
Ground	0.89	0.89	0.89	0.89
Machined or cold-drawn	0.84	0.71	0.66	0.63
Hot-rolled	0.70	0.50	0.39	0.31
As forged	0.54	0.36	0.27	0.20

Size Factor k_b

$$k_b = \begin{cases} 0.869d^{-0.097} & 0.3 \text{ in} < d \leq 10 \text{ in} \\ 1 & d \leq 0.3 \text{ in or } d \leq 8 \text{ mm} \\ 1.189d^{-0.097} & 8 \text{ mm} < d < 250 \text{ mm} \end{cases}$$

Reliability Factor k_c

Reliability, %	k_c
50	1.00
90	0.89
95	0.87
99	0.81

Temperature Factor k_d

Temperature		k_d
°F	°C	
840	450	1.00
940	482	0.71
1,020	550	0.42

Notch Sensitivity q

S_{ut}, kips	Notch radius r, in			
	0.02	0.06	0.10	0.14
60	0.56	0.70	0.74	0.78
100	0.68	0.79	0.83	0.85
150	0.80	0.90	0.91	0.92
200	0.90	0.95	0.96	0.96

See Table 8.2.45 for fatigue stress concentration factors for plain press fits.

A general representation of material strengths (Marin, Design for Fatigue Loading, *Mach. Des.* 29, no. 4, Feb. 21, 1957, pp. 128–131, and series of the same title) is given as

$$\left(\frac{S_a}{S_e} \right)^m + \left(\frac{KS_m}{S_{ut}} \right)^p = 1$$

where S_a = variable portion of material strength; S_m = mean portion of material strength; S_e = adjusted endurance limit; S_{ut} = ultimate strength. Table 8.2.46 lists the constants m, K, and P for various failure criteria.

For purposes of design, safety factors are introduced into the equation resulting in:

$$\left(\frac{\sigma'_{a,p}}{S_e/n_{se}} \right)^m + \left(\frac{K\sigma'_{m,p}}{S_{ut}/n_{ut}} \right)^P = 1$$

where

$$\sigma'_{a,p} = \frac{1}{\pi d^3} \sqrt{(32n_{Ma}M_a)^2 + 3(16n_{\tau a}T_a)^2}$$

$$\sigma'_{m,p} = \frac{1}{\pi d^3} \sqrt{(32n_{Mm}M_m)^2 + 3(16n_{\tau m}T_m)^2}$$

and n_{ij} = safety factor pertaining to a particular stress (that is, n_a = safety factor for amplitude shear stress).

Stiffness of shafting may become important where critical speeds, vibration, etc., may occur. Also, the lack of sufficient stiffness in shafts may give rise to bearing troubles. Critical speeds of shafts in torsion or bending and shaft deflections may be calculated using the methods of Sec. 5. For shafts of variable diameter see Spotts, "Design of Machine Elements," Prentice-Hall. In order to avoid trouble where sleeve bearings are used, the angular deflections at the bearings in general must be kept within certain limits. One rule is to make the shaft deflection over the bearing width equal to a small fraction of the oil-film thickness. Note that since stiffness is proportional to the modulus of elasticity, alloy-steel shafts are no stiffer than carbon-steel shafts of the same diameter.

Crankshafts For calculating the torsional stiffness of crankshafts, the formulas given in Sec. 5 may be used.

Table 8.2.44a Limits of Fits
(Dimensions in mm)

Preferred hole basis clearance fits

Basic size		Loose-running			Free-running			Close-running			Sliding			Locational clearance		
		Hole H11	Shaft c11	Fit	Hole H9	Shaft d9	Fit	Hole H8	Shaft f7	Fit	Hole H7	g6	Fit	Hole H7	Shaft h6	Fit
1	max	1.060	0.940	0.180	1.025	0.980	0.070	1.014	0.994	0.030	1.010	0.998	0.018	1.010	1.000	0.016
	min	1.000	0.880	0.060	1.000	0.955	0.020	1.000	0.984	0.006	1.000	0.992	0.002	1.000	0.994	0.000
1.2	max	1.260	1.140	0.180	1.225	1.180	0.070	1.214	1.194	0.030	1.210	1.198	0.018	1.210	1.200	0.016
	min	1.200	1.080	0.060	1.200	1.155	0.020	1.200	1.184	0.006	1.200	1.192	0.002	1.200	1.194	0.000
1.6	max	1.660	1.540	0.180	1.625	1.580	0.070	1.614	1.594	0.030	1.610	1.598	0.018	1.610	1.600	0.016
	min	1.600	1.480	0.060	1.600	1.555	0.020	1.600	1.584	0.006	1.600	1.592	0.002	1.600	1.594	0.000
2	max	2.060	1.940	0.180	2.025	1.980	0.070	2.014	1.994	0.030	2.010	1.998	0.018	2.010	2.000	0.016
	min	2.000	1.880	0.060	2.000	1.955	0.020	2.000	1.984	0.006	2.000	1.992	0.002	2.000	1.994	0.000
2.5	max	2.560	2.440	0.180	2.525	2.480	0.070	2.514	2.494	0.030	2.510	2.498	0.018	2.510	2.500	0.016
	min	2.500	2.380	0.060	2.500	2.455	0.020	2.500	2.484	0.006	2.500	2.492	0.002	2.500	2.494	0.000
3	max	3.060	2.940	0.180	3.025	2.980	0.070	3.014	2.994	0.030	3.010	2.998	0.018	3.010	3.000	0.016
	min	3.000	2.880	0.060	3.000	2.955	0.020	3.000	2.984	0.006	3.000	2.992	0.002	3.000	2.994	0.000
4	max	4.075	3.930	0.220	4.030	3.970	0.090	4.018	3.990	0.040	4.012	3.996	0.024	4.012	4.000	0.020
	min	4.000	3.855	0.070	4.000	3.940	0.030	4.000	3.978	0.010	4.000	3.988	0.004	4.000	3.992	0.000
5	max	5.075	4.930	0.220	5.030	4.970	0.090	5.018	4.990	0.040	5.012	4.996	0.024	5.012	5.000	0.020
	min	5.000	4.855	0.070	5.000	4.940	0.030	5.000	4.978	0.010	5.000	4.988	0.004	5.000	4.992	0.000
6	max	6.075	5.930	0.220	6.030	5.970	0.090	6.018	5.990	0.040	6.012	5.996	0.024	6.012	6.000	0.020
	min	6.000	5.855	0.070	6.000	5.940	0.030	6.000	5.978	0.010	6.000	5.988	0.004	6.000	5.992	0.000
8	max	8.090	7.920	0.260	8.036	7.960	0.112	8.022	7.987	0.050	8.015	7.995	0.029	8.015	8.000	0.024
	min	8.000	7.830	0.080	8.000	7.924	0.040	8.000	7.972	0.013	8.000	7.986	0.005	8.000	7.991	0.000
10	max	10.090	9.920	0.260	10.036	9.960	0.112	10.022	9.987	0.050	10.015	9.995	0.029	10.015	10.000	0.024
	min	10.000	9.830	0.080	10.000	9.924	0.040	10.000	9.972	0.013	10.000	9.986	0.005	10.000	9.991	0.000
12	max	12.110	11.905	0.315	12.043	11.950	0.136	12.027	11.984	0.061	12.018	11.994	0.035	12.018	12.000	0.029
	min	12.000	11.795	0.095	12.000	11.907	0.050	12.000	11.966	0.016	12.000	11.983	0.006	12.000	11.989	0.000
16	max	16.110	15.905	0.315	16.043	15.950	0.136	16.027	15.984	0.061	16.018	15.994	0.035	16.018	16.000	0.029
	min	16.000	15.795	0.095	16.000	15.907	0.050	16.000	15.966	0.016	16.000	15.983	0.006	16.000	15.989	0.000
20	max	20.130	19.890	0.370	20.052	19.935	0.169	20.033	19.980	0.074	20.021	19.993	0.041	20.021	20.000	0.034
	min	20.000	19.760	0.110	20.000	19.883	0.065	20.000	19.959	0.020	20.000	19.980	0.007	20.000	19.987	0.000
25	max	25.130	24.890	0.370	25.052	24.935	0.169	25.033	24.980	0.074	25.021	24.993	0.041	25.021	25.000	0.034
	min	25.000	24.760	0.110	25.000	24.883	0.065	25.000	24.959	0.020	25.000	24.980	0.007	25.000	24.987	0.000
30	max	30.130	29.890	0.370	30.052	29.935	0.169	30.033	29.980	0.074	30.021	29.993	0.041	30.021	30.000	0.034
	min	30.000	29.760	0.110	30.000	29.883	0.065	30.000	29.959	0.020	30.000	29.980	0.007	30.000	29.987	0.000
40	max	40.160	39.880	0.440	40.062	39.920	0.204	40.039	39.975	0.089	40.025	39.991	0.050	40.025	40.000	0.041
	min	40.000	39.720	0.120	40.000	39.858	0.080	40.000	39.950	0.025	40.000	39.975	0.009	40.000	39.984	0.000
50	max	50.160	49.870	0.450	50.062	49.920	0.204	50.039	49.975	0.089	50.025	49.991	0.050	50.025	50.000	0.041
	min	50.000	49.710	0.130	50.000	49.858	0.080	50.000	49.950	0.025	50.000	49.975	0.009	50.000	49.984	0.000
60	max	60.190	59.860	0.520	60.074	59.900	0.248	60.046	59.970	0.106	60.030	59.990	0.059	60.030	60.000	0.049
	min	60.000	59.670	0.140	60.000	59.826	0.100	60.000	59.940	0.030	60.000	59.971	0.010	60.000	59.981	0.000
80	max	80.190	79.850	0.530	80.074	79.900	0.248	80.046	79.970	0.106	80.030	79.990	0.059	80.030	80.000	0.049
	min	80.000	79.660	0.150	80.000	79.826	0.100	80.000	79.940	0.030	80.000	79.971	0.010	80.000	79.981	0.000
100	max	100.220	99.830	0.610	100.087	99.880	0.294	100.054	99.964	0.125	100.035	99.988	0.069	100.035	100.000	0.057
	min	100.000	99.610	0.170	100.000	99.793	0.120	100.000	99.929	0.036	100.000	99.966	0.012	100.000	99.978	0.000
120	max	120.220	119.820	0.620	120.087	119.880	0.294	120.054	119.964	0.125	120.035	119.988	0.069	120.035	120.000	0.057
	min	120.000	119.600	0.180	120.000	119.793	0.120	120.000	119.929	0.036	120.000	119.966	0.012	120.000	119.978	0.000
160	max	160.250	159.790	0.710	160.100	159.855	0.345	160.063	159.957	0.146	160.040	159.986	0.079	160.040	160.000	0.065
	min	160.000	159.540	0.210	160.000	159.755	0.145	160.000	159.917	0.043	160.000	159.961	0.014	160.000	159.975	0.000

Table 8.2.44b Limits of Fits
(Dimensions in mm)

Preferred hole basis transition and interference fits

Basic size		Locational transition Hole H7	Shaft k6	Fit	Hole H7	Shaft n6	Fit	Locational interference Hole H7	Shaft p6	Fit	Medium drive Hole H7	Shaft s6	Fit	Force Hole H7	Shaft u6	Fit
1	max	1.010	1.006	0.010	1.010	1.010	0.006	1.010	1.012	0.004	1.010	1.020	−0.004	1.010	1.024	−0.008
	min	1.000	1.000	−0.006	1.000	1.004	−0.010	1.000	1.006	−0.012	1.000	1.014	−0.020	1.000	1.018	−0.024
1.2	max	1.210	1.206	0.010	1.210	1.210	0.006	1.210	1.212	0.004	1.210	1.220	−0.004	1.210	1.224	−0.008
	min	1.200	1.200	−0.006	1.200	1.204	−0.010	1.200	1.206	−0.012	1.200	1.214	−0.020	1.200	1.218	−0.024
1.6	max	1.610	1.606	0.010	1.610	1.610	0.006	1.610	1.612	0.004	1.610	1.620	−0.004	1.610	1.624	−0.008
	min	1.600	1.600	−0.006	1.600	1.604	−0.010	1.600	1.606	−0.012	1.600	1.614	−0.020	1.600	1.618	−0.024
2	max	2.010	2.006	0.010	2.010	2.010	0.006	2.010	2.012	0.004	2.010	2.020	−0.004	2.013	2.024	−0.008
	min	2.000	2.000	−0.006	2.000	2.004	−0.010	2.000	2.006	−0.012	2.000	2.014	−0.020	2.000	2.018	−0.024
2.5	max	2.510	2.506	0.010	2.510	2.510	0.006	2.510	2.512	0.004	2.510	2.520	−0.004	2.513	2.524	−0.008
	min	2.500	2.500	−0.006	2.500	2.504	−0.010	2.500	2.506	−0.012	2.500	2.514	−0.020	2.500	2.518	−0.024
3	max	3.010	3.006	0.010	3.010	3.010	0.006	3.010	3.012	0.004	3.010	3.020	−0.004	3.013	3.024	−0.008
	min	3.000	3.000	−0.006	3.000	3.004	−0.010	3.000	3.006	−0.012	3.000	3.014	−0.020	3.003	3.018	−0.024
4	max	4.012	4.009	0.011	4.012	4.016	0.004	4.012	4.020	0.000	4.012	4.027	−0.007	4.012	4.031	−0.011
	min	4.000	4.001	−0.009	4.000	4.008	−0.016	4.000	4.012	−0.020	4.000	4.019	−0.027	4.000	4.023	−0.031
5	max	5.012	5.009	0.011	5.012	5.016	0.004	5.012	5.020	0.000	5.012	5.027	−0.007	5.012	5.031	−0.011
	min	5.000	5.001	−0.009	5.000	5.008	−0.016	5.000	5.012	−0.020	5.000	5.019	−0.027	5.000	5.023	−0.031
6	max	6.012	6.009	0.011	6.012	6.016	0.004	6.012	6.020	0.000	6.012	6.027	−0.007	6.012	6.031	−0.011
	min	6.000	6.001	−0.009	6.000	6.008	−0.016	6.000	6.012	−0.020	6.000	6.019	−0.027	6.000	6.023	−0.031
8	max	8.015	8.010	0.014	8.015	8.019	0.005	8.015	8.024	0.000	8.015	8.032	−0.008	8.015	8.037	−0.013
	min	8.000	8.001	−0.010	8.000	8.010	−0.019	8.000	8.015	−0.024	8.000	8.023	−0.032	8.000	8.028	−0.037
10	max	10.015	10.010	0.014	10.015	10.019	0.005	10.015	10.024	0.000	10.015	10.032	−0.008	10.015	10.037	−0.013
	min	10.000	10.001	−0.010	10.000	10.010	−0.019	10.000	10.015	−0.024	10.000	10.023	−0.032	10.000	10.028	−0.037
12	max	12.018	12.012	0.017	12.018	12.023	0.006	12.018	12.029	0.000	12.018	12.039	−0.010	12.018	12.044	−0.015
	min	12.000	12.001	−0.012	12.000	12.012	−0.023	12.000	12.018	−0.029	12.000	12.028	−0.039	12.000	12.033	−0.044
16	max	16.018	16.012	0.017	16.018	16.023	0.006	16.018	16.029	0.000	16.018	16.039	−0.010	16.018	16.044	−0.015
	min	16.000	16.001	−0.012	16.000	16.012	−0.023	16.000	16.018	−0.029	16.000	16.028	−0.039	16.000	16.033	−0.044
20	max	20.021	20.015	0.019	20.021	20.028	0.006	20.021	20.035	−0.001	20.021	20.048	−0.014	20.021	20.054	−0.020
	min	20.000	20.002	−0.015	20.000	20.015	−0.028	20.000	20.022	−0.035	20.000	20.035	−0.048	20.000	20.041	−0.054
25	max	25.021	25.015	0.019	25.021	25.028	0.006	25.021	25.035	−0.001	25.021	25.048	−0.014	25.021	25.061	−0.027
	min	25.000	25.002	−0.015	25.000	25.015	−0.028	25.000	25.022	−0.035	25.000	25.035	−0.048	25.000	25.048	−0.061
30	max	30.021	30.015	0.019	30.021	30.028	0.006	30.021	30.035	−0.001	30.021	30.048	−0.014	30.021	30.061	−0.027
	min	30.000	30.002	−0.015	30.000	30.015	−0.028	30.000	30.022	−0.035	30.000	30.035	−0.048	30.000	30.048	−0.061
40	max	40.025	40.018	0.023	40.025	40.033	0.008	40.025	40.042	−0.001	40.025	40.059	−0.018	40.025	40.076	−0.035
	min	40.000	40.002	−0.018	40.000	40.017	−0.033	40.000	40.026	−0.042	40.000	40.043	−0.059	40.000	40.060	−0.076
50	max	50.025	50.018	0.023	50.025	50.033	0.008	50.025	50.042	−0.001	50.025	50.059	−0.018	50.025	50.086	−0.045
	min	50.000	50.002	−0.018	50.000	50.017	−0.033	50.000	50.026	−0.042	50.000	50.043	−0.059	50.000	50.070	−0.086
60	max	60.030	60.021	0.028	60.030	60.039	0.010	60.030	60.051	−0.002	60.030	60.072	−0.023	60.030	60.106	−0.057
	min	60.000	60.002	−0.021	60.000	60.020	−0.039	60.000	60.032	−0.051	60.000	60.053	−0.072	60.000	60.087	−0.106
80	max	80.030	80.021	0.028	80.030	80.039	0.010	80.030	80.051	−0.002	80.030	80.078	−0.029	80.030	80.121	−0.072
	min	80.000	80.002	−0.021	80.000	80.020	−0.039	80.000	80.032	−0.051	80.000	80.059	−0.078	80.000	80.102	−0.121
100	max	100.035	100.025	0.032	100.035	100.045	0.012	100.035	100.059	−0.002	100.035	100.093	−0.036	100.035	100.146	−0.089
	min	100.000	100.003	−0.025	100.000	100.023	−0.045	100.000	100.037	−0.059	100.000	100.071	−0.093	100.000	100.124	−0.146
120	max	120.035	120.025	0.032	120.035	120.045	0.012	120.035	120.059	−0.002	120.035	120.101	−0.044	120.035	120.166	−0.109
	min	120.000	120.003	−0.025	120.000	120.023	−0.045	120.000	120.037	−0.059	120.000	120.079	−0.101	120.000	120.144	−0.166
160	max	160.040	160.028	0.037	160.040	160.052	0.013	160.040	160.068	−0.003	160.040	160.125	−0.060	160.040	160.215	−0.150
	min	160.000	160.003	−0.028	160.000	160.027	−0.052	160.000	160.043	−0.068	160.000	160.100	−0.125	160.000	160.190	−0.215

SOURCE: ANSI B4.2-1978 (R94), reprinted by permission.

Table 8.2.45 K_f Values for Plain Press Fits
Obtained from fatigue tests in bending)

Shaft material	Shaft diam, in	Collar or hub material	K_f	Remarks
0.42% carbon steel	1⅝	0.42% carbon steel	2.0	No external reaction through collar
0.45% carbon axle steel	2	Ni-Cr-Mo steel (case-hardened)	2.3	No external reaction through collar
0.45% carbon axle steel	2	Ni-Cr-Mo steel (case-hardened)	2.9	External reaction taken through collar
Cr-Ni-Mo steel (310 Brinell)	2	Ni-Cr-Mo steel (case-hardened)	3.9	External reaction taken through collar
2.6% Ni steel (57,000 lb/in² fatigue limit)	2	Ni-Cr-Mo steel (case-hardened)	3.3–3.8	External reaction taken through collar
Same, heat-treated to 253 Brinell	2	Ni-Cr-Mo steel (case-hardened)	3.0	External reaction taken through collar

Marine-engine shafts and **diesel-engine crankshafts** should be designed not only for strength but for avoidance of critical speed. (See Applied Mechanics, *Trans. ASME,* **50,** no. 8, for methods of calculating critical speeds of diesel engines.)

Table 8.2.46 Constants for Use in
$(S_a/S_e)^m + (KS_m/S_{ut})^P = 1$

Failure theory	K	P	m
Soderberg	S_{ut}/S_y	1	1
Bagci	S_{ut}/S_y	4	1
Modified Goodman	1	1	1
Gerber parabolic	1	2	1
Kececioglu	1	2	m†
Quadratic (elliptic)	1	2	2

$$\dagger m = \begin{cases} 0.8914 \text{ UNS G } 10180 \ H_B = 130 \\ 0.9266 \text{ UNS G } 10380 \ H_B = 164 \\ 1.0176 \text{ UNS G } 41300 \ H_B = 207 \\ 0.9685 \text{ UNS G } 43400 \ H_B = 233 \end{cases}$$

PULLEYS, SHEAVES, AND FLYWHEELS

Arms of pulleys, sheaves, and flywheels are subjected to stresses due to condition of founding, to details of construction (such as split or solid), and to conditions of service, which are difficult to analyze. For these reasons, no accurate stress relations can be established, and the following formulas must be understood to be only approximately correct. It has been established experimentally by Benjamin (*Am. Mach.,* Sept. 22, 1898) that thin-rim pulleys do not distribute equal loads to the several pulley arms. For these, it will be safe to assume the tangential force on the pulley rim as acting on half of the number of arms. Pulleys with comparatively thick rims, such as engine band wheels, have all the arms taking the load. Furthermore, while the stress action in the arms is similar to that in a beam fixed at both ends, the amount of restraint at the rim depending on the rim's elasticity, it may nevertheless be assumed for purposes of design that cantilever action is predominant. The bending moment at the hub in arms of thin-rim pulleys will be $M = PL/(\frac{1}{2}N)$, where M = bending moment, in·lb; P = tangential load on the rim, lb; L = length of the arm, in; and N = number of arms. For thick-rim pulleys and flywheels, $M = PL/N$.

For arms of elliptical section having a width of two times the thickness, where E = **width of arm section** at the rim, in, and s_t = intensity of tensile stress, lb/in²

$$E = \sqrt[3]{40PL/(s_tN)} \text{ (thin rim)} = \sqrt[3]{20PL/(s_tN)} \text{ (thick rim)}$$

For single-thickness belts, P may be taken as $50B$ lb and for double-thickness belts $P = 75B$ lb, where B is the width of pulley face, in. Then $E = k \times \sqrt[3]{BL/(s_tN)}$, where k has the following values: for thin rim, single belt, 13; thin rim, double belt, 15; thick rim, single belt, 10; thick rim, double belt, 12. For cast iron of good quality, s_t due to bending may be taken at 1,500 to 2,000. The arm section at the rim may be made from ⅔ to ¾ the dimensions at the hub.

For high-speed pulleys and flywheels, it becomes necessary to check the arm for tension due to rim expansion. It will be safe to assume that each arm is in tension due to one-half the centrifugal force of that portion of the rim which it supports. That is, $T = As_t = Wv^2/(2NgR)$, lb, where T = tension in arm, lb; N = number of arms; v = speed of rim, ft/s; R = radius of pulley, ft; A = area of arm section, in²; W = weight of pulley rim, lb; and s_t = intensity of tensile stress in arm section, lb/in², whence $s_t = WRn^2/(6,800NA)$, where n = r/min of pulley.

Arms of flywheels having heavy rims may be subjected to severe stress action due to the inertia of the rim at sudden load changes. There being no means of predicting the probable maximum to which the inertia may rise, it will be safe to make the arms equal in strength to ¾ of the shaft strength in torsion. Accordingly, for elliptical arm sections,

$$N \times 0.5E^3s_t = \frac{3}{4} \times 2s_s d^3 \quad \text{or} \quad E = 1.4d\sqrt[3]{s_s/(s_tN)}$$

For steel shafts with $s_s = 8,000$ and cast-iron arms with $s = 1,500$,

$$E = 2.4d/\sqrt[3]{N} = 1.3d \text{ (for 6 arms)} = 1.2d \text{ (for 8 arms)}$$

where $2E$ = width of elliptical arm section at hub, in (thickness = E), and d = shaft diameter, in.

Rims of belted pulleys cast whole may have the following proportions (see Fig. 8.2.86):

$$t_2 = \frac{3}{4}h + 0.005D \qquad t_1 = 2t_2 + C \qquad W = \frac{9}{8}B \text{ to } \frac{5}{4}B$$

where h = belt thickness, $C = \frac{1}{24}W$, and B = belt width, all in inches.

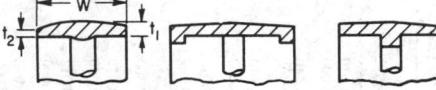

Fig. 8.2.86 Rims for belted pulleys.

Engine band wheels, flywheels, and pulleys run at **high speeds** are subjected to the following stress actions in the rim:

Considering the rim as a free ring, i.e., without arm restraint, and made of cast iron or steel, $s_t = v^2/10$ (approx), where s_t = intensity of tensile stress, lb/in², and v = rim speed, ft/s. For beam action between the arms of a solid rim, $M = Pl/12$ (approx), where M = bending moment in rim, in·lb; P = centrifugal force of that portion of rim between arms, lb, and l = length of rim between arms, in; from which $s_t = WR^2n^2/(450N^2Z)$, where W = weight of entire rim, lb; R = radius of wheel, ft; n = r/min of wheel; and Z = section modulus of rim section, in³. In case the rim section is of the forms shown in Fig. 8.2.86, care must be taken that the flanges do not reduce the section modulus from that of the rectangular section. For **split rims** fastened with bolts the stress analysis is as follows:

Let w = weight of rim portion, lb (with length L, in) lb; w_1 = weight of lug, lb; L_1 = lever arm of lug, in; and s_t = intensity of tensile stress lb/in² in rim section joining arm. Then $s_t = 0.00034n^2R(w_1L_1 + wL/2)/Z$, where n = r/min of wheel; R = wheel radius, ft; and Z = section modulus of rim section, in^3. The above equation gives the value of s_t for bending when the bolts are loose, which is the worst possible condition that may arise. On this basis of analysis, s_t should not be greater than 8,000 lb/in^2. The stress due to bending in addition to the stress due to rim expansion as analyzed previously will be the probable maximum intensity of stress for which the rim should be checked for strength. The flange bolts, because of their position, do not materially relieve the bending action. In case a tie rod leads from the flange to the hub, it will be *safe* to consider it as an additional factor of safety. When the tie rod is kept tight, it very materially strengthens the rim.

A more accurate method for calculating maximum stresses due to centrifugal force in flywheels with arms cast integral with the rim is given by Timoshenko, "Strength of Materials," Pt. II, 1941, p. 98. More exact equations for calculating stresses in the arms of flywheels and pulleys due to a combination of belt pull, centrifugal force, and changes in velocity are given by Heusinger, *Forschung,* 1938, p. 197. In both treatments, shrinkage stresses in the arms due to casting are neglected.

Large flywheels for high rim speeds and severe working conditions (as for rolling-mill service) have been made from flat-rolled steel plates with holes bored for the shaft. A group of such plates may be welded together by circumferential welds to form a large flywheel. By this means, the welds do not carry direct centrifugal loads, but serve merely to hold the parts in position. Flywheels up to 15-ft diam for rolling-mill service have been constructed in this way.

BELT DRIVES

Flat-Belt Drives

The primary drawback of flat belts is their reliance on belt tension to produce frictional grip over the pulleys. Such high tension can shorten bearing life. Also, tracking may be a problem. However, flat belts, being thin, are not subject to centrifugal loads and so work well over small pulleys at high speeds in ranges exceeding 9,000 ft/min. In light service flat belts can make effective clutching drives. Flat-belt drives have efficiencies of about 90 percent, which compares favorably to geared drives. Flat belts are also quiet and can absorb torsional vibration readily.

Leather belting has an ultimate tensile strength ranging from 3,000 to 5,000 lb/in^2. Average values of breaking strength of good oak-tanned belting (determined by Benjamin) are as follows: single (double) in solid leather 900 (1,400); at riveted joint 600 (1,200); at laced joint 350 lb/in of width. Well-made cemented joints have strengths equal to the belt, leather-laced and riveted joints about one-third to two-thirds as strong, and wire-laced joints about 85 to 95 percent as strong.

Rubber belting is made from fabric or cord impregnated and bound together by vulcanized rubber compounds. The fabric or cord may be of cotton or rayon. Nylon cord and steel cord or cable are also available. Advantages are high tensile strength, strength to hold metal fasteners satisfactorily, and resistance to deterioration by moisture. The best rubber fabric construction for most types of service is made with hard or tight-woven fabric with a "skim coat" or thin layer of rubber between plies. The cord type of construction allows the use of smaller pulley diameters than the fabric type, and also develops less stretch in service. It must be used in the endless form, except in cases where the oil-field type of clamp may be used.

Initial tensions in rubber belts run from 15 to 25 lb/ply/in width. A common rule is to cut belts 1 percent less than the minimum tape-line measurement around the pulleys. For heavy loads, a 1½ percent allowance is usually required, although, because of shrinkage, less initial tension is required for wet or damp conditions. Initial tensions of 25 lb/ply/in may overload shafts or bearings. Maximum safe tight-side tensions for rubber belts are as follows:

Duck weight, oz	28	32	32.66	34.66	36
Tension, lb/ply/in width	25	28	30	32	35

Centrifugal forces at high speeds require higher tight-side tensions to carry rated horsepower.

Rubber belting may be bought in endless form or made endless in the field by means of a vulcanized splice produced by a portable electric vulcanizer. For endless belts the drive should provide take-up of 2 to 4 percent to allow for length variation as received and for stretch in service. The amount of take-up will vary with the type of belt used. For certain drives, it is possible to use endless belts with no provision for take-up, but this involves a heavier belt and a higher initial unit tension than would be the case otherwise. Ultimate tensile strength of rubber belting varies from 280 to 600 lb or more per inch width per ply. The weight varies from 0.02 to 0.03 or more lb/in width/ply. Belts with steel reinforcement are considerably heavier. For horsepower ratings of rubber belts, see Table 8.2.47c.

Arrangements for Belt Drives   In belt drives, the centerline of the belt advancing on the pulley should lie in a plane passing through the midsection of the pulley at right angles to the shaft. Shafts inclined to each other require connections as shown in Fig. 8.2.87a. In case guide pulleys are needed their positions can be determined as shown in Fig. 8.2.87a, b, and c. In Fig. 8.2.87d the center circles of the two pulleys to be connected are set in correct relative position in two planes, *a* being the angle between the planes (= supplement of angle between shafts). If any two points as *E* and *F* are assumed on the line of intersection *MN* of the planes, and tangents *EG, EH, FJ,* and *FK* are drawn from them to the circles, the center circles of the guide pulleys must be so arranged that these tangents are also tangents to them, as shown. In other words, the middle planes of the guide pulleys must lie in the planes *GEH* and *JFK.*

Table 8.2.47a   Service Factors S

Application	Squirrel-cage ac motor		Wound rotor ac motor (slip ring)	Single-phase capacitor motor	DC shunt-wound motor	Diesel engine, 4 or more cyl, above 700 r/min
	Normal torque, line start	High torque				
Agitators	1.0–1.2	1.2–1.4	1.2			
Compressors	1.2–1.4		1.4	1.2	1.2	1.2
Belt conveyors (ore, coal, sand)	—	1.4	—	—	1.2	
Screw conveyors	—	1.8	—	—	1.6	
Crushing machinery	—	1.6	1.4	—	—	1.4–1.6
Fans, centrifugal	1.2	—	1.4	—	1.4	1.4
Fans, propeller	1.4	2.0	1.6	—	1.6	1.6
Generators and exciters	1.2	—	—	—	1.2	2.0
Line shafts	1.4	—	1.4	1.4	1.4	1.6
Machine tools	1.0–1.2	—	1.2–1.4	1.0	1.0–1.2	
Pumps, centrifugal	1.2	1.4	1.4	1.2	1.2	
Pumps, reciprocating	1.2–1.4	—	1.4–1.6	—	—	1.8–2.0

Table 8.2.47b Arc of Contact Factor _K_—Rubber Belts

Arc of contact, deg	140	160	180	200	220
Factor _K_	0.82	0.93	1.00	1.06	1.12

When these conditions are met, the belts will run in either direction on the pulleys.

To avoid the necessity of taking up the **slack** in belts which have become stretched and permanently lengthened, a **belt tightener** such as shown in Fig. 8.2.88 may be employed. It should be placed on the slack side of the belt and nearer the driving pulley than the driven pulley. Pivoted motor drives may also be used to maintain belt tightness with minimum initial tension.

Length of Belt for a Given Drive The length of an **open belt** for a given drive is equal to $L = 2C + 1.57(D + d) + (D - d)^2/(4C)$, where L = length of belt, in; D = diam of large pulley, in; d = diam of small pulley, in; and C = distance between pulley centers, in. Center

distance C is given by $C = 0.25b + 0.25\sqrt{b^2 - 2(D - d)^2}$, where $b = L - 1.57(D + d)$. When a **crossed belt** is used, the length in $L = 2C + 1.57(D + d) + (D + d)^2/(4C)$.

Step or Cone Pulleys For belts operating on step pulleys, the pulley diameters must be such that the belt will fit over any pair with equal tightness. With **crossed belts**, it will be apparent from the equation for length of belt that the sum of the pulley diameters need only be constant in order that the belt may fit with equal tightness on each pair of pulleys. With open belts, the length is a function of both the sum and the difference of the pulley diameters; hence no direct solution of the problem is possible, but a graphical approach can be of use.

A graphical method devised by Smith (*Trans. ASME*, 10) is shown in Fig. 8.2.89. Let A and B be the centers of any pair of pulleys in the set, the diameters of which are known or assumed. Bisect AB in C, and draw CD at right angles to AB. Take $CD = 0.314$ times the center distance L, and draw a circle tangent to the belt line EF. The belt line of any other pair of pulleys in the set will then be tangent to this circle. If the angle

Table 8.2.47c Horsepower Ratings of Rubber Belts
(hp/in of belt width for 180° wrap)

						Belt speed, ft/min						
	Ply	500	1,000	1,500	2,000	2,500	3,000	4,000	5,000	6,000	7,000	8,000
32-oz fabric	3	0.7	1.4	2.1	2.7	3.3	3.9	4.9	5.6	6.0		
	4	0.9	1.9	2.8	3.6	4.4	5.2	6.5	7.4	7.9		
	5	1.2	2.3	3.4	4.5	5.5	6.5	8.1	9.2	9.8		
	6	1.4	2.8	4.1	5.4	6.6	7.8	9.6	11.0	11.7		
	7	1.6	3.2	4.7	6.2	7.7	9.0	11.2	12.8	13.6		
	8	1.8	3.6	5.3	7.0	8.7	10.2	12.7	14.6	15.5		
32-oz hard fabric	3	0.7	1.5	2.2	2.9	3.5	4.1	5.1	5.8	6.2	6.1	5.5
	4	1.0	2.0	3.0	3.9	4.7	5.5	6.8	7.8	8.3	8.1	7.3
	5	1.3	2.5	3.7	4.9	5.9	6.9	8.5	9.8	10.3	9.1	9.0
	6	1.5	3.0	4.5	5.9	7.1	8.3	10.2	11.7	12.3	12.1	10.7
	7	1.7	3.5	5.2	6.9	8.3	9.7	11.9	13.6	14.3	14.1	12.4
	8	1.9	4.0	5.9	7.9	9.5	11.1	13.6	15.5	16.3	16.0	14.1
	9	2.1	4.5	6.6	8.9	10.6	12.4	15.3	17.4	18.3	17.9	15.8
	10	2.3	5.0	7.3	9.8	11.7	13.7	17.0	19.3	20.3	19.8	17.5
No. 70 rayon cord	3	1.6	3.1	4.6	6.0	7.3	8.6	10.6	12.0	12.7	12.3	10.7
	4	2.1	4.1	6.1	8.0	9.8	11.5	14.5	16.6	17.8	17.8	16.4
	5	2.6	5.1	7.6	10.1	12.3	14.5	18.3	21.1	23.0	23.5	22.2
	6	3.1	6.2	9.2	12.1	14.8	17.5	22.1	25.7	28.1	28.9	27.9
	7	3.6	7.2	10.7	14.1	17.4	20.4	26.0	30.3	33.2	34.5	33.7
	8	4.1	8.2	12.2	16.2	19.9	23.4	29.8	34.8	38.4	40.0	39.4

Table 8.2.47d Minimum Pulley Diameters—Rubber Belts, in

						Belt speed, ft/min						
	Ply	500	1,000	1,500	2,000	2,500	3,000	4,000	5,000	6,000	7,000	8,000
32-oz fabric	3	4	4	4	4	5	5	5	6	6		
	4	4	5	6	6	7	7	8	9	10		
	5	6	7	9	10	10	11	12	13	14		
	6	9	10	11	13	14	14	16	18	19		
	7	13	14	16	17	18	19	21	22	24		
	8	18	19	21	22	23	24	25	27	29		
32-oz hard fabric	3	3	3	3	4	4	4	4	5	5	6	7
	4	4	4	5	5	6	6	7	7	8	9	12
	5	5	6	7	8	8	9	10	11	12	13	16
	6	6	8	10	11	11	12	13	15	16	18	21
	7	10	12	14	15	15	16	17	19	20	22	26
	8	14	16	17	18	19	20	21	23	24	27	31
	9	18	20	21	22	23	24	25	27	28	31	36
	10	22	24	25	26	27	28	29	31	33	35	41
No. 70 rayon cord	3	5	6	7	7	8	8	9	10	11	12	13
	4	7	8	9	9	10	11	12	12	14	15	17
	5	9	10	11	12	13	13	15	16	17	19	21
	6	13	14	15	16	16	17	18	19	21	23	25
	7	16	17	18	19	20	21	22	23	24	26	29
	8	19	20	22	23	23	24	25	26	28	30	33

EF makes with *AB* is greater than 18°, draw a tangent to the circle *D*, making an angle of 18° with *AB*; and from a center on *CD* distant 0.298*L* above *C*, draw an arc tangent to this 18° line. All belt lines with angles greater than 18° will be tangent to this last drawn arc.

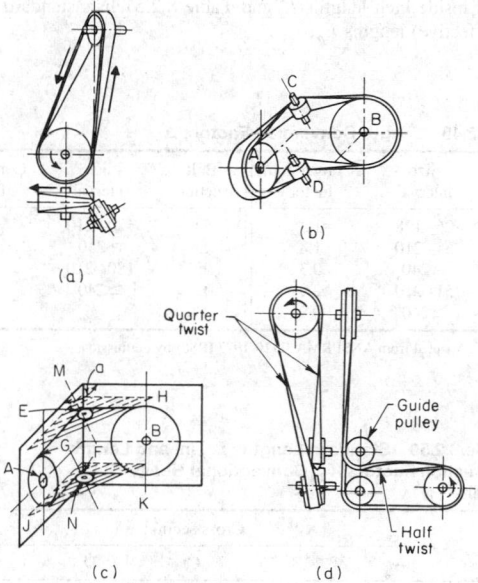

Fig. 8.2.87 Arrangements for flat-belt drives.

A very slight error in a graphical solution drawn to any scale much under full size will introduce an error seriously affecting the equality of belt tensions on the various pairs of pulleys in the set, and where much power is to be transmitted it is advisable to calculate the pulley diameters from the following **formulas** derived from Burmester's graphical method ("Lehrbuch der Mechanik").

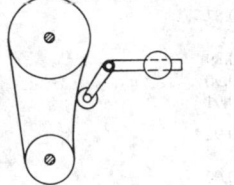

Fig. 8.2.88 Belt tightener.

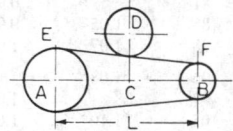

Fig. 8.2.89 Symbols for cone pulley graphical method.

Let D_1 and D_2 be, respectively, the diameters of the smaller and larger pulleys of a pair, $n = D_2/D_1$, and l = distance between shaft centers, all in inches. Also let $m = 1.58114l - D_0$, where D_0 = diam of both pulleys for a speed ratio $n = 1$. Then $(D_1 + m)^2 + (nD_1 + m)^2 = 5l^2$. First settle on values of D_0, l, and n, and then substitute in the equation and solve for D_1. The diameter D_2 of the other pulley of the pair will then be nD_1. The values are correct to the fourth decimal place.

The speeds given by cone pulleys should increase in a **geometric ratio**; i.e., each speed should be multiplied by a constant a in order to obtain the next higher speed. Let n_1 and n_2 be, respectively, the lowest and highest speeds (r/min) desired and k the number of speed changes. Then $a = \sqrt[k-1]{n_2/n_1}$. In practice, a ranges from 1.25 up to 1.75 and even 2. The ideal value for a in machine-tool practice, according to Carl G. Barth, would be 1.189. In the example below, this would mean the use of 18 speeds instead of 8.

EXAMPLE. Let $n_1 = 16$, $n_2 = 400$, and $k = 8$, to be obtained with four pairs of pulleys and a back gear. From formula, $a = \sqrt[7]{25} = 1.584$, whence speeds will be

16, $(16 \times 1.584 =)$ 25.34, $(25.34 \times 1.584 =)$ 40.14, and similarly 63.57, 100.7, 159.5, 252.6, and 400. The first four speeds are with the back gear in; hence the back-gear ratio must be $100.7 \div 16 = 6.29$.

Transmission of Power by Flat Belts The theory of flat-belt drives takes into account changes in belt tension caused by friction forces between belt and pulley, which, in turn, cause belt elongation or contraction, thus inducing relative movement between belt and pulley. The transmission of power is a complex affair. A lengthy mathematical presentation can be found in Firbank, "Mechanics of the Flat Belt Drive," ASME Paper 72-PTG-21. A simpler, more conventional analysis used for many years yields highly serviceable designs.

The turning force (tangential) on the rim of a pulley driven by a flat belt is equal to $T_1 - T_2$, where T_1 and T_2 are, respectively, the tensions in the driving (tight) side and following (slack) side of the belt. (For the relations of T_1 and T_2 at low peripheral speeds, see Sec. 3.) Log $(T_1/T_2) = 0.0076fa$ when the effect of centrifugal force is neglected and $T_1/T_2 = 10^{0.0076fa}$. Figure 8.2.90 gives values of this function. When the speeds are high, however, the relations of T_1 to T_2 are modified by centrifugal stresses in the belt, in which case log $(T_1/T_2) = 0.0076f(1 - x)a$, where f = coefficient of friction between the belt and pulley surface, a = angle of wrap, and $x = 12wv^2/(gt)$ in which w =

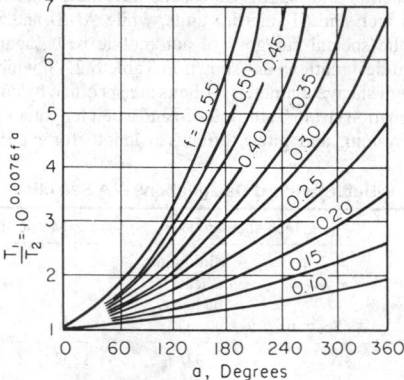

Fig. 8.2.90 Values of $10^{0.0076fa}$.

weight of 1 in³ of belt material, lb; v = belt speed, ft/s; $g = 32.2$ ft/s²; and t = allowable working tension, lb/in². Values of x for leather belting (with $w = 0.035$ and $t = 300$) are as follows:

u	30	40	50	60	70	
x	0.039	0.070	0.118	0.157	0.214	

uv	80	90	100	110	120	130
x	0.279	0.352	0.435	0.526	0.626	0.735

Researches by Barth (*Trans. ASME,* 1909) seem to show that f is a function of the belt velocity, varying according to the formula $f = 0.54 - 140/(500 + V)$ for leather belts on iron pulleys, where V = belt velocity for ft/min. For practical design, however, the following values of f may be used: for leather belts on cast-iron pulleys, $f = 0.30$; on wooden pulleys, $f = 0.45$; on paper pulleys, $f = 0.55$. The treatment of belts with belt dressing, pulleys with cork inserts, and dampness are all factors which greatly modify these values, tending to make them higher.

The **arc of contact** on the smaller of two pulleys connected by an open belt, in degrees, is approximately equal to $180 - 60(D - d)/l$, where D and d are the larger and smaller pulley diameters and l the distance between their shaft centers, all in inches. This formula gives an error not exceeding 0.5 percent.

Selecting a Belt Selecting an appropriate belt involves calculating horsepower per inch of belt width as follows:

$$\text{hp/in} = (\text{demanded hp} \times S)/(K \times W)$$

where demanded hp = horsepower required by the job at hand; S = service factor; K = arc factor; W = proposed belt width (determined from pulley width). One enters a belt manufacturer's catalog with *hp/in*, *belt speed*, and *small pulley diameter*, then selects that belt which has a matching maximum hp/in rating. See Table 8.2.47*a, b, c,* and *d* for typical values of S, K, hp/in ratings, and minimum pulley diameters.

V-Belt Drives

V-belt drives are widely used in power transmission, despite the fact that they may range in efficiency from about 70 to 96 percent. Such drives consist essentially of endless belts of trapezoidal cross section which ride in V-shaped pulley grooves (see Fig. 8.2.93*a*). The belts are formed of cord and fabric, impregnated with rubber, the cord material being cotton, rayon, synthetic, or steel. V-belt drives are quiet, able to absorb shock and operate at low bearing pressures. A V belt should ride with the top surface approximately flush with the top of the pulley groove; clearance should be present between the belt base and the base of the groove so that the belt rides on the groove walls. The friction between belt and groove walls is greatly enhanced beyond normal values because sheave groove angles are made somewhat less than belt-section angles, causing the belt to wedge itself into the groove. See Table 8.2.56*a* for standard groove dimensions of sheaves.

The cross section and lengths of V belts have been standardized by ANSI in both inch and SI (metric) units, while ANSI and SAE have standardized the special category of automobile belts, again in both units. Standard designations are shown in Table 8.2.48, which also includes minimum sheave diameters. V belts are specified by combining a standard designation (from Table 8.2.48) and a belt length; inside length for the inch system, and pitch (effective) length for metric system.

Table 8.2.48   V-Belt Standard Designations—A Selection

	Inch standard		Metric standard	
Type	Section	Minimum sheave diameter, in	Section	Minimum sheave diameter, mm
Heavy-duty	A	3.0	13 C	80
	B	5.4	16 C	140
	C	9.0	22 C	214
	D	13.0	32 C	355
	E	21.04		
Automotive	0.25	2.25	6 A	57
	0.315	2.25	8 A	57
	0.380	2.40	10 A	61
	0.440	2.75	11 A	70
	0.500	3.00	13 A	76
	$1\frac{1}{16}$	3.00	15 A	76
	$\frac{3}{4}$	3.00	17 A	76
	$\frac{7}{8}$	3.50	20 A	89
	1.0	4.00	23 A	102
Heavy-duty narrow	3 V	2.65		
	5 V	7.1		
	8 V	12.3		
Notched narrow	3 VX	2.2		
	5 VX	4.4		
Light-duty	2 L	0.8		
	3 L	1.5		
	4 L	2.5		
	5 L	3.5		
Synchronous belts	MXL			
	XL			
	L			
	H			
	XH			
	XXH			

NOTE: The use of smaller sheaves than minimum will tend to shorten belt life.
SOURCE: Compiled from ANSI/RMA IP-20, 21, 22, 23, 24, 25, 26; ANSI/SAE J636C.

Sheaves are specified by their pitch diameters, which are used for velocity ratio calculations in which case inside belt lengths must be converted to pitch lengths for computational purposes. Pitch lengths are calculated by adding a conversion factor to inside length (i.e., $L_p = L_s + \Delta$). See Table 8.2.49 for conversion factors. Table 8.2.50 lists standard inside inch lengths L_s and Table 8.2.51 lists standard metric pitch (effective) lengths L_p.

Table 8.2.49   Length Conversion Factors Δ

Belt section	Size interval	Conversion factor	Belt section	Size interval	Conversion factor
A	26–128	1.3	D	120–210	3.3
B	35–210	1.8	D	≥240	0.8
B	≥240	0.3	E	180–210	4.5
C	51–210	2.9	E	≥240	1.0
C	≥240	0.9			

SOURCE: Adapted from ANSI/RMA IP-20-1977 (R88) by permission.

Table 8.2.50   Standard Lengths L_s, in, and Length Correction Factors K_2: Conventional Heavy-Duty V Belts

	Cross section				
L_s	A	B	C	D	E
26	0.78				
31	0.82				
35	0.85	0.80			
38	0.87	0.82			
42	0.89	0.84			
46	0.91	0.86			
51	0.93	0.88	0.80		
55	0.95	0.89			
60	0.97	0.91	0.83		
68	1.00	0.94	0.85		
75	1.02	0.96	0.87		
80	1.04				
81		0.98	0.89		
85	1.05	0.99	0.90		
90	1.07	1.00	0.91		
96	1.08		0.92		
97		1.02			
105	1.10	1.03	0.94		
112	1.12	1.05	0.95		
120	1.13	1.06	0.96	0.88	
128	1.15	1.08	0.98	0.89	
144		1.10	1.00	0.91	
158		1.12	1.02	0.93	
173		1.14	1.04	0.94	
180		1.15	1.05	0.95	0.92
195		1.17	1.06	0.96	0.93
210		1.18	1.07	0.98	0.95
240		1.22	1.10	1.00	0.97
270		1.24	1.13	1.02	0.99
300		1.27	1.15	1.04	1.01
330			1.17	1.06	1.03
360			1.18	1.07	1.04
390			1.20	1.09	1.06
420			1.21	1.10	1.07
480				1.13	1.09
540				1.15	1.11
600				1.17	1.13
660				1.18	1.15

SOURCE: ANSI/RMA IP-20-1977 (R88), reproduced by permission.

Table 8.2.51   Standard Pitch Lengths L_p (Metric Units) and Length Correction Factors K_2

13C		16C		22C		32C	
L_p	K_2	L_p	K_2	L_p	K_2	L_p	K_2
710	0.83	960	0.81	1,400	0.83	3,190	0.89
750	0.84	1,040	0.83	1,500	0.85	3,390	0.90
800	0.86	1,090	0.84	1,630	0.86	3,800	0.92
850	0.88	1,120	0.85	1,830	0.89	4,160	0.94
900	0.89	1,190	0.86	1,900	0.90	4,250	0.94
950	0.90	1,250	0.87	2,000	0.91	4,540	0.95
1,000	0.92	1,320	0.88	2,160	0.92	4,720	0.96
1,075	0.93	1,400	0.90	2,260	0.93	5,100	0.98
1,120	0.94	1,500	0.91	2,390	0.94	5,480	0.99
1,150	0.95	1,600	0.92	2,540	0.96	5,800	1.00
1,230	0.97	1,700	0.94	2,650	0.96	6,180	1.01
1,300	0.98	1,800	0.95	2,800	0.98	6,560	1.02
1,400	1.00	1,900	0.96	3,030	0.99	6,940	1.03
1,500	1.02	1,980	0.97	3,150	1.00	7,330	1.04
1,585	1.03	2,110	0.99	3,350	1.01	8,090	1.06
1,710	1.05	2,240	1.00	3,550	1.02	8,470	1.07
1,790	1.06	2,360	1.01	3,760	1.04	8,850	1.08
1,865	1.07	2,500	1.02	4,120	1.06	9,240	1.09
1,965	1.08	2,620	1.03	4,220	1.06	10,000	1.10
2,120	1.10	2,820	1.05	4,500	1.07	10,760	1.11
2,220	1.11	2,920	1.06	4,680	1.08	11,530	1.13
2,350	1.13	3,130	1.07	5,060	1.10	12,290	1.14
2,500	1.14	3,330	1.09	5,440	1.11		
2,600	1.15	3,530	1.10	5,770	1.13		
2,730	1.17	3,740	1.11	6,150	1.14		
2,910	1.18	4,090	1.13	6,540	1.15		
3,110	1.20	4,200	1.14	6,920	1.16		
3,310	1.21	4,480	1.15	7,300	1.17		
		4,650	1.16	7,680	1.18		
		5,040	1.18	8,060	1.19		
		5,300	1.19	8,440	1.20		
		5,760	1.21	8,820	1.21		
		6,140	1.23	9,200	1.22		
		6,520	1.24				
		6,910	1.25				
		7,290	1.26				
		7,670	1.27				

SOURCE: ANSI/RMA IP-20-1988 revised, reproduced by permission.

For given large and small sheave diameters and center-to-center distance, the needed V-belt length can be computed from

$$L_p = 2C + 1.57(D + d) + \frac{(D - d)^2}{4C}$$

$$C = \frac{K + \sqrt{K^2 - 32(D - d)^2}}{16} \qquad K = 4L_p - 6.28\,(D + d)$$

where C = center-to-center distance; D = pitch diameter of large sheave; d = pitch diameter of small sheave; L_p = pitch (effective) length.

Arc of contact on the smaller sheave (degrees) is approximately

$$\theta = 180 - \frac{60(D - d)}{C}$$

Transmission of Power by V Belts   Unfortunately there is no theory or mathematical analysis that is able to explain all experimental results reliably. Empirical formulations based on experimental results, however, do provide very serviceable design procedures, and together with data published in V-belt manufacturers' catalogs provide the engineer with the necessary V-belt selection tools.

For satisfactory performance under most conditions, ANSI provides the following empirical single V-belt power-rating formulation (inch and metric units) for 180° arc of contact and average belt length:

$$H_r = \left(C_1 - \frac{C_2}{d} - C_3(r - d)^2 - C_4 \log rd \right) rd + C_2 r \left(1 - \frac{1}{K_A} \right)$$

where H_r = rated horsepower for inch units (rated power kW for metric units); C_1, C_2, C_3, C_4 = constants from Table 8.2.53; r = r/min of high-speed shaft times 10^{-3}; K_A = speed ratio factor from Table 8.2.52; d = pitch diameter of small sheave, in (mm).

Selecting a Belt   Selecting an appropriate belt involves calculating horsepower per belt as follows:

$$NH_r = (\text{demanded hp} \times K_s)/(K_1 K_2)$$

where H_r = hp/belt rating, either from ANSI formulation above, or from manufacturer's catalog (see Table 8.2.55); demanded hp = horsepower required by the job at hand; K_s = service factor accounting for driver and driven machine characteristics regarding such things as shock, torque level, and torque uniformity (see Table 8.2.54); K_1 = angle of contact correction factor (see Fig. 8.2.91a); K_2 = length correction factor (see Tables 8.2.50 and 8.2.51); N = number of belts. See Fig. 8.2.91b for selection of V-belt cross section.

V band belts, effectively joined V belts, serve the function of multiple single V belts (see Fig. 8.2.93b).

Long center distances are not recommended for V belts because excess slack-side vibration shortens belt life. In general $D \leqq C \leqq 3(D + d)$. If longer center distances are needed, then link-type V belts can be used effectively.

Since belt-drive capacity is normally limited by slippage of the smaller sheave, V-belt drives can sometimes be used with a flat, larger pulley rather than with a grooved sheave, with little loss in capacity. For instance the flat surface of the flywheel in a large punch press can serve such a purpose. The practical range of application is when speed ratio is over $3:1$, and center distance is equal to or slightly less than the diameter of the large pulley.

Table 8.2.52   Approximate Speed-Ratio Factor K_A for Use in Power-Rating Formulation

D/d range	K_A	D/d range	K_A
1.00–1.01	1.0000	1.15–1.20	1.0586
1.02–1.04	1.0112	1.21–1.27	1.0711
1.05–1.07	1.0226	1.28–1.39	1.0840
1.08–1.10	1.0344	1.40–1.64	1.0972
1.11–1.14	1.0463	Over 1.64	1.1106

SOURCE: Adapted from ANSI/RMA IP-20-1977 (R88), by permission.

Table 8.2.53   Constants C_1, C_2, C_3, C_4 for Use in Power-Rating Formulation

Belt section	C_1	C_2	C_3	C_4
		Inch		
A	0.8542	1.342	2.436×10^{-4}	0.1703
B	1.506	3.520	4.193×10^{-4}	0.2931
C	2.786	9.788	7.460×10^{-4}	0.5214
D	5.922	34.72	1.522×10^{-3}	1.064
E	8.642	66.32	2.192×10^{-3}	1.532
		Metric		
13C	0.03316	1.088	1.161×10^{-8}	5.238×10^{-3}
16C	0.05185	2.273	1.759×10^{-8}	7.934×10^{-3}
22C	0.1002	7.040	3.326×10^{-8}	1.500×10^{-2}
32C	0.2205	26.62	7.037×10^{-8}	3.174×10^{-2}

SOURCE: Compiled from ANSI/RMA IP-20-1977 (R88), by permission.

Table 8.2.54   Approximate Service Factor K_s for V-Belt Drives

Power source torque	Load			
	Uniform	Light shock	Medium shock	Heavy shock
Average or normal	1.0–1.2	1.1–1.3	1.2–1.4	1.3–1.5
Nonuniform or heavy	1.1–1.3	1.2–1.4	1.4–1.6	1.5–1.8

SOURCE: Adapted from ANSI/RMA IP-20-1977 (R88), by permission.

Table 8.2.55 Horsepower Ratings of V Belts

Belt section	Speed of faster shaft, r/min	Rated horsepower per belt for small sheave pitch diameter, in							Additional horsepower per belt for speed ratio				
		2.60	3.00	3.40	3.80	4.20	4.60	5.00	1.02–1.04	1.08–1.10	1.15–1.20	1.28–1.39	1.65–over
A	200	0.20	0.27	0.33	0.40	0.46	0.52	0.59	0.00	0.01	0.01	0.02	0.03
	800	0.59	0.82	1.04	1.27	1.49	1.70	1.92	0.01	0.04	0.06	0.08	0.11
	1,400	0.87	1.25	1.61	1.97	2.32	2.67	3.01	0.02	0.06	0.10	0.15	0.19
	2,000	1.09	1.59	2.08	2.56	3.02	3.47	3.91	0.03	0.09	0.15	0.21	0.27
	2,600	1.25	1.87	2.47	3.04	3.59	4.12	4.61	0.04	0.12	0.19	0.27	0.35
	3,200	1.37	2.08	2.76	3.41	4.01	4.57	5.09	0.05	0.14	0.24	0.33	0.43
	3,800	1.43	2.23	2.97	3.65	4.27	4.83	5.32	0.06	0.17	0.28	0.40	0.51
	4,400	1.44	2.29	3.07	3.76	4.36	4.86*	5.26*	0.07	0.20	0.33	0.46	0.59
	5,000	1.39	2.28	3.05	3.71	4.24*	4.48*	4.64*	0.07	0.22	0.37	0.52	0.65
	5,600	1.29	2.17	2.92	3.50*				0.08	0.25	0.42	0.58	0.75
	6,200	1.11	1.98	2.65*					0.09	0.28	0.46	0.64	0.83
	6,800	0.87	1.68*	2.24*					0.10	0.30	0.51	0.71	0.91
	7,400	0.56	1.28*						0.11	0.33	0.55	0.77	0.99
	7,800	0.31*							0.12	0.35	0.58	0.81	1.04
		4.60	5.20	5.80	6.40	7.00	7.60	8.00					
B	200	0.69	0.86	1.02	1.18	1.34	1.50	1.61	0.01	0.02	0.04	0.05	0.07
	600	1.68	2.12	2.56	2.99	3.41	3.83	4.11	0.02	0.07	0.12	0.16	0.21
	1,000	2.47	3.16	3.84	4.50	5.14	5.78	6.20	0.04	0.12	0.19	0.27	0.35
	1,400	3.13	4.03	4.91	5.76	6.59	7.39	7.91	0.05	0.16	0.27	0.38	0.49
	1,800	3.67	4.75	5.79	6.79	7.74	8.64	9.21	0.07	0.21	0.35	0.49	0.63
	2,200	4.08	5.31	6.47	7.55	8.56	9.48	10.05	0.09	0.26	0.43	0.60	0.77
	2,600	4.36	5.69	6.91	8.01	9.01	9.87	10.36	0.10	0.30	0.51	0.71	0.91
	3,000	4.51	5.89	7.11	8.17	9.04	9.73*		0.12	0.35	0.58	0.82	1.05
	3,400	4.51	5.88	7.03	7.95*				0.13	0.40	0.66	0.93	1.19
	3,800	4.34	5.64	6.65*					0.15	0.44	0.74	1.04	1.33
	4,200	4.01	5.17*						0.16	0.49	0.82	1.15	1.47
	4,600	3.48							0.18	0.54	0.90	1.25	1.61
	5,000	2.76*							0.19	0.59	0.97	1.36	1.75
		7.00	8.00	9.00	10.00	11.00	12.00	13.00					
C	100	1.03	1.29	1.55	1.81	2.06	2.31	2.56	0.01	0.03	0.05	0.08	0.10
	400	3.22	4.13	5.04	5.93	6.80	7.67	8.53	0.04	0.13	0.22	0.30	0.39
	800	5.46	7.11	8.73	10.31	11.84	13.34	14.79	0.09	0.26	0.43	0.61	0.78
	1,000	6.37	8.35	10.26	12.11	13.89	15.60	17.24	0.11	0.33	0.54	0.76	0.97
	1,400	7.83	10.32	12.68	14.89	16.94	18.83	20.55	0.15	0.46	0.76	1.06	1.36
	1,800	8.76	11.58	14.13	16.40	18.37	20.01	21.31*	0.20	0.59	0.98	1.37	1.75
	2,000	9.01	11.90	14.44	16.61	18.37	19.70*		0.22	0.65	1.08	1.52	1.95
	2,400	9.04	11.87	14.14					0.26	0.78	1.30	1.82	2.34
	2,800	8.38	10.85*						0.30	0.91	1.52	2.12	2.73
	3,000	7.76							0.33	0.98	1.63	2.28	2.92
	3,200	6.44*							0.36	1.07	1.79	2.50	3.22

* For footnote see end of table on next page.

In general the V-belting of skew shafts is discouraged because of the decrease in life of the belts, but where design demands such arrangements, special deep-groove sheaves are used. In such cases center distances should comply with the following:

For 90° turn $\qquad C_{min} = 5.5(D + W)$
For 45° turn $\qquad C_{min} = 4.0(D + W)$
For 30° turn $\qquad C_{min} = 3.0(D + W)$

where W = width of group of individual belts. Selected values of W are shown in Table 8.2.56d.

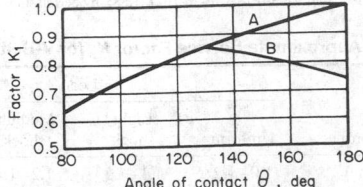

Fig. 8.2.91a Angle-of-contact correction factor, where A = grooved sheave to grooved pulley distance (V to V) and B = grooved sheave to flat-face pulley distance (V to flat).

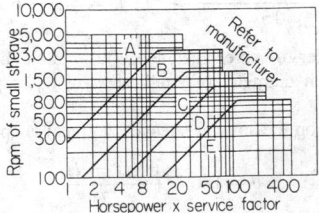

Fig. 8.2.91b V-belt section for required horsepower ratings. Letters A, B, C, D, E refer to belt cross section. (See Table 8.2.54 for service factor.)

Figure 8.2.92 shows a 90° turn arrangement, from which it can be seen that the horizontal shaft should lie some distance Z higher than the center of the vertical-shaft sheave. Table 8.2.56c lists the values of Z for various center distances in a 90° turn arrangement.

Cogged V belts have cogs molded integrally on the underside of the belt (Fig. 8.2.94a). Sheaves can be up to 25 percent smaller in diameter with cogged belts because of the greater flexibility inherent in the cogged construction. An extension of the cogged belt mating with a sheave or pulley notched at the same pitch as the cogs leads to a drive particularly useful for timing purposes.

Table 8.2.55 Horsepower Ratings of V Belts (*Continued*)

Belt section	Speed of faster shaft, r/min	Rated horsepower per belt for small sheave pitch diameter, in							Additional horsepower per belt for speed ratio				
		12.00	14.00	16.00	18.00	20.00	22.00	24.00	1.02–1.04	1.08–1.10	1.15–1.20	1.28–1.39	1.65 over
D	50	1.96	2.52	3.08	3.64	4.18	4.73	5.27	0.02	0.06	0.10	0.13	0.17
	200	6.28	8.27	10.24	12.17	14.08	15.97	17.83	0.08	0.23	0.38	0.54	0.69
	400	10.89	14.55	18.12	21.61	25.02	28.35	31.58	0.15	0.46	0.77	1.08	1.38
	600	14.67	19.75	24.64	29.33	33.82	38.10	42.15	0.23	0.69	1.15	1.61	2.07
	800	17.70	23.91	29.75	35.21	40.24	44.83	48.94	0.31	0.92	1.54	2.15	2.77
	1,000	19.93	26.94	33.30	38.96	43.86	47.93	51.12	0.38	1.15	1.92	2.69	3.46
	1,200	21.32	28.71	35.05	40.24	44.18*			0.46	1.39	2.31	3.23	4.15
	1,400	21.76	29.05	34.76*					0.54	1.62	2.69	3.77	4.84
	1,600	21.16	27.81*						0.62	1.85	3.08	4.30	5.53
	1,800	19.41							0.69	2.08	3.46	4.84	6.22
	1,950	17.28*							0.75	2.25	3.75	5.25	6.74
		18.00	21.00	24.00	27.00	30.00	33.00	36.00					
E	50	4.52	5.72	6.91	8.08	9.23	10.38	11.52	0.04	0.11	0.18	0.26	0.33
	100	8.21	10.46	12.68	14.87	17.04	19.19	21.31	0.07	0.22	0.37	0.51	0.66
	200	14.68	18.86	22.97	27.00	30.96	34.86	38.68	0.15	0.44	0.73	1.03	1.32
	300	20.37	26.29	32.05	37.67	43.13	48.43	53.58	0.22	0.66	1.10	1.54	1.98
	400	25.42	32.87	40.05	46.95	53.55	59.84	65.82	0.29	0.88	1.47	2.06	2.64
	500	29.86	38.62	46.92	54.74	62.05	68.81	75.00	0.37	1.10	1.84	2.57	3.30
	600	33.68	43.47	52.55	60.87	68.36	74.97	80.63	0.44	1.32	2.20	3.08	3.96
	700	36.84	47.36	56.83	65.15	72.22	77.93*		0.51	1.54	2.57	3.60	4.62
	800	39.29	50.20	59.61	67.36	73.30*			0.59	1.76	2.94	4.11	5.28
	900	40.97	51.89	60.73					0.66	1.98	3.30	4.63	5.94
	1,000	41.84	52.32	60.04*					0.73	2.21	3.67	5.14	6.60
	1,100	41.81	51.40*						0.81	2.43	4.04	5.65	7.26
	1,200	40.83							0.88	2.65	4.41	6.17	7.93
	1,300	38.84*							0.95	2.87	4.77	6.68	8.59

* Rim speed above 6,000 ft/min. Special sheaves may be necessary.
SOURCE: Compiled from ANSI/RMA IP-20-1988 revised, by permission.

Ribbed V belts are really flat belts molded integrally with longitudinal ribbing on the underside (Fig. 8.2.94*b*). Traction is provided principally by friction between the ribs and sheave grooves rather than by wedging action between the two, as in conventional V-belt operation. The flat

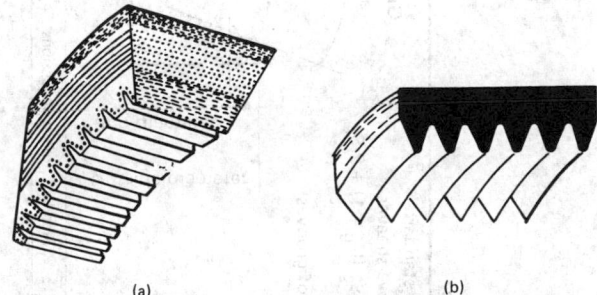

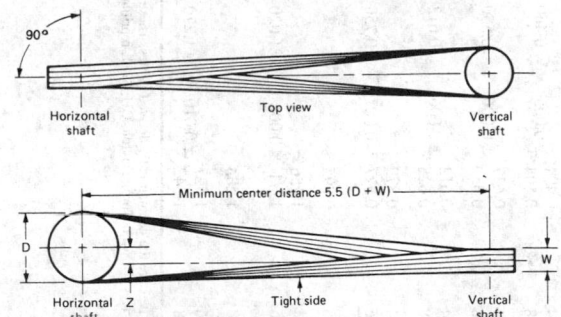

Fig. 8.2.92 Quarter-turn drive for V belts.

upper portion transmits the tensile belt loads. Ribbed belts serve well when substituted for multiple V-belt drives and for all practical purposes eliminate the necessity for belt-matching in multiple V-belt drives.

Fig. 8.2.93 V- and V-band belt cross section.

Fig. 8.2.94 Special V belts. (*a*) Cogged V belt; (*b*) ribbed V belt.

Adjustable Motor Bases

To maintain proper belt tensions on short center distances, an adjustable motor base is often used. Figure 8.2.95 shows an embodiment of such a base made by the Automatic Motor Base Co., in which adjustment for proper belt tension is made by turning a screw which opens or closes the center distance between pulleys, as required. The carriage portion of the base is spring loaded so that after the initial adjustment for belt tension

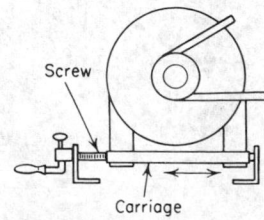

Fig. 8.2.95 Adjustable motor base.

Table 8.2.56a Standard Sheave Groove Dimensions, in

Face width of standard and deep groove sheaves

Face width = $S_g (N_g - 1) + 2 S_e$

where:

N_g = number of grooves

Cross section	Outside diameter range	Groove angle $\alpha \pm 0.33°$	b_g	h_g min	R_B min	$d_B \pm 0.0005$	$S_g \pm 0.025$	S_e	Pitch diameter range	Minimum recommended pitch diameter	$2a$
A	Up through 5.65	34	0.494 ±0.005	0.460	0.148	0.4375 (7/16)	0.625	0.375 +0.090 −0.062	Up through 5.40	3.0	0.250
	Over 5.65	38	0.504		0.149				Over 5.40		
B	Up through 7.35	34	0.637 ±0.006	0.550	0.189	0.5625 (9/16)	0.750	0.500 +0.120 −0.065	Up through 7.00	5.4	0.350
	Over 7.35	38	0.650		0.190				Over 7.00		
A/B	Up through 7.35	34	0.612 ±0.006	0.612	0.230	0.5625 (9/16)	0.750	0.500 +0.120 −0.065		A = 3.0	A = 0.620*
	Over 7.35	38	0.625		0.226					B = 5.4	B = 0.280
C	Up through 8.39	34	0.879	0.750	0.274	0.7812 (25/32)	1.000	0.688 +0.160 −0.70	Up through 7.99	9.0	0.400
	Over 8.39 to and incl. 12.40	36	0.877 ±0.007		0.276				Over 7.99 to and incl. 12.00		
	Over 12.40	38	0.895		0.277				Over 12.00		
D	Up through 13.59	34	1.256	1.020	0.410	1.1250 (1 1/8)	1.438	+0.220 −0.80	Up through 12.99	13.0	0.600
	Over 13.59 to and incl. 17.60	36	1271 ±0.008		0.410				Over 12.99 to and incl. 17.00		
	Over 17.60	38	1.283		0.411				Over 17.00		
E	Up through 24.80	36	1.527 ±0.010	1.270	0.476	1.3438 (1 11/32)	1.750	1.125 +0.280 −0.090	Up through 24.00	21.0	0.800
	Over 24.80	38	1.542		0.477				Over 24.00		

Standard groove dimensions

Drive design factors

* The a values shown for the A/B combination sheaves are the geometrically derived values. These values may be different than those shown in manufacturers' catalogs.

SOURCE: "Dayco Engineering Guide for V-Belt Drives," Dayco Corp., Dayton, OH, 1981, reprinted by permission.

Table 8.2.56b   Classical Deep Groove Sheave Dimensions, in

Cross section	Outside diameter range	Deep groove dimensions						
		Groove angle $\alpha \pm 0.33°$	b_g	h_g min	$2a$	$S_g \pm 0.025$	S_e	
A	Up through 5.96	34	0.539 ±0.005	0.615	0.560	0.750	0.438	$+0.090$ -0.062
	Over 5.96	38	0.611					
B	Up through 7.71	34	0.747 ±0.006	0.730	0.710	0.875	0.562	$+0.120$ -0.065
	Over 7.71	38	0.774					
C	Up through 9.00	34	1.066 +0.007	1.055	1.010	1.250	0.812	$+0.160$ -0.070
	Over 9.00 to and incl. 13.01	36	1.085					
	Over 13.01	38	1.105					
D	Up through 14.42	34	1.513	1.435	1.430	1.750	1.062	$+0.220$ -0.080
	Over 14.42 to and incl. 18.43	36	1.541 ±0.008					
	Over 18.43	38	1.569					
E	Up through 25.69	36	1.816 +0.010	1.715	1.690	2.062	1.312	$+0.280$ -0.090
	Over 25.69	38	1.849					

SOURCE: "Dayco Engineering Guide for V-Belt Drives," Dayco Corp., Dayton, OH, 1981, reprinted by permission.

Table 8.2.56c   Z Dimensions for Quarter-Turn Drives, in

Center distance	3V, 5V, 8V Z dimension	A B, C, D Z dimension
20		0.2
30		0.2
40		0.4
50		0.4
60	0.2	0.5
80	0.3	0.5
100	0.4	1.0
120	0.6	1.5
140	0.9	2.0
160	1.2	2.5
180	1.5	3.5
200	1.8	4.0
220	2.2	5.0
240	2.6	6.0

SOURCE: "Dayco Engineering Guide for V-Belt Drives," Dayco Corp., Dayton, OH, 1981, reprinted by permission.

Table 8.2.56d   Width W of Set of Belts Using Deep-Groove Sheaves, in

No. of belts	V-belt cross section						
	3V	5V	8V	A	B	C	D
1	0.4	0.6	1.0	0.5	0.7	0.9	1.3
2	0.9	1.4	2.3	1.3	1.6	2.2	3.1
3	1.4	2.2	3.6	2.0	2.5	3.4	4.8
4	1.9	3.0	4.0	2.8	3.3	4.7	6.6
5	2.4	3.8	6.2	3.5	4.2	5.9	8.3
6	2.9	4.7	7.6	4.3	5.1	7.2	10.1
7	3.4	5.5	8.9	5.0	6.0	8.4	11.8
8	3.9	6.3	10.2	5.8	6.8	9.7	13.6
9	4.4	7.1	11.5	6.5	7.7	10.9	15.3
10	4.9	7.9	12.8	7.3	8.6	12.2	17.1

SOURCE: "Dayco Engineering Guide for V-Belt Drives," Dayco Corp., Dayton, OH, 1981, reprinted by permission.

has been made by the screw, the spring will compensate for a normal amount of stretch in the belts. When there is more stretch than can be accommodated by the spring, the screw is turned to provide the necessary belt tensions. The carriage can be moved while the unit is in operation, and the motor base is provided for vertical as well as horizontal mounting.

CHAIN DRIVES

Roller-Chain Drives

The advantages of finished steel roller chains are high efficiency (around 98 to 99 percent), no slippage, no initial tension required, and chains may travel in either direction. The basic construction of roller chains is shown in Fig. 8.2.96 and Table 8.2.57.

The shorter the pitch, the higher the permissible operating speed of roller chains. Horsepower capacity in excess of that provided by a single chain may be had by the use of multiple chains, which are essentially parallel single chains assembled on pins common to all strands. Because of its lightness in relation to tensile strength, the effect of centrifugal pull does not need to be considered; even at the unusual speed of 6,000 ft/min, this pull is only 3 percent of the ultimate tensile strength.

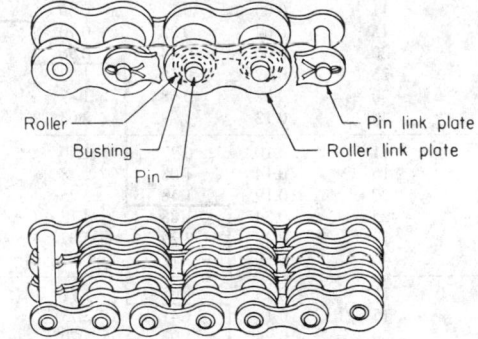

Fig. 8.2.96 Roller chain construction.

Sprocket wheels with fewer than 16 teeth may be used for relatively slow speeds, but 18 to 24 teeth are desirable for high-speed service. Sprockets with fewer than 25 teeth, running at speeds above 500 or 600 r/min, should be heat-treated to give a tough wear-resistant surface testing between 35 and 45 on the Rockwell C hardness scale.

If the speed ratio requires the larger sprocket to have as many as 128 teeth, or more than eight times the number on the smaller sprocket, it is usually better, with few exceptions, to make the desired reduction in two or more steps. The ANSI tooth form ASME B29.1 M-1993 allows roller chain to adjust itself to a larger pitch circle as the pitch of the chain elongates owing to natural wear in the pin-bushing joints. The greater the number of teeth, the sooner the chain will ride out too near the ends of the teeth.

Idler sprockets may be used on either side of the standard roller chain,

Table 8.2.57 Roller-Chain Data and Dimensions, in*

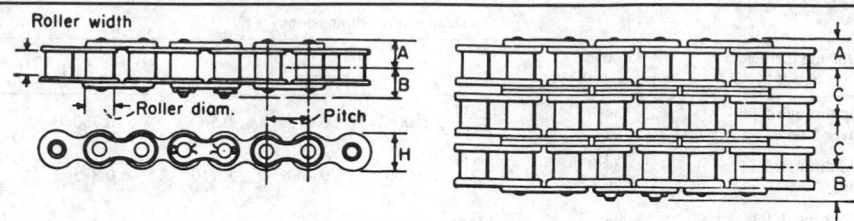

ANSI chain no.	ISO chain no.	Roller				Roller link plate		Dimension			Tensile strength per strand, lb	Recommended max speed r/min		
		Pitch	Width	Diam	Pin diam	Thickness	Height H	A	B	C		12 teeth	18 teeth	24 teeth
25	04C-1	¼	⅛	0.130	0.091	0.030	0.230	0.150	0.190	0.252	925	5,000	7,000	7,000
35	06C-1	⅜	3/16	0.200	0.141	0.050	0.344	0.224	0.290	0.399	2,100	2,380	3,780	4,200
41	085	½	¼	0.306	0.141	0.050	0.383	0.256	0.315	—	2,000	1,750	2,725	2,850
40	08A-1	½	5/16	0.312	0.156	0.060	0.452	0.313	0.358	0.566	3,700	1,800	2,830	3,000
50	10A-1	⅝	⅜	0.400	0.200	0.080	0.594	0.384	0.462	0.713	6,100	1,300	2,030	2,200
60	12A-1	¾	½	0.469	0.234	0.094	0.679	0.493	0.567	0.897	8,500	1,025	1,615	1,700
80	16A-1	1	⅝	0.625	0.312	0.125	0.903	0.643	0.762	1.153	14,500	650	1,015	1,100
100	20A-1	1¼	¾	0.750	0.375	0.156	1.128	0.780	0.910	1.408	24,000	450	730	850
120	24A-1	1½	1	0.875	0.437	0.187	1.354	0.977	1.123	1.789	34,000	350	565	650
140	28A-1	1¾	1	1.000	0.500	0.218	1.647	1.054	1.219	1.924	46,000	260	415	500
160	32A-1	2	1¼	1.125	0.562	0.250	1.900	1.250	1.433	2.305	58,000	225	360	420
180		2¼	1 13/32	1.406	0.687	0.281	2.140	1.421	1.770	2.592	76,000	180	290	330
200	40A-1	2½	1½	1.562	0.781	0.312	2.275	1.533	1.850	2.817	95,000	170	260	300
240	48A-1	3	1⅞	1.875	0.937	0.375	2.850	1.722	2.200	3.458	135,000	120	190	210

* For conversion to metric units (mm) multiply table values by 25.4.

Table 8.2.58 Selected Values of Horsepower Ratings of Roller Chains

ANSI no. and pitch, in	Number of teeth in small socket	Small sprocket, r/min										
		50	500	1,200	1,800	2,500	3,000	4,000	5,000	6,000	8,000	10,000
25	11	0.03	0.23	0.50	0.73	0.98	1.15	1.38	0.99	0.75	0.49	0.35
¼	15	0.04	0.32	0.70	1.01	1.36	1.61	2.08	1.57	1.20	0.78	0.56
	20	0.06	0.44	0.96	1.38	1.86	2.19	2.84	2.42	1.84	1.201	0.86
	25	0.07	0.56	1.22	1.76	2.37	2.79	3.61	3.38	2.57	1.67	1.20
	30	0.08	0.68	1.49	2.15	2.88	3.40	4.40	4.45	3.38	2.20	1.57
	40	0.12	.092	2.03	2.93	3.93	4.64	6.00	6.85	5.21	3.38	2.42
35	11	0.10	0.77	1.70	2.45	3.30	2.94	1.91	1.37	1.04	0.67	0.48
⅜	15	0.14	1.08	2.38	3.43	4.61	4.68	3.04	2.17	1.65	1.07	0.77
	20	0.19	1.48	3.25	4.68	6.29	7.20	4.68	3.35	2.55	1.65	1.18
	25	0.24	1.88	4.13	5.95	8.00	9.43	6.54	4.68	3.56	2.31	1.65
	30	0.29	2.29	5.03	7.25	9.74	11.5	8.59	6.15	4.68	3.04	2.17
	40	0.39	3.12	6.87	9.89	13.3	15.7	13.2	9.47	7.20	4.68	—
41	11	0.13	1.01	1.71	0.93	(0.58)	0.43	0.28	0.20	0.15	0.10	
½	15	0.18	1.41	2.73	1.49	(0.76)	0.69	0.45	0.32	0.24	0.16	
	20	0.24	1.92	4.20	2.29	(1.41)	1.06	0.69	0.49	0.38		
	25	0.31	2.45	5.38	3.20	(1.97)	1.49	0.96	0.69	0.53		
	30	0.38	2.98	6.55	4.20	(2.58)	1.95	1.27	0.91	0.69		
	40	0.51	4.07	8.94	6.47	(3.97)	3.01	1.95	1.40			
40	11	0.23	1.83	4.03	4.66	(3.56)	2.17	1.41	1.01	0.77	0.50	
½	15	0.32	2.56	5.64	7.43	(4.56)	3.45	2.24	1.60	1.22	0.79	
	20	0.44	3.50	7.69	11.1	(7.03)	5.31	3.45	2.47	1.88		
	25	0.56	4.45	9.78	14.1	(9.83)	7.43	4.82	3.45	2.63		
	30	0.68	5.42	11.9	17.2	(12.9)	9.76	6.34	4.54	3.45		
	40	0.93	7.39	16.3	23.4	(19.9)	15.0	9.76	6.99			
50	11	0.45	3.57	7.85	5.58	(3.43)	2.59	1.68	1.41	1.20	0.92	
⅝	15	0.63	4.99	11.0	8.88	(5.46)	4.13	2.68	2.25	1.92		
	20	0.86	6.80	15.0	13.7	(8.40)	6.35	4.13	3.46	2.95		
	25	1.09	8.66	19.0	19.1	(11.7)	8.88	5.77	4.83			
	30	1.33	10.5	23.2	25.1	(15.4)	11.7	7.58				
	40	1.81	14.4	31.6	38.7	(23.7)	18.0					

Table 8.2.58 Selected Values of Horsepower Ratings of Roller Chains *(Continued)*

ANSI no. and pitch, in	Number of teeth in small sprocket	Small sprocket, r/min: 10	50	100	200	500	700	1,000	1,400	2,000	2,700	4,000
60 ¾	11	0.18	0.77	1.44	2.69	6.13	8.30	11.4	9.41	5.51	(3.75)	1.95
	15	0.25	1.08	2.01	3.76	8.57	11.6	16.0	15.0	8.77	(6.18)	3.10
	20	0.35	1.47	2.75	5.13	11.7	15.8	21.8	23.1	13.5	(9.20)	
	25	0.44	1.87	3.50	6.52	14.9	20.1	27.8	32.2	18.9	(12.9)	
	30	0.54	2.28	4.26	7.94	18.1	24.5	33.8	42.4	24.8	(16.9)	
	40	0.73	3.11	5.81	10.8	24.7	33.5	46.1	62.5	38.2		
80 1	11	0.42	1.80	3.36	6.28	14.3	19.4	19.6	11.8	6.93	4.42	
	15	0.59	2.52	4.70	8.77	20.0	27.1	31.2	18.9	11.0	7.04	
	20	0.81	3.44	6.41	12.0	27.3	37.0	48.1	29.0	17.0		
	25	1.03	4.37	8.16	15.2	34.7	47.0	64.8	40.6	23.8		
	30	1.25	5.33	9.94	18.5	42.3	57.3	78.9	53.3	31.2		
	40	1.71	7.27	13.6	25.3	57.7	78.1	108	82.1	48.1		
100 1¼	11	0.81	3.45	6.44	12.0	27.4	37.1	23.4	14.2	8.29		
	15	1.13	4.83	9.01	16.8	38.3	51.9	37.3	22.5	13.2		
	20	1.55	6.58	12.3	22.9	52.3	70.8	57.5	34.7	20.3		
	25	1.97	8.38	15.5	29.2	66.6	90.1	80.3	48.5	28.4		
	30	2.40	10.2	19.0	35.5	81.0	110	106	63.7	10.0		
	40	3.27	13.9	26.0	48.5	111	150	163	98.1			
120 1½	11	1.37	5.83	10.9	20.3	46.3	46.3	27.1	16.4	9.59		
	15	1.91	8.15	15.2	28.4	64.7	73.8	43.2	26.1			
	20	2.61	11.1	20.7	38.7	88.3	114	66.5	40.1			
	25	3.32	14.1	26.4	49.3	112	152	92.9	56.1			
	30	4.05	17.2	32.1	60.0	137	185	122	73.8			
	40	5.52	23.5	43.9	81.8	187	253	188	59.5			
140 1¾	11	2.12	9.02	16.8	31.4	71.6	52.4	30.7	18.5			
	15	2.96	12.6	23.5	43.9	100	83.4	48.9	29.5			
	20	4.04	17.2	32.1	59.9	137	128	75.2	45.4			
	25	5.14	21.9	40.8	76.2	174	180	105	63.5			
	30	6.26	26.7	49.7	92.8	212	236	138				
	40	8.54	36.4	67.9	127	289	363	213				
160 2	11	3.07	13.1	24.4	45.6	96.6	58.3	34.1				
	15	4.30	18.3	34.1	63.7	145	92.8	54.4				
	20	5.86	25.0	46.4	86.9	198	143	83.7				
	25	7.40	31.8	59.3	111	252	200	117				
	30	9.08	38.7	72.2	125	307	263	154				
	40	12.4	52.8	98.5	184	419	404					
180 2¼	11	4.24	18.1	33.7	62.9	106	64.1	37.5				
	15	5.93	25.3	47.1	88.0	169	102	59.7				
	20	8.10	34.5	64.3	120	260	157	92.0				
	25	10.3	43.9	81.8	153	348	220					
	30	12.5	53.4	99.6	186	424	289					
	40	17.1	72.9	136	254	579	398					
200 2½	11	5.64	24.0	44.8	83.5	115						
	15	7.88	33.5	62.6	117	184						
	20	10.7	45.8	85.4	159	283						
	25	13.7	58.2	109	203	396						
240 3	11	9.08	38.6	72.1	135							
	15	12.7	54.0	101	188							
	20	17.3	73.7	138	257							
	25	22.0	93.8	175	327							

NOTE: The sections separated by heavy lines denote the method of lubrication as follows: type A (left section), manual; type B (middle section), bath or disk; type C (right section), oil stream.
SOURCE: Supplementary section of ANSI B29.1-1975 (R93), adapted by permission.

to take up slack, to guide the chain around obstructions, to change the direction of rotation of a driven shaft, or to provide more wrap on another sprocket. Idlers should not run faster than the speeds recommended as maximum for other sprockets with the same number of teeth. It is desirable that idlers have at least two teeth in mesh with the chain, and it is advisable, though not necessary, to have an idler contact the idle span of chain.

Horsepower ratings are based upon the number of teeth and the rotative speed of the smaller sprocket, either driver or follower. The pin-bushing bearing area, as it affects allowable working load, is the important factor for medium and higher speeds. For extremely slow speeds, the chain selection may be based upon the ultimate tensile strength of the chain. For chain speeds of 25 ft/min and less, the chain pull should be not more than ⅕ of the ultimate tensile strength; for 50 ft/min, ⅙; for

100 ft/min, $\frac{1}{7}$; for 150 ft/min, $\frac{1}{8}$; for 200 ft/min, $\frac{1}{9}$; and for 250 ft/min, $\frac{1}{10}$ of the ultimate tensile strength.

Ratings for **multiple-strand chains** are proportional to the number of strands. The recommended numbers of strands for multiple chains are 2, 3, 4, 6, 8, 10, 12, 16, 20, and 24, with the maximum overall width in any case limited to 24 in.

The **horsepower ratings** in Table 8.2.58 are modified by the **service factors** in Table 8.2.59. Thus for a drive having a nominal rating of 3 hp, subject to heavy shock, abnormal conditions, 24-h/day operation, the chain rating obtained from Table 8.2.58 should be at least $3 \times 1.7 = 5.1$ hp.

Table 8.2.59 Service Factors for Roller Chains

	Load		
Power source	Smooth	Moderate shock	Heavy shock
Internal combustion engine with hydraulic drive	1.0	1.2	1.4
Electric motor or turbine	1.0	1.3	1.5
Internal combustion engine with mechanical drive	1.2	1.4	1.7

SOURCE: ANSI B29.1-1975, adapted by permission.

Chain-Length Calculations Referring to Fig. 8.2.97, L = length of chain, in; P = pitch of chain, in; R and r = pitch radii of large and small sprockets, respectively, in; D = center distance, in; A = tangent length, in; a = angle between tangent and centerline; N and n = number

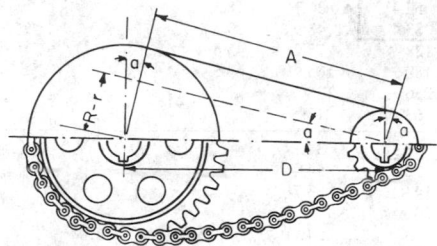

Fig. 8.2.97 Symbols for chain length calculations.

of teeth on larger and smaller sprockets, respectively; $180 + 2a$ and $180 - 2a$ are angles of contact on larger and smaller sprockets, respectively, deg.

$$a = \sin^{-1}[(R - r)/D] \qquad A = D \cos a$$
$$L = NP(180 + 2a)/360 + nP(180 - 2a)/360 + 2D \cos a$$

If L_p = length of chain in pitches and D_p = center distance in pitches,

$$L_p = (N + n)/2 + a(N - n)/180 + 2D_p \cos a$$

Avoiding the use of trigonometric tables,

$$L_p = 2C + (N + n)/2 + K(N - n)^2/C$$

where C is the center distance in pitches and K is a variable depending upon the value of $(N - n)/C$. Values of K are as follows:

$(N - n)/C$	0.1	1.0	2.0	3.0
K	0.02533	0.02538	0.02555	0.02584

$(N - n)/C$	4.0	5.0	6.0
K	0.02631	0.02704	0.02828

Formulas for chain length on multisprocket drives are cumbersome except when all sprockets are the same size and on the same side of the chain. For this condition, the chain length in pitches is equal to the sum of the consecutive center distances in pitches plus the number of teeth on one sprocket.

Actual chain lengths should be in even numbers of pitches. When necessary, an odd number of pitches may be secured by the use of an offset link, but such links should be avoided if possible. An offset link is one pitch; half roller link at one end and half pin link at the other end. If center distances are to be nonadjustable, they should be selected to give an initial snug fit for an even number of pitches of chain. For the average application, a center distance equal to 40 ± 10 pitches of chain represents good practice.

There should be at least 120° of wrap in the arc of contact on a power sprocket. For ratios of 3 : 1 or less, the wrap will be 120° or more for any center distance or number of teeth. To secure a wrap of 120° or more, for ratios greater than 3 : 1, the center distance must be not less than the difference between the pitch diameters of the two sprockets.

Sprocket Diameters N = number of teeth; P = pitch of chain, in; D = diameter of roller, in. The pitch of a standard roller chain is measured from the center of a pin to the center of an adjacent pin.

$$\text{Pitch diam} = P/\sin \frac{180}{N}$$

$$\text{Bottom diam} = \text{pitch diam} - D$$

$$\text{Outside diam} = P \left(0.6 + \cot \frac{180}{N} \right)$$

$$\text{Caliper diam} = \left(\text{pitch diam} \times \cos \frac{90}{N} \right) - D$$

The exact bottom diameter cannot be measured for an odd number of teeth, but it can be checked by measuring the distance (caliper diameter) between bottoms of the two tooth spaces nearest opposite to each other. Bottom and caliper diameters must not be oversize—all tolerances must be negative. ANSI negative tolerance = $0.003 + 0.001P\sqrt{N}$ in.

Design of Sprocket Teeth for Roller Chains The **section profile** for the teeth of roller chain sprockets, recommended by ANSI, has the proportions shown in Fig. 8.2.98. Let P = chain pitch; W = chain width (length of roller); n = number of strands of multiple chain; M = overall width of tooth profile section; H = nominal thickness of the link plate, all in inches. Referring to Fig. 8.2.98, $T = 0.93W - 0.006$, for

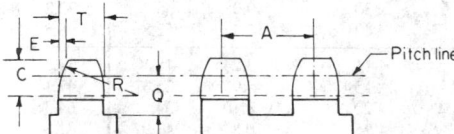

Fig. 8.2.98 Sprocket tooth sections.

single-strand chain; $= 0.90W - 0.006$, for double- and triple-strand chains; $= 0.88W - 0.006$, for quadruple- or quintuple-strand chains; and $= 0.86W - 0.006$, for sextuple-strand chain and over. $C = 0.5P$. $E = \frac{1}{8}P$. $R(\text{min}) = 1.063P$. $Q = 0.5P$. $A = W + 4.15H + 0.003$. $M = A(n - 1) + T$.

Inverted-tooth (silent) chain drives have a typical tooth form shown in Fig. 8.2.99. Such chains should be operated in an oil-retaining casing with provisions for lubrication. The use of offset links and chains with an uneven number of pitches should be avoided.

Horsepower ratings per inch of silent chain width, given in ANSI B29.2-1957 (R1971), for various chain pitches and speeds, are shown in

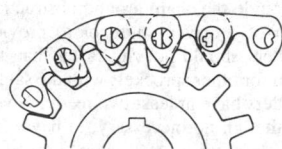

Fig. 8.2.99 Inverted tooth (silent-chain) drive.

Table 8.2.60 Horsepower Rating per Inch of Chain Width, Silent-Chain Drive (Small Pitch)

Pitch, in	No. of teeth in small sprocket	Small sprocket, r/min											
		500	600	700	800	900	1,200	1,800	2,000	3,500	5,000	7,000	9,000
3/16	21	0.41	0.48	0.55	0.62	0.68	0.87	1.22	1.33	2.03	2.58	3.12	3.35
	25	0.49	0.58	0.66	0.74	0.82	1.05	1.47	1.60	2.45	3.13	3.80	4.10
	29	0.57	0.67	0.76	0.86	0.95	1.21	1.70	1.85	2.83	3.61	4.40	4.72
	33	0.64	0.75	0.86	0.97	1.07	1.37	1.90	2.08	3.17	4.02	4.85	—
	37	0.71	0.84	0.96	1.08	1.19	1.52	2.11	2.30	3.48	4.39	5.24	—
	45	0.86	1.02	1.15	1.30	1.43	1.83	2.53	2.75	4.15	5.21	—	—
Type*		I						II			III		

Pitch, in	No. of teeth in small sprocket	Small sprocket, r/min											
		100	500	1,000	1,200	1,500	1,800	2,000	2,500	3,000	3,500	4,000	5,000	6,000
3/8	21	0.58	2.8	5.1	6.0	7.3	8.3	9.0	10	11	12	12	12	10
	25	0.69	3.3	6.1	7.3	8.8	10	11	13	14	15	5	15	14
	29	0.80	3.8	7.3	8.5	10	12	13	15	16	18	19	19	18
	33	0.90	4.4	8.3	9.8	12	14	15	18	19	21	21	21	20
	37	1.0	4.9	9.1	11	14	15	16	20	21	24	24	24	—
	45	1.3	6.0	11	13	16	19	20	24	26	28	29	—	—
Type*		I		II				III					

Pitch, in	No. of teeth in small sprocket	Small sprocket, r/min										
		100	500	700	1,000	1,200	1,800	2,000	2,500	3,000	3,500	4,000
1/2	21	1.0	5.0	6.3	8.8	10	14	14	15	16	16	—
	25	1.2	5.0	7.5	10	13	16	18	20	21	21	20
	29	1.4	6.3	8.8	13	14	19	21	24	25	25	25
	33	1.6	7.5	10	14	16	23	24	28	29	30	29
	37	1.9	8.8	11	16	19	25	26	30	33	33	—
	45	2.5	10	14	19	23	30	30	36	39	—	—
Type*		I		II			III				

Pitch, in	No of teeth in small sprocket	Small sprocket, r/min									
		100	500	700	1,000	1,200	1,800	2,000	2,500	3,000	3,500
5/8	21	1.6	7.5	10	13	15	19	20	20	20	—
	25	1.9	8.8	11	16	19	24	25	26	26	24
	29	2.1	10	14	19	21	28	30	31	31	29
	33	2.5	11	16	21	25	33	34	36	36	34
	37	2.8	13	18	24	28	36	39	43	41	—
	45	3.4	16	21	29	34	44	46	—	—	—
Type*		I		II		III				

Pitch, in	No. of teeth in small sprocket	Small sprocket, r/min								
		100	500	700	1,000	1,200	1,500	1,800	2,000	2,500
3/4	21	2.3	10	14	18	20	23	24	25	24
	25	2.8	13	16	21	25	29	31	31	30
	29	3.1	15	20	26	30	34	36	38	38
	33	3.6	16	23	30	34	39	43	44	44
	37	4.0	19	25	34	39	44	48	49	49
	45	4.9	23	30	40	46	53	56	58	—
Type*		I		II		III			

* Type I: manual, brush, or oil cup. Type II: bath or disk. Type III: circulating pump.
NOTE: For best results, smaller sprocket should have at least 21 teeth.
SOURCE: Adapted from ANSI B29.2M-1982.

Tables 8.2.60 and 8.2.61. These ratings are based on ideal drive conditions with relatively little shock or load variation, an average life of 20,000 h being assumed. In utilizing the horsepower ratings of the tables, the nominal horsepower of the drive should be multiplied by a service factor depending on the application. Maximum, or worst-case scenario, service factors are listed in Table 8.2.62.

For details on lubrication, sprocket dimensions, etc., see ANSI B29.2M-1957(R71). In utilizing Tables 8.2.60 and 8.2.61 (for a complete set of values, see ANSI B29.2M-1982) the required chain width is obtained by dividing the design horsepower by the horsepower ratings given. For calculating silent-chain lengths, the procedure for roller-chain drives may be used.

Table 8.2.61   Horsepower Rating per Inch of Chain Width, Silent-Chain Drive (Large Pitch)

Pitch, in	No. of teeth in small sprocket	Small sprocket, r/min										
		100	200	300	400	500	700	1,000	1,200	1,500	1,800	2,000
1	21	3.8	7.5	11	15	18	23	29	31	33	33	—
	25	5.0	8.8	14	18	21	28	35	39	41	41	41
	29	5.0	11	16	20	25	33	11	46	50	51	50
	33	6.3	13	18	24	29	38	49	54	59	59	58
	37	6.8	14	20	26	33	43	54	60	65	66	—
	45	8.8	16	25	31	39	51	65	71	76	—	—
Type*		I			II				III			

Pitch, in	No. of teeth in small sprocket	Small sprocket, r/min										
		100	200	300	400	500	600	700	800	1,000	1,200	1,500
$1\frac{1}{4}$	21	6.3	11	18	23	26	30	33	36	40	41	—
	25	7.5	14	20	26	31	36	40	44	50	53	53
	29	8.6	16	24	31	38	43	48	53	59	63	64
	33	9.9	19	28	35	43	49	55	60	69	73	74
	37	11	21	30	40	48	55	63	68	76	81	—
	45	13	26	38	49	59	68	75	81	91	—	—
Type*		I			II				III			

Pitch, in	No. of teeth in small sprocket	Small sprocket, r/min										
		100	200	300	400	500	600	700	800	900	1,000	1,200
$1\frac{1}{2}$	21	8.8	16	24	30	36	40	44	46	49	49	—
	25	10	20	29	38	44	50	55	59	61	65	64
	29	13	24	34	44	51	59	65	70	74	75	76
	33	14	28	39	50	59	68	75	80	85	88	89
	37	16	30	44	59	66	76	84	90	96	99	—
	45	19	38	54	68	81	93	101	108	113	—	—
Type*		I		II			III					

Pitch, in	No. of teeth in small sprocket	Small sprocket, r/min								
		100	200	300	400	500	600	700	800	900
2	21	16	29	40	50	53	63	65	—	—
	25	18	35	49	61	70	78	83	85	85
	29	21	41	58	73	84	93	99	103	103
	33	25	46	66	83	96	106	114	118	118
	37	28	53	75	72	110	124	128	131	—
	45	34	64	90	113	131	144	151	—	—
Type*		I		II			III			

* Type I: manual, brush, or oil cup. Type II: bath or disk. Type III: circulating pump.
NOTE: For best results, small sprocket should have at least 21 teeth.
SOURCE: Adapted from ANSI B29.2-1982.

Table 8.2.62   Service Factors* for Silent-Chain Drives

Application	Fluid-coupled engine or electric motor		Engine with straight mechanical drive		Torque converter drives		Application	Fluid-coupled engine or electric motor		Engine with straight mechanical drive		Torque converter drives	
	10 h	24 h	10 h	24 h	10 h	24 h		10 h	24 h	10 h	24 h	10 h	24 h
Agitators	1.1	1.4	1.3	1.6	1.5	1.8	Line shafts	1.6	1.9	1.8	2.1	2.0	2.3
Brick and clay machinery	1.4	1.7	1.6	1.9	1.8	2.1	Machine tools	1.4	1.7	—	—	—	—
Centrifuges	1.4	1.7	1.6	1.9	1.8	2.1	Mills—ball, hardinge, roller, etc.	1.6	1.9	1.8	2.1	—	—
Compressors	1.6	1.9	1.8	2.1	2.0	2.3	Mixer—concrete, liquid	1.6	1.9	1.8	2.1	2.0	2.3
Conveyors	1.6	1.9	1.8	2.1	2.0	2.3	Oil field machinery	1.6	1.9	1.8	2.1	2.0	2.3
Cranes and hoists	1.4	1.7	1.6	1.9	1.8	2.1	Oil refinery equipment	1.5	1.8	1.7	2.0	1.9	2.2
Crushing machinery	1.6	1.9	1.8	2.1	2.0	2.3	Paper machinery	1.5	1.8	1.7	2.0	1.9	2.2
Dredges	1.6	1.9	1.8	2.1	2.0	2.3	Printing machinery	1.5	2.0	1.4	1.7	1.6	1.9
Elevators	1.4	1.7	1.6	1.9	1.8	2.1	Pumps	1.6	1.9	1.8	2.1	2.0	2.3
Fans and blowers	1.5	1.8	1.7	2.0	1.9	2.2	Rubber plant machinery	1.5	1.8	1.7	2.0	1.9	2.2
Mills—flour, feed, cereal	1.4	1.7	1.6	1.9	1.8	2.1	Rubber mill equipment	1.6	1.9	1.8	2.1	2.0	2.3
Generator and excitors	1.2	1.5	1.4	1.7	1.6	1.9	Screens	1.5	1.8	1.7	2.0	1.9	2.2
Laundry machinery	1.2	1.5	1.4	1.7	1.6	1.9	Steel plants	1.3	1.6	1.5	1.8	1.7	2.0
							Textile machinery	1.1	1.4	—	—	—	—

* The values shown are for the maximum worst-case scenario for each application. The table was assembled from ANSI B29.2M-1982, by permission. Because the listings above are maximum values, overdesign may result when they are used. Consult the ANSI tables for specific design values.

ROTARY AND RECIPROCATING ELEMENTS

Slider Crank Mechanism

Kinematics   The slider crank mechanism is widely used in automobile engines, punch presses, feeding mechanisms, etc. For such mechanisms, displacements, velocities, and accelerations of the parts are important design parameters. The basic mechanism is shown in Fig. 8.2.100. The slider displacement x is given by

$$x = r \cos \theta + l \sqrt{1 - [(r/l) \sin \theta]^2}$$

where r = crank length, l = connecting-rod length, θ = crank angle measured from top dead center position. Slider velocity is given by

$$\dot{x} = V = -r\omega \left(\sin \theta + \frac{r \sin 2\theta}{2l \cos \beta} \right)$$

where ω = instantaneous angular velocity of the crank at position θ and

$$\cos \beta = \sqrt{1 - (r/l)^2 \sin^2 \theta}$$

Slider acceleration is given by

$$a = \ddot{x} = -r\alpha \left(\sin \theta + \frac{r}{l} \frac{\sin 2\theta}{2 \cos \beta} \right)$$
$$- r\omega^2 \left(\cos \theta + \frac{r}{l} \frac{\cos 2\theta}{\cos \beta} + \frac{r}{l^3} \frac{\sin^2 2\theta}{4 \cos^3 \beta} \right)$$

where α = instantaneous angular acceleration of the crank at position θ.

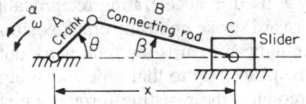

Fig. 8.2.100   Basic slider-crank mechanism.

Forces   Neglecting gravity effects, the forces in a mechanism arise from those produced by input and output forces or torques (herewith called static components). Such forces may be produced by driving motors, shaft loads, expanding cylinder gases, etc. The net forces on the various links cause accelerations of the mechanism's masses, and can be thought of as dynamic components. The static components must be borne by the various links, thus giving rise to internal stresses in those parts. The supporting bearings and slide surfaces also feel the effects of these components, as do the support frames. Stresses are also induced by the dynamic components in the links, and such components cause shaking forces and shaking moments in the support frame.

By building onto the basic mechanism appropriately designed countermasses, the support frame and its bearings can be relieved of a significant portion of the dynamic component effects, sometimes called inertia effects. The augmented mechanism is then considered to be "balanced." The static components cannot be relieved by any means, so that the support frame and its bearings must be designed to carry safely the static component forces and not be overstressed. Figure 8.2.101 illustrates a common, simple form of approximate balancing. Sizing of the countermass is somewhat complicated because the total

Fig. 8.2.101   "Balanced" slider-crank mechanism where T = center of mass, S = center of mass of crank A, and Q = center of mass of connecting rod B.

inertia effect is the vector sum of the separate link inertias, which change in magnitude and direction at each position of the crank. The countermass D is sized to contain sufficient mass to completely balance the crank plus an additional mass (effective mass) to "balance" the other links (connecting rod and slider). In simple form, the satisfying

condition is approximately

$$M_D e\omega^2 = M_A a\omega^2 + M_{\text{eff}} e\omega^2$$

where a = distance from crankshaft to center of mass of crank (note that most often a is approximately equal to the crank radius r); e = distance from crankshaft to center of mass of countermass; M_{eff} = additional mass of countermass to "balance" connecting rod and slider.

From one-half to two-thirds the slider mass (e.g., $\frac{1}{2}M_c \leq M_{\text{eff}} \leq \frac{2}{3}M_c$) is usually added to the countermass for a single-cylinder engine. For critical field work, single and multiple slider crank mechanisms are dynamically balanced by experimental means.

Forces and Torques   Figure 8.2.102a shows an exploded view of the slider crank mechanism and the various forces and torques on the links (neglecting gravity and weight effects). Inertial effects are shown as broken-line vectors and are manipulated in the same manner as the actual or real forces. The inertial effects are also known as **D'Alembert forces**. The meanings are as follows: $F_1 = -M_D e\omega^2$, parallel to crank A; $F_2 = -M_a e\alpha$, perpendicular to crank A; $F_3 = (z/l)F_9$; F_4 = as found in Fig. 8.2.102b; $F_5 = F_2 - F_7$; $F_6 = F_1 - F_8$; $F_7 = -M_A r\omega^2$, parallel to crank A; $F_8 = -M_A r\alpha$, perpendicular to crank A; $F_9 = -M_B \times$ absolute acceleration of Q, where acceleration of point Q can be found

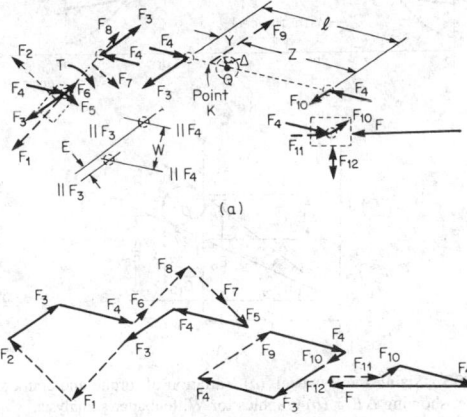

Fig. 8.2.102   (a) Forces and torques; (b) force polygons.

graphically by constructing an acceleration polygon of the mechanism (see for example Shigley, "Kinematic Analysis of Mechanisms," McGraw-Hill); $F_{10} = (y/l)F_9$; $F_{11} = -M_c \times$ absolute acceleration of slider $\ddot{x}$; F_{12} = normal wall force (neglecting friction); F = external force on slider's face, where the vector sum $F + F_4 + F_{10} + F_{11} + F_{12} = 0$; T = external crankshaft torque, where the algebraic sum $T + F_3 E + F_4 W + F_2 e + F_7 a = 0$ (note that signs must reflect direction of torque); $T_t = F_3 E + F_w W$ = transmitted torque; K = the effective location of force F_9. The distance Δ of force F_9 from the center of mass of connecting rod B (see Fig. 8.2.102a) is given by $\Delta = J_{B(\text{cm})} \times$ angular acceleration of link $B/M_B \times$ absolute acceleration of point Q. Figure 8.2.102b shows the force polygons of the separate links of the mechanism.

Flywheel

One can surmise that both F and T may be functions of crank angle θ. Even if one or the other were deliberately kept constant, the remaining one would still be a function of θ. If a steady-speed crank is desired (ω = constant and α = zero), then external crankshaft torque T must be constantly adjusted to equal transmitted torque T_t. In such a situation a motor at the crankshaft would suffer fatigue effects. In a combustion engine the crankshaft would deliver a fluctuating torque to its load.

Inserting a flywheel at the crankshaft allows the peak and valley excursions of ω to be considerably reduced because of the flywheel's ability to absorb energy over periods when $T > T_t$ and to deliver back into the system such excess energy when $T < T_t$. Figure 8.2.103a illustrates the above concepts, also showing that over one cycle of a repeated event the excess ($+$) energies and the deficient ($-$) energies are equal. The greatest crank speed change tends to occur across a single large positive loop, as illustrated in Fig. 8.2.103a.

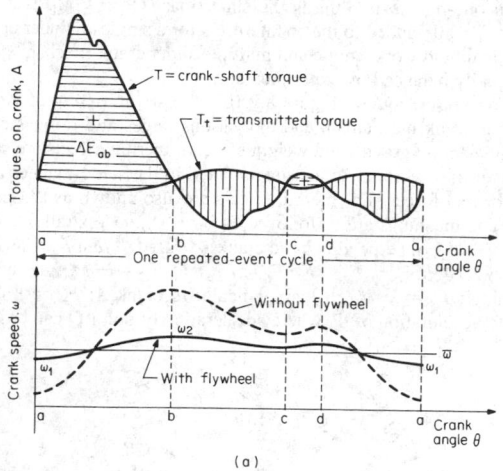

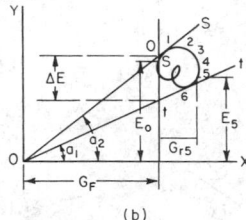

Fig. 8.2.103 Sizing the flywheel. (a) Variation of torque and crank speed vs. crank angle, showing ΔE_{ab}; (b) graphics for Wittenbauer's analysis.

Sizing the Flywheel For the single largest energy change we can write

$$\Delta E_{ab} \int_a^b (T - T_t)\, d\theta = \frac{J_0}{2}(\omega_2^2 - \omega_1^2)$$

$$= \frac{J_0}{2}(\omega_2 - \omega_1)(\omega_2 + \omega_1)$$

where J_0 = flywheel moment of inertia plus effective mechanism moment of inertia.

Define $\overline{\omega} \doteq (\omega_2 + \omega_1)/2$ and C_s = coefficient of speed fluctuation = $(\omega_2 - \omega_1)/\overline{\omega}$. Hence $\Delta E_{ab} = J_0 C_s \overline{\omega}^2$. Acceptable values of C_s are:

Pumps	0.03–0.05
Machine tools	0.025–0.03
Looms	0.025
Paper mills	0.025
Spinning mills	0.015
Crusher	0.02
Electric generators, ac	0.003
Electric generators, dc	0.002

Evaluating ΔE_{ab} involves finding the integral

$$\int_a^b (T - T_1)\, d\theta$$

which can be done graphically or by a numerical technique such as Simpson's rule.

Wittenbauer's Analysis for Flywheel Performance This method does not involve more computation work than the one described above, but it is more accurate where the reciprocating parts are comparatively heavy. Wittenbauer's method avoids the inaccuracy resulting from the evaluation of the inertia forces on the reciprocating parts on the basis of the uniform nominal speed of rotation for the engine.

Let the crankpin velocity be represented by v_r, and the velocity of any moving masses (m_1, m_2, m_3, etc.) at any instant of phase be represented, respectively, by v_1, v_2, v_3, etc. The kinetic energy of the entire engine system of moving masses may then be expressed as

$$E = \tfrac{1}{2}(m_1 v_1^2 + m_2 v_2^2 + m_3 v_3^2 + \cdots) = \tfrac{1}{2} M_r v_r^2$$

or, the single reduced mass M_r at the crankpin which possesses the equivalent kinetic energy is

$$M_r = [m_1(v_1/v_r)^2 + m_2(v_2/v_r)^2 + m_3(v_3/v_r)^2 + \cdots]$$

In an engine mechanism, sufficiently accurate values of M_r can be obtained if the weight of the connecting rod is divided between the crankpin and the wrist pin so as to retain the center of gravity of the rod in its true position; usually 0.55 to 0.65 of the weight of the connecting rod should be placed on the crankpin, and 0.45 to 0.35 of the weight on the wrist pin. M_r is a variable in engine mechanisms on account of the reciprocating parts and should be found for a number of crank positions. It should include all moving masses except the flywheel.

The total energy E used in accelerating reciprocating parts from the beginning of the forward stroke up to any crank position can be obtained by finding from the indicator cards the total work done in the cylinder (on both sides of the piston) up to that time and subtracting from it the work done in overcoming the resisting torque, which may usually be assumed constant. The mean energy of the moving masses is $E_0 = \tfrac{1}{2} M_r v_r^2$.

In Fig. 8.2.103b, the reduced weights of the moving masses $G_F + G_{r5}$ are plotted on the X axis corresponding to different crank positions. $G_F = gM_F$ is the reduced flywheel weight and $G_{r5} = gM_{r5}$ is the sum of the other reduced weights. Against each of these abscissas is plotted the energy E available for acceleration measured from the beginning of the forward stroke. The curve $O123456$ is the locus of these plotted points.

The diagram possesses the following property: Any straight line drawn from the origin O to any point in the curve is a measure of the velocity of the moving masses; tangents bounding the diagram measure the limits of velocity between which the crankpin will operate. The maximum linear velocity of the crankpin in feet per second is $v_2 = \sqrt{2g} \tan a_2$, and the minimum velocity is $v_1 = \sqrt{2g} \tan a_1$. Any desired change in v_1 and v_2 may be accomplished by changing the value of G_F, which means a change in the flywheel weight or a change in the flywheel weight reduced to the crankpin. As G_F is very large compared with G_r and the point 0 cannot be located on the diagram unless a very large drawing is made, the tangents are best formed by direct calculation:

$$\tan \alpha_2 = \frac{v_r^2}{2g}(1 + k) \qquad \tan \alpha_1 = \frac{v_r^2}{2g}(1 - k)$$

where k is the coefficient of velocity fluctuation. The two tangents ss and tt to the curve $O123456$, thus drawn, cut a distance ΔE and on the ordinate E_0. The reduced flywheel weight is then found to be

$$G_F(\Delta E)g/(v_r^2 k)$$

SPRINGS

It is assumed in the following formulas that the springs are in no case stressed beyond the elastic limit (i.e., that they are perfectly elastic) and that they are subject to Hooke's law.

Notation

P = safe load, lb
f = deflection for a given load P, in
l = length of spring, in
V = volume of spring, in^3
S_s = safe stress (due to bending), lb/in^2
S_v = safe shearing stress, lb/in^2
U = resilience, in·lb

For sheet metal and wire gages, ferrous and nonferrous, see Table 8.2.76 and metal suppliers' catalogs.

The **work** in inch-pounds performed in **deflecting a spring** from 0 to f (spring duty) is $U = Pf/2 = s_s^2 V/(CE)$. This is based upon the assumption that the deflection is proportional to the load, and C is a constant dependent upon the shape of the springs.

The **time of vibration** T (in seconds) **of a spring** (weight not considered) is equal to that of a simple circular pendulum whose length l_0 equals the deflection f (in feet that is produced in the spring by the load P. $T = \pi\sqrt{l_0/g}$, where g = acceleration of gravity, ft/s^2.

Springs Subjected to Bending

1. **Rectangular plate spring** (Fig. 8.2.104).

$$P = bh^2 S_s/(6l) \qquad I = bh^3/12 \qquad U = Pf/2 = VS_s^2/(18E)$$
$$f = Pl^3/(3EI) = 4Pl^3/(bh^3 E) = 2l^2 S_s/(3hE)$$

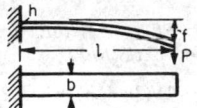

Fig. 8.2.104 Rectangular plate spring.

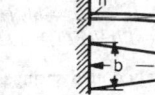

Fig. 8.2.105 Triangular plate spring.

2. **Triangular plate spring** (Fig. 8.2.105). The elastic curve is a circular arc.

$$P = bh^2 S_s/(6l) \qquad I = bh^3/12 \qquad U = Pf/2 = S_s^2 V/(6E)$$
$$f = Pl^3/(2EI) = 6Pl^3/(bh^3 E) = l^2 S_s/(hE)$$

3. **Rectangular plate spring with end tapered** in the form of a cubic parabola (Fig. 8.2.106). The elastic curve is a circular arc; P, I, and f same for triangular plate spring (Fig. 8.2.105); $U = Pf/2 = S_s^2 V/(9E)$.

The strength and deflection of **single-leaf flat springs** of various forms are given (Bruce, *Am. Mach.*, July 19, 1900) by the formulas $h = al^2/f$ and $b = cPl/h^2$. The volume of the spring is given by $V = vlbh$. The values of constants a and c and the resilience in inch-pounds per cubic inch are given in Table 8.2.63, in terms of the safe stress S_s. Values of v are given also.

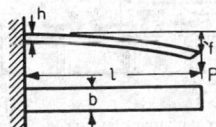

Fig. 8.2.106 Rectangular plate spring: tapered end.

4. **Compound (leaf or laminated) springs.** If several springs of rectangular section are combined, the resulting compound spring should (1) form a beam of uniform strength that (2) does not open between the joints while bending (i.e., elastic curve must be a circular arc). Only the type immediately following meets both requirements, the others meeting only the second requirement.

5. **Laminated triangular plate spring** (Fig. 8.2.107). If the triangular plate spring shown at I is cut into an even number (= $2n$) of strips of equal width (in this case eight strips of width $b/2$), and these strips are combined, a laminated spring will be formed whose carrying capacity will equal that of the original uncut spring; or $P = nbh^2 S_s/(6l)$; $n = 6Pl/(bh^2 S_s)$.

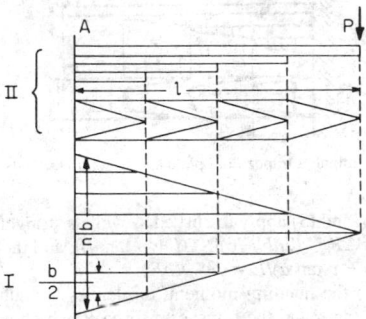

Fig. 8.2.107 Laminated triangular plate spring.

Table 8.2.63 Strength and Deflection of Single-Leaf Flat Springs

Plans and elevations of springs	a	U	v	Plans and elevations of springs	a	U	v
	Load applied at end of spring; $c = 6/S_s$				Load applied at center of spring; $c = 6/4S_s$		
	$\dfrac{S_s}{E}$	$\dfrac{S_s^2}{6E}$	$\dfrac{1}{2}$		$\dfrac{S_s}{4E}$	$\dfrac{S_s^2}{6E}$	$\dfrac{1}{2}$
	$\dfrac{4S_s}{3E}$	$\dfrac{S_s^2}{6E}$	$\dfrac{2}{3}$		$\dfrac{0.87S_s}{4E}$	$\dfrac{0.70S_s^2}{6E}$	$\dfrac{5}{8}$
Parabolic arc	$\dfrac{2S_s}{3E}$	$\dfrac{0.33S_s^2}{6E}$	1		$\dfrac{S_s}{3E}$	$\dfrac{S_s^2}{6E}$	$\dfrac{2}{3}$
	$\dfrac{0.87S_s}{E}$	$\dfrac{0.70S_s^2}{6E}$	$\dfrac{5}{8}$	Parabolic arcs	$\dfrac{1.09S_s}{4E}$	$\dfrac{0.725S_s^2}{6E}$	$\dfrac{3}{4}$
	$\dfrac{1.09S_s}{E}$	$\dfrac{0.725S_s^2}{6E}$	$\dfrac{3}{4}$		$\dfrac{S_s}{6E}$	$\dfrac{0.33S_s^2}{6E}$	1

6. **Laminated rectangular plate spring with lead ends tapered** in the form of a cubical parabola (Fig. 8.2.108); see case 3.

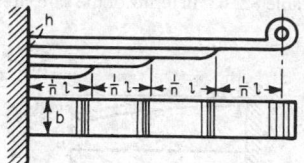

Fig. 8.2.108 Laminated rectangular plate spring with leaf end tapered.

7. **Laminated trapezoidal plate spring with leaf ends tapered** (Fig. 8.2.109). The ends of the leaves are trapezoidal and are tapered according to the formula

$$z = \frac{h}{\sqrt[3]{1 + (b_1/b)(a/x - 1)}}$$

8. **Semielliptic springs** (for locomotives, trucks, etc.). Referring to Fig. 8.2.110, the load $2P$ (lb) acting on the spring center band produces a tensional stress $P/\cos a$ in each of the inclined shackle links. This is resolved into the vertical force P and the horizontal force $P \tan a$, which together produce a bending moment $M = P(l + p \tan a)$. The equations

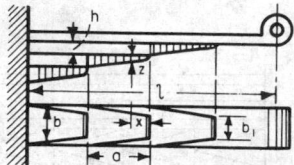

Fig. 8.2.109 Laminated trapezoidal plate spring with leaf ends tapered.

given in (1), (2), and (3) apply to curved as well as straight springs. The bearing force $= 2P = (2nbh^2/6)[S_s/(l + p \tan a)]$, and the deflection $= [6l^2/(nbh^3)]P(l + p \tan a)/E = l^2S_2/(hE)$.

In addition to the bending moment, the leaves are subjected to the tension force $P \tan a$ and the transverse force P, which produce in the upper leaf an additional stress $S = P \tan a/(bh)$, as well as a transverse shearing stress.

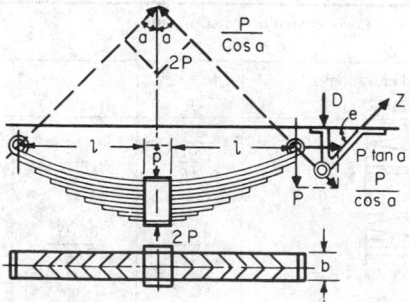

Fig. 8.2.110 Semielliptic springs.

In determining the number of leaves n in a given spring, allowance should be made for an excess load on the spring caused by the vibration. This is usually made by decreasing the allowable stress about 15 percent.

The foregoing does not take account of initial stresses caused by the band. For more detailed information, see Wahl, "Mechanical Springs," Penton.

9. **Elliptic springs.** Safe load $P = nbh^2S_s/(6l)$, where $l = \frac{1}{2}$ distance between bolt eyes (less $\frac{1}{2}$ length of center band, where used); deflection $f = 4l^2S_sK/(hE)$, where

$$K = \frac{1}{(1-r)^3}\left[\frac{1-r^2}{2} - 2r(1-r) - r^2 \ln r\right]$$

r being the number of full-length leaves ÷ total number (n) of leaves in the spring. All dimensions in inches. For semielliptic springs, the deflection is only half as great. Safe load $= nbh^2S_s/(3l)$. (Peddle, *Am. Mach.*, Apr. 17, 1913.)

Coiled Springs In these, the load is applied as a couple Pr which turns the spring while winding or holds it in place when wound up. If the spindle is not to be subjected to bending moment, P must be replaced by two equal and opposite forces ($P/2$) acting at the circumference of a circle of radius r. The formulas are the same in both cases. The springs are assumed to be fixed at one end and free at the other. The bending moment acting on the section of least resistance is always Pr. The length of the straightened spring $= l$. See Benjamin and French, Experiments on Helical Springs, *Trans. ASME*, **23**, p. 298.

For **heavy closely coiled helical springs** the usual formulas are inaccurate and result in stresses greatly in excess of those assumed. See Wahl, Stresses in Heavy Closely-Coiled Helical Springs, *Trans. ASME*, 1929. In springs 10 to 12 and 15 to 18, the quantity k is unity for lighter springs and has the stated values (supplied by Wahl) for heavy closely coiled springs.

10. **Spiral coiled springs of rectangular cross section** (Fig. 8.2.111).

$$P = bh^2S_s/(6rk) \qquad I = bh^3/12 \qquad U = Pf/2 = S_s^2V/(6Ek^2)$$
$$f = ra = Plr^2/(EI) = 12Plr^2/(Ebh^3) = 2rlS_s/(hEk)$$

For heavy closely coiled springs, $k = (3c - 1)/(3c - 3)$, where $c = 2R/h$ and R is the minimum radius of curvature at the center of the spiral.

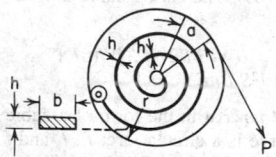
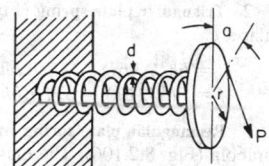

Fig. 8.2.111 Spiral coiled spring: rectangular cross section.

Fig. 8.2.112 Cylindrical helical spring: circular cross section.

11. **Cylindrical helical spring of circular cross section** (Fig. 8.2.112).

$$P = \pi d^3S_s/(32rk) \qquad I = \pi d^4/64 \qquad U = Pf/2 = S_s^2V/(8Ek^2)$$
$$f = ra = Plr^2/(EI) = 64Plr^2/(\pi Ed^4) = 2rlS_s/(dEk)$$

For heavy closely coiled springs, $k = (4c - 1)/(4c - 4)$, where $c = 2r/d$.

12. **Cylindrical helical spring of rectangular cross section** (Fig. 8.2.113).

$$P = bh^2S_s/(6rk) \qquad I = bh^3/12 \qquad U = Pf/2 = S_s^2V/(8Ek^2)$$
$$f = ra = Plr^2/(EI) = 12Plr^2/(Ebh^3) = 2rlS_s/(hEk)$$

For heavy closely coiled springs, $k = (3c - 1)/(3c - 3)$, where $c = 2r/h$.

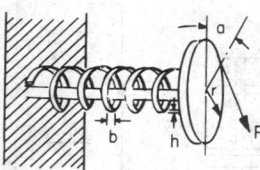

Fig. 8.2.113 Cylindrical helical spring: rectangular cross section.

Springs Subjected to Torsion

The statements made concerning coiled springs subjected to bending apply also to springs 13 and 14.

13. **Straight bar spring of circular cross section** (Fig. 8.2.114).

$$P = \pi d^3 S_v/(16r) = 0.1963 d^3 S_v/r \qquad U = Pf/2 = S_v^2 V/(4G)$$
$$f = ra = 32r^2 lP/(\pi d^4 G) = 2rlS_v/(dG)$$

14. **Straight bar spring of rectangular cross section** (Fig. 8.2.115).

$$P = 2b^2 h S_v/(9r) \qquad K = b/h$$
$$U = Pf/2 = 4S_v^2 V(K^2 + 1)/(45G) \qquad \text{max when } K = 1$$
$$f = ra = 3.6r^2 lP(b^2 + h^2)/(b^3 h^3 G) = 0.8rlS_v(b^2 + h^2)/(bh^2 G)$$

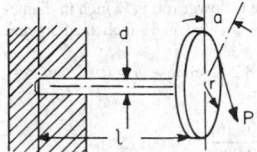

Fig. 8.2.114   Straight bar spring: circular cross section.

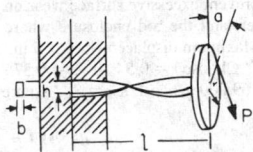

Fig. 8.2.115   Straight bar spring: rectangular cross section.

Springs Loaded Axially in Either Tension or Compression

NOTE.   For springs 15 to 18, r = mean radius of coil; n = number of coils.

15. **Cylindrical helical spring of circular cross section** (Fig. 8.2.116).

$$P = \pi d^3 S_v/(16rk) = 0.1963 d^3 S_v/(rk)$$
$$U = Pf/2 = S_v^2 V/(4Gk^2)$$
$$f = 64nr^3 P/(d^4 G) = 4\pi nr^2 S_v/(dGk)$$

For heavy closely coiled springs, $k = (4c - 1)/(4c - 4) + 0.615/c$, where $c = 2r/d$.

Fig. 8.2.116   Cylindrical helical spring: circular cross section.

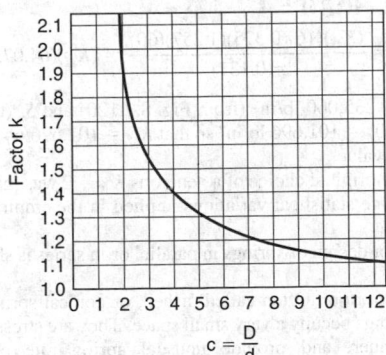

Fig. 8.2.117   Wahl correction factor.

16. **Cylindrical helical spring of rectangular cross section** (Fig. 8.2.118).

$$P = 2b^2 h S_v/(9rk) \qquad K = b/h$$
$$U = Pf/2 = 4S_v^2 V(K^2 + 1)/(45Gk^2) \qquad \text{max when } K = 1$$
$$f = 7.2\pi nr^3 P(b^2 + h^2)/(b^3 h^3 G) = 1.6\pi nr^2 S_v(b^2 + h^2)/(bh^2 Gk)$$

For heavy closely coiled springs, $k = (4c - 1)/(4c - 4) + 0.615/c$, where $c = 2r/b$.

Fig. 8.2.118   Cylindrical helical spring: rectangular cross section.

17. **Conical helical spring of circular cross section** (Fig. 8.2.119a).

$$l = \text{length of developed spring}$$
$$d = \text{diameter of wire}$$
$$r = \text{maximum mean radius of coil}$$
$$P = \pi d^3 S_v/(16rk) = 0.1963 d^3 S_v/(rk)$$
$$U = Pf/2 = S_v^2 V/(8Gk^2)$$
$$f = 16r^2 lP/(\pi d^4 G) = 16nr^3 P/(d^4 G)$$
$$= rlS_v/(dGk) = \pi nr^2 S_v/(dGk)$$

For heavy closely coiled springs, $k = (4c - 1)/(4c - 4) + 0.615/c$, where $c = 2r/d$.

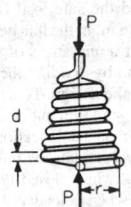

Fig. 8.2.119a   Conical helical spring: circular cross section.

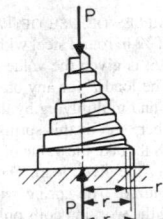

Fig. 8.2.119b   Conical helical spring: rectangular cross section.

18. **Conical helical spring of rectangular cross section** (Fig. 8.2.119b).

$$b = \text{small dimension of section}$$
$$d = \text{large dimension of section}$$
$$r = \text{maximum mean radius of coil}$$
$$K = b/h \; (\leq 1) \qquad P = 2b^2 h S_v/(9rk)$$
$$U = Pf/2 = 2S_v^2 V(K^2 + 1)/(45Gk^2) \qquad \text{max when } K = 1$$
$$f = 1.8r^2 lP(b^2 + h^2)/(b^3 h^3 G) = 1.8\pi nr^3 P(b^2 + h^2)/(b^3 h^3 G)$$
$$= 0.4rlS_v(b^2 + h^2)/(bh^2 Gk) = 0.4\pi nr^2 S_v(b^2 + h^2)/(bh^2 Gk)$$

For heavy closed coiled springs, $k = (4c - 1)/(4c - 4) + 0.615c$, where $c = 2r/r_o - r_i$.

19. **Truncated conical springs** (17 and 18). The formulas under 17 and 18 apply for truncated springs. In calculating deflection f, however, it is necessary to substitute $r_1^2 + r_2^2$ for r^2, and $\pi n(r_1 + r_2)$ for πnr, r_1 and r_2 being, respectively, the greatest and least mean radii of the coils.

NOTE.   The preceding formulas for various forms of coiled springs are sufficiently accurate when the cross-section dimensions are small in comparison with the radius of the coil, and for small pitch. Springs 15 to 19 are for either tension or compression but formulas for springs 17 and 18 are good for compression only until the largest coil flattens out; then r becomes a variable, depending on the load.

Design of Helical Springs

When sizing a new spring, one must consider the spring's available working space and the loads and deflections the spring must experience. Refinements dictated by temperature, corrosion, reliability, cost, etc. may also enter design considerations. The two basic formulas of load and deflection (see item 15, Fig. 8.2.116) contain eight variables (f, P, d, S, r, k, n, G) which prevent one from being able to use a one-step solution. For instance, if f and P are known and S and G are chosen, there still remain d, r, k, and n to be found.

A variety of solution approaches are available, including: (1) slide-rule-like devices available from spring manufacturers, (2) nomographic methods (Chironis, "Spring Design and Application," McGraw-Hill; Tsai, Speedy Design of Helical Compression Springs by Nomography Method, *J. of Eng. for Industry,* Feb. 1975), (3) table methods (Carlson, "Spring Designer's Handbook," Marcel Dekker), (4) formula method (ibid.), and (5) computer programs and subroutines.

Design by Tables

Safe working loads and deflections of cylindrical helical springs of round steel wire in tension or compression are given in Table 8.2.64. The table is based on the formulas given for spring 15. d = diameter of steel wire, in; D = pitch diameter (center to center of wire), in; P = safe working load for given unit stress, lb; f = deflection of 1 coil for safe working load, in.

The table is based on the values of unit stress indicated, and $G = 12{,}500{,}000$. For any other value of unit stress, divide the tabular value by the unit stress used in the table and multiply by the unit stress to be used in the design. For any other value of G, multiply the value of f in the table by 12,500,000 and divide by the value of G chosen. For **square steel wire**, multiply values of P by 1.06, and values of f by 0.75. For **round brass wire**, take $S_s = 10{,}000$ to 20,000, and multiply values of f by 2 (Howe).

EXAMPLES OF USE OF TABLE 8.2.64. 1. Required the safe load (P) for a spring of ⅜-in round steel with a pitch diameter (D) of 3½ in. In the line headed D, under 3½, is given the value of P, or 678 lb. This is for a unit stress of 115,000 lb/in². The load P for any other unit stress may be found by dividing the 678 by 115,000 and multiplying by the unit stress to be used in the design. To determine the number of coils this spring would need to compress (say) 6 in under a load of (say) 678 lb, take the value of f under 678, or 0.938, which is the deflection of one coil under the given load. Therefore, 6/0.938 = 6.4, say 7, equals the number of coils required. The spring will therefore be 2⅝ in long when closed (7 × ⅜), counting the working coils only, and must be 8⅝ in long when unloaded. Whether there is an extra coil at one end which does not deflect will depend upon the details of the particular design. The deflection in the above example is for a unit stress of 115,000 lb/in². The rule is, divide the deflection by 115,000 and multiply by the unit stress to be used in the design.

2. A ⁷⁄₁₆-in steel spring of 3½-in OD has its coils in close contact. How much can it be extended without exceeding the limit of safety? The maximum safe load for this spring is found to be 1,074 lb, and the deflection of one coil under this load is 0.810 in. This is for a unit stress of 115,000 lb/in². Therefore, 0.810 is the greatest admissible opening between any two coils. In this way, it is possible to ascertain whether or not a spring is overloaded, without knowledge of the load carried.

Design by Formula

A design formula can be constructed by equating calculated stress s_v (from load formula, Fig. 8.2.117) to an allowable working stress in torsion:

$$s_v = \sigma_{max} = \frac{16rP}{\pi d^3}\left(\frac{4c-1}{4c-4}+\frac{0.615}{c}\right) = \frac{16rPk}{\pi d^3} = \frac{S_v}{K_{sf}}$$

where P = axial load on spring, lb; $r = D/2$ = mean radius of coil, in; D = mean diameter of coil, in (outside diameter minus wire diameter); d = wire diameter, in; σ_{max} = torsional stress, lb/in²; K_{sf} = safety factor and $c = D/d$. Note that the expression in parentheses [$(4c-1)(4c-4) + 0.615/c$] is the **Wahl correction factor** k, which accounts for the added stresses in the coils due to curvature and direct shear. See Fig. 8.2.117. Values of S_v, yield point in shear from standard tests, are strongly dependent on d, hence the availability of S_v in the literature is limited. However, an empirical relationship between S_{uT} and d is available (see Shigley, "Mechanical Engineering Design," McGraw-Hill, 4th ed., p. 452). Using also the approximate relations $S_y = 0.75S_{ut}$

and $S_v = 0.577S_y$ results in the following relationship:

$$S_v = \frac{0.43A}{d^m}$$

where A and m are constants (see Table 8.2.65).

Substituting S_v and rearranging yield the following useful formula:

$$d^{3-m} = \frac{K_{sf}16rPk}{\pi 0.43A}$$

EXAMPLE. A slow-speed follower is kept in contact with its cam by means of a helical compression spring, in which the minimum spring force desired is 20 lb to assure firm contact, while maximum spring force is not to exceed 60 lb to prevent excessive surface wear on the cam. The follower rod is ¼ inch in diameter, and the rod enclosure where the spring is located is ⅞ inch in diameter. Maximum displacement is 1.5 in.

Choose $r = 0.5 \times 0.75 = 0.375$. From Table 8.2.65 choose $m = 0.167$ and $A = 169{,}000$. Choose $K_{sf} = 2$. Assume $k = 1$ to start.

$$d^{3-0.167} = \frac{(2)(0.375)(60)(16)}{(\pi)(0.43)(169{,}000)} = 0.00315$$

$$d = (0.00315)^{0.35298} = 0.131 \text{ in}$$

$$\text{Spring constant} = \frac{\Delta P}{\Delta f} = \frac{40}{1.5} = 26.667$$

and from the deflection formula, the number of active turns

$$n = \frac{(0.131)^4(11{,}500{,}000)}{(26.667)(64)(0.375)^3} = 37.6$$

For squared and ground ends add two dead coils, so that

$$n_{total} = 40$$
$$H = \text{solid height} = (40)(0.131) = 5.24 \text{ in}$$
$$f_0 = \text{displacement from zero to maximum load}$$
$$= 60/26.667 = 2.249$$
$$L_0 = \text{approximate free length} = 5.24 + 2.249 = 7.49 \text{ in}$$

NOTE. Some clearance should be added between coils so that at maximum load the spring is not closed to its solid height. Also, the nearest commercial stock size should be selected, and recalculations made on this stock size for S_v, remembering to include k at this juncture. If recalculated S_v is satisfactory as compared with published values, the design is retained, otherwise enough iterations are performed to arrive at a satisfactory result. Figure 8.2.120 shows a plot of S_{uT} versus d. To convert S_{uT} to S_v, multiply S_{uT} by 0.43.

$$c = 2(0.375/0.131) = 5.73$$

$$k = \frac{4(5.73)-1}{4(5.73)-4} + \frac{0.615}{5.73} = 1.257$$

$$S_v = \frac{(K_{sf})(16)(0.375)(1.257)(60)}{\pi(0.131)^3} = (K_{sf})(64{,}072)$$

Now $S_{uT} = 235{,}000$ lb/in² (from Fig. 8.2.120) and S_v (tabulated) = $(0.43)(235{,}000) = 101{,}696$ lb/in² so that $K_{sf} = 101{,}696/64{,}072 = 1.59$, a satisfactory value.

NOTE. The original choice of a generous $K_{sf} = 2$ was made to hedge against the large statistical variations implied in the empirical formula $S_v = 0.43A/d^m$.

The basis for design of springs in parallel or in series is shown in Fig. 8.2.121.

Belleville Springs Often called dished, or conical spring, washers, Belleville springs occupy a very small space. They are stressed in a very complex manner, and provide unusual spring rate curves (Fig. 8.2.122a). These springs are nonlinear, but for some proportions, they behave with approximately linear characteristics in a limited range. Likewise, for some proportions they can be used through a spectrum of spring rates, from positive to flat and then through a negative region. The snap-through action, shown at point A in Fig. 8.2.122b, can be useful in particular applications requiring reversal of spring rates. These

Table 8.2.64 Safe Working Loads _P_ and Deflections _f_ of Cylindrical Helical Steel Springs of Circular Cross Section
(For closely coiled springs, divide given load and deflection values by the curvature factor _k_.)

Allowable unit stress, lb/in²	Diam, in	D	5/32	3/16	1/4	5/16	3/8	7/16	1/2	5/8	3/4	7/8	1	1 1/8	1 1/4	1 3/8	1 1/2	1 5/8	1 3/4	1 7/8	2	2 1/4	2 1/2
150,000	0.035	P	16.2	13.4	10.0	8.10	6.66	5.75	4.96	4.05	3.39												
		f	.026	.037	.067	.105	.149	.200	.276	.420	.608												
	0.041	P	26.2	21.6	16.2	13.0	10.8	9.27	8.10	6.52	5.35	4.57											
		f	.023	.032	.057	.089	.128	.175	.229	.362	.512	.697											
	0.047	P	39.1	32.6	24.5	19.6	16.4	13.9	12.3	9.80	8.10	6.92	6.14										
		f	.019	.028	.049	.078	.112	.153	.200	.311	.449	.610	.800										
	0.054	P	59.4	49.6	37.2	29.7	24.6	21.2	18.5	14.7	12.4	10.5	9.25	8.23									
		f	.016	.024	.043	.067	.098	.133	.174	.273	.390	.532	.695	.878									
	0.062	P		74.9	56.1	44.9	37.3	32.0	28.0	22.4	18.6	16.1	13.9	12.5	11.2								
		f		.021	.037	.058	.084	.115	.151	.235	.340	.460	.605	.760	.947								
	0.063	P		78.2	58.7	46.9	39.2	33.9	29.4	23.5	19.6	16.8	14.7	13.2	11.9								
		f		.020	.037	.057	.083	.113	.148	.233	.335	.445	.591	.748	.930								
	0.072	P		117.	80.7	70.0	58.7	50.2	43.6	35.2	29.0	25.0	21.9	19.5	17.5	16.0							
		f		.018	.032	.050	.077	.100	.130	.203	.294	.405	.521	.652	.802	.986							
	0.080	P			121	96.6	80.5	69.1	60.4	48.2	40.2	34.6	30.1	26.7	24.2	22.1	20.2						
		f			.029	.045	.065	.090	.117	.183	.262	.359	.470	.593	.735	.886	1.105						
140,000	0.092	P				136	113	97.6	85.5	68.9	57.3	48.8	42.6	37.8	34.5	31.3	28.6						
		f				.037	.053	.072	.098	.148	.214	.291	.388	.481	.596	.720	.854						
	0.093	P					118	99.5	89.0	71.2	59.1	50.9	44.3	39.6	35.7	32.3	29.6	27.3					
		f					.052	.071	.093	.146	.211	.286	.376	.473	.585	.707	.841	.986					
	0.105	P				204	170	147	127	102	85.4	73.0	63.4	56.6	51.1	46.3	42.6	38.9					
		f				.032	.047	.064	.083	.122	.188	.256	.336	.425	.512	.632	.755	.880					
	0.120	P				303	253	217	190	152	126	108	95.2	84.2	76.2	69.2	63.5	58.5	54.3				
		f				.028	.041	.055	.073	.114	.174	.223	.296	.368	.449	.551	.657	.768	.893				
	0.125	P					286	245	214	171	143	121	107	95.5	85.2	78.0	71.5	65.8	60.8	57.2			
		f					.039	.053	.069	.109	.169	.213	.278	.353	.437	.528	.626	.731	.855	.981			
	0.135	P					359	309	270	217	171	154	135	120	108	98.7	90.2	82.7	76.2	71.8	67.5		
		f					.036	.049	.064	.106	.145	.198	.260	.327	.399	.486	.581	.680	.791	.908	1.04		
	0.148	P						408	356	285	237	207	178	158	142	130	118	109	102	95.0	89.0		
		f						.045	.059	.092	.132	.180	.236	.293	.370	.448	.530	.620	.723	.828	.945		

Pitch diameter D, in

Table 8.2.64   Safe Working Loads P and Deflections f of Cylindrical Helical Steel Springs of Circular Cross Section   (Continued)

Allowable unit stress, lb/in²	Diam, in	D	7/16	1/2	5/8	3/4	7/8	1	1⅛	1¼	1⅜	1½	1⅝	1¾	1⅞	2	2¼	2½	2¾	3	3½	4	4½	5
140,000	0.156	P	480	418	330	270	234	208	185	167	152	139	128	119	111	104	92.7							
		f	0.42	.056	.087	.125	.171	.223	.282	.350	.422	.509	.588	.685	.785	.896	1.12							
	0.162	P		468	376	311	276	234	207	187	170	156	143	134	125	117	103							
		f		.054	.085	.121	.165	.216	.273	.338	.409	.488	.566	.663	.757	.863	1.09							
	0.177	P		608	487	406	347	305	270	243	223	205	187	174	163	152	135	122						
		f		.049	.077	.115	.151	.198	.251	.311	.375	.447	.522	.606	.695	.793	1.00	1.24						
125,000	0.187	P		642	522	426	367	320	284	256	233	213	197	183	170	160	142	128						
		f		.041	.065	.093	.127	.166	.210	.260	.314	.373	.487	.510	.584	.665	.832	1.04						
	0.192	P		696	556	465	396	348	309	278	254	233	214	199	186	174	154	139	126					
		f		.040	.063	.091	.124	.160	.205	.252	.308	.366	.428	.499	.571	.652	.825	1.02	1.23					
	0.207	P			694	579	495	432	385	346	315	288	266	247	232	216	192	173	158	144				
		f			.059	.085	.115	.151	.191	.236	.286	.342	.396	.462	.570	.607	.757	.943	1.11	1.36				
	0.218	P			812	678	580	509	452	408	360	339	310	291	270	255	225	204	185	169				
		f			.055	.080	.109	.142	.180	.223	.269	.321	.374	.437	.488	.570	.710	.891	1.08	1.28				
	0.225	P			895	746	640	560	498	447	407	372	345	320	299	280	248	224	203	187				
		f			.054	.078	.106	.138	.175	.213	.262	.312	.372	.425	.486	.565	.691	.866	1.05	1.24				
	0.244	P			1120	950	811	711	632	570	517	475	438	406	381	356	316	284	259	237				
		f			.049	.071	.098	.138	.161	.200	.240	.287	.336	.391	.449	.537	.646	.800	.965	1.14				
	0.250	P				1027	880	760	685	617	560	513	476	440	410	385	342	308	281	266	222			
		f				.070	.095	.131	.157	.191	.236	.281	.328	.385	.439	.524	.624	.780	.946	1.12	1.53			

Pitch diameter D, in

0.263	P	256	298	326	359	400	448	478	501	551	598	652	717	795	895	1125	1195
	f	1.44	1.06	.896	.740	.592	.475	.416	.363	.312	.266	.224	.183	.149	.118	.089	.066
0.281	P	310	362	395	437	482	543	580	620	665	724	794	863	969	1087	1240	1450
	f	1.36	1.02	.840	.692	.562	.443	.390	.340	.292	.250	.209	.172	.140	.111	.085	.062
0.283	P	317	370	402	439	492	564	592	634	682	740	805	886	985	1110	1264	
	f	1.35	.990	.883	.690	.559	.440	.386	.338	.289	.246	.207	.169	.139	.111	.084	
0.312	P	343	392	460	500	550	610	687	733	775	845	915	1000	1100	1220	1376	1575
	f	1.47	1.12	.829	.697	.577	.467	.368	.322	.283	.242	.207	.174	.144	.116	.092	.070
0.331	P	410	468	545	594	653	725	818	870	932	1000	1090	1187	1316	1455	1636	
	f	1.30	1.05	.770	.654	.541	.437	.346	.343	.264	.227	.194	.163	.135	.109	.088	
0.341	P	454	520	608	661	728	808	910	970	1040	1120	1214	1325	1452	1620	1820	
	f	1.32	1.02	.745	.625	.522	.413	.330	.293	.256	.218	.186	.156	.127	.105	.082	
0.362	P	535	612	714	778	858	950	1070	1147	1220	1318	1430	1560	1714	1910	2140	
	f	1.26	.965	.713	.598	.495	.400	.317	.273	.243	.207	.177	.149	.123	.100	.079	
0.375	P	528	592	678	790	860	950	1058	1185	1265	1354	1458	1580	1780	1940	2110	
	f	1.54	1.22	.938	.688	.579	.478	.382	.308	.268	.234	.201	.172	.144	.117	.079	
0.393	P	670	682	780	910	990	1092	1212	1365	1458	1560	1680	1820	1984	2180	2430	
	f	1.47	1.16	.890	.657	.550	.457	.369	.292	.256	.223	.195	.164	.137	.114	.092	
0.406	P	666	750	855	1000	1090	1200	1330	1500	1600	1710	1840	2000	2170	2400		
	f	1.43	1.13	.867	.640	.525	.444	.353	.284	.248	.217	.168	.159	.134	.108		
0.430	P	800	900	1028	1200	1308	1440	1598	1798	1918	2050	2210	2400	2610	2875		
	f	1.35	1.06	.815	.600	.503	.418	.338	.267	.234	.204	.175	.150	.126	.104		
0.437	P	750	835	940	1074	1250	1365	1500	1665	1800	2000	2140	2310	2500	2730	3000	
	f	1.64	1.33	1.05	.810	.593	.490	.412	.327	.264	.231	.201	.173	.148	.124	.100	

115.000

Table 8.2.64 Safe Working Loads P and Deflections f of Cylindrical Helical Steel Springs of Circular Cross Section (Continued)

Allowable unit stress, lb/in²	Diam, in	D	1¼	1⅜	1½	1⅝	1¾	1⅞	2	2¼	2½	2¾	3	3½	4	4½	5	5½	6
110,000	0.460	P		3065	2800	2580	2400	2230	2100	1865	1680	1530	1400	1200	1058	952	840		
		f		.112	.134	.157	.183	.209	.239	.303	.374	.447	.536	.729	.956	1.21	1.49		
	0.468	P		3265	2940	2725	2530	2375	2210	1970	1770	1610	1472	1265	1110	935	885		
		f		.111	.132	.154	.182	.206	.235	.295	.368	.444	.530	.720	.943	1.19	1.47		
	0.490	P		3675	3270	3115	2890	2710	2535	2245	2025	1840	1690	1445	1268	1125	1015	920	
		f		.106	.126	.148	.172	.196	.225	.284	.351	.424	.506	.688	.900	1.13	1.40	1.70	
	0.500	P			3610	3320	3090	2890	2710	2410	2160	1970	1810	1550	1352	1205	1082	985	
		f			.123	.144	.168	.192	.220	.274	.347	.415	.495	.672	.880	1.11	1.37	1.65	
	0.562	P				4700	4390	4090	3830	3420	3080	2790	2565	2190	1913	1710	1535	1395	1280
		f				.128	.149	.175	.195	.248	.306	.372	.440	.596	.782	.990	1.22	1.47	1.75
	0.625	P					6100	5600	5260	4660	4210	3825	3505	3000	2630	2340	2110	1913	1750
		f					.134	.154	.176	.218	.275	.328	.397	.538	.705	.875	1.05	1.33	1.58
100,000	0.687	P							6325	5660	5090	4630	4250	3625	3195	2825	2560	2330	2125
		f							.145	.183	.228	.274	.3278	.443	.580	.733	.908	1.00	1.30
	0.750	P								7400	6640	6030	5540	4745	4150	3690	3325	3025	2770
		f								.178	.218	.252	.299	.402	.532	.671	.832	1.00	1.19
	0.812	P									8420	7660	7000	6000	5260	4675	4200	3825	3500
		f									.192	.232	.276	.376	.490	.620	.766	.880	1.10
	0.875	P									10830	9550	8700	7500	6560	5740	5250	4770	4730
		f									.179	.218	.257	.348	.456	.577	.712	.860	1.02
90,000	0.937	P										10600	9700	8400	7160	6470	5810	5290	4850
		f										.179	.217	.290	.383	.480	.591	.715	.855
	1.000	P											11780	10100	8800	7850	7050	6330	5870
		f											.206	.276	.360	.454	.561	.680	.803
	1,125	P												14400	12600	11230	10100	9200	8400
		f												.244	.320	.405	.496	.600	.718
	1,250	P												24700	18200	15300	13250	12540	11500
		f												.260	.287	.364	.442	.545	.648
80,000	1.375	P													20400	18100	16150	14850	13600
		f													.280	.294	.364	.440	.522

Pitch diameter *D*, in

Table 8.2.65   Constants for Use in $S_v = 0.43A/d^m$

Material	Size range, in	Size range, mm	Exponent m	Constant A kips	Constant A MPa
Music wire*	0.004–0.250	0.10–6.55	0.146	196	2,170
Oil-tempered wire†	0.020–0.500	0.50–12	0.186	149	1,880
Hard-drawn wire‡	0.028–0.500	0.70–12	0.192	136	1,750
Chrome vanadium§	0.032–0.437	0.80–12	0.167	169	2,000
Chrome silicon¶	0.063–0.375	1.6–10	0.112	202	2,000

* Surface is smooth, free from defects, and has a bright, lustrous finish.
† Has a slight heat-treating scale which must be removed before plating.
‡ Surface is smooth and bright, with no visible marks.
§ Aircraft-quality tempered wire; can also be obtained annealed.
¶ Tempered to Rockwell C49 but may also be obtained untempered.
SOURCE: Adapted from "Mechanical Engineering Design," Shigley, McGraw-Hill, 1983 by permission.

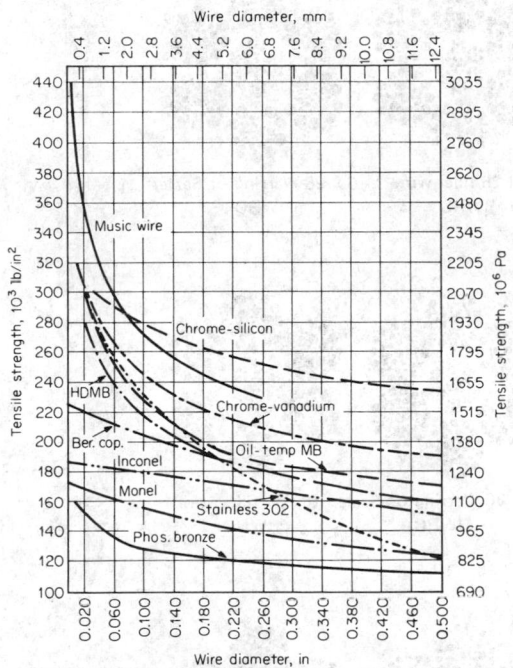

Fig. 8.2.120 Minimum tensile strength for the most popular spring materials, spring-quality wire. *(Reproduced from Carlson, "Spring Designer's Handbook," Marcel Dekker, by permission.)*

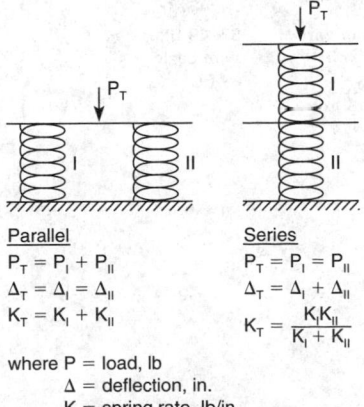

Parallel
$P_T = P_I + P_{II}$
$\Delta_T = \Delta_I = \Delta_{II}$
$K_T = K_I + K_{II}$

Series
$P_T = P_I = P_{II}$
$\Delta_T = \Delta_I + \Delta_{II}$
$K_T = \dfrac{K_I K_{II}}{K_I + K_{II}}$

where P = load, lb
Δ = deflection, in.
K = spring rate, lb/in.

Fig. 8.2.121 Springs in parallel and in series.

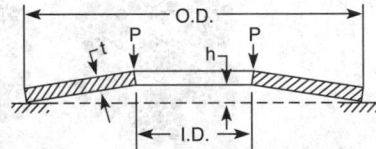

Fig. 8.2.122a Sectional view of Belleville spring.

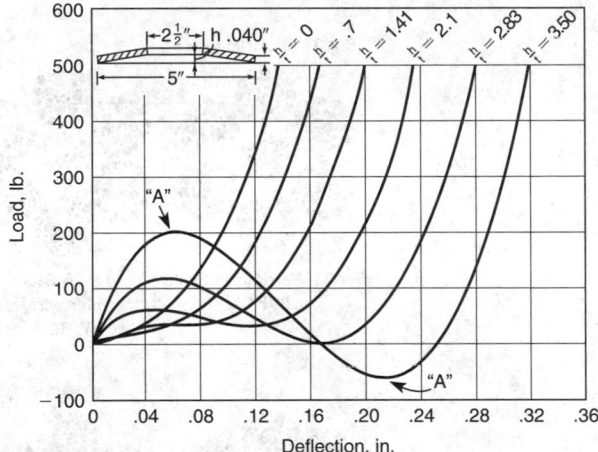

Fig. 8.2.122b Load deflection curves for a family of Belleville springs. *(Associated Spring Corp.)*

springs are used for very high and special spring rates. They are extremely sensitive to slight variations in their geometry. A wide range are available commercially.

WIRE ROPE

When power source and load are located at extreme distances from one another, or loads are very large, the use of wire rope is suggested. Design and use decisions pertaining to wire rope rest with the user, but manufacturers generally will help users toward appropriate choices. The following material, based on the Committee of Wire Rope Producers, "Wire Rope User's Manual," 2d ed., 1981, may be used as an initial guide in selecting a rope.

Wire rope is composed of (1) wires to form a strand, (2) strands wound helically around a core, and (3) a core. Classification of wire ropes is made by giving the number of strands, number of minor strands in a major strand (if any), and nominal number of wires per strand. For example 6 × 7 rope means 6 strands with a nominal 7 wires per strand (in this case no minor strands, hence no middle number). A nominal value simply represents a range. A nominal value of 7 can mean anywhere from 3 to 14, of which no more than 9 are outside wires. A full

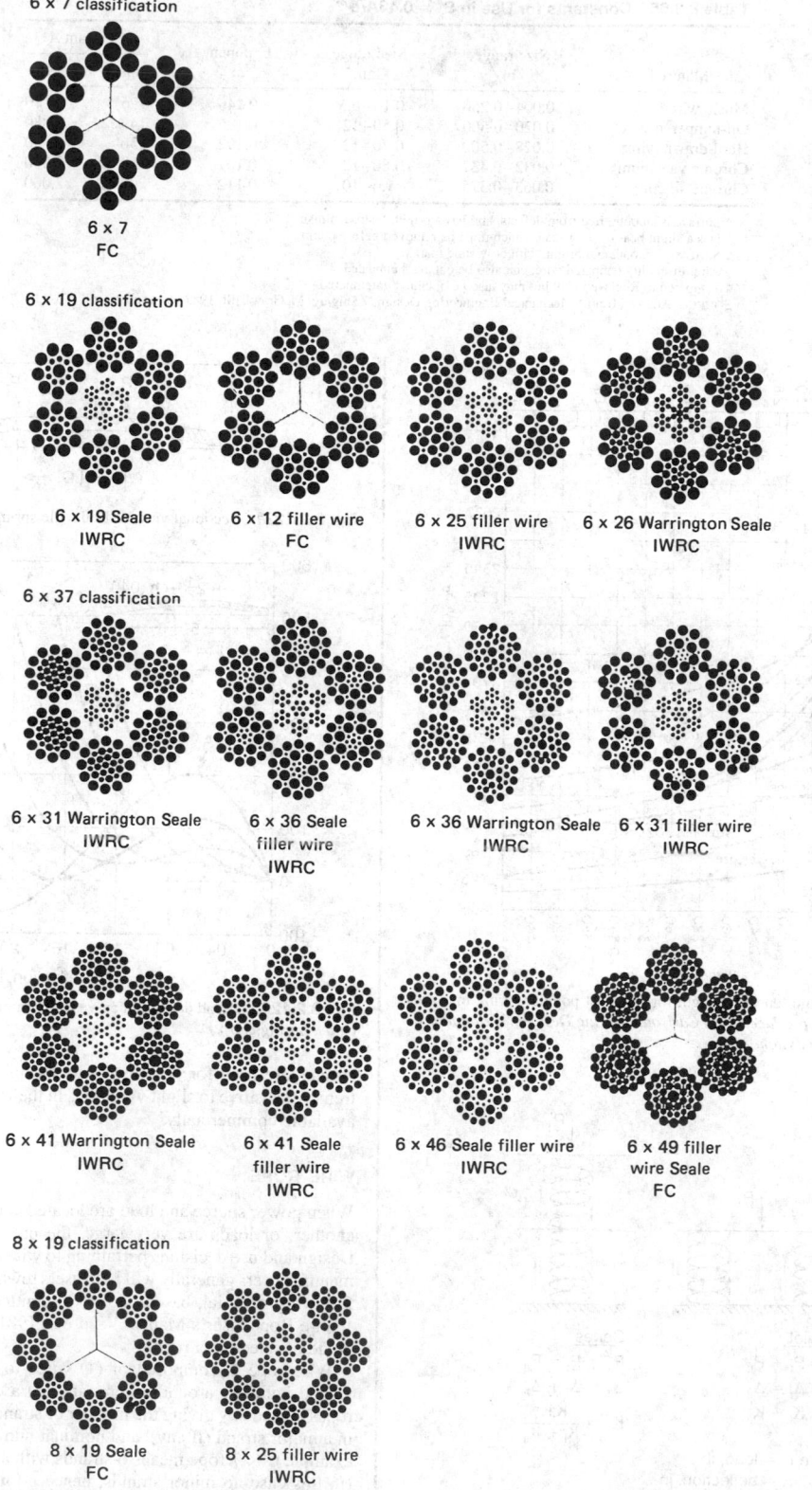

Fig. 8.2.123 Cross sections of some commonly used wire rope construction. *(Reproduced from "Wire Rope User's Manual," AISI, by permission.)*

rope description will also include length, size (diameter), whether wire is preformed or not prior to winding, direction of lay (right or left, indicating the direction in which strands are laid around the core), grade of rope (which reflects wire strength), and core. The most widely used classifications are: 6×7, 6×19, 6×37, 6×61, 6×91, 6×127, 8×19, 18×7, 19×7. Some special constructions are: 3×7 (guardrail rope); 3×19 (slusher), 6×12 (running rope); 6×24 and 6×30 (hawsers); 6×42 and $6 \times 6 \times 7$ (tiller rope); $6 \times 3 \times 19$ (spring lay); 5×19 and 6×19 (marlin clad); $6 \times 25B$, $6 \times 27H$, and $6 \times 30G$ (flattened strand). The diameter of a rope is the circle which just contains the rope. The right-regular lay (in which the wire is twisted in one direction to form the strands and the strands are twisted in the opposite direction to form the rope) is most common. Regular-lay ropes do not kink or untwist and handle easily. Lang-lay ropes (in which wires and strands are twisted in the same direction) are more resistant to abrasive wear and fatigue failure.

Cross sections of some commonly used wire rope are shown in Fig. 8.2.123. Figure 8.2.124 shows rotation-resistant ropes, and Fig. 8.2.125 shows some special-purpose constructions.

The core provides support for the strands under normal bending and loading. Core materials include fibers (hard vegetable or synthetic) or steel (either a strand or an independent wire rope). Most common core designations are: fiber core (FC), independent wire-rope core (IWRC), and wire-strand core (WSC). Lubricated fiber cores can provide lubrication to the wire, but add no real strength and cannot be used in high temperature environments. Wire-strand or wire-rope cores add from 7 to 10 percent to strength, but under nonstationary usage tend to wear from interface friction with the outside strands. Great flexibility can be achieved when wire rope is used as strands. Such construction is very pliable and friction resistant. Some manufacturers will provide plastic coatings (nylon, Teflon, vinyl, etc.) upon request. Such coatings help provide resistance to abrasion, corrosion, and loss of lubricant. *Crushing* refers to rope damage caused by excessive pressures against drum or sheave, improper groove size, and multiple layers on drum or sheave. Consult wire rope manufacturers in doubtful situations.

Wire-rope materials and their strengths are reflected as grades. These are: traction steel (TS), mild plow steel (MPS), plow steel (PS), improved plow steel (IPS), and extra improved plow (EIP). The plow steel strength curve forms the basis for calculating the strength of all steel rope wires. American manufacturers use color coding on their ropes to identify particular grades.

The grades most commonly available and tabulated are IPS and EIP. Two specialized categories, where selection requires extraordinary attention, are elevator and rotation-resistant ropes.

Elevator rope can be obtained in four principal grades: iron, traction steel, high-strength steel, and extra-high-strength steel.

Bronze rope has limited use; iron rope is used mostly for older existing equipment.

Selection of Wire Rope

Appraisal of the following is the key to choosing the rope best suited to the job: resistance to breaking, resistance to bending fatigue, resistance to vibrational fatigue, resistance to abrasion, resistance to crushing, and reserve strength. Along with these must be an appropriate choice of safety factor, which in turn requires careful consideration of all loads, acceleration-deceleration, shocks, rope speed, rope attachments, sheave

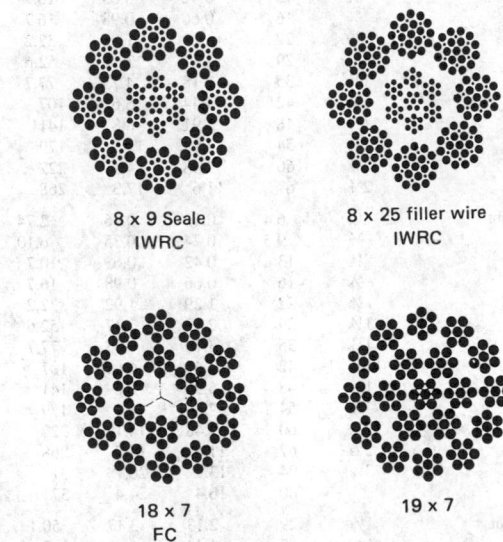

Fig. 8.2.124 Cross section of some rotation-resistant wire ropes. (*Reproduced from "Wire Rope User's Manual," AISI, by permission.*)

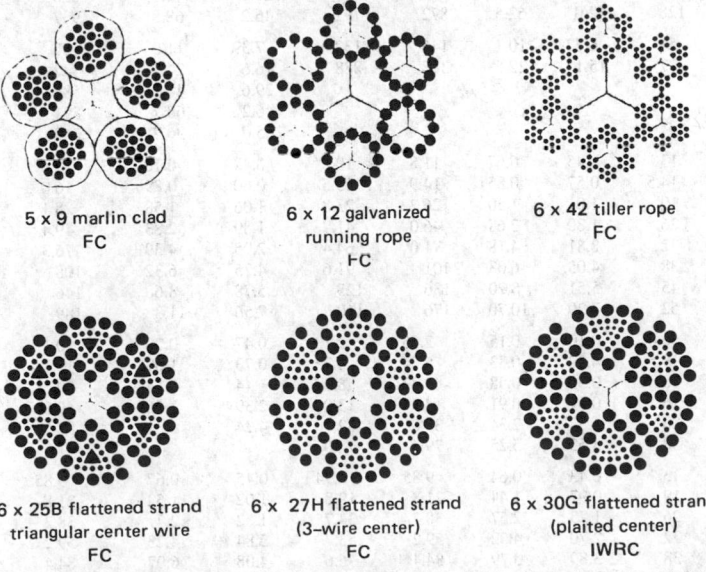

Fig. 8.2.125 Some special constructions. (*Reproduced from "Wire Rope User's Manual," AISI, by permission.*)

Table 8.2.66 Selected Values of Nominal Strengths of Wire Rope

			Fiber core				IWRC				
	Nominal diameter		Approximate mass		Nominal strength IPS		Approximate mass		Nominal strength		
									IPS		EIP
Classification	in	mm	lb/ft	kg/m	tons	t	lb/ft	kg/m	tons	t	tons	t
6 × 7 Bright	¼	6.4	0.09	0.14	2.64	2.4	0.10	0.15	2.84	2.58		
(uncoated)	⅜	9.5	0.21	0.31	5.86	5.32	0.23	0.34	6.30	5.72		
	½	13	0.38	0.57	10.3	9.35	0.42	0.63	11.1	10.1		
	⅝	16	0.59	0.88	15.9	14.4	0.65	0.97	17.1	15.5		
	⅞	22	1.15	1.71	30.7	27.9	1.27	1.89	33.0	29.9		
	1⅛	29	1.90	2.83	49.8	45.2	2.09	3.11	53.5	48.5		
	1⅜	35	2.82	4.23	73.1	66.3	3.12	4.64	78.6	71.3		
6 × 19 Bright	¼	6.4	0.11	0.16	2.74	2.49	0.12	0.17	2.94	2.67	3.40	3.08
(uncoated)	⅜	9.5	0.24	0.35	6.10	5.53	0.26	0.39	6.56	5.95	7.55	6.85
	½	13	0.42	0.63	10.7	9.71	0.46	0.68	11.5	10.4	13.3	12.1
	⅝	16	0.66	0.98	16.7	15.1	0.72	1.07	17.7	16.2	20.6	18.7
	⅞	22	1.29	1.92	32.2	29.2	1.42	2.11	34.6	31.4	39.8	36.1
	1⅛	29	2.13	3.17	52.6	47.7	2.34	3.48	56.5	51.3	65.0	59.0
	1⅜	35	3.18	4.73	77.7	70.5	3.5	5.21	83.5	75.7	96.0	87.1
	1⅝	42	4.44	6.61	107	97.1	4.88	7.26	115	104	132	120
	1⅞	48	5.91	8.8	141	128	6.5	9.67	152	138	174	158
	2⅛	54	7.59	11.3	179	162	8.35	12.4	192	174	221	200
	2⅜	60	9.48	14.1	222	201	10.4	15.5	239	217	274	249
	2⅝	67	11.6	17.3	268	243	12.8	19.0	288	261	331	300
6 × 37 Bright	¼	6.4	0.11	0.16	2.74	2.49	0.12	0.17	2.94	2.67	3.4	3.08
(uncoated)	⅜	9.5	0.24	0.35	6.10	5.53	0.26	0.39	6.56	5.95	7.55	6.85
	½	13	0.42	0.63	10.7	9.71	0.46	0.68	11.5	10.4	13.3	12.1
	⅝	16	0.66	0.98	16.7	15.1	0.72	1.07	17.9	16.2	20.6	18.7
	⅞	22	1.29	1.92	32.2	29.2	1.42	2.11	34.6	31.4	39.5	36.1
	1⅛	29	2.13	3.17	52.6	47.7	2.34	3.48	56.5	51.3	65.0	59.0
	1⅜	35	3.18	4.73	77.7	70.5	3.50	5.21	83.5	75.7	96.0	87.1
	1⅝	42	4.44	6.61	107	97.1	4.88	7.26	115	104	132	120
	1⅞	48	5.91	8.8	141	128	6.5	9.67	152	138	174	158
	2⅛	54	7.59	11.3	179	162	8.35	12.4	192	174	221	200
	2⅜	60	9.48	14.1	222	201	10.4	15.5	239	217	274	249
	2⅞	67	11.6	17.3	268	243	12.8	19.0	288	261	331	300
	3⅛	74	13.9	20.7	317	287	15.3	22.8	341	309	392	356
		80	16.4	24.4	371	336	18.0	26.8	399	362	458	415
6 × 61 Bright	1⅛	29	2.13	3.17	50.1	45.4	2.34	3.48	53.9	48.9	61.9	56.2
(uncoated)	1⅝	42	4.44	6.61	103	93.4	4.88	7.62	111	101	127	115
	2	52	6.77	10.1	154	140	7.39	11.0	165	150	190	172
	2⅝	67	11.6	17.3	260	236	12.8	18.3	279	253	321	291
	3	77	15.1	22.5	335	304	16.6	24.7	360	327	414	376
	4	103	26.9	40.0	577	523	29.6	44.1	620	562	713	647
	5	128	42.0	62.5	872	791	46.2	68.8	937	850	1,078	978
6 × 91 Bright	2	51	6.77	10.1	146	132	7.39	11.0	157	142	181	164
(uncoated)	3	77	15.1	22.5	318	288	16.6	24.7	342	310	393	357
	4						29.6	44.1	589	534	677	614
	5						46.2	68.7	891	808	1,024	929
	6						65.0	96.7	1,240	1,125	1,426	1,294
6 × 25B	½	13	0.45	0.67	11.8	10.8	0.47	0.70	12.6	11.4	14	12.7
6 × 27H	9⁄16	14.5	0.57	0.85	14.9	13.5	0.60	0.89	16.0	14.5	17.6	16
6 × 30G	¾	19	1.01	1.50	26.2	23.8	1.06	1.58	28.1	25.5	31	28.1
Flattened strand bright	1	26	1.80	2.68	46.0	41.7	1.89	2.83	49.4	44.8	54.4	49.4
(uncoated)	1¼	32	2.81	4.18	71.0	64.4	2.95	4.39	76.3	60.2	84	76.2
	1½	38	4.05	6.03	101	91.6	4.25	6.32	108	98	119	108
	1¾	45	5.51	8.20	136	123	5.78	8.60	146	132	161	146
	2	52	7.20	10.70	176	160	7.56	11.3	189	171	207	188
8 × 19 Bright	¼	6.4	0.10	0.15	2.35	2.13	0.47	0.70	10.1	9.16	11.6	10.5
(uncoated)	⅜	9.5	0.22	0.33	5.24	4.75	0.73	1.09	15.7	14.2	18.1	16.4
	½	13	0.39	0.58	9.23	8.37	1.44	2.14	30.5	27.7	35.0	31.8
	⅝	16	0.61	0.91	14.3	13.0	2.39	3.56	49.8	45.2	57.3	51.7
	1	26	1.57	2.34	36.0	32.7	4.24	6.31	87.3	79.2	100.0	90.7
	1½	38	3.53	5.25	79.4	72.0						
18 × 7	½	13	0.43	0.64	9.85	8.94	0.45	0.67	9.85	8.94	10.8	9.8
Rotation resistant, bright	¾	19	0.97	1.44	21.8	19.8	1.02	1.52	21.8	19.8	24.0	21.8
(uncoated)	1	26	1.73	2.57	38.3	34.7	1.82	2.71	38.3	34.7	42.2	38.3
	1¼	32	2.70	4.02	59.2	53.7	2.84	4.23	59.2	53.7	65.1	59.1
	1½	38	3.89	5.79	84.4	76.6	4.08	6.07	84.4	76.6	92.8	84.2

SOURCE: ''Wire Rope User's Manual,'' AISI, adapted by permission.

arrangements as well as their number and size, corrosive and/or abrasive environment, length of rope, etc. An approximate selection formula can be written as:

$$\text{DSL} = \frac{(\text{NS})K_b}{K_{sf}}$$

where DSL (demanded static load) = known or dead load **plus** additional loads caused by sudden starts or stops, shocks, bearing friction, etc., tons; NS (nominal strength) = published test strengths, tons (see Table 8.2.66); K_b = a factor to account for the reduction in nominal strength due to bending when a rope passes over a curved surface such as a stationary sheave or pin (see Fig. 8.2.126); K_{sf} = safety factor. (For average operation use $K_{sf} = 5$. If there is danger to human life or other critical situations, use $8 \le K_{sf} \le 12$. For instance, for elevators moving at 50 ft/min, $K_{sf} = 8$, while for those moving at 1,500 ft/min, $K_{sf} = 12$.)

Having made a tentative selection of a rope based on the demanded static load, one considers next the wear life of the rope. A loaded rope

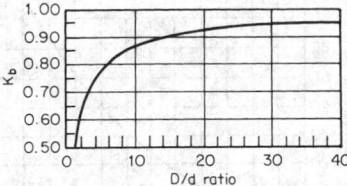

Fig. 8.2.126 Values of K_{bend} vs. D/d ratios (D = sheave diameter, d = rope diameter), based on standard test data for 6 × 9 and 6 × 17 class ropes. (*Compiled from "Wire Rope User's Manual," AISI, by permission.*)

bent over a sheave stretches elastically and so rubs against the sheave, causing wear of both members. Drum or sheave size is of paramount importance at this point.

Sizing of Drums or Sheaves

Diameters of drums or sheaves in wire rope applications are controlled by two main considerations: (1) the radial pressure between rope and groove and (2) degree of curvature imposed on the rope by the drum or sheave size.

Radial pressures can be calculated from $p = 2T/(Dd)$, where p = unit radial pressure, lb/in²; T = rope load, lb; D = tread diameter of drum or sheave, in; d = nominal diameter of rope, in. Table 8.2.67 lists suggested allowable radial bearing pressures of ropes on various sheave materials.

All wire ropes operating over drums or sheaves are subjected to cyclical stresses, causing shortened rope life because of fatigue. Fatigue resistance or relative service life is a function of the ratio D/d. Adverse

effects also arise out of relative motion between strands during passage around the drum or sheave. Additional adverse effects can be traced to poor match between rope and groove size, and to lack of rope lubrication. Table 8.2.68 lists suggested and minimum sheave and drum ratios for various rope construction. Table 8.2.69 lists relative bending life factors; Figure 8.2.127 shows a plot of relative rope service life versus D/d. Table 8.2.70 lists minimum drum (sheave) groove dimensions. Periodic groove inspection is recommended, and worn or corrugated grooves should be remachined or the drum replaced, depending on severity of damage.

Seizing and Cutting Wire Rope   Before a wire rope is cut, seizings (bindings) must be applied on either side of the cut to prevent rope distortion and flattening or loosened strands. Normally, for preformed ropes, one seizing on each side of the cut is sufficient, but for ropes that

Table 8.2.68   Sheave and Drum Ratios

Construction*	Suggested	Minimum
6 × 7	72	42
19 × 7 or 18 × 7 Rotation-resistant	51	34
6 × 19 S	51	34
6 × 25 B flattened strand	45	30
6 × 27 H flattened strand	45	30
6 × 30 G flattened strand	45	30
6 × 21 FW	45	30
6 × 26 WS	45	30
6 × 25 FW	39	26
6 × 31 WS	39	26
6 × 37 SFW	39	26
6 × 36 WS	35	23
6 × 43 FWS	35	23
6 × 41 WS	32	21
6 × 41 SFW	32	21
6 × 49 SWS	32	21
6 × 46 SFW	28	18
6 × 46 WS	28	18
8 × 19 S	41	27
8 × 25 FW	32	21
6 × 42 Tiller	21	14

* WS—Warrington Seale; FWS—Filler Wire Seale; SFW—Seale Filler Wire; SWS—Seale Warrington Seale; S—Seale; FW—Filler Wire.

† D = tread diameter of sheave; d = nominal diameter of rope. To find any tread diameter from this table, the diameter for the rope construction to be used is multiplied by its nominal diameter d. For example, the minimum sheave tread diameter for a ½-in 6 × 21 FW rope would be ½ in (nominal diameter) × 30 (minimum ratio), or 15 in.

NOTE: These values are for reasonable service. Other values are permitted by various standards such as ANSI, API, PCSA, HMI, CMAA, etc. Similar values affect rope life.

SOURCE: "Wire Rope User's Manual," AISI, reproduced by permission.

Table 8.2.67   Suggested Allowable Radial Bearing Pressures of Ropes on Various Sheave Materials

Material	Regular lay rope, lb/in²				Lang lay rope, lb/in²			Flattened strand lang lay, lb/in²	Remarks
	6 × 7	6 × 19	6 × 37	8 × 19	6 × 7	6 × 19	6 × 37		
Wood	150	250	300	350	165	275	330	400	On end grain of beech, hickory, gum.
Cast iron	300	480	585	680	350	550	660	800	Based on minimum Brinell hardness of 125.
Carbon-steel casting	550	900	1,075	1,260	600	1,000	1,180	1,450	30–40 carbon. Based on minimum Brinell hardness of 160.
Chilled cast iron	650	1,100	1,325	1,550	715	1,210	1,450	1,780	Not advised unless surface is uniform in hardness.
Manganese steel	1,470	2,400	3,000	3,500	1,650	2,750	3,300	4,000	Grooves must be ground and sheaves balanced for high-speed service.

SOURCE: "Wire Rope User's Manual," AISI, reproduced by permission.

Table 8.2.69   Relative Bending Life Factors

Rope construction	Factor	Rope construction	Factor
6 × 7	0.61	6 × 36 WS	1.16
19 × 7 or 18 × 7	0.67	6 × 43 FWS	1.16
Rotation-resistant	0.81	6 × 41 WS	1.30
6 × 19 S	0.90	6 × 41 SFW	1.30
6 × 25 B flattened strand	0.90	6 × 49 SWS	1.30
6 × 27 H flattened strand	0.90	6 × 43 FW (2 op)	1.41
6 × 30 G flattened strand	0.89	6 × 46 SFW	1.41
6 × 21 FW	0.89	6 × 46 WS	1.41
6 × 26 WS	1.00	8 × 19 S	1.00
6 × 25 FW	1.00	8 × 25 FW	1.25
6 × 31 WS	1.00	6 × 42 Tiller	2.00
6 × 37 SFW			

SOURCE: "Wire Rope User's Manual," AISI, reproduced by permission.

Table 8.2.70   Minimum Sheave- and Drum-Groove Dimensions*

Nominal rope diameter		Groove radius			
		New		Worn	
in	nm	in	mm	in	mm
¼	6.4	0.135	3.43	.129	3.28
⁵⁄₁₆	8.0	0.167	4.24	.160	4.06
⅜	9.5	0.201	5.11	.190	4.83
⁷⁄₁₆	11	0.234	5.94	.220	5.59
½	13	0.271	6.88	.256	6.50
⁹⁄₁₆	14.5	0.303	7.70	.288	7.32
⅝	16	0.334	8.48	.320	8.13
¾	19	0.401	10.19	.380	9.65
⅞	22	0.468	11.89	.440	11.18
1	26	0.543	13.79	.513	13.03
1⅛	29	0.605	15.37	.577	14.66
1¼	32	0.669	16.99	.639	16.23
1⅜	35	0.736	18.69	.699	17.75
1½	38	0.803	20.40	.759	19.28
1⅝	42	0.876	22.25	.833	21.16
1¾	45	0.939	23.85	.897	22.78
1⅞	48	1.003	25.48	.959	24.36
2	52	1.085	27.56	1.025	26.04
2⅛	54	1.137	28.88	1.079	27.41
2¼	58	1.210	30.73	1.153	29.29
2⅜	60	1.271	32.28	1.199	30.45
2½	64	1.338	33.99	1.279	32.49
2⅝	67	1.404	35.66	1.339	34.01
2¾	71	1.481	37.62	1.409	35.79
2⅞	74	1.544	39.22	1.473	37.41
3	77	1.607	40.82	1.538	39.07
3⅛	80	1.664	42.27	1.598	40.59
3¼	83	1.731	43.97	1.658	42.11
3⅜	87	1.807	45.90	1.730	43.94
3½	90	1.869	47.47	1.794	45.57
3¾	96	1.997	50.72	1.918	48.72
4	103	2.139	54.33	2.050	52.07
4¼	109	2.264	57.51	2.178	55.32
4½	115	2.396	60.86	2.298	58.37
4¾	122	2.534	64.36	2.434	61.82
5	128	2.663	67.64	2.557	64.95
5¼	135	2.804	71.22	2.691	68.35
5½	141	2.929	74.40	2.817	71.55
5¾	148	3.074	78.08	2.947	74.85
6	154	3.198	81.23	3.075	78.11

* Values given are applicable to grooves in sheaves and drums; they are not generally suitable for pitch design since this may involve other factors. Further, the dimensions do not apply to traction-type elevators; in this circumstance, drum- and sheave-groove tolerances should conform to the elevator manufacturer's specifications. Modern drum design embraces extensive considerations beyond the scope of this publication. It should also be noted that drum grooves are now produced with a number of oversize dimensions and pitches applicable to certain service requirements.
SOURCE: "Wire Rope User's Manual," AISI, reproduced by permission.

are not preformed a minimum of two seizings on each side is recommended, and these should be spaced six rope diameters apart (see Fig. 8.2.128). Seizings should be made of soft or annealed wire or strand, and the width of the seizing should never be less than the diameter of the rope being seized. Table 8.2.71 lists suggested seizing wire diameters.

Wire Rope Fittings or Terminations   End terminations allow forces to be transferred from rope to machine, or load to rope, etc. Figure 8.2.129 illustrates the most commonly used end fittings or terminations. Not all terminations will develop full strength. In fact, if all of the rope elements are not held securely, the individual strands will sustain unequal loads causing unequal wear among them, thus shortening the effective rope service life. Socketing allows an end fitting which reduces the chances of unequal strand loading.

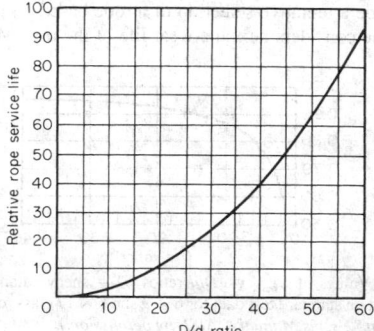

Fig. 8.2.127   Service life curves for various *D/d* ratios. Note that this curve takes into account only bending and tensile stresses. *(Reproduced from "Wire Rope User's Manual," AISI, by permission.)*

Wire rope manufacturers have developed a recommended procedure for socketing. A tight wire serving band is placed where the socket base will be, and the wires are unlaid, straightened, and "broomed" out. Fiber core is cut close to the serving band and removed, wires are cleaned with a solvent such as SC-methyl chloroform, and brushed to remove dirt and grease. If additional cleaning is done with muriatic acid this must be followed by a neutralizing rinse (if possible, ultrasonic cleaning is preferred). The wires are dipped in flux, the socket is positioned, zinc (spelter) is poured and allowed to set, the serving band is removed, and the rope lubricated.

A somewhat similar procedure is used in thermoset resin socketing.

Socketed terminations generally are able to develop 100 percent of nominal strength.

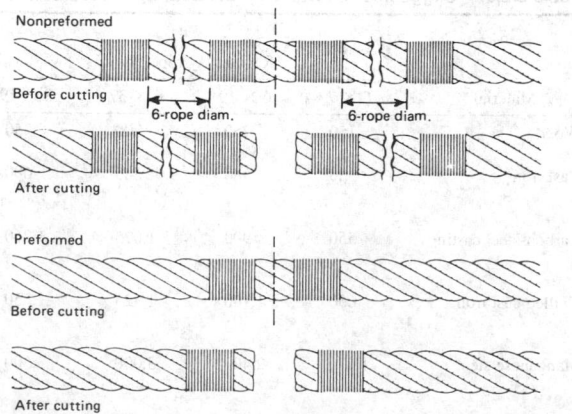

Fig. 8.2.128   Seizings. *(Reproduced from "Wire Rope User's Manual," AISI, by permission.)*

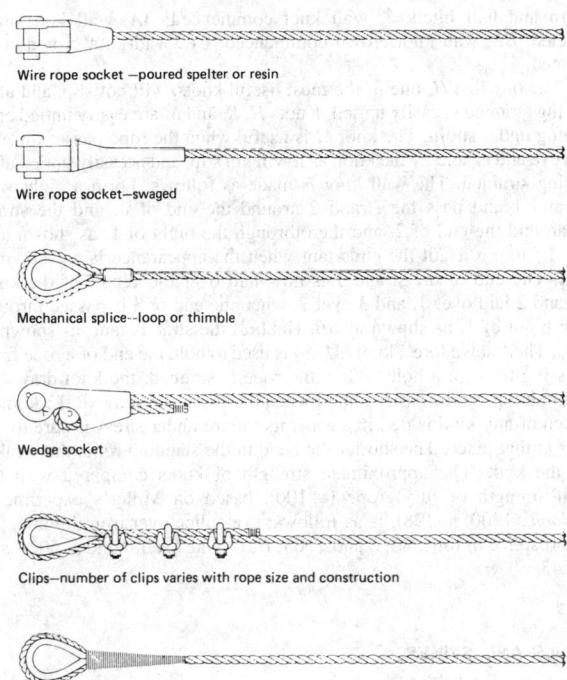

Wire rope socket —poured spelter or resin

Wire rope socket—swaged

Mechanical splice—loop or thimble

Wedge socket

Clips—number of clips varies with rope size and construction

Loop or thimble splice—hand tucked

Fig. 8.2.129 End fittings, or terminations, showing the six most commonly used. *(Reproduced from "Wire Rope User's Manual," AISI, by permission.)*

Table 8.2.71 Seizing*

Rope diameter		Suggested seizing wire diameter†	
in	mm	in	mm
1/8 – 5/16	3.5 – 8.0	0.032	0.813
3/8 – 9/16	9.4 – 14.5	0.048	1.21
5/8 – 15/16	16.0 – 24.0	0.063	1.60
1 – 1 5/16	26.0 – 33.0	0.080	2.03
1 3/8 – 1 11/16	35.0 – 43.0	0.104	2.64
1 3/4 and larger	45.0 and larger	0.124	3.15

* Length of the seizing should not be less than the rope diameter.
† The diameter of seizing wire for elevator ropes is usually somewhat smaller than that shown in this table. Consult the wire rope manufacturer for specific size recommendations. Soft annealed seizing strand may also be used.
SOURCE: "Wire Rope User's Manual," AISI, reproduced by permission.

FIBER LINES

The breaking strength of various fiber lines is given in Table 8.2.72.

Knots, Hitches, and Bends

No two parts of a knot which would move in the same direction if the rope were to slip should lie alongside of and touching each other. The knots shown in Fig. 8.2.130 are known by the following names:

A, bight of a rope; *B*, simple or overhand knot; *C*, figure 8 knot; *D*, double knot; *E*, boat knot; *F*, bowline, first step; *G*, bowline, second step; *H*, bowline, completed; *I*, square or reef knot; *J*, sheet bend or weaver's knot; *K*, sheet bend with a toggle; *L*, carrick bend;

Table 8.2.72 Breaking Strength of Fiber Lines, Lb

Size, in												
Diam	Cir.	Manila	Composite	Sisal	Sisal mixed	Sisal hemp	Agave or jute	Nylon	Dacron	Poly-ethylene	Poly-propylene (mono-filament)	Esterlon (polyester)
3/16	5/8	450	—	360	340	310	270	1,000	850	700	800	720
1/4	3/4	600	—	480	450	420	360	1,500	1,380	1,200	1,200	1,150
5/16	1	1,000	—	800	750	700	600	2,500	2,150	1,750	2,100	1,750
3/8	1 1/8	1,350	—	1,080	1,010	950	810	3,500	3,000	2,500	3,100	2,450
7/16	1 1/4	1,750	—	1,400	1,310	1,230	1,050	4,800	4,500	3,400	3,700	3,400
1/2	1 1/2	2,650	—	2,120	1,990	1,850	1,590	6,200	5,500	4,100	4,200	4,400
9/16	1 3/4	3,450	—	2,760	2,590	2,410	2,070	8,300	7,300	4,600	5,100	5,700
5/8	2	4,400	—	3,520	3,300	3,080	2,640	10,500	9,500	5,200	5,800	7,300
3/4	2 1/4	5,400	—	4,320	4,050	3,780	3,240	14,000	12,500	7,400	8,200	9,500
13/16	2 1/2	6,500	—	5,200	4,880	4,550	3,900	17,000	15,000	8,900	9,800	11,500
7/8	2 3/4	7,700	—	—	—	—	—	20,000	17,500	10,400	11,500	13,500
1	3	9,000	—	7,200	6,750	6,300	5,400	24,000	20,000	12,600	14,000	16,500
1 1/16	3 1/4	10,500	—	8,400	7,870	7,350	6,300	28,000	22,500	14,500	16,100	19,000
1 1/8	3 1/2	12,000	—	9,600	9,000	8,400	7,200	32,000	25,000	16,500	18,300	21,500
1 1/4	3 3/4	13,500	—	10,800	10,120	9,450	8,100	36,500	28,500	18,600	21,000	24,300
1 5/16	4	15,000	—	12,000	11,250	10,500	9,000	42,000	32,000	21,200	24,000	28,000
1 1/2	4 1/2	18,500	16,600	14,800	13,900	12,950	11,100	51,000	41,000	26,700	30,000	34,500
1 5/8	5	22,500	20,300	18,000	16,900	15,800	13,500	62,000	50,000	32,700	36,500	41,500
1 3/4	5 1/2	26,500	23,800	21,200	19,900	18,500	15,900	77,500	61,000	39,500	44,000	51,000
2	6	31,000	27,900	24,800	23,200	21,700	18,600	90,000	72,000	47,700	53,000	61,000
2 1/8	6 1/2	36,000	—	—	—	—	—	105,000	81,000	55,800	62,000	70,200
2 1/4	7	41,000	36,900	32,800	30,800	28,700	—	125,000	96,000	63,000	70,000	81,000
2 1/2	7 1/2	46,500	—	—	—	—	—	138,000	110,000	72,500	80,500	92,000
2 5/8	8	52,000	46,800	41,600	39,000	36,400	—	154,000	125,000	81,000	90,000	103,000
2 7/8	8 1/2	58,000	—	—	—	—	—	173,000	140,000	92,000	100,000	116,000
3	9	64,000	57,500	51,200	48,000	44,800	—	195,000	155,000	103,000	116,000	130,000
3 1/4	10	77,000	69,300	61,600	57,800	53,900	—	238,000	190,000	123,000	137,000	160,000
3 1/2	11	91,000	—	—	—	—	—	288,000	230,000	146,000	162,000	195,000
4	12	105,000	94,500	84,000	78,800	73,500	—	342,000	275,000	171,000	190,000	230,000

Breaking strength is the maximum load the line will hold at the time of breaking. The working load of a line is one-fourth to one-fifth of the breaking strength.
SOURCE: Adapted, by permission of the U.S. Naval Institute, Annapolis, MD, and Wall Rope Works, Inc., New York, NY.

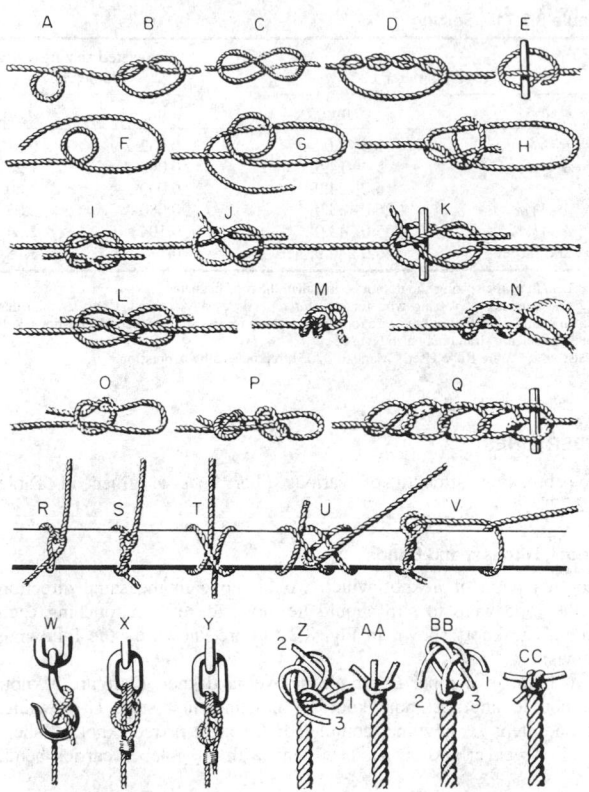

Fig. 8.2.130 Knots, hitches, and bends.

M, "stevedore" knot completed; N, "stevedore" knot commenced; O, slip knot; P, Flemish loop; Q, chain knot with toggle; R, half hitch; S, timber hitch; T, clove hitch; U, rolling hitch; V, timber hitch and half hitch; W, blackwall hitch; X, fisherman's bend; Y, round

turn and half hitch; Z, wall knot commenced; AA, wall knot completed; BB, wall-knot crown commenced; CC, wall-knot crown completed.

The bowline H, one of the most useful knots, will not slip, and after being strained is easily untied. Knots H, K, and M are easily untied after being under strain. The knot M is useful when the rope passes through an eye and is held by the knot, as it will not slip, and is easily untied after being strained. The wall knot is made as follows: Form a bight with strand 1 and pass the strand 2 around the end of it, and the strand 3 around the end of 2, and then through the bight of 1, as shown at Z in the figure. Haul the ends taut when the appearance is as shown in AA. The end of the strand 1 is now laid over the center of the knot, strand 2 laid over 1, and 3 over 2, when the end of 3 is passed through the bight of 1, as shown at BB. Haul all the strands taut, as shown at CC. The "stevedore" knot (M, N) is used to hold the end of a rope from passing through a hole. When the rope is strained, the knot draws up tight, but it can be easily untied when the strain is removed. If a knot or hitch of any kind is tied in a rope, its failure under stress is sure to occur at that place. The shorter the bend in the standing rope, the weaker is the knot. The approximate strength of knots compared with the full strength of (dry) rope (= 100), based on Miller's experiments (*Mach.*, 1900, p. 198), is as follows: eye splice over iron thimble, 90; short splice in rope, 80; S and Y, 65; H, O, and T, 60; I and J, 50; B and P, 45.

NAILS AND SPIKES

Nails are either **wire nails** of circular cross section and constant diameter or **cut nails** of rectangular cross section with taper from head to point. The larger sizes are called **spikes**. The length of the nail is expressed in the "penny" system, the equivalents in inches being given in Tables 8.2.73 to 8.2.75. The letter d is the accepted symbol for penny. A keg of nails weighs 100 lb. **Heavy hinge nails** or **track nails** with countersunk heads have chisel points unless diamond points are specified. **Plasterboard nails** are smooth with circumferential grooves and have diamond points. **Spikes** are made either with flat heads and diamond points or with oval heads and chisel points.

Table 8.2.73 Wire Nails for Special Purposes
(Steel wire gage)

Length, in	Barrel nails		Barbed roofing nails		Barbed dowel nails	
	Gage	No. per lb	Gage	No. per lb	Gage	No. per lb
⅝	15½	1,570	—	—	8	394
¾	15½	1,315	13	729	8	306
⅞	14½	854	12	478	8	250
1	14½	750	12	416	8	212
1⅛	14½	607	12	368	8	183
1¼	14	539	11	250	8	16
1⅜	13	386	11	228	8	145
1½	13	355	10	167	8	131
1¾	—	—	10	143		
2	—	—	9	104		

Length, in	Clout nails		Slating nails		Fine nails		
	Gage	No. per lb	Gage	No. per lb	Length, in	Gage	No. per lb
¾	15	999					
⅞	14	733					
1	14	648	12	425	1	16½	1,280
1⅛	14	580	10½	229			
1¼	13	398	—	—	1	17	1,492
1⅜	13	365					
1½	13	336	10½	190	1⅛	15	757
1¾	—	—	10	144	1⅛	16	984
2	—	—	9	104			

Table 8.2.74   Wire Nails and Spikes
(Steel wire gage)

Size of nail	Length, in	Casing nails		Finishing nails		Clinch nails		Shingle nails	
		Gage	No. per lb	Gage	No. per lb	Gage	No. per lb	Gage	No. per lb
2d	1	15½	940	16½	1.473	14	723	13	434
3d	1¼	14½	588	15½	880	13	432	12	271
4d	1½	14	453	15	634	12	273	12	233
5d	1¾	14	389	15	535	12	234	12	203
6d	2	12½	223	13	288	11	158		
7d	2¼	12½	200	13	254	11	140		
8d	2½	11½	136	12½	196	10	101		
9d	2¾	11½	124	12½	178	10	91.4		
10d	3	10½	90	11½	124	9	70		
12d	3¼	10½	83	11½	113	9	64.1		
16d	3½	10	69	11	93	8	50		
20d	4	9	51	10	65	7	36.4		
30d	4½	9	45						
40d	5	8	37						

Size of nail	Length, in	Boat nails				Hinge nails				Flooring nails	
		Heavy		Light		Heavy		Light			
		Diam, in	No. per lb	Diam, in	No. per lb	Diam, in	No. per lb	Diam, in	No. per lb	Gage	No. per lb
4d	1½	¼	47	3/16	82	¼	53	3/16	90		
6d	2	¼	36	3/16	62	¼	39	3/16	66	11	168
8d	2½	¼	29	3/16	50	¼	31	3/16	53	10	105
10d	3	⅜	11	¼	24	⅜	12	¼	25	9	72
12d	3¼	⅜	10.4	¼	22	⅜	11	¼	23	8	56
16d	3½	⅜	9.6	¼	20	⅜	10	¼	22	7	44
20d	4	⅜	8	¼	18	⅜	8	¼	19	6	32

Size of nail	Length, in	Common wire nails and brads		Barbed car nails				Spikes			
				Heavy		Light					
		Gage	No. per lb	Gage	No. per lb	Gage	No. per lb	Size	Length, in	Gage	Approx no. per lb
2d	1	15	847	—	—	—	—	10d	3	6	43
3d	1¼	14	548	—	—	—	—	12d	3¼	6	39
4d	1½	12½	294	10	179	12	284	16d	3½	5	31
5d	1¾	12½	254	9	124	10	152	20d	4	4	23
6d	2	11½	167	9	108	10	132	30d	4½	3	18
7d	2¼	11½	150	8	80	9	95	40d	5	2	14
8d	2½	10¼	101	8	72	9	88	50d	5½	1	11
9d	2¾	10¼	92	7	55	8	65	50d	6	1	10
10d	3	9	66	7	50	8	59	—	7	5/16 in.	7
12d	3¼	9	61	6	39	7	46	—	8	⅜	4.1
16d	3½	8	47	6	36	7	43	—	9	⅜	3.7
20d	4	6	30	5	27	6	32	—	10	⅜	3.3
30d	4½	5	23	5	24	6	28	—	12	⅜	2.7
40d	5	4	18	4	18	5	22				
50d	5½	3	14	3	14	4	17				
60d	6	2	11	3	13	4	15				

Table 8.2.75 Cut Steel Nails and Spikes
(Sizes, lengths, and approximate number per lb)

Size	Length, in	Common	Clinch	Finishing	Casing and box	Fencing	Spikes	Barrel	Slating	Tobacco	Brads	Shingle
2d	1	740	400	1,100	—	—	—	450	340	—	—	—
3d	1¼	460	260	880	—	—	—	280	280	—	—	—
4d	1½	280	180	530	420	—	—	190	220	—	—	—
5d	1¾	210	125	350	300	100	—	—	180	130	—	—
6d	2	160	100	300	210	80	—	—	—	97	120	—
7d	2¼	120	80	210	180	60	—	—	—	85	94	90
8d	2½	88	68	168	130	52	—	—	—	68	74	72
9d	2¾	73	52	130	107	38	—	—	—	58	62	60
10d	3	60	48	104	88	26	—	—	—	48	50	—
12d	3¼	46	40	96	70	20	—	—	—	—	40	—
16d	3½	33	34	86	52	18	17	—	—	—	27	—
20d	4	23	24	76	38	16	14	—	—	—	—	—
25d	4¼	20	—	—	—	—	11	—	—	—	—	—
30d	4½	16½	—	—	30	—	9	—	—	—	—	—
40d	5	12	—	—	26	—	7½	—	—	—	—	—
50d	5½	10	—	—	20	—	6	—	—	—	—	—
60d	6	8	—	—	16	—	5½	—	—	—	—	—
—	6½	—	—	—	—	—	5	—	—	—	—	—
—	7	—	—	—	—	—	—	—	—	—	—	—

WIRE AND SHEET-METAL GAGES

In the metal industries, the word *gage* has been used in various systems, or scales, for expressing the thickness or weight per unit area of thin plates, sheet, and strip, or the diameters of rods and wire. Specific diameters, thicknesses, or weights per square foot have been or are denoted in gage systems by certain numerals followed by the word *gage,* for example, no. 12 gage, or simply 12 gage. Gage numbers for flat rolled products have been used only in connection with thin materials (Table 8.2.76). Heavier and thicker, flat rolled materials are usually designated by thickness in English or metric units.

There is considerable danger of confusion in the use of gage number in both foreign and domestic trade, which can be avoided by specifying thickness or diameter in inches or millimeters.

DRILL SIZES

See Table 8.2.77.

Table 8.2.76 Comparison of Standard Gages*
Thickness of diameter, in

Gage no.	BWG; Stubs Iron Wire	AWG; B&S	U.S. Steel Wire; Am. Steel & Wire; Washburn & Moen; Steel Wire	Galv. sheet steel	Manufacturers' standard
0000000	—	—	0.4900	—	—
000000	—	0.580000	0.4615	—	—
00000	—	0.516500	0.4305	—	—
0000	0.454	0.460000	0.3938	—	—
000	0.425	0.409642	0.3625	—	—
00	0.380	0.364796	0.3310	—	—
0	0.340	0.324861	0.3065	—	—
1	0.300	0.289297	0.2830	—	—
2	0.284	0.257627	0.2625	—	—
3	0.259	0.229423	0.2437	—	0.2391
4	0.238	0.204307	0.2253	—	0.2242
5	0.220	0.181940	0.2070	—	0.2092
6	0.203	0.162023	0.1920	—	0.1943
7	0.180	0.144285	0.1770	—	0.1793
8	0.165	0.128490	0.1620	0.1681	0.1644
9	0.148	0.114423	0.1483	0.1532	0.1495
10	0.134	0.101897	0.1350	0.1382	0.1345
11	0.120	0.090742	0.1205	0.1233	0.1196
12	0.109	0.080808	0.1055	0.1084	0.1046
13	0.095	0.071962	0.0915	0.0934	0.0897
14	0.083	0.064084	0.0800	0.0785	0.0747
15	0.072	0.057068	0.0720	0.0710	0.0673
16	0.065	0.050821	0.0625	0.0635	0.0598
17	0.058	0.045257	0.0540	0.0575	0.0538
18	0.049	0.040303	0.0475	0.0516	0.0478
19	0.042	0.035890	0.0410	0.0456	0.0418
20	0.035	0.031961	0.0348	0.0396	0.0359
21	0.032	0.028462	0.03175	0.0366	0.0329
22	0.028	0.025346	0.0286	0.0336	0.0299
23	0.025	0.022572	0.0258	0.0306	0.0269
24	0.022	0.020101	0.0230	0.0276	0.0239
25	0.020	0.017900	0.0204	0.0247	0.0209
26	0.018	0.015941	0.0181	0.0217	0.0179
27	0.016	0.014195	0.0173	0.0202	0.0164
28	0.014	0.012641	0.0162	0.0187	0.0149
29	0.013	0.011257	0.0150	0.0172	0.0135
30	0.012	0.010025	0.0140	0.0157	0.0120
31	0.010	0.008928	0.0132	0.0142	0.0105
32	0.009	0.007950	0.0128	0.0134	0.0097
33	0.008	0.007080	0.0118	—	0.0090
34	0.007	0.006305	0.0104	—	0.0082
35	0.005	0.005615	0.0095	—	0.0075
36	0.004	0.005000	0.0090	—	0.0067
37	—	0.004453	0.0085	—	0.0064
38	—	0.003965	0.0080	—	0.0060
39	—	0.003531	0.0075	—	—
40	—	0.003144	0.0070	—	—

* Principal uses—BWG: strips, bands, hoops, and wire; AWG or B&S: nonferrous sheets, rod, and wire; U.S. Steel Wire: steel wire except music wire; manufacturers' standard: uncoated steel sheets.

Table 8.2.77 Diameters of Small Drills

Number, letter, metric, and fractional drills in order of size

No.	Ltr	mm	in	Diam, in
		0.10		0.0039
		0.15		0.0059
		0.20		0.0079
		0.25		0.0098
		0.30		0.0118
80				0.0135
		0.35		0.0137
79				0.0145
			1/64	0.0156
		0.40		0.0157
78				0.0160
		0.45		0.0177
77				0.0180
		0.50		0.0197
76				0.0200
75				0.0210
		0.55		0.0216
74				0.0225
		0.60		0.0236
73				0.0240
72				0.0250
		0.65		0.0255
71				0.0260
		0.70		0.0275
70				0.0280
69				0.0292
		0.75		0.0295
68				0.0310
			1/32	0.0312
		0.80		0.0314
67				0.0320
66				0.0330
		0.85		0.0334
65				0.0350
		0.90		0.0354
64				0.0360
63				0.0370
		0.95		0.0374
62				0.0380
61				0.0390
		1.00		0.0393
60				0.0400
59				0.0410
		1.05		0.0413
58				0.0420
57				0.0430
		1.10		0.0433
		1.15		0.0452
56				0.0465
			3/64	0.0468
		1.20		0.0472
		1.25		0.0492
		1.30		0.0512
55				0.0520
		1.35		0.0531
54				0.0550
		1.40		0.0551
		1.45		0.0570
		1.50		0.0590
53				0.0595
		1.55		0.0610
			1/16	0.0625
		1.60		0.0630
52				0.0635
		1.65		0.0649
		1.70		0.0669
51				0.0670
		1.75		0.0688
50				0.0700
		1.80		0.0709
		1.85		0.0728
49				0.0730
		1.90		0.0748

No.	Ltr	mm	in	Diam, in
48				0.0760
		1.95		0.0767
			5/64	0.0781
47				0.0785
		2.00		0.0787
		2.05		0.0807
46				0.0810
45				0.0820
		2.10		0.0827
		2.15		0.0846
44				0.0860
		2.20		0.0866
		2.25		0.0885
43				0.0890
		2.30		0.0906
		2.35		0.0925
42				0.0935
			3/32	0.0937
		2.40		0.0945
41				0.0960
		2.45		0.0964
40				0.0980
		2.50		0.0984
39				0.0995
38				0.1015
		2.60		0.1024
37				0.1040
		2.70		0.1063
36				0.1065
		2.75		0.1082
			7/64	0.1093
35				0.1100
		2.80		0.1102
34				0.1110
33				0.1130
		2.90		0.1142
32				0.1160
		3.00		0.1181
31				0.1200
		3.10		0.1220
			1/8	0.1250
		3.20		0.1260
		3.25		0.1279
30				0.1285
		3.30		0.1299
		3.40		0.1339
29				0.1360
		3.50		0.1378
28				0.1405
			9/64	0.1406
		3.60		0.1417
27				0.1440
		3.70		0.1457
26				0.1470
		3.75		0.1476
25				0.1495
		3.80		0.1496
24				0.1520
		3.90		0.1535
23				0.1540
			5/32	0.1562
22				0.1570
		4.00		0.1575
21				0.1590
20				0.1610
		4.10		0.1614
		4.2		0.1654
19				0.1660
		4.25		0.1673
		4.30		0.1693
18				0.1695
			11/64	0.1718
17				0.1730

No.	Ltr	mm	in	Diam, in
		4.40		0.1732
16				0.1770
		4.50		0.1772
15				0.1800
		4.60		0.1811
14				0.1820
13				0.1850
		4.70		0.1850
		4.75		0.1850
			3/16	0.1875
		4.80		0.1890
12				0.1890
11				0.1910
		4.90		0.1920
10				0.1935
9				0.1960
		5.00		0.1968
8				0.1990
		5.10		0.2008
7				0.2010
			13/64	0.2031
6				0.2040
		5.20		0.2047
5				0.2055
		5.25		0.2066
		5.30		0.2087
4				0.2090
		5.40		0.2126
3				0.2130
		5.50		0.2165
			7/32	0.2187
		5.60		0.2205
2				0.2210
		5.70		0.2244
		5.75		0.2263
1				0.2280
		5.80		0.2283
		5.90		0.2323
	A			0.2340
			15/64	0.2340
		6.00		0.2362
	B			0.2380
		6.10		0.2402
	C			0.2420
		6.20		0.2441
	D			0.2460
		6.25		0.2460
		6.30		0.2480
	E		1/4	0.2500
		6.40		0.2519
		6.50		0.2559
	F			0.2570
		6.60		0.2598
	G			0.2610
		6.70		0.2637
			17/64	0.2656
		6.75		0.2657
	H			0.2660
		6.80		0.2677
		6.90		0.2717
	I			0.2720
		7.00		0.2756
	J			0.2770
		7.10		0.2795
	K			0.2810
			9/32	0.2812
		7.20		0.2835
		7.25		0.2854
		7.30		0.2874
	L			0.2900
		7.40		0.2913
	M			0.2950
		7.50		0.2953

Table 8.2.77   Diameters of Small Drills   (Continued)

No.	Ltr	mm	in	Diam, in	No.	Ltr	mm	in	Diam, in	No.	Ltr	mm	in	Diam, in
			19/64	0.2968			9.70		0.3819			17.00		0.6693
		7.60		0.2992			9.75		0.3838				43/64	0.6718
	N			0.3020			9.80		0.3858				11/16	0.6875
		7.70		0.3031		W			0.3860			17.50		0.6890
		7.75		0.3051			9.90		0.3898				45/64	0.7031
		7.80		0.3071				25/64	0.3906			18.00		0.7087
		7.90		0.3110			10.00		0.3937				23/32	0.7187
			5/16	0.3125		X			0.3970			18.50		0.7283
		8.00		0.3150		Y			0.4040				47/64	0.7374
	O			0.3160				13/32	0.4062			19.00		0.7480
		8.10		0.3189		Z			0.4130				3/4	0.7500
		8.20		0.3228			10.50		0.4134				49/64	0.7656
	P			0.3230				27/64	0.4218			19.50		0.7677
		8.25		0.3248			11.00		0.4331				25/32	0.7812
		8.30		0.3268				7/16	0.4375			20.00		0.7874
			21/64	0.3281			11.50		0.4528				51/64	0.7968
		8.40		0.3307				29/64	0.4531			20.50		0.8070
	Q			0.3320				15/32	0.4687				13/16	0.8125
		8.50		0.3346			12.00		0.4724			21.00		0.8267
		8.60		0.3386				31/64	0.4843				53/64	0.8281
	R			0.3390			12.50		0.4921				27/32	0.8437
		8.70		0.3425			1/2		0.5000			21.50		0.8464
			11/32	0.3437			13.00		0.5118				55/64	0.8593
		8.75		0.3444				33/64	0.5156			22.00		0.8661
		8.80		0.3464				17/32	0.5312				7/8	0.8750
	S			0.3480			13.50		0.5315			22.50		0.8858
		8.90		0.3504				35/64	0.5468				57/64	0.8906
		9.00		0.3543			14.00		0.5512			23.00		0.9055
	T			0.3580				9/16	0.5625				29/32	0.9062
		9.10		0.3583			14.50		0.5708				59/64	0.9218
			23/64	0.3593				37/64	0.5781			23.50		0.9251
		9.20		0.3622			15.00		0.5905				15/16	0.9375
		9.25		0.3641				19/32	0.5937			24.00		0.9448
		9.30		0.3661				39/64	0.6093				61/64	0.9531
	U			0.3680			15.50		0.6102			24.50		0.9646
		9.40		0.3701				5/8	0.6250				31/32	0.9687
		9.50		0.3740			16.00		0.6299			25.00		0.9842
			3/8	0.3750				41/64	0.6406				63/64	0.9843
	V			0.3770			16.50		0.6496				1.0	1.0000
		9.60		0.3780				21/32	0.6562			25.50		1.0039

SOURCE: Adapted from Colvin and Stanley, "American Machinists' Handbook," 8th ed., McGraw-Hill, New York, 1945.

8.3   GEARING
by George W. Michalec

REFERENCES: Buckingham, "Manual of Gear Design," Industrial Press. Cunningham, Noncircular Gears, Mach. Des., Feb. 19, 1957. Cunningham and Cunningham, Rediscovering the Noncircular Gear, Mach. Des., Nov. 1, 1973. Dudley, "Gear Handbook," McGraw-Hill. Dudley, "Handbook of Practical Gear Design," McGraw-Hill. Michalec, "Precision Gearing: Theory and Practice," Wiley. Shigely, "Engineering Design," McGraw-Hill. AGMA Standards. "Gleason Bevel and Hypoid Gear Design," Gleason Works, Rochester. "Handbook of Gears: Inch and Metric" and "Elements of Metric Gear Technology," Designatronics, New Hyde Park, NY. Adams, "Plastics Gearing: Selection and Application," Marcel Dekker.

Notation

a = addendum
b = dedendum
B = backlash, linear measure along pitch circle
c = clearance
C = center distance
d = pitch diam of pinion
d_b = base circle diam of pinion
d_o = outside diam of pinion
d_r = root diam of pinion
D = pitch diameter of gear
D_P = pitch diam of pinion
D_G = pitch diam of gear
D_o = outside diam of gear
D_b = base circle diam of gear
D_t = throat diam of wormgear
F = face width
h_k = working depth
h_t = whole depth
$\text{inv } \phi$ = involute function ($\tan \phi - \phi$)

l = lead (advance of worm or helical gear in 1 rev)
$l_p(l_G)$ = lead of pinion (gear) in helical gears
L = lead of worm in one revolution
m = module
m_G = gear ratio ($m_G = N_G/N_P$)
m_p = contact ratio (of profiles)
M = measurement of over pins
$n_P(n_G)$ = speed of pinion (gear), r/min
$N_P(N_G)$ = number of teeth in pinion (gear)
n_w = number of threads in worm
p = circular pitch
p_b = base pitch
p_n = normal circular pitch of helical gear, worm mesh
P_d = diametral pitch
P_{dn} = normal diametral pitch
R = pitch radius
R_c = radial distance from center of gear to center of measuring pin
$R_P(R_G)$ = pitch radius of pinion (gear)
R_T = testing radius when rolled on a variable-center-distance inspection fixture
s = stress
t = tooth thickness
t_n = normal circular tooth thickness
$T_P(T_G)$ = formative number of teeth in pinion (gear) (in bevel gears)
v = pitch line velocity
X = correction factor for profile shift
α = addendum angle of bevel gear
γ = pitch angle of bevel pinion
γ_R = face angle at root of bevel pinion tooth
γ_o = face angle at tip of bevel pinion tooth
Γ = pitch angle of bevel gear
Γ_R = face angle at root of bevel gear tooth
Γ_{g_o} = face angle at tip of bevel gear tooth
δ = dedendum angle of bevel gear
$\overline{\Delta C}$ = relatively small change in center distance C
ϕ = pressure angle
ϕ_n = normal pressure angle
ψ = helix or spiral angle
$\psi_P(\psi_G)$ = helix angle of teeth in pinion (gear)
Σ = shaft angle of meshed bevel pair

BASIC GEAR DATA

Gear Types Gears are grouped in accordance with tooth forms, shaft arrangement, pitch, and quality. Tooth forms and shaft arrangements are:

Tooth form	Shaft arrangement
Spur	Parallel
Helical	Parallel or skew
Worm	Skew
Bevel	Intersecting
Hypoid	Skew

Pitch definitions (see Fig. 8.3.1). **Diametral pitch** P_d is the ratio of number of teeth in the gear to the diameter of the pitch circle D measured in inches, $P_d = N/D$. **Circular pitch** p is the linear measure in inches along the pitch circle between corresponding points of adjacent teeth. From these definitions, $P_d p = \pi$. The **base pitch** p_b is the distance along the line of action between successive involute tooth surfaces. The base and circular pitches are related as $p_b = p \cos \phi$, where ϕ = the pressure angle.

Pitch circle is the imaginary circle that rolls without slippage with a pitch circle of a mating gear. The pitch (circle) diameter equals $D = N/P_d = Np/\pi$. The basic relation between P_d and p is $P_d p = \pi$.

Tooth size is related to pitch. In terms of diametral pitch P_d, the relationship is inverse; i.e., large P_d implies a small tooth, and small P_d

implies a large tooth. Conversely, there is a direct relationship between tooth size and circular pitch p. A small tooth has a small p, but a large tooth has a large p. (See Fig. 8.3.1b.) In terms of P_d, coarse teeth comprise P_d less than 20; fine teeth comprise P_d of 20 and higher. (See Fig. 8.3.1b.) **Quality** of gear teeth is classified as commercial, precision, and ultraprecision.

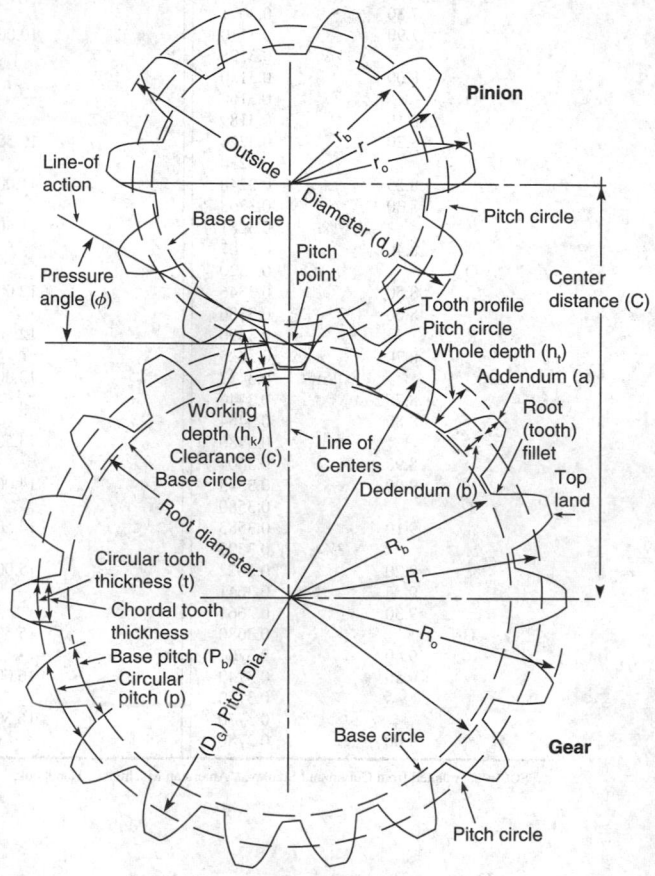

Fig. 8.3.1a Basic gear geometry and nomenclature.

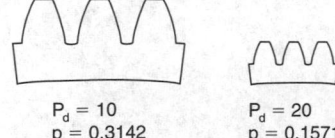

$P_d = 10$	$P_d = 20$
$p = 0.3142$	$p = 0.1571$

Fig. 8.3.1b Comparison of pitch and tooth size.

Pressure angle ϕ for all gear types is the acute angle between the common normal to the profiles at the contact point and the common pitch plane. For spur gears it is simply the acute angle formed by the common tangent between base circles of mating gears and a normal to the line of centers. For standard gears, pressure angles of 14½°, 20°, and 25° have been adopted by ANSI and the gear industry (see Fig. 8.3.1a). The 20° pressure angle is most widely used because of its versatility. The higher pressure angle 25° provides higher strength for highly loaded gears. Although 14½° appears in standards, and in past decades was extensively used, it is used much less than 20°. The 14½° standard is still used for replacement gears in old design equipment, in applications where backlash is critical, and where advantage can be taken of lower backlash with change in center distance.

The **base circle** (or **base cylinder**) is the circle from which the involute tooth profiles are generated. The relationship between the base-circle and pitch-circle diameter is $D_b = D \cos \phi$.

Tooth proportions are established by the addendum, dedendum, working depth, clearance, tooth circular thickness, and pressure angle (see Fig. 8.3.1). In addition, gear face width F establishes thickness of the gear measured parallel to the gear axis.

For involute teeth, proportions have been standardized by ANSI and AGMA into a limited number of systems using a basic rack for specification (see Fig. 8.3.2 and Table 8.3.1). Dimensions for the basic rack are normalized for diametral pitch = 1. Dimensions for a specific pitch are

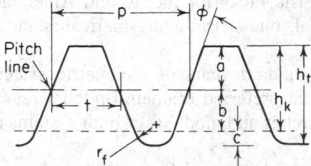

Fig. 8.3.2 Basic rack for involute gear systems. a = addendum; b = dedendum; c = clearance; h_k = working depth; h_t = whole depth; p = circular pitch; r_f = fillet radius; t = tooth thickness; ϕ = pressure angle.

obtained by dividing by the pitch. Standards for basic involute spur, helical and face gear designs, and noninvolute bevel and wormgear designs are listed in Table 8.3.2.

Gear ratio (or **mesh ratio**) m_G is the ratio of number of teeth in a meshed pair, expressed as a number greater than 1; $m_G = N_G/N_P$, where the pinion is the member having the lesser number of teeth. For spur and parallel-shaft helical gears, the base circle ratio must be identical to the gear ratio. The speed ratio of gears is inversely proportionate to their numbers of teeth. Only for standard spur and parallel-shaft helical gears is the pitch diameter ratio equal to the gear ratio and inversely proportionate to the speed ratio.

Metric Gears—Tooth Proportions and Standards

Metric gearing not only is based upon different units of length measure but also involves its own unique design standard. This means that metric gears and American-standard-inch diametral-pitch gears are not interchangeable.

In the metric system the *module m* is analogous to pitch and is defined as

$$m = \frac{D}{N} = \text{mm of pitch diameter per tooth}$$

Table 8.3.2 Gear System Standards

Gear type	ANSI/AGMA no.	Title
Spur and helical	201.02	Tooth Proportions for Coarse-Pitch Involute Spur Gears
Spur and helical	1003-G93	Tooth Proportions for Fine-Pitch Spur and Helical Gearing
Spur and helical	370.01	Design Manual for Fine-Pitch Gearing
Bevel gears	2005-B88	Design Manual for Bevel Gears (Straight, Zerol, Spiral, and Hypoid)
Worm gearing	6022-C93	Design of General Industrial Coarse-Pitch Cylindrical Worm Gearing
Worm gearing	6030-C87	Design of Industrial Double-Enveloping Worm Gearing
Face gears	203.03	Fine-Pitch on Center-Face Gears for 20-Degree Involute Spur Pinions

Note that, for the module to have proper units, the pitch diameter must be in millimeters.

The **metric module** was developed in a number of versions that differ in minor ways. The German DIN standard is widely used in Europe and other parts of the world. The Japanese have their own version defined in JS standards. Deviations among these and other national standards are minor, differing only as to dedendum size and root radii. The differences have been resolved by the new unified module standard promoted by the International Standards Organization (ISO). This unified version (Fig. 8.3.3) conforms to the new SI in all respects. All major industrial

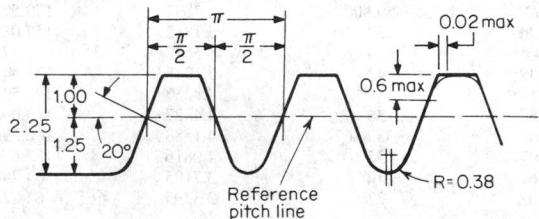

Fig. 8.3.3 The ISO basic rack for metric module gears.

countries on the metric system have shifted to this ISO standard, which also is the basis for American metric gearing. Table 8.3.3 lists pertinent current ISO metric standards.

Table 8.3.1 Tooth Proportions of Basic Rack for Standard Involute Gear Systems

Tooth parameter (of basic rack)	Symbol, Figs. 8.3.1a and 8.3.2	Tooth proportions for various standard systems					
		1 Full-depth involute, 14½°	2 Full-depth involute, 20°	3 Stub involute, 20°	4 Coarse pitch involute spur gears, 20°	5 Coarse pitch involute spur gears, 25°	6 Fine-pitch involute, 20°
1. System sponsors		ANSI and AGMA	ANSI	ANSI and AGMA	AGMA	AGMA	ANSI and AGMA
2. Pressure angle	ϕ	14½°	20°	20°	20°	25°	20°
3. Addendum	a	$1/P_d$	$1/P_d$	$0.8/P_d$	$1.000/P_d$	$1.000/P_d$	$1.000/P_d$
4. Min dedendum	b	$1.157/P_d$	$1.157/P_d$	$1/P_d$	$1.250/P_d$	$1.250/P_d$	$1.200/P_d + 0.002$
5. Min whole depth	h_t	$2.157/P_d$	$2.157/P_d$	$1.8/P_d$	$2.250/P_d$	$2.250/P_d$	$2.2002/P_d$ 0.002 in
6. Working depth	h_k	$2/P_d$	$2/P_d$	$1.6/P_d$	$2.000/P_d$	$2.000/P_d$	$2.000/P_d$
7. Min clearance	h_c	$0.157/P_d$	$0.157/P_d$	$0.200/P_d$	$0.250/P_d$	$0.250/P_d$	$0.200/P_d + 0.002$ in
8. Basic circular tooth thickness on pitch line	t	$1.5708/P_d$	$1.5708/P_d$	$1.5708/P_d$	$\pi/(2P_d)$	$\pi/(2P_d)$	$1.5708/P_d$
9. Fillet radius in basic rack	r_f	1⅓ × clearance	1½ × clearance	Not standardized	$0.300/P_d$	$0.300/P_d$	Not standardized
10. Diametral pitch range		Not specified	Not specified	Not specified	19.99 and coarser	19.99 and coarser	20 and finer
11. Governing standard: ANSI		B6.1	B6.1	B6.1			
AGMA		201.02		201.02	201.02	201.02	1,003–G93

Table 8.3.3   ISO Metric Gearing Standards

ISO 53:1974	Cylindrical gears for general and heavy engineering—Basic rack
ISO 54:1977	Cylindrical gears for general and heavy engineering—Modules and diametral pitches
ISO 677:1976	Straight bevel gears for general and heavy engineering—Basic rack
ISO 678:1976	Straight bevel gears for general and heavy engineering—Modules and diametral pitches
ISO 701:1976	International gear notation—symbols for geometric data
ISO 1122-1:1983	Glossary of gear terms—Part 1: Geometric definitions
ISO 1328:1975	Parallel involute gears—ISO system of accuracy
ISO 1340:1976	Cylindrical gears—Information to be given to the manufacturer by the purchaser in order to obtain the gear required
ISO 1341:1976	Straight bevel gears—Information to be given to the manufacturer by the purchaser in order to obtain the gear required
ISO 2203:1973	Technical drawings—Conventional representation of gears

Tooth proportions for standard spur and helical gears are given in terms of the basic rack. Dimensions, in millimeters, are normalized for module $m = 1$. Corresponding values for other modules are obtained by multiplying each dimension by the value of the specific module m. Major tooth parameters are described by this standard:

Tooth form: Straight-sided and full-depth, forming the basis of a family of full-depth interchangeable gears.

Pressure angle: 20°, conforming to worldwide acceptance.

Addendum: Equal to module m, which corresponds to the American practice of $1/P_d$ = addendum.

Dedendum: Equal to 12.5m, which corresponds to the American practice of $1.25/P_d$ = dedendum.

Root radius: Slightly greater than American standards specifications.

Tip radius: A maximum is specified, whereas American standards do not specify. In practice, U.S. manufacturers can specify a tip radius as near zero as possible.

Note that the basic racks for metric and American inch gears are essentially identical, but metric and American standard gears are not interchangeable.

The preferred standard gears of the metric system are not interchangeable with the preferred diametral-pitch sizes. Table 8.3.4 lists commonly used pitches and modules of both systems (preferred values are boldface).

Metric gear use in the United States, although expanding, is still a small percentage of total gearing. Continuing industry conversions and imported equipment replacement gearing are building an increasing demand for metric gearing. The reference list cites a domestic source of stock metric gears of relatively small size, in medium and fine pitches. Large diameter coarse pitch metric gears are made to order by many gear fabricators.

Table 8.3.4   Metric and American Gear Equivalents

Diametral pitch P_d	Module m	Circular pitch		Circular tooth thickness		Addendum	
		in	mm	in	mm	in	mm
1/2	50.8000	6.2832	159.593	3.1416	79.7965	2.0000	50.8000
0.5080	**50**	6.1842	157.080	3.0921	78.5398	1.9685	50
0.5644	**45**	5.5658	141.372	2.7850	70.6858	1.7730	45
0.6048	**42**	5.1948	131.947	2.5964	65.9734	1.6529	42
0.6513	**39**	4.8237	122.522	2.4129	61.2610	1.5361	39
0.7056	**36**	4.4527	113.097	2.2249	56.5487	1.4164	36
3/4	33.8667	4.1888	106.396	2.0943	53.1977	1.3333	33.8667
0.7697	**33**	4.0816	103.673	2.0400	51.8363	1.2987	33
0.8467	**30**	3.7105	94.248	1.8545	47.1239	1.1806	30
0.9407	**27**	3.3395	84.823	1.6693	42.4115	1.0627	27
1	25.4000	3.1416	79.800	1.5708	39.8984	1.0000	25.4001
1.0583	**24**	2.9685	75.398	1.4847	37.6991	0.9452	24
1.1546	**22**	2.7210	69.115	1.3600	34.5575	0.8658	22
1.2700	**20**	2.4737	62.832	1.2368	31.4159	0.7874	20
1.4111	**18**	2.2263	56.548	1.1132	28.2743	0.7087	18
1.5	16.9333	2.0944	53.198	1.0472	26.5988	0.6667	16.933
1.5875	**16**	1.9790	50.267	0.9894	25.1327	0.6299	16
1.8143	**14**	1.7316	43.983	0.8658	21.9911	0.5512	14
2	12.7000	1.5708	39.898	0.7854	19.949	0.5000	12.7000
2.1167	**12**	1.4842	37.699	0.7420	18.8496	0.4724	12
2.5	10.1600	1.2566	31.918	0.6283	15.9593	0.4000	10.1600
2.5400	**10**	1.2368	31.415	0.6184	15.7080	0.3937	10
2.8222	**9**	1.1132	28.275	0.5565	14.1372	0.3543	9
3	8.4667	1.0472	26.599	0.5235	13.2995	0.3333	8.4667
3.1416	8.0851	1.0000	25.400	0.5000	12.7000	0.3183	0.0851
3.1750	**8**	0.9895	25.133	0.4948	12.5664	0.3150	8.00
3.5	7.2571	0.8976	22.799	0.4488	11.3994	0.2857	7.2571
3.6286	**7**	0.8658	21.991	0.4329	10.9956	0.2756	7.000
3.9078	**6.5**	0.8039	20.420	0.4020	10.2101	0.2559	6.5
4	6.3500	0.7854	19.949	0.3927	9.9746	0.2500	6.3500
4.2333	**6**	0.7421	18.850	0.3710	9.4248	0.2362	6.0000
4.6182	**5.5**	0.6803	17.279	0.3401	8.6394	0.2165	5.5
5	5.0801	0.6283	15.959	0.3142	7.9794	0.2000	5.080
5.0802	**5**	0.6184	15.707	0.3092	7.8537	0.1968	5.000
5.3474	**4.75**	0.5875	14.923	0.2938	7.4612	0.1870	4.750
5.6444	**4.5**	0.5566	14.138	0.2783	7.0688	0.1772	4.500
6	4.2333	0.5236	13.299	0.2618	6.6497	0.1667	4.233
6.3500	**4**	0.4947	12.565	0.2473	6.2827	0.1575	4.000
6.7733	**3.75**	0.4638	11.781	0.2319	5.8903	0.1476	3.750
7	3.6286	0.4488	11.399	0.2244	5.6998	0.1429	3.629

Table 8.3.4   Metric and American Gear Equivalents   (Continued)

Diametral pitch P_d	Module m	Circular pitch		Circular tooth thickness		Addendum	
		in	mm	in	mm	in	mm
7.2571	**3.5**	0.4329	10.996	0.2164	5.4979	0.1378	3.500
7.8154	**3.25**	0.4020	10.211	0.2010	5.1054	0.1279	3.250
8	3.1750	0.3927	9.974	0.1964	4.9886	0.1250	3.175
8.4667	**3**	0.3711	9.426	0.1855	4.7130	0.1181	3.000
9	2.8222	0.3491	8.867	0.1745	4.4323	0.1111	2.822
9.2364	**2.75**	0.3401	8.639	0.1700	4.3193	0.1082	2.750
10	2.5400	0.3142	7.981	0.1571	3.9903	0.1000	2.540
10.1600	**2.50**	0.3092	7.854	0.1546	3.9268	0.0984	2.500
11	2.3091	0.2856	7.254	0.1428	3.6271	0.0909	2.309
11.2889	**2.25**	0.2783	7.069	0.1391	3.5344	0.0886	2.250
12	2.1167	0.2618	6.646	0.1309	3.3325	0.0833	2.117
12.7000	**2**	0.2474	6.284	0.1236	3.1420	0.0787	2.000
13	1.9538	0.2417	6.139	0.1208	3.0696	0.0769	1.954
14	1.8143	0.2244	5.700	0.1122	2.8500	0.0714	1.814
14.5143	**1.75**	0.2164	5.497	0.1082	2.7489	0.0689	1.750
15	1.6933	0.2094	5.319	0.1047	2.6599	0.0667	1.693
16	1.5875	0.1964	4.986	0.0982	2.4936	0.0625	1.587
16.9333	**1.5**	0.1855	4.712	0.0927	2.3562	0.0591	1.500
18	1.4111	0.1745	4.432	0.0873	2.2166	0.0556	1.411
20	1.2700	0.1571	3.990	0.0785	1.9949	0.0500	1.270
20.3200	**1.25**	0.1546	3.927	0.0773	1.9635	0.0492	1.250
22	1.1545	0.1428	3.627	0.0714	1.8136	0.0455	1.155
24	1.0583	0.1309	3.325	0.0655	1.6624	0.0417	1.058
25.4000	1	0.1237	3.142	0.0618	1.5708	0.0394	1.000
28	0.90701	0.1122	2.850	0.0561	1.4249	0.0357	0.9071
28.2222	**0.9**	0.1113	2.827	0.0556	1.4137	0.0354	0.9000
30	0.84667	0.1047	2.659	0.0524	1.3329	0.0333	0.8467
31.7500	**0.8**	0.0989	2.513	0.04945	1.2566	0.0315	0.8000
32	0.79375	0.0982	2.494	0.04909	1.2468	0.0313	0.7937
33.8667	**0.75**	0.0928	2.357	0.04638	1.1781	0.0295	0.7500
36	0.70556	0.0873	2.217	0.04363	1.1083	0.0278	0.7056
36.2857	**0.7**	0.0865	2.199	0.04325	1.0996	0.0276	0.7000
40	0.63500	0.0785	1.994	0.03927	0.9975	0.0250	0.6350
42.3333	**0.6**	0.0742	1.885	0.03710	0.9423	0.0236	0.6000
44	0.57727	0.0714	1.814	0.03570	0.9068	0.0227	0.5773
48	0.52917	0.0655	1.661	0.03272	0.8311	0.0208	0.5292
50	0.50800	0.0628	1.595	0.03141	0.7976	0.0200	0.5080
50.8000	**0.5**	0.06184	1.5707	0.03092	0.7854	0.0197	0.5000
63.5000	**0.4**	0.04947	1.2565	0.02473	0.6283	0.0157	0.4000
64	0.39688	0.04909	1.2469	0.02454	0.6234	0.0156	0.3969
67,7333	0.375	0.04638	1.1781	0.02319	0.5890	0.0148	0.3750
72	0.35278	0.04363	1.1082	0.02182	0.5541	0.0139	0.3528
72.5714	0.35	0.04329	1.0996	0.02164	0.5498	0.0138	0.3500
78.1538	0.325	0.04020	1.0211	0.02010	0.5105	0.0128	0.3250
80	0.31750	0.03927	0.9975	0.01964	0.4987	0.0125	0.3175
84.6667	**0.3**	0.03711	0.9426	0.01856	0.4713	0.0118	0.3000
92.3636	0.275	0.03401	0.8639	0.01700	0.4319	0.0108	0.2750
96	0.26458	0.03272	0.8311	0.01636	0.4156	0.0104	0.2646
101.6000	0.25	0.03092	0.7854	0.01546	0.3927	0.00984	0.2500
120	0.21167	0.02618	0.6650	0.01309	0.3325	0.00833	0.2117
125	0.20320	0.02513	0.6383	0.01256	0.3192	0.00800	0.2032
127.0000	**0.2**	0.02474	0.6284	0.01237	0.3142	0.00787	0.2000
150	0.16933	0.02094	0.5319	0.01047	0.2659	0.00667	0.1693
169.3333	0.15	0.01855	0.4712	0.00928	0.2356	0.00591	0.1500
180	0.14111	0.01745	0.4432	0.00873	0.2216	0.00555	0.1411
200	0.12700	0.01571	0.3990	0.00786	0.1995	0.00500	0.1270
203.2000	0.125	0.01546	0.3927	0.00773	0.1963	0.00492	0.1250

FUNDAMENTAL RELATIONSHIPS OF SPUR AND HELICAL GEARS

Center distance is the distance between axes of mating gears and is determined from $C = (n_G + N_P)/(2P_d)$, or $C = (D_G + D_P)/2$. Deviation from ideal center distance of involute gears is not detrimental to proper (conjugate) gear action which is one of the prime superiority features of the involute tooth form.

Contact Ratio   Referring to the top part of Fig. 8.3.4 and assuming no tip relief, the pinion engages in the gear at a, where the outside circle of the gear tooth intersects the line of action ac. For the usual spur gear and pinion combinations there will be two pairs of teeth theoretically in contact at engagement (a gear tooth and its mating pinion tooth considered as a pair). This will continue until the pair ahead (bottom part of Fig. 8.3.4) disengages at c, where the outside circle of the pinion intersects the line of action ac, the movement along the line of action being *ab*. After disengagement the pair behind will be the only pair in contact until another pair engages, the movement along the line of action for single-pair contact being *bd*. Two pairs are theoretically in contact during the remaining intervals, *ab* + *dc*.

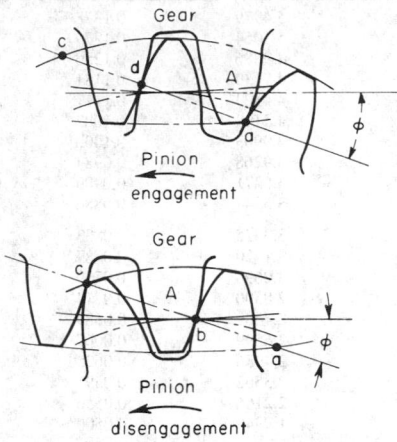

Fig. 8.3.4 Contact conditions at engagement and disengagement.

Contact ratio expresses the average number of pairs of teeth theoretically in contact and is obtained numerically by dividing the length of the line of action by the normal pitch. For full-depth teeth, without undercutting, the contact ratio is $m_p = (\sqrt{D_0^2 - D_b^2} + \sqrt{d_0^2 - d_b^2} - 2C \sin \phi)/(2p \cos \phi)$. The result will be a mixed number with the integer portion the number of pairs of teeth always in contact and carrying load, and the decimal portion the amount of time an additional pair of teeth are engaged and share load. As an example, for m_p between 1 and 2:

Load is carried by one pair, $(2 - m_p)/m_p$ of the time.

Load is carried by two pairs, $2(m_p - 1)/m_p$ of the time.

In Figs. 8.3.5 to 8.3.7, contact ratios are given for standard generated gears, the lower part of Figs. 8.3.5 and 8.3.6 representing the effect of undercutting.

These charts are applicable to both standard dimetral pitch gears made in accordance with American standards and also standard metric gears that have an addendum of one module.

Tooth Thickness For standard gears, the tooth thickness *t* of mating gears is equal, where $t = p/2 = \pi/(2P_d)$ measured linearly along the arc of the pitch circle. The tooth thickness t_1 at any radial point of the tooth (at diameter D_1) can be calculated from the known thickness *t* at the pitch radius $D/2$ by the relationship $t_1 = t(D_1/D) - D_1$ (inv ϕ_1 − inv ϕ), where inv $\phi = \tan \phi - \phi$ = involute function. Units for ϕ must be radians. Tables of values for inv ϕ from 0 to 45° can be found in the references (Buckingham and Dudley).

Over-plus measurements (spur gears) are another means of deriving tooth thickness. If cylindrical pins are inserted in tooth spaces diametrically opposite one another (or nearest space for an odd number of teeth) (Fig. 8.3.8), the tooth thickness can be derived from the measurement *M* as follows:

$$t = D(\pi/N + \text{inv } \phi_1 - \text{inv } \phi - d_w/D_b)$$
$$\cos \phi_1 = (D \cos \phi)/2R_c$$
$$R_c = (M - d_w)/2 \qquad \text{for even number of teeth}$$
$$R_c = (M - d_w)/[2 \cos (90/N)] \qquad \text{for odd number of teeth}$$

where d_w = pin diameter, R_c = distance from gear center to center of pin, and *M* = measurement over pins.

For the reverse situation, the over-pins measurement *M* can be found for a given tooth thickness *t* at diameter *D* and pressure angle ϕ by the following: inv $\phi_1 = t/D + \text{inv } \phi + d_w/(D \cos \phi) - \pi/N$, $M = D \cos \phi/\cos \phi_1 + d_w$ (for even number of teeth), $M = (D \cos \phi/\cos \phi_1) \cos (90°/N) + d_w$ (for odd number of teeth).

Table values of over-pins measures (see Dudley and Van Keuren) facilitate measurements for all standard gears including those with slight departures from standard. (For correlation with tooth thickness and testing radius, see Michalec, *Product Eng.*, May 1957, and "Precision Gearing: Theory and Practice," Wiley.)

Testing radius R_T is another means of determining tooth thickness and refers to the effective pitch radius of the gear when rolled intimately with a master gear of known size calibration. (See Michalec, *Product Eng.*, Nov. 1956, and "Precision Gearing: Theory and Practice," op. cit.) For standard design gears the testing radius equals the pitch radius. The testing radius may be corrected for small departures Δt from ideal tooth thickness by the relationship, $R_T = R + \Delta t/2 \tan \phi$, where $\Delta t = t_1 - t$ and is positive and negative respectively for thicker and thinner tooth thicknesses than standard value *t*.

Backlash *B* is the amount by which the width of a tooth space exceeds the thickness of the engaging tooth measured on the pitch circle. Backlash does not adversely affect proper gear function except for lost mo-

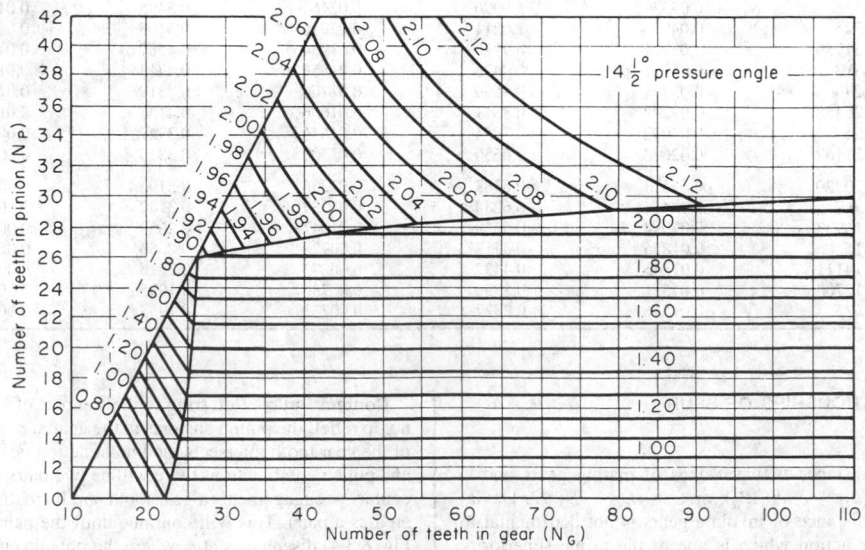

Fig. 8.3.5 Contact ratio, spur gear pairs—full depth, standard generated teeth, 14½° pressure angle.

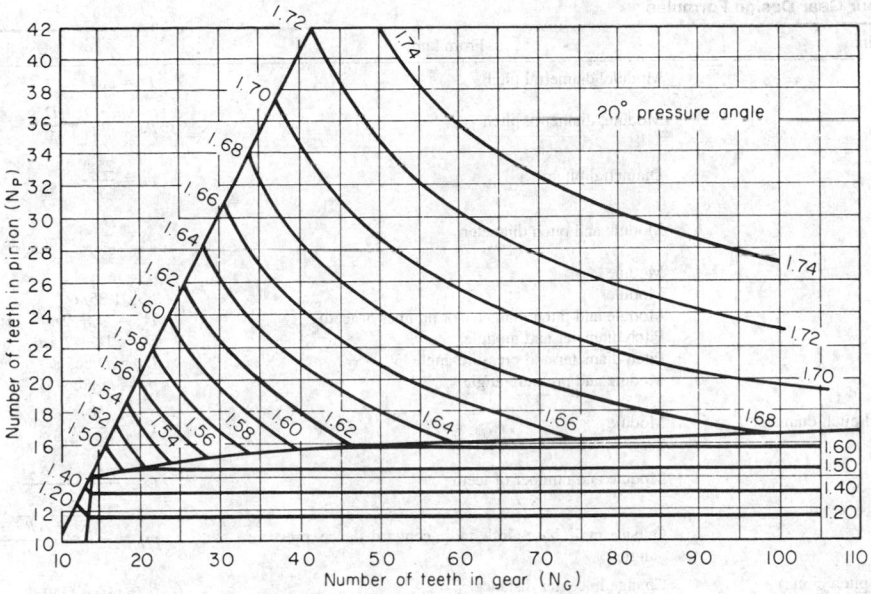

Fig. 8.3.6   Contact ratio, spur gear pairs—full-depth standard generated teeth, 20° pressure angle.

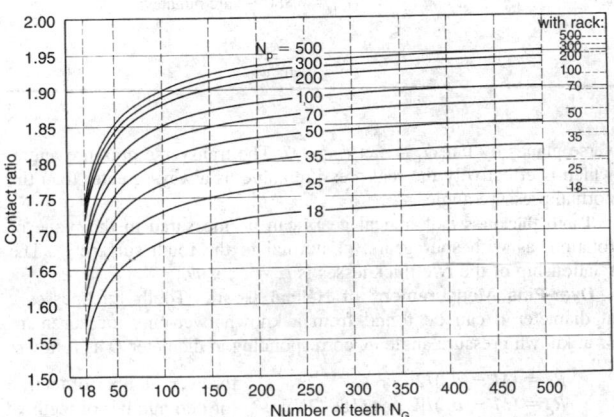

Fig. 8.3.7   Contact ratio for large numbers of teeth—spur gear pairs, full-depth standard teeth, 20° pressure angle. (*Data by R. Feeney and T. Wall.*)

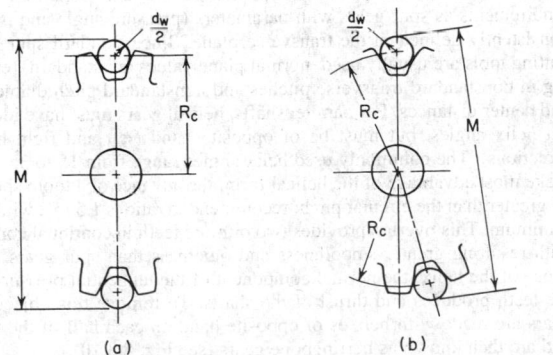

Fig. 8.3.8   Geometry of over-pins measurements (*a*) for an even number of teeth and (*b*) for an odd number of teeth.

tion upon reversal of gear rotation. Backlash inevitably occurs because of necessary fabrication tolerances on tooth thickness and center distance plus need for clearance to accommodate lubricant and thermal expansion. Proper backlash can be introduced by a specified amount of tooth thinning or slight increase in center distance. The relationship between small change in center distance $\overline{\Delta C}$ and backlash is $B = 2\, \overline{\Delta C} \tan \phi$ (see Michalec, "Precision Gearing: Theory and Practice").

Total composite error (tolerance) is a measure of gear quality in terms of the net sum of irregularity of its testing radius R_T due to pitch-circle runout and tooth-to-tooth variations (see Michalec, op. cit.).

Tooth-to-tooth composite error (tolerance) is the variation of testing radius R_T between adjacent teeth caused by tooth spacing, thickness, and profile deviations (see Michalec, op. cit.).

Profile shifted gears have tooth thicknesses that are significantly different from nominal standard value; excluded are deviations caused by normal allowances and tolerances. They are also known as **modified gears, long and short addendum gears**, and **enlarged gears**. They are produced by cutting the teeth with standard cutters at enlarged or reduced outside diameters. The result is a relative shift of the two families of involutes forming the tooth profiles, simultaneously with a shift of the tooth radially outward or inward (see Fig. 8.3.9). Calculation of operating conditions and tooth parameters are

$$C_1 = \frac{(C \cos \phi)}{\cos \phi_1}$$

$$\text{inv } \phi_1 = \text{inv } \phi + \frac{N_P(t'_G + t'_P) - \pi D_P}{D_P(N_P + N_G)}$$

$$t'_G = t + 2X_G \tan \phi$$
$$t'_P = t + 2X_P \tan \phi$$
$$D'_G = (N_G/P_d) + 2X_G$$
$$D'_P = (N_P/P_d) + 2X_P$$
$$D'_o = D_G + (2/P_d)$$
$$d'_o = D_P + (2/P_d)$$

where ϕ = standard pressure angle, ϕ_1 = operating pressure angle, C = standard center distance = $(N_G + N_P)/2P_d$, C_1 = operating center distance, X_G = profile shift correction of gear, and X_P = profile shift correction of pinion. The quantity X is positive for enlarged gears and negative for thinned gears.

Table 8.3.5 Metric Spur Gear Design Formulas

To obtain:	From known	Use this formula*
Pitch diameter D	Module; diametral pitch	$D = mN$
Circular pitch p_c	Module; diametral pitch	$p_c = m\pi = \dfrac{D}{N}\pi = \dfrac{\pi}{P}$
Module m	Diametral pitch	$m = \dfrac{25.4}{P}$
No. of teeth N	Module and pitch diameter	$N = \dfrac{D}{m}$
Addendum a	Module	$a = m$
Dedendum b	Module	$b = 1.25m$
Outside diameter D_o	Module and pitch diameter or number of teeth	$D_o = D + 2m = m(N + 2)$
Root diameter D_r	Pitch diameter and module	$D_r = D - 2.5m$
Base circle diameter D_b	Pitch diameter and pressure angle ϕ	$D_b = D \cos\phi$
Base pitch p_b	Module and pressure angle	$p_b = m\pi \cos\phi$
Tooth thickness at standard pitch diameter T_{std}	Module	$T_{std} = \dfrac{\pi}{2}m$
Center distance C	Module and number of teeth	$C = \dfrac{m(N_1 + N_2)}{2}$
Contact ratio m_p	Outside radii, base-circle radii, center distance, pressure angle	$m_p = \dfrac{\sqrt{{}_1R_o^2 - {}_1R_b^2} + \sqrt{{}_2R_o^2 - {}_2R_b^2} - C\sin\phi}{m\pi\cos\phi}$
Backlash (linear) B (along pitch circle)	Change in center distance	$B = 2(\Delta C)\tan\phi$
Backlash (linear) B (along pitch circle)	Change in tooth thickness, T	$B = \Delta T$
Backlash (linear) (along line of action) B_{LA}	Linear backlash (along pitch circle)	$B_{LA} = B\cos\phi$
Backlash (angular) B_a	Linear backlash (along pitch circle)	$B_a = 6{,}880\,\dfrac{B}{D}$ (arc minutes)
Min. number teeth for no undercutting, N_c	Pressure angle	$N_c = \dfrac{2}{\sin^2\phi}$

* All linear dimensions in millimeters.

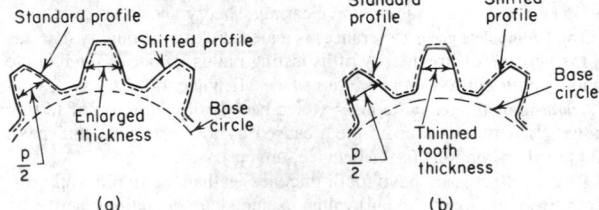

Fig. 8.3.9 Geometry of profile-shifted teeth. (*a*) Enlarged case; (*b*) thinned tooth thickness case.

Metric Module Gear Design Equations Basic design equations for spur gearing utilizing the metric module are listed in Table 8.3.5. (See Designatronics, "Elements of Metric Gear Technology.")

HELICAL GEARS

Helical gears divide into two general applications: for driving parallel shafts and for driving skew shafts (mostly at right angles), the latter often referred to as *crossed-axis* helical gears. The helical tooth form may be imagined as consisting of an infinite number of staggered laminar spur gears, resulting in the curved cylindrical helix.

Pitch of helical gears is definable in two planes. The diametral and circular pitches measured in the plane of rotation (transverse) are defined as for spur gears. However, pitches measured normal to the tooth are related by the cosine of the helix angle; thus normal diametral pitch = $P_{dn} = P_d/\cos\psi$, normal circular pitch = $p_n = p\cos\psi$, and $P_{dn}p_n = \pi$. **Axial pitch** is the distance between corresponding sides of adjacent teeth measured parallel to the gear axis and is calculated as $p_a = p\cot\psi$.

Pressure angle of helical gears is definable in the normal and trans-

verse planes by $\tan\phi_n = \tan\phi\cos\psi$. The transverse pressure angle, which is effectively the real pressure angle, is always greater than the normal pressure angle.

Tooth thickness t of helical gears can be measured in the plane of rotation, as with spur gears, or normal to the tooth surface t_n. The relationship of the two thicknesses is $t_n = t\cos\psi$.

Over-Pins Measurement of Helical Gears Tooth thicknesses t at diameter d can be found from a known over-pins measurement M at known pressure angle ϕ, corresponding to diameter D as follows:

$$R_c = (M - d_w)/2 \qquad \text{for even number of teeth}$$
$$R_c = (M - d_w)/[2\cos(180/2N)] \qquad \text{for odd number of teeth}$$
$$\cos\phi_1 = (D\cos\phi)/2R_c$$
$$\tan\phi_n = \tan\phi\cos\psi$$
$$\cos\psi_b = \sin\phi_n/\sin\phi$$
$$t = D[\pi/N + \operatorname{inv}\phi_1 - \operatorname{inv}\phi - d_w/(D\cos\phi\cos\psi_b)]$$

Parallel-shaft helical gears must conform to the same conditions and requirements as spur gears with parameters (pressure angle and pitch) consistently defined in the transverse plane. Since standard spur gear cutting tools are usually used, normal plane values are standard, resulting in nonstandard transverse pitches and nonstandard pitch diameters and center distances. For parallel shafts, helical gears must have identical helix angles, but must be of opposite hand (left and right helix directions). The commonly used helix angles range from 15 to 35°. To make most advantage of the helical form, the advance of a tooth should be greater than the circular pitch; recommended ratio is 1.5 to 2 with 1.1 minimum. This overlap provides two or more teeth in continual contact with resulting greater smoothness and quietness than spur gears. Because of the helix, the normal component of the tangential pressure on the teeth produces end thrust of the shafts. To remove this objection, gears are made with helixes of opposite hand on each half of the face and are then known as herringbone gears (see Fig. 8.3.10).

Crossed-axis helical gears, also called *spiral* or *screw gears* (Fig. 8.3.11), are a simple type of involute gear used for connecting nonparallel, nonintersecting shafts. Contact is point and there is considerably more sliding than with parallel-axis helicals, which limits the load capacity. The individual gear of this mesh is identical in form and specification to a parallel-shaft helical gear. Crossed-axis helicals can connect

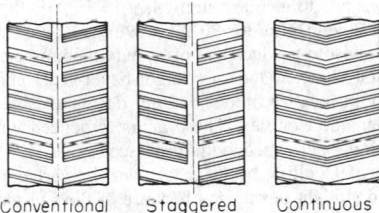

Fig. 8.3.10 Herringbone gears.

any shaft angle Σ, although 90° is prevalent. Usually, the helix angles will be of the same hand, although for some extreme cases it is possible to have opposite hands, particularly if the shaft angle is small.

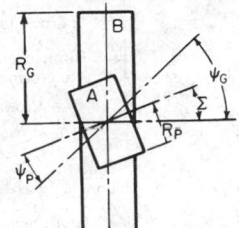

Fig. 8.3.11 Crossed-axis helical gears.

Helical Gear Calculations For parallel shafts the center distance is a function of the helix angle as well as the number of teeth, that is, $C = (N_G + N_P)/(2P_{dn} \cos \psi)$. This offers a powerful method of gearing shafts at any specified center distance to a specified velocity ratio. For crossed-axis helicals the problem of connecting a pair of shafts for any velocity ratio admits of a number of solutions, since both the pitch radii and the helix angles contribute to establishing the velocity ratio. The formulas given in Tables 8.3.6 and 8.3.7 are of assistance in calculations. The notation used in these tables is as follows:

$N_P(N_G)$ = number of teeth in pinion (gear)
$D_P(D_G)$ = pitch diam of pinion (gear)
$p_P(p_G)$ = circular pitch of pinion (gear)
p = circular pitch in plane of rotation for both gears
P_d = diametral pitch in plane of rotation for both gears
p_n = normal circular pitch for both gears
P_{dn} = normal diametral pitch for both gears
ψ_G = tooth helix angle of gear
ψ_P = tooth helix angle of pinion
$l_P(l_G)$ = lead of pinion (gear)
= lead of tooth helix
$n_P(n_G)$ = r/min of pinion (gear)
Σ = angle between shafts in plan
C = center distance

Table 8.3.6 Helical Gears on Parallel Shafts

To find:	Formula
Center distance C	$\dfrac{N_G + N_P}{2P_{dn} \cos \psi}$
Pitch diameter D	$\dfrac{N}{P_d} = \dfrac{N}{P_{dn} \cos \psi}$
Normal diametral pitch P_{dn}	$\dfrac{P_d}{\cos \psi}$
Normal circular pitch p_n	$p \cos \psi$
Pressure angle ϕ	$\tan^{-1} \dfrac{\tan \phi_n}{\cos \psi}$
Contact ratio m_p	$\dfrac{\sqrt{_G D_o^2 - _G D_b^2} + \sqrt{_P D_o^2 - _P D_b^2} + 2C \sin \phi}{2p \cos \phi} + \dfrac{F \sin \psi}{p_n}$
Velocity ratio m_G	$\dfrac{N_G}{N_P} = \dfrac{D_G}{D_P}$

Table 8.3.7 Crossed Helical Gears on Skew Shafts

To find:	Formula
Center distance C	$\dfrac{P_n}{2\pi} \left(\dfrac{N_G}{\cos \psi_G} + \dfrac{N_G}{\cos \psi_P} \right)$
Pitch diameter D_G, D_P	$D_G = \dfrac{N_G}{P_{dG}} = \dfrac{N_G}{P_{dn} \cos \psi_G} = \dfrac{N_G p_n}{\pi \cos \psi_G}$
	$D_P = \dfrac{N_P}{P_{dP}} = \dfrac{N_P}{P_{dn} \cos \psi_P} = \dfrac{N_P p_n}{\pi \cos \psi_P}$
Gear ratio m_G	$\dfrac{N_G}{N_P} = \dfrac{D_G \cos \psi_G}{D_P \cos \psi_P}$
Shaft angle Σ	$\psi_G + \psi_P$

NONSPUR GEAR TYPES*

Bevel gears are used to connect two intersecting shafts in any given speed ratio. The tooth shapes may be designed in any of the shapes shown in Fig. 8.3.12. A special type of gear known as a **hypoid** was developed by Gleason Works for the automotive industry (see *Jour. SAE*, **18**, no. 6). Although similar in appearance to a spiral bevel, it is not a true bevel gear. The basic pitch rolling surfaces are hyperbolas of revolution. Because a ''spherical involute'' tooth form has a curved crown tooth (the basic tool for generating all bevel gears), Gleason used a straight-sided crown tooth which resulted in bevel gears differing slightly from involute form. Because of the figure 8 shape of the complete theoretical tooth contact path, the tooth form has been called ''octoid.'' Straight-sided bevel gears made by reciprocating cutters are of this type. Later, when curved teeth became widely used (spiral and Zerol), practical limitations of such cutters resulted in introduction of the ''spherical'' tooth form which is now the basis of all curved tooth bevel gears. (For details see Gleason's publication, ''Guide to Bevel Gears.'') Gleason Works also developed the generated tooth form

* In the following text relating to bevel gearing, all tables and figures have been extracted from Gleason Works publications, with permission.

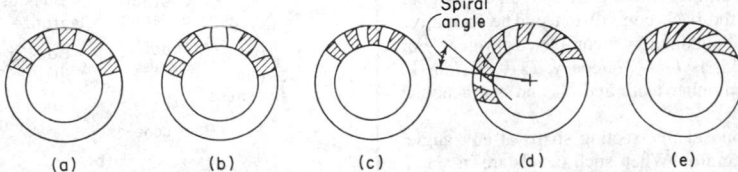

Fig. 8.3.12 Bevel gear types. (*a*) Old-type straight teeth; (*b*) modern Coniflex straight teeth (exaggerated crowning); (*c*) Zerol teeth; (*d*) spiral teeth; (*e*) hypoid teeth.

Revecycle and the nongenerated tooth forms Formate and Helixform, used principally for mass production of hypoid gears for the automotive industry.

Referring to Fig. 8.3.13, we see that the pitch surfaces of bevel gears are frustums of cones whose vertices are at the intersection of the axes; the essential elements and definitions follow.

Addendum angle α: The angle between elements of the face cone and pitch cone.

Back angle: The angle between an element of the back cone and a plane of rotation. It is equal to the pitch angle.

Back cone: The angle of a cone whose elements are tangent to a sphere containing a trace of the pitch circle.

Back-cone distance: The distance along an element of the back cone from the apex to the pitch circle.

Cone distance A_o: The distance from the end of the tooth (heel) to the pitch apex.

Crown: The sharp corner forming the outside diameter.

Crown-to-back: The distance from the outside diameter edge (crown) to the rear of the gear.

Dedendum angle δ: The angle between elements of the root cone and pitch cone.

Face angle γ_o: The angle between an element of the face cone and its axis.

Face width F: The length of teeth along the cone distance.

Front angle: The angle between an element of the front cone and a plane of rotation.

Generating mounting surface, GMS: The diameter and/or plane of rotation surface or shaft center which is used for locating the gear blank during fabrication of the gear teeth.

Heel: The portion of a bevel gear tooth near the outer end.

Mounting distance, MD: For assembled bevel gears, the distance from the crossing point of the axes to the registering surface, measured along the gear axis. Ideally, it should be identical to the pitch apex to back.

Mounting surface, MS: The diameter and/or plane of rotation surface which is used for locating the gear in the application assembly.

Octoid: The mathematical form of the bevel tooth profile. Closely resembles a spherical involute but is fundamentally different.

Pitch angle Γ: The angle formed between an element of the pitch cone and the bevel gear axis. It is the half angle of the pitch.

Pitch apex to back: The distance along the axis from apex of pitch cone to a locating registering surface on back.

Registering surface, RS: The surface in the plane of rotation which locates the gear blank axially in the generating machine and the gear in application. These are usually identical surfaces, but not necessarily so.

Root angle γ_R: The angle formed between a tooth root element and the axis of the bevel gear.

Shaft angle Σ: The angle between mating bevel-gear axes; also, the sum of the two pitch angles.

Spiral angle ψ: The angle between the tooth trace and an element of the pitch cone, corresponding to helix angle in helical gears. The spiral angle is understood to be at the mean cone distance.

Toe: The portion of a bevel tooth near the inner end.

Bevel gears are described by the parameter dimensions at the large end (heel) of the teeth. Pitch, pitch diameter, and tooth dimensions, such as addendum are measurements at this point. At the large end of the gear, the tooth profiles will approximate those generated on a spur gear pitch circle of radius equal to the back cone distance. The formative number of teeth is equal to that contained by a complete spur gear. For pinion and gear, respectively, this is $T_P = N_P/\cos \gamma$; $T_G = N_G/\cos \Gamma$, where T_P and T_G = formative number teeth and N_P and N_G = actual number teeth.

Although bevel gears can connect intersecting shafts at any angle, most applications are for right angles. When such bevels are in a 1 : 1 ratio, they are called **mitre gears.** Bevels connecting shafts other than 90° are called **angular bevel gears.** The speeds of the shafts of bevel gears are

determined by $n_P/n_G = \sin \Gamma/\sin \gamma$, where $n_P(n_G)$ = r/min of pinion (gear), and $\gamma(\Gamma)$ = pitch angle of pinion (gear).

All standard bevel gear designs in the United States are in accordance with the **Gleason bevel gear system.** This employs a basic pressure angle of 20° with long and short addendums for ratios other than 1 : 1 to avoid undercut pinions and to increase strength.

20° Straight Bevel Gears for 90° Shaft Angle Since straight bevel gears are the easiest to produce and offer maximum precision, they are frequently a first choice. Modern straight-bevel-gears generators produce a tooth with localized tooth bearing designated by the Gleason registered tradename Coniflex. These gears, produced with a circular cutter, have a slightly crowned tooth form (see Fig. 8.3.12b). Because of the superiority of Coniflex bevel gears over the earlier reciprocating cutter produced straight bevels and because of their faster production, they are the standards for all bevel gears. The design parameters of Fig. 8.3.13 are calculated by the formulas of Table 8.3.8. Backlash data are given in Table 8.3.9.

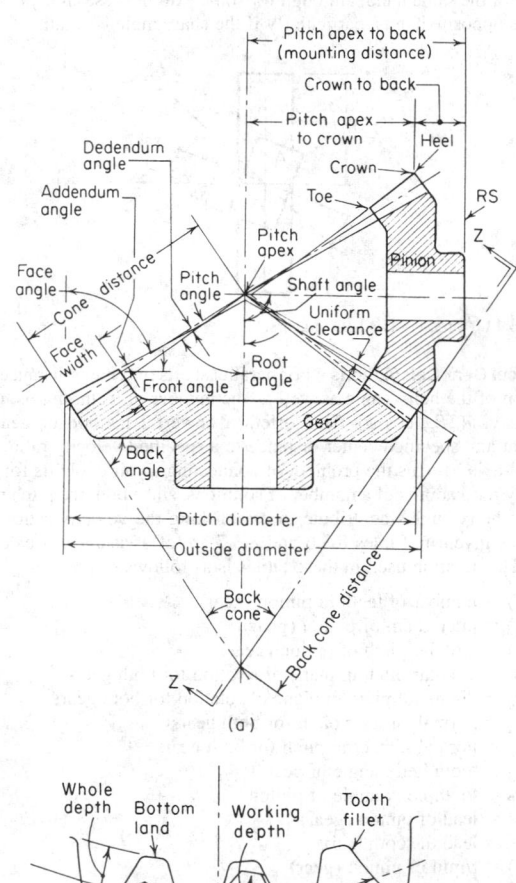

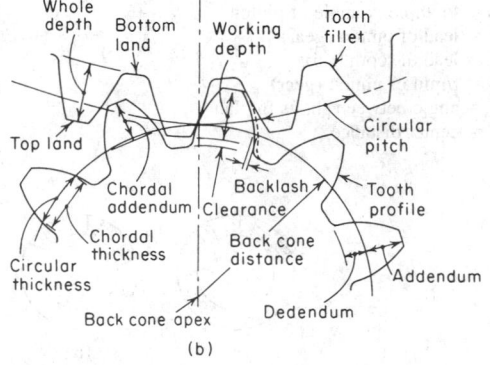

Fig. 8.3.13 Geometry of bevel gear nomenclature. (*a*) Section through axes; (*b*) view along axis Z/Z.

Table 8.3.8 Straight Bevel Gear Dimensions*
(All linear dimensions in inches)

1. Number of pinion teeth†	n	5. Working depth	$h_k = \dfrac{2.000}{P_d}$
2. Number of gear teeth†	N	6. Whole depth	$h_i = \dfrac{2.188}{P_d} + 0.002$
3. Diametral pitch	P_d	7. Pressure angle	ϕ
4. Face width	F	8. Shaft angle	Σ

	Pinion	Gear
9. Pitch diameter	$d = \dfrac{n}{P_d}$	$D = \dfrac{N}{P_d}$
10. Pitch angle	$\gamma = \tan^{-1}\dfrac{n}{N}$	$\Gamma = 90° - \gamma$
11. Outer cone distance	$A_O = \dfrac{D}{2\sin\Gamma}$	
12. Circular pitch	$p = \dfrac{3.1416}{P_d}$	
13. Addendum	$a_{OP} = h_k - a_{OG}$	$a_{OG} = \dfrac{0.540}{P_d} + \dfrac{0.460}{P_d(N/n)^2}$
14. Dedendum‡	$b_{OP} = \dfrac{2.188}{P_d} - a_{OP}$	$b_{OG} = \dfrac{2.188}{P_d} - a_{OG}$
15. Clearance	$c = h_t - h_k$	
16. Dedendum angle	$\delta_P = \tan^{-1}\dfrac{b_{OP}}{A_O}$	$\delta_G = \tan^{-1}\dfrac{b_{OG}}{A_O}$
17. Face angle of blank	$\gamma_O = \gamma + \delta_G$	$\Gamma_O = \Gamma + \delta_P$
18. Root angle	$\gamma_R = \gamma - \delta_P$	$\Gamma_R = \Gamma - \delta_G$
19. Outside diameter	$d_O = d + 2a_{OP}\cos\gamma$	$D_O = D + 2a_{OG}\cos\Gamma$
20. Pitch apex to crown	$x_O = \dfrac{D}{2} - a_{OP}\sin\gamma$	$X_O = \dfrac{d}{2} - a_{OG}\sin\Gamma$
21. Circular thickness	$t = p - T$	$T = \dfrac{p}{2} - (a_{OP} - a_{OG})\tan\phi - \dfrac{K}{P_d}$ (see Fig. 8.3.14)
22. Backlash	B See Table 8.3.9	
23. Chordal thickness	$t_C = t - \dfrac{t^2}{6d^2} - \dfrac{B}{2}$	$T_C = T - \dfrac{T^2}{6D^2} - \dfrac{B}{2}$
24. Chordal addendum	$a_{CP} = a_{OP} + \dfrac{t^2\cos\gamma}{4d}$	$a_{CG} = a_{OG} + \dfrac{T^2\cos\Gamma}{4D}$
25. Tooth angle	$\dfrac{3,438}{A_O}\left(\dfrac{t}{2} + b_{OP}\tan\phi\right)$ minutes	$\dfrac{3,438}{A_O}\left(\dfrac{T}{2} + b_{OG}\tan\phi\right)$ minutes
26. Limit-point width (L.F.)	$W_{LOP} = (T - 2b_{OP}\tan\phi) - 0.0015$	$W_{LOG} = (t - 2b_{OG}\tan\phi) - 0.0015$
27. Limit-point width (S.E.)	$W_{LiP} = \dfrac{A_O - F}{A_O}(T - 2b_{OP}\tan\phi) - 0.0015$	$W_{LiG} = \dfrac{A_O - F}{A_O}(t - 2b_{OG}\tan\phi) - 0.0015$
28. Tool-point width	$W = W_{LiP} -$ stock allowance	$W = W_{LiG} -$ stock allowance

* Abstracted from "Gleason Straight Bevel Gear Design," Tables 8.3.8 and 8.3.9 and Fig. 8.3.4. Gleason Works, Inc.
† Numbers of teeth; ratios with 16 or more teeth in pinion: 15/17 and higher; 14/20 and higher; 13/31 and higher. These can be cut with 20° pressure angle without undercut.
‡ The actual dedendum will be 0.002 in greater than calculated.

Table 8.3.9 Recommended Normal Backlash for Bevel Gear Meshes*

P_d	Backlash range	P_d	Backlash range
1.00–1.25	0.020–0.030	3.50–4.00	0.007–0.009
1.25–1.50	0.018–0.026	4–5	0.006–0.008
1.50–1.75	0.016–0.022	5–6	0.005–0.007
1.75–2.00	0.014–0.018	6–8	0.004–0.006
2.00–2.50	0.012–0.016	8–10	0.003–0.005
2.50–3.00	0.010–0.013	10–12	0.002–0.004
3.00–3.50	0.008–0.011	Finer than 12	0.001–0.003

* The table gives the recommended normal backlash for gears assembled ready to run. Because of manufacturing tolerances and changes resulting from heat treatment, it is frequently necessary to reduce the theoretical tooth thickness by slightly more than the tabulated backlash in order to obtain the correct backlash in assembly. In case of choice, use the smaller backlash tolerances.

Angular straight bevel gears connect shaft angles other than 90° (larger or smaller), and the formulas of Table 8.3.8 are not entirely applicable, as shown in the following:

Item 8, shaft angle, is the specified non-90° shaft angle.

Item 10, pitch angles. Shaft angle Σ less than 90°, $\tan \gamma = \sin \Sigma/(N/n + \cos \Sigma)$; shaft angle Σ greater than 90°, $\tan \gamma = \sin (180 - \Sigma)/[N/n - \cos (180° - \Sigma)]$.

For all shaft angles, $\sin \gamma/\sin \Gamma = n/N$; $\Gamma = \Sigma - \gamma$.

Item 13, addendum, requires calculation of the equivalent 90° bevel gear ratio m_{90}, $m_{90} = [N \cos \gamma/(n \cos \Gamma)]^{1/2}$. The value m_{90} is used as the ratio N/n when applying the formula for addendum. The quantity under the radical is always the absolute value and is therefore always positive.

Item 20, pitch apex to crown, $x_o = A_0 \cos \gamma - a_{op} \sin \gamma$, $X_o = A_0 \cos \Gamma - a_{oG} \sin \Gamma$.

Item 21, circular thickness, except for high ratios, K may be zero.

Spiral Bevel Gears for 90° Shaft Angle   The spiral curved teeth produce additional overlapping tooth action which results in smoother gear action, lower noise, and higher load capacity. The spiral angle has been standardized by Gleason at 35°. Design parameters are calculated by formulas of Table 8.3.10.

Angular Spiral Bevel Gears   Several items deviate from the formulas of Table 8.3.10 in the same manner as angular straight bevel gears. Therefore, the same formulas apply for the deviating items with only the following exception:

Item 21, circular thickness, the value of K in Fig. 8.3.15 must be determined from the equivalent 90° bevel ratio (m_{90}) and the equivalent 90° bevel pinion. The latter is computed as $n_{90} = n \sin \Gamma_{90}/\cos \gamma$, where $\tan \Gamma_{90} = m_{90}$.

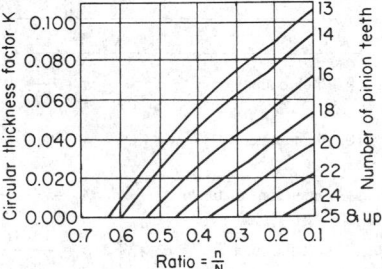

Fig. 8.3.14   Circular thickness factor for straight bevel gears.

Table 8.3.10   Spiral Bevel Gear Dimensions
(All linear dimensions in inches)

1. Number of pinion teeth	n		5. Working depth	$h_k = \dfrac{1.700}{P_d}$
2. Number of gear teeth	N		6. Whole depth	$h_i = \dfrac{1.888}{P_d}$
3. Diametral pitch	P_d		7. Pressure angle	ϕ
4. Face width	F		8. Shaft angle	Σ

	Pinion	Gear
9. Pitch diameter	$d = \dfrac{n}{P_d}$	$D = \dfrac{N}{P_d}$
10. Pitch angle	$\gamma = \tan^{-1} \dfrac{n}{N}$	$1 = 90° - \gamma$
11. Outer cone distance	$A_O = \dfrac{D}{2 \sin \Gamma}$	
12. Circular pitch	$p = \dfrac{3.1416}{P_d}$	
13. Addendum	$A_{OP} = h_k - a_{OG}$	$a_{OG} = \dfrac{0.460}{P_d} + \dfrac{0.390}{P_d(N/n)^2}$
14. Dedendum	$b_{OP} = h_t - a_{OP}$	$b_{OG} = h_t - a_{OG}$
15. Clearance	$c = h_t - h_k$	
16. Dedendum angle	$\delta_P = \tan^{-1} \dfrac{b_{OP}}{A_O}$	$\delta_G = \tan^{-1} \dfrac{b_{OG}}{A_O}$
17. Face angle of blank	$\gamma_O = \gamma + \delta_G$	$\Gamma_O = \Gamma + \delta_P$
18. Root angle	$\gamma_R = \gamma - \delta_P$	$\Gamma_R = \Gamma - \delta_G$
19. Outside diameter	$d_o = d + 2a_{OP} \cos \gamma$	$D_o = D + 2a_{OG} \cos \Gamma$
20. Pitch apex to crown	$x_O = \dfrac{D}{2} - a_{OP} \sin \gamma$	$X_O = \dfrac{d}{2} - a_{OG} \sin \Gamma$
21. Circular thickness	$t = p - T$	$T = \dfrac{p}{2}(a_{OP} - a_{OG})\dfrac{\tan \phi}{\cos \phi} - \dfrac{K}{P_d}$ (see Fig. 8.3.15)
22. Backlash	See Table 8.3.9	
23. Hand of spiral	Left or right	Right or left
24. Spiral angle	35°	
25. Driving member	Pinion or gear	
26. Direction of rotation	Clockwise or counterclockwise	

SOURCE: Gleason, "Spiral Bevel Gear System."

The **zerol bevel gear** is a special case of a spiral bevel gear and is limited to special applications. Design and fabrication details can be obtained from Gleason Works.

Hypoid gears are special and are essentially limited to automotive applications.

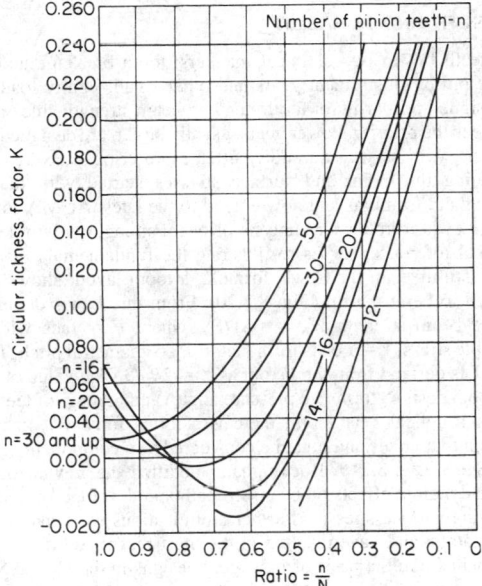

Fig. 8.3.15 Circular thickness factors for spiral bevel gears with 20° pressure angle and 35° spiral angle. Left-hand pinion driving clockwise or right-hand pinion driving counterclockwise.

WORMGEARS AND WORMS

Worm gearing is used for obtaining large speed reductions between nonintersecting shafts making an angle of 90° with each other. If a wormgear such as shown in Fig. 8.3.16 engages a straight worm, as shown in Fig. 8.3.17, the combination is known as **single enveloping worm gearing**. If a wormgear of the kind shown in Fig. 8.3.16 engages a worm as shown in Fig. 8.3.18, the combination is known as **double enveloping worm gearing**.

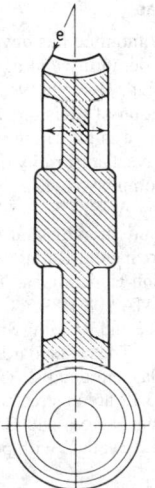

Fig. 8.3.16 Single enveloping worm gearing.

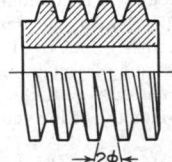

Fig. 8.3.17 Straight worm.

With worm gearing, the **velocity ratio** is the ratio between the number of teeth on the wormgear and the number of threads on the worm. Thus, a 30-tooth wormgear meshing with a single threaded worm will have a velocity ratio of 1 : 30; that is, the worm must make 30 rv in order to revolve the wormgear once. For a double threaded worm, there will be 15 rv of the worm to one of the wormgear, etc. High-velocity ratios are thus obtained with relatively small wormgears.

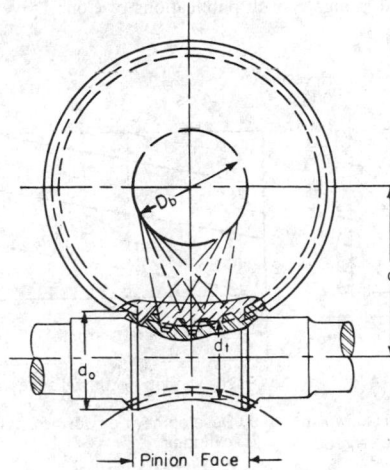

Fig. 8.3.18 Double enveloping worm gearing.

Tooth proportions of the worm in the central section (Fig. 8.3.17) follow standard rack designs, such as 14½, 20, and 25°. The mating wormgear is cut conjugate for a unique worm size and center distance. The geometry and related design equations for a straight-sided cylindrical worm are best seen from a development of the pitch plane (Fig. 8.3.19).

$$D_w = \text{pitch diameter of worm} = \frac{n_w p_n}{\pi \sin \lambda}$$

$$p_n = p \cos \lambda = \frac{\pi D_w}{n_w} \sin \lambda$$

$$L = \text{lead of worm} = n_w p$$

$$D_g = \text{pitch diameter of wormgear}$$

$$= \frac{N_g}{P_d} = \frac{p N_g}{\pi} = \frac{P_n N_g}{\pi \cos \lambda}$$

$$C = \text{center distance}$$

$$= \frac{D_w + D_g}{2} = \frac{p_n}{2\pi} \left(\frac{N_g}{\cos \lambda} + \frac{n_w}{\sin \lambda} \right)$$

where n_w = number of threads in worm; N_g = number of teeth in wormgear; Z = velocity ratio = N_g/n_w.

The pitch diameter of the wormgear is established by the number of teeth, which in turn comes from the desired gear ratio. The pitch diameter of the worm is somewhat arbitrary. The lead must match the wormgear's circular pitch, which can be satisfied by an infinite number of worm diameters; but for a fixed lead value, each worm diameter has a unique lead angle. AGMA offers a design formula that provides near optimized geometry:

$$D_w = \frac{C^{0.875}}{2.2}$$

where C = center distance. Wormgear face width is also somewhat arbitrary. Generally it will be ⅗ to ⅔ of the worm's outside diameter.

Worm mesh nonreversibility, a unique feature of some designs, occurs because of the large amount of sliding in this type of gearing. For a given coefficient of friction there is a critical value of lead angle below which the mesh is nonreversible. This is generally 10° and lower but is related to the materials and lubricant. Most single thread worm meshes are in this category. This locking feature can be a disadvantage or in some designs can be put to advantage.

Double enveloping worm gearing is special in both design and fabrication. Application is primarily where a high load capacity in small space is desired. Currently, there is only one source of manufacture in the United States: Cone Drive Division of Ex-Cello Corp. For design details and load ratings consult publications of Cone Drive and AGMA Standards.

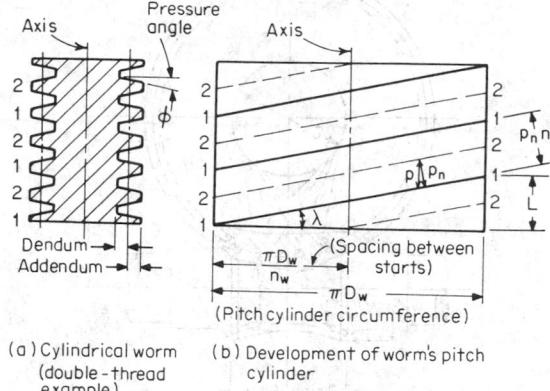

(a) Cylindrical worm (double-thread example)

(b) Development of worm's pitch cylinder

Fig. 8.3.19 Cylindrical worm geometry and design parameters.

Other Gear Types

Gears for special purposes include the following (details are to be found in the references):

Spiroid (Illinois Tool Works) gears, used to connect skew shafts, resemble a hypoid-type bevel gear but in performance are more like worm meshes. They offer very high ratios and a large contact ratio resulting in high strength. The **Helicon** (Illinois Tool Works) gear is a variation in which the pinion is not tapered, and ratios under 10:1 are feasible.

Beveloid (Vinco Corp.) gears are tapered involute gears which can couple intersecting shafts, skew shafts, and parallel shafts.

Face gears have teeth cut on the rotating face plane of the gear and mate with standard involute spur gears. They can connect intersecting or nonparallel, nonintersecting shafts.

Noncircular gears or **function gears** are used for special motions or as elements of analog computers. They can be made with elliptical, logarithmic, spiral, and other functions. See Cunningham references; also, Cunningham Industries, Inc., Stamford, CT.

DESIGN STANDARDS

In addition to the ANSI and AGMA standards on basic tooth proportions, the AGMA sponsors a large number of national standards dealing with gear design, specification, and inspection. (Consult AGMA, 1500 King St., Arlington, VA 22314, for details.) Helpful general references are AGMA, "Gear Handbook," 390.03 and ANSI/AGMA, "Gear Classification and Inspection Handbook," 2000-A88, which establish a system of quality classes for all gear sizes and pitches, ranging from crude coarse commercial gears to the highest orders of fine and coarse ultra-precision gears.

There are 13 quality classes, numbered from 3 through 15 in ascending quality. Tolerances are given for key functional parameters: runout, pitch, profile, lead, total composite error, tooth-to-tooth composite error, and tooth thickness. Also, tooth thickness tolerances and recommended mesh backlash are included. These are related to diametral pitch and pitch diameter in recognition of fabrication achievability. Data are available for spur, helical, herringbone, bevel, and worm gear-

ing; and spur and helical racks. Special sections cover gear applications and suggested quality number; gear materials and treatments; and standard procedure for identifying quality, material, and other pertinent parameters. These data are too extensive for inclusion in this handbook, and the reader is referred to the cited AGMA references.

STRENGTH AND DURABILITY

Gear teeth fail in two classical manners: tooth breakage and surface fatigue pitting. Instrument gears and other small, lightly loaded gears are designed primarily for tooth-bending beam strength since minimizing size is the priority. Power gears, usually larger, are designed for both strengths, with surface durability often more critical. Expressions for calculating the beam and surface stresses started with the Lewis-Buckingham formulas and now extend to the latest AGMA formulas.

The Lewis formula for analysis of beam strength, now relegated to historical reference, serves to illustrate the fundamentals that current formulas utilize. In the Lewis formula, a tooth layout shows the load assumed to be at the tip (Fig. 8.3.20). From this Lewis demonstrated that the beam strength $W_b = FSY/P_d$, where F = face width; S = allowable stress; Y = Lewis form factor; P_d = diametral pitch. The form factor Y is derived from the layout as $Y = 2P_d/3$. The value of Y varies with tooth design (form and pressure angle) and number of teeth. In the case of a helical gear tooth, there is a thrust force W_{th} in the axial direction that arises and must be considered as a component of bearing load. See Fig. 8.3.21b. Buckingham modified the Lewis formula to include dynamic effects on beam strength and developed equations for evaluating surface stresses. Further modifications were made by other investigators, and have resulted in the most recent AGMA rating formulas which are the basis of most gear designs in the United States.

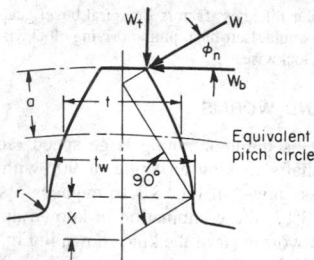

Fig. 8.3.20 Layout for beam strength (Lewis formula).

AGMA Strength and Durability Rating Formulas

For many decades the AGMA Gear Rating Committee has developed and provided tooth beam strength and surface durability (pitting resistance) formulas suitable for modern gear design. Over the years, the formulas have gone through a continual evolution of revision and improvement. The intent is to provide a common basis for rating various gear types for differing applications and thus have a uniformity of practice within the gear industry. This has been accomplished via a series of standards, many of which have been adopted by ANSI.

The latest standards for rating bending beam strength and pitting resistance are ANSI/AGMA 2001-C95, "Fundamental Rating Factors and Calculation Methods for Involute, Spur, and Helical Gear Teeth" (available in English and metric units) and AGMA 908-B89, "Geometry Factors for Determining the Pitting Resistance and Bending Strength of Spur, Helical, and Herringbone Gear Teeth." These standards have replaced AGMA 218.01 with improved formulas and details.

The rating formulas in Tables 8.3.11 and 8.3.12 are abstracted from ANSI/AGMA 2001-B88, "Fundamental Rating Factors and Calculation Methods for Involute Spur and Helical Gear Teeth," with permission.

Overload factor K_o is intended to account for an occasional load in excess of the nominal design load W_t. It can be established from experience with the particular application. Otherwise use $K_o = 1$.

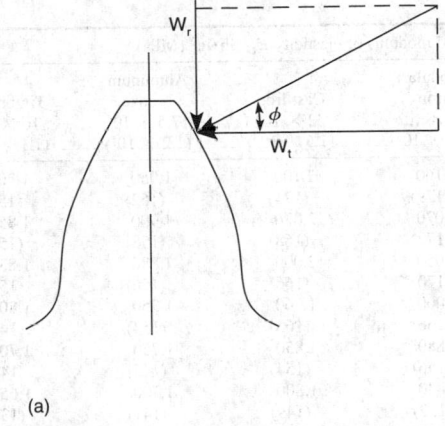

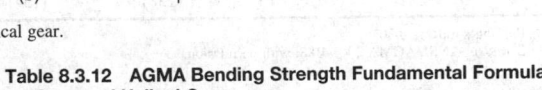

Fig. 8.3.21   Forces on spur and helical teeth. (*a*) Spur gear; (*b*) helical gear.

Table 8.3.11   AGMA Pitting Resistance Formula for Spur and Helical Gears
(See Note 1 below.)

$$s_c = C_p \sqrt{W_t K_o K_v K_s \frac{K_m}{dF} \frac{C_f}{I}}$$

where s_c = contact stress number, lb/in²
C_p = elastic coefficient,* (lb/in²)$^{0.5}$ (see text and Table 8.3.13)
W_t = transmitted tangential load, lb
K_o = overload factor (see text)
K_v = dynamic factor (see Fig. 8.3.22)
K_s = size factor (see text)
K_m = load distribution factor (see text and Table 8.3.14)
C_f = surface condition factor for pitting resistance (see text)
F = net face width of narrowest member, in
I = geometry factor for pitting resistance (see text and Figs. 8.3.23 and 8.3.24)
d = operating pitch diameter of pinion, in

$$= \frac{2C}{m_G + 1} \quad \text{for external gears}$$

$$= \frac{2C}{m_G - 1} \quad \text{for internal gears}$$

where C = operating center distance, in
m_G = gear ratio (never less than 1.0)

Allowable contact stress number s_{ac}

$$s_c \leq \frac{s_{ac} Z_N C_H}{S_H K_T K_R}$$

where s_{ac} = allowable contact stress number, lb/in² (see Tables 8.3.15 and 8.3.16; Fig. 8.3.34)
Z_N = stress cycle factor for pitting resistance (see Fig. 8.3.35)
C_H = hardness ratio factor for pitting resistance (see text and Figs. 8.3.36 and 8.3.37)
S_H = safety factor for pitting (see text)
K_T = temperature factor (see text)
K_R = reliability factor (see Table 8.3.19)

* **Elastic coefficient C_p** can be calculated from the following equation when the paired materials in the pinion-gear set are not listed in Table 8.3.13:

$$C_p = \sqrt{\frac{1}{\pi[(1 - \mu_P^2)/E_P + (1 - \mu_G^2)/E_G]}}$$

where $\mu_P(\mu_G)$ = Poisson's ratio for pinion (gear)
$E_P(E_G)$ = modulus of elasticity for pinion (gear), lb/in²

Note 1: If the rating is calculated on the basis of uniform load, the transmitted tangential load is

$$W_t = \frac{33,000P}{v_t} = \frac{2T}{d} = \frac{126,000P}{n_p d}$$

where P = transmitted power, hp
T = transmitted pinion torque, lb · in
v_t = pitch line velocity at operating pitch diameter, ft/min $= \dfrac{\pi n_p d}{12}$

Table 8.3.12   AGMA Bending Strength Fundamental Formula for Spur and Helical Gears
(See Note 1 in Table 8.3.11.)

$$s_t = W_t K_o K_v K_s \frac{P_d}{F} \frac{K_m K_B}{J}$$

where s_t = bending stress number, lb/in?
K_B = rim thickness factor (see Fig. 8.3.38)
J = geometry factor for bending strength (see text and Figs. 8.3.25 to 8.3.31)
P_d = transverse diametral pitch, in^{-1}*;
P_{dn} for helical gears

$$P_d = \frac{\pi}{p_x \tan \psi_s} = P_{dn} \cos \psi_s \quad \text{for helical gears}$$

where P_{dn} = normal diametral pitch, in^{-1}
p_x = axial pitch, in
ψ_s = helix angle at standard pitch diameter

$$\psi_s = \arcsin \frac{\pi}{p_x P_{dn}}$$

Allowable bending stress numbers s_{at}

$$s_t \leq \frac{s_{at} Y_N}{S_F K_T K_R}$$

where s_{at} = allowable bending stress number, lb/in² (see Tables 8.3.17 and 8.3.18 and Figs. 8.3.39 to 8.3.41)
Y_N = stress cycle factor for bending strength (see Fig. 8.3.42)
S_F = safety factor for bending strength (see text)

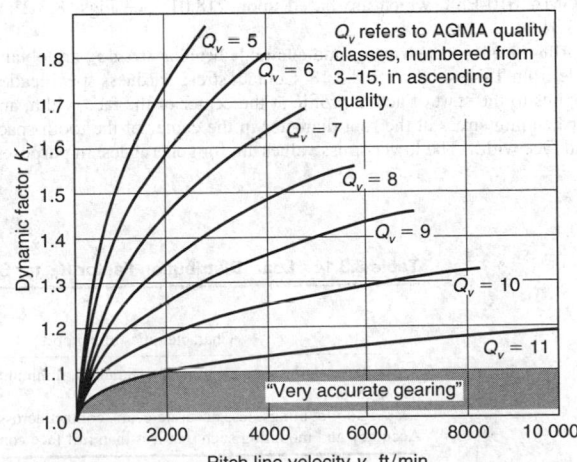

Fig. 8.3.22   Dynamic factor K_v. (*Source: ANSI/AGMA 2001-C95, with permission.*)

Table 8.3.13   Elastic Coefficient C_p

Pinion material	Pinion modulus of elasticity E_P, lb/in^2 (MPa)	Gear material and modulus of elasticity E_G, lb/in^2 (MPa)					
		Steel 30×10^6 (2×10^5)	Malleable iron 25×10^6 (1.7×10^5)	Nodular iron 24×10^6 (1.7×10^5)	Cast iron 22×10^6 (1.5×10^5)	Aluminum bronze 17.5×10^6 (1.2×10^5)	Tin bronze 16×10^6 (1.1×10^5)
Steel	30×10^6 (2×10^5)	2,300 (191)	2,180 (181)	2,160 (179)	2,100 (174)	1,950 (162)	1,900 (158)
Malleable iron	25×10^6 (1.7×10^5)	2,180 (181)	2,090 (174)	2,070 (172)	2,020 (168)	1,900 (158)	1,850 (154)
Nodular iron	24×10^6 (1.7×10^5)	2,160 (179)	2,070 (172)	2,050 (170)	2,000 (166)	1,880 (156)	1,830 (152)
Cast iron	22×10^6 (1.5×10^5)	2,100 (174)	2,020 (168)	2,000 (166)	1,960 (163)	1,850 (154)	1,800 (149)
Aluminum bronze	17.5×10^6 (1.2×10^5)	1,950 (162)	1,900 (158)	1,880 (156)	1,850 (154)	1,750 (145)	1,700 (141)
Tin bronze	16×10^6 (1.1×10^5)	1,900 (158)	1,850 (154)	1,830 (152)	1,800 (149)	1,700 (141)	1,650 (137)

Poisson's ratio = 0.30.
SOURCE: ANSI/AGMA 2001-B88; with permission.

Size factor K_s is intended to factor in material nonuniformity due to tooth size, diameter, face width, etc. AGMA has not established factors for general gearing; use $K_s = 1$ unless there is information to warrant using a larger value.

Load distribution factor K_m reflects the nonuniform loading along the lines of contact due to gear errors, installation errors, and deflections. Analytical and empirical methods for evaluating this factor are presented in ANSI/AGMA 2001-C95 but are too extensive to include here. Alternately, if appropriate for the application, K_m can be extrapolated from values given in Table 8.3.14.

Surface condition factor C_f is affected by the manufacturing method (cutting, shaving, grinding, shotpeening, etc.). Standard factors have not been established by AGMA. Use $C_f = 1$ unless experience can establish confidence for a larger value.

Geometry factors I and J relate to the shape of the tooth at the point of contact, the most heavily loaded point. AGMA 908-B89 (Information Sheet, Geometry Factors for Determining the Pitting Resistance and Bending Strength for Spur, Helical and Herringbone Gear Teeth) presents detailed procedures for calculating these factors. The standard also includes a collection of tabular values for a wide range of gear tooth designs, but they are too voluminous to be reproduced here in their entirety. Earlier compact graphs of I and J values in AGMA 218.01 are still valid. They are presented here along with several curves from AGMA 610-E88, which are based upon 218.01. See Figs. 8.3.23 to 8.3.31.

Allowable contact stress s_{ac} and **allowable bending stress s_{at}** are obtainable from Tables 8.3.15 to 8.3.18. Contact stress hardness specification applies to the start of active profile at the center of the face width, and for bending stress at the root diameter in the center of the tooth space and face width. The lower stress values are for general design purposes; the upper values are for high-quality materials and high-quality control. (See ANSI/AGMA 2001-C95, tables 7 through 10, regarding detailed metallurgical specifications; stress grades 1, 2, and 3; and type A and B hardness patterns.)

For **reversing loads,** allowable bending stress values, s_{at} are to be reduced to 70 percent. If the rim thickness cannot adequately support the load, an additional derating factor K_B is to be applied. See Fig. 8.3.38.

Hardness ratio factor C_H applies when the pinion is substantially harder than the gear, and it results in work hardening of the gear and increasing its capacity. Factor C_H applies to only the gear, not the pinion. See Figs. 8.3.36 and 8.3.37.

Safety factors S_H and S_F are defined by AGMA as factors beyond K_O and K_R; they are used in connection with extraordinary risks, human or economic. The values of these factors are left to the designer's judgment as she or he assesses all design inputs and the consequences of possible failure.

Temperature factor $K_T = 1$ when gears operate with oil temperature not exceeding 250°F.

Reliability factor K_R accounts for statistical distribution of material failures. Typically, material strength ratings (Tables 8.3.13, and 8.3.15 to 8.3.18; Fig. 8.3.34 and 8.3.42) are based on probability of one failure in 100 at 10^7 cycles. Table 8.3.19 lists reliability factors that may be used to modify the allowable stresses and the probability of failure.

Strength and Durability of Bevel, Worm, and Other Gear Types

For bevel gears, consult the referenced Gleason publications; for worm gearing refer to AGMA standards; for other special types, refer to Dudley, "Gear Handbook."

Table 8.3.14   Load-Distribution Factor K_m for Spur Gears*

Characteristics of support	Face width, in			
	0–2	6	9	16 up
Accurate mountings, small bearing clearances, minimum deflection, precision gears	1.3	1.4	1.5	1.8
Less rigid mountings, less accurate gears, contact across full face	1.6	1.7	1.8	2.2
Accuracy and mounting such that less than full face contact exists		Over 2.2		

* An approximate guide only. See ANSI/AGMA 2001-C95 for derivation of more exact values.
SOURCE: Darle W. Dudley, "Gear Handbook," McGraw-Hill, New York, 1962.

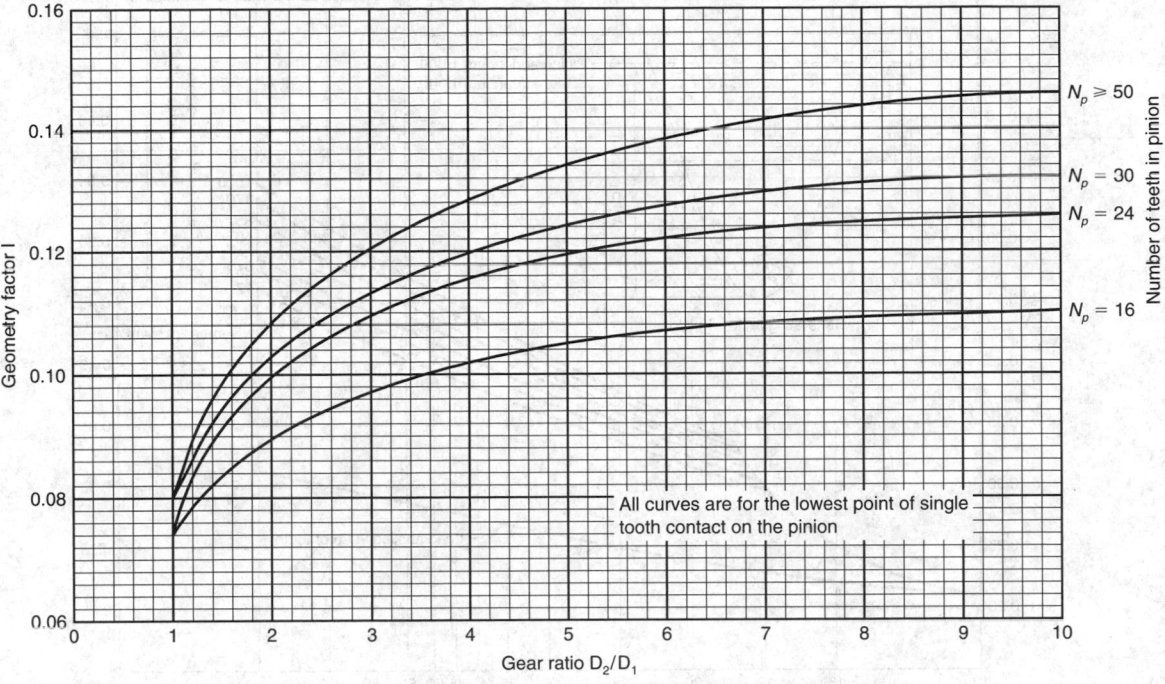

Fig. 8.3.23   Geometry factor *I* for 20° full-depth standard spur gears. (*Source: ANSI/AGMA 2018-01, with permission.*)

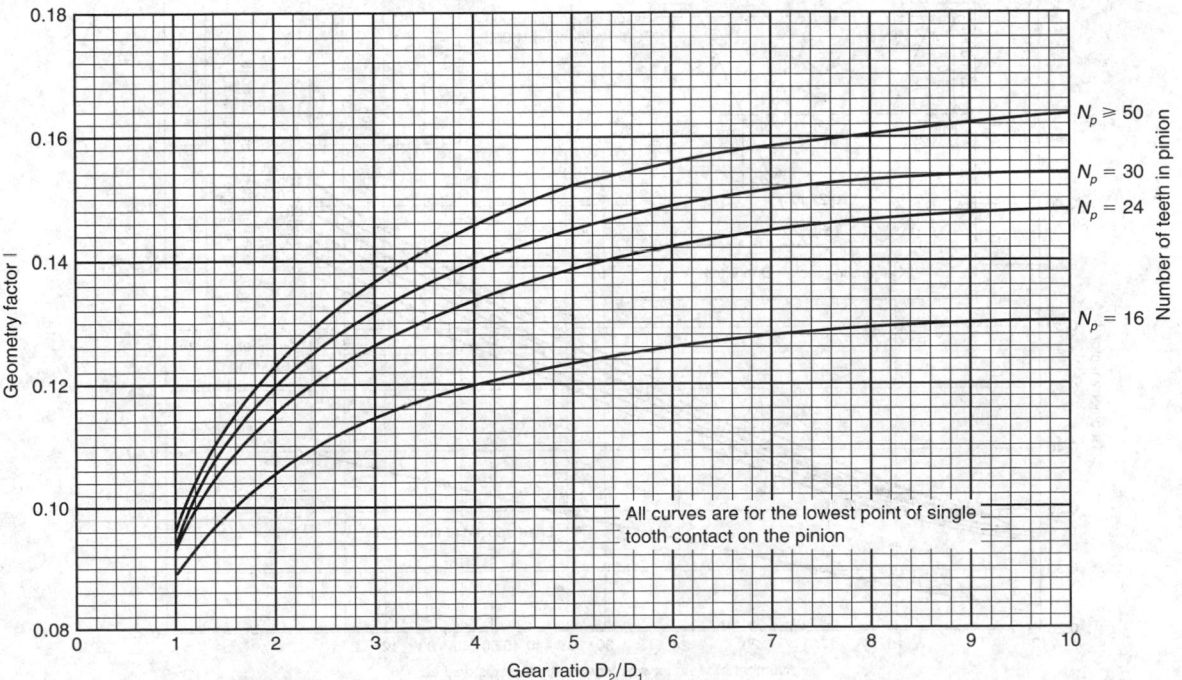

Fig. 8.3.24   Geometry factor *I* for 25° full-depth standard spur gears. (*Source: ANSI/AGMA 2018-01, with permission.*)

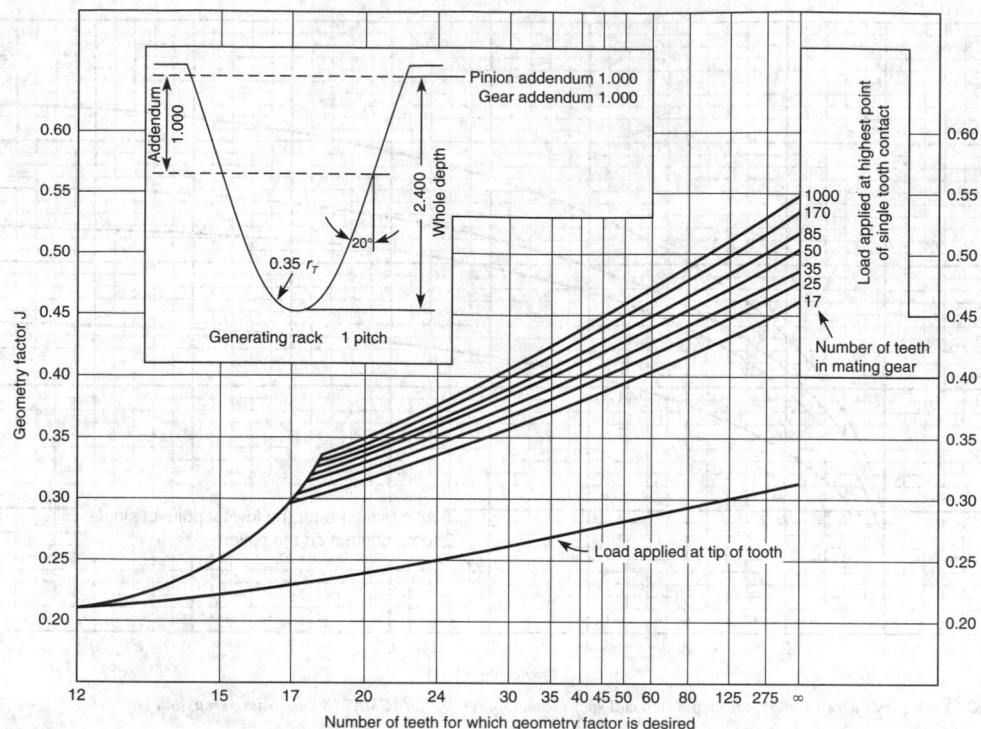

Fig. 8.3.25 Geometry factor *J* for 20° standard addendum spur gears. (*Source: ANSI/AGMA 2018-01, with permission.*)

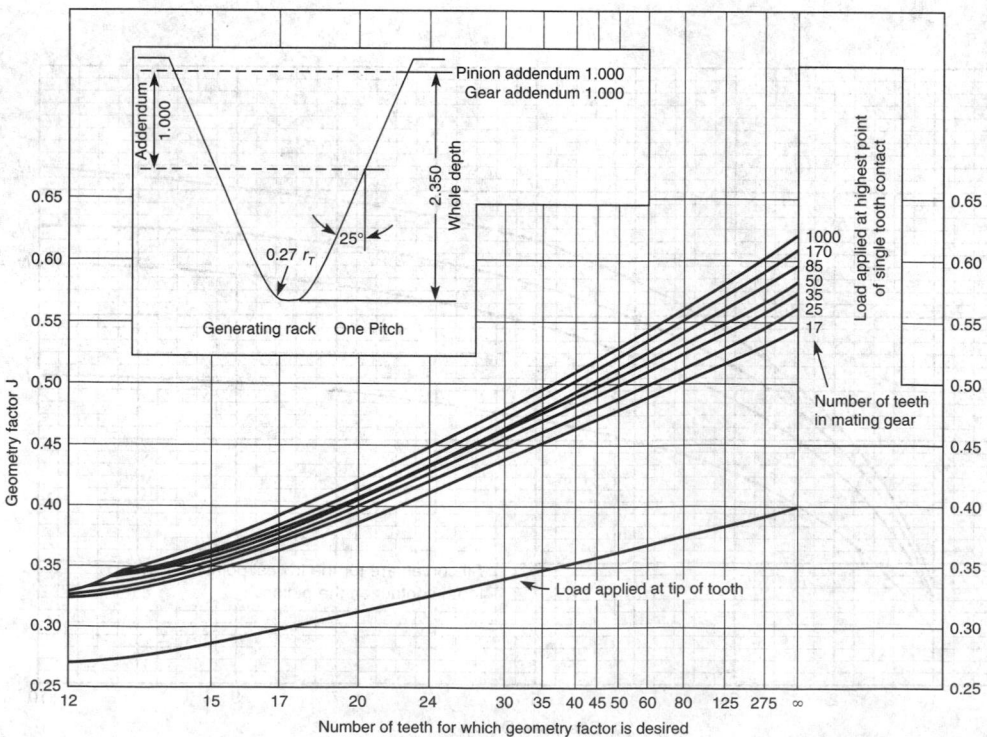

Fig. 8.3.26 Geometry factor *J* for 25° standard addendum spur gears. (*Source: ANSI/AGMA 2018-01, with permission.*)

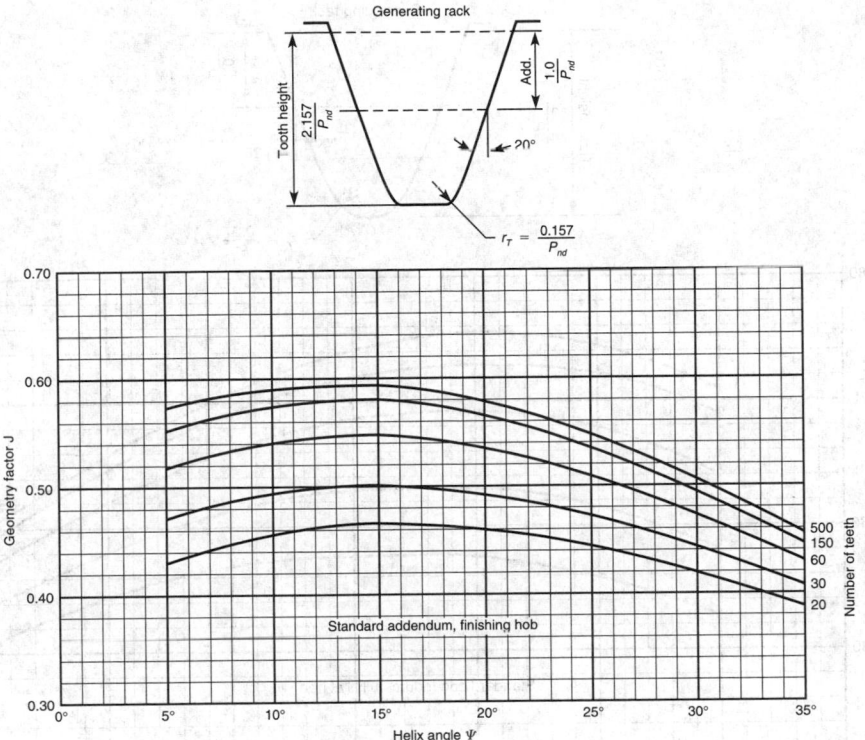

Fig. 8.3.27 Geometry factor J for 20° normal pressure angle helical gears. (Standard addendum, finishing hob.) (*Source: ANSI/AGMA 2018-01, with permission.*)

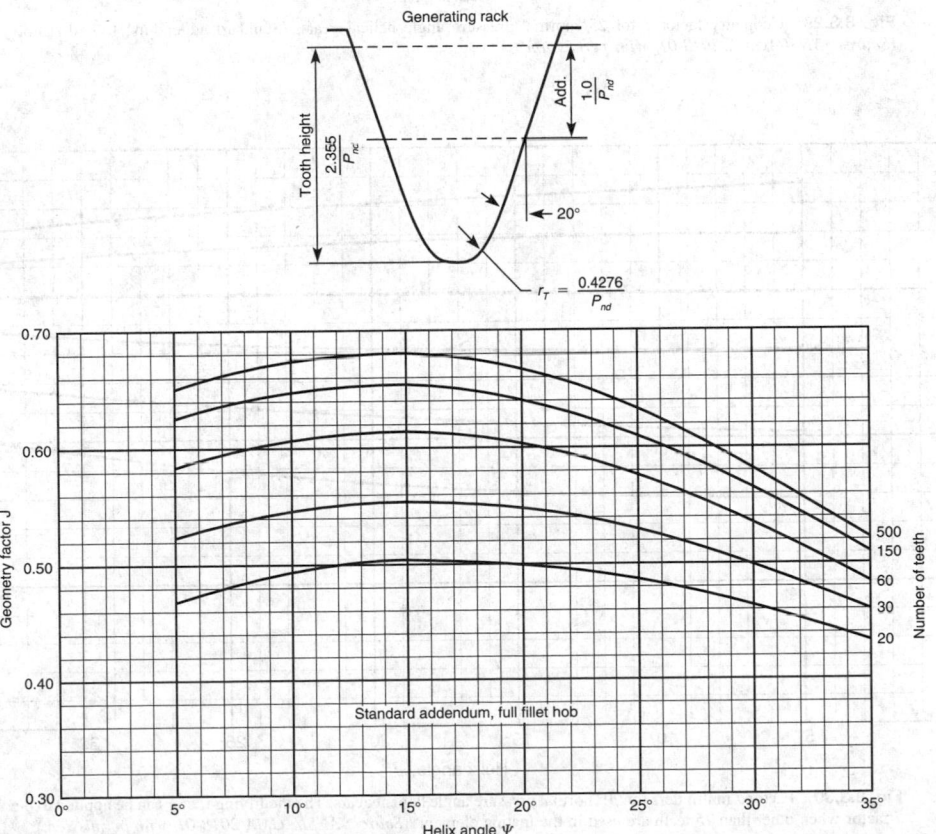

Fig. 8.3.28 Geometry factor J for 20° normal pressure angle helical gears. (Standard addendum, full fillet hob.) (*Source: ANSI/AGMA 2018-01, with permission.*)

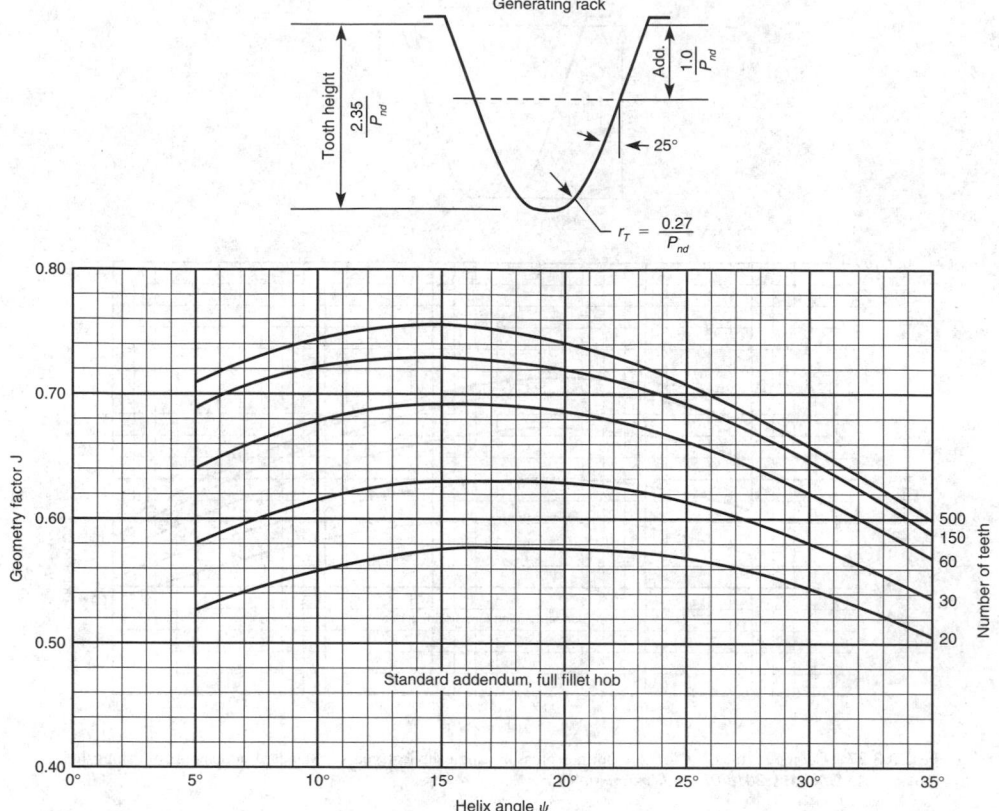

Fig. 8.3.29 Geometry factor J for 25° normal pressure angle helical gears. (Standard addendum, full fillet hob.) (*Source: ANSI/AGMA 2018-01, with permission.*)

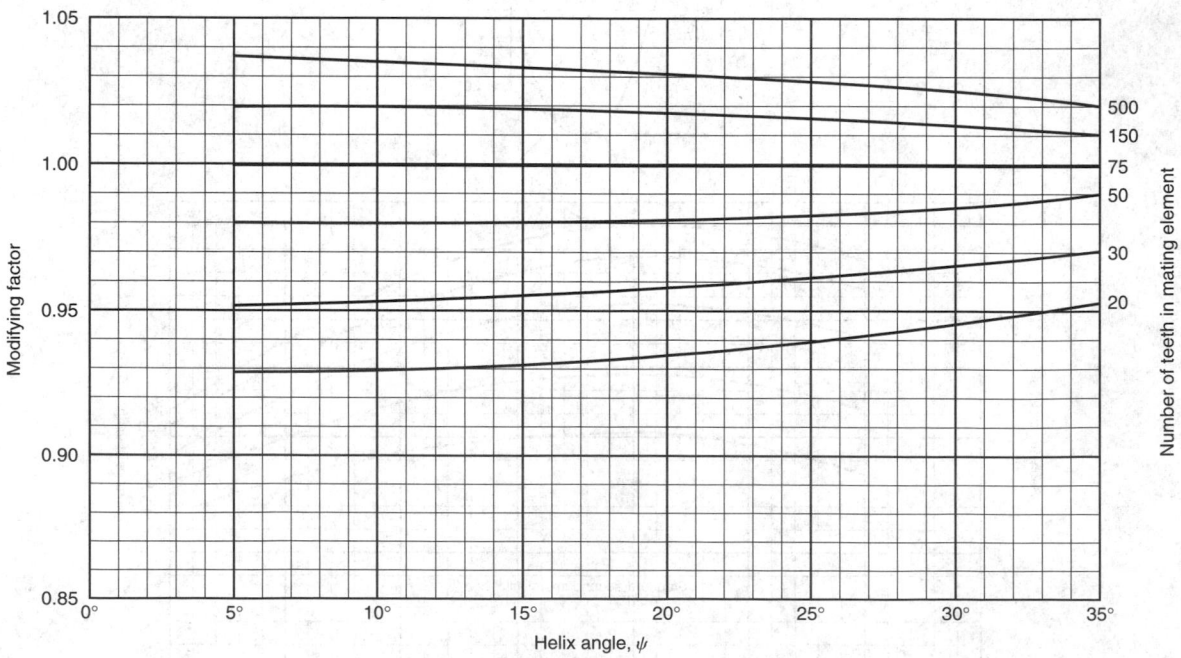

Fig. 8.3.30 Factor J multipliers for 20° normal pressure angle helical gears. The modifying factor can be applied to the J factor when other than 75 teeth are used in the mating element. (*Source: ANSI/AGMA 2018-01, with permission.*)

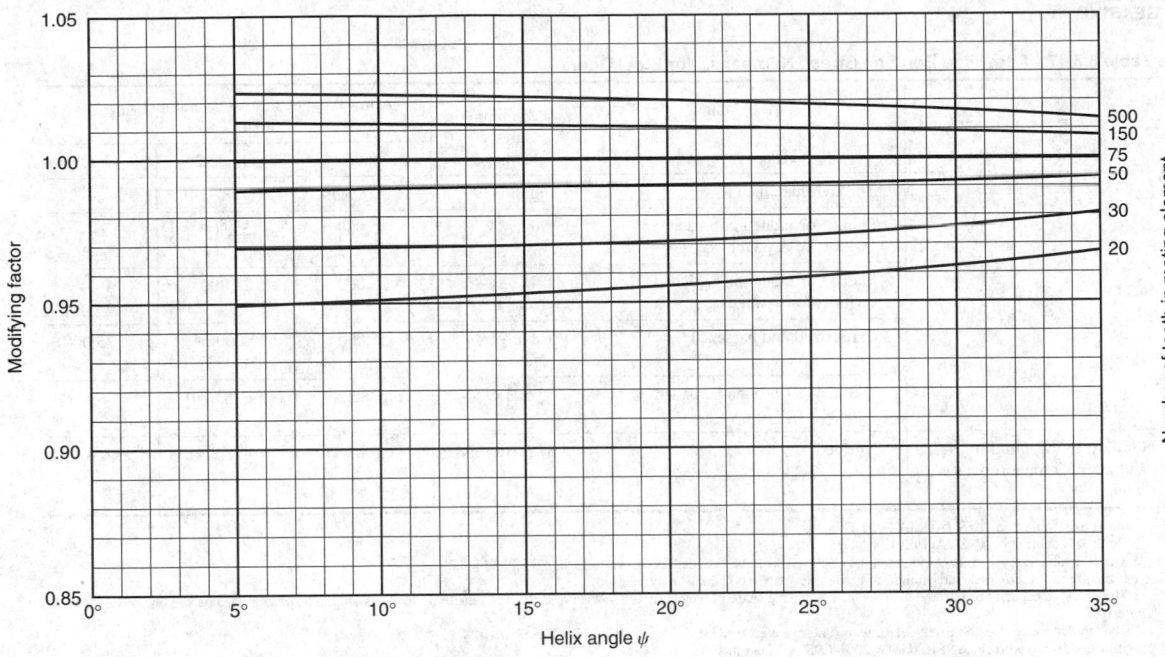

Fig. 8.3.31 Factor *J* multipliers for 25° normal pressure angle helical gears. The modifying factor can be applied to the *J* factor when other than 75 teeth are used in the mating element. (*Source: ANSI/AGMA 6010-E88, with permission.*)

Table 8.3.15 Allowable Contact Stress Number s_{ac} for Steel Gears

Material designation	Heat treatment	Minimum surface hardness*	Allowable contact stress number s_{ac}, lb/in²		
			Grade 1	Grade 2	Grade 3
Steel	Through-hardened*	Fig. 8.3.34	Fig. 8.3.34	Fig. 8.3.34	—
	Flame-* or induction-hardened*	50 HRC	170,000	190,000	—
		54 HRC	175,000	195,000	—
	Carburized and hardened*	Table 9 Note 2	180,000	225,000	275,000
	Nitrided* (through-hardened steels)	83.5 HR15N	150,000	163,000	175,000
		84.5 HR15N	155,000	168,000	180,000
2.5% Chrome (no aluminum)	Nitrided*	87.5 HR15N	155,000	172,000	189,000
Nitralloy 135M	Nitrided*	90.0 HR15N	170,000	183,000	195,000
Nitralloy N	Nitrided*	90.0 HR15N	172,000	188,000	205,000
2.5% Chrome (no aluminum)	Nitrided*	90.0 HR15N	176,000	196,000	216,000

Note 1: Table 9 and Tables 7, 8, and 10 cited in Tables 8.3.16 to 8.3.18 are in ANSI/AGMA 2001-95
* The allowable-stress numbers indicated may be used with the case depths shown in Figs. 8.3.32 and 8.3.33.
SOURCE: Abstracted from ANSI/AGMA 2001-C95, with permission.

Table 8.3.16 Allowable Contact Stress Number s_{ac} for Iron and Bronze Gears

Material	Material designation*	Heat treatment	Typical minimum surface hardness	Allowable contact stress number s_{ac}, lb/in²
ASTM A48 gray cast iron	Class 20	As cast	—	50,000–60,000
	Class 30	As cast	174 HB	65,000–75,000
	Class 40	As cast	201 HB	75,000–85,000
ASTM A536 ductile (nodular) iron	Grade 60-40-18	Annealed	140 HB	77,000–92,000
	Grade 80-55-06	Quenched & tempered	179 HB	77,000–92,000
	Grade 100-70-03	Quenched & tempered	229 HB	92,000–112,000
	Grade 120-90-02	Quenched & tempered	269 HB	103,000–126,000
Bronze	—	Sand-cast	Minimum tensile strength 40,000 lb/in²	30,000
	ASTM B-148 Alloy 954	Heat-treated	Minimum tensile strength 90,000 lb/in²	65,000

* See ANSI/AGMA 2004–B89, "Gear Materials and Heat Treatment Manual."
SOURCE: Abstracted from ANSI/AGMA 2001-C95, with permission.

Table 8.3.17 Allowable Bending Stress Number s_{at} for Steel Gears

Material designation	Heat treatment	Minimum surface hardness	Allowable bending stress number s_{at}, lb/in^2		
			Grade 1	Grade 2	Grade 3
Steel	Through-hardened	Fig. 8.3.39	Fig. 8.3.39	Fig. 8.3.39	—
	Flame* or induction-hardened* with type A pattern	Table 8§	45,000	55,000	—
	Flame* or induction hardened* with type B pattern	Table 8§ Note 1	22,000	22,000	—
	Carburized and hardened*	Table 9§ Note 2	55,000	65,000 or 70,000†	75,000
	Nitrided*‡ (through-hardened steels)	83.5 HR15N	Fig. 8.3.40	Fig. 8.3.40	—
Nitralloy 135M, nitralloy N, and 2.5% Chrome (no aluminum)	Nitrided‡	87.5 HR15N	Fig. 8.3.41	Fig. 8.3.41	Fig. 8.3.41

Note 1: See Table 8 in ANSI/AGMA 2001-C95.
Note 2: See Table 9 in ANSI/AGMA 2001-C95.
* The allowable-stress numbers indicated may be used with the case depths shown in Figs. 8.3.32 and 8.3.33.
† If bainite and microcracks are limited to grade 3 levels, 70,000 lb/in^2 may be used.
‡ The overload capacity of nitrided gears is low. Since the shape of the effective $S-N$ curve is flat, the sensitivity to shock should be investigated before one proceeds with the design.
§ The tabular material is too extensive to record here. Refer to ANSI/AGMA 2001-C95, tables 7 to 10.
SOURCE: Abstracted from ANSI/AGMA 2001-C95, with permission.

Table 8.3.18 Allowable Bending Stress Number s_{at} for Iron and Bronze Gears

Material	Material designation*	Heat treatment	Typical minimum surface hardness	Allowable bending stress number s_{at}, lb/in^2
ASTM A48 gray cast iron	Class 20	As cast	—	5,000
	Class 30	As cast	174 HB	8,500
	Class 40	As cast	201 HB	13,000
ASTM A536 ductile (nodular) iron	Grade 60-40-18	Annealed	140 HB	22,000–33,000
	Grade 80-55-06	Quenched & tempered	179 HB	22,000–33,000
	Grade 100-70-03	Quenched & tempered	229 HB	27,000–40,000
	Grade 120-90-02	Quenched & tempered	269 HB	31,000–44,000
Bronze		Sand-cast	Minimum tensile strength 40,000 lb/in^2	5,700
	ASTM B-148 Alloy 954	Heat-treated	Minimum tensile strength 90,000 lb/in^2	23,600

* See ANSI/AGMA 2004-B89, "Gear Materials and Heat Treatment Manual."
SOURCE: Abstracted from ANSI/AGMA 2001-C95, with permission.

Table 8.3.19 Reliability Factors K_R

Requirements of application	K_R*
Fewer than one failure in 10,000	1.50
Fewer than one failure in 1,000	1.25
Fewer than one failure in 100	1.00
Fewer than one failure in 10	0.85†
Fewer than one failure in 2	0.70†,‡

* Tooth breakage is sometimes considered a greater hazard than pitting. In such cases a greater value of K_R is selected for bending.
† At this value, plastic flow might occur rather than pitting.
‡ From test data extrapolation.
SOURCE: Abstracted from ANSI/AGMA 2001-C95, with permission.

GEAR MATERIALS

(See Tables 8.3.15 to 8.3.18.)

Metals

Plain carbon steels are most widely used as the most economical; similarly, cast iron is used for large units or intricate body shapes. Heat-treated carbon and alloy steels are used for the more severe load- and wear-resistant applications. Pinions are usually made harder to equalize wear. Strongest and most wear-resistant gears are a combination of heat-treated high-alloy steel cores with case-hardened teeth. (See Dudley, "Gear Handbook," chap. 10.) Bronze is particularly recommended for wormgears and crossed helical gears. Stainless steels are limited to special corrosion-resistant environment applications. Aluminum alloys are used for light-duty instrument gears and airborne lightweight requirements.

Sintered powdered metals technology offers commercial high-quality gearing of high strength at very economical production costs. Die-cast gears for light-duty special applications are suitable for many products.

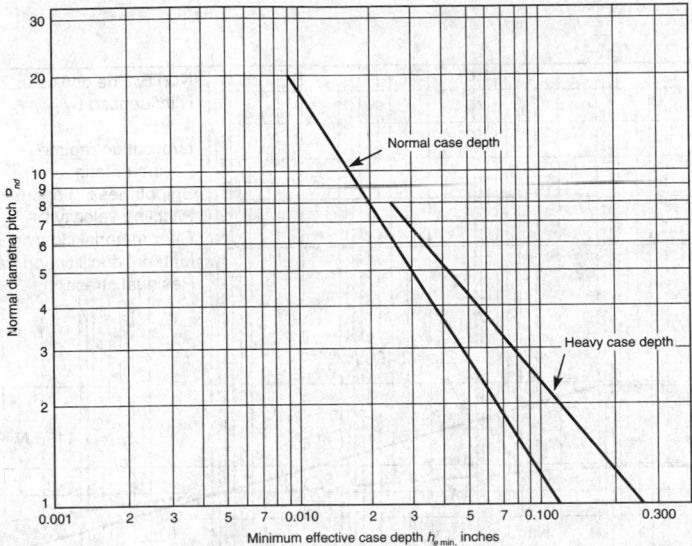

Fig. 8.3.32 Minimum effective case depth for carburized gears $h_{e\,min}$. Effective case depth is defined as depth of case with minimun hardness of 50 RC. Total case depth to core carbon is approximately 1.5 times the effective case depth. (*Source: ANSI/AGMA 2001-C95, with permission.*)

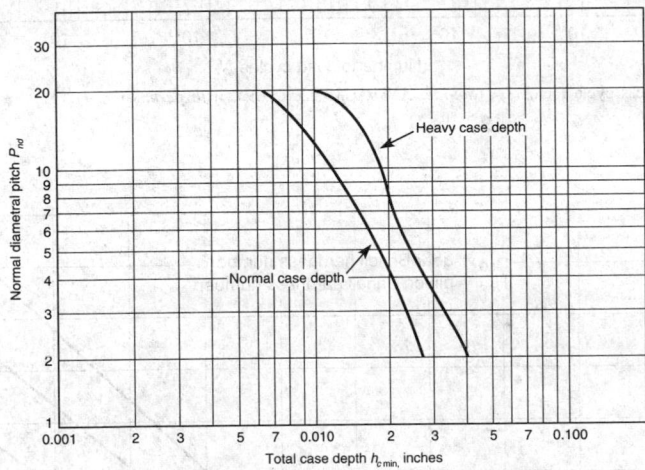

Fig. 8.3.33 Minimum total case depth for nitrided gears $h_{c\,min}$. (*Source: ANSI/AGMA 2001-C95, with permission.*)

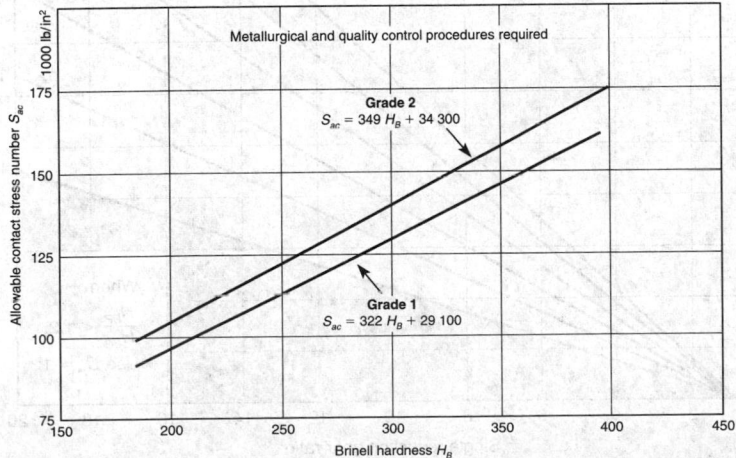

Fig. 8.3.34 Allowable contact stress number for through-hardened steel gears s_{ac}. (*Source: ANSI/AGMA 2001-C95, with permission.*)

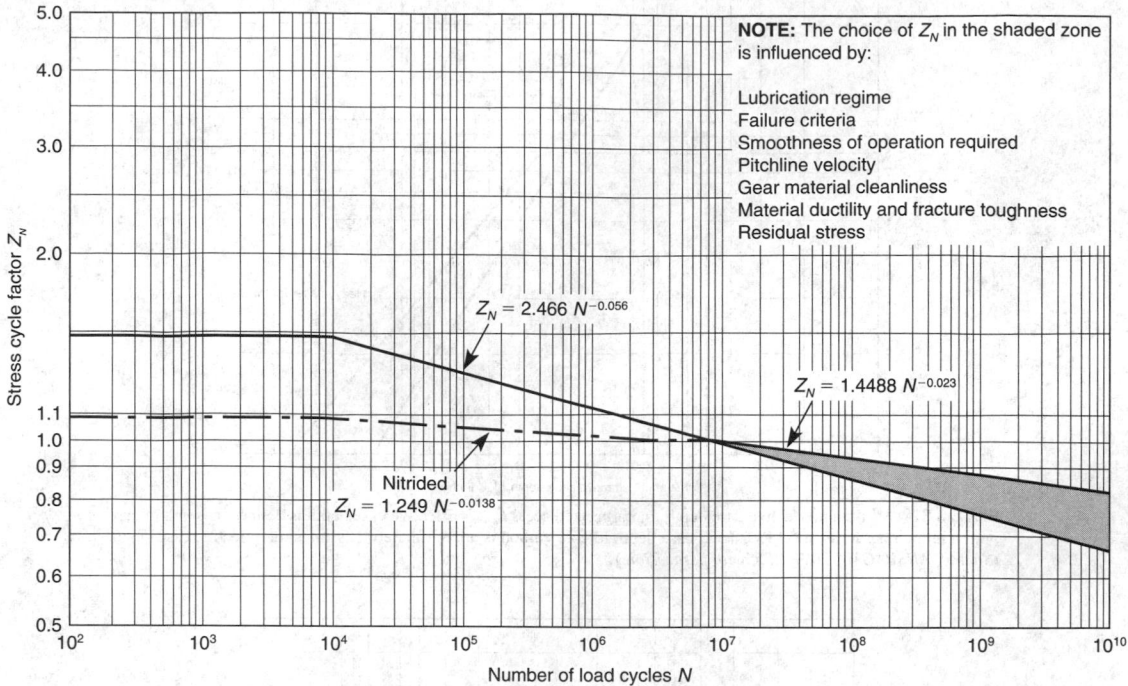

Fig. 8.3.35 Pitting resistance stress cycle factor Z_N. (*Source: ANSI/AGMA 2001-C95, with permission.*)

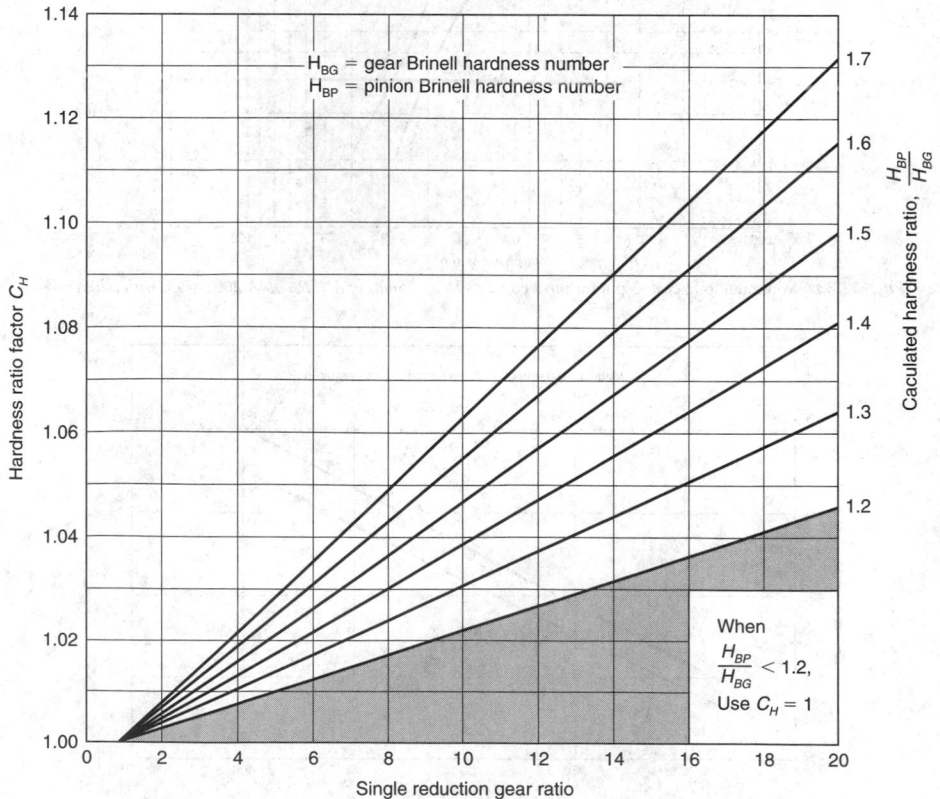

Fig. 8.3.36 Hardness ratio factor C_H (through-hardened). (*Source: ANSI/AGMA 2001-C95, with permission.*)

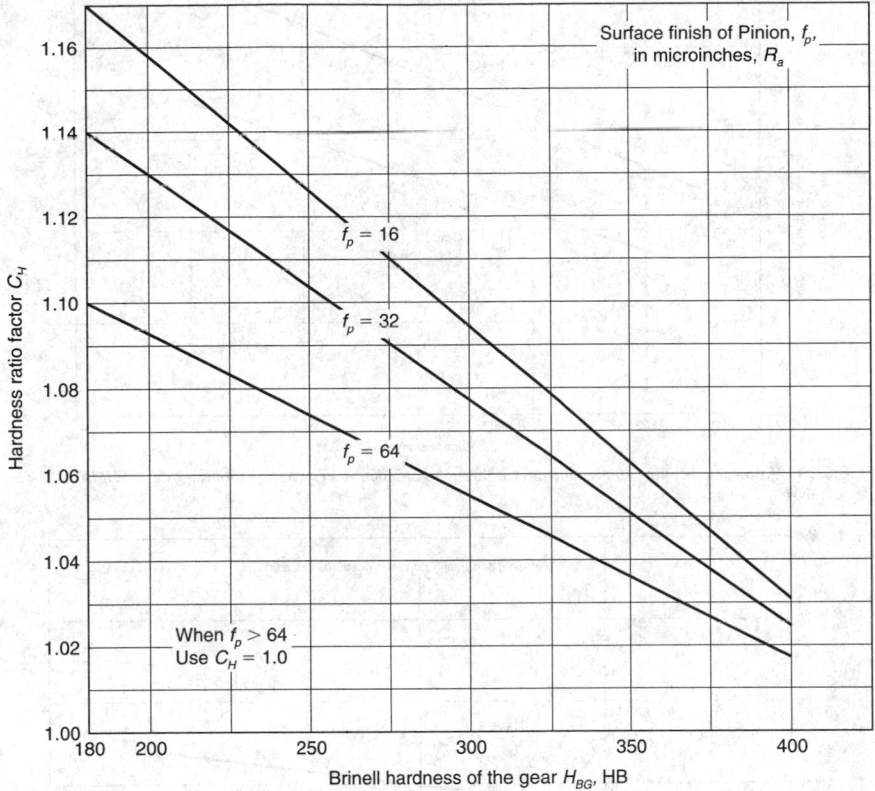

Fig. 8.3.37 Hardness ratio factor C_H (surface-hardened pinions). (*Source: ANSI/AGMA 2001-C95, with permission.*)

For power gear applications, heat treatment is an important part of complete and proper design and specification. Heat treatment descriptions and specification tolerances are given in the reference cited below.

Precision gears of the small device and instrument types often require protective coatings, particularly for aircraft, marine, space, and military applications. There is a wide choice of chemical and electroplate coatings offering a variety of properties and protection.

For pertinent properties and details of the above special materials and protective coatings see Michalec, "Precision Gearing," chap. 9, Wiley.

Plastics

In recent decades, various forms of nonmetallic gears have displaced metal gears in particular applications. Most plastics can be hobbed or shaped by the same methods used for metallic gears. However, high-strength composite plastics suitable for good-quality gear molding have become available, along with the development of economical high-speed injection molding machines and improved methods for producing accurate gear molds.

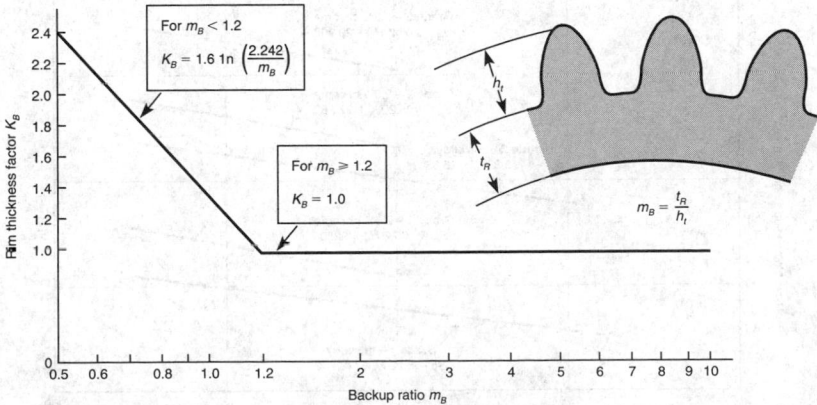

Fig. 8.3.38 Rim thickness factor K_B. (*Source: ANSI/AGMA 2001-C95, with permission.*)

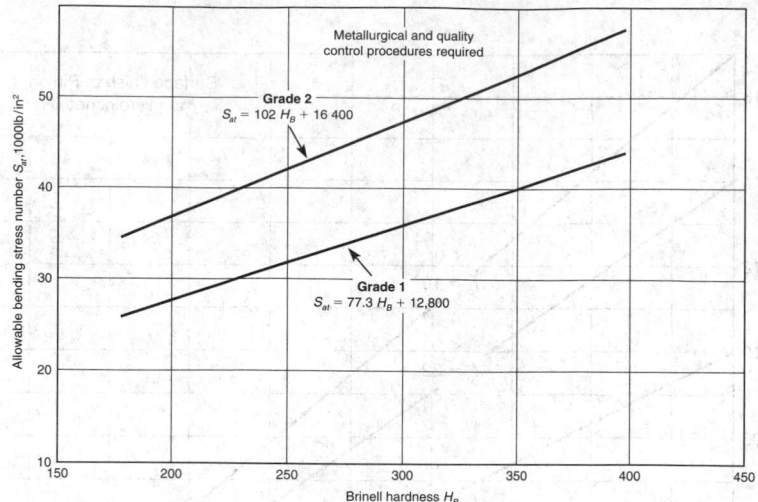

Fig. 8.3.39 Allowable bending stress numbers for through-hardened steel gears s_{at}. (*Source: ANSI/AGMA 2001-C95, with permission.*)

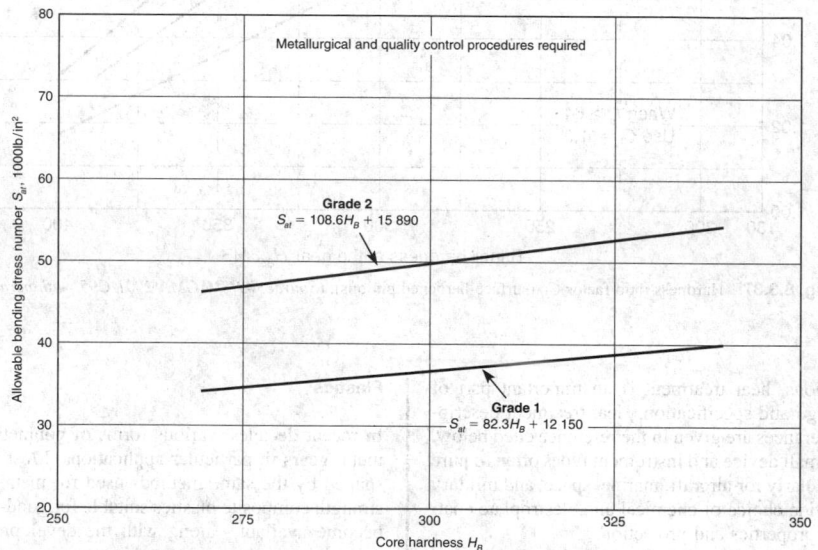

Fig. 8.3.40 Allowable bending stress numbers s_{at} for nitrided through-hardened steel gears (i.e., AISI 4140 and 4340). (*Source: ANSI/AGMA 2001-C95, with permission.*)

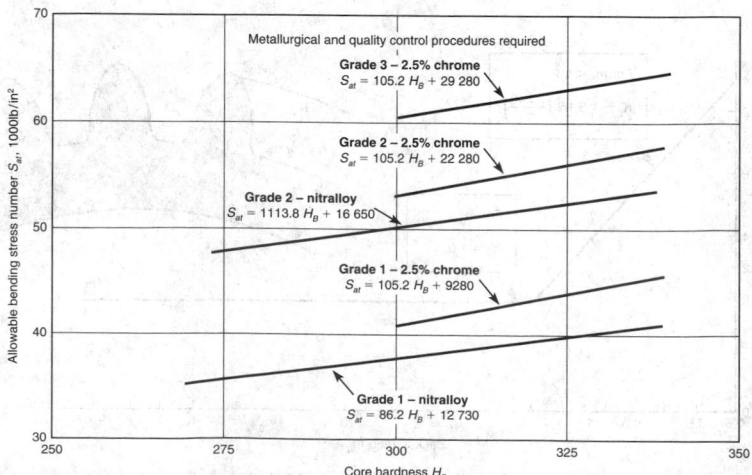

Fig. 8.3.41 Allowable bending stress numbers for nitrided steel gears s_{at}. (*Source: ANSI/AGMA 2001-C95, with permission.*)

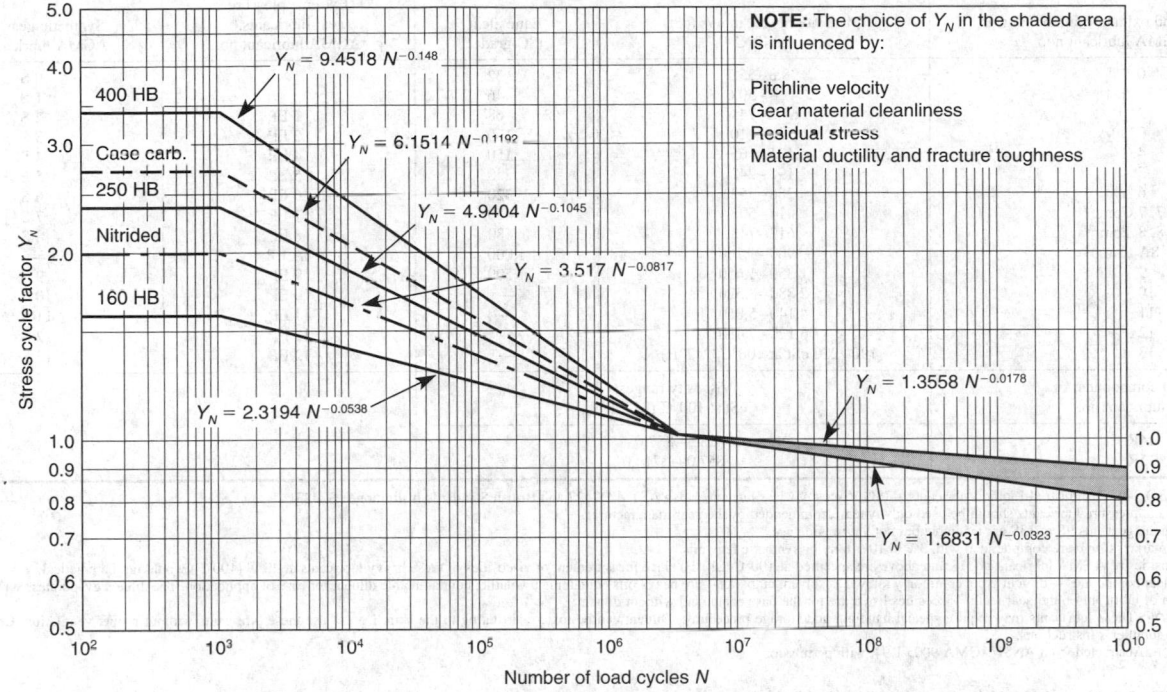

Fig. 8.3.42 Bending strength stress cycle factor Y_N. (*Source: ANSI/AGMA 2001-C95, with permission.*)

The most significant features and advantages of plastic gear materials are:

Cost-effectiveness of injection molding process
Wide choice of characteristics: mechanical strength, density, friction, corrosion resistance, etc.
One-step production; no preliminary or secondary operations
Uniformity of parts
Ability to integrate special shapes, etc., into gear body
Elimination of machining operations
Capability to mold with metallic insert hubs, if required, for more precise bore diameter or body stability
Capability to mold with internal solid lubricants
Ability to operate without lubrication
Quietness of operation
Consistent with trend toward greater use of plastic housings and components

Plastic gears do have limitations relative to metal gears. The most significant are:

Less load-carrying capacity
Cannot be molded to as high accuracy as machined metal gears
Much larger coefficient of expansion compared to metals
Less environmentally stable with regard to temperature and water absorption
Can be negatively affected by some chemicals and lubricants
Initial high cost in mold manufacture to achieve proper tooth geometry accuracies
Narrower range of temperature operation, generally less than 250°F and not lower than 0°F

For further information about plastic gear materials and achievable precision, consult the cited reference: Michalec, "Precision Gearing: Theory and Practice."

For a comprehensive presentation of gear molding practices, design, plastic materials, and strength and durability of plastic gears, consult the cited reference: Designatronics, "Handbook of Gears: Inch and Metric," pp. T131–T158.

GEAR LUBRICATION

Proper lubrication is important to prevention of premature wear of tooth surfaces. In the basic action of involute tooth profiles there is a significant sliding component along with rolling action. In worm gearing sliding is the predominant consideration. Thus, a lubricant is essential for all gearing subject to measurable loadings, and even for lightly or negligibly loaded instrument gearing it is needed to reduce friction. Excellent oils and greases are available for high unit load, high speed gearing. See Secs. 6.11 and 8.4; also consult lubricant suppliers for recommendations and latest available high-quality lubricants with special-purpose additives.

General and specific information for lubrication of gearing is found in ANSI/AGMA 9005-D94, "Industrial Gear Lubrication," which covers open and enclosed gearing of all types. AGMA lists a family of lubricants in accordance with viscosities, numbered 1 through 13, with a cross-reference to equivalent ISO grades. See Table 8.3.20. For AGMA's lubrication recommendations for open and closed gearing related to pitch line speed and various types of lubrication systems, refer to Tables 8.3.21 to 8.3.23. For worm gearing and other information, see ANSI/AGMA 9005-D94.

Information in Table 8.3.24 may be used as a quick guide to gear lubricants and their sources for general-purpose instrument and medium-size gearing. Lubricant suppliers should be consulted for specific high-demand applications.

Often gear performance can be enhanced by special additives to the oil. For this purpose, colloidal additives of graphite, molybdenum disulfide (MoS_2), and Teflon are very effective. These additives are particularly helpful to reduce friction and prevent wear; they are also very beneficial to reduce the rate of wear once it has begun, and thus they prolong gear life. The colloidal additive combines chemically with the metal surface material, resulting in a tenacious layer of combined material interposed between the base metals of the meshing teeth. The size of the colloidal additives is on the order of 2 μm, sufficiently fine not to interfere with the proper operation of the lubricant system filters. Table 8.3.25 lists some commercial colloidal additives and their sources of supply.

Plastic gears are often operated without any external lubrication, and

Table 8.3.20 Viscosity Ranges for AGMA Lubricants

Rust- and oxidation-inhibited gear oils, AGMA lubricant no.	Viscosity range[a] mm²/s (cSt) at 40°C	Equivalent ISO grade[a]	Extreme pressure gear lubricants,[b] AGMA lubricant no.	Synthetic gear oils,[c] AGMA lubricant no.
0	28.8–35.2	32		0 S
1	41.4–50.6	46		1 S
2	61.2–74.8	68	2 EP	2 S
3	90–110	100	3 EP	3 S
4	135–165	150	4 EP	4 S
5	198–242	220	5 EP	5 S
6	288–352	320	6 EP	6 S
7, 7 Comp[d]	414–506	460	7 EP	7 S
8, 8 Comp[d]	612–748	680	8 EP	8 S
8A Comp[d]	900–1,100	1,000	8A EP	—
9	1,350–1,650	1,500	9 EP	9 S
10	2,880–3,520	—	10 EP	10 S
11	4,140–5,060	—	11 EP	11 S
12	6,120–7,480	—	12 EP	12 S
13	190–220 cSt at 100°C (212°F)[e]	—	13 EP	13 S

Residual compounds,[f] AGMA lubricant no.	Viscosity ranges[e] cSt at 100°C (212°F)			
14R	428.5–857.0			
15R	857.0–1,714.0			

[a] Per ISO 3448, ''Industrial Liquid Lubricants—ISO Viscosity Classification,'' also ASTM D2422 and British Standards Institution B.S. 4231.
[b] Extreme-pressure lubricants should be used only when recommended by the gear manufacturer.
[c] Synthetic gear oils 9S to 13S are available but not yet in wide use.
[d] Oils marked *Comp* are compounded with 3% to 10% fatty or synthetic fatty oils.
[e] Viscosities of AGMA lubricant no. 13 and above are specified at 100°C (210°F) since measurement of viscosities of these heavy lubricants at 40°C (100°F) would not be practical.
[f] Residual compounds—diluent type, commonly known as solvent cutbacks—are heavy oils containing a volatile, nonflammable diluent for ease of application. The diluent evaporates, leaving a thick film of lubricant on the gear teeth. Viscosities listed are for the base compound without diluent.
CAUTION: These lubricants may require special handling and storage procedures. Diluent can be toxic or irritating to the skin. Do not use these lubricants without proper ventilation. Consult lubricant supplier's instructions.
SOURCE: Abstracted from ANSI/AGMA 9005-D94, with permission.

Table 8.3.21 AGMA Lubricant Number Guidelines for Open Gearing (Continuous Method of Application)[a,b]

		Pressure lubrication		Splash lubrication		Idler immersion
		Pitch line velocity		Pitch line velocity		Pitch line velocity
Ambient temperature[c] °C (°F)	Character of operation	Under 5 m/s (1,000 ft/min)	Over 5 m/s (1,000 ft/min)	Under 5 m/s (1,000 ft/min)	5–10 m/s (1,000–3,000 ft/min)	Up to 1.5 m/s (300 ft/min)
−10 to 15[d] (15–60)	Continuous	5 or 5 EP	4 or 4 EP	5 or 5 EP	4 or 4 EP	8–9 8 EP–9 EP
	Reversing or frequent ''start/stop''	5 or 5 EP	4 or 4 EP	7 or 7 EP	6 or 6 EP	8–9 8 EP–9 EP
10 to 50[d] (50–125)	Continuous	7 or 7 EP	6 or 6 EP	7 or 7 EP	8–9[f] 8 EP–9 EP	11 or 11 EP
	Reversing or frequent ''start/stop''	7 or 7 EP	6 or 6 EP	9–10[e] 9 EP–10 EP		

[a] AGMA lubricant numbers listed above refer to gear lubricants shown in Table 8.3.20. Physical and performance specifications are shown in Tables 1 and 2 of ANSI/ASMA 9005-D94. Although both R & O and EP oils are listed, the EP is preferred. Synthetic oils in the corresponding viscosity grades may be substituted where deemed acceptable by the gear manufacturer.
[b] Does not apply to worm gearing.
[c] Temperature in vicinity of the operating gears.
[d] When ambient temperatures approach the lower end of the given range, lubrication systems must be equipped with suitable heating units for proper circulation of lubricant and prevention of channeling. Check with lubricant and pump suppliers.
[e] When ambient temperature remains between 30°C (90°F) and 50°C (125°F) at all times, use 10 or 10 EP.
[f] When ambient temperature remains between 30°C (90°F) and 50°C (125°F) at all times, use 9 or 9 EP.
SOURCE: Abstracted from ANSI/AGMA 9005-D94, with permission.

Table 8.3.22 AGMA Lubricant Number Guidelines for Open Gearing Intermittent Applications[a,b,c]

Gear pitch line velocity does not exceed 7.5 m/s (1,500 ft/min)

	Intermittent spray systems[e]			Gravity feed or forced-drip method[g]	
Ambient temperature,[d] °C (°F)	R&O or EP lubricant	Synthetic lubricant	Residual compound[f]	R&O or EP lubricant	Synthetic lubricant
−10–15 (15–60)	11 or 11 EP	11 S	14 R	11 or 11 EP	11 S
5–40 (40–100)	12 or 12 EP	12 S	15 R	12 or 12 EP	12 S
20–50 (70–125)	13 or 13 EP	13 S	15 R	13 or 13 EP	13 S

[a] AGMA viscosity number guidelines listed above refer to gear oils shown in Table 8.3.20.
[b] Does not apply to worm gearing.
[c] Feeder must be capable of handling lubricant selected.
[d] Ambient temperature is temperature in vicinity of gears.
[e] Special compounds and certain greases are sometimes used in mechanical spray systems to lubricate open gearing. Consult gear manufacturer and spray system manufacturer before proceeding.
[f] Diluents must be used to facilitate flow through applicators.
[g] EP oils are preferred, but may not be available in some grades.
SOURCE: Abstracted from ANSI/AGMA 9005-D94, with permission.

Table 8.3.23   AGMA Lubricant Number Guidelines for Enclosed Helical, Herringbone, Straight Bevel, Spiral Bevel, and Spur Gear Drives[a]

| Pitch line velocity[b,c] of final reduction stage | AGMA lubricant numbers,[a,d,e] ambient temperature °C (°F)[f,g] | | | |
	−40 to −10 (−40 to +14)	−10 to +10 (14 to 50)	10 to 35 (50 to 95)	35 to 55 (95 to 131)
Less than 5 m/s (1,000 ft/min)[h]	3 S	4	6	8
5–15 m/s (1,000–3,000 ft/min)	3 S	3	5	7
15–25 m/s (3,000–5,000 ft/min)	2 S	2	4	6
Above 25 m/s (5,000 ft/min)[h]	0 S	0	2	3

[a] AGMA lubricant numbers listed above refer to R&O and synthetic gear oil shown in Table 8.3.20. Physical and performance specifications are shown in Tables 1 and 3 of ANSI/AGMA 9005-D94. EP or synthetic gear lubricants in the corresponding viscosity grades may be substituted where deemed acceptable by the gear drive manufacturer.

[b] Special considerations may be necessary at speeds above 40 m/s (8,000 ft/min). Consult gear drive manufacturer for specific recommendations.

[c] Pitch line velocity replaces previous standards' center distance as the gear drive parameter for lubricant selection.

[d] Variations in operating conditions such as surface roughness, temperature rise, loading, speed, etc., may necessitate use of a lubricant of one grade higher or lower. Contact gear drive manufacturer for specific recommendations.

[e] Drives incorporating wet clutches or overrunning clutches as backstopping devices should be referred to the gear manufacturer, as certain types of lubricants may adversely affect clutch performance.

[f] For ambient temperatures outside the ranges shown, consult the gear manufacturer.

[g] Pour point of lubricant selected should be at least 5°C (9°F) lower than the expected minimum ambient starting temperature. If the ambient starting temperature approaches lubricant pour point, oil sump heaters may be required to facilitate starting and ensure proper lubrication (see 5.1.6 in ANSI/AGMA 9005-D94).

[h] At the extreme upper and lower pitch line velocity ranges, special consideration should be given to all drive components, including bearing and seals, to ensure their proper performance.

SOURCE: Abstracted from ANSI/AGMA 9005-D94, with permission.

Table 8.3.24   Typical Gear Lubricants

| Lubricant type | Military specification | Useful temp range, °F | Commercial source and specification (a partial listing) | | Remarks and applications |
			Source	Identification	
Oils					
Petroleum	MIL-L-644B	−10 to 250	Exxon Corporation	#4035 or Unvis P-48	Good general-purpose lubricant for all quality gears having a narrow range of operating temperature
			Franklin Oil and Gas Co.	L-499B	
			Royal Lubricants Co.	Royco 380	
			Texaco	1692 Low Temp. Oil	
Diester	MIL-L-6085A	−67 to 350	Anderson Oil Co.	Windsor Lube I-245X	General-purpose, low-starting torque, and stable over a wide temperature range. Particularly suited for precision instrument gears and small machinery gears
			Eclipse Pioneer Div., Bendix	Pioneer P-10	
			Shell Oil Co.	AeroShell Fluid 12	
			E. F. Houghton and Co.	Cosmolubric 270	
Diester	MIL-L-7808C	−67 to 400	Sinclair Refining Co.	Aircraft Turbo S Oil	Suitable for oil spay or mist system at high temperature. Particularly suitable for high-speed power gears
			Socony Mobil Oil Co.	Avrex S Turbo 251	
			Bray Oil Co.	Brayco 880	
			Exxon Corporation	Exxon Turbine Oil 15	
Silicone		−75 to 350	Dow-Corning Corp.	DC200	Rated for low-starting torque and lightly loaded instrument gears
Silicone		−100 to 600	General Electric Co.	Versilube 81644	Best load carrier of silicone oils with widest temperature range. Applicable to power gears requiring wide temperature ranges
Greases					
Diester oil-lithium soap	MIL-G-7421A	−100 to 200	Royal Lubricants Co.	Royco 21	For moderately loaded gears requiring starting torques at low temperatures
			Texaco	Low Temp. No. 1888	
Diester oil-lithium soap	MIL-G-3278A	−67 to 250	Exxon Corporation	Beacon 325	General-purpose light grease for precision instrument gears, and generally lightly loaded gears
			Shell Oil Co.	AeroShell Grease II	
			Sinclair Refining Co.	Sinclair 3278 Grease	
			Bray Oil Co.	Braycote 678	
Petroleum oil-sodium soap	MIL-L-3545	−20 to 300	Exxon Corporation	Andok 260	A high-temperature lubricant for high speed and high loads
			Standard Oil Co. of Calif.	RPM Aviation Grease #2	
Mineral oil-sodium soap	—	−25 to 250	Exxon Corporation	Andok C	Stiff grease that channels readily. Suitable for high speeds and highly loaded gears

Table 8.3.25 Solid Oil Additives

Lubricant type	Temperature range, °F	Source	Identification	Remarks
Colloidal graphite	Up to 1,000	Acheson Colloids Co.	SLA 1275	Good load capacity, excellent temperature resistance
Colloidal MoS$_2$	Up to 750	Acheson Colloids Co.	SLA 1286	Good antiwear
Colloidal Teflon	Up to 575	Acheson Colloids Co.	SLA 1612	Low coefficient of friction

they will provide long service life if the plastic chosen is correct for the application. Plastics manufacturers and their publications can be consulted for guidance. Alternatively, many plastic gear materials can be molded with internal solid lubricants, such as MoS$_2$, Teflon, and graphite.

GEAR INSPECTION AND QUALITY CONTROL

Gear performance is not only related to the design, but also depends upon obtaining the specified quality. Details of gear inspection and control of subtle problems relating to quality are given in Michalec, "Precision Gearing," Chap. 11.

COMPUTER MODELING AND CALCULATIONS

A feature of the latest AGMA rating standard is that the graphs, including those presented here, are accompanied by equations which allow application of computer-aided design. Gear design equations and strength and durability rating equations have been computer modeled by many gear manufacturers, users, and university researchers. Numerous software programs, including integrated CAD/CAM, are available from these places, and from computer system suppliers and specialty software houses. It is not necessary for gear designers, purchasers, and fabricators to create their own computer programs.

With regard to gear tooth strength and durability ratings, many custom gear house designers and fabricators offer their own computer modeling which incorporates modifications of AGMA formulas based upon experiences from a wide range of applications.

The following organizations offer software programs for design and gear ratings according to methods outlined in AGMA publications: Fairfield Manufacturing Company Gear Software; Geartech Software, Inc.; PC Gears; Universal Technical Systems, Inc. For details and current listings, refer to AGMA's latest "Catalog of Technical Publications."

8.4 FLUID FILM BEARINGS
by Vittorio (Rino) Castelli

REFERENCES: "General Conference on Lubrication and Lubricants," ASME. Fuller, "Theory and Practice of Lubrication for Engineers," 2d ed., Wiley. Booser, "Handbook of Lubrication, Theory and Design," vol. 2, CRC Press. Barwell, "Bearing Systems, Principles and Practice," Oxford Univ. Press. Cameron, "Principles of Lubrication," Longmans Greene. "Proceedings," Second International Symposium on Gas Lubrication, ASME. Gross, "Fluid-Film Lubrication," Wiley. Gunter, "Dynamic Stability of Rotor-Bearing Systems," NASA SP-113, Government Printing Office.

Plain bearings, according to their function, may be

Journal bearings, cylindrical, carrying a rotating shaft and a radial load

Thrust bearings, the function of which is to prevent axial motion of a rotating shaft

Guide bearings, to guide a machine element in its translational motion, usually without rotation of the element

In exceptional cases of design, or with a complete **failure of lubrication,** a bearing may run dry. The coefficient of friction is then between 0.25 and 0.40, depending on the materials of the rubbing surfaces. With the **bearing barely greasy,** or when the bearing is well lubricated but the speed of rotation is very slow, boundary lubrication takes place. The coefficient of friction may vary from 0.08 to 0.14. This condition occurs also in any bearing when the shaft is starting from rest if the bearing is not equipped with an oil lift.

Semifluid, or **mixed,** lubrication exists between the journal and bearing when the conditions are not such as to form a load-carrying fluid film and thus separate the surfaces. Semifluid lubrication takes place at comparatively low speed, with intermittent or oscillating motion, heavy load, insufficient oil supply to the bearing (wick or waste-lubrication, drop-feed lubrication). Semifluid lubrication may also exist in thrust bearings with fixed parallel-thrust collars, in guide bearings of machine tools, in bearings with copious lubrication where the shaft is bent or the bearing is misaligned, or where the bearing surface is interrupted by improperly arranged oil grooves. The coefficient of friction in such bearings may range from 0.02 to 0.08 (Fuller, Mixed Friction Conditions in Lubrication, *Lubrication Eng.,* 1954).

Fluid or **complete lubrication,** when the rubbing surfaces are completely separated by a fluid film, provides the lowest friction losses and prevents wear. A certain amount of oil must be fed to the oil film in order to compensate for end leakage and maintain its carrying capacity. Such lubrication can be provided under pressure from a pump or gravity tank, by automatic lubricating devices in self-contained bearings (oil rings or oil disks), or by submersion in an oil bath (thrust bearings for vertical shafts).

Notation

R = radius of bearing, length
r = radius of journal, length
$c = mr = R - r$ = radial clearance, length
W = bearing load, force
μ = viscosity = force × time/length^2
Z = viscosity, centipoise (cP); 1 cP = 1.57×10^{-7} lb · s/in^2 (0.001 N · s/m)
β = angle between load and entering edge of oil film
η = coefficient for side leakage of oil
ν = kinematic viscosity = μ/ρ, length^2/time
R_e = Reynolds number = umr/ν
P_a = absolute ambient pressure, force/area
$P = W/(ld)$ = unit pressure, lb/in^2
N = speed of journal, r/min
m = clearance ratio (diametral clearance/diameter)
F = friction force, force
A = operating characteristic of plain cylindrical bearing
P' = alternate operating characteristic of plain cylindrical bearing
h_0 = minimum film thickness, length
ε = eccentricity ratio, or ratio of eccentricity to radial clearance
e = eccentricity = distance between journal and bearing centers, length
f = coefficient of friction
f' = friction factor = $F/(\pi rl\rho u^2)$
l = length of bearing, length
$d = 2r$ = diameter of journal, length

K_f = friction factor of plain cylindrical bearing
t_w = temperature of bearing wall
t_0 = temperature of air
t_1 = temperature of oil film
u = surface speed, length/time
ω = angular velocity, rad/time
ρ = mass density, mass/length^3
Λ = bearing compressibility parameter = $6\mu\omega r^2/(P_a c^2)$

INCOMPRESSIBLE AND COMPRESSIBLE LUBRICATION

Depending on the fluid employed and the pressure regime, the fluid density may or may not vary appreciably from the ambient value in the load-carrying film. Typically, oils, water, and liquid metals can be considered incompressible, while gases exhibit compressibility effects even at modest loads. The difference comes from the fact that, in incompressible lubricants, fluid flow rates are linearly proportional to pressure differences, whereas for compressible lubricants the mass flow rates are proportional to the difference of some power of the pressure. This is because the pressure affects the fluid density. The bearing behavior is somewhat dissimilar. In incompressible lubrication, gage pressures can be used and the value of the ambient pressure has no effect on the load-carrying capacity, which is linearly related to viscosity and speed. This is not true in compressible lubrication, where the value of ambient pressure has a direct effect on the load-carrying capacity which, in turn, increases with viscosity and speed, but only up to a limit dependent on the bearing geometry. In what follows, incompressible lubrication is treated first and compressible lubrication second.

Incompressible (Plain Cylindrical Journal Bearings)

Fluid lubrication in plain cylindrical bearings depends on the viscosity of the lubricant, the speed of the bearing components, the geometry of the film, and possible external sources of pressurized lubricant. The oil is entrained by the journal into the film by the action of the viscosity which, if the passage is convergent, causes the creation of a pressure field, resulting in a force sufficient to float the journal and carry the load applied to it.

The **minimum film thickness** h_0 determines the closest approach of the journal and bearing surfaces (Fig. 8.4.1). The allowable closest approach depends on the finish of these surfaces and on the rigidity of the journal and bearing structures. In practice, $h_0 = 0.00075$ in (0.019 mm) is common in electric motors and generators of medium speed, with

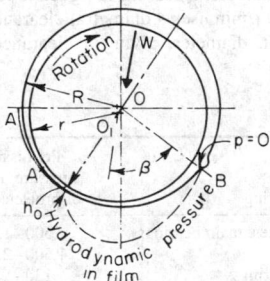

Fig. 8.4.1 Journal bearing with perfect lubrication.

steel shafts in babbitted bearings; $h_0 = 0.003$ in (0.076 mm) to 0.005 in (0.127 mm) for large steel shafts running at high speed in babbitted bearings (turbogenerators, fans), with pressure oil-supply for lubrication; $h_0 = 0.0001$ in (0.0025 mm) to 0.0002 in (0.005 mm) in automotive and aviation engines, with very fine finish of the surfaces.

Figure 8.4.2 gives the relationship between ε and the load-carrying coefficient A for a plain cylindrical journal. The operating characteristic

of the bearing is

$$A = (132/\eta)(1,000m)^2[P/(ZN)]$$

In Fig. 8.4.1, β is the angle between the direction of the load W and the entering edge of the load-carrying oil film, in degrees. The entering edge is at the place where the hydrodynamic pressure is equal or nearly equal to the atmospheric pressure and may be at the location of the

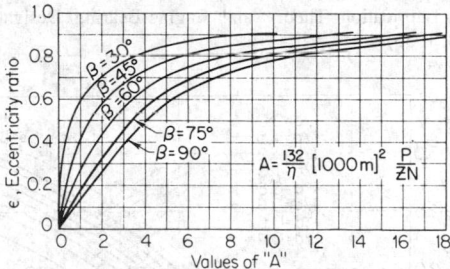

Fig. 8.4.2 Eccentricity ratio for a plain cylindrical journal.

oil-distributing groove B, or at the end of the machined recess pocket as at AA. For **complete bearings,** i.e., when the inner surface of the bearing is not interrupted by grooves, β may be taken as 90°. The reason for this assumption is the fact that, where the film diverges, the bearing pumping action tends to generate negative pressure, which liquids cannot sustain. The film **cavitates;** i.e., it breaks up in regions of fluid intermixed with either air or fluid vapor, while the pressure does not deviate substantially from ambient. For a 120° bearing with a central load, β may be taken as 60°.

The coefficient η corrects for side leakage. There is a loss of load-carrying capacity caused by the drop in the hydrodynamic pressure p in the oil film from the midsection of the bearing toward its ends; $p = 0$ at the ends. The value of η depends on the length-diameter ratio l/d and ε, the eccentricity ratio. Values of η are given in Fig. 8.4.3.

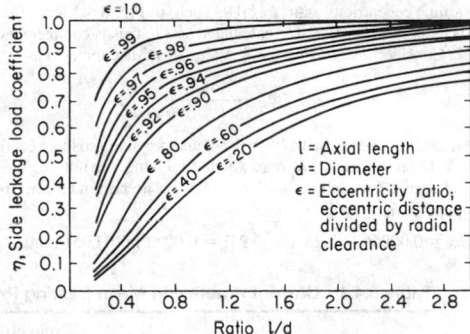

Fig. 8.4.3

EXAMPLE 1.   A generator bearing, 6 in diam by 9 in long, carries a vertical downward load of 8,650 lb; $N = 720$ r/min. The diametral clearance of the bearing is 0.012 in; the bearing is split on its horizontal diameter, and the lower half is relieved 40° down on each side, for oil distribution along journal; the bearing arc is therefore 100°; with the load vertical, $\beta = 50°$; bearing temperature 160°F. The absolute viscosity of the oil in the film is 12 centipoises (medium turbine oil). $P = W/ld = 160$ lb/in^2; $\mu = 12 \times 1.45 \times 10^{-7} = 17.4 \times 10^{-7}$ lb · s/in^2. The solution is one of trial and error. By using Fig. 8.4.3 in conjunction with Fig. 8.4.2, only a few trials are necessary to obtain the answer. As a first trial assume $\varepsilon = 0.85$. For an l/d ratio of 1.5 in Fig. 8.4.3, η, the end-leakage factor, will be 0.77. Compute A using this value of η. $m = 0.012/6 = 0.002$.

$$A = \frac{132}{0.77}(2)^2\frac{160}{12 \times 720} = 12.7$$

Enter Fig. 8.4.2 with this value of a and at $\beta = 50°$, and find that $\varepsilon = 0.9$. This value is larger than the initial assumption for ε. As a second trial, $\varepsilon = 0.88$. Then $\eta = 0.8$, $A = 12.2$, and $\varepsilon = 0.89$. This is a sufficiently close check. The minimum film thickness is $h_0 = mr(1 - \varepsilon) = 0.002 \times 3 \times 0.12 = 0.0007$ in (0.01778 mm).

For severe operating conditions the value of A may exceed 18, the limit of Fig. 8.4.2. For complete journal bearings under extreme operating conditions, Fig. 8.4.4 should be used. The ordinate is P', defined as shown. The curves are drawn for various values of l/d instead of values of β as in Fig. 8.4.2. Values of ε may thus be obtained directly (Dennison, Film-Lubrication Theory and Engine-Bearing Design, *Trans. ASME*, **58**, 1936).

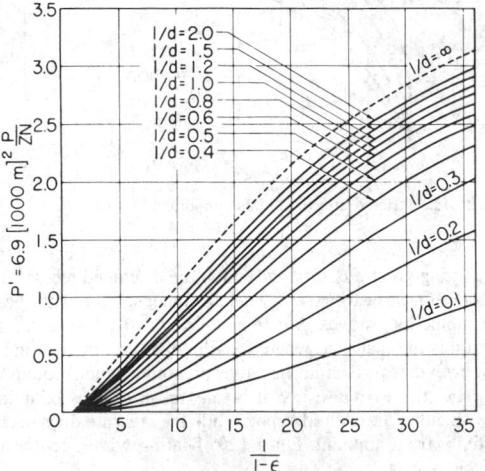

Fig. 8.4.4 Load-carrying parameter in terms of eccentricity.

EXAMPLE 2. A 360° journal bearing 2½ in diam and 3⅞ in long carries a steady load of 3,875 lb. Speed $N = 500$ r/min; diametral clearance, 0.0064 in; average viscosity of the oil in the film, 23.4 centipoises (SAE 20 light motor oil at 105°F). $P = 3,875/(2.5 \times 3.875) = 400$ lb/in². Value of $m = 0.0064/2.5 = 0.00256$. Value of $l/d = 1.55$. First, attempt to use Figs. 8.4.2 and 8.4.3 in this solution. Assume eccentricity ratio ε is 0.9. Then, in Fig. 8.4.3, with $l/d = 1.55$, value of η is determined as 0.8. A is calculated as 37. This is completely off scale in Fig. 8.4.2. Consider instead Fig. 8.4.4. Value of P' is computed as

$$P' = 6.9(2.56)^2 \, \frac{400}{23.4 \times 500} = 1.54$$

In Fig. 8.4.4, enter the curves with $P' = 1.54$, and move left to intersect the curve for $l/d = 1.5$. Drop downward to read a value for $1/(1 - \varepsilon)$ of 16. Then $\frac{1}{16} = 1 - \varepsilon$, or the eccentricity ratio $\varepsilon = \frac{15}{16}$, or 0.94. The minimum film thickness, as in Example 1 is $h_0 = mr(1 - \varepsilon)$, or

$$h_0 = 0.00256 \times 1.25(1 - 0.94) = 0.0002 \text{ in (0.0051 mm)}$$

Allowable mean bearing pressures in bearings with fluid film lubrication are given in Table 8.4.1. If the load maintains the same magnitude and direction when the journal is at rest (heavily loaded shafts, heavy gears), the mean bearing pressure should be somewhat less than when bearings are loaded only when running.

For internal-combustion-engine bearing design, Etchells and Underwood (*Mach. Des.*, Sept. 1942) list the following maximum design pressures for bearing alloys, pounds per square inch of projected area: lead-base babbitt (75 to 85 percent lead, 4 to 10 percent tin, 9 to 15 percent antimony) 600 to 800; tin-base babbitt (0.35 to 0.6 percent lead, 86 to 90 percent tin, 4 to 9 percent antimony, 4 to 6 percent copper) 800 to 1,000; cadmium-base alloy (0.4 to 0.75 percent copper, 97 percent cadmium, 1 to 1.5 percent nickel, 0.5 to 1.0 percent silver) 1,200 to 1,500; copper-lead alloy (45 percent lead, 55 percent copper) 2,000 to 3,000; copper-lead (25 percent lead, 3 percent tin, 72 percent copper) 3,000 to 4,000; silver (0.5 to 1.0 percent lead on surface, 99 percent silver) 5,000 up. The above pressures are based on fatigue life of 500 h at 300°F bearing temperature, and a bearing metal thickness 0.01 to 0.015 in for lead-, tin-, and cadmium-base metals and 0.25 in for copper, lead, and silver. At lower temperatures the life will be greatly extended.

Much higher pressures are encountered in rolling element bearings, such as ball and roller bearings, and gears. In these situations, the formation of fluid films capable of preventing contact between surface asperities is aided by the increase of viscosity with pressure, as exhibited by most lubricating oils. The relation is typically exponential, $\mu = \mu_0 e^{\alpha p}$, where α is the so-called pressure coefficient of viscosity.

Length-diameter ratios are usually chosen between $l/d = 1$ and $l/d = 2$, although many engine bearings are designed with $l/d = 0.5$, or even less. In shorter bearings, the carrying capacity of the oil film is greatly impaired by the effect of side leakage. Longer bearings are used to restrain the shaft from vibration, as in line shafts, or to position the shaft accurately, as in machine tools. In power machines, the tendency is toward shorter bearings. Typical values are as follows: turbogenerators, 0.8 to 1.5; gasoline and diesel engines for main and crankpin bearings, 0.4 to 1.0, with most values between 0.5 and 0.8; generators and motors, 1.5 to 2.0; ordinary shafting, heavy, with fixed bearings, 2 to 3; light, with self-aligning bearings, 3 to 4; machine-tool bearings, 2 to 4; railroad journal bearings, 1.2 to 1.8.

For the **clearance between journal and bearing** see Fits in Sec. 8. Medium fits may be used for journals running at speeds under 600 r/min, and free fits for speeds over 600 r/min. Kingsbury suggests for these journals a diametral clearance $= 0.002 + 0.001d$ in. In journals running at high speed, diametral clearance $= 0.002d$ should be used in order to lower the friction losses in the bearing. All units are in inches. The most satisfactory clearance should, of course, be based on a complete bearing analysis which includes both load-carrying capacity and heat generation due to friction. For example, a bearing designed to run at the extremely high speed of 50,000 r/min uses a diametral clearance of 0.0025 in for a journal with 0.8-in diameter, giving a clearance ratio, clearance/diameter, of 0.00316.

Table 8.4.1 Current Practice in Mean Bearing Pressures

Type of bearing	Permissible pressure, lb/in², of projected area	Type of bearing	Permissible pressure, lb/in², of projected area
Diesel engines, main bearings	800–1,500	Automotive gasoline engines, main bearings	500–1,000
Crankpin	1,000–2,000	Crankpin	1,500–2,500
Wrist pin	1,800–2,000	Air compressors, main bearings	120– 240
Electric motor bearings	100– 200	Crankpin	240– 400
Marine diesel engines, main bearings	400– 600	Crosshead pin	400– 800
Crankpin	1,000–1,400	Aircraft engine crankpin	700–2,000
Marine line-shaft bearings	25– 35	Centrifugal pumps	80– 100
Steam engines, main bearings	150– 500	Generators, low or medium speed	90– 140
Crankpin	800–1,500	Roll-neck bearings	1,500–2,500
Crosshead pin	1,000–1,800	Locomotive crankpins	1,500–1,900
Flywheel bearings	200– 250	Railway-car axle bearings	300– 350
Marine steam engine, main bearings	275– 500	Miscellaneous ordinary bearings	80– 150
Crankpin	400– 600	Light line shaft	15– 25
Steam turbines and reduction gears	100– 220	Heavy line shaft	100– 150

For high-speed internal-combustion-engine bearings using forced-feed lubrication, medium fits are used. Federal-Mogul recommends the following diametral clearances in inches per inch of shaft diameter for insert-type bearings: tin-base and high-lead babbitts, 0.0005; cadmium-silver-copper, 0.0008; copper-lead, 0.001.

The dependence of the **coefficient of friction** for journal bearings on the bearing clearance, lubricant viscosity, rotational speed, and loading pressure, as reported by McKee and others, is shown in Sec. 3. A plot of the coefficient of friction against the parameter ZN/P is a convenient method for showing this relationship. ZN/P is a parameter based on mixed units. Z is the viscosity in centipoise, N is r/min, P is the mean pressure on the bearing due to the load, pounds per square inch of projected area, and m is the clearance ratio. Values of ZN/P greater than about 30 indicate fluid film conditions in the bearings. If the viscosity of the lubricant becomes lower or if there is a reduction in rotational speed or an increase in load, the value of ZN/P will become smaller until the coefficient of friction reaches a minimum value. Any further reduction in ZN/P will produce breakdown of the oil film, marking the transition from fluid film lubrication with complete separation of the moving surfaces to semifluid or mixed lubrication, where there is partial contact. As soon as semifluid conditions are initiated, there will be a sharp increase in the coefficient of friction. The critical value of ZN/P, where this transition takes place, will be lowest for a rigid bearing and shaft with finely finished surfaces.

Figure 8.4.5 shows a generalization of the relationship between the coefficient of friction for a journal bearing and the parameter ZN/P,

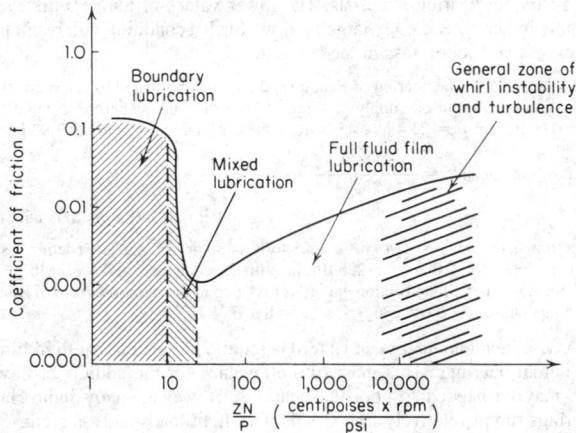

Fig. 8.4.5 Various zones of possible lubrication for a journal bearing.

indicating the various possible lubrication regimes that may be expected. For optimum design, a value of ZN/P somewhere between 30 and 300 would be recommended, but, in any case, the determination of minimum film thickness h_0 should be the deciding parameter. For extremely large values of ZN/P, resulting from high speeds and low loads,

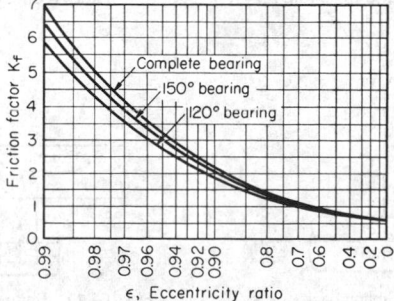

Fig. 8.4.6 Variation of the friction factor of a bearing with eccentricity ratio.

whirl instability may be developed. (See material on gas-lubricated bearings in this section.) With large values of ZN/P and a lubricant having a low kinematic viscosity, turbulent conditions may develop in the bearing clearance.

The friction force in plain journal bearings may be estimated by the use of the expression $F = K_f \mu Nrl/m$, where μ is in lb·s/in² units. The value of K_f depends upon the magnitude of ε and the type of bearing. Figure 8.4.6 shows values of K_f for a complete bearing, a 150° partial bearing, and a 120° partial bearing, assuming that the clearance space is at all times filled with lubricant. Note that F is the friction force at the surface of the bearing. Consequently, the friction torque is obtained by multiplying F by the bearing radius.

EXAMPLE 3.   As an illustration of the use of Fig. 8.4.6, determine the friction force in the bearing of Example 2. This is a complete journal bearing 2½-in diam by 3⅞ in. The value of ε was determined as 0.94. From Fig. 8.4.6, $K_f = 2.8$. Then

$$F = \frac{2.8 \times 23.4 \times 1.45 \times 10^{-7} \times 500 \times 1.25 \times 3.875}{0.00256}$$

$$= 8.97 \text{ lb } (4.08 \text{ kg})$$

The coefficient of friction $F/W = 8.97/3875 = 0.00231$. The mechanical loss in the bearing is $FV/33,000$ hp, where V is the peripheral velocity of the journal, ft/min.

$$\text{Friction hp} = (8.97 \times 500 \times \pi \times 2.5)/(33,000 \times 12)$$
$$= 0.089 \text{ hp } (66.37 \text{ W})$$

Departure from laminarity in the fluid film of a journal bearing will increase the friction loss. Figure 8.4.7 (Smith and Fuller, Journal Bearing Operation at Super-laminar Speeds, *Trans. ASME*, **78**, 1956) shows test results for such bearings, expressed in terms of a Reynolds number for the fluid film, $R_e = umr/\nu$. Laminar conditions hold up to an R_e of about 1,000. Friction may be calculated for laminar flow by using Fig. 8.4.6 or the left branch of the curve in Fig. 8.4.7, where $f' = 2/R_e$, and which applies to low values of the eccentricity ratio ($K_f = 0.66$). The values from Fig. 8.4.7 may be converted to friction torque T by the use of the expression $T = f' \pi \rho u^2 r^2 l$, where ρ is the mass density of the lubricant. In Fig. 8.4.7, a transition region spans values of the Reynolds number from 1,000 to 1,600. Here, two types of flow instability can occur. Usually, the first is due to **Taylor vortices** which are wrapped in

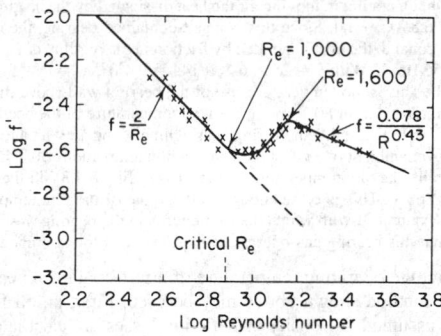

Fig. 8.4.7 Friction f' as a function of the Reynolds number for an unloaded journal bearing with $l/d = 1$. (*Smith and Fuller.*)

regular circumferential structures, each of which occupies the entire clearance. The onset of this phenomenon takes place at a value of the Reynolds number exceeding the threshold $R_e = 41.1(r/c)^{1/2}$. The second instability is due to turbulence, occurring at $R_e > 2,000$.

EXAMPLE 4.   A journal bearing is 4.5 in diameter by 4.5 in long. Speed 22,000 r/min. $mr = 0.002$ in. Viscosity μ, 1 cP (water) $= 1.45 \times 10^{-7}$ lb·s/in²; mass density $\rho = 62.4/1,728 \times 386 = 9.35 \times 10^{-5}$ lb·s²/in⁴; $\nu = \mu/\rho = 1.45 \times 10^{-7}/9.35 \times 10^{-5} = 0.155 \times 10^{-2}$ in²/s; $u = 22,000 \times 2\pi \times 2.25/60 = 5,180$ in/s; $R_e = 5,180 \times 0.002/0.155 \times 10^{-2} = 6,680$. This would indicate turbulence in the film. Value of f' is then $0.078/6,680^{0.43} = 0.078/44.2 = 1.765 \times 10^{-3}$. Friction torque $T = 1.765 \times 10^{-3} \times \pi \times 9.35 \times 10^{-5} \times 5,180^2 \times 2.25^2 \times 4.5$, $T = 317.5$ in·lb. Friction horsepower $= 2\pi TN/12 \times 33,000 = 2\pi \times 317.5 \times 22,000/12 \times 33,000$, FHP $= 111$ (82.77 kW).

In self-contained bearings (electric motor, line shaft, etc.) without external oil or water cooling, the **heat dissipation** is equal to the heat generated by friction in the bearing.

The heat dissipated from the outside bearing wall to the surrounding air is governed by the laws of heat transfer $Q = hS(t_w - t_0)$, where S is the surface area from which the heat is convected, Q is the rate of energy flow; t_w and t_0 are the temperatures of the wall and ambient air, respectively; and h is the heat convection coefficient, which has values from 2.2 Btu/(h·ft²·°F) for still air to 6.5 Btu/(h·ft²·°F) for air moving at 500 ft/min. Calculations of heat loss are extremely important due to the strong temperature dependence of the viscosity of most oils.

The temperature of the oil film will be higher than the temperature of the bearing wall. Typical ranges of values according to Karelitz (*Trans. ASME,* **64**, 1942), Pearce (*Trans. ASME,* **62**, 1940), and Needs (*Trans. ASME,* **68**, 1948) for self-contained bearings with oil bath, oil ring, and waste-packed lubrication are shown in Fig. 8.4.8.

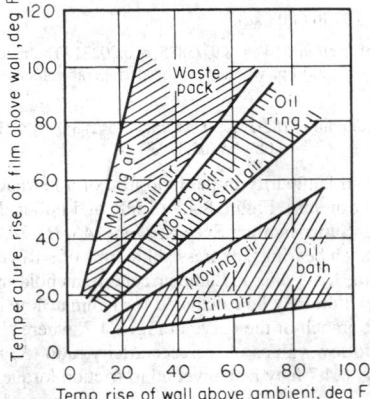

Fig. 8.4.8 Temperature rise of the film.

EXAMPLE 5.   The frictional loss for the generator bearing of Example 1, computed by the method outlined in Example 3, is 0.925 hp with $\varepsilon = 0.88$, $K_f = 1.6$, and $F = 27$ lb. Operating in moving air the heat dissipated by the bearing housing will be $L = 6.5S(t_w - t_0)$. Since this is a self-contained bearing, the heat dissipated is also equal to the heat generated by friction in the oil film, or $L = 0.925 \times 2,545 = 2,355$ Btu/h. With $S = 25 \times 6 \times 9/144 = 9.4$ ft², $t_w = t_0 = 2,355/6.5 \times 9.4 = 38.5$°F. This is the temperature rise of the bearing wall above the ambient room temperature. For an 80°F room, the wall temperature of the bearing would be about 118°F. In Fig. 8.4.8 an oil-ring bearing in moving air with a temperature rise of wall over ambient of 38°F should have a film temperature 50°F higher than that of the wall. The film temperature on the basis of Fig. 8.4.8 will then be 80 + 38 + 50, or 168°F. This is close enough to the value of the film temperature of 160°F from Example 1, with which the friction loss in the bearing was computed, to indicate that this bearing can operate without the need for external cooling.

To predict the operating temperature of a self-contained bearing, the cut-and-try method shown above may be used. First, an oil-film temperature is assumed. Viscosity and friction losses are calculated. Then the temperature rise of the wall over ambient is computed so as to dissipate to the atmosphere an amount of heat equal to the friction loss. Lastly from Fig. 8.4.8 the corresponding oil-film temperature is estimated and compared to the value that was originally assumed. A few adjustments of the assumed film temperature will produce satisfactory agreement and indicate the leveling-off temperature of the bearing. Self-contained bearings have been built with diameters of 3, 8, and 24 in (7.62, 20.32, and 60.96 cm) to operate at shaft speeds of 3,600, 1,000, and 200 r/min, respectively. These designs indicate a rough limit for bearings with no external cooling. The highest bearing temperature permissible with normal lubricants is about 210°F (100°C).

The temperature of automotive-type bearings is held within safe limits by using a **pressure-feed oil supply**. Sufficient lubricant is forced through the bearing to act as a coolant and prevent overheating. One widely used practice is to place a circumferential groove at the center of

the bearing to which the oil supply is fed. This is effective as far as cooling is concerned but has the disadvantage of interrupting the active length of the bearing and lowering its l/d ratio (see Fig. 8.4.9). The axial flow through each side of the bearing is given by

$$Q_1 = \frac{\Delta P m^3 r^4 \pi}{6\mu b}\left(1 + \frac{3}{2}\varepsilon^2\right)$$

where b is the effective axial length of the half bearing and ΔP is the difference between the oil pressure in the circumferential groove and

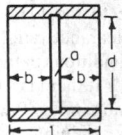

Fig. 8.4.9 Bearing with central circumferential groove.

the pressure at the ends of the bearing. The value of the last term in this equation will vary from 1.0 for a concentric shaft and bearing indicated by $\varepsilon = 0$ to a value of 2.5 for the extreme case of the shaft touching the bearing wall, indicated when $\varepsilon = 1$. Most of the heat caused by friction in the bearing is carried away by the circulating oil. Permissible temperature rises for this type of bearing may range from 15 to 50°F (8 to 28°C). In extreme cases a rise of 100°F (55°C) can be tolerated for high-strength bearing materials. The lower values of temperature rise usually indicate needlessly large oil flow. Such a condition will result in an excessive friction loss in the bearing.

EXAMPLE 6.   The bearing of Examples 2 and 3 is lubricated by a circumferential groove with an oil supply pressure of 30 lb/in² and, as before, $\varepsilon = 0.94$, $m = 0.0026$, and $\mu = 23.4 \times 1.45 \times 10^{-7}$ lb·s/in². Length b is about 1.93 in.

$$Q_1 \text{ flow out one side} = \frac{30 \times 0.0026^3 \times 1.25^4 \times \pi}{6 \times 23.4 \times 1.45 \times 10^{-7} \times 1.93}$$
$$\times [1 + 3/2(0.94)^2] = 0.240 \text{ in}^3/\text{s} \ (3.93 \text{ cm}^3/\text{s})$$

Total flow (two sides) = 0.48 in³/s = 53 lb/h for sp gr = 0.85. The friction loss from Example 3 = 0.089 hp = 226 Btu/h. With a specific heat of 0.5 Btu/(lb·°F) and assuming that all the friction energy is given up to the oil in the form of heat, the temperature rise $\Delta t = 226/0.5 \times 53 = 8.5$°F (4.72°C).

A definite **minimum rate of oil feed** is required to maintain a fluid film in journal bearings. This makes no allowance for the additional flow that may be needed to cool the bearings. However, many industrial bearings run at relatively low speeds with light loads and, as a consequence, additional oil flow to provide cooling is not necessary. But if a fluid film is desired, a definite minimum amount of lubricant is required. If the volume of lubricant fed to the bearing is less than this minimum requirement, there will not be a complete fluid film in the bearing. Friction will rise, wear will become greater, and the satisfactory service life of such a bearing will be reduced. This minimum lubricant supply can be evaluated by using the equation

$$Q_M = K_M urml$$

where Q_M is the flow rate and K_M is approximately 0.006.

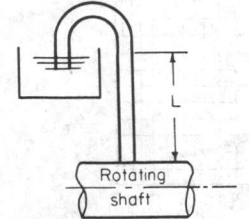

Fig. 8.4.10 Siphon wick.

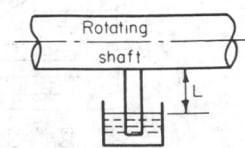

Fig. 8.4.11 Bottom wick.

EXAMPLE 7. The minimum feed rate for a journal bearing 2⅛-in diam by 2⅛ in long will be determined. Diametral clearance is 0.0045 in; speed, 1,230 r/min; load, 40 lb/in² based on projected area. $u = 1{,}230 \times \pi \times 2.125 = 10{,}220$ in/min, $r = 1.062$ in, $m = 0.0045/2.125 = 0.00212$, $l = 2.125$ in. Substituting,

$$Q_M = 0.006 \times 10{,}220 \times 1.062 \times 0.00212 \times 2.125$$
$$= 0.28 \text{ in}^3/\text{min}$$

(Fuller and Sternlicht, Preliminary Investigation of Minimum Lubricant Requirements of Journal Bearings, *Trans. ASME*, **78**, 1956.)

Many bearings are supplied with oil at low rates of feed by **felts, wicks,** and **drop-feed oilers.** Wicks can supply substantial rates of feed if they are properly designed. The two basic types of wick feed are siphon wicks, as shown in Fig. 8.4.10, and bottom wicks, as shown in Fig.

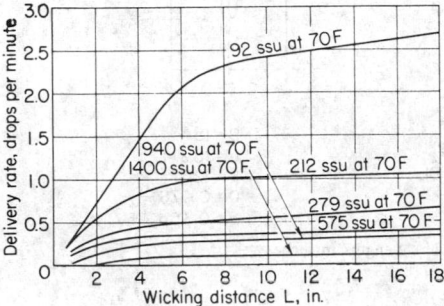

Fig. 8.4.12 Oil delivery with siphon wick (Fig. 8.4.10).

8.4.11. Data on oil delivery for these wicks are shown in Figs. 8.4.12 and 8.4.13. The data, from the American Felt Co., are for SAE Fl felts, based on a cross-sectional area of 0.1 in². The flow rate is indicated in drops per minute. One drop equals 0.0026 in³ or 0.043 cm³.

EXAMPLE 8. If it is desired to deliver 12.5 drops/min to a journal bearing, and if the viscosity of the oil is 212 s Saybolt Universal at 70°F, and if *L*, Fig. 8.4.10, is 5 in, what size of round wick would be required? From Fig. 8.4.12, for the stated conditions the delivery rate would be 0.9 drop/min for an area of 0.1 in². If 12.5 drops/min is needed, this would mean an area of 12.5 divided by 0.9 and multiplied by 0.1, or 1.4 in². For a round wick this would mean a diameter of 1⅜ in (3.49 cm).

If a **bottom wick** is considered with $L = 4$ in, Fig. 8.4.11, then in Fig. 8.4.13 the delivery rate using the same oil would be 1.6 drops/min; and if 12.5 drops/min is required, the area would be 12.5 divided by 1.6 and multiplied by 0.1, or 0.78 in². This would mean a bottom wick of 1 in diam if it is round (2.54 cm).

When journal **bearings** are **started, stopped,** or **reversed,** or whenever conditions are such that the operating value of *ZN/P* falls below the critical value for that bearing, the oil film will be ruptured and metal-to-metal contact will increase friction and cause wear. This condition can be eliminated by using a **hydrostatic oil lift.** High-pressure oil is introduced to the area between the bottom of the journal and the bearing (Fig. 8.4.14). If the pressure and quantity of flow are great enough, the shaft, whether it is rotating or not, will be raised and supported by an oil film. Neglecting axial flow, which is small, the flow up one side is

$$Q_1 = \frac{W r m^3}{A \mu} \quad \text{in}^2/\text{s}$$

and the inlet pressure required, $P_o = \mu Q_1 B/(b r^2 m^3)$, where *b* is the axial length of the high-pressure recess. Values of *A* and *B* are dimensionless factors which represent geometric effects and are given in the following table as a function of ε:

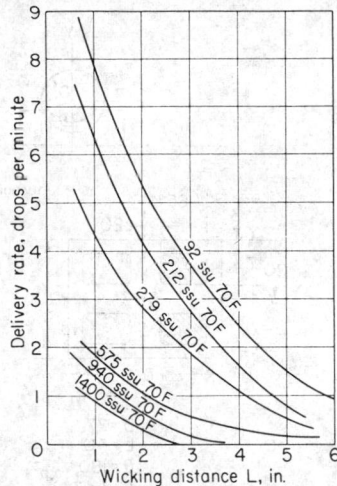

Fig. 8.4.13 Oil delivery with bottom wick (Fig. 8.4.11).

Current practice is to make the total area of the high-pressure recess in a bearing 2½ to 5 percent of the projected area *ld* of the bearing. It is generally desirable to use a check valve in the supply line to the oil lift so that, when the journal builds up a hydrodynamic oil-film pressure, reverse flow of oil in the supply line will be prevented.

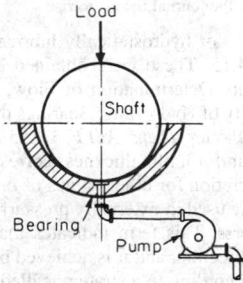

Fig. 8.4.14 Diagram of oil lift.

EXAMPLE 9. A 4.000-in-diam journal rests in a bearing of 4.012-in-diam. SAE 30 oil at 100°F (105 cP) is supplied under pressure to a groove at the lowest point in the bearing. Length of bearing, 6 in, length of groove, 3 in, load on bearing, 3,600 lb. What inlet pressure and oil flow are needed to raise the journal 0.004 in?

$$h_0 = mr(1 - \varepsilon)$$
$$0.004 = 0.006(1 - \varepsilon)$$
$$\varepsilon = 0.333$$

From the table, $A = 44.5$, $B = 42$.

$$Q_1 = \frac{3{,}600 \times 2}{44.5 \times 105 \times 1.45 \times 10^{-7}} (0.003)^3$$
$$= 0.287 \text{ in}^3/\text{s, one side (4.70 cm}^3/\text{s)}$$

Flow from both sides = $(0.287 \times 2) \times {}^{60}\!/_{231} = 0.149$ gal/min (0.564 l/min).
Oil supply pressure is

$$P_o = \frac{105 \times 1.45 \times 10^{-7} \times 0.287 \times 42}{3 \times 4} \times \frac{1}{0.003^3} = 566 \text{ lb/in}^2$$

ε	0	0.1	0.2	0.3	0.4	0.5	0.6	0.7	0.75	0.8	0.85	0.9
A	24.0	28.1	33.8	41.6	53.3	72.0	105	173	237	360	613	1,320
B	18.9	23.2	29.0	38.2	52.7	77.9	128	246	344	634	1,260	3,360

ε	0.91	0.92	0.93	0.94	0.95	0.96	0.97	0.98	0.99
A	1,620	2,070	2,620	3,530	5,040	7,800	13,700	30,600	121,000
B	4,340	5,810	8,040	11,800	18,400	32,100	65,300	179,000	348,000

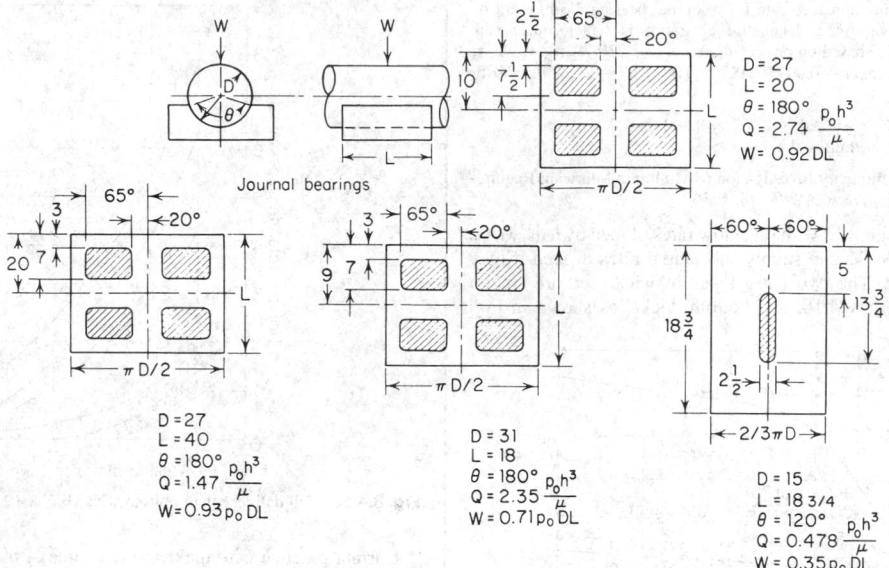

Fig. 8.4.15 Load-carrying capacity and flow for journal bearings *(Loeb)*. Lengths in inches.

An adjustable constant-volume pump or a spur-gear pump with a capacity of about 1,000 lb/in² (6.894 kN/m²) should be used to allow for pressure that may be built up in the line before the journal begins to rise.

Other configurations for hydrostatically lubricated journal bearings are shown in Fig. 8.4.15. These were obtained by means of electric analog solutions (Loeb, Determination of Flow, Film Thickness and Load-Carrying Capacity of Hydrostatic Bearings through the Use of the Electric Analog Field Plotter, *Trans. ASLE,* **1**, 1958). The data from Fig. 8.4.15 are exact for a uniform film thickness corresponding to $\varepsilon = 0$ but may be used with discretion for other values of ε.

Multiple recesses are used in externally pressurized bearings in order to provide local **stiffness**. This term indicates that the bearing resists shaft motions in any direction, and it is achieved by properly arranging the feeding network according to a strategy called **compensation**. Three main types are employed: orifice (and its variant, inherent), capillary, and fixed flow rates. In the first two, the idea is to insert a hydraulic resistance in each of the recess feeding lines and to use a single pump to feed all recesses. The flow rate q through orifices varies with the square root of the pressure drop Δp

$$q \propto \sqrt{\Delta p}$$

while for capillary tubes the relation is linear:

$$q = \frac{\pi \, \Delta p \, d^4}{64 \, l_1 \mu}$$

The general rule of thumb in designing orifices or capillary restrictors is to generate a pressure drop approximately equal to that taking place through the bearing, i.e., from the recesses to the ambient. The recess geometry and distribution, on the other hand, are designed so that $W = 0.5 p_{recess} \, DL$. Thus, the pump supply pressure is 4 times the average

bearing pressure. The bearing stiffness is usually equal to $K = 0.5 p_{recess} \, DL/c$.

The third method of compensation consists of forcing the same amount of flow to reach each recess regardless of clearance distribution. This can be achieved either by using separate pumps for each recess or by using a hydraulic device called a *flow divider*. With recess distributions as indicated above, the pump pressure need only be double the average bearing pressure; thus, this method of compensation leads to half the power dissipation of the other two. It is commonly used in large machinery, where power consumption must be limited. The polar axis bearings of the 200-in Hale telescope on Mount Palomar were the first large-scale demonstration of this technique. The azimuth axis thrust bearing of the 270-ft-diameter Goldstone radio telescope is probably the largest example of this type of bearing.

ELEMENTS OF JOURNAL BEARINGS

Typical dimensions of solid and split **bronze bushings** are given in Table 8.4.2.

Bronze bushings made from hard-drawn sheets and rolled into cylindrical shape are made with a wall thickness of only ¹⁄₃₂ in for bearings up to ½ in diam and with a wall thickness of ¹⁄₁₆ in for bearings from 1 in diam up. The wall thickness of these bearings depends chiefly upon the strength of the material which supports them. Bushings of this type are pressed into place, and the bearing surface is finished by burnishing with a slightly tapered bar to a mirror finish. The allowable bearing pressures may exceed those of cast bronze shown in Table 8.4.1 by 10 to 20 percent.

Babbitt linings in larger bearings are generally employed in thickness of ⅛ in or over and must be provided with sufficient anchorage in the

Table 8.4.2 Wall Thickness of Bronze Bushings, in

	Diam of journal, in						
	¼	¼–½	½–1	1–1½	1½–2½	2½–4	4–5½
Solid bushing, normal	¹⁄₁₆	³⁄₃₂	⅛	³⁄₁₆	¼	⅜	½
Split bushing, normal	³⁄₃₂	⅛	⁵⁄₃₂	⁷⁄₃₂	⁵⁄₁₆	¹⁵⁄₃₂	⅝
Solid bushing, thin	¹⁄₁₆	³⁄₃₂	³⁄₃₂	⅛	³⁄₁₆	¼	⅜
Split bushing, thin	¹⁄₁₆	³⁄₃₂	⅛	³⁄₁₆	¼	⅜	½

supporting shell. The anchors take the form of dovetailed grooves or holes drilled in the shell and counterbored from the outside.

Improved conditions are obtained by sweating or bonding the babbitt to the shell by tinning the latter, using potassium chlorate as flux. Tin-base babbitts and other low-strength materials evidence some yielding when subjected to heavy pressures. This tendency may be alleviated by the use of a thinner layer of the bearing material, fused either to a bronze or to a steel shell. This improves the fatigue life of the bearing material. Standard bearing inserts of this type are available in tin-base babbitts, high-lead babbitts, cadmium alloys, and copper-lead mixtures in diameters up to about 6 in (15.24 cm) (Fig. 8.4.16). A few materials can be obtained in sizes up to 8 in (20.32 cm). Some types are available with flanges or with other special features. The bearing lining may vary from about 0.001 in (0.025 mm) to 0.1 in (2.5 mm) in thickness depending upon the size of the bearing.

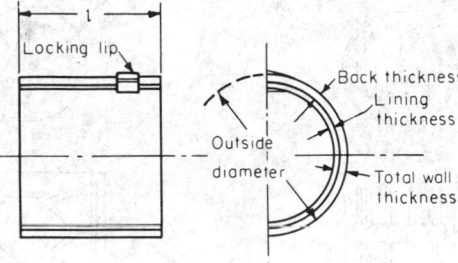

Fig. 8.4.16   Bearing insert.

Figure 8.4.17 shows the principal types of bonded babbitt linings. Figure 8.4.17a is for normal operating conditions. Figure 8.4.17b is for more severe operating conditions.

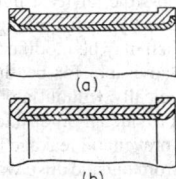

Fig. 8.4.17

General practice for the **thickness of babbitt lining and shells** is as follows: Fig. 8.4.18, $b = \frac{1}{32}d + \frac{1}{8}$ in, $S = 0.18d$ for bronze or steel $= 0.2d$ for cast iron; Fig. 8.4.18a, $t = b/2 + \frac{1}{16}$ in, $W = 1.8t$, $W_1 = 2.2t$.

Solid bronze or steel bushings, when pressed into the bearing housing, must be finished after pressing in. Light press fits and securing by

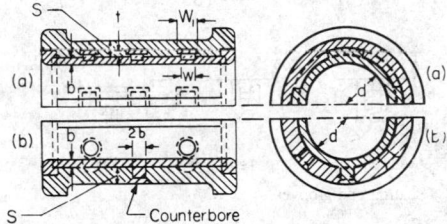

Fig. 8.4.18

setscrews or keys are preferable to heavy press fits and no keying, since heavy pressure, especially in thin-walled bushings, will set up stresses which will release themselves if bearings should run hot in service and will result in closing in on the journal and scoring when cooling.

Uniform Load Distribution Misalignment between journal and bearing should never be so great as to cause metallic contact. The max-

imum allowable inclination α of the shaft to the bearing is given by $\tan \alpha = md/l$.

Whenever the deflection angle of the bearing installation is greater than α, either the bearing length should be reduced or, if that is not feasible, the bearing should be mounted on a spherical seat to permit self-alignment.

Oil grooves are of two kinds, axial and circumferential; the former distribute the oil lengthwise in the bearing; the latter distribute it around the shaft at the oil hole, and also collect and return oil which would

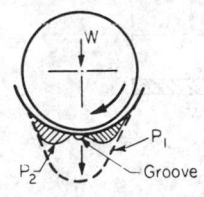

Fig. 8.4.19

otherwise be forced out at the ends of the bearing. Grooves have often been put into bearings indiscriminatingly, with the result that they scrape off the oil and interrupt the film.

In Fig. 8.4.19, W is the resultant force or load, pounds, on the bearing or journal. The radial ordinates P_1, to the dotted curve, show the pressures, lb/in², of the journal on the oil film due to the load when there is no axial groove, while the ordinates P_2, to the solid curve, show the pressures with an incorrectly located groove. Since there is no oil pressure near the groove, the permissible load W must be reduced or the film will be ruptured.

Groove dimensions (Fig. 8.4.20) are given by the following relations: $a = \frac{1}{3}$ wall thickness; $W_o = 2.5a$; $W_d = 3a$; $c = 0.5W_d$; $f = \frac{1}{16}$ in to $0.5W_d$.

In order to maintain the oil film, **the axial distributing groove should be placed in the unloaded sector** of the bearing. The location of grooves in a variety of cases is shown in Figs. 8.4.21 to 8.4.30.

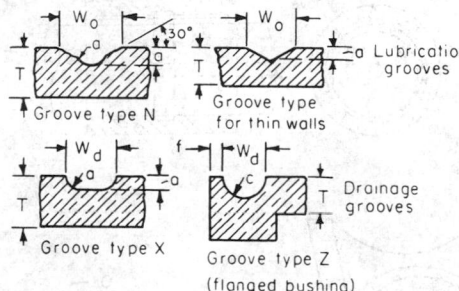

Fig. 8.4.20   Lubrication and drainage grooves.

Horizontal Bearings, Rotational Motion

DIRECTION OF LOAD KNOWN AND CONSTANT

Load downward or inside the lower 60° segment as in the case of ring-oiling bearings (Fig. 8.4.21).

Load at an angle more than 45° to the vertical centerline (Fig. 8.4.22).

In force- or drop-feed oiling, the oil inlet may be anywhere within the no-load sector (Fig. 8.4.23).

Oil can be introduced through the center of the revolving shaft (Fig. 8.4.24).

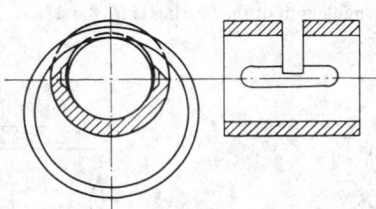

Fig. 8.4.21

Where oil-ring electric-motor bearings will be subjected by the purchaser to belt loads varying from vertical downward to horizontal, a continuous type of oil groove developed by General Electric Co. has proved very successful (Fig. 8.4.25). There are no critical spots with this groove because only a small percentage of the babbitt surface is removed along any axial line.

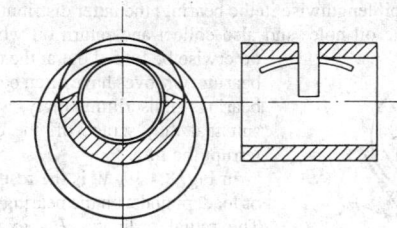

Fig. 8.4.22

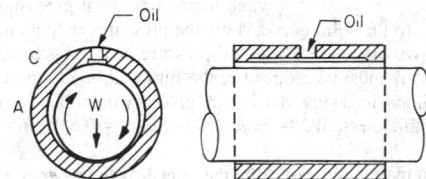

Fig. 8.4.23

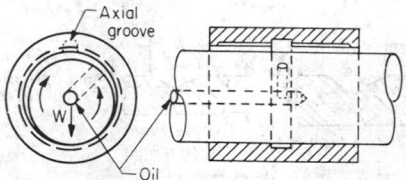

Fig. 8.4.24

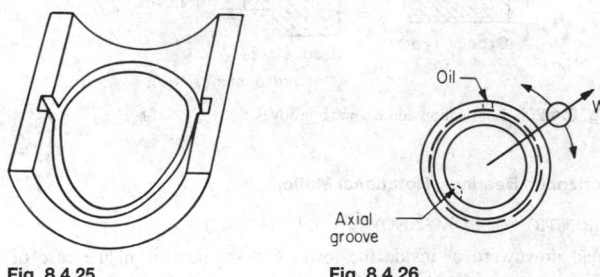

Fig. 8.4.25   **Fig. 8.4.26**

ROTATING LOAD

For rotating shafts, a circumferential groove at the middle of the bearing and an axial groove on the no-load side (Fig. 8.4.26).

For stationary shafts and rotating bearings, a circumferential groove in the bearing and an axial groove on the no-load side. The oil hole is in the shaft at the midlength of the bearing (Fig. 8.4.27).

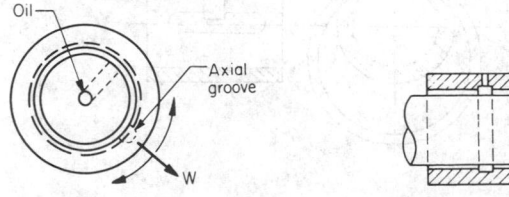

Fig. 8.4.27   **Fig. 8.4.28**

LOAD DIRECTION UNCERTAIN

Oil-ring bearings (Figs. 8.4.21 and 8.4.22) may be used, although they have defects under certain load directions. With forced or drop feed, the oil hole enters a circumferential groove at the middle of the bearing and the axial groove is omitted (Fig. 8.4.28). Arrangements for introducing oil through the rotating shaft can be made.

Bearings with Oscillatory Motion

DIRECTION OF LOAD CONSTANT

No oil film can be built up owing to the small sliding velocity, and boundary lubrication will exist. Axial grooves in the loaded sector distribute the lubricant to all parts of the bearing and avoid dry spots (Fig. 8.4.29).

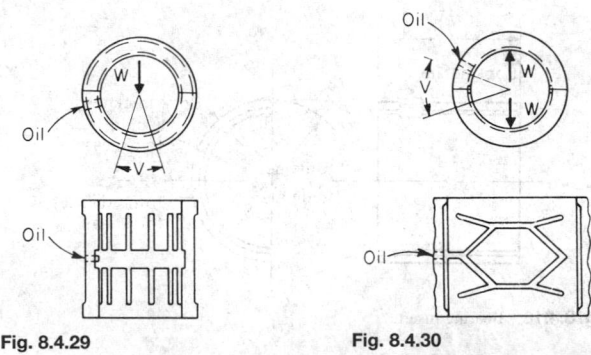

Fig. 8.4.29   **Fig. 8.4.30**

LOAD DIRECTION REVERSED DURING OSCILLATION

Fluid film lubrication is possible, at least during part of the motion, owing to the vacuum caused by shaft moving back and forth. Figure 8.4.30 shows grooving which may be modified to suit local conditions. This arrangement is also advisable for bearings under a load which reverses in direction periodically without any rotation of the bearing. The lubrication may then provide an oil cushion to soften shocks.

Bearing seals are used to prevent oil leakage from the bearing housing and to protect the bearing from outside dust, water, vapors, etc. A drainage groove at the end of the bearing is effective to divert the oil passing through the bearing back into the oil well (Fig. 8.4.31a). The drain holes at the bottom of the groove must be ample for passage of the oil flow.

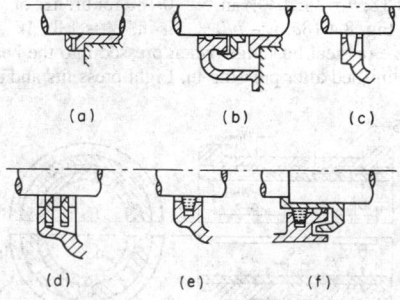

Fig. 8.4.31 Sealing end grooves.

An oil thrower mounted on the shaft is shown in Fig. 8.4.31b. The bearing housing may be provided with a single (Fig. 8.4.31c) or double collecting groove, or with brass or aluminum strip scrapers (Fig. 8.4.31d), to collect the oil creeping along the shaft.

For protection from dust, etc., felt packing rings are often used (Fig. 8.4.31e). The felt ring is soaked in oil to prevent charring by friction heat. In severe cases, additional protection by a labyrinth runner is very effective (Fig. 8.4.31f).

Standard seals are available for oil and grease retention as shown in

Fig. 8.4.32*a*, *b*, and *c*. The seal material that is pressed against the rotating shaft is typically made of synthetic rubber, which is satisfactory for temperatures as high as about 250°F (121°C). Figure 8.4.32*a* shows the seal material pressed against the shaft by a series of flexible fingers

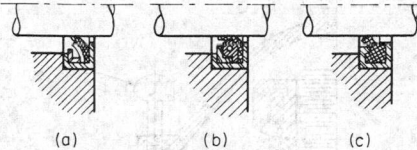

Fig. 8.4.32   Seals for oil and grease retention.

or leaf springs. In Fig. 8.4.32*b* a helical garter spring provides the gripping force. In Fig. 8.4.32*c* the rubber acts as its own spring.

Types of bearings are shown in Figs. 8.4.33 to 8.4.38. They include the principal methods of lubrication and types of construction.

Oiless bearings is the accepted term for self-lubricating bearings containing lubricants in solid or liquid form in their material. Graphite, molybdenum disulfide, and Teflon are used as solid lubricants in one group, and another group consists of porous structures (wood, metal), containing oil, grease, or wax.

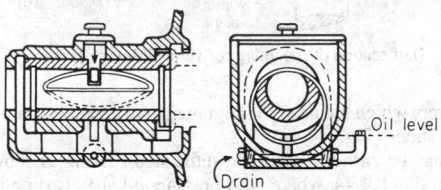

Oil grooves: Full lines for load in direction of arrow. Full and dotted lines for indeterminate load

Fig. 8.4.33   Ring-oiled bearing solid bushing.

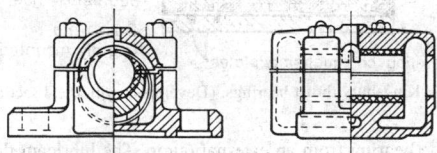

Fig. 8.4.34   Rigid ring-oiling pillow block. (*Link Belt Co.*)

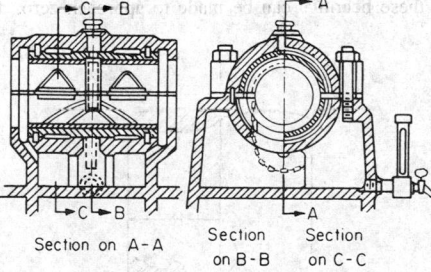

Section on A-A   Section on B-B   Section on C-C

Fig. 8.4.35   Split bearing with one chain. Main crankshaft bearing; vertical oil engine.

Graphite-lubricated bearings (bridge bearings, sheaves, trolley wheels, high-temperature applications) consist generally of cast bearing bronze as a supporting structure containing various overlapping designs of grooves which are filled with graphite. The graphite is mixed with a binder, and the plastic mass is pressed into the cavities to the hardness of a lead pencil; 45 percent of the bearing area may be graphite.

Porous-metal bearings, compressed from metal powders and sintered, contain up to 35 percent of liquid lubricant. See ASTM B202-45T for sintered bronze and iron bearings, and also Army and Navy Specification AN-B-7G. The porous metal generally consists of a 90-10 copper-

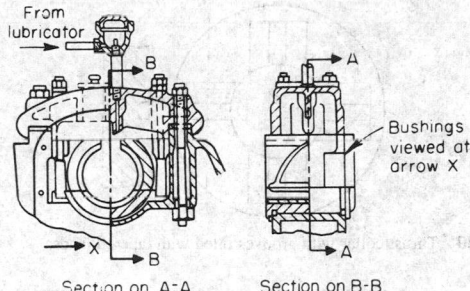

Section on A-A   Section on B-B

Fig. 8.4.36   Crankshaft main bearing. Horizontal engine with drop-feed lubrication.

tin bronze with 1½ percent graphite. These bearings do not require oil grooves since capillarity distributes the oil and maintains an oil film. If additional lubrication from an oil well should be provided, oil will be absorbed through the porous wall as required. For high temperatures where oil will carburize, a higher percentage of graphite (6 to 15 percent) is used.

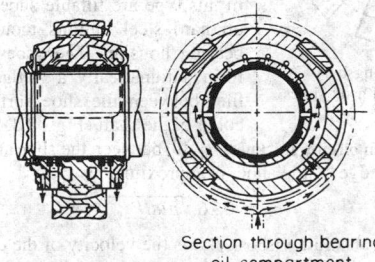

Section through bearing oil compartment

Fig. 8.4.37

Porous-metal bearings are used, where plain metal bearings are impractical because of lack of space, cost, or inaccessibility for lubrication, as in automotive generators and motors, hand power tools, vacuum cleaner motors, and the like.

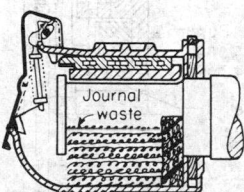

Journal waste

Fig. 8.4.38

THRUST BEARINGS

At low speeds, shaft shoulders or collars bear against flat bearing rings. The lubrication may be semifluid, and the friction is comparatively high.

For hardened-steel collars on bronze rings, with intermittent service, pressures up to 2,000 lb/in² (13,790 kN/m²) are permissible; for continuous low-speed operation, 1,500 lb/in² (10,341 kN/m²); for steel collars on babbitted rings, 200 lb/in² (1,378.8 kN/m²). In multicollar thrust bearings, the values are reduced considerably because of the difficulty in distributing the load evenly between the several collars.

The performance of the bearing thrust rings is much improved by the introduction of **grooves** with tapered lands as shown in Fig. 8.4.39. The lands extend on either side of the groove. The taper angle of the lands is very slight, so that a pressure oil film is formed between the bearing ring

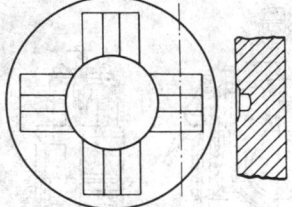

Fig. 8.4.39 Thrust collar with grooves fitted with tapered lands.

and the collar of the shaft. It is generally known that slightly tapered radial grooves will develop a hydrodynamic load-carrying film, when formed in the manner of Fig. 8.4.39. The taper angle should be on the order of 0.5°. Alternatively, a shallow recessed area that is a couple of film thicknesses deep can be used in place of the taper.

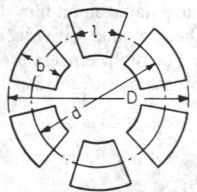

Fig. 8.4.40 Kingsbury thrust bearing with six shoes.

For high speeds or where low friction losses and a low wear rate are essential, **pivoted segmental thrust bearings** are used (Kingsbury thrust bearing, or Michell bearing in Europe). The bearing members in this type are tiltable shoes which rest on hard steel buttons mounted on the bearing housing. The shoes are free to form automatically a wedge-shaped oil film between the shoe surface and the collar of the shaft (Figs. 8.4.40 to 8.4.42).

The **minimum oil-film thickness** h_0, in, between the shoe and the collar, at the trailing edge of the shoe, is approximately

$$h_0 = 0.26 \sqrt{\mu u l / P_{avg}}$$

where μ is the absolute viscosity; u is the velocity of the collar, on the mean diam; l is the length of a shoe, at the mean diam of the collar, in the direction of sliding motion; P_{avg} is the average load on the shoes. As indicated in Fig. 8.4.40, $b = l$, approximately. The standard thrust bearings have six shoes. Load-carrying capacities of Kingsbury thrust bearings are given in Table 8.4.3.

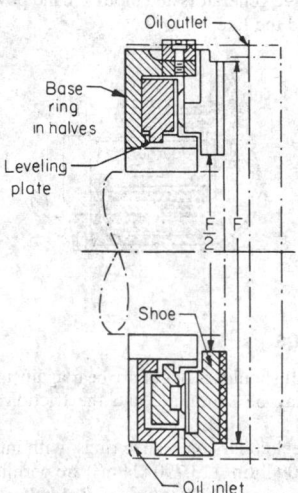

Fig. 8.4.41 Left half of six-shoe self-aligning equalizing horizontal thrust bearing for load in either axial direction.

The coefficient of friction in Kingsbury thrust bearings, referred to the mean diameter of the shoes, is approximately $f = 11.7 h_0/l$, where h_0 is computed as shown above. Figures 8.4.41 and 8.4.42 show typical pivoted segmental thrust bearings. They usually embody a system of

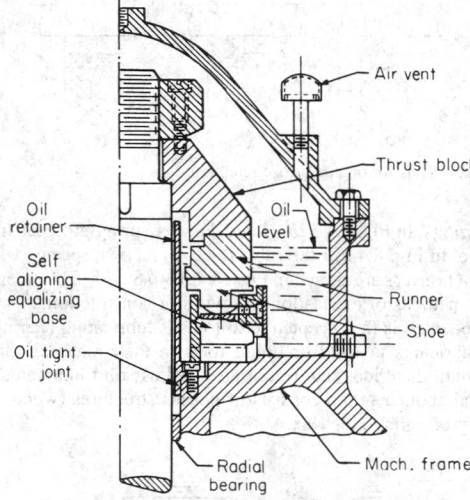

Fig. 8.4.42 Half section of mounting for vertical thrust bearing.

rocking levers which are used for alignment and equalization of load on the several shoes (Fig. 8.4.43).

Thrust may be carried on a hydrostatic step bearing as shown schematically in Fig. 8.4.44, where high-pressure oil at P_o is supplied at the

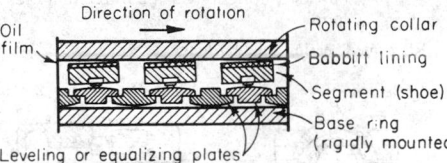

Fig. 8.4.43 Kingsbury thrust bearings. (Developed cylindrical sections.)

center of the bearing from an external pump. The lubricant flows radially outward through the annulus of depth h_0 and escapes at the periphery of the shaft at some pressure P_1 which is usually at atmospheric pressure. An oil film will be present whether the shaft rotates or not. Friction in these bearings can be made to approach zero, depending

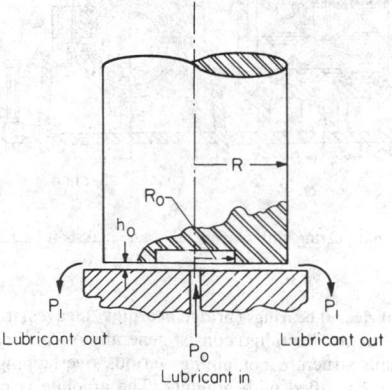

Fig. 8.4.44 Hydrostatic step bearing.

Table 8.4.3   Capacities of Six-Shoe Standard-Duty Horizontal and Vertical Thrust Bearings
(Based on viscosity of 150 s Saybolt at operating temperatures. Capacities given may be increased from 10 to 25% if viscosity is increased in same proportion)

Bearing size, in	Area, in²	Speed, r/min 100	200	400	800	1,800	3,600	Bearing size, in	Area, in²	Speed, r/min 100	150	200	300	500	700
		Safe load, 10³ lb								Safe load, 10³ lb					
5	12.5	1.44	1.7	2.0	2.4	2.9	3.5	19	180	40.00	44.0	48.0	53.0	60.0	65.0
6	18.0	2.30	2.7	3.2	3.8	4.6	5.5	21	220	51.00	57.0	61.0	68.0	77.0	84.0
7	24.5	3.30	3.9	4.7	5.6	6.8	8.0	23	264	65.00	72.0	77.0	85.0	97.0	105.0
8	32.0	4.60	5.5	6.6	7.8	9.6	11.4	25	312	80.00	88.0	95.0	105.0	119.0	123.0
9	40.5	6.20	7.4	8.8	10.4	13.0	15.0	27	364	97.00	107.0	115.0	127.0	144.0	146.0
10½	55.1	9.20	10.8	13.0	15.4	19.0	22.0	29	420	116.00	128.0	137.0	152.0	168.0	168.0
12	72.0	12.80	15.2	18.0	21.0	26.0	29.0	31	480	137.00	151.0	162.0	180.0	192.0	192.0
13½	91.1	17.20	20.0	24.0	29.0	35.0	36.0	33	544	160.00	177.0	189.0	210.0	220.0	220.0
15	112.5	22.00	26.0	32.0	37.0	45.0		37	684	215.00	235.0	250.0	275.0	275.0	
17	144.5	30.00	36.0	43.0	51.0	58.0		41	840	275.00	305.0	325.0	335.0	335.0	
								45	1012	345.00	385.0	405.0	405.0		

upon the rotational velocity and the viscosity of the lubricant film. Figure 8.4.45 shows the step bearing of a vertical turbogenerator. The load-carrying capacity is

$$W = \frac{P_o \pi}{2} \frac{R^2 - R_o^2}{\ln (R/R_o)}$$

This equation is valid even when the recess is eliminated, in which case R_o becomes the radius of the inlet oil supply pipe. The volume flow rate of lubricant and the clearance are related thus:

$$Q = P_o \pi h_0^3 [6\mu \ln(R/R_o)]$$

The friction power loss in the bearing is

$$H_f = \frac{\pi \mu \omega^2 (R^4 - R_o^4)}{2h_0}$$

The pumping power loss in forcing the lubricant through the bearing is $H_p = Q(P_o - P_1)/\eta$, where η is the efficiency of the pump.

EXAMPLE 10.   A typical 5,000-kW vertical turbogenerator has a thrust load of about 101,000 lb; outside diameter of bearing, 16 in; diam of recess, 10 in; pump efficiency, 0.5; speed, 750 r/min. Substituting these values,

$$101,000 = \frac{P_o \pi}{2} \left[\frac{8^2 - 5^2}{\ln (8/5)} \right]$$

or

$$P_o = 774 \text{ lb/in}^2$$

In practice, about 825 lb/in² is used on this step bearing to provide some margin of safety. Film thickness in the bearing should be from 0.001 to 0.01 in to protect the surfaces from metal-to-metal contact and allow passage of harmful grit that may

find its way into the system. The film thickness determines the oil flow for a given viscosity and pressure. With $h_0 = 0.006$ in (0.1524 mm) and SAE 20 oil at 130°F (29 centipoises), $Q = 8.25 \times \pi \times (0.006)^3/6 \times 29 \times 1.45 \times 10^{-7} \times 0.470 = 46.8$ in³/s (766.91 cm³/s). Flow $= 46.8 \times 60/231 = 12.15$ gal/min (45.99 L/min). The horsepower lost due to friction, $H_f = 750^2 \times 29 \times 1.45(8^4 - 5^4)/383,000 \times 0.006 = 3.58$ hp (2.669 kW). The horsepower lost due to pumping with pump efficiency of 0.5, $H_p = 46.8 \times 825/6,600 \times 0.5 = 11.7$ hp (8.725 kW). The total energy lost $= 11.7 + 3.58 = 15.28$ hp (11.39 kW).

Evaluation of these equations for other film thicknesses will show that the minimum lost energy will occur between $h_0 = 0.004$ and $h_0 = 0.006$ in (0.1016 and 0.1524 mm). The coefficient of friction corresponding to an energy loss of 15.28 hp in the above example is 0.002.

Other configurations for hydrostatically lubricated thrust bearings are shown in Fig. 8.4.46 from Loeb. They may be used directly to obtain the value of load-carrying capacity W and flow rate Q.

LINEAR SLIDING BEARINGS

All sliding bearings (Fig. 8.4.47), to wear true, must have the sliding parts of nearly equal lengths. Bearings made in this way will be found not to wear out of true. Oiling is accomplished in several ways, an acceptable method being that shown in Fig. 8.4.48. Short slides in many machine tools are lubricated by oil pads or direct oil application. The weight of the table and work and thrust of the tool cause wear on the bottom and sides of the guides. To compensate for the wear in both directions, bearings are sometimes made V-shaped, as shown in Fig. 8.4.49.

Simpler sliding bearings in machine tools are made with provision for adjustment (as shown in Fig. 8.4.50) of which there are many modifications. Recent applications involving hydrostatic lubrication on machine-tool ways have been very successful.

GAS-LUBRICATED BEARINGS

The fluid-film calculations included in Examples 1 through 10 have assumed that oil (or, in one case, water) was the lubricant. Actually, almost any **process fluid** may be used if proper recognition is given to the viscosity, corrosive action, change in state (where a liquid is close to its boiling point), toxicity, and in the case of a gas, its compressibility. Fluid-film journal and thrust bearings have run successfully, for example, on water, kerosene, gasoline, acid, liquid refrigerants, mercury, molten metals, and a wide variety of gases.

The previous equations for load-carrying capacity, film thickness, friction, and flow may be used for process liquids, but for gases, proper recognition must be made of the compressibility effects.

Because of the great value of gas-lubricated bearings for special applications, and to demonstrate the methods for handling the compressibility action, an introduction to the **design** of **gas-lubricated bearings** follows.

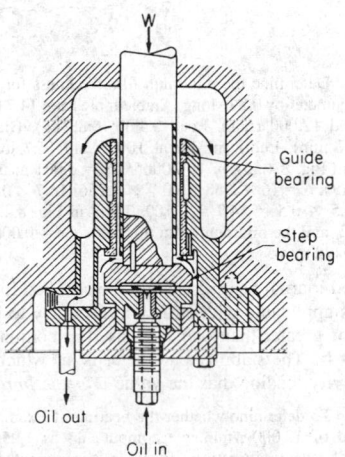

W

Guide bearing

Step bearing

Oil out

Oil in

Fig. 8.4.45   Step bearing of a vertical turbogenerator.

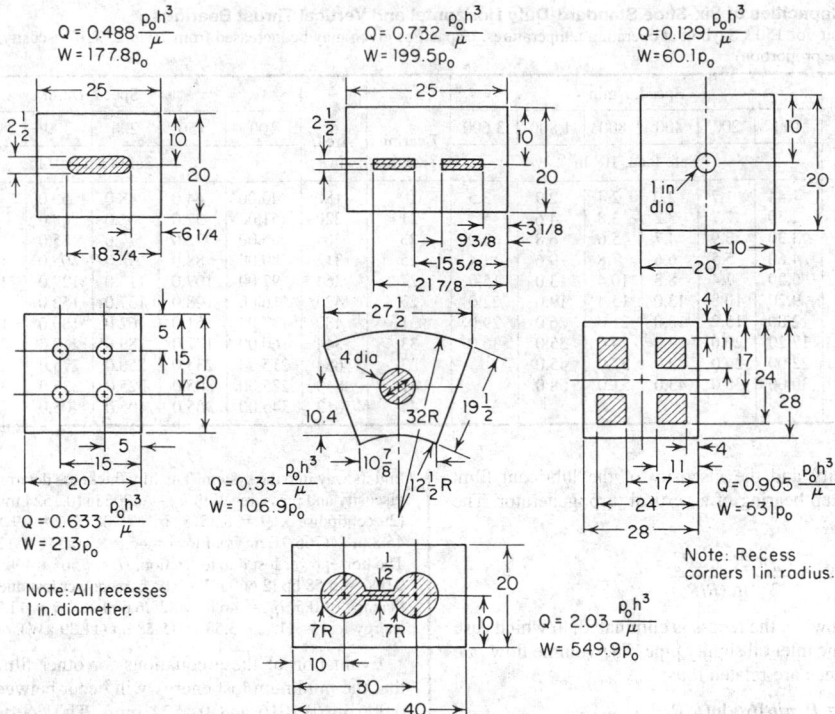

Fig. 8.4.46 Load-carrying capacity and flow for several flat thrust bearings *(Loeb)*. Lengths in inches.

Naturally, if the change in pressure within the bearing clearance is small compared to ambient pressure, the compressibility effect will be likewise small, and lubrication equations based on liquids may be used. A **compressibility parameter** Λ indicates the extent of this action. For hydrodynamic journal bearings it has the form $\Lambda = 6\mu\omega/(P_a m^2)$. For values of Λ less than one, the previous equations of this section for journal bearings may be used. For values of Λ greater than one, compressibility effects are included through the use of Figs. 8.4.51 to 8.4.54. (Data from Elrod and Burgdorfer, Proceedings First International Symposium on Gas-lubricated Bearings, 1959, and Raimondi, *Trans. ASLE*, vol. IV, 1961.)

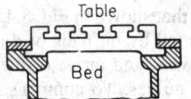

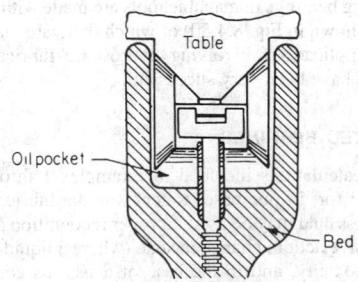

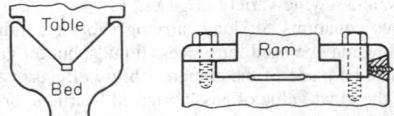

Fig. 8.4.47

Fig. 8.4.48

Fig. 8.4.49

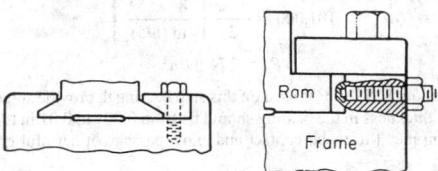

Fig. 8.4.50

EXAMPLE 11. Determine the minimum film thickness for a journal bearing 0.5 in (1.27 cm) diameter by 0.5 in long. Ambient pressure 14.7 lb/in² abs (101.34 kN/m² abs). Speed 12,000 r/min. Load 0.4 lb (0.88 kg). Diametral clearance 0.0005 in (0.0127 mm). Lubricant, air at 100°F and 14.7 lb/in² abs (2.68 × 10^{-9} lb·s/in² from Fig. 8.4.55). $m = 0.0005/0.5 = 0.001$ in/in. $\omega = 12,000 \times 2\pi/60 = 1,256$ rad/s, $\Lambda = (6 \times 2.68 \times 10^{-9} \times 1,256)/14.7 \times 0.001^2 = 1.37$, and $W/(dlP_a) = 0.4/0.5 \times 0.5 \times 14.7 = 0.109$. Then, in Fig. 8.4.53 ($l/d = 1$), we find that $\varepsilon = 0.22$, and the minimum film thickness $h_0 = 0.00025(1 - 0.22) = 0.000195$ in (0.00495 mm).

Gas-lubricated journal bearings should be checked for **whirl stability.** Figure 8.4.56 is applicable with sufficient accuracy to bearings where l/d is equal to or greater than one. It is used in conjunction with Fig. 8.4.51 for $l/d = \infty$. The stability parameter is ω_1^* which, for a bearing having only gravity loading, has the value $\omega_1^* = \omega\sqrt{mr/g}$.

EXAMPLE 12. To determine whether the bearing of Example 11 is stable at the running speed of 12,000 r/min, we compute ω_1^* as $1,256\sqrt{0.00025/386} = 1.015$. The value of eccentricity ratio ε_0 for $l/d = \infty$ is computed from Fig. 8.4.51

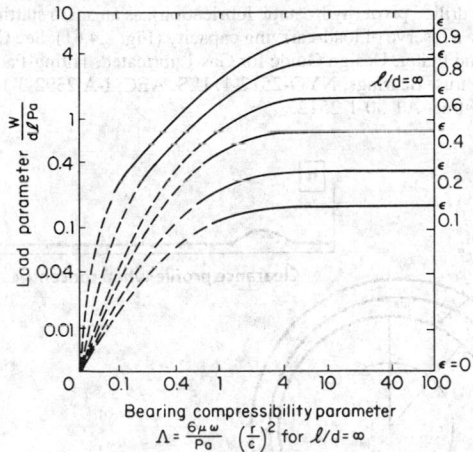

Fig. 8.4.51

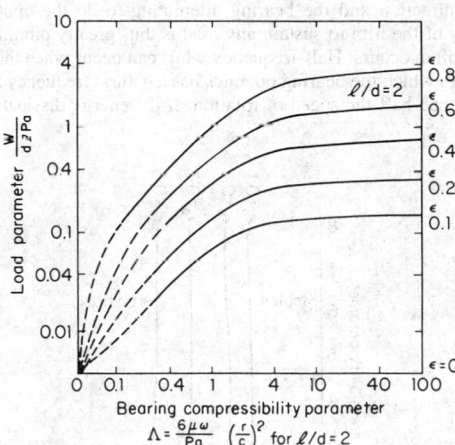

Fig. 8.4.52

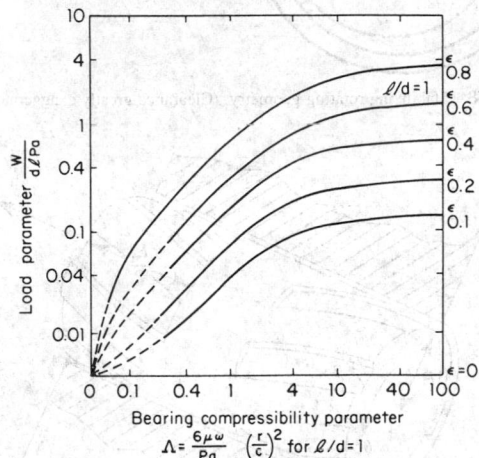

Fig. 8.4.53

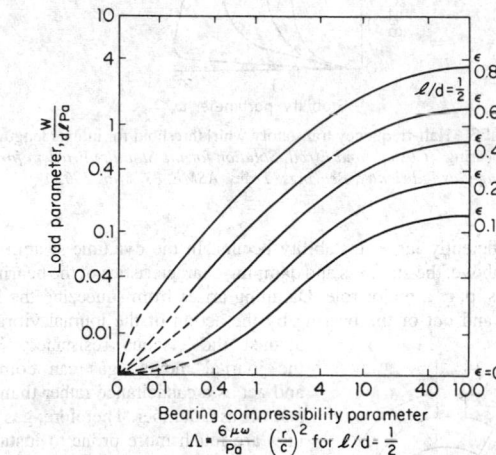

Fig. 8.4.54

Figs. 8.4.51–8.4.54 Theoretical load-carrying parameter vs. compressibility parameter for a full journal bearing: $l/d = \infty$. (Fig. 8.4.51), $l/d = 2$ (Fig. 8.4.52), $l/d = 1$ (Fig. 8.4.53), and $l/d = 0.5$ (Fig. 8.4.54). (*Elrod and Raimondi.*)

on the basis of W being the load per inch of bearing length. Thus $W(\text{lb/in}) = 0.4/0.5 = 0.8$ lb/in. For Fig. 8.4.51, $W/(dlP_a) = 0.218$ and in Fig. 8.4.51, we determine $\varepsilon_0 = 0.18$. Then (in Fig. 8.4.56), for $\omega_t^* = 1.015$ and $\Lambda = 1.37$, we find the intersection at about where a curve for $\varepsilon_0 = 0.18$ would be found. The bearing should just be stable. An intersection point on the ε_0 line or to the left should represent a stable condition. An intersection point to the right of the appropriate ε_0 line would predict an unstable condition.

The plain cylindrical journal bearing (360°) is the least stable of possible bearing designs. Control of "half-frequency" whirl has been achieved and the threshold of instability has been raised through modification of the geometry of the plain bearing. The simplest modification is the insertion of axial grooves. Bearings with three or four such grooves have been successful, but lose much stiffness.

A typical three-groove (three-sector) bearing is shown in Fig. 8.4.57.

Half-frequency whirl indicates a dynamic instability in which the journal orbits at approximately one-half of the shaft rotational speed, which coincides with the average speed of the lubricant in the film. If one looked at the bearing geometry from a coordinate system rotating at this whirl speed, one would see the journal attempting to pump lubricant

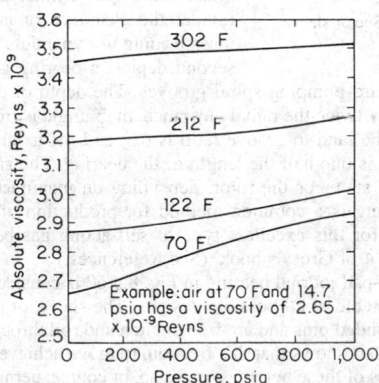

Fig. 8.4.55 Absolute viscosity of air. (*Iwaski, Sci. Rpts., Research Inst., Tohuku Univ., Ser. A; Kestin and Pliarczyk, Trans. ASME, 56, 1954.*) Reyns = 1.45×10^{-7} cP.

in one direction and the bearing attempting to do the opposite. The capacity of the film to sustain any load is thus greatly diminished, and failure often occurs. Half-frequency whirl can occur when the dynamic system in which the bearing operates has a natural frequency at approximately one-half the speed of rotation. If the energy dissipation rate is

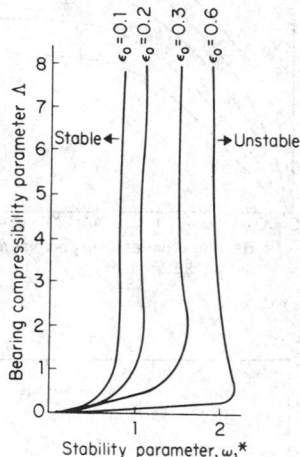

Fig. 8.4.56 Half-frequency translatory whirl threshold for infinite length, 360° journal bearing. *(Castelli and Elrod, Solution for the Stability Problem for 360°, Self Acting Gas Lubricated Bearings, Trans. ASME, 87, Mar. 1965.)*

not sufficiently large, instability occurs. In the dynamic system mentioned above, the stiffness and damping characteristics of the bearing, or bearings, play a major role. Damping arises from squeezing the lubricant in and out of the bearing by the action of the journal vibrations

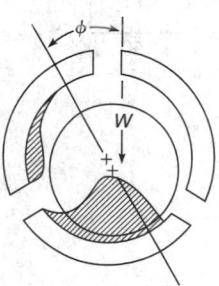

Fig. 8.4.57 Sector sleeve bearing.

against the viscous resistance. When the journal moves, gas can compress and act as a capacitance rather than flow and act as a damper. Therefore, gas bearings are much more prone to instability than are liquid-lubricated ones.

Aside from avoiding resonance conditions, two often successful techniques for whirl prevention are shown in Figs. 8.4.58 and 8.4.59. The first depicts a complete journal bearing with a rotating geometric artifact causing a synchronous disturbance. This artifact can be a variation in the clearance or an asymmetric mass leading to dynamic unbalance. The second depicts a bearing geometry with shallow, inward-pumping spiral grooves. The depth of the grooves is approximately twice the radial clearance, their angle is from 25° to 30° to the axis, the land-to-groove ratio is 0.5, and the axial extent of the grooved area is one-half the length of the bearing. The grooves can be on either the stator or the rotor, depending on manufacturing convenience. Etching is a common method for production of the grooves. Design data for this excellent type of self-acting gas bearing can be found in sec. 4 of Gross's book. (See References.)

The tilting-pad journal bearing in Fig. 8.4.60 is considered to be one of the most stable of all possible designs. The shoes, or pads, are supported on rounded pins and are free to pitch and roll through very small angles. Analysis shows that this freedom to move achieves the stability characteristics of these bearings, and also, of course, permits them to be self-aligning. Three-, four-, and five-shoe configurations are often used. Figure 8.4.60 shows an early design used for machine tool grinding spindles.

When applied to gas lubrication, pressurized gas may be supplied

through a drilled pivot (hydrostatic lubrication), as an aid in starting and to provide a reserve of load-carrying capacity (Fig. 8.4.61). See Gunter, Hinkle, and Fuller, Design Guide for Gas-Lubricated, Tilting-Pad Journal and Thrust Bearings, NYO-2512-1, U.S. AEC, I-A 2392-3-1, Nov. 1964, Contract AT 30-1-2512.

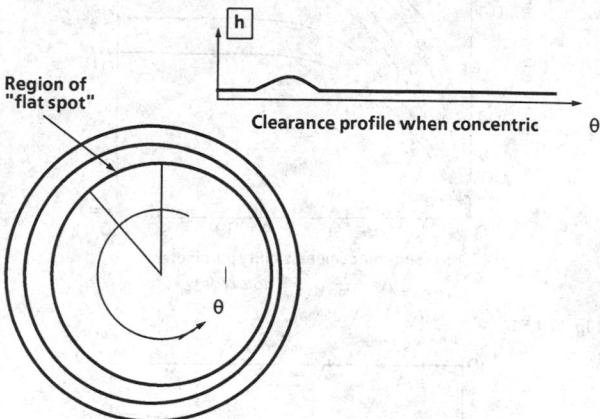

Fig. 8.4.58 Stabilizing rotating geometry. (Clearance greatly exaggerated.)

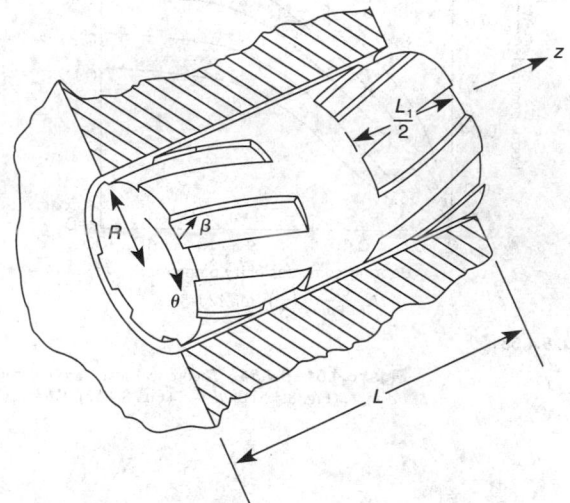

Fig. 8.4.59 Spiral groove bearing shown as outward-pumping. (Clearance is greatly exaggerated.)

A bearing design development that is simple, inexpensive, and very stable is the foil bearing. It is also very tolerant of thermal distortions and possible loss of clearance resulting from elevated temperatures. Probably the most widely used of several designs is shown in Fig. 8.4.62. It consists of overlapping metal shims, anchored at the base end like a cantilever beam. The "free" ends deflect and are able to automatically form their own operating clearance. These bearings are widely used in aircraft cabin cooling and for auxiliary power supply systems (Suriano, Dayton, and Woessner, Test Experience with Turbine-end Foil Bearing Equipped Gas Turbine Engines, ASME Paper 83-GT-73, 1983).

Thrust bearings of the tilting-pad variety are less susceptible to compressibility effects and may be considered as liquid-lubricated for values of Λ (suitable for thrust bearings) less than about 30. $\Lambda = 6u\mu l/(h_0^2 P_a)$ where l is the length of the shoe in the direction of sliding and U is the linear velocity at the mean radius. However, the shoes

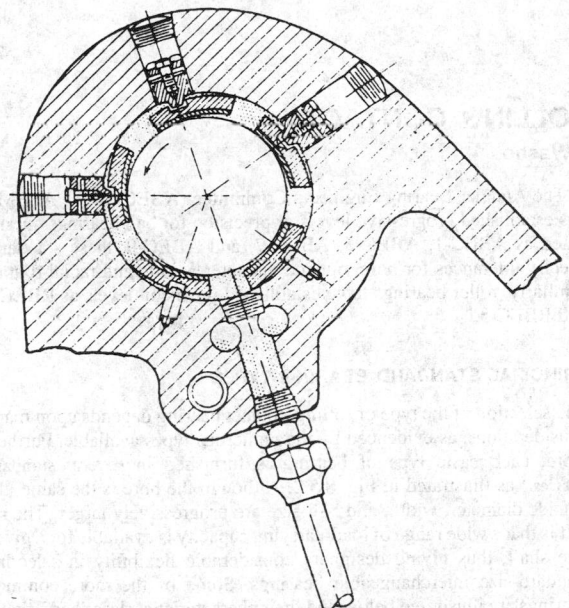

Fig. 8.4.60   Filmatic bearing. *(Courtesy Cincinnati Milacron Corp.)*

should not be made flat for gas operation but should have a crowned contour (see Fig. 8.4.63). (Gross, "Gas Film Lubrication," Wiley.) An approximate value for the crown is to make $\delta = \frac{3}{4}h_0$. The tilting-pad bearing design is probably the most common gas bearing presently in existence. Every hard-disk computer memory since the early 1960s has had its read-write heads supported by self-acting tilting-pad sliders. Hundreds of millions of such units, called *flying heads,* have been manufactured to date. Some designs employ the crown geometry while,

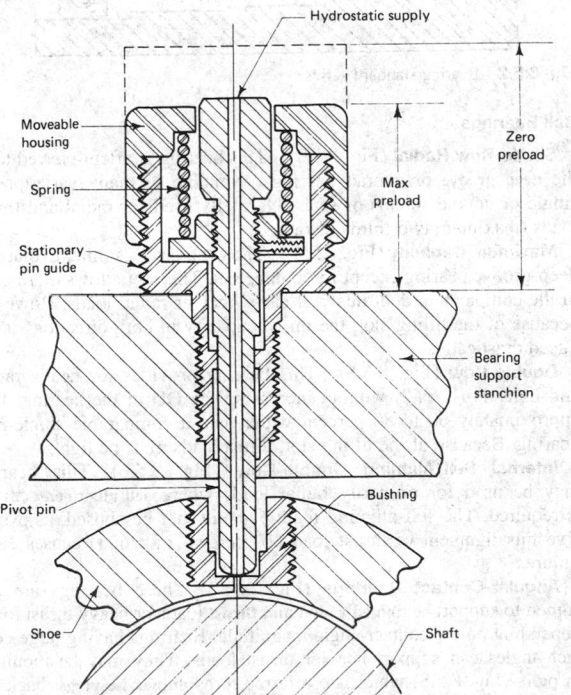

Fig. 8.4.61   Cross-sectional view, spring-mounted pivot assembly. *(Courtesy of The Franklin Institute Research Labs.)*

most commonly, heads with flat multiple sliders with straight ramps in their forward sections are used. The reason for the multiple thin sliders is the achievement of maximum damping possible. The typical minimum film heights have decreased steadily through the years from 1 μm

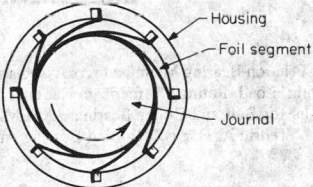

Fig. 8.4.62   Bending-dominated segments foil bearing.

(40 millionths of an inch) 25 years ago to less than 0.2 μm (8 millionths of an inch) currently (1995). This trend is driven by the achievement of the higher and higher recording densities possible at lower flying heights. Design of these devices is done rather precisely from first principles by means of special simulation programs. At these low clearances, allowance must be made for the finiteness of the **molecular mean free path,** which represents the mean distance that a gas molecule must travel between collisions. This effect manifests itself in a lowering of viscosity and wall shear resistance.

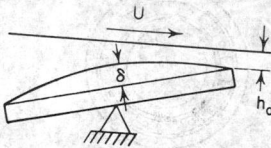

Fig. 8.4.63   Schematic of tilting-pad shoe, showing crown height δ.

Gas-lubricated hydrostatic bearings, unlike liquid-lubricated bearings, cannot be designed on the basis of fixed flow rate. They are designed instead to have a pressure loss produced by an **orifice restrictor** in the supply line. Such throttling enables the bearing to have load-carrying capacity and stiffness. For maximum stiffness the pressure drop in the orifice may be about one-half of the manifold supply pressure. For a circular thrust bearing with a single circular orifice, the load-carrying capacity is given with sufficient accuracy by the equation previously used for liquids (see Fig. 8.4.44). $W = (P_R - P_a/2)[R^2 - R_0^2/\ln (R/R_0)]$, where P_R is the recess pressure, lb/in^2 abs. The flow volume, however, is given by $Q_0 = \pi h_0^3/[6\mu \ln (R/R_0)](P_0^2 - P_1^2)/2P_0$. Q_0 and P_0 refer to recess conditions, and Q_1 and P_1 refer to ambient conditions. Pressures are absolute.

EXAMPLE 13.   A circular thrust bearing 6 in (15.24 cm) diameter with a recess 2 in (5.08 cm) diameter has a film thickness of $h_0 = 0.0015$ in (0.0381 mm). $P_0 = 30$ lb/in^2 gage or 44.7 lb/in^2 abs (308.16 kN/m^2). P_1 is room pressure, 14.7 lb/in^2 abs (101.34 kN/m^2 abs). Depth of recess is 0.02 in. Applied load is 375 lb. $Q_0 = (\pi \times 0.0015^3)/(6 \times 2.68 \times 10^{-9} \ln 3)(44.7^2 - 14.7^2)/(2 \times 44.7)$, $Q_0 = 12.3$ in^3/s (201.6 cm^3/s) at recess pressure. Converted to free air, $Q_1 = Q_0(P_0/P_1)$ with isothermal expansion, $Q_1 = 12.3(44.7/14.7) = 37.4$ in^3/s (612.87 cm^3/s), or $Q_1 = 37.4 \times 60 = 2,244$ in^3/min (36.77 L/min). Actual measured flow = 2,440 in^3/min (39.98 L/min).

Externally pressurized gas bearings are not as easily designed as liquid-lubricated ones. Whenever a volume larger than approximately that of the film is present between the restrictor and the film, a phenomenon known as **air hammer** or **pneumatic instability** can take place. Therefore, in practical terms, recesses cannot be used and orifice restrictors must be obtained by the smallest flow cross-section at the very entrance to the film; this area is equal to the perimeter of the inlet holes multiplied by the local height of the film. This technique is called **inherent compensation.** Unfortunately, as one can readily see, the area of the restrictors is smaller where the film is smaller; thus, the stiffness is lower than that obtainable by incompressible lubrication. Design data are available in Sec. 5 of Gross's book (see References).

8.5 BEARINGS WITH ROLLING CONTACT

by Michael W. Washo

REFERENCES: Anti-Friction Bearing Manufacturers Association, Inc. (AFBMA), Method of Evaluating Load Ratings. American National Standards Institute (ANSI), Load Ratings for Ball and Roller Bearings. AFBMA, "Mounting Ball and Roller Bearings." Tedric A. Harris, "Rolling Bearing Analysis."

COMPONENTS AND SPECIFICATIONS

Rolling-contact bearings are designed to support and locate rotating shafts or parts in machines. They transfer loads between rotating and stationary members and permit relatively free rotation with a minimum of friction. They consist of **rolling elements (balls or rollers)** between an **outer** and **inner ring. Cages** are used to space the rolling elements from each other. Figure 8.5.1 illustrates the common terminology used in describing rolling-contact bearings.

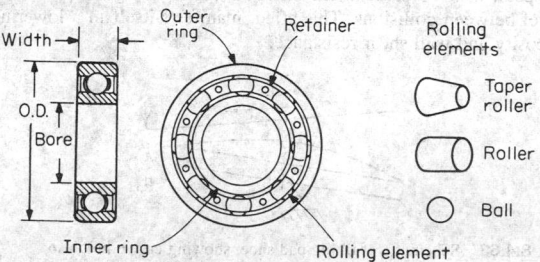

Fig. 8.5.1 Radial contact bearing terminology.

Rings The inner and outer rings of a rolling-contact bearing are normally made of SAE 52100 steel, hardened to Rockwell C 60 to 67. The rolling-element raceways are accurately ground in the rings to a very fine finish (16 μin or less).

Rings are available for special purposes in such materials as stainless steel, ceramics, and plastic. These materials are used in applications where corrosion is a problem.

Rolling Elements Normally the rolling elements, balls or rollers, are made of the same material and finished like the rings. Other rolling-element materials, such as stainless steel, ceramics, Monel, and plastics, are used in conjunction with various ring materials where corrosion is a factor.

Cages Cages, sometimes called separators or retainers, are used to space the rolling elements from each other. Cages are furnished in a wide variety of materials and construction. Pressed-steel cages, riveted or clinched and filled nylon, are most common. Solid machined cages are used where greater strength or higher speeds are required. They are fabricated from bronze or phenolic-type materials. At high speeds, the phenolic type operates more quietly with a minimum amount of friction. Bearings without cages are referred to as full-complement.

A wide variety of rolling-contact bearings are normally manufactured to standard boundary dimensions (bore, outside diameter, width) and tolerances which have been standardized by the AFBMA. All bearing manufacturers conform to these standards, thereby permitting interchangeability. ANSI has for the most part adopted these and published them jointly as AFBMA/ANSI standards as follows:

Title	Standard	Title	Standard
Terminology	1	Ball Standards	10
Gaging Practice	4	Roller Load Ratings	11
Mounting Dimensions	7	Instrument Bearings	12
Mounting Accessories	8.2	Vibration and Noise	13
Ball Load Ratings	9	Basic Boundary Dimensions	20

The Annular Bearing Engineers Committee (ABEC) of the AFBMA has established progressive levels of precision for ball bearings. Designated as ABEC-1, ABEC-5, ABEC-7, and ABEC-9, these standards specify tolerances for bore, outside diameter, width, and radial runout. Similarly, roller bearings have established precision levels as RBEC-1 and RBEC-5.

PRINCIPAL STANDARD BEARING TYPES

The selection of the type of rolling-contact bearing depends upon many considerations, as evidenced by the numerous types available. Furthermore, each basic type of bearing is furnished in several **standard "series"** as illustrated in Fig. 8.5.2. Although the bore is the same, the outside diameter, width, and ball size are progressively larger. The result is that a wide range of load-carrying capacity is available for a given size shaft, thus giving designers considerable flexibility in selecting standard-size interchangeable bearings. Some of the more common bearings are illustrated below and their characteristics described briefly.

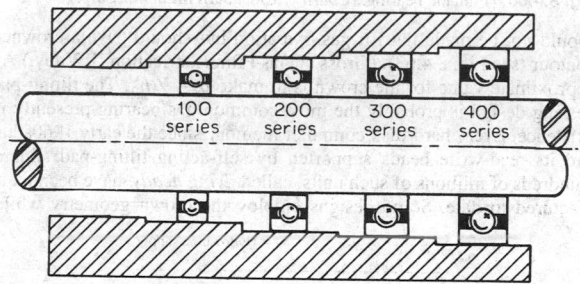

Fig. 8.5.2 Bearing standard series.

Ball Bearings

Single-Row Radial (Fig. 8.5.3) This bearing is often referred to as the **deep groove** or conrad bearing. Available in many variations—single or double shields or seals. Normally used for radial and thrust loads (maximum two-thirds of radial).

Maximum Capacity (Fig. 8.5.4) The geometry is similar to that of a deep-groove bearing except for a **filling slot.** This slot allows more balls in the complement and thus will carry heavier radial loads. However, because of the filling slot, the thrust capacity in both directions is reduced drastically.

Double-Row (Fig. 8.5.5) This bearing provides for heavy radial and light thrust loads without increasing the OD of the bearing. It is approximately 60 to 80 percent wider than a comparable single-row bearing. Because of the filling slot, thrust loads must be light.

Internal Self-Aligning Double-Row (Fig. 8.5.6) This bearing may be used for primarily radial loads where **self-alignment** ($\pm 4°$) is required. The self-aligning feature should not be abused, as excessive misalignment or thrust load (10 percent of radial) causes early failure.

Angular-Contact Bearings (Fig. 8.5.7) These bearings are designed to support **combined radial and thrust** loads or heavy thrust loads depending on the contact-angle magnitude. Bearings having large contact angles can support heavier thrust loads. They may be mounted in pairs (Fig. 8.5.8) which are referred to as **duplex bearings:** back-to-back, tandem, or face-to-face. These bearings (ABEC-7 or ABEC-9) may be preloaded to minimize axial movement and deflection of the shaft.

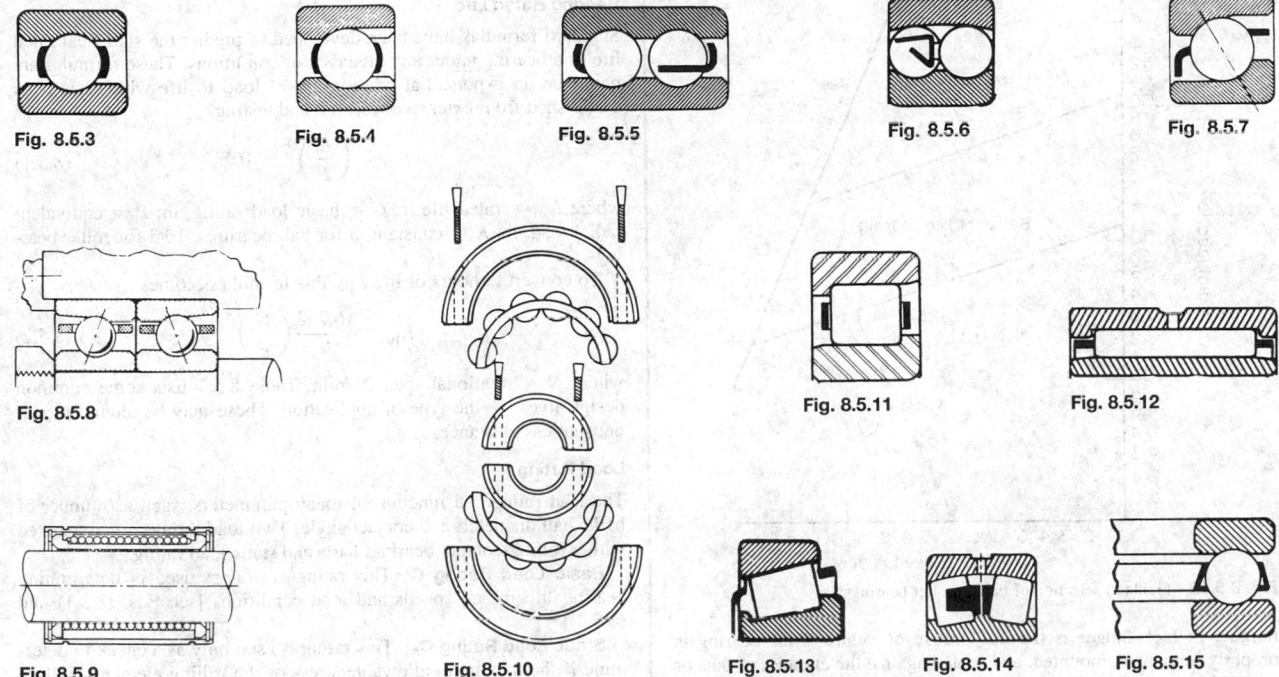

Fig. 8.5.3   Fig. 8.5.4   Fig. 8.5.5   Fig. 8.5.6   Fig. 8.5.7

Fig. 8.5.8   Fig. 8.5.11   Fig. 8.5.12

Fig. 8.5.9   Fig. 8.5.10   Fig. 8.5.13   Fig. 8.5.14   Fig. 8.5.15

Ball Bushings (Fig. 8.5.9)   This type of bearing is used for linear motions on hardened shafts (Rockwell C 58 to 64). Some types can be used for linear and rotary motion.

Split-Type Ball Bearing (Fig. 8.5.10)   This type of ball or roller bearing has split inner ring, outer ring, and cage. They are assembled by screws. This feature is expensive but useful where it is difficult to install or remove a solid bearing.

Roller Bearings

Cylindrical Roller (Fig. 8.5.11)   These bearings utilize cylinders with approximate length/diameter ratio ranging from 1:1 to 1:3 as rolling elements. Normally used for heavy radial loads. Especially useful for free axial movement of the shaft. Highest speed limits for roller bearings.

Needle Bearings (Fig. 8.5.12)   These bearings have rollers whose length is at least 4 times their diameter. They are most useful where space is a factor and are available with or without inner race. If shaft is used as inner race, it must be hardened and ground. Full-complement type is used for high loads, oscillating, or slow speeds. Cage type should be used for rotational motion. They cannot support thrust loads.

Tapered-Roller (Fig. 8.5.13)   These bearings are used for heavy radial and thrust loads. The bearing is designed so that all elements in the rolling surface and the raceways intersect at a common point on the axis: thus **true rolling** is obtained. Where maximum system rigidity is required, the bearings can be adjusted for a preload. They are available in double row.

Spherical-Roller (Fig. 8.5.14)   These bearings are excellent for heavy radial loads and moderate thrust. Their internal self-aligning feature is useful in many applications such as HVAC fans.

Thrust Bearings

Ball Thrust Bearing (Fig. 8.5.15)   It may be used for low-speed applications where other bearings carry the radial load. These bearings are made with shields, as well as the open type.

Straight-Roller Thrust Bearing (Fig. 8.5.16)   These bearings are made of a series of short rollers to minimize the skidding, which causes

twisting, of the rollers. They may be used for moderate speeds and loads.

Tapered-Roller Thrust (Fig. 8.5.17)   It eliminates the skidding that takes place with straight rollers but causes a thrust load between the ends of the rollers and the shoulder on the race. Thus speeds are limited because the roller end and race flange are in sliding contact.

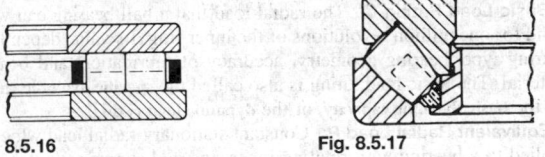

Fig. 8.5.16   Fig. 8.5.17

Selection of Ball or Roller Bearing

Selection of the type of rolling-element bearing is a function of many factors, such as load, speed, misalignment sensitivity, space limitations, and desire for precise shaft positioning. However, to determine if a ball or roller bearing should be selected, the following **general rules** apply:

1. Ball bearings function on theoretical point contact. Thus they are suited for higher speeds and lighter loads than roller bearings.

2. Roller bearings are generally more expensive except in larger sizes. Since they function theoretically on line contact, they will carry heavy loads, including shock, more satisfactorily, but are limited in speed.

Use Fig. 8.5.18 as a general guide to determine if a ball or roller bearing should be selected. This figure is based on a rated life of 30,000 h.

ROLLING-CONTACT BEARINGS' LIFE, LOAD, AND SPEED RELATIONSHIPS

An accurate knowledge of the load-carrying capacity and expected life is essential in the proper selection of ball and roller bearings. Bearings that are subject to millions of different stress applications fail owing to

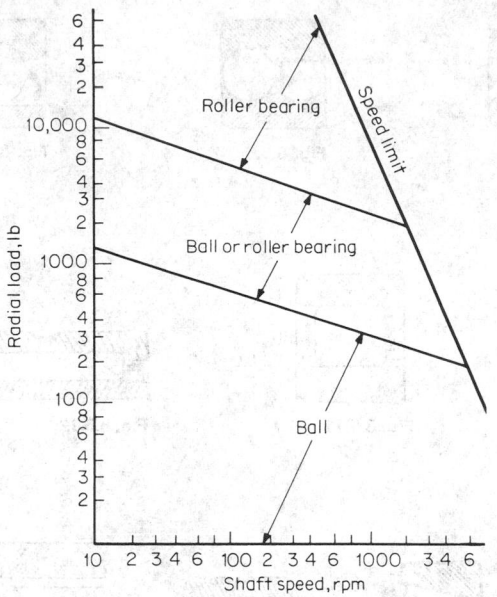

Fig. 8.5.18 Guide to selection of ball or roller bearings.

fatigue. In fact, **fatigue** is the only cause of **failure** if the bearing is properly lubricated, mounted, and sealed against the entrance of dust or dirt and is maintained in this condition. For this reason, the **life** of an individual bearing is defined as the total number of revolutions or hours at a given constant speed at which a bearing runs before the first evidence of fatigue develops.

Definitions

Rated Life L_{10} The number of revolutions or hours at a given constant speed that 90 percent of an apparently identical group of bearings will complete or exceed before the first evidence of fatigue develops; i.e., 10 out of 100 bearings will fail before rated life. The names **Minimum life** and L_{10} **life** are also used to mean rated life.

Basic Load Rating C The radial load that a ball bearing can withstand for one million revolutions of the inner ring. Its value depends on bearing type, bearing geometry, accuracy of fabrication, and bearing material. The basic load rating is also called the **specific dynamic capacity,** the **basic dynamic capacity,** or the **dynamic load rating**.

Equivalent Radial Load P Constant stationary radial load which, if applied to a bearing with rotating inner ring and stationary outer ring, would give the same life as that which the bearing will attain under the actual conditions of load and rotation.

Static Load Rating C_0 Static radial load which produces a maximum contact stress of 580,000 lb/in² (4,000 MPa).

Static Equivalent Load P_0 Static radial load, if applied, which produces a maximum contact stress equal in magnitude to the maximum contact stress in the actual condition of loading.

Bearing Rated Life

Standard formulas have been developed to predict the statistical rated life of a bearing under any given set of conditions. These formulas are based on an exponential relationship of load to life which has been established from extensive research and testing.

$$L_{10} = \left(\frac{C}{P}\right)^K \times 10^6 \qquad (8.5.1)$$

where L_{10} = rated life, r; C = basic load rating, lb; P = equivalent radial load, lb; K = constant, 3 for ball bearings, 10/3 for roller bearings.

To convert to hours of life L_{10}, this formula becomes

$$L_{10} = \frac{16,700}{N}\left(\frac{C}{P}\right)^K \qquad (8.5.2)$$

where N = rotational speed, r/min. Table 8.5.1 lists some common design lives vs. the type of application. These may be altered to suit unusual circumstances.

Load Rating

The **load rating** is a function of many parameters, such as number of balls, ball diameter, and contact angle. Two load ratings are associated with a rolling-contact bearing: **basic** and **static** load rating.

Basic Load Rating C This rating is *always* used in determining bearing life for all speeds and load conditions [see Eqs. (8.5.1) and (8.5.2)].

Static Load Rating C_0 This rating is used only as a check to determine if the maximum allowable stress of the rolling elements will be exceeded. It is *never* used to calculate bearing life.

Values for C and C_0 are readily attainable in any bearing manufacturer's catalog as a function of size and bearing type. Table 8.5.2 lists the basic and static load ratings for some common sizes and types of bearings.

Equivalent Load

There are two **equivalent-load** formulas. Bearings operating with some finite speed use the equivalent radial load P in conjunction with C [Eq. (8.5.1)] to calculate bearing life. The static equivalent load is used in comparison with C_0 in applications when a bearing is highly loaded in a static mode.

Equivalent Radial Load P All bearing loads are converted to an equivalent radial load. Equation (8.5.3) is the general formula used for both ball and roller bearings.

$$P = XR + YT \qquad (8.5.3)$$

where P = equivalent radial loads, lb; R = radial load, lb; T = thrust (axial) load, lb; X and Y = radial and thrust factors (Table 8.5.3). The empirical X and Y factors in Eq. (8.5.3) depend upon the geometry, loads, and bearing type. Average X and Y factors can be obtained from Table 8.5.3. Two values of X and Y are listed. The set $X_1\,Y_1$ or $X_2\,Y_2$ giving the largest equivalent load should always be used.

Static Equivalent Load P_0 The static equivalent load may be compared directly to the static load rating C_0. If P_0 is greater than the C_0

Table 8.5.1 Design-Life Guide

Application	Design life, h, L_{10}	Application	Design life, h, L_{10}
Agricultural equipment	3,000–6,000	Domestic appliances	1,000–2,000
Aircraft engines	1,000–3,000	Electric motors:	
Aircraft jet engines	1,500–4,000	Domestic	1,000–2,000
Automotive:		Industrial	20,000–30,000
Bus, car	2,000–5,000	Elevator	8,000–15,000
Trucks	1,500–2,500	Fans:	
Blowers:	20,000–30,000	Industrial	8,000–15,000
Continuous 8-h service	20,000–40,000	Mine ventilation	40,000–50,000
Continuous 24-h service	40,000–60,000	Gearing units (multipurpose)	8,000–15,000
Continuous 24-h service (extreme reliability)	100,000–200,000	Intermittent service	8,000–15,000
Compressors	40,000–60,000	Paper machines	50,000–60,000
Conveyors	20,000–40,000	Pumps	40,000–60,000

Table 8.5.2 Approximate Basic and Static Load Ratings vs. Types and Sizes
(Ratings are in pounds) 1 lb = 4,448 N

Bearing bore, mm	Ball single-row 200 series		Ball single-row 300 series		Ball double-row 200 series		Roller cylindrical 300 series		Roller spherical 22200 series	
	C_0	C	C_0	C	C_0	C	C_0	C	C_0	C
10	600	1,040	850	1,430	800	1,210	1,020	1,960		
12	680	1,180	1,040	1,650	1,250	1,820	1,350	2,540		
15	780	1,330	1,220	1,970	1,430	2,030	1,520	2,820		
17	1,000	1,660	1,470	2,340	1,840	2,510	2,070	3,700		
20	1,390	2,220	1,760	2,730	2,540	3,480	2,560	4,490		
25	1,560	2,420	2,350	3,550	2,858	3,780	3,720	6,360		
30	2,250	3,360	3,120	4,600	4,110	5,140	5,070	8,460		
35	3,070	4,430	4,020	5,770	5,600	6,700	6,400	10,400		
40	3,520	5,040	5,020	7,060	6,430	7,680	7,930	12,500	11,800	15,200
45	4,000	5,660	6,130	8,430	7,320	8,620	9,310	14,700	12,600	15,900
50	4,450	6,070	8,010	10,750	8,130	9,220	11,600	17,900	13,600	16,800
55	5,630	7,500	9,400	12,410	10,300	11,400	12,600	19,100	16,500	20,300
60	6,950	9,070	10,902	14,179	12,700	13,800	15,200	22,800	20,800	25,200
65	7,660	9,900	12,516	16,051	14,000	15,000	19,900	29,000	25,500	30,200
70	8,410	10,714	14,240	18,030	15,400	16,300	21,400	30,800	27,500	31,900
75	9,190	11,610	16,080	19,600	16,900	17,300	23,200	32,900	29,100	33,100
80	10,010	12,550	18,020	21,230	18,300	19,100	27,000	38,100	32,100	36,800
85	11,750	14,490	20,080	22,880	19,500	19,700	30,900	43,300	38,200	43,200
90	13,630	16,540	22,250	24,580	22,100	22,600	35,200	48,800	44,500	49,800
95	15,650	18,740	24,530	26,300	28,600	28,600	39,500	54,200	48,800	54,700
100	17,800	21,130	29,430	29,940	32,500	32,100	44,700	60,800	55,700	61,900
110	20,100	23,000	32,040	31,800	30,500	30,700	53,200	70,500	72,000	78,400

rating, permanent deformation of the rolling element will occur. Calculate P_0 as follows:

$$P_0 = X_0 R = Y_0 T \qquad (8.5.4)$$

where P_0 = static equivalent load, lb; X_0 = radial factor (see Table 8.5.4); Y_0 = thrust factor (see Table 8.5.4); R = radial load, lb; T = thrust (axial) load, lb. If a load higher than the basic static load rating is

Table 8.5.3 Radial and Thrust Factors

Bearing type	X_1	Y_1	X_2	Y_2
Single-row ball	1.0	0.0	0.56	1.40
Double-row ball	1.0	0.75	0.63	1.25
Cylindrical roller	1.0	0.0	1.0	0.0
Spherical roller	1.0	2.5	0.67	3.7

imposed while rotating, the deformation is distributed evenly and no practical impairment occurs until the deformation becomes quite large. Some equipment operates with loads greatly exceeding the static capacity, such as bearings supporting artillery (twice static capacity), or aircraft control pulleys (four times static capacity). The load which will fracture a bearing is approximately eight times the static load rating.

Table 8.5.4 Radial and Thrust Factors

Type of bearing	X_0	Y_0
Single-row ball	0.6	0.5
Double-row ball	0.6	0.5
Spherical roller, 22200 series	1.0	2.9
Cylindrical roller	1.0	0.0

Oscillating loads, where the motion is such that the rolling element rotates less than half a revolution, approach static load conditions. This type of load is conducive to rapid **false brinelling** and requires special lubrication techniques.

Required Capacity

The basic load rating C is very useful in the selection of the type and size of bearing. By calculating the required capacity needed for a bearing in a certain application and comparing this with known capacities, a bearing can be selected. To calculate the required capacity, the following formula can be used:

$$C_r = \frac{P(L_{10}N)^{1/K}}{Z} \qquad (8.5.5)$$

where C = required capacity, lb; L_{10} = rated life, h; P = equivalent radial load, lb; K = constant, 3 for ball bearings, 10/3 for roller bearings; Z = constant, 25.6 for ball bearings, 18.5 for roller bearings; N = rotation speed, r/min.

LIFE ADJUSTMENT FACTORS

Modifications to Eq. (8.5.2) can be made, based on a better understanding of causes of fatigue. Influencing factors include
  1. Reliability factors for survival rates greater than 90 percent
  2. Improved raw materials and manufacturing processes for ball bearing rings and balls.
  3. The beneficial effects of elastrohydrodynamic lubricant films

Equation (8.5.2) can be rewritten to reflect these influencing factors:

$$L_{10} \text{ modified} = A_1 A_2 A_3 \frac{16,700}{N} \left(\frac{C}{P}\right)^K \qquad (8.5.6)$$

where A_1 = statistical life reliability factor for a chosen survival rate, A_2 = life-modifying factor reflecting bearing material type and condition, and A_3 = elastohydrodynamic lubricant film factor.

Factor A_1

Reliability factors listed in Table 8.5.5 represent rates from 90 to 99 percent. Using rates other than 90 percent will yield larger bearings for equivalent loads and, therefore, will increase costs. Rates higher than 90 percent should be used only when absolutely necessary.

Table 8.5.5 Reliability Factor A_1 for Various Survival Rates

Survival rate, %	Bearing life notation	Reliability factor A_1
90	L_{10}	1.00
95	L_5	0.62
96	L_4	0.53
97	L_3	0.44
98	L_2	0.33
99	L_1	0.21

Factor A_2

While not formally recognized by AFBMA, estimated A_2 factors are commonly used as represented by the values in Table 8.5.6. The main considerations in establishing A_2 values are the material type, melting procedure, mechanical working and grain orientation, and hardness.

Table 8.5.6 Life-Modifying Factor A_2

Material	Factor A_2
A1S1 440C, Air Melted	0.025
SAE 52100, Vacuum Processed	1.0
M50 VIM-VAR Melted	2.0

Factor A_3

This factor is based on elastohydrodynamic lubricant film calculations which relate film thickness and surface finish to fatigue life. A factor of 1 to 3 indicates adequate lubrication, with 1 being the minimum value for which the fatigue formula can still be applied. As A_3 goes from 1 to 3, the life expectancy will increase proportionately, with 3 being the largest value for A_3 that is meaningful. If A_3 is less than 1, poor lubrication conditions are presumed. Calculations for A_3 are beyond the scope of this section.

Speed Limits

Many factors combine to determine the limiting speeds of ball and roller bearings. It depends on several factors, like bearing size, inner- or outer-ring rotation, contacting seals, radial clearance and tolerances, operating loads, type of cage and cage material, temperature, and type of lubrication. A convenient check on speed limits can be made from a dn value. The dn value is a direct function of size and speed and is dependent on type of lubrication. It is calculated by multiplying the bore in millimeters (mm) by the speed in r/min.

$$dn = \text{bore (mm)} \times \text{speed (r/min)} \qquad (8.5.7)$$

A guide for dn values is listed in Table 8.5.7. When these values are exceeded, bearing life is shortened. The values are only a guide for approaching difficulties and can be exceeded by special bearings, lubrication, and application.

Table 8.5.7 dn Values vs. Bearing Types

Bearing type	Series	Max dn value	
		Grease	Oil
Single-row ball	100, 200, 300, 400, 30, in	200,000	300,000
Double-row ball	200, 300	160,000	220,000
Cylindrical roller	200, 300	150,000	200,000
Spherical roller	22200	120,000	170,000

Friction

One of the assets of rolling-contact bearings is their low friction. The **coefficient of friction** varies appreciably with the type of bearing, load, speed, lubrication, and sealing element. For rough calculations the fol-

lowing coefficients can be used for normal operating conditions and favorable lubrication:

Single-row ball bearings	0.0015
Roller bearings	0.0018

Excess grease, contact seals, etc., will increase these values, and allowances should be made.

PROCEDURE FOR DETERMINING SIZE, LIFE, AND BEARING TYPE

Basically, three common situations may be encountered in the analysis of a bearing system; bearing-size selection, bearing-type selection, and bearing-life determination. Each of these problems requires the following conditions to be known; radial load, thrust load, and speed. The static load capacity is not considered in the following procedures but should be analyzed if the bearing rotational speed is slow or if the bearing is idle for a period of time.

Bearing Size Selection

Known type and series:
1. Select desired design life (Table 8.5.1).
2. Calculate equivalent radial load P [Eq. (8.5.3)].
3. Calculate required capacity C_r [Eq. (8.5.5)].
4. Compare C_r with capacities C in Table 8.5.2. Select first bore size having a capacity C greater than C_r.
5. Check bearing speed limit [Eq. (8.5.7)].

Bearing-Type Selection

Known bore size and life:
1. Select ball or roller bearing (Fig. 8.5.18).
2. Calculate equivalent load P [Eq. (8.5.3)] for various bearing types (conrad, spherical, etc.).
3. Calculate C_r [(Eq. (8.5.5)].
4. Compare C_r with capacities C in Table 8.5.3, and select the type that has a capacity equal to or greater than C_r.
5. Check bearing speed limit [Eq. (8.5.7)].

Bearing-Life Determination

Known bearing size:
1. Select ball or roller bearing (Fig. 8.5.18).
2. Calculate equivalent radial load P [Eq. (8.5.3)].
3. Select basic load rating C from Table 8.5.3.
4. Calculate rated life L_{10} [Eq. (8.5.1) or (8.5.2)].
5. Check calculated life with design life.

BEARING CLOSURES

Rolling-element bearings are made with a wide variety of **closures**. Basically, they are open, shielded, or sealed (Figs. 8.5.19 and 8.5.20). **Shielded bearings** have a small clearance between the stationary shield and rotating ring. This provides reasonable exclusion of dirt without an

Fig. 8.5.19

Fig. 8.5.20

increase in friction. **Sealed bearings** have a flexible lip (usually synthetic rubber) in contact with the inner ring. Friction is increased, but more effective retention of lubricant and exclusion of dirt is obtained. Seals should not be used to seal a fluid head or at high speeds.

BEARING MOUNTING

Correct **mounting** of a rolling-contact bearing is essential to obtain its rated life. Many types of mounting methods are available. The selection of the proper method is a function of the accuracy, speed, load, and cost of the application. The most common and best method of bearing retention is a press fit against a shaft shoulder secured with a locknut. End caps are used to secure the bearing against the housing shoulder (Fig. 8.5.21). Retaining rings are also used to fix a bearing on a shaft or in a housing (Fig. 8.5.22). Each shaft assembly normally must provide for expansion by allowing one end to float. This can be accomplished by

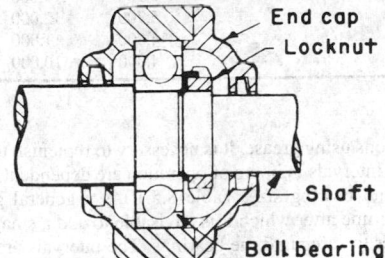

Fig. 8.5.21

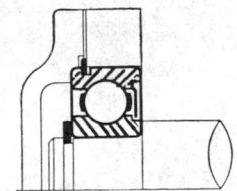

Fig. 8.5.22

allowing the bearing to expand linearly in the housing or by using a straight roller bearing on one end. Care must be exercised when designing a **floating installation** because it requires a slip fit. An excessively loose fit will cause the bearing to spin on the shaft or in the housing.

Table 8.5.8 lists shaft and housing tolerances for press fits with ABEC 1 precision applications (pumps, gear reducers, electric motors, etc.) and ABEC 7 precision applications (grinding spindles, etc.).

Table 8.5.8   Shaft and Housing Tolerances for Press Fit

Bearing bore, mm	Shaft tolerances, in, ABEC 1 precision	Bearing bore, mm	Shaft tolerances, in, ABEC 7 precision
4–6	+ 0.0000 − 0.0002	4–30	+ 0.00000 − 0.00015
7–17	+ 0.0000 − 0.0003	35–50	+ 0.0000 − 0.0002
20–50	+ 0.0000 − 0.0004	55–80	+ 0.0000 − 0.0003
55–80	+ 0.0000 − 0.0005	85–120	+ 0.00000 − 0.00035
85–120	+ 0.0000 − 0.0006		

Bearing OD, mm	Housing tolerances, in, ABEC 1 precision	Bearing OD, mm	Housing tolerances, in, ABEC 7 precision
16–30	+ 0.0008 − 0.0000	16–80	+ 0.0002 − 0.0000
32–47	+ 0.0010 − 0.0000	85–120	+ 0.0003 − 0.0000
52–80	+ 0.0012 − 0.0000	125–225	+ 0.0004 − 0.0000
85–120	+ 0.0014 − 0.0000		
125–180	+ 0.0016 − 0.0000		

Mounted units such as **ball-bearing pillow blocks** (Fig. 8.5.23) are frequently used for fans and conveyors. Three common methods are used to attach the bearing to the shaft; setscrew, eccentric locking collar, and taper-sleeve adapter.

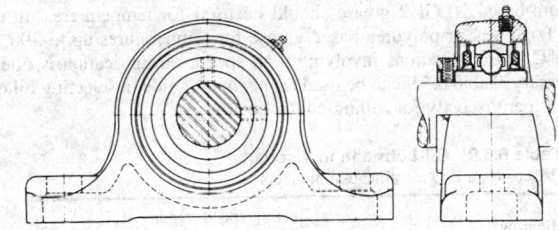

Fig. 8.5.23

Setscrew   Figure 8.5.24 illustrates the use of an extended inner-ring bearing held to the shaft with a setscrew. This is a simple method and is suitable only for lightly loaded bearings.

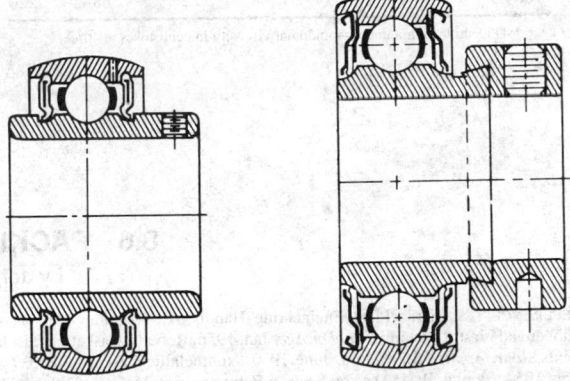

Fig. 8.5.24   Fig. 8.5.25

Eccentric Locking Collar   Figure 8.5.25 illustrates the use of an extended inner-ring bearing held to the shaft with an eccentric collar. This method tends to keep the shaft centered in the bearing more concentrically than the setscrew method. It is suitable for light to moderate loads.

Taper-Sleeve Adapter   Figure 8.5.26 illustrates the use of a taper-sleeve adapter to mount the bearing on the shaft. It provides uniform concentric contact between the shaft and bearing bore. However, skill is required to tighten the locking nut enough to keep the sleeve from spinning on the shaft and yet not so tight that the inner race of the bearing is expanded to the point where the clearance is removed from the bearing. It is very difficult to obtain the correct setting with light-series bearings. They are excellent for heavy-duty spherical roller bearings.

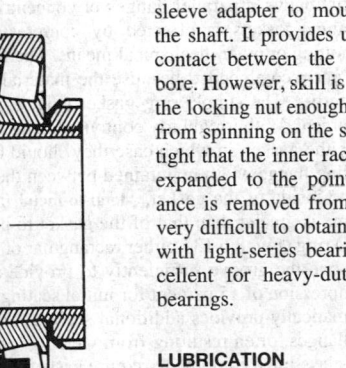

Fig. 8.5.26

LUBRICATION

Rolling-contact bearings need a **fluid lubricant** to obtain or exceed their rated life. In the absence of high-temperature environment, only a small amount of lubricant is required for excellent performance. Excess lubricant will cause heating of the bearing and accelerate the deterioration of the lubricant. Optimum lubrication of rolling-contact bearings can be predicted by elastohydrodynamic theory (EHD). It has been shown that film thickness is sensitive to bearing speed of operation and lubricant viscosity properties and, moreover, that the film thickness is virtually insensitive to load.

Grease is commonly used for lubrication of rolling-contact bearings because of its convenience and minimum maintenance. A high-quality lithium-based NLGI 2 grease should be used for temperatures up to 180°F (82°C), or polyurea-based grease for temperatures up to 300°F (150°C). In applications involving high speed, oil lubrication is often necessary. Table 8.5.9 can be used as a general guide in selecting oil of the proper viscosity for rolling-contact bearings.

Table 8.5.9 Oil-Lubrication Viscosity
(Viscosity in ISO identification numbers*)

Bearing bore, mm	Bearing speed, r/min				
	10,000	3,600	1,800	600	50
4–7	68	150	220		
10–20	32	68	150	220	460
25–45	10	32	68	150	320
50–70	7	22	68	150	320
75–90	3	10	22	68	220
100	3	7	22	68	220

* ISO identification number = midpoint viscosity in centistokes at 40°C.

Table 8.5.10 Ball-Bearing Grease Relubrication Intervals
(Hours of operation)

Bearing bore, mm	Bearing speed, r/min				
	5,000	3,600	1,750	1,000	200
10	8,700	12,000	25,000	44,000	220,000
20	5,500	8,000	17,000	30,000	150,000
30	4,000	6,000	13,000	24,000	127,000
40	2,800	4,500	11,000	20,000	111,000
50		3,500	9,300	18,000	97,000
60		2,600	8,000	16,000	88,000
70			6,700	14,000	81,000
80			5,700	12,000	75,000
90			4,800	11,000	70,000
100			4,000	10,000	66,000

In applications using grease, it is necessary to replenish the lubricant. Relubrication intervals in hours of operation are dependent on temperature, speed, and bearing size. Table 8.5.10 is a general guide which represents the time after which it is advisable to add a small amount of grease in order to safeguard the bearings. The intervals are valid up to 160°F (71°C) and should be divided by 2 for cylindrical roller bearings and by 10 for spherical roller bearings.

8.6 PACKINGS AND SEALS
by John W. Wood, Jr.

REFERENCES: Staniar, "Plant Engineering Handbook," McGraw-Hill. Thorn, Rubber and Plastic Packings, *Rubber Age,* Jan. 1956. Roberts, Gaskets and Bolted Joints, *Jour. Applied Mechanics,* June 1950. Nonmetallic Gaskets, *Mach. Des.,* Nov. 1954. Elonka, Basic Data on Seals, a *Power* reprint, McGraw-Hill. Fluidtec Engineered Products, Training Manuals.

Packings are materials used to control or stop leakage of fluids (liquids and/or gases) or solid dry products through mechanical clearances when the contained material is under static or dynamic pressure.

Gaskets are compressible materials installed in static clearances which normally exist between parallel flanges or concentric cylinders. Sealing of flat flange gaskets is effected by compressive loading achieved through bolting or other mechanical means. The full face gasket (Fig. 8.6.1) is not recommended because the material outside the bolt holes is ineffective. The simple ring gasket (Fig. 8.6.2) is more efficient and economical. With irregularly contoured flanges, bolt holes may serve to locate the gasket, in which case they should be placed in lobes with full sealing flange width maintained between the inner edge of the holes and the inside of the gasket. **Metal-to-metal** fits require a recess whose volume is greater than that of the gasket to be used. The gasket, such as an O ring (Fig. 8.6.13), either rectangular or round cross section, extends above the groove sufficiently to provide a minimum cross-sectional compression of 15 percent for initial seating. In service, the fluid load automatically provides additional sealing force. **Warped, wavy,** or irregular flanges, often resulting from welding, other fabrication, or as found in glass-lined equipment, require gaskets that are softer or thicker than normal in order to compensate for surface imperfections. Excessive thickness or volume of gasket material, even though the gasket is installed in a groove, must be avoided to prevent distortion or "mushrooming," which will result in inadequate loading. Tongue and groove joints (Fig. 8.6.4) confine the gasket material and may adapt to the extra thickness, within limits.

In addition to the types (Figs. 8.6.5 to 8.6.7) shown, as defined in the table (Fig. 8.6.37), there are the machined metal profile gasket (Fig. 8.6.8) and solid metal designs in flat, round, and either octagonal or oval API ring joint gaskets for extreme pressures and temperatures to seal against steam, oil, and gases. These types have very low compressibilities, and their behavior depends on their cross sections. The envelope gasket (Fig. 8.6.3), usually polytetrafluoroethylene with a variety of cores, is particularly useful for extremely corrosive or noncontaminating service under average pressure.

Cylindrical or **concentric** gasketing uses a retaining gland follower and is mechanically loaded, e.g., the standard mechanical joint for cast-iron pipe (Fig. 8.6.10) or the condenser tube-sheet ferrule (Fig. 8.6.11). Cup-shaped gaskets are designed to be self-tightening under pressure (Fig. 8.6.12). The O ring (Fig. 8.6.13) located in an annular groove and precompressed as in the grooved flange, is a self-energized gasket. A cylindrical ring with internal single lip or double lips, also automatic in action, is quite common in pipe joints.

Beyond these types are many specialty gaskets designed for specific or proprietary use, e.g., a seal for a removable drumhead.

The compressibility of various gasketing materials is shown in Fig. 8.6.37, and their common usage is listed in Table 8.6.1. Beyond rubber are many elastomeric materials generally similar in mechanical behavior but varying as to temperature limits and fluid compatibility (see Sec. 6).

The **proper design** of a gasketed joint requires flange rigidity to avoid distortion, surface finish commensurate with gasket type and good sealing pressure, and adequate bolt loading. The load must seat the gasket, i.e., cause the material to flow into and fill flange irregularities. It must seal sufficiently that the residual fluid pressure on the gasket exceeds the pressure of the fluid being contained. These values, known respectively as the *seating load y* in lb/in^2 and the *gasket factor m,* vary with gasket material and thickness. The ASME Code for Unfired Pressure Vessels, section VIII, gives sufficient detail for typical joint design and tabulates values for y and m for various gasketing materials.

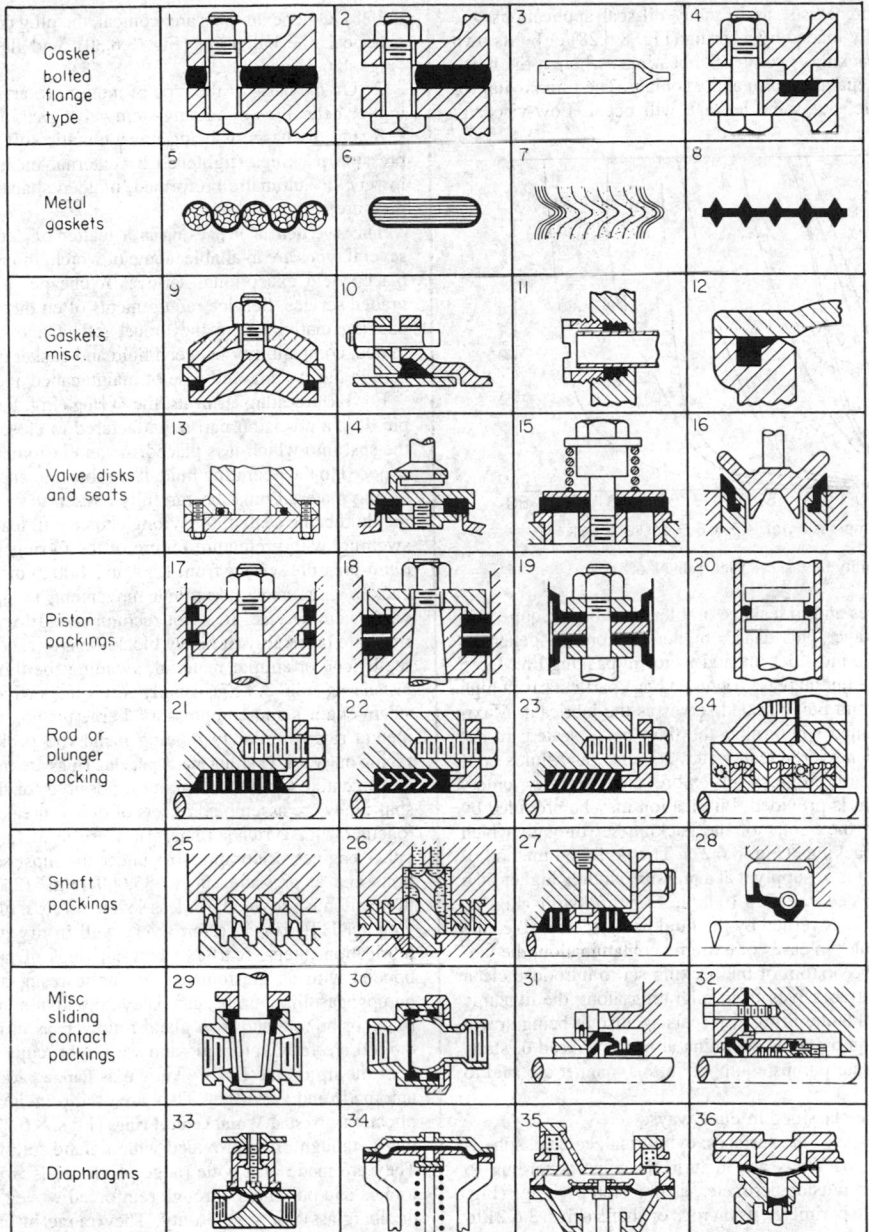

Fig. 8.6.1–8.6.36 Packings.

High bolt loading is desirable for tight and enduring gasket joints, but it must not crush the gasket material. Crushing-strength values, which will vary with thickness and temperature, can be obtained from the gasket manufacturers. Consistent with the condition of the flanges, the thinner the gasket, the more efficient the joint.

Data on the design of **O-ring joints** are available from suppliers. The nominal pressure limit for O rings, based on typical mechanical clearances, is 1,500 lb/in² (10 MN/m²) without backup rings and 3,000 lb/in² (20 MN/m²) with backup rings. If clearances can be eliminated, as in a flanged joint with close metal-to-metal contact, no limit can be set. Other self-energizing joints, such as the boiler hand-hole plate (Fig. 8.6.9), need only sufficient load to effect an initial seal.

Valve disks are specialized gaskets designed for joints that are frequently broken and reseated. Disks for globe valves (Fig. 8.6.14) are usually encased in a disk holder with a swivel mounting, which ensures precise reseating without abrasion during the closing and opening cycles. They are made of firm rubber for bib washers, hard rubber and phenolics for more severe service, and plastics such as nylon and polytetrafluoroethylene. Pump valves (Fig. 8.6.15) are described in Sec. 14. Rubber **valve seats** are used with metal valve disks on some pumps, e.g., the rotary drilling pump valve (Fig. 8.6.16). Plastics are also used for seats, notably in ball valves.

Dynamic packings include all packings that operate on moving surfaces. To retain fluid under pressure, they are subjected to the hydraulic load. When no pressure exists, as in many oil-seal applications, the packing is mechanically loaded by a spring (Fig. 8.6.28) or by its own resiliency. Dynamic packings operate like bearings, wherein the lubricant serves as both a separating film and a coolant. The film is vital for satisfactory service life, but some leakage will occur. Low-viscosity

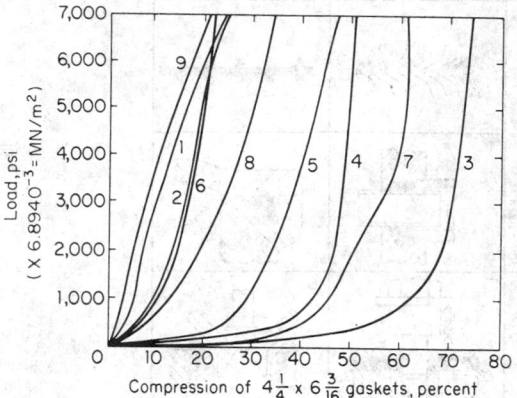

Fig. 8.6.37   Compressibility of gaskets. (See Table 8.6.1.)

fluids and high pressures add to leakage problems, as both require thin films to minimize leakage. This causes higher friction and generates heat, which is the single most detrimental factor in packing life. Deep packings reduce leakage but increase frictional heat, particularly at high speeds. Normally the fluid being sealed serves as the lubricant. Maximum efficiency is attained when oil is the fluid being sealed; in decreasing order of efficiency are clean water, solvents, and fluids containing solids. These are progressively more unsatisfactory unless supplemental lubrication is provided. Lubrication may be provided by using a lantern ring in the center of the packing set through which lubricant is fed to the packings (Fig. 8.6.27). The preferred method of introducing the lubricant is to supply it at a pressure slightly higher than that of the fluid being sealed, say, 5 to 10 lb/in^2 (3.5 to 7 kN/m^2) higher. The choice of lubricant is governed by the fluid being sealed, since the two should be compatible. In cases of extreme contamination, the lantern ring is moved to the bottom of the packing set to introduce clean lubricant and to prevent abrasives from migrating along the dynamic sealing surface. Use of a lantern ring also seals against air being drawn into the system when equipment is operating at negative head or starts under vacuum. Centrifugal pumps equipped in this manner are said to have a **water seal**.

Dynamic packings are classified in three ways:

1. On the basis of shape of the surfaces: cylindrical, conical, spherical, or flat. Cylindrical packings are in turn classified according to whether they pack on the outside perimeter, as in piston packings (Fig. 8.6.17 to 8.6.20) or inside perimeter, as on rods or shafts (Figs. 8.6.21 to

8.6.28). Other examples are: conical, the plug cock lining (Fig. 8.6.29); spherical, the ball joint (Fig. 8.6.30); and flat, the mechanical seals (Figs. 8.6.31 and 8.6.32).

2. On the basis of the type of motion: rotary, oscillating, reciprocating, or helical (as in a rising-stem valve packing).

3. On the basis of being nonautomatic soft or jamb packings/compression packings (tightened by external means, usually a gland follower); or automatic preformed, molded shapes (self-tightening under pressure).

The **selection of a packing** is a matter of economics. In most cases several types are available, some of which, though expensive in the first place, yield exceptional service. A cheaper packing could yield degraded service. Service requirements often dictate the final choice of a packing material, and they must reflect and balance the fluid being sealed, compatibility between fluid and gasket material, operating pressure in the system, and ease of maintenance and replacement.

For **reciprocating elements**, the **O ring** (Fig. 8.6.20) is extremely simple. It is a precision part manufactured to close tolerances, as must be the seat into which it is placed. As an elastomeric material completely exposed to the operating fluid, it is subject to chemical degradation. The O-ring material must be carefully chosen to ensure compatibility with the fluid being sealed; the wrong choice will lead to either shrinkage or swelling, with premature failure of the O ring. It is best suited to medium-pressure service from 1,500 to 3,000 lb/in^2 (10 to 20 MN/m^2) with backup rings and intermittent movement, as in hydraulic cylinder or valve stem service. It is not recommended for pump service. Backup rings are preferably of heavy blocklike cross section in either tetrafluoroethylene or similar material, avoiding the thin spiral type. The split **piston ring** (Fig. 8.6.17), usually cast iron, is widely used in gas, oil, and steam engines and compressors. Large pistons frequently employ segmental rings similar to floating metal rod packings (Fig. 8.6.24) but facing outward. **Floating metal packing rings** are made of numerous radial or tangential segments, making it possible for them to contract on the shaft; they are assembled in sets of two to break the joints and are held together with garter springs. They are used for steam, gas, or air, in either engines or compressors under the most severe operating conditions and at pressures up to 35,000 lb/in^2 (241 MPa). Normally no oil lubrication is provided; for less severe service, filled polytetrafluoroethylene (PTFE) rings perform very well in dry gases without auxiliary lubrication. Step-, scarf-, or butt-cut rings of laminated cotton fabric, bonded with an elastomer or phenolic resin, are employed in water pumps, gasoline pumps, etc. They may float similar to cast iron piston rings, or be retained by a gland follower, as in Fig. 8.6.18. **Cups** (Fig. 8.6.19) are fully automatic and very tight; cups in their inverted form, with the lip on the ID, are known as **flange packings** and are also fully automatic and very tight. They are used principally for slow-speed applications. **Nested V and conical rings** (Figs. 8.6.22 and 8.6.23) are automatic, though often provided with a gland follower to effect initial fit. They are made of a wide range of materials from homogeneous elastomers and polymers, through reinforced woven fibers (cotton, aramid, or fiberglass) for severe duty. They range in hardness from soft and

Table 8.6.1   Common Usage of Gasketing Materials*

No.	Type	Service principally for:	Thickness tested, in ($\times 25.4 = $ mm)
1	Sheet rubber	Water	1/16
2	Cloth inserted sheet	Water	1/16
3	Cork composition	Oil, low-pressure	1/8
4	Gasket paper	Oil, low-pressure	1/16
5	Rubberized asbestos cloth (Fig. 8.6.9)	Hot water (boiler manholes, etc.)	1/4
6	Compressed asbestos sheet	All services up to 750°F (400°C)	1/16
7	Corrugated sheet metal with filling (Fig. 8.6.5)	Steam, oil at high temperatures	1/4
8	Metal jacket over asbestos center (Fig. 8.6.6)	Steam, oil at high temperatures	1/8
9	Spirally wound steel strip with intervening asbestos layers (Fig. 8.6.7)	Steam, oil at high temperatures	3/16

* Asbestos bearing material is found generally in older equipment; current, new, and/or replacement parts are compounded of other materials suitable to the service application. Fiberglass is a common substitute for asbestos in these applications.

flexible to semirigid. Use of multiple rings allows them to be of the cut or split type for ease of installation and replacement. **Soft or jamb packings** are best suited for rod or plunger service, since an adjustable gland follower (Fig. 8.6.21) is required. They are normally formed in rectangular section with a butt joint staggered from ring to ring at installation. Many materials are employed, such as braided flax saturated with wax or viscous lubricants for water and aqueous solutions; braided fiberglass similarly treated or often impregnated with PTFE/graphite suspensoid for more severe service; laminated rubberized cotton fabric for hot water, low-pressure steam, and ammonia; rolled rubberized fiberglass or aramid fabric for steam; and rolled or twisted metal foil for high-temperature and high-pressure conditions. Packings containing woven or braided fibers are also made from wire-inserted yarns to gain additional strength. For pipe expansion joints, see Sec. 8.

Rotary shafts are generally packed with adjustable soft packings, with the notable exception of the mechanical seals (Figs. 8.6.31 and 8.6.32); where pressures are low, nested V or conical styles may be used. At zero or negligible pressures, the oil seal, a spring-loaded flange packing (Fig. 8.6.28), is very widely used. Where some leakage can be tolerated, the labyrinth (Fig. 8.6.25) and controlled-gap seals are used, particularly on high-speed equipment such as steam and gas turbines. **Soft packings** are of the same general type as those used for reciprocating service, with the fiber braid lubricated with grease and graphite or with polytetra-fluoroethylene fibers and suspensoid. Aramid, carbon, and graphite fibers filled with various lubricants and reinforcements are used at higher speeds and fluid pressures. Fiber braid with PTFE suspensoid is widely applied on valve stems operating below 500°F (260°C) and on centrifugal pumps. This material is an insulator, however, and results in high heat buildup on the dynamic surface; a better choice lies in use of a packing with better heat-transfer characteristics, such as one containing carbon or graphite. For continuous rotary service, **automatic packings** are best restricted to low pressure because their tightness under high pressure results in overheating. For intermittent service, as on valve stems, they are excellent.

Oil seals (Fig. 8.6.28) are unique flange packings having an elastomer lip generally bonded to a metal cup which is press-fitted into a smooth cylindrical bore. Basically, an oil seal is a flange packing with a flexible lip and a narrow contact area about 1/16 in (1.6 mm) wide which, under pressure, causes extreme local heating and wear. They are recommended only for nonpressure service and perform best in good lubricating media. To accommodate shaft runout up to 0.020 in (0.5 mm) depending on the rotating speed, the lip is spring-loaded with a coil spring or a finger spring. Coil springs are safer inasmuch as they are molded into the elastomer and are less likely to become dislodged and cause shaft damage. Since the lip is completely exposed to the sealed fluid, particular care should be taken to ensure compatibility between the elastomer and the fluid. Temperature is another operating condition which must be taken into consideration when one is using oil seals.

Mechanical, Rotary, or End Face Seals

The greatest advancements in the design of end face mechanical seals have come about in response to environmental regulations; requirements to minimize energy consumption and operating costs; safety; and concerns over loss of the product which is being sealed. The application of seals to replace packing in rotary equipment has increased dramatically and continues.

All end face mechanical seals (Figs. 8.6.31 and 8.6.32) consist of four parts: a stationary flat face, a rotating flat face, secondary sealing elements (usually elastomeric), and a flexible loading device. The assembled seal is placed and effects proper leak control. The two flat-face seal rings (one stationary, one rotating) rub and create the primary seal. Normally, the flat seal rings have different hardness values, and the soft one is narrower than the hard one. Secondary sealing elements prevent leakage between the rotating shaft and the rotating seal ring, and they block the leakage path around the outside of the stationary seal face. They also serve as gaskets between the assembled parts (i.e., gland plate and housing). The flexible loading device usually consists of one or more springs which press the flat seal rings together. Spring loading

ensures a seal when there is little or no hydraulic pressure available to press the faces together and helps maintain constant pressure between the faces as the soft (sacrificial) face wears down. The springs also act as vibration dampers to mitigate against the intrusion of transmitted vibrations, which may affect the efficient operation of the seal assembly.

Types of End Face Mechanical Seals

1. *Inside-mounted.* The seal head is mounted inside the stuffing box (Fig. 8.6.38a).

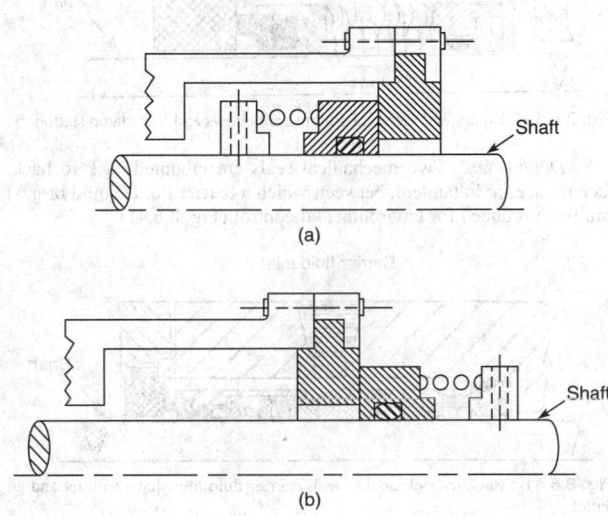

(a)

(b)

Fig. 8.6.38 Rotary end face seal. (*a*) Inside the seal chamber/stuffing box; (*b*) outside the seal chamber/stuffing box.

2. *Outside-mounted.* The seal head is mounted outside the stuffing box (Fig. 8.6.38b).

3. *Unbalanced seal.* The full hydraulic pressure in the seal chamber is transmitted to the seal faces (Fig. 8.6.39a).

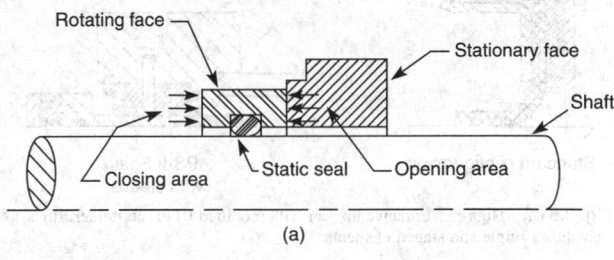

(a)

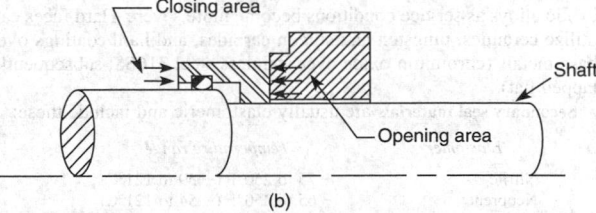

(b)

Fig. 8.6.39 Rotary end face seals showing (*a*) unbalanced and (*b*) balanced configurations.

4. *Balanced.* The seal elements are designed to reduce the hydraulic forces transmitted to the seal faces (Fig. 8.6.39b). A complete balance is not practical.

5. *Rotary seal.* In this design, the springs rotate with the shaft.

6. *Stationary seal.* In this design, the springs do not rotate with the shaft.

7. *Metal bellows.* Welded or formed metal bellows exert a spring load; there is no dynamic secondary seal element (Fig. 8.6.40).

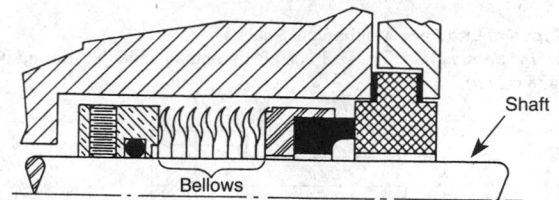

Fig. 8.6.40 Rotary end face seal with metal bellows and "t" clamp stationary.

8. *Double seal.* Two mechanical seals are mounted back to back, face to face, or in tandem, between which a barrier fluid (liquid or gas) can be introduced for environmental control (Fig. 8.6.41).

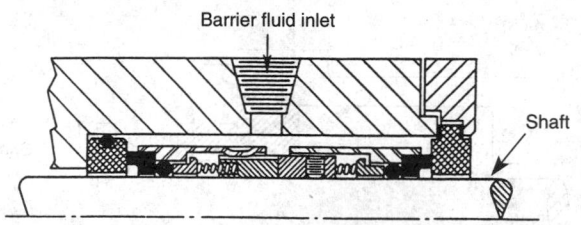

Fig. 8.6.41 Back-to-back double seal. Barrier fluid must have an inlet and an outlet.

End face mechanical seal materials must satisfy a number of design requirements, including chemical compatibility between the sealed fluid and the seal materials, ability of the seal materials to remain serviceable under the worst operating conditions, and ability to provide a reasonably long life in service at the operating conditions. Mating faces of the seals can be made from ordinary materials like bronze and PTFE

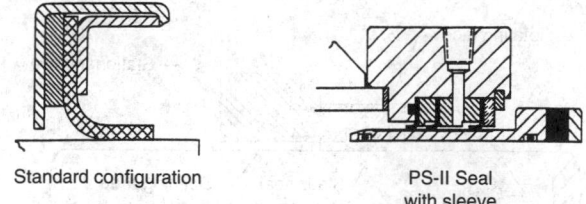

Fig. 8.6.42 High-performance lip seal with modified PTFE elastomer. Illustration shows single and staged elements.

for mild service on up to carbon, carbides, stainless steels, and other exotic alloys as service conditions become more severe. Hard faces can utilize ceramics, tungsten and silicon carbides, and hard coatings over base metals (chromium oxide over stainless steel 316SS, subsequently lapped flat).

Secondary seal materials are usually elastomeric and include these:

Elastomer	Temperature range
Nitrile	−75 to 250°F (−59 to 121°C)
Neoprene	−65 to 250°F (−54 to 121°C)
Viton	−40 to 450°F (−40 to 232°C)
Ethylene propylene	−65 to 300°F (−59 to 149°C)
Kalrez	32 to 550°F (−0 to 288°C)

Environmental controls as applied to seals are special techniques used to control the environment in which the seal operates. Some examples include discharge return recirculation, suction return, flush from an outside source, flush, and quench and drain. The controlling techniques often require supplementary equipment such as heaters, coolers, pumping rings, external circulation devices, various special solenoids and valves compatible with the working environment, and so on. Sealed products that solidify, vaporize, abrade, or carry contaminants are successfully sealed with end face seals augmented by environmental controls.

High Performance Lip Seals The nature of some sealed products is such that end face mechanical seals are not applicable. In many difficult instances of that type, sealing can be achieved with high-performance modified PTFE lip seals (Fig. 8.6.42). Gylon is such a material which can serve in seals operating over a wide range of pressures, temperatures, and rotating speeds. It is particularly useful to seal against dry products, viscous resins, heavy slurries, salting solutions, and products which tend to solidify on seal faces. Dry running is possible under some circumstances. Unlike conventional lip seal material, modified PTFE lip seals in multiples can operate from high vacuums (10^{-3} inHg) up to 10 bar (150 lb/in^2), within a temperature range of −130 to +500°F (−90 to +260°C), and exhibit excellent compatibility with a wide range of sealed fluids. Manufacturers' literature will provide data showing the effect of temperature and rotating speed on the permissible operating pressure.

For extremely high speeds, where it is desirable to eliminate all rubbing contact, the **labyrinth seal** (Fig. 8.6.25) is chosen. This seal is not fluid-tight but restricts serious flow by means of a torturous path and induced turbulence. It is widely used on steam turbines (Sec. 9.4). Where no leakage is permissible, a liquid seal based on the U-tube principle (Fig. 8.6.26) may be used. The natural weight of the liquid is amplified by centrifugal force so that under high rotating speed a fair pressure differential can be sealed. Another noncontacting seal is the **controlled gap seal** which is being used on gas turbines where pressure differentials are not excessive and a small amount of leakage can be tolerated. The seal consists of a ring with a shaft clearance in the range of 0.0005 to 0.0015 in (0.013 to 0.038 mm) and is made of exotic heat-resisting materials capable of maintaining that clearance at all operating temperatures. Usually one end of the ring is faced to form an axial seal against the inside of its housing.

Diaphragms are a form of dynamic packing but include the requirements of a gasket where they are gripped or held in position. In service they are leakless, although generally limited in travel. By literally rolling one cylinder inside another, considerable increase in travel is possible. This type is often called a bellows, and a simple application is the mechanical seal suspension shown in Fig. 8.6.31. In the diaphragm valve (Fig. 8.6.33) the diaphragm replaces both the conventional stem packing and valve disk. **Diaphragms** of fabric such as cotton or nylon (except friable materials such as glass) covered with an elastomer suitable for the fluids and temperatures involved are used in pumps (fuel pump, Fig. 8.6.35) and in motors (Fig. 8.6.34) to operate valves, switches, and other controls. Correctly designed diaphragms are made with slack to permit a natural rolling action. Flat sheet stock should be used only where limited travel is desired. An unusual application is shown in Fig. 8.6.36, where the diaphragm is under balanced fluid pressure on both sides and is unstressed. Thin sheet metal, usually with concentric corrugations, is used where movement is limited and long life is desired. Where considerable movement is involved, the possibility of fatigue must be considered.

PTFE and Glycon diaphragms are used with chemically aggressive fluids. Experience shows that PTFE has a tendency toward cold flow, which leads to leaking at the clamp areas; Gylon has proved more dimensionally stable and serviceable.

8.7 PIPE, PIPE FITTINGS, AND VALVES
by Helmut Thielsch

REFERENCES: M. L. Nayyar, "Piping Handbook," McGraw-Hill. ANSI Code for Power Piping. ASTM Specifications. Tube Turns Division, Natural Cylinder Gas Co., catalogs. Crane Co., catalogs and bulletins. Grinnell Co., Inc., "Piping Design and Engineering." M. W. Kellogg Co., "Design of Piping Systems," Wiley. United States Steel Co., catalogs and bulletins.

EDITOR'S NOTE: The several piping standards listed in this section are subject to continuing periodic review and/or modification. It is suggested that the reader make inquiry to the issuing organizations (see Table 8.7.1) as to the currency of a given standard as listed.

PIPING STANDARDS

Codes for various piping services have been developed by nationally recognized engineering societies, standardization bodies, and trade associations. The sound engineering practices incorporated in these codes generally cover minimum safety requirements for the selection of materials, dimensions, design, fabrication, erection, and testing of piping systems. By means of interpretation and revision these codes continually reflect the knowledge gained through experience, testing, and research.

Generally, piping codes form the basis for many state and municipal safety laws. Compliance with a code which has attained this status is mandatory for all systems included within the jurisdiction. Although some of today's piping installations are not within the scope of any mandatory code, it is advisable to comply with the applicable code in the interests of safety and as a basis for contract negotiations. Contracts with various agencies of the federal government are regulated by federal specifications or rules. These often do not have a direct connection with the codes enumerated below.

The reader is cautioned that the **piping standards** are changing more often than in previous years. Although the formulas and other data provided are in accordance with the code rules in effect at the time of publication, it must be recognized that code rules may change, and piping engineering and design work performed in accordance with information contained herein does not provide complete assurance that all extant code requirements have been met. The reader is urged to become familiar with the specific code edition and addenda applicable in a particular project, for they may contain mandatory requirements applicable to the particular project.

The **ASME Boiler and Pressure Vessel Code** is mandatory in many cities, states, and provinces in the United States and Canada. Local application of this code into law is not uniform, making it necessary to investigate the city or state laws which have jurisdiction over the installation in question. Compliance with this code is required in all locations to qualify for insurance approval.

Section I: "Power Boilers" concerns all piping connections to power boilers or superheaters including the first stop valve on single boilers, or including the second stop valve for cross-connected multiple-boiler installations. Section I refers to ASME B31.1 which contains rules for design and construction of "boiler external piping." "Boiler external piping" is under the jurisdiction of Section I and requires inspection and code stamping in accordance with Section I even though the rules for its design and construction are contained in the ASME Code for Pressure Piping, section B31.1.

Section II "Material Specifications" provides detailed specifications of the materials which are acceptable under this code. (These specifications generally are identical to the corresponding ASTM Standards.)

Section III: "Nuclear Components" includes all nuclear piping. It is the responsibility of the designer to determine whether or not a particular piping system is "nuclear" piping, since Section III makes this determination the responsibility of the designer. In general, piping whose failure could result in the release of radiation which would endanger the public or plant personnel is considered "nuclear" piping.

Section VIII: "Unfired Pressure Vessels" concerns piping only to the extent of the flanged or threaded connections to the pressure vessel, except that the entire section will apply in those special cases where unfired pressure vessels are made from pipe and fittings.

Section IX: "Welding and Brazing Qualifications" establishes the minimum requirements for ASME Code welding.

Section XI: "Rules for Inservice Inspection of Nuclear Power Plant Components" contains rules for the examination and repair of components throughout the life of the plant.

The ASME Code for Pressure Piping B31 is, at present, a nonmandatory code in the United States except where U.S. state legislative bodies and Canadian provinces have adopted this code as a legal requirement. The minimum safety requirements of these codes have been accepted by the industry as a standard for all piping outside the jurisdiction of other codes. The piping systems covered by the separate sections of this code are listed below:

Power Piping	B31.1
Fuel Gas Piping	B31.2
Chemical Plant and Petroleum Refinery Piping	B31.3
Liquid Petroleum Transportation Piping Systems	B31.4
Refrigeration Piping	B31.5
Gas Transmission and Distribution Piping Systems	B31.8
Building Service Piping	B31.9

Several other engineering societies and trade associations have also issued standards covering piping. Foremost among these is the American Society for Testing and Materials (ASTM), the American National Standards Institute (ANSI), the American Water Works Association (AWWA), the American Petroleum Institute (API), and the Manufacturers Standardization Society of the Valve and Fitting Industry (MSS).

Additional piping specifications have been issued by the American Welding Society (AWS), the Pipe Fabrication Institute (PFI), the National Fire Protection Association (NFPA), the Copper Development Association (CDA), the Plastics Pipe Institute (PPI), and several others.

The piping standards issued by the ASTM are most commonly referred to in specifications covering piping for power plants, chemical plants, refineries, pulp and paper mills, and other industrial plants. The large majority of ASTM Standards has also been issued by the ASME in Section II of the ASME Boiler and Pressure Vessel Code. The same specification numbers are applied by the ASME as were originally assigned by the ASTM.

The ANSI formerly prepared the various standards of the B31 Code for Pressure Piping. These standards are now issued by the ASME. The ANSI, however, continues to prepare and issue various standards covering pipe fittings, flanges, and other piping components. Note that ASME B16 prepares and issues standards for fittings, flanges, etc.

The AWWA has issued various standards for waterworks applications. The majority of these involve ductile iron pipe, ductile iron and cast iron pipe fittings, etc.

The MSS has prepared various standards for valves, hangers, and fittings, generally involving the lower range of pressures and temperatures.

Table 8.7.1 gives the most **commonly used piping standards** and the organizations from which the standards are available.

Table 8.7.1 Commonly Used Piping Standards

ASTM Specifications	ASTM Specifications (*Cont.*)	ASTM Specifications (*Cont.*)	ASTM Specifications (*Cont.*)
*A 20-82	*A 516-82	*C 361-78	*F 492-77
*A 36-81a	*A 522-81	*C 582-68 (R1974)	*F 493-80
*A 47-77	*A 524-80	*C 599-70 (R1977)	*F 546-77
*A 48-76	*A 530-82	*D 1503-73 (R1978)	*F 599-78
*A 53-81a	*A 537-82	*D 1527-77	
*A 105-82	*A 553-82	D 1600-80	SAE Specifications
*A 106-82	*A 579-79	D 1694-79	
*A 120-82	*A 571-82	*D 1785-76	*J 513f-1977
*A 126-73 (R1980)	*A 587-78	*D 2104-74	*J 514j-1980
*A 134-80	*A 645-82	*D 2235-81	*J 518c-1972
*A 135-79	*A 658-82	*D 2239-74	
*A 139-74 (R1980)	*A 671-80	*D 2241-80	ASNT Standard
*A 167-81a	*A 672-81	*D 2282-77	
*A 179-79b	*A 675-82	*D 2310-80	SNT TC-1A-1980
*A 181-81	*A 691-81	*D 2412-77	*A3.0-1980
*A 182-82	*B 12-76a	*D 2446-73 (R1978)	*A5.1-1981
*A 193-82	*B 21-81	*D 2447-74	*A5.4-1981
*A 194-82	*B 26-82b	*D 2464-76	*A5.5-1981
*A 197-79	*B 42-82	*D 2465-73 (R1979)	
*A 202-82	*B 43-80	*D 2466-78	Copper Development Assn.
*A 203-82	*B 61-82a	*D 2467-76a	
*A 204-82	*B 62-82a	*D 2468-80	Copper Tube Handbook
*A 211-75a (R1980)	*B 68-80	*D 2469-76	
*A 216-82	*B 75-81a	*D 2513-80a	EJMA Standards, Current
*A 217-81	*B 88-81	D 2517-81	
*A 225-82	*B 96-82	*D 2560-80	*American National Standards
*A 234-82	*B 98-82	*D 2564-80	
*A 240-82b	*B 127-80a	*D 2609-74	A21.14-1979
*A 263-82	*B 148-82	*D 2657-79	A21.52-1981
*A 264-82	*B 150-82a	*D 2662-81	A58.1-1972
*A 265-81a	*B 152-82	*D 2666-81a	B1.1-1974
*A 268-82a	*B 160-81	*D 2672-80	B1.20.3-1976
*A 269-82	*B 161-81	*D 2683-80	B1.20.1-1983
*A 276-82a	*B 162-80	D 2737-74	B1.20.7-1966 (R1983)
*A 278-75 (R1980)	*B 164-81	*D 2740-80	B16.1-1975
*A 283-81	*B 165-81	*D 2837-76	B16.3-1977
*A 285-82	*B 166-81	*D 2846-80	B16.4-1977
*A 299-82	*B 167-81	*D 2855-80	B16.5-1981
*A 302-82	*B 168-80a	*D 2992-71 (R1977)	B16.9-1978
*A 307-82a	*B 169-82	*D 2996-81	B16.9a-1981
*A 312-82	*B 209-82b	*D 2997-71 (R1977)	B16.10-1973
*A 320-82	*B 210-82a	D 3000-73	B16.11-1980
*A 325-82	*B 211-82b	D 3035-74	B16.14-1977
*A 333-82	*B 221-82a	D 3036-73	B16.15-1978
*A 334-79	*B 241-82a	D 3139-77	B16.18-1978
*A 335-81a	*B 247-82a	*D 3140-72 (R1977)	B16.20-1973
*A 338-61 (R1977)	*B 283-81	D 3197-73	B16.21-1978
*A 350-81a	*B 333-77	D 3287-73	B16.22-1980
*A 351-82	*B 335-77	*D 3309-80a	B16.24-1979
*A 352-82	*B 337-78	*E 112-82	B16.25-1979
*A 353-82	*B 345-82	*E 114-75 (R1981)	B16.26-1975
*A 354-82b	*B 361-81	*E 125-63 (R1980)	B16.28-1978
*A 358-81	*B 366-81	*E 142-77	B16.33-1981
*A 369-79a	*B 402-81	*E 155-79	B16.34-1981
*A 370-82	*B 407-77	*E 165-80	B16.36-1975 (A1979)
*A 376-81	*B 443-82	*E 186-81	B16.38-1978
A 377-79	*B 444-82	*E 272-75 (R1979)	B16.39-1977
*A 381-81	*B 446-82	*E 280-81	B16.42-1979
*A 387-82	*B 466-82a	*E 310-75 (R1979)	B18.2.1-1981
*A 395-80	*B 467-82	*E 446-81	B18.2.2-1972
*A 403-82	*B 574-77	*E 709-80	B36.10-1979
*A 409-81a	*B 575-77	*F 336-78	B36.19-1976
*A 420-81a	*B 581-81	F 423-75	B46.1-1978
*A 426-80	*B 582-81	*F 437-77	B93.11-1981
*A 430-79	*B 584-82	*F 438-77	
*A 437-82	*B 619-81	*F 439-77	PPI Technical Reports
*A 451-80	*B 620-77	*F 441-77	
*A 452-79	*B 621-77	*F 442-77	TR-21-1973
*A 453-80	*C 14-78	*F 443-77	
*A 494-81	*C 296-78	*F 491-77	Pipe Fabrication Institute
*A 515-82	*C 301-79		
			ES-7-1962 (R1980)
			Federal Specification
			WW-P-421D. Sept. 1976

NOTE: Footnotes appear at the end of the table.

Table 8.7.1 Commonly Used Piping Standards (*Continued*)

API Standards†	API Standards (*Cont.*)	MSS Standard Practices (*Cont.*)	MSS Standard Practices (*Cont.*)
5B, 10th ed., 1979 (A1982)	*C151-1976	SP-9-1984	SP-82-1976 (R81)
5L, 32d ed., 1982	*C200-1980	SP-25-1978 (R83)	SP-83-1976
5LE, 2d ed., 1976	*C207-1978	SP-42-1985	SP-85-1985
5LP, 4th ed., 1976	C208-1959	SP-43-1982	SP-86-1981
5LR, 4th ed., 1976	C300-1974	SP-44-1985	SP-87-1987
5LS, 12th ed., 1982	*C301-1972 (A1974)	SP-45-1982	SP-88-1978
5LX, 24th ed., 1982	C302-1974	SP-51-1982	SP-89-1985
*526, 2d ed., 1969 (R1977)	C400-1977	SP-53-1985	SP-90-1980
593, 2d ed., 1981	*C402-1977	SP-55-1985	SP-91-1984
594, 2d ed., 1977	*C500-1980	*SP-58-1983	SP-92-1982
595, 2d ed., 1979	*C504-1980	SP-60-1982	SP-93-1982
597, 3d ed., 1981	*C600-1982	SP-61-1985	SP-94-1983
599, 2d ed., 1978	C900-1975	SP-65-1983	
600, 8th ed., 1981		SP-67-1985	CGA
601, 5th ed., 1982	**ASME Codes**	SP-68-1984	
602, 4th ed., 1978		SP-69-1983	G-4.1-1977
603, 3d ed., 1977	*ASME Boiler and Pressure	SP-70-1984	
604, 4th ed., 1981	Vessel Code, 1980 ed.	SP-71-1984	NACE
*605, 3d ed., 1980	*Section V, incl. addenda through	SP-72-1970	
606, 1st ed., 1976	W82	SP-73-1982	Corrosion Data Survey
609, 2d ed., 1978	*Section VIII, Division 1	SP-75-1983	
*C101-1967 (R1977)	*Section VIII, Division 2	SP-77-1984	NBS
*C110-1977	*Section IX, incl. addenda through	SP-78-1977	
*C111-1980	W82	SP-79-1980	PS 15-69
*C115-1975		SP-80-1979	NFPA Specifications, current
*C150-1976	**MSS Standard Practices**	SP-81-1981	Uniform Building Code, current
	SP-6-1985		Aluminum Assn.

The referenced standards are available from the listed organizations:

		Standards sources	
Alum. Assn.	Aluminum Association 900 19th St., NW, Washington, DC 20006 202 862-5100	MSS	Manufacturers Standardization Society of the Valve and Fittings Industry 127 Park Street, NE, Vienna, VA 22180 703 281-6613
ANSI	American National Standards Institute, Inc. 11 West 42d St., New York, NY 10036 212 642-4900	NACE	National Association of Corrosion Engineers P.O. Box 986 Katy, TX 77450 713 492-0535
API	American Petroleum Institute 1220 L Street, NW, Washington, DC 20005-8029 202 682-8000	NIST	National Institute of Standards and Technology (U.S. Dept. of Commerce): Publications available from Superintendent of Documents
ASME	The American Society of Mechanical Engineers 345 East 47th Street, New York, NY 10017 212 705-7722		United States Government Printing Office Washington, DC 20402 202 541-3000
ASNT	American Society for Nondestructive Testing 3200 Riverside Drive, Columbus, OH 43221 614 488-7921	NFPA	National Fire Protection Association P.O. Box 9101 1 Batterymarch Park, Quincy, MA 02269-9101 617 770-3000
ASTM	American Society for Testing and Materials 1916 Race Street, Philadelphia, PA 19103 215 299-5400	PFI	Pipe Fabrication Institute Box 173, Lenore Avenue, Springdale, PA 15144-1518 412 274-4722
AWWA	American Water Works Association 6666 W. Quincy Avenue, Denver, CO 80235 303 794-7711	PPI	Plastics Pipe Institute 65 Madison Avenue, Morristown, NJ 07960-6078 No telephone listed
AWS	American Welding Society 2501 N.W. 7th Street, Miami, FL 33125 305 642-7090	SAE	Society of Automotive Engineers 400 Commonwealth Drive Warrendale, PA 15096
CDA	Copper Development Association 260 Madison Avenue, New York, NY 10016 212 251-7234	UBC	412 776-4841 Uniform Building Code
(a) CGA	Compressed Gas Association 1235 Jefferson Davis Highway Arlington, VA 22202		International Conference of Building Officials 5360 South Workman Mill Road Whittier, CA 90601 213 699-0541
(a) EJMA	Expansion Joint Manufacturers Association 25 North Broadway, North Tarrytown, NY 10591 914 382-0040		
Fed. Spec.	Federal Specification: Superintendent of Documents United States Government Printing Office Washington, DC 20402 202 541-3000		

* Indicates that the standard has been approved as an American National Standard by the American National Standards Institute.
† Including supplements to these API Standards through spring 1981.
NOTE: The issue date shown immediately following the hyphen after the number of the standard (e.g., B16.9-1978, C207-1978, and A 47-77) is the effective date of the issue (edition) of the Standard. Any additional number shown following the issue date and prefixed by the letter R is the latest date of reaffirmation [e.g., C101-1967 (R1977)]. Any edition number prefixed by the letter A is the date of the latest addenda accepted [e.g., B16.36-1975 (A1979)].

PIPING, PIPE, AND TUBING

The term **piping** generally is broadly applied to pipe, fittings, valves, and other components that convey liquids, gases, slurries, etc.

The term **pipe** is applied to tubular products of dimensions and materials commonly used for pipelines and connections, formerly designated as **iron pipe size** (IPS). The outside diameter of all weights and kinds of IPS pipe is of necessity the same for a given pipe size on account of threading. Nevertheless, the large majority of pipe is furnished unthreaded with butt-weld ends.

The word **tube** (or **tubing**) is generally applied to tubular products as utilized in boilers, heat exchangers, instrumentation, and in the machine, aircraft, automotive, and related industries.

Pipe and Tube Products—General

Commercial pipe and tube products are grouped into various classifications generally based on the application or use and not on the manufacturing method. Most tubular products fall into one of three very broad classifications: (1) pipe, (2) pressure tubes, and (3) mechanical tubes. Each classification falls into various subgroupings, which may have been defined and standardized differently by the different trade or user groups. The same standard materials specifications may apply to several of the (user) classifications. For example, ASTM A120 or A53 pipe may be used for applications representing refrigeration, pressure, and nipple service.

Cost considerations enter into the selection of specific piping materials. In some sizes, prices of pipe made to different materials specifications may vary, whereas in other sizes, they may be identical.

Within the broad **use classifications** listed above, the **production method classifications** are also recognized. These are primarily (1) seamless wrought pipe, (2) seamless cast pipe, and (3) seam-welded pipe or tubes. The large variety of single and combination pipe- or tube-forming methods can produce different characteristics and properties in essentially identical pipe materials. In addition, the final finishing can result in hot-finished or cold-finished products. Cold-finishing may be accomplished by reducing or by expanding. Heat treatments may also affect the properties of the finished product.

Piping

On the basis of user classification, the more commonly used types of pipe are tabulated in Table 8.7.2. This listing ignores method of manufacture, size range, wall thickness, and finish, for which the different user groups may have developed different standard requirements.

Table 8.7.2 Major Pipe Classification and Examples of Applications

Identification of pipe	Uses
Standard	Mechanical (structural) service pipe, low-pressure service pipe, refrigeration (ice-machine) pipe, ice-rink pipe, dry-kiln pipe
Pressure	Liquid, gas, or vapor service pipe, service for elevated temperature or pressure, or both
Line	Threaded or plain end, gas, oil, and steam pipe
Water well	Reamed and drifted, water-well casing, drive pipe, driven well pipe, pump pipe, turbine-pump pipe
Oil country tubular goods	Casing, well tubing, drill pipe
Other pipe	Conduit, piles, nipple pipe, sprinkler pipe, bedstead tubing

Standard Pipe Mechanical service pipe is produced in three classes of wall thickness—standard weight, extra strong, and double extra strong. It is available as welded or seamless pipe of ordinary finish and dimensional tolerances, produced in sizes up to 12-in nominal OD. This pipe is used for structural and mechanical purposes. Certain applications have other requirements for size, surface finish, or straightness.

Refrigeration Pipe This pipe is also known as ice-machine pipe or ammonia pipe. It may be butt-welded, lap-welded, electric-resistance-welded, or seamless and is intended for use as a conveyor of refrigerants. This pipe is suitable for coiling, bending, and welding. The sizes commonly used range from ¾ to 2 in. The piping is produced in random and double random lengths in standard line pipe sizes and weights. Double random lengths are used as ice-rink pipe. It can be produced with plain ends, with threaded ends only, or with threaded ends and line pipe couplings, as desired.

Dry-Kiln Pipe This pipe is butt-welded, electric-resistance-welded, or seamless pipe for use in the lumber industry. It is produced in standard-weight pipe sizes of ¾, 1, and 1¼ in. Joints are designed to permit subsequent "makeup" after expansion has occurred. Dry-kiln pipe is commonly produced with threaded ends and couplings and in random lengths.

Pressure Pipe Pressure pipe is used for conveying fluids or gases at normal, subzero, or elevated temperatures and/or pressures. It generally is not subjected to external heat application. The range of sizes is ⅛-in nominal size to 36-in actual OD. It is produced in various wall thicknesses. Pressure piping is furnished in random lengths, with threaded or plain ends, as required. Pressure pipe generally is hydrostatically tested at the mill.

Line Pipe Line pipe is seamless or welded pipe produced in sizes from ⅛-in nominal OD to 48-in actual OD. It is used principally for conveying gas, oil, or water. Line pipe is produced with ends which are plain, threaded, beveled, grooved, flanged, or expanded, as required for various types of mechanical couplers, or for welded joints. When threaded ends and couplings are required, recessed couplings are normally supplied.

Water-Well Pipe Water-well pipe is welded or seamless steel pipe used for conveying water for municipal and industrial applications. Pipelines for such purposes involve flow mains, transmission mains, force mains, water mains, or distribution mains. The mains are generally laid underground. Sizes range from ⅛- to 106-in OD in a variety of wall thicknesses. Pipe is produced with ends suitably prepared for mechanical couplers, with plain ends beveled for welding, with ends fitted with butt straps for field welding, or with bell-and-spigot joints with rubber gaskets for field joining. Pipe is produced in double random lengths of about 40 ft, single random lengths of about 20 ft, or in definite cut lengths, as specified. Wall thicknesses vary from 0.068 in for ⅛-in nominal OD to 1.00 in for 106-in actual OD.

When required, water-well pipe is produced with a specified coating or lining or both. For example, cement-mortar lining and coatings are extensively used.

Oil Country Goods Casing is used as a structural retainer for the walls of oil or gas wells. It is also used to exclude undesirable fluids, and to confine and conduct oil or gas from productive subsurface strata to the ground level. Casing is produced in sizes 4½- to 20-in OD. Size designations refer to actual outside diameter and weight per foot. Ends are commonly threaded and furnished with couplings. When required, the ends are prepared to accommodate other types of joints.

Drill Pipe Drill pipe is used to transmit power by rotary motion from ground level to a rotary drilling tool below the surface and also to convey flushing media to the cutting face of the tool. Drill pipe is produced in sizes 2⅜- to 6⅝-in OD. Size designations refer to actual outside diameter and weight per foot. Drill pipe is generally upset, either internally or externally, or both, and is furnished with threaded ends and couplings, threaded only, or prepared to accommodate other types of joints.

Tubing is used within the casing of oil wells to conduct oil to ground level. It is produced in sizes 1.050- to 4.500-in OD in several weights per foot. Ends are threaded and fitted with couplings and may or may not be upset externally.

Other Pipe Classifications **Rigid conduit pipe** is welded or seamless pipe intended especially for the protection of electrical wiring systems. Conduit pipe is not subjected to hydrostatic tests unless so specified. It is furnished in standard-weight pipe sizes from ¼- to 6-in OD in 10-ft lengths,* with plain ends or with threaded ends and couplings, as specified.

Piling pipe is welded or seamless pipe for use as piles, where the cylinder section acts as a permanent load-carrying member or where it acts as a shell to form cast-in-place concrete piles. Specifications provide for the choice of three grades by minimum tensile strength, in which the sizes listed are 8⅝- to 24-in OD in a variety of wall thicknesses and in two length ranges. Ends are plain or beveled for welding.

Nipple pipe is standard-weight, extra-strong, or double-extra-strong welded or seamless pipe produced for the manufacture of pipe nipples. Standard-weight pipe with threaded ends is also used in sprinkler systems. Nipple pipe is commonly produced in random lengths with plain ends in nominal sizes ⅛- to 12-in OD. Close OD tolerances, sound welds, good threading properties, and surface cleanliness are essential in this product. It is commonly coated with oil or zinc and well protected in shipment. When reference is made to ASTM Specifications for this application, Specification A120 is generally used for diameters to 5-in OD and A53 for diameters of 5 in and over.

Standard Pipe Sizes   Standard pressure, line, and other pipe with plain ends for welding or with threaded ends is standardized in two ranges. Diameters of 12 in and less have a nominal size which represents approximately that of the inside diameter of standard-weight pipe. The nominal outside diameter is standard, regardless of weight. Increase in wall thickness results in a decrease of the inside diameter.

The standardization of pipe sizes over 12 in is based on the actual outside diameter, the wall thickness, and the weight per foot.

The principal dimensions, weights, and characteristics of commercial piping materials are summarized in Table 8.7.3.

The weights of butt-welding elbows, tees, and laterals and flanges are given in Tables 8.7.4 to 8.7.9 for several common pipe sizes. The weights of reducing fittings are approximately the same as for full-size fittings.

The weights of welding reducers are for one size reduction and are thus only approximately correct for other reductions.

Hot-finished or cold-drawn seamless low-alloy steel tubes generally are process-annealed at temperatures between 1,200 and 1,350°F.

Austenitic stainless-steel tubes are usually annealed at temperatures between 1,800 and 2,100°F, with specific temperatures varying somewhat with each grade. This is generally followed by pickling, unless bright-annealing was done.

Mechanical Tubing

Unlike pipe and pressure tubes, mechanical tubing is generally classified by the method of manufacture and the degree of finish. Examples of classifications are "seamless hot-finished," "cold-drawn welded," "flash-in-grade," etc.

Seamless Tubes   Seamless tubes are available as either hot- or cold-finished. They are normally made in sizes from 0.187-in OD to 10.750-in OD.

Dimensions for hot-finished mechanical tubes are provided in Table 8.7.11. Dimensions for cold-finished tubes are listed in Table 8.7.12.

Welded Tubes   Welded tubes generally are produced by electric resistance methods. Where required, the welding flash is removed with a cutting tool. Industry practice normally recognizes a number of finish conditions which are summarized in Table 8.7.13.

Flash-in Type Tubing   This tubing is generally limited to applications where nothing is inserted in the tube.

Flash-Controlled Tubing   This tubing is used where moderate control of the inside diameter is required. Generally, the outside and inside diameters are specified.

For special materials, the equations listed below for weights of tubes and weights of contents of tubes are helpful.

$$\text{Weight of tube, lb/ft} = F \times 10.68 \times T \times D - T$$

where T = wall thickness, in; D = outside diameter, in; F = relative weight factor.

The weight of tube calculation is based on low-carbon steel weighing 0.2833 lb/in³ and is extended to other materials through the factor F.

Relative weight factor F

Aluminum	0.35
Brass	1.12
Cast-iron	0.91
Copper	1.14
Ferritic stainless steel	1.02
Steel	1.00
Wrought iron	0.98

$$\text{Weight of contents of tube, lb/ft} = G \times 0.3405 \times (D - 2T)^2$$

where G = specific gravity of contents; T = tube wall thickness, in; D = tube outside diameter, in.

The weight per foot of steel pipe is subject to the tolerances listed in Table 8.7.10.

The designation *sink-draw tubes* is specified where close control over the outer diameter is required with normal tolerance applying to the wall thickness. Smoothness of the inside surface is not controlled, except that the flash is generally controlled to a height of 0.005 or 0.010 in maximum.

Mandrel-drawn tubes usually are normalized after welding by passing the tubes through a continuous atmosphere-controlled furnace. After descaling, the tubes are cold-drawn through a die with a mandrel on the inside of the tube. These tubes provide maximum control over surface finish, outside or inside diameters, and wall thickness. The normalizing heat treatment removes the effects of welding and provides a uniform microstructure around the tube circumference.

The different finish classifications may result in substantial differences in the mechanical properties of the steel material.

Typical examples for low-carbon steel material are given in Table 8.7.14. Differences in carbon content and other chemistry, heat treatment, etc., may significantly change these typical values.

Other Tubing Types   Among other tube classifications are sanitary tubing usually made of 18% Cr-8% Ni stainless steel and available as seamless or welded tubing. This tubing is used extensively in the dairy, beverage, and food industries. Sanitary tubing is generally available in sizes from 1- to 4-in OD. It may be furnished either hot- or cold-finished. The tubes are normally annealed at temperatures above 1,900°F.

Some welded tube is also produced by fusion-welding methods utilizing either the inert-gas tungsten-arc-welding or gas-shielded consumable metal-arc-welding process. This tubing is generally more expensive than the resistance-welded types.

The butt-welded cold-finished tubes are made from hot-rolled or cold-rolled strip and fusion-welded. This tubing is usually furnished as sink-drawn or mandrel-drawn.

Butt-welded tubing is made in heavier wall thicknesses than the resistance-welded tube.

Several tubing materials used in the automobile industry are covered by specifications of the Society of Automotive Engineers, "SAE Handbook."

Pressure Tubing

Pressure-tube applications commonly involve external heat applications, as in boilers or superheaters.

Pressure tubing is produced to the actual outside diameter and minimum wall or average wall thickness specified by the purchaser. Pressure tubing may be hot- or cold-finished.

The wall thickness is normally given in decimal parts of an inch rather than as a fraction or gage number. When gage numbers are given without reference to a gage system, Birmingham wire gage (BWG) is implied.

Pressure tubing is usually made from steel produced by the open-hearth, basic oxygen, or electric-furnace processes.

---

* Although some specifications of rigid conduit pipe list lengths to 20 ft, the National Electric Code, 1965, limits lengths to 10 ft.

Table 8.7.3 Properties of Commercial Steel Pipe

Nominal pipe size, outside diameter, in	Schedule number* a	b	c	Wall thickness, in	Inside diameter, in	Inside area, in^2	Metal area, in^2	Outside surface, ft^2/ft	Inside surface, ft^2/ft	Weight, lb/ft	Weight of water, lb/ft	Moment of inertia, in^4	Section modulus, in^3	Radius of gyration, in
1/8 0.405			10S	0.049	0.307	0.0740	0.0548	0.106	0.0804	0.186	0.0321	0.00088	0.00437	0.1271
	40	Std	40S	0.068	0.269	0.0568	0.0720	0.106	0.0705	0.245	0.0246	0.00106	0.00525	0.1215
	80	XS	80S	0.095	0.215	0.0364	0.0925	0.106	0.0563	0.315	0.0157	0.00122	0.00600	0.1146
1/4 0.540			10S	0.065	0.410	0.1320	0.0970	0.141	0.1073	0.330	0.0572	0.00279	0.01032	0.1694
	40	Std	40S	0.088	0.364	0.1041	0.1250	0.141	0.0955	0.425	0.0451	0.00331	0.01230	0.1628
	80	XS	80S	0.119	0.302	0.0716	0.1574	0.141	0.0794	0.535	0.0310	0.00378	0.01395	0.1547
3/8 0.675			10S	0.065	0.545	0.2333	0.1246	0.177	0.1427	0.423	0.1011	0.00586	0.01737	0.2169
	40	Std	40S	0.091	0.493	0.1910	0.1670	0.177	0.1295	0.568	0.0827	0.00730	0.02160	0.2090
	80	XS	80S	0.126	0.423	0.1405	0.2173	0.177	0.1106	0.739	0.0609	0.00862	0.02554	0.1991
1/2 0.840			5S	0.065	0.710	0.396	0.1582	0.220	0.1859	0.538	0.171	0.01197	0.0285	0.2750
			10S	0.083	0.674	0.357	0.1974	0.220	0.1765	0.671	0.1547	0.01431	0.0341	0.2692
	40	Std	40S	0.109	0.622	0.304	0.2503	0.220	0.1628	0.851	0.1316	0.01710	0.0407	0.2613
	80	XS	80S	0.147	0.546	0.2340	0.320	0.220	0.1433	1.088	0.1013	0.02010	0.0478	0.2505
	160			0.187	0.466	0.1706	0.383	0.220	0.1220	1.304	0.0740	0.02213	0.0527	0.2402
		XXS		0.294	0.252	0.0499	0.504	0.220	0.0660	1.714	0.0216	0.02425	0.0577	0.2192
3/4 1.050			5S	0.065	0.920	0.655	0.2011	0.275	0.2409	0.684	0.2882	0.02451	0.0467	0.349
			10S	0.083	0.884	0.614	0.2521	0.275	0.2314	0.857	0.2661	0.02970	0.0566	0.343
	40	Std	40S	0.113	0.824	0.533	0.333	0.275	0.2157	1.131	0.2301	0.0370	0.0706	0.334
	80	XS	80S	0.154	0.742	0.432	0.435	0.275	0.1943	1.474	0.1875	0.0448	0.0853	0.321
	160			0.218	0.614	0.2961	0.570	0.275	0.1607	1.937	0.1284	0.0527	0.1004	0.304
		XXS		0.308	0.434	0.1479	0.718	0.275	0.1137	2.441	0.0641	0.0579	0.1104	0.2840
1 1.315			5S	0.065	1.185	1.103	0.2553	0.344	0.310	0.868	0.478	0.0500	0.0760	0.443
			10S	0.109	1.097	0.945	0.413	0.344	0.2872	1.404	0.409	0.0757	0.1151	0.428
	40	Std	40S	0.133	1.049	0.864	0.494	0.344	0.2746	1.679	0.374	0.0874	0.1329	0.421
	80	XS	80S	0.179	0.957	0.719	0.639	0.344	0.2520	2.172	0.311	0.1056	0.1606	0.407
	160			0.250	0.815	0.522	0.836	0.344	0.2134	2.844	0.2261	0.1252	0.1903	0.387
		XXS		0.358	0.599	0.2818	1.076	0.344	0.1570	3.659	0.1221	0.1405	0.2137	0.361
1 1/4 1.660			5S	0.065	1.530	1.839	0.326	0.434	0.401	1.107	0.797	0.1038	0.1250	0.564
			10S	0.109	1.442	1.633	0.531	0.434	0.378	1.805	0.707	0.1605	0.1934	0.550
	40	Std	40S	0.140	1.380	1.496	0.669	0.434	0.361	2.273	0.648	0.1948	0.2346	0.540
	80	XS	80S	0.191	1.278	1.283	0.881	0.434	0.335	2.997	0.555	0.2418	0.2913	0.524
	160			0.250	1.160	1.057	1.107	0.434	0.304	3.765	0.458	0.2839	0.342	0.506
		XXS		0.382	0.896	0.631	1.534	0.434	0.2346	5.214	0.2732	0.341	0.411	0.472
1 1/2 1.900			5S	0.065	1.770	2.461	0.375	0.497	0.463	1.274	1.067	0.1580	0.1663	0.649
			10S	0.109	1.682	2.222	0.613	0.497	0.440	2.085	0.962	0.2469	0.2599	0.634
	40	Std	40S	0.145	1.610	2.036	0.799	0.497	0.421	2.718	0.882	0.320	0.326	0.623
	80	XS	80S	0.200	1.500	1.767	1.068	0.497	0.393	3.631	0.765	0.391	0.412	0.605
	160			0.281	1.338	1.406	1.429	0.497	0.350	4.859	0.608	0.483	0.508	0.581
		XXS		0.400	1.100	0.950	1.855	0.497	0.288	6.408	0.412	0.568	0.598	0.549
				0.525	0.850	0.567	2.267	0.497	0.223	7.710	0.246	0.6140	0.6470	0.5200
				0.650	0.600	0.283	2.551	0.497	0.157	8.678	0.123	0.6340	0.6670	0.4980

Nominal Size	OD	Sched. No.	Desig.	Sched. S	t	ID									
2	2.375			5S	0.065	2.245	3.96	0.472	0.622	1.604	0.588	1.716	0.315	0.2652	0.817
				10S	0.109	2.157	3.65	0.776	0.622	2.638	0.555	1.582	0.499	0.420	0.802
		40	Std	40S	0.154	2.067	3.36	1.075	0.622	3.653	0.541	1.455	0.666	0.561	0.787
		80	XS	80S	0.218	1.939	2.953	1.477	0.622	5.022	0.508	1.280	0.868	0.731	0.766
		160			0.343	1.689	2.240	2.190	0.622	7.444	0.442	0.971	1.163	0.979	0.729
			XXS		0.436	1.503	1.774	2.656	0.622	9.029	0.393	0.769	1.312	1.104	0.703
					0.562	1.251	1.229	3.199	0.622	10.882	0.328	0.533	1.442	1.2140	0.6710
					0.687	1.001	0.787	3.641	0.622	12.385	0.262	0.341	1.5130	1.2740	0.6440
2½	2.875			5S	0.083	2.709	5.76	0.728	0.753	2.475	0.709	2.499	0.710	0.494	0.988
				10S	0.120	2.635	5.45	1.039	0.753	3.531	0.690	2.361	0.988	0.687	0.975
		40	Std	40S	0.203	2.469	4.79	1.704	0.753	5.793	0.646	2.076	1.530	1.064	0.947
		80	XS	80S	0.276	2.323	4.24	2.254	0.753	7.661	0.608	1.837	1.925	1.339	0.924
					0.375	2.125	3.55	2.945	0.753	10.01	0.556	1.535	2.353	1.637	0.894
		160			0.552	1.771	2.464	4.03	0.753	13.70	0.464	1.067	2.872	1.998	0.844
			XXS		0.675	1.525	1.826	4.663	0.753	15.860	0.399	0.792	3.0890	2.1490	0.8140
					0.800	1.275	1.276	5.212	0.753	17.729	0.334	0.554	3.2250	2.2430	0.7860
3	3.500			5S	0.083	3.334	8.73	0.891	0.916	3.03	0.873	3.78	1.301	0.744	1.208
				10S	0.120	3.260	8.35	1.274	0.916	4.33	0.853	3.61	1.822	1.041	1.196
		40	Std	40S	0.216	3.068	7.39	2.228	0.916	7.58	0.803	3.20	3.02	1.724	1.164
		80	XS	80S	0.300	2.900	6.61	3.02	0.916	10.25	0.759	2.864	3.90	2.226	1.136
		160			0.437	2.626	5.42	4.21	0.916	14.32	0.687	2.348	5.03	2.876	1.094
			XXS		0.600	2.300	4.15	5.47	0.916	18.58	0.602	1.801	5.99	3.43	1.047
					0.725	2.050	3.299	6.317	0.916	21.487	0.537	1.431	6.5010	3.7150	1.0140
					0.850	1.800	2.543	7.073	0.916	24.057	0.471	1.103	6.8530	3.9160	0.9840
3½	4.000			5S	0.083	3.834	11.55	1.021	1.047	3.47	1.004	5.01	1.960	0.980	1.385
				10S	0.120	3.760	11.10	1.463	1.047	4.97	0.984	4.81	2.756	1.378	1.372
		40	Std	40S	0.226	3.548	9.89	2.680	1.047	9.11	0.929	4.28	4.79	2.394	1.337
		80	XS	80S	0.318	3.364	8.89	3.68	1.047	12.51	0.881	3.85	6.28	3.14	1.307
			XXS		0.636	2.728	5.845	6.721	1.047	22.850	0.716	2.530	9.8480	4.9240	1.2100
4	4.500			5S	0.083	4.334	14.75	1.152	1.178	3.92	1.135	6.40	2.811	1.249	1.562
				10S	0.120	4.260	14.25	1.651	1.178	5.61	1.115	6.17	3.96	1.762	1.549
					0.188	4.124	13.357	2.547	1.178	8.560	1.082	5.800	5.8500	2.6000	1.5250
		40	Std	40S	0.237	4.026	12.73	3.17	1.178	10.79	1.054	5.51	7.23	3.21	1.510
		80	XS	80S	0.337	3.826	11.50	4.41	1.178	14.98	1.002	4.98	9.61	4.27	1.477
		120			0.437	3.626	10.33	5.58	1.178	18.96	0.949	4.48	11.65	5.18	1.445
					0.500	3.500	9.621	6.283	1.178	21.360	0.915	4.160	12.77710	5.6760	1.4250
		160			0.531	3.438	9.28	6.62	1.178	22.51	0.900	4.02	13.27	5.90	1.416
			XXS		0.674	3.152	7.80	8.10	1.178	27.54	0.825	3.38	15.29	6.79	1.374
					0.800	2.900	6.602	9.294	1.178	31.613	0.759	2.864	16.6610	7.4050	1.3380
					0.925	2.650	5.513	10.384	1.178	35.318	0.694	2.391	17.7130	7.8720	1.3060
5	5.563			5S	0.109	5.345	22.44	1.868	1.456	6.35	1.399	9.73	6.95	2.498	1.929
				10S	0.134	5.295	22.02	2.285	1.456	7.77	1.386	9.53	8.43	3.03	1.920
		40	Std	40S	0.258	5.047	20.01	4.30	1.456	14.62	1.321	8.66	15.17	5.45	1.878
		80	XS	80S	0.375	4.813	18.19	6.11	1.456	20.78	1.260	7.89	20.68	7.43	1.839
		120			0.500	4.563	16.35	7.95	1.456	27.04	1.195	7.09	25.74	9.25	1.799
		160			0.625	4.313	14.61	9.70	1.456	32.96	1.129	6.33	30.0	10.80	1.760
			XXS		0.750	4.063	12.97	11.34	1.456	38.55	1.064	5.62	33.6	12.10	1.722
					0.875	3.813	11.413	12.880	1.456	43.810	0.998	4.951	36.6450	13.1750	1.6860
					1.000	3.563	9.966	14.328	1.456	47.734	0.933	4.232	39.1110	14.0610	1.6520

NOTE: See footnotes at end of table.

Table 8.7.3 Properties of Commercial Steel Pipe (Continued)

Nominal pipe size, outside diameter, in	Schedule number*			Wall thickness, in	Inside diameter, in	Inside area, in^2	Metal area, in^2	Outside surface, ft^2/ft	Inside surface, ft^2/ft	Weight, lb/ft†	Weight of water, lb/ft	Moment of inertia, in^4	Section modulus, in^3	Radius of gyration, in
	a	b	c											
6 / 6.625			5S	0.109	6.407	32.2	2.231	1.734	1.677	5.37	13.98	11.85	3.58	2.304
			10S	0.134	6.357	31.7	2.733	1.734	1.664	9.29	13.74	14.40	4.35	2.295
				0.219	6.187	30.100	4.410	1.734	1.620	15.020	13.100	22.6600	6.8400	2.2700
	40	Std	40S	0.280	6.065	28.89	5.58	1.734	1.588	18.97	12.51	28.14	8.50	2.245
	80	XS	80S	0.432	5.761	26.07	8.40	1.734	1.508	28.57	11.29	40.5	12.23	2.195
	120			0.562	5.501	23.77	10.70	1.734	1.440	36.39	10.30	49.6	14.98	2.153
	160			0.718	5.189	21.15	13.33	1.734	1.358	45.30	9.16	59.0	17.81	2.104
		XXS		0.864	4.897	18.83	15.64	1.734	1.282	53.16	8.17	66.3	20.03	2.060
				1.000	4.625	16.792	17.662	1.734	1.211	60.076	7.284	72.1190	21.7720	2.0200
				1.125	4.375	15.025	19.429	1.734	1.145	66.084	6.517	76.5970	23.1240	1.9850
8 / 8.625			5S	0.109	8.407	55.5	2.916	2.258	2.201	9.91	24.07	26.45	6.13	3.01
			10S	0.148	8.329	54.5	3.94	2.258	2.180	13.40	23.59	35.4	8.21	3.00
				0.219	8.187	52.630	5.800	2.258	2.150	19.640	22.900	51.3200	11.9000	2.9700
	20			0.250	8.125	51.8	6.58	2.258	2.127	22.36	22.48	57.7	13.39	2.962
	30			0.277	8.071	51.2	7.26	2.258	2.113	24.70	22.18	63.4	14.69	2.953
	40	Std	40S	0.322	7.981	50.0	8.40	2.258	2.089	28.55	21.69	72.5	16.81	2.938
	60			0.406	7.813	47.9	10.48	2.258	2.045	35.64	20.79	88.8	20.58	2.909
	80	XS	80S	0.500	7.625	45.7	12.76	2.258	1.996	43.39	19.80	105.7	24.52	2.878
	100			0.593	7.439	43.5	14.96	2.258	1.948	50.87	18.84	121.4	28.14	2.847
	120			0.718	7.189	40.6	17.84	2.258	1.882	60.63	17.60	140.6	32.6	2.807
	140			0.812	7.001	38.5	19.93	2.258	1.833	67.76	16.69	153.8	35.7	2.777
		XXS		0.875	6.875	37.1	21.30	2.258	1.800	72.42	16.09	162.0	37.6	2.757
	160			0.906	6.813	36.5	21.97	2.258	1.784	74.69	15.80	165.9	38.5	2.748
				1.000	6.625	34.454	23.942	2.258	1.734	81.437	14.945	177.1320	41.0740	2.7190
				1.125	6.375	31.903	26.494	2.258	1.669	90.114	13.838	190.6210	44.2020	2.6810
10 / 10.750			5S	0.134	10.482	86.3	4.52	2.815	2.744	15.15	37.4	63.7	11.85	3.75
			10S	0.165	10.420	85.3	5.49	2.815	2.728	18.70	36.9	76.9	14.30	3.74
				0.219	10.312	83.52	7.24	2.815	2.70	24.63	36.2	100.46	18.69	3.72
	20			0.250	10.250	82.5	8.26	2.815	2.683	28.04	35.8	113.7	21.16	3.71
	30			0.307	10.136	80.7	10.07	2.185	2.654	34.24	35.0	137.5	25.57	3.69
	40	Std	40S	0.365	10.020	78.9	11.91	2.815	2.623	40.48	34.1	160.8	29.90	3.67
	60	XS	80S	0.500	9.750	74.7	16.10	2.815	2.553	54.74	32.3	212.0	39.4	3.63
	80			0.593	9.564	71.8	18.92	2.815	2.504	64.33	31.1	244.9	45.6	3.60
	100			0.718	9.314	68.1	22.63	2.815	2.438	76.93	29.5	286.2	53.2	3.56
	120			0.843	9.064	64.5	26.24	2/815	2.373	89.20	28.0	324	60.3	3.52
				0.875	9.000	63.62	27.14	2.815	2.36	92.28	27.6	333.46	62.04	3.50
	140			1.000	8.750	60.1	30.6	2.815	2.091	104.13	26.1	368	68.4	3.47
	160			1.125	8.500	56.7	34.0	2.815	2.225	115.65	24.6	399	74.3	3.43
				1.250	8.250	53.45	37.31	2.815	2.16	126.832	23.2	428.17	79.66	3.39
				1.500	7.750	47.15	43.57	2.815	2.03	148.19	20.5	478.59	89.04	3.31

NPS / OD	Sched.	Ident.	t	ID									
12 / 12.750		5S	0.156	12.438	121.4	6.17	3.34	3.26	20.99	52.7	122.2	19.20	4.45
		10S	0.180	12.390	120.6	7.11	3.34	3.24	24.20	52.2	140.5	22.03	4.44
	20		0.250	12.250	117.9	9.84	3.34	3.21	33.38	51.1	191.9	30.1	4.42
	30		0.330	12.090	114.8	12.88	3.34	3.17	43.77	49.7	248.5	39.0	4.39
		Std, 40S	0.375	12.000	113.1	14.58	3.34	3.14	49.56	49.0	279.3	43.8	4.38
	40		0.406	11.938	111.9	15.74	3.34	3.13	53.53	48.5	300	47.1	4.37
		XS, 80S	0.500	11.750	108.4	19.24	3.34	3.08	65.42	47.0	362	56.7	4.33
	60		0.562	11.626	106.2	21.52	3.34	3.04	73.16	46.0	401	62.8	4.31
	80		0.687	11.376	101.6	26.04	3.34	2.978	88.51	44.0	475	74.5	4.27
			0.750	11.250	99.40	28.27	3.34	2.94	96.2	43.1	510.7	80.1	4.25
	100		0.843	11.064	96.1	31.5	3.34	2.897	107.20	41.6	562	88.1	4.22
			0.875	11.000	95.00	32.64	3.34	2.88	110.9	41.1	578.5	90.7	4.21
	120	XXS	1.000	10.750	90.8	36.9	3.34	2.814	125.49	39.3	642	100.7	4.17
	140		1.125	10.500	86.6	41.1	3.34	2.749	139.68	37.5	701	109.9	4.13
			1.250	10.250	82.50	45.16	3.34	2.68	153.6	35.8	755.5	118.5	4.09
	160		1.312	10.126	80.5	47.1	3.34	2.651	160.27	34.9	781	122.6	4.07
14 / 14.000		5S	0.156	13.688	147.20	6.78	3.67	3.58	23.0	63.7	162.6	23.2	4.90
		10S	0.188	13.624	145.80	8.16	3.67	3.57	27.7	63.1	194.6	27.8	4.88
			0.210	13.580	144.80	9.10	3.67	3.55	30.9	62.8	216.2	30.9	4.87
			0.219	13.562	144.50	9.48	3.67	3.55	32.2	62.6	225.1	32.2	4.86
	10		0.250	13.500	143.1	10.80	3.67	3.53	36.71	62.1	255.4	36.5	4.85
			0.281	13.438	141.80	12.11	3.67	3.52	41.2	61.5	285.2	40.7	4.84
	20		0.312	13.376	140.5	13.42	3.67	3.50	45.68	60.9	314	44.9	4.83
			0.344	13.312	139.20	14.76	3.67	3.48	50.2	60.3	344.3	49.2	4.82
	30	Std	0.375	13.250	137.9	16.05	3.67	3.47	54.57	59.7	373	53.3	4.80
	40		0.437	13.126	135.3	18.62	3.67	3.44	63.37	58.7	429	61.2	4.79
			0.469	13.062	134.00	19.94	3.67	3.42	67.8	58.0	456.8	64.3	4.78
		XS	0.500	13.000	132.7	21.21	3.67	3.40	72.09	57.5	484	69.1	4.74
	60		0.593	12.814	129.0	24.98	3.67	3.35	84.91	55.9	562	83.3	4.73
			0.625	12.750	127.7	26.26	3.67	3.34	89.28	55.3	589	84.1	4.69
	80		0.750	12.500	122.7	31.2	3.67	3.27	106.13	53.2	687	98.2	4.63
	100		0.937	12.126	115.5	38.5	3.67	3.17	130.73	50.0	825	117.8	4.58
	120		1.093	11.814	109.6	44.3	3.67	3.09	150.67	47.5	930	132.8	4.53
	140		1.250	11.500	103.9	50.1	3.67	3.01	170.22	45.0	1017	145.8	4.48
	160		1.406	11.188	98.3	55.6	3.67	2.929	189.12	42.6	1127	159.6	
16 / 16.000		5S	0.165	15.670	192.90	8.21	4.19	4.10	28	83.5	257	32.2	5.60
		10S	0.188	15.624	191.70	9.34	4.19	4.09	32	83.0	292	35.5	5.59
	10		0.250	15.500	188.7	12.37	4.19	4.06	42.05	81.8	384	43.0	5.57
	20		0.312	15.376	185.7	15.38	4.19	4.03	52.36	80.5	473	59.2	5.55
	30	Std	0.375	15.250	182.6	18.41	4.19	3.99	62.58	79.1	562	70.3	5.53
	40	XS	0.500	15.000	176.7	24.35	4.19	3.93	82.77	76.5	732	91.5	5.48
	60		0.656	14.688	169.4	31.6	4.19	3.85	107.50	73.4	933	116.6	5.43
	80		0.843	14.314	160.9	40.1	4.19	3.75	136.46	69.7	1157	144.6	5.37
	100		1.031	13.938	152.6	48.5	4.19	3.65	164.83	66.1	1365	170.6	5.30
	120		1.218	13.564	144.5	56.6	4.19	3.55	192.29	62.6	1556	194.5	5.24
	140		1.437	13.126	135.3	65.7	4.19	3.44	223.64	58.6	1750	220.0	5.17
	160		1.593	12.814	129.0	72.1	4.19	3.35	245.11	55.9	1894	236.7	5.12

Table 8.7.3 Properties of Commercial Steel Pipe (Continued)

Nominal pipe size, outside diameter, in	Schedule number* a	b	c	Wall thickness, in	Inside diameter, in	Inside area, in²	Metal area, in²	Outside surface, ft²/ft	Inside surface, ft²/ft	Weight, lb/ft†	Weight of water, lb/ft	Moment of inertia, in⁴	Section modulus, in³	Radius of gyration, in
18 *18.000*			5S	0.165	17.670	245.20	9.24	4.71	4.63	31	106.2	368	40.8	6.31
			10S	0.188	17.624	243.90	10.52	4.71	4.61	36	105.7	417	46.4	6.30
	10			0.250	17.500	240.5	13.94	4.71	4.58	47.39	104.3	549	61.0	6.28
	20			0.312	17.376	237.1	17.34	4.71	4.55	59.03	102.8	678	75.5	6.25
		Std		0.375	17.250	233.7	20.76	4.71	4.52	70.59	101.2	807	89.6	6.23
	30			0.437	17.126	230.4	24.11	4.71	4.48	82.06	99.9	931	103.4	6.21
		XS		0.500	17.00	227.0	27.49	4.71	4.45	93.45	98.4	1053	117.0	6.19
	40			0.562	16.876	223.7	30.8	4.71	4.42	104.75	97.0	1172	130.2	6.17
	60			0.750	16.500	213.8	40.6	4.71	4.32	138.17	92.7	1515	168.3	6.10
	80			0.937	16.126	204.2	50.2	4.71	4.22	170.75	88.5	1834	203.8	6.04
	100			1.156	15.688	193.3	61.2	4.71	4.11	207.96	83.7	2180	242.2	5.97
	120			1.375	15.250	182.6	71.8	4.71	3.99	244.14	79.2	2499	277.6	5.90
	140			1.562	14.876	173.8	80.7	4.71	3.89	274.23	75.3	2750	306	5.84
	160			1.781	14.438	163.7	90.7	4.71	3.78	308.51	71.0	3020	336	5.77
20 *20.000*			5S	0.188	19.634	302.40	11.70	5.24	5.14	40	131.0	574	57.4	7.00
			10S	0.218	19.564	300.60	13.55	5.24	5.12	46	130.2	663	66.3	6.99
	10			0.250	19.500	298.6	15.51	5.24	5.11	52.73	129.5	757	75.7	6.98
	20	Std		0.375	19.250	291.0	23.12	5.24	5.04	78.60	126.0	1114	111.4	6.94
	30	XS		0.500	19.000	283.5	30.6	5.24	4.97	104.13	122.8	1457	145.7	6.90
	40			0.593	18.814	278.0	36.2	5.24	4.93	122.91	120.4	1704	170.4	6.86
	60			0.812	18.376	265.2	48.9	5.24	4.81	166.40	115.0	2257	225.7	6.79
				0.875	18.250	261.6	52.6	5.24	4.78	178.73	113.4	2409	240.9	6.77
	80			1.031	17.938	252.7	61.4	5.24	4.70	208.87	109.4	2772	277.2	6.72
	100			1.281	17.438	238.8	75.3	5.24	4.57	256.10	103.4	3320	332	6.63
	120			1.500	17.00	227.0	87.2	5.24	4.45	296.37	98.3	3760	376	6.56
	140			1.750	16.500	213.8	100.3	5.24	4.32	341	92.6	4200	422	6.48
	160			1.968	16.064	202.7	111.5	5.24	4.21	379.01	87.9	4650	459	6.41
22 *22.000*			5S	0.188	21.624	367.3	12.88	5.76	5.66	44	159.1	766	69.7	7.71
			10S	0.218	21.564	365.1	14.92	5.76	5.65	51	158.2	885	80.4	7.70
	10			0.250	21.500	363.1	17.18	5.76	5.63	58	157.4	1010	91.8	7.69
	20	Std		0.375	21.250	354.7	25.48	5.76	5.56	87	153.7	1490	135.4	7.65
	30	XS		0.500	21.000	346.4	33.77	5.76	5.50	115	150.2	1953	177.5	7.61
				0.625	20.750	338.2	41.97	5.76	5.43	143	146.6	2400	218.2	7.56
	60			0.750	20.500	330.1	50.07	5.76	5.37	170	143.1	2829	257.2	7.52
	80			0.875	20.250	322.1	58.07	5.76	5.30	197	139.6	3245	295.0	7.47
	100			1.125	19.750	306.4	73.78	5.76	5.17	251	132.8	4029	366.3	7.39
	120			1.375	19.250	291.0	89.09	5.76	5.04	303	126.2	4758	432.6	7.31
	140			1.625	18.750	276.1	104.02	5.76	4.91	354	119.6	5432	493.8	7.23
	160			1.875	18.250	261.6	118.55	5.76	4.78	403	113.3	6054	550.3	7.15
				2.125	17.750	247.4	132.68	5.76	4.65	451	107.2	6626	602.4	7.07

24	24.000		5S	0.218	23.564	436.1	16.29	6.28	6.17	55	188.9	1152	96.0	8.41
		10		0.250	23.500	434	18.65	6.28	6.15	63.41	188.0	1316	109.6	8.40
		20	Std	0.375	23.250	425	27.83	6.28	6.09	94.62	183.8	1943	161.9	8.35
			XS	0.500	23.000	415	36.9	6.28	6.02	125.49	180.1	2550	212.5	8.31
		30		0.562	22.876	411	41.4	6.28	5.99	140.80	178.1	2840	237.0	8.29
				0.625	22.750	406	45.9	6.28	5.96	156.03	176.2	3140	261.4	8.27
		40		0.687	22.626	402	50.3	6.28	5.92	171.17	174.3	3420	285.2	8.25
				0.750	22.500	398	54.8	6.28	5.89	186.24	172.4	3710	309	8.22
				0.875	22.250	388.6	63.54	6.28	5.83	216	168.6	4256	354.7	8.18
		60		0.968	22.064	382	70.0	6.28	5.78	238.11	165.8	4650	388	8.15
		80		1.218	21.564	365	87.2	6.28	5.65	296.36	158.3	5670	473	8.07
		100		1.531	20.938	344	108.1	6.28	5.48	367.40	149.3	6850	571	7.96
		120		1.812	20.376	326	126.3	6.28	5.33	429.39	141.4	7830	652	7.87
		140		2.062	19.876	310	142.1	6.28	5.20	483.13	134.5	8630	719	7.79
		160		2.343	19.314	293	159.4	6.28	5.06	541.94	127.0	9460	788	7.70
26	26.000			0.250	25.500	510.7	19.85	6.81	6.68	67	221.4	1646	126.6	9.10
		10		0.312	25.376	505.8	25.18	6.81	6.64	86	219.2	2076	159.7	9.08
			Std	0.375	25.250	500.7	30.19	6.81	6.61	103	217.1	2478	190.6	9.06
		20	XS	0.500	25.000	490.9	40.06	6.81	6.54	136	212.8	3259	250.7	9.02
				0.625	24.750	481.1	49.82	6.81	6.48	169	208.6	4013	308.7	8.98
				0.750	24.500	471.4	59.49	6.81	6.41	202	204.4	4744	364.9	8.93
				0.875	24.250	461.9	69.07	6.81	6.35	235	200.2	5458	419.9	8.89
				1.000	24.000	452.4	78.54	6.81	6.28	267	196.1	6149	473.0	8.85
				1.125	23.750	443.0	87.91	6.81	6.22	299	192.1	6813	524.1	8.80
28	28.000		5S	0.250	27.500	594.0	21.80	7.33	7.20	74	257.3	2098	149.8	9.81
		10	10S	0.312	27.376	588.6	27.14	7.33	7.17	92	255.0	2601	185.8	9.79
			Std	0.375	27.250	583.2	32.54	7.33	7.13	111	252.6	3105	221.8	9.77
		20	XS	0.500	27.000	572.6	43.20	7.33	7.07	147	248.0	4085	291.8	9.72
		30		0.625	26.750	562.0	53.75	7.33	7.00	183	243.4	5038	359.8	9.68
				0.750	26.500	551.6	64.21	7.33	6.94	218	238.9	5964	426.0	9.64
				0.875	26.250	541.2	74.56	7.33	6.87	253	234.4	6865	490.3	9.60
				1.000	26.000	530.9	84.82	7.33	6.81	288	230.0	7740	552.8	9.55
				1.125	25.750	520.8	94.98	7.33	6.74	323	225.6	8590	613.6	9.51
30	30.000		5S	0.250	29.500	683.4	23.37	7.85	7.72	79	296.3	2585	172.3	10.52
		10	10S	0.312	29.376	677.8	29.19	7.85	7.69	99	293.7	3201	213.4	10.50
			Std	0.375	29.250	672.0	34.90	7.85	7.66	119	291.2	3823	254.8	10.48
		20	XS	0.500	29.000	660.5	46.34	7.85	7.59	158	286.2	5033	335.5	10.43
		30		0.625	28.750	649.2	57.68	7.85	7.53	196	281.3	6213	414.2	10.39
		40		0.750	28.500	637.9	68.92	7.85	7.46	234	276.6	7371	491.4	10.34
				0.875	28.250	626.7	80.06	7.85	7.39	272	271.8	8494	566.2	10.30
				1.000	28.000	615.7	91.11	7.85	7.33	310	267.0	9591	639.4	10.26
				1.125	27.750	604.7	102.05	7.85	7.26	347	262.2	10653	710.2	10.22
32	32.000			0.250	31.500	779.2	24.93	8.38	8.25	85	337.8	3141	196.3	11.22
		10		0.312	31.376	773.2	31.02	8.38	8.21	106	335.2	3891	243.2	11.20
			Std	0.375	31.250	766.9	37.25	8.38	8.18	127	332.5	4656	291.0	11.18
		20	XS	0.500	31.000	754.7	49.48	8.38	8.11	168	327.2	6140	383.8	11.14
		30		0.625	30.750	742.5	61.59	8.38	8.05	209	321.9	7578	473.6	11.09
		40		0.688	30.624	736.6	67.68	8.38	8.02	230	319.0	8298	518.6	11.07
				0.750	30.500	730.5	73.63	8.38	7.98	250	316.7	8990	561.9	11.05
				0.875	30.250	718.3	85.52	8.38	7.92	291	311.6	10372	648.2	11.01
				1.000	30.000	706.8	97.38	8.38	7.85	331	306.4	11680	730.0	10.95
				1.125	29.750	694.7	109.0	8.38	7.79	371	301.3	13023	814.0	10.92

NOTE: See footnotes at end of table.

Table 8.7.3 Properties of Commercial Steel Pipe (Continued)

Nominal pipe size, outside diameter, in	Schedule number* a	b	c	Wall thickness, in	Inside diameter, in	Inside area, in²	Metal area, in²	Outside surface, ft²/ft	Inside surface, ft²/ft	Weight, lb/ft†	Weight of water, lb/ft	Moment of inertia, in⁴	Section modulus, in³	Radius of gyration, in
34 34.000	10			0.250	33.500	881.2	26.50	8.90	8.77	90	382.0	3773	221.9	11.93
		Std		0.312	33.376	874.9	32.99	8.90	8.74	112	379.3	4680	275.3	11.91
	20	XS		0.375	33.250	867.8	39.61	8.90	8.70	135	376.2	5597	329.2	11.89
	30			0.500	33.000	855.3	52.62	8.90	8.64	179	370.8	7385	434.4	11.85
	40			0.625	32.750	841.9	65.53	8.90	8.57	223	365.0	9124	536.7	11.80
				0.688	32.624	835.9	72.00	8.90	8.54	245	362.1	9992	587.8	11.78
				0.750	32.500	829.3	78.34	8.90	8.51	266	359.5	10829	637.0	11.76
				0.875	32.250	816.4	91.01	8.90	8.44	310	354.1	12501	735.4	11.72
				1.000	32.000	804.2	103.67	8.90	8.38	353	348.6	14114	830.2	11.67
				1.125	31.750	791.3	116.13	8.90	8.31	395	343.2	15719	924.7	11.63
36 36.000	10			0.250	35.500	989.7	28.11	9.42	9.29	96	429.1	4491	249.5	12.64
				0.312	35.376	982.2	34.95	9.42	9.26	119	426.1	5565	309.1	12.62
	20	Std		0.375	35.250	975.8	42.01	9.42	9.23	143	423.1	6664	370.2	12.59
	30	XS		0.500	35.000	962.1	55.76	9.42	9.16	190	417.1	8785	488.1	12.55
				0.625	34.750	948.3	69.50	9.42	9.10	236	411.1	10872	604.0	12.51
	40			0.750	34.500	934.7	83.01	9.42	9.03	282	405.3	12898	716.5	12.46
				0.875	34.250	920.6	96.50	9.42	8.97	328	399.4	14903	827.9	12.42
				1.000	34.000	907.9	109.96	9.42	8.90	374	393.6	16851	936.2	12.38
				1.125	33.750	894.2	123.19	9.42	8.89	419	387.9	18763	1042.4	12.34
42 42.000		Std		0.250	41.500	1352.6	32.82	10.99	10.86	112	586.4	7126	339.3	14.73
	20	XS		0.375	41.250	1336.3	49.08	10.99	10.80	167	579.3	10627	506.1	14.71
	30			0.500	41.000	1320.2	65.18	10.99	10.73	222	572.3	14037	668.4	14.67
	40			0.625	40.750	1304.1	81.28	10.99	10.67	276	565.4	17373	827.3	14.62
				0.750	40.500	1288.2	97.23	10.99	10.60	330	558.4	20689	985.2	14.59
				1.000	40.000	1256.6	128.81	10.99	10.47	438	544.8	27080	1289.5	14.50
				1.250	39.500	1225.3	160.03	10.99	10.34	544	531.2	33233	1582.5	14.41
				1.500	39.000	1194.5	190.85	10.99	10.21	649	517.9	39181	1865.7	14.33

NOTE: The following formulas are used in the computation of the values shown in the table:

Weight of pipe per foot (pounds) $= 10.6802t(D - t)$
Weight of water per foot (pounds) $= 0.3405d^2$
Square feet outside surface per foot $= 0.2618D$
Square feet inside surface per foot $= 0.2618d$
Inside area (square inches) $= 0.785d^2$
Area of metal (square inches) $= 0.785(D^2 - d^2)$
Moment of inertia (inches⁴) $= 0.0491(D^4 - d^4)$
$= A_m R_g^2$

Section modulus (inches³) $= \dfrac{0.0982(D^4 - d^4)}{D}$

Radius of gyration (inches) $= 0.25\sqrt{D^2 + d^2}$

where A_m = area of metal, in²; d = inside diameter, in; D = outside diameter, in; R_g = radius of gyration, in; t = pipe wall thickness, in.
* Schedule numbers: Standard-weight pipe and schedule 40 are the same in all sizes through 10 in; from 12 in through 24 in, standard weight pipe has a wall thickness of ⅜ in. Extra-strong-weight pipe and schedule 80 are the same in all sizes through 8 in; from 8 in through 24 in, extra-strong-weight pipe has a wall thickness of ½ in. Double-extra-strong-weight pipe has no corresponding schedule number.
  a: ANSI B36.10 steel pipe schedule numbers
  b: ANSI B36.10 steel pipe nominal wall thickness designation
  c: ANSI B36.19 stainless steel pipe schedule numbers
† The ferritic steels may be about 5% less and the austenitic stainless steels about 2% greater than the values shown in this table, which are based on weights for carbon steel.

Table 8.7.4 Weight of Standard Pipe Fittings and Materials, 3-in Size (3.500-in OD)

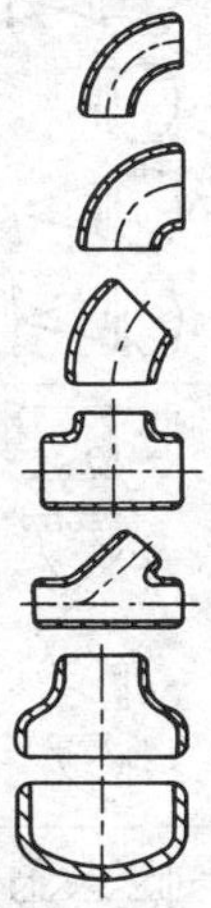

Schedule no.	40	80	160	XXS
Wall designation	Std	XS		
Thickness, in	0.216	0.300	0.438	0.600
Pipe, lb/ft	7.58	10.25	14.32	18.58
Water, lb/ft	3.20	2.86	2.35	1.80

Welding fittings

	40 Std		80 XS		160		XXS	
L.R. 90° elbow	4.6	0.8	6.1	0.8	8.4	0.8	10.7	0.8
S.R. 90° elbow	3	0.5	4	0.5				
L.R. 45° elbow	2.4	0.3	3.2	0.3	4.4	0.3	5.4	0.3
Tee	7.4	0.8	9.5	0.8	12.2	0.8	14.8	0.8
Lateral	13	1.8	19	1.8				
Reducer	2.2	0.3	2.9	0.3	3.7	0.3	4.7	0.3
Cap	1.4	0.5	1.8	0.5	3.5	0.5	3.7	0.5

Insulation

Temperature range, °F	100–199	200–299	300–399	400–499	500–599	600–699	700–799	800–899	900–999	1,000–1,099	1,100–1,200
85% magnesia calcium silicate — Nom. thick., in	1	1	1½	2	2	2½	3	3	3	3½	3½
lb/ft	1.25	1.25	2.08	3.01	3.01	4.07	5.24	5.24	5.24	6.65	6.65
Combination — Nom. thick., in						2½	3	3	3	3½	3½
lb/ft						5.07	6.94	6.94	6.94	9.17	9.17
*Asbestos fiber-sodium silicate — Nom. thick., in	1	1	1	1½	1½	2	2	3	3	3½	3½
lb/ft	1.61	1.61	1.61	2.74	2.74	3.98	3.98	6.99	6.99	8.99	8.99

Table 8.7.4 Weight of Standard Pipe Fittings and Materials, 3-in Size (3.500-in OD) (Continued)

		Cast iron			Steel					
								Pressure rating, lb/in²		
		125	250	150	300	400	600	900	1,500	2,500
Flanges	Screwed or slip-on	9 / 1.5	17 / 1.5	9 / 1.5	17 / 1.5	20 / 1.5	20 / 1.5	37 / 1.5	61 / 1.5	102 / 1.5
	Welding neck			11 / 1.5	19 / 1.5	27 / 1.5	27 / 1.5	38 / 1.5	61 / 1.5	113 / 1.5
	Lap joint			9 / 1.5	17 / 1.5	19 / 1.5	19 / 1.5	36 / 1.5	60 / 1.5	99 / 1.5
	Blind	10 / 1.5	19 / 1.5	10 / 1.5	20 / 1.5	24 / 1.5	24 / 1.5	38 / 1.5	61 / 1.5	105 / 1.5
Flanged fittings	S.R. 90° elbow	26 / 3.9	46 / 4	32 / 3.9	53 / 4		67 / 4.1	98 / 4.3	150 / 4.6	
	L.R. 90° elbow	30 / 4.3	50 / 4.3	40 / 4.3	63 / 4.3					
	45° elbow	22 / 3.5	41 / 3.6	28 / 3.5	46 / 3.6		60 / 3.8	93 / 3.9	135 / 4	
	Tee	39 / 5.9	67 / 6	52 / 5.9	81 / 6		102 / 6.2	151 / 6.5	238 / 6.9	
Valves	Flanged bonnet gate	66 / 7	112 / 7.4	70 / 4	125 / 4.4		155 / 4.8	260 / 5	410 / 5.5	
	Flanged bonnet globe or angle	56 / 7.2	121 / 7.6	60 / 4.3	95 / 4.5		155 / 4.8	225 / 5	495 / 5.5	
	Flanged bonnet check	46 / 7.2	100 / 7.6	60 / 4.3	70 / 4.4		120 / 4.8	150 / 4.9	440 / 5.8	
	Pressure seal bonnet–gate							208 / 3	235 / 3.2	
	Pressure seal bonnet–globe							135 / 2.5	180 / 3	

NOTES: Boldface type is weight in pounds. Lightface type beneath weight is weight factor for insulation.
Insulation thicknesses and weights are based on average conditions and do not constitute a recommendation for specific thicknesses of materials. Insulation weights are based on 85% magnesia and hydrous calcium silicate at 11 lb/ft³. The listed thicknesses and weights of combination covering are the sums of the inner layer of diatomaceous earth at 21 lb/ft³ and the outer layer at 11 lb/ft³.
Insulation weights include allowances for wire, cement, canvas, bands and paint, but no special surface finishes.
To find the weight of covering on flanges, valves or fittings, multiply the weight factor by the weight per foot of covering used on straight pipe.
Valve weights are approximate. When possible, obtain weights from the manufacturer.
Cast-iron valve weights are for flanged end valves; steel weights for welding end valves.
All flanged fitting, flanged valve and flange weights include the proportional weight of bolts or studs to make up all joints.
* 16 lb/ft³ density. (Existing installations only.)

Table 8.7.5 Weights of Standard Pipe Fittings and Materials, 4-in Size (4.500-in OD)

Schedule no.	40	80	120	160	XXS
Wall designation	Std	XS			XXS
Thickness, in	0.237	0.337	0.438	0.531	0.674
Pipe, lb/ft	10.79	14.98	18.96	22.51	27.54
Water, lb/ft	5.51	4.98	4.48	4.02	3.38
L.R. 90° elbow	**8.7** / 1	**11.9** / 1		**17.6** / 1	**21** / 1
S.R. 90° elbow	**5.8** / 0.7	**7.9** / 0.7			
L.R. 45° elbow	**4.3** / 0.4	**5.9** / 0.4		**8.5** / 0.4	**10.1** / 0.4
Tee	**12.6** / 1	**16.4** / 1		**23** / 1	**27** / 1
Lateral	**21** / 2.1	**33** / 2.1			
Reducer	**3.6** / 0.3	**4.9** / .3		**6.6** / .3	**8.2** / 0.3
Cap	**2.6** / 0.6	**3.4** / 0.6		**6.5** / 0.6	**6.7** / 0.6

Welding fittings

Insulation		Temperature range, °F	100–199	200–299	300–399	400–499	500–599	600–699	700–799	800–899	900–999	1,000–1,099	1,100–1,200
85% magnesia calcium silicate		Nom. thick., in	1		1½	2	2½	2½	3	3	3½	3½	4
		lb/ft	1.62		2.55	3.61	4.66	4.66	6.07	6.07	7.48	7.48	9.10
Combination		Nom. thick., in						2½	3	3	3½	3½	3½
		lb/ft						6.07	8.30	8.30	10.6	10.6	10.6
Asbestos fiber-sodium silicate		Nom. thick., in	1	1	1	1½	1½	2	2	3	3	3½	3½
		lb/ft	2.04	2.04	2.04	3.28	3.28	4.70	4.70	8.29	8.29	10.25	10.25

Table 8.7.5 Weights of Standard Pipe Fittings and Materials, 4-in Size (4.500-in OD) (Continued)

	Cast iron		Steel						
	125	250	150	300	400	600	900	1,500	2,500
Flanges									
Screwed or slip-on	16 / 1.5	26 / 1.5	15 / 1.5	26 / 1.5	32 / 1.5	43 / 1.5	66 / 1.5	90 / 1.5	158 / 1.5
Welding neck			17 / 1.5	29 / 1.5	41 / 1.5	48 / 1.5	64 / 1.5	90 / 1.5	177 / 1.5
Lap joint			15 / 1.5	26 / 1.5	31 / 1.5	42 / 1.5	64 / 1.5	92 / 1.5	153 / 1.5
Blind	18 / 1.5	29 / 1.5	19 / 1.5	31 / 1.5	39 / 1.5	47 / 1.5	67 / 1.5	90 / 1.5	164 / 1.5
Flanged fittings									
S.R. 90° elbow	45 / 4.1	72 / 4.2	59 / 4.1	85 / 4.2	99 / 4.3	128 / 4.4	185 / 4.5	254 / 4.8	
L.R. 90° elbow	52 / 4.5	79 / 4.5	72 / 4.5	98 / 4.5					
45° elbow	40 / 3.7	65 / 3.8	51 / 3.7	78 / 3.8	82 / 3.9	119 / 4	170 / 4.1	214 / 4.2	
Tee	70 / 6.1	109 / 6.2	86 / 6.1	121 / 6.3	153 / 6.4	187 / 6.6	262 / 6.8	386 / 7.2	
Valves									
Flanged bonnet gate	109 / 7.2	188 / 7.5	100 / 4.2	175 / 4.5	195 / 5	255 / 5.1	455 / 5.4	735 / 6	
Flanged bonnet globe or angle	97 / 7.4	177 / 7.8	95 / 4.3	145 / 4.8	215 / 5	230 / 5.1	415 / 5.5	800 / 6	
Flanged bonnet check	80 / 7.4	146 / 7.8	80 / 4.3	105 / 4.5	160 / 4.8	195 / 5	320 / 5.6	780 / 6	
Pressure seal bonnet–gate						215 / 2.8	380 / 3	520 / 4	
Pressure seal bonnet–globe							240 / 2.7	290 / 3	

NOTE: See footnotes to Table 8.7.4.

Table 8.7.6 Weights of Standard Pipe Fittings and Materials, 8-in Size (8.625-in OD)

Schedule no.	20	30	40	60	80	100	120	140	XXS	160
Wall designation			Std		XS				XXS	
Thickness, in	0.250	0.277	0.322	0.406	0.500	0.598	0.718	0.812	0.875	0.906
Pipe, lb/ft	22.36	24.70	28.55	35.64	43.4	50.9	60.6	67.8	72.4	74.7
Water, lb/ft	22.48	22.18	21.69	20.79	19.8	18.8	17.6	16.7	16.1	15.8

Welding fittings

Fitting	40	80	XXS	160
L.R. 90° elbow	**46** / 2	**69** / 2	**114** / 2	**117** / 2
 S.R. 90° elbow	**31** / 1.3	**46** / 1.3		
 L.R. 45° elbow	**23** / 0.8	**34** / 0.8	**55** / 0.8	**56** / 0.8
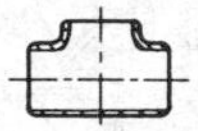 Tee	**54** / 1.8	**76** / 1.8	**118** / 1.8	**120** / 1.8
 Lateral	**76** / 3.8	**140** / 3.8		
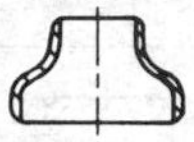 Reducer	**13.9** / 0.5	**20** / 0.5	**32** / 0.5	**33** / 0.5
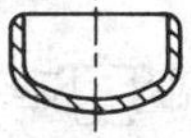 Cap	**11.3** / 1	**16.3** / 1	**31** / 1	**32** / 1

Insulation

Temperature range, °F		100–199	200–299	300–399	400–499	500–599	600–699	700–799	800–899	900–999	1,000–1,099	1,100–1,200
85% magnesia calcium silicate	Nom. thick., in	1½	1½	2	2	2½	3	3½	3½	4	4	4½
	lb/ft	4.13	4.13	5.64	5.64	7.85	9.48	11.5	11.5	13.8	13.8	16.0
Combination	Nom. thick., in						3	3½	3½	4	4	4½
	lb/ft						12.9	16.2	16.2	20.4	20.4	23.8
Asbestos fiber-sodium silicate	Nom. thick., in	1½	1½	1½	1½	1½	2½	2½	3½	3½	4½	4½
	lb/ft	5.38	5.38	5.38	5.38	5.38	10.60	10.60	15.85	15.85	20.85	20.85

Table 8.7.6 Weights of Standard Pipe Fittings and Materials, 8-in Size (8.625-in OD) (Continued)

		Pressure rating, lb/in²										
		Cast iron				Steel						
		125	250	150	300	400	600	900	1,500	2,500
Flanges	Screwed or slip-on	34	64	33	67	82	135	207	319	601
		1.5	1.5	1.5	1.5	1.5	1.5	1.5	1.5	1.5
	Welding neck			42	76	104	137	222	334	692
				1.5	1.5	1.5	1.5	1.5	1.5	1.5
	Lap joint	1.5	1.5	33	67	79	132	223	347	587
			1.5	1.5	1.5	1.5	1.5	1.5	1.5	1.5
	Blind	45	83	48	90	115	159	232	363	649
		1.5	1.5	1.5	1.5	1.5	1.5	1.5	1.5	1.5
Flanged fittings	S.R. 90° elbow	117	201	157	238	310	435	639	995	
		4.5	4.7	4.5	4.7	5	5.2	5.4	5.7	
	L.R. 90° elbow	152	236	202	283					
		5.3	5.3	5.3	5.3					
	45° elbow	101	171	127	203	215	360	507	870	
		3.9	4	3.9	4	4.1	4.4	4.5	4.8	
	Tee	175	304	230	337	445	610	978	1,465	
		6.8	7.1	6.8	7.1	7.5	7.8	8.1	8.6	
Valves	Flanged bonnet gate	251	583	305	505	730	960	1,180	2,740	
		7.5	8.1	4.5	5.1	6	6.3	6.6	7	
	Flanged bonnet globe or angle	317	554	475	505	610	1,130	1,160	2,865	
		8.4	8.6	5.4	5.5	5.9	6.3	6.3	7	
	Flanged bonnet check	302	454	235	310	475	725	1,140	2,075	
		8.4	8.6	5.2	5.3	5.6	6	6.4	7	
	Pressure seal bonnet—gate						925	1,185	2,345	
							4.5	4.7	5.5	
	Pressure seal bonnet—globe							1,550	1,680	
								4	5	

NOTE: See footnotes to Table 8.7.4.

Table 8.7.7 Weights of Standard Pipe Fittings and Materials, 12-in Size (12.750-in OD)

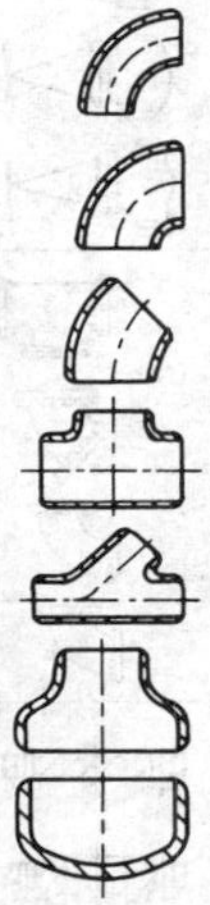

Schedule no.	20	30	Std	40	XS	60	80	100	120	140	160
Wall designation			Std		XS						
Thickness, in	0.250	0.330	0.375	0.406	0.500	0.562	0.687	0.843	1.000	1.125	1.312
Pipe, lb/ft	33.38	43.8	49.6	53.5	65.4	73.2	88.5	107.2	125.5	139.7	160.3
Water, lb/ft	51.10	49.7	49.0	48.5	47.0	46.0	44.0	41.6	39.3	37.5	34.9

Welding fittings

Fitting	20	30	Std	40	XS	60	80	100	120	140	160
L.R. 90° elbow			119 / 3		157 / 3						375 / 3
S.R. 90° elbow			80 / 2		104 / 2						
L.R. 45° elbow			60 / 1.3		78 / 1.3						181 / 1.3
Tee			132 / 2.5		167 / 2.5						360 / 2.5
Lateral			180 / 5.4		273 / 5.4						
Reducer			33 / 0.7		44 / 0.7						94 / 0.7
Cap			30 / 1.5		38 / 1.5						89 / 1.5

Insulation

Temperature range, °F		100–199	200–299	300–399	400–499	500–599	600–699	700–799	800–899	900–999	1,000–1,099	1,100–1,200
85% magnesia calcium silicate	Nom. thick., in	1½	1½	2	2½	3	3	3½	4	4	4½	4½
	lb/ft	6.04	6.04	8.13	10.5	12.7	12.7	15.1	17.9	17.9	20.4	20.4
Combination	Nom. thick., in			1½	1½	1½	3	3½	4	4	4½	4½
	lb/ft			5.22	5.22	5.22	17.7	21.9	26.7	26.7	31.1	31.1
Asbestos fiber-sodium silicate	Nom. thick., in	1½	1½	1½	1½	1½	2½	2½	4	4	5	5
	lb/ft	5.22	5.22	5.22	5.22	5.22	14.20	14.20	24.64	24.64	32.40	32.40

Table 8.7.7 Weights of Standard Pipe Fittings and Materials, 12-in Size (12.750-in OD) (Continued)

Pressure rating, lbf/in²

		Cast iron		Steel						
		125	250	150	300	400	600	900	1,500	2,500
Flanges	Screwed or slip-on	71 / 1.5	137 / 1.5	72 / 1.5	140 / 1.5	164 / 1.5	261 / 1.5	388 / 1.5	820 / 1.5	1,611 / 1.5
	Welding neck			88 / 1.5	163 / 1.5	212 / 1.5	272 / 1.5	434 / 1.5	843 / 1.5	1,919 / 1.5
	Lap joint			72 / 1.5	164 / 1.5	187 / 1.5	286 / 1.5	433 / 1.5	902 / 1.5	1,573 / 1.5
	Blind	96 / 1.5	177 / 1.5	118 / 1.5	209 / 1.5	261 / 1.5	341 / 1.5	475 / 1.5	928 / 1.5	1,755 / 1.5
Flanged fittings	S.R. 90° elbow	265 / 5	453 / 5.2	345 / 5	509 / 5.2	669 / 5.5	815 / 5.8	1,474 / 6.2		
	L.R. 90° elbow	375 / 6.2	553 / 6.2	485 / 6.2	624 / 6.2			1,598 / 6.2		
	45° elbow	235 / 4.3	383 / 4.3	282 / 4.3	414 / 4.3	469 / 4.5	705 / 4.7	1,124 / 4.8		
	Tee	403 / 7.5	684 / 7.8	513 / 7.5	754 / 7.8	943 / 8.3	1,361 / 8.7	1,928 / 9.3		
Valves	Flanged bonnet gate	687 / 7.8	1,298 / 8.5	635 / 4	1,015 / 5	1,420 / 5.5	2,155 / 7	2,770 / 7.2	4,650 / 8	
	Flanged bonnet globe or angle	808 / 9.4	1,200 / 9.5	710 / 5	1,410 / 5.5					
	Flanged bonnet check	674 / 9.4	1,160 / 9.5	560 / 6	720 / 6.5					
	Pressure seal bonnet—gate						1,410 / 7.2	2,600 / 8	3,370 / 8	
	Pressure seal bonnet—globe						1,975 / 5.5	2,560 / 6	4,515 / 7	

NOTE: See footnotes to Table 8.7.4.

Table 8.7.8 Weights of Standard Pipe Fittings and Materials, 24-in Size (24-in OD)

Schedule no.	10	20		30	40	60	80	100	120	140	160
Wall designation		Std	XS	30	40	60	80	100	120	140	160
Thickness, in	0.250	0.375	0.500	0.562	0.687	0.968	1.218	1.531	1.812	2.062	2.343
Pipe, lb/ft	63.4	94.6	125.5	140.8	171.2	238.1	296.4	367.4	429.4	483.1	541.9
Water, lb/ft	188.0	183.8	180.1	178.1	174.3	165.8	158.3	149.3	141.4	134.5	127.0

Welding fittings

	Std	XS
L.R. 90° elbow	458 / 6	606 / 6
 S.R. 90° elbow	305 / 3.7	404 / 3.7
 L.R. 45° elbow	229 / 2.5	302 / 2.5
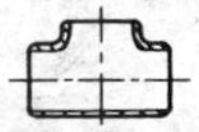 Tee	445 / 4.9	563 / 4.9
 Lateral	544 / 10	882 / 10
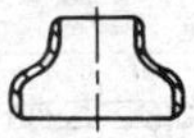 Reducer	167 / 1.7	220 / 1.7
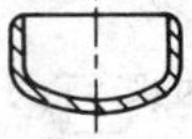 Cap	102 / 2.8	134 / 2.8

Insulation

Temperature range, °F	100–199	200–299	300–399	400–499	500–599	600–699	700–799	800–899	900–999	1,000–1,099	1,100–1,200
85% magnesia calcium silicate Nom. thick., in lb/ft	1½ 10.0	1½ 10.0	2 13.4	2½ 17.0	3 21.0	3 21.0	3½ 24.8	4 28.7	4 28.7	4½ 32.9	4½ 32.9
Combination Nom. thick., in lb/ft						30 29.2	3½ 36.0	4 43.1	4 43.1	4½ 50.6	4½ 50.6
Asbestos fiber-sodium silicate Nom. thick., in lb/ft	1½ 13.55	1½ 13.55	1½ 13.55	2 18.44	2 18.44	3 28.38	3 28.38	4½ 45.06	4½ 45.06	5 50.97	5 50.97

Table 8.7.8 Weights of Standard Pipe Fittings and Materials, 24-in Size (24-in OD) (Continued)

		Cast iron		Steel						
		125	250	150	300	400	600	900	1,500	2,500
Flanges	Screwed or slip-on	255 / 1.5		245 / 1.5	577 / 1.5	676 / 1.5	1,056 / 1.5	1,823 / 1.5	3,378 / 1.5	
	Welding neck			295 / 1.5	632 / 1.5	777 / 1.5	1,157 / 1.5	2,450 / 1.5	4,153 / 1.5	
	Lap joint			295 / 1.5	617 / 1.5	752 / 1.5	1,046 / 1.5	2,002 / 1.5	3,478 / 1.5	
	Blind	405 / 1.5	757 / 1.5	446 / 1.5	841 / 1.5	1,073 / 1.5	1,355 / 1.5	2,442 / 1.5		
Flanged fittings	S.R. 90° elbow	1,231 / 6.7	2,014 / 6.8	1,671 / 6.7	2,174 / 6.8	2,474 / 7.1	3,506 / 7.6	6,155 / 8.1		
	L.R. 90° elbow	1,711 / 8.7	2,644 / 8.7	1,821 / 8.7	2,874 / 8.7					
	45° elbow	871 / 4.8	1,604 / 5	1,121 / 4.8	1,634 / 5	1,974 / 5.1	2,831 / 5.5	5,124 / 6		
	Tee	1,836 / 10	3,061 / 10.2	2,276 / 10	3,161 / 10.2	3,811 / 10.6	5,184 / 11.4	9,387 / 12.1		
Valves	Flanged bonnet gate	3,062 / 8.5	6,484 / 9.8	2,500 / 5	4,675 / 7	6,995 / 8.7	8,020 / 9.5			
	Flanged bonnet globe or angle									
	Flanged bonnet check	2,956 / 12								
	Pressure seal bonnet–gate									
	Pressure seal bonnet–globe									

Pressure rating, lb/in²

NOTE: See footnotes to Table 8.7.4.

Table 8.7.9 Weights of Standard Pipe Fittings and Materials, 36-in Size (36-in OD)

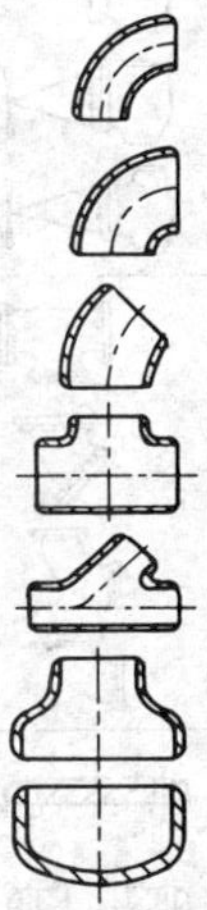

Schedule no.	10	Std	20 XS	30	40
Wall designation		Std	XS		
Thickness, in	0.312	0.375	0.500	0.625	0.750
Pipe, lb/ft	119.1	142.7	189.6	236.1	282.4
Water, lb/ft	425.9	422.6	416.6	411.0	405.1
L.R. 90° elbow		1,040 / 12	1,380 / 12		
S.R. 90° elbow		692 / 5	913 / 5		
L.R. 45° elbow		518 / 4.8	686 / 4.8		
Tee		1,294 / 9.5	1,610 / 9.5		
Lateral					
Reducer		340 / 3.6	360 / 3.6		
Cap		175 / 6	235 / 6		

Welding fittings

Temperature range, °F	100–199	200–299	300–399	400–499	500–599	600–699	700–799	800–899	900–999	1,000–1,099	1,100–1,200
85% magnesia calcium silicate — Nom. thick, in	1½	1½	2	2½	3	3½	4	4½	5	5	6
lb/ft	14.2	14.2	19.2	24.2	29.5	34.8	40.3	45.9	51.7	51.7	63.5
Combination — Nom. thick, in						3½	4½	5½	6	6½	7
lb/ft						49.8	69.3	89.7	100.2	111.0	122.0
Asbestos fiber-sodium silicate — Nom. thick, in	3	3	3	3	3	3	3	4½	4½	5	5
lb/ft	40.84	40.84	40.84	40.84	40.84	40.84	40.84	55.83	55.83	71.48	71.48

Insulation

Table 8.7.9 Weights of Standard Pipe Fittings and Materials, 36-in Size (36-in OD) (Continued)

					Pressure rating, lb/in²						
		Cast iron		Steel							
		125	250	150	300	400	600	900	1,500	2,500	
Flanges	Screwed or slip-on			480 1.5	1,200 1.5	1,325 1.5	1,699 1.5	3,350 1.5			
	Welding neck			520 1.5	1,300 1.5	1,475 1.5	1,750 1.5	3,450 1.5			
	Lap joint										
	Blind			1,125 1.5	2,275 1.5	2,525 1.5	2,950 1.5	4,900 1.5			
Flanged fittings	S.R. 90° elbow										
	L.R. 90° elbow										
	45° elbow										
	Tee										
Valves	Flanged bonnet gate										
	Flanged bonnet globe or angle										
	Flanged bonnet check										
	Pressure seal bonnet–gate										
	Pressure seal bonnet–globe										

NOTE: See footnotes to Table 8.7.4.

Table 8.7.10   Weight Tolerances of Steel Piping

Specification	Size	Tolerance, %	
ASTM A53	Std wt	+ 5	− 5
	XS wt	+ 5	− 5
ASTM A120	XXS wt	+ 10	− 10
ASTM A106	Sch 10–120	+ 6.5	− 3.5
	Sch 140–160	+ 10	− 3.5
ASTM A335	12 in and under	+ 6.5	− 3.5
	Over 12 in	+ 10	− 5
ASTM A312 ASTM A376	12 in and under	+ 6.5	− 3.5
API 5L	Std. wt Reg wt XS wt XXS wt	+ 10	− 3.5
	Spec. plain end pipe	+ 10	− 5

Seamless pressure tubing may be either hot-finished or cold-drawn. Cold-drawn steel tubing is frequently process-annealed at temperatures above 1,200°F. To ensure quality, maximum hardness values are frequently specified. For example, in ASME Specification SA192,* Specification for Seamless Carbon Steel Boiler Tubes for High-Pressure Service, the following maximum hardness values are given.

Boiler tubes	Brinell hardness no. (tubes 0.200 in and over in wall thickness)	Rockwell hardness no. (tubes less than 0.200 in in wall thickness)
Hot-finished tubes	137	B77
Cold-drawn (normalized) tubes	125	B72

Piping Fabrication Methods Commercial steel pipe is furnished most commonly as seamless pipe. Seamless pipe can be produced by (1) piercing, (2) hollow forging, and (3) forging, turning, and boring. Commercial pipe is also produced by welding involving (1) electric resistance welding, (2) electric fusion welding, or (3) submerged arc welding. Some mills are prepared to extrude small-diameter pipe and tubing in a variety of geometric shapes or to cast large-diameter steel pipe.

Electric-Fusion Welding Flat plate, known as **skelp,** is prepared in proper width and thickness for the desired pipe inside and outside diameters. It is then charged into an electric furnace and, when the proper welding temperature has been reached, is drawn through a funnel-

* Specifications for boiler tubes generally reference the ASME Standards such as SA178, SA192, SA210, etc.

shaped die so shaped that the plate is gradually formed into the shape of a tube, with the edges of the plate being forced squarely together and fused. The formed pipe then passes through a series of rolls in which it is sized or drawn to final dimensions.

Electric-Resistance Welding For pipes or tubes sized 4-in (10.2-cm) OD and under, strip is fed onto a set of forming rolls which consists of horizontal and vertical rollers so placed as to gradually form the flat strip into a tube. The tube form then passes to the welding electrodes. The electrodes are copper disks connected to the secondary of a revolving transformer assembly. The copper-disk electrodes make contact on each side of the seam of the tube form; a flow of current takes place across the seam, and temperature is raised to the welding point. Outside flash is removed by a cutting tool as the tube leaves the electrodes; inside flash is removed either by an air hammer or by passing a mandrel through the welded tube after the tube has been cooled.

Submerged-Arc Electric Welding This process is used for pipes from 24-in to 36-in (61.0- to 9.1.4-cm) OD. Flat plate is first pressed into a U and later into an O shape. The O shape is placed in an automatic welder and backed up on the inside by a water-cooled copper shoe. Two electrodes in close proximity are used. The electrodes are not in actual contact with the pipe. The current passes from one electrode through a granular flux and across the gap in the pipe to the second electrode. The high temperature of the arc heats the edges of the plate; a welding rod placed just over the seam is thereby melted and metal is deposited in the groove. After the outside weld has been made, the pipe is conveyed to an inside welder where a similar operation is carried on, except that no backup shoe is needed.

Seamless Tubing and Pipe A heated billet is brought into contact with tapered revolving rolls in such a way that the billet is pulled into the space allowed between the rolls. A piercing mandrel is placed in this space; the soft center of the billet makes it possible for the rolls to draw the billet over the mandrel, producing a hollow shell. When the billet has entirely passed over the mandrel, it is in the form of a thick-walled seamless tube. The heavy-walled tube is then passed to a rolling mill which reduces the tube to pipe of proper outside diameter and wall thickness.

The method of fabrication described above is limited as to diameter and thickness. For seamless alloy tubes and for heavy-wall carbon-steel tubes or pipe, a process known as **cupping and drawing** is frequently used. A circular flat plate of proper diameter and thickness is heated, placed in a hydraulic press, and pressed by a ram through a die. The cup so formed is reheated and pressed through a smaller die, thus elongating the cup so that it becomes a short cylinder with one closed end. This short cylinder is then placed in a horizontal drawbench, and with reheating as necessary, is pushed by a ram through dies of successively smaller diameters until the desired outer diameter is reached.

Forged, Turned, and Bored Tubing In this process, the ingot is heated and forged to a rough cylindrical shape, oversize in both diameter and length. The forging is then placed in a lathe and the outside turned down to the desired outer diameter. Rough ends are then removed so that the finally desired length is obtained. The cylinder is then placed in a boring mill, and the inside bored out until the desired wall

Table 8.7.11   Diameter and Wall-Thickness Tolerances for Seamless Hot-finished Mechanical Tubing of Carbon and Alloy Steel (AISI)*

Specified size, OD, in	Ratio of wall thickness to OD	OD tolerance		Wall thicknesses tolerance, %							
				0.109 in and under		Over 0.109 to 0.172 in, incl.		Over 0.172 to 0.203 in, incl.		Over 0.203 in	
		Over	Under	Over	Under	Over	Under	Over	Under	Over	Under
Under 3	All wall thicknesses	0.023	0.023	16.5	16.5	15	15	14	14	12.5	12.5
3–5½, excl.	All wall thicknesses	0.031	0.031	16.5	16.5	15	15	14	14	12.5	12.5
5½–8 excl.	All wall thicknesses	0.047	0.047					14	14	12.5	12.5
8–10¾, incl	5% and over	0.047	0.047							12.5	12.5
8–10¾, incl.	Under 5%	0.063	0.063							12.5	12.5

* The common range of sizes of hot-finished tubes is 1½ to and including 10¾ in outside diameter with wall thickness not less than 0.095 in (no. 13 BWG) or 3% or more of the outside diameter. For sizes under 1½ or over 10¾ in outside diameter, the tolerances are commonly negotiated between the purchaser and producer.
SOURCE: AISI, ''Steel Products Manual.''

Table 8.7.12 Diameter and Wall-Thickness Tolerances for Seamless Cold-Worked Mechanical Tubing of Carbon and Alloy Steel (AISI)*

Size, OD, in	Unannealed or finish-annealed				Soft annealed or normalized				Quenched and tempered				Wall thickness all conditions, %	
	OD, in		ID, in		OD, in		ID, in		OD, in		ID, in			
	Over	Under	Over	Under	Over	Under	Over	Under	Over	Under	Over	Under	Over	Under
3/16–1/2, excl.††	0.004	0			0.006	0.002			0.010	0.010			15	15
1/2–1 1/2, excl.††‡§¶	0.005	0	0	0.005	0.008	0.002	0.002	0.008	0.015	0.015	0.015	0.015	10	10
1 1/2–3 1/2, excl.††‡§¶	0.010	0	0.005	0.010	0.015	0.005	0.005	0.015	0.030	0.030	0.030	0.030	10	10
3 1/2–5 1/2, excl.§¶	0.015	0	0.035	0.035	0.023	0.007	0.015	0.025	0.045	0.045	0.045	0.045	10	10
5 1/2–8, excl.¶ wall less than 5% OD	0.030	0.030	0.025	0.025	0.060	0.060	0.070	0.070					10	10
5 1/2–8, excl., wall from 5 to 7.5% OD	0.020	0.020	0.015	0.030	0.040	0.040	0.050	0.050						
5 1/2–8, excl.§ wall over 7.5% OD	0.030	0			0.045	0.015	0.037	0.053						
8–10 3/4, incl.,¶ wall less than 5% OD	0.045	0.045	0.050	0.050									10	10
8–10 3/4, incl., wall from 5 to 75% OD	0.035	0.035	0.040	0.040										
8–10 3/4, incl.,§ wall over 7.5% OD	0.045	0.045	0.015	0.040										

* For tolerances closer than those indicated, availability, and applicable tolerances for tubing less than 3/16 in OD or larger than 10 3/4-in OD, the producer should be consulted.

† For those tubes with inside diameter less than 1/2 in (or less than 5/8 in when the wall thickness is more than 20% of the outside diameter), which are not commonly drawn over a mandrel. Note § is not applicable. Unless otherwise agreed upon by the purchaser and producer, the wall thickness may vary 15% over and under that specified, and the inside diameter is governed by the outside diameter and wall-thickness tolerances shown.

‡ For tubes with inside diameter less than 1/2 in (or less than 5/8 in when the wall thickness is more than 20% of the outside diameter), which can be produced by the rod or bar mandrel process, the tolerances are as shown in the above table except that the wall-thickness tolerances are 10% over and under the specified wall thickness.

§ Many tubes with inside diameter less than 50% of outside diameter, or with wall thickness over 1 1/4 in, or weighing more than 90 lb/ft are difficult to draw over a mandrel. Unless otherwise agreed upon by the purchaser and producer the inside diameter may vary over or under by an amount equal to 10% of the wall thickness and the wall thickness may vary 12 1/2% over and under that specified.

¶ Tubing having a wall thickness less than 3% of the outside diameter cannot be straightened properly without a certain amount of distortion. Consequently, such tubes, while having an average outside diameter and inside diameter within the tolerances shown in the above table, require an ovality tolerance of 0.5% over and under nominal outside diameter, this being in addition to the tolerances indicated in the above table.

SOURCE: AISI, "Steel Products Manual."

Table 8.7.13 Finish Classifications Normally Recognized as Welded Mechanical Tubing

	Type	Condition
A	Flash-in, hot-rolled	Made by longitudinally forming and welding hot-rolled strip; the interior welding flash remains in place
B	Flash-in, cold-rolled	Made by longitudinally forming and welding cold-rolled strip; the interior flash remains in place
C	Flash controlled–0.010 in max, hot-rolled	Same as A, except that the interior flash is partially removed so that the height remaining does not exceed 0.010 in
D	Flash controlled 0.010 in max, cold-rolled	Same as C, except that cold-rolled strip has been used
E	Flash controlled–0.005 in max, hot-rolled	Same as C, except that flash height on tube inside does not exceed 0.005 in
F	Flash controlled–0.005 in max, cold-rolled	Same as E, except that cold-rolled strip has been used
G	Sink-drawn, hot-rolled	Tubing with flash controlled to 0.010 in max, which has been descaled and cold-drawn through a die to cold-finish the exterior surface
H	Sink-drawn, cold-rolled	Tubing with flash controlled to 0.010 in max, which has been cold-drawn through a die to cold-finish the exterior surface
I	Mandrel-drawn	Cold-rolled tubing with the interior flash removed and drawn through a die and over a mandrel to cold-finish the exterior and interior surfaces

thickness is secured. Because of the relatively high cost, this process is now rarely used.

Hollow-Forged Pipe and Tubing In this process, ingots are cast and their ends cropped; then they are placed in a furnace and heated to a specified temperature. The heated ingot is placed in a press where it is pierced. This hollow cylinder, open at one end, is then descaled and drawn over a mandrel on a horizontal drawbench. The closed end is then burned off, and the hollow forging is chemically descaled. Following this, the forging is straightened, placed in a lathe, and the outer diameter machined to a true dimension. The inside is dressed to remove scale, but no machining is done on the inside.

Carbon-steel piping is most frequently used as manufactured in accordance with ASTM Specifications A106 and A53 (or ASME Specifications SA106 and SA53). The chemical compositions of these two materials are identical except for the deoxidation practice which applies to the A106 pipe. Both are subjected to physical tests, but those for A106 are more rigorous. A53 and A106 are made in grades A and B; grade B has higher strength properties but is less ductile and, for this reason, grade A is permitted only for cold bending or close coiling. When carbon steel is intended for use in welded construction at temperatures in excess of 775°F (413°C), consideration should be given to the possibility of graphite formation.

Chromium-molybdenum steel has been used for temperatures up to 1,100°F (593°C). In the small diameters, the material is usually available in the seamless construction; because of the inability of the seamless mills to fabricate large-diameter and heavy-walled pipe, it may be necessary to resort to the more expensive hollow-forged or forged-and-bored piping for higher pressures and temperatures. The material for a high-temperature piping system should be selected after a careful review of technical and economic considerations; the following is intended only as being indicative of recent and current practice.

For temperatures up to 1,000°F, 1¼% Cr–½% Mo (A335, grade P11) is used. For temperatures from 950 to 1050°F, 2¼% Cr–1% Mo (A335, P22) generally is used. Where there is a combination of high temperatures and erosive action, 5% Cr–½% Mo (A335, grade 5) or other more highly chromium-molybdenum or chromium stainless steels have been used.

Stainless-steel piping is available in a variety of compositions, most popular of which are ASTM A213, grade TP304 (16% Cr–8% Ni), and ASTM A213, grade TP316 (18% Cr–12% Ni and 3% Mo). For high-temperature service, type 34 stainless steels are used (18% Cr–8% Ni and stabilized with columbium). This material may be used up to 1,200°F (649°C); particular care must be given to choice of welding filler metal to avoid brittleness in the welds.

The permissible stress values for a large variety of piping materials at low and elevated temperatures are provided in Table 8.7.15.

PIPE FITTINGS

The various major piping materials are also produced in the form of standard fittings. Among the more widely used are wrought-steel fittings, welded-steel fittings, cast-steel fittings, cast-iron fittings, ductile-iron fittings, malleable-iron fittings, brass and copper fittings, aluminum fittings, etc. Other major nonferrous piping materials are also produced in the form of cast and wrought fittings.

Cast-iron, ductile-iron, and malleable-iron fittings are made by conventional founding methods for a variety of joints including bell-and-spigot, flanged, and mechanical (gland-type), or other proprietary joint designs.

Schedule Designations Over 100 years ago piping was designated as standard, extra-strong, and double extra-strong. There was no provision for thin-walled pipe, and no intervening standard thicknesses between the three schedules, which covered too great a spread to be economical without intermediate weights. Table 8.7.3 lists piping as a function of the schedule number which is given, approximately, by the following relationship: Schedule no. = 1,000 $P/(SE)$, where P is operating pressure, lb/in^2 gage, S is allowable stress, lb/in^2 (Table 8.7.15), and E is the quality factor (Tables 8.7.16 and 8.7.17).

Commercial sizes of steel pipe are known by their nominal inside diameter (ID) from ⅛ in (0.3175 cm) to 12 in (30.5 cm). Above 12-in ID, pipe is usually known by its outside diameter (OD). All classes of pipe of a given nominal size have the same OD, the extra thickness for different weights being on the inside.

Thickness of Pipe The following notes, covering power piping systems, have been abstracted from Part 2 of the Code for Power Piping (ASME B31.1).

For inspection purposes, the minimum thickness of pipe wall to be used for piping at different pressures and for temperatures not exceeding those for the various materials listed in Table 8.7.15 shall be determined by the formula

$$t_m = \frac{PD}{2(SE + Py)} + A \qquad (8.7.1)$$

where t_m = minimum pipe-wall thickness, in, allowable on inspection; P = maximum internal service pressure, lb/in^2 gage (plus water-

Table 8.7.14 Typical Mechanical Properties of Resistance-Welded Mechanical Carbon-Steel Tubing

Type	Yield strength, lb/in^2	Tensile strength, lb/in^2	Elongation, %	Hardness, Rockwell B
Flash-in or flash-controlled, hot-rolled	48,000	58,000	29	68
Flash-in or flash-controlled, cold-rolled	68,000	76,000	17	84
Normalized	34,000	52,000	39	61
Sink-drawn	73,000	76,000	20	84
Mandrel-drawn	80,000	83,000	15	86

Table 8.7.15   Basic Allowable Stresses in Tension for Metals

Material	Spec. no.	P no. (5)†	Grade	Notes	Min temp. (6)
Iron					
Centrifugally cast pipe	A 377			(8) (9) (48)	−20
Castings					
Gray	A 48		20	(8) (9) (48)	−20
Gray	A 48		25	(8) (9) (48)	−20
Gray	A 48		30	(8) (9) (48)	−20
Gray	A 48		35	(8) (9) (48)	−20
Gray	A 48		40	(8) (9) (48)	−20
Gray	A 48		45	(8) (9) (48)	−20
Gray	A 48		50	(8) (9) (48)	−20
Gray	A 48		55	(8) (9) (48)	−20
Gray	A 48		60	(8) (9) (48)	−20
Gray	A 278		70	(8) (9) (53)	−20
Gray	A 278		80	(8) (9) (53)	−20
Cupola malleable	A 197			(8) (9)	−20
Malleable	A 47		32510	(8) (9)	−20
Malleable	A 47		35018	(8) (9)	−20
Ductile	A 395			(8) (9)	−20
Ferritic ductile	A 395			(8) (9)	−20
Austenitic ductile	A 571		Type D-2M	(9)	−20

Materials	Spec. no.	P no. (5)	Grade	Notes	Min temp. (6)	SMTS, ksi	SMYS, ksi
Carbon steel							
Pipes and tubes							
	A 120	1		(8)	−20		
A 285 gr. A	A 672	1	A45	(57) (59) (67)	−20	45	24
	A53	1	Type F	(8)	−20	45	25
Butt weld	API 5L	1	A25				
Smls & ERW	API 5L	1	A25	(57) (59)	−20	45	25
	A 179	1		(57) (59)	−20	47	26
	A 135	1	A	(57) (59)	−20	48	30
A 285 gr. B	A 672	1	A50	(57) (59) (67)	−20	50	27
A 285 gr. C	A 134	1		(8) (57)	−20	55	30
A 285 gr. C	A 671	1	CA55	(59) (67)	−20	55	30
A 516 gr. 60	A 671	1	CC60	(57) (67)	−20	60	32
A 515 gr. 60	A 672	1	B60	(57) (67)	−20	60	32
	A 135	1	B	(57) (59)	−20	60	35
	A 53	1	B	(57) (59)	−20		
	A 106	1	B	(57)	−20	60	35
	A 139	1	D	(8)	−20	60	46
(> ⅜ in thick)	A 381	SP3	Y48	(51)	−20	62	48
(> ⅜ in thick)	A 381	SP3	Y50	(51)	−20	64	50
A 515 gr. 65	A 672	1	B65	(57) (67)	−20	65	35
(> ⅜ in thick)	A 381	SP3	Y52	(51)	−20	66	52
(≤ ⅜ in thick)	A 381	SP3	Y48	(51)	−20	67	48
A 516 gr. 70	A 671	1	CC70	(57) (67)	−20	70	38
	A 106	1	C	(57)	−20	70	40
(≤ ⅜ in thick)	A 381	SP3	Y52	(51)	−20	72	52
A 299 (> 1 in thick)	A 672	1	N75	(57) (67)	−20	75	40
A 299 (≤ 1 in thick)	A 672	1	N75	(57) (67)	−20	75	42

NOTE: Footnotes appear at the end of the table.

SMTS,‡ ksi	SMYS,‡ ksi	Basic allowable stress S, ksi (1), at metal temperature, °F (7)						
		Min temp. to 100	200	300	400	500	600	650
		4.0	4.0	4.0	4.0			
20		2.0	2.0	2.0	2.0 ‖			
25		2.5	2.5	2.5	2.5 ‖			
30		3.0	3.0	3.0	3.0 ‖			
35		3.5	3.5	3.5	3.5 ‖			
40		4.0	4.0	4.0	4.0 ‖			
45		4.5	4.5	4.5	4.5 ‖			
50		5.0	5.0	5.0	5.0 ‖			
55		5.5	5.5	5.5	5.5 ‖			
60		6.0	6.0	6.0	6.0 ‖			
70		7.0	7.0	7.0	7.0	7.0	7.0	7.0 ‖
80		8.0	8.0	8.0	8.0	8.0	8.0	8.0 ‖
40	30	‖ 8.0	8.0	8.0	8.0	8.0	8.0	8.0 ‖
50	32	10.0	10.0	10.0	10.0	10.0	10.0	10.0
53	35	10.6	10.6	10.6	10.6	10.6	10.6	10.6 ‖
60	40	‖ 20.0	19.0	17.9	16.9	15.9	14.9	14.1 ‖
65	30	20.0						

Basic allowable stress S, ksi (1), at metal temperature, °F (7)															
Min temp. to 100	200	300	400	500	600	650	700	750	800	850	900	950	1,000	1,050	1,100
‖ 12.0	11.4														
15.0	14.6	14.2	13.7 ‖	13.0	11.8	11.6	11.5	10.3	9.0	7.8	6.5	4.5	2.5	1.6	1.0
‖ 15.0	15.0	14.5	13.8												
15.0	15.0	14.5	13.8 ‖												
15.7	15.0	14.2	13.5	12.8	12.1	11.8	11.5	10.6	9.2	7.9	6.5	4.5	2.5	1.6	1.0
16.0	16.0	16.0	16.0	16.0	14.8	14.5	14.4	10.7	9.3	7.9	6.5	4.5	2.5	1.6	1.0
16.7	16.4	16.0	15.4	14.6	13.3	13.1	13.0	11.2	9.6	8.1	6.5	4.5	2.5	1.6	1.0
18.3	18.3	17.7	17.2	16.2	14.8	14.5	14.4	12.0	10.2	8.3	6.5				
18.3	18.3	17.7	17.2	16.2	14.8	14.5	14.4	12.1	10.2	8.4	6.5	4.5	2.5	1.6	1.0
20.0	19.5	18.9	18.3	17.3	15.8	15.5	15.4	13.0	10.8	8.7	6.5	4.5	2.5		
20.0	19.5	18.9	18.3	17.3	15.8	15.5	15.4	13.0	10.8	8.7	6.5	4.5	2.5	1.6	1.0
20.0	20.0	20.0	20.0	18.9	17.3	17.0	16.5	13.0	10.8	8.7	6.5	4.5	2.5		
20.0	20.0	20.0	20.0	18.9	17.3	17.0	16.5	13.0	10.8	8.7	6.5	4.5	2.5	1.6	1.0
20.0	20.0	20.0													
20.6	19.7	18.7	17.8	16.9	16.0	15.5									
21.3	20.3	19.3	18.4	17.4	16.5	16.0									
21.7	21.3	20.7	20.0	18.9	17.3	17.0	16.8	13.9	11.4	9.0	6.5	4.5	2.5	1.6	1.0
22.0	21.0	19.9	18.9	17.9	17.0	16.5									
22.3	21.3	20.2	19.2	18.2	17.3	16.7									
23.3	23.1	22.5	21.7	20.5	18.7	18.4	18.3	14.8	12.0	9.3	6.5	4.5	2.5		
23.3	23.3	23.3	22.9	21.6	19.7	19.4	19.2	14.8	12.0						
24.0	22.9	21.7	20.7	19.6	18.6	18.0									
25.0	24.4	23.7	22.9	21.6	19.7	19.4	19.2	15.7	12.6	9.5	6.5	4.5	2.5	1.6	1.0
25.0	25.0	24.8	24.0	22.7	20.7	20.4	20.2								

Table 8.7.15 Basic Allowable Stresses in Tension for Metals (*Continued*)

Materials	Spec. no.	P no. (5)	Grade	Notes	Min temp. (6)	SMTS, ksi	SMYS, ksi
Carbon steel (*Cont.*)							
Pipes (structural grade)							
A 283 gr. A	A 134	1		(8)	−20	45	24
A 570 gr. 30	A 134	1		(8)	−20	49	30
A 283 gr. B	A 134	1		(8)	−20	50	27
A 570 gr. 33	A 134	1		(8)	−20	52	33
A 570 gr. 36	A 134	1		(8)	−20	53	36
A 570 gr. 40	A 134	1		(8)	−20	55	40
A 36	A 134	1		(8)	−20	58	36
A 570 gr. 45	A 134	1		(8)	−20	60	45
A 570 gr. 50	A 134	1		(8)	−20	65	50
Forgings and fittings							
	A 350	1	LF-1	(9) (57) (59)	−20	60	30
	A 105	1		(9) (57) (59)	−20	70	36
Castings							
	A 216	1	WCA	(9) (57)	−20	60	30
	A 352	1	LCB	(9) (57)	−50	65	35
	A 216	1	WCC	(9) (57)	−20	70	40

Material	Spec. no.	P no. (5)	Grade	Notes	Min temp. (6)
Low and intermediate alloy steel					
Pipes					
C–½Mo	A 335	3	P1	(58)	−20
½Cr–½Mo	A 691	3	½Cr	(11) (67)	−20
A 387 gr. 2 cl. 1					
1Cr–½Mo	A 691	4	1Cr	(11) (67)	−20
A 387 gr. 12 cl. 1					
1½Si–½Mo	A 335	3	P15		−20
7Cr–½Mo	A 426	5	CP7	(10)	−20
3Cr–Mo	A 426	5	CP21	(10)	−20
¾C–¾Ni–Cu–Al	A 333	4	4		−150
2Cr–½Mo	A 369	4	FP3b		−20
7Cr–½Mo	A 335	5	P7		−20
1¼Cr–½Mo	A 335	4	P11		−20
5Cr–½Mo	A 691	5	5Cr	(11) (67)	
A 387 gr. 5 cl. 1					
5Cr–½Mo–Si	A 335	5	P5b		−20
9Cr–1Mo	A 335	5	F9		−20
3Cr–1Mo	A 335	5	P21		−20
2¼Cr–1Mo	A 691	5	2¼Cr	(11) (67)	−20
A 387 gr. 22 cl. 1					
2¼Cr–1Mo	A 335	5	P22		−20
2Ni–1Cu	A 333	9A	9		−100
C–Mo A 204 gr. B	A 672	3	L65	(11) (58) (67)	−20
2¼Ni	A 333	9A	7		−100
3½Ni	A 333	9B	3		−150
2¼Ni A 203 gr. B	A 671	9A	CF70	(11) (65) (67)	−20
C–Mo A 204 gr. B	A 672	3	L70	(11) (58) (67)	−20
1¼Cr–½Mo	A 426	4	CP11	(10)	−20
C–Mo A 204 gr. C	A 672	3	L75	(11) (58) (67)	−20
12¾Cr	A 426	6	CPCA-15	(10)	−20
5Cr–½Mo	A 426	5	CP5	(10)	−20
9Ni	A 333	11A-SG1	8	(47)	−320
Forgings and fittings					
C–½Mo	A 234	3	WP1	(58)	−20
1Cr–½Mo	A 234	4	WP12		−20
1¼Cr–½Mo	A 234	4	WP11		−20

Basic allowable stress S, ksi (1), at metal temperature, °F (7)															
Min temp. to 100	200	300	400	500	600	650	700	750	800	850	900	950	1,000	1,050	1,100
13.7	13.0	12.4													
15.0	15.0	15.0	15.0												
15.3	14.4	13.9													
15.9	15.9	15.9	15.9												
16.3	16.3	16.3	16.3												
16.9	16.9	16.9	16.9												
17.6	16.8	16.8	16.8												
18.4	18.4	18.4	18.4												
19.9	19.9	19.9	19.9												
20.0	18.3	17.7	17.2	16.2	14.8	14.5	14.4	12.9	10.8	8.6	6.5	4.5	2.5	1.6	1.0
23.3	21.9	21.3	20.6	19.4	17.8	17.4	17.3	14.8	12.0	9.3	6.5	4.5	2.5	1.6	1.0
20.0	18.3	17.7	17.2	16.2	14.8	14.5	14.4	13.0	10.8	8.6	6.5	4.5	2.5	1.6	1.0
21.7	21.3	20.7	20.0	18.9	17.3	17.0	16.8	13.8	11.4	8.9	6.5	4.5	2.5	1.6	1.0
23.3	23.3	23.3	22.9	21.6	19.7	19.4	19.2	14.8	12.0	9.3	6.5	4.5	2.5		

		Basic allowable stress S, ksi (1), at metal temperature, °F (7)							
SMTS, ksi	SMYS ksi	Min temp. to 100	200	400	600	800	1,000	1,100	1,200
55	30	18.3	18.3	16.9	15.7	13.5	4.8		
55	33	18.5	18.3	18.3	17.3	13.8	5.9		
55	33	18.3	18.3	18.3	17.3	15.9	6.6	2.6	1.0
60	30	18.8	18.2	17.0	15.9	14.4	6.3	2.4	
60	30	18.8	17.9	16.2	14.5	12.5	5.0	2.5	1.2
60	30	18.8	18.1	16.8	15.5	13.9	7.0	4.0	1.5
60	35	20.0	19.1	17.3	15.5				
60	30	20.0	18.5	16.4	15.7	13.5	6.2	2.6	1.0
60	30	20.0	18.1	17.2	16.8	12.8	5.8	2.9	1.2
60	30	20.0	18.7	17.5	16.7	15.0	7.8	4.0	1.2
60	30	20.0	18.1	17.2	16.8	12.8	5.8	2.9	1.3
60	30	20.0	18.1	17.2	16.8	12.8	7.4	3.3	1.5
60	30	20.0	18.7	17.5	16.7	15.0	7.0	4.0	1.5
60	30	20.0	18.5	17.9	17.9	15.2	7.8	4.2	1.6
60	30	20.0	18.5	17.9	17.9	15.2	7.8	4.2	2.0
63	46	21.0							
65	37	21.7	21.7	20.7	19.3	15.8	4.8		
65	35	21.7	19.6	18.7	16.8	11.4	2.5	1.0	
65	35	21.7	19.6	18.7	16.8	11.4	2.5	1.0	
70	40	23.3							
70	40	23.3	23.3	22.5	20.9	17.5	4.8		
70	40	23.3	23.3	23.3	22.3	15.0	7.8	4.0	1.2
75	43	25.0	25.0	24.1	22.5	18.8	4.8		
90	65	30.0							
90	60	30.0	28.0	24.1	20.1	14.5	5.6	3.1	1.3
100	75	31.7	31.7						
55	30	18.3	18.3	16.9	15.7	13.5	4.8		
60	30	20.0	18.7	17.5	16.7	15.0	7.5	2.8	1.0
60	30	20.0	18.7	17.5	16.7	15.0	7.8	4.0	1.2

Table 8.7.15   Basic Allowable Stresses in Tension for Metals (*Continued*)

Material	Spec. no.	P no. (5)†	Grade	Notes	Min temp. (6)
Low and intermediate alloy steel (*Cont.*)					
Forgings and fittings (*Cont.*)					
5Cr–½Mo	A 234	5	WP5		−20
9Cr–½Mo	A 234	5	WP9		−20
3½Ni	A 350	9B	LF3	(9)	−150
C–½Mo	A 182	3	F1	(9) (58)	−20
½Cr–½Mo	A 182	3	F2	(9)	−20
1¼Cr–½Mo	A 182	4	F11	(9)	−20
5Cr–½Mo	A 182	5	F5	(9)	−20
13Cr	A 182	6	F6a Cl.1		−20
3Cr–1Mo	A 182	5	F21	(9)	−20
13Cr	A 182		F6a Cl.2		−20
5Cr–½Mo	A 182	5	F5a	(9)	−20
9Ni	A 420	11A-SG1	WPL8	(47)	−320
13Cr	A 182		F6a Cl.3		−20
13Cr–½Mo	A 182		F6b		
Castings					
C–½Mo	A 217	3	WC1	(9) (58)	−20
2½ Ni	A 352	9A	LC2	(9)	−100
Ni–Cr–1Mo	A 217	4	WC5	(9)	−20
2¼Cr–1Mo	A 217	5	WC9	(9)	−20
12Cr	A 217	6	CA15	(9) (35)	−20
5Cr–½Mo	A 217	5	C5	(9)	−20

Material	Spec. no.	P no. (5)	Grade	Notes	Min temp. (6)	SMTS, ksi	SMYS, ksi
Stainless steel (4) (40)							
Pipes and tubes							
18CR–8Ni pipe	A 312	8	TP304L		−425	70	25
Type 304L A 240	A 358	8	304L	(36)	−425	70	25
16Cr–12Ni–2Mo pipe	A 312	8	TP316L		−325	70	25
23Cr–13Ni	A 451	8	CPH8	(26) (28) (35)	−325	65	28
18Cr–Ti tube	A 268	7	TP430TI	(35) (49)	−20	60	40
16Cr–8Ni–2Mo pipe	A 376	8	16–8–2H	(26) (31) (35)	−325	75	30
12Cr–Al tube	A 268	7	TP405	(35)	−20	60	30
16Cr tube	A 268	7	TP430	(35) (49)	−20	60	35
Type 310S A 240	A 358	8	310S	(28) (31) (35) (36)	−325	75	30
18Cr–10Ni–Ti	A 312	8	TP321	(28)	−325	75	30
Type 321 A 240	A 358	8	321	(28) (30) (36)	−325	75	30
Type 309S A 240	A 358	8	309S	(28) (31) (35) (36)	−325	75	30
18Cr–8Ni	A 451	8	CPF8	(26) (28)	−425	70	30
Type 347 A 240	A 358	8	347	(28) (30) (36)	−425	75	30
Type 348 A 240	A 358	8	348	(28) (30) (36)	−325		
18 Cr–10Ni–Cb pipe	A 409	8	TP348	(28) (30) (36)	−325		
Type 310S A 240	A 358	8	310S	(28) (29) (31) (35) (36)	−325	75	30
18Cr–10Ni–Ti pipe	A 312	8	TP321H		−325	75	30
Type 316 A 240	A 358	8	316	(26) (28) (31) (36)	−325		
16Cr–12Ni–2Mo pipe	A 376	8	TP316	(26) (28) (31) (35)	−325	75	30
18Cr–13Ni–3Mo pipe	A 312	8	TP317	(26) (28)	−325		
16Cr–12Ni–2Mo pipe	A 430	8	FP316H	(26) (31) (36)	−325	70	30
16Cr–12Ni–2Mo pipe	A 312	8	TP316H	(26)	−325	75	30
18Cr–10Ni–Cb pipe	A 430	8	FP347H	(30) (36)	−325	70	30
18Cr–10Ni–Cb pipe	A 312	8	TP348H		−325	75	30
18Cr–8Ni pipe	A 430	8	FP304	(26) (31) (36)	−425	70	30
Type 304 A 240	A 358	8	304	(26) (28) (31) (36)	−425	75	30
						70	30
18Cr–8Ni pipe	A 312	8	TP304H	(26)	−325	75	30
18Cr–8Ni	A 452	8	TP304H	(26)	−325	75	30

SMTS, ksi	SMYS, ksi	Basic allowable stress S, ksi (1), at metal temperature, °F (7)							
		Min temp. to 100	200	400	600	800	1,000	1,100	1,200
60	30	20.0	18.1	17.2	16.8	12.8	5.8	2.9	1.3
60	30	20.0	18.1	17.2	16.8	12.8	7.4	3.3	1.5
70	37.5	23.3							
70	40	23.3	23.3	22.5	20.9	17.5	4.8		
70	40	23.3	23.3	22.5	20.9	17.5	5.9		
70	40	23.3	23.3	22.5	20.9	19.2	6.9	2.8	1.2
70	40	23.3	23.3	22.4	22.0	14.8	5.8	2.9	1.3
70	40								
75	45	25.0	25.0	24.1	23.8	22.5	6.8	3.2	1.3
85	55	28.3	28.3	27.2	26.1	23.2			
90	65	30.0	29.9	28.9	28.3	19.1	5.8	2.9	1.3
110	75	31.7	31.7						
110	85								
110–135	90								
65	35	21.7	21.5	19.7	18.3	15.8	4.8		
70	40	23.3	17.5	17.5	17.5				
70	40	23.3	23.3	22.5	20.9	17.5	6.9	2.8	
70	40	23.3	23.3	22.5	22.4	21.0	7.6	4.4	1.3
90	65	30.0	21.5	20.0	18.8	16.8	5.0	2.3	1.0
90	60	30.0	29.9	28.9	28.3	19.1	5.8	2.9	1.3

Basic allowable stress S, ksi (1), at metal temperature, °F (7)								
Min temp. to 100	200	400	600	800	1,000	1,200	1,400	1,500
16.7	16.7	15.8	14.0	13.0	7.8	3.2	1.1	0.9
16.7	16.7	15.5	13.5	12.4	11.2	6.4	1.8	1.0
18.7	18	18.7	18.0	16.3	10.4	3.7	1.3	0.8
20.0								
20.0								
20.0	18.4	17.4	16.8	11.1	4.0			
20.0	20.0	19.2	18.5	11.1	6.5	1.7		
20.0	20.0	20.0	19.2	17.5	11.0	2.5	0.4	0.2
20.0	20.0	18.6	16.4	15.5	13.8	3.6	0.8	0.3
20.0	20.0	20.0	19.2	17.5	10.5	3.8	1.3	0.7
20.0	20.0	17.5	15.7	14.8	10.8	4.4	1.3	0.8
20.0	20.0	20.0	19.3	18.3	14.0	4.4	1.2	0.8
20.0	20.0	20.0	19.2	17.5	11.0	6.0	1.6	0.8
20.0	20.0	18.6	16.4	15.5	14.0	5.4	1.9	1.1
20.0	20.0	19.3	17.0	15.9	15.3	7.4	2.3	1.3
20.0	20.0	19.3	17.0	15.9	15.3	7.4	2.3	1.3
20.0	20.0	19.2	18.3	18.2	18.0	7.9	2.5	1.3
20.0	20.0	20.0	19.3	18.3	18.0	7.9	2.5	1.3
20.0	20.0	18.7	16.4	15.2	13.8	6.0	2.3	1.4
20.0	20.0	18.7	16.4	15.2	13.8	6.0	2.3	1.4
20.0	20.0	18.7	16.5	15.1	13.8	6.0	2.2	1.4

Table 8.7.15 Basic Allowable Stresses in Tension for Metals (*Continued*)

Material	Spec. no.	P no. (5)	Grade	Notes	Min temp. (6)	SMTS, ksi	SMYS, ksi
Stainless steel (4) (40) (*Cont.*)							
Pipes and tubes (*Cont.*)							
18Cr–10Ni–Mo	A 451	8	CPF8M	(26) (28)	−425	70	30
27Cr tube	A 268	10E	TP446	(35)	−20	70	40
26Cr–3Ni–1Mo tube	A 268	10E	TP329	(35)	−20	90	70
Forgings and Fittings							
18Cr–8Ni	A 182	8	F304L	(9)	−425	65	25
16Cr–12Ni–2Mo	A 182	8	F316L	(9)	−325	65	25
20Ni–8Cr	A 182	8	F10	(9) (26) (28) (39)	−325	80	30
25Cr–20Ni	A 182	8	F310	(9) (28) (35) (39)	−425	75	30
18Cr–10Ni–Ti	A 182	8	F321	(9) (21) (28)	−325	75	30
25Cr–20Ni	A 182	8	F310	(9) (28) (29) (35) (39)	−425	75	30
18Cr–10Ni–Cb	A 182	8	F347	(9) (21) (28)	−425	75	30
18Cr–10Ni–Ti	A 182	8	F321H	(9) (21)	−325	75	30
16Cr–12Ni–2Mo	A 182	8	F316H	(9) (21) (26)	−325	75	30
18Cr–10Ni–Cb	A 182	8	F347H	(9) (21)	−325	75	30
16Cr–12Ni–2Mo	A 403	8	WP316	(26) (28) (31) (32) (37)	−325	75	30
18Cr–8Ni	A 182	8	F304	(9) (21) (26) (28)	−425	75	30
18Cr–8Ni	A 403	8	WP304H	(26) (31) (32) (37)	−325	75	30
Castings							
28Ni–20Cr–2Mo–3Cb	A 351	8	CN7M	(9) (30)	−325	62	25
35Ni–15Cr–Mo	A 351	8	HT30	(36) (39)	−325	65	28
15Cr–15Ni–2Mo–Cb	A 351	8	CF10MC	(9) (30)	−325	70	30
18Cr–8Ni	A 351	8	CF3	(9)	−425	70	30
18Cr–8Ni	A 351	8	CF8	(9) (26) (27) (31)	−425	70	30
25Cr–13Ni	A 351	8	CH10	(9) (27) (31) (35)	−325	70	30
18Cr–10Ni–2Mo	A 351	8	CF8M	(9) (26) (27) (30)	−425	70	30
25Cr–20Ni	A 351	8	HK40	(35) (36) (39)	−325	62	35
18Cr–8Ni	A 351	8	CF3A	(9) (26)	−425	70	35

Material	Spec. no.	P no. (5)	Grade	Class§	Notes	Size range, in	Min temp. (6)
Nickel and nickel alloy (4)							
Pipes and tubes							
Low C Ni	B 161	41	201 (N02201)	Ann.		>5 OD	−325
Ni	B 161	41	200 (N02200)	Ann.		>5 OD	−325
Ni–Cu	B 165	42	400 (N04400)	Ann.		>5 OD	−325
Ni–Cr–Fe	B 167	43	600 (N06600)	H.F. or H.F. Ann.		>5 OD	−325
Ni–Cu	B 165	42	400 (N04400)	Ann.		≤5 OD	−325
Ni–Cr–Fe	B 167	43	600 (N06600)	H.F. or H.F. Ann.		≤5 OD	−325
Ni–Fe–Cr	B 407	45	800 (N08800)	C.D. Ann.	(61)		−325
Ni–Cr–Fe–Mo–Cu	B 619	45	G1 (N06007)	Sol. Ann.			−325
Cr–Ni–Fe–Mo–Cu–Cb	B 464	45	20Cb (N08020)	Ann.			−325
Ni–Cr–Fe–Mo–Cu	B 622	45	G (N06007)	Sol. Ann.			−325
Ni–Cr–Mo–Fe	B 619		X (N06002)	Sol. Ann.			−325
Ni–Mo–Cr	B 619	44	C-276 (N10276)	Sol. Ann.			−325
Ni–Mo–Cr	B 622	44	C-276 (N10276)	Sol. Ann.			−325
Ni–Mo	B 619	44	B-2 (N10665)	Sol. Ann.			−325
Ni–Mo	B 619		B (N10001)	Sol. Ann.			−325
Ni–Cr–Mo–Cb	B 444	43	625 (N06625)	Ann.	(64)		−325
Ni–Mo	B 622	44	B-2 (N10655)	Sol. Ann.			−325
Forgings and fittings							
Low C Ni	B 160	41	201 (N02201)	Ann.	(9)	All	−325
Ni	B 160	41	200 (N02200)	H.F.	(9)	All	−325
Ni–Cu	B 164	42	400 (N04400)	Ann. Forg.	(9) (13)	All	−325
	B 366	43	WPNC1 (N06600)		(32)	All	−325

Basic allowable stress S, ksi (1), at metal temperature, °F (7)								
Min temp. to 100	200	400	600	800	1,000	1,200	1,400	1,500
20.0	20.0	19.4	17.1	15.5	15.4	6.8	2.3	1.4
23.3	23.3	20.4	18.4	16.2	4.5			
30.0								
16.7	16.7	15.8	14.0	13.0	7.8	3.2	1.1	0.9
16.7	16.7	15.5	13.5	12.4	11.2	6.4	1.8	1.0
20.0								
20.0	20.0	20.0	19.2	17.5	11.0	2.5	0.4	0.2
20.0	20.0	18.6	16.4	15.5	13.8	3.6	0.8	0.3
20.0	20.0	20.0	19.2	17.5	11.0	6.0	1.6	0.8
20.0	20.0	20.0	19.3	18.3	14.0	4.4	1.2	0.8
20.0	20.0	18.6	16.4	15.5	14.0	5.4	1.9	1.1
20.0	20.0	19.3	17.0	15.9	15.3	7.4	2.3	1.3
20.0	20.0	20.0	19.3	18.3	18.0	7.9	2.5	1.3
20.0	20.0	19.3	17.0	15.9	15.3	7.4	2.3	1.3
20.0	20.0	18.7	16.4	15.2	13.8	6.0	2.3	1.4
20.0	20.0	18.7	16.4	15.2	13.8	6.0	2.3	1.4
16.6								
18.6								
20.0								
20.0	20.0	17.6	15.6	14.7				
20.0	20.0	17.6	15.6	14.7	10.7	4.5	1.4	0.7
20.0	20.0	20.0	19.2	17.5	10.5	3.7	1.2	0.7
20.0	20.0	19.4	17.1	15.6	13.1	6.7	2.4	1.5
20.6								
23.2								

		Basic allowable stress S, ksi (1), at metal temperature, °F (7)								
SMTS, ksi	SMYS, ksi	Min temp. to 100	200	400	600	800	1,000	1,200	1,400	1,500
50	10	6.7	6.4	6.2	6.2	5.9	3.0	1.2		
55	12	8.0	8.0	8.0	8.0					
70	25	16.7	14.7	13.2	13.2	12.7				
75	25	16.7	16.7	16.7	16.7	16.7	7.0	2.0		
70	28	18.7	16.4	14.8	14.8	14.2				
80	30	20.0	20.0	20.0	20.0	20.0	7.0	2.0		
75	30	20.0	20.0	20.0	20.0	20.0	17.6	6.6	1.1	0.8
90	35	22.5	22.5	21.9	21.1	20.5	18.9			
85	35	23.3	21.3	20.6	20.3	19.2				
90	35	23.5	23.3	23.3	22.7	21.8				
100	40	26.6	24.1	22.9	21.1	19.8	18.6	11.3		
100	41	27.3	27.3	27.3	25.4	22.9	21.8			
100	41	27.5	27.3	27.3	25.4	23.0				
110	51	29.1	28.9	28.9	28.9	28.9				
100	45	30.0	30.0	30.0	30.0	27.6				
120	60	30.0	30.0	28.2	26.4	26.0	26.0	13.2		
110	51	34.2	34.0	34.0	34.0	34.0				
50	10	6.6	6.4	6.2	6.2	5.9	3.0	1.2		
60	15	10.0	10.0	10.0	8.3					
70	25	16.6	14.6	13.2	13.1	12.7				
75	25	16.7	16.7	16.7	16.7	16.7	7.0	2.0		

Table 8.7.15 Basic Allowable Stresses in Tension for Metals (*Continued*)

Material	Spec. no.	P no. (5)	Grade	Class§	Notes	Size range, in	Min temp. (6)
Nickel and nickel alloy (4) (*Cont.*)							
Forgings and fittings (*Cont.*)							
Ni–Cr–Fe	B 166	43	600 (N06600)	H.F.	(9) (13)	All	− 325
			WPHX				
Ni–Cr–Mo–Fe	B 366		(N06002)		(32)		− 325
Castings							
Ni–Mo–Cr	A 494		CW-12M-1		(9) (44)		− 325
Ni–Mo–Cr	A 494		CW-12M-2		(9) (44)		− 325

Material	Spec. no.	P no. (5)	Grade	Notes	Min temp. (6)
Unalloyed titanium					
Pipes and tubes					
C.P.	B 337	51	1	(17)	− 75
C.P.	B 337	51	2	(17)	− 75
C.P.	B 337	52	3	(17)	− 75

Material	Spec. no.	P no. (5) (46)	Class	Temper	Size range, in	Notes
Copper and copper alloy						
Pipes and tubes						
Cu tube	B 75	31	C10200, C12000, C12200, C14200	Annealed		(14)
Cu tube	B 88	31	C10200, C12000, C12200	Annealed		(14) (24)
Red brass	B 43	32	C23000	Annealed		(14)
Cu–Ni 90/10	B 467	34	C70600	Annealed	> 4.5 OD	(14)
Cu–Ni 90/10	B 467	34	C70600	Annealed	≤ 4.5 OD	(14)
Cu–Ni 70/30	B 467	34	C71500	Annealed	> 4.5 OD	(14)
Cu	B 42	31	C10200, C1200, C12200	Drawn	NPS 2½ thru 12	(14) (34)
Cu tube	B 88	31	C10200, C12000, C12200	Drawn		(14) (24) (34)
Cu–Ni 70/30	B 466	34	C71500	Annealed		(14)
Forgings						
Cu	B 283	a	C11000			(9) (14)
High Si bronze (A)	B 283	33	C65500			(9) (14)
Forging brass	B 283	a	C37700			(9) (14)
Leaded naval brass	B 283	a	C48500			(9) (14)
Naval brass	B 283	32	C46400			(9) (14)
Mn–bronze (A)	B 283	32	C67500			(9) (14)
Al–Si bronze	B 283	35	C63900			(9) (14)
Castings						
Composition bronze	B 62	a	C83600			(9)
Leaded Ni–bronze	B 584	a	C97600			(9)
Leaded Sn–bronze	B 584	a	C92200			(9)
Steam bronze	B 61	a	C92200			(9)
Sn–bronze	B 584	b	C90300			(9)
Leaded Mn–bronze	B 584	a	C86400			(9)
No. 1 Mn–bronze	B 584	b	C86500			(9)
Al–bronze	B 584	b	C95200			(9)
Si–Al–bronze	B 584	b	C95600			(9)

SMTS, ksi	SMYS, ksi	Basic allowable stress S, ksi (1), at metal temperature, °F (7)								
		Min temp. to 100	200	400	600	800	1,000	1,200	1,400	1,500
85	35	23.3	21.2	21.2	21.2	20.4	14.5	5.5		
95	35	23.3	23.3	22.9	21.1	19.8	19.3	11.3		
72	46	24.0	16.4	16.4	16.4					
72	46	24.0	16.4	16.4	16.4	15.5	14.6			

SMTS, ksi	SMYS, ksi	Basic allowable stress S, ksi (1), at metal temperature, °F (7)					
		Min temp. to 100	200	300	400	500	600
35	25	11.7	9.7	7.7	6.4	5.3	4.2
50	40	16.7	16.7	12.3	9.8	8.0	7.3
65	55	21.7	19.0	15.6	12.3	9.9	8.0

Min temp. (6)	Specified min strength, ksi		Basic allowable stress S, ksi (1), at metal temperature, °F (7)						
	Tensile	Yield	Min temp. to 100	200	300	400	500	600	700
−325	30	9	6.0	4.8	4.7	3.0	0.8		
−325	30	9	6.0	5.9	5.0	2.5	0.8		
−325	40	12	8.0	8.0	8.0	5.0			
−325	38	13	8.7	8.1	7.8	7.5	7.2	6.0	
−325	40	15	10.0	9.5	8.9	8.5	8.0	6.0	
−325	45	15	10.0	9.5	9.1	8.6	8.2	8.0	7.8
−325	36	30	12.0	9.0	8.7	8.2			
−325	36	30	12.0	8.7	8.0	2.5	0.8		
−325	50	18	12.0	11.3	10.8	10.3	9.9	9.6	9.4
−325	33	11	7.3	6.5	5.0	2.5	0.8		
−325	52	18	12.0	10.0	10.0	2.0			
−325	58	23	15.3	12.0	10.5	2.0			
−325	62	24	16.0	15.0	13.0	2.0			
−325	64	26	17.3	15.3	13.0	2.0			
−325	72	34	22.7	12.0	10.5	2.0			
−325	83	41	27.3	17.3	17.1	16.8			
−325	30	14	9.4	9.4	9.1	8.6			
−325	40	17	10.0	7.3	6.3				
−325	34	16	10.6	10.6	10.6	10.3			
−325	34	16	10.6	10.6	10.6	10.3	9.0		
−325	40	18	12.0	9.5	8.5	7.0			
−325	60	20	13.3	12.0	10.5				
−325	65	25	16.6	13.4	10.5				
−325	65	25	16.6	16.1	15.5	14.5	10.0		
−325	60	28	18.8						

Table 8.7.15 Basic Allowable Stresses in Tension for Metals (*Continued*)

Material	Spec. no.	P no. (5) (46)	Class	Temper	Size range, in	Notes
Copper and copper alloy (*Cont.*)						
Al–bronze	B 584	b	C95400			(9)
Mn–bronze	B 584	a	C86700			(9)
High-strength Mn–bronze	B 584	b	C86300			(9)

Material	Spec. no.	P no. (5)	Grade	Temper	Notes	Size or thickness limitations, in
Aluminum alloy						
Seamless pipes and tubes						
	B 210, B 241, B 345	21	1060	0, H112	(14) (33)	
	B 241	21	1100	0, H112	(14) (33)	
	B 210, B 241, B 345	21	3003	H18	(14) (33)	
	B 210, B 241, B 345	21	Alclad 3002	0, H112	(14) (33)	
	B 210	21	Alclad 3003	H18	(14) (33)	
	B 210, B 241	22	5052	0	(14)	
	B 210, B 241, B 345	25	5083	0, H112	(33)	
	B 210, B 241, B 345	25	5086	0, H112	(33)	
	B 210	25	5086	H34	(33)	
	B 210	22	5154	H34	(33)	
	B 241	22	5454	0, H112	(33)	
	B 210, B 241	25	5456	0, H112	(33)	
	B 241	22	5652	0, H112	(33)	
	B 210	23	6061	T4	(33)	
	B 241, B 345	23	6061	T4	(33) (63)	
	B 241, B 345	23	6061	T6	(33)	Pipe < NPS 1
	B 241, B 345	23	6061	T6	(33) (63)	Pipe ≥ NPS 1 all tube
	B 241, B 345	23	6063	T4	(33)	≤ 0.500
	B 241, B 345	23	6063	T5	(33)	≤ 0.500
	B 241, B 345	23	6063	T6	(33)	
	B 210, B 241, B 345	23	6063	T4, T5, T6 Wld.		
Structural tubes						
	B 221	21	1100	0, H112	(14) (33)	
	B 221	21	3003	0, H112	(14) (33)	
	B 221	22	5052	0	(14)	
	B 221	25	5083	0		
	B 221	25	5086	0		
	B 221	22	5454	0		
	B 221	23	6061	T4	(33) (63)	
	B 221	23	6061	T4, T6 Wld.	(22) (63)	
	B 221	23	6063	T4	(33)	≤ 0.500
	B 221	23	6063	T5	(33)	≤ 0.500
	B 221	23	6063	T6	(33)	
	B 221	23	6063	T4, T5, T6 Wld.		
Forgings and fittings						
	B 247	25	5083	0, H112	(9) (33)	
	B 247	23	6061	T6	(9) (33) (45)	
	B 361	21	WP3003	0, H112	(13) (14) (23) (32) (33)	
	B 361	22	WP5154	0, H112	(14) (23) (32) (33)	
	B 361	23	WP6061	T4	(13) (14) (23) (32) (33)	
	B 361	23	WP6061	T4, T6 Wld.	(22) (23) (32)	
	B 361	23	WP6063	T4	(13) (14) (23) (32) (33)	
	B 361	23	WP6063	T4, T6 Wld.	(23) (32)	
Castings						
	B 26		443.0	F	(9) (43)	
	B 26		356.0	T6	(9) (43)	

SOURCE: Adapted from ASME B31.3-1984 with permission.
† Numbers in parentheses refer to notes at end of table. All specifications are ASTM unless noted otherwise.
‡ In the table, SMTS = standard minimum tensile stress, SMYS = standard minimum yield stress.
§ Abbreviations in class column: Ann. = annealed, C.D. = cold drawn, Forg. = forged, H.F. = hot-finished, H.R. = hot-rolled, Plt. = plate, R. = Rolled, Rel. = relieved, Sol. = solution, and Str. = stress.

Min temp. (6)	Specified min strength, ksi		Basic allowable stress S, ksi (1), at metal temperature, °F (7)						
	Tensile	Yield	Min temp. to 100	200	300	400	500	600	700
−325	75	30	20.0	18.0	16.3	14.8	11.0		
−325	80	32	21.3	15.3	10.5				
−325	110	60	36.6	19.0	10.5				

Min temp. (6)	SMTS, ksi	SMYS, ksi	Basic allowable stress S, ksi (1), at metal temperature, °F (7)						
			Min temp. to 100	150	200	250	300	350	400
−452	8.5	2	1.7	1.7	1.6	1.5	1.3	1.1	0.8
−452	11	3	2.0	2.0	2.0	1.9	1.7	1.3	1.0
−452	27	24	9.0	9.0	8.9	6.3	5.4	3.5	2.5
−452	13.5	4.5	3.0	3.0	3.0	2.8	2.2	1.6	1.3
−452	26	23	8.1	8.1	8.0	5.7	4.9	3.2	2.2
−452	25	10	6.7	6.7	6.7	6.2	5.6	4.1	2.3
−452	39	16	10.7	10.7					
−452	35	14	9.3	9.3					
−452	44	34	14.7	14.7					
−452	39	29	13.0	13.0					
−452	31	12	8.0	8.0	8.0	7.4	5.5	4.1	3.0
−452	41	19	12.7	12.7					
−452	25	10	6.7	6.7	6.7	6.2	5.6	4.1	2.3
−452	30	16	10.0	10.0	10.0	9.8	9.2	7.9	5.6
−452	26	16	8.7	8.7	8.7	8.5	8.0	7.9	5.6
−452	42	35							
−452	38	35	12.7	12.7	12.7	12.1	10.6	7.9	5.6
−452	19	10	6.7	6.7	6.7	6.7	6.7	3.4	2.0
−452	22	16	7.3	7.3	7.2	6.8	6.1	3.4	2.0
−452	30	25	10.0	10.0	9.8	9.0	6.6	3.4	2.0
−452	17		5.7	5.7	5.7	5.6	5.2	3.0	2.0
−452	11	3	2.0	2.0	2.0	1.9	1.7	1.3	1.0
−452	14	5	3.3	3.3	3.3	3.1	2.4	1.8	1.4
−452	25	10	6.7	6.7	6.7	6.2	5.6	4.1	2.3
−452	39	16	10.7	10.7					
−452	35	14	9.3	9.3					
−452	31	12	8.0	8.0	8.0	7.4	5.5	4.1	3.0
−452	26	16	8.7	8.7	8.7	8.5	8.0	7.7	5.3
−452	24		8.0	8.0	8.0	7.9	7.4	6.1	4.3
−452	19	10	6.4	6.4	6.4	6.4	6.4	3.4	2.0
−452	22	16	7.3	7.3	7.2	6.8	6.1	3.4	2.0
−452	30	25	10.0	10.0	9.8	9.0	6.6	3.4	2.0
−452	17		5.7	5.7	5.7	5.6	5.2	3.0	2.0
−452	39	16	10.7	10.7					
−452	38	35	12.7	12.7	12.7	12.1	10.6	7.9	5.6
−452	14	5	3.3	3.3	3.3	3.1	2.4	1.8	1.4
−452	30	11	7.3	7.3					
−452	26	16	8.7	8.7	8.7	8.5	8.0	7.7	5.6
−452	24		8.0	8.0	8.0	7.9	7.4	6.1	4.3
−452	18	9	6.0	6.0	6.0	6.0	6.0	3.4	2.0
−452	17		5.7	5.7	5.7	5.6	5.2	3.0	2.0
−452	17	6	4.0	4.0	4.0	4.0	4.0	4.0	3.0
−452	30	20	10.0	10.0	10.0	8.4			

Table 8.7.15 Basic Allowable Stresses in Tension for Metals (*Continued*)

NOTES: These notes are requirements of the Code. Those marked with an asterisk (*) restate requirements found in the text of the Code. The other notes are limitations or special requirements applicable to particular materials. At this time, metric equivalents have not been provided in the stress tables for metals. (P-number groupings, tables, and appendixes cited in these notes for Table 8.7.15 will be found in the basic reference ASME B31.3.)

(1)* The stress values in Table A-1 and the design stress values in Table A-2 are basic allowable stresses in tension in accordance with 302.3.1(a). For pressure design, the stress values from Table A-1 are multiplied by the appropriate quality factor E (E_c from Table A-1A, or E_j from Table A-1B). Stress values in shear and bearing are stated in 302.3.1(b); those in compression in 302.3.1(c).

(2)* The quality factors for castings E_c in Table A-1A are basic factors in accordance with 302.3.3(b). The quality factors for longitudinal weld joints E_j in Table A-1B are basic factors in accordance with 302.3.4(a). See 302.3.3(c) and 302.3.4(b) for enhancement of quality factors. See also 302.3.1(a), footnote 9.

(3)* This casting quality factor can be enhanced by supplementary examination in accordance with 302.3.3(c) and Table 302.3.3C. The higher factor from Table 302.3.3C may be substituted for this factor in pressure design equations.

(4)* In shaded areas, stress values printed in *italics* exceed two-thirds of the expected yield strength at temperature. All other stress values in shaded areas are equal to 90% of expected yield strength at temperature. See 302.3.2(d)(4) and 302.3.2(d) [Note (3)].

(5)* See 327.5.2 for description of P-number groupings.

(6)* The minimum temperature shown is that design minimum temperature for which the material is normally suitable without impact testing other than that required by the material specification. However, the use of a material at a design minimum temperature below − 20°F (− 29°C) is established by rules elsewhere in this Code, including any necessary impact test requirements.

(7)* A single bar (|) in these stress tables indicates there are conditions other than stress which affect usage above or below the temperature, as described in other referenced notes. A double bar (||) after a tabled stress indicates that use of the material is prohibited above that temperature. A double bar (||) before the stress value for "Min. temp. to 100°F" indicates that the use of the material is prohibited below the listed minimum temperature. At temperatures where there are no stress values, the material may be used in accordance with 323.2 unless prohibited by a double bar (||).

(8)* There are restrictions on the use of this material in the text of the Code.

(9)* Pressure-temperature ratings of cast and forged parts as published in standards referenced in this Code section may be used for parts meeting requirements of these standards. Allowable stresses for castings and forgings, where listed, are for use in design of special components not furnished in accordance with such standards.

(10)* These casting quality factors are applicable only when proper supplementary examination has been specified (see 302.3.3).

(11) For use under this Code, radiography shall be performed after heat treatment.

(12)* Certain forms of this material, as stated in Table 323.2.2, must be impact-tested to qualify for service below − 20°F (− 29°C). Alternatively, if provisions for impact testing are included in the material specification as supplementary requirements and are invoked, the material may be used down to the temperature at which the test was conducted in accordance with the specification.

(13) Properties of this material vary with thickness or size. Stresses are based on minimum properties for the thickness listed.

(14) For use in Code piping at the stated stress values, the required minimum tensile and yield properties must be verified by tensile test at the mill. If such tests are not required by the material specification, they shall be specified in the purchase order.

(15) These stress values are established from a consideration of strength only and will be satisfactory for average service. For bolted joints where freedom from leakage over a long period of time without retightening is required, lower stress values may be necessary as determined from the flexibility of the flange and bolts and corresponding relaxation properties.

(16) This joint factor shall be applied to stress values in Table A-1 unless the longitudinal welded joint has been 100% radiographed as required by the specification for fabricated welds.

(17)* Filler metal shall not be used in the manufacture of this pipe or tube.

(18)* This specification does not include requirements for 100% radiographic inspection. If this higher joint factor is to be used, the material shall be purchased to the special requirements of Table 327.4.1A for longitudinal butt welds with 100% radiography in accordance with Table 302.3.4.

(19)* This specification includes requirements for random radiographic inspection for mill quality control. If the 0.90 joint factor is to be used, the welds shall meet the requirements of Table 327.4.1A for longitudinal butt welds with spot radiography in accordance with Table 302.3.4. This shall be a matter of special agreement between purchaser and manufacturer.

(20) For pipe sizes NPS 8 and larger and for wall thicknesses of Schedule 140 or heavier, the minimum specified tensile strength is 70.0 ksi (483 MPa).

(21) For material thickness greater than 5 in (125 mm), the minimum specified tensile strength shall be 70.0 ksi (483 MPa).

(22) The minimum tensile strength for weld (qualification) and stress values shown shall be multiplied by 0.90 for pipe having an outside diameter less than 2 in (51 mm) and a D/t value less than 15. This requirement may be waived if it can be shown that the welding procedure to be used will consistently produce welds that meet the listed minimum tensile strengths of 24.0 ksi (165 MPa).

(23) Stress values apply only to fittings made from seamless material conforming to ASTM B 210 or B 241. Otherwise, the value of factor E_j shall be selected from Table 302.3.4 for the appropriate construction.

(24) Yield strengths listed are not included in the material specifications. The value shown is based on yield strengths of materials with similar characteristics.

(26) These unstabilized grades of stainless steel have increasing tendency to intergranular carbide precipitation as the carbon content increases above 0.03%.

(27) For temperatures above 800°F (425°C), these stress values apply only when the carbon content is 0.04% or higher.

(28) For temperatures above 1,000°F (538°C), these stress values apply only when the carbon content is 0.04% or higher.

(29) The higher stress values at 1,050°F (566°C) and above for this material shall be used only when the steel has an austenitic micrograin size no. 6 or less (coarser grain) as defined in ASTM E 112. Otherwise, the lower stress values shall be used.

(30) For temperatures above 1,000°F (538°C), these stress values may be used only if the material has been heat-treated at a temperature of 2,000°F (1,090°C) minimum.

(31) For temperatures above 1,000°F (538°C), these stress values may be used only if the material has been heat-treated by heating to a minimum temperature of 1,000°F (1,040°C) and quenching in water or rapidly cooling by other means.

(32) Stress values shown are for the lowest-strength base material permitted by the specification to be used in the manufacture of this grade of fitting. If a higher strength base material is used, the higher stress values for that material may be used in design.

(33) For welded construction with work-hardened grades, use the stress values for annealed material; for welded construction with precipitation hardened grades, use the special stress values for welded construction given in the table.

(34) After use above the temperature indicated by a single bar (|), use at a lower temperature shall be based on the stress values allowed for the annealed condition of the material.

(35) These steels are intended for use at high temperatures; however, they may have low ductility and/or low impact properties at room temperature after being used above the temperature indicated by the single bar (|).

(36) The specification permits this material to be furnished without solution heat treatment or with other than a solution heat treatment. When the material has not been solution heat-treated, the minimum temperature shall be − 20°F (− 29°C) unless the material is impact tested per 323.3.

(37) Impact requirements for seamless fittings shall be governed by those listed in this table for the particular basic material specification in the grades permitted (A 312, A 240, and A 182). When A 276 materials are used in the manufacture of these fittings, the notes, minimum temperatures, and allowable stresses for comparable grades of A 240 materials shall apply.

(38) For use at temperatures below − 20°F through − 50°F (− 29°C through − 45°C), this material must be quenched and tempered.

(39) This material when used below − 20°F (− 29°C) requires impact testing if the carbon content is above 0.10%.

(40) The stress values for austenitic stainless steels in this table may not be applicable if the material has been given a final heat treatment other than that required by the material specification and any overriding requirements of this Code called for by note (30) or (31).

(41) Design stresses for the cold-drawn temper are based on hot-rolled properties until required data on cold-drawn are submitted.

(42) This is a product specification. No design stresses are necessary. Limitations on metal temperature for materials covered by this specification are:

	°F	°C		°F	°C
Grades 1 and 2	− 20 to 900	− 29 to 480	Grade 6	− 20 to 800	− 29 to 425
Grade 2H	− 50 to 1,100	− 45 to 595	Grade 8FA [see note (39)]	− 20 to 800	− 29 to 425
Grade 3	− 20 to 1,100	− 29 to 595	Grades 8MA and 8TA	− 325 to 1,500	− 198 to 815
Grade 4	− 150 to 1,100	− 100 to 595	Grades 8A and 8CA	− 425 to 1,500	− 254 to 815

(43)* The stress values given for this material are not applicable when either welding or thermal cutting is employed [see 323.4.2(c)].

(44) This material shall not be welded.

(45) Stress values shown are applicable for "die" forgings only.

(46) The letter "a" indicates alloys which are not recommended for welding and which, if welded, must be individually qualified. The letter "b" indicates copper-base alloys which must be individually qualified.

(47) If no welding is employed in fabrication of piping from these materials, the stress values may be increased to 33.3 ksi (230 MPa).

(48) The stress value to be used for this gray cast iron material at its upper temperature limit of 450°F (232°C) is the same as that shown in the 400°F (204°C) column.

(49) If the chemical composition of this grade is such as to render it hardenable, qualification under P no. 6 is required.

(50) This material is grouped in P no. 7 because its hardenability is low.

(51) Special P-numbers SP-1, SP-2, and SP-3 of carbon steels are not included in P no. 1 because of possible high-carbon, high-manganese combination which would require special consideration in qualification. Qualification of any high carbon, high manganese grade may be extended to other grades in its group.

(52) Copper-silicon alloys are not always suitable when exposed to certain media and high temperature, particularly above 212°F (100°C). The user should satisfy himself that the alloy selected is satisfactory for the service for which it is to be used.

(53) Stress relief heat treatment is required for service above 450°F (232°C).

Table 8.7.15   Basic Allowable Stresses in Tension for Metals (*Continued*)

(54) The maximum operating temperature is arbitrarily set at 500°F (260°C) because harder temper adversely affects design stress in the creep rupture temperature ranges.

(55) Pipe produced to this specification is not intended for high-temperature service. The stress values apply to either nonexpanded or cold-expanded material in the as-rolled, normalized, or normalized and tempered condition.

(56) Because of thermal instability, this material is not recommended for service above 800°F (425°C).

(57)* Conversion of carbides to graphite may occur after prolonged exposure to temperatures over 800°F (425°C) (see App. F).

(58)* Conversion of carbides to graphite may occur after prolonged exposure to temperatures over 875°F (468°C) (see App. F).

(59)* For temperatures above 900°F (480°C), consider the advantages of killed steel (see App. F).

(60) For all design temperatures, the maximum hardness shall be Rockwell C35 immediately under the thread roots. The hardness shall be taken on a flat area at least ⅛ in (3 mm) across, prepared by removing threads. No more material than necessary shall be removed to prepare the area. Hardness determination shall be made at the same frequency as tensile tests.

(61) Annealed at approximately 1,800°F (980°C).

(62) Annealed at approximately 2,000°F (1,150°C).

(63) For stress-relieved tempers (T351, T3510, T3511, T451, T4510, T4511, T651, T6510, T6511), stress values for material in the listed temper shall be used.

(64) The minimum tensile strength of the reduced section tensile specimen in accordance with QW-462.1 of BPV Code, Section IX, shall not be less than 110.0 ksi (758 MPa).

(65) The minimum temperature shown is for the heaviest wall permissible by the specification. The minimum temperature for lighter walls shall be as shown in the following tabulation:

| Spec. and grade | Temp. for plate thicknesses shown | | | | | |
| | °F | | | °C | | |
	1 in max	2 in max	Over 2 to 3 in	25 mm max	50 mm max	Over 50 to 76 mm
A 203, A	−90	−90	−75	−68	−68	−60
A 203, B	−90	−90	−75	−68	−68	−60
A 203, D	−150	−150	−125	−101	−101	−87
A 203, E	−150	−150	−125	−101	−101	−87

(66) Stress values shown are 90% of those for the corresponding core material.

(67) For use under this Code, the heat treatment requirements for pipe manufactured to ASTM A 671, A 672, and A 691 shall be as required by 331 for the particular material being used. In some cases, 331 does not require heat treatment. In these cases, if the user requires no radiography, the designation shall be class 13. If 100% radiography is required, it shall be class 12. If heat treatment is required by the plate thickness or by the engineering design, the designation shall be class 23 (no radiography) or class 22 (100% radiography).

(68) The tension test specimen from plate 0.500 in (12.7 mm) and thicker is machined from the core and does not include the cladding alloy; therefore, the stress values listed are those for materials less than 0.500 in (12.7 mm).

Table 8.7.16   Basic Casting Quality Factors E_c

Abstracted from ASME B31.3 (1984) with permission

Spec. no.	Description	E_c (2)*	Notes*
Iron			
FS-WW-P421c	Centrifugally cast pipe	1.00	
A 377	Centrifugally cast pipe	1.00	
A 47	Malleable iron castings	1.00	
A 48	Gray iron castings	1.00	
A 126	Gray iron castings	1.00	
A 197	Cupola malleable iron castings	1.00	
A 278	Gray iron castings	1.00	
A 338	Malleable iron castings	1.00	
A 395	Ductile and ferritic ductile iron castings	0.80	(3)
A 571	Austenitic ductile iron castings	0.80	(3)
Carbon steel			
A 216	Carbon steel castings	0.80	(3)
A 352	Ferritic steel castings	0.80	(3)
Low and intermediate alloy steel			
A 426	Centrifugally cast pipe	1.00	(10)
A 217	Martensitic stainless and alloy castings	0.80	(3)
A 352	Ferritic steel castings	0.80	(3)
Stainless steel			
A 451	Centrifugally cast pipe	0.90	(3) (10)
A 452	Centrifugally cast pipe	0.85	(3)
A 351	Austenitic steel castings	0.80	(3)
Copper and copper alloy			
B 61	Steam bronze castings	0.80	(3)
B 62	Composition bronze castings	0.80	(3)
B 148	Al–bronze and Si–Al–bronze castings	0.80	(3)
B 584	Copper alloy castings	0.80	(3)
Nickel and nickel alloy			
A 494	Nickel and nickel alloy castings	0.80	(3)
Aluminum alloy			
B 26, temper F	Aluminum alloy castings	1.00	(10)
B 26, temper T6, T71	Aluminum alloy castings	0.80	(3)

* See Notes at end of Table 8.7.15.

Table 8.7.17   Basic Quality Factors for Longitudinal Weld Joints in Pipes, Tubes, and Fittings, E_j

Spec. no.	Class material	Description	E_j (2)*	Notes*
Carbon steel				
API 5L		Seamless	1.00	
		Electric resistance welded	0.85	
		Electric fusion welded, double butt, straight or spiral	0.85	
		Furnace butt welded	0.60	
A 53	Type S	Seamless	1.00	
	Type E	Electric resistance welded	0.85	
	Type F	Furnace butt welded	0.60	
A 105		Forgings and fittings	1.00	
A 106		Seamless	1.00	
A 120		Seamless	1.00	
		Electric resistance welded	0.85	
		Furnace butt welded	0.60	
A 134		Electric fusion welded, single butt, straight or spiral	0.80	
A 135		Electric resistance welded	0.85	
A 139		Electric fusion welded, straight or spiral	0.80	
A 179		Seamless	1.00	
A 181		Forgings and fittings	1.00	
A 211		Spiral welded	0.75	
A 234		Seamless and welded fittings	1.00	
A 333		Seamless	1.00	
		Electric resistance welded	0.85	
A 334		Seamless	1.00	
A 350		Forgings and fittings	1.00	
A 369		Seamless	1.00	
A 381		Electric fusion welded, 100% radiograph	1.00	(18)
		Electric fusion welded, spot radiograph	0.90	(19)
		Electric fusion welded, as manufactured	0.85	
A 420		Welded fittings, 100% radiograph	1.00	
A 524		Seamless	1.00	
A 587		Electric resistance welded	0.85	
A 671	12, 22	Electric fusion welded, 100% radiograph	1.00	
	13, 23	Electric fusion welded, double butt	0.85	

Table 8.7.17 Basic Quality Factors for Longitudinal Weld Joints in Pipes, Tubes, and Fittings, E_j (Continued)

Spec. no.	Class material	Description	E_j (2)*	Notes*
A 672	12, 22	Electric fusion welded, 100% radiograph	1.00	
	13, 23	Electric fusion welded, double butt	0.85	
A 691	12, 22	Electric fusion welded, 100% radiograph	1.00	
	13, 23	Electric fusion welded, double butt	0.85	
Low and intermediate alloy steel				
A 182		Forgings and fittings	1.00	
A 234		Seamless and welded fittings	1.00	
A 333		Seamless	1.00	
		Electric resistance welded	0.85	
A 334		Seamless	1.00	
A 335		Seamless	1.00	
A 350		Forgings and fittings	1.00	
A 369		Seamless	1.00	
A 420		Welded fittings, 100% radiograph	1.00	
A 671	12, 22	Electric fusion welded, 100% radiograph	1.00	
	13, 23	Electric fusion welded, double butt	0.85	
A 672	12, 22	Electric fusion welded, 100% radiograph	1.00	
	13, 23	Electric fusion welded, double butt	0.85	
A 691	12, 22	Electric fusion welded, 100% radiograph	1.00	
	13, 23	Electric fusion welded, double butt	0.85	
Stainless steel				
A 182		Forgings and fittings	1.00	
A 268		Seamless	1.00	
		Electric fusion welded, double butt	0.85	
		Electric fusion welded, single butt	0.80	
A 269		Seamless	1.00	
		Electric fusion welded, double butt	0.85	
		Electric fusion welded, single butt	0.80	
A 312		Seamless	1.00	
		Electric fusion welded, double butt	0.85	
		Electric fusion welded, single butt	0.80	
A 358	1, 3, 4	Electric fusion welded, 100% radiograph	1.00	
	5	Electric fusion welded, spot radiograph	0.90	
	2	Electric fusion welded, double butt	0.85	
A 376		Seamless	1.00	
A 403		Seamless fitting	1.00	
		Welded fitting, 100% radiograph	1.00	(16)
		Welded fitting, double butt	0.85	
		Welded fitting, single butt	0.80	
A 409		Electric fusion welded, double butt	0.85	
		Electric fusion welded, single butt	0.80	
A 430		Seamless	1.00	
Copper and copper alloy				
B 42		Seamless	1.00	
B 43		Seamless	1.00	
B 68		Seamless	1.00	
B 75		Seamless	1.00	
B 88		Seamless	1.00	
B 466		Seamless	1.00	

Spec. no.	Class material	Description	E_j (2)*	Notes*
Copper and copper alloy (Cont.)				
B 467		Electric resistance welded	0.85	
		Electric fusion welded, double butt	0.85	(16)
		Electric fusion welded, single butt	0.80	(16)
Nickel and nickel alloy				
B 160		Forgings and fittings	1.00	
B 161		Seamless	1.00	
B 164		Forgings and fittings	1.00	
B 165		Seamless	1.00	
B 166		Forgings and fittings	1.00	
B 167		Seamless	1.00	
B 366		Seamless and welded fittings	1.00	(16)
B 407		Seamless	1.00	
B 444		Seamless	1.00	
B 619		Electric resistance welded	0.85	
		Electric fusion welded, double butt	0.85	
		Electric fusion welded, single butt	0.80	
Unalloyed titanium				
B 337		Seamless	1.00	
		Electric fusion welded, double butt	0.85	
Aluminum alloy				
B 210		Seamless	1.00	
B 241		Seamless	1.00	
B 247		Forgings and fittings	1.00	
B 345		Seamless	1.00	
B 361		Seamless fittings	1.00	

* See Notes at end of Table 8.7.15.
SOURCE: Abstracted from ASME B31.3 (1984) with permission.

hammer allowance in case of cast-iron conveying liquids); D = OD of pipe, in; S = maximum allowable stress in material due to internal pressure, lb/in²; E = quality factor, y = a coefficient, values for which are listed in Table 8.7.18; A = allowance for threading, mechanical strength, and corrosion, in, with values of A listed in Table 8.7.19.

The thickness of ductile-iron pipe conveying liquid may be taken from Table 8.7.24, using the pressure class next higher than the maximum internal service pressure in pounds per square inch. Where

Table 8.7.18 Values of y
(Interpolate for intermediate values) (ASME B31.1)

	Temp, °F (°C)					
	900 (482) and below	950 (510)	1,000 (538)	1,050 (566)	1,100 (593)	1,150 (621) and above
Ferritic steels	0.4	0.5	0.7	0.7	0.7	0.7
Austenitic steels	0.4	0.4	0.4	0.4	0.5	0.7

Table 8.7.19 Values of A
(ASME B31.1)

Type of pipe	A, in
Cast-iron pipe, centrifugally cast	0.14
Cast-iron, pit-cast	0.18
Threaded-steel, wrought-iron, or nonferrous pipe:	
⅜ in and smaller	0.05
½ in and larger	Depth of thread
Grooved-steel, wrought-iron or nonferrous pipe	Depth of groove
Plain-end steel, wrought-iron or tube:	
1 in and smaller	0.05
1¼ in and larger	0.065
Plain-end nonferrous pipe or tube	0.000

ductile-iron pipe is used for steam service, the thickness should be calculated by Eq. (8.7.1).

Plain-end pipe includes pipe joined by flared compression couplings, lapped joints, and by welding, i.e., by any method that does not reduce the wall thickness of the pipe at the joint.

Physical and Chemical Properties of Pipes, Tubes, etc. The design of piping for operation above 750°F (399°C) presents many problems not encountered at lower temperatures. For the properties of steel applicable to high-temperature service (as well as to ordinary service) for pipes, tubes, fittings, bolting material, etc., see Sec. 6. For a discussion of creep properties, see Sec. 5.

Piping of thickness designed in accordance with Eq. (8.7.1) may be used for any combination of pressure and temperature for which S and E values are listed in Tables 8.7.15 to 8.7.17. The following summarizes piping industry practice.

Steam Pressures above 250 lb/in² (1,724 kPa), and Not above 2,500 lb/in² (17,238 kPa), Temperatures Not above 1,100°F (593°C) For pressures in excess of 100 lb/in² (690 kPa), the pipe may be seamless steel (A106), (A312), (A335), or (A376); or electric-fusion-welded steel (A691); or forged-and-bored steel (A369); or automatic-welded steel (A312). For pressures between 250 and 600 lb/in² (9,224 and 22,137 N/m²) the pipe may be seamless steel (A106) or (A53); electric-fusion-welded steel (A155); electric-resistance-welded steel (A135) or (A53). For pressures of 250 lb/in² (1,724 kPa) and lower and for service up to 750°F (399°C), any of the following may be used: electric-fusion-welded steel (A134) or (A139); electric-resistance-welded steel (A135); seamless or welded steel (A53). Grade A seamless pipe (A106) or (A53); or grade A electric-welded pipe (A53), (A135), or (A139) is used for close coiling or cold bending. Pipe permissible for services specified may be used for temperatures higher than 750°F (399°C), unless otherwise prohibited, if the S and E values of Tables 8.7.15 to 8.7.17 are used when calculating the required wall thickness.

Because of several failures in seam-welded 1¼% Cr–½% Mo or 2¼% Cr–1% Mo piping produced to ASTM Specification A155 operating at temperatures above 950°F, a preference has developed for seamless piping in these applications. Nevertheless, in general, seam-welded piping has provided entirely satisfactory service in high-temperature applications above 950°F.

Valves and fittings must have flange openings or welded ends, and valves must have external stem threads. Valves must be of cast or forged steel or may be fabricated from plate and pipe. Valves of nonferrous materials are generally cast or forged. Forged and cast-steel threaded valves and fittings may be used up to 300 lb/in² and 500°F for 3 (2) [1½] in pipe, and pressure from 250 to 400 (400 to 600) [600 to 2,500] lb/in². Malleable-iron threaded fittings (300 lb/in²) may be used for pressures not greater than 300 lb/in² and temperatures not over 500°F. Valves 8 in and larger should have the bypass of at least ¾ in, commercial size.* Welded fittings may be used of the same material and thickness as the pipe to which they are to be connected.

Steam Pressures from 125 to 250 lb/in² (862 to 1,724 kPa), Temperature Not above 450°F (232°C) Pipe may be electric-fusion-welded steel (A134 or A139). Copper and brass may be used if the temperature does not exceed 406°F. Cast iron may also be used. For close coiling or cold bending, grade A seamless steel (A53); or grade A electric-welded steel (A53), (A135), or (A139) is suitable. Pipe permissible for this service may be used for temperatures above 450°F (232°C) if the proper S and E are used in calculating the pipe-wall thickness.

Valves below 3 in may have inside stem screws. Stop valves 8 in and over must be bypassed. Bodies, bonnets, and yokes are of cast iron, malleable iron, steel, bronze, brass, or Monel. Flanged-steel fittings must conform to the class 300 ANSI Standard B16.5; if of cast iron, to the class 250 ANSI Standard B16.1; or, for threaded fittings, to the ANSI Standard B16.4. Malleable-iron threaded fittings must conform to the class 300 ANSI B16.3 standard, except that the class 150

* See Manufacturers Standardization Society SP-45 for recommended size of bypass valves.

ANSI Standard B16.3 may be used for pressures not greater than 150 lb/in². Welded fittings may be used.

Steam Pressures from 25 to 125 lb/in² (172 to 802 kPa) Temperatures Not above 450°F Pipe may be of steel, ductile iron, copper, or brass; valve bodies of cast iron, malleable iron, ductile iron, steel, or brass. Fittings are of class 125 or class 150 American Standard cast iron with screwed or flanged ends, or of ductile or malleable iron with screwed ends.

Steam Pressures 25 lb/in² (172 kPa) and Less, Temperature up to 450°F Pipe may be of steel, brass, copper, or cast iron. Flanged fittings conform to the class 25 ANSI Standard B16.1. Screwed fittings are of the class 125 ANSI Standard B16.4 or of the class 150 ANSI Standard B16.3 for malleable iron, or conform to B16.15 for cast bronze. Welded-steel fittings are extensively used.

Pipe coils are made from any of the commercial sizes of iron, steel, brass, and copper pipe and tubing. Limiting center-to-center dimensions, to which pipe coils can be fabricated in sizes ¾ to 2 in, are given in Table 8.7.20. Steel tubing cannot be bent to the absolute limits of brass or copper.

Table 8.7.20   Center-to-Center Dimensions of Pipe Coils

Nominal pipe size, in	Recommended and advisable minimum, in	
	Schedule 40	Schedule 80
¾	3½	2½
1	4	3
1¼	5	4
1½	6	5
2	8	6

Seamless mechanical tubing is obtainable in outside diameters ranging from ¼ to 10¾ in and in wall thickness from 20 gage to 2 in (0.091 to 5.08 cm). Oval, square, rectangular, and other special shapes can be obtained in various sizes and wall thicknesses. Mechanical tubing is available either hot-finished or cold-drawn, but is furnished principally cold-drawn. It is readily adaptable to varied treatment by expansion, cupping, tapering, swaging, flanging, coiling, welding, and similar manipulations. Typical of the many uses are aircraft tubing, automobile axle housings, driveshafts, drive-shaft housings, tie rods, steering columns, piston rods and pins, gear rings, roller-bearing cases and cones, cylinders for various purposes, machine parts, sleeves, bushings, spacers, surgical instruments, and hypodermic needles. Table 8.7.21 lists weights and dimensions of round seamless-steel tubing for sizes that have by common usage become standard. Detailed information on mechanical tubing for any particular applications can be obtained from manufacturers.

Dimensions and weights of **condenser** and **heat-exchanger tubes** are given in Table 8.7.22 and of **boiler** tubes in Table 8.7.23.

Spiral Pipe Spiral pipe is strong lightweight steel pipe with a single continuous welded helical seam from end to end stiffening it throughout. It is listed in sizes 6- to 42-in ID (15.24- to 106.7-cm), in various thicknesses, and in lengths up to 40 ft (12.19 m). It is used for high- and low-pressure water lines, vacuum lines, exhaust-steam lines, low-pressure air lines, sand and gravel slurry conveying and similar services. It is also used extensively by the petroleum industry, for oil and gas lines, for low-pressure steam lines, etc.

Spiral pipe may be asphalt-coated or galvanized. The pipe is designed for special joints, flanges, and lightweight fittings, but the ANSI flanges and fittings can be furnished, if desired.

The **sleeve-type coupling** illustrated in Fig. 8.7.1 is particularly suitable for plain-end pipe and is widely used. A gasket is used to make a tight joint. Advantages of this coupling are low cost, the use of unskilled labor in making the connections, and the fact that small changes in alignment and grade can be made with regular straight lengths of pipe by a movement in the coupling. This type of coupling is used extensively in long oil lines.

Table 8.7.21 Approx Weight of Round Seamless Cold-Finished Carbon-Steel Mechanical Tubing, lb/ft*
(Carbon 0.25% max. Standard sizes for warehouse stocks random lengths. United States Steel Corporation)

Wall thickness		OD, in																	
in	g or in	3⁄8	1⁄2	5⁄8	3⁄4	7⁄8	1	1 1⁄8	1 1⁄4	1 3⁄8	1 1⁄2	1 5⁄8	1 3⁄4	1 7⁄8	2	2 1⁄8	2 1⁄4	2 3⁄8	2 1⁄2
0.035	20 g	0.127	0.174	0.221	0.267	0.314	0.361	0.407	0.454	—	0.548								
0.049	18 g	0.171	0.236	0.301	0.367	0.432	0.498	0.563	0.629	0.694	0.759	0.825	0.890	—	1.02				
0.058	17 g	—	0.274	0.351	0.429	—	0.584												
0.065	16 g	0.215	0.302	0.389	0.476	0.562	0.649	0.736	0.823	0.909	0.996	1.08	1.17	1.26	1.34	1.43	1.52	1.60	1.69
0.083	14 g	—	0.370	0.480	0.591	0.702	0.813	0.924	1.03	1.15	1.26	—	—	—	1.70				
0.095	13 g	—	0.411	0.538	0.665	0.791	0.918	1.05	1.17	1.30	1.43	1.55	1.68	1.81	1.93	2.06	2.19	2.31	2.44
0.109	12 g	—	0.455	0.601	0.746	0.892	1.04	1.18	1.33	—	1.62								
0.120	11 g	—	0.487	0.647	0.807	0.968	1.13	1.29	1.45	1.61	1.77	1.93	2.09	2.25	2.41	2.57	2.73	2.89	3.05
0.134	10 g	—	—	0.703	0.882	—	1.24	1.42	1.60	1.78	1.96	2.13	2.31	—	2.67				
0.156	5⁄32	—	—	0.781	0.990	1.20	1.41	1.61	1.82	2.03	2.24	2.45	2.66	2.86	3.07	3.28	3.49	3.70	3.91
0.188	3⁄16	—	—	—	1.13	1.38	1.63	1.88	2.13	2.38	2.63	2.89	3.14	3.39	3.64	3.89	4.14	4.39	4.64
0.219	7⁄32	—	—	—	—	1.53	1.83	2.12	2.41	2.70	3.00	3.29	3.58	3.87	4.17	4.46	4.75	5.04	5.34
0.250	1⁄4	—	—	—	1.34	1.67	2.00	2.34	2.67	3.00	3.34	3.67	4.01	4.34	4.67	5.01	5.34	5.67	6.01
0.281	9⁄32	—	—	—	—	—	—	—	2.91	—	3.66	4.03	4.41	4.78	5.16	—	5.91	6.28	6.66
0.313	5⁄16	—	—	—	—	—	—	—	3.13	3.55	3.97	4.39	4.80	5.22	5.64	6.06	6.48	6.89	7.31
0.375	3⁄8	—	—	—	—	—	—	—	3.50	4.01	4.51	5.01	5.51	6.01	6.51	7.01	7.51	8.01	8.51
0.438	7⁄16	—	—	—	—	—	—	—	—	—	4.97	5.53	6.14	6.72	7.31	7.89	8.48	9.06	9.65
0.500	1⁄2	—	—	—	—	—	—	—	—	—	5.34	6.01	6.68	7.34	8.01	8.68	9.35	10.0	10.7
0.625	5⁄8	—	—	—	—	—	—	—	—	—	—	—	—	—	9.18	—	10.8	11.7	12.5

* Other standard sizes, in certain standard wall thicknesses, vary by 1⁄8-in increments for 2 1⁄2 to 3 1⁄2 in; by 1⁄4-in increments from 3 1⁄2 to 7 1⁄2 in; 1⁄2-in increments from 7 1⁄2 to 10 1⁄2 in OD. There are also standard sizes for every 1⁄16 in from 3⁄8 to 1 5⁄8 in OD.

To obtain weights in kg/m, multiply tabular values shown by 1.42.

Table 8.7.22 Steel Condenser and Heat-Exchanger Tubes
(Dimensions and weights. United States Steel Corporation)

		Avg wall			Min wall		
OD, in	Thickness, in	ID, in	Area of metal, in²*	Weight per ft, lb†	ID, in	Area of metal, in²*	Weight per ft, lb†
1⁄2	0.035	0.430	0.0511	0.1738	0.423	0.0558	0.1898
	0.050	0.400	0.0707	0.2403	0.390	0.0769	0.2614
	0.065	0.370	0.0888	0.3020	0.357	0.0963	0.3272
5⁄8	0.035	0.555	0.0649	0.2205	0.548	0.0709	0.2412
	0.050	0.525	0.0903	0.3071	0.515	0.0985	0.3348
	0.065	0.495	0.1144	0.3888	0.482	0.1243	0.4227
	0.085	0.455	0.1442	0.4902	0.438	0.1561	0.5308
3⁄4	0.050	0.650	0.1100	0.3738	0.640	0.1201	0.4082
	0.065	0.620	0.1399	0.4755	0.607	0.1524	0.5181
	0.085	0.580	0.1776	0.6037	0.563	0.1928	0.6556
	0.095	0.560	0.1955	0.6646	0.541	0.2119	0.7204
7⁄8	0.050	0.775	0.1296	0.4406	0.765	0.1417	0.4817
	0.065	0.745	0.1654	0.5623	0.732	0.1805	0.6136
	0.085	0.705	0.2110	0.7172	0.688	0.2296	0.7804
	0.095	0.685	0.2328	0.7914	0.666	0.2530	0.8599
1	0.050	0.900	0.1492	0.5073	0.890	0.1633	0.5551
	0.065	0.870	0.1909	0.6491	0.857	0.2086	0.7090
	0.085	0.830	0.2443	0.8306	0.813	0.2663	0.9052
	0.095	0.810	0.2701	0.9182	0.791	0.2940	0.9994
1 1⁄4	0.050	1.150	0.1885	0.6408	1.140	0.2065	0.7020
	0.065	1.120	0.2420	0.8226	1.107	0.2647	0.8999
	0.085	1.080	0.3111	1.058	1.163	0.3397	1.155
	0.095	1.060	0.3447	1.172	1.041	0.3761	1.278
	0.105	1.040	0.3777	1.284	1.019	0.4117	1.399
1 1⁄2	0.050	1.400	0.2278	0.7743	1.390	0.2497	0.8488
	0.065	1.370	0.2930	0.9962	1.357	0.3209	1.091
	0.085	1.330	0.3779	1.285	1.313	0.4053	1.378
	0.095	1.310	0.4193	1.426	1.291	0.4581	1.557
	0.105	1.290	0.4602	1.564	1.269	0.5024	1.708
1 3⁄4	0.065	1.620	0.3441	1.170	1.606	0.3803	1.293
	0.085	1.580	0.4446	1.512	1.561	0.4907	1.668
	0.095	1.560	0.4939	1.679	1.539	0.5448	1.852
	0.105	1.540	0.5426	1.845	1.517	0.5902	2.007
	0.120	1.510	0.6145	2.089	1.484	0.6766	2.300
2	0.065	1.870	0.3951	1.343	1.856	0.4370	1.486
	0.085	1.830	0.5114	1.738	1.811	0.5649	1.920
	0.095	1.810	0.5685	1.933	1.789	0.6275	2.133
	0.105	1.790	0.6251	2.125	1.767	0.6896	2.344
	0.120	1.760	0.7087	2.409	1.734	0.7812	2.656

* Multiply values shown by 0.0645 to obtain areas in cm².
† Multiply values shown by 1.42 to obtain weights in kg/m.

Table 8.7.23 Seamless-Steel Boiler Tubes

Outside diam, in	Thickness BWG	Thickness in	Mfg.* wt, lb/ft	Outside diam, in	Thickness BWG	Thickness in	Mfg.* wt, lb/ft	Outside diam, in	Thickness BWG	Thickness in	Mfg.* wt, lb/ft
1	13	0.095	1.037	2½	12	0.109	3.171	4½	10	0.134	7.103
	12	0.109	1.168		11	0.120	3.457		9	0.148	7.817
	11	0.120	1.263		10	0.134	3.835		8	0.165	8.702
	10	0.134	1.384		9	0.148	4.207		7	0.180	9.447
1¼	13	0.095	1.323	2¾	12	0.109	3.504	5	9	0.148	8.720
	12	0.109	1.502		11	0.120	3.823		8	0.165	9.711
	11	0.120	1.628		10	0.134	4.244		7	0.180	10.550
	10	0.134	1.793		9	0.148	4.658		6	0.203	11.810
1½	13	0.095	1.619	3	12	0.109	3.838	5½	9	0.148	9.622
	12	0.109	1.836		11	0.120	4.189		8	0.165	10.720
	11	0.120	1.994		10	0.134	4.652		7	0.180	11.650
	10	0.134	2.201		9	0.148	5.110		6	0.203	13.050
1¾	13	0.095	1.910	3¼	11	0.120	4.555	6	7	0.180	12.750
	12	0.109	2.169		10	0.134	5.061		6	0.203	14.290
	11	0.120	2.360		9	0.148	5.061		5	0.220	15.410
	10	0.134	2.610		8	0.165	6.179		4	0.238	16.640
2	13	0.095	2.201	3½	11	0.120	4.921				
	12	0.109	2.503		10	0.134	5.469				
	11	0.120	2.726		9	0.148	6.012				
	10	0.034	3.018		8	0.065	6.683				
2¼	13	0.095	2.492	4	10	0.134	6.286				
	12	0.109	2.837		9	0.148	6.915				
	11	0.120	3.092		8	0.165	7.693				
	10	0.134	3.427		7	0.180	8.347				

* Multiply values shown by 1.42 to obtain weights in kg/m.
SOURCE: United States Steel Corporation.

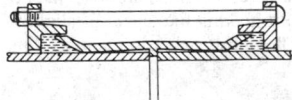

Fig. 8.7.1 Sleeve type, plain end coupling.

CAST-IRON AND DUCTILE-IRON PIPE

Cast-iron pipe was extensively produced since before the turn of the century until approximately 1960. Even though cast-iron pipe generally has given excellent service, the manufacture of cast-iron pipe was discontinued at that time.

Since then, **ductile-iron pipe** has replaced cast-iron pipe because of its improved ductility. Ductile-iron pipe is now extensively utilized for water, gas, sewage, culverts, drains, etc. It is produced in a wide range of sizes for varying pressures. Ductile-iron pipe is particularly adaptable to underground and submerged service because of its comparatively superior corrosion resistance compared to steel pipe. Nevertheless, steel pipe, when properly coated and wrapped, can also provide adequate resistance to corrosion when placed in certain soils.

Pipe fittings are available as cast-iron pipe fittings and ductile-iron pipe fittings. Both types of fittings are used with ductile-iron pipe.

Ductile-iron pipe may be obtained in various thicknesses and weights with (1) flanges cast on, (2) ends threaded for screwed-on flanges, (3) ends prepared for mechanical joint, (4) ends grooved or shouldered for patented coupling, (5) one end bell, other end spigot, and (6) one end hub, other end spigot. **Bell-and-spigot** ends are most popular for underground work; hub-and-spigot ends are most frequently used for sewage systems in enclosed spaces. Spigot-end joints are prepared by tightly tamping in hemp or jute at the bottom of the recess with a yarning iron and then pouring in molten lead; the lead, when cooled, is caulked in tightly with a caulking iron and makes a gastight joint. For exposed piping, flanged ends are used, the joints being made up with gaskets. Flanged pipe has superior strength and tightness of the joint and is used where pipelines can be well supported. The bell-and-spigot joint possesses greater flexibility and provides for expansion and contraction. It is therefore suitable for water pipe and is largely used for that purpose.

Figure 8.7.2 shows a typical form of this joint for ordinary pressures. Figure 8.7.3 shows one form of this joint for ordinary pressures. Figure 8.7.3 shows one form of **mechanical joint** suitable for water, gas, or oil. Other forms of joint, plain-end pipe with couplings, and threaded pipe also are manufactured. Cast-iron and ductile-iron pipe, fittings, and valves have been found unsuitable for superheated steam service. The Code for Pressure Piping, B31.1 (Power Piping), states that cast iron or ductile-iron pipe may be used for steam service not over 250 lb/in^2 or 406°F (1,724 kPa or 208°C) provided that it meets the requirements as dictated by Eq. (8.7.1).

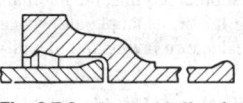

Fig. 8.7.2 Standard bell-and-spigot joint.

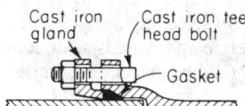

Fig. 8.7.3 Mechanical joint.

Wall thicknesses for the various conditions which ductile-iron pipe is designed to meet are determined in accordance with the requirements of ANSI 21.50 (AWWA C150).

Ductile-iron pipe is made by **centrifugal casting,** in which molten iron is admitted to the interior of a sand-lined or metal-lined mold, the mold being rotated at high speeds so that the molten metal is thrown by centrifugal force against the lining. ANSI specifications have been prepared for the various combinations of fabrication procedure and intended end use.

Table 8.7.24 lists thicknesses and weight data for centrifugally cast ductile-iron pipe intended for use with water or other liquids.

The employment of ductile-iron pipe for gas supply and distribution is second in importance only to its use for carrying water. Bell-and-spigot gas pipe is similar in design to bell-and-spigot water pipe (Fig. 8.7.2). For flanged gas pipe, the class 25 ANSI B16.1 Standard flanges are approved for maximum gas pressures of 25 lb/in^2 (172 kPa). The class 125 ANSI B16.1 Standard flanges are approved for gas pressures of 125 lb/in^2 (862 kPa), up to 4 in nominal pipe size; 100 lb/in^2 (689 kPa), 6 to 12 in; and 80 lb/in^2 (552 kPa), 16 to 48 in. The type of joint shown in Fig. 8.7.2 is also widely used for gas.

Table 8.7.24   Standard Weights and Thicknesses of Ductile-Iron Bell-and-Spigot Pipe for Water

Nominal size, in	Class 50, 50 lb/in^2 (345 kPa) 115-ft (35.05-m) head			Class 100, 100 lb/in^2 (690 kPa) 231-ft (70.41-m) head			Class 150, 150 lb/in^2 (1,034 kPa) 346-ft (105.5-m) head			Approx lb lead per joint 2 in thick	Approx lb hemp or jute per joint
	Thickness, in	OD, in	Wt, lb per avg ft	Thickness, in	OD, in	Wt, lb per avg ft	Thickness, in	OD, in	Wt, lb per avg ft		
3	0.32	3.96	12.4	0.32	3.96	12.4	0.32	3.96	12.4	6.2	0.17
4	0.35	4.80	16.5	0.35	4.80	16.5	0.35	4.80	16.5	7.5	0.21
6	0.38	6.90	25.9	0.38	6.90	25.9	0.38	6.90	25.9	10.3	0.31
8	0.41	9.05	37.0	0.41	9.05	37.0	0.41	9.05	37.0	13.3	0.44
10	0.44	11.10	49.1	0.44	11.10	49.1	0.44	11.10	49.1	16.0	0.53
12	0.48	13.20	63.7	0.48	13.20	63.7	0.48	13.20	63.7	19.0	0.61
14	0.48	15.30	74.6	0.51	15.30	78.8	0.51	15.65	80.7	22.0	0.81
16	0.54	17.40	95.2	0.54	17.40	95.2	0.54	17.80	97.5	30.0	0.94
18	0.54	19.50	107.6	0.58	19.50	114.8	0.58	19.92	117.2	33.8	1.00
20	0.57	21.60	125.9	0.62	21.60	135.9	0.62	22.06	138.9	37.0	1.25
24	0.63	25.80	166.0	0.68	25.80	178.1	0.73	26.32	194.0	44.0	1.50
30	0.79	32.00	257.6	0.79	32.00	257.6	0.85	32.00	275.4	54.3	2.06
36	0.87	38.30	340.9	0.87	38.30	340.9	0.94	38.30	365.9	64.8	3.00
42	0.97	44.50	442.0	0.97	44.50	442.0	1.05	44.50	475.3	75.3	3.62
48	1.06	50.80	551.6	1.06	50.80	551.6	1.14	50.80	589.6	85.5	4.37

Pipe weights indicated are approximate and include allowance for bell based on a 16-ft laying length. Calculations are for pipe laid without blocks, on flat-bottom trench, with tamped backfill under 5 ft of cover. Thicknesses given above include allowance for water hammer and factory tolerance.
To obtain weights in kg/m, multiply values shown in lb per avg ft by 1.42.
SOURCE: Condensed from Table 8.2 of Specification AWWA C 108-70.

Flexible-Joint Pipe   The necessity for crossing streams and other waterways and of laying pipelines into them has led to the development of various forms of flexible-joint pipe adapted to laying under water, which are capable of motion through several degrees without leakage. Figure 8.7.4 shows one style of such joint which has an adjustment of about 15° in standard sizes.

Fig. 8.7.4   Flexible joint. (See Table 8.7.25 for dimensions.)

In selecting the thickness of a pipe for a submerged line, the internal-service pressure is seldom the determining factor, as ample allowance should be made to minimize the risk of breakage in laying and to with-

stand external shocks from floating ice or other objects. The dimensions and weights given in Table 8.7.25 are typical of those listed by several manufacturers.

"Universal" pipe (Fig. 8.7.5) is ductile-iron pipe with hub-and-spigot ends, the contact surfaces of which are machined on a taper, giving an iron-to-iron joint. By making the tapers of slightly different pitch, the joint provides for flexibility while remaining tight. Two bolts to the

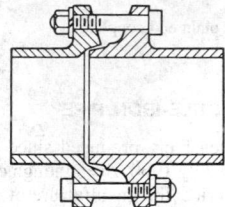

Fig. 8.7.5   Universal ductile-iron pipe and joint.

Table 8.7.25   Dimensions and Weights of Flexible-Joint Pipe*
(Dimensions refer to Fig. 8.7.4)

Nominal diam, in	Class	Dimensions, in			Bolts			Average metal thickness, in	Weight† of pipe, incl. bell, lb per 12-ft length
		A	B	C	No. required	Size, in	Length, in		
4	B	12.13	9.75	5.00	8	0.75	4.50	0.45	290
4	C	12.13	9.75	5.00	8	0.75	4.50	0.48	305
4	D	12.13	9.75	5.00	8	0.75	4.50	0.52	325
6	B	14.25	11.75	7.10	12	0.75	4.50	0.48	440
6	C	14.25	11.75	7.10	12	0.75	4.50	0.51	460
6	D	14.25	11.75	7.10	12	0.75	4.50	0.55	490
8	B	17.25	14.75	9.30	12	0.75	5.25	0.51	635
8	C	17.25	14.75	9.30	12	0.75	5.25	0.56	680
8	D	17.25	14.75	9.30	12	0.75	5.25	0.60	720
10	B	20.56	18.00	11.40	16	0.75	5.25	0.57	905
10	C	20.56	18.00	11.40	16	0.75	5.25	0.62	965
10	D	20.56	18.00	11.40	16	0.75	5.25	0.68	1,035
12	B	23.75	21.00	13.50	16	0.75	6.25	0.62	1,200
12	C	23.75	21.00	13.50	16	0.75	6.25	0.68	1,280
12	D	23.75	21.00	13.50	16	0.75	6.25	0.75	1,375

* United States Pipe and Foundry Co.
† Weights do not include follower rings, bolts, or gaskets. For sizes above 12 in, see manufacturers' catalogs. Multiply weights shown by 38.7 to obtain weight in kg per m or by 0.4536 to obtain weight in kg per 12-ft length.

Table 8.7.26   Standard Weights and Thicknesses of Universal Ductile-Iron Pipe
(Central Foundry Co.)

Nominal ID, in	Class 150, 150 lb/in² (1,034 kPa)			Class 250, 250 lb/in² (1,724 kPa)		
	Approx thickness, in	Estimated wt, lb per*		Approx thickness, in	Estimated wt, lb per*	
		ft	6-ft length		ft	6-ft length
2	0.25	7	42	0.31	8	48
3	0.30	11¼	67½	0.34	12½	75
4	0.32	16½	99	0.37	18	108
6	0.36	26.6	160	0.43	30	180
8	0.39	38½	231	0.47	44¼	265½
10	0.43	53½	321	0.50	60½	363
12	0.47	69	414	0.53	77½	465
14	0.50	87	522	0.565	98½	591
16	0.53	106	636	0.60	121	726

ID, in	2	3	4	6	8	10	12	14	16
Bolt sizes, in	½ × 4	½ × 4¼	⅝ × 5	¾ × 6	⅞ × 6¾	1 × 7½	1 × 8	1⅛ × 9	1¼ × 9½

* Multiply wts. by 0.4536 to obtain wt. in kilograms.

joint are sufficient, except for pressures above 175 lb/in² (1,207 kPa). Universal ductile-iron pipe is used largely for carrying gas and water and is suitable for all pressures and services. The pipe is tested with hydrostatic pressure of 300 to 500 lb/in² (2,067 to 3,448 kPa). All universal pipe and special castings of a given diameter and of any class are interchangeable with those of a different class. Standard laying lengths are 6 ft (1.83 m). Thicknesses and weights of standard types up to 16 in are given in Table 8.7.26. Information on other types and sizes and on fittings may be obtained from pipe producers.

Fittings for Ductile-Iron Water Pipe   Flanged fittings of the dimensions of the ANSI standard for steam are not often used with ductile-iron water pipe. The longer fittings of the AWWA are generally preferred because of low friction loss. The dimensions of the flanged fittings of this class conform very closely to the dimensions of the bell-and-spigot fittings of the AWWA. The flange thicknesses and drillings conform to those of the ANSI standards. These fittings, both flange and bell-and-spigot type, are made in a great variety of forms

known as "standard special fittings." For dimensions and weights, see manufacturers' catalogs or standard specifications of the AWWA.

Cast-iron soil pipe and fittings are of the hub-and-spigot form, similar in design to the pipe shown in Fig. 8.7.2. Tapped openings and pipe plugs are threaded in accordance with the taper pipe thread requirements of ANSI B1.20.1-1983.

The ANSI standard, Threaded Cast-Iron Pipe for Drainage, Vent, and Waste Services, ANSI A40.5-1943, covers two types of pipes having threaded joints in nominal pipe sizes 1¼ to 12 in and in lengths 5 to 27 ft. One type has external threads on both ends; the other type has external threads on one end and an internal threaded drainage hub on the other end.

PIPES AND TUBES OF NONFERROUS MATERIALS

Brass tubing is commercially available in the form known as yellow brass, an alloy consisting of approximately 65 percent copper and 35

Table 8.7.27   Sizes and Weights of SPS Copper and 85 Red Brass Pipe*

Standard size, in	Nominal dimensions, in			Theoretical areas based on nominal dimensions			Theoretical weight, lb/ft	Allowable internal pressure, lb/in²			
	Outside diameter	Inside diameter	Wall thickness	Cross-sectional area of bore, in²	External surface, ft²/lin. ft	Internal surface, ft²/lin. ft		100°F, S = 6,000	200°F, S = 4,800	300°F, S = 4,700	400°F, S = 3,000
¼	0.540	0.410	0.065	0.132	0.141	0.107	0.376	1,500	1,200	1,180	740
⅜	0.675	0.545	0.065	0.233	0.177	0.143	0.483	1,170	940	910	580
½	0.840	0.710	0.065	0.396	0.220	0.186	0.613	920	740	720	460
¾	1.050	0.920	0.065	0.665	0.275	0.241	0.780	730	580	560	360
1	1.315	1.185	0.065	1.10	0.344	0.310	0.989	580	460	450	290
1¼	1.660	1.530	0.065	1.84	0.435	0.401	1.26	450	360	350	220
1½	1.900	1.770	0.065	2.46	0.497	0.464	1.45	400	310	300	190
2	2.375	2.245	0.065	3.96	0.622	0.588	1.83	300	240	240	140
2½	2.875	2.745	0.065	5.92	0.753	0.719	2.22	250	200	190	120
3	3.500	3.334	0.083	8.73	0.916	0.874	3.45	260	210	210	130
3½	4.000	3.810	0.095	11.4	1.05	0.998	4.52	270	210	210	130
4	4.500	4.286	0.107	14.4	1.18	1.12	5.72	270	210	210	130
5	5.562	5.298	0.132	22.0	1.46	1.39	8.73	270	210	210	130
6	6.625	6.309	0.158	31.3	1.73	1.65	12.4	270	210	210	130
8	8.625	8.215	0.205	53.0	2.26	2.15	21.0	270	210	210	130
10	10.750	10.238	0.256	82.3	2.81	2.68	32.7	270	210	210	130
12	12.750	12.124	0.313	115.	3.34	3.18	47.4	280	220	220	130

* The values in the table above are based on the formula in The Code for Pressure Piping, ANSI B31:

$$t_m = \frac{PD}{2S + 0.8P} + C \quad \text{or when } C \text{ is } O \quad P = \frac{2St_m}{D - 0.8t_m}$$

where t_m = minimum pipe wall thickness, in; P = maximum rated internal working pressure, lb/in²; D = outside diameter of pipe, in; S = allowable stress in material due to internal pressure, at operating temperature, lb/in²; C = allowance for threading, mechanical strength and/or corrosion, in. The allowable internal pressures apply to the pipe itself after brazing.
   SOURCE: Copper Development Association.

Table 8.7.28 Allowable Internal Pressures, lb/in², for Temperatures up to 400°F for Red Brass and Copper Pipe*

Standard size, in	Pipe with threaded ends								Pipe with plain ends for use with welded, brazed, or soldered fittings							
	Red brass				Copper				Red brass				Copper			
	100°F S=8,000	200°F S=8,000	300°F S=8,000	400°F S=5,000	100°F S=6,000	200°F S=4,800	300°F S=4,700	400°F S=3,000	100°F S=8,000	200°F S=8,000	300°F S=8,000	400°F S=5,000	100°F S=6,000	200°F S=4,800	300°F S=4,700	400°F S=3,000
Regular																
⅛	370	370	370	230	280	210	210	130	2,640	2,640	2,640	1,650	1,980	1,570	1,540	980
¼	870	870	870	550	650	510	510	320	2,610	2,610	2,610	1,640	1,960	1,560	1,530	970
⅜	890	890	890	570	670	530	510	320	2,280	2,280	2,280	1,440	1,710	1,350	1,330	840
½	900	900	900	570	670	530	520	320	2,160	2,160	2,160	1,350	1,620	1,280	1,260	800
¾	810	810	810	520	610	480	470	300	1,800	1,800	1,800	1,130	1,350	1,070	1,050	670
1	630	630	630	400	480	370	370	230	1,580	1,580	1,580	1,000	1,190	940	920	590
1¼	690	690	690	430	520	410	400	250	1,440	1,440	1,440	900	1,080	890	840	530
1½	630	630	630	400	480	370	370	230	1,280	1,280	1,280	800	960	750	740	470
2	540	540	540	350	410	320	310	190	1,050	1,050	1,050	670	790	620	610	380
2½	450	450	450	280	340	260	250	160	1,040	1,040	1,040	650	780	620	610	380
3	510	510	510	320	380	300	290	180	1,000	1,000	1,000	630	750	590	580	370
3½	570	570	570	370	430	330	330	200	1,000	1,000	1,000	630	750	590	580	370
4	510	510	510	320	380	300	290	180	880	880	880	550	660	530	520	320
5	410	410	410	270	310	230	240	140	710	710	710	450	540	420	410	260
6	340	340	340	220	260	190	200	120	600	600	600	380	450	350	340	220
8	360	360	360	230	270	210	210	130	560	560	560	350	420	320	320	200
10	360	360	360	230	270	210	210	130	520	520	520	330	390	300	300	190
12	320	320	320	200	240	190	200	120	450	450	450	280	340	260	250	160
Extra-Strong																
⅛	1,960	1,960	1,960	1,240	1,470	1,160	1,140	720	4,630	4,630	4,630	2,900	3,470	2,750	2,710	1,730
¼	2,210	2,210	2,210	1,340	1,660	1,310	1,290	820	4,200	4,200	4,200	2,640	3,150	2,500	2,460	1,570
⅜	1,840	1,840	1,840	1,150	1,380	1,090	1,070	680	3,360	3,360	3,360	2,100	2,520	2,000	1,960	1,250
½	1,760	1,760	1,760	1,100	1,320	1,040	1,030	660	3,130	3,130	3,130	1,970	2,350	1,860	1,830	1,160
¾	1,510	1,510	1,510	950	1,140	900	880	560	2,560	2,560	2,560	1,600	1,920	1,520	1,500	960
1	1,340	1,340	1,340	850	1,010	790	780	490	2,360	2,360	2,360	1,490	1,770	1,400	1,370	880
1¼	1,160	1,160	1,160	730	880	690	680	430	1,950	1,950	1,950	1,220	1,460	1,160	1,140	720
1½	1,090	1,090	1,090	680	820	650	640	410	1,770	1,770	1,770	1,120	1,330	1,050	1,030	660
2	1,000	1,000	1,000	630	750	590	580	360	1,520	1,520	1,520	950	1,140	910	890	590
2½	970	970	970	620	730	580	560	360	1,600	1,600	1,600	1,000	1,200	950	930	590
3	910	910	910	570	680	530	530	340	1,420	1,420	1,420	900	1,070	840	830	530
3½	860	860	860	550	650	510	500	310	1,300	1,300	1,300	820	980	790	760	480
4	840	840	840	530	630	490	480	300	1,080	1,080	1,080	680	810	640	630	400
5	770	770	770	480	580	450	450	290	1,060	1,060	1,060	670	800	620	610	380
6	800	800	800	500	600	470	460	300	910	910	910	570	680	540	540	340
8	710	710	710	450	530	430	410	260	710	710	710	450	540	440	420	240
10	550	550	550	350	420	340	330	200								

* The values above are based on the formula in The Code for Pressure Piping, ANSI B31:

$$t_m = \frac{PD}{2S + 0.8P} + C \qquad \text{or} \qquad P = \frac{2S(t_m - C)}{D - 0.8(t_m - C)}$$

$$\text{or when } C \text{ is O} \qquad P = \frac{2St_m}{D - 0.8t_m}$$

where t_m = minimum pipe wall thickness, in; P = maximum rated internal working pressure, lb/in²; D = outside diameter of pipe, in; S = allowable stress in annealed material due to internal pressure, at operating temperature, lb/in²; C = allowance for threading, mechanical strength and/or corrosion, in. C equals 0.05 for sizes ⅜ in and smaller and the depth of the thread for all other sizes, and equals O for pipe with plain ends. These allowable internal pressures apply to the pipe itself and do not take into account the limitations which may be imposed by the type of joint and the joining material.

SOURCE: Copper Development Association.

percent zinc, and is used principally for ornamental work and hand railings, It has a density of approximately 0.3 lb/in³ (8.31 g/cm³), the exact density being dependent upon the specific chemical composition. **Brass piping** is most frequently furnished as red brass, an alloy consisting of approximately 85 percent copper and 15 percent zinc. This alloy, having a density of about 0.32 lb/in³ (8.86 g/cm³), has been found to be structurally superior to the yellow brasses and is used where the fluid being conveyed has corrosive properties.

Copper is available either as pipe or as tubing. In the form of piping, it has the same outer diameter as that of standard steel pipe (Tables 8.7.27 and 8.7.28). As tubing, it is used for a variety of purposes, such as for compressed-air instrumentation lines, hydraulic control lines around machinery, domestic oil-burner and heating systems, and for general plumbing purposes. **Copper tubing** (Table 8.7.29) is furnished in 12-ft (3.7-m) and 20-ft (6.1-m) straight lengths or in coils of 100-ft (30.5-m) length. **Type K** tubing, in coils, is used for underground work where the minimum number of joints, combined with greater thickness of type K tubing, is of distinct advantage. **Type L** tubing, usually in straight lengths, is used as a principal piping material for plumbing systems in homes and buildings; this is largely due to the economy of installation made possible by the use of soldered fittings. Copper deteriorates rapidly under high temperatures and repeated stresses. At a temperature of 360°F (182°C) its strength is reduced 15 percent, and on this account it should never be used for high steam pressures and temperatures.

Commercial sizes of **aluminum tubing** are listed by the manufacturers in even outside diameters and in wall thicknesses conforming to Stubs gage. Aluminum pipe is available as listed in Table 8.7.30. To obtain the approximate weight per foot of aluminum pipe or tubing, a weight of 0.098 lb/in³ (2.71 g/cm³) may be used.

Lead pipe is supplied in straight lengths, in coils, or in reels.

Block tin is a term used in the metal trade to refer to products made wholly from strictly pure high-grade tin. While tin pipe has many and varied uses, its most important applications are in types of equipment handling liquids intended for human consumption. Tin pipe does not corrode, and therefore does not contaminate most of the liquids passing through it.

Plastic pipes and **tubes** are available in a wide range of diameters and thicknesses, with Table 8.7.31 generally representative. The plastic used is resistant to attack by many chemicals, light in weight, flexible, and available in coiled form so that installation time is low. It is used for a variety of purposes including drainage, irrigation, sewage, and for conveying chemical solutions or waters that would attack metal piping.

Plastic pipe used in gas service is listed in Table 8.7.32. Caution should be used in selection of plastic piping insofar as service temperature is concerned: e.g., polyethylene is suitable for a maximum temperature of 120°F (49°C). Table 8.7.33 lists corrosion-resistance data for polyethylene plastic piping.

Pipes with Special Linings For use in lines through which are passed solutions containing more or less free acid or other corrosive agents, standard pipe, valves, and fittings may be **lead-lined, tin-lined,** or **rubber-lined,** to resist corrosive action. This lining prolongs the life of the pipe and also gives it additional strength. For mine service in coal districts where the drainage water is more or less impregnated with sulfur or free sulfuric acid, **wood-lined pipe** and fittings are sometimes used. For special service, **seamless-copper-lined pipe** is also used. The **cement lining** of ductile-iron and steel pipe for water and other services is advantageous because of its protection against unusual destructive agencies and its ability to prevent tuberculation. Standard **hard-rubber pipe** and fittings have been developed for working pressures of 50 lb/in² (345 kPa) at normal temperatures. Standard sizes run from ¼- to 4-in diam (0.635- to 10.2-cm), in 10-ft (3.05-m) lengths. For temperature above 120°F (49°C), the use of hard-rubber-lined steel pipe is recommended. This pipe is suitable for conveying strong acids and chemicals.

VITRIFIED, WOODEN-STAVE, AND CONCRETE PIPE

Vitrified pipe is used extensively for drains and sewerage systems. Burnt-clay tile, being rendered impervious to water by glazing, is by far the best material for sewage purposes as it is not attacked by acids. Dimensions are given in Table 8.7.34 (see also Fig. 8.7.6). For sizes larger than 36 in (91.4 cm) and other data, refer to the publications of the Clay Products Assoc., Chicago (see ASTM Standard C-700).

Wood-stave pipe (Fig. 8.7.7) is used to a large extent for municipal water supply, outfall sewers, mining, irrigation, and various other uses providing for the transportation of water. The water carried may be hot, cold, or acid. It is made either untreated or creosoted by a vacuum and pressure process. This process uses 8 lb of creosote per cubic foot of wood treated. The untreated pipe is most used where the pipe is constantly full of water, and the wood therefore completely saturated, although in many such instances the creosoted wood is used to give assurance of permanence. (See also Sec. 6.)

Wood-stave pipe is made in two types: machine-banded pipe and continuous-stave pipe. **Machine-banded pipe** is banded with wire and is

Table 8.7.29 Sizes and Weights of Copper Tubes

Nominal size, in	OD, in, types* K, L	ID, in Type K	ID, in Type L	Wall thickness, in Type K	Wall thickness, in Type L	Permissible variation of mean OD, in Types K, L Annealed	Permissible variation of mean OD, in Types K, L Hard-drawn	Weight,† lb/ft Type K	Weight,† lb/ft Type L
⅜	0.500	0.402	0.430	0.049	0.035	0.0025	0.001	0.269	0.198
½	0.625	0.527	0.545	0.049	0.040	0.0025	0.001	0.344	0.285
⅝	0.750	0.652	0.666	0.049	0.042	0.0025	0.001	0.418	0.362
¾	0.875	0.745	0.785	0.065	0.045	0.003	0.001	0.641	0.455
1	1.125	0.995	1.025	0.065	0.050	0.0035	0.0015	0.839	0.655
1¼	1.375	1.245	1.265	0.065	0.055	0.004	0.0015	1.04	0.884
1½	1.625	1.481	1.505	0.072	0.060	0.0045	0.002	1.36	1.14
2	2.125	1.959	1.985	0.083	0.070	0.005	0.002	2.06	1.75
2½	2.625	2.435	2.465	0.095	0.080	0.005	0.002	2.93	2.48
3	3.125	2.907	2.945	0.109	0.090	0.005	0.002	4.00	3.33
3½	3.625	3.385	3.425	0.120	0.100	0.005	0.002	5.12	4.29
4	4.125	3.857	3.905	0.134	0.110	0.005	0.002	6.51	5.38
5	5.125	4.805	4.875	0.160	0.125	0.005	0.002	9.67	7.61
6	6.125	5.741	5.845	0.192	0.140	0.005	0.002	13.9	10.2
8	8.125	7.583	7.725	0.271	0.200	0.006	0.003	25.9	19.3
10	10.125	9.449	9.625	0.338	0.250	0.008	0.004	40.3	30.1
12	12.125	11.315	11.565	0.405	0.280	0.008	0.004	57.8	40.4

* Type K recommended for underground service and general plumbing. Type L suitable for interior plumbing and other services.
† Multiply these values by 1.48 to obtain weight in kg/m.
SOURCE: American Brass Co.

Table 8.7.30   Aluminum Piping

Nominal pipe size, in	Schedule no.	OD, in	ID, in	Wall thickness, in	Weight per linear foot, lb, plain ends†	Cross-sectional wall area, in²	Inside cross-sectional area, in²	Moment of inertia, in⁴	Section modulus, in³	Radius of gyration, in
⅛	40‡	0.405	0.269	0.068	0.085	0.0720	0.0568	0.0011	0.0053	0.1215
	80§	0.405	0.215	0.095	0.109	0.0925	0.0363	0.0012	0.0060	0.1146
¼	40‡	0.540	0.364	0.088	0.147	0.1250	0.1041	0.0033	0.0123	0.1628
	80§	0.540	0.302	0.119	0.185	0.1574	0.0716	0.0038	0.0139	0.1547
⅜	40‡	0.675	0.493	0.091	0.196	0.1670	0.1909	0.0073	0.0216	0.2090
	80§	0.675	0.423	0.126	0.256	0.2173	0.1405	0.0086	0.0255	0.1991
½	40‡	0.840	0.622	0.109	0.294	0.2503	0.3039	0.0171	0.0407	0.2613
	80§	0.840	0.546	0.147	0.376	0.3200	0.2341	0.0201	0.0478	0.2505
¾	10	1.050	0.884	0.083	0.297	0.2521	0.6138	0.0297	0.0566	0.3432
	40‡	1.050	0.824	0.113	0.391	0.3326	0.5333	0.0370	0.0705	0.3337
	80§	1.050	0.742	0.154	0.510	0.4335	0.4324	0.0448	0.0853	0.3214
1	5	1.315	1.185	0.065	0.300	0.2553	1.103	0.0500	0.0760	0.4425
	10	1.315	1.097	0.109	0.486	0.4130	0.9452	0.0757	0.1151	0.4382
	40‡	1.315	1.049	0.133	0.581	0.4939	0.8643	0.0873	0.1328	0.4205
	80§	1.315	0.957	0.179	0.751	0.6388	0.7193	0.1056	0.1606	0.4066
1¼	5	1.660	1.530	0.065	0.383	0.3257	1.839	0.1037	0.1250	0.5644
	10	1.660	1.442	0.109	0.625	0.5311	1.633	0.1605	0.1934	0.5497
	40‡	1.660	1.380	0.140	0.786	0.6685	1.496	0.1947	0.2346	0.5397
	80§	1.660	1.278	0.191	1.037	0.8815	1.283	0.2418	0.2913	0.5238
1½	5	1.900	1.770	0.065	0.441	0.3747	2.461	0.1579	0.1662	0.6492
	10	1.900	1.682	0.109	0.721	0.6133	2.222	0.2468	0.2598	0.6344
	40‡	1.900	1.610	0.145	0.940	0.7995	2.036	0.3099	0.3262	0.6226
	80§	1.900	1.500	0.200	1.256	1.068	1.767	0.3912	0.4118	0.6052
2	5	2.375	2.245	0.065	0.555	0.4717	3.958	0.3149	0.2652	0.8170
	10	2.375	2.157	0.109	0.913	0.7760	3.654	0.4992	0.4204	0.8021
	40‡	2.375	2.067	0.154	1.264	1.074	3.356	0.6657	0.5606	0.7871
	80§	2.375	1.939	0.218	1.737	1.477	2.953	0.8679	0.7309	0.7665
2½	5	2.875	2.709	0.083	0.856	0.7280	5.764	0.7100	0.4939	0.9876
	10	2.875	2.635	0.120	1.221	1.039	5.453	0.9873	0.6868	0.9750
	40‡	2.875	2.469	0.203	2.004	1.704	4.788	1.530	1.064	0.9474
	80§	2.875	2.323	0.276	2.650	2.254	4.238	1.924	1.339	0.9241
3	5	3.500	3.334	0.083	1.048	0.8910	8.730	1.301	0.7435	1.208
	10	3.500	3.260	0.120	1.498	1.274	8.346	1.822	1.041	1.196
	40‡	3.500	3.068	0.216	2.621	2.228	7.393	3.017	1.724	1.164
	80§	3.500	2.900	0.300	3.547	3.016	6.605	3.894	2.225	1.136
3½	5	4.000	3.834	0.083	1.201	1.021	11.55	1.960	0.9799	1.385
	10	4.000	3.760	0.120	1.720	1.463	11.10	2.755	1.378	1.372
	40‡	4.000	3.548	0.226	3.151	2.680	9.887	4.788	2.394	1.337
	80§	4.000	3.364	0.318	4.326	3.678	8.888	6.281	3.140	1.307
4	5	4.500	4.334	0.083	1.354	1.152	14.75	2.810	1.249	1.562
	10	4.500	4.260	0.120	1.942	1.651	14.25	3.963	1.761	1.549
	40‡	4.500	4.026	0.237	3.733	3.174	12.73	7.232	3.214	1.510
	80§	4.500	3.826	0.337	5.183	4.407	11.50	9.611	4.272	1.477
5	40‡	5.563	5.047	0.258	5.057	4.300	20.01	15.16	5.451	1.878
	80§	5.563	4.813	0.375	7.188	6.112	18.19	20.67	7.432	1.839
6	40‡	6.625	6.065	0.280	6.564	5.581	28.89	28.14	8.496	2.246
	80§	6.625	5.761	0.432	9.884	8.405	26.07	40.49	12.22	2.195
8	30	8.625	8.071	0.277	8.543	7.265	51.16	63.35	14.69	2.953
	40‡	8.625	7.981	0.322	9.878	8.399	50.03	72.49	16.81	2.938
	80§	8.625	7.625	0.500	15.01	12.76	45.66	105.7	24.51	2.878
10		10.750	10.192	0.279	10.79	9.178	81.59	125.9	23.42	3.704
	30	10.750	10.136	0.307	11.84	10.07	80.69	137.4	25.57	3.694
	40‡	10.750	10.020	0.365	14.00	11.91	78.85	160.7	29.90	3.674
	80§	10.750	9.750	0.500	18.93	16.10	74.66	211.9	39.43	3.628
12	30	12.750	12.090	0.330	15.14	12.88	114.8	248.5	38.97	4.393
	‡	12.750	12.000	0.375	17.14	14.58	113.1	279.3	43.81	4.377
	§	12.750	11.750	0.500	22.63	19.24	108.4	361.5	56.71	4.335
Construction pipe										
		2.00	1.900	0.050	0.3602	0.3063	2.835	0.1457	0.1457	0.6897
		3.00	2.900	0.050	0.5449	0.4634	6.605	0.5042	0.3361	1.043
		4.00	3.900	0.050	0.7297	0.6205	11.95	1.210	0.6051	1.397
		5.00	4.896	0.052	0.9506	0.8083	18.83	2.474	0.9896	1.749
		6.00	5.876	0.062	1.360	1.157	2.712	5.098	1.699	2.100
		7.00	6.856	0.072	1.843	1.567	36.92	9.403	2.687	2.450
		8.00	7.812	0.094	2.745	2.335	47.93	18.24	4.561	2.795

* Aluminum Co. of America.
† Weights calculated for 6061 and 6063. For 3003 multiply by 1.010.
‡ Also designated as standard pipe.
§ Also designated as extra-heavy or extra-strong pipe. All calculations based on nominal dimensions.

Table 8.7.31   Commerical Sizes (IPS) and Weights of Polyvinyl Chloride (PVC) Pipe*

Nominal size, in	Schedule	Wall thickness,* in	OD, in	ID, in	Theoretical weight,† lb/ft	Calculated min bursting pressure, lb/in²	
						Note 1	Note 2
¼	40	0.088	0.540	0.364	0.076	2,490	1,950
	80	0.119	0.540	0.302	0.096	3,620	2,830
½	40	0.109	0.840	0.622	0.153	1,910	1,490
	80	0.147	0.840	0.546	0.185	2,720	2,120
¾	40	0.113	1.050	0.824	0.203	1,540	1,210
	80	0.154	1.050	0.742	0.265	2,200	1,720
1	40	0.133	1.315	1.049	0.305	1,440	1,130
	80	0.179	1.315	0.957	0.385	2,020	1,580
1¼	40	0.140	1.660	1.380	0.409	1,180	920
	80	0.191	1.660	1.278	0.550	1,660	1,300
1½	40	0.145	1.900	1.610	0.489	1,060	830
	80	0.200	1.900	1.500	0.653	1,510	1,180
2	40	0.154	2.375	2.067	0.640	890	690
	80	0.218	2.375	1.939	0.910	1,290	1,010
3	40	0.216	3.500	3.068	1.380	840	660
	80	0.300	3.500	2.900	1.845	1,200	940
4	40	0.237	4.500	4.026	1.965	710	560
	80	0.337	4.500	3.826	2.710	1,040	810

* Thicknesses listed are minimum values. Tolerance is generally $-0 + 10\%$.
† These representative values are not specified in ASTM D1785-85.
NOTES:
1. Materials are PVC 1120, 1220, and 4116. A fiber stress of 6,400 lb/in² was used in bursting pressure calculations.
2. Materials are PVC 2112, 2116, and 2120. A fiber stress of 5,000 lb/in² was used in bursting pressure calculations.
SOURCE: Abstracted from ASTM Specification D1785-85.

Table 8.7.32   Dimensions of Plastic Pipe for Gas Service (AGA Requirements)

Nominal size, in	Nominal OD, in	Nominal ID, in	Sleeved weight, lb	Max working pressure at 73°F, lb/in²
½	0.625	0.500	0.054	250
¾	0.875	0.750	0.079	175
1	1.125	1.000	0.103	125
1¼	1.375	1.250	0.127	110
1½	1.625	1.500	0.152	90
1¾	1.875	1.750	0.177	80
2	2.125	2.000	0.200	75
2½	2.660	2.500	0.320	75
3	3.190	3.000	0.457	75
4	4.250	4.000	0.800	75

made with wood or metal collars, or with inserted joints. **Continuous-stave pipe** is manufactured in units consisting of staves, bands, and shoes, shipped in knocked-down form, and constructed in the trench. In building this type of pipe, the staves are laid so as to break joints and the completed pipe is without joints. Continuous-stave pipe is banded with individual bands, ranging in size from ⅜ to 1 in (0.95 to 2.54 cm), depending upon the size of the pipe. A factor of safety of 4 is maintained in the band, based on an ultimate strength of 60,000 lb/in² (414 MPa) of cross section. The maximum pressure to which a continuous-stave pipe may be subjected depends upon the size of the pipe. The head for small pipes may run as high as 400 ft (121.9 m) while in the largest sizes the head would be less than 200 ft (61 m).

Machine-banded pipe is made for pressures of 50 to 400 ft (15.2 to 121.9 m). The staves are made from redwood or Douglas-fir lumber, dried and carefully selected. The inside and outside of the staves are dressed to conform to the circumferential lines, and the edges of the staves dressed to conform to the radial lines.

Wooden pipe is largely built in western sections of the United States, close to the natural lumber market. The sizes of machine-banded pipe range from 2 to 24 in (5.08 to 61 cm), and of the continuous-stave pipe from 6- to 20-ft (15.2-cm to 6.1-m) inside diameter.

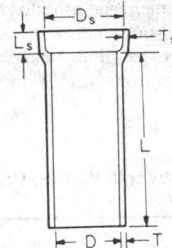

Fig. 8.7.6   Joint for vitrified pipe.

Pipe made from **plywood** is molded in lengths up to 11 ft (3.35 m) in diameters 3 in (7.62 cm) and up, and in wall thicknesses to specifications. Tubes made from fiber, by a molding process, are obtainable in a variety of sizes and lengths.

Concrete pipe is an important factor in sewer, conduit, railroad, culvert, and water-pipe construction. The pipe, as usually made, is constructed of concrete reinforced longitudinally with bars and transversely with wire mesh or steel bands. It is made in sections of definite length, with the longitudinal reinforcement so disposed as to provide for the interlocking of one section with another, and so formed that when these are locked together and cemented they form a continuous line of pipe free from leakage or seepage. Various forms of joints are used, all capable of taking care of expansion. Figure 8.7.8 shows one type of construction. Concrete pipe is manufactured in a great variety of diameters, thicknesses, and lengths to suit almost any requirement arising in practice. (See also Sec. 6.)

Table 8.7.33   Corrosion-Resistance Data, Polyethylene Pipe

Reagent	Performance at: 75°F (24°C)	Performance at: 120°F (49°C)	Reagent	Performance at: 75°F (24°C)	Performance at: 120°F (49°C)	Reagent	Performance at: 75° (24°C)	Performance at: 120°F (49°C)
Acetic acid, glacial*	F	NG	Ferrous sulfate, 15 percent aq.	E	E	Potassium chloride, saturated	E	E
Acetic acid, 10 percent*	E	E	Fluorine	E	NG	Potassium dichromate	E	E
Acetone	NG	NG	Fluosilicic acid, concentrated	E	F	Potassium hydroxide	E	E
Ammonia, dry gas	E	E	Formaldehyde, 40 percent	E	E	Potassium nitrate	E	E
Ammonium hydroxide, 10 percent	E	E	Formic acid, 50 percent	E	E	Potassium permanganate	E	E
Ammonium hydroxide, 28 percent	E	E	Furfuryl alcohol	NG	NG	Silicic acid	E	E
Amyl acetate	NG	NG	Gasoline	NG	NG	Silver nitrate	E	E
Aniline	E	F	Hydrobromic acid	E	E	Sodium benzoate	E	E
Benzene	NG	NG	Hydrochloric acid, 10 percent	E	E	Sodium bisulfite	E	E
Bromine	NG	NG	Hydrochloric acid, 37 percent	E	E	Sodium carbonate, concentrated	E	E
Butyraldehyde	E	G	Hydrofluoric acid, 48 percent	E	E	Sodium chloride, saturated solution	E	E
Calcium chloride, saturated	E	E	Hydrofluoric acid, 75 percent	E	F	Sodium hydroxide, 10 percent	E	E
Calcium hydroxide	E	E	Hydrogen peroxide, 30 percent	E	G	Sodium hydroxide, 50 percent	E	E
Calcium hypochlorite	E	E	Hydrogen peroxide, 90 percent	G	NG	Sodium sulfate	E	E
Carbon disulfide	NG	NG	Lactic acid, 90 percent	E	E	Stannic chloride, saturated	E	E
Carbon tetrachloride	NG	NG	Linseed oil	E	E	Stearic acid, 100 percent	E	E
Carbonic acid	E	E	Lubricating oil	NG	NG	Sulfuric acid, 10 percent	E	E
Chlorine, dry gas	F	NG	Magnesium chloride	E	E	Sulfuric acid, 30 percent	E	E
Chlorine, liquid	NG	NG	Magnesium sulfate	E	E	Sulfuric acid, 60 percent	E	F
Chlorosulfonic acid	NG	NG	Methyl bromide	NG		Sulfuric acid, 98 percent	F	NG
Citric acid, saturated	E	E	Methyl isobutyl ketone	F	NG	Tannic acid	E	E
Copper sulfate	E	E	Nitric acid, 10 percent	E	E	Toluene	NG	NG
Cyclohexanone	NG	NG	Nitric acid, 30–50 percent	E	E	Trichlorobenzene	F	NF
Diethylene glycol	E	E	Nitric acid, 70 percent	E	E	Trichloroethylene	NG	NG
Dioxane	E	G	Oleic acid	F	NG	Vinegar	E	E
Ethyl acetate	F	NG	Phosphoric acid, 30 percent	E	E	Xylene	NG	NG
Ethyl alcohol, 35 percent	NG	NG	Phosphoric acid, 90 percent	E	NG	Zinc chloride	E	E
Ethyl butyrate	F	NG	Photographic developer	E	E	Zinc sulfate	E	E
Ethylene dichloride	NG	NG	Potassium borate	E	E			
Ferric chloride	E	E	Potassium carbonate	E	E			

Corrosion-resistance data given in this table based on laboratory tests conducted by the manufacturers of the materials covered, and are indicative only of the conditions under which the tests were made. This information may be considered as a basis for recommendation, but not as a guarantee. Materials should be tested under actual service to determine suitability for a particular purpose.
E = excellent, G = good, F = fair, NG = not good.
* Polyethylene is permeable to acetic acid.

Asbestos-cement pipe, known by the trade name **Transite** pipe in this country, was developed initially in Europe. It was widely used in the United States for many years in a large variety of services. It was made of a mixture of portland cement and asbestos fiber, was highly resistant to corrosion, and provided outstanding service in mine drainage systems, waterworks systems, gas lines, and sewerage systems. It was manufactured in diameters from 3 to 36 in (7.6 to 91 cm), in stock lengths of 13 ft (4 m), and in pressure classes of 50, 100, 150, and 200 lb/in² (349, 689, 1,034, and 1,379 kPa). While it is no longer manufactured, it is still in place in some of the above-cited applications. Repairs to existing Transite piping generally devolves into replacing it with pipe made of a material compatible with its service, often with fiberglass-reinforced pipe, plastic pipe, or some form of metal pipe.

Table 8.7.34   Standard-Strength Vitrified Clay Pipe
(Dimensions refer to Fig. 8.7.6)

Size D, in	Laying length L: Nominal, ft	Laying length L: Limit of minus variation,* in/ft length	Max difference in length of two opposite sides, in	OD of barrel, in Min	OD of barrel, in Max	ID of socket ½ in above base D_s, in Min	ID of socket ½ in above base D_s, in Max	Depth of socket L_s, in Nominal	Depth of socket L_s, in Min	Thickness of barrel T, in Nominal	Thickness of barrel T, in Min	Thickness of socket at ½ in from outer end T_s, in Nominal	Thickness of socket at ½ in from outer end T_s, in Min
4	2, 2½, 3	¼	5⁄16	4⅞	5⅛	5¾	6⅛	1¾	1½	½	7⁄16	7⁄16	⅜
6	2, 2½, 3	¼	⅜	7 1⁄16	7 7⁄16	8 3⁄16	8⅝	2½	2	⅝	9⁄16	½	7⁄16
8	2, 2½, 3	¼	7⁄16	9¼	9¾	10½	11	2½	2¼	¾	11⁄16	9⁄16	½
10	2, 2½, 3	¼	7⁄16	11½	12	12¾	13¼	2⅝	2⅜	⅞	13⁄16	⅝	9⁄16
12	2, 2½, 3	¼	7⁄16	13¾	14 5⁄16	15⅛	15¾	2¾	2½	1	15⁄16	¾	11⁄16
15	3, 4	¼	½	17 3⁄16	17 13⁄16	18⅝	19¼	2⅞	2⅝	1¼	1⅛	15⁄16	⅞
18	3, 4	¼	½	20⅝	21 7⁄16	22¼	23	3	2¾	1½	1⅜	1⅛	1 1⁄16
21	3, 4	¼	9⁄16	24⅛	25	25⅞	26¾	3¼	3	1¾	1⅝	1 5⁄16	1 3⁄16
24	3, 4	⅜	9⁄16	27½	28½	29⅜	30⅜	3⅜	3⅛	2	1⅞	1½	1⅜
27	3, 4	⅜	9⁄16	31	32⅛	33	34⅛	3½	3¼	2¼	2⅛	1 11⁄16	1 9⁄16
30	3, 4	⅜	⅝	34⅜	35⅝	36½	37¾	3⅝	3⅜	2½	2⅜	1⅞	1¾
33	3, 4	⅜	⅝	37⅝	38 15⁄16	39⅞	41¼	3¾	3½	2⅝	2½	2	1 13⁄16
36	3, 4	⅜	11⁄16	40¾	42¼	43¼	44¾	4	3¾	2¾	2⅝	2 1⁄16	1⅞

When ordering standard-strength vitrified-clay pipe, give the size of pipe (ID) and the laying strength wanted, and refer to ASTM Specification C-700. Standard lengths of pipe shown meet normal practice in various sections of the country. Manufacturers' stocks include those lengths conforming to local practice.
* There is no limit for plus variation.

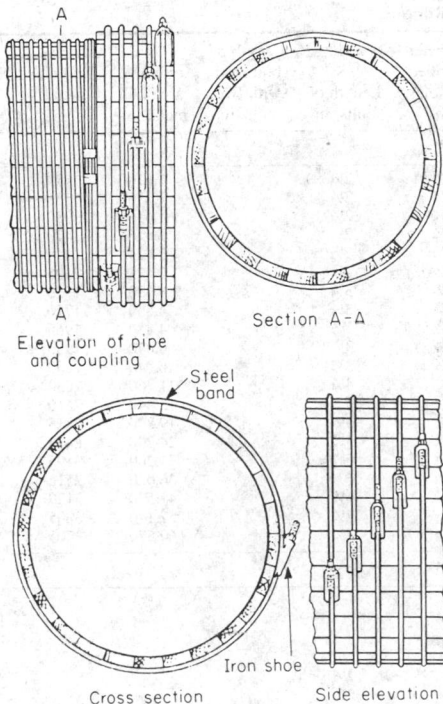

Elevation of pipe
and coupling

Section A-A

Cross section

Side elevation

Steel band

Iron shoe

Fig. 8.7.7   Wood-stave pipes.

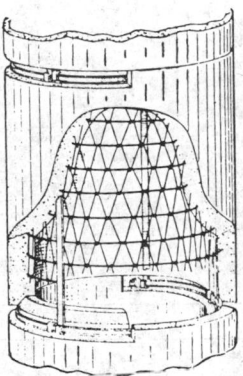

Fig. 8.7.8   Reinforced-concrete pipe.

FITTINGS FOR STEEL PIPE

American National Standard Cast-Iron Pipe Flanges and Flanged Fittings for Maximum Working Saturated Steam Pressure of 25, 125, and 250 lb/in² (172, 862, and 1,724 kPa)

INTRODUCTORY NOTES

Sizes   The sizes of the fittings in the following tables are nominal pipe sizes. In the class 25 standard, the nominal pipe size is the same as the port diameter of the fittings, for *all* sizes. In the class 125 and class 250 standards the nominal pipe size is the same as the port diameter of fittings for pipe having inside diameters of 12 in (30.5 cm) and smaller. For pipe 14 in (35.6 cm) and larger, the corresponding outside diameter of the pipe is given, and consequently the fittings will have a smaller port diameter.

Pressure Rating   In the class 25 standard the sizes 36 in (91.4 cm) and smaller may also be used for maximum nonshock working hydraulic pressures of 43 lb/in² gage (296 kPa) or a maximum gas pressure of 25 lb/in² gage (172 kPa), at or near the ordinary range of air temperatures. In the class 125 standard, the sizes 12 in (30.5 cm) and smaller may also be used for maximum nonshock working hydraulic pressure of 175 lb/in² gage (1,207 kPa) at or near the ordinary range of air temperatures. In the class 250 standard, the sizes 12 in and smaller may be used for maximum nonshock working hydraulic pressures of 400 lb/in² gage (2,758 kPa) at or near the ordinary range of air temperatures.

Facing   All class 25 and class 125 cast-iron flanges and flanged fittings are plain faced, i.e., without projection or raised face. All class 250 cast-iron flanges and flanged fittings have a raised face ¹⁄₁₆ (0.16 cm) high, of the diameters given in Table 8.7.35. The raised face is included in the minimum flange thickness and center-to-face dimensions.

An **inspection limit** of ± ¹⁄₃₂ in (0.08 cm) is allowed on all center-to-contact-surface dimensions for sizes up to and including 10 in (25.4 cm) and ± ¹⁄₈ in on sizes larger than 10 in.

Dimensions   In the class 25 standard, the flange diameters, bolt circles, and number of bolts are the same as in the class 125 ANSI Standard B16.1-1975, with a reduction in the thickness of flanges and bolt diameters, thereby maintaining interchangeability between the two standards.

The center-to-face and face-to-face dimensions of class 25 standard fittings are the same as for the class 125 standard.

Bolting   Drilling templates are in multiples of four, so that fittings may be made to face in any quarter. **Bolt-holes** straddle the centerline. For bolts smaller than 1¾ in (4.45 cm) the bolt-holes are drilled ⅛ in larger in diameter than the nominal size of the bolt. Holes for bolts 1¾ in and larger are drilled ¼ in (0.64 cm) larger than nominal diameter of bolts. **Bolts** of steel are with standard "rough square heads" and the nuts are of steel with standard "rough hexagonal" dimensions; all as given in the American Standard on Wrench Head Bolts and Nuts and Wrench Openings of the National Screw Thread Commission. For bolts, 1¾-in (4.45-cm) diam and larger, bolt studs with a nut on each end are recommended.

Hexagonal nuts for pipe sizes 1 to 48 in (2.54 to 122 cm) in the class 125 standard and 1 to 16 in (40.6 cm) in the class 250 standard can be conveniently pulled up with open wrenches of minimum design of heads. Hexagonal nuts for pipe sizes 48 to 96 in (244 cm) in the class 125 standard and 18 to 48 in (45.7 to 122 cm) in the class 250 standard can be conveniently pulled up with box wrenches.

Spot Facing   The bolt-holes of classes 25, 125, and 250 cast-iron flanges and flanged fittings are not spot-faced for ordinary service. When required, the flanges and fittings in sizes 30 in (76.2 cm) and larger may be spot-faced or back-faced to the minimum thickness of flange with a plus tolerance of ⅛ in (0.32 cm).

Reducing Fittings   Reducing elbows and side-outlet elbows carry same dimensions center to face as straight-size elbows corresponding to the size of the larger opening.

Tees, side-outlet tees, crosses, and laterals sizes 16 in (40.6 cm) and smaller, reducing on the outlet or branch, have the same dimension center to face and face to face as straight-size fittings corresponding to the size of the larger opening. Sizes 18 in (45.7 cm) and larger, reducing on the outlet or branch, are made in two lengths depending on the size of the outlet or branch as given in the tables of dimensions.

Tees, crosses, and laterals, reducing on the run only, have the same dimensions center to face and face to face as straight-size fittings corresponding to the size of the larger opening.

Reducers and eccentric reducers for all reduction have the same face-to-face dimensions for the larger opening.

Special double-branch elbows whether straight or reducing have the same dimension center to face as straight-size elbows corresponding to the size of the larger opening.

Side-outlet elbows and side-outlet tees have all openings on intersecting centerlines.

Table 8.7.35 Templates for Drilling Cast-Iron Pipe Flanges, Flanged Valves, and Fittings*

Nominal pipe size, in	Diameter of flange, in	Thickness of flange (minimum), in	Diameter of raised face, in	Diameter of bolt circle, in	Number of bolts	Diameter of bolts, in	Diameter of drilled bolt-holes, in	Length of bolts, in	Length of bolt stud with two nuts, in	Total effective area bolt metal, in²	Stress in bolt metal, lb/in²†	Size of ring gasket, in
Class 25 standard (922 N/m²)												
4	9	¾		7½	8	⅝	¾	2¼		1.616	570	4 × 6⅞
5	10	¾		8½	8	⅝	¾	2¼		1.616	750	5 × 7⅞
6	11	¾		9½	8	⅝	¾	2¼		1.616	930	6 × 8¾
8	13½	¾		11¾	8	⅝	¾	2¼		1.616	1470	8 × 11
10	16	⅞		14¼	12	⅝	¾	2½		2.424	1440	10 × 13⅜
12	19	1		17	12	⅝	¾	2¾		2.424	2195	12 × 16⅛
14	21	1⅛		18¾	12	¾	⅞	3¼		3.620	1750	14 × 18
16	23½	1⅛		21¼	16	¾	⅞	3¼		4.830	1710	16 × 20½
18	25	1¼		22¾	16	¾	⅞	3½		4.830	1965	18 × 22
20	27½	1¼		25	20	¾	⅞	3½		6.040	1920	20 × 24½
24	32	1⅜		29½	20	¾	⅞	3¾		6.040	2690	24 × 28¾
30	38¾	1½		36	28	⅞	1	4¼		11.760	2030	30 × 35⅛
36	46	1⅝		42¾	32	⅞	1	5		13.440	2610	36 × 41⅞
42	53	1¾		49½	36	1	1⅛	5¼		19.800	2315	42 × 48½
48	59½	2		56	44	1	1⅛	5½		24.200	2475	48 × 55
54	66¼	2¼		62¾	44	1	1⅛	5¾		24.200	3195	54 × 61¾
60	73	2¼		69¼	52	1⅛	1¼	6		36.020	2515	60 × 68⅛
72	86½	2½		82½	60	1⅛	1¼	6¼		41.570	3120	72 × 81⅜
84	99¾	2¾		95½	64	1¼	1⅜	7¼		57.140	3005	84 × 94¼
96	113¼	3		108½	68	1¼	1⅜	7¾		60.570	3705	96 × 107¼
Class 125 standard (4,612 N/m²)												
1	4¼	7⁄16		3⅛	4	½	⅝	1¾				1 × 2⅝
1¼	4⅝	½		3½	4	½	⅝	2				1¼ × 3
1½	5	9⁄16		3⅞	4	½	⅝	2				1½ × 3⅜
2	6	⅝		4¾	4	⅝	¾	2¼				2 × 4⅛
2½	7	11⁄16		5½	4	⅝	¾	2½				2½ × 4⅞
3	7½	¾		6	4	⅝	¾	2½				3 × 5⅜
3½	8½	13⁄16		7	8	⅝	¾	2¾				3½ × 6⅜
4	9	15⁄16		7½	8	⅝	¾	3				4 × 6⅞
5	10	15⁄16		8½	8	¾	⅞	3				5 × 7¾
6	11	1		9½	8	¾	⅞	3¼				6 × 8¾
8	13½	1⅛		11¾	8	¾	⅞	3½				8 × 11
10	16	1³⁄₁₆		14¼	12	⅞	1	3¼				10 × 13⅜
12	19	1¼		17	12	⅞	1	3¾				12 × 16⅛
14 OD	21	1⅜		18¾	12	1	1⅛	4¼				14 × 17¾
16 OD	23½	1⁷⁄₁₆		21¼	16	1	1⅛	4½				16 × 20¼
18 OD	25	1⁹⁄₁₆		22¾	16	1⅛	1¼	4¾				18 × 21⅝
20 OD	27½	1¹¹⁄₁₆		25	20	1⅛	1¼	5				20 × 23⅞
24 OD	32	1⅞		29½	20	1¼	1⅜	5½				24 × 28¼
30 OD	38¾	2⅛		36	28	1¼	1⅜	6¼				30 × 34⅝
36 OD	46	2⅜		42¾	32	1½	1⅝	7				36 × 41¼
42 OD	53	2⅝		49½	36	1½	1⅝	7½				42 × 48
48 OD	59½	2¾		56	44	1½	1⅝	7¾				48 × 54½
54 OD	66¼	3		62¾	44	1¾	2	8½	10½			54 × 61
60 OD	73	3⅛		69¼	52	1¾	2	8¾	11			60 × 67½
72 OD	86½	3½		82½	60	1¾	2	9½	12			72 × 80¾
84 OD	99¾	3⅞		95½	64	2	2¼	10½	13			84 × 93½
96 OD	113¼	4¼		108½	68	2¼	2½	11½	14½			96 × 106¼
Class 250 standard (9,224 N/m²)												
1	4⅞	11⁄16	2¹¹⁄₁₆	3½	4	⅝	¾	2¼				1 × 2⅞
1¼	5¼	¾	3¹⁄₁₆	3⅞	4	⅝	¾	2½				1¼ × 3¼
1½	6⅛	13⁄16	3⁹⁄₁₆	4½	4	¾	⅞	2½				1½ × 3¾
2	6½	⅞	4³⁄₁₆	5	8	⅝	¾	2½				2 × 4⅜
2½	7½	1	4¹⁵⁄₁₆	5⅞	8	¾	⅞	3				2½ × 5⅛
3	8¼	1⅛	5¹¹⁄₁₆	6⅝	8	¾	⅞	3¼				3 × 5⅞
3½	9	1³⁄₁₆	6⁵⁄₁₆	7¼	8	¾	⅞	3¼				3½ × 6½
4	10	1¼	6¹⁵⁄₁₆	7⅞	8	¾	⅞	3½				4 × 7⅛
5	11	1⅜	8⁵⁄₁₆	9¼	8	¾	⅞	3¾				5 × 8½
6	12½	1⁷⁄₁₆	9¹¹⁄₁₆	10⅝	12	¾	⅞	3¾				6 × 9⅞
8	15	1⅝	11¹⁵⁄₁₆	13	12	⅞	1	4¼				8 × 12⅛
10	17½	1⅞	14¹⁄₁₆	15¼	16	1	1⅛	5				10 × 14¼
12	20½	2	16⁷⁄₁₆	17¾	16	1⅛	1¼	5½				12 × 16⅝
14 OD	23	2⅛	18¹⁵⁄₁₆	20¼	20	1⅛	1¼	5¾				13¼ × 19⅛
16 OD	25½	2¼	21¹⁄₁₆	22½	20	1¼	1⅜	6				15¼ × 21¼
18 OD	28	2⅜	23⁵⁄₁₆	24¾	24	1¼	1⅜	6¼				17 × 23½
20 OD	30½	2½	25⁹⁄₁₆	27	24	1¼	1⅜	6½				19 × 25¾
24 OD	36	2¾	30⁵⁄₁₆	32	24	1½	1⅝	7½	9½			23 × 30½
30 OD	43	3	37³⁄₁₆	39¼	28	1¾	2	8¼	10½			29 × 37½
36 OD	50	3⅜	43¹¹⁄₁₆	46	32	2	2¼	9¼	11½			34½ × 44
42 OD	57	3¹¹⁄₁₆	50⁷⁄₁₆	52¾	36	2	2¼	9¾	12			40¼ × 50¾
48 OD	65	4	58⁷⁄₁₆	60¾	40	2	2¼	10½	13			46 × 58¾

* To obtain linear dimensions in cm, multiply tabular values in in by 2.54. To obtain areas in cm², multiply tabular values in in² by 6.45. To obtain stress in N/m², multiply values in lb/in² by 36.895.

† The stress shown is that of internal pressure assumed to act only on a circular area equal in diameter to the outside diameter of the ring gasket covering the flange to the inside of the bolts for the 25-lb standard.

SOURCE: ANSI B16.1-1975.

Elbows Special degree elbows ranging from 1 to 45° have the same center-to-face dimensions given for 45° elbows, and those over 45° and up to 90° have the same center-to-face dimensions given for 90° elbows. The angle designation of an elbow is its deflection from straight-line flow and is the angle between its flange faces.

Threaded Companion Flanges Threaded companion flanges in the class 25 standard should not be thinner than those in the class 125 standard on sizes 24 in (61 cm) and smaller. Other types of flanges may have thicknesses as given in Table 8.7.35.

Laterals Laterals (Y branches) both straight and reducing sizes 8 in and larger are reinforced to compensate for the inherent weakness in the casting design.

The American National Standard B16.1-1975 covers also dimensions (not included in the tables) of base elbows and base tees and anchorage bases for straight tees and reducing tees.

American National Standard cast-iron pipe flanges and flanged fittings are available for maximum nonshock working hydraulic pressure of 800 lb/in² gage (5,516 kPa) at ordinary air temperatures.

Assembly of Flanged Joints The optimum degree of tightening occurs when a stress of 30,000 lb/in² (207 MPa) is uniformly reached in each flange stud or bolt. For a modulus of elasticity of 30,000,000 lb/in² (207 GPa) a stress of 30,000,000 lb/in² occurs when the elongation, determined with a dial indicator or micrometer, is 0.001 in/in of stud length measured between centers of nuts. Uniform tension in flange bolts may also be obtained by use of a torque wrench; bearing surfaces of nuts must have a good machine finish, and threads must be properly lubricated for reliable results with a torque wrench. The following torque values have been found to give 30,000 lb/in² stress in studs:

Study diam, in	Threads per inch	Torque, lb · ft*
⅝	11	89
¾	10	107
⅞	9	162
1	8	244
1⅛	8	322
1¼	8	410
1⅜	8	510
1½	8	615

* Multiply by 0.149 to obtain torque in kilogram-metres.

American National Standard Steel Pipe Flanges and Flanged Fittings

INTRODUCTORY NOTES

Pressure Ratings and Tests These standards are known as "American Class 150, 300, 400, 600, 900, 1,500, and 2,500 Steel Flange Standards" (ANSI B16.5-1981), said pressure designation being the recommended rating at the temperatures given in Table 8.7.35. This table shows recommended ratings for various temperatures. For other tables, refer to ANSI B16.5-1981.

Sizes The size of the fittings and companion flanges in the tables is identified by the corresponding nominal pipe size. For pipe NPS 14 (35.6 cm) and larger, the corresponding outside diameter of the pipe is given.

Materials The flanged fittings and flanges should be either steel castings or steel forgings of the grade complying with the ASTM specifications recommended under these standards for the various pressure-

Table 8.7.36 Pressure-Temperature Ratings for Carbon-Steel Flanges and Flanged Fittings, Working Pressure in lb/in² Gage

Temperature, °F	Class 150 Material group* 1.1	1.2	1.4	Class 300 Material group* 1.1	1.2	1.4	Class 1500 Material group* 1.1	1.2	1.4
	Carbon steel Norm.	High	Low	Carbon steel Norm.	High	Low	Carbon steel Norm.	High	Low
−20–100	285	290	235	740	750	620	3,705	3,750	3,085
200	260	260	215	675	750	560	3,375	3,750	2,810
300	230	230	210	655	730	550	3,280	3,640	2,735
400				635	705	530	3,170	3,530	2,645
500				600	665	500	2,995	3,325	2,490
600				550	605	455	2,735	3,025	2,285
650				535	590	450	2,685	2,940	2,245
700				535	570	450	2,665	2,840	2,245
750				505	505	445	2,520	2,520	2,210
800				410	410	370	2,060	2,060	1,850
850					270			1,340	
900					170			860	
950					105			515	
1,000					50			260	

Material group	Materials (spec. grade)	See notes
1.1	A105, A181-II, A216-WCB, A515-70	†, ¶
	A516-70	†, §
	A350-LF2, A537-C1.1	‡
1.2	A203-B, A203-E, A216-WCC	†, ¶
	A350-LF3, A352-LC2, A352-LC3	‡
1.4	A181-I, A515-60	†, ¶
	A516-60	†, §
	A350-LF1	‡

* Ratings shown apply to other material groups where columns dividing lines are omitted.
† Permissible but not recommended for prolonged use above about 800°F.
‡ Not to be used over 650°F.
§ Not to be used over 850°F.
¶ Not to be used over 1,000°F.
SOURCE: Abstracted from ASME B16.5-81 with permission.

temperature ratings for which these standards are designed. A few of these characteristics from ANSI B16.5-1981 are given in Table 8.7.36.

Bolting material including nuts and washers are based on a high-grade product equal to that given in ASTM Standard Specifications for Alloy-Steel Bolting Material for High Temperature Service (Table 8.7.37) and with physical and chemical requirements in accordance with the tables given under ANSI B16.5-1981. **Commercial steel bolts should not be used at steam pressures over 250** lb/in^2 (1,724 kPa) and temperatures over 450°F (232°C). Nuts should be of carbon or alloy steel. Washers when used under nuts should be of forged or rolled carbon steel.

Bolting Drilling templates are in multiples of four, so that fittings may be made to face in any quarter. Bolt-holes straddle the centerlines. Bolt-holes are drilled ⅛ in (0.32 cm) larger in diameter than the nominal size of bolt. Bolts or bolt studs threaded at both ends may be used and should be equipped with cold-punched or cold-pressed semifinished nuts of American National Standard rough dimensions, chamfered and trimmed.

All bolts and bolt studs having diameters 1 in and smaller, and the corresponding nuts are threaded with the American National Standard screw thread, coarse thread series, medium fit, class 3, while those bolts and bolt studs whose diameters are 1⅛ in (2.86 cm) and larger have special threads of the American National form whose pitch is ⅛ in (8 threads per inch). It is recommended that these special threads be allowed a pitch-diameter tolerance of − 0.006 in and a lead tolerance of ± 0.002 in.

Bolt studs with a nut at each end are recommended for high-temperature service.

The allowable working fiber stress, considering internal allowable working pressure only, in bolting material for valve bonnet flanges, cleanout flanges, etc., is not to exceed 9,000 lb/in^2 (62 MPa) assuming the pressure to act upon an area circumscribed by the periphery of the outside of the contact surface.

Metal Thickness Minimum metal thicknesses specified in the tables are based on an allowable fiber stress of 7,000 lb/in^2 (48 MPa), using the modified Barlow formula of the ASME Boiler Construction Code for cylindrical sections and adding 50 percent to the thickness thus determined to compensate for the shape of the fittings. The minimum commercial casting thickness is considered to be ¼ in; therefore, the standards do not show thicknesses less than this. The minimum thickness in these standards means the minimum thickness in any part of the finished casting.

The modified Barlow formula is as follows: For pipes having nominal diameters of ¼ to 5 in, $P + 125 = 2S(t - 0.065)/D$. For pipes of nominal diameters over 5 in, $P = 2S(t - 0.1)/D$, where P is the working pressure, lb/in^2, t is the thickness of wall of pipe, in; D is the actual outside diameter of pipe, in; and S is 7,000 lb/in^2 (48 MPa).

Ring Joints The dimensions used for ring and groove joint facings were developed by a committee of the API. The corresponding dimensions and ring numbers incorporated in the ANSI Standard are identical with those given in API Standard 5G. The dimension for the depth of groove is added to the basic flange thickness, which makes it necessary to include separate tables of dimensions for fittings having the ring joint facing.

Fitting Dimensions An inspection limit of ± ¹⁄₃₂ in is allowed on all center-to-contact surface dimensions for sizes up to and including 10 in, and ± ¹⁄₁₆ in on sizes larger than NPS 10. An inspection limit of ± ¹⁄₁₆ in is allowed on all contact-surface to contact-surface dimensions for sizes up to and including NPS 10, and ± ⅛ in, on sizes larger than NPS 10.

When elbows having longer radii than specified in the standards are required, the use of pipe bends is recommended.

Laterals The 45° laterals of the larger sizes may require additional reinforcement to compensate for the inherent weakness in this shape of casting.

Valve Dimensions Face-to-face and end-to-end dimensions of ferrous valves for the various pressures are in accordance with the requirements of ANSI B16.10-1973 for ferrous flanged valves.

Reducing Fittings Reducing fittings have the same center-to-flange edge dimensions as those of straight-size fittings of the largest opening.

Side-Outlet Fittings All side-outlet fittings have all openings on the intersecting centerlines.

Welding Neck Flanges The materials, facings, spot facings, etc., conform to the requirements given for other flanges, with the additional provision that the carbon content of the steel shall not exceed 0.35 percent.

Templates for drilling and center to contact-surface dimensions of the American Standard class 150 steel flanges and flanged fittings are the same as for the American Standard class 125 cast-iron flanged fitting standard.

Templates for drilling and center to contact-surface dimensions of the ANSI class 300 steel flanges and flanged fittings are the same as for the ANSI class 600 steel flanged fitting standard for sizes NPS ½ to 1½ (Table 8.7.38); and the same as for the ANSI class 250 cast-iron flanged fitting standard for sizes NPS 2 to NPS 24.

Flanged Pipe Joints

The usual form of pipe joint is that made up by bolting together flanges cast or forged integral with the pipe or fitting, threaded flanges, loose flanges on pipes with lapped ends, and flanges arranged for welding. These forms are illustrated in Tables 8.7.39 and 8.7.40 and in Fig. 8.7.9. The threaded joint is satisfactory for low and medium steam pressures. The lapped joint is permitted in the same sizes and service ratings as for joints with integral flanges. It is extensively used in high-class work. With the ring joint a higher pressure can be maintained with the same total bolt stress than is possible with the flat gasket type of joint. The welded joint eliminates possibility of leakage between flange and pipe. It is very successful in lines subject to high temperatures and pressures and heavy expansion strains. The welding-neck flange is available in the various pipe sizes. Specific requirements covering the application of all the types of joints in common use are outlined in the Code for Power Piping (ASME B31.1 and B31.3).

Facing of Flanges Various styles of finish are used on the faces of flanges, for the purpose of the retention of the gasket used to make a tight joint. Those in general use are as follows (see Table 8.7.39): plain straight face, plain face corrugated or scored, male and female, tongue and groove, and raised face.

The **plain straight face** has the entire face of the flange faced straight across and may be used with either a full face or ring gasket. The **plain face, serrated or V-grooved,** is a plain face upon which concentric grooves have been cut with either a round-nose or V-shaped tool. This finish is sometimes of advantage when the service demands an exceptionally thick, loosely woven fibrous or soft metallic gasket, because the roughening of the faces of the flanges tends to keep the gasket from blowing out. The **male-and-female** facing consists of a recess in one flange and a corresponding raised face or projection on the other mating flange extending from the inside of the pipe nearly to the inside of the bolt holes. In the **tongue-and-groove** facing, the tongue or raised face and the groove or recess are narrow rings located between the bolt holes and the port. The male-and-female and the tongue-and-groove facing have been extensively used, particularly on hydraulic lines. To a more limited extent they have been used also on high-pressure steam lines. Both these types, however, have in common several objectionable features from the standpoint of manufacture, erection, and maintenance. These objections are removed by the use of the **raised-face** facing, which consists of a high narrow raised ring on each of the mating flanges, whose inside diameters is the same as that of the pipe or port. It is particularly recommended for high-pressure steam and hydraulic lines. **Gaskets** used in this type of joint are either soft fibrous material or soft metal and extend from the inside of the pipe to the bolt holes. Only the small portion in contact with the narrow raised face is subjected to the compressive effect of the bolts.

The following advantages are claimed for the raised-face type of facing: all mating of flanges has been eliminated; any valve or fitting may be removed from the line without springing the line apart; the gasket is automatically centered by its outer edge coming in contact with the bolts; the outside edges of the flanges are far enough apart to make it possible to determine whether the joint has been properly made.

Table 8.7.37 Mechanical Properties of Alloy Steel Bolting Material for High-Temperature Service

Ferritic steels

Grade	Diameter, in (mm)	Minimum tempering temperature,* °F (°C)	Tensile strength, min ksi (MPa)	Yield strength, min 0.2% offset, ksi (MPa)	Elongation in 2 in (50.8 mm), min. %	Reduction of area, min %	Hardness, max
B5, 4 to 6% chromium	Up to 4 (101.6), incl.	1,100 (593)	100 (690)	80 (550)	16	50	
B6, 13% chromium	Up to 4 (101.6), incl.	1,100 (593)	110 (760)	85 (585)	15	50	
B6X, 13% chromium	Up to 4 (101.6), incl.	1,100 (593)	90 (620)	70 (485)	16	50	26 HRC
B7, chromium-molybdenum	2½ (63.5) and under	1,100 (593)	125 (860)	105 (720)	16	50	
	Over 2½ to 4 (63.5 to 101.6)	1,100 (593)	115 (790)	95 (655)	16	50	
	Over 4 to 7 (101.6 to 117.8)	1,100 (593)	100 (690)	75 (515)	18	50	
B7M,† chromium-molybdenum	2½ (63.5) and under	1,150 (620)	100 (690)	80 (550)	18	50	235 HB or 99 HRB
B16, chromium-molybdenum-vanadium	2½ (63.5) and under	1,200 (650)	125 (860)	105 (720)	18	50	
	Over 2½ to 4 (63.5 to 101.6)	1,200 (650)	110 (760)	95 (655)	17	45	
	Over 4 to 7 (101.6 to 177.8)	1,200 (650)	100 (690)	85 (585)	16	45	

Austenitic steels

Class and grade, diameter, in (mm)	Heat treatment‡	Tensile strength, min ksi (MPa)	Yield strength, min 0.2% offset, ksi (MPa)	Elongation in 2 in (50.8 mm), min %	Reduction of area, min %	Hardness, max
Class 1: B8, B8C, B8M, B8P, B8T, B8LN, B8MLN, all diameters	Carbide solution treated	75 (515)	30 (205)	30	50	223 HB or 96 HRB
Class 1A: B8A, B8CA, B8MA, B8PA, B8TA, B8LNA, B8MLNA, B8NA, B8MNA, all diameters	Carbide solution treated in finished condition	75 (515)	30 (205)	30	50	192 HB or 90 HRB
Class 1B: B8N, B8MN, all diameters	Carbide solution treated	80 (550)	35 (240)	30	40	223 HB or 96 HRB
Class 1C: B8R, all diameters	Carbide solution treated	100 (690)	55 (380)	35	55	271 HB or 28 HRC
B8RA, all diameters	Carbide solution treated in finished condition	100 (690)	55 (380)	35	55	271 HB or 28 HRC
B8S, all diameters	Carbide solution treated	95 (655)	50 (345)	35	55	271 HB or 28 HRC
B8SA, all diameters	Carbide solution treated in finished condition	95 (655)	50 (345)	35	55	271 HB or 28 HRC
Class 2: B8, B8C, B8P, B8T, B8N, ¾ (19.05) and under	Carbide solution treated and strain-hardened	125 (860)	100 (690)	12	35	321 HB or 35 HRC
Over ¾ to 1 (19.05 to 25.4) incl.		115 (790)	80 (550)	15	35	321 HB or 35 HRC
Over 1 to 1¼ (25.4 to 31.6) incl.		105 (720)	65 (450)	20	35	321 HB or 35 HRC
Over 1¼ to 1½ (31.6 to 37.9) incl.		100 (690)	50 (345)	28	45	321 HB or 35 HRC
Class 2: B8M, B8MN, ¾ (19.05) and under	Carbide solution treated and strain-hardened	110 (760)	95 (655)	15	45	321 HB or 35 HRC
Over ¾ to 1 (19.05 to 25.4) incl.		100 (690)	80 (550)	20	45	321 HB or 35 HRC
Over 1 to 1¼ (25.4 to 31.6) incl.		95 (655)	65 (450)	25	45	321 HB or 35 HRC
Over 1¼ to 1½ (31.6 to 37.9) incl.		90 (620)	50 (345)	30	45	321 HB or 35 HRC

* For sizes ¾ in. in diameter and smaller, a maximum hardness of 241 HB (100 HRB) is permitted.
† To meet the tensile requirements, the Brinell hardness shall be over 201 HB (94 HRB).
‡ Class 1 is solution treated. Class 1A is solution treated in the finished condition for corrosion resistance: heat treatment is critical due to physical property requirement. Class 2 is solution-treated and strain-hardened. Austenitic steels in the strain-hardened condition may not show uniform properties throughout the section particularly in sizes over ¾ in (19.05 mm) in diameter.

Table 8.7.38 Templates for Drilling, American National Standard Steel Pipe Flanges and Flanged Fittings (ANSI B16.5-1981)
(All dimensions in inches)

	Class 400 standard					Class 600 standard					Class 900 standard					Class 1,500 standard				
Nominal pipe size	Outside diam of flange	Thickness of flange, minimum	Diam of bolt circle	Number of bolts	Size of bolts	Outside diam of flange	Thickness of flange, minimum	Diam of bolt circle	Number of bolts	Size of bolts	Outside diam of flange	Thickness of flange, minimum	Diam of bolt circle	Number of bolts	Size of bolts	Outside diam of flange	Thickness of flange, minimum	Diam of bolt circle	Number of bolts	Size of bolts
½	For sizes below 4 in use dimensions of 600-lb fittings					3¾	9/16	2⅝	4	½	For sizes below 3 in use dimensions of 1,500-lb fittings					4¾	⅞	3¼	4	¾
¾						4⅝	⅝	3¼	4	⅝						5⅛	1	3½	4	¾
1						4⅞	11/16	3½	4	⅝						5⅞	1⅛	4	4	1⅞
1¼	—	—	—	—	—	5¼	13/16	3⅞	4	⅝						6¼	1⅛	4⅜	4	⅞
1½	—	—	—	—	—	6⅛	⅞	4½	4	¾	—	—	—	—	—	7	1¼	4⅞	4	1
2	—	—	—	—	—	6½	1	5	8	⅝	—	—	—	—	—	8½	1½	6½	8	⅞
2½	—	—	—	—	—	7½	1⅛	5⅞	8	¾	—	—	—	—	—	9⅝	1⅝	7½	8	1
3	—	—	—	—	—	8¼	1¼	6⅝	8	¾	9½	1½	7½	8	⅞	10½	1⅞	8	8	1⅛
3½	—	—	—	—	—	9	1⅜	7¼	8	⅞										
4	10	1⅜	7⅞	8	⅞	10¾	1½	8½	8	⅞	11½	1¾	9¼	8	1⅛	12¼	2⅛	9½	8	1¼
5	11	1½	9¼	8	⅞	13	1¾	10½	8	1	13¾	2	11	8	1¼	14¾	2⅞	11½	8	1½
6	12½	1⅝	10⅝	12	⅞	14	1⅞	11½	12	1	15	2 3/16	12½	12	1⅛	15½	3¼	12½	12	1⅜
8	15	1⅞	13	12	1	16½	2 3/16	13¾	12	1⅛	18½	2½	15½	12	1⅜	19	3⅝	15½	12	1⅝
10	17½	2⅛	15¼	16	1⅛	20	2½	17	16	1¼	21½	2¾	18½	16	1⅜	23	4¼	19	12	1⅞
12	20½	2¼	17¾	16	1¼	22	2⅝	19¼	20	1¼	24	3⅛	21	20	1⅜	26½	4⅞	22½	16	2
14 OD	23	2⅜	20¼	20	1¼	23¾	2¾	20¾	20	1⅜	25¼	3⅜	22	20	1½	29½	5¼	25	16	2¼
16 OD	25½	2½	22½	20	1⅜	27	3	23¾	20	1½	27¾	3½	24¼	20	1⅝	32½	5¾	27¾	16	2½
18 OD	28	2⅝	24¾	24	1⅜	29¼	3¼	25¾	20	1⅝	31	4	27	20	1⅞	36	6⅜	30½	16	2¾
20 OD	30½	2¾	27	24	1½	32	3½	28½	24	1⅝	33¾	4¼	29½	20	2	38¾	7	32¾	16	3
24 OD	36	3	32	24	1¾	37	4	33	24	1⅞	41	5½	35½	20	2½	46	8	39	16	3½

Table 8.7.39   Facing Dimensions for the American Class 150, 300, 400, 600, 900, 1,500, and 2,500 Steel Flanges (ANSI B16.5-1981)

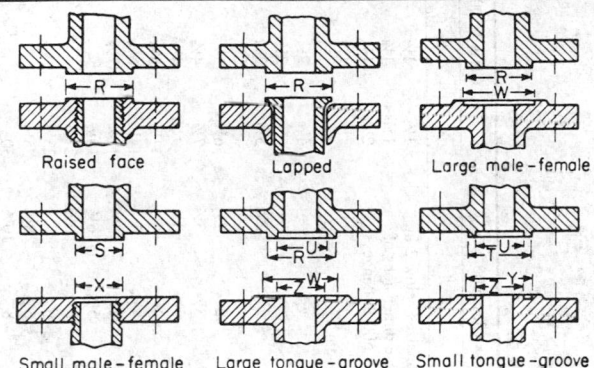

Raised face   Lapped   Large male–female   Small male–female   Large tongue–groove   Small tongue–groove

| Nominal pipe size, in | Outside diameter, in | | | ID of large and small tongue, U | Outside diameter, in | | | ID of large and small groove, Z | Height, in | | Depth of groove or female companion flanges |
	Raised face, lapped, large male, and large tongue, R	Small male, S	Small tongue, T		Large female and large groove, W	Small female, X	Small groove, Y		Raised face, 150 and 300 lb std*	Raised face, large and small male and tongue 400-, 600-, 900-, 1,500- and 2,500-lb studs	
$\frac{1}{2}$	$1\frac{3}{8}$	$\frac{23}{32}$	$1\frac{3}{8}$	1	$1\frac{7}{16}$	$\frac{25}{32}$	$1\frac{7}{16}$	$\frac{15}{16}$	$\frac{1}{16}$	$\frac{1}{4}$	$\frac{3}{16}$
$\frac{3}{4}$	$1\frac{11}{16}$	$\frac{15}{16}$	$1\frac{11}{16}$	$1\frac{5}{16}$	$1\frac{3}{4}$	1	$1\frac{3}{4}$	$1\frac{1}{4}$	$\frac{1}{16}$	$\frac{1}{4}$	$\frac{3}{16}$
1	2	$1\frac{3}{16}$	$1\frac{7}{8}$	$1\frac{1}{2}$	$2\frac{1}{16}$	$1\frac{1}{4}$	$1\frac{15}{16}$	$1\frac{7}{16}$	$\frac{1}{16}$	$\frac{1}{4}$	$\frac{3}{16}$
$1\frac{1}{4}$	$2\frac{1}{2}$	$1\frac{1}{2}$	$2\frac{1}{4}$	$1\frac{7}{8}$	$2\frac{9}{16}$	$1\frac{9}{16}$	$2\frac{5}{16}$	$1\frac{13}{16}$	$\frac{1}{16}$	$\frac{1}{4}$	$\frac{3}{16}$
$1\frac{1}{2}$	$2\frac{7}{8}$	$1\frac{3}{4}$	$2\frac{1}{2}$	$2\frac{1}{8}$	$2\frac{15}{16}$	$1\frac{13}{16}$	$2\frac{9}{16}$	$2\frac{1}{16}$	$\frac{1}{16}$	$\frac{1}{4}$	$\frac{3}{16}$
2	$3\frac{5}{8}$	$2\frac{1}{4}$	$3\frac{1}{4}$	$2\frac{7}{8}$	$3\frac{11}{16}$	$2\frac{5}{16}$	$3\frac{5}{16}$	$2\frac{13}{16}$	$\frac{1}{16}$	$\frac{1}{4}$	$\frac{3}{12}$
$2\frac{1}{2}$	$4\frac{1}{8}$	$2\frac{11}{16}$	$3\frac{3}{4}$	$3\frac{3}{8}$	$4\frac{3}{16}$	$2\frac{3}{4}$	$3\frac{13}{12}$	$3\frac{5}{16}$	$\frac{1}{16}$	$\frac{1}{4}$	$\frac{3}{16}$
3	5	$3\frac{5}{16}$	$4\frac{5}{8}$	$4\frac{1}{4}$	$5\frac{1}{16}$	$3\frac{3}{8}$	$4\frac{11}{16}$	$4\frac{3}{16}$	$\frac{1}{16}$	$\frac{1}{4}$	$\frac{3}{16}$
$3\frac{1}{2}$	$5\frac{1}{2}$	$3\frac{13}{16}$	$5\frac{1}{8}$	$4\frac{3}{4}$	$5\frac{9}{16}$	$3\frac{7}{8}$	$5\frac{3}{16}$	$4\frac{11}{16}$	$\frac{1}{16}$	$\frac{1}{4}$	$\frac{3}{16}$
4	$6\frac{3}{16}$	$4\frac{5}{16}$	$5\frac{11}{16}$	$5\frac{3}{16}$	$6\frac{1}{4}$	$4\frac{3}{8}$	$5\frac{3}{4}$	$5\frac{1}{8}$	$\frac{1}{16}$	$\frac{1}{4}$	$\frac{3}{16}$
5	$7\frac{5}{16}$	$5\frac{3}{8}$	$6\frac{13}{16}$	$6\frac{5}{16}$	$7\frac{3}{8}$	$5\frac{7}{16}$	$6\frac{7}{8}$	$6\frac{1}{4}$	$\frac{1}{16}$	$\frac{1}{4}$	$\frac{3}{16}$
6	$8\frac{1}{2}$	$6\frac{3}{8}$	8	$7\frac{1}{2}$	$8\frac{9}{16}$	$6\frac{7}{16}$	$8\frac{1}{16}$	$7\frac{7}{16}$	$\frac{1}{16}$	$\frac{1}{4}$	$\frac{3}{16}$
8	$10\frac{5}{8}$	$8\frac{3}{8}$	10	$9\frac{3}{8}$	$10\frac{11}{16}$	$8\frac{7}{16}$	$10\frac{1}{16}$	$9\frac{5}{16}$	$\frac{1}{16}$	$\frac{1}{4}$	$\frac{3}{16}$
10	$12\frac{3}{4}$	$10\frac{1}{2}$	12	$11\frac{1}{4}$	$12\frac{13}{16}$	$10\frac{9}{16}$	$12\frac{1}{16}$	$11\frac{3}{16}$	$\frac{1}{16}$	$\frac{1}{4}$	$\frac{3}{16}$
12	15	$12\frac{1}{2}$	$14\frac{1}{4}$	$13\frac{1}{2}$	$15\frac{1}{16}$	$12\frac{9}{16}$	$14\frac{5}{16}$	$13\frac{7}{16}$	$\frac{1}{16}$	$\frac{1}{4}$	$\frac{3}{16}$
14 OD	$16\frac{1}{4}$	$13\frac{3}{4}$	$15\frac{1}{2}$	$14\frac{3}{4}$	$16\frac{5}{16}$	$13\frac{13}{16}$	$15\frac{9}{16}$	$14\frac{11}{16}$	$\frac{1}{16}$	$\frac{1}{4}$	$\frac{3}{16}$
16 OD	$18\frac{1}{2}$	$15\frac{3}{4}$	$17\frac{5}{8}$	$16\frac{3}{4}$	$18\frac{9}{16}$	$15\frac{13}{16}$	$17\frac{11}{16}$	$16\frac{11}{16}$	$\frac{1}{16}$	$\frac{1}{4}$	$\frac{3}{16}$
18 OD	21	$17\frac{3}{4}$	$20\frac{1}{8}$	$19\frac{1}{4}$	$21\frac{1}{16}$	$17\frac{13}{16}$	$20\frac{3}{16}$	$19\frac{3}{16}$	$\frac{1}{16}$	$\frac{1}{4}$	$\frac{3}{16}$
20 OD	23	$19\frac{3}{4}$	22	21	$23\frac{1}{16}$	$19\frac{13}{16}$	$22\frac{1}{16}$	$20\frac{15}{16}$	$\frac{1}{16}$	$\frac{1}{4}$	$\frac{3}{16}$
24 OD	$27\frac{1}{4}$	$23\frac{3}{4}$	$26\frac{1}{4}$	$25\frac{1}{4}$	$27\frac{5}{16}$	$23\frac{13}{16}$	$26\frac{5}{16}$	$25\frac{3}{16}$	$\frac{1}{16}$	$\frac{1}{4}$	$\frac{3}{16}$

* Included in the minimum flange thickness dimensions. A $\frac{1}{16}$-in raised face is also permitted on the class 400, 600, 900, 1,500, and 2,500 flange standards, but it must be added to the minimum flange thicknesses. Regular facing for class 400, 600, 900, 1,500, and 2,500 flange standards is a $\frac{1}{4}$-in raised face not included in minimum flange thickness dimensions.
A tolerance of $\frac{1}{64}$ in is allowed on the inside and outside diameters of all facings.
Gaskets for male-female and tongue-groove joints should cover the bottom of the recess with minimum clearance taking into account the tolerances stated above.

Unions may be classified as **screw** and **flange**. Typical designs are shown in Fig. 8.7.10, where at the top left is represented a female screw union of the gasket type, at the top right a female screw union having a brass to iron seat that is noncorrosive and a ground joint that eliminates the need for a gasket, and at the bottom a flange union of the gasket type. As in the case of other pipe fittings, unions and union fittings are available in the various pipe sizes and in materials and designs suitable for any service conditions. Very large flange unions can be made by bolting together two screwed companion flanges.

Threaded Fittings

Threaded fittings are made of cast iron, malleable iron, cast steel, forged steel, or brass. Plain standard fittings are generally used for low-pressure gas and water, as in house plumbing and railing work, while the beaded fitting is the standard steam, air, gas, or oil fitting. Screwed fittings are supplied with a large factor of safety. The questions of strength involve much more than the pressure from within the pipe which induces a comparatively low stress in the material. The greater strains come from expansion, contraction, weight of piping, settling, water hammer, etc. Dimensions of cast-iron and malleable-iron screwed fittings of the American National Standard are given in Tables 8.7.41 and 8.7.42.

The dimensions of ferrous plugs, bushings, locknuts, and caps with pipe threads are covered by ANSI B16.14-1983. The dimensions of pipe plugs from this standard are given in Table 8.7.43.

The normal **amount of thread engagement** necessary to make a tight

Table 8.7.40 Dimensions of American National Standard Companion Flanges (ANSI B16.5-1981)*

Threaded flange and Lapped flange, showing dimensions X, Y, Z.

Nominal pipe size	Class 150 X	Class 150 Y	Class 150 Z	Class 300 X	Class 300 Y	Class 300 Z	Class 400 X	Class 400 Y	Class 400 Z	Class 600 X	Class 600 Y	Class 600 Z	Class 900 X	Class 900 Y	Class 900 Z	Class 1,500 X	Class 1,500 Y	Class 1,500 Z
1/2	1 3/16	5/8	5/8	1 1/2	7/8	7/8	For sizes below 4 in, use dimensions of 600-lb flanges			1 1/2	7/8	7/8	For sizes below 3 in, use dimensions of 1,500-lb flanges			1 1/2	1 1/4	1 1/4
3/4	1 1/2	5/8	5/8	1 7/8	1	1				1 7/8	1	1				1 3/4	1 3/8	1 3/8
1	1 15/16	11/16	11/16	2 1/8	1 1/16	1 1/16				2 1/8	1 1/8	1 1/16				2 1/16	1 5/8	1 5/8
1 1/4	2 5/16	13/16	13/16	2 1/2	1 1/16	1 1/16				2 1/2	1 1/8	1 1/8				2 1/2	1 5/8	1 5/8
1 1/2	2 9/16	7/8	7/8	2 3/4	1 3/16	1 3/16				2 3/4	1 1/4	1 1/4				2 3/4	1 3/4	1 3/4
2	3 3/16	1	1	3 5/16	1 5/16	1 1/2				3 5/16	1 7/16	1 7/16				4 1/4	2 1/4	2 1/4
2 1/2	3 9/16	1 1/8	1 1/8	3 15/16	1 1/2	1 5/8				3 15/16	1 5/8	1 5/8				4 7/8	2 1/2	2 1/2
3	4 1/4	1 3/16	1 3/16	4 5/8	1 11/16	1 11/16				4 5/8	1 13/16	1 13/16				5 1/4	2 7/8	2 7/8
3 1/2	4 13/16	1 1/4	1 1/4	5 1/4	1 3/4	1 3/4				5 1/4	1 15/16	1 15/16	5	2 1/8	2 7/8	—	—	—
4	5 5/16	1 5/16	1 5/16	5 3/4	1 7/8	1 7/8	5 1/4	2	2	6	2 1/8	2 1/8	6 1/4	2 3/4	2 3/4	6 3/8	3 9/16	3 9/16
5	6 7/16	1 7/16	1 7/16	7	2	2	7	2 1/4	2 1/8	7 7/16	2 3/8	2 3/8	7 1/2	3 1/8	3 1/8	7 3/8	4 1/4	4 1/4
6	7 9/16	1 9/16	1 9/16	8 1/8	2 1/16	2 1/8	8 1/8	2 1/4	2 1/4	8 3/8	2 5/8	2 5/8	9 1/4	3 3/8	3 3/8	9	4 11/16	4 11/16
8	9 11/16	1 3/4	1 3/4	10 1/4	2 7/16	2 11/16	10 1/4	2 11/16	2 11/16	10 3/4	3	3	11 3/4	4	4 1/2	11 1/2	5 5/8	5 5/8
10	12	1 15/16	1 15/16	12 5/8	2 5/8	3 3/4	12 5/8	2 7/8	4	13 1/2	3 3/8	4 3/8	14 1/2	4 1/4	5	14 1/2	6 1/4	7
12	14 3/8	2 3/16	2 3/16	14 3/4	2 7/8	4 3/8	14 3/4	3 1/8	4 1/4	15 3/4	3 5/8	4 5/8	16 1/2	4 5/8	5 5/8	17 3/4	7 3/8	8 3/8
14 OD	15 3/4	2 1/4	3 1/8	16 3/4	3	4 3/4	16 3/4	3 3/8	4 3/4	17	3 11/16	5	17 3/4	5 1/8	6 1/8	19 1/2	—	9 1/2
16 OD	18	2 1/2	3 7/16	19	3 1/4	5	19	3 11/16	5	19 1/2	4 3/8	5 3/8	20	5 1/4	6 1/2	21 3/4	—	10 1/4
18 OD	19 7/8	2 11/16	3 13/16	21	3 1/2	5 5/8	21	3 7/8	5 5/8	21 1/2	4 5/8	6	22 1/4	6	7 1/2	23 3/4	—	10 7/8
20 OD	22	2 7/8	4 1/16	23 1/8	3 3/4	5 1/2	23 3/8	4	5 3/8	24	5	6 1/4	24 1/2	6 1/2	8 1/4	25 3/4	—	11 1/2
24 OD	26 1/8	3 1/4	4 3/8	27 7/8	4 3/16	6	27 7/8	4 1/2	6 1/4	28 1/4	5 1/2	7 1/4	29 1/2	8	10 1/2	30	—	13

* Other dimensions are given in Tables 8.7.38 and 8.7.39. Finished bore on lapped flange to be such as method of attachment of pipe requires.

Table 8.7.41   Dimensions of ANSI Class 150 Standard Malleable-Iron Threaded Fittings*
(All dimensions in inches)

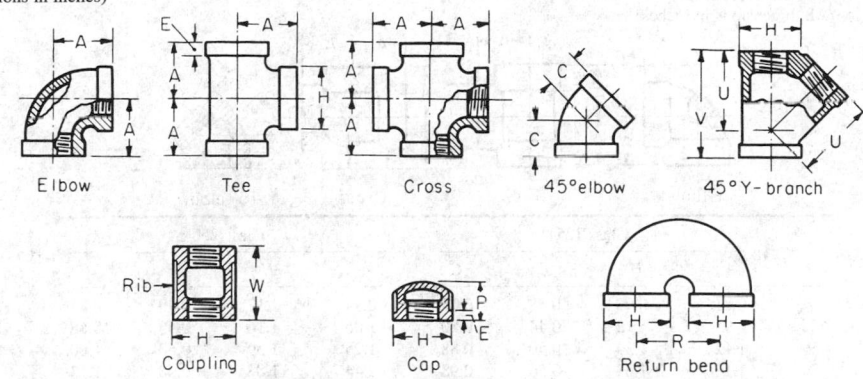

Elbow   Tee   Cross   45°elbow   45°Y-branch

Coupling   Cap   Return bend

Size	A	H	E	C	V	U	W	P	R Close	R Medium	R Open
⅛	0.69	0.693	0.200				0.96				
¼	0.81	0.844	0.215	0.73			1.06				
⅜	0.95	1.015	0.230	0.80	1.93	1.43	1.16				
½	1.12	1.197	0.249	0.88	2.32	1.71	1.34	0.87	1.000	1.25	1.50
¾	1.31	1.458	0.273	0.98	2.77	2.05	1.52	0.97	1.250	1.50	2.00
1	1.50	1.771	0.302	1.12	3.28	2.43	1.67	1.16	1.500	1.875	2.50
1¼	1.75	2.153	0.341	1.29	3.94	2.92	1.93	1.28	1.750	2.25	3.00
1½	1.94	2.427	0.368	1.43	4.38	3.28	2.15	1.33	2.188	2.50	3.50
2	2.25	2.963	0.422	1.68	5.17	3.93	2.53	1.45	2.625	3.00	4.00
2½	2.70	3.589	0.478	1.95	6.25	4.73	2.88	1.70			4.50
3	3.08	4.285	0.548	2.17	7.26	5.55	3.18	1.80			5.00
3½	3.42	4.843	0.604	2.39			3.43	1.90			
4	3.79	5.401	0.661	2.61	8.98	6.97	3.69	2.08			
5	4.50	6.583	0.780	3.05				2.32			
6	5.13	7.767	0.900	3.46				2.55			

* The complete standard (ANSI B16.3-1977) covers also reducing couplings, elbows, tees, crosses, and service or street elbows and tees.
SOURCE: ANSI B16.3-1977.

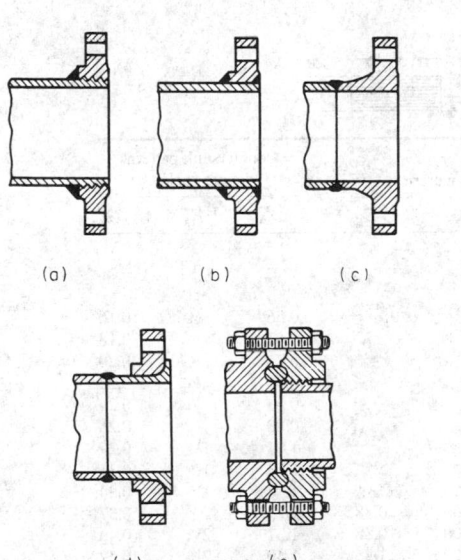

(a)   (b)   (c)

(d)   (e)

Fig. 8.7.9   Welded flange joints and ring joint. (*a*) Forged steel, screwed flange, back-welded and refaced; (*b*) forged steel, slip-on welding flange, welded front and back, refaced; (*c*) forged steel, welding neck flange, butt-welded to pipe; (*d*) lap-welding nipple, butt-welded to pipe; (*e*) ring joint.

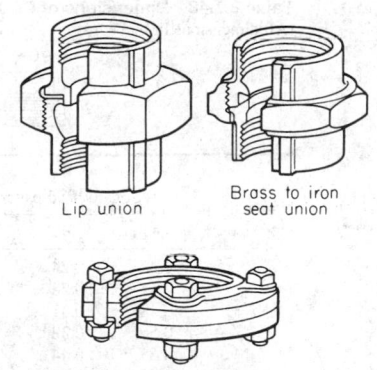

Lip union   Brass to iron seat union

Flange union

Fig. 8.7.10   Types of pipe unions.

joint for ANSI Standard pipe thread joints as recommended by Crane Co. is as follows:

Size of pipe, in	⅛	¼	⅜	½	¾	1	1¼	1½	2
Length of thread, in	¼	⅜	⅜	½	⁹⁄₁₆	¹¹⁄₁₆	¹¹⁄₁₆	¹¹⁄₁₆	¾

Size of pipe, in	2½	3	3½	4	5	6	8	10	12
Length of thread, in	¹⁵⁄₁₆	1	1¹⁄₁₆	1⅛	1¼	1⁵⁄₁₆	1⁷⁄₁₆	1⅝	1¾

Table 8.7.42 Dimensions of Class 125 and Class 250 Standard Cast-Iron Threaded Fittings*
(All dimensions in inches)

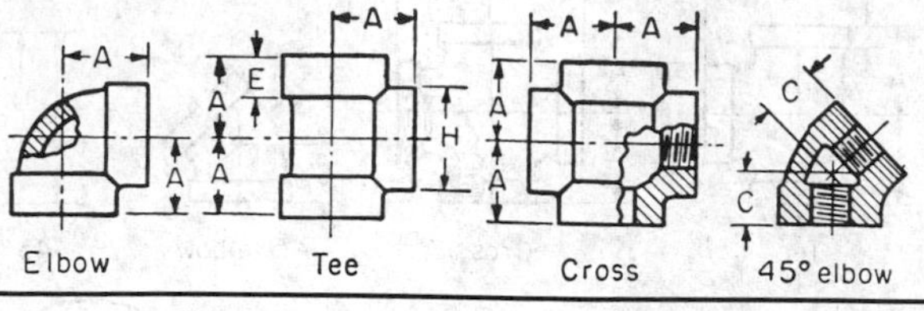

Size	Class 125 A	Class 125 H	Class 125 E	Class 125 C	Class 250 A	Class 250 H	Class 250 E	Class 250 C
¼	0.81	0.93	0.38	0.73	0.94	1.17	0.49	0.81
⅜	0.95	1.12	0.44	0.80	1.06	1.36	0.55	0.88
½	1.12	1.34	0.50	0.88	1.25	1.59	0.60	1.00
¾	1.13	1.63	0.56	0.98	1.44	1.88	0.68	1.13
1	1.50	1.95	0.62	1.12	1.63	2.24	0.76	1.31
1¼	1.75	2.39	0.69	1.29	1.94	2.73	0.88	1.50
1½	1.94	2.68	0.75	1.43	2.13	3.07	0.97	1.69
2	2.25	3.28	0.84	1.68	2.50	3.74	1.12	2.00
2½	2.70	3.86	0.94	1.95	2.94	4.60	1.30	2.25
3	3.08	4.62	1.00	2.17	3.38	5.36	1.40	2.50
3½	3.42	5.20	1.06	2.39	3.75	5.98	1.49	2.63
4	3.79	5.79	1.12	2.61	4.13	6.61	1.57	2.81
5	4.50	7.05	1.18	3.05	4.88	7.92	1.74	3.19
6	5.13	8.28	1.28	3.46	5.63	9.24	1.91	3.50
8	6.56	10.63	1.47	4.28	7.00	11.73	2.24	4.31
10	8.08	13.12	1.68	5.16	8.63	14.37	2.58	5.19
12	9.50	15.47	1.88	5.97	10.00	16.84	2.91	6.00

* This applies to elbows and tees only.
The class 125 standard covers also reducing elbows and tees. The class 250 standard covers only the straight sizes.
SOURCE: ANSI B16.4-1971.

Table 8.7.43 Dimensions of Class 125, 150, and 250 Pipe Plugs*
(All dimensions in inches)

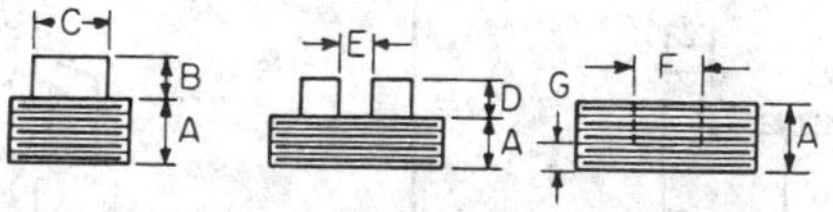

Nominal pipe size	Square-head pattern A	Square-head pattern B	Square-head pattern C	Slotted pattern A	Slotted pattern D	Slotted pattern E	Countersunk pattern† (square sockets) A	Countersunk pattern† (square sockets) F	Countersunk pattern† (square sockets) G
⅛	0.37	0.24	9/32						
¼	0.44	0.28	⅜						
⅜	0.48	0.31	7/16						
½	0.56	0.38	9/16				0.56	⅜	0.16
¾	0.63	0.44	⅝				0.63	½	0.18
1	0.75	0.50	13/16				0.75	½	0.20
1¼	0.80	0.56	15/16				0.80	¾	0.22
1½	0.83	0.62	1⅛				0.83	¾	0.24
2	0.88	0.68	1 5/16				0.88	⅞	0.26
2½	1.07	0.74	1½				1.07	1⅛	0.29
3	1.13	0.80	1 11/16				1.13	1⅜	0.31
3½	1.18	0.86	1⅞				1.18	1½	0.34
4				1.22	1.00	0.88	1.22	2	0.37
5				1.31	1.00	0.88	1.31	2¼	0.46
6				1.40	1.25	1.25	1.40	2½	0.52
8				1.57	1.38	1.50			

* The material of (ANSI B16.14-1983) is to be cast iron, malleable iron, or steel, for use in connection with fittings covered by the American National Standard class 125 cast-iron threaded fittings (ANSI B16.4) and the American National Standard class 150 malleable-iron screwed fittings (ANSI B16.3).
† Hexagon sockets (sizes ⅛ to 1 in) have dimensions to fit regular wrenches used with hexagon socket setscrews.
SOURCE: ANSI B16.14-1983.

The Manufacturers' Standardization Society of Valve and Fitting Industry (MSS) has standardized malleable-iron and brass threaded fittings for several pressures.

Cast-bronze threaded fittings are made in both the class 125 and 250 standards. They are used for any water pipe where bad water makes steel pipe undesirable. Bronze fittings may be had in iron pipe sizes. Forged-steel threaded fittings are made for cold water or oil-working pressures up to 6,000 lb/in^2 (41.4 MPa) hydrostatic. The ANSI has approved standard B16.26-1983 for cast copper alloy fittings for flared copper tubes for maximum cold-water service pressure of 175 lb/in^2 (1,207 kPa).

Railing Fittings Fittings of special construction and of tighter weight than standard steam, gas, and water pipe fittings are widely used for hand railings around areaways, on stairs, for office enclosures with gates, and for permanent ladders. Railing fittings are made in various styles, generally globe-shaped in body, with ends reduced to take thread and recessed to cover all threads. They are furnished in malleable iron, black and galvanized, and in brass.

Special railing-fitting joints are available, such as the slip-and-screwed joint, where the post connection is screwed and the rim of the fitting is so made that the rail will slip into the fitting and allow for an angular variation of several degrees, being fastened by pins which are riveted over and filed smooth. The flush-joint stair-rail fitting is another special style of fitting which provides a hand rail with even surfaces at the joints.

Drainage fittings, as shown in the figures accompanying Table 8.7.44, have no pockets for the lodgment of solids, and the length of the thread chamber is such that when the pipe is threaded to the American National Standard dimensions, the end of the pipe will practically touch the shoulder when screwed in. They are especially adapted to plumbing work and vacuum-cleaning pipe installations. Dimensions in Table 8.7.44 conform to ANSI Standard B16.12-1983, Cast-Iron Threaded Drainage Fittings.

The development of standards for **cast-iron long-turn sprinkler fittings** was begun by the National Fire Protection Assoc. in 1914 with a study of the peculiar needs of fittings intended for fire-protection purposes. These fittings (screwed and flanged) are rated at 175 and 250 lb/in^2 (1,207 and 1,724 kPa).

American National Standard Air Gaps and Backflow Preventers in Plumbing Systems, ANSI A40.4-1942 and A40.6-1943, was prepared to establish minimum requirements for plumbing, including water-supply distributing systems, drainage and venting systems, fixtures, apparatus, and devices, and the standardization of plumbing equipment in general.

Ammonia valves and fittings must provide a high margin of safety against accidents. Flanged valves and fittings have tongue-and-groove faces to assure tightness at the joints and against blowing out gaskets. Gaskets are compressed asbestos sheet. Threaded valves and fittings have long threads and are recessed so that the joints may be soldered. These valves and fittings are made of malleable iron, ductile iron, ferrosteel, or forged steel; depending on the size and style. Valves are all iron, with steel stems, and have special lead disk faces or steel disks. Copper or brass must not be used in their construction. Flanged valves are generally interchangeable with flanged fittings. All valves and fittings for ammonia are tested to 300 lb/in^2 (2,069 kPa) air pressure under water. For dimensions of valves, fittings, and specialties for ammonia, refer to manufacturers' catalogs.

Soldered-Joint Fittings The American standard for these fittings

Table 8.7.44 Dimensions of American National Standard Cast-Iron Threaded Drainage Fittings
(All dimensions in inches)

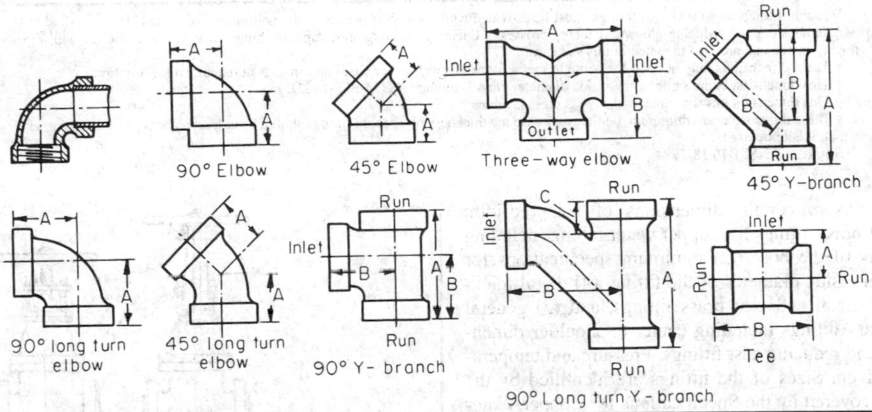

Size, in	90° elbows*	45° elbows*	90° long-turn elbows	45° long-turn elbows	Three-way elbows†		Tees*		90° Y branches		90° long-turn Y branches			45° Y branches*	
	A	A	A	A	A	B	A	B	A	B	A	B	C	A	B
1¼	1¾	1⁵⁄₁₆	2¼	1¾	4½	2¼	1¾	3½	3¾	2¼	4¾	3⅝	1⅛	5	3¼
1½	1¹⁵⁄₁₆	1⁷⁄₁₆	2½	1⅞	5	2½	1¹⁵⁄₁₆	3⅞	4¼	2½	5⅜	4⅛	1¼	5½	3⅝
2	2¼	1¹¹⁄₁₆	3¹⁄₁₆	2¼	6⅛	3¹⁄₁₆	2¼	4½	5³⁄₁₆	3¹⁄₁₆	7	5¼	1¾	6½	4⅜
2½	2¹¹⁄₁₆	1¹⁵⁄₁₆	3¹¹⁄₁₆	2⅝	7⅝	3¹¹⁄₁₆	2¹¹⁄₁₆	5⅜	6⁵⁄₁₆	3¹¹⁄₁₆	8¼	6¼	2	7⅞	5⅜
3	3¹⁄₁₆	2³⁄₁₆	4¼	2¹⁵⁄₁₆	8½	4¼	3¹⁄₁₆	6⅛	7¼	4¼	9⅞	7½	2⅜	9	6³⁄₁₆
4	3¹³⁄₁₆	2⅝	5³⁄₁₆	3½	10⅜	5³⁄₁₆	3¹³⁄₁₆	7⅝	8¾	5³⁄₁₆	13	9⅞	3⅛	10⅞	7¹¹⁄₁₆
5	4½	3¹⁄₁₆	6⅛	4⅛	12¼	6⅛	4½	9	10⁵⁄₁₆	6⅛	15¾	12¼	3½	12¹⁵⁄₁₆	9¾
6	5⅛	3⁷⁄₁₆	7⅛	4⅞	14¼	7⅛	5⅛	10¼	11¹⁵⁄₁₆	7⅛	18¾	14⅝	4⅛	14⅞	10¾

* Same as adopted for Class 125 Cast-iron Threaded Fittings, ANSI B16.4-1983.
† Three-way elbows have same dimensions as 90° long-radius elbows.
Double Y branches have the same dimensions as single Y branches.
Other fittings which are available are as follows: 5⅝, 11¼, and 60° elbows; basin tees and crosses; double 90° Y branches; double 90° long-turn Y branches; 45° double Y branches; S traps; half S traps; offsets, couplings, increasers, and reducing sizes.
SOURCE: ANSI B16.2-1983.

Table 8.7.45 Soldered-Joint Fittings—Dimensions of Elbows, Tees, and Crosses
(All dimensions in inches)

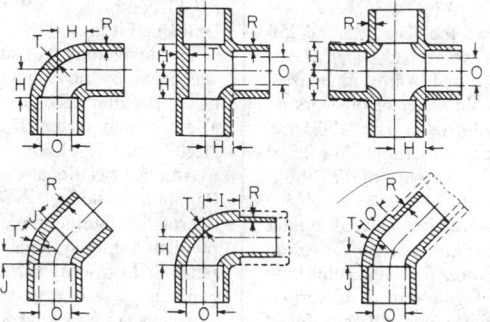

Nominal size	Cast brass†							Wrought metal
	H^*	I	J	Q	$O‡$	T	R	$(T$ and $R)§,¶$
¼	¼	⅜	³⁄₁₆	¼	0.31	0.08	0.048	0.030
⅜	⁵⁄₁₆	⁷⁄₁₆	³⁄₁₆	⁵⁄₁₆	0.43	0.08	0.048	0.035
½	⁷⁄₁₆	⁹⁄₁₆	³⁄₁₆	⁵⁄₁₆	0.54	0.09	0.054	0.040
¾	⁹⁄₁₆	¹¹⁄₁₆	¼	⅜	0.78	0.10	0.060	0.045
1	¾	⅞	⁵⁄₁₆	⁷⁄₁₆	1.02	0.11	0.066	0.050
1¼	⅞	1	⁷⁄₁₆	⁹⁄₁₆	1.26	0.12	0.072	0.055
1½	1	1⅛	½	⅝	1.50	0.13	0.078	0.060
2	1¼	1⅜	⁹⁄₁₆	¾	1.98	0.15	0.090	0.070
2½	1½	1⅝	⅝	⅞	2.46	0.17	0.102	0.080
3	1¾	1⅞	¾	1	2.94	0.19	0.114	0.090
3½	2	2⅛	⅞	1⅛	3.42	0.20	0.120	0.100
4	2¼	2⅜	¹⁵⁄₁₆	1¼	3.90	0.22	0.132	0.110
5	3⅛		1⁷⁄₁₆		4.87	0.28	0.168	0.125
6	3⅝		1⅝		5.84	0.34	0.204	0.140

Wrought fittings as well as cast fittings, must be provided with a shoulder or stop at the bottom end of socket.

 * Dimensions for reducing elbows, reducing crosses, reducing tees, couplings, caps, bushings, adapters, and fittings with pipe thread on one end are also included in this standard.

 † These dimensions may be used for cast wrought metal fittings as well as for cast brass fittings at manufacturer's option.

 ‡ This dimension is the same as the inside diameter class L tubing (ASTM B88-1983).

 § This dimension has the same thickness as class L tubing.

 ¶ These dimensions are minimum, but in every case the thickness of wrought fittings should be at least as heavy as the tubing with which it is to be used.

 SOURCE: ANSI B16.18-1984.

(ANSI B16.18-1984) covers certain dimensions of soldered-joint wrought metal and cast brass fittings for copper water tubing including (1) detailed dimensions of the bore, (2) minimum specifications for materials, (3) minimum inside diameter of the fitting, (4) metal thickness for both wrought metal and cast brass fittings, and (5) general dimensions for cast brass fittings including center-to-shoulder dimensions for both straight and reducing cast fittings. Pressure and temperature ratings are also given. Sizes of the fittings are identified by the nominal tubing size as covered by the Specifications for Copper Water Tube (ASTM B88-1983). Dimensions of some of the fittings from this standard are given in Table 8.7.45.

Valves

The face-to-face dimensions of ferrous flanged and welding end valves are given in ANSI B16.10-1973. The types covered are:

Wedge-Gate Valves Cast iron, for 125-, 175-, and 250-lb/in² (862, 1,207, and 1,724 kPa) steam service pressure and 800-lb/in² (5,516 kPa) hydraulic pressure, and steel, for 150-, 300-, 400-, 600-, 900-, and 1,500-lb/in² (1,034-, 2,068-, 2,758-, 4,137-, 6,206-, and 10,343-kPa) steam service pressures (see Fig. 8.7.11).

Double-Disk Gate Valves Cast iron, for 125- and 250-lb/in² (862- and 1,724-kPa) steam service pressure and 800-lb/in² (5,516-kPa) hydraulic pressure.

Globe and Angle Valves Cast iron, for 125- and 250-lb/in² (862- and 1,724-kPa) steam service pressure, and steel, for 150-, 300-, 400-, 600-, 900-, 1,500-, and 2,500-lb/in² (862-, 2,068-, 2,758-, 4,137-, 6,206-, 10,343-, and 17,238-kPa) steam service pressures (see Fig. 8.7.12).

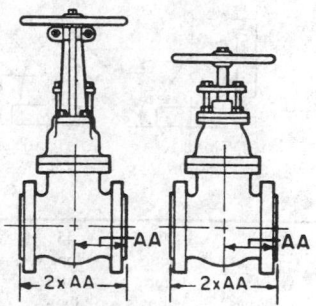

Fig. 8.7.11 Wedge gate valves.

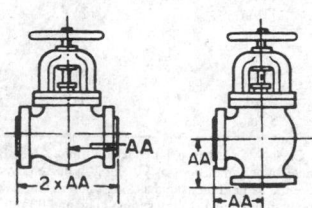

Fig. 8.7.12 Globe valve and angle valve.

Swing-Check Valves   Cast iron, for 125- and 250-lb/in² (862- and 1,724-kPa) steam service pressure and 800 lb/in² (5,516 kPa) hydraulic pressure, and steel, for 150-, 300-, 400-, and 600-lb/in² (1,034-, 2,068-, 2,758-, and 4,137-kPa) steam service pressures.

Except for ring-joint facings to the face-to-face dimension for flanged valves is the distance between the faces of the connecting end flanges upon which the gaskets are actually compressed, i.e., the "contact surfaces."

All flanges for class 125 cast-iron valves are plain-faced. The facings of the class 250 cast-iron, and the class 150 and 300 steel valves have a 1/16-in raised face which is included in the contact-surface to contact-surface dimensions. The contact-surface to contact-surface dimensions of steel valves for class 400 and higher pressures and for cast-iron valves for class 800 hydraulic pressure include a ¼-in raised face.

The end-to-end dimensions for **welding-end** valves for sizes NPS 1 to 8 are the same as the contact-surface to contact-surface dimensions given in the tables for steel valves. For details of welding bevel see ANSI B16.10-1973 and Fig. 8.7.15.

A plus or minus **tolerance** of 1/16 in is allowed on all face-to-face dimensions of valves NPS 10 and smaller, and a tolerance of 1/8 on sizes NPS 12 and larger.

Cocks   The ordinary plug cock operated by a handle or wrench is a form of valve in comparatively small sizes suitable for ordinary service only. The ASME Code for Pressure Piping requires that where cocks are used for high-temperature service they shall be so designed as to prevent galling, either by making the plugs of different material from the body of the cock or by treating the plugs to ensure different physical properties. By means of special design features that eliminate the tendency to leak and stick, the plug-cock type of valve has become available in large sizes and for severe service conditions. Sizes are listed as high as 30 in and are gear-operated in the larger sizes. For further details, refer to manufacturers' catalogs.

Expansion and Flexibility

Piping systems must be designed so that they (1) will not fail because of excessive stresses, (2) will not produce excessive thrusts or moments at connected equipment, or (3) will not leak at joints because of expansion of the pipe. Flexibility is provided by changes of direction in the piping through the use of bends or loops, or provision may be made to absorb thermal strains by use of expansion joints. All, or portions, of the pipe may be corrugated to improve flexibility; in many systems, however, sufficient change is provided by the geometry of the layout to make unnecessary the use of either expansion joints or corrugated sections of piping. Proper cold springing is beneficial in assisting the piping system to attain its most favorable condition. Because of plastic flow of the piping material, hot stresses tend to decrease with time while cold stresses tend to increase with time; their sum, called the stress range, remains substantially constant. For this reason no credit is warranted with regard to stresses; for calculation of forces and moments, the effect of cold spring is recognized by use of a cold-spring factor varying from 0 to 1 for cold spring varying from 0 to 100 percent.

The allowable stress range S_A is calculated by

$$S_A = f(1.25S_c + 0.25S_h)$$

where S_c and S_h are the S values for the minimum cold and maximum hot conditions, respectively, as given in Table 8.7.15. The stress-reduction factor f is a function of the number of hot-to-cold-to-hot (full) temperature cycles anticipated over the life of the plant, as follows:

Total no. of full temp cycles over expected life	Stress-reduction factor
7,000 and less	1.0
14,000 and less	0.9
22,000 and less	0.8
45,000 and less	0.7
100,000 and less	0.6
Over 100,000	0.5

The bending and torsional stresses calculated (see paragraph 119.6.4 of ANSI B31.1.0-1983) are used to determine the maximum computed expansion stress $S_E = \sqrt{S_b^2 + 4S_t^2}$, where S_b and S_t are bending and torsional stresses, respectively. S_E must not exceed the allowable stress range S_A.

In recent years, many principal high-temperature steam lines have either been analyzed, tested in a model-testing machine, or both. No rigid rule is stipulated for the requirement of analysis or model test; however, the Code for Pressure Piping suggests that when the following criterion is not satisfied, need for an analysis is indicated: $DY/(L - U)^2 \leq 0.03$, where D is the nominal pipe size, in; Y is the resultant of movements to be absorbed by pipeline, in; U is the length of straight line joining the anchor points, ft; and L is the length of the developed line axis, ft.

Expansion Joints for Steam Pipelines   In many instances it may be economical to care for thermal expansion by use of expansion joints. For low-pressure steam lines, the use of packed expansion joints may be feasible; experience has indicated that packed joints are difficult to maintain when used on high-pressure lines. Figure 8.7.13 shows a type of joint that has been successfully used for high-pressure, high-

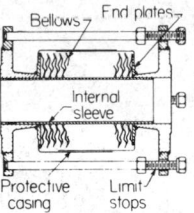

Fig. 8.7.13   Expansion joint for steam line. *(Croll-Reynolds, Inc.)*

temperature service. The bellows is designed to take either axial, lateral, or combined axial and lateral deflections. The internal sleeve guides movement of the joint and also protects the flexible bellows from direct contact with the fluid being handled. Face-to-face dimensions, as well as permissible axial and lateral deflections, are indicated in Table 8.7.46.

Where large lateral deflections are to be absorbed, two expansion joints separated by a length of pipe as shown in Fig. 8.7.14 may be used. With such an arrangement, the lateral deflection permissible with one joint only may be increased many times. Tie rods, as shown, should always be installed to protect the joint against overtravel and externally to guide movement of the joint.

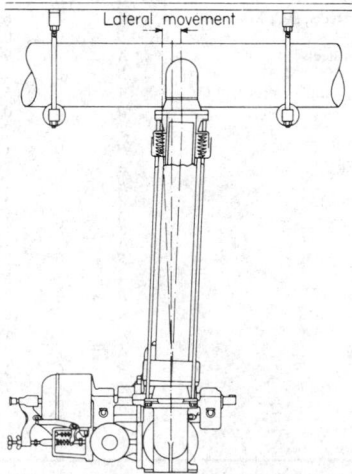

Fig. 8.7.14   Arrangement of expansion joints for large lateral deflection. *(Croll-Reynolds, Inc.)*

Table 8.7.46 Dimensions of Expansion Joints*

Pipe size	Pressure series, lb/in² gage	Face-to-face dimensions,† axial movements of 1 in	2 in	3 in	Lateral movements‡ equivalent to axial movements of 1 in	2 in	3 in
4	150	8½	11	15½	1/16	1/4	9/16
	300	12	17	24	1/8	1/2	1
	600	17½	26½		1/4	1	
	900	31½			11/16		
6	150	9½	12	16½	1/16	3/16	3/8
	300	13	18	25	3/32	13/32	15/16
	600	18½	27½		7/32	27/32	
	900	33½			1/2		
8	150	10½	13	17½	1/32	3/16	3/8
	300	14	19	26	3/32	3/8	27/32
	600	20	29		3/16	3/4	
	900	35½			15/32		
10	150	10½	13	17½	1/32	1/8	5/16
	300	14	19	26	1/16	5/16	23/32
	600	21½	30½		5/32	5/8	
	900	37			3/8		
12	150	22½	14	18½	1/32	1/8	1/4
	300	15	20	27	1/16	1/4	19/32
	600	21½	30½		1/8	1/2	
	900	38½			5/16		
14	100	12½	15½		1/32	1/8	
	150	15	20		1/16	7/32	
	300	20½			3/32		
16	100	12½	15½		1/32	3/32	
	150	15	20		1/32	3/16	
	300	20½			3/32		
18	100	13½	16½		1/32	3/32	
	150	16	21		1/32	3/16	
	300	21½			3/32		
20	100	14½	16½			3/32	
	150	16	21		1/32	3/16	
	300	21½			3/32		
24	100	14½	17½			1/16	
	150	17	22		1/32	5/32	
	300	22½			1/16		
30	100	9½	12½			1/16	
	150	12	17		1/32	1/8	
	300	19½			1/16		

* Croll-Reynolds, Inc.
† For welding ends, add 4 in to face-to-face dimension shown.
‡ Consult manufacturer for permissible combined axial and lateral deflection.

Table 8.7.47 Thermal Expansion Data

Material	Coefficient	Temp range: 70°F (21°C) to: 70 (21)	200 (93)	300 (149)	400 (205)	500 (260)	600 (316)	700 (371)	800 (427)	900 (482)	1,000 (538)	1,100 (593)	1,200 (649)	1,300 (705)	1,400 (760)
Carbon steel: carbon-moly steel low-chrome steels (through 3% Cr)	*A*		6.38	6.60	6.82	7.02	7.23	7.44	7.65	7.84	7.97	8.12	8.19	8.28	8.36
	B	0	0.99	1.82	2.70	3.62	4.60	5.63	6.70	7.81	8.89	10.04	11.10	12.22	13.34
Intermediate alloy steels: 5 Cr Mo-9 Cr Mo	*A*		6.04	6.19	6.34	6.50	6.66	6.80	6.96	7.10	7.22	7.32	7.41	7.49	7.55
	B	0	0.94	1.71	2.50	3.35	4.24	5.14	6.10	7.07	8.06	9.05	10.00	11.06	12.05
Austenitic stainless steels	*A*		9.34	9.47	9.59	9.70	9.82	9.92	10.05	10.16	10.29	10.39	10.48	10.54	10.60
	B	0	1.46	2.61	3.80	5.01	6.24	7.50	8.80	10.12	11.48	12.84	14.20	15.56	16.92
Straight chromium stainless steels: 12 Cr, 17 Cr, and 27 Cr	*A*		5.50	5.66	5.81	5.96	6.13	6.26	6.39	6.52	6.63	6.72	6.78	6.85	6.90
	B	0	0.86	1.56	2.30	3.08	3.90	4.73	5.60	6.49	7.40	8.31	9.20	10.11	11.01
25 Cr-20 Ni	*A*		7.76	7.92	8.08	8.22	8.38	8.52	8.68	8.81	8.02	9.00	9.08	9.12	9.18
	B	0	1.21	2.18	3.20	4.24	5.33	6.44	7.60	8.78	9.95	11.12	12.31	13.46	14.65
Monel 67: Ni-30 Cu	*A*		7.84	8.02	8.20	8.40	8.58	8.78	8.96	9.16	9.34	9.52	9.70	9.88	10.04
	B	0	1.22	2.21	3.25	4.33	5.46	6.64	7.85	9.12	10.42	11.77	13.15	14.58	16.02
Monel 66: Ni-29 CuAl	*A*		7.48	7.68	7.90	8.09	8.30	8.50	8.70	8.90	9.10	9.30	9.50	9.70	9.89
	B	0	1.17	2.12	3.13	4.17	5.28	6.43	7.62	8.86	10.16	11.50	13.00	14.32	15.78
Aluminum	*A*		12.95	13.28	13.60	13.90	14.20								
	B	0	2.00	3.66	5.39	7.17	9.03								
Gray cast iron	*A*		5.75	5.93	6.10	6.28	6.47	6.65	6.83	7.00	7.19				
	B	0	0.90	1.64	2.42	3.24	4.11	5.03	5.98	6.97	8.02				
Bronze	*A*		10.03	10.12	10.23	10.32	10.44	10.52	10.62	10.72	10.80	10.90	11.00		
	B	0	1.56	2.79	4.05	5.33	6.64	7.95	9.30	10.68	12.05	13.47	14.92		
Brass	*A*		9.76	10.00	10.23	10.47	10.69	10.92	11.16	11.40	11.63	11.85	12.09		
	B	0	1.52	2.76	4.05	5.40	6.80	8.26	9.78	11.35	12.98	14.65	16.39		
Wrought iron	*A*		7.32	7.48	7.61	7.73	7.88	8.01	8.13	8.29	8.39				
	B	0	1.14	2.06	3.01	3.99	5.01	6.06	7.12	8.26	9.36				
Copper-nickel (70/30)	*A*		8.54	8.71	8.90										
	B	0	1.33	2.40	3.52										

A = mean coefficient of thermal expansion × 10^6, in/(in · °F) in going from 70°F (21°C) to indicated temperature.
B = linear thermal expansion, in/100 ft in going from 70°F (21°C) to indicated temperature.
Multiply values of *A* shown by 1.8 to obtain coefficient of expansions in cm/(cm · °C).
Multiply values of *B* shown by 8.33 to obtain linear expansion in cm per 100 m.
SOURCE: ANSI B31.1-1983.

Table 8.7.47, extracted from the Code for Pressure Piping, lists thermal-expansion data for both ferrous and nonferrous piping. For expansion at temperatures intermediate between those shown, straight-line interpolation is permitted.

The **rubber expansion joint** has become an established part of pipeline equipment. Its special field of application is on low-pressure and vacuum lines in condenser applications, etc., and it is recommended for pressures up to 25 lb/in² (172 kPa) gage where the maximum temperature does not exceed 250°F. Standard joints for pressure installations are reinforced to withstand working pressures up to 125 lb/in² (862 kPa) gage and temperatures up to 200°F. Joints are available in all standard pipe sizes.

Welding in Power-Plant Piping

(For dimensions of welding fittings see Tables 8.7.48 to 8.7.51; for welding techniques see also Sec. 13.3.)

The majority of main-cycle and service steel piping in modern steam power plants is of welded construction. Steel pipe of NPS 2 and smaller is generally socket-welded; larger-size piping is usually butt-welded. Frequently, depending on location and scheduling, piping larger than NPS 2 is prefabricated; smaller piping is shipped to the construction site in random lengths and is fabricated concurrently with installation. Small-sized chromium-molybdenum piping requiring bending is frequently also shop-fabricated so as to avoid high field preheat, welding, and stress-relieving costs. It is desirable to schedule shipment of

Table 8.7.48 Dimensions of Long-Radius 90° Butt-Welding Elbows
(Standard weight—ANSI B16.9-1978, ASTM A234)
(All dimensions in inches)

Nominal pipe size	OD	ID	Wall thickness	Center to face	Pipe schedule numbers	Approx wt, lb
2½	2.875	2.469	0.203	3¾	40	2.92
3	3.500	3.068	0.216	4½	40	4.58
3½	4.000	3.548	0.226	5¼	40	6.43
4	4.500	4.026	0.237	6	40	8.70
5	5.563	5.047	0.258	7½	40	14.7
6	6.625	6.065	0.280	9	40	22.9
8	8.625	7.981	0.322	12	40	46.0
10	10.750	10.020	0.365	15	40	81.5
12	12.750	12.000	0.375	18	ST*	119
14	14.000	13.250	0.375	21	30	154
16	16.000	15.250	0.375	24	30	201
18	18.000	17.250	0.375	27	ST*	256
20	20.000	19.250	0.375	30	20	317
22	22.000	21.250	0.375	33	ST*	385
24	24.000	23.250	0.375	36	20	458
26	26.000	25.250	0.375	39	ST*	539
30	30.000	29.250	0.375	45	ST*	720
34	34.000	33.250	0.375	51	ST*	926
36	36.000	35.250	0.375	54	ST*	1,040
42	42.000	41.250	0.375	63	ST*	1,420

* Standard weight.

Table 8.7.49 Dimensions of Straight Butt-Welding Tees
(Standard weight—ANSI B16.9-1978, ASTM A234)
(Dimensions in inches)

Nominal pipe size	OD	ID	Wall thickness	Center to end	Pipe schedule numbers	Approx wt, lb
2½	2.875	2.469	0.203	3	40	5.21
3	3.500	3.068	0.216	3⅜	40	7.44
3½	4.000	3.548	0.226	3¾	40	9.85
4	4.500	4.026	0.237	4⅛	40	12.6
5	5.563	5.047	0.258	4⅞	40	19.8
6	6.625	6.065	0.280	5⅝	40	29.3
8	8.625	7.981	0.322	7	40	53.7
10	10.750	10.020	0.365	8½	40	91.2
12	12.750	12.000	0.375	10	ST*	132
14	14.000	13.250	0.375	11	30	172
16	16.000	15.250	0.375	12	30	219
18	18.000	17.250	0.375	13½	ST*	282
20	20.000	19.250	0.375	15	20	354
22	22.000	21.250	0.375	16½	ST*	437
24	24.000	23.250	0.375	17	20	493
26	26.000	25.250	0.375	19½	ST*	634
30	30.000	29.250	0.375	22	ST*	855
34	34.000	33.250	0.375	25	ST*	1,136
36	36.000	35.250	0.375	26½	ST*	1,294

* Standard weight.

Table 8.7.50   Dimensions of Long-Radius 45° Butt-Welding Elbows
(Standard weight—ANSI B16.9-1978, ASTM A234)
(Dimensions in inches)

Nominal pipe size	OD	ID	Wall thickness	Center to face	Radius	Pipe schedule numbers	Approx wt, lb
2½	2.875	2.469	0.203	1¾	3¾	40	1.64
3	3.500	3.068	0.216	2	4½	40	2.43
3½	4.000	3.548	0.226	2¼	5¼	40	3.29
4	4.500	4.026	0.237	2½	6	40	4.31
5	5.563	5.047	0.258	3⅛	7½	40	7.30
6	6.625	6.065	0.280	3¾	9	40	11.3
8	8.625	7.981	0.322	5	12	40	22.8
10	10.750	10.020	0.365	6¼	15	40	40.4
12	12.750	12.000	0.375	7½	18	ST*	59.5
14	14.000	13.250	0.375	8¾	21	30	76.5
16	16.000	15.250	0.375	10	24	30	100
18	18.000	17.250	0.375	11¼	27	ST*	128
20	20.000	19.250	0.375	12½	30	20	158
22	22.000	21.250	0.375	13½	33	ST*	192
24	24.000	23.250	0.375	15	36	20	229
26	26.000	25.250	0.375	16	39	ST*	269
30	30.000	29.250	0.375	18½	45	ST*	358
34	34.000	33.250	0.375	21	51	ST*	463
36	36.000	35.250	0.375	22¼	54	ST*	518
42	42.000	41.250	0.375	26	63	ST*	707

* Standard weight.

hangers so that they will be available at the job site upon arrival of the prefabricated piping; this avoids the expense of providing, installing, and later removing temporary hangers and supports. Aside from the economy of welded construction, it is a virtual necessity in high-pressure, high-temperature work because of danger of leakage if joints are flanged.

Shop welds are frequently made by automatic or semiautomatic sub-merged-arc or inert-gas shielded-arc processes; field welds are gener-

Table 8.7.51   Dimensions of Concentric and Eccentric Butt-Welding Reducers
(Standard weight—ANSI B16.9-1978, ASTM A234)
(Dimensions in inches)

Nominal pipe size	Length	Approx wt, lb	Nominal pipe size	Length	Approx wt, lb	Nominal pipe size	Length	Approx wt, lb	
2½ × 1	3½	1.30	8 × 6	6	13.4	22 × 20	20	157	
2½ × 1¼	3½	1.47	8 × 6	6	13.9	24 × 16	20	160	
2½ × 1½	3½	1.51	10 × 4	7	21.1	24 × 18	20	163	
2½ × 2	3½	1.60	10 × 5	7	21.8	24 × 20	20	167	
3 × 1¼	3½	1.70	10 × 6	7	22.3	26 × 18	24	200	
3 × 1½	3½	1.89	10 × 8	7	23.2	26 × 20	24	200	
3 × 2	3½	2.00	12 × 5	8	30.5	26 × 22	24	200	
3 × 2½	3½	2.16	12 × 6	8	31.1	26 × 24	24	200	
3½ × 1¼	4	2.35	12 × 8	8	32.1	30 × 20	24	220	
3½ × 1½	4	2.52	12 × 10	8	33.4	30 × 24	24	220	
3½ × 2	4	2.71	14 × 6	13	55.8	30 × 26	24	220	
3½ × 2½	4	2.96	14 × 8	13	57.2	30 × 28	24	220	
3½ × 3	4	3.05	14 × 10	13	60.4				
4 × 1½	4	2.73	14 × 12	13	63.4			Conc.	Ecc.
4 × 2	4	3.17	16 × 8	14	70.2	34 × 24	24	270	229
4 × 2½	4	3.34	16 × 10	14	72.9	34 × 26	24	270	237
4 × 3	4	3.50	16 × 12	14	75.6	34 × 30	24	270	253
4 × 3½	4	3.61	16 × 14	14	77.5	34 × 32	24	270	261
5 × 2	5	5.05	18 × 10	15	86.9	36 × 24	24	340	237
5 × 2½	5	5.52	18 × 12	15	89.2	36 × 26	24	340	245
5 × 3	5	5.73	18 × 14	15	90.9	36 × 30	24	340	261
5 × 3½	5	5.86	18 × 16	15	94.0	36 × 32	24	340	269
5 × 4	5	5.99	20 × 12	20	134	36 × 34	24	340	277
6 × 2½	5½	7.61	20 × 14	20	135	42 × 24	24	260	
6 × 3	5½	8.00	20 × 16	20	138	42 × 26	24	270	
6 × 3½	5½	8.14	20 × 18	20	142	42 × 30	24	285	
6 × 4	5½	8.19	22 × 14	20	148	42 × 32	24	295	
6 × 5	5½	8.65	22 × 16	20	151	42 × 34	24	300	
8 × 3½	6	12.8	22 × 18	20	154	42 × 36	24	310	
8 × 4	6	13.1							

ally of the manual type and may be done by the shielded metal-arc and/or inert-gas metal-arc processes. Welding in power piping systems, whether in the shop or at the job site, must be done by welders who have qualified under provisions of the Code for Pressure Piping or the ASME Boiler and Pressure Vessel Code.

End Preparation for Butt Welds   Figure 8.7.15 shows the end preparation recommended (not required) for piping whose wall thickness is ¾ in or less, and Fig. 8.7.16 shows that required for piping with wall thickness above ¾ in. During the welding process, to avoid entrance of welding material into the pipe, backing rings may be used as shown in Fig. 8.7.17a, b, and c.* Note that thick-walled pipes (over ¾ in) are taper-bored on the inside in order that they may receive a tapered, machined backing ring.

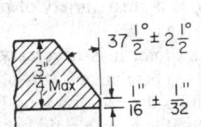

Fig. 8.7.15 Recommended end preparation for pipe wall thickness of ¾ in or less.

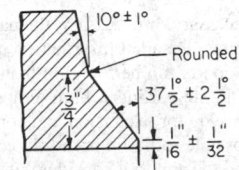

Fig. 8.7.16 Recommended end preparation for pipe wall thickness greater than ¾ in.

Preheating   Prior to start of welding, many materials require preheat to a specified temperature: preheat may be done by electrical-resistance or induction heating or by ring-type gas burners placed concentrically with the pipe. The preheat temperature is measured by indicating crayons or by thermocouple pyrometers and must be maintained during the welding operation. Table 1, Appendix D, of the Code for Pressure Piping lists materials used in piping systems and the appropriate temperatures for preheat. In general, the following is indicative of the intent only; for specific instances, the Code must be consulted.

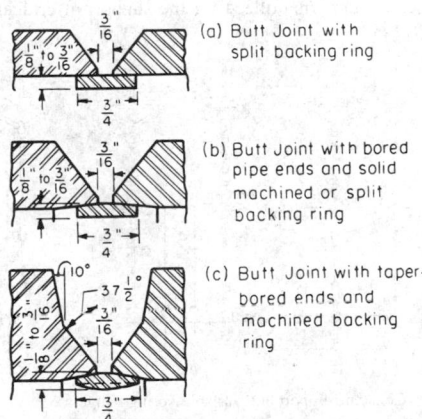

(a) Butt joint with split backing ring

(b) Butt joint with bored pipe ends and solid machined or split backing ring

(c) Butt joint with taper-bored ends and machined backing ring

Fig. 8.7.17 Recommended backing ring types. (a) Butt joint with split backing ring; (b) butt joint with bored pipe ends and solid machined or split backing ring; (c) butt joint with taper-bore ends and machined backing ring.

Carbon steel and **wrought iron** should be preheated to a "hand-hot" condition if the ambient temperature at time of field installation is 32°F (0°C) or less: carbon steels which have minimum tensile properties of 70,000 lb/in² (483 MPa) or higher should be preheated to 250°F (121°C); under other conditions, preheat is not mandatory, but some purchasers insist that the contractor preheat heavy-walled piping such as boiler feed.

* Consumable inserts are also available. They are recommended for installation in piping systems which require a smooth, unobstructed interior surface.

Low-alloy steels with a chromium content not exceeding ¾ percent and low-alloy steels with a total alloy content not exceeding 2 percent are required to be preheated to a minimum temperature of 300°F (149°C).

Alloy steels with a chromium content between ¾ and 2 percent and low-alloy steels with a total alloy content not exceeding 2¾ percent require preheating to 375°F (191°C) minimum. Those with a total alloy content greater than 2¾ percent but not exceeding 10 percent require preheating to a temperature of 450°F (232°C) minimum.

High-alloy steels containing the martensitic phase require preheating to 450°F (232°C) minimum; preheating is a matter of agreement between the purchaser and contractor in the case of welding high-alloy ferritic steels (ASTM A240 and A268). The possible advantages of preheat have not been established in the case of welding high-alloy austenitic steels, and for this reason the Code for Pressure Piping states that preheat is optional for these materials.

Welding procedure varies with material and welding process. In general, the pipe ends must be cleaned of oil or grease, and excessive amounts of scale or rust should be removed. The size and type of welding rod must be stated; the number of layers or passes is determined by the thickness of the pieces being joined. All slag or flux remaining on any bead of welding must be removed before laying down the next successive bead; any cracks or blowholes that appear on the surface of any bead must be chipped or ground away before the next bead of weld material is deposited. Throughout the welding process, it is essential that the minimum specified preheat temperature be maintained.

Stress Relieving   Welded joints in all carbon-steel material whose thickness is ¾ in (1.91 cm) or greater must be stress-relieved at a temperature of 1,100°F (593°C) or over for a period of time proportioned on the basis of at least 1 h/in of pipe-wall thickness (but in no case less than ½ h) and then allowed to cool slowly (generally under a blanket) and uniformly. No stress relief is required for joints in carbon-steel piping whose wall thickness is less than ¾ in.

Welded joints in alloy steels with a wall thickness of ½ in (1.27 cm) or greater, having a chromium content not exceeding ¾ percent, and low-alloy steels with a total alloy content not exceeding 2 percent require stress-relieving at a temperature of 1,200°F (649°C) or over for a period of time proportioned on the basis of at least 1 h/in (0.4 h/cm) of wall thickness, but in no case less than ½ h.

Welded joints in alloy steels having a chromium content exceeding ¾ percent, or a total alloy content exceeding 2 percent, except high-alloy ferritic (ASTM A240, A268) and austenitic steels, regardless of wall thickness, require stress relief at a temperature of 1,200°F or over for a period of time proportioned on the basis of at least 1 h/in of wall thickness, but in no case less than ½ h. Stress relief of high-alloy ferritic steel (A240, A268) and austenitic steels is not required but may be performed as agreed upon by purchaser and contractor. In welds between austenitic and ferritic materials, stress relieving is optional and, if used, shall be a matter of agreement between the purchaser and contractor. Because of the difference between the coefficients of thermal expansion of the two dissimilar materials, careful consideration should be given to the selection of a heat treatment, if any, that will be beneficial to the welded joint.

Graphitization is precipitation of carbon at the grain boundaries in the heat-affected zone during the welding process. Such a phenomenon occurs when some metals operate at high temperatures for extended periods. It has been observed particularly in carbon-molybdenum steels that operate at 900°F (482°C) or higher.

Graphitization does not generally occur in carbon-molybdenum steels with over 1 percent molybdenum. It also has generally not occurred in the chromium-molybdenum low-alloy steels operated at temperatures between 900°F (482°C) and 1,050°F (566°C). Where graphitization has occurred, the two most commonly used methods for rehabilitation of the pipe are (1) gouging out the heat-affected zone of the weld deposit and rewelding the area with electrodes depositing carbon-molybdenum weld metal, followed by a stabilization heat treatment at 1,300°F (704°C) for 4 h, or (2) solution annealing the weld joints at 1,800°F (982°C), followed by a stabilization heat treatment.

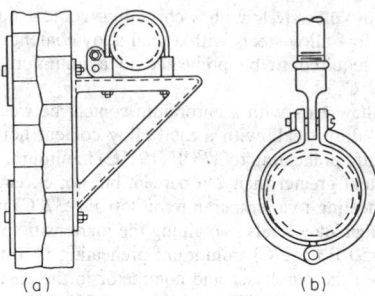 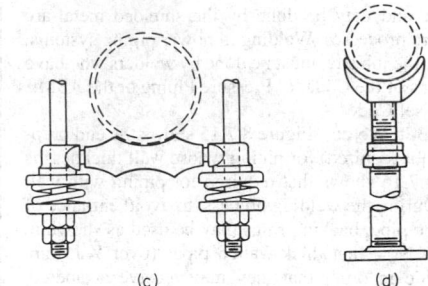

Fig. 8.7.18 Methods of supporting pipes.

Pipe Supports

The Code for Pressure Piping includes many types of supports and gives directions for their application. A proper pipe support must have a strong rigid base properly supported, and an adjustable roll construction which will maintain the alignment in any direction. It is important to avoid friction caused by the movement of the pipe in the support and to have all parts of sufficient strength to maintain alignment at all times. Wire hangers, band iron hangers, wooden hangers, hangers made from small pipe, and hangers having one vertical pipe support do not maintain alignment.

The direction of expansion in a pipe run can be predetermined by anchoring one end, both ends, or the middle. **Anchors** must be firmly fastened to a rigid and heavy part of the power-plant structure, and must also be securely fastened to the pipe; otherwise the equipment for absorbing expansion is useless, and severe stresses may be thrown on parts of the piping system. Some methods of support are shown in Figs. 8.7.18 and 8.7.19. Welded steel brackets (Fig. 8.7.18a) are available in light, medium, and heavy weights. Many types of supports can be mounted on these brackets, such as the anchor chair shown on the bracket at (a), pipe roller supports of the type at (c), pipe roll stands of various types such as shown in Fig. 8.7.19, pipe seats, etc. Figure 8.7.18b illustrates one of the many types of adjustable ring hangers in use. The split ring hanger can be applied after the pipeline is in place. At (c) in Fig. 8.7.18 is shown a spring cushion pipe roll hanger recommended for service where constant support* is required and compensation must be made for movement of the piping. The springs provide an efficient means of absorbing the vibration. Figure 8.7.18d shows one of the many types of pipe saddle supports available. Figure 8.7.19 shows a cast-iron pipe roll stand designed for cases where vertical adjustment is not necessary but where provision must be made for expansion and contraction of the pipeline. Several designs of such stands with provision for vertical adjustment and of the same general dimensions are also available. One type of cast-iron roll and plate, illustrated in Fig. 8.7.19, provides for expansion and contraction where vertical adjustment is not

* The support afforded by the hanger of Fig. 8.7.18c is constant only in the sense that some degree of support is always present. It might be more appropriately termed a variable-support device.

necessary. If necessary, the baseplate can be raised or lowered by use of shims. Detailed information and dimensions of a great variety of pipe supports can be found in manufacturers' catalogs.

In supporting a high-temperature piping system, it is necessary to provide for expansion and contraction due to cyclic changes. It is often possible to find a point of zero movement along the run of a long line and to support a considerable portion of the total load by a rigid hanger or support of the type shown in Figs. 8.7.18 and 8.7.19. However, for other portions of the run, some form of spring support is often indicated. For relatively light lines, which are not subjected to excessive movements from hot to cold positions, a variable spring hanger will frequently suffice; for heavy lines, or those in which expansion movements are great, it is advisable to use constant support of counterweighted hangers so that transfer of weight to other hangers or equipment connections is prevented. Parts (a) and (b) of Fig. 8.7.20 indicate, respectively, a horizontal and vertical run of piping supported by a **constant-support hanger**. Figure 8.7.20c and 8.7.21a indicate horizontal runs supported by **variable-spring hangers**. Figure 8.7.21b shows a riser supported by a variable spring beneath a base elbow. Figure 8.7.21c indicates a **sway brace** that is used to control vibration and undesirable movement in a piping system.

The principal supports utilized for the support of critical piping in-

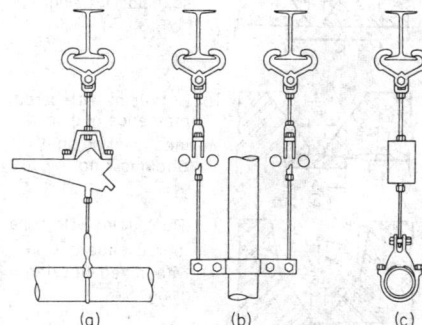

Fig. 8.7.20 Constant support and variable-spring hangers.

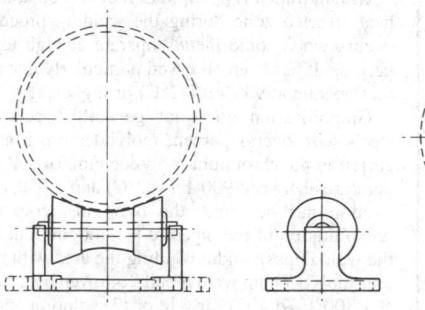 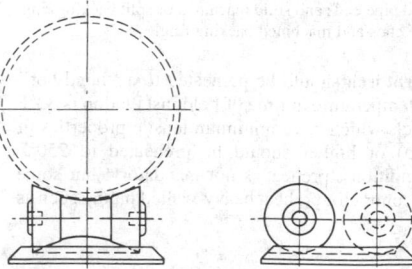

Fig. 8.7.19 Pipe supports on cast-iron rolls.

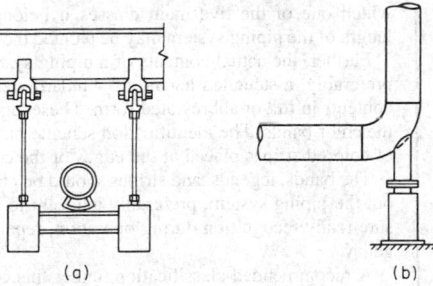

Fig. 8.7.21 Spring hangers and sway brace.

volve constant-support hangers, variable-spring hangers, rigid hangers, and restraints.

Constant-Support Hangers This type of hanger provides a constant supporting force for the piping system throughout its full range of vertical pipe movement. This is accomplished through use of a spring coil working in conjunction with a lever in such a way that the spring force times its distance to the lever pivot is always equal to the pipe load times its distance to the lever pivot. This type of support is "thermally invisible," as the supporting force equals the pipe weight throughout its entire expansion or contraction cycles.

These hangers are used on systems or at locations where stresses are considered critical. Pipe weight reactions or transfer of loads are not imposed in the system or connections with this type of device.

As the load is considered constant with the unit in travel, readings from inspections are based on travel. The readings are taken from the position of an indicator and its relation to the numbers on the travel scale. The scale is divided into 10 divisions. H or high on the scale is equal to (0.0), M or midway on the scale is equal to (5.0), and L or low on the scale is equal to 10.

The design settings were obtained from the position of the factory-installed buttons that are placed adjacent to the travel scale.

"Perfect" readings would be if the indicator were to line up with the white button (cold) and the red (hot); however, this is rarely the case. Generally, readings are considered acceptable and not noteworthy as long as they reflect movement consistent with design in both direction and length. A general rule is that when the hot setting is higher than the cold setting, then movement is down from cold to hot. If the cold setting is higher, movement is up.

The following terms are normally applied to these devices:

Actual travel: Anticipated movement of the pipe from design. The hot and cold position stickers are a function of this movement.

Total travel: The maximum movement a support can accept without danger of topping or bottoming out. The scale from H (0.0) to L (10.0) is a function of this.

Topped out: The indicator is above the high point and in contact with the end of the slot. This condition means the support is unloaded or is in the process of unloading.

Bottomed out: The indicator is below the (10.0) point and in contact with the end of the travel slot; the support is overloaded.

Variable-Spring Hangers These devices are installed at locations where stresses are not considered to be critical or where movement and economics permit their use.

The inherent characteristics of a variable are such that the supporting force varies with the spring's deflection. Movement of the pipe causes the spring to extend or compress. This results in a change or variance in the supporting force. Since the weight of the pipe is the same in either condition, hot or cold, the variation results in pipe weight transfer to equipment and adjacent hangers and consequently additional stresses in the piping system. The effects of this variation are usually considered during the original design.

In addition, as it is desirable to support the actual weight of the pipe when the system is hot, when the stresses tend to become most critical, the hot load is the dead weight of the pipe. The cold load is actually under- or oversupporting the pipe, depending on the movement from cold to hot.

The general rule for determining movement is similar to that of constant supports. If the hot load is higher than the cold, then pipe movement is down from cold to hot. If the cold load is higher, then movement is up.

Unlike constant supports, the readings from variables are measured in pounds. The readings are taken by noting the position of the indicator relative to a load scale that is adjacent to the travel slot.

The **distance between supports** will vary with the kind of piping and the number of valves and fittings. Supports should be provided near changes in direction, branch lines, and particularly near valves. The weight of piping must not be carried through valve bodies. In establishing the location of pipe supports, the designer should be guided by two requirements: (1) the horizontal span must not be so long that sag in the pipe will impose an excessive stress in the pipe wall and (2) the pipeline must be pitched downward so that the outlet of each span is lower than maximum sag in the span. Otherwise entrapped water can result in severe water hammer and pipe swings, particularly during plant start-up of steam piping.

Fabrication and installation practices are provided in MSS Standard Practice SP-89. Table 8.7.52 lists spacing for standard-weight pipe supports.

Pipe Insulation

(see Secs. 4 and 6 for heat-transmission data.)

The value of a steam-pipe covering is measured by its ability to reduce heat losses. This might range from 50 percent for small, low-temperature lines to 90 percent for large, high-temperature lines. Many **pipe-insulating materials** are available: 85 percent magnesia, foam glass, calcium silicate, and various forms of diatomaceous earths. Some of these materials are suited for relatively low temperatures only, others are best suited for high temperatures, and still others are suitable over a considerable temperature range.

Pipe insulation is applied in molded sections 3 ft long. For high-temperature work, the insulation is applied in at least two layers with the

Table 8.7.52 Maximum Spacing of Pipe Supports at 750°F (399°C)*

Nominal pipe size, in	1	1½	2	3	4	6	8	10	12	14	16	18	20	24
Maximum span, ft	7	9	10	12	14	17	19	22	23	25	27	28	30	32
Maximum span, m	2.13	2.74	3.05	3.66	4.27	5.18	5.79	6.71	7.01	7.62	8.23	8.53	9.14	9.75

* This tabulation assumes that concentrated loads, such as valves and flanges, are separately supported. Spacing is based on a combined bending and shear stress of 1,500 lb/in² when pipe is filled with water; under this condition, sag in pipeline between supports will be approximately 0.1 in.

joints staggered so as to prevent a direct channel for heat loss. Because of its maximum-temperature limitation of about 600°F (316°C), 85 percent magnesia is used as the second layer with a high-temperature-resistant material placed in direct contact with the pipe. The molded insulation is fastened securely in place with copper or galvanized wire and is then given a surface finish; indoor pipes are first sheathed with resin paper and covered with canvas, either pasted or sewed; outdoor pipes may be weather-protected by a coating of asphaltic-type waterproofing compound, they may be sheathed and canvased and then given a weatherproof surface, so they may be encased in metallic (steel or aluminum) jackets.

The heat loss from an insulated pipe appears in three phases: heat passes by *conduction* through the metallic pipe walls and through the insulating material; it then is dissipated from the outdoor surface of the insulation by *convection* and by *radiation*. Extremely accurate calculations must also take into account the temperature drop by convection through the film on the inside surface of the pipe. The task of accurately calculating heat losses is somewhat tedious, since the convection and radiation losses are related to the surface temperature (outside of insulation), which is unknown until conduction losses are balanced against surface losses.

For combined convection and radiation coefficients for bare pipes, and all necessary formulas to permit trial-and-error calculations, see Sec. 4. Insulation manufacturers publish data which give heat losses for wide ranges of pipe size and temperature.

Identification of Piping

The American National Standards Institute has approved a **Scheme for the Identification of Piping Systems** (ANSI A13.1-1981). This scheme is limited to the identification of piping systems in industrial plants, not including pipes buried in the ground, and electric conduits. Fittings, valves, and pipe coverings are included, but not supports, brackets, or other accessories.

Classification by Color All piping systems are classified by the nature of the material carried. Each piping system is placed, by the nature of its contents, in the following classifications:

Class	*Color*
F—Fire-protection equipment	Red
D—Dangerous materials	Yellow (or orange)
S—Safe materials	Green (or the achromatic colors, white, black, gray, or aluminum)
P—Protective materials	Bright blue
V—Extra valuable materials	Deep purple

Method of Identification At conspicuous places throughout a piping system, color bands should be painted on the pipes to designate to which one of the five main classes it belongs. If desired, the entire length of the piping system may be painted the main classification color.

Further, the actual contents of a piping system may be indicated by, preferably, a stenciled legend of standard size giving the name of the contents in full or abbreviated form. These legends should be placed on the color bands. The identification scheme may be extended by the use of colored stripes placed at the edges of the colored bands.

The bands, legends, and stripes should be placed at intervals throughout the piping system, preferably adjacent to valves and fittings to ensure ready recognition during operation, repairs, and at times of emergency.

A recommended classification, under this color scheme of materials carried in pipes, includes, as dangerous, combustible gases and oils, hot water and steam above atmospheric pressure; as safe, compressed air, cold water, and steam under vacuum.

Pressure Hose

Hose with durable rubber lining may be obtained to withstand any needed pressure. If the rubber compound is properly made, the life of a hose will be 7 to 10 years, while a cheaper hose, lined with inferior material, will probably not last more than 3 or 4 years. See also Secs. 3 and 12.

American National Fire-Hose Coupling Screw Thread (ANSI B1.20-7-1966) This standard is intended to cover the threaded part of fire-hose couplings, hydrant outlets, standpipe connections, and at other special fittings on fire lines, where fittings of the nominal diameters given in Table 8.7.53 are used. It also includes the limiting dimensions of the field inspection gages. The American National Standard form of thread must be used.

Table 8.7.53 Dimensions of Standard Fire-Hose Couplings
(All dimensions in inches. Letters refer to Fig. 8.7.22)

Inside diam, *C*	Diam of thread, *D*	No. of threads per inch	*L*	*I*	*H*	*J*	*T*
2½	3¹⁄₁₆	7½	1	¼	¹⁵⁄₁₆	³⁄₁₆	¹¹⁄₁₆
3	3⅝	6	1⅛	⁵⁄₁₆	1¹⁄₁₆	¼	¹³⁄₁₆
3½	4¼	6	1⅛	⁵⁄₁₆	1¹⁄₁₆	¼	¹³⁄₁₆
4½	5¾	4	1¼	⁷⁄₁₆	1³⁄₁₆	⅜	¹⁵⁄₁₆

SOURCE: ANSI B1.20.7-1966.

American National Standard Hose-Coupling Screw Threads (ANSI B1.20.7-1966) These standards apply to the threaded parts of hose couplings, valves, nozzles, and all other fittings used in direct connection with hose intended for fire protection or for domestic, industrial, or general service in nominal sizes given in Table 8.7.54. The American

Table 8.7.54 Dimensions of Standard Hose Couplings
(All dimensions in inches. Letters refer to Fig. 8.7.22)

Service and nominal size	Inside diam, *C*	Diam of thread, *D*	No. of threads per inch	*L*	*I*	*H*	*T*
Garden:							
½, ⅝, ¾	²⁵⁄₃₂	1¹⁄₁₆	11½	⁹⁄₁₆	⅛	¹⁷⁄₃₂	⅜
Chemical:							
¾, 1	1¹⁄₃₂	1⅜	8	⅝	⁵⁄₃₂	¹⁹⁄₃₂	¹⁵⁄₃₂
Fire:							
1½	1¹⁷⁄₃₂	2	9	⅝	⁵⁄₃₂	¹⁹⁄₃₂	¹⁵⁄₃₂
Other connections:							
½	¹⁷⁄₃₂	¹³⁄₁₆	14	½	⅛	¹⁵⁄₃₂	⁵⁄₃₂
¾	²⁵⁄₃₂	1¹⁄₃₂	14	⁹⁄₁₆	⅛	¹⁷⁄₃₂	⅜
1	1¹⁄₃₂	1⁹⁄₃₂	11½	⁹⁄₁₆	⁵⁄₃₂	¹⁷⁄₃₂	⅜
1¼	1⁹⁄₃₂	1⅝	11½	⅝	⁵⁄₃₂	¹⁹⁄₃₂	¹⁵⁄₃₂
1½	1¹⁷⁄₃₂	1⅞	11½	⅝	⁵⁄₃₂	¹⁹⁄₃₂	¹⁵⁄₃₂
2	2¹⁄₃₂	2¹¹⁄₃₂	11½	¾	³⁄₁₆	²³⁄₃₂	¹⁹⁄₃₂

SOURCE: ANSI B1.20.7-1966.

National Standard thread form is used. This coupling is similar in design to the fire-hose couplings illustrated in Fig. 8.7.22.

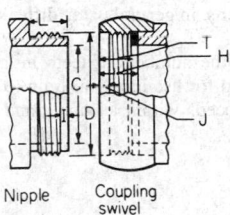

Fig. 8.7.22 Typical form of standard coupling.

Flexible metal hose and tubing are available for a wide range of conditions of temperature, pressure, vibration, and corrosion, and are made in two basic constructions, corrugated or interlocked, and in either bronze or steel.

The corrugated type (Fig. 8.7.23) may have either annular or helical corrugated formations, usually covered with metal braid, and is adapted to high-pressure high-temperature leak-proof service. Some typical applications include diesel-engine exhaust hose, reciprocating flexible connections, loading and unloading hose, saturated and superheated steam lines, lubricating lines, gas and oil lines, vibration connections, etc.

The interlocked type is made in several ways; the fully interlocked type is illustrated in Fig. 8.7.24. Typical applications include wiring conduit, cable armor, decorative wiring covering, dust-collective tubing, grease and oil connections, flexible spouts, and moderate-pressure oil lines.

Standard couplings and fittings can be attached to flexible metal hose or tubing by various methods such as brazing or welding. Each type of

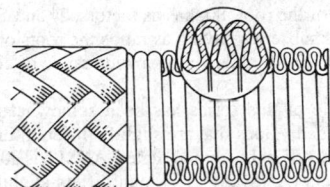

Fig. 8.7.23   Flexible metal hose.

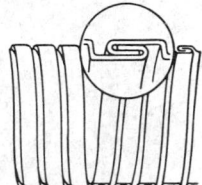

Fig. 8.7.24   Interlocked flexible metal hose.

hose construction has limits of service use and proved application usages. Information and recommendations as to the type and size to use under any given conditions should be obtained from the manufacturers.

8.8   PREFERRED NUMBERS
by C. H. Berry

REFERENCES: Hirshfeld and Berry, Size Standardization by Preferred Numbers, *Mech. Eng.*, Dec. 1922. Schlink, A New Tool for Standardizers, *Am. Mach.*, July 12, 1923. ANSI Standard Z17.1.

Many manufactured articles are made in several sizes which may be designated by some dimension, speed, capacity, or other feature. Each such series of products may be paralleled by a series of numbers.

It is generally agreed that such number series should be **geometric progressions**; i.e., each term should be a fixed percentage larger than the preceding. A geometric series provides small steps for small numbers, large steps for large numbers, and this best meets most requirements. The small steps in the diameter of the numbered twist drills would be absurd in drills of 1 in diameter and larger.

In the case of sized objects that are used principally as raw material, e.g., steel rod, an arithmetic progression may be preferable because it tends to reduce the cost of machining. It is desirable to be able to buy raw material a fixed amount (rather than a fixed percentage) larger than the finished article.

Preferred numbers is the name given to various series proposed for general use. These are either geometric progressions or approximations thereto. A geometric series is defined by one term and the ratio of each term to the preceding one. On the choice of these elements for a preferred number series, there is as yet no general agreement. The same value would hardly be satisfactory for all cases. The idea of preferred numbers is to provide a master series from which terms can be chosen to suit any needs. This would ultimately lead to a comprehensive plan in all fields of manufacture, so that, for example, the sizes of shafting would be in accord with the sizes of bearings, and indeed with all manner of cylindrical machine elements.

An advantage of a geometric series is that if linear dimensions are chosen in the series, areas, volumes, and other functions of powers of dimensions are also members of the same series.

In one of the most carefully considered systems of preferred numbers the base term is 1, and the ratio is $\sqrt[80]{10}$. In this series, the 81st term is 10, and accordingly the series from 10 to 100 or from 0.01 to 0.1, or, in general, from 10^n to 10^{n+1} is identical with the series from 1 to 10 with the decimal point shifted. This series will rarely be used in full; some will choose alternate terms, some every fourth, fifth, tenth, or twentieth

Table 8.8.1   Basic Series of Preferred Numbers: R 80 Series

1.00	1.80	3.15	5.60
1.03	1.85	3.25	5.80
1.06	1.90	3.35	6.00
1.09	1.95	3.45	6.15
1.12	2.00	3.55	6.30
1.15	2.06	3.65	6.50
1.18	2.12	3.75	6.70
1.22	2.18	3.87	6.90
1.25	2.24	4.00	7.10
1.28	2.30	4.12	7.30
1.32	2.36	4.25	7.50
1.36	2.43	4.37	7.75
1.40	2.50	4.50	8.00
1.45	2.58	4.62	8.25
1.50	2.65	4.75	8.50
1.55	2.72	4.87	8.75
1.60	2.80	5.00	9.00
1.65	2.90	5.15	9.25
1.70	3.00	5.30	9.50
1.75	3.07	5.45	9.75

SOURCE: American National Standard Preferred Numbers Z17.1, reproduced with permission of ANSI.

term. The index of the root, 80, has as factors, 2^4 and 5, so that the series readily yields subseries having as ratios the roots of 10 with indices 2, 4, 8, 16, 5, 10, 20, 40, thus giving a wide range of choice. See Table 8.8.1.

The strict logic of this series has been somewhat impaired by the adoption of rounded values that are slightly different in the 1-to-10 and 10-to-100 intervals. For the United States, ANSI has adopted a Table of Preferred Numbers (ANSI Z17.1) which differs slightly from the system described in the preceding paragraph.

Another type of series is the **semigeometric series** consisting of a basic geometric series with 1 as the base term and a ratio of 2, giving a series . . . ⅛, ¼, ½, 1, 2, 4, ... Between consecutive terms are inserted arithmetic series of 2, 4, 8, or 16 terms, in general using different numbers of terms in different intervals.

A similar procedure is used to establish the numbers of teeth in a prescribed number of gears that are intended for use in a gear train to provide a stepped gradation of rotational speeds within an upper and lower bound.

Power Generation

BY

EZRA S. KRENDEL *Emeritus Professor of Operations Research and Statistics, University of Pennsylvania*

R. RAMAKUMAR *Professor of Electrical Engineering, Oklahoma State University*

C. P. BUTTERFIELD *Chief Engineer, Wind Technology Division, National Renewable Energy Laboratory*

ERICH A. FARBER *Distinguished Service Professor Emeritus; Director Emeritus Solar Energy and Energy Conversion Laboratory, University of Florida*

KENNETH A. PHAIR *Senior Mechanical Engineer, Stone and Webster Engineering Corp.*

SHERWOOD B. MENKES *Professor of Mechanical Engineering, Emeritus, The City College, The City University of New York*

JOSEPH C. DELIBERT *Retired Executive, The Babcock & Wilcox Co.*

FREDERICK G. BAILY *Consulting Engineer; formerly Technical Coordinator, Thermodynamics and Applications Engineering, General Electric Co.*

WILLIAM J. BOW *Director (Retired), Heat Transfer Products Dept., Foster and Wheeler Energy Corp.*

DONALD E. BOLT *Engineering Manager, Heat Transfer Products Dept., Foster and Wheeler Energy Corp.*

DENNIS N. ASSANIS *Professor of Mechanical Engineering, University of Michigan*

CLAUS BORGNAKKE *Associate Professor of Mechanical Engineering, University of Michigan*

DAVID E. COLE *Director, Office for the Study of Automotive Transportation, Transportation Research Institute, University of Michigan*

D. J. PATTERSON *Professor of Mechanical Engineering, Emeritus, University of Michigan*

JOHN H. LEWIS *Technical Staff, Pratt & Whitney, Division of United Technologies Corp.; Adjunct Associate Professor, Hartford Graduate Center, Rensselaer Polytechnic Institute*

ALBERT H. REINHARDT *Technical Staff, Pratt & Whitney, Division of United Technologies Corp.*

LOUIS H. RODDIS, JR. *Late Consulting Engineer, Charleston, SC*

DANIEL J. GARNER *Senior Program Manager, Institute of Nuclear Power Operations*

JOHN E. GRAY *ERCI, International*

EDWIN E. KINTNER *GPU Nuclear Corp.*

NUNZIO J. PALLADINO *Dean Emeritus, College of Engineering, Pennsylvania State University*

GEORGE SEGE *Technical Assistant to the Director, Office of Nuclear Regulatory Research, U.S. Nuclear Regulatory Commission*

PAUL E. NORIAN *Special Assistant, Regulatory Applications, Office of Nuclear Regulatory Research, U.S. Nuclear Regulatory Commission*

ROBERT D. STEELE *Manager, Turbine and Rehabilitation Design, Voith Hydro, Inc.*

9.1 SOURCES OF ENERGY

Contributors are shown at the head of each category.

REFERENCES: Latest available published data from the following: "Reserves of Crude Oil, Natural Gas Liquids, and Natural Gas in the U.S. and Canada," American Petroleum Institute. "Annual Statistical Review—Petroleum Industry Statistics," American Petroleum Institute. Worldwide Issue, *Oil & Gas Jour.* annually. "Potential Supply of Natural Gas in the U.S.," Mineral Resources Institute. Colorado School of Mines. Coal resources in the United States, *U.S. Geol. Surv. Bull.* 1412. Geological Estimates of Undiscovered Recoverable Oil and Gas Resources in the U.S., *U.S. Geol. Surv. Circ.* 725, United Nations Statistical Office, "Statistical Yearbook," New York, U.N. Department of Economic and Social Affairs. Bureau of Mines, Metals, Minerals, and Fuels, vol. I of "Minerals Yearbook" published annually. "Coal Data," National Coal Association. "International Coal," National Coal Association. Annual Technical Literature Data Base, *Power,* McGraw-Hill.

INTRODUCTION
Staff Contribution

Global energy requirements are supplied primarily by fossil fuels, nuclear fuels, and hydroelectric sources; about 1 to 2 percent of global requirements are supplied from other miscellaneous sources. In the United States in 1994, total domestic power requirements were supplied approximately as follows: 70 percent fossil fuel (of which coal accounted for 58 percent), 20 percent nuclear fuel, 10 percent from hydroelectric sources, and less than 1 percent from all other sources. In spite of the large increase in nuclear generated power, both in the United States and globally, coal continues to be the major fuel consumed.

In the United States, new power plants constructed at this time are designed to consume fossil fuels—primarily coal, many with gas, and a few with petroleum. The situation with regard to nuclear power is complicated by a number of circumstances; see Sec. 9.8. Energy statistics and accompanying data stay current for a short time. Bear in mind that when quantities of known reserves of fuel of all types are stated, there is implied the significant matter of whether they are, indeed, producible in a given economic climate. Estimates for additional reserves remaining to be discovered are available, by and large, only for the United States. At any given time, the situation with regard to estimates of recoverable fuel sources is subject to wide swings whose source is manifold: national and international politics, environmental concerns, significant progress in energy conservation, unsettled political and social conditions in locations within which reside much of the world reserves of fossil fuels, economic impact of financing, effects of inflation, and so on. The references cited, in their most current form, will provide the reader with realistic and authoritative compilations of data.

Fossil Fuels

Petroleum Proved reserves of crude oil and natural gas liquids in the United States, based upon estimated discovered quantities which geological and engineering data demonstrate with reasonable certainty to be recoverable in future years from presently known reservoirs under existing economic and operating conditions, are published annually by the American Petroleum Institute. Estimates of additional remaining producible reserves which will be discovered, proved, and produced in the future from the total original oil in place, are derived by *U.S. Geol. Surv. Circ.* 725 from present and projected conditions in the industry.

Estimates of proved crude oil reserves in all countries of the world are published by *Oil and Gas Journal.* New discoveries are continually adding to and changing proved reserves in many parts of the world, and these estimates are indicative of producible quantities.

Natural Gas Proved reserves of natural gas in the United States, based upon the same definition as for crude oil and natural gas liquids, are estimated annually by the American Gas Association. The estimated total additional potential supply remaining to be discovered is prepared by the Potential Gas Committee, sponsored by the Potential Gas Agency, Colorado School of Mines Foundation, Inc.

Estimates of proved reserves of natural gas in all countries of the world are published by *Oil and Gas Journal.* As with crude oil, large additional natural gas reserves are currently being discovered and developed in Alaska, the arctic regions, offshore areas, northern Africa, and other locations remote from consuming markets. Valid estimates of additional probable remaining reserves in the world are not available.

Coal (See also Secs. 7.1 and 7.2.) Authoritative information about reserves of coal is presented in *Geol. Surv. Bull.* 1412, Coal Resources of the United States. Remaining U.S. proved reserves (1974) of bituminous, subbituminous, lignite, and anthracite have been estimated by mapping and exploration of areas with 0 to 3,000-ft overburden. The U.S. Geological Survey (USGS) estimates probable additional resources in unmapped and unexplored areas with 0 to 3,000-ft overburden and in areas with 3,000- to 6,000-ft overburden. Slightly more than one-half of the proved reserves are considered producible (at this time) because of favorable depth of overburden and thickness of coal strata. Approximately 30 percent of all ranks of coal are commerically available in beds less than 1,000 ft deep. The USGS estimates that about 65 percent contains less than 1 percent sulfur; most of the low-sulfur coals are located west of the Mississippi. *USGS Bull.* 1412 also estimates global coal resources, but in view of the questionable validity of much of the global data, it can but offer gross approximations. (See Sec. 7.1.)

Shale Oil The portion of total U.S. reserves of oil from oil shale, measured or proved, considered minable and amenable to processing is estimated to be over 150 billion bbl (30 billion m^3), based upon grades averaging 30 gal/ton in beds at least 100 ft thick (*USGS Bull.* 1412). Most oil shale occurs in Colorado. No commercial production is expected for many years. World reserves occur largely in the United States and Brazil, with small quantities elsewhere.

Tar Sands Large deposits are in the Athabasca area of northern Alberta, Canada, estimated capable of producing 100 to 300 billion bbl (15.9 to 47.7 billion m^3) of oil. About 6.3 billion bbl (1.0 billion m^3) has been proved economically recoverable within the radius of the present large mining and recovery plant in Athabasca. Commercial quantities of oil have been produced there since the 1960s. Sizable deposits are lo-

Table 9.1.1 Major U.S. Coal-Producing Locations

Anthracite and semianthracite
Pennsylvania

Bituminous coal
Illinois
West Virginia
Kentucky
Colorado
Pennsylvania
Ohio
Indiana
Missouri

Subbituminous coal
Montana
Alaska
Wyoming
New Mexico

Lignite
North Dakota
Montana

cated elsewhere; they have not been exploited to date, meaningful data for them are not available, and there is no report of those other deposits having been worked. (See Sec. 7.1.)

Nuclear Fuels

Uranium Reserves of uranium in the United States are reported by the Department of Energy (DOE). The proved reserves, usually presented in terms of quantity of U_3O_8, refer to ore deposits (concentrations of 0.01 percent, or 0.0016 oz/lb ore, are viable) of grade, quantity, and geological configuration that can be mined and processed profitably with existing technology. *Estimated additional resources* refer to uranium surmised to occur in unexplored extensions of known deposits or in undiscovered deposits in known uranium districts, and which are expected to be discovered and economically exploitable in the given price range. The total of these uranium reserves would yield about 3,000 tons of U_3O_8. United States uranium resources are located mainly in New Mexico, Wyoming, and Colorado.

Thorium Total known resources of thorium, the availability of which is considered reasonably assured, are estimated in the millions of tons of thorium oxide. Additional actual reserves will increase in response to the demand and concomitant market price. Most of the larger known resources are in India and Brazil. There seems to be little prospect of significant requirements for thorium as a nuclear fuel in the near future.

Hydroelectric Power

Hydroelectric and Pumped Storage for Electric Generation Although most available sites for economical production of hydroelectric energy have been developed, some additional hydroelectric capacity will be provided at new sites or by additions at existing plants. Increased pumped storage capacity will be limited by the availability of suitable sites and a dependable supply of economical pumping energy. The flexibility of operation of a pumped storage plant in meeting sudden load changes and its ability to provide high inertia spinning reserve at low operating cost are additional benefits that can weigh heavily in favor of this type of installation, particularly in the future if (when) the proportion of nuclear capacity in service increases. At this time, hydro and pumped storage account for about 10 percent of electricity generated by all sources of energy in the United States.

World installed hydropower capacity presently is located about 40 percent in North America and 40 percent in Europe.

ALTERNATIVE ENERGY, RENEWABLE ENERGY, AND ENERGY CONVERSION: AN INTRODUCTION
Staff Contribution

REFERENCES: AAAS, *Science*. Hottell and Howard, "New Energy Technology—Some Facts and Assessments," MIT Press. Fisher, "Energy Crises in Perspective," Wiley-Interscience. Hammond, Metz, and Maugh, "Energy and the Future," AAAS.

Many sources of raw energy have been proposed or used for the generation of power. Only a few sources—fossil fuels, nuclear fission, and elevated water—are dominant in practical applications today.

A more complete list of sources would include fossil fuels (coal, petroleum, natural gas); nuclear (fission and fusion); wood and vegetation; elevated water supply; solar; winds; tides; waves; geothermal; muscles (human, animal); industrial, agricultural, and domestic wastes; atmospheric electricity; oceanic thermal gradients; oceanic currents. There are others.

Historically, wood, muscles, elevated water, and wind were prominent. These sources were superseded in the industrial era by fossil fuels, with nuclear energy the most recent addition. This dominance rests in the suitability of the thermal sources for practical stationary and transportation power plants. Features of acceptability include reliability, flexibility, portability, maneuverability, size, bulk, weight, efficiency, economy, maintenance, and costs. The plant for transportation service

must be self-contained. For stationary service there is wider latitude for choice.

The dominant end product, especially for stationary applications, is electricity, because of its favorable distribution and control features. However, there is no practical way of storing electric energy. Electricity must be generated at the instant of its use. Reliability and continuity of service consequently dictate the need for reserve, alternate, and interconnection supports. Pumped storage, coal piles, and tanks of liquid and gaseous fuels, e.g., offer the necessary continuity, flexibility, and reliability.

Raw energy sources, other than fuels (fossil and nuclear) and elevated water, are particularly deficient in this storage aspect. For example, wind power is best for jobs that can wait for the wind, e.g., pumping water or grinding grain. Solar power, to avoid foul weather and the darkness of night, could call for desert locations or extraterrestrial satellites.

Despite such limitations an energy-intensive society can expect to see increasing efforts to harness many of the raw energy sources cited. Several of these topics are treated in the following pages to show the factual and technical progress that has been made to adapt sources to practicality.

MUSCLE-GENERATED POWER
by Ezra S. Krendel, Amended by Staff

REFERENCES: Whitt and Wilson, "Bicycling Science," 2d ed., MIT Press. Harrison, Maximizing Human Power Output by Suitable Selection of Motion Cycle and Load, *Human Factors,* **12**, 1970. Krendel, Design Requirements for Man-Generated Power, *Ergonomics,* **3**, 1960. Wilkie, Man as a Source of Mechanical Power, *Ergonomics,* **3**, 1960. Brody, "Bioenergetics and Growth," Reinhold.

The use of human muscles to generate work will be examined from two points of view. The first is that of measuring the energy expended in gross, long duration physical activities such as marching, forestry work, freight handling, and factory work. The second is that of determining the useful mechanical work which can be performed by specified muscle groups for brief or extended periods of time in well defined work situations, such as pedaling or cranking.

Labor

Over an 8-h day for a 48-h week, a useful norm for a 35-year-old laborer for total power expenditure, including basal metabolism energy, is 0.49 hp (366 W). Of this total expenditure, approximately 0.1 hp (75 W) is available for useful work. A 20-year-old man can generate about 15 percent more power than this norm, and a 60-year-old man about 20 percent less. The total energy or power expenditure is needed for determining nutritional requirements for classes of labor. A rule of thumb for power developed by European males can be expressed as a function of age and duration of effort in minutes for work lasting from 4 to about 480 min, assuming that 20 percent of the total output is useful power.

Age, years	Useful horsepower (t in min)
20	hp $= 0.40 - 0.10 \log t$
35	hp $= 0.35 - 0.09 \log t$
60	hp $= 0.30 - 0.08 \log t$

For a well-trained man, useful power production by pedaling, hand cranking, or a combination of the two for working durations of from 20 to 120 s may be summarized as follows (t is in seconds):

Arms and legs	hp $= 4.4t^{-0.40}$
Legs only	hp $= 2.8t^{-0.40}$
Arms only	hp $= 1.5t^{-0.40}$

There are examples of well-trained athletes generating between 1.5 and 2 hp for efforts of 5 to 20 s, using both arms and legs to generate power.

For pedaling efforts of from 1 to about 100 min, the useful power generated may be expressed as hp = 0.53 − 0.13 log t (t is in minutes).

Work scheduling, either as rhythmic work activity or with rest stops for recuperation, the temperature and humidity of the environment, and the detailed nature of the laborer's diet are factors which influence ability to generate and maintain the above nominal power values. These considerations should be factored in for specific work situations.

Steady State and Transient

When a human and a passive mechanism are working together to generate power, the following conditions obtain: Energy is available both from stores residing in the muscles [a total usefully available energy of about 0.6 hp · min (27 kJ), usually applied in transient bursts of activity] and from the oxidation of foods (for producing steady state power). For an aerobic transient activity, energy production depends on the mass of muscle which can be brought into effective contact with the power transmission mechanism. For example, bicycle pedaling is an effective use of a large muscle mass. For steady-state activity, assuming adequate food for fuel energy, power generated depends on the oxygen supply and the efficiency with which oxygenated blood can be transported to the muscles as well as on the muscle mass.

The **physiological limit,** determined by oxygen-respiration capacity, for steady-state useful mechanical power generation is between 0.4 and 0.54 hp (300 and 400 W), depending on the man's physical condition.

Useful power production may be achieved by such methods as rowing, cranking, or pedaling. The **highest values** of human-generated horsepower using robust subjects have been achieved using a rowing assembly which restrained nonuseful motions of the torso and major limbs. Under these conditions up to 2 hp (1,500 W) was generated over intervals of 0.6 s, and averages of about 1 hp (750 W) were generated over 2 min.

In order to approach an optimal **conversion efficiency** (mechanical work/food energy) of 25 percent, a mechanism would be required to store and to transmit energy from the body muscle masses when they were operating at optimal efficiency. This condition occurs when the force exerted by the muscle is about one-half of its maximum and the speed of muscle movement one-quarter of its maximum. Data on both force and speed for a given set of muscles are best measured in situ. Optimal conversion efficiency and maximum output power do not occur together.

Examples of High Output

Data for human-generated power come from measurements of subjects with different kinds of training, skill, body builds, diets, and motivation using a variety of mechanical devices such as bicycles, ergonometers, and variations on rowing machines. For strong, healthy young men, aggregations of such data for power produced in an interval t of 10 to 120 s can be approximated as follows:

$$hp = 2.5t^{-0.40}$$

For world-class athletes this becomes hp = 0.25 + 2.5$t^{-0.40}$. These values can be exceeded for bursts of power of less than 10 s. For long-term efforts of from 2 to about 200 min, the aggregated data for useful power generated by strong, healthy young men can be approximated as follows (t in minutes):

$$hp = 0.50 − 0.13 \log t$$

For world-class athletes this becomes hp = 0.65 − 0.13 log t. The pilot of the Gossamer Albatross, who flew 22 mi from England to France in 2 h 55 min on August 12, 1979 entirely by pedal-generated power, sustained an output of about ⅓ hp (250 W) during the flight.

Maximum power output occurs at a load impedance of 5 to 10 times the size of the human being's source impedance.

Brody has developed detailed nomograms for determining the energetic cost of muscular work by farm animals; these nomograms are useful for precise cost-effectiveness comparisons between animal and mechanical power generation methods. A 1,500- to 1,900-lb horse can work continuously for up to 10 h/day at a rate of 1 hp, or equivalently

pull 10 percent of its body weight for a total of 20 mi/day, and retain its vigor to an advanced age. Brody's work allows the following approximations for estimating the useful power output of work animals of varying sizes: The ratio of the power exerted in maximal energy production for a few seconds to the maximum steady-state power maintained for 5 to 30 min to the power produced in sustained heavy work over a 6- to 10-h day is approximately 25 : 4 : 1. For any one of these conditions, it has been found that, for healthy, mature specimens,

$$hp_{animal} = hp_{man}(\text{mass of animal/mass of man})^{0.73}$$

Thus, from the previously given horsepower magnitudes for men, one can compute the power generated by ponies, horses, bullocks, or elephants under the specified working conditions.

WIND POWER
by R. Ramakumar and C. P. Butterfield

REFERENCES: NREL technical information at Internet address: http://gopher.nrel.gov.70. AWEA information at Internet address: awea.wind-net@notes.igc.apc.org. Hansen and Butterfield, Aerodynamics of Horizontal-Axis Wind Turbines, *Ann. Rev. Fluid Mech.*, **25**, 1993, pp. 115–149. Touryan, Strickland, and Berg, "Electric Power from Vertical-Axis Wind Turbines," *J. Propulsion,* **3**, no. 6, 1987. Betz, "Introduction to the Theory of Flow Machines," Pergamon, New York. Eldridge, "Wind Machines," 2d ed., Van Nostrand Reinhold, New York. Glauert, "Aerodynamic Theory," Durand, ed., 6, div. L, p. 324, Springer, Berlin, 1935. Richardson and McNerney, Wind Energy Systems, *Proc. IEEE,* **81**, no. 3, Mar. 1993, pp. 378–389. Elliott et al., "Wind Energy Resource Atlas," Wilson and Lissaman, Applied Aerodynamics of Wind Power Machines, Oregon State University Report, 1974. Eggleston and Stoddard, *Wind Turbine Engineering Design,* New York, Van Nostrand Reinhold. Spera, *Wind Turbine Technology,* ASME Press, New York. Gipe, "Wind Power for Home and Business," Chelsea Green Publishing Company. Ramakumar et al., Economic Aspects of Advanced Energy Technologies, *Proc. IEEE,* **81,** no. 3, Mar. 1993, pp. 318–332.

Wind is one of the oldest widely used sources of energy. Although its use is many centuries old, it has not been a dominant factor in the energy picture of developed countries for the past 50 years because of the abundance of fossil fuels. Recently, the realization that fossil fuels are in limited supply has awakened the need to develop wind power with modern technology on a large scale. Consequently, there has been a tremendous resurgence of effort in wind power in just the past few years. The state of knowledge is rapidly increasing, and the reader is referred to the current literature and the NREL Internet address cited above for information on the latest technology. Wind energy is one of the lowest-cost forms of renewable energy. In 1995, more than 1,700 MW of wind energy capacity was operating in California, generating enough energy to supply a city the size of San Francisco with all its energy needs. European capacity was almost the same. For the latest status on worldwide use of wind energy, the reader is referred to the American Wind Energy Association (AWEA) at the Internet address cited above. The fundamental principles of wind power technology do not change and are discussed here.

Wind Turbines   The essential ingredient in a **wind energy conversion system (WECS)** is the wind turbine, traditionally called the windmill. The predominant configurations are **horizontal-axis propeller turbines (HAWTs)** and **vertical-axis wind turbines (VAWTs),** the latter most often termed **Darrieus rotors.** In the performance analysis of wind turbines, the propeller devices were studied first, and their analysis set the current conventions for the evaluation of all turbines.

General Momentum Theory for Horizontal-Axis Turbines   Conventional analysis of horizontal-axis turbines begins with an axial momentum balance originated by Rankine using the control volume depicted in Fig. 9.1.1. The turbine is represented by a porous disk of area A which extracts energy from the air passing through it by reducing its pressure: on the upstream side the pressure has been raised above atmospheric by the slowing airstream; on the downstream side pressure is lower, and atmospheric pressure will be recovered by further slowing of the stream. V is original wind speed, decelerated to $V(1 − a)$ at the turbine disk, and to $V(1 - 2a)$ in the wake of the turbine (a is called the interference factor). Momentum analysis predicts the axial thrust on the turbine of radius R to be

$$T = 2\pi R^2 \rho V^2 a (1 - a) \tag{9.1.1}$$

where air density, ρ, equals 0.00237 lbf·s²/ft⁴ (or 1.221 kg/m³) at sea-level standard-atmosphere conditions.

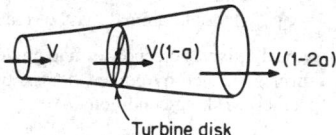

Fig. 9.1.1   Control volume.

Application of the mechanical energy equation to the control volume depicted in Fig. 9.1.1 yields the prediction of power to the turbine of

$$P = 2\pi R^2 \rho V^3 a (1 - a)^2 \tag{9.1.2}$$

This power can be nondimensionalized with the energy flux E in the upstream wind covering an area equal to the rotor disk, i.e.,

$$E = \tfrac{1}{2}\rho V^3 \pi R^2 \tag{9.1.3}$$

The resulting power coefficient is

$$C_p = \frac{P}{E} = 4a(1 - a)^2 \tag{9.1.4}$$

This power coefficient has a theoretical maximum at $a = \frac{1}{3}$ of $C_p = 0.593$. This result was first predicted by Betz and shows that the load placed on a windmill must be optimized to obtain the best power output: If the load is too small (small a), too much of the power is carried off with the wake; if the load is too large (large a), the flow is excessively obstructed and most of the approaching wind passes around the turbine.

This derivation includes some important assumptions which limit its accuracy and applicability. In particular, the portion of the kinetic energy in the swirl component of the wake is neglected. Partial accounting for the rotation in the wake has been included in the analysis of Glauert with the resulting prediction of ideal power coefficient as a function of turbine tip speed ratio $X = \Omega R/V$ (where Ω is the angular velocity of the turbine) shown in Fig. 9.1.2. Clearly, the swirl is made up of wasted kinetic energy and is largest for a high-torque, low-speed turbine. Actual farm, multiblade, and two- or three-bladed turbines show somewhat lower than ideal performance because of drag effects neglected in ideal flow analysis, but the high-speed two- or three-bladed turbines do tend to yield higher efficiency than low-speed multiblade windmills.

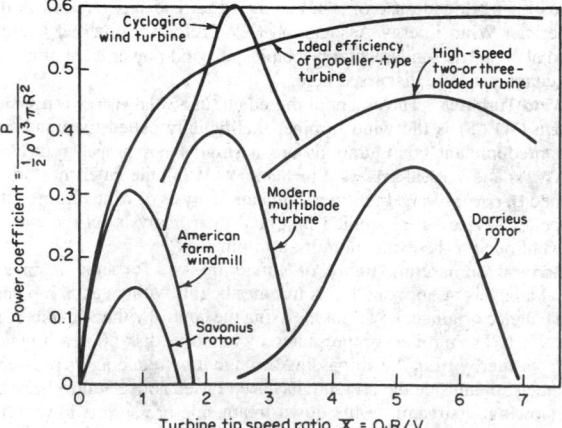

Fig. 9.1.2   Performance curves for wind turbines.

Blade Element Theory for Horizontal-Axis Turbines   Wilson and Eggleston describe blade element theory as a mechanism for analyzing the relationship between the individual airfoil properties and the interference factor a, the power produced P, and the axial thrust T of the turbine. Rather than the stream tube of Fig. 9.1.1, the control volume consists of the annular ring bounded by streamlines depicted in Fig. 9.1.3. It is assumed that the flow in each annular ring is independent of the flow in all other rings.

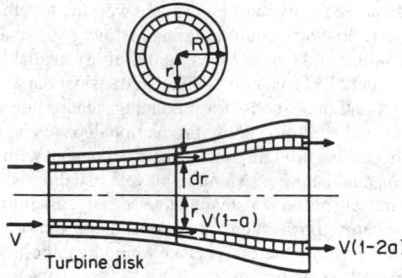

Fig. 9.1.3   Annular ring control volume.

A schematic of the velocity and force vector diagrams is given in Fig. 9.1.4. The turbine is defined by the number B of its blades, by the variation of chord c, by the variation in blade angle θ, and by the shape of blade sections $a' = \omega/(2\Omega)$, where ω is the angular velocity of the air just behind the turbine and Ω is the turbine angular velocity. Also W is the velocity of the wind relative to the airfoil. Note that the angle ϕ will be different for each blade element, since the velocity of the blade is a function of the radius. In order to keep the local flow angle of attack $\alpha = \theta - \phi$ at a suitable value, it will generally be necessary to construct twisted blades, varying θ with the radius. The sectional lift and drag coefficients C_L and C_D are obtained from empirical airfoil data and are unique functions of the local flow angle of attack $\alpha = \theta - \varphi$ and the local Reynolds number of the flow. The entire calculation requires trial-and-error procedures to obtain the axial interference factor a and the angular velocity fraction a'. It can, however, be reduced to programs for small computers.

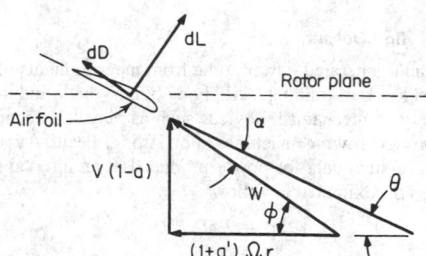

Fig. 9.1.4   Velocity and force vector diagrams.

A typical solution for steady-state operation of a two- or three-bladed wind axis turbine is shown in Fig. 9.1.2. When optimized, these turbines run at high tip speed ratios. The curve shown in Fig. 9.1.2 for the two- or three-bladed wind turbine is for constant blade pitch angle. These turbines typically have pitch change mechanisms which are used to feather the blades in extreme wind conditions. In some instances the blade pitch is continuously controlled to assist the turbine to maintain constant speed and appropriate output. Turbines with continuous pitch control typically have flatter, more desirable operating curves than the one depicted in Fig. 9.1.2.

The traditional U.S. farm **windmill** has a large number of blades with a high solidity ratio σ. (σ is the ratio of area of the blades to swept area of the turbine πR^2.) It operates at slower speed with a lower power coefficient than high-speed turbines and is primarily designed for good starting torque.

The curves depicted in Fig. 9.1.2 representing the performance of high- and low-speed wind axis turbines are theoretically predicted performance curves which have been experimentally confirmed.

Vertical-Axis Turbines The Darrieus rotor looks somewhat like an eggbeater (Fig. 9.1.5). The blades are high-performance symmetric airfoils formed into a gentle curve to minimize the bending stresses in the blades. There are usually two or three blades in a turbine, and as shown in Fig. 9.1.2, the turbines operate efficiently at high speed. Wilson shows that VAWT performance analysis also takes advantage of the same momentum principles as the horizontal axis wind turbines. However, the blade element momentum analysis becomes much more complicated (see Touryan et al.).

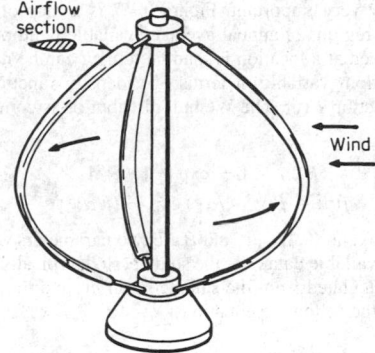

Fig. 9.1.5 Darrieus rotor.

Care must be taken not to overemphasize the aerodynamic efficiency of wind turbine configurations. The most important criterion in evaluating WECSs is the power produced on a per-unit-cost basis.

Drag Devices Rotors utilizing drag rather than lift have been constructed since antiquity, even though they are bulky and limited to low coefficients of performance. The Savonius rotor is a modern variation of these devices; in practice it is limited to small sizes. Eldridge describes the history and theory of this type of windmill.

Augmentation Occasionally, the use of structures designed to concentrate and equalize the wind at the turbine is proposed. For its size, the most effective of these has been a short diffuser (hollow cone) placed around and downwind of a wind turbine. The disadvantage of such augmentation devices is the cost of the bulky static structures required.

Rotor Configuration Trends Hansen and Butterfield describe some trends in turbine configurations which have developed from 1975 to 1995. Although no single configuration has emerged which is clearly superior, HAWTs have been more widely used than VAWTs. Only about 3 percent of turbines installed to date are VAWTs.

HAWT rotors are generally classified according to rotor orientation (upwind or downwind of the tower), blade articulation (rigid or teetering), and number of blades (generally two or three). **Downwind turbines** were favored initially in the United States, but the trend has been toward greater use of **upwind turbines** with a current split between upwind (55 percent) and downwind (45 percent) configurations.

Downwind orientation allows blades to deflect away from the tower when thrust loading increases. Coning can also be easily introduced to decrease mean blade loads by balancing aerodynamic loads with centrifugal loads. Figure 9.1.6 shows typical upwind and downwind configurations along with definitions for blade coning and yaw orientation.

Free yaw, or passive orientation with the wind direction, is also possible with downwind configurations, but yaw must be actively controlled with upwind configurations. Free-yaw systems rely on rotor thrust loads and blade moments to orient the turbine. Net yaw moments for rigid rotors are sensitive to inflow asymmetry caused by turbulence, wind shear, and vertical wind. These are in addition to the moments caused by changes in wind direction which are commonly, though often incorrectly, considered the dominant cause of yaw loads.

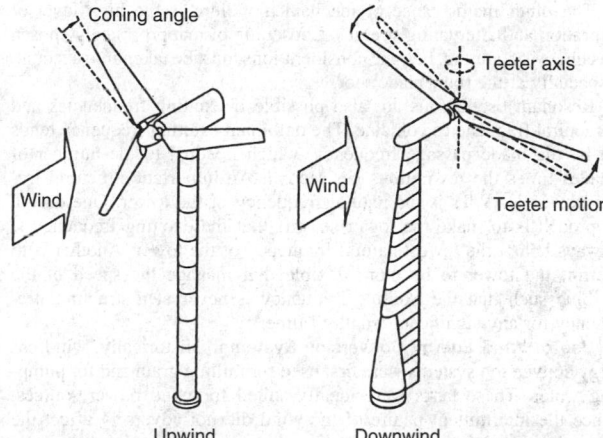

Fig. 9.1.6 HAWT configurations. *(Courtesy of Atlantic Orient Corp.)*

Some early downwind turbine designs developed a reputation for generating subaudible noise as the blades passed through the tower shadow (tower wake). Most downwind turbines operating today have greater tower clearances and lower tip speeds, which result in negligible infrasound emissions (Kelley and McKenna, 1985).

Blade Articulation Several different rotor blade articulations have been tested. Only two have survived—the three-blade, rigid rotor and the two-blade, teetered rotor. The rigid, three-blade rotor attaches the blade to a hub by using a stiff cantilevered joint. The first bending natural frequency of such a blade is typically greater than twice the rotor rotation speed $2p$. Cyclic loads on rigid blades are generally higher than on teetering blades of the same diameter. Richardson and McNerney describe a 33-m, 300-kW turbine currently under development which reflects a mature version of this configuration.

Teetered, two-blade rotors use relatively stiff blades rigidly connected to a hub, but the hub is attached to the main drive shaft through a **teeter hinge.** This type rotor is commonly used in tail rotors and some main rotors on helicopters. Two-blade rotors usually require teeter hinges or flexible root connections to reduce dynamic loading resulting from nonaxisymmetric mass moments of inertia. In normal operation, the cyclic loads on the teetering rotor are low, but there is risk of teeter-stop bumping ("mast bumping" in helicopter terminology) that can greatly increase dynamic loads in unusual situations.

Number of Blades Most two-blade rotors operating today use teetering hinges, but all three-blade rotors use rigid root connections. For small turbines (smaller than 50-ft diameter) rigid, three-blade rotors are inexpensive and simple and have the lowest system cost. As the turbines become larger, blade weight (and hence cost) increases in proportion to the third power of the rotor diameter, whereas power output increases only as the square of the diameter. This makes it cost-effective to reduce the number of blades to two and to add the complexity of a teeter hinge or flex beams to reduce blade loads. In the midscale rotor size (15 to 30 m), it is difficult to determine whether three rigidly mounted blades or two teetered blades are more cost-effective. In many cases, the choice between two- and three-blade rotors has been driven by designers' lack of experience and the potential risk of high development cost rather than by technical and economic merit. Currently 10 percent of the turbines installed are two-bladed, yet approximately 60 percent of all new designs being considered in the United States are two-blade, teetered rotors.

Design Problems A key design consideration is survival in severe storms. Various systems for furling the rotor, feathering the blades, or braking the shaft have been employed; failure of these systems in a high wind has been known to cause severe damage to the turbine.

A different, but related, consideration is the control of the turbine after a loss of electrical load, which also could cause severe overspeeding and catastrophic failure.

The other major cause of mechanical failure is the high level of vibration and alternating stresses. Loosening of inappropriately chosen fasteners is common. Fatigue considerations must be taken into account, especially at the rotor blade root.

Resonant oscillations are also possible if exciting frequencies and structural frequencies coincide. The dominant exciting frequency tends to be the blade passage frequency, which is equal to the number of blades times the revolutions per second. An important structural frequency in HAWTs is the natural frequency of the tower. One design approach is to make the tower so stiff that the exciting frequency is always below the lowest natural frequency of the tower. Another is to permit the tower to be more flexible, but manage the speed of the turbine such that the exciting frequency is never at a structural frequency for any significant length of time.

Use of Wind Energy Conversion Systems Historically, wind energy conversion systems were first used for milling grain and for pumping water. These tasks were ideally suited for wind power sources, since the intermittent nature of the wind did not adversely affect the operation.

The largest impact of wind power on the energy picture in the developed countries of the world is expected to be in the generation of electric power. In most cases, this involves feeding power into the power grid, and requires induction or synchronous generators. These generators require that the rotor turn at a constant speed. Wind turbines operate more efficiently (aerodynamically speaking) if they turn at an optimum ratio of tip speed to wind speed. Thus the use of variable-speed operation, using power electronics to obtain constant-frequency utility-grade ac power, has become attractive. Richardson describes the modern use of variable speed in wind turbines.

Gipe explains that in remote locations, where the power grid is not accessible and the first few units of electric energy may be very valuable, dc generation with storage and/or wind and diesel "village power systems" have been used. These systems are now being optimized to supply stable, constant-frequency ac electric energy.

Power in the Wind Since wind is air in motion, the power in wind can be expressed as

$$P_w = \tfrac{1}{2}\rho V^3 A \tag{9.1.5}$$

where P_w = power, W; A = reference area, m²; V = wind speed, m/s; ρ = air density, kg/m³. Since V appears to the third power, the wind speed is clearly very important. Figure 9.1.7 is a map of the United States showing regions of annual average available wind power.

The wind speed at a location is random; thus it can be modeled as a continuous random variable in terms of a density function $f(v)$ or a distribution function $F(v)$. The Weibull distribution is commonly used to model wind:

$$F(v) = 1 - \exp\left[-(v/\alpha)^\beta\right] \tag{9.1.6}$$

$$f(v) = \beta(v^{\beta-1}/\alpha^\beta) \exp\left[-(v/\alpha)^\beta\right] \tag{9.1.7}$$

In Eqs. (9.1.6) and (9.1.7), α and β are two parameters which can be adjusted to fit available data over the study period, typically one month. They can be calculated from the sample mean m_v and the sample variance σ_v^2 using the following equations:

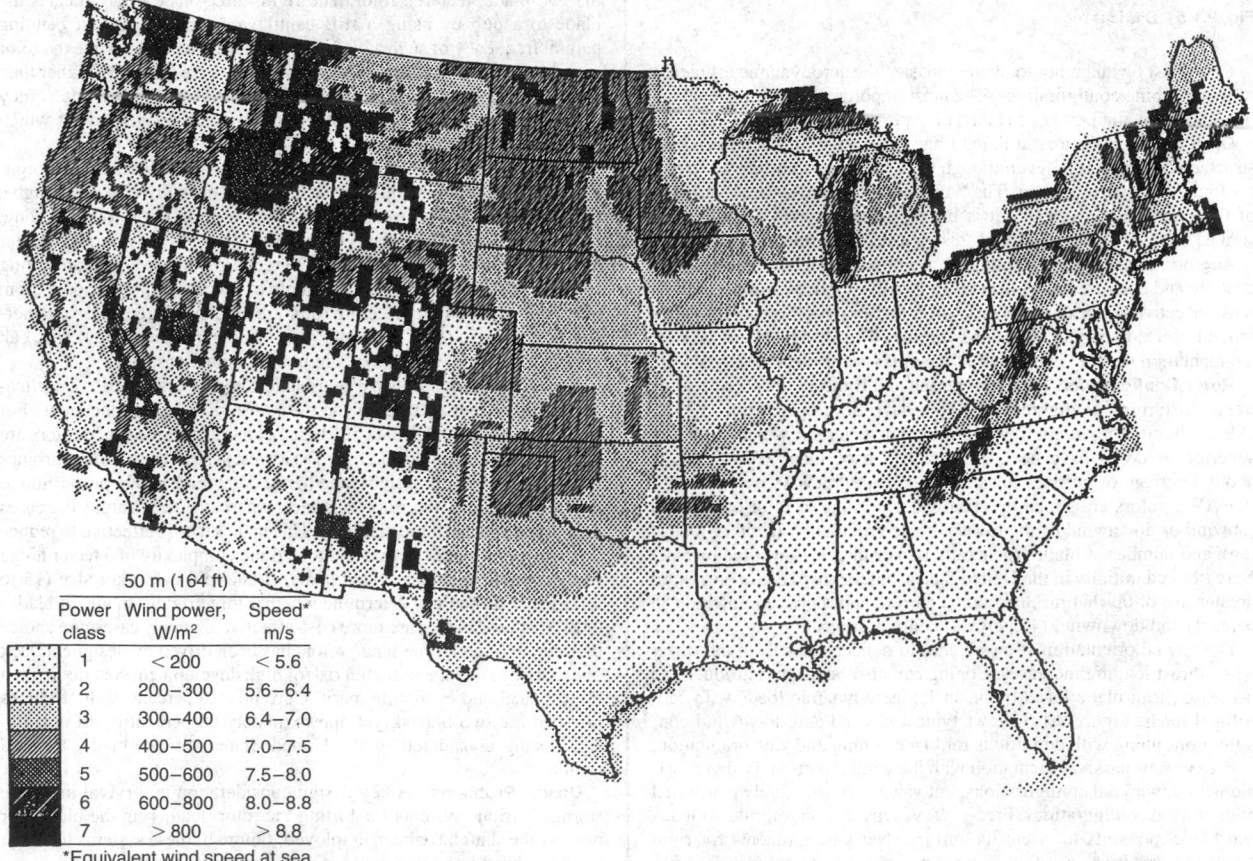

50 m (164 ft)

Power class	Wind power, W/m²	Speed* m/s
1	< 200	< 5.6
2	200–300	5.6–6.4
3	300–400	6.4–7.0
4	400–500	7.0–7.5
5	500–600	7.5–8.0
6	600–800	8.0–8.8
7	> 800	> 8.8

*Equivalent wind speed at sea level for a Rayleigh distribution.

Fig. 9.1.7 Gridded map of annual average wind energy resource estimates in the contiguous United States. Grid cells are ¼° latitude by ⅓° longitude.

$$m_v = \alpha\Gamma(1 + 1/\beta) \tag{9.1.8}$$
$$(\sigma_v/m_v)^2 = \Gamma(1 + 2/\beta)/\Gamma^2(1 + 1/\beta) - 1 \tag{9.1.9}$$

Typically, the sample mean is the only piece of information readily available for many potential sites. In such cases, a knowledge of the variability of the wind speed can be used to select an appropriate value for β, which can be used in (9.1.8) to obtain α. A good compromise value for β is about 4 for wind regimes with low variances.

In addition to α and β, several other parameters are used to characterize wind regimes. Some of the important ones are listed below.

$$\text{Mean cubed wind speed} = \langle v^3\rangle$$
$$= \int_0^\infty v^3 f(v)\,dv \tag{9.1.10}$$
$$\text{Cube factor } K_c = (\langle v^3\rangle)^{1/3}/m_v \tag{9.1.11}$$
$$= \alpha^3\Gamma(1 + 3/\beta)$$
$$\text{for Weibull model}$$
$$\text{Average power density} = P_{av} = \tfrac{1}{2}\rho\langle v^3\rangle \quad \text{W/m}^2 \tag{9.1.12}$$
$$\text{Energy pattern factor} = K_{ep} = \langle v^3\rangle/m_v^3 = K_c^3 \tag{9.1.13}$$

Values of K_{ep} range from 1.5 to 3 for typical wind regimes.

The annual average available wind power for the contiguous United States is shown in Fig. 9.1.7. The values shown must be regarded as averages over large areas. The possibility of finding small pockets of sites with excellent wind regimes because of special terrain anywhere in the country should not be overlooked. The variability of the wind can also be shown in terms of a speed duration curve. Figure 9.1.8 shows the wind speed duration curve for Plum Brook, OH, for 1972.

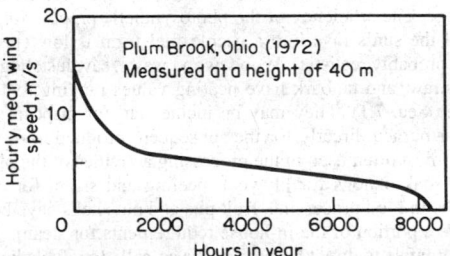

Fig. 9.1.8 Wind variability at Plum Brook, OH (1972).

Wind speed varies with the height above ground level (Fig. 9.1.9). Anemometers are usually located at a height of 10 m above ground level. The long-term average wind speed at height h above ground can be expressed in terms of the average wind speed at 10-m height using a one-seventh power law:

$$(v/v_{10\,m}) = (h/10)^{1/7} \tag{9.1.14}$$

The power $\tfrac{1}{7}$ in the power law equation above depends on surface roughness and other terrain-related factors and can range from 0.1 to 0.3. The value of $\tfrac{1}{7}$ should be regarded as a compromise value in the absence of other information regarding the terrain. Clearly, it is advantageous to construct an adequately high support tower for a wind energy conversion system.

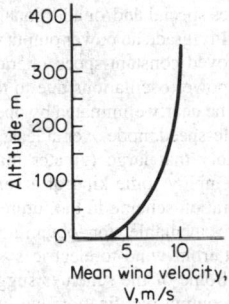

Fig. 9.1.9 Typical variation of mean wind velocity with height.

Table 9.1.2 shows average and peak wind velocities at locations within the continental United States.

Wind to Electric Power Conversion The ease with which wind energy can be converted to rotary mechanical energy and the maturity of electromechanical energy converters and solid-state power conditioning equipment clearly point to wind-to-electric conversion as the most promising approach to harnessing wind power in usable form.

The electric power output of a wind-to-electric conversion system can be expressed as

$$P_e = \tfrac{1}{2}\rho\eta_g\eta_m\eta_p A C_p V^3 \tag{9.1.15}$$

Table 9.1.2 Wind Velocities in the United States

Station	Avg velocity, mi/h	Prevailing direction	Fastest mile	Station	Avg velocity, mi/h	Prevailing direction	Fastest mile
Albany, N.Y.	9.0	S	71	Louisville, Ky.	8.7	S	68
Albuquerque, N.M.	8.8	SE	90	Memphis, Tenn.	9.9	S	57
Atlanta, Ga.	9.8	NW	70	Miami, Fla.	12.6	—	132
Boise, Idaho	9.6	SE	61	Minneapolis, Minn.	11.2	SE	92
Boston, Mass.	11.8	SW	87	Mt. Washington, N.H.	36.9	W	150
Bismarck, N. Dak.	10.8	NW	72	New Orleans, La.	7.7	—	98
Buffalo, N.Y.	14.6	SW	91	New York, N.Y.	14.6	NW	113
Burlington, Vt.	10.1	S	72	Oklahoma City, Okla.	14.6	SSE	87
Chattanooga, Tenn.	6.7	—	82	Omaha, Neb.	9.5	SSE	109
Cheyenne, Wyo.	11.5	W	75	Pensacola, Fla.	10.1	NE	114
Chicago, Ill.	10.7	SSW	87	Philadelphia, Pa.	10.1	NW	88
Cincinnati, Ohio	7.5	SW	49	Pittsburgh, Pa.	10.4	WSW	73
Cleveland, Ohio	12.7	S	78	Portland, Maine	8.4	N	76
Denver, Colo.	7.5	S	65	Portland, Ore.	6.8	NW	57
Des Moines, Iowa	10.1	NW	76	Rochester, N.Y.	9.1	SW	73
Detroit, Mich.	10.6	NW	95	St. Louis, Mo.	11.0	S	91
Duluth, Minn.	12.4	NW	75	Salt Lake City, Utah	8.8	SE	71
El Paso, Tex.	9.3	N	70	San Diego, Calif.	6.4	WNW	53
Galveston, Tex.	10.8	—	91	San Francisco, Calif.	10.5	WNW	62
Helena, Mont.	7.9	W	73	Savannah, Ga.	9.0	NNE	90
Kansas City, Mo.	10.0	SSW	72	Spokane, Wash.	6.7	SSW	56
Knoxville, Tenn.	6.7	NE	71	Washington, D.C.	7.1	NW	62

U.S. Weather Bureau records of the average wind velocity, and fastest mile, at selected stations. The period of record ranges from 6 to 84 years, ending 1954. No correction for height of station above ground.

where P_e = electric power output, W; η_g and η_m = efficiencies of the electric generator and the mechanical interface, respectively; η_ρ = efficiency of the power conditioning equipment (if employed). The product of these efficiencies and the coefficient of performance (Fig. 9.1.2) usually will be in the range of 20 to 35 percent.

The electrical equipment needed for wind-to-electric conversion depends, above all, on whether the aeroturbine is operated in the constant-speed, nearly constant-speed, or variable-speed mode. With constant-speed and nearly constant-speed operation, the power coefficient C_p in Eq. (9.1.15) becomes a function of wind speed. If variable-speed mode is used, it is possible to operate the turbine at a constant optimum C_p over a range of wind speeds, thus extracting a larger fraction of the energy in the wind.

Synchronous and induction generators are ideally suited for constant-speed and nearly constant-speed operation, respectively. Variable-speed operation requires special and/or additional electrical hardware if constant-frequency utility-grade ac power output is desired. Most of the early prototypes employed constant-speed operation and synchronous generators. However, power oscillations due to tower interference and wind shear effects can be nearly eliminated by operating the turbine and the generator in variable-speed mode over at least some limited range of speeds. It appears likely that large (greater than 100-kW) wind-to-electric systems may employ some kind of a variable-speed constant-frequency power generation scheme in the future.

Several options are available for obtaining constant frequency utility-grade ac output from wind-to-electric systems operated in the variable-speed mode. Some of the schemes suggested are: permanent magnet alternator with output rectification and inversion, dc generator feeding a line commutated (synchronous) or force commutated inverter, ac commutator generator, ac-dc-ac link, field modulated generator system, and slip ring induction machines operated as generators with rotor power conditioning. The last type is also known as a double output induction generator or simply a doubly fed machine. In general, the simpler the electrical generation scheme, the poorer the quality of the constant-frequency ac output. For example, synchronous inverters are very simple, are economical, and have been popular in small (less than 50 kW) commercial units; however, they have power quality and harmonic injection problems, and they absorb (on the average) more reactive voltamperes from the utility line than the watts they deliver. The latter is also a problem with simple induction generators. Schemes such as the field modulated generator system and doubly fed machine deliver excellent power quality, but at a higher cost for the hardware.

Economics Costs of wind energy systems are often divided into two categories: annualized fixed costs and operation and maintenance (O&M) costs. Annualized fixed costs are comprised primarily of the cost of capital required to purchase and install the turbines. In addition, they include certain fixed costs such as taxes and insurance. O&M costs include scheduled and unscheduled maintenance and the levelized cost for major equipment overhauls.

The initial capital cost of a wind turbine system includes the cost of the turbine, installation, and balance of plant. Turbine costs are often expressed in terms of nameplate rating ($/kW). In 1995, utility-grade turbines cost on the order of $800 per kilowatt. Installation and balance of plant costs add approximately 20 percent. The cost of capital varies, but (in 1995) was estimated at 8 percent per annum for wind energy projects. Other fixed costs were estimated at around 3 percent of the installed turbine cost. The **fixed charge rate (FCR)**, combined capital and other fixed costs, was approximately 11 percent per annum. O&M costs in modern wind farms are around $0.01 per kilowatthour (1995).

In addition to capital and O&M costs, an economic assessment of wind energy systems must account for system performance. A commonly used parameter that describes the production of useful energy by wind and other energy systems is the **capacity factor C,** also called the **plant factor** or **load factor.** It is the ratio of the annual energy produced (AEP) to the energy that would be produced if the turbine operated at full-rated output throughout the year:

$$C_f = \frac{\text{AEP}}{8,760\,P_R} \qquad (9.1.16)$$

where AEP is in kWh, 8,760 is the number of hours in 1 year, and P_R is the unit's nameplate rating in kW.

In order of decreasing importance, C_f is affected by the average power available in the wind, speed vs. duration curve of the wind regime, efficiency of the turbine, and reliability of the turbine. Variable-speed turbines which tend to have low cut-in speeds and high efficiency in low winds exhibit better capacity factors than constant-speed turbines. Modern utility-grade turbines at good sites (class 4) can achieve capacity factors in the range of 25 to 30 percent.

The combination of cost and performance can be used to calculate the cost of energy (COE) as follows:

$$\text{COE} = \frac{\text{FCR} \times \text{ICC}}{8,760\,C_f} + (\text{O\&M}) \qquad (9.1.17)$$

where FCR is the fixed charge rate for the cost of capital and for other fixed charges such as taxes and insurance, ICC is the installed capital cost of the turbine and balance of plant in dollars per kilowatt. This method is useful to estimate the cost of energy for different technologies or sites. However, for investment decisions, more detailed analyses that include the effects of various investment strategies, tax incentives, and environmental factors should be performed. Ramakumar et al. discuss the economic aspects of advanced energy technologies, including wind energy systems.

POWER FROM VEGETATION AND WOOD
Staff Contribution

Vegetation offers, by photosynthesis, a natural process for the storage of solar energy. The efficiency of the photosynthetic process for the conversion of the sun's rays into a usable fuel form is low (less than 2 percent is probably realistic). Wood, wood waste, sawdust, hogged fuel, bagasse, straw, and tanbark have heating values ranging to 10,000 ± Btu/lb (see Sec. 7.1). They may be incinerated for disposal as waste material or burned directly for the subsequent production of steam or hot water, most often used in the processing activities of the plant, e.g., hot water soak of logs for plywood peeling and steam for drying in paper mills. In food processing, fruit pits and nut shells have been used to generate a portion of the in-house requirements for steam.

The alternative to direct burning of the so-called biofuels lies in their possible conversion to gaseous fuel by gasification at high temperature in the presence of air. Pyrolitic treatment can render biofuels to fractions of liquids and gases that have thermal value. In both cases, the solid residue remaining also has some thermal value which can be utilized in normal combustion.

Tree farming, with controlled growth and cutting, proposes to balance harvesting plans to load demands; e.g., Szego and Kemp (*Chem. Tech.,* May 1973) project a 400-mi^2 "energy plantation" to serve a 400-MW steam electric plant. Such proposals would utilize proved steam power plant cycles and equipment for novel breeding, growing, harvesting, preparation, and combustion of vegetation. (See also Sec. 7.1.)

The **photosynthesis process** is basic to all agricultural practice. The human animal has long known how to convert grain to alcohol. It can be said that as long as we can grow green stuff we should be able to harness some of the sun's energy. The prohibition era in the United States saw many efforts to use the alcohol production capacity of the nation to offer alcohol as a substitute or supplementary fuel for internal combustion engines. Ethanol (C_2H_5OH) and methanol (CH_3OH) have properties that are basically attractive for internal combustion engines, to wit, smokeless combustion, high volatility, high octane ratings, high compression ratios ($R_v > 10$). Heating values are 9,600 Btu/lb for methanol and 12,800 Btu/lb for ethanol. On a volume basis these translate, respectively, to 63,000 and 85,000 Btu/gal for methanol and ethanol. Gasoline, by comparison, has 126,000 Btu/gal (20,700 Btu/lb). (See Sec. 7 for values.) The blending of ethanol and methanol with gasolines (9 ± gasoline to 1 ± alcohol) has been used particularly in Europe since the 1930s as a suitable internal combustion engine fuel. The miscibility of the lighter alcohols with water and gasoline introduces corrosion problems for engine parts and lowers the octane number. Higher-carbon alcohols (e.g., butyl) which are immiscible with water are possible blending substitutes, but their availability and cost are not presently attractive. Such properties as flash point would introduce further problems. While these constitute some of the unsolved technical problems, the basic principle of harnessing the sun's energy through vegetation will continue as a provocative challenge not only in the field of power generation but also as a solution for the perennial farm problem of waste disposal.

SOLAR ENERGY
Erich A. Farber

REFERENCES: Daniels, ''Direct Use of the Sun's Energy,'' Yale. ASHRAE, ''Solar Energy Use for Heating and Cooling of Buildings.'' Duffie and Beckman, ''Solar Engineering of Therma Processes,'' Wiley. Lunde, ''Solar Thermal Engineering,'' McGraw-Hill. Kreider and Kreith, ''Solar Heating and Cooling,'' McGraw-Hill. ASHRAE, ''Handbook of Fundamentals, 1993,'' ''Handbook of Applications, 1982.'' Yellott, Selective Reflectance, *Trans. ASHRAE*, 69, 1963. Hay, Natural Air Conditioning with Roof Ponds and Movable Insulation, *Trans. ASHRAE*, 75, part I, p. 165, 1969. Backus, ''Solar Cells,'' IEEE. Bliss, Atmospheric Radiation near the Surface of the Ground, *Solar Energy*, 5, no. 3, 1961. Yellott, Passive and Hybrid Cooling Research, ''Advances in Solar Energy— 1982,'' American Solar Energy Soc., Boulder. Haddock, Solar Energy Collection, Concentration, and Thermal Conversion—A Review, *Energy Sources*, 13, 1991, pp. 461–482. ''Proceedings of the Intersociety Energy Conversion Conferences,'' annual (with the 29th in 1994), The University of Florida, ''Erich A. Farber Archives.'' ''European Directory of Renewable Energy, Suppliers and Services 1994,'' James & James Ltd., London, England.

Notation

A, R, T = subscripts denoting absorbed, reflected, and transmitted solar radiation
C = concentration ratio
c = subscript denoting collector cover
c_p = specific heat of fluid, Btu/(lb·°F)
I_{DN} = direct normal solar intensity, Btu/(ft²·h)
I_d = diffuse radiation, Btu/(ft²·h)
I_o = radiation intensity beyond earth's atmosphere, Btu/(ft²·h)
I_r = reflected solar radiation, Btu/(ft²·h)
I_{SC} = solar constant; normal incidence intensity at average earth-sun distance, Btu/(ft²·h)
I_t = total solar radiation, Btu/(ft²·h)
L = latitude, deg
m = air mass
o, i = subscripts denoting outgoing and incoming fluid conditions
q = rate of heat flux, Btu/(ft²·h)
q_I = heat flow through insulation, Btu/(ft²·h)
T_p = temperature of absorbing surface, °R
U = overall coefficient of heat transfer, Btu/(ft²·h·°F)
w_f = flow rate of collecting fluid, lb/(h·ft²)
ϕ = solar azimuth, deg from south
α, ρ, τ = absorptance, reflectance, and transmittance for solar radiation
β = solar altitude, deg
δ = solar declination, deg
ε = emittance for long-wave radiation
γ = wall-solar azimuth, deg
λ = unit of wavelength, μm
Σ = angle of tilt from horizontal, deg
θ = incident angle, deg, from perpendicular to surface

Introduction and Scope

The sun exerts forces upon the earth and radiates solar energy produced within the sun by nuclear fusion. A small fraction of that energy is intercepted by the earth and is converted by nature to heat, winds, ocean currents, waves, tides; makes plants grow, some of which over millions of years produced fossil fuels (oil, coal, and gas); and creates biomass which can be burned to generate heat and/or power. Solar energy is implicit in many subject areas treated elsewhere in the Handbook; only the more direct uses such as water heating, space heating and cooling, swimming pool heating, solar distillation, solar drying and cooking, solar furnaces, solar engines, solar electricity generation, and solar assisted transportation will be treated here. The total field is widely termed *alternative or renewable sources of energy and their conversion*.

Solar Energy Utilization Solar energy reaches the earth's surface as shortwave electromagnetic radiation in the wavelength band between 0.3 and 3.0 μm; its peak spectral sensitivity occurs at 0.48 μm (Fig. 9.1.10). Total solar radiation intensity on a horizontal surface at sea level varies from zero at sunrise and sunset to a noon maximum which can reach 340 Btu/(ft²·h) (1,070 W/m²) on clear summer days. This inexhaustible source of energy, despite its variability in magnitude and

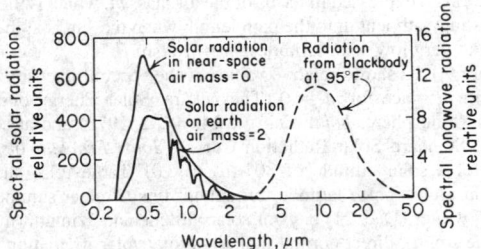

Fig. 9.1.10 Spectral distribution of solar radiation and radiation emitted by blackbody at 95°F (35°C).

direction, can be used in three major processes (Daniels, ''Direct Use of the Sun's Energy,'' Yale; Zarem and Erway, ''Introduction to the Utilization of Solar Energy,'' McGraw-Hill): (1) **Heliothermal,** in which the sun's radiation is absorbed and converted into heat which can then be used for many purposes, such as evaporating seawater to produce salt or distilling it into potable water; heating domestic hot water supplies; house heating by warm air or hot water; cooling by absorption refrigeration; cooking; generating electricity by vapor cycles and thermoelectric processes; attaining temperatures as high as 6,500°F (3,600°C) in solar furnaces. (2) **Heliochemical,** in which the shorter wavelengths can cause chemical reactions, can sustain growth of plants and animals, can convert carbon dioxide to oxygen by photosynthesis, can cause degradation and fading of fabrics, plastics, and paint, can be used to detoxify toxic waste, and can increase the rate of chemical reactions. (3) **Helioelectrical,** in which part of the energy between 0.33 and 1.3 μm can be converted directly to electricity by photovoltaic cells. Silicon solar batteries have become the standard power sources for communication satellites, orbiting laboratories, and space probes. Their use for terrestrial power generation is currently under intensive study, with primary emphasis upon cost reduction. Other methods include thermoelectric, thermionic, and photoelectromagnetic processes and the use of very small antennas in arrays for the conversion of solar energy to electricity (Antenna Solar Energy to Electricity Converter/ASETEC, Air Force Report, AF C FO 8635-83-C-0136, Task 85-6, Nov. 1988).

Solar Radiation Intensity In space at the average earth-sun distance, 92.957 million mi (150 million km), solar radiation intensity on a surface normal to the sun's rays is 434.6 ± 1 Btu/(ft²·h) (1,370 ± 3 W/m²). This quantity, called the solar constant I_{SC}, undergoes small (±1 percent) periodic variations which affect primarily the shortwave portion of the spectrum (Abbott, in Moon, Standard Polar Radiation Curves, *Jour. Franklin Inst.*, Nov. 1940). Recent measurements using satellites give essentially the same results. Since the earth-sun distance varies throughout the year, the intensity beyond the earth's atmosphere I_o also varies by ± 3.3 percent (Table 9.1.3). The great seasonal variations in terrestrial solar radiation intensity are due to the earth's tilted axis, which causes the solar declination δ (the angle between the earth's equatorial plane and earth-sun line) to change from 0° on Mar. 21 and Sept. 21 to −23.5° on Dec. 21 and +23.5° on June 21.

In passing through the earth's atmosphere, the sun's radiation is partially and selectively absorbed, scattered, and reflected by water vapor

Table 9.1.3 Annual Variation in Solar Declination and Solar Radiation Intensity beyond the Earth's Atmosphere

Date	Jan. 1	Feb. 1 Nov. 10	Mar. 1 Oct. 13	Apr. 1 Sept. 12	May 1 Aug. 12	June 1 July 12	July 1
Declination, deg	−23.0	−17.1	−7.7	+4.4	+15.0	+22.0	+23.1
Ratio, I_o/I_{sc}	1.033	1.029	1.017	1.000	0.983	0.971	0.967
Intensity I_o, Btu/(ft² · h) (W/m²)	449 1,416	447 1,409	442 1,393	435 1,370	427 1,347	422 1,331	420 1,324

and ozone, air molecules, natural dust, clouds, and artificial pollutants. Some of the scattered and reflected energy reaches the earth as diffuse or sky radiation I_d.

The intensity of the direct normal radiation I_{DN} depends upon the clarity and the amount of precipitable moisture in the atmosphere and the length of the atmospheric path, which is determined by the solar altitude β and expressed in terms of the air mass m, which is the ratio of the existing path length to the path length when the sun is at the zenith. Except at very low solar altitudes, $m = 1.0/\sin \beta$.

Figure 9.1.10 shows relative values of the spectral intensity of solar radiation in space for $m = 0$ (Thekaekara, Solar Energy outside the Earth's Atmosphere, *Solar Energy,* 14, no. 2, 1973) and at sea level (Moon, Standard Solar Radiation Curves, *Jour. Franklin Inst.,* Nov. 1940) for a solar altitude of 30° ($m = 2.0$). Table 9.1.4 shows the variation at 40° north latitude throughout typical clear summer (June 21) and winter (Dec. 21) days of solar altitude and azimuth (measured from the south), direct normal radiation, total solar irradiation of horizontal and vertical south-facing surfaces.

The total solar irradiation reaching a terrestrial surface is the sum of the direct, diffuse, and reflected components: $I_t = I_{DN} \cos \theta + I_d + I_r$, where θ is the incident angle between the sun's rays and a line perpendicular to the receiving surface and, I_r is the shortwave radiation reflected from adjacent surfaces.

Direct beam solar radiation intensity is measured by **pyroheliometers** with collimating tubes to exclude all but the direct rays from their sensors, which may use calorimetric, thermoelectric, or photovoltaic means to produce a response proportional to the irradiation rate. Similar but uncollimated instruments called **pyranometers** are used to measure the total radiation from sun and sky; when their sensors are shaded from the sun's direct rays, they also can measure the diffuse component.

Incident Angle Determination The incident angle θ affects both the direct solar intensity and the solar optical properties of the irradiated surface. For a flat surface tilted at an angle Σ from the horizontal, $\cos \theta = \cos \beta \cos \gamma \sin \Sigma + \sin \beta \cos \Sigma$. For vertical surfaces, $\Sigma = 90°$; so $\cos \theta = \cos \beta \cos \gamma$; for horizontal surfaces, $\Sigma = 0°$ and $\theta = 90° − \beta$. (See ASHRAE, "Handbook of Fundamentals," for values of solar altitude, azimuth, and direct normal radiation throughout the year for 0 to 56° north latitude.)

Solar Optical Properties of Transparent Materials When solar radiation with total intensity I_t falls on a transparent material, part of the energy is reflected, part is absorbed, and the remainder is transmitted. At any instant,

$$I_t = q_\tau + q_A + q_R = I_t(\tau + \alpha + \rho)$$

The sum of the solar optical properties τ, α, and ρ must equal unity, but the individual values depend upon the incident angle and wavelength of the radiation, the composition of the material, and the nature of any coatings which may be applied to the surfaces.

For uncoated single-strength (³⁄₃₂-in or 2.4-mm) clear window glass (Fig. 9.1.11), solar transmittance at normal incidence ($\theta = 0°$) is approximately 0.90, but the transmittance for longwave thermal radiation (5 μm) is virtually zero. Thus glass acts as a "heat trap" by admitting solar radiation readily but retaining most of the heat produced by the absorbed sunshine. This "greenhouse effect," which is also exhibited but to a lesser degree by some plastic films (see Whillier, Plastic Covers for Solar Collectors, *Solar Energy,* 7, no. 3, 1964), is the basis for most heliothermal processes. Heat absorbing glass [¼ in (6.3 mm) thick (Fig. 9.1.11)], which absorbs more than 50 percent of the incident solar radiation, is widely used by architects to reduce the heat and glare admitted through unshaded windows. Reflective coatings (Yellott, Selective Reflectance, *Trans. ASHRAE,* 69, 1963) have been developed to serve similar purposes.

For all types of glass, transmittance falls and reflectance rises as θ exceeds about 30°. Absorptance increases somewhat owing to the increased path length and then drops off sharply toward zero as θ exceeds 60°.

Absorptance and Emittance of Opaque Surfaces Opaque materials absorb or reflect all the incident sunshine. The absorptance α for solar radiation and the emittance ε for longwave radiation at the temperature of the receiving surface are particularly important in heliotechnology. For a true blackbody, the absorptance and emittance are equal and do not change with wavelength. Most real surfaces have reflectances and absorptances which vary with wavelength (Fig. 9.1.12). Aluminum foil has a consistently low absorptance and high reflectance over the entire spectrum from 0.25 to 25 μm, while black paint has a high absorptance and low reflectance. White paint, however, has low

Table 9.1.4 Solar Altitude and Azimuth, Direct Normal Radiation, and Total Solar Radiation on Horizontal and Vertical South-Facing Surfaces, June 21 and Dec. 21, for 40° North Latitude

June 21, declination = +23.45°							
Time: A.M.; P.M.	6; 6	7; 5	8; 4	9; 3	10; 2	11; 1	12; 12
Solar altitude, deg	14.8	26.0	37.4	48.8	59.8	69.2	73.5
Solar azimuth, deg	108.4	99.7	90.7	80.2	65.8	41.9	0.0
Direct normal irradiation, Btu/(ft² · h)	154	215	246	262	272	276	278
Total irradiation, Btu/(ft² · h)							
On horizontal surface	60	123	182	233	273	296	304
On vertical south surface	10	14	16	47	74	92	98
Dec. 21, declination = −23.45°							
Time: A.M.; P.M.	6; 6	7; 5	8; 4	9; 3	10; 2	11; 1	12; 12
Solar altitude, deg			5.5	14.0	20.7	25.0	26.6
Solar azimuth, deg			53.0	41.9	29.4	15.2	0.0
Direct normal irradiation, Btu/(ft² · h)			88	217	261	279	284
Total irradiation, Btu/(ft² · h)							
On horizontal surface			14	65	107	119	143
On vertical south surface			56	163	221	252	263

Values adapted from ASHRAE, "Handbook of Applications," 1982.

shortwave (solar) absorptance, but beyond 3 μm its absorptance and reflectance are virtually the same as for black paint.

Solar collectors require a high α/ε ratio, while surfaces which should remain cool, such as rooftops or space vehicles, should have low ratios

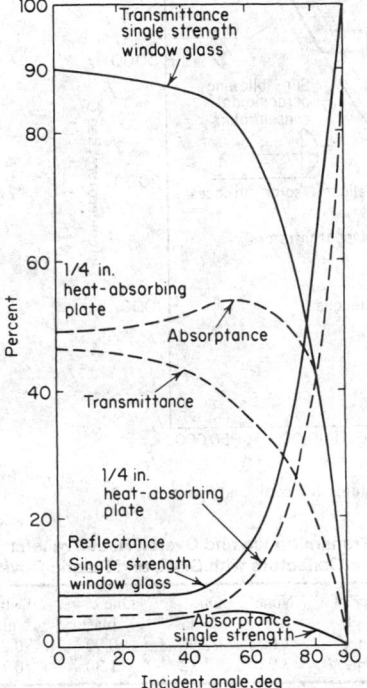

Fig. 9.1.11 Variation with incident angle of solar optical properties of ³⁄₃₂-in (2.4-mm) clear glass and ¼-in (6.3-mm) heat-absorbing glass.

since their objective usually is to absorb as little solar radiation and emit as much longwave radiation as possible. Special surface treatments have been developed (see ASHRAE, "Solar Energy Use for Heating and Cooling of Buildings," 1977) for which the ratio α/ε is above 7.0, making them suitable for solar collectors; others with ratios as low as 0.15 are useful as heat rejectors for space applications (see Table 9.1.5). In addition, the absorptance can be changed and controlled by paint

Table 9.1.5 Solar Absorptance, Longwave Emittance, and Radiation Ratio for Typical Surfaces

Surface or material	Shortwave (solar) absorptance α	Longwave emittance ε	Radiation ratio, α/ε
Flat, oil-based paints:			
Black	0.90	0.90	1.00
Red	0.74	0.90	0.82
Green	0.50	0.90	0.55
Aluminum	0.45	0.90	0.50
White	0.25	0.90	0.28
Whitewash on galvanized iron	0.22	0.90	0.25
Building materials:			
Asbestos slate	0.81	0.96	0.84
Tar paper, black	0.93	0.93	1.00
Brick, red	0.55	0.92	0.59
Concrete	0.60	0.88	0.68
Sand, dry	0.82	0.90	0.92
Glass	0.04–0.70	0.84	
Metals:			
Copper, polished	0.18	0.04	4.50
Copper, oxidized	0.64	0.60–0.90	1.03–0.71
Aluminum, polished	0.30	0.05	6.00
Selective surfaces:			
Tabor, electrolytic	0.90	0.12	7.50
Silicon cell, uncoated	0.94	0.30	3.13
Black cupric oxide on copper	0.91	0.16	5.67

composition (grain material and size, binder, thickness etc.) and surface configuration, both large and small.

Equilibrium Temperatures for Concentrating Collectors When a surface is irradiated, its temperature rises until the rate of solar radiation absorption equals the rate at which heat is removed from the surface. If no heat is intentionally removed, the maximum temperature which can be attained by a blackbody ($\alpha = \varepsilon$) is found from $I_{DN}C\alpha = 0.1713\varepsilon(T_p/100)^4$, where C is the concentration ratio. Figure 9.1.13 shows the variation of blackbody equilibrium temperatures for earth and near space where $I_{DN} = 320$ and $I_o = 435$ Btu/ft²·h (1,000 and 1,370 W/m²).

For flat plate collectors, $C = 1.0$; so their maximum attainable temperatures are below 212°F (100°C) unless a selective surface is used with $\alpha/\varepsilon > 1.0$, or both radiation and convection loss are suppressed by the use of multiple glass cover plates. Only the direct component of the total solar radiation can be concentrated, and concentrating collectors must follow the sun's apparent motion across the sky or use heliostats

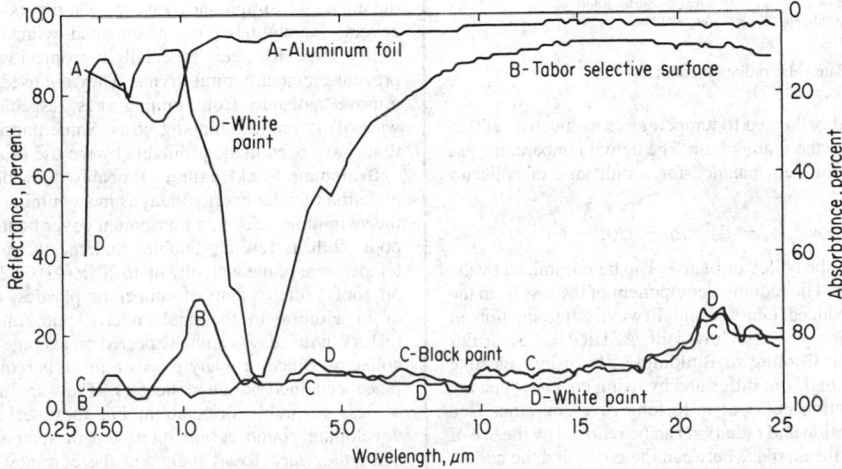

Fig. 9.1.12 Variation with wavelength of reflectance and absorptance for opaque surfaces.

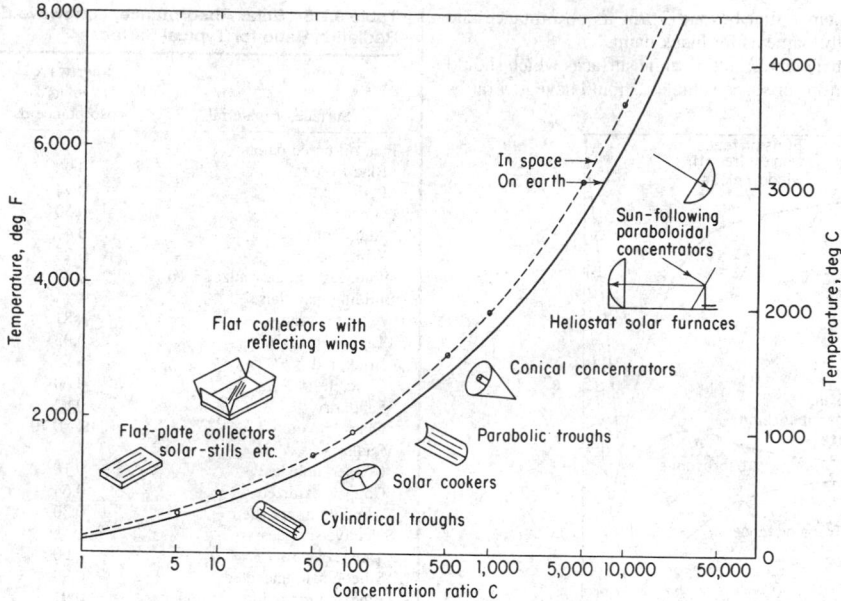

Fig. 9.1.13 Variation with concentration ratio of equilibrium temperatures for earth and space.

which serve the same function. Diffuse radiation cannot be concentrated effectively.

Flat-Plate Collectors Direct, diffuse, and reflected solar radiation can be collected and converted into heat by flat-plate collectors (Fig. 9.1.14). These generally use blackened metal plates which are finned, tubed, or otherwise provided with passages through which water, air, or

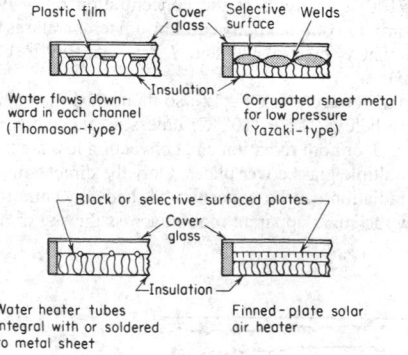

Fig. 9.1.14 Typical flat-plate solar radiation collectors.

other fluids may flow and be heated to temperatures as much as 100 to 150°F (55 to 86°C) above the ambient air. The actual temperature rise may be estimated from the heat balance for a unit area of collector surface:

$$q_A = I_t \tau_c \alpha_p = w_f c_p (t_o - T_i) \div q_I + U(t_p - t_a)$$

The loss from the back of the collector plate q_I can be minimized by the use of adequate insulation. The radiation component of the loss from the upper surface can be reduced (Zarem and Erway, "Introduction to the Utilization of Solar Energy," McGraw-Hill; ASHRAE, "Solar Energy Use for Heating and Cooling of Buildings") by using selective reflectance coatings with high α/ε ratios and by using covers which are transparent to solar radiation but opaque to longwave emissions (see Table 9.1.6). Both convection and radiation can be reduced by the use of honeycomb structures in the airspace between the cover and the collector plate (see Hollands, Honeycomb Devices in Flat-Plate Solar Collec-

Table 9.1.6 Transmittance and Overall Heat-Transfer Coefficients for Collectors with Glass and Plastic Covers

Type and number of covers	None	One glass	One plastic	Two glass	Two plastic
Solar transmittance	1.00	0.90	0.92	0.81	0.85
Overall coefficient U	3.90	1.12	1.30	0.71	0.87

tors, *Solar Energy,* **9**, no. 3, 1965). For the transparent covers, glass is best since it lets through the shortwave solar radiation but stops the longwave radiation given off by the collector plate. Plastics do not have this trapping characteristic, do have a shorter life, may lose their transparency, and outgas when heated. The fumes may condense on other surfaces, forming a film to reduce the collector performance.

Applications of Heliotechnology

Solar Drying Probably the largest use of solar energy over the centuries—the drying of agricultural crops, evaporation of ocean and salt lake ponds for salt production, etc.—has led more recently to more efficient dryers. The newer dryers prevent rain and dew from rewetting the materials. Simple inexpensive solar dryers—essentially transparent covers—sometimes are supplemented by air heaters providing hot air to dry crops, fish, etc., especially in tropical regions where daily rains prevent efficient natural drying. They are used for curing wood and to remove moisture from mining ores (especially if they have been washed) to reduce shipping costs. Some uneconomical mining operations have been made profitable by the use of solar drying.

Swimming Pool Heating Probably the widest U.S. commercial application of solar energy today is in swimming pool heating. To extend the swimming season, a transparent cover floating on the surface of the pool, with as few air bubbles underneath as possible, will raise the temperature of the water by up to 20°F (11°C). Flat-plate type of heaters on roofs, often made of rubber or plastics, can be used instead of or in addition to the pool cover. Approximately 1,000 Btu/ft² · day (3.1 kWh/m² · day) can be expected on average from a reasonably good collector. Since in many places a fence is required around a pool, the fence can incorporate collectors. They are not as efficient because of less favorable orientation, but they serve a dual purpose. The developing countries are interested in solar swimming pool heating since they lack fossil fuels and the currency to buy them, but want to attract tourists with modern conveniences.

Solar Ponds If water in ponds or reservoirs contains salts in solution, the warmer layers will have higher concentrations and, being heavier, will sink to the bottom. The hot water on the bottom is insulated against heat losses by the cooler layers above. Heat can be extracted from these ponds for power generation. Large solar ponds have been used in the Mideast along the Dead and Red seas, along the Salton Sea; a number of artificial ponds have been established elsewhere. The inexpensive extraction of heat at over 200°F (94°C) still requires improvement. Solar ponds are, effectively, large inexpensive solar collectors. Their heat can be used to power vapor engines and turbines; these, in turn, drive electric generators.

Solar Stills Covering swimming pools, ponds, or basins with airtight covers (glass or plastic) will condense the water vapors on the underside of the covers. The condensate produced by the solar energy can be collected in troughs as distilled water. Deep-basin stills have a water depth of several feet (between approximately 0.5 and 1.5 m) and require renewal only every few months. Shallow-basin stills have a water depth of about 0.5 to 2.0 in (approximately 1 to 5 cm) and have to be fed and flushed out frequently. Solar stills can be designed to also collect rainwater (in Florida this can double the freshwater production). If the supply water is not contaminated and only too high in solids content, it can be mixed with the distilled water to increase the actual output. This is often done for farm animal water and water used for irrigation.

The glass-covered roof-type solar still (Fig. 9.1.15a) is in wide use in arid areas for the production of drinking water from salty or brackish sources. The sun's rays enter through the cover glasses, warm the water, and thus produce vapor which condenses on the inner surface of the cover. The water droplets coalesce and flow downward into the discharge troughs, while the remaining brine is periodically replaced with a new supply of nonpotable water. Daily yield ranges from 0.4 lb/ft² (2 kg/m²) of water surface in winter to 1.0 lb/ft² (5 kg/m²) in summer (Daniels, "Direct Use of the Sun's Energy," Yale, p. 174).

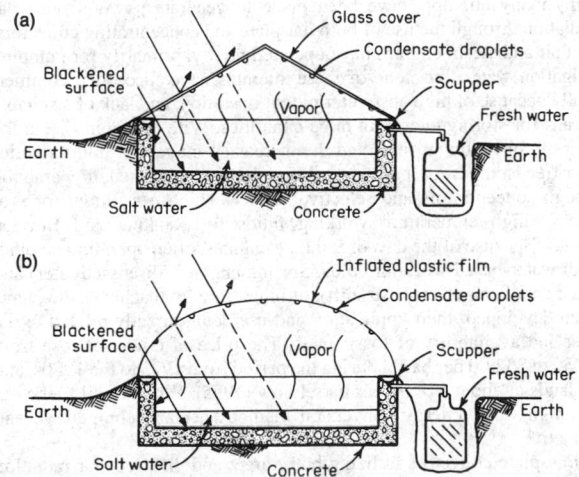

Fig. 9.1.15 Shallow-basin horizontal-surface solar stills. (*a*) Glass-covered roof type; (*b*) inflated plastic type.

Inflated plastic films (Fig. 9.1.15*b*) have also been used to cover solar stills, but their greatest success has been achieved in controlled-environment greenhouses where the vapor which transpires from plant leaves is condensed and reused at the plant roots. Stills made of inflatable plastic also are equipment in survival kits, on lifeboats, etc. Wicks in some of the solar stills can improve their performance. Most plastics have to be surface-treated for this application to produce film condensation (for good solar transmission) rather than dropwise condensation, which reflects a considerable fraction of the impinging solar radiation.

Solar Water Heaters These can be the pan (batch) type—a tank or basin with transparent cover—or tube collector type, described previously. The simple thermosiphon solar water heater (Fig. 9.1.16) with a glass-covered flat-plate collector is used in thousands of homes all over the world, but mainly in Australia, Japan, Israel, North Africa, and Central and South America. Under favorable climatic conditions (abun-

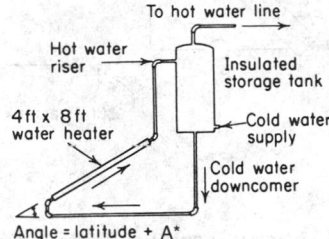

Fig. 9.1.16 Thermosiphon type of water heater. (*The collector angle *A* depends on the latitude. More favorable orientation toward the winter sun when days are shorter can produce the same amount of hot water all year round. In Florida, *A* is 10°.)

dant sunshine and moderate winter temperatures) they can produce 30 to 50 gal (110 to 190 L) of water at temperatures up to 160°F (70°C) in summer and 120°F (50°C) in winter. Auxiliary electric heaters are often used to produce higher temperatures during unfavorable winter weather.

In the United States and Europe, solar collectors are usually placed on the roof, with the hot water storage tank lower. This requires a small circulating pump. The pump is controlled by a timer, a temperature sensor in the collector, or a differential temperature controller. Ideally, the pump runs when heat can be added to the water in the tank.

To protect the system from freezing, the collectors are drained, manually or automatically. The system can also be designed with a primary circuit containing antifreeze and effecting heat transfer with a heat exchanger in the tank, or by use of a double-walled tank. Another method for protection is a dual system, in which the water drains from the collectors when the pump stops. This is the preferred method for large systems.

Auxiliary heaters, usually electric, are used often in solar water heaters to handle overloads and unusually bad weather conditions.

Solar House Heating and Cooling House heating can be accomplished in temperate climates by collecting solar radiation with flat-plate devices (Fig. 9.1.14) mounted on south-facing roofs or walls (in the northern hemisphere). Water or air, warmed by solar radiation, can be used in conventional heating systems, with small auxiliary fuel-burning apparatus available for use during protracted cloudy periods. Excess heat collected during the day can be stored for use at night in insulated tanks of hot water or beds of heated gravel. Heat-of-fusion storage systems, which use salts that melt and freeze at moderate temperatures, may also be used to improve heat storage capacity per unit volume.

Solar air conditioning and refrigeration can be done with absorption systems supplied with moderately high-temperature (200°F or 93°C) working fluids from high-performance good flat-plate collectors. The economics of solar energy utilization for domestic purposes become much more favorable when the same collection and storage apparatus can be used for both summer cooling and winter heating. One such system (see Hay, Natural Air Conditioning with Roof Ponds and Movable Insulation, *Trans. ASHRAE,* **75**, part I, 1969, p. 165) uses a combination of shallow ponds of water on horizontal rooftops with panels of insulation which may be moved readily to cover or uncover the water surfaces. During the winter, the ponds are uncovered during the day to absorb solar radiation and covered at night to retain the absorbed heat. The house is warmed by radiation from metallic ceiling panels which are in thermal contact with the roof ponds. During the summer, the ponds are covered at sunrise to shield them from the daytime sun, and uncovered at sunset to enable them to dissipate heat by radiation, convection, and evaporation to the sky.

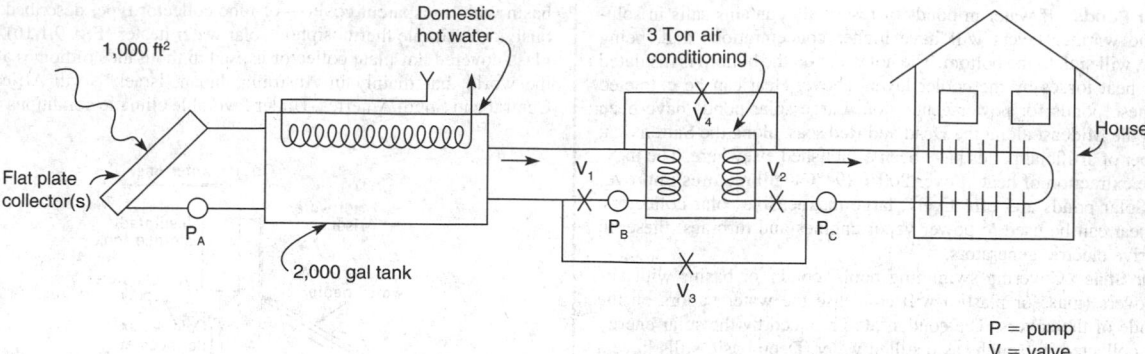

Fig. 9.1.17 Schematic of typical solar house heating, cooling, and hot water system.

Figure 9.1.17 is a schematic of a more typical active solar house heating, cooling, and hot water system. Flat-plate collectors, roughly 1,000 ft² (93 m²) to provide both 3 tons of air conditioning and hot water, properly oriented on the roof (they can actually be the roof), supply hot water to a storage tank (usually buried) of about 2,000 gal (about 7,570-L) capacity. A heat-exchanger coil, submerged in and near the top of the tank, provides domestic hot water. When heat is needed, the thermostat orders the following: Valves V_1 and V_2 close, V_3 and V_4 open, and pump P_C circulates hot water through the house as long as required. When cooling is required, the thermostat orders the following: Valves V_1 and V_2 open, V_3 and V_4 close, and pumps P_B and P_C feed hot water to the air conditioner, which, in turn, produces chilled water for circulation through the house as long as needed. This system can be used over a wide range of latitudes, and only the heating and cooling duty cycles will change (i.e., more heating in the north and more cooling in the south). The proper orientation of the collectors will depend upon the duty cycles. In the northern hemisphere, the collectors will face south. When used mostly for heating, they are inclined to the horizontal by latitude plus up to 20°. When they are used mostly for cooling, the inclination will be latitude minus up to 10°. The basic idea is to orient the collectors more favorably toward the winter sun when mostly heating and more favorably toward the summer sun when mostly cooling. The collectors can be made adjustable at extra cost. Air can be used instead of water with rock bin storage and blowers instead of pumps. Blowers require more energy to run.

The air conditioner can be a conventional type, driven by solar engines or solar electricity, or preferably an absorption system (e.g., low-temperature NH_3/H_2O), a jet air conditioning system, a liquid or solid dessicant system, or other direct energy conversion systems described later (in "Direct Energy Conversion").

Solar cooking utilizes (1) a sun-following broiler-type device with a metallized parabolic reflector and a grid in the focal area where cooking pots can be placed; (2) an oven-type cooker comprising an insulated box with glass covers over an open end which is pointed toward the sun. When reflecting wings are used to increase the solar input, temperatures as high as 400°F (204°C) are reached at midday. Large solar cookers for community cooking in third world country villages can be floated on water and thus easily adjusted to point at the sun. For cooking when the sun does not shine, oils or other fluids can be heated to a high temperature, 800°F (around 425°C), with a solar concentrator when the sun shines, and then stored. A range similar to an electric range but with the hot oil flowing through the coils, at adjustable rates, is then used to cook with solar energy 24 h/day.

Solar Furnaces Precise paraboloid concentrators can focus the sun's rays upon small areas, and if suitable receivers are used, temperatures up to 6,500°F (3,600°C) can be attained. The concentrator must be able to follow the sun, either through movement of the paraboloidal reflector itself (Fig. 9.1.13) or by the use of a heliostat which tracks the sun and reflects the rays along a horizontal or vertical axis into the concentrator. This pure, noncontaminating heat can be used to produce highly purified materials through zone refining, in a vacuum or controlled atmosphere. This allows us to grow crystals of high-temperature materials, crystals not existing in nature, or to do simple things such as determining the melting points of exotic materials. Other methods of heating contaminate these materials before they melt. With solar energy the materials can be sealed in a glass or plastic bulb, and the solar energy can be concentrated through the glass or plastic onto the target. The glass or plastic is not heated appreciably since the energy is not highly concentrated when it passes through it.

Solar furnaces are used in high-temperature research, can simulate the effects of nuclear blasts on materials, and at the largest solar furnace in the world (France) produce considerable quantities of highly purified materials for industry.

Power from Solar Energy During the past century (Zarem and Erway, "Introduction to the Utilization of Solar Energy," McGraw-Hill) many attempts have been made to generate power from solar radiation through the use of both flat-plate and concentrating collectors. Hot air and steam engines have operated briefly, primarily for pumping irrigation water, but none of these attempts have succeeded commercially because of high cost, intermittent operation, and lack of a suitable means for storing energy in large quantities. With the rapid rise in the cost of conventional fuels and the increasing interest in finding pollution-free sources of power, attention has again turned to parabolic trough concentrators and selective surfaces (high α/ε ratios) for producing high-temperature working fluids for Rankine and Brayton cycles. Because of the cost of Rankine engines, often operating on other than water-based working fluids, Stirling engines (discussed later) and other small engines (phase shift), turbines, gravity machines, etc., have been developed, their application and use being directly related to cost and availability of fossil fuels. The price of crude oil rose from $2.50 to $32.00 per barrel during the period from 1973 to the 1980s, and it stands at about $18.00 per barrel now (1995). When fossil fuel costs are high, solar energy conversion methods become competitive and attractive.

Flat-plate collectors utilizing both direct and diffuse solar radiation can be used to drive small, low-temperature Stirling engines, which operate off the available hot water; in turn, the engines can circulate water from the buried storage tank through the collectors on the roof. These engines have low efficiency, but that is not quite so important since solar energy is free. Low efficiency generally implies larger, and thus more expensive, equipment. The application cited above is ideal for small circulating electric pumps powered by solar cells.

Concentrating collector systems, having higher conversion efficiencies, can only utilize the direct portion of the solar radiation and in most cases need tracking mechanisms, adding to the cost. A 10-MW plant was built and had been operating in Barstow, CA until recently. That plant was used both for feasibility studies and to gather valuable operating experience and data. Although intrinsically attractive by virtue of zero fuel cost, solar-powered central plants of this type are not quite yet state of the art. Proposed plants of this and competitive types require abundant sunshine most of the time; obviously, they are not very effective on cloudy days, and their siting would appear to be circumscribed to desertlike areas.

Direct Conversion of Solar Radiation to Electricity Photovoltaic cells made from silicon, cadmium sulfide, gallium arsenide, and other semiconductors (see "Solar Cells," National Academy of Science, Washington, 1972) can convert solar radiation directly to electricity without the intervention of thermal cycles. Of primary importance today are the silicon solar batteries which are used in large numbers to provide power for space probes, orbiting laboratories, and communication satellites. Their extremely high cost and relatively low efficiency have thus far made them noncompetitive with conventional power sources for large-scale terrestrial applications, but intensive research is currently underway to reduce their production cost and to improve their efficiency. Generation of power from solar radiation on the earth's surface encounters the inherent problems of intermittent availability and relatively low intensity. At the maximum noon intensity of 340 Btu/(ft$^2 \cdot$ h) (1,080 W/m^2) and 100 percent energy conversion, 10 ft^2 (1.1 m^2) of collection area would produce 1 thermal kilowatt, but with a conversion efficiency of 10 percent, the area required for an electrical kilowatt approaches 100 ft^2 (9.3 m^2). Thus very large collection areas are essential, regardless of what method of conversion may be employed. However, the total amount of solar radiation falling on the arid southwestern section of the United States is great enough to supply all the nation's electrical needs, provided that the necessary advances are made in collection, conversion, and storage of the unending supply of energy from the sun.

Solar Transportation Presently the best application of solar energy to transportation seems to be electric vehicles, although solar-produced hydrogen could be used in hydrogen-propelled cars. Since there is not enough surface area on these vehicles to collect the solar energy needed for effective propulsion, storage is needed. Vehicles have been designed and built so that batteries are charged by solar energy. Charging is by Rankine engine-, Stirling engine-, or other engine-driven generators or by photovoltaic panels. A number of utilities, government agencies, municipalities, and universities have electric vehicles, cars, trucks, or buses, the batteries in which are charged by solar energy. For general use a nationwide pollution-free system is proposed, with solar battery charging stations replacing filling stations. They would provide, for a fee, charged batteries in exchange for discharged ones. A design objective to help implement this concept would require that the change of batteries be effected quickly and safely. Electric cars with top speeds of over 65 mi/h (88 km/h) and a range of 200 mi (320 km) have been built. Regenerative braking (the motor becomes a generator when slowing down, charging the batteries), especially in urban driving, can increase the range by up to 25 percent.

Closure Our inherited energy savings, in the form of fossil fuels, cannot last forever; indeed, an energy income must be part of the overall picture. That income, in the form of solar energy, will have to assume a larger role in the future. Fossil fuels will definitely fade from the picture at some time; it behooves us to plan now for the benefit of future generations.

For the successful application of solar energy, as with any other source of energy, each potential use must be analyzed carefully and the following criteria must be met:

1. Use the minimum amount of energy to do the task (efficiency).
2. Use the best overall energy source available.
3. The end result must be feasible and workable.
4. Cost must be reasonable.
5. The end result must fit the lifestyle and habits of the user.

Schemes which have failed in the past have violated one or more of these criteria.

In utilizing conventional fossil fuels, the energy conversion equipment is a capital cost to which must be added the periodic cost of fuel. Solar energy conversion systems, likewise, represent a capital cost, but there is no periodic cost for fuel. To be competitive, solar capital costs must be less than the total cost of a conventional plant (capital cost plus fuel). There exist circumstances where this is, indeed, the case. Financing will continue to look favorably on such investments.

There are several reasons why solar energy conversion has not had a wider impact, especially in the fossil-fuel-rich countries: lack of awareness of the long-term problems associated with fossil fuel consumption; the fact that solar energy conversion equipment is not as available as would be desirable; the current continuing supply of fossil fuels at very competitive prices; and so forth. There will come a time, however, when the bank of fossil fuels will have been exhausted; solar energy conversion looms large in the future.

A significant consideration with regard to fossil fuels is the realization that they constitute an irreplaceable source of raw materials which ought really to be husbanded for their greater utility as feedstocks for medicines, fertilizers, petrochemicals, etc. Their utility in serving these purposes overrides their convenient use as cheap fossil fuels burned for their energy content alone.

GEOTHERMAL POWER
by Kenneth A. Phair

REFERENCES: Assessment of Geothermal Resources of the United States—1978, *U.S. Geol. Surv. Circ. 790*, 1979, "Geothermal Resources Council Transactions," vol. 17, Geothermal Resources Council, Davis, CA. Getting the Most out of Geothermal Power, *Mech. Eng.*, publication of ASME, Sept. 1994. "Geothermal Program Review XII," U.S. Department of Energy, DOE/GO 10094-005, 1994.

Geothermal energy is a naturally occurring, semirenewable source of thermal energy. Thermal energy within the earth approaches the surface in many different geologic formations. Volcanic eruptions, geysers, fumaroles, hot springs, and mud pots are visual indications of geothermal energy.

Significant **geothermal reserves** exist in many parts of the world. The U.S. Geological Survey, in *Circ. 790*, has estimated that in the United States alone there is the potential for 23,000 megawatts (MW) of electric power generation for 30 years from recoverable hydrothermal (liquid- or steam-dominated) geothermal energy. Undiscovered reserves may add significantly to this total. Many of the known resources can be developed using current technology to generate electric power and for various direct uses. For other reserves, technical breakthroughs are necessary before this energy source can be fully developed.

Power generation from geothermal energy is cost-competitive with most combustion-based power generation technologies. In a broader picture, geothermal power generation offers additional benefits to society by producing significantly less carbon dioxide and other pollutants per kilowatt-hour than combustion-based technologies.

Electric power was first generated from geothermal energy in 1904. Active worldwide development of geothermal resources began in earnest in 1960 and continues. In 1993, the capacity of geothermal power plants worldwide exceeded 6,000 MW. Table 9.1.7 lists the installed capacity by country.

Table 9.1.7 Worldwide Geothermal Capacity, 1993

Country	Capacity installed, MW
United States	2,913
Philippines	894
Mexico	700
Italy	545
New Zealand	295
Japan	270
Indonesia	142
El Salvador	95
Nicaragua	70
Iceland	45
Kenya	45
Others	65
Total	6,079

SOURCE: From "Geothermal Resources Council Transactions," vol. 17, Geothermal Resources Council, Davis, CA, 1993.

Geothermal power plants are typically found in areas with "recently" active volcanoes and continuing seismic activity. In 1994 and 1995, significant additional geothermal power generation facilities were installed in Indonesia and the Philippines. Many countries not included in Table 9.1.7 also have significant geothermal resources that are not yet developed.

Although geothermal energy is a renewable resource, economic development of geothermal resources usually extracts energy from the reservoir at a much higher rate than natural recharge can replenish it. Therefore, facilities that use geothermal energy should be designed for high efficiency to obtain maximum benefit from the resource.

Geothermal Resources Geothermal resources may be described as hydrothermal, hot dry rock, geopressured, or magma.

Hydrothermal resources contain hot water, steam, or a mixture of water and steam. These fluids transport thermal energy from the reservoir to the surface. Reservoir pressures are usually sufficient to deliver the fluids to the surface at useful pressures, although some liquid-dominated resources require downhole pumps for fluid production. Hydrothermal resources may be geologically closed or open systems. In a closed system, the reservoir fluids are contained within an essentially impermeable boundary. Communication and fluid transport within the reservoir occur through fractures in the reservoir rock. There is little, if any, natural replenishment of fluids from outside the reservoir boundary. An open system allows influx of cold subsurface fluids into the reservoir as the reservoir pressure decreases. Hydrothermal reservoirs have been found at depths ranging from 400 ft (122 m) to over 10,000 ft (3,050 m).

Hot dry rock resources are geologic formations that have high heat content but do not contain meteoric or magmatic waters to transport thermal energy. Thus water must be injected to carry the energy to the surface. The difficulty in recovering a sufficient percentage of the injected water and the limited thermal conductivity of rock have hindered development of hot dry rock resources. Because of the vast amount of energy in these resources, additional research and development is justified to evaluate whether it is technically exploitable. In 1994 research and development of hot dry rock resources was proceeding in Australia, France, Japan, and the United States.

Geopressured resources are liquid-dominated resources at unusually high pressure. They occur between 5,000 and 20,000 ft (1,500 and 6,100 m), contain water that varies widely in salinity and dissolved minerals, and usually contain a significant amount of dissolved gas. Pressures in such reservoirs vary from about 3,000 to 14,000 lb/in² gage (21 to 96 MPa) with temperatures between 140°F (60°C) and 360°F (182°C). The largest geopressured zones in the United States exist beneath the continental shelf in the Gulf of Mexico, near the Texas, Louisiana, and Mississippi coasts. Other zones of lesser extent are scattered throughout the United States.

Magma resources occur as formations of molten rock that have temperatures as high as 1,300°F (700°C). In most regions in the continental United States, such resources occur at depths of 100,000 ft (30,500 m) or more. However, in the vicinity of current or recent volcanic activity, magma chambers are believed to be within 20,000 ft (6,100 m) of the surface. The Department of Energy initiated a magma energy research program in 1975, and an exploratory well was drilled in the Long Valley Caldera in California in 1989. The drilling program was planned in four phases to reach a final depth of 20,000 ft (6,100 m) or a temperature of 900°F (500°C), whichever is reached first. The first phases have been completed, but the deep, and most significant, drilling remains to be done.

Exploration Technology Geothermal sites historically have been identified from obvious surface manifestations such as hot springs, fumaroles, and geysers. Some discoveries have been made accidentally during exploring or drilling for other natural resources. This approach has been replaced by more scientific prospecting methods that appraise the extent, as well as the physical and thermodynamic properties, of the reservoir. Modern methods include geological studies involving aerial, surface, and subsurface investigations (including remote infrared sensing) and geochemical analyses which provide a guide for selecting specific drilling sites. Geophysical methods include drilling, measuring the temperature gradient in the drill hole, and measuring the thermal conductivity of rock samples taken at various depths.

Resource Development Extraction of fluids from a geothermal resource entails drilling large-diameter production wells into the reservoir formation. Bottom hole temperatures in hydrothermal wells and hot dry rock formations can exceed 450°F (232°C). Geopressured resources have lower temperatures but offer energy in the form of fluids at unusually high pressures that frequently contain significant amounts of dissolved combustible gases. Although research and development projects continue to seek ways to efficiently extract and use the energy contained in hot dry rock, geopressured, and magma resources, virtually all current geothermal power plants operate on hydrothermal resources.

Production Facilities For most projects, a number of wells drilled into different regions of the reservoir are connected to an aboveground piping system. This system delivers the geothermal fluid to the power plant. As with any fluid flow system, the geothermal reservoir, wells, and production facilities operate with a specific flow vs. pressure relationship. Fig. 9.1.18 shows a typical steam deliverability curve for a 110-MW geothermal power plant.

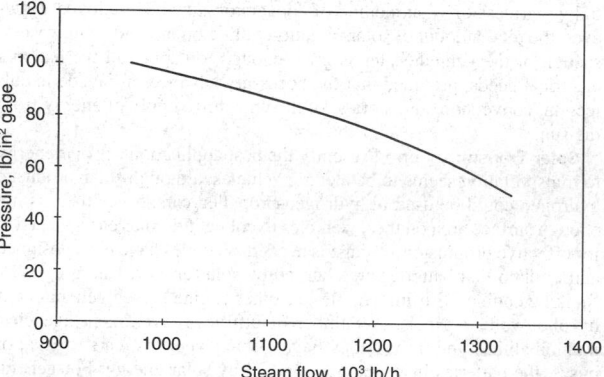

Fig. 9.1.18 Typical deliverability curve; steam flow to power plant [1,000 lb/h = 453 kg/h)] vs. turbine inlet pressure [100 lb/in² gage (689 kPa gage)].

Resource permeability; the number, depth, and size of the wells; and the surface equipment and piping arrangement all contribute to make the deliverability curve different for each power plant. Production of geothermal fluids over time results in declining deliverability. For one-half or more of the operating life of a reservoir, the deliverability can usually be held constant by drilling additional production wells into other regions of the resource. As the resource matures, this technique ceases to provide additional production. The deliverability curve begins to change shape and slope as deliverability declines. The power plant design must be matched to the deliverability curve if maximum generation from the resource is to be achieved.

Geothermal Power Plants A steam-cycle geothermal power plant is very much like a conventional fossil-fueled power plant, but without a boiler. There are, however, significant differences. The turbines, condensers, noncondensable gas removal systems, and materials used to fabricate the equipment are designed for the specific geothermal application. With geothermal steam delivered to the power plant at approximately 100 lb/in² gage (689 kPa), only the low-pressure sections of a conventional turbine generator are used. Additionally, the geothermal turbine must operate with steam that is far from pure. Chemicals and compounds in solid, liquid, and gaseous phases are transported with the steam to the power plant. At the power plant, the steam passes through a separator that removes water droplets and particulates before it is delivered to the turbine. Geothermal turbines are of conventional design with special materials and design enhancements to improve reliability in geothermal service. Turbine rotors, blades, and diaphragms operate in a wet, corrosive, and erosive environment. High-alloy steels, stainless steels, and titanium provide improved durability and reliability. Still, frequent overhauls are necessary to maintain reliability and performance. The high moisture content and the corrosive nature of the condensed steam require effective moisture removal techniques in the later (low-pressure) stages of the turbine. Scale formation on rotating and stationary parts of the turbine occurs frequently. Water washing of the turbine at low-load operation is sometimes used between major overhauls to remove scale.

Most geothermal power plants use direct-contact condensers. Only when control of hydrogen sulfide emissions has been required or anticipated have surface condensers been used. Surface condensers in geothermal service are subject to fouling on both sides of the tubes. Power plants in The Geysers in northern California use conservative cleanliness factors to account for the expected tube-side and shell-side fouling. Some plants have installed on-line tube-cleaning systems to combat tube-side fouling on a continuous basis, whereas other plants mechanically clean the condenser tubes to restore lost performance.

Noncondensable gas is transported with the steam from the geothermal resource. The gas is primarily carbon dioxide but contains lesser amounts of hydrogen sulfide, ammonia, methane, nitrogen, and other gases. Noncondensable gas content can range from 0.1 percent to more than 5 percent of the steam. The makeup and quantity of noncondensable gas vary not only from resource to resource but also from well to well within a resource. The noncondensable gas removal system for a geothermal power plant is substantially larger than the same system for a conventional power plant. The equipment that removes and compresses the noncondensable gas from the condenser is one of the largest consumers of auxiliary power in the facility, requiring up to 15 percent of the thermal energy delivered to the power plant. A typical system uses two stages of compression. The first stage is a steam jet ejector. The second stage may be another steam jet ejector, a liquid ring vacuum pump, or a centrifugal compressor. The choice of equipment selected for the second stage is influenced by project economics and the amount of gas to be compressed.

The chemicals and compounds in geothermal fluids are highly corrosive to the materials normally used for power plant equipment and facilities. The chemical content of geothermal fluids is unique to each resource; therefore, each resource must be evaluated separately to determine suitable materials for system components. Carbon steel usually will degrade at alarmingly high rates when exposed to geothermal fluids. Corrosion-resistant materials such as stainless steel may perform satisfactorily, but may experience rapid, unpredictable local failures depending upon the composition of the geothermal fluid. Based on experience with a number of geothermal resources:

Carbon steel with a corrosion allowance is usually suitable for transporting dry geothermal steam.

Geothermal condensate and cooling water usually require corrosion-resistant piping and equipment.

Because noncondensable gas is also corrosive, special materials are usually required.

Copper is extremely vulnerable to attack from the atmosphere surrounding a geothermal power plant. Therefore, copper wire and electrical components should be protected with tin plating and isolated from the corrosive atmosphere.

Within the context of these generalities, the fluids at each resource must be evaluated before construction materials are chosen.

The steam Rankine cycle used in fossil-fueled power plants is also used in geothermal power plants. In addition, a number of plants operate with binary cycles. Combined cycles also find application in geothermal power plants. The basic cycles are shown in Fig. 9.1.19.

The direct steam cycle shown in Fig. 9.1.19a is typical of power plants at The Geysers in northern California, the world's largest geothermal field. Steam from geothermal production wells is delivered to power plants through steam-gathering pipelines. The wells are up to 1 mi or more from the power plant. The number of wells required to supply steam to the power plant varies with the geothermal resource as well as the size of the power plant. The 55-MW power plants in The Geysers receive steam from between 8 and 23 production wells.

A flash steam cycle for a liquid-dominated resource is shown in

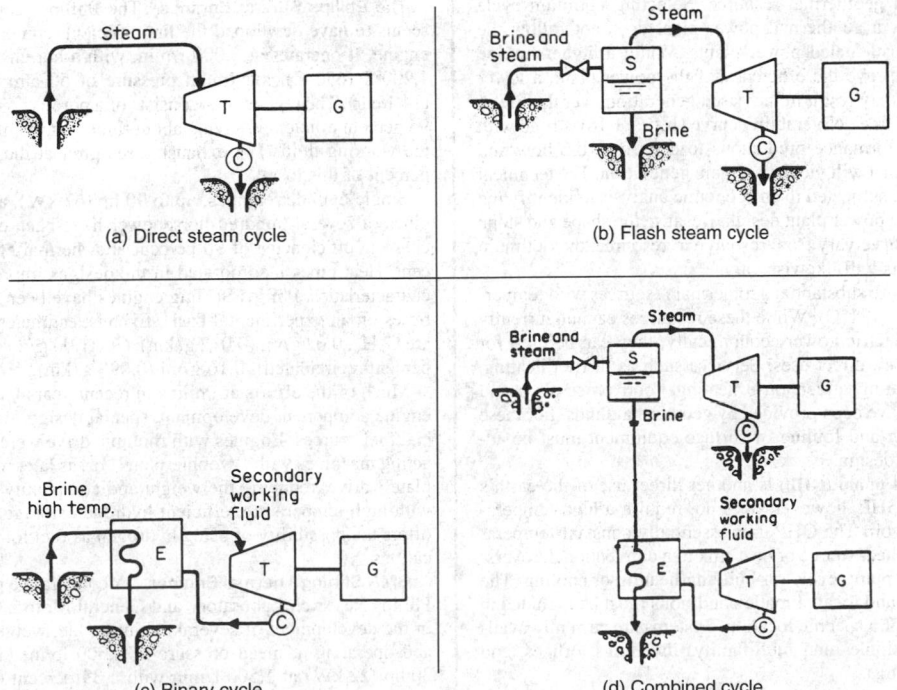

(a) Direct steam cycle

(b) Flash steam cycle

(c) Binary cycle

(d) Combined cycle

Fig. 9.1.19   Geothermal power cycles. T = turbine, G = generator, C = condenser, S = separator, E = heat exchanger.

Fig. 9.1.19b. Geothermal brine or a mixture of brine and steam is delivered to a flash vessel at the power plant by either natural circulation or pumps in the production wells. At the entrance to the flash vessel, the pressure is reduced to produce flash steam, which then is delivered to the turbine. This cycle has been used at power plants in California, Nevada, Utah, and many other locations around the world. Increased thermal efficiency is available from the use of a second, lower-pressure flash to extract more energy from the geothermal fluid. However, this technique must be approached carefully as dissolved solids in the geothermal fluids will concentrate and may precipitate as more steam is flashed from the fluid. The solids also tend to form scale at lower temperatures, resulting in clogged turbine nozzles and rapid buildup in equipment and piping to unacceptable levels.

A binary cycle is the economic choice for hydrothermal resources with temperatures below approximately 330°F (166°C). A binary cycle uses a secondary heat-transfer fluid instead of steam in the power generation equipment. A typical binary cycle is shown in Fig. 9.1.19c. Binary cycles can be used to generate electric power from resources with temperatures as low as 250°F (121°C). The binary cycle shown in Fig. 9.1.19c uses isobutane as the heat-transfer fluid. It is representative of units of about 10-MW capacity. Many small modular units of 1- or 2-MW capacity use pentane as the binary fluid. Heat from geothermal brine vaporizes the binary fluid in the brine heat exchanger. The binary fluid vapor drives a turbine generator. The turbine exhaust vapor is delivered to an air-cooled condenser where the vapor is condensed. Liquid binary fluid drains to an accumulator vessel before being pumped back to the brine heat exchangers to repeat the cycle. Binary-cycle geothermal plants are in operation in several countries. In the United States, they are located in California, Nevada, Utah, and Hawaii.

A geothermal combined cycle is shown in Fig. 9.1.19d. Just as combustion-based power plants have achieved improved efficiencies by using combined cycles, geothermal combined cycles also show improved efficiencies. Some new power plants in the Philippines use a combination of steam and binary cycles to extract more useful energy from the geothermal resource. Existing steam-cycle plants can be modified with a binary bottoming cycle to improve efficiency.

Cycle optimization is critically important to maximize the power generation potential of a geothermal resource. Selecting optimum cycle design parameters for a geothermal power plant does not follow the practices used for fossil-fueled power plants. While a higher turbine inlet pressure will improve the efficiency of the power plant, a lower turbine inlet pressure may result in increased generation over the life of the resource. The resource deliverability curve (Fig. 9.1.18) is used with turbine and cycle performance predictions to determine the flow and turbine inlet pressure that will yield maximum generation. The technical optimum must then be subjected to an economic analysis to identify the best parameters for the power plant design. Because the shape and slope of the deliverability curve vary from resource to resource, the optimum turbine inlet conditions will likewise vary.

Direct Use There are substantial geothermal resources with temperatures less than 250°F (121°C). While these resources cannot currently be used to generate electric power economically, they can be used for various low-temperature direct uses. Services such as district heating, industrial process heating, greenhouse heating, food processing, and aquaculture farming have been provided by geothermal fluids. For these applications, corrosion and fouling of surface equipment must be addressed in the system design.

The geothermal heat pump (GHP) is another direct use of the earth's thermal energy. The GHP, however, does not require a high-temperature geothermal reservoir. The GHP uses essentially constant-temperature groundwater as a heat source or heat sink in a conventional, reversible, water-to-air heat pump cycle for building heating or cooling. The ground, groundwater, and local climatic conditions must be included in the design of a GHP for a specific location. Systems are currently available for residential (single- and multifamily) dwellings, offices, and small industrial buildings.

Environmental Considerations Geothermal fluids contain many chemicals and compounds in solid, liquid, and gaseous phases. For both environmental protection and resource conservation, spent geothermal liquids are returned to the reservoir in injection wells. This limits the release of compounds to the environment to a small amount of liquid lost as drift from the cooling tower and noncondensable gases. Problems with arsenic and boron contamination have been encountered in the immediate vicinity of cooling towers at geothermal power plants. The noncondensable gases, composed primarily of carbon dioxide, usually also contain hydrogen sulfide. Along with its noxious odor, hydrogen sulfide is hazardous to human and animal life. Although many geothermal power plants do not currently control the release of hydrogen sulfide, others use process systems to oxidize the hydrogen sulfide to less toxic compounds. A number of the process systems produce 99.9 percent pure sulfur that can be sold as a by-product. Using geothermal energy for power generation and other direct applications provides environmental benefits. Carbon dioxide released from a geothermal power plant is approximately 90 percent less than the amount released from a combustion-based power plant of equal size, and they create little, if any, liquid or solid waste.

STIRLING (HOT AIR) ENGINES
by Erich A. Farber

Hot air engines, frequently referred to as Stirling engines, are heat engines with regenerative features in which air; other gases such as H_2, He, N_2; or even vapors are used as working fluids, operating, theoretically at least, on the Stirling or Ericsson cycle (see Sec. 4.1) or modifications of them. While the earlier engines of this type were bulky, slow, and low in efficiency, a number of new developments have addressed these deficiencies. Stirling engines are multifuel engines and have been driven by solid, liquid, or gaseous fuels, and in some cases with solar energy. They can be reciprocating or rotary, include special features, run quietly, are relatively simple in construction (no valves, no electrical systems), and if used with solar energy, produce no waste products. (See Walker, "Stirling Engines," Clarendon Press, Oxford; *Proceedings,* 19th Intersociety Energy Conversion Engineering Conference, Aug. 1984, San Francisco.)

The Philips Stirling Engines The Philips Laboratory (in Holland) seems to have developed the first efficient, compact hot air or Stirling engine. It operates at 3,000 r/min, with a hot chamber temperature of 1,200°F (650°C), maximum pressure of 50 atm, and mep of 14 atm (14.1 bar). The regenerator consists of a porous coil of thin wires having 95 percent efficiency, saving about three-fourths of the heat required by the working fluid. The exhaust gases preheat the air, saving about 70 percent of this loss.

Single-cylinder engines, up to 90 hp (67 kW), and multicylinder engines of several hundred horsepower have been constructed with mechanical efficiencies of 90 percent and thermal efficiencies of 40 percent. Heat pipes incorporated in the designs improve the heat transfer characteristics. Philips Stirling engines have been installed in clean-air buses on an experimental basis. Exhaust estimates for an 1,800-kg car are C_xH_y, 0.02 g/mi (0.012 g/km); CO, 1.00 g/mi (0.62 g/km); NO (25 percent recirculated), 0.16 g/mi (0.099 g/km).

Much of the efforts at Philips in recent years have gone into Stirling engine component development, special design features, and even special fuel sources. Engines with rhombic drive were replaced by double-acting machines with "wobble-plate" or, as later referred to, as "swash plate" drive, reducing the weight and complexity of the design. Work with high temperature, efficient hydrogen storage in metallic hydrides offers the possibility of using hydrogen as fuel for transportation applications.

GMR Stirling Thermal Engines A cooperative program between the Philips Research Laboratory and General Motors Corporation resulted in the development of several engines. One, weighing 450 lb (200 kg) and operating at mean pressure of 1,500 lb/in² (103.4 bar), produces 30 hp (22 kW) at 1,500 r/min with a 39 percent efficiency and 40 hp (30 kW) at 2,500 r/min with a 33.3 percent efficiency. Another weighing 127 lb (57 kg) and operating at a mean pressure of 1,000 lb/in²

(6.9 MN/m²), produces 6 hp (4.5 kW) at 2,400 r/min with a 29.6 percent efficiency and 8.63 hp (6.4 kW) at 3,600 r/min with a 26.4 percent efficiency.

One such engine was used for a portable Stirling engine electric generator set; another was installed in a Stirling engine electric hybrid car. A 360 hp (265 kW) marine engine was delivered to the U.S. Navy. Another 400-hp (295-kW) engine with special control features was built and tested, and could reverse its direction of rotation almost instantaneously.

Ford-Philips Stirling Engine Development In 1972, Ford Motor Company and Philips entered into a joint development program and developed Stirling engines which were installed experimentally in then-current automobile models.

MAN/MWM Stirling Engines The German company Entwicklungsgruppe Stirlingmotor MAN/MWM, in cooperation with Philips, developed a single acting engine with rhombic drive which developed 30 hp (22 kW) at 1500 r/min and formed the basic test unit for a four-cylinder 120-hp (88-kW) engine. Some double acting engines have been developed. In cooperation with the Battelle Institut, Frankfurt, a 15-kVA Stirling engine hydroelectric generator was developed. It operated at 3,000 r/min, pressurized with helium, with an efficiency of 25 percent.

United Stirling Engines United Stirling AB (Sweden) in cooperation with Philips developed Stirling engines for boats, including those of the Swedish Royal Navy, and engines for buses. One generated 200 hp (145 kW) at 3,000 r/min and a mean helium pressure of 220 atm (22.3 MN/m²).

Internally Focusing Regenerative Gas Engines These engines, conceived at the Solar Energy Laboratory of the University of Wisconsin, use solar energy, concentrated by a parabolic reflector and directed through a quartz dome upon an internal absorber. This reduces the heat losses, since the engine has no external high-temperature heat transfer surfaces. A small working model of this engine has been built at Battelle Memorial Institute and was demonstrated driving a small fan.

Fractional-Horsepower Solar Hot Air Engines The Solar Energy and Energy Conversion Laboratory of the University of Florida has developed small (¼ to ⅓ hp; 0.186 to 0.25 kW) solar hot air engines (some of them converted lawnmower engines). The actual power output of the engines is determined by the size of the solar concentrator rather than by the engine. Some of these engines are self-supercharging to increase power. Water injection, self-acting, increased power by 19 percent. The average speed of the closed-cycle engines is about 500 r/min; average conversion efficiency is about 9 percent. Open-cycle engines separate the heating process from the working cycle, allowing the design of high-speed or low-speed engines as desired. Any heat source can be used with these engines, such as solar, wood, farm wastes, etc. They are simple, rugged, and designed for possible use in developing countries.

The Stirling Engine for Space Power General Motors Corporation, under contract to the U.S. Air Force Aeronautical Systems Command, adapted the GMR Stirling engine to possible space applications. A 3-kW engine was built utilizing NaK heated to 1,250°F (677°C) as a heat source and water at 150°F (66°C) as the cooling medium. The engine is pressurized to a mean pressure of 1,500 lb/in² (103.4 bar), giving an efficiency of 27 percent at 2,500 r/min. The weight of this solar energy conversion system is 550 lb (249 kg). Chemical, nuclear, or other energy sources can also be used.

Free-Piston Stirling Engines The free-piston Stirling engines, pioneered principally by William Beale, consist of displacer and power pistons, coupled by springs, inside one cylinder. They are relatively simple, self-starting, and if pressurized, can be hermetically sealed. The power piston can be coupled to a pump piston since the motion is reciprocating. Single- and double-acting engines have been designed, built, and demonstrated for water pumping and electricity generation. Some of the engines are presently under evaluation by the U.S. Agency for International Development for possible use in developing countries. They can use alternative energy sources, principally solar energy.

Closed-Environment Stirling Engines A number of Stirling engines have been developed to utilize energy sources which do not require coupling to the external environment. Such engines can be powered by specially prepared fuel sources or by stored energy.

Artificial-Heart Stirling Engines Considerable interest has been shown in the possible use of Stirling engines either to assist weakened hearts or to replace them if they have been damaged beyond repair. The program is supported by the National Heart Institute and has involved many organizations (e.g., Philips, Westinghouse, Aerojet-General, McDonnell-Douglas, University of California, Washington State University). Most of the engines are powered by nuclear fuel sources.

Low-Temperature Stirling Engines In many applications, low-temperature sources such as exhaust gases from combustion, waste steam, and hot water from solar collectors are available. Several groups (University of Florida, University of Wisconsin, Zagreb University in former Yugoslavia, etc.) are working on the development of low-temperature Stirling engines. Models for demonstration have been built and their performance has been evaluated.

Heat Pump and Cryocooler Stirling Engines A Stirling engine can be driven by any mechanical source or by another Stirling engine, and when so motored becomes a heat pump or cooler, depending upon the effects desired and utilized. Special duplex designs for Stirling engines lend themselves especially well for these applications. A number of private companies and public laboratories are involved in this development. Philips manufactured small cryocoolers in the past and sold them throughout the world.

Liquid Piston Stirling Engines Liquid piston engines are extremely simple. The basic liquid piston Stirling engine consists of two U tubes. A pipe connects the two ends of one of the U tubes with one end of the other. The unconnected end of the second U tube is left open. Both U tubes are filled with liquid, thus forming the liquid pistons. The closed U tube liquid acts as the displacer and the other as the power unit. The section of the connecting pipe between the displacer U tube ends contains the regenerator. This engine is referred to as the basic **Fluidyne**. Even though these engines have been around for a long time much development work is still needed. Their efficiencies are still extremely low.

Closure During their history, Stirling engines have experienced periods of high interest and rapid development. Stirling Engines for Energy Conversion in Solar Energy Units (Trukhov and Tursunbaev, *Geliotekhnika*, **29**, no. 2, 1993, pp. 27–31) summarizes the performance of 15 Stirling engines. With supply temperatures of about 600°C, their output varies from 0.55 to 43.2 kW, their speed varies from 833 to 4,000 r/min, and their conversion efficiencies vary from 12.5 to 30.3 percent. Development work continues on some problem areas (seals, hydrogen embrittlement, weight, etc.). Interest in the potential of these engines remains high, as indicated by an average of about 50 Stirling engine papers presented and published yearly in each of the 1991 (26th) through 1994 (29th) "Proceedings of the Intersociety Energy Conversion Conferences."

POWER FROM THE TIDES
Staff Contribution

REFERENCES: The Rance Estuary Tidal Power Project, *Pub. Util. Ftly.*, Dec. 3, 1964. *Mech. Eng.*, Ap. 1984. ASCE Symposium, 1987: Tidal Power. Gray and Gashus, "Tidal Power," Department of Commerce, NOAA, Water for Energy, *Proc. 3d Intl. Symp.*, 1986.

The tides are a renewable source of energy originating in the gravitational pull of the moon and sun, coupled with the rotation of the earth. The consequent portion of the earth's rotation is a mean ocean tide of 2 ± ft (0.6 m). The seashore periodic variation of the tides averages 12 h 25 min.

Tidal power is derivable from the large periodic variations in tidal flows and water levels in certain oceanic coastal basins. Suitable configurations of the continental shelves and of the coastal profiles result in reflection and resonance that amplify normally small bulges to ranges as high as 50 ± ft (15 ± m).

Principal tidal-power sites include the North Sea [12 ft (3.6 m) average tidal range]; the Irish Sea [22 ft (6.7 m)]; the west coast of India [23 ft (7 m)]; the Kimberly coast of western Australia [40 ft (12 m)]; San Jose Bay on the east coast of the Argentine [23 ft (7 m)]; the Kislaya Guba (Kisgalobskaia Bay) near the White Sea (no data); St. Michel (including the Rance estuary) on the Brittany coast of France [26 ft (8 m)]; the Bristol Channel (Severn) in England [32 ft (9.8 m)]; the Bay of Fundy (including the Chignecto Bay between New Brunswick and Nova Scotia and the Minas Basin in Nova Scotia) [40 ft (12 m)]; Passamaquoddy Bay between Maine and New Brunswick [18 ft (5.5 m)].

The harnessing of the tides reaches back into ancient history. Tidal mills, typically with undershot water wheels, were used in New England raceway estuaries, with reversible features for ebb and flood conditions. These power applications were suitable for purposes such as grinding grain, but their number and size were small. In recent times the unique tidal ranges to 50 ± ft (15 ± m) have prompted many studies, proposals, and projects for most of the regions cited above. The attraction for the utilization of tides to generate electric power lies in the facts that there results no air or thermal pollution, the source is effectively inexhaustible, and the construction work related to the tidal power plant is relatively benign in its environmental impact. Despite these efforts for the generation of electricity, there are only four tidal power developments in actual service (1995)—the Rance estuary in France (240,000 kW), the Kislaya Guba in Russia (400 kW), the Bay of Fundy in Canada (20,000 kW), and a small pilot plant in Kiangshia, China.

Developments take one of two general forms: **single-basin** or **multiple-basin**. A single-basin project, such as the Rance, has a dam, sluices, locks, and generating units in a structure separating a tidal basin from the sea. Water is trapped in the basin after a high tide. As the water level outside the basin falls with the tide, flow from the basin through turbines generates power. Power also may be generated when a basin emptied during a low tide is refilled on a rising tide. Numerous variations in operation are possible, depending on tide conditions and the relationship between the tide cycles and the load cycles. Pumping into or out of the basin increases the availability of the installed capacity for peak load service. A multiple-basin development, such as projected for Chignecto or Passamaquoddy, generally has the power house between two basins. Sluices between the sea and the basins are so arranged that one basin is filled twice a day on high tide and the other emptied twice a day on low tide. Power output can be made continuous.

The amount of **energy available** from a tidal development is proportional to the basin area and to the square of the tidal range. Head variations are large in tidal projects during generating cycles and on a daily, monthly, and annual basis owing to various cosmic factors. Intermittent power, as from all single-basin plans and from two-basin plans with low-capacity factor, implies that the output can best be utilized as peaking capacity. Because of low heads, particularly toward the end of any generating cycle when pools have been drawn down, the cost of adding generating units only to tidal projects is well over the total installation cost of alternative peaking capacity. To be economically competitive with alternative capacity, the tidal projects must produce enough energy to pay the power-plant costs and also to pay for the dams and other costs, such as general site development, transmission, operation, and replacements.

The **risks and uncertainties** involved in designing, pricing, building, and operating capital-intensive tidal works, and the technological developments in alternative types of generating capacity, have tended to defeat tidal developments. Civil works are too extensive; transmission distances to load centers are too great; the required scale of development is too large for existing loads; the coordination of system demands and tidal generation requires interconnections for economic loading; the ultimate capacity of all the world's tidal potential is practically insignificant to meet the world's demands for electricity. In addition, the matter of interrupting tidal action and storage of tidewaters in large basins for extended periods of time raises the possibility of saltwater infiltration of adjacent underground fresh water supplies which, in many cases, are the source of drinking water for the contiguous areas.

UTILIZATION OF ENERGY OF THE WAVES
Staff Contribution

According to Albert W. Stahl, USN (*Trans, ASME,* **13,** p. 438), the total energy of a series of **trochoidal deep-sea waves** may be expressed as follows: hp per ft of breadth of wave = $0.0329 \times H^2 \sqrt{L}(1 - 4.935 H^2 / L^2)$, where H = height of wave, ft, and L = length of wave between successive crests, ft. For example, with $L = 25$ ft and $L/H = 50$, hp = 0.04; with $L = 100$ ft and $L/H = 10$, hp = 31.3. Not much more than a quarter of the total energy of such waves would probably be available after reaching shallow water, and apparatus rugged enough for this purpose would doubtless be unable to utilize more than a third of this amount. **Wave motors** brought out from time to time have depended for their operation largely on the lifting power of the waves.

Gravity waves may be only a few feet high yet develop as much as 50 kW/ft of wavefront. Historical wave motors utilize (1) the kinetic energy of the waves by a device such as a paddle wheel or turbine or (2) the potential energy from devices such as a series of floats or by impoundment of water above sea level. Few devices proposed utilize both forms of energy. Jacobs (*Power Eng.,* Sept. 1956) has analyzed the periodic fluctuation of **"seiching"** of the water level of harbors or basins where, with a resonant port, a 1,000-ft wavefront might be used to achieve a liquid piston effect for the compression of air, the air to be subsequently used in an air turbine.

The principle of using an oscillating column of displaced air has been employed for many years in buoys and at lighthouses, where the wave-actuated rise and fall of the column of air actuated sound horns. A wave-actuated air turbine and electric generator have been operating on the Norwegian coast to study feasibility and to gather operating data. Another similar unit has been emplaced recently off the Scottish coast, and while serving to provide operational data, it also feeds about 2 MW of power into the local grid. If the results are favorable, this particular type of unit may be expanded at this site or emplaced at other similar sites.

A variation of capture of sea wave energy is to cause waves to spill over a low dam into a reservoir, whence water is conducted through water turbines as it flows back to sea level. Any attempt to channel significant quantities of water by this method would require either a natural location or one in which large concrete structures (much of them under water) are emplaced in the form of guides and dams. The capital expenses implicit in this scheme would be enormous.

The extraction of sea wave energy is attractive because not only is the source of energy free, but also it is nonpolluting. Most probably, the capture of wave energy for beneficial transformation to electric power may be economically effected in isolated parts of the world where there are no viable alternatives. Remote island locations are candidates for such installations.

UTILIZATION OF HEAT ENERGY OF THE SEA
Staff Contribution

REFERENCES: Claude and Boucherot, *Compt. rend.,* 183, 1926, pp. 929–933. *The Engineer,* 1926, p. 584. Anderson and Anderson, *Mech. Eng.,* Apr. 1966. Othmer and Roels, Power, Fresh Water, and Food from Cold, Deep Sea Water, *Science,* Oct. 12, 1973. Roe and Othmer, *Mech. Eng.,* May 1971. Veziroglu, "Alternate Energy Sources: An International Compendium." Department of Commerce, NOAA, Water for Energy, *Proc. 3d Intl. Symp.,* 1986.

Deep seawater, e.g., at 1-mi (1.6-km) depth in some tropical regions, may be as much as 50°F (28°C) colder than the surface water. This difference in temperature is a fundamental challenge to the power engineer, as it offers a potential for the conversion of heat into work. The Carnot cycle (see Sec. 4.1) specifies the limits of conversion efficiency. Typically, with a heat source surface temperature T_1 = 85°F (545°R, 29.4°C, 303 K), and a heat sink temperature T_2 50° lower, or T_2 = 35°F

(495°R, 1.7°C, 275 K), the ideal Carnot cycle thermal efficiency = $(T_1 - T_2)/T_1 = [(545 - 495)/545]100 = 9.2$ percent.

Some units in experimental or pilot operation have demonstrated actual thermal efficiency in the range of 2 to 3 percent. These efficiencies, both ideal and actual, are far lower than those obtained with fossil or nuclear fuel-burning plants. The fundamental attraction of **ocean thermal energy conversion (OTEC)** is the vast quantity of seawater exhibiting sufficient difference in temperature between shallow and deep layers. In reality, the sea essentially represents a limitless store of solar energy which manifests itself in the warmth of seawater, especially in the top, shallow layers. Water temperatures fluctuate very little over time; thus the thermodynamic properties are relatively constant. In addition, the thermal energy is available on a 24-h basis, can be harnessed to serve a plant on land or offshore, provides ''free'' fuel, and results in a nonpolluting recycled effluent.

Consider the extent of the ocean between latitudes of 30°S and 30°N, and ascribe a temperature difference between shallow and very deep waters of 20°F. The theoretical energy content comes to about 8×10^{21} Btu. Of this enormous amount of raw energy, the actual amount posited for eventual recovery in the best of circumstances via an OTEC system is a miniscule percentage of that total. The challenge is to develop practical machinery to harness thermal energy of the sea in a competitive way.

The concept was put forward first in the nineteenth century, and it has been reduced to practice in several experimental or pilot plants in the past decades. There are three basic conversion schemes: closed Rankine cycle, open Rankine cycle, and mist cycle. In the closed cycle, warm surface water is pumped through a heat exchanger (boiler) which transfers heat to a low-temperature, high-vapor-pressure working fluid (e.g., ammonia). The working fluid vaporizes and expands, drives a turbine, and is subsequently cooled by cold deep water in another heat exchanger (condenser). The heat exchangers are large; the turbine, likewise, is large, by virtue of the low pressure of the working fluid flowing through; enormous quantities of water are pumped through the system. In the open cycle, seawater itself is the working fluid. Steam is generated by flash evaporation of warm surface water in an evacuated chamber (boiler), flows through a turbine, and is cooled by pumped cold deep water in a direct-contact condenser. The reduction of heat-transfer barriers between working fluid and seawater increases the overall system efficiency and requires smaller volumes of seawater than are used in the closed cycle. Introduction of a closed-cycle heat exchanger into the open-cycle scheme results in a slightly reduced thermal efficiency, but provides a valuable by-product in the form of freshwater, which is suitable for human and animal consumption and may be used to irrigate vegetation. A plant of this last type operates in Hawaii and generates 210 kW of electric power, with a 50-kW net surplus power available after supplying all pumping and other house power. This plant continues to provide operational and feasibility data for further development.

The mist cycle mimics the natural cycle which converts evaporated seawater to rain which is collected and impounded and ultimately flows through a hydraulic turbine to generate power.

Problems encountered in the implementation of any OTEC system revolve on material selection, corrosion, maintenance, and significant fouling of equipment and heat-transfer surfaces by marine flora and fauna. Although development of OTEC systems will continue, with a view to their application in fairly restricted locales, there is no prospect that the systems will make any significant impact on power generation in the foreseeable future.

POWER FROM HYDROGEN
Staff Contribution

REFERENCES: Stewart and Edeskuty, Alternate Fuels for Transportation, *Mech. Eng.,* June 1974. Winshe et al., Hydrogen: Its Future Role in the Nation's Energy Economy, *Science,* **180,** 1973.

Hydrogen offers many attractive properties for use as fuel in a power plant. Fundamentally it is a ''clean'' fuel, smokeless in combustion with no particulate products, and if burned with oxygen, water vapor is the sole end product. If, however, it is burned with air, some of the nitrogen may combine at elevated temperature to form NO_x, a troublesome contaminant. If carbon is present as a fuel constituent, or if it can be picked up from a source such as a lubricant, the carbon introduces further contaminant potentials, e.g., carbon monoxide and cyanogens.

Basically the potential cleanliness of combustion is supported by other properties that make hydrogen a significant fuel, to wit, prevalence as a chemical element, calorific value, ignition temperature, explosibility limits, diffusivity, flame emissivity, flame velocity, ignition energy, and quenching distance.

Hydrogen offers a unique calorific value of 61,000 Btu/lb (140,000 kJ/kg). With a specific volume of 190 ft^3/lb (12 m^3/kg) this translates to 319 Btu/ft^3 (12,000 kJ/m^3) at normal pressure and temperature, 14.7 lb/in^2 absolute and 32°F (1 bar at 0°C). These figures, particularly on the volume basis, introduce many practical problems because hydrogen, with a critical point of -400°F (33 K) at 12.8 atm, is a gas at all normal, reasonable temperatures. When compared with alternative fuels, results are as shown in Table 9.1.8.

These figures demonstrate the volumetric deficiency of gaseous hydrogen. High-pressure storage (50 to 100 atm) is a dubious substitute for the gasoline tank of an automobile. Liquefaction calls for cryogenic elements (Secs. 11 and 19). Chemical compounds, metallic hydrides, hydrazene, and alcohols are potential alternates, but practicality and cost are presently disadvantageous.

Hydrogen has been used to power a number of different vehicles. Its use as a rocket fuel is well documented; in that application, cost is no concern. Experimental use in automotive and other commercial vehicles with slightly altered internal combustion engines has not advanced beyond very early stages. Aircraft jet engines have been powered successfully for short flight times on an experimental basis, and it is conjectured that the first successful commercial application of hydrogen as a source of power will be as fuel for jet aircraft early in the twenty-first century.

In spite of the demonstrated thermodynamic advantages inherent in

Table 9.1.8 Bulk and Calorific Power of Selected Fuels (Approximate and Comparative)

Fuel	State	Sp. wt., lb/ft^3	Sp. gr.	Btu/lb	Btu/ft^3	Btu/gal
Hydrogen	Gas (NTP)	0.0052	0.07	61,000	320	(40)
Natural gas	Gas (NTP)	0.042	0.67	24,000	1,000	(130)
Gasoline (reg., 90 oct.)	Liquid	46	0.72	20,500	950,000	125,000
Ethanol (99 oct.)	Liquid	49	0.79	12,800	620,000	82,000
Methanol (98 oct.)	Liquid	49	0.79	9,600	480,000	64,000
Hydrogen	Liquid (36°R, 14.7 lb/in^2 abs)	4.4	0.07	56,000	240,000	32,000
Coal	Piled	50	0.8	12,000	600,000	80,000

hydrogen as a source of power, in the current competitive economic market for fuels, hydrogen still faces daunting problems because of its high cost and difficulties related to its efficient storage and transportation.

DIRECT ENERGY CONVERSION
by Erich A. Farber

REFERENCES: Kaye and Welsh, "Direct Conversion of Heat to Electricity," Wiley. Chang, "Energy Conversion," Prentice-Hall. Shive, "Properties, Physics and Design of Semiconductor Devices," Van Nostrand. Bredt, Thermoelectric Power Generation, *Power Eng.*, Feb.–Apr. 1963. Wilson, Conversion of Heat to Electricity by Thermionic Conversion, *Jour. Appl. Phys.*, Apr. 1959. Angrist, "Direct Energy Conversion," Allyn & Bacon, Harris and Moore, Combustion—MHD Power Generation for Central Stations, *IEEE Trans. Power Apparatus and Systems*, **90**, 1971. Roberts, Energy Sources and Conversion Techniques, *Am. Scientist*, Jan.–Feb. 1973. Poule, Fuel Cells: Today and Tomorrow, *Heating, Piping, and Air Conditioning*, Sept. 1970. Fraas, "Engineering Evaluation of Energy Systems," McGraw-Hill. Kattani, "Direct Energy Conversion, Addison-Wesley. Commercialization of Fuel Cell Technology, *Mech. Eng.*, Sept. 1992, p. 82. The Power of Thermionic Energy Conversion, *Mech. Eng.*, Sept. 1993, p. 78. Fuel Cells Turn Up the Heat, *Mech. Eng.*, Dec. 1994, p. 62. "Proceedings of the Intersociety Energy Conversion Conferences," published yearly with the 29th in 1994.

In contrast to the conventional thermal cycle for the conversion of heat into electricity are several more direct methods of converting thermal and chemical energy into electric power. The methods which seem to have the greatest potential possibilities are thermoelectric, thermionic, magnetohydrodynamic (MHD), fuel cell, and photovoltaic. The principles of operation of these processes have long been known, but technological and economic obstacles have limited their use. New applications, materials, and technology now provide increased impetus to the development of these processes.

Thermoelectric Generation

Thermoelectric generation is based on the phenomenon, discovered by **Seebeck** in 1821, that current is produced in a closed circuit of two dissimilar metals if the two junctions are maintained at different temperatures, as in thermocouples for measuring temperature. A thermoelectric generator is a low-voltage, dc device. To obtain higher voltages, the elements must be stacked. Typical thermocouples produce potentials on the order of 50 to 70 μV/°C and power at efficiencies on the order of 1 percent.

Certain semiconductors have thermoelectric properties superior to conductor materials, with resultant improved efficiency. The criterion for evaluating material characteristics for thermoelectric generation is the **figure of merit Z**, measured in (°C)$^{-1}$ and defined as $Z = S^2/(\rho K)$, where S = Seebeck coefficient, V/°C; ρ = electrical resistivity, $\Omega \cdot$ cm; K = thermal conductivity, W/(°C $\cdot$ cm).

An ideal thermoelectric material would have a high Seebeck coefficient, low electrical resistivity, and low thermal conductivity. Unfortunately, materials with low electrical resistivity have a high thermal conductivity since both properties are dependent, to some extent, on the number of free electrons in the material. The maximum conversion efficiency of a thermoelectric generator is a function of the figure of merit, the hot junction temperature, and the temperature difference between the hot and cold junctions.

In some types of thermoelectric materials, the voltage difference between the hot and cold junctions results from the flow of negatively charged electrons (n type, hot junction positive), whereas in other types, the voltage difference between the cold and hot junctions results from the flow of positively charged voids vacated by electrons (p type, cold junction positive). Since the voltage output of a typical semiconductor thermoelectric couple is low (about 100 to 300 μV/°C temperature difference between the hot and cold junctions), it is advantageous to use both p- and n-type materials in constructing a thermoelectric generator. The two types of materials make it possible to connect the thermojunctions in series electrically and in parallel thermally (Fig. 9.1.20).

Typical semiconductor thermoelectric materials are compounds and alloys of lead, selenium, tellurium, antimony, bismuth, germanium, tin, manganese, cobalt, and silicon. To these materials, minute quantities of

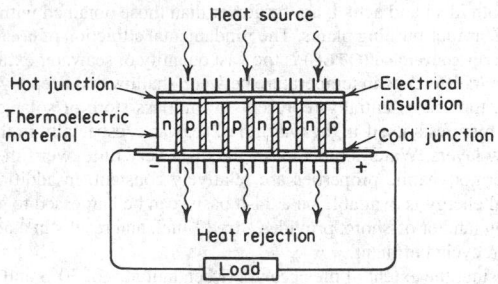

Fig. 9.1.20

"dopants" such as boron, phosphorus, sodium, and iodine are sometimes added to improve properties. Typical Z values for the more commonly used thermoelectric materials are in the range of 0.5×10^{-3} to 3.0×10^{-3} (°C)$^{-1}$. The onset of deleterious thermochemical effects at elevated temperatures, such as sublimation or reaction, limit the materials' application. Bismuth telluride alloys, which have the highest Z values, cannot be used beyond a hot-side temperature of about 300°C without encountering undue degradation. Silicon-germanium alloys have high-temperature capability up to 1,000°C that can take advantage of higher Carnot efficiencies. However, these alloys possess low Z values. Optimized designs of thermoelectric junctions using semiconductor materials have resulted in experimental conversion efficiencies as high as 13 percent; however, the efficiency of practical thermoelectric generators is lower, e.g., 4 to 9 percent. Materials which have higher figures of merit (2 or 3×10^{-3}) and which are capable of operating at higher temperatures (800 to 1,000°C) are required for an appreciable improvement in efficiency.

Thermoelectric-generation technology has matured considerably through its application to nuclear power systems for space vehicles where modules as large as 500 W have been used. It is also used in terrestrial applications such as gas pipeline cathodic protection and power for microwave repeater stations. Development work continues, but the use of this technology is expected to be limited to special cases where power source selection criteria other than efficiency and first cost will dominate.

Thermoelectric Cooling

The **Peltier effect,** discovered in 1834, is the inverse of the Seebeck effect. It involves the heating or cooling of the junction of two thermoelectric materials by passing current through the junction. The effectiveness of the thermojunction as a cooling device has been greatly increased by the application of semiconductor thermoelectric materials. Typical applications of thermoelectric coolers include electronic circuit cooling, small-capacity ice makers, and dew-point hygrometers, small refrigerators, freezers, portable coolers or heaters, etc. These devices make it possible to preserve vaccines, medicines, etc., in remote areas and in third world countries during disasters or military conflicts.

Thermionic Generation

Thermionic generation, proposed by **Schlicter** in 1915, uses a thermionic converter (Fig. 9.1.21), which is a vacuum or gas-filled device with a hot electron "emitter" (cathode) and a cold electron "collector" (anode) in or as part of a suitable gastight enclosure, with electrical connections to the anode and cathode, and with means for heating the cathode and cooling the anode. A thermionic generator is a low-voltage dc device.

Figure 9.1.22 is a plot of the electron energy at various places in the converter. The abscissa is cathode-anode spacing, and the ordinate is electron energy. The base line corresponds to the energy of the electrons in the cathode before heating. Heating the cathode imparts sufficient energy to some of the electrons to lift them over the **work function barrier** (retaining force) at the surface of the cathode into the interelectrode space. (The lower the work function, the easier it is for an electron

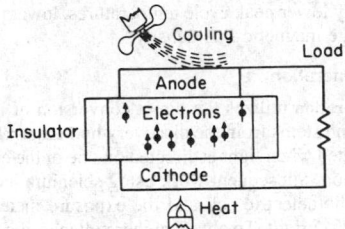

Fig. 9.1.21

to escape from the surface of the cathode.) If it is assumed that the electrons can follow path *a* to the anode with only a small loss of energy, they will "drop down" the work function barrier as they join the electrons in the anode still retaining some of their potential energy (**Fermi level**), which is available to cause an electric current to flow in the external circuit. The work function of the anode should be as small as possible. The anode should be maintained at a lower temperature to prevent anode emission or back current. This pattern presumes that the electrons could follow path *a* from the cathode to the anode with little interference. Since, however, electrons are charged particles, those in the space between the cathode and anode form a space charge barrier, as shown by *b*. This space charge barrier limits the electrons emitted from the cathode. Space charge formation can be reduced by close spacing of the cathode and anode surfaces or by the introduction of a suitable gas atmosphere that can be ionized by heating and thus neutralize the space charge. In vacuum-type thermionic converters, the spacing between cathode and anode must be less than 0.02 mm to get as many as 10 percent of the electrons over to the cathode and to achieve an efficiency of 4 to 5 percent. In gas-filled converters, the negative electron space charge is neutralized by positive ions. Cesium vapor is used for this purpose. At low pressure, it will also lower the work function of the

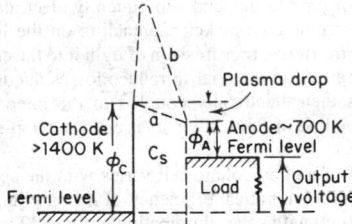

Fig. 9.1.22

anode, and at high pressure, it can, in addition, be used to adjust the work function of the cathode. Efficiencies as high as 17 percent have been obtained with gas-filled converters operating at a cathode temperature of 1,900°C (2,173 K). The output voltage is 1 to 2 V, so the units must be connected in series for reasonable utilization voltages.

Thermionic development results have been encouraging, but major technical challenges remain to be resolved before reliable, long-life converters become available. Effort has been focused on problem areas such as the limited life of emitter materials, leaktightness of the converter, and dimensional stability of the converter gap. Studies of thermionic converters incorporated into nuclear reactors and as a topping cycle for fossil fuel fired steam generators as well as for space and solar applications have received the greatest attention.

Fuel Cells

The **fuel cell** is an electrochemical device in which electric energy is generated by chemical reaction without altering the basic components (electrodes and electrolyte) of the cell itself. It is a low-voltage, dc device. To obtain higher voltages, the elements must be stacked. The fact that electrode and electrolyte are invariant distinguishes the fuel cell from the primary cell and storage battery. The fuel cell dates back to 1839, when **Grove** demonstrated that the electrolysis of water could be reversed using platinum electrodes. The fuel cell is unique in that it converts chemical energy to electric energy without an intermediate conversion to heat energy; its efficiency is therefore independent of the thermodynamic limitation of the Carnot cycle. In practical units, however, its efficiency is comparable with the efficiency of Carnot limited engines.

Figure 9.1.23 is a simplified version of a hydrogen or hydrocarbon fuel cell with air or oxygen as the other reactant. The fuel is supplied to the anode, where it is ionized, freeing electrons, which flow in the external circuit, and hydrogen ions, which pass through the electrolyte to the cathode, which is supplied with oxygen. The oxygen is ionized by

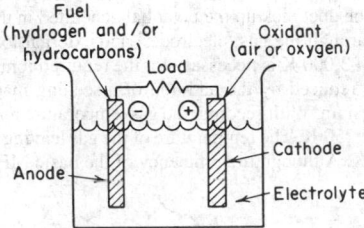

Fig. 9.1.23

electrons flowing into the cathode from the external circuit. The ionized oxygen and hydrogen ions react to form water. Electrodes for this type of cell are usually porous and impregnated with a catalyst. In a simple cell of this type, chemical and catalytic action take place only at the line (**notable surface of action**) where the electrolyte, gas, and electrode meet. One of the objectives in designing a practical fuel cell is to increase the notable surface of action. This has been accomplished in a number of ways, but usually by the creation of porous electrodes within which, in the case of gas diffusion electrodes, the fuel and oxidant in gaseous state can come in contact with the electrolyte at many sites. If the electrolyte is a liquid, a delicate balance must be achieved in which surface tension and density of the liquid must be considered and gas pressure and electrode pore size must be chosen to hold their interface inside the electrode. If the gas pressure is too high, the electrolyte is excluded from the electrode, gas leaks into the electrolyte, and ion flow stops; if the gas pressure is too low, drowning of the electrode occurs and electron flow stops.

Fuel cells may be classified broadly by operating temperature level, type of electrolyte, and type of fuel. Low-temperature (less than 150°C) fuel cells are characterized by the need for good and expensive catalysts, such as platinum and relatively simple fuel, such as hydrogen. High temperatures (500 to 1,000°C) offer the potential for use of hydrocarbon fuels and lower-cost catalysts. Electrolytes may be either acidic or alkaline in liquid, solid, or solid-liquid composite form. In one type of fuel cell, the electrolyte is a solid polymer.

Low-temperature fuel cells of the hydrogen-oxygen type, one a solid-polymer electrolyte type, and the other using free KOH as an electrolyte, have been successfully applied in generating systems for U.S. space vehicles. High-temperature fuel cell development has been primarily in molten carbonate cells (500 to 700°C) and solid-electrolyte (zirconia) cells (1,100°C), but no significant practical applications have resulted.

Considerable study and development work has been done toward the application of fuel cell generating systems to bulk utility power systems. Low-temperature cells of the phosphoric acid matrix and solid-polymer electrolyte types using petroleum fuels and air have been considered. Cell efficiencies of about 50 percent have been achieved; but with losses in the fuel reformers and electrical inverters, the overall system efficiency becomes of the order of 37 percent. In this application fuel cells have environmental advantages, such as low noise, low atmospheric emissions, and low heat rejection requirements. Additional development work is necessary to overcome the disadvantage of high catalyst costs and requirements for expensive fuels. More than 50 phosphoric acid fuel cell units, having a capacity of 200 kW, are in use. Companies in Canada, Germany, and the United States have demonstrated fuel cells in passenger bus propulsion systems. They are cooperating now on the development of proton exchange fuel cells. Other U.S. and Japanese companies are developing power plants for transportation.

Magnetohydrodynamic Generation

Magnetohydrodynamic generation utilizes the movement of electrically conducting gas through a magnetic field. Normally, it results in a high-voltage, dc output, but it can be designed to provide alternating current. In the simple open-cycle MHD generator (Fig. 9.1.24), hot, partially ionized, compressed gas, which is the product of combustion, is expanded in a duct and forced through a strong magnetic field. Electrodes in the sides of the duct pick up the potential generated in the gas, so that current flows through the gas, electrodes, and external load. Temperature in excess of 3,000 K is necessary for the required ionization of gas, but this can be reduced by the addition of a seeding material such as potassium or cesium. With seeding, the gas temperature may be reduced to the order of 2,750 K. The temperature of the gas leaving the generator is about 2,250 K. Although the efficiency of the basic MHD channel is

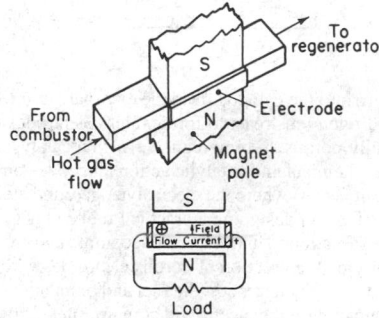

Fig. 9.1.24

on the order of 70 percent, only a portion of the available thermal energy can be removed in the channel. The remainder of the energy contained in the hot exhaust gas must be removed by a more conventional steam cycle. In this combined-cycle plant, the exhaust gas from the MHD generator is passed sequentially through an air preheater, the steam superheater and boiler, and an economizer and stack gas cooler. The air preheater is necessary to raise the temperature of incoming combustion air to some 1,900 K in order to obtain the initial gas temperature of 2,750 K.

The potential improvement in efficiency from the use of MHD generator in a combined-cycle plant is in the order of 15 to 30 percent. An overall steam plant efficiency of 38 percent could be raised to some 45 to 54 percent. Contrasted to other methods for direct conversion, MHD generation appears best suited to large blocks of power. For example, an MHD generator 75 m long with an average magnetic field of 5 T (attained by means of a superconductive magnet) would have a net output of about 1,000 MW dc at 5 to 10 kV. Typically, this would provide topping energy for a steam plant of about 500 MW.

Although the MHD topping cycle offers the highest peak cycle temperature and thermodynamic cycle efficiency of any system that has been studied, none of the generators tested have yielded enough efficiency to account for even half of the power required to supply oxidant to the combustor. Serious materials problems have also been experienced, with severe erosion, corrosion, and thermal stresses in the electrodes and insulators. Slow progress in the solution of these and other difficulties diminishes the prospect for a viable MHD system in the foreseeable future.

Closed-cycle MHD generators are also under study for bulk power generation. They are of two types: first, one in which the working fluid is an inert gas such as argon seeded with cesium; and second, the liquid-metal type in which the working fluid is a helium-sodium mixture. Closed-cycle MHD generation offers the potential for high efficiency with considerably lower peak cycle temperatures, lower pressure ratios, and lower average magnetic-flux density.

Photovoltaic Generation

Photovoltaic generation utilizes the direct conversion of light energy to electric energy and stems from the discovery by **Becquerel** in 1839 that a voltage is generated when light is directed on one of the electrodes in an electrolyte solution. Subsequent work using selenium led to the development of the photoelectric cell and the exposure meter. It results in low-voltage direct current. To obtain higher voltages, the elements must be stacked.

Photovoltaic effect is the generation of electric potential by the ionization by light energy (photons) of the area at or near the p-n junction of a semiconductor. The p-n junction constitutes a one-way potential barrier which permits the passage of photon-generated $(-)$ electrons from the p to the n material and $(+)$ "holes" from the n to the p material. The resulting excess of $(-)$ electrons in the n material and $(+)$ holes in the p material produces a voltage at the terminals comparable to the junction potential.

Solar cells have been a useful source of electricity since about 1960 and have enjoyed widespread use for small amounts of electric energy in remote locations. They have proved particularly well suited for use in spacecraft; in fact, much of the effort in photovoltaic R&D has been funded by the space program. Solar cells have been applied also to remote weather monitoring and recording stations (some of them equipped with transmitters to send the data to collecting stations), traffic control devices, buoys, channel markers, navigational beacons, etc. They can also be used for applications such as battery chargers for cars, boats, flashlights, tools, watches, calculators, and emergency radios. Cost reduction through the use of polycrystalline or thin-film techniques constitutes a major development effort.

A commercially available solar cell is constructed of a 0.3-mm-thick silicon wafer (2 × 2 cm or 2 × 6 cm) that is doped with boron to give it a p-type characteristic. It is then diffused with phosphorus to a depth of about 10^{-4} cm (n-type layer), and subsequently electrically contacted with titanium-silver or gold-nickel. Contacting on the light exposure side is limited to maximize transmission of light into the cell. The cell is coated with antireflection material to reduce losses due to light reflection. The cell is then electrically coupled to the intercell circuit by soldering. The method of fabricating solar cells is complex and expensive.

The efficiency of a photovoltaic cell varies with the spectrum of the light. The maximum theoretical efficiency of a single-junction, single-transition silicon cell with solar illumination is about 22 percent. Actual cell performance has been realized at 12 to 15 percent efficiency at 0.6 V open circuit and 0.02 W/cm². Advanced cells using materials such as gallium arsenide and cadmium telluride offer maximum theoretical efficiencies above 25 percent. Further cell efficiency improvement is being investigated by concentrating the sunlight with a Fresnel lens or parabolic mirror and by selecting the light spectrum.

Factors reducing the efficiency of conversion of light energy into electricity using solar cells include: (1) the fact that only a certain bandwidth of the solar spectrum can be effective, (2) structural defects and chemical impurities within the materials, (3) reflection of incident light, and (4) cell internal resistance. These factors lower the overall efficiency of solar arrays to about 6 to 8 percent. Another drawback is the intermittent nature of the solar source, which necessitates the use of an energy storage facility.

It is commonly agreed that substantial investments will be necessary to make solar cell energy conversion economically competitive with terrestrial fossil fuel fired or nuclear power plants. Space applications remain a practical use of this technology, since long life and reliability override cost considerations.

Other Energy Converters

Each of the converters described in the following section has characteristics which makes it suitable for specific tasks. Some produce low-voltage direct current and must be stacked for higher voltage output. Other converters produce high-voltage direct or alternating current depending upon their design. High voltages can be used to drive Klystron or X-ray tubes, or similar equipment, and can be transmitted without step-up transformers. Some medium- or low-temperature converters can be used in low-grade energy (heat) applications; in medical practice, e.g., body heat can drive heart pacemakers, artificial heart pumps, internal medicine dispensing devices, and organ monitoring equipment.

Electrohydrodynamic Converters When positive ions are transported by neutral hot gases against an electric field, high potential differences result. The charges produced by the ions do work when allowed to flow through a load. These devices are also called **electrogas-dynamic (EGD)** converters. If the gases are allowed to condense, producing small liquid droplets, the devices are often referred to as **aerosol EGD** converters. A number of different basic designs exist.

Van der Graaff Converter This device operates on the same principle as the EHD converter, except that a belt is used instead of hot gases to transport the ions against an electric field. Very high potential differences are produced, which may be utilized in high-energy particle accelerators, atom smashers, artificial lightning generators, and the like.

Ferroelectric Converters Certain materials exhibit a rapid change in their dielectric constant k around their Curie temperature. The voltage produced by a charge is the charge divided by the capacitance, or $V = Q/C$. The variation between capacitance and dielectric constant is expressed by $C_h = (k_h/k_c)C_c$, where c = cold and h = hot.

A capacitor containing ferroelectric material is charged when the capacitance is high. When the temperature is changed to lower the dielectric constant, the capacitance is lowered, with an accompanying increase in voltage. The charge is dissipated through a load when the voltage is high, resulting in the performance of work. Barium titenate, e.g., can produce a fivefold voltage swing between temperatures of 100 and 120°C. Thermocycling could be produced by the sun on a spinning satellite.

Ferromagnetic Converters Ferromagnetic material is used to complete the magnetic circuit of a permanent magnet. The ferromagnetic material is heat-cycled through its Curie point, producing flux changes in a coil wrapped around the magnet. The operating conditions can be selected by the Curie temperature of the ferromagnetic material. Gadolinium, e.g., has a Curie point near room temperature.

Piezoelectric Converters When axisymmetric crystals are compressed parallel to their polar axes, they become polarized; i.e., positive charges are generated on one side, and negative charges are generated on the other side of the crystal. The induced compressive stresses can be produced mechanically or by heating the crystal. The resulting potential difference will do work when allowed to flow through a load. In a reverse process, imposing a potential difference between the ends of the crystal will result in a compressive stress within the crystal. The imposition of alternating current will result in controlled oscillations useful in sonar, ultrasound equipment, and the like.

Pyroelectric Converters Some materials become electrically polarized when heated, and the conversion of heat to electricity can be utilized as it is in piezoelectric converters.

Bioenergetic Converters Energy requirements for medical devices used to monitor or control the performance of human organs (heart, brain, etc.) range from a few microwatts to a few watts. In many cases, body heat is sufficient to operate an energy converter which, in turn, will power the device.

Nernst Effect Converters When heat flows through certain semiconductors exposed to a magnetic field perpendicular to the direction of heat flow, an electric potential difference will be induced along a third orthogonal axis. This conversion of heat to electricity can be utilized to do work. In a reverse fashion, crossing a magnetic and an electric field will produce a temperature difference (Ettinghausen effect, the reverse of the Nernst effect). This reverse conversion is useful in electric heating, cooling, and refrigeration.

Thermophotovoltaic Converters A radiant heat source surrounded by photovoltaic cells will result in the radiant energy being converted to electricity. Source radiation and photovoltaic cell characteristics can be controlled to operate anywhere in the spectrum.

Photoelectromagnetic Converters When certain semiconductors (Cu_2O, for example) are placed in a tangential magnetic field and illuminated by visible light, there will result an electric potential difference along an axis orthogonal to the other two axes. The resulting flow of electric current can be used to do work.

Magnetothermoelectric Converters A magnetic field applied to certain thermoelectric semiconductors produces electric potential differences, useful in power generation.

Superconducting Converters The phase transition in a superconductor can be utilized similarly to a ferroelectric converter. Thermal cycling of the superconductor material will produce alternating current in the coil surrounding it. An idealized analysis for niobium at 8 K yields a conversion efficiency of about 44 percent.

Magnetostrictive Converters Changes in dimensions of materials in a magnetic field produce electric potential differences, thus converting mechanical energy to electricity. The effects can be reversed by combining an electric field with a magnetic field to produce dimensional changes.

Electron Convection Converters When a liquid is heated (sometimes to the boiling point), electrons and neutral atoms are emitted from the liquid surface. The flow of vapor transports the electrons upward, where they are collected on screens. The vapor condenses and recycles into the liquid pool. High electric potential differences can be produced in this manner between the screens and the liquid pool, allowing the subsequent flow of current to do work. The process is similar to that for EGD converters.

Electrokinetic Converters Certain fluids flowing through capillary tubes due to pressure gradients produce an electric potential difference between the ends of the capillaries, converting flow (kinetic) energy to electricity.

Particle-Collecting Converters When an alpha, beta, or gamma particle emitter is surrounded by a collector surface, an electric potential difference is produced between the emitter and the collector. Biased screen grids can improve the performance.

EHD Water Drop Converter Two separate streams of water coming from the same reservoir, in falling, are allowed to break up into droplets. At the breakup points, each stream is surrounded by a short metal cylinder. Each cylinder is connected electrically to a screen at the bottom of the opposite stream. High potential differences are produced between the two metal cylinders.

Photogalvanic Converters Photochemical reactions often produced by solar radiation (especially at the shorter wavelengths) can be used to generate electricity. Concentration of solar energy can increase the power of the converters considerably. The actual processes are similar to those in fuel cells.

The field of instrumentation provides other techniques which could become useful as energy conversion devices. While many of the methods cited and described above are not economically competitive with conventional conversion methods in current use, some are adapted to unique situations where the matter of cost becomes inconsequential. Certainly, it is expected that as progress is made in the field of energy conversion, certain techniques will be refined to greater practicality, and others will be developed.

FLYWHEEL ENERGY STORAGE
by Sherwood B. Menkes

REFERENCES: The Oerlikon Electrogyro: Its Development and Application for Omnibus Service, *Auto Eng.*, Dec. 1955. Beams, Magnetic Bearings, *SAE Automotive Congress Proc.*, Jan. 1964. Beachley and Frank, Electric and Electric-Hybrid Cars, SAE paper 730619, Mar. 1973. Clerk, The Utilization of Flywheel Energy, SAE Paper 711A, June 1963. Lawson and Hellman, Design and Testing of High Energy Density Flywheels for Application to Flywheel/Heat Engine Hybrid Drives, SAE Paper 719150, Aug. 1971. Post and Post, Flywheels, *Sci. Am.*,

Dec. 1973. Rabenhorst, Primary Energy Storage and the Super Flywheel, Johns Hopkins University, TG 1081, Oct. 1970. Weber and Menkes, Flywheel Energy Propulsion and the Electric Vehicle, Paper 7458, Electric Vehicle Symposium, Feb. 1974. Ashley, UT and BMW Collaborate on Flywheel Systems, *Mech. Eng.,* Aug. 1994. Olszewski, Eisenhaure, Beachley, and Kirk, On the Fly or under Pressure, *Mech. Eng.,* June 1988. Ashley, Flywheels Put a New Spin on Electric Vehicles, *Mech. Eng.,* Oct. 1993. Flywheels Are Back, *Compressed Air Mag.,* June 1993.

For many years a **flywheel** has been defined as a heavy wheel which is used to oppose and moderate by its inertia any fluctuation of speed in the machinery with which it revolves. Shafts in many different kinds of machinery are subjected to torque loading that is not uniform throughout a work cycle. By utilizing a flywheel, the designer can incorporate a smaller driving motor, and achieve a smoother operation.

Until recently, design of flywheels has not posed any serious difficulties, for work cycles have been relatively short, and the flywheel functioned solely to regulate speed. The kinetic energy has been relatively small. Concentration of material in a massive rim provides the maximum moment of inertia for a given amount of material.

Within the last few years, as a result of concern about fuel shortages and environmental pollution, suggestions have been made to utilize unconventional energy sources. Accordingly, there is much interest in the use of flywheels to store large amounts of kinetic energy. *Thus the flywheel is proposed as a major storage device, rather than as a means to effect speed regulation.*

As Post and Post observed, old concepts often reappear in technology as our needs change. Flywheels were probably first used as energy storage devices in the potter's wheel, perhaps 5,000 years ago. The spindle was vertical; there was a head, on which clay was placed, and a separate flywheel below. The flywheel was used to store enough energy to turn rapidly and for a long time.

A power plant is designed to operate most efficiently under a set of stated conditions. When it is necessary to operate the plant at off-design conditions, efficiency decreases, often quite severely. If the plant is operated only at high efficiency, and the excess energy is stored until it is needed, fuel is conserved. In addition, certain sources of energy (solar radiation, wind, etc.) become attractive provided that we can deal effectively with the question of storing energy thus *freely available* until such time as it can be used.

The flywheel is an attractive energy storage concept for several reasons: (1) it is simple; (2) it is possible to store and abstract energy readily, either by mechanical means or by using electric motors and generators; (3) high power rates are practicable; (4) there is no stringent limitation on the number of charge and discharge cycles that can be used; (5) reliability promises to be high; and (6) maintenance costs should be low.

Modern flywheel technology is in its **infancy.** The first symposium on the state of the art was held in November 1975. Any specific application will require consideration of technical alternatives and a cost analysis. The following must be evaluated in each case: (1) how much energy can be stored per unit weight or volume of flywheel material, which in turn controls (2) the size flywheel required, (3) relative importance of friction losses and associated inefficiency, (4) system safety, and (5) nature of controls and systems needed to provide the proper interface between source of energy and the demand for it.

A uniform flat disk with a central hole was suggested to replace the massive rim flywheel, but the resultant dynamic stress distribution limits its use.

Improved stress distribution (for an **isotropic material**) can be effected by thickening the flywheel toward the center and making it possible to achieve a constant tangential stress distribution. The energy density capability of a flywheel in which constraints other than those due to stress considerations are removed can be calculated from

$$T = K_s \sigma / \rho$$

in which T is the specific energy, K_s a flywheel shape factor, σ the material working stress, and ρ the material density. For a solid metal wheel, the ideal shape is one in which K_s is unity. Lawson reports that

Lockheed has achieved a shape factor of 0.832; such a wheel constructed of *maraging steel* results in a T value of 52 Wh/kg.

The parameter T is useful to compare candidate energy storage concepts. Table 9.1.9, prepared by Weber and Menkes as part of a feasibility study of a flywheel powered local-duty automobile, indicates the range of possible values. Note the inclusion of the *Oerlikon gyrobus,* the first vehicular application of flywheel energy storage. Advanced anisotropic materials offer great promise as flywheel materials, and many organizations are now engaged in the design and development of fiber composite flywheels. These high-strength fibers, which include E glass, PRD − 49 (Kevlar), S glass, fused silica, and others, dictate radical changes in design concepts.

The size range for suggested flywheels is considerable, as are the recommended speed and energy capacity. Some applications are discussed below.

Central Stations Long-range energy storage in central stations is accomplished by storing fuel (coal, oil, or gas), using a hydro reservoir and, more recently, cryogenic tanks. The basic problem is brought about by **highly fluctuating power demand.** A typical electric utility load cycle has a peak on weekdays nearly double the demand at night, while there are no comparable peak demands on weekends.

Considerations of economy and efficiency make it attractive to increase the base capacity of the central station, to generate and store excess energy when it is available, and to draw on the stored energy when it is needed. One technique, in limited use, employs a pumped hydrostorage installation. The principal advantage there is that while the potential energy of fluid is stored at a higher elevation, there is no continuous loss of energy; this cannot be said for flywheel energy storage. Furthermore, pumped hydrostorage systems are completely safe and make use of existing technology both to store and to abstract the energy.

Unfortunately, *severe geographical constraints* limit the use of pumped hydrostorage as a *universal solution.*

Flywheels offer a good alternative to *pumped hydrostorage,* on the grounds of (1) compactness, (2) high power density, (3) reliability and low maintenance, (4) unlimited cycle life, and (5) good thermal compatibility with the environment. Several technological advances must be achieved, however, before flywheel energy storage becomes cost-effective. These *necessary* improvements are: (1) development of low-cost, high-energy-density composite rotors, (2) development of very low-friction bearing systems, and (3) development of improved motor generator systems and controls.

The first two factors are self-evident; the last is not. A generator must extract energy from a constantly decelerating flywheel, and then feed it into a power network at constant voltage and frequency. The generator must either invert the variable-frequency input to the desired frequency

Table 9.1.9 Energy Density T for Various Storage Elements

Storage element	Wh/kg
Internal combustion engine system	550*
Electrochemical storage:	
Lead-acid	18–33
Nickel-cadmium	26–40
Silver-zinc	66–132
Zinc-air (experimental)	110–176
Sodium-sulfur (experimental)	154–220
Lithium-halide (experimental)	220
Flywheels:	
Gyreacta transmission	0.7†
Oerlikon gyrobus	7.0†
4340 steel	26
Maraging steel	55
Advanced anisotropic materials	190–870
Hydraulic accumulator	7–15
Natural elastic band	9

* Based on specific fuel consumption of 0.5 lb/(bhp · h) and engine weight equal to that of gasoline carried.
† Systems actually operated.

or use some other scheme to accomplish the same result. Several systems are being developed which will do this.

Transportation Applications Ground transport vehicles are powered, by and large, exclusively by internal combustion engines. In passenger vehicles in particular, the thermal efficiency of the cycle is of the order of 10 to 15 percent. The waste of fossil fuel distillates and the concomitant problem of air pollution are well documented. Accordingly, it is attractive to consider the possibility of generating electricity at a remote site, and *providing on-board energy storage.* Under certain circumstances, an auxiliary supply can be maintained external to the vehicle (as in a third rail), but for reasonable route flexibility, a self-contained store of energy is required.

A number of suggestions have been made which are in various stages of development. At one extreme is an *all-electric local-duty vehicle;* at the other is a *hybrid heat engine and flywheel energy storage* without electric energy utilization at all. An intermediate arrangement would use a *heat engine, a flywheel, and an electric traction motor* drive system.

In the **all-electric vehicle,** major design problems include (1) development of a passive bearing system with an ultra-low-energy drain, (2) an increase in energy density capability of flywheels to provide reasonable *range* and *speed,* (3) a design safe enough to withstand collisions, and (4) development of a compact and efficient motor generator unit.

In the **heat engine flywheel hybrid** with entirely mechanical means of using flywheel energy, no new technology is needed. Such a vehicle can make fairly impressive gains in fuel economy, especially by means of *regenerative dynamic braking.* The major difference between this and conventional vehicles lies in the need for a *continuously variable transmission unit* coupled to the flywheel.

A **modification of the all-electric** vehicle would require the addition of a small heat engine, perhaps 25 percent the size of those now in use. This heat engine can be operated at maximum efficiency, with the storage element being used to supply energy for acceleration. The driver could switch to all-electric mode for urban driving or short trips.

Greater attention is being paid to hybrid vehicles utilizing flywheels in conjunction with either an all-electric vehicle or a combined internal combustion/electric battery drive vehicle. Together with continuing attention given to flywheel material and construction, efforts are being made to introduce electric drive vehicles for passenger automobiles within the next several years in several states. Those efforts are accompanied by continued advances in high-strength composite materials for flywheels, frictionless magnetic bearings, high-efficiency motor generators, and continued miniaturization of power components and control electronics. The reference to Olszewski et al. is particularly instructive as to recent data and design details which have met with some success in the continuing development work in this generic area.

Regenerative dynamic braking is in use in the New York subway system; a **flywheel trolley coach** was developed for the city of San Francisco.

The output from an **exotic source**—sun or wind—is cyclical in nature. Exploitation of this type of energy source, especially for generating electric power, must be accompanied by suitable "flywheel" energy storage devices. Toward that end, rotating flywheels may hold promise for small units adaptable to residential use, especially in remote areas.

9.2 STEAM BOILERS
by Joseph C. Delibert

REFERENCES: The Babcock & Wilcox Co., "Steam—Its Generation and Use." Combustion Engineering, Inc., "Combustion—Fossil Power Systems." Staniar, "Plant Engineering Handbook," McGraw-Hill. Powell, "Water Conditioning for Industry," McGraw-Hill. "Boiler and Pressure Vessel Code." "Power Test Code for Steam Generating Units," American Society of Mechanical Engineers. Also, *Proceedings* of the ASME, the Joint Power Generation Conference, and the American Power Conference.

FUELS AVAILABLE FOR STEAM GENERATION

(See also Secs. 7, 9.1, and 9.8.)

A large variety of materials and heat sources can be used for steam generation. In the absence of other considerations, boilers are designed to use the most economical fuel or combinations of fuels available. These include natural gas, residual oil, and coal. Some typical solid fuels are: anthracite coal, bituminous and subbituminous coal, coke breeze, fluid petroleum coke (4 to 5 percent volatile), lignite, low-temperature fluid coal char (with auxiliary fuel), petroleum coke (9 to 14 percent volatile; with auxiliary fuel), pipeline slurry, wood and bark, and bagasse and other agricultural wastes. Other less common energy sources, many suitable for steam generation, are described in Sec. 9.1.

Sources of fuel nearest the plant are generally favored, but the spread of oil and gas pipelines throughout the United States and improved efficiency of coal transportation (unit trains, pipeline slurry) have greatly altered regional use patterns. Restrictions by the Environmental Protection Agency (EPA) on air and water pollution (Sec. 18) and the potential for interruption of oil supplies must also be considered. Many industries use their by-products for steam generation. Paper mills burn the black liquor from the cooking of wood pulp. Steel mill boilers may be fired with blast furnace and coke oven gases. Oil refineries burn lean CO gas, a waste product, in combination with a richer gas. The increased cost of basic fuels has caused many other industries to consider the use of their own combustible or heat-bearing by-products. The use of municipal wastes for steam generation is also accelerating (Sec. 7.4).

EFFECT OF FUEL ON BOILER DESIGN

Fuel is the governing factor in boiler design. Clean natural gas leads to the simplest design. If it is the only fuel, the boiler can be relatively small and compact. When solid or liquid fuels are to be used, the boilers will be larger because of the need to provide the required furnace volume for combustion and to accommodate ash and slag. Also, unless the fuel is low in sulfur content, the resultant combustion products will contribute to air pollution and must be reduced to approved levels before release. Equipment for this purpose can be large and expensive (Secs. 17.5 and 18). Some manufacturers offer **fluidized-bed firing** which uses a mixture of fine coal and limestone to trap most of the sulfur compounds in the ash. The equipment is not yet in general use.

Depending on the type of firing, considerable amounts of fly ash may be carried to the stack. Fly ash collectors and electrostatic precipitators are usually required to meet governmental requirements (Sec. 18). Oxides of nitrogen can also contribute to air pollution. These are formed from both fuel-bound nitrogen and the nitrogen contained in the combustion air. Nitric oxide omission can be controlled primarily by equipment design and operating techniques without appreciable impact on costs (Secs. 4, 7, and 18).

SLAG AND ASH

Boilers fired with pulverized coal can be designed for either dry-ash or slag-tap operation. The dry-ash type is particularly suited for coals with high ash-fusion temperatures. The ash impinging on the water-cooled furnace walls can be readily removed. The slag-tap furnace uses coals having low ash-fusion temperatures and is designed to have high temperatures near the furnace floor, thus keeping the ash molten for tapping.

When sintered or fused, ash forms deposits on furnace walls, boiler surfaces, and superheater tubes, thus reducing heat absorption, increasing draft loss and possibly causing overheating of tubes. Two general types of slag deposition can occur on furnace walls and convection surfaces. **Slagging** takes place when molten or partially fused ash particles entrained in the gas strike a wall or tube surface, become chilled, and solidify. Coals with low ash-fusion temperatures [i.e., those that are plastic or semimolten at temperatures less than 2,000°F (1,093°C)] have a high potential for slagging. Although normally confined to the furnace area, slagging can occur in the convection sections if proper design and operating parameters are not observed.

Fouling occurs when the volatile constituents in the ash condense on fly-ash particles, convection tubes, and existing ash deposits, at temperatures which keep the volatile constituents liquid and allow them to react chemically to form bonded deposits.

Slagging and fouling characteristics can be evaluated from the chemical composition of the ash by empirically determined relationships. The amount and the chemical and physical characteristics of coal ash vary over a wide range, not only from mine to mine, but also from different parts of the same mine. Thus, in design of boilers or selection of new coal sources for existing units, it is essential to have a thorough knowledge of the coal ash characteristics. Considerable data have been accumulated over the years, much of it based on eastern coals. Western coals are being used more extensively, and new criteria, often at variance with eastern experience, are being developed (Heil and Durrant, *Proc. JPGC,* 1978).

Coal ash may be classified as **eastern** or **lignitic**. By definition, if MgO + CaO is greater than Fe_2O_3, the ash is lignitic. If it is smaller, the ash is bituminous. This is important because subbituminous coal can have a lignitic type ash. Ash from eastern coals generally falls in the bituminous category while that from the west tends to be lignitic. Some of the parameters used to evaluate the effect of an ash on furnace slagging and deposition on both furnace walls and convection surfaces include:

Ash fusion temperatures
Iron content and ferritic percentage

$$\frac{(Fe_2O_3) \times 100}{Fe_2O_3 + 1.11FeO + 1.43Fe}$$

Silica ratio

$$\frac{(SiO_2) \times 100}{SiO_2 + Fe_2O_3 + CaO + MgO}$$

Base/acid ratio B/A

$$\frac{Fe_2O_3 + CaO + MgO + Na_2O + K_2O}{SiO_2 + Al_2O_3 + TiO_2}$$

Dolomite percentage

$$\frac{(CaO + MgO) \times 100}{Fe_2O_3 + CaO + MgO + Na_2O + K_2O}$$

Viscosity
Sintered strength

The preferred procedure for determining ash fusion temperatures is outlined in ASTM Standard D-1857, which defines and provides procedures to determine **initial deformation temperature (IDT), softening temperature (ST), hemispherical temperature (HT),** and **fluid temperature (FT).**

Iron has a significant effect on the behavior of coal ash. In the completely oxidized form it tends to raise fusion temperatures; in the lesser oxidized form it tends to lower them (Fig. 9.2.1). The iron content and its degree of oxidation also have a great influence on the viscosity, which increases with ferritic percentage. Liquid slag is not troublesome

as long as it remains a true liquid with a viscosity below 250 poise. The most troublesome form is plastic slag which is arbitrarily defined to exist in the region where the viscosity is 250 to 10,000 poise.

Slagging Indices The most accurate indicator of potential slagging for eastern or western coals is the viscosity-temperature relationship of

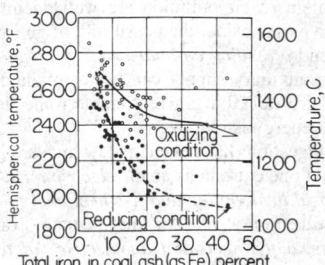

Fig. 9.2.1 Influence of iron on hemispherical temperature of ash.

the ash (Moore and Ehrler, *Proc. ASME,* WAM, 1973). Since viscosity measurements are costly and time-consuming, means have been developed to calculate furnace slagging potential from chemical analyses. For *eastern bituminous coals* the index is $(B/A)S$, where S is the percent sulfur, as S, on the dry-coal basis. The potential of ash with an index less than 0.6 is low; 0.6 to 2.0, medium; 2.0 to 2.6, high; above 2.6, severe.

For *lignitic-type ash,* slagging indices are based on fusion temperatures: [max HT + 4(min IDT)]/S, where the temperatures are the highest and the lowest reducing or oxidizing temperatures. Indices less than 2,100°F (1,149°C) are classed as severe slagging; 2,100 to 2,250°F (1,149 to 1,232°C), high; 2,250 to 2,450°F (1,232 to 1,343°C), medium.

Fouling Indices The volatile constituents of the ash (i.e., Na_2SO_4 or $CaSO_4 \cdot Na_2SO_4$) cause fouling and can be used as an indication of the fouling potential of a given coal. Two factors that affect fouling are deposit hardness and rate of deposition. Ash fusion temperatures bear little relation to the tendency to form bonded deposits (Fig. 9.2.2). Deposit hardness is affected by the chemical composition, temperature,

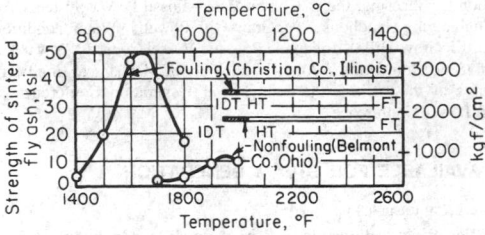

Fig. 9.2.2 Comparative sintered strengths and ash fusion temperatures for a fouling and a nonfouling coal.

and, to some extent, time. The rate of deposition is dependent on the volatile constituents and the amount of ash in the coal. Sintering strength of the ash, as determined in the laboratory, is an indication of how hard a deposit might become at different temperatures and has been used to predict fouling potential. Since these tests are expensive, the sintering strength has been related to the chemical composition (Fig. 9.2.3). (See Attig and Duzy, *Proc. Amer. Pwr. Conf.,* 1969). For *eastern coals* the **chemical index** is $(B/A) \times Na_2O_2$ where Na_2O is the weight percent in the ash prepared in accordance with ASTM D-271. For an index less than 0.2, fouling potential is low; 0.2 to 0.5, medium; 0.5 to 1.0 high; above 1.0, severe.

For **lignitic ash,** Na_2O is the determinant. Fouling tendency is low to medium for less than 3 percent, high for 3 to 6 percent, and severe for over 6 percent.

Additives such as dolomite, lime and magnesia are effective in reducing the sintered strength of ash (Fig. 9.2.4). Dolomite is also effective in neutralizing the acid in the flue gas and eliminating condensation and subsequent plugging in the cold end of air heaters.

The ash content of residual **fuel oil** seldom exceeds 0.2 percent but this relatively small amount is capable of causing severe problems of deposits on tubes and corrosion. To predict the effect of oil ash on slagging

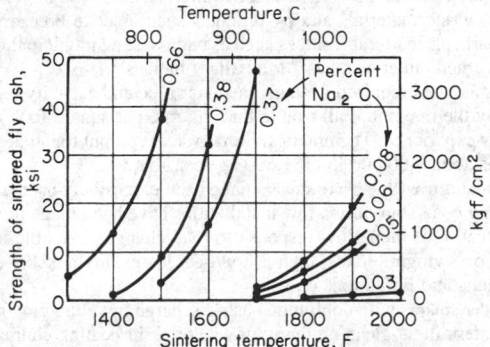

Fig. 9.2.3   Effect of sodium oxide content of coal on the sintered strength of fly ash.

and tube bank fouling, several variables are considered, including (1) ash content, (2) ash analysis, (3) melting and freezing temperatures of the ash, and (4) the total sulfur content. When oil is burned, complex chemical reactions occur, resulting in the formation of various oxides,

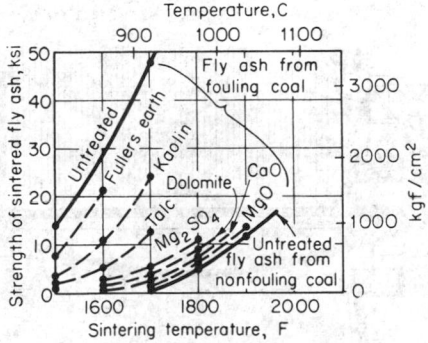

Fig. 9.2.4   Effect of additives on the sintered strength of fly ash (1 part additive to 4 parts fly ash).

Table 9.2.1   Analyses of Ash from Heavy Fuel Oil

	Analysis, %		
	Troublefree fuel oil	Troublesome fuel oils	
Ferric oxide, Fe_2O_3	56	8	6
Silica, SiO_2	25	9	5
Alumina, Al_2O_3	6	4	1
Lime, CaO	3	10	1
Magnesia, MgO	9	3	2
Vandium pentoxide, V_2O_5		1	39
Alkali sulfates	1	65	46

	Melting point in air	
Constituent	°F	°C
V_2O_5	1,274	690
$NaSO_4$	1,625	885
$MgSO_4$	2,165	1,185
$CaSO_4$	2,640	1,449
Fe_2O_3	2,850	1,566

vanadates, and sulfates. Many of these have melting points between 480 and 1,250°F (249 and 677°C), falling within the range of tube metal temperatures in the furnaces and superheaters of oil-fired boilers. As indicated in the analyses of Table 9.2.1, some oil ashes with high alkali sulfates can be troublesome. Dolomite added in quantities equal to the weight of the ash can be used to produce a softer slag which can be removed easily by soot blowing. In some installations air heater corrosion and pluggage and acid stack discharge are also minimized by dolomite additives.

SOOT BLOWER SYSTEMS

While slagging and fouling of coal- and oil-fired boilers can be minimized by proper design and operation, auxiliary equipment for cleaning furnace walls and removing deposits from convection surfaces must be supplied to maintain capacity and efficiency. Steam or air jets from the soot blower nozzles dislodge the dry or sintered ash and slag, which then fall into hoppers or travel along with the gaseous products of combustion to the removal equipment.

Types of soot blowers vary with their location in the boiler unit, the severity of ash or slag conditions, and the arrangement of the heat-absorbing surfaces.

Furnace walls are generally cleaned with **wall blowers** (Fig. 9.2.5) which project a nozzle assembly into the furnace for blowing and then retract it behind the wall tubes for protection after operation.

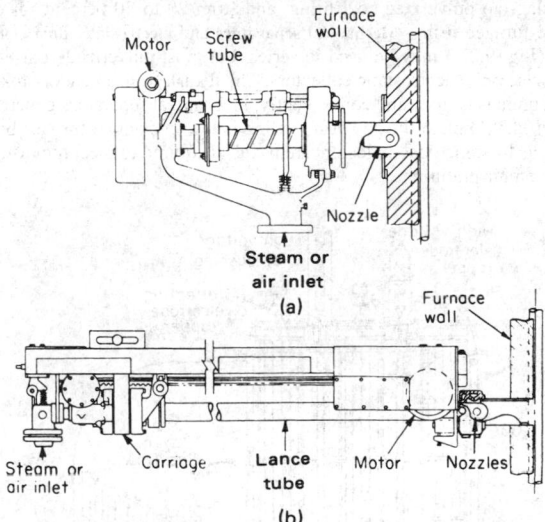

Fig. 9.2.5   Retracting soot blowers. (*a*) Furnace wall blower; (*b*) long-lance blower.

Tube banks in high-gas-temperature zones, such as slag screens, superheaters, and reheaters, where slag or sintered ash may accumulate, are generally cleaned by **long-lance retracting-type blowers** (Fig. 9.2.5). The lance, which rotates or oscillates as it advances into the boiler, is fitted with large nozzles to supply a powerful cleaning action and is retracted from the boiler for protection when it is not operating.

Tube banks located in low-gas temperature zones, including the economizer and boiler sections, where uncooled metals have satisfactory life and ash removal is easier, usually can be cleaned by **multiple-nozzle rotating-type soot blowers** (Fig. 9.2.6). However, long-lance retracting-type blowers may be necessary for very wide boilers, for extended cleaning ranges, or where the ash tends to pack or cake.

Soot blowers for air heaters generally are arranged to blow through the plate or tube assemblies with single- or multiple-nozzle elements moving in an arc or straight-line motion. These blowers may also supply water for washing the air-heater surface.

Additives to soften oil slag for easier removal by soot blowers may be introduced as a slurry spray through long-lance retractable blowers or through separate spraying equipment.

Automatic controls for soot blower systems often are used and can be arranged to operate the blowers in a prescribed sequence at time intervals adjusted to blower-cleaning requirements or to receive signals from the blower unit's instruments and controls so as to operate the soot blowers selectively in the various heat-absorbing sections in order to maintain the required cleanliness and heat absorption.

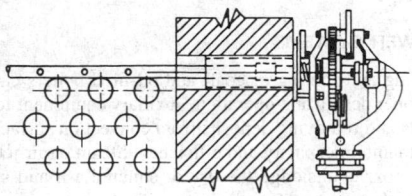

Fig. 9.2.6 Rotating soot blower with multiple nozzles.

ASH AND SLAG REMOVAL

Dust collectors (see also Sec. 18) are required for all large coal-burning blower units in order to reduce atmospheric pollution. The amount of ash entrained in the flue gas varies from about 80 percent of the ash in the coal for dry-ash pulverized-coal firing to approximately 50 percent for slag-tap pulverized-coal firing, and from 20 to 30 percent for cyclone-furnace firing. **Mechanical separators** and **electrostatic dust collectors** (Fig. 9.2.7) may be used in series, but most pulverized-coal-fired units use only electrostatic collectors. The fly ash from spreader-stoker-fired units is coarse, and consequently, mechanical separators generally are used. Although the gas-dust loading is low in cyclone-furnace boilers, electrostatic dust collectors are usually required to meet the restrictions on air pollution.

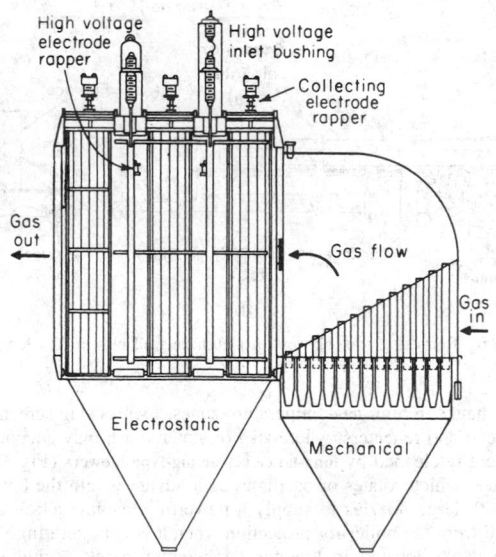

Fig. 9.2.7 Mechanical and electrostatic dust collectors in series.

The **bottom ash** recovered from the ashpits of chain-grate and spreader-type stokers is usually sold for cement-block aggregate. **Slag** from slag-tap furnaces can be used as a black granular coating for asbestos roofing shingles, as a mixture containing slag, fly ash, lime, and water for Poz-o-pac roads, or as an antiskid material for icy roads.

Fly ash presents disposal problems because of its low density and the consequent large volume which must be handled. It is not suitable for fill materials unless quickly covered. However, it can be utilized as an

admixture, replacing 20 to 30 percent of portland cement, and as a lightweight aggregate after sintering.

STOKERS

Almost any coal can be burned successfully on some type of stoker. In addition, waste materials and by-products such as coke breeze, wood waste, bark, agricultural residues such as bagasse, and municipal wastes can be burned either as a base or auxiliary fuel.

The grate area required for a given stoker type and capacity is determined by the maximum allowable burning rate per square foot, established by experience. The practical limit to steam output for stoker-fired boilers is about 400,000 lb (181,600 kg) per hour.

Chain- and **traveling-grate stokers** have been extensively used to burn noncoking coals, but only a few installations have been made in recent years because of their slow response to load changes, possible loss of ignition on swinging loads, high ashpit heat losses, high excess-air requirements, and limitations on size.

Spreader stokers with continuous-ash-discharge traveling grates (Fig. 9.2.8), intermittent-cleaning dump grates, or reciprocating continuous-cleaning grates are capable of burning all types of bituminous and lignitic coals. The fines are burned in suspension, and the larger fuel particles are burned on the grate. The use of a thin, fast-burning fuel bed provides rapid response to variations in load. Rotating mechanical feeding and distributing devices are generally used with spreader stokers. These stokers operate with low excess air and high efficiencies when the carbon in the fly ash is reinjected above the grate. However, relatively low gas velocities through the boiler are necessary to prevent fly-ash erosion, and fly-ash collectors should be used to reduce air pollution.

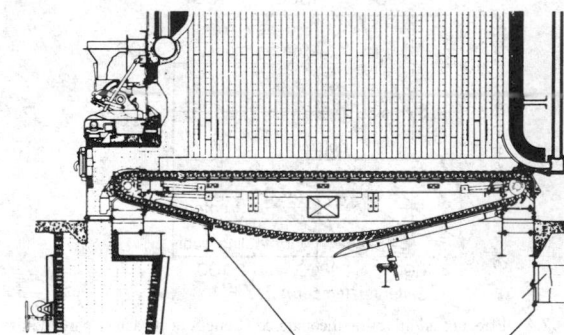

Fig. 9.2.8 Spreader stoker with traveling grate.

Single- or double-retort **underfeed stokers** with side ash dump and multiple-retort underfeed stokers with rear ash discharge are well suited for the burning of coking coals. These stokers operate best at steady loads. Both the ashpit heat loss and the maintenance are high.

PULVERIZERS

Pulverized coal firing is rarely used for boilers of less than 100,000 lb (45 t*) per hour steaming capacity since the use of stokers is more economical for those capacities. Most installations use the direct-fired system in which the coal and air pass directly from the pulverizers to the burners, and the desired firing rate is regulated by the rate of pulverization. Some types of direct-fired pulverizers are capable of grinding as much as 100 tons (91 t) per hour (Fig. 9.2.9). **Primary air** enters the pulverizer at temperatures that may run 650°F (343°C) or higher, depending on the amount of moisture in the coal and the type of pulverizer. The pulverizer provides the active mixing necessary for drying. The percentage of volatile matter in the fuel has a direct bearing on the

* 1 metric ton (t) = 1,000 kg = 2,205 lb.

recommended primary-air-fuel temperature for combustion. The generally accepted safe values for pulverizer exit fuel-air temperatures are:

Fuel	Exit temperature	
	°F	°C
Lignite	120–140	49–60
High-volatile bituminous	150	66
Low-volatile bituminous	150–175	66–79
Anthracite	175–212	79–100

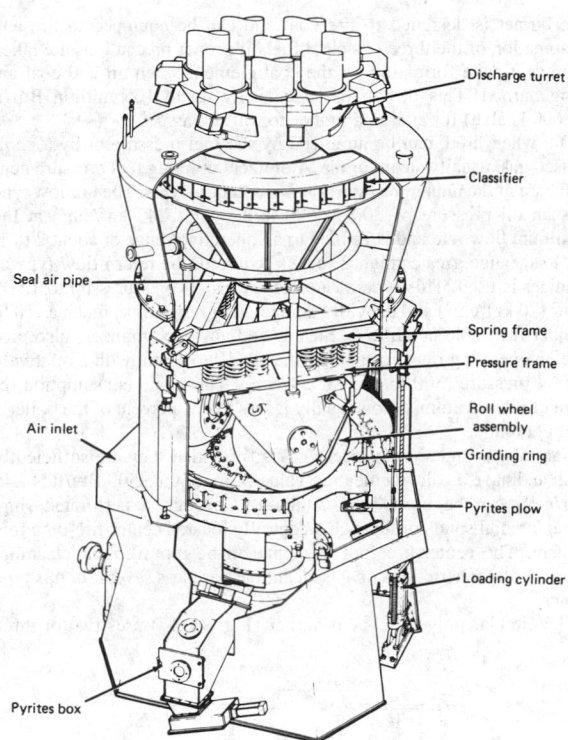

Fig. 9.2.9 Slow-speed pulverizer, roll-and-race type.

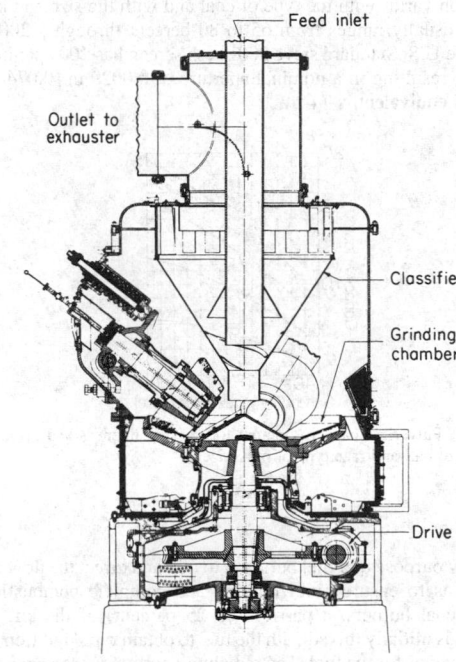

Fig. 9.2.11 Medium-speed pulverizer, roller type.

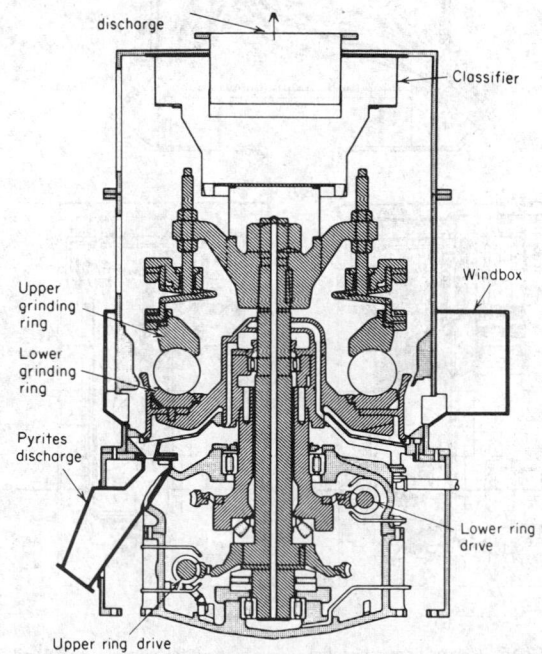

Fig. 9.2.10 Medium-speed pulverizer, contrarotation ball-and-race type.

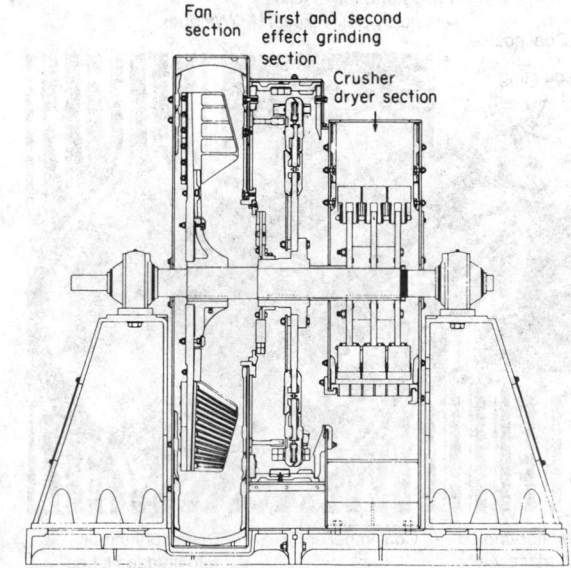

Fig. 9.2.12 High-speed pulverizer, attrition type.

The three principal types of pulverizers may be classified as *slow speed* (below 75 r/min), *medium speed* (75–225 r/min), and *high speed* (above 225 r/min). Figure 9.2.9 shows a slow-speed pulverizer using the roll-and-race principle. Medium-speed pulverizers are generally of either the ball-and-race type (Fig. 9.2.10) or the bowl-and-roller type (Fig. 9.2.11). High-speed pulverizers are usually of the attrition type (Fig. 9.2.12).

When a variety of coal is to be used, the pulverizer should be sized for the coal that gives the highest **base capacity,** which is the desired capacity divided by the *capacity factor,* a function of the *grindability* of the coal and the *fineness* required (Fig. 9.2.13). Capacities are established by testing with coals of different grindability. The required fineness of pulverization varies with the type of coal and with the size and kind of furnace. It usually ranges from 65 to 80 percent through a 200-mesh screen. [The U.S. Standard sieve 200-mesh screen has 200 openings per linear inch, resulting in a nominal aperture of 0.0029 in (0.074 mm).] The ASTM equivalent is 74 μm.

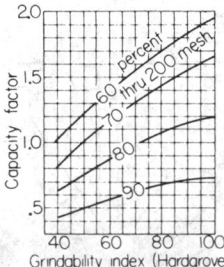

Fig. 9.2.13 Pulverizer capacity factors for varying fineness and grindability; medium-speed, ball-and-race type of pulverizers.

BURNERS

The primary purpose of a fuel burner is to mix and direct the flow of fuel and air so as to ensure rapid ignition and complete combustion. In pulverized-coal burners, a part (15 to 25 percent) of the air, called **primary air,** is initially mixed with the fuel to obtain rapid ignition and to act as a conveyor for the fuel. The remaining portion, or **secondary air,** is introduced through registers in the windbox (Fig. 9.2.14). This circular-

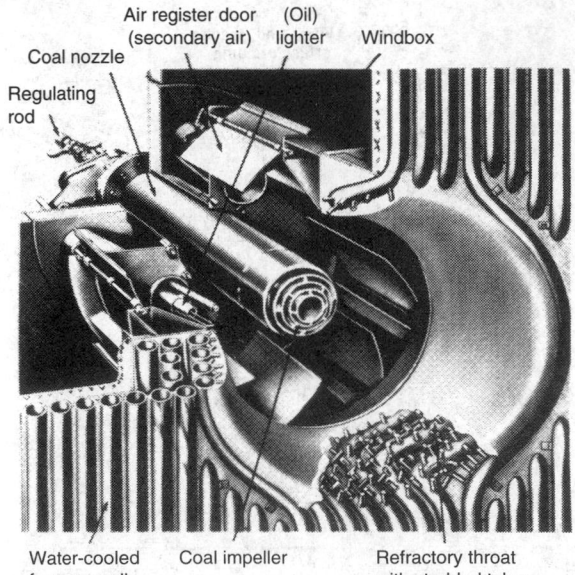

Fig. 9.2.14 Circular burner for pulverized coal, oil, or gas.

type burner is designed to fire coal and can be equipped to fire any combination of the three principal fuels if proper precautions are taken to prevent coke formation on the coal element when oil and coal are being burned. This design has a capacity up to 165 million Btu/h (41,600 kcal/h) for coal and higher for oil or gas.

Oil, when fired, can be atomized by the fuel pressure or by a compressed gas, usually steam or air. Atomizers utilizing fuel pressure generally are of the **uniflow or return-flow mechanical types.** The uniflow type uses an oil pressure of 300 to 600 lb/in² (21 to 42 kgf/cm²) at the maximum flow rate and is limited to an operating range of about 2 to 1. If a load range greater than 2 to 1 is required, the return-flow type of atomizer is used. This type of atomizer uses oil pressures up to 1,000 lb/in² (70 kgf/cm²) and provides an operating range of as much as 10 to 1 under favorable conditions. Steam- and air-type atomizers also provide an operating range of approximately 10 to 1, but with a relatively low oil pressure (300 lb/in²; 21 kgf/cm²). The steam consumption required for good atomization usually is less than 1 percent of the boiler's steam output.

Natural gas and some process gases [provided they are sufficiently clean and have a calorific heating value of more than 500 Btu/ft³ (4.45 kcal/m³)] can be burned by admission through a perforated ring, through radial spuds or through a centrally located center-fire-type fuel element. The center-fire fuel element can be removed for cleaning; consequently, restrictions on gas cleanliness are less severe for this type burner.

The circular pulverized-coal burner (Fig. 9.2.14) uses two or three

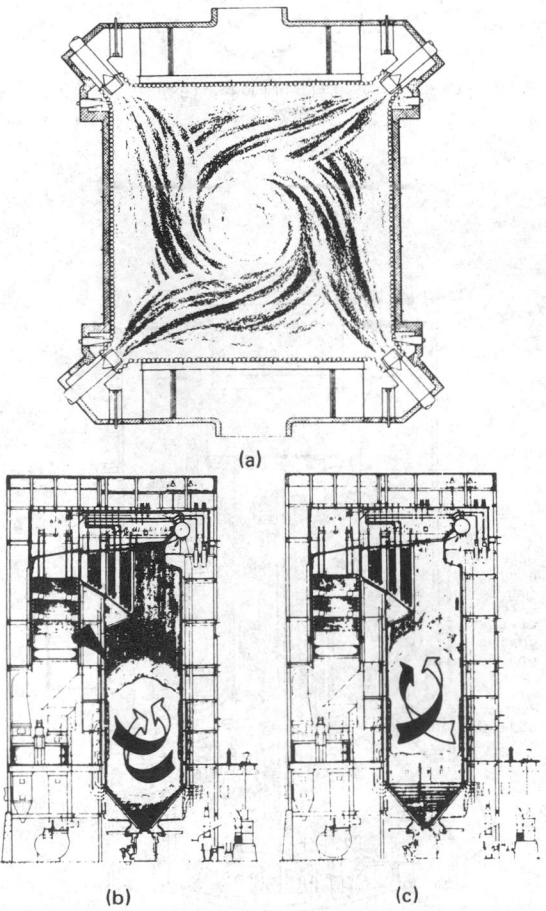

Fig. 9.2.15 Corner fired tangential tilting burners for pulverized coal, oil, or gas. (*a*) Plan section; (*b*) burners tilted down; (*c*) burners tilted up.

fuel nozzles and provides the excellent ignition characteristics of the circular burner. Gas, when fired in these burners, is introduced through fixed-spud-type elements located in the burner throat, and return-flow or steam- or compressed-air-type atomizers are used for firing oil.

When corner-fired burners (Fig. 9.2.15) are used, the mixing of fuel and combustion air takes place in the furnace (Fig. 9.2.15a). Oil and gas also can be fired in these burners by inserting fuel elements in the corner ports adjacent to the pulverized-coal nozzles. The burner tips can be tilted, as shown in Fig. 9.2.15b and c, to control the steam temperature.

A properly designed pulverized-coal installation should operate satisfactorily over a range of 2 or 3 to 1 without the need of auxiliary fuel to maintain ignition and without increasing or decreasing the number of burners in service. If some of the pulverizers and burners on large units are taken out of service as the load decreases, it is possible to operate at ratings down to one-sixth of full-load steam flow without the use of auxiliary fuel.

Blast-furnace and coke-oven gas burners are usually of the circular type, in which the gas is introduced either through a centrally positioned nozzle or through an annular port surrounding the coal nozzle, or of the intertube type, where the fuel and air ports are alternated across the width of the burner.

CYCLONE FURNACES

The cyclone furnace is designed to burn low-ash fusion coals and to retain most of the coal ash in the slag, which is then tapped from the furnace, thus preventing the passage of the ash through the heat-absorbing surfaces. The coal, crushed to 4-mesh size, is admitted with primary air in a tangential manner to the primary burner (Fig. 9.2.16). The finer particles burn in suspension while the coarser particles are thrown by centrifugal force to the outer wall of the cyclone furnace. The wall surface with its sticky coating of molten slag retains most of the particles of coal until they burn and leave their molten ash on the wall. The molten ash drains into the boiler furnace and then, through an opening in the boiler furnace floor, into the slag-collecting tank.

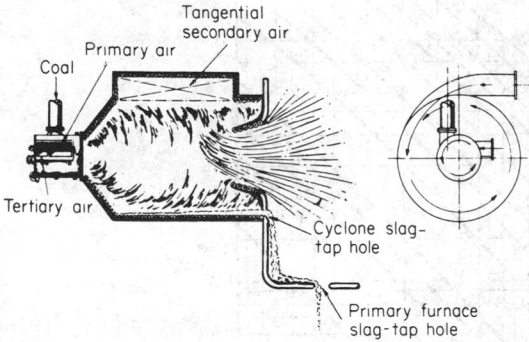

Fig. 9.2.16 Cyclone furnace.

The secondary air, which is admitted tangentially at the top of the cyclone, vigorously scrubs the coal particles on the wall, and combustion is completed at a firing rate of about ½ million Btu/(ft³ · h) [4,450 kcal/(m³ · h)]. The primary-furnace walls (consisting of fully studded tubes) also are wetted with molten ash and help to catch ash particles that are not retained in the cyclone furnaces. Gas, when available, can be burned by injection through openings at the bottom of the secondary-air ports. Oil is burned by spraying it axially into the cyclone through the primary burner or by firing it tangentially through an oil element located in the secondary-air port.

Figure 9.2.24 shows a boiler fired with twenty-three 10-ft-diam cyclone furnaces.

UNBURNED COMBUSTIBLE LOSS

The unburned combustible loss in the fly ash from pulverized-coal firing varies with the furnace-heat liberation, type of furnace cooling, use

of slag-tap or dry-ash removal, volatility and fineness of the coal, excess air, and type of burner (see Fig. 9.2.17). There is practically no combustible in the fluid slag from slag-tap furnaces. Although the hopper refuse from dry-ash furnaces usually is low in combustible, the combustible may be appreciable in some cases. The fly ash from cyclone-furnace boilers has a very low combustible content, varying from an equivalent 0.03 percent efficiency loss when burning Illinois coal to 0.15 percent for Ohio coal.

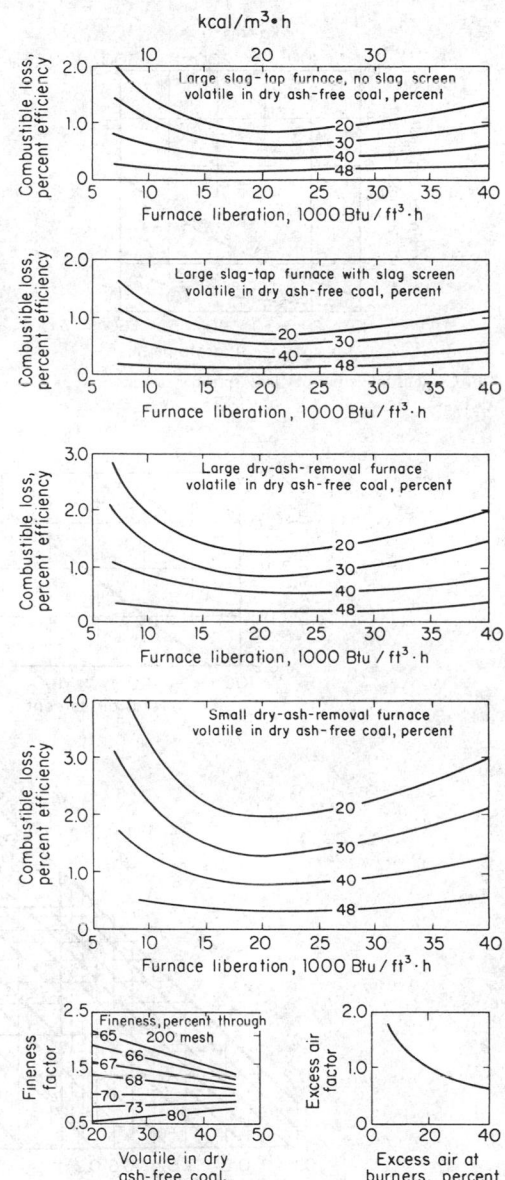

Fig. 9.2.17 Principal factors affecting combustible loss for pulverized coal firing in various types of furnaces.

Fly-ash combustible from spreader stokers varies widely with the rating, size, and type of coal burned. The combustible carryover at high rates of firing is relatively large, but reinjection of the fly ash is common and the loss in efficiency can be reduced as shown in Fig. 9.2.18.

The combustible loss in the fly ash from solid fuels is determined by withdrawing a representative sample of fly ash and flue gas from the boiler outlet flue, or stack, at the same velocity in the sampling tip as the gas velocity in the flue (see ASME Test Code). The rates of flue-gas flow and fly-ash collection are measured, and the dust loading in pounds per 1,000 lb of flue gas is then calculated. The combustible in fly ash also can be measured. The data in Fig. 9.2.19 can be used for rapid determination of efficiency loss when the dust loading and amount of combustible are known.

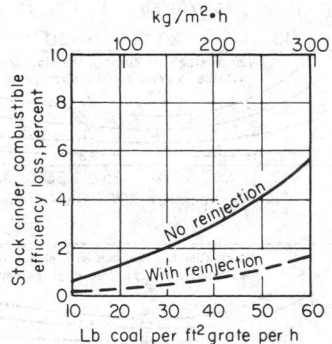

Fig. 9.2.18 Combustible efficiency loss, spreader stoker.

BOILER TYPES

The greater safety of **water-tube boilers** was recognized more than 100 years ago, and water-tube boilers have generally superseded the fire-tube type except in special cases such as small package-boiler designs and waste-heat boiler designs for medium- and low-pressure applications.

Water-tube boilers are available in a wide range of capacities—from as low as 5,000 lb (2.3 t) to as high as 9,000,000 lb (4,082 t) of steam per hour. By coordinating the various components—boilers, furnaces, fuel burners, fans, and controls—boiler manufacturers have produced a broad series of standardized and economical steam generating units, with capacities up to 550,000 lb (249 t) of steam per hour, burning oil or natural gas. For capacities up to 200,000 lb (90.7 t) per hour most units can be shipped by rail or truck as a completely shop-assembled package (Fig. 9.2.20). For larger units, barges or selected ocean vessels may be used if the site is suitably located. Beyond the limits for shop assembly, greatest economy is obtained by using modularized sections and shipping large assembled components. The two-drum-type boiler shown in Fig. 9.2.21, which comes in capacities up to 1,200,000 lb (544 t) per hour, is one example of modularized design.

Boilers utilizing banks of tubes directly connected to the steam and water drums are, in general, limited to a maximum steam pressure of 1,650 lb/in² (116 kgf/cm²), since the wide tube spacing required to maintain high drum-ligament efficiency reduces the effectiveness of heat absorption.

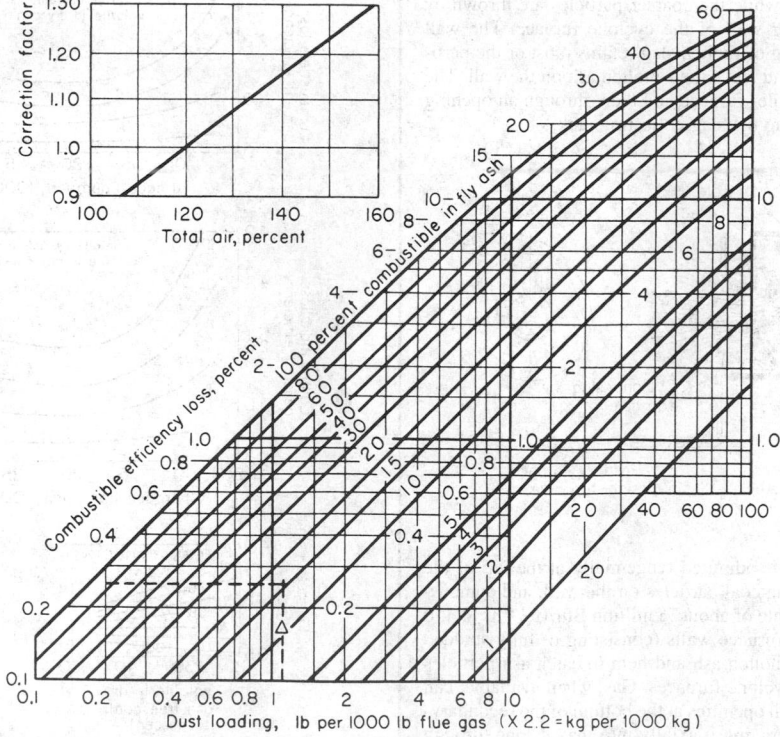

Fig. 9.2.19 Chart for determining combustible efficiency loss in fly ash. Heat loss in percent from combustible in fly ash is

$$\left[\frac{DL(C/100)(14,600)}{10,000,000/[7.6(TA/100) + 0.9] + [DL(C/100)(14,600)]} \right] \times 100$$

where DL = dust loading, lb/1,000 lb flue gas (× 2.2 = kg/1,000 kg); C = combustible in fly ash, percent; TA = total air, percent.

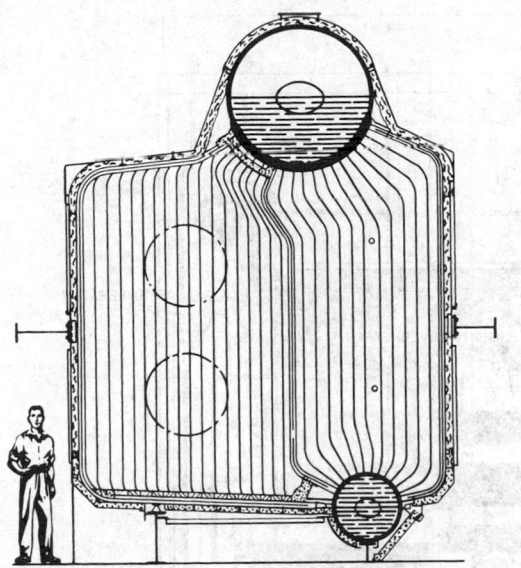

Fig. 9.2.20   Water-tube package boiler, single-pass gas.

Many designs of high-pressure, high-temperature, utility-type boilers, ranging from capacities of 500,000 to 9,000,000 lb (230 to 4,100 t) of steam per hour, are used, but they can generally be classed as radiant-type boilers. In radiant boilers, little or no steam is generated by convection-heat-absorbing surfaces since virtually all the steam is generated in the tubes forming the furnace enclosure walls from heat radiated to these tubes from the hot combustion gases. Figure 9.2.22 illustrates a large oil- and/or gas-fired natural-circulation type radiant-boiler unit,

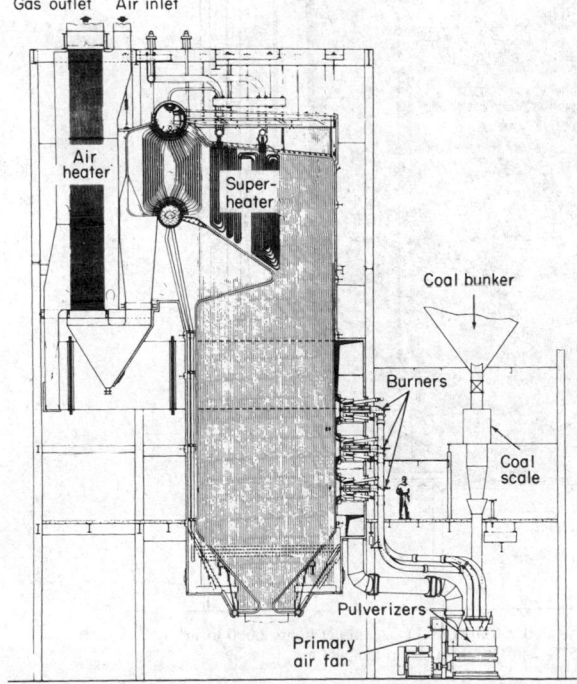

Fig. 9.2.21   Two-drum type of boiler designed for pulverized-coal firing.

and Fig. 9.2.23 shows a tangentially pulverized coal-fired, supercritical-pressure, combined-circulation type of boiler. The cyclone furnace forced-flow, once-through unit shown in Fig. 9.2.24 is known as a universal pressure boiler, since it can be designed for operation at pressures either above or below the critical pressure (3,206 lb/in²; 225 kgf/cm²).

Drum-type natural or assisted-circulation boilers are restricted to a maximum pressure of about 2,600 lb/in² (183 kgf/cm²) at the superheater outlet because of circulation and steam separation characteristics. However, once-through forced-flow-type boilers are not restricted to any pressure plateau by circulation limits.

In once-through forced-flow boilers, the water generally flows from the economizer to the furnace-wall tubes, then to the gas-convection-pass enclosure tubes and the primary superheater. Usually, the transition to the vapor phase (if operation is below the critical pressure) begins in the furnace circuits and, depending upon the operating conditions and the design, is completed either in the gas-convection-pass enclosure or in the primary superheater. The steam from the primary superheater passes to the secondary (and possibly to a tertiary) superheater. One or more reheaters are provided to reheat the low-pressure steam.

In addition to boilers for the conversion of energy in conventional fuels (coal, oil, and gas) to steam for power, heating or process use, many boilers have been developed for special requirements.

Waste-heat and exhaust-gas boilers utilize the sensible heat in the gas to generate steam. In the recovery of heat from the gas, water-tube boilers, often in conjunction with superheaters and economizers, are generally used; but fire-tube boilers may be used for cooling process or other gases when the containment of pressurized gas is a factor and the steam requirements are small.

High-temperature-water boilers provide hot water, under pressure, for space heating of large areas. Water is circulated at pressures up to 450 lb/in² (31.6 kgf/cm²) through the generator and the heating system. The water leaves the generator at subsaturated temperatures ranging up to 400°F (200°C). The boilers usually incorporate a water-cooled furnace and convection-gas-pass enclosure, with the convection-heat-absorbing surface arranged in sections similar to those of an economizer. Sizes generally range up to 60 million Btu/h (15,120 kcal/h) for package units, and field-erected units can be designed for much higher capacities.

The exhaust CO gas from oil refinery fluid catalytic-cracking units is used as the fuel for carbon monoxide boilers. Generally, a cylindrical furnace is used to contain the pressurized gas and the CO burners are arranged tangentially to increase the residence time of the gas in the furnace. The furnace walls are water cooled and tubes are refractory covered to promote ignition. Conventional-type gas and/or burners are provided for startup, for continuous pilots, and for the generation of steam when the cracker is out of service.

Recovery-type boilers are designed specifically for the recovery of chemicals in the spent cooking liquors from kraft, sulfite, soda, and other papermaking processes. The liquor is fired in a water-cooled furnace, either in suspension or in a smelt bed on the furnace floor. The chemicals, depending upon the process, are recovered from the smelt or the flue gas in a form which permits economical conversion for reuse.

FURNACES

A furnace is an enclosure provided for the combustion of fuel. The enclosure confines the products of combustion and is capable of withstanding the high temperatures developed and the pressures used. Its dimension and geometry are adapted to the rate of heat release, the type of fuel, and the method of firing so as to promote complete burning of combustible and provide suitable disposal of the ash.

Water-cooled furnaces are used with most boiler units and for all types of fuel and methods of firing. Water cooling of the furnace walls reduces the transfer of heat to the structural members and, consequently, their temperature can be limited to that which will meet the requirements of strength and resistance to oxidation. Water-cooled tube con-

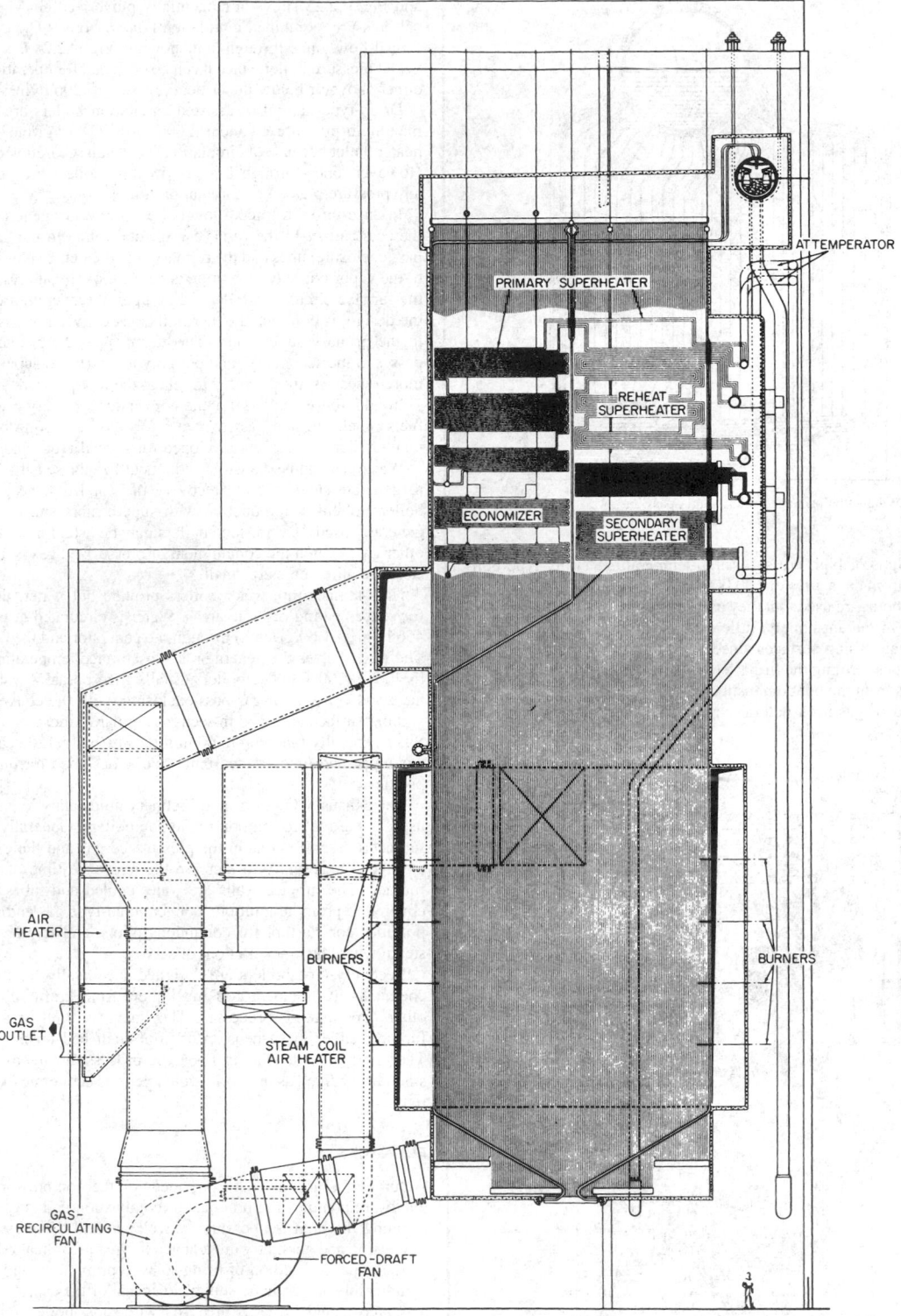

Fig. 9.2.22 Natural circulation radiant boiler, oil- and gas-fired; 4,200,000 lb (1,900 t) steam per hour; 2,600 lb/in² (183 kgf/cm²) pressure; 1,005°F (540°C) steam temperature; 1,005°F (540°C) reheat steam temperature.

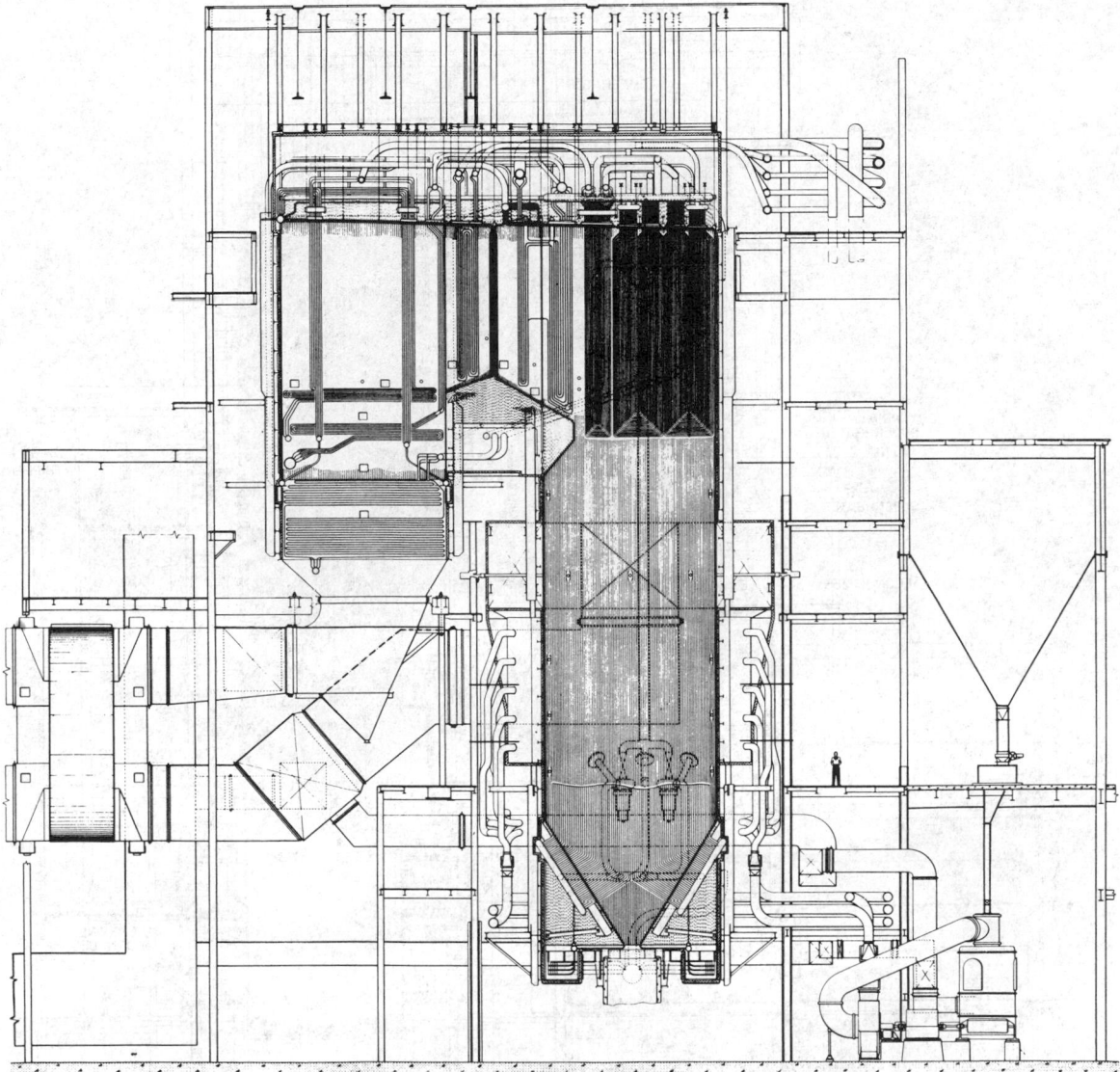

Fig. 9.2.23 Once-through boiler with combined circulation; twin pressurized furnaces; pulverized coal tangentially fired; 6,400,000 lb (2,900 t) steam per hour; 3,650 lb/in² (257 kgf/cm²) pressure; 1,003°F (539°C) steam temperature; 1,003°F (539°C) reheat steam temperature.

structions facilitate large furnace dimensions and optimum arrangements of roofs, hoppers, arches, and mountings for burners; and the use of tubular screens, platens, or division walls to increase the amount of heat-absorbing surface in the combustion zone. The use of water-cooled furnaces reduces the external heat losses; these losses for conventional-type furnaces are shown in Fig.9.2.37.

Heat-absorbing surfaces in the furnace receive heat from the products of combustion and, consequently, contribute directly to steam generation while lowering the furnace exit-gas temperature. The principal mechanisms of heat transfer take place simultaneously. These include intersolid radiation from the fuel bed or fuel particles, nonluminous radiation from the products of combustion, convection heat transfer from the furnace gases, and heat conduction through deposits and tube metals. (See also Sec. 4.) The absorption effectiveness of the furnace surfaces is influenced by the deposits of ash or slag.

Furnaces vary in shape and size, in the location and spacing of burners, in the disposition of heat-absorbing surface, and in the arrangement of arches and hoppers. Flame shape and length affect the geometry of radiation and the rate and distribution of heat absorption by the water-cooled surfaces.

Analytical solutions of the transfer of heat in the furnaces of steam-generating units are extremely complex, and it is most difficult to calculate furnace outlet-gas temperature by theoretical methods. Nevertheless, the furnace outlet-gas temperature must be accurately predicted, since this temperature determines the design of the remainder of the boiler unit, particularly that of the superheater and reheater. The calculations must therefore be based upon test results supplemented by data accumulated from operating experience and by judgments predicated upon knowledge of the principles of heat transfer and the characteristics of fuels and slags.

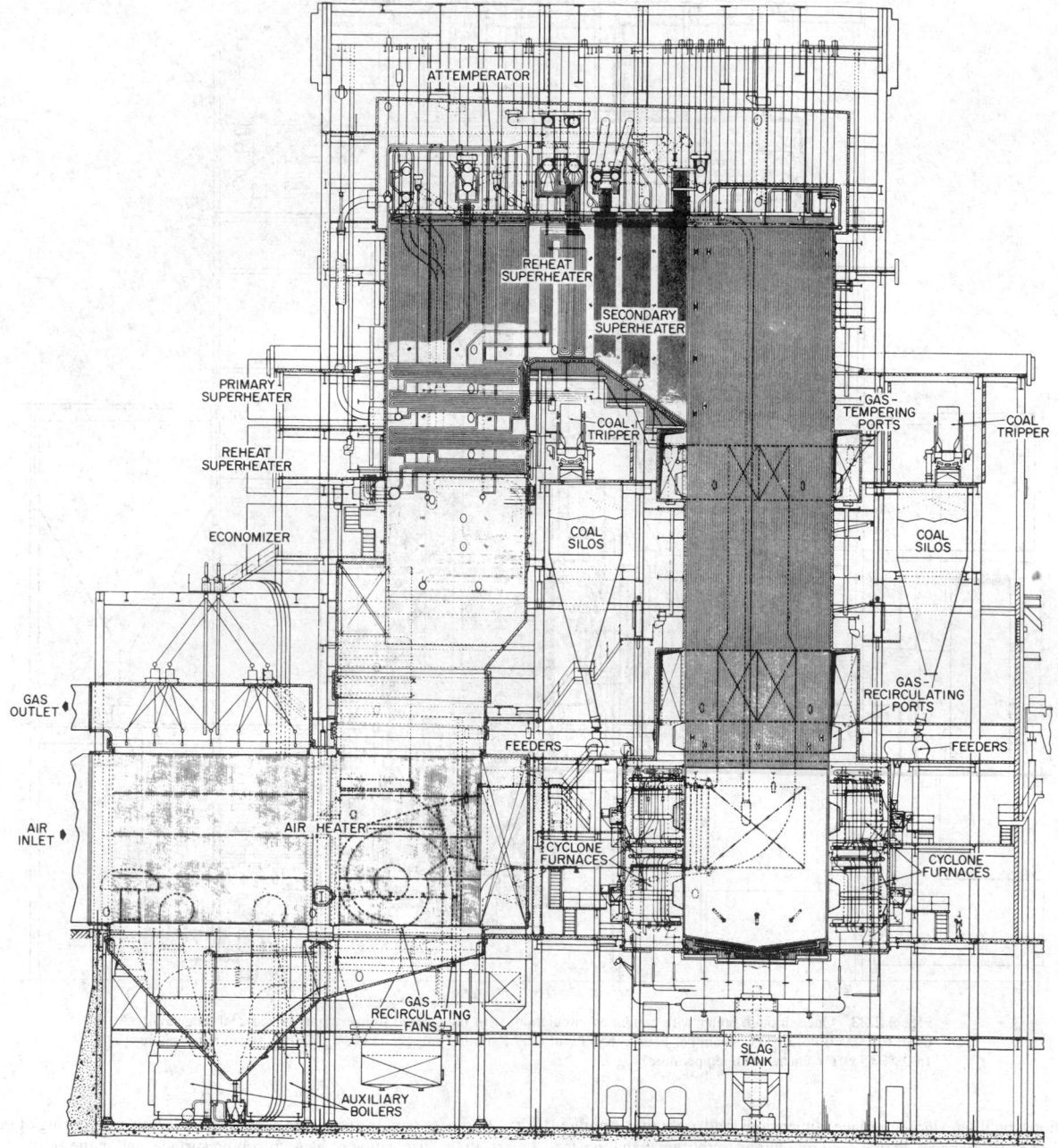

Fig. 9.2.24   Universal Pressure boiler; pressurized cyclone furnace, coal-fired; 8,000,000 lb (3,640 t) steam per hour; 3,650 lb/in² (257 kgf/cm²) pressure; 1,003°F (539°C) steam temperature; 1,003°F (539°C) reheat steam temperature.

The curves in Figs. 9.2.25 and 9.2.26 show the gas temperatures at the furnace outlet of typical boiler units when firing coal, oil, and gas. The furnace exit-gas temperatures vary considerably with coal firing because of the insulating effect of ash and slag deposits on the heat-absorbing surfaces. The amount of surface is the major factor in overall furnace heat absorption, and the heat released and available for absorption per hour per square foot of effective heat-absorbing surface is therefore a satisfactory basis for correlation. The heat released and available for absorption is the sum of the calorific heat content of the fuel fired and the sensible heat of the combustion air, less the sum of the heat unavailable owing to the unburned portion of the fuel and latent heat of the water vapor formed from moisture in the fuel and the combustion of hydrogen.

Furnace-wall tubes often are pitched on close centers to obtain maxi-

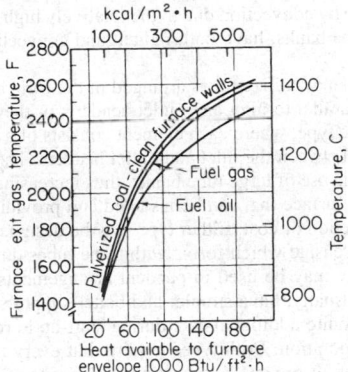

Fig. 9.2.25 Approximate gas temperatures at water-cooled furnace outlet with different fuels.

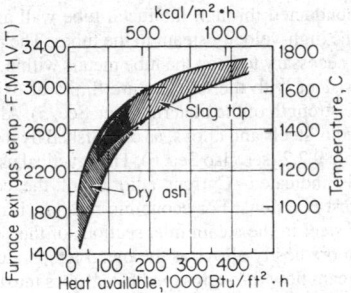

Fig. 9.2.26 General range of furnace exit gas temperatures, pulverized-coal firing. (MHVT = multiple-shield, high-velocity thermocouple.)

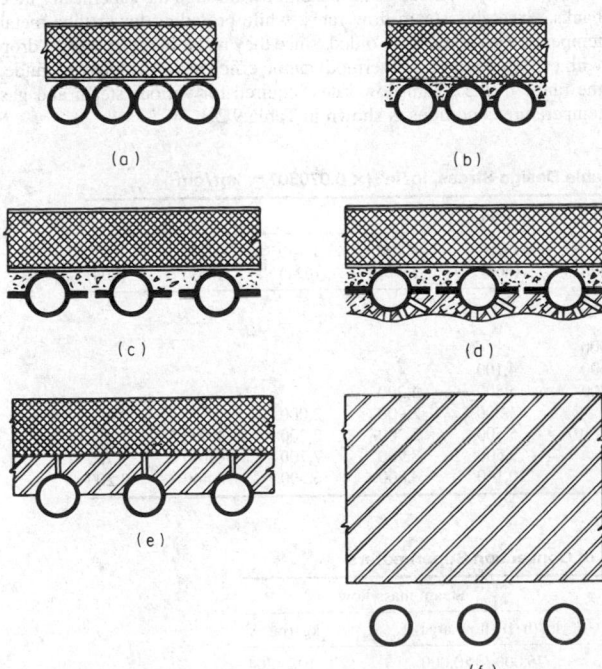

Fig. 9.2.27 Water-cooled furnace wall construction types. (*a*) Tangent tube wall; (*b*) welded membrane tube wall; (*c*) flat studs welded to sides of tubes; (*d*) full stud tube wall, refractory-covered; (*e*) tube and tile wall; (*f*) tubes spaced from refractory wall.

mum heat absorption and to facilitate ash removal. The arrangement takes the form of the so-called tangent tube construction (Fig. 9.2.27) wherein the adjacent tubes are almost touching with only a small clearance provided for erection purposes. However, most boilers now use **membrane tube walls** in which a steel bar, or membrane, is welded between adjacent tubes. This construction facilitates the fabrication of water-cooled walls in large shop-assembled tube panels. Less effective cooling is obtained, at a lower cost, by placing the tubes on wider spacing and using extended metal surface in the form of flat studs welded to the tubes. If even less cooling is desired, the tube spacing can be increased and refractory installed between or behind the tubes to form the wall enclosure.

Additional furnace cooling in the form of tubular platens, division walls, or wide-spaced tubular screens may be used. In high heat input zones, the tubes can be protected by refractory coverings anchored to the tubes by studs. Peak heat-absorption rates of furnace-wall tubes in the combustion zone may, in some designs, approximate 200,000 Btu/h · ft^2 [542 kcal/(h · m^2)] of projected surface, but the average heat absorption rate for the furnace is considerably lower (Fig. 9.2.28).

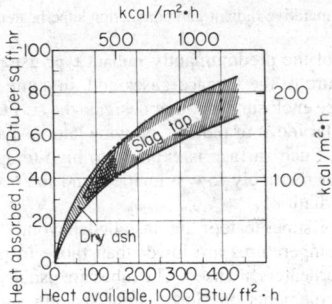

Fig. 9.2.28 General range of average heat absorption rate in water-cooled pulverized-coal-fired furnaces.

Furnace walls must be adequately supported with provision for thermal expansion and with reinforcing buckstays to withstand the lateral forces caused by the difference between the furnace pressure and the surrounding atmosphere. The furnace enclosure must prevent air infiltration when the furnace is operated under suction and gas leakage when the furnace is operated at pressures above atmospheric.

SUPERHEATERS AND REHEATERS

The addition of heat to steam after evaporation, or change of state, is accompanied by an increase in the temperature and the enthalpy of the fluid. The heat is added to the steam in boiler components called superheaters and reheaters, which are comprised of tubular elements exposed to the high-temperature gaseous products of combustion.

The advantages of superheat and reheat in power generation result from thermodynamic gain in the Rankine cycle (see Sec. 4) and from the reduction of heat losses due to moisture in the low-pressure stages of the turbine. With high steam pressures and temperatures, more useful energy is available, but the advances to high steam temperature often are restricted by the strength and the oxidation resistance of the steel and the ferrous alloys currently available and economically practical for use in boiler pressure-part and turbine-blade constructions.

The term superheating is applied to the higher-pressure steam and the term reheating to the lower-pressure steam which has given up some of its energy during expansion in the high-pressure turbine. With high initial steam pressure, one or more stages of reheating may be employed to improve the thermal efficiency.

Separately fired superheaters may be used, but superheaters usually are installed as an integral part of the steam-generating unit and broadly classified as **radiant** or **convection** types, depending upon the predominant method of heat transfer to the heat-absorbing surfaces.

The quantity of heat absorbed and the amount of superheat attained are dependent upon the size, location, and arrangement of the heat-absorbing surfaces; the temperature differentials between the gas, the tube metal, and the steam; and the heat-transfer coefficients. Steam-temperature characteristics of radiant- and convection-type super-heaters are shown in Fig. 9.2.29, as well as the effect of using a combination of these types to produce a more uniform steam temperature over a wide operating range.

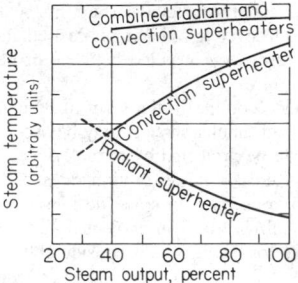

Fig. 9.2.29 Comparative radiant and convection superheater characteristics.

Superheaters of the predominantly **radiant type** usually are arranged for direct exposure to the furnace gases and, in some designs, form a part of the furnace enclosure. In other designs, the surface is arranged in the form of tubular loops or platens, on wide lateral spacing, extending into the furnace. Such surface is exposed to high-temperature furnace gases traveling at relatively low velocities, and the transfer of heat is principally by radiation.

Convection-type superheaters are installed beyond the furnace exit where the gas temperatures are lower than those in the zones where radiant-type superheaters are used. The tubes are usually arranged in the form of parallel elements on close lateral spacing and in tube banks extending partially or completely across the width of the gas stream, with the gas flowing through the relatively narrow spaces between the tubes. High rates of gas flow and, thus, high convection heat-transfer rates are obtained at the expense of gas-pressure drop through the tube bank.

Superheaters, shielded from the furnace combustion zone by arches or wide-spaced screens of steam-generating tubes, which receive heat by radiation from the high-temperature gases in cavities or intertube spaces and also by convection due to the relatively high rate of gas flow through the tube banks, have both radiant and convection characteristics.

Superheaters may utilize tubes arranged in the form of hairpin loops connected in parallel to inlet and outlet headers; or they may be of the continuous-tube type, where each element consists of a number of tube loops in series between the inlet and outlet headers. The latter arrangement permits the use of large tube banks, thus increasing the amount of heat-absorbing surface that can be installed and providing economy of space and reduction of cost. Either type may be designed for the drainage of the condensate which forms within the tubes during outages of the unit, or they may be used in pendent arrangements which are not drainable but, usually, have simpler and better supports. Nondrainable superheaters require additional care during start-up to remove the condensate by evaporation. Both types require that every tube have sufficient steam flow to prevent overheating during operation. In start-up, when flow is insufficient, gas temperatures entering the superheater must be controlled to limit tube metal temperatures to safe values for the material used.

The heat transferred from high-temperature gases by radiation and convection is conducted through the metal tube wall and imparted by convection to the high-velocity steam in the tubes. The removal of heat by the steam is necessary to keep the tube metals within a safe temperature range consistent with the temperature limits of oxidation and the creep or rupture strength of the materials (see Sec. 5). Allowable design stresses for various steels and alloys, as established by the ASME Code, are listed in Table 9.2.2 (see also Sec. 6). The practical use limit for each material also is indicated. Current editions of the code should be checked for latest revisions. For economic reasons, it is customary to use low-carbon steel in the steam inlet sections of the superheater and, progressively, more costly alloys as the metal temperatures increase.

The rate of steam flow through superheater tubes must be sufficiently high to keep the metal temperatures within a safe operating range and to ensure good distribution of flow through all the elements connected in parallel circuits. This can be accomplished by arrangements which provide for multiple passes of the steam flow through the superheater tube banks. Excessive steam-flow rates, while providing lower tube-metal temperatures, should be avoided, since they result in high pressure drop with consequent loss of thermodynamic efficiency. As a general guide, the range of the steam-flow rates required for various steam and gas temperature conditions is shown in Table 9.2.3.

Table 9.2.2 Superheater and Reheater Tubes—Maximum Allowable Design Stress, lb/in² (× 0.070307 = kgf/cm²)

Material	ASME spec. no. and type	900 (482)	950 (510)	1,000 (538)	1,050 (566)	1,100 (593)	1,150 (621)	1,200 (649)	1,300 (704)
					Temp, °F (°C)				
Carbon steel	SA210, grade C	5,000	3,000						
Carbon moly	SA209, T1a	13,600	8,200						
Croloy ½	SA213, T2	12,800	9,200	5,900					
Croloy 1¼	SA213, T11	13,100	11,000	6,600	4,100				
Croloy 2¼	SA213, T22		11,000	7,800	5,800	4,200			
Croloy 5	SA213, T5				4,200	2,900	2,000		
Croloy 9	SA213, T9				5,500	3,300	2,200	1,500	
Croloy 304H	SA213, TP 304H				9,500	8,900	7,700	6,100	3,700
Croloy 32H	SA213, TP 321H				10,100	8,800	6,900	5,400	3,200

SOURCE: ASME Code, 1983.

Table 9.2.3 Range of Steam Mass Flow Values of Convection Superheaters

Steam	Gas	lb/(h·ft² flow area)	kg/(m²·s)
	Temp, °F (°C)		Steam mass flow
Less than 750 (399)	1200 (649)	75,000–150,000	102–204
700–800 (371–426)	1600 (871)	250,000–350,000	340–475
800–900 (426–482)	2400 (1316)	400,000–500,000	545–680
900–1,000 (482–538)	2400 (1316)	500,000–600,000	680–816
1,000–1,100 (538–593)	2400 (1316)	700,000 and higher	950

The spacing of the tubular elements in the tube bank and, consequently, the rate of gas flow and convection heat transfer are governed primarily by the types of fuel fired, draft-loss considerations, and the fouling and erosive characteristics of fuel ash carried in the gas stream. With clean gases, or in the low-gas-temperature zones of coal-fired units, a gas flow rate of about 6,000 lb/(h · ft²) [8.2 kg/(m² · s)] of free-flow area is generally within economic limits. In the higher-gas-temperature zones, 1,600 to 2,300°F (871 to 1,260°C), the adherence and the accumulation of ash deposits can reduce the gas-flow area and, in some cases, may completely bridge the space between tubes. Thus, as gas temperatures increase, it is customary to increase the tube spacings in the tube banks to avoid excessive draft loss and to facilitate ash removal.

ECONOMIZERS

Economizers remove heat from the moderately low-temperature combustion gases after the gases leave the steam-generating and superheating/reheating sections of the boiler unit. Economizers are, in effect, feedwater heaters which receive water from the boiler feed pumps and deliver it at a higher temperature to the steam generator. Economizers are used instead of additional steam-generating surface, since the feedwater and, consequently, the heat-receiving surface are at temperatures below the saturated-steam temperature. Thus, the gases can be cooled to lower temperature levels for greater heat recovery and economy.

Economizers are forced-flow, once-through, convection heat-transfer devices, usually consisting of steel tubes, to which feedwater is supplied at a pressure above that in the steam-generating section and at a rate corresponding to the steam output of the boiler unit. They are classed as horizontal- or vertical-tube types, according to geometrical arrangement; as longitudinal or crossflow, depending upon the direction of gas flow with respect to the tubes; as parallel or counterflow, with respect to the relative direction of gas and water flow; as steaming or nonsteaming, depending on the thermal performance, as return-bend or continuous-tube, depending upon the details of design; and as bare-tube or extended-surface, according to the type of heat-absorbing surface. Staggered or in-line tube arrangements may be used. The arrangement of tubes affects the gas flow through the tube bank, the draft loss, the heat-transfer characteristics, and the ease of cleaning.

The size of an economizer is governed by economic considerations involving the cost of fuel, the comparative cost and thermal performance of alternate steam-generating or air-heater surface, the feedwater temperature, and the desired exit-gas temperature. In many cases, it is more economical to use both an economizer and an air heater.

The temperatures of the economizer tube metals generally approximate those of the water flowing within the tubes, and thus with low feedwater temperatures, condensation and external corrosion are encountered in those locations where the tube-metal temperature is below that of the acid or water dew point of the gas (see Fig. 9.2.30). Internal

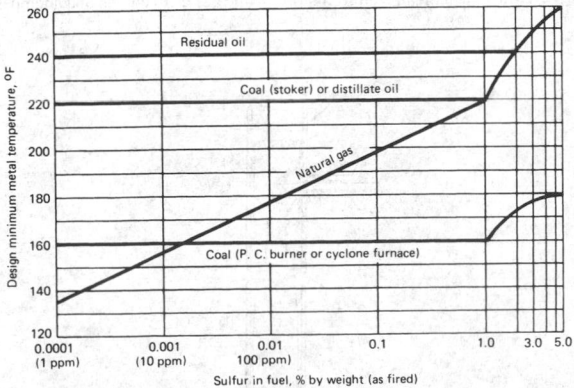

Fig. 9.2.30 Limiting metal temperatures to avoid external corrosion in economizers or air heaters when fuel containing sulfur is burned.

corrosion and pitting also may be experienced if the feedwater contains more than 0.007 ppm of dissolved oxygen. Therefore, it is imperative to maintain feedwater temperatures above the dew-point temperature of the gas and to provide suitable deaeration of the feedwater for the removal of oxygen.

AIR HEATERS

Air heaters, like economizers, remove heat from the relatively low-temperature combustion gases. The temperature of the inlet air is less than that of the water to the economizer, and thus it is possible to reduce the temperature of the gaseous products of combustion further before they are discharged to the stack.

The heat recovered from the combustion gases is recycled to the furnace with the combustion air and, when added to the thermal energy released from the fuel, is available for absorption by the steam-generating unit, with a resultant gain in overall thermal efficiency. The use of preheated combustion air accelerates ignition and promotes rapid and complete burning of the fuel.

Air heaters are usually classed as **recuperative** or **regenerative**. Both types utilize the convection transfer of heat from the gas stream to a metal or other solid surface and the convection transfer of heat from this surface to the air. In recuperative air heaters, exemplified by the tubular or plate types (Fig. 9.2.31), the stationary metal parts form a separating boundary between the heating and cooling fluids, and the heat passes by

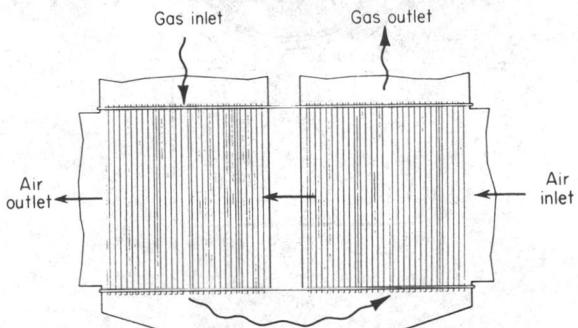

Fig. 9.2.31 Recuperative-type tubular air heaters, two gas passes, single air pass.

conduction through the metal wall. There are two commonly used types of regenerative air heaters (Fig. 9.2.32). In Fig. 9.2.32a, the heat transfer members are moved alternately through the gas and air streams undergoing successive heating and cooling cycles and transferring heat by the thermal-storage capacity of the members. The other type of regenerative air heater (Fig. 9.2.32b) has stationary elements, and the alternate flow of gas and air is controlled by rotating the inlet and outlet connections.

Recuperative and regenerative air heaters may be arranged either vertically or horizontally and for either parallel or counterflow of the gas and air. The gases are usually passed through the tubes of tubular air heaters to facilitate cleaning, although in some designs, particularly for marine installations, the air flows through the tubes.

Improved heat transfer and better utilization of the heat-absorbing surfaces are obtained with a counterflow of the gases and the use of small flow channels. Regenerative type air heaters readily lend themselves to these two principles and thus offer high capability in minimum space. However, regenerative air heaters have the disadvantages of air leakage into the gas stream and the transport of fly ash into the combustion air system. Tubular-type recuperative air heaters do not encounter these problems.

The products of combustion from most fuels contain a high percentage of water vapor, and thus condensation will be experienced in air heaters if the exposed metal surfaces are cooled below the dew point of the gas. Minute concentrations of sulfur trioxide in the gases, originating from the combustion of sulfur and varying with the sulfur content of the fuel and the method of firing, combine with the water vapor in the combustion gases to form sulfuric acid, which may condense on the metal surfaces at acid-dew-point temperatures as high as 250 to 300°F (121 to 149°C), well above the water dew point (Huge and Piotter, *Trans. ASME,* 1955). Such condensation leads to corrosion and/or the fouling of the gas-flow area. It is most likely to occur during the winter when the entering-air temperature is low, and at low operating loads or in localized sections at the cold-air inlet if there is poor distribution of the air or the gas flowing through the air heater. Corrosion and fouling can be prevented by the use of auxiliary steam-heated air heaters located ahead of the air inlet, by recirculating heated air from the outlet duct, or by bypassing a portion of the cold air to reduce the airflow through the air heater. Both recuperative and regenerative air heaters often are designed with separate corrosion sections arranged to facilitate the replacement of the vulnerable cold-end portions.

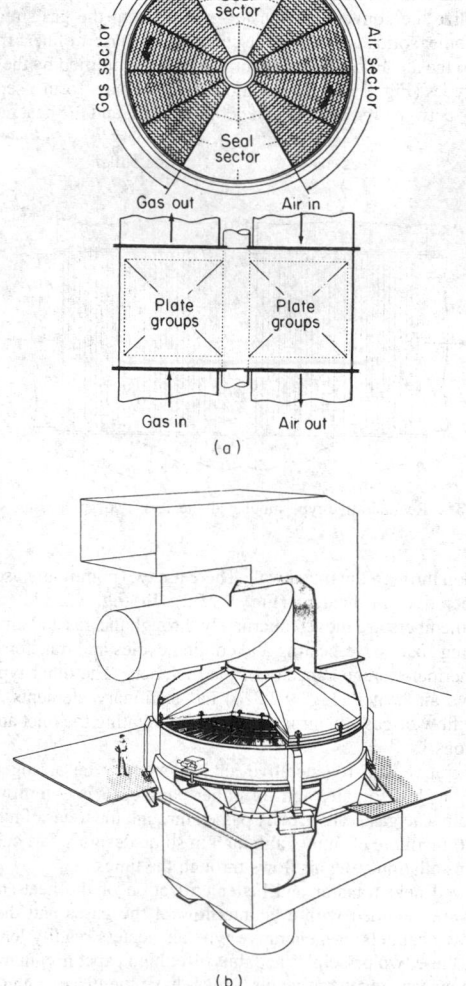

Fig. 9.2.32 Two designs of rotary regenerative air heaters.

STEAM TEMPERATURE, ADJUSTMENT AND CONTROL

The control of steam temperatures is vital to the life of high-temperature equipment and to the economy of power generation. Actual, or operat-

ing, steam temperatures below the design temperature reduce the thermodynamic efficiency and increase fuel cost, and temperatures above the design temperature reduce the margins of reserve in the strength of tubes, headers, piping, valves, and turbine elements. Further, sudden or extreme temperature variations may cause destructive stresses, particularly in rotating equipment.

It is sometimes necessary, because of the complexities involved in the design evaluation of heat-transfer rates and fuel characteristics, to modify installed equipment to obtain the required steam temperatures. Such changes might involve the installation of baffles for the distribution of gas through the superheater and the removal, or addition, of tubular elements in the superheater or in those components preceding the superheater which affect the temperature of the gas to the superheater. Therefore, it is desirable, and usually essential, to provide some means of controlling steam temperature to compensate for the variations in fuel, heat transfer, and surface cleanliness conditions encountered during operation. These may include (1) damper control of the gases to the superheater and/or reheater; (2) recirculation of the low-temperature gaseous products of combustion to the furnace to change the relative amounts of heat absorbed in the furnace, in the superheater, and/or in the reheater; (3) selective use of burners at different elevations in the furnace or the use of tilting burners to change the location of the combustion zone with respect to the furnace heat-absorbing surface; (4) attemperation or controlled cooling of the steam at superheater inlet, at superheater outlet, or between the primary and secondary stages of the superheater; (5) control of the firing rate in divided furnaces; and (6) control of the firing rate relative to the pumping rate of water in forced-flow once-through boilers.

The speed of response differs for the various methods, and the control of steam temperature by gas bypass or flame position is slower than that by spray-water attemperation. The operating controls for these methods can be arranged for manual, automatic, or combination adjustment, and the use of more than one method often facilitates the maintaining of constant steam temperature over a wider range of boiler load (Fig. 9.2.33).

The attemperation of superheated steam by direct-contact water spray (Fig. 9.2.34) results in an equivalent increase in high-pressure steam generation without thermal loss. Spray attemperation requires the use of

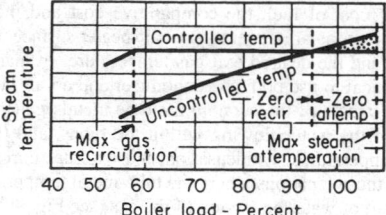

Fig. 9.2.33 Steam temperature control by flue gas recirculation and attemperation.

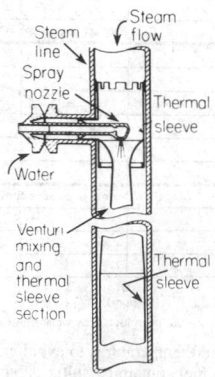

Fig. 9.2.34 Spray-type direct-contact attemperator.

essentially pure water, such as condensate, to avoid impurities in the steam. Submerged-type attemperators, when used, are generally restricted to relatively low-pressure boilers operating with steam temperatures of 850°F (454°C) or less. Usually, spray attemperators are not used for the control of reheat-steam temperature since their use reduces the overall thermal-cycle efficiency. They are, however, often installed for the emergency control of reheat-steam temperatures.

Ash and slag deposits on superheater and reheater surfaces reduce heat transfer and lower steam temperatures. Similar deposits on the furnace walls and steam-generating surface ahead of the superheater and/or reheater also reduce heat transfer to those surfaces, resulting in higher-temperature gas to the superheater and reheater and, consequently, increased steam temperatures. Thus the control of surface cleanliness is an important factor in the control of steam temperature.

Increased excess air results in higher steam temperatures because of reduction in furnace radiant-heat absorption, the greater amount of gas, and the increased convection heat transfer in the superheater and/or reheater. Operation with feedwater temperatures below that anticipated also results in increased steam temperatures because of the greater firing rate required to maintain steam generation.

OPERATING CONTROLS

(See also Sec. 16.)

The need for operating instruments and manual or automatic controls varies with the size and type of equipment, method of firing, and proficiency of operating personnel.

Safe operation and efficient performance require information relative to the (1) water level in the boiler drum; (2) burner performance; (3) steam and feedwater pressures; (4) superheated and reheated steam temperatures; (5) pressures of the gas and air entering and leaving principal components; (6) feedwater and boiler-water chemical conditions and particle carryover; (7) operation of feed pumps, fans, and fuel-burning and fuel-preparation equipment; (8) relationship of the actual combustion air passing through the furnace to that theoretically required for the fuel fired; (9) temperatures of the fuel, water, gas, and air entering and leaving the principal components of the boiler unit; and (10) fuel, feedwater, steam, and air flows so as to monitor operating conditions continuously and to make such adjustments as might be necessary.

Control of the various functions to maintain the desired operating conditions may be accomplished on small-capacity boilers by the manual adjustment of valves, dampers, and motor speeds. Most oil- and gas-fired package boilers are equipped with automatic controls to purge the furnace, to start and stop the burners, and to maintain the required steam pressure and water level. The operating requirements of utility and large industrial boilers dictate the use of automatic controls for the major variables, such as feedwater flow, firing rate, and steam temperature. The type of boiler and its components generally establishes the basic mode of control. Analog controls of either the pneumatic or the electric type are available. Digital control is being used more extensively.

Sequence controls often are applied in the start-up of utility boilers to program the furnace purge, burner light-off, and burner control. Interlocks are essential to ensure the proper starting and firing sequence and to alarm or automatically shut down the unit in the event of the failure of essential auxiliaries.

BOILER CIRCULATION

Adequate circulation in the steam-generating section of a boiler is required to prevent overheating of the heat-absorbing surfaces, and it may be provided naturally by gravitational forces, mechanically by pumps, or by a combination of both methods.

Natural circulation is produced by the difference in the densities of the water in the unheated downcomers and the steam water mixture in the heated steam generating tubes. This density differential provides a large circulating force (curve A, Fig. 9.2.35). The downcomers and the heated circuits are so designed that the friction, or resistance to flow,

through the system balances the circulating force at the desired total circulating flow.

The forced-recirculation or assisted-circulation type boiler uses a steam drum similar to that used with natural-circulation boilers. The

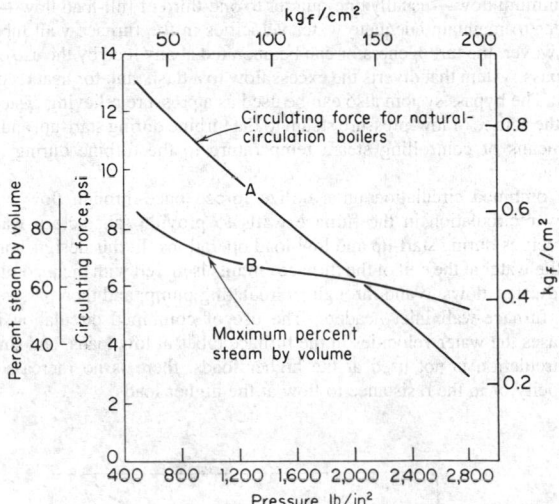

Fig. 9.2.35 Maximum percent steam by volume and circulating force for natural-circulation boilers.

supply of water to the furnace walls and the boiler surfaces flows from this drum to a circulating pump, which supplies the pressure necessary to force the water through the water-steam mixture circuits and then back to the drum, where the steam and water are separated. The total quantity of water pumped usually is 4 to 6 times the amount of steam evaporated, as shown in Fig. 9.2.36. The recirculating pump produces a differential pressure of 30 to 40 lb/in^2 (2.1 to 2.8 kgf/cm^2), and the power required is equivalent to about 0.5 percent of the heat input to the boiler. Resistor orifices are required at the entrance to each tube, or circuit, to control flow distribution.

In assisted-circulation designs, the velocities or flows are independent of boiler rating, thus facilitating the use of smaller connecting piping and, sometimes, smaller-diameter furnace-wall tubing than that used with natural-circulation units. Both drum-type natural- and assisted-circulation boilers operate with essentially saturated steam temperatures in all parts of the steam-evaporating sections, and they can be used for drum operating pressures ranging up to approximately 2,800 lb/in^2 (197 kgf/cm^2).

In natural- and assisted-circulation boilers, it is essential to wet the inside surfaces continually with water of the two-phase water-steam mixture to prevent overheating these heat-absorbing surfaces.

Satisfactory cooling of the heat-absorbing steam-generating surfaces is dependent upon the pressure, heat flux, water-steam mass velocity, percent steam by weight (quality), tube diameter, and the tube's internal geometry. The heat flux is one of the most predominant of these parameters, and the rate of furnace heat absorption at the maximum firing rate generally dictates design considerations.

In forced-flow once-through boilers, the water from the feed supply is pumped to the inlet of the heat-absorbing circuits. Evaporation, or change of state, takes place along the length of the circuit, and when evaporation is completed the steam is superheated. These units do not require steam or water drums and, in most cases, use relatively small-diameter tubes. The boilers can be started rapidly owing to the elimination of the drums and the reduced amount of metal. The water flow to the unit is the same as the steam output (Fig. 9.2.36), and fluid velocities greater than those needed for natural- or assisted-circulation units must be used at full load so as to maintain adequate velocities at the low loads and, thus, satisfactory tube-metal temperatures at all loads.

The transition from a liquid to a vapor at, or above, the critical steam pressure of 3,206 lb/in² (225 kgf/cm²) is dependent upon temperature and takes place without a change in density. Thus, separation of steam and water is impossible and forced-flow once-through boilers must be used.

Forced-flow once-through boilers must be operated above a specified minimum flow—usually one-quarter to one-third of full-load flow—in order to maintain adequate water velocities in the furnace-wall tubes. However, the turbogenerator can be operated at any load by the use of a bypass system that diverts the excess flow to a flash tank for heat recovery. The bypass system also can be used as a pressure relieving system, as the source of low-pressure steam to the turbine during start-up, and as a means of controlling steam temperature to the turbine during hot restarts.

Combined circulation units utilize forced once-through flow with flow recirculation in the furnace walls to provide satisfactory water velocities during start-up and low-load operations. In this design, some of the water at the exit of the furnace circuits is mixed with the incoming feedwater, flows to and through a circulating pump, and then passes to the furnace-wall inlet headers. The use of combined circulation increases the water velocities in the furnace tubes at low loads, and since recirculation is not used at the higher loads, there is no increase in velocity or in the resistance to flow at the higher loads.

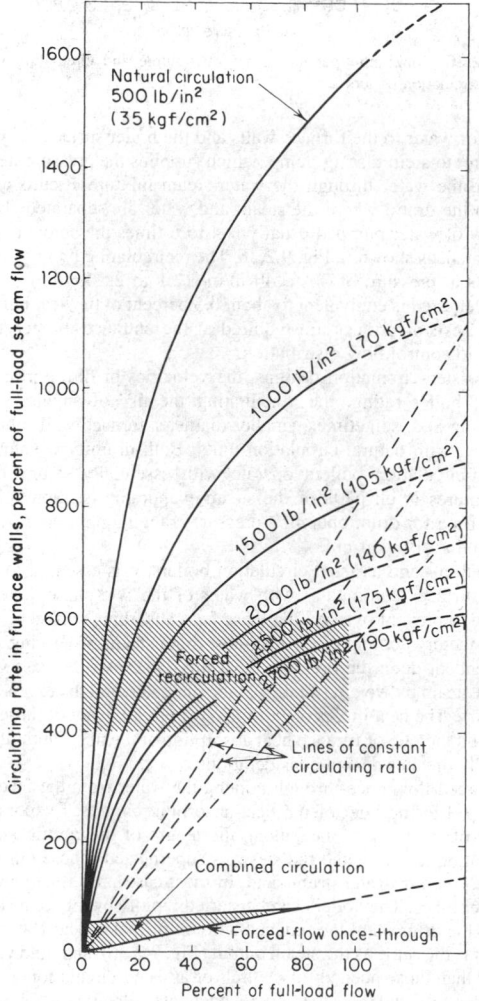

Fig. 9.2.36 Circulating flow for natural- and forced-circulation systems.

FLOW OF GAS THROUGH BOILER UNIT

During the combustion of fuel and the transfer of heat to the heat-absorbing surfaces, it is necessary to maintain sufficient pressure to overcome the resistance to flow imposed by the burning equipment, tube banks, directional turns, and the flues and dampers in the system. The resistance to air and gas flows depends upon the arrangement of the equipment and varies with the rates of flow and the temperatures of the air and gas.

The term draft denotes the difference between atmospheric pressure and some lower pressure existing in the furnace or gas passages of a steam-generating unit. Draft loss is defined as the difference in static pressure of a gas between two points in a system, both of which are below atmospheric pressure, and is the result of the resistance to flow. These terms originated with the use of the so-called natural-draft units in which the pressure differentials are obtained from a chimney or stack which produces static pressures throughout the boiler setting that are below that of the atmosphere. The terms are rather loosely applied to modern boilers using induced draft and/or forced draft, mechanically produced by fans, in which the pressures throughout the boiler unit may be well above atmospheric pressure.

Forced-draft fans, handling cold, clean air, provide the most economical source of energy to produce flow through high-capacity units (see Sec. 14.5). Induced-draft fans, handling hot flue gases, require more power and are subject to fly-ash erosion. However, they facilitate operation by providing a draft in the boiler setting and thus prevent the outward leakage of gas through joints or crevices in the boiler enclosure. As a result of advances in furnace and boiler-setting designs to eliminate gas leakage, modern units often are built for positive-pressure operation, thus eliminating the need for induced-draft fans. Such units are generally referred to as pressure-fired units, while those using induced-draft fans are classed as suction units. When both forced- and induced-draft fans are used, the boilers are designated as balanced-draft units.

The pressure drop throughout the unit is caused by the fluid friction in the gas stream and the shock losses at the turns or the contractions and enlargements of sections. It can be calculated as a function of the fluid mass flow and fluid properties in accordance with the principles of fluid flow (see Sec. 3). It is essential in the design of a boiler to determine the sum of all component resistances in the flow system at the maximum load in order to establish the fan requirements. It is customary to specify test-block static head, temperature, and capacity requirements of the fan in excess of those calculated so as to allow for departure from ideal flow conditions and to provide a satisfactory margin of reserve.

Stack effect is caused by the difference in densities resulting from the difference in the temperatures of two vertical columns of gas. In a chimney, or stack, the stack effect is due to the difference between the confined hot gas and the cooler surrounding air and the equal static pressure at the top or free outlet of the stack. The stack effect, which varies with the height and the mean temperature of the columns, can be calculated from the data in Table 9.2.4. The effect is the static draft produced by a stack, at sea level, with no gas flow. When flow occurs, a portion of the stack effect is used to establish gas velocity and the remainder is used to overcome the resistance of the connected system, including the dampers and the stack. The limit of natural-draft capacity is reached when these forces are in balance with the dampers in a wide-open position. Stack performance may be favorably or adversely affected by external factors such as the wind and the atmospheric conditions. The available draft varies directly with the barometric pressure for altitudes above sea level.

Stack effects also exist within the boiler setting and are most pronounced in tall units with vertical gas passes. The individual gas columns within the setting may aid the head produced by the fan or chimney if the flow is upward or may reduce it if the flow is downward. The net stack effect, and its overall influence on the performance of the fan, may be calculated from the data in Table 9.2.4, taking into account the positive or negative effects. The relationship between local static pressures and the atmospheric pressure is most important, since gas may

Table 9.2.4   Stack Effect of Pressure Difference, in of Water for 1 ft of Vertical Height (mm of Water for Each 1 m of Vertical Height)
Barometer = 29.92 inHg (759.97 mmHg)

Avg temp in flue, °F (°C)	Air temp outside flue, °F (°C)			
	40°F (5°C)	60°F (15°C)	80°F (25°C)	100°F (35°C)
250 (125)	0.0041 inH₂O/ft (0.346 mmH₂O/m)	0.0035 (0.303)	0.0030 (0.262)	0.0025 (0.224)
500 (250)	0.0070 (0.563)	0.0064 (0.520)	0.0058 (0.480)	0.0053 (0.442)
1,000 (500)	0.0098 (0.788)	0.0092 (0.774)	0.0086 (0.708)	0.0081 (0.665)
1,500 (750)	0.0111 (0.902)	0.0106 (0.858)	0.0100 (0.818)	0.0095 (0.780)
2,000 (1,000)	0.0120 (0.972)	0.0114 (0.925)	0.0108 (0.887)	0.0103 (0.850)
2,500 (1,250)	0.0125 (1.018)	0.0119 (0.975)	0.0114 (0.934)	0.0109 (0.896)

blow into the room through an open inspection door at the top of a furnace, even though a strong draft, or negative pressure, exists at some lower elevation.

PERFORMANCE

Steam-generating units are designed for specific operating conditions and are generally sold with a guarantee of performance. The boiler rating is usually specified and guaranteed in terms of steam output (lb/h) at a given pressure and temperature at full load or maximum continuous operation. When the steam is reheated, the rating includes this requirement in terms of the quantity of reheat steam at stated inlet and outlet steam pressures and temperatures.

Generally, either the efficiency or the gas temperature leaving the unit is guaranteed at a specified rate of operation, and the draft loss and the quality or purity of the steam also may be guaranteed at this rate. When component equipment such as stokers, pulverizers, burners, and air heaters are supplied by different manufacturers, the performance of the individual components is usually guaranteed by the various manufacturers and then, in turn, guaranteed by the prime contractor.

Anticipated-performance data for several rates of operation may be given to the purchaser in addition to the guaranteed-performance data. Guarantees may be demonstrated by acceptance tests, conducted in accordance with established codes, as agreed upon by the parties to the contract. However, acceptance tests are more difficult to perform as unit size and capacity increase, and overall performance usually is determined from the operating data. Guarantees of materials and the quality of manufacture and erection are usually considered separately from those pertaining to operating performance.

Heat balances account for the thermal energy entering the system in terms of its ultimate useful heat absorption or thermal loss. Methods of measuring and calculating the quantities involved in heat balances are presented in the ASME Power Test Code for Stationary Steam Generating Units.

The heat input is predicated upon the hourly firing rate, the calorific heating value of the fuel, and any additional heat supplied from an outside source. Heat in the preheated combustion air obtained from an air heater integral with the boiler unit is not considered in the determination of heat input, since this heat is recycled within the system.

The heat absorption in a boiler is calculated from the rate of steam output and the increase in fluid enthalpy from feedwater conditions to that at the superheater outlet. The amount of heat absorbed by the steam passing through the reheater, if used, is added to the heat absorbed in the boiler, economizer, and superheater. The total heat absorption also must take into account any steam generated which bypasses the superheater. Usually, the heat absorption is determined on an hourly basis.

In its simplest form,

$$\text{Efficiency (percent)} = [(\text{heat absorbed, Btu/h (cal/h)}/ \\ (\text{heat input, Btu/h (cal/h)}] \times 100$$

Both the heat input and the heat absorption may be very large quantities. Therefore, unless elaborate precautions are taken in the sampling and the measurement of fuel and steam quantities, it is difficult to obtain test data having the degree of accuracy required to determine the actual efficiency of the boiler unit. For this reason, boiler efficiency usually is established from the heat losses, since each of the thermal losses is a relatively small percentage of the heat entering the system and reasonable errors in measurement will not appreciably affect the final result.

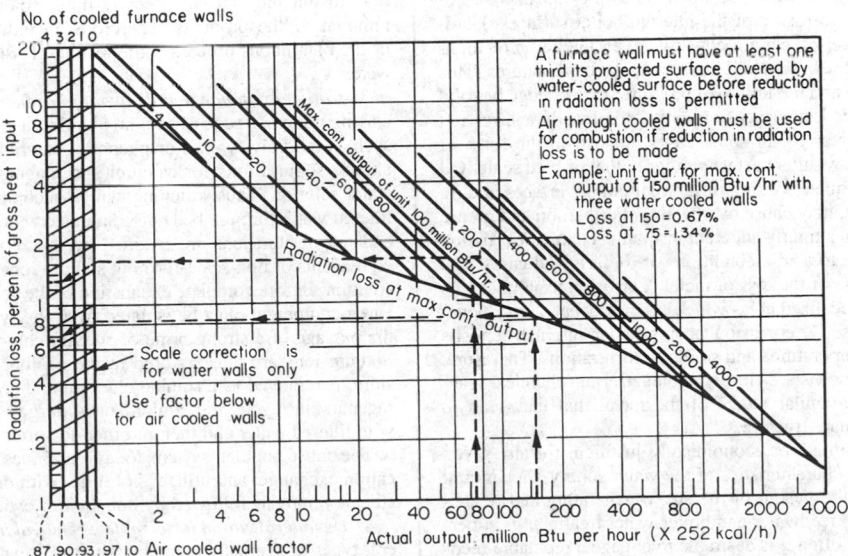

Fig. 9.2.37   External heat loss from boiler setting (ABMA).

The principal thermal losses are those due to the sensible heat in the gases leaving the unit, latent-heat losses associated with the evaporation of fuel moisture and formation of water vapor resulting from burning of hydrogen in the fuel, unburned-combustible loss, loss from the boiler setting or enclosure due to external convection and radiation, and ashpit loss. The first two losses can be derived from the fuel analysis, the exit-gas temperature, and the analysis of the flue gas (see Sec. 4). The unburned-combustible loss can be established by a qualitative and quantitative sampling of refuse and fly ash. The radiation from the boiler setting can be estimated in detail, but it also can be approximated from Fig. 9.2.37. The heat loss to the ashpit of large units can be determined by measuring the quantity of quenching water evaporated from the furnace ash hopper or the slag tank; and for small units the ashpit loss is included in the heat loss from the boiler setting. The sum of these heat losses, expressed as a percentage of the total heat input, is the total measurable loss. The anticipated or guaranteed performance data include a tolerance for the so-called manufacturer's margin (unaccountable losses) in the order of 0.5 to 0.75 percent, depending upon the type of fuel fired. The thermal efficiency of the unit is established by subtracting the sum of all these losses from 100 percent.

Tests of the individual components of the boiler unit, such as the furnace, superheater, economizer, or air heater, may be conducted for the determination of heat-transfer and gas-flow characteristics, for comparisons with other units, or to facilitate changes in operating procedures or equipment. Available ASME codes delineate such tests.

WATER TREATMENT AND STEAM PURIFICATION

(Also see Secs. 6.5 and 6.10.)

Purity of the water used in steam generation and of the steam leaving the unit is of paramount importance. The following discusses the general problem and then reviews the four areas of concern: raw water, feedwater, boiler water, and steam purification.

General

With few exceptions, the waters found in nature are not suitable for use as boiler feedwater; but they can be used after proper treatment. In essence, this entails the removal from the raw water of those constituents which are known to be harmful; supplementary treatment, within the boiler or connected system, of residual impurities to convert them into harmless forms; and systematic removal, by blowdown of boiler-water concentrates, to prevent excessive accumulation of solids within the boiler unit.

The ultimate purpose of water treatment is to prevent deposits of scale or sludge on and corrosion of the internal boiler surfaces. Hard-scale formations, formed by certain constituents in the zones of high-heat input, retard the flow of heat and raise the metal temperatures. This can lead to overheating and the failure of pressure parts. Sludge or solid particles normally carried in suspension may settle locally and restrict the flow of cooling water or, in some cases, deposit in the form of insulating layers with a resultant effect similar to that of hard scale. Oil and grease prevent adequate wetting of internal boiler surfaces and, in areas of high heat input, may cause overheating. Further, oil and grease may carbonize and form a tightly adherent insulating coating. Corrosion (see also Sec. 6.5), due to acidic conditions or to dissolved gases, can weaken a boiler because of the loss of metal. Corrosion usually occurs as cavities and pits in localized areas which, as they deepen, may penetrate the metal. Frequently, corrosion occurs under internal deposits because of elevated temperatures and solids concentrations. Therefore, corrosion and solids deposits are closely related. Certain chemical reactions produce an intergranular attack of the metal that may lead to embrittlement and ultimate fracture.

The treatment best suited, or economically justified, for any given plant depends upon the characteristics of the water supply, the amount of makeup water, and the design of the steam-generating and related equipment. Usually, the feedwater and boiler-water treatment is supervised by a chemist, and often it is desirable to engage a reputable feedwater specialist to prescribe specific procedures. However, the results obtained depend upon the diligence and integrity of routine sampling and the control carried out by plant personnel.

Raw Water

All natural waters contain impurities, many of which may be harmful in boiler operation. These impurities originate from the earth and the atmosphere (or from municipal and industrial wastes) and are broadly classed as suspended or dissolved organic and inorganic matter and as dissolved gases.

The concentration of the impurities is customarily expressed in terms of the parts by weight of the constituent per million parts of water (ppm). This is equivalent to the percentage of concentration multiplied by 10^4 and has the advantage of using positive whole numbers for small concentrations. However, for exceedingly small concentrations, especially those involving gases, the quantities are sometimes expressed as parts per billion (ppb). Concentrations also may be expressed in terms of the number of grains per gallon, but in boiler practice this has generally been superseded by the gravimetric relation, ppm. One grain per gallon is equal to 17.1 ppm.

The treatment of raw water for makeup and boiler feedwater involves one or more of the following:

1. *The removal of suspended solids.* Large particles in the water are removed by settling and decantation or by filtering through screens, fabrics, or beds of granular material. Small particles which settle slowly or colloidal particles which do not settle can be removed by coagulation using floc-forming chemicals, such as alum or ferrous sulfate, to trap the particles in the floc. The floc is then removed by settling or filtration. The solids can be removed intermittently or on a continuous basis.

2. *Chemical treatment for removal of hardness.* Calcium, magnesium, and silica are the principal scale-forming impurities in water and, if present in the boiler water, may form compounds whose solubility decreases with an increase in temperature.

In the lime-soda process for softening water, lime (calcium hydroxide) reacts with the soluble calcium and magnesium bicarbonates to form precipitates of calcium carbonate and magnesium hydroxide which can be removed as sludge. The soda ash (sodium carbonate) reacts with the scale-forming calcium sulfate and magnesium sulfate, and precipitates the calcium and magnesium as insoluble carbonates. Both reactions produce sodium sulfate, a soluble and non-scale-forming compound. When the hot-lime-soda process is carried out at temperatures of 200 to 250°F (93 to 121°C), the reactions are accelerated and some of the silica may be removed.

The reactions, as in all chemical processes, tend to approach equilibrium but they are affected by time, completeness of mixing, and removal of the products. Therefore, in either the intermittent batch or the continuous process some residual hardness is left in the treated water.

3. *Cation exchange for removal of hardness.* Certain naturally occurring minerals, such as sodium aluminum silicate, or synthetic resins, such as the polystyrenes or phenolic materials, have the ability to exchange sodium ions for calcium and magnesium ions if present in a water solution. Thus softening can be accomplished by passing raw or filtered water through beds of granulated zeolite particles. The calcium and magnesium ions are retained by the zeolite material, while their equivalents of non-scale-forming sodium ions are released to the water solution. Before complete exhaustion of the sodium is reached, the softening equipment must be isolated from the system and regenerated by the passage of a strong brine of sodium chloride through the softener. Sodium ions are thus restored to the zeolite, and calcium, or magnesium, is removed as a soluble chloride and drained to waste. After the regenerating cycle, the equipment is purged of the brine by flushing with filtered water and then returned to softening service.

The most popular system today combines chemical treatment and cation exchange, and utilizes hot lime (with or without magnesium for silicate removal) followed by hot sodium-cation exchange.

4. *Demineralization for complete removal of dissolved solids.* Several types of synthetic organic resins are capable of selectively removing undesirable cations or anions from water solutions by their ex-

change for hydrogen or hydroxyl ions. When used in combination, as separate or mixed beds of small-sized beads or particles through which the water flows, they can produce an effluent that is virtually free of mineral solutes and satisfactory for boiler feedwater. The cation exchanger is regenerated by acid which restores hydrogen ions to the resin in exchange for the calcium, sodium, or other metallic cations removed from the water. The anion exchanger is regenerated by the use of caustic soda, or another appropriate base, which restores the hydroxyl ions in exchange for the chloride, sulfate, or other negative chemical radicals previously removed from the water. The hydrogen and hydroxyl ions released from the resins during the heating process combine to form pure water. The greatest effectiveness is attained by a mixed-bed arrangement of resins, since the interchange of cation and anion components proceeds in minute increments and with less probability of the escape of unexchanged ions. With individual regeneration, the mixed resins are separated hydraulically because of differences in specific gravity. The resins can then be sluiced to external regeneration facilities or regenerated in place. The resins must be remixed before the demineralizer is returned to service.

5. *Evaporation.* Essentially pure water can be obtained by the evaporation of raw water and the collection of the distillate. Evaporation leaves the soluble constituents as concentrates in the residual water which can be removed by blowdown, or as scale on the heat-absorbing surfaces which can be mechanically removed. There may be some contamination in the distillate because of the carryover of water particles with the vapor and the reabsorption of noncondensable gases.

Feedwater

Boiler feedwater may consist of condensate, treated water, or a mixture of both. Usually, there is only a small amount of dissolved and suspended solids as a result of the treatment and, generally, the removal of additional solids is not required. However, any dissolved gases present must be removed to prevent corrosion in the boiler and the preboiler system.

When condensate is used as the feedwater to a boiler, water treatment is minimized, since it is required only for the small amounts of raw water that may leak into the system and the makeup water needed (usually ½ to 3 percent) to replace the loss of steam and condensate from the system. However, in industrial plants using a large portion of the steam generated for process work, the makeup-water requirements may be 90 to 100 percent of the total feedwater flow. Such plants require a considerable amount of water treatment.

Dissolved oxygen is, perhaps, the greatest factor in the corrosion of steel surfaces in contact with water. It may be present in the makeup water or the feedwater because of previous contacts with atmospheric air, or it may be added to the water by the leakage of air into the system through low-pressure-pump seals, storage tanks, etc.

Oxygen may be partially removed (to a residual of 0.2 to 0.3 ppm) by heating the water to boiling temperature in open type feedwater heaters. Tray or spray-type deaerating heaters are more effective in removing oxygen (to residuals of 0.02 to 0.04 ppm), and the amount of oxygen in the water can be reduced to 0.007 ppm or less by the use of multistage deaerators arranged for the countercurrent scavenging of noncondensable gases. (See Sec. 9.5.)

It is customary to supplement feedwater deaeration by adding a scavenging agent, such as sodium sulfite or hydrazine, to effect the complete removal of residual oxygen. Sodium sulfite combines with oxygen to form sodium sulfate, but it should not be used at operating pressures in excess of 1,800 lb/in² (127 kgf/cm²), since it decomposes to corrosive products at high temperatures. Thus hydrazine should be used at high pressures.

In boiler plants using high-purity feedwater, corrosion may be experienced in the condensate piping and the preboiler system because of dissolved gases such as carbon dioxide, sulfur dioxide, or hydrogen sulfide in the water. These gases may originate from the atmosphere or from constituents in the boiler water. They are released in the steam generators, intimately mixed with the outgoing steam, and with the exception of those partially removed by the vacuum pumps, returned to

solution in the condenser. These gases in the condenser produce an acidic reaction leading to corrosion, even in the absence of dissolved oxygen. The corrosion products in the preboiler cycle often are carried into the boiler and may deposit on the heat-absorbing surfaces, with resultant overheating of these surfaces.

A small amount of alkaline boiler water is sometimes recirculated to the feedwater heaters to raise the pH of the feedwater and thus prevent corrosion in the preboiler system. However, this procedure may precipitate sludge in the feedwater piping if appreciable hardness is present in the boiler water.

The pH of the feedwater can be increased by the addition of ammonia or volatile amines, such as morpholine or cyclohexylamine. Generally, these compounds are added as early as possible in the preboiler system. This procedure prevents corrosion in the early stages of moisture formation in the turbine and the condenser, as well as in the entire condensate-return system.

Filming amines also can be used and are generally introduced to the system through chemical pumps in the feedwater or steam lines. These materials do not change the pH of the fluid but protect against corrosion by forming a monomolecular coating on the metal surfaces. However, caution must be exercised since excessive use of filming amines has been known to agglomerate boiler sludge and produce strongly adherent internal deposits.

Boiler Water

In boilers, water is converted into steam and the steam leaves the boiler drum in a relatively pure state. Impurities, other than the gases which enter with the feedwater, are thus retained and concentrated in the boiler water. High concentrations of foam-producing solids in the boiler water contribute to particle and water carryover and contamination of the steam. Chemical and solubility changes also take place in the boiler, particularly as temperature is increased.

Boiler water is treated internally to prevent corrosion, the fouling of heat-absorbing surfaces, and the contamination of steam. Internal treatment also aids in maintaining water conditions within satisfactory limits. The internal treatment requires the introduction of chemicals in suitable amounts to react with the residual impurities in the feedwater.

Corrosion in boilers is prevented or minimized by maintaining alkaline boiler water. This condition may be expressed in terms of pH or as total alkalinity.

Acid or alkaline reactions of aqueous solutions are due to the presence of free or excess hydrogen (H^+) or hydroxyl (OH^-) ions, and the strength of the reaction varies with the concentration or activity of the excess ions. Some compounds enter into solution without dissociation while others dissociate partially or completely into ions carrying positive or negative electrical charges. If such ionizable compounds contribute hydrogen (H^+) ions to the solution (e.g., HCl), they add to the strength of its acid reaction; if they contribute hydroxyl (OH^-) ions (e.g., NaOH), they add to its alkaline or base reaction. When the ions of many different compounds are present, as is the usual case with boiler waters, their interaction or buffering affects the resulting concentration of the specific ions, and the solution tends to approach a balance or equilibrium in accordance with the principles of chemical mass action. It is therefore possible by the addition of some compounds which in themselves contain neither hydrogen nor hydroxyl components to suppress or release these ions from other constituent solutes and thereby change the acidity of alkalinity of the solution.

The pH value of a solution, which designates its acidity or alkalinity, refers to a logarithmic scale proposed by Sorenson in 1909. The symbol p is derived from the German word *Potenz,* meaning power or exponent; and the symbol H represents the hydrogen-ion concentrations. Thus, by definition, the pH value is equal to the logarithm of the reciprocal of the hydrogen-ion concentration measured in gram-moles per litre.

Pure water, which may be considered as composed primarily of molecular H_2O, exhibits a slight degree of dissociation to hydrogen ($+$) and hydroxyl ($-$) ions in the equilibrium amounts, at room temperature, of 0.0000001 gmol each per litre of water. It thus has the somewhat unusual capability of reacting, under proper conditions, as a weak acid or as a weak base and is said to be amphoteric. The H^+ and OH^- ions are in exact balance, and the water is electrically neutral.

The equation which expresses the equilibrium dissociation of water and also applies to water solutions is

$$H^+OH^- = K^{H_2O} = 10^{-14} \quad \text{at } 25°C$$

H^+ and OH^- represent the respective concentrations of the ionized hydrogen and hydroxyl groups, and the dissociation product K^{H_2O} is found by experimental methods to be $1/10^{14}$ at 25°C.

Since the product of the two concentrations is a constant, some OH^- ions are present even in a highly acidic solution and some H^+ ions are present in a basic solution, and the relationship of these factors can be determined from the measurement of either term. Thus, in the case of neutral water, the value of each is 10^{-7}, or 0.0000001, gmol/L.

In a solution having a hydrogen-ion concentration of 10^{-3} gmol/L, the corresponding hydroxyl-ion concentration is 10^{-11}, as a result of the dominating influence of the solvent, which is present in great excess and maintains the product equilibrium. Although either factor in the equation can be used, the conventional scale is based upon the measurement of the hydrogen ion.

Numerical values of this relationship extend over an extremely wide range and can best be expressed in terms of logarithms or exponents; thus:

$$\log H^+ + \log OH^- = -14$$
$$-\log H^+ - \log OH^- = 14$$
$$\log (1/H^+) \log = 14 - (1/OH^-) = pH$$

The term pH is used to represent $\log(1/H^+)$ and is therefore the logarithm of the reciprocal of the hydrogen-ion concentration (more properly, the hydrogen-ion activity which is equal to the concentration multiplied by an activity factor that approaches unity in dilute solutions).

For neutral water, the pH = $\log (1/H^+)$ = $\log (1/0.0000001)$ = $\log (1/10^{-7})$ = 7.0.

For an acid solution in which H^+ exceeds OH^-, say $H^+ = 10^{-3}$, the pH = $\log (1/10^{-3})$ = 3.0.

For an alkaline solution in which OH^- exceeds H^+, say $OH^- = 10^{-3}$, the pH = $\log (1/10^{-11})$ = 11.0.

In practical terms, the pH scale extends from 0 to 14, as shown in Table 9.2.5. The value 7.0, corresponding to pure water, is considered the neutral point; values below 7.0 are increasingly acidic, and values above 7.0 are increasingly alkaline. Since the pH scale is logarithmic, a change from one number to the next in series is equivalent to a change of ten times the activity. Beyond the range of the scale, the strength of acid or alkaline solutions is expressed in terms of normality or percentage of concentration.

The pH of a water sample can be determined accurately by the measurement of its electrical potential. It also can be approximated by indicators that change color in certain pH ranges as the result of their reaction with the solution. The pH of boiler water usually is maintained within the range of 10.2 to 11.5 for boilers operating at pressures compatible with an 1,800 lb/in² (127 kgf/cm²) turbine throttle pressure. Above these pressures, mixed-bed demineralizers are generally used to treat the makeup water, the boiler-water treatment is low in solids, and the pH ranges from 9.0 to 10.0.

In forced-flow once-through units the recommended pH range is 8.8 to 9.2 for preboiler systems using copper alloys. If the preboiler system does not incorporate the use of copper alloys, the recommended pH is 9.2 to 9.5.

Total alkalinity (expressed in ppm) is a measure of all reactives that have the ability to neutralize acids. It is determined by titrating a water sample with a standard acid, and it is frequently expressed as equivalent calcium carbonate, which has a molecular weight of 100. Total alkalinity, as determined in this manner, is not exactly the same as the pH measurement of alkalinity because of the buffering action which occurs in complex solutions. However, it is widely used as a reference and in the case of low-pressure boilers where higher concentrations of greater diversity of solids can be tolerated, it often is more satisfactory than the measurement of pH as an index of boiler-water conditions.

The elimination of hardness in boiler water is necessary to prevent scale. Hardness can be removed by introducing one of the forms of sodium or potassium phosphate and thoroughly mixing it with the boiler water. The residual calcium ions entering with the feedwater are precipitated as an insoluble phosphate sludge and the magnesium is precipitated as a nonadherent magnesium hydroxide, if the alkalinity is maintained at a pH of 10 or higher. A lower pH may result in the formation of magnesium phosphate, an adherent type of sludge. Routine control facilitates the adjustment of the pH by the addition of sodium hydroxide, or its equivalent, and the maintenance of a moderate excess of phosphate ions in the boiler water.

Early methods of internal treatment employed the use of soda ash for hardness removal. However, the hydrolysis of soda ash at the temperatures encountered with high operating pressures releases carbon dioxide into the steam, making it difficult to maintain an excess of carbonate and promoting corrosion in the condensate system. In some services, sodium carbonate in combination with the hydroxides and phosphates of sodium is used for hardness removal. A phosphate sludge is preferred since it is less adherent and more easily kept in suspension.

Silica may enter the system in the form of soluble compounds or as finely divided particles which are not removed by filtration. It dissolves in alkaline boiler waters and will, with unreacted calcium or magnesium hardness in the water, form an adherent scale. Under some conditions, it may produce complex scale-forming silicates with soluble or colloidal iron oxide (acmite) or alumina (analcite). The crystalline matrix of these deposits tends to trap sludge particles and contributes to the accumulation of scale on heated surfaces.

Silica also is soluble in steam, and its solubility increases rapidly at temperatures above 500°F (260°C). Thus it can be transported in a vapor phase into the turbine and deposited on the turbine blading. This characteristic necessitates the limiting of the silica concentration in the boiler water in order to avoid turbine deposits, and the limits, varying with operating pressure, range from about 10 ppm at 1,000 lb/in² (70 kgf/cm²) to 0.3 ppm at 2,500 lb/in² (176 kgf/cm²).

Silica is partially removed from raw water by the hot lime-soda softening process and can be completely removed by the evaporation of the makeup water. Soluble silica can be removed by demineralization, but in colloidal forms it may pass through the treating beds. The silica concentration in the boiler water can be controlled by blowdown.

Many operators of industrial boilers use the Chelant methods of water treatment. Chelants react with the residual divalent metal ions (calcium, magnesium, and iron) in the boiler water to form soluble complexes. The resultant soluble complexes are removed by use of continuous blowdown. One of the most popular methods uses the sodium salt of ethylenediaminetetraacetic acid (Na_4 EDTA). Chelant methods of treatment have been used in boilers operating at pressures as high as 1,500 lb/in² (105 kgf/cm²).

Table 9.2.5 Relationship of pH and Hydrogen-Ion Concentration

	pH	Hydrogen-ion concentration, gmol/L	
	0	1.0	10^0
	1	0.1	10^{-1}
	2	0.01	10^{-2}
Acid range	3	0.001	10^{-3}
	4	0.000,1	10^{-4}
	5	0.000,01	10^{-5}
	6	0.000,001	10^{-6}
Neutral	7	0.000,000,1	10^{-7}
	8	0.000,000,01	10^{-8}
	9	0.000,000,001	10^{-9}
	10	0.000,000,000,1	10^{-10}
Alkaline range	11	0.000,000,000,01	10^{-11}
	12	0.000,000,000,001	10^{-12}
	13	0.000,000,000,000,1	10^{-13}
	14	0.000,000,000,000,01	10^{-14}

Table 9.2.6   Recommended Limits of Boiler-Water Concentration (ABMA)

Pressure at outlet of steam-generating unit, lb/in² gage (× 0.07037 = kgf/cm²)	Total solids, ppm	Total alkalinity, ppm	Suspended solids, ppm
0–300	3,500	700	300
301–450	3,000	600	250
451–600	2,500	500	150
601–750	2,000	400	100
751–900	1,500	300	60
901–1,000	1,250	250	40
1,001–1,500	1,000	200	20
1,501–2,000	750	150	10
2,001 and higher	500	100	5

The recommended limits of boiler-water concentration, as defined in the ABMA manual, are listed in Table 9.2.6. These data are not applicable to forced-flow once-through boilers. The total solids content can be determined by weighing the residue of a water sample which has been evaporated by dryness. The dissolved-solids content can be determined in a similar manner from a filtered sample, but for immediate determinations and control purposes, it can be quickly approximated by an electrical-conductivity measurement and the use of conversion factors previously established by comparisons with gravimetric determinations.

Solids concentration also can be controlled by intermittent or continuous blowdown. The amount of blowdown and the time interval between blows should be coordinated with operation and should consider or anticipate load changes, water conditioning, and chemical treatment.

In forced-flow once-through boilers, the impurities entering with the feedwater must leave with the steam or be deposited within the unit. Thus, such units require high-purity feedwater and the control of corrosion by volatile bases, such as ammonia, which will prevent or minimize deposits in the boiler unit or the turbine. Raw-water leakage into the system must be prevented, and the makeup must be evaporated or demineralized water.

When sampling water from high-pressure, high-temperature sources, cooling is required to prevent the flashing or selective loss of water vapor at atmospheric pressure. The approved methods for water sampling and analysis are discussed in the *Annual Book of ASTM Standards*, Part 23.

STEAM PURIFICATION

In drum-type boilers, using either natural or assisted circulation, a mixture of steam and water is delivered to the upper or steam drum where separation of the steam and water takes place and a water level is maintained. The water is recycled through downcomers to the heat-absorbing circuits, and the steam is discharged from the top of the drum for use as saturated steam or the supply to the superheater.

Separation of the steam and water by buoyant or gravitational force requires a relatively large cross-sectional area and, consequently, a low fluid velocity within the drum as well as an effective difference in the fluid densities, which decreases as pressure is increased. Steam entrainment in the recycled water impedes circulation; and water entrainment in the outlet steam transports dissolved or suspended particulate matter into the superheater, steam piping, and turbine, where particle deposition can cause overheating of the tubes or flow obstructions in the turbine blading with subsequent loss of capacity, efficiency, and dynamic balance.

Gravity separation of steam and water may be satisfactory in low-pressure, low-duty boilers. This type of separation can be augmented by the use of baffles which utilize a change in direction to throw out the water droplets, or by dry pipes which impose a pressure drop that promotes evaporation of the moisture and reduces the tendency of solids to deposit in the superheater.

In high-pressure boiler units, particularly those employing high evaporative ratings, a part of the circulating head can be utilized to provide a separating force, many times greater than that of gravity, in centrifugal separating devices such as the cyclone steam separators shown in Fig. 9.2.38. These separators deliver steam-free water to the drum and downcomers, and discharge steam with a minimum of water entrainment. Secondary steam and water separation is accomplished by passing the steam at a low velocity through sinuously shaped passages between closely spaced corrugated plates, which provide a large surface area for intercepting entrained boiler-water particles. In modern high-capacity boilers, the steam leaving the steam drum contains less than 0.1 ppm of total solids.

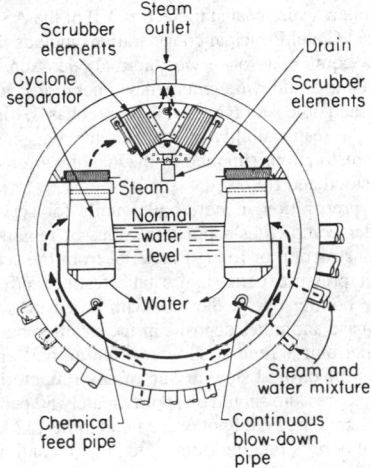

Fig. 9.2.38   Drum internal arrangement with cyclone steam separators and scrubber elements.

Mechanical steam separators do not prevent the transport of silica in a vapor solution. The amount of silica dissolved in the steam is dependent upon its concentration in the boiler water, and for a given concentration, the ratio of silica in the steam to the silica in the water increases rapidly with an increase in the operating pressure (see Fig. 9.2.39). Silica can be removed by steam washers which provide a large surface area for contact with the relatively pure feedwater and which reabsorb the silica and return it to the boiler-water system. Turbine deposits can be practically eliminated and the requirements of boiler blowdown materially reduced by the use of steam washers for steam purification. Steam washers are used to best advantage in medium-pressure boilers operating with large amounts of makeup water, particularly if the makeup water contains silica in an insoluble form

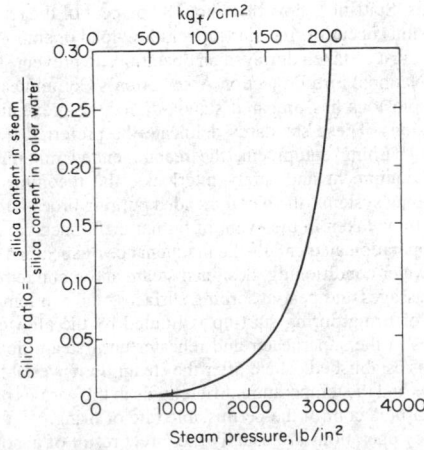

Fig. 9.2.39   Equilibrium relationship of silica ratio and operating pressure for a given concentration of silica in boiler water.

Impurities in the water entering a forced-flow once-through boiler leave with the steam unless they are deposited in the boiler unit. Therefore, the equivalent of steam purification must be accomplished by treatment of the feedwater before the water enters the boiler so as to prevent the accumulation of deposits in the boiler unit and the turbine. In the treatment, most of the feedwater, including the steam condensate, is passed through a mixed-bed demineralizer that removes suspended as well as dissolved impurities.

CARE OF BOILERS

The care of boilers is delineated in Section VII of the ASME Boiler and Pressure Vessel Code. Principal considerations include the initial preparation of new equipment for service; normal operation, including routine start-up and shutdown; emergency operations; inspection and maintenance; and idle storage. In all these phases, the handling of equipment is the responsibility of the operator, but recommendations and operating instructions supplied by the manufacturer should be thoroughly understood and followed.

The initial preparation of new boiler units for service, or of old equipment after completing major alterations or repairs, involves the removal of construction or foreign material from the setting and from the interior of pressure parts, hydrostatic testing and inspection for leaks, and the boiling out of the unit with a caustic solution for the removal of grease and other deposits in the steam-generating pressure parts. The boiler unit is fired at a low rate during boil-out. This procedure facilitates the desired slow drying of any refractories used in the setting. Boil-out pressure should be approximately 50 percent of normal operating pressure but should not exceed 600 lb/in^2 (42 kgf/cm^2). During the boil-out period, ranging from 12 to 36 h, the unit is blown down periodically through all the blowdown connections so as to eject any sediment removed from the surfaces. The boil-out often is supplemented, particularly in high-pressure boilers, by inhibited-acid cleaning for the removal of mill scale.

It is general practice, following the boil-out, to reduce the concentration of boil-out chemicals to a satisfactory level for operation by blowing down and replenishing the amount of water blown down with fresh water. The pressure is then raised to test and set the safety valves to code requirements; the superheater and the steam piping are blown out to remove any foreign material; and the boiler is placed on the line for a period of low-load operation, during which the auxiliary equipment, controls, and interlocks are test-operated. After these operations, it is advisable to shut down, cool, and drain the boiler unit prior to a thorough internal and external inspection and any adjustments or modifications required to the equipment.

Normal operation involves the orderly start-up and shut-down of equipment and operation, under controlled conditions, to meet plant requirements. Statistics show that about 80 percent of the recorded furnace explosions occur during start-up and low-load operation, and particular care must be taken during such operations to prevent such explosions. The National Fire Protection Association's Committee on Boiler Furnace Explosions has prepared standards for the prevention of furnace explosions. These standards delineate the preferred sequence of starting fuel-burning equipment, the recommended minimum flame-monitoring equipment and safety interlocks, the recommended fuel-transport piping systems, the recommended purging procedures, and the procedures to be taken in the event of burner or furnace flame-out.

Normal operation also entails the maintenance of specified feedwater and boiler water conditioning, designed steam and metal temperatures, clean gas passages and heat-absorbing surfaces, and ash removal.

The rate of firing during start-up is limited by the allowable metal temperatures in the superheater and reheater until steam flow through the turbine is established. Then, after the steam flow is established, the temperatures and the temperature differentials in the various parts of the boiler and turbine control the permissible rate of firing.

Emergency operations are usually the direct result of abnormal conditions such as the failure of the feedwater supply, the rupture of a pressure part, the interruption of the fuel supply, the loss of air, or a burner flame-out. Automatic safety interlocks usually are installed which trip the fuel supply and shut down the unit if these or other hazardous conditions are experienced. Abnormal operating conditions, which might become hazardous if allowed to persist, such as low (fuel-rich) or high (air-rich) air-fuel ratios or the failure of essential auxiliaries, require appropriate action to correct operating conditions. An operator who cannot correct an abnormal condition must determine whether operation can continue and, if not, must shut down the unit in the proper manner or activate the emergency trip through the automatic interlock system.

Inspection and maintenance should be performed during regularly scheduled outages. A list of the known items requiring repair or maintenance should be prepared before the outage and should be supplemented by any additional items noted in thorough inspection of the boiler and auxiliary equipment during the outage. A major item on the work list should be the maintenance of internal and external cleanliness. External cleaning is usually accomplished by water washing or air lancing. The internal surfaces of small boiler units are usually mechanically cleaned, but the large-capacity units are generally chemically cleaned. Under competent supervision, which can be obtained from several firms specializing in chemical cleaning, this method can be used with complete confidence for boilers of all sizes. The chemical-cleaning solution normally is composed of a 3 to 5 percent solution of hydrochloric acid, wetting and complexing agents for the removal of silica or other hard-to-remove deposits (such as iron and copper oxides), and a suitable inhibitor to prevent excessive chemical attack on the pressure parts of the boiler unit. Hydrochloric acid, however, should not be used for cleaning stainless-steel surfaces, since it can cause stress-corrosion cracking. Thus other organic or inorganic acids are used depending upon the type of material to be cleaned and the composition of the deposit to be removed.

The chemical cleaning of drum-type boilers involves filling the boiler [which has previously been uniformly heated to a temperature of about 175°F (79°C)] with the cleaning solution at 150 to 160°F (65 to 71°C) and allowing it to soak for 6 to 8 h or until samples of the solution show no appreciable further reduction in acid strength. The boiler should never be fired while it contains an acid solution, and open lights or other ignition sources must be prohibited in the area to avoid the ignition of the explosive gases, usually hydrogen, evolved during the cleaning operation. The unit is then drained and flushed several times, preferably under a nitrogen blanket, with neutral or slightly acidic water to remove the loosened deposits and to displace the acid solution and any corrosive gases. The flushing is followed by boil-out with an alkaline solution to neutralize any residual acid and to passivate the surfaces. The unit is then flushed with clean water to remove the remaining loose deposits, and it is inspected before being returned to service.

When chemically cleaning forced-flow once-through units, the solvent is continuously circulated through the unit for 4 to 6 h. Generally, the solvent is an inhibited solution of hydroxyacetic-formic acids at a temperature of 200°F (93°C).

Boiler units removed from service for long periods of time may be stored wet or dry. It is practically impossible to drain and dry modern high-pressure utility boilers completely with their complex furnace and superheater circuitry. Thus wet lay-up of the unit with water treated with 10 ppm ammonia and 200 ppm hydrazine is the best means of protection for both short- and long-term lay-up. If, however, dry storage is utilized, the system must be kept dry; and low humidity within the pressure parts and setting can be maintained by the use of trays of moisture-absorbing materials, such as silica gel or lime, which must be replenished at intervals to retain their effectiveness. When using either wet or dry storage, the system should be pressurized to a few pounds above atmospheric pressure with nitrogen gas.

CODES

The ASME Boiler and Pressure Vessel Code, initiated in 1914 and supplemented by continuing revisions, contains the basic rules for the safe design, the construction, and the materials for steam generating units. Its legal status depends upon its adoption by state or municipal authority. The Code is administered by the National Board of Boiler and Pressure Vessel Inspectors. This organization also has established the ''Recommended Rules for Repairs by Fusion Welding to Power Boilers and Unfired Pressure Vessels.'' The adoption of both codes has been widespread, and they form the basis for the pertinent legal requirements in all but a few localities throughout the country. The National Bureau of Casualty and Surety Underwriters' book titled ''Synopsis of Boiler Rules and Regulations'' lists the states and the communities having laws which govern the installation and the operation of steam boilers.

NUCLEAR BOILERS

(See also Sec. 9.8.)

Nuclear power boilers are usually identified by the primary fluid used as the reactor coolant. A number of coolants and system design concepts have been studied, but only three basic coolants have been used in U.S. power plants—liquid metals, gases, and water.

Most of the nuclear systems for commercial power production use water in some form as the primary coolant, principally the pressurized-water-reactor (PWR) and the boiling-water-reactor (BWR) systems. The preference for water as the reactor coolant is due to the fact that its physical, chemical, and thermodynamic characteristics are well known and materials and equipment are available for its handling and containment.

The steam-generating unit shown in Fig. 9.2.40 is typical of those used in PWR installations. In this design, hot primary fluid enters one side of the divided primary head, passes through the U-type tubes, and leaves through the primary outlet nozzle. Boiling takes place on the outside, or secondary side, of the tubes, and the steam-water mixture passes upward through the riser section and then the steam and water separators. The steam is discharged from the scrubbers into the outlet connection. The separated water flows downward, mixes with the incoming feedwater, and is then circulated downward through the annulus around the tube bundle to reenter the tube bundle at the bottom.

Generally, in units of this type, stainless-steel or Inconel tubes are used to minimize corrosion, the structural components are of carbon or

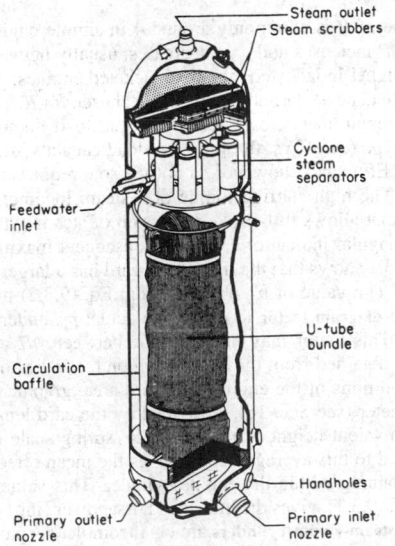

Fig. 9.2.40 Nuclear power boiler.

low-alloy steel, and the primary head and primary tube-sheet faces are completely clad with stainless steel or Inconel.

Design considerations are similar to those for fossil-fuel boiler units, but particular emphasis is placed upon possible hazards, codes and specifications, and system economics.

1. **Hazards.** In the design of nuclear systems, great consideration must be given to the damage and the loss of life that could result from an accident, and also the psychological effect of such an accident. Potential dangers that must be considered by the boiler designer include the following:

Radioactivity. The primary fluid and any materials transported in the primary passages may become radioactive from neutron bombardment. This radioactivity can hold at a high level for varying lengths of time, and since the radioactive products may be deposited anywhere in the primary system, the boiler designer must design the equipment for minimum exposure of personnel to radiation.

Chemical poisons. Some fission products are lethal, not only because they are radioactive but because they are chemically poisonous by both ingestion and inhalation. Personnel must be protected against exposure to these poisons.

Chemical reactions. In some nuclear systems, possible contact between the primary and secondary fluids could cause strong chemical reactions. Sodium and water, for example, react violently, yet sodium has been selected as the primary fluid in some systems because of its excellent nuclear properties and good heat-transfer and heat-transport characteristics. Therefore, steam generators in which sodium is used for the primary fluid must be designed so that direct contact between the sodium and the water is unlikely and so that the effects of such a contact, if one should occur, are minimized.

2. **Codes and specifications.** The requirements for safe design and fabrication are delineated in the ASME Boiler and Pressure Vessel Code, augmented by Special Code Cases. The special cases applying specifically to nuclear components establish design, inspection, and fabrication rules which are more stringent than those usually required for other types of equipment. The construction of commercial nuclear components is governed by the ASME Code, Section III, ''Nuclear Power Plant Components'' while that of equipment for military use is governed by special military specifications.

3 **Economics.** Basic design considerations of PWR and the BWR systems differ greatly from those of fossil-fuel systems. The temperature difference that produces the flow or transfer of heat between the primary and secondary water is dependent upon the difference in the pressures of the two fluids, and the hotter primary fluid is maintained at the higher pressure. Therefore, an optimized steam-generator design using a minimum amount of heat-absorbing surface should have the boiling secondary fluid on the outside of the tubes since a tube, or cylinder, can withstand greater internal pressures. This contrasts with fossil-fuel-fired water-tube boiler designs in which the boiling fluid is contained within the tubes.

Both internal and external heat-transfer coefficients are high in nuclear steam generators, and consequently, in most cases the tube metal and the fouling film coefficients control the overall rate of heat transfer. Thus tube diameters should be small and tube walls as thin as practical. Further, the water must be conditioned to minimize scale and sludge formations and, thus, fouling coefficients.

Primary-water systems are designed for relatively high pressures 1,200 to 2,500 lb/in^2 (84 to 176 kgf/cm^2), and extremely compact systems are economically justified in the efforts to minimize the size and the weight of the steam generator, reactor, and other pressure vessels. Fluid-temperature differences are necessarily small when operating with the highest practical secondary-steam pressure; this tends to increase surface requirements and, consequently, the size of the steam generator.

9.3 STEAM ENGINES
Staff Contribution

REFERENCES: Ewing, "The Steam Engine and Other Engines," Cambridge. Ripper, "Steam Engine Theory and Practice," Longmans. Heck, "The Steam Engine and Turbine," Van Nostrand. Allen, "Uniflow, Back Pressure, and Steam Extraction Engines," Pitman. Peabody, "Valve Gears for Steam Engines," Wiley. Zeuner, "Treatise on Valve Gears," Spon. Spangler, "Valve Gears," Wiley. Dalby, "Valves and Valve Gear Mechanisms," Longmans.

EDITOR'S NOTE: Although largely relegated to history and nostalgia, small steam engines are still manufactured in the United States, primarily as replacement items. Steam locomotives in the United States cater only to the tourist trade; they are kept in repair and in service mainly with parts manufactured locally. The remaining steam locomotives in India are in the last stage of being phased out of service; all are scrapped and replaced by diesel, electric, or diesel-electric locomotives. China is still engaged in the manufacture of steam locomotives, and has based its railroad system on their continued use.

The reciprocating steam engine was the heart of the early industrial era. It dominated power generation for stationary and transportation service for more than a century until the development of the steam turbine and the internal-combustion engine. The mechanisms were numerous (see Patent Office listings), but practicality essentially standardized the positive-displacement, double-acting, piston and cylinder design in a vertical or horizontal configuration. These engines were heavy, cast-iron structures, e.g., 50 to 100 lb/hp; they had low piston speed (600 to 1,200 ft/min); long stroke (up to 6 ft); low turning speeds (50 to 500 r/min); steam conditions less than 300 lb/in² (dry saturated, or 100 to 200°F superheat); noncondensing or condensing (25 ± in Hg vacuum); sized from children's toys to 25,000 hp. Diversity of valve gear was an inventor's paradise with many suitable for reversible operation, as in locomotive, rolling-mill, and ship applications. Most of the machine elements known today, such as cylinders, piston rods, crossheads, connecting rods, crankshafts, flywheels, and governors, were developed in steam engines.

These engines utilize the expansive power of steam. Theoretically, the more the steam can be expanded in the engine cylinder, the better will be the economy. Practical losses, which occur in every steam engine, limit the expansion ratio and result in a minimum steam rate for a definite degree of expansion. Cylinder dimensions preclude the practical utilization of high-vacuum conditions because of the high specific volume (e.g., 339 ft³/lb dry and saturated at 2 inHg abs). The steam turbine can effectively utilize maximum vacuum.

WORK AND DIMENSIONS OF THE STEAM ENGINE

Thermodynamic principles define the limits of the conversion efficiency of heat into work (see Rankine and Carnot cycles, Sec. 4). In fact, the historical development of thermodynamic principles was primarily aimed in the nineteenth century at defining the thermal performance of steam engines and steam power plants. It can be said that the steam engine did more for thermodynamics than thermodynamics did for the steam engine. Steam tables and steam charts (e.g., temperature-entropy; enthalpy-entropy; and pressure-volume) are essential to evaluate theoretical and actual equipment performance.

The pressure-volume diagram, or indicator card, is of primary significance in both the design and operation of the reciprocating piston and cylinder mechanism—not only steam engines, but also internal-combustion engines, air compressors, and pumps. The utility of the p-v diagram is often lost in attempts to improve fluid-dynamic reciprocating mechanisms. The work and power are the consequence of the difference in pressure on the two sides of the piston, expressed as mean effective pressure (mep or p_m). This is applied to the cylinder dimensions and rotating speed in the "plan" equation

$$\text{hp} = \frac{p_m L a n}{33,000} \qquad (9.3.1)$$

where p_m = mep, lb/in²; L = stroke, ft; a = effective piston area, in²; n = number of cycles completed per min; 33,000 = mechanical equivalent of horsepower, ft · lb/min, by definition.

A typical steam-engine indicator card is shown in Fig. 9.3.1, identifying the established nomenclature for the events of the cycle. Superimposed is a theoretical diagram, with expansion, but without clearance or compression. The terminal pressure controls the steam, or water, rate. The term "back pressure" is used for pressures above the atmosphere, while "condenser pressure" is used when engines operate with negative exhaust pressure. The volume ratio $v_j/v_h = R$ is called the **ratio of expansion.**

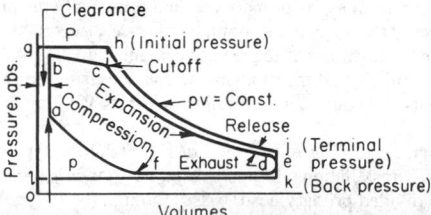

Fig. 9.3.1 Typical steam engine indicator diagram.

The average ordinate, or **mean effective pressure** p_m [applicable in Eq. (9.3.1)] for the ideal cycle diagram $ghjkl$ is

$$p_m = P[(1 + \log_e R)/R] - p \qquad (9.3.2)$$

The expansion phase h to j of the cycle is *logarithmic* where pv = constant. It is not isothermal but reasonably approximates expansion (and compression) in actual engines where condensation and evaporation on cylinder walls, heads, pistons, and valves are cyclically recurrent.

The value of R is commonly around 4 in simple engines. It should increase as P increases and as p decreases, usually between 3 and 5. It should be higher in jacketed than in unjacketed engines. The efficiency of the engine depends largely on the value chosen for R. This is dictated by service requirements, e.g., load variation, load fluctuation, engine governing type (cutoff vs. throttle), overload capacity, steam cost, and economics. Efficiency, however, is not the sole requirement for all installations. The high starting torque of a steam locomotive dictates a valve gear that allows full-stroke admission of steam with zero expansion, a rectangular indica tor card with consequent maximum p_m.

Figure 9.3.1 shows that the theoretical card has a larger area than the actual card. The value of p_m obtained from Eq. (9.3.2) must be multiplied by the **diagram factor** to obtain the actual p_m under the assumed conditions. This factor may have a value between 0.7 and 0.95. The actual p_m is obtained from the card drawn on the indicator drum under running conditions of the engine. The card area, graphically measured by a planimeter (see Sec. 16), is divided by the card length to get the average equivalent height of the card. The spring scale of the instrument, applied to this average height, gives the mean effective pressure actually obtaining within the engine cylinder. This value, when introduced in Eq. (9.3.1), gives the indicated horsepower (ihp) of the engine.

Losses in steam-engine cylinders are (1) incomplete expansion; (2) initial condensation; (3) throttling, affected by valve and port area; and (4) radiation, which can be considered constant. The point of best steam economy occurs with the p_m at which the total of all losses is a minimum.

Mechanical Efficiency and Shaft Output

$$\text{Brake horsepower (bhp)} = \text{ihp} \times \text{mechanical efficiency} \qquad (9.3.3)$$
$$\text{Friction horsepower} = \text{ihp} - \text{bhp} \qquad (9.3.4)$$

bhp	25	50	75	100
Friction hp	6	6	6	6
ihp	31	56	81	106
Mechanical efficiency, %	81	89.5	92.5	94

Figure 9.3.2 shows the effect of mechanical efficiency and generator efficiency on the steam rate of an engine-generator set, both as to relative magnitude and as to location of the minimum values.

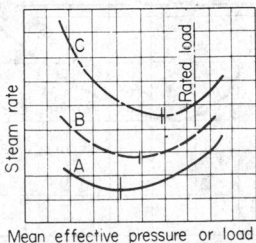

Fig. 9.3.2 Engine steam rate curves (*a*) at the steam cylinder, lb/(ihp·h); (*b*) at the engine shaft, lb/(bhp·h); (*c*) at the generator, lb/kWh.

Engine economy may be improved by a number of methods: (1) **separation of inlet and outlet ports;** (2) **steam jackets,** applied to cylinders and heads to keep surfaces hot and dry; (3) **multiple expansion.** Condensation losses are related to the temperature difference existing in the cylinder. Cylinders in series reduce this temperature difference and allow more complete overall expansion of the steam. Two, three, and four cylinders in series, as in **compound-, triple-,** and **quadruple-expansion** engines were common constructions, but the successive improvement is smaller for each additional stage. (4) **Superheating** gives the vapor the properties of a gas, reduces cylinder condensation, and necessitates decisive changes in engine design. The overall improvement in performance and water rate is so substantial that superheat is prevalent in practice. (5) The **uniflow** arrangement was the last great improvement in design (Figs. 9.3.3 and 9.3.4). Its high economy results from the high temperature of the residual steam at the end of compression. This temperature, aided by jackets, is higher than the live steam temperature, so that initial condensation is reduced to a negligible amount. For **condensing operation** the

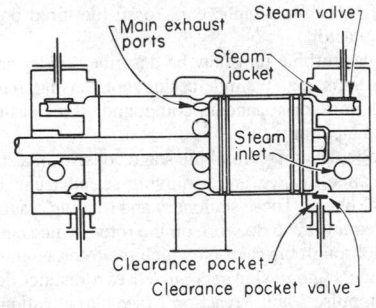

Fig. 9.3.3 Condensing uniflow engine cylinder, with clearance pockets.

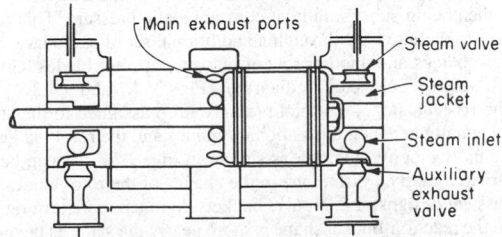

Fig. 9.3.4 Noncondensing uniflow engine cylinder, with auxiliary exhaust valves.

engines are built with steam valves only (two-valve type), Fig. 9.3.3. Clearance pockets are provided, with either hand-operated or automatic valves to permit operation with atmospheric exhaust or against back pressure. **Noncondensing uniflow** engines have, in addition, auxiliary exhaust valves to reduce the otherwise large clearance required (four-valve type). If back pressure is variable, exhaust-valve *gears* are designed to change the length of compression while the engine is in operation (Fig. 9.3.4).

Engine Steam (Water) Rates The efficiency of steam engines has generally been expressed in terms of pounds of steam per horsepower-hour. The term water rate was frequently used because of the convenient accuracy in weighing liquid water in the condensing plant. Figure 9.3.5 shows, on a percentage basis, two types of performance curves, one in

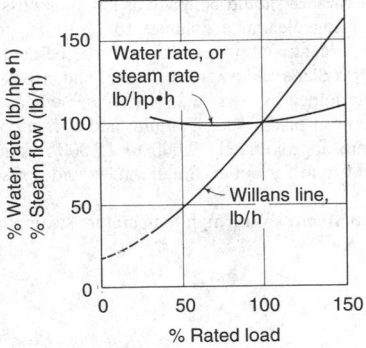

Fig. 9.3.5 Actual steam (water) rate, lb/(hp·h), and Willans line, lb/h, for a noncondensing uniflow engine, percentage basis.

lb/hph and the other in lb/h. The latter is identified as the "total steam" or **Willans line** and is straight for an engine with fixed cutoff and variable initial pressure. Figure 9.3.6 reflects the impact of steam pressure and superheat on the steam consumption of a condensing uniflow engine.

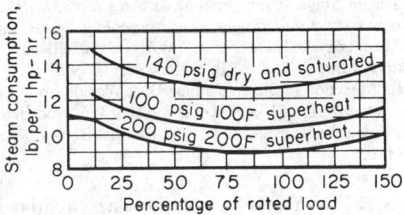

Fig. 9.3.6 Steam consumption of uniflow engine with 27.5-in vacuum.

The **Rankine cycle** (see Sec. 4) is the accepted thermodynamic standard for comparing the performance of steam prime movers (engines and turbines). It is predicated on complete isentropic (reversible adiabatic) expansion of the steam from initial to back pressure. It is shown on the *p-v* basis in Fig. 9.3.7. There is no compression or clearance. The water rate of this cycle is most conveniently calculated by use of the Mollier chart (Sec. 4) where

$$\text{Rankine steam rate} = 2{,}545/(h_1 - h_2) \qquad \text{lb/hph} \qquad (9.3.5)$$

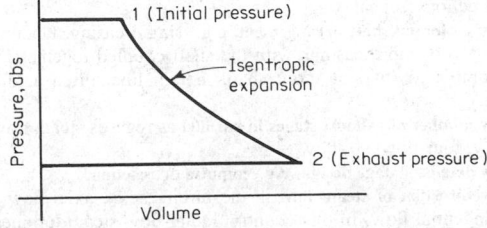

Fig. 9.3.7 Rankine cycle, *p-v* diagram.

and h_1 and h_2 are the initial and exhaust enthalpies, Btu/lb (constant entropy). The lowest point of the actual steam-rate curve (Fig. 9.3.6) is usually referred to as the Rankine rate, and the ratio of Rankine rate to actual rate is the **Rankine efficiency ratio (RER)**. This ratio may vary from 0.5 to 0.9, depending on the type of engine.

When clearance, compression, and incomplete expansion are introduced, the methods of Sec. 4 should be used to evaluate steam rate. This involves steam tables and charts to find the net work of the indicator card from the positive and negative areas of the several component phases. Figure 9.3.8 shows a theoretical steam-rate curve, computed by such methods, together with the actual test curve for the engine.

Engine Details Valves and valve-gear types range from the simplest D slide to gridiron, double-slide (Meyer or Rider), rocking, piston, releasing and nonreleasing Corliss, and poppet.

Volumetric clearance should be made as small as possible. Slide and piston valves bring clearance volumes to 12 to 15 percent, Corliss valves 6 to 10 percent, and poppet valves 4 to 8 percent.

Valve and Port Sizes Flow area of valves and cross section of ports are usually determined by port area $= AS/C$ in^2, where A is net piston area, in^2, S is mean piston speed, ft/min, and C is a constant, ft/min. Values of C are approximately 9,000 to 15,000 ft/min for inlet and 6,000 to 7,000 ft/min for outlet. Small valves and ports represent lost work.

Superheated Steam With high-temperature steam, the cylinder and parts design must allow for free expansion. Poppet and piston-valve cylinders can easily meet these requirements. The orthodox type of Corliss cylinder, however, is not suitable for highly superheated steam.

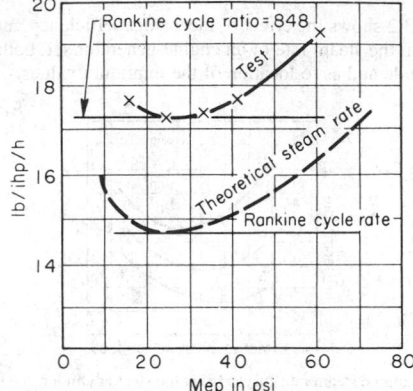

Fig. 9.3.8 Theoretical and test steam rate curves. Uniflow engine 20 × 24 at 200 r/min; throttle conditions 150 lb/in^2 saturated steam; exhaust to atmosphere.

9.4 STEAM TURBINES
by Frederick G. Baily

REFERENCES: Stodola, "Steam and Gas Turbines," trans. by L. C. Loewenstein, McGraw-Hill. Bartlett, "Steam Turbine Performance and Economics," McGraw-Hill. Warren, Development of Steam Turbines for Main Propulsion of High-Powered Combatant Ships, *Trans. Soc. Naval Architects Marine Engrs.,* 1946. Newman, Modern Extraction Turbines, *Power Plant Eng.,* Jan.–Apr., 1945. Salisbury, "Steam Turbines and Their Cycles," Wiley. Campbell and Heckman, Tangential Vibration of Steam Turbine Buckets, *Trans. ASME,* **47,** 1925, pp. 643–671. Campbell, Protection of Steam Turbine Disk Wheels from Axial Vibration, *Trans. ASME,* **46,** 1924, pp. 31–139. Deak and Baird, A Procedure for Calculating the Packet Frequencies of Steam Turbine Exhaust Blades, *Trans. ASME,* **85,** series A, Oct. 1963.

Steam turbines have established a wide usefulness as prime movers, and are manufactured in many different forms and arrangements. They are used to drive many different types of apparatus, e.g., electric generators, pumps, compressors, and for driving ship propellers, through suitable gears. When designed for variable-speed operation, a turbine may be operated over a considerable speed range, which may be of advantage in many applications. Steam turbines range in output capacity from a few horsepower to over 1,300 MW. The largest ones are used for generator drive in central power stations.

Turbines are classified descriptively in various ways.

1. **By steam supply and exhaust conditions,** e.g., condensing, noncondensing, automatic extraction, mixed pressure (in which steam is supplied from more than one source at more than one pressure), regenerative extractions, reheat.

2. **By casing or shaft arrangement,** e.g., single casing, tandem compound (two or more casings with the shaft coupled together in line), cross-compound (two or more shafts not in line, often at different speeds).

3. **By number of exhaust stages in parallel** as regards steam flow, e.g., two-flow, four-flow, six-flow.

4. **By details of stage design,** e.g., **impulse** or **reaction**.

5. **By direction of steam flow** in the turbine, e.g., axial flow, radial flow, tangential flow. In this country, radial-flow steam turbines have not been used; there are quite a few such machines abroad. Axial-flow units predominate; some small turbines in this country operate on the tangential-flow principle.

6. **Whether single-stage or multistage.** Small turbines, or those designed for small energy drop, may have only one stage; larger units are always multistage.

7. **By type of driven apparatus,** e.g., generator, mechanical, or ship drive.

8. **By nature of steam supply,** e.g., fossil-fuel-fired boiler, or light-water nuclear reactor.

Any particular turbine unit may be described under one or more of these classifications, e.g., a single-casing, condensing, regenerative extraction fossil unit, or a tandem-compound, three-casing, four-flow steam-reheat nuclear unit.

Turbine-Stage Design A turbine stage consists of a **stationary set of blades,** often called **nozzles,** and a moving set adjacent thereto, called **buckets,** or **rotor blades.** These stationary and rotating blades act together to allow the steam flow to do work on the rotor, which can be transmitted to the **load** through the **shaft** on which the rotor assembly is carried. Classical turbine-stage design recognized two distinct designs of turbine stage, "impulse" and "reaction" (see classification 4 above). In the **impulse** stage, the total pressure drop for the stage is taken across the nozzles or stationary element, the flow through the buckets or rotor blades then being substantially at constant static pressure. This may be extended to include flow through an additional set of stationary "intermediate" blades and another row of buckets, or rotor blades (Curtis or two row stages). See velocity diagrams, Figs. 9.4.1 and 9.4.2.

In the **reaction** stage, the total pressure drop assigned to the stage is divided equally between the stationary blades and the rotor blades, giving rise to a velocity diagram, as shown in Fig. 9.4.3. As can be seen, there arises a marked difference in the shapes of the rotor blades in the two classical designs; the impulse buckets do much more turning of the stream; the reaction-bucket shape is more nearly the same as the nozzle-blade shape.

Fluid flow theory recognizes that only in rare cases can an axial-flow turbine stage be either pure impulse or pure reaction. The annulus following the nozzle exit is filled with steam flowing with a high tangential velocity, i.e., a vortex, confined between inner and outer boundaries,

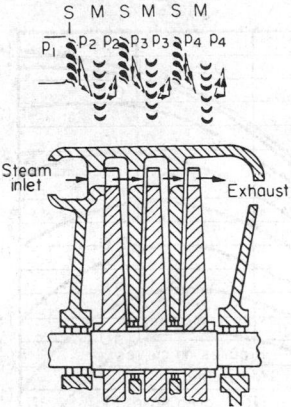

Fig. 9.4.1 Impulse turbine with single-velocity stages.

and for equilibrium to exist, there must be a gradient in static pressure from a lower-than-average value at the inner boundary to a higher-than-average value at the outer boundary. The amount of this depends upon the boundary **radius ratio** R_{outer}/R_{inner}. Thus only for a radius ratio near

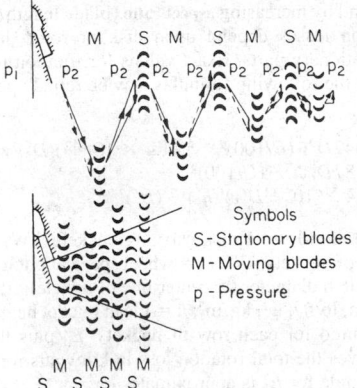

Symbols
S - Stationary blades
M - Moving blades
p - Pressure

Fig. 9.4.2 Impulse turbine with multivelocity stages.

1.0 (small-height blades) can it be said that any one pressure condition exists for the stage. All axial-flow turbine stages of larger radius tend to be more nearly impulse at the inner diameter and more nearly reaction at the outer diameter.

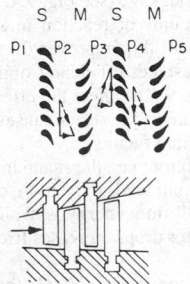

Fig. 9.4.3 Reaction turbine.

The designers of multistage steam turbines have continued to refine the efficiency of their steam paths over the years. For example, the 1950s saw the broad introduction of **twisted free-vortex staging.** Vane profiles have been improved since the 1950s by drawing on developments made in their aerodynamic laboratories and those of others. Complex computational fluid-dynamics computer codes based on true three-dimensional formulations of the inviscid Euler and viscous Navier-Stokes equations, and the supercomputers on which to run them, have recently become available. Manufacturing technology advancement permits the creation of steam-path components of almost any desired three-dimensional configuration. The result has been the introduction of **controlled-vortex staging** in the 1990s, in which the flow is biased toward the more efficient midsection of the blade, and secondary flow and profile losses are minimized by refinement in nozzle and bucket profiles.

Differences in basic mechanical construction of axial-flow turbine stages exist. Generally, the reaction type of turbine has continued as in the past with a "drum rotor" and stationary blades fixed in the casing, while the impulse type continues as a "diaphragm-and-wheel" construction. However, the mechanical construction bears no fixed relation to the degree of impulse or reaction adopted in the blading design. Designers employ the mechanical construction which they deem suitable for best reliability and efficiency.

An impulse stage, when employed for the first expansion, permits nozzle group control, i.e., steam admission to each group, opening and closing, successively, in response to load changes. This improves the efficiency at low loads through the reduction of the loss due to throttling.

A greater enthalpy drop can be employed per stage with the impulse element, particularly in the case of multivelocity elements, thus reducing the number of stages in a turbine. This is of special importance in the first expansion if the nozzle chamber is not integral with the turbine casing, so that the casing is not subjected to the initial steam conditions. If turbine elements of equal blade speed could have equal efficiency, one 2 row impulse element would equal four 1 row impulse elements or 16 rows (8 pairs) of reaction elements.

General Advantages of Steam Turbines Compared with reciprocating engines steam turbines require less floor space, lighter foundations, and less attendance; have a lower lubricating-oil consumption, with no internal lubrication, the exhaust steam being free from oil; have no reciprocating masses with their resulting vibrations; have uniform torque; have no rubbing parts excepting the bearings; have great overload capacity, great reliability, low maintenance cost, and excellent regulation; are capable of operating with higher steam temperature and of expanding to lower exhaust pressure than the reciprocating steam engine. Their efficiencies may be as good as steam engines for small powers, and much better at large capacities. Single units can be built of greater capacity than can any other type of prime mover. Small turbines cost about the same as reciprocating engines; larger turbines cost much less than corresponding sizes of reciprocating engines, and they can be built in capacities never reached by reciprocating engines. Combustion-gas turbines possess many of the advantages of steam turbines but are not available in ratings much exceeding 175 and 225 MW for 60- and 50-Hz service, respectively.

STEAM FLOW THROUGH NOZZLES AND BUCKETS IN IMPULSE TURBINES

Nozzles For the general treatment of the flow of steam and for maximum weight of flow of saturated steam, see Sec. 4.

The **theoretical work** obtainable from the expansion of 1 lb of steam is equal to the enthalpy drop in isentropic expansion $h_1 - h_{s2}$, in Btu/lb, and the spouting velocity in 223.8 $\sqrt{h - h_{s2}}$ ft/s (m/s = 44.7 $\sqrt{h'_1 - h'_{s2}}$, where h' is in kJ/kg). The actual expansion is not isentropic but follows a path such as $h_1 h_2$ on the enthalpy-entropy diagram (Fig. 9.4.4), and the available work becomes $h_1 - h_2$.

The **nozzle efficiency** is $(h_1 - h_2)/(h_1 - h_{s2})$.

The required **throat area** of the nozzle is $A_t = W v_t / V_t$, and the **mouth** area is $A_m = W v_m / V_m$, where v is specific volume, V is velocity, and the subscripts relate to throat and mouth, respectively.

In the case of a subcritical pressure ratio, throat and mouth conditions will coincide; with a supercritical pressure ratio, the mouth area will be

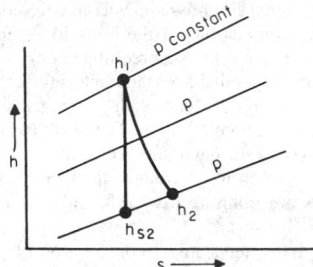

Fig. 9.4.4 Enthalpy-entropy diagram.

larger than the throat area. For **nozzle velocity coefficients** based on tests, see Keenan and Kraft, *Trans. ASME,* **71,** 1949, pp. 773–787. The **bucket-velocity coefficient** is the ratio of the average exit velocity from the bucket divided by the velocity equivalent of the total energy available to the bucket, i.e., the sum of inlet-velocity energy and pressure-drop energy. Typical values of tests on impulse buckets are shown in Fig. 9.4.5, when the bucket inlet angle is 3 to 5° larger than the exit angle. These are for subsonic flow. Reaction buckets can have the same coefficients as nozzles. For intermediate cases between pure impulse and pure reaction, interpolate between the values for these limiting cases.

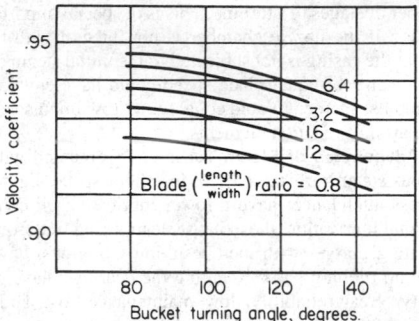

Fig. 9.4.5 Impulse-bucket velocity coefficients.

If, in the velocity diagram for a usual turbine stage, V_1 = actual nozzle exit velocity, V_2 = bucket relative entrance velocity, V_3 = bucket relative exit velocity, V_4 = absolute leaving velocity, V_0 = velocity corresponding to total stage available energy, then the **diagram efficiency** is $(V_1^2 - V_2^2 + V_3^2 - V_4^2)/V_0^2$.

Diagram efficiencies calculated with nozzle and bucket coefficients given above will be higher than the efficiencies derived by turbine tests, because of losses not existing in stationary tests of nozzles and buckets.

For supersonic bucket velocities, the impulse bucket-velocity coefficient is lower by the factors in Table 9.4.1.

Turbine Stage Efficiency Single-row stages of short blade length have relatively lower efficiency, owing to inner and outer sidewall losses; stages with longer blades are therefore higher in efficiency. Figure 9.4.6 shows typical values of stage efficiency for single-row stages, plotted against the wheel-speed steam-speed ratio, with pitch diam, inches/nozzle area, square inches, as a parameter. These curves reflect the net total of losses: (1) friction losses in nozzles (stationary

Table 9.4.1 Impulse Bucket-Velocity Coefficient Factors

Mach no.	<	1.0	1.2	1.3	1.5	1.75	2.0
Factor		1.0	0.997	0.995	0.978	0.928	0.816

blades); (2) friction losses in buckets (rotor blades); (3) rotation loss of rotor; (4) leakage loss between inner circumference of stationary element and rotor; (5) leakage loss between tip of rotor blades and casing; (6) moisture and supersaturation losses, if steam is wet (not included in curves of Fig. 9.4.6).

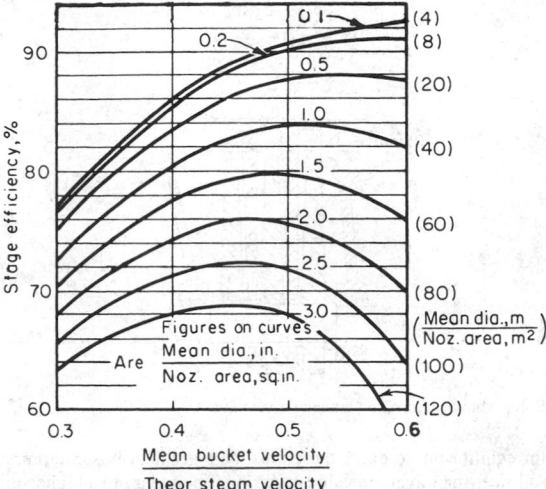

Fig. 9.4.6 Turbine single-row stage efficiency.

Nozzles and bucket friction losses are minimized by good aerodynamic design and by increasing **aspect ratio** (blade length/steam passage width). Rotation losses depend upon disk or rotor dimensions and surrounding stationary parts. Exact values for rotation loss depend on several factors; the following formulas may be relied upon for all usual purposes:

$$L_d = 0.042 D^2 w (U/100)^{2.9} = (1.4 \times 10^{-12})(D')^2 w'(U')^{2.9}$$
$$L_b = 0.187 Dwh^{1.25}(U/100)^{2.9}$$
$$= 3.34 \times 10^{-11} D'w'(h')^{1.25}(U')^{2.9}$$

where L_d = rotation loss of disk carrying buckets, kW; L_b = rotation loss of one row of buckets, kW; U = wheel speed at pitch diameter, ft/s; U', m/s; D = pitch diameter (at center line of nozzle), ft; D', mm; w = density of steam, lb/ft^3; w', kg/m^3; h = mean bucket height, in; h', mm. L_b must be figured for each row of buckets. L_d plus the sum of the values of L_b gives the total rotation loss in kilowatts for dry saturated steam. The formula for L_b is approximate.

Leakage loss of steam between inner circumference of stationary element and rotor is minimized by maintaining minimum practical clearance and by use of labyrinth packings (see Figs. 9.4.14 and 9.4.15 and accompanying text). Leakage loss of steam between tip of rotor blades and casing is similar to that through labyrinths between shaft and stationary parts; the magnitude depends upon the clearance area and the amount of reaction; other things being equal, the larger the percentage of reaction, the larger the leakage (see Fig. 9.4.7). Thus designers often employ considerable amounts of reaction in stages with long blades where such losses are small; this improves the net efficiency. In stages with short blades, the best net efficiency obtains with near impulse design. The curves in Fig. 9.4.6 reflect the effects of these practices.

The presence of moisture in the steam causes extra losses. These are probably mainly due to three factors:

1. Effect of supersaturation; i.e., the steam in expanding rapidly does not remain in equilibrium but tends to be more or less supercooled; thus less than theoretical equilibrium energy is available.

2. The presence of water drops increases friction losses in the steam itself.

3. Water drops tend to move more slowly than the vapor; they strike the rotor blades at unfavorable velocities and exert a braking effect.

Figure 9.4.8 gives correction factors which may be applied to the values from Fig. 9.4.6 to arrive at stage efficiencies in the "wet" region. Curves are identified by initial superheat, °F, or by initial quality, percent.

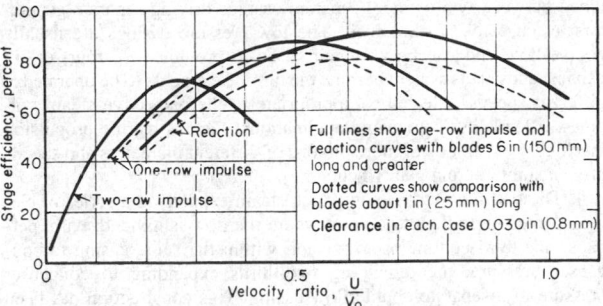

Fig. 9.4.7 Variation of the efficiency of turbine elements with velocity ratio.

Two-row stages, with one set of nozzles, have lower basic efficiency than single-row stages; they are useful for the first, or governing, stage in small to medium units. They are no longer employed in large central station designs. The approximate relative efficiency level of these stages is shown in Fig. 9.4.7.

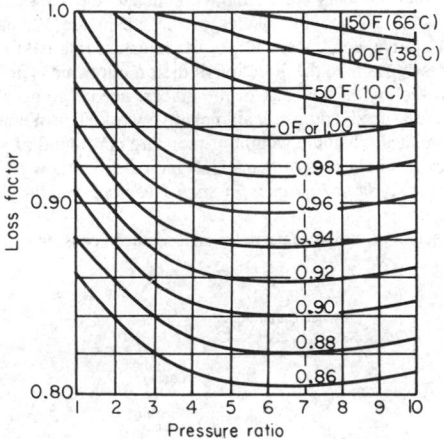

Fig. 9.4.8 Supersaturation and moisture loss.

The losses occurring in a turbine stage are partially recoverable in succeeding stages in a multistage turbine because the energy available to succeeding stages is increased above that resulting from isentropic expansion. The amount of this factor depends upon the temperature at which the loss takes place, and the turbine exhaust temperature. Values for the reheat or energy-regain factor are shown in Fig. 9.4.9.

Turbine Steam-Flow Requirements The steam flow required by a turbine is related to its power output and steam conditions by the following expressions:

Flow, lb/h = (TSR/efficiency) × power output, hp or kW

where TSR = theoretical steam rate, lb/(hp · h) = 2,545/available energy, Btu/lb, or TSR, lb/kWh = 3,412.14/available energy, Btu/lb.

Values of TSR are given in tables or may be calculated. In case of mixed-pressure or extraction turbines, the various sections of the turbine where flows are added or subtracted must be treated separately.

Turbine Steam-Path Design The basic quantities required for this are the steam conditions (i.e., inlet pressure and temperature and exhaust pressure), the required flow (see above), and the turbine speed. The latter is often fixed by the requirements of the driven machine; if not, the choice of speed by the designer is based upon experience, and

factors such as space or weight limitations, efficiency requirements, stress limits, or required exhaust area. Often several preliminary layouts are needed to arrive at the best design. If it appears that a single stage will suffice, the problem is simple, since the entire available energy is allotted to the one stage. If a multistage machine is required, the total available energy must be divided properly between the various stages.

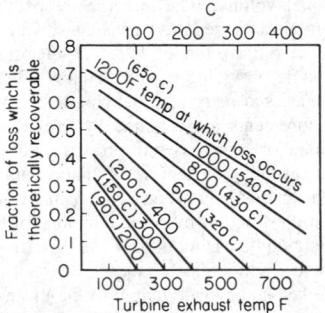

Fig. 9.4.9 Energy regain chart.

For a given wheel-speed/steam-speed ratio, the energy to be allotted to the stage is directly proportional to the square of the product (r/min × D), D being the rotor mean diameter. If available energy AE is in Btu/lb, D, in, and velocity ratio, 0.50, then

$$AE = (r/min × D)^2/(6.57 × 10^8)$$

The sum of the AE values per stage for all the stages must equal the total energy on the turbine, which is greater than the isentropic available energy by the amount of the reheat factor.

Having determined the energy to be assigned to each stage, the steam pressure in each stage is fixed, and this, together with the enthalpy determined from the efficiency of the machine from inlet, determines the steam specific volume. The velocity ratio and the degree of reaction decided upon fix the velocity diagram, from which the velocities through the stationary and rotating blades are determined. With the flow Q, lb/h, the velocity V, ft/s, and the volume v, ft³/lb, determined, theoretical areas required are given by

$$A = Qv/(25V) \quad in^2$$

The actual area required will be larger than theoretical because of friction losses; a value of 0.98 for nozzle flow coefficient is reasonable.

There is no fixed criterion for the number of stages to be used in a steam-turbine design, and experienced designers differ in the number they will choose for any particular design. The stage velocity ratio, the mean diameter, and often the r/min are subject to judgment, bearing in mind the general relationships shown in Figs. 9.4.6, 9.4.8, and 9.4.9. Cross-compound arrangements are possible, allowing higher speed for the high-pressure section, where steam volume flow is smaller, and a lower speed for the low-pressure section, where the final exhaust area desired may require long blades on a large diameter.

A detailed calculation of the efficiencies and energy outputs of each stage can be summed up to the total "internal used energy" of the turbine, which, when divided by the isentropic available energy, results in an "**internal efficiency**." Then, account must be taken of other losses to arrive at the turbine overall efficiency. These losses are

1. Exhaust loss, i.e., the kinetic energy corresponding to the absolute velocity of the steam leaving the last stage, plus the pressure drop through the exhaust connection to the turbine outlet flange, where exhaust pressure is by custom measured as a static pressure.
2. Pressure drops through interconnecting piping if turbine has more than one casing
3. Shaft end-packing leakages
4. Valve-stem leakages, if any
5. Inlet-valve and intermediate-valve pressure-drop losses
6. Bearing, oil-pump, and coupling power losses

7. If the turbine drives a generator or gear, the losses of these elements

LOW-PRESSURE ELEMENTS OF TURBINES

In condensing turbines expanding to high vacuum, the steam increases in specific volume as it passes through the stages, exhausting at about 1,000 times the inlet volume. The increase in specific volume from stage to stage is greatest in the latter few stages, which are commonly designed for pressure ratios of about 2 : 1. Considerations of efficiency and economics dictate providing reasonably low interstage steam velocity in these stages, and reasonable leaving loss from the last-stage blades. These requirements are satisfied by providing a steam path whose cross-sectional area increases in proportion to the specific volume. The area increase may be achieved by increasing blading length, by increasing the mean diameter of the steam path, by arranging the last expansions in multiple flow, or by some combination of two or more. The relatively small single casing unit of Fig. 9.4.10 is provided with increasing area by the first two techniques.

The ratio of the last-stage blade height to the mean diameter of the steam path may be as high as 0.35 in the last stage of large central station units. With such a ratio, there is a variation in blade velocity between the inner and the outer radii of the steam path that cannot be properly satisfied by any one steam velocity. Losses incidental to this may be minimized or eliminated by providing blades with warped surfaces, the sections at the inner radius partaking of impulse form with relatively small inlet angles, and those at the outer radius of reaction form with large inlet angles. The longest last-stage blades in service today have tips whose speed exceeds sonic velocity. Many are of transonic design employing subsonic profiles toward the inner radius, blending into supersonic profiles at the tip.

Among the methods of obtaining increased low-pressure blade areas through multiple flowing are:

1. A single-casing turbine with the last elements arranged for double flow.

2. The steam expansion divided between two casings, coupled in tandem, and driving a single main generator, the low-pressure casing being arranged for double flow. This arrangement is commonly extended to provide tandem-compound turbines with four and six exhaust ends. Figure 9.4.11 illustrates the application of a double-flow low-pressure casing to a unit of medium rating.

3. Cross-compound turbines, in which the steam expansion is divided between two or more separate casings driving separate generators, electrically synchronized. The low-pressure casings are usually double-flow and can be arranged so as to provide two, four, or six exhaust ends. This system permits the turbine elements to be operated at different speeds, selected as appropriate to the respective steam volumes. It lends itself to geared applications, such as marine propulsion machinery, when two or more pinions of different diameters and speeds drive a single-output gear wheel.

4. Divided-flow turbines in which steam expands in a series of elements in a single flow to a point where the flow is divided, with, perhaps, one-third continuing expansion within the same casing to condenser pressure, the remaining two-thirds expanding to condenser pressure in a separate double-flow casing. This construction has been used to provide triple-flow exhaust ends.

The **leaving loss** at the exit from the last row of blades is $V_{c2}^2/50,100$ Btu/lb of steam, where V_{c2} is the absolute terminal velocity in feet per second [or $(V_{c2}')^2/2,000$ kJ/kg, where V_{c2}' is in m/s]. The presence of moisture in the steam causes **moisture loss**. The acceleration of moisture particles is less as the density of the steam decreases; hence the difference between the velocities of the particles and the steam increases. As indicated in Fig. 9.4.12, with steam velocity, V_{s1} and moisture velocity V_m leaving the stationary blades and with bucket velocity V_b, the velocities relative to the moving blades of the steam are V_{c1} and of the moisture V_{m1}. The component V_{m2} of the moisture relative to the moving blades is opposite to the direction of their motion and is proportional to the force acting on the back of the blades, needed to accelerate the water to blade speed, and results in negative work. This negative work can be calculated when the weight of moisture per pound of steam and V_m are known. The results of many tests indicate that the *efficiency of a stage is reduced about 1 percent for each 1 percent moisture present in the steam.*

The presence of moisture particles will result in erosion of the blades

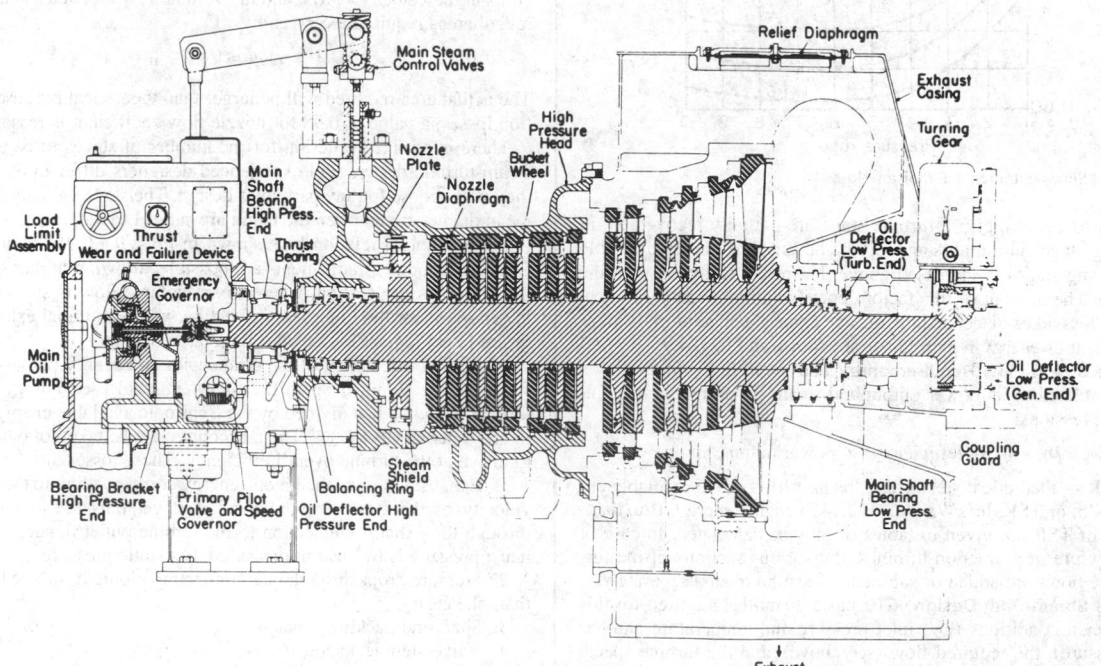

Fig. 9.4.10 Cross section of a multistage impulse condensing turbine rated at 30,000 kW. (*General Electric Co.*)

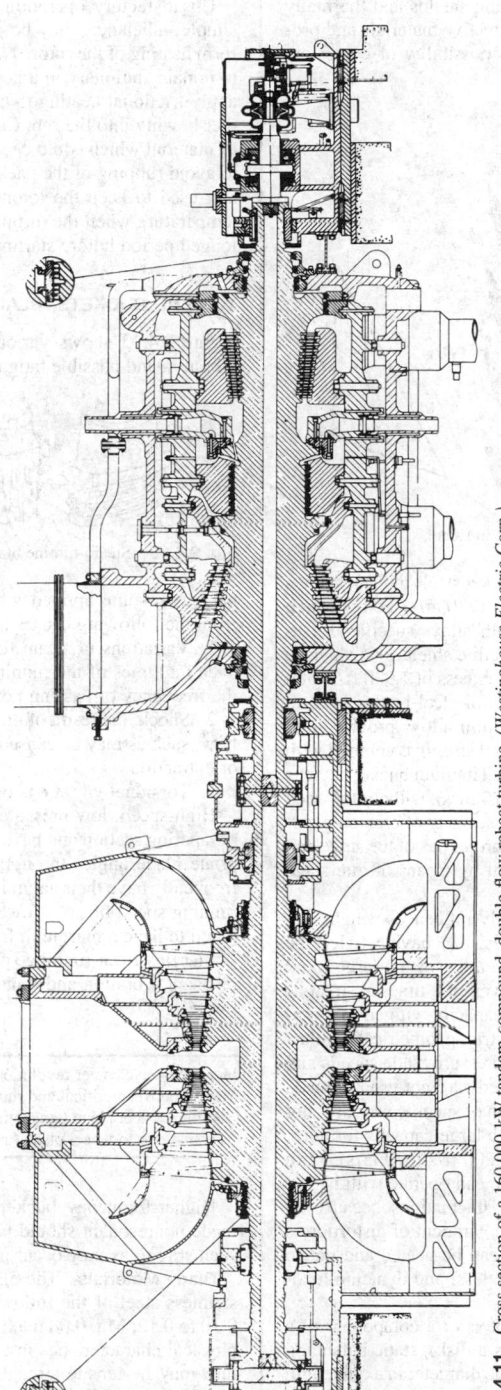

Fig. 9.4.11 Cross section of a 160,000-kW tandem-compound, double-flow reheat turbine. (*Westinghouse Electric Corp.*)

along their inlet edges if V_{m1} is too large—unless the moisture content is very small. The rate of erosion is reduced by using materials of inherently high erosion resistance, or protective shielding of hardened materials along the inlet edge. Attached Stellite shields and thermally hardened edges are commonly employed. Protective materials and processes are carefully chosen to minimize the possibility of corrosion.

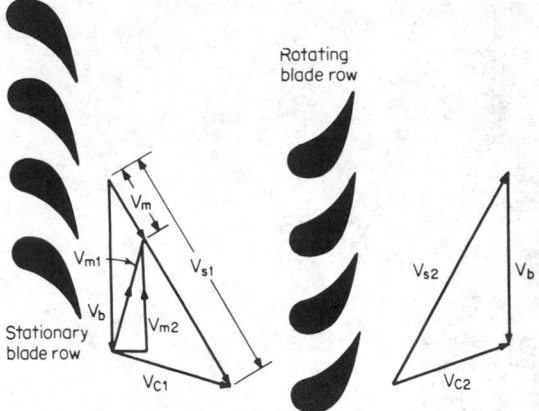

Fig. 9.4.12   Velocity of steam and of the moisture in steam.

Experience with the usual 12-chrome alloy bucket steels indicates that a threshold velocity exists at about $V_b = 900$ ft/s (270 m/s), below which impact erosion does not normally occur. With alloys specifically selected for their erosion resistance and with Stellite shields, satisfactory service has been obtained with values of V_b in excess of 1,900 ft/s (580 m/s). Centrifugal stress limits the application of steel blades to a tip velocity of about 2,000 ft/s (610 m/s). Titanium alloys provide high strength, low density (and hence low centrifugal stress), combined with excellent inherent moisture erosion resistance. Titanium bucket designs are available with a tip speed of 2,200 ft/s (670 m/s), without the need for separate erosion protection shielding. The severity of erosion penetration is dependent upon the thermodynamic properties of the stage and the effectiveness of reducing moisture content by means of interstage collection and drainage as well.

Rotative Speed   The speed selected greatly influences weight and cost. With two geometrically similar turbines, one having twice the linear dimensions of the other, the steampath areas of the larger, and hence its capacity, would be 4 times and its weight 8 times as great as those of the smaller. The weight per unit of capacity with similar machines increases inversely as the speed with strict geometrical similarity. For this reason, the highest possible low-pressure blade speeds and r/min are selected. Machines of different speeds are not usually made strictly geometrically similar, and the reduction of specific weight is not so rapid as the above rule would indicate. With large-capacity turbines, blade speeds, at the outer radius, can reach 2,200 ft/s (670 m/s). With high speeds and small dimensions, the turbine can operate with higher steam temperature and greater temperature fluctuations because of lighter casing walls and less mass of rotor; the amount of distortion is less with more uniform heating; the turbine can be heated and put in service more quickly; space requirements are less; and dynamic loadings on foundations are less.

Balancing   (see Secs. 3 and 5)   With a rotor, or a component of a rotor, of relatively short axial length (such as a disk), static balancing may suffice. Single bodies of more than half the diameter in axial length are usually dynamically balanced by the use of balancing machines. The balancing may be done at less than the running speed, since a bladed turbine rotor cannot be rotated in air at a speed approaching the running speed, or at high speed in an evacuated spin facility, or with a combination of both. Balance at full speed may not be satisfactory unless the balance weights are applied at points diametrically opposite the errors in balance, so that balance corrections must frequently be provided along

the length of the rotor. Five balance planes are commonly used for the rotors of large central-station turbines, of which three are in the rotor body between bearings and two are in the overhung couplings.

Unsatisfactory operation of turbine rotors in service, resembling a simple unbalance, may be caused by nonuniform material or nonuniform heating of the rotor. The latter may be caused by permitting a rotor to remain stationary in a hot casing, or by packing rubbing which may apply frictional heating to the "high" side of the rotor, leading to further bowing into the rub. Care must be taken to see that nonuniformity of material which could cause rotor distortion with heat is avoided, and to avoid rubbing of the packings on the rotor. Turning gears are generally used to keep the rotor turning at low speed to maintain uniform temperature when the turbine is shut down and cooling, and for a prolonged period before starting.

TURBINE BUCKETS, BLADING, AND PARTS

Figure 9.4.13 shows various blade fastenings. Blades are subject to vibration and possible fatigue fracture if their natural frequency is reso-

Fig. 9.4.13   Steam-turbine blade fastenings.

nant with some applied vibration force. Forced vibrations may arise from the following causes (see also Sec. 5):

1. Variations in steam forces. The blade frequencies should not be even multiples of the running speed, nor should they be resonant with the frequency of passing nozzle partitions or exhaust-hood struts.

2. Shock, the result of blades being subjected to discontinuous steam flow, such as may be caused by incomplete peripheral steam admission or extraction.

3. Torsional vibrations of the rotor.

High-speed, low-pressure blades of condensing turbines are usually of tapering section and have a warped surface in order to provide appropriate blade angles throughout their length. Long blades of this type frequently have their natural frequency between three and four times the running speed or even lower. Such blades should always be specially tuned to have a margin in frequency away from running-speed stimuli.

Margins from running speed to assure freedom from fatigue due to resonant vibration and transverse to the plane of the wheel are as follows:

Frequency, cycles per revolution	2	3	4	5
Margin between critical and running speeds, %:				
Within wheel plane (tangential)	15	10	5	5
Transverse to wheel plane (axial)	20	10	10	5

Higher-frequency buckets whose frequencies cannot assuredly be made nonresonant should be designed with adequate strength to resist such stimuli as may occur under service conditions.

Blade Materials   The material in most general use is a low-carbon stainless steel of the following composition: Cr, 12 to 14 percent; C, 0.10 to 0.12; Mn, 0.08 max; P, 0.03 max; S, 0.05 max; Si, 0.25 max. Its physical characteristics in the heat-treated condition at room temperature may be tensile strength, 100,000 lb/in^2 (690 MPa); yield point, 80,000 lb/in^2 (550 MPa); elongation, 21 percent; reduction of area, 60 percent. For the higher-temperature blades, particularly on large machines, it is practice to use alloyed chrome steel to achieve the required strength and oxidation-erosion resistance (see also Sec. 6).

Rotor Materials   Since steam turbines operate at high speeds, rotor materials must be of very high integrity and of basically high strength. In addition, the material should be "tough" at the temperatures at which it is to be highly stressed. A measure of this toughness may be obtained by running Charpy notch impact tests at various temperatures (see Sec. 5). In large modern machines the rotor forgings are almost exclusively made of steel melted in basic electric furnaces and vacuum poured to achieve freedom from internal defects. Turbine rotor forgings are usually made with small amounts of alloying elements such as Ni, Cr, V, or Mo.

Casing and Bolting Materials High-temperature and -pressure casings are almost always made of castings in order to achieve the complicated shapes required by these components. The alloy compositions used are selected so as to provide good weldability and castability as well as good physical properties. Low-temperature and -pressure casings are usually fabricated from steel plate. Bolts are made of forged or rolled materials.

The practical use of higher steam pressures and temperatures is limited by the strength and cost of available materials.

Leakage Metallic labyrinth packings are employed to (1) reduce internal steam leakage from stage to stage, (2) prevent steam from escaping the turbine from elevated-pressure ends, and (3) prevent air from leaking into the turbine at subatmospheric-pressure shaft ends. Interstage packings usually employ single rings with multiple teeth. End packings are arranged in multiple rings. At the high-pressure end, the leakage of steam past the first few rings may be carried to a lower-pressure stage of the turbine so that the outer rings need only prevent the leakage of low-pressure steam to atmosphere. The annulus above the last ring, or group of rings, is connected to a packing exhauster which maintains a pressure slightly below atmospheric. In consequence, no steam leaks out along the shaft, while a small amount of air is drawn past the final packing to the exhauster. Vacuum packings are provided with an inner annulus which is supplied with steam above atmospheric pressure. Steam flows inward from that annulus to supply the leakage toward the vacuum end, and outward toward a packing-exhauster connection. Thus steam is prevented from escaping along the shaft, while air is prevented from being drawn into the turbine.

Some turbines use carbon end packings or water seals. Carbon packings consist of one or more rings of pure carbon made in sections of 90 or 120° and held toward the shaft with small clearances by means of springs. The springs should have an axial component of force to hold the rings against the side of the box.

Labyrinths dependent upon radial clearances are shown in Fig. 9.4.14. In the design with "high and low" teeth, heavy teeth are cut on the turbine shaft, and thin teeth are part of a renewable packing ring which is made in segments backed and held inward by flat springs. The key indicated in the figure prevents turning of the segments. These types require that the rotor remain sensibly concentric with the stator but do not require a close axial adjustment.

Labyrinths dependent upon axial clearances are shown in Fig. 9.4.15. These require the maintenance of a close axial adjustment of the rotor.

The flow through a labyrinth may be approximately determined by the formula

$$W = 25KA \sqrt{\frac{(P_1/V_1)[1 - (P_2/P_1)]}{N - \ln (P_2/P_1)}}$$

where W = mass flow of steam, lb/h; K = experimentally determined coefficient; A = area through packing clearance space, in²; P_1 = initial

pressure, lb/in² abs; V_1 = initial specific volume of steam, ft³/lb; P_2 = final pressure, lb/in² abs; N = number of throttlings.

The value of K for interlocking labyrinths where the flow velocity is effectively destroyed between throttlings is approximately 50 and is independent of clearance for usual clearance values.

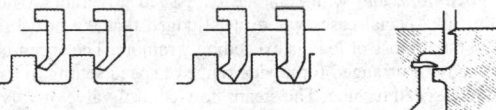

Fig. 9.4.15 Labyrinths with axial clearance.

For noninterlocking labyrinths, i.e., stationary teeth against a straight cylindrical shaft, the value of K varies with the ratio of tooth spacing to radial clearance, being about 120 for tooth spacing of five times the radial clearance, reducing to approximately 50 for a tooth spacing fifty times the clearance.

Turning Gears Large turbines are equipped with turning gears to rotate the rotors slowly during warming up, cooling off, and particularly during shutdown periods of several days when it may be necessary to start the turbine again on short notice. The object is to maintain the shaft or rotor at an approximately uniform temperature circumferentially, so as to maintain straightness and preserve the balance. Turning gears permit an appreciable reduction in starting time, particularly following a relatively short shutdown.

It is seldom necessary to use high-pressure oil to lift the journals off their bearings when using a turning gear. A low-pressure motor-driven oil pump is used which floods the bearings with about half their usual flow of oil. The turning gear is made powerful enough to start the rotor and rotate it at low speed.

High-Temperature Bolting The bolting of high-pressure, high-temperature joints, particularly turbine-shell or valve-bonnet joints, is very exacting. It is worthwhile to taper the threads of either the male or the female element so that the engagement of the threads throughout the length of the engaged thread portion will give approximately uniform bearing. The reliability of taper-threaded bolts is superior to that of parallel-threaded bolts.

Thrust bearings must usually be designed to carry axial rotor thrust in either direction, with sufficient margin to take care of unusual operating conditions. Thrust runners may be machined solid on the shaft or can be a separate piece shrunk on and secured from endwise motion. The stationary bearing surfaces may be of the pivoted-shoe type, or made in solid plates with babbitt or other bearing-material facing, with grooves for oil supply and "lands" to carry the thrust load.

Axial thrust on the turbine rotor is caused by pressure and velocity differences across the rotor blades, pressure differences from one side to the other on wheels or rotor bodies, and pressure differences across shaft labyrinths which have steps in diameter. The net thrust is the sum of all these effects, some of which may be in one direction, and some in the opposite direction. Rotor-blade and wheel or body thrust are usually in the direction of steam flow. It is usual to balance this thrust either partially or completely by proper choice of shaft packing diameters and pressure differences so that the net thrust is not too large. The thrust

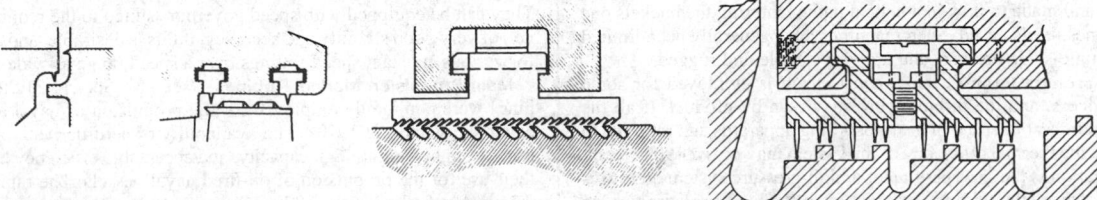

Fig. 9.4.14 Labyrinths with radial clearance.

bearing must be made large enough so that it is not overloaded by the net thrust. In this respect it is necessary to foresee all operating conditions which may influence the net thrust and allow for these; in addition some margin must be allowed for abnormal or unforeseen circumstances which may occur in service. (See also Sec. 8.)

Controls   Steam turbines are nearly always equipped with speed-control governors, and with separate overspeed governors. The only exceptions are special cases where it is judged that the possibility of overspeed due to loss of load is exceedingly remote. The speed-control governor may be arranged for a wide range of speed setting in the case of a variable-speed turbine. The steam flow-control valve or valves are operated by this governor, usually through a hydraulic relay mechanism. The overspeed governor is usually of an overisochronous type, arranged to trip at 10 percent over normal full speed (on some small turbines, 15 percent), actuating quick-closing stop valves to shut off the steam supply to the turbine. Speed-control governing systems are usually designed so that the overspeed-governing system is not brought into action on sudden loss of full load.

On automatic extraction machines, the speed governor must be correlated with the extraction-pressure controlling-valve system.

On fossil-reheat turbines, because of the large stored steam volume in the reheater and piping, and on nuclear turbines, because of the moisture separator/steam reheater and piping volumes, it is necessary to protect the turbine from overspeed on sudden loss of load by shutting off this stored steam ahead of the lower-pressure stages. It is done by intermediate intercept valves, actuated by a governor set slightly higher than the speed-control governor. An intermediate stop valve actuated by the overspeed trip is usually added in series for additional protection.

Speed-governing systems can be supplied of various sensitivities and speed ranges, to suit the requirements of the driven apparatus. Both mechanical-hydraulic and electrohydraulic systems are employed. Analog electrohydraulic systems were introduced during the 1960s as electronic components of sufficient reliability for turbine service became available. Digital systems were introduced in the 1980s and have become the standard control technology. Modern systems control transient thermal stress and the expenditure of low-cycle fatigue life of high-temperature components, in addition to controlling speed and load. (See "Steam Temperature—Starting and Loading.")

Steam turbines are usually provided with a supervisory instrumentation system. That for a large central station unit may process several hundred channels of information, providing alarms and/or trips for abnormal parameters. Data may be stored on paper charts, on magnetic media, or in computer memory depending upon the design of the system. Displays frequently employ color cathode-ray tubes.

INDUSTRIAL AND AUXILIARY TURBINES

Low-capacity turbines are employed for services such as engine-room auxiliaries and small generating sets. Usually they comprise a single turbine element. Their efficiency may be less than that of a corresponding reciprocating engine, but they are employed because of their compactness and because they require no internal lubrication. The exhaust steam is free from oil and grease and is available for heating purposes. They are frequently coupled to the driven machine by means of speed-reducing gears. Turbines of this type are usually of the axial-flow type, but the **tangential helical-flow** turbine, in which the steam is directed tangentially and radially inward by nozzles against buckets milled in the wheel rim and made to flow in a helical path reentering the buckets one or more times, is also used. Such machines have generally been limited to small single-stage designs, and are very simple and rugged.

In **back-pressure turbines,** the exhaust steam is employed for some heating process, and the turbine work may be a by-product. If all the exhaust steam is condensed in heat-absorbing apparatus and returned to the system, the thermal efficiency of the system may be over 90 percent. One application is the **superposition** of a high-pressure system on lower-pressure power units, with the exhaust from the high-pressure turbine power units, with the exhaust from the high-pressure turbine going to the low-pressure steam mains. By this device, an old power station can

be rehabilitated and its capacity increased. Two methods of operation are in use: (1) with constant intermediate pressure as when the lower-pressure power units operate also with steam from existing lower-pressure boilers and (2) with variable intermediate pressure as when the low-pressure units receive steam only from the back-pressure turbine.

Boiler-feed-pump-drive turbines have been used extensively as part of the power-plant system, especially for large, high-pressure plants where the required feed-pump power may amount to 4 percent of the gross plant output, and for large nuclear units. The turbine and pump can be matched as to rotative speed. These turbines are variously integrated into the main cycle. The most common present practice is to use condensing, nonextracting turbines supplied with steam in the range of 150 to 200 lb/in^2 abs (1,000 to 1,400 kPa) taken from the exhaust of the intermediate sections of fossil turbines, or from the inlet of the low-pressure sections of nuclear units. These turbines normally have a connection to the main steam supply for starting and low-load operation. Similar auxiliary turbines are frequently used to drive the forced- or induced-draft fans of large fossil-fuel-fired boilers.

With **extraction turbines,** partly expanded steam is extracted for external process use at one or more points. The turbines may be either condensing or noncondensing. Extraction turbines are usually designed to sustain full rated output, with or without extraction, and are provided with automatic regulating mechanisms to deliver steam from the extraction points at constant pressure, as long as there is sufficient power load to permit the necessary flow. The use of such extraction turbines, particularly with high initial pressures in connection with many industrial processes requiring moderate- or low-pressure steam, results frequently in a high efficiency of power production, i.e., the only heat required in such a plant over and above that to provide the required process steam is the heat equivalent of the power generated by the steam before extraction. This means that such power can be produced at nearly 100 percent thermal efficiency.

Figure 9.4.16 illustrates a typical double-automatic, condensing extraction turbine, providing two controlled extraction pressures. In this case, the unit is equipped with internal spool valves at both extraction points. Grid and poppet-type valves have been used for this purpose. The extraction-stage valves are under the control of an extraction-pressure-actuated governor; they determine the flow to the subsequent stages of the turbine and maintain the pressure in the extraction stage. The operation of the valves is by means of a pilot valve controlling the admission of high-pressure fluid to an actuating cylinder, which, in turn, opens or closes the valves to the nozzle ports to the succeeding stage. Extractions of this kind are called pressure-controlled extractions, and the pressure is maintained practically constant over a wide range if the load is sufficient to permit the required steam flow.

Mechanical-drive turbines are commonly applied where moderate to high power and/or precise speed control of the driven machine are needed. Typical applications include the powering of papermaking machines and the driving of fluid compressors in petrochemical plants. So many sizes and types are available from the manufacturers, and they have been adapted to so many applications, that it is impossible to give here more than a general description. These turbines are commonly built in sizes from a few to several thousand horsepower. If the speed of the driven machine is low, a reduction gear may be used in order to reduce the size and cost of the driving turbine and to improve its efficiency. Mechanical-drive turbines have a wide range of application, being adaptable for any steam conditions and a wide variety of speeds. They can be equipped with speed governors suited to the requirements, i.e., of very good stability and accuracy, if this is desirable, and arranged for various constant speed settings over a speed range as wide as 10 : 1.

Main Propulsion Marine Turbines   (see also Sec. 11.3) Steam turbines were commonly employed for the propulsion of naval and merchant ships into the 1970s. The availability of aero-derivative gas turbines of light weight, high capacity, and acceptable efficiency has led to their use for the propulsion of oil-fired naval vessels. The rapid rise in the price of fuel oil in the 1970s led merchant ship operators away from steam propulsion to the use of the more efficient diesel engine. Steam turbines continue to be used for the propulsion of all nuclear-powered

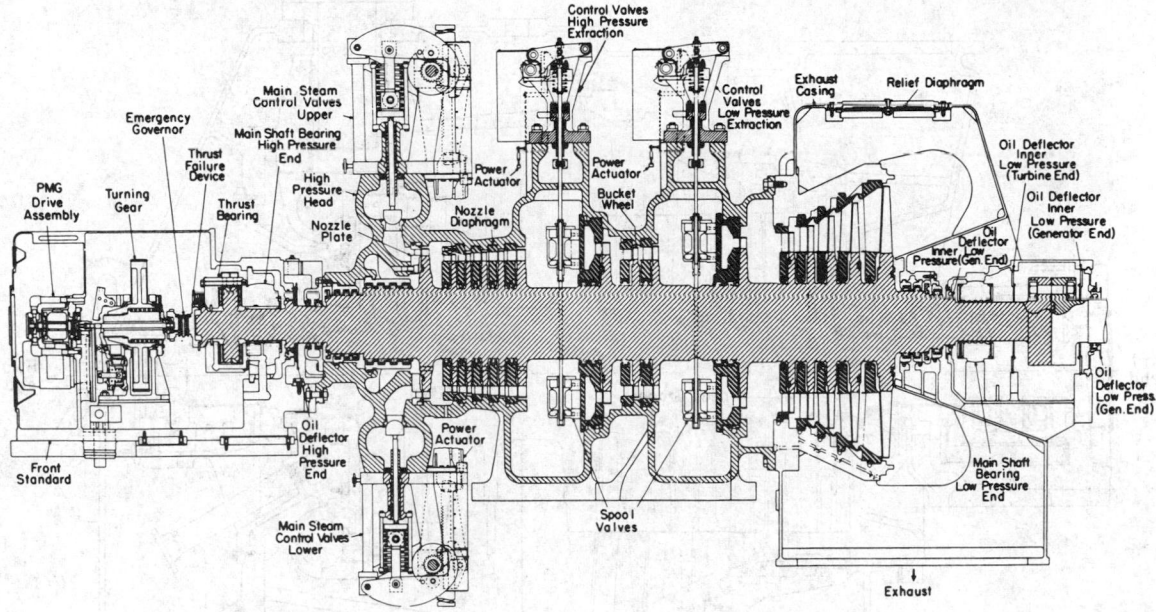

Fig. 9.4.16 Double automatic condensing extracting turbine, 25,000 kW. *(General Electric Co.)*

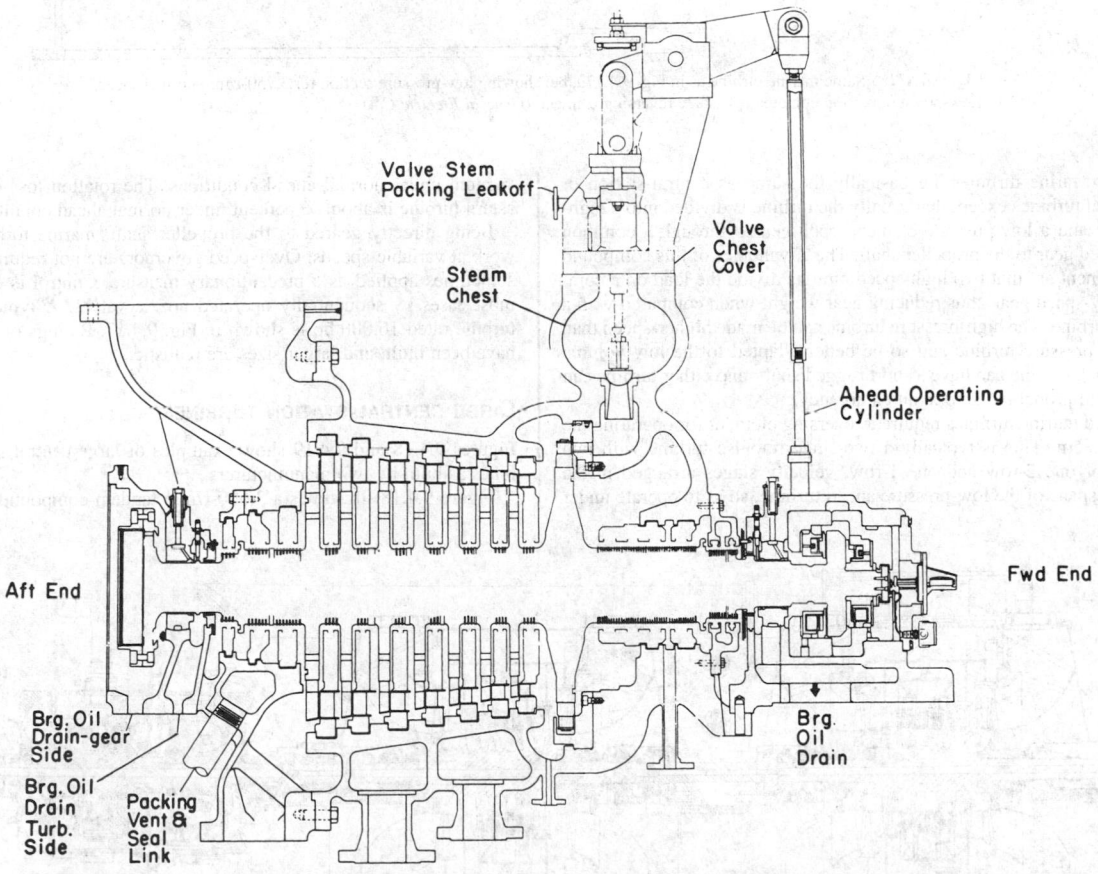

Fig. 9.4.17a A 16,000-hp cross-compound marine turbine designed for steam conditions of 600 lb/in² gage, 850°F, 1½ inHg abs. High-pressure section for 6,550-r/min normal speed. Low-pressure section in Fig. 9.4.17b. *(General Electric Co.)*

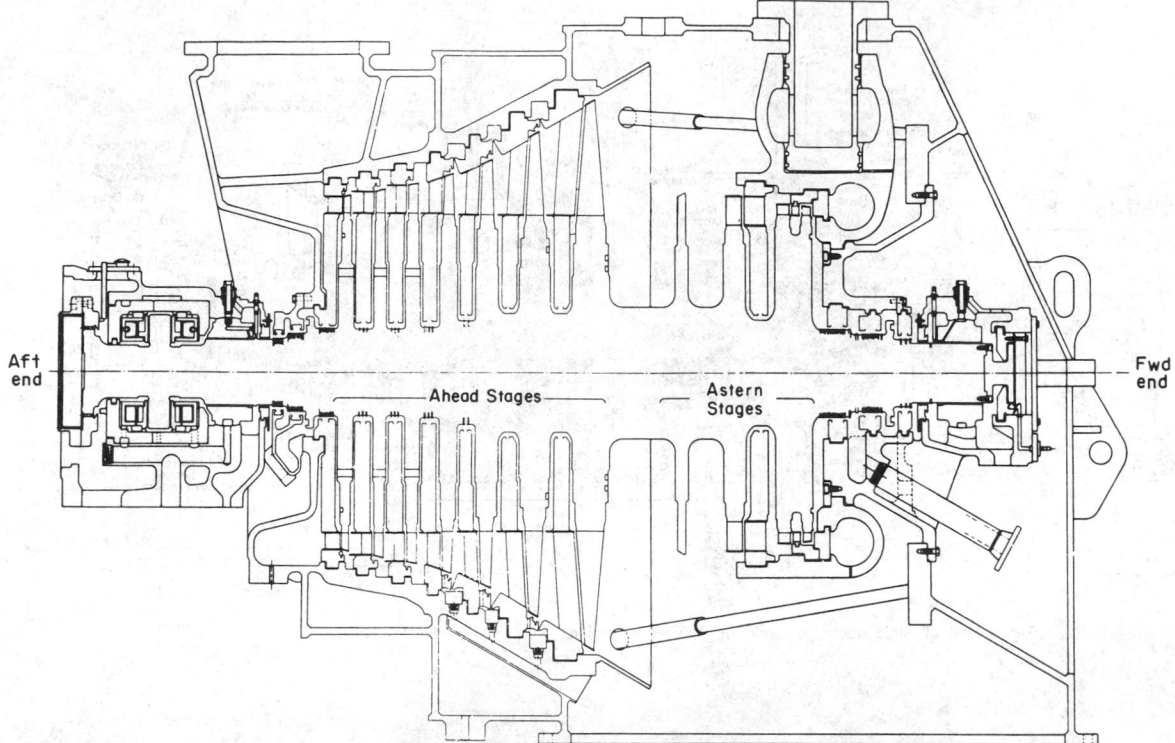

Fig. 9.4.17b Same marine turbine as in Fig. 9.4.17a, but showing low-pressure section for 3,750-r/min normal speed. Low-pressure section contains two-stage reversing element. *(General Electric Co.)*

vessels. Marine turbines are basically the same as central-station or industrial turbines except that usually the turbine is divided into a high-pressure and a low-pressure element, each geared through a common low-speed gear to the propeller shaft. The advantages of this compound arrangement are that two high-speed pinions divide the load on a common low-speed gear, thus reducing gear weight when compared with a single turbine. The high-pressure turbine can be made higher-speed than the low-pressure turbine and so be better adapted to the low volume flow. Each turbine can have a short rugged shaft, and either turbine can be used to propel the ship in an emergency.

Geared marine turbines require a reversing element for operating the vessel astern. This is typically a two-stage impulse turbine with two 2-row, or one 2-row and one 1-row, velocity stages arranged in the exhaust space of the low-pressure ahead turbine, so as to operate under

vacuum under normal ahead conditions. The rotation loss of such an astern turbine is about ½ percent under normal ahead conditions.

Being directly geared to the propeller shaft, marine turbines must work at variable speeds. Overspeed governors are not required but are sometimes applied as a precautionary measure. Control is effected in most cases by sequentially operated nozzle valves. A typical marine turbine rated 16,000 hp is shown in Fig. 9.4.17. Ratings of 70,000 hp have been built, and larger sizes are realistic.

LARGE CENTRAL-STATION TURBINES

Figures 9.4.18 and 9.4.19 show examples of large central-station turbines as built by two manufacturers.

Figure 9.4.18 illustrates a 3,600-r/min tandem-compound four-flow

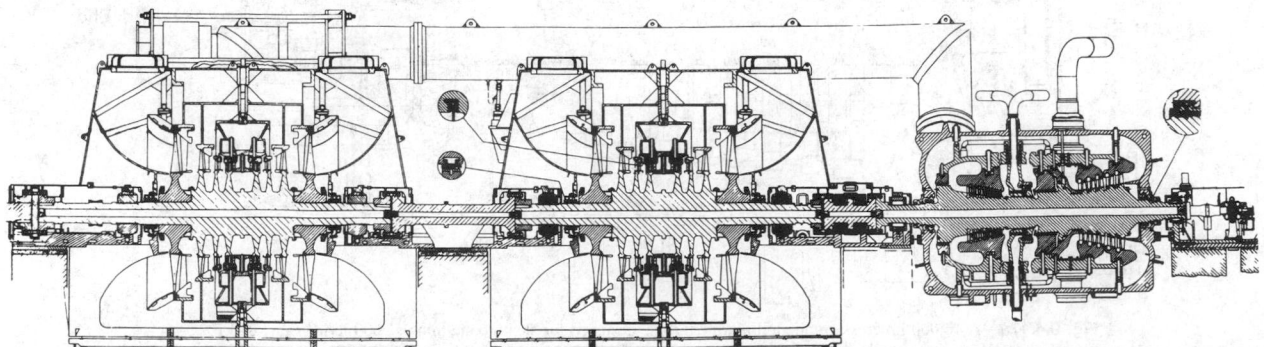

Fig. 9.4.18 Cross section of a 500-MW tandem-compound, quadruple-flow 3,600-r/min reheat turbine. *(Westinghouse Electric Corp.)*

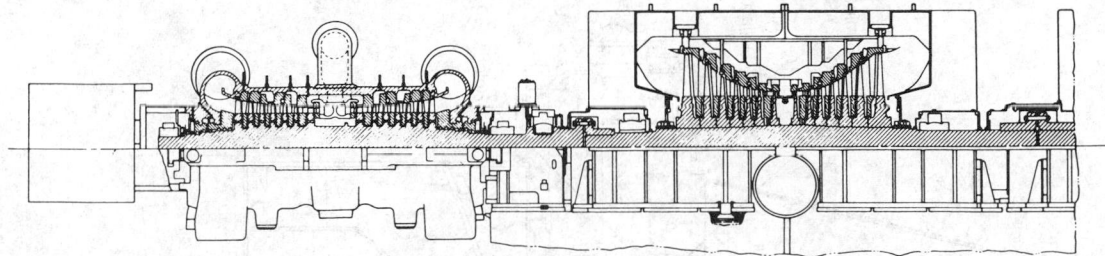

Fig. 9.4.19   Cross section of a 1,300-MW class tandem-compound, six-flow 1,800-r/min turbine for steam from light-water nuclear reactors, showing one of the three low-pressure sections. *(General Electric Co.)*

unit rated 500 MW. It is a single-reheat fossil unit for nominal inlet-steam conditions of 2,400 lb/in² gage (16,650 kPa) 1,000°F (538°C), reheat to 1,000°F (538°C). The right-hand casing is a combined high-pressure and reheat section. Steam flows from the left center to the right through the impulse-type governing stage, then reverses, flowing to the left through the nine reaction-type high-pressure stages, and exhausts from the casing to the reheat section of the boiler. Reheated steam reenters that casing at its right center, flowing to the right through five reheat stages, turning once more to flow to the left between the inner and outer casings, finally exhausting up and to the left to the two double-flow low-pressure casings on the left end of the unit. The inlet stop and control valves, the reheat stop and intercepter valves, and the generator are not shown.

Four-flow units of this general type employ last-stage rotor blades from 25 to 40 in. (600 to 900 mm) long and in ratings up to about 700 MW. Significantly higher ratings require dividing the functions of the combined casing into a separate single-flow high-pressure casing and a separate two-flow reheat casing, for a total of four casings. The use of the long titanium last-stage buckets makes this configuration suitable for ratings up to 1,200 MW. In cases requiring additional exhaust area, a third double-flow low-pressure element can be employed for a total of five casings. Tandem-compound 3,600- and 3,000-r/min units are commonly offered for ratings up to about 1,200 MW. Somewhat larger ratings can be accommodated by 3,600/3,600- or 3,600/1,800-r/min cross-compound units, but economic considerations makes their application infrequent.

Figure 9.4.19 illustrates an 1,800-r/min tandem-compound six-flow turbine designed for steam from light-water nuclear reactors. Reactors of both the boiling and pressurized-water types raise steam at 1,000 lb/in² abs (7,000 kPa) approximately with little or no initial superheat, so that the initial temperature is about 545°F (285°C), with a fraction of 1 percent of moisture frequently present. The poorer steam conditions result in higher steam rates than seen by fossil turbines. The lower initial pressure causes larger initial specific volume. In consequence, a typical nuclear turbine must accommodate 2.5 to 4 times the initial volume flow, and about 1½ times the exhaust volume flow of a fossil unit of the same rating. These considerations and the fact that the low temperature of the steam results in high moisture content in the expansion lead to the choice of 1,800 r/min. In halving the speed, diameters are less than doubled, balancing the advantages of larger steam-path area to accommodate large flow, while reducing velocities to minimize the occurrence of impact moisture erosion. The shortened energy range due to the lower initial conditions requires only two kinds of casings, high pressure and low pressure, compared with the three needed by fossil-reheat units.

Referring to Fig. 9.4.19, the steam enters the double-flow nozzle boxes of the high-pressure section, to the left, through stop and control valves which are not shown. It flows in both directions through the impulse-type stages, exhausting through four connections on each end of the shell. At the exhaust, the pressure is reduced to 200 lb/in² abs (1,400 kPa) approximately, and the moisture content is increased to 12 percent. That moisture poses an erosion risk and a performance loss to the low-pressure section following. It is current practice to dry the steam in an external moisture separator, frequently combined with one or two stages of steam-to-steam reheating, before admission to the low-

pressure casings. Figure 9.4.20 is a cross section through a combination moisture separator and two-stage steam reheater. The exhaust from the high-pressure turbine enters the shell at the bottom, flowing upward through the inclined corrugated-plate moisture-separating elements, which remove essentially all the entrained water. It continues upward, flowing over the tubes of the first-stage bundle, which are supplied with

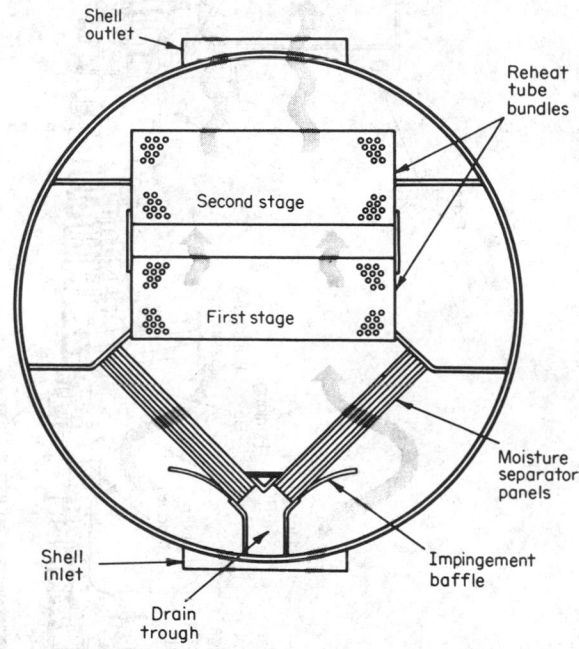

Fig. 9.4.20   Cross section of a combination moisture separator and two-stage steam reheater for use with nuclear reactor turbines.

steam extracted from the high-pressure turbine, approaching to within 20 to 50°F (11 to 28°C) of that temperature. Next, it flows over the tubes of the second stage bundle, which are supplied with initial steam, approaching to within 20 to 50°F (11 to 28°C) of that temperature, or to 495 to 525°F (257 to 274°C). The steam leaves the vessel at the top and is admitted to the low-pressure sections of Fig. 9.4.19, through stop and intercept valves, not shown. The last-stage blade length is in the range of 38 to 52 in (960 to 1,320 mm).

Units of the type described are built to ratings of approximately 1,300 MW with larger sizes available. Similar four-flow units are employed for ratings up to approximately 1,000 MW.

Other types of nuclear reactors, such as the high-temperature gas-cooled reactor and the liquid-metal-cooled fast-breeder reactor, produce steam conditions at temperature and pressure levels comparable with fossil-fuel-fired boilers, leading to the use of 3,600-r/min units similar to fossil practice.

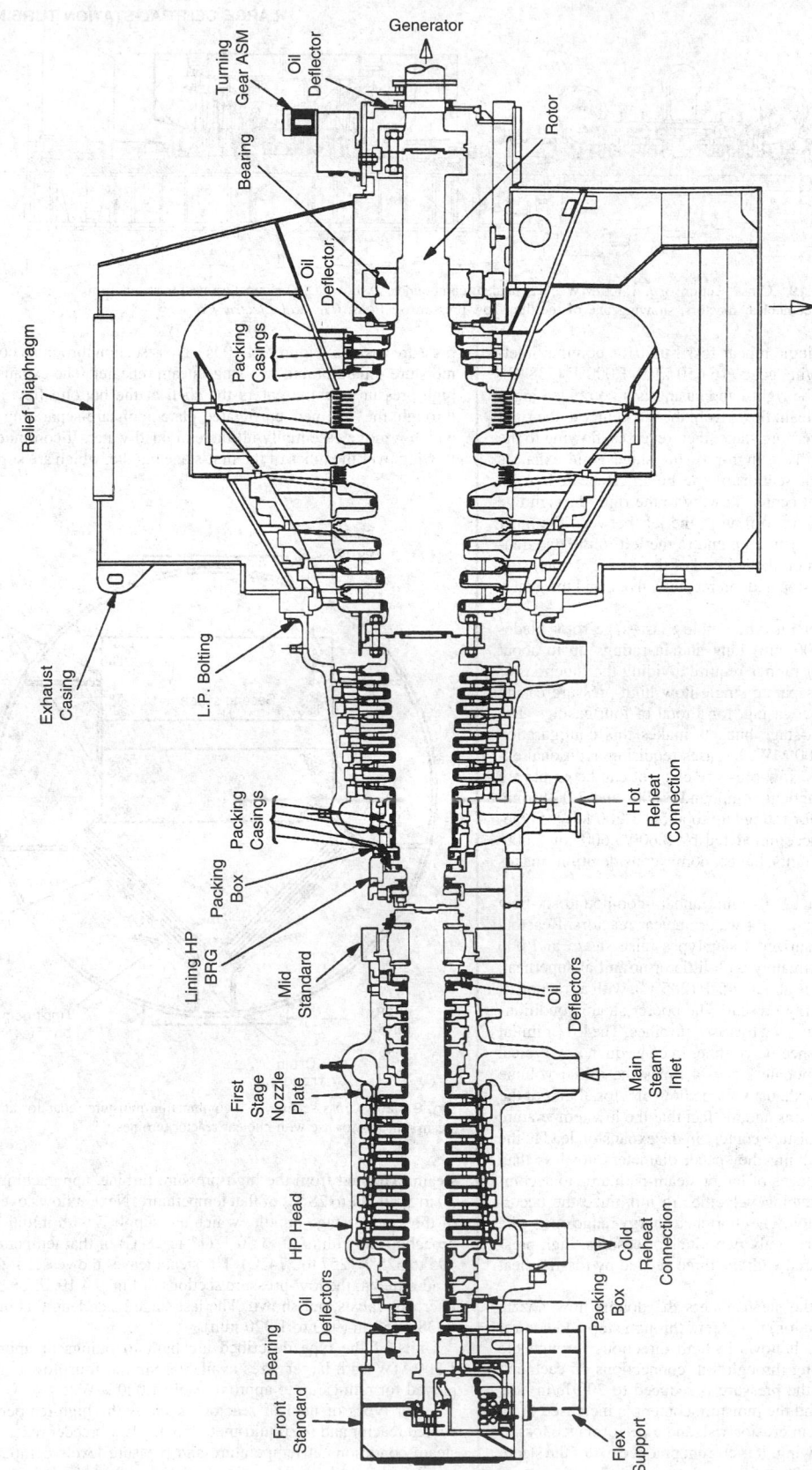

Fig. 9.4.21 Two-casing reheat combined-cycle turbine with single-flow down exhaust. (*General Electric Co.*)

STEAM TURBINES FOR COMBINED CYCLES

(See also Sec. 9.7)

Modern combustion **gas turbines (GTs)** have ratings approaching 250 MW. Considerable sensible heat is available in the gas-turbine exhaust-gas flow which ranges from 1,000 to 1,100°F (530 to 600°C) in temperature. In the **combined cycle,** the gas-turbine exhaust is directed to an unfired steam boiler called a **heat-recovery steam generator (HRSG).** The steam generated in the HRSG is admitted to a steam turbine, whose electrical output is approximately one-half that of the gas turbine. The resulting **combined thermal efficiency** can range from 50 to 55 percent, better than that of any other available power cycle. High efficiency combined with the clean exhaust of gas turbines burning natural gas has created a broad field of application for combined cycles and the associated steam turbines.

Combined cycles can be arranged in several ways. In the *single-shaft* configuration, the gas and steam turbines are coupled together, driving a single generator. Units rated 352 MW have been manufactured for 50-Hz power systems, in which the gas turbine contributes 223 MW while the steam turbine generates 129 MW. The *multiple-shaft* arrangement uses a separate generator for each steam turbine and gas turbine. Each gas turbine has its own HRSG. The steam output from two or more GT/HRSG pairs can be manifolded to a single steam turbine. For example, a typical application for 60-Hz power systems combines two 164-MW gas turbines with one 188-MW steam turbine for a combined output of 516 MW. The steam conditions are 1,400 lb/in² gage (9,760 kPa), 1,000°F (538°C) with reheat in the HRSGs back to 1,000°F (538°C).

Steam turbines in combined-cycle service operate following the gas turbine(s), with their admission valves wide open accepting the full instantaneous output of the HRSG(s). The valves are located away from the turbine casings and are used only for speed control upon starting and for overspeed protection in the event of loss of load. The result is a very simple symmetric casing arrangement which reduces transient thermal stress and helps the steam turbine follow the potentially rapid changes in gas-turbine output. These features can be seen in Fig. 9.4.21. A two-casing reheat unit with single-flow exhaust is illustrated.

STEAM-TURBINE PERFORMANCE

The ideal **steam rate** (steam consumption, lb/kWh) of a simple turbine cycle is $3,412.14/(h_1 - h_{s2})$, where h is in Btu/lb [or kg/kWs = $1/(h_1' - h_{s2}')$, h' in kJ/kg]. (See Fig. 9.4.4.)

The actual steam rate is $3,412.14/[\eta_t(h_1 - h_{s2})]$, where η_t is the **engine efficiency** of the turbine only, inclusive of mechanical losses {or kg/kWs = $1/[\eta_t(h_1' - h_{s2}')]$}.

The actual steam rate of turbine and generator is $3,412.14/\eta_e(h_1 - h_{s2})]$, where η_e is the engine efficiency including all mechanical and electrical losses {or kg/kWs = $1/[\eta_e(h_1' - h_{s2}')]$}.

The enthalpy of the steam leaving the turbine elements h_2 is

$$h_2 = h_1 - \eta_s(h_1 - h_{s2}) = (Wh_1 - 3,412.14P_g)/W$$

where η_s is the engine efficiency of the steam path, inclusive of leakages and losses but exclusive of mechanical and electrical losses; W is the steam flow, lb/h; and the gross output P_g is the net output plus mechanical and electrical losses, kW [or $h_2' = h_1' - \eta_s(h_1' - h_{s_2}') = (W'h_1' - P_g)/W'$, where W' is flow in kg/s].

Table 9.4.2 gives steam rates for the ideal simple turbine cycle through a wide range of operating conditions. The performance of a turbine is usually expressed as a steam rate in the case of machines having no extraction or admission of steam between inlet and exhaust, which is generally true of small units and most noncondensing turbines.

Table 9.4.2 Theoretical Steam Rates for Typical Steam Conditions, lb/kWh*

	Initial pressure, lb/in² gage															
	150	250	400	600	600	850	850	900	900	1,200	1,250	1,250	1,450	1,450	1,800	2,400
	Initial temp, °F															
	365.9	500	650	750	825	825	900	825	900	825	900	950	825	950	1000	1000
	Initial Superheat, °F															
	0	94.0	201.9	261.2	336.2	297.8	372.8	291.1	366.1	256.3	326.1	376.1	232.0	357.0	377.9	337.0
	Initial enthalpy, Btu/lb															
Exhaust pressure	1,195.5	1,261.8	1,334.9	1,379.6	1,421.4	1,410.6	1,453.5	1,408.4	1,451.6	1,394.7	1,438.4	1,468.1	1,382.7	1,461.2	1,480.1	1,460.4
inHg abs																
2.0	10.52	9.070	7.831	7.083	6.761	6.580	6.282	6.555	6.256	6.451	6.133	5.944	6.408	5.900	5.668	5.633
2.5	10.88	9.343	8.037	7.251	6.916	6.723	6.415	6.696	6.388	6.584	6.256	6.061	6.536	6.014	5.773	5.733
3.0	11.20	9.582	8.217	7.396	7.052	6.847	6.530	6.819	6.502	6.699	6.362	6.162	6.648	6.112	5.862	5.819
4.0	11.76	9.996	8.524	7.644	7.282	7.058	6.726	7.026	6.694	6.894	6.541	6.332	6.835	6.277	6.013	5.963
lb/in² gage																
5	21.69	16.57	13.01	11.05	10.42	9.838	9.288	9.755	9.209	9.397	8.820	8.491	9.218	8.351	7.874	7.713
10	23.97	17.90	13.83	11.64	10.95	10.30	9.705	10.202	9.617	9.797	9.180	8.830	9.593	8.673	8.158	7.975
20	28.63	20.44	15.33	12.68	11.90	11.10	10.43	10.982	10.327	10.490	9.801	9.415	10.240	9.227	8.642	8.421
30	33.69	22.95	16.73	13.63	12.75	11.80	11.08	11.67	10.952	11.095	10.341	9.922	10.801	9.704	9.057	8.799
40	39.39	25.52	18.08	14.51	13.54	12.46	11.66	12.304	11.52	11.646	10.831	10.380	11.309	10.134	9.427	9.136
50	46.00	28.21	19.42	15.36	14.30	13.07	12.22	12.90	12.06	12.16	11.284	10.804	11.779	10.531	9.767	9.442
60	53.90	31.07	20.76	16.18	15.05	13.66	12.74	13.47	12.57	12.64	11.71	11.20	12.22	10.90	10.08	9.727
75	69.4	35.77	22.81	17.40	16.16	14.50	13.51	14.28	13.30	13.34	12.32	11.77	12.85	11.43	10.53	10.12
80	75.9	37.47	23.51	17.80	16.54	14.78	13.77	14.55	13.55	13.56	12.52	11.95	13.05	11.60	10.67	10.25
100		45.21	26.46	19.43	18.05	15.86	14.77	15.59	14.50	14.42	13.27	12.65	13.83	12.24	11.21	10.73
125		57.88	30.59	21.56	20.03	17.22	16.04	16.87	15.70	15.46	14.17	13.51	14.76	13.01	11.84	11.28
150		76.5	35.40	23.83	22.14	18.61	17.33	18.18	16.91	16.47	15.06	14.35	15.65	13.75	12.44	11.80
160		86.8	37.57	24.79	23.03	19.17	17.85	18.71	17.41	16.88	15.41	14.69	16.00	14.05	12.68	12.00
175			41.16	26.29	24.43	20.04	18.66	19.52	18.16	17.48	15.94	15.20	16.52	14.49	13.03	12.29
200			48.24	29.00	26.95	21.53	20.05	20.91	19.45	18.48	16.84	16.05	17.39	15.23	13.62	12.77
250			69.1	35.40	32.89	24.78	23.08	23.90	22.24	20.57	18.68	17.81	19.11	16.73	14.78	13.69
300				43.72	40.62	28.50	26.53	27.27	25.37	22.79	20.62	19.66	20.89	18.28	15.95	14.59
400				72.2	67.0	38.05	35.43	35.71	33.22	27.82	24.99	23.82	24.74	21.64	18.39	16.41
425				84.2	78.3	41.08	38.26	38.33	35.65	29.24	26.21	24.98	25.78	22.55	19.03	16.87
600						78.5	73.1	68.11	63.4	42.10	37.03	35.30	34.50	30.16	24.06	20.29

* From Theoretical Steam Rate Tables—compatible with the 1967 ASME Steam Tables, ASME 1969.

Turbines having automatic pressure-controlled extractions or admissions of steam between inlet and exhaust usually have their performance expressed by a chart showing required throttle flow vs. load for varying amounts of steam extracted or admitted at specified conditions.

Turbines working on regenerative and/or reheat cycles, condensing, usually have performance expressed as a heat rate, based upon a carefully specified heat cycle arrangement. This is usually illustrated by a diagram that defines all the surrounding conditions. See, for example, Fig. 9.4.26, actual cycle.

The above methods for expressing turbine performance are more satisfactory for most application than the use of turbine "engine efficiencies." However, it is useful to know the general range of turbine efficiency realized in practice. The engine efficiency of a turbine depends mainly upon the flow areas and diameter of stages, the average velocity ratio, as can be deduced from Fig. 9.4.6, 9.4.7, and 9.4.9, the number of turbine stages, and the steam conditions. With so many variables, it is not possible to do more than show a general picture of efficiency as a function of rating, as in Fig. 9.4.22, for multistage condensing turbines. Noncondensing turbines will usually have similar efficiency levels; automatic extraction turbines will generally be slightly lower because of extra losses in the control-stage sections.

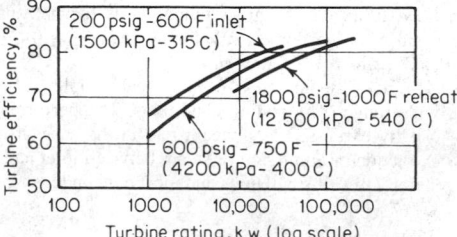

Fig. 9.4.22 Turbine efficiencies vs. capacity.

Approximate steam rates for turbines operating without auxiliary admissions or extractions of steam between inlet and exhaust may be estimated for any turbine rating by dividing theoretical steam rate, corresponding to inlet steam pressure and temperature and exhaust pressure, by the appropriate turbine efficiency from Fig. 9.4.22.

A short method for calculating extraction-turbine performance is illustrated by the following example:

Assume a 12,500-kW automatic-extraction-condensing unit operating at 10,000 kW, with 175,000 lb/h extraction for process at 150 psig with no extraction for feedwater heating, and throttle steam conditions of 850 lb/in², 825°F, exhaust at 2 inHg abs.

PROCEDURE. Find theoretical steam rates (TSR) from Table 9.4.2 or steam charts; TSR$_1$ for 850 lb/in², 825°F, 2 inHg is 6.58 lb/kWh; TSR$_2$ for 850 lb/in², 825°F to 150 lb/in² is 18.61 lb/kWh.

Turbine-generator efficiency from Table 9.4.3, single autoextraction at 80 percent rating (10,000 kW on a 12,500-kW unit), is 78 percent. Efficiency correction for autoextraction (see Table 9.4.3) is 0.92. Then actual steam rate (ASR) is TSR/(efficiency × correction); ASR$_1$ = 6.58/78% × 0.92 = 9.17 lb/kWh; ASR$_2$ = 18.61/78% × 0.92 = 25.9 lb/kWh; kW generation from extraction flow = extraction flow/ASR$_2$ = 175,000/25.9 = 6,760 kW.

kW to be generated by condenser flow = 10,000 − 6,760 = 3,240. Condenser steam flow required is 3,240 × 9.17 = 29,700 lb/h, or (say) 30,000 lb/h. Total steam flow to throttle then is 175,000 + 30,000 = 205,000 lb/h.

Figure 9.4.23 gives correction factors for speed which differ from 3,600 r/min and is representative of units designed for about 4 inHg abs exhaust pressure.

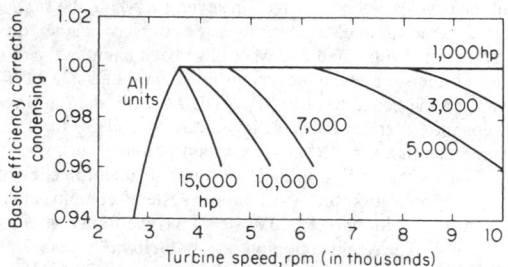

Fig. 9.4.23 Correction factor for condensing mechanical-drive turbines.

Mechanical-Drive Turbines Table 9.4.3 and Figs. 9.4.22 and 9.4.23 provide efficiency-estimating data for typical condensing turbines, primarily for 3600-r/min generator drive. Figure 9.4.24 shows approximate values for turbine efficiency to be expected for noncondensing mechanical-drive units designed for a broad range of horsepower rating, speed, and inlet steam conditions.

Large Central-Station Turbine-Generators

REFERENCES: Spencer, Cotton, and Cannon, A Method for Predicting the Performance of Steam Turbine-Generators . . . 16,500 kW and Larger, *Trans. ASME*, ser. A, Oct. 1963. Baily, Cotton, and Spencer, Predicting the Performance of Large Steam Turbine-Generators Operating with Saturated and Low Superheat Steam Conditions, *Proc. Amer. Power Conf.*, 1967; discussion of foregoing, *Combustion*, Sept. 1967. Spencer and Booth, Heat Rate Performance of Nuclear Steam Turbine-Generators, *Proc. Amer. Power Conf.*, 1968. Baily, Booth, Cotton, and Miller, "Predicting the Performance of 1800-rpm Large Steam Turbine-Generators Operating with Light Water-Cooled Reactors," General Electric publication GET-6020, 1973. "Heat Rates for Fossil Reheat Cycles Using General Electric Steam Turbine-Generators 150,000 kW and Larger," General Electric publication GET-2050C, 1974.

Table 9.4.3 Basic Efficiency for Steam Turbines, Straight Condensing at Rated Load*

kW capacity	Equivalent mechanical drive, hp	Initial steam conditions (gage pressure and temp.)				
		250 lb/in² 500°F	400 lb/in² 650°F	600 lb/in² 750°F	800 lb/in² 825°F	1,250 lb/in² 900°F
875	1,200	63	63	62		
1,875	2,600	76	67	66		
2,500	3,500	69	69	68		
5,000	6,900		74	73	73	
7,500	10,300		76	75	75	
12,500	17,200		78	78	78	77
15,625	21,500		79	79	79	77
20,000	27,100		79	80	79	79

* Efficiency correction factors, mechanical drive and auto extraction-condensing turbines—multiply basic efficiencies by:

	At 80% rating	At 100% rating
Single autoextraction-condensing	0.92	0.96
Double autoextraction-condensing	0.88	0.92

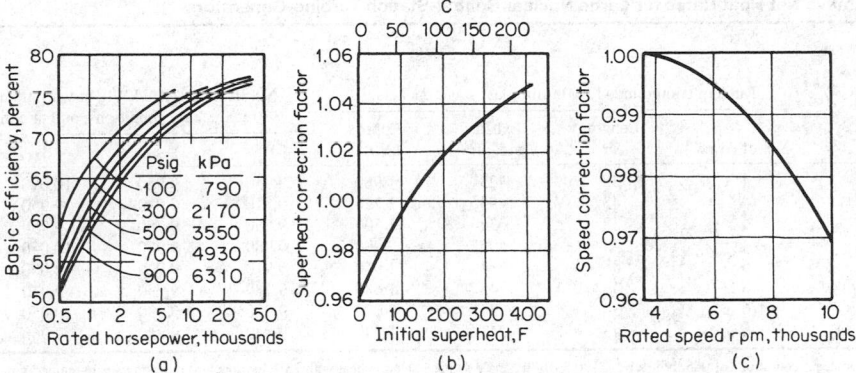

Fig. 9.4.24 Mechanical-drive turbine efficiencies. (*a*) Basic efficiency, 3,600 r/min. Figures on curve are inlet steam pressure in lb/in² gage. (*b*) Superheat correction factor. (*c*) Rated-load speed-correction factor.

The performance of central-station turbine-generators is generally expressed as **heat rate**, Btu/kWh, the ratio of the heat added to the cycle in Btu/h, to generation, in kW. Heat rate may be converted to **thermal efficiency** using the relationship, Efficiency = 3,412.14/heat rate (or 1/heat rate expressed in kJ/kWs). Heat rates are calculated by the preparation of a **heat balance**, which considers steam conditions, steam flow, turbine-expansion efficiency, packing leaking losses, exhaust loss at the end of the low-pressure expansion (perhaps other casings as well), mechanical losses, electrical losses associated with the generator, moisture separation and reheat if present, and extraction for feedwater heating. **Gross heat rate** is calculated without consideration of the power consumed by the boiler feed pump. **Net heat rate** does consider pump power and is higher (poorer) than gross by a factor related to the initial steam pressure. If the pump is driven by an auxiliary turbine, as is the present usual practice, net heat rate is the natural result of the heat-balance calculation, and gross heat rate has little meaning. **Net station heat rate** considers the auxiliary power required by the rest of the power-plant equipment, and the boiler efficiency of fossil plants. It is generally about 3 percent higher than net heat rate in nuclear plants (3 percent auxiliary power, 100 percent ''boiler'' efficiency), and about 16 percent higher than net heat in the case of a coal-fired plant (4 percent auxiliary power, 90 percent boiler efficiency).

A typical current value of net station heat rate for the fossil steam conditions is 9,000 Btu/kWh (2.64 kJ/kWs), equivalent to a thermal efficiency of 38 percent. A typical net station heat rate for a light-water nuclear-reactor plant is about 10,100 Btu/kWh (2.96 kJ/kWs), or about 34 percent thermal efficiency.

Table 9.4.4 lists representative net heat rates for large fossil turbines

of today's types and steam conditions. Steam pressures in excess of 3,500 lb/in² gage (24,200 kPa) and initial and reheat temperatures in excess of 1,000°F (538°C) were frequently employed in the past. However, a number of operating and economic considerations have led to near standardization on the single reheat cycle with initial pressure of 2,400 or 3,500 lb/in² gage (16,650 or 24,240 kPa), with initial and reheat temperature of 1,000°F (538°C).

Table 9.4.5 lists some representative net heat rates for large nuclear turbines for service with steam from boiling-water reactors (BWR), at 950 lb/in² gage (6,650 kPa), ½ percent initial moisture. Values for other light-water reactors may be approximated by reducing heat rate by 1 percent for each 100 lb/in² (690 kPa) pressure increase, reducing heat rate by 0.15 percent for reducing initial moisture to 0 percent, reducing heat rate by 0.3 percent for each 10°F (6°C) of initial superheat provided.

Reheating with Regenerative Cycle

REFERENCES: Reynolds, Reheating in Steam Turbines, *Trans. ASME,* **71,** 1949, p. 701. Harris and White, Development in Resuperheating in Steam Power Plants, *Trans. ASME,* **71,** 1949, p. 685.

Reheating is currently used on all new large fossil central-station turbines. It is accomplished by taking the steam from the turbine after partial expansion, reheating it in a separate section of the boiler, and returning it to the next lower-pressure section of the turbine. Reheating results in lowering of the turbine heat rate by approximately 5 percent; the exact improvement is dependent on several factors. Roughly speaking, 40 percent of the improvement comes from having added heat to

---

Table 9.4.4   Representative Net Heat Rates for Large Fossil Central-Station Turbine-Generators

| Nominal rating, MW, at 1.5 inHg abs | Steam conditions | | | Tandem compound, 3,600 r/min, last-stage buckets | | | | | Net heat rate, Btu/kWh, at rated load and steam conditions, and at exhaust pressure, inHg abs | | | | |
	Throttle pressure lb/in² gage	Temp, °F	Reheat temp, °F	No. of rows	Length, in	Exhaust area, ft²	Approx kW/ft²	Boiler feed-pump drive	1.5	2	3	4	5
150	1,800	1,000	1,000	2	26	82	1,820	Motor	8,010	8,060	8,230	8,440	8,630
235	1,800	1,000	1,000	2	26	82	2,860	Motor	8,240	8,240	8,290	8,380	8,500
250	1,800	1,000	1,000	2	30	111	2,250	Motor	8,080	8,100	8,220	8,400	8,620
250	1,800	1,000	1,000	2	30	111	2,250	Turbine	8,030	8,060	8,200	8,390	8,610
250	2,400	1,000	1,000	2	30	111	2,250	Turbine	7,850	7,890	8,030	8,240	8,450
500	2,400	1,000	1,000	4	30	222	2,250	Turbine	7,790	7,830	7,970	8,170	8,370
700	2,400	1,000	1,000	4	33.5	264	2,650	Turbine	7,860	7,870	7,970	8,130	8,320
1,000	2,400	1,000	1,000	6	30	334	3,000	Turbine	7,920	7,930	8,000	8,100	8,250
500	3,500	1,000	1,000	4	30	222	2,250	Turbine	7,620	7,660	7,820	8,030	8,220
700	3,500	1,000	1,000	4	33.5	264	2,650	Turbine	7,670	7,690	7,810	7,980	8,170
1,000	3,500	1,000	1,000	6	30	334	3,000	Turbine	7,710	7,730	7,810	7,940	8,090
1,100	3,500	1,000	1,000	6	33.5	397	2,770	Turbine	7,680	7,700	7,810	7,960	8,140

Table 9.4.5 Representative Net Heat Rates for Large Nuclear Central-Station Turbine-Generators

Warranted reactor thermal power, MWt	Nominal turbine rating, MWe at 2 inHg abs	Tandem compound, 1,800 r/min, last-stage buckets				Net heat rate, Btu/kWh, at warranted reactor thermal power, at rated steam conditions, and at exhaust pressure, inHg abs				
		No. of rows	Length, in	Exhaust area, ft^2	Approx kW/ft^2	1.5	2	3	4	5
2,440	840	4	38	423	1,980	9,950	9,950	10,090	10,190	10,410
2,440	850	4	43	495	1,720	9,810	9,820	9,950	10,170	10,440
2,890	1,010	6	38	634	1,590	9,750	9,780	9,950	10,200	10,480
2,890	990	4	43	495	2,000	9,980	9,980	10,050	10,200	10,410
3,580	1,230	6	38	634	1,940	9,910	9,920	10,000	10,170	10,380
3,580	1,250	6	43	743	1,680	9,780	9,790	9,930	10,160	10,430
3,830	1,310	6	38	634	2,070	9,990	9,990	10,050	10,190	10,390
3,830	1,330	6	43	743	1,790	9,840	9,850	9,960	10,170	10,420

All units boiling-water reactor steam conditions of 965 lb/in^2 abs, 1,190.8 Btu/lb, and two stage steam reheat with 25°F approach to reheating steam temperature.

the cycle at a higher-than-average temperature (thermodynamic gain), and the remaining 60 percent comes from improvement in turbine efficiency due to reduced moisture loss and increased reheat factor.

Reheating can theoretically be done any number of times, but because of extra cost of apparatus and piping, and the steam pressure drops required in practice (8 to 10 percent of the reheat pressure), the economic gains diminish rapidly with more than one reheating (see Fig. 9.4.25). In a few cases, two reheatings are employed. (See also Sec. 4.)

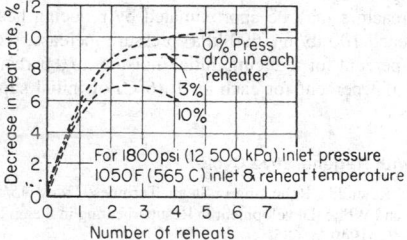

Fig. 9.4.25 Approximate gains due to reheating.

The throttle and condenser steam-flow rates for a given turbine output are reduced approximately 17 and 13 percent, respectively, by reheating once to the initial temperature, as compared with no reheat with the same initial steam conditions.

The maximum gain in heat rate from one reheating with a fixed-percentage pressure drop through the reheating system occurs when the reheat pressure is about 0.15 of the initial pressure. In practice, however, the reheat pressure is higher, 0.20 to 0.30 times initial pressure, because of the extra cost of larger piping, valves, etc., required for lower reheater pressures owing to the larger steam volume.

Regenerative Feedwater Heating
(See also Sec. 4.)

The heat consumption of a turbine may be reduced by heating the condensate (feedwater) in stages by the condensation of steam extracted at various points from the turbine. This is shown diagrammatically for an ideal cycle and a more practical cycle in Fig. 9.4.26. The difference between the two is in the use of mixing heaters in the ideal cycle with each discharge pumped back while the practical cycle has closed heaters with cascaded drains in the upper and pumped drains in the lowest heater, together with some pressure drop between turbine and heaters and a terminal temperature difference between saturated-steam temperature in the heater and feedwater temperature coming out. Usually the difference between such an ideal and a practical cycle is about 1½ percent. A deaerating type of contact heater with no terminal difference may be substituted for one of the closed heaters as is shown in

Fig. 9.4.26. Other variations are the use of (1) all open-contact heaters, or (2) drain coolers to reduce the loss due to cascading the drips, or (3) a desuperheating section on the top heater to get a higher final feed temperature, thereby approaching most closely to the ideal cycle.

Figures 9.4.27 and 9.4.28 and Table 9.4.6 supply data on the results of regenerative heating based on the ideal cycle of Fig. 9.4.26. Figure 9.4.27 shows the reduction in heat rate for various initial pressures and

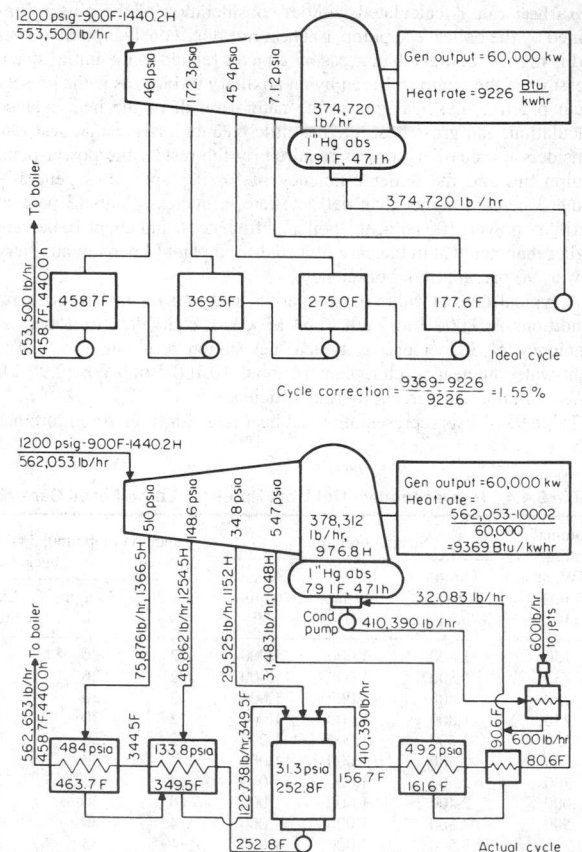

Fig. 9.4.26 Comparison of ideal and actual cycles for regenerative feedwater heating.

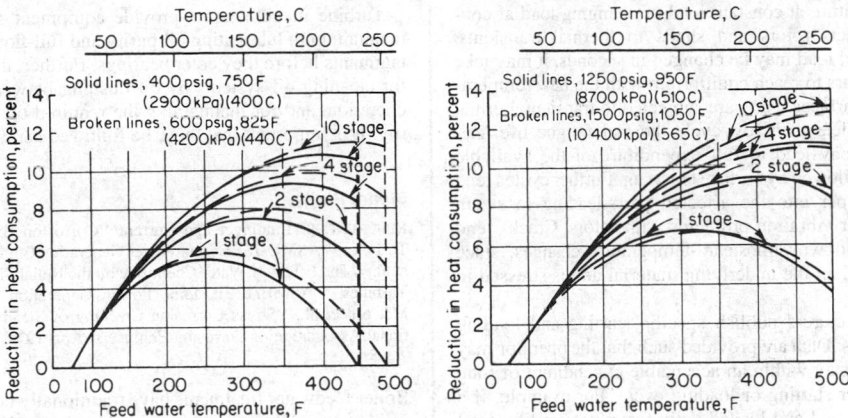

Fig. 9.4.27 Reduction in heat rate by use of ideal regenerative cycle, with 1-inHg back pressure.

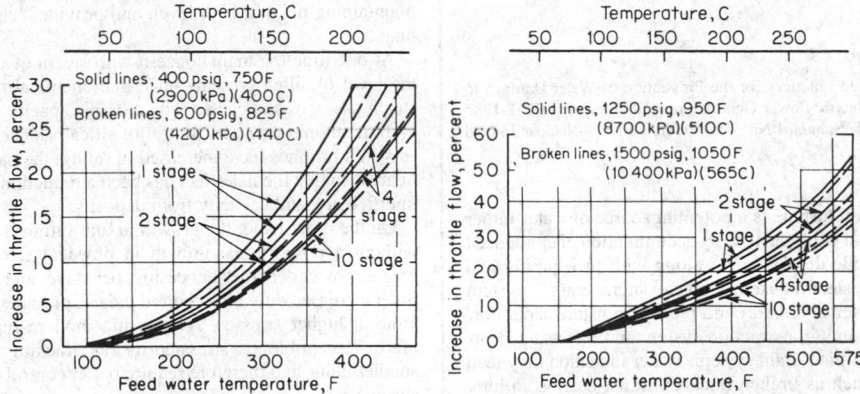

Fig. 9.4.28 Increase in steam flow necessary to maintain the same power output when using the ideal regenerative cycle, with 1-inHg back pressure.

temperatures at 1 inHg abs (3.4 kPa) exhaust pressure, for various feed-water temperatures and number of heaters. The increase in throttle flow necessary to maintain the same power output when extracting steam for feedwater heating is shown in Fig. 9.4.28.

INSTALLATION, OPERATION, AND MAINTENANCE CONSIDERATIONS

Steam turbines are capable of long life and high reliability with relatively little maintenance, if proper attention is paid to their installation, operation, and preventive maintenance. This section considers four areas proved to be of particular importance by operating experience.

Steam Temperature—Starting and Loading

REFERENCES: Mora et al., "Design and Operation of Large Fossil-Fueled Steam Turbines Engaged in Cyclic Duty," ASME/IEEE Joint Power Generation Conference, Oct. 1979. Spencer and Timo, Starting and Loading of Large Steam Turbines, *Proc. Amer. Power Conf.*, 1974. Ipsen and Timo, The Design of Turbines for Frequent Starting, *Proc. Amer. Power Conf.*, 1969. Timo and Sarney, "The Operation of Large Steam Turbines to Limit Cyclic Shell Cracking," ASME Paper 67-WA/PWR-4, 1967.

Table 9.4.6 Total Steam Bled, Percent of Throttle Flow

		Steam pressure and temperature							
		400 lb/in² gage (2,900 kPa)		600 lb/in² gage 825°F (4,200 kPa, 440°C)		1,250 lb/in² gage 950°F (8,700 kPa, 510°C)		1,500 lb/in² gage 1,050°F 10,400 kPa, 565°C)	
		Stages of feedwater heating							
Final feed temp									
°F	°C	2	10	2	10	2	10	2	10
150	65	7.0	7.1	6.9	7.0				
200	93	11.4	11.8	11.3	11.6	11.2	11.5	10.7	11.0
250	121	15.6	16.2	15.5	16.0	15.5	15.9	12.6	13.1
300	149	19.6	20.6	19.5	20.4	19.5	20.3	18.8	19.5
350	177	23.6	24.8	23.5	24.6	23.6	24.6	20.7	21.6
400	204	27.1	29.0	27.1	28.8	27.4	28.9	26.4	27.8
450	232			30.2	32.5	31.1	33.2	30.0	32.0
500	260					35.0	37.4	33.9	36.1

Changing steam temperature at constant load or changing load at constant temperature subjects rotors and shells to thermal transients. Whereas temperature and load may be changed in seconds, it may take heavy metal sections hours to reach equilibrium with the new temperatures imposed on their surfaces. Parts are subjected to transient thermal stresses which may deplete their low-cycle thermal fatigue life. Repeated thermal cycles may lead to full expenditure of the available fatigue life of the part, followed by surface cracking. Further cycles tend to drive the cracks deeper into the affected part, leading to steam leakage through shells or vibration problems with rotors. Cracks tend to be driven deeper by downward steam-temperature changes, since the surface chills faster than the underlying material and is stressed in tension.

Steam-turbine manufacturers publish specific starting and loading instructions for their units. Data are provided such that the operator may select loading rates that stay within an acceptable expenditure of total low-cycle fatigue life per starting or loading cycle. For example, if a unit is expected to be started and loaded daily for a 30-year life, total cycles will be about 10,000, and it would be desirable to avoid exceeding 0.01 percent life expenditure per daily cycle.

Water-Induction Damage

REFERENCES: Recommended Practices for the Prevention of Water Damage to Steam Turbines Used for Electric Power Generation, ANSI/ASME TDP-1-1985 Fossil Fueled Plants. ASME Standard No. TWPDS-1, Part 2—Nuclear Fueled Plants, Apr. 1973.

Any connection to a steam turbine is a potential source of water either by induction from external equipment or by accumulation of condensed steam. The sources include the following along with their piping and drains: main and reheat steam systems; reheat attemperating system (fossil units); bypass systems, crossaround piping, moisture separator/reheater system (nuclear units); extraction system and feedwater heaters; steam-seal system; turbine-drain system. Water induction may lead to steam-path damage such as broken buckets, thrust-bearing failure, rotor bowing, and shell distortion, which may be indicated by abnormal vibration or differential expansion, or inability to turn the rotor on turning gear.

Water induction may be prevented by proper system installation, provision of protective and indicating devices, and periodic testing, inspection, and maintenance. Detailed recommendations are given in the ANSI/ASME and ASME standards and in manufacturers' instructions.

Lubricating-Oil and Hydraulic-Fluid Purity

Most steam turbines are provided with a lubricating-oil system consisting of reservoir, pumps, coolers, and piping to provide the thrust and journal bearings with a generous supply of oil at the proper temperature and viscosity. Some units also use the lube-oil system as a source of fluid power for control devices and steam-valve actuation. It is most important to assure the cleanliness and purity of the lube oil at all times to avoid bearing and journal damage, or control-system malfunction. Bearings have failed because of oil starvation caused by clogged lines. At least one overspeed failure has resulted from the silting of control devices with rust caused by the entry of water into the oil system.

Units employing an electrohydraulic control system frequently use a synthetic fire-resistant fluid for the high-pressure control hydraulics, separate from the petroleum oil used in the bearing lubrication system. High-pressure hydraulic systems employing synthetic fluid offer advantages in size reduction, speed of response, and fire safety. However, because of the need for very close clearances between small parts at high pressures, and because of the poorer rust-preventing properties of the fluid, cleanliness is of even greater importance than with oil-based systems.

Turbine manufacturers provide equipment such as conditioners to maintain bulk lubricating oil purity and full-flow filters to remove contaminants before they enter bearings. Further, they provide instructions for cleaning oil systems by oil flushing between installation and first operation, and for maintaining the required oil and hydraulic-fluid purity. It is important that these be followed carefully.

Steam Purity

REFERENCES: Lindinger and Curren, "Corrosion Experience in Large Steam Turbines," ASME/IEEE Joint Power Generation Conference, Oct. 1981. Bussert et al., "The Effect of Water Chemistry on the Reliability of Modern Large Steam Turbines," ASME/IEEE Joint Power Generation Conference, Sept. 1978. McCord et al., "Stress Corrosion Cracking of Steam Turbine Materials," National Association of Corrosion Engineers, Apr. 1975.

Boiler feedwater treatments have traditionally been designed to remove solids that might clog steam passages in the boiler, remove salts that could cause scaling of tube surfaces and interfere with heat transfer, prevent corrosion of tube surfaces by reducing oxygen content and by maintaining pH at a high level, and provide "clean" steam to the turbine.

At one time the main concern with steam quality in the turbine was the level of silica present, since that chemical tends to deposit in the steam passages and causes reduction in capacity and efficiency. In general, the monitoring and control of silica has been well developed, and as steam turbines have increased in rating, the passages have increased in area, so that the net effect has been a reduction in the extent of losses in efficiency and capacity from deposits.

On the other hand, the growth in unit ratings has been accomplished without a proportional growth in physical size, and has resulted in greater power densities per casing, per stage, and per pound of material. Such increased duty has required the use of higher-strength alloys operating at higher stresses. As a result, modern turbine components are more susceptible to stress-corrosion cracking than those in older, smaller units, and therefore require better control of contaminants in the steam.

The **feedwater treatment** in most fossil fuel stations is designed to provide a sufficiently low level of contaminants so that stress-corrosion cracking of turbine components should not be a problem. Both the "zero solids" and the "coordinated phosphate" treatments can be controlled to provide steam of acceptable chemistry. Unfortunately, there are a number of situations in which undesirable chemicals can be introduced in the steam in spite of well-intentioned "normal" water-control practices. For example, the composition of coatings put on turbine components for corrosion protection during shipment, storage, and installation must be controlled. The chemistry of solutions used for the removal of coatings during installation, and the methods used, must be carefully regulated. The turbine must be protected during the chemical cleaning of related components such as the boiler, condenser, and feedwater heaters. Critical components have been damaged by fumes from cleaning. The feedwater system must be designed so that only water of high purity is used for boiler desuperheater sprays, and for the turbine exhaust-hood sprays used to limit temperature during light-load operation. Condensate demineralizers must be operated and regenerated so as to ensure that they do not introduce the harmful chemical they are intended to remove. In the event of a leak into the condenser of impure cooling water, it is important to avoid changing the feedwater treatment in such a way that the turbine is subjected to harmful contaminants introduced to protect other station components.

In each of these undesirable instances, the average concentration of chemicals in the steam can be quite low, but high local concentrations can be developed through several mechanisms. For example, dilute solutions can enter crevices not washed by flowing steam; as water evaporates on heating, the concentration of the solution wetting the surfaces tends to increase. In the case of expansion-joint bellows, chemicals contained in steam condensing on shutdown or cold start-up tend to be trapped and concentrated in the bottom of the bellows convolutions.

Succeeding cycles can lead to dry residue or to concentrated solutions during operating conditions which provide moisture. In the case of the steam path, as the expansion crosses into the moisture region, the first droplets of water condensed from the steam will tend to contain most of the contaminants. Concentration-enhancement factors of 100 to 1,000 can be achieved. In modern reheat turbines the early-moisture region occurs in one of the later few stages of the low-pressure sec-tion, and at light loads can occur on the most highly stressed last-stage buckets.

Extreme care must be exercised in protecting turbines from chemical contamination during installation, operation, and maintenance. Any de-viation from sound feedwater treatment practice during condenser leaks should be done with the full realization that damage to the turbine may result.

9.5 POWER PLANT HEAT EXCHANGERS
by William J. Bow, assisted by Donald E. Bolt

NOTE: Standards for this industry retain USCS units except as indicated in the text.

SURFACE CONDENSERS

REFERENCE: Heat Exchange Institute Standards for Steam Surface Condensers.

The power plant surface condenser is attached to the low-pressure ex-haust of a steam turbine (see Figs. 9.5.1 and 9.5.2). Its purposes are (1) to produce a vacuum or desired back pressure at the turbine exhaust for the improvement of plant heat rate, (2) to condense turbine exhaust steam for reuse in the closed cycle, (3) to deaerate the condensate, and (4) to accept heater drains, makeup water, steam drains, and start-up and emergency drains.

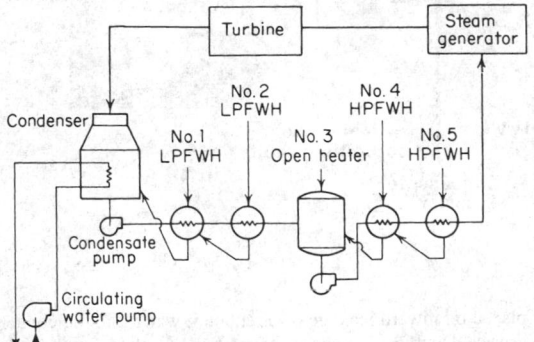

Fig. 9.5.1 Equipment arrangement, schematic.

An economical turbine back pressure is from 1.0 to 3.5 inHg abs. The factors involved in establishing this pressure are involved and will not be discussed here.

An equipment diagram of a closed power plant cycle is shown in Fig. 9.5.1.

For a condenser to deaerate the condensate, it must remove oxygen and other noncondensable gases to an acceptable level compatible with material selection and/or chemical treatment of the feedwater (conden-sate). Depending on materials and treatment, the dissolved O_2 level must normally be kept below 0.005 cm^3/L for turbine units operating with high-pressure and -temperature steam.

Deaeration in a condenser is accomplished by applying Henry's law, which states that the concentration of the dissolved gas in a solution is directly proportional to the partial pressure of that gas in the free space above the condensate level in the hot well, with the exception of those gases (e.g., $CO_2 + NH_3$) which unite chemically with the solvent. In a condenser droplets of condensate are continually scrubbed with steam, liberating the O_2 and permitting it to flow to the low-pressure air-re-moval section, where it is discharged to the atmosphere by the air-re-moval equipment.

To remove the last traces of O_2 from the condensate, an ammonia compound such as hydrazine is normally added. Free ammonia is liber-ated in the cycle and is either removed with the noncondensables as a gas or is condensed and retained in the condensate, depending on the detailed design of the condenser air-removal section. If the ammonia is concentrated as a liquid, it can be very corrosive to certain copper base materials.

Most condenser manufacturers have **tube-bundle configurations** unique to their design philosophy. Basically, pressure losses from tur-bine exhaust to the air offtake are kept to a minimum and tubes are arranged to promote good heat-transfer rates. Small condensers are usu-ally cylindrical, whereas large ones are rectangular for better utilization of space. Most turbines exhaust downward into the condenser, but con-densers are also built to accommodate side as well as axial exhaust turbines.

Because of the inherent strength of cylindrical shapes as opposed to flat plates, condenser **water boxes** are generally made with curved sur-faces. This has come about because of the increased pressure resulting from cooling towers, which in turn, are the result of environmental influences. With a cooling tower, pressures are in the 60 to 80 lb/in^2 range, whereas with water from lakes, rivers, etc., where a siphon sys-tem can be employed, water-box design pressures are in the 20 to 30 lb/in^2 range.

As a general rule, **tube selection** is based on economics; 18 BWG admiralty metal has been satisfactory for freshwater service and 90-10 copper-nickel material likewise for seawater. The current trend is to use 22 BWG titanium or one of the new specially formulated stainless-steel tube materials for this application. Material prices fluctuate greatly, and selection can be influenced by first cost. Lost revenue due to downtime caused by tube leaks or other causes, particularly in larger units, can usually justify the use of more exotic and expensive materials.

Low-pressure feedwater heaters are frequently located in the steam-inlet neck of a condenser. This is done to minimize pressure drop in the extraction steam piping and to utilize floor space surrounding the con-denser better.

A sufficient number of **tube supports** must be provided within the condenser so that the tubes will not vibrate excessively, which will cause tubes to rub or crack circumferentially. During low-water-tem-perature operation, the steam entering the condenser will often reach sonic velocities, causing severe **flow-induced vibration** and ultimate tube failures if the tube support system is inadaquate.

Where once-through boiler or nuclear steam generators are used, it is imperative to dispose of large quantities of steam during starting and stopping of a turbine unit. The condenser, because of its large volume, has been used as a convenient dumping place for this steam. Means must be provided within the condenser to accommodate the high-energy steam without damage to condenser tubing, structural members, or the low-pressure end of the turbine.

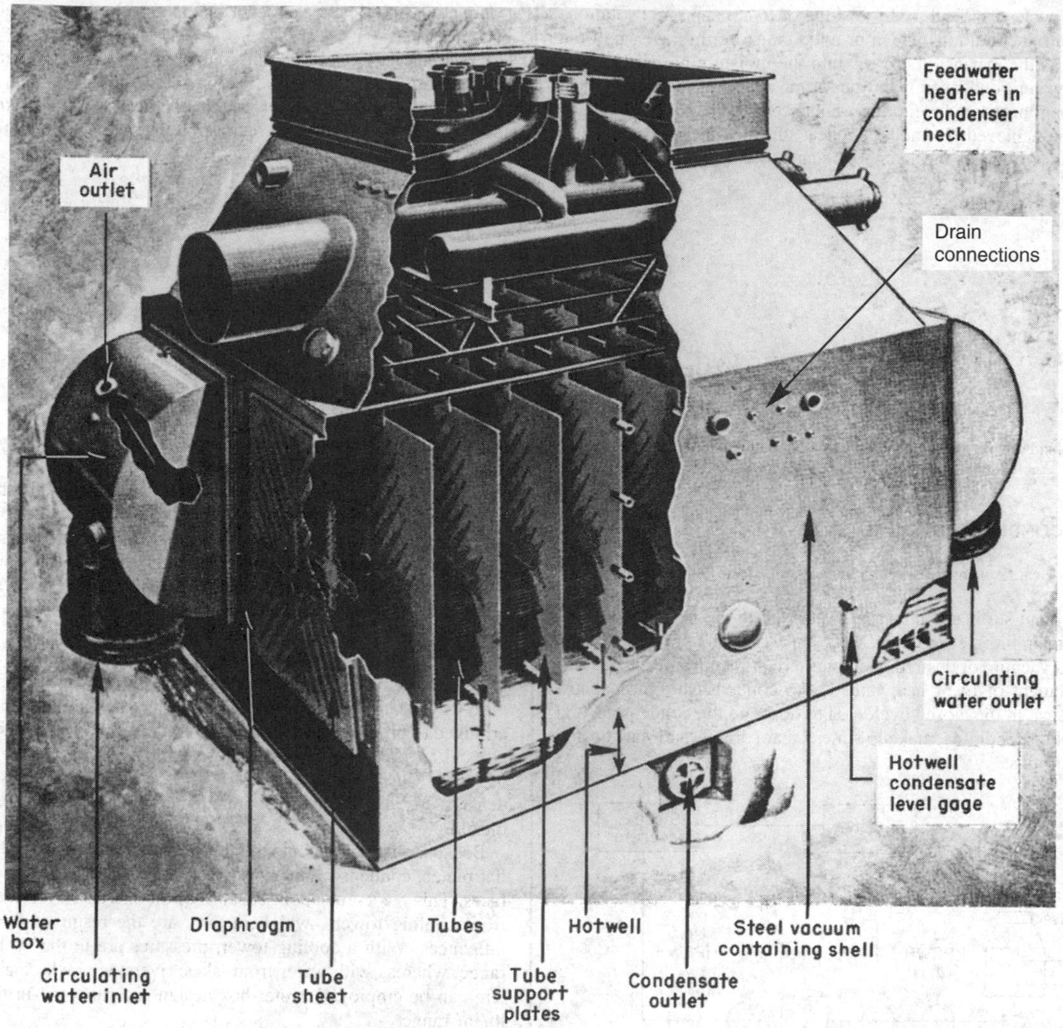

Fig. 9.5.2 One-pass rectangular surface condenser.

A major factor in condenser performance is **tube cleanliness.** Tubes can be plugged with leaves, marine life, and other debris deposited on the face of the tube sheets. By valving, various arrangements for back washing, or flushing, are effected to remove these materials.

The inside surfaces of the tubes can also become fouled by scaling, corrosion products, silt, products of biofouling such as slimes, or a combination of these. To clean the inside surfaces of the tubes, it is necessary to force brushes or plugs through them individually by the use of water or air jets. To do this, the affected portion of the condenser must be shut down. As an alternative, recirculating sponge rubber ball or brush systems are available that can be used during normal operation. To remove scale formations, sponge rubber balls with abrasive coatings are effective, but their continued use will shorten tube life.

Movement due to **temperature differences** between turbine and condenser is usually accommodated by a stainless-steel bellow or rubber-belt-type **expansion joint.** For small units the condenser may be supported on springs and rigidly connected to the turbine. To accommodate differential expansion between condenser shell and tubes, a flexible diaphragm or other expansion element may be installed.

The **tube-to-tube sheet joint** is usually rolled but has been welded in certain installations. Proper material selection must be made regardless of whether the joint is rolled or welded. Currently, considerable emphasis is placed on inward leakage of circulating water into the condenser steam space. Double tube sheets have been supplied for some large central-station condensers for nuclear units. They are standard in the Navy's nuclear-powered ships and submarines to minimize this source of contamination.

Condensers for some large turbine units having two or three low-pressure ends are designed for **multipressure application.** Usually there will be a gain, in the form of either heat-rate improvement or a reduction in capital investment, with a multipressure condenser. A rather complex analysis must be made for each application before a decision can be made.

The current version of ''Standards for Steam Surface Condensers'' is published by the Heat Exchange Institute in Cleveland, OH. These are available to the public and provide guidance in solving most structural and sizing problems related to surface condensers.

Performance Calculations and Sizing

Notation

S = condenser tube surface area, ft^2
C_c = cleanliness factor
C_1 = heat-transfer-rate constant
C_m = material and gage factor

C_t = temperature correction factor
c_p = specific heat, Btu/lb, °F
G = circulating-water quantity, gal/min
h = enthalpy, Btu/lb
h_r = heat rejected by steam, Btu/lb
L = length of water travel, ft
NPSH = net positive suction head, ft
OD = tube outside diameter, in
Q = heat transferred, Btu/h
R = temperature rise $(t_o - t_i)$, °F
TDH = total dynamic head, ft
t = tube thickness, in
TTD = terminal temperature difference = $t_s - t_o$
t_i = inlet-water temperature, °F
t_o = outlet-water temperature, °F
t_s = saturation temperature in condenser, °F
U_o = overall heat-transfer rate, Btu/(ft² · h · °F)
V = water velocity, ft/s
W_s = steam to be condensed, lb/h
Δt_m = logarithmic mean temperature difference, °F

In **sizing** a condenser, the steam flow and heat rejected to the condenser are obtained from the turbine heat balance. Table 9.5.1 gives representative steam flows; heat rejected to the condenser is approximately 950 Btu/lb of steam for nonreheat turbines and 975 Btu/lb for reheat machines. Figure 9.5.3 shows approximate average water temperatures for the United States; local water temperature should be used, when known. The number of passes is usually dictated by the plant arrangement, with total length of water travel and tube diameter dictated by economic considerations. Normally, small-diameter tubes, single-pass condensers are used where water is plentiful, and larger-diameter tubes, two-pass condensers when water is scarce. The vacuum, or back pressure, is determined by economic evaluation, but Table 9.5.2 gives normal recommended values for average water temperatures. Table 9.5.3 is a pressure-temperature conversion table.

Table 9.5.1   Steam Flow to Condensers*

Turbine throttle conditions	Weight flow, lb/kWh
600 lb/in², 825°F	7.08
850 lb/in², 900°F	6.45
1,250 lb/in², 950°F	5.80
1,450 lb/in², 1,000°F	5.44
1,450 lb/in², 1,000°F, 1,000°F	4.70

* Approximate values for unit sizes to 100,000 kW and exhaust at 1.0 inHg abs.
NOTE: For exhaust pressures other than 1.0 inHg abs, multiply tabular values by 1.02 (1.04)[1.06] for 1.5 (2.0)[2.5] in.

Table 9.5.2   Normal Condenser Pressures and Circulating-Water Temperatures

Water inlet temp t_i, °F	Normal back pressure, inHg abs
55	1.0
70	1.5
80	2.0
85	2.5
90	3.0
95	3.5

A **cleanliness factor** is applied to the heat-transfer rate of new, clean tubes to allow for gradual decrease by fouling. A standarized cleanliness factor of 85 percent is frequently used, but this can often be misleading and even erroneous. The fouling is attributable to (1) sedimentation, (2) scaling, (3) steam-side deposits, (4) corrosion, and (5) biological growth. The fouling is correctly determined by use; in some cases the cleanliness factor may be 90 percent, whereas in other cases it may never rise above 75 percent.

Velocities normally used are: for clean water, 7.0 ft/s; for very clean water with cooling towers, 8.0 ft/s; and for seawater, with entrained sand, as low as 6.0 ft/s to minimize erosion. Prevalent velocities are 6.5 ft/s with aluminum-brass tubes, 7.0 ft/s with admiralty metal, and

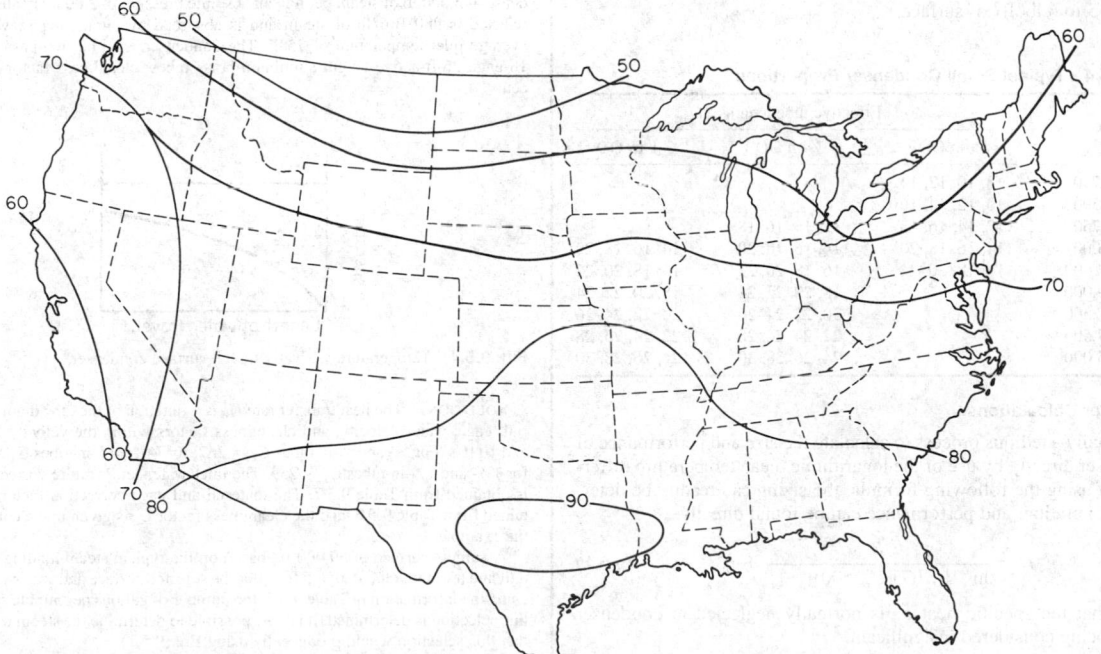

Fig. 9.5.3   Average inlet temperature of circulating water, °F, United States.

Table 9.5.3   Pressure-Temperature Conversion Table for Water*

Abs press, inHg	Sat temp, °F	Abs press, inHg	Sat temp, °F	Abs press, inHg	Sat temp, °F	Abs press, inHg	Sat temp, °F
0.20	34.57	1.35	88.36	2.50	108.71	8.00	152.24
0.25	40.23	1.40	89.51	2.55	109.38	9.00	157.09
0.30	44.96	1.45	90.64	2.60	110.06	10.00	161.49
0.35	49.06	1.50	91.72	2.65	110.72	11.00	165.54
0.40	52.64	1.55	92.77	2.70	111.37	12.00	169.28
0.45	55.89	1.60	93.81	2.75	112.01	13.00	172.78
0.50	58.80	1.65	94.80	2.80	112.63	14.00	176.05
0.55	61.48	1.70	95.78	2.85	113.25	15.00	179.14
0.60	63.96	1.75	96.73	2.90	113.86	16.00	182.05
0.65	66.26	1.80	97.65	2.95	114.45	17.00	184.82
0.70	68.41	1.85	98.56	3.00	115.06	18.00	187.45
0.75	70.43	1.90	99.43	3.20	117.35	19.00	189.96
0.80	72.32	1.95	100.30	3.40	119.51	20.00	192.37
0.85	74.13	2.00	101.14	3.60	121.57	21.00	194.68
0.90	75.84	2.05	101.96	3.80	123.53	22.00	196.90
0.95	77.48	2.10	102.77	4.00	125.43	23.00	199.03
1.00	79.03	2.15	103.56	4.20	127.21	24.00	201.09
1.05	80.53	2.20	104.33	4.40	128.94	25.00	203.08
1.10	81.96	2.25	105.09	4.60	130.61	26.00	205.00
1.15	83.33	2.30	105.85	4.80	132.20	27.00	206.87
1.20	84.64	2.35	106.58	5.00	133.76	28.00	208.67
1.25	85.93	2.40	107.30	6.00	140.78	29.00	210.43
1.30	87.17	2.45	108.01	7.00	146.86	29.921	212.00

* See also Sec. 4 for further data.

8 + ft/s with stainless steel or titanium. **Water-temperature rise** is about 10°F for single-pass condensers and 15°F for two-pass condensers, with a minimum 5° terminal temperature difference (TTD).

Approximate general rules for condensers serving turbines rated up to 100 MW. The **surface area**, ft², is equal to the steam flow, lb/h, divided by 10 for single-pass condensers and by 7.5 for two-pass condensers. **Circulating-water quantity**, gal/min, is equal to the area, ft², for a two-pass condenser and is twice the area for a single-pass condenser. Condenser **proportions** are given in Table 9.5.4. Empty **weight** of an installed condenser is 5 to 6 lb/ft² of surface.

Table 9.5.4   Typical Small Condenser Proportions

Surface area, ft²	Effective tube lengths, ft		
	¾-in OD	⅞-in OD	1-in OD
1,000–1,750	8, 10, 12, 14		
2,000–2,500	10, 12, 14, 16		
2,750–4,750	12, 14, 16, 18	12, 14, 16, 18	
5,000–7,000	14, 16, 18, 20	14, 16, 18, 20	14, 16, 18, 20
7,250–14,000	16, 18, 20, 22	16, 18, 20, 22	16, 18, 20, 22
15,000–19,000		18, 20, 22, 24	18, 20, 22, 24
20,000–27,500		20, 22, 24, 26	20, 22, 24, 26
30,000–47,500		22, 24, 26, 28	22, 24, 26, 28
50,000–75,000		24, 26, 28, 30	24, 26, 28, 30

Condenser Calculations

It is normally a tedious process to calculate the size and performance of a condenser directly by use of the logarithmic mean temperature difference. But using the following formula, the sizing can readily be determined by iteration, and performance can be found directly.

$$t_s = \frac{R}{\ln^{-1}[(S/G)(U_o/500)] - 1} + t_o \qquad (9.5.1)$$

Note that the specific heat c_p is normally neglected in condenser work, it being considered insignificant.

The basic diagram is shown in Fig. 9.5.4, and the applicable equations are

$$Q = U_o S \, \Delta t_m \qquad (9.5.2)$$
$$Q = 500 G c_p (t_o - t_i) \qquad (9.5.3)$$
$$R = W_s h_r / (500 G) \qquad (9.5.4)$$
$$U_o = C_1 \sqrt{V} C_t C_m C_c \qquad (9.5.5)$$

EXAMPLE.   Calculate the surface and circulating-water requirements to condense 445,000 lb of steam per h to an absolute pressure of 2.00 inHg abs. The heat rejected is 980 Btu/lb of steam and is absorbed by circulating seawater at an average inlet temperature of 75°F. The condenser is to be two-pass with ⅞-in 18-gage 26-ft-active-length aluminum-brass tubes; cleanliness factor = 85 percent.

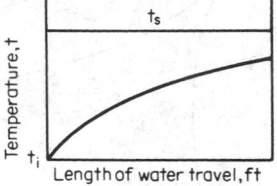

Fig. 9.5.4   Temperature vs. water travel, surface condenser.

SOLUTION.   The heat transfer rate U_o is a function of the tube diameter, material, gage, water velocity, and cleanliness factors within the velocity limits of 3 and 10 ft/s. For ⅝- and ¾-in tubes $C_1 = 267$; for ⅞- and 1-in tubes $C_1 = 263$, and for 1⅛- and 1¼-in tubes $C_1 = 259$. The inlet-water-temperature correction factor is obtained from Table 9.5.7. The material and gage correction factor C_m is obtained from Table 9.5.6, and the cleanliness factor C_c is given in the conditions of the example.

Assume a surface of 59,000 ft² based on the typical steam loading for such a unit and a water velocity of 6.5 ft/s. For the selected surface and water velocity, by using the information in Table 9.5.5, the number of gallons per minute required by this selection is determined. It is now possible to determine the steam temperature that this selection would produce, by using Eq. (9.5.1).

The results show a steam temperature of 99°F, indicating that a smaller surface is required to attain a 2-inHg abs condenser pressure. Try a smaller unit until a surface with no more than three significant figures will yield a steam temperature of no more than 101.14°F. The surface arrived at is 54,400 ft² with 43,900 gal/min of cooling water.

Table 9.5.5 Tube Characteristics

OD and ext. surface, ft²/ft	BWG no.	ID, in	Water, gal/min @ 1 ft/s velocity
⅝-in OD 0.1636	16	0.495	0.600
	18	0.527	0.680
	20	0.555	0.754
	22	0.569	0.793
¾-in OD 0.1963	16	0.620	0.941
	18	0.652	1.041
	20	0.680	1.132
	22	0.694	1.179
⅞-in OD 0.2291	16	0.745	1.359
	18	0.777	1.478
	20	0.805	1.586
	22	0.819	1.642
1-in OD 0.2618	16	0.870	1.853
	18	0.902	1.992
	20	0.930	2.117
	22	0.944	2.182

Table 9.5.6 Material and Gage Factor

Tube material	C_m						
	BWG no.						
	24	22	20	18	16	14	12
Admiralty metal	1.03	1.02	1.01	1.00	0.98	0.96	0.93
Arsenical copper	1.04	1.03	1.03	1.02	1.01	1.00	0.98
Copper iron 194	1.04	1.04	1.03	1.03	1.02	1.01	1.00
Aluminum brass	1.02	1.02	1.01	0.99	0.97	0.95	0.92
Aluminum bronze	1.02	1.01	1.00	0.98	0.96	0.93	0.89
90–10 Cu-Ni	0.99	0.98	0.96	0.93	0.89	0.85	0.80
70–30 Cu-Ni	0.97	0.95	0.92	0.88	0.83	0.78	0.71
Cold-rolled carbon steel	1.00	0.98	0.97	0.93	0.89	0.85	0.80
Stainless steel type 304/316	0.90	0.86	0.82	0.75	0.69	0.62	0.54
Titanium	0.94	0.91	0.88	0.82	0.77	0.71	0.63

Table 9.5.7 Inlet-Water-Temperature Correction Factors

Inlet temp t_i, °F	Correction factor C_t
40	0.740
50	0.830
60	0.920
70	1.000
80	1.045
90	1.075
100	1.100

If this condenser is single-shell with one main turbine exhaust inlet and has no auxiliary turbine exhausts entering it (such as from a turbine-driven boiler feed pump), then the air-removal equipment must have a minimum capacity of 10 scfm of air at 1-inHg abs suction pressure. (See Table 9.5.8.)

Performance curves (Fig. 9.5.5) are drawn for a condenser to show the back pressure for various condenser steam loads and inlet-water temperatures maintaining a minimum TTD of 5°. The zero-load back pressure and the cutoff pressure are shown in Fig. 9.5.6, where the cutoff

Table 9.5.8 Venting-Equipment Capacities*

Effective steam flow each main exhaust opening, lb/h	One condenser shell		
		Total number of exhaust openings	
		1	2
Up to 25,000	scfm†	3.0	4.0
	Dry air, lb/h	13.5	18.0
	Water vapor, lb/h	29.7	39.6
	Total mixture, lb/h	43.2	57.6
25,001–50,000	scfm†	4.0	5.0
	Dry air, lb/h	18.0	22.5
	Water vapor, lb/h	39.6	49.5
	Total mixture, lb/h	57.6	72.0
50,001–100,000	scfm†	5.0	7.5
	Dry air, lb/h	22.5	33.8
	Water vapor, lb/h	49.5	74.4
	Total mixture, lb/h	72.0	108.2
100,001–250,000	scfm†	7.5	12.5
	Dry air, lb/h	33.8	56.2
	Water vapor, lb/h	74.4	123.6
	Total mixture, lb/h	108.2	179.8
250,001–500,000	scfm†	10.0	15.0
	Dry air, lb/h	45.0	67.5
	Water vapor, lb/h	99.0	148.5
	Total mixture, lb/h	144.0	216.0
500,001–1,000,000	scfm†	12.5	20.0
	Dry air, lb/h	56.2	90.0
	Water vapor, lb/h	123.6	198.0
	Total mixture, lb/h	179.8	288.0
1,000,001–2,000,000	scfm†	15.0	25.0
	Dry air, lb/h	67.5	112.5
	Water vapor, lb/h	148.5	247.5
	Total mixture, lb/h	216.0	360.0

* Heat Exchange Institute.
† 14.7 lb/in² abs at 70°F.
NOTE: These tables are based on air leakage only and the air-vapor mixture at 1 inHg abs and 71.5°F.

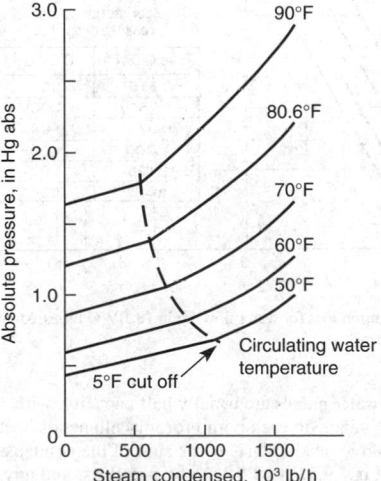

Fig. 9.5.5 Representative performance curves for a 157,500-ft² one-pass surface condenser, 1-in, 18-gage, 37.8-ft-active-length aluminum-brass tubes. Performance is based on 85 percent clean tubes, 221,400 gal/min cooling water at a velocity of 7 ft/s, and 968 Btu/lb rejected to the cooling water.

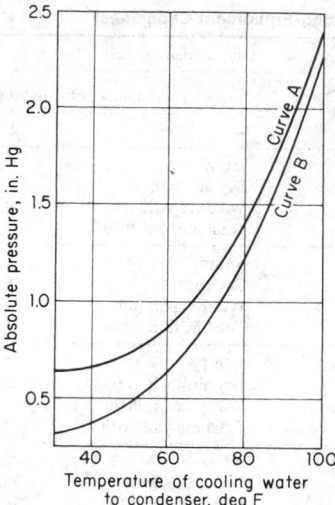

Fig. 9.5.6 Cutoff and zero load vacuums. Curve *A*, cutoff (except where the absolute pressure is limited to 5°F TTD); curve *B*, zero load.

limitation is set by the air-removal equipment. A combined turbine-condenser performance curve is sometimes drawn in which heat rejected vs. back pressure for a given turbine load is superimposed on the condenser performance.

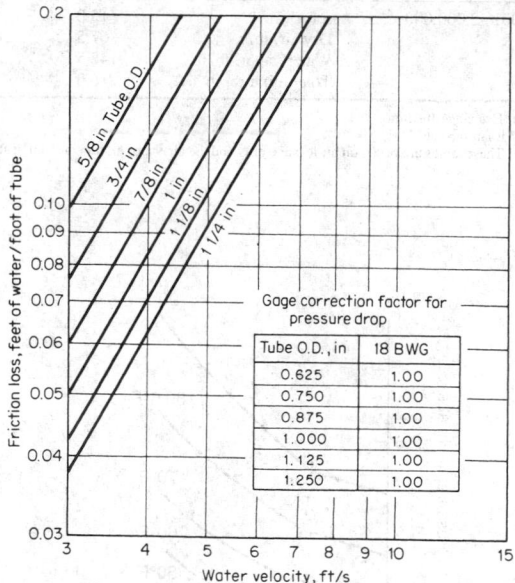

Fig. 9.5.7 Friction loss for water flowing in 18 BWG tubes, low velocity. *(Heat Exchange Institute.)*

Circulating-water pumps are usually half capacity, with 3 or 4 percent additional allowance on each pump for miscellaneous heat exchangers. The total dynamic head (TDH) is the sum of the condenser and water-box friction (Figs. 9.5.7 to 9.5.11), the pipe loss, and any unrecovered static head. Normal heads are about 20 ft; for cooling-tower installations, 60 ft. (See also Sec. 14 and Fig. 9.5.12.)

Condensate pumps are usually full capacity, determined by condenser flow plus any heater drains dumped into the condenser and 5 percent

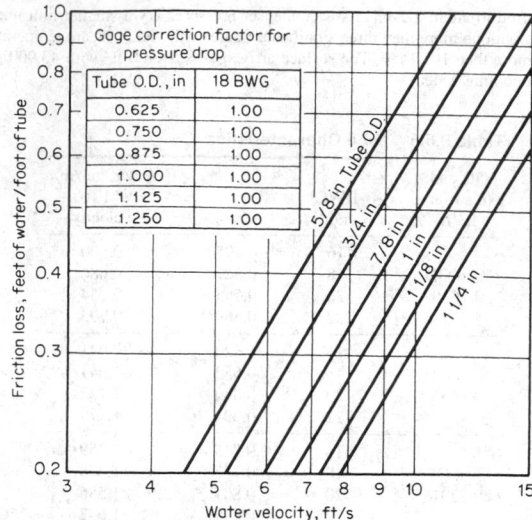

Fig. 9.5.8 Friction loss for water flowing in 18 BWG tubes, high velocity. *(Heat Exchange Institute.)*

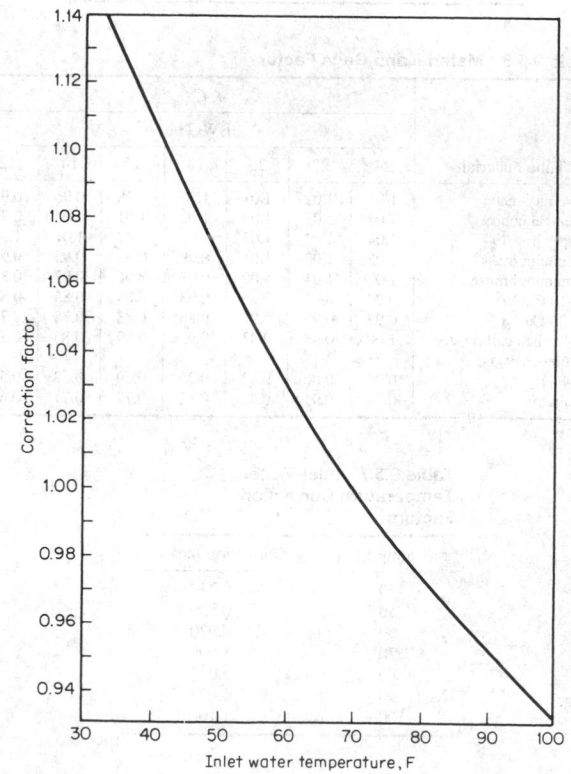

Fig. 9.5.9 Temperature correction for friction losses in tubes (see Figs. 9.5.7 and 9.5.8). *(Heat Exchange Institute.)*

additional margin. Vertical-pit-type designs are commonly used because of low net-positive-suction-head (NPSH) requirements with water at the boiling point. TDH ranges from a normal value of 100 ft on small plants to 800 ft on larger plants. (See Sec. 14.)

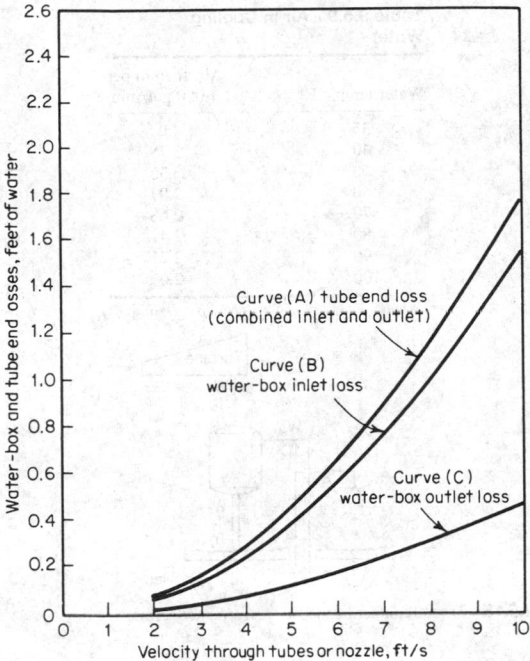

Fig. 9.5.10 Water-box and tube end losses, single-pass condensers. *(Heat Exchange Institute.)*

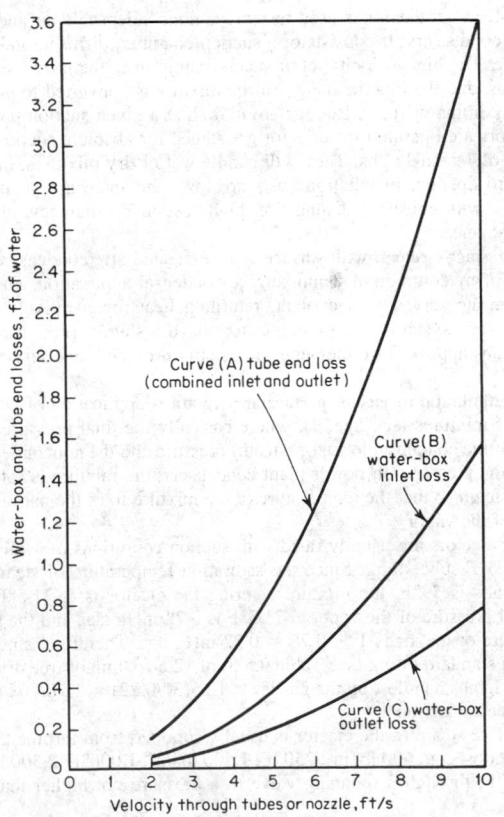

Fig. 9.5.11 Water-box and tube end losses, two-pass condensers. *(Heat Exchange Institute.)*

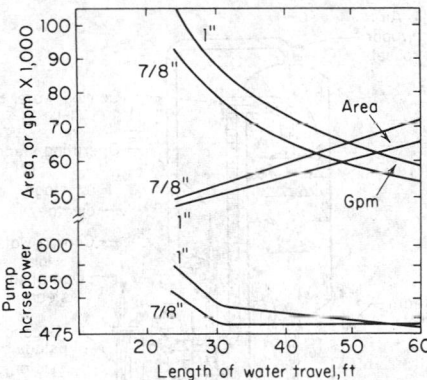

Fig. 9.5.12 Condenser configuration selection for a given heat load, $W_s =$ 477,355 lb/h; $h = 975$ Btu/lb; $t_i = 70.0°F$; 1.5 inHg abs; 18-gage admiralty metal tubes; 7.0-ft/s water velocity; 85 percent cleanliness factor.

AIR-COOLED CONDENSERS

The air-cooled condenser is used where adequate water is not available in sufficient quantities to permit the use of conventional cooling towers. The water lost by evaporation and cooling tower blow-down is approximately equal to the steam condensed from the turbine exhaust.

In an air-cooled condenser, the steam is condensed inside tubes while cooling air flows over fins on the external surfaces. These tubes are arranged in the area where packing or fill would be arranged in a conventional cooling tower, and propeller fans (see Sec. 14) supply air across the tube surfaces.

Aluminum is generally the material of choice for tubing because of its excellent heat transfer properties and availability. Items to be considered in design are the overall heat-transfer coefficient, steam-side pressure drop, uniformity of steam distribution, noncondensible gas concentration for air removal, and, of course, provision for freeze protection. Turbine design back pressures are normally in the range of 5 in Hga when served by air-cooled condensers.

DIRECT-CONTACT CONDENSERS

REFERENCES: Heat Exchange Institute Standards for Direct Contact and Low Level Condensers.

When (1) low investment is desired and (2) condensate recovery is not a factor, direct-contact condensers are effective. They are relatively simple to build and operate, are limited to sizes less than 250,000 lb of steam per h, and are built in three types: (1) barometric, (2) low-level, and (3) jet.

Figure 9.5.13 shows a self-supported counterflow **barometric condenser** with tail pipe, hot well, and air ejector. Steam and cooling water flow in opposite directions, with the coldest water for final condensation and cooling of noncondensables. The air pump must handle that part of the air disengaged from the cooling water as well as air leakage. The head required to pump water is pipe friction plus static head, minus 75 percent approx of the design vacuum. The barometric condenser is usually placed outdoors, and the water leg must be more than 34 ft high.

The **low-level condenser** substitutes a pump for the water leg of the barometric condenser to remove the liquid from the vacuum space. The **jet condenser** utilizes the aspirating effect of a jet for the entrainment of noncondensables and the consequent elimination of a separate air pump.

In the usual direct-contact condenser, where steam and raw circulating water are mixed, the recovery of pure condensate is precluded; greater feedwater makeup is necessary, and poorer vacuums are attained than with surface condensers. Direct-contact-condenser installations are not found in large plants, but there is some recent interest in their use with dry cooling towers.

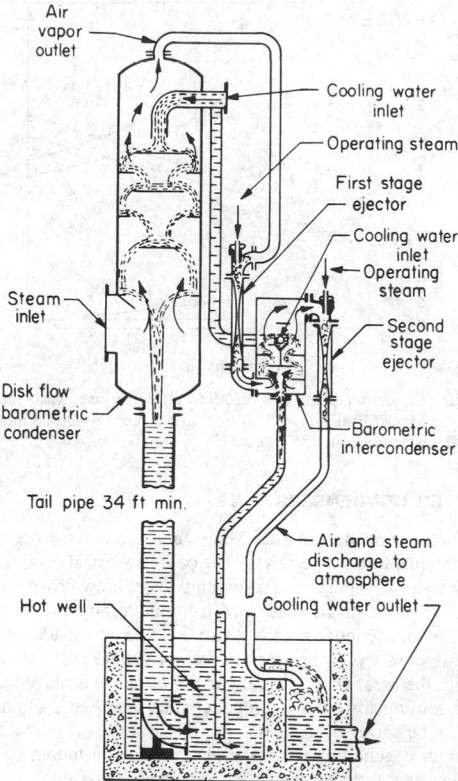

Fig. 9.5.13 Counterflow barometric condenser with two-stage air ejector. *(Ingersoll-Rand Co.)*

Table 9.5.9 gives values of expected air content of cooling waters; Fig. 9.5.14 gives typical performance curves; Fig. 9.5.15 defines flows W, temperatures t, and enthalpies h needed for the basic heat-balance equation $W_s(h_s - h_2) = W_w(t_2 - t_1)$. The latent heat $(h_s - h_2)$ is frequently taken as 950 Btu/lb of steam.

Shell materials are usually steel plate; with dirty or corrosive water, bronze, stainless steel, or linings of ceramic, plastic, or rubber may be used.

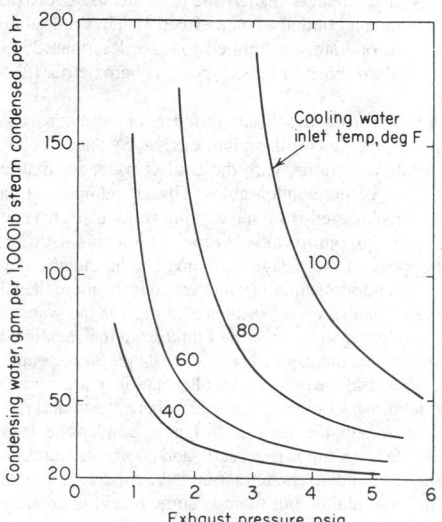

Fig. 9.5.14 Representative performance curves for a direct-contact condenser.

Table 9.5.9 Air in Cooling Water

Water temp, °F	Air, ft³/min per 1,000 gal/min
35	4.0
40	3.78
50	3.35
60	2.97
70	2.68
80	2.41
90	2.21
100	2.0

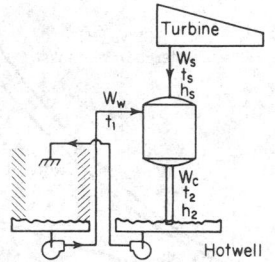

Fig. 9.5.15 Direct-contact condenser performance terms.

AIR EJECTORS

REFERENCE: Heat Exchange Institute, Standards for Steam Jet Ejectors.

The steam-jet ejector is used to remove noncondensable air and gases from condensers. It consists of a suction chamber, diffuser, and steam nozzle. The high-velocity jet of steam issuing from the nozzle entrains the gas, and the kinetic energy of the mixture is converted to pressure energy in the diffuser. Ratings are in lb/h at a given suction pressure. Ejectors are operated in series or are staged for absolute suction pressures of $1 \pm$ inHg abs. They will handle wet or dry mixtures, they are easy to operate, installation costs are low, and there are no moving parts—with consequent long life, high sustained efficiency, and low maintenance.

Two-stage ejectors, with surface-type inter- and aftercondensers (Fig. 9.5.16), are common in steam-surface-condenser application. The main condensate serves as a coolant, returning heat regeneratively to the boiler feed system. Two-stage ejectors have a shutoff pressure of 0.5 inHg abs approx. Two elements are usually provided, one serving as a spare.

Computation of **ejector performance** requires application of Dalton's law of mixtures (see Sec. 4), where basically the total pressure is the sum of the condensable vapor (steam) pressure and the noncondensable gas (air) pressure. In power plant condensers, the mixture is saturated with steam so that the temperature of the mixture fixes the partial pressure of the vapor.

Air ejectors are usually rated with suction conditions of 1 inHg abs and 7.5°F subcooling. Since the saturation temperature of steam at 1 inHg abs is 79°F, the mixture entering the ejector is at 71.5°F. The partial pressure of the vapor at 71.5°F is 0.78 inHg abs, and the partial pressure of the air is $1 - 0.78 = 0.22$ inHg abs. Therefore, an ejector with a standard rating (see Table 9.5.8) of 12.5 ft³/min of free dry air at 30 inHg abs handles, by the gas laws, $12.5(30/0.22) = 1,704$ ft³/min at the ejector suction.

Motive steam to the ejector is usually supplied from turbine throttle conditions, e.g., 600 lb/in², 750°F; 1,500 lb/in², 1,000°F; 2,500 lb/in², 1,050°F, through a reducing valve to a pressure not higher than 600 lb/in².

The **inter- and aftercondenser,** one- or two-pass, is a small heat exchanger which condenses the motive steam and allows the noncondensable gases, such as O_2 and CO_2 to be expelled to the atmosphere. However, the highly soluble ammonia gas may be returned to the feedwater system. The water-box pressure of tube-side pressure must be designed for the condensate-pump shutoff head. The pressure drop through the inter- and aftercondensers varies from 1 to 10 ft of water, with 3 ft as a reasonable average.

Normal **materials** used are: nozzle, stainless steel type 303; steam

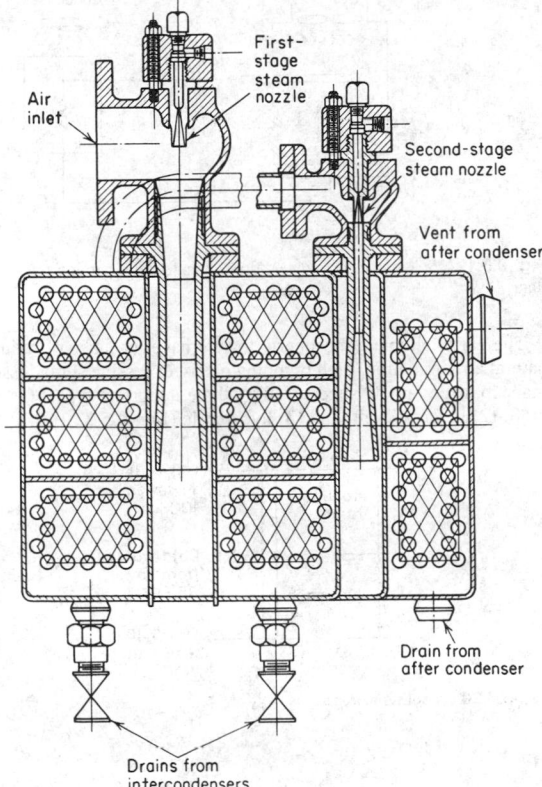

Fig. 9.5.16 Two-stage air ejector with surface-type inter- and aftercondensers. *(Westinghouse Electric Corp.)*

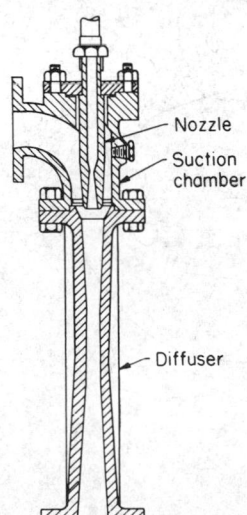

Fig. 9.5.17 Single-stage air ejector.

chest, steel; inlet-air chamber, cast steel; diffuser, aluminum bronze; inter- and aftercondenser shell, steel; inter- and aftercondenser tubes, stainless steel type 304; inter- and after-condenser tube sheet, carbon steel.

Hogging and Priming Ejectors Single-stage steam-jet ejectors (Fig. 9.5.17) are used for evacuating, or **hogging**, the air from the steam side of a surface condenser, in 15 or 20 min time, so that steam flow can be established and the condenser brought on the line. These ejectors are usually noncondensing and exhaust to the atmosphere. Ratings (Table 9.5.10) are typically in lb/h of free dry air at 70°F, suction conditions of 15 inHg abs, shutoff pressure about 2 inHg abs, motive steam pressures and temperatures as with multistage ejectors.

Priming ejectors are also single-stage noncondensing units and are used to withdraw air from the water side and to fill the condenser with circulating water during start-up periods. The hogging ejector may also serve this priming function by a suitable system of piping and valves.

Materials normally employed are: for the nozzle, stainless steel, 18-8 chrome nickel; steam chest, steel; air-inlet chamber and diffuser, cast steel.

Table 9.5.10   Hogger Capacities*

Total steam condensed, lb/h	scfm†—dry air at 10 inHg abs design suction pressure
Up to 100,000	50
100,001–250,000	100
250,001–500,000	200
500,001–1,000,000	350
1,000,001–2,000,000	700
2,000,001–3,000,000	1,050
3,000,001–4,000,000	1,400
4,000,001–5,000,000	1,750
5,000,001–6,000,000	2,100
6,000,001–7,000,000	2,450
7,000,001–8,000,000	2,800
8,000,001–9,000,000	3,150
9,000,001–10,000,000	3,500

* Heat Exchange Institute.
† 14.7 lb/in² abs at 70°F.
NOTE: In the range of 500,000 lb/h steam condensed and above, the table provides evacuation of the air in the condenser and low-pressure turbine from atmospheric pressure to 10 inHg abs in about 30 min if the volume of condenser and low-pressure turbine is assumed to be 26 ft³/(1,000 lb/h) of steam condensed.

VACUUM PUMPS

REFERENCE: Woodman, Rotary or Steam-Jet Air Pumps, *Power*, Aug. 1948.

Mechanical vacuum pumps, used for hogging, holding, and priming, are of several types for power-plant service: (1) reciprocating—piston, diaphragm; (2) rotary—sliding-vane, oval-water- seal, eccentric-rotor. (See Sec. 14 for details.) Multistage pumps usually need intercooling. A silencer is generally provided on the air discharge. Normally, two half-capacity pumps are installed, both operating during hogging and one adequate for holding service. The relative capabilities of a vacuum pump and hogging and holding ejectors are illustrated in Fig. 9.5.18 (note the considerable hogging capacity of the vacuum pump on start-up).

Advantages of vacuum pumps compared to steam-jet air ejectors include (1) system independent of a steam supply; (2) start-up when steam is not available, as on a once-through boiler system; (3) system capable of completely automatic operation; (4) high-pressure steam piping eliminated; (5) recycling of noncondensables and ammonia eliminated; (6) quieter operation. Disadvantages include (1) damage by water entering the intake (except when pumps having a liquid rotating seal); (2) higher initial cost; (3) higher maintenance cost. Operating costs are approximately equal. Various combination arrangements are found in practice, e.g., an air-operated air-ejector first stage followed by

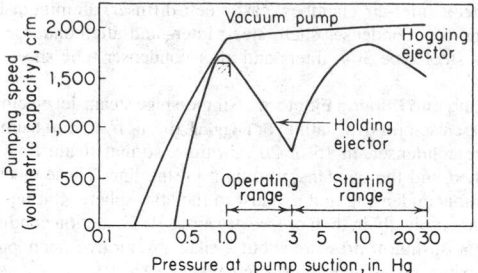

Fig. 9.5.18 Relative capabilities of air-removal pumps.

a vacuum-pump second stage, with consequent savings in motor-horse-power and space requirements.

COOLING TOWERS

REFERENCES: "Evaluated Weather Data for Cooling Equipment Design," Fluor Products Co. Cooling Towers, *Power*, Mar. 1963. Baker and Shryock, A Comprehensive Approach to the Analysis of Cooling Tower Performance, *Trans. ASME*, 1961.

The prevalent type of cooling tower (see Figs. 9.5.19 and 9.5.22) dissipates heat by the evaporation of some of the water sprayed into the air circulated through the tower. It is used where water is in limited supply, where temperature pollution of natural water bodies is to be avoided, where water conservation is to be effected, or where otherwise polluted sources must be avoided. Figure 9.5.20 illustrates the functional cycle and the significant basic terms (see also Sec. 4).

Wet-bulb temperature, for design purposes, should not exceed the maximum expected value more than 5 percent of the time during summer (June to September). (See Fig. 9.5.21.)

Approach is the difference in temperature between the cold water leaving the tower and the ambient wet bulb.

Cooling range is the difference in temperature between the hot water entering and the cold water leaving the tower.

Drift is the water lost as mist or droplets entrained by the circulating air and discharged to the atmosphere. It is in addition to the evaporative loss and is minimized by good design.

Makeup is the water required to replace total losses by evaporation, drift, blowdown, and small leaks.

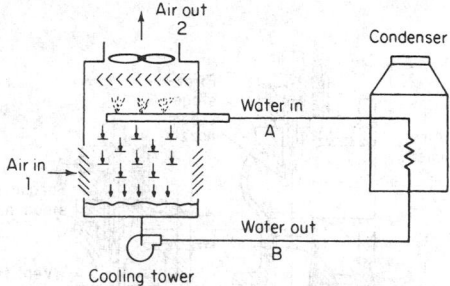

Fig. 9.5.19 Schematic for induced-draft, counterflow, cooling tower installation.

The early, simple atmospheric towers have, because of reliance on natural air circulation, high pumping heads, excessive spray losses, and makeup, been largely superseded by three important types: (1) forced-draft, (2) induced-draft, and (3) hyperbolic (Fig. 9.5.22).

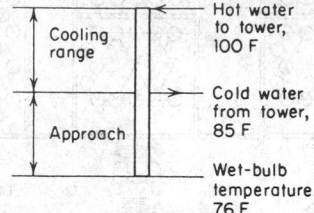

Fig. 9.5.20 Cooling tower terms.

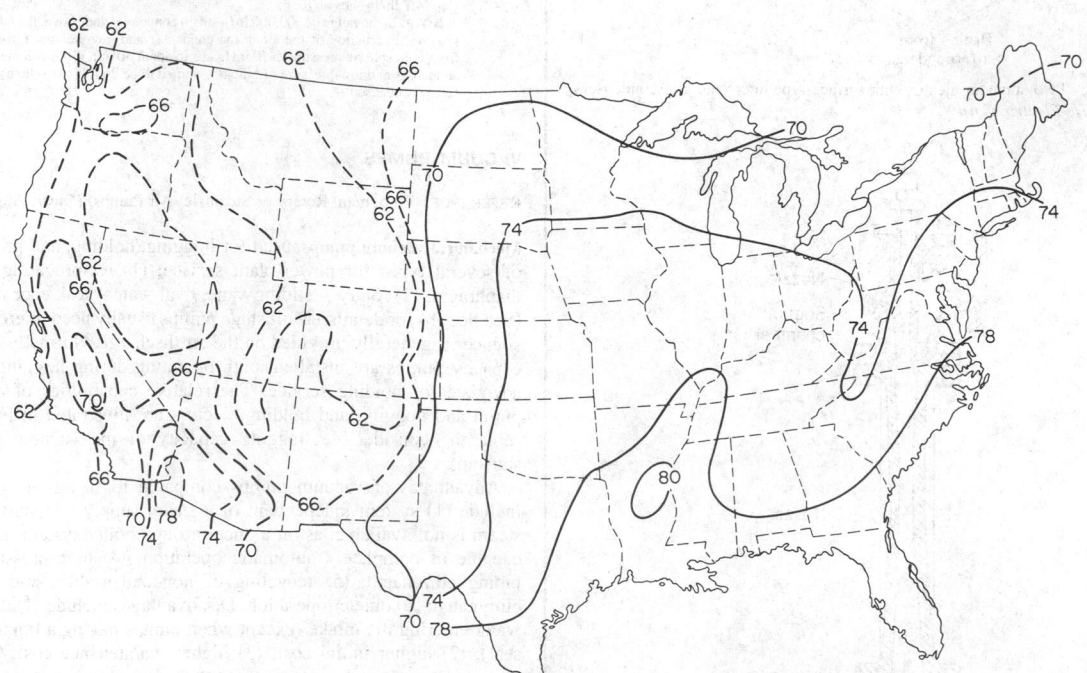

Fig. 9.5.21 Wet-bulb temperature, °F, isoclines at the 5 percent level, United States. (*Adapted from Fluor Products Co., by permission.*)

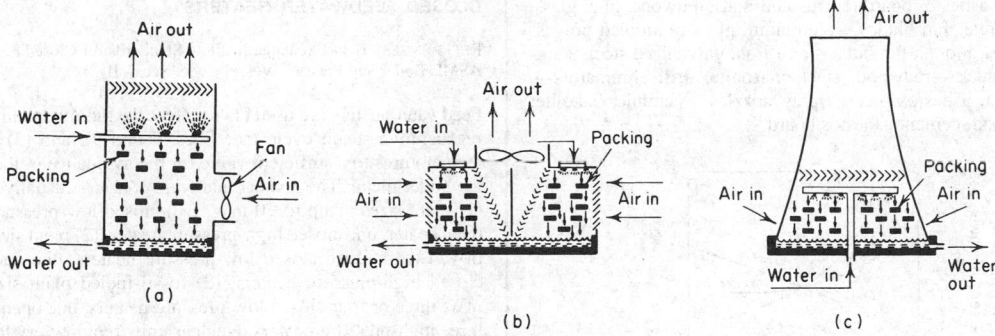

Fig. 9.5.22 Cooling tower types. (*a*) Forced draft; (*b*) crossflow-induced draft; (*c*) hyperbolic.

The **mechanical forced-draft tower** (Fig. 9.5.22*a*) gives (1) a controllable air supply from fans conveniently located for inspection and maintenance at ground level, and (2) reduced water-pumping head. Nonuniform distribution of air over the ground area of the tower cell, recirculation of vapor from the tower discharge to the tower inlet, with its deleterious effects and fan icing in cold weather, and limitations on the physical diameter of fans are all problems with forced-draft towers.

The **induced-draft tower** (Fig. 9.5.22*b*) is prevalent in U.S. practice. The fan is mounted on the top (discharge) of the cell, with consequent improved air distribution within the cell; drift eliminators reduce makeup requirements; spray nozzles, downspouts, splash plates, and splash bars ensure ample evaporative surface for the water, with maximum volumetric heat-transfer rates. In the **counterflow** design, air is introduced beneath the cell fill, but in the **crossflow** design (Fig. 9.5.22*b*), air is introduced at the sides of the fill.

The **hyperbolic tower** (Fig. 9.5.22*c*) utilizes the chimney effect (height, 300 ± ft) for natural circulation. It has been favored in large European installations where the prevalent lower dry-bulb ambient temperature gives a greater difference in density between entrance to and exit from the tower. The wide approach keeps the unit size within practical bounds, and the savings in fan power support the higher investment. Recent installations in the United States attest to its attractiveness.

Precautions are necessary to avoid freezing troubles in cold weather, fire hazards with intermittent operations, corrosion, scale, and microbiological growth problems. **Location** must avoid recirculation: for tower lengths to 250 ft, the long-axis orientation should be parallel to the prevailing wind; towers longer than 250 ft should be arranged broadside to the prevailing wind. The tower should be isolated as much as possible; adjacent heat sources compel the specification of higher design wet-bulb temperatures.

Performance Calculations

(See Fig. 9.5.19.)

Applicable equations are

$$W_{a1}h_{a1} + W_{v1}h_{v1} + W_{wA}h_{fA} = W_{a2}h_{a2} + W_{v2}h_{v2} + W_{wB}h_{fB} \quad (9.5.6)$$
$$W_{wB} = W_{wA} - (W_{v2} - W_{v1}) \qquad W_{a1} - W_{a2} \quad (9.5.7)$$
$$W_{wA}(h_{fA} - h_{fB}) = W_{a2}h_{a2} + W_{v2}h_{v2}$$
$$\qquad - (W_{a1}h_{a1} + W_{v1}h_{v1}) - (W_{v2} - W_{v1})h_{fB} \quad (9.5.8)$$
$$h_{fA} - h_{fB} = t_{wA} - t_{wB} \quad (9.5.9)$$

and

$$W_a h_a + W_v h_v$$
$$= \text{total heat, from psychrometric chart} \quad \text{(See Sec. 4.)} \quad (9.5.10)$$
$$W_{wA}(t_{wA} - t_{wB}) = \text{total heat at 2}$$
$$\qquad - \text{total heat at 1} - (W_{v2} - W_{v1})h_{fB} \quad (9.5.11)$$

where W = flow, lb/h; a = air; w = water; v = vapor; h = enthalpy, Btu/lb; f = fluid; t = temperature, °F; 1, 2, A, B = locations.

EXAMPLE. Given: flow = 80,000; wet bulb = 76°F; dry bulb = 86°F; range (also condenser rise) = 100 − 85 = 15°F; approach = 9°F. Find: eco-

nomical tower size, circulating-pump power, fan capacity and power, and makeup water.

SOLUTION. (1) *Economical tower size:* A study must be made to tie in with condenser. Heat to be dissipated is constant. Vary rise (range), approach, and then compare capital costs of cooling tower, circulating pump, and condenser with operating costs for fan and pump power. (2) *Circulating-water-pump power;* bhp = gal/min × TDH/3,960 × eff. TDH = condenser friction + pipe friction + static head. (3) *Fan capacity and power:* If air leaves tower saturated at 95°F, then from psychrometric charts (Sec. 4),

$$\text{Total heat} = 1 \text{ lb dry air } C_p(t_a - 0°) + W_v h_g \quad \text{at } t_a \quad (9.5.12)$$

For air and vapor at 1, vapor = 120 grains/lb dry air. By Eqs. (9.5.10) and (9.5.12), total heat = $1 \times 0.24(86 - 0) + (120/7,000)1,099.2 = 39.5$ Btu/lb dry air. For air and vapor at 2, vapor = 256 grains/lb dry air. Total heat = 63.1 per lb dry air. Sp vol (air and vapor) = 14.80 ft³/lb dry air. By Eq. (9.5.11), $W_{wA}(100 - 85) = 63.1 - 39.5 - (256 - 120)\ ^{63}/_{7,000} = 1.50$ lb water/lb dry air. 80,000 gal/min/(1.50 × 500) = 26.6×10^6 lb dry air/h, or $26.6 \times 10^6 \times 14.80/60 = 6.58 \times 10^6$ ft³/min. With test data on static-pressure requirements for the tower and with the fan characteristics, the horsepower to drive the fan can be calculated. (4) *Makeup:* $W_{v2} - W_{v1} = (256 - 120)\ 26.6 \times 10^6/7,000 = 515,000$ lb/h.

Normally, a cooling tower is purchased for only one guarantee point. It is well, however, to have **performance** curves (Fig. 9.5.23) showing operation for various wet-bulb temperatures and cooling ranges. The **investment** for a cooling tower is essentially a matter of water flow and is

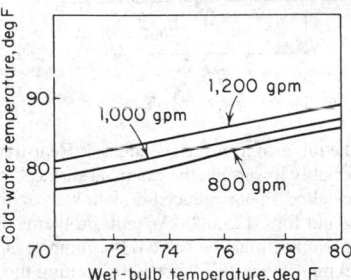

Fig. 9.5.23 Cooling tower performance. Design for 85°F cold water, 95°F hot water, 78°F wet-bulb temperature, 10° range; 1,000-gal/min cell.

influenced by approach, range, and wet bulb (Fig. 9.5.24). The **cost evaluation** should include consideration of tower frame and fill, fans, motors, basin and pump pit, pump head, fan horsepower, freight, labor, and erection. In the choice of a tower for a power plant, there should be coordinated study and evaluation of the turbine and condenser for best overall economy.

The **height** of a field-erected induced-draft tower, from basin curb to fan deck, ranges from 8 to 50 ft; widths vary from 6 to 60 ft; lengths from 8 to 500 ft; fan-stack height between 2 and 15 ft.

Materials used are: frame—redwood (treated or untreated), Douglas fir (treated), steel (galvanized), concrete; fill—red wood, plastics (polyethylene, polypropylene), cement asbestos board (extruded);

casing—cement asbestos board (corrugated, slat), redwood, fiber glass, aluminum, concrete; fan blades—aluminum, glass-reinforced polyester, stainless steel, monel; fan hubs—cast iron, galvanized steel, stainless steel; fan stack—redwood, steel, masonite, drift eliminators—redwood, cement, asbestos board; spray nozzles—ceramic, bakelite; louvers—redwood, cement asbestos board.

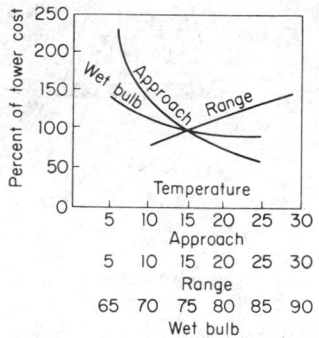

Fig. 9.5.24 Comparative cooling tower costs. *(Foster Wheeler Corp.)*

DRY COOLING TOWERS, WITH DIRECT-CONTACT CONDENSERS

REFERENCE: Ritchings and Lotz, Economics of Closed vs. Open Cooling Water Cycles, *Power Eng.,* May and June 1963.

The thermodynamic requirement for a heat sink can generally best be met with a natural body of water such as a river, lake, or ocean. When water supplies are limited, a **"wet" cooling tower** (water losses less than 2 or 3 percent) may be used for reclamation. A **"dry" cooling tower** will further reduce this loss. Figure 9.5.25 shows an application with a direct-contact condenser. The finned surfaces of the dry cooling tower are

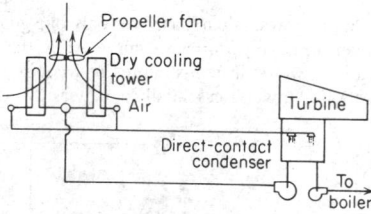

Fig. 9.5.25 Dry cooling tower system.

substituted for the tubes of a surface condenser. Heat-transfer rates (see Sec. 4) are appreciably lower, but the extended surface, relatively low in cost, may be installed in an induced-draft tower or in a hyperbolic tower, as in England for a 125,000-kW unit. Problems of recirculation, freezing, and air leakage must be solved. If a minor evaporative loss is permissible, peaking capacity is obtained by wetting the heat-exchange surface in the tower.

SPRAY PONDS

If land area is available, a spray pond may serve as an alternative reclamation system. Pipelines are typically arranged on 50-ft centers, with five nozzles (50 ± gal/min each) every 12 ft of pipeline. Cooling is limited by the relatively short period of contact between spray and air. Drift, especially with adverse wind conditions, can make losses higher than on towers. Generally, if cooling efficiencies above 50 percent are required, a tower is favored, where efficiency = $(t_h - t_c) \times 100/(t_h - t_{wb})$, where the temperature t subscripts are h = hot water, c = cold water, wb = wet bulb of the ambient air. (See Wright and Kirsopp, Pond Surface Cooling of a Chemical Plant Cooling Water, *Proc. Am. Power Conf.,* 1960.)

CLOSED FEEDWATER HEATERS

REFERENCES: Heat Exchange Institute Standards for Closed Feedwater Heaters. ASME Boiler and Pressure Vessel Code, Sec. VIII.

Feedwater heaters are used (1) to effect the thermodynamic gains of the regenerative steam cycle (see Secs. 4 and 9), and (2) to raise water temperatures to a sufficient value for the avoidance of thermal shock to the boiler metal. The **number and types of heaters** usually employed are: (1) plant sizes of up to 70 mW, two closed low-pressure heaters, one open heater, one closed high-pressure heater; (2) plant sizes of 75 to 300 mW, two or three closed low-pressure heaters, one open heater, two closed high-pressure heaters; (3) fossil-fueled plant sizes above 300 mW, three or four closed low-pressure heaters, one open heater, two or three high-pressure heaters. Nuclear units require very large feedwater flows, necessitating double or triple parallel strings of heaters. There are generally five or six low-pressure and one high-pressure heater (no open heaters) in each string.

Minimum heater cost prevails with minimum restrictive specifications, e.g., horizontal, two-pass, high water velocity (10 ft/s at 60°F), no length limits. Overall heater length is limited by maximum available tube lengths of 100 ft for copper alloys, admiralty metal, and copper-nickel and 85 ft for Monel. With U-tube construction, this results in heater lengths of about 48 and 40 ft, respectively. A general rule, to ensure good steam distribution, is that the length in feet shall not exceed the shell diameter in inches plus 2; i.e., with a 30-in diam shell, the length should not exceed 32 ft. **Pressure drop** through the tubes must be economically evaluated as it varies approximately with the square of the water velocity.

If a **length restriction** is imposed, the designer may have to substitute a four-pass arrangement for the two-pass design, with consequent large-diameter shell and water chamber, heavier walls, more tube holes to be drilled, more tubes to be installed, and a cost increase. If a **pressure-drop restriction** is imposed, a lower water velocity results, with more tubes, larger shell and chamber diameter, and more surface because of the lower heat-transfer rate. Vertical heaters, with appropriate construction details, are also higher in cost.

The **condensing heater** (Fig. 9.5.26), like a surface condenser, performs at constant saturation temperature outside the tubes with no condensate subcooling. Feedwater is generally heated to within 5°F of the

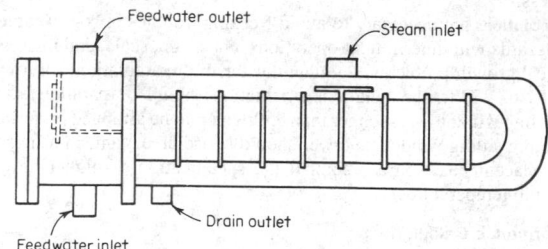

Fig. 9.5.26 Basic heater section for condensing heater.

saturation temperature. The addition of a **drain-cooling section** (Fig. 9.5.27), with a cover or enveloping baffle around the tubes of the inlet pass, can reduce the drip temperature to within 10 or 15°F of the incoming water. When the steam is at sufficient pressure (above 125 lb/in² abs) and contains enough superheat (more than 250°), a **desuperheating**

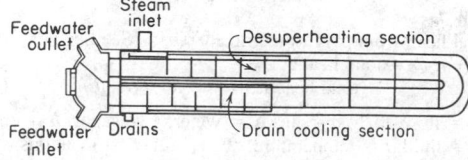

Fig. 9.5.27 Basic heater section with drain cooling and desuperheater sections added.

section (Fig. 9.5.27), using an enveloping baffle around the tubes at feedwater outlet, can raise the water temperature above the saturation temperature of the entering steam. Drain coolers and desuperheating sections improve overall plant heat rate by reduction of energy degradation. Subcooling of drains reduces flashing, erosion, vibration, and noise when drains are dropped to a lower-pressure heater. Special attention should be given to drains flashed to the main condenser, as this is a direct thermodynamic loss. As an alternate, these drains can be pumped forward into the feedwater line.

Heat-transfer rates for a condensing feedwater heater range, overall, between 500 and 900 Btu/(h · ft² · °F). (See Figs. 9.5.28 and 9.5.29.)

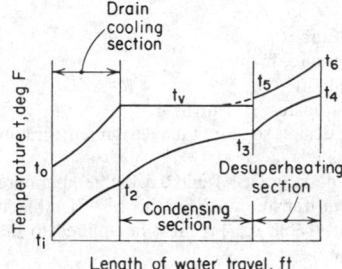

Fig. 9.5.28 Temperature vs. water travel, closed feed heater.

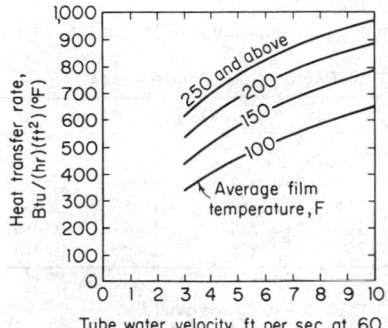

Fig. 9.5.29 Approximate condenser heat-transfer rates with 18 BWG admiralty metal tubes. Average film temperature = $t_s - 0.8\delta t_m$.

Heat-transfer rates for drain-cooling and desuperheating sections must be separately evaluated, with overall ranges from 300 to 500 on the former and 80 to 140 on the latter. (See Figs. 9.5.30 and 9.5.31.) Normally, the three sections are included in the same shell (Fig. 9.5.27); when the cooler becomes too large, as with the lowest-pressure heater, a separate shell-and-tube exchanger may be justified.

Low-pressure heaters (upstream of the boiler feed pump) are usually designed for tube-side pressure of less than 90 lb/in² gage. This is based on the shutoff head of the condensate pump plus 10 percent. For pressures up to 300 to 400 lb/in² gage, a bolted cover with flange and gasket

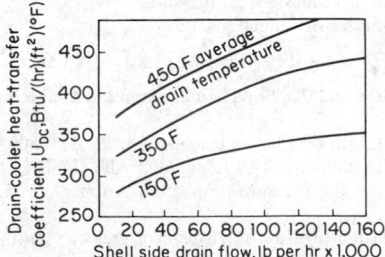

Fig. 9.5.30 Approximate drain-cooler heat-transfer rates.

(Fig. 9.5.32a) is used, while higher pressures usually justify a welded joint to eliminate the gasket (Fig. 9.5.32b and c). A manway is utilized for access to the tube sheet. Tubes are generally rolled into the tube sheet.

High-pressure heaters (downstream of the boiler feed pump) are designed for shutoff head of the pump plus 10 percent, resulting in about

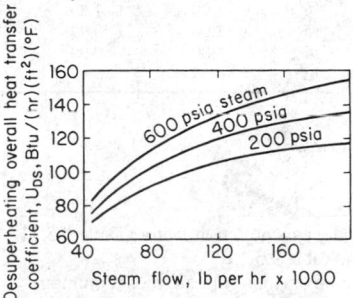

Fig. 9.5.31 Approximate desuperheater section heat-transfer rates.

1,500 lb/in² gage for a nuclear plant and up to 5,000 lb/in² gage for fossil-fuel plants. Hemispherical head construction (Fig. 9.5.32c) and welded tube-to-tube sheet joints are almost exclusively used. Since the cost of the tubes represents the largest portion of the heater cost, material selection is important (see Table 9.5.11). Depending on the desired water-chemistry requirements, low-pressure heaters usually use admiralty metal, 90-10 Cu-Ni, stainless or carbon steel. High-pressure heaters require the use of the higher-strength alloys of 70-30 Cu-Ni, monel, stainless, or carbon steel. The thinnest wall possible [$t = Pd/(2s + 0.8p)$] or the industry minimum standard of 18 BWG (20 BWG for stainless steel) is utilized.

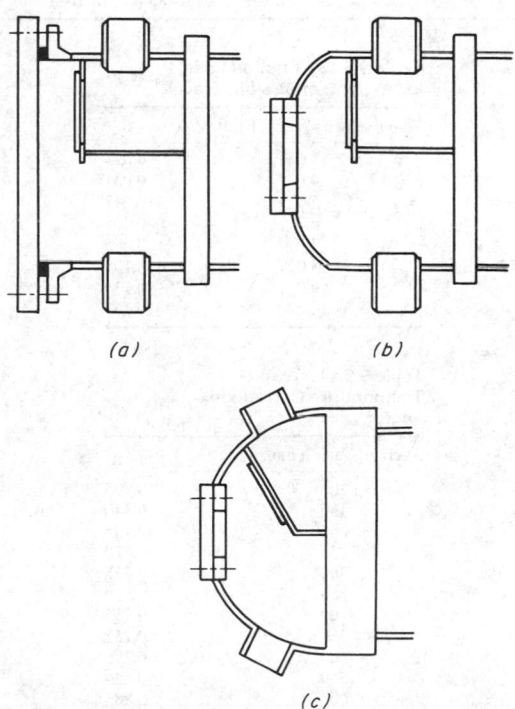

Fig. 9.5.32 Feedwater heater channels. (a) Bolted; (b) dished; (c) hemispherical.

Table 9.5.11 Material and Gage Correction Factor C_m
(For $\frac{5}{8}$- to 1-in tubes)

BWG no.	Tube material							
	Ars-Cu	Admiralty	90-10 Cu-Ni	80-20 Cu-Ni	70-30 Cu-Ni	Monel	18-8 SS	Carbon steel
18	1.00	1.00	0.97	0.95	0.92	0.89	0.85	0.96
17	1.00	1.00	0.94	0.91	0.87	0.85	0.80	0.92
16	1.00	1.00	0.91	0.88	0.84	0.82	0.77	0.89
15	1.00	0.99	0.89	0.86	0.82	0.79	0.74	0.87
14	1.00	0.96	0.85	0.82	0.77	0.75	0.70	0.83
13	0.98	0.93	0.81	0.78	0.73	0.70	0.65	0.79

Performance Calculations

Notation

A = heat-transfer surface, ft^2
C_m = material and gage correction factor, Table 9.5.11
D = tube ID, in, Table 9.5.12
d = OD of tube, in ($\frac{5}{8}$ and $\frac{3}{4}$ in are most common)
F_1 = friction factor, Table 9.5.13
F_2 = water-temperature correction factor, Table 9.5.14
h = enthalpy, Btu/lb
k = tube diameter and gage factor, Table 9.5.15
L = active length of tubes per pass, ft
N = number of passes

n = number of tubes
P = design pressure, lb/in^2
p = pressure, lb/in^2
Q = total heat transferred, Btu/h
S = allowable design stress at tube design temperature, lb/in^2, from ASME Code
t = wall thickness, in (see Table 9.5.16); temperature, °F
U_o = overall heat-transfer rate, Btu/(h · ft^2 · °F) (the material correction factor, Table 9.5.11, must be applied to the overall heat-transfer rate)
V = water velocity, ft/s
W = steam flow, lb/h
W_w = feedwater flow, lb/h
Δp = pressure drop, lb/in^2
Δt_m = logarithmic mean temperature difference, °F (Fig. 9.5.28)

Table 9.5.12 Feedwater-Heater Tube Constants

Tube BWG	$\frac{1}{8}$-in OD		$\frac{3}{4}$-in OD		$\frac{7}{8}$-in OD	
	D	$D^{1.24}$	D	$D^{1.24}$	D	$D^{1.24}$
18	0.527	0.452	0.652	0.589	0.777	0.731
17	0.509	0.433	0.634	0.567	0.759	0.710
16	0.495	0.418	0.620	0.554	0.745	0.695
15	0.481	0.404	0.606	0.538	0.731	0.678
14	0.458	0.380	0.584	0.514	0.709	0.652
13	—	—	0.560	0.488	0.685	0.625

Table 9.5.15 Tube Diameter and Gage Factor k

Diam, in	BWG no.				
	20	18	17	16	15
$\frac{5}{8}$	0.217	0.2407	0.2580	0.2728	0.289
$\frac{3}{4}$	0.1732	0.1887	0.1995	0.2089	0.218
$\frac{7}{8}$	0.1445	0.1550	0.1624	0.1686	0.1748
1	0.1238	0.1314	0.1369	0.1413	0.1458

Table 9.5.13 Friction Loss, lb/in^2 per ft of Straight Travel

Water velocity (at 60°F), ft/s	F_1
3.0	0.018
4.0	0.031
5.0	0.047
6.0	0.065
7.0	0.085
8.0	0.108
9.0	0.134
10.0	0.162

Table 9.5.16 Tube-Wall Thickness

Gage, BWG	Thickness, in
15	0.072
16	0.065
17	0.058
18	0.049
19	0.042
20	0.035
22	0.028

Table 9.5.14 Water-Temperature Correction Factor*

Average water temp, °F	F_2
100	0.905
150	0.840
200	0.795
250	0.770
300	0.755
350	0.750
400	0.755
450	0.757
500	0.795
550	0.830
600	0.885

* Average water temperature = $t_s - \Delta t_m$.

Heat-balance data will provide the major parameters of steam pressure, feedwater flows, and temperatures. The choice of tube diameter and material must be made first. Tube thickness is calculated by

$$t = Pd/(2S + 0.8P)$$

Test pressure is 1.5 times design pressure.

Condensing-heater-size equations are

$$Q = UA\,\Delta t_m = W_w(h_3 - h_2) = W_s(\Delta h_s)$$

(See Figs. 9.5.28 and 9.5.29.) The minimum terminal temperature difference is 2°F.

The active length of tubes is calculated from $L = 500AV/(NW_wk)$, where maximum velocity is 10 ft/s at 60°F; the number of tubes, from $n = 3.82A/(Ld)$; the pressure drop, from $\Delta p = (L + 5.5D)F_1F_2N/D^{1.24}$.

The shell design pressure is 20 percent greater than the maximum operating pressure, rounded to the higher 25 lb/in^2 as a minimum.

A **drain-cooling section,** when added, is treated as a separate heat exchanger where the quantities t_v, t_o, t_i, W_s, and W_w are known; t_2 is the only unknown and is to be calculated from a heat balance on the section or $Q = W_s(h_v - h_o) = W_w(h_2 - h_1)$. Compute Δt_m; find the area of the drain-cooler section A_{dc} from $Q = U_o A_{dc} \, \Delta t_m$, with overall U_o from Fig. 9.5.30.

A **desuperheating section** is similarly calculated with t_v, t_6, t_4, p_6, h_6, h_4, W_s known. Allow approximately 1 percent pressure drop through the section to get p_v, and 80 to 90 percent desuperheating (30°F superheat entering condensing section) to obtain t_5 and h_5. By a heat balance on section, calculate h_3 from $Q = W_s(h_6 - h_5) = W_w(h_4 - h_3)$. With equivalent t_3, calculate Δt_m and substitute in $Q = U_o A \, \Delta t_m$ for area of desuperheating section (overall U_o from Fig. 9.5.31).

Materials

Heaters are designed in accordance with the Heat Exchange Institute Standards for Closed Feedwater Heaters and the ASME Boiler and Pressure Vessel Code, Sec. VIII, using the following materials:

Tubes:

Material		Max temp, °F	
		Rolled joint	Welded joint
Admiralty metal		350	
90-10 Cu-Ni	ASME SB-395	400	450
80-20 Cu-Ni		400	475
70-30 Cu-Ni		400	525
70-30 Cu-Ni, stress-relieved		400	525
Monel, ASME SB-163		400	600
Carbon steel, ASME SA-556, grades A, B, and C		350	650
Stainless steel, ASME SA-249; TP 304 (welded)		350	650

Shell, nozzles: carbon-steel pipe and plant, ASME SA-515 grade 70; SA-106 grade B.

Channels, or heads, channel covers: carbon-steel plate, forgings, and pipe, ASME SA-515 grade 70; SA-105 grade II; SA-106 grade B.

Baffles and tube supports: steel plate, ASME SA-283 grade C.

Tube sheets: carbon-steel plate and forgings, ASME SA-515 grade 70; SA-105 grade II.

Carbon steel is used for temperatures up to 800°F. Chromium-molybdenum steel (ASME SA-387 grade C) is used for higher temperatures.

OPEN, DEAERATING, AND DIRECT-CONTACT HEATERS

REFERENCE: Heat Exchange Institute Standards for Deaerators and Deaerating Heaters.

Deaerators or deaerating heaters serve (1) to degasify feedwater and thus reduce equipment corrosion (see Sec. 6), (2) to heat feedwater regeneratively and improve thermodynamic efficiency (see Sec. 9), and (3) to provide storage, positive submergence, and surge protection on the boiler feed-pump suction (see Sec. 14).

Removal of oxygen and carbon dioxide from boiler feedwater and process water at elevated temperature is essential for adequate conditioning. In some power-plant applications, a well-designed surface condenser gives adequate deaeration and the accompanying exclusive use of closed feedwater heaters.

A modern **deaerator** will, by mechanical action, reduce O_2 content of effluent to less than 0.005 cm³/L and CO_2 content to a negligible amount. Water must (1) be heated to and kept at saturation temperature, as the gas solubility is zero at the boiling point of the liquid, and (2) be mechanically agitated by spraying or cascading over trays for effective scrubbing, release, and removal of gases. Gases must be swept away by an adequate supply of steam. Since the water is heated to saturation conditions, the terminal temperature difference is zero with maximum improvement in associated turbine heat rate. Extremely low partial gas pressures, dictated by Henry's law, call for large volumes of scrubbing steam. **Vent condensers** of the shell-and-tube or direct-contact types, cooled by incoming feed, serve to recover heat and water before release of gases to the atmosphere.

The **tray-type deaerator** (Fig. 9.5.33) is prevalent. While it has some tendency to scale, it will operate at wide load conditions and is practically independent of water-inlet temperature. Trays can be loaded to some 10,000 lb/(ft² · h), and the deaerator seldom exceeds 8 ft in height.

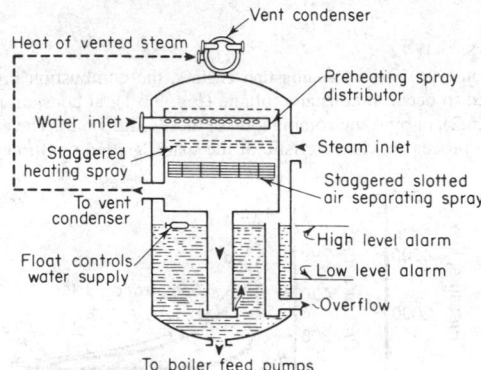

Fig. 9.5.33   Tray-type deaerating heater section.

The **spray type** uses a high-velocity steam jet to atomize and scrub the preheated water. Applications are (1) marine service, where it is unaffected by ship roll and pitch, and (2) industrial plants where operating pressures are stable. It requires a temperature gradient, e.g., 50°F minimum, to produce the fine sprays and vacuums with the cold water required.

Materials

Open-feed heaters are designed to the ASME Boiler and Pressure Vessel Code. Materials used are: shell—steel, ASME SA-285 grade C and SA-515 grade 70; trays—stainless steel, type 304; baffles in vent condenser—stainless steel, type 304; spray valves—stainless steel, type 304.

EVAPORATORS

For general treatment of evaporators used in the preparation of makeup water, see Sec. 6.

9.6 INTERNAL COMBUSTION ENGINES
by D. Assanis, C. Borgnakke, D. E. Cole, and D. J. Patterson

REFERENCES: Lichty "Internal-Combustion Engines," McGraw-Hill. Obert, "Internal-Combustion Engines and Air Pollution," Harper & Row. Taylor and Taylor, "The Internal-Combustion Engine," International Textbook. Vincent, "Supercharging the Internal-Combustion Engine," McGraw-Hill. Crouse, "Automotive Engine Design," McGraw-Hill. Patterson and Henein, "Emissions from Combustion Engines," Ann Arbor Science. Starkman, "Combustion Generated Air Pollution," Plenum. *Trans. ASME Trans. SAE Proc. I Mech E*, Automobile Division, *Automotive Inds.* Ferguson, "Internal Combustion Engines," Wiley. Heywood, "Internal Combustion Engine Fundamentals," McGraw-Hill. Blair, "The Basic Design of Two-Stroke Engines," SAE. Watson and Janota, "Turbocharging the Internal Combustion Engine," Wiley. Benson and Whitehouse, "Internal Combustion Engines," Pergamon.

GENERAL FEATURES

Cycles

(See also Sec. 4.1)

In piston-type **internal combustion engines,** the combustion process is assumed to occur at constant volume (Fig. 9.6.1), at constant pressure (Fig. 9.6.2), or by some combination of these (Fig. 9.6.3). The constant-volume process is characteristic of the **spark ignition** or **Otto cycle;** the

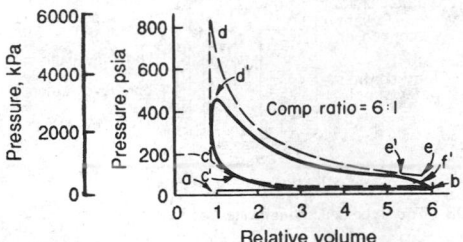

Fig. 9.6.1 Otto cycle (constant-volume combustion, no excess air). Process of the ideal cycle: *a-b,* suction stroke, admission of the charge; *b-c,* compression stroke; at *c,* ignition of the compressed charge; *d-e,* expansion; at *e,* exhaust valve opens; *b-a,* exhaust stroke.

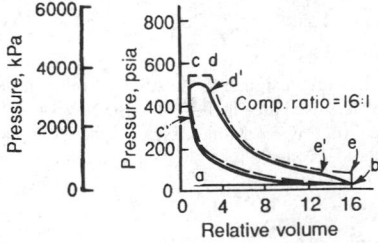

Fig. 9.6.2 Diesel cycle (constant-pressure combustion, 50 percent excess air). Processes of the ideal cycle: *a-b,* suction stroke, admission of air; *b-c,* compression of air; *c-d,* injection and combustion of the fuel; *d-e,* expansion; at *e,* exhaust valve opens; *b-a,* exhaust stroke.

constant pressure is found only in the slow-speed **compression ignition** or **diesel cycle;** with both processes, the cycle is sometimes called a **mixed, combination,** or **limited pressure cycle.** Actually, the indicator card obtained from a high-speed, mixed-cycle, compression ignition engine may be similar to that obtained from a spark ignition engine. The fundamental differences are the methods of mixing the air and fuel (before compression in the Otto cycle, and usually near the end of compression in the diesel cycle) and the methods of ignition.

The nominal **compression ratio** (usually specified) is the displacement

plus clearance volume divided by the clearance volume. The actual compression ratio is appreciably less than the nominal value because of late intake valve or port closing.

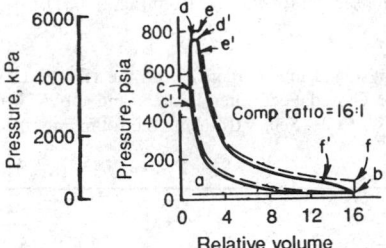

Fig. 9.6.3 Mixed cycle (constant-volume and constant-pressure combustion, 50 percent excess air). Processes of the ideal cycle: *a-b,* suction stroke, admission of air; *b-c,* compression of air; *c-d,* ignition and constant-volume combustion; *e-f,* expansion; at *f,* exhaust begins; *b-a,* exhaust stroke. Broken lines represent ideal cycle; solid lines show characteristics of actual cycle. Primes attached to letters show actual locations of events, though some lower-compression-ratio and multiple-chamber engines may require starting aids.

The **compression pressure** may be estimated from the relation $p = r_a^{1.33}$, where p_m is the intake manifold pressure and r_a is the actual compression ratio.

The actual compression pressure can be determined with a compression pressure gage that traps the gate by means of a check valve, thus indicating the maximum pressure under motoring conditions.

Spark ignition engines use volatile liquids or gases as fuel; have compression ratios between 6 : 1 and 12 : 1 (limited by combustion knock or the fuel/air mixture) and compression pressures from below 150 to above 300 lb/in² (1,030 to 2,060 kPa); use carburetors, gas mixing valves, or fuel injection systems; and operate on the Otto cycle. Gasoline is the fuel commonly used in airplane, automobile, small marine, small stationary, and tractor engines. Commercial gases such as blast furnace gas, coal gas, coke oven gas, carbureted water gas, producer gas, and natural gas are used in large stationary engines. The mixtures used generally are near chemically correct. Combustion pressures are usually 3.5 to 5 times the compression pressures. Load and speed are usually controlled by throttling the charge; piston speeds above 3,000 ft/min (15 m/s) are permissible.

Advantages are low first cost, low specific weight, low cranking effort required, wide variation obtainable in speed and load, high mechanical efficiency, and fairly low specific fuel consumption at high compression ratios and wide-open throttle.

Compression ignition engines use liquid fuels of low volatility varying from fuel oil and distillates to crude oil (see Sec. 7.1), have compression ratios between 11.5 : 1 and 22 : 1 and compression pressures 400 to 700 lb/in² (2,760 to 4,830 kPa), and operate on the diesel or mixed cycle. Generally no ignition devices are used, although some lower-compression-ratio and multiple-chamber engines may require starting aids. Load and speed are controlled by varying the fuel quantity injected.

The **dual-fuel engine** is a diesel engine with a compression ratio which is too low to result in ignition, at the desired time, of the gas/air mixture inducted into the cylinder. A pilot (small) injection of liquid fuel with good ignition quality is used to initiate the combustion process.

Advantages are low specific fuel consumption, high thermal efficiency at partial loads, possibly lower fuel cost, no preignition, low CO and hydrocarbon emission at low and moderate loads, suitability for two-stroke operation, and excellent durability. The lower-compression engines are of simpler construction (usually valveless two-stroke cycle), are more lightweight, and have lower first cost, lower operating expense, and higher mechanical efficiency than the higher-compression engines.

The **four-stroke cycle** (four-cycle) engine requires four piston strokes or two crankshaft revolutions per cycle (Figs. 9.6.1 and 9.6.2). This engine cycle is used almost exclusively in automobile, tractor, and aircraft engines of all types and sizes and in engines of other classifications with the exception of most outboard engines. Small four-cycle engines are always single-acting (combustion on only one side of the piston). The pumping loss, indicated by the area between the exhaust and the induction curves (Fig. 9.6.12), is more than the negative loop of the indicator card and at wide-open throttle depends on valve openings and speed and amounts to 3 to 7 percent of the engine indicated power. Throttling increases the pumping loss and reduces the positive indicator card area.

Advantages compared with two-cycle crankcase compression spark ignition engines include the wider variation in speed and load, cooler pistons, common crankcase in multicylinder engines, good lubrication secured more easily, no fuel loss during exhaust, lower specific fuel consumption lower pumping losses, less exhaust dilution, lower hydrocarbon emissions, positive-displacement inlet and exhaust processes, and easier power regulation.

The **two-stroke cycle** (two-cycle) engine requires two piston strokes or only one revolution for each cycle. Exhaust ports in the cylinder wall are uncovered by the piston, or exhaust valves in the cylinder head are opened near the end (at 60 to 88 percent) of the expansion stroke, permitting the escape of exhaust gases and reducing the pressure in the cylinder (Fig. 9.6.4). The charge of air or combustible mixture flows into and is compressed in a separate crankcase compartment for each cylinder, by a compressor or a blower to slightly above atmospheric pressure. Intake ports are uncovered by the piston or intake valves and

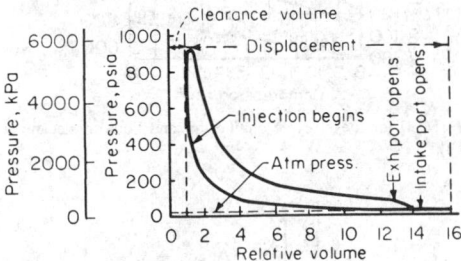

Fig. 9.6.4   Indicator card for two-cycle diesel engine (16:1 compression ratio).

are opened soon after the opening of the exhaust, and the compressed charge flows into the cylinder, expelling most of the exhaust products, some charge escaping with the exhaust. Blowers or compressors, driven either mechanically or by an exhaust turbine, are often used to supply sufficient air to scavenge the cylinder and clearance space and, in some cases, to supercharge the engine. Although small engines use only ports and crankcase compression (Fig. 9.6.5), large engines usually have separate compressors or blowers and either ports or both valves and ports. With crankcase compression, low scavenging of the cylinder occurs, and the volumetric efficiency is low (30 to 70 percent). The crankcase compression work amounts to 7 to 12 percent of the total work, but scavenging blowers with displacements 20 to 80 percent greater than the piston displacement may require as high as 30 percent of the indicated work in high-speed engines.

Advantages compared with four-cycle engines include the 50 to 80 percent greater power output per unit piston displacement and same speed (depending on scavenging) twice at many power impulses per cylinder per revolution, low cost for valveless designs employing crankcase compression, low NO_x emissions (spark ignited), and light weight.

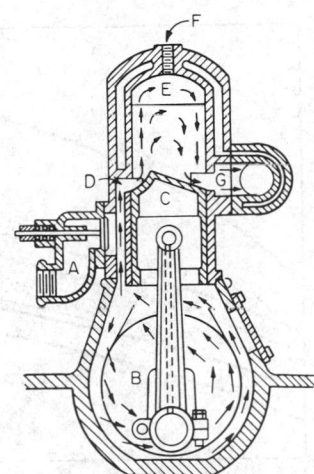

Fig. 9.6.5   Small two-cycle gasoline engine with crankcase compression.

ANALYSIS OF ENGINE PROCESS

Cycle Analysis   The theoretical indicated thermal efficiency η_a of the air standard Otto cycle depends only on the volumetric compression ratio r_c (Sec. 4.1):

$$\eta = 1 - \left(\frac{1}{r_c}\right)^{0.4} \tag{9.6.1}$$

For the air standard diesel cycle (constant-pressure combustion),

$$\eta_a = 1 - \frac{(r_d^{1.4} - 1)}{1.4 r_c^{0.4}(r_d - 1)} \tag{9.6.2}$$

where r_d is the volumetric expansion ratio during constant-pressure combustion.

However, the diesel engine is more closely approximated by the limited pressure combustion (dual) cycle for which the thermal efficiency is given by

$$\eta_a = 1 - \frac{r_p r_d^{1.4} - 1}{r_c^{0.4}[r_p - 1 + 1.4 r_p(r_d - 1)]} \tag{9.6.3}$$

where r_p is the pressure ratio during the constant-volume heat addition.

The efficiencies indicated by the air standard analysis are much higher than can be attained, principally because of variable specific heats, dissociation, and heat and time losses. The fuel-air cycle analysis takes into consideration variable fuel/air mixture properties and dissociation of combustion products and results in lower, more realistic values for efficiencies (Figs. 9.6.6 and 9.6.7). Real cycle analyses include, in addition, combustion losses, heat losses, and leakages, resulting in thermal efficiencies approximately 80 percent of those predicted by the fuel-air cycle analyses.

The **mean effective pressure** (mep) is equal to net work divided by volumetric displacement. Horsepower follows from

$$hp = P_{mep}\, SA\, rpm/K \tag{9.6.4}$$

where P_{mep} = mean effective pressure, lb/in^2 (kPa); S = stroke, ft (m); A = total piston area, in^2 (m^2); rpm = number of cycles completed per min; K = 33,000 (0.4566 for SI units).

Increasing the compression ratio and the air-fuel ratio increases the thermal efficiency of the Otto cycle (Fig. 9.6.6). Increasing the compression ratio and the air/fuel ratio also increases the thermal efficiency of the diesel cycle with constant-pressure combustion (Fig. 9.6.7). Mean effective pressures depend on charge input and thermal efficiency, both of which depend on compression ratio and air and fuel supply (Figs. 9.6.8 and 9.6.9).

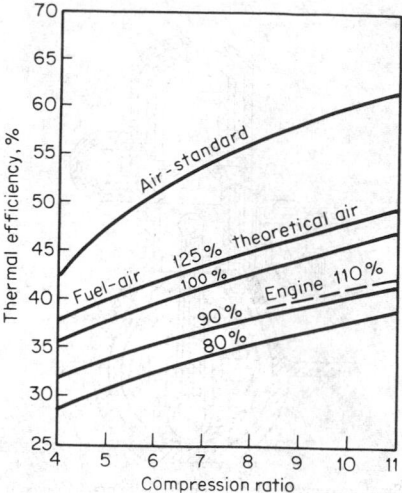

Fig. 9.6.6 Indicated thermal efficiencies of various Otto cycle analyses and engine test results.

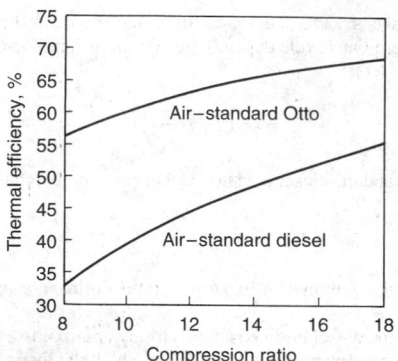

Fig. 9.6.7 Indicated thermal efficiency of the air standard diesel cycle with constant-pressure combustion.

Deviations from Ideal Processes

Changes in atmospheric pressure and temperature change power output although they do not appreciably affect thermal efficiency. The standard correction formulas adopted by SAE for brake power output are, for spark ignition engines,

$$\text{bhp}_s = \text{bhp}_o \left(\frac{P_s}{P_o} \right) \sqrt{\frac{T_o}{T_s}}$$

and for diesel engines,

$$\text{bhp}_s = \text{bhp}_o \left(\frac{P_s}{P_o} \right) \left(\frac{T_o}{T_s} \right)^{0.7}$$

where subscripts o and s indicate observed and standard conditions, respectively, P_o being the dry barometric pressure.

Pumping work (area EFGAE, Fig. 9.6.10) is required to overcome the resistance to flow of fresh charge into and exhaust products out of the cylinder. It varies from almost 0 for an engine running at slow speed with wide-open throttle to about 3 lb/in² (21 kPa) mep at high speed, and to 10 lb/in² with idle throttle position. The negative loop (area AXFGA, Fig. 9.6.10) in a normally aspirated engine represents about 60 percent of the pumping work.

Volumetric efficiency is the ratio of the volume of air (for a liquid-fuel engine) and charge (for a gas engine) actually admitted, measured at atmospheric pressure and temperature, to the displacement volume. High manifold and cylinder temperatures and resistance to flow reduce volumetric efficiency. High-output aircraft engines have maximum vol-

umetric efficiencies of 85 to 90 percent at rated speeds. Supercharged engines can have volumetric efficiencies well above 100 percent or can be designed to maintain high volumetric efficiencies with an increase in speed.

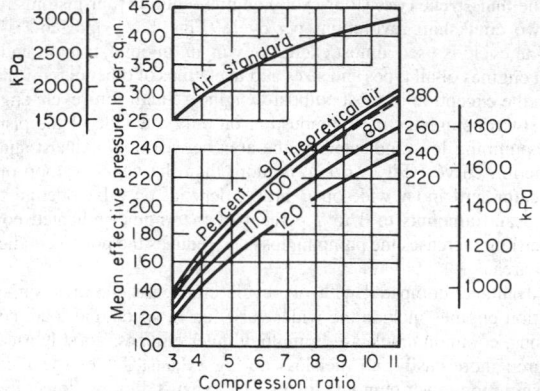

Fig. 9.6.8 MEP for Otto cycle engine [air and liquid iso-octane supplied at 60°F (16°C)].

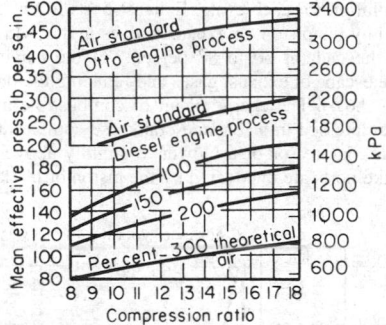

Fig. 9.6.9 MEP for diesel cycle engine [air and liquid dodecane supplied at 60°F (16°C)].

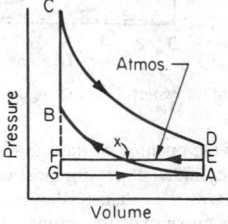

Fig. 9.6.10 Throttled Otto cycle.

Volumetric efficiencies of automobile engines (Fig. 9.6.11) diminish appreciably as piston speeds exceed 1,200 to 1,600 ft/min (6 to 8 m/s), decreasing to approximately 60 percent at top speeds of 2,500 to 3,200 ft/min (13 to 16 m/s). Volumetric efficiency varies appreciably with valve timing.

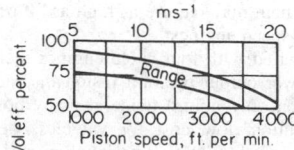

Fig. 9.6.11 Volumetric efficiencies of normally aspirated engines with poppet valves.

Table 9.6.1   Relative Ideal Volumetric Efficiencies with Liquid Fuels
(Air and liquid fuel at 60°F)

Fuel	Chemically correct air-fuel ratio	Heat required per lb of fuel for complete evaporation, Btu/lb (J/gm)	Percent evap with 200 Btu/lb fuel	Suction temperature, °F (°C)		Relative volumetric efficiency	
				Complete evaporation	200 Btu/lb fuel	Complete evaporation	200 Btu/lb fuel
Gasoline	15.11	330 (768)	81	106 (41)	81 (27)	100	100
Benzene	13.26	149 (347)	100+	50 (10)	68 (20)	111	103
Ethyl alcohol	8.99	425 (989)	55	71 (22)	55 (13)	107	105
Methyl alcohol	6.46	519 (1,207)	49	68 (20)	45 (7)	108	107

Two-cycle crankcase compression engines have volumetric efficiencies of about 60 percent at low speeds and 50 percent or below at high speeds.

Fuels with high latent heats and low boiling temperatures reduce the charge temperature and result in high volumetric efficiencies (Table 9.6.1).

Air cleaners, governors, and carburetors increase the resistance to flow and decrease the volumetric efficiency.

The **compression stroke** begins with heat transfer from the cylinder walls to the charge but ends with heat transfer from the charge to the walls. The net result is a heat loss of about 0.5 to 1.5 percent of the heat value of the charge. Heat loss during compression lowers the compression curve and work, but lowers the expansion work a greater amount, thus slightly reducing the net work output.

The mean value of the exponent n in the approximate equation $pv^n = $ constant for the polytropic compression of a correct mixture of octane and air is about 1.33. For the diesel cycle in which the gas under compression is mainly air, the exponent is about 1.38 for a 15 : 1 compression ratio. Heat transfer into the charge increases the exponent, but leakage lowers its value.

The **combustion process** begins before the end of the compression stroke and ends after the beginning of the expansion stroke. Rassweiler, Withrow, and Cornelius (*Trans. SAE,* **34,** 1939) found a maximum variation of 30 percent in combustion time for a single-cylinder engine while variations of both maximum cylinder pressure and rate of pressure rise of about 30 percent were found by Patterson (SAE paper 660129) in a V-8 engine. These effects are thought to be due primarily to cyclic variations in swirl and turbulence at the spark plug, but can arise also from variations in air/fuel ratio and homogeneity of the mixture. Poor distribution in multicylinder engines also causes variations between cylinders. It has been shown that the loss due to the actual finite combustion time is 3 to 6 percent of the ideal efficiency with constant-volume combustion. Any energy released before or after top-center piston position has an availability corresponding to the expansion ratio for the piston position at which the energy is released. Heat loss during the combustion process amounts to 5 to 7 percent of the heat supply of the charge at rated outputs and speeds for automotive engines.

Flame Travel   Flame begins at the spark plug in the spark ignition engine or at various points in the combustion chamber of the compression ignition engine and travels in all directions through the mixture. At the end of combustion, the first part of the charge to burn has reached a higher temperature than the last portion to burn, except in the constant-pressure process. Rassweiler and Withrow (*Trans. SAE,* **30,** 1935, p. 125) observed temperature differences of over 400°F (200°C) (non-knocking) and over 600°F (315°C) (knocking) in a nonturbulent combustion chamber.

The combustion of part of the charge compresses the remainder. Thus with 30 percent of the mass burned, the volume of the unburned portion will be about 35 percent of the total volume. This increases the temperature of the last fraction to burn in a constant-volume process by about 500°F (280°C) above its temperature at the beginning of combustion.

The mean adiabatic exponent for the **expansion process** is about 1.22, varying from about 1.20 at the beginning to 1.25 at the end of the process. Heat transfer and leakage increase the exponent, an average value of 1.33 being found by Rassweiler and Withrow (*Trans. SAE,* **33,** 1938, p. 185) in an L-head 2⅞- by 4¾-in (7.3- by 12.1-cm) cylinder. The heat transfer during the expansion stroke amounts to 8 to 12 percent of the heat value of the charge. Because of heat losses and real gas effects, the thermal efficiency of the engine is decreased by about 40 percent.

Blowdown begins at 80 to 90 percent of the expansion stroke. This reduces the work about 1 to 2 percent. High gas velocities [1,200 to 1,500 ft/s (360 to 460 m/s)] are attained, and heat transfer, including that through exhaust port walls, amounts to 10 to 20 percent of the heat value of the charge, but represents practically no loss of work or availability.

The **exhaust process** is the discharge of some of the gases from the cylinder by the piston. The exhaust pressure is usually above atmospheric (Fig. 9.6.12) but may run below atmospheric for part of the stroke. Heat loss during exhaust amounts to 3 to 5 percent of the heat

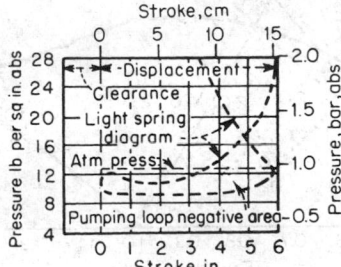

Fig. 9.6.12   Pumping loop of Otto cycle engine.

value of the charge but does not represent any loss of work or availability. At full load about 80 percent of the gases in the cylinder escape during blowdown, and about 15 percent are pushed out of the cylinder during exhaust, the remainder being left in the clearance space at the end of the exhaust stroke. At partial load, the blowdown pulse is smaller and the residual fraction higher.

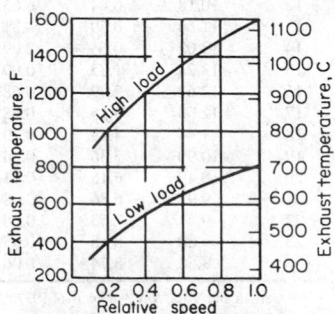

Fig. 9.6.13   Exhaust temperatures for Otto cycle engines.

Exhaust gas temperatures vary with speed and load (Figs. 9.6.13 and 9.6.14), high loads and speeds resulting in the highest temperatures. The first gases escaping during release are at the highest temperatures.

Exhaust Gas Analysis The composition of exhaust varies depending upon the equivalence ratio and the hydrogen/carbon ratio of the fuel. Figure 9.6.15 shows the calculated composition including water

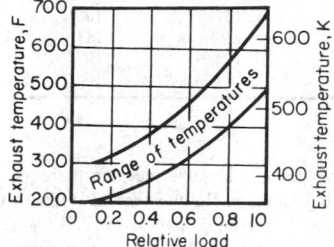

Fig. 9.6.14 Exhaust temperatures for diesel cycle engines.

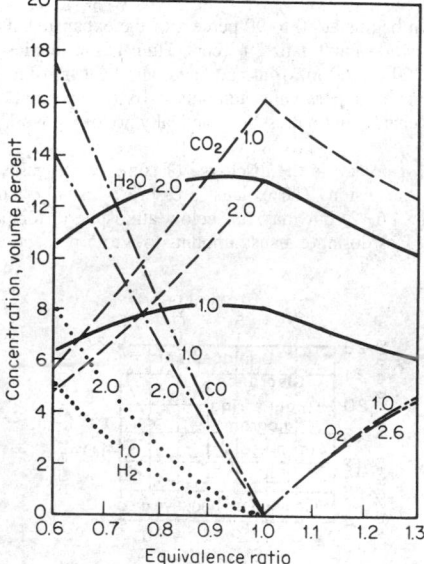

Fig. 9.6.15 Theoretical mole percent of principal combustion products of hydrocarbon fuels for fuel hydrogen/carbon ratios.

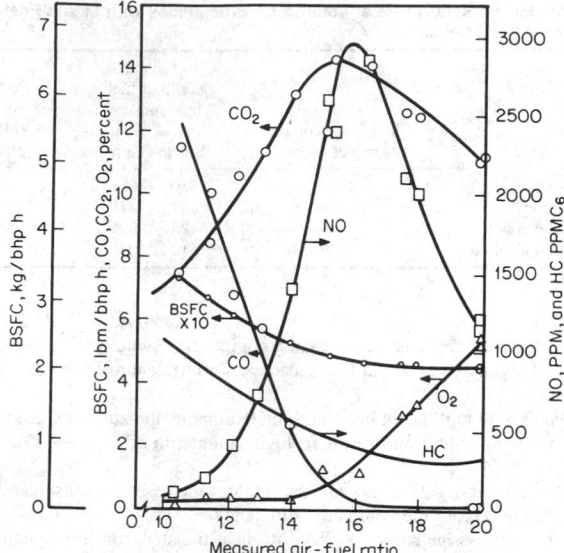

Fig. 9.6.16 Gasoline engine basic specific fuel consumption and emissions vs. air/fuel ratio.

(wet basis) for complete combustion. Charts of theoretical exhaust composition for various fuels have been prepared by D'Alleva (*General Motors Res. Rept.,* **372,** 1960) and Eltinge (*SAE Trans.,* 1968). In practice, nonequilibrium products, namely, unburned hydrocarbons and nitrogen oxide, appear in trace quantities. Figure 9.6.16 shows measured exhaust composition on a dry basis for a fuel with a hydrogen/carbon ratio of 1.85.

Equations for calculating the air/fuel ratio from exhaust CO, CO_2, HC, and O_2 have been developed by Spindt (*SAE Trans.,* 1966) and others. Such determinations and those for calculation of remaining but unmeasured exhaust species depend upon carbon, hydrogen, oxygen, and nitrogen balances employing measured values for some exhaust constituents (Obert, "Internal Combustion Engines").

Table 9.6.2 summarizes the principal exhaust gas constituents and combustion efficiency. HC and NO were not measured.

U.S. AUTOMOBILE ENGINES

The size (displacement) of U.S. automobile engines varies from 61 in³ (1.0 L) to 488 in³ (8 L). Rated horsepower and displacement have

Table 9.6.2 Summary of Exhaust Gas Constituents and Combustion Efficiency*

| Air | Percent by volume | | | | | | $\dfrac{H_2O}{CO_2}$ | $\dfrac{H_2O}{Fuel}$ | Combustion† |
Fuel	CO_2	O_2	CO	H_2	N_2	H_2O			eff., percent
11	8.76	0.15	9.14	4.66	77.08	13.78	1.57	0.972	66.7
12	10.18	0.44	6.65	3.39	79.13	13.93	1.37	1.043	73.8
13	11.60	0.59	4.31	2.20	81.09	14.16	1.22	1.122	81.5
14	13.02	0.63	2.09	1.07	82.99	14.46	1.11	1.205	89.6
15	13.23	1.35	0.99	0.50	83.72	14.09	1.06	1.247	93.8
16	12.62	2.49	0.68	0.35	83.65	13.30	1.05	1.256	94.8
17	12.00	3.55	0.48	0.25	83.51	12.54	1.05	1.261	95.5
18	11.45	4.49	0.30	0.16	83.39	11.88	1.04	1.267	96.2
19	10.90	5.36	0.20	0.10	83.23	11.25	1.03	1.269	96.5
20	10.40	6.15	0.11	0.06	83.07	10.68	1.03	1.272	96.9
21	9.92	6.86	0.08	0.04	82.90	10.16	1.03	1.271	96.9
22	9.44	7.55	0.06	0.03	82.71	9.65	1.02	1.268	96.8
23	9.00	8.18	0.05	0.03	82.53	9.19	1.02	1.266	96.7
24	8.60	8.74	0.06	0.03	82.37	8.78	1.02	1.264	96.6

* Gerrish and Voss, *NACA Rept.* 616, 1937.
† Low-heat-value basis.

trended downward while horsepower per unit of displacement has generally increased during the period from 1984 to 1994 (Fig. 9.6.17). Most U.S. automobiles are spark-ignited and are fueled with gasoline.

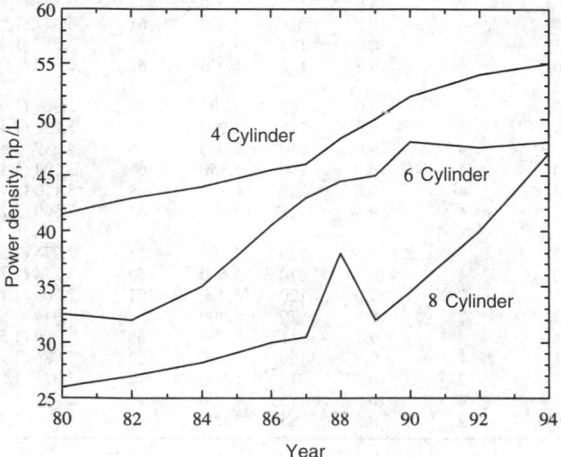

Fig. 9.6.17 Trend of average U.S. automobile power density.

Emission control has brought about (1) reduced compression ratios, (2) improved valves, (3) more precise and accurate fuel metering (carburetion and fuel injection), (4) controlled inlet air temperatures, and (5) electronically controlled fuel management (most with closed-loop feedback to maintain a chemically correct mixture), exhaust gas recirculation, and ignition timing. Current advertised horsepower is based on net rather than gross horsepower. Net horsepower is measured with a fully equipped engine, including all parts needed to perform its intended function unaided, such as fuel pump, oil pump, intake air system, and exhaust system.

Valve timing, particularly the closure of intake valves (approximately 50° crankshaft rotation after bottom dead center), is selected to give a rising volumetric efficiency and torque as the rotation speed is increased. At high speeds, restrictions in the intake system, developed primarily by the carburetor venturi and the intake valves, begin to throttle the engine. The 1994 base engines developed maximum torque at approximately 2,600 r/min. Despite reduced torque at higher speeds, the power peaks at 4,200 to 5,200 r/min. Higher powers were obtained with four-barrel (compound) carburetors, electronic fuel injection, and other modifications.

Engines have been produced with various numbers of cylinders over the years (from 2 to 16), but the most common engines in modern cars are in-line four-cylinder designs (Fig. 9.6.18) for smaller cars and V-6 and some V-8 designs for larger cars. All V-8 engines have an angle of 90° between the two banks of four cylinders, whereas the V-6 engines have an angle of either 60° or 90° between two banks of three cylinders. Smaller engines, in terms of both cylinder number and displacement, provide improved fuel economy, which is particularly important because of rigid federal corporate average fuel economy (CAFE) standards. Neither four- nor six-cylinder engines operate as smoothly as the V-8 because of less effective dynamic balancing and greater torque fluctuations. With the rapid growth in the use of transverse engines with front drive, and smaller and lighter vehicle structures with sloping aerodynamic front hoods, packaging constraints on automobile engines are severe.

All modern engines employ compact, rigid structures. Most cylinder blocks and heads are made of cast iron, although use of aluminum is increasing, particularly in cylinder heads. Four-cylinder engines use three or five crankshaft bearings; the V-6 uses four, and the V-8 uses five. Cylinder bore is usually greater than piston stroke.

Liquid cooling passages are provided around the bore of each cylinder as well as at the hot regions of the combustion chambers and exhaust ports in the head. Modern pistons are almost always made of aluminum. Crankshafts and connecting rods are usually made of cast iron or forged steel. Four-cylinder engines and some six-cylinder engines use a sepa-

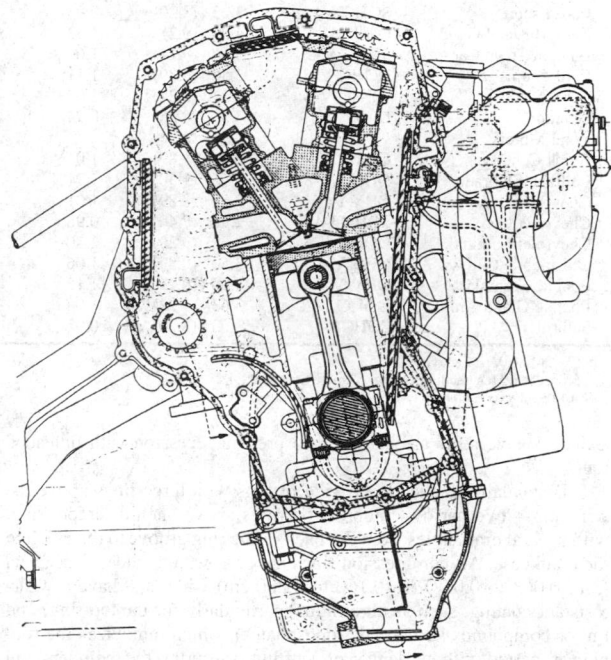

Fig. 9.6.18 Oldsmobile Division Quad 4 with four valves per cylinder (1987).

rate crank throw (eccentric portion of the crankshaft) for each connecting rod, whereas in V-8s, connecting rods for corresponding cylinders on each side of the V are side by side on the same crank throw.

Detachable cylinder heads contain the inlet and exhaust valves (overhead valves). Most engines employ a single exhaust and inlet valve, although there is a growing trend toward designs with three or four valves per cylinder (Table 9.6.3). Many four-cylinder engines use a belt-driven overhead camshaft, although most V-6s and V-8s still have a camshaft in the valley between the cylinder banks. Exhaust gas is directed through short passages to the exhaust collector or manifold.

Fuel and air are delivered to the head by a compact inlet manifold. Fuel management is provided by either a carburetor or an electronic fuel injection system which may be either multipoint (one injector for each inlet port) or single-point (throttle body injection). In most cases, feedback from an exhaust gas composition sensor is used to provide precise control of air/fuel ratio and optimize performance of the emission control system. Electronic ignition virtually eliminates the need for periodic ignition system maintenance. Manufacturing precision has been improved considerably, leading to improved overall quality, reduced friction, and greater reliability.

Quiet operation is essential in automobile engines and requires close manufacturing control of clearances. Most engines have hydraulic valve lifters and means of rotating the exhaust valves, at least, to prolong life.

Table 9.6.3 Selected U.S. Automobile Engines

Make and model	Type and no.*	Bore, in (cm)	B/S ratio	Displacement, in³ (L)	No. of valves per cyl.	Comp. ratio	bhp	r/min	bhp/L	Car weight, lb (kg)
Chrysler Corp.										
Chrysler LeBaron	SOHC V-6	3.55 (9.02)	1.19	181 (3.0)	2	8.9	141	5,000	47	3,122 (1,416)
Chrysler New Yorker	SOHC V-6	3.78 (9.60)	1.19	215 (3.5)	4	10.5	214	5,800	61	3,457 (1,568)
Dodge Intrepid	OHV V-6	3.66 (9.30)	1.15	201 (3.3)	2	8.9	153	5,300	46	3,217 (1,459)
Plymouth Neon	SOHC I-4	3.45 (8.76)	1.06	122 (2.0)	4	9.8	132	6,000	66	2,320 (1,052)
Ford Motor Co.										
Ford Escort	SOHC I-4	3.23 (8.20)	0.93	114 (1.9)	2	9.0	88	4,400	46	2,304 (1,045)
Ford Mustang GT	OHV V-8	4.00 (10.2)	1.33	302 (5.0)	2	9.0	215	4,200	43	3,258 (1,478)
Lincoln Town Car	SOHC V-8	3.60 (9.14)	1.00	281 (4.6)	4	9.0	210	4,600	46	4,039 (1,832)
Ford Taurus†	DOHC V-6	3.50 (8.89)	1.11	182 (3.0)	4	10.0	200	5,750	67	3,104 (1,408)
Tarus SHO†	DOHC V-8	3.24 (8.24)	1.04	207 (3.4)	4	10.0	235	6,100	69	3,150 (1,429)
Contour†	DOHC I-4	3.30 (8.38)	0.94	121 (2.0)	4	9.6	125	5,500	63	2,500 (1,135)
General Motors										
Buick Century	OHV I-4	3.50 (8.89)	1.01	133 (2.2)	2	9.0	120	5,200	55	2,974 (1,349)
Cadillac Eldorado	DOHC V-8	3.66 (9.30)	1.11	279 (4.6)	4	10.3	270	5,600	59	3,774 (1,712)
Chevrolet Cavalier	OHV I-4	3.50 (8.89)	1.01	133 (2.2)	2	9.0	120	5,200	55	2,509 (1,138)
Chevrolet Geo†	SOHC I-4	2.91 (7.40)	0.98	79 (1.3)	2	9.5	70	5,500	54	1,949 (884)
Chevrolet Lumina†	DOHC V-6	3.62 (9.20)	1.10	207 (3.4)	4	9.7	215	5,200	63	3,309 (1,501)
Oldsmobile Cutlass	OHV V-6	3.50 (8.89)	1.06	191 (3.1)	2	9.5	160	5,200	52	3,086 (1,400)
Pontiac Formula	OHV V-8	4.00 (10.2)	1.15	350 (5.7)	2	10.5	275	5,000	48	3,425 (1,553)
Pontiac Grand Am	DOHC I-4	3.63 (9.22)	1.08	138 (2.3)	4	10.0	175	6,200	76	2,822 (1,280)
Saturn	SOHC I-4	3.23 (8.20)	0.91	116 (1.9)	2	9.3	85	5,000	45	2,314 (1,050)

* I = in-line, V = vee type.
† From 1996 Manufacturers Motor Vehicle Specifications.
SOURCE: *Wards Automotive*, 1994; *Auto Ind.*, July 1994.

Exhaust valve seats are hardened to prevent recession with unleaded fuel.

It is standard practice to provide pistons with three rings above the wrist pin—two narrow compression rings above one oil scraper ring, with several drain holes from the back of the ring groove to return oil to the crankcase. All compression rings are narrow, are made of cast iron from 0.063 to 0.0787 in (0.160 to 0.200 cm) wide, and have a wear-resistant coating, such as chromium (particularly for the top ring) and tin, or compounds (iron oxide, phosphates), which minimizes the running-in period without danger of scuffing or scoring the cylinders. Oil scraper rings are of steel or cast iron, generally are a little less than 3/16 in (0.476 cm) wide, and are provided with internal expander springs to increase pressure against the cylinder walls. Many oil scraper rings, especially those of steel, also use wear-resistant coatings.

Liquid coolant under a pressure of 15 lb/in² (103 kPa) is standard for U.S. cars. Coolant jacket temperatures are controlled by thermostats which open at approximately 195°F (90.6°C), permitting coolant from the cylinder heads to flow to the radiator. All engines either are designed with coolant all around each cylinder or are "siamesed" and with jackets the full length of the piston travel.

FOREIGN AUTOMOBILE ENGINES

Foreign automobile engines sold in the United States are increasingly similar to engines produced in the United States. (See Table 9.6.4.) These similarities include size and basic design features. They are mostly four-stroke, Otto cycle engines; the rest are diesels. All engines are liquid-cooled. Engines produced for areas of the world other than

Table 9.6.4 Selected Foreign Automobile Engines

Make and model	Type and no.*	Bore, in (cm)	B/S ratio	Displacement, in³ (L)	No. of valves per cyl.	Comp. ratio	bhp	r/min	bhp/L	Car weight, lb (kg)
Audi 90 and 100	DOHC V-6	3.25 (8.26)	0.96	169 (2.8)	4	10.3	172	5,500	61	3,400 (1,542)
BMW 325i	DOHC I-6	3.31 (8.41)	1.12	152 (2.5)	4	10.5	189	5,900	76	3,087 (1,400)
Chrysler Colt	SOHC I-4	3.19 (8.10)	0.91	112 (1.8)	4	9.5	113	6,000	63	2,195 (996)
Ford Aspire	SOHC I-4	2.79 (7.09)	0.85	81 (1.3)	2	9.7	63	5,000	48	2,004 (909)
Honda Corp.:										
Accord	SOHC I-4	3.35 (8.50)	0.90	132 (2.2)	4	8.8	130	5,300	59	2,800 (1,270)
Civic	SOHC I-4	2.95 (7.49)	0.89	91 (1.5)	4	9.2	102	5,900	68	2,224 (1,009)
Mazda 626	DOHC I-4	3.30 (8.38)	0.92	122 (2.0)	4	9.0	118	5,500	59	2,606 (1,182)
Mercedes 220	DOHC I-4	3.54 (9.0)	1.04	134 (2.2)	4	9.8	110	5,500	50	3,150 (1,429)
Nissan Centra	DOHC I-4	2.99 (7.60)	0.86	97 (1.6)	4	9.5	110	6,000	69	2,346 (1,064)
Porsche 911	SOHC H-6	3.94 (10.0)	1.31	220 (3.6)	2	11.3	247	6,100	69	3,031 (1,375)
Saab 900	DOHC I-4	3.54 (9.0)	1.00	140 (2.3)	4	10.5	155	4,300	67	2,950 (1,338)
Subaru Legacy	SOHC H-4	3.82 (9.70)	1.29	135 (2.2)	2	9.5	130	5,600	59	2,825 (1,280)
Toyota:										
Camry	DOHC I-4	3.43 (8.71)	0.96	134 (2.2)	4	9.5	130	5,400	59	3,086 (1,400)
Tercel	SOHC I-4	2.87 (7.29)	0.84	89 (1.5)	3	9.3	82	5,200	55	1,975 (896)
VW Jetta	SOHC I-4	3.25 (8.26)	0.89	121 (2.0)	4	10.0	115	5,400	58	2,647 (1,200)
Volvo 850	DOHC I-5	3.32 (8.43)	0.92	149 (2.4)	4	10.5	168	6,200	70	3,280 (1,488)

* I = in-line, V = vee type.
SOURCE: *Ward's Automotive*, 1994; *Auto Ind.*, July 1994.

the United States exhibit a broader range of characteristics. Very small cars with engines of less than 40 in³ (0.66 L) and 35 hp are found in some areas. Engines produced for the U.S. market are slightly larger, on average, in terms of piston displacement, than those used elsewhere. Some European engines, however, have very high specific output in terms of horsepower per unit displacement because of taxation based on engine displacement, higher vehicle speed limits, and less severe pollution requirements. Many countries tax larger engines at a higher rate than smaller engines.

Most engines are normally aspirated, although the use of turbocharging is increasing. Displacements range from three cylinders, 61 in³ (1.0 L); four cylinders, 82 to 156 in³ (1.35 to 2.6 L); five-cylinders, 109 to 183 in³ (1.8 to 3.0 L); six cylinders, 152 to 258 in³ (2.5 to 4.26 L); eight cylinders, 149 to 412 in³ (2.46 to 6.8 L).

The maximum horsepower of foreign automobile engines ranges from approximately 50 to over 400 hp, equivalent to the range for U.S. engines. For a number of years, power has decreased because of emphasis on fuel economy and emission controls, but power levels have once more increased in response to consumer demand.

A number of foreign engines were considered "high tech" engines, with such features as three and four valves per cylinder, aluminum cylinder heads and blocks, and lightweight precision components. These features are now common in all cars produced worldwide. See Tables 9.6.3 and 9.6.4.

TRUCK AND BUS ENGINES

Truck and bus engines (Figs. 9.6.19 and 9.6.20) are similar to automobile engines but, in general, are larger, having 300- to 1,000-in³ (4.6- to 16-L) displacement, are more rugged, and run at lower speeds. Both Otto and diesel cycles are used, with the diesel predominant in the larger sizes. The two-cycle blower scavenged diesel engine with or without a turbocharger is in common use (Fig. 9.6.20). Water cooling is universally used. The gasoline engines are naturally aspirated and have compression ratios slightly lower than automobile engines but are valved

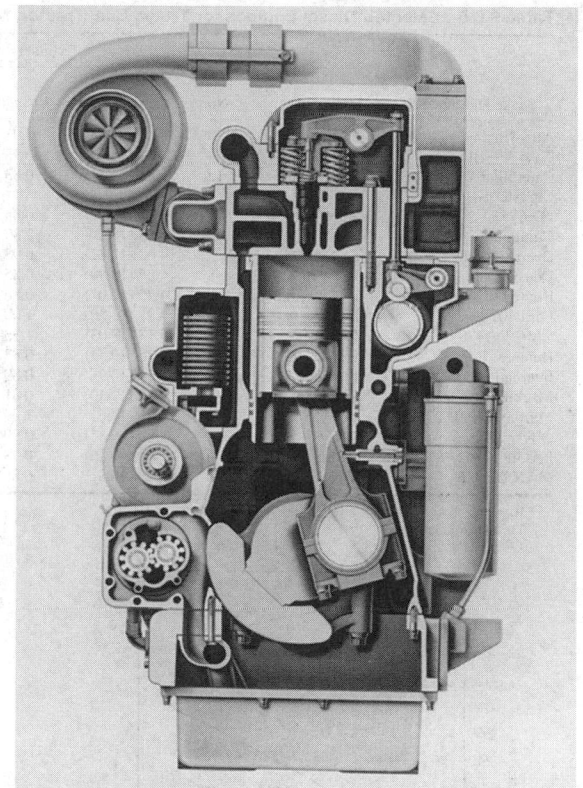

Fig. 9.6.19   Cummins model KT 450-hp turbocharged, four-stroke diesel engine.

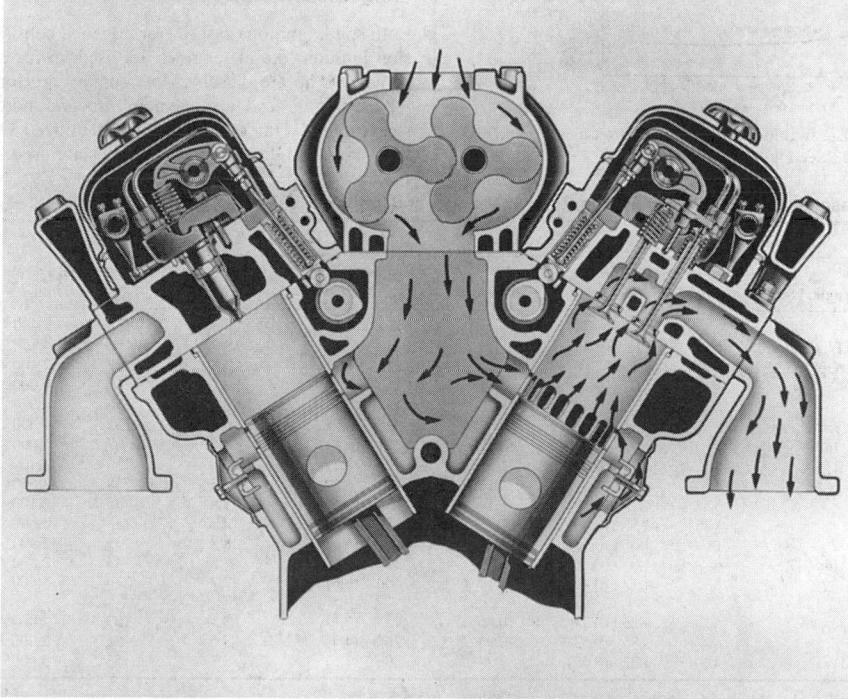

Fig. 9.6.20   Blower-scavenged Detroit diesel model 8V-71N two-stroke diesel engine.

Table 9.6.5   Selected Diesel Engines for Truck, Bus, Tractor, Marine, and Industrial Uses

Make and model	Designed for*	No.	Bore, in (cm)	B/S ratio	Displacement, in³ (cm³)	Compression ratio	Max continuous bhp	at r/min	Max torque, lb·ft (N·m)	at r/min
Allis-Chalmers 2800	Tr, Ind	6	3.88 (9.85)	0.91	301 (4,932)	16.25	68	2,200	225 (305)	1,400
Case 336 BD	Tr, Ind	4	4.63 (11.76)	0.93	336 (5,506)	16.5	72	2,200	281 (381)	1,400
Caterpillar D 348	T, Ind	12	5.41 (13.74)	0.83	1,786 (29,267)	16.5	725	1,800	2,840 (3,850)	1,500
Chrysler Nissan IN675	Ind	6	4.33 (10.99)	0.84	452 (7,407)	16	168	1,800	664 (900)	1,400
Cummins V-785-C	Ind	8	5.50 (14.0)	1.33	785 (12,864)	15.9	220	2,300	560 (759)	1,600
Deere 3164 D	Tr, Ind	3	4.02 (10.21)	0.93	164 (2,687)	16.3†	40	2,200	122 (165)	1,500
Detroit Diesel 8V-71T	T, Tr, Ind	8	4.25 (10.80)	0.85	568 (9,308)	17†	350	2,100	965 (1,312)	1,600
Deutz F6L-714	T, B, Tr, Ind	6	4.75 (12.07)	0.86	579 (9,488)	17.8	117	2,000	376 (510)	1,300
Ford 401-D	Tr	6	4.41 (11.20)	1.00	401 (6,571)	16.5	108	2,300	284 (385)	1,400
GMC DH478	T, B	6	5.13 (13.03)	1.33	478 (7,833)	17.5	155	2,800	337 (457)	2,000
Hercules D-3000	Tr, Ind	6	3.75 (9.53)	0.83	298 (4,883)	17.5	85	2,400	244 (331)	1,400
International DT-407	Tr, Ind	6	4.33 (10.99)	0.89	407 (6,670)	17†	130	2,400	393 (533)	1,800
Mack END 673E	T	6	4.88 (12.40)	0.81	672 (11,012)	16.11	168	2,100	540 (732)	1,400
Murphy MP-321	Ind	6	5.75 (14.61)	1.09	1,013 (16,600)	16	207	1,400	946 (1,283)	800
Oliver 1855	Tr	6	3.88 (9.85)	0.89	310 (5,080)	16	98	2,400	298 (404)	1,600
Perkins 6-354	Tr, Ind	6	3.88 (9.85)	0.78	354 (5,801)	16	88	2,250	288 (390)	1,400
MAN DO826	B	6	4.25 (10.8)	0.86	419 (6,871)	16	270	2,400	710 (960)	1,500

* B = bus; T = truck; Ind = industrial; Tr = tractor.
† Turbocharged.

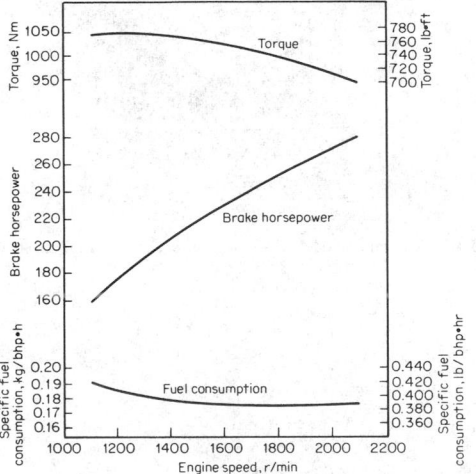

Fig. 9.6.21 Performance data for blower-scavenged Detroit diesel model 8V-71N two-stroke diesel engine. (See Fig. 9.6.20.)

similarly, have similar performance curves, and attain maximum brake horsepower (bhp) (55 to above 300) at a mean piston speed averaging about 2,600 ft/min (13 m/s). Maximum torque is obtained at about 1,500 r/min, appreciably lower than for automobile engines. Diesel engines, which are often supercharged and/or turbocharged, have piston speeds ranging from 1,500 to 2,100 ft/min (7.6 to 11 m/s) and operate at somewhat lower speeds than gasoline engines. Truck and bus gasoline engines can operate at speeds slightly higher than those at which maximum power is attained. Some diesel engines can be operated near the maximum power speed. Most are operated below (Fig. 9.6.21).

Truck and bus engines are usually built with four or six cylinders in line or with eight cylinders in V-type construction. The bore/stroke ratio ranges from about 0.75 to 1.4, the latter engines having the larger values (see Tables 9.6.5 and 9.6.6).

TRACTOR ENGINES

Both spark ignition (gasoline or liquefied petroleum gas) and compression ignition (diesel) engines are used in tractors. Almost all engines above 100 hp are diesels. Most engines are four-cycle, water-cooled, with valve-in-head arrangement. Integral bore, dry-sleeve, or wet-sleeve cylinder liner arrangements are utilized. One-, two-, three-, four-, five-, six-, and eight-cylinder engines are used. All engines except the

Table 9.6.6   Selected Gasoline Engines for Truck, Bus, Tractor, Marine, and Industrial Uses

Make and model	Designed for*	No.	Bore, in (cm)	B/S ratio	Displacement, in³ (L)	Compression ratio	Rating† bhp	at r/min	Max torque, lb·ft (N·m)	at r/min
Allis-Chalmers G-153	Ind	4	3.44 (8.74)	0.83	153 (2.5)	8	52	2,400	132 (179)	1,460
Jeep Cherokee	T, Tr, Gp, Ind	I-6	3.88 (9.53)	1.13	242 (4.0)	8.8	190	4,750	215 (292)	1,400
Case 201 G	Tr, Gp, Ind	4	4.0 (10.16)	—	251 (4.1)	7.5	87	2,400	221 (300)	1,400
Chevrolet 350	Gp	V-8	4.0 (10.16)	1.15	350 (5.7)	9.1	210	4,000	—	
Chevrolet 454	T	8	4.25 (10.79)	1.06	454 (7.4)	8.3	250	4,000	365 (495)	2,800
Chrysler H-225	M, Ind	6	3.41 (8.66)	0.83	225 (3.7)	8.4	132	4,000	203 (275)	2,000
Continental R 821-46	Tr, Ind	4	3.03 (7.70)	0.92	95.5 (1.6)	8.6	67.2	5,000	84 (114)	2,400
Diamond Reo 8-250	T, B	8	4.24 (10.79)	1.03	468 (7.7)	7.5	250	3,400	420 (569)	2,400
Dodge CT 900	T	8	4.19 (10.64)	1.12	413 (6.8)	7.5	238	3,600	355 (481)	2,000
Ford 158-G	Tr	3	4.20 (10.67)	1.10	158 (2.6)	7.8	45.7	2,100	120 (163)	1,350
Ford Aerostar	Gp	V-6	3.5 (8.89)	1.11	182 (3.0)	9.2	135	4,600	—	
Ford F150	T	I-6	4.0 (10.16)	1.00	300 (4.9)	8.8	145	3,400	—	
Hercules G-3400	Tr, Ind	6	4.0 (10.16)	0.89	339 (5.6)	7.5	143	2,800	311 (422)	1,400
International V5-401	T	8	4.13 (10.49)	1.10	400 (6.6)	7.7	226	3,600	355 (481)	2,000
Minneapolis-Moline HD 425	Tr	6	4.25 (10.79)	0.85	425 (7.0)	8.1	133	2,000	378 (512)	800
Oliver 1655	Tr	6	3.75 (9.53)	0.94	265 (4.4)	8.5	84	2,200	220 (298)	1,300
Volkswagen 124A	Ind	4	3.38 (8.59)	1.24	96.7 (1.6)	7.7	53	3,600	75.5 (102)	2,500

* B = bus; T = truck; Tr = tractor; M = marine; Gp = general purpose; Ind = industrial.
† hp without accessories.

six-cylinder are usually in-line in order to minimize the width of the hood in the interest of driver visibility. Tractor applications have a high load factor, are relatively low-speed, and deliver high low-speed torque (Fig. 9.6.22). Future off-road engines are expected to be emission-controlled.

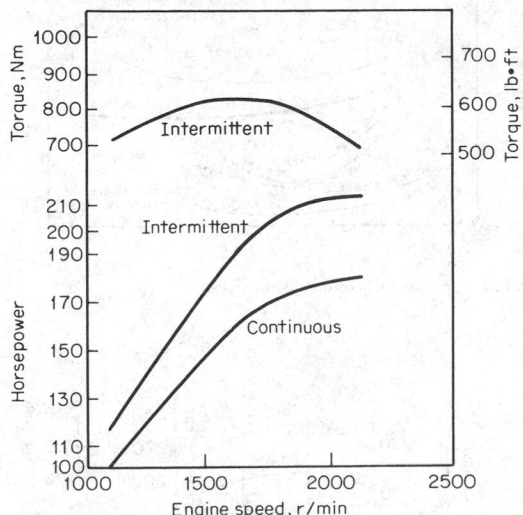

Fig. 9.6.22 Performance of John Deere model 6351 A, a diesel engine.

STATIONARY ENGINES

Stationary engines may be either Otto or diesel cycle, use either liquid or gaseous fuel, and use the two- or four-stroke cycle. The power output ranges up to about 48,000 bhp per engine. Piston rods, which permit the crankcase to be isolated from blowby gases and the bottom face of the piston to function as an air pump, are commonly used. Nearly all engines are now single-acting. In the larger sizes, cylinders are built up of several parts, cylinder lines are used, and long pistons for single-acting diesel engines usually have about five piston rings. The valves are usually massive and in the larger sizes may be cooled. The exhaust pipe is sometimes water-cooled, and the pistons for large-bore engines are always liquid-cooled with oil or water. In general, there is much less integral construction (Fig. 9.6.23) in the large stationary engine than in the automobile type.

An interesting medium-speed, four-stroke diesel engine, V-type design with rotating pistons, is illustrated in Fig. 9.6.23. The symmetric piston is cooled with lubricating oil, and besides the reciprocating motion, it performs a slow rotating motion around its longitudinal axis. At each stroke, a new oil-wetted portion of the piston is always in contact with the pressure side of the liner, thus minimizing wear, oil consumption, and local overheating of the liner resulting from blowby.

Stationary engines usually operate at constant speed and are governed by throttling the charge of the Otto engine and by varying the amount of fuel injected into the diesel engine. Compression ratios as low as 6:1 are used with spark ignition engines and as low as 12:1 with compression ignition engines. Stationary engines are usually built with 1 to 12 cylinders, with cylinders in a vertical line, or with a V-type arrangement having up to 16 or more cylinders.

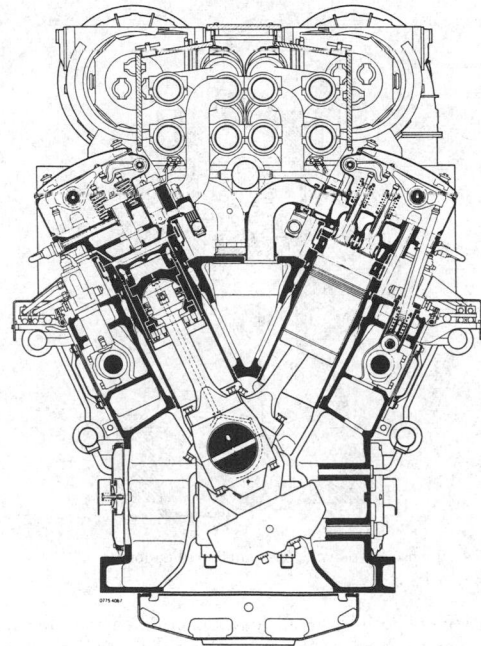

Fig. 9.6.23 Cross section of Sulzer V-type ZV-40 engine. Bore = 400 mm; stroke = 480 mm.

MARINE ENGINES

(See also Sec. 11.3)

Marine engines may be either Otto or diesel cycle and may be normally aspirated or supercharged. Outboard engines range in power from less than 1 to more than 200 bhp and are generally of the two-cycle gasoline type. Modified automotive gasoline engines are also used in larger boats. Diesel engines range in power from below 20 to 48,000 bhp and may be either two- or four-cycle. Both automotive and stationary types are used in marine work; consequently, wide variations in specific weights are found. (See Tables 9.6.5, 9.6.6, and 9.6.7.)

The outboard two-cycle engines (see Table 9.6.7) have crankcase compression (Fig. 9.6.24) and may have two ports, three ports, or two

Table 9.6.7 Selected Outboard Engines

Make and model	Cylinders Type and no.*	Bore, in (cm)	B/S ratio	Displacement, in³ (cm³)	Rating, bhp at r/min		Weight, lb (kg)
Evinrude Yachtwin 6	I-2	1.937 (4.92)	1.14	10 (164)	6	5,000	56 (25.5)
Evinrude 25	I-2	3.000 (7.62)	1.33	31.8 (521)	25	5,000	117 (53.6)
Evinrude 90	V-4	3.500 (8.89)	1.35	99.6 (1,632)	90	5,000	301 (136)
Yamaha Force 9.9	I-2	2.250 (5.71)	1.16	14.2 (232)	9.9	5,050	91 (41.5)
Yamaha Force C85	I-3	3.313 (8.41)	1.18	69.6 (1,140)	85	5,000	248 (113)
Johnson 15	I-2	2.375 (5.56)	1.35	15.6 (255)	15	5,500	77 (35)
Johnson 225	V-6	3.685 (8.89)	1.29	183.3 (3,000)	225	5,500	470 (212)
Mercury 8	I-2	2.13 (5.4)	1.20	12.8 (210)	8	5,000	69 (31)
Mercury 100	I-4	3.5 (8.9)	1.19	113 (1,848)	100	5,000	348 (158)
Mercury 200	V-6	3.5 (8.9)	1.32	153 (2,507)	200	5,500	422 (192)

* I = in-line, V = vee type.
SOURCE: Manufacturer's publications.

ports with a rotary or reed crankcase inlet valve for the highest outputs (Fig. 9.6.25). Otto four-cycle engines have either L-head or valve-in-head construction, and diesel four-cycle engines are invariably of valve-in-head construction.

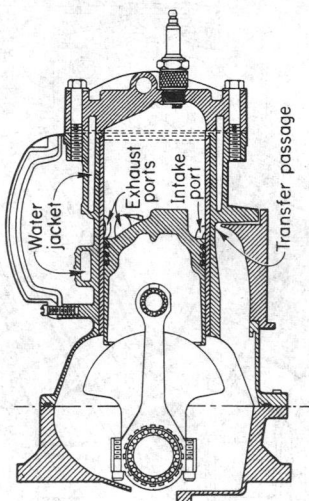

Fig. 9.6.24 Johnson two-cycle, 30-hp outboard motor.

Regular gasoline is required for Otto cycle engines, the fuel and lubricating oil being mixed for outboard two-cycle engines. A comparison of a number of fuel-consumption curves for various types of engine shows the large low-speed marine diesel engine as having the lowest specific fuel consumption (Fig. 9.6.26).

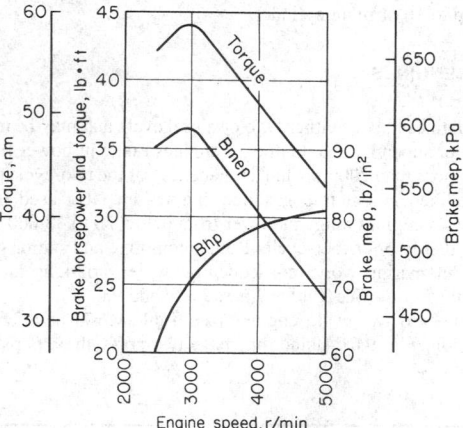

Fig. 9.6.25 Performance of Johnson outboard motor.

Low-speed engines are directly connected to the propellers, and high-speed engines are connected through reduction gearing. Outboard engines have maximum speeds of 4,500 to 6,000 r/min with the piston speeds ranging from 1,000 to about 2,500 ft/min (5.0 to 12.7 m/s).

Maximum engine speed is obtained at the intersection of the propeller/horsepower and the maximum horsepower curves (Fig. 9.6.27). This may occur well below the speed for maximum horsepower or beyond this speed. The effect of throttling to reduce speed is to increase the specific fuel consumption considerably.

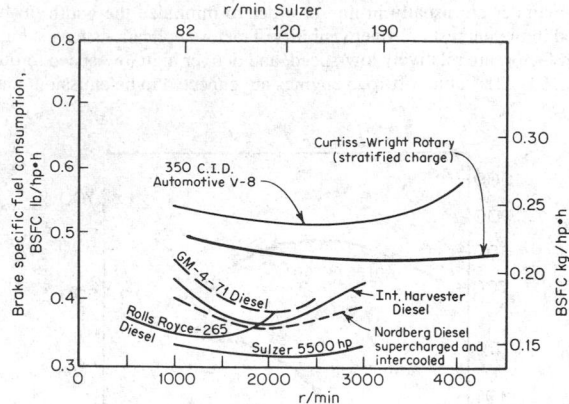

Fig. 9.6.26 Specific fuel consumption curves.

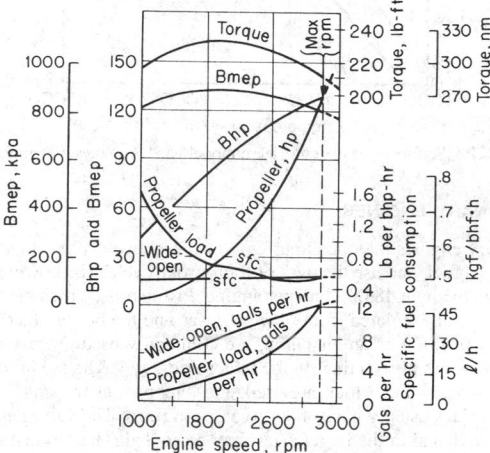

Fig. 9.6.27 Performance of six-cylinder $3\frac{7}{16} \times 4\frac{3}{4}$ in (8.73 × 12.07 cm) Chrysler marine engine with 7 : 1 compression ratio.

SMALL INDUSTRIAL, UTILITY, AND RECREATIONAL VEHICLE GASOLINE ENGINES

The Otto cycle engine with two or fewer cylinders is predominant in this class (Table 9.6.8). Fuel economy is generally not a major concern, whereas simplicity and performance are important. For reciprocating engines, the L-head design is predominant. Both high- and low-speed engines are used, depending on the application. Engines of this type generally employ air cooling of the cylinder and charge cooling of the interior. Water or liquid cooling is used on a few high-output small engines. Higher-mep air-cooled engines used forced fan cooling. Two-stroke engines are often used in applications where minimum weight and cost are important.

Carburetion, ignition systems, and other auxiliary equipment are usually less sophisticated than those on larger engines. A given engine is often used in a wide range of applications.

Engines used in motorcycles and in all-terrain vehicles are generally very sophisticated and constructed with great precision. Features such as overhead valves and overhead cams are common. The specific output of these engines is often high as measured by power per unit of piston displacement.

Table 9.6.8 Selected Small Gasoline Engines

Make and model	Cycles	Cylinders						Valve location*	Continuous rating, bhp at r/min	Max torque, lb·ft (N·m), at r/min	Weight, lb (kg)
		Type	No.	Bore, in (cm)	B/S ratio	Displacement, in³ (cm³)	Compression ratio				
Briggs & Stratton 100900	4	V	1	2.5 (6.50)	1.17	10.43 (171)	6.2	L	3.40 3,600	5.9 (8.0) 3,100	30.5 (13.9)
Briggs & Stratton 80400	4	H	1	2.38 (6.04)	1.36	7.75 (127)	6.2	L	2.55 3,600	4.6 (6.2) 3,100	25.25 (11.9)
Chrysler 500	2	H, V	1	2.00 (6.45)	1.26	5.00 (82)	6.5		3.33 7,000	3 (4.1) 5,000	8.75 (3.9)
Homelite XL-123	2	H	1	1.81 (4.60)	1.31	3.60 (59)	8.3				13.0 (5.9)
Kohler MV18	4	H, V	2	3.12 (7.92)	1.13	42.2 (692)	6	L	18 3,600	26 (35) 2,600	130 (59)
O & R 13A	2	V	1	1.25 (3.18)	1.15	1.34 (22)	9		1.0 6,300	0.88 (1.2) 5,200	3.75 (1.70)
Onan AI	4	V	2	2.75 (6.99)	1.10	14.90 (244)	6.25	L	3.86 3,600	8 (10.8) 2,100	150 (68.2)
Rockwell ZF-400	2	V	2	2.56 (6.50)	1.00	24.29 (398)	7.5		24.0 6,250	28 (38) 6,250	62 (28.2)
Tecumseh LAV 35-1A	4	V	1	2.50 (6.35)	1.36	9.06 (149)	—	L	2.27 2,400	5.25 (7.12) 3,100	21.5 (9.77)
Tecumseh HH160	4	H	1	3.5 (8.89)	1.22	27.66 (453)	8.7	L	16.0 3,600	25 (33.9) 2,400	84 (38.2)
WE400 Wisconsin TRA-10D	4	V	1	3.3 (8.4)	1.20	23.7 (388)	6.5	L	10.0 3,400	16.5 (22.4) 2,800	80 (36.4)
Fichtel & Sachs KM 914	4	H	1	—	—	18.5 (303)	8	RC	20 5,000	—	56 (25.5)

* l = valve in head; L = L head; RC = rotary combustion.
SOURCE: *Auto. Ind.*, **48**, no. 7, Apr. 1, 1973, pp. 104–106.

Table 9.6.9 Selected Medium-Size Diesel Engines*

Make and model	Cylinder arrangement†	Cycles	Cylinders				Compression ratio	Max rating, bhp. at r/min	Weight, lb (kg)
			No.	Bore, in (cm)	B/S ratio	Displacement, in³ (L)			
Alco 251	V	4	12	9.0 (22.86)	0.86	8,016 (131.4)	11.5	3,150 1,000	35,581 (16,152)
DeLaval GVB-16	V	4	16	13.0 (33.0)	0.87	31,856 (522.0)	—	2,900 1,000	141,000 (64,005)
Electro-Motive 12-645	V	2	12	9.125 (22.0)	0.92	7,740 (128.4)	16	2,950 950	28,600 (11,350)
Fairbanks-Morse 12-38D 1/8	OP	2	12	8.13 (20.65)	0.81	12,443 (203.9)	16.1	3,840 900	45,000 (20,430)
White-Superior 40-VX-12	V	4	12	10.0 (25.4)	0.95	9,896 (162.2)	—	1,800 1,000	—

* "Diesel and Gas Turbine Catalog," 1994. All engines are turbocharged.
† V = V type; OP = opposed.

LOCOMOTIVE ENGINES

Diesel engines are used for locomotive service because of high thermal efficiency (overall diesel brake thermal efficiency of about 38 to 40 percent with the net locomotive traction efficiency of about 32 percent for a diesel locomotive, compared with 6 percent for a steam locomotive), high availability for service, elimination of destructive pounding of rails, and elimination of shutdown fuel losses. The locomotive diesels are usually lighter, have lower displacement per cylinder, but higher outputs per cubic inch of displacement than the corresponding stationary diesel engine. These engines are built with 4 to 20 cylinders. Usually one engine is used per locomotive, and one or more locomotives are used per train. On an experimental basis, locomotive diesel engines have used coal-water slurries and natural gas as fuel. The convenience and low cost of conventional liquid diesel fuel will cause liquid fuels to predominate into the foreseeable future. (See Table. 9.6.9.)

AIRCRAFT ENGINES

Small aircraft piston engines are air-cooled and invariably four-stroke. All types are naturally aspirated or supercharged. These engines have the lowest specific weight of piston engines (Table 9.6.10). Aircraft engines always have valve-in-head construction; they also have low specific fuel consumption, and the guaranteed minimum may be as low as 0.40 lb (0.18 kg)/(bhp · h). Maximum cruising bmep ranges from 100 to 150 lb/in^2 (690 to 1,030 kPa), while takeoff bmep ranges from 120 to 200 lb/in^2 (830 to 1,380 kPa), depending principally on the compression ratio, fuel, and supercharge.

Aircraft engines are valved to obtain high mean effective pressure at rated speed. High-output engines usually operate with compression ratios and supercharging levels that prohibit full-throttle operation at or near sea level with the specified fuel. Such engines can be operated at full throttle at or above a critical altitude, beyond which the power decreases with an increase in altitude. Below the critical altitude, the power decreases with a decrease in altitude, at constant intake-manifold pressure, because of the increase in exhaust back pressure and ambient temperature. Turbocharged engines, however, maintain essentially constant-power with constant manifold pressure from sea level to a critical altitude, which can be as high as 20,000 ft (6,100 m).

The performance of an aircraft engine varies with speed, manifold pressure, and altitude. At any given speed and with a fixed supercharger or blower ratio (supercharger r/min to engine r/min) and wide-open throttle, the bhp of an engine varies almost linearly with the density of the atmosphere. This characteristic makes it possible to estimate the power curves at altitude if the sea-level output is known. The power at 20,000-ft (6,100-m) altitude is about 50 percent of the sea-level wide-open-throttle output for any given speed.

Low specific fuel consumption is obtained by operating with reduced r/min for the desired power with a lean mixture setting, the optimum output at this point being about one-half to two-thirds of the normal rated output of the engine.

Aircraft engine cylinders are limited to a maximum diameter of about 6 in (0.15 m) by piston cooling and knocking difficulties that arise with large cylinders, high compression ratios, and high intake manifold pressures. The bore/stroke ratio is commonly above 1.0. The larger piston engine sizes are being replaced by turbine-type engines.

Piston aircraft engines are mostly of the horizontally opposed design (desirable profile) with an even number of cylinders, usually four, six, or eight. They are equipped with a dual ignition system, resulting in more reliable combustion.

WANKEL (ROTARY) ENGINES

The Wankel (rotary) SI engine is used in the Mazda RX series. A triangular rotor rotates on an eccentric shaft inside an epitrochoidal housing. The rotor tips are in constant contact with the housing and form three working chambers. The operating cycle (Fig. 9.6.28) follows the four-stroke cycle principle, although the output shaft rotates at 3 times rotor speed, providing one power stroke per crankshaft revolution (as in the single-cylinder two-cycle). A fixed gear meshes with an internal gear in the rotor to maintain the proper relationship between the rotor and housing. Housing cooling is by either water or air and rotor cooling by oil or fuel/air charge. Auxiliary equipment (fuel system, ignition system, etc.) is similar to that used on piston engines. A major problem is the sealing grid (90° intersection of seals and single-line seal at apexes of rotor), and this tends to produce high hydrocarbon emissions.

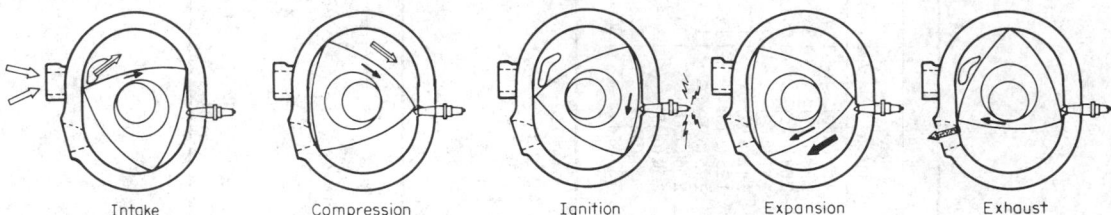

Intake   Compression   Ignition   Expansion   Exhaust

Fig. 9.6.28   Operating cycle of Wankel rotating combustion engine.

Table 9.6.10   Selected Piston-Type Aircraft Engines

Make and model	No.	Arrangements*	Bore, in (cm)	Stroke, in (cm)	Displacement, in^3 (cm^3)	Comp. ratio	bhp‡	r/min	At sea level (SL) or altitude, ft	bmep at max bhp, lb/in^2 (kPa)	Engine, dry, lb (kg)	Fuel octane no. required
Avco Lycoming												
0-320-E	4	H	5.13 (13.03)	3.88 (9.86)	319.8 (5,240.6)	7.0	150	2,700	SL	138 (952)	244 (110.4)	80/87
I0-360-A	4	H	5.13 (13.03)	4.38 (9.86)	361.0 (5,915.7)	8.7	200	2,700	SL	163 (1,120)	293 (132.6)	100/130
TIGO-541-E	6	H	5.13 (13.03)	4.38 (9.86)	541.5 (8,873.6)	7.3	425	3,200	15,000	195 (1,340)	700 (316.8)	100/130
Teledyne Continental												
0-200	4	H	4.06 (10.31)	3.88 (9.86)	201 (3,293.8)	7.1	100	2,750	SL	143 (986)	190 (86.0)	80/87
I0-470	6	H	5.0 (12.7)	4.0 (10.16)	471 (7,718.3)	8.6	260	2,600	SL	166 (1,140)	465 (210.5)	100/130
GTSIO-520-F	6	H	5.25 (13.34)	4.0 (10.16)	520 (8,521.3)	7.5	435	3,400	19,000	195 (1,340)	647 (292.8)	100/130

* H = horizontal (production of radial and vertical cylinder arrangements has been discontinued).
† METO = maximum except during takeoff.
‡ Takeoff power is usually higher than the recommended maximum continuous power.
§ Weights are for dry engines without hub or starter.

Table 9.6.11 Data on Fuel Properties

Fuel	Formula (phase)	Molecular weight	Specific gravity (density* kg/m³)	Heat of vaporization, kJ/kg†	Specific heat Liquid, kJ/(kg·K)	Specific heat Vapor c_p, kJ/(kg·K)	Higher heating value, MJ/kg	Lower heating value, MJ/kg	LHV of stoich. mixture, MJ/kg	$(A/F)_s$	$(F/A)_s$	Fuel octane rating RON	Fuel octane rating MON
Practical fuels													
Gasoline	$C_nH_{1.87n}(l)$	~110	0.72–0.78	305	2.4	~1.7	47.3	44.0	2.83	14.6	0.0685	92–98	80–90
Light diesel	$C_nH_{1.8n}(l)$	~170	0.84–0.88	270	2.2	~1.7	44.8	42.5	2.74	14.5	0.0690	—	—
Heavy diesel	$C_nH_{1.7n}(l)$	~200	0.82–0.95	230	1.9	~1.7	43.8	41.4	2.76	14.4	0.0697	—	—
Natural gas‡	$C_nH_{3.8n}N_{0.1n}(g)$	~18	(~0.79†)	—	—	~2	50	45	2.9	14.5	0.069	—	—
Pure hydrocarbons													
Methane	$CH_4(g)$	16.04	(0.72*)	509	0.63	2.2	55.5	50.0	2.72	17.23	0.0580	120	120
Propane	$C_3H_8(g)$	44.10	0.51 (2.0*)	426	2.5	1.6	50.4	46.4	2.75	15.67	0.0638	112	97
Isooctane	$C_8H_{18}(l)$	114.23	0.692	308	2.1	1.63	47.8	44.3	2.75	15.13	0.0661	100	100
Cetane	$C_{16}H_{34}(l)$	226.44	0.773	358	—	1.6	47.3	44.0	2.78	14.82	0.0675	—	—
Benzene	$C_6H_6(l)$	78.11	0.879	433	1.72	1.1	41.9	40.2	2.82	13.27	0.0753	—	115
Toluene	$C_7H_8(l)$	92.14	0.867	412	1.68	1.1	42.5	40.6	2.79	13.50	0.0741	120	109
Alcohols													
Methanol	$CH_3OH(l)$	32.04	0.792	1,103	2.6	1.72	22.7	20.0	2.68	6.47	0.155	106	92
Ethanol	$C_2H_6(l)$	46.07	0.785	840	2.5	1.93	29.7	26.9	2.69	9.00	0.111	107	89
Other fuels													
Carbon	$C(s)$	12.01	~2‡	—	—	—	33.8	33.8	2.70	11.51	0.0869	—	—
Carbon monoxide	$CO(g)$	28.01	(1.25*)	—	—	1.05	10.1	10.1	2.91	2.467	0.405	—	—
Hydrogen	$H_2(g)$	2.015	(0.090*)	—	—	1.44	142.0	120.0	3.40	34.3	0.0292	—	—

(l) = liquid phase; (g) = gaseous phase; (s) = solid phase; RON = research octane number; MON = motor octane number.
* Density in kg/m³ at 0°C and 1 atm.
† At 1 atm and 25°C for liquid fuels; at 1 atm and boiling temperature for gaseous fuels.
‡ Typical values.

SOURCES: E. M. Goodger, "Hydrocarbon Fuels: Production, Properties and Performance of Liquids and Gases," Macmillan, London, 1975. E. F. Obert, "Internal Combustion Engines and Air Pollution," Intext Educational Publishers, 1973. C. F. Taylor, "The Internal Combustion Engine in Theory and Practice," vol. 1, MIT Press, 1966. J. W. Rose and J. R. Cooper (eds.), "Technical Data on Fuel," 7th ed., British National Committee, World Energy Conference, London, 1977.

Advantages include small size, low weight, simple and fewer components (compared with piston-type four-cycle engines), superior breathing, no valves (rotor and its seals serve as valves), lower friction, low-octane fuel requirement, low NO emissions, and no reciprocating unbalance.

FUELS

Fuels for internal combustion engines are predominantly petroleum products. Natural and manufactured gases are also used in limited quantities. Data on properties of fuels which are commonly used for internal combustion engines are summarized in Table 9.6.11. Additional details on fuels are given in Sec. 7.1.

Gasoline

Gasoline consists of various amounts of many hydrocarbons, each having its vapor, pressure, and temperature characteristics. Also included are small amounts of additives such as knock suppressers, deposit modifiers, antioxidants, metal deactivators, antirust agents, anti-icing agents, detergents, upper cylinder lubricants, and dyes. Different grades of gasoline are marketed on the basis of octane number. The overall volatility of a gasoline sample is specified in U.S. industry by the ASTM Distillation Test (D86) and the ASTM Vapor Pressure Test (D323, Reid Method). The distillation test records the increasing temperature of the vapor in the neck of a flask vs. percent evaporated as 100 mL of fuel is gradually heated and distilled. The vapor pressure test gives the pressure in a split-chamber bomb which contains the fuel sample and room air in thermal equilibrium with a 100°F (311 K) temperature bath. The pressure, corrected to standard initial bomb air conditions, gives the **Reid vapor pressure** (**RVP**) of the gasoline. Typical results for summer- and winter-grade gasolines are given in Table 9.6.12.

In general, volatility is specified by giving the initial boiling point (IBP); the 10, 50, and 90 percent evaporation points; and the endpoint (EP) temperatures as well as the Reid vapor pressure. These data indicate the effects of fuel volatility on engine performance, e.g., ease of starting, warm-up, and fuel economy (Figs. 9.6.29 and 9.6.30.)

Reformulated Gasoline (RFG) The composition of gasoline is known to affect the exhaust and evaporative emissions. The C_4 and C_5 paraffins and olefins (photochemically active) are the main contributors to evaporative losses. Reformulated gasolines have much reduced amounts, and thus lower vapor pressures (Reid vapor pressure maximum of 48 kPa) and a maximum of 6 percent by volume olefins. Benzene, a toxic compound, is limited to a maximum of 0.85 percent by volume, while aromatics (photochemically active, deposits) as a class are limited to 25 percent by volume. Oxygenated fuel components (methanol, ethanol, etc.) are added at a level of 1.8 to 2.2 percent by weight. This reduces CO somewhat, especially from older vehicles without feedback fuel controls. For the federal RFG standards, sulfur is limited to 130 ppm; this lowers acid emissions and lengthens engine life. Further, the ASTM 90 percent distillation temperature is limited to 165°C. This minimizes deposits and leads to less enrichment required upon cold starting. Such a fuel may give fewer miles per gallon. The California Air Resources Board (CARB) applies even more stringent limits, with sulfur held to 30 ppm and the 90 percent distillation temperature limited to 143°C.

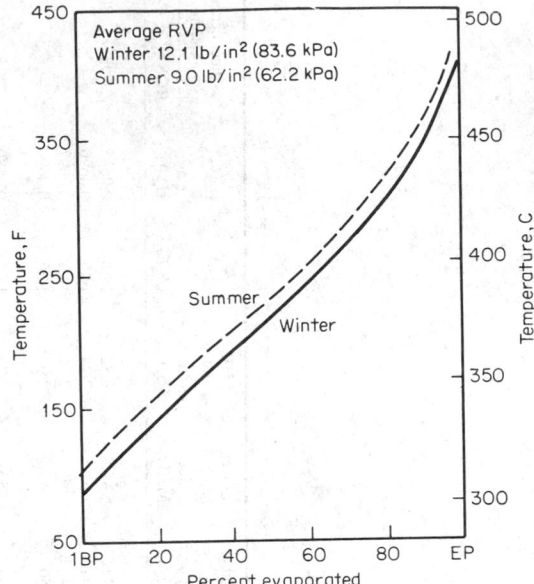

Fig. 9.6.29 Distillation curves for typical winter and summer gasolines. (*Ethyl Corp.*)

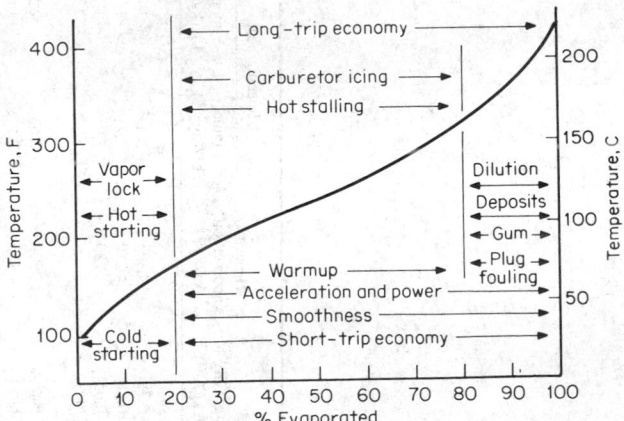

Fig. 9.6.30 Effects of volatility characteristics on engine performance.

Diesel Fuels Automotive and railroad diesel fuels are derived from petroleum refinery products which are referred to as **middle distillates;** the latter have a higher boiling range than gasoline. The properties of commercial distillate diesel fuels depend on the refinery practices employed and the nature of the crude oil from which they are derived, and thus vary substantially. Such fuels generally boil over a range between 163 and 371°C (325 and 700°F). Their makeup can represent various

Table 9.6.12 Average Volatility Characteristics of U.S. Motor Gasolines

	Summer				Winter			
	Premium		Regular		Premium		Regular	
IBP temp, °F (°C)	93	(34)	94	(34)	82	(28)	82	(28)
10% temp, °F (°C)	126	(52)	125	(52)	107	(42)	107	(42)
50% temp, °F (°C)	216	(102)	207	(97)	207	(97)	198	(92)
90% temp, °F (°C)	321	(160)	338	(170)	316	(157)	333	(167)
Endpoint temp, °F (°C)	402	(205)	414	(212)	397	(203)	407	(209)
RVP, lb/in² (kPa)	8.9	(61.3)	8.8	(60.6)	12.3	(84.8)	12.1	(83.4)

Table 9.6.13 Detailed Requirements for Diesel Fuel Oils[a,b,i] (ASTM D975)

Grade of diesel fuel oil	Flash point, °C (°F) Min	Cloud point, °C (°F) Max	Water and sediment, vol % Max	Carbon residue on 10 percent residuum, % Max	Ash, wt % Max	Distillation temperatures, °C (°F), at 90% point Min	Distillation temperatures, °C (°F), at 90% point Max	Viscosity kinematic, cSt (mm²/s), at 40°C Min	Viscosity kinematic, cSt (mm²/s), at 40°C Max	Viscosity, Saybolt, SUS at 100°F Min	Viscosity, Saybolt, SUS at 100°F Max	Sulfur,[d] wt % Max	Copper strip corrosion Max	Cetane no. Min
No. 1-D—a volatile distillate fuel oil for engines in service requiring frequent speed and load changes	38 (100)	b	0.05	0.15	0.01	—	288 (550)	1.3	2.4	—	34.4	0.50	No. 3	40[f]
No. 2-D—a distillate fuel oil of lower volatility for engines in industrial and heavy mobile service	52 (125)	b	0.05	0.35	0.01	282[c] (540)	338 (640)	1.9	4.1	32.6	40.1	0.50	No. 3	40[f]
No. 4-D—a fuel oil for low- and medium-speed engines	55 (130)	b	0.05	—	0.10	—	—	5.5	24.0	45.0	125.0	2.0	—	30[f]

[a] To meet special operating conditions, modifications of individual limiting requirements may be agreed upon between purchaser, seller, and manufacturer.

[b] It is unrealistic to specify low-temperature properties that will ensure satisfactory operation on a broad basis. Satisfactory operation should be achieved in most cases if the cloud point (or wax appearance point) is specified at 6°C (10°F) above the tenth percentile minimum ambient temperature for the area in which the fuel will be used. The tenth percentile minimum ambient temperatures for the months of October, November, December, January, February, and March are given in ASTM D 975 in maps for the 48 contiguous states and Alaska. This guidance is of a general nature; some equipment designs, use of flow improver additives, fuel properties, and/or operations may allow higher or require lower cloud point fuels. Appropriate low-temperature operability properties should be agreed on between the fuel supplier and purchaser for the intended use and expected ambient temperatures.

[c] When cloud point less than −12°C (10°F) is specified, the minimum viscosity shall be 1.7 cSt and the 90% point shall be waived.

[d] In countries outside the U.S., other sulfur limits may apply.

[e] Where cetane number by method D 613 is not available, method D 976 may be used as an approximation. Where there is disagreement, method D 613 shall be the preferred method.

[f] Low atmospheric temperatures as well as engine operation at high altitudes may require system use of fuels with higher cetane ratings.

[g] Millimeter squared per second (official SI unit) (Note: 1 cSt = 1 mm²/s).

[h] The values in SI units are to be regarded as the standard. The values in U.S. Customary system units are for information only.

[i] Nothing in this specification shall preclude observance of federal, state, or local regulations which may be more restrictive.

combinations of volatility, ignition quality, viscosity, sulfur level, gravity, and other characteristics. Additives may be used to enhance specific properties. The SAE recommended practices are described in SAE J313 (March 1992). Permissible limits of significant diesel fuel properties are summarized in ASTM's classification chart D975, reproduced as Table 9.6.13.

Alternative Fuels   Other than gasoline and diesel fuels, alternative fuels used in various parts of the world include propane, frequently referred to as LPG; hydrous ethyl alcohol (ethanol), particularly in Brazil; gasohol, a blend of 10 percent by volume of gasoline and 90 percent by volume of denatured anhydrous ethanol. Other possible fuels include compressed or liquefied natural gas (CNG or LNG)—primarily methane, other alcohols—particularly methanol, which is also used in racing, and biomass fuels. Details can be found in SAE J1297.

GAS EXCHANGE PROCESSES

Intake Manifolds   The intake manifold distributes the air (diesel) or the air and fuel (Otto) to various cylinders of a multicylinder engine. Intake manifolds are **streamlined** (always with diesel), the **rake type,** or a combination of both. The sections may be either circular or rectangular. In some cases, the circular sections are combined with a flat **floor** or bottom to provide more surface on which the liquid fuel may spread and be vaporized. **Dams** are also used to prevent liquid fuel from flowing along the manifold floor. In carbureted engines, the updraft manifold should have a **riser velocity** of not less than 40 ft/s (12 m/s) to entrain or lift the liquid fuel particles from the carburetor to the **distributor** section of the manifold. This determines the minimum wide-open throttle speed for satisfactory operation. The downdraft manifold permits larger areas and lower velocities between the carburetor and manifold.

The overlapping of the intake valve periods produces complicated disturbances in the intake manifold and has resulted in the use of dual manifolds in some engines with six or more cylinders. V-8 automobile engines have complicated dual manifolds, integrally cast, with each manifold usually feeding two cylinders of each bank. Port fuel injection permits extensive manifold tuning because the wall wetting produced by the carburetor is eliminated. Rotary engines have a simple intake system requirement. A separate carburetor or fuel injector may be used to feed each chamber.

Preheating the intake air and controlling its temperature are desirable in that they reduce air density variation and thereby allow better control of the mixture ratio for economy and emission control. In carbureted engines, carburetor icing is thereby minimized and the need for an exhaust manifold crossover valve reduced. However, air preheating is undesirable in that it reduces volumetric efficiency, heats the carburetor, and evaporates fuel in the float chamber, and may increase intake system coking. While preheating the fuel is also undesirable in that the light fuel fractions will be lost by vaporization, many intake manifolds for liquid fuels have a **hot spot** which receives heat from the exhaust (usually thermostatically controlled) and on which the fuel particles impinge upon leaving the manifold riser. The dew points or condensation temperatures for fuel in an air/fuel mixture depend on the fuel, air/fuel ratio, and total pressure (Table 9.6.14), and they may be determined from equilibrium air distillation data.

Exhaust Manifolds   In addition to adequate flow area, the most important consideration in exhaust manifold design is expansion. Small multicylinder engines employ a one-piece exhaust manifold. This is anchored at the center of the engine, and elongated holes are provided at other points for expansion. Large engines employ multipiece manifolds with expansion joints. They are usually water-cooled in marine applications to prevent strain and for safety reasons. In turbocharged engines, exhaust manifolds are usually insulated to conserve energy.

Manifold flow area should be large enough to avoid local back pressure buildup during exhaust and in connection with the total exhaust system should not raise back pressure excessively at maximum flow. A flow area of 0.7 times the intake manifold area may be used for four-cycle engines. For two-cycle engines, the exhaust flow area usually exceeds the intake flow area. Overlapping of the exhaust valve periods in multicylinder engines makes multiple manifolds desirable.

Mufflers (see Sec. 12.6)   In order to be efficient as sound silencers, exhaust mufflers must decrease the exhaust gas velocity and either absorb the sound waves or cancel them by interference with other waves from the same source. Mufflers should have volumes 6 to 8 times the piston displacement and may contain baffles with or without holes. Mufflers that cancel sound waves by interference usually break the waves into two parts which follow different paths and meet again, out of phase, before leaving the muffler (Davis et al., NACA Rept. 1192).

Exhaust **muffle pits** are used with large engines, are usually made of concrete, have a volume about 20 times the piston displacement, and may be open or provided with baffling or partly filled with loose stone through which the gases must pass. A stack is usually provided for the escape of the gases, and a drain for the discharge of the condensed water from the exhaust products.

Exhaust **back pressure** should be kept to a minimum since an increase of 1 lb/in² (6.9 kPa) in back pressure decreases the maximum power output about 2 percent, about 1 percent being due to more exhaust work and the balance of the effect of increased clearance gas pressure on volumetric efficiency.

Supercharging

Supercharging increases the amount of charge per cycle (above that of the normally aspirated engine) by increasing its pressure and hence its density. Supercharging permits more fuel to be burned and is a practical means to increase engine power, but it increases mechanical and thermal stresses in the engine. For aircraft piston engines, it is a means to high-power output for takeoff and to offset the rare atmosphere at high altitude. For diesel engines supercharging results in smaller and lighter power plants. It is especially practical for diesel engines because supercharging does not add to the fuel quality required, and because diesel engines do not use combustion air as effectively as spark ignited engines.

Three alternative methods can be used to achieve supercharging. In **mechanical supercharging,** a positive-displacement type of Roots blower or compressor is geared directly to the engine shaft and thus is driven by engine power. This type is desirable for variable-speed engines (diesel railcar) where high torque is required at various speeds, since the pressure and capacity characteristics of this type do not decrease with speed. In **turbocharging,** exhaust energy drives a turbine, which, in turn, provides the necessary work to drive the turbocharger's compressor, mounted on the same shaft as the turbine (Fig. 9.6.31). The centrifugal compressor is particularly adaptable to aircraft engines because of high capacity, small size, and low weight. It is also used with truck diesel

Table 9.6.14   Condensation Temperatures of Liquid Fuels, °F (°C)

		ASTM distillation				Condensation temp.			
						Atmospheric pressure		Pressure, ⅔ atmospheric	
						Air/fuel ratio		Air/fuel ratio	
Fuel	Deg API	Initial point	50 percent	90 percent	Endpoint	12:1	15:1	12:1	15:1
Gasoline	60	100 (38)	249 (121)	354 (179)	400 (204)	118 (48)	114 (46)	111 (44)	107 (42)
Kerosene	43	350 (177)	450 (232)	510 (266)	550 (288)	204 (96)	200 (93)	201 (94)	197 (92)

engines to improve the volumetric efficiency and power at high engine speeds. Centrifugal blowers may have one or two stages with cooling between stages or after the single or last stage. The rotors run at maximum speeds of 15,000 to 70,000 r/min. Finally, in **pressure wave supercharging,** wave action and resonant tuning in the intake and exhaust systems are used to compress the charge and thereby boost volumetric efficiency.

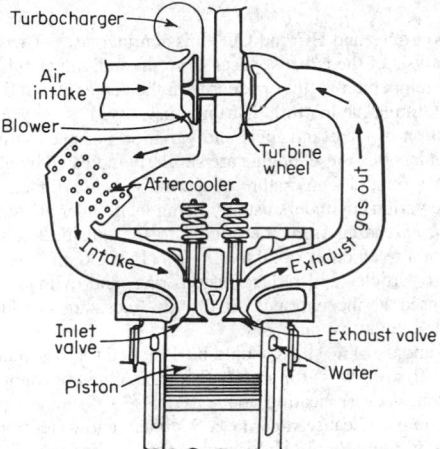

Fig. 9.6.31 Schematic of exhaust turbosupercharger installation on diesel engine cylinder.

Supercharging a spark ignition, premixed-charge engine which has the optimum compression ratio for the available fuel causes combustion knock and necessitates lowering the compression ratio, enriching the mixture, or increasing the octane number of the fuel. Supercharging a diesel engine increases the maximum pressure and necessitates lowering the compression ratio or changing the fuel injection timing, if the limiting pressure is already attained without supercharging. The use of variable engine compression ratio together with supercharging is very desirable but difficult and costly.

Turbocharging of diesel engines is usual for most applications, including truck, locomotive, and marine uses. The exhaust gas turbine

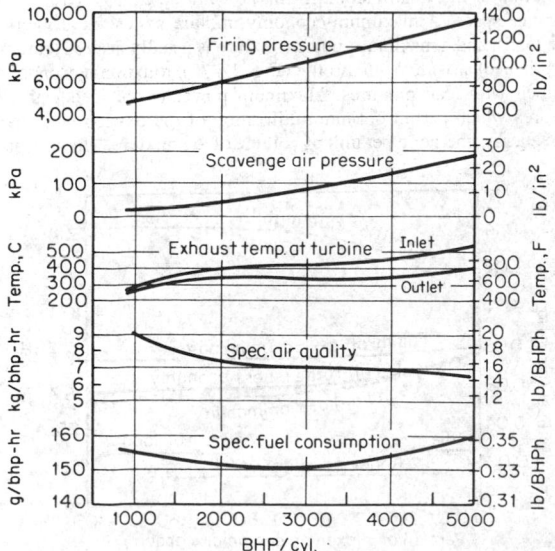

Fig. 9.6.32 Performance curves of a turbocharged diesel engine. *(Sulzer.)*

provides more complete expansion of the combustion gases than is practical in the engine cylinder and contributes to improved thermal efficiency. Higher output also increases the engine mechanical efficiency (Buchi, *ASME Trans.,* **59,** pp. 85–96). Supercharging assists with NO_x control without a fuel economy penalty. The fuel economy and other performance characteristics of a supercharged Sulzer marine diesel engine of 30-in (760-mm) cylinder diameter are shown in Fig. 9.6.32.

Scavenging Two-Stroke Cycle Engines

Crankcase compression for each cylinder or an engine driven blower may be used for scavenging and charging the cylinders of a two-stroke cycle engine (Fig. 9.6.33). With crankcase compression, used on small gasoline outboard marine engines, a third port uncovered by the piston near the top of the stroke, a rotary valve, or an automatic poppet valve may be used. The top of the piston is shaped to deflect the gases entering the cylinder and to prevent short-circuiting to the exhaust port. A "dead" spot of gases may remain in the lower center of the cylinder or in the upper corners.

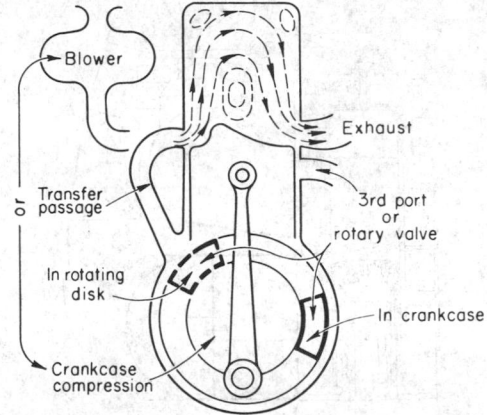

Fig. 9.6.33 Common scavenging systems.

The intake may be inclined and tangential, giving the entering charge an upward helical motion (Fig. 9.6.34). This motion keeps the entering charge near the cylinder walls, forcing the exhaust gases to the center of the cylinder and down to the exhaust port. Some engines employ loop scavenging (Fig. 9.6.35).

Fig. 9.6.34 Hesselman helical-loop system.

Fig. 9.6.35 M-A-N loop scavenging.

Some large diesel engines, such as the General Motors engine shown in Fig. 9.6.20, have poppet exhaust valves in the cylinder head, eliminating the hot exhaust ports from the cylinder walls. Intake ports can then completely surround the cylinder. The resulting flow pattern is referred to as **uniflow scavenging.** Typical blower pressures and related conditions are as shown in Table 9.6.15 for a General Motors two-stroke diesel engine.

Opposed-piston engines have intake and exhaust ports located in the cylinder walls at opposite ends of the cylinder. A blower forces the

Table 9.6.15 General Motors Two-Stroke Cycle Diesel Engine

Engine type	Air-box pressure at 500 to 2,300 r/min		Max bmep*		Max imep*	
	inHg	cmHg	lb/in²	kPa	lb/in²	kPa
Naturally aspirated	1–9	2.54–22.86	101	697	129	890
Turbocharged	1.5–40	3.81–101.62	136	938	163	1,125

* At 2,300 r/min.

entering charge through the intake ports which extend around the entire circumference of the cylinder.

The two-stroke cycle Sulzer diesel engine (Fig. 9.6.36) makes use of a crosshead and piston rod design. The air in the space between the piston and the diaphragm carrying the piston rod stuffing box is displaced by the descending piston and is forced into the cylinder through the intake ports as they are uncovered by the piston. This aids the

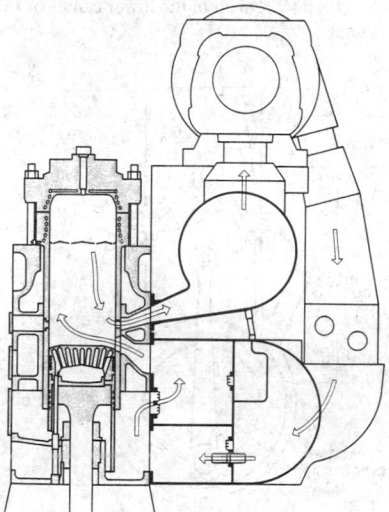

Fig. 9.6.36 A Sulzer series RL two-stroke crosshead diesel engine.

scavenging of the cylinder. Supercharging is provided by exhaust gas turbine-driven blowers. Air from these blowers passes through intercoolers and is admitted through automatic one-way valves to the scavenging chambers. Electric motor-driven air blowers are sometimes used on large engines to assist scavenging at idle and low power.

The crosshead design, with its stuffing box, also prevents combustion products from contaminating the crankcase oil system and is commonly used on large engines. This RL design (Fig. 9.6.36) is used for cylinder diameters of 560, 660, 760, and 1,900 mm. The largest cylinder, with a stroke of 1,900 mm, delivers a normal output of 4,000 bhp at 102 r/min corresponding to a bmep of 1,430 kPa (207 lb/in²) and a piston speed of 6,480 m/s (1,300 ft/min). Figure 9.6.32 shows the performance curves of this type up to an overload of 5,000 bhp per cylinder. Smaller cylinders produce correspondingly lower outputs with roughly the same mep and piston speed.

FUEL-AIR MIXTURE PREPARATION

Mixture Distribution Perfect distribution of fuel-air mixture consists of equal distribution of both air and fuel to the various cylinders of a multicylinder engine for each cycle. The compression pressure of each cylinder is an indication of the air distribution if all cylinders have identical valve timing and compression ratios and no leakage. Cylinders receiving lean mixtures develop less power than those receiving correspondingly rich mixtures. Poor distribution can be masked by enriching the mixture, and high-power output attained at the cost of high fuel

consumption rates and HC and CO emission increases. The CO, CO_2, and O_2 analysis of the exhaust gases from the individual cylinders and spark plug temperatures (thermocouple in the center electrode) may be used to determine the mixture distribution.

Distribution of load (and fuel) and variation in injection timing between cylinders of a diesel engine are similarly indicated by the exhaust gas composition and temperature of the individual cylinders, the mixtures in the various cylinders usually being lean under all conditions.

Air lines, extending from the engine to the outside air, should be designed for air velocities of 50 to 100 ft/s (15 to 30 m/s). Air cleaners remove dirt particles and reduce piston, ring, and cylinder wear. Air silencers, used at the entrance to the air lines, are combined with cleaners for automotive purposes.

Gasoline and **diesel fuel lines** should be designed for maximum velocities under 1 ft/s (0.3 m/s) for gravity feed. Gasoline lines should be in a cool location, because heating above 100°F (38°C) may vaporize the lighter fractions and cause **vapor lock**. Sediment and water traps as well as strainers or filters should be located in the fuel line ahead of the fuel injector or carburetor, fuel pump, or injection pump.

Gas lines should be designed for gas velocities of 30 to 60 ft/s (9 to 18 m/s). A pressure drop of 0.07 to 0.12 in (1.8 to 3 mm) of water should be allowed per 100 ft (30 m) of pipe. Close calculation of the **size of gas pipe** for industrial gas installations is undesirable, because of the reduction of the normal pipe cross section by tar and dust deposits. A large-diameter storage chamber should be located in the pipe near the engine. The gas line should be free of traps and as free as possible from changes in direction of flow, and it should be pitched slightly from both ends toward a low point in the center, where a sealed drain should be located. All water seals in the gas line should be as cool as possible, to minimize evaporation and the breaking of the seal.

Mixture Preparation in Spark Ignition Engines

A rich mixture is required for idling and small throttle openings because of dilution of the small incoming charge with the exhaust gases in the clearance space. A maximum-economy mixture is desired at intermediate loads, and a maximum-power mixture is usually desired (automotive practice) at wide-open throttle (Fig. 9.6.37). Maximum economy is obtained with lean mixtures. Maximum power is obtained with rich mixtures which permit optimum utilization of the oxygen with maximum energy liberation per unit of volume of the mixture. Mixture ratios

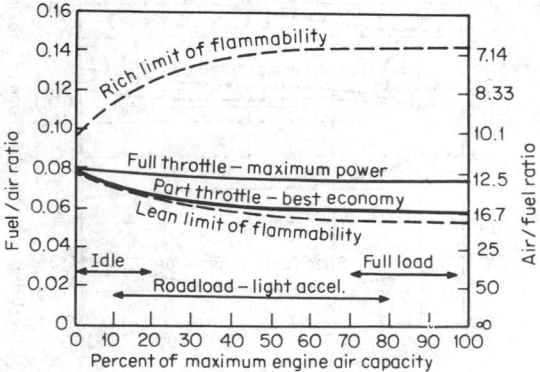

Fig. 9.6.37 Mixture ratios for spark ignition engines.

for minimum specific fuel consumption and for maximum power are also indicated on Fig. 9.6.37. The rich mixture which provides maximum power also results in large amounts of carbon monoxide and unburned fuel in the exhaust.

For many years, carburetion has been used to meter, atomize (if liquid), and mix the fuel with the air flowing to the engine. This method is still in use for mixing gaseous fuels with air, for small engines such as those used in household and off-road applications, for piston-type air-cooled engines, and for older passenger cars. However, except for parts of the world with less stringent emission standards, carburetion has been replaced by fuel injection for automotive applications. The two methods of fuel-air mixture preparation are described below.

Carburetion

The basic principle of carburetion is illustrated in Fig. 9.6.38. A float maintains a constant fuel level in a float chamber which is vented to the atmosphere or air entrance of the carburetor. The float chamber fuel level is slightly below the outlet of the fuel jet. The engine air flows through a venturi tube which has a throat diameter of 0.75 to 0.85 of the diameter of the carburetor bore. The reduction in pressure at the venturi

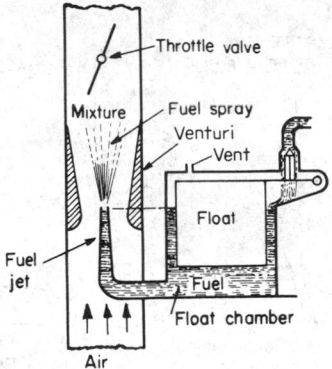

Fig. 9.6.38   Principle of carburetion.

throat causes fuel to flow from the float chamber through the main fuel jet into the airstream. Fuel atomization is accomplished by the velocity difference between the air and fuel. Vaporization of the fuel continues while flowing to the engine cylinder but is usually not completed in the intake manifold at wide-open throttle.

The metering characteristic of a carburetor is the weight ratio of fuel to air supplied over the operating range of airflow. These carburetor flow characteristics are commonly shown on a graph, similar to Fig. 9.6.37, which gives the fuel/air ratios at idle, along the "road-load curve" for an automobile engine, and at wide-open throttle. Although computations can be made to establish the fuel/air ratio under specified conditions, estimates are approximate because air is usually bled into the fuel channels to help control the fuel flow. Also, airflow through the venturi is pulsating in nature and makes both airflow and fuel flow somewhat unpredictable. Actual carburetor characteristics approach the desirable characteristics indicated in Fig. 9.6.37.

When the throttle is opened quickly, there is a momentary lag of the fuel flow in relation to the airflow due to the formation of a liquid film on the intake manifold walls. This necessitates the use of some device (accelerating pump or well) to enrich the mixture during sudden accelerations and thereby avoid engine backfire. At idle, the depression of the throat of the venturi is insufficient to overcome the viscosity of the fuel, thus requiring special idle circuits to overcome this problem and to enrich the otherwise very dilute, high-residual cylinder contents. Provisions are also made for mixture enrichment at high loads. In the 1980s, with the enforcement of more stringent emissions standards and the introduction of closed-loop control of the fuel/air ratio to ensure operation in a narrow window about the stoichiometric ratio, the electronic carburetor was introduced (Fig. 9.6.39). In this device, the flow area

for fuel metering is computer-controlled by a pulse-width-modulated solenoid.

Carburetors for piston-type air-cooled aircraft engines provide fuel/air ratios that vary from 0.11 at idle to 0.085 for cruising rich (cruising lean being about 0.072). As the full-power condition is approached, the

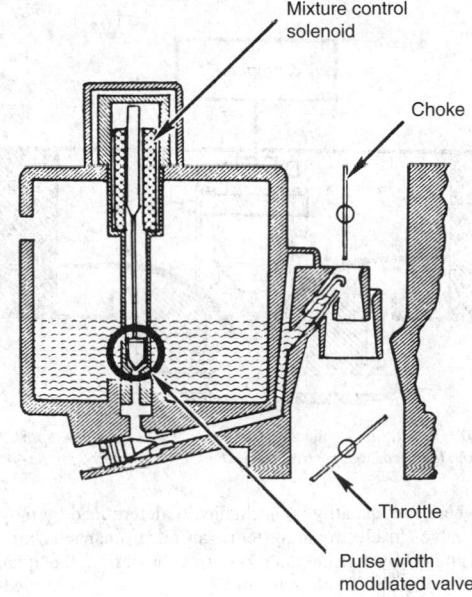

Fig. 9.6.39   Electronic carburetor. *(Amann, "The Automotive Spark-Ignition Engine—An Historical Perspective," ASME-ICE vol. 8, 1989, pp. 33–45.)*

mixture richness is increased to aid cylinder cooling, becoming about 0.108 at maximum output. Aircraft carburetors must compensate for the effect of altitude on the air density and the resulting change in the mixture ratio. Compensation for the enrichment that occurs with increasing altitude is usually provided by reducing the area of the main fuel-metering orifice. This is accomplished by moving a tapered needle in the orifice, which may be done manually or automatically by means of an air-pressure-sensitive bellows. A carburetor providing a 0.067 fuel/air ratio at sea level would provide a fuel/air ratio of about 0.091 at an altitude of 20,000 ft (6,100 m).

Fuel Injection in Spark Ignition Engines   A majority of new automobile and truck engines and some small marine and utility engines (future) employ fuel injection. Fuel is injected into the supercharger, intake manifold, intake valve ports, or combustion chambers of spark ignition engines. Fuel injection consists of an in-tank pump, common-rail recirculating supply, and injection nozzles. With single-point, central fuel injection above the throttle, as typified by the throttle body injector of Fig. 9.6.40, one or two injectors with injection pressure of about 70 kPa are used for multicylinder engines. This system provides an electronically controlled injector at the intake manifold entrance where the carburetor was formerly positioned. With multipoint fuel injection, an injector is positioned in each port, and pressures from 400 to 700 kPa are used to avoid vapor formation in the tip located in the hot spit-back region just upstream of the intake valve. In certain cases, an additional injector is used to supplement the fuel flow during starting.

The advantages of fuel injection over carburetion are the more uniform distribution of fuel between the engine cylinders, the reduction of combustion knock caused by nonuniform fuel distribution, the elimination of heat to the intake manifold to obtain the desirable fuel vaporization (thus obtaining higher-power outputs because of higher charge density), more rapid throttle response, and the use of less volatile fuel. Carburetor and induction system icing are eliminated. With an injection system, it is easier to cut off the fuel flow during deceleration (coasting) than is the case with the carburetor with its fuel-wetted intake manifold.

Note that throttle body injection encounters problems associated with the slower transport of fuel than air during transients, as with carburetors.

Fuel injection systems may have mechanical or electronic control of fuel flow, with the electronic type being predominant. In mechanical systems, such as the Bosch K-Jetronic, fuel is continuously injected into

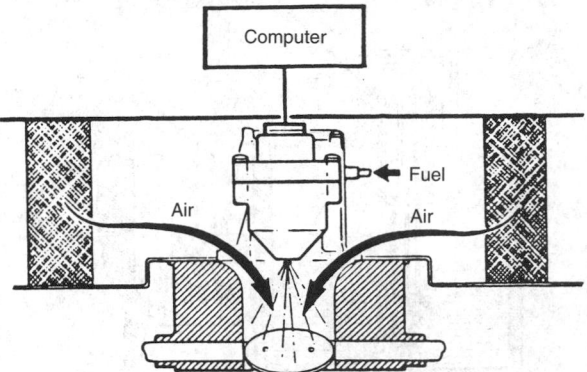

Fig. 9.6.40 Throttle body injector. *(Amann, "The Automotive Spark-Ignition Engine—An Historical Perspective," ASME-ICE vol. 8, 1989, pp. 33–45.)*

the port where fuel quantity is mechanically determined by movement of an air valve. In electronic systems, an electromagnetic injector is pulsed on and off by the computer to supply the desired fuel quantity in proportion to the air. The **injector pulse** width may be trimmed by an exhaust sensor signal to achieve the desired mixture strength. The airflow must be sensed to establish the correct fuel/air ratio for various engine conditions. This can be done by sensing the airflow with a venturi or equivalent, a hot-film, hot-wire, or other air anemometer (such as in the L-Jetronic system), or by sensing the principal engine variables which control engine air consumption. The latter is referred to as a **speed-density** control system. A system of this type is shown schematically in Fig. 9.6.41. Engine speed and manifold pressure and air temperature are used to calculate the engine airflow from prior dynamometer data, a method which has proved less accurate in practice.

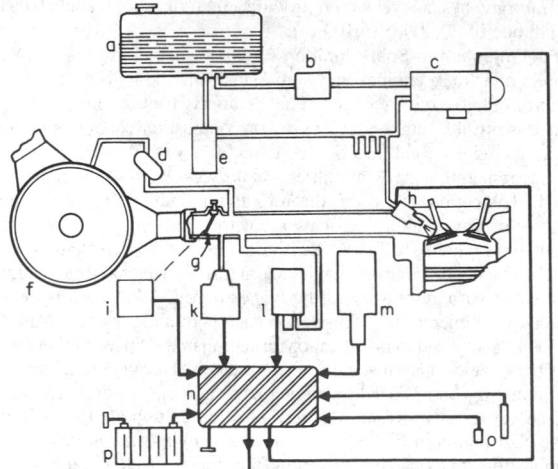

Fig. 9.6.41 Schematic of electronic gasoline injection system for an automobile. (*a*) Gasoline tank; (*b*) fuel filter; (*c*) fuel supply pump; (*d*) auxiliary idling air valve; (*e*) fuel pressure regulator; (*f*) air filter; (*g*) throttle plate; (*h*) fuel injection valve; (*i*) throttle position switch; (*k*) full load pressure switch; (*l*) manifold pressure sensor; (*m*) ignition distributor with trigger contacts; (*n*) electronic control unit; (*o*) temperature sensor; (*p*) battery. (*Robert Bosch.*)

For central fuel injection designs, the injector opening angle may be independent of engine cylinder events (asynchronous). For port injection designs, the opening is normally synchronized with cylinder events to open before the intake valve opens. This minimizes axial mixture stratification which occurs when the injector supplies fuel when the intake valve is open. Such axial stratification may be used advantageously to create a lean-burning engine (Toyota). Injection of fuel into the intake port usually occurs early in the intake stroke, to provide the maximum time for fuel evaporation and mixing to occur. Injection into the intake port provides good mixing as the fuel and air are drawn at high velocity through the intake valve passages, and it isolates the injector from combustion gases at high pressure and temperature. Direct injection into the cylinder occurs during intake or early compression, and it usually results in charge stratification.

Spray targeting in port fuel injection is especially important for fuel evaporation, wall wetting, and engine emissions. Typically when the spray is directed at the center of the intake valve, toward its valve stem base, hydrocarbon and CO emissions are minimized (Fig. 9.6.42). Injector design, spray angle, and droplet size distribution are important. Quader (1989) also showed that targeting the fuel spray at the intake valve centerline toward the valve head resulted in the lowest smoke and

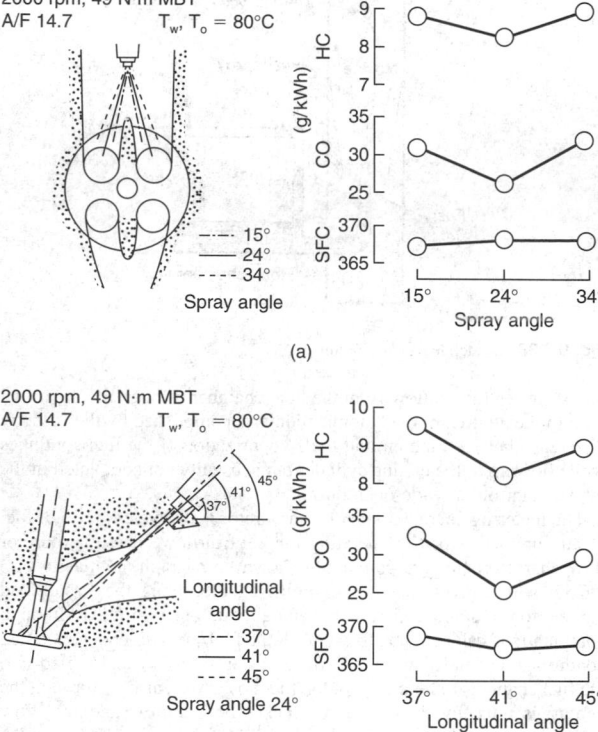

Fig. 9.6.42 Effects of spray angle and injection targeting on engine emissions. (*a*) Effect of dual stream separation angle; (*b*) effect of injection direction. (*Iwata et al., SAE paper 870125, 1987.*)

HC emissions under high-speed and heavy-load conditions. Some designs are less affected by deposits. Air-assisted fuel sprays provide smaller droplets and may be an advantage in both port and direct cylinder injection. Air-assisted direct cylinder injection has shown promise for improved economy and lower emissions in two-stroke, spark ignition, orbital combustion process (OCP) engines.

Fuel Injection in Diesel Engines The fuel injection system for diesel engines usually consists of a pump, fuel line, and nozzle. This system provides control of the beginning, rate and duration of injection, and desired fuel quantity per injection, and atomizes and partially distributes the fuel in the combustion chamber. Methods of fuel injection include pump or unit injectors and common-rail systems. In the **pump injection system** (Fig. 9.6.43), a pump forces a desired quantity of fuel through a high-pressure fuel line and an atomizing nozzle into the combustion chamber. Unit injectors combine the plunger pump and spray nozzle, thus eliminating the high-pressure fuel line. Injection pressures usually range from 1,500 to 7,000 lb/in² (10,300 to 48,000 kPa) but may run above 30,000 lb/in² (206,000 kPa) at high injection rates. Fuel quantity

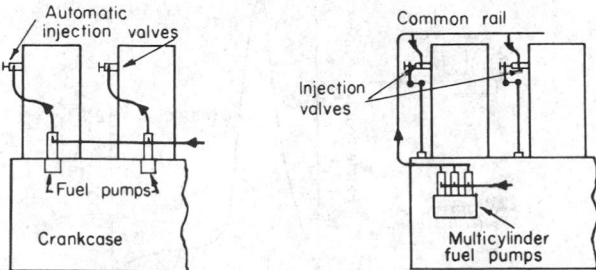

Fig. 9.6.43 Pump injection.   **Fig. 9.6.44** Common rail.

control with this type of injection is usually affected by controlling a bypass valve in the pump or the pump discharge line or by varying the time of closure or opening of the fuel-pump inlet port. A separate fuel-pump is commonly required for each engine cylinder, although rotating plunger pumps serving from two to eight cylinders per plunger are in common use, especially on small engines. In the **common-rail system** (Fig. 9.6.44), a constant pressure is maintained in the fuel discharge line, and the discharge of fuel to the engine cylinder is controlled by the lift, timing, and duration of opening of the fuel valve. Considerable capacity of the discharge line is necessary in this case, a triplex pump being considered desirable to supply fuel at a constant rate.

Pumps Fuel pumps for pump injection systems require rugged construction, are usually actuated by a cam, and have constant mechanical stroke. The pump plunger and barrel (Fig. 9.6.45) are a very close lapped fit to minimize leakage. The inlet port is uncovered near the

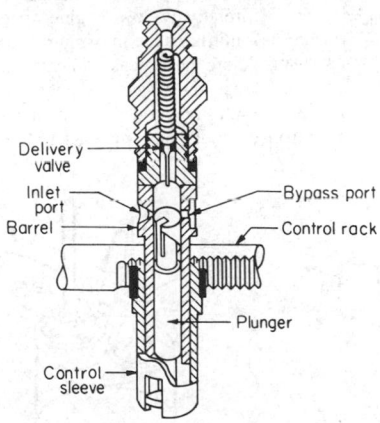

Fig. 9.6.45 Bosch plunger, barrel, and delivery valve assembly.

bottom of the plunger stroke, the low vapor pressure in the pump cylinder causing fuel flow into the barrel. On the upward plunger stroke, fuel flows past the discharge valve and out of the spring-loaded nozzle. Fuel spray from the nozzle continues until the pump bypass port is uncovered. The pressure then subsides, the nozzle closes, and the pump delivery valve closes to maintain a desired residual line pressure until the next injection. Fuel quantity control is obtained by turning the pump plunger in the barrel, which varies the time of uncovering the bypass

port by the plunger helix. This in turn varies the duration of the effective plunger stroke and the amount of fuel injected. The high pressures which occur in these injection systems, together with the compressibility of the fuel oil, give rise to transient pressure wave phenomena in the connecting lines which change injection characteristics.

Fuel Lines The tubing connecting the injection pump with the spray nozzle must be thick-walled, of smooth and uniform bore, and made of ductile material to permit bending. Tubes range in size from ¼-in (0.64-cm) OD and ¹⁄₁₆-in (0.16-cm) ID to ½-in (1.3-cm) OD and ³⁄₁₆-in (0.48-cm) ID. Fuel lines should be the same length for each cylinder, to obtain the same transient hydraulic characteristics. Pressure waves due to plunger motion and nozzle efflux characteristics are developed, travel back and forth, and create disturbances during the injection process (De Juhasz, *Trans. ASME, Oil and Gas Power,* **60**, p. 2; Bolt et al., *SAE Trans.,* 1971, paper 710569), making short lines desirable.

Fuel Injection Nozzles Injection nozzles are usually hydraulically operated, having a differential-diameter valve (Fig. 9.4.46), held on its seat by a spring, which opens when the fuel line pressure is increased by the injection pump. The valve closing pressure is always lower than the opening pressure, depending on the relative effective valve areas exposed to fuel pressure. The valves may be either the pintle or multiple-orifice type (Fig. 9.6.46). The former has a single orifice through which the end of the valve protrudes, and it is used mostly with precombustion

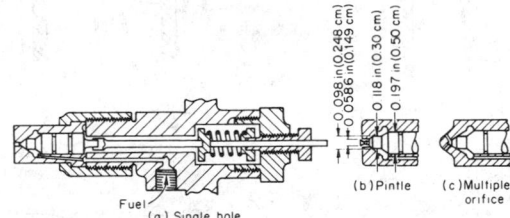

Fig. 9.6.46 Solid injection closed nozzle.

chambers. A pintle may be designed to restrict the fuel flow as it opens; this results in a throttling nozzle. The pintle nozzle provides a hollow conical spray with an included angle which varies from 4° to 60° with various pintle sizes. The multiple-orifice type of nozzle (Fig. 9.6.46c) is used principally with the open combustion chamber. The resulting fuel spray consists of a core and surrounding envelope of atomized fuel.

Diesel Fuel Injection System Characteristics At any given setting, most fuel injection systems deliver more fuel per cycle with a decrease in speed, except under the idling condition of very low speed and small fuel delivery. The rising fuel delivery curve with decrease in speed at full load contributes to the good torque rise and lugging ability of the high-speed diesel. At the idle condition, decrease in speed reduces fuel delivery and causes the engine to run still more slowly, etc. Thus the diesel engine is usually unstable and requires a governor for idling as well as at other loads.

COMBUSTION CHAMBERS

Spark Ignition Engines

Most combustion chambers for four-cycle engines, especially automotive, have an open design and employ overhead valves actuated by either pushrods or overhead cams. This design provides fast burning of the charge because of the concentrated combustion volume and minimizes surface area and thus heat losses and wall quenching. Figure 9.6.47 shows classical, two-valve chamber designs. In each case, the spark plug location and chamber shape yield a relation between flame front area and flame travel similar to that in Fig. 9.6.48. For a given engine, the maximum rate of pressure development is proportional to the peak flame-front area viewed at TDC. For the chambers of Fig. 9.6.47a, b, and d, virtually all the volume is in the head. This design allows large valves and minimizes piston inertia. The hemispherical or domed head of Fig. 9.6.47*b* allows very large valves and with a domed piston provides a combustion chamber of normal compression ratio. Aircraft engines usually have this chamber shape. The chamber of Fig. 9.6.47*c*, used by Jaguar in its V-12 design, is in the piston bowl. The bathtub chamber of Fig. 9.6.47*d* is very compact for good fuel economy. To suppress knock, each chamber has the spark close to the exhaust valve and a well-cooled, quench region in the end gas far from the spark plug. While all these chambers have a squish area to generate turbulence, compression-induced turbulence is low and mixture motion depends on intake-induced swirl, which is generally quite considerable.

More recently, pent-roof heads with four valves per cylinder and a centrally located spark plug have become very popular. This arrangement gives minimum flame travel distance and hence fast burn, as well as large valve flow area and thus high volumetric efficiency. In such

designs, turbulence has been generated through the complex interaction of the two inlet flows which produce a tumbling motion. Other fast-burn chambers have been designed by Nissan and Ricardo. In the Nissan NAPS-Z combustion system, there are twin spark plugs and an induction system producing a comparatively high level of axial swirl (Fig. 9.6.49). In Ricardo's high-ratio compact chamber (HRCC), the combustion chamber is centered on the exhaust valve. Chamber compactness, high turbulence, and lean burning allow this design to operate at compression ratios exceeding 11 without knock.

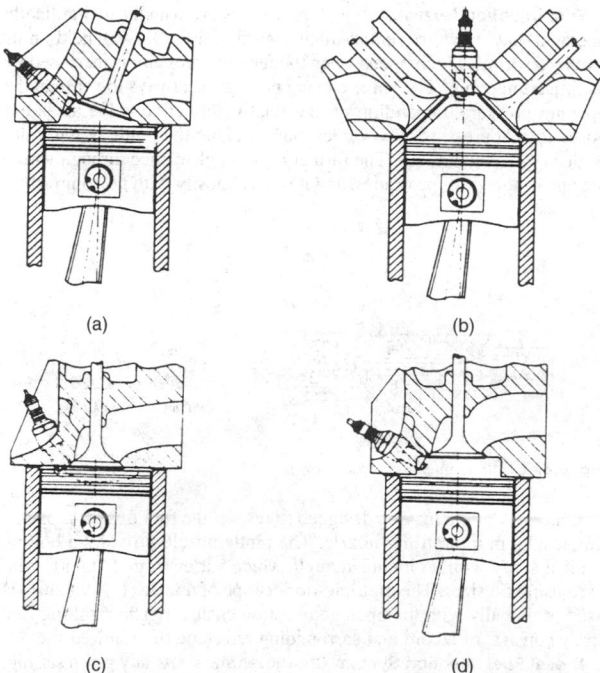

Fig. 9.6.47 Classical combustion chambers for spark-ignition engines; (*a*) Wedge; (*b*) hemispherical head; (*c*) bowl in piston; (*d*) bathtub head. (*Courtesy: R. Stone, SAE, 1995.*)

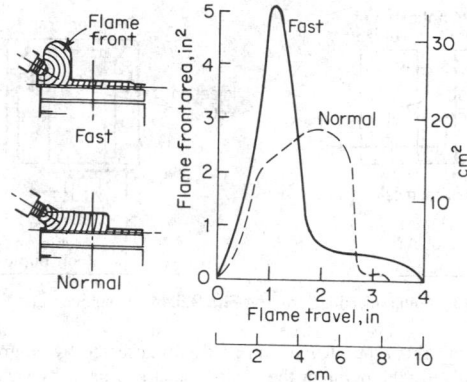

Fig. 9.6.48 Flame-front area curves for fast- and normal-burn combustion chambers.

Where cost or engine height is a principal concern, a design such as the L-head (Fig. 9.6.50), in which the valves are actuated from below directly by the cam, is commonly used. Spark plug location can compensate to some extent for the slow burn rate of the elongated chamber. The L-head has a relatively high surface area, and thus greater heat loss and wall quenching, and a small space for valves.

Fig. 9.6.50 L-head engine.

Combustion Roughness The high rate of pressure rise during combustion at relatively low engine speeds (1,000 r/ min) and the maximum combustion pressure near peak indicated torque engine speeds subject the engine mechanism to severe stresses which may result in combustion roughness noise unless the pressure rate is controlled. Andon and Marks (*SAE Trans.* **72,** 1964,

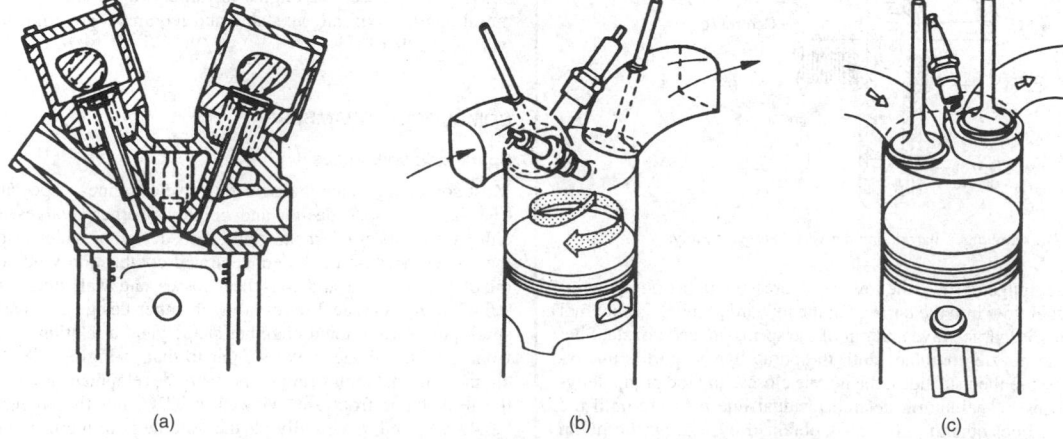

Fig. 9.6.49 Modern fast-burn combustion chambers. (*a*) Four-valve pent-roof; (*b*) NAPS-Z with twin spark plugs and high axial swirl; (*c*) high-ratio compact chamber (HRCC). (*Collins and Stokes, SAE, 1983.*)

p. 636) found that spark plug location, combustion chamber shape, and degree of turbulence are the principal factors that can be varied to control combustion roughness and the resulting noise. In a given engine, the maximum combustion rate of pressure rise can be reduced by retarding the spark timing or decreasing the volumetric efficiency.

A rate of pressure rise of 30 lb/in² (207 kPa) per degree of crank travel results in near-maximum efficiency, and higher rates of pressure rise usually result in combustion roughness noise. Increasing the rigidity of the crankshaft increases the natural frequency of vibration and reduces the deflection of the structure caused by combustion and inertial forces and thereby reduces roughness noise.

Die-cast aluminum transmits roughness noise more readily than cast iron. Structurally stiff compact engines such as the V-block are better than the in-line engines in this regard.

Compression Ignition Engines

Compression ignition engines have valve-in-head construction if operating with four cycles. Two types of combustion chamber arrangements are used: a single chamber, referred to as the *open* or *direct* injection type, and a two-chamber, referred to as a *prechamber* or *divided-chamber* type.

Open Combustion Chambers (Direct Injection) Fuel is injected under high pressure, usually through a multiple-orifice nozzle, directly into the clearance space or chamber between the piston and the cylinder head. The piston head is usually conformed (Figs. 9.6.51 and 9.6.52) to fit the fuel spray, and swirl moves the air into the fuel spray. Air swirl is accomplished by intake port design, by shrouding the intake valve, or, in the two-stroke cycle engine, by using tangential intake ports. High turbulence is accomplished by having the piston closely approach part

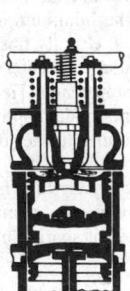

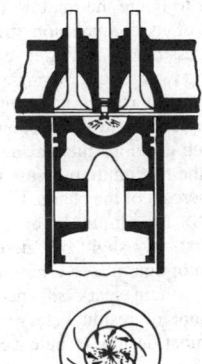

Fig. 9.6.51 Open combustion chamber with swirl produced by intake ports.

Fig. 9.6.52 Open combustion chamber with high turbulence and some swirl.

of the cylinder head. This forces the gases out of the small clearances and agitates the mixture.

Precombustion chambers are divided into two parts, the major volume being between the piston and the cylinder head and connected by a small passageway to the minor volume located in the cylinder head (Fig. 9.6.53). Fuel is injected into only the smaller chamber, and except under light loads, partial combustion occurs and discharges the burning mixture into the large chamber in which combustion is completed. This

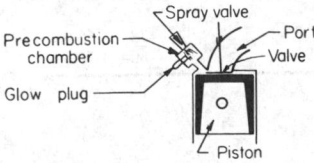

Fig. 9.6.53 Precombustion chamber.

type of combustion chamber produces smooth combustion but has fairly high fluid friction and heat transfer losses.

Turbulence or **divided combustion chambers** are modifications of the precombustion chamber, having the major chamber in the cylinder head and usually only a small clearance space between piston and cylinder head (Fig. 9.6.54). The passage between the two chambers is considerably larger than that in the precombustion chamber. Close piston clearance produces high turbulence in the prechamber and promotes rapid combustion. Part of the prechamber containing the transfer passage commonly is thermally insulated. The chamber (Fig. 9.6.53) is typical of those used in automotive diesel engines for passenger automobile use. An electrically heated hot-wire glow plug is used to assist cold starting, together with compression ratios up to 24 : 1.

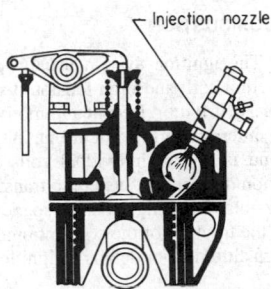

Fig. 9.6.54 Ricardo divided combustion chamber.

A **stratified-charge spark-ignited engine** has potential for burning lean mixtures for improved fuel economy and reduced emissions. In theory, an overall lean mixture is formed in the combustion chamber, while controlled to be slightly richer than stoichiometric in the vicinity of the spark plug at the time of ignition. The richer portion is thus easily ignited, and this in turn ignites the remaining lean mixture. Some of the benefits claimed are (1) lower flame temperature, (2) improved cycle efficiency, (3) less heat loss, (4) less dissociation, and (5) less pumping loss. In addition, less NO_x would be expected because of the lower combustion temperatures; likewise, less unburned hydrocarbon and CO emissions would result because of the overall excess of oxygen.

For effective stratification, very careful development has proved to be necessary to obtain complete combustion of fuel under the wide range of speed and load condition required of an automotive type of engine. Three rather distinct means for accomplishing the stratified charge condition have been explored.

1. A single combustion chamber with a well-controlled rotating air motion is used. This arrangement is illustrated (Fig. 9.6.55) by the Texaco combustion process (TCP), patented in 1949. Ford Motor Company's programmed combustion (PROCO) is a similar, single-chamber concept.

2. There is a prechamber or two-chamber system. This is illustrated by Fig. 9.6.56, which shows the general arrangement of the Honda compound-vortex, controlled-combustion (CVCC) system.

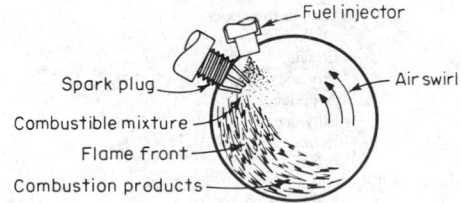

Fig. 9.6.55 Texaco combustion process (TCP).

3. There is axially stratified charge. By turning on the port fuel injector toward the end of the filling process, a rich mixture is produced near the spark plug, while the overall mixture may be lean. This method is known as the *Toyota lean-burn system*.

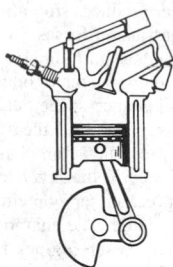

Fig. 9.6.56 Honda CVCC combustion process.

SPARK IGNITION COMBUSTION

Ignition System The ignition system usually consists of a 12-V storage battery, induction coil, and high-tension distributor, or a high-tension magneto with a distributor. Recent automotive practice employs electronic means to interrupt primary current in place of a contact breaker, and the trend is toward individual coils for each plug. For aircraft, the low-tension distribution system to transformer spark plugs, or other apparatus for obtaining a high-tension spark at the plug, appears to eliminate some of the troubles of the conventional ignition system in aircraft engines at high altitude. For details on ignition systems, see Sec. 15.1.

The usual firing order for in-line engines is 1, 3, 4, 2, or 1, 2, 4, 3 for four cylinders; 1, 5, 3, 6, 2, 4 (U.S. automobile engines) or 1, 4, 2, 6, 3, 5 for six cylinders in line; and 1, 6, 5, 4, 3, 2 for V-6 engines. A common procedure for V-8 engines is to number the cylinders from front to rear, with the odd numbers in the left bank, as viewed from the driver's seat. For this arrangement a typical firing order is 1, 8, 4, 3, 6, 5, 7, 2. Radial engines have firing orders that skip every other cylinder. For a nine-cylinder, single-row engine the order is 1, 3, 5, 7, 9, 2, 4, 6, 8.

Spark plug gaps vary from 0.020 to 0.080 in (0.5 to 2.0 mm). The smaller gaps require 4,000 to 8,000 V while the larger gaps require 10,000 to 34,000 V. In wedge-shaped, two-valve combustion chambers, spark plugs were typically located near the exhaust valve (hottest spot) so that the flame progressed toward the cooler part of the combustion chamber; this arrangement permitted higher compression ratios without combustion knock. In modern, four-valve, pent-roof combustion chambers, a central spark plug location results in minimum combustion time and good tolerance for dilution. Dual ignition (two spark plugs located opposite each other) also reduces combustion time and results in a slight gain in efficiency (usually less than 1 percent) compared with single ignition with optimum spark advance. Rotary (Wankel) engines employ one or two spark plugs per rotor. Single plugs are centrally located either above or below the trochoid waist. In modern, direct injection, stratified-charge, lean-burn engines, optimum placement of the spark plug is crucial to provide a combustible mixture near its vicinity.

Flame Speed Flame speeds in spark ignition engines are low immediately after ignition, attain maximum values when about one-half the combustion chamber has been inflamed, and decrease toward the end of the piston. Mean flame speeds are a maximum for mixtures usually 10 to 20 percent richer than the chemically correct air/fuel ratio, and they vary with fuel, engine speed, and turbulence.

At 900-ft/min (4.5-m/s) mean piston speed, mean flame speeds with maximum-power gasoline-air mixtures and optimum spark advance range from about 50 to 100 ft/s (15 to 30 m/s). Mean flame speeds under the same mixture, spark, and speed conditions in the CFR Waukesha engine are estimated at 55 (70) [80] ft/s (17, 21, 24 m/s) for a 4.6 : 1 (6.5 : 1) [8 : 1] compression ratio, while with ethyl alcohol–air mixtures the corresponding mean flame speeds are 50 (65) [75] ft/s (15, 20, 23 m/s).

The mean flame speed in an automotive engine under the foregoing conditions is estimated from the optimum spark advance (approximately five-ninths of combustion time) to be 90 ft/s (27 m/s) for a 6.5 : 1 compression ratio. In the same engine, mean flame speeds with optimum conditions vary from about 60 ft/s (18 m/s) at 500 ft/min (2.5 m/s) to about 170 ft/s (0.9 m/s) at 3,000 ft/min (15 m/s) mean piston speed. The higher flame speeds in the automotive engine at comparable piston speeds are due to greater turbulence than that obtained with the usual flat valve-in-head (CFR) design.

Spark Advance Since combustion requires finite time, it is necessary to ignite the mixture in the cylinder before the piston reaches the end of its compression stroke. This should distribute the combustion process before and after top center in order to obtain maximum torque and power. Spark advance is measured by the number of degrees the crankshaft rotates between the time of the spark and the end of the compression stroke. **Optimum spark advance** is the minimum advance which develops the **maximum brake torque** (**MBT**). Usually this causes (1) the maximum pressure to peak at about 16° after top center and (2) 50 percent of the charge to be burned by 10° after top center (Heywood, 1988). The torque lost by a spark timing earlier or later than optimum is at first only slight but increases rapidly as the timing further deviates from optimum (see Fig. 9.6.57).

Optimum spark advance depends principally on air-fuel mixture, amount of residual gas, emission control requirements, combustion chamber design, turbulence, engine speed, number of spark plugs, and

Table 9.6.16 Ignition Temperatures of Air-Gas Mixtures at Atmospheric Pressure

Gas	Chemical symbol	With ignition lag*		Instantaneous† ignition	
		% gas in mixture	Avg temp, °F (K)	% gas in mixture	°F (K)
Hydrogen	H_2	8–24	1,130 (893)	10	1,377 (1,020)
Carbon monoxide	CO	13–47	1,204 (924)	50	1,708 (1,204)
Methane	CH_4	5–38	1,202‡ (923)	25	>1,832 (1,273)
Ethane	C_2H_6	§	968‡ (793)		
Propane	C_3H_8	§	914‡ (763)		
Acetylene	C_2H_2	4–22	804 (702)		
Ethylene	C_2H_4	6–19	1,004 (813)	10	1,832 (1,273)
Benzene	C_6H_6			50	1,943 (1,335)
Ether	$C_4H_{10}O$			50	1,892 (1,306)
Gasoline:					
92 octane					734 (663¶)
100 octane					804 (1,702¶)

* Dixon and Coward, *Trans. Chem. Soc.*, **95**, 1909, p. 514.
† David, *Trans. Chem. Soc.*, **111**, 1917, p. 1003.
‡ Minimum value, since spread in values was appreciable.
§ Composition not reported.
¶ Jones, *U.S. Bur. Mines Bull.*, 1946.

Table 9.6.17   Ignition Temperatures at Various Air Pressure, °F (K)

Gas	Air pressure, atm								
	1	3	5	7	10	15	20	25	30
Hydrogen*	1,134 (885)	1,132 (884)	1,112 (873)	1,100 (866)					
Methane*	1,343 (1,001)	1,269 (960)	1,215 (930)	1,175 (908)					
Gasoline†			590 (583)		480 (522)	420 (489)			
Kerosine†			670 (628)		490 (528)	430 (494)	400 (478)	385 (469)	
Gas oil†			580 (578)		500 (533)	450 (505)	435 (497)	415 (486)	
Machine oil†			710 (650)		610 (594)	550 (561)	520 (544)		
Benzene†			685 (636)		585 (580)	545 (558)	530 (550)	515 (541)	500 (533)
Cylinder oil†			825 (714)		710 (650)	645 (614)	620 (600)	595 (586)	572 (573)
Benzol†					1,095 (864)	970 (794)	930 (772)	905 (758)	880 (744)

* Bone and Townsend, "Flame and Combustion in Gases, p. 69.
† Tausz and Schulte, *Zeit. V.D.I.,* 1924, p. 574.

spark plug location. The optimum spark advance is also affected by knock considerations, as described later. Maximum power air/fuel ratios require the minimum spark advance. Low-speed engines require 10° to 20° spark advance, and high-speed automotive engines require 30° to 40° spark advance. Racing engines require still more. Full load requires less spark advance than part throttle. Spark advance in contemporary automotive engines is based on computer-stored values and is a function of speed, load, and other engine operating variables.

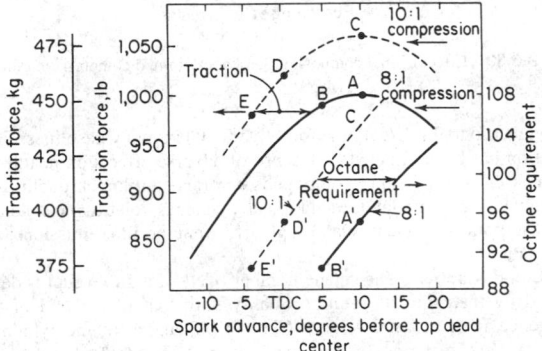

Fig. 9.6.57   Effect of spark advance on traction force and octane requirement.

Autoignition   The autoignition temperature of an air-fuel mixture is the lowest temperature at which chemical reaction proceeds at a rate sufficient to result eventually (long time lag) in inflammation. This temperature depends principally on the air-fuel mixture, fuel properties, mixture pressure, and test apparatus. Subjecting a mixture to a temperature higher than the autoignition temperature results in inflammation after a shorter time lag, there being a temperature for each mixture which results in practically instantaneous ignition (Table 9.6.16). An increase in pressure decreases the ignition temperature of an air-fuel mixture (Table 9.6.17). See CRC Report on Project CF-37-53.

Compression ignition engines, such as diesel engines, rely on autoignition for initiation of combustion. However, in spark ignition engines, autoignition of the gas ahead of the flame front is undesirable as it leads to combustion knock. This abnormal combustion phenomenon is described in greater detail in the following section.

COMBUSTION KNOCK

Combustion knock in the internal combustion engine is produced by the spontaneous combustion or autoignition of an appreciable portion of the charge (Rassweiler and Withrow, *Trans, SAE,* **31**, 1936, p. 297), which results in an extremely rapid local pressure rise and sharp metallic knock. In the spark ignition engine, this occurs when the last fraction of the charge to burn is compressed appreciably above its autoignition temperature by the fraction that has burned. However, if the lag in time (ignition lag) between the attainment of this temperature and the spon-

taneous local appearance of the flame is increased by the use of a knock-suppressor blending agent or additive, then the flame from the spark plug will be permitted to travel through the unburned fraction before it can knock. Thus knocking tendency depends on the autoignition temperature, ignition lag, and flame speed of the air-fuel mixture.

The tendency of a spark ignition engine to knock increases almost directly with spark advance (Fig. 9.6.57). Prior to the introduction of exhaust emission control requirements, it was general practice in automobile engine design to use a spark advance a few degrees less than optimum at wide-open throttle. This materially reduces the tendency to knock with only a negligible loss of torque. At part-throttle operation, where combustion knock may not be a problem, the spark is advanced closer to optimum. However, in many cases a compromise must be made between spark advance for good torque and spark retard for emission control.

In the compression ignition engine, combustion knock occurs when the ignition lag is comparatively long, thus permitting the evaporation of a considerable portion of the injected fuel, which suddenly inflames and produces the characteristic knock at the initiation of combustion. Thus knock suppression requires long ignition lags (high octane, low cetane) in spark ignition engines and short ignition lags (high cetane, low octane) in compression ignition engines.

Knock Rating of Fuels

For detailed information regarding methods, apparatus, reference fuels, and suppliers of apparatus and reference fuels, see ASTM Manual of Engine Test Methods for Rating Fuels, current edition.

Octane Number

The octane number (ON) of a fuel is the percentage by volume of iso-octane (2,2,4 trimethylpentane) in a mixture of iso-octane and normal heptane which matches the unknown fuel in knock tendency when compared by a specific procedure in the ASTM CFR knock-testing engine. A fuel of high octane will have a low cetane number and vice versa (Fig. 9.6.58).

The **performance number** scale for aviation gasolines (Fig. 9.6.58) is designed to relate fuel rating to average knock-limited performance (imep). A gasoline of the 100/130 grade indicates that in the aviation test (lean mixture, normally aspirated) the fuel is equivalent to iso-octane in imep while in the supercharge test (rich mixture) it permits an imep of 130 percent of that of iso-octane.

There are four methods—**research, motor, aviation** (lean mixture), and **supercharge** (rich mixture) (ASTM D 908, 357, 614, and 909, respectively)—in general use in the rating of gasolines according to knocking tendency. These methods vary principally in the engine operating conditions and the means for indicating knock intensity (see ASTM manual).

Engine conditions, as indicated by the air or mixture temperature and by the coolant temperature, are less severe in the research method than in the other methods. Thus, some fuels rate lower by the motor method than by the research method. The sensitivity of a fuel is defined quantitatively as the difference between the research and motor octane numbers. Although most gasolines rate lower by the motor method than by the research method, straight-run gasolines have about the same rating by both methods. With present fuels and engines, the average of the research and motor method numbers provides the best correlation with road ratings.

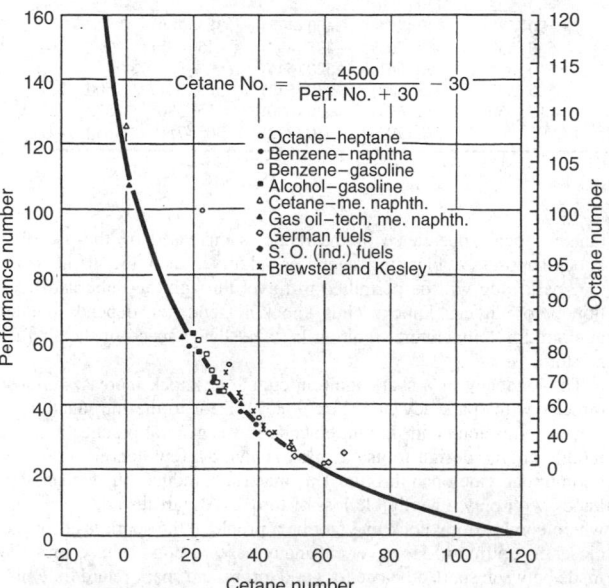

Fig. 9.6.58 Relationship between antiknock scales (octane number and performance number) and cetane number. (*Brewster and Kesley, SAE, 1963.*)

The use of the aviation and supercharge methods for aviation gasolines results in two performance numbers (such as 100/130). This grading indicates that the fuel is less sensitive than iso-octane to the difference in engine severity in the two test methods.

Road ratings of gasoline are determined by means of borderline knock curves for the test fuel and various blends of the reference fuels at various spark advances. The car is accelerated with wide-open throttle, and the speed at which steady knock ceases is recorded as the knock die-out speed (CRC procedure E-2). A plot of these data on a spark-advance vs. knock die-out speed chart (Fig. 9.6.59) makes possible the determination of the octane number rating of the test fuel over the desired speed range.

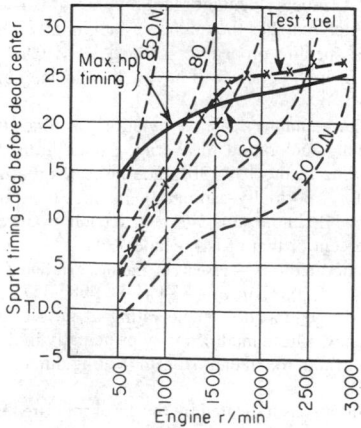

Fig. 9.6.59 Road octane rating.

Mechanical octane number is the resulting decrease in fuel octane requirements brought about by mechanical changes in engine design or control. Efficiency design variables include ignition control, combustion chamber design, cooling, valve timing, fuel system, and engine-transmission combinations.

Knock Suppression The antiknock characteristics of a fuel can be improved by blending in either another fuel of better antiknock characteristics or a knock-suppressor fuel additive (Fig. 9.6.60). Tetraethyllead (TEL) and tetramethyllead (TML) are the most effective knock suppressors (Table 9.6.18) and have been used extensively in small

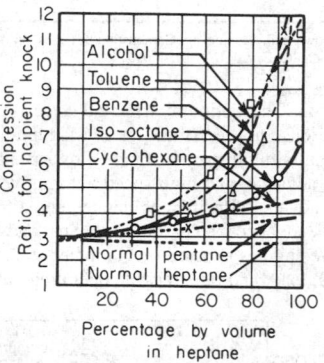

Fig. 9.6.60 Effect of fuel composition on knock-limited compression ratio.

amounts (up to 3 mL/gal of automotive fuels) in many gasolines. Their addition has been eliminated because of adverse effects on health and exhaust emissions. Other compounds can raise antiknock quality, but the much larger amount needed typically changes volatility and stoichiometric point significantly and may affect engine life, emissions, and lubrication.

Cetane Number The cetane number (CN) of a diesel fuel is determined by matching its ignition quality with that of a blend of two reference fuels, normal cetane (CN = 100) and heptamethylnonane (CN = 15). The comparison is made in an ASTM CFR diesel engine (Waukesha Motor Co.) using a specified procedure (ASTM D 613 or CRC-F-S). Cetane number for the reference blend is calculated by

$$CN = (\text{vol \% } n\text{-cetane}) + 0.15(\text{vol \% heptamethylnonane})$$

The **ignition quality** of a fuel for a compression ignition engine is indicated by the time delay between beginning of injection and the start of rapid pressure rise caused by combustion. The compression ratio in the test engine is adjusted to result in a 13° crank-angle delay for the unknown fuel when injection begins at 13° before top center. The mixture of reference fuels which has the same lag under the same conditions indicates the cetane number of the fuel.

The factors that tend to aggravate knocking in a spark ignition gaso-

Table 9.6.18 Relative Effect of Antiknock Compounds

Compound	Chemical symbol	Weight for a given effect, g
Tetraethyllead (TEL)	$Pb(C_2H_6)_4$	0.0295
Aniline	$C_6H_5NH_2$	1.0000
Ethyl iodide	C_2H_5I	1.55
Ethyl alcohol	C_2H_5OH	4.75
Xylene	$C_6H_4(CH_3)_2$	8.0
Toluene	$C_6H_5CH_3$	8.8
Benzene	C_6H_6	9.8

SOURCE: "International Critical Tables," vol. 2, pp. 162–163.

Table 9.6.19   Antiknock Indices (ASTM D4814)

Unleaded fuel* (for vehicles that can or must use unleaded fuel)

Antiknock index (RON + MON)/2	Application
87	Designed to meet antiknock requirements of most 1971 and later model vehicles
89	Satisfies vehicles with somewhat higher antiknock requirements
91 +	Satisfies vehicles with high antiknock requirements

Leaded fuel* (for vehicles that can or must use leaded fuel)

88	For most vehicles that were designed to operate on leaded fuel

* Unleaded fuel having an antiknock index of at least 87 should also have a minimum motor octane number of 82 in order to adequately protect those vehicles that are sensitive to motor octane quality.
SOURCE: Abstracted from ASTM D4814, with permission.

line engine tend to suppress it in a diesel engine. Thus, octane has a low, while heptane has a high, cetane number. High-CN fuels burn more smoothly and start cold engines more readily than low-CN fuels.

Cetane number has been correlated with the mid-boiling point and the API gravity of the fuel, the relation for computing the approximate cetane number being (*Jour. Inst. Petroleum*, **30**, 1944, p. 193)

$$CN = 175.4(\log \text{mid-boiling pt, °F}) + 1.98(\text{API gr}) - 496$$

The deviation from engine test CN for 579 fuels was less than ± 5 CN for 94 percent of the fuels (see also Sec. 7.1).

Ignition accelerators increase the cetane number of diesel fuels, but the susceptibility of the fuel to the accelerator decreases with an increase in the amount of the accelerator (Fig. 9.6.61). Commercial amyl

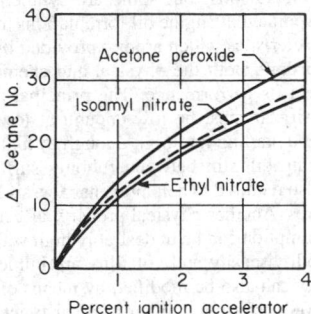

Fig. 9.6.61   Improvement in CN with ignition accelerators. *(Bogen and Wilson.)*

nitrate is slightly less effective than the pure isoamyl nitrate (Bogen and Wilson, *Petroleum Refiner*, July 1944).

Permissible Compression Ratios   Data obtained on single-cylinder test engines and multicylinder automotive engines show wide variations in the compression ratios producing incipient knocking of fuels of various octane numbers. High compression ratios in engines using fuels of a given octane number do not necessarily indicate excellent design and high performance, since valve timing, manifolding, and porting may act as throttling devices, or spark timing may be retarded. Average compression ratios vary considerably depending on the design and fuel used. An increase in cylinder size usually necessitates a lower compression ratio (Fig. 9.6.62), which may be due in some cases to the use of lower grades of fuels with the larger spark ignition engines. The larger diesel engines require less compression to obtain the desired temperatures since the heat transfer effect is less than in the smaller engine.

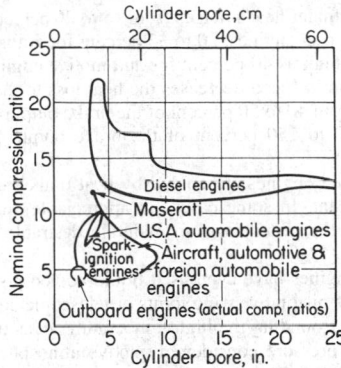

Fig. 9.6.62   Range of compression ratios as a function of cylinder bore.

OUTPUT CONTROL

Speed and **load regulation** are usually obtained in four-stroke cycle gasoline engines by throttling the charge or by quantitative governing. The air/fuel ratio may vary some with load, depending on the engine requirements. Intake throttling reduces the pressure in the cylinder at the end of the intake stroke and thereby reduces the mass of charge trapped per cycle. While throttling is the simplest means for regulation and provides a stable idling condition, it increases the pumping work and consequently decreases the efficiency of the engine. Presently, there is a trend toward using variable valve timings and variable valve lift as a means of controlling load without increasing pumping losses. In direct injection, stratified-charge engines, load control may be accomplished through control of the amount of fuel injected per cylinder.

Power control of carbureted, spark-ignited, two-stroke cycle engines is difficult because intake throttling reduces or prevents cylinder scavenging. Qualitative governing is used in some cases with constant-speed gas engines. The fuel is throttled while the air is unrestricted. Qualitative governing is limited by the lean limit of flammability. In addition, lean mixtures burn more slowly and may lower the thermal efficiency of the engine. Advancing the ignition timing, as the mixture is made leaner, reduces this loss of efficiency.

Governing by variation of ignition timing, as used on some small two-stroke cycle gasoline engines, is undesirable since this is regulation by variation of thermal efficiency and may result in overheating the engine. Variation of ignition timing is required for each change in speed and load to obtain optimum results.

Power control or regulation of diesel engines is obtained by varying the amount of fuel injected per cycle, which provides increasing richness of the fuel-air mixture as the engine load is increased. Maximum specific fuel consumption usually occurs at about 85 percent of the maximum load.

Diesel engines require a speed governor to provide a stable idle. Mechanical, hydraulic, and electronic speed-control governors are used on diesel engines. These control the fuel pump injection quantity and provide speed regulations at all operating conditions.

COOLING SYSTEMS

Engine cylinders must be cooled to maintain a lubricant film on the cylinder walls and other sliding surfaces. The cylinder heads, pistons, and exhaust valves are cooled to prevent combustion knock or destruction of these parts due to overheating. The lubricant must be cooled to maintain the desirable viscosity under operating conditions. Liquid or air cooling systems are used. Pistons, exhaust valves, and lubricants in comparatively small or low-duty engines are sufficiently cooled by contact with other engine parts or the lubricant between them and do not require separate cooling systems.

Liquid Cooling   The heat removed by the cooling liquid from the cylinders and cylinder heads ranges from 15 to 20 percent of the energy input for large diesel engines, 20 to 35 percent for automotive engines, and may run as high as 40 percent for automotive engines at one-third load. An increase in speed decreases the heat loss to the coolant. Heat loss can range from 40 to 50 percent of the brake output for large diesel engines and 100 to 150 percent of the brake output for automotive engines.

Pistons in small engines are cooled by heat transfer to the cylinder walls and lubricant. In some cases, an appreciable quantity of oil is directed against the piston head to maintain desirable head temperatures.

Some large engines have even used liquid-cooled pistons. Liquid is usually fed to the piston by swing joints or telescopic connections. The coolant jacket surrounding the high-temperature areas of the engine is maintained at a pressure considerably above atmospheric—10 to 15 lb/in^2 (70 to 100 kPa). This permits operation at higher coolant temperatures and/or decreases the tendency for local boiling in the region of "hot spots," which could lead to component failures. The average percentage distribution of the coolant to the various jackets is approximately 60 percent to the cylinder jacket and 40 to the cylinder-head jacket with uncooled piston; 50 to the cylinder jacket, 25 to the cylinder-head jacket, and 25 to the piston and piston rod with liquid-cooled piston.

The cooling system may be noncirculating, in which case cold water is supplied, flows through the water jacket, and is wasted. With a temperature rise in large engines of 90°F (32°C), from 2 to 4 gal/(bhp · h) [7.6 to 15 L/(bhp · h)] of coolant must be supplied. In some engines, coolant outlet temperatures as low as 120°F (49°C) must be maintained, which, with high inlet temperatures, may increase the required quantity of coolant to as much as 10 gal/(bhp · h) (38 L)/(bhp · h). Coolant is circulated at a rate of 25 to 50 gal/(bhp · h) [95 to 190 L/(bhp · h)] with a temperature rise of 10 to 20°F (5.6 to 11°C) while flowing through the water jacket.

Natural circulation (**thermosiphon**), with low velocities, requiring large connections, may be used if the coolant forms a complete circuit. In this case, hot coolant rises in the engine jacket, flows to the radiator, is cooled, descends, and flows back to the engine jacket.

Small stationary engines are often **hopper-cooled**. This is an **evaporative cooling system** requiring about 4 to 6 lb/(bhp · h) [1.8 to 2.7 kg/(bhp · h)] of makeup water. The ASTM CFR knock-testing engines are cooled by evaporation, the vapor being condensed by a **reflex condenser** (copper coil with cold water flowing through) and returned to the cylinder jacket.

In high-output engines, the coolant is directed at the hottest spot, usually the exhaust valve seat, with either an external or internal inlet manifold from a common cylinder block jacket; otherwise vapor bubbles will form, cling to the surface, and cause overheating. In vertical engines, the coolant usually flows upward, around the cylinder barrel into the cylinder head jacket, and to the outlet. Higher output and reduced knocking tendency have resulted from directing the entering coolant at the hot spots in the cylinder head and then letting the coolant flow downward around the cylinders.

The **size of piping** required for the inlet to the jacket may be calculated for 3-ft/s (1-m/s) coolant velocity, and for the discharge 2 ft/s (0.6 m/s) for large engines. In engines with recirculation, these values are increased up to 10 to 15 ft/s (3 or 4.5 m/s), and the size of the inlet line to the circulating pump is usually larger than the outlet from the coolant jacket. It is desirable to have an open or visible outlet from each separately cooled part, showing the flow of coolant. To ensure totally filled systems which are as air-free as possible, overflow bottles may be used. Water is an outstanding liquid coolant. However, where freeze protection is necessary or where higher jacket temperatures are desired, other cooling media are generally used. For example, light-duty engines typically use a mixture of water and ethylene glycol. (See Sec. 6.8.)

Air Cooling Many small industrial engines and most motorcycle and aircraft engines are air-cooled. Even here, however, higher specific output engines are increasingly liquid-cooled. Air cooling eliminates the necessity for water or other liquid cooling media, coolant jackets, pumps, radiators, and coolant connections, but necessitates single-cylinder construction, finning, baffles, and, in some cases, blowers. Air-cooled engines also tend to be more noisy since the combustion chamber walls radiate sound and the engine structure is less rigid. The permissible compression ratio and output of air-cooled engines depend on the efficiency of the cooling of the cylinder head, exhaust valves, and seats. Long fins [1 to 2 in (25 to 50 mm) depending on cylinder size], closely spaced [0.10 to 0.20 in (2.5. to 5 mm)], with baffles directing the air at high velocity at the hot spots, have made high-output aircraft engines possible at the expense of considerable air resistance or drag. Properly designed cooling ducts may make use of the heat added to the cooling air to obtain a jet effect, thereby reducing drag.

Oil Cooling Shearing of various oil films by moving parts and oil contact with hot parts of an engine cause a rise in oil temperature until energy input to the oil equals the energy transferred by the oil to the cooler parts that it contacts. Oil coolers are required for high-duty engines to maintain oil temperatures of 200°F (94°C) or under. Temperatures of 250°F (120°C) are not uncommon in automobile engines on hot days: 300°F (150°C) is considered too high, particularly for oils which decompose rapidly under conditions causing the high oil temperatures. The desired oil temperature is maintained by circulating the oil through a radiator or cooler to an oil sump and then through the engine. Transmission oil cooling is also accomplished by placing plate coolers in the end tanks of radiators or with separate radiators.

LUBRICATION
(See also Secs. 6.10 and 8.4)

General A suitable liquid lubricant is vital to the satisfactory operation of an internal combustion engine. It prevents excessive wear and deposit accumulation and removes heat from the areas of relatively high temperature within the engine.

Most engine oils are composed of base oils and additives. The base oils are usually mineral oils, but some are synthetic oils. Chemical additives are incorporated in engine oil formulations to impart desirable performance characteristics which are not provided by base oils alone.

Physical Properties Both the physical and chemical properties of an engine oil affect its performance. The principal physical property involved is viscosity. It must be low enough at low temperatures to permit cranking and starting and high enough at high temperatures to provide an adequate oil film between rubbing surfaces. Sufficiently high viscosity is also required to help prevent excessive oil consumption at high temperatures. Another physical property affecting consumption is volatility. Consumption can be undesirably high with an oil of excessive volatility. Both viscosity and volatility are influenced by base oil selection. Viscosity can also be modified by means of additives.

Chemical additives perform a number of other functions in an engine oil. For example, oxidation inhibitors reduce oxidative and thermal degradation, which can lead to varnish and sludge deposition and to excessive thickening. Detergent-dispersant additives suspend insoluble products formed during engine operation, so they can be removed from the engine when the used oil is drained. Antiwear agents help protect rubbing surfaces, especially those subject to boundary lubrication. Antifoam agents prevent foam and air entrainment and their adverse effects on oil pressure and heat transfer. Rust inhibitors eliminate the formation of corrosion products on ferrous surfaces and reduce corrosive wear. Friction-reducing additives lower boundary friction.

SAE Classifications Engine oils are classified by the Society of Automotive Engineers in two general groups, viscosity (SAE J300 standard) and performance (SAE J183 standard). Viscosity is measured at both 0°F (−18°C) and 210°F (98.9°C). SAE 5 W, 10 W, 15 W, 20 W, and 25 W viscosity numbers are related to 0°F (−18°C) measurements; SAE 20, 30, 40, 50, and 60 to 210°F (98.9°C) measurements. Multigrade oils, such as SAE 5W-20, SAE 10W-30, and SAE20W-40, are those which have viscosity characteristics meeting requirements at both temperatures. Multigrade oils can be used throughout the year, reduce friction and wear, and provide better lubrication and fuel economy during a cold start; e.g., the viscosity of an SAE 20W-40 oil will be less than that of an SAE 40 oil at ambient conditions.

The SAE performance classification includes engine oils ranging in performance quality from that of a straight mineral oil plus a small amount of pour-point and/or foam depressants, to those required in severe passenger car service and heavy-duty truck and off-highway operation. Letter designations such as SG apply to active test techniques for oils for gasoline engines existing since 1988; designations CD and CE apply to oils for naturally aspirated and turbocharged diesel engine service; and designation CD-II applies to oils for two-stroke cycle diesel engines. Performance of these oils is evaluated in both single- and multi-cylinder engine tests. Factors measured include rust, wear, deposit formation, oil consumption, oil thickening, and fuel economy.

REFERENCES: Most of the foregoing information on engine lubrication is covered in greater detail, and references are provided in the "SAE Handbook." See also the list of references in Sec. 6.11.

AIR POLLUTION
(See also Sec. 18.1)

General Relationship

Air pollution from automobile engines (smog) was first detected about 1942 in Los Angeles, CA (see Haagen-Smit, *Sci. Amer.*, **210**, no. 1, 1964, p. 25). Smog arises from sunlight-induced photochemical reactions between nitrogen dioxide and the several hundred hydrocarbons in the atmosphere (see Caplan, SAE Pt. 12, p. 20). Undesirable products of the reactions include ozone, aldehydes, and peroxyacylnitrates (PAN). These are highly oxidizing in nature and cause eye and throat irritation. Visibility-decreasing nitrogen dioxide and aerosols are also formed.

According to an EPA study, transportation engine emissions comprised 42 percent by weight of all U.S. artificial air pollutants in 1980. At that time, virtually all the transportation CO, about one-half of the hydrocarbons and oxygenates, and about one-third of the nitrogen oxides came from gasoline engines. Diesel engines account for particulates. Table 9.6.20 illustrates the situation extant in 1980 and addresses the five categories of air pollutants listed which were attributed to all transportation sources and highway vehicles. Even more current statistics will show trends only, inasmuch as there is a sizable time lag between the collection and publication of data. Suffice it to say that improvements have been noted in many parts of the country and have resulted from improved engine design, universal use of catalytic converters in new automobiles, improved fuel technology, and so forth.

Locally, artificial pollutants can exceed those from natural sources and can become dominant. However, emitted mass does not reflect pollution severity. For example, 1 unit mass of oxides of sulfur may equal 30 units of carbon monoxide or more in terms of health effects.

Emissions from internal combustion engines include those from blowby, evaporation, and exhaust. These can vary considerably in amount and composition depending on engine type, design, and condition; fuel-system type; fuel volatility; and engine operating point. For an automobile without emission control it is estimated that of the organic emissions, 20 to 25 percent arise from blowby, 60 percent from the exhaust, and the balance from evaporative losses primarily from the fuel tank and to a lesser extent from the carburetor (if used). All other nonhydrocarbon emissions emanate from the exhaust.

At least 200 hydrocarbon (HC) and other organic compounds have been identified in exhaust. Some of those compounds, such as the olefins, are referred to as **reactive** since they react rapidly and to a great extent in the atmosphere to form smog products. Others such as the paraffins are virtually unreactive. Methane is deemed unreactive. Of several reactivity scales, the Carter scale (Table 9.6.21) has become most widely used. It ranks nonmethane organic gases (NMOGs) on a scale from 0 to 11 g ozone/g NMOG. The reactivity of a mixture is computed as the sum of the grams of ozone from individual species, with lower levels typical of the larger molecules. A value of 3.42 g NMOG/mi is representative of non-catalyst-equipped cars.

Table 9.6.21   Carter Scale of Reactivity

Nonmethane organic gas	g ozone/g NMOG (approx. range)
Methane	0
C_2-C_5 paraffins	0–1.5
Acetylene	0
Benzene	0.42
C_6 + C_8 paraffins	2–0.2
Toluene (and higher aromatics)	2–10
Alcohols	0.5–2.5
1-alkenes	9–1
Diolefins	11–9
Aldehydes	3–7
Internally double-bonded olefins	10–3

SOURCE: Carter and Atkinson, *Env. Sc. Tech.* **23**, 1989, p. 864.

Atmospheric oxides of nitrogen (commonly termed NO_x) are a mixture of nitrogen oxide (NO) and nitrogen dioxide (NO_2). Oxides of sulfur (SO_z) arising from fuel sulfur are a mixture of SO_2, SO_3, and SO_4. Sulfuric acid may be a product in some oxidizing catalytic exhaust converters.

Emission quantities are most often reported as mole percent for the principal exhaust constituents (CO, H_2O, CO_2, H_2, O_2, N_2) and as moles per million moles (ppm) for trace constituents (HC, NO). Parts per million may be assumed to be equal to volume percent and ppm by volume. Hydrocarbons are reported commonly as either ppm equivalent hexane (C_6H_{14}) or as ppm carbon (C_1). One ppm C_6 is equivalent to 6 ppm C. Other reporting parameters are (1) specific emission rate, g pollutant/kWh or g pollutant/vehicle mi (km), and (2) emission index, g pollutant/1,000 g fuel.

Emission Sources

Homogenous Combustion and Spark Ignition Engines ORGANIC GAS EMISSIONS. Emissions of unburned hydrocarbon and other organic gases can be generated from several sources, as shown in Fig. 9.6.63. The largest source is thought to be flow during compression of un-

Table 9.6.20   Estimated Total Annual U.S. Emissions from Artificial Sources (1980)*

	Carbon monoxide	Hydrocarbons	Sulfur oxides	Nitrogen oxides	Particulates
Total, Tg/yr	85.4	21.8	23.7	20.7	7.8
All transportation, %	81	36	3.8	44	18
Highway vehicles, %	72	29	1.7	32	14

* Current data from the EPA and other sources will provide an updated version of these values. Rapid changes result from greater use of catalytic converters (universal in all spark ignition automotive engines), weather conditions, seasonal variations, etc.

SOURCE: EPA Rept. 450/4-82-001, 1982.

burned fuel-air mixture into the crevice volume between the piston and liner above the top ring; the crevice mixture returns to the combustion during expansion at a time when it is too late to burn. A similar mechanism is caused by the absorption of fuel vapor into oil layers and combustion chamber deposits during compression, and subsequent desorption during expansion and exhaust.

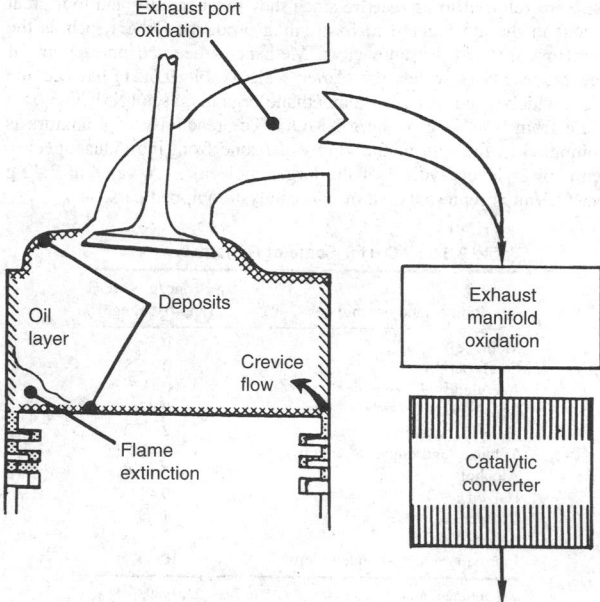

Fig. 9.6.63 Emission sources of unburned hydrocarbon and other organic gases. *(Amann, "The Automotive Spark-Ignition Engine—An Historical Perspective," ASME-ICE vol. 8, 1989, pp. 33–45.)*

Emissions also are generated from flame extinction or wall quenching due to the inability of the flame to propagate totally to a surface or into a crevice. Daniel ("Sixth Symposium on Combustion," Reinhold, 1957, p. 886) described the result of wall quenching to be a visibly dark layer of unburned and partially burned fuel-air mixture left adjacent to all engine combustion chamber surfaces. Subsequent studies showed that the layer adjacent to single walls is largely burned prior to being exhausted. Thus the organic emission from the engine is mainly due to crevice quenching, of which the top piston land clearance is the major source.

Under engine operating conditions where cylinder exhaust residual is high, incomplete flame propagation may leave a portion of the combustion chamber contents unburned. This can produce organic emissions of several thousand ppm C_6. When residual exceeds 15 percent by weight, some incomplete combustion may occur. In some stratified-charge, direct-injection designs, bulk quenching as the flame comes in contact with cooler or leaner gases may also be present.

Wankel engine organic emissions arise from the above sources, but significant additional quantities may arise from mixture leakage through the leading rotor apex seal directly into the exhaust.

In two-stroke engines, where intake pressure exceeds exhaust during the scavenging or overlap period, raw-mixture loss to the exhaust creates very high organic emissions. Emissions from such two-cycle engines may be 10 times higher than those from four-cycle engines. Note also that most two-stroke designs utilize once-through lubrication; i.e., the oil is mixed with the fuel, typically in a fuel/oil ratio of 50:1. Hence, increased deposits and oil films generate organic emissions by physical and chemical storage and release of fuel components.

Any unburned hydrocarbons and other organic gases escaping the cylinder during exhaust can still oxidize in the exhaust temperature and manifold if the exhaust temperature is high enough for the available

residence time and if there is sufficient oxygen present. In addition, where combustion is close to stoichiometric, oxidation of unburned gases is accomplished in catalysts (see later).

CARBON MONOXIDE EMISSIONS.   The principal source of exhaust carbon monoxide (CO) is rich-mixture combustion. Figure 9.6.16 shows exhaust composition as mixture strength is varied. Cycle-to-cycle and cylinder-to-cylinder maldistribution of the fuel-air mixture, especially during transients, increases CO. Additional CO may arise from incomplete combustion of organics in an exhaust treatment device.

OXIDES OF NITROGEN EMISSIONS.   The principal oxide of nitrogen produced inside the engine cylinder is nitrogen oxide (NO). However, since the residence time of gases within the reaction zone is very short, NO forms primarily in postflame combustion gases which do not attain chemical equilibrium. The formation depends primarily on peak combustion gas temperature and to a lesser extent on oxygen content. Once formed, NO tends to be frozen and persists in high, nonequilibrium concentrations during expansion and exhaust. Nitrogen oxide variation with mixture strength is shown in Fig. 9.6.16. Other important variables that affect NO emissions include the amount of exhaust gas recirculation (EGR) and spark timing. When mixtures are more than 30 percent lean or where air is injected into the exhaust, about 5 percent of NO may be oxidized to NO_2 prior to leaving the exhaust pipe. Typically NO and NO_2 emissions are measured together with a chemiluminescent analyzer, and their sum is reported as NO_x emissions.

ALDEHYDE EMISSIONS.   Aldehydes are incomplete combustion products of the oxidation of organics. They arise from the same wall quenching and incomplete combustion mechanisms as the unburned organic emissions themselves, but in addition may be formed in low-temperature exhaust system oxidation reactions. Typically gasoline engine exhaust contains 100 ppm C aldehyde, less for rich and more for lean mixtures. Methanol fuel tends to produce high emission levels of formaldehyde.

PARTICULATE EMISSIONS.   Solid-particle emissions from gasoline engines consist primarily of carbon, metals, and metal oxides. Metals arise from fuel and lubricant additives such as lead, phosphorus, zinc, and barium, as well as engine wear and corrosion. Sulfuric acid particles may be emitted as a result of reactions involving SO_2 over platinum metal catalysts. SO_2 is formed during combustion from fuel sulfur. Liquid aerosols of heavy hydrocarbons are emitted not only during starting, but also during running. Of these, polynuclear aromatics (PNAs) are health concerns.

Heterogeneous Combustion and Diesel Engines   The hydrocarbon, carbon monoxide, and nitrogen oxide emission formation in heterogeneous combustion processes in diesel engines derives from the same basic mechanisms as in the spark ignition engine. However, the relative importance of each is different. In the diesel engine, the emission formation in combustion can be considered to arise from a broad distribution of individual fuel-air elements, some of which burn rich and others lean. Patterson and Henein (Emissions from Combustion Engines and Their Control, *Ann Arbor Sci.*, 1972, p. 260) have suggested the various emission sources in the combustion of a fuel spray injected into swirling air, as shown in Fig. 9.6.64. In addition, fuel impingement on chamber surfaces contributes to unburned hydrocarbons and carbon smoke particles. Hydrocarbon emissions from diesels contain many

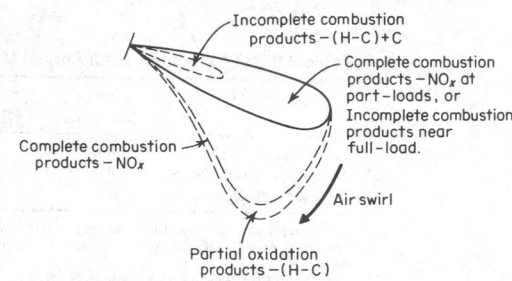

Fig. 9.6.64   Emission formation in a fuel spray injected in swirling air.

heavy fuel types. Normally, heated sampling lines are required to avoid HC condensation during sample analysis.

The amount of carbon, carbon monoxide, hydrocarbon, and nitrogen oxide emissions from diesel engines varies considerably with engine and fuel system design as well as with the operating point. Moreover, because of the high air mass flow rate at partial load, emissions from diesel and gasoline engines cannot be compared directly except on a specific emission basis. Diesel emissions as mole percent or ppm may be corrected for dilution to an equivalence ratio of unity for comparison with spark ignition engine data. On a corrected basis, diesel engines emit in the range of 50 to 500 ppm C_6 and NO_x up to 3,000 ppm. Diesel engines which run lean tend to show higher NO_2/NO ratios (between 10 and 30 percent), especially at light loads when cooler regions quench any NO_2 formed in the reaction zone (Hiliard and Wheeler, SAE 790691, *SAE Trans.*, **88,** 1979). Carbon monoxide emission from the diesel engine is very low, typically under 1,000 ppm (0.1 percent).

Diesel engines emit considerable amounts of carbon smoke and odor constituents. Most carbon arises from pyrolysis in rich combustion regions. Spindt, Barnes, and Somers (SAE paper 710605, 1971) have identified low concentrations of many potential oxidation products and certain fuel fractions as being the odor-producing compounds. Human-panel techniques are commonly used to measure odor.

Strategies for Controlling Spark Ignition Engine Emissions: Effects of Engine Design Variables

Combustion Chamber Shape The shape of the combustion chamber affects organic emissions through a change in surface area and thus heat loss. Surface area relative to the entire clearance volume is an indicator of the volume fraction of organic emissions. This may be stated as

$$\text{Organic, ppm} \propto \frac{A_s}{V_c}$$

where A_s is the chamber surface area and V_c the clearance volume at TDC. According to Sheffler (SAE, Pt. 12, p. 60), the following change the surface/volume ratio, thereby changing hydrocarbon emissions: chamber shape, bore/stroke ratio, compression ratio, and cylinder displacement. In automotive reciprocating engines, surface/volume ratios range from 5 to 8 in²/in³ (2 to 3 cm²/cm³). Wankel rotary engines have surface/volume ratios about 50 percent higher. In well-designed engines, combustion chamber shape has little, if any, effect on CO. A fast-burning chamber increases NO because of an increase in peak temperature.

Ignition System Under most running conditions, improvements in spark energy, rise time, and spark duration produce little improvement in emissions. However, a high-energy, long-duration spark in conjunction with extended electrodes and wider spark gap has been found to minimize the increase in organic emissions due to misfire, as mixtures are leaned well beyond what is chemically correct. (See Quader, *SAE Trans.*, **85,** 1976.)

Exhaust back pressure and valve overlap affect organics and NO through changes in the exhaust residual. Increasing overlap or back pressure lowers NO_x and may lower organic emissions. At very light loads, excessive residual leads to incomplete combustion, thereby increasing organic emissions.

Induction Systems and Fuel-Air Mixture Preparation Optimum mixture preparation and distribution are a key to minimum emissions. With carburetors, high-velocity manifolds, increased manifold heating, and staged throttles have shown emission reductions (see Bartholomew, SAE Pt. 12, 1971). Port fuel injection should be well-targeted on the intake valve to minimize hydrocarbon and soot emission. (See Kosowski, SAE paper 850294, 1985.)

Effects of Operating Variables

Mixture ratio and ignition timing are the most important operating variables affecting emissions. Figure 9.6.16 shows the effect of the mixture ratio on major emission amounts. Retarding ignition timing from the best-efficiency setting reduces HC and NO_x emissions significantly. Excessive retard increases HC and CO emissions. Increasing engine speed reduces HC emissions as ppm. NO_x emissions increase as load increases. Increasing coolant temperature normally reduces HC emissions and increases NO_x emissions.

Limiting fuel enrichment during warm-up is critical to low emissions, especially without exhaust treatment devices. With exhaust treatment, emissions may be reduced to virtually zero once the system is warmed up. Thus the only significant emissions from such systems are those generated during starting and warm-up.

Exhaust Treatment Devices

Catalytic or noncatalytic exhaust treatment devices may be used to clean up remaining exhaust emissions. Thermal reactors are noncatalytic devices which rely on homogeneous bulk gas reactions to oxidize CO and HC. Commonly they appear to be enlarged exhaust manifolds, but they may have internal baffling. In thermal reactors, NO_x is unaffected. Reactions are enhanced by increased exhaust temperature (reduced compression ratio, retarded timing) or by increasing exhaust combustibles (rich mixtures). Typically, temperatures of 1,500°F (800°C) or more are required for peak efficiency. Usually the engine is run rich to give 1 percent CO, and air is injected into the exhaust. A typical thermal reactor residence time is 100 ms. Required reactor volume is commonly 1.5 times engine displacement. Maximum allowable reactor temperature is governed by materials and is typically 1,750°F (950°C). Thermal reactors are seldom used because the required engine calibration produces low fuel economy.

Catalytic systems are capable of reducing NO_x as well as oxidizing CO and HC. However, a reducing environment for NO_x treatment is required which necessitates a richer than chemically correct engine mixture ratio. A two-bed converter may be used in which air is injected into the second stage to oxidize CO and HC. It is highly efficient but results in lower fuel economy. Single-stage, three-way catalysts (TWCs) are widely used, but they require extremely precise fuel control to be effective. As illustrated in Fig. 9.6.65, only in the vicinity of stoichiometric ratio is the efficiency high for all three pollutants. Such TWC systems employ either a zirconia or a titanium oxide exhaust oxygen sensor (EGO, HEGO, or UEGO) and a feedback fuel system to maintain the required mixture strength near stoichiometric. Although a

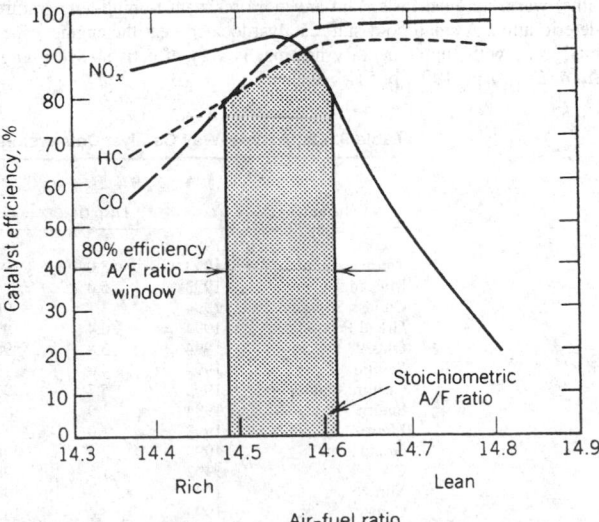

Fig. 9.6.65 Conversion efficiency of three-way catalyst (TWC) as a function of air/fuel ratio. (*Kummer, Prog. Energy Combust. Sci.,* **6,** *1981, pp. 177–199. Reprinted with permission.*)

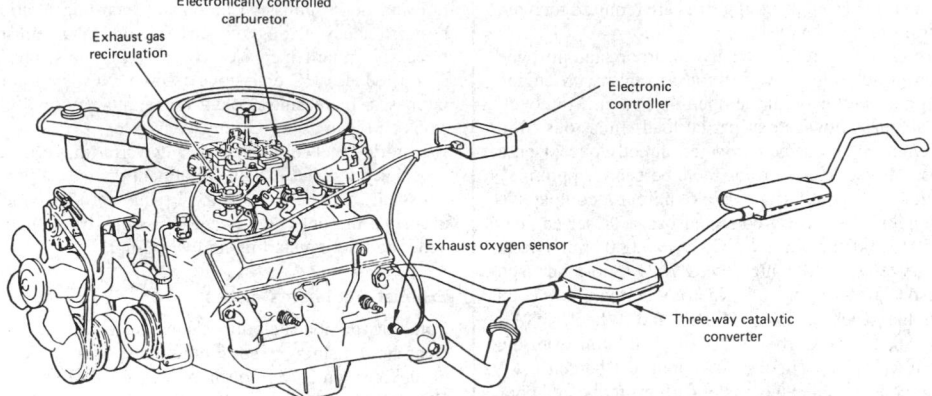

Fig. 9.6.66 Closed-loop three-way catalyst (TWC) system.

carburetor is shown in Fig. 9.6.66, the feedback concept is similar for both central port and multiport fuel injection. Table 9.6.22, which lists average hot stabilized conversion efficiencies during an FTP cycle, demonstrates that many modern vehicles achieve average conversion efficiencies of over 90 percent for all three pollutants once the catalyst has reached normal operating temperatures. Efficiencies are considerably lower on the high-speed acceleration cycle.

Catalyst support beds may be of the pellet or honeycomb (monolith) type. Materials for reducing catalysts include rhodium, Monel, and ruthenium. Materials for oxidizing catalysts include platinum and palladium. The amount of active catalyst material required may be as low as 0.05 troy oz (1.4 g) per vehicle system. In 1996, catalyst volume is about 50 to 100 percent of the engine displacement. The weight of the catalyst bed typically is about 2.2 to 5 lb (1 to 2.3 kg) excluding the metal container. Many aged catalysts can oxidize at 85 percent or higher efficiency at temperatures as low as 800 to 1,000°F (430 to 540°C). Above 1,500°F (820°C), they deteriorate rapidly. Compared with thermal reactors, catalytic converters warm up slowly because of the larger mass of material. Catalytic converters may be located farther from the engine than thermal reactors, since converters operate efficiently at lower temperatures. However, this slows warm-up. Monolithic converters, which can normally be operated hotter, can be smaller and can be placed closer to the engine. Catalysts with metal substrates (monoliths) warm up even faster and are more resistant to high-temperature deterioration. A small cold-start catalyst located near the engine, especially one with supplementary heating, is very effective. (See Laing, *Auto. Eng.*, Apr. 1994, p. 51.)

Strategies for Controlling Diesel Emissions: Engine Design and Operating Variables

Horrocks (*Proc. Inst. Mech. Engr.*, **208**, 1994, pp. 289–297) has summarized the strategies for controlling particulate and HC emissions as follows: (1) optimization of fuel injection and air motion; (2) good atomization of the fuel spray, particularly at idle, at partial load, and during transients; (3) precise control of fuel injection timing, particularly during transients; (4) minimization of parasitic losses in the combustion chamber; (5) low sac volume or valve cover orifice nozzles for direct injection; (6) minimization of contribution of lubricating oil; (7) central injector and combustion chamber for high-speed, direct injection engines; (8) modulated EGR; (9) electronic controls; (10) rapid warm-up of engine.

Control of NO_x emissions from diesels presents additional challenges since suppression of NO formation tends to cause an increase in particulates. Nevertheless, the following strategies are frequently used (Horrocks, op. cit.): (1) injection retard; (2) injection rate shaping; (3) intercooling of turbocharged engines; (4) centralized combustion for high-speed, direct injection; (5) modulated EGR; (6) electronic controls.

Exhaust Treatment Devices

Since diesel engines burn lean overall, the exhaust will always contain excess oxygen. Hence the reduction of NO_x with conventional three-way catalysts is not feasible. Currently, NO_x is removed from the diesel exhaust by either selective catalytic reduction (e.g., Feeley et al., SAE-

Table 9.6.22 Three-Way Catalyst Conversion Efficiency, Percent

Vehicle	Year	Disp (L)	Hot, stabilized FTP			High-speed cycle		
			HC	CO	NO_x	HC	CO	NO_x
Escort	1993	1.9	—	—	—	89.3	41.2	57.8
Taurus	1993	3.0	94.4	90.2	69.9	90.8	56.5	81.5
Mustang	1993	5.0	96.3	97.4	71.9	96.6	89.4	73.9
Cutlass Supreme	1994	3.1	98.3	91.5	84.7	96.0	70.8	84.1
Grand Prix	1994	3.4	97.5	95.1	88.3	88.0	60.2	76.1
Olds 98	1994	3.8	96.7	83.9	94.0	96.9	76.4	92.1
Seville	1994	4.6	97.3	87.6	76.6	93.1	60.1	74.7
Custom Cruiser	1992	5.7	97.1	87.4	90.2	92.9	47.0	62.6
Saturn	1994	1.9	—	—	—	91.6	43.2	91.1
Metro	1993	1.0	—	—	—	85.3	23.0	70.1
Grand Am	1993	2.3	98.3	92.1	93.9	93.0	63.9	98.0
Civic	1992	1.5	98.5	92.8	95.2	86.7	57.6	97.2
Mirage	1993	1.8	98.8	96.4	98.9	94.4	83.6	99.0
Camry	1993	3.0	95.7	92.8	98.4	95.1	79.5	96.3
Corolla	1993	1.8	91.8	86.3	94.9	85.3	66.9	92.3
420 SEL	1993	4.2	99.3	94.4	97.7	99.3	90.9	96.5

SOURCE: German, SAE paper 950812.

Table 9.6.23 Federal Gasoline and Diesel Engine Vehicle Emission Standards (EPA 664200)*†

Year(s)	Test procedure	Nonmethane organic gases, g/mi	Nonmethane hydrocarbons, g/mi	Carbon monoxide, g/mi	Spark-ignited oxides of nitrogen, g/mi	Diesel oxides of nitrogen, g/mi	Particulates, g/mi	Cold carbon monoxide, g/mi	Evaporative losses, g/test
Automobiles: 50,000-mi schedule (100,000-mi); phased in; light-duty trucks up to 3,750-lb loaded vehicle weight (LVW) (*Fed Reg*, v. 56, no. 108, June 5, 1991)									
1995–2002	FTP-tier I	0.257 (0.319)	0.25 (0.31)	3.4 (4.2)	0.4 (0.6)	1.0 (1.25)	0.08 (0.1)	10	2
2003 and later	FTP-tier II	(0.0128)	(0.0125)	(1.7)	(0.2)	(0.4)	(0.1)	10	Not specified
Light-duty trucks between 3,751- and 5,750-lb LVW, phased in (*Fed Reg*, **56**, no. 108, June 5, 1991)									
1995–2002	FTP-tier I		0.32 (0.40)	4.4 (5.5)	0.7 (0.976)	— (0.97)	0.08 (0.1)		2
Light-duty trucks over 5,750-lb LVW, phased in (*Fed Reg*, **56**, no. 108, June 5, 1991)									
1995–2002	FTP-tier I		0.39 (0.56)	5.0 (7.3)	1.1 (1.53)	— (1.53)	— (0.12)		2

		Total hydrocarbons		Carbon monoxide	Oxides of nitrogen	Hydrocarbons + oxides of nitrogen	Particulates		Accel/Lug Pea %
Heavy-duty diesel, emission in g/(bhp·h)									
1994	EPA Trans.	1.3		15.5	5.0		0.10 (0.05 buses 1996)		20/15/50
1998	EPA Trans.	1.3		15.5	4.0		0.10 (0.05 buses)		20/15/50
Off-highway diesels 37 kW and greater, emission in g/kWh (*Fed. Reg.* **59**, no. 116, June 17, 1994); phase-in 1996 to 2000									
1996–2000									
I. 130 kW or more		1.3		11.4	9.2		0.54		20/15/50
II. 75–130 kW; 8 mode					9.2				20/15/50
III. 37–75 kW					9.2				20/15/50
Motorcycles (50 cm³ and over), emission in g/km (*Fed. Reg.* **42**, no. 3, June 28, 1979)									
1980 and later	FTP	5		12					0.6
Small utility engines, emission in g/kWh (*Fed. Reg.*, **60**, no. 127, July 3, 1995)									
1997 and later (SAE cycle J-1088)									
I. Nonhandheld under 225 cm³				469	16.1				
II. Nonhandheld 225 cm³ and over				469	13.4				
III. Handheld under 20 cm³		295		805	5.36				
IV. Handheld 20–50 cm³		241		805	5.36				
V. Handheld 50 cm³ and over		1,561		603	5.36				
SI outboard and personal watercraft. (*Fed. Reg.* **59**, No. 216, Nov. 9, 1994)									
1998–2006	ICOMIA	Phase in		400	6				

Standard calls for average 75% reduction in HC emission from 1990 baseline levels
Baseline HC, g/kWh = $151 + 557/\text{Power}^{0.9}$
Final = 0.25 (baseline HC)

* Numbers in parentheses refer to 100,000 mi standard.
† Refer to the standard, which is periodically updated, for further details of terms, etc.

paper 950747) or catalyzed thermal decomposition of NO into O_2 and N_2. In addition, lean NO_x catalysts are under development.

Diesel particulate traps have been developed which employ ceramic or metal filters (see Wade et al., SAE 830083, 1983). Thermal and catalytic regeneration can burn out the material stored. Particulate standards of 0.2 g/mi may necessitate such traps. Both fuel sulfur and aromatic content contribute strongly to particulates. Catalysts have been developed for diesels which are very effective in oxidizing the organic portion of the particulate (see Ball and Stack, SAE paper 902110, 1990).

Emission Regulations

Light-Duty Passenger Cars and Trucks Emissions are measured on a light-duty driving cycle termed the **Federal Test Procedure (FTP)**. This involves various accelerations, decelerations, and cruise models on a chassis dynamometer. The vehicle is started after a 12-h soak in a 60 to 86°F (16 to 30°C) ambient temperature. Top cycle speed is 56.7 mi/h (91.3 km/h). Test procedures, fuel, and instrumentation are detailed in the Code of Federal Regulations (70, parts 85 and 86, 1986) for cars, trucks, and diesel engines.

In the FTP, the exhaust is introduced into a constant-volume sampler (CVS) in which ambient air is mixed with raw exhaust. Emissions are calculated in grams per mile from average concentrations and total sampler volume flow. In theory, the constant-volume sampler yields true mass emission results. Three sample bags have been used to segregate emissions during the key cycle portions of cold start, hot restart, and cruise. The emission concentrations from each bag are weighed, and the sums are used for certification purposes. Particulate emissions from diesels are measured by weighing filters. Summaries of light-duty, passenger car, and truck standards can be found in Table 9.6.23 (Federal) and Table 9.6.24 (California).

In order to meet the **ultra-low emission vehicle (ULEV)** standards in the cold phase of the FTP-75 test cycle, the time period where the standard

Table 9.6.24   California Air Resources Board Exhaust Emission (CARB) Standards for Passenger Cars (Phased in), 50,000-mi (100,000-mi) Standards

CARB	Effective date	Nonmethane organic gases, g/mi	Carbon monoxide, g/mi	Oxides nitrogen, g/mi
1993 base	1993	0.25 (0.31)	3.4 (4.2)	0.4
TLEV	1995	0.125 (0.156)	3.4 (4.2)	0.4 (0.6)
LEV	1995	0.075 (0.090)	3.4 (4.2)	0.2 (0.3)
ULEV	1995	0.040 (0.055)	1.7 (2.1)	0.2 (0.3)
ZEV	1998	0.000 (0.000)	0.0 (0.0)	0.0 (0.0)

TLEV = transitional low-emission vehicle; LEV = low-emission vehicle; ULEV = ultra-low-emission vehicle; ZEV = zero-emission vehicle.

three-way catalyst has low efficiency (light-off time) needs to be reduced drastically. Exhaust gas ignition (run rich, pump air into the exhaust, and use ignition device such as glow plug to ignite mixture in exhaust), electric-heated catalysts, burner-heated catalysts (Achleitner et al., SAE paper 950479), stainless-steel or double-wall exhaust pipes for insulation, and insulated catalysts are some of the enabling technologies currently being explored. In addition, more sophisticated transient fuel strategies and closed-loop control are employed.

Heavy-Duty Engines Emissions from heavy-duty diesel engines are determined in transient engine dynamometer tests. This involves starts, stops, and speed/load changes. Gaseous and particulate emissions are determined in g/(bhp · h). These standards are shown in Table 9.6.23.

Additional Emission Control The California Air Resources Board (CARB) and the EPA have set standards for a variety of off-highway and marine engines. Some of these are given in Table 9.6.24. Future standards may include a wider group of off-highway engines, locomotives, and light aircraft.

9.7 GAS TURBINES
by John H. Lewis and Albert H. Reinhardt

REFERENCES: Kerrebrock, ''Aircraft Engines and Gas Turbines,'' MIT Press. Koff, Spanning the Globe with Jet Propulsion, 1991, AIAA paper 2987. ''Turbine Engine Directory,'' Flight Int., June 8–14, 1994. Hill and Peterson, ''Mechanics and Thermodynamics of Propulsion,'' Addison-Wesley. Bejan, ''Advanced Engineering Thermodynamics,'' Wiley. Horlock, ''Combined Cycle Powerplants,'' Pergamon. ''Gas Turbine World Handbook,'' published annually, Pequot Publishing, Inc.

INTRODUCTION

The **gas turbine** is a turbine-type engine which operates with gas (usually air) as distinguished from steam or water as the working substance. In its most basic configuration, the gas-turbine engine consists of a compressor to draw in and compress the gas, a combustor or heat source to add energy to the compressed gas, and a turbine to extract power from the heated gas flow. Practical applications of gas turbines first occurred from 1939 to 1941. In 1939, the firm of Brown Boveri of Switzerland used a gas turbine to generate electricity. Also in 1939, the first flight of an aircraft powered by a gas turbine developed by Hans von Ohain took place in Germany. Another aircraft gas turbine was developed by Frank Whittle, who powered an aircraft in 1941 in England. From these early applications, the gas turbine has been developed to the point where today it is the most important aircraft power plant in use. In addition to its aviation applications, the gas turbine is a significant power plant for electric power generation, pipeline pumping power, and marine propulsion. It is a key component of combined-cycle power plants and cogeneration plants (see Fig. 9.7.18 and Sec. 9.4).

Depending upon its development roots and applications, the gas tur-

bine is referred to by many names. In aviation, it is called a jet engine, a turbojet, a jet turbine, a turboprop, a turbofan, and a fanjet. In land and marine applications it is called a gas turbine, a turboshaft, a combustion turbine, and an industrial turbine. In the industrial world, engines generally fall into two categories. **Heavy-duty** or **frame machines** refer to gas turbines designed and built strictly for ground applications. **Aeroderivative machines**, as the name implies, are gas turbines originally designed for aircraft use and adapted for industrial applications. A typical aeroderivative is shown in Fig. 9.7.1. Future trends to marry heavy industrial and aerospace technologies will blur this distinction between the two types.

Gas turbines offer significant advantages compared with other types of engines because they are compact, lightweight, reliable, and efficient. They are capable of rapid start-up, follow transient loading well, and can be operated remotely or left unattended. They have a long service life, long service intervals, and low maintenance costs. Cooling fluids are usually not required. Their compact size high power-to-weight ratio, and high reliability make gas turbines the ideal power plant for aircraft and for portable power plants. In industrial applications, the engine's small footprint allows relatively small foundations and enclosure buildings.

FUELS

Gas turbines use a wide variety of **gaseous and liquid fuels**. For ground and marine applications, natural gas and light distillate oils are most commonly used, but gasified coal and heavy residual oils can also be

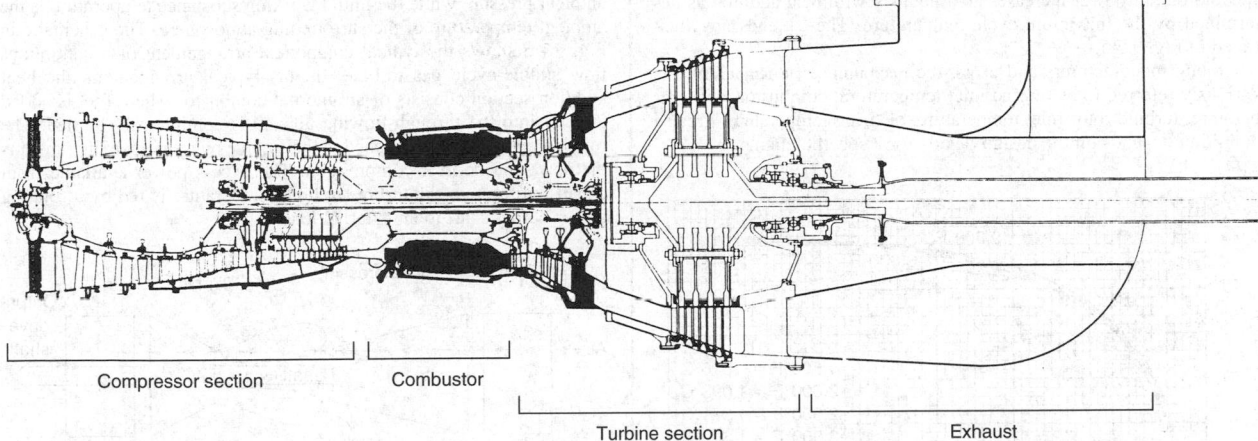

Fig. 9.7.1 Cross section of Pratt and Whitney FT8 gas turbine derived from the JT8D commercial aircraft engine.

used. When fuels that contain significant amounts of sulfur, salt, vanadium, and other metals are used, the hot-section components such as the combustor and turbine must be designed to withstand the corrosive effects of these impurities. For aircraft applications, specific kerosine-type fuels have been developed that control important qualities of the fuel, to obtain maximum performance.

THERMODYNAMIC CYCLE BASIS

The gas turbine operates on the principle of the **Brayton** (or **Joule**) **cycle**. The working substance, usually ambient air, is compressed adiabatically to high pressure where heat is added at constant pressure to elevate the temperature, at which point it is expanded, also adiabatically, back to the original pressure. Because of the expansion from a higher temperature, the work extracted in the expansion process exceeds the work required for compression by an amount that equals the net output work of the cycle. The cycle is completed at the original pressure with heat rejection in the exhaust. See Fig. 9.7.2. At the end of the expansion process, the working substance temperature is higher than its initial temperature by an amount that corresponds to the waste energy to be dissipated at constant pressure to the surroundings. Unique to the Brayton cycle are the heat addition and heat rejection processes at (ideally) constant pressure. The constant-pressure process lends itself to continuous flow of the working substance. When coupled with compressors and turbines which operate on the rotating turbomachinery principle

1-2 Adiabatic compression

2-3 Constant pressure heat addition

3-4 Adiabatic expansion

4-1 Constant pressure heat rejection

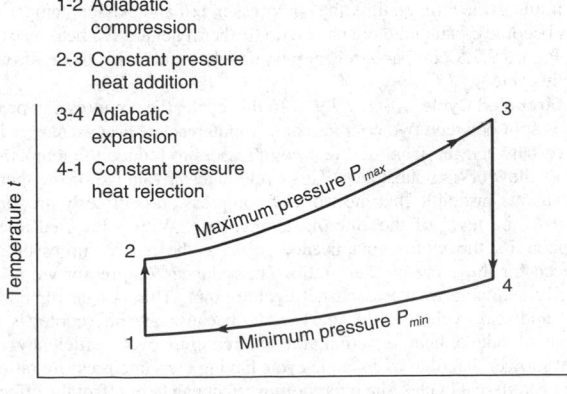

Fig. 9.7.2 Brayton cycle temperature vs. entropy diagram. Ideal cycle.

(see Sec. 14), the entire gas turbine acts as a **continuous flow engine**. The continuously flowing working substance (air) results in very high power density. The gas turbine's intrinsic compactness favors its use over other engine types in many power generation applications.

Temperature-entropy diagrams describe the Brayton cycle. Figure 9.7.2 shows the ideal case where all component efficiencies are 100 percent. Figure 9.7.3 shows the real cycle with nonideal components, including friction losses in the ducting between the compressor and turbine.

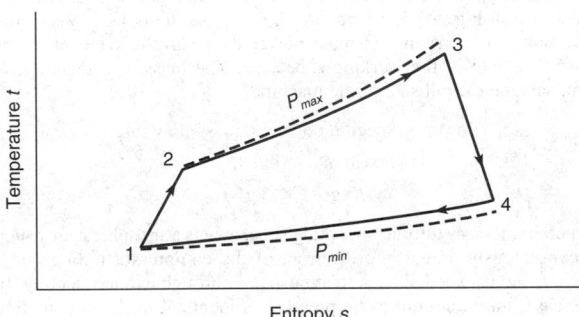

Fig. 9.7.3 Brayton cycle temperature vs. entropy diagram. Real cycle reflects component inefficiencies and pressure losses.

The **ideal Brayton cycle efficiency** η is related to the cycle pressure ratio by the following expression, where P_{max}/P_{min} = pressure ratio, R = gas constant, and c_p (the latter being specific heat at constant pressure properties of the working substance):

$$\eta \text{ Brayton, ideal} = 1 - 1/(P_{max}/P_{min})^{R/C_p} \qquad (9.7.1)$$

Thus with ideal components, the gas-turbine efficiency will be increasingly high as its characteristics maximum/minimum pressure ratio continues to increase approaching 100 percent when the pressure ratio becomes infinitely large. With real components, however, when the pressure ratio increases, the component inefficiencies become larger contributors to exhaust waste heat. The net result is that efficiency reaches a peak at some level and then falls off. The point where this peaking occurs is directly related to the quantity of cycle heat addition, which, in turn, is related to the maximum cycle temperature. As illustrated in Fig. 9.7.4, the peak shifts to a higher efficiency level at high pressure ratio as the turbine temperature increases. At the same time, the specific power output has a similar trend of increasing up to a peak and then falling off as the pressure ratio increases, while the amount of specific output power increases continuously with heat addition as determined by the maximum cycle temperature. These trends are illustrated in Fig. 9.7.5.

Among manufacturers and users, the **maximum cycle temperature** is variously referred to as turbine inlet temperature, combustor exit temperature, turbine rotor inlet temperature, or firing temperature. Precise definition depends on the design details for a specific engine.

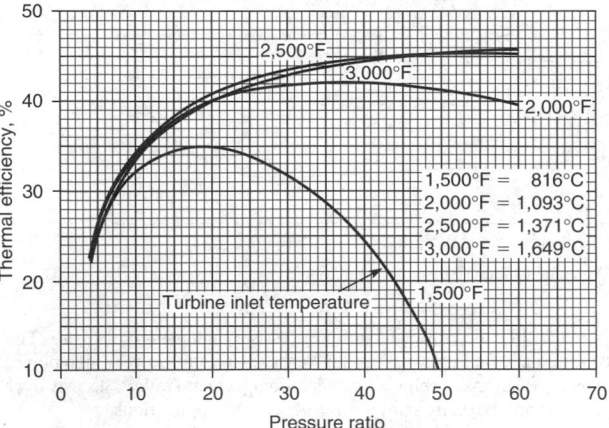

Fig. 9.7.4   Representative Brayton cycle variation of thermal efficiency with pressure ratio at different turbine inlet temperatures.

Specific power refers to output per unit mass flow of the working substance, expressed in hp/(lb·s) or its electrical equivalent kW/(kg·s). As shown in Fig. 9.7.5, a given cycle fixes specific power. When specific power is constant, absolute power then varies directly with the mass flow rate of the working substance. The mass flow quantity, in turn, sets the overall size of the machine

$$\text{Power} = \text{specific power} \times \text{mass flow rate}$$
$$= \text{hp/(lb·s)} \times \text{(lb/s)}$$
$$= \text{kW/(kg·s)} \times \text{(kg/s)}$$

Generally, the gas turbine's working substance is atmospheric air which is continuously drawn into the front of the compressor through inlet ducting and is expelled back to atmosphere through exhaust ducting. In this case, the minimum cycle pressure is identical to the surrounding

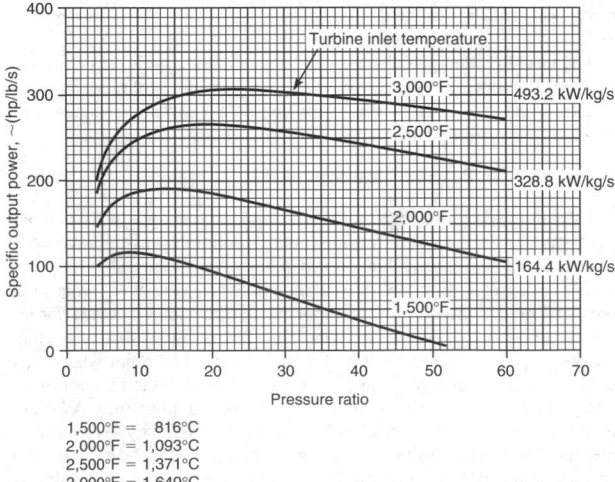

1,500°F =   816°C
2,000°F = 1,093°C
2,500°F = 1,371°C
3,000°F = 1,649°C

Fig. 9.7.5   Representative Brayton cycle variation of specific output power with different pressure ratios and turbine inlet temperatures.

ambient pressure while the initial working substance temperature is the ambient temperature of the surrounding atmosphere. The schematic in Fig. 9.7.6 shows the typical component arrangement of a continuous flow simple-cycle gas turbine. In this typical arrangement, the heat addition section consists of an internal combustor where fuel is added directly into the through-flowing air and is continuously burned. The rotating compressor is attached to the turbine by a shaft which transmits the power needed for compression. The excess power available from expansion through the turbine at high temperature is fed by a rotating power output shaft to an external load.

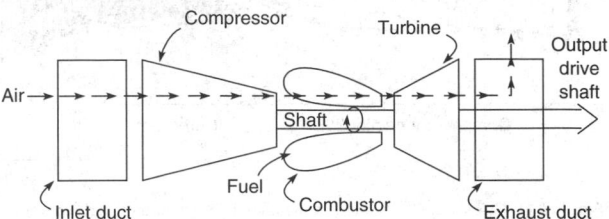

Fig. 9.7.6   Simple Brayton cycle gas-turbine component arrangement.

BRAYTON CYCLE VARIATIONS

Regenerative Cycle   (Fig. 9.7.7)   In this cycle, air exiting the compressor is directed through a heat exchanger, or **regenerator** before being introduced to the combustion section. Then the heated air exhausting from the turbine passes through the other side of the regenerator. By this

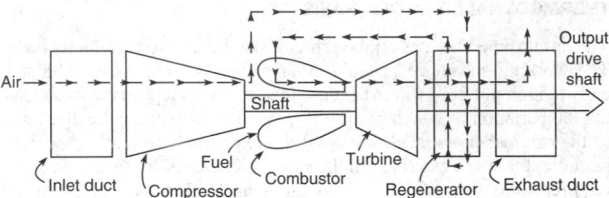

Fig. 9.7.7   Regenerative Brayton cycle gas-turbine component arrangement.

means, some of the exhaust heat is recovered and used as an additional source of energy to heat the high-pressure air, instead of being dissipated as waste heat in the low-pressure exhaust. The effect of regeneration is to increase cycle efficiency, but it is only effective at low levels of pressure ratio. This is because as the pressure ratio increases, the compressor exit air temperature increases and the turbine exhaust temperature is lowered, thereby reducing the regenerator temperature difference necessary for heat transfer. At some point, depending on the maximum cycle temperature, the compressor exit and exhaust temperatures become equal and there can be no further regenerative heat recovery. Figure 9.7.8 compares regenerative cycle trends with those of the simple cycle.

Intercooled Cycle   (Fig. 9.7.9)   In this cycle, the compression process is split between two compressors. An **intercooling heat exchanger** is placed between the front and rear compressors to reduce the temperature of the working substance. This cycle takes advantage of the thermodynamic principle that the work of compression is directly proportional to the level of the entering temperature. With a lowered inlet temperature, the work input needed to drive the rear compressor is reduced for the same pressure ratio. The reduced compressor work is directly available as increased net cycle output. This is true for both ideal and real cycles. In the ideal cycle, because the intercooler is a source of added heat rejection, the intercooling cycle efficiency is always lower than that of a simple cycle having the same pressure ratio. However, in real cycles, the intercooling effect can help offset the effect of real component efficiencies. In some circumstances, intercooling improves both specific power output and efficiency over simple cycle levels. Theoretically, the maximum power output occurs when the total

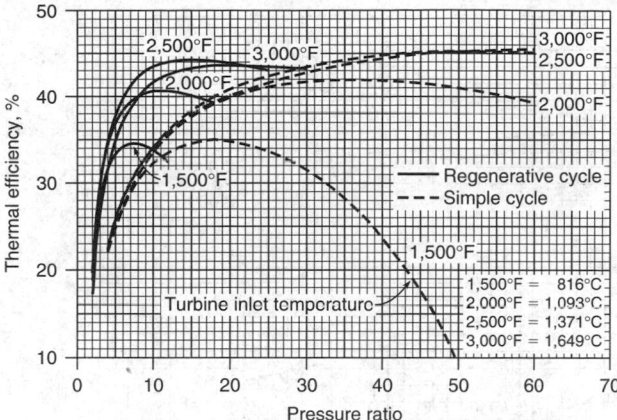

Fig. 9.7.8 Regenerative Brayton cycle variation of thermal efficiency with pressure ratio at different turbine inlet temperatures.

pressure ratio is split evenly between front and rear compressors. However, maximum cycle efficiency occurs at a split which depends on the turbine inlet temperature as well as the efficiency levels of compressors and turbines. Typically, peak efficiency occurs when intercooling takes

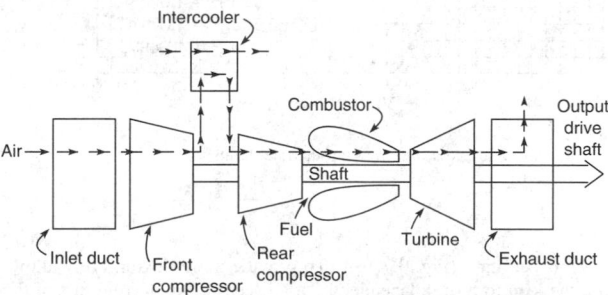

Fig. 9.7.9 Intercooled Brayton cycle gas-turbine component arrangement.

place well in front of the midpoint of the compression process. Figures 9.7.10 and 9.7.11 compare intercooled thermal efficiency and specific power output trends for pressure-ratio splits optimized for power output and for pressure-ratio splits optimized for efficiency.

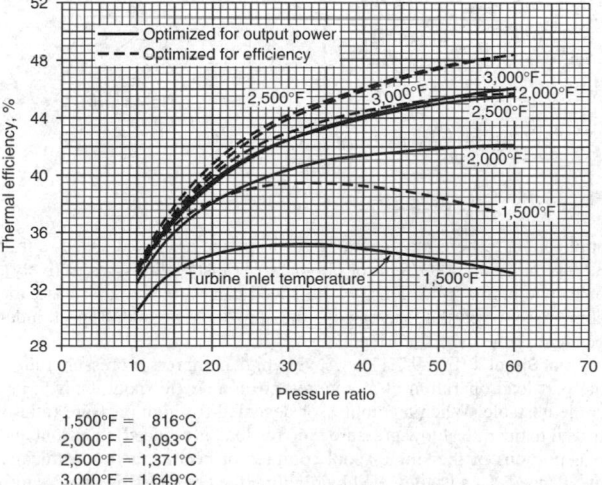

1,500°F = 816°C
2,000°F = 1,093°C
2,500°F = 1,371°C
3,000°F = 1,649°C

Fig. 9.7.10 Intercooled Brayton cycle variation of thermal efficiency with pressure ratios at different turbine inlet temperatures, optimized for output power and for efficiency.

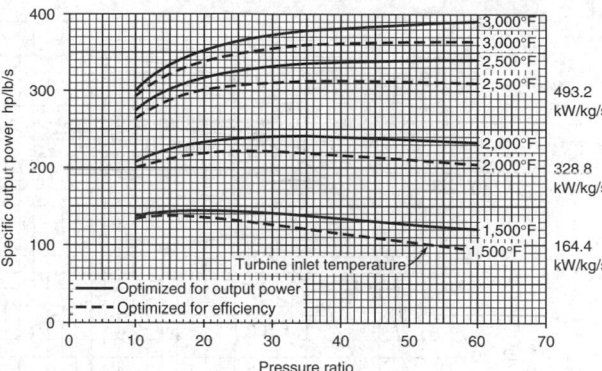

Fig. 9.7.11 Intercooled Brayton cycle variation of specific output power with pressure ratio at different turbine inlet temperatures, optimized for output power and for efficiency.

Reheat Cycle (Fig. 9.7.12) In this cycle, the turbine expansion process is interrupted at some intermediate point before it reaches minimum (or ambient) pressure in the exhaust. At this point, additional heat is added to the working substance by combustion at constant pressure in a second combustor. This takes advantage of the basic thermodynamic principle that the work of any expansion process increases in direct

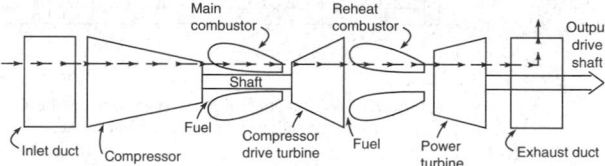

Fig. 9.7.12 Reheat Brayton cycle gas-turbine component arrangement.

proportion to the temperature entering the turbine. (This is the converse of the principle that applies to compression, as cited above for the intercooled process.) The elevated temperature entering the second turbine increases its work output; this additional work appears directly as greater power on the output shaft. Since the extra work is more than offset by the added heat energy of the second combustor, the result is also reduced cycle efficiency.

Associated with the **reheat cycle** is an increase in exhaust temperature which ordinarily represents higher exhaust waste heat. However, in the case of a **combined cycle,** the higher exhaust temperature is a source of waste heat recovery that can be used to generate steam. The higher exhaust temperature can improve cycle efficiency of the steam system to the extent that the **overall system efficiency** can be improved. Reheat thus lends itself to use with exhaust waste heat recovery systems.

Intercooled and Regenerative Cycle This cycle (Fig. 9.7.13) has both an intercooler in the middle of the compression process and a regenerator to capture exhaust waste heat. The intercooling results in a

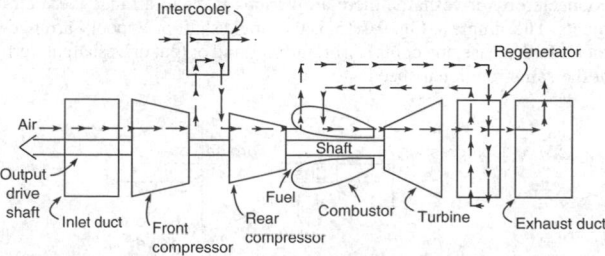

Fig. 9.7.13 Intercooled regenerative Brayton cycle gas-turbine component arrangement.

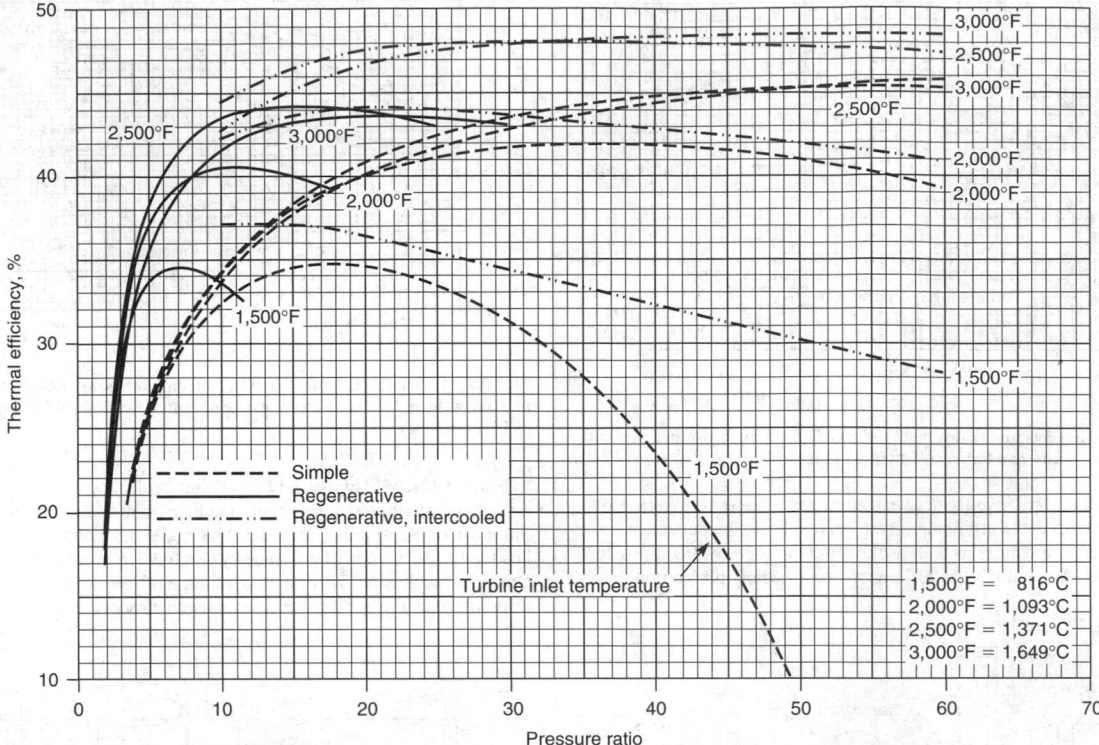

Fig. 9.7.14 Variation of Brayton cycle thermal efficiency with pressure ratio and at different turbine inlet temperatures, for regenerative cycle and for simple cycle.

lower temperature leaving the rear compressor. This circumvents the temperature difference limitation reached in regeneration. That is, at increasingly high pressure-ratio levels, the turbine exhaust temperature remains sufficiently higher than the compressor exit temperature to sustain heat exchange and thus waste heat recovery. Overall, the cycle efficiency of the intercooled regenerative cycle is superior to that of the simple Brayton cycle and all its common variants over a wide range of cycle pressure-ratio levels. Figure 9.7.14 shows variation of efficiency for an intercooled regenerative cycle compared to simple and pure regenerative cycles.

In actual practice, the choice of a simple cycle or one of its variants depends on overall economics applied to a specific use. The extra cycle benefits must trade favorably against the additional cost and complexity of the added heat exchangers or combustors and the penalties of the added ducting in between.

CONFIGURATION VARIATIONS

Since the gas turbine is made up of rotating compressors and turbines connected by driveshafts, there are various ways to arrange these elements. The simplest, Fig. 9.7.15, is the single shaft or "spool" arrangement. In this case, the compressor, turbine, and output driveshaft all turn at the same rotational speed.

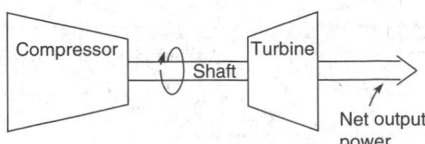

Fig. 9.7.15 Single-spool gas-turbine component arrangement.

Free Turbine (Fig. 9.7.16)   To provide more flexibility by allowing the load to turn independently at a speed different from that of the compressor and its drive turbine, a **free turbine** configuration connects the compressor and its turbine on one driveshaft, while hot air continues to expand through a separate turbine which delivers net output power. The free turbine finds use either in electric generator power

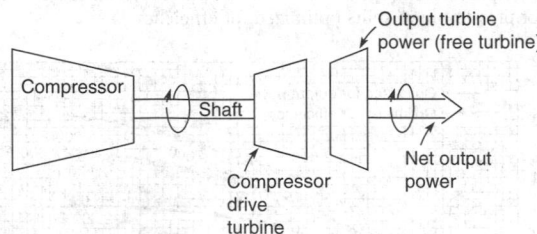

Fig. 9.7.16 Free turbine gas-turbine component arrangement.

applications, where the generator must operate at a fixed speed (either 3,600 r/min or 3,000 r/min for 60- or 50-Hz installations, respectively), or in pumping applications, where speed can vary. In either case, the compressor or partial load can operate at its most favorable speed independent of the driveshaft load.

Dual Spool (Fig. 9.7.17)   At very high compressor pressure ratios, the part load operation of a compressor on a single spool tends to become unstable. When a compressor designed for high pressure ratio is forced to operate at low pressure ratio (as load is reduced), the front and rear portions of the single-spool compressor become aerodynamically unbalanced. This is alleviated by dividing the compression process into two or more spools on separate shafts driven by separate turbines. With this arrangement, the front compressor can slow down more rapidly than the one behind, thus maintaining a more stable overall aerodynamic balance. Conventionally, to avoid unnecessary turning and ducting of the air, the spools are arranged in a straight-through alignment. This is made possible by **concentric shafting**. The shaft connecting the second compressor and its drive turbine (commonly known as the *high-pressure spool*) is made hollow to accommodate the driveshaft which connects the first or front compressor with its drive turbine at the rear (commonly known as the *low-pressure spool*).

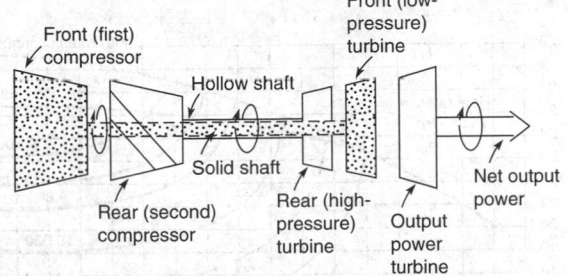

Fig. 9.7.17 Dual-spool gas-turbine component arrangement.

Closed vs. Open System In circumstances where heat addition by internal combustion is not practical, the open cycle using atmospheric air as the working substance may be replaced by another working substance continuously flowing in a **closed circuit**. An example of such a power plant would be the helium-cooled nuclear reactor. As the name implies, helium under high pressure replaces air. With the working substance completely contained, the minimum cycle pressure can be made much higher than ambient. Even such a low-molecular-weight gas as helium can be compressed into a low volume and thus keep the overall power plant relatively compact. Helium lends itself to the nuclear reactor application by virtue of its being almost completely chemically inert, which prevents radioactive transport.

External Combustion The gas turbine can maintain its continuous flow feature without internal combustion; such is the case with the direct use of coal as a fuel. Early attempts to use coal directly in internally fired gas turbines always resulted in severe and rapid damage to the turbine materials. Associated with the coal burning are not only the erosive particulates but also highly corrosive impurities such as sulfur. As an alternative, gas-turbine power plants using indirect firing of coal are being considered. Coal, on a Btu/lb basis, is much cheaper than any other available fossil fuel. Conversely, indirect heat addition through a heat exchanger not only adds expense to the power plant but also wastes energy because of heat exchanger effectiveness limitations. The overall economics of a specific application will dictate power plant choice.

WASTE HEAT RECOVERY SYSTEMS

A characteristic of gas-turbine engines is the incentive to operate at as high a ''firing'' (turbine inlet) temperature as the prevailing technology will allow. This incentive comes from the direct benefit to both specific output power and cycle efficiency, as illustrated in Figs. 9.7.4 and 9.7.5. Associated with the high maximum temperature is a high exhaust temperature which, if not utilized, represents waste heat dissipated to the atmosphere. Schemes to capture this high-temperature waste heat are prevalent in industrial applications of the gas turbine.

Combined Cycle This term commonly refers to the scheme that places a boiler and superheater directly behind the gas turbine to generate steam and produce additional power in a steam turbine (Rankine cycle). A simplified version of a combined cycle is shown in Fig. 9.7.18. Figure 9.7.19 illustrates the combined-cycle process on a temperature vs. entropy diagram. The steam portion is sometimes called the **bottoming** cycle. Early combined cycles had **fired** as well as **unfired** versions. The supplementary firing was required because of the high sensitivity of the Rankine cycle to maximum superheat temperature, as

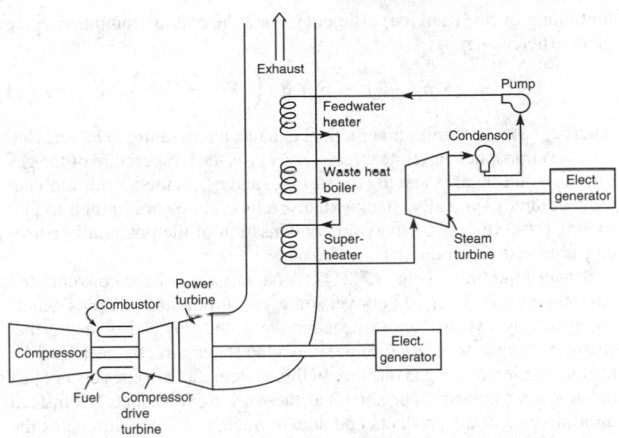

Fig. 9.7.18 Schematic of combined-cycle heat recovery system.

1-2 Feedwater heating
2-3 Boiling
3-4 Superheating
4-5 Expansion
5-1 Condensing

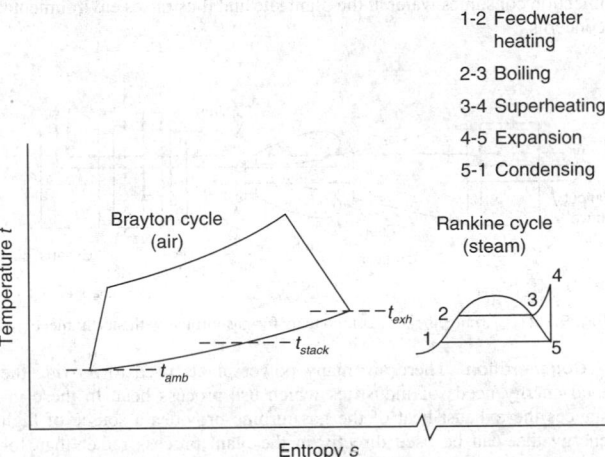

Fig. 9.7.19 Combined-cycle temperature vs. entropy diagram.

shown in Fig. 9.7.20. When the gas-turbine exhaust temperature is too low to sustain a good steam cycle efficiency, the extra combustion in the gas-turbine exhaust could result in an overall system efficiency improvement. Modern gas turbines, with their increasingly high turbine inlet and exhaust temperatures, can approach or exceed 60 percent in **overall combined-cycle plant efficiency** without supplementary firing.

If η_{gt} is the gas-turbine (Brayton) cycle efficiency and η_{st} is the steam

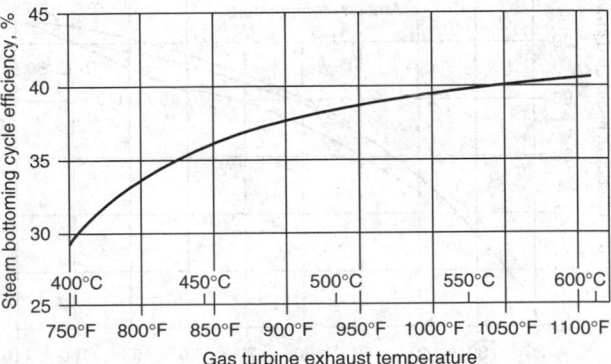

Fig. 9.7.20 Variation of typical steam system heat recovery with turbine exhaust temperature.

bottoming cycle (Rankine) efficiency, then the overall combined-cycle plant efficiency η_{cc} is:

$$\eta_{cc} = \eta_{gt} + (1 - \eta_{gt})\,\eta_{st}\left(\frac{t_{exh} - t_{stack}}{t_{exh} - t_{amb}}\right) \qquad (9.7.2)$$

where t_{stack} is the minimum practical exhaust temperature to be released into the atmosphere after heat recovery, t_{exh} is the temperature of the gas turbine exhaust gases leaving the turbine, and t_{amb} is the outside ambient temperature. (Typically, stack exhaust temperatures are limited to 200 to 300°F (93 to 150°C) to avoid condensation of the potentially corrosive combustion products.)

Steam Injection (Fig. 9.7.21) As a variation of the conventional combined cycle described above, some gas-turbine installations generate steam by exhaust waste heat recovery and, instead of using that steam in a separate steam turbine, inject the steam directly back into the turbine section of the gas turbine. In this system, the turbine acts not just on air (plus combustion products) as the working substance, but instead on a mixture of air products and steam. Such a scheme eliminates the need for additional equipment and is thus less expensive than the normal combined cycle. This system has a drawback in that direct steam injection consumes water at the plant site and thus raises environmental concerns.

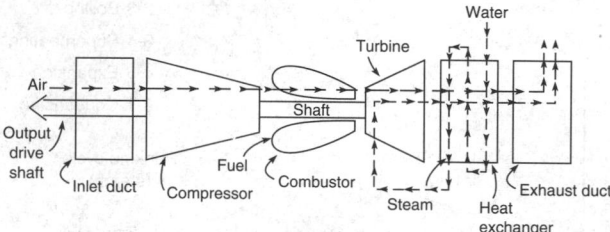

Fig. 9.7.21 Arrangement of components for gas turbine with steam injection.

Cogeneration There are many power plants used to provide the total energy needs of industries which use process heat. In these instances the exhaust heat of the gas turbine provides a source of high energy that can be used directly in the plant process rather than for producing more power. This is particularly true in the chemical industry, where many operational **total energy plants** can be found. In addition, some of or all the exhaust waste heat can be used for **district heating** where, especially in colder climates, steam or warm water generated from the gas-turbine exhaust waste heat can be circulated around the community to heat homes and buildings. Use of gas-turbine exhaust for heating is especially suited to provide the total energy needs of moderate-size to large institutions such as schools, hospitals, or commercial complexes like shopping malls.

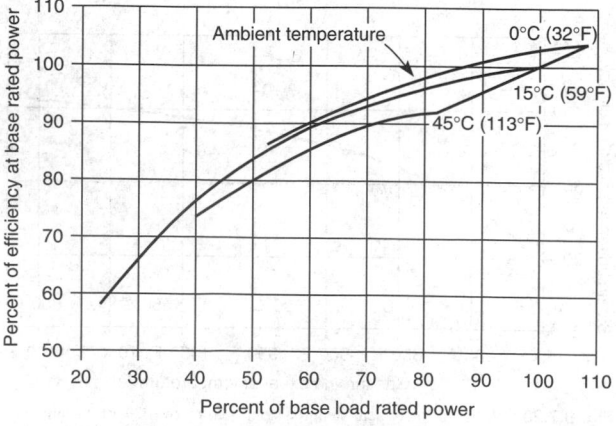

Fig. 9.7.22 Typical partial-load performance characteristics for fixed output speed.

OPERATING CHARACTERISTICS

Partial Load Gas turbines respond rapidly to throttle variations. They are capable of accelerating from idle to full power in minutes or, in the case of the lightweight flight engine designs, seconds. They can be started in a matter of minutes and, via automatic controls, can operate for long periods of time unattended. Partial-load fuel efficiency remains high. Typical partial-load performance characteristics are shown in Fig. 9.7.22 for fixed output rotating speed (r/min) and in Fig. 9.7.23 for a variable-speed (free-turbine) drive.

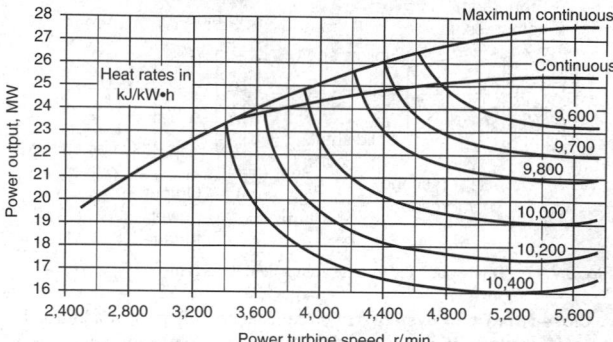

Fig. 9.7.23 Typical partial-load performance characteristics for variable output speed in a free-turbine drive.

Ambient Conditions Gas-turbine operation is sensitive to ambient conditions. Being directly dependent on mass flow for a fixed cycle, power output varies directly with air density and thus ambient pressure. Variations in ambient temperature affect the thermodynamic cycle and engine operating characteristics. Normally, any particular gas-turbine model will have an associated **power rating curve** which varies as shown in Fig. 9.7.24.

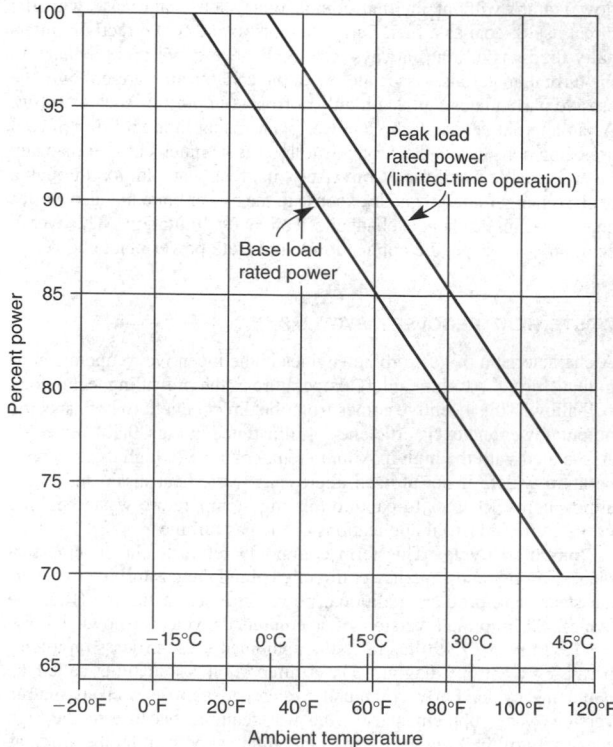

Fig. 9.7.24 Typical variation of output power with ambient temperature.

GAS-TURBINE COMPONENTS

The primary components of a gas turbine consist of a **compressor,** a **combustor** and a **turbine**. Other elements of a gas turbine, depending on the installation, are the electronic control, inlet to the compressor, exhaust nozzle, power turbine, regenerators, and intercoolers. The component that poses the greatest aerodynamic and thermodynamic design challenge, as well as having the greatest influence on engine performance, is the compressor.

Compressors   Compressors can be centrifugal flow, axial flow, or a combination centrifugal and axial flow. In a **centrifugal flow compressor,** the airflow enters at the hub and is then turned from axial to radial by the compressor rotor. Early gas turbines and some modern small engines employ centrifugal compressors. In an **axial flow compressor,** air flows in an axial direction through a series of rotating blades and stationary vanes which are concentric with the axis of rotation. All large gas turbines employ multistage axial flow compressors because of their higher efficiency and capacity. Figure 9.7.25 illustrates the rotor, stators, and assembly of a multistage axial flow compressor. To achieve high pressure ratios and maintain stability at all operating conditions, **dual-spool** or three-spool arrangements are employed, which permit each spool to operate at optimum speed. Interstage bleeds, discharge bleeds, and variable-angle vanes are also employed to avoid surge. Compression ratios in any one spool can exceed 20 : 1 with spools of 10 or more stages, and polytropic efficiency can be as high as 92 percent.

Combustors   A number of arrangements are used for the **combustion chamber** including side-by-side can annular combustors or a single annular combustion chamber (Fig. 9.7.26). Fuel is introduced through a manifold and an arrangement of multiple fuel nozzles. The **combustor** must maintain high combustion efficiency, and low levels of pollutant emissions and smoke, have a minimum pressure loss, maintain stable combustion over a wide range of operating conditions, and produce a uniform temperature in the hot gases distributed to the turbine. The combustor must be designed to utilize some of the airflow to atomize the fuel to achieve efficient combustion, generate turbulent mixing needed to produce a uniform exit temperature, and cool the metal parts of the combustor for durability at high operating temperature. Burner efficiency usually exceeds 99 percent except at low power in the idle region.

Turbines   The **turbine** consists of one or more stages of blades and interstage guide vanes located immediately to the rear of the combustor section. Figure 9.7.27 illustrates the rotors or wheels, sets of stationary

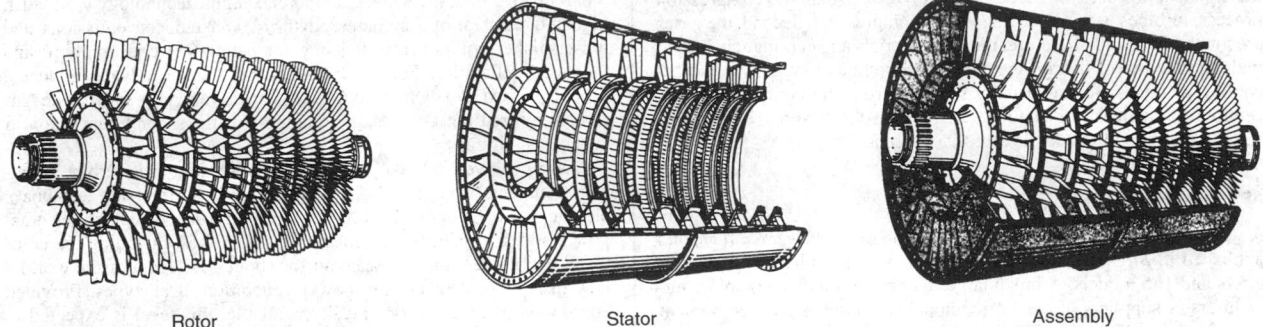

Rotor   Stator   Assembly

Fig. 9.7.25   Components and assembly of axial flow compressor.

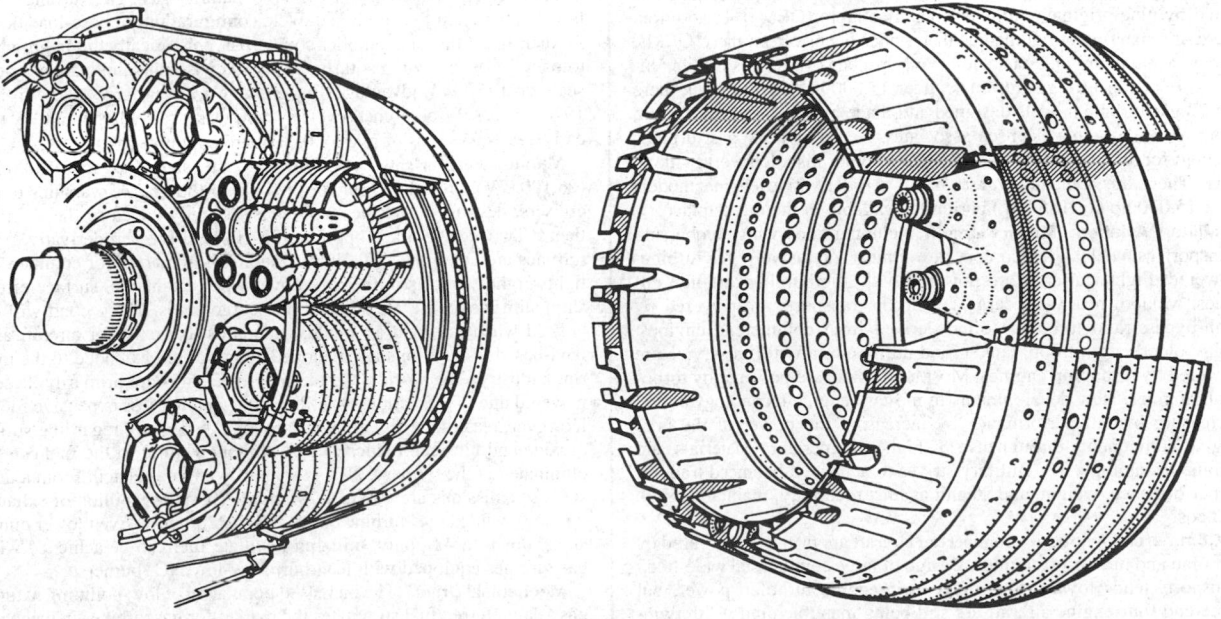

Can annular combustion chamber   Annular combustion chamber

Fig. 9.7.26   Combustion chamber (combustor) configurations.

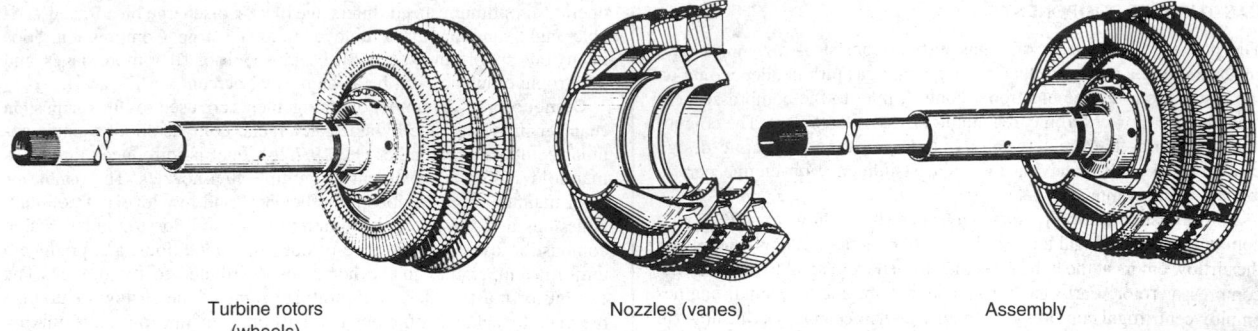

Turbine rotors (wheels)   Nozzles (vanes)   Assembly

Fig. 9.7.27 Turbine elements and assembly.

vanes or nozzles, and their assembly. The turbine extracts kinetic energy from the heated gas stream and converts it to shaft power to drive the compressors and accessories. In the case of a turboprop, turbofan, or turboshaft, power is also extracted to drive a propeller, a fan, or an output shaft (see Sec. 11.5). The arrangement of the turbine is similar to that of the compressor with two major differences. First, since the expansion process is aerodynamically easier than the compression process, turbines have fewer stages than compressors. Second, the gases are much hotter and, therefore, require materials and cooling schemes to maintain the life of the turbine parts. High-temperature alloys, internal cooling passages (for air usually supplied from the compressor), and ceramic thermal barrier coatings are used in the turbine. The normal range of turbine efficiency is 90 to 95 percent.

APPLICATIONS

A worldwide directory of currently available **gas-turbine aircraft engines** published in *Flight International* (June 8–14, 1994) lists 37 manufacturers and 166 models. A much larger number of companies are manufacturers or suppliers of parts or components to the original equipment manufacturers. More than 700 model designations are now or have been in service.

For **industrial gas turbines** above 15,000 hp (11,185 kW), there are 28 gas-turbine models designed, manufactured, and currently being marketed by nine original equipment manufacturers (OEMs). In addition, there are many other companies that, under license from these OEMs, either manufacture and package or only package these gas-turbine engines. For industrial gas turbines under 15,000 hp (11,185 kW), there are 73 gas-turbine models designed and made by 17 companies; again, there are many other licensee associates. Aeroderivative gas turbines account for almost two-thirds of the gas turbines installed worldwide to date. There are 4 manufacturers of 10 aeroderivative engine models over 15,000 hp (11,185 kW) currently available in the marketplace.

Military Aviation **Military aircraft,** including fixed-wing combat and transport as well as helicopters, are almost exclusively gas-turbine-powered. Fighter aircraft are powered with afterburning turbofan engines. Modern heavy-lift cargo and troop transports are powered by high-bypass-ratio turbofans often derived from commercial engines. Many older cargo, troop, tanker, and utility aircraft still in service are powered by turboprop engines. Most helicopters are powered by turboshaft engines. New developments in military engine technology aim at increasing overall performance by increasing the power of the core engine components through use of higher-temperature materials and turbine technology. The military strives to achieve a balanced tradeoff of performance, weight, and stealth against reliability, maintainability, and cost.

Commercial Aviation **Commercial aircraft** are primarily powered by turbofan and turboprop engines. Design of large commercial wide-body transports tends toward long-range twin-engine turbofan power, but three- and four-engine aircraft are still being manufactured. With twin-engine aircraft approaching the size and range of four-engine aircraft, thrust requirements are ever increasing. Turbofan engines in the 60,000- to 90,000-lb (267- to 400-kN) thrust range have been certified, and versions to 100,000 lb (444.8 kN) are planned. Medium-size commercial transports are also turbofan-powered, with engines in the 20,000- to 40,000-lb (89- to 178-kN) thrust range. Small regional or commuter aircraft are primarily powered by turboprop engines from 500 to 3,000 hp (372 to 2,237 kW). Commercial engine technology is aimed at higher bypass ratios for increased thrust and reduced emissions and noise; higher component efficiency for better fuel consumption; and increased reliability. Future developments include advanced ducted propulsors (ADPs) which couple a gear-driven prop fan to the gas turbine, geared fan engines, and engines for the second generation of supersonic aircraft.

Electric Power Of the three generic industrial gas-turbine application markets, i.e., electric, mechanical, and marine (ship and boat), electric power generation comprises roughly 90 percent of the total power added to generating capacity. Gas turbines, including the combined-cycle steam turbine, account for about 40 percent of the world's new installations for electric power generation of all types. Provided appropriate fuels are economically available, this level is expected to increase to over 50 percent. Historically, gas turbines were used primarily for peak duty. Because of marked increases in ease of operation, reliability, thermal efficiency (approaching 60+ percent) as well as the availability of relatively inexpensive natural gas, gas turbines are largely utilized in the combined-cycle configuration for base-load duty. As their fossil fuel steam boiler counterparts historically did in growing to over 1 GW in size, gas turbines are likewise pursuing economy-of-scale cost ($/kW) advantages by growing to very large sizes. In the 1970s, a 75-MW combined-cycle system was a fairly large unit; by the early 1990s, systems of over 400 MW are common.

Marine **Aeroderivative gas turbines,** in sizes up to 35,000 hp (26,100 kW), provide power for the bulk of the surface warships (e.g., cruisers, destroyers, frigates, corvettes) recently added to the world's fleets. Two new market opportunities for marine aeroderivative gas turbines are emerging. The first is the growing market for commercial lightweight, very fast (40+ kn), novel-type (catamaran, surface effect ship, slim monohull, etc.) ferries and cargo ships. The second will be created when the rigid air quality standards that exist for aircraft and land-based installations in developed countries are extended to the marine industry. The world's cargo and cruise ships are primarily diesel-powered due to their ability to utilize poor-quality, less expensive fuels. However, resulting effluent levels are high, especially regarding sulfur dioxide and nitrogen oxides (NO_x). Although sulfur in the fuel can be eliminated at the refinery, NO_x is primarily engine-related. From a diesel NO_x emissions are 600 to 1,700 ppmv (parts per million of exhaust volume), while a gas turbine emits 200 to 220 ppmv. Even lower emissions, down to 42 ppmv, utilizing distillate fuel can be achieved with gas turbines equipped with lean-burn, dry-low NO_x burners.

Mechanical Drive The increased popularity of low-polluting natural gas as a preferred fuel stimulates the market for gas turbines in mechanical drive applications. Mechanical drive gas turbines, up to 40,000 hp

(29,800 kW), are largely aeroderivative and are found in such applications as powering natural gas pipeline compressors, providing gas-lift compressor power for enhanced oil well recovery in offshore platforms and land-based installations, and powering compressors for natural gas gathering, liquefied natural gas production, and various processing plants.

Small Engines Gas turbines in smaller sizes [less than 1,000 hp (746 kW)] are used in several fields of application. Such engines usually feature centrifugal compressors and are all manufactured by specialized companies. In the ground transportation field, automobile and truck manufacturers have experimented since 1950 with gas turbines, primarily of the low-pressure-ratio (less than 10 : 1) regenerative type.

While several prototypes have been tried over the years, displacement of the widespread, well entrenched automobile piston engine has not taken place, with one exception: The U.S. Army has found the gas turbine to be well suited as a tank engine. As with the automobile engine, gas turbines have been experimentally tried in railroad trains, but have yet to make inroads to replace the diesel engine.

Small gas turbines are used as a compact, simple, and reliable power source on board aircraft for start-ups and auxiliary power. They are known as auxiliary power units or APUs. Similar engines find use as quick-start standby units to drive generators for emergency power. Finally, very small jet engines provide propulsion for remotely controlled aircraft with surveillance or cruise missile military missions.

9.8 NUCLEAR POWER

by Louis H. Roddis, Jr., Daniel J. Garner, John E. Gray, Edwin E. Kintner, Nunzio J. Palladino, supplemented by George Sege and Paul E. Norian of the NRC

REFERENCES: Etherington, "Nuclear Engineering Handbook," McGraw-Hill. Hogerton, "The Atomic Energy Deskbook," Reinhold. Kramer, "Nuclear Energy—What It Is and How It Acts," Technical Publishing Co. Blatz, "Radiation Hygiene Handbook," McGraw-Hill. Ellis, "Nuclear Technology for Engineers," McGraw-Hill. Glasstone, "Source Book on Atomic Engineering," Van Nostrand. Murray, "Nuclear Reactor Physics," Prentice-Hall. Annual reports of the International Atomic Energy Agency, Vienna. "Nuclear News," LaGrange Park, IL. Cole, "Understanding Nuclear Power," Gower. Glasstone and Sesonake, "Nuclear Reactor Engineering," 3d ed.; Murray and Powell, "Understanding Radioactive Waste," 3d ed.; Nero, "Guidebook to Nuclear Reactors," 1979; Rahn et al., "A Guide to Nuclear Power Technology," 1984.

NOTE: The material related to nuclear power plant safety and licensing was revised and amended by personnel of the Nuclear Regulatory Commission (NRC) on official time. That material, therefore, is in the public domain and not covered by copyright.

FISSION AND FUSION ENERGY

Nuclear power is the energy derived from the fission (or splitting) of the nuclei of heavy elements, such as uranium or thorium, or from the fusion (or combining) of the nuclei of light elements, such as deuterium or tritium. Particles set in motion by these processes yield heat energy virtually instantaneously. The amount of energy released per atom exceeds by a factor of several million the amount of energy obtainable per atom in a chemical reaction, such as the burning of fossil fuels. While control of the fusion process is still surrounded by tremendous technical problems, considerable success has been achieved in utilizing heat energy produced by fission for power generation, propulsion, industrial production, and scientific experiments.

NUCLEAR PHYSICS

A nuclear reaction occurs when the particles making up the nucleus of an atom are rearranged—as distinct from a chemical reaction, which is caused by changes in the electron structure surrounding the nucleus.

Fundamental (elementary) particles are, theoretically, the irreducible constituents of the material world. More than 20 particles were regarded as elementary in 1972, the exact number being dependent on how they are classified. New experimental techniques lead to discovery of more particles; advances in theory indicate that some are compounds of particles. For example, atoms and atomic nuclei were at one time believed to be noncomposite. Except for the electron and proton, all the fundamental particles are unstable.

An **electron** carries one unit of negatively charged electricity [defined as 1.6×10^{-19} coulomb (C)] and has a mass at rest equal to 1/1,836 times the mass of the hydrogen atom. Electrons emitted by radioactive atoms are called **beta** (β) **rays** and have energies up to several MeV (million electron volts).

A **proton** carries one unit of positively charged electricity equal in magnitude to that of the electron and has a mass, at rest, of 1.0072764 atomic mass units [u when 1 u (1.66057×10^{-27} kg) is defined as $\frac{1}{12}$ the mass of a neutral atom of the most abundant isotope of carbon]. It is identical to the nucleus of the hydrogen atom. Protons are not emitted spontaneously from atomic nuclei.

A **neutron** has a mass approximately equal to that of the proton but lacks an electric charge. It cannot be detected by subjecting it to electric or magnetic fields, nor can its presence be shown by its passage through cloud chambers. The presence of neutrons can be shown only by their interaction with other particles.

An **atom**, which is the basic unit of any chemical element, consists of a central nucleus surrounded by planetary electrons. The number of its electrons determines its chemical characteristics and in electrically neutral atoms equals the positive charge on its nucleus. The nuclei of all atoms except the hydrogen atom are composed of protons and neutrons. The nucleus of the hydrogen atom consists of a single proton, which gives it an atomic number of 1. This number is designated by the letter Z. The number of neutrons in the nucleus is represented by the letter N. From this it follows that the mass number A, which represents the atomic weight, is equal to $N + Z$. In 1977, 105 elements were known. All with atomic numbers above 92 (uranium) must be produced artificially and have a short half-life.

Isotopes are elements which have identical chemical characteristics but different atomic weights, i.e., certain atoms of the same element have different numbers of neutrons in their nuclei. There are considerably more than a thousand isotopes of the known elements. Only 320 of these exist in nature, of which approximately 40 are unstable and radioactive.

For example, hydrogen has three isotopes. One is ordinary hydrogen with a proton nucleus. Another is deuterium, whose nucleus consists of a proton and a neutron. The third is tritium, which consists of one proton and two neutrons. At the heavy end of the natural periodic table is uranium, with 92 protons. When it has 146 neutrons, it is $_{92}U^{238}$, which comprises 99.28 percent of natural uranium. The remainder of natural uranium consists of 0.006 percent U^{234} and 0.71 percent U^{235}. The latter isotope, with 92 protons and 143 neutrons, is the only one of the three which is readily fissionable and is the one utilized in most of the nuclear power reactors operating in the United States. It is separated from the natural element in the enrichment desired by a gaseous diffusion process in government facilities.

Energy and mass are used interchangeably in nuclear physics, and the total energy of a nucleus can be determined by measurement of its exact mass m, which is related to its energy E by the Einstein equation $E = mc^2$, where c is the velocity of light, 3×10^{10} cm/s. This law of equivalence of mass and energy unifies two long-accepted laws: the conservation of energy and the conservation of matter. Singly, the two older laws must be regarded as high-order approximations, adequate in all engineering fields except atomic energy. The exact mass m of a nucleus differs by only a few hundredths of a percent from an integral number A of proton masses, but this small deviation is of significance in nuclear theory, since it measures the difference in energy between the nucleus and its separate components, i.e., its binding energy. The energies of nuclides are recorded in the tables in terms of their masses.

Physical atomic weights do not quite equal chemical atomic weights because the oxygen used in the chemical determinations is a mixture of isotopes, whereas in mass spectrographic measurements the single isotope C^{12} is used for reference. The relationship is: Physical atomic weight = 1.000280 × chemical atomic weight.

Planck's Constant   Since radiation has the properties of both waves and particles, it is theorized that in the emission or absorption of energy by atoms or molecules, the process takes place by steps, each step being the emission or absorption of an **individual quantity of electromagnetic energy,** which is considered the elementary quantum of action. It is known as **Planck's constant** and has the value $h = 6.624 \times 10^{-27}$ erg·s. It is used in virtually all quantum relationships, including Schrödinger's equation and Heisenberg's uncertainty principle, and in the formula for the energy levels of atomic hydrogen. The standard notion is $\hbar = h/2\pi$, where $\hbar$ signifies the angular momentum of an orbiting electron.

Relativistic Mass   When moving particles approach the speed of light, the approximation of kinetic energy, $\Sigma \frac{1}{2} m_0 v^2$, fails; it then becomes necessary to use the equation $E = mc^2$ and to use the relativistic mass, a function that increases with velocity according to the law $m - m_0/\sqrt{1 - (v/c)^2}$, where m_0 is the rest mass (i.e., the mass of newtonian mechanics) and v is the coordinate velocity of the particle.

Binding energy is the work required to disintegrate an atom completely into Z protons and $A - Z$ neutrons. It is a measure of the total kinetic and potential energy of the nucleons in the nucleus and may be calculated from the equations

$$B = C^2 \Delta \quad \text{and} \quad \Delta = Z m_H + (A - Z) m_n - m$$

where m is the mass of the nuclide, m_H the mass of the hydrogen atom, and m_n the mass of the neutron, all on the physical atomic-mass scale (u). The quantity of Δ is known as the **mass defect**. The energy involved is usually stated in MeV.

Fusion   Figure 9.8.1 shows that if two deuterium (H^2) nuclei react to produce one He^3 nucleus plus one neutron, the total binding energy of the four particles (two neutrons plus two protons) involved in the reaction is increased from approximately 4.4 to 7.6 MeV, releasing 3.2 MeV for the reaction. Such a process is called **fusion**. The deuterium-deuterium fusion reaction may also produce an H^3 plus an H^1 nucleus, with a release of 4.0 MeV.

For fusion to occur, the two nuclei must approach each other with exceptionally high kinetic energy in order to overcome their electrostatic repulsion; and since kinetic energy is proportional to absolute temperature, effective fusion can occur only at extremely high temperatures. For fusion to continue as a chain reaction, it is further necessary that nuclear collisions be sufficiently frequent to maintain the high temperature in spite of heat radiation. The condition necessary for frequent collisions is high pressure.

A light-element chain reaction is the principal source of heat in the sun, hydrogen being converted to helium as the net result of a multistage process. The temperature and pressure at which this reaction is sustained at the center of the sun are 35 million °F and 1.5×10^{11} lb/in² respectively.

A deuterium-deuterium reaction can be sustained at temperatures and pressures which are considerably lower than those in the solar reactions but still fantastically high by terrestrial standards. Achievement of a controlled light-element chain reaction for power generation by fusion depends on producing high particle velocity and high density locally by electromagnetic means, without development of correspondingly high temperatures and pressures at the container walls.

Fission is the division of an atomic nucleus into parts of comparable mass, either naturally (spontaneously) or under bombardment (induced) with neutrons, α particles, γ rays, deuterons, or protons. In elements with $Z > 90$, fission can be produced by neutrons of low or moderate energy.

Fission is important because in the process neutrons are emitted that may, under certain conditions, be utilized to produce further fissions, thus leading to a self-sustaining, or ''chain,'' reaction. It is thus possible to ''burn'' uranium to obtain an energy release per atom approximately 10^8 greater than from a chemical fuel. The energy release in the fission of U^{235} is

	MeV
Kinetic energy of fission products	168
Kinetic energy of fission neutrons	5
Energy of γ rays	10
Energy of β rays	5
Energy of neutrons	11
	199

The total of 199 MeV is equivalent to 3.2×10^{-11} W·s, which indicates that it requires the fissioning of 1.12×10^{17} atoms to release a kilowatthour of energy. This number of atoms forms only a speck of matter six thousands of an inch in diameter.

Radioactivity occurs when an unstable nucleus undergoes atomic disintegration by emitting α, β, or $\beta +$ particles, γ or X-ray electromagnetic radiation.

The **alpha (α) particle** is identical with the nucleus of a helium atom. The emission of an α particle creates a new nucleus, with Z reduced by 2 and A decreased by 4 mass units.

Beta (β) decay involves emission of an electron from the nucleus of an atom, thereby increasing Z by unity.

The **gamma (γ) ray** is electromagnetic radiation originating in the nucleus—as differentiated from X-rays, which usually are less energetic and arise from energy adjustments as electrons move between orbital shells outside the nucleus. Gamma rays are emitted instantaneously when a neutron is captured in a nucleus and are frequently emitted following ejection of a particle from a radioactive nucleus.

The **positron ($\beta +$)** is a particle whose mass is the same as that of an electron and whose charge is equal in magnitude but opposite in sign. Positron emission decreases Z by unity.

Electron capture (EC) results when an unstable atom decays by capturing an orbital electron in the nucleus, resulting also in a decrease of Z by unity. This capture produces a vacancy in the orbital shell which is filled by an electron moving from an outer shell, giving the rise to X-ray characteristics.

Half-life is the time required for half the original nuclei in a sample of an isotope to decay; it is given as $0.693/\lambda$ s. Half-lives of the radioactive nuclides vary from 10^{-7} s to 10^{10} years.

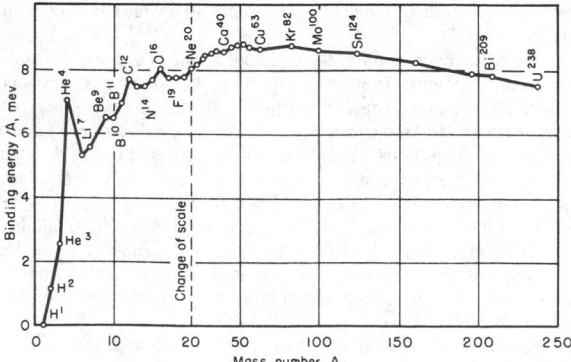

Fig. 9.8.1   Binding energy per nuclear particle for stable nuclei.

UTILIZATION OF FISSION ENERGY

Nonpower Types

Nuclear reactors having a purpose other than the generation of usable power may be classified into two general types: research and test reactors and weapons material production reactors.

The latter category of **production reactors** includes both light- and heavy-water and gas-cooled reactors of high thermal power (comparable in thermal output to large power reactors). Their fuel cycle is adjusted to produce plutonium and tritium of weapons quality for nuclear weapons programs. Most production reactors operate with outlet temperatures near or below the boiling point of water, so the thermal heat is wasted to the environment. A few "dual-purpose" production reactors operate at temperatures which allow the coupling of useful electric power output facilities.

Teaching reactors have the characteristics of very little excess reactivity, very low power capability (a few to a few thousand watts of heat energy), and relatively easy access to the core. Teaching reactors are used mainly for the purpose of instruction in the behavior of a nuclear reactor. With a teaching reactor novice reactor operators may experience the response of reactor instrumentation as reactivity is inserted and criticality is approached, "just critical" behavior, and power increases or decreases.

Teaching reactors are generally owned and operated by universities. In addition to being used for operator training, they are used in a variety of nuclear-engineering instructional experiments. Typical experiments include measurements of neutron-flux distribution; measurement of neutronic characteristics such as generation time, leakage fraction, thermal-utilization factor, resonance escape probability, and other characteristics affecting criticality and kinetics; and measurement of parameters affecting control such as control-rod worth and reactivity feedback coefficients. Although the neutron flux available from a teaching reactor is relatively small, it is nevertheless significant and can be used for limited production of radioactive isotopes.

Research reactors generally have a power capability of 10 thermal MW approximately. They are for the purpose of providing an intense radiation source, both neutrons and gamma rays. A research reactor is usually designed to have a high leakage flux in order to facilitate its use as a radiation source.

The neutrons from a research reactor have a variety of uses including neutron activation analysis, neutron radiography, neutron diffractometry, and production of radioactive isotopes. The gamma rays are applied in materials testing and in studies using the interaction of gamma rays with the nuclei of various elements.

In **neutron activation analysis,** an unknown sample is irradiated in a calibrated neutron flux, and the constituents of the sample become radioactive through neutron absorption. The radiation from the sample is then analyzed. Since the radiation from a specific radioactive isotope is unique to that isotope, a qualitative and quantitative determination of the constituents of the sample is possible. Extremely small quantities of radioactivity can be detected; therefore, neutron activation analysis is a very sensitive analytical tool.

Neutron radiography is a technique similar to X-radiography. The perturbation of a collimated beam of neutrons that has passed through the radiographed item is determined by film exposure or other means. Unlike x-rays that serve to outline dense, highly absorptive materials, neutron radiography serves to outline light materials that preferentially scatter neutrons from the incident beam.

Neutron diffractometry is a technique used to determine the molecular and crystalline structure of materials. The analysis is accomplished through a precise determination of the diffraction pattern from a collimated monoenergetic beam of neutrons that has interacted with the sample.

Certain radioactive isotopes are useful and effective in the **treatment of some diseases.** Isotopes that do not occur naturally can be produced in large quantities by exposure of the precursor element to the neutron fluxes available from research or test reactors.

Test reactors generally have a power capability ranging from 25 to 100 thermal MW. They are used largely for the purpose of providing an experimental radiation environment similar to that which exists in the core of a power reactor. They are also used for large scale radioactive isotope production.

Power Cycles

Although the energy of fission appears as kinetic energy (85 percent) and photon energy of γ rays, there is no practical method of converting this energy directly into useful work on a large scale. Therefore, a nuclear reactor must be treated as a heat source which differs from a chemical heat source in that no oxygen is required and the heat does not have to be removed from gaseous combustion products which possess poor heat-transfer properties.

In the large power capacities associated with nuclear power plants, the **Rankine vapor cycle** is the one in use. The Rankine cycle appears to be best suited for nuclear power because (1) the maximum practicable operating temperature for a reactor corresponds to or is lower than the temperature used in the conventional Rankine vapor cycle; (2) the problem of containment of radioactivity is better solved by this cycle than by other cycles; (3) reactors are most economical in large sizes, as are Rankine engines.

Three variations of the Rankine cycle (Fig. 9.8.2) are being used. The **direct cycle** is the least complicated and thermodynamically the most desirable. Its disadvantages are the facts that the boiling process does not permit the high power densities which are attainable with liquid cooling and that radioactive steam is carried into the turbine and condenser. The **indirect cycle** gives greater power density and eliminates radioactivity in the turbine, although it too produces steam at lower

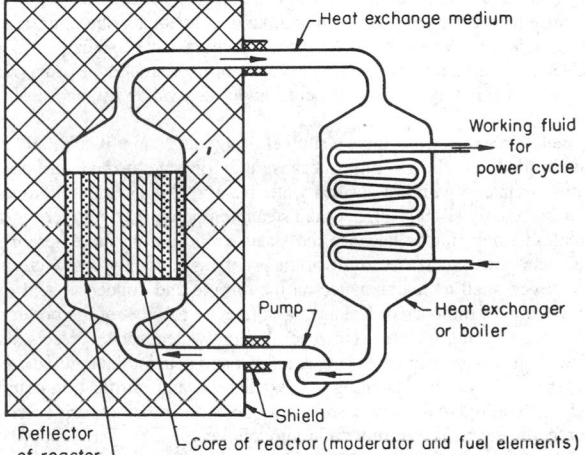

Fig. 9.8.2 Plant cycles—coupling with a working fluid. (*Amorosi, Selection of Reactors, "Nuclear Engineering Handbook," McGraw-Hill, 1957.*)

pressures and temperatures than those considered most efficient for modern turbines. Both concepts are highly developed and commercially available. Plants utilizing sodium or other liquid metal as the heat-exchange medium use an intermediate-link system to isolate the water system from radioactive sodium, but the ability to operate at high temperature with liquid-metal coolants permits the generation of high-temperature steam to utilize the present state of steam turbine technology.

The only other cycle which has been given consideration is the **Brayton cycle**. This cycle is being examined with renewed interest because of the rapid advances in gas-cooled reactor technology and the availability of large gas turbines suitable for reliable utility industry service. The use of an open cycle using air has been discarded because of the spread of activation products from the turbine exhaust and the corrosion problems in the reactor core and fuel elements. The use of a closed Brayton cycle with the latest high-temperature gas-cooled reactors (HTGR) has potential advantages in that the gas temperatures can match the latest gas-turbine designs and a very compact plant can be built.

The ability of a nuclear reactor to produce high temperatures not limited by chemical reaction equilibrium has generated interest in advanced concepts such as magnetohydrodynamics and thermionic generators. These systems are limited by the availability of suitable materials of construction and appear destined to remain in the research phase until suitable materials are developed.

Power Reactor Types

During the early years in the development of power reactors, a large number of possible reactor designs were investigated. In the present state of the industry, the designs which have gained the greatest acceptance all are solid fuel in the form of uranium dioxide. The moderator may be either graphite or water and the coolant may be water, gas, or liquid sodium. Experimental power reactors using a solution of uranium compounds in water or molten salt have been built but have not assumed industrial importance because of many unsolved corrosion problems. Light-water (moderated and cooled) reactors (LWRs), comprised of the boiling water reactor (BWR) and pressurized water reactor (PWR), represent about 99 percent of the operating reactors in the United States. A high-temperature gas-cooled reactor (HTGR) is in operation in a small plant. The liquid-metal-cooled fast breeder reactor (LMFBR) concept is employed in two operating test reactors in the United States, but has been successfully implemented in full-scale operating plants in France and Russia. For the efficient utilization of the energy represented by the world's uranium reserves, a successful breeder reactor is desirable since the present generation of reactors are all consumers of only the naturally occurring fissionable isotope of uranium, U^{235}.

For utilization of natural uranium as a fuel, D_2O is substituted for H_2O as the coolant moderator. The reactor must be refueled more frequently than enriched reactors to maintain an adequate operating reactivity margin. However, D_2O reactors such as the Canadian designs (CANDU) utilize on-line refueling to minimize shutdown time. The success of this type of reactor depends on the availability and cost of D_2O.

Boiling water reactors have a simpler design and can utilize relatively thin-walled vessels and pipes because they operate at moderate pressures compared with pressurized-water reactors. Fuel cladding temperatures are only slightly higher than steam temperatures, and there is an inherent safety factor because the steam-void volume increases on a transient power increase. Some of the problems caused by radioactivity carryover, such as maintenance on the turbine and condenser and the prevention of radioactive leakage, are offset by the cost of the boiler in pressurized water reactors. However, boiling water reactor vessels, despite their lower design pressure, are larger and heavier than pressurized water vessels of similar rating and extensive water purification equipment is required to remove corrosion products and other impurities from the feedwater before introduction into the reactor.

Limitations on the **power density** imposed by exit voids (i.e., the vapor volume of the exiting steam-water mixture from the reactor core) are a disadvantage in that low-power density contributes to high fuel inventory charges. Load changes are accomplished by steam bypass control or rods since adjustment of the turbine throttle and consequent reduction in steam flow will cause an increase in pressure in the reactor, which in turn will collapse the steam bubbles and increase reactivity. The control system functions to maintain constant reactor pressure, and reactivity control is achieved in part by varying the recirculation rate in the reactor. The net decomposition of the water into O and H is much greater than in a pressurized water reactor.

Pressurized Water Reactors Properties of water and steam which led to their predominance as general-purpose heat-transfer media have also caused their widespread application as a reactor coolant. A major disadvantage of water results from its relatively high vapor pressure. However, this can be partially overcome by allowing boiling in the reactor. Thermal efficiencies up to 36 percent are possible.

The use of H_2O as a coolant and moderator is based on well-developed technology which indicates that the ultimate size (or capacity) will be dictated primarily by heat transfer requirements. The average heat flux q/A is around 200,000 Btu/(h·ft²) (630,000 W/m²), with a maximum of 600,000 Btu/(h·ft²) approximately (1,890,000 W/m²).

The **light-water** (H_2O) **reactors** require slightly enriched fuel. However, the cost of enrichment is economically justified because the increased power density reduces inventory charges. To provide sufficient neutron moderation, the H_2O/U volume ratio is kept slightly above 2 : 1.

The use of oxide fuel minimizes corrosion. Design problems are reasonably well understood, although costly structural provisions must be made to load and unload fuel because of the high pressures. Fuel enrichment usually runs between 1.5 and 4.5 percent, depending on the alloy used for corrosion resistance.

Control rods, burnable neutron absorbers, and soluble boron in the coolant are used to control excess reactivity, achieve high fuel burnup, and maintain power distribution within the design limits.

Gas-cooled reactors offer low fuel and operating costs because of their ability to utilize natural uranium as fuel. This advantage is offset by higher capital costs caused by the large system size, in turn resulting from the lower power density usually used. More advanced designs such as the high-temperature gas-cooled reactor and its variants can use low enriched uranium and are able to use higher temperatures and the required structural material for such temperatures. These newer gas-cooled designs are much more compact than the natural-uranium reactors which they are beginning to supplant.

Carbon dioxide is the coolant usually used in natural uranium reactors, as it is nontoxic, only mildly radioactive, nonflammable, noncontaminating in case of leakage, and relatively inexpensive. Helium gives indication of being an ideal coolant, but its limited availability and high cost make it attractive in a power reactor only where the inventory can be kept small and its superior performance at high temperatures can be utilized. Helium, being inert, does not chemically react with graphite and the reactor structure at high temperatures as does carbon dioxide.

The latest gas-cooled reactor designs are prestressed-concrete vessels which contain the core, gas circulators, and steam generators. The inclusion of as much equipment and piping as possible inside the vessel structure greatly simplifies the control of gas leakage.

Fast breeder reactors operate at extremely high power densities (as they are unmoderated) and are liquid-metal-cooled. Their attractiveness stems from an ability to "breed," i.e., produce more fuel than they consume. Aside from savings in fuel cost, this characteristic is believed necessary to conserve the world's supply of energy.

Two conflicting definitions of a breeder reactor are in common use: (1) a nuclear reactor that produces more fissionable material than it consumes, regardless of type of fuel used; (2) a nuclear reactor that produces the same species of material as it consumes, regardless of the net gain or loss. The first definition has wider acceptance in the United States.

The selection of a coolant for a breeder reactor is restricted by the requirement that it not act as a moderator. This requirement rules out the use of water, although the attainment of some breeding in a water-cooled reactor has been investigated. The choice of coolant is limited to helium or liquid metal.

Choice of a liquid-metal coolant is based mainly on nuclear properties and on the engineering difficulties associated with melting temperatures and corrosion effects. Those of prime interest are sodium, sodium-potassium alloy, bismuth, lead, and lead-bismuth alloy. Separated Li^7 would be very attractive were it not for its high cost. Sodium has been selected as the preferred coolant because of its availability, good nuclear properties, and the short half-life of its induced radioactivity. However, in the presence of oxygen, it burns and can react violently with water.

Advantages of operating in the fast neutron spectrum include (1) structural material for the reactor core can be selected without consideration of neutron absorption; (2) very little excess reactivity is needed to compensate for fission product buildup; and (3) reprocessed fuel in which fission product removal may not be complete can be used.

The disadvantages are the increased damage to structural materials by the high fast flux and the control problems resulting from the short neutron lifetime.

Fuel and Waste Cycle

The steps and processes in a typical fuel cycle for light-water reactors are described below. Figure 9.8.3 illustrates the relationship of these steps and processes.

Natural Uranium Uranium occurs naturally in ores in very low concentrations, less than ½ percent by weight. In certain very unusual cases, the uranium content in the ore, called ore grade, may be as high as 10 percent, but these occurrences are extremely rare.

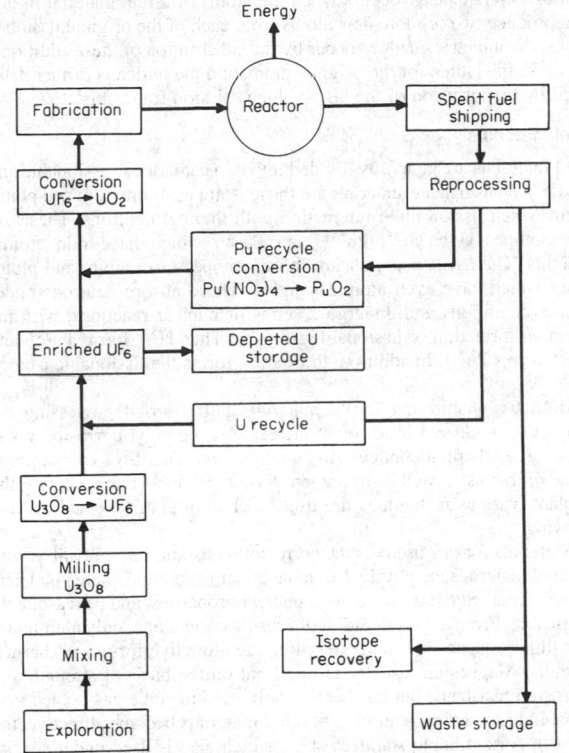

Fig. 9.8.3 Nuclear fuel cycle.

Uranium ores are mined by conventional techniques and then crushed and ground in a mill, which has special equipment for the recovery and purification of the ore into uranium concentrates. The product is in the form of a powder called yellow cake, which has the chemical form of either ammonium or sodium diuranate, both of which are yellow, hence the name.

Conversion of U_3O_8 to UF_6 The process of enriching the U^{235} isotope (from 0.711 percent to the 2 or 3 percent required by water reactors) requires that the uranium be in gaseous form. The gas UF_6 is used; so the natural U_3O_8 must be converted to natural UF_6.

Enrichment alters the relative amounts of U^{235} and U^{238} in uranium, raising the U^{235} concentration from 0.711 percent to the percentage required for use in a reactor. By so doing, the input (feed) stream of natural UF_6 is broken into two output (product plus tails) streams, one enriched and one depleted.

The process of enrichment makes use of the mass difference between U^{235} and U^{238} isotopes. In a gaseous form, the average energies of all UF_6 molecules are equal whether the U atom is U^{235} or U^{238}. Since the energies are equal, the average velocity of the lighter $U^{235}F_6$ molecule is greater than that of the $U^{238}F_6$. Because of this greater velocity and, hence, a greater momentum, the $U^{235}F_6$ molecules can diffuse through a barrier more easily. Gas diffusion plants work on this principle.

In the **gas centrifuge** process, gaseous UF_6 is fed into centrifuges which rotate at very high speed. The heavier U^{238}-bearing molecules drift to the outside wall, while those containing the U^{235} atoms accumu-

late close to the axis of the machine. The separation process is enhanced by longitudinal gas circulation along the wall and near the axis, in currents counter to each other. This also allows end scoops to collect the enriched and depleted streams. Centrifuges are connected in cascade to obtain specific enrichment assay levels and the cascades are operated in parallel to obtain adequate amounts of enriched product. A plant will contain several thousand stages. Electric power demands for the centrifuge enrichment process are typically 5 to 10 percent those of the diffusion process for similar enrichment and amount. The term **separative work unit (SWU)** is used to express the quantity. Centrifuge plants are in use in European countries, but the United States continues to use the more energy-intensive diffusion process, after abandoning a major centrifuge plant half-built.

There are a number of other possible separation processes for uranium (or for that matter, other isotopes of any material). These include various thermal separation processes and plasma processes, activated either by electronic or laser energy, which work on atomic or molecular ions with separation by magnetic field. The lighter isotope will have a slightly smaller radius of curvature and can be caught in a pocket separate from the heavier. Several of these processes have been proven at the experimental level, and some have been used to produce significant amounts of material in the past. The United States, when it abandoned the centrifuge plant, authorized further work on the **atomic vapor laser isotope separation (AVLIS)** process to carry it to full-scale development. This process utilizes vaporized uranium metal, the U^{235} and U^{238} atoms of which are selectively energized to different ionized states by laser energy, and then separated in a magnetic field, all at high temperature. This is potentially even more energy-efficient than the centrifuge process.

Fabrication The gaseous enriched UF_6 that comes out of the enrichment plant is delivered in standard cylindrical gas bottles. The UF_6 is then converted to UO_2, the fuel form is which it is used. The UO_2 is in the form of a powder which is then milled to provide suitable sintering properties. The powder is formed into pellets by cold pressing, sintered, and ground to final dimensions. The pellets are loaded into a zirconium alloy tube (cladding) which is backfilled with helium and sealed at each end with a welded-end plug. The fuel rods are assembled into complete fuel assemblies. The number of fuel rods in an assembly varies from 50 to 200 or more, depending on the reactor type and specific design.

Spent Fuel Storage and Transportation Several options are available for managing spent fuel after its discharge from the reactor. It may be stored at the reactor site for a minimum time of 3 months or longer and then shipped in a shielded cask to a reprocessing plant for recovery of residual fuel valves. This procedure is used in most power reactors, except in the United States. Alternatively, it may be stored at reactor water pools, or in an away-from-reactor storage facility for an extended period of time, and then shipped to an interim dry storage site, possibly after fuel mechanical compaction, or sent to a geological repository for long-term disposal. This latter procedure is mandated by current U.S. law.

Reprocessing and Reconversion Spent nuclear fuel has substantial residual value in its uranium and plutonium, and possibly in other fission products or transuranic elements. The plutonium and other transuranics are the longer-lived radioactive components. The spent fuel is mechanically chopped and dissolved in acid, and the solution processed through subsequent steps to purify the plutonium, uranium, and any other desired product. The remaining fission wastes remain in the high-level waste streams. The plutonium product is usually in the form of a nitrate, and the uranium, either directly or in a subsequent process as UF_6.

Uranium and Plutonium Recycle As described above, the components of value in the spent fuel are the remaining uranium, now of lower enrichment, and the plutonium generated through neutron capture in U^{238}. The uranium can either be reenriched or blended with new enriched uranium and used to make new fuel elements. The plutonium can also be used either alone or mixed with uranium in making new fuel elements. A breeder reactor, by definition, will reuse the plutonium as fuel. Water reactors are nearly all designed to reuse fuel, but this is being done only outside the United States. By law the United States has mandated the use of a once-through cycle, without reprocessing the spent fuel. The economics of the two approaches depend on the relative cost of newly mined uranium.

Waste Disposal Disposal of **low-level radioactive wastes,** from power plants or other radioactive material uses, such as medical is usually done by dry burial in a dedicated site.

High-level waste in the form of solid spent fuel containing transuranics as well as high-level fission products, or reprocessing plant high-level waste converted to a solid form as a glass or ceramic, can be disposed of in a mined geologic repository to ensure long-term isolation of the waste from the environment. Current U.S. policy calls for acceptance of commercial high-level waste by the Department of Energy for disposal in mined repositories beginning in 1998, in accord with the Nuclear Waste Policy Act of 1982. At this time (1996), it seems likely that this goal will remain unmet; rather, it is expected that such repositories will not be ready to accommodate these wastes until sometime in the early twenty-first century. The electric power producing utilities pay a fee on the basis of kilowatthours generated for this service. Isolation will be by the use of both engineered (waste package and container) and geologic (geologic media, such as salt, basalt, or granite) barriers. Other countries, after separating the uranium and plutonium, generally plan to store the waste for extended periods of time before final disposal. Key technical areas in implementation include the development of highly stable materials for encapsulating the waste (e.g., glass, ceramic, corrosion-resistant metals), methodologies for predicting repository performance over extended time periods, and repository closure techniques. Demonstration of the ability to safely dispose of waste has become critical to public acceptance of nuclear power in many countries.

PROPERTIES OF MATERIALS
(See also Table 9.8.1.)

Materials used in nuclear reactors can be divided into six basic categories: fuel, radiation, control, coolant, structural, and shielding. While the nature of the materials may be different, they must be physically and chemically stable and compatible with their environment for a sufficient period to enable the reactor to operate predictably, stably, and economically under the particular set of parameters imposed by the design. (See also Sec. 6.)

In addition to the environmental conditions to which materials may ordinarily be exposed, reactor materials must also sustain the extraordinary bombardment of X-rays (photons) and other atomic particles which are products of nuclear fission. In their passage through matter, these photons and atomic particles may inflict severe damage, since the energy of these particles is spent mainly in ejecting or pulling electrons from their orbits, i.e., producing **ionization.** Gamma photons and neutrons are highly penetrating, whereas fission fragments have a path under 0.001 in the solids and liquids (but cause intense ionization over that path).

The **behavior of materials** following breakage of a bond by ionization may differ markedly. The fragments of simple, highly stable molecules such as water (H_2O) or carbon dioxide (CO_2) will most probably recombine. However, it is highly probable that the parts of complex **organic compounds** such as lubricants and electrical insulation will suffer permanent fragmentation and/or recombination in a different manner to produce new molecules.

Metals, on the other hand, are essentially unaffected by ionization since the probability of displacement of excited atoms in a metal lattice is small and the atoms will merely revert to their original status on ridding themselves of their excess energy. The atoms of **nonconducting crystals** and glasses are similarly immobile, and most of their properties are insensitive to the ionization. However, since these crystals are nonconducting, some of the electrons displaced by ionization become trapped in lattice vacancies, producing changes of optical properties (such as darkening of glass by gamma irradiation) and change of electrical resistivity.

On direct collision, **fast neutrons** may displace atoms from a chemical compound without affording opportunity for immediate recombination. Similarly, **energetic neutrons** may displace atoms from their normal lattice positions in a metal or other crystalline material and push them into abnormal interstitial positions in the lattice, generally producing an effect on metals similar to work hardening. The hardness and tensile strength of the metal will increase, but the metal becomes more brittle. **Thermal neutrons** affect the properties of metals only in those cases where the probability of absorption is high enough to cause significant transmutation to other elements.

Most severe damage occurs when the atoms of a fuel material fission and produce two or more new atoms from each of the original fissioned atoms. Volumetric changes occur by the substitution of these additional atoms in the lattice for the original atom, and the lattice is further damaged by the intrusion of the high-velocity fission fragments.

Fuel Materials

Fuel materials are generally divided into two categories: fissionable and fertile. The **fissionable materials** are those isotopes of uranium and plutonium which fission upon interaction with thermal neutrons. These are the isotopes U^{233} U^{235}, Pu^{239}, and Pu^{241}, which have odd atomic weights. The **fertile materials** are those isotopes of uranium and plutonium which have even atomic weights. These absorb neutrons under proper conditions and undergo a series of nuclear reactions, with the eventual formation of fissionable isotopes. Thus U^{238} forms Pu^{239}, and Pu^{240} forms Pu^{241}. In addition, thorium^{232} forms the fissionable isotope U^{233}.

Both fissionable and fertile materials fall short of possessing the **properties considered ideal** for a nuclear fuel, namely, corrosion resistance, good thermal conductivity, strength, and ductility. To overcome these problems as well as to prevent fission products from entering the coolant, various techniques are used, including plating, cladding, and alloying.

Materials for this use must be corrosion-resistant, be compatible with the fuel materials, be physically stable under irradiation, have good heat transfer characteristics, have good nuclear properties, and be reasonably fabricable. From a consideration of cross section alone, only aluminum, beryllium, magnesium, and zirconium are attractive for use in thermal reactors. Magnesium usually is ruled out on the basis of strength and corrosion resistance but has been widely used in some gas-cooled systems. In fast reactors, a number of other materials become attractive, the most important being stainless steel, which also is used under special conditions in thermal reactors. Consideration also is being given to ceramic materials, BeO, Be_2C, SiC, and $MoSi_2$, which are very attractive in all phases except resistance to thermal shock.

Moderating material must be capable of reducing neutron energy very rapidly and should have good strength at high temperatures, corrosion resistance, thermal and radiation stability, and reasonable cost. Such properties are not available in a single material. Graphite is the most widely used solid moderator. Attention also is being given to beryllium and its compounds, deuterium, oxygen, and hydrogen.

The nuclear and physical properties considered desirable for **fuel diluents and cladding** are, in general, desirable for structural materials. For the latter, however, more emphasis is placed on strength and corrosion resistance than on nuclear properties. Stainless steel, aluminum, zirconium, molybdenum, titanium, niobium, and their alloys are used. As operating temperatures continue to rise, ceramics and cermets will have to be developed for this purpose.

Materials used for **control purposes** must have a high cross section for absorption of neutrons, adequate strength, low mass to permit movement, good corrosion resistance, chemical and dimensional stability under heat and irradiation, and low cost. Boron, cadmium, and hafnium are given the most consideration in various metallic and ceramic forms and as compounds dissolved in the coolant.

Coolant Properties

The choice of a coolant also is a compromise, since no known material possesses all the **desirable characteristics.** These are good heat transfer coefficient and good heat capacity on a volume basis, low absorption

Table 9.8.1 General Properties of Reactor Materials before Irradiation*†

Materials	Density, g/cm^3	Melting point, °C	Specific heat cal/(mol·°C)	Thermal conductivity† cal/(s·cm·°C)	Coefficient of thermal expansion, 10^{-6} per °C	Ultimate tensile strength† 10^3 lb/in²	Modulus of elasticity‡ 10^4 lb/in²	Capture cross sections, 0.025 eV (σ_c barns)
Fuels and fertile materials								
Uranium, normal	19.3	1,133 ± 1	6.649	100°C, 0.063	25–125°C, 45.8	56	24	7.68
Thorium, normal	11.71	1,690 ± 10	100°C, 6.59	100°C, 0.090	30–100°C, 11.5	30.6	10.6	7.56
Plutonium^{239}	19.60	623 ± 7		100°C, 0.061		125	13–16	1.026
Uranium oxide (UO_2)	10.96	2,500–2,600	15.38	100°C, 0.022	22–363°C, 9.3–10.8	17.7	21	24.3
Al-16 weight % U alloy		~650		200°C, 0.42	20–100°C, 20		10.6	
Zr-4 weight % U alloy	6.72	1,840 ± 25		100°C, 0.033	190–300°C, 6.66	63	14.4	
Diluents and cladding materials								
Zirconium	6.50	1,845 ± 25	6.31	50°C, 0.05	5.82	30–38	13.8	0.18
Zircaloy 2	6.55	1,820 ± 25		<0.04	20–100°C, 5.2 ± 0.6	65	14	
Aluminum (28)	2.694	600.2	100 C, 6.07	0.53	20–100°C, 23.8	13	10.3	0.26
Magnesium	1.74	650	25 C, 6.08	18°C, 0.376	49°C, 26	Annealed, 27	Annealed, 6.5	0.069
$MoSi_2$	6.24	1,870		370°C, 0.088	0–1,500°C, 5.1	27°C, 22		
Solid Moderators								
Graphite	2.27	Sublimes at 3,650 ± 25°C	2.066	100°C, 0.37–0.48	20–250°C, 1.9–4.0	0.500–2.4	0.5–1.2	0.0032
Be_2C	6.24	1,870	100°C, 13.69	30°C, 0.02	25–200°C, 7.7		400°C, 23	
Beryllium	1.847	1,315	100°C, 7.7	100°C, 0.34	25–100°C, 11.54	29	44	0.01
Sintered BeO†	2.2–2.8	2,550 ± 25		200°C, 0.19	25–100°C, 5.5	400°C, 15		
SiC	3.21	Decomposes at 2,500°C	327°C, 10.06	400°C, 0.060	0–1,700°C, 4.4	1,700°C, 50	400°C, 39	0.00072
Structural materials								
Stainless steel (347)	8.027	1,427	100°C, 0.12‡	100°C, 0.037	20–100°C, 16.5	85	29	3
Inconel X	8.51	1,395–1,425	25–1,100°C, 0.109‡	0–100°C, 0.036	20–100°C, 11.5	Annealed, 100–120		4.1
Molybdenum	10.2	2,622 ± 10	100°C, 6.24	0°C, 0.32	5.1	67	48	2.7
Titanium (commercial purity)	4.507	1,690	0–500°C, 6.61	25°C, 0.41	25°C, 8.5	85–100	15.5	5.8
Niobium	8.57	2,415	0°C, 6.01			1,200°C, 14.7		1.1
Control materials								
Cadmium	8.694	321	25–321°C, 6.19	0.22	20–100°C, 31.8	10.3	7.1–10	2,450
Hafnium	13.36	2,130 ± 15	25–2,227°C, 6.16		0–100°C, 5.9	67.5	14	105

* All properties change with irradiation at room temperature except when specified.

† Thermal conductivity, strength, and the modulus of elasticity vary directly with the density; however, the coefficient of thermal expansion appears unaffected.

‡ Cal/(g · °C).

SOURCES: H. A. Seller, Properties of Reactor Materials; "Nuclear Engineering Handbook," McGraw-Hill, 1958; "Reactor Physics Constants," ANL 5800, 1963.

Table 9.8.2 Properties of Nuclear Coolants*†

	Water	Heavy water	Diphenyl	Dowtherm-A	Santowax R
			Aqueous and organic liquids		
Density, lb/ft³	42.4 (600°F)	46.2 (600°F)	46.4 (675°F)	46.6 (675°F)	52.29 (675°F)
Viscosity, lb/(h · ft)	0.20 (600°F)	0.220 (600°F)	0.353 (675°F)	0.7986 (675°F)	0.630 (675°F)
Melting point, °F	32	39	156	53.2	293
Boiling point, °F	212	214	49	496	687–784
Heat capacity, Btu/(lb · °F)	1.54 (600°F)	1.68 (600°F)	0.637 (675°F)	0.675 (675°F)	0.539 (675°F)
Thermal conductivity,	0.294 (600°F)	0.294 (600°F)	0.0763 (675°F)	0.098 (675°F)	0.064 (675°F)
Capture cross section σ_a, 0.025 eV (barns)	0.66	0.0011	3.3	3.3	4.6

	Hydrogen	Helium	CO_2	Air	Nitrogen	Steam	Neon
				Gases			
Density, lb/ft³	0.00185	0.00375	0.041	0.0275	0.026	0.0167	0.0190
Viscosity, lb/(h · ft)	0.042	0.094	0.081	0.089	0.086	0.069	0.1445
Heat capacity, Btu/(lb · °F)	3.625	1.245	0.283	0.263	0.268	0.518	0.246
Thermal conductivity, Btu/(h · ft · °F)	0.224	0.159	0.0328	0.0337	0.0327	0.0341	0.0536
Prandtl no.	0.660	0.735	0.6988	0.6898	0.7097	1.048	0.6632

	Gallium	Lithium	Potassium	Rubidium	Sodium	Tin	Na (56%) K (44%)	Pb (44.5%) Bi (55.5%)
				Liquid metals				
Density, lb/ft³	359	29.9	44.6	84.4	51.2	421	48.8	625.5
Viscosity, lb/(h · ft)	1.836	1.188	0.3816	0.415	0.5040	2.736	0.435	2.88
Melting point, °F	85.86	354	147	102	208	449	66.2	257
Boiling point, °F	3,601	2,403	1,400	1,270	1,621	4,118	1,518	3,038
Heat capacity, Btu/(lb · ft · °F)	0.082	1.0	0.180	0.0877	0.3005	0.0639	0.2484	0.035
Thermal conductivity, Btu/(h · ft · °F)	18.0	18.0	21.15	13.2	37.7	19.0	16.35	8.05
Capture cross section σ_a, 0.025 eV (barns)	2.77	71.0†	1.97	0.70	0.505	0.60		0.094

* At 1,000°F unless otherwise specified. Adapted from L. Green, Reactor Coolant Properties, *Nucleonics*, **19**, no. 11, which contains a detailed bibliography.
† Thermodynamic properties of a variety of other specific materials are listed also in Secs. 4.1, 4.2, and 6.1.

cross section, low vapor pressure at the operating temperature, minimum of chemical reaction with the material contacted by the coolant, good resistance to forming γ emitters with long half-lives, stability under irradiation and high temperatures, and workable melting points.

Coolants are divided into three broad **classes:** aqueous and organic liquids, liquid metals, and gases. **Water** has found the widest application, largely because it is readily available and economical, has fairly good heat transfer properties, and offers no corrosion problems which cannot be handled. Its most serious drawback is its high vapor pressure. **Heavy water** has superior qualifications except for cost, which is high. Diphenyl and other **organic liquids** have demonstrated an ability to stand up under irradiation but are handicapped by a tendency to deposit decomposition products on the fuel element plates. See Table 9.8.2.

Liquid metals are attractive because they provide high heat-transfer rates and do not require pressurizing to operate at high temperatures. Sodium is most attractive because of its high boiling point and high heat conductivity, but it reacts violently with water and has poor lubricating qualities. Lithium 7 better meets qualifications but is harder to obtain and is costly.

Gas coolants offset their advantage of being inert at elevated temperatures with poor heat transfer qualities. They must be operated at high pressures to be useful in power reactors. Sodium hydroxide and yellow phosphorus are among other materials considered as coolants, but their acceptability has not been demonstrated.

FISSION REACTOR DESIGN

Neutron Balance The main objective of reactor design is to achieve neutron balance during reactor operation. At steady state, the number of neutrons produced during fission is equal to the neutrons lost by absorption and leakage processes. The net rate of change of neutron density n (neutrons per cubic centimeter) is given by the neutron conservation equation:

$$\frac{\partial n}{\partial t} = \text{production} - \text{absorption} - \text{leakage}$$

If $\partial n/\partial t$ is positive, the reactor is said to be supercritical; if it is zero, the reactor is at steady state and is said to be critical; if it is negative, the reactor is subcritical.

Cross Section The microscopic cross section σ for a particular nuclear process is the effective target area of the nucleus with which a neutron must interact to produce the given reaction. The unit of σ is barn (1 barn = 10^{-24} cm²). Neutron cross sections (absorption, scattering, and fission) constitute the basic nuclear data for reactor design. Absorption cross sections for thermal neutrons range from 0.00046 barn for deuterium to 3.3×10^6 barns for xenon[135]. Moderating atoms for the slowing down of fast neutrons have low atomic mass, small absorption, and large scattering cross sections. Examples are hydrogen, deuterium beryllium, and carbon. Table 9.8.3 gives thermal neutron cross sections in barns for selected elements (based upon the naturally occurring combinations of isotopes).

Multiplication Factor The number of neutrons produced in any one generation in a reactor for each neutron produced in a previous generation is called the multiplication factor k. If $k = 1$, the reactor is critical; if $k < 1$, it is subcritical; and if $k > 1$, it is supercritical. For an infinite thermal reactor, multiplication factor $k_\infty = \eta f p \varepsilon$. η is the number of fission neutrons produced per neutron absorption in fuel, f is the thermal utilization factor (neutron absorption rate in fuel/total absorption rate), ε is the fast-fission factor, and p is the resonance capture escape probability. The leakage of neutrons from the reactor core is accounted for by multiplying k_∞ by the factor $1/(1 + M^2 B^2)$, where M^2 is the migration area (sum of the squares of the diffusion length L and the slowing-down

Table 9.8.3   Absorption (σ_a) and Scattering (σ_s) Cross Section for Thermal Neutrons, in barns

Element	σ_a	σ_s	Element	σ_a	σ_s	Element	σ_a	σ_s
Aluminum	0.23	1.4	Helium	0.007	0.8	Potassium	1.97	1.5
Beryllium	0.01	7	Hydrogen	0.33	38	Plutonium[239]	1.026	9.6
Bismuth	0.032	9	Iron	2.53	11	Sodium	0.505	4.0
Boron	755	4	Lead	0.17	11	Thorium	7.56	12.6
Cadmium	2,450	7	Magnesium	0.063	3.6	Tin	0.6	4
Carbon	0.0032	4.8	Molybdenum	2.7	7	Titanium	5.6	4
Chromium	2.9	3.0	Nickel	4.6	17.5	Uranium	7.68	8.3
Copper	3.7	7.2	Niobium	1.1	5	Zirconium	0.18	8
Hafnium	105	8	Oxygen	< 0.0002	4.2			

NOTE: Individual isotopes have markedly different cross section. For example, U[235] has total σ_a = 682 barns and fission cross section σ_f = 580 barns.

length τ) and B^2 is the geometric buckling ($B^2 = 0$ for an infinite reactor).

Breeding Ratio   If the ratio of a number of fissile nuclides produced by the capture of neutrons by the fertile nuclides to the number of fissile nuclides consumed is greater than unity, it is called the **breeding ratio;** if it is less than unity, it is called the conversion ratio. For breeding, the value of η (defined above) must be greater than 2, as one neutron is required to maintain the chain reaction by fission, another neutron maintains breeding by its capture by the fertile nuclide, and the remaining to compensate for the loss of neutrons due to absorption and leakage. Neutron economy favors the breeding of Pu[239] fissile nuclides from fast neutron capture by U[238] fertile nuclides. This breeding process is the basis of LMFBR (liquid-metal fast breeder reactor). On the other hand, thermal breeders are based upon the breeding of uranium 233 fissile nuclides from thorium 232 fertile nuclides by the capture of a thermal neutron.

Diffusion Theory   Multienergy group diffusion theory is presently the basis of reactor design. For a critical reactor the neutron balance equation reduces to

$$D_g \nabla^2 \phi_g - \sum R_g \phi_g + S_g = 0$$

where ϕ_g and S_g are the gth energy group flux and the slowing-down neutron source; D_g and R_g are the diffusion coefficient and removal cross section. The latter are called group constants, which are obtained by the analysis of the $\Sigma R_g \phi_g$ basic fuel lattice. The solution of the above equations provides the reactor multiplication factor, reactor critical size, neutron fluxes, power distributions, isotopics of elements, and fuel-cycle length.

Control   Extra fuel is required for depletion to compensate fuel lost in producing energy, to account for the loss of reactivity due to the buildup of fission products, and to account for the change in reactivity due to changes in fuel and moderator temperatures. The control of excess reactivity is undertaken by the control rods of strongly absorbing materials having large absorption cross sections (such as boron, hafnium, or silver-indium-cadmium) and the burnable poison rods containing boron or gadolinium. In PWR, soluble boron is also employed. Control rods are inserted to shut down the reactor and are withdrawn to make the reactor critical.

A fraction of prompt neutrons (0.0065) are delayed neutrons. The presence of the delayed neutrons makes control of the reactor easier.

Thermal

Sources of Heat   The energy produced by fissioning appears in several forms, each of which eventually degrades to heat. To provide for removal of this heat, the core, or active portion of the reactor, usually consists of uranium bearing fuel elements with passages around these elements for flow of coolant. The rate of coolant flow and the temperature of the coolant must be such as to permit removal of all the heat generated in the elements without exceeding allowable materials properties.

Hot Spot and Hot Channel Factors   The problem of heat removal is complicated by the fact that heat is not generated uniformly throughout the core. The geometric configuration of most power reactors is approximately represented by a finite cylinder of height H and radius R. Diffu-

sion theory predicts the following thermal neutron flux pattern for an unreflected, unrodded, and homogenized cylindrical reactor as a function of axial Z and radial r variables:

$$\phi(z, r) = \phi_0 \left(\cos \frac{\pi Z}{H} \right) J_0 \frac{2.405 r}{R}$$

where ϕ_0 is peak flux and J_0 is the Bessel function.

In addition, the power generation is influenced by mechanical factors such as variables in fuel element dimensions, fuel concentrations, and flow. These nuclear and mechanical factors are generally called hot spot factors. Thus the peak power density of heat flux in the core is greater than the average by an amount indicated by the hot spot factor. In addition, hot channel factors are identified which indicate how much greater the temperature rise of coolant is in the hot channel than in the average channel. As an example of how such factors are used, consider a reactor in which the axial heat generation pattern is a cosine and the total flow passes uniformly once through all channels of the core; the maximum fuel element surface temperature T_{sm} depends on the hot channel and hot spot factors $F_{\Delta T}$ and F_θ and on the three temperatures—the coolant temperature at the core inlet T_1, the coolant temperature rise in the core ΔT, and the average film temperature drop in the core θ_a. (F_θ includes the axial peak-to-average of $\pi/2$.)

$$T_{sm} = T_1 + \tfrac{1}{2} F_{\Delta T} \Delta T + \tfrac{1}{2} \sqrt{F_{\Delta T}^2 \Delta T^2 + 4 F_\theta^2 \theta_a^2}$$

Design Criteria   The maximum fuel element surface temperature T_{sm} is of interest because it may be limited by corrosion effects or by a desire to avoid film boiling, a condition which lowers the heat transfer from the fuel to the coolant. However, other limiting conditions also exist. In addition the internal fuel element temperature during all anticipated reactor transients is generally limited to a maximum value (generally below the melting point of the fuel) or to a maximum average value.

Coolant temperature rise generally is limited indirectly by fuel temperature limits. Occasionally, however, thermal stresses within the fuel element may impose a direct limit. In pressurized water reactors, the coolant temperature is limited not to exceed the saturation temperature of the fluid; thus it is more significant to speak of the maximum enthalpy rise, which is related directly to the steam quality of the fluid leaving the hot channel.

Density changes within the coolant may be limited to provide control stability in reactors that use boiling heat transfer.

Coolant velocity usually is limited to a maximum value by erosion or by vibration resonances, and sometimes is required to exceed a minimum value because of crud deposition. The burnout condition is determined in a given fuel element geometry for a particular combination of fuel element heat flux, primary coolant flow, and primary coolant enthalpy. Steady-state, local, or bulk boiling may have to be limited, even when burnout is not a danger, because of pitting of fuel element surfaces or crud deposition on the fuel surfaces. The burnout problem is most acute during reactor transients; hence steady state burnout generally is not limiting.

Mechanical

The mechanical design of nuclear reactors is influenced by three factors not encountered with other apparatus: the need to control criticality, the effects of irradiation, and the high power density produced in the core.

Control of Criticality The following requirements are introduced by the need to control criticality:

1. Not only must components perform reliably, but also if they fail to function as designed, the reactor will shut down. Small parts located elsewhere in the core should be so designed that if they fail they do not interfere with the action of control rods in shutting down the reactor.

2. Emergency shutdown mechanisms must be fast-acting devices to reduce the time delay in initiating shutdown.

3. During normal operation, components of the core must be rigid enough so that deflections due to mechanical or thermal stresses do not introduce unpredicted criticality variations.

Effects of Irradiation Neutron and gamma irradiations cause the following effects, which must be considered in the design of reactor systems and reactor components:

1. Radiation effects on properties of materials—particularly on the brittle fracture characteristics of the reactor vessel, on the relaxation characteristics of bolts, and on crevice corrosion.

2. Heat generation brought about by attenuation of radiation within components materials—probably insofar as such heat influences heat removal requirements, thermal stresses, and thermal distortions of parts.

3. Induced radioactivity in corrosion and wear products that could be carried by the coolant and deposited in such regions as pumps and valves.

High Power Density A nuclear reactor requires a large amount of heat transfer surface for removal of the high power per unit volume which can be produced. Thus the reactor consists of many small, closely packed fuel elements. Mechanical tolerances are often appreciable fractions of the dimensions and spacing of these elements and therefore influence thermal performance. The influence of such tolerances on performance must be evaluated. Parts also must be rigid and vibration-free in the face of the high fluid flow rates required for power removal. The high power density may also lead to large thermal stresses in fuel elements.

Furthermore, moving parts may have to operate without benefit of organic lubricants because such lubricants tend to sludge under irradiation, and if deposited on heat surfaces could interfere with removal of heat from the core.

Residual radioactivity also influences programs and facilities for refueling of the core and for periodic maintenance of components in the reactor system. In addition, personnel exposure by the above radiation requires adequate shielding.

Shielding To protect reactor operating personnel against damaging biological effects of neutrons and gamma rays, shielding is required around a nuclear reactor. Neutron and gamma ray fluxes in the range of 10^{13} to 10^{14} must be attenuated to 10^{3} particles/(cm$^2 \cdot$ s) to meet the tolerance radiation level.

To attenuate gamma rays, which interact primarily with the orbital electrons of atoms, a material with a high atomic number containing a high density of these electrons is required. Examples are lead, tungsten, depleted uranium, or concrete containing high-Z elements in the form of scrap or heavy ore.

To attenuate neutrons, they must be slowed down and then absorbed. Hydrogenous materials such as water, concrete, or polyethylene are excellent moderators. The slowed neutrons must be absorbed without producing high-energy-capture gamma rays by using boron 10. Therefore, some gamma shielding outside the neutron shield is generally required.

Electrical

Neutron level instruments, located adjacent to but outside the core or core vessel, are used for the basic control of nuclear reactors. Thermocouples, flowmeters, pressure taps, and fission chambers are used inside the core for power calibration of nuclear instruments and for determining the distribution of power and/or temperature within the core; in turn, the power distribution is used to compute heat fluxes and fuel temperatures for comparison with limiting values.

Neutron level instruments utilize the fact that neutrons can produce ionizing particles to indicate their presence. The two most common processes used for this purpose are the (n,α) reaction with boron and the fission reaction. The resulting ion pair is used to produce ionization of a gas in a tube containing a central anode and outer electrode. The electric discharge produced by the ionization is used for measurement purposes. The detailed design features vary widely with the range of the instrument and the type of reactor in which it is to be used.

Different neutron level instruments are employed during the reactor operation at various power levels, because of the wide range of sensitivity required to monitor neutron fluxes from start-up to full-power condition. In power reactors at least three instrument ranges are used with the overlap between the adjacent ranges. These are (1) source range, (2) intermediate range, and (3) power range. During the reactor start-up (source range) the BF$_3$ proportional counters are used to measure neutron flux in the range of 10^{-1} to 5×10^4 neutrons/(cm$^2 \cdot$ s). The compensated ionization chambers are employed to monitor neutron flux in the intermediate power range of 2.5×10^2 to 2.5×10^{10} neutrons/(cm$^2 \cdot$ s). In the power range, where neutron flux ranges from 2.5×10^7 to 2.5×10^{10} neutrons/(cm$^2 \cdot$ s), uncompensated ionization chambers are employed.

Compensated ionization chambers are designed to count neutrons against a background of gamma radiation. Uncompensated chambers are useful for providing axial power distribution.

All the above instruments are located outside the core and monitor neutron flux, hence reactor power. The latter is also obtained from the measure of ΔT (change in temperature) across the reactor core from the temperature measuring instruments located in the primary loop of the reactor. However, these instruments do not provide local flux peaking and local anomalous coolant temperature situations. In-core instrumentation must be used for this purpose.

In-core instrumentation is employed to provide information on neutron flux and fuel-assembly-coolant outlet temperature at a few selected locations in the reactor core. Thermocouples measure the coolant outlet temperature and thus provide the rise in local coolant temperature, a measure of the increase in local reactor power. Miniature, fixed, and movable fission chambers containing U$_3$O$_8$ which is 90 to 95 percent enriched U^{235} constitute neutron flux detectors. The experimental data obtained during the reactor operation from the in-core instrumentation provide the check against the reactor design parameters such as the assembly distributions and hot channel factors.

In addition to thermocouples and fission chambers, other in-core instruments such as pressure, displacement, and strain transducers are also desirable. None of these instruments except thermocouples have operated satisfactorily over a long period. The installation of in-core instrumentation is vital to the safe and economic operation of the reactor.

NUCLEAR POWER PLANT ECONOMICS

Decisions by electric utilities as to what type of generating plants to order to meet future base load requirements are based largely, though by no means wholly, on a comparison of the costs for which the alternative types can produce electricity over their lifetimes. Other factors, which have become relatively more important in recent years, include adequacy and reliability of fuel supply, including dependence on foreign sources; environmental and safety considerations; availability of suitable sites; public acceptance as influenced particularly by environmental opposition; the costs, uncertainties, and delays involved in government regulation; lead times required from initial decision to get plants on line; and plant reliability and availability.

As a practical matter in the United States, utilities seeking to add to their base load capacity have the choice only of plants utilizing nuclear fuel or one of the fossil fuels; coal, gas, or oil. Most usable hydrosites have already been developed.

The discussion of nuclear generating costs here is directed to light water reactors of either the pressurized water reactor (PWR) or boiling water reactor (BWR) types. Although other types are in use (e.g., high-temperature gas-cooled) or under development (e.g., fast breeder), the overwhelming preponderance of reactors in commercial use are, and will for some time continue to be, light-water types.

It is customary to compare generating costs for nuclear and fossil fueled plants in three broad categories: (1) fixed costs on capital investment, (2) fuel costs, and (3) operating, maintenance, and insurance costs.

Fixed Costs on Capital Investment

In standard accounting practice, the capital investment in a nuclear power plant is considered to be the total cost of building the plant and placing it in commercial operation. Accounting classification systems employed by both NRC and FPC provide separately for "direct" and "indirect" costs. Direct costs are associated on an item-by-item basis with land and land rights, the equipment and structures which comprise the complete power plant, and coolant and moderator materials. Equipment and structures are customarily subdivided into reactor plant and equipment, turbine plant and equipment, other electrical equipment, equipment and facilities of a general nature, and transmission plant. Indirect costs consist mainly of expenses which apply to all portions of the physical plant, including engineering and design costs, construction facilities, taxes, interest during construction, and the owner's administrative and overhead costs.

The total of these items tends to be higher for nuclear than for fossil fueled plants largely because of (1) the need for containment shells, shielding, instrumentation, and other measures to contain radioactivity and ensure safety; (2) the greater complexity, hence greater costs, of reactor equipment compared to conventional boiler equipment; and (3) the lower turbine steam temperatures and pressures, and the lower thermal efficiencies of nuclear plants compared to conventional plants, requiring the use of larger, hence costlier, turbines. The discrepancy is wider in warm climates, which permit fossil fuel plants to utilize outdoor (unenclosed) construction.

Mitigating factors which prevent the construction cost discrepancy between nuclear and coal plants, in particular, from being still larger are (1) the lower cost of fuel handling equipment in nuclear plants and (2) the need in coal plants but not in nuclear plants, for smoke control and ash removal equipment.

Unit construction costs (expressed in dollars per kilowatt) for both nuclear and fossil fueled plants are lower at any given time as plants increase in size because of various economies of scale. The decreases are sharper in the case of nuclear plants, largely because of the prominence of certain minimum and relatively fixed costs to ensure safety. The existence of these inescapable costs places small nuclear plants at a particularly great economic disadvantage.

Adding a nuclear unit to a generating station which already has one or more nuclear units is less costly than constructing a first nuclear unit of otherwise identical characteristics. The principal savings in multiple-unit plants result from use of a developed site; reduced engineering and design efforts; and use of joint facilities and equipment such as cooling water intake and discharge facilities, control rooms, warehouses, shops, offices, roads, fuel handling and storage facilities, and temporary construction facilities.

The unit capital costs of nuclear plants often can be reduced substantially over their lifetimes by using improved fuel cores to achieve outputs higher than the original rating, which is usually conservative. To take advantage of this opportunity, allowance for higher output must be made at the outset in the capacity of turbine generators. Conversely, should it be necessary to derate plants to take account of safety uncertainties such as the effectiveness of emergency core-cooling systems, the effect would be to increase unit capital costs.

Total construction costs of nuclear plants have risen rapidly since about 1963, when it was believed that a large light-water plant could be built for $125 per kilowatt or less. Plants now are being estimated to cost in excess of $2,000 per kilowatt. The principal factors making for such increases in cost have been inflation, stretch-outs of plant schedules (adding to interest during construction), lower labor productivity, increases in the scope of projects to add safety and environmental features, and more stringent quality control.

Some of these factors may continue to push unit construction costs upward in the future. Counterbalancing factors tending toward a moderation in unit costs include continued increases in plant size, the likelihood of greater standardization of systems and components, a larger sales volume over which to spread manufacturing overhead costs, and improved field assembly and construction methods. One should take note of the fact that the historical tendency in the industry has been to underestimate construction costs by a wide margin.

The fixed cost component of generating costs (mills per kilowatt-hour) can be obtained from construction costs (dollars per kilowatt) by applying an annual fixed charge rate and an annual plant capacity factor.

Annual fixed charges cover all costs that are in direct proportion to the initial capital investment and are independent of the extent to which the facilities represented by this investment are used. A simple procedure for evaluation of a plant, which approximates the levelized annual charges obtained through utility accounting procedures, is to use constant annual fixed charges, i.e., to apply a constant fixed charge rate each year to the initial investment.

The fixed charge rate for a utility will vary because of a number of factors including amount of state and local taxes, plant life and accounting method for depreciation purposes, and debt-equity composition of the utility's financial structure. A representative range of fixed charge rates for depreciating capital of an investor-owned utility illustrate the components and relative importance of these items, it being understood that the actual components and range will vary, depending on the time of implementation and the then-current financial climate.

Component	Fixed charge range, %
Return on investment	7.0–9.0
Depreciation	1.0–1.3
Interim replacements of short-lived assets	0.3–0.5
Property insurance	0.3–0.4
Federal income taxes	2.0–2.8
State and local taxes	2.9–3.5
	13.5–17.5

A lower fixed charge rate would be applied to the nondepreciating capital (land and land rights and working capital) because depreciation, interim replacements, and property insurance are eliminated and associated adjustments are made in taxes. Not included in annual fixed charges are relatively constant fixed costs, such as nuclear liability insurance or plant staffing expense, which are only indirectly related to plant investment.

The annual capacity factor is calculated by dividing the kilowatthours generated by the plant in a year by the kilowatthours that would have been generated had the plant operated at full capacity throughout all the hours in the year (8.760 except in leap years).

The conversion from capital to generating costs may be illustrated as follows: Assume a plant of 1,000,000-kW capacity costing $2 billion, with annual fixed charges of 14 percent and operating at 80 percent capacity factor. The construction-cost contribution to generating costs would then be 40 mills/kWh, calculated as follows:

$$\frac{\$2,000,000,000 \times 14\%}{1,000,000 \text{ kW} \times 8,760 \text{ h} \times 80\%} = 40 \text{ mills/kWh}$$

Fuel Costs

The economic advantage of nuclear power plants lies essentially in fuel costs which are substantially lower than those of fossil fueled plants.

The fuel cost advantage is implicit in the compactness of nuclear fuel: 685 lb of uranium, if fully consumed, contains the energy equivalent of 1.7 million tons of coal, 7.2 million bbl of oil, or 32 billion ft^3 of natural gas. It has not yet proved possible in any water cooled reactor to consume more than a small fraction of the uranium used in nuclear fuel; yet this fraction is high enough to yield the cost advantage noted above.

Determining nuclear fuel costs is a far more complicated matter than determining fossil fuel costs because of the need to take account of a complex sequence of events which begins long before the uranium fuel is inserted in the reactor and ends long after the spent fuel has been removed from the reactor. The entire sequence, which is called the nuclear fuel cycle, includes mining and milling of uranium, refining of uranium and conversion to uranium hexafluoride, enrichment of the uranium in the isotope uranium 235, conversion of enriched uranium to fuel material, fabrication of reactor fuel elements, use of fuel elements in nuclear power plants, recovery and marketing of by-product plutonium (not in the United States), chemical reprocessing of spent fuel to obtain reusuable fuel material (not in the United States), and disposal of radioactive wastes.

Some principal items of nuclear fuel cost and possibilities of change include the following:

Fabrication The cost of fuel element fabrication is taken to include all the steps necessary to change a starting material (UF_6 in the case of enriched uranium elements) into a usable element. These steps involve chemical conversion to a powder or metal, metallurgical and mechanical processing to form and clad the elements, inspection, testing, and scrap recovery. Fabrication costs are expected to continue declining as greater quantities are handled, processing becomes more efficient, and fuel design is improved.

Recycling of Bred Plutonium Uranium-fueled light-water reactors in normal operation produce a certain amount of fissionable plutonium 239 through capture of neutrons in the nonfissionable isotope uranium 238. The separation is done by well-established chemical processes, the end products being uranium dioxide and plutonium dioxide. The plan to recycle spent nuclear fuel resulted in several reprocessing plants being built in the United States, but only one was ever operated. Widespread public concern over proliferation of nuclear weapons and the possible use of plutonium by terrorists and others resulted in a ban on reprocessing in the late 1970s. The ban was lifted in 1980, but since that time there has been no economic incentive to embark on an extended program of industrial reprocessing of nuclear fuel. This situation is entirely different elsewhere; reprocessing is currently a going operation in France, Russia, Great Britain, and elsewhere. The confluence of many factors will foster increased reprocessing of spent nuclear fuels, among which will be the matter of depletion of fossil fuels, accumulated pollution products from the combustion thereof, growth of power requirements worldwide, and so on. Certainly, at such future time as breeder reactors may become viable, they will provide an ideal place to utilize plutonium derived from reprocessed spent nuclear fuel produced in light-water reactors.

Shipping Charge The costs of transporting irradiated fuel elements from a reactor site to a chemical processing plant include charges for freight, shipping casks, handling, and insurance. The shape of the fuel element (having an effect on the size of cask required), the irradiation level, the cooling time required, the shipping distance, the route and type of transport, and the regulatory requirements are all factors which affect the cost.

Enrichment Enrichment in the fissionable isotope uranium 235 is accomplished at present (in the United States) only in three DOE-owned plants. This is the only part of the nuclear power industry not yet in private hands.

Burnup The amount of heat that can be obtained from a given weight of nuclear fuel before it is discharged from the reactor is known as the fuel's burnup. Higher burnups are expected as a result of technological improvements. This would have the consequences of reducing the amount of fuel that must be fabricated, reprocessed, and shipped per unit of electrical output. Cost reductions should be achieved as a result, but they may be limited by the fact that fuel elements capable of achieving the higher burnups may be more expensive to fabricate and process and may require higher enrichment.

Chemical Reprocessing and Waste Disposal The cost of chemical reprocessing, less any income from the plutonium or uranium recovered, is usually a net credit for foreign reactors. For U.S. and foreign reactors, the cost of waste disposal is an appreciable cost factor. Under U.S. law, that cost is assessed against the user utility as the power is generated, even though the disposal will take place decades later.

Decommissioning The cost of decommissioning the nuclear power plant when it is eventually closed down is treated differently in different countries. In the United States, these costs are required to be guaranteed by various accounting practices, which in effect accumulate a reserve for costs to be accrued in the future. Reactors decommissioned and dismantled are utilized as case studies to accumulate data on the costs incurred thereby. Other reactors that have been shut down have been defueled, and either stored under guard, or entombed in concrete, with ultimate decommissioning postponed to the future.

Operating and Maintenance Costs

The main cost items included in this category are station labor, training of replacement personnel, expendable materials, and supplies other than fuel, maintenance operations, and insurance.

The novelty and complexity of nuclear plants, a conservative approach to health and safety problems, and the need for specialized knowledge are all factors requiring a more skilled staff in nuclear plants and a longer period of training. Problems of coal and ash handling and maintenance of tubes and related boiler components require a larger working force in coal plants. While insurance costs have been higher for nuclear plants, it is expected that a continuance of safe operating experience will eventually equalize this item.

Numbers of operating personnel per generating unit are less for a plant having two or more nuclear units than for a single-unit plant. Increasing the size of a nuclear unit does not materially affect the number of people required to operate it. For these two reasons, one can expect declining personnel requirements per kilowatt of capacity at such time as the nuclear power industry expands present plant capacity and/ or adds new plants to the overall inventory.

Total Generating Costs

Although many uncertainties becloud cost estimates for both nuclear and fossil fuel plants, in the 1970s, comparisons tended to favor nuclear power in many parts of the United States on either economic or noneconomic bases, or both. The continued and expanded use of fossil fuels in power plants is driven by several considerations, not the least of which is the current (1995) apparant plentiful supply at reasonable prices. In looking to the future, however, there are certain problems which are recognized and which must be addressed in the matter of fossil fuels. Domestic supplies of low-sulfur oil are diminishing, and dependence on imports makes for uncertain supply and is always subject to sharp price increases. Domestic coal is plentiful, but most of it, especially in the eastern United States, has a relatively high sulfur content which requires extensive and expensive steps to be taken to desulfurize it or to treat the effluent gases to meet local air pollution standards. Vast quantities of natural gas have come on the market in the last decade, but it is recognized that this source, too, is finite. Oil and natural gas have required deeper and deeper wells, in locations which add to the difficulties of extraction (i.e., offshore platforms in deep waters) and shipment.

The large-scale plans for expansion of the nuclear power plant base in the United States were dramatically scaled back in the 1980s, for a number of reasons. Nationwide, actual and forecasted growth of demand was markedly lower; the regulatory climate and length of time required from inception to the plant's going on-line were such that construction financing presented onerous economic problems which largely offset the erstwhile advantages to be gained from the use of nuclear fuel; and there was denial of retraction of plant certification by local authorities.

In the past several years, efforts have been made to reduce the time required from planning to final operation of nuclear plants. To that end, the industry has designed a number of standard size and type nuclear reactors which, once approved by the NRC, will not have to undergo the long-time, one-of-a-kind review for each proposed installation. Although no plants are on order in the United States, the major U.S. fabricators and suppliers of nuclear reactors are engaged, either solely or in partnership, with the growing nuclear energy programs in many other countries. It is expected that the store of expertise gained thereby will be available for application in the United States at such future time when, as expected, nuclear power plants once more will serve to meet growing demands. At this time (1995), approximately 20 percent of the electric power generated in the United States originates in nuclear reactors. Outside the United States, there is great activity in the emplacement and operation of an expanded nuclear power base in the electric power generating industries.

Effect on Fuel Resources

The emergence of nuclear power as a viable energy source comes at a time when additional environmentally acceptable sources of energy are sorely needed, both in the United States and in other countries, to meet continued rapid increases in demand. As indicated above, domestic production of both natural gas and oil appears to be approaching its peak and may decline in the future. Dependence on foreign sources entails risks with regard to dependability of supply and the U.S. balance of payments. New technology is needed in order to be able to mine and use this country's extensive coal reserves in an environmentally and economically acceptable manner.

Uranium supplies are limited and, if utilized only in light-water receptors, would rapidly diminish. Breeder reactors, now under active development outside the United States, can utilize most (60 percent or more) of our uranium (and/or thorium) resources rather than just a minute percentage.

A practically unlimited source of energy is available from the deuterium in the world's oceans should development work now underway in several countries lead to production of useful power from controlled fusion reactions. The technical problems are formidable, however, and the harnessing of fusion as an energy source in large power generating plants is not expected until the latter half of the twenty-first century.

NUCLEAR POWER PLANT SAFETY
Revised and amended by George Sege

Through the end of 1993, over 1800 reactor-years of operation had been accumulated by commercial nuclear power plants in the United States, and about 6,900 worldwide, not including military or shipboard reactors (International Atomic Energy Agency, "Nuclear Power Reactors of the World," 1994). This operational history has been accompanied by an impressive safety record with respect to health and safety impacts on the public, yet accidents have occurred. The most disastrous occurred in 1986 at the Chernobyl nuclear power plant complex located in the then U.S.S.R., resulting in loss of life and long-term radioactive contamination over a large area in the vicinity of the plant as well as transport of short-lived radioactive contamination dictated by the prevailing local weather patterns. Ultimate disposition of the destroyed reactor, now encased in an added containment structure ("sarcophagus"), remains a concern. In the United States, two of the worst commercial accidents occurred at the Brown's Ferry facility in 1975 and the Three Mile Island (TMI) facility in 1979. In the Brown's Ferry accident, a fire burned through hundreds of electrical power and control cables, threatening the plant's capability to be shut down and maintained in a safe condition. There was ultimately no damage to the reactor and no off-site radioactive releases occurred. At Three Mile Island significant reactor damage occurred through a sequence of events that began with a stuck-open primary system relief valve. Radioactivity was released to the environment, mostly in the form of noble gas fission products. However, estimated exposures to members of the public were found by a Federal interagency working group to be exceedingly small, even for the maximally exposed individual. The TMI reactor vessel did not fail despite relocation of approximately 19 tons of molten core materials. Although these accidents had minimal public impact, safety systems and plant operators were severely challenged during the events.

Commercial reactors have inherent safety features that prevent them from exploding like bombs. In a fission bomb, the fissioning of the fuel is caused by neutrons whose energy is about the same as the level at which they are produced by fission of the material in the previous generations. As a result, the average neutron lifetime is very small, and this allows the supercritical chain reaction to increase the power level to incredibly high levels before the device explosively destructs itself. In a power reactor, light-water or gas-cooled, enrichments of the fuel is very small (a few percent). As a result, essentially all fissioning of the fuel is caused by neutrons that have been reduced greatly in energy by the moderator (water or carbon). This results in a long average neutron lifetime compared to that in a bomb. If the chain reaction is allowed to increase in an uncontrolled fashion, this long lifetime will prevent the power of the reactor from rising to very high levels before the inherent design characteristics of negative temperature and void coefficients cause the reaction to shut down. While the Chernobyl reactor had design deficiencies in this regard, the violence of the accident was very far removed from what would result from detonation of a nuclear bomb.

There are two fundamental inherent safety features that contribute to the controllability of commercial power reactors. The first of these is the **Doppler effect**. As the power level of a reactor increases rapidly, the temperature of the fuel increases very rapidly. This rise in temperature causes the low-energy resonant peaks in the microscopic cross section to broaden, effectively creating a larger probability that neutrons will suffer nonfission captures by the U^{238} in the fuel. The second inherent safety feature is caused by the physical densities of materials, particularly the moderator, that decrease as the temperature increases. This results in higher probabilities that neutrons will escape from the reactor rather than cause fission in the fuel. Both of these inherent safety features feed back to the chain reaction in a negative way such that an uncontrolled increase in power level tends to shut the reactor down. This is called a **negative temperature coefficient**. Steam voids in a water moderator have a similar effect.

In spite of the inherent safety features, a reactor still provides a potential hazard to plant workers and members of the public in the vicinity of the plant. This is because of the generation of highly radioactive fission products within the fuel. The essential safety challenge is to assure that these fission products are kept confined at all times.

Nuclear power plant safety is based on a defense-in-depth philosophy. The defense-in-depth relies not only upon the inherent safety features, but also on high quality standards that provide a high degree of assurance that accidents are not likely to occur, and that if they do occur, the radioactivity will be contained locally.

High quality begins with the selection of highly qualified personnel to design, construct, and operate a plant. Each applicant for a license to build a nuclear plant must institute a quality assurance program to be applied to design, fabrication, construction, and testing of the structures, systems, and components of the facility. An applicant for a license to operate a plant must also provide for appropriate managerial and administrative controls to be used to assure safe operation. The function of the quality assurance program is to provide a high degree of confidence that the structures, systems, and components in the plant will perform as intended during the service life of the plant.

Nevertheless, accidents can occur, and other safety features are designed to keep fission products confined in the event of a broad range of accidents that could conceivably occur. Two categories of anticipated accidents are undercooling events and positive reactivity addition events. An undercooling event can be caused by such things as ruptures in the reactor cooling system, loss of flow through the reactor core, or loss of feedwater flow. Positive reactivity addition events are the result of such things as inadvertent control-rod withdrawals or dissolved boron dilution, main steam line breaks (pressurized water reactor), or overfeeding of a steam generator.

There are various types of engineered safety features found in today's commercial reactors. They can be grouped into four major areas: (1) reactor protection system, (2) emergency core-cooling systems, (3) containment, and (4) emergency power supplies. The following discussion of these features is generally applicable to light-water-cooled reactors, pressurized water and boiling water reactors, but the same general features are applicable to other types.

The purpose of the **reactor protection system** is to rapidly insert the control rods into the reactor upon receipt of a signal indicative of a malfunction that potentially threatens the integrity of the core. For example, a low reactor coolant system pressure is indicative of a loss-of-coolant accident. High temperatures in the reactor coolant system or the core indicate the possibility of a loss of coolant or loss of flow through the core. The reactor protection system processes the various input signals and compares them to predetermined set points. If a set point is reached or exceeded, the system is designed to automatically insert the control rods into the reactor. The control rods absorb neutrons and quickly reduce the fission rate in the core. This rapidly drops the power level in the reactor and aids in prevention of damage to the fuel.

In the event of a loss-of-coolant accident, it is not sufficient to shut the reactor down in order to prevent damage to the fuel. Decay heat produced from the radioactive decay of fission products (initially some 4 percent of the previous operating power, but decreasing rapidly with time after stopping the chain reaction) would cause damage to the fuel if the coolant inventory losses were not made up by an injection system. **Emergency core-cooling systems** are designed to permit injection of coolant at high reactor coolant system pressures that exist during small ruptures as well as at low reactor coolant system pressures resulting from large breaks. Redundancy is provided to assure that equipment failures do not result in the loss of capability to keep the core covered with water. Large sources of water are stored to meet the long-term cooling requirements to remove the decay heat. In the event that a pipe rupture results in the depletion of the stored injection water, provisions are made to allow recirculation of water from the emergency sumps in the containment building back through the injection pumps. In time the rate of heat generation drops to the point where forced-circulation cooling is no longer required. After the Three Mile Island core damage event the system was cooling by natural convection to the surroundings at atmospheric pressure after about 3 years.

Containment features provide an important function essential to the defense-in-depth philosophy. There are actually three levels in the containment scheme. The first level is the fuel cladding, which not only mechanically holds the fuel pellets in the core, but also provides a boundary to prevent migration of fission products to the coolant. The second level is the reactor cooling system pressure boundary. This consists of the reactor vessel system piping, and other components designed to withstand pressures in the order of hundreds of atmospheres without failure. Parts of these systems that are outside containment in normal operation, as in a boiling water reactor, are isolated at the containment building boundary in case of an accident. Thus even if the fuel cladding should fail, the pressure boundary is likely to keep the fission products contained. In the event that the accident would cause both of these boundaries to fail, a third level is provided in the form of a containment building. The typical containment building consists of a steel shell surrounded by a concrete structure subjected to precompression by steel cables. Thick steel only, reinforced concrete, and combinations have all been used in containment schemes, most of which are designed to withstand 5- to 10-atm pressure, based on containing the total energy available during an accident. The destroyed Chernobyl reactor did not have a containment building; it had only confinement provisions for small local pipe breaks.

Emergency power supplies also exhibit redundancy and defense-in-depth features. The power requirements for auxiliary systems within a power plant are normally supplied from the reactor output itself through the main electrical generator and auxiliary station transformer. In the event of an accident, the reactor would not be available to produce useful power; thus the preferred power source for supplying emergency electrical loads is from the transmission lines from other power plants. At least two off-site power supplies are required. In case of loss of these sources, redundant on-site sources are also required, usually powered by quick-start diesel generators. Battery systems are also needed to provide certain direct-current loads and some alternating-current loads through inverters. Control, computer, and instrument supplies are typically so served.

Protection is provided not only for events originating in the plant but also for external events. Plants are designed to withstand safely the worst anticipated earthquakes, floods, and winds. Containment and other safety features of nuclear power plants generally offer significant protection against consequences of accidents exceeding the severity of design-basis accidents. In recent years, experimental work and probabilistic studies have been carried out to determine the extent of such margins and discover residual vulnerabilities that would then become the basis for possible further safety improvements, generally on a plant-specific basis.

As the operating years of plants have accumulated, there has been increasing attention to detecting and counteracting age-related degradation of plant materials and equipment. A notable example is the varying degrees of radiation-induced embrittlement of reactor vessel materials, which if unchecked should increase vulnerability to overcooling accidents (pressurized thermal shock).

Another development in recent years has been the increasing use of probabilistic safety assessments as a partial guide to plant safety actions as well as safety regulation. In such assessments, **fault trees** are used to trace malfunctions through all antecedent events to the initiating causes and **event trees** to trace undesired events through all their consequences to ultimate outcomes. Known or estimated probabilities are assigned at each step in the chain of causation, to derive probabilities of sequences of safety (or risk) interest.

The various inherent and engineered safety features, along with the high quality standards which are required in the design, construction, and operation of a nuclear power plant, make accidental releases of radioactivity to the environment extremely unlikely. Nonetheless, additional defense-in-depth is provided in the form of trained operators and procedural adherence. As a final feature of the defense-in-depth philosophy, emergency planning is required that establishes the protective measures that would be taken to protect workers and members of the public against possible radioactive releases. Such emergency measures are required to be practiced at least annually in the United States.

Further assurance of nuclear plant safety is provided by the regulatory requirements summarized in the next section.

NUCLEAR POWER PLANT LICENSING
Revised and amended by Paul E. Norian

NOTE. The following describes the process in the United States. Nearly all countries have similar processes, but the degree of public participation and appeal opportunities may differ in significant degree, generally resulting in less complex and time-consuming processes in other countries, especially those with nationally owned industries and utilities.

Licensing of commercial power reactors is conducted in accordance with the Atomic Energy Act of 1954, as amended. The applicable regulations appear in Title 10 of the Code of Federal Regulations (10 CFR). Licensing is currently conducted by the NRC in a two-step process involving the issuance of a construction permit (CP) and then an operating license (OL) when the plant is completed. The process begins with the filing by a utility of an application for the construction permit and is followed by safety, environmental, security, and antitrust reviews by the NRC staff. These reviews are conducted in parallel. Following the staff reviews, and report to the NRC, another review is conducted by the Advisory Committee on Reactor Safeguards (ACRS), a body independent of the NRC, which gives a formal opinion letter to the NRC. Finally a mandatory public hearing is held by a three-member Atomic Safety and Licensing Board (ASLB), which reports to the NRC. If all issues are resolved satisfactorily, the staff is allowed by the NRC to issue the construction permit. The applicant may then begin construction, provided other local, state, and federal requirements are met for a variety of nonnuclear matters such as water discharge, building construction, worker safety, etc. The total number of permits required may exceed one hundred.

Several years before the plant construction is completed the applicant files an application to the NRC for an operating license. The process that follows is similar to that for the construction permit except that the public hearing is not mandatory during this step. However, a hearing may be requested by affected members of the public or by the commission (NRC). In practice this usually occurs on request from some member of the public.

Major features of the licensing process, are summarized below.

Preliminary Safety Analysis Report (PSAR) The PSAR is the primary component of the application for a CP. It generally comprises 10 or more volumes of information concerning site characteristics, plant design features, operating features, and documentation of compliance with NRC regulations.

Final Safety Analysis Report (FSAR)   The FSAR accompanies the application for an operating license. This document provides more detailed information than the PSAR about changes in design, requests for information from the NRC staff during the CP phase, changes in regulations since the issuance of the CP, and conformance with the additional regulations that are only applicable at the OL stage.

NRC Staff and ACRS Safety Reviews   The applicant's safety analysis reports provide the basis for the safety reviews. The staff reviews the reports against standard NRC acceptance criteria derived from the regulations and documents the results of these reviews in **Safety Evaluation Reports (SERs)** at both steps in the licensing process. Clarifications are usually required from the applicant both in writing and in meetings. Following completion of the SERs, the ACRS independently completes its reviews. These ACRS reviews typically include many meetings with the NRC staff and applicant. At the conclusion of the ACRS review, a letter is sent to the chairman of the NRC providing the ACRS recommendation as to whether the CP or OL should be issued. Supplements to the SERs are issued by the staff that incorporate appropriate changes as a result of the ACRS recommendations.

Environmental Review (ER)   This review is held at both the CP and OL steps pursuant to the National Environmental Policy Act of 1969. The staff review is based on a Safety Evaluation Report (SER) prepared by the applicant. Upon completion of the staff's analysis, a draft Environmental Statement is prepared and distributed for comment to the various federal, state, and local agencies and to members of the public. Comments are taken into account in the preparation of a final Environmental Statement.

Antitrust Reviews   This review is conducted by the NRC and the attorney general (Department of Justice) to ensure that operation of the facility will not be in violation of the antitrust laws.

Public Hearings   Separate hearings may be held on safety and environmental matters, or the issues may be consolidated into a single hearing. If an antitrust hearing is required, it is always held separately. Hearings are normally held close to the reactor site to allow local public participation.

License Issuance and Commission Reviews   Since 1981 it has been NRC practice for the staff to initially issue an operating license restricting operations of the plant to power levels not exceeding 5 percent of the full power rating. This low power license is issued after the completion of the safety, environmental, and antitrust reviews, and after completion of the public hearings, and resolution of any appeals. The exception to this is that public hearings on off-site emergency planning need not be completed prior to the issuance of a low-power license. The low-power license permits plant heat-up and reactor system testing, which typically takes 2 or 3 months. After the major remaining safety issues are resolved and the staff has confidence that the applicant can operate the plant safely, the staff brings the matter before the five-member commission (NRC) for review. Following this review and a favorable finding by the commission, the staff is allowed to issue a license amendment authorizing full-power operation of the plant. Federal courts may review the commission decision on appeal from interested parties, with or without a stay of execution of the license as the court may decide. Following the term of the full-power license, the license can be renewed for a period up to 20 years in accordance with the requirements established by the NRC.

The NRC has issued a safety goal policy statement that established goals that broadly define an acceptable level of radiological risk that might be imposed on the public as a result of nuclear power plant operation. To implement the goals, numerical objectives on core damage frequency and containment performance have been established. The safety goals are among the considerations underlying NRC's generic regulatory actions, but do not supplant NRC regulations and are not applied to individual plants.

Continuing Review   A licensed reactor is subject to continual oversight by the NRC throughout its operating life to assure that regulations and license conditions are observed. The NRC maintains an active program for evaluating and resolving generic safety issues, i.e., issues relevant to all plants or to a class of plants. Plant or procedural modifications are required if needed for adequate safety or justified on a safety-cost tradeoff basis. Licensees are required to monitor the performance or condition of certain structures, systems, and components, and the effectiveness of maintenance, to provide reasonable assurance that capability to perform their intended safety functions will be retained.

Reactor licensees are also required to perform a probabilistic risk assessment, called an **individual plant examination,** for their facilities. These risk assessment analyses are a result of the NRC policy regarding assessment of beyond-design-basis accidents (severe accidents) and are intended to identify any plant vulnerabilities needing attention.

On-site resident inspectors carry out a continuous inspection program and are augmented by specialized inspectors based in the NRC regional offices or headquarters. Significant events are subjected to thorough investigations to determine causes and to assure corrective actions. Should a safety hazard ever be presented at a facility, the NRC is empowered to take any necessary enforcement action, including the shutdown of the plant.

Standardized Design   Standard plant designs can lead to greater assurance of safety, reduced costs through quality production, and less licensing review time by NRC. Under NRC's procedures, a standard design is reviewed using bounding site conditions for the range of sites at which the plant might be located. For safety considerations which do not involve site-specific issues, the licensing review is therefore conducted only once, even though many applicants may reference the design. The NRC is in the process of certifying several standard nuclear power plant designs, and it strongly encourages future applicants for licenses to reference them in their applications. The operation of multiple plants having essentially the same design features will provide a substantial safety benefit because experience gained at one facility may be applied to other like facilities. This should lead to more effective maintenance of key safety components and improved reliability of plant operations. Standardization would also stimulate programs of quality control and lead to better and faster training of operators and workers. It appears likely that when U.S. utilities do begin ordering nuclear plants again, the standard plant concept will be used, as has successfully been done in other countries.

Operator Licenses   In addition to issuing operating licenses to the owners of nuclear power plants, the NRC also licenses persons who actually manipulate the controls of a reactor, at two responsibility levels, reactor operator and senior reactor operator. The licensing process involves training, experience requirements, and examinations, both written and practical, to verify an acceptable degree of knowledge on the part of the candidate before the license is granted or renewed. Because of the high cost of using an operating reactor for training and demonstration of practical facility in operation, many utilities use elaborate replica simulators of the control-room panels and systems of the power plant for operator training.

Decommissioning   Five years before an operating license is to expire, licensees are required to inform the NRC for approval of their program for funding the management of spent fuel until disposal in a permanent repository. No later than 1 year before the license is to expire, the licensee must file an application to terminate its operating license, together with a detailed plan for decommissioning. With license renewal, decommissioning follows the period of extended operation.

Radioactive Waste   The nuclear waste generated at nuclear power plants is generally classified as low-level waste, mixed waste, and spent fuel, although these plants are not significant generators of mixed waste. The low-level and mixed wastes can be shipped off-site to regional storage facilitates, if available. If not available, the low-level and mixed waste can be safely stored on-site until a facility is available. There is currently no permanent repository for storage of spent fuel, although the Department of Energy is working to establish one. Spent fuel is stored in spent fuel storage pools at each facility, and when they are filled, additional storage can be provided on-site by dry storage in independent spent fuel storage installations (ISFSIs). The spent fuel has normally been allowed to cool for at least 5 years before it is placed in an ISFSI; these facilities are certified for 20 years, can be recertified, if needed, and have a design life of 50 years.

OTHER POWER APPLICATIONS

Nuclear Ships The earliest and most successful of the efforts to develop nuclear power for ship propulsion were those by the U.S. AEC and U.S. Navy. These initial efforts toward this end predated and contributed to the development of large electric power generators. The U.S. Navy launched the *USS Nautilus* in 1954 and followed it with about 125 other operational nuclear submarines and 13 operational surface vessels. Additional submarines and surface vessels have been committed by the U.S. Navy. Nuclear submarines and ships are also being operated by the British, Russian, and French and Chinese navies.

Although a ship can be propelled by diesel engines, steam engines, gas turbines, and other devices, the steam turbine has emerged as the principal modern marine propulsion unit. When a nuclear reactor of a type developed for generating stations is used to produce steam, the combustion process is eliminated. Such a propulsion plant is especially advantageous for a submarine, since it is then completely independent of the earth's atmosphere. The U.S. Navy's nuclear ships and submarines all employ the pressurized water type of nuclear reactor.

On commercial ships, the advantages of nuclear power are related to higher-speed operation over long run routes. Experience indicates that an increase in speed usually produces substantial increases in cargoes, thereby increasing a ship's load factor. In conventional ships, the power required for higher speed is increased by the third power of the speed ratio. This means larger engines and greater fuel storage capacity. On the other hand, a nuclear ship can operate continuously at the maximum level of its hull form without penalty in payload. Such advantages are especially desirable for icebreakers, large bulk cargo vessels, and high-speed long-distance passenger service. Efforts in the 1960s and 1970s to build commercial nuclear-powered ships resulted in the N.S. *Savannah* in the United States, the N.S. *Otto Hahn* in West Germany, the N.S. *Mutsu* in Japan, and the icebreaker *Lenin* in the Soviet Union. All proved to be failures. The first two were technical successes, but economic failures. The *Savannah* is a museum, and the *Otto Hahn* had her power plant removed and replaced by diesel propulsion units. The *Mutsu* had a radiation shield leak, never operated, and was eventually broken up. The *Lenin* had her nuclear power plant replaced by a more modern one, and with three larger nuclear-powered icebreakers and a large supply ship, also nuclear-powered, operates in the Russian Arctic. Economics of a state-owned ship are difficult to assess, but these are the only unarmed nuclear-powered surface ships in operation.

Other Vehicles The use of nuclear power in airplanes, automobiles, and railroad trains has been deterred by the requirements for heavy shielding to protect against radioactivity and collisions.

Remote Power Sources Small nuclear power plants have been built and operated by the U.S. Corps of Engineers in such remote regions as the Greenland ice cap and the Antarctic ice sheet. A barge-mounted plant was also built and operated in the Panama Canal Zone, but was scrapped.

Isotopic Power Generators Isotopic power generators are direct conversion devices that convert the energy of decaying radioisotopes to electricity, using direct thermoelectric or thermionic devices, or thermal conversion systems. (See Sec. 9.1.) The first isotopic power generator, completed in 1959, produced 2.5 W of power and used polonium 210 as fuel. The first commercial isotopic power generator, fueled by strontium 90, became available in 1966, and produced 25 W of power. There are now eight radioisotopes available in this area (see Table 9.8.4). Plutonium 238 was used in the long-range space probes Voyager 1 and 2.

Space Power Sources Most satellites and space vehicles use solar and battery power sources. However for very high powers, into the multikilowatt region, and for very deep space probes where solar energy is not available, nuclear sources (either radioisotopes or reactor heat systems) are required. Several dozen such systems have been deployed since 1970. Most are isotopic, but the United States orbited one reactor system in 1965. The Soviet Union has put several reactor-powered systems into orbit. On at least two occasions such U.S.S.R. systems have decayed back into the earth's atmosphere, and in one instance radioactive parts landed in Canada. In 1984, the United States reactivated development programs for high-energy reactor-powered systems for space use.

NUCLEAR FUSION

Development of nuclear fusion reactors for the production of useful energy will require significant advances from the present state of the art in both plasma physics and engineering. A large fusion effort, especially in physics research, has been carried on since 1956 in many industrial nations, notably the Soviet Union, Japan, the United States, and several in western Europe; considerable progress has been made.

There are a number of possible **fuel cycles** involving the fusion of light ions into heavier atoms, the most promising of which are shown in Table 9.8.5. The most accessible of these is the deuterium-tritium (DT) reaction, which involves fusion of two hydrogen isotopes. Even this lowest-temperature reaction of the possible fusion cycles has a thresh-

Table 9.8.5   Possible Fusion Fuel Cycles

Reactor equation	Approx threshold plasma temp*	Approx avg energy gain per fusion
$D + T \rightarrow {}^4He\ (3.5\ MeV) + n(14.1\ MeV)$	10 keV	1,800
$D + D \begin{cases} {}^3He\ (0.82\ MeV) + n(2.45\ MeV) \\ T\ (1.01\ MeV) + p(3.02\ MeV) \end{cases}$	50 keV	70
$D + {}^3He \rightarrow {}^4He\ (3.6\ MeV) + p(14.7\ MeV)$	100 keV	180

* 1 keV = 11,000,000 K approximately.

Table 9.8.4   Radioisotopic Materials

Radioisotope	Half-life, yr	Typical power density,* W/cm^3	Radiation present†	Shielding required
Strontium^{90}	28	1.4	Beta and bremsstrahlung	Heavy
Cesium^{137}	30	0.21	Beta and gamma	Heavy
Cerium^{144}	0.78	24.5	Beta and gamma	Heavy
Promethium^{147}	2.7	1.8	Beta	Minor
Polonium^{210}	0.38	1,210	Alpha	Minor
Plutonium^{238}	89	3.5	Alpha	Minor
Curium^{242}	0.45	1,150	Alpha	Minor
Curium^{244}	18	26.4	Alpha and neutron	Moderate

* Reduced below theoretical maxima in order to account for expected isotopic impurities.
† Of dominant importance in heat generation and for consideration of radiation shielding.
SOURCE: AEC TID-20079.

old temperature of about 50 million K, and a practical reactor would require operating temperatures of about 100 million K. At these temperatures the ionized gas or plasma cannot be contained in vessels of conventional materials.

Fortunately, ionized plasmas can be contained by magnetic or inertial forces on earth. (The level of gravitation forces which produce fusion in the sun and stars cannot be achieved on earth.) Research seeking how this can be done at high enough combinations of temperatures, pressures, and densities to obtain practical net energy output is the main effort of fusion programs worldwide. This research is concentrated in three major approaches:

1. **Toroidal sytems,** especially of the tokamak variety, in which very high currents are circulated through an endless plasma confined by toroidal (doughnut-shaped) magnetic systems of various physical cross sections and magnetic shaped fields.

2. **Magnetic mirror systems,** in which the ends of a solenoidal (linear) magnetic field are "plugged" by a combination of electrostatic and magnetic forces to make the plasma reflect back and forth.

3. **Inertial systems,** in which small pellets of fusionable material are rapidly compressed by beams of laser-produced light, electrons, heavy ions, or other compressive forces. This is the principle employed in nuclear fusion weapons.

For practical energy production, a DT plasma should have a temperature of about 100 million K. This is usually expressed as equivalent ion temperature in electron volts (1 eV = 11,000 K approx.). Coincidentally with this temperature the product of density (in ions per cubic centimeter) and confinement time (in seconds) should reach 10^{14}. These conditions have been achieved experimentally.

Three large tokamak installations (TFTR in the United States, JET in Europe, and JT-60 in Japan) have begun experimental operations. Each of these machines is expected to create plasma conditions approximating those required in practical power producing reactors, although they are not in themselves net power producing devices. Likewise, a large magnetic mirror experiment is under construction at Lawrence Livermore National Laboratory in the United States to determine if similar plasma conditions can be achieved in a mirror configuration. Also at Livermore a very large laser device is being used to attempt to establish similar conditions by inertial means, using very tiny pellets filled with DT mixtures. All these devices are experimental.

After practical reactor conditions have been achieved in a confined plasma, much engineering development associated with removing the energy and converting it into useful form will be required. Developmental work on fusion engineering continues, but formidable obstacles lie ahead, especially in the requirements for materials capable of surviving the environment of burning plasmas for long periods of time.

Although scientific feasibility seems ensured and is expected to be demonstrated in the late 1990s, the very considerable engineering problems yet unresolved put the operation of a demonstration fusion power reactor well into the twenty-first century, with any significant effect on world energy supplies occurring at some future time. Since uranium resources appear adequate for many decades, the essentially unlimited fuel supplies for fusion (from the oceans) may not be needed until then.

9.9 HYDRAULIC TURBINES
by Robert D. Steele

REFERENCES: "Water Power Engineering," McGraw-Hill. Creager and Justin, "Hydroelectric Handbook," Wiley. Daugherty and Ingersoll, "Fluid Mechanics," McGraw-Hill. Hammitt, "Cavitation and Multiphase Flow Phenomena," McGraw-Hill. Kovalev, "Hydroturbines, Design and Construction," Russian translated by U.S. Department of the Interior and National Science Foundation. Mosonyi, "Water Power Development," Hungarian Academy of Science.

GENERAL

Notation

B = width of distributor or height of wicket gates, in (mm)
D = diameter of runner, in (mm)
D_l = diameter of runner, at centerline of guide case, in (mm)
D_{th} = throat diameter of runner, in (mm)
D_r = relative diameter of runner, in (mm) (= entrance diameter for low-speed Francis turbines; throat diameter for high-speed Francis turbines; bore of throat ring at centerline of blade for propeller turbines)
D_d = diameter top of draft tube, in (mm)
D_p = pitch diameter of impulse-turbine runner, in (mm)
d = jet diameter of impulse turbine, in (mm)
e = overall efficiency of turbine = $e_h e_m$
e_h = hydraulic efficiency (including draft tube)
e_m = mechanical efficiency
g = acceleration of gravity, ft/s² (m/s²)
H = net effective head, ft (m)
b = head change due to load change, ft (m)
n = r/min
n_l = r/min at 1 ft (m) head = $n/\sqrt{H}$
n_s = specific speed = $n\sqrt{P}/H^{5/4}$
P = horsepower = 0.746 kW
P_1 = horsepower (kW) at 1 ft (m) head = $P/H^{3/2}$
Q = discharge, ft³/s (m³/s)

Q_1 = discharge at 1 ft (m) head = $Q/\sqrt{H}$, ft³/s (m³/s)
t = thickness, in (mm), or time, s
u = circumferential velocity of a point on runner, ft/s (m/s)
V = absolute velocity of water, ft/s (m/s)
V_u = tangential components of absolute velocity = $V \cos \alpha$
V_r = component of V in radial plane
v = velocity of water relative to runner, ft/s (m/s)
w = weight of water lb/ft³ (kg/m³)
z = number of runner buckets (blades for propeller turbines)
α = angle between V and u (measured between positive directions)
β = angle between v and u (measured between positive directions)
θ = angle of water deflection relative to bucket
ϕ = peripheral coefficient = $[\pi Dn/(720\sqrt{2gH})][\pi Dn/(60{,}000\sqrt{2gH})]$
σ = cavitation coefficient

Fundamental Formulas   The actual power of a hydraulic turbine is the theoretical power P_t multiplied by the turbine efficiency e.

Power in U.S. Customary System (USCS) Units

$$P = eP_t = \frac{eHQw}{550}$$

In the power equation above, **power** is calculated in horsepower. In the following text, when power P is used in stated equations, it is in horsepower.

Power in Metric Units

$$P = eP_t = \frac{eHQw \times .746}{76.04} = \frac{eHQw}{101.93}$$

The metric equation stated above for power calculates power in kilowatts. In the following text, where power P is used in metric equations (shown in parentheses), the power stated in the equations is in kilowatts. [The customary unit of horsepower in the United States should not be confused with the term *metric horsepower* (see the conversion and equivalency tables of this handbook).]

The laws of proportionality for homologous turbines are:

For constant runner diam	For constant head	For variable diam and head
$P \propto H^{3/2}$	$P \propto D^2$	$P \propto D^2 H^{3/2}$
$n \propto H^{1/2}$	$n \propto 1/D$	$n \propto H^{1/2}/D$
$Q \propto H^{1/2}$	$Q \propto D^2$	$Q \propto D^2 H^{1/2}$

Nomenclature The nomenclature used throughout this section is based on NEMA's standards publication HT1-1957, "Hydraulic Turbines, Governors and Accessory Equipment," which also contains definitions.

Types of Turbines Three characteristic types of hydraulic turbines are now in general use: the **Francis reaction** type (Figs. 9.9.1 and 9.9.2);

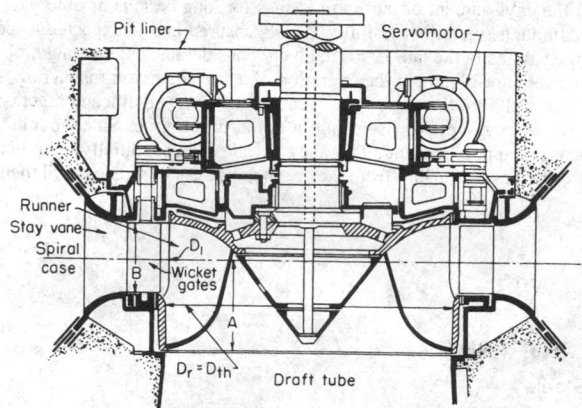

Fig. 9.9.1 Medium-head Francis turbine.

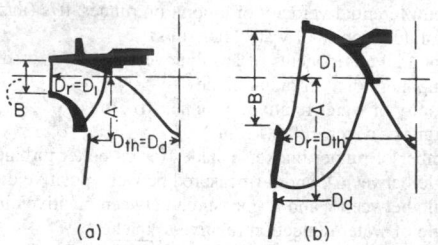

Fig. 9.9.2 Typical profile of Francis runners. (*a*) Low specific speed; (*b*) high specific speed.

the **propeller reaction** type (Fig. 9.9.3) and the **impulse** type (Fig. 9.9.12). The propeller type may be further divided into fixed and adjustable blade types. All three types have in common a stationary guide case (or nozzle in the case of the impulse type) in which the static head is transformed partly, or wholly, into velocity, and a revolving part, the runner.

In the guide case of the impulse turbine, the static head is completely transformed into velocity, so that air surrounds both the jet issuing from the nozzle and the runner.

In the guide case of the reaction types, the static head is only partly transformed into velocity, leaving an overpressure between the guide case and the runner. This overpressure causes an acceleration of the relative velocity of the water passing through the runner, the discharge area of which is smaller than the entrance area. Except when operating vented at low loads, the water passages are completely filled with water from the intake to the end of the draft tube.

Selection of Turbine Type and Casing Impulse turbines receive

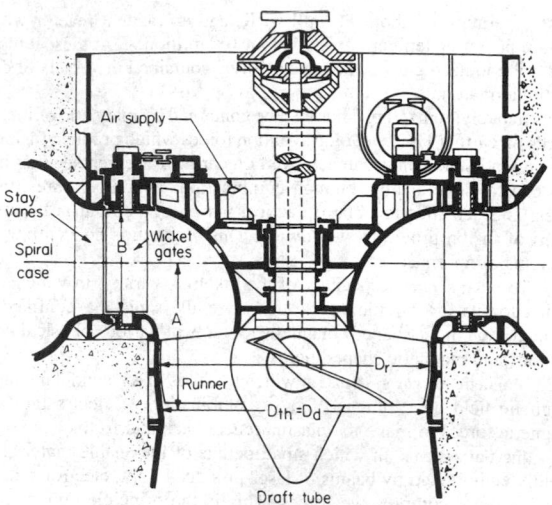

Fig. 9.9.3 Adjustable-blade Kaplan turbine.

their water supply directly from a pipeline. Francis and propeller reaction types are set in a case of concrete or metal. While the limits of head to which the impulse and reaction types are adopted may be fairly well defined in practice, as roughly outlined in Table 9.9.1, there is no definite line where the application of one type ends and the other begins. The choice between impulse and reaction types depends upon the size of the unit as well as upon the head and other considerations.

Specific Speed The common basis of comparison between turbine runners of different types and between runners of the same type but different design and characteristics is termed *specific speed* n_s. This is the relationship between the speed of a runner at the point of highest efficiency and the maximum power output at this speed regardless of size. However, since both power and speed vary with head, specific speed is defined as the relationship between the speed n_1 and power P_1 at 1 ft (m) head. Subscript 1 denotes that the value is reduced by the similarity law to a 1 ft (m) head basis. Since $n \propto 1/D$ and $P \propto D^2$, the product $n_1\sqrt{P_1}$ remains a constant regardless of the size of the runner and is designated the specific speed of the runner. The term *specific speed* for this relationship stems from the fact that $n_1\sqrt{P_1}$ also is the value of the speed in r/min at best efficiency which the runner would have if operated under 1 ft (m) head, the runner being of such size as to develop 1 hp (kW) ($P_1 = 1$).

Since $n \propto \sqrt{H}$ and $P \propto H^{3/2}$, the specific speed of any runner operating under head H will be $n_s = n\sqrt{P}/H^{5/4}$, where n is the best efficiency speed and P the maximum power output at this speed, all at head H.

Selecting the Speed Hydraulic turbines are usually connected to ac generators. The turbine speed must agree with one of the synchronous speeds required for the system frequency. The prevailing frequency for most systems in the United States is 60 Hz. Synchronous speeds are determined by the formula $n = 120 \times$ frequency/number of poles in generator. The number of poles must be even.

The speed should be as high as practicable, as the higher the speed the less expensive will be the turbine and generator and the more efficient will be the generator.

A convenient way of determining the highest practicable speed is by the relation of the specific speed to the head. For Francis turbines this may be taken as $n_s = 900/\sqrt{H}$ ($1,900/\sqrt{H}$); for propeller-type turbines as $n_s = 1,000/\sqrt{H}$ ($2,100/\sqrt{H}$). The adoption of these values of specific speed is predicated on the use of a reasonable setting of the unit with reference to tailwater level, i.e., selection of proper cavitation coefficient.

Number of Units From the standpoint of reducing the number of auxiliaries and the amount of associated equipment and also reducing initial and maintenance costs for the entire plant, the number of units should be kept to a minimum. Also the larger the unit, the higher the

Table 9.9.1   General Arrangements of Turbine Installations and Usual Head Limits Employed

Type	Setting	Construction	No. of runners	Usual head limits for direct-connected units	
				ft	m
Reaction turbines, 5 to 1,000 ft (1.5 to 300 m) head	Axial flow	Vertical or horizontal or slanted	1	5–60	1.5–20
	Encased	Concrete vertical	1	15–130	5–40
		Cast or welded plate steel. Stainless steel. Vertical or horizontal	1 or 2	30–1,600	10–500
Impulse turbines, 500 to 6,000 ft (150 to 1,800 m) head	Encased	Horizontal (1–2 nozzles)	2	500–6,000	150–1,800
		Vertical (1–6 nozzles)	1	500–6,000	150–1,800

efficiency and generally the lower the cost per unit power output. However, other considerations, such as flexibility of operation, higher efficiency operation during low-load demands, and minimum loss of capacity during shutdown for repair or maintenance, might dictate the use of multiple units where one unit would be feasible in terms of physical size. For some projects, the physical size of the unit has been limited to the maximum size runner that could be shipped in one piece; this is largely due to the extra manufacturing costs involved in furnishing split runners. However, since split runners present no serious mechanical difficulties, the tendency in recent years has been to disregard this limitation.

REACTION TURBINES

Francis   Figure 9.9.1 shows a Francis-type reaction turbine for medium head. The runner consists of a relatively large number of shrouded buckets. Movable wicket gates with axes parallel to the turbine shaft control the flow. This type of turbine is normally used for heads ranging from 75 to 1,600 ft (25 to 500 m). Specific speeds vary from 15 to 100 (50 to 400).

Propeller Turbines, Fixed and Adjustable Blade   The propeller turbine has a runner which is normally provided with from 3 to 10 unshrouded blades, either fixed or adjustable. This type of turbine usually is used for heads from 5 ft (1.5 m) up to 120 ft (40 m), although in a few cases they have been used for heads up to 200 ft (60 m). The higher the head, the greater the number of blades used. Specific speeds vary from 80 to 250 (300 to 1,000). Propellers have very steep efficiency vs. power curves (4 and 5, Fig. 9.9.5). Adjustable-blade propeller runners are used to produce a flat efficiency curve over a wide range of power (6, Fig. 9.9.5) and to produce considerably more power beyond the maximum efficiency point than can be obtained with a fixed-blade runner of equal diameter. For fixed-blade runners, the blade angle is usually set between 16 and 28°, where maximum efficiency occurs. For adjustable blade runners, the blade angle may vary from −10° min to +45° max. Normally, automatic oil-pressure-operated blades are used in a **Kaplan turbine** (Fig. 9.9.3). The blades are adjusted by means of an oil-operated piston located either within the main shaft or in the runner hub. The oil is admitted to and discharged from the piston by means of a distributor head either on top of the generator shaft or surrounding the main shaft. The oil pressure is supplied from the governor oil pressure system. The controls are so arranged that the blade tilt varies automatically with the wicket gate opening to produce a maximum efficiency envelope curve (Fig. 9.9.9). A large operating-cylinder capacity is required. The capacity S, ft·lb (J) may be approximated from the formula $S = 20 P n_s^{1/4}/\sqrt{H}$ $(14 P n_s^{1/4}/\sqrt{H})$. If antifriction bearings are used in the hub instead of bronze bearings for the blade trunnions, this formula becomes $S = 10 P n_s^{1/4}/\sqrt{H}$ $(7 P n_s^{1/4}/\sqrt{H})$. However, since the hydraulic unbalance on the blades varies with speed, gate opening, and the location of the blade pivot axis, the oil-operated cylinder size should be calculated from model test data.

Axial-Flow Turbines   Axial-flow turbines use the propeller type runner with either fixed or adjustable blades. Their characteristic feature

is the straight-through, or nearly straight-through, water passageway from intake to discharge. The shaft is, therefore, either horizontal, vertical, or inclined (Fig. 9.9.4). The spiral or semispiral case and elbow draft tube, which require substantial widths and depth of excavation, are eliminated. Therefore, with a reduction in height and area of the powerhouse and the turbine's suitability for location directly within the dam, an overall construction-cost savings may be obtained for the power plant part of the project compared with conventional vertical units. This may make it possible to build or redevelop power plants for low heads or in small sizes that have previously been considered uneconomical.

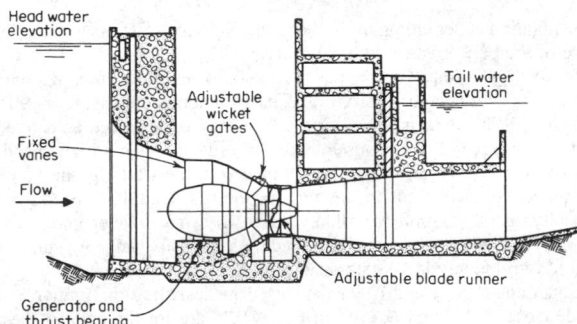

Fig. 9.9.4   Axial-flow bulb turbine.

In recent years, axial-flow turbines have been installed for use with tidal power since they can be arranged to operate with water flowing in either direction and, when required, to pump with water flowing in either direction.

There are four general types of axial-flow turbines. The first type has the generator rotor mounted around the periphery of the turbine runner. This type of turbine and generator arrangement has been called a **rim** turbine. The second type is the **bulb** turbine, where the generator is directly connected to the turbine at a submerged elevation. The bulb turbine has the generator enclosed in a streamlined, watertight housing located in the water passageway on either the upstream or downstream side of the runner (Fig. 9.9.4). The third type of axial-flow turbine is the **pit** turbine. In the pit turbine arrangement, a high-speed generator is driven from the high-speed output shaft of a speed increaser, and the low-speed input torque to the speed increaser is supplied from the low-speed shaft of the turbine. The generator of an axial-flow pit turbine can be mounted in line with the turbine, or in smaller units the generator may be mounted inclined or vertical.

The fourth type of axial-flow turbine is the **tube** type, which has the generator located outside the water passages. The construction characteristic of this turbine is the slight bend of the water passageway that permits the extension of the turbine shaft to the outside of the water passageway. The tube turbine is typically arranged so that the generator is on the downstream side of the turbine, but it could be constructed to have the generator on the upstream side. The shaft from the turbine to the generator of a tube turbine can be inclined to reduce excavation, thereby raising the generator higher above tailwater elevations. In some cases, a gear-type speed increaser is used to reduce the combined equipment cost and generator size and weight.

Axial-flow turbines are suitable for heads up to at least 90 ft (30 m) with basically the same limitations that apply to the conventional Kaplan or other propeller-type turbines. Maximum unit capacity is limited, however, by maximum practical speed increaser torques and maximum practical horizontal generator capacities. This appears to be approximately 100 MW at present. Either fixed-position or movable wicket gates can be used. Power and efficiency performance is comparable to conventional vertical shaft propeller turbines.

Design The water entering the runner is given a whirl or tangential component by the guide vanes. This whirl is taken out by the runner so that, at the design point (point of best efficiency), the water leaves the runner without appreciable whirl. According to Euler's theorem, the power of a turbine in steady motion equals the angular velocity multiplied by the change in angular momentum experienced by the mass of water flowing in a unit time in its passage through the turbine. This is expressed in the following equations:

$$P = 62.4Q(u_1V_{u1} - u_2V_{u2})/(550g)$$
$$= QHe_n/8.81 \tag{9.9.1}$$

and $$u_1V_{u1} - u_2V_{u2} = e_hgH$$

$$P = 1,000Q(u_1v_{u1} - u_2v_{u2})/(101.93g)$$
$$= QHe_hw/101.93 \tag{9.9.2}$$

(See notation at beginning of this section.) Subscript 1 refers to the inlet edge of the bucket and 2 to the discharge edge.

In practice, e_h may be taken as 0.92 to 0.94, and the term u_2V_{u2} may be omitted as it becomes zero for discharge in a radial plane, $\alpha_2 = 90°$.

The right hand member of Eq. (9.9.2) may be regarded as a fixed value for a given head. Consequently, in order to have a high-speed-type runner (u_1, large) the whirl V_{u1} placed in the water by the guide case must be small. This means that for the best load, low-speed low-capacity (small n_s) runners will have relatively small gate openings and short wicket gates, whereas high-speed high-capacity (large n_s) runners will have relatively large gate openings and long wicket gates.

Equations (9.9.1) and (9.9.2) are used in designing the runner and guide case. The P and Q used in Eq. (9.9.1) are for the point of best efficiency and not for the rated values. For specific speeds up to about 60 (250) the power at best efficiency may be taken as 85 percent of the rated power. Above that specific speed the point of best efficiency occurs at higher loads as indicated in Fig. 9.9.5. Adjustable-blade propeller turbine-runner blades are laid out for an intermediate load of 75 to 80 percent of the rated load. The actual design of a runner should be worked out by an experienced designer on the basis of coordinated test data.

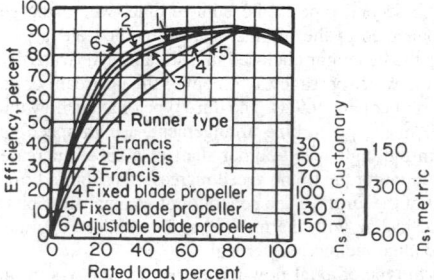

Fig. 9.9.5 Variation of efficiency vs. load for reaction turbines.

The general proportions of a normal line of turbine runners may be determined as follows for any combination of net head H and rated power P. (Rated power is usually considered to be 95 percent of full gate power.)

First select the specific speed appropriate to the head (see "Specific Speed" above). Next calculate n from the formula $n_s = n\sqrt{P}/H^{5/4}$. If $n/H^{5/4}$ is not a synchronous speed, select the nearest synchronous speed and calculate the corresponding n_s. This value of n_s may be used in Fig. 9.9.6 to determine the peripheral coefficient ϕ for D_r and D_1. Then the diameter may be determined from the formula

$$\phi = \pi Dn/720\sqrt{2gH} \ [\ = \pi Dn/(60,000\sqrt{2gH})]$$

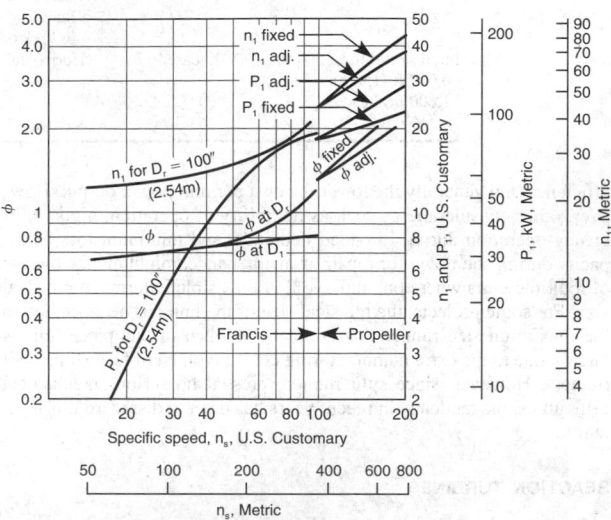

Fig. 9.9.6 Turbine characteristics and specific speed.

Other proportions of the runner may be determined from Fig. 9.9.7, which shows the ratio to D_r of width of distributor B; diameter at top of draft tube D_d; and centerline of distributor to top of draft tube A. See Fig. 9.9.1 for D_r, B, and A.

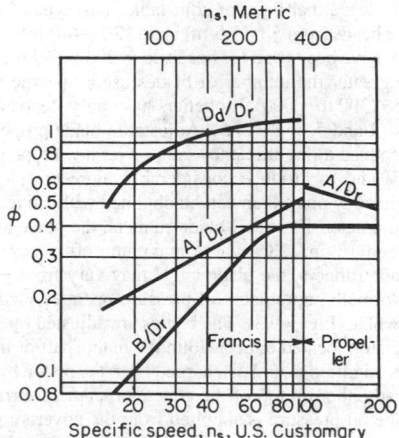

Fig. 9.9.7 Turbine characteristics and specific speed.

The values of unit power P_1, and unit speed n_1, for $D_r = 100$ in (2,540 mm), which correspond to the specific speed of the runner may be determined from Fig. 9.9.6. The following formulas may then be used: $P = P_1H^{3/2}(D_r/100)^2 \ [\ = P_1H^{3/2}(D_r/(2,540)^2]$ and $n = n_1\sqrt{H}$ $(100/D_r) \ [n_1\sqrt{H} \ (2,540/D_r)]$. These formulas may be used to verify the diameter calculated from the peripheral coefficient, and will prove useful in selecting runner sizes for units which must meet power requirements other than rated power at design head. In such cases the values of n_1 and P_1 from Fig. 9.9.6 should be used in conjunction with Fig. 9.9.8.

For preliminary powerhouse layouts these ratios of diameter to D_r may be used for all specific speeds: gate circle, 1.20; stay ring, 1.65; pit liner, 1.50; unit spacing, 3.5 to 4.5.

The clearance space between the discharge tips of the wicket gates and the entrance edge of the runner buckets is treated as a free vortex space in which the tangential component of the absolute velocity of

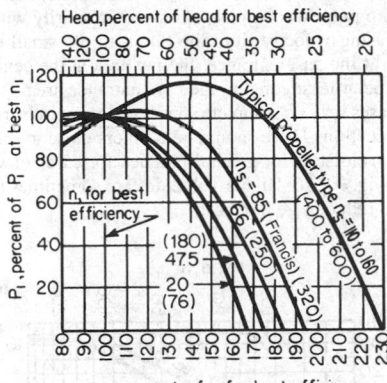

Fig. 9.9.8 Variation of unit power with unit speed and specific speed.

discharge from the gates (V_u) increases in inverse proportion to the radius while the component in a radial plane (V_r) increases in inverse proportion to the area.

To obtain the best efficiency, the water must enter the runner without shock and leave it with as little velocity as possible; i.e., the entrance angles of the buckets must be approximately in line with the relative direction of the entering flow and the direction of the absolute discharge velocity should be approximately in a radial plane (axial in the case of propeller runners). This condition prevails only at the most efficient load. Above that load the water enters the draft tube with reverse whirl and below that load with a forward whirl.

The shape of the buckets should be kept as simple as possible, consistent with meeting the proper angles. The curvature of the bucket along a flow line is made greatest near the entrance and reduces as the discharge edge is approached, somewhat the shape of a portion of a parabola.

Turbine Characteristics Speed, power, and efficiency characteristics vary widely with specific speed as shown in Figs. 9.9.5 and 9.9.6. The variation of unit power P_1 with the unit speed (as a percent of normal basis) is shown in Fig. 9.9.8 for typical specific speeds ranging from 20 to 160 (70 to 600). For low specific speed Francis-type turbines, P_1 decreases as n_1 increases, as when operating at heads below normal. This is due to the centrifugal effect, which tends to decrease the flow as n_1 increases. This characteristic of low specific speed Francis runners makes them less suitable for operation under extreme variations in head. With the higher specific speeds and particularly with the propeller type, P_1 increases as n_1 increases above normal. This makes the higher specific speeds particularly suitable for operation at widely varying heads.

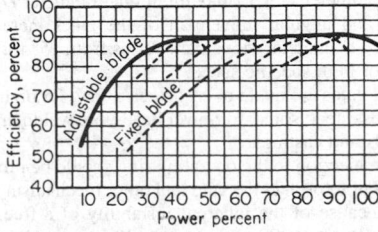

Fig. 9.9.9 Variation of efficiency vs. power for fixed- and adjustable-blade propeller turbines.

Figure 9.9.9 shows the advantage of blade adjustment for propeller-type turbines. By angular adjustment of the blades as the load changes, the peaked power-efficiency curve of the fixed blade propeller is transformed into a flat power-efficiency curve, and the maximum power output is considerably increased.

Runaway Speed If a turbine runner is allowed to revolve freely without load and with the wicket gates wide open, it will overspeed to a value called the *runaway speed*. The runaway speed, at normal head, varies with the specific speed and for Francis turbines ranges from 170 percent (normal speed = 100 percent) at low specific speed [n_s = 20 (76)] to some 195 percent at high specific speed [n_s = 100 (383)]. For propeller turbines, the runaway speed varies with blade angle—the steeper the blade angle, the lower the runaway speed. For fixed-blade propellers with the blades set from 16 to 28°, where maximum efficiency is usually obtained, the runaway speed ranges approx from 255 to 180 percent, respectively. For adjustable blade turbines, where the minimum blade angle is sometimes as low as 6 to 16° in order to obtain high efficiency at part load, the maximum possible runaway speed will be about 290 to 270 percent, respectively. However, with adjustable-blade propeller turbines, there is from the standpoint of efficiency an optimum relationship between runner-blade angle and wicket gate opening, usually controlled by a cam in the operating mechanism; the higher the gate opening, the steeper the blade angle. Thus the combination of wide-open gate and minimum design blade angle can only occur in the so-called "off-cam" position, which is an extremely rare possibility. In some units, this maximum possible off-cam runaway speed is reduced by limiting the minimum design blade angle to 16°. Another method is the use of a runaway speed limiter. There are several designs in use. One of the most frequently used is a valve designed to open by centrifugal force at speeds of 120 percent (approximately). The open valve bypasses the runner blade servomotor piston, thus equalizing the oil pressure on both sides of this piston. If the runner blades are designed to open because of hydraulic unbalance at overspeeds, they will then go to a higher blade angle under off-cam overspeed conditions. The off-cam runaway speed thus can be limited to some 185 percent if the runner is designed for a maximum blade angle of 28 to 40°.

For all turbines, if the maximum head is higher than the normal head, the runaway speed will be increased in proportion to the square root of the head. Therefore, runaway speeds should be based on the maximum operating head rather than on the normal head. Any runaway speed above 180 percent can increase appreciably the cost of the generator.

Turbine Thrust The water passing through the turbine creates a thrust load, that is carried by a thrust bearing, typically supplied with the generator. For vertical units, the thrust T consists of the hydraulic thrust T_h plus the weight of the turbine runner and shaft T_w, lb (kg). Units have been constructed with the thrust bearing supporting the total weight of rotating components of the turbine and generator, plus the hydraulic thrust, located on the turbine head cover.

For horizontal or inclined units, the thrust T consists of the hydraulic thrust T_h and a component of the weight T_w of the rotating components, if the unit is inclined. The thrust bearing can be located in the generator, in a speed increaser, in the hydraulic turbine, or in a separate thrust bearing housing.

The hydraulic thrust may be obtained by using a **thrust coefficient K_t** multiplied by the weight, lb (kg), of a circular column of water whose diameter is the diameter of the runner D_r, in (mm), and whose height is equal to the maximum head H_m, ft (m), under which the turbine must operate, or

$$T_h = \frac{K_t D_r^2 H_m}{2.94} \quad \text{lb} \qquad \left(T_h = \frac{K_t D_r^2 H_m}{1{,}273} \quad \text{kg} \right)$$

The thrust coefficient for Francis turbines varies with the specific speed and may be taken as approximately $K_t = n_s/250$ ($K_t = n_s/954$). In using this value of K_t, it is assumed that the runner seals are properly arranged and that the crown of the runner inside the seals is properly drained; otherwise, the thrust may be much higher.

For propeller turbines, with either fixed or adjustable blades, the thrust coefficient K_t may be taken as 0.90 for all specific speeds.

Reaction Turbine Elements

Runner and Wearing Rings The number of buckets z for Francis runners ranges from about 21 for low specific speed to 12 for high specific speed. The approximate number may be taken as $z = 55/n_s^{1/3}$ ($86/n_s^{1/3}$).

Materials that can readily be repaired by welding with either carbon-steel or stainless-steel electrodes are selected for the construction of runners. Most runners are made of cast carbon steel, cast stainless steel, formed carbon steel plate, formed stainless-steel plates, or a combination of carbon steel and stainless steel. The majority of new equipment is supplied manufactured of stainless steel, using a combination of martensitic and austenitic stainless-steel casting, forgings, and plates. The use of high-strength stainless steel for runners is increasing for high heads and for conditions where cavitation pitting may be a design consideration.

The number of blades for propeller-type turbines ranges from 10 for low specific speeds to 3 for high specific speeds. Propeller runner blades are usually made of cast carbon steel, formed carbon steel, formed stainless steel, or cast stainless steel. If they are made of cast steel or formed carbon plate steel, the surfaces over which pitting may be expected are overlay-welded with stainless steel before finishing to final contour. In place of overlay welding, solid stainless-steel inserts are sometimes welded into propeller blades, and stainless trim is added to the periphery of the blades.

The functions of the runner seals for Francis turbines are to prevent excessive leakage loss and thus improve the efficiency and control the hydraulic thrust.

Rolled, forged, and cast steels make excellent wearing rings for low- and medium-head units. To prevent seizure in case of contact the rotating ring material should differ from that of the stationary rings. Stainless steel, bronze, or steel with bronze inserts make excellent wearing rings. For extremely high heads, stainless steel should be used to prevent undue wear and erosion.

Seal clearances are made as small as practicable to reduce leakage, particularly on high-head units. Larger clearances are required with water-lubricated main bearings, subject to considerable wear before being readjusted.

Main Shaft and Bearings The main shaft, for a typical vertical hydraulic turbine unit, must be rigid and should be made of a medium-grade carbon-steel forging. The size of the shaft is determined by limiting the torsional stress from 3,000 to 6,000 lb/in² (20 to 41 N/mm²).

The shafting system of a horizontal or an inclined hydraulic turbine unit is subject to the torsional stress produced by the torque transmitted and the bending stress produced by the weight of the runner, the shaft, and the components that are attached to the shafting system. The design of a shaft for a horizontal unit must be subjected to review by experienced mechanical engineering designers to ensure acceptable stress levels, to account for points of stress concentration, and to determine the overall safety margin of a design. The allowable stress levels in a horizontal shaft must be evaluated after the fatigue properties of the shafting material and the points of stress concentration are taken into consideration. Should the bearing be designed to be water-lubricated, a stainless-steel sleeve is installed on the shaft to provide a bearing surface for the guide bearing. A stainless-steel sleeve is typically placed on the shaft in the area of a packing box or stuffing box.

Vertical hydraulic turbines are usually provided with one main bearing located in the head cover, as near the runner as practicable. The main turbine bearing is generally a babbitted oil bearing, supplied in halves, or as independently adjustable segmented babbitted shoes. The oil is supplied to the bearing by an independent low-pressure oiling system. Self-lubricated babbitted shell or shoe bearings, containing an oil reservoir, and pumping grooves in the babbitt are sometimes used.

Water-lubricated bearings of wood, lignum vitae, rubber, or special composition are sometimes preferred, particularly for small and medium-size propeller-type turbines where the bearing is located at the bottom of the head cover cone, where the packing box and the oil-lubri- cated bearing would be inaccessible, and where it would be difficult to avoid water contamination. With the water-lubricated bearing, the packing box is placed above the bearing.

Spiral Case The spiral case must be proportioned so as to cause relatively low friction losses, as well as to prevent eddying which would travel into the runner and affect its efficiency. The cases generally are made of (1) metal: cast steel, cast iron, or steel plate, and (2) concrete. Metal cases are made as complete spirals, customarily with uniform or slightly increasing velocity from the throat to the small end. When a valve is used at the case entrance, the net area at the centerline of the valve should be at least equal to the case entrance area.

Concrete cases may be complete spirals or semispirals, rectangular or oval in cross section. There should be no piers close to the turbine.

Rated load velocities for general practice for cases of various types are shown in Fig. 9.9.10. Higher velocities are sometimes used with the larger units to reduce size and cost.

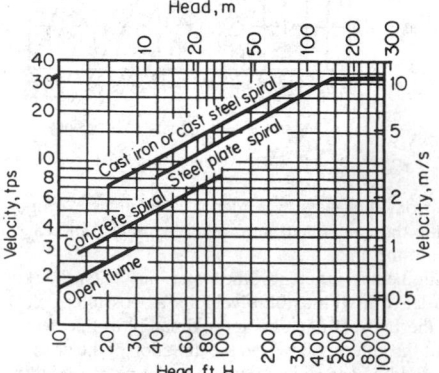

Fig. 9.9.10 Case velocities.

Stay Ring That part of the guide apparatus between the spiral case and the wicket gates, containing stationary stay vanes, is called the stay ring. The water is accelerated within this space as it approaches the gates. The number of stay vanes employed is usually made equal to or one-half the number of gates. They are placed at the angle which will cause the least obstruction to the flow.

The stay ring is cast integral with cast cases and is made separately of welded or cast steel for concrete or steel plate spiral cases. It should be a continuous ring to facilitate erection, and very rigid, because it serves as a foundation for the rest of the turbine and generator.

Wicket Gates and Operating Mechanism Wicket gates control the power and speed of the turbine. The number of gates m ranges from 12 for small units to 28 for large units. The overall dimensions of the turbine decrease as the number of gates increases.

To prevent interference between the gates and the runner buckets, which may cause noise and vibration, the discharge tips of the fully open gates should be kept well away from the inlet edges of the runner buckets of Francis turbines, the radial clearance being large enough to prevent the gate tips from overhanging the curved part of the discharge ring. Gate tip diameter = 1.1 (maximum inlet diameter D_r).

The height and angular movement of the gates increase with the specific speed. The angular movement varies from 15° for low specific speed to 80° for high specific speed.

Most wicket gates are made of cast steel; a few, for higher heads, are made of stainless steel; some, for lower heads, are weldments built from rolled materials and castings.

Each gate connection to the operating ring should be provided with a breaking element to protect the gate and other mechanism in case of an obstruction. Because of the inherent instability of a free gate and its tendency to flutter, it is advisable to install gate-restraining devices to hold the gate if a breaking element fails. This device also avoids potential resonant flutter with the natural frequencies in the system. Each gate should also be provided with stops to prevent it from striking the runner or reversing after the breaking element fails.

The capacity G, ft·lb (J), of the oil pressure cylinder or cylinders (sometimes known as governor capacity) may be computed from the formula

$$G = 18 Pn_s^{1/4}/\sqrt{H} \qquad (13 Pn_s^{1/4}/\sqrt{H})$$

Some units now provide individual servomotors on each wicket gate. This synchronized system allows the hydraulic control system to dampen a tendency to gate flutter.

One or more vacuum breakers or air valves are installed in the head cover to admit air to the runner or draft tube, to improve efficiency at low gate openings or to alleviate draft tube vortex cavitation. An air valve is also necessary on propeller-type units to break the vacuum under the head cover and to help prevent the runner from "screwing up" in the water when the gates are suddenly closed. The air valve is piped to the outside of the powerhouse above the floodwater level.

Draft Tubes The draft tube may serve the double purpose of (1) allowing the turbine to be set above tailwater level, without loss of head, to facilitate inspection and maintenance, and (2) regaining, by diffuser action, the major portion of the kinetic energy delivered to it from the runner.

At rated load the velocity at the upstream end of the tube for modern units ranges from 24 to 40 ft/s (7 to 12 m/s), representing from 9 to 25 ft (3 to 8 m) head. As the specific speed is increased and the head reduced, it becomes increasingly more important to have an efficient draft tube. Good practice limits the velocity at the discharge end of the tube to 5 to 10 ft/s (2 to 3 m/s), representing less than 1 ft (0.3 m) velocity head loss.

The elbow type of tube is now used with most vertical turbine installations. With this type the vertical portion begins with a conical section which gradually flattens in the elbow section and then discharges horizontally through substantially rectangular sections to the tailrace. Most of the regain of energy takes place in the vertical portion, very little in the elbow section which is shaped to deliver the water to the horizontal portion so that the regain may be efficiently completed. Figure 9.9.11 shows proportions of a good elbow tube. One or two vertical piers are placed in the horizontal portion of the tube for structural reasons.

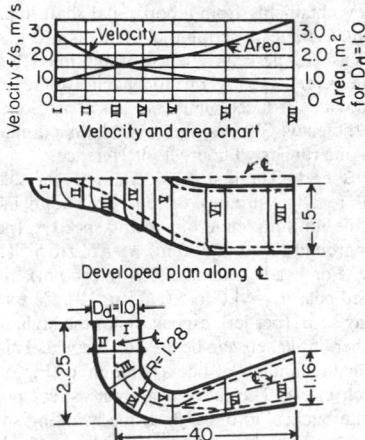

Fig. 9.9.11 Elbow draft tube layout.

Most tubes are made of concrete with a steel plate lining extending from the upper end to a point where the velocity has been sufficiently reduced [say 20 ft/s (7 m/s)] to prevent erosion of the concrete. Pier noses are also lined where necessary to prevent erosion and for structural reasons.

IMPULSE TURBINES

Impulse turbines are utilized when the head is too high for the practical use of Francis turbines, which is normally a head exceeding 1,600 ft (500 m). Impulse turbines are also sometimes used for heads below 1,600 ft (500 m) where excessive erosion due to foreign materials in the water presents a problem. The main disadvantage of impulse turbines, especially for low heads, is their low specific speed. In the past, this was overcome on conventional horizontal shaft units by the use of two runners or two jets per runner. In recent years, the vertical shaft multijet impulse turbine (Fig. 9.9.12) has become popular.

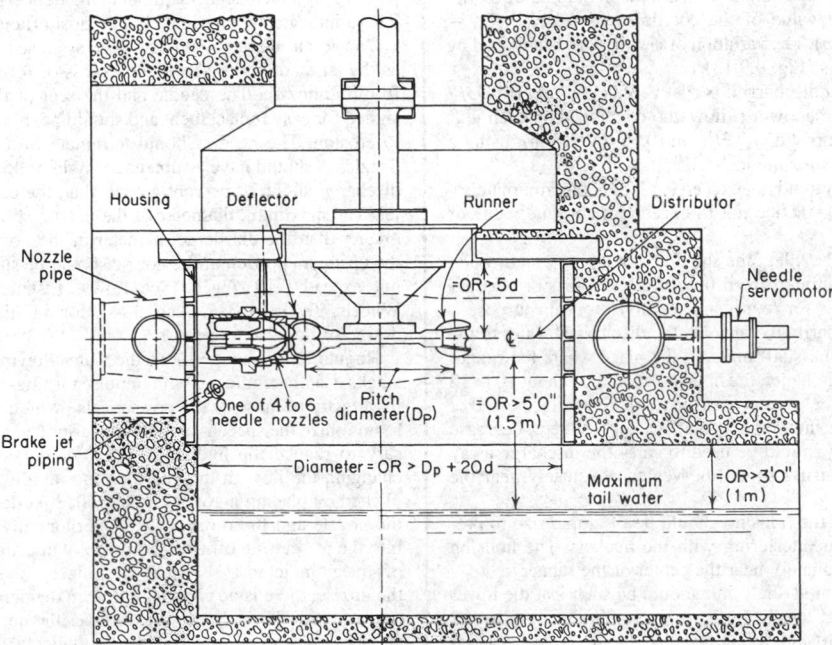

Fig. 9.9.12 Vertical-shaft multijet impulse turbine installation.

The efficiency obtainable from a horizontal shaft impulse turbine is about 90 percent. Field tests on multijet vertical units have shown efficiencies as high as 91.5 percent. The use of multiple jets on the vertical units reduces the percentage loss due to windage of the runner. The unit can be operated with a reduced number of jets at part load, thus increasing part load efficiency. Six jets are about the maximum number that can be used on one runner without jet interference.

Selection of Speed   The first consideration is the selection of suitable n_s. As with reaction turbines, there is a relation which determines approximately the limiting value of specific speed n_s (per jet) for any head. For heads around 1,000 ft (300 m), n_s = 5.0 to 5.5 (19 to 21) gives high efficiency. For heads around 2,000 ft (600 m), maximum efficiency is attained near n_s = 4.0 to 5.0 (15 to 19). Care has to be exercised in selecting an n_s (per jet), especially for the higher heads, so that the proper number of buckets can be used on the wheel disk. The higher the n_s, the fewer the number of buckets required (Fig. 9.9.13) but the smaller the pitch circle. The latter can create stress problems in the attachment of the buckets to the disk. If the resulting speed n is quite low for the power to be developed, the n_s of the unit and consequently the speed n can be increased by increasing the number of runners or the number of jets per runner. The n_s of the unit will then be n_s (per jet) times the square root of the number of jets.

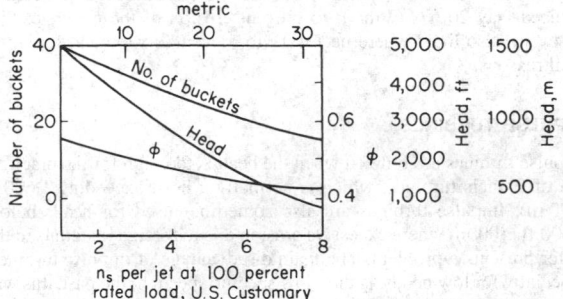

Fig. 9.9.13   Impulse turbine characteristics and proportions.

Basic Dimensions   The pitch diameter D_p of the runner is twice the distance from the center of the runner to the axial centerline of the jet and is determined by the value of the coefficient ϕ selected, $D_p = 1,840\phi\sqrt{H}/n$ ($84,578\phi\sqrt{H}/n$). The variation of ϕ with n_s is established by experience and model tests (Fig. 9.9.13).

The quantity of water Q discharged per jet is $550P/(wHe)$ [$101.93P/(wHe)$], where P is the horsepower (kilowatts) developed by each jet.

The diameter of the jet is $4.85\sqrt{Q}/H^{1/4}$ in ($0.55\sqrt{Q}/H^{1/4}$ m), using a velocity coefficient of 0.97 for the jet.

The **velocity** in the inlet should not exceed $0.12\sqrt{2gH}$ (approximately), and it is considered good practice not to exceed an absolute value of 40 ft/s (12 m/s).

The **valve** employed in the inlet pipe should be of a type which, when open, permits a smooth, uninterrupted flow, free of eddies or obstructions. Modern preference is for rotary sphere valves for this purpose.

The **housing** serves primarily to carry off the discharged water to the tail pit below. On horizontal shaft units, at the place where the runner receives the discharge from the jet, the housing should be about 10 to 12 times the jet diameter. At the place where the runner has been cleared of the discharge, the housing should be as narrow as possible to decrease windage. Suitable baffling should be used to carry the discharge away from the buckets. The housing should be vented adequately near the center of the runner.

For vertical-shaft units, the housing should be of ample size to prevent discharge water from interfering with the buckets. The housing should also be vented adequately near the center of the runner.

The **setting** of the horizontal-shaft unit should be such that the lower edges of the buckets are at least 3 ft (1 m) above maximum tailwater elevation. For vertical shaft units, this distance should be at least 5 ft (1.5 m). Impulse turbine discharge entrains large amounts of air, and unless this is completely replaced, a vacuum will be produced in the

housing which will draw the tailwater up and drown out the runner. Thus the roof of the discharge tunnel or passageway, for both horizontal- and vertical-shaft units, should be at least 3 ft (1 m) above the maximum operating tailwater elevation to permit the free circulation of air to the runner (Fig. 9.9.12).

In cases where extremely high tailwater occurs for a short period, compressed air can be used to depress the tailwater.

Experience has shown that the ft^3/s (m^3/s) of air (at the discharge pressure of the compressor) required to maintain the depressed tailwater is about 15 percent of the ft/3/s (m^3/s) of water (Q) discharged by the turbine.

Runner   The force F exerted by a stream upon a stationary bucket (Fig. 9.9.14) is

$$F = MV(1 - \cos\theta) = wQV(1 - \cos\theta)/g$$

where M = mass per s and θ = angle in deg through which the water is turned relative to the bucket (see section Y-Y, Fig. 9.9.14). If the bucket is moving in the direction of the stream with a velocity u, $F = wQ(V - u)(1 - \cos\theta)/g$. The work done by the stream = $Fu = wQn (V - u)(1 - \cos\theta)/g$. When θ is greater than 90°, the cosine becomes negative. Thus the closer the angle θ is to 180°, the greater the amount of work done with the same Q and V. However, in order to prevent the water from striking the succeeding bucket, θ should never be 180° (see section Y-Y, Fig. 9.9.14). Also, since the water discharged from the bucket must have some velocity, the value of u is made somewhat less than $V/2$.

The angle β_1 at the centerline of the bucket should be calculated from the velocity triangle (Fig. 9.9.14). The value of β_1 is greatest as the entrance lip E enters the jet. The angle of the underside of the bucket should be greater than β_1 to avoid pitting.

In terms of the jet diameter which will give maximum guaranteed or rated capacity, the approximate proportions of the buckets are $B = 3d$, $L = 2.6d$, $D = 0.85d$ (Fig. 9.9.14). The contour of the entrance lip E is important, since the manner in which this lip comes in contact with the jet affects the path of the water passing to the preceding bucket. The size of the bucket with relation to the jet diameter also determines the percentage of full load at which maximum efficiency occurs. The larger the bucket, the higher this percentage of load.

The number of buckets on a runner should be such that no water can pass through the buckets without being deflected by the buckets. Figure 9.9.13 indicated approximately the most efficient number.

The **needle nozzle** should be placed as close to the buckets as possible, as the jet tends to lose its compactness of form shortly after emerging from the nozzle. The needle and the seat of the nozzle should be designed for easy replacement and should be of a material highly resistant to erosion. The needle should terminate in a cone of 30 to 45°. The nozzle tip should have a subtended angle of 45 to 60° and a diameter at discharge about 20 percent greater than the calculated diameter of the jet. The maximum diameter of the needle should be about 15 percent greater than the discharge diameter of the nozzle tip. The diameter of the upstream portion of the nozzle should be such that the velocity does not exceed $0.10\sqrt{2gH}$. (See Lowy, Efficiency Analysis of Pelton Wheels, *Trans. ASME*, Aug. 1944, for additional information on the design of impulse turbines.)

Regulation   The inertia of the water flowing through the long penstocks usually employed with impulse turbines prohibits rapid reduction in velocity because of the pressure rise which would occur. Therefore, to minimize the speed rise following a sudden load rejection, it is necessary to reduce the hydraulic power delivered to the runner without changing the flow in the penstock too rapidly. This is usually accomplished by placing a governor controlled jet deflector between the needle nozzle and the runner. The governor moves this deflector rapidly into the jet, cutting off the load. It is not unusual for the deflector to cut off the entire jet in 1½ s. Since the deflector acts on the jet after it leaves the nozzle, there is no change of flow in the penstock; hence, there is no pressure rise. The governor then moves the needle at a permissible rate (in terms of pressure rise), with simultaneous automatic withdrawal of the deflector. The jet is finally reduced the necessary amount to correspond to the reduced load. The needle must also move slowly in the

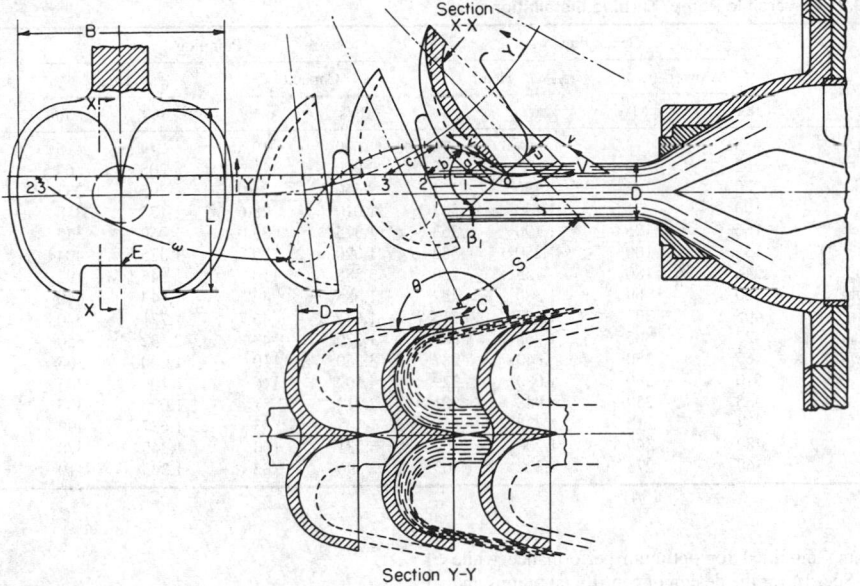

Fig. 9.9.14   Impulse turbine bucket design.

opening direction for oncoming loads to avoid penstock collapse because of large pressure drops.

   Runaway Speed   The runaway speed for impulse turbines ranges from 160 to 190 percent of normal speed, depending upon the specific speed of the runner; the higher the specific speed, the higher the runaway speed.

REVERSIBLE PUMP/TURBINES

In this type of project, surplus low-value energy from either hydro or thermal plants is used to pump water during off-peak periods to an elevated reservoir, where it becomes available for generating high-value peaking energy. While separate pumps and hydraulic turbines of conventional design can and have been used for this purpose, the development of single reversible pump/turbines has made many pumped storage projects economically feasible.

Figure 9.9.15 is a cross section of a reversible pump/turbine, showing that, with the exception of the runner, it is essentially a conventional turbine. The wicket gates must be designed for flow in both directions. A few units have been built without movable wicket gates. The runner is

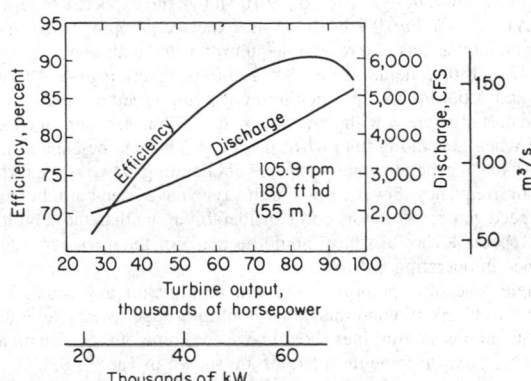

Fig. 9.9.16   Generating.

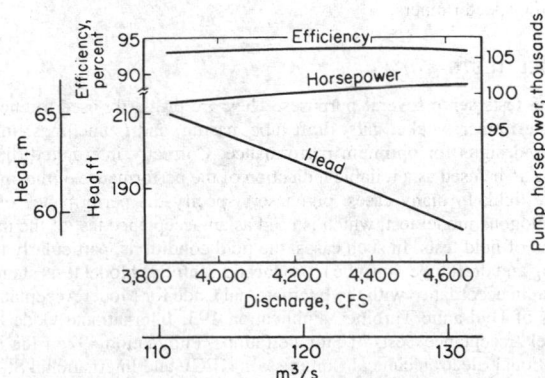

Fig. 9.9.17   Pumping.

Figs. 9.9.16 and 9.9.17   Typical performance curves for a pump turbine.

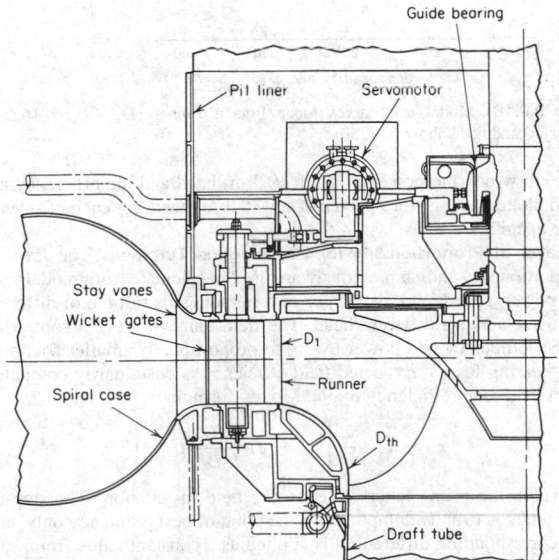

Fig. 9.9.15   Reversible pump turbine.

Table 9.9.2   Reversible Pump/Turbine Installations

| Plant | Generating | | | | Pumping | | | | Year in operation |
| | Power | | Head | | Capacity | | Head | | |
	1,000 hp	MW	ft	m	ft³/s	m³/s	ft	m	
Pedreira "A"	5.2	3.9	49	15	690	19.5	49	15	1939
Hiwassee	62	46	190	58	3,900	111	205	62.5	1956
Taum Sauk	220	165	790	241	2,650	75.0	765	233	1963
Cruachan	100	74.6	1,130	343	1,010	28.6	1,175	358	1966
Cabin Creek	166	125	1,190	363	1,375	39.0	1,060	323	1967
Villarino	135	100	1,250	382	1,020	29	1,325	404	1969
Kisenyami	240	180	755	220	3,880	110	645	197	1969
Drakensburg	360	269	1,380	422	1,685	42	1,443	440	1977
Dinorwic	405	302	1,682	513	1,647	46.7	1,771	540	1975
Bajina Basta	422	385	1,968	600	1,520	43	2,132	650	1976
Raccoon Mt.	465	350	940	287	3,850	110	1,000	305	1974
Bath Co.	510	380	1,080	329	4,100	116	1,100	335	1983
Palmiet	335	250	988	301	2,012	57	1,000	305	1988
La Muela	284	212	1,725	525	1,165	33	1,637	499	1990
Bad Creek	482	360	1,200	367	3,531	100	1,240	378	1991
Mingtan	368	275	1,322	403	2,896	82	1,345	410	1992

essentially a pump runner modified for optimum performance while generating power. (See Sec. 14 for the design of centrifugal pumps and their characteristics.) Conventional turbine runners, because of their short blades, are not suited for the pumping cavitation requirements.

The reversible pump/turbines have certain fundamental performance characteristics which are inherent in the design. The relationship between pumping and generating performance for a given specific speed is more or less fixed and can be modified only to a minor degree by alterations in the design. Figure 9.9.16 shows the expected generating performance, and Fig. 9.9.17 shows the expected pumping performance based on model tests of a reversible pump/turbine with a specific speed $n_s = 47.3$ (180) generating and $n_s = 2,600$ (50) pumping (see "Centrifugal and Axial Pumps" for definition of pump specific speed).

The best efficiency, with reversible pump/turbines, occurs at a lower speed when generating than when pumping. This can be compensated for by using a generator/motor capable of operating at two speeds with a constant frequency. Several units of this type have been built, but since two-speed generator/motors cost considerably more than those built for single speed, a careful study should be made of the advantages to be obtained in operating at two speeds.

Single reversible pump/turbines can be built for any heads up to 2,000 ft (600 m). Beyond this, either multiple-stage reversible units or separate pumps and turbines should be used. Some of the more notable reversible pump/turbine installations are shown in Table 9.9.2.

Runaway Speed   The runaway speed for pump/turbines is considerably lower than that for conventional turbines. It ranges from 150 percent of normal for low specific speed runners to 175 percent for high specific speed runners.

MODEL TESTS

Model tests serve several purposes. They are primarily used to check turbine runner, wicket gate, draft tube, casing, and (sometimes) inlet works designs for optimum performance. Correctly interpreted, they may also be used as a reliable indication of the performance of the units in the field. In many cases, purchasers specify the performance of a homologous model test, which is used as an acceptance test of the unit in lieu of field tests. In such cases, the field conditions, particularly the casing and draft tube, must be reproduced faithfully. Model tests should be run in accordance with the International Code for Model Acceptance Tests of Hydraulic Turbines, Publication 193; International Code for Model Acceptance Tests of Storage Pumps, Publication 497 of the International Electrotechnical Commission (IEC); and International Standard Publication, IEC 995, that supplements IEC 193 and 497.

Figure 9.9.18 shows typical model test results of a reaction-type tur-

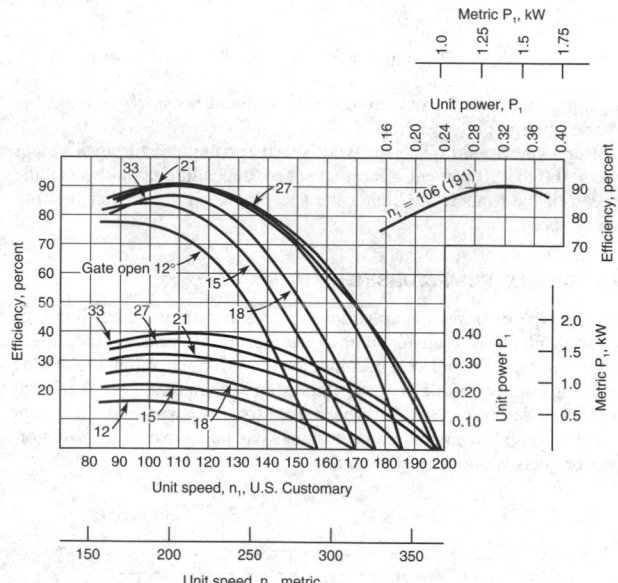

Fig. 9.9.18   Model test curves for a Francis runner ($D_r = D_{th} = 16.2$ in (0.41 m) and $n_s =$ about 65 (250).

bine, in which the power P at 1 ft (0.3 m) head and the efficiency are each plotted against the speed n at 1 ft (0.3 m) head, for each of several gate openings.

Laws of Proportionality for Homologous Turbines   The laws of proportionality shown previously are used to calculate from model tests the power, speed, and discharges of a homologous turbine of different diameter under a different head. The field unit will have a somewhat higher efficiency and power owing to proportionally smaller frictional and bearing losses. Expected field efficiency is customarily computed from the model efficiency by the Moody formula

$$c' = 1 - \frac{2}{3}\left[(1 - e)\left(\frac{D}{D'}\right)^{1/5}\right]$$

in which the prime letters refer to the field installation. The step-up efficiency is usually computed for the point of best efficiency only, and the corresponding differential is applied as a constant value from, say, half load to full load.

CAVITATION

Cavitation occurs when the pressure at any point in the flowing water drops below the vapor pressure of the water.

The relationship which produces cavitation is between vapor pressure, barometric pressure, setting of the runner with respect to tailwater, and net effective head on the turbine and is expressed by the **Thoma cavitation coefficient** $\sigma = (H_b - H_v - H_s)/H$, where H_b = barometric head, ft (m) of water; H_v = vapor pressure of water, abs; H_s = elevation, ft (m) of the runner above tailwater, measured at the centerline of the distributor of a Francis runner and at the centerline of the blades of a propeller runner (if the runner is submerged, H_s becomes negative); and H = total or net effective head, ft (m), on the turbine. The setting of the runner depends upon the value of σ, which varies with specific speed n_s of the runner and the individual characteristics of a particular runner design. In practice, the model of the proposed runner is first tested with relatively high back pressure (H, small or negative). Then the back pressure is reduced in increments until the breaking point, as indicated by a drop in power, efficiency, and discharge, is reached. This breaking point is designated as the critical σ and will vary with gate opening and speed and, on propeller turbines, with blade angle. Consequently, σ must be determined for a range of limiting conditions.

In the absence of cavitation tests, the value of σ should not be lower than $\sigma = n_s^{3/2}/2,000$ ($n_s^{3/2}/15,000$) for Francis and propeller runners and $\sigma = n_s^2/25,000$ ($n_s^2/350,000$) for adjustable-blade propeller runners.

The value of σ at which a plant operates, depending largely upon the setting of the runner with respect to tailwater, is called the **plant** σ. To avoid excessive cavitation, the plant σ should not exceed the critical σ. The greater this margin, the less possibility of cavitation during operation. For a general discussion of cavitation phenomena, see Knapp, Recent Investigations of the Mechanics of Cavitation and Cavitation Damage, *Trans. ASME*, Oct. 1955. Laboratory tests and experience have shown that materials having a high resistance to cavitation erosion (pitting) and suitable for use in hydraulic turbines are the stainless steels and aluminum bronzes, especially when used as welding overlays. (See Rheingans, "Resistance of Various Materials to Cavitation Damage," ASME Report of 1956 Cavitation Symposium.)

SPEED REGULATION

(See also Sec. 16.)

Regulation is accomplished by changing the flow of water to the turbine. The flow is controlled by the wicket gates of reaction turbines and by the needle valve or jet deflector of impulse turbines. The governor moves the gates or needle in response to speed changes resulting from load or head changes.

A schematic diagram of a classic governor is shown in Fig. 9.9.19. The parts consist of a speed-responsive device, a power element that changes the gates or needle position, and a follow-up or compensating device that prevents hunting.

The **speed-responsive device** originally was a pair of spring-loaded flyballs mounted directly on the turbine shaft, or driven from the shaft by belt or gears, or driven by an electric motor that receives its power either from the bus line or from an independent generator driven from the main turbine generator shaft.

The **power element** consists of oil-operated power cylinders or servomotors which operate the turbine gates or needle. Oil pumps and a pressure tank or accumulator maintain a supply of oil. A valve operated by the speed sensor controls the flow of oil to the servomotors or acts as a pilot valve controlling a larger relay valve, which in turn controls the oil to the servomotors. The pump capacity is usually 3 servomotor volumes per minute. The capacity of the pressure tank is generally made 20 times the servomotor volume, allowing for 8 volumes of oil and 12 volumes of air. The velocity of oil in the pipelines is kept below 15 ft/s (5 m/s). When using individual servomotors on each gate, these values are usually increased 50 percent to balance the usual damping effect of the common massive two-servomotor system.

The **follow-up** or **compensating device** connects the power piston of the servomotor to the control valve, usually through a dashpot, and causes the motion of gates or needle to stop when they have moved sufficiently to compensate for the load change.

The time for a full stroke or traverse of the governor is controlled by the rate of flow of oil to the servomotors; most governors have provisions for varying this time. The gate opening changes at a uniform rate

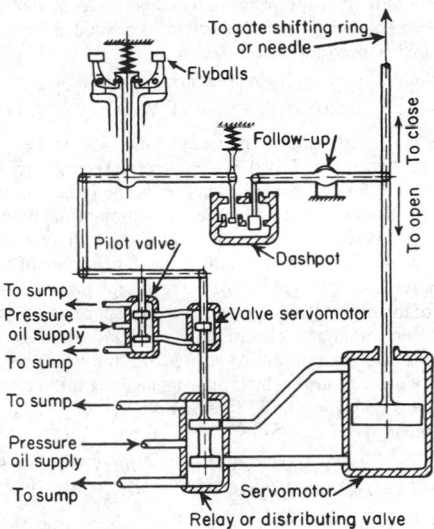

Fig. 9.9.19 Schematic diagram of governor.

over the major portion of the stroke and at a somewhat slower rate at the ends of the stroke. The governor *dead time,* or the elapsed time from the initial speed change to the first movement of the gates, is usually less than 0.2 s.

Electric Governor In recent years, the functional requirements placed upon the hydraulic-turbine governor have increased to the point where electrical control of the hydraulic turbine is attractive in view of the simplicity with which electrical signals can be manipulated. The basic elements of an electric-hydraulic governing system are (1) permanent magnet generator (PMG) or equivalent for measurement of turbine speed and the means for transmittal of such speed signals to the electrical portion of the governor; (2) an electric circuit sensitive to speed variations about some adjustable reference point; (3) amplifying circuits to convert speed reference changes, speed error signals, and auxiliary signals into a useful electric current; (4) an electrohydraulic transducer to transform the electric current into a hydraulic output signal; (5) hydraulic amplifying equipment to deliver suitable power and the desired signal to the gate servomotors as a function of the output of the electrohydraulic transducer; (6) power supplies for the electric and hydraulic portions of the control. For further particulars of the electric governor, see Leum, Electric Governors for Hydro Turbines, ASME Paper 62-WA165.

Electronic Governors With the advent of minicomputers, microprocessors, and programmable controllers, many of the traditional mechanical functions in the hydraulic turbine governor have been replaced electronically. While the basic elements of the governor remain the same, the use of electronics in the feedback sensing and control circuits has greatly reduced the size and mechanical complexity of these units. The basic elements of the modern electronic governoring system are: (1) Inductive pickups or permanent magnetic generators or equivalent to measure the turbine speed and to transmit the information to the processor. (2) Electronic circuitry to establish the speed-control function of the governor (integrated circuits or digital programming). Basically the circuitry consists of a proportional, an integrator, and a derivative function. (3) A servo valve or equivalent to translate the electronic signal from the processor to a hydraulic signal. (4) Where required, a means to hydraulically amplify the output from the servo valve so that a suitable power level can be delivered to the gate and blade servomotors. (5) Auxiliary equipment as required in the form of power supplies, etc. to support the system.

Speed Regulation Requirements Usually, a sufficient measure of the regulation provided is the maximum speed rise resulting from sudden rejection of full load, as from the breaker tripping. A maximum speed rise of 30 percent of normal speed for this condition is a common limitation.

Speed Rise Following Load Reduction For sudden load reductions, the approximate speed rise is

$$n_t/n = [1 + 1,620,000T_tP_t(1 + h/H)^{3/2}/WR^2n^2]^{1/2} \quad \text{(USCS)}$$
$$= [1 + 365,000T_tP_t(1 + h/H)^{3/2}/GD^2n^2]^{1/2} \quad \text{(SI)}$$

where n_t is the speed, rpm, at the end of time T_t; n is the speed, rpm, before the load decrease; T_t is the time interval, seconds, for the governor to adjust the flow to the new load; P_t is the reduction in load, hp (kW); h is the head rise caused by the retardation of the flow, ft (m); H is the net effective head before the load change, ft (m); WR^2 is the product of the revolving parts weight, lb, and the square of their radius of gyration, ft; GD^2 is the product of the revolving parts weight, kg, and the square of their diameter of gyration, m. For approximate values of h, see below. Very rapid gate closure produces a reduction of pressure in the draft tube and the possibility of breaking the water column, with subsequent violent resurge which may damage the turbines.

Speed Drop Following Load Increase For sudden load increases, the approximate speed drop is

$$n_t/n = [1 - 1,620,000T_tP_t/WR^2n^2(1 - h/H)^{3/2}]^{1/2} \quad \text{(USCS)}$$
$$= [1 - 365,000T_tP_t/GD^2n^2(1 - h/H)^{3/2}]^{1/2} \quad \text{(SI)}$$

where P_t is the actual load increase and h is the head drop caused by the increase of the flow. If the speed drop is to be determined for a given increase in gate opening, the governor time T_t for making this increase and the normal change in load for the change in gate opening, under constant head H, can be used in the following formula:

$$n_t/n = [1 - 1,620,000T_tP_t(1 - h/H)^{3/2}/WR^2n^2]^{1/2} \quad \text{(USCS)}$$
$$= [1 - 365,000T_tP_t(1 - h/H)^{3/2}/GD^2n^2]^{1/2} \quad \text{(SI)}$$

The actual change in load, however, will be $P_t(1 - h/H)^{3/2}$.

For derivation of the above speed variation formulas and for a more accurate determination, see Strowger and Kerr, Speed Changes of Hydraulic Turbines for Sudden Changes of Load, *Trans. ASME,* 1926; and Rich, "Hydraulic Transients," McGraw-Hill.

Water Hammer in Penstocks If a gate movement is considered as a series of instantaneous movements with a very small interval between each movement, the pressure variation in the penstock following the gate movement will be the effect of a series of pressure waves, each caused by one of the instantaneous small gate movements. For a steel penstock, the velocity of the pressure wave $a = 4,660/\sqrt{1 + d/100t}$ $(1,420/\sqrt{1 + d/100t})$, where d is the penstock diameter, in (m), and t is the penstock wall thickness, in (m). The pressure change at any point along the penstock at any time after the start of the gate movement may be calculated by summing up the effect of the individual pressure waves. See "Symposium on Water Hammer," ASME, 1933; and Parmakian, "Water Hammer Analysis," Dover.

Approximate formulas (De Sparre) for the increase in pressure h, ft (m), following gate closure, are given below. They are quite accurate for pressure rises not exceeding 50 percent of the initial pressure, which includes most practical cases.

$h = aV/g$	for $K < 1, N < 1$
$h = aV/\{g[N + K(N - 1)]\}$	for $K < 1, N > 1$
$h = aV/[g(2N - K)]$	for $K > 1, N > 1$

where $K = aV/(2gH)$; $N = aT/(2L)$. V and H are the penstock velocity, ft/s (m/s), and head, ft (m), prior to closure; L is the penstock length, ft (m), and T is the time of gate closure. For full load rejection, T may be taken as 85 percent of the total gate traversing time to allow for nonuniform gate motion.

For pressure drop following a complete gate opening, the following formula (S. Logan Kerr) may be used with T not less than $2L/a$:

$$h = \frac{aV}{g}\left(\frac{-K + \sqrt{K^2 + N^2}}{N^2}\right) = \text{pressure drop, ft (m)}$$

Pressure variations exceeding 40 percent rise and about 25 percent drop should be avoided. When the control, directly by the governor, causes undesirable pressure variations, a surge tank, a pressure regulator, or a jet deflector may be used. A **surge tank** is a standpipe with an atmospheric tank, attached to the penstock as close as possible to the casing inlet. The tank provides a reservoir and expansion chamber for the water demand or the water rejection following sudden gate movements, so that sudden accelerations or decelerations of the flow in the penstock are avoided.

Pressure regulators may be of either the water-wasting or water-saving type. The **water-wasting type** is a synchronous bypass, generally attached to the turbine casing. It is operated directly from the governor, or the gate mechanism of the turbine, and wastes such an amount as to keep the total water discharge equal at all times to the full-load discharge of the turbine. The bypass is a needle nozzle or a mushroom-shaped disk valve or a cone valve which opens and is partly balanced hydraulically by a piston under pipeline pressure. The **water-saving type** permits the regulator to open upon rapid closure of the turbine gates, and then close slowly, so that the total water discharge is gradually reduced and finally limited to that through the turbine, adjusted for the new load.

AUXILIARIES

Valves and head gates are provided for shutting off the water to each turbine for safety, for ease of maintenance, and to reduce water leakage losses. Motor-operated steel head gates are generally used for low- and medium-head plants with concrete scroll cases, although a few still use stop logs. The butterfly type of valve placed close to the turbine casing is suitable for medium- and high-head units with metal casings of circular inlet diameter. Butterfly valves of 8-ft (2.5 m) diameter for 1,000-ft (305 m) head and 27-ft (8.2 m) diameter for 100-ft (30 m) head have been built. The new flow through valve or biplane valve is becoming more popular for this application. In recent years, rotary sphere valves have replaced gate valves for high heads and where the loss through the butterfly valve is excessive because of the obstruction to flow by the valve disk.

COMPUTER-AIDED DESIGN

The continued development, refinement, and use of proven high-technology computer-aided design and analysis tools help create new generations of designs. Design objectives today are to produce equipment which has characteristics of reliability, easy servicing, operational flexibility, and high performance. The success of this approach is made possible by integrating design history and field experience into hydraulic, mechanical, and operational computer design and analysis systems.

TURBINE TESTS

Field testing of hydraulic turbines to determine the absolute efficiency and output involves careful and accurate measurement of the power available in the water supplied to the turbine [water hp (kW)] and the turbine output [developed hp (kW)] e = developed hp (kW)/water hp (kW) = $550P/(wQH)$ [$101.93P/(QHw)$]. The tests should be conducted in accordance with the International Code for Field Acceptance Tests of Hydraulic Turbines and/or Storage Pumps, IEC Publications 41 and 198.

Because of the difficulties and costs involved in making accurate measurements of horsepower, net head, and discharge in the field, there has been a trend in recent years to dispense with the field test, especially where a laboratory test on a homologous model turbine is available. Instead, an index test is made on the unit in the field, which measures the turbine output and relative discharge under various conditions. Index tests should be conducted in accordance with the International Code for Field Acceptance Tests of Hydraulic Turbines, IEC Publication 41.

Section 10

Materials Handling

by

VINCENT M. ALTAMURO *President, VMA Inc., Toms River, NJ*
ERNST K. H. MARBURG *Manager, Product Standards and Service, Columbus McKinnon Corp.*

10.1 MATERIAL HOLDING, FEEDING, AND METERING

by Vincent M. Altamuro

FACTORS AND CONSIDERATIONS

The movement of material from the place where it is to the place where it is needed can be time consuming, expensive, and troublesome. The material can be damaged or lost in transit. It is important, therefore, that it be done smoothly, directly, with the proper equipment and so that it is under control at all times. The several factors that must be known when a material handling system is designed include:

1. Form of material at point of origin, e.g., liquid, granular, sheets, etc.
2. Characteristics of the material, e.g., fragile, radioactive, oily, etc.
3. Original position of the material, e.g., under the earth, in cartons, etc.
4. Flow demands, e.g., amount needed, continuous or intermittent, timing, etc.
5. Final position, where material is needed, e.g., distance, elevation differences, etc.
6. In-transit conditions, e.g., transocean, jungle, city traffic, in-plant, etc., and any hazards, perils, special events or situations that could occur during transit
7. Handling equipment available, e.g., devices, prices, reliability, maintenance needs, etc.
8. Form and position needed at destination
9. Integration with other equipment and systems
10. Degree of control required

Other factors to be considered include:

1. Labor skills available
2. Degree of mechanization desired
3. Capital available
4. Return on investment
5. Expected life of installation

Since material handling adds expense, but not value, it should be reduced as much as possible with respect to time, distance, frequency, and overall cost. A straight steady flow of material is usually most efficient. The use of mechanical equipment rather than humans is usually, but not always, desirable — depending upon the duration of the job, frequency of trips, load factors, and characteristics of the material. When equipment is used, maximizing its utilization, using the correct equipment, proper maintenance, and safety are important considerations.

The proper material handling equipment can be selected by analyzing the material, the route it must take from point of origin to destination, and knowing what equipment is available.

FLOW ANALYSIS

The path materials take through a plant or process can be shown on a flowchart and floor plan. See, Figs. 10.1.1 and 10.1.2. These show not only distances and destinations, but also time required, loads, and other information to suggest which type of handling equipment will be needed.

Description of operation	Hdlg.	Move	Insp.	Temp. stor.	Perm. stor.	People req.	Dist. moved	Time req.	Equipment used
Unload from truck to receiving dock	○	○	☐	△	△	1	10 ft	2 min. 1 0s	Hand truck
Inspect and tag	○	○	☐	△	△	1		30s	
Move to raw material storage	○	○	☐	△	△	1	40 ft	1 min. 10s	Hand truck
Raw material storage	○	○	☐	△	△				
Move to machine	○	○	☐	△	△	1	120 ft.	1 min. 30s	Hand truck
Set in machine	○	○	☐	△	△	1		10 min.	Manual handling
Place finished stock on pallet	○	○	☐	△	△	1		12 min 10s	Manual handling
Move to inspection and test	○	○	☐	△	△	1	110 ft	1 min.	Pallet truck
Move to packing	○	○	☐	△	△	1	60 ft	1 min. 25s	Pallet truck
Move to finished goods storage	○	○	☐	△	△	1	70 ft	1 min. 10s	Pallet truck
Finished goods storage	○	○	☐	△	△				
Move to shipping dock	○	○	☐	△	△	1	40 ft	1 min. 10s	Pallet truck
Load on trucks	○	○	☐	△	△	1	10 ft	2 min. 30s	Manual handling
Total manpower travel and time						11	460 ft	34 min. 45s	

Fig. 10.1.1 Material flowchart.

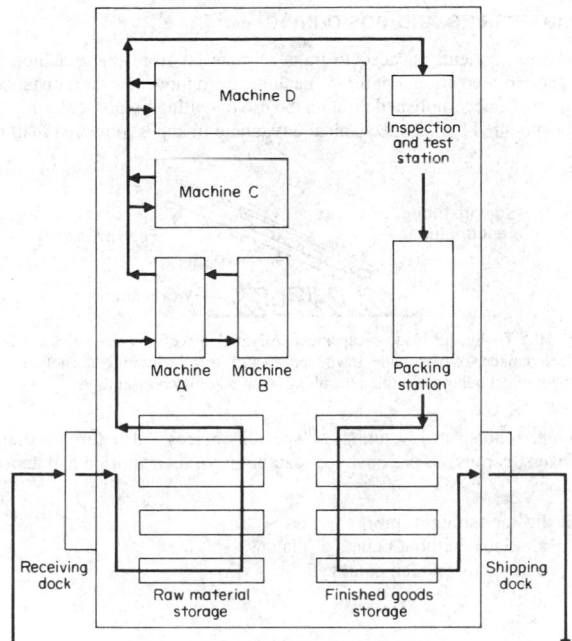

Fig. 10.1.2   Material flow floor plan.

CLASSIFICATIONS

Material handling may be divided into classifications or actions related to the stage of the process. For some materials the first stage is its existence in its natural state in nature, e.g., in the earth or ocean. In other cases the first stage would be its receipt at the factory's receiving dock. Subsequent material handling actions might be:

1. Holding, feeding, metering
2. Transferring, positioning
3. Lifting, hoisting, elevating
4. Dragging, pulling, pushing
5. Loading, carrying, excavating
6. Conveyor moving and handling
7. Automatic guided vehicle transporting
8. Robot manipulating
9. Identifying, sorting, controlling
10. Storing, warehousing
11. Order picking, packing
12. Loading, shipping

FORMS OF MATERIAL

Material may be in solid, liquid, or gaseous form. Solid material may be in end-product shapes or in intermediate forms of slabs, sheets, bars, wire, etc., which may be rigid, soft, or amorphous. Further, each item may be unique or all may be the same, allowing for indiscriminate selection and handling. Items may be segregated according to some characteristic or all may be commingled in a mixed container. They might have to be handled as discrete items, or portions of a bulk supply may be handled. If bulk, as coal or powder, their size, size distribution, flowability, angle of repose, abrasiveness, contaminants, weight, etc., must be known. In short, before any material can be held, fed, sorted, and transported, all important characteristics of its form must be known.

HOLDING DEVICES

Devices used to hold material should be selected on the basis of the form of the material, what is to be done to it while being held (e.g.,

heated, mixed, macerated, dyed, etc.), and how it is to be fed out of the device. Some such devices are tanks, bins, hoppers, reels, spools, bobbins, trays, racks, magazines, tubes, and totes. Such devices may be stationary—either with or without moving parts—or moving, e.g., vibrating, rotating, oscillating, or jogging.

The material holding system is the first stage of a material handling system. Efforts should be made to assure that the holding devices never become empty, lest the entire system's flow stops. They can be refilled manually or automatically and have signals to call for refills at carefully calculated trigger points.

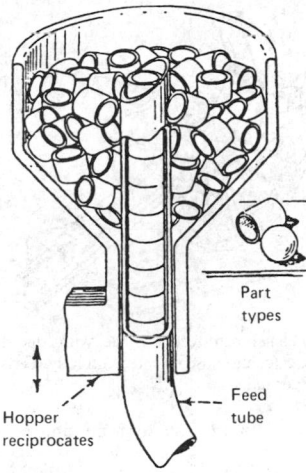

Fig. 10.1.3   Reciprocating feeder. Either hopper or tube reciprocates to obtain, orient, and feed items.

MATERIAL FEEDING AND METERING MODES

Material can be fed from its holding devices to processing areas in several modes:

1. Randomly or by selection of individual items
2. Pretested or untested
3. Linked together or separated
4. Oriented or in any orientation
5. Continuous flow or interruptible flow
6. Specified feed rate or any rate of flow
7. Live (powered) or unpowered action
8. With specific spaces between items or not

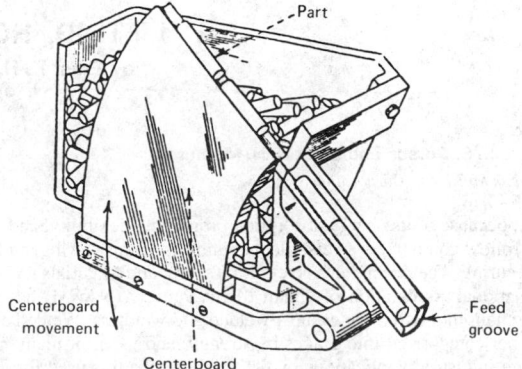

Fig. 10.1.4   Centerboard feeder. Board, with edge shaped to match part, dips and then rises through parts to select part, which then slides into feed groove.

FEEDING AND METERING DEVICES

The feeding and metering devices selected depend on the above factors and the properties of the material. Some, but not all, possible devices include:

1. Pumps, dispensers, applicators
2. Feedscrews, reciprocating rams, pistons
3. Oscillating blades, sweeps, arms
4. Belts, chains, rotaries, turntables
5. Vibratory bowl feeders, shakers
6. Escapements, mechanisms, pick-and-place devices

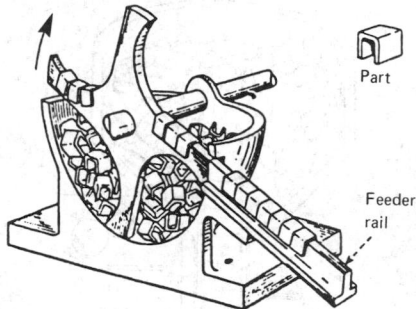

Fig. 10.1.5   Rotary centerboard feeder. Blade, with edge shaped to match part, rotates through parts, catching those oriented correctly. Parts slide on contour of blade to exit onto feeder rail.

Figures 10.1.3 to 10.1.8 illustrate some possible feeding and metering devices.

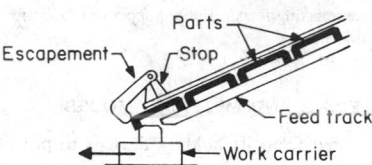

Fig. 10.1.6   Escapement feeder. One of several variations. Escapement is weighted or spring-loaded to stop part from leaving track until it is snared by the work carrier. Slope of track feeds next part into position.

TRANSFERRING AND POSITIONING

There is frequently a need to transfer material from one machine to another, to reposition it within a machine, or to move it a short distance. This can be accomplished through the use of ceiling-, wall-, column-, or floor-mounted hoists, mechanical-advantage linkages, powered manip-

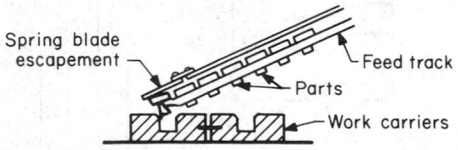

Fig. 10.1.7   Spring blade escapement. Advancing work carrier strips off one part. Remainder of parts slide down feed track. Work carrier may be another part of the product being assembled, resulting in an automated assembly.

ulators, robots, airflow units, or special devices. Also turrets, dials, indexers, carousels, or conveyors can be used; their motion and action can be:

1. In-line (straight line)
   a. Plain (without pallets or platens)
   b. Pallet or platen equipped

Fig. 10.1.8   Lead screw feeder. Parts stack up behind the screw. Rotation of screw feeds parts with desired separation and timing.

2. Rotary (dial, turret, or table)
   a. Horizontal
   b. Vertical
3. Shuttle table
4. Trunnion
5. Indexing or continuous motion
6. Synchronous or nonsynchronous movement
7. Single- or multistation

10.2   LIFTING, HOISTING, AND ELEVATING
by Ernst K. H. Marburg and Associates

CHAIN
by Joseph S. Dorson and Ernst K. H. Marburg
Columbus McKinnon Corporation

Chain, because of its energy absorption properties, flexibility, and ability to follow contours, is a versatile medium for lifting, towing, pulling, and securing. The American Society for Testing and Materials (ASTM) and National Association of Chain Manufacturers (NACM) have divided chain into two main groups: welded and weldless. Welded chain is further categorized into grades by strength and used for many high-strength industrial applications, while weldless chain is used for light-duty low-strength industrial and commercial applications. The grade designations recognized by ASTM and NACM for welded steel chain are 30, 43, 70, and 80. These grade numbers relate the mean stress in newtons per square millimeter (N/mm²) at the minimum breaking load.

As an example, the mean stress of Grade 30 chain at its minimum breaking load is 300 N/mm² while those for Grades 80, 70, and 43 are 800, 700, and 430 N/mm² respectively.

Graded welded steel chain is further divided by material and use. Grades 30, 43, and 70 are welded carbon-steel chains intended for a variety of uses in the trucking, construction, agricultural, and lumber industries, while Grade 80 is welded alloy-steel chain used primarily for overhead lifting (sling) applications.

As a matter of convenience when discussing graded steel chains, chain designers have developed an expression which contains a constant for each given grade of chain regardless of the material size. This expression is $M = N \times (\text{dia.})^2 \times 2000$ where M is the chain minimum ultimate strength in pounds; N is a constant: 91 for Grade 80, 80 for Grade 70, 49 for Grade 43, and 34 for Grade 30. Note that N has units of lb/in² and dia. is the actual material stock diameter in inches.

Table 10.2.1   Grade 80 Alloy Chain Mechanical and Dimensional Requirements

Nominal chain size		Material diameter		Working load limit, max		Proof test,* min		Minimum breaking force*		Inside length, max		Inside width, min to max	
in	mm	in	mm	lb	kg	lb	kN	lb	kN	in	mm	in	mm
7/32	5.5	0.217	5.5	2,100	970	4,300	19.0	8,500	38.0	0.693	17.6	0.270 to 0.325	6.87 to 8.25
9/32	7.0	0.276	7.0	3,500	1,570	7,000	30.8	13,800	61.6	0.900	22.9	0.344 to 0.430	8.75 to 10.92
3/8	10.0	0.394	10.0	7,100	3,200	14,200	63.0	28,300	126.0	1.260	32.0	0.492 to 0.591	12.50 to 15.00
1/2	13.0	0.512	13.0	12,000	5,400	23,900	107.0	47,700	214.0	1.640	41.6	0.640 to 0.768	16.25 to 19.50
5/8	16.0	0.630	16.0	18,100	8,200	36,200	161.0	72,300	322.0	2.020	51.2	0.787 to 0.945	20.00 to 24.00
3/4	20.0	0.787	20.0	28,300	12,800	56,500	252.0	113,000	504.0	2.520	64.0	0.984 to 1.180	25.00 to 30.00
7/8	22.0	0.866	22.0	34,200	15,500	68,400	305.0	136,700	610.0	2.770	70.4	1.080 to 1.300	27.50 to 33.00
1	26.0	1.024	26.0	47,700	21,600	95,400	425.0	191,000	850.0	3.280	83.2	1.280 to 1.540	32.50 to 39.00
1¼	32.0	1.260	32.0	72,300	32,800	144,600	644.0	289,300	1288.0	4.030	102.4	1.580 to 1.890	40.00 to 48.00

* The proof test and minimum breaking force loads *shall not* be used as criteria for service or design purposes.
SOURCE: ASTM and NACM; copyright material reprinted with permission.

Note that the foregoing expression can also be stated as chain ultimate strength = $N \times (\text{dia.})^2 \times$ (short tons) where short tons = 2,000. Therefore, knowing the value of the constant N and the stock diameter of the chain, the minimum ultimate strength can be determined.

As an example, the ultimate strength of ½-in Grade 80 chain can be verified as follows by using the information contained in Table 10.2.1:

$$M = (91 \text{ lb/in}^2)\,(0.512 \text{ in})^2 \times 2,000 = 47,700 \text{ lb } (214 \text{ kN})$$

While graded steel chains are not classified as calibrated chain, since links are manufactured to accepted commercial tolerances, they are sometimes used for power transmission in pocket wheel applications. In most such applications, however, power transmission chain is special in that link dimensions are held to very close tolerances and the chain is processed to achieve special wear and strength properties. More in-depth discussions of chain, chain fittings, power transmission, and special chain follow in this section.

Grade 80 Welded Alloy-Steel Chain (Sling Chains)

The ASTM and NACM specifications for alloy-steel chain (sling chain) has evolved over a period of many years dating back to the mid-1950s. The load and strength requirements for slings and sling chain have unified and are currently specified at Grade 80 levels in the ASME standard for slings and the NACM and ASTM specifications for alloy steel chain. Table 10.2.1 relates the ASTM and NACM Grade 80 specifications. As noted, the working load limit or maximum load to be applied during use is approximately 25 percent of the minimum breaking force. Because the endurance limit is about 18 percent of the breaking strength, applications which call for a high number of load cycles near the rated load should be reconsidered to maintain the load within endurance limits. As an alternative, a larger-size chain with a higher working load limit could be applied.

Alloy steel chain manufactured to the Grade 80 specification varies in hardness among producers from about 360 to 430 Brinell. This range has narrowed, on average, in recent years from a prior range of 250 to 450 Brinell which was common in the 1960s. The current values equate to a material tensile strength range of 175,000 to 215,000 lb/in² or an average chain tensile strength of 116,000 lb/in² (800 N/mm²) to 145,000 lb/in² (1000 N/mm²). Thus the tensile strengths expressed for Grade 80 chain provide a breaking strength range of $91 \times (\text{diam})^2$ to $114 \times (\text{diam})^2$ short tons. The approximate conversion is $1.0 \times (\text{diam})^2$ short tons = 8.78 N/mm². Thus, Grade 80 chain has a minimum breaking strength of $91 \times (\text{diam})^2$ short tons.

As is the case with chain strength, chain dimensions have become standardized in the interest of interchangeability of chain and chain fittings. Thus, the dimensions presented in Table 10.2.1 are intended to assure interchangeability of products between manufacturers.

Present ASTM and NACM chain specifications call for a minimum elongation of 15 percent at rupture. This figure was established in 1924 as an attempt to guarantee against brittle fracture and to give visible warning of overloading. For low-hardness carbon-steel chains (under 200 Brinell) in use prior to the advent of hardened alloy steel, the requirement for sling chains to have 15 percent minimum elongation at rupture provided some usefulness as an overload warning signal. However, since high-quality Grade 80 alloy steel chain of 360 Brinell or higher has replaced carbon-steel chain in high-strength applications, elongation as an indicator of overload is not so apparent. Grade 80 chain achieves about 50 percent of its total elongation during the final 10 percent of loading just prior to rupture. For this reason, only about 50 percent of the total elongation is useful as a visual indicator of overloading. Thus, it is increasingly important that sling chain and slings of optimum strength, toughness, and durability be properly selected, applied, and maintained as advised in applicable standards.

Total deformation (plastic plus elastic) at failure is important as one of the determinants of energy-absorption capability and therefore of impact resistance. However, its companion determinant—breaking strength—is equally important. The only practical way to measure their composite effect is by actual impact tests on actual chain samples—not by tests on Charpy or other prepared specimens of material. Special impact testers equipped with high-speed measuring and integrating devices have been developed for this purpose.

General-Purpose Welded Carbon-Steel Chains

Grade 30 (proof coil) chain is made from low-carbon steel containing about 0.08 percent carbon. It has a minimum breaking strength of approximately $34 \times (\text{diam})^2$ short tons at 125 Brinell, while its working load limit is 25 percent of its minimum breaking force. Applications include load securement, guard rail chain, and boat mooring. It is not for use in critical or lifting applications. Table 10.2.2 gives pertinent ASTM-NACM size and load data.

Grade 43 (high test) chain is made from medium carbon steel having a carbon content of 0.15 to 0.22 percent. It is typically heat-treated to a minimum breaking strength of $49 \times (\text{diam})^2$ short tons at a hardness of 200 Brinell. Table 10.2.3 presents the ASTM-NACM specifications for Grade 43 chain. As shown, the working load limit is approximately 35 percent of the minimum breaking force. The major use of this chain grade is load securement, although it also finds use for towing and dragging in the logging industry.

Grade 70 (transport) chain was previously designated *binding chain* by NACM. This is a high-strength chain which finds use as a load securement medium on log and cargo transport trucks. It may be made from carbon or low-alloy steel with boron or manganese added in some cases to provide a uniformly hardened cross section. Carbon content is in the range of 0.22 to 0.30 percent. The chain is heat-treated to a minimum breaking force of $80 \times (\text{diam})^2$ short tons with the working load limit at 25 percent of the minimum breaking force. Although this chain approaches the strength of Grade 80, it is not recommended for overhead lifting because of the greater energy absorption capability and toughness demanded of sling chain. Table 10.2.4 presents the ASTM-NACM specifications for Grade 70 chain.

Table 10.2.2 Grade 30 Proof Coil Chain
Note: Not to be used in overhead lifting applications.

Nominal chain size		Material diameter		Working load limit, max		Proof test,* min		Minimum breaking force*		Inside length, max		Inside width, min	
in	mm	in	mm	lb	kg	lb	kN	lb	kN	in	mm	in	mm
1/8	4.0	0.156	4.0	375	170	800	3.6	1,600	7.1	0.94	23.9	0.25	6.4
3/16	5.5	0.217	5.5	800	365	1,600	7.2	3,200	14.3	0.98	24.8	0.30	7.7
1/4	7.0	0.276	7.0	1,300	580	2,600	11.6	5,200	23.1	1.24	31.5	0.38	9.8
5/16	8.0	0.315	8.0	1,900	860	3,400	15.1	6,800	30.2	1.29	32.8	0.44	11.2
3/8	10.0	0.394	10.0	2,650	1200	5,300	23.6	10,600	47.1	1.38	35.0	0.55	14.0
1/2	13.0	0.512	13.0	4,500	2030	8,950	39.8	17,900	79.6	1.79	45.5	0.72	18.2
5/8	16.0	0.630	16.0	6,900	3130	13,600	60.3	27,200	120.6	2.20	56.0	0.79	20.0
3/4	20.0	0.787	20.0	10,600	4800	21,200	94.3	42,400	188.5	2.76	70.0	0.98	25.0
7/8	22.0	0.866	22.0	12,800	5810	25,600	114.1	51,200	228.1	3.03	77.0	1.08	27.5

* The proof test and minimum breaking force loads *shall not* be used as criteria for service or design purposes.
SOURCE: ASTM and NACM; copyright material reprinted with permission.

Table 10.2.3 Grade 43 High Test Chain
Note: Not to be used in overhead lifting applications.

Nominal chain size		Material diameter		Working load limit, max		Proof test,* min		Minimum breaking force*		Inside length, max		Inside width, min	
in	mm	in	mm	lb	kg	lb	kN	lb	kN	in	mm	in	mm
1/4	7.0	0.276	7.0	2,600	1,180	3,750	16.6	7,500	33.1	1.24	31.5	0.38	9.8
5/16	8.0	0.315	8.0	3,900	1,770	4,900	21.6	9,700	43.2	1.29	32.8	0.44	11.2
3/8	10.0	0.394	10.0	5,400	2,450	7,600	33.8	15,200	67.6	1.38	35.0	0.55	14.0
1/2	13.0	0.512	13.0	9,200	4,170	12,900	57.1	25,700	114.2	1.79	45.5	0.72	18.2
5/8	16.0	0.630	16.0	11,500	5,220	19,500	86.5	38,900	172.9	2.20	56.0	0.79	20.0
3/4	20.0	0.787	20.0	16,200	7,350	30,400	130.1	60,700	270.2	2.76	70.0	0.98	25.0

* The proof test and minimum breaking force loads *shall not* be used as criteria for service or design purposes.
SOURCE: ASTM and NACM; copyright material reprinted with permission.

Table 10.2.4 Grade 70 Transport Chain
Note: Not to be used in overhead lifting applications.

Nominal chain size		Material diameter		Working load limit, max		Proof test,* min		Minimum breaking force*		Inside length, max		Inside width, min	
in	mm	in	mm	lb	kg	lb	kN	lb	kN	in	mm	in	mm
1/4	7.0	0.276	7.0	3,150	1,430	6,100	27.0	12,100	53.9	1.24	31.5	0.38	9.8
5/16	8.7	0.343	8.7	4,700	2,130	9,400	41.8	18,800	83.5	1.32	33.5	0.48	12.2
3/8	10.0	0.394	10.0	6,600	2,990	12,400	55.0	24,700	110.0	1.38	35.0	0.55	14.0
7/16	11.9	0.468	11.9	8,750	3,970	17,500	77.7	35,000	155.4	1.64	41.6	0.65	16.6
1/2	13.0	0.512	13.0	11,300	5,130	20,900	92.9	41,800	185.8	1.79	45.5	0.72	18.2
5/8	16.0	0.630	16.0	15,800	7,170	31,700	140.8	63,300	281.5	2.20	56.0	0.79	20.0
3/4	20.0	0.787	20.0	24,700	11,200	49,500	219.9	98,900	439.8	2.76	70.0	0.98	25.0

* The proof test and minimum breaking force loads *shall not* be used as criteria for service or design purposes.
SOURCE: ASTM and NACM; copyright material reprinted with permission.

Chain Strength

Welded chain links are complex, statically indeterminate structures subjected to combinations of bending, shear, and tension under normal axial load. When a link is loaded, the maximum tensile stress normal to the section occurs in the outside fiber at the end of the link [see Fig. 10.2.1], which shows the stress distribution in a 3/8-in (10-mm) Grade 80 chain link with rated load, determined by finite element analysis (FEA). In Fig. 10.2.1, stress distribution is obtained from FEA by assuming that peak stress is below the elastic limit of the material. If the material is ductile and the peak stress from linear static FEA exceeds the elastic limit, plastic deformation will occur, resulting in a distribution of stresses over a larger area and a reduction of peak stress.

It is interesting to note that the highest stress in the chain link is compressive and occurs at the point of loading on the inside end of the link. Conversely, traveling around the curved portion of the link, the compressive and tensile stresses reverse and the outside of the link barrels are in compression. Since the outside of the barrels of the link, unlike the link ends, are unprotected, they are vulnerable to damage

such as wear from abrasion, nicks, and gouges. Fortunately, since the damage occurs in an area of compressive stress, their potentially harmful effects are reduced.

Maximum shear stress, as indicated in Fig. 10.2.1, occurs on a plane approximately 45° to the x axis of the link where the bending moment is zero. Chains such as Grade 80 with low to medium hardness (BHN 400 or less) normally break when overloaded in this region. As hardness increases, there is a tendency for the fracture mode to shift from one of shear as described above to one of tension bending through the end of the link where the maximum tensile stress occurs. A member made of ductile material and subjected to uniaxial stress fails only after extensive plastic deformation of the material. The yielding results in deformation, causing a reduction in the link's cross section until final failure occurs in shear.

Combined stresses referred to initially in this section result in higher stresses than those computed by considering the load applied uniformly in simple tension across both barrels. For example, a 3/8-in (10-mm) Grade 80 alloy chain with a BHN of 360 has a material strength of

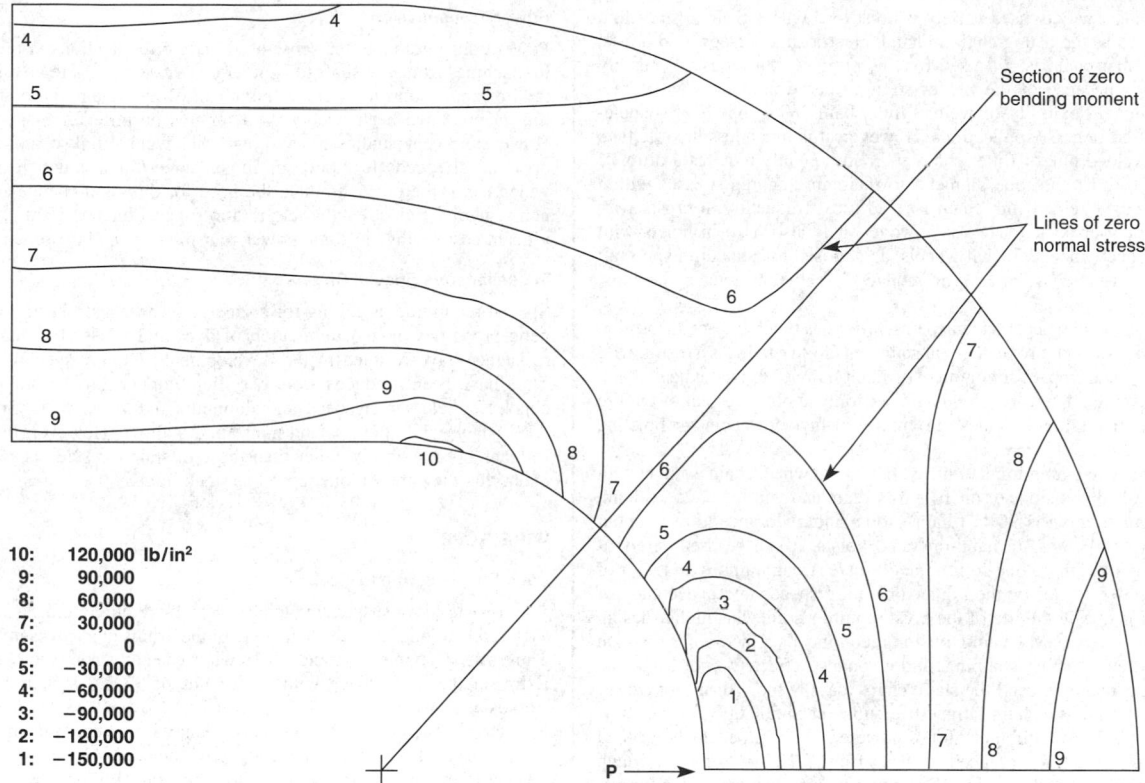

Fig. 10.2.1   Stress distribution in a ⅜-in (10-mm) Grade 80 chain link at rated load.

10:	**120,000 lb/in²**
9:	90,000
8:	60,000
7:	30,000
6:	0
5:	−30,000
4:	−60,000
3:	−90,000
2:	−120,000
1:	−150,000

180,000 lb/in² (1240.9 MN/m²). Based on a specified ultimate strength of the chain of 91 × (diam)² tons, which is 28,300 lb (12,800 kg), the average stress in each barrel of the chain would be 116,000 lb/in² (799.7 MN/m²). The actual breaking strength, however, because of the effects of the combined stresses, is magnified by a factor of 180,000/116,000 = 1.5 over the average stress in each barrel.

Most chain manufacturers now manufacture a through-hardened commercial alloy sling chain, discussed in this section. Although designed primarily for sling service, it is used in many original equipment applications. Typical strength of this Grade 80 chain is 91 × (diam)² tons with a BHN of 360 and a minimum elongation of 15 percent. This combination of strength, wear resistance, and energy absorption characteristics makes it ideal for applications where the utmost dependability is required under rugged conditions. Typical inside width and pitch dimensions are 1.25 × diam and 3.20 × diam.

Chain End Fittings

Most industrial chains must be equipped with some type of end fittings. These usually consist of oblong links, pear-shaped links, or rings (called masters) on one end to fit over a crane hook and some variety of hook or enlarged link at the other end to engage the load. The hooks are normally drop-forged from carbon or alloy steel and subsequently heat-treated. They are designed (using curve-beam theories) to be compatible in strength with the chain for which they are recommended. Master links or rings must fit over the rather large section thicknesses of crane hooks, and they therefore must have large inside dimensions. For this reason they are subjected to higher bending moments, requiring larger section diameters than those on links with a narrower configuration.

CM Chain Division of Columbus McKinnon recommends that oblong master links be designed with an inside width of 3.5 × diam and an inside length of 7.0 × diam, where diam is the section diameter in inches. Using a material with a yield strength of 100,000 lb/in², the diameter can be calculated from the relationship diam = (WLL)/20,000¹ᐟ², where WLL is the working load limit in pounds.

Rings require a somewhat greater section diameter than oblong links to withstand the same load without deforming because of the higher bending moment. **Master rings** with an inside diameter of 4 × diam can be sized from the relationship diam = (WLL)/15,000¹ᐟ² in. Rings require a 15 percent larger section diameter and a 33 percent greater inside width than oblong links for the same load-carrying capacity, and the use of the less bulky oblong links is therefore preferred.

Pear-shaped master links are not nearly as widely in demand as oblong master links. Some users, however, prefer them for special applications. They are less versatile than oblong links and can be inadvertently reversed, leading to undesirable bending stresses in the narrow end due to jamming around the thick saddle of the crane hook.

Hooks and **end links** are attached to sling chains by means of coupling links, either welded or mechanical. Welded couplers require special equipment and skill to produce quality and reliability compatible with the other components of a sling and, consequently, must be installed by sling chain manufacturers in their plants. The advent of reliable mechanical couplers such as **Hammerloks** (CM Chain Division), made from high-strength alloy forgings, have enabled users to repair and alter the sling in the field. With such attachments, customized slings can be assembled by users from component parts carried in local distributor stocks.

Welded-Link Wheel Chains

These differ from sling chains and general-purpose chains in two principal aspects: (1) they are precisely calibrated to function with pocket or sprocket wheels and (2) they are usually provided with considerably higher surface hardness to provide adequate wear life.

The most widely used variety is hoist load wheel chain, a short link style used as the lifting chain in hand, electric, and air-powered hoists. Some roller chain is still used for this purpose, but its use is steadily declining because of the higher strength per weight ratio and three-dimensional flexibility of welded link chain. Wire rope is also sometimes used in hoist applications. However, it is much less flexible than either welded link or roller chain and consequently requires a drum of 20 to 50 times the rope diameter to maintain bending stresses within safe limits. Welded link chain is used most frequently in hoists with three- or four-pocket wheels; however, it is also used in hoists with multipocket wheels. Pocket wheels, as opposed to cable drums, permit the use of much smaller gear reductions and thus reduce the hoist weight, bulk, and cost.

Welded load wheel chain also has 3 times as much impact absorption capability as a wire rope of equal static tensile strength. As a result, wire rope of equal impact strength costs much more than load wheel chain. Since average wire rope life in a hoist is only about 5 percent of the life of chain, overall economics are greatly in favor of chain for hoisting purposes.

To achieve maximum flexibility, hoist load wheel chain is made with link inside dimensions of pitch = 3 × diam and width = 1.25 × diam. Breaking strength is 90 × (diam)^2 tons, and endurance limit is about 18 × (diam)^2 tons. In hand-operated hoists, it can be used safely at working loads up to one-fourth the ultimate strength (a design factor of 4). In powered hoists, the higher operating speeds and expectancy of more lifts during the life of the hoist require a somewhat higher design factor of safety. CM Chain Division recommends a factor of 7 for such units where starting, stopping, and resonance effects do not cause the dynamic load to exceed the static chain load by more than 25 percent. For this condition, chain fatigue life will exceed 500,000 lifts, the normal maximum requirement for a powered hoist. Standard hoist load wheel chains, now produced in sizes from 0.125-in (3-mm) through 0.562-in (14-mm) diameter, meet requirements for working load limits up to approximately 5 or 6 short tons (5 long tons) for hand-operated hoists and 3 short tons (2.5 long tons) for power-operated hoists. More complete technical information on this special chain product and on chain-hoist design considerations is available from CM Chain Division.

Another widespread use of welded link chain is in conveyor applications where high loads and moderate speeds are encountered. As contrasted to link-wheel chain used in hoists, conveyor chain requires a longer pitch. For example, the pitch of hoist chain is approximately 3.0 × diam and conveyor chain is 3.5 × diam. The longer-pitch chain, in addition to being more economical, accommodates the use of shackle connectors between sections of the chain to which flight bars are attached.

CM Corporation, for example, manufactures conveyor chains in four metric sizes and in two different thermal treatments (Table 10.2.5). These chains are precisely calibrated to operate in pocket wheels. They may have a high surface hardness (500 Brinell) to provide adequate wear life for ash conveyors in power generating plants and to provide high strength and good wear properties for drive chain applications. Also, the long-pitch chain is used extensively in longwall mining operations to move coal in the mines.

Table 10.2.5 Conveyor Chain Specifications

Chain size, mm	Link pitch, mm	Minimum breaking force, 1000 lb (454 kg)	
		Case-hardened	Through-hardened
14	50	36.5	50.0
18	64	65.0	92.2
22	86	75.0	121.5
26	92	90.0	169.8

SOURCE: Columbus McKinnon Corporation.

Power Transmission

Power transmission, in the sense used to describe the transfer of power in machines from one shaft to another one close by, is a relatively new field of application for welded-link chain. Roller chain and other pin-link chains have been widely used for this purpose for many years. However, recent studies have shown that welded-link chain can be operated at speeds to 3,000 ft/min. Its three-dimensional flexibility, which can sometimes eliminate the need for direction-change components, and its high strength-weight ratio suggest the possibility for significant cost savings in some power-transmission applications.

Miscellaneous Special Chains

Special requirements calling for corrosion or heat resistance, nonmagnetic properties, noncontamination of dyes and foodstuffs, and spark resistance have resulted in the development of many special chains. They have been produced from beryllium copper, bronze, monel, Inconel, Hadfield's manganese, and aluminum and from a wide variety of AISI analyses of stainless and nonstainless alloys. However, the need for such special chains, other than those of stainless steel, is so infrequent that they are seldom carried as stock items.

WIRE ROPE

Load Suspension and Haulage

Wire rope used for suspending loads is usually required to be as flexible as possible to minimize the diameters of the drums or sheaves involved. Thus, a rope having six strands of 19 wires each on a hemp core is used. Extra pliable ropes made with six strands of 37 wires each or eight strands of 19 wires each on a hemp core are also available but are much less durable because of the finer individual wires used. Hoisting ropes are constructed with the relative twist of the wires in the strands the reverse of the twist of the strands about the core (Fig. 10.2.2).

Based upon the service to be expected, special attention should be given to the ratio of drum and sheave diameters to cable diameter. For example, hoists having moderate duty cycle may have a ratio of diameters as low as 20 : 1, but for extensive duty or where there is need for

Fig. 10.2.2 Haulage rope.

great safety, such as in elevators, this should be at least 45 : 1 or larger. Often the need for storing enough cable to obtain sufficient lift length may require a larger drum diameter than would otherwise be needed.

Haulage ropes are of the same construction as **suspension ropes** or are of the **lang-lay** type shown in Fig. 10.2.3, with the twist of the wire and strand in the same direction. This lang-lay construction increases the wear resistance of the rope, but it tends to untwist and should not be used where the load is in free suspension. Lang-lay rope is difficult to splice. By preforming individual wires and strands before laying up, secondary stresses due to bending are reduced and longer life is obtained. Preformed ropes have less tendency to kink and are easier to handle.

Fig. 10.2.3 Lang-lay rope.

Hoisting and haulage ropes should be frequently **greased** to minimize wear and to prevent corrosion; either a special commercial lubricant or boiled linseed oil may be used on ropes subjected to atmospheric action, and a tacky petroleum and graphite on hoisting ropes in wet places. Crude oil or other lubricants having an acid or basic characteristic should not be used because of corrosive action on both wire and sisal core. To ensure penetration, lubricant can be applied hot, or a volatile solvent can be used. Ropes should be inspected frequently for broken wires and excessive wear.

Track Cables

Cables used as **tracks to support loads suspended on trolleys** are either the locked-coil type for longest life (Fig. 10.2.4) or round-wire track strand.

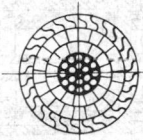

Fig. 10.2.4   Locked-coil track strand wire rope. *(United States Steel.)*

The strength of the locked-coil type is given in Table 10.2.6. This type of wire minimizes the impact loads on the outer wires which result from the rolling of the trolley.

Fittings

Fittings for ropes are attached at the ends by (1) passing the rope around a minimum-radius thimble and then *(a)* attaching the rope to itself with rope clips (approximately 80 percent efficient), *(b)* splicing the rope to itself (80 to 95 percent efficient), *(c)* attaching the rope to itself by a metal ring which is swaged or crimped on (90 to 95 percent efficient), (2) using a fitting part of which is a steel tube which is pressed or swaged over the rope (90 to 100 percent efficient), (3) using zinc to embed the end of the rope in a fitting having a socket to receive it (100 percent efficient).

Drums

Drums are made with smooth surfaces on hand-powered hoists and on power hoists subject to light-duty operation. Medium-and heavy-duty drums are normally grooved. Drums can be welded or cast, depending upon the quantity to be manufactured, since cast drums are economical when mass-produced. Large drums frequently have separate shells welded to the spider or end plate. Drum shells may be made from steel plates which are bent to a cylindrical shape and welded to the end plates with welded hubs before they are grooved for the rope. Steel-plate shells

Table 10.2.6   Locked-Coil Track Strand Wire Rope

Diameter		Special grade		Standard grade		Weight	
in	mm	Short ton	tonne	Short ton	tonne	lb/ft	kg/m
¾	19.1	31.5	28.6	25	22.7	1.41	2.10
⅞	22.2	41.5	37.6	32	29.0	1.92	2.86
1	25.4	52.5	47.6	42	38.1	2.50	3.72
1⅛	28.6	66.0	59.9	54	49.0	3.16	4.70
1¼	31.8	81.0	73.5	65	59.0	3.91	5.82
1⅜	34.9	100.0	90.7	78	70.8	4.73	7.04
1½	38.1	120.5	109.3	93	84.4	5.63	8.38
1⅝	41.3	140.0	127.0	108	98.0	6.60	9.82
1¾	44.5	165.0	150	125	113.4	7.66	11.4
1⅞	47.6	187.5	170	138	125.2	8.79	13.1
2	50.8	215	195	158	143	10.00	14.9
2¼	57.2	280	254			12.50	18.6
2½	63.5	345	313			15.2	22.6
2¾	69.9	420	381			18.3	27.2
3	76.2	500	454			22.2	33.0
3¼	82.6	580	526			25.6	38.1
3½	88.9	690	626			29.9	44.5
3¾	95.3	785	712			33.9	50.4
4	101.6	880	798			38.4	57.1

SOURCE: United States Steel Corp.

are stronger than cast shells, better balanced, and free from hidden initial defects. The thickness can be less, thus reducing the inertia of the rotating drum and the resulting acceleration-peak loads. Conical and cylindroconical drums are frequently used on large mine hoists. Faces

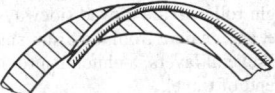

Fig. 10.2.5   Hole for anchoring rope on drums.

of drums for medium and heavy duty are made wide enough to hold the rope in one layer plus two to four holding turns. The hole for attachment of the rope should be as shown in Fig. 10.2.5 to prevent excessive bending; this method of anchoring is normally done on cast drums which have a limited face width. Figure 10.2.6 shows an alternate,

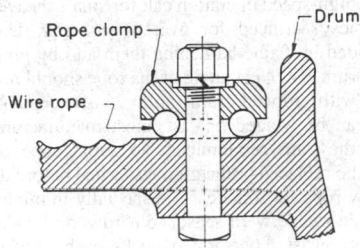

Fig. 10.2.6   Rope-anchoring attachment. *(McDowell Wellman.)*

preferred method of anchoring the rope on welded and cast drums when space is not a problem. The pitch diameter of the drum should be at least 20 times the rope diameter in order to obtain reasonable life for both drum and rope. Long life requires 45 to 60 times the rope diameter.

Where there is side draft on the rope, movable **idlers** are provided to align the rope and groove. The idlers may be moved parallel to the face

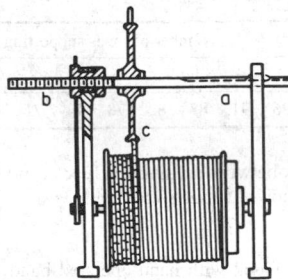

Fig. 10.2.7   Hoisting drum.

of the drum by the side pressure of the rope or may be driven positively sideways, thus eliminating friction and increasing the life of rope. In Fig. 10.2.7, idler sheave *c* revolves between fixed collars on shaft *a*, which is connected to the drum shaft by sprocket and chain. The shaft is prevented from rotating by a feather key. On sprocket *b*, a nut is held

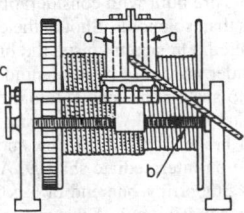

Fig. 10.2.8   Hoisting drum.

from moving sideways by flanges; the shaft *a* with sheave *c* moves in the direction of its axis. In an alternative construction, the idler shaft is threaded but held stationary, and the sheave hub is a nut. The sheave is turned by the friction of the rope, which causes it to travel back and forth. Figure 10.2.8 shows a construction used when the side draft is excessive. The upright rollers *a* are moved sideways by screw *b*, which is driven by sprocket and chain *c* from the drum shaft. For winding the rope on the drum in several layers, a clutch is provided which reverses the direction at the end of travel.

Sheaves

Sheaves should be grooved to fit the rope as closely as possible in order to prevent the rope from assuming an oval or elliptical shape under heavy load. They should be balanced and properly aligned to prevent swaying of the rope and abrasion against the sheave flanges. Sheaves and drums should be as large as possible to obtain maximum rope life, but factors such as weight of machinery for easy transport, minimizing headroom, and high-speed operation call for small sheaves. Hence rope life is sometimes sacrificed for overall economy. Undue wear on sheaves is avoided by flame-hardening them and by properly aligning them with the drum. The fleet angle of the rope should not exceed $1\frac{1}{2}°$, but sometimes with grooved drums up to $2°$ is acceptable. Sheaves of any diameter can be welded or cast; most manufacturers make cast sheaves since they are economical. To avoid rope damage, worn sheaves should be replaced or the grooves turned before the sheaves are used with a new rope. In some cases, especially in mines, the grooves are lined with renewable, well-seasoned hardwood blocks.

Tackle blocks consist of one or two blocks, each carrying one or more sheaves. A single-sheave block (generally used to change the direction of a lead line and frequently arranged for easy removal of the loop of rope) is called a **snatch block**. Blocks are made for both manila and wire rope. Those for manila rope are usually made with wooden cheeks to prevent chafing of the rope and have sheaves of smaller diameter than wire-rope blocks for the same size of rope. Heavy hoisting is almost universally done with wire-rope blocks. Tests by the American Bridge Co. found the following approximate efficiencies for well-designed and properly maintained ¾-in (19.1-mm) wire-rope tackle.

Number of ropes supporting load											
1	2	3	4	5	6	7	8	9	10	11	12
Efficiency, %	86	96	91	87	82	78	74	71	68	65	62	59

Each snatch block between the hoisting blocks and the hoist or winch will have an efficiency of about 86 percent.

Brakes

Small hoists are provided with hand-operated band brakes or electrically operated disk brakes; larger hoists have mechanically operated post brakes. On electric hoists, it is usual to apply the brake with a weight or spring and to remove it by a solenoid. Where controlled rate of application of a brake is desirable, a **thrustor** is frequently used. This is a self-contained unit consisting of a vertical motor, a centrifugal pump in a piston, and a cylinder filled with oil. Starting the motor raises the piston and connected counterweight. Stopping the motor permits the weight to fall. An adjustable range of several seconds in falling sets the brake slowly. Thrustors are built with considerably greater load capacities and stroke lengths than solenoids. Should the current fail, the brakes are automatically applied. On steam hoists, the brake is taken off by a steam piston and cylinder instead of by a solenoid. On overhead cranes, load brakes are used to sustain the load automatically at any point and occasionally to regulate the speed when lowering.

In one type of **load brake** (Fig. 10.2.9), the motor *A* drives drum *B* through the load brake on intermediate shaft *D*. A spider *C* is keyed to shaft *D*, its inner end supporting one end of a coiled bronze spring of square section *E*. The opposite end of the spring is fixed in flange *F*, which is loosely fitted to shaft *D* and directly attached to pinion *G*. Any

relative angular motion of flange *F* and spider *C* alters the closeness of the coiling of spring *E*, consequently altering its outside diameter (considered as a drum). This outer surface is one of the friction surfaces of the brake, the other being provided by the internal face of drum *H*, which revolves loosely on shaft *D* at one end and on flange *F* at the

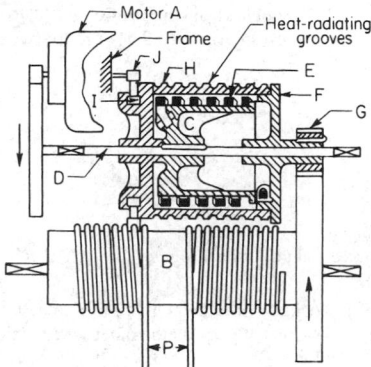

Fig. 10.2.9 Load brake.

other. Drum *H* is restrained from moving in one direction by ratchet *I* and pawls *J*. The exterior of drum *H* is grooved for heat dissipation.

The action of the **load brake** in Fig. 10.2.9 is as follows: When hoisting, the brake revolves as shown by the arrow. Pawls *J* permit drum *H* to revolve; consequently, the whole mechanism is locked and revolves as one piece. When stopping the load, the downward pull of the load reacts to drive drum *H* against pawl *J*. Flange *F* therefore moves slightly in an angular direction relative to spider *C*, and spring *E* consequently untwists until it grips the interior of drum *H*, thus locking the load. The action is such that the grip is slightly more than necessary to hold the load. Reversing the motor for lowering the load drives the interior of the brake surface against drum *H* so that the power consumed is the amount necessary to overcome the excess holding power of the brake over the load reaction.

Load brakes of the **disk** and **cone** types are also used and embody the same principle of pawl locks and differential action. The choice of brake type should be based on considerations of smoothness of working and lack of chatter, as well as on the power requirements for lowering at different values of load within the range of the crane.

In **regenerative braking,** the motor, when overhauled, acts as a generator to pump current back into the line.

HOLDING MECHANISMS

Lifting Tongs

Figure 10.2.10 shows the type of tongs used for lifting plates. The cams *a* grip the plate when the chains tighten, preventing slipping. Self-

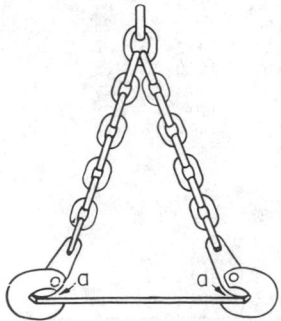

Fig. 10.2.10 Lifting tongs.

closing tongs (Fig. 10.2.11) are used for handling logs, manure, straw, etc. The rope *a* is attached to and makes several turns around drum *b*; chains *c* are attached to the bucket head *e* and to drum *b*. When power is applied to rope *a*, drum *b* is revolved, winding itself upon chains *c* and closing the tongs. To open, slack off on rope *a*, holding tongs on rope *d*, attached to head *e*.

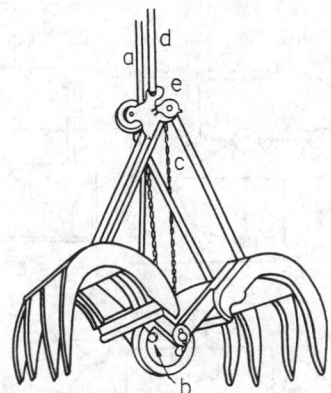

Fig. 10.2.11   Lifting tongs.

Lifting Magnets

Lifting magnets are materials-handling devices used for handling pig iron, scalp iron, castings, billets, tubes, rails, plates, skull-cracker balls, and other magnetic material.

At temperatures above dull-red heat, ordinary magnetic materials lose their magnetic properties, while certain stainless and high-manganese steels are nonmagnetic even at normal temperatures. Such materials cannot be handled by lifting magnets.

Most lifting magnets are not designed for continuous operation but for operation with the normal off times generally associated with materials-handling applications. In addition, attention should be given to operation of magnets on high-temperature loads so as to keep the magnet-coil temperature within the design limits of the insulation. Lifting magnets can be used for underwater operation when they are supplied with watertight cases and specially designed lead connections.

To obtain optimum magnet performance, a suitable **magnetic controller** should be used. It is necessary in most cases to provide means to reverse the current in the magnet in order to release the materials efficiently from the magnet. The controller should also have protective features to absorb the stored energy of the magnet during its discharge, especially when the dc power requirements for a magnet are supplied from a **rectifier-type** power supply.

Lifting magnets can be classified as follows:

1. **Circular Magnets**. This configuration makes the most efficient use of materials, is extremely rugged, and is best suited for general lifting applications. Recently this category of magnets has been subdivided into two distinct types: a relatively large diameter magnet, which is especially efficient on low-permeability material such as scrap; and a magnet which, for equal weight, has a smaller diameter but a much deeper magnetic field, making it more suitable for high-permeability loads typically found in steel mills. Dings Elektrolift series of 8, 13, and 18 in diameters (203, 330, and 457 mm) is designed for machine-shop usage from 2,500 to 13,500 lb (1,130 to 6,120 kg).

2. **Rectangular Magnets**. Many types of materials such as rails, beams, and plates can be handled more efficiently with a rectangular magnet. The tendency of these materials to pull away from the face of the magnet because of deflection rather than total weight is a limiting factor when these magnets are applied. Two or more small rectangular magnets mounted on a spreader beam usually give a better lifting performance than a single large magnet of equal weight. This is especially true in the case of thin plates.

3. **Specialty Magnets**. Loads such as coiled steel present unique prob-

lems when lifted by a magnet. Special magnets are available for loads of this type, but the scope of this text does not permit detailed discussion.

Construction features of a typical **circular magnet** are shown in Fig. 10.2.12. A winding of either strap aluminum or copper is located inside the cast-steel case. Strap material with insulating tape between turns,

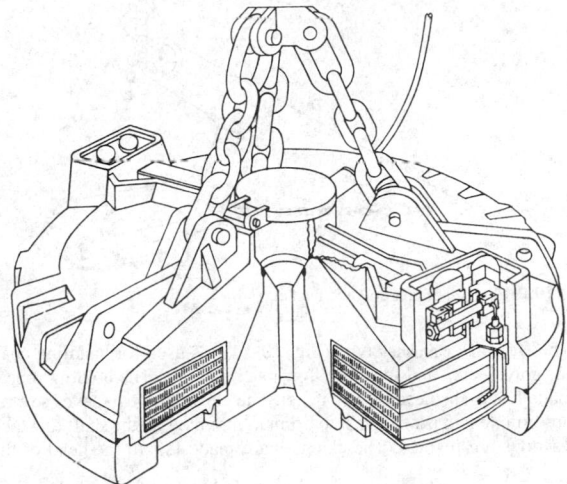

Fig. 10.2.12   Circular lifting magnet. *(Dings.)*

rather than insulated wire, is used for windings since it permits a greater number of turns in the same space. Recent advances in the processing of anodized aluminum have permitted the elimination of the turn-to-turn insulation, thereby allowing an increase in the number of turns, which results in increasing lifting capacity. A nonmagnetic manganese-steel bottom plate is welded to the case to make the coil cavity watertight. The center and outer pole shoes are made so that they can be replaced, since they will wear in severe service.

Rectangular-magnet construction features are shown in Fig. 10.2.13, which illustrates a typical plate-handling magnet. Coils for this type of

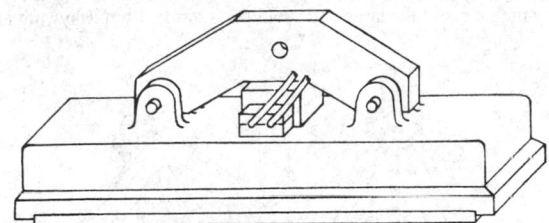

Fig. 10.2.13   Rectangular lifting magnet. *(Dings.)*

magnet are generally wound with wire and formed to fit over the center pole. A manganese bottom plate is used to hold the coils in place and seal the coil cavity. The external flux path is from the center pole to each of the outer poles.

Buckets and Scrapers

One type of **Williams self-filling dragline bucket** (Fig. 10.2.14) consists of a bowl with the top and digging end open. The shackles at *c* are so attached that the cutting edge will penetrate the material. When filled, the bucket is raised by keeping the line *a* taut and lifting on the fall line *d*. In this position, the bucket is carried to the dumping position, where by slacking off on line *a*, it is dumped. Line *b* holds up line *a* and the bridle chains while dumping. Such buckets are built in sizes from ¾ yd³ to as large as 85 yd³ (0.6 to 65 m³), weighing 188,000 lb (85,000 kg).

Figure 10.2.15 shows an **open-bottom-type scraper**. Inhauling on cable *a* causes the sloping bottom plate to dig and load until upward pressure of the material against the top prevents further loading. The scraper continues its forward haul to the dumping point. Pulling on cable *b* deposits the load and returns the scraper to the excavation point. Scrapers are made in sizes from ⅓ to 20 yd³ capacity (0.25 to 15 m³).

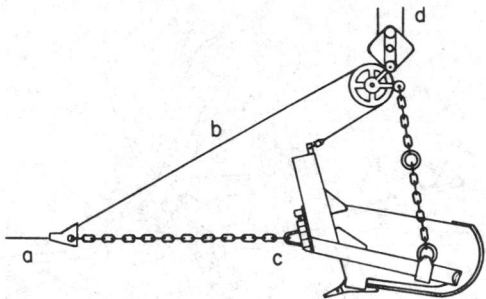

Fig. 10.2.14 Self-filling dragline bucket.

The **Hayward clamshell** type (Fig. 10.2.16) is used for handling coal, sand, gravel, etc., and for other flowable materials. The holding rope *a* is made fast to the head of the bucket. The closing rope *b* makes several wraps around and is made fast to drum *d*, mounted on the shaft to which the scoops are pivoted. The chains *c* are made fast to the head of the

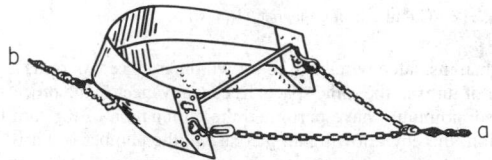

Fig. 10.2.15 Scraper bucket. *(Sauerman.)*

bucket and to the smaller diameter of drum *d*. When power is applied to rope *b*, it causes the drum to wind itself up on chain *c*, raising the drum and closing the bucket. To dump, hold rope *a* and slack off rope *b*. The digging power of the bucket is determined by its weight and by the ratio of the diameters of the large and small parts of the drum. Hayward also

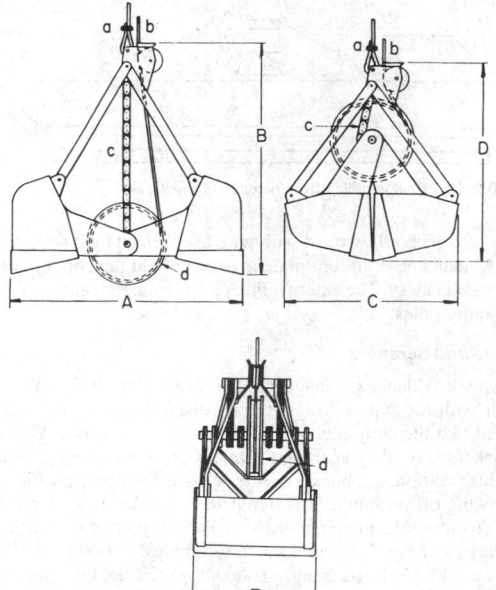

Fig. 10.2.16 Hayward grab bucket.

has an **orange-peel** bucket that operates like the clamshell type (Fig. 10.2.16) but has four blades pivoted to close.

The **Hayward electrohydraulic** single-rope type (Fig. 10.2.17) is used not only for handling ore and sand but also where it is desirable to hook the bucket on a derrick or crane hook. The bucket requires only to be hung from the crane or derrick hook by the eye and an electric line plugged in. No ropes are fed into the bucket and no time-consuming

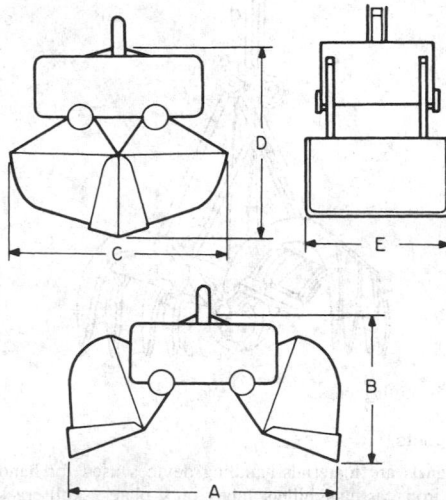

Fig. 10.2.17 Hayward electrohydraulic single-rope-type grab bucket.

shifting of lines is required—thus facilitating changing from magnet to bucket. The bowls are suspended from the head, which contains a complete electrohydraulic power unit consisting of an ac or dc motor, hydraulic pump, directional valve, sump, filters, and cylinders. When current is turned on, the motor drives the pump which provides system pressure. Energizing the solenoids in the directional valve directs the fluid flow to either end of the cylinders, which are attached to the blade arms, thus opening or closing the bucket. The system uses a pressure-compensated variable-displacement pump which enters a no-delivery mode when system pressure is reached during the closing or opening action, thus avoiding any heat generation and reducing the load on the motor while maintaining system pressure. The 1 yd³ (0.76 m³) bucket is provided with a 15-hp (11.2-kW) motor and 15 gal/min (0.95 l/s) pump. The 2 and 3 yd³ (1.5 to 2.3 m³) buckets have motors of 20 to 25 hp (15 to 19 kW), according to duty.

HOISTS

Hand-Chain Hoists

Hand chain hoists are portable lifting devices suspended from a hook and operated by pulling on a hand chain. There are two types currently in common use, the high-speed, high-efficiency hand hoist (Fig. 10.2.18) and the economy hand hoist. The economy hand hoist is similar in appearance to the high-efficiency hand hoist except the handwheel diameter is smaller. This smaller diameter accounts for a smaller headroom dimension for the economy hoist. High-speed hoists have mechanical efficiency as high as 80 percent, employ Weston self-energizing brakes for load control, and can incorporate load-limiting devices which prevent the lifting of excessive overloads. They require less hand chain pull and less chain overhaul for movement of a given load at the expense of the larger head size when compared to economy hoists. As noted, this larger head size also creates a minimum headroom dimension somewhat greater than that for the economy hoists. High-speed hoists find application in production operations where low operating effort and long life are important. Economy hoists find application in construction and shop use where more compact size and lower unit cost compensate for the higher operating effort and lower efficiency.

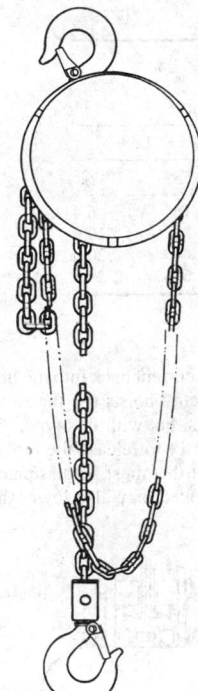

Fig. 10.2.18 High-speed hand hoist. (*CM Hoist Division, Columbus McKinnon.*)

Both hoist types function in a similar manner. The hand chain operates over a handwheel which is connected via the brake mechanism and gear train to a pocket wheel over which the load chain travels. The brake is disengaged during hoisting by a one-way ratchet mechanism. In lowering, the hand chain is pulled continuously in a reverse direction to overcome brake torque, thus allowing the load to descend.

Although a majority of hand chain hoist applications are fixed hook-suspended applications, sometimes a hand hoist is mounted to a trolley to permit horizontal movement of the load, as in a forklift battery-changing operation. This is accomplished by directly attaching the hoist hook to the trolley load bar, or by using an integrated trolley hoist which combines a high-speed hoist with a trolley to save headroom (Fig. 10.2.19).

Hand chain hoists are available in capacities to 50 tons (45 tonnes). Tables 10.2.7 and 10.2.8 provide data for high-speed and economy hoists in readily available capacities to 10 tons (9 tonnes), respectively.

Pullers or Come-Alongs

Pullers, or come-alongs, are lever-operated chain or wire rope hoists (Fig. 10.2.20). Because they are smaller in size and lower in weight than equivalent ca-

pacity hand chain hoists, they find use in applications where the travel (take-up) distance is short and the lever is within easy reach of the operator. They may be used for lifting or pulling at any angle as long as the load is applied in a straight line between hooks. There are two types of lever-operated chain or wire rope hoists currently in use: long handle and short handle. The long-handle variety, because of direct drive or numerically lower gear ratios, allows a greater load chain take-up per handle stroke than the short-handle version. Conversely, the short-

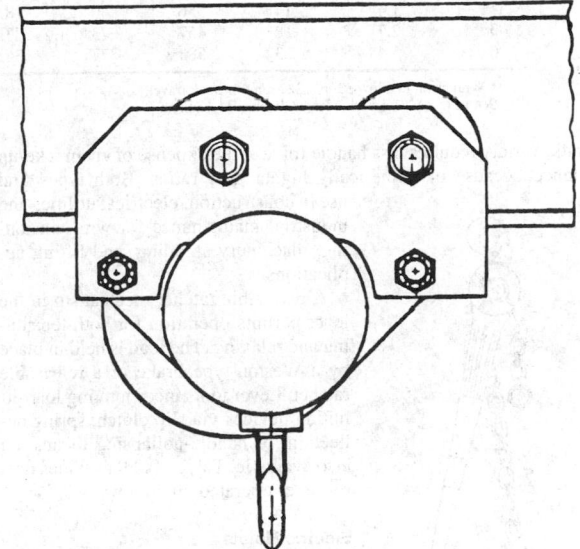

Fig. 10.2.19 Trolley hoist, shown with hand chain removed for clarity. (*CM Hoist Division, Columbus McKinnon.*)

Table 10.2.7   Typical Data for High-Speed Hand Chain Hoists with 8-ft Chain Lift

Capacity		Hand chain pull to lift rated capacity load		Retracted distance between hooks (headroom)		Net weight		Hand chain overhaul to lift load 1 ft	
short tons	tonnes	lb	kg	in	mm	lb	kg	ft	m
¼	0.23	23	10.4	13	330.2	33	15.0	22.5	6.9
½	0.45	46	20.9	13	330.2	33	15.0	22.5	6.9
1	0.91	69	31.3	14	355.6	36	16.3	30	9.1
1½	1.36	80	36.3	17.5	444.5	59	26.8	40.5	12.3
2	1.81	83	37.6	17.5	444.5	60	27.2	52	15.8
3	2.7	85	38.6	21.5	546.1	84	38.1	81	24.7
4	3.6	88	39.9	21.5	546.1	91	41.3	104	31.7
5	4.5	75	34.0	24.5	622.3	122	55.3	156	47.5
6	5.4	90	40.8	25.5	647.7	127	57.6	156	47.5
8	7.3	89	40.3	35.5	901.7	207	93.9	208	63.4
10	9.1	95	43.1	35.5	901.7	219	99.3	260	79.2

SOURCE: CM Hoist Division, Columbus McKinnon.

Table 10.2.8   Typical Data for Economy Hand Chain Hoist with 8-ft Chain Lift

Capacity		Hand chain pull to lift rated capacity load		Retracted distance between hooks (headroom)		Net weight		Hand chain overhaul to lift load 1 ft	
short tons	tonnes	lb	kg	in	mm	lb	kg	ft	m
½	0.45	45	20	10	254	15	7	32	9.8
1	0.91	74	34	12	305	22	10	39	11.9
2	1.81	70	32	17	432	53	24	77	23.5
3	2.7	54	25	22	559	69	31	154	46.9
5	4.5	88	40	24	610	74	34	154	46.9
10	9.1	90	41	32	813	139	63	308	93.9

SOURCE: CM Hoist Division, Columbus McKinnon.

Table 10.2.9   Typical Data for Short- and Long-Handle Lever-Operated Chain Hoists

Capacity		Retracted distance between hooks (headroom)		Handle pull to lift rated capacity				Net weight*			
				Short		Long		Short		Long	
short tons	tonnes	in	mm	lb	kg	lb	kg	lb	kg	lb	kg
¾	0.68	11	279	35	15.8	58	26.3	18	8.2	14	6.4
1½	1.4	14	356	40	18.1	83	37.6	27	12.2	24	10.9
3	2.7	18	457	73	33.1	95	43.1	45	20.4	44	20.0
6	5.4	23	584	77	35.0	96	43.5	66	30.0	65	29.5

* With chain to allow 52 in (1322 mm) of hook travel.
SOURCE: CM Hoist Division, Columbus McKinnon.

handle variety requires less handle force at the expense of chain take-up distance because of numerically higher gear ratios. Both types find use in construction, electrical utilities and industrial maintenance for wire tensioning, machinery skidding, and lifting applications.

A reversible ratchet mechanism in the lever permits operation for both tensioning and relaxing. The load is held in place by a Weston-type brake or a releasable ratchet. Lever tools incorporating load-limiting devices via slip clutch, spring deflection, or handle-collapsing means are also available. Table 10.2.9 provides data for lever operated chain hoists.

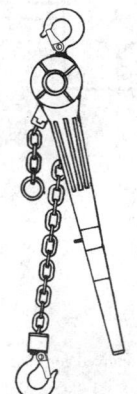

Fig. 10.2.20   Puller or come-along hoist. *(CM Hoist Division, Columbus McKinnon.)*

Electric Hoists

Electric hoists are used for repetitive or high-speed lifting. Two types are available: chain (see Fig. 10.2.21) (both link and roller), in capacities to 15 tons (13.6 tonnes); wire rope (see Fig. 10.2.22), in ratings to 25 tons (22.7 tonnes). The typical hoist has a drum or sprocket centered in the frame, with the motor and gearing at opposite ends, the motor shaft passing through or alongside the drum or sprocket. Many have an integral load-limiting device to prevent the lifting of gross overloads.

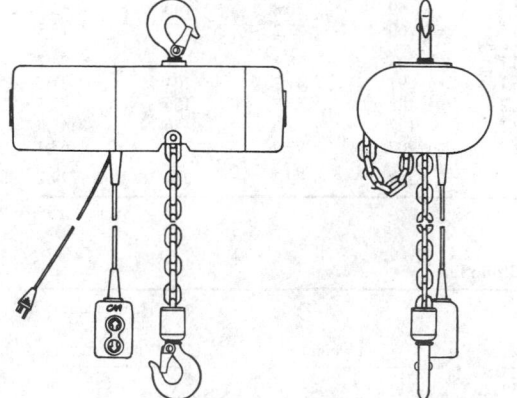

Fig. 10.2.21   Electric chain hoist. *(CM Hoist Division, Columbus McKinnon.)*

Electric hoists are equipped with at least two independent braking means. An electrically released brake causes spring-loaded disk brake plates to engage when current is off. When the hoist motor is activated, a solenoid overcomes the springs to release the brake. In the lowering direction, the motor acts as a generator, putting current back into the line and controlling the lowering speed. Some electric hoists use the same type of Weston brake as is used in hand hoists, but with this type, the motor must drive the load downward so as to try to release the brake. This type of brake generates considerable heat that must be dissipated—usually through an oil bath. The heat generated may also lower the

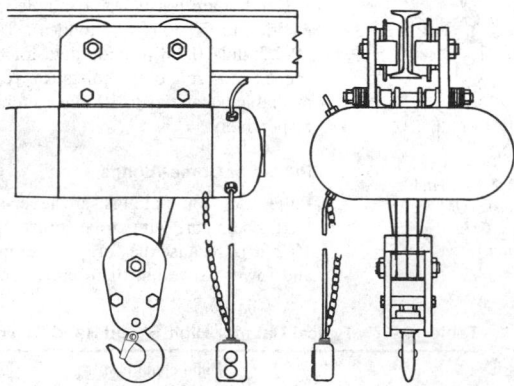

Fig. 10.2.22   Electric wire-rope hoist. *(CM Hoist Division, Columbus McKinnon.)*

useful-duty cycle of the hoist. If the Weston brake is used, an additional auxiliary hand-released or electrically released friction brake must be provided since the Weston brake will not act in the raising direction. All electric hoists have upper-limit devices; lower-limit devices are standard on chain hoists, optional on wire-rope hoists. Control is usually by push button: pendant ropes from the controller are obsolescent. Control is "deadman" type, the hoist stopping instantly upon release. Modern multiple-speed and variable-speed ac controls have made dc hoists obsolete. Single-phase ac hoists are available to 1 hp, polyphase in all sizes.

The hoist may be suspended by an integral hook or bolt-type lug or may be attached to a trolley rolling on an I beam or monorail. The trolley may be plain (push type), geared (operated by a hand chain), or motor-driven; the latter types are essential for heavier loads. Table 10.2.10 gives data for electric **chain hoists**, and Table 10.2.11 for **wire-rope hoists**.

Air Hoists

Air hoists are similar to electric hoists except that air motors are used. Hoists with roller chain are available to 1 ton capacity, with link chain to 3 short tons (2.7 tonnes) and with wire rope to 15 short tons (13.6 tonnes) capacity. The motor may be of the rotary-vane or piston type. The piston motor is more costly but provides the best starting and low-speed performance and is preferred for larger-capacity hoists. A brake, interlocked with the controls, automatically holds the load in neutral; control movement releases the brake, either mechanically or by air

Table 10.2.10   Typical Data for Electric Chain Hoists*

Capacity		Lifting speed		Retracted distance between hooks (headroom)		Net weight		Motor	
short tons	tonnes	ft/min	m/min	in	mm	lb	kg	hp	kW
⅛	0.11	32	10	15	381	62	28	¼	0.19
⅛	0.11	60	18	15	381	68	31	½	0.37
¼	0.23	16	5	15	381	62	28	¼	0.19
¼	0.23	32	10	15	381	68	31	½	0.37
½	0.45	8	2.5	18	457	73	33	¼	0.19
½	0.45	16	5	15	381	68	31	½	0.37
½	0.45	32	10	16	406	114	52	1	0.75
½	0.45	64	20	16	406	121	55	2	1.49
1	.91	8	2.5	18	457	79	36	½	0.37
1	.91	16	5	16	406	114	52	1	0.75
1	.91	32	10	16	406	121	55	2	1.49
2	1.81	8	2.5	23	584	139	63	1	0.75
2	1.81	16	5	23	584	146	66	2	1.49
3	2.7	5.5	1.5	25	635	163	74	1	0.75
3	2.7	11	3.3	25	635	170	77	2	1.49
5	4.5	10	3	37	940	668	303	5	3.73
5	4.5	24	7	37	940	684	310	7½	5.59
7½	6.8	7	2	40	1016	929	421	5	3.73
7½	6.8	16	5	40	1016	957	434	7½	5.59
10	9.1	7	2	42	1067	945	429	5	3.73
10	9.1	13	4	42	1067	961	436	7½	5.59
15	13.6	4	1.2	48	1219	1155	524	5	3.73
15	13.6	8	2.5	48	1219	1167	529	7½	5.59

* Up to and including 3-ton hook suspended with 10-ft lift. 5- through 15-ton plain trolley suspended with 20-ft lift. Headroom distance is beam to high hook.
SOURCE: CM Hoist Division, Columbus McKinnon Corporation.

Table 10.2.11   Typical Data for Electric Wire Rope Hoists*

Capacity		Lifting speed		Beam to high hook distance		Net weight		Motor	
short tons	tonnes	ft/min	m/min	in	mm	lb	kg	hp	kW
½	0.45	60	18	27	686	430	195	4.5	3.36
¾	0.68	37	11	27	686	430	195	4.5	3.36
1	0.91	30	9	25	635	465	211	4.5	3.36
1	0.91	37	11	27	686	435	197	4.5	3.36
1	0.91	60	18	27	686	445	202	4.5	3.36
1½	1.36	18	5.5	25	635	465	211	4.5	3.36
1½	1.36	30	9	25	635	465	211	4.5	3.36
1½	1.36	37	11	27	686	445	202	4.5	3.36
2	1.81	18	5.5	25	635	465	211	4.5	3.36
2	1.81	30	9	25	635	480	218	4.5	3.36
3	2.7	18	5.5	25	635	560	254	4.5	3.36
5	4.5	13	4	30	762	710	322	4.5	3.36
7½	6.8	15	4.6	36	914	1100	499	7.5	5.59
10	9.1	13.5	4.1	40	1016	1785	810	10	7.46
15	13.6	13	4	49	1245	2510	1139	15	11.19

* Up to and including 5-ton with plain trolley. 7½- to 15-ton with motor-driven trolley.
SOURCE: CM Hoist Division, Columbus McKinnon.

pressure. Some air hoists also include a Weston-type load brake. The hoist may be suspended by a hook, lug, or trolley; the latter may be plain, geared, or air-motor-driven. Horizontal movement is limited to about 25 ft (7.6 m) because of the air hose, although a runway system is available with a series of normally closed ports that are opened by a special trolley to supply air to the hoist.

Air hoists provide **infinitely variable speed,** according to the movement of the control valve. Very high speeds are possible with light loads. When severely overloaded, the air motor stalls without damage. Air hoists are smaller and lighter than electric hoists of equal capacity and can be operated in explosive atmospheres. They are more expensive than electric hoists, require mufflers for reasonably quiet operation, and normally are fitted with automatic lubricators in the air supply.

Jacks

Jacks are portable, hand-operated devices for moving heavy loads through short distances. There are three types in common use: screw jacks, rack-and-lever jacks, and hydraulic jacks. Bell-bottom **screw jacks** (Fig. 10.2.23) are available in capacities to 24 tons and lifting ranges to 14 in. The screw is rotated by a bar inserted in holes in the screw head or by a ratchet lever fitted to the head. Geared bridge jacks will lift up to 50 short tons (45 tonnes). A lever ratchet mechanism turns a bevel pinion; an internal thread in the gear raises the nonrotating screw. **Rack-and-lever jacks** (Fig. 10.2.24) consist of a cast-steel or malleable-iron housing in which the lever pivots. The rack toothed bar passes through the hollow housing; the load may be lifted either on the top or on a toe

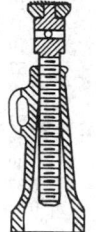
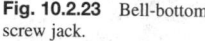
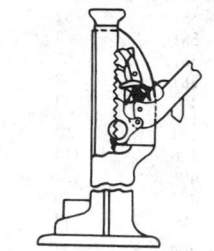

Fig. 10.2.23 Bell-bottom screw jack.

Fig. 10.2.24 Rack-and-lever jack.

extending from the bottom of the bar. The lever pawl may be biased either to raise or to lower the bar, the housing pawl holding the load on the return lever stroke. Rack-and-lever jacks to 20 short tons (18 tonnes) are direct-acting. Lever-operated geared jacks range up to 35 short tons (32 tonnes). Lifting heights to 18 in (0.46 m) are provided. **Track jacks** are rack-and-lever jacks which may be tripped to release the load. They are used for railroad-track work but not for industrial service where the tripping features might be hazardous. **Hydraulic jacks** (Fig. 10.2.25) consist of a cylinder, a piston, and a lever-operated pump. Capacities to 100 short tons (91 tonnes) and lifting heights to 22 in (0.56 m) are available. Jacks 25 short tons (22.7 tonnes) and larger may be provided with two pumps, the second pump being a high-speed unit for rapid travel at partial load.

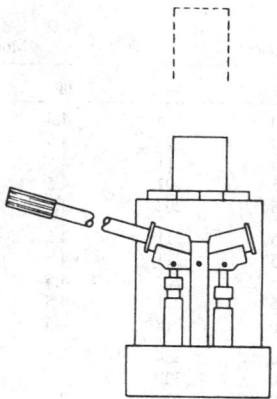

Fig. 10.2.25 Hydraulic jack.

MINE HOISTS AND SKIPS
by Burt Garofab

Pittston Corporation

There are two types of hoists, the drum hoist and the friction hoist. On a **drum hoist,** the rope is attached to the drum and is wound around and stored on the drum. A drum hoist may be single-drum or double-drum. There are also several different configurations of both the single-drum and the double-drum hoist. A drum hoist may be further divided as either unbalanced, partially balanced, or fully balanced.

On a **friction hoist,** which is often referred to as a **Koepe hoist,** the rope is not wound around the drum, but rather passes over the drum. A friction hoist may utilize a single rope or multiple ropes.

The operation of hoists may be controlled manually, automatically, or semiautomatically. There are also arrangements that use combinations of the different control types.

There are currently three types of grooving being used on new drums for drum hoists. They are helical grooving, counterbalance (Lebus) grooving, and antisynchronous grooving. Helical grooving is used primarily for applications where only a single layer of rope is wound around the drum. Counterbalance and antisynchronous grooving are used for applications where multiple layers of rope are wound around the drum.

Mine hoists are generally divided into (1) **Metal-mine hoists** (e.g., iron, copper, zinc, salt, gypsum, silver, gold, ores) and (2) **coal-mine hoists.** These classes subdivide into main hoists (for handling ores or coal) and hoists for men, timbers, and supplies. They are designed for operating (1) mine shafts, vertical and inclined, balanced and unbalanced; and (2) slopes, balanced and unbalanced. When an empty cage or platform descends while the loaded cage or platform ascends, as when both cables are wound on a single drum, the machine is referred to as a **balanced hoist.** Most medium- and large-sized hoists normally operate in balance, as the tonnage obtained for a given load and rope speed is about double that for an unbalanced hoist and the power consumption per ton hoisted is lower. Balanced operation can also be obtained by a counterweight. The counterweight (approximately equal to all the dead loads plus one-half the live load) is usually installed in guides within a single shaft compartment. The average depth of ore mines is about 2,000 ft (610 m), and that of coal mines is close to 500 ft (152 m). Most ore-mine hoists are of the double-drum type, normally hoisting in balance, each drum being provided with a friction clutch for changing the relative positions of the two skips when operating from various levels.

Hoists for coal mines are principally of the keyed-drum type, for operating in balance from one level. For high rope speeds in shallow shafts, it is generally advantageous to use **combined cylindrical and conical drums.** The cylindroconical drum places the maximum rope pull (weight of rope and loaded skip) on the small diameter so that during the acceleration period of the cycle, the weight of the opposing skip is offering the greatest counterbalance torque, reducing motor peak loads and slightly reducing the power consumption. The peak reduction obtained becomes greater when shafts are shallower and hoisting speeds are higher, provided that the proportion between the smaller and the large diameters is increased as these conditions increase. By varying the ratio of diameters and the distribution of rope on the drum profiles with respect to the periods of acceleration, retardation, and constant speed, static and dynamic torques can be modified to produce the most economical power consumption and the minimum size of motor. The conical drum is not applicable for multiple-level operation, and except in very special cases, only a single layer of rope can be used.

Skip hoists for industrial purposes such as power-plant fuel handling and blast-furnace charging are similar to shallow-lift slow-speed coal-mine hoists in that they operate from a single level. Speeds of 100 to 400 ft/min (0.5 to 2 m/s) are usual. For blast-furnace charging with combined bucket and load weights up to 31,000 lb (14,000 kg) and a speed of 500 ft/min (2.5 m/s), modern plants consist of straight-drum geared engines, frequently with Ward Leonard control.

Industrial skip hoists may be specified where the lift is too high for a bucket elevator, where the lumps are too large for elevator buckets, or where the material is pulverized and extremely abrasive or actively corrosive. For high lifts having a vertical or nearly vertical path, the skip with supporting structure usually costs less than a bucket elevator or an inclined-belt conveyor with bridge. Typical paths are shown in Fig. 10.2.26, paths *C* and *D* being suitable when the load is received through a track hopper.

The skip may be **manually loaded** direct from a dump car or **automatically loaded** by a pivoted chute, which is actuated by the bucket and which, when upturned, serves as a cutoff gate (Fig. 10.2.27).

For small capacities, the skip can be manually loaded with semiautomatic control. When the bucket has been filled, the operator pushes the start button and the bucket ascends, dumps, and returns to loading position. With automatic loading and larger capacity, the skip may have full automatic control. For economy, the bucket is counterbalanced by a weight, usually equaling the weight of the empty bucket plus half the load. For large capacity, a balanced skip in which one bucket rises as the other descends may be used. High-speed skips usually have automatic slowdown (two-speed motor) as the bucket nears the loading and discharge points.

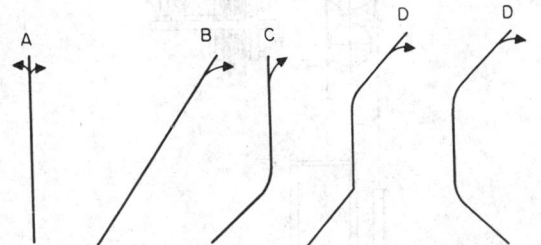

Fig. 10.2.26 Typical paths for skip hoists. (*a*) Vertical with discharge to either side; (*b*) straight inclined run; (*c*) incline and vertical; (*d*) incline, vertical, incline.

There are various types of wire ropes used in hoisting. Hoist ropes can be categorized into three main types, round strand, flattened strand, and locked coil. **Round strand rope** is used on drum hoists in applications when a single layer of rope is wound on the drum. **Flattened strand rope** is used on drum hoists when multiple layers of rope are wound on the drum. Flattened strand rope can also be used on friction hoists. **Locked coil rope** is used on friction hoists.

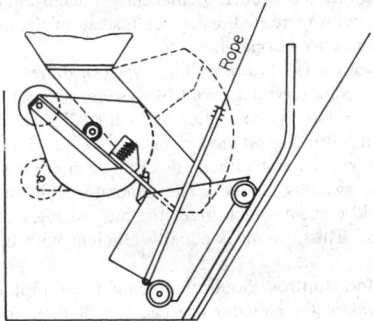

Fig. 10.2.27 Automatic loader (in loading position).

Plow steel and improved plow steel are the most commonly used grades; the latter is used where the service is severe. Some state mining regulations require higher factors of safety than the usual hoisting requirements. The working capacity of new ropes is usually computed by using the minimum breaking strength given in the manufacturer's tables and the following factors of safety: rope lengths of 500 ft (152 m) or less, minimum factor 8; 500 to 1,000 ft (152 to 305 m), 7; 1,000 to 2,000 ft (305 to 610 m), 6; 2,000 to 3,000 ft (610 to 914 m), 5; 3,000 ft (914 m) or over, 4½. These are gross factors between the rated minimum breaking strength of the rope and the maximum static pull due to suspended load plus rope weight. The net factor, which should be used in dealing with large capacities and great depth, must take into consideration stresses due to acceleration and bending around the drum, together with suitable allowances for shock. With 6-by-19 wire construction, the pitch diameter of the drums is generally not less than sixty times the diameter of the rope. Drums are made of either cast iron or cast steel machine-grooved to suit the size of rope (see above). In large hoists, a lifting device is installed at the free rope end of the drum to assist the rope in doubling back over the first layer.

Brakes   There are three main types of brakes used on hoists: the jaw brake, the parallel motion brake, and disk brakes. Brake control is ac-

complished through air or hydraulics. Brake shoes are steel with attached friction material surfaces.

Hoist Motors   Determining the proper size of motor for driving a hoist calls for setting up a definite cycle of duty based upon the required daily or hourly tonnage.

The **permissible hoisting speed** for mine hoists largely depends upon the depth of the shaft; the greater the depth, the higher the allowable speed. Conservative maximum hoisting speeds, as recommended by *Bu. Mines Bull.* 75, are as follows:

Depth of shaft		Hoisting speed	
ft	m	ft/min	m/s
500 or less	150 or less	1,200	6
500–1,000	150–300	1,600	8
1,000–2,000	300–600	2,000	10
2,000–3,000	600–900	3,000	15

High hoisting speeds call for rapid **acceleration** and **retardation**. For small hoists, the rate of acceleration may be made as low as 0.5 ft/s^2 (0.15 m/s^2). An average value of 3 ft/s^2 is adopted for large hoists with fairly high speeds. Exceptional cases may require up to 6 ft/s^2 (1.9 m/s^2). The speed should also be considered with regard to the weight of the material to be hoisted per trip. The question of whether the load should be increased and the speed reduced or vice versa is controlled by local conditions, mining laws, and practical experience. The rest period assigned to the duty cycle, i.e., the requisite time for loading at the bottom and unloading at the top, is dependent upon the equipment employed. With skips loaded from underground ore-storage hoppers, 5 to 6 s is the minimum that can be assumed. Unless special or automatic provision is made, the loading time should be taken as 8 to 10 s minimum. When the (1) hoisting speed, (2) weight of skip or cage, (3) weight of load, (4) periods of acceleration, (5) retardation, and (6) time for loading have been decided upon, the next step is to ascertain the "root-mean-square" equivalent continuous load-heating effect on the motor, taking into account rope and load weights, acceleration and deceleration of all hanging and rotating masses, and friction of sheaves, machines, etc. The friction load is usually taken as constant throughout the running period of the cycle. The overall efficiency of the mechanical parts of a single-reduction-geared hoist averages 80 percent; that of a first-motion hoist is closer to 85 percent. The motor selected must have sufficient starting torque to meet the temporary peaks of any cycle, including, in the case of balanced hoists, the requirements of trips out of balance.

Electrical equipment for driving mine hoists is of four classes:

1. *Direct-current motors* with resistance control for small hoists, usually series-wound but occasionally compound-wound in conjunction with dynamic braking control.

2. *Alternating-current slip-ring-type motor* with secondary resistance.

3. *Ward Leonard system of control* for higher efficiency, particularly on short lifts at high rope speeds, where the rheostatic losses during acceleration and retardation represent a large proportion of the net work done during the cycle; for accuracy of speeds, with high-speed hoists; and for equalization of power demands. Complete control of the speed from standstill to maximum is obtained for all values of load from maximum positive to maximum negative. The lowering of unbalanced loads without the use of brakes is as readily accomplished as hoisting.

4. The *Ilgner Ward Leonard system* consists of a flywheel directly connected to a Ward Leonard motor-generator set and a device for automatically varying the speed through the secondary rheostatic control of the slip-ring induction motor driving the set. This form of equipment is used under conditions that prohibit the carrying of heavy loads or where power is purchased under heavy reservation charges for peak loads. It limits the power taken from the supply circuit to a certain predetermined value; whatever is required in excess of this value is produced by the energy given up by the flywheel as its speed is reduced.

ELEVATORS, DUMBWAITERS, ESCALATORS
by Louis Bialy
Otis Elevator Corporation

The advent of the safety elevator changed the concept of the city by making high-rise buildings possible. Elevators are widely used to transport passengers and freight vertically or at an incline in buildings and structures. Elevators are broadly classified as low-rise, medium-rise, and high-rise units. Low-rise elevators typically serve buildings with between 2 and 7 floors, medium-rise elevators serve buildings with between 5 and 20 floors, while high-rise elevators serve buildings with more than 15 floors. The speed of the elevator is indexed to the rise of the building so that the overall flight time from bottom to top, or vice versa, is approximately the same. A typical **flight time** is about one minute. Typical speeds for low-rise elevators are up to 200 ft/min (1.0 m/s). Typical speeds for medium rise elevators are up to 400 ft/min (2.0 m/s). High rise elevators typically travel at speeds of up to 1,800 ft/min (9.0 m/s).

Low rise elevators are usually oil-power hydraulic devices. The simplest version consists of a hydraulic jack buried in the ground beneath the elevator car. The jack is approximately centrally located beneath the car and the ram or plunger is connected to the platform or structure which supports the car. The car is guided by guiderails which cover the full rise of the elevator hoistway. Guideshoes or guiderollers typically guide the elevator as it ascends and descends the hoistway. The hydraulic cylinder is equipped with a cylinder head which houses the seals and bearing rings which locate the ram. The ram typically runs clear of the inner cylinder wall. Buried cylinders are typically protected against corrosion.

Hydraulic power is usually supplied by a screw-type positive-displacement pump driven by an induction motor. It is common for the motor and pump to be coaxially mounted and submersed in the hydraulic reservoir. Operating pressures are typically 300 to 600 lb/in² (2 to 4 MPa). A hydraulic control valve controls the flow of oil to the hydraulic jack and hence the speed of the elevator. In the down direction, the pump is not powered, and the elevator speed is controlled by bleeding fluid through the control valve.

Another manifestation of the **hydraulic elevator** is called the *holeless hydraulic* elevator. These typically have one or more hydraulic jacks mounted vertically alongside the elevator car, the plunger being either directly attached to the car or connected to the car by steel wire cables (also known as *wire ropes*). Elevators of the latter type are known as *roped hydraulic elevators*. These elevators are often roped 1 : 2 so that the elevator moves at twice the speed of the hydraulic ram. The rise of the elevator is also twice the stroke of the ram. Holeless hydraulic elevators with the ram directly connected to the elevator car may have single-stage jacks or multistage telescopic jacks, depending on the required rise.

Medium- and high-rise elevators are typically traction-driven units; i.e., the rotary motion of the drive sheave is transmitted to the steel wire cables or ropes via friction. The elevators are typically counterweighted so that the motor and drive only need overcome the unbalanced load. The **counterweight** mass is typically the car mass plus approximately half the duty load (load of passengers or freight). With high-rise elevators, the weight of the rope is neutralized by compensating chains hung from the underside of the counterweight and looped to the bottom of the car. If the compensation weight matches the suspension rope weight, then irrespective of the position of the car, there will be no imbalance due to rope weight. The drive sheave is usually grooved to guide the rope and enhance the traction. Roping ratios may be 1 : 1 or 2 : 1. With 1 : 1 ratios the car speed is the same as the rope speed. With 2 : 1 ratios the elevator speed is half of the rope speed.

Medium-rise elevators are typically driven by geared machines which transmit the motor power to the drive sheave. Gear reduction ratios are typically in the range from 12 : 1 to 30 : 1.

Right-angle worm reduction gear sets are most common; however, helical gears are becoming more acceptable because of their higher operating efficiency and low wear characteristics.

High-rise elevators are typically driven by gearless machines, which provide the smoothest and most precise performance of all elevators (see Fig. 10.2.28).

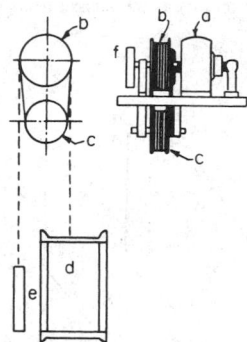

Fig. 10.2.28 Electric traction elevator.

Motors Both dc and ac motors are used to power the driving sheave. Direct current motors have the advantages of good starting torque and ease of speed control. Elevator motors are obliged to develop double rated torque at 125 percent rated current and have frequent starts, stops, reversals, and runs at constant speed. With sparkless commutation under all conditions, commercial motors cannot as a rule meet these requirements. For constant voltage, cumulative-compound motors with heavy series fields are used. The series fields are gradually short-circuited as the motor comes up to speed, after which the motor operates shunt. The shunt field is excited permanently, the current being reduced to a low value with increased resistance instead of the circuit opening when the motor is not in operation.

Drive Control for DC Motors Two types of drives are used for dc motors: Ward Leonard (voltage control) systems using motor-generator sets, and thyristor [silicon controlled rectifier (SCR)] drives. With voltage control, an individual dc generator is used for each elevator. The generator may be driven by either a dc or an ac motor (see Ward Leonard and Ilgner systems, above). The generator voltage is controlled through its field current which gives the highest rates of acceleration and retardation. This system is equally efficient with either dc or ac supply.

AC Motor and Control Squirrel-cage induction motors are becoming more prevalent for elevator use because of their ruggedness and simplicity. The absence of brushes is considered a distinct advantage. The motors are typically three-phase machines driven by variable-voltage, variable frequency (VVVF) drives. The VVVF current is provided by an inverter. High-current-capacity power transistors modulate the power to the motor. Motion of the rotor or the elevator is often monitored by devices such as optional encoders which provide a feedback to the control system for closed loop control.

Elevators installed in the United States are usually required to comply with the ANSI/ASME A17.1 Safety Code or other applicable local codes. The A17.1 code imposes specific safety requirements pertaining to the mechanical, structural, and electrical integrity of the elevator. Examples of **code requirements** include: factors of safety for mechanical and structural elements; the number and type of wire ropes to be used; the means of determining the load capacity of the elevator; the requirements for speed governors and independent safety devices to arrest the car should the car descend at excessive speeds; the requirements for brakes to hold the stationary car with loads of up to 125 percent of rated load; the requirements for buffers at the bottom of the elevator shaft to decelerate the elevator or counterweight should they descend beyond the lowest landing; the type and motion of elevator and hoistway doors; the requirement that doors not open if the elevator is not at a landing; the requirement that door motion cease or reverse if a passenger or obstruction is in their path and that the kinetic energy of the door motion be limited so as to minimize the impact with a passenger or obstruction should they fail to stop in time. The code also requires electrical protective devices which stop the car should it transcend either upper or lower terminal landings, remove power to the motor, and cause the brake to apply should the speed exceed governor settings, etc. The A17.1 code also provides special requirements for elevators located in high seismic risk zones.

Loads   Table 10.2.12 shows the rated load of passenger elevators.

Table 10.2.12   Rated Load of Passenger Elevators

Rated Capacity,	lb	2,500	3,500	4,500	6,000	9,000	12,000
	kg	1,135	1,590	2,045	2,725	4,085	5,450
Net inside area,	ft²	29.1	38.0	46.2	57.7	80.5	103.0
Platform area,	m²	2.7	3.5	4.3	5.4	7.5	9.6

Efficiencies and Energy Dissipation per Car Mile   The overall efficiency and hence the energy dissipation varies greatly from elevator to elevator, depending on the design, hoistway conditions, loading conditions, acceleration and deceleration profiles, number of stops, etc.

For a load of 2,500 lb (1,130 kg) for **geared traction elevators** driven at 350 ft/min (1.75 m/s) by SCR-controlled dc motor or VVVF-controlled ac motor, typical values are as follows: efficiency 60 percent, energy dissipation 4.5 kWh (16 MJ).

Under same conditions for **gearless traction elevators** traveling at 700 ft/min (3.5 m/s), the efficiency may be 70 percent and the energy dissipation 4 kWh (14.5 MJ).

Note that efficiencies are based on net load, i.e., full load minus overbalance.

Car Mileage and Stops (per elevator, 8-h day)

Office buildings, local elevators: intensive service, 12 to 20 car miles (19 to 32 car kilometers), making about 150 regular stops per mile

Express elevators: 20 to 40 car miles (32 to 64 car kilometers), making about 75 to 100 regular stops per mile

Department store elevators: 4 to 8 car miles (6 to 13 car kilometers), making about 350 regular stops per mile

Operational Control of Elevators   Most modern passenger elevators are on fully automatic group collective control. Each group of elevators is controlled by a dispatching system which assigns specific elevators to answer specific hall calls. Modern dispatch systems are microprocessor-based or have some equivalent means of processing information so that each call registered for an elevator is answered in a timely manner. The determination as to which elevator is assigned to answer a specific call is a complex process which in the case of microprocessor-controlled dispatch systems requires the use of sophisticated algorithms which emulate the building dynamics. Some dispatch systems use advanced decision-making processes based on *artificial intelligence* such as *expert systems* and *fuzzy logic* to optimize group collective service in buildings. With better dispatching systems it is possible to achieve excellent elevator service in buildings with the minimum number of elevators, or the efficient service of taller buildings without devoting more space to elevators.

The operational control systems also provide signals for door opening and closing.

Dumbwaiters follow the general design philosophy of elevators except that code requirements are somewhat more relaxed. For example, roller-link chain can be used for support instead of wire rope. Moreover, a single steel wire rope can be used instead of the mandatory minimum of three for traction-type elevators and two for drum-type elevators. Dumbwaiters are not intended for the transportation of people.

Escalators have the advantages of continuity of motion, great capacity, and small amounts of space occupied and current consumed for each passenger carried. Escalators are built with step widths of 24, 32, and 40 in (610, 810, and 1015 mm). The angle of incline is 30° from the horizontal, and the speed is 90 to 120 ft/min (0.45 to 0.6 m/s).

Approximate average handling capacities for escalators traveling at 90 ft/min are 2,000, 2,300, and 4,000 passengers per hour for 24-in, 32-in, and 40-in (610-, 810-, and 1015-mm) escalator step sizes respectively. Higher-speed escalators have proportionally higher passenger-carrying capacity.

Escalators are equipped with a handrail on each side, mounted on the balustrade. The handrail moves at the same speed as the escalator. Escalator steps are so arranged that as they approach the upper and lower landings they recede so that they are substantially level with the floor for safe embarkation and disembarkation of the escalator.

10.3   DRAGGING, PULLING, AND PUSHING
by Harold V. Hawkins
revised by Ernst K. H. Marburg and Associates

HOISTS, PULLERS, AND WINCHES

Many of the fundamental portable lifting mechanisms such as hoists or pullers (see above) can also be used forcefully to drag or pull materials. In addition, nonmobile versions, called **winches**, utilizing hoisting drums can also be used.

LOCOMOTIVE HAULAGE, COAL MINES
by Burt Garofab
Pittston Corporation

The haulage system of an underground mine is used to transport personnel and material between the face and the portal. It can be subdivided into face haulage and main haulage. The main haulage system extends from the end of the face haulage system to the outside.

Rubber-tired haulage at the coal face was introduced in 1935 and received further impetus with the introduction of the rubber-tired shuttle car in 1938. Crawler-mounted loaders and rubber-tired universal coal cutters completed the equipment needed for complete off-track mining. This off-track mining caused a revolution in face haulage, since it eliminated the expense of laying track in the rooms and advancing the track as the face of the coal advanced. It also made gathering locomotives of the cable-reel, crab-reel, and battery types obsolete. Practically all the gathering duty is now performed by rubber-tired shuttle cars, chain conveyors, extensible belt conveyors, and other methods involving off-track equipment. Some mines eliminated track completely by having belt conveyors carry the coal to the outside.

Most coal mines today utilize a combination of haulage systems. Personnel and supplies are transported by either rail or rubber-tired haulage. Coal is transported by belt or rail, predominantly belt.

Locomotives used in rail haulage may be trolley wire powered, battery powered, or diesel powered. Battery and diesel powered locomotives are commonly used to transport supplies in mines which utilize belt as the main coal haulage system. Trolley-wire-powered locomotives are used when rail is the main coal haulage system. The sleek, streamlined, fast, and easy-riding **Jeffrey** eight-wheel four-motor locomotives are available in 27-, 37-, and 50-short ton (24-, 34-, and 45-tonne) sizes. This type of locomotive has two four-wheel trucks, each having two motors. The trucks, having Pullman or longitudinal-type equalizers with snubbers, will go around a 50 ft (15 m) radius curve, have short overhang, and provide a very easy ride. This construction is easy on the track, with consequent low track-maintenance cost. Speed at full load ranges from 10 to 12.5 mi/h (4.5 to 5.6 m/s), and the maximum safe speed is approximately 30 mi/h (13.4 m/s).

The **eight-wheel locomotive** usually has a box frame; series-wound motors with single-reduction spur gearing; 10 steps of straight parallel, full electropneumatic contactor control; dynamic braking; 32-V battery-operated control and headlights, with the battery charged automatically from the trolley; straight air brakes; air-operated sanders; air horn; automatic couplers; one trolley pole; two headlights at each end; and blowers to ventilate the traction motors. The equipment is located so that it is easily accessible for repair and maintenance.

The eight-wheel type of locomotive has, to a great extent, superseded the tandem type, consisting of two four-wheel two-motor locomotives coupled together and controlled from the cab of one of the units of the tandem.

The older Jeffrey **four-wheel-type locomotive** is available in 11-, 15-, 20-, and 27-short ton (10-, 13.6-, 18-, 24.5-tonne) nominal weights. The 20- and 27-short ton (18- and 24.5-tonne) sizes have electrical equipment very much like that of the eight-wheel-type locomotive. Speeds are also comparable. The 15-short ton (13.6-tonne) locomotive usually has semielectropneumatic contactor control, rather than full electropneumatic control, and dynamic braking but usually does not have the 32-V battery-operated control. The 11-short ton (10-tonne) locomotive has manual control, with manual brakes and sanders, although contactor control, air brakes, blowers, etc., are optional.

Locomotive motors have a horsepower rating on the basis of 1 h at 75°C above an ambient temperature of 40°C. Sizes range from a total of 100 hp (75 kW) for the 11-ton (10-tonne) to a total of 720 hp for the 50-ton.

The following formulas are recommended by the Jeffrey Mining Machinery Co. to **determine the weight of a locomotive** required to haul a load. Table 10.3.1 gives haulage capacities of various weights of locomotives on grades up to 5 percent. The tabulation of haulage capacities shows how drastically the tons of trailing load decrease as the grade increases. For example, a 50-ton locomotive can haul 1,250 tons trailing load on the level but only 167 tons up a 5 percent grade.

The following formulas are based on the use of steel-tired or rolled-steel wheels on clean, dry rail.

Weight of locomotive required on level track:

$$W = L(R + A)/(0.3 \times 2{,}000 - A)$$

Weight of locomotive required to haul train up the grade:

$$W = L(R + G)/(0.25 \times 2{,}000 - G)$$

Weight of locomotive necessary to start train on the grade:

$$W = L(R + G + A)/(0.30 \times 2{,}000 - G - A)$$

where W is the weight in tons of locomotive required; R is the frictional resistance of the cars in pounds per ton and is taken as 20 lb for cars with antifriction bearings and 30 lb for plain-bearing cars; L is the weight of the load in tons; A is the acceleration resistance [this is 100 for 1 mi/(h·s) and is usually taken at 20 for less than 10 mi/h or at 30 from 10 to 12 mi/h, corresponding to an acceleration of 0.2 or 0.3 mi/(h·s)]; G is the grade resistance in pounds per ton or 20 lb/ton for each percent of grade (25 percent is the running adhesion of the locomotive, 30 percent is the starting adhesion using sand); 2,000 is the factor to give adhesion in pounds per ton.

Where the grade is in favor of the load:

$$W = L(G - R)/(0.20 \times 2{,}000 - G)$$

To brake the train to a stop on grade:

$$W = L(G + B - R)/(0.20 \times 2{,}000 - G - B)$$

where B is the braking (or decelerating) effort in pounds per ton and equals 100 lb/ton for a braking rate of 1 mi/(h·s) or 20 lb/ton for a braking rate of 0.2 mi/(h·s) or 30 lb for a braking rate of 0.3 mi/(h·s). The adhesion is taken from a safety standpoint as 20 percent. It is not advisable to rely on using sand to increase the adhesion, since the sandboxes may be empty when sand is needed.

Time in seconds to brake the train to stop:

$$s = \frac{\text{mi/h (start)} - \text{mi/h (finish)}}{\text{deceleration in mi/(h·s)}}$$

Distance in feet to brake the train to a stop:

$$\text{ft} = [\text{mi/h (start)} - \text{mi/h (finish)}] \times s \times 1.46/2$$

Storage-battery locomotives are used for hauling muck cars in tunnel construction where it is inconvenient to install trolley wires and bond the track as the tunnel advances. They are also used to some extent in metal mines and in mines of countries where trolley locomotives are not permitted. They are often used in coal mines in the United States for hauling supplies. Their first cost is frequently less than that for a trolley installation. They also possess many of the advantages of the trolley locomotive and eliminate the danger and obstruction of the trolley wire. Storage-battery locomotives are limited by the energy that is stored in the battery and should not be used on steep grades or where large, continuous overloads are required. Best results are obtained where light

Table 10.3.1 Haulage Capacities of Locomotives with Steel-Tired or Rolled-Steel Wheels*

Grade		Weight of locomotive, tons†					
		11	15	20	27	37	50
Level	Drawbar pull, lb	5,500	7,500	10,000	13,500	18,500	25,000
	Haulage capacity, gross tons	275	375	500	675	925	1,250
1%	Drawbar pull, lb	5,280	7,200	9,600	12,960	17,760	24,000
	Haulage capacity, gross tons	132	180	240	324	444	600
2%	Drawbar pull, lb	5,260	6,900	9,200	12,420	17,020	23,000
	Haulage capacity, gross tons	88	115	153	207	284	384
3%	Drawbar pull, lb	4,840	6,600	8,800	11,880	16,280	22,000
	Haulage capacity, gross tons	67	82	110	149	204	275
4%	Drawbar pull, lb	4,620	6,300	8,400	11,340	15,540	21,000
	Haulage capacity, gross tons	46	63	84	113	155	210
5%	Drawbar pull, lb	4,400	6,000	8,000	10,800	14,800	20,000
	Haulage capacity, gross tons	31	50	67	90	123	167

* Jeffrey Mining Machinery Co.
† Haulage capacities are based on 20 lb/ton rolling friction, which is conservative for roller-bearing cars. Multiply lb by 0.45 to get kg and tons by 0.91 to get tonnes.

and medium loads are to be handled intermittently over short distances with a grade of not over 3 percent against the load.

The general construction and mechanical features are similar to those of the four-wheel trolley type, with battery boxes located either on top of the locomotive or between the side frames, according to the height available. The motors are rugged, with high efficiency. Storage-battery locomotives for coal mines are generally of the explosion-tested type approved by the Bureau of Mines for use in gaseous mines.

The **battery** usually has sufficient capacity to last a single shift. For two- or three-shift operation, an extra battery box with battery is required so that one battery can be charging while the other is working on the locomotive. Motor-generator sets or rectifiers are used for charging the batteries. The overall efficiency of the battery, motor, and gearing is approximately 63 percent. The speed varies from 3 to 7 mi/h, the average being 3½ to 4½ mi/h. Battery locomotives are available in sizes from 2 to 50 tons. They are usually manufactured to suit individual requirements, since the sizes of motors and battery are determined by the amount of work that the locomotive has to do in an 8-h shift.

INDUSTRIAL CARS

Various types of narrow-gage industrial cars are used for handling bulk and package materials inside and outside of buildings. Those used for bulk material are usually of the dumping type, the form of the car being determined by the duty. They are either pushed by workers or drawn by mules, locomotives, or cable. The **rocker side-dump car** (Fig. 10.3.1) consists of a truck on which is mounted a V-shaped steel body supported on rockers so that it can be tipped to either side, discharging material. This type is mainly used on construction work. Capacities vary from ⅔ to 5 tons for track gages of 18, 20, 24, 30, 36, and 56½ in. In the **gable-bottom car** (Fig. 10.3.2), the side doors a are hinged at the top and controlled by levers b and c, which lock the doors when closed. Since this type of car discharges material close to the ground on both sides of

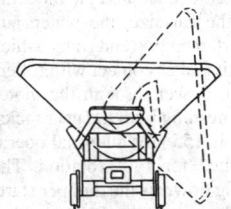

Fig. 10.3.1 Rocker side-dump car.

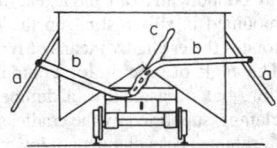

Fig. 10.3.2 Gable-bottom car.

the track simultaneously, it is used mainly on trestles. Capacities vary from 29 to 270 ft³ for track gages of 24, 36, 40, and 50 in. The **scoop dumping car** (Fig. 10.3.3) consists of a scoop-shaped steel body pivoted at a on turntable b, which is carried by the truck. The latch c holds the body in a horizontal position, being released by chain d attached to handle e. Since the body is mounted on a turntable, the car is used for service where it is desirable to discharge material at any point in the

circle. This car is made with capacities from 12 to 27 ft³ to suit local requirements. The **hopper-bottom car** (Fig. 10.3.4) consists of a hopper on wheels, the bottom opening being controlled by door a, which is operated by chain b winding on shaft c. The shaft is provided with handwheel and ratchet and pawl. The type of door or gate controlling the bottom opening varies with different materials.

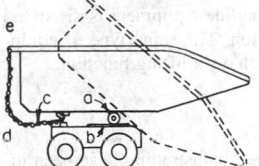
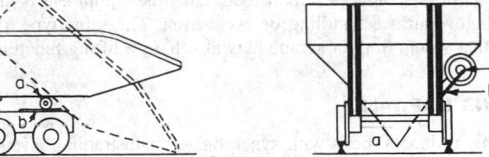

Fig. 10.3.3 Scoop dumping car.   **Fig. 10.3.4** Hopper-bottom car.

The **box-body dump car** (Fig. 10.3.5) consists of a rectangular body pivoted on the trucks at a and held in horizontal position by chains b. The side doors of the car are attached to levers so that the door is automatically raised when the body of the car is tilted to its dumping position. The cars can be dumped to either side. On the large sizes, where rapid dumping is required, dumping is accomplished by compressed air. This type of car is primarily used in excavation and quarry work, being loaded by power shovels. The greater load is placed on the side on which the car will dump, so that dumping is automatic when the operator releases the chain or latch. The car bodies may be steel or steel-lined wood. **Mine cars** are usually of the four-wheel type, with low bodies, the doors being at one end, and pivoted at the top with latch at

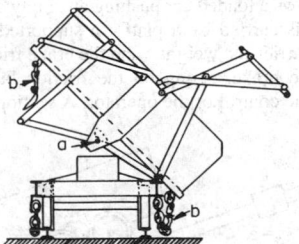

Fig. 10.3.5 Box-body dump car.

the bottom. Industrial **tracks** are made with rails from 12 to 45 lb/yd (6.0 to 22 kg/m) and gages from 24 in to 4 ft 8½ in (0.6 to 1.44 m). Either steel or wooden ties are used. Owing to its lighter weight, the steel tie is preferred where tracks are frequently moved, the track being made up in sections. Industrial cars are frequently built with one wheel attached to the axle and the other wheel loose to enable the car to turn on short-radius tracks. Capacities vary from 4 to 50 yd³ (3.1 to 38 m³) for track gages of 36 to 56½ in (0.9 to 1.44 m), with cars having weights from 6,900 to 80,300 lb (3,100 to 36,000 kg). The frictional resistance per ton (2,000 lb) (8,900 N) for different types of mine-car bearings are given in Table 10.3.2.

Table 10.3.2 Frictional Resistance of Mine Car Bearings

	Drawbar pull					
	Level track		2% grade		4% grade	
Types of bearings	lb/short ton	N/tonne	lb/short ton	N/tonne	lb/short ton	N/tonne
Spiral roller	13	58	15	67	46	205
Solid roller	14	62	18	80	53	236
Self-oiling	22	98	31	138		
Babbitted, old style	24	107	40	178		

DOZERS, DRAGLINES

The dual capability of some equipment, such as **dozers** and **draglines**, suggests that it should be mentioned as prime machinery in the area of materials handling by dragging, pulling, or pushing. Dozers are described in the discussion on earthmoving equipment since their basic frames are also used for power shovels and backhoes. In addition, dozers perform the auxiliary function of pushing carryall earthmovers to assist them in scraping up their load. Dragline equipment is discussed with below-surface handling or excavation. The same type of equipment that would drag or scrape may also have a lifting function.

MOVING SIDEWALKS

Moving horizontal belts with synchronized balustrading have been introduced to expedite the movement of passengers to or from railroad trains in depots or planes at airports (see belt conveyors). A necessary feature is the need to prevent the clothing of anyone (e.g., a child) sitting on the moving walk from being caught in the mechanism at the end of the walk. Use of a comblike stationary end fence protruding down into longitudinal slots in the belt is an effective preventive.

CAR-UNLOADING MACHINERY

Four types of devices are in common use for unloading material from all types of open-top cars: crossover and horn dumps, used to unload mine cars with swinging end doors; rotary car dumps, for mine cars without doors; and tipping car dumps, for unloading standard-gage cars where large unloading capacity is required.

Crossover Dump   Figure 10.3.6 shows a car in the act of dumping. Figure 10.3.7 shows a loaded car pushing an empty car off the dump. A section of track is carried by a platform supported on rockers a. An extension bar b carries the weight c and the brake friction bar d. A hand lever controls the brake, acting on the friction bar and placing the dumping under the control of the operator. A section of track e in front

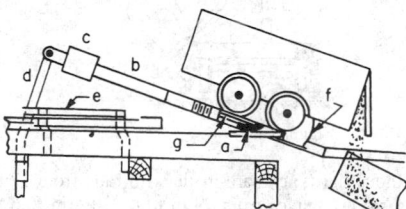

Fig. 10.3.6   Crossover dump: car unloading.

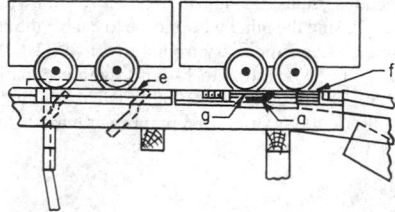

Fig. 10.3.7   Crossover dump: empty car being pushed away.

of the dump is pivoted on a parallel motion and counterbalanced so that it is normally raised. The loaded car depresses the rails e and, through levers, pivots the horns f around the shafts g, releasing the empty car. The loaded car strikes the empty car, starting it down the inclined track. After the loaded car has passed the rails e, the springs return the horns f so that they stop the loaded car in the position to dump. Buffer springs on the shaft g absorb the shock of stopping the car. Since the center of gravity of the loaded car is forward of the rockers, the car will dump automatically under control of the brake. No power is required for this dump, and one operator can dump three or four cars per minute.

Rotary Gravity Dump   (Fig. 10.3.8)   This consists of a steel cylinder supported by a shaft a, its three compartments carrying three tracks. The loaded car b is to one side of the center and causes the cylinder to rotate, the material rolling to the chute beneath. The band brake c, with counterweight d, is operated by lever e, putting the dumping under control of the operator. No power is required; one operator can dump two or three cars per minute.

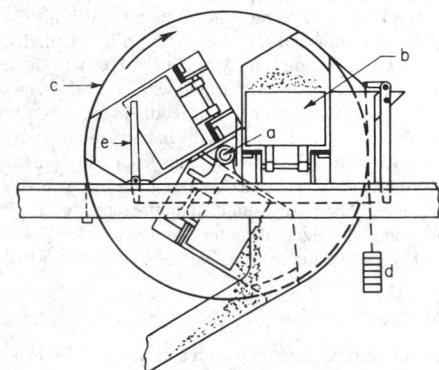

Fig. 10.3.8   Rotary gravity dump.

Rotary-power dumpers are also built to take any size of open-top railroad car and are frequently used in power plants, coke plants, ports, and ore mines to dump coal, coke, ore, bauxite, and other bulk material. They are mainly of two types: (1) single barrel, and (2) tandem.

The McDowell-Wellman Engineering Co. dumper consists of a revolving cradle supporting a platen (with rails in line with the approach and runoff car tracks in the upright position), which carries the car to be dumped. A blocking on the dumping side supports the side of the car as the cradle starts rotating. Normally, the platen is movable and the blocking is fixed, but in some cases the platen is fixed and the blocking movable. Where there is no variation in the car size, the platen and blocking are both fixed. The cradle is supported on two end rings, which are bound with a rail and a gear rack. The rail makes contact with rollers mounted in sills resting on the foundation. Power through the motor rotates the cradle by means of two pinions meshing with the gear racks. The angle of rotation for a complete dump is 155° for a normal operation, but occasionally, a dumper is designed for 180° rotation. The clamps, supported by the cradle, start moving down as the dumper starts to rotate. These clamps are lowered, locked, released, and raised either by a gravity-powered mechanism or by hydraulic cylinders.

With the advent of the **unit-train system,** the investment and operating costs for a dumper have been reduced considerably. The design of dumpers for unit train has improved and results in fewer maintenance problems. The use of rotary couplers on unit train eliminates uncoupling of cars while dumping because the center of the rotary coupler is in line with the center of rotation of the dumper.

Car Shakers   As alternatives to rotating or tilting the car, several types of car shakers are used to hasten the discharge of the load. Usually the shaker is a heavy yoke equipped with an unbalanced pulley rotated at 2,000 r/min by a 20-hp (15-kW) motor. The yoke rests upon the car top sides, and the load is actively vibrated and rapidly discharged. While a car shaker provides a discharge rate about half that of a rotary dumper, the smaller investment is advantageous.

Car Positioner   As the popularity of unit-train systems consisting of rail cars connected by rotary couplers has increased, more rotary dumping stations have been equipped with an automatic train positioner developed by McDowell-Wellman Engineering Company. This device consists of a carriage moving parallel to the railroad track actuated by either hydraulic cylinders or wire rope driven by a winch, which carries an arm that rotates in a vertical plane to engage the coupling between the cars. The machines are available in many sizes, the largest of which are capable of indexing 200-car trains in one- or two-car increments through the rotary dumper. These machines or similar ones are also available from FMC/Materials Handling System Division, Heyl and Patterson, Inc., and Whiting Corp.

10.4 LOADING, CARRYING, AND EXCAVATING
by Ernst K. H. Marburg

CONTAINERIZATION

The proper packaging of material to assist in handling can significantly minimize the handling cost and can also have a marked influence on the type of handling equipment employed. For example, partial carload lots of liquid or granular material may be shipped in rigid or nonrigid containers equipped with proper lugs to facilitate in-transit handling. Heavy-duty rubberized containers that are inert to most cargo are available for repeated use in shipping partial carloads. The nonrigid container reduces return shipping costs, since it can be collapsed to reduce space. Disposable light-weight corrugated-cardboard shipping containers for small and medium-sized packages both protect the cargo and permit stacking to economize on space requirements. The type of container to be used should be planned or considered when the handling mechanism is selected.

SURFACE HANDLING
by Colin K. Larsen
Blue Giant Equipment Co.

Lift Trucks and Palletized Loads

The basis of all efficient handling, storage, and movement of unitized goods is the cube concept. Building a cube enables a large quantity of unit goods to be handled and stored at one time. This provides greater efficiency by increasing the volume of goods movement for a given amount of work. Examples of cube-facilitating devices include pallets, skids, slip sheets, bins, drums, and crates.

The most widely applied cube device is the pallet. A **pallet** is a low platform, usually constructed of wood, incorporating openings for the forks of a lift truck to enter. Such openings are designed to enable a lift truck to pick up and transport the pallet containing the cubed goods.

Lift truck is a loose term for a family of pallet handling/transporting machines. Such machines range from manually propelled low-lift devices (Fig. 10.4.1) to internal combustion and electric powered ride-on high-lift devices (Fig. 10.4.2). While some machines are substitutes in terms of function, each serves its own niche when viewed in terms of individual budgets and applications.

Pallet trucks are low-lift machines designed to raise loaded pallets sufficiently off the ground to enable the truck to transport the pallet horizontally. Pallet trucks are available as manually operated and propelled models that incorporate a hydraulic pump and handle assembly (Fig. 10.4.1). This pump and handle assembly enables the operator to raise the truck forks, and push/pull the load. Standard manual pallet trucks are available in lifting capacities from 4,500 to 5,500 lb (2,045 to 2,500 kg), with customer manufactured models to 8,000 lb (3,636 kg). While available in a variety of fork sizes, by far the most common is 27 in wide × 48 in long (686 mm × 1,219 mm). This size accommodates the most common pallet sizes of 40 in × 48 in (1,016 mm ×

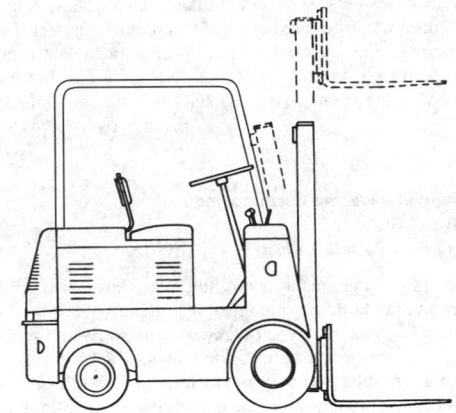

Fig. 10.4.2 Lift truck powered by an internal-combustion engine.

1,219 mm) and 48 in × 48 in (1,219 mm × 1,219 mm). Pallet trucks are also available in motorized versions, equipped with dc electric motors to electrically raise and transport. The power supply for these trucks is an on-board lead-acid traction battery that is rechargeable when the truck is not in use. Control of these trucks is through a set of lift, lower, speed, and direction controls fitted into the steering handle assembly. Powered pallet trucks are available in walk and ride models. Capacities range from 4,000 to 10,000 lb (1,818 to 4,545 kg), with forks up to 96 in (2,438 mm) long. The longer fork models are designed to allow the truck to transport two pallets, lined up end to end.

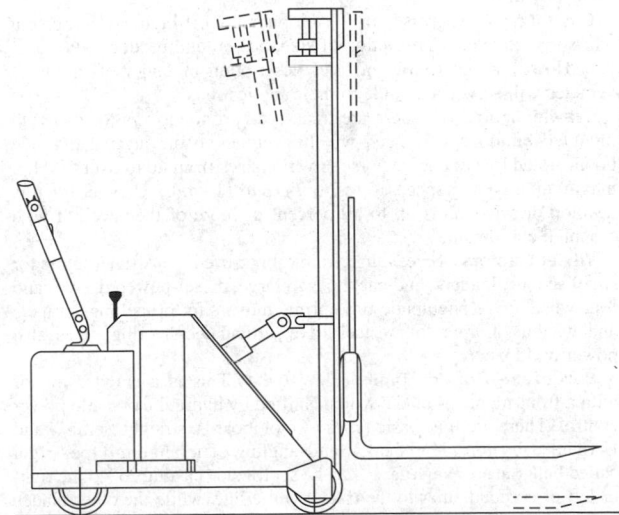

Fig. 10.4.3 Counterbalanced electric-battery-powered pallet truck. (*Blue Giant.*)

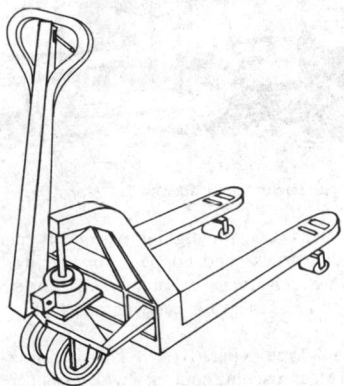

Fig. 10.4.1 Manually operated and propelled pallet truck.

Stackers, as the name implies, are high lift machines designed to raise and stack loaded pallets in addition to providing horizontal transportation. Stackers are separated into two classes: straddle and counterbalanced. **Straddle stackers** are equipped with legs which straddle the pallet and provide load and truck stability. The use of the straddle leg system results in a very compact chassis which requires minimal aisle space for turning. This design, however, does have its trade-offs insomuch as the straddles limit the truck's usage to smooth level floors. The limited leg underclearance inherent in these machines prohibits their use on dock levelers for loading/unloading transport trucks. Straddle stackers are available from 1,000 to 4,000 lb (455 to 1,818 kg) capacity with lift heights to 16 ft (4,877 mm). **Counterbalanced stackers** utilize a counterweight system in lieu of straddle legs for load and vehicle stability (Fig. 10.4.3). The absence of straddle legs results in a chassis with increased underclearance which can be used on ramps, including dock levelers. The counterbalanced chassis, however, is longer than its straddle counterpart, and this requires greater aisle space for maneuvering. For materials handling operations that require one machine to perform a multitude of tasks, and are flexible in floor layout of storage areas, the counterbalanced stacker is the recommended machine.

Off-Highway Vehicles and Earthmoving Equipment

by Darrold E. Roen, John Deere and Co.

The movement of large quantities of bulk materials, earth, gravel, and broken rock in road building, mining, construction, quarrying, and land clearing may be handled by **off-highway vehicles.** Such vehicles are mounted on large pneumatic tires or on crawler tracks if heavy pulling and pushing are required on poor or steep terrain. Width and weight of the rubber-tired equipment often exceed highway legal limits, and use of grouser tracks on highways is prohibited. A wide range of working tool attachments, which can be added (and removed) without modification to the basic machine, are available to enhance the efficiency and versatility of the equipment.

Proper selection of size and type of equipment depends on the amount, kind, and density of the material to be moved in a specified time and on the distances, direction, and steepness of grades, footing for traction, and altitude above sea level. Time cycles and pay loads for production per hour can then be estimated from manufacturers' performance data and job experience. This production per hour, together with the corresponding owning, operating, and labor costs per hour, enables selection by favorable cost per cubic yard, ton, or other pay unit.

Current rapid progress in the development of off-highway equipment will soon make any description of size, power, and productivity obsolete. However, the following brief description of major off-highway vehicles will serve as a guide to their applications.

Crawler Tractors These are track-type prime movers for use with mounted bulldozers, rippers, winches, cranes, cable layers, and side booms rated by net engine horsepower in sizes from 40 to over 500 hp; maximum traveling speeds, 5 to 7 mi/h (8 to 11 km/h). Crawler tractors develop drawbar pulls up to 90 percent or more of their weight with mounted equipment.

Wheel Tractors Sizes range from rubber-tired industrial tractors for small scoops, loaders, and backhoes to large, diesel-powered, two- and four-wheel drive pneumatic-tired prime movers for propelling scrapers and wagons. Large, four-wheel-drive, articulated-steering types also power bulldozers.

Bulldozer—Crawler Type (Fig. 10.4.4) This is a crawler tractor with a front-mounted blade, which is lifted by hydraulic or cable power control. There are four basic types of moldboards: straight, semi-U and U (named by top-view shape), and angling. The angling type, often called **bullgrader** or **angledozer,** can be set for side casting 25° to the right or left of perpendicular to the tractor centerline, while the other blades can be tipped forward or back through about 10° for different digging

conditions. All blades can be tilted for ditching, with hydraulic-power tilt available for all blades.

APPLICATION. This is the best machine for pioneering access roads, for boulder and tree removal, and for short-haul earthmoving in rough terrain. It push-loads self-propelled scrapers and is often used with a

Fig. 10.4.4 Crawler tractor with dozer blade. *(John Deere.)*

rear-mounted ripper to loosen firm or hard materials, including rock, for scraper loading. U blades drift 15 to 20 percent more loose material than straight blades but have poor digging ability. Angling blades expedite sidehill benching and backfilling of trenches. Loose-material capacity of straight blades varies approximately as the blade height squared, multiplied by length. Average capacity of digging blades is about 1 yd³ loose measure per 30 *net* hp rating of the crawler tractor. Payload is 60 to 90 percent of loose measure, depending on material swell variations.

Bulldozer—Wheel Type This is a four-wheel-drive, rubber-tired tractor, generally of the hydraulic articulated-steering type, with front-mounted blade that can be hydraulically raised, lowered, tipped, and tilted. Its operating weights range to 150,000 lb, with up to 700 hp, and its traveling speeds range from stall to about 20 mi/h for pushing and mobility.

APPLICATION. It is excellent for push-loading self-propelled scrapers, for grading the cut, spreading and compacting the fill, and for drifting loose materials on firm or sandy ground for distances up to 500 ft. Useful tractive effort on firm earth surfaces is limited to about 60 percent of weight, as compared with 90 percent for crawler dozers.

Loader—Crawler Type (Fig. 10.4.5) This is a track-type prime mover with front-mounted bucket that can be raised, dumped, lowered, and tipped by power control. Capacities range from 0.7 to 5.0 yd³ (0.5 to 3.8 m³), SAE rated. It is also available with grapples for pulpwood, logs, and lumber.

Fig. 10.4.5 Crawler tractor with loader bucket. *(John Deere.)*

APPLICATION. It is used for digging basements, pools, ponds, and ditches; for loading trucks and hoppers; for placing, spreading, and compacting earth over garbage in sanitary fills; for stripping sod; for removing steel-mill slag; and for carrying and loading pulpwood and logs.

Loader—Wheel Type (Fig. 10.4.6) This is a four-wheel, rubber-tired, articulated-steer machine equipped with a front-mounted, hydraulic-powered bucket and loader linkage that loads material into the bucket through forward motion of the machine and which lifts, transports, and discharges the material. The machine is commonly referred to as a *four-wheel-drive loader tractor*. Bucket sizes range from ½ yd³ (0.4 m³) to more than 20 yd³ (15 m³), SAE rated capacity. The addition

Fig. 10.4.6   Four-wheel-drive loader. (*John Deere.*)

of a quick coupler to the loader linkage permits convenient interchange of buckets and other working tool attachments, adding versatility to the loader. Rigid-frame machines with variations and combinations of front/rear/skid steer, front/rear drive, and front/rear engine are also used in various applications.

APPLICATION.   Four-wheel-drive loaders are used primarily in construction, aggregate, and utility industries. Typical operations include truck loading, filling hoppers, trenching and backfilling, land clearing, and snow removal.

Backhoe Loader   (Fig. 10.4.7) This is a self-propelled, highly mobile machine with a main frame to support and accommodate both the rear-mounted backhoe and front-mounted loader. The machine was designed with the intention that the backhoe will normally remain in place when the machine is being used as a loader and vice versa. The backhoe digs, lifts, swings, and discharges material while the machine is stationary. When used in the loader mode, the machine loads material into the bucket through forward motion of the machine and lifts, transports, and discharges the material. Backhoe loaders are categorized according to digging depth of the backhoe. Backhoe loader types include variations of front/rear/articulated and all-wheel steer and rear/four-wheel drive.

Fig. 10.4.7   Backhoe loader. (*John Deere.*)

APPLICATION.   Backhoe loaders are used primarily for trenching and backfilling operations in the construction and utility industries. Quick couplers for the loader and backhoe are available which quickly interchange the working tool attachments, thus expanding machine capabilities. Backhoe loader mobility allows the unit to be driven to nearby job sites, thus minimizing the need to load and haul the machine.

Scrapers   (Fig. 10.4.8) This is a self-propelled machine, having a cutting edge positioned between the front and rear axles which loads, transports, discharges, and spreads material. Tractor scrapers include

open-bowl and self-loading types, with multiple steer and drive axle variations. Scraper rear wheels may also be driven by a separate rear-mounted engine which minimizes the need for a push tractor. Scraper ratings are provided in cubic yard struck/heaped capacities. Payload capacities depend on loadability and swell of materials but approximate the struck capacity. Crawler tractor-drawn, four-wheel rubber-tired scrapers have traditionally been used in a similar manner—normally in situations with shorter haul distances or under tractive and terrain conditions that are unsuitable for faster, self-propelled scrapers.

Fig. 10.4.8   Two-axle articulated self-propelled elevating scraper. (*John Deere.*)

APPLICATION.   Scrapers are used for high-speed earth moving, primarily in road building and other construction work where there is a need to move larger volumes of material relatively short distances. The convenient control of the cutting edge height allows for accurate control of the grade in either a cut or fill mode. The loaded weight of the scraper can contribute to compaction of fill material. All-wheel-drive units can also load each other through a push-pull type of attachment. Two-axle, four-wheel types have the best maneuverability; however, the three-axle type is sometimes preferred for operator comfort on longer, higher-speed hauls.

Motor Grader   (Fig. 10.4.9) This is a six-wheel, articulated-frame self-propelled machine characterized by a long wheelbase and mid-mounted blade. The blade can be hydraulically positioned by rotation about a vertical axis—pitching fore/aft, shifting laterally, and independently raising each end—in the work process of cutting, moving, and spreading material, to rough- or finish-grade requirements. Motor graders range in size to 60,000 lb (27,000 kg) and 275 hp (205 kW) with typical transport speeds in the 25-mi/h (40-km/h) range. Rigid-frame machines with various combinations of four/six wheels, two/four-wheel drive, front/rear-wheel steer are used as dictated by the operating requirements.

Fig. 10.4.9   Six-wheel articulated-frame motor grader. (*John Deere.*)

APPLICATION.   Motor graders are the machine of choice for building paved and unpaved roads. The long wheelbase in conjunction with the midmounted blade and precise hydraulic controls allows the unit to finish-grade road beds within 0.25 in (6 mm) prior to paving. The weight, power, and blade maneuverability enable the unit to perform all the necessary work, including creating the initial road shape, cutting the ditches, finishing the bank slopes, and spreading the gravel. The motor grader is also a cost-effective and vital part of any road maintenance fleet.

Table 10.4.1 Typical Monorail Trolley Dimensions

Capacity, short tons	I-beam range (depth), in	Wheel-tread diam, in	Net weight, lb	*B*,* in	*C*, in	*H*, in	*M*, in	*N*, in	Min beam radius,* in
½	5–10	3½	32	3¼	4⅛	9⅞	2⅜	6¼	21
1	5–10	3½	32	3¼	4⅛	9⅞	2⅜	6¼	21
1½	6–10	4	52	3⅞	4⅝	11⅜	$2\frac{15}{16}$	$7\frac{7}{16}$	30
2	6–10	4	52	3⅞	4⅝	11⅜	$2\frac{15}{16}$	$7\frac{7}{16}$	30
3	8–15	5	88	$4\frac{7}{16}$	5⅜	13½	$2\frac{13}{16}$	$7\frac{15}{16}$	42
4	8–15	5	88	$4\frac{7}{16}$	5⅜	13½	$2\frac{13}{16}$	$7\frac{15}{16}$	42
5	10–18	6	137	$5\frac{3}{16}$	$6\frac{3}{16}$	15⅜	$3\frac{5}{16}$	10⅛	48
6	10–18	6	137	$5\frac{3}{16}$	$6\frac{3}{16}$	15⅜	$3\frac{5}{16}$	10⅛	48
8	12–24	8	279	5½	$7\frac{7}{16}$	21⅜	$4\frac{3}{16}$	13¾	60
10	12–24	8	279	5½	$7\frac{7}{16}$	21⅜	$4\frac{3}{16}$	13¾	60

Metric values, multiply tons by 907 for kg, inches by 25.4 for mm, and lb by 0.45 for kg.
* These dimensions are given for minimum beam.
SOURCE: CM Hoist Division, Columbus McKinnon.

Excavator (Fig. 10.4.10) This is a mobile machine which is propelled by either crawler track or rubber-tired undercarriage, with the unique feature being an upper structure that is capable of continuous rotation and a wide working range. The unit digs, elevates, swings, and dumps material by action of the boom, arm, or telescoping boom and bucket. Excavators include the hoe type (digging tool cuts toward the machine and generally downward) and the shovel type (digging tool cuts away from the machine and generally upward). Weight of the machines ranges from 17,600 lb (8 tonnes) to 1,378,000 lb (626 tonnes) with power ratings from 65 to 3644 hp (48.5 to 2719 kW).

Fig. 10.4.10 Excavator with tracked undercarriage. *(John Deere.)*

APPLICATION. The typical attachment for the unit is the bucket, which is used for trenching in the placement of pipe and other underground utilities, digging basements or water retention ponds, maintaining slopes, and mass excavation. Other specialized attachments include hydraulic hammers and compactors, thumbs, clamshells, grapples, and long-reach front ends which expand the capabilities of the excavator.

Dumpers A dumper is a self-propelled machine, having an open cargo body, which is designed to transport and dump or spread material. Loading is accomplished by means external to the dumper. Types are generally categorized into rear, side, or bottom dump with multiple variations of front/articulated steer, two to five axles, and front/rear/center/multiaxle drive.

APPLICATION. Dumpers are used for hauling and dumping blasted rock, ore, earth, sand, gravel, coal, and other hard and abrasive materials in road and dam construction and in quarries and mines. The units are capable of 30 to 40 mi/h (50 to 65 km/h) when loaded (depending on terrain/slopes), which makes the dumper an excellent choice for longer haul distances.

Owning and Operating Costs These include depreciation; interest, insurance, taxes; parts, labor, repairs, and tires; fuel, lubricant, filters, hydraulic-system oil, and other operating supplies. This is reduced to cost per hour over a service life of 4 to 6 years of 2,000 h each—average 5 years, 10,000 h. Owning and operating costs of diesel-powered bulldozers and scrapers, excluding operator's wages, average 3 to 4 times the delivered price in 10,000 h.

ABOVE-SURFACE HANDLING

Monorails

Materials can be carried on light, rigid trackage, as described for overhead conveyors (see below). Trolleys are supported by structural I beams, H beams, or I-beam-like rails with special flat flanges to improve rolling characteristics of the wheels. Size of wheels and smoothness of tread are important in reducing rolling resistance. Figure 10.4.11 shows a typical rigid trolley for traversing short-radius track curves. Typical dimensions for both types are given in Table 10.4.1. These trolleys may be plain, with geared handwheel and hand chain, or motor-drive. For very low headroom, the trolley can be built into the hoist; this is known as a **trolley hoist**.

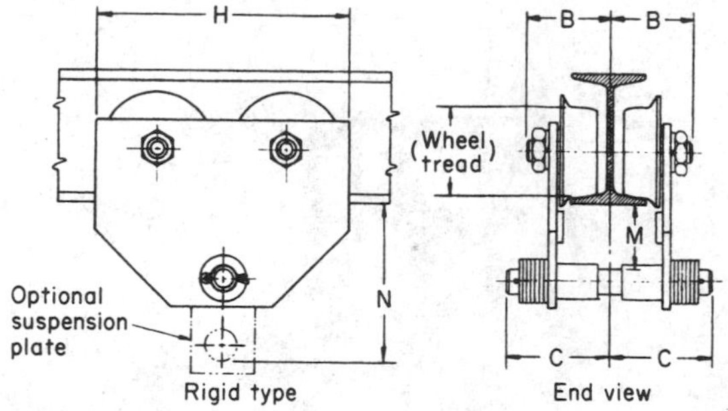

Fig. 10.4.11 Monorail trolley. *(CM Hoist Division, Columbus McKinnon.)*

Overhead Traveling Cranes

by Alger Anderson, Lift-Tech International, Inc.

An **overhead traveling crane** is a vehicle for lifting, transporting, and lowering loads. It consists of a bridge supporting a hoisting unit and is equipped with wheels for operating on an elevated runway or track. The hoisting unit may be fixed relative to the bridge but is usually supported on wheels, permitting it to traverse the length of the bridge.

The motions of the crane—hoisting, trolley traversing, and bridging—may be powered by hand, electricity, air, hydraulics, or a combination of these. Hand-powered cranes are generally built in capacities under 50 tons (45 tonnes) and are used for infrequent service where slow speeds are acceptable. Pneumatic cranes are used where electricity would be hazardous or where advantage can be taken of existing air supply. Electric cranes are the most common overhead type and can be built to capacities of 500 tons (454 tonnes) or more and to spans of 150 ft (46 m) and over.

Single-Girder Cranes   (Fig. 10.4.12) In its simplest form, this consists of an I beam *a* supported by four wheels *b*. The trolley *c* traveling on the lower flanges carries the chain hoist, forming the lifting unit. The crane is moved by hand chain *d* turning sprocket wheel *e*, which is keyed to shaft *f*. The pinions on shaft *f* mesh with gears *g*, keyed to the axles of two wheels. An underslung construction may also be used, with pairs of wheels at each corner which ride on the lower flange of I-beam rails. Single-girder cranes may be hand-powered by pendant hand chains or electric-powered as controlled by a pendant push-button station.

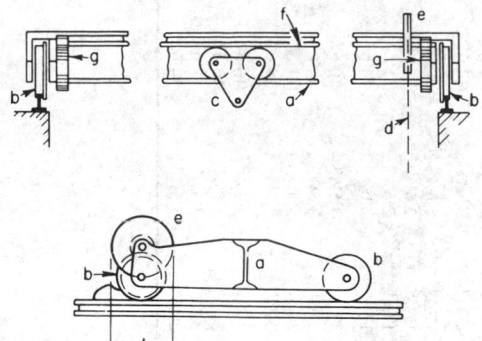

Fig. 10.4.12   Hand-powered crane.

Electric Traveling Crane, Double-Girder Type   (Fig. 10.4.13) This consists of two bridge girders *a*, on the top of which are rails on which travels the self-contained hoisting unit *b*, called the trolley. The girders are supported at the ends by trucks with two or four wheels, according to the size of the crane. The crane is moved along the track by motor *c*, through shaft *d* and gearing to the truck wheels. Suspended from the girders on one side is the operator's cab *e*, containing the controller, or master switches, master hydraulic brake cylinder, warning device, etc. The bridge girders for small cranes are of the I-beam type, but on the

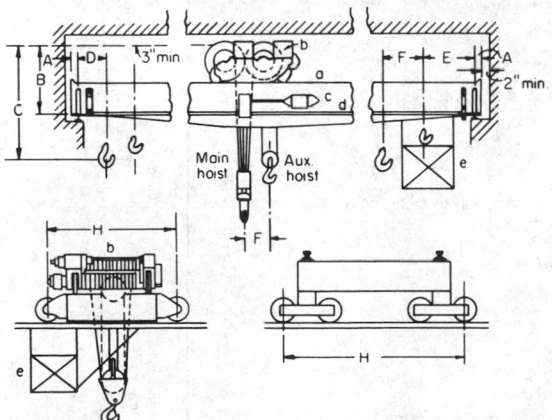

Fig. 10.4.13   Electric traveling crane.

longer spans, box girders are used to give torsional and lateral stiffness. The girders are rigidly attached to the truck end framing, which carries the double-flanged wheels for supporting the bridge. The end frames project over the rails so that in case of a broken wheel or axle, the frame

will rest on the rail, preventing the crane from dropping. One wheel axle on each truck is fitted with gears for driving the crane or is coupled directly to the shaft which transmits power from the gear reducer. On a cab-operated bridge, a brake, usually hydraulic, is applied to the motor shaft to stop the crane. Floor-operated cranes generally utilize spring engaged, electrically released brakes.

The trolley consists of a frame which carries the hoisting machinery and is supported on wheels for movement along the bridge rails. The wheels are coupled to the trolley traverse motor through suitable gear reduction. Trolleys are frequently equipped with a second set of hoisting machinery to provide dual lifting means or an auxiliary of smaller capacity. The hoisting machinery consists of motor, motor brake, load brake, gear reduction, and rope drum. Wire rope winding in helical grooves on the drum is reeved over sheaves in the upper block and lower hook block for additional mechanical advantage. Limit switches are provided to stop the motors when limits of travel are reached. Current is brought to the crane by sliding or rolling collectors in contact with conductors attached to or parallel with the runway and preferably located at the cab end of the crane. Current from the runway conductors and cab is carried to the trolley in a like manner from conductors mounted parallel to the bridge girder. Festooned multiconductor cables are also used to supply current to crane or trolley.

Electric cranes are built for either alternating or direct current, with the former predominating. The motors for both kinds of current are designed particularly for crane service. Direct-current motors are usually series-wound, and ac motors are generally of the wound-rotor or two-speed squirrel-cage type. The usual ac voltages are 230 and 460, the most common being 460. Variable-frequency drives (VFDs) are used with ac squirrel cage motors to provide precise control of the load over a wide range of speeds. Cranes and hoists equipped with VFDs are capable of delicate positioning and swift acceleration of loads to the maximum speed. Standard, inexpensive squirrel cage motors may be used with VFDs to provide high-performance control of all crane motions. The **capacities** and other dimensions for standard electric cranes are given in Table 10.4.2.

Gantry Cranes

Gantry cranes are modifications of traveling cranes and are generally used outdoors where it is not convenient to erect on overhead runway. The bridge (Fig. 10.4.14) is carried at the ends by the legs *a*, supported by trucks with wheels so that the crane can travel. As with the traveling crane, the bridge carries a hoisting unit; a cover to protect the machinery from the weather is often used. The crane is driven by motor *b* through a gear reduction to shaft *c*, which drives the vertical shafts *d* through bevel gears. Bevel- and spur-gear reductions connect the axles of the wheels with shafts *d*. Many gantry cranes are built without the cross shaft, employing separate motors, brakes, and gear reducers at each end of the crane. Gantry cranes are made in the same sizes as standard traveling cranes.

Special-Purpose Overhead Traveling Cranes

A wide variety can be built to meet special conditions or handling requirements; examples are stacker cranes to move material into and out of racks, wall cranes using a runway on only one side of a building, circular running or pivoting cranes, and semi-gantries. Load-weighing arrangements can be incorporated, as well as special load-handling devices such as lifting beams, grapples, buckets, forks, and vacuum grips.

Rotary Cranes and Derricks

Rotary cranes are used for lifting material and moving it to points covered by a boom pivoted to a fixed or movable structure. **Derricks** are used outdoors (e.g., in quarries and for construction work), being built so that they can be easily moved. Pillar cranes are always fixed and are used for light, infrequent service. Jib cranes are used in manufacturing plants. Locomotive cranes mounted on car wheels are used to handle loads by hook or bulk material by means of tubs, grab buckets, or magnets. Wrecking cranes are of the same general type as locomotive cranes and are used for handling heavy loads on railroads.

Table 10.4.2 Dimensions, Loads, and Speeds for Industrial-Type Cranes[a,e,f]

Capacity main hoist, tons, 2,000 lb	Span, ft	Std. lift, main hoist, ft[c]	Std. hoist speed, ft/min[b]	Dimensions, refer to Fig. 10.4.13							Max load per wheel, lb[d]	Runway rail, lb/yd	X, in	No. of bridge wheels
				A	*B min*	*C*	*D*	*E*[c]	*F*	*H*				
6	40	36	31	5⅞″	3′9″	3′¼″	29⅞″	31⅞″		8′0″	9,470	25	12	4
	60	53	31	5⅞″	3′9″	3′¼″	29⅞″	31⅞″		9′6″	12,410	25	12	4
	80	86	31	5⅞″	3′9″	3′¼″	29⅞″	31⅞″		11′6″	15,440	25	12	4
	100	118	31	5⅞″	3′9″	3′¼″	29⅞″	31⅞″		14′6″	19,590	40	12	4
10	40	36	23	5⅞″	3′9″	3′¼″	29⅞″	31⅞″		8′0″	15,280	25	12	4
	60	53	23	5⅞″	3′9″	3′¼″	29⅞″	31⅞″		9′6″	18,080	40	12	4
	80	86	23	5⅞″	3′9″	3′¼″	29⅞″	31⅞″		11′6″	20,540	40	12	4
	100	118	23	5⅞″	3′9″	3′¼″	29⅞″	31⅞″		14′6″	24,660	60	12	4
16	40	28	19	5⅞″	3′9″	4′0″	28⅛″	38½″		8′0″	20,950	60	12	4
	60	40	19	5⅞″	3′9″	4′0″	28⅛″	38½″		9′6″	23,680	60	12	4
	80	62	19	5⅞″	3′9″	4′0″	28⅛″	38½″		11′6″	26,140	60	12	4
	100	84	19	5⅞″	3′10″	4′0″	28⅛″	38½″		14′6″	30,560	80	12	4
20	40	22	14	5⅞″	3′9″	4′0″	40⅝″	28⅝″	2′2″	8′0″	26,350	60	18	4
5 ton aux	60	30	14	5⅞″	3′9″	4′0″	40⅝″	28⅝″	2′2″	9′6″	30,000	80	18	4
	80	46	14	6⅜″	3′11″	4′0″	40⅛″	28⅛″	2′2″	11′6″	33,370	60	18	4
	100	63	14	6⅜″	3′11″	4′0″	40⅛″	28⅛″	2′2″	14′6″	38,070	80	18	4
30	40	22	9	6⅜″	3′11″	4′6″	41⅞″	28⅞″	1′11½″	9′6″	37,300	80	24	4
5 ton aux	60	22	9	6⅜″	3′11″	4′6″	41⅞″	28⅞″	1′11½″	9′6″	40,050	135	24	4
	80	34	9	7⅞″	3′11″	4′6″	40⅜″	27⅜″	1′11½″	11′6″	44,680	80	24	4
	100	48	9	7⅞″	3′11″	4′6″	40⅜″	27⅜″	1′11½″	14′6″	51,000	135	24	4
40	40	25	7	7⅞″	5′5″	6′6″	46⅞″	17⅝″	2′6¾″	9′6″	48,630	135	24	4
5 ton aux	60	25	7	7⅞″	5′5″	6′6″	46⅞″	17⅝″	2′6¾″	9′6″	52,860	135	24	4
	80	40	7	7⅞″	5′5″	6′6″	46⅞″	17⅝″	2′6¾″	11′6″	59,040	135	24	4
	100	54	7	7⅞″	5′5″	6′6″	46⅞″	17⅝″	2′6¾″	14′6″	65,200	135	24	4
50	40	25	5	7⅞″	6′10″	7′6″	56⅜″	29¼″	3′5¼″	10′6″	55,810	135	24	4
10 ton aux	60	25	5	7⅞″	6′7″	7′6″	56⅜″	29¼″	3′5¼″	10′6″	62,050	135	24	4
	80	32	5	9″	6′7″	7′6″	55¼″	28⅛″	3′5¼″	11′6″	68,790	135	24	4
	100	46	5	9″	6′7″	7′6″	55¼″	28⅛″	3′5¼″	14′6″	76,160	176	24	4
60	40	25	4	9″	6′8″	7′6″	55¼″	28⅛″	3′5¼″	10′6″	65,970	135	24	4
10 ton aux	60	25	4	9″	6′8″	7′6″	55¼″	28⅛″	3′5¼″	10′6″	72,330	135	24	4
	80	32	4	9″	6′8″	7′6″	55¼″	28⅛″	3′5¼″	11′6″	79,230	175	24	4
	100	46	4	9″	6′10″	7′6″	55¼″	28⅛″	3′5¼″	14′6″	89,250	175	24	4

[a] Lift-Tech International, Inc.
[b] Trolley speeds 50–70 ft/min and bridge speed 100–150 ft/min.
[c] For each 10 ft extra lift, increase *H* by *X*.
[d] Direct loads, no impact.
[e] These figures should be used for preliminary work only as the data varies among manufacturers.
[f] Multiply ft by 0.30 for m, ft/min by 0.0051 for m/s, in by 25.4 for mm, lb by 0.45 for kg, lb/yd by 0.50 for kg/m.

Derricks are made with either wood or steel masts and booms, are of the guyed or stiff-leg type, and are either hand-slewed or power-swung with a bull wheel. Figure 10.4.15 shows a guyed wooden derrick of the bull-wheel type. The mast *a* is carried at the foot by pivot *k* and at the

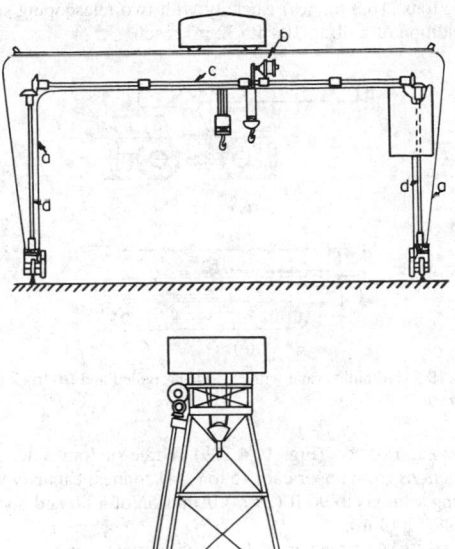

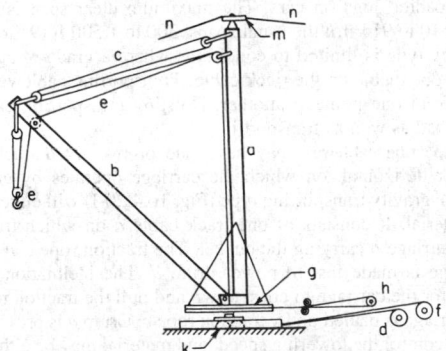

Fig. 10.4.14   Gantry crane.

top by pivot *m*, held by rope guys *n*. The boom *b* is pivoted at the lower end of the mast. The rope *c*, passing over sheaves at the top of the mast and at the end of the boom and through the pivot *k*, is made fast to drum *d* and varies the angle of the boom. The hoisting rope *e*, from which the load is suspended, is made fast to drum *f*. The bull wheel *g* is attached to the mast and swings the derrick by a rope made fast to the bull wheel

Fig. 10.4.15   Guyed wooden derrick.

and passing around the reversible drum *h*. In derricks of the self-slewing type, the engine is mounted on a platform attached to the mast and the derrick is swung by a pinion meshing with a gear attached to the foundation. Either the bull-wheel or the self-slewing type may be made of steel or wood construction and may be of the guyed or stiff-leg type. Figure 10.4.16 shows a **column jib crane,** consisting of pivoted post *a* and carrying boom *b*, on which travels either an electric or a hand hoist *c*. The post *a* is attached to building column *d* so that it can swing through approximately 270°. Cranes of this type are rapidly being replaced by such other methods of handling materials as the mobile lift truck or the

automotive-type crane. Column jib cranes are built with radii up to 20 ft (6 m) and for loads up to 5 tons (4.5 tonnes). Yard jib cranes are generally designed to meet special conditions.

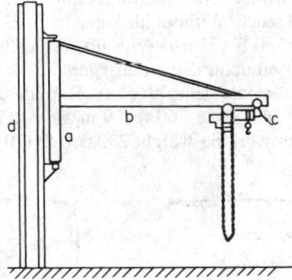

Fig. 10.4.16   Column jib crane.

Locomotive Cranes

The locomotive crane (Fig. 10.4.17) is self-propelled and provided with trucks, brakes, automatic couplers, fittings, and clearances which will permit it to be used or hauled in a train; it can function as a complete unit on any railroad. Locomotive cranes are of the rotating-deck type, consisting of a hinged boom attached to the machinery deck, which is turntable-mounted and operated either by mechanical rotating clutches

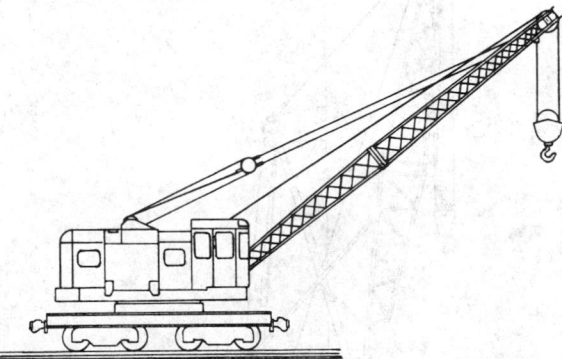

Fig. 10.4.17   Locomotive crane. *(American Hoist and Derrick.)*

or by a separate electric or hydraulic swing motor. The boom is operated by powered topping line, with a direct-geared hoisting mechanism to raise and lower it. Power to operate the machinery is deck-mounted, and the machinery deck is completely housed. The crane may be powered by internal-combustion engine or electric motor. The combination of internal-combustion engine, generator, and electric motor makes up the power arrangement for the diesel-electric locomotive crane. Another power arrangement is made up of internal-combustion engines driving hydraulic pumps for hydraulically powered swing and travel mechanisms. The car body and machinery deck are ballasted, thereby adding stability to the crane when it is rotated under load. The basic boom is generally 50 ft (15 m) in length; however, booms range to 130 ft (40 m) in length. Locomotive cranes are so designed that power-shovel, pile-driver, hook, bucket, or magnet attachments can be installed and the crane used in such service. Locomotive cranes are used most extensively in railroad work, steel mills, and scrap yards. The cranes usually have sufficient propelling power not only for the crane itself, but also for switching service and hauling cars.

Truck Cranes

The advent of the truck crane has changed significantly the methods of lifting and placing heavy items such as concrete buckets, logs, pipe, and bridge or building members. Truck cranes can, without assistance, be rapidly equipped with accessory booms to reach to 260 ft (79 m) vertically—or 180 ft (55 m) vertically with 170 ft of horizontal reach.

Mechanical Models

TOWER CRANE. (Fig. 10.4.18a and b). Has vertical and horizontal members together with a boom and jib. Permits location close to building with horizontal reach. Without jib, capacity to 27 tons (24.5 tonnes). With jib, reach to 180 ft (55 m). With jib, vertically to 190 ft (58 m). Lifting capacity based upon using outriggers.

CONVENTIONAL CRANE. (Fig. 10.4.18c). With boom and with or without jib. Boom plus jib to 260 ft (79 m) and 125 tons (113 tonnes) capacity. Maximum working weight 230,000 lb (104,000 kg).

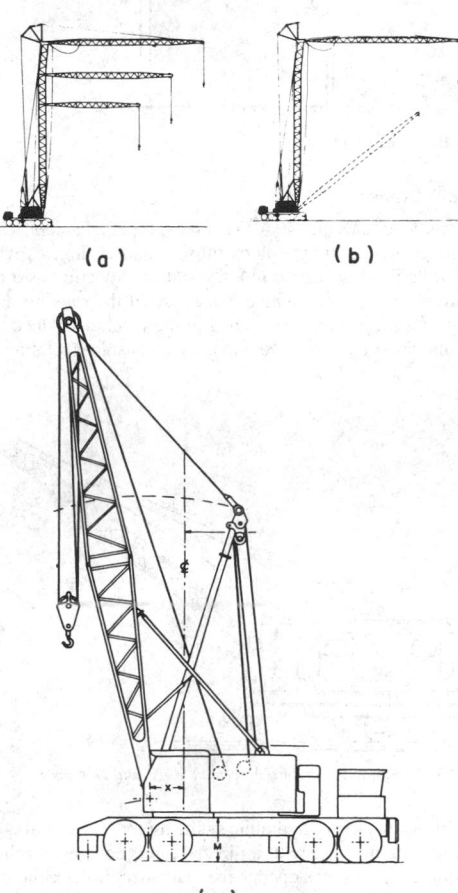

Fig. 10.4.18 Mechanical tower crane with vertical and horizontal extensions. *(a)* Tower working heights; *(b)* normal crane position; *(c)* crane with conventional boom. *(FMC Corp.)*

Table 10.4.3   Conventional Crane* Capacity and Limits of Operation

Boom				On outriggers	
Length, ft	Radius, ft	Angle, deg	Point height, ft	Rear lb	Side lb
30	11	81.0	33.5	250,000	250,000
30	25	46.3	24.5	123,300	123,300
60	16	80.7	63.1	145,100	145,100
60	50	39.7	41.0	52,200	50,100
90	20	81.3	92.9	131,600	131,600
90	80	31.5	49.8	30,500	26,000
180	40	79.2	180.6	46,700	46,700
180	170	21.6	68.9	8,300	6,200
230	50	79.0	229.6	21,000	21,000
230	220	18.9	77.5	2,900	1,800

* Multiply ft by 0.30 for m, lb by 0.45 for kg.
SOURCE: FMC Corporation.

GENERAL CHARACTERISTICS. (Table 10.4.3). 8 × 4 drive wheels with air brakes on all eight wheels. Power hydraulic steering. Removable-pin-connected counterweights front or rear removable for roadability.

Hydraulic Models

SELF-PROPELLED. (Fig. 10.4.19a). Short wheelbase, two axle, single cab; 18½ tons (16.8 tonnes) capacity with two telescoping sections to 64 ft. Addition of a jib to 104 ft (32 m) reach.

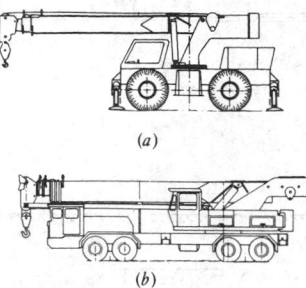

(a)

(b)

Fig. 10.4.19 Hydraulic crane with *(a)* self-propelled and *(b)* truck-type bases. *(FMC Corp.)*

HYDRAULIC TRUCK. (Fig. 10.4.19b). Three or four axle, two cabs, crane functions from upper cab; 45 tons (41 tonnes) capacity with three telescoping sections to 96 ft (29 m). Addition of a jib and boom extension to 142 ft (43 m).

GENERAL CHARACTERISTICS. Hydraulic extensions save setup time and provide job-to-job mobility. Equipment such as this comes under Commercial Standard specification CS90-58, ''Power Cranes and Shovels.'' Similar equipment, called **utility cranes**, without the highway-truck-type cabs, is also available.

Cableways

Cableways are **aerial hoisting and conveying devices** using suspended steel cable for their tracks, the loads being suspended from carriages and moved by gravity or power. The most common uses are transporting material from open pits and quarries to the surface; handling construction material in the building of dams, docks, and other structures where the construction of tracks across rivers or valleys would be uneconomical; and loading logs on cars. The maximum clear **span** is 2,000 to 3,000 ft (610 to 914 m); the usual spans, 300 to 1,500 ft (91 to 457 m). The gravity type is limited to conditions where a grade of at least 20 percent is obtainable on the track cable. Transporting cableways move the load from one point to another. Hoisting transporting cableways hoist the load as well as transport it.

A **transporting cableway** may have one or two fixed track cables, inclined or horizontal, on which the carriage operates by gravity or power. The gravity transporting type (Fig. 10.4.20-I) will either raise or lower material. It consists of one track cable a on which travels the wheeled carriage b carrying the bucket. The traction rope c attached to the carriage is made fast to power drum d. The inclination must be sufficient for the carriage to coast down and pull the traction rope after it. The carriage is hauled up by traction rope c. Drum d is provided with a brake to control the lowering speed, and material may be either raised or lowered. When it is not possible to obtain sufficient fall to operate the load by gravity, traction rope c (Fig. 10.4.20-II) is made endless so that carriage b is drawn in either direction by power drum d. Another type of inclined cableway, shown in Fig. 10.4.20-III, consists of two track cables aa, with an endless traction rope c, driven and controlled by drum d. When material is being lowered, the loaded bucket b raises the empty carriage bb, the speed being controlled by the brake on the drum. When material is being raised, the drum is driven by power, the descending empty carriage assisting the engine in raising the loaded carriage. This type has twice the capacity of that shown in Fig. 10.4.20-I.

A **hoisting and conveying cableway** (Fig. 10.4.20-IV) hoists the material at any point under the track cable and transports it to any other point. It consists of a track cable a and carriage b, moved by the endless traction rope c and by power drum d. The hoisting of the load is accomplished by power drum e through fall rope f, which raises the fall block g suspended from the carriage. The fall-rope carriers h support the fall rope; otherwise, the weight of this sagging rope would prevent fall block g from lowering when without load. Where it is possible to obtain a minimum inclination of 20° on the track cable, the traction-rope drum d is provided with a brake and is not power-driven. The carriage then

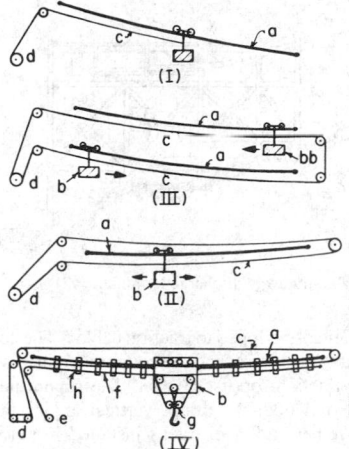

Fig. 10.4.20 Cableways.

descends by gravity, pulling the fall and traction ropes to the desired point. Brakes are applied to drum d, stopping the carrier. The fall block is lowered, loaded, and raised. If the load is to be carried up the incline, the carriage is hauled up by the fall rope. With this type, the friction of the carriage must be greater than that of the fall block or the load will run down. A novel development is the use of self-filling grab buckets operated from the carriages of cableways, which are lowered, automatically filled, hoisted, carried to dumping position, and discharged.

The **carriage speed** is 300 to 1,400 ft/min (1.5 to 7.1 m/s) [in special cases, up to 1,800 ft/min (9.1 m/s)]; average hoisting speed is 100 to 700 ft/min. The **average loads** for coal and earth are 1 to 5 tons (0.9 to 4.5 tonnes); for rock from quarries, 5 to 20 tons; for concrete, to 12 yd³ (9.1 m³) at 50 tons.

The **deflection of track cables** with their maximum gross loads at midspan is usually taken as 5½ to 6 percent of the span. Let S = span between supports, ft; L = one-half the span, ft; w = weight of rope, lb/ft; P = total concentrated load on rope, lb; h = deflection, ft; H = horizontal tension in rope, lb. Then $h = (wL + P)L/2H$; $P = (2h - wL^2)/L = (8hH - wS^2)/2S$.

For track cables, a **factor of safety** of at least 4 is advised, though this may be as low as 3 for locked smooth-coil strands that use outer wires of high ultimate strength. For traction and fall ropes, the sum of the load and bending stress should be well within the elastic limit of the rope or, for general hoisting, about two-thirds the elastic limit (which is taken as 65 percent of the breaking strength). Let P = load on the rope, lb; A = area of metal in rope section, in²; $E = 29{,}500{,}000$; R = radius of curvature of hoisting drum or sheave, whichever is smaller, in; d = diameter of individual wires in rope, in (for six-strand 19-wire rope, d = $\frac{1}{15}$ rope diam; for six-strand 7-wire rope, d = $\frac{1}{9}$ rope diam). Then load stress per in² = $T_1 = P/A$, and bending stress per in² = $T_b = Ed/2R$. The radius of curvature of saddles, sheaves, and driving drums is thus important to fatigue life of the cable. In determining the horsepower required, the load on the traction ropes or on the fall ropes will govern, depending upon the degree of inclination.

Cable Tramways

Cable tramways are aerial conveying devices using suspended cables, carriages, and buckets for transporting material over level or mountainous country or across rivers, valleys, or hills (they transport but do not hoist). They are used for handling small quantities over long distances,

and their construction cost is insignificant compared with the construction costs of railroads and bridges. Five types are in use:

Monocable, or Single-Rope, Saddle-Clip Tramway Operates on grades to 50 percent gravity grip or on higher grades with spring grip and has capacity of 250 tons/h (63 kg/s) in each direction and speeds to 500 ft/min. Single section lengths to 16 miles without intermediate stations or tension points. Can operate in multiple sections without transshipment to any desired length [monocables to 170 miles (274 km) over jungle terrain are practical]. Loads automatically leave the carrying moving rope and travel by overhead rail at angle stations and transfer points between sections with no detaching or attaching device required. Main rope constantly passes through stations for inspection and oiling. Cars are light and safe for passenger transportation.

Single-Rope Fixed-Clip Tramway Endless rope traveling at low speed, having buckets or carriers fixed to the rope at intervals. Rope passes around horizontal sheaves at each terminal and is provided with a driving gear and constant tension device.

Bicable, or Double-Rope, Tramway Standing track cable and a moving endless hauling or traction rope traveling up to 500 ft/min (2.5 m/s). Used on excessively steep grades. A detacher and attacher is required to open and close the car grip on the traction rope at stations. Track cable is usually in sections of 6,000 to 7,000 ft (1,830 to 2,130 m) and counterweighted because of friction of stiff cable over tower saddles.

Jigback, or Two-Bucket, Reversing Tramway Usually applied to hillside operations for mine workings so that on steep slopes loaded bucket will pull unloaded one up as loaded one descends under control of a brake. Loads to 10 tons (9 tonnes) are carried using a pair of track cables and an endless traction rope fixed to the buckets.

To-and-Fro, or Single-Bucket, Reversing Tramway A single track rope and a single traction rope operated on a winding or hoist drum. Suitable for light loads to 3 tons (2.7 tonnes) for intermittent working on a hillside, similar to a hoisting and conveying cableway without the hoisting feature.

The **monocable tramway** (Fig. 10.4.21) consists of an endless cable a passing over horizontal sheaves d and e at the ends and supported at intervals by towers. This cable is moved continuously, and it both supports and propels carriages b and c. The carriages either are attached permanently to the cable (as in the **single-rope fixed clip tramway**), in

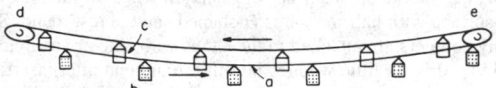

Fig. 10.4.21 Single-rope cable tramway.

which case they must be loaded and dumped while in motion, or are attached by friction grips so that they may be connected automatically or by hand at the loading and dumping points. When the tramway is lowering material from a higher to a lower level, the grade is frequently sufficient for the loaded buckets b to raise the empty buckets c, operating the tramway by gravity, the speed being controlled by a brake on grip wheel d.

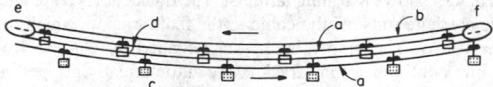

Fig. 10.4.22 Double-rope cable tramway.

The **bicable tramway** (Fig. 10.4.22) consists of two stationary track cables a, on which the wheeled carriages c and d travel. The endless traction rope b propels the carriages, being attached by friction grips. Figure 10.4.23 shows the arrangement of the **overhead type**. The track cable a is supported at intervals by towers b, which carry the saddles c in which the track cable rests. Each tower also carries the sheave d for supporting traction rope e. The self-dumping bucket f is suspended from carriage g. The grip h, which attaches the carriage to traction rope e, is controlled by lever *k*. In the **underhung type,** shown in Fig. 10.4.24, track cable *a* is carried above traction rope *e*. Saddle *c* on top of the tower supports the track cable, and sheave *d* supports the traction cable. The sheave is provided with a rope guard *m*. The lever *h*, with a roller on the end, automatically attaches and detaches the grip by coming in contact

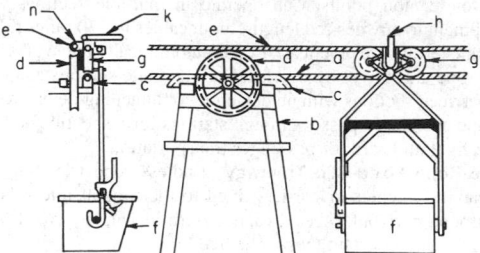

Fig. 10.4.23 Overhead-type double-rope cable tramway.

with guides at the loading and dumping points. The carriages move in only one direction on each track. On steep downgrades, special hydraulic speed controllers are used to fix the speed of the carriages.

The **track cables** are of the special locked-joint smooth-coil, or tramway, type. Nearly all wire rope is made of plow steel, with the old cast-steel type no longer being in use. The track cable is usually pro-

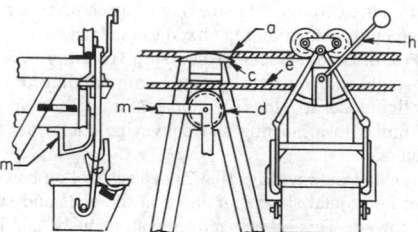

Fig. 10.4.24 Underhung-type double-rope cable tramway.

vided with a smooth outer surface of Z-shaped wires for full lock type or with a surface with half the wires *H*-shaped and the rest round. Special tramway couplers are attached in the shops with zinc or are attached in the field by driving little wedges into the strand end after inserting the end into the coupler. The second type of coupling is known as a **dry socket** and, though convenient for field installation, is not held in as high regard for developing full cable strength. The usual spans for level ground are 200 to 300 ft (61 to 91 m). One end of the track cable is anchored; the other end is counterweighted to one-quarter the breaking strength of the rope so that the horizontal tension is a known quantity. The **traction ropes** are made six-strand 7-wire or six-strand 19-wire, of cast or plow steel on hemp core. The maximum diameter is 1 in, which limits the length of the sections. The traction rope is endless and is driven by a drum at one end, passing over a counterweighted sheave at the other end.

Fig. 10.4.25 shows a **loading terminal.** The track cables *a* are anchored at *b*. The carriage runs off the cable to the fixed track *c*, which makes a 180° bend at *d*. The empty buckets are loaded by chute *e* from the loading bin, continue around track *c*, are automatically gripped to traction cable *f*, and pass on the track cable *a*. Traction cable *f* passes around

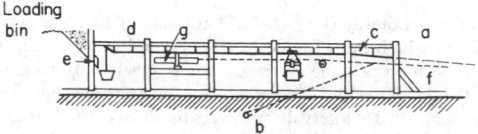

Fig. 10.4.25 Cable-tramway loading terminal.

and is driven by drum *g*. When the carriages are permanently attached to the traction cable, they are loaded by a moving hopper, which is automatically picked up by the carriage and carried with it a short distance while the bucket is being filled. Figure 10.4.26 shows a **discharge terminal.** The carriage rolls off from the track cable *a* to the fixed track *c*, being automatically ungripped. It is pushed around the 180° bend of track *c*, discharging into the bin underneath and continuing on track *c*

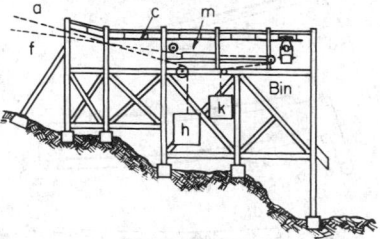

Fig. 10.4.26 Cable-tramway discharge terminal.

until it is automatically gripped to traction cable *f*. The counterweights *h* are attached to track cables *a*, and the counterweight *k* is attached to the carriage of the traction-rope sheave *m*. The **supporting towers** are A frames of steel or wood. At abrupt vertical angles the supports are placed close together and steel tracks installed in place of the cable. Spacing of towers will depend upon the capacity of the track cables and sheaves and upon the terrain as well as the bucket spacing.

Stress In Ropes (Roebling) The deflection for track cables of tramways is taken as one-fortieth to one-fiftieth of the span to reduce the grade at the towers. Let S = span between supports, ft; h = deflection, ft; P = gross weight of buckets and carriages, lb; Z = distance between buckets, ft; W_1 = total load per ft of rope, lb; H = horizontal tension of rope, lb. The formulas given for cableways then apply. When several buckets come in the span at the same time, special treatment is required for each span. For large capacities, the buckets are spaced close together, the load may be assumed to be uniformly distributed, and the live load per linear foot of span = P/Z. Then $H = W_1 S^2/8h$, where W_1 = (weight of rope per ft) + (P/Z). When the buckets are not spaced closely, the equilibrium curve can be plotted with known horizontal tension and vertical reactions at points of support.

For figuring the traction rope, t_0 = tension on counterweight rope, lb; t_1, t_2, t_3, t_4 = tensions, lb, at points shown in Fig. 10.4.27; n = number of carriers in motion; a = angle subtended between the line connecting the tower supports and the horizontal; W_1 = weight of each loaded carrier, lb; W_2 = weight of each empty carrier, lb; w = weight of

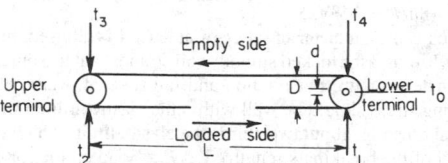

Fig. 10.4.27 Diagram showing traction rope tensions.

traction rope, lb/ft; L = length of tramway of each grade *a*, ft; D = diameter of end sheave, ft; d = diameter of shaft of sheave, ft; f_1 = 0.015 = coefficient of friction of shaft; f_2 = 0.025 = rolling friction of carriage wheels. Then, if the loads descend, the maximum stress on the loaded side of the traction rope is

$$t_2 = t_1 + \Sigma(Lw \sin a + \tfrac{1}{2}nW_1 \sin a) - f_2\Sigma(Lw \cos a + \tfrac{1}{2}nW_1 \cos a)$$

where $t_1 = \tfrac{1}{2}t_0[1 - f_1(d/D)]$. If the load ascends, there are two cases: (1) driving power located at the lower terminal, (2) driving power at the upper terminal. If the line has no reverse grades, it will operate by gravity at a 10 percent incline to 10 tons/h capacity and at a 4 percent grade for 80 tons/h. The preceding formula will determine whether it will operate by gravity.

The **power required** or developed by tramways is as follows: Let $V =$ velocity of traction rope, ft/min; $P =$ gross weight of loaded carriage, lb; $p =$ weight of empty carriage, lb; $N =$ number of carriages on one track cable; $P/50 =$ friction of loaded carriage; $p/50 =$ friction of empty carriage; $W =$ weight of moving parts, lb; $E =$ length of tramway divided by difference in levels between terminals, ft. Then, power required is

$$hp = \frac{NV}{33,000}\left(\frac{P-p}{E} \pm \frac{P+p}{50}\right) \pm 0.0000001\,WV$$

Where power is developed by tramways, use 80 instead of 50 under $P + p$.

BELOW-SURFACE HANDLING (EXCAVATION)

Power Shovels

Power shovels stand upon the bottom of the pit being dug and dig above this level. Small machines are used for road grading, basement excavation, clay mining, and trench digging; larger sizes are used in quarries, mines, and heavy construction; and the largest are used for removing overburden in opencut mining of coal and ore. The uses for these machines may be divided into two groups: (1) **loading,** where sturdy machines with comparatively short working ranges are used to excavate material and load it for transportation; (2) **stripping,** where a machine of very great dumping and digging reaches is used to both excavate the material and transport it to the dump or wastepile. The **full-revolving shovel,** which is the only type built at the present (having entirely displaced the old railroad shovel), is usually composed of a crawler-mounted truck frame with a center pintle and roller track upon which the revolving frame can rotate. The revolving frame carries the swing and hoisting machinery and supports, by means of a socket at the lower end and cable guys at the upper end, a boom carrying guides for the dipper handles and machinery to thrust the dipper into the material being dug.

Figure 10.4.28 shows a full-revolving shovel. The dipper a, of cast or plate steel, is provided with special wear-resisting teeth. It is pulled through the material by a steel cable b wrapped on a main drum c. Gasoline engines are used almost exclusively in the small sizes, and diesel, diesel-electric, or electric power units, with Ward Leonard control, in the large machines. The commonly used sizes are from ½ to 5 yd³ (0.4 to 3.8 m³) capacity, but special machines for coal-mine stripping are built with buckets holding up to 33 yd³ (25 m³) or even more. The very large machines are not suited for quarry or heavy rock work. Sizes up to 5 yd³ (3.8 m³) are known as **quarry machines.** Stripping shovels are crawler-mounted, with double-tread crawlers under each of the four corners and with power means for keeping the turntable level when traveling over uneven ground. The crowd motion consists of a

chain which, through the rack-and-pinion mechanism, forces the dipper into the material as the dipper is hoisted and withdraws it on its downward swing. On the larger sizes, a separate engine or motor is mounted on the boom for crowding. A separate engine working through a pinion and horizontal gear g swings the entire frame and machinery to bring the

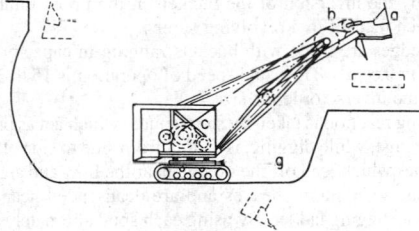

Fig. 10.4.28   Revolving power shovel.

dipper into position for dumping and to return it to a new digging position. Dumping is accomplished by releasing the hinged dipper bottom, which drops upon the pulling of a latch. With gasoline-engine or diesel-engine drives, there is only one prime mover, the power for all operations being taken off by means of clutches.

Practically all **power shovels** are readily **converted** for operation as **dragline excavators,** or **cranes.** The changes necessary are very simple in the case of the small machines; in the case of the larger machines, the installation of extra drums, shafts, and gears is required, in addition to the boom and bucket change.

The **telescoping boom,** hydraulically operated excavator shown in Fig. 10.4.29 is a versatile machine that can be quickly converted from the rotating-boom power shovel shown in Fig. 10.4.29a to one with a crane boom (Fig. 10.4.29b) or backhoe shovel boom (Fig. 10.4.29c). It can dig ditches reaching to 22 ft (6.7 m) horizontally and 9 ft 6 in (2.9 m) below grade; it can cut slopes, rip, scrape, dig to a depth of 12 ft 6 in (3.8 m), and load to a height of 11 ft 2 in (3.4 m). It is completely hydraulic in all powerized functions.

Dredges

Placer dredges are used for the mining of gold, platinum, and tin from placer deposits. The usual maximum digging depth of most existing dredges is 65 to 70 ft (20 to 21 m), but one dredge is digging to 125 ft (38 m). The dredge usually works with a bank above the water of 8 to 20 ft (2.4 to 6 m). Sometimes hydraulic jets are employed to break down these banks ahead of the dredge. The excavated material is deposited astern, and as the dredge advances, the pond in which the dredge floats is carried along with it.

The digging element consists of a chain of closely connected buckets passing over an idler tumbler and an upper or driving tumbler. The chain is mounted on a structural-steel ladder which carries a series of

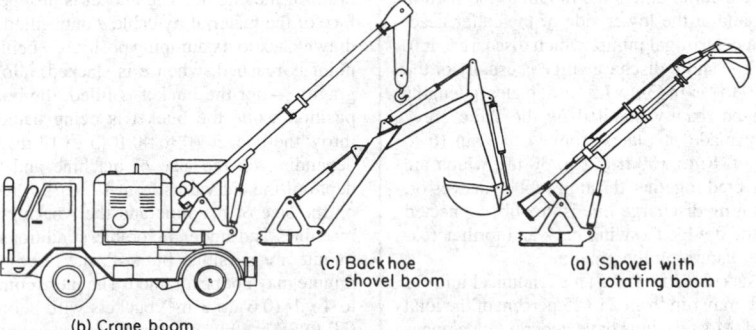

(c) Backhoe shovel boom

(a) Shovel with rotating boom

(b) Crane boom

Fig. 10.4.29   Hydraulically operated excavator. *(Link-Belt.)*

rollers to provide a bearing track for the chain of buckets. The upper tumbler is placed 10 to 40 ft (3 to 13 m) above the deck, depending upon the size of the dredge. Its fore-and-aft location is about 65 percent of the length of the ladder from the bow of the dredge. The ladder operates through a well in the hull, which extends from the bow practically to the upper-tumbler center. The material excavated by the buckets is dumped by the inversion of the buckets at the upper tumbler into a hopper, which feeds it to a revolving screen.

Placer dredges are made with buckets ranging in capacity from 2 to 20 ft³ (0.06 to 0.6 m³). The usual speed of operation is 15 to 30 buckets per min, in the inverse order of size.

The digging reaction is taken by stern spuds, which act as pivots upon which the dredge, while digging, is swung from side to side of the cut by swinging lines which lead off the dredge near the bow and are anchored ashore or pass over shore sheaves and are dead-ended near the lower tumbler on the digging ladder. By using each spud alternately as a pivot, the dredge is fed forward into the bank.

Elevator dredges, of which dredges are a special classification, are used principally for the excavation of sand and gravel beds from rivers, lakes, or ocean deposits. Since this type of dredge is not as a rule required to cut its own flotation, the bow corners of the hull may be made square and the digging ladder need not extend beyond the bow. The bucket chain may be of the close-connected placer-dredge type or of the open-connected type with one or more links between the buckets. The dredge is more of an elevator than a digging type, and for this reason the buckets may be flatter across the front and much lighter than the placer-dredge bucket.

The excavated material is usually fed to one or more revolving screens for classification and grading to the various commercial sizes of sand and gravel. Sometimes it is delivered to sumps or settling tanks in the hull, where the silt or mud is washed off by an overflow. Secondary elevators raise the material to a sufficient height to spout it by gravity or to load it by belt conveyors to the scows.

Hydraulic dredges are used most extensively in river and harbor work, where extremely heavy digging is not encountered and spoil areas are available within a reasonable radius of the dredge. The radius may vary from a few hundred feet to a mile or more, and with the aid of booster pumps in the pipeline, hydraulic dredges have pumped material through distances in excess of 2 mi (3.2 km), at the same time elevating it more than 100 ft (30 m). This type of dredge is also used for sand-and-gravel-plant operations and for land-reclamation work. Levees and dams can be built with hydraulic dredges. The usual maximum digging depth is about 50 ft (16 m). Hydraulic dredges are reclaiming copper stamp-mill tailings from a depth of 115 ft (35 m) below the water, and a depth of 165 ft (50 m) has been reached in a land-reclamation job.

The usual type of hydraulic dredge has a **digging ladder** suspended from the bow at an angle of 45° for the maximum digging depth. This ladder carries the suction pipe and cutter, with its driving machinery, and the swinging-line sheaves. The cutter head may have applied to it 25 to 1,000 hp (3.7 to 746 kW). The 20-in (0.5-m) dredge, which is the standard, general-purpose machine, has a cutter drive of about 300 hp. The usual operating speed of the cutter is 5 to 20 r/min.

The material excavated by the cutter enters the mouth of the suction pipe, which is located within and at the lower side of the cutter head. The material is sucked up by a centrifugal pump, which discharges it to the dump through a pipeline. The shore discharge pipe is usually of the telescopic type, made of No. 10 to ³/₁₀-in (3- to 7.5-mm) plates in lengths of 16 ft (5 m) so that it can be readily handled by the shore crew. Floating pipelines are usually made of plates from ¼ to ½ in (6 to 13 mm) thick and in lengths of 40 to 100 ft (12 to 30 m), which are floated on pontoons and connected together through rubber sleeves or, preferably, ball joints. The floating discharge line is flexibly connected to the hull in order to permit the dredge to swing back and forth across the cut while working without disturbing the pipeline.

Pump efficiency is usually sacrificed to make an economical unit for the handling of material, which may run from 2 to 25 percent of the total volume of the mixture pumped. Most designs have generous clearances and will permit the passage of stone which is 70 percent of the pipeline diameter. The pump efficiencies vary widely but in general may run from 50 to 70 percent.

Commercial dredges vary in size and discharge-pipe diameters from 12 to 30 in (0.3 to 0.8 m). Smaller or larger dredges are usually special-purpose machines. A number of 36-in (0.9-m) dredges are used to maintain the channel of the Mississippi River. The power applied to pumps varies from 100 to 3,000 hp (75 to 2,200 kW). The modern 20-in (0.5-m) commercial dredge has about 1,350 bhp (1,007 kW) applied to the pump.

Diesel dredges are built for direct-connected or electric drives, and modern steam dredges have direct-turbine or turboelectric drives. The steam turbine and the dc electric motor have the advantage that they are capable of developing full rating at reduced speeds.

Within its scope, the hydraulic dredge can work more economically than any other excavating machine or combination of machines.

Dragline Excavators

Dragline excavators are typically used for digging open cuts, drainage ditches, canals, sand, and gravel pits, where the material is to be moved 20 to 1,000 ft (6 to 305 m) before dumping. They cannot handle rock unless the rock is blasted. Since they are provided with long booms and mounted on turntables, permitting them to swing through a full circle, these excavators can deposit material directly on the spoil bank farther from the point of excavation than any other type of machine. Whereas a shovel stands below the level of the material it is digging, a dragline excavator stands above and can be used to excavate material under water.

Figure 10.4.30 shows a self-contained dragline mounted on crawler treads. The drive is almost exclusively gasoline in the small sizes and diesel, diesel-electric, or electric, frequently with Ward Leonard control, in the large sizes. The boom a is pivoted at its lower end to the turntable, the outer end being supported by cables b. so that it can be raised or lowered to the desired angle. The scraper bucket c is supported

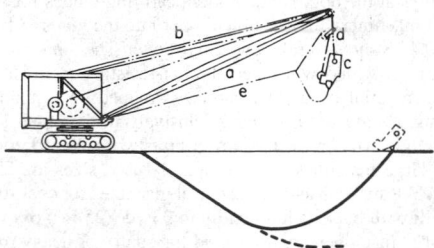

Fig. 10.4.30 Dragline excavator.

by cable d, which is attached to a bail on the bucket, passes over a sheave at the head of the boom, and is made fast to the engine. A second cable e is attached to the front of the bucket and made fast to the second drum of the engine. The bucket is dropped and dragged along the surface of the material by cable e until filled. It is then hoisted by cable d, drawn back to its dumping position, e being kept tight until the dumping point is reached, when e is slacked, allowing the bucket to dump by gravity. After the bucket is filled, the boom is swung to the dumping position while the bucket is being hauled up. A good operator can throw the bucket 10 to 40 ft (3 to 12 m) beyond the end of the boom, depending on the size of machine and the working conditions. The depth of the cut varies from 12 to 75 ft (3.7 to 23 m), again depending on the size of machine and the working conditions. With the smaller machines and under favorable conditions, two or even three trips per minute are possible; but with the largest machines, even one trip per minute may not be attained. The more common sizes are for handing ¾- to 4-yd³ (0.6- to 3-m³) buckets with boom lengths up to 100 or 125 ft (30 to 38 m), but machines have been built to handle an 8-yd³ (6-m³) bucket with a boom length of 200 ft (60 m). The same machine can handle a 12-yd³ (9-m³) bucket with the boom shortened to 165 ft (50 m).

Stackline Cableways

Used widely in sand-and-gravel plants, the **slackline cableway** employs an open-ended dragline bucket suspended from a carrier (Fig. 10.4.31) which runs upon a track cable. It will dig, elevate, and convey materials in one continuous operation.

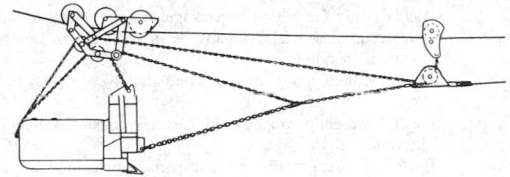

Fig. 10.4.31   Slackline-cable bucket and trolley.

Figure 10.4.32 shows a typical slackline-cableway operation. The bucket and carrier is *a*; *b* is the track cable, inclined to return the bucket and carrier by gravity; *c* is a tension cable for raising or lowering the track cable; *d* is the load cable; and *e* is a power unit with two friction drums having variable speeds. A mast or tower *f* is used to support guide

and tension blocks at the high end of the track cable; a movable tail tower *g* supports the lower end of the track cable. The bucket is raised and lowered by tensioning or slacking off the track cable. The bucket is loaded, after lowering, by pulling on the load cable. The loaded bucket, after raising, is conveyed at high speed to the dumping point and is

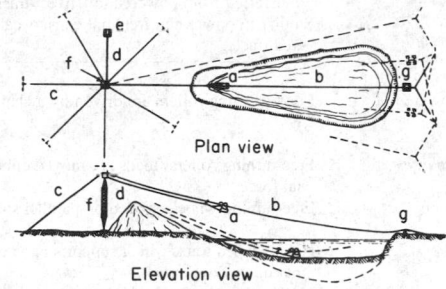

Plan view

Elevation view

Fig. 10.4.32   Slackline-cable plant. *(Sauerman.)*

returned at a still higher speed by gravity to the digging point. The cableway can be operated in radial lines from a mast or in parallel lines between two moving towers. It will not dig rock unless the rock is blasted. The depth of digging may vary from 5 to 100 ft.

10.5   CONVEYOR MOVING AND HANDLING
by Vincent M. Altamuro

OVERHEAD CONVEYORS
by Ivan L. Ross
Acco Chain Conveyor Division

Conveyors are primarily horizontal-movement, fixed-path, constant-speed material handling systems. However, they often contain inclined sections to change the elevation of the material as it is moving, switches to permit alternate paths and "power-and-free" capabilities to allow the temporary slowing, stopping, or accumulating of material.

Conveyors are used, not only for transporting material, but also for in-process storage. They may be straight, curved, closed-loop, irreversible, or reversible. Some types of conveyors are:

air blower	monorail
apron	pneumatic tube
belt	power-and-free
bucket	roller
car-on-track	screw
carousel	skate wheel
chain	slat
flight	spiral
hydraulic	towline
magnetic	trolley

Tables 10.5.1 and 10.5.2 describe and compare some of these.

Conveyors are often used as integral components of assembly systems. They bring the correct material, at the required rate, to each worker and then to the next operator in the assembly sequence. Figure 10.5.1 shows ways conveyors can be used in assembly.

Overhead conveyor systems are defined in two general classifications: the basic trolley conveyor and the power-and-free conveyor, each of which serves a definite purpose.

Trolley conveyors, often referred to as overhead power conveyors, consist of a series of trolleys or wheels supported from or within an overhead track and connected by an endless propelling means, such as chain, cable, or other linkages. Individual loads are usually suspended from the trolleys or wheels (Fig. 10.5.2). Trolley conveyors are utilized for transportation or storage of loads suspended from one conveyor which follows a single fixed path. They are normally used in applications where a balanced, continuous production is required. Track sections range from lightweight "tee" members or tubular sections, to medium- and heavy-duty I-beam sections. The combinations and sizes of trolley-propelling means and track sections are numerous. Normally this type of conveyor is continually in motion at a selected speed to suit its function.

Power-and-free conveyor systems consist of at least one power conveyor, but usually more, where the individual loads are suspended from one or more free trolleys (not permanently connected to the propelling means) which are conveyor-propelled through all or part of the system. Additional portions of the system may have manual or gravity means of propelling the trolleys.

Worldwide industrial, institutional, and warehousing requirements of in-process and finished products have affected the considerable growth and development of power-and-free conveyors. Endless varieties of size, style, color, and all imaginable product combinations have extended the use of power-and-free conveyors. The power-and-free system combines the advantages of continuously driven chains with the versatile traffic system exemplified by traditional monorail unpowered systems. Thus, high-density load-transportation capabilities are coupled with complex traffic patterns and in-process or work-station requirements to enable production requirements to be met with a minimum of manual handling or transferring. Automatic dispatch systems for coding and programming of the load routing are generally used.

Trolley Conveyors

The load-carrying member of a trolley conveyor is the trolley or series of wheels. The wheels are sized and spaced as a function of the imposed

Table 10.5.1 Types of Conveyors Used in Factories

Type of conveyor	Description	Features/limitations
Overhead		
Power-and-free	Carriers hold parts and move on overhead track. With inverted designs, track is attached to factory floor. Carriers can be transferred from powered to "free" track.	Accumulation; flexible routing; live storage; can hold parts while production steps are performed; can travel around corners and up inclines.
Trolley	Similar to power and free, but carriers cannot move off powered track.	Same as power and free, but flexible routing is not possible without additional equipment
Above-floor		
Roller, powered	Load-carrying rollers are driven by a chain or belt.	Handles only flat-bottomed packages, boxes, pallets or parts; can accumulate loads; can also be suspended from ceiling for overhead handling.
Roller, gravity	Free-turning rollers; loads are moved either by gravity or manual force.	When inclined, loads advance automatically.
Skate wheel	Free-turning wheels spaced on parallel shafts.	For lightweight packages and boxes; less expensive than gravity rollers.
Spiral tower	Helix-shaped track which supports parts or small "pallets" that move down track.	Buffer storage; provides surge of parts to machine tools when needed.
Magnetic	Metal belt conveyor with magnetized bed.	Handles ferromagnetic parts, or separates ferromagnetic parts from nonferrous scrap.
Pneumatic	Air pressure propels cylindrical containers through metal tubes.	Moves loads quickly; can be used overhead to free floor space.
Car-on-track	Platforms powered by rotating shaft move along track.	Good for flexible manufacturing systems; precise positioning of platforms; flexible routing.
In-floor		
Towline	Carts are advanced by a chain in the floor.	High throughput; flexible routing possible by incorporating spurs in towline; tow carts can be manually removed from track; carts can travel around corners.

SOURCE: *Modern Materials Handling,* Aug. 5, 1983, p. 55.

Table 10.5.2 Transportation Equipment Features

Equipment	Most practical travel distance	Automatic load or unload	Typical load	Throughput rate	Travel path	Typical application
Conveyors						
Belt	Short to medium	No	Cases, small parts	High	Fixed	Take-away from picking, sorting
Chain	Short to medium	Yes	Unit	High	Fixed	Deliver to and from automatic storage and retrieval systems
Roller	Short to long	Yes	Unit, case	High	Fixed	Unit: same as chain Case: sorting, delivery between pick station
Towline	Medium to long	Yes	Carts	High	Fixed	Delivery to and from shipping or receiving
Wire-guided vehicles						
Cart	Short to long	Yes	Unit	Low	Fixed	Delivery between two drop points
Pallet truck	Short to long	Yes	Unit	Low	Fixed	Delivery between two drop points
Tractor train	Short to long	Yes	Unit	Medium	Fixed	Delivery to multiple drop points
Operator-guided vehicles						
Walkie pallet	Short	No	Unit	Low	Flexible	Short hauls at shipping and receiving
Rider pallet	Medium	No	Unit	Low	Flexible	Dock-to-warehouse deliveries
Tractor train	Short to long	No	Unit	Medium	Flexible	Delivery to multiple drop points

SOURCE: *Modern Materials Handling,* Feb. 22, 1980, p. 106.

load, the propelling means, and the track capability. The load hanger (carrier) is attached to the conveyor and generally remains attached unless manually removed. However, in a few installations, the load hanger is transferred to and from the conveyor automatically.

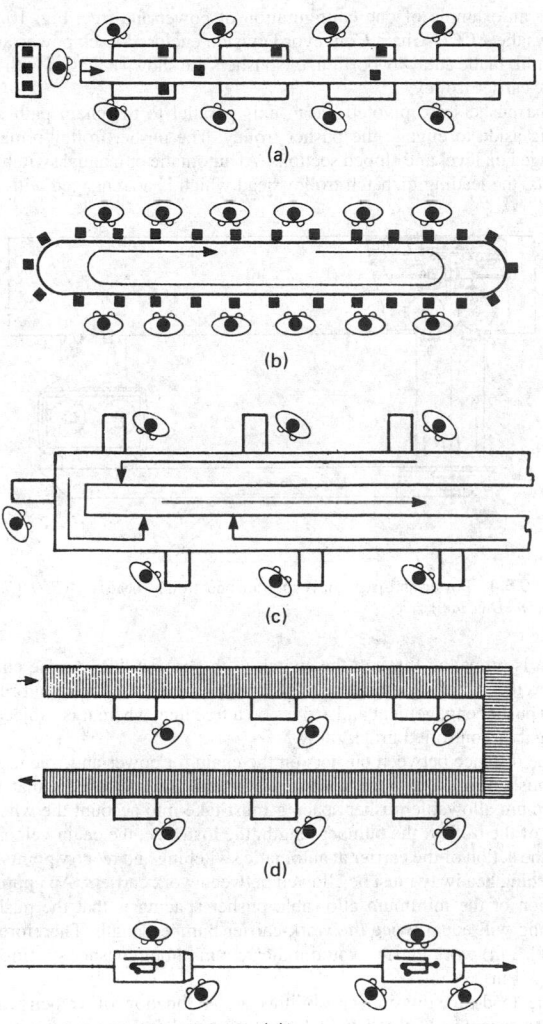

Fig. 10.5.1 Conveyors used in assembly operations. (*a*) Belt conveyor, with diverters at each station; (*b*) carousel circulates assembly materials to workers; (*c*) multiple-path conveyors allow several products to be built at once; (*d*) gravity roller conveyors; (*e*) towline conveyor, moving assemblies from one group of assemblers to another. (*Modern Materials Handling, Nov. 1979, page 116.*)

The trolley conveyor can employ any chain length consistent with allowable propelling means and drive(s) capability. The track layout always involves horizontal turns and commonly has vertical inclines and declines.

When a dimensional layout, load spacing, weights of moving loads, function, and load and unload points are determined, the chain or cable pull can be calculated. Manufacturer's data should be used for frictional values. A classical point-to-point analysis should be made, using the most unfavorable loading condition.

In the absence of precise data, the following formulas can be used to find the approximate drive effort:

$$\text{Max drive effort, lb} = A + B + C$$
$$\text{Net drive effort, lb} = A + B + C - D$$

$$\text{Max drive effort, N} = (A + B + C)\,9.81$$
$$\text{Net drive effort, N} = (A + B + C - D)\,9.81$$

where $A = fw$ [where w = total weight of chain, carriers, and live load, lb (kg) and f = coefficient of friction]; $B = wS$ [where w = average carrier load per ft (m), lb/ft (kg/m) and S = total vertical rise, ft (m)]; $C = 0.017f(A + B)N$ (where N = sum of all horizontal and vertical curves, deg); and $D = w'S'$ [where w' = average carrier load per ft (m), lb/ft (kg/m) and s' = total vertical drop, ft (m)].

For conveyors with antifriction wheels, with clean operating condition, the coefficient of friction f may be 0.13.

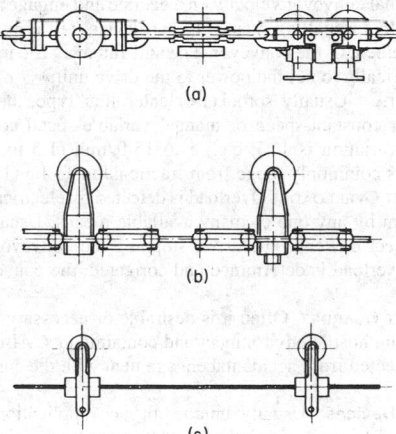

Fig. 10.5.2 Typical conveyor chains or cable.

Where drive calculations indicate that the allowable chain or cable tension may be exceeded, multiple-drive units are used. When multiple drives are used for constant speed, high-slip motors or fluid couplings are commonly used. If variable speed is required with multiple drives, it is common to use direct-current motors with direct-current supply and controls. For both constant and variable speed, drives are balanced to share drive effort. Other than those mentioned, many various methods of balancing are available for use.

Complexities of overhead conveyors, particularly with varying loads on inclines and declines, as well as other influencing factors (e.g., conveyor length, environment, lubrication, and each manufacturer's design recommendations), usually require detail engineering analysis to select the proper number of drives and their location. Particular care is also required to locate the take-up properly.

The following components or devices are used on trolley-conveyor applications:

Trolley Assembly Wheels and their attachment portion to the propelling means (chain or cable) are adapted to particular applications, depending upon loading, duty cycle, environment, and manufacturer's design.

Carrier Attachments These are made in three main styles: (1) enclosed tubular type, where the wheels and propelling means are carried inside; (2) semienclosed tubular type, where the wheels are enclosed and the propelling means is external; and (3) open-tee or I-beam type, where the wheels and propelling means are carried externally.

Sprocket or Traction Wheel Turn Any arc of horizontal turn is available. Standards usually vary in increments of 15° from 15 to 180°.

Roller Turns Any arc of horizontal turn is available. Standards usually vary in increments of 15° from 15 to 180°.

Track Turns These are horizontal track bends without sprockets, traction wheels, or rollers; they are normally used on enclosed-track conveyors where the propelling means is fitted with horizontal guide wheels.

Track Hangers, Brackets, and Bracing These conform to track size and shape, spaced at intervals consistent with allowable track stress and deflection applied by loading and chain or cable tensions.

Track Expansion Joints For use in variable ambient conditions, such as ovens, these are also applied in many instances where conveyor track crosses building expansion joints.

Chain Take-Up Unit Required to compensate for chain wear and/or variable ambient conditions, this unit may be traction-wheel, sprocket, roller, or track-turn type. Adjustment is maintained by screw, screw spring, counterweight, or air cylinder.

Incline and Decline Safety Devices An "anti-back-up" device will ratchet into a trolley or the propelling means in case of unexpected reversal of a conveyor on an incline. An "anti-runaway" device will sense abnormal conveyor velocity on a decline and engage a ratchet into a trolley or the propelling means. Either device will arrest the uncontrolled movement of the conveyor. The anti-runaway is commonly connected electrically to cut the power to the drive unit.

DRIVE UNIT. Usually sprocket or caterpillar type, these units are available for constant-speed or manual variable-speed control. Common speed variation is 1 : 3; e.g., 5 to 15 ft/min (1.5 to 4.5 m/min). Drive motors commonly range from fractional to 15 hp (11 kW).

DRIVE UNIT OVERLOAD. Overload is detected by electrical, torque, or pull detection by any one of many available means. Usually overload will disconnect power to the drive, stopping the conveyor. When the reason for overload is determined and corrected, the conveyor may be restarted.

EQUIPMENT GUARDS. Often it is desirable or necessary to guard the conveyor from hostile environment and contaminants. Also employees must be protected from accidental engagement with the conveyor components.

Transfer Devices Usually unique to each application, automatic part or carrier loading, unloading, and transfer devices are available. With growth in the use of power-and-free, carrier transfer devices have become rare.

Power-and-Free Conveyor

The power-and-free conveyor has the highest potential application wherever there is a requirement for other than a single fixed-path flow (trolley conveyor). Power-and-free conveyors may have any number of automatic or manual switch points. A system will permit scheduled transit and delivery of work to the next assigned station automatically. Accumulation (storage) areas are designed to accommodate in-process inventory between operations.

The components and chain-pull calculation discussed for powered overhead conveyors are basically applicable to the power-and-free conveyor. Addition of a secondary free track surface is provided for the work carrier to traverse. This free track is usually disposed directly below the power rail but is sometimes found alongside the power rail. (This arrangement is often referred to as a "side pusher" or "drop finger.") The power-and-free rails are joined by brackets for rail (free-

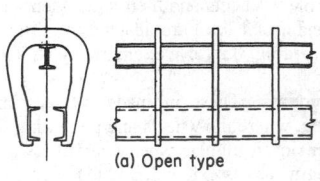

(a) Open type

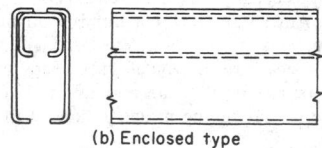

(b) Enclosed type

Fig. 10.5.3 Conveyor tracks.

track) continuity. The power chain is fitted with pushers to engage the work-carrier trolley. Track sections are available in numerous configurations for both the power portion and the free portion. Sections will be enclosed, semienclosed, or open in any combinations. Two of the most common types are shown in Fig. 10.5.3.

As an example of one configuration of power-and-free, Fig. 10.5.4 shows the ACCO Chain Conveyor Division enclosed-track power-and-free rail. In the cutaway portion, the pushers are shown engaged with the work-carrier trolley.

The pushers are pivoted on an axis parallel to the chain path and swing aside to engage the pusher trolley. The pusher trolley remains engaged on level and sloped sections. At automatic or manual switching points, the leading dispatch trolley head which is not engaged with the

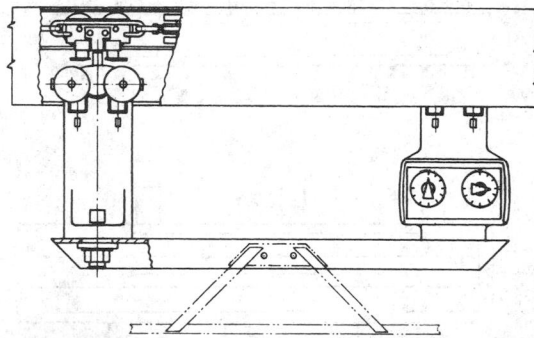

Fig. 10.5.4 Power-and-free conveyor rail and trolley heads. (*ACCO Chain Conveyor Division.*)

chain is propelled through the switch to the branch line. As the chain passes the switching point, the pusher trolley departs to the right or left from pusher engagement and arrives on a free line, where it is subject to manual or controlled gravity flow.

The distance between pushers on the chain for power-and-free use is established in accordance with conventional practice, except that the minimum allowable pusher spacing must take into account the wheelbase of the trolley, the bumper length, the load size, the chain velocity, and the action of the carrier at automatic switching and reentry points. A switching headway must be allowed between work carriers. An approximation of the minimum allowable pusher spacing is that the pusher spacing will equal twice the work-carrier bumper length. Therefore, a 4-ft (1.2-m) work carrier would indicate a minimum pusher spacing of 8 ft (2.4 m).

The load-transmission capabilities are a function of velocity and pusher spacing. At the 8-ft (2.4-m) pusher spacing and a velocity of 40 ft/min (0.22 m/s), five pushers per minute are made available. The load-transmission capability is five loads per minute, or 300 loads per hour.

Method of Automatic Switching from a Powered Line to a Free Line Power-and-free work carriers are usually switched automatically. To do this, it is necessary to have a code device on the work carrier and a decoding (reading) device along the track in advance of the track switch. Figure 10.5.5 shows the equipment relationship. On each carrier, the free trolley carries the code selection, manually or automatically introduced, which identifies it for a particular destination or routing. As the free trolley passes the reading station, the trolley intelligence is decoded and compared with a preset station code and its current knowledge of the switch position and branch-line condition; a decision is then reached which results in correct positioning of the rail switch.

In Fig. 10.5.5, the equipment illustrated includes a transistorized readout station *a*, which supplies 12 V direct current at stainless-steel code brushes *b*. The code brushes are "matched" by contacts on the encoded trolley head *c*. When a trolley is in register and matches the code-brush positioning, it will be allowed to enter branch line *d* if the line is not full. If carrier *e* is to be entered, the input signal is rectified and amplified so as to drive a power relay at junction box *f*, which in turn actuates solenoid *g* to operate track switch *h* to the branch-line position. A memory circuit is established in the station, indicating that a full-line condition exists. This condition is maintained until the pusher pin of the switched carrier clears reset switch *j*.

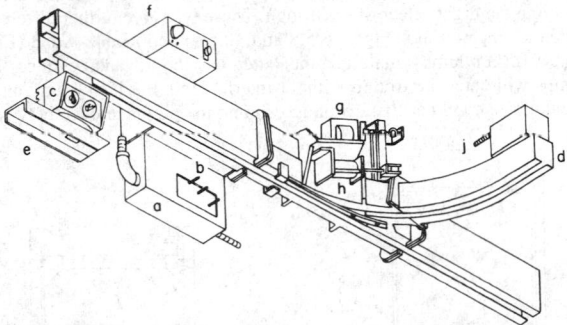

Fig. 10.5.5 Switch mechanism exiting trolley from power rail to free rail. *(ACCO Chain Conveyor Division.)*

The situations which can be handled by the decoding stations are as follows: (1) A trolley with a matching code is in register; there is space in the branch line. The carrier is allowed to enter the branch line. (2) A trolley with a matching code is in register; there is no space in the branch line. The carrier is automatically recirculated on the powered system and will continue to test its assigned destination until it can be accommodated. (3) A trolley with a matching code is in register; there is no space in the branch line. The powered conveyor is automatically stopped, and visual and/or audible signals are started. The conveyor can be automatically restarted when the full-line condition is cleared, and the waiting carrier will be allowed to enter. (4) A trolley with a nonmatching code is in register; in this case, the decoding station always returns the track switch to the main-line position, if necessary, and bypasses the carrier.

Use and Control of Carriers on Free and Gravity Lines Free and gravity lines are used as follows: (1) To connect multiple power-and-free conveyors, thus making systems easily extensible and permitting different conveyor designs for particular use. (2) To connect two auxiliary devices such as vertical conveyors and drop-lift stations. (3) As manned or automatic workstations. In this case, the size of the station depends on the number of carriers processed at one time, with additional space provision for arriving and departing carriers. (Considerable knowledge of work-rate standards and production-schedule requirements is needed for accurate sizing.) (4) As manned or automatic storage lines, especially for handling production imbalance for later consumption.

Nonmanned automatic free lines require that the carrier be controlled in conformance with desired conveyor function and with regard to the commodity being handled. The two principal ways to control carriers in nonmanned free rails are (1) to slope the free rail so that all carriers will start from rest and use incremental spot retarders to check velocity, and (2) to install horizontal or sloped rails and use auxiliary power conveyor(s) to accumulate carriers arriving in the line.

In manned free lines, it is usual to have slope at the automatic arrival and automatic departure sections only. These sections are designed for automatic accumulation of a finite number of carriers, and retarding or feeding devices may be used. Throughout the remainder of the manned station, the carrier is propelled by hand.

Method of Automatic Switching from a Free Line to a Powered Line Power-and-free work carriers can be reentered into the powered lines either manually or automatically. The carrier must be integrated with traffic already on the powered line and must be entered so that it will engage with a pusher on the powered chain. Figure 10.5.6 illustrates a typical method of automatic reentry. The carrier *a* is held at a rest on a slope in the demand position by the electric trolley stop *b*. The demand to enter enables the sensing switches *c* and *d* mounted on the powered rail to test for the availability of a pusher. When a pusher that is not propelling a load is sensed, all conditions are met and the carrier is released by the electric trolley stop such that it arrives in the pickup position in advance of the pusher. A retarding or choke device can be used to keep the entering carrier from overrunning the next switch position or another carrier in transit. The chain pusher engages the pusher trolley and departs the carrier. The track switch *f* can remain in the branch-line position until a carrier on the main line *g* would cause the track switch to reset. Automatic reentry control ensures that no opportunity to use a pusher is overlooked and does not require the time of an operator.

Power-and-Free Conveyor Components All components used on trolley conveyors are applied to power-and-free conveyors. Listed below are a few of the various components unique to power-and-free systems.

TRACK SWITCH. This is used for diverting work carriers either automatically or manually from one line or path to another. Any one system may have both. Switching may be either to the right or to the left. Automatically, stops are usually operated pneumatically or electromechanically. Track switches are also used to merge two lines into one.

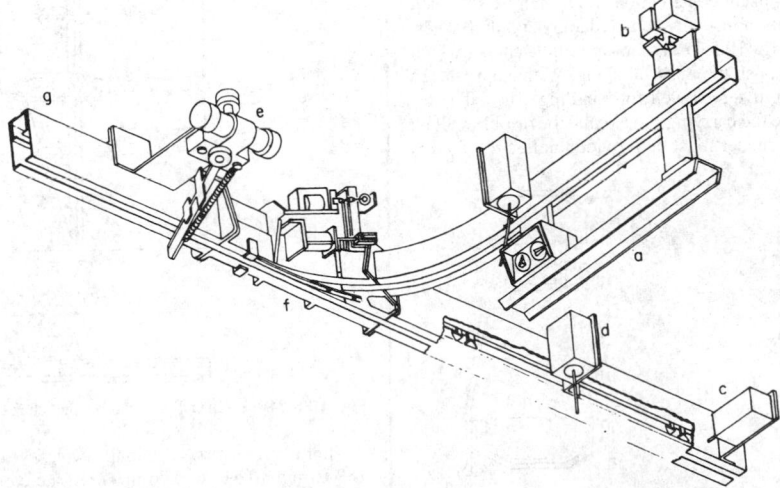

Fig. 10.5.6 Switch mechanism reentering trolley from free rail to power rail. *(ACCO Chain Conveyor Division.)*

TROLLEY STOPS. Used to stop work carriers, these operate either automatically or manually on a free track section or on a powered section. Automatically, stops are usually operated pneumatically or electromechanically.

STORAGE. Portions or spurs of power-and-free conveyors are usually dedicated to the storage (accumulation) of work carriers. Unique to the design, type, or application, storage may be accomplished on (1) level hand-pushed lines, (2) gravity sloped lines (usually with overspeed-control retarders), (3) power lines with spring-loaded pusher dogs, or (4) powered lines with automatic accumulating free trolleys.

INCLINES AND DECLINES. As in the case of trolley conveyors, vertical inclines and declines are common to power-and-free. In addition to safety devices used on trolley conveyors, similar devices may be applied to free trolleys.

LOAD BARS AND CARRIERS. The design of load bars, bumpers, swivel devices, index devices, hooks, or carriers is developed at the time of the initial power-and-free investigation. The system will see only the carrier, and all details of system design are a function of its design. How the commodity is being handled on or in the carrier is carefully considered to facilitate its use throughout the system, manually, automatically, or both.

VERTICAL CONVEYOR SECTIONS. Vertical conveyor sections are often used as an accessory to power-and-free. For practical purposes, the vertical conveyor can be divided into two classes of devices:

DROP (LIFT) SECTION. This device is used to drop (or lift) the work carrier vertically to a predetermined level in lieu of vertical inclines or declines. One common reason for its use is to conserve space. The unit may be powered by a cylinder or hoist, depending on the travel distance, cycle time, and load. One example of the use of a drop (lift) section is to receive a carrier on a high level and lower it to an operations level. The lower level may be a load-unload station or processing station. Automatic safety stops are used to close open rail ends.

INTERFLOOR VERTICAL CONVEYOR. When used for interfloor service and long lifts, the vertical conveyor may be powered by high-speed hoists or elevating machines. In any case, the carriers are automatically transferred to and from the lift, and the dispatch control on the carrier can instruct the machine as to the destination of the carrier. Machines can be equipped with a variety of speeds and operating characteristics. Multiple carriers may be handled, and priority-call control systems can be fitted to suit individual requirements.

NONCARRYING CONVEYORS

Flight Conveyors **Flight conveyors** are used for moving granular, lumpy, or pulverized materials along a horizontal path or on an incline seldom greater than about 40°. Their principal application is in handling coal. The flight conveyor of usual construction should not be specified for a material that is actively abrasive, such as damp sand and ashes. The **drag-chain conveyor** (Fig. 10.5.7) has an open-link chain, which serves, instead of flights, to push the material along. With a hard-faced concrete or cast-iron trough, it serves well for handling ashes. The return run is, if possible, above the carrying run, so that the dribble will be back into the loaded run. A feeder must be provided unless the feed is

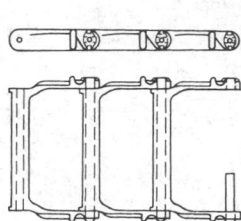

Fig. 10.5.7 Drag chain.

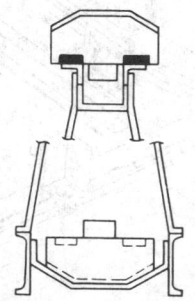

Fig. 10.5.8 Single-strand scraper-flight conveyor.

otherwise controlled, as from a tandem elevator or conveyor. As the feeder is interlocked, either mechanically or electrically, the feed stops if the conveyor stops. Flight conveyors may be classified as **scraper type** (Fig. 10.5.8), in which the element (chain and flights) rests on the trough; **suspended-flight type** (Fig. 10.5.9), in which the flights are carried clear of the trough by shoes resting on guides; and **suspended-chain type** (Fig. 10.5.10), in which the chain rests on guides, again carrying the flights clear of the trough. These types are further differentiated as **single-strand** (Figs. 10.5.8 and 10.5.9) and **double-strand** (Fig. 10.5.10). For lumpy material, the latter has the advantage since the lumps will enter the trough without interference. For heavy duty also, the double strand has the advantage, in that the pull is divided between

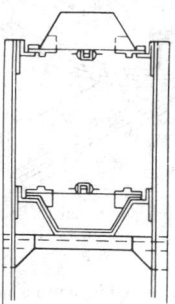

Fig. 10.5.9 Single-strand suspended-flight conveyor.

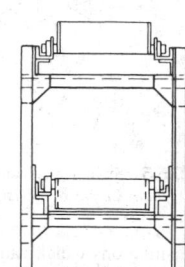

Fig. 10.5.10 Double-strand roller-chain flight conveyor.

two chains. A special type for simultaneous handling of several materials may have the trough divided by longitudinal partitions. The material having the greatest coefficient of friction is then carried, if possible, in the central zone to equalize chain wear and stretch.

Improvements in the welding and carburizing of welded-link chain have made possible its use in flight conveyors, offering several significant advantages, including economy and flexibility in all directions. Figure 10.5.11 shows a typical scraper-type flight cast from malleable iron incorporated onto a slotted conveyor bed. The small amount of fines that fall through the slot are returned to the top of the bed by the returning flights. Figure 10.5.12 shows a double-chain scraper conveyor in which the ends of the flights ride in a restrictive channel. These types of flight conveyors are driven by pocket wheels.

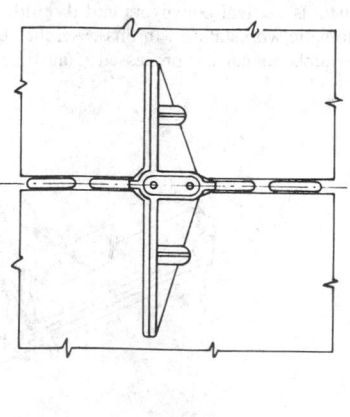

Fig. 10.5.11 Scraper flight with welded chain.

Flight conveyors of small capacity operate usually at 100 to 150 ft/min (0.51 to 0.76 m/s). Large-capacity conveyors operate at 100 ft/min (0.51 m/s) or slower; their long-pitch chains hammer heavily against the drive-sprocket teeth or pocket wheels at higher speeds. A conveyor steeply inclined should have closely spaced flights so that the material will not avalanche over the tops of the flights. The capacity of a given conveyor diminishes as the angle of slope increases. For the heaviest duty, hardened-face rollers at the articulations are essential.

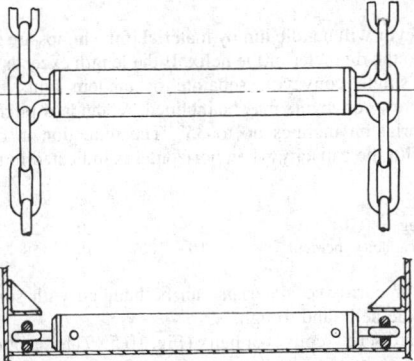

Fig. 10.5.12 Scraper flight using parallel welded chains.

Cautions in Flight-Conveyor Selections With abrasive material, the trough design should provide for renewal of the bottom plates without disturbing the side plates. If the conveyor is inclined and will reverse when halted under load, a solenoid brake or other automatic backstop should be provided. Chains may not wear or stretch equally. In a double-strand conveyor, it may be necessary to shift sections of chain from one side to the other to even up the lengths.

Intermediate slide gates should be set to open in the opposite direction to the movement of material in the conveyor.

The **continuous-flow conveyor** serves as a conveyor, as an elevator, or as a combination of the two. It is a slow-speed machine in which the material moves as a continuous core within a duct. Except with the **Redler** conveyor, the element is formed by a single strand of chain with closely spaced impellers, somewhat resembling the flights of a flight conveyor.

The **Bulk flo** (Fig. 10.5.13) has peaked flights designed to facilitate the outflow of the load at the point of discharge. The load, moved by a positive push of the flights, tends to provide self-clearing action at the end of a run, leaving only a slight residue.

The **Redler** (Fig. 10.5.14) has skeletonized or U-shaped impellers which move the material in which they are submerged because the resistance to slip through the element is greater than the drag against the walls of the duct.

Materials for which the continuous-flow conveyor is suited are listed below in groups of increasing difficulty. The constant C is used in the power equations below, and in Fig. 10.5.16.

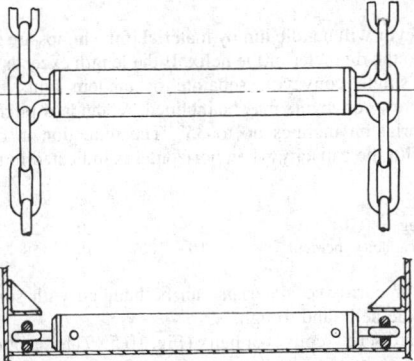

Fig. 10.5.13 Bulk-flow continuous-flow elevator.

$C = 1$: clean coal, flaxseed, graphite, soybeans, copra, soap flakes

$C = 1.2$: beans, slack coal, sawdust, wheat, wood chips (dry), flour

$C = 1.5$: salt, wood chips (wet), starch

$C = 2$: clays, fly ash, lime (pebble), sugar (granular), soda ash, zinc oxide

$C = 2.5$: alum, borax, cork (ground), limestone (pulverized)

Among the materials for which special construction is advised are bauxite, brown sugar, hog fuel, wet coal, shelled corn, foundry dust, cement, bug dust, hot brewers' grains. The machine should not be specified for use with ashes, bagasse, carbon-black pellets, sand and gravel, sewage

sludge, and crushed stone. Fabrication from corrosion-resistant materials such as brass, monel, or stainless steel may be necessary for use with some corrosive materials.

Where a single **runaround conveyor** is required with multiple feed points and some recirculation of excess load, the Redler serves. The U-frame flights do not squeeze the loads, as they resume parallelism after separating when rounding the terminal wheels. As an elevator, this machine will also handle sluggish materials that do not flow out readily. A pusher plate opposite the discharge chute can be employed to enter

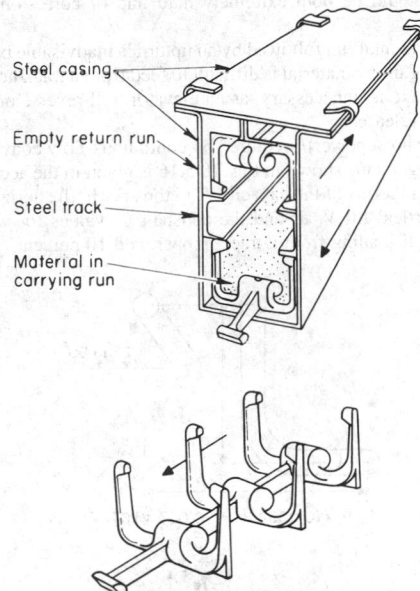

Fig. 10.5.14 Redler U-type continuous-flow conveyor.

between the legs of the U flights to push out such material. When horizontal or inclined, continuous-flow elevators such as the Redler are normally self-cleaning. When vertical, the Redler type can be made self-cleaning (except with sticky materials) by use of special flights.

Continuous-flow conveyors and elevators do not require a feeder (Fig. 10.5.15). They are self-loading to capacity and will not overload, even though there are several open- or uncontrolled-feed openings, since the duct fills at the first opening and automatically prevents the entrance of additional material at subsequent openings. Some special care may be required with free-flowing material.

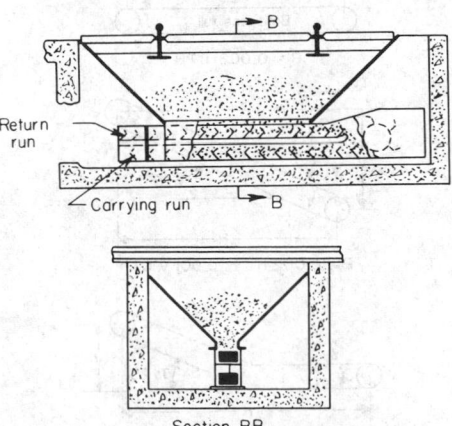

Fig. 10.5.15 Shallow-track hopper for continuous-flow conveyor with feed to return run.

The duct is easily insulated by sheets of asbestos cement or similar material to reduce cooling in transit. As the duct is completely sealed, there is no updraft where the lift is high. The material is protected from exposure and contamination or contact with lubricants. The handling capacity for horizontal or inclined lengths (nearly to the angle of repose of the material) approximates 100 percent of the volume swept through by the movement of the element. For steeper inclines or elevators, it is between 50 and 90 percent.

If the material is somewhat abrasive, as with wet bituminous coal, the duct should be of corrosion-resistant steel, of extra thickness, and the chain pins should be both extremely hard and of corrosion-resistant material.

A long horizontal run followed by an upturn is inadvisable because of radial thrust. Lumpy material is difficult to feed from a track hopper. An automatic brake is unnecessary, as an elevator will reverse only a few inches when released.

The motor horsepower P required by continuous-flow conveyors for the five arrangements shown in Fig. 10.5.16 is given in the accompanying formulas in terms of the capacity T, in tons per h; the horizontal run H, ft; the vertical lift V, ft; and the constant C, values for which are given above. If loading from a track hopper, add 10 percent.

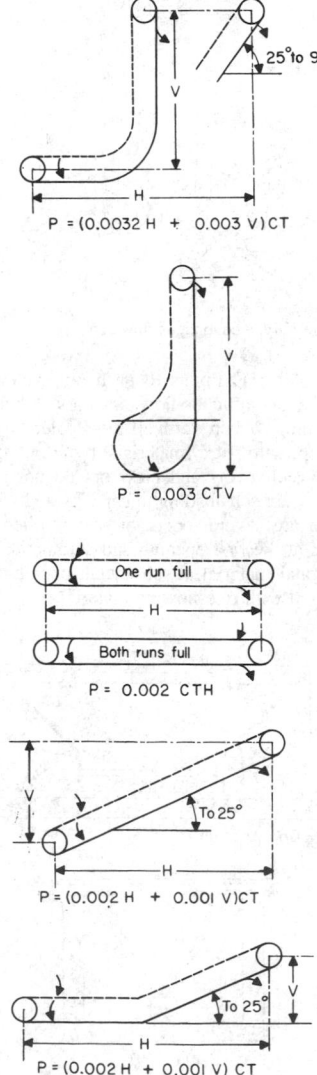

Fig. 10.5.16 Continuous-flow-conveyor arrangements.

Screw Conveyors

The screw, or spiral, conveyor is used quite widely for pulverized or granular, noncorrosive, nonabrasive materials when the required capacity is moderate, when the distance is not more than about 200 ft (61 m), and when the path is not too steep. It usually costs substantially less than any other type of conveyor and is readily made dusttight by a simple cover plate.

The conveyor will handle lumpy material if the lumps are not large in proportion to the diameter of the helix. If the length exceeds that advisable for a single conveyor, separate or tandem units are readily arranged. Screw conveyors may be inclined. A standard-pitch helix will handle material on inclines up to 35°. The reduction in capacity as compared with the capacity when horizontal is indicated in the following table:

Inclination, deg	10	15	20	25	30	35
Reduction in capacity, percent	10	26	45	58	70	78

Abrasive or corrosive materials can be handled with suitable construction of the helix and trough.

The standard screw-conveyor helix (Fig. 10.5.17) has a pitch approximately equal to its outside diameter. Other forms are used for special cases.

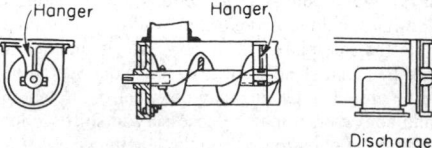

Fig. 10.5.17 Spiral conveyor.

Short-pitch screws are advisable for inclines above 29°.

Variable-pitch screws, with short pitch at the feed end, automatically control the flow to the conveyor so that the load is correctly proportioned for the length beyond the feed point. With a short section either of shorter pitch or of smaller diameter, the conveyor is self-loading to capacity and does not require a feeder.

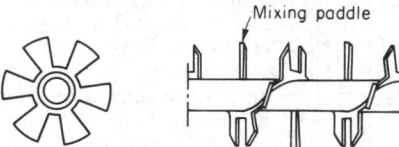

Fig. 10.5.18 Cut-flight conveyor.

Cut flights (Fig. 10.5.18) are used for conveying and mixing cereals, grains, and other light materials.

Ribbon screws (Fig. 10.5.19) are used for wet and sticky materials, such as molasses, hot tar, and asphalt, which might otherwise build up on the spindle.

Paddle screws are used primarily for mixing such materials as mortar and bitulithic paving mixtures. One typical application is to churn ashes and water to eliminate dust.

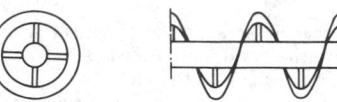

Fig. 10.5.19 Ribbon conveyor.

Standard constructions have a plain or galvanized-steel helix and trough. For abrasives and corrosives such as wet ashes, both helix and trough may be of hard-faced cast iron. For simple abrasives, the outer edge of the helix may be faced with a renewable strip of Stellite or

Table 10.5.3   Capacities and Speed of Spiral Conveyors

Group	Max percent of cross section occupied by the material	Max density of material, lb/ft³ (kg/m³)	Max r/min for diameters	
			6 in (152 mm)	20 in (508 mm)
1	45	50 (800)	170	110
2	38	50 (800)	120	75
3	31	75 (1,200)	90	60
4	25	100 (1,600)	70	50
5	12½		30	25

Group 1 includes light materials such as barley, beans, brewers grains (dry), coal (pulv.), corn meal, cottonseed meal, flaxseed, flour, malt, oats, rice, wheat. The value of the factor F is 0.5.

Group 2 includes fines and granular materials. The values of F are alum (pulv.), 0.6, coal (slack or fines), 0.9; coffee beans, 0.4; sawdust, 0.7; soda ash (light), 0.7; soybeans, 0.5; fly ash, 0.4.

Group 3 includes materials with small lumps mixed with fines. Values of F are alum, 1.4; ashes (dry), 4.0, borax, 0.7; brewers grains (wet), 0.6; cottonseed, 0.9; salt, course or fine, 1.2; soda ash (heavy), 0.7.

Group 4 includes semiabrasive materials, fines, granular and small lumps. Values of F are acid phosphate (dry), 1.4; bauxite (dry), 1.8; cement (dry), 1.4; clay, 2.0; fuller's earth, 2.0; lead salts, 1.0; limestone screenings, 2.0; sugar (raw), 1.0; white lead, 1.0; sulfur (lumpy), 0.8; zinc oxide, 1.0.

Group 5 includes abrasive lumpy materials which must be kept from contact with hanger bearings. Values of F are wet ashes, 5.0; flue dirt, 4.0; quartz (pulv.), 2.5; silica sand, 2.0; sewage sludge (wet and sandy), 6.0.

similar extremely hard material. For food products, aluminum, bronze, monel metal, or stainless steel is suitable but expensive.

Table 10.5.3 gives the capacities, allowable speeds, percentages of helix loading for five groups of materials, and the factor F used in estimating the power requirement.

Table 10.5.4 gives the handling capacities for standard-pitch screw conveyors in each of the five groups of materials when the conveyors are operating at the maximum advised speeds and in the horizontal position. The capacity at any lower speed is in the ratio of the speeds.

Power Requirements   The power requirements for horizontal screw conveyors of standard design and pitch are determined by the Link-Belt Co. by the formula that follows. Additional allowances should be made for inclined conveyors, for starting under load, and for materials that tend to stick or pack in the trough, as with cement.

$$H = \text{hp at conveyor head shaft} = (ALN + CWLF) \times 10^{-6}$$

where A = factor for size of conveyor (see Table 10.5.5); C = quantity of material, ft³/h; L = length of conveyor, ft; F = factor for material (see Table 10.5.3); N = r/min of conveyor; W = density of material, lb/ft³.

The motor size depends on the efficiency E of the drive (usually close to 90 percent); a further allowance G, depending on the horsepower, is made:

H	1	1–2	2–4	4–5	5
G	2	1.5	1.25	1.1	1.0

$$\text{Motor hp} = HG/E$$

When the material is distributed into a bunker, the conveyor has an open-bottom trough to discharge progressively over the crest of the pile so formed. This trough reduces the capacity and increases the required power, since the material drags over the material instead of over a polished trough.

If the material contains unbreakable lumps, the helix should clear the trough by at least the diameter of the average lump. For a given capacity, a conveyor of larger size and slower speed is preferable to a conveyor of minimum size and maximum speed. For large capacities

and lengths, the alternatives—a flight conveyor or a belt conveyor—should receive consideration.

EXAMPLES. 1. Slack coal 50 lb/ft³ (800 kg/m³); desired capacity 50 tons/h (2,000 ft³/h) (45 tonnes/h); conveyor length, 60 ft (18 m); 14-in (0.36-m) conveyor at 80 r/min. F for slack coal = 0.9 (group 2).

$$H = (255 \times 60 \times 80 + 2,000 \times 50 \times 60 \times 0.9)/1,000,000 = 6.6$$
$$\text{Motor hp} = (6.6 \times 1.0)/0.90 = 7.3 \quad (5.4 \text{ kW})$$

Use 7½-hp motor.

2. Limestone screenings, 90 lb/ft³; desired capacity, 10 tons/h (222 ft³/h); conveyor length, 50 ft; 9-in conveyor at 50 r/min. F for limestone screenings = 2.0 (group 4).

$$H = (96 \times 50 \times 50 + 222 \times 90 \times 50 \times 2.0)/1,000,000 = 2.24$$
$$\text{Motor hp} = (2.24 \times 1.25)/0.90 = 2.8$$

Use 3-hp motor.

Chutes

Bulk Material   If the material is fragile and cannot be set through a simple **vertical chute**, a **retarding chute** may be specified. Figure 10.5.20 shows a ladder chute in which the material trickles over shelves instead of falling freely. If it is necessary to minimize breakage when material is fed from a bin, a vertical box chute with flap doors opening inward, as shown in Fig. 10.5.21, permits the material to flow downward only from the top surface and eliminates the degradation that results from a converging flow from the bottom of the mass.

Straight inclined chutes for coal should have a slope of 40 to 45°. If it is found that the coal accelerates objectionably, the chute may be provided with cross angles over which the material cascades at reduced speed (Fig. 10.5.22).

Lumpy material such as coke and large coal, difficult to control when flowing from a bin, can be handled by a chain-controlled feeder chute with a screen of heavy endless chains hung on a sprocket shaft (Fig. 10.5.23). The weight of the chain curtain holds the material in the chute. When a feed is desired, the sprocket shaft is revolved slowly, either manually or by a motorized reducer.

Unit Loads   Mechanical handling of unit loads, such as boxes, barrels, packages, castings, crates, and palletized loads, calls for methods and

Table 10.5.4   Screw-Conveyor Capacities
(ft³/h)

Group	Conveyor size, in*							
	6	9	10	12	14	16	18	20
1	350	1,100	1,600	2,500	4,000	5,500	7,600	10,000
2	220	700	950	1,600	2,400	3,400	4,500	6,000
3	150	460	620	1,100	1,600	2,200	3,200	4,000
4	90	300	400	650	1,000	1,500	2,000	2,600
5	20	68	90	160	240	350	500	650

*Multiply by 25.4 to obtain mm.

Table 10.5.5 Factor *A*
(Self-lubricating bronze bearings assumed)

Diam of conveyor, in	6	9	10	12	14	16	18	20	24
mm	152	229	254	305	356	406	457	508	610
Factor *A*	54	96	114	171	255	336	414	510	690

mechanisms entirely different from those adapted to the movement of bulk materials.

Fig. 10.5.20 Ladder chute.

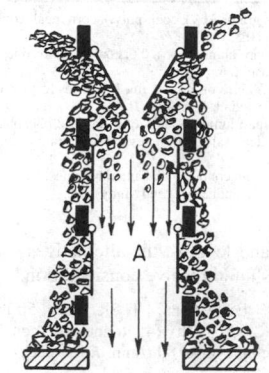

Fig. 10.5.21 Box chute with flap doors. Chute is always full up to discharging point.

Spiral chutes are adapted for the direct lowering of unit loads of various shapes, sizes, and weights, so long as their slide characteristics do not vary widely. If they do vary, care must be exercised to see that items accelerating on the selected helix pitch do not crush or damage those ahead.

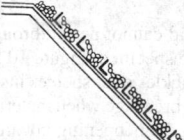

Fig. 10.5.22 Inclined chute with cross angles.

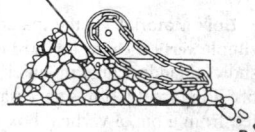

Fig. 10.5.23 Chain-controlled feeder chute.

A spiral chute may extend through several floors, e.g., for lowering parcels in department stores to a basement shipping department. The opening at each floor must be provided with automatic closure doors, and the design must be approved by the Board of Fire Underwriters.

At the discharge end, it is usual to extend the chute plate horizontally to a length in which the loads can come to rest. A tandem gravity roll conveyor may be advisable for distribution of the loads.

The **sheet-metal spiral** (Fig. 10.5.24) has a fixed blade and can be

furnished in varying diameters and pitches, both of which determine the maximum size of package that can be handled. These chutes may have receive and discharge points at any desired floors. There are certain kinds of commodities, such as those made of metal or bound with wire or metal bands, that cannot be handled satisfactorily unless the spiral chute is designed to handle only that particular commodity. Sheet-metal spirals can be built with double or triple blades, all mounted on the same standpipe. Another form of sheet-metal spiral is the open-core type, which is especially adaptable for handling long and narrow articles or bulky classes of merchandise or for use where the spiral must wind around an existing column or pass through floors in locations limited by beams or girders that cannot be conveniently cut or moved. For handling bread or other food products, it is customary to have the spiral tread made from monel metal or aluminum.

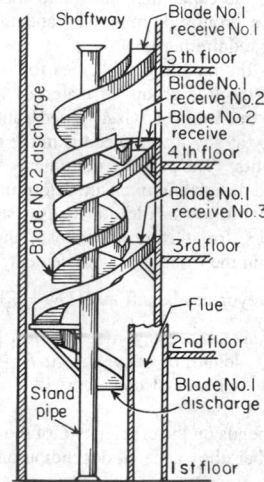

Fig. 10.5.24 Metal spiral chute.

CARRYING CONVEYORS

Apron Conveyors

Apron conveyors are specified for granular or lumpy materials. Since the load is carried and not dragged, less power is required than for screw or scraper conveyors. Apron conveyors may have stationary skirt or side plates to permit increased depth of material on the apron, e.g., when used as a feeder for taking material from a track hopper (Fig. 10.5.25)

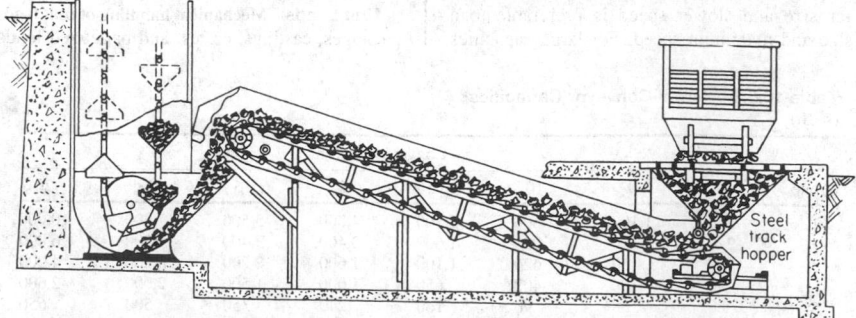

Fig. 10.5.25 Track hopper and apron feeder supplying a gravity-discharge bucket-elevator boot.

with controlled rate of feed. They are not often specified if the length is great, since other types of conveyor are substantially lower in cost. Sizes of lumps are limited by the width of the pans and the ability of the conveyor to withstand the impact of loading. Only end discharge is possible. The apron conveyor (Fig. 10.5.26) consists of two strands of roller chain separated by overlapping apron plates, which form the carrying surface, with sides 2 to 6 in (51 to 152 mm) high. The chains are driven by sprockets at one end, take-ups being provided at the other end. The conveyors always pull the material toward the driving end. For light duty, flangeless rollers on flat rails are used; for heavy duty, single-flanged rollers and T rails are used. Apron conveyors may be run without feeders, provided that the opening of the feeding hopper is made sufficiently narrow to prevent material from spilling over the sides of the conveyor after passing from the opening. When used as a conveyor, the **speed** should not exceed 60 ft/min (0.30 m/s); when used as a feeder, 30 ft/min (0.15 m/s). Table 10.5.6 gives the capacities of apron feeders with material weighing 50 lb/ft³ (800 kg/m³) at a speed of 10 ft/min (0.05 m/s).

Chain pull for horizontal-apron conveyor:

$$2LF(W + W_1)$$

Chain pull for inclined-apron conveyor:

$$L(W + W_1)(F \cos \theta + \sin \theta) + WL(F \cos \theta - \sin \theta)$$

where L = conveyor length, ft; W = weight of chain and pans per ft, lb; W_1 = weight of material per ft of conveyor, lb; θ = angle of inclination, deg; F = coefficient of rolling friction, usually 0.1 for plain roller bearings or 0.05 for antifriction bearings.

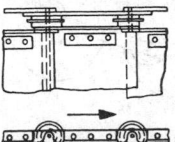

Fig. 10.5.26 Apron conveyor.

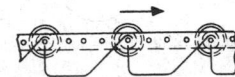

Fig. 10.5.27 Open-top carrier.

Bucket Conveyors and Elevators

Open-top bucket carriers (Fig. 10.5.27) are similar to apron conveyors, except that dished or bucket-shaped receptacles take the place of the flat or corrugated apron plates used on the apron conveyor. The carriers will operate on steeper inclines than apron conveyors (up to 70°), as the buckets prevent material from sliding back. Neither sides extending above the tops of buckets nor skirtboards are necessary. **Speed,** when loaded by a feeder, = 60 ft/min (max)(0.30 m/s) and when dragging the load from a hopper or bin, $\leqq$ 30 ft/min. The **capacity** should be calculated on the basis of the buckets being three-fourths full, the angle of inclination of the conveyor determining the loading condition of the bucket.

V-bucket carriers are used for elevating and conveying nonabrasive materials, principally coal when it must be elevated and conveyed with one piece of apparatus. The length and height lifted are limited by the strength of the chains and seldom exceed 75 ft (22.9 m). These carriers can operate on any incline and can discharge at any point on the horizontal run. The size of lumps carried is limited by the size and spacing of the buckets. The carrier consists of two strands of roller chain separated by V-shaped steel buckets. Figure 10.5.28 shows the most common form, where material is received on the lower horizontal run, elevated, and discharged through openings in the bottom of the trough of the upper horizontal run. The material is scraped along the horizontal trough of the conveyor, as in a flight conveyor. The steel guard plates *a*

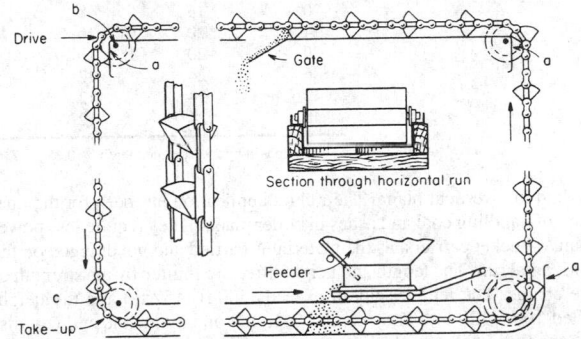

Fig. 10.5.28 V-bucket carrier.

at the right prevent spillage at the bends. Figure 10.5.29 shows a different form, where material is dug by the elevator from a boot, elevated vertically, scraped along the horizontal run, and discharged through gates in the bottom of the trough. Figure 10.5.30 shows a variation of the type shown in Fig. 10.5.29, requiring one less bend in the conveyor. The troughs are of steel or steel-lined wood. When feeding material to the horizontal run, it is advisable to use an automatic feeder driven by

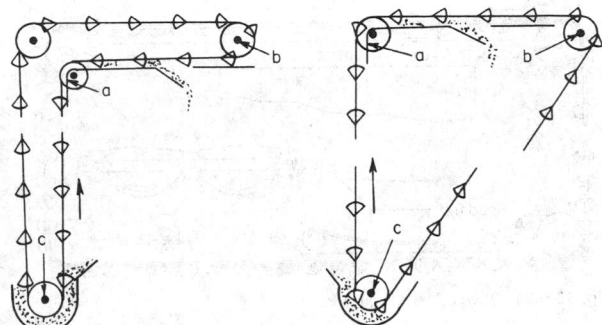

Fig. 10.5.29 and 10.5.30 V-bucket carriers.

power from one of the bend shafts to prevent overloading. Should the buckets of this type of conveyor be overloaded, they will spill on the vertical section. The drive is located at *b*, with take-up at *c*. The speed should not exceed 100 ft/min (0.51 m/s) when large material is being handled, but when material is small, speed may be increased to 125 ft/min (0.64 m/s). The best results are obtained when speeds are kept low. Table 10.5.7 gives the **capacities** and **weights** based on an even and continuous feed.

Pivoted-bucket carriers are used primarily where the path is a run-

Table 10.5.6 Capacities of Apron Conveyors

Width between skirt plates		Capacity, 50 lb/ft³ (800 kg/m³) material at 10 ft/min (0.05 m/s) speed			
		Depth of load, in (mm)			
in	mm	12 (305)	16 (406)	20 (508)	24 (610)
24	610	22 (559)	30 (762)		
30	762	26 (660)	37 (940)	47 (1,194)	56 (1,422)
36	914	34 (864)	45 (1,143)	56 (1,422)	67 (1,702)
42	1,067	39 (991)	52 (1,321)	65 (1,651)	79 (2,007)

Table 10.5.7   Capacities and Weights of V-Bucket Carriers*

Buckets				Capacity, tons of coal per hour at 100 ft/min	Weight per ft of chains and buckets, lb
Length, in	Width, in	Depth, in	Spacing, in		
12	12	6	18	29	36
16	12	6	18	32	40
20	15	8	24	43	55
24	20	10	24	100	65
30	20	10	24	126	70
36	24	12	30	172	94
42	24	12	30	200	105
48	24	12	36	192	150

* Multiply in by 25.4 for mm, ton/h by 0.25 for kg/s or by 0.91 for tonnes/h.

around in a vertical plane. Their chief application has been for the dual duty of handling coal and ashes in boiler plants. They require less power than V-bucket carriers, as the material is carried and not dragged on the horizontal run. The length and height lifted are limited by the strength of the chains. The length seldom exceeds 500 ft (152 m) and the height lifted 100 ft (30 m). They can be operated on any incline and can discharge at any point on the horizontal run. The size of lumps is limited by the size of buckets. The maintenance cost is extremely low. Many carrier installations are still in operation after 40 years of service. Other applications are for hot clinker, granulated and pulverized chemicals, cement, and stone.

The carrier consists of two strands of roller chain, with flanged rollers, between which are pivoted buckets, usually of malleable iron. The drive (Fig. 10.5.31) is located at *a* or *a'*, the take-up at *b*. The material is fed to the buckets by a feeder at any point along the lower horizontal run, is elevated, and is discharged on the upper horizontal

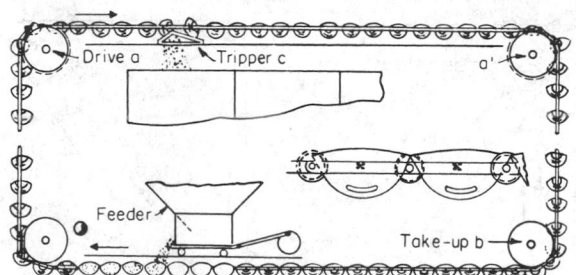

Fig. 10.5.31   Pivoted-bucket carrier.

run. The tripper *c*, mounted on wheels so that it can be moved to the desired dumping position, engages the cams on the buckets and tips them until the material runs out. The buckets always remain vertical except when tripped. The chain rollers run on T rails on the horizontal sections and between guides on the vertical runs. Speeds range from 30 to 60 ft/min (0.15 to 0.30 m/s).

After dumping, the overlapping bucket lips are in the wrong position to round the far corner; after rounding the take-up wheels, the lap is wrong for making the upturn. The Link-Belt **Peck** carrier eliminates this by suspending the buckets from trunnions attached to rearward cantilever extensions of the inner links (Fig. 10.5.32). As the chain rounds

the turns, the buckets swing in a larger-radius curve, automatically unlatch, and then lap correctly as they enter the straight run.

The pivoted-bucket carrier requires little attention beyond periodic lubrication and adjustment of take-ups. For the dual service of coal and ash handling, its only competitor is the skip hoist.

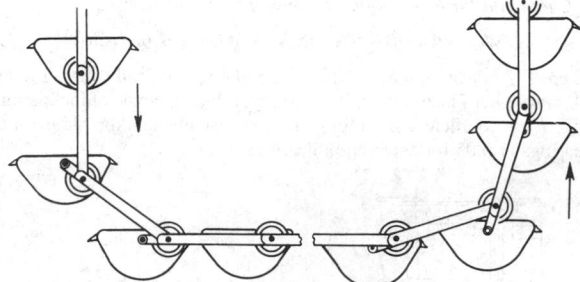

Fig. 10.5.32   Link-Belt Peck carrier buckets.

Table 10.5.8 shows the capacities of pivoted-bucket carriers with materials weighing 50 lb/ft³ (800 kg/m³), with carriers operating at 40 to 50 ft/min (0.20 to 0.25 m/s), and with buckets loaded to 80 percent capacity.

Bucket elevators are of two types: (1) chain-and-bucket, where the buckets are attached to one or two chains; and (2) belt-and-bucket, where the buckets are attached to canvas or rubber belts. Either type may be vertical or inclined and may have continuous or noncontinuous buckets. Bucket elevators are used to elevate any bulk material that will not adhere to the bucket. Belt-and-bucket elevators are particularly well adapted to handling abrasive materials which would produce excessive wear on chains. Chain-and-bucket elevators are frequently used with perforated buckets when handling wet material, to drain off surplus water. The length of elevators is limited by the strength of the chains or belts. They may be built up to 100 ft (30 m) long, but they average 25 to 75 ft (7.6 to 23 m). Inclined-belt elevators operate best on an angle of about 30° to the vertical. At greater angles, the sag of the return belt is excessive, as it cannot be supported by rollers between the head and foot pulleys. This applies also to single-strand chain elevators. Double-strand chain elevators, however, if provided with roller chain, can run on an angle, as both the upper and return chains are supported by rails.

Table 10.5.8   Capacities of Pivoted-Bucket Carriers with Coal or Similar Materials Weighing 50 lb/ft³ (800 kg/m³) at Speeds Noted

Bucket pitch × width		Capacity of coal		Speed	
in	mm	Short ton/h	tonne/h	ft/min	m/s
24 × 18	610 × 457	35–45	32–41	40–50	0.20–0.25
24 × 24	610 × 610	50–60	45–54	40–50	0.20–0.25
24 × 30	610 × 762	60–75	54–68	40–50	0.20–0.25
24 × 36	610 × 914	70–90	63–82	40–50	0.20–0.25

The size of lumps is limited by the size and spacing of the buckets and by the speed of the elevator.

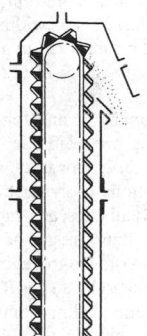

Fig. 10.5.33   Continuous bucket.

Continuous-bucket elevators (Fig. 10.5.33 and Table 10.5.9) usually operate at 100 ft/min (0.51 m/s) or less and are single- or double-strand. The contents of each bucket discharge over the back of the preceding bucket. For maximum capacity and a large proportion of lumps, the buckets extend rearward behind the chain runs. The elevator is then called a supercapacity elevator (Fig. 10.5.34 and Table 10.5.10).

Gravity-discharge elevators operate at 100 ft/min (0.51 m/s) or less and are double-strand, with spaced V buckets. The path may be an L, an inverted L, or a runaround in a vertical plane (Fig. 10.5.28). Along the horizontal run, the buckets function as pushers within a trough. An elevator with a tandem flight conveyor costs less. For a runaround path, the pivoted-bucket carrier requires less power and has lower maintenance costs.

Table 10.5.9   Continuous Bucket Elevators*

Bucket size in	Max lump size, in		Capacity with 50 lb material at 100 ft/min tons/h
	All lumps	10% lumps	
10 × 5 × 8	¾	2½	17
10 × 7 × 12	1	3	21
12 × 7 × 12	1	3	25
14 × 7 × 12	1	3	30
14 × 8 × 12	1¼	4	36
16 × 8 × 12	1½	4½	42
18 × 8 × 12	1½	4½	46

* Multiply in by 25.4 for mm, lb by 0.45 for kg, tons by 0.91 for tonnes.

As bucket elevators have no **feed control,** an interlocked feeder is desirable for a gravity flow. Some types scoop up the load as the buckets round the foot end and can take care of momentary surges by spilling the excess back into the boot. The continuous-bucket elevator, however, must be loaded after the buckets line up for the lift, i.e., when the gaps between buckets have closed.

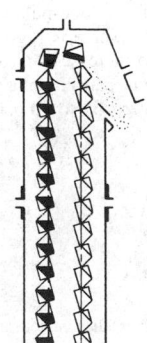

Fig. 10.5.34
Supercapacity bucket.

Belt-and-bucket elevators are advantageous for grain, cereals, glass batch, clay, coke breeze, sand, and other abrasives if the temperature is not high enough to scorch the belt [below 250°F (121°C) for natural rubber].

Elevator casings usually are sectional and dusttight, either of ³⁄₁₆-in (4.8-mm) sheet steel or, better, of aluminum. If the elevator has considerable height, its cross section must be sufficiently large to prevent sway contact between buckets and casing. Chain guides extending the length of both runs may be provided to control sway and to prevent piling up of the element, at the boot, should the chain break. **Caution:** Indoor high elevators may develop considerable updraft tending to sweep up light, pulverized material. Provision to neutralize the pressure differential at the top and bottom may be essential. Figure 10.5.35 shows the **cast-iron boot** used with centrifugal-discharge and V-bucket chain elevators and belt elevators. Figure 10.5.36 shows the general form of a belt-and-bucket elevator with struc-

Table 10.5.10   Supercapacity Elevators*
(Link-Belt Co.)

Bucket, in length × width × depth	Max lump size, large lumps not more than 20%, in	Capacity with 50 lb material tons/h
16 × 12 × 18	8	115
20 × 12 × 18	8	145
24 × 12 × 18	8	175
30 × 12 × 18	8	215
24 × 17 × 24	10	230
36 × 17 × 24	10	345

* Multiply in by 25.4 for mm, lb by 0.45 for kg, tons by 0.91 for tonnes.

tural-steel boot and casing. Elevators of this type must be run at sufficient speed to throw the discharging material clear of the bucket ahead.

Capacity   Elevators are rated for capacity with the buckets 75 percent loaded. The buckets must be large enough to accommodate the lumps, even though the capacity is small.

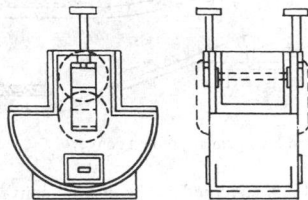

Fig. 10.5.35   Cast-iron boot.

Power Requirements   The motor horsepower for the continuous-bucket and supercapacity elevators can be approximated as

$$\text{Motor hp} = (2 \times \text{tons/h} \times \text{lift, ft})/1{,}000$$

The motor horsepower of gravity-discharge elevators can be approximated by using the same formula for the lift and adding for the horsepower of the horizontal run the power as estimated for a flight conveyor. For a vertical runaround path, add a similar allowance for the lower horizontal run.

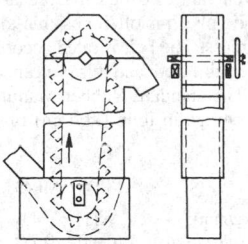

Fig. 10.5.36   Structural-steel boot and casing.

Belt Conveyors

The belt conveyor is a heavy-duty conveyor available for transporting large tonnages over paths beyond the range of any other type of mechanical conveyor. The capacity may be several thousand tons per hour, and the distance several miles. It may be horizontal or inclined upward or downward, or it may be a combination of these, as outlined in Fig. 10.5.37. The limit of incline is reached when the material tends to slip on the belt surface. There are special belts with molded designs to assist in keeping material from slipping on inclines. They will handle pulverized, granular, or lumpy material. Special compounds are available if material is hot or oily.

In its simplest form, the conveyor consists of a head or drive pulley, a take-up pulley, an endless belt, and carrying and return idlers. The spacing of the carrying idlers varies with the width and loading of the belt and usually is 5 ft (1.5 m) or less. Return idlers are spaced on 10-ft (3.0-m) centers or slightly less with wide belts. Sealed antifriction idler bearings are used almost exclusively, with pressure-lubrication fittings requiring attention about once a year.

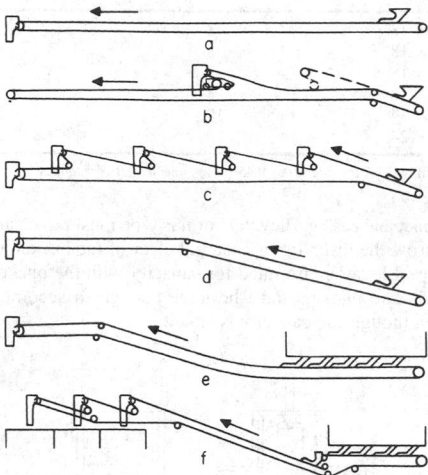

Fig. 10.5.37 Typical arrangements of belt conveyors.

Belts Belt width is governed by the desired conveyor capacity and maximum lump size. The standard rubber belt construction (Fig. 10.5.38) has several plies of square woven cotton duck or synthetic fabric such as rayon, nylon, or polyester cemented together with a rubber compound called ''friction'' and covered both top and bottom with rubber to resist abrasion and keep out moisture. Top cover thickness is

Fig. 10.5.38 Rubber-covered conveyor belt.

determined by the severity of the job and varies from $\frac{1}{16}$ to $\frac{3}{4}$ in (1.6 to 19 mm). The bottom cover is usually $\frac{1}{16}$ in (1.6 mm). By placing a layer of loosely woven fabric, called the breaker strip, between the cover and outside fabric ply, it is often possible to double the adhesion of the cover to the carcass. The belt is rated according to the tension to which it may safely be subjected, and this is a function of the length and lift of the conveyor. The standard Rubber Manufacturers Association (RMA) multiple ply ratings in lb/in (kg/mm) of width per ply are as follows:

RMA multiple ply No.	MP35	MP43	MP50	MP60	MP70
Permissible pull, lb/in (kg/mm) of belt width per ply, vulcanized splice	35 (0.63)	43 (0.77)	50 (0.89)	60 (1.07)	70 (1.25)
Permissible pull, lb/in (kg/mm) of belt width per ply, mechanical splice	27 (0.48)	33 (0.60)	40 (0.71)	45 (0.80)	55 (0.98)

Thus, for a pull of 4,200 lb (1,905 kg) and 24-in (0.61-m) belt width, a five-ply MP35 could be used with a vulcanized splice or a five-ply MP50 could be used with a mechanical splice.

High-Strength Belts For belt conveyors of extremely great length, a greater strength per inch of belt width is available now through the use of improved weaving techniques that provide straight-warp synthetic fabric to support the tensile forces (Uniflex by Uniroyal, Inc., or Flexseal by B. F. Goodrich). Strengths go to 1,500 lb/in width tension rating. They are available in most cover and bottom combinations and have good bonding to carcass. The number of plies is reduced to two instead of as many as eight so as to give excellent flexibility. Widths to 60 in are available. Other conventional high-strength fabric belts are available to

somewhat lower tensile ratings of 90 (1.61), 120 (2.14), 155 (2.77), 195 (3.48), and 240 (4.29) lb/in (kg/mm) per ply ratings.

The B. F. Goodrich Company has developed a steel-cable-reinforced belt rated 700 to 4,400 lb/in (12.5 to 78.6 kg/mm) of belt width. The belt has parallel brass-plated 7 by 19 steel airplane cables ranging from $\frac{5}{32}$ to $\frac{3}{8}$ in (4.0 to 9.5 mm) diameter placed on $\frac{1}{2}$- to $\frac{3}{4}$-in (12.7- to 19.0-mm) centers.

These belts are used for long single-length conveyors and for high-lift, extremely heavy duty service, e.g., for taking ore from deep open pits, thus providing an alternative to a spiraling railway or truck route.

Synthetic rubber is in use for belts. Combinations of synthetic and natural rubbers have been found satisfactory. Synthetics are superior under special circumstances, e.g., neoprene for flame resistance and resistance to petroleum-based oils, Buna N for resistance to vegetable, animal, and petroleum oils, and butyl for resistance to heat (per RMA).

Belt Life With lumpy material, the impact at the loading point may be destructive. Heavy lumps, such as ore and rock, cut through the protective cover and expose the carcass. The impact shock is reduced by making the belt supports flexible. This can be done by the use of idlers with cushion or pneumatic tires (Fig. 10.5.39) or by supporting the idlers on rubber mountings. Chuting the load vertically against the belt should be avoided. Where possible, the load should be given a movement in the direction of belt travel. When the material is a mixture of lumps and fines, the fines should be screened through to form a cushion for the lumps. Other destructive factors are belt overstressing, belts

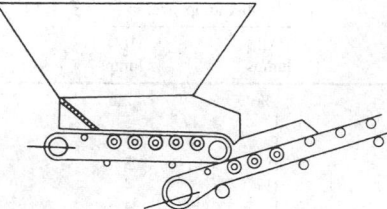

Fig. 10.5.39 Pneumatic-tired idlers applied to belt feeder at loading point of belt conveyor.

running out of line and rubbing against supports, broken idlers, and failure to clean the belt surface thoroughly before it comes in contact with the snub and tripper pulleys. Introduction of a 180° twist in the return belt (B. F. Goodrich Co.) at both head and tail ends can be used to keep the clean side of the belt against the return idlers and to prevent buildup. Using one 180° twist causes both sides of the belt to wear evenly. For each twist, 1 ft of length/in of belt width is required.

Idler Pulleys Troughing idlers are usually of the three-pulley type (Fig. 10.5.40), with the troughing pulleys at 20°. There is a growing tendency toward the use of 35 and 45° idlers to increase the volume capacity of a belt; 35° idlers will increase the volume capacity of a given belt 25 to 35 percent over 20° idlers, and 45° idlers, 35 to 40 percent.

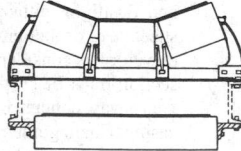

Fig. 10.5.40 Standard assembly of three-pulley troughing idler and return idler.

The bearings, either roller or ball type, are protected by felt or labyrinth grease seals against the infiltration of abrasive dust. A belt running out of line may be brought into alignment by shifting slightly forward one end or the other with a few idler sets. **Self-aligning idlers** (Fig. 10.5.41) should be spaced not more than 75 ft (23 m) apart. These call attention to the necessity of lining up the belt and should not serve as continuing correctives.

Drive Belt slip on the conveyor-drive pulley is destructive. There is little difference in tendency to slip between a bare pulley and a rubber-lagged pulley when the belt is clean and dry. A wet belt will adhere to a lagged pulley much better, especially if the lagging is grooved. Heavy-duty conveyors exposed to the possibility of wetting the belt are gener-

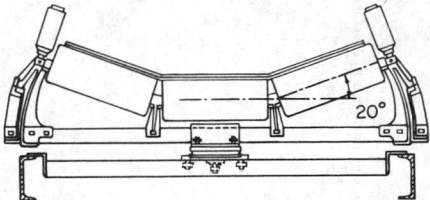

Fig. 10.5.41 Self-aligning idler.

ally driven by a head pulley lagged with a ½-in (12.7-mm) rubber belt and with ¼- by ¼-in (6.4- by 6.4-mm) grooves spaced ½ in (12.7 mm) apart and, preferably, diagonally as a herringbone gear. A snub pulley can be employed to increase the arc of contact on the head pulley, and since the pulley is in contact with the dirty side of the belt, a belt cleaner is essential. The belt cleaner may be a high-speed bristle brush, a spiral rubber wiper (resembling an elongated worm pinion), circular disks mounted slantwise on a shaft to give a wiping effect when rotated, or a scraper. Damp deposits such as clay or semifrozen coal dirt are best removed by multiple diagonal scrapers of stainless steel.

A belt conveyor should be emptied after each run to avoid a heavy starting strain. The motor should have high starting torque, moderate starting-current inrush, and good characteristics when operating under full load. The double-squirrel-cage ac motor fulfills these requirements.

Heavy-Duty Belt-Conveyor Drives For extremely heavy duty, it is essential that the drive torque be built up slowly, or serious belt damage will occur. The hydraulic clutch, derived from the hydraulic automobile clutch, serves nicely. The best drive developed to date is the **dynamic clutch** (Fig. 10.5.42). This has a magnetized rotor on the extended motor shaft, revolving within an iron ring keyed to the reduction gearing of the conveyor. The energizing current is automatically built up over a period that may extend to 2 min, and the increasing magnetic pull on the ring builds up the belt speed.

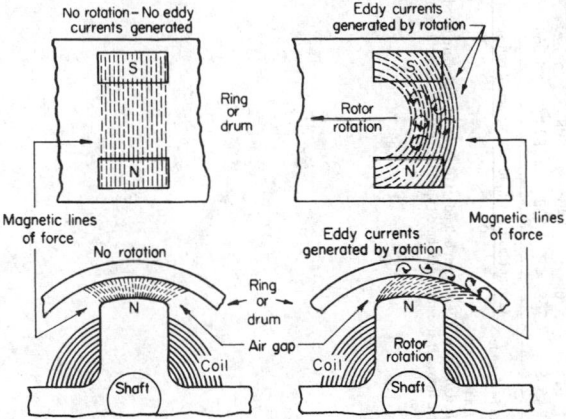

Fig. 10.5.42 Operating principle of the dynamic clutch.

Take-ups For short conveyors, a screw take-up is satisfactory. For long conveyors, the expansion and contraction of the belt with temperature changes and the necessity of occasional cutting and resplicing make a weighted gravity take-up preferable, especially if a vulcanized splice is used. The take-up should, if possible, be located where the slack first occurs, usually just back of the drive except in a conveyor inclined

downward (retarding conveyor), when the take-up is located at the downhill end.

Trippers The load may be removed from the belt by a diagonal or V plow, but a tripper that snubs the belt backward is standard equipment. Trippers may be (1) stationary, (2) manually propelled by crank, or (3) propelled by power from one of the snubbing pulleys or by an independent motor. The discharge may be chuted to either side or back to the belt by a deflector plate. When the tripper must move back to the load-receiving end of the conveyor, it is usual to incline the belt for about 15 ft (4.6 m) to match the slope up to the tripper top pulley. As the lower tripper snub pulleys are in contact with the dirty side of the belt, a cleaner must be provided between the pulleys. A scraper in light contact with the face of the pulley may be advisable.

Belt Slope The slopes (in degrees) given in Table 10.5.11 are the maximum permissible angles for various materials.

Table 10.5.11 Maximum Belt Slopes for Various Materials, Degrees

Coal: anthracite, sized; mined, 50 mesh and under; or mined and sized	16
Coal, bituminous, mined, run of mine	18
Coal: bituminous, stripping, not cleaned; or lignite	22
Earth, as excavated, dry	20
Earth, wet, containing clay	23
Gravel, bank run	20
Gravel, dry, sharp	15–17
Gravel, pebbles	12
Grain, distillery, spent, dry	15
Sand, bank, damp	20–22
Sand, bank, dry	16–18
Wood Chips	27

SOURCE: Uniroyal, Inc.

Determination of Motor Horsepower The power required to drive a belt conveyor is the sum of the powers required (1) to move the empty belt, (2) to move the load horizontally, and (3) to lift the load if the conveyor is inclined upward. If (3) is larger than the other two, an automatic brake must be provided to hold the conveyor if the current fails. A solenoid brake is usual. The power required to move the empty belt is given by Fig. 10.5.43. The power to move 100 tons/h horizontally is given by the formula $hp = 0.4 + 0.00325L$, where L is the distance between centers, ft. For other capacities the horsepower is proportional.

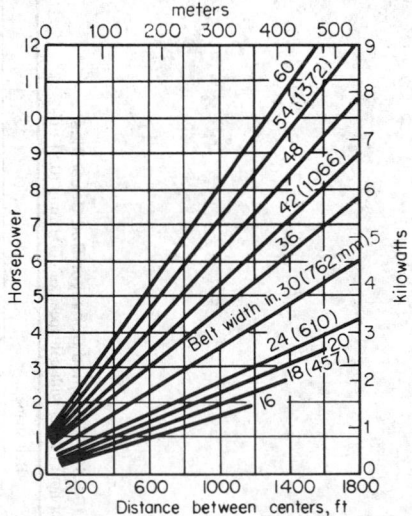

Fig. 10.5.43 Horsepower required to move belt conveyor empty at 100 ft/min (0.15 m/s).

Table 10.5.12 Troughed Conveyor-Belt Capacities*

Tons per hour (TPH) with belt speed of 100 ft/min (0.51 m/s).

Belt width, in	Equal length, 3 roll										Long center, 3 roll						
	20°			35°			45°			CED‡	35°			45%			CED‡
SCA†	0°	10°	30°	0°	10°	30°	0°	10°	30°		0°	10°	30°	0°	10°	30°	
12	10	14	24	16	20	28	19	29	35	.770							
24	52	74	120	83	102	143	98	115	150	1.050	82	101	141	96	113	149	1.05
30	86	121	195	137	167	232	161	188	244	1.095	111	144	212	133	163	225	1.12
42	177	249	400	282	345	476	332	386	500	1.130	179	248	394	216	281	417	1.22
60	375	526	843	598	729	1,003	702	815	1,053	1.187	266	417	734	324	468	772	1.35
72	548	768	1,232	874	1,064	1,464	1,026	1,190	1,535	1.205	291	516	987	356	573	1,030	1.44

* Tons per hour (TPH) = value from table $\times \dfrac{\text{(actual material wt., lb/ft}^3)}{100} \times \dfrac{\text{(actual belt speed, ft/min)}}{100}$

† Surcharge angle (see Fig 10.5.44).

‡ Capacity calculated for standard distance of load from belt edge: $(0.55\,b + 0.9)$, where b = belt width, inches. For constant 2-in edge distance (CED) multiply by CED constant as given in this table. For metric units multiply in by 25.4 for mm; tons per hour by 0.91 for tonne per hour; ft/min by 0.0051 for m/s.

For slumping materials (very free flowing), use capacities based upon 2-in CED. This includes dry silica sand, dry aerated portland cement, wet concrete, etc., with surcharge angle 5° or less.

Table 10.5.13   Minimum Belt Width for Lumps

Belt width,	in	12	18	24	30	42	60	72
	mm	305	457	610	762	1,067	1,524	1,829
Sized material,	in	2	4	5	6	8	12	14
	mm	51	102	127	152	203	305	356
Unsized material,	in	4	6	8	10	14	20	24
	mm	102	152	203	254	356	508	610

SOURCE Uniroyal, Inc.

Table 10.5.14   Maximum Belts Speeds, ft/min, for Various Materials*

Width of belt, in	Light or free-flowing materials, grains, dry sand, etc.	Moderately free-flowing sand, gravel, fine stone, etc.	Lump coal, coarse stone, crushed ore	Heavy sharp lumpy materials, heavy ores, lump coke
12–14	400	250		
16–18	500	300	250	
20–24	600	400	350	250
30–36	750	500	400	300
42–60	850	550	450	350

* Multiply in by 25.4 for mm, ft/min by 0.005 for m/s.

The capacity in tons per hour for materials of various weights per cubic foot is given by Table 10.5.12. Table 10.5.13 gives minimum belt widths for lumps of various sizes. Table 10.5.14 gives advisable maximum belt speeds for various belt widths and materials. The speed should ensure full-capacity loading so that the entire belt width is utilized.

Drive Calculations   From the standpoint of the application of power to the belt, a conveyor is identical with a power belt. The determining factors are the coefficient of friction between the drive pulley and the belt, the tension in the belt, and the arc of contact between the pulley and the belt. The arc of the contact is increased up to about 240° by using a

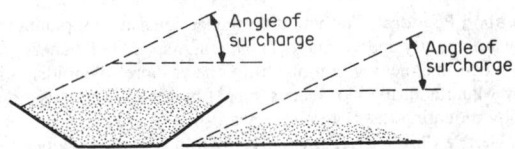

Fig. 10.5.44   (a) Troughed belt; (b) flat belt. (Uniroyal, Inc.)

snub pulley and up to 410° by using two pulleys geared together or driven by separate motors and having the belt wrapped around them in the form of a letter S. The resistance to be overcome is the sum of all the frictional resistances throughout the length of the conveyor plus, in the case of a rising conveyor, the resistance due to lifting the load. The sum of the conveyor and load resistances determines the working pull that has to be transmitted to the belt at the drive pulley. The total pull is increased by the slack-side tension necessary to keep the belt from slipping on the pulley. Other factors adding to the belt pull are the component of the weight of the belt if the conveyor is inclined and a take-up pull to keep the belt from sagging between the idlers at the loading point. These, however, do not add to the working pull. The maximum belt pull determines the length of conveyor that can be used. If part of the conveyor runs downgrade, the load on it will reduce the working pull. In moderate-length conveyors, stresses due to acceleration or deceleration are safely carried by the factor of safety used for belt-life calculations.

The total or maximum tension T_{max} must be known to specify a suitable belt. The effective tension T_e is the difference between tight-side tension and slack-side tension, or $T_e = T_1 - T_2$. The coefficient of friction between rubber and steel is 0.25; with a lagged pulley, between rubber and rubber, it is 0.55 for ideal conditions but should be taken as 0.35 to allow for loss due to dirty conditions.

Except for extremely heavy belt pulls, the tandem drive is seldom used since it is costly; the lagged-and-grooved drive pulley is used for most industrial installations.

For a belt with 6,000-lb (26,700-N) max tension running on a bare

pulley drive with 180° wrap (Table 10.5.15), $T = 1.85T_e = 6,000$ lb; $T_e = 3,200$ lb (14,200 N). Such a belt, 30 in wide, might be a five-ply MP50, a reduced-ply belt rated at 200 lb/in, or a steel-cable belt with 5/32-in (4-mm) cables spaced on 0.650-in (16.5-mm) centers. The last is the most costly.

Table 10.5.15   Ratio T_1 to T_e for Various Arcs of Contact with Bare Pulleys and Lagged Pulleys
(Coefficients of friction 0.25 and 0.35, respectively)

Belt wrap, deg	180	200	210	215	220	240
Bare pulley	1.85	1.72	1.67	1.64	1.62	1.54
Lagged pulley	1.50	1.42	1.38	1.36	1.35	1.30

In an inclined belt with single pulley drive, the T_{max} is lowest if the drive is at the head end and increases as the drive shifts toward the foot end.

Allowance for Tripper   The belt lifts about 5 ft to pass over the top snub pulley of the tripper. Allowance should be made for this lift in determining the power requirement of the conveyor. If the tripper is propelled by the belt, an allowance of 1 hp (0.75 kW) for a 16-in (406-mm) belt, 3 hp (2.2 kW) for a 36-in (914-mm) belt, or 7 hp (5.2 kW) for a 60-in (1,524-mm) belt is ample. If a rotary cleaning brush is driven from one of the snub shafts, an allowance should be made which is approximately the same as that for the propulsion of the tripper.

Magnetic pulleys are frequently used as head pulleys on belt conveyors to remove tramp iron, such as stray nuts or bolts, before crushing; to concentrate magnetic ores, such as magnetic or nickeliferous pyrrhotite, from nonmagnetic material; and to reclaim iron from foundry refuse. A chute or hopper automatically receives the extracted material as it is drawn down through the other nonmagnetic material, drawn around the end pulley on the belt, and finally released as the belt leaves the pulley. Light-duty permanent-magnet types [for pulleys 12 to 24 in (305 to 610 mm) in diameter] will separate material through a 2-in (51-mm) layer on the belt. Heavy-duty permanent-magnet units (12 to 24 in in diameter) will separate material if the belt carries over 2 in of material or if the magnetic content is very fine. Even larger units are available for special applications. So effective and powerful are the permanent-magnet types that **electromagnetic pulleys** are available only in the larger sizes, from 18 to 48 in in diameter. The permanent-magnet type requires no slip rings, external power, or upkeep.

Overhead magnetic separators (Dings), both electromagnetic and Ceramox permanent-magnet types, for suspension above a belt conveyor are also available for all commercial belt widths to pull magnetic material from burden as thick as 40 in and at belt speeds to 750 ft/min. These may or may not be equipped with a separately encompassing belt to be self-cleaning. Wattages vary from 1,600 to 17,000. The permanent-magnet type requires no electric power and have nonvarying magnet strength. An alternate type of belt protection is to use a Ferro Guard Detector (Dings) to stop belt motion if iron is detected.

Trippers of the fixed or movable type are used for discharging material between the ends of a belt conveyor. A **self-propelling tripper** consists of two pulleys, over which the belt passes, the material being discharged into the chute as the belt bends around the upper pulley. The pulleys are mounted on a frame carried by four wheels and power-driven. A lever on the frame and stops alongside the rails enable the tripper, taking power from the conveyor belt, to move automatically between the stops, thus distributing the material. Rail clamps are provided to hold the tripper in a fixed position when discharging. Motor-driven trippers are used when it is desirable to move the tripper independently of the conveyor belt. Fixed trippers have their pulleys mounted on the conveyor framework instead of on a movable carriage.

Shuttle conveyors are frequently used in place of trippers for distributing materials. They consist of a reversible belt conveyor mounted upon a movable frame and discharging over either end.

Belt-Conveyor Arrangements Figure 10.5.37 shows **typical arrangements of belt conveyors.** *a* is a level conveyor receiving material at one end and discharging at the other. *b* shows a level conveyor with traveling tripper. The receiving end of the conveyor is depressed so that the belt will not be lifted against the chute when the tripper is at its extreme loading end. *c* is a level conveyor with fixed trippers. *d* shows an inclined end combined with a level section. *e* is a combination of level conveyor, vertical curve, and horizontal section. The radius of the vertical curve depends upon the weight of the belt and the tension in the belt at the point of tangency. This must be figured in each case and is found by the formula: min radius, ft = belt tension at lowest point of curve divided by weight per ft of belt. The belt weight should be for the worn belt with not over 1/16-in (1.6-mm) top cover. *f* is a combination of level conveyor receiving material from a bin, a fixed dump, and inclined section, and a series of fixed trippers.

Portable conveyors are widely used around retail coal yards, small power plants, and at coal mines for storing coal and reclaiming it for loading into trucks or cars. They are also used for handling other bulk materials. They consist of a short section of chain or belt conveyors mounted on large wheels or crawler treads and powered with a gasoline engine or electric motor. They vary in length from 20 to 90 ft and can handle up to 250 tons/h (63 kg/s) of coal. For capacities greater than what two people can shovel onto a belt, some form of power loader is necessary.

Sectional-belt conveyors have come into wide use in coal mines for bringing the coal from the face and loading it into cars in the entry. They consist of short sections (6 ft or more) of light frame of special low-type construction. The sections are designed for ease of connecting and disconnecting for transfer from one part of the mine to another. They are built in lengths up to 1,000 ft (305 m) or more under favorable conditions and can handle 125 tons/h (32 kg/s) of coal.

Sliding-belt conveyors use belts sliding on decks instead of troughed belts carried on rollers. Sliding belts are used in the shipping rooms of department stores for handling miscellaneous parcels, in post offices for handling mail bags, in chemical plants for miscellaneous light waste, etc. The decking preferably is of maple strips. If of steel, the deck should be perforated at intervals to relieve the vacuum effect between the bottom of the belt and the deck. Cotton or balata belts are best. The speed should be low. The return run may be carried on 4-in straight-face idlers. The power requirement is greater than with idler rollers.

The **oscillating conveyor** is a horizontal trough varying in width from 12 to 48 in (305 to 1,219 mm), mounted on rearward-inclined cantilever supports, and driven from an eccentric at 400 to 500 r/min. The effect is to "bounce" the material along at about 50 ft/min (0.25 m/s) with minimum wear on the trough. The conveyor is adapted to abrasive or hot fragmentary materials, such as scrap metals, castings, or metal chips. The trough bottom may be a screen plate to cull fine material, as when cleaning sand from castings, or the trough may have louvers and a ventilating hood to cool the moving material. These oscillating conveyors may have unit lengths up to 100 ft (30 m). Capacities range from a few tons to 100 tons/h (25 kg/s) with high efficiency and low maintenance.

Feeders

When material is drawn from a hopper or bin to a conveyor, an **automatic feeder** should be used (unless the material is dry and free-running, e.g., grain). The satisfactory operation of any conveyor depends on the material being fed to it in an even and continuous stream. The automatic feeder not only ensures a constant and controlled feed, irrespective of the size of material, but saves the expense of an operator who would otherwise be required at the feeding point. Figure 10.5.45 shows a **reciprocating-plate feeder,** consisting of a plate mounted on four wheels and forming the bottom of the hopper. When the plate is moved forward, it carries the material with it; when moved back, the plate is withdrawn from under the material, allowing it to fall into the chute. The plate is moved by connecting rods from cranks or eccentrics. The capacity of this feeder is determined by the length and number of strokes, width of plate, and location of the adjustable gate. The number of strokes should not exceed 60 to 70 per min. When used under a track hopper, the material remaining on the plate may freeze in winter, as this type of feeder is not self-clearing.

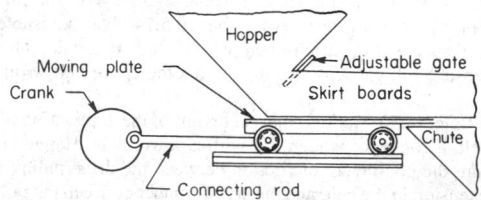

Fig. 10.5.45 Reciprocating-plate feeder.

Vibrating Feeders The vibrating feeder consists of a plate inclined downward slightly and vibrated (1) by a high-speed unbalanced pulley, (2) by electromagnetic vibrations from one or more solenoids, as in the Jeffrey Manufacturing Co. feeder, or (3) by the slower pulsations secured by mounting the plate on rearward-inclined leaf springs.

The **electric vibrating feeder** (Fig. 10.5.46) operates magnetically with a large number of short strokes (7,200 per min from an alternating current in the small sizes and 3,600 from a pulsating direct current in the larger sizes). It is built to feed from a few pounds per minute to 1,250 tons/h (315 kg/s) and will handle any material that does not adhere to the pan. It is self-cleaning, instantaneously adjustable for capacity, and controllable from any point near or remote. It is usually supported from above with spring shock absorbers *a* in each hanger, but it be may supported from below with similar springs in the supports. A modified form can be set to feed a weighed **constant amount** hourly for process control.

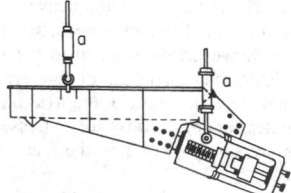

Fig. 10.5.46 Electric vibrating feeder.

Roller Conveyors

The principle involved in **gravity roller conveyors** is the control of motion due to gravity by interposing an antifriction trackage set at a definite grade. Roller conveyors are used in the movement of all sorts of package goods with smooth surfaces which are sufficiently rigid to prevent sagging between rollers—in warehouses, brickyards, building-supply yards, department stores, post offices, and the manufacturing and shipping departments of industrial manufacturers.

The **rollers vary in diameter** and strength from 1 in, with a capacity of 5 lb (2.3 kg) per roller, up to 4 in (102 mm), with a capacity of 1,800 lb (816 kg) per roller. The heavier rollers are generally used in foundries and steel mills for moving large molds, castings, or stacks of sheet steel. The small roller is used for handling small, light objects. The spacing of the rollers in the frames varies with the size and weight of the objects to be moved. Three rollers should be in contact with the package to prevent hobbling. The grade of fall required to move the object varies from 1½ to 7 percent, depending on the weight and character of the material in contact with the rollers.

Figure 10.5.47 shows a typical cross section of a roller conveyor. Curved sections are similar in construction to straight sections, except that in the majority of cases multiple rollers (Fig. 10.5.48) are used to keep the package properly lined up and in the center of the conveyor.

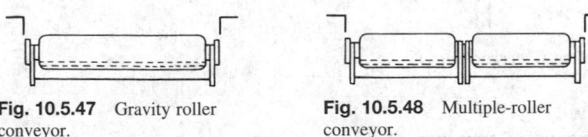

Fig. 10.5.47   Gravity roller conveyor.

Fig. 10.5.48   Multiple-roller conveyor.

Figure 10.5.49 illustrates a **wheel conveyor,** used for handling bundles of shingles, fruit boxes, bundles of fiber cartons, and large, light cases. The wheels are of ball-bearing type, bolted to flat-bar or angle-frame rails.

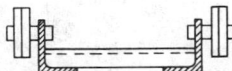

Fig. 10.5.49   Wheel-type conveyor.

When an installation involves a **trunk line with several tributary runs,** a simple two-arm deflector at each junction point holds back the item on one run until the item on the other has cleared. **Power-operated** roller conveyors permit handling up an incline. Usually the rolls are driven by sprockets on the spindle ends. An alternative of a smooth deck and pusher flights should be considered as costing less and permitting steeper inclines.

Platform conveyors are single- or double-strand conveyors (Fig. 10.5.50) with plates of steel or hardwood forming a continuous platform on which the loads are placed. They are adapted to handling heavy drums or barrels and miscellaneous freight.

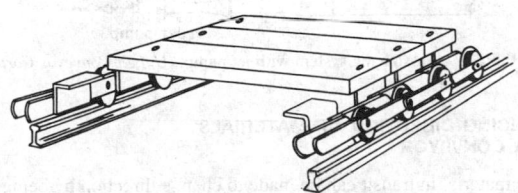

Fig. 10.5.50   Double-strand platform conveyor.

Pneumatic Conveyors

The pneumatic conveyor transports dry, free-flowing, granular material in suspension within a pipe or duct by means of a high-velocity airstream or by the energy of expanding compressed air within a comparatively dense column of fluidized or aerated material. Principal uses are (1) dust collection; (2) conveying soft materials, such as grain, dry foodstuff (flour and feeds), chemicals (soda ash, lime, salt cake), wood chips, carbon black, and sawdust; (3) conveying hard materials, such as fly ash, cement, silica metallic ores, and phosphate. The need in processing of bulk-transporting plastic pellets, powders, and flour under contamination-free conditions has increased the use of pneumatic conveying.

Dust Collection   All pipes should be as straight and short as possible, and bends, if necessary, should have a radius of at least three diameters of the pipe. Pipes should be proportioned to keep down friction losses and yet maintain the air velocities that will prevent settling of the material. Frequent cleanout openings must be provided. Branch connections should go into the sides of the main and deliver the incoming stream as nearly as possible in the direction of flow of the main stream. Sudden changes in diameter should be avoided to prevent eddy losses.

When **vertical runs** are short in proportion to the horizontal runs, the size of the riser is locally restricted, thereby increasing the air velocity and producing sufficient lifting power to elevate the material. If the vertical pipes are comparatively long, they are not restricted, but the necessary lifting power is secured by increased velocity and suction throughout the entire system.

The area of the main at any point should be 20 to 25 percent in excess of the sums of the branches entering it between the point in question and the dead end of the main. Floor sweepers, if equipped with efficient blast gates, need not be included in computing the main area. The diameter of the connecting pipe from machine to main and the section required at each hood are determined by experience. The sum of the volumes of each branch gives the total volume to be handled by the fan.

Fan Suction   The maintained resistance at the fan is composed of (1) suctions of the various hoods, which must be chosen from experience, (2) collector loss, and (3) loss due to pipe friction.

The **pipe loss** for any machine is the sum of the losses in the corresponding branch and in the main from that branch to the fan. For each elbow, add a length equal to 10 diameters of straight pipe. The total loss in the system, or static pressure required at the fan, is equal to the sum of (1), (2), and (3).

For conveying soft materials, a fan is used to create a suction. The suspended material is collected at the terminal point by a separator upstream from the fan. The material may be moved from one location to another or may be unloaded from barge or rail car. Required conveying velocity ranges from 2,000 ft/min (10.2 m/s) for light materials, such as sawdust, to 3,000 to 4,000 ft/min (15.2 to 20.3 m/s) for medium-weight materials, such as grain. Since abrasion is no problem, steel pipe or galvanized-metal ducts are satisfactory. Unnecessary bends and fittings should be avoided to minimize power consumption.

For conveying hard materials, a water-jet exhauster or steam exhauster is used on suction systems, and a positive-displacement blower on pressure systems. A mechanical exhauster may also be used on suction systems if there is a bag filter or air washer ahead of the exhauster. The largest tonnage of hard material handled is fly ash. A single coal-fired, steam-electric plant may collect more than 1,000 tons (907 tonnes) of fly ash per day. Fly ash can be conveyed several miles pneumatically at 30 tons (27 tonnes) or more per h using a pressure conveyor. Another high-tonnage material conveyed pneumatically is cement. Individual transfer conveyors may handle several hundred tons per hour. Hard materials are usually also heavy and abrasive. Required conveying velocities vary from 4,000 to 5,000 ft/min (20.3 to 25.4 m/s). Heavy cast-iron or alloy pipe and fittings are required to prevent excessive wear.

Vacuum pneumatic ash-handling systems have the airflow induced by steam- or water-jet exhausters, or by mechanical blowers. Cyclone-type Nuveyor receivers collect the ash for storage in a dry silo. A typical Nuveyor system is shown in Fig. 10.5.51. The conveying velocity is dependent upon material handled in the system. Fly ash is handled at approximately 3,800 ft/min (19.3 m/s) and capacity up to 60 tons/h (15.1 kg/s). Positive-pressure pneumatic ash systems are becoming more common because of higher capacities required. These systems can convey fly ash up to 1½ mi (2.4 km) and capacities of 100 tons/h (25.2 kg/s) for shorter systems.

The **power requirement** for pneumatic conveyors is much greater than for a mechanical conveyor of equal capacity, but the duct can be led along practically any path. There are no moving parts and no risk of injury to the attendant. The vacuum-cleaner action provides dustless

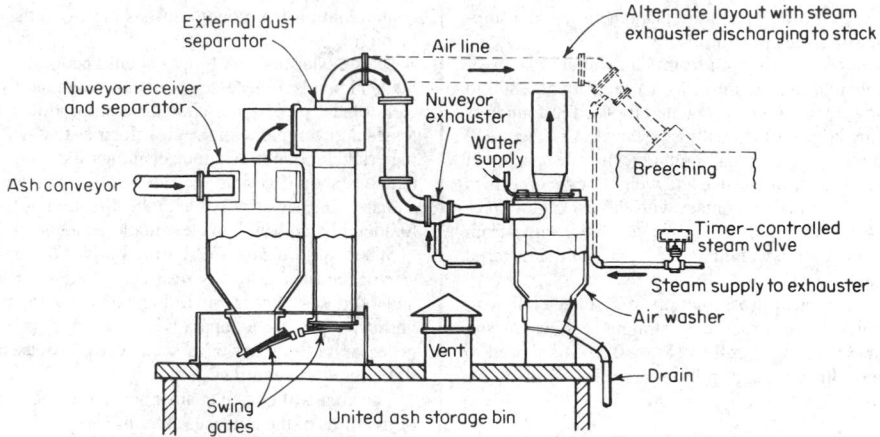

Fig. 10.5.51   Nuveyor pneumatic ash-handling system. (*United Conveyor Corp.*)

operation, sometimes important when pulverized material is unloaded from boxcars through flexible hose and nozzle. A few materials build up a state-electric charge which may introduce an explosion risk. Sulfur is an outstanding example. Sticky materials tend to pack at the elbows and are unsuitable for pneumatic handling.

The performance of a pneumatic conveyor cannot be predicted with the accuracy usual with the various types of mechanical conveyors and elevators. It is necessary to rely on the advice of experienced engineers.

The **Fuller-Kinyon system** for transporting dry pulverized material consists of a motor- or engine-driven pump, a source of compressed air for fluidizing the material, a conduit or pipe-line, distributing valves (operated manually, electropneumatically, or by motor), and electric bin-level indicators (high-level, low-level or both). The impeller is a specially designed differential-pitch screw normally turning at 1,200 r/min. The material enters the feed hopper and is compressed in the decreasing pitch of the screw flights. At the discharge end of the screw, the mass is introduced through a check valve to a mixing chamber, where it is aerated by the introduction of compressed air. The fluidized material is conveyed in the transport line by the continuing action of the impeller screw and the energy of expanding air. Practical distance of transportation by the system depends upon the material to be handled. Cement has been handled in this manner for distances up to a mile. The most important field of application is the handling of portland cement. For this material, the Fuller-Kinyon pump is used for such operations as moving both raw material and finished cement within the cement-manufacturing process; loading out; unloading ships, barges, and railway cars; and transferring from storage to mixer plant on large construction jobs.

The **Airslide** (registered trademark of the Fuller Company) **conveyor system** is an air-activated gravity-type conveyor using low-pressure air to aerate or fluidize pulverized material to a degree which will permit it to flow on a slight incline by the force of gravity. The conveyor comprises an enclosed trough, the bottom of which has an inclined air-permeable surface. Beneath this surface is a plenum chamber through which air is introduced at low pressure, depending upon the application. Various control devices for controlling and diverting material flow and for controlling air supply may be provided as part of complete systems. For normal conveying applications, the air is supplied by an appropriate fan; for operation under a head of material (as in a storage bin), the air is supplied by a rotary positive blower. The Airslide conveyor is widely used for horizontal conveying, discharge of storage bins, and special railway-car and truck-trailer transport, as well as in stationary blow-tank-type conveying systems. An important feature of this conveyor is low power requirement.

Hydraulic Conveyors

Hydraulic conveyors are used for handling boiler-plant ash or slag from an ash hopper or slag tank located under the furnace. The material is flushed from the hopper to a grinder, which discharges to a jet pump or a mechanical pump for conveying to a disposal area or a dewatering bin (Fig. 10.5.52). Water requirements average 1 gal/lb of ash.

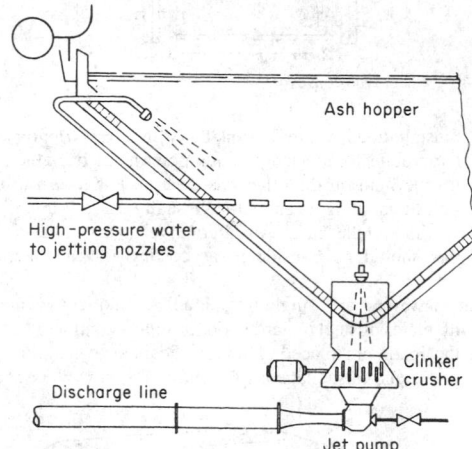

Fig. 10.5.52   Ash-sluicing system with jet pump. (*United Conveyor Corp.*)

CHANGING DIRECTION OF MATERIALS ON A CONVEYOR

Some material in transit can be made to change direction by being bent around a curve. Metal being rolled, wire, strip, material on webs, and the like, can be guided to new directions by channels, rollers, wheels, pulleys, etc.

Some conveyors can be given curvatures, such as in overhead monorails, tow car grooves, tracks, pneumatic tube bends, etc. Other, straight sections of conveyors can be joined by curved sections to accomplish directional changes. Figures 10.5.53 to 10.5.56 show some examples of these. In some cases, the simple intersection, or overlapping, at an angle of two conveyors can result in a direction change. See Figs. 10.5.57 and 10.5.58. When not all of the material on the first section of conveyor is to be diverted, sweeps or switches may be used. Figure 10.5.59 shows how fixed diverters set at decreasing heights can direct boxes of certain sizes onto other conveyors. Switches allow the material to be sent in one direction at one time and in another at other times. In Fig. 10.5.60 the diverter can swing to one side or the other. In Fig. 10.5.61, the switching section has a dual set of wheels, with the set being made higher at the time carrying the material in their angled direction. Pushers, as shown in Fig. 10.5.62 can selectively divert material, as can air blasts for lightweight items (Fig. 10.5.63) and tipping sections (Fig. 10.5.64).

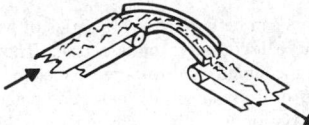

Fig. 10.5.53   Fixed curve turn section. Joins two conveyors and changes direction and level of material flow.

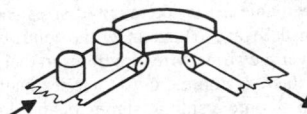

Fig. 10.5.54   Fixed curve turn section. Joins two conveyors and changes direction of material flow. Turn section has no power. Incoming items push those ahead of them around curve. Wheels, balls, rollers, etc. may be added to turn section to reduce friction of dead plate section shown.

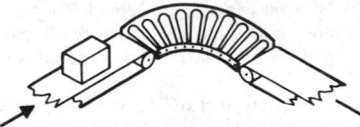

Fig. 10.5.55   Fixed curve turn section. Uses tapered rollers, skate wheels, balls, or belt. May be level or inclined, "dead" or powered.

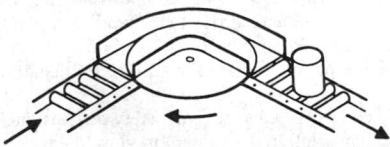

Fig. 10.5.56   Fixed curve turn section. Disk receives item from incoming conveyor section, rotates it, and directs it onto outgoing conveyor section.

Fig. 10.5.57   Simple intersection. Incoming conveyor section overlaps outgoing section and merely dumps material onto lower conveyor.

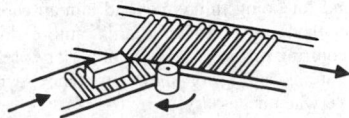

Fig. 10.5.58   Simple intersection. Both incoming and outgoing sections are powered. Rotating post serves as a guide.

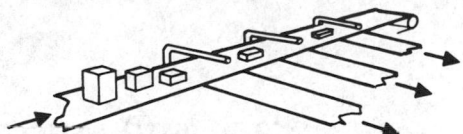

Fig. 10.5.59   Fixed diverters. Diverters do not move. They are preset at heights to catch and divert items of given heights.

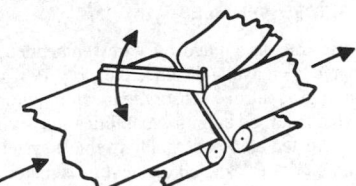

Fig. 10.5.60   Moving diverter. Material on incoming conveyor section can be sent to either outgoing section by pivoting the diverter to one side or the other.

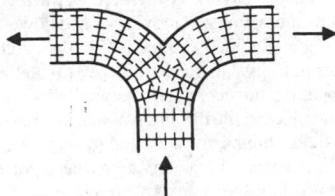

Fig. 10.5.61   Wheel switch. Fixed wheels carry material on incoming conveyor to outgoing conveyor until movable set of wheels is raised above the fixed set. Then material is carried in the other direction.

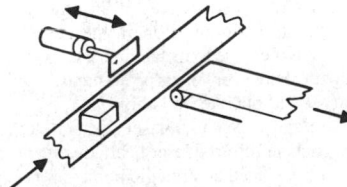

Fig. 10.5.62   Pusher diverter. When triggered, the pusher advances and moves item onto another conveyor. A movable sweep can also be used.

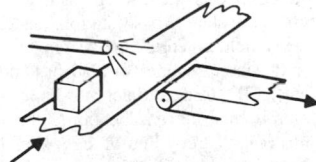

Fig. 10.5.63   Air diverter. When activated, a jet of air pushes a lightweight item onto another conveyor.

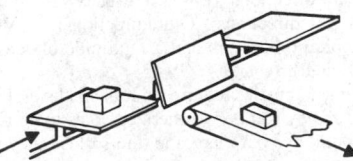

Fig. 10.5.64   Tilting section. Sections of main conveyor move along, level, until activated to tilt at chosen location to dump material onto another conveyor.

10.6 AUTOMATIC GUIDED VEHICLES AND ROBOTS

by Vincent M. Altamuro

AUTOMATIC GUIDED VEHICLES

Driverless towing tractors guided by wires embedded into or affixed onto the floor have been available since the early 1950s. Currently, the addition of computer controls, sensors that can monitor remote conditions, real-time feedback, switching capabilities and a whole new family of vehicles have created automatic guided vehicle systems (AGVs) that compete with industrial trucks and conveyors as material handling devices.

Most AGVs have only horizontal motion capabilities. Any vertical motion is limited. Fork lift trucks usually have more vertical motion capabilities than do standard AGVs. Automatic storage and retrieval systems (AS/RS) usually have very high rise vertical capabilities, with horizontal motions limited only to moving to and down the proper aisle. Power for the automatic guided vehicle is usually a battery, like that of the electric industrial truck. Guidance may be provided several ways. In *electrical* or *magnetic* guidance, a wire network is usually embedded in a narrow, shallow slot cut in the floor along the possible paths. The electromagnetic field emitted by the conductor wire is sensed by devices on-board the vehicle which activate a steering mechanism, causing the vehicle to follow the wire. An *optical* guidance system has a similar steering mechanism, but the path is detected by a photosensor picking out the bright path (paint, tape, or fluorescent chemicals) previously laid down. The embedded wire system seems more popular in factories, as it is in a protective groove, whereas the optical tape or painted line can get dirty and/or damaged. In offices, where the AGV may be used to pick up and deliver mail, the optical system may be preferred, as it is less expensive to install and less likely to deteriorate in such an environment. However, in carpeted areas, a wire can easily be laid under the carpet and operate invisibly.

In a **laser beam** guidance system, a laser scanner on the vehicle transmits and receives back an infrared laser beam that reads passive retroreflective targets that are placed at strategic points on x and y coordinates in the facility. The vehicle's on-board computers take the locations and distances of the targets and calculate the vehicle's position by triangulation. The locations of the loads to be picked up, the destinations at which they are to be dropped off, and the paths the vehicle is to travel are transmitted from the system's base station. Instructions are converted to inputs to the vehicle's steering and driving motors. A variation of the system is for the laser scanner to read bar codes or radio-frequency identification (RFID) targets to get more information regarding their mission than merely their current location. Other, less used, guidance methods are **infrared,** whereby line-of-sight signals are sent to the vehicle, and **dead reckoning,** whereby a vehicle is programmed to traverse a certain path and then turned loose.

The directions that an AGV can travel may be classified as unidirectional (one way), bidirectional (forward or backward along its path), or omnidirectional (all directions). Omnidirectional AGVs with five or more on-board microprocessors and a multitude of sensors are sometimes called **self-guided vehicles** (SGVs).

Automatic guided vehicles require smooth and level floors in order to operate properly. They can be weatherized so as to be able to run outdoors between buildings but there, too, the surface they travel on must be smooth and level.

AGV equipment can be categorized as:

1. Driverless tractors
2. Guided pallet trucks
3. Unit load transporters and platform carriers
4. Assembly or tool bed robot transporters

Driverless tractors can be used to tow a series of powerless material handling carts, like a locomotive pulls a train. They can be routed, stopped, coupled, uncoupled, and restarted either manually, by a programmed sequence, or from a central control computer. They are suited to low-volume, heavy or irregularly shaped loads which have to be moved over longer distances than would be economical for a conveyor.

Guided pallet trucks, like the conventional manually operated trucks they replace, are available in a wide range of sizes and configurations. In operation, they usually are loaded under the control of a person, who then drives them over the guide wire, programs in their desired destination, then switches them to automatic. At their destination, they can turn off the main guidepath onto a siding, automatically drop off their load, then continue back onto the main guidepath. The use of guided pallet trucks reduces the need for conventional manually operated trucks and their operators.

Unit load transporters and platform carriers are designed so as to carry their loads directly on their flat or specially contoured surface, rather than on forks or on carts towed behind. They can either carry material or work-in-process from workstation to workstation or they can be workstations themselves and process the material while they transport it.

The assembly or tool bed type of AGV is used to carry either work-in-process or tooling to machines. It may also be used to carry equipment for an entire process step — a machine plus its tooling — to large, heavy, or immovable products.

Robot transporters are used to make robots mobile. A robot can be fitted atop the transporter and carried to the work. Further, the robot can process the work as the transporter carries it along to the next station, thereby combining productive work with material handling and transportation.

Most AGVs and SGVs have several safety devices, including flashing in-motion lights, infrared scanners to slow them down when approaching an obstacle, sound warnings and alarms, stop-on-impact bumpers, speed regulators, and the like.

ROBOTS

A robot is a machine constructed as an assemblage of links joined so that they can be articulated into desired positions by a reprogrammable controller and precision actuators to perform a variety of tasks. Robots range from simple devices to very complex and "intelligent" systems by virtue of added sensors, computers, and special features. See Figure 10.6.1 for the possible components of a robotic system.

Robots, being programmable multijointed machines, fall between humans and fixed-purpose machines in their utility. In general, they are most useful in replacing humans in jobs that are dangerous, dirty, dull, or demeaning, and for doing things beyond human capabilities. They are usually better than conventional "hard" automation in short-run jobs and those requiring different tasks on a variety of products. They are ideal for operations calling for high consistency, cycle after cycle, yet are changeable when necessary.

There are several hundred types and models of robots. They are available in a wide range of shapes, sizes, speeds, load capacities, and other characteristics. Care must be taken to select a robot to match the requirements of the tasks to be done. One way to classify them is by their intended application. In general, there are industrial, commercial, laboratory, mobile, military, security, service, hobby, home, and personal robots. While they are programmable to do a wide variety of tasks, most are limited to one or a few categories of capabilities, based on their construction and specifications. Thus, within the classification of industrial robots, there are those that can paint but not assemble products, and of those that can assemble products, some specialize in assembling very small electronic components while others make automobiles.

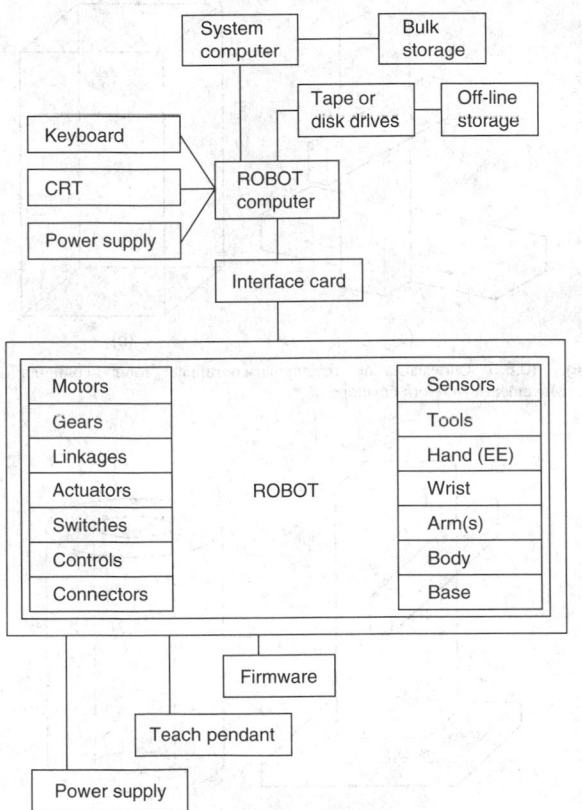

Fig. 10.6.1 Robotic system schematic. (*Robotics Research Division, VMA, Inc.*)

Some of the common uses of industrial robots include: loading and unloading machines, transferring work from one machine to another, assembling, cutting, drilling, grinding, deburring, welding, gluing, painting, inspecting, testing, packing, palletizing, and many others.

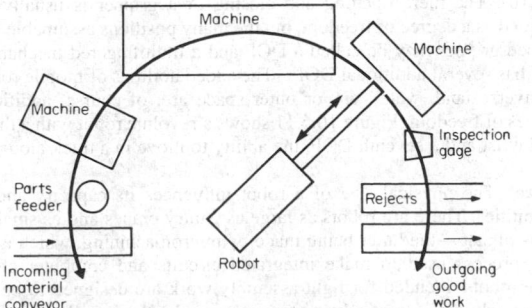

Fig. 10.6.2 Robot work cell. Robot is positioned to service a number of machines clustered about it.

Rather than have a robot perform only one task, it is sometimes possible to select a model and its tooling so that it can be positioned to do several jobs. Figure 10.6.2 shows a robot set in a work cell so that it can service the incoming material, part feeders, several machines, inspection gages, reject bins, and outgoing product conveyors arranged in an economical cluster around it. Robots are also used in chemical mixing and measuring; bomb disassembly; agriculture; logging; fisheries; department stores; amusement parks; schools; prisons; hospitals; nursing homes; nuclear power plants; space vehicles; underwater; surveillance; police, fire, and sanitation departments; inside the body; and an increasing number of innovative places.

Robot Anatomy

All robots have certain basic sections, and some have additional parts that give them added capabilities. All robots must have a power source, either mechanical, electrical, electromechanical, pneumatic, hydraulic, nuclear, solar, or a combination of these. They all must also have a means of converting the source of power to usable power and transmitting it: motors, pumps, and other actuators, and gears, cams, drives, bars, belts, chains, and the like. To do useful work, most robots have an assemblage of links arranged in a configuration best-suited to the tasks it is to do and the space within which it is to do them. In some robots this is called the *manipulator,* and the links and the joints connecting them are sometimes referred to as the robot's *body, shoulder, arm,* and *wrist.* A robot may have no, one, or multiple arms. Multiarmed robots must have control coordination protocols to prevent collisions, as must robots that work very close to other robots. Some robots have an immovable base that connects the manipulator to the floor, wall, column, or ceiling. In others the base is attached to, or part of, an additional section that gives it mobility by virtue of components such as wheels, tracks, rollers, "legs and feet," or other devices.

All robots need an intelligence and control section that can range from a very simple and relatively limited type up to a very complex and sophisticated system that has the capability of continuously interacting with varying conditions and changes in its environment. Such advanced control systems would include computers of various types, a multitude of microprocessors, memory media, many input/output ports, absolute and/or incremental encoders, and whatever type and number of sensors may be required to make it able to accomplish its mission. The robot may be required to sense the positions of its several links, how well it is doing its mission, the state of its environment, and other events and conditions. In some cases, the sensors are built into the robots; in others they are handled as an adjunct capability—such as machine vision, voice recognition, and sonic ranging—attached and integrated into the overall robotic system.

Other items that may be built into the robots or handled as attachments are the **tooling**—the things that interface the robot's arm or wrist and the workpiece or object to be manipulated and which accomplish the intended task. These are called the robot's *end effectors* or *end-of-arm tooling* (EOAT) and may be very simple or so complex that they become a major part of the total cost of the robotic system. A gripper may be more than a crude open or closed clamp (See Fig. 10.6.3), jaws equipped with sensors (See Figure 10.6.4), or a multifingered hand complete with several types of miniature sensors and capable of articulations that even the human hand cannot do. They may be binary or servoed. They may be purchased from stock or custom-designed to do specific tasks. They may be powered by the same type of energy as the basic robot or they may have a different source of power. A hydraulically powered robot may have pneumatically powered grippers and tooling, for example, making it a hybrid system. Most end effectors and EOAT are easily changeable so as to enable the robot to do more tasks.

Another part of the anatomy of many robots is a safety feature. Many robots are fast, heavy, and powerful, and therefore a source of danger. Safety devices can be both built into the basic robot and added as an adjunct to the installation so as to reduce the chances that the robot will do harm to people, other equipment, the tooling, the product, and itself.

The American National Standards Institute, New York City, issued ANSI/RIA R 15.06-1986 that is supported as a robot safety standard by the Robotic Industries Association, Dearborn, MI.

Finally, ways of programming robots, communicating with them, and monitoring their performance are needed. Some of the devices used for these purposes are teach boxes on pendants, keyboards, panels, voice recognition equipment, and speech synthesis modules.

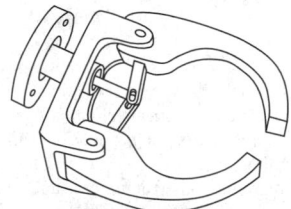

Fig. 10.6.3 A simple nonservo, no-sensor robot gripper.

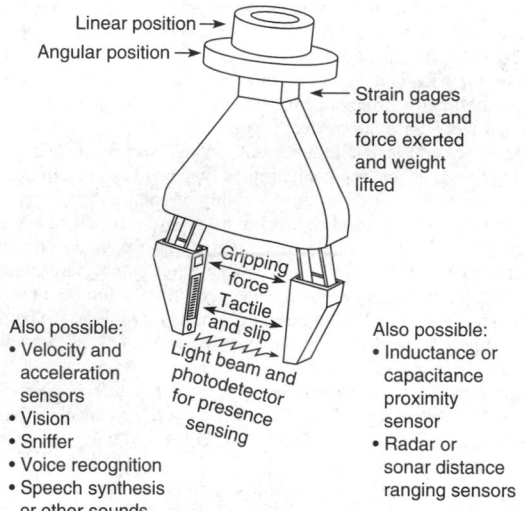

Linear position →
Angular position →

Strain gages for torque and force exerted and weight lifted

Gripping force
Tactile and slip
Light beam and photodetector for presence sensing

Also possible:
• Velocity and acceleration sensors
• Vision
• Sniffer
• Voice recognition
• Speech synthesis or other sounds

Also possible:
• Inductance or capacitance proximity sensor
• Radar or sonar distance ranging sensors

Fig. 10.6.4 A complex servoed, multisensor robot gripper.

Robot Specifications

While most robots are made "to stock" and sold from inventory, many are made "to order" to the specifications required by the intended application.

Configuration The several links of a robot's manipulator may be joined to move in various combinations of ways relative to one another. A joint may turn about its axis (rotational axis) or it may translate along its axis (linear axis). Each particular arrangement permits the robot's control point (usually located at the center of its wrist flange, at the center of its gripper jaws, or at the tip of its tool) to move in a different way and to reach points in a different area. When a robot's three movements are along translating joints, the configuration is called *cartesian* or *rectangular* (Fig. 10.6.5). When one of the three joints is made a rotating joint, the configuration is called *cylindrical* (Fig. 10.6.6). When two of the three joints are rotational, the configuration is called *polar* or

spherical (Fig. 10.6.7). And if all three joints are rotational, the robot is said to be of the *revolute* or *jointed-arm* configuration (Fig. 10.6.8). The selective compliance assembly robot arm (SCARA) is a special type of revolute robot in which the joint axes lie in the vertical plane rather than in the horizontal (Fig. 10.6.9).

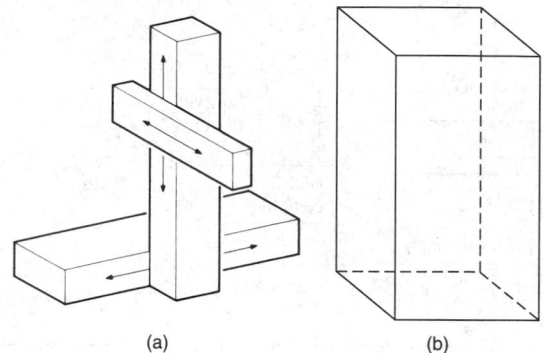

(a)   (b)

Fig. 10.6.5 Cartesian- or rectangular-coordinate robot configuration. (*a*) Movements; (*b*) work envelope.

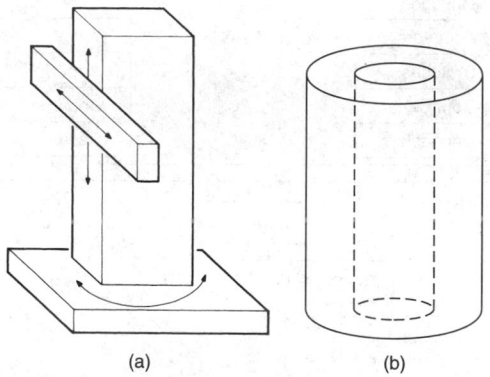

(a)   (b)

Fig. 10.6.6 Cylindrical-coordinate robot configuration. (*a*) Movements; (*b*) work envelope.

Articulations A robot may be described by the number of articulations, or jointed movements, it is capable of doing. The joints described above that establish its configuration give the typical robot three degrees of freedom (DOF) and at least three more may be obtained from the motions of its wrist—these being called *roll, pitch,* and *yaw* (Fig. 10.6.10). The mere opening and closing of a gripper is usually not regarded as a degree of freedom, but the many positions assumable by a servoed gripper may be called a DOF and a multifingered mechanical hand has several additional DOFs. The added abilities of mobile robots to traverse land, water, air, or outer space are, of course, additional degrees of freedom. Figure 10.6.11 shows a revolute robot with a three-DOF wrist and a seventh DOF, the ability to move in a track along the floor.

Size The physical size of a robot influences its capacity and its capabilities. There are robots as large as gantry cranes and as small as grains of salt—the latter being made by micromachining, which is the same process used to make integrated circuits and computer chips. Some robots intended for light assembly work are designed to be approximately the size of a human so as to make the installation of the robot into the space vacated by the replaced human as easy and least disruptive as possible.

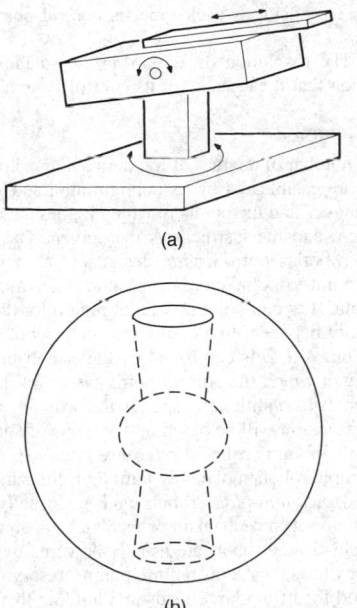

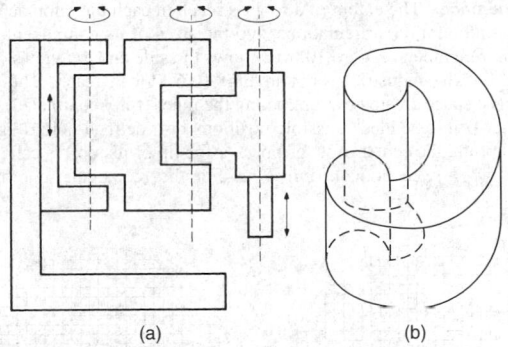

Fig. 10.6.9 SCARA coordinate robot configuration. (*a*) Movements; (*b*) work envelope.

Fig. 10.6.7 Polar- or spherical-coordinate robot configuration. (*a*) Movements; (*b*) work envelope.

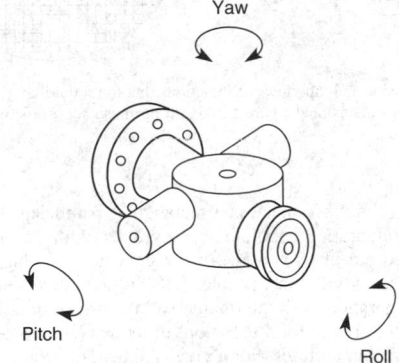

Fig. 10.6.10 A robot wrist able to move in three ways: roll, pitch, and yaw.

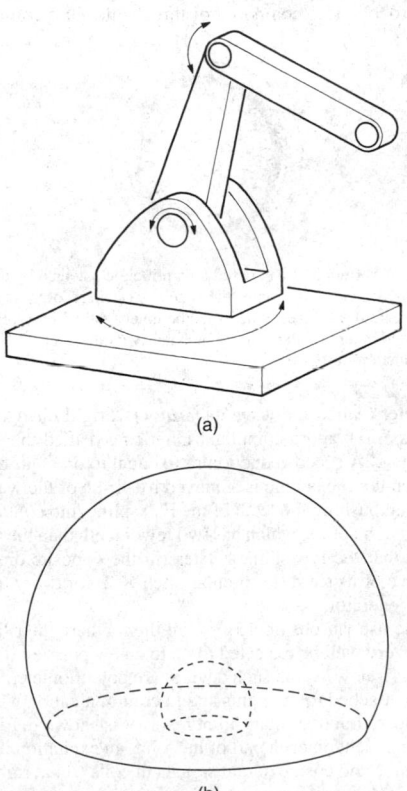

Fig. 10.6.8 Revolute– or jointed-arm–coordinate robot configuration. (*a*) Movements; (*b*) work envelope.

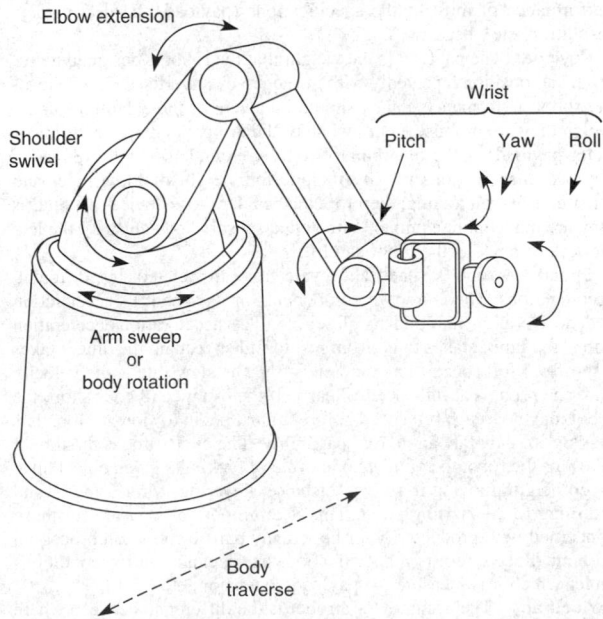

Fig. 10.6.11 A typical industrial robot with three basic degrees of freedom, plus three more in its wrist, and a seventh in its ability to move back and forth along the floor.

Workspace The extent of a robot's reach in each direction depends on its configuration, articulations, and the sizes of its component links and other members. Figure 10.6.12 shows the side and top views of the points that the industrial robot in Fig. 10.6.11 can reach. The solid geometric space created by subtracting the inner (fully contracted) from the outer (full extended) possible positions of a defined point (e.g., its control point, the center of its gripper, or the tip of its tool) is called the robot's *workspace* or *work envelope*. For the rectangular-coordinate

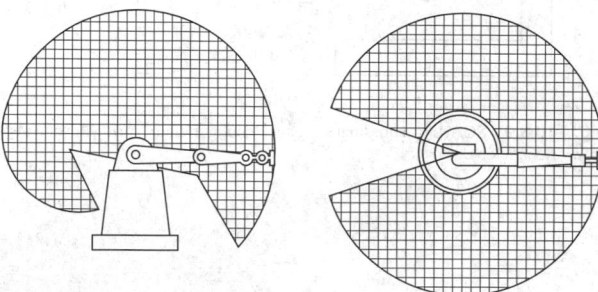

Fig. 10.6.12 Side and top views of the points that the robot in Fig. 10.6.11 can reach. The three-dimensional figure generated by these accessible end points is shown in Fig. 10.6.8*b*.

configuration robot, this space is a rectangular parallelepiped; for the cylindrical-configuration robot, it is a cylinder with an inaccessible space hollowed out of its center; for the polar and revolute robots, it consists of spheres with inaccessible spots at their centers; and for the SCARA configuration, it is the unique solid shown in the illustration. See Fig. 10.6.5 to 10.6.9 for illustrations of the work envelope of robots of the various configurations. For a given application, one would select a robot with a work envelope slightly larger than the space required for the job to be done. Selecting a robot too much larger than is needed could increase the costs, control problems, and safety concerns. Many robots have what is called a *sweet spot:* that smaller space where the performance of most of the specifications (payload, speed, accuracy, resolution, etc.) peaks.

Payload The payload is the weight that the robot is designed to lift, hold, and position repeatedly with the accuracy advertised. People reading robot performance claims should be aware of the conditions under which they are promised, e.g., whether the gripper is empty or at maximum payload, or the manipulator is fully extended or fully retracted. Payload specifications usually include the weight of the gripper and other end-of-arm tooling, therefore the heavier those devices become as they are made more complex with added sensors and actuators, the less workpiece payload the robot can lift.

Speed A robot's speed and cycle time are related, but different, specifications. There are two components of its speed: its acceleration and deceleration rate and its slew rate. The acceleration/deceleration rate is the time it takes to go from rest to full speed and the time it takes to go from full speed to a complete stop. The slew rate is the velocity once the robot is at full speed. Clearly, if the distance to go is short, the robot may not reach full speed before it must begin to slow to stop, thus making its slew rate an unused capability. The cycle time is the time it takes for the robot to complete one cycle of picking a given object up at given height, moving it a given distance, lowering it, releasing it, and returning to the starting point. This is an important measure for those concerned with productivity, or how many parts a robot can process in an hour. Speed, too, is a specification that depends on the conditions under which it is measured—payload, extension, etc.

Accuracy The accuracy of a robot is the difference between where its control point goes and where it is instructed or programmed to go.

Repeatability The repeatability or precision of a robot is the var-

iance in successive positions each time its control point returns to a taught position.

Resolution The resolution of a robot is the smallest incremental change in position that it can make or its control system can measure.

Robot Motion Control

While a robot's number of degrees of freedom and the dimensions of its work envelope are established by its configuration and size, the paths along which it moves and the points to which it goes are established by its control system and the instructions it is given. The motion paths generated by a robot's controller are designated as point-to-point or continuous. In point-to-point motion, the controller moves the robot from starting point *A* to end point *B* without regard for the path it takes or of any points in between. In a continuous-path robot controller, the path in going from *A* to *B* is controlled by the establishment of *via* or *way points* through which the control point passes on its way to each end point. If there is a conditional branch in the program at an end point such that the next action will be based on the value of some variable at that point, then the point is called a *reference point*.

Some very simple robots move only until their links hit preset stops. These are called *pick-and-place* robots and are usually powered by pneumatics with no servomechanism or feedback capabilities.

The more sophisticated robots are usually powered by hydraulics or electricity. The hydraulic systems require pumps, reservoirs, and cylinders and are good for lifting large loads and holding them steady. The electric robots use stepper motors or servomotors and are suited for quick movements and high-precision work. Both types of robots can use harmonic drives (Fig. 10.6.13), gears, or other mechanisms to reduce the speed of the actuators.

The **harmonic drive** is based on a principle patented by the Harmonic Drive unit of the Emhart Division of the Black & Decker Corp. It permits dramatic speed reductions, facilitating the use of high-speed actuators in robots. It is composed of three concentric components: an

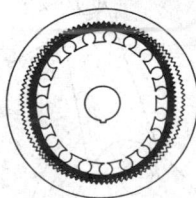

Fig. 10.6.13 A harmonic drive, based on a principle patented by the Harmonic Drive unit of the Emhart Division of the Black & Decker Corp. As the elliptical center wave generator revolves, it deforms the inner toothed Flexspline into contact with the fixed outer circular spline which has two more teeth. This results in a high speed-reduction ratio.

elliptical center (called the *wave generator*), a rigid outer ring (called the *circular spline*), and a compliant inner ring (called the *Flexspline*) between the two. A speed-reduction ratio equal to one-half of the number of teeth on the Flexspline is achieved by virtue of the wave generator rotating and pushing the teeth of the Flexspline into contact with the stationary circular spline, which has two fewer teeth than the Flexspline. The result is that the Flexspline rotates (in the opposite direction) the distance of two of its teeth (its circular pitch × 2) for every full rotation of the wave generator.

Robots also use various devices to tell them where their links are. A robot cannot very well be expected to go to a new position if it does not know where it is now. Some such devices are potentiometers, resolvers, and encoders attached to the joints and in communication with the controller. The operation of a rotary-joint resolver is shown in Fig. 10.6.14. With one component on each part of the joint, an excitation input in one part induces sine and cosine output signals in coils (set at right angles in the other part) whose differences are functions of the amount of angular offset between the two. Those angular displacement waveforms are sent to an analog-to-digital converter for output to the robot's motion controller.

Encoders may be of the contact or noncontact type. The contact type has a metal brush for each band of data, and the noncontact type either reflects light to a photodetector or lets the light pass or not pass to a photodetector on its opposite side, depending on its pattern of light/dark

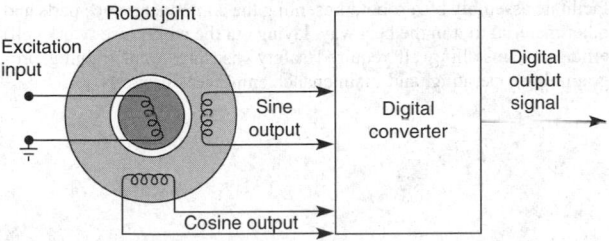

Fig. 10.6.14 Rotary-joint resolver. Resolver components on the robot's joint send analog angular displacement signals to a digital converter for interpretation and output to the robot's controller.

or clear/solid marks. Both types of encoders may be linear or rotary—that is, shaped like a bar or a disk—to match the type of joint to which they are attached. The four basic parts of a rotary optical encoder (the light sources, the coding disk, the photodetectors, and the signal processing electronics) are all contained in a small, compact, sealed package.

Encoders are also classified by whether they are incremental or absolute. Figure 10.6.15 shows two types of incremental optical encoder disks. They have equally spaced divisions so that the output of their photodetectors is a series of square waves that can be counted to tell the amount of rotational movement of the joint to which they are attached. Their resolution depends on how many distinct marks can be painted or etched on a disk of a given size. Disks with over 2,500 segments are available, such that the counter recording 25 impulses would know that the joint rotated 1/100 of a circle, reported as either degrees or radians, depending on the type of controller. It can also know the speed of that movement if it times the intervals between incoming pulses. It will not be able to tell the direction of the rotation if it uses a single-track, or tachometer-type, disk, as shown in Fig. 10.6.15a. A quadrature device, as shown in Fig. 10.6.15b, has two output channels, A and B, which are offset 90°, thus allowing the direction of movement to be determined by seeing whether the signals from A lead or lag those from B.

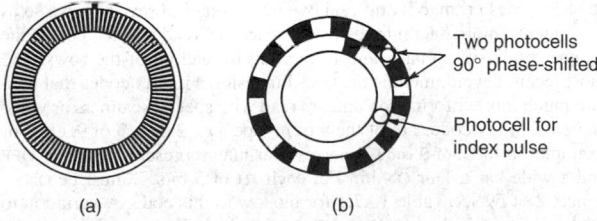

Fig. 10.6.15 Two types of incremental optical encoder disks. Type a, a single track incremental disk, has a spoke pattern that shutters the light onto a photodetector. The resulting triangular waveform is electronically converted to a square-wave output, and the pulses are counted to determine the amount of angular movement of the joint to which the encoder components are attached. Type b, a quadrature incremental disk, uses two light sources and photodetectors to determine the direction of movement.

The major limitation of incremental encoders is that they do not tell the controller where they are to begin with. Therefore, robots using them must be sent "home," to a known or initial position of each of its links, as the first action of their programs and after each switch-off or interruption of power. Absolute encoders solve this problem, but are

more complex and more expensive. They have a light source, a photodetector, and two electrical wires for each track or concentric ring of marks. Each mark or clear spot in each ring represents a bit of data, with the whole word being made up by the number of rings on the disk. Figure 10.6.16 shows 4-bit-word and 8-bit-word disks. Each position of the joint is identified by a specific unique word. All of the tracks in a segment of the disk are read together to input a complete coded word (address) for that position. The codes mostly used are natural binary, binary-coded decimal (BCD), and Gray. Gray code has the advantage that the state (ON/OFF or high/low) of only one track changes for each position change of the joint, making error detection easier.

Fig. 10.6.16 Two types of absolute optical encoder disks. Type a is for 4-bit-word systems, and type b is for 8-bit-word systems. Absolute encoders require a light source and photodetector set for each bit of word length.

All robot controllers require a number of input and output channels for communication and control. The number of I/O ports a robotic system has is one measure of its capabilities. Another measure of a robot's sophistication is the number and type of microprocessors and microcomputers in its system. Some have none; some have a microprocessor at each joint, as well as one dedicated to mathematical computations, one to safety, several to vision and other sensory subsystems, several to internal mapping and navigation (if they are mobile robots), and a master controller micro- or minicomputer.

Programming Robots

The methods of programming robots range from very elementary to advanced techniques. The method used depends on the robot, its controller, and the task to be performed. At the most basic level, some robots can be programmed by setting end stops, switches, pegs in holes, cams, wires, and so forth, on rotating drums, patch boards, control panels, or the robot's links themselves. Such robots are usually the pick-and-place type, where the robot can stop only at preset end points and nowhere in between, the number of accessible points being 2^N, where N is the robot's number of degrees of freedom. The more advanced robots can be programmed to go to any point within their workspace and to execute other commands by using a teach box on a tether (called a *teach pendant*) or a remote control unit. Such devices have buttons, switches, and sometimes a joystick to instruct the robot to do the desired things in the desired sequence. Errors can be corrected by simply overwriting the mistake in the controller's memory. The program can be tested at a slow speed, then, when perfected, can be run at full speed by merely turning a switch on the teach box. This is called the *walk-through* method of programming.

Another way some robots can be programmed is to physically take hold of it (or a remote representation of it) and lead it through the desired sequence of motions. The controller records the motions when in *teach mode* and then repeats them when switched to *run mode*. This is called the *lead-through* method of programming. The foregoing are all *teach-and-repeat* programming methods and are done on-line, that is, on a specific robot as it is in place in an installation but not doing other work while it is being programmed.

More advanced programming involves the use of a keyboard to type in textual language instructions and data. This method can be done on line or off line (away from and detached from the robot, such that it can be working on another task while the program is being created). There are many such programming languages available, but most are limited to use on one or only a few robots. Most of these languages are explicit, meaning that each of their instructions gives the robot a specific command to do a small step (e.g., open gripper) as part of a larger task. The more advanced higher-level implicit languages, however, give instructions at the task level (e.g., assemble 200 of product XYZ), and the robot's computer knows what sequence of basic steps it must execute to do that. These are also called *object-level* or *world-modeling languages* and *model-based systems*. Other ways of instructing a robot include spoken commands and feedback from a wide array of sensors, including vision systems.

Using Robots

The successful utilization of robots involves more than selecting and installing the right robot and programming it correctly. For an assembly robot, for example, care must be taken to design the product so as to facilitate assembly by a robot, presenting the component piece parts and other material to it in the best way, laying out the workplace (work cell) efficiently, installing all required safety measures, and training programming, operating, and maintenance employees properly.

10.7 MATERIAL STORAGE AND WAREHOUSING
by Vincent M. Altamuro

The warehouse handling of material is often more expensive than in-process handling, as it frequently requires large amounts of space, expensive equipment, labor, and computers for control. Warehousing activities, facilities, equipment, and personnel are needed at both ends of the process—at the beginning as receiving and raw materials and purchased parts storage, and at the end as finished goods storage and shipping. These functions are aided by various subsystems and equipment—some simple and inexpensive, some elaborate and expensive.

The rapid and accurate identification of materials is essential. This can be done using only human senses, humans aided by devices, or entirely automated. Bar codes have become an accepted and reliable means of identifying material and inputting that data into an information and control system.

Material can be held, stacked, and transported in simple devices, such as shelves, racks, bins, boxes, baskets, tote pans, pallets and skids, or in complex and expensive computer-controlled systems, such as automated storage and retrieval systems.

IDENTIFICATION AND CONTROL OF MATERIALS

Materials must be identified, either by humans or by automated sensing devices, so as to:
1. Measure presence or movement
2. Qualify and quantify characteristics of interest
3. Monitor ongoing conditions so as to feed back corrective actions
4. Trigger proper marking devices
5. Actuate sortation or classification mechanisms
6. Input computing and control systems, update data bases, and prepare analyses and summary reports

To accomplish the above, the material must have, or be given, a unique code symbol, mark, or special feature that can be sensed and identified. If such a coded symbol is to be added to the material, it must be:
1. Easily and economically produced
2. Easily and economically read
3. Able to have many unique permutations—flexible
4. Compact—sized to the package or product
5. Error resistant—low chance of misreading, reliable
6. Durable
7. Digital—compatible with computer data formats

The codes can be read by contact or noncontact means, by moving or stationary sensors, and by humans or devices.

There are several technologies used in the identification and control of materials. The most important of these are:
1. Bar codes
2. Radio-frequency identification (RFID)
3. Smart cards
4. Machine vision
5. Voice recognition
6. Magnetic stripes
7. Optional character recognition (OCR)

All can be used for automatic data collection (ADC) purposes and as part of a larger electronic data interchange (EDI), which is the paperless, computer-to-computer, intercompany, and international communications and information exchange using a common data language. The dominant data format standard in the world is Edifact (Electronic Data Interchange for Administration, Commerce, and Transportation).

Bar codes are one simple and effective means of identifying and controlling material. Bar codes are machine-readable patterns of alternating, parallel, rectangular bars and spaces of various widths whose combinations represent numbers, letters, or other information as determined by the particular symbology, or language, employed. Several types of bar codes are available. While multicolor bar codes are possible, simple black-and-white codes dominate because a great number of permutations are available by altering their widths, presence, and sequences. Some codes are limited to numeric information, but others can encode complete alphanumeric plus special symbols character sets. Most codes are binary digital codes, some with an extra parity bit to catch errors. In each bar code, there is a unique sequence set of bars and spaces to represent each number, letter, or symbol. One-dimensional, or linear, codes print all of the bars and spaces in one row. Two-dimensional stacked codes arrange sections above one another in several rows so as to condense more data into a smaller space. To read stacked codes, the scanner must sweep back and forth so as to read all of the rows. The most recent development is the two-dimensional matrix codes that contain much more information and are read with special scanners or video cameras. Many codes are of the n of m type, i.e., 2 of 5, 3 of 9, etc. For example, in the **2-of-5 code**, a narrow bar may represent a 0 bit or OFF and a wide bar a 1 or ON bit. For each set of 5 bits, 2 must be ON—hence, 2 of 5. See Table 10.7.1 for the key to this code. A variation to the basic 2 of 5 is the **Interleaved 2-of-5**, in which the spaces between the bars also can be either narrow or wide, permitting the reading of a set of bars, a set of spaces, a set of bars, a set of spaces, etc., and yielding more information in less space. See Fig. 10.7.1. Both the 2-of-5 and the Interleaved 2-of-5 codes have ten numbers in their character sets.

Code 39, another symbology, permits alphanumeric information to be encoded by allowing each symbol to have 9 modules (locations) along the bar, 3 of which must be ON. Each character comprises five bars and four spaces. Code 39 has 43 characters in its set (1-10, A-Z, six symbols, and one start/stop signal). Each bar or space in a bar code is called an element. In Code 39, each element is one of two widths, referred to as *wide* and *narrow*. As in other bar codes, the narrowest element is referred to as the X dimension and the ratio of the widest to narrowest element widths is referred to as the N of the code. For each code, X and

Table 10.7.1   Bar Code 2-of-5 and I 2/5 Key

	Code				
Character	1	2	4	7	P
0	0	0	1	1	0
1	1	0	0	0	1
2	0	1	0	0	1
3	1	1	0	0	0
4	0	0	1	0	1
5	1	0	1	0	0
6	0	1	1	0	0
7	0	0	0	1	1
8	1	0	0	1	0
9	0	1	0	1	0

Wide bars and spaces = 1
Narrow bars and spaces = 0

N are constants. These values are used to calculate length of the bar code labels. The Y dimension of a bar code is the length (or height) of its bars and spaces. It influences the permissible angle at which a label may be scanned without missing any of the pattern. See Table 10.7.2 for the Code 39 key. The patterns for the Code 39 bars and spaces are designed in such a way that changing or misreading a single bit in any of them

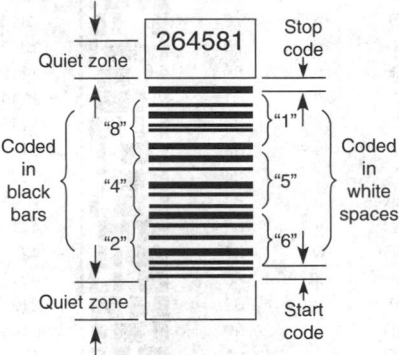

Fig. 10.7.1   A sample of a Uniform Symbology Specification Interleaved 2-of-5 bar code (also called I 2/5 or ITF).

results in an illegal code word. The bars have only an even number of wide elements and the spaces have only an odd number of wide elements. The code is called *self-checking* because that design provides for the immediate detection of single printing or reading errors. See Fig. 10.7.2. Code 39 was developed by Intermec (Everett, WA), as was **Code 93,** which requires less space by permitting more sizes of bars and spaces. Code 93 has 128 characters in its set (the full ASCII set) and permits a maximum of about 14 characters per inch, versus Code 39's maximum of about 9. Code 39 is *discrete,* meaning that each character is printed independently of the other characters and is separated from the characters on both sides of it by a space (called an intercharacter gap) that is not part of the data. Code 93 is a *continuous* code, meaning that there are no intercharacter gaps in it and all spaces are parts of the data symbols.

Code 128, a high-density alphanumeric symbology, has 128 characters and a maximum density of about 18 characters per inch (cpi) for numeric data and about 9 cpi for alphanumeric data. A Code 128 character comprises six elements: three bars and three spaces, each of four possible widths, or modules. Each character consists of eleven $1X$ wide modules, where X is the width of the narrowest bar or space. The sum of the number of bar modules in any character is always even (even parity), while the sum of the space modules is always odd (odd parity), thus permitting self-checking of the characters. See Table 10.7.3 for the

Table 10.7.2   Uniform Symbology Specification Code 39 Symbol Character Set

Char.	Encodation Pattern	bsbsbsbsb	ASCII
0		000110100	48
1		100100001	49
2		001100001	50
3		101100000	51
4		000110001	52
5		100110000	53
6		001110000	54
7		000100101	55
8		100100100	56
9		001100100	57
A		100001001	65
B		001001001	66
C		101001000	67
D		000011001	68
E		100011000	69
F		001011000	70
G		000001101	71
H		100001100	72
I		001001100	73
J		000011100	74
K		100000011	75
L		001000011	76
M		101000010	77
N		000010011	78
O		100010010	79
P		001010010	80
Q		000000111	81
R		100000110	82
S		001000110	83
T		000010110	84
U		110000001	85
V		011000001	86
W		111000000	87
X		010010001	88
Y		110010000	89
Z		011010000	90
–		010000101	45
.		110000100	46
SPACE		011000100	32
$		010101000	36
/		010100010	47
+		010001010	43
%		000101010	37
S/S		010010100	none

Note: "S/S" denotes special Code 39 Start and Stop Character.

Note: In the columns headed "b" and "s" 1 is used to represent a wide element and 0 is used to represent a narrow element.

SOURCE: Uniform Symbology Specification Code 39, Table 2: Code 39 Symbol Character Set, © Copyright 1993 Automatic Identification Manufacturers, Inc.

Code 128 key. Code 128 is a continuous, variable-length, bidirectional code.

The **Universal Product Code** (U.P.C.) is a 12-digit bar code adopted by the U.S. grocery industry in 1973. See Figure 10.7.3 for a sample of the U.P.C. standard symbol. Five of the six digits in front of the two long bars in the middle [called the *center guard pattern* (CGP)] identify the item's manufacturer and five of the six digits after the CGP identifies the specific product. There are several variations of the U.P.C., all controlled by the Uniform Code Council, located in Dayton, OH. One variation is an 8-digit code that identifies both the manufacturer and the specific product with one number. The U.P.C. uses prefix and suffix digits separated from the main number. The prefix digit is the *number system character.* See Table 10.7.4 for the key to these prefix numbers. The suffix is a ''modulo-10'' check digit. It serves to confirm that the prior 11 digits were read correctly. See Figure 10.7.4 for the method by

Table 10.7.3   Uniform Symbology Specification Code 128 Symbol Character Set

Value	Code set A	Code set B	Code set C	Encodation Pattern	bsbsbs
0	SP	SP	00		212222
1	!	!	01		222122
2	"	"	02		222221
3	#	#	03		121223
4	$	$	04		121322
5	%	%	05		131222
6	&	&	06		122213
7	'	'	07		122312
8	(	(	08		132212
9	)	)	09		221213
10	*	*	10		221312
11	+	+	11		231212
12	,	,	12		112232
13	-	-	13		122132
14	.	.	14		122231
15	/	/	15		113222
16	0	0	16		123122
17	1	1	17		123221
18	2	2	18		223211
19	3	3	19		221132
20	4	4	20		221231
21	5	5	21		213212
22	6	6	22		223112
23	7	7	23		312131
24	8	8	24		311222
25	9	9	25		321122
26	:	:	26		321221
27	;	;	27		312212
28	<	<	28		322112
29	=	=	29		322211
30	>	>	30		212123
31	?	?	31		212321
32	@	@	32		232121
33	A	A	33		111323
34	B	B	34		131123
35	C	C	35		131321
36	D	D	36		112313
37	E	E	37		132113
38	F	F	38		132311
39	G	G	39		211313
40	H	H	40		231113
41	I	I	41		231311
42	J	J	42		112133
43	K	K	43		112331
44	L	L	44		132131
45	M	M	45		113123
46	N	N	46		113321
47	O	O	47		133121
48	P	P	48		313121
49	Q	Q	49		211331
50	R	R	50		231131
51	S	S	51		213113
52	T	T	52		213311
53	U	U	53		213131
54	V	V	54		311123
55	W	W	55		311321

Value	Code set A	Code set B	Code set C	Encodation Pattern	bsbsbs
56	X	X	56		331121
57	Y	Y	57		312113
58	Z	Z	58		312311
59	[	[	59		332111
60	\	\	60		314111
61	]	]	61		221411
62	^	^	62		431111
63	_	_	63		111224
64	NUL	`	64		111422
65	SOH	a	65		121124
66	STX	b	66		121421
67	ETX	c	67		141122
68	EOT	d	68		141221
69	ENQ	e	69		112214
70	ACK	f	70		112412
71	BEL	g	71		122114
72	BS	h	72		122411
73	HT	i	73		142112
74	LF	j	74		142211
75	VT	k	75		241211
76	FF	l	76		221114
77	CR	m	77		413111
78	SO	n	78		241112
79	SI	o	79		134111
80	DLE	p	80		111242
81	DC1	q	81		121142
82	DC2	r	82		121241
83	DC3	s	83		114212
84	DC4	t	84		124112
85	NAK	u	85		124211
86	SYN	v	86		411212
87	ETB	w	87		421112
88	CAN	x	88		421211
89	EM	y	89		212141
90	SUB	z	90		214121
91	ESC	{	91		412121
92	FS			92		111143
93	GS	}	93		111341
94	RS	~	94		131141
95	US	DEL	95		114113
96	FNC 3	FNC 3	96		114311
97	FNC 2	FNC 2	97		411113
98	SHIFT	SHIFT	98		411311
99	CODE C	CODE C	99		113141
100	CODE B	FNC 4	CODE B		114131
101	FNC 4	CODE A	CODE A		311141
102	FNC 1	FNC 1	FNC 1		411131

					bsbsbs
103	START A				211412
104	START B				211214
105	START C				211232

			bsbsbsb
	STOP		2331112

Note: The numeric values in the "b" and "s" columns represent the number of modules in each of the symbol characters' bars and spaces.

Note: Dashed line indicates leading edge of adjacent character.

SOURCE: Uniform Symbology Specification Code 128, Table 2: Code 128 Symbol Character Set, © Copyright 1993 Automatic Identification Manufacturers, Inc.

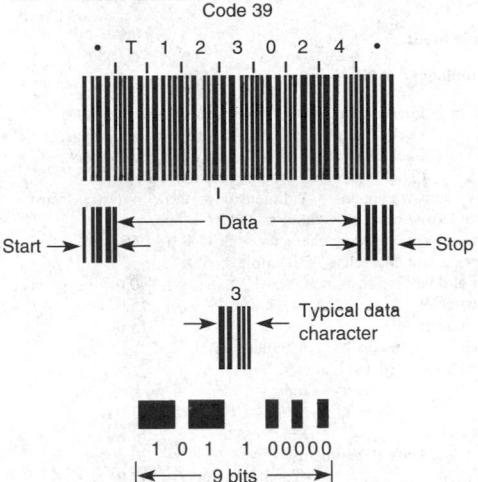

Code 39

Fig. 10.7.2 A sample of Uniform Symbology Specification Code 39 bar code.

which the check digit is calculated. Each U.P.C. character comprises two bars and two spaces, each of which may be one, two, three, or four modules wide, such that the entire character consumes a total of seven modules of space. See Table 10.7.5 for the key to the code. It will be noted that the codes to the left of the CGP all start with a zero (a space)

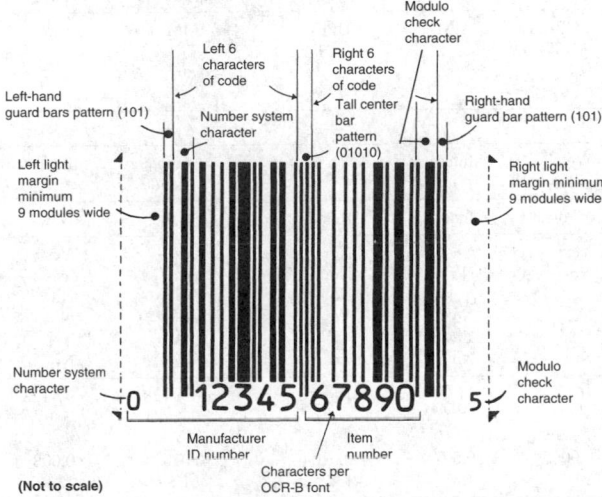

Fig. 10.7.3 A sample of the Universal Product Code Standard Symbol. (*U.P.C. Symbol Specification Manual, January 1986, Copyright© 1986 Uniform Code Council, Inc. The Uniform Code Council prohibits the commercial use of their copyrighted material without prior written permission.*)

and end with a one (a bar) while those to the right of the CGP all start with a one and end with a zero. This permits the code to be scanned from either left to right or right to left (that is, it is omnidirectional) and allows the computer to determine the direction and whether it needs to be reversed before use by the system. The U.P.C. label also has the number printed in human-readable form along the bottom.

When integrated into a complete system, the U.P.C. can serve as the input data point for supermarket pricings, checkouts, inventory control, reordering stock, employee productivity checks, cash control, special promotions, sales analyses, and management reports. Its use increases checkout speed and reduces errors. In some stores, it is combined with a speech synthesizer that voices the description and price of each item scanned. It is a vital component in a fully automated store.

Table 10.7.4   U.P.C. Prefix Number System Character Key
The human-readable character identifying the encoded number system will be shown in the left-hand margin of the symbol as per Fig. 10.7.3.

Number system character	Specified use
0	Regular U.P.C. codes (Versions A and E)
2	Random weight items, such as meat and produce, symbol-marked at store level (Version A)
3	National Drug Code and National Health Related Items Code in current 10-digit code length (Version A). Note that the symbol is not affected by the various internal structures possible with the NDC or HRI codes.
4	For use without code format restriction with check digit protection for in-store marking of nonfood items (Version A).
5	For use on coupons (Version A).
6, 7	Regular U.P.C. codes (Version A).
1, 8, 9	Reserved for uses unidentified at this time.

SOURCE: U.P.C. Symbol Specification Manual, January 1986, © Copyright 1986 Uniform Code Council, Inc. The Uniform Code Council prohibits the commercial use of their copyrighted material without prior written permission.

The following example will illustrate the calculation of the check character for the symbol shown in Fig. 10.7.3. Note that the code shown in Fig. 10.7.3 is in concurrent number system 0.

Step 1: Starting at the left, sum all the characters in the *odd* positions (that is, first from the left, third from the left, and so on), *starting with the number system character.*

(For the example, $0 + 2 + 4 + 6 + 8 + 0 = 20$.)

Step 2: Multiply the sum obtained in Step 1 by 3.

(The product for the example is 60.)

Step 3: Again starting at the left, sum all the characters in the *even* positions.

(For the example, $1 + 3 + 5 + 7 + 9 = 25$.)

Step 4: Add the product of step 2 to the sum of step 3.

(For the example, the sum is 85.)

Step 5: The modulo-10 check character value is the smallest number which when added to the sum of step 4 produces a multiple of 10.

(In the example, the check character value is 5.)

The human-readable character identifying the encoded check character will be shown in the right-hand margin of the symbol as in Fig. 10.7.3.

Fig. 10.7.4 U.P.C. check character calculation method. (*U.P.C. Symbol Specification Manual, January 1986, Copyright© 1986 Uniform Code Council, Inc. The Uniform Code Council prohibits the commercial use of their copyrighted material without prior written permission.*)

Over the years, nearly 50 different one-dimensional bar code symbologies have been introduced. Around 20 found their way into common use at one time or another, with six or seven being most important. These are the Interleaved 2-of-5, Code 39, Code 93, Code 128, Codabar, Code 11, and the U.P.C. Table 10.7.6 compares some of the characteristics of these codes. The choice depends on the use by others in the same industry, the need for intercompany uniformity, the type and amount of data to be coded, whether mere numeric or alphanumeric plus special symbols will be needed, the space available on the item, and intracompany compatibility requirements.

Some bar codes are limited to a fixed amount of data. Others can be extended to accommodate additional data and the length of their label is variable. Figure 10.7.5 shows a sample calculation of the length of a Code 39 label.

Table 10.7.5 U.P.C. Encodation Key

Encodation for U.P.C. characters, number system character, and modulo check character.

Decimal value	Left characters (Odd parity—O)	Right characters (Even parity—E)
0	0001101	1110010
1	0011001	1100110
2	0010011	1101100
3	0111101	1000010
4	0100011	1011100
5	0110001	1001110
6	0101111	1010000
7	0111011	1000100
8	0110111	1001000
9	0001011	1110100

The encodation for the left and right halves of the regular symbol, including UPC characters, number system character and modulo check character, is given in this following chart, which is applicable to version A. Note that the left-hand characters always use an odd number (3 or 5) of modules to make up the dark bars, whereas the right-hand characters always use an even number (2 or 4). This provides an "odd" and "even" parity encodation for each character and is important in creating, scanning and decoding a symbol.

SOURCE: U.P.C. Symbol Specification Manual, January 1986, © Copyright 1986 Uniform Code Council, Inc. The Uniform Code Council prohibits the commercial use of their copyrighted material without prior written permission.

Example used:

Symbology: Code 39

System: Each bar or space = one element
Each character = 9 elements (5 bars and 4 spaces)
Each character = 3 wide and 6 narrow elements
(That is, 3 of 9 elements must be wide)
Width of narrow element = X dimension = 0.015 in (0.0381 cm)
Wide to narrow ratio = N value = 3:1 = 3
Length segments per data character = 3(3) + 6 = 15
Number of data characters in example = 6

Length of data section = 6 × 5	= 90.0	segments of length
Start character	= 15.0	
Stop character	= 15.0	

Intercharacter gaps (at 1 segment per gap):
No. of gaps = (data characters − 1)
+ start + stop
= (6 − 1) + 1 + 1 = 7   = 7.0
Quiet zones
(2, each at least 10 segments)   = 20.0
Total segments = 147.0
Times X, length of each segment   × 0.015
= 2.205 in (5.6007 cm)

Fig. 10.7.5 Calculation of a bar code label length.

Table 10.7.6 Comparison of Bar Code Symbologies

	Interleaved	Code 39	Codabar	Code 11	U.P.C./EAN	Code 128	Code 93
Date of inception	1972	1974	1972	1977	1973	1981	1982
Industry-standard specification	AIM ANSI UPCC AIAG	AIM ANSI AIAG HIBC	CCBBA ANSI	AIM	UPCC IAN	AIM	AIM
Government support		DOD					
Corporate sponsors	Computer Identics	Intermec	Welch Allyn	Intermec		Computer Identics	Intermec
Most-prominent application area	Industry	Industry	Medical	Industry	Retail	New	New
Variable length	No[a]	Yes	Yes	Yes	No	Yes	Yes
Alphanumeric	No	Yes	No	No	No	Yes	Yes
Discrete	No	Yes	Yes	Yes	No	No	No
Self-checking	Yes	Yes	Yes	No	Yes	Yes	No
Constant character width	Yes	Yes	Yes[b]	Yes	Yes	Yes	Yes
Simple structure (two element widths)	Yes	Yes	No	No	No	No	No
Number of data characters in set	10	43/128	16	11	10	103/128	47/128
Density[c]:							
Units per character	7–9	13–16	12	8–10	7	11	9
Smaller nominal bar, in	0.0075	0.0075	0.0065	0.0075	0.0104	0.010	0.008
Maximum characters/inch	17.8	9.4	10	15	13.7	9.1	13.9
Specified print tolerance at maximum density, in							
Bar width	0.0018	0.0017	0.0015	0.0017	0.0014	0.0010	0.0022
Edge to edge					0.0015	0.0014	0.0013
Pitch					0.0030	0.0029	0.0013
Does print tolerance leave more than half of the total tolerance for the scanner?	Yes	Yes	No	Yes	No	No	Yes
Data security[d]	High	High	High	High	Moderate	High	High

[a] Interleaved 2 of 5 is fundamentally a fixed-length code.

[b] Using the standard dimensions Codabar has constant character width. With a variant set of dimensions, width is not constant for all characters.

[c] Density calculations for interleaved 2 of 5, Code 39, and Code 11 are based on a wide-to-narrow ratio of 2.25:1. Units per character for these symbols are shown as a range corresponding to wide/narrow ratios from 2:1 to 3:1. A unit in Codabar is taken to be the average of narrow bars and narrow spaces, giving about 12 units per character.

[d] High data security corresponds to less than 1 substitution error per several million characters scanned using reasonably good-quality printed symbols. Moderate data security corresponds to 1 substitution error per few hundred thousand characters scanned. These values assume no external check digits other than those specified as part of the symbology and no file-lookup protection or other system safeguards.

[e] AIM = Automatic Identification Mfrs Inc (Pittsburgh)
ANSI = American National Standards Institute (New York)
AIAG = Automative Industry Action Group (Southfield, MI)
IAN = Intl Article Numbering Assn (Belgium)
DOD = Dept of Defense
CCBBA = Committee for Commonality in Blood Banking (Arlington, VA)
UPCC = Uniform Product Code Council Inc (Dayton, OH)

SOURCE: Intermec—A Litton Company, as published in the March, 1987 issue of *American Machinist* magazine, Penton Publishing Co., Inc.

The foregoing one-dimensional, linear bar codes are sometimes called *license plates* because all they can do is contain enough information to identify an item, thereby permitting locating more information stored in a host computer. Two-dimensional stacked codes contain more information. The stacked codes, such as **Code 49,** developed by Dr. David Allais at Intermec, **Code 16K,** invented by Ted Williams, of Laserlight Systems, Inc., Dedham, MA, and others are really one-dimensional codes in which miniaturized linear codes are printed in multiple rows, or tiers. They are useful on small products, such as electronic components, single-dose medicine packages, and jewelry, where there is not enough space for a one-dimensional code. To read them, a scanner must traverse each row in sequence. Code 49 (see Fig. 10.7.6) can arrange all 128 ASCII encoded characters on up to eight stacked rows. Figure 10.7.7 shows a comparison of the space required for the same amount (30 characters) of alphanumeric information in three bar code symbologies: Code 39, Code 93, and Code 49. Code 16K (Fig. 10.7.8), which stands for 128 squared, shrinks and stacks linear codes in from 2

Fig. 10.7.6 Code 49. *(Uniform Symbology Specification Code 49, Copyright© 1993 Automatic Identification Manufacturers, Inc.)*

to 16 rows. It offers high-data-density encoding of the full 128-character ASCII set and double-density encoding of numerical data strings. It can fit 40 characters of data in the same space as eight would take in Code 128, and it can be printed and read with standard printing and scanning equipment.

Code 39

Code 94

Code 49

Fig. 10.7.7 A comparison of the space required by three bar codes. Each bar code contains the same 30 alphanumeric characters.

Two-dimensional codes allow extra information regarding an item's description, date and place of manufacture, expiration date, lot number, package size, contents, tax code, price, etc. to be encoded, rather than merely its identity. Two-dimensional matrix codes permit the encoding of very large amounts of information. Read by video cameras or special laser scanners and decoders, they can contain a complete data file about

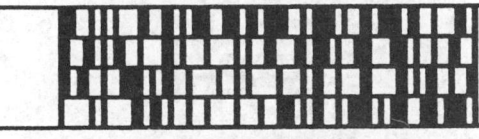

Abcd-1234567890-wxyZ

Fig. 10.7.8 Code 16K with human-readable interpretation. *(Uniform Symbology Specification Code 16K, Copyright© 1993 Automatic Identification Manufacturers, Inc.)*

the product to which they are affixed. In many cases, this eliminates the need to access the memory of a host computer. That the data file travels with (on) the item is also an advantage. Their complex patterns and interpretation algorithms permit the reading of complete files even when part of the pattern is damaged or missing.

Code PDF417, developed by Symbol Technologies Inc., Bohemia, NY (and put in the public domain, as are most codes), is a two-dimensional bar code symbology that can record large amounts of data, as well as digitized images such a fingerprints and photographs. To read the high-density PDF417 code, Symbol Technologies Inc. has developed a rastering laser scanner. This scanner is available in both a hand-held and fixed-mount version. By design, PDF417 is also scannable using a wide range of technologies, as long as specialized decoding algorithms are used. See Fig. 10.7.9 for an illustration of PDF417's ability to compress Abraham Lincoln's Gettysburg Address into a small space. The minimal unit that can represent data in PDF417 is called a *codeword*. It comprises 17 equal-length modules. These are grouped to form 4 bars and 4 spaces, each from one to six modules wide, so that they total 17 modules. Each unique codeword pattern is given one of 929 values. Further, to utilize different data compression algorithms based on the

Fig. 10.7.9 A sample of PDF417 Encoding: Abraham Lincoln's Gettysburg Address. *(Symbol Technologies, Inc.)*

data type (i.e., ASCII versus binary) the system is multimodal, so that a codeword value may have different meanings depending on which of the various modes is being used. The modes are switched automatically during the encoding process to create the PDF417 code and then again during the decoding process. The mode-switching instructions are carried in special codewords as part of the total pattern. Thus, the value of a scanned codeword is first determined (by low-level decoding), then its meaning is determined by virtue of which mode is in effect (by high-level decoding). Code PDF417 is able to detect and correct errors because of its sophisticated error-correcting algorithms. Scanning at angles is made possible by giving successive rows of the stacked code different identifiers, thereby permitting the electronic "stitching" of the individual codewords scanned. In addition, representing the data by using codewords with numerical values that must be interpreted by a decoder permits the mathematical correction of reading errors. At its simplest level, if two successive codewords have values of, say, 20 and 52, another codeword could be made their sum, a 72. If the first codeword were read, but the second missed or damaged, the system would subtract 20 from 72 and get the value of the second codeword. More complex high-order simultaneous polynomial equations and algorithms are also used. The PDF417 pattern serves as a portable data file that provides local access to large amounts of information about an item (or person) when access to a host computer is not possible or economically feasible. PDF417 provides a low-cost paper-based method of transmitting machine-readable data between systems. It also has use in WORM (written once and read many times) applications. The size of the label and its aspect ratio (the shape, or relationship of height to width) are variable so as to suit the user's needs.

Data Matrix, a code developed by International Data Matrix Inc., Clearwater, FL, is a true matrix code. It may be square or rectangular, and can be enlarged or shrunk to whatever size [from a 0.001-in (0.254-mm), or smaller, to a 14-in (36-cm), or larger, square] suits the user and application. Figure 10.7.10 shows the code in three sizes, but with the same contents. Its structure is uniform—square cells all the same size for a given pattern. To fit more user data into a matrix of a specific size, all the internal square cells are reduced to a size that will accommodate

0123456789 0123456789

0123456789 0123456789

0123456789
0123456789

0123456789
ABCDEFGHIJKLMN
OPQRSTUVWXYZ

0123456789ABCD
EFGHIJKLMNOPQ
RSTUVWXYZabcd
efghijklmnopqrstuv
wxyz

0123456789ABCD
EFGHIJKLMNOPQ
RSTUVWXYZabcd
efghijklmnopqrstuv
wxyz!@#$%^&*()_
+{}:"<>?~',./;'[]-=

0123456789012345 6789012345678901234567890123456789

Fig. 10.7.10   Three sizes of Data Matrix patterns, all with the same contents. (*International Data Matrix, Inc.*)

the amount of data to be encoded. Figure 10.7.11 shows five patterns, all the same size, containing different amounts of data. As a user selects additional data, the density of the matrix naturally increases. A pattern can contain from 1 to 2000 characters of data. Two adjacent sides are always solid, while the remaining two sides are always composed of alternating light and dark cells. This signature serves to have the video [charge-coupled device (CCD)] camera locate and identify the symbol. It also provides for the determination of its angular orientation. That capability permits a robot to rotate the part to which the code is affixed to the proper orientation for assembly, test, or packaging. Further, the information it contains can be program instructions to the robot. The labels can be printed in any combination of colors, as long as their difference provides a 20 percent or more contrast ratio. For aesthetic or security reasons, it can be printed in invisible ink, visible only when illuminated with ultraviolet light. The labels can be created by any printing method (laser etch, chemical etch, dot matrix, dot peen, thermal transfer, ink jet, pad printer, photolithography, and others) and on any type of computer printer through the use of a software interface driver provided by the company. The pattern's binary code symbology is read with standard CCD video cameras attached to a company-made controller linked to the user's computer system with standard interfaces and a Data Matrix Command Set. The algorithms for error correction that it uses create data values that are cumulative and predictable so that a missing or damaged character can be deduced or verified from the data values immediately preceding and following it.

Code One (see Fig. 10.7.12), invented by Ted Williams, who also created Code 128 and Code 16K, is also a two-dimensional checker-

0123456789012345678901234567890123456789 0123456789
0123456789012345678901234567890123456789 0123456789
0123456789012345678901234567890123456789 0123456789
0123456789012345678901234567890123456789 0123456789
0123456789012345678901234567890123456789 0123456789
0123456789012345678901234567890123456789 0123456789
0123456789012345678901234567890123456789 0123456789
0123456789012345678901234567890123456789 0123456789
0123456789012345678901234567890123456789 0123456789
0123456789012345678901234567890123456789 0123456789

Fig. 10.7.11   Five Data Matrix patterns, all the same size but containing different amounts of information. (*International Data Matrix, Inc.*)

board or matrix type code read by a two-dimensional imaging device, such as a CCD video camera, rather than a conventional scanner. It can be read at any angle. It can encode the full ASCII 256-character set in addition to four function characters and a pad character. It can encode from 6 to over 2000 characters of data in a very small space. If more data is needed, additional patterns can be linked together.

123456789012345678 9012

Fig. 10.7.12   Code One. (*Laserlight Systems, Inc.*)

Figure 10.7.13 shows a comparison of the sizes of four codes—Code 128, Code 16K, PDF-417, and Code One—to illustrate how much smaller Code One is than the others. Code One can fit 1000 characters into a 1-in (2.54-cm) square of its checkerboard-like pattern. All Code One patterns have a finder design at their center. These patterns serve several purposes. They permit the video camera to find the label and determine its position and angular orientation. They eliminate the need for space-consuming quiet zones. And, because the internal patterns are different for each of the versions (sizes, capacities, and configurations) of the code, they tell the reader what version it is seeing. There are 10 different versions and 14 sizes of the symbol. The smallest contains 40 checkerboard-square type bits and the largest has 16,320. Code 1H, the largest version, can encode 3,550 numeric or 2,218 alphanumeric characters, while correcting up to 2240 bit errors. The data is encoded in rectangular tile groups, each composed of 8-bit squares arranged in a four-wide by two-high array. Within each tile, each square encodes one bit of data, with a white square representing a zero and a black square representing a one. The squares are used to create the 8-bit bytes that are the symbol's characters. The most significant bit in the character pattern is in the top left corner of the rectangle, proceeding left to right through the upper row and then the lower row, to the least significant bit in the bottom right corner of the pattern. The symbol characters are ordered left to right and then top to bottom, so that the first character is in the top left corner of the symbol and the last character is in the bottom right corner.

The fact that the patterns for its characters are arrays of spots whose images are captured by a camera, rather than the row of bars that one-dimensional and stacked two-dimensional codes have, frees it from the Y-dimension scanning angle restraint of those codes and makes it less susceptible to misreads and no-reads. The size of each square data bit is controlled only by the smallest optically resolvable or printable spot.

Bar codes may be printed directly on the item or its packaging or on a label that is affixed thereon. Some are scribed on with a laser. In many cases, they are printed on the product or package at the same time other printing or production operations are being performed, making the cost of adding them negligible. Labels may be produced in house or ordered from an outside supplier.

In addition to the bar code symbology, an automatic identification system also requires a scanner or a video camera. A scanner is an electrooptical sensor which comprises a light source [typically a light-emitting diode (LED), laser diode, or helium-neon-laser tube], a light detector [typically a photodiode, photo–integrated circuit (photo-IC), phototransistor, or charge-coupled device], an analog signal amplifier, and (for digital scanners) an analog-to-digital converter. The system also needs a microprocessor-based decoder, a data communication link to a computer, and one or more output or actuation devices. See Fig. 10.7.14.

The scanner may be hand-held or mounted in a fixed position, but something must move relative to the other, either the entire scanner, the scanner's oscillating beam of light, or the package or item containing the bar code. For moving targets, strobe lights are sometimes used to freeze the image while it is being read.

The four basic categories of scanners are: hand-held fixed beam, hand-held moving beam, stationary fixed beam, and stationary moving beam. Some scanners cast crosshatched patterns on the object so as to catch the bar code regardless of its position. Others create holographic laser patterns that wrap a field of light beams around irregularly shaped articles. Some systems require contact between the scanner and the bar code, but most are of the noncontact type. A read-and-beep system emits a sound when a successful reading has been made. A specialized type of scanner, a slot scanner, reads bar codes on badges, cards, envelopes, documents, and the like as they are inserted in the slot and swiped.

In operation, the spot or aperture of the scanner beam is directed by optics onto the bar code and reflected back onto a photodetector. The relative motion of the scanner beam across the bar code results in an analog response pattern by the photodetector as a function of the relative reflectivities of the dark bars and light spaces. A sufficient contrast between the bars, spaces, and background color is required. Some scanners see red bars as white. Others see aluminum as black, requiring the spaces against such backgrounds to be made white.

The system anticipates an incoming legitimate signal by virtue of first receiving a zero input from the "quiet zone"—a space of a certain width (generally at least 10 times the code's X dimension) that is clear of all marks and which precedes the start character and follows the stop character. The first pattern of bars and spaces (a character) that the scanner sees is the start/stop code. These not only tell the system that data codes are coming next and then that the data section has ended, but in bidirectional systems (those that can be scanned from front to back or from back to front) they also indicate that direction. In such bidirectional systems, the data is read in and then reversed electronically as necessary before use. The analog signals pass through the scanner's signal conditioning circuitry, were they are amplified, converted into digital form, tested to see whether they are acceptable, and, if so, passed to the decoder section microprocessor for algorithmic interpretation.

The decoder determines which elements are bars and which are spaces, determines the width (number of modules) of each element, determines whether the code has been read frontward or backward and reverses the signal if needed, makes a parity check, verifies other criteria, converts the signal into a computer language character—the most common of which is asynchronous American Standard Code for Information Interchange (ASCII)—and sends it, via a data communication link, to the computer. Virtually all automatic identification systems are computer-based. The bar codes read become input data to the computer, which then updates its stored information and/or triggers an output.

Other types of automatic identification systems include radio-frequency identification, smart cards, machine vision, voice recognition, magnetic stripes, and optical character recognition.

Radio-frequency identification has many advantages over the other identification, control, and data-collection methods. It is immune to factory noise, heat, cold, and harsh environments. Its tags can be covered with dirt, grease, paint, etc. and still operate. It does not require line of sight from the reader, as do bar codes, machine vision, and OCR. It

Code 128

Code 16K

PDF-417

Code One

Fig. 10.7.13 Comparative sizes of four symbols encoding the same data with the same x dimension. (*Laserlight Systems, Inc.*)

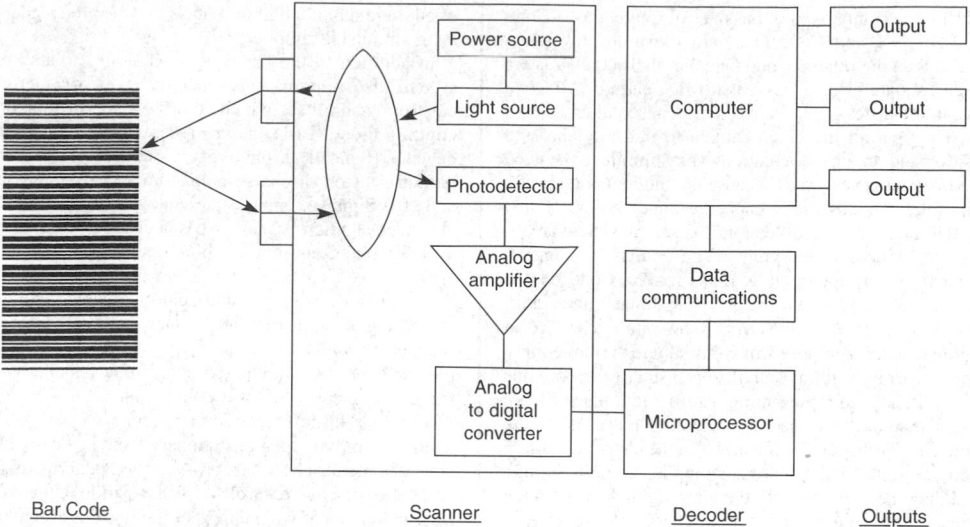

Fig. 10.7.14 Components of a bar code scanner.

can be read from a distance, typically 15 ft (4.6 m). Its tags can both receive and send signals. Many types have higher data capacities than one-dimensional bar codes, magnetic stripes, and OCR. Some tags can read/write and process data, much like smart cards. However, RFID also has some disadvantages, compared to other methods. The tags are not human-readable and their cost for a given amount of data capacity is higher than that of some other methods. The RFID codes are currently closed—that is, a user's code is not readable by suppliers or customers—and there are no uniform data identifiers (specific numeric or alphanumeric codes that precede the data code and that identify the data's industry, category, or use) like the U.P.C. has.

An RFID system consists of a transceiver and a tag, with a host computer and peripheral devices possible. The transceiver has a coil or antenna, a microprocessor, and electronic circuitry, and is called a *reader* or *scanner*. The tag can also have an antenna coil, a nonvolatile type of memory, transponder electronic circuitry, and control logic—all encapsulated for protection.

The system's operations are very simple. The reader transmits an RF signal to the tag, which is usually on a product, tote bin, person, or other convenient carrier. This signal tells the tag that a reader is present and wants a response. The activated tag responds by broadcasting a coded signal on a different frequency. The reader receives that signal and processes it. Its output may be as simple as emitting a sound for the unauthorized removal of an item containing the tag (retail theft, library books, and the like), flashing a green light and/or opening a door (employee ID badges), displaying information on a human-readable display, directing a robot or automatic guided vehicle (AGV), sending a signal to a host computer, etc. The signals from the reader to the tags can also provide energy to power the tag's circuitry and supply the digital clock pulses for the tag's logic section.

The tags may be passive or active. Passive tags use power contained in the signal from the reader. The normal range for these is 18 in (0.46 m) or less. Active tags contain a battery for independent power and usually operate in the 3- to 15-ft (0.9- to 4.6-m) range. Tags programmed when made are said to be *factory programmed* and are read-only devices. Tags programmed by plugging them into, or placing them very near, a programmer are called *field-programmed* devices. Tags whose program or data can be changed while they are in their operating locations are called *in-use programmed* devices. The simplest tags contain only one bit of code. They are typically used in retail stores to detect shoplifting and are called *electronic article surveillance* (EAS), *presence sensing,* or *Level I* tags. The next level up are the tags that, like bar codes, identify an object and serve as an input code to a database in a host computer. These "electronic license plates" are called *Level II* tags and typically have a capacity of between 8 and 128 bits. The third level devices are called *Level III,* or *transaction/routing* tags. They hold up to 512 bits so that the item to which they are attached may be described as well as identified. The next level tags, *Level IV,* are called *portable databases,* and, like matrix bar codes, carry large amounts of information in text form, coded in ASCII. The highest current level tags incorporate microprocessors, making them capable of data processing and decision making, and are the RF equivalent of smart cards.

Smart cards, or memory cards, are the size of credit cards but twice as thick. They are composed of self-contained circuit boards, stacked memory chips, a microprocessor, a battery, and input/output means, all laminated between two sheets of plastic which usually contain identifiers and instructions. They are usually read by inserting them into slots and onto pins of readers. The readers can be free-standing devices or components of production and inventory control systems, computer systems, instruments, machines, products, maintenance records, calibration instruments, vending machines, telephones, automatic teller machines, road toll booths, doors and gates, hospital and patient records, etc. One smart card can hold the complete data file of a person or thing and that data can be used as read and/or updated. Cards with up to 64-Mbytes of flash memory are available. They are made by tape-automated bonding of 20 or more blocks of two-layered memory chips onto both sides of a printed-circuit board.

Machine vision can be used to:

1. Detect the presence or absence of an entire object or a feature of it, and sort, count, and measure.

2. Find, identify, and determine the orientation of an object so that a robot can go to it and pick it up.

3. Assure that the robot or any other machine performed the task it was supposed to do, and inspect the object.

4. Track the path of a seam so that a robot can weld it.

5. Capture the image of a matrix-type bar code for data input or a scene for robotic navigation (obstacle avoidance and path finding).

In operation, the machine-vision system's processor digitizes the image that the camera sees into an array of small square cells (called picture elements, or pixels). It classifies each pixel as light or dark or a relative value between those two extremes. The system's processor then employs methods such as pixel counting, edge detection, and connectivity analysis to get a pattern it can match (a process called *windowing and feature analysis*) to a library of images in its memory, triggering a specific output for each particular match. Rather than match the entire image to a complete stored pattern, some systems use only common-feature tests, such as total area, number of holes, perimeter length, minimum and maximum radii, and centroid position. The resolution of a machine vision system is its field of view (the area of the image it captures) divided by its number of pixels (e.g., $256 \times 256 = 65,536$, $600 \times 800 = 480,000$). The system can evaluate each pixel on either a binary or a gray scale. Under the binary method, a threshold of darkness is set such that pixels lighter than it are assigned a value of 1 and those darker than it are given a value of 0 for computer processing. Under the gray scale method, several intensity levels are established, the number depending on the size of the computer's words. Four bits per pixel permit 16 levels of gray, 6 bits allow 64 classifications, and 8-bit systems can classify each pixel into any of 256 values, but at the cost of more computer memory, processing time, and expense. As the cost of vision systems drops, their use in automatic identification increases. Their value in reading matrix-type bar codes and in robotics can justify their cost.

Voice recognition systems process patterns received and match them to a computer's library of templates much like machine vision does. Memory maps of sound pattern templates are created either by the system's manufacturer or user, or both. A *speaker-dependent* system recognizes the voice of only one or a few people. It is taught to recognize their words by having the people speak them several times while the system is in the teach mode. The system creates a speech template for each word and stores it in memory. The system's other mode is the recognition mode, during which it matches new input templates to those in its memory. *Speaker-independent* systems can recognize anyone's voice, but can handle many fewer words than speaker-dependent systems because they must be able to recognize many template variations for each word. The number of words that each system can recognize is constantly being increased.

In operation, a person speaks into a microphone. The sound is amplified and filtered and fed into an analyzer, thence to a digitizer. Then a synchronizer encodes the sound by separating its pattern into equal slices of time and sends its frequency components to a classifier where it is compared to the speech templates stored in memory. If there is a match, the computer produces an output to whatever device is attached to the system. Some systems can respond with machine-generated speech. The combination of voice recognition and machine speech is called *speech processing* or *voice technology*. Products that can "hear" and "talk" are called *conversant products*. Actual systems are more complicated and sophisticated than the foregoing simplified description. The templates, for example, instead of having a series of single data points define them, have numerical fields based on statistical probabilities. This gives the system more tolerant template-to-template variations. Further, and to speed response, neural networks, fuzzy logic, and associative memory techniques are used.

Some software uses expert systems that apply grammatical and context rules to the inputs to anticipate that a word is a verb, noun, number, etc. so as to limit memory search and thereby save time. Some systems are adaptive, in that they constantly retrain themselves by modifying the templates in memory to conform to permanent differences in the speakers' templates. Other systems permit the insertion by each user of an individual module which loads the memory with that person's particular speech templates. This increases the capacity, speed, and accuracy of the system and adds a new dimension: security against use by an unauthorized person. Like fingerprints, a person's voice prints are unique, making voice recognition a security gate to entry to a computer, facility, or other entity. Some systems can recognize thousands of words with almost perfect accuracy. Some new products, such as the VCP 200 chip, of Voice Control Products, Inc., New York City, are priced low enough, by eliminating the digital signal processor (DSP) and limiting them to the recognition of only 8 or 10 words (commands), to permit their economic inclusion in products such as toys, cameras, and appliances.

A major use of voice recognition equipment is in industry where it permits assemblers, inspectors, testers, sorters, and packers to work with both hands while simultaneously inputting data into a computer directly from the source. They are particularly useful where keyboards are not suitable data-entry devices. They can eliminate paperwork, du-

plication, and copying errors. They can be made to cause the printing of a bar code, shipping label, or the like. They can also be used to instruct robots—by surgeons describing an operation, for example—and in many other applications.

Magnetic stripes encode data on magnetic material in the form of a strip or stripe on a card, label, or the item itself. An advantage to the method is that the encoded data can be changed as required.

Optical character recognition uses a video camera to read numbers, letters, words, signs, and symbols on packages, labels, or the item itself. It is simpler and less expensive than machine vision but more limited.

The industry's trade group is the Automatic Identification Manufacturers, Inc. (AIM), located in Pittsburgh.

STORAGE EQUIPMENT

The functions of storage and handling devices are to permit:

The greatest use of the space available, stacking high, using the "cube" of the room, rather than just floor area

Multiple layers of stacked items, regardless of their sizes, shapes, and fragility

"Unit load" handling (the movement of many items of material each time one container is moved)

Protection and control of the material

Shelves and Racks   Multilevel, compartmentalized storage is possible by using shelves, racks, and related equipment. These can be either prefabricated standard size and shape designs or modular units or components that can be assembled to suit the needs and space available. Variations to conventional equipment include units that slide on floor rails (for denser storage until needed), units mounted on carousels to give access to only the material wanted, and units with inclined shelves in which the material rolls or slides forward to where it is needed.

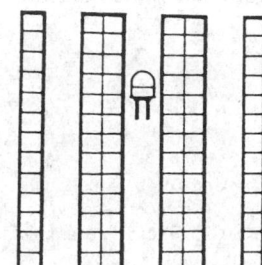

Fig. 10.7.15   Pallet racks. (*Modern Materials Handling, Feb. 22, 1980, p. 85.*)

Figure 10.7.15 shows a typical arrangement of pallet racks providing access to every load in storage. Storage density is low because of the large amount of aisle space. Figure 10.7.16 shows multiple-depth pallet racks with proportionally smaller aisle requirement. Figure 10.7.17 shows cantilever racks for holding long items. Figure 10.7.18 shows flow-through racks. The racks are loaded on one side and emptied from the other. The lanes are pitched so that the loads advance as they are

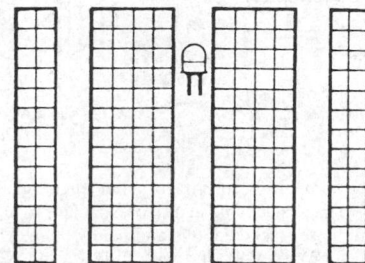

Fig. 10.7.16   Multiple-depth pallet racks. (*Modern Materials Handling, Feb. 22, 1980, p. 85.*)

removed. Figure 10.7.19 shows mobile racks. Here the racks are stored next to one another. When material is required, the racks are separated and become accessible. Figure 10.7.20 shows racks where the picker passes between the racks. The loads are supported by arms attached to

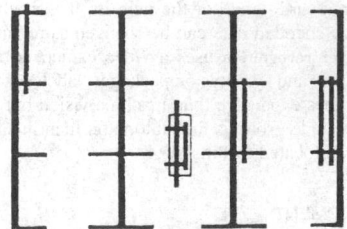

Fig. 10.7.17 Cantilever racks. *(Modern Materials Handling, Feb. 22, 1980, p. 85.)*

uprights. Figure 10.7.21 shows block storage requiring no rack. Density is very high. The product must be self-supporting or in stacking frames. The product should be stored only a short time unless it has a very long shelf life.

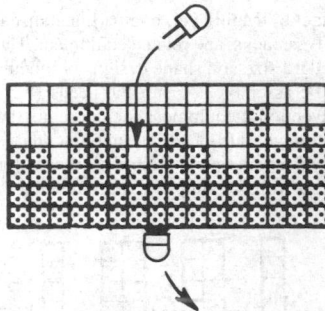

Fig. 10.7.18 Flow-through racks. *(Modern Materials Handling, Feb. 22, 1980, p. 85.)*

Bins, Boxes, Baskets, and Totes   Small, bulk, odd-shaped, or fragile material is often placed in containers to facilitate unit handling. These bins, boxes, baskets, and tote pans are available in many sizes, strength grades, and configurations. They may be in one piece or with hinged flaps for ease of loading and unloading. They may sit flat on the

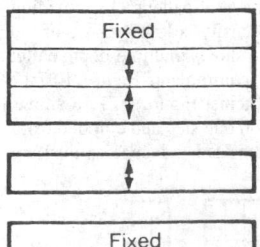

Fig. 10.7.19 Mobile racks and shelving. *(Modern Materials Handling, Feb. 22, 1980, p. 85.)*

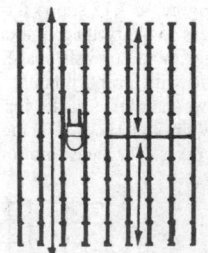

Fig. 10.7.20 Drive-in drive-through racks. *(Modern Materials Handling, Feb. 22, 1980, p. 85.)*

floor or be raised to permit a forklift or pallet truck to get under them. Most are designed to be stackable in a stable manner, such that they nest or interlock with those stacked above and below them. Many are sized to fit (in combination) securely and with little wasted space in trucks, ships, between building columns, etc.

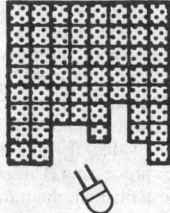

Fig. 10.7.21 Block storage. *(Modern Materials Handling, Feb. 22, 1980, p. 85.)*

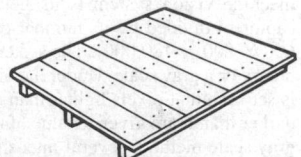

Fig. 10.7.22 Single-faced pallet.

Pallets and Skids   Pallets are flat, horizontal structures, usually made of wood, used as platforms on which material is placed so that it is unitized, off the ground, stackable, and ready to be picked up and moved by a forklift truck or the like. They can be single-faced (Fig. 10.7.22), double-faced, nonreversible (Fig. 10.7.23) double-faced, reversible (Fig. 10.7.24), or solid (slip pallets). Pallets with overhanging stringers are called *wing pallets,* single or double (Figs. 10.7.25 and 10.7.26).

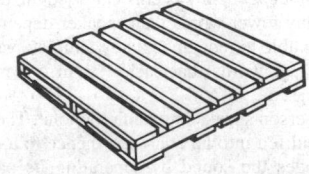

Fig. 10.7.23 Double-faced pallet.

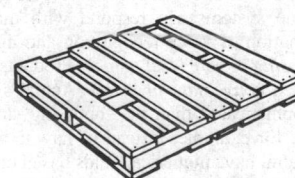

Fig. 10.7.24 Double-faced reversible pallet.

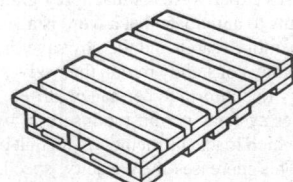

Fig. 10.7.25 Single-wing pallet.

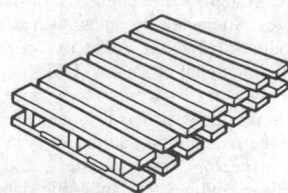

Fig. 10.7.26 Double-wing pallet.

Skids typically stand higher off the ground than do pallets and stand on legs or stringers. They can have metal legs and framing for extra strength and life (Fig. 10.7.27), be all steel and have boxes, stacking alignment tabs, hoisting eyelets, etc. (Fig. 10.7.28).

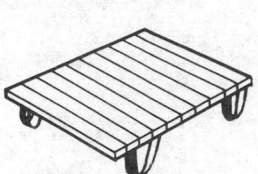
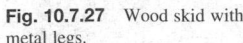

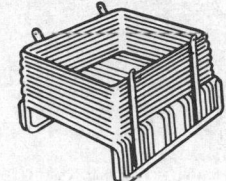

Fig. 10.7.27   Wood skid with metal legs.

Fig. 10.7.28   Steel skid box.

AUTOMATED STORAGE/RETRIEVAL SYSTEMS

An automated storage/retrieval system (AS/RS) is a high-rise, high-density material handling system for storing, transferring, and controlling inventory in raw materials, work-in-process, and finished goods stages. It comprises:

1. Structure
2. Aisle stacker cranes or storage-retrieval machines and their associated transfer devices
3. Controls

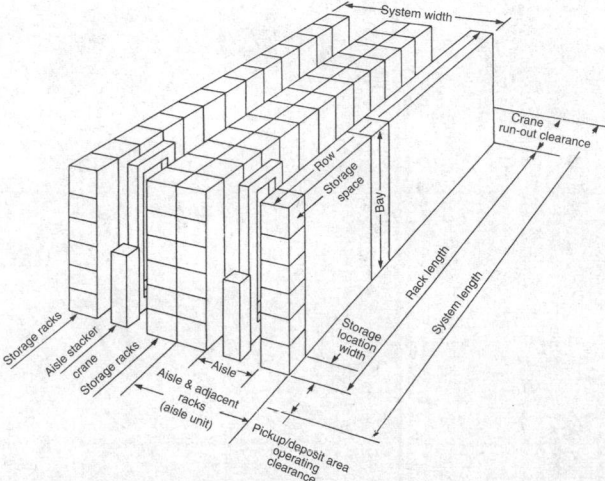

Fig. 10.7.29   Schematic diagram of an automated storage and retrieval system (AS/RS) structure. (*Harnischfeger Corp.*)

The AS/RS structure (see Fig. 10.7.29) is a network of steel members assembled to form an array of storage spaces, arranged in bays, rows, and aisles. These structures typically can be 60 to 100 ft (18 to 30 m) high, 40 to 60 ft (12 to 18 m) wide and 300 to 400 ft (91 to 121 m) long. Usually, they are erected before their protective building, which might be merely a "skin" and roof wrapped around them.

The AS/RS may contain only one aisle stacker crane which is transferred between aisles or it may have an individual crane for each aisle. The function of the stacker crane storage-retrieval machine is to move down the proper aisle to the proper bay, then elevate to the proper storage space and then, via its shuttle table, move laterally into the space to deposit or fetch a load of material.

AS/RS controls are usually computer-based. Loads to be stored need not be given a storage address. The computer can find a suitable location, direct the S/R machine to it, and remember it for retrieving the load from storage.

ORDER PICKING

Material is often processed or manufactured in large lot sizes for more economical production. It is then stored until needed. If the entire batch is not needed out of inventory at one time, order picking is required. Usually, to fill a complete order, a few items or cartons of several different materials must be located, picked, packed in one container, address labeled, loaded, and delivered intact and on time.

In picking material, a person can go to the material, the material can be brought to the person doing the packing, or the process can be automated, such that it is released and routed to a central point without a human order picker.

When the order picking is to be done by a human, it is important to have the material stored so that it can be located and picked rapidly, safely, accurately, and with the least effort. The most frequently picked items should be placed in the most readily accessible positions. Gravity racks, with their shelves sloped forward so that as one item is picked, those behind it slide or roll forward into position for ease of the next picking, are used when possible.

LOADING DOCK DESIGN

In addition to major systems and equipment such as automatic guided vehicles and automated storage/retrieval systems, a material handling capability also requires many auxiliary or support devices, such as the various types of forklift truck attachments, slings, hooks, shipping cartons, and packing equipment and supplies. Dock boards to bridge the gap between the building and railroad cars or trucks may be one-piece portable plates of steel or mechanically or hydraulically operated leveling mechanisms. Such devices are required because the trucks that deliver and take away material come in a wide variety of sizes. The shipping-receiving area of the building must be designed and equipped to handle these truck size variations. The docks may be flush, recessed, open, closed, side loading, saw tooth staggered, straight-in, or turn-around.

Transportation

BY

JOHN T. BENEDICT *Retired Standards Engineer and Consultant, Society of Automotive Engineers*

V. TERREY HAWTHORNE *Vice President, Engineering and Technical Services, American Steel Foundries*

KEITH L. HAWTHORNE *Senior Assistant Vice President, Transportation Technology Center, Association of American Railroads*

MICHAEL C. TRACY *Captain, U.S. Navy*

MICHAEL W. M. JENKINS *Professor, Aerospace Design, Georgia Institute of Technology*

SANFORD FLEETER *Professor of Mechanical Engineering and Director, Thermal Sciences and Propulsion Center, School of Mechanical Engineering, Purdue University*

AARON COHEN *Retired Center Director, Lyndon B. Johnson Space Center, NASA and Zachry Professor, Texas A&M University*

G. DAVID BOUNDS *Senior Engineer, PanEnergy Corp.*

11.1 AUTOMOTIVE ENGINEERING
by John T. Benedict

REFERENCES: "Motor Vehicle Facts and Figures, 1995," American Automobile Manufacturers Association. "Fundamentals of Automatic Transmissions and Transaxles," Chrysler Corp. "Year Book," The Tire and Rim Association, Inc. "Automobile Tires," The Goodyear Tire and Rubber Co. "Fundamentals of Vehicle Dynamics," GMI Engineering and Management Institute. "Vehicle Performance and Economy Prediction," GMI Engineering and Management Institute. Various publications of the Society of Automotive Engineers, Inc. (SAE) including: "Tire Rolling Losses," Proceedings P-74; "Automotive Aerodynamics," PT-78; "Driveshaft Design Manual," AE-7; "Design for Fuel Economy," SP-452; Bosch, "Automotive Handbook"; Limpert, "Brake Design and Safety"; Fitch, "Motor Truck Engineering Handbook"; "Truck Systems Design Handbook," PT-41; "Antilock Systems for Air-Braked Vehicles," SP-789; "Vehicle Dynamics, Braking and Steering," SP-801; "Heavy-Duty Drivetrains," SP-868; "Transmission and Driveline Developments for Trucks," SP-893; "Design and Performance of Climate Control Systems," SP-916; "Vehicle Suspension and Steering Systems," SP-952; "ABS/TCS and Brake Technology," SP-953; "Automotive Transmissions and Drivelines," SP-965; "Automotive Body Panel and Bumper System Materials and Design," SP-902; "Light Truck Design and Process Innovation," SP-1005.

GENERAL

(See also Sec. 9.6, "Internal Combustion Engines.")

In the United States the **automobile** is the **dominant mode** of personal transportation. Approximately nine of every ten people commute to work in a private motor vehicle. Public transportation is the means of transportation to work for 5 percent of the work force. More than 90 percent of households have a motor vehicle. Nearly two-thirds have two or more vehicles.

U.S. Federal Highway Administration data, illustrated in Fig. 11.1.1, shows that cars, vans, station wagons, or pickup trucks were used for more than **90 percent** of all trips. In 1993, **1.72 trillion passenger-miles** were traveled by car and truck. Automobile and truck usage accounted for 80.8 percent of intercity passenger-miles, and, when bus usage is added, the figure increases to 82 percent. Intercity motor carriers of freight handled 29 percent of the freight ton-miles; while 37 percent was carried by railroad. Pipelines and inland waterways, respectively, accounted for 19 percent and 15 percent of the freight shipments.

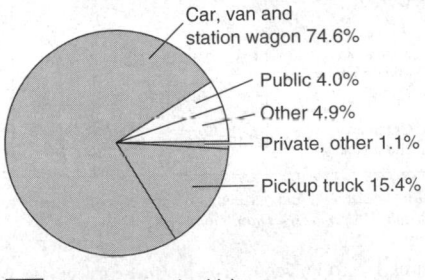

Car, van and
station wagon 74.6%

Public 4.0%

Other 4.9%

Private, other 1.1%

Pickup truck 15.4%

Privately owned vehicles

Fig. 11.1.1 Personal trips grouped by mode of transportation. *("Motor Vehicle Facts & Figures.")*

In 1994, **147 million cars, 48 million trucks,** and **676,000 buses** were registered in the United States. Included were 9 million new cars, of which 1.7 million were imported. The average age of cars was 8.4 years. More than 15 million cars were at least 15 years old.

Sales statistics for the 1994 model year reflect continued popularity of small and midsize cars, which accounted for the five top-selling makes. However, the most notable 1994 sales trend was seen in the continued rapid rise of **truck** sales. Total sales of the top five pickup trucks, sport-utility vehicles, and minivans (which are classified as trucks) surpassed total sales of the top five automobiles.

In 1994, 9 million passenger **cars** were sold in the United States. **Truck** sales rose to 6.4 million, constituting 42 percent of the 15.4 million total vehicle sales in the United States. The two top-selling nameplates were Ford and Chevrolet pickup trucks, whose sales exceeded any of the passenger car nameplates.

Average **dimensions** of the three top-selling 1994 passenger cars produced by the "big three" U.S. manufacturers were: wheelbase, 107 in (2,718 mm); length, 190 in (4,826 mm); width, 71 in (1,803 mm); tread, 58 in (1,473 mm); height, 54.5 in (1,384 mm); and turning diameter, 38 ft (11.6 m). Weight (mass) of a typical compact size car was 3,145 lb (1,427 kg).

Characteristics of cars purchased in 1994 are further described by their **optional equipment** and accessories: engine, four-cylinder, 46 percent; six-cylinder, 39 percent; eight-cylinder, 14 percent. Additional percentages include: automatic transmission, 88; power steering, 93; antilock brakes, 56; and air conditioning, 94. **Front-wheel drive** (FWD) accounted for nearly 90 percent of the vehicle totals.

TRACTION REQUIRED

The total resistance, which determines the traction force and power (road load horsepower) required for steady motion of a vehicle on a level road, is the sum of: (1) air resistance and (2) friction resistance. **Road load horsepower,** therefore, can be divided into two general parts; aerodynamic horsepower, which includes all aerodynamic losses (both internal and external to the vehicle), and mechanical horsepower or rolling resistance horsepower, which includes drivetrain power losses from the engine to the driving wheels, the wheel bearing losses of front and rear wheels, and the power losses in the four tires. The **rolling resistance** and power consumption of the tires is such a dominant factor that, for a first-order approximation, the frictional loss and the power consumed by the vehicle's equipment and accessories may be disregarded.

Tire rolling resistance, as reported by Hunt, Walter, and Hall (Conference Proceedings, P-74, SAE) was about 1 percent of the load carried at low speeds and increased to about 1.5 percent at 60 mi/h (96.6 km/h). For modern radial-ply passenger car tires, these values are about 1.2 to 1.4 percent at 30 mi/h (48.3 km/h), increasing to 1.6 to 1.8 percent at 70 mi/h (112.7 km/h).

Greater tire deflection, caused by deviation from recommended loads and air pressures (see Table 11.1.1) increases tire resistance. Low temperatures do likewise. Figure 11.1.2 shows the dependence of rolling resistance on inflation pressure for an FR78-14 tire tested at 1,280 lb

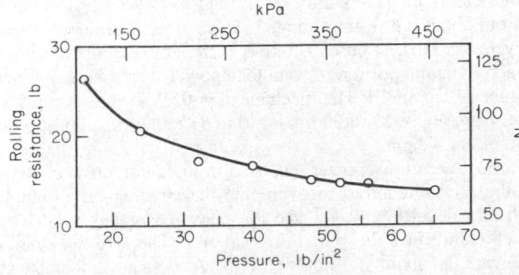

Fig. 11.1.2 Dependence of rolling resistance on inflation pressure for FR78-14 tire, 1,280-lb load and 60-mi/h speed. *("Tire Rolling Losses.")*

Table 11.1.1 Passenger-Car Tire Inflation Pressures and Load Limits

Tire size designation	Cold inflation pressure, lb/in² (kPa)							Diam., in (mm)	r/ mi
	17 (120)	20 (140)	23 (160)	26 (180)	29 (200)	32 (220)	35 (240)		
13-in nominal wheel diameter—80 series									
P135/80*13	540 (245)	584 (265)	628 (285)	661 (300)	694 (315)	728 (330)	761 (345)	21.4 (544)	965
P145/80*13	606 (275)	661 (300)	705 (320)	750 (340)	783 (355)	827 (375)	860 (390)	22.1 (561)	948
P155/80*13	683 (310)	739 (335)	783 (355)	838 (380)	882 (400)	926 (420)	959 (435)	22.7 (577)	922
P165/80*13	761 (345)	816 (370)	871 (395)	926 (420)	981 (445)	1025 (465)	1069 (485)	23.3 (592)	897
P175/80*13	838 (380)	904 (410)	970 (440)	1025 (465)	1080 (490)	1135 (515)	1179 (535)	23.9 (607)	873
P185/80*13	915 (415)	992 (450)	1058 (480)	1124 (510)	1190 (540)	1246 (565)	1301 (590)	24.6 (625)	851
13-in nominal wheel diameter—70 series									
P175/70*13	739 (335)	794 (360)	849 (385)	893 (405)	948 (430)	992 (450)	1036 (470)	22.6 (574)	925
P185/70*13	805 (365)	871 (395)	926 (420)	992 (450)	1036 (470)	1091 (495)	1135 (515)	23.2 (589)	903
P195/70*13	882 (400)	948 (430)	1014 (460)	1080 (490)	1135 (515)	1190 (540)	1246 (565)	23.8 (605)	878
P205/70*13	959 (435)	1036 (470)	1113 (505)	1179 (535)	1235 (560)	1301 (590)	1356 (615)	24.3 (617)	862

Left side label: Standard tire load, lb (kg)

* Space for R, B, or D tire-type designation.
SOURCE: The Tire and Rim Association, Inc.

(581 kg) load and 60 mi/h (96.6 km/h) speed. A tire's rolling resistance is fairly constant from 25 to 60 mi/h (40 to 97 km/h), hence power consumption is a direct function of vehicle speed and load carried by the tire.

There is general agreement that, at 45 mi/h (72.4 km/h), in 70°F (21°C) air, a run of approximately 20 min is necessary to reach **temperature equilibrium** in the tires. This is a significant factor, since tire rolling resistance typically decreases by about 25 percent during the first 10 min of operation.

Aerodynamic drag force is a function of a car's shape, size, and speed. Air resistance varies closely with the square of car speed and has a value in the range of 50 to 85 lbf (223 to 378 N) at 50 mi/h (80 km/h). The total car air drag and mechanical (friction) resistance at 50 mi/h varies from 90 to 200 lbf (400 to 890 N), and a road load power requirement of 15 to 20 hp (11 to 15 kW) at 50 mi/h is typical.

Drag force is defined by the equation $D = C'_D (\frac{1}{2} \rho V^2) A$, where A is the car's frontal cross-sectional area, ρ is the air density, V is car speed, and C'_D is a nondimensional drag coefficient determined by the vehicle's shape. Frontal area of automobiles varies from about 18 to 23 ft² (1.7 to 2.1 m²). Contemporary cars have drag coefficients ranging from 0.27 to 0.55. For comparison, the **drag coefficients** and **power requirements** for various automobile body shapes are shown in Fig. 11.1.3. Typically, drag coefficients for heavy-duty trucks and truck/trailers (not shown) range from about 0.6 to more than 1. Other drag coefficients include: motorcycles, 0.5 to 0.7; buses, 0.6 to 0.7; streamlined buses, 0.3 to 0.4.

The aerodynamic portion of road load power increases as a function of the cube of car speed. The mechanical portion increases at a slower rate and, from about 25 to 60 mi/h (40 to 97 km/h), is almost a direct function of car weight.

The term "aero horsepower" is used in the automotive industry to denote the power required to overcome the air drag on a vehicle at 50 mi/h on a level highway. Aero horsepower is equal to 0.81 times drag coefficient times the vehicle frontal area. The **aero horsepower** of midsize cars is about 7.7 hp (5.7 kW). A typical subcompact car is 30 percent lower, at 5.3 hp (4 kW). By contrast, a **heavy-duty truck** has an aero horsepower as high as 100 hp (75 kW) at 50 mi/h.

	Drag coefficient	Drag power in kW, average values for A = 2m² at various speeds		
		40 km/h	120 km/h	160 km/h
 Open convertible	0.5 ... 0.7	1	27	63
 Station wagon (2-box)	0.5 ... 0.6	0.91	24	58
 Conventional form (3-box)	0.4 ... 0.55	0.78	21	50
 Wedge shape, headlights & bumpers integrated in body	0.3 ... 0.4	0.58	16	37
 Optimum streamlining	0.15 ... 0.20	0.29	7.8	18

Fig. 11.1.3 Drag coefficient and aerodynamic power requirements for various body shapes. *(Bosch, "Automotive Handbook," SAE.)*

A reduction of 1 aero horsepower is the **fuel-efficiency equivalent** of taking 300 lb (135 kg) of weight out of a car. Figure 11.1.4 shows a typical relationship between aerodynamic and mechanical horsepower for vehicle constant-speed operation (Kelly and Holcombe: "Automotive Aerodynamics," PT-16, SAE). Variations in the body, drivetrain, and tires alter the shapes of the curves—but equal road load power for aerodynamic and mechanical requirements at 50 to 55 mi/h is typical. At 55 mi/h (88 km/h), a car expends about half of its power overcoming air drag. For speeds above 55 mi/h, there is a 2 to 3 mi/gal (0.9 to 1.3 km/L) decrease in fuel economy for each 10 mi/h (16 km/h) increase in speed, depending on power plant, driveline components, and drag coefficient. At about 55 mi/h and up, any percentage reduction of the vehicle's aerodynamic drag makes a decrease in fuel consumption of one-half or more of that same percentage. For example, for a typical automobile, a 10 percent reduction in aerodynamic drag yields a 5 percent reduction in fuel consumption at 55 mi/h.

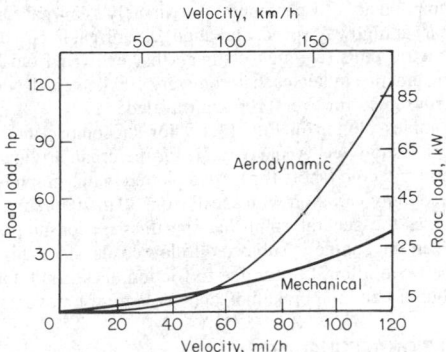

Fig. 11.1.4 Vehicle road load horsepower requirements. (Four-door sedan: frontal area 22 ft²; weight 3,675 lb; C_D = 0.45.) (*"Automotive Aerodynamics,"* PT-78.)

Average values of **traction requirements** for several large cars with average weight (including two passengers and luggage) of 4,000 lb (1,814 kg) are shown in Fig. 11.1.5. Curves A, R, and T represent the air, rolling, and total resistance, respectively, on a level road, with no wind. Curves T′, parallel to curve T, represent the displacement of the latter for gravity effects on the grades indicated, the additional traction being equal to the car weight (4,000 lb) times the percent grade. Curve E shows the traction available in high gear in this average car. The intersection of the "traction available" curve with any of the constant-gradient curves indicates the top speed that may be attained on a given grade. For example, this hypothetical large car should negotiate a 12.5 percent grade in high gear at 60 mi/h (97 km/h). Top speed on a level grade would be about 100 mi/h (161 km/h).

Effects of Transmission Gear Ratios Constant-horsepower parabolas, which apply to any vehicle, are shown as light curves in Fig. 11.1.5.

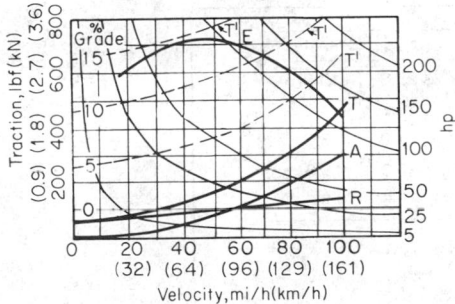

Fig. 11.1.5 Traction available and traction required for a typical large automobile.

Except for small effects of friction losses, changes in gear ratio move points of a curve for traction available from an engine along these constant-power parabolas to traction values multiplied by the change in gear reduction. Such shifting of the traction values available from engines follows gear changes in the axle as well as in the gear box.

Acceleration The difference between traction force available from power generated by an engine and the force required for constant speed on a given grade may be used for acceleration [acceleration = 21.9(100) × (surplus traction force/total effective car mass)], where acceleration is in mi/h · s, force is in lbf, and mass is in lbm.

Car weight must include a factor for the rotating parts of the engine, which may be of considerable magnitude when a high gear ratio is used. It is common for about one-third or more of the torque developed at the engine crankshaft to be absorbed in accelerating the engine, drivetrain and its rotating masses, and the road wheels. For this instance, this means that the "effective mass" equals 1.33 × actual mass.

The **effective mass** of engine rotating parts increases as the square of the engine revolutions per mile and, for a typical example, may equal or approach the car mass at a gear ratio giving about 12,000 engine rev/min. Since the effective mass of engine rotating parts increases with the square of the gear reduction, and the traction force increases directly, there is an optimum gear reduction for maximum acceleration.

The maximum acceleration rate possible is limited by the friction between the driving tires and the road surface. The coefficient of friction is dependent on vehicle speed, tire condition, and road conditions. At 50 mi/h (80 km/h) on a dry roadway, the coefficient of friction is about 1.0; but, when the roadway is wet, this drops to 0.4 or lower, depending on the amount of water and the polish of the surface ("Bosch Automotive Handbook," 3d English ed., SAE, 1993).

For a car with 2,000 lb (907 kg) weight on the driving wheels, the limiting traction force would be about 2,000 lbf (8,900 N) on a dry pavement and less than half this value on a wet pavement. From the equation given previously, the theoretical **maximum accelerations** possible under these two conditions for a car of 4,000 lb (1,814 kg) would be equivalent to a speed change of from 0 to 60 mi/h (97 km/h) in about 5.5 and 11 s, respectively. For 0 to 60 mi/h acceleration, a rough approximation of the time also is given by the empirical equation (Campbell, "The Sports Car," Robert Bentley, Inc., 1978): $t = (2W/T)^{0.6}$, where t = time, s; W = weight, lb; T = maximum engine torque, lb · ft.

FUEL CONSUMPTION

Because motor vehicles consume more than 25 percent of the nation's gasoline fuel and it is in the national interest to conserve energy supplies, the corporate average fuel economy (CAFE) of cars and trucks was regulated in 1975. Calculated according to production and sales level of a company's various models, the CAFE for cars was mandated to rise from 18 mi/gal in 1978 to 27 mi/gal in 1985 and thereafter. For light trucks, from 17.2 mi/gal in 1979 to 20.6 in 1995, and 20.7 in 1996. The upward trend in fuel economy is shown graphically in Figs. 11.1.6 (cars) and 11.1.7 (light trucks and vans).

Vehicle **design-related factors** that affect fuel economy are: the vehicle's purpose; performance goals; size; weight; aerodynamic drag; engine type, size, output, and brake specific fuel consumption; transmis-

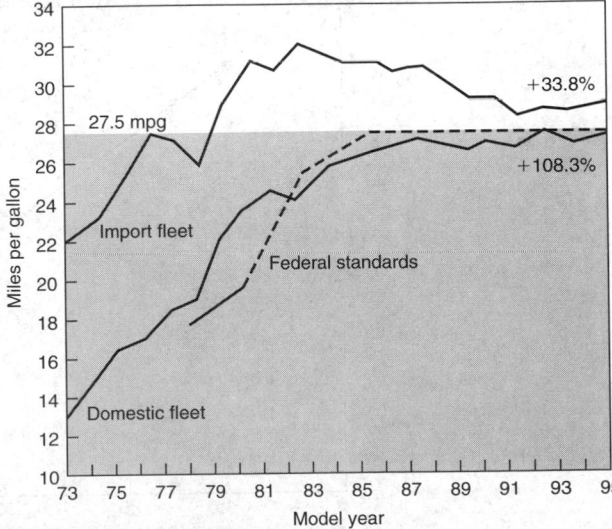

Fig. 11.1.6 Passenger car corporate average fuel economy. (*"Motor Vehicle Facts & Figures."*)

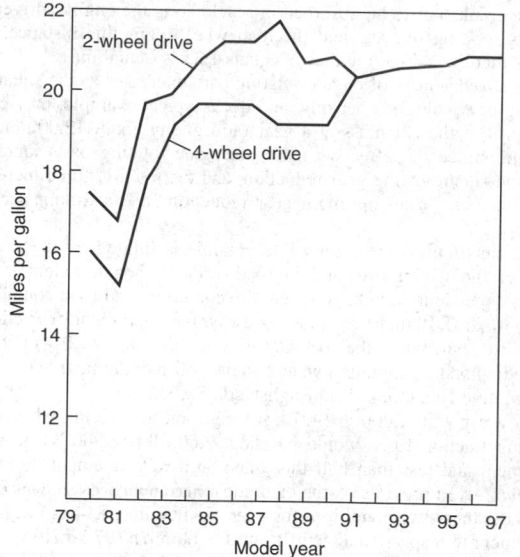

Fig. 11.1.7 U.S. federal light truck fuel economy standards. (*"Motor Vehicle Facts & Figures."*)

sion type; axle ratio; tire construction; and federal standards for fuel economy, emissions, and safety.

The principal **customer-** or **owner-related factors** include: driving pattern; trip length and number of stops; driving technique, especially acceleration, speed, and braking; vehicle maintenance; accessory operation; vehicle loading; terrain; and weather. Several popular accessories affect fuel economy as follows: automatic transmission, 2 to 3 mi/gal; air conditioning, 1 to 3 mi/gal; power steering, about ½ mi/gal; and power brakes, negligible.

Figure 11.1.8 illustrates how fuel economy is affected by design changes such as axle ratio, engine displacement, and vehicle weight. **Weight** is a key determinant of fuel economy. For a rough rule-of-thumb, it may be estimated that the addition of 300 lb weight increases fuel consumption about 1 percent (approximately ⅓ mi/gal for a typical compact car) at highway speed, and about 0.8 mi/gal in city driving. In terms of metric units, a rough estimate that evolved from European engineering practice indicates that, for every 100 kg vehicle weight, 1 L of fuel is consumed for every 100 km traveled.

The examples plotted in Fig. 11.1.9 for subcompact and intermediate-size cars show **fuel economy range** for the urban driving cycle for engine warm and cold (short-trip). Also plotted in this graph is the road load fuel economy variation with speed from 30 to 70 mi/h. This graph also illustrates the general point that specific fuel consumption is the greatest when the engine is subjected to low loads, since this is where the ratio between idling losses (due to friction, leaks, and nonuniform fuel distribution) and the brake horsepower is most unfavorable.

TRANSMISSION MECHANISMS

Friction clutches are either (1) the single-disk type (Fig. 11.1.10), connecting the engine to a manual transmission, or (2) the hydraulically operated multiple-disk type (Fig. 11.1.11, schematic), for control of the various planetary-gear changes in automatic transmissions. In (1), the area of the friction facing is usually based on a pressure of 30 lb/in² (206.9 kPa), and the torque rating on a friction coefficient of 0.25. The clutch is held in engagement by several coiled springs or a diaphragm spring and is disengaged by means of a pedal with such leverage that 30 to 40 lb (133 to 178 N) will overcome the clutch springs.

Fluid couplings between the engine and transmission formerly were used to provide a smooth drive by the flow of oil between the flat radial blades in two adjacent toroidal casings (Fig. 11.1.12a, schematic). The difference in centrifugal force between the mass of oil contained in each toroid, when either is running at a speed higher than the other, causes a flow of oil from the periphery of the faster one to the slower one. Since this mass of oil is also rotating around the shaft at the speed of the

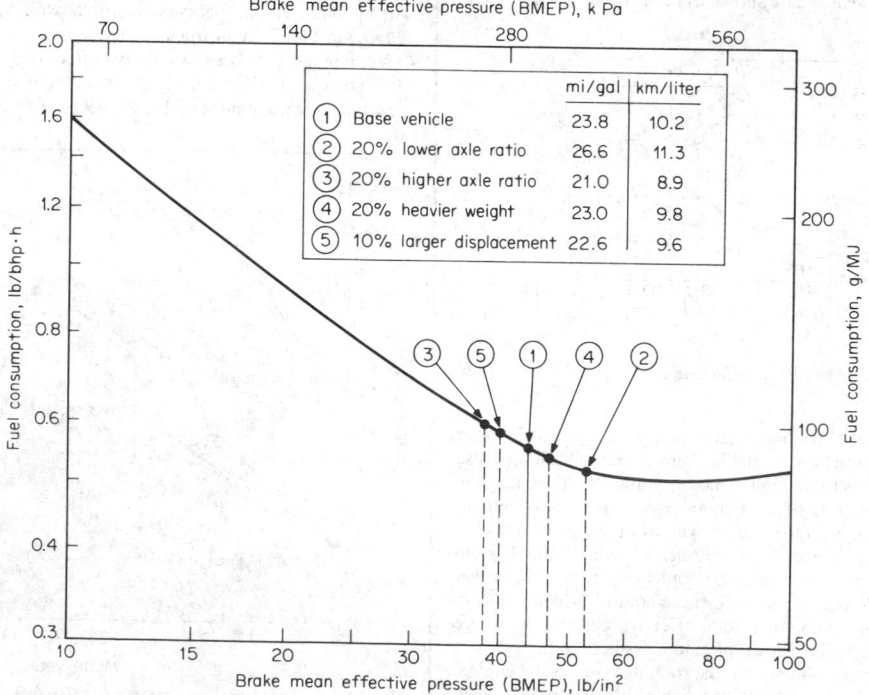

Fig. 11.1.8 Effect of vehicle design changes on road load 55-mi/h fuel consumption. (*Chrysler Corp.*)

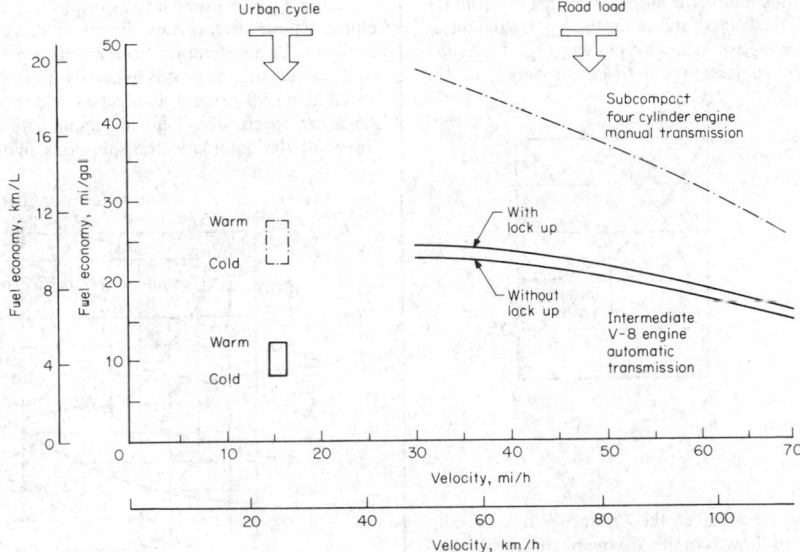

Fig. 11.1.9   Vehicle fuel consumption increases with speed, for subcompact and midsize cars. (*Chrysler Corp.*)

driving torus, its impact on the blades of the slower torus develops a torque on the latter. The developed torque is equal to, and cannot exceed, the torque of the driving torus. In this respect it is similar to a slipping friction clutch. The driven member must always run at a lower speed, though at high rotative speeds, and when the torque demand is small, the slip may be only 2 or 3 percent. The **stalled torque** increases with the square of the engine speed, so that very little is developed when idling. Since torque may be transmitted in either direction, depending only on which member is rotating at the higher speed, the engine may be used as a brake as with friction clutches, and the car may be started by pushing.

Torque-converter couplings (Fig. 11.1.13*a*) have largely replaced fluid couplings because the torque transmitted can be increased at high slippage. The circulation of oil between the driving, or higher-speed, torus (the pump) and the driven or lower-speed, torus (the turbine) results from the difference in centrifugal force developed in these two units, just as in the fluid coupling. With the torque converter, however, the turbine blades are given a curvature so that an additional torque is developed by the reaction of a backward-spinning mass of oil as it leaves the turbine. Stationary, or stator, blades are interposed between the turbine and the pump to change the direction of the oil spin. The entrance angle of the stator blades required for tangential flow varies

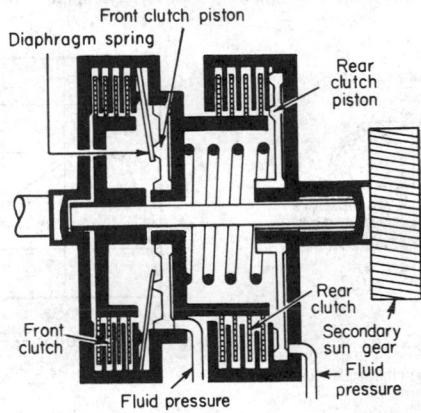

Fig. 11.1.11   Schematic of two hydraulically operated multiple-disk clutches in an automatic transmission. (*Ford Motor Co.*)

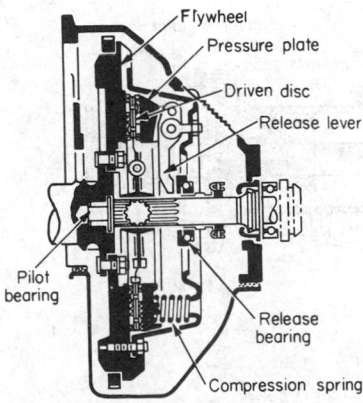

Fig. 11.1.10   Single-plate dry-disk friction clutch.

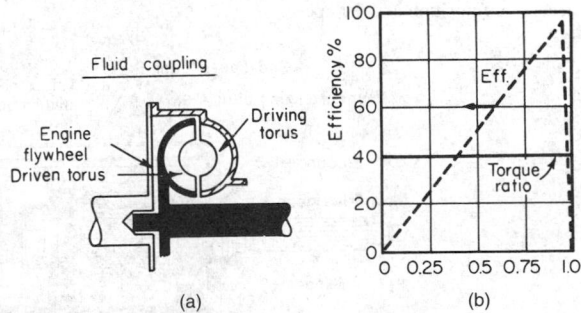

Fig. 11.1.12   Fluid coupling: (*a*) section; (*b*) characteristics.

widely with the slip ratio. For a given blade angle there is a hydraulic shock loss at any slip ratio greater or less than that which provides tangential flow. This is reflected in the rapid fall of the efficiency curve in Fig. 11.1.14*b* on each side of the maximum. The essential parts of a torque converter with its stationary stator are shown in Fig. 11.1.14*a*. A **stalled-torque** multiplication of 2.0 to 2.7 is used in various designs.

When the torque ratio is almost unity, the slip is such that oil from the turbine starts to impinge on the back of the stator blades. By mounting the stator assembly on a **sprag,** or one-way clutch (Fig. 11.1.13*a*), it remains stationary while subject to the reversing action of the

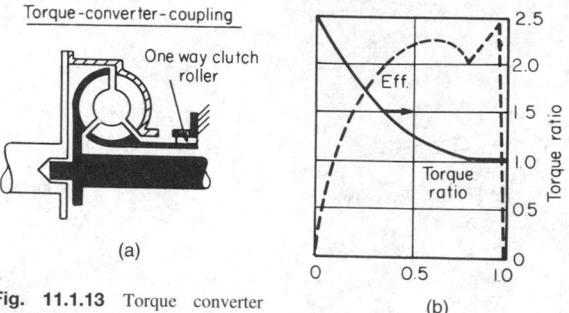

Fig. 11.1.13 Torque converter coupling: (*a*) section; (*b*) characteristics.

backward-spinning oil mass as it leaves the turbine. When the slip reaches the point where the oil flow from the turbine begins to spin forward, the stator is free to turn with it. When the slip is further reduced, the unit acts as a fluid coupling with improved efficiency (Fig. 11.1.13*b*). Such a unit is a **fluid torque-converter coupling.** Some designs

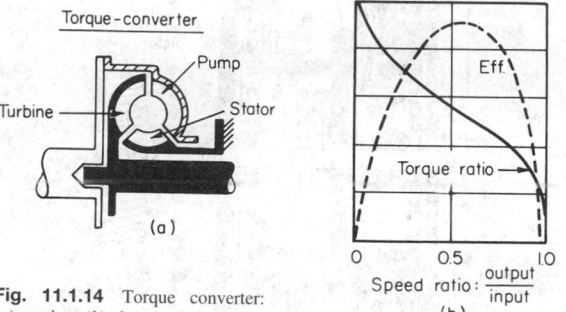

Fig. 11.1.14 Torque converter: (*a*) section; (*b*) characteristics.

eliminate slippage by the inclusion of a friction clutch which carries the load when a predetermined car speed is reached. These clutches are hydraulically operated, the engagement being controlled automatically by accelerator position and car speed.

Figure 11.1.15 compares a torque-converter coupling to a friction clutch on **car performance** in direct drive. The increase in traction available for acceleration from a standing start substantiates its public acceptance. An axle gear is generally used which gives a propeller shaft speed about 90 percent as great as with a manual transmission at the same car speed. The gain in engine efficiency compensates for the losses of the automatic transmissions under steady cruising speeds.

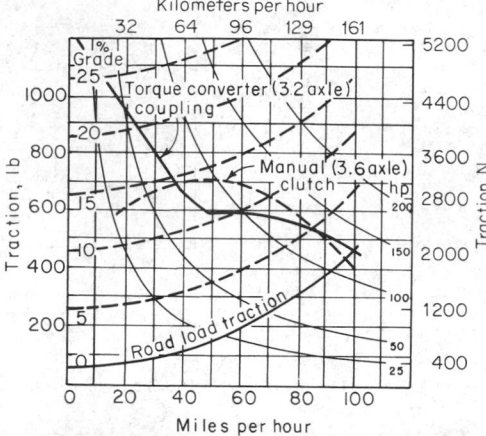

Fig. 11.1.15 Comparative traction available in the performance of a fluid torque-converter coupling and a friction clutch.

However, there may be a considerable **loss of power during acceleration** unless supplemented by such modifications as one or more auxiliary gear ratios or variable-angle stator vanes. Various design modifications of torque-converter couplings have been introduced by different manufacturers, such as two or more stators, each independently mounted on one way clutches, and variable pitch angles for the stator blades. These provide compromises in blade angles for the development of rapid acceleration without sacrifice of high efficiency while cruising.

Manual transmissions installed as standard equipment on American cars have three, four, or five forward speeds, including direct drive, and one reverse. These speeds are obtained by sliding either one of two gears along a splined shaft to bring it into mesh with a corresponding gear on a countershaft which is, in turn, driven by a pair of gears in constant mesh. Helical gears are used to minimize noise. A **"synchromesh"** device (Fig. 11.1.16), acting as a friction clutch, brings the gears to be meshed approximately to the correct speed just before meshing and minimizes **"clashing,"** even with inexperienced drivers. Gear

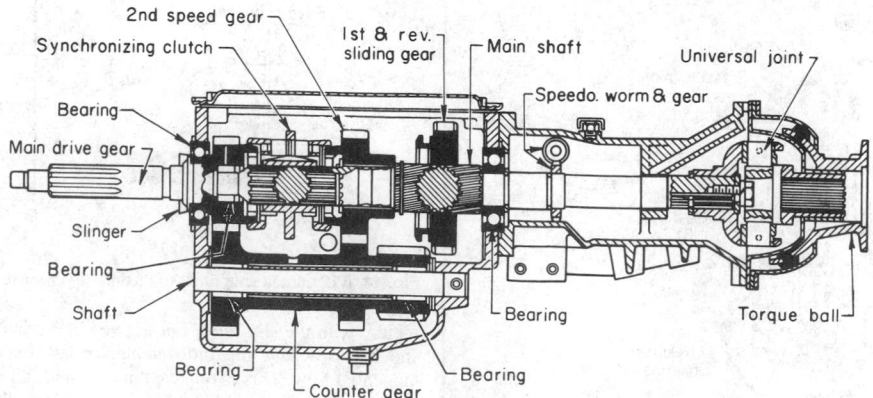

Fig. 11.1.16 Three-speed synchromesh transmission. *(Buick.)*

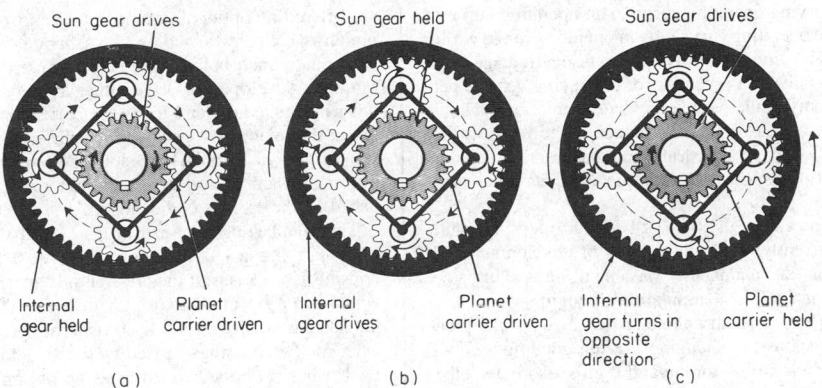

Sun gear drives   Sun gear held   Sun gear drives

Internal   Planet   Internal   Planet   Internal   Planet
gear held   carrier driven   gear drives   carrier driven   gear turns in   carrier held
                                                                                          opposite
                                                                                          direction
(a)   (b)   (c)

Fig. 11.1.17 Planetary gear action: (*a*) Large speed reduction: ratio = 1 + (internal gear diam.)/(sun gear diam.) = 3.33 for example shown. (*b*) Small speed reduction: ratio = 1 + (sun gear diam.)/(internal gear diam.) = 1.428. (*c*) Reverse gear ratio = (internal gear diam.)/(sun gear diam.) = −2.33.

changes are generally in geometric ratios. Transmission ratios average about 2.76 in first gear, 1.64 in second, 1.0 in third or direct drive, and 3.24 in reverse. The shift lever is generally located on the steering column. Four-speed transmissions usually have the shift lever on the floor. Average gear ratios are about 2.67 in first, 1.93 in second, 1.45 in third, and 1.0 in fourth or direct drive.

Overdrives have been available for some cars equipped with manual transmissions. These are supplemental planetary gear units with three planetary pinions driven around a stationary sun gear. The surrounding internal gear is coupled to the propeller shaft, which thus turns faster than the engine. The gear ratio is selected to permit the engine to slow down to about 70 percent of the propeller-shaft speed and operate with less noise and friction. These units automatically come into action when the driver momentarily releases the accelerator pedal at a car speed above 25 to 28 mi/h (40 to 45 km/h).

AUTOMATIC TRANSMISSIONS

Automatic transmissions commonly use torque-converter couplings with planetary-gear units that can supply one or two gear reductions and reverse, depending on the design, by simultaneously engaging or locking various elements of planetary systems (Fig. 11.1.17). Automatic control is provided by disk clutches or brake bands which lock the various elements, operated by oil pressure as regulated by governors at car speeds where shifts are made from one speed to another.

A schematic of a representative automatic transmission, combining a three-element torque converter and a compound planetary gear is shown in Fig. 11.1.18. The speed reductions and reverse are provided by a compound planetary system consisting of two simple systems in series. The two sun gears are an integral unit with the same number of teeth,

and the forward internal gear carries the planets of the rear unit. The internal gears of both systems are all of the same size, and consequently all planets are of equal size. This arrangement, together with three clutches, two brake bands, and suitable one-way sprags, makes possible three forward gear or torque ratios, plus direct drive and reverse.

Many automatic transmissions use a **lockup clutch** to improve performance and fuel economy. The torque converter is used for power and smoothness while accelerating in first and second gears until road speed reaches about 40 mi/h (64 km/h). Then, after the transmission upshifts from second to third gear, the clutch automatically locks up the torque converter so there is a direct mechanical drive through the transmission. Normal slippage in the converter is eliminated, engine speed is reduced, and fuel economy is improved. The lockup clutch disengages automatically during part-throttle or full-throttle downshifts and, when the vehicle is slowed, to a speed slightly below the lockup speed.

Approximately two-thirds of 1995 cars have **electronically controlled** automatic transmissions. Combined electronic-hydraulic units for control of automatic transmissions are, increasingly, superseding systems that rely solely on hydraulic control. Hydraulic actuation is retained for the clutches, while **electronic modules** assume control functions for gear selection and for modulating pressure in accordance with torque flow. **Sensors** monitor load, selector-lever position, program, and kick-down switch positions, along with rotational speed at both the engine and transmission shafts. The control unit processes these data to produce control signals for the transmission. **Advantages** include: diverse shift programs, smooth shifts, flexibility for various vehicles, simplified hydraulics, and elimination of one-way clutches.

Worldwide, automatic transmission engineering practice includes some evolving design development and growing production acceptance of **continuously variable transmissions** (CVT). The CVT can convert the

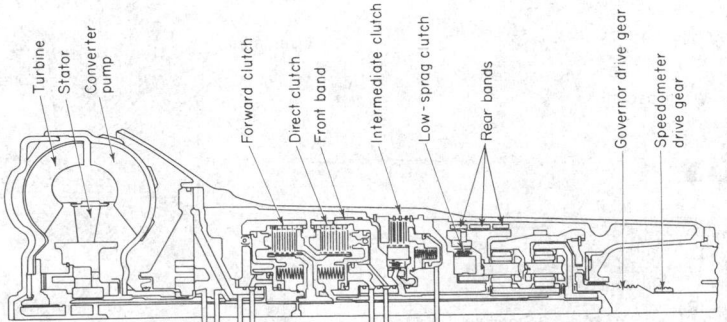

Turbine   Stator   Converter pump   Forward clutch   Direct clutch   Front band   Intermediate clutch   Low-sprag clutch   Rear bands   Governor drive gear   Speedometer drive gear

Fig. 11.1.18 Three-element torque converter and planetary gear. (*General Motors Corp.*)

engine's continuously varying operating curve to an operating curve of its own, and every engine operating curve into an operating range within the field of potential driving conditions. The **theoretical advantage** (over fixed-ratio transmissions) lies in a potential for enhancing vehicle performance and fuel economy while reducing exhaust emissions. This is done by maintaining the engine in a performance range for best fuel economy. There are, however, **practical limitations** and considerations that constrain the full exploitation of the CVT's theoretical capabilities.

The CVT can operate mechanically (belt or friction roller), hydraulically, or electrically. Currently, the highest level of development has been attained with **mechanical continuously variable** designs using steel belts. A general feature of CVT is manipulation of engine speed, with the objective of maintaining constant engine speed; or optimizing engine speed changes in response to changing driving conditions. CVT **developmental activity** includes: high-torque and high-speed belts, electronic control for line pressure and engine speed, torque converters with electronically controlled lockup clutch, and roller vane pumps with electronically operated flow control valve.

FINAL DRIVE

The **differential** is a unit attached to the ring gear (Figs. 11.1.19 and 11.1.20) which equalizes the traction of both wheels and permits one wheel to turn faster than the other, as needed on curves. Each axle is driven by a bevel gear meshing with pinions on a cross-shaft pinion pin

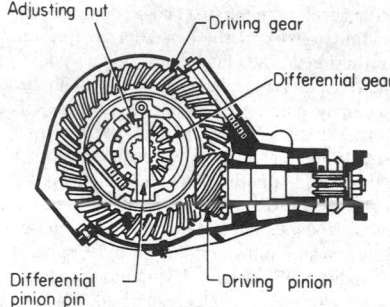

Fig. 11.1.19 Rear-axle hypoid gearing.

secured to the differential case. The case also carries the ring gear. An undesirable feature of the conventional differential is that no more traction may be developed on one wheel than on the other. If one wheel slips on ice, there is no traction to move the car. **Limited-slip differentials**

are offered as optional equipment on most cars. One design has four pinions which are carried on two separate cross shafts at right angles to each other, each being driven by V-shaped notches in the carrier. As torque is developed to drive either axle, one pinion cross shaft or the other moves axially and locks the corresponding disk-clutch plates between that axle drive gear and the differential housing. In another design, similar disk clutches are locked by spring pressure, which prevents differential action until a differential torque greater than the limit established by the springs is developed.

The **semifloating rear axle** (Fig. 11.1.20) used on many rear-wheel drive cars has a bearing for each drive axle at the outer end of the housing as well as near the differential carrier, with the full load on each wheel taken by the drive axles in combined bending and shear. The **full-floating axle,** generally used on commercial vehicles, supports each wheel on two bearings carried by the axle housing or an extension to it. Each wheel is bolted to a flange on one of the axle shafts. The axle shafts carry none of the vehicle weight and may be withdrawn without jacking up the wheel.

The **front-wheel drive** (FWD) cars commonly use a **transaxle** that combines a torque converter, automatic three-speed or four-speed transmission, final drive gearing, and differential into a compact drive system, such as illustrated in the cutaway view shown in Fig. 11.1.21. Typically, the torque converter, transaxle unit, and differential are housed in an integral aluminum die casting. The differential oil sump is separate from the transaxle sump. The torque converter is attached to the crankshaft through a flexible driving plate. The converter is cooled by circulating the transaxle fluid through an oil-to-water type cooler, located in the radiator side tank.

Engine torque is transmitted to the torque converter through the input shaft to multiple disk clutches in the transaxle. The power flow depends on the application of the clutches and bands. As illustrated in Fig. 11.1.22, the transaxle consists of two multiple-disk clutches, an overrunning clutch, two servos, a hydraulic accumulator, two bands, and two planetary gear sets, to provide three or four forward ratios and a reverse ratio.

The common sun gear of the planetary gear sets is connected to the front clutch by a driving shell that is splined to the sun gear and to the front clutch retainer. The hydraulic system consists of an oil pump and a single valve body that contains all of the valves except the governor valves.

Output torque from the main centerline is delivered through helical gears to the transfer shaft. This gear set is a factor in the final drive (axle) ratio. The shaft also carries the governor and the parking sprag. An integral helical gear on the transfer shaft drives the differential ring gear. In a representative FWD vehicle, the final drive gearing is completed with either of two gear sets to produce overall ratios of 3.48, 3.22, and 2.78.

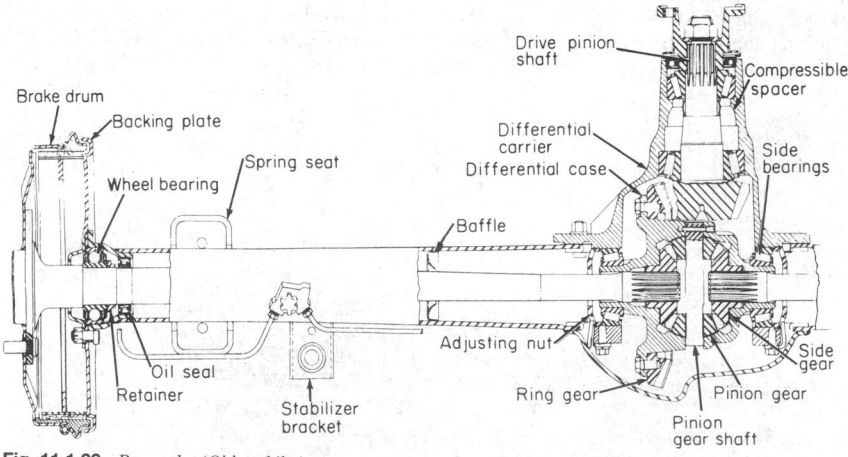

Fig. 11.1.20 Rear axle. *(Oldsmobile.)*

Advances introduced in 1995 include an **electronically controlled four-speed** automatic transaxle with **nonsynchronous** shifting that allows independent movement of two gear sets at one time and smooths torque demand and coasting down-shifts.

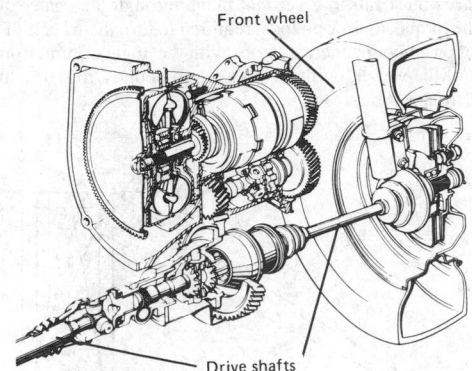

Fig. 11.1.21   Automatic transaxle used on front-wheel-drive car. *(Chrysler Corp.)*

Various types of **four-wheel drive** (4WD) systems have been developed to improve driving performance and vehicle stability on roads with a low friction coefficient surface. Most systems distribute driving force evenly (50:50) to the front and rear wheels. Driving performance, stability, and control near the limits of tire adhesion are improved, especially on low-friction surfaces. However, because of the improved levels of stability and control. the driver may be unaware that the vehicle is approaching a critical limit. The **advanced technology** in the 4WD field now extends to the ''intelligent'' four-wheel drive system. This system is designed to distribute driving force to front and rear wheels at varying, optimal ratios (instead of equal front/rear sharing of driving force) according to driving conditions and the critical limit of vehicle dynamics. The result is improved balancing of stability considerations along with cornering performance and critical limit predictability.

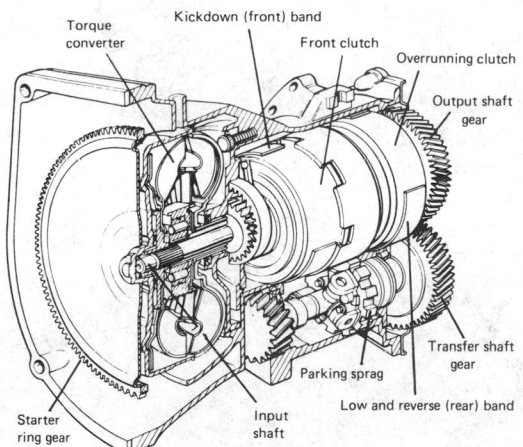

Fig. 11.1.22   Cross-section view of typical transaxle. *(Chrysler Corp.)*

SUSPENSIONS

Rear Suspensions   Torque reactions may be taken through longitudinal leaf springs, as in the **Hotchkiss drive,** or through radius rods when coil springs are used. Some designs in the past used a torque tube around the propeller shaft, bolted to the axle housing, with universal joints for both at the forward ends. Most contemporary rear suspension designs use either coil springs (Fig. 11.1.23) or leaf springs. Spring stiffness at the rear wheels ranges from 85 lb/in (15 N/mm) to about 160 lb/in (28 N/mm). Shock absorbers, to dampen road shock and vibration, are used on all cars.

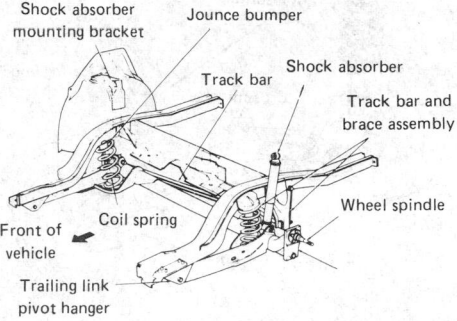

Fig. 11.1.23   Trailing-arm type rear suspension with coil springs, used on some front-wheel-drive cars. *(Chrysler Corp.)*

Front-Wheel Suspensions   Independent front-wheel suspensions are used on all cars. Rear-wheel drive cars typically use the short-and-long-arm (SLA) design, with the steering knuckle held directly between the wishbones by spherical joints (Fig. 11.1.24). The upper wishbone is shorter than the lower, to allow the springs to deflect without lateral movement of the tire at the point of ground contact.

A modification of the conventional suspension consists of sloping the upper wishbones down toward the rear, so that the steering spindle is given more ''caster'' when the front springs are compressed. This geometry causes the torque produced from braking at the front wheels to develop a couple on the inclined wishbones, which tends to raise the front of the car frame. By suitable proportioning of the parts it is possible by this means to reduce ''nose diving'' of the car when the brakes are applied. The load on these wishbones is generally taken by coil springs acting on the lower wishbone or by torsion-bar springs mounted longitudinally.

Figure 11.1.25 shows a representative application of the **spring strut (McPherson) system** for the front suspension of front-wheel drive cars. The lower end of the telescoping shock absorber (strut) is mounted within a coil spring and connected to the steering knuckle. The upper end is anchored to the car body structure.

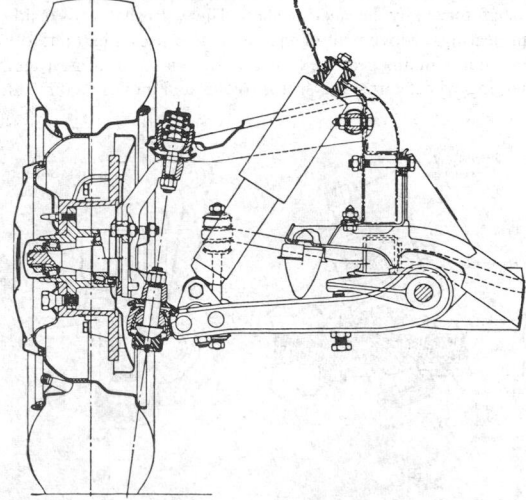

Fig. 11.1.24   Front-wheel suspension for rear-wheel-drive car.

Suspension system state of the art, with **electronic control** modules, includes user-selected dynamic tailoring of suspension characteristics and a continuously variable road-sensing suspension that senses wheel motion and other parameters.

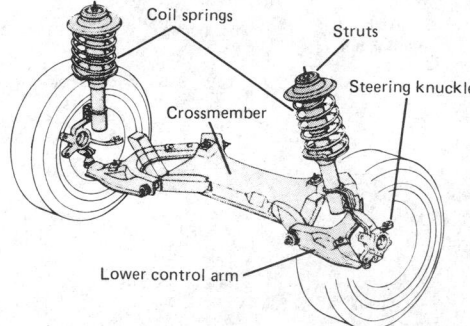

Fig. 11.1.25 Typical McPherson strut, coil spring front suspension for front-wheel-drive car. *(Chrysler Corp.)*

WHEEL ALIGNMENT

Caster is the angle, in side elevation, between the steering axis and the vertical. It is considered positive when the upper end of the steering axis is inclined rearward. Manufacturer's specifications vary considerably, with the range from 1½ to − 2¼°. **Camber**, the inclination of the wheel plane from the vertical, is positive when the wheel leans outward, and varies from 1 to − ½° with many preferring 0°. **Toe-in** of a pair of wheels is the difference in transverse distance between the wheel planes taken at the extreme rear and front points of the tire treads. It is limited to ¼ in (6.4 mm), with ⅛ in (3.2 mm) or less generally preferred.

STEERING

The force applied to the steering wheel is generally multiplied through a **worm-and-roller**, a **recirculating-ball**, or a **rack-and-pinion** type of steering gear. The overall ratios are such that 20° to 33° rotation of the steering wheel results in 1° turn of the front wheels for manual steering and 17.6° to 25.0° for power steering. The **responsiveness** of rack-and-pinion steering (Fig. 11.1.26a) results from the basic design, in which one pinion is attached directly to the steering shaft. This gear meshes with the rack, which directly extends the linkage to turn the front wheels. By comparison, a recirculating-ball steering shaft (Fig. 11.1.26b) includes a wormshaft turned by the steering shaft. The wormshaft rolls inside a set of ball bearings. Movement of the bearings causes a ball nut to move. The ball nut contains gear teeth that mesh to a sector of gear teeth. As the steering wheel turns, the sector rotates and moves a connection of

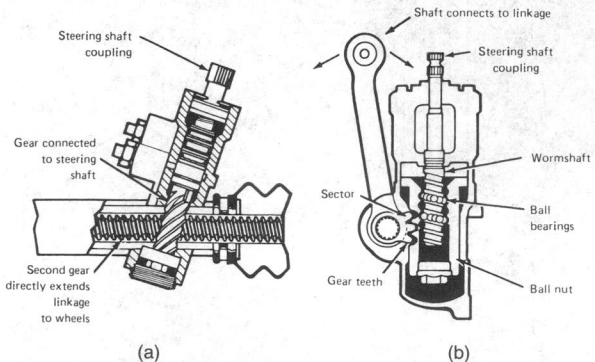

Fig. 11.1.26 Two commonly used types of steering gears. (a) Rack-and-pinion; (b) recirculating ball.

parallel links to turn the wheels of the car. Both steering gear systems operate satisfactorily. The rack-and-pinion is more direct.

Figure 11.1.27 illustrates the geometry of the prevalent **Ackermann steering gear** layout. To avoid slippage of the wheels when turning a curve of radius *r*, the point of intersection *M* for the projected front-wheel axes must fall in a vertical plane through the center of the rear axle. The torque, in foot-pounds, required to turn the wheels of a vehicle standing on smooth concrete varies with the angle of turn, from about 6 percent of the weight on the front axle, in pounds, to start a turn, to 17 percent for a 30° turn.

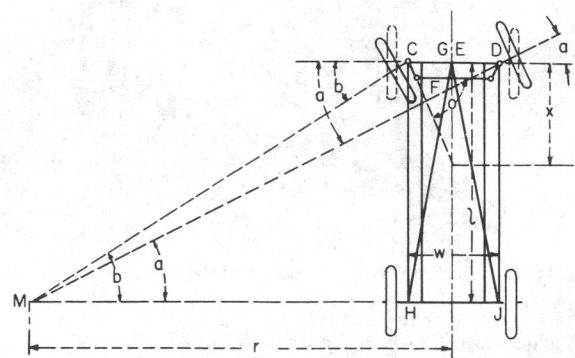

Fig. 11.1.27 Geometry of the Ackermann steering gear.

Power-Assisted Steering

Power steering is a steering control system in which an auxiliary power source assists the driver by providing the major force required to direct the road wheels of the car. The **principal components** of the power steering system (Fig. 11.1.28) are: power steering gear, with servo valve; oil pump, with flow control and relief valves; reserve tank; hydraulic tubing; and oil cooler. These components operate in conjunction with the car's steering wheel, linkage system, and steered wheels. The physical

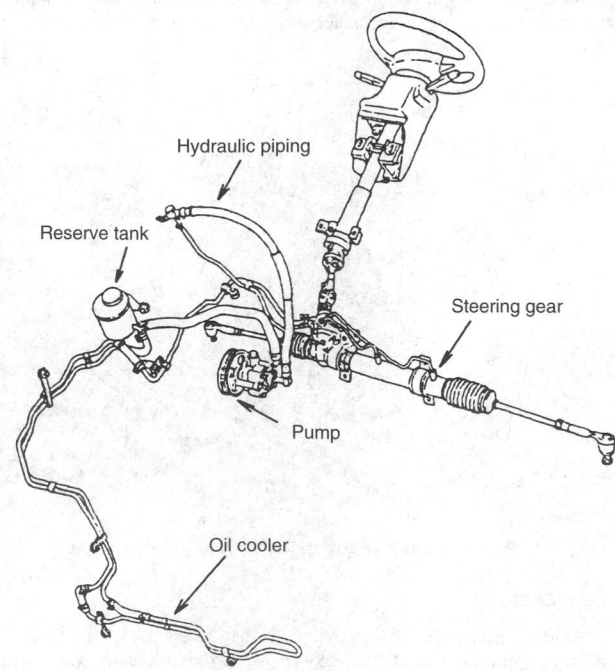

Fig. 11.1.28 Principal components of power steering systems with variable displacement pump. *(Reprinted with permission from SAE, SP-952, ©1993, Society of Automotive Engineers, Inc.)*

effort required to steer an automobile, especially when parking, is appreciably lessened by the power-assisted steering device. This permits reduction in the gear ratio between the steering wheel and the car wheels from some 30 to 15, with consequent reduction in the number of turns of the steering wheel for the complete movement of the front wheels from extreme right to left from 5.5 to 3. Power-assisted steering has been offered for many years on U.S. cars as standard or optional equipment. Public acceptance is such that 88 percent of the cars sold in 1993 were so equipped.

All systems provide (1) steering control in case of failure of the hydraulic-power assistance, and (2) a "feel of the road," by which the driver's effort on the steering wheel is proportional to the force needed to turn the front wheels and by which the tendency of a car to straighten out from a turn or the drag of a soft front tire may be felt at the steering wheel.

Power assistance is effected by hydraulic pressure from an engine-drive pump, acting on a piston in the steering linkage. The piston and its cylinder are incorporated in the steering-gear housing. Oil pressure on the piston is controlled by a valve, such as the balanced spool valve of Fig. 11.1.29.

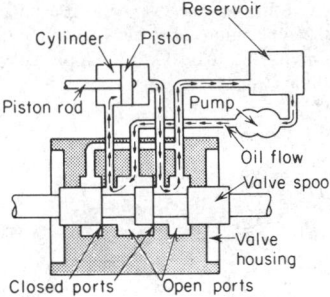

Fig. 11.1.29 Control valve positioned for full-turn power steering assistance using maximum pump pressure.

When the spool is moved slightly to the right, lands on the spool restrict the return of oil from the pump through both return circuits, thus building up delivery pressure. Since the pump delivery is still open to the left end of the power cylinder while the right end is open to the pump suction, a force is developed to move the piston to the right. The greater the restriction imposed on the return of oil to the pump, the greater will be the pressure and the resulting force on the piston.

The spool is centered to the neutral position by suitable **centering springs**. These provide an increasing effort on the steering wheel for increasing steering angle. Although they aid in straightening out from a turn, they do not give the driver a feel of the force required to provide the steering direction. **Hydraulic reaction** against the spool, which is felt at the steering wheel and is proportional to the force developed by the steering gear, is developed by subjecting the ends of the spool to the oil pressure on either side of the power piston.

The valve is held in its neutral position by **preloading** the centering springs. Steering effort at the wheel overcomes this preload. During normal, straight-highway driving, the steering effort is less than the preload and there is no hydraulic assistance; the steering gear is freely reversible, and the driver can "feel the road" and correct for elements such as road camber and crosswinds. The caster action of the front wheels straightens the path of the car when it is coming out of a turn. Any steering effort greater than the preload of the centering springs allows the spool movement to develop a steering assistance proportional to the steering effort and to correspondingly reduce, but not eliminate, the road reactions and shocks felt by the driver.

Oil pumps for power-assisted steering gears are generally driven from the engine by belts, though in some instances they have been driven at higher speeds directly from the electric generator. A typical unit delivers 1.75 gal/min (6.62 L/min) at engine idling speed, at any pressure up to 1,200 lb/in² (8.3 MPa) as may be required while parking.

Three types of **rotary pumps** for the high pressures required are shown in Fig. 11.1.30 (see also Sec. 14.1). Centrifugal force holds the sliding elements against a cam-shaped or eccentric case at high speeds. At low speeds, the sliding elements are held against the case—in design a by springs and in design c by oil pressure admitted to the base of the vanes. The double cam of design c, in addition to doubling the normal volumetric displacement, provides for balancing the oil pressure on each side of the rotor and on the bearings. The cam is contoured for uniform acceleration.

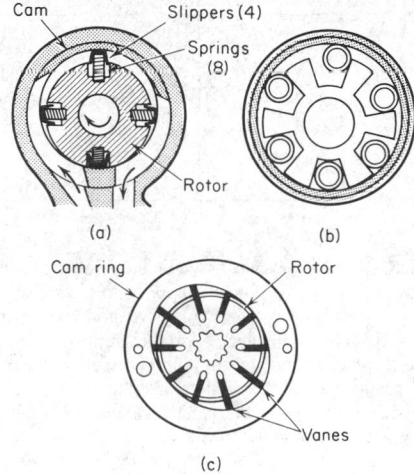

Fig. 11.1.30 Rotary pump types: (a) Chrysler; (b) Ford (Eaton); (c) General Motors (Saginaw).

Variable displacement vane pumps (instead of fixed displacement) are used to raise the efficiency of power steering systems. The variable displacement design **reduces power consumption** by curtailing the surplus oil flow at the middle and high revolution speeds of the steering apparatus. The amount of oil that is pumped is matched to the requirements of the system in its various operating stages.

BRAKES

Stopping distance—the distance traveled by a vehicle after an obstacle has been spotted until the vehicle is brought to a halt—is the sum of the distances traveled during the reaction time and the braking time.

The braking ratio z, usually expressed as a percentage, is the ratio between braking deceleration and the acceleration due to gravity ($g = 32.2$ ft/s² or 9.8 m/s²). The upper and lower braking ratio values are limited by static friction between tire and road surface and the legally prescribed values for stopping distances.

The **reaction time** is the time that elapses between the driver's perception of an object and commencement of action to apply the brakes. This time is not constant; it varies from 0.3 to 1.7 s, depending on personal and external factors. For a reaction time of 1 s, Table 11.1.2 gives stopping distances for various speeds and values of braking ratio (deceleration rates).

The **maximum retarding force** that can be applied to a vehicle through its wheels is limited by the friction between the tires and the road, equal to the coefficient of friction times the vehicle weight. With a coefficient of 1.0, which is about the maximum for dry pavement, this force can equal the car weight and can develop a retardation of 1.0 g. In this instance, stopping distance $S = V^2/29.9$, where V is in mi/h and S is expressed in feet. For metric units, where S is meters and V is km/h, the equation is $S = V^2/254$.

For typical vehicle, tire, and road conditions, with a 0.4 coefficient of friction, 0.4 g deceleration rate, and a reaction time of 1 s, the following is a rule-of-thumb equation for stopping distance; $S \approx (V/10)^2 + (3V/10)$, where S is stopping distance in meters and V is the speed in km/h.

Table 11.1.2   Stopping Distances (Calculated)

Braking ratio z, %	Driving speed before applying brakes, mi/h (km/h)									
	12 (20)	25 (40)	31 (50)	37 (60)	43 (70)	50 (80)	56 (90)	62 (100)	69 (110)	75 (120)
	Reaction distance traveled in 1 s (no braking), ft (m)									
	18 (5.6)	36 (11)	46 (14)	56 (17)	62 (19)	72 (22)	82 (25)	92 (28)	102 (31)	108 (33)
	Stopping distance (reaction + braking), ft (m)									
30	36 (11)	105 (32)	151 (46)	207 (63)	269 (82)	344 (105)	427 (130)	509 (155)	607 (185)	705 (215)
50	29 (8.7)	75 (23)	108 (33)	148 (45)	187 (57)	233 (71)	285 (87)	344 (105)	410 (125)	476 (145)
70	26 (7.8)	66 (20)	92 (28)	121 (37)	151 (46)	187 (57)	230 (70)	272 (83)	318 (97)	360 (110)
90	24 (7.3)	53 (18)	82 (25)	105 (32)	131 (40)	164 (50)	197 (60)	233 (71)	272 (83)	312 (95)

SOURCE: Bosch, "Automotive Handbook," SAE.

The **automobile's brake system** is based on the principles of hydraulics. Hydraulic action begins when force is applied to the brake pedal. This force creates pressure in the master cylinder, either directly or through a power booster. It serves to displace hydraulic fluid stored in the master cylinder. The displaced fluid transmits the pressure through the fluid-filled brake lines to the wheel cylinders that actuate the brake shoe (or pad) mechanisms. Actuation of these mechanisms forces the brake pads and linings against the rotors (front wheels) or drums (rear wheels) to stop the wheels.

All automobiles have two independent systems of brakes for safety. One is generally a **parking brake** and is rarely used to stop a car from speed, though it should be able to. The brake manually operates on the rear wheels through cables or mechanical linkage from an auxiliary foot lever (or a hand pull); it is held on by a ratchet until released by some means such as a push button or a lever.

The main system, or **service brakes,** on all U.S. cars is hydraulically operated, with equalized pressure to all four wheels, except with disk brakes on front wheels, where a proportioning valve is used to permit increased pressure to the disk calipers. Rubber seals preclude the use of petroleum products; hydraulic fluids are generally mixtures of glycols with inhibitors. Figure 11.1.31 shows the **split system,** for improved

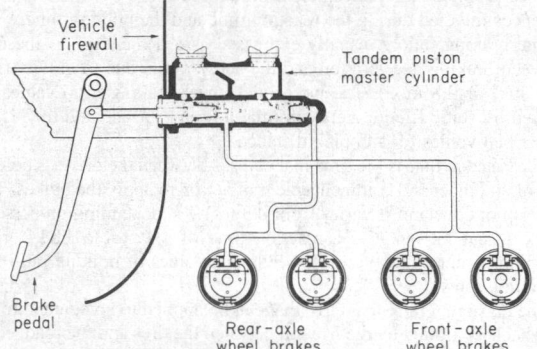

Fig. 11.1.31   "Split" hydraulic brake system.

safety, with two independent master cylinders in tandem, each actuating half the brakes, either front or rear or one front and the opposite rear. Failure of either hydraulic section allows stopping of the car by brakes on two wheels.

Figure 11.1.32 shows the customary design of a **brake dual master cylinder** by which the brake shoes are applied in the conventional internal-expanding brakes (Fig. 11.1.33). When the brakes are released, a spring in the master cylinder returns the pistons to their original positions. This uncovers the compensating ports, permitting brake fluid to enter from the reservoir or to escape from the wheel cylinders after brake application. The check valve facilitates the maintenance of 8 to 16 lb/in² (55 to 110 kPa) line pressure to prevent the entrance of air into the system.

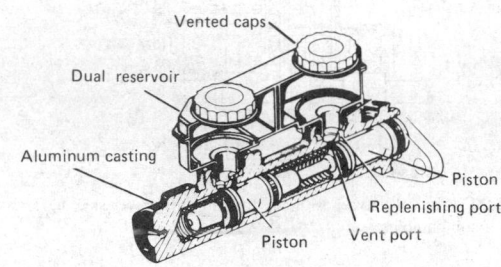

Fig. 11.1.32   Typical design of a dual master cylinder for a split brake system.

Three **types of internal-expanding brakes** (Fig. 11.1.33) have been accepted in service. All are **self-energizing,** where the drum rotation increases the applying force supplied by the wheel cylinder.

With the **trailing shoe** (Fig. 11.1.33a), friction is opposed to the actuating force. The resulting deenergizing of this shoe causes it to do about one-third the work of the leading shoe. Its tendency to lock or squeal is less, and the length and position of the lining are not so critical. The type of brake shown in Fig. 11.1.33a, with one leading and one trailing shoe, formerly was used for the rear wheels. The braking work and wear of the two shoes can be equalized by use of a larger bore for that half of the wheel cylinder which operates the trailing shoe.

The design shown in Fig. 11.1.33b has two leading shoes, each actuated by a single-piston wheel cylinder and each self-energizing. This design has been used for the front wheels where the Fig. 11.1.33a design was used for the rear wheels.

Figure 11.1.33c shows the **Bendix Duo-Servo design,** used on many cars, in which the self-energizing action of **two leading shoes** is much increased by turning them "**in series**"; the braking force developed by the primary shoe becomes the actuating force for the secondary shoe. The action reverses with rotation.

Adjustment for lining wear is effected automatically on most cars. If sufficient wear has developed, a linkage may turn the notched wheel on the adjusting screw (Fig. 11.1.33c) by movement of the primary shoe relative to the anchor pin when the brake is applied with the car moving in reverse. On other designs, adjustment is by linkage between the hand brake and the adjusting wheel.

Brake drums are designed to be as large as practicable in order to develop the necessary torque with the minimum application effort and

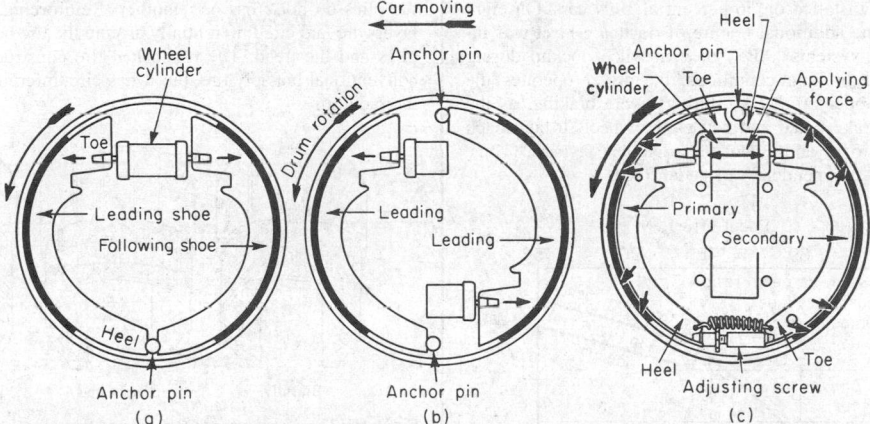

Fig. 11.1.33   Three types of internal-expanding brakes.

to limit the temperature developed in dissipating the heat of friction. The 14- and 15-in (36- and 38-cm) wheel-rim diameters limit the drum diameters to 10 to 12 in (25 to 30 cm), and the 13-in (33-cm) rims limit the diameters to 8 to 9½ in (20 to 24 cm). Drum widths limit unit pressures between the linings and the drums to 16 to 23 lb (7.2 to 10.4 kg) of car weight per in². Drum **friction surfaces** are usually cast iron or iron alloy. **Drum brake shoes** and **disk brake caliper pads** are lined with compounds of resin, metal powder, solid lubricant, abrasives, organic and inorganic fillers, and fibers. Environmental concerns led to the development of **asbestos-free** brake system friction materials. These materials take into account the full range of brake performance requirements, including the mechanism of brake noise and the causes for brake judder (abnormal vibration), and squeal. Current **nonasbestos,** nonsteel friction materials for brake linings and pads include those with main fibers of carbon and aramid plastic. Secondary fibers are copper and ceramic. Friction coefficients range from 0.3 to 0.4.

Where identical brakes are used on front and rear wheels, the rear-wheel cylinders are smaller, so that about 40 to 45 percent of total braking force is developed at the rear wheels. With the split system (Fig. 11.1.31), a smaller master cylinder for the rear brakes gives a similar division. Master cylinders are about 1 in (25.4 mm) in diameter, and other parts of the brake system are so proportioned that a 100-lb (445-N) brake-pedal force develops 600- to 1,200-lb/in² (4.1- to 8.3-MPa) fluid pressure. Air in a hydraulic system makes the brakes feel spongy, and it must be bled wherever it accumulates, as at each wheel cylinder.

Caliper disk brakes (Fig. 11.1.34) offer better heat dissipation by direct contact with moving air; they are not self-energizing, so that there is

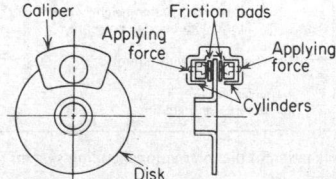

Fig. 11.1.34   Caliper disk brake (schematic).

less drop in the friction coefficient with temperature rise of the brake pads. Contrarily, the absence of self-energization requires higher hydraulic-system pressures and consequent power boosters on heavier cars. **Wear** of the friction pads is normally greater because of the smaller area of contact and the greater exposure to road dirt. The pads are consequently made thicker than the linings of drum brakes, and automatic retraction is incorporated in the hydraulic cylinders.

Power-assisted brakes relieve the driver of much physical effort in retarding or stopping a car. They are either standard or optional equip-

ment on virtually all car models. The supplemental force is developed on a diaphragm by vacuum from the engine intake manifold, either mechanically to the master cylinder or hydraulically, to boost (1) the force between the pedal and the master cylinder, or (2) the hydraulic pressure between the master cylinder and the brakes. Common characteristics are (1) a braking force which is related to pedal pressure so that the driver can feel a pedal reaction proportional to the force applied, and (2) ability to apply the brakes in the absence of the supplemental power.

Figure 11.1.35 illustrates a passenger-car **vacuum-suspended type** of power brake, where vacuum exists on both sides of the main power element when the brakes are released. In the released position, there is contact between the valve plunger and the poppet; thus the port is closed between the power cylinder and the atmosphere.

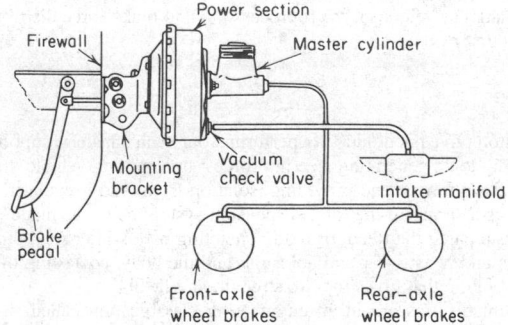

Fig. 11.1.35   Power-assisted brake installation.

Physical effort applied to the brake pedal moves the valve operating rod toward the master-cylinder section. Initial movement of this rod closes the port between the poppet and the power piston. This closes the vacuum passage and brings the valve plunger into contact with the resilient reaction disk. Additional movement of the valve rod then separates the valve plunger from the poppet, thus opening the atmospheric port and admitting air to the control chamber. Air pressure in this chamber depends upon the amount of **physical effort** applied to the pedal. The pressure differences between the two sides of the power piston cause it to move toward the master cylinder, closing the vacuum port and transferring its force through the reaction disk to the hydraulic piston of the master cylinder. This force tends to extrude the reaction disk against the valve plunger and react against the valve operating rod, thus reducing the pedal effort required. An inherent feature of the vacuum-suspended type of power brake is the existence of vacuum, without an additional reservoir, for at least one brake stop after the engine is stopped. Figure 11.1.36 shows the relationship between pedal effort and hydraulic line pressure.

Antilock brakes were installed on 56 percent of 1994 cars. On eight percent of these cars, the additional feature of **traction control** was included. Antilock brake systems (ABS) prevent wheel lockup during braking. Under normal braking conditions, the driver operates the brakes as usual. On slippery roads, or during severe braking, as the driver pushes on the brake pedal and causes the wheels to approach lockup, the ABS takes over and **modulates brake line pressure**. Thus, braking force is applied independently of pedal force.

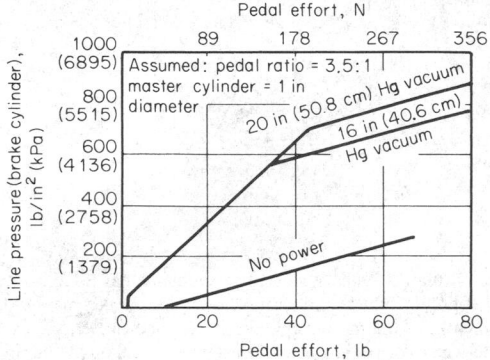

Fig. 11.1.36 Performance chart of a typical power-assisted brake.

Traction control functions as an extension of antilock brake systems, with which it shares numerous components. Traction control helps maintain directional stability along with good traction. An ABS system with traction control uses brake applications to **control slip**. The amount of force transferred when starting off or during deceleration or acceleration is a function of the amount of slip between the tire and the road. The combined antilock and traction control systems make extensive use of automotive **electronics** including elements such as sensors (to measure wheel angular velocity), microprocessors, and brake force distribution controllers.

TIRES

The automotive pneumatic **tire** performs four **main functions**: supporting a moving load; generating steering forces; transmitting vehicle driving and braking forces; and providing isolation from road irregularities by acting as a spring in the total suspension system. A tire is made up of two basic parts: the tread, or road-contacting part, which must provide traction and resist wear and abrasion; and the body, consisting of rubberized fabric that gives the tire strength and flexibility.

Pneumatic tires are in an engineering classification called "**tensile structures**." Other examples of tensile structures are bicycle wheels (spokes in tension, rim in compression); sailboat sails (fabric in tension, air in compression); and prestressed concrete (tendons in tension, concrete in compression).

Tensile structures contain a compression member to provide a tensile preload in the tensile member. In tires, the **cords** are the tensile members and the **compressed air** is the compression member. The common **misunderstanding** is that the tire uses air pressure beneath the rigid wheel to lift it from the flattened tire. Actually, this is invalid; load support must come through the tire casing structure and enter the rim through the tire bead.

The three **principal types** of automobile and truck tires are the crossply or bias-ply, the radial-ply, and the bias-belted (Fig. 11.1.37). In the **bias-ply** tire design, the cords in each layer of fabric run at an angle from one bead (rim edge) to the opposite bead. To balance the tire strength symmetrically across the tread center, another layer is added at an opposing angle of 30 to 38°, producing a two-ply tire. The addition of two more criss-crossed plies make a four-ply tire.

In the **radial tire** (used on most automobiles), cords run straight across from bead to bead. The second layer of cords runs the same way; the plies do not cross one another. Reinforcing belts of two or more layers are laid circumferentially around the tire between the radial body plies and the tread. The **bias-belted** tire construction is similar to the conventional bias-ply tire, but it has circumferential belts like those in the radial tire.

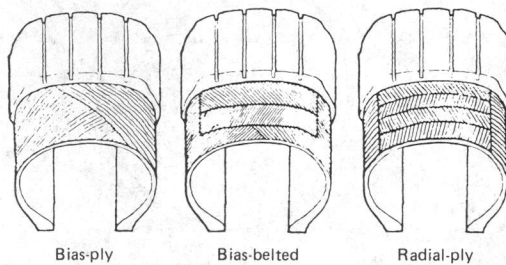

Fig. 11.1.37 Three types of tire construction.

Of the three types, the radial-ply offers the longest tread life, the best traction, the coolest running, the highest gasoline mileage, and the greatest resistance to road hazards. The bias-ply tire provides a softer, quieter ride and is the least expensive. The bias-belted tire design is intermediate between the good-quality bias-ply and the radial tire. It has a longer tread life and is cooler running than the bias-ply, and it gives a smoother low-speed ride than the radial tire. Figure 11.1.38 explains the coding system used for **metric tire size** designation. See Table 11.1.1 for **inflation pressures** and **load limits** for 13-in tires.

Various means for continuous measurement and remote display of car and truck **tire pressures** are well established in vehicle engineering practice. The new designs for tire pressure monitoring systems (on 1996 model vehicles) use **wheel-mounted sensors**. The module containing the sensor, a 6-V lithium battery, and a radio transmitter, is mounted on the wheel rim inside the tire. Once a minute, the tire pressure signal is transmitted to a dash-mounted receiver. The encoded signal is translated and displayed for each wheel. In one U.S. car application, the monitor is installed when optional **run-flat** tires are ordered. Such tires typically allow driving to continue for up to 200 miles at 55 mi/h with zero air pressure. The tires perform so well that there is a possibility that the driver will not be aware of the tire pressure loss, hence the need for monitoring.

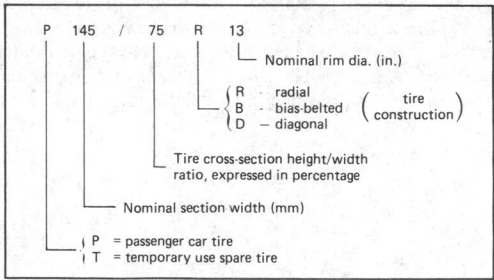

Fig. 11.1.38 Explanation of the international coding system for metric tire size (P series) designations.

AIR CONDITIONING AND HEATING

(See also Sec. 12.4.)

Automobiles are generally **ventilated** through an opening near the windshield with a plenum chamber to separate rain from air. Airflow developed by car motion is augmented, especially at low speed, by a variable-speed, electrically driven blower. When **heat** is required, this air is passed through a finned core served by the engine-jacket water. Core design typically calls for delivery of 20,000 Btu/h (6 kW) and 125 ft³/min (0.1 m³/s) airflow at 130°F (55°C) with 0°F (−18°C) ambient. Car **temperature** is controlled by (1) mixing ambient with heated air, (2) mixing heated with recirculated air, or (3) variation of blower speed. Provision is always made to direct heated air against the interior of the windshield to prevent formation of ice or fog. Figure 11.1.39 illustrates schematically a three-speed blower that drives fresh air through (1) a radiator core or (2) a bypass. The degree and direction of air heating are further regulated by the doors.

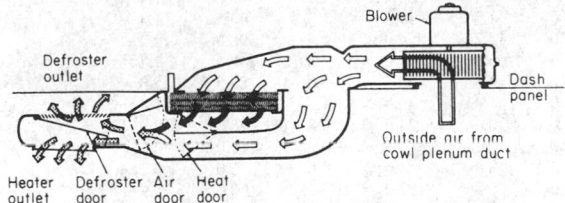

Fig. 11.1.39   Heater airflow diagram.

In 1994, approximately 94 percent of cars built in the United States were equipped with air-conditioning systems. The **refrigeration capacity** of a typical system is 18,000 Btu/h, or 1.5 tons, at 25 mi/h. Cooling capacity increases with car speed. In a **"cool-down" test,** beginning at a test point with 110°F ambient temperature and bright sunshine, the car is "soaked" until its interior temperature has leveled off (at about 140°F). The car is then started and run at 25 mi/h, with interior car temperature checked at up to 48 locations throughout the passenger compartment.

After 10 min of operation, the average car temperature has dropped to the range of 80 to 90°F (27 to 32°C). However, in terms of passenger comfort, at 2 min after starting, the air discharged from the outlets is at about 70°F—and, at 10 minutes, the discharge air is at 55°F. And, by adjusting the outlets, the cool air is directed onto the front seat occupants as desired.

Figure 11.1.40 shows schematically a **combined air-heating and air-cooling system;** various dampers control the proportions of fresh and recirculated air to the heater or evaporator core; air temperature is controlled by a thermostat, which switches the compressor on and off through a magnetic clutch. Electronic control systems eliminate manual changeover and thermostatically actuate the heating and cooling functions.

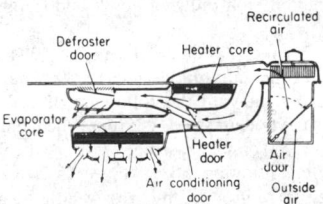

Fig. 11.1.40   Combined heater and air conditioner. *(Chevrolet.)*

General acceptance of **ozone depletion** attributable, in part, to **chlorinated refrigerants** in vehicular air-conditioning systems, led to legislation and regulations that brought about significant changes in air conditioning system design. These systems must be safe and environmentally acceptable, while performing the full range of vehicle cooling and heating requirements.

Regulations banning the use of chlorofluorocarbons (CFs) have been adopted to protect the planet's ozone layer from these chemicals. In the mid-1990s, automotive air conditioning systems commonly use HFC-134a refrigerant, which has thermodynamic properties similar to the formerly used CFC-12. Extensive developmental activity and engineering performance testing shows that cooling capacity and compressor power consumption for the two refrigerants are **closely comparable.** Figure 11.1.41 presents a schematic drawing of a representative air conditioning system using HFC-134a refrigerant and a variable displacement compressor.

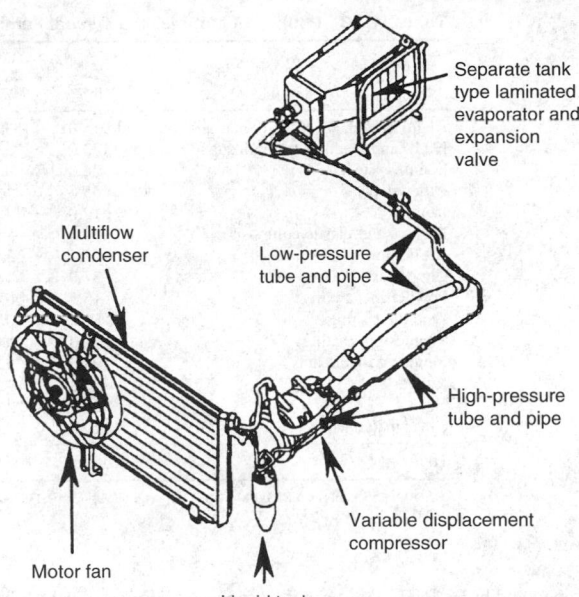

Fig. 11.1.41   Schematic of air conditioning system using HFC-134a refrigerant. *(Reprinted with permission from SAE, SP-916, ©1992, Society of Automotive Engineers, Inc.)*

BODY STRUCTURE

Vehicle design layout can be characterized generally as front-engine, rear-wheel drive (large cars); front-engine, front-wheel drive (small and midsize cars); four-wheel drive (utility vehicles); and midengine, rear-wheel drive (sports cars). The two most commonly used basic body constructions are the unit construction and the body-and-frame. As illustrated in Figs. 11.1.42 and 11.1.43, a car of the **body-and-frame** design has a body that is bolted to a separate frame. Most of the suspension,

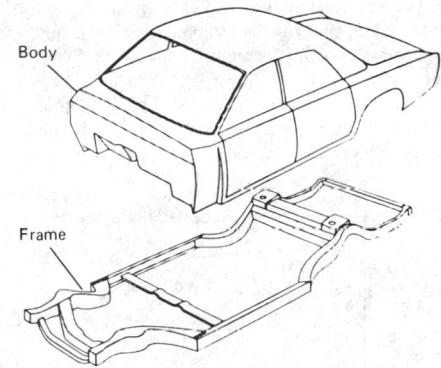

Fig. 11.1.42   Separate body and frame design.

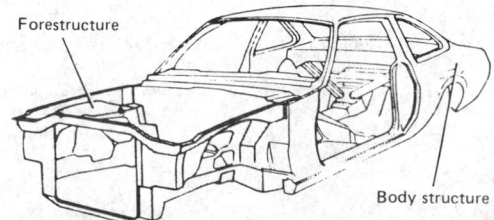

Fig. 11.1.43   Unit construction design for automobiles.

Table 11.1.3   Pounds of Material in a Typical Family Vehicle, 1978–1995

Material	1978 lb	1978 %	1985 lb	1985 %	1990 lb	1990 %	1995 lb	1995 %
Regular steel, sheet, strip, bar, and rod	1,915.0	53.6	1,481.5	46.5	1,405.0	44.7	1,398.0	43.6
High- and medium-strength steel	133.0	3.7	217.5	6.8	238.0	7.6	279.5	8.7
Stainless steel	26.0	0.7	29.0	0.9	34.0	1.1	46.0	1.4
Other steels	55.0	1.5	54.5	1.7	39.5	1.3	43.5	1.4
Iron	512.0	14.3	468.0	14.7	454.0	14.5	398.5	12.4
Plastics and plastic composites	180.0	5.0	211.5	6.6	229.0	7.3	246.5	7.7
Aluminum	112.5	3.2	138.0	4.3	158.5	5.0	187.5	5.8
Copper and brass	37.0	1.0	44.0	1.4	48.5	1.5	43.5	1.4
Powder metal parts	15.5	0.4	19.0	0.6	24.0	0.8	28.0	0.9
Zinc die castings	31.0	0.9	18.0	0.6	18.5	0.6	16.0	0.5
Magnesium castings	1.0	0.0	2.5	0.0	3.0	0.0	5.0	0.2
Fluids and lubricants	198.0	5.5	184.0	5.8	182.0	5.8	190.0	5.9
Rubber	146.5	4.1	136.0	4.3	136.5	4.3	136.0	4.2
Glass	86.5	2.4	85.0	2.7	86.5	2.8	91.5	2.9
Other materials	120.5	3.4	99.0	3.1	83.5	2.7	98.5	3.1
Total	3,569.5	100.0	3,187.5	100.0	3,140.5	100.0	3,208.0	100.0

SOURCE: "American Metal Market," copyright 1995 Capital Cities/ABC Inc. and "Motor Vehicle Facts & Figures," AAMA.
NOTE: 1 lb = 0.45 kg.

bumper, and brake loads are transmitted to the car's frame. A car of the **unit construction** or unitized design, commonly used for small and midsize cars, utilizes the body structure to react to stresses and loads. Unit bodies are complex structures consisting of stamped sheet metal sections that are welded together, forming a framework to which an outer skin is attached.

MATERIALS

(See also Sec. 6.)

In the 1990s, the trend toward smaller, lighter, more fuel-efficient, more corrosion-resistant cars has led to growth in the use of such materials as high-strength steel, galvanized steel, aluminum, and a variety of plastics materials. Table 11.1.3 provides a **breakdown of materials usage** beginning with 1978 and including a 1995 projection. In this table, the "other" material category includes sound deadeners and sealers, paint and corrosion-protective dip, ceramics, cotton, cardboard, and miscellaneous other materials.

Motor vehicle design and materials specification are greatly influenced by life-cycle requirements that extend to **recycling** the material from scrapped vehicles. In the mid 1990s, 94 percent of scrapped vehicles are processed for recycling. Approximately 75 percent of an automobile's material content is recycled. Figure 11.1.44 shows a general breakdown for materials disposition from **recycled vehicles**.

TRUCKS

Since the beginning of the motor vehicle industry at the turn of the century, **trucks** have been an important segment of the industry. Many of the first motor vehicles were trucks, and closely paralleled automobile design technology. In the United States, in 1992, a total of nearly **60 million trucks** were in use. More than 50 million (of the total) were **light trucks** weighing 6,000 lb or less; about 5 million were light trucks weighing between 6,000 and 10,000 lb; and 700,000 were light trucks between 10,000 and 14,000 lb. The remaining approximately 3.3 million ranged in **medium** and **heavy** sizes from 14,001 to 60,000 lb and larger. In 1994 the average age of trucks in use was 8.4 years. About 29 percent of U.S. intercity ton-miles of goods and freight are transported by truck.

The breakdown of 6.4 million **truck sales** in 1994 was: 6.1 million light trucks, 167,000 medium-duty, and 186,000 heavy-duty. In 1994, light trucks were bought with the following percentages of optional equipment: automatic transmission, 78; four-wheel antilock brakes, 32; rear antilock brakes, 54; power steering, 97; four-wheel drive, 34; diesel engine, 4; and air conditioning, 86.

Trucks, generally, are grouped by **gross vehicle weight (GVW)** rating into light-duty (0 to 14,000 lb), medium-duty (14,001, to 33,000 lb), and heavy-duty (over 33,000 lb) categories. To determine GVW classification, truck capacities are further broken down into **classes** 1 to 8, as listed in Table 11.1.4.

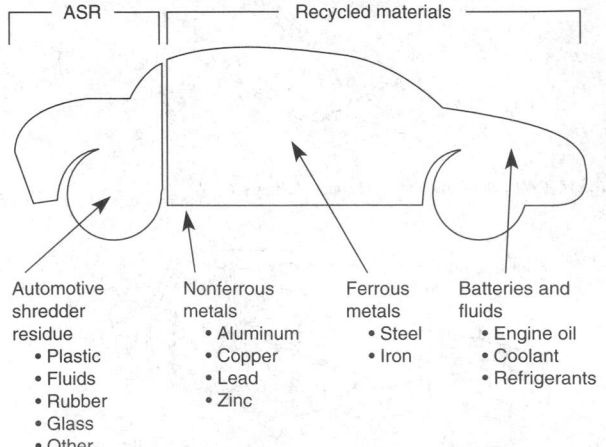

Fig. 11.1.44 Materials disposition from recycled vehicles. (*"Motor Vehicle Facts & Figures,"* AAMA.)

Table 11.1.4   Gross Vehicle Weight Classification

GVW group	Class	Weight, kg (max)
6000 lb or less	1	2722
6001–10,000 lb	2	4536
10,001–14,000 lb	3	6350
14,001–16,000 lb	4	7258
16,001–19,500 lb	5	8845
19,501–26,000 lb	6	11,794
26,001–33,000 lb	7	14,469
33,001 lb and over	8	14,970

Truck Service Operations

Truck usage, purpose, or "vocation" is categorizd by type of **service operation** as follows:

Class A Service: operation of motor vehicles on smooth, hard-surfaced highways in **level** country, where transmission gears are used only to accelerate the vehicle and payload from rest. Fast axle gear ratios are used.

Class B Service: operation of motor vehicles on smooth, hard-surfaced highways in **hilly** country where numerous grades are encountered. Requires intermittent use of transmission for short periods to surmount grades. The intermediate range of axle gear ratios is used in this class of service.

Class C Service: operation of motor vehicles for dump truck service, where the vehicle operates on loose **dirt, sand,** or **muddy roads** that offer high rolling resistance. This class of service also covers operation under conditions where the transmission is used for long periods in overcoming bad road conditions on long mountain grades. Vehicle design for this class of service usually calls for the slow range of axle gear ratios.

Class D Service: operation of motor vehicles in connection with **semitrailers or trailers.** This type of service is confined to smooth, hard-surfaced highways. In this service, the gross vehicle weight is the gross train weight, that is, the total weight of truck, trailer, and load.

Truck weight classifications are used to differentiate truck sizes. To fully define the vehicle, engine type and truck mission or purpose need to be known. Figure 11.1.45 illustrates medium- and heavy-duty **truck classifications** by gross weight and vocation. Figure 11.1.46 shows life expectancy within the weight classes.

Class 8 (33,001 lb or over)

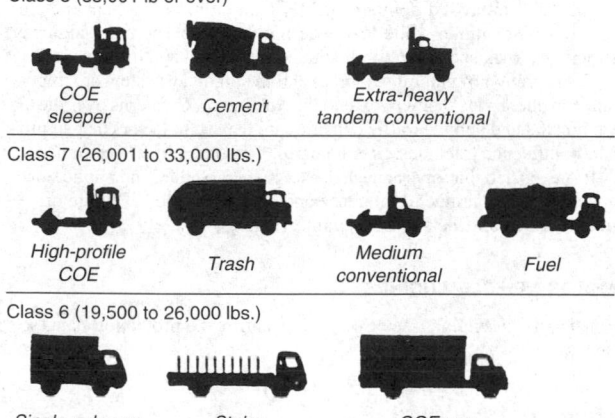

Class 7 (26,001 to 33,000 lbs.)

Class 6 (19,500 to 26,000 lbs.)

Fig. 11.1.45 Medium- and heavy-duty truck classification by gross weight and vehicle usage. (*Reprinted with permission from SAE, SP-868, ©1991, Society of Automotive Engineers, Inc.*)

Truck Engineering

A heavy-duty **truck drivetrain** is treated as an engineered, integral system that starts at the engine flywheel housing, transmission mounting surface, and flywheel pilot bearing and ends at the rear axle suspension mounting points. The most important information needed is the end-user description, including a customer/fleet profile. The following items provide a framework for starting and focusing the **truck engineering process:**

Current vehicle specifications
Gross cargo weight and percent of time at gross cargo weight
What is hauled, where it is hauled
Competition, including driveline
Future trends and changes

Past experience
Trade-in cycle

System characteristics that dictate design include:
1. Basic vehicle performance: startability, gradeability, acceleration, fuel economy, trip time
2. Reliability and expected system life
3. Noise emission—pass-by
4. Ergonomics—controls, in-cab noise, vibration
5. Dynamics—vibration related to noise, safety, or early component failure
6. Mechanical mounting and attachment of system components
7. Maintenance and serviceability

Performance is a basic requirement in designing a truck or truck-trailer. Broadly, performance relates to a truck's ability to move a load economically under varying conditions of operation. The ability of a truck to move a load depends on the engine, transmission, axle ratio, and tire matched to the expected gross weight, road conditions, and road speed. The **engine** must provide adequate power to maintain the desired road speed with the specified load; the **transmission** must embody a range of speed selection to permit the engine to operate in the optimum range; **axle ratio** must be compatible with required road speed.

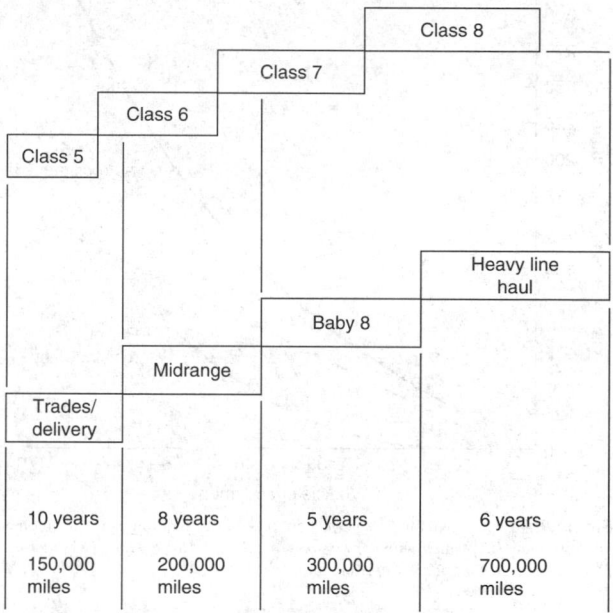

Fig. 11.1.46 Truck life expectancy in medium- and heavy-duty service. (*Reprinted with permission from SAE, SP-868, ©1991, Society of Automotive Engineers, Inc.*)

Power requirements for a given vehicle must be known before an engine and driveline can be selected. **Power required** is the sum of air, rolling, and grade resistances. Assuming the road surface remains constant, the grade resistance may be considered as increased rolling resistance. The **total resistance** and the power required can be determined when the following factors are known:

Gross vehicle weight
Maximum frontal cross-section area
Drag coefficient
Vehicle speed
Road surface and grade

Truck **top speed** is design-controlled using one or more of three approaches:
1. **Gear limited:** the engine provides adequate reserve power to reach governed maximum rotational speed while driving the fully laden

vehicle over level ground. In this case, maximum truck speed is proportional to the maximum governed engine speed, cruise gear ratios, and tire rolling radius. Reserve power is provided to permit the truck to maintain the desired **geared speed** when slight grades or head winds are encountered.

2. **Road speed governed:** applied and installed as a means to limit vehicle top speed, without sacrificing performance when the truck is operating at speeds below maximum. Safety is a principal purpose of a road speed governor. Governing also contributes to engine life and fuel economy through reduced engine speed while cruising at the maximum **governed truck speed.** Electronic fuel systems provide increased engine-drivetrain-vehicle control.

3. **Horsepower limited:** the horsepower to propel the truck over level ground must overcome rolling resistance and air resistance. Rolling resistance increases linearly with vehicle speed. Air resistance increases as the cube of vehicle speed.

Because the power required to overcome air resistance increases as the **cube of vehicle speed,** aerodynamic drag or wind resistance is a

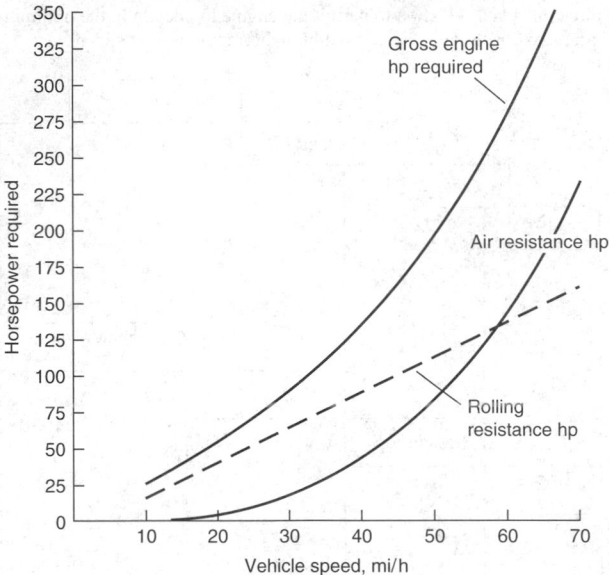

Fig. 11.1.47 Aerodynamic drag effect on truck power requirement to overcome air resistance. *(Reprinted with permission from SAE, "Motor Truck Engineering Handbook," ©1994, Society of Automotive Engineers, Inc.)*

significant consideration in truck engineering and truck operations. For heavy-duty trucks, including truck-trailer combinations, frontal area ranges from 50 to 102 ft^2 (4.6 to 9.5 m^2). Drag coefficients range from about 0.6 to 1.0. The **drag coefficient** for a representative Class 8 truck-trailer is about 0.7.

Published data and testing indicate that **fuel savings** from aerodynamic improvements are equivalent to fuel savings attained by chassis weight reduction and require fewer services to maintain. Approximately half the driving force reaching the powered wheels is needed to overcome air resistance at cruising speed, as shown in Fig. 11.1.47. This is similar, in order of magnitude, to the aerodynamic effects discussed earlier for automobiles. For trucks, a reduction of approximately 10 percent in drag results in a fuel savings of about 5 percent, or nearly 1000 gallons per year.

Truck Noise Emissions

Commercial vehicle manufacturers are required to comply with a U.S. federal law restricting the **pass-by noise** emitted from a new production vehicle. A maximum sound pressure level of 80 dBA is allowed when subjecting a vehicle to the pass-by test described in SAE J366, "Exterior Sound Level for Heavy Trucks and Buses." This procedure measures the maximum **exterior sound level** the vehicle produces under a prescribed condition where the vehicle does not exceed 35 mi/h. Tire noise becomes a significant factor at higher speeds, so the test is designed specifically to include all noise except tires. Another procedure, SAE J57, "Truck Noise Control," is used to measure tire noise.

Truck Design Constraints and Prospects

Historically, **commercial trucks** replaced the horse and wagon. Dirt roads evolved into a paved national highway system. Early truck developments focused on increased load-carrying capacity, higher highway speeds, and improved component life. Currently, component life improvements continue while loads are limited to conserve the highway structure. Truck speeds are limited to conserve fuel and improve safety.

Conservation of natural resources and the highway system are important considerations that will extend into the future. Concerns over the air we breathe and vehicle traffic densities are **important factors** that significantly influence truck design engineering and operational performance.

It is expected that **environmental, energy conservation,** and public **safety** concerns will continue to influence and direct the focus of future drivetrain and overall truck vehicle engineering.

MOTOR VEHICLE ENGINES

For details on engines, see Sec. 9. For antifreeze-protection data, see Sec. 6.

11.2 RAILWAY ENGINEERING
by V. Terrey Hawthorne and Keith L. Hawthorne
(in collaboration with David G. Blaine, E. Thomas Harley, Charles M. Smith, John A. Elkins, and A. John Peters)

REFERENCES: *Proceedings,* Association of American Railroads (AAR), Mechanical Division. *Proceedings,* American Railway Engineering Association (AREA). *Proceedings,* American Society of Mechanical Engineers (ASME). "The Car and Locomotive Cyclopedia," Simmons-Boardman. J. K. Tuthill, "High Speed Freight Train Resistance," *University of Illinois Engr. Bull.,* **376**, 1948. A. I. Totten, "Resistance of Light Weight Passenger Trains," *Railway Age,* **103**, Jul. 17, 1937. J. H. Armstrong, "The Railroad—What It Is, What It Does," Simmons-Boardman, 1978. Publications of the International Government-Industry Research Program on Track Train Dynamics, Chicago. Publications of AAR Research and Test Department, Chicago. Max Ephraim, Jr., "The AEM-7 —A New High Speed, Light Weight Electric Passenger Locomotive," ASME RT Division, 82-RT-7, 1982. W. J. Davis, Jr., *General Electric Review,* October 1926. R. A. Allen, Conference on the Economics and Performance of Freight Car Trucks, October 1983. "Engineering and Design of Railway Brake Systems," The Air Brake Association, Sept. 1975. Code of Federal Regulations, Title 40, Protection of Environment. Code of Federal Regulations, Title 49, Transportation.

DIESEL-ELECTRIC LOCOMOTIVES

Diesel-electric locomotives and electric locomotives are classified by wheel arrangement; letters represent the number of adjacent driving axles in a rigid truck (A for one axle, B for two axles, C for three axles, etc.). Idler axles between drivers are designated by numerals. A plus sign indicates articulated trucks or motive power units. A minus sign indicates separate swivel trucks not articulated. This nomenclature is fully explained in Standard S-523 issued by the Association of American Railroads (AAR). Virtually all modern locomotives are of either B-B or C-C configuration.

The high efficiency of the diesel engine is an important factor in its selection as a prime mover for traction. This efficiency at full or part load makes it ideally suited to the variable service requirements found in daily railroad operations. The diesel engine is a constant-torque machine that cannot be started under load and hence requires a variably coupled transmission arrangement. The electric transmission system allows it to make use of its full rated power output at low track speeds for starting and efficient hauling of heavy trains. Examples of the various diesel-electric locomotive types in service are shown in Table 11.2.1. A typical diesel-electric locomotive is shown in Fig. 11.2.1.

Most of the diesel-electric locomotives have a dc generator or rectified alternator coupled directly to the diesel engine crankshaft. The generator/alternator is electrically connected to dc series traction motors having nose suspension mountings. Some recent locomotives utilize gate turn-off (GTO) inverters and ac traction motors to obtain the benefits of increased adhesion and increased horsepower per axle. The tooth ratio of the axle-mounted gears to the motor pinion gears which they engage is determined by the speed range and is related to the type of service. A high ratio is used for freight where high tractive efforts and low speeds are required, whereas high-speed passenger locomotives have a lower ratio.

Diesel Engines Most new diesel-electric locomotives are equipped with either a V-type, two strokes per cycle, or V-type, four strokes per cycle, engine. Engines range from 8 to 20 cylinders each. Output power ranges from 1,000 hp to over 4,400 hp for a single engine application.

Two-cycle engines are aspirated with either a gear-driven blower or a turbocharger. Because these engines lack an intake stroke for natural aspiration, the turbocharger is gear driven at lower engine speeds. At higher engine speeds, when the exhaust gases contain enough energy to drive the turbocharger, an overriding clutch disengages the gear train. Free-running turbochargers are used on four-cycle engines, as at lower speeds the engines are aspirated by the intake stroke. Two-cycle engines utilize a unit fuel injector which pumps, meters, and atomizes the fuel. The four-cycle engines use separate metering pumps and atomizing injectors.

The engine control governor is an electrohydraulic device used to regulate the speed and horsepower output of the diesel engine. It is a self-contained unit mounted on the engine and driven from one of the engine camshafts. It is equipped with its own oil supply and pressure pump.

The governor contains four solenoids which are actuated individually or in combination from the 74-V auxiliary generator/battery supply by a series of switches actuated by the engineer's throttle. There are eight power positions of the throttle, each corresponding to a specific value of engine rpm and horsepower. The governor maintains the predetermined value of engine rpm through a mechanical linkage to the engine fuel racks which control the amount of fuel metered to the cylinders.

One or more centrifugal pumps, gear-driven from the engine crankshaft, force water through passages in the cylinder heads and liners. The water temperature is automatically controlled by regulating shutter and fan operation, which in turn controls the passage of air through the cooling radiators, or by bypassing the water around the radiators. The fans (one to four per engine) may be motor driven or mechanically driven by the engine crankshaft. If mechanical drive is used, current practice is to drive the fans through a clutch, since it is wasteful of energy to operate the fans when cooling is not required.

The lubricating oil system supplies clean oil at the proper temperature and pressure to the various bearing surfaces of the engine, such as crankshaft, camshaft, wrist pins, and cylinder walls, and supplies oil to the heads of the pistons to remove excess heat. One or more gear-type pumps driven from the crankshaft are used to move the oil from the crankcase through strainers to the bearings and the piston cooling passages, after which the oil runs by gravity to the crankcase. A heat exchanger is located in the system to permit all the oil to pass through it at some time during a complete cycle. The oil is cooled by passing engine cooling water through the other side of the exchanger. Paper

Table 11.2.1   Locomotives in Service in North America

Locomotive*		Service	Arrangement	Weight min./max. (1,000 lb)	Starting† for min./max. weight (1,000 lb)	Tractive effort At continuous‡ speed (mi/h)	Number of cylinders	Horsepower rating (r/min)
Builder	Model							
Bombardier	HR-412	General purpose	B-B	240/280	60/70	60,400 (10.5)	12	2,700 (1,050)
Bombardier	HR-616	General purpose	C-C	380/420	95/105	90,600 (10.0)	16	3,450 (1,000)
Bombardier	LRC	Passenger	B-B	252 nominal	63	19,200 (42.5)	16	3,725 (1,050)
EMD	SW-1001	Switching	B-B	230/240	58/60	41,700 (6.7)	8	1,000 (900)
EMD	MP15	Multipurpose	B-B	248/278	62/69	46,822 (9.3)	12	1,500 (900)
EMD	GP40-2	General purpose	B-B	256/278	64/69	54,700 (11.1)	16	3,000 (900)
EMD	SD40-2	General purpose	C-C	368/420	92/105	82,100 (11.0)	16	3,000 (900)
EMD	GP50	General purpose	B-B	260/278	65/69	64,200 (9.8)	16	3,500 (950)
EMD	SD50	General purpose	C-C	368/420	92/105	96,300 (9.8)	16	3,500 (950)
EMD	F40PH-2	Passenger	B-B	260 nominal	65	38,240 (16.1)	16	3,000 (900)
GE	B18-7	General purpose	B-B	231/268	58/67	61,000 (8.4)	8	1,800 (1,050)
GE	B30-7A	General purpose	B-B	253/280	63/70	64,600 (12.0)	12	3,000 (1,050)
GE	C30-7A	General purpose	C-C	359/420	90/105	96,900 (8.8)	12	3,000 (1,050)
GE	B36-7	General purpose	B-B	260/280	65/70	64,600 (12.0)	16	3,600 (1,050)
GE	C36-7	General purpose	C-C	367/420	92/105	96,900 (11.0)	16	3,600 (1,050)
EMD	AEM-7	Passenger	B-B	201 nominal	50	33,500 (10.0)	NA¶	7,000 §
EMD	GM6C	Freight	C-C	365 nominal	91	88,000 (11.0)	NA	6,000 §
EMD	GM10B	Freight	B-B-B	390 nominal	97	100,000 (5.0)	NA	10,000 §
GE	E60C	General purpose	C-C	364 nominal	91	82,000 (22.0)	NA	6,000 §
GE	E60CP	Passenger	C-C	366 nominal	91	34,000 (55.0)	NA	6,000 §
GE	E25B	Freight	B-B	280 nominal	70	55,000 (15.0)	NA	2,500 §

* Engines: Bombardier—model 251, 4-cycle, V-type, 9 × 10½ in cylinders.
    EMD—model 645E, 2-cycle, V-type, 9 9/16 × 10 in cylinders.
    GE—model 7FDL, 4-cycle, V-type, 9 × 10½ in cylinders.
† Starting tractive effort at 25% adhesion.
‡ Continuous tractive effort for smallest pinion (maximum).
§ Electric locomotive horsepower expressed as diesel-electric equivalent (input to generator).
¶ Not applicable.

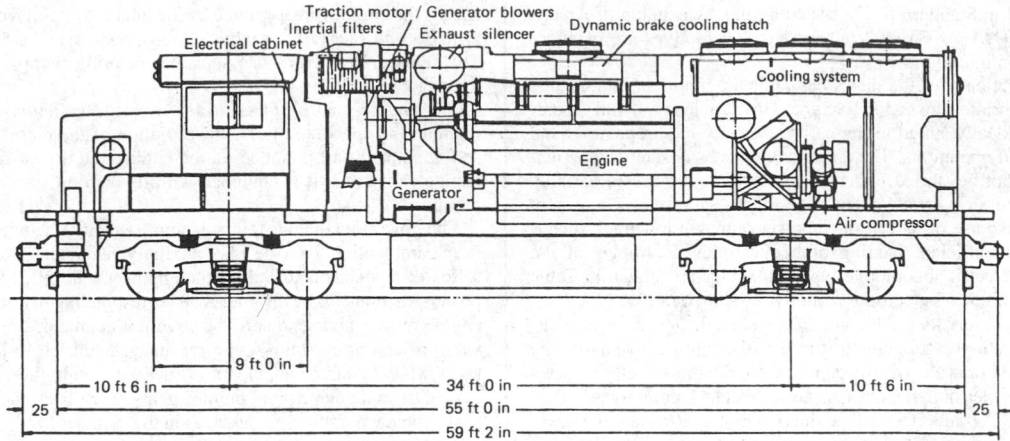

Fig. 11.2.1 Locomotive general arrangement. (*Courtesy of M. Ephraim, ASME, Rail Transportation Division.*)

element cartridge filters in series with the oil flow are used to remove any fine impurities in the oil before it enters the engine.

Electric Transmission Equipment A main generator or three-phase ac alternator is directly coupled to the diesel engine crankshaft. Alternator output is rectified through a full-wave bridge to minimize ripple to a level acceptable for operation of series field dc motors or for input to the solid-state inverter system for ac motors.

Generator power output is controlled by (1) varying engine speed through movement of the engineer's controller and (2) controlling the flow of current in its battery field or in the field of a separate exciter generator. The shunt and differential fields (if used) are designed to maintain a constant generator power output for a given engine speed as the propulsion system and voltage vary. They do not completely accomplish this, thus the battery field or separately excited field must be controlled by the load regulator to provide the final adjustment in excitation to load the engine properly. This field is also automatically deenergized to reduce or remove the load, when certain undesirable conditions occur, to prevent damage to the power plant or other traction equipment.

The engine main generator or alternator power plant functions at any throttle setting as a constant-horsepower source of energy. Therefore, the main generator voltage must be controlled to provide constant power output for each specific throttle position under varying conditions of locomotive speed, train resistance, atmospheric pressure, and quality of fuel. The load regulator, which is an integral part of the governor, accomplishes this within the maximum safe values of main-generator voltage and current. For example, when the locomotive experiences an increase in track gradient with a consequent reduction in speed, traction-motor counter-emf decreases, causing a change in traction-motor and main-generator current. Because this alters the load demand on the engine, the speed of the engine tends to change to compensate. As the speed changes, the governor begins to reposition the fuel racks, but at the same time a pilot valve in the governor directs hydraulic pressure into a load regulator vane motor which changes the resistance value of the load regulator rheostat in series with the main-generator excitation circuit. This alters the main-generator excitation current and consequently, main-generator voltage and returns the value of main-generator power output to normal. Engine fuel racks return to normal, consistent with constant values of engine speed.

The load regulator is effective within maximum and minimum limit values of the rheostat. Beyond these limits the power output of the engine is reduced. However, protective devices in the main generator excitation circuit limit the voltage output to ensure that values of current and voltage in the traction motor circuits are within safe limits.

Auxiliary Generating Apparatus Dc power for battery charging, lighting, control, and cab heaters is furnished by a separate generator, geared to the main engine. Voltage output is regulated within 1 percent

of 74 V over the full range of engine speeds. Auxiliary alternators with full-wave bridge rectifiers are also utilized for this application.

Traction motor blowers are mounted above the locomotive underframe. Air from the centrifugal blower housings is carried through the underframe and into the motor housings through flexible ducts. Other designs are being developed to vary the air output with cooling requirements to conserve parasitic energy demands.

Traction motors are nose-suspended from the truck frame and bearing-suspended from the axle (Fig. 11.2.2). The traction motors employ

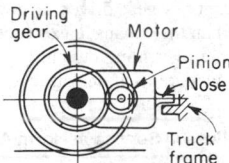

Fig. 11.2.2 Axle-hung traction motor for diesel-electric and electric locomotives.

series-exciting (main) and commutating field poles. The current in the series field is reversed to change locomotive direction and may be partially shunted through resistors to reduce counter-emf as locomotive speed is increased. Newer locomotives dispense with field shunting to improve commutation. Early locomotive designs also required motor connection changes (series, series/parallel, parallel) referred to as transition, to maintain motor current as speed increased.

AC traction motors employ a variable-frequency supply derived from a computer-controlled solid-state inverter system, fed from the rectified alternator output. Locomotive direction is controlled by reversing the sequence of the three-phase supply.

The development of the modern traction alternator with its high output current has resulted in a trend toward traction motors that are permanently connected in parallel. DC motor armature shafts are equipped with grease-lubricated roller bearings, while ac traction-motor shafts are supplied with a grease-lubricated roller bearing at the free end and an oil-lubricated bearing at the pinion drive end. Traction-motor support bearings are usually of the plain sleeve type with lubricant wells and spring-loaded felt wicks which maintain constant contact with the axle surface.

Electrical Controls In the conventional dc propulsion system, electropneumatic or electromagnetic contactors are used to make and break the circuits between traction motors and the main generator. They are equipped with interlocks for various control-circuit functions. Similar contactors are used for other power and excitation circuits of lower power (amperage). An electropneumatic or electric-motor-operated cam switch, consisting of a two-position drum with copper segments moving between spring-loaded fingers, is generally used to reverse traction-motor field current ("reverser") or to set up the circuits for dynamic braking. This switch is not designed to operate under load. On some dc locomotives these functions have been accomplished with a system of contactors. In the ac propulsion systems, the power-control contactors are totally eliminated since their function is performed by the solid-state switching devices in the inverters.

Cabs In order to promote uniformity and safety, the AAR has issued standards for many locomotive cab features, such as cab seats, water cooler and toilet installation, and flooring. Locomotive cab noise levels are prescribed by CFR 49§229.121.

Propulsion control circuits transmit the engineer's movements of the throttle lever, reverse lever, and transition or dynamic-brake control lever in the controlling unit to the power-producing equipment of each unit operating in multiple in the locomotive consist. Before power is applied, all reversers must move to provide proper motor connections for the directional movement desired, taking into account the direction in which each locomotive unit is headed. Power contactors complete the circuits between generators and motors. For dc propulsion systems excitation circuits then function to provide the proper main-generator field current while the engine speed increases to correspond to the engineer's throttle position. In ac propulsion systems, all power circuits are controlled by computerized switching of the inverter. To provide for multiple-unit operation, the control circuits are connected between locomotive units by jumper cables. The AAR has issued Standard S-512 covering standard dimensions and contact identification for 27-point control jumpers used between diesel-electric locomotive units.

Wheel slip is detected by some form of sensing equipment connected electrically to the motor circuits or mechanically to the axles. When slipping occurs on some units, relays automatically reduce main generator excitation until slipping ceases, whereupon power is gradually reapplied. On newer units, an electronic system senses small changes in motor current and reduces motor current before a slip occurs. An advanced system recently introduced adjusts wheel creep to maximize wheel-to-rail adhesion. Wheel speed is compared to ground speed, which is accurately measured by radar. A warning light and/or buzzer in the operating cab informs the engineer, who, if the slip condition persists, must then notch back on the throttle.

Batteries Lead-acid storage batteries of 280 or 420 A · h capacity are usually used for starting the diesel engine. Thirty-two cells on each locomotive unit are used to provide 64 V to the system. (See also Sec. 15.) The batteries are charged from the 74-V power supply.

Air Brake System The "independent" brake valve handle at the engineer's position controls air pressure supplied from the locomotive reservoirs to the brake cylinders on the locomotive itself. The "automatic" brake valve handle (Fig. 11.2.3) controls the air pressure in the

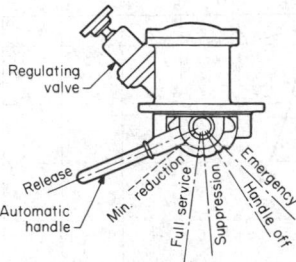

Fig. 11.2.3 Automatic-brake valve-handle positions for 26-L brake equipment.

brake pipe to the train. The AAR has issued Standard S-529, "Specifications for New Locomotive Standard Air Brake Equipment."

Compressed air for braking and for various pneumatic controls on the locomotive is supplied by a two-stage three-cylinder compressor, usually connected directly or through a clutch to the engine crankshaft. An unloader or the clutch is activated to maintain a pressure of approximately 130 to 140 lb/in² (896 to 965 kPa) in the main reservoirs. When

charging an empty trainline with the locomotive at standstill (maximum compressor demand), the engineer may increase engine (and compressor) speed without feeding power to the traction motors.

Dynamic Braking On most locomotives, dynamic brakes supplement the air brake system. The traction motors are used as generators to convert the kinetic energy of the locomotive and train into electrical energy, which is dissipated through resistance grids located near the locomotive roof. Motor-driven blowers, designed to utilize some of this braking energy, force cooler outside air over the grids and out through roof hatches. By directing a generous and evenly distributed airstream over the grids, their physical size is reduced in keeping with the relatively small space available in the locomotive. On some locomotives, resistor-grid cooling is accomplished by an engine-driven radiator/braking fan, but energy conservation is causing this arrangement to be replaced by motor-driven fans which can be powered in response to need using the parasitic power generated by dynamic braking itself.

By means of a cam-switch reverser, the traction motors are connected to the resistance grids; the motor fields are usually connected in series across the main generator to supply the necessary high excitation current. The magnitude of the braking force is set by controlling the traction motor excitation and the resistance of the grids. Conventional dynamic braking is not usually effective below 10 mi/h (16 km/h), but it is very useful at 20 to 30 mi/h (32 to 48 km/h). Some locomotives are equipped with "extended range" dynamic braking which enables dynamic braking to be used at speeds as low as 3 mi/h (5 km/h) by shunting out grid resistance (both conventional and extended range shown on Fig. 11.2.4). Dynamic braking is now controlled according to the "ta-

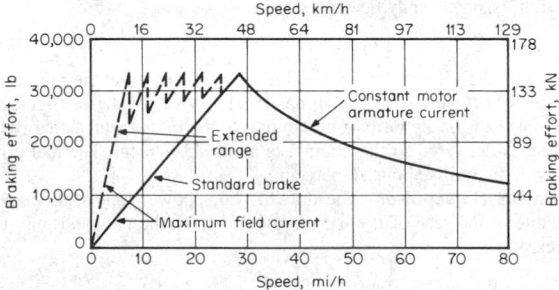

Fig. 11.2.4 Dynamic-braking effort versus speed. (*Electro-Motive Division, General Motors Corp.*)

pered" system, although the "flat" system has been used in the past. Dynamic braking control requirements are specified by AAR Standard S-518. Dynamic braking is especially advantageous on long grades where accelerated brake shoe wear and potential thermal damage to wheels could otherwise be problems. Other advantages are relieving crews from setting up air brake retainers on freight cars and smoother control of train speed. Dynamic brake grids can also be used for a self-contained load-test feature which permits a standing locomotive to be tested for power output. On locomotives equipped with ac traction motors a constant dynamic braking (flat-top) force can be achieved from the horsepower limit down to 2 mi/h (3 km/h).

Performance

Engine Indicated Horsepower The horsepower delivered at the diesel locomotive drawbar is the end result of a series of subtractions from the original indicated horsepower of the engine, made to take into account efficiency of transmission equipment and losses due to the power requirements of various auxiliaries. The formula for the engine's indicated horsepower is

$$\text{ihp} = PLAN/33{,}000 \qquad (11.2.1)$$

where P = mean effective pressure in the cylinder, lb/in²; L = length of piston stroke, ft; A = piston area, in²; and N = total number of cycles completed per min. Factor P is governed by the overall condition of the engine, quality of fuel, rate of fuel injection, completeness of combustion, compression ratio, etc. Factors L and A are fixed with design of engine. Factor N is a function of engine speed, number of working chambers, and strokes needed to complete a cycle. (See also Secs. 4 and 9.)

Engine Brake Horsepower In order to calculate the horsepower delivered to the crankshaft coupling (main-generator connection), frictional losses in bearings and gears must be subtracted from the **indicated horsepower** (ihp). Some power is also used to drive lubricating-oil pumps, governor, water pump, scavenging blower, and other auxiliary devices. The resultant horsepower at the coupling is **brake horsepower** (bhp).

Rail Horsepower A portion of the engine bhp is transmitted mechanically via couplings or gears to operate the traction motor blowers, air brake compressor, auxiliary generator, and radiator cooling fan generator or alternator. Part of the auxiliary generator electric output is used to run some of the auxiliaries. The remainder of the engine bhp transmitted to the main generator or main alternator for traction purposes must be multiplied by generator efficiency (usually about 91 percent), and the result again multiplied by the efficiency of the traction motors (including power circuits) and gearing to develop rail horsepower. Power output of the main generator for traction may be expressed as

$$\text{Watts}_{\text{traction}} = E_g \times I_m \qquad (11.2.2)$$

where E_g is the main-generator voltage and I_m is the traction motor current in amperes, multiplied by the number of parallel paths or the dc link current in the case of an ac traction system.

Rail horsepower may be expressed as

$$\text{hp}_{\text{rail}} = V \times \text{TE}/375 \qquad (11.2.3)$$

where V is the velocity, mi/h, and TE is tractive effort at the rail, lb.

Thermal Efficiency The thermal efficiency of the diesel engine at the crankshaft, or the ratio of bhp output to the rate at which energy of the fuel is delivered to the engine, is about 33 percent. Thermal efficiency at the rail is about 26 percent.

Drawbar Horsepower The drawbar horsepower represents power available at the rear of the locomotive to move the cars and may be expressed as

$$\begin{aligned}\text{hp}_{\text{drawbar}} = \text{hp}_{\text{rail}} \\ - \text{ locomotive running resistance} \times V/375 \quad (11.2.4)\end{aligned}$$

where V is the speed in mi/h. Running resistance calculations are discussed later under Vehicle/Track Interaction. Theoretically, therefore, drawbar horsepower available is power output of the diesel engine less parasitic losses described above.

Speed-Tractive Effort At full throttle the losses vary somewhat at different values of speed and tractive efforts, but a curve of tractive effort plotted against speed is nearly hyperbolic. Figure 11.2.5 is a typical speed-tractive effort curve for a 3,500 hp (2,600 kW) freight locomotive. The diesel-electric locomotive has full horsepower available over the entire speed range (within limits of adhesion described below). The reduction in power as continuous speed is approached is referred to as "power matching." This allows the mixing of units with varied power per axle at the same continuous speed.

Adhesion In Figure 11.2.6 the maximum value of tractive effort represents the usually achievable level just before the wheels slip under average rail conditions. Adhesion is usually expressed as a percentage with the nominal level being 25 percent. This means that a force equal to 25 percent of the total locomotive weight on drivers is available as tractive effort. Actually, at the point of wheel slip, adhesion will vary widely with rail conditions, from as low as 5 percent to as high as 35 percent or more. Adhesion is severely reduced by lubricants which spread as thin films in the presence of moisture on running surfaces. Adhesion can be increased with sand applied to the rails from the locomotive sanding system. More recent wheel slip systems permit wheel creep (very slow controlled slip) to achieve greater levels of tractive effort. Higher adhesion levels are available from ac traction motors; for

example, 45 percent at start-up and low speed with a nominal value of 35 percent.

Traction Motor Characteristics Motor torque is a function of armature current and field flux (which is a function of field current). Since

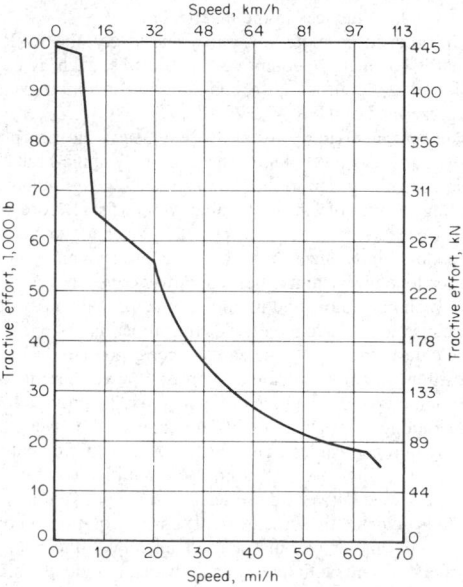

Fig. 11.2.5 Tractive effort versus speed.

traction motors are series connected, armature and field current are the same (except when field shunting circuits are introduced), and therefore tractive effort is a function solely of motor current. Figure 11.2.7 presents a group of traction motor characteristic curves with tractive effort, speed and efficiency plotted against motor current for full field (FF) and at 35 (FS1) and 55 (FS2) percent field shunting. Wheel diameter and gear ratio must be specified when plotting torque in terms of tractive effort. (See also Sec. 15.)

Traction motors are usually rated in terms of maximum continuous current. This represents the current at which the heating effect due to electrical losses in the armature and field windings is sufficient to raise the temperature of the motor to its maximum safe limit when cooling air

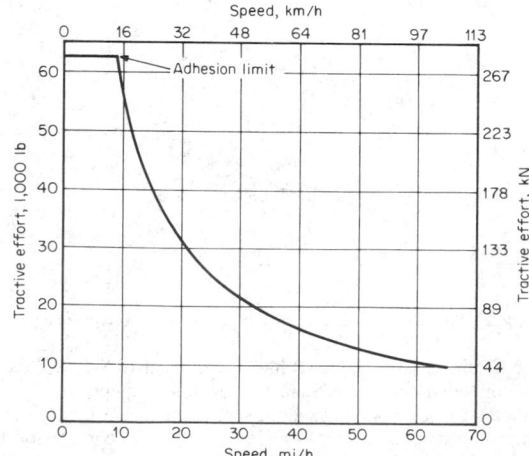

Fig. 11.2.6 Typical tractive-effort versus speed characteristics. (*Electro-Motive Division, General Motors Corp.*)

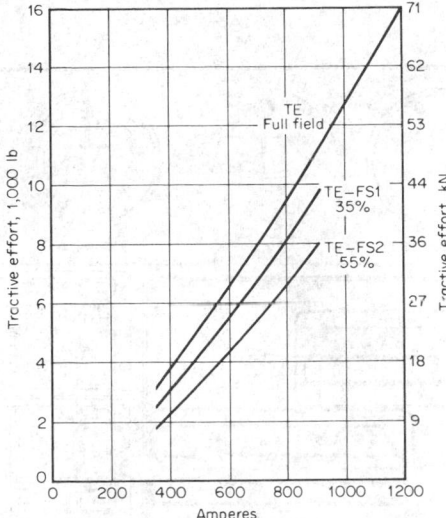

Fig. 11.2.7 Traction-motor characteristics. (*Electro-Motive Division, General Motors Corp.*)

at maximum expected ambient temperature is forced through it at the prescribed rate by the blowers. Continuous operation at this current level ideally allows the motor to operate at its maximum safe power level, with waste heat generated equal to heat dissipated. The tractive effort corresponding to this current is usually somewhat lower than that allowed by adhesion at very low speeds. Higher current values may be permitted for short periods of time (as when starting). These ratings are specified in time intervals and are posted on or near the load meter (ammeter) dial in the cab.

Maximum Speed Traction motors are also rated in terms of their maximum safe speed in r/min, which in turn limits locomotive speed. The gear ratio and wheel diameter are directly related to speed as well as the maximum tractive effort and the minimum speed at which full horsepower can be developed at the continuous rating of the motors. Maximum locomotive speed may be expressed as follows:

$$(\text{mi/h})_{\text{max}} = \frac{\text{wheel diameter (in)} \times \text{maximum motor r/min}}{\text{gear ratio} \times 336} \quad (11.2.5)$$

where the gear ratio is the number of teeth on the motor gear mounted on the axle divided by the number of teeth on the pinion mounted on the armature shaft.

Locomotive Compatibility The AAR has developed two standards in an effort to improve compatibility between locomotives of different model, manufacture, and ownership: a standard 27-point control system (Standard S-512, Table 11.2.2) and a standard control stand (Standard S-523, Fig. 11.2.8). These features allow locomotives to be combined in multiple-unit service on all railroads irrespective of model or manufacturer.

Energy Conservation Efforts to improve efficiency and fuel economy have resulted in major changes in the prime movers, including more efficient turbocharging, fuel injection, and combustion. The auxiliary (parasitic) power demands have also been reduced with improvements which include fans and blowers which only move the air required by the immediate demand, air compressors that declutch when unloaded, and a selective low-speed idle. Fuel-saver switches permit dropping trailing locomotive units off the line when less than maximum power is required, and allow the remaining units to operate at maximum efficiency.

ELECTRIC LOCOMOTIVES

Electric locomotives are presently in limited use in North America. Freight locomotives are in dedicated service primarily for coal or min-

Table 11.2.2 Standard Dimensions and Contact Identification of 27-Point Control Plug and Receptacle for Diesel-Electric Locomotives

Receptacle point	Function	Code	Wire size, AWG
1	Power reduction setup, if used	(PRS)	14
2	Alarm signal	SG	14
3	Engine speed	DV	14
4*	Negative	N	14 or 10
5	Emergency sanding	ES	14
6	Generator field	GF	12
7	Engine speed	CV	14
8	Forward	FO	12
9	Reverse	RE	12
10	Wheel slip	WS	14
11	Spare		14
12	Engine speed	BV	14
13	Positive control	PC	12
14	Spare		14
15	Engine speed	AV	14
16	Engine run	ER	14
17	Dynamic brake	B	14
18	Unit selector circuit	US	12
19	2d negative, if used	(NN)	12
20	Brake warning light	BW	14
21	Dynamic brake	BG	14
22	Compressor	CC	14
23	Sanding	SA	14
24	Brake control/power reduction control	BC/PRC	14
25	Headlight	HL	12
26	Separator blowdown/ remote reset	SV/RR	14
27	Boiler shutdown	BS	14

* Receptacle point 4—AWG wire size 12 is "standard" and AWG wire size 10 is "Alternate standard" at customer's request. A dab of white paint in the cover latch cavity must be added for ready identification of a no. 10 wire present in a no. 4 cavity.

eral hauling. Electric passenger locomotives are used in high-density service in the northeastern United States and a combination electric passenger and freight locomotive has been produced for use in Mexico.

Electric locomotives draw power from overhead catenary systems. While earlier systems used either direct current up to 3,000 V or single-phase alternating current at 11,000 V, 25 Hz, the newer systems use 25,000 or 50,000 V at 60 Hz. The higher voltage levels can only be used where clearances permit. Three-phase supply was tried briefly in this country and overseas about 50 years ago, but was abandoned because of the complexity of the required double catenary.

While the older dc locomotives used resistance control, ac locomotives have used a variety of systems including Scott-connected transformers; series ac motors; motor generators and dc motors; ignitrons and dc motors; silicon thyristors and dc motors; and, more recently, chopper control in experimental locomotives. Examples of the various electric locomotives in service are shown in Table 11.2.1.

An electric locomotive used in high-speed passenger service is shown in Fig. 11.2.9. High short-time ratings (Figs. 11.2.10 and 11.2.11) render electric locomotives suitable for passenger service where high acceleration rates and high speeds are combined to meet demanding schedules.

The modern electric locomotive in Fig. 11.2.9 obtains power for the main circuit and motor control from the catenary through a pantograph. A motor-operated switch provides for the transformer change from series to parallel connection of the primary windings to give a constant secondary voltage with either 25 kV or 12.5 kV primary supply. The converters for armature current consist of two asymmetric type bridges for each motor. The traction-motor fields are each separately fed from a one-way connected-field converter.

Identical control modules separate the control of motors on each truck. Motor sets are therefore connected to the same transformer wind-

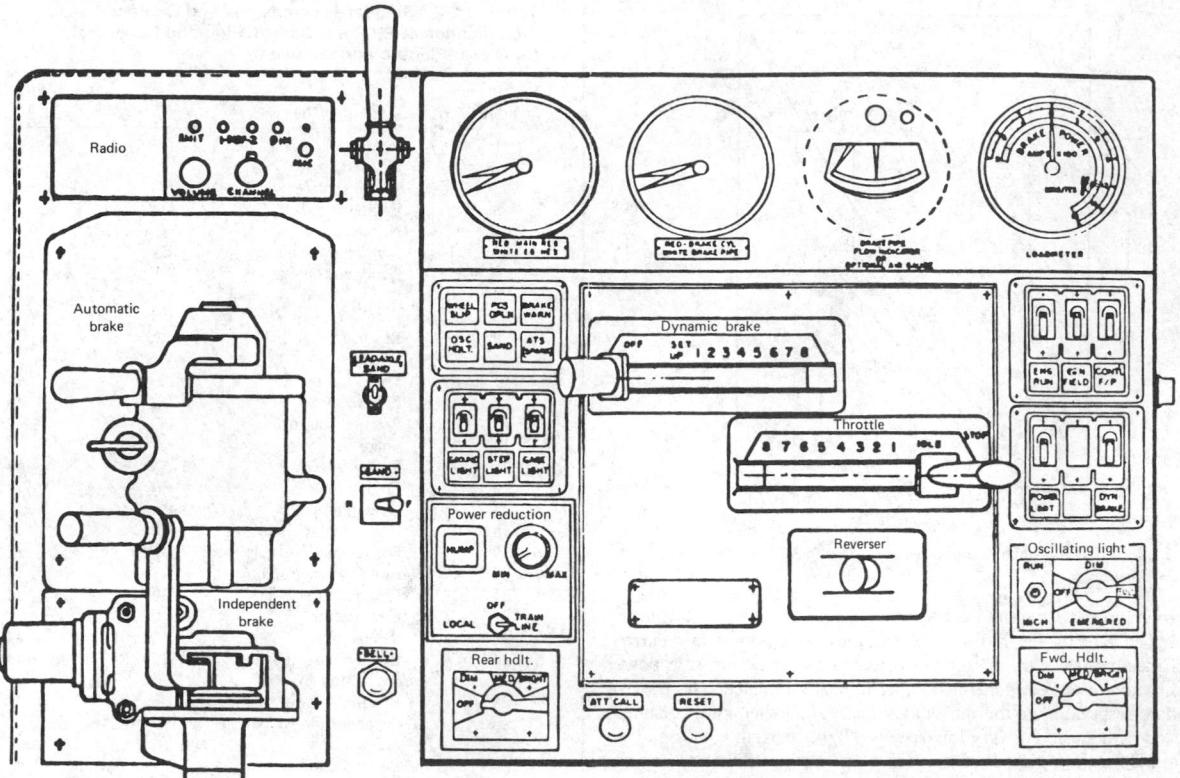

Fig. 11.2.8 AAR standard control stand.

ing. Wheel slip correction is also modularized, utilizing one module for each two-motor truck set. This correction is made with a complementary wheel-slip detection and correction system. A magnetic pickup speed signal is used for the basic wheel slip. Correction is enhanced by a magnetoelastic transducer used to measure force swings in the traction-motor reaction rods. This system provides the final limit correction.

To optimize the utilization of available adhesion, the wheel-slip control modules operate independently to allow the motor modules to receive different current references depending on their respective adhesion conditions.

All auxiliary machines, air compressor, traction motor blower, cooling fans, etc., are driven by three-phase 400-V, 60-Hz induction motors powered by a static inverter which has a rating of 175 kVA at a 0.8 power factor. When cooling requirements are reduced, the control system automatically reduces the voltage and frequency supplied to the blower motors to the required level. As a backup, the system can be powered by the static converter used for the head-end power requirements of the passenger cars. This converter has a 500-kW, 480-V, three-phase, 60-Hz output capacity and has a built-in overload capacity of 10 percent for half of any 1-h period.

Dynamic brake resistors are roof-mounted and are cooled by ambient

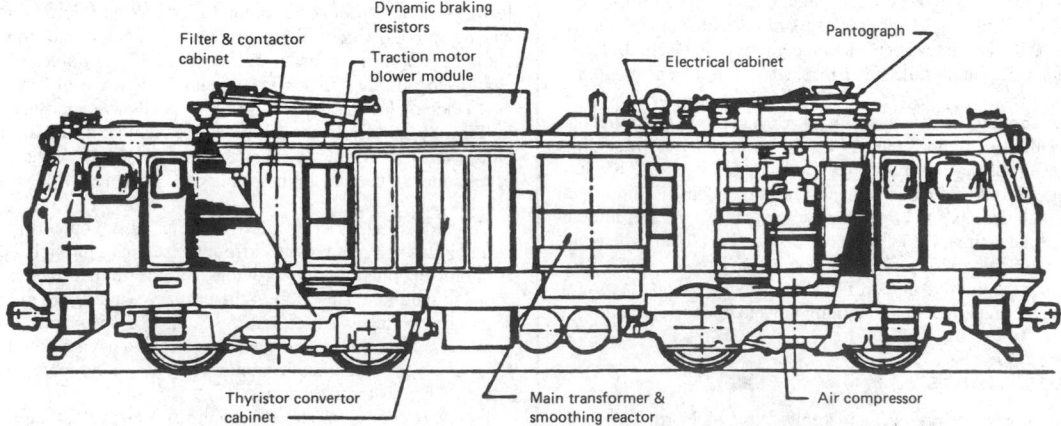

Fig. 11.2.9 Locomotive general arrangement. (*Courtesy of M. Ephraim, ASME, Rail Transportation Division.*)

airflow induced by locomotive motion. The dynamic brake capacity relative to train speed is shown in Fig. 11.2.12.

Many electric locomotives utilize traction motors identical to those used on diesel-electric locomotives. Some, however, use frame sus-

pended motors with quill-drive systems (Fig. 11.2.13). In these systems, torque is transmitted from the traction motor by a splined coupling to a quill shaft and rubber coupling to the gear unit. This transmission allows for greater relative movement between the traction motor (truck frame) and gear (wheel axle) and reduces unsprung weight.

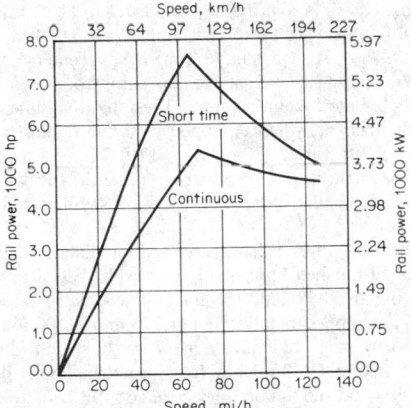

Fig. 11.2.10 Speed-horsepower characteristics (half-worn wheels). [Gear ratio = 85/36; wheel diameter = 50 in (1,270 mm); ambient temperature − 60°F (15.5°C).] *(Courtesy of M. Ephraim, ASME, Rail Transportation Division.)*

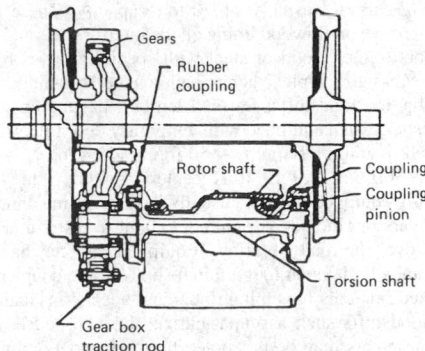

Fig. 11.2.13 Frame-suspended locomotive transmission. *(Courtesy of M. Ephraim, ASME, Rail Transportation Division.)*

FREIGHT CARS

Freight Car Types

Freight cars are designed and constructed for either general service or specific ladings. The Association of American Railroads (AAR) has established design and maintenance criteria to assure the interchangeability of freight cars on all North American railroads. Freight cars which do not conform to AAR standards have been placed in service by special agreement of the railroads over which the cars operate. The AAR "Manual of Standards and Recommended Practices" specifies dimensional limits, weights, and other design criteria for cars which may be freely interchanged between North American railroads. This manual is revised annually by the AAR. Many of the standards are reproduced in the "Car and Locomotive Cyclopedia," which is revised about every 4 years. Safety appliances (such as end ladders, sill steps, and uncoupling levers), braking equipment, and certain car design features and maintenance practices must comply with the Safety Appliances Act, the Power Brake Law, and the Freight Car Safety Standards covered by the Federal Railroad Administration's (FRA) Code of Federal Regulations (CFR Title 49). Maintenance practices are discussed in the Field Manual of the AAR Interchange Rules and pricing information is provided in the companion Office Manual.

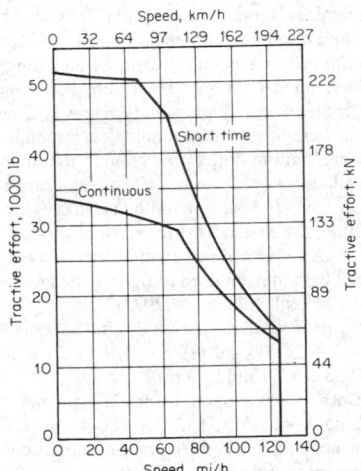

Fig. 11.2.11 Tractive effort versus speed (half-worn wheels). [Voltage = 11.0 kV at 25 Hz; gear ratio = 85/36; wheel diameter = 50 in (1,270 mm).] *(Courtesy of M. Ephraim, ASME, Rail Transportation Division.)*

The AAR identifies most cars by nominal capacity, type, and in some cases there are restrictions as to type of load (i.e., food, automobile parts, coil steel, etc.). Most modern cars have nominal capacities of either 70 or 100 tons.* A 100-ton car is limited to a maximum gross rail load of 263,000 lb (119.3 t*) with 6½ by 12 in (165 by 305 mm) axle journals, four pairs of 36-in (915-mm) wheels and a 70-in (1,778 mm) rigid wheel base. Some 100-ton cars are rated at 286,000 lb and are handled by individual railroads on specific routes or by mutual agreement between the handing railroads. A 70-ton car is limited to a maximum gross rail load of 220,000 lb (99.8 t) with 6 by 11 in (152 by 280 mm) axle journals, four pairs of 33-in (838-mm) wheels with trucks having a 66-in (1,676-mm) rigid wheel base.

On some special cars where height limitations are critical, 28-in (711-mm) wheels are used with 70-ton axles and bearings. In these cases wheel loads are restricted to 22,400 lb (10.2 t). Some special cars are equipped with 38-in (965-mm) wheels in four wheel trucks having 7 by 14 in (178 by 356 mm) axle journals and 72-in (1,829-mm) wheel base. This application is prevalent in articulated double-stack (two-high) container cars. Interchange of these very heavy cars is by mutual agreement between the operating railroads involved.

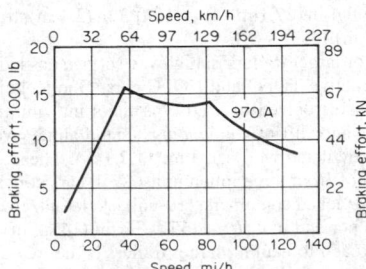

Fig. 11.2.12 Dynamic-brake performance. *(Courtesy of M. Ephraim, ASME, Rail Transportation Division.)*

* 1 ton = 1 short ton = 2,000 lb; 1 t = 1 metric ton = 1,000 kg = 2205 lb.

The following are the most common car types in service in North America. Dimensions given are for typical cars; actual measurements may vary.

Box Cars (Fig. 11.2.14a) There are six popular box-car types.

1. Unequipped box cars may have sliding doors or plug doors. Plug doors provide a tight seal from weather and a smooth interior. Unequipped box cars are usually of 70-ton capacity. These cars have tongue-and-groove or plywood lining on the interior sides and ends with nailable floors (either wood or steel with special grooves for locking nails). The cars carry typical general merchandise lading: packaged, canned, or bottled foodstuffs; finished lumber; bagged or boxed bulk commodities; or, when equipped with temporary door fillers, bulk commodities such as grain. [70 ton: L = 50 ft 6 in (15.4 m), H = 11 ft 0 in (3.4 m), W = 9 ft 6 in (2.9 m), truck centers = 40 ft 10 in (12.4 m).]

2. Specially equipped box cars usually have the same dimensions as unequipped cars but include special equipment to restrain lading from impacts and over-the-road vibrations. Equipped cars may have hydraulic-cushion units to dampen longitudinal shock at the couplers.

3. Insulated box cars have plug doors and special insulation. These cars carry foodstuffs such as unpasteurized beer, produce, and dairy products. These cars may be precooled by the shipper and maintain a heat loss rate equivalent to 1°F (0.55°C) per day.

4. Refrigerated box cars are used where transit times are longer. These cars are equipped with diesel-powered refrigeration units and are primarily used to carry produce. These are often 100-ton cars [L = 52 ft 6 in (16.0 m), H = 10 ft 6 in (3.2 m), truck centers = 42 ft 11 in (13.1 m)].

5. "All door" box cars have doors which open the full length of the car for loading package lumber products such as plywood and gypsum board.

6. High-cubic-capacity box cars with an inside volume of 10,000 ft^3 (283 m^3) have been designed for light-density lading.

Box-car door widths vary from 6 to 10 ft for single-door cars and 16 to 20 ft (4.9 to 6.1 m) for double-door cars. "All door" cars have clear doorway openings in excess of 25 ft (7.6 m). The floor height above rail for an empty uninsulated box car is approximately 44 in (1,120 mm) and for an empty insulated box car approximately 48 in (1,220 mm). The floor height of a loaded car can be as much as 3 in (76 mm) lower than the empty car.

Covered hopper cars (Fig. 11.2.14b) are used to haul bulk commodities which must be protected from the environment. Modern covered hopper cars are typically 100-ton cars with roof hatches for loading and from two to six bottom outlets for discharge. Cars used for dense commodities, such as fertilizer or cement, have two bottom outlets, round roof hatches, and volumes of 3,000 to 4,000 ft^3 (84.9 to 113.2 m^3). [100 ton: L = 39 ft 3 in (12.0 m). H = 14 ft 10 in (4.5 m), truck centers = 26 ft 2 in (8.0 m).] Cars used for grain service (corn, wheat, rye, etc.) have three or four bottom outlets, longitudinal trough roof hatches, and volumes of from 4,000 to 5,000 ft^3 (113 to 142 m^3). Cars used for hauling plastic pellets have four to six bottom outlets (for pneumatic unloading with a vacuum system), round roof hatches, and volumes of from 5,000 to 6,000 ft^3 (142 to 170 m^3). [100 ton: L = 65 ft 7 in (20.0 m), H = 15 ft 5 in (4.7 m), truck centers = 54 ft 0 in (16.5 m).]

Open-top hopper cars (Fig. 11.2.14c) are used for hauling bulk commodities such as coal, ore, or wood chips. A typical 100-ton coal hopper car will vary in volume depending on the light weight of the car and density of the coal to be hauled. Volumes range from 3,900 to 4,800 ft^3 (110 to 136 m^3). Cars may have three or four manually operated or 12 to 16 automatically operated bottom hoppers. Some cars are equipped with rotating couplers on one end to allow rotary dumping without uncoupling. [100 ton: L = 53 ft 1/2 in (16.2 m), H = 12 ft 8½ in (3.9 m), truck centers = 40 ft 6 in (12.3 m).] Cars intended for aggregate or ore service have smaller volumes for the more dense commodity. These cars typically have two manual bottom outlets. [100 ton: L = 40 ft 8 in (12.4 m), H = 11 ft 10⅞ in (3.6 m), truck centers = 29 ft 9 in (9.1 m).] Hopper cars used for chip services are configured for low-density loads. Volumes range from 6,500 to 7,500 ft^3 (184 to 212 m^3).

High-side gondola cars (Fig. 11.2.14d) are open-top cars typically used to haul coal or wood chips. These cars are similar to open-top hopper cars in volume but require a rotary coupler on one end for rotary dumping to discharge lading since they do not have bottom outlets. Rotary-dump coal gondolas are usually used in dedicated, unit-train service between a coal mine and an electric power plant. The length over coupler pulling faces is approximately 53 ft 1 in (16.2 m) to suit the standard coal dumper. [100 ton: L = 50 ft 5½ in (15.4 m), H = 11 ft 9 in (3.6 m), W = 10 ft 5⅜ in (3.2 m), truck centers = 50 ft 4 in (15.3 m).] Wood chip cars are used to haul chips from sawmills to paper mills or particle-board manufacturers. These high-volume cars are either rotary-dumped or, when equipped with end doors, end-dumped. [100 ton: L = 62 ft 3 in (19.0 m), H = 12 ft 7 in (3.8 m), W = 10 ft 5⅜ in (3.2 m), truck centers = 50 ft 4 in (15.3 m).] Rotary-dump aggregate or ore cars, called "ore jennies," have smaller volumes for the high-density load.

Bulkhead flat cars (Fig. 11.2.14e) are used for hauling such commodities as packaged finished lumber, pipe, or with special inward canted floors, pulpwood. Both 70- and 100-ton bulkhead flats are used. Typical deck heights are approximately 50 in (1,270 mm). [100 ton: L = 61 ft ¾ in (18.6 m), W = 10 ft 1 in (3.1 m), H = 11 ft 0 in (3.4 m), truck centers = 55 ft 0 in (16.8 m).] A few special bulkhead flat cars designed for pulpwood and lumber service have a full-height, longitudinal divider from bulkhead to bulkhead.

Tank cars (Fig 11.2.14f) are used for liquids, compressed gases, and other ladings, such as sulfur, which can be loaded and unloaded in a molten state. Nonhazardous liquids such as corn syrup, crude oil, and mineral spring water are carried in nonpressure cars. Cars used to haul hazardous substances such as liquefied petroleum gas (LPG), vinyl chloride, and anhydrous ammonia are regulated by the U.S. Department of Transportation. Newer and earlier-built retrofitted cars equipped for hazardous commodities have safety features including safety valves, specially designed top and bottom "shelf" couplers which increase the interlocking effect between couplers and decrease the danger of disengagement due to derailment, head shields on the ends of the tank to prevent puncturing, bottom-outlet protection if bottom outlets are used, and thermal insulation to reduce the risk of rupturing in a fire. These features resulted from industry-government studies in the RPI-AAR Tank Car Safety Research and Test Program.

Cars for sulfur and other viscous-liquid service have heating coils on the shell so that steam may be used to liquefy the lading for discharge. [Pressure cars, 100 ton: volume = 20,000 gal (75.7 m^3), L = 59 ft 11¾ in (18.3 m), truck centers = 49 ft 0¼ in (14.9 m).] [Nonpressure, 100 ton, volume = 21,000 gal (79.5 m^3), L = 51 ft 3¼ in (15.6 m), truck centers = 38 ft 11¼ in (11.9 m).]

Intermodal Cars Conventional 89-ft (27.1-m) intermodal flat cars are equipped to haul one 45-ft (13.7 m) and one 40-ft (12.2 m) trailer with or without nose-mounted refrigeration units, two 45-ft (13.7 m) trailers (dry vans), or combinations of containers from 20 to 40 ft (6.1 to 12.2 m) in length. Hitches to support trailer fifth wheels may be fixed for trailer-only cars or retractable for conversion to haul containers or for driving trailers onto the cars where circus loading is required. Trailer hauling service (Fig. 11.2.14g) is called TOFC (trailer on flat car). Container service (Fig. 11.2.14h) is called COFC (container on flat car). [70 ton: L = 89 ft 4 in (27.1 m), W = 10 ft 3 in (3.1 m), truck centers = 64 ft 0 in (19.5 m).]

Introduction of larger trailers in highway service has led to the development of alternative TOFC and COFC cars. One technique involves articulation of skeletonized or well-type units into multiunit cars for lift-on loading and lift-off unloading (stand-alone well car, Fig. 11.2.14n and articulated well car, Fig. 11.2.14o). These cars are typically composed of from three to ten units. Well cars consist of a center well for double-stacked containers. [10-unit, skeletonized car; L = 46 ft 6⅜ in (14.2 m) per end unit, L = 465 ft 3½ in (141.8 m).]

Another approach to hauling larger trailers is the two-axle skeletonized car. These cars, used either singly or in multiple combinations, can haul a single trailer from 40 to 48 ft long (12.2 to 14.6 m) with nose-

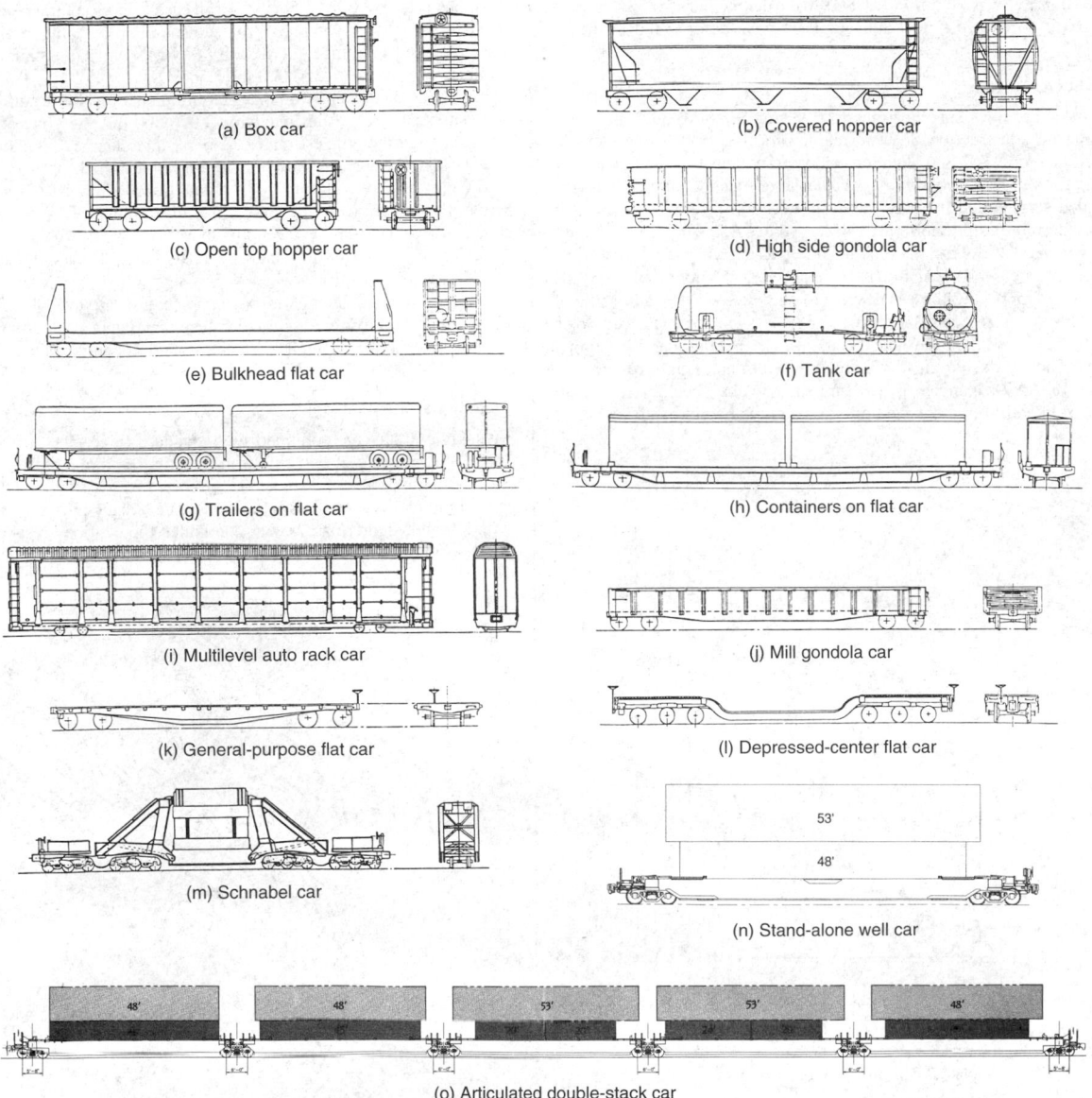

(a) Box car

(b) Covered hopper car

(c) Open top hopper car

(d) High side gondola car

(e) Bulkhead flat car

(f) Tank car

(g) Trailers on flat car

(h) Containers on flat car

(i) Multilevel auto rack car

(j) Mill gondola car

(k) General-purpose flat car

(l) Depressed-center flat car

(m) Schnabel car

(n) Stand-alone well car

(o) Articulated double-stack car

Fig. 11.2.14 Typical freight cars.

mounted refrigeration unit and 36- or 42-in (914- to 1070-mm) spacing between the kingpin and the front of the trailer.

Bilevel and trilevel **auto rack cars** (Fig. 11.2.14*i*) are used to haul finished automobiles and other vehicles. Most recent designs of these cars feature fully enclosed racks to provide security against theft and vandalism. [70 ton: $L = 89$ ft 4 in (27.2 m), $H = 18$ ft 11 in (5.8 m), $W = 10$ ft 7 in (3.2 m), truck centers = 64 ft 0 in (19.5 m).]

Mill gondolas (Fig. 11.2.14*j*) are 70-ton or 100-ton open-top cars principally used to haul pipe, structural steel, scrap metal, and when specially equipped, coils of aluminum or tinplate and other steel materials. [100 ton: $L = 52$ ft 6 in (16.0 m), $W = 9$ ft 6 in (2.9 m), $H = 4$ ft 6 in (1.4 m), truck centers = 43 ft 6 in (13.3 m).]

General-purpose or **machinery flat cars** (Fig. 11.2.14*k*) are 70-ton or 100-ton cars used to haul machinery such as farm equipment and highway tractors. These cars usually have wood decks for nailing lading-restraint dunnage. Some heavy-duty six-axle cars are used for hauling off-highway vehicles such as army tanks and mining machinery. [100 ton, four axle: $L = 60$ ft 0 in (18.3 m), $H = 3$ ft 9 in (1.1 m), truck centers = 42 ft 6 in (13.0 m).] [200 ton, 8 axle: $L = 44$ ft 4 in (13.5 m), $H = 4$ ft 0 in (1.2 m), truck centers = 33 ft 9 in (10.3 m).]

Depressed-center flat cars (Fig. 11.2.14*l*) are used for hauling transformers and other heavy, large materials which require special clearance considerations. Depressed-center flat cars may have four-, six-, or dual four-axle trucks with span bolsters, depending on weight requirements.

Schnabel cars (Fig. 11.2.14*m*) are special cars for transformers and nuclear components. With these cars the load itself provides the center section of the car structure during shipment. Some Schnabel cars are equipped with hydraulic controls to lower the load for height restrictions and shift the load laterally for wayside restrictions. [472 ton: $L =$

22 ft 10 in to 37 ft 10 in (7.0 to 11.5 m), truck centers = 55 ft 6 in to 70 ft 6 in (16.9 to 21.5 m).] Schnabel cars must be operated in special trains.

Freight Car Design

The AAR provides specifications to cover minimum requirements for design and construction of new freight cars. Experience has demonstrated that the AAR Specifications alone do not ensure an adequate car design for all service conditions. The designer must be familiar with the intended service and increase the design criteria for the particular car above the minimum criteria provided by the AAR. The AAR requirements include stress calculations for the load-carrying members of the car and physical tests which may be required at the option of the AAR committee approving the car design. In some cases, it is advisable to operate an instrumented prototype car in service to detect problems which might result from unexpected track or train-handling input forces. The car design must comply with width and height restrictions shown in AAR clearance plates furnished in the specifications, Fig. 11.2.15. In addition, there are limitations on the height of the center of

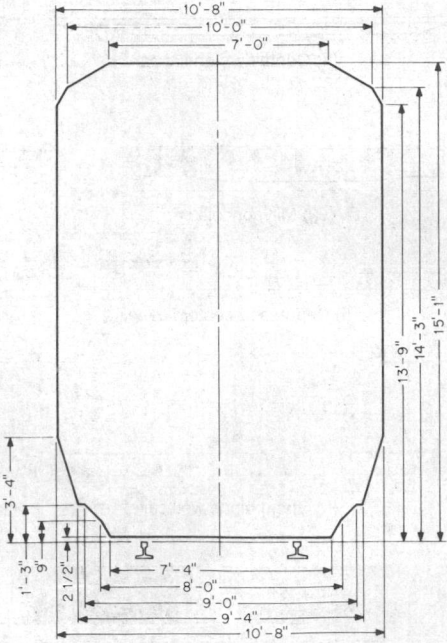

Fig. 11.2.15   AAR plate B equipment-clearance diagram.

gravity of the loaded car and on the vertical and horizontal curving capability allowed by the clearance provided at the coupler. The AAR provides a method of calculating the minimum radius curve which the car design can negotiate when coupled to another car of the same type or to a standard AAR "base" car. In the case of horizontal curves, the requirements are based on the length over the pulling faces of the car.

In the application for approval of a new or untried type of car, the Car Construction Committee of the AAR may require either additional calculations or tests to assess the design's ability to meet the AAR minimum requirements. These tests might consist of a static compression test of 1,000,000 lb (4.4 MN), a static vertical test applied at the coupler, and impact tests simulating yard impact conditions.

Freight cars are designed to withstand single-ended impact or coupling loads based upon the type of cushioning provided in the car design. Conventional friction, elastomer, or combination draft gears or short-travel hydraulic cushion units which provide less than 6 in (152 mm) of travel require a structure capable of withstanding a 1,250,000-lb (5.56-MN) impact load. For cars with hydraulic units that provide greater than 14 in (356 mm) of travel, the required design im-

pact load is 600,000 lb (2.7 MN). In all cases, the structural connections to the car must be capable of withstanding a static compressive (squeeze) end load of 1,000,000 lb (4.44 MN) or a dynamic (impact) compressive load of 1,250,000 lb (5.56 MN).

The AAR has recently adopted requirements for unit trains of high-utilization cars to be designed for 3,000,000 mi (4.8 Gm) of service based upon fatigue life estimates. General-interchange cars which accumulate less mileage in their life should be designed for 1,000,000 mi (1.6 Gm) of service. Road environment spectra for various locations within the car are being developed for different car designs for use in this analysis. The fatigue strengths of various welded connections are provided in the AAR "Manual of Standards and Recommended Practices," Sec. C, Part II.

Many of the design equations and procedures are available from the AAR. Important information on car design and approval testing is contained in the AAR "Manual of Standards and Recommended Practice," Sec. C-II M-1001, Chap. XI.

Freight-Car Suspension

Most freight cars are equipped with standard three-piece trucks (Fig. 11.2.16) consisting of two side-frame castings and one bolster casting. Side-frame and bolster designs are subjected to both static and fatigue test requirements specified by the AAR. The bolster casting is equipped with a female centerplate bowl upon which the car body rests and side bearings located [generally 25 in (635 mm)] each side of the centerline.

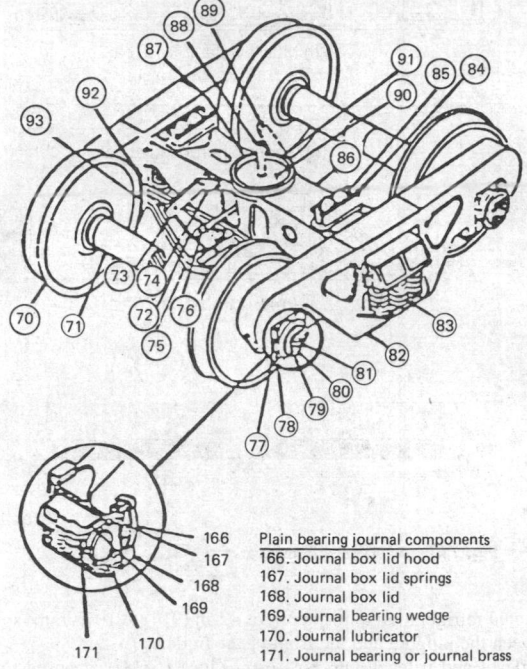

Plain bearing journal components
166. Journal box lid hood
167. Journal box lid springs
168. Journal box lid
169. Journal bearing wedge
170. Journal lubricator
171. Journal bearing or journal brass

Unit beam roller bearing truck components	
70. Wheel	82. Truck side frame
71. Axle	83. Truck springs
72. Truck dead lever	84. Truck side bearing
73. Dead lever fulcrum	85. Side bearing roller
74. Dead lever fulcrum bracket	86. Truck bolster
75. Brake beam	87. Truck center plate cast
76. Bottom rod	88. Truck live lever
77. Roller bearing adapter	89. Center pin
78. Roller bearing assembly	90. Horizontal wear plate
79. End cap	91. Vertical wear plate
80. End cap retaining bolt	92. Brake shoe key
81. Locking plate	93. Brake shoe

Fig. 11.2.16   Unit-beam roller-bearing truck with inset showing plain-bearing journal. *(AAR Research and Test Department.)*

In most cases, the side bearings have clearance to the car body and are equipped with either flat sliding plates or rollers. In some cases, constant-contact side bearings or centerplate extension pads provide a resilient material between the car body and the truck bolster.

The centerplate arrangement consists of various styles of wear plates or friction materials and a vertical loose or locked pin between the truck centerplate and the car body.

Truck springs nested into the bottom of the side-frame opening support the end of the truck bolster. Requirements for spring designs and the grouping of springs are generally specified by the AAR. Historically, the damping provided within the spring group has utilized a combination of springs and friction wedges. In addition to friction wedges, in more recent years, some cars have been equipped with hydraulic damping devices which parallel the spring group.

A few trucks with a "steering" feature which includes an interconnection between axles to increase the lateral interaxle stiffness and decrease the interaxle yaw stiffness. Increased lateral stiffness improves the lateral stability and decreased yaw stiffness improves the curving characteristics.

Freight-Car Wheel-Set Design

A freight-car wheel set consists of wheels, axle, and bearings. Cast- and wrought-steel wheels are used on freight cars in North America (AAR "Manual of Standards and Recommended Practice," Sec. G). Freight-car wheels are subjected to thermal loads from braking, as well as mechanical loads at the wheel-rail interface. Experience with thermal damage to wheels has led to the introduction of "low stress" or curved plate wheels (Fig. 11.2.17). These wheels are less susceptible to the

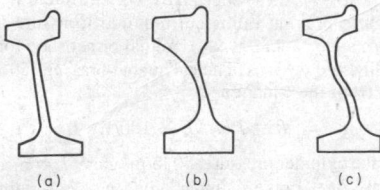

Fig. 11.2.17 Wheel-plate designs. (*a*) Flat plate; (*b*) parabolic plate; (*c*) S-curved plate.

development of circumferential tensile residual stresses which render the wheel vulnerable to sudden failure if a flange or rim crack occurs. New wheel designs introduced for interchange service must be evaluated using a finite-element technique employing both thermal and mechanical loads (AAR S-660).

Freight-car wheels range in diameter from 28 to 38 in (711 to 965 mm) depending on car weight, Table 11.2.3. The old AAR standard tread profile (Fig. 11.2.18*a*) has been replaced with the AAR-1B (Fig. 11.2.18*c*) profile which represents a worn profile to minimize early tread loss due to wear and provides a stable profile over the life of the tread. Several variant wheel tread profiles, including the AAR-1B, were developed from the basic Heumann design, Fig. 11.2.18*b*. One of these was for application in Canada and provided increasing conicity into the

Table 11.2.3   Wheel and Journal Sizes of Eight-Wheel Cars

Nominal car capacity, ton	Maximum gross weight, lb	Journal (bearing) size, in	Wheel diameter, in
50	177,000	5½ × 10	33
*	179,200	6 × 11	28
70	220,000	6 × 11	33
100	263,000‡	6½ × 12	36
125†	315,000	7 × 12	38

* Limited by wheel rating.
† Not approved for free interchange.
‡ 286,000 in special cases.

throat of the flange, a design feature of the Heumann profile. This reduces curving resistance and increases wheel wear life.

Wheels are also specified by chemistry and heat treatment. Low-stress wheel designs of Classes B and C are required for freight cars. Class B wheels have a carbon content of 0.57 to 0.67 percent and are rim-quenched. Class C wheels have a carbon content of 0.67 to 0.77 percent and are also rim-quenched. Rim quenching provides a hardened running surface for a long wear life. Lower carbon levels than those in Class B may be used where thermal cracking is experienced but freight-car equipment generally does not require their use.

Axles used in interchange service are solid steel forgings with raised wheel seats. Axles are specified by journal size for different car capacities, Table 11.2.3.

Most freight car bearings are grease-lubricated, tapered-roller bearings (see Sec. 8). The latest bearing designs eliminate the need for periodic field lubrication.

Wheels are mounted and secured on axles with an interference fit. Bearings are mounted with an interference fit and retained by an end cap bolted to the end of the axle. Wheels and bearings for cars in interchange service must be mounted by an AAR-inspected and -approved facility.

Special Features

Many components are available to enhance the usefulness of freight cars. In most cases, the design or performance of the component is specified by the AAR.

Coupler Cushioning   Switching of cars in a classification yard can result in relatively high coupler forces at the time of the impact between the moving and standing cars. Nominal coupling speeds of 4 mi/h (6.4 km/h) or less are sometimes exceeded, with lading damage a possible result. Conventional cars are equipped with an AAR-approved draft gear, usually a friction-spring energy-absorbing device, mounted between the coupler and the car body. The rated capacity of draft gears ranges between 20,000 ft·lb (27.1 kJ) for earlier units to over 65,000 ft·lb (88.1 kJ) for later designs. Impact forces of 1,250,000 lb (5.56 MN) can be expected when a moving 100-ton car strikes a string of standing cars at 8 to 10 mi/h (12.8 to 16 km/h). Hydraulic cushioning devices are available which will reduce the impact force to 500,000 lb (2.22 MN) at impact speeds of 12 to 14 mi/h (19 to 22 km/h). These

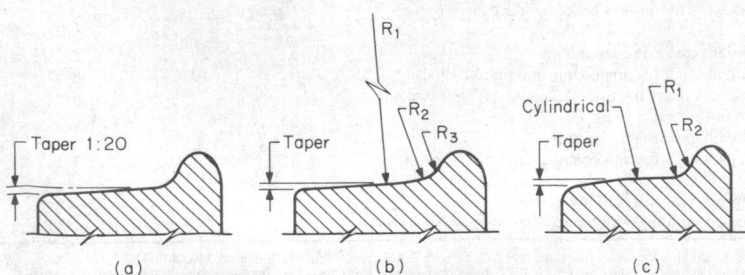

Fig. 11.2.18   Wheel-tread designs. (*a*) Obsolete standard AAR; (*b*) Heumann, (*c*) new AAR-1B.

devices may be either mounted at each end of the car (end-of-car devices) or mounted in a long beam which extends from coupler to coupler (sliding centersill devices).

Lading Restraint Many forms of lading restraint are available, from tie-down chains for automobiles on rack cars to movable bulkheads for boxcars. Most load-restraining devices are specified by the AAR "Manual of Standards and Recommended Practices" and approved car loading arrangements are specified in AAR "Loading Rules," a multivolume publication for enclosed, open-top, and TOFC and COFC cars.

Covered Hopper Car Discharge Gates The majority of covered hopper cars are equipped with rack-and-pinion slide gates which allow the lading to discharge by gravity onto conveyor equipment located between the rails. These gates can be operated manually with a simple bar or a torque-multiplying wrench or mechanically with an impact or hydraulic wrench. Many special covered hopper cars are equipped with discharge gates with nozzles and metering devices for vacuum or pneumatic unloading.

Coupling Systems A majority of freight cars are connected with AAR standard couplers. A specification has been developed to permit the use of alternative coupling systems such as articulated connectors, drawbars, and rotary-dump couplers.

Freight-Train Braking

The retarding forces acting on a railway train are rolling and mechanical resistance, aerodynamic drag, curvature and grade, plus that force resulting from the friction of the brake shoes rubbing on the wheel treads. On locomotives so equipped, the dynamic or rheostatic brake using the traction motors as generators can provide all or a portion of the retarding force to control train speed on down grades and sometimes during retardation and service stops.

Quick-action automatic air brakes of the type specified by the AAR are the common standard in North America. With the automatic air brake, the brake pipe extends through every vehicle in the train, connected by hoses between each locomotive unit and car. The front and rear brake pipe angle cocks are closed.

Air pressure is provided by compressors on the locomotive units to the main reservoirs, usually at 130 to 150 lb/in² (900 to 965 kPa). (Pressure values are gage pressures.) The engineer's automatic brake valve in release position provides air to the brake pipe on freight trains at reduced pressure, usually at 75, 80, 85, or 90 lb/in² (520, 550, 585, or 620 kPa) depending on the type of service, train weight, descending grades, and speeds that a train will operate. In passenger service, brake pipe pressure is usually 90 or 100 lb/in² (620 or 760 kPa).

When brake pipe pressure is increased, the control valve allows the reservoir capacity on each car and locomotive to be charged and at the same time connects the brake cylinders to exhaust. Brake pipe pressure is reduced when the engineer's brake valve is placed in a service position, and the control valve cuts off the charging function and allows the reservoir air on each car to flow into the brake cylinder. This moves the piston and, through a system of levers and rods, pushes the brake shoes against the wheel treads.

When the engineer's automatic brake valve is placed in the emergency position the brake pipe pressure (BP) is reduced very rapidly. The control valves on each car move to the emergency-application position and rapidly open a large vent valve, exhausting brake pipe pressure to atmosphere. This will serially propagate the emergency application through the train at from 900 to 950 ft/s (280 to 290 m/s). With the control valve in the emergency position both auxiliary- and emergency-reservoir volumes (pressures) equalize with the brake cylinder, and higher brake cylinder pressure (BCP) results, building up at a faster rate than in service applications.

The foregoing briefly describes the functions of the fundamental automatic air brake based on the functions of the control valve. AAR-approved brake equipment is required on all freight cars used in interchange service. The functions of the control valve have been refined to permit the handling of longer trains by more uniform brake performance. Important improvements in this design have been (1) reduction of the time required to apply the brakes on the last car of a train, (2) more uniform and faster release of the brakes, and (3) availability of emergency application with brake pipe pressure greater than 40 lb/in² (275 kPa).

The braking ratio of a car is defined as the ratio of brake shoe (normal) force to the car's rated gross weight. Two types of brake shoes, high-friction composition and high-phosphorus cast iron, are in use in interchange service. As these shoes have very different friction characteristics, different braking ratios are required to assure uniform train braking performance (Table 11.2.4). Actual or net shoe forces are measured with calibrated devices. The calculated braking ratio R (nominal) is determined from the equation

$$R = PLANE \times 100/W \qquad (11.2.6)$$

where P = brake cylinder pressure, 50 lb/in² gage; L = mechanical ratio of brake levers; A = brake cylinder area, in²; N = number of brake cylinders; E = brake rigging efficiency = $E_r \times E_b \times E_c$; and W = car weight (lb).

To estimate rigging efficiency consider each pinned joint and horizontal sliding joint as a 0.01 loss of efficiency; i.e., in a system with 20 pinned and horizontal sliding joints, $E_r = 0.80$. For unit-type (hangerless) brake beams $E_b = 0.90$ and for the brake cylinder $E_c = 0.95$, giving overall efficiency of 0.684 or 68.4 percent.

Table 11.2.4 Braking Ratios, AAR Standard S-401

Type of brake rigging and shoes	With 50 lb/in² brake cylinder pressure			Hand brake*
	Percent of gross rail load		Maximum percent of light weight	Minimum percent of gross rail load
	Min.	Max.		
Conventional body-mounted brake rigging or truck-mounted brake rigging using levers to transmit brake cylinder force to the brake shoes				
Cars equipped with cast iron brake shoes	13	20	53	13
Cars equipped with high-friction composition brake shoes	6.5	10	30	11
Direct-acting brake cylinders not using levers to transmit brake cylinder force to the brake shoes				
Cars equipped with cast iron brake shoes				
Cars equipped with high-friction composition brake shoes	6.5	10	33	11
Cabooses†				
Cabooses equipped with cast iron brake shoes			35–45	
Cabooses equipped with high-friction composition brake shoes			18–23	

* Hand brake force applied at the horizontal hand brake chain with AAR certified or AAR approved hand brake.
† Effective for cabooses ordered new after July 1, 1982, hand brake ratios for cabooses to the same as lightweight ratios for cabooses.
NOTE: Above braking ratios also apply to cars equipped with empty and load brake equipment.

The total retarding force in pounds per ton may be taken as:

$$F = (PLef/W) = F_g G \qquad (11.2.7)$$

where P = total brake-cylinder piston force, lbf; L = multiplying ratio of the leverage between cylinder pistons and wheel treads; ef = product of the coefficient of brake shoe friction and brake rigging efficiency; W = loaded weight of vehicle, tons; F_g = gravity force, 20 lb/ton/percent grade; G = ascending grade, percent.

Stopping distance can be found by adding the distance covered during the time the brakes are fully applied to the distance covered during the equivalent instantaneous application time.

$$S = \frac{0.0334 \, V_2^2}{\left[\dfrac{W_n B_n (p_a/p_n) ef}{W_a}\right] + \left(\dfrac{R}{2000}\right) \pm (G)}$$
$$+ \, 1.467 \, t_1 \left[V_1 - \left(\frac{R + 2000G}{91.1}\right) \frac{t_1}{2} \right] \qquad (11.2.8)$$

where S = stopping distance, ft; V_1 = initial speed when brake applied, mi/h; V_2 = speed, mi/h, at time t_1; W_a = actual vehicle or train weight, lb; W_n = weight on which braking ratio B_n is based, lb (see table for values of W_n for freight cars):

Capacity, ton	W_n, 1,000 lb
50	177
70	220
100	263
125	315

(for passenger cars and locomotives: W_n is based on empty or ready-to-run weight); B_n = braking ratio (total brake shoe force at stated brake cylinder, lb/in², divided by W_n); p_n = brake cylinder pressure on which B_n is based, usually 50 lb/in², p_a = full brake cylinder pressure, t_1 to stop; e = overall rigging and cylinder efficiency, decimal; f = typical friction of brake shoes, see below; R = total resistance, mechanical plus aerodynamic and curve resistance, lb/ton; G = grade in decimal, + upgrade, − downgrade; t_1 = equivalent instantaneous application time, s.

Equivalent instantaneous application time is that time on a curve of average brake cylinder buildup versus time for a train or car where the area above the buildup curve is equal to the area below the curve. A straight-line buildup curve starting at zero time would have a t_1 of half the total buildup time.

The friction coefficient f varies with the speed; it is usually lower at high speed. To a lesser extent, it varies with brake shoe force and with the material of the wheel and shoe. For stops below 60 mi/h (97 km/h), a conservative figure for a high-friction composition brake shoe on steel wheels is approximately

$$ef = 0.30 \qquad (11.2.9)$$

In the case of high-phosphorus iron shoes, this figure must be reduced by approximately 50 percent.

P_n is based on 50 lb/in² (345 kPa) air pressure in the cylinder; 80 lb/in² is a typical value for the brake pipe pressure of a fully charged freight train. This will give a 50-lb/in² (345 kPa) brake cylinder pressure during a full service application on AB equipment, and a 60 lb/in² (415 kPa) brake cylinder pressure with an emergency application.

To prevent wheel sliding, $F_R \leq \phi W$, where F_R = retarding force at the wheel rims resisting rotation of any pair of connected wheels, lb; ϕ = coefficient of wheel-rail adhesion or friction (a decimal); and W = weight upon a pair of wheels, lb. Actual or adhesive weight on wheels when the vehicle is in motion is affected by weight transfer (force transmitted to the trucks and axles by the inertia of the car body through the truck center plates), center of gravity, and vertical oscilla-

tion of body weight upon truck springs. The value of ϕ varies with speed as shown in Fig. 11.2.19.

The relationship between the required coefficient ϕ_1 of wheel-rail adhesion to prevent wheel sliding and rate of retardation A in miles per hour per second may be expressed by $A = 21.95\phi_1$.

Test Devices   Special devices have been developed for testing brake components and cars on a repair track.

End-of-Train Devices   To eliminate the requirement for a caboose car crew at the end of the train, special electronic devices have been developed to transmit the end-of-train brake-pipe pressure by telemetry to the locomotive operator.

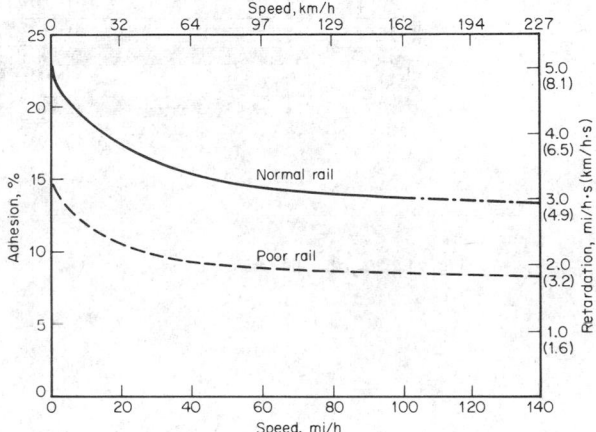

Fig. 11.2.19   Typical wheel-rail adhesion. Track has jointed rails. *(Air Brake Association.)*

PASSENGER EQUIPMENT

During the past two decades most main-line or long-haul passenger service in North America has become a function of government agencies, i.e., Amtrak in the United States and Via Rail in Canada. Equipment for intraurban service is divided into three major categories: commuter rail, full-scale rapid transit, and light rail transit, depending upon the characteristics of the service. Commuter rail equipment operates on a conventional railroad right-of-way usually intermixed with other long-haul passenger and freight traffic. Full-scale rapid transit operates on a dedicated right-of-way which is commonly in subways or on elevated sections. Light-rail transit (LRT) utilizing light rail vehicles (LRV) is derived from the older trolley or streetcar concepts and may operate in any combination of surface subway or elevated dedicated rights-of-way, semireserved surface rights-of-way with grade crossing, or intermixed with other traffic on surface streets. In a few cases LRT shares the trackage with freight operations which are usually conducted at night on a noninterference basis.

Since main-line and commuter rail equipment operates over a conventional railroad right-of-way, the structural design is heavy to provide the "crashworthiness" of the vehicles in the possible event of a collision with freight equipment or a grade-crossing accident with an automobile or a truck. Although transit vehicles are designed to stringent structural criteria, the requirements are somewhat less severe to compensate for the fact that freight equipment does not usually occupy the same track and there are usually no highway grade crossings. Minimum weight is particularly important for transit vehicles so that demanding schedules over lines with close station spacing can be met with minimum energy consumption.

Main-Line Passenger Equipment   In recent years the design of main-line passenger equipment has been controlled by specifications provided by the operating authority. All of the newer cars provided for Amtrak have had stainless steel structural components. These cars have been designed to be locomotive-hauled and to use a locomotive 480-V

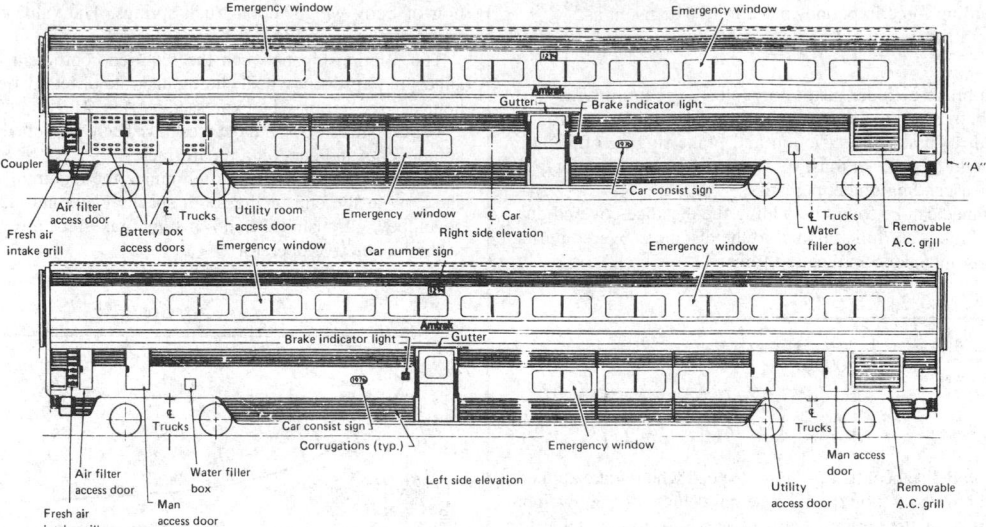

Fig. 11.2.20 Main-line passenger car.

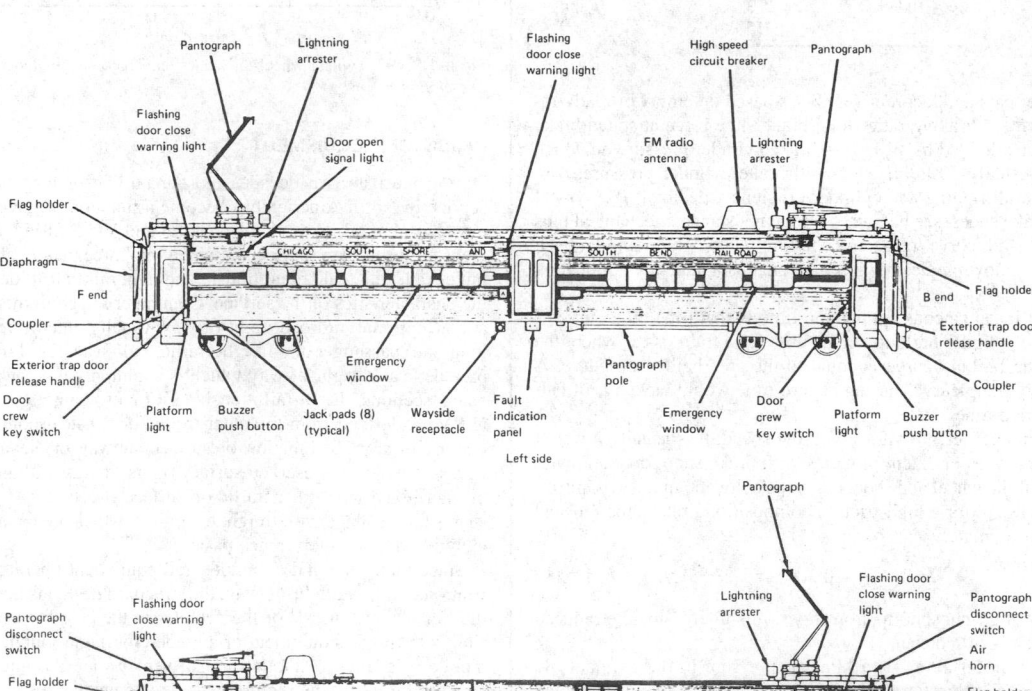

Fig. 11.2.21 Commuter rail car.

three-phase power supply for heating, ventilation, air conditioning, food car services, and other control and auxiliary power requirements. Figure 11.2.20 shows an Amtrak coach car of the Superliner class.

A few earlier high-speed main-line passenger cars (Metroliners) are equipped with special "hook"-type couplers which allow automatic connection of the pneumatic system as the cars are mechanically coupled together. During the past 10 years all of the major new Amtrak equipment has been equipped with the H-type coupler, a passenger version of the AAR standard coupler.

Trucks for passenger equipment are designed to provide a superior ride as compared to freight-car trucks. As a result, passenger trucks include a form of "primary" suspension to isolate the wheel set from the frame of the truck. A softer "secondary" suspension is provided to isolate the truck from the car body. In most cases, the primary suspension uses either coil springs, elliptical springs, or elastomeric components. The secondary suspension generally utilizes either large coil springs or pneumatic springs with special leveling valves to control the height of the car body. Hydraulic dampers are also applied to improve the vertical and lateral ride quality.

Commuter Rail Passenger Equipment Commuter rail equipment can be either locomotive-hauled or self-propelled (Fig. 11.2.21). Some locomotive-hauled equipment is arranged for "push-pull" service. This configuration permits the train to be operated with the locomotive either pushing or pulling the train. For push-pull service some of the passenger cars must be equipped with cabs to allow the engineer to operate the train from the end opposite the locomotive during the push operation. All cars must have control trainlines to connect the lead (cab) car to the trailing locomotive. Locomotive-hauled commuter rail cars use AAR-type H couplers. Self-propelled (single or multiple-unit) cars sometimes use other coupler designs which can be coupled to an AAR coupler with an adaptor.

Full-Scale Rapid Transit Equipment The equipment is used on traditional subway/elevated properties in such cities as Boston, New York, Philadelphia (Fig. 11.2.22), and Chicago in a semiautomatic mode of operation over dedicated rights-of-way which are constrained by limiting civil features. State-of-the-art subway-elevated properties include such cities as Washington, Atlanta, Miami, and San Francisco, where the equipment provides highly automated modes of operation on rights-of-way with generous civil alignments.

The cars can operate bidirectionally in multiple with as many as 12 cars controlled from the leading cab. They are electrically propelled,

usually from a dc third rail which makes contact with a shoe insulated from and supported by the frame of the truck. Occasionally, roof-mounted pantographs are used. Voltages range from 600 to 1000 V dc.

The cars range from 48 to 75 ft (14.6 to 22.9 m) over the anti-climbers, the longer cars being used on the newer properties. Passenger seating varies from 40 to 80 seats per car depending upon length and the local policy (or preference) regarding seated to standee ratio. Older properties require negotiation of curves as sharp as 50 ft (15.2 m) minimum radius with speeds up to only 50 mi/h (80 km/h) on tangent track, while newer properties usually have no less than 125 ft (38.1 m) minimum radius curves with speeds up to 75 mi/h (120 km/h) on tangent track.

All properties operate on standard-gage track with the exception of the 5 ft 6 in (1.7 m) San Francisco Bay Area Rapid Transit (BART), the 4 ft 10⅞ in (1.5 m) Toronto Transit Subways, and the 5 ft 2½ in (1.6 m) Philadelphia Southeastern Pennsylvania Transportation Authority (SEPTA) Market-Frankford line. Grades seldom exceed 3 percent, and 1.5 to 2.0 percent is the desired maximum.

Typically, newer properties require maximum acceleration rates of between 2.5 and 3.0 mi/(h·s) [4.0 to 4.8 km/(h·s)] as nearly independent of passenger loads as possible from 0 to approximately 20 mi/h (32 km/h). Depending upon the selection of motors and gearing, this rate falls off as speed is increased, but rates can generally be controlled at a variety of levels between zero and maximum.

Deceleration is typically accomplished by a blended dynamic and electropneumatic friction brake, although a few properties use an electrically controlled hydraulic friction brake. Either of these systems usually provides a maximum braking rate of between 3.0 and 3.5 mi/(h·s) [4.8 and 5.6 km/(h·s)] and is made as independent of passenger loads as possible by a load-weighing system which adjusts braking effort to suit passenger loads.

Dynamic braking is now generally used as the primary stopping mode and is effective from maximum speed down to 10 mi/h (16 km/h) with friction braking supplementation as the characteristic dynamic fade occurs. The friction brakes provide the final stopping forces. Emergency braking rates depend upon line constraints, car subsystems, and other factors, but generally rely on the maximum retardation force that can be provided by the dynamic and friction brakes within the limits of available wheel-to-rail adhesion.

Acceleration and braking on modern properties are controlled by a single master controller handle which has power positions in one direc-

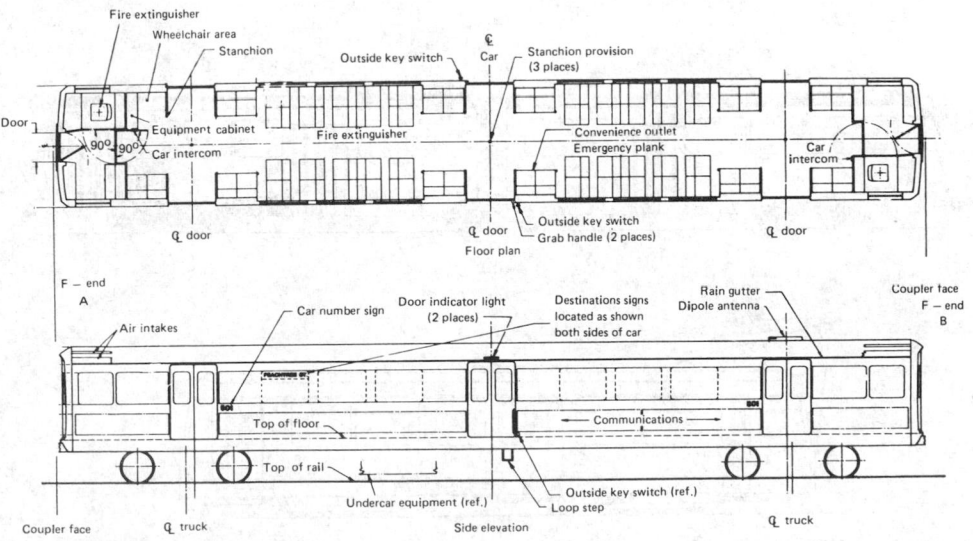

Fig. 11.2.22 Full-scale transit car.

tion and a coasting or neutral center position and braking positions in the opposite direction. The number of positions depends upon the property policy or preference and the control subsystems on the car. Control elements include motor-current sensors and some or all of the following: speed sensors, rate sensors, and load-weighing sensors. Signals are processed by an electronic control unit (ECU) which processes them and provides control functions to the propulsion and braking systems. The propulsion systems currently include pilot-motor-operated cams which actuate switches to control resistance steps, or chopper systems which electronically provide the desired voltages and currents at the motors.

Most modern cars are equipped with two trucks, each of which carries two series-wound dc traction motors. A few use motors with separately excited fields. Monomotor trucks are not used on full-scale rapid transit equipment. The motors are rated between 100 and 200 shaft hp at between 300 and 750 V dc, depending upon line voltage, series or series-parallel connections (electronic or electromechanical control), and performance level desired. The motors, gear units (right angle or parallel), and axles are joined variously through flexible couplings. Electronic-inverter control drives with ac motors have been employed in recent applications in Europe and North America.

In most applications, dynamic braking utilizes the traction motors as generators to dissipate energy through the on-board resistors also used in acceleration. It is expected that regenerative braking will become more common since energy can be returned to the line for use by other cars. Theoretically, 35 to 50 percent of the energy can be returned, but the present state-of-the-art is limited to a practical 20 percent on properties with large numbers of cars on close headways.

Bodies are made of welded stainless steel or **low-alloy high-tensile** (LAHT) steel of a design which carries structural loads (load bearing). Earlier problems with aluminum, primarily electrolytic action among dissimilar metals and welding techniques, have been resolved and aluminum is used on a significant number of new cars.

Trucks may be cast or fabricated steel with the frames and journal bearings either inside or outside the wheels. Axles are carried in roller-bearing journals connected to the frames so as to be able to move vertically against a variety of types of primary spring restraint. Metal springs or air bags are used as the secondary suspension between the trucks and the car body. Most wheels are solid cast or wrought steel. Resilient wheels have been tested but are not in general service on full-scale transit equipment.

All full-scale rapid transit systems use high-level loading platforms to speed passenger flow. This adds to the civil construction costs, but is necessary to achieve the required level of service.

Light Rail Transit Equipment This equipment differs only slightly from full-scale rapid transit equipment since the term ''light rail'' means only that more severe civil restraints are acceptable to achieve about 80 percent of the capability at about 50 percent of the cost of full-scale rapid transit.

The cars are called light rail vehicles (Fig. 11.2.23 shows typical **LRVs**) and are used on a few remaining streetcar systems such as those in Boston, Philadelphia, and Toronto on city streets and on state-of-the-art subway, surface, and elevated systems such as those in Edmonton, Calgary, San Diego, and Portland (OR) in semiautomated modes over partially or wholly reserved rights-of-way.

As a practical matter, the LRV's track, signal systems, and power systems utilize the same subsystems as full-scale rapid transit and are not ''lighter'' in a physical sense.

The LRVs are designed to operate bidirectionally in multiple with up to four cars controlled from the leading cab. The LRV's are electrically propelled from an overhead contact wire often of catenary design which makes contact with a pantograph on the roof. For wayside safety reasons third-rail pickup is not used. Voltages range from 550 to 750 V dc.

The cars range from 60 to 65 ft (18.3 to 19.8 m) over the anticlimbers for single cars and from 70 to 90 ft (21.3 to 27.4 m) for articulated cars. The choice is determined by passenger volumes and civil constraints.

Articulated cars have been used in railroad and transit applications since the 1920s, but have found favor in light rail applications because of their ability to increase passenger loads in a longer car which can negotiate relatively tight-radius curves. These cars require the additional mechanical complexity of the articulated or hinged center section carried on a third truck which fits between two car-body sections.

Passenger seating varies from 50 to 80 seats per car depending upon length, placement of seats in the articulation and the policy regarding seated/standee ratio. Older systems require negotiation of curves down to 30 ft (9.1 m) minimum radius with speeds up to only 40 mi/h (65 km/h) on tangent track while newer systems usually have no less than 75 ft (22.9 m) minimum radius curves with speeds up to 65 mi/h (104 km/h) on tangent track.

The newer properties all use standard gage; however, existing older systems include 4 ft 10⅞ in (1,495 mm) and 5 ft 2½ in (1,587 mm) gages. Grades have reached 12 percent but 6 percent is now considered the maximum and 5 percent is preferred.

Typically, newer properties require maximum acceleration rates of

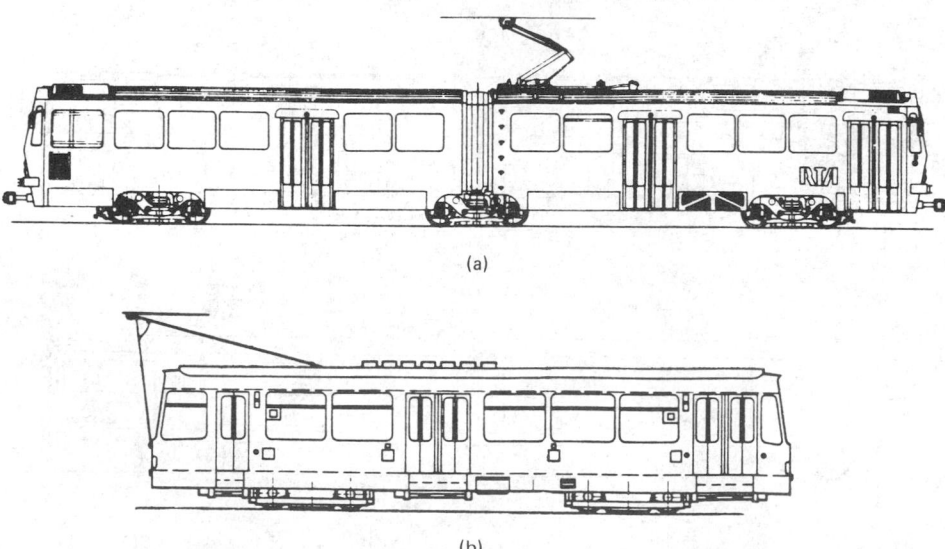

(a)

(b)

Fig. 11.2.23 Light-rail vehicle (LRV). (*a*) Articulated; (*b*) nonarticulated.

between 3.0 to 3.5 mi/(h·s) [4.8 to 5.6 km/(h·s)] as nearly independent of passenger load as possible from 0 to approximately 20 mi/h (32 km/h). Depending upon the selection of motors and gearing, this rate falls off as speed is increased, but the rates can generally be controlled at a variety of levels.

Unlike most full-scale rapid transit cars, LRVs incorporate three braking modes: dynamic, friction, and track brake which typically provide maximum service braking at between 3.0 and 3.5 mi/(h·s) [4.8 to 5.6 km/(h·s)] and 6.0 mi/(h·s) [9.6 km/(h·s)] maximum emergency braking rates. The dynamic and friction brakes are usually blended, but a variety of techniques exist. The track brake is intended to be used primarily for emergency conditions and may or may not be controlled with the other braking systems.

The friction brakes are almost exclusively disk brakes since LRVs use resilient wheels which can be damaged by tread-brake heat buildup. No single consistent pattern exists for the actuation mechanism. All-electric, all-pneumatic, electropneumatic, electrohydraulic, and electropneumatic over hydraulic are in common use.

Dynamic braking is generally used as the primary mode and is effective from maximum speed down to about 5 mi/h (8 km/h) with friction braking supplementation as the characteristic dynamic fade occurs.

As with full-scale rapid transit, the emergency braking rates depend upon line constraints, car-control subsystems selected, and other factors, but the use of track brakes means that higher braking rates can be achieved because the wheel-to-rail adhesion is not the limiting factor.

Acceleration and braking on modern properties are usually controlled by a single master controller handle which has power positions in one direction and a coasting or neutral center position and braking positions in the opposite direction. A few properties use foot-pedal control with a ''deadman'' pedal operated by the left foot, a brake pedal operated by the right foot, and an accelerator (power) pedal also operated by the right foot. In either case, the number of positions depends upon property policy and control subsystems on the car.

Control elements include motor-current sensors and some or all of the following: speed sensors, rate sensors, and load-weighing sensors. Signals from these sensors are processed by an electronic control unit (ECU) which provides control functions to the propulsion and braking systems.

The propulsion systems currently include pilot-motor-operated cams which actuate switches or electronically controlled unit switches to control resistance steps. Chopper systems which electronically provide the desired voltages and currents at the motors are also used.

Most modern LRVs are equipped with two powered trucks. In articulated designs (Fig. 11.2.23a), the third (center) truck is left unpowered but usually has friction and track brake capability. Some European designs use three powered trucks but the additional cost and complexity have not been found necessary in North America.

Unlike full-scale rapid transit, there are three major dc-motor configurations in use: the traditional series-wound motors used in bimotor trucks, the European-derived monomotor, and a hybrid monomotor with a separately excited field—the latter in chopper-control version only.

The bimotor designs are rated between 100 and 125 shaft hp per motor at between 300 and 750 V dc, depending upon line voltage and series or series-parallel control schemes (electronic or electromechanical control). The monomotor designs are rated between 225 and 250 shaft hp per motor at between 300 and 750 V dc (electronic or electromechanical control).

The motors, gear units (right angle or parallel), and axles are joined variously through flexible couplings. In the case of the monomotor, it is supported in the center of the truck, and right-angle gearboxes are mounted on either end of the motor. Commonly, the axle goes through the gearbox and connection is made with a flexible coupling arrangement. Electronic inverter control drives with ac motors have been applied in recent conversions and new equipment.

Dynamic braking is achieved in the same manner as full-scale rapid transit.

Unlike full-scale rapid transit, LRV bodies are made only of welded LAHT steel and are of a load-bearing design. Because of the semireserved right-of-way, the risk of collision damage with automotive vehicles is greater than with full-scale rapid transit and the LAHT steel has been found to be easier to repair than stainless steel or aluminum. Although the LAHT steel requires painting, this can be an asset since the painting can be done in a highly decorative manner pleasing to the public and appropriate to themes desired by the cities.

Trucks may be cast or fabricated steel with either inside or outside frames. Axles are carried in roller-bearing journals which are usually resiliently coupled to the frames with elastomeric springs as a primary suspension. Both vertical and a limited amount of horizontal movement occur. Since tight curve radii are common, the frames are usually connected to concentric circular ball-bearing rings, which, in turn, are connected to the car body. Air bags, solid elastomeric springs, or metal springs are used as a secondary suspension. Resilient wheels are used on virtually all LRVs.

With a few exceptions, LRVs have low-level loading doors and steps which minimize station platform costs.

TRACK

Gage   The gage of railway track is the distance between the inner sides of the rail heads measured 5/8 in (15.9 mm) below the top of rail. North American railways are laid with a nominal gage of 4 ft 8½ in (1,435 mm), which is known as **standard gage**. Rail wear causes an increase in gage. On sharp curves [over 8°, see Eq. (11.2.10)], it has been the practice to widen the gage to reduce rail wear.

Track Structure   The basic track structure is composed of six major elements: rail, tie plates, fasteners, cross ties, ballast, and subgrade (Fig. 11.2.24).

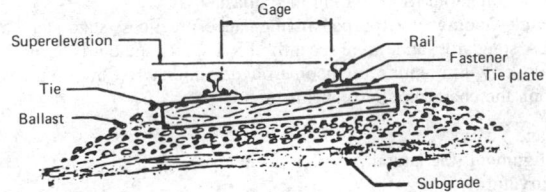

Fig. 11.2.24   Track structure.

Rail   The Association of American Railroads has defined standards for the cross section of rails of various weights per yard. The **AREA standards** provide for rail weights varying from 90 to 140 lb/yd. A few railroads use special sections of their own design. The standard length in use is 39 ft (11.9 m) although some mills are now rolling 78 ft (23.8 m) or longer lengths.

The prevailing weight of rails used in main-line track is 112 to 140 lb/yd. Secondary and branch lines are generally laid with 75 to 100 lb/yd rail which has been previously used and partly worn in main-line service. In the past, rails were connected using bolted joint bars. Current practice is the use of **continuous welded rail** (CWR). In general, rails are first welded into lengths up to 20 rails or more, then laid, and later field welded to eliminate conventional joints between rail strings and other track appliances. Electric-flash, gas-fusion, and Thermit welding processes are used. CWR requires the use of more rail anchors than bolted rail to clamp the base of the rail and position it against the sides of cross ties in order to prevent rail movement. This is necessary to restrain longitudinal forces which result from thermal expansion and contraction due to ambient temperature changes. Failure to adequately restrain the rail can result in track buckling either longitudinally, laterally, or both, in extremely hot weather or in rail ''pull aparts'' in extremely cold weather. Care must be exercised to install the rail at temperatures which do not approach high or low extremes.

A variety of heat-treating and alloying processes have been applied to produce rail that is more resistant to wear and less susceptible to fatigue damage. Comprehensive studies indicate that gage face lubrication can dramatically reduce the wear rate of rail in curves (see publications of AAR Research and Test Department). This results in rail fatigue becoming the dominant cause for replacement at some longer life.

Tie Plates Rail is supported on tie plates which restrain the rail laterally, distribute the vehicle loads onto the cross tie, and position (cant) the rail for optimum vehicle performance. Tie plates cant the rail toward the gage side. Cants vary from 1:14 to 1:40, depending on service conditions and operating speeds.

Fasteners Rail and tie plates are fixed to the cross ties by fasteners. Conventional construction with wood cross ties utilizes steel spikes driven into the tie through holes in the tie plate. In some cases, screw-type fasteners are used to provide more resistance to vertical forces. Elastic, clip-type fasteners are used on concrete and, at times, on wooden cross ties to provide a uniform, resilient attachment and longitudinal restraint in lieu of anchors.

Cross Ties To retain gage and provide a further distribution of vehicle loads to the ballast, lateral cross ties are used. The majority of cross ties used in North America are wood. However, concrete cross ties are being used more and more frequently in areas where track surface and lateral stability are difficult to maintain and in high-speed passenger corridors.

Ballast Ballast serves to further distribute vehicle loads to the subgrade, restrain vertical and lateral displacement of the track structure, and provide for drainage. Typical main-line ballast materials are limestone, trap rock, granite, or slag. Lightly built secondary lines use gravel, cinders, or sand.

Subgrade The subgrade serves as the interface between the ballast and the native soil. Subgrade material is typically compacted or stabilized soil. Instability of the subgrade due to groundwater permeation is a concern in some areas. In some difficult locations, a semipermeable geotextile fabric is used to separate the ballast and subgrade and prevent subgrade material from fouling the ballast.

Track Geometry Railroad maintenance practices, state standards, AREA standards, and, more recently, FRA safety standards (CFR Title 49) specify limitations on geometric deviations for track structure. Among the characteristics considered are:

Gage
Alignment (the lateral position of the rails)
Curvature
Easement (the rate of change from tangent to curve)
Superelevation
Runoff (the rate of change in superelevation)
Cross level (the relative height of opposite rails)
Twist (the change in relative position of the rails about a longitudinal axis)

FRA standards specify maximum operating speeds by class of track ranging from Class I (poorest acceptable) to Class 6 (excellent). Track geometry is a significant consideration in vehicle suspension design.

Curvature The curvature of track is designated in terms of "degrees of curvature." The degree of curve is the number of degrees of central angle subtended by a chord of 100-ft length (measured on the track centerline). Equation (11.2.10) gives the approximate radius, R, in feet for ordinary railway curves.

$$R = \frac{5730}{\text{degrees per 100-ft chord}} \qquad (11.2.10)$$

On important main lines where trains are operated at relatively high speed, the curves are ordinarily not sharper than 6 or 8°. In mountainous territory, in rare instances, main-line curves as sharp as 18° occur. Most diesel and electric locomotives are designed to traverse curves up to 21°. Most uncoupled cars will pass considerably sharper curves, the limiting factor in this case being the clearance of the truck components to car body structural members or equipment or the flexibility of the brake connections to the trucks (AAR "Manual of Standards and Recommended Practices," Sec. C).

Clearances The AREA standard clearance diagrams provide for a clear height of 22 ft (6.7 m) above the tops of the rail heads and for a width of 16 ft (4.9 m) on bridges and 15 ft (4.6 m) elsewhere. Where conditions require the minimum clearance, as in tunnels, these dimensions are much reduced. For tracks entering buildings, an opening 12 ft (3.7 m) wide and 17 ft (5.2 m) high will ordinarily suffice to pass the largest locomotives and cars. A standard clearance diagram for railway bridges (Fig. 11.2.25) represents the AAR recommendation for new construction. Since some clearances are dictated by state authorities, for specific clearance limitations on a particular railroad, refer to the official publication "Railway Line Clearances," Railway Equipment and Publishing Co.

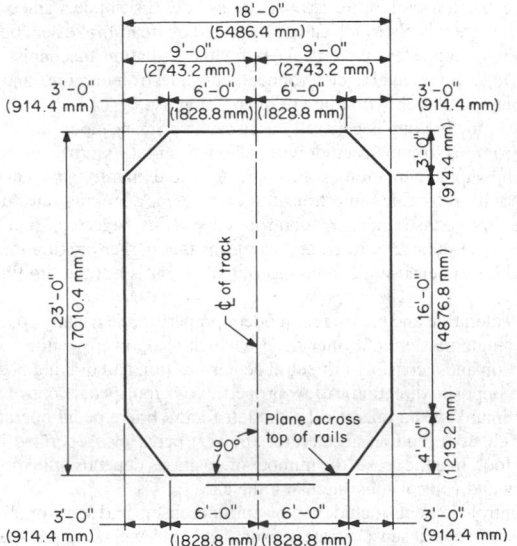

Fig. 11.2.25 Wayside clearance diagram. (AREA "Manual for Railway Engineering—Fixed Properties.")

Track Spacing The distance between centers of main-line track varies between 12 and 14 ft (3.6 and 4.3 m). Twelve-foot spacing exists, however, only on old construction. The prevailing spacing, 14 ft 0 in (4.3 m), has the implied endorsement of the AREA and is required by some states.

Turnouts Railway turnouts (switches) are described by frog number. The frog number is the cotangent of the frog angle, which is the ratio of the lateral displacement of the diverging track per unit of longitudinal distance along the nondiverging track. Typical turnouts in yard and industry trackage range from number 7 to number 10. For main-line service, typical turnouts range from number 10 to number 20. Generally, the higher the turnout number, the higher the allowable speed for a train traversing the diverging track. As a general rule, the allowable speed (in mi/h) is twice the turnout number.

Noise Allowable noise levels from railroad operation are prescribed by CFR40 §201.

VEHICLE-TRACK INTERACTION

Train Resistance The resistance to a train in motion along the track is of prime interest, as it is reflected directly in locomotive energy (fuel) requirements. This resistance is expressed in terms of "pounds per ton" of train weight. **Gross train resistance** is that force which must be overcome by the locomotives at the driving-wheel-rail interface. **Trailing train resistance** must be overcome at the locomotive rear drawbar.

There are two classes of resistance which must be overcome: "inherent" and "incidental." **Inherent resistance** includes the rolling resistance of bearings and wheels and aerodynamic resistance due to motion through still air. It may be considered equal to the force necessary to maintain motion at constant speed on level tangent track in still air. **Incidental resistance** includes resistance due to grade, curvature, wind, and vehicle dynamics.

Inherent Resistance Of the elements of inherent resistance, at low speeds rolling resistance is dominant, but at high speeds aerodynamic resistance is the predominant factor. Attempts to differentiate and evaluate the various elements through the speed range are a continuing part of industry research programs to reduce train resistance. At very high speeds, the effect of air resistance can be approximately determined. This is an aid to studies in its reduction by means of cowling and fairing. The resistance of a car moving in still air on straight, level track increases parabolically with speed. Because the aerodynamic resistance is independent of car weight, the resistance in pounds per ton decreases as the weight of the car increases. The total resistance in pounds per ton of a 100-ton car is much less than twice as great as that of a 50-ton car under similar conditions. WIth known conditions of speed and car weight, inherent resistance can be predicted with reasonable accuracy. Knowledge of track conditions will permit further refining of the estimate, but for very rough track or extremely cold ambient temperatures, generous allowances must be made. Under such conditions normal resistance may be doubled. A formula proposed by Davis (*General Electric Review*, Oct. 1926) and revised by Tuthill "High Speed Freight Train Resistance," *Univ. of Illinois Engr. Bull.*, **376**, 1948) has been used extensively for inherent freight-train resistances at speeds up to 40 mi/h:

$$R = 1.3W + 29n + 0.045WV + 0.0005 AV^2 \qquad (11.2.11)$$

where R = train resistance, lb/car; W = weight per car, tons; V = speed, mi/h; n = total number of axles; A = cross-sectional area, ft².

With freight-train speeds of 50 to 70 mi/h (80 to 112 km/h), it has been found that actual resistance values fall considerably below calculations based on the above formula. Several modifications of the Davis equation have been developed for more specific applications. All of these equations apply to cars trailing locomotives.

1. Davis equation modified by J. K. Tuthill (*University of Illinois Engr. Bull.*, **376**, 1948):

$$R = 1.3W + 29n + 0.045WV + 0.045V^2 \qquad (11.2.12)$$

Note: In the Tuthill modification, the equation is augmented by a matrix of coefficients when the velocity exceeds 40 mi/h.

2. Modified by Canadian National Railway:

$$R = 0.6W + 20n + 0.01WV + 0.07V^2 \qquad (11.2.13)$$

3. Modified by Canadian National Railway and Erie-Lackawanna Railroad for trailers and containers on flat cars:

$$R = 0.6W + 20n + 0.01WV + 0.2V^2 \qquad (11.2.14)$$

Other modifications of the Davis equation have been developed for passenger cars by Totten ("Resistance of Light Weight Passenger Trains," *Railway Age*, **103**, Jul. 17, 1937). These formulas are for passenger cars pulled by a locomotive and do not include head-end air resistance.

1. Modified by Totten for streamlined passenger cars:

$$R = 1.3W + 29n + 0.045WV$$
$$+ [0.00005 + 0.060725 (L/100)^{0.88}]V^2 \qquad (11.2.15)$$

2. Modified by Totten for nonstreamlined passengers cars

$$R = 1.3W + 29n + 0.045WV$$
$$+ [0.0005 + 0.1085 (L/100)^{0.7}]V^2 \qquad (11.2.16)$$

where L = car length in feet.

Aerodynamic and Wind Resistance Wind-tunnel testing has indicated a significant effect on train resistance due to vehicle spacing, open tops of hopper and gondola cars, open boxcar doors, vertical side reinforcements on railway cars and intermodal trailers, and protruding appurtenances on cars. These effects can cause significant increases in train resistance at higher speeds. For example, spacing of intermodal trailers or containers greater than approximately 6 ft apart can result in a new frontal area to be considered in determining train resistance. Frontal or cornering ambient wind conditions can also have an adverse effect on train resistance which is increased with discontinuities along the length of the train.

Curve Resistance Resistance due to track curvature varies with speed and degree of curvature. The behavior of railroad equipment in curve negotiation is the subject of several ongoing AAR studies. Lubrication of the rail gage face or wheel flanges is common practice for reducing friction forces and the resulting wheel and rail wear. Recent studies, in fact, indicate that flange and/or gage face lubrication can significantly reduce train resistance on tangent track as well (Allen, "Conference on the Economics and Performance of Freight Car Trucks," October 1983). In addition, a variety of special trucks (wheel assemblies) which reduce curve resistance by allowing axles to steer toward a radial position in curves have been developed. For general estimates of car resistance and locomotive hauling capacity on dry (unlubricated) rail with conventional trucks, speed and gage relief may be ignored and a figure of 0.8 lb/ton per degree of curvature used.

Grade resistance depends only on the angle of ascent or descent and relates only to the gravitational forces acting on the vehicle. It equates to 20 lb/ton for each "percent of grade" or 0.379 lb/ton for each foot per mile rise.

Acceleration Resistance The force (tractive effort) required to accelerate the train is the sum of the forces required for translation acceleration and that required to produce rotational acceleration of the wheels about their axle centers. A translatory acceleration of 1 mi/h·s (1.6 km/h·s) is produced by a force of 91.1 lb/ton. The rotary acceleration requirement adds 6 to 12 percent, so that the total is nearly 100 lb/ton (the figure commonly used) for each mile per hour per second. If greater accuracy is required, the following expression is used:

$$R_a = A(91.05W + 36.36n) \qquad (11.2.17)$$

where R_a = the total accelerating force, lb; A = acceleration, mi/h·s; W = weight of train, tons; n = number of axles.

Acceleration and Distance If in a distance of S ft the speed of a car or train changes from V_1 to V_2 mi/h, the force required to produce acceleration (or deceleration if the speed is reduced) is

$$R_a = 74(V_2^2 - V_1^2)/S \qquad (11.2.18)$$

The coefficient 74 corresponds to the use of 100 lb/ton above. This formula is useful in the calculation of the energy required to climb a grade with the assistance of stored energy. In any train-resistance calculation or analysis, assumptions with regard to acceleration will generally submerge all other variables; e.g., an acceleration of 0.1 mi/h·s (0.16 km/h·s) requires more tractive force than that required to overcome inherent resistance for any car at moderate speeds.

Starting Resistance Most railway cars are equipped with roller bearings with a starting force of 5 or 6 lb/ton.

Vehicle Suspension Design The primary consideration in the design of the vehicle suspension system is to isolate track input from the vehicle car body and lading. In addition, there are a few specific areas of instability which railway suspension systems must address, see AAR "Manual of Standards and Recommended Practice," Sec. C-II-M-1001, Chap. XI.

Harmonic roll is the tendency of a freight car with a high center of gravity to rotate about its longitudinal axis (parallel to the track). This instability is excited by passing over staggered low rail joints at a speed which causes the frequency of the input for each joint to match the natural roll frequency of the car. Unfortunately, in many car designs, this occurs for loaded cars at 12 to 18 mi/h (19.2 to 28.8 km/h), a common speed for trains moving in yards or on branch lines where tracks are not well maintained.

This adverse behavior is more noticeable in cars with truck centers approximately the same as the rail length. The effect of harmonic roll can be mitigated by improved track surface and by damping in the truck suspension.

Pitch and bounce are the tendencies of the vehicle to either translate vertically up and down (bounce), or rotate (pitch) about a horizontal axis perpendicular to the centerline of track. This response is also excited by low track joints and can be relieved by increased truck damping.

Yaw is the tendency of the freight car to rotate about its axis vertical to the centerline of track. Yaw responses are usually related to truck hunting.

Truck hunting is an instability inherent in the design of the truck and dependent upon the stiffness parameters of the truck and the wheel conicity (tread profile). The instability is observed as a "parallelogramming" of the truck components at a frequency of 2 to 3 Hz causing the car body to yaw or translate laterally. This response is excited by the effect of the natural frequency of the "gravitational stiffness" of the wheel set when the speed of the vehicle approaches the kinematic velocity of the wheel set. This problem is discussed in analytic work available from the AAR Research and Test Department.

Superelevation As a train passes around a curve, there is a tendency for the cars to tip toward the outside of the curve due to the centrifugal force acting on the center of gravity of the car body (Fig. 11.2.26a). To compensate for this effect, the outside rail is superelevated, or raised, relative to the inside rail (Fig. 11.2.26b). The amount of superelevation for a particular curve is based upon the radius of the curve and the operating speed of the train. The "balance" or equilibrium speed for a given curve is that speed at which the centrifugal force on the car matches the component of gravity force resulting from the superelevation between the amount required for high-speed trains and the amount required for slower-operating trains. The FRA allows a railroad to operate with 3 in of outward unbalance, or at the speed at which equilibrium would exist if the superelevation were 3 in greater. The maximum superelevation is usually 6 in.

Longitudinal Train Action Longitudinal train (slack) action is a term associated with the dynamic action between individual cars in a train. An example would be the effect of starting a long train in which the couplers between each car had been compressed (i.e., bunched up). As the locomotive begins to pull the train, the slack between the locomotive and the first car must be traversed before the first car begins to accelerate. Next the slack between the first and second car must be traversed before the second car beings to accelerate, and so on. Before the last car in a long train begins to move, the locomotive and the moving cars may be traveling at a rate of several miles per hour. This effect can result in coupler forces sufficient to cause the train to break in two.

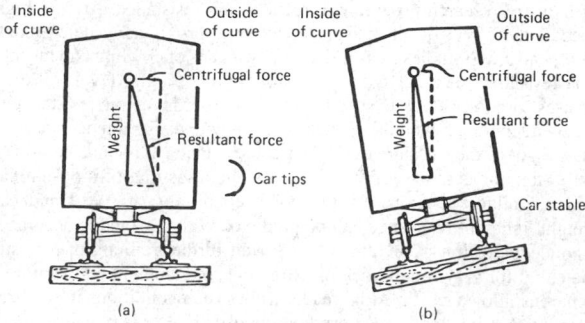

Fig. 11.2.26 Effect of superelevation on center of gravity of car body.

Longitudinal train action is also induced by serial braking, undulating grades, or by braking on varying grades. The Track Train Dynamics Program has published guidelines titled "Track Train Dynamics to Improve Freight Train Performance" which explain the causes of undesirable train action and how to minimize the effects. Analysis of the forces developed by longitudinal train action requires the application of the Davis equation to represent the resistance of each vehicle based upon its velocity and location on a grade or curve. Also, the longitudinal stiffness of each car and the tractive effort of the locomotive must be considered in equations which model the kinematic response of each vehicle in the train. Computer programs are available from the AAR to assist in the analysis of longitudinal train action.

11.3 MARINE ENGINEERING
by Michael C. Tracy

REFERENCES: Harrington, "Marine Engineering," SNAME, 1992. Lewis, "Principles of Naval Architecture," SNAME, 1988/89. Myers, "Handbook of Ocean and Underwater Engineering," McGraw-Hill, 1969. "Rules for Building and Classing Steel Vessels," American Bureau of Shipping. Rawson and Tupper, "Basic Ship Theory," American Elsevier, 1968. Gillmer and Johnson, "Introduction to Naval Architecture," Naval Institute, 1982. Barnaby, "Basic Naval Architecture," Hutchinson, London, 1967. *Jour. Inst. Environ. Sci.* Taggert, "Ship Design and Construction," SNAME, 1980. *Trans. Soc. Naval Architects and Marine Engrs.*, SNAME. *Naval Engrs. Jour.*, Am. Soc. of Naval Engrs., ASNE. *Trans. Royal Inst. of Naval Architects*, RINA. Figures and examples herein credited to SNAME have been included by permission of The Society of Naval Architects and Marine Engineers.

Marine engineering is an integration of many engineering disciplines directed to the development and design of systems of transport, warfare, exploration, and natural-resource retrieval which have one thing in common: operation in or on a body of water. Marine engineers are responsible for the engineering systems required to propel, work, or fight ships. They are responsible for the main propulsion plant; the powering and mechanization aspects of ship functions such as steering, anchoring, cargo handling, heating, ventilation, air conditioning, etc.; and other related requirements. They usually have joint responsibility with naval architects in areas of propulsor design; hull vibration excited by the propeller or main propulsion plant; noise reduction and shock hardening, in fact, dynamic response of structures or machinery in general; cargo-handling pumping systems; and environmental control and habitability. Marine engineering is a distinct multidiscipline and characteristically a dynamic, continuously advancing technology.

THE MARINE ENVIRONMENT

Marine engineers must be familiar with their environment so that they may understand fuel and power requirements, vibration effects, and propulsion-plant strength considerations. The outstanding characteristic of the open ocean is its irregularity in storm winds as well as under relatively calm conditions. The **irregular sea** can be described by statistical mathematics based on the superposition of a large number of regular waves having different lengths, directions, and amplitudes. The characteristics of idealized regular waves are fundamental for the description and understanding of realistic, irregular seas. Actual sea states consist of a combination of many sizes of waves often running in different directions, and sometimes momentarily superimposing into an exceptionally large wave.

The effects of the marine environment also vary with water depth. As a ship passes from deep to shallow water, there is an appreciable change in the potential flow around the hull and a change in the wave pattern produced. Additionally, silt, sea life, and bottom growth may affect seawater systems or foul heat exchangers.

MARINE VEHICLES

The platform is additionally a part of marine engineers' environment. Ships are supported by a buoyant force equal to the weight of the volume of water displaced. For surface ships, this weight is equal to the total weight of the structure, machinery, outfit, fuel, stores, crew, and useful load. The principal sources of resistance to propulsion are skin friction and the energy lost to surface waves generated by moving in the interface between air and water. Minimization of one or both of these sources of resistance has generally been a primary objective in the design of marine vehicles.

Displacement Hull Forms

Displacement hull forms are the familiar monohull, the catamaran, and the submarine. The moderate-to-full-displacement **monohull** form provides the best possible combinations of high-payload-carrying ability, economical powering characteristics, and good seakeeping qualities. A more slender hull form achieves a significant reduction in wave-making resistance, hence increased speed; however, it is limited in its ability to carry topside weight because of the low transverse stability of its narrow beam. The **catamaran** provides a solution to the problem of low transverse stability. It is increasingly popular in sailing yachts and is under development for high-speed passenger ferries and research and small support ships. Sailing catamarans, with their superior transverse stability permitting large sail-plane area, gain a speed advantage over monohull craft of comparable size. A powered catamaran has the advantage of increased deck space and relatively low roll angles over a monohull ship. The **submarine,** operating at depths which preclude the formation of surface waves, experiences significant reductions in resistance compared to a well-designed surface ship of equal displacement.

Planing Hull Forms

The planing hull form, although most commonly used for yachts and racing craft, is used increasingly in small, fast commercial craft and in coastal patrol craft. The weight of the planing hull ship is partially borne by the dynamic lift of the water streaming against a relatively flat or V-shaped bottom. The effective displacement is hence reduced below that of a ship supported statically, with significant reduction in wave-making resistance at higher speeds.

High-Performance Ships

In a search for high performance and higher speeds in rougher seas, several advanced concepts to minimize wave-making resistance have been investigated. These concepts have been or are being developed in hydrofoil craft, surface-effect vehicles, and small water-plane-area twin hull (SWATH) forms (Fig. 11.3.1). The **hydrofoil craft** has a planing hull that is raised clear of the water through dynamic lift generated by an underwater foil system. The **surface-effect vehicles** ride on a cushion of compressed air generated and maintained in the space between the ve-

hicle and the surface over which it hovers or moves. The most practical vehicles employ a peripheral-jet principle, with flexible skirts for obstacle or wave clearance. A rigid sidewall craft, achieving some lift from hydrodynamic effects, is more adaptable to marine construction techniques. The **SWATH** gains the advantages of the catamaran, twin displacement hulls, with the further advantage of minimized wave-making resistance and wave-induced motions achieved by submarine-shaped hulls beneath the surface and the small water-plane area of the supporting struts at the air-water interface.

SEAWORTHINESS

Seaworthiness is the quality of a marine vehicle being fit to accomplish its intended mission. In meeting their responsibilities to produce seaworthy vehicles, marine engineers must have a basic understanding of the effects of the marine environment with regard to the vehicle's (1) **structure,** (2) **stability** and **motions,** and (3) **resistance** and **powering** requirements.

Units and Definitions

The introduction of SI units to the marine engineering field presents somewhat of a revolutionary change. Displacement, for instance, is a force and therefore is expressed in newtons (N) or meganewtons (MN); what for many years has been known as a 10,000-ton ship therefore becomes a 99.64-MN ship. To assist in the change, examples and data have been included in both USCS and SI units.

The **displacement** Δ is the weight of the water displaced by the immersed part of the vehicle. It is equal (1) to the buoyant force exerted on the vehicle and (2) to the weight of the vehicle (in equilibrium) and everything on board. Displacement is expressed in long tons equal to 1.01605 metric tons or 2,240 lb (1 lb = 4.448 N). The specific weight of seawater averages about 64 lb/ft; hence, the displacement in seawater is measured by the displaced volume ∇ divided by 35. In fresh water, divide by 35.9. (Specific weight of seawater = 10,053 N/m³; 1 MN = 99.47 m³; $\Delta = \nabla/99.47$ MN; in fresh water 1 MN = 102 m³.)

Two measurements of a merchant ship's earning capacity that are of significant importance to its design and operation are deadweight and tonnage. The **deadweight** of a ship is the weight of cargo, stores, fuel, water, personnel, and effects that the ship can carry when loaded to a specific load draft. Deadweight is the difference between the load displacement, at the minimum permitted freeboard, and the light displacement, which comprises hull weight and machinery. Deadweight is expressed in long tons (2,240 lb each) or MN. The volume of a ship is expressed in tons of 100 ft³ (2.83 m³) each and is referred to as its **tonnage.** Charges for berthing, docking, passage through canals and locks, and for many other facilities are based on tonnage. **Gross tonnage** is based on cubic capacity below the tonnage deck, generally the uppermost complete deck, plus allowances for certain compartments above,

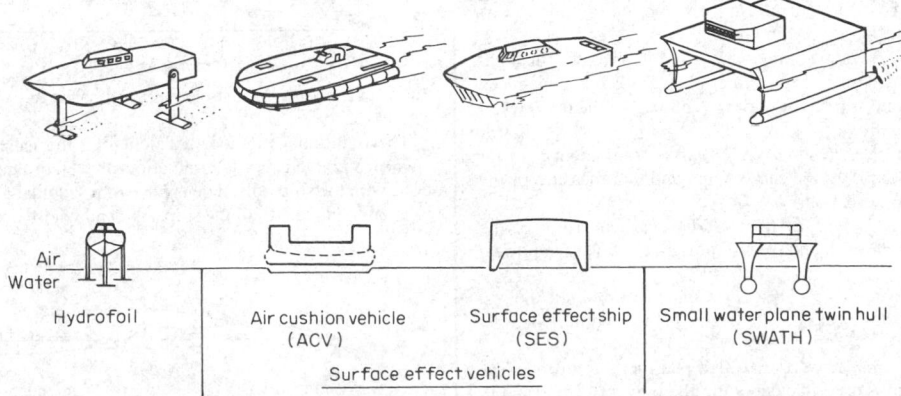

Fig. 11.3.1 Family of high-performance ships.

Table 11.3.1 Coefficients of Form

Type of vessel	$\dfrac{v}{\sqrt{gL}}$	C_B	C_M	C_P	C_{WP}	$\dfrac{\Delta}{(L/100)^3}$
Great Lakes ore ships	0.116–0.128	0.85–0.87	0.99–0.995	0.86–0.88	0.89–0.92	70–95
Slow ocean freighters	0.134–0.149	0.77–0.82	0.99–0.995	0.78–0.83	0.85–0.88	180–200
Moderate-speed freighters	0.164–0.223	0.67–0.76	0.98–0.99	0.68–0.78	0.78–0.84	165–195
Fast passenger liners	0.208–0.313	0.56–0.65	0.94–0.985	0.59–0.67	0.71–0.76	75–105
Battleships	0.173–0.380	0.59–0.62	0.99–0.996	0.60–0.62	0.69–0.71	86–144
Destroyer, cruisers	0.536–0.744	0.44–0.53	0.72–0.83	0.62–0.71	0.67–0.73	40–65
Tugs	0.268–0.357	0.45–0.53	0.71–0.83	0.61–0.66	0.71–0.77	200–420

which are used for cargo, passengers, crew, and navigating equipment. Deduction of spaces for propulsion machinery, crew quarters, and other prescribed volumes from the gross tonnage leaves the **net tonnage.**

The dimensions of a ship may refer to the molded body (or form defined by the outside of the frames), to general outside or overall dimensions, or to dimensions on which the determination of tonnage or of classification is based. There are thus (1) molded dimensions, (2) overall dimensions, (3) tonnage dimensions, and (4) classification dimensions. The published rules and regulations of the classification societies and the U.S. Coast Guard should be consulted for detailed information.

The **designed load waterline** (DWL) is the waterline at which a ship would float freely, at rest in still water, in its normally loaded or designed condition. The keel line of most ships is parallel to DWL. Some keel lines are designed to slope downward toward the stern, termed **designed drag.**

A vertical line through the intersection of DWL and the foreside of the stem is called the **forward perpendicular, FP.** A vertical line through the intersection of DWL with the afterside of the straight portion of the rudder post, with the afterside of the stern contour, or with the centerline of the rudder stock (depending upon stern configuration), is called the **after perpendicular, AP.**

The **length** on the designed load waterline, L_{WL}, is the length measured at the DWL, which, because of the stern configuration, may be equal to the **length between perpendiculars,** L_{pp}. The **classification society length** is commonly noted as L_{pp}. The extreme length of the ship is the **length overall,** L_{OA}.

The **molded beam** B is the extreme breadth of the molded form. The extreme or overall breadth is occasionally used, referring to the extreme transverse dimension taken to the outside of the plating.

The **draft** T (molded) is the distance from the top of the keel plate or bar keel to the load waterline. It may refer to draft amidships, forward, or aft.

Trim is the longitudinal inclination of the ship usually expressed as the difference between the draft forward, T_F, and the draft aft, T_A.

Coefficients of Form Assume the following notation: L = length on waterline; B = beam; T = draft; ∇ = volume of displacement; A_{WP} = area of water plane; A_M = area of midship section, up to draft T; v = speed in ft/s (m/s); and V = speed in knots.

Block coefficient, $C_B = \nabla/LBT$, may vary from about 0.38, for high-powered yachts and destroyers, to greater than 0.90 for slow-speed seagoing cargo ships and is a measure of the fullness of the underwater body.

Midship section coefficient, $C_M = A_M/BT$, varies from about 0.75 for tugs or trawlers to about 0.99 for cargo ships and is a measure of the fullness of the maximum section.

Prismatic coefficient, $C_P = \nabla/LA_M = C_B/C_M$, is a measure of the fullness of the ends of the hull, and is an important parameter in powering estimates.

Water-plane coefficient, $C_{WP} = A_{WP}/LB$, ranges from about 0.67 to 0.95, is a measure of the fullness of the water plane, and may be estimated by $C_{WP} \approx \frac{2}{3}C_B + \frac{1}{3}$.

Displacement/length ratio, $\Delta/L = \Delta/(L/100)^3$, is a measure of the slenderness of the hull, and is used in calculating the power of ships and in recording the resistance data of models.

A similar coefficient is the **volumetric coefficient,** $C_V = \nabla/(L/10)^3$, which is commonly used as this measure.

Table 11.3.1 presents typical values of the coefficients with representative values for Froude number = $v/\sqrt{gL}$.

Structure

The structure of a ship is a complex assembly of small pieces of material. Common hull structural materials for small boats are wood, aluminum, and fiberglass-reinforced plastic. Large ships are nearly always constructed of steel. Past practices of using aluminum in superstructures are usually nowadays limited to *KG*-critical ship designs.

The **analysis** of a ship structure is accomplished through the following simplified steps: (1) Assume that the ship behaves like a box-shaped girder supported on a simple wave system; (2) estimate the loads acting on the ship, using simplified assumptions regarding weight and buoyancy distribution; (3) calculate the static shear forces and bending moments; (4) analyze the resulting stresses; and (5) iterate the design until the stresses are acceptable. The maximum longitudinal bending stresses which result from such simplified loading assumptions are used as an indicator of the maximum stress that will be developed, and an approximate factor of safety is introduced to allow for stresses induced from other types of external loading, from local loadings, from stress concentration, and from material fatigue over the life of the ship.

Weight, buoyancy, and load curves (Figs. 11.3.2 and 11.3.3) are developed for the ship for the determination of shear force and bending moment. Several extreme conditions of loading may be analyzed. The weight curves include the weights of the hull, superstructure, rudder, and castings, forgings, masts, booms, all machinery and accessories, solid ballast, anchors, chains, cargo, fuel, supplies, passengers, and baggage. Each individual weight is distributed over a length equal to the distance between frames at the location of that particular weight.

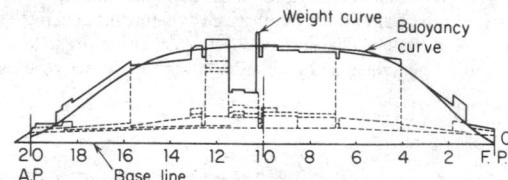

Fig. 11.3.2 Representative buoyancy and weight curves for a ship in still water.

The traditional method of calculating the **maximum design bending moment** involved the determination of weight and buoyancy distributions with the ship poised on a wave of trochoidal form whose length L_w is equal to the length of the ship, L_{pp}, and whose height $H_w = L_w/20$. To

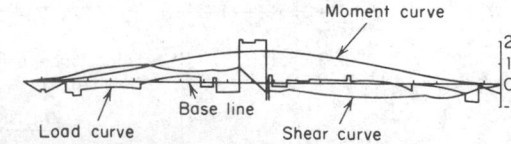

Fig. 11.3.3 Representative moment and net-load curves for a ship in still water.

more accurately model waves whose lengths exceed 300 ft, $H_w = 1.1\sqrt{L}$ is widely used (including standard use by the U.S. Navy), and others have been suggested and applied. With the wave crest amidships, the ship is in a **hogging** condition (Fig. 11.3.4a); the deck is in tension and the bottom shell in compression. With the trough amidships, the ship is

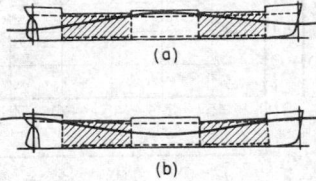

Fig. 11.3.4 Hogging (a) and sagging (b) conditions of a vessel. *(From "Principles of Naval Architecture," SNAME, 1967.)*

in a **sagging** condition (Fig. 11.3.4b); the deck is in compression and the bottom in tension. For normal cargo ships with the machinery amidships, the hogging condition produces the highest bending moment, whereas for normal tankers and ore carriers with the machinery aft, the sagging condition gives the highest bending moment.

The artificiality of the above assumption should be apparent; nevertheless, it has been an extremely useful one for many decades, and a great deal of ship data have been accumulated upon which to base refinements. Many advances have been made in the theoretical prediction of the actual loads a ship is likely to experience in a realistic, confused seaway over its life span. Today it is possible to predict reliably the bending moments and shear forces a ship will experience over a short term in irregular seas. Long-term prediction techniques use a probabilistic approach which associates load severity and expected periodicity to establish the design load.

In determining the **design section modulus** Z which the continuous longitudinal material in the midship section must meet, an **allowable bending stress** σ_{all} must be introduced into the bending stress equation. Based on past experience, an appropriate choice of such an allowable stress is $1.19\,L^{1/3}$ tons/in^2 ($27.31\,L^{1/3}$ MN/m^2) with L in ft (m).

For ships under 200 ft (61 m) in length, strength requirements are dictated more on the basis of locally induced stresses than longitudinal bending stresses. This may be accounted for by reducing the allowable stress based on the above equation or by reducing the K values indicated by the trend in Table 11.3.2 for the calculation of the ship's bending moment. For aluminum construction, Z must be twice that obtained for steel construction. Minimum statutory values of section modulus are published by the U.S. Coast Guard in "Load Line Regulations." Shipbuilding classification societies such as the American Bureau of Shipping have section modulus standards somewhat greater.

Table 11.3.2 Constants for Bending Moment Approximations

Type	Δ tons (MN)	L ft (m)	$K*$
Tankers	35,000–150,000 (349–1495)	600–900 (183–274)	35–41
High-speed cargo passenger	20,000–40,000 (199–399)	500–700 (152–213)	29–36
Great Lakes ore carrier	28,000–32,000 (279–319)	500–600 (152–183)	54–67
Destroyers	3,500–7,000 (35–70)	400–500 (122–152)	23–30
Destroyer escorts	1,600–4,000 (16–40)	300–450 (91–137)	22–26
Trawlers	180–1,600 (1.79–16)	100–200 (30–61)	12–18
Crew boats	65–275 (0.65–2.74)	80–130 (24–40)	10–16

* Based on allowable stress of $\sigma_{all.} = 1.19\sqrt[3]{L}$ tons/in^2 ($27.31\sqrt[3]{L}$ MN/m^2) with L in ft (m).

The **maximum bending stress** at each section can be computed by the equation

$$\sigma = M/Z$$

where σ = maximum bending stress, lb/in^2 (N/m^2) or tons/in^2 (MN/m^2); M = maximum bending moment, ft·lb (N·m) or ft·tons; Z = section modulus, in^2·ft (m^3); $Z = I/y$, where I = minimum vertical moment of inertia of section, in^2·ft^2 (m^4), and y = maximum distances from neutral axis to bottom and strength deck, ft (m).

The **maximum shear stress** can be computed by the equation

$$\tau = Vac/It$$

where τ = horizontal shear stress, lb/in^2 (N/m^2); V = shear force, lb (N); ac = moment of area above shear plane under consideration taken about neutral axis, ft^3 (m^3); I = vertical moment of inertia of section, in^2·ft^2 (m^4); t = thickness of material at shear plane, ft (m).

For an approximation, the **bending moment** of a ship may be computed by the equation

$$M = \Delta L/K \qquad \text{ft·tons (MN·m)}$$

where Δ = displacement, tons (MN); L = length, ft (m); and K is as listed in Table 11.3.2.

The above formulas have been successfully applied in the past as a convenient tool for the initial assessment of strength requirements for various designs. At present, computer programs are routinely used in ship structural design. Empirical formulas for the individual calculation of both still water and wave-induced bending moments have been derived by the classification societies. They are based on ship length, breadth, block coefficient, and effective wave height (for the wave-induced bending moment). Similarly, permissible bending stresses are calculated from formulas as a function of ship length and its service environment. Finite element techniques in the form of commercially available computer software packages combined with statistical reliability methods are being applied with increasing frequency.

For ships, the maximum longitudinal bending stresses occur in the vicinity of the midship section at the deck edge and in the bottom plating. Maximum shear stresses occur in the shell plating in the vicinity of the quarter points at the neutral axis. For long, slender girders, such as ships, the maximum shear stress is small compared with the maximum bending stress.

The structure of a ship consists of a grillage of stiffened plating supported by longitudinals, longitudinal girders, transverse beams, transverse frames, and web frames. Since the primary stress system in the hull arises from longitudinal bending, it follows that the longitudinally continuous structural elements are the most effective in carrying and distributing this stress system. The strength deck, particularly at the side, and the keel and turn of bilge are highly stressed regions. The shell plating, particularly deck and bottom plating, carry the major part of the stress. Other **key longitudinal elements**, as shown in Fig. 11.3.5, are longitudinal deck girders, main deck stringer plate, gunwale angle, sheer strake, bilge strake, inner bottom margin plate, garboard strake, flat plate keel, center vertical keel, and rider plate.

Transverse elements include deck beams and transverse frames and web frames which serve to support and transmit vertical and transverse loads and resist hydrostatic pressure.

Good structural design minimizes the structural weight while providing adequate strength, minimizes interference with ship function, provides for effective continuity of the structure, facilitates stress flow around deck openings and other stress obstacles, and avoids square corner discontinuities in the plating and other stress concentration "hot spots."

A longitudinally framed ship is one which has closely spaced longitudinal structural elements and widely spaced transverse elements. A transversely framed ship has closely spaced transverse elements and widely spaced longitudinal elements. Longitudinal framing systems are generally more efficient structurally, but because of the deep web frames supporting the longitudinals, it is less efficient in the use of internal space than the transverse framing system. Where interruptions of open internal spaces are unimportant, as in tankers and bulk carriers, longitudinal framing is universally used. However, modern practice tends increasingly toward longitudinal framing in other types of ships also.

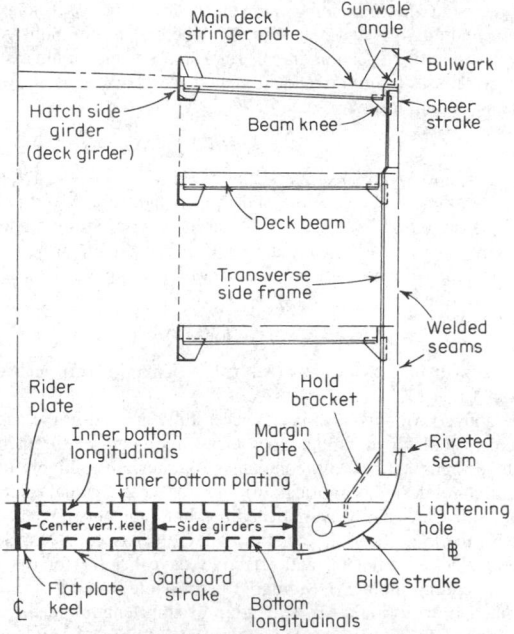

Fig. 11.3.5 Key structural elements.

Stability

A ship afloat is in vertical equilibrium when the force of **gravity**, acting at the ship's **center of gravity** G, is equal, opposite, and in the same vertical line as the force of **buoyancy**, acting upward at the **center of buoyancy** B. Figure 11.3.6 shows that an upsetting, transverse couple

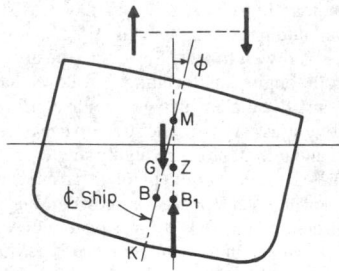

Fig. 11.3.6 Ship stability.

acting on the ship causes it to rotate about a longitudinal axis, taking a list ϕ. G does not change; however, B moves to B_1, the centroid of the new underwater volume. The resulting couple, created by the transverse separation of the two forces ($\overline{GZ}$), opposes the upsetting couple, thereby righting the ship. A static stability curve (Fig. 11.3.7), consisting of values for the righting arm $\overline{GZ}$ plotted against angles of inclination (ϕ), gives a graphic representation of the static stability of the ship. The primary indicator for the safety of a ship is the maximum righting arm it develops and the angle heel at which this righting arm occurs.

For small angles of inclination ($\phi < 10°$), centers of buoyancy follow a locus whose instantaneous center of curvature M is known as the transverse **metacenter**. When M lies above G ($\overline{GM} > 0$), the resulting gravity-buoyancy couple will right the ship; the ship has positive stability. When G and M are coincident, the ship has neutral stability. When G

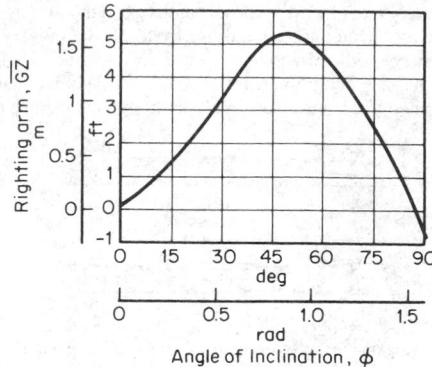

Fig. 11.3.7 Static stability curve.

is above M ($\overline{GM} < 0$), negative stability results. Hence $\overline{GM}$, known as the **transverse metacentric height,** is an indication of **initial stability** of a ship.

The transverse **metacentric radius** $\overline{BM}$ and the vertical location of the center of buoyancy are determined by the design of the ship and can be calculated. Once the vertical location of the ship's center of gravity is known, then $\overline{GM}$ can be found. The vertical center of gravity of practically all ships varies with the condition of loading and must be determined either by a careful calculation or by an inclining experiment.

Minimum values of $\overline{GM}$ ranging from 1.5 to 3.5 ft (0.46 to 1.07 m), corresponding to small and large seagoing ships, respectively, have been accepted in the past. The $\overline{GM}$ of passenger ships should be under 6 percent of the beam to ensure a reasonable (comfortable) rolling period. This relatively low value is offset by the generally large range of stability due to the high freeboard of passenger ships.

The question of **longitudinal stability** affects the trim of a ship. As in the transverse case, a longitudinal metacenter exists, and a longitudinal metacentric height $\overline{GM}_L$ can be determined. The moment to alter trim 1 in, $\overline{MT1}$, is computed by $\overline{MT1} = \Delta \overline{GM}_L/12L$, ft·tons/in and moment to alter trim 1 m is $\overline{MT1} = 10^6 \, \Delta \overline{GM}_L/L$, N·m/m. Displacement Δ is in tons and MN, respectively.

The **location of a ship's center of gravity** changes as small weights are shifted within the system. The vertical, transverse, or longitudinal component of movement of the center of gravity is computed by $\overline{GG}_1 = w \times d/\Delta$, where w is the small weight, d is the distance the weight is shifted in a component direction, and Δ is the displacement of the ship, which includes w.

Vertical changes in the location of G caused by **weight addition** or **removal** can be calculated by

$$KG_1 = \frac{\Delta \overline{KG} \pm w \overline{K}g}{\Delta_1}$$

where $\Delta_1 = \Delta \pm w$ and $\overline{K}g$ is the height of the center of gravity of w above the keel.

The free surface of the liquid in fuel oil, lubricating oil, and water storage and service tanks is deleterious to ship stability. The weight shift of the liquid as the ship heels can be represented as a virtual rise in G, hence a reduction in $\overline{GM}$ and the ship's initial stability. This virtual rise, called the **free surface effect**, is calculated by the expression

$$GG_v = \gamma_i i/\gamma_w \nabla$$

where γ_i, γ_w = specific gravities of liquid in tank and sea, respectively; i = moment of inertia of free surface area about its longitudinal centroidal axis; and ∇ = volume of displacement of ship. The ship designer can minimize this effect by designing long, narrow, deep tanks or by using baffles.

Seakeeping

The seakeeping qualities of a ship determine its ability to function normally in a seaway and are based on its motions in waves. These motions

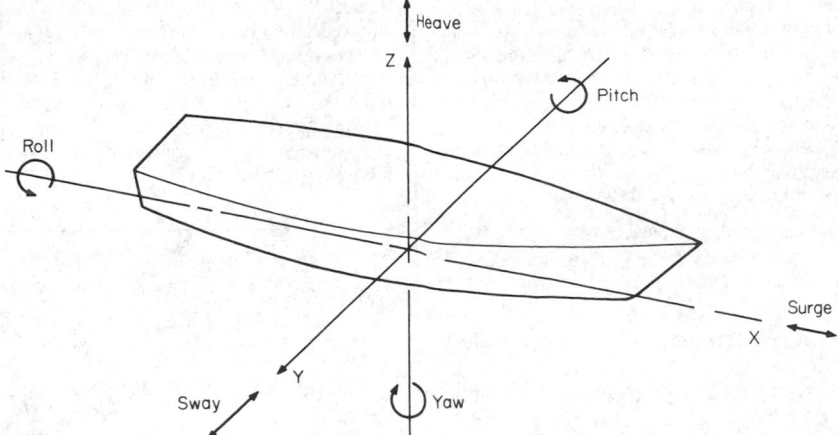

Fig. 11.3.8   Conventional ship coordinate axes and ship motions.

in turn directly affect the practices of marine engineers. The motion of a floating object has six degrees of freedom. Figure 11.3.8 shows the conventional ship coordinate axes and ship motions. Oscillatory movement along the x axis is called **surge**; along the y axis, **sway**; and along the z axis, **heave**. Rotation about x is called **roll**; about y, **pitch**; and about z, **yaw**.

Rolling has a major effect on crew comfort and on the structural and bearing requirements for machinery and its foundations. The natural period of roll of a ship is $T = 2\pi K/\sqrt{g\overline{GM}}$, where K is the radius of gyration of virtual ship mass about a longitudinal axis through G. K varies from 0.4 to 0.5 of the beam, depending on the ship depth and transverse distribution of weights.

Angular acceleration of roll, if large, has a very distressing effect on crew, passengers, machinery, and structure. This can be minimized by increasing the period or by decreasing the roll amplitude. Maximum angular acceleration of roll is

$$d^2\phi/dt^2 = -4\pi^2(\phi_{max}/T^2)$$

where ϕ_{max} is the maximum roll amplitude. The period of the roll can be increased effectively by decreasing $\overline{GM}$; hence the lowest value of $\overline{GM}$ compatible with all stability criteria should be sought.

Pitching is in many respects analogous to rolling. The natural period of pitching, bow up to bow down, can be found by using the same expression used for the rolling period, with the longitudinal radius of gyration K_L substituted for K. A good approximation is $K_L = L/4$, where L is the length of the ship. The natural period of pitching is usually between one-third and one-half the natural period of roll. A by-product of pitching is **slamming,** the reentry of bow sections into the sea with heavy force.

Heaving and **yawing** are the other two principal rigid-body motions caused by the sea. The amplitude of heave associated with head seas, which generate pitching, may be as much as 15 ft (4.57 m); that arising from beam seas, which induce rolling, can be even greater. Yawing is started by unequal forces acting on a ship as it quarters into a sea. Once a ship begins to yaw, it behaves as it would at the start of a rudder-actuated turn. As the ship travels in a direction oblique to its plane of symmetry, forces are generated which force it into heel angles independent of sea-induced rolling.

Analysis of ship movements in a seaway is performed by using probabilistic and statistical techniques. The seaway is defined by a mathematically modeled wave energy density spectrum based on data gathered by oceanographers. This wave spectrum is statistically calculated for various sea routes and weather conditions. A ship response amplitude operator or transfer function is determined by linearly superimposing a ship's responses to varying regular waves, both experimentally and theoretically. The wave spectrum multiplied by the ship response

transfer function yields the ship response spectrum. The ship response spectrum is an energy density spectrum from which the statistical character of ship motions can be predicted. (See Edward V. Lewis, Motions in Waves, "Principles of Naval Architecture," SNAME, 1989.)

Resistance and Powering

Resistance to ship motion through the water is the aggregate of (1) wave-making, (2) frictional, (3) pressure or form, and (4) air resistances.

Wave-making resistance is primarily a function of Froude number, $Fr = v/\sqrt{gL}$, where v = ship speed, ft/s (m/s); g = acceleration of gravity, ft/s^2 (m/s^2); and L = ship length, ft (m). In many instances, the dimensional speed-length ratio $V/\sqrt{L}$ is used for convenience, where V is given in knots. A ship makes at least two distinct wave patterns, one from the bow and the other at the stern. There also may be other patterns caused by abrupt changes in section. These patterns combine to form the total wave system for the ship. At various speeds there is mutual cancellation and reinforcement of these patterns. Thus, a plot of total resistance of the ship versus Fr or $V/\sqrt{L}$ is not smooth but shows humps and hollows corresponding to the wave cancellation or reinforcement. Normal procedure is to design the operating speed of a ship to fall at one of the low points in the resistance curve.

Frictional resistance is a function of Reynolds number (see Sec. 3). Because of the size of a ship, the Reynolds number is large and the flow is always turbulent.

Pressure or **form resistance** is a viscosity effect but is different from frictional resistance. The principal observed effects are boundary-layer separation and eddying near the stern.

It is usual practice to combine the wave-making and pressure resistances into one term, called the **residuary resistance,** assumed to be a function of Froude number. The combination, although not strictly legitimate, is practical because the pressure resistance is usually only 2 to 3 percent of the total resistance. The frictional resistance is then the only term which is considered to be a function of Reynolds number and can be calculated. Based on an analysis of the water resistance of flat, smooth planes, Schoenherr gives the formula

$$R_f = 0.5\rho S v^2 C_f$$

where R_f = frictional resistance, lb (N); ρ = mass density, lb/s$^2 \cdot$ ft^4 (kg/m^3); S = wetted surface area, ft^2 (m^2); v = velocity, ft/s (m/s); and C_f is the frictional coefficient computed from the ITTC formula

$$C_f = \frac{0.075}{(\log_{10} Re - 2)^2}$$

and Re = Reynolds number = vL/ν.

Through towing tests of ship models at a series of speeds for which Froude numbers are equal between the model and the ship, **total model resistance** (R_{tm}) is determined. Residuary resistance (R_{rm}) for the model is obtained by subtracting the frictional resistance (R_{fm}). By **Froude's law of comparison**, the residuary resistance of the ship (R_{rs}) is equal to R_{rm} multiplied by the ratio of ship displacement to model displacement. Total ship resistance (R_{ts}) then is equal to the sum of R_{rs}, the calculated ship frictional resistance (R_{fs}), and a correlation allowance that allows for the roughness of ship's hull opposed to the smooth hull of a model.

$$R_{tm} = \text{measured resistance}$$
$$R_{rm} = R_{tm} - R_{fm}$$
$$R_{rs} = R_{rm}(\Delta s/\Delta m)$$
$$R_{ts} = R_{rs} + R_{fs} + R_a$$

A nominal value of 0.0004 is generally used as the correlation allowance coefficient C_a. $R_a = 0.5\rho S v^2 C_a$.

From R_{ts}, the **total effective power** P_E required to propel the ship can be determined:

$$P_E = \frac{R_{ts}v}{550} \quad \text{ehp} \quad \left(P_E = \frac{R_{ts}v}{1,000} \quad \text{kW}_E\right)$$

where R_{ts} = total ship resistance, lb (N); v = velocity, ft/s (m/s); and $P_E = P_S \times P.C.$, where P_S is the shaft power (see Propulsion Systems) and $P.C.$ is the propulsive coefficient, a factor which takes into consideration mechanical losses, propeller efficiency, and the flow interaction with the hull. $P.C.$ = 0.45 to 0.53 for high-speed craft; 0.50 to 0.60, for tugs and trawlers; 0.55 to 0.65, for destroyers; and 0.63 to 0.72, for merchant ships.

Figure 11.3.9 illustrates the specific effective power for various displacement hull forms and planing craft over their appropriate speed regimes. Figure 11.3.10 shows the general trend of specific resistance versus Froude number and may be used for coarse powering estimates.

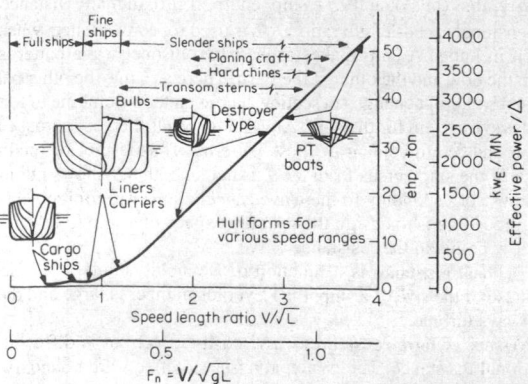

Fig. 11.3.9 Specific effective power for various speed-length regimes. (*From "Handbook of Ocean and Underwater Engineering," McGraw-Hill, 1969.*)

In early design stages, the hull form is not yet defined, and model testing is not feasible. Reasonably accurate power calculations are made by using preplotted series model test data of similar hull form, such as from Taylor's Standard Series and the Series 60. The Standard Series data were originally compiled by Admiral David W. Taylor and were based on model tests of a series of uniformly varied twin screw, cruiser hull forms of similar geometry. Revised Taylor's Series data are available in "A Reanalysis of the Original Test Data for the Taylor Standard Series," Taylor Model Basin, Rept. no. 806.

Air Resistance The air resistance of ships in a windless sea is only a few percent of the water resistance. However, the effect of wind can have a significant impact on drag. The wind resistance parallel to the ship's axis is roughly 30 percent greater when the wind direction is about 30° ($\pi/6$ rad) off the bow vice dead ahead, since the projected above-water area is greater. Wind resistance is approximated by $R_A =$

$0.002B^2V_R^2$ lb ($0.36B^2V_R^2$ N), where B = ship's beam, ft (m) and V_R = ship speed relative to the wind, in knots (m/s).

Powering of Small Craft The American Boat and Yacht Council, Inc. (ABYC) provides the following guide for determining the maximum safe brake power for pleasure craft.

Compute a length-width factor by multiplying the overall boat length in feet by the overall stern width in feet (widest part of stern excluding fins and sheer). For SI units, use length and width in metres times 10.76.

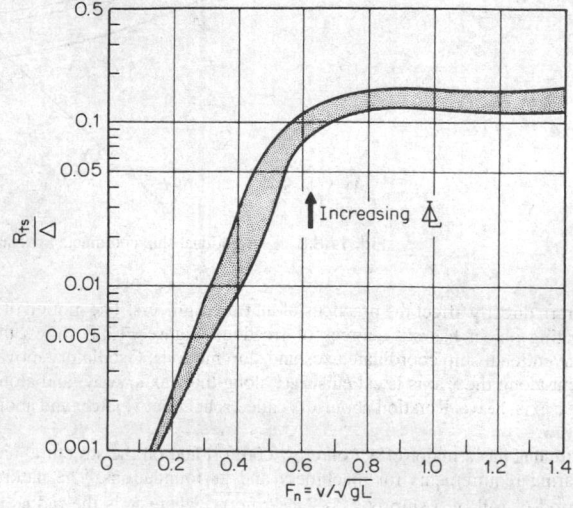

Fig. 11.3.10 General trends of specific resistance versus Froude number.

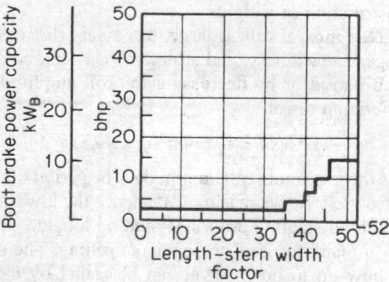

Fig. 11.3.11 Boat brake power capacity for length-width factor under 52. (*From "Safety Standards for Small Craft," ABYC, 1974.*)

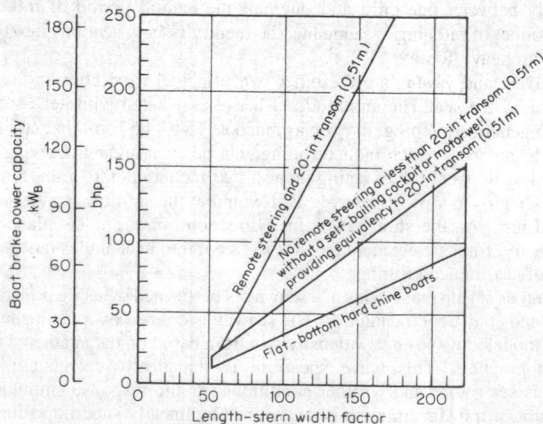

Fig. 11.3.12 Boat brake power capacity for length-width factor over 52. (*From "Safety Standards for Small Craft," ABYC, 1974.*)

Locate the factor in Fig. 11.3.11 or Fig. 11.3.12 and read the boat brake power capacity.

For powering canoes, the maximum should be 3.0 bhp (2.2 kW$_B$) for lengths under 15 ft (4.6 m); 5 bhp (3.7 kW$_B$) for 15 to 18 ft (4.6 to 5.5 m); and 7.5 bhp (5.6 kW$_B$) for over 18 ft (5.5 m).

ENGINEERING CONSTRAINTS

The constraints affecting marine engineering design are too numerous, and some too obvious, to include in this section. Three significant categories, however, are discussed. The geometry of the hull forms immediately suggests **physical constraints**. The interaction of the vehicle with the marine environment suggests **dynamic constraints,** particularly vibration, noise, and shock. The broad topic of **environmental protection** is one of the foremost engineering constraints of today, having a very pronounced effect on the operating systems of a marine vehicle.

Physical Constraints

Formerly, tonnage laws in effect made it economically desirable for the propulsion machinery spaces of a merchant ship to exceed 13 percent of the gross tonnage of the ship so that 32 percent of the gross tonnage could be deducted in computing net tonnage. In most design configurations, however, a great effort is made to minimize the space required for the propulsion plant in order to maximize that available to the mission or the money-making aspects of the ship.

Specifically, **space** is of extreme importance as each component of support equipment is selected. Each component must fit into the master compact arrangement scheme to provide the most efficient operation and maintenance by engineering personnel.

Weight constraints for a main propulsion plant vary with the application. In a tanker where cargo capacity is limited by draft restrictions, the weight of machinery represents lost cargo. Cargo ships, on the other hand, rarely operate at full load draft. Additionally, the low weight of propulsion machinery somewhat improves inherent cargo ship stability deficiencies. The weight of each component of equipment is constrained by structural support and shock resistance considerations. Naval shipboard equipment, in general, is carefully analyzed to effect weight reduction.

Dynamic Constraints

Dynamic effects, principally mechanical vibration, noise, shock, and ship motions, are considered in determining the dynamic characteristics of a ship and the dynamic requirements for equipment.

Vibration (See Sec. 9.) Vibration analyses are especially important in the design of the propulsion shafting system and its relation to the excitation forces resulting from the propeller. The main propulsion **shafting** can vibrate in longitudinal, torsional, and lateral modes. Modes of **hull vibration** may be vertical, horizontal, torsional, or longitudinal; may occur separately or coupled; may be excited by synchronization with periodic harmonics of the propeller forces acting either through the shafting, by the propeller force field interacting with the hull afterbody, or both; and may also be set up by unbalanced harmonic forces from the main machinery, or, in some cases, by impact excitation from slamming or periodic-wave encounter.

It is most important to reduce the excitation forces at the source. Very objectionable and serious vibrations may occur when the frequency of the exciting force coincides with one of the hull or shafting-system natural frequencies.

Vibratory forces generated by the propeller are (1) alternating pressure forces on the hull due to the alternating hydrodynamic pressure fields of the propeller blades; (2) alternating propeller-shaft bearing forces, primarily caused by wake irregularities; and (3) alternating forces transmitted throughout the shafting system, also primarily caused by wake irregularities.

The most effective means to ensure a satisfactory level of vibration is to provide adequate clearance between the propeller and the hull surface and to select the propeller revolutions or number of blades to avoid synchronism. Replacement of a four-blade propeller, for instance, by a three- or five-blade, or vice versa, may bring about a reduction in vibration. Singing propellers are due to the vibration of the propeller blade edge about the blade body, producing a disagreeable noise or hum. Chamfering the edge is sometimes helpful.

Vibration due to variations in **engine torque reaction** is hard to overcome. In ships with large diesel engines, the torque reaction tends to produce an athwartship motion of the upper ends of the engines. The motion is increased if the engine foundation strength is inadequate.

Foundations must be designed to take both the thrust and torque of the shaft and be sufficiently rigid to maintain alignment when the ship's hull is working in heavy seas. Engines designed for foundations with three or four points of support are relatively insensitive to minor working of foundations. Considerations should be given to using flexible couplings in cases when alignment cannot be assured.

Torsional vibration frequency of the prime mover–shafting-propeller system should be carefully computed, and necessary steps taken to ensure that its natural frequency is clear of the frequency of the main-unit or propeller heavy-torque variations; serious failures have occurred. Problems due to resonance of engine torsional vibration (unbalanced forces) and foundations or hull structure are usually found after ship trials, and correction consists of local stiffening of the hull structure. If possible, engines should be located at the nodes of hull vibrations.

Although more rare, a **longitudinal vibration** of the propulsion shafting has occurred when the natural frequency of the shafting agreed with that of a pulsating axial force, such as propeller thrust variation.

Other forces inducing vibrations may be vertical inertia forces due to the acceleration of reciprocating parts of an unbalanced reciprocating engine or pump; longitudinal inertia forces due to reciprocating parts or an unbalanced crankshaft creating unbalanced rocking moments; and horizontal and vertical components of centrifugal forces developed by unbalanced rotating parts. Rotating parts can be balanced.

Noise The noise characteristics of shipboard systems are increasingly important, particularly in naval combatant submarines where remaining undetected is essential, and also from a human-factors point of view on all ships. Achieving significant reduction in machinery noise level can be costly. Therefore desired noise levels should be analyzed. Each operating system and each piece of rotating or reciprocating machinery installed aboard a submarine are subjected to intensive airborne (noise) and structureborne (mechanical) vibration analyses and tests. Depending on the noise attenuation required, similar tests and analyses are also conducted for all surface ships, military and merchant.

Shock In naval combatant ships, shock loading due to noncontact underwater explosions is a major design parameter. Methods of qualifying equipment as ''shock-resistant'' might include ''static'' shock analysis, ''dynamic'' shock analysis, physical shock tests, or a combination.

Motions Marine lubricating systems are specifically distinguished by the necessity of including list, trim, roll, and pitch as design criteria. The American Bureau of Shipping requires satisfactory functioning of lubricating systems when the vessel is permanently inclined to an angle of 15° (0.26 rad) athwartship and 5° (0.09 rad) fore and aft. In addition, electric-generator bearings must not spill oil under a momentary roll of 22½° (0.39 rad). Military specifications cite the same permanent trim and list as for surface ships but add 45° (0.79 rad) roll and 10° (0.17 rad) pitch requirements. For submarines, a requirement of 30° (0.52 rad) trim, 15° (0.26 rad) list, 60° (1.05 rad) roll, and 10° (0.17 rad) pitch is imposed.

Environmental Constraints

The **Refuse Act of 1899** (33 U.S.C. 407) prohibits discharge of any refuse material from marine vehicles into navigable waters of the United States or into their tributaries or onto their banks if the refuse is likely to wash into navigable waters. The term **refuse** includes nearly any substance, but the law contains provisions for sewage. The Environmental Protection Agency (EPA) may grant permits for the discharge of refuse into navigable waters.

The **Oil Pollution Act of 1961** (33 U.S.C.1001-1015) prohibits the discharge of oil or oily mixtures (over 100 mg/L) from vehicles generally within 50 mi of land; some exclusions, however, are granted. (The Oil

Pollution Act of 1924 was repealed in 1970 because of supersession by subsequent legislation.)

The **Port and Waterways Safety Act of 1972** (PL92-340) grants the U.S. Coast Guard authority to establish and operate mandatory vehicle traffic control systems. The control system must consist of a VHF radio for ship-to-shore communications, as a minimum. The Act, in effect, also extends the provisions of the Tank Vessel Act, which protects against hazards to life and property, to include **protection of the marine environment.** Regulations stemming from this Act govern standards of tanker design, construction, alteration, repair, maintenance, and operation.

The most substantive marine environmental protection legislation is the **Federal Water Pollution Control Act (FWPCA)**, 1948, as amended (33 U.S.C. 466 et seq.). The 1972 and 1978 Amendments contain provisions which greatly expand federal authority to deal with **marine pollution,** providing authority for control of pollution by oil and hazardous substances other than oil, and for the assessment of penalties. **Navigable waters** are now defined as ''. . . the waters of the United States, including territorial seas.''

''Hazardous substances other than oil'' and harmful quantities of these substances are defined by the EPA in regulations. The Coast Guard and EPA must ensure removal of such discharges, costs borne by the responsible vehicle owner or operator if liable. **Penalties** now may be assessed by the Coast Guard for any discharge. The person in charge of a vehicle is to notify the appropriate U.S. government agency of any discharge upon knowledge of it.

The Coast Guard regulations establish **removal procedures** in coastal areas, contain regulations to **prevent discharges** from vehicles and transfer facilities, and regulations governing the **inspection of cargo vessels** to prevent hazardous discharges. The EPA regulations govern inland areas and non-transportation-related facilities.

Section 312 of the FWPCA as amended in 1972 deals directly with **discharge of sewage.** The EPA must issue standards for marine sanitation devices, and the Coast Guard must issue regulations for implementing those standards. In June, 1972, EPA published standards that now prohibit discharge of any sewage from vessels, whether it is treated or not. Federal law now prohibits, on all inland waters, operation with marine sanitation devices which have not been securely sealed or otherwise made visually inoperative so as to prevent overboard discharge.

The **Marine Plastic Pollution Control Act of 1987** (PL100-220) prohibits the disposal of plastics at sea within U.S. waters.

The **Oil Pollution Act of 1990** (PL101-380) provides extensive legislation on liability and includes measures that impact the future use of single-hull tankers.

PROPULSION SYSTEMS

(See Sec. 9 for component details.)

The basic operating requirement for the main propulsion system is to propel the vehicle at the required sustained speed for the range or endurance required and to provide suitable maneuvering capabilities. In meeting this basic requirement, the marine propulsion system integrates the power generator/prime mover, the transmission system, the propulsor, and other shipboard systems with the ship's hull or vehicle platform. Figure 11.3.13 shows propulsion system alternatives with the most popular drives for fixed-pitch and controllable-pitch propellers.

Definitions for Propulsion Systems

Brake power P_B, bhp (kW$_B$), is the power delivered by the output coupling of a prime mover before passing through speed-reducing and transmission devices and with all continuously operating engine-driven auxiliaries in use.

Shaft power P_S, shp (kW$_S$), is the net power supplied to the propeller shafting after passing through all reduction gears or other transmission devices and after power for all attached auxiliaries has been taken off. Shaft power is measured in the shafting within the ship by a torsionmeter as close to the propeller or stern tube as possible.

Delivered power P_D, dhp (kW$_D$), is the power actually delivered to the propeller, somewhat less than P_S because of the power losses in the

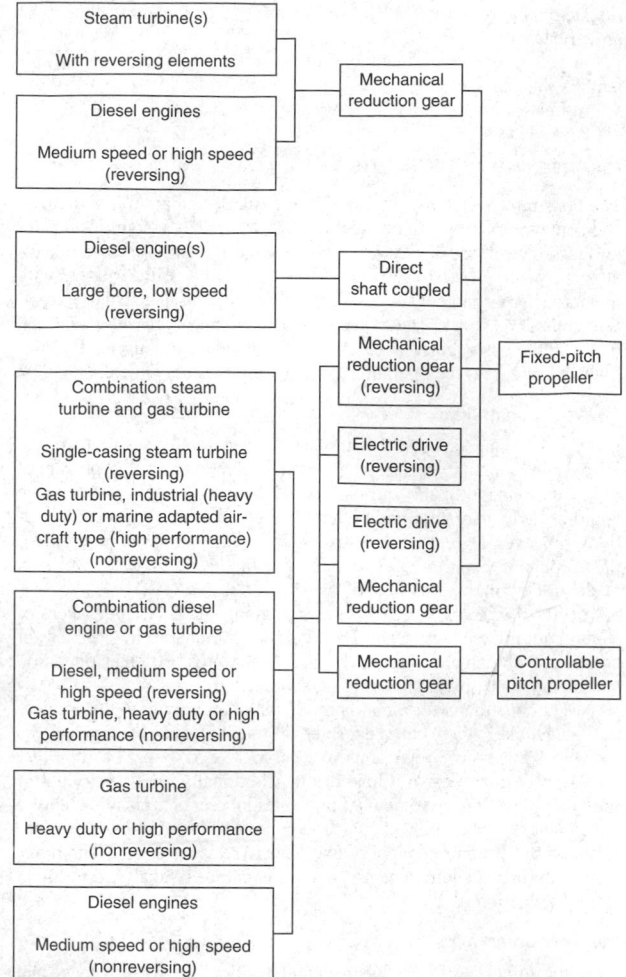

Fig. 11.3.13 Alternatives in the selection of a main propulsion plant. *(From ''Marine Engineering,'' SNAME, 1992.)*

stern tube and other bearings between the point of measurement and the propeller. Sometimes called **propeller power,** it is also equal to the effective power P_E, ehp (kW$_E$), plus the power losses in the propeller and the losses in the interaction between the propeller and the ship.

Normal shaft power or **normal power** is the power at which a marine vehicle is designed to run most of its service life.

Maximum shaft power is the highest power for which propulsion machinery is designed to operate continuously.

Service speed is the actual speed maintained by a vehicle at load draft on its normal route and under average weather and sea conditions typical of that route and with average fouling of bottom.

Designed service speed is the speed expected on trials in fair weather at load draft, with clean bottom, when machinery is developing a specified fraction of maximum shaft power. This fraction may vary but is of order of 0.8.

MAIN PROPULSION PLANTS

Steam Plants

The basic steam propulsion plant contains main boilers, steam turbines, a condensate system, a feedwater system, and numerous auxiliary components necessary for the plant to function. A heat balance calculation, the basic tool for determining the effect of various configurations on plant thermal efficiency, is demonstrated for the basic steam cycle. Both fossil-fuel and nuclear energy sources are successfully employed for marine applications.

Main Boilers (See Sec. 9.) The pressures and temperatures achieved in steam-generating equipment have increased steadily over recent years, permitting either a higher-power installation for a given space or a reduction in the size and weight of a given propulsion plant.

The **trend in steam pressures** and **temperatures** has been an increase from 600 psig (4.14 MN/m^2) and 850°F (454°C) during World War II to 1,200 psig (8.27 MN/m^2) and 950°F (510°C) for naval combatants in and since the postwar era. For merchant ships, the progression has been from 400 psig (2.76 MN/m^2) and 750°F (399°C) gradually up to 850 psig (5.86 MN/m^2) and 850°F (454°C) in the 1960s, with some boilers at 1,500 psig (10.34 MN/m^2) and 1000°F (538°C) in the 1970s.

The **quantity of steam** produced by a marine boiler ranges from approximately 1,500 lb/h (680 kg/h) in small auxiliary boilers to over 400,000 lb/h (181,500 kg/h) in large main propulsion boilers. Outputs of 750,000 lb/h (340,200 kg/h) or more per boiler are practical for high-power installations.

Most marine boilers are **oil-fired.** Compared with other fuels, oil is easily loaded aboard ship, stored, and introduced into the furnace, and does not require the ash-handling facilities required for coal firing.

Gas-fired boilers are used primarily on power or drill barges which are fixed in location and can be supplied from shore (normally classed in the Ocean Engineering category). At sea, tankers designed to carry liquefied natural gas (LNG) may use the natural boil-off from their cargo gas tanks as a supplemental fuel (**dual fuel system**). The cargo gas boil-off is collected and pumped to the boilers where it is burned in conjunction with oil. The quantity of boil-off available is a function of the ambient sea and air temperatures, the ship's motion, and the cargo loading; thus, it may vary from day to day.

For economy of space, weight, and cost and for ease of operation, the trend in **boiler installations** is for fewer boilers of high capacity rather than a large number of boilers of lower capacity. The minimum installation is usually two boilers, to ensure propulsion if one boiler is lost; one boiler per shaft for twin-screw ships. Some large ocean-going ships operate on single boilers, requiring exceptional reliability in boiler design and operation.

Combustion systems include forced-draft fans or blowers, the fuel oil service system, burners, and combustion controls. Operation and maintenance of the combustion system are extremely important to the efficiency and reliability of the plant. The best combustion with the least possible excess air should be attained.

Main Turbines (See Sec. 9.) Single-expansion (i.e., single casing) marine steam turbines are fairly common at lower powers, usually not exceeding 4,000 to 6,000 shp (2,983 to 4,474 kW$_S$). Above that power range, the turbines are usually double-expansion (cross-compound) with high- and low-pressure turbines each driving the main reduction gear through its own pinion. The low-pressure turbine normally contains the reversing turbine in its exhaust end. The main condenser is either underslung and supported from the turbine, or the condenser is carried on the foundations, with the low-pressure turbine supported on the condenser.

The inherent **advantages** of the steam turbine have favored its use over the reciprocating steam engine for all large, modern marine steam propulsion plants. Turbines are not size-limited, and their high temperatures and pressures are accommodated safely. Rotary motion is simpler than reciprocating motion; hence unbalanced forces can be eliminated in the turbine. The turbine can efficiently utilize a low exhaust pressure; it is lightweight and requires minimum space and low maintenance.

The **reheat cycle** is the best and most economical means available to improve turbine efficiencies and fuel rates in marine steam propulsion plants. In the reheat cycle, steam is withdrawn from the turbine after partial expansion and is passed through a heat exchanger where its temperature is raised; it is then readmitted to the turbine for expansion to condenser pressure. Marine reheat plants have more modest steam conditions than land-based applications because of lower power ratings and a greater reliability requirement for ship safety.

The reheat cycle is applied mostly in high-powered units above 25,000 shp (18,642 kW$_S$) and offers the maximum economical thermal efficiency that can be provided by a steam plant. Reheat cycles are not used in naval vessels because the improvement in efficiency does not warrant the additional complexity; hence a trade-off is made.

Turbine foundations must have adequate rigidity to avoid vibration conditions. This is particularly important with respect to periodic variations in propeller thrust which may excite longitudinal vibrations in the propulsion system.

Condensate System (See Sec. 9.) The condensate system provides the means by which feedwater for the boilers is recovered and returned to the feedwater system. The major components of the condensate system of a marine propulsion plant are the **main condenser,** the **condensate pumps,** and the **deaerating feed tank** or **heater.** Both single-pass and two-pass condenser designs are used; the single-pass design, however, allows somewhat simpler construction and lower water velocities. The single-pass condenser is also adaptable to scoop circulation, as opposed to pump circulation, which is practical for higher-speed ships. The deaerating feedwater heater is supplied from the condensate pumps, which take suction from the condenser hot well, together with condensate drains from steam piping and various heaters.

The deaerating feed heater is normally maintained at about 35 psig (0.241 MN/m^2) and 280°F (138°C) by auxiliary exhaust and turbine extraction steam. The condensate is sprayed into the steam atmosphere at the top of the heater, and the heated feedwater is pumped from the bottom by the feed booster or main feed pumps. In addition to removing oxygen or air, the heater also acts as a surge tank to meet varying demands during maneuvering conditions. Since the feed pumps take suction where the water is almost saturated, the heater must be located 30 to 50 ft (9.14 to 15.24 m) above the pumps in order to provide enough positive suction head to prevent flashing from pressure fluctuations during sudden plant load change.

Feedwater System The feedwater system comprises the pumps, piping, and controls necessary to transport feedwater to the boiler or steam generator, to raise water pressure above boiler pressure, and to control flow of feedwater to the boiler. **Main feed pumps** are so vital that they are usually installed in duplicate, providing a standby pump capable of feeding the boilers at full load. Auxiliary steam-turbine-driven centrifugal pumps are usually selected. A typical naval installation consists of three main feed booster pumps and three main feed pumps for each propulsion plant. Two of the booster pumps are turbine-driven and one is electric. Additionally, an electric-motor-driven emergency booster pump is usually provided. The total capacity of the main feed pumps must be 150 percent of the boiler requirement at full power plus the required recirculation capacity. Reliable feedwater regulators are important.

Steam Plant—Nuclear (See Sec. 9.) The compact nature of the energy source is the most significant characteristic of nuclear power for marine application. The fission of one gram of uranium per day produces about one megawatt of power. (One pound produces 608,579 horsepower.) In other terms, the fission of 1 lb of uranium is equivalent to the combustion of about 86 tons (87,380 kg) of 18,500 Btu/lb (43.03 MJ/kg) fuel oil. This characteristic permits utilization of large power plants on board ship without the necessity for frequent refueling or large bunker storage. Economic studies, however, show that nuclear power, as presently developed, is best suited for military purposes, where the advantages of high power and endurance override the pure economic considerations. As the physical size of nuclear propulsion plants is reduced, their economic attractiveness for commercial marine application will increase.

Major differences between the nuclear power and fossil-fuel plants are: (1) The **safety aspect** of the nuclear reactor system—operating personnel must be shielded from fission product radiation, hence the size and weight of the reactor are increased and maintenance and reliability become more complicated; (2) the steam produced by a pressurized-water reactor plant is saturated and, because of the **high moisture content** in the turbine steam path, the turbine design requires more careful attention; (3) the steam pressure provided by a pressurized-water reactor plant varies with output, the maximum pressure occurring at no load and decreasing approximately linearly with load to a minimum at full power; hence, **blade stresses** in a nuclear turbine increase more rapidly with a decrease in power than in a conventional turbine. Special attention must be given to the design of the control stage of a nuclear turbine. The *N.S. Savannah* was designed for 700 psig (4.83 MN/m²) at no load and 400 psig (2.76 MN/m²) at full power.

Steam Plant—Heat Balance The heat-balance calculation is the basic analysis tool for determining the effect of various steam cycles on the thermal efficiency of the plant and for determining the quantities of steam and feedwater flow.

The thermal arrangement of a **simple steam cycle,** illustrated in Fig. 11.3.14, and the following simplified analysis are taken from "Marine Engineering," SNAME, 1992:

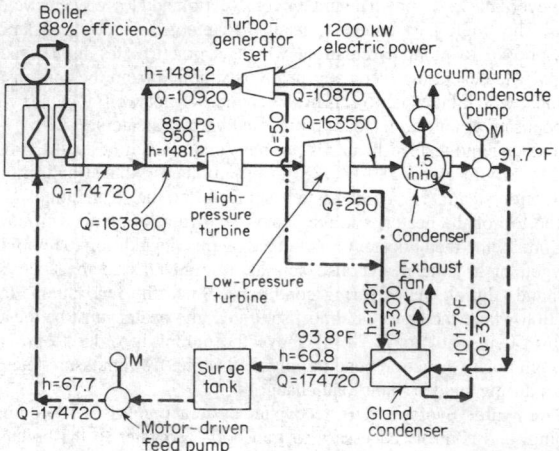

Fig. 11.3.14 Simple steam cycle. *(From "Marine Engineering," SNAME, 1992.)*

The unit is assumed to develop 30,000 shp (22,371 kW$_S$). The steam rate of the main propulsion turbines is 5.46 lb/(shp·h) [3.32 kg/(kW$_S$·h)] with throttle conditions of 850 psig (5.86 MN/m²) and 950°F (510°C) and with the turbine exhausting to the condenser at 1.5 inHg abs (5,065 N/m²). To develop 30,000 shp (22,371 kW$_S$), the throttle flow must be 163,800 lb/h (74,298 kg/h).

The generator load is estimated to be 1,200 kW, and the turbogenerator is thus rated at 1,500 kW with a steam flow of 10,920 lb/h (4,953 kg/h) for steam conditions of 850 psig (5.86 MN/m²) and 950°F (510°C) and a 1.5 inHg abs (5,065 N/m²) back pressure. The total steam flow is therefore 174,720 lb/h (79,251 kg/h).

Tracing the steam and water flow through the cycle, the flow exhausting from the main turbine is 163,800 − 250 = 163,550 lb/h (74,298 − 113 = 74,185 kg/h), 250 lb/h (113 kg/h) to the gland condenser, and from the auxiliary turbine is 10,870 lb/h (4,930 kg/h), 50 lb/h (23 kg/h) to the gland condenser. The two gland leak-off flows return from the gland leak-off condenser to the main condenser. The condensate flow leaving the main condenser totals 174,720 lb/h (79,251 kg/h).

It is customary to allow a 1°F (0.556°C) hot-well depression in the condensate temperature. Thus, at 1.5 inHg abs (5,065 N/m²) the condenser saturation temperature is 91.7°F (33.2°C) and the condensate is 90.7°F (32.6°C). Entering the gland condenser there is a total energy flow of 174,720 × 59.7 = 10,430,784 Btu/h (11,004,894,349 J/h). The gland condenser receives gland steam at 1,282 Btu/lb (2,979,561 J/kg) and drains at a 10°F (5.6°C) terminal difference or 101.7°F (38.6°C). This adds a total of 300 × (1,281 − 69.7) or 363,390 Btu/h (383,390,985 J/h) to the condensate, making a total of 10,794,174 Btu/h (11,388,285,334 J/h) entering the surge tank. The feed leaves the surge tank at the same enthalpy with which it enters.

The feed pump puts an amount of heat into the feedwater equal to the total power of the pump, less any friction in the drive system. This friction work can be neglected but the heat from the power input is a significant quantity. The power input is the total pump head in feet of feedwater, times the quantity pumped in pounds per hour, divided by the mechanical equivalent of heat and the efficiency. Thus,

Heat equivalent of feed pump work

$$= \frac{144 \Delta P v_f Q}{778 E} \quad \text{Btu/h} \quad \left(\frac{\Delta P v_f Q}{E} \quad \text{J/h} \right)$$

where ΔP = pressure change, lb/in² (N/m²); v_f = specific volume of fluid, ft³/lb (m³/kg); Q = mass rate of flow, lb/h (kg/h); and E = efficiency of pump.

Assuming the feed pump raises the pressure from 15 to 1,015 lb/in² abs (103,421 to 6,998,178 N/m²), the specific volume of the water is 0.0161 ft³/lb (0.001005 m³/kg), and the pump efficiency is 50 percent, the heat equivalent of the feed pump work is 1,041,313 Btu/h (1,098,626,868 J/h). This addition of heat gives a total of 11,835,487 Btu/h (12,486,912,202 J/h) entering the boiler. Assuming no leakage of steam, the steam leaves the boiler at 1,481.2 Btu/lb (3,445,219 J/kg), with a total thermal energy flow of 1,481.2 × 174,720 = 258,795,264 Btu/h (273,039,355,300 J/h). The difference between this total heat and that entering [258,795,264 − 11,835,487 = 246,959,777 Btu/h (260,552,443,098 J/h)] is the net heat added in the boiler by the fuel. With a boiler efficiency of 88 percent and a fuel having a higher heating value of 18,500 Btu/lb (43,030,353 J/kg), the quantity of fuel burned is

Fuel flow rate = 246,959,777/(18,500)(0.88) = 15,170 lb/h

or 260,552,443,098/(43,030,353)(0.88) = 6,881 kg/h

The specific fuel rate is the fuel flow rate divided by the net shaft power [15,170/30,000 = 0.506 lb/(shp·h) or 6881/22,371 = 0.308 kg/(kW$_S$·h)]. The heat rate is the quantity of heat expended to produce one horsepower per hour (one kilowatt per hour) and is calculated by dividing the net heat added to the plant, per hour, by the power produced.

Heat rate

= (15,170 lb/h)(18,500 Btu/lb)/30,000 shp = 9,335 Btu/(shp·h)

or 296,091,858,933/22,371 = 13,235,522 J/(kW$_S$·h)

This simple cycle omits many details that must necessarily be included in an actual steam plant. A continuation of the example, developing the details of a complete analysis, is given in "Marine Engineering," SNAME, 1992. A heat balance is usually carried through several iterations until the desired level of accuracy is achieved. The first heat balance may be done from approximate data given in the SNAME Technical and Research Publication No. 3-11, "Recommended Practices for Preparing Marine Steam Power Plant Heat Balances," and then updated as equipment data are known.

Diesel Engines
(See Sec. 9.)

While steam plants are custom-designed, diesel engines and gas turbines are selected from commercial sources at discrete powers. Diesel engines are referred to as being high, medium, or low speed. Low-speed diesels are generally categorized as those with engine speeds less than about 300 r/min, high-speed in excess of 1,200 r/min. Low-speed marine diesel engines are directly coupled to the propeller shaft. Unlike steam turbines and gas turbines, which require special reversing provisions, most direct-drive diesel engines are readily reversible.

Slow-speed diesel engines are well-suited to marine propulsion. Although larger, heavier, and initially more expensive than higher-speed engines, they generally have lower fuel, operating, and maintenance costs. Slow-speed engine parts take longer to wear to the same percentage of their original dimension than high-speed engine parts. Large-bore, low-speed diesel engines have inherently better combustion performance with lower-grade diesel fuels. However, a well-designed high-speed engine which is not overloaded can give equally good service as a slow-speed engine.

Medium- and High-Speed Diesels The number of medium- and high-speed diesel engines used in marine applications is relatively small compared to the total number of such engines produced. The medium- and high-speed marine engine of today is almost universally an adaptation of engines built in quantities for service in automotive and stationary applications. The automotive field contributes high-speed engine applications in the 400-hp (298-kW) range, with engine speeds commonly varying from 1,800 to 3,000 r/min, depending on whether use is continuous or intermittent. Diesels in the 1,200- to 1,800-r/min range are typical of off-highway equipment engines at powers from 500 to upward of 1,200 hp (373 to 895 kW). Medium-speed diesel engines in units from 6 to 20 cylinders are available at ratings up to and exceeding 8,000 hp (5,966 kW) from both locomotive and stationary engine manufacturers. These applications are not all-inclusive, but only examples of the wide variety of diesel ratings commercially available today.

Some of the engines were designed with **marine applications** in mind, and others require some modification in external engine hardware. The changes are those needed to suit the engine to the marine environment, meaning salt-laden air, high humidity, use of corrosive seawater for cooling, and operating from a pitching and rolling platform. It also may mean an installation made in confined spaces. In order to adapt to this environment, the prime requisite of the marine diesel engine is the ability to resist corrosion. Because marine engines may be installed with their crankshafts at an angle to the horizontal and because they are subjected to more motion than in many other applications, changes are also necessary in the lubricating oil system. The air intake to a marine engine may not be dust-free and dirt-free when operating in harbors, inland waters, or close offshore; therefore, it is as important to provide a good air cleaner as in any automotive or stationary installation.

Diesel engines are utilized in all types of marine vehicles, both in the merchant marine and in the navies of the world. The power range in which diesel engines have been used has increased directly with the availability of higher-power engines. The line of demarcation in horsepower between what is normally assigned to diesel and to steam has continually moved upward, as has the power installed in ships in general.

Small high-speed engines are commonly used in pleasure boats where the owner is safety-conscious and wants to avoid the use of gasoline. For the same reason, small boats in use in the Navy are usually powered by diesel engines, although the gas turbine is being used in special boats where high speed for short periods of time is the prime requirement. Going a little higher in power, diesels are used in many kinds of workboats such as fishing boats, tugs, ferries, dredges, river towboats and pushers, and smaller types of cargo ships and tankers. They are used in the naval counterparts of these ships and, in addition, for military craft such as minesweepers, landing ships, patrol and escort ships, amphibious vehicles, tenders, submarines, and special ships such as salvage and rescue ships and icebreakers. In the nuclear-powered submarine, the diesel is relegated to emergency generator-set use; however, it is still the best way to power a nonnuclear submarine when not operating on the batteries.

Diesel engines are used either singly or in multiple to drive propeller shafts. For all but high-speed boats, the rpm of the modern diesel is too fast to drive the propeller directly with efficiency, and some means of speed reduction, either mechanical or electrical, is necessary. If a single engine of the power required for a given application is available, then a decision must be made whether it or several smaller engines should be used. The decision may be dictated by the available space. The diesel power plant is flexible in adapting to specific space requirements. When more than one engine is geared to the propeller shaft, the gear serves as both a speed reducer and combining gear. The same series of engines could be used in an electric-drive propulsion system, with even greater flexibility. Each engine drives its own generator and may be located independently of other engines and the propeller shaft.

Low-Speed, Direct-Coupled Diesels Of the more usual prime mover selections, only low-speed diesel engines are directly coupled to the propeller shaft. This is due to the low rpm required for efficient pro-

peller operation and the high rpm inherent with other types of prime movers.

A rigid hull foundation, with a high resistance to vertical, athwartship, and fore-and-aft deflections, is required for the low-speed diesel. The engine room must be designed with sufficient overhaul space and with access for large and heavy replacement parts to be lifted by cranes.

Because of the low-frequency noise generated by low-speed engines, the operating platform often can be located at the engine itself. Special control rooms are often preferred as the noise level in the control room can be significantly less than that in the engine room.

Electric power may be produced by a generator mounted directly on the line shafting. Operation of the entire plant may be automatic and remotely controlled from the bridge. The engine room is often completely unattended for 16 h a day.

Gas Turbines
(See Sec. 9.)

The gas turbine has developed since World War II to join the steam turbine and the diesel engine as alternative prime movers for various shipboard applications.

In gas turbines, the efficiency of the components is extremely important since the compressor power is very high compared with its counterpart in competitive thermodynamic cycles. For example, a typical marine propulsion gas turbine rated at 20,000 bhp (14,914 kW_B) might require a 30,000 hp (22,371 kW) compressor and, therefore, 50,000 hp (37,285 kW) in turbine power to balance the cycle.

The basic **advantages** of the gas turbine for marine applications are its simplicity and light weight. As an internal-combustion engine, it is a self-contained power plant in one package with a minimum number of large supporting auxiliaries. It has the ability to start and go on line very quickly. Having no large masses that require slow heating, the time required for a gas turbine to reach full speed and accept the load is limited almost entirely by the rate at which energy can be supplied to accelerate the rotating components to speed. A further feature of the gas turbine is its low personnel requirement and ready adaptability to automation.

The relative simplicity of the gas turbine has enabled it to attain outstanding records of reliability and maintainability when used for aircraft propulsion and in industrial service. The same level of reliability and maintainability is being achieved in marine service if the unit is properly applied and installed.

Marine units derived from aircraft engines usually have the gas generator section, comprising the compressor and its turbine, arranged to be removed and replaced as a unit. Maintenance on the power turbine, which usually has the smallest part of the total maintenance requirements, is performed aboard ship. Because of their light weight, small gas turbines used for auxiliary power or the propulsion of small boats, can also be readily removed for maintenance.

Units designed specifically for marine use and those derived from industrial gas turbines are usually maintained and overhauled in place. Since they are somewhat larger and heavier than the aviation-type units, removal and replacement are not as readily accomplished. For this reason, they usually have split casings and other provisions for easy access and maintenance. The work can be performed by the usual ship repair forces.

Both **single-shaft** and **split-shaft** gas turbines can be used in marine service. Single-shaft units are most commonly used for generator drives. When used for main propulsion, where the propeller must operate over a very wide speed range, the single-shaft unit must have a controllable-pitch propeller or some equivalent variable-speed transmission, such as an electric drive, because of its limited speed range and poor acceleration characteristics. A multishaft unit is normally used for main propulsion, with the usual arrangement being a split-shaft unit with an independent variable-speed power turbine; the power turbine and propeller can be stopped, if necessary, and the gas producer kept in operation for rapid load pickup. The use of variable-area nozzles on the power turbine increases flexibility by enabling the compressor to be maintained at or near full speed and the airflow at low-power turbine speeds. Nearly full power is available by adding fuel, without waiting for the compressor to accelerate and increase the airflow. Where low-load economy is of importance, the controls can be arranged to reduce the compressor speed at low loads and maintain the maximum turbine inlet and/or exhaust temperature for best efficiency. Since a gas turbine inherently has a poor part-load fuel rate performance, this variable-area nozzle feature can be very advantageous.

The **physical arrangement** of the various components, i.e., compressors, combustion systems, and turbines, that make up the gas turbine is influenced by the thermodynamic factors (i.e., the turbine connected to a compressor must develop enough power to drive it), by mechanical considerations (i.e., shafts must have adequate bearings, seals, etc.), and also by the necessity to conduct the very high air and gas flows to and from the various components with minimum pressure losses.

In marine applications, the gas turbines usually **cannot be mounted rigidly** to the ship's structure. Normal movement and distortions of the hull when under way would cause distortions and misalignment in the turbine and cause internal rubs or bearing and/or structural failure. The turbine components can be mounted on a subbase built up of structural sections of sufficient rigidity to maintain the gas-turbine alignment when properly supported by the ship's hull.

Since the gas turbine is a high-speed machine with output shaft speeds ranging from about 3,600 r/min for large machines up to 100,000 r/min for very small machines (approximately 25,000 r/min is an upper limit for units suitable for the propulsion of small boats), a **reduction gear is necessary** to reduce the speed to the range suitable for a propeller. Smaller units suitable for boats or driving auxiliary units, such as generators in larger vessels, frequently have a reduction gear integral with the unit. Larger units normally require a separate reduction gear, usually of the double- or triple-reduction type.

A gas turbine, in common with all turbine machinery, is **not inherently reversible** and must be reversed by external means. Electric drives offer ready reversing but are usually ruled out on the basis of weight, cost, and to some extent efficiency. From a practical standpoint there are two alternatives, a reversing gear or a controllable pitch (CRP) propeller. Both have been used successfully in gas-turbine-driven ships, with the CRP favored.

Combined Propulsion Plants

In some shipboard applications, diesel engines, gas turbines, and steam turbines can be employed effectively in various combinations. The prime movers may be combined either mechanically, thermodynamically, or both. The leveling out of specific fuel consumption over the operating speed range is the goal of most combined systems.

The gas turbine is a very flexible power plant and consequently figures in most possible combinations which include combined diesel and gas-turbine plants (**CODAG**); combined gas- and steam-turbine plants (**COGAS**); and combined gas-turbine and gas-turbine plants (**COGAG**). In these cycles, gas turbines and other engines or gas turbines of two different sizes or types are combined in one plant to give optimum performance over a very wide range of power and speed. In addition, combinations of diesel or gas-turbine plants (**CODOG**), or gas-turbine or gas-turbine plants (**COGOG**), where one plant is a diesel or small gas turbine for use at low or cruising powers and the other a large gas turbine which operates alone at high powers, are also possibilities. Even the combination of a small nuclear plant and a gas turbine plant (**CONAG**) has undergone feasibility studies.

COGAS Gas and steam turbines are connected to a common reduction gear, but thermodynamically are either independent or combined. Both gas- and steam-turbine drives are required to develop full power. In a thermodynamically independent plant, the boost power is furnished by a lightweight gas turbine. This combination produces a significant reduction in machinery weight.

In the thermodynamically combined cycle, commonly called STAG (steam and gas turbine) or RACER (Rankine cycle energy recovery), energy is recovered from the exhaust of the gas turbine and is used to augment the main propulsion system through a steam turbine. The prin-

cipal advantage gained by the thermodynamic interconnection is the potential for improved overall efficiency and resulting fuel savings. In this arrangement, the gas turbine discharges to a heat-recovery boiler where a large quantity of heat in the exhaust gases is used to generate steam. The boiler supplies the steam turbine that is geared to the propeller. The steam turbine may be coupled to provide part of the power for the gas-turbine compressor. The gas-turbine may be used to provide additional propulsion power, or its exhaust may be recovered to supply heat for various ship's services (Fig. 11.3.15).

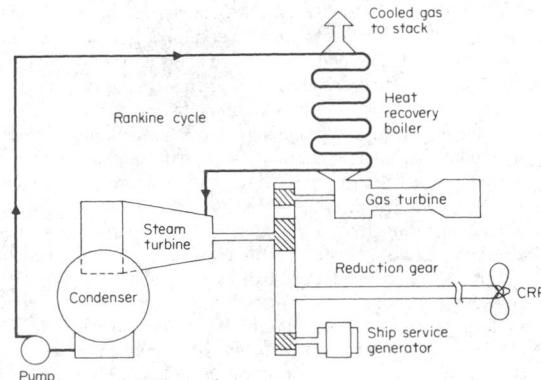

Fig. 11.3.15 COGAS system schematic.

Combined propulsion plants have been used in several applications in the past. **CODOG** plants have serviced some Navy patrol gunboats and the Coast Guard's Hamilton class cutters. Both **COGOG** and **COGAS** plants have been used in foreign navies, primarily for destroyer-type vessels.

PROPULSORS

The force to propel a marine vehicle arises from the rate of change of momentum induced in either the water or air. Since the force produced is directly proportional to the mass density of the fluid used, the reasonable choice is to induce the momentum change in water. If air were used, either the cross-sectional area of the jet must be large or the

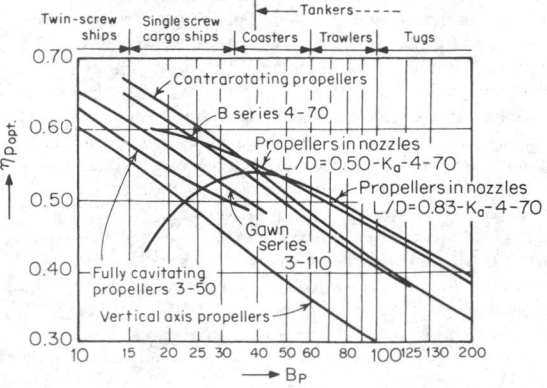

Fig. 11.3.16 Comparison of optimum efficiency values for different types of propulsors. (*From "Marine Engineering," SNAME, 1992.*)

velocity must be high. A variety of propulsors are used to generate this stream of water aft relative to the vehicle: screw propellers, controllable-pitch propellers, water jets, vertical-axis propellers, and other thrust devices. Figure 11.3.16 indicates the type of propulsor which provides the best efficiency for a given vehicle type.

Screw Propellers

The screw propeller may be regarded as part of a helicoidal surface which, as it rotates, appears to "screw" its way through the water, driving water aft and the vehicle forward. A propeller is termed "**right-handed**" if it turns clockwise (viewed from aft) when producing ahead thrust; if counterclockwise, "**left-handed.**" In a twin-screw installation, the starboard propeller is normally right-handed and the port propeller left-handed. The surface of the propeller blade facing aft, which experiences the increase in pressure, producing thrust, is the **face** of the blade; the forward surface is the **back.** The face is commonly constructed as a true helical surface of constant pitch; the back is not a helical surface. A **true helical surface** is generated by a line rotated about an axis normal to itself and advancing in the direction of this axis at constant speed. The distance the line advances in one revolution is the **pitch.** For simple propellers, the pitch is constant on the face; but in practice, it is common for large propellers to have a reduced pitch toward the hub and less usually toward the tip. The pitch at 0.7 times the maximum radius is usually a representative mean pitch; maximum lift is generated at that approximate point. Pitch may be expressed as a dimensionless ratio, P/D.

The **shapes** of blade outlines and sections vary greatly according to the type of ship for which the propeller is intended and to the designer's ideas. Figure 11.3.17 shows a typical design and defines many of the terms in common use. The **projected area** is the area of the projection of the propeller contour on a plane normal to the shaft, and the **developed area** is the total face area of all the blades. If the variation of helical cord length is known, then the true blade area, called **expanded area,** can be obtained graphically or analytically by integration.

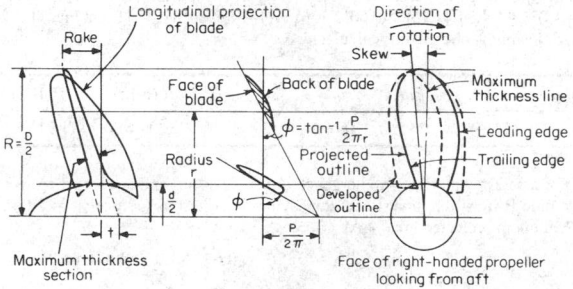

Fig. 11.3.17 Typical propeller drawing. *(From "Principles of Naval Architecture," SNAME, 1988.)*

Diameter = D Pitch ratio = P/D Pitch = P
Blade thickness ratio = t/D No. of blades = 4
Pitch angle = ϕ
Disk area = area of tip circle = $\pi D^2/4 = A_O$
Developed area of blades, outside hub $- A_D$
Developed area ratio = DAR = A_D/A_O
Projected area of blades (on transverse plane) outside hub = A_P
Projected area ratio = PAR = A_P/A_O
Blade width ratio = BWR = (max. blade width)/D
Mean width ratio = MWR = $[A_D$/length of blades (outside hub)]/D

Consider a section of the propeller blade at a radius r with a pitch angle ϕ and pitch P working in an unyielding medium; in one revolution of the propeller it will advance a distance P. Turning n revolutions in unit time it will advance a distance $P \times n$ in that time. In a real fluid, there will be a certain amount of yielding when the propeller is developing thrust and the screw will not advance $P \times n$, but some smaller distance. The difference between $P \times n$ and that smaller distance is called the **slip velocity. Real slip ratio** is defined in Fig. 11.3.18.

A wake or a frictional belt of water accompanies every hull in motion; its velocity varies as the ship's speed, hull shape, the distance from the ship's side and from the bow, and the condition of the hull surface. For ordinary propeller design the **wake velocity** is a fraction w of the ship's speed. Wake velocity = wV. The **wake fraction** w for a ship may be obtained from Fig. 11.3.19. The velocity of the ship relative to the ship's wake at the stern is $V_A = (1 - w)V$.

The **apparent slip ratio** S_A is given by

$$S_A = (Pn - V)/Pn = 1 - V/Pn$$

Although real slip ratio, which requires knowledge of the wake fraction, is a real guide to ship performance, the apparent slip ratio requires only ship speed, revolutions, and pitch to calculate and is therefore often recorded in ships' logs.

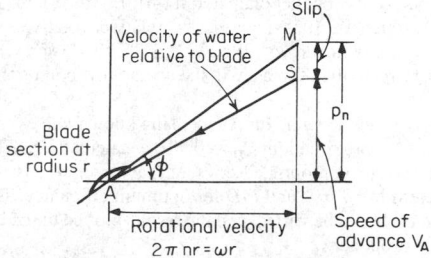

Fig. 11.3.18 Definition of slip. *(From "Principles of Naval Architecture," SNAME, 1988.)*

$$\tan \varphi = Pn/2\pi nr = P/2\pi r$$
$$\text{Real slip ratio} = S_R = MS/ML = (Pn - V_A)/Pn = 1 - V_A/Pn$$

For P in ft (m), n in r/min, and V in knots, $S_A = (Pn - 101.3V)/Pn$; $[S_A = (Pn - 30.9V)/Pn]$.

Propeller Design The design of a marine propeller is usually carried out by one of two methods. In the first, the design is based upon charts giving the results of open-water tests on a series of model propellers. These cover variations in a number of the design parameters such as pitch ratio, blade area, number of blades, and section shapes. A propeller which conforms with the characteristics of any particular series can be rapidly designed and drawn to suit the required ship conditions.

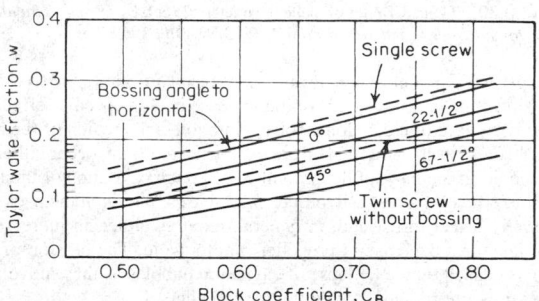

Fig. 11.3.19 Wake fractions for single- and twin-screw models. *(From "Principles of Naval Architecture," SNAME, 1988.)*

The second method is used in cases where a propeller is heavily loaded and liable to cavitation or has to work in a very uneven wake pattern; it is based on purely theoretical calculations. Basically this involves finding the chord width, section shape, pitch, and efficiency at a number of radii to suit the average circumferential wake values and give optimum efficiency and protection from cavitation. By integration of the resulting thrust and torque-loading curves over the blades, the thrust, torque, and efficiency for the whole propeller can be found.

Using the first method and **Taylor's propeller** and **advance coefficients** B_p and δ, a convenient practical design and an initial estimate of propeller size can be obtained

$$B_p = \frac{n(P_D)^{0.5}}{(V_A)^{2.5}} \qquad \left[\frac{1.158n(P_D)^{0.5}}{(V_A)^{2.5}}\right]$$

$$\delta = \frac{nD}{V_A} \qquad \left(\frac{3.281nD}{V_A}\right)$$

$$\eta_O = \frac{TV_A}{325.7P_D\eta_R} \qquad \left(\frac{TV_A}{1,942.5P_D\eta_R}\right)$$

where B_p = Taylor's propeller coefficient; δ = Taylor's advance coefficient; n = r/min; P_D = delivered power, dhp (kW$_D$); V_A = speed of advance, knots; D = propeller diameter, ft (m); T = thrust, lb (N); η_O = open-water propeller efficiency; η_R = relative rotative efficiency ($0.95 < \eta_R < 1.0$, twin-screw; $1.0 < \eta_R < 1.05$, single-screw), a factor which corrects η_O to the efficiency in the actual flow conditions behind the ship.

Assume a reasonable value for n and, using known values for P_D and V_A (for a useful approximation, $P_D = 0.98P_S$), calculate B_p. Then enter Fig. 11.3.20, or an appropriate series of Taylor or Troost propeller charts, to determine δ, η_O, and P/D for optimum efficiency. The charts and parameters should be varied, and the results plotted to recognize the most suitable propeller.

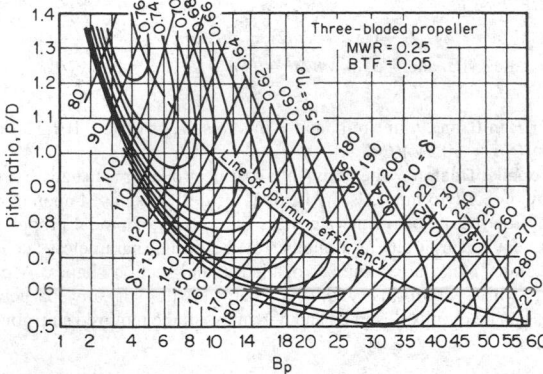

Fig. 11.3.20 Typical Taylor propeller characteristic curves. *(From "Handbook of Ocean and Underwater Engineering," McGraw-Hill, 1969.)*

Propeller cavitation, when severe, may result in marked increase in rpm, slip, and shaft power with little increase in ship speed or effective power. As cavitation develops, noise, vibration, and erosion of the propeller blades, struts, and rudders are experienced. It may occur either on the face or on the back of the propeller. Although cavitation of the face has little effect on thrust and torque, extensive cavitation of the back can materially affect thrust and, in general, requires either an increase in blade area or a decrease in propeller rpm to avoid. The erosion of the backs is caused by the collapse of cavitation bubbles as they move into higher pressure regions toward the trailing edge.

Avoidance of cavitation is an important requirement in propeller design and selection. The Netherlands Ship Model Basin suggests the following **criterion** for the minimum blade area required **to avoid cavitation:**

$$A_p^2 = \frac{T^2}{1,360(p_0 - p_v)^{1.5}V_A} \quad \text{or} \quad \left[\frac{T^2}{5.44(p_0 - p_v)^{1.5}V_A}\right]$$

where A_p = projected area of blades, ft^2 (m^2); T = thrust, lb (N); p_0 = pressure at screw centerline, due to water head plus atmosphere, lb/in^2 (N/m^2); p_v = water-vapor pressure, lb/in^2 (N/m^2); and V_A = speed of advance, knots (1 knot = 0.515 m/s); and

$$p_0 - p_v = 14.45 + 0.45h \qquad \text{lb/in}^2$$
$$\text{or} \qquad (p_0 - p_v = 99,629 + 10,179h \qquad \text{N/m}^2)$$

where h = head of water at screw centerline, ft (m).

A four-blade propeller of 0.97 times the diameter of the three-blade, the same pitch ratio, and four-thirds the area will absorb the **same power** at the same rpm as the three-blade propeller. Similarly, a two-blade propeller of 5 percent greater diameter is approximately equivalent to a three-blade unit. Figure 11.3.20 shows that propeller efficiency increases with decrease in value of the propeller coefficient B_p. A low value of B_p in a slow-speed ship calls for a large-diameter (optimum) propeller. It is frequently necessary to limit the diameter of the propeller and accept the accompanying loss in efficiency.

The number of propeller blades is usually three or four. Four blades are commonly used with single-screw merchant ships. Recently five, and even six, blades have been used to reduce vibration. Alternatively, highly skewed blades reduce vibration by creating a smoother interaction between the propeller and the ship's wake. Highly loaded propellers of fast ships and naval vessels call for large blade area, preferably three to five blades, to reduce blade interference and vibration.

Propeller fore-and-aft clearance from the stern frame of large single-screw ships at 70 percent of propeller radius should be greater than 18 in. The stern frame should be streamlined. The clearance from the propeller tips to the shell plating of twin-screw vessels ranges from about 20 in for low-powered ships to 4 or 5 ft for high-powered ships. U.S. Navy practice generally calls for a propeller tip clearance of $0.25D$. Many high-speed motorboats have a tip clearance of only several inches. The immersion of the propeller tips should be sufficient to prevent the drawing in of air.

EXAMPLE. Consider a three-blade propeller for a single-screw installation with P_s = 2,950 shp (2,200 kW$_S$), V = 17 knots, and C_B = 0.52. Find the propeller diameter D and efficiency η_O for n = 160 r/min.

From Fig. 11.3.19, the wake fraction w = 0.165. $V_A = V(1 - 0.165)$ = 14.19 knots. $P_D = 0.98P_s$ = 2,891 dhp (2,156 kW$_D$). At 160 r/min, $B_p = 160(2,891)^{0.5}/(14.19)^{2.5}$ = 11.34. [$1.158 \times 160(2,156)^{0.5}/(14.19)^{2.5}$ = 11.34.] From Fig. 11.3.20, the following is obtained or calculated:

r/min	B_p	P/d	η_O	δ	D, ft (m)	P, ft (m)
160	11.34	0.95	0.69	139	12.33 (3.76)	11.71 (3.57)

For a screw centerline submergence of 9 ft (2.74 m) and a selected projected area ratio PAR = 0.3, investigate the cavitation criterion. Assume η_R = 1.0. Minimum projected area:

$$T = 325.7\,\eta_O\eta_R\,P_D/V_A = 325.7 \times 0.69 \times 1.0 \times 2,891/14.19$$
$$= 45,786 \text{ lb}$$
$$(T = 1942.5 \times 0.69 \times 2,156/14.19 = 203,646 \text{ N})$$
$$p_0 - p_v = 14.45 + 0.45 (9) = 18.5 \text{ lb/in}^2$$
$$(p_0 - p_v = 99,629 + 10,179 (2.74) = 127,519 \text{ N/m}^2)$$
$$A_p^2 = (45,786)^2/1,360 (18.5)^{1.5}(14.19) = 1,365, A_p$$
$$= 36.95 \text{ ft}^2 \text{ minimum}$$
$$A_p^2 = (203,646)^2/5.44 (127,519)^{1.5}(14.19) = 11.79, A_p$$
$$= 3.43 \text{ m}^2 \text{ minimum}$$

Actual projected area: PAR = A_p/A_0 = 0.3, $A_p = 0.3A_0 = 0.3\pi (12.33)^2/4$ = 35.82 ft^2 (3.33 m^2).

Since the selected PAR does not meet the minimum cavitation criterion, either the blade width can be increased or a four-blade propeller can be adopted. To absorb the same power at the same rpm, a four-blade propeller of about 0.97 (12.33) = 11.96 ft (3.65 m) diameter with the same blade shape would have an A_p = 35.82 × ⅓ × $(11.96/12.33)^2$ = 44.94 ft^2 (4.17 m^2), or about 25 percent increase. The pitch = 0.95 (11.96) = 11.36 ft (3.46 m). The real and apparent slip ratios for the four-blade propeller are:

$$S_R = 1 - (101.3)(14.19)/(11.36)(160) = 0.209$$
$$S_A = 1 - (30.9)(17.0)/(3.46)(160) = 0.051$$

that is, 20.9 and 5.1 percent (S_A calculated using SI values).

Since the projected area in the three-blade propeller of this example is only slightly under the minimum, a three-blade propeller with increased blade width would be the more appropriate choice.

Controllable-Pitch Propellers

Controllable-pitch propellers are screw propellers in which the blades are separately mounted on the hub, each on an axis, and in which the pitch of the blades can be changed, and even reversed, while the propeller is turning. The pitch is changed by means of an internal mechanism consisting essentially of hydraulic pistons in the hub acting on crossheads. Controllable-pitch propellers are most suitable for vehicles which must meet different operating conditions, such as tugs, trawlers, ferries, minesweepers, and landing craft. As the propeller pitch is changed, the engine can still run at its most efficient speed. Maneuvering is more rapid since the pitch can be changed more rapidly than could the shaft revolutions. By use of controllable-pitch propellers, neither reversing mechanisms are necessary in reciprocating engines, nor astern turbines in turbine-powered vehicles, especially important in gas-turbine installations. Except for the larger hub needed to house the pitch-changing mechanism, the controllable-pitch propeller can be made almost as efficient as the solid, fixed-blade propeller. The newest application is in the U.S. Navy's DDG 51 class destroyer, developing approximately 50,000 shp (24,785 kW_S) per shaft.

Water Jet

This method usually consists of an impeller or pump inside the hull, which draws water from outside, accelerates it, and discharges it astern as a jet at a higher velocity. It is a reaction device like the propeller but in which the moving parts are contained inside the hull, desirable for shallow-water operating conditions and maneuverability. The overall efficiency is lower than that of the screw propeller of diameter equal to the jet orifice diameter, principally because of inlet and ducting losses. Other disadvantages include the loss of volume to the ducting and impeller and the danger of fouling of the impeller. Water jets have been used in several of the U.S. Navy's hydrofoil and surface-effect vehicles.

Vertical-Axis Propellers

There are two types of vertical-axis propulsor systems consisting of one or two vertical-axis rotors located underwater at the stern. Rotor disks are flush with the shell plating and have five to eight streamline, spade-like, vertical impeller blades fitted near the periphery of the disks. The blades feather during rotation of the disk to produce a maximum thrust effect in any direction desired. In the **Kirsten-Boeing** system the blades are interlocked by gears so that each blade makes a half revolution about its axis for each revolution of the disk. The blades of the **Voith-Schneider** system make a complete revolution about their own axis for each revolution of the disk. A bevel gear must be used to transmit power from the conventional horizontal drive shaft to the horizontal disk; therefore, limitations exist on the maximum power that can be transmitted. Although the propulsor is 30 to 40 percent less efficient than the screw propeller, it has obvious maneuverability advantages. Propulsors of this type have also been used at the bow to assist in maneuvering.

Other Thrust Devices

Pump Jet   In a pump-jet arrangement, the rotating impeller is external to the hull with fixed guide vanes either ahead and/or astern of it; the whole unit is enclosed in a duct or long shroud ring. The duct diameter increases from the entrance to the impeller so that the velocity falls and the pressure increases. Thus the impeller diameter is larger, thrust loading less, and the efficiency higher; the incidence of cavitation and noise is delayed. A penalty is paid, however, for the resistance of the duct.

Kort Nozzles   In the Kort nozzle system, the screw propeller operates in a ring or nozzle attached to the hull at the top. The longitudinal sections are of airfoil shape, and the length of the nozzle is generally about one-half its diameter. Unlike the pump-jet shroud ring, the Kort nozzle entrance is much bigger than the propeller, drawing in more water than the open propeller and achieving greater thrust. Because of the acceleration of the water into the nozzle, the pressure inside is less; hence a forward thrust is exerted on the nozzle and the hull. The greatest advantage is in a tug, pulling at rest. The free-running speed is usually less with the nozzle than without. In some tugs and rivercraft, the whole nozzle is pivoted and becomes a very efficient steering mechanism.

Tandem and Contrarotating Propellers   Two or more propellers arranged on the same shaft are used to divide the increased loading factor when the diameter of a propeller is restricted. Propellers turning in the same direction are termed **tandem**, and in opposite directions,

contrarotating. In tandem, the rotational energy in the race from the forward propeller is augmented by the after one. Contrarotating propellers work on coaxial, contrary-turning shafts so that the after propeller may regain the rotational energy from the forward one. The after propeller is of smaller diameter to fit the contracting race and has a pitch designed for proper power absorption. Such propellers have been used for years on torpedoes to prevent the torpedo body from rotating. Hydrodynamically, the advantages of contrarotating propellers are increased propulsive efficiency, improved vibration characteristics, and higher blade frequency. Disadvantages are the complicated gearing, coaxial shafting, and sealing problems.

Supercavitating Propellers   When cavitation covers the entire back of a propeller blade, an increase of rpm cannot further reduce the pressure at the back but the pressure on the face continues to increase as does the total thrust, though at a slower rate than before cavitation began. Advantages of such fully cavitating propellers are an absence of back erosion and less vibration. Although the characteristics of such propellers were determined by trial and error, they have long been used on high-speed racing motor boats. The blade section design must ensure clean separation of flow at the leading and trailing edges and provide good lift-drag ratios for high efficiency. Introducing air to the back of the blades (**ventilated propeller**) will ensure full cavitation and also enables use at lower speeds.

Partially Submerged Propellers   The appendage drag presented by a propeller supported below high-speed craft, such as planing boats, hydrofoil craft, and surface-effect ships, led to the development of partially submerged propellers. Although vibration and strength problems, arising from the cyclic loading and unloading of the blades as they enter and emerge from the air-water interface, remain to be solved, it has been demonstrated that efficiencies in partially submerged operation, comparable to fully submerged noncavitating operation, can be achieved. The performance of these propellers must be considered over a wide range of submergences.

Outboard Gasoline Engine   Outboard gasoline engines of 1 to 300 bhp (0.75 to 225 kW_B), combining steering and propulsion, are popular for small pleasure craft. Maximum speeds of 3,000 to 6,000 r/min are typical, as driven by the propeller horsepower characteristics (see Sec. 9).

PROPULSION TRANSMISSION

In modern ships, only large-bore, slow-speed diesel engines are directly connected to the propeller shaft. Transmission devices such as mechanical speed-reducing gears or electric drives (generator/motor transmissions) are required to convert the relatively high rpm of a compact, economical prime mover to the relatively low propeller rpm necessary for a high propulsive efficiency. In the case of steam turbines, medium- and high-speed diesel engines, and gas turbines, speed reduction gears are used. Gear ratios vary from relatively low values for medium-speed diesels up to approximately 50:1 for a compact turbine design. Where the prime mover is unidirectional, the drive mechanism must also include a reversing mechanism.

Reduction Gears

Speed reduction is usually obtained with reduction gears. The simplest arrangement of a marine reduction gear is the **single-reduction**, single-input type, i.e., one pinion meshing with a gear (Fig. 11.3.21). This ar-

Fig. 11.3.21 Single-reduction, single-input gear. *(From "Marine Engineering," SNAME, 1992.)*

rangement is used for connecting a propeller to a diesel engine or to an electric motor but is not used for propelling equipment with a turbine drive.

The usual arrangement for turbine-driven ships is the **double-reduction,** double-input, articulated type of reduction gear (Fig. 11.3.22). The two input pinions are driven by the two elements of a cross-compound turbine. The term *articulated* applies because a flexible coupling is generally provided between the first reduction or primary gear wheel and the second reduction or secondary pinion.

The **locked-train** type of double-reduction gear has become standard for high-powered naval ships and, because it minimizes the total weight and size of the assembly, is gaining increased popularity for higher-powered merchant ships (Fig. 11.3.23).

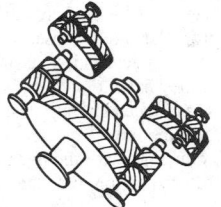

Fig.11.3.22 Double-reduction, double-input, articulated gear. *(From ''Marine Engineering,'' SNAME, 1992.)*

Fig.11.3.23 Double-reduction, double-input, locked-train gear. *(From ''Marine Engineering,'' SNAME, 1992.)*

Electric Drive

Electric propulsion drive is the principal alternative to direct- or geared-drive systems. The prime mover drives a generator or alternator which in turn drives a propulsion motor which is either coupled directly to the propeller or drives the propeller through a low-ratio reduction gear.

Among its **advantages** are the ease and convenience by which propeller speed and direction are controllable; the freedom of installation arrangement offered by the electrical connection between the generator and the propulsion motor; the flexibility of power use when not used for propulsion; the convenience of coupling several prime movers to the propeller without mechanical clutches or couplings; the relative simplicity of controls required to provide reverse propeller rotation when the prime mover is unidirectional; and the speed reduction that can be provided between the generator and the motor, hence between the prime mover and the propeller, without mechanical speed-reducing means.

The disadvantages are the inherently higher first cost, increased weight and space, and the higher transmission losses of the system. The advent of **superconductive electrical machinery** suitable for ship propulsion, however, has indicated order-of-magnitude savings in weight, greatly reduced volume, and the distinct possibility of lower costs. Superconductivity, a phenomenon occurring in some materials at temperatures near absolute zero, is characterized by the almost complete disappearance of electrical resistance. Research and development studies

have shown that size and weight reductions by factors of 5 or more are possible. With successful development of superconductive electrical machinery and resolution of associated engineering problems, monohulled craft, such as destroyers, can benefit greatly from the location flexibility and maneuverability capabilities of superconductive electric propulsion, but the advantages of such a system will be realized to the greatest extent in the new high-performance ships where mechanical-drive arrangement is extremely complex.

In general, electric propulsion drives are employed in marine vehicles requiring a high degree of maneuverability such as ferries, icebreakers, and tugs; in those requiring large amounts of special-purpose power such as self-unloaders, fireboats, and large tankers; in those utilizing nonreversing, high-speed, and multiple prime movers; and in deep-submergence vehicles. Diesel-electric drive is ideally adapted to bridge control.

Reversing

Reversing may be accomplished by stopping and reversing a reversible engine, as in the case of many reciprocating engines, or by adding reversing elements in the prime mover, as in the case of steam turbines. It is impracticable to provide reversing elements in gas turbines; hence a reversing capability must be provided in the transmission system or in the propulsor itself. Reversing reduction gears for such transmissions are available up to quite substantial powers, and CRP propellers are also used with diesel or gas-turbine drives. Electrical drives provide reversing by dynamic braking and energizing the electric motor in the reverse direction.

Marine Shafting

A main propulsion shafting system consists of the equipment necessary to transmit the rotative power from the main propulsion engines to the propulsor; support the propulsor; transmit the thrust developed by the propulsor to the ship's hull; safely withstand transient operating loads (e.g., high-speed maneuvers, quick reversals); be free of deleterious modes of vibration; and provide reliable operation.

Figure 11.3.24 is a shafting arrangement typical of multishaft ships and single-shaft ships with transom sterns. The shafting must extend outboard a sufficient distance for adequate clearance between the propeller and the hull. Figure 11.3.25 is typical of single-screw merchant ships.

The shafting located inside the ship is **line shafting.** The outboard section to which the propeller is secured is the **propeller shaft** or **tail shaft.** The section passing through the stern tube is the **stern tube shaft** unless the propeller is supported by it. If there is a section of shafting between the propeller and stern tube shafts, it is an **intermediate shaft.**

The principal dimensions, kind of material, and material tests are specified by the classification societies for marine propulsion shafting. The American Bureau of Shipping Rules for Steel Vessels specifies the minimum diameter of propulsion shafting, tail shaft bearing lengths, bearing liner and flange coupling thicknesses, and minimum fitted bolt diameters.

Tail or propeller shafts are of larger diameter than the line or tunnel shafting because of the propeller-weight-induced bending moment;

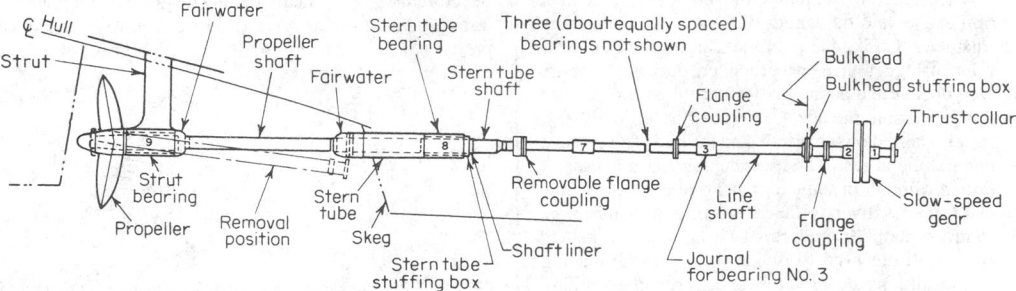

Fig. 11.3.24 Shafting arrangement with strut bearings. *(From ''Marine Engineering,'' SNAME, 1992.)*

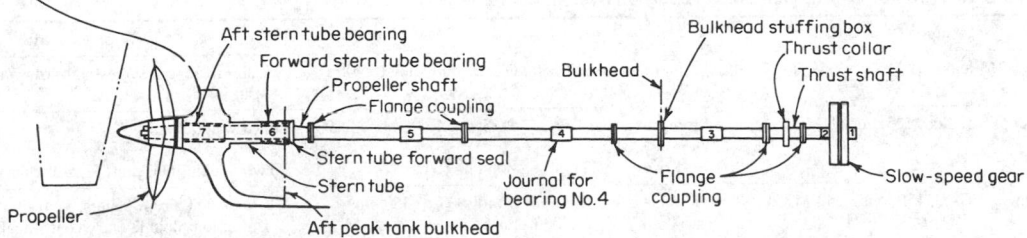

Fig. 11.3.25 Shafting arrangement without strut bearings. *(From "Marine Engineering," SNAME, 1992.)*

also, inspection during service is possible only when the vessel is dry-docked.

Propellers are fitted to shafts over 6 in (0.1524 m) diameter by a taper fit. Aft of the propeller, the shaft is reduced to 0.58 to 0.68 of its specified diameter under the liner and is fitted with a relatively fine thread, nut, and keeper. Propeller torque is taken by a long flat key, the width and total depth of which are, respectively, about 0.21 and 0.11 times the shaft diameter under the liner. The forward end of the key slot in the shaft should be tapered off in depth and terminate well aft of the large end of the taper to avoid a high concentrated stress at the end of the keyway. Every effort is made to keep saltwater out of contact with the steel tail shaft so as to prevent failure from corrosion fatigue.

A cast-iron, cast-steel, or wrought-steel-weldment **stern tube** is rigidly fastened to the hull. In single-screw ships it supports the propeller shaft and is normally provided with a packing gland at the forward end as seawater lubricates the stern-tube bearings. The aft stern-tube bearing is made four diameters in length; the forward stern-tube bearing is much shorter. A bronze liner, ¾ to 1 in (0.0191 to 0.0254 m) thick, is shrunk on the propeller shaft to protect it from corrosion and to give a good journal surface. Lignum-vitae inserts, arranged with the grain on end for medium and large shafts, are normally used for stern-tube bearings; the average bearing pressure is about 30 lb/in² (206,842 N/m²). Rubber and phenolic compound bearings, having longitudinal grooves, are also extensively used. White-metal oil-lubricated bearings and, in Germany, roller-type stern bearings have been fitted in a few installations.

The line or tunnel shafting is laid out, in long equal lengths, so that, in the case of a single-screw ship, withdrawal of the tail shaft inboard for inspection every several years will require only the removal of the next inboard length of shafting. For large outboard or water-exposed shafts of twin-screw ships, a bearing spacing up to 30 shaft diameters has been used. A greater ratio prevails for steel shafts of power boats. Bearings inside the hull are spaced closer—normally under 15 diameters if the shaft is more than 6 in (0.1524 m) diameter. Usually, only the bottom half of the bearing is completely white-metaled; oil-wick lubrication is common, but oil-ring lubrication is used in high-class installations.

Bearings are used to support the shafting in essentially a straight line between the main propulsion engine and the desired location of the propeller. Bearings inside the ship are most popularly known as line-shaft bearings, steady bearings, or spring bearings. Bearings which support outboard sections of shafting are stern-tube bearings if they are located in the stern tube and strut bearings when located in struts.

The propeller thrust is transmitted to the hull by means of a main thrust bearing. The main thrust bearing may be located either forward or aft of the slow-speed gear.

HIGH-PERFORMANCE SHIP SYSTEMS

High-performance ships have spawned new hull forms and propulsion systems which were relatively unknown just a generation ago. Because of their unique characteristics and requirements, conventional propulsors are rarely adequate. The prime thrust devices range from those using water, to water-air mixtures, to large air propellers.

The total propulsion power required by these vehicles is presented in general terms in the Gabrielli and Von Karman plot of Fig. 11.3.26. Basically, there are those vehicles obtaining lift by displacement and those obtaining lift from a wing or foil moving through the fluid. In order for marine vehicles to maintain a reasonable lift-to-drag ratio over a moderate range of speed, the propulsion system must provide increased thrust at lower speeds to overcome low-speed wave drag. The weight of any vehicle is converted into total drag and, therefore, total required propulsion and lift power.

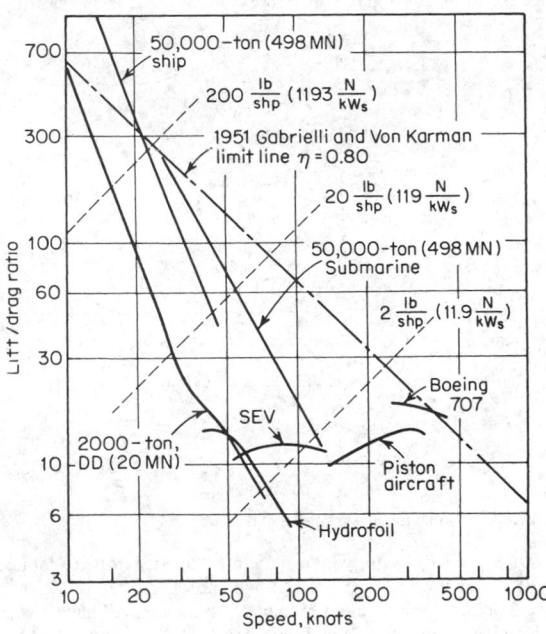

Fig. 11.3.26 Lift-drag ratio versus calm weather speed for various vehicles.

Lightweight, compact, and efficient propulsion systems to satisfy extreme weight sensitivity are achieved over a wide range of power by the marinized aircraft gas-turbine engine. Marinization involves addition of a power turbine unit, incorporation of air inlet filtration to provide salt-free air, addition of silencing equipment on both inlet and exhaust ducts, and modification of some components for resistance to salt corrosion. The shaft power delivered is then distributed to the propulsion and lift devices by lightweight shaft and gear power transmission systems.

A comparison of high-performance ship propulsion system characteristics is given in Table 11.3.3.

Hydrofoil Craft

Under foilborne conditions, the support of the hydrofoil craft is derived from the dynamic lift generated by the foil system, and the craft hull is lifted clear of the water. Special problems relating to operation and directly related to the conditions presented by waves, high fluid density, surface piercing of foils and struts, stability, control, cavitation, and ventilation of foils have given rise to foil system configurations defined in Fig. 11.3.27.

Table 11.3.3   U.S. Navy High-Performance Ship Propulsion System Characteristics

Ship	Engine	Propulsor/lift
Hydrofoil, PHM, 215 tons (2.1 MN), 50 knots	Foilborne: one LM-2500 GT, 22,000 bhp (15,405 kW$_B$), 3,600 r/min	Foilborne: one 2-stage axial-flow waterjet, twin aft strut inlets
Surface-effect ship, SES-200, 195 tons (1.94 MN), 30 + knots	Propulsion: two 16V-149TI diesel, 1,600 bhp (1,193 kW$_B$)	Two waterjet units
	Lift: two 8V-92TI diesel, 800 bhp (597 kW$_B$)	Two double-inlet centrifugal lift fans
Air-cushion vehicle, LCAC, 150 tons (1.49 MN), 50 knots	Propulsion and lift: four TF-40B GT, 3,350 bhp (2,498 kW$_B$), 15,400 r/min	Two ducted-air propellers, 12 ft (3.66 m) dia., 1,250 r/min; four double-entry centrifugal lift fans, 1,900 r/min
SWATH, SSP, 190 tons (1.89 MN), 25 knots	Propulsion: two T-64 GT, 3,200 bhp (2,386 kW$_B$), 1,000 r/min at output of attached gearbox	Two subcavitating controllable-pitch propellers, 450 r/min

The hydrofoil craft has three modes of operation: **hullborne, takeoff,** and **foilborne.** The thrust and drag forces involved in hydrofoil operation are shown in Fig. 11.3.28. Design must be a compromise between the varied requirements of these operating conditions. This may be difficult,

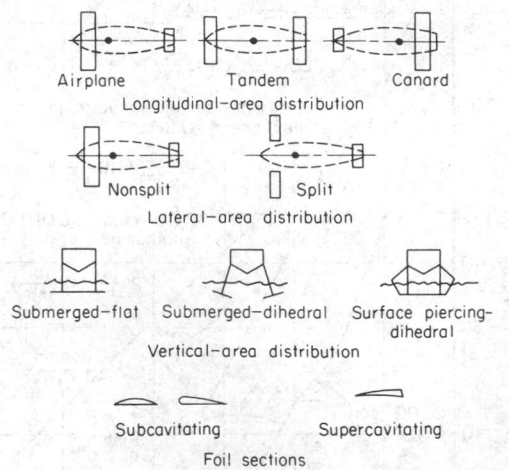

Fig. 11.3.27   Some foil arrangements and sections.

as with propeller design, where maximum thrust is often required at takeoff speed (normally about one-half the operating speed). A propeller designed for maximum efficiency at takeoff will have a smaller efficiency at design speeds, and vice versa. This is complicated by the

need for supercavitating propellers at operating speeds over 50 knots. Two sets of propellers may be needed in such a case, one for hullborne operation and the other for foilborne operation. The selection of a surface-piercing or a fully submerged foil design is determined from specified operating conditions and the permissible degree of complexity. A **surface-piercing system** is inherently stable, as a deviation from the equilibrium position of the craft results in corrective lift forces. **Fully submerged foil** designs, on the other hand, provide the best lift-drag ratio but are not stable by themselves and require a depth-control system for satisfactory operation. Such a control system, however, permits operation of the submerged-foil design in higher waves.

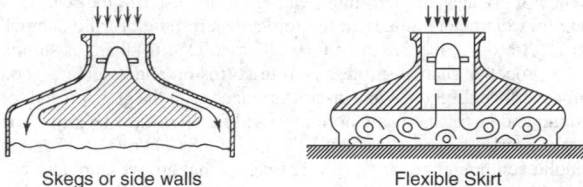

Skegs or side walls   Flexible Skirt

Fig. 11.3.29   Two configurations of the surface-effect principle.

Surface-Effect Vehicles (SEV)

The air-cushion-supported **surface-effect ship (SES)** and **air-cushion vehicle (ACV)** show great potential for high-speed and amphibious operation. Basically, the SEV rides on a cushion of air generated and maintained in the space between the vehicle and the surface over which it moves or hovers. In the SES, the cushion is captured by rigid sidewalls

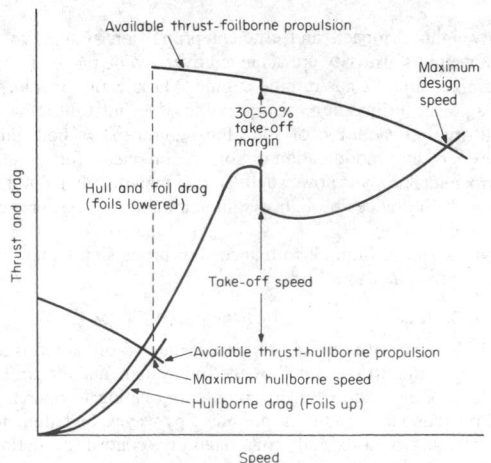

Fig. 11.3.28   Thrust and drag forces for a typical hydrofoil.

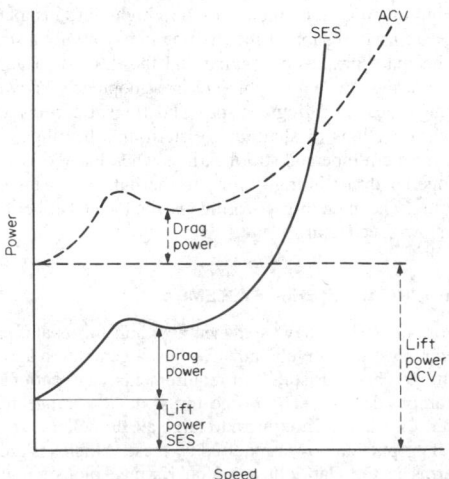

Fig. 11.3.30   Comparison of power requirements: SES versus ACV.

and flexible skirts fore and aft. The ACV has a flexible seal or skirt all around. Figure 11.3.29 illustrates two configurations of the surface-effect principle.

Since the ACV must rise above either land or water, the propulsors use only air for thrust, requiring a greater portion of the total power allocated to lift than in the SES design as shown in Fig. 11.3.30.

While SES technology has greatly improved the capability of amphibious landing vehicles, it is rapidly being developed for high-speed large cargo vessels.

Small Water-Plane Area Twin Hull

The **SWATH,** configured to be relatively insensitive to seaway motions, gains its buoyant lift from two parallel submerged, submarinelike hulls which support the main hull platform through relatively small vertical struts. In this configuration, a propulsion arrangement choice must be with regard to the location of the main engines: lower hull with the propulsor or upper hull with a suitable transmission system. In the U.S. Navy's semisubmerged platform (SSP), power from two independently operated gas turbines is transmitted down each aft strut by a group of three wide-belt chain drives with speed reduction to deliver 450 r/min to the propellers. By contrast, the 3,400-ton (33.87-MN), 11-knot T-AGOS 19 series SWATH vessels employ diesel-electric drive, with both motors and propulsors in the lower hull.

CARGO SHIPS

The development of containerization in the 1950s sparked a growth in ship size. More efficient cargo-handling methods and the desire to reduce transit time led to the emergence of the high-speed container vessel.

The Container Vessel

Because of low fuel costs in the 1950s and 1960s, early container vessel design focused on higher speed and container capacity. Rising fuel costs shifted emphasis toward ship characteristics which enhanced reduced operating costs, particularly as affected by fuel usage and trade route capacity. By the early 1980s, ships were generally sized for TEU (twenty-foot-equivalent-unit) capacity less than 2,500 and speeds of about 20 knots, lower if fuel was significant in the overall operating costs. Largely a function of the scale of trade route capacity, the weighted average full containership size is on the order of 1,000 TEU. The larger the trade capacity, the larger the containerships operating the route. However, many trade route scales do not support even moderately sized containership service.

The **on deck** and **under deck** container capacity varies with the type of vessel, whether a **lift on/lift off** vessel, a bulk carrier equipped for containers, or a **roll on/roll off** vessel. Extensive studies have been conducted to optimize the hull form and improve the operating efficiency for the modern containership.

11.4   AERONAUTICS
by M. W. M. Jenkins

REFERENCES: National Advisory Committee for Aeronautics, *Technical Reports* (designated *NACA-TR* with number), *Technical Notes* (*NACA-TN* with number), and *Technical Memoranda* (*NACA-TM* with number). British Aeronautical Research Committee *Reports and Memoranda* (designated *Br. ARC-R & M* with number). *Ergebnisse der Aerodynamischen Versuchsanstalt zu Göttingen.* Diehl, "Engineering Aerodynamics," Ronald. Reid, "Applied Wing Theory," McGraw-Hill. Durand, "Aerodynamic Theory," Springer. Prandtl-Tietjens, "Fundamentals of Hydro- and Aeromechanics," and "Applied Hydro- and Aeromechanics," McGraw-Hill. Goldstein, "Modern Developments in Fluid Dynamics," Oxford. Millikan, "Aerodynamics of the Airplane," Wiley. Von Mises, "Theory of Flight," McGraw-Hill. Hoerner, "Aerodynamic Drag," Hoerner. Glauert, "The Elements of Airfoil and Airscrew Theory," Cambridge. Milne-Thompson, "Theoretical Hydrodynamics," Macmillan. Munk, "Fluid Dynamics for Aircraft Designers," and "Principles of Aerodynamics," Ronald. Abraham, "Structural Design of Missiles and Spacecraft," McGraw-Hill. "U.S. Standard Atmosphere, 1962," U.S. Government Printing Office.

DEFINITIONS

Aeronautics is the science and art of flight and includes the operation of heavier-than-air aircraft. An **aircraft** is any weight-carrying device designed to be supported by the air, either by buoyancy or by dynamic action. An **airplane** is a mechanically driven fixed-wing aircraft, heavier than air, which is supported by the dynamic reaction of the air against its wings. A **helicopter** is a kind of aircraft lifted and moved by a large propeller mounted horizontally above the fuselage. It differs from the **autogiro** in that this propeller is turned by motor power and there is no auxiliary propeller for forward motion. A **ground-effect machine (GEM)** is a heavier-than-air surface vehicle which operates in close proximity to the earth's surface (over land or water), never touching except at rest, being separated from the surface by a cushion or film of air, however thin, and depending entirely upon aerodynamic forces for propulsion and control.

Aerodynamics is the branch of aeronautics that treats of the motion of air and other gaseous fluids and of the forces acting on solids in motion relative to such fluids. Aerodynamics falls into velocity ranges, depending upon whether the velocity is below or above the local speed of sound in the fluid. The velocity range below the local speed of sound is called the **subsonic** regime. Where the velocity is above the local speed of sound, the flow is said to be **supersonic**. The term **transonic** refers to flows in which both subsonic and supersonic regions are present. The **hypersonic** regime is that speed range usually in excess of five times the speed of sound.

STANDARD ATMOSPHERE

The standard atmosphere of Table 11.4.1 is a revised U.S. Standard Atmosphere, adapted by the United States Committee on Extension to the Standard Atmosphere (COESA) in 1962.

The values given up to about 65,000 ft are designated as **standard**. The region from 65,000 to 105,000 ft is designated **proposed standard**. U.S. Standard Atmosphere, 1962, gives data out to 2,320,000 ft; however, the region from 105,000 to 295,000 ft is designated **tentative**, and that portion of the atmosphere above 295,000 ft is termed **speculative**.

The assumed sea-level conditions are: pressure, $p_0 = 29.91$ in (760 mm Hg) = 2,116.22 lb/ft^2; mass density, $\rho_0 = 0.002378$ slugs/ft^3 (0.001225 g/cm^3); $T_0 = 59°F$ (15°C).

UPPER ATMOSPHERE

High-altitude atmospheric data have been obtained directly from balloons, sounding rockets, and satellites and indirectly from observations of meteors, aurora, radio waves, light absorption, and sound effects. At relatively low altitudes, the earth's atmosphere is, for aerodynamic purposes, a uniform gas. Above 250,000 ft, day and night standards differ because of dissociation of oxygen by solar radiation. This difference in density is as high as 35 percent, but it is usually aerodynamically negligible above 250,000 ft since forces here will become less than 0.05 percent of their sea-level value for the same velocity.

Temperature profile of the COESA atmosphere is given in Fig. 11.4.1. From these data, other properties of the atmosphere can be calculated.

Table 11.4.1   U.S. Standard Atmosphere, 1962

Altitude h, ft*	Temp T, °F†	Pressure ratio, p/p_0	Density ratio, ρ/ρ_0	$(\rho_0/\rho)^{0.5}$	Speed of sound V_s, ft/s‡
0	59.00	1.0000	1.0000	1.000	1,116
5,000	41.17	0.8320	0.8617	1.077	1,097
10,000	23.34	0.6877	0.7385	1.164	1,077
15,000	5.51	0.5643	0.6292	1.261	1,057
20,000	−12.62	0.4595	0.5328	1.370	1,036
25,000	−30.15	0.3711	0.4481	1.494	1,015
30,000	−47.99	0.2970	0.3741	1.635	995
35,000	−65.82	0.2353	0.3099	1.796	973
36,089	−69.70	0.2234	0.2971	1.835	968
40,000	−69.70	0.1851	0.2462	2.016	968
45,000	−69.70	0.1455	0.1936	2.273	968
50,000	−69.70	0.1145	0.1522	2.563	968
55,000	−69.70	0.09001	0.1197	2.890	968
60,000	−69.70	0.07078	0.09414	3.259	968
65,000	−69.70	0.05566	0.07403	3.675	968
65,800	−69.70	0.05356	0.07123	3.747	968
70,000	−67.30	0.04380	0.05789	4.156	971
75,000	−64.55	0.03452	0.04532	4.697	974
80,000	−61.81	0.02725	0.03553	5.305	977
85,000	−59.07	0.02155	0.02790	5.986	981
90,000	−56.32	0.01707	0.02195	6.970	984
95,000	−53.58	0.01354	0.01730	7.600	988
100,000	−50.84	0.01076	0.01365	8.559	991

* × 0.3048 = metres.
† × (°F − 32)/1.8 = °C.
‡ × 0.3048 = m/s.

Pressure and Density For all practical purposes, both decrease exponentially with altitude. The variations in T (Fig. 11.4.1) are accompanied by slight inflections in the curves of p and ρ, but the deviations from a mean curve are far less than the scatter in test data from various sources. For altitudes above 100,000 ft, pressure and density may be approximated by

$$\log p_0/p = 0.00001910h + 0.0140$$
$$\log \rho_0/\rho = 0.00001890h - 0.1000$$

where h is in ft. At $h = 320,000$ ft, or about 60 mi (96 km), p and ρ are about one-millionth of the sea-level values.

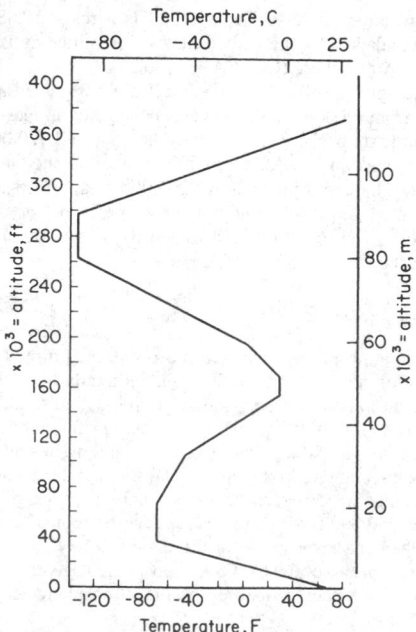

Fig. 11.4.1   Temperature as a function of altitude. *(COESA.)*

Speed in aeronautics may be given in knots (now standard for USAF, USN, and FAA), miles per hour, feet per second, or metres per second. The international nautical mile = 6,076.1155 ft (1,852 m). The basic relations are: 1 knot = 0.5144 m/s = 1.6877 ft/s = 1.1508 mi/h. For additional conversion factors see Sec. 1. Higher speeds are often given as **Mach number,** or the ratio of the particular speed to the speed of sound in the surrounding air. For additional data see "Supersonic and Hypersonic Aerodynamics" below.

Axes   The forces and moments acting on an airplane (and the resultant velocity components) are referred to a set of three mutually perpendicular axes having the origin at the airplane center of gravity (cg) and moving with the airplane. Three sets of axes are in use. The basic difference in these is in the direction taken for the longitudinal, or x, axis, as follows:

Wind axes: The x axis lies in the direction of the relative wind. This is the system most commonly used, and the one used in this section. It is shown on Fig. 11.4.2.

Body axes: The x axis is fixed in the body, usually parallel to the thrust line.

Stability axes: The x axis points in the initial direction of flight. This results in a simplification of the equations of motion.

Absolute Coefficients   Aerodynamic force and moment data are usually presented in the form of absolute coefficients. Examples of force coefficients are: lift $C_L = L/qS$; drag $C_D = D/qS$; side force $C_Y = Y/qS$; where q is the dynamic pressure $\frac{1}{2}\rho V^2$, ρ = air mass-density, and V = air speed. Examples of moment coefficients are $C_m = M/qSc$; roll $C_l = L/qSb$; and yaw $C_n = N/qSb$; where c = mean wing chord and b = wing span.

Section Coefficients   NACA basic test data on wing sections are usually given in the form of section lift coefficient C_l and section drag coefficient C_d. These apply directly to an infinite aspect ratio or to two-dimensional flow, but aspect-ratio and lift-distribution corrections are necessary in applying to a finite wing. For a wing having an elliptical lift distribution, $C_L = (\pi/4)C_l$, where C_l is the centerline value of the local lift coefficient.

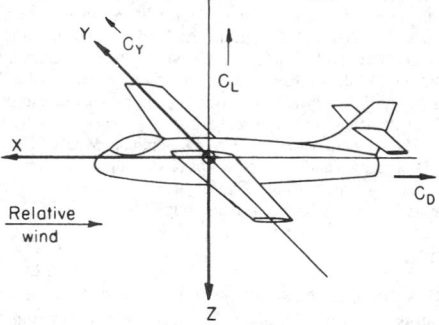

Positive moments: X→Y, Y→Z, Z→X

Fig. 11.4.2   Wind axes.

SUBSONIC AERODYNAMIC FORCES

When an airfoil is moved through the air, the motion produces a pressure at every point of the airfoil which acts normal to the surface. In addition, a frictional force tangential to the surface opposes the motion. The sum of these pressure and frictional forces gives the **resultant force** R acting on the body. The point at which the resultant force acts is defined as the **center of pressure**, c.p. The resultant force R will, in general, be inclined to the airfoil and the relative wind velocity V. It is resolved (Fig. 11.4.3) either along wind axes into

L = **lift** = component normal to V
D = **drag** or resistance = component along V, or along
    body axes into

N = **normal force** = component perpendicular to airfoil chord
T = **tangential force** = component along airfoil chord

Instead of specifying the center of pressure, it is convenient to specify the moment of the air forces about the so-called *aerodynamic center*, a.c. This point lies at a distance a (about a quarter-chord length) back of the leading edge of the airfoil and is defined as the point about which the moment of the air forces remains constant when the angle of attack α is changed. Such a point exists for every airfoil. The force R acting at the c.p. is equivalent to the same force acting at the a.c. plus a moment equal to that force times the distance between the c.p. and the a.c. (see Fig. 11.4.3). The location of the aerodynamic center, in terms of the chord c and the section thickness t, is given for NACA 4- and 5-digit series airfoils approximately by

$$a/c = 0.25 - 0.40(t/c)^2$$

The distance $x_{c.p.}$ from the leading edge to the center of pressure expressed as a fraction of the chord is, in terms of the moment $M_{c/4}$ about the quarter-chord point,

$$M_{c/4} = (\tfrac{1}{4} - x_{c.p.}) \cdot cN$$

$$x_{c.p.} = \frac{1}{4} - \frac{M_{c/4}}{cN}$$

From dimensional analysis it can be seen that the **air force on a body** of length dimension l moving with velocity V through air of density ρ can be expressed as

$$F = \varphi \cdot \rho V^2 l^2$$

where φ is a coefficient that depends upon all the dimensionless factors of the problem. In the case of a wing these are:

1. **Angle of attack** α, the inclination between the chord line and the velocity V
2. **Aspect ratio** $A = b/c$, where b is the span and c the mean chord of the wing
3. **Reynolds number** $\mathrm{Re} = \rho Vl/\mu$, where μ is the coefficient of dynamic viscosity of air
4. **Mach number** V/V_s, where V_s is the velocity of sound
5. Relative surface roughness
6. Relative turbulence

The dependence of the force coefficient φ upon α and A can be theoretically determined; the variation of φ with the other parameters must be established experimentally, i.e., by model tests.

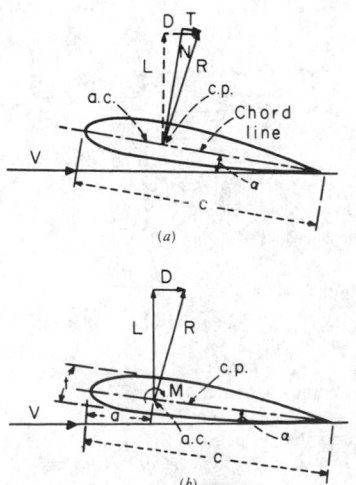

Fig. 11.4.3 Forces acting on an airfoil. (*a*) Actual forces; (*b*) equivalent forces through the aerodynamic center plus a moment.

AIRFOILS

Applying Bernoulli's equation to the flow around a body, if p represents the **static pressure**, i.e., the atmospheric pressure in the undisturbed air, and if V is superposed as in Fig. 11.4.4 and if p_1, V_1 represent the pressure and velocity at any point 1 at the surface of the body,

$$p + \tfrac{1}{2}\rho V^2 = p_1 + \tfrac{1}{2}\rho V_1^2$$

The maximum pressure occurs at a point s on the body at which the velocity is zero. Such a point is defined as the **stagnation point**. The maximum pressure increase occurs at this point and is

$$p_s - p = \tfrac{1}{2}\rho V^2$$

This is called the stagnation pressure, or **dynamic pressure**, and is denoted by q. It is customary to express all aerodynamic forces in terms of $\tfrac{1}{2}\rho V^2$, hence

$$F = \tfrac{1}{2}\varphi\rho V^2 l^2 = q\varphi l^2$$

For the case of a wing, the forces and moments are expressed as

$$\textbf{Lift} = L = C_L \tfrac{1}{2}\rho V^2 S = C_L qS$$
$$\textbf{Drag} = D = C_D \tfrac{1}{2}\rho V^2 S = C_D qS$$
$$\textbf{Moment} = M = C_M \tfrac{1}{2}\rho V^2 Sc = C_M qSc$$

where S = wing area and c = wing chord.

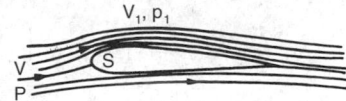

Fig. 11.4.4 Fluid flow around an airfoil.

The lift produced can be determined from the intensity of the circulatory flow or **circulation** Γ by the relation $L' = \rho V\Gamma$, where L' is the lift per unit width of wing. In a **wing of infinite aspect ratio**, the flow is two-dimensional and the lift reaction is at right angles to the line of relative motion. The **lift coefficient** is a function of the angle of attack α, and by mathematical analysis $C_L = 2\pi \sin\alpha$, which for small angles becomes $C_L = 2\pi\alpha$. Experiments show that $C_L = 2\pi\eta(\alpha - \alpha_0)$, where α_0 is the angle of attack corresponding to zero lift. $\eta \approx 1 - 0.64(t/c)$, where t is airfoil thickness.

In a **wing of finite aspect ratio**, the circulation flow around the wing creates a strong vortex trailing downstream from each wing tip (Fig.

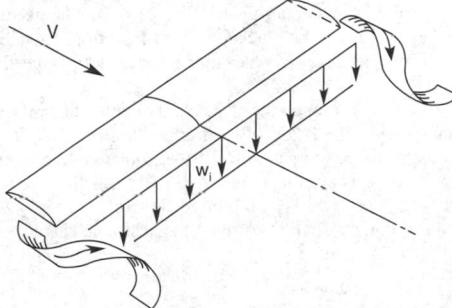

Fig. 11.4.5 Vortex formation at wing tips.

11.4.5). The effect of this is that the direction of the resultant velocity at the wing is tilted downward by an **induced angle of attack** $\alpha_i = w_i/V$ (Fig. 11.4.6). If friction is neglected, the resultant force R is now tilted to the rear by this same angle α_i. The lift force L is approximately the same as R. In addition, there is also a drag component D_i, called the **induced drag**, given by

$$D_i = L \tan\alpha_i = Lw_i/V$$

For a given geometrical angle of attack α the **effective angle of attack** has been reduced by α_i, and thus

$$C_L = 2\pi\eta(\alpha - \alpha_0 - \alpha_i)$$

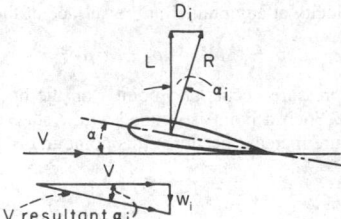

Fig. 11.4.6 Induced angle of attack.

According to the Lanchester-Prandtl theory, for **wings having an elliptical lift distribution**

$$\alpha_i = w_i/V = C_L/\pi A \qquad \text{rad}$$
$$D_i = L^2/\pi b^2 q \quad \text{and} \quad C_{Di} = D_i/qS = C_L^2/\pi A$$

where b = span of wing and $A = b^2/S$ = aspect ratio of wing. These results also apply fairly well to wings not differing much from the elliptical shape. For **square tips**, correction factors are required:

$$\left.\begin{array}{l} \alpha_i = \dfrac{C_L}{\pi A}(1 + \tau) \quad \text{where } \tau = 0.05 + 0.02A \\[2mm] C_{Di} = \dfrac{C_L^2}{\pi A}(1 + \sigma) \quad \text{where } \sigma = 0.01A - 0.01 \end{array}\right\} \text{ for } A < 12$$

These formulas are the basis for transforming the characteristics of rectangular wings from an **aspect ratio** A_1 to an aspect ratio A_2:

$$\alpha_2 = \alpha_1 + \frac{57.3C_L}{\pi}\left(\frac{1 + \tau_2}{A_2} - \frac{1 + \tau_1}{A_1}\right) \quad \text{deg}$$

$$C_{D2} = C_{D1} + \frac{C_L^2}{\pi}\left(\frac{1 + \sigma_2}{A_2} - \frac{1 + \sigma_1}{A_1}\right)$$

For elliptical wings the values of τ and σ are zero. Most wing-section data are given in terms of aspect ratios 6 and ∞. For other values, the preceding formulas must be used.

Characteristics of Airfoils Airfoil characteristics are expressed in terms of the dimensionless coefficients C_L, C_D, and C_M and the angle of attack α. The NACA presents the results for wings of aspect ratio 6 and also corrected to aspect ratio ∞ (see Fig. 11.4.10).

The lift coefficient C_L is a linear function of the angle of attack up to a critical angle called the **stalling angle** (Fig. 11.4.7). The **maximum lift coefficient** C_{Lmax} which can be reached is one of the important characteristics of a wing because it determines the landing speed of the airplane.

The **drag of a wing** is made up of two components: the **profile drag** D_0 and the **induced drag** D_i. The profile drag is due principally to surface friction. At aspect ratio ∞ or at zero lift the induced drag is zero, and thus the entire drag is profile drag. In Fig. 11.4.8 the coefficient of induced drag $C_{Di} = C_L^2/\pi A$ is also plotted. The difference $C_D - C_{Di} = C_{D0}$, the profile-drag coefficient. Among the desirable characteristics of

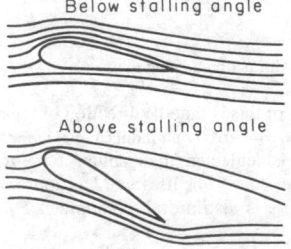

Fig. 11.4.7 Stalling angle of an airfoil.

an airfoil is a small value of the minimum profile-drag coefficient and a large value of C_L/C_D.

The moment characteristics of a wing are obtainable from the curve of the center of pressure as a function of α or by the **moment coefficient taken about the aerodynamic center** $C_{Ma.c.}$ as a function of C_L. A forward motion of the c.p. as α is increased corresponds to an unstable wing section. This instability is undesirable because it requires a large download on the tail to counteract it.

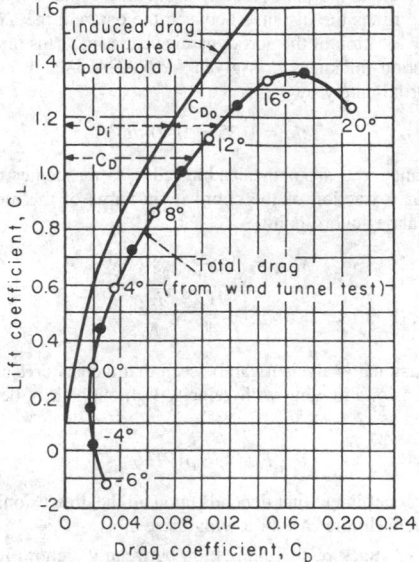

Fig. 11.4.8 Polar-diagram plot of airfoil data.

The characteristics of a wing section (Fig. 11.4.9) are determined principally by its **mean camber line**, i.e., the curvature of the median line of the profile, and secondly by the thickness distribution along the chord. In the NACA system of designation, when a four-digit number is used such as 2412, the significance is always first digit = maximum camber in percent of chord, second digit = location of position of maximum camber in tenths of chord measured from the leading edge (that is, 4 stands for 40 percent), and the last two figures indicate the maximum thickness in percent of chord. The NACA five-digit system is explained in *NACA-TR* 610.

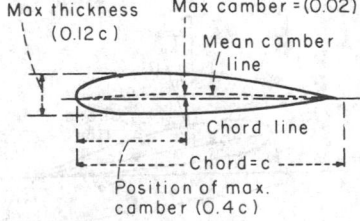

Fig. 11.4.9 Characteristics of a wing section.

Selection of Wing Section

In selecting a wing section for a particular airplane the following factors are generally considered:

1. Maximum lift coefficient C_{Lmax}
2. Minimum drag coefficient C_{Dmin}
3. Moment coefficient at zero lift C_{m0}
4. Maximum value of the ratio C_L/C_D

For certain special cases it is necessary to consider one or more factors from the following group:

5. Value of C_L for maximum C_L/C_D
6. Value of C_L for minimum profile drag

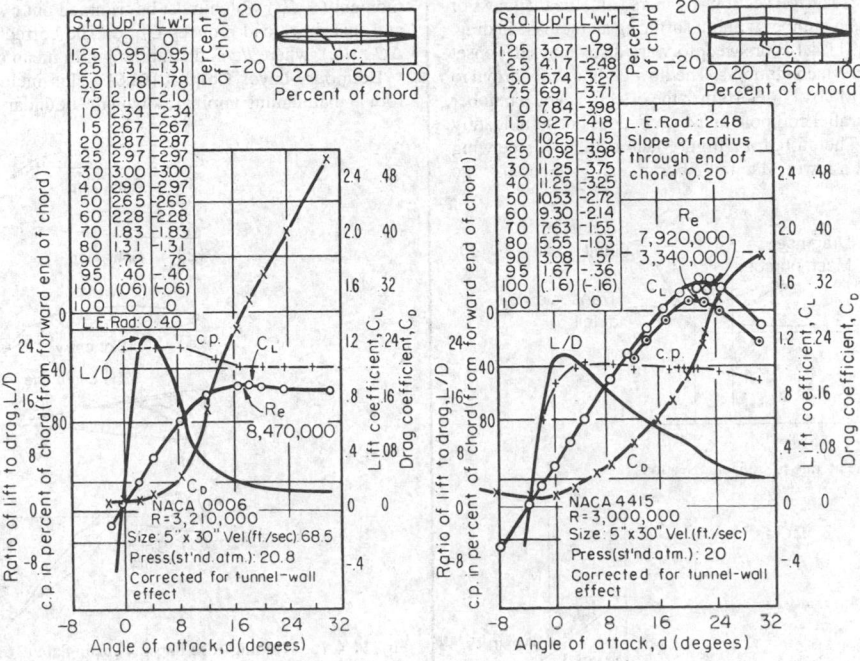

Fig. 11.4.10 Properties of an airfoil section.

7. Maximum value of $C_L^{3/2}/C_D$
8. Maximum value of $C_L^{1/2}/C_D$
9. Type of lift curve peak (stall characteristics)
10. Drag divergence Mach number M_{DD}

Characteristics of airfoil sections are given in *NACA-TR* 586, 647, 669, 708, and 824. Figure 11.4.10 gives data on two typical sections: 0006 is often used for tail surfaces, and 4415 is especially suitable for the wings of subsonic airplanes. For Mach numbers greater than 0.6 thin wings must be used.

The dimensionless coefficients C_L and C_D are functions of the **Reynolds number** $\text{Re} = \rho V l/\mu$. For wings, the characteristic length l is taken to be the chord. In standard air at sea level

$$\text{Re} = 6,378 V_{ft/s} \cdot l_{ft} = 9,354 V_{mi/h} \cdot l_{ft}$$

For other heights, multiply by the coefficient $K = (\rho/\mu)/(\rho_0/\mu_0)$. Values of K are as follows:

variation of the drag coefficient for a smooth flat plate for laminar and turbulent flow, respectively, and also for the transition region.

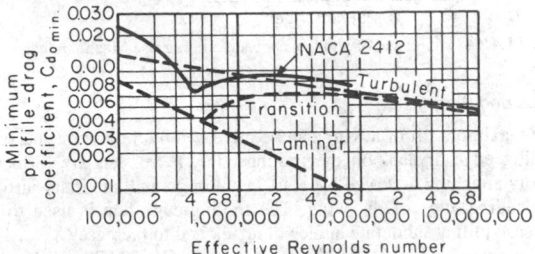

Fig. 11.4.12 Scale effect on minimum profile-drag coefficient.

Altitude, ft*	0	5,000	10,000	15,000	20,000	25,000	30,000	40,000	50,000	60,000
K	1.000	0.884	0.779	0.683	0.595	0.515	0.443	0.300	0.186	0.116

* $\times$ 0.305 = m.

The variation of C_L and C_D with the Reynolds number is known as **scale effect.** These are shown in Figs. 11.4.11 and 11.4.12 for some typical airfoils. Figure 11.4.12 shows in dashed lines also the theoretical

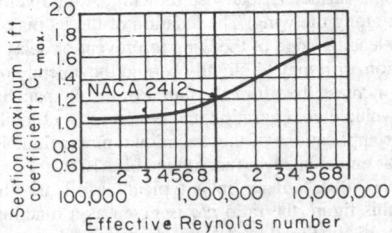

Fig. 11.4.11 Scale effect on section maximum lift coefficient.

In addition to the Reynolds number, airfoil characteristics also depend upon the **Mach number** (see "Supersonic and Hypersonic Aerodynamics" below).

Laminar-flow Wings NACA developed a series of **low-drag wings** in which the distribution of thickness along the chord is so selected as to maintain **laminar flow** over as much of the wing surface as possible.

The low-drag wing under controlled conditions of surface smoothness may have drag coefficients about 30 percent lower than those obtained on normal conventional wings. Low-drag airfoils are so sensitive to roughness in any form that the full advantage of laminar flow is unobtainable.

Transonic Airfoils At airplane Mach numbers of about M = 0.75, normal airfoil sections begin to show a greater drag, which increases sharply as the speed of sound is approached. Airfoil sections known as **transonic airfoils** have been developed which significantly delay this **drag rise** to Mach numbers of 0.9 or greater (Fig. 11.4.13). Advantage may also be taken of these airfoil sections by increasing the thickness of the airfoil at a given Mach number without suffering an increase in drag.

Wing sweepback (Fig. 11.4.14) or sweepforward also delays the onset of the drag rise to higher Mach numbers. The flow component normal to the wing leading edge, $M \cos \Gamma$, determines the effective Mach number felt by the wing. The parallel component, $M \sin \Gamma$, does not significantly influence the drag rise. Therefore for a flight Mach number M, the wing senses it is operating at a lower Mach number given by $M \cos \Gamma$.

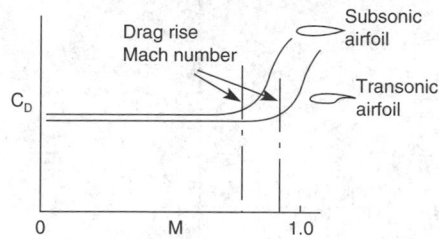

Fig. 11.4.13 Drag rise delay due to transonic section.

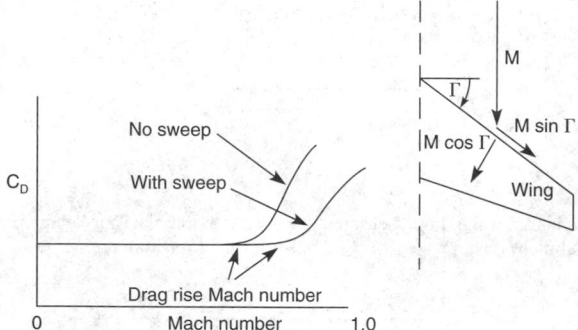

Fig. 11.4.14 Influence of wing sweepback on drag rise Mach number.

Flaps and Slots

The maximum lift of a wing can be increased by the use of **slots** on the leading edge or **flaps** on the trailing edge. **Fixed slots** are formed by rigidly attaching a curved sheet of metal or a small auxiliary airfoil to the leading edge of the wing. The trailing-edge flap is used to give increased lift at moderate angles of attack and to increase C_{Lmax}. Theoretical analyses of the effects are given in *NACA-TR* 938 and *Br. ARC-R&M* 1095. Flight experience indicates that the four types shown in Fig. 11.4.15 have special advantages over other known types. Values of C_{Lmax} for these types are shown in Table 11.4.2. All flaps cause a diving

Table 11.4.2 Flaps*

Type of flap	Section C_{Lmax} (approx)	Finite wing C_{Lmax} (approx)	Remarks
Plain	2.4	1.9	$C_{f/c} = 0.20$. Sensitive to leakage between flap and wing.
Split	2.6	2.0	$C_{f/c} = 0.20$. Simplest type of flap.
Single slotted	2.8 3.0	2.2 2.4	$C_{f/c} = 0.25$. Shape of slot and location of deflected flap are critical.
Double slotted	3.0 3.4	2.4 2.7	$C_{f/c} = 0.25$. Shape of slot and location of deflected flap are critical.

* The data is for wing thickness $t/c = 0.12$ or 0.15. Lift increment is dependent on the leading-edge radius of the wing.

moment $(-C_m)$ which must be trimmed out by a download on the tail, and this download reduces C_{Lmax}. The correction is ΔC_{Lmax} (trim) $= \Delta C_{mo}(l/c)$, where l/c is the tail length in mean chords.

Boundary-Layer Control (BLC) This includes numerous schemes for (1) maintaining laminar flow in the boundary layer in the flow over a

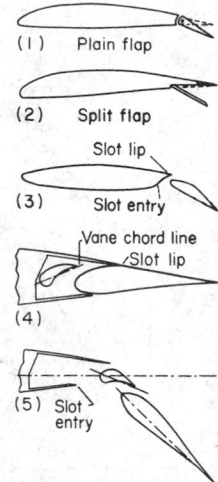

Fig. 11.4.15 Trailing-edge flaps: (1) plain flap; (2) split flap; (3) single-slotted flap; (4) double-slotted flap (retracted); (5) double-slotted flap (extended).

wing or (2) preventing flow separation. Schemes in the first category try to obtain the lower frictional drag of laminar flow either by providing favorable pressure gradients, as in the NACA "low-drag" wings, or by removing part of the boundary layer. The boundary-layer thickness can be partially controlled by the use of suction applied either to spanwise slots or to porous areas. The flow so obtained approximates the laminar skin-friction drag coefficients (see Fig. 11.4.12). Schemes in the second category try to delay or improve the stall. Examples are the leading-edge slot, slotted flaps, and various forms of suction or blowing applied through transverse slots. The slotted flap is a highly effective form of boundary-layer control that delays flow separation on the flap.

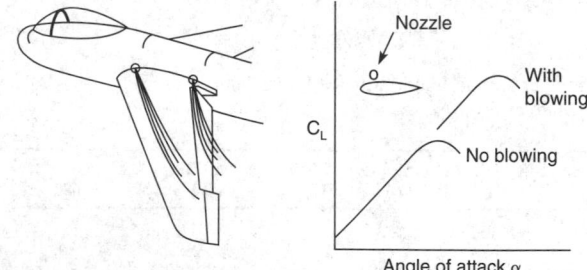

Fig. 11.4.16 Influence of spanwise blowing.

Spanwise Blowing Another technique for increasing the lift from an aerodynamic surface is spanwise blowing. A high-velocity jet of air is blown out along the wing. The location of the jet may be parallel to and near the leading edge of the wing to prevent or delay leading-edge flow separation, or it may be slightly behind the juncture of the trailing-edge flap to increase the lift of the flap. Power for spanwise blowing is sensitive to vehicle configuration and works most efficiently on swept wings. Other applications of this technique may include local blowing on ailerons or empennage surfaces (Fig. 11.4.16).

The **pressure distributions** on a typical airfoil are shown in Fig. 11.4.17. In this figure the ratio p/q is given as a function of distance along the chord. In Fig. 11.4.18 the effect of addition of an external airfoil flap is shown.

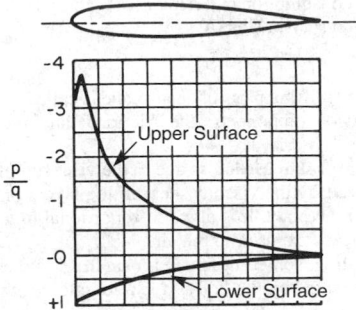

Fig. 11.4.17 Pressure distribution over a typical airfoil.

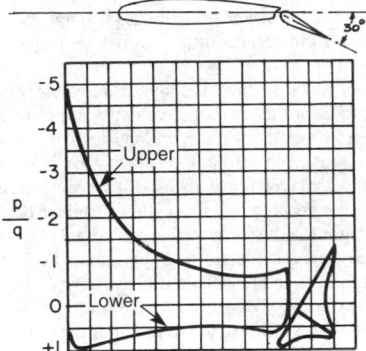

Fig. 11.4.18 Pressure distribution on an airfoil-flap combination.

Airplane Performance

In **horizontal flight** the lift of the wings must be equal to the weight of the airplane, or $L = W$, where W is the gross weight, lb (kg). This equation determines the minimum horizontal speed of an airplane since

$$W = C_{Lmax} \tfrac{1}{2}\rho V_{min}^2 S \quad \text{or} \quad V_{min} = \sqrt{2W/C_{Lmax}\rho S}$$

This corresponds to the **landing speed** without power, or the **stalling speed.**

In uniform horizontal flight the propeller thrust T must be exactly equal to the drag of the airplane, or $T = D$. Then TV = power available from the propeller and DV = power required for overcoming the drag of the airplane; for horizontal flight $TV = DV$. If the airplane climbs at a rate dh/dt, where h = height, then an additional power = $W\,(dh/dt)$ is required to increase the potential energy of the airplane. The equilibrium condition becomes

$$TV = DV + W(dh/dt)$$

If the **thrust power available** P_{Ta} is measured in hp, D and W in lb, and V and dh/dt in ft/s,

$$P_{Ta} = \frac{1}{550}DV + \frac{1}{550}W(dh/dt) \qquad (11.4.1)$$

The thrust horsepower available $P_{Ta} = \eta P_a$, where P_a = **available engine horsepower** and η is the propeller efficiency. P_{Ta} is determined as a function of V from the engine and propeller characteristics. The **thrust horsepower required** to overcome the drag D is $P_{Tr} = DV/550$. The **drag** is the sum of the wing-profile drag D_0, the wing induced drag D_i, and the parasite drag of the other airplane parts D_p. D_0 can be obtained from wing-profile tests corrected to full-scale Reynolds number, D_i from the induced drag formula $D_i = L^2/\pi q b^2$, and D_p by summation of the parasite drag components due to fuselage, tail surfaces, landing gear, etc.

From Eq. (11.4.1) the performance of an airplane can be obtained either graphically or analytically. In the graphical determination the curves of power available and power required are plotted for a fixed

altitude (Fig. 11.4.19). If $P_{Ta} > P_{Tr}$, the excess horsepower can be used either for horizontal acceleration to a higher speed or for climbing. The maximum speed V_{max} occurs when $P_{Ta} = P_{Tr}$. The rate of climb is determined by the equation $dh/dt = 550(P_{Ta} - P_{Tr})/W$. To calculate the

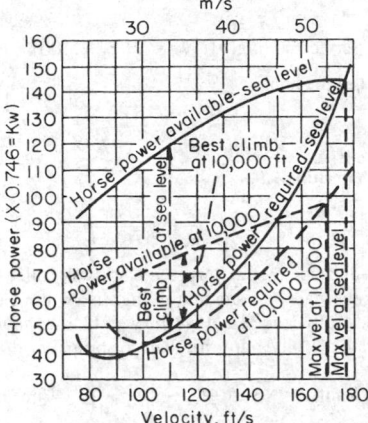

Fig. 11.4.19 Construction for determining airplane performance.

maximum velocity and rate of climb for another altitude, another set of curves of P_{Ta} and P_{Tr} versus V is constructed. At any altitude at which the air density is ρ and for a given angle of attack (corresponding to an unchanged lift/drag ratio), $L/D = L_0/D_0$. As the lift is equal to the weight, $L = L_0 = W$, therefore $D = D_0$ and

$$P_{Tr}/P_{Tr0} = VD/V_0 D_0 = V/V_0$$

But $C_L = C_{L0}$ and $C_L \cdot \tfrac{1}{2}\rho V^2 S = C_{L0}\tfrac{1}{2}\rho_0 V_0^2 S$. Therefore $V/V_0 = \sqrt{\rho_0/\rho}$. From this relation, new curves of P_{Tr} versus V may be constructed for various altitudes by multiplying both the ordinates and abscissas of the original curve by $\sqrt{\rho_0/\rho}$.

Figure 11.4.19 shows a performance chart for a 2,100-lb airplane. The maximum velocity at 10,000 ft altitude is 168.5 ft/s. At sea level $(P_{Ta} - P_{Tr})_{max} = 71$ and the rate of climb is $71 \times 33,000/2,100 = 1,120$ ft/min. At 10,000 ft altitude the rate is $30 \times 33,000/2,100 = 470$ ft/min.

When the curves of power required and power available become tangent to each other, there is only one speed at which the airplane can fly level and the rate of climb is zero. The corresponding altitude is the **absolute ceiling** H. It can be determined from the curve of maximum rate of climb as a function of altitude when this is approximated by a straight

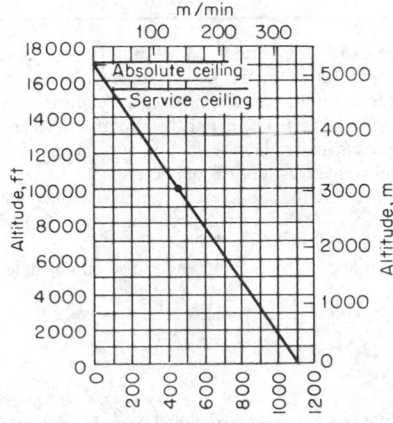

Fig. 11.4.20 Characteristic rate-of-climb curves.

line (Fig. 11.4.20). The **service ceiling** h_s is defined as the altitude at which the rate of climb is 100 ft/min. For a linear decrease of rate of climb the following approximations hold:

Absolute ceiling: $\quad H = r_0 h/(r_0 - r)$

where r_0 is the rate of climb at sea level and r is the rate of climb at altitude h.

Service ceiling: $\quad h_s = H(r_0 - 100)/r_0$

Altitude climbed in t min,

$$h = H(1 - e^k)$$

where $k = -r_0 t/H$.

Time to climb to altitude h,

$$t = 2.303 \frac{H}{r_0} \log \frac{H}{H - h}$$

The maximum distance that an airplane can fly is called its **range**, and the length of time that it can remain flying, its **endurance**. If W_0 is the weight in pounds fully loaded and W_1 the weight after having consumed its fuel at the rate of C lb/(bhp · h), then for propeller-driven aircraft,

$$\text{Range} = 863.5 \frac{\eta}{C} \cdot \frac{L}{D} \cdot \log \frac{W_0}{W_1} \quad \text{mi}$$

$$\text{Endurance} = 750 \frac{\sqrt{W}}{V_c} \cdot \frac{\eta}{C} \cdot \frac{L}{D} \left(\frac{1}{\sqrt{W_1}} - \frac{1}{\sqrt{W_0}} \right) \quad \text{hours}$$

where η is the average propulsive efficiency and $L/D = C_L/C_D$ is the ratio of lift to drag. V_c is the cruising speed, mi/h, at any gross weight W, lb. See Table 11.4.3 for dimensions and performance of selected airplanes.

Power Available The maximum efficiency η_m and the diameter of a propeller to absorb a given power at a given speed and rpm are found from a propeller-performance curve. The thrust horsepower at maximum speed P_{Tm} is found from $P_{Tm} = \eta_m P_m$. The **thrust horsepower at any speed** can be approximately determined from the ratios given in the following table:

V/V_{max}		0.20	0.30	0.40	0.50	0.60	0.70	0.80	0.90	1.00	1.10
P_T/P_{Tm}	Fixed-pitch propeller	0.29	0.44	0.57	0.68	0.77	0.84	0.90	0.96	1.00	1.03
	constant-rpm propeller	0.47	0.62	0.74	0.82	0.88	0.93	0.97	0.99	1.00	1.00

The **brake horsepower** of an engine decreases with increase of altitude. The variation of P_T with altitude depends on engine and propeller characteristics with average values as follows:

h, ft	0	5,000	10,000	15,000	20,000	25,000
P_T/P_{T0} (fixed pitch)	1.00	0.82	0.66	0.52	0.41	0.30
P_T/P_{T0} (controllable pitch)	1.00	0.85	0.71	0.59	0.48	0.38

Performance with Jet Thrust In all cases where jet thrust constitutes either a part or all of the power source, it is necessary to use graphical methods, plotting thrust and drag as functions of speed at each altitude to be investigated. The major thrust corrections are those due to losses in (1) the air intake, (2) the ducting system, and (3) the tailpipe. (See Jet Propulsion and Aircraft Propellers.)

An excess thrust T, lb, at a speed V, mi/h, is equivalent to a thrust horsepower $P_T = TV/375$. The corresponding rate of climb is $dh/dt = 33,000 P_T/W = 88 TV/W$.

For jet-driven aircraft, range and endurance can be determined from

$$\text{Range} = 1.9285 \left(\frac{1}{\sqrt{\rho S}} \right) \cdot \frac{1}{C_t} \cdot \frac{C_L^{1/2}}{C_D} \cdot (W_0^{1/2} - W_1^{1/2}) \quad \text{mi}$$

and

$$\text{Endurance} = \frac{1}{C_t} \cdot \frac{C_L}{C_D} \cdot \log \left(\frac{W_0}{W_1} \right) \quad \text{hours}$$

where C_t is fuel consumption in lb/(lb thrust) · hour, S is wing area in ft², and ρ is the air density in slugs/ft³.

Parasite Drag

The drag of the nonlifting parts of an airplane is called the **parasite drag**. It consists of two components: the frictional and the eddy-making drag.

Frictional drag, or **skin friction**, is due to the viscosity of the fluid. It is the force produced by the viscous shear in the layers of fluid immediately adjacent to the body. It is always proportional to the wetted area, i.e., the total surface exposed to the air.

Eddy-making drag, sometimes called **form drag**, is due to the disturbance or wake created by the body. It is a function of the shape of the body.

The total drag of a body may be composed of the two components in any proportion, varying from almost pure skin friction for a plate edgewise or a good streamline form to 100 percent form drag for a flat plate normal to the wind. A **steamline form** is a shape having very low form drag. Such a form creates little disturbance in moving through a fluid.

When air flows past a surface, the layer immediately adjacent to the surface adheres to it, or the tangential velocity at the surface is zero. In the transition region near the surface, which is called the **boundary layer**, the velocity increases from zero to the velocity of the stream. When the flow in the boundary layer proceeds as if it were made up of laminae sliding smoothly over each other, it is called a **laminar boundary layer**. If there are also irregular motions in the layers normal to the surface, it is a **turbulent boundary layer**. Under normal conditions the flow is laminar at low Reynolds numbers and turbulent at high Re with a transition range of values of Re extending between 5×10^5 and 5×10^7. The profile-drag coefficients corresponding to these conditions are shown in Fig. 11.4.12.

For laminar flow the friction-drag coefficient is practically independent of surface roughness and for a flat plate is given by the Blasius equation,

$$C_{DF} = 2.656/\sqrt{Re}$$

The turbulent boundary layer is thicker and produces a greater frictional drag. For a smooth flat plate with a turbulent boundary layer

$$C_{DF} = 0.91/\log Re^{2.58}$$

For rough surfaces the drag coefficients are increased (see von Kármán, *Jour. Aeronaut. Sci.,* **1,** no. 1, 1934). The drag coefficients as given above are based on *projected area* of a double-surfaced plane. If *wetted* area is used, the coefficients must be divided by 2. The frictional drag is $D_F = C_{DF} qS$, where S is the *projected area*, or $D_F = \frac{1}{2} C_{DF} qA$, where A is the wetted area.

Values of C_{DF} for double-surfaced planes may be estimated from the following tabulation:

Laminar flow (Blasius equation):

Re	10	10^2	10^3	10^4	10^5	10^6
C_{DF}	0.838	0.265	0.0838	0.0265	0.0084	0.0265

Turbulent flow:

Re	10^5	10^6	10^7	10^8	10^9	10^{10}
C_{DF}	0.0148	0.0089	0.0060	0.0043	0.00315	0.0024

These are double-surface values which facilitate direct comparison with wing-drag coefficients based on projected area. For calculations involving wetted area, use one-half of double-surface coefficients. In-

Table 11.4.3   Principal Dimensions and Performance of Typical Airplanes*

	Aero Commander	BAC	Beech	Bell	Boeing	Boeing	Lockheed	Douglas	McDonnell Douglas	Piper	Sikorsky
Designation model no.	680 F-P	One-Eleven	Bonanza S35	Ranger 47J2-A	747-200B	727	1011-1	DC-8 Series 5D	DC-10-10	Super Cub Pa 18-150	S-58
Type	Executive transport	Jet transport	Business plane	Executive helicopter	Passenger	Jet transport	Passenger	Jet transport	Passenger	Utility plane	Helicopter
No. passengers	5–7	63	4–6	4	374–500	70–114	256–400	116–176	255–380	2	12–18
Cargo capacity, lb	2,815	N.A.	270	1,120		8,550		20,850		50	
Span, ft	49.5	83.5	33.4	37.0†	195.6	108.6	155.3	142.3	155.3	35.3	56.0†
Overall length, ft	35.1	92.1	26.4	43.3	231.3	134.3	178.6	150.5	181.4	22.5	65.8
Overall height, ft	14.5	23.8	6.5	9.3	63.4	34.0	55.3	42.3	58	6.7	14.3
Wing area, ft²	255	980	181	Rotor	5,500	1,650	3,456	2,868	3,861	178.5	Rotor
Weight empty, lb‡	5,185	58,000	1,885	1,730	361,216	86,000	234,275	124,529	231,779	930	7,900
Weight gross, lb	8,000	73,800	3,300	2,850	775,000	142,000	430,000	310,000	430,000	1,750	13,000
Power plant	(2) Lyc IGSO-540B1A	(2) RR Spey 2	Con IO-520-B	Lyc VO540	JT9D-3, -7	(3) PW JT8D-1	RB-211-22B	(4) PW JT3D-3	CF6-6D	Lyc O-320	WR R1820
High speed, knots	252	469	184	91	528	539	530	521	530	113	106
Cruise speed, knots	220	441	178	81	478	513	473	472	473	100	84
Landing speed, knots	91	N.A.	54	0	140	121	125	133	138	37	0
Range, nautical miles	1,310	1,700	1,145	260	5,748	2,320	2,500	8,600	2,110	460	247

* ft × 0.305 = m; ft² × 0.0929 = m²; lb m × 0.454 = kg.
† Rotor diameter.
‡ Operating weight empty (includes trapped oil).
N.A. Not available.

terpolations in the foregoing tables must allow for the logarithmic functions; i.e., the variation in C_{DF} is not linear with Re.

Drag Coefficients of Various Bodies

For **bodies with sharp edges** the drag coefficients are almost independent of the Reynolds number, for most of the resistance is due to the difference in pressure on the front and rear surfaces. Table 11.4.4 gives $C_D = D/qS$, where S is the maximum cross section perpendicular to the wind.

For **rounded bodies** such as **spheres, cylinders,** and **ellipsoids** the drag coefficient depends markedly upon the Reynolds number, the surface roughness, and the degree of turbulence in the airstream. A sphere and a cylinder, for instance, experience a sudden reduction in C_D as the Reynolds number exceeds a certain critical value. The reason is that at low speeds (small Re) the flow in the boundary layer adjacent to the body is laminar and the flow separates at about 83° from the front (Fig. 11.4.21). A wide wake thus gives a large drag. At higher speeds (large Re) the boundary layer becomes turbulent, gets additional energy from the outside flow, and does not separate on the front side of the sphere. The drag coefficient is reduced from about 0.47 to about 0.08 at a critical Reynolds number of about 400,000 in free air. Turbulence in the airstream reduces the value of the critical Reynolds number (Fig. 11.4.22). The Reynolds number at which the sphere drag $C_D = 0.3$ is

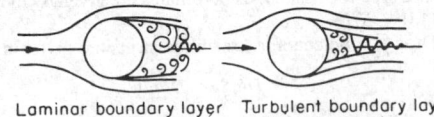

Fig. 11.4.21 Boundary layer of a sphere.

Laminar boundary layer (early separation— wide wake)   Turbulent boundary layer (later separation— narrow wake)

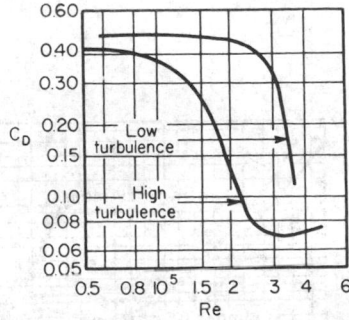

Fig. 11.4.22 Drag coefficients of a sphere as a function of Reynolds number and of turbulence.

taken as a criterion of the amount of turbulence in the airstream of wind tunnels.

Cylinders The drag coefficient of a cylinder with its axis normal to the wind is given as a function of Reynolds number in Fig. 11.4.23. Cylinder drag is sensitive to both Reynolds and Mach number. Figure 11.4.23 gives the Reynolds number effect for M = 0.35. The increase in C_D because of Mach number is approximately:

M	0.35	0.4	0.6	0.8	1.0
C_D increase, percent	0	2	20	50	70

(See *NACA-TN* 2960.)

Streamline Forms The drag of a streamline body of revolution depends to a very marked extent on the Reynolds number. The difference between extreme types at a given Reynolds number is of the same

Table 11.4.4   Drag Coefficients

Object	Proportions	Attitude	C_D
Rectangular plate, sides a and b	$\dfrac{a}{b} = $ 1 4 8 12.5 25 50 ∞		1.16 1.17 1.23 1.34 1.57 1.76 2.00
Two disks, spaced a distance l apart	$\dfrac{l}{d} = $ 1 1.5 2 3		0.93 0.78 1.04 1.52
Cylinder	$\dfrac{l}{d} = $ 1 2 4 7		0.91 0.85 0.87 0.99
Circular disk			1.11
Hemispherical cup, open back			0.41
Hemispherical cup, open front, parachute			1.35
Cone, closed base			$\alpha = 60°$, 0.51 $\alpha = 30°$, 0.34

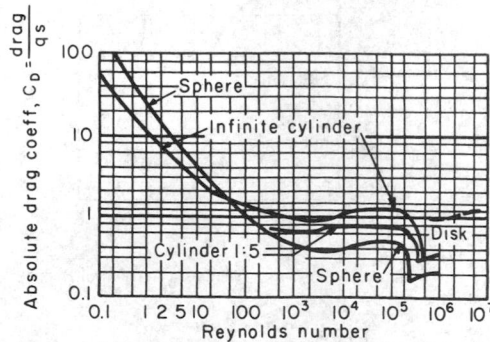

Fig. 11.4.23 Drag coefficients of cylinders and spheres.

order as the change in C_D for a given form for values of Re from 10^6 to 10^7.

Tests reported in *NACA-TN* 614 indicate that the shape for minimum drag should have a fairly sharp nose and tail. At Re = 6.6×10^6 the best forms for a fineness ratio (ratio of length to diameter) 5 have a drag

$$D = 0.040qA = 0.0175qV^{2/3}$$

where A = max cross-sectional area and V = volume. This value is equivalent to about 1.0 lb/ft² at 100 mi/h. For Re > 5×10^6, the drag coefficients vary approximately as Re$^{0.15}$.

Minimum drag on the basis of cross-sectional or frontal area is obtained with a fineness ratio of the order of 2 to 3. Minimum drag on the basis of contained volume is obtained with a fineness ratio of the order of 4 to 6 (see *NACA-TR* 291).

The following table gives the ordinates for good streamline shapes: the **Navy strut,** a two-dimensional shape; and the Class C **airships,** a three-dimensional shape. Streamline shapes for high Mach numbers have a high fineness ratio.

Percent length		1.25	2.50	4.00	7.50	10.0	12.50
Percent of max ordinate	Navy strut	26	37.1	52.50	63.00	72.0	78.50
	Class C airship	20	33.5	52.60	65.80	75.8	83.50

Percent length		20	40	60	80	90
Percent of max ordinate	Navy strut	91.1	99.5	86.1	56.2	33.8
	Class C airship	94.7	99.0	88.5	60.5	49.3

Test data on the *RM*-10 shape are given in *NACA-TR* 1160. This shape is a parabolic-arc type for which the coordinates are given by the equation $r_x = X/7.5(1.0 - X/L)$.

Struts   Drag coefficients for streamline struts are given in the form

$$C_D = D/qS_f = D/qdl$$

where S_f is the projected frontal area; d is the thickness, ft; and l is the length, ft. The variation of C_D with Re for a Navy No. 1 strut of fineness ratio 3 is as follows:

$Re \times 10^{-5}$	0.75	1.0	1.25	1.5	2.0	3.0	4.0
C_D	0.114	0.102	0.093	0.088	0.085	0.077	0.073

Wing-Profile Drag For accurate basic values of profile drag it is necessary to refer to test data on the wing section employed. In the absence of test data, the following approximate values may be used:

Average t/c		0.10	0.12	0.14	0.16
Basic	C_{D0}	0.0058	0.0060	0.0063	0.0067
	D_0/S	0.148	0.154	0.161	0.171
Best wing	C_{D0}	0.0078	0.0080	0.0083	0.0087
	D_0/S	0.20	0.205	0.212	0.223
Average roughness	C_{D0}	0.0098	0.0100	0.0103	0.0107
	D_0/S	0.25	0.256	0.264	0.274

The "basic" values are for a perfectly smooth wing of infinite aspect ratio, "best wing" values are for smooth full flush-riveted construction, and "average-roughness" values are for flush-riveted leading-edge with brazier head rivets back of the 20 percent chord point. C_{D0} is the profile-drag coefficient ($D_0 = C_{D0}qS$). The values of D_0/S are drags in lb/ft^2 of projected wing area at 100 mi/h in standard air ($q_0 = 25.58$ lb/ft^2). t/c is the maximum wing thickness as a fraction of the chord.

Faired values of NACA data on **symmetrical sections** at Re = 8×10^6 give the variation of minimum C_{D0} and C_{DA} with thickness as follows:

NACA section	0006	0009	0012	0015	0018	0021	0025	0030	0035
Min C_{D0}	0.0051	0.0056	0.0061	0.0067	0.0073	0.0080	0.0089	0.0103	0.0120
C_{DA} (see below)	0.085	0.062	0.050	0.045	0.041	0.038	0.036	0.034	0.034

NOTE: C_{DA} is the drag coefficient based on the frontal area. The 00 section is also suitable for use in struts or fairings. (See *NACA-TR* 628, 647, 669, 708.)

Drag of Tail Surfaces Owing to joints, control balances, and interference effects, the drag of a control surface is much higher than that of the basic section. The average value is about 0.40 lb/ft^2 at 100 mi/h in standard density.

Streamline Wire The drag coefficient of a standard streamline wire of lenticular or elliptical cross section is given as a function of Reynolds number in Fig. 11.4.24.

The drag coefficients of **wires and cables** are also shown in Fig. 11.4.24.

Elliptic Cylinders

Fineness ratio	C_D when Re is			
	3×10^4	6×10^4	1×10^5	2×10^5
1:1	1.20	1.22	1.22	1.23
2:1	0.62	0.57	0.46	0.35
4:1	0.32	0.32	0.30	0.24
8:1	0.27	0.23	0.22	0.21

The above data are for M < 0.4. For additional data see *NACA-TR* 619.

C_L and C_D for **inclined wires** are as follows:

Angle of attack, deg	0	15	30	45	60	75	90
C_L	0	0.09	0.25	0.39	0.42	0.27	0
C_D	0.01	0.05	0.17	0.46	0.77	1.01	1.12

C_L and C_D are based on the area $S = LD$, where L is the length and D the diam, ft.

(Additional data on prismatic cylinders are given in *NACA-TR* 619 and *NACA-TN* 3038.)

Engine Drag, Nacelle Drag The drag of an uncowled air-cooled engine is approximately $D = 0.050d^2(V/100)^2$, where d is the overall diameter of the engine, in, and V is the speed, mi/h. The average drag

in pounds at 100 mi/h for a fixed-slot NACA-type cowl is as follows:

Engine diam, in	40	44	48	52	56
Drag, lb at 100 mi/h	28	34	40	47	54

If adjustable **cowl flaps** or controlled airflow is used, the engine drag need not vary as the square of the speed; it may remain substantially constant. Theoretically, it is possible to cool an engine by forced cooling at an expenditure of about 2 percent of the engine power.

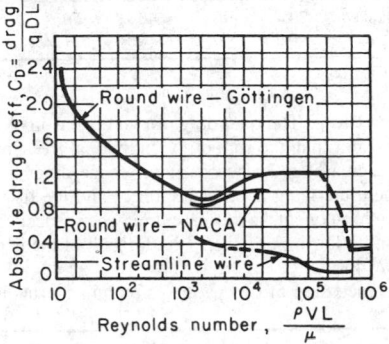

Fig. 11.4.24 Drag coefficients of wires.

The average drag coefficient of a nacelle is $C_D = 0.20$, based on frontal area. This is equivalent to about 5.0 lb/ft^2 at 100 mi/h. A very clean form may be as low as 3 lb/ft^2 at 100 mi/h. The drag of a pure streamline form would be of the order of 1 lb/ft^2 at 100 mi/h (see *NACA-TR* 313, 314, and 415).

Fuselage Drag Owing to various projections and irregularities in the surface, the drag of an average airplane fuselage is considerably greater than the drag of a pure streamline form. Since the increase is due in effect to a substantially constant drag increment, the drag per unit of cross-sectional area tends to decrease with increase in the size of the fuselage. At 100 mi/h the drag in lb/ft^2 will be approximately as follows:

A, ft^2	10	15	20	40	60
D/A, average	5.5	4.5	4.0	3.3	3.0
D/A, lower limit	4.0	3.3	3.0	2.4	2.3

Cockpit enclosures if properly designed and blended into fuselage lines, do not appreciably increase the fuselage drag coefficient. Sharp junctures must be avoided. The best fairing radius is approximately 25 percent of the enclosure height. Length of tail fairing should be 4 × height (see *NACA-TR* 730).

Seaplane Floats, Flying-Boat Hulls The drag of a seaplane float is between 3 and 6 lb/ft^2 at 100 mi/h, depending on the lines. The step accounts for 5 to 10 percent of the drag.

The drag of a flying-boat hull is comparable to, but slightly higher than, the drag of a fuselage of the same cross-sectional area. Average values at 100 mi/h are as follows:

A, ft^2	40	60	80	100
D/A, lb/ft^2	4.5	4.3	4.1	4.0

Wire Mesh Measurements on square pieces of exposed wire mesh have given the following results:

Percent area blocked	100	80	60	50	40	30	20
Percent flat plate, C_D	100	92	77	60	43	26	10

The pressure drop through a screen in a tube is given by

$$\Delta p = C_D q = \frac{1}{2} C_D \rho V^2$$

where C_D is a function of the percent area blocked as follows:

Percent area blocked	20	30	40	50	60	70
C_D	0.20	0.45	0.90	1.60	3.40	7.20

(See *Br. ARC-R&M* 1469.)

Interference Drag The total drag of two objects in close proximity is generally greater than the sum of the individual free-flow drags. The increase (or decrease) is known as interference drag. It is especially important where the wing is one component and the fuselage or a nacelle the other component. In this case it may be reduced to a minimum by an appropriate fairing or "fillet" in the path of the expanding flow (see *NACA-TN* 460).

The interference drag of two parallel streamline struts is as follows:

Spacing/thickness	5	4	3	2	1.5	1.25
D/D_0	1.00	1.06	1.12	1.25	2.25	2.87

The interference drag of a strut intersection with a flat surface is a function of the intersection angle θ. The drag increase, expressed as an equivalent length increase ΔL in diameters, is as follows:

θ, deg	90	70	60	50	40	30	20
ΔL	0	1.5	2.5	4.0	6.0	9.5	14.0

Nonintrusive flow field measurements about airfoils can be made by laser Doppler velocimeter (LDV) techniques. A typical LDV measurement system setup is shown in Fig. 11.4.25. A seeded airflow generates the necessary reflective particles to permit two-dimensional flow directions and velocities to be accurately recorded.

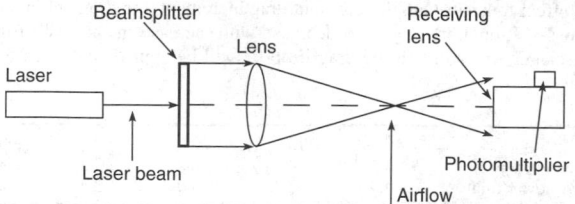

Fig. 11.4.25 Typical laser doppler velocimeter measurement system.

STABILITY AND CONTROL

Control An airplane is controlled in flight by imposing yawing, pitching, and rolling moments by use of rudders, elevators, and ailerons. This is known as the **three-control** system and is in almost universal use, although it is possible to dispense with either the ailerons or the rudders and so obtain a two-control system.

Controllability is separate and distinct from stability in physical significance, but not necessarily so in flight. An airplane may be made automatically stable by gyroscopic or other devices that actuate the controls mechanically; it is inherently stable if, on disturbance from any cause, the aerodynamic forces and moments induced always tend to return the airplane to its original attitude.

Stability may be either static or dynamic. An airplane is **statically stable** if the moments tend to return it to the original attitude. It is **dynamically stable** if the oscillations produced by the static stability are rapidly damped out. If it is **statically unstable,** any departure tends to increase, there being an upsetting moment instead of a restoring moment. If it is **dynamically unstable,** the oscillations due to static stability tend to increase in amplitude with time. Static stability or a condition of stable equilibrium is necessary to obtain dynamic stability, but static stability does not ensure dynamic stability. Too much static stability may cause dynamic instability if damping is inadequate.

The common method of getting **static longitudinal stability** is by means of a tail surface. A wing alone has a lift force L acting at the a.c. and a moment M_0 about the a.c. (Fig. 11.4.26). The moment is measured

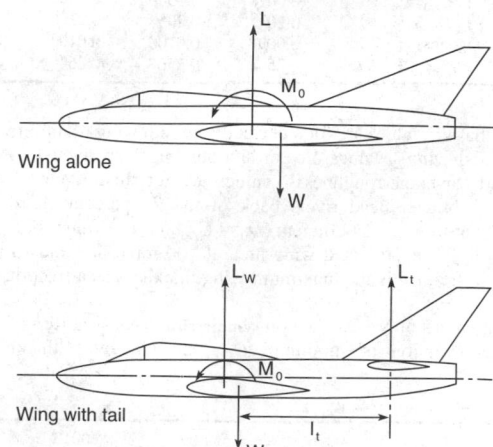

Fig. 11.4.26 Longitudinal stability of an airplane.

positively as airplane nose-up. Most wings alone are "unstable" (positive lift-curve slope), though possessing a small negative moment (diving moment). If there were no other air forces acting, this could be balanced by putting the center of gravity of the airplane behind the a.c. a distance δ such that $W\delta = M_0$. This arrangement would be in equilibrium but would be unstable. Upon a small increase in the angle of attack such as by a gust, the lift force will be increased. This gives a moment tending to increase further the angle of attack. The curve of the moment of the air forces on a wing about the center of gravity thus starts from the value M_0 and becomes positive for larger angles of attack. The curve is shown in Fig. 11.4.27. The positive slope of M versus α thus corresponds to an unstable moment. When a horizontal tail surface is added, a lift force also acts upon the tail. This gives a diving moment

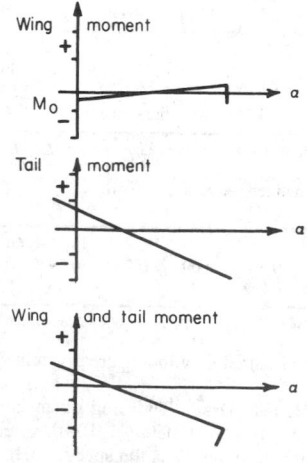

Fig. 11.4.27 Longitudinal stability. Wing and foil moments.

$M_t = -L_t \cdot l_t$. Upon an increase in the angle of attack α, the diving tail moment is increased more rapidly than the stalling moment of the wing. Thus, the combination is stable. The resultant curve of pitching moment against angle of attack has a small negative slope. Too steep a slope would indicate longitudinal stiffness and difficulty in control.

Rolling (or **banking**) does not produce any lateral shift in the center of the lift, so that there is no restoring moment as in pitch. However, when banked, the airplane sideslips toward the low wing. A fin placed above the center of gravity gives a lateral restoring moment that can correct the roll and stop the slip. The same effect can be obtained by **dihedral**, i.e., by raising the wing tips to give a transverse V. An effective dihedral of 1 to 3° on each side is generally required to obtain stability in roll. In low-wing monoplanes, 2° effective dihedral may require 8° or more of geometrical dihedral owing to interference between the wing and fuselage.

In a **yaw** or **slip** the line of action of the lateral force depends on the size of the effective vertical fin area. Insufficient fin surface aft will allow the skid or slip to increase. Too much fin surface aft will swing the nose of the plane around into a tight spiral. Sound design demands sufficient vertical tail surface for adequate directional control and then enough dihedral to provide lateral stability.

When moderate positive effective dihedral is present, the airplane will possess **static lateral stability,** and a low wing will come up automatically with very little yaw. If the dihedral is too great, the airplane may roll considerably in gusts, but there is little danger of the amplitude ever becoming excessive. **Dynamic lateral stability** is not assured by static stability in roll and yaw but requires that these be properly proportioned to the damping in roll and yaw.

Spiral instability is the result of too much fin surface and insufficient dihedral.

HELICOPTERS

REFERENCES: *Br. ARC-R & M* 1111, 1127, 1132, 1157, 1730, and 1859. *NACA-TM* 827, 836, 858. *NACA-TN* 626, 835, 1192, 3323, 3236. *NACA-TR* 434, 515, 905, 1078. *NACA Wartime Reports* L-97, L-101, L-110. NACA, "Conference on Helicopters," May, 1954. Gessow and Myers, "Aerodynamics of the Helicopter," Macmillan.

Helicopters derive lift, propulsion force, and control effect from adjustments in the blade angles of the rotor system. At least two rotors are required, and these may be arranged in any form that permits control over the reaction torque. The common arrangements are main lift rotor; auxiliary torque-control or tail rotor at 90°; two main rotors side by side; two main rotors fore and aft; and two main rotors, coaxial and oppositely rotating.

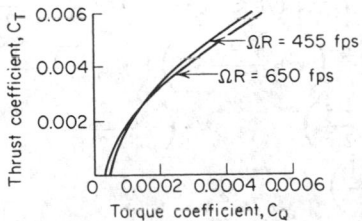

Fig. 11.4.28 Static thrust performance of NACA 8-H-12 blades.

The helicopter rotor is an actuator disk or momentum device that follows the same general laws as a propeller. In calculating the rotor performance, the major variables concerned are diameter D, ft (radius R); tip speed, $V_t = \Omega R$, ft/s; angular velocity of rotor $\Omega = 2\pi n$, rad/s; and rotor solidity q (= ratio of blade area/disk area). The rotor performance is usually stated in terms of coefficients similar to propeller coefficients. The rotor coefficients are:

$$\text{Thrust coefficient } C_T = T/\rho(\Omega R)^2 \pi R^2$$
$$\text{Torque coefficient } C_Q = Q/\rho(\Omega R)^2 \pi R^3$$
$$\text{Torque } Q = 5{,}250 \ \text{bhp/rpm}$$

Hovering The **hovering flight** condition may be calculated from basic rotor data given in Fig. 11.4.28. These data are taken from full-scale rotor tests. Surface-contour accuracy can reduce the total torque coefficient 6 to 7 percent. The power required for hovering flight is greatly reduced near the ground. Observed flight-test data from various sources are plotted in Fig. 11.4.29. The ordinates are heights above the ground measured in rotor diameters.

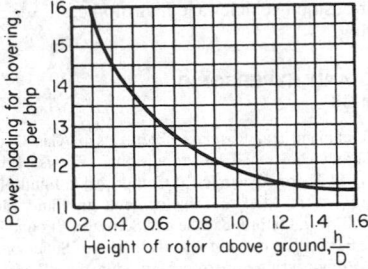

Fig. 11.4.29 Observed ground effect on Sikorsky-type helicopters.

Effect of Gross Weight on Rate of Climb Figure 11.4.30 from Talkin (*NACA-TN* 1192) shows the rate of climb that can be obtained by reducing the load on a helicopter that will just hover. Conversely, given the rate of climb with a given load, the curves determine the increase in load that will reduce the rate of climb to zero; i.e., they determine the maximum load for which hovering is possible.

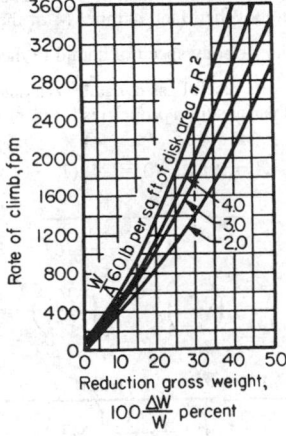

Fig. 11.4.30 Effect of gross weight on the rate of climb of a helicopter.

Performance with Forward Speed The performance of a typical single-main-rotor-type helicopter may be shown by a curve of C_{TR}/C_{QR} plotted against $\mu = v/\pi nD$, where v is the speed of the helicopter, as in Fig. 11.4.31. This curve includes the parasite-drag effects which are

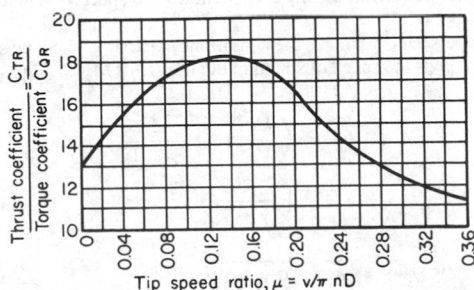

Fig. 11.4.31 Performance curve for a single-main-rotor-type helicopter.

appreciable at the higher values of μ. It is only a rough approximation to the experimentally determined values. C_{TR} may be calculated to determine C_{QR}, from which Q is obtained.

The performance of rotors at forward speeds involves a number of variables. For more complete treatment, see *NACA-TN* 1192 and *NACA Wartime Report* L-110.

GROUND-EFFECT MACHINES (GEM)

For data on air-cushion vehicles and hydrofoil craft, see Sec. 11.3 "Marine Engineering."

SUPERSONIC AND HYPERSONIC AERODYNAMICS

REFERENCES: Liepmann and Roshko, "Elements of Gas Dynamics," Wiley. "High Speed Aerodynamics and Jet Propulsion" 12 vols., Princeton. Shapiro, "The Dynamics and Thermodynamics of Compressible Fluid Flow," Vols. I, II, Ronald. Howarth (ed.), "Modern Developments in Fluid Mechanics—High Speed Flow," Vols. I, II, Oxford. Kuethe and Schetzer, "Foundations of Aerodynamics," Wiley. Ferri, "Elements of Aerodynamics of Supersonic Flows," Macmillan. Bonney, "Engineering Supersonic Aerodynamics," McGraw-Hill.

The effect of the compressibility of a fluid upon its motion is determined primarily by the **Mach number** M.

$$M = V/V_s \tag{11.4.2}$$

where V = speed of the fluid and V_s = speed of sound in the fluid which for air is $49.1\sqrt{T}$, where T is in °R and V_s in ft/s. The Mach number varies with position in the fluid, and the compressibility effect likewise varies from point to point.

If a *body moves through the atmosphere*, the overall compressibility effects are a function of the Mach number $\overline{M}$ of the body, defined as

$$\overline{M} = \text{velocity of body/speed of sound in the atmosphere}$$

Table 11.4.5 lists the useful **gas dynamics relations** between velocity, Mach number, and various fluid properties for isentropic flow. (See also Sec. 4, "Heat.")

$$V^2 = 2(h_0 - h) = 2C_p(T_0 - T)$$

$$\frac{V^2}{V_{s0}^2} = \frac{M^2}{1 + \dfrac{\gamma - 1}{2}M^2}$$

$$\left(\frac{p}{p_0}\right)^{(\gamma-1)/\gamma} = \frac{T}{T_0} = \left(\frac{\rho}{\rho_0}\right)^{\gamma-1} = \left(\frac{V_s}{V_{s0}}\right)^2 = \frac{1}{1 + \dfrac{\gamma - 1}{2}M^2}$$

$$\frac{\frac{1}{2}\rho V^2}{p_0} = \frac{(\gamma/2)M^2}{\left(1 + \dfrac{\gamma - 1}{2}M^2\right)^{\gamma/(\gamma-1)}} \tag{11.4.3}$$

$$\left(\frac{A}{A^*}\right)^2 = \frac{1}{M^2}\left[\frac{2}{\gamma + 1}\left(1 + \frac{\gamma - 1}{2}M^2\right)\right]^{(\gamma+1)/(\gamma-1)}$$

where h = enthalpy of the fluid, A = stream-tube cross section normal to the velocity, and the subscript $_0$ denotes the isentropic stagnation condition reached by the stream when stopped frictionlessly and adiabatically (hence isentropically). V_{s0} denotes the speed of sound in that medium.

The superscript $*$ denotes the conditions occurring when the speed of the fluid equals the speed of sound in the fluid. For M = 1, Eqs. (11.4.2) become

$$\frac{T^*}{T_0} = \left(\frac{V_s^*}{V_{s0}}\right)^2 = \frac{2}{\gamma + 1}$$

$$\frac{p^*}{p_0} = \left(\frac{2}{\gamma + 1}\right)^{\gamma/(\gamma-1)} \tag{11.4.4}$$

$$\frac{\rho^*}{\rho_0} = \left(\frac{2}{\gamma + 1}\right)^{1/(\gamma-1)}$$

For air $p^* = 0.52828p_0$; $\rho^* = 0.63394\rho_0$; $T^* = 0.83333T_0$.

For **subsonic regions** flow behaves similarly to the familiar hydraulics or incompressible aerodynamics: In particular, an increase of velocity is associated with a decrease of stream-tube area, and friction causes a pressure drop in a tube. There are no regions where the local flow velocity exceeds sonic speed. For **supersonic regions,** an increase of velocity is associated with an increase of stream-tube area, and friction causes a pressure rise in a tube. M = 1 is the dividing line between these two regions. For **transonic flows,** there are regions where the local flow exceeds sonic velocity, and this mixed-flow region requires special analysis. For **supersonic flows,** the entire flow field, with the exception of the regime near stagnation areas, has a velocity higher than the speed of sound.

The **hypersonic regime** is that range of very high supersonic speeds (usually taken as M > 5) where even a very streamlined body causes disturbance velocities comparable to the speed of sound, and stagnation temperatures can become so high that the gas molecules dissociate and become ionized.

Shock waves may occur at locally supersonic speeds. At low velocities the fluctuations that occur in the motion of a body (at the start of the motion or during flight) propagate away from the body at essentially the speed of sound. At higher subsonic speeds, fluctuations still propagate at the speed of sound in the fluid away from the body in all directions, but now the waves cannot get so far ahead; therefore, the force coefficients increase with an increase in the Mach number. When the Mach number of a body exceeds unity, the fluctuations instead of traveling away from the body in all directions are actually left behind by the body. These cases are illustrated in Fig. 11.4.32. As the supersonic speed is increased, the fluctuations are left farther behind and the force coefficients again decrease.

Where, for any reason, waves form an envelope, as at M > 1 in Fig. 11.4.32, a wave of finite pressure jump may result.

The speed of sound is higher in a higher-temperature region. For a compression wave, disturbances in the compressed (hence high-temperature) fluid will propagate faster than and overtake disturbances in the lower-temperature region. In this way shock waves are formed.

For a **stationary normal shock wave,** the fluid velocities V_1 before and V_2 after the shock are related by

$$V_1 V_2 = (V_s^*)^2 \tag{11.4.5}$$

Other properties are given by

$$p_2 - p_1 = \rho_1 V_1(V_1 - V_2) \tag{11.4.6}$$

$$\frac{T_2}{T_1} = \left(\frac{V_{s2}}{V_{s1}}\right)^2$$

$$= \left(\frac{2}{\gamma + 1}\right)^2 \frac{1}{M_1^2}\left(1 + \frac{\gamma - 1}{2}M_1^2\right)\left(\gamma M_1^2 - \frac{\gamma - 1}{2}\right) \tag{11.4.7}$$

$$M_2^2 = \left(1 + \frac{\gamma - 1}{2}M_1^2\right)\left(\gamma M_1^2 - \frac{\gamma - 1}{2}\right)^{-1} \tag{11.4.8}$$

$$\frac{p_{20}}{p_1} = \left(\frac{\gamma + 1}{2}\right)^{(\gamma+1)/(\gamma-1)} M^2 \left(\gamma - \frac{\gamma - 1}{2M_1^2}\right)^{-1/(\gamma-1)} \tag{11.4.9a}$$

$$\frac{p_{20}}{p_{10}} = \left(\frac{\gamma + 1}{2}\right)^{(\gamma+1)/(\gamma-1)} M^2$$

$$\times \left(1 + \frac{\gamma - 1}{2}M^2\right)^{-\gamma/(\gamma-1)}\left(\gamma - \frac{\gamma - 1}{2M^2}\right)^{-1/(\gamma-1)} \tag{11.4.9b}$$

The subscript 20 refers to the stagnation condition after the shock, and the subscript 10 refers to the stagnation condition ahead of the shock.

In a normal shock M, V, and p_0 decrease; p, ρ, T, and s increase, whereas T_0 remains unchanged. These relations are given numerically in Table 11.4.6.

If the shock is moving, these same relations apply *relative to the shock*. In particular, if the shock advances into stationary fluid, it does so at the speed V_1 which is always greater than the speed of sound in the stationary fluid by an amount dependent upon the shock strength.

Table 11.4.5 Isentropic Gas Dynamics Relations

M	$\dfrac{p}{p_0}$	$\dfrac{V}{V_{s0}}$	$\dfrac{A}{A^*}$	$\dfrac{\frac{1}{2}\rho V^2}{p_0}$	$\dfrac{\rho V}{\rho_0 V_{s0}}$	$\dfrac{\rho}{\rho_0}$	$\dfrac{T}{T_0}$	$\dfrac{V_s}{V_{s0}}$
0.00	1.000	0.000	∞	0.000	0.000	1.000	1.000	1.000
0.05	0.998	0.050	11.59	0.002	0.050	0.999	0.999	1.000
0.10	0.993	0.100	5.82	0.007	0.099	0.995	0.998	0.999
0.15	0.984	0.150	3.91	0.016	0.148	0.989	0.996	0.998
0.20	0.972	0.199	2.964	0.027	0.195	0.980	0.992	0.996
0.25	0.957	0.248	2.403	0.042	0.241	0.969	0.988	0.994
0.30	0.939	0.297	2.035	0.059	0.284	0.956	0.982	0.991
0.35	0.919	0.346	1.778	0.079	0.325	0.941	0.976	0.988
0.40	0.896	0.394	1.590	0.100	0.364	0.924	0.969	0.984
0.45	0.870	0.441	1.449	0.123	0.399	0.906	0.961	0.980
0.50	0.843	0.488	1.340	0.148	0.432	0.885	0.952	0.976
0.55	0.814	0.534	1.255	0.172	0.461	0.863	0.943	0.971
0.60	0.784	0.580	1.188	0.198	0.487	0.840	0.933	0.966
0.65	0.753	0.624	1.136	0.223	0.510	0.816	0.922	0.960
0.70	0.721	0.668	1.094	0.247	0.529	0.792	0.911	0.954
0.75	0.689	0.711	1.062	0.271	0.545	0.766	0.899	0.948
0.80	0.656	0.753	1.038	0.294	0.557	0.740	0.887	0.942
0.85	0.624	0.795	1.021	0.315	0.567	0.714	0.874	0.935
0.90	0.591	0.835	1.009	0.335	0.574	0.687	0.861	0.928
0.95	0.559	0.874	1.002	0.353	0.577	0.660	0.847	0.920
1.00	0.528	0.913	1.000	0.370	0.579	0.634	0.833	0.913
1.05	0.498	0.950	1.002	0.384	0.578	0.608	0.819	0.905
1.10	0.468	0.987	1.008	0.397	0.574	0.582	0.805	0.897
1.15	0.440	1.023	1.018	0.407	0.569	0.556	0.791	0.889
1.20	0.412	1.057	1.030	0.416	0.562	0.531	0.776	0.881
1.25	0.386	1.091	1.047	0.422	0.553	0.507	0.762	0.873
1.30	0.361	1.124	1.066	0.427	0.543	0.483	0.747	0.865
1.35	0.337	1.156	1.089	0.430	0.531	0.460	0.733	0.856
1.40	0.314	1.187	1.115	0.431	0.519	0.437	0.718	0.848
1.45	0.293	1.217	1.144	0.431	0.506	0.416	0.704	0.839
1.50	0.272	1.246	1.176	0.429	0.492	0.395	0.690	0.830
1.55	0.253	1.274	1.212	0.426	0.478	0.375	0.675	0.822
1.60	0.235	1.301	1.250	0.422	0.463	0.356	0.661	0.813
1.65	0.218	1.328	1.292	0.416	0.443	0.337	0.647	0.805
1.70	0.203	1.353	1.338	0.410	0.433	0.320	0.634	0.796
1.75	0.188	1.378	1.387	0.403	0.417	0.303	0.620	0.788
1.80	0.174	1.402	1.439	0.395	0.402	0.287	0.607	0.779
1.85	1.161	1.425	1.495	0.386	0.387	0.272	0.594	0.770
1.90	0.149	1.448	1.555	0.377	0.372	0.257	0.581	0.762
1.95	0.138	1.470	1.619	0.368	0.357	0.243	0.568	0.754
2.00	0.128	1.491	1.688	0.358	0.343	0.230	0.556	0.745
2.50	0.059	1.667	2.637	0.256	0.219	0.132	0.444	0.667
3.00	0.027	1.793	4.235	0.172	0.137	0.076	0.357	0.598
3.50	0.013	1.884	6.790	0.112	0.085	0.045	0.290	0.538
4.00	0.007	1.952	10.72	0.074	0.054	0.028	0.238	0.488
4.50	0.003	2.003	16.56	0.049	0.035	0.017	0.198	0.445
5.00	0.002	2.041	25.00	0.033	0.023	0.011	0.167	0.408
10.00	0.00002	2.182	536.00	0.002	0.001	0.005	0.048	0.218

SOURCE: Emmons, "Gas Dynamics Tables for Air," Dover Publications, Inc., 1947.

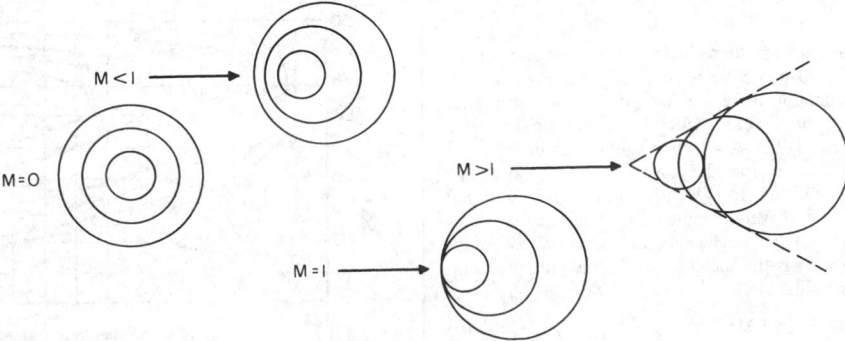

Fig. 11.4.32 Propagation of sound waves in moving streams.

Table 11.4.6 Normal Shock Relations

M	p_2/p_1	p_{20}/p_1	p_{20}/p_{10}	M_2	V_{s2}/V_{s1}	V_2/V_1	T_2/T_1	ρ_2/ρ_1
1.00	1.000	1.893	1.000	1.000	1.000	1.000	1.000	1.000
1.05	1.120	2.008	1.000	0.953	1.016	0.923	1.033	1.084
1.10	1.245	2.133	0.999	0.912	1.032	0.855	1.065	1.169
1.15	1.376	2.266	0.997	0.875	1.047	0.797	1.097	1.255
1.20	1.513	2.408	0.993	0.842	1.062	0.745	1.128	1.342
1.25	1.656	2.557	0.987	0.813	1.077	0.700	1.159	1.429
1.30	1.805	2.714	0.979	0.786	1.091	0.660	1.191	1.516
1.35	1.960	2.878	0.970	0.762	1.106	0.624	1.223	1.603
1.40	2.120	3.049	0.958	0.740	1.120	0.592	1.255	1.690
1.45	2.286	3.228	0.945	0.720	1.135	0.563	1.287	1.776
1.50	2.458	3.413	0.930	0.701	1.149	0.537	1.320	1.862
1.55	2.636	3.607	0.913	0.684	1.164	0.514	1.354	1.947
1.60	2.820	3.805	0.895	0.668	1.178	0.492	1.388	2.032
1.65	3.010	4.011	0.876	0.654	1.193	0.473	1.423	2.115
1.70	3.205	4.224	0.856	0.641	1.208	0.455	1.458	2.198
1.75	3.406	4.443	0.835	0.628	1.223	0.439	1.495	2.279
1.80	3.613	4.670	0.813	0.617	1.238	0.424	1.532	2.359
1.85	3.826	4.902	0.790	0.606	1.253	0.410	1.573	2.438
1.90	4.045	5.142	0.767	0.596	1.268	0.398	1.608	2.516
1.95	4.270	5.389	0.744	0.586	1.284	0.386	1.647	2.592
2.00	4.500	5.640	0.721	0.577	1.299	0.375	1.688	2.667
2.50	7.125	8.526	0.499	0.513	1.462	0.300	2.138	3.333
3.00	10.333	12.061	0.328	0.475	1.637	0.259	2.679	3.857
3.50	14.125	16.242	0.213	0.451	1.821	0.235	3.315	4.261
4.00	18.500	21.068	0.139	0.435	2.012	0.219	4.047	4.571
4.50	23.458	26.539	0.092	0.424	2.208	0.208	4.875	4.812
5.00	29.000	32.653	0.062	0.415	2.408	0.200	5.800	5.000
10.00	116.500	129.220	0.003	0.388	4.515	0.175	20.388	5.714

SOURCE: Emmons, "Gas Dynamics Tables for Air," Dover Publications, Inc., 1947.

A **shock wave** may be **oblique** to a supersonic stream (see Fig. 11.4.33). If so, it behaves exactly like a normal shock to the normal component of the stream. The tangential component is left unchanged. Thus the resultant velocity not only drops abruptly in magnitude but also changes discontinuously in direction. Figure 11.4.34 gives the relations between M_1, M_2, θ_w, and δ.

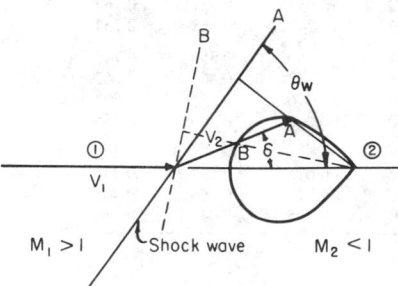

Fig. 11.4.33 Shock polar.

If, for a given supersonic stream, the velocities following all possible oblique shocks are plotted, a **shock polar** is obtained, as in Fig. 11.4.33. For any given stream deflection there are two possible shock angles. The **supersonic flow past a wedge** can, in principle, occur with either the strong shock B or the weak shock A attached to the leading edge; in practice only the weak shock occurs. The exact flow about the wedge may be computed by Eqs. (11.4.5) to (11.4.9) or more easily with the help of Fig. 11.4.34. For small wedge angles, the shock angle differs only slightly from $\sin^{-1}(1/M)$, the **Mach angle**, and the velocity component normal to this **Mach wave** is the speed of sound; the pressure jump is small and is given approximately by

$$p - p_1 = (\gamma p_1 \overline{M}_1^2 / \sqrt{\overline{M}_1^2 - 1})\delta \qquad (11.4.10)$$

where δ is the wedge semiangle.

Exact solutions also exist for **supersonic flow past a cone.** Above a certain supersonic Mach number, a conical shock wave is attached to the apex of the cone. Figures 11.4.35 and 11.4.36 show these exact relations; they are so accurate that they are often used to determine the Mach number of a stream by measuring the shock-wave angle on a cone of known angle. For small cone angles the shock differs only slightly from the **Mach cone**, i.e., a cone whose semiapex angle is the Mach angle; the pressure on the cone is then given approximately by

$$p - p_1 = \gamma p_1 \overline{M}_1^2 \delta^2 (\ln 2/\delta \sqrt{\overline{M}_1^2 - 1}) \qquad (11.4.11)$$

where δ is the cone semiangle.

A **nozzle** consisting of a single contraction will produce at its exit a jet of any velocity from $M = 0$ to $M = 1$ by a proper adjustment of the pressure ratio. For use as a subsonic wind-tunnel nozzle where a uniform parallel gas stream is desired, it is only necessary to connect the supply section to the parallel-walled or open-jet test section by a smooth gently curving wall. If the radius of curvature of the wall is nowhere less than the largest test-section cross-sectional dimension, no flow separation will occur and a good test gas stream will result.

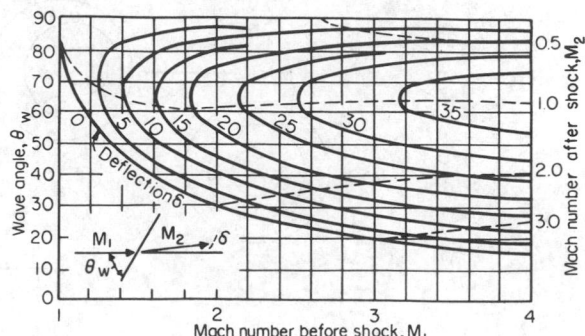

Fig. 11.4.34 Oblique shock wave relations.

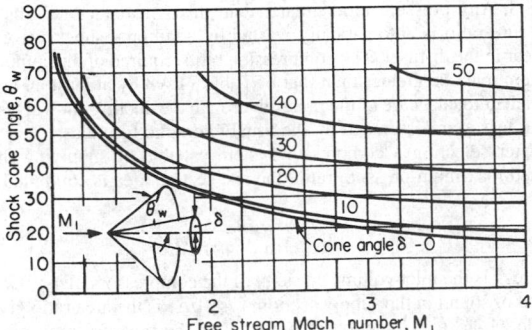

Fig. 11.4.35 Wave angles for supersonic flow around cones.

When a **converging nozzle** connects two chambers with the pressure drop beyond the critical $[(p/p_0) < (p^*/p_0)]$, the Mach number at the exit of the nozzle will be 1; the pressure ratio from the supply section to the nozzle exit will be critical; and all additional expansion will take place outside the nozzle.

A nozzle designed to supply a supersonic jet at its exit must converge to a minimum section and diverge again. The area ratio from the minimum section to the exit is given in the column headed A/A^* in Table 11.4.5. A converging-diverging nozzle with a pressure ratio $p/p_0 =$ (gas pressure)/(stagnation pressure), Table 11.4.5, gradually falling from unity to zero will produce shock-free flow for all exit Mach numbers from zero to the subsonic Mach number corresponding to its area ratio. From this Mach number to the supersonic Mach number corresponding to the given area ratio, there will be shock waves in the nozzle. For all smaller pressure ratios, the Mach number at the nozzle exit will not change, but additional expansion to higher velocities will occur outside of the nozzle.

To obtain a uniform parallel shock-free supersonic stream, the converging section of the nozzle can be designed as for a simple converging nozzle. The diverging or supersonic portion must be designed to produce and then cancel the expansion waves. A series of designs are given

in Table 11.4.7. These nozzles would perform as designed if it were not for the growth of the boundary layer. Experience indicates that these nozzles give a good first approximation to a uniform parallel supersonic stream but at a somewhat lower Mach number.

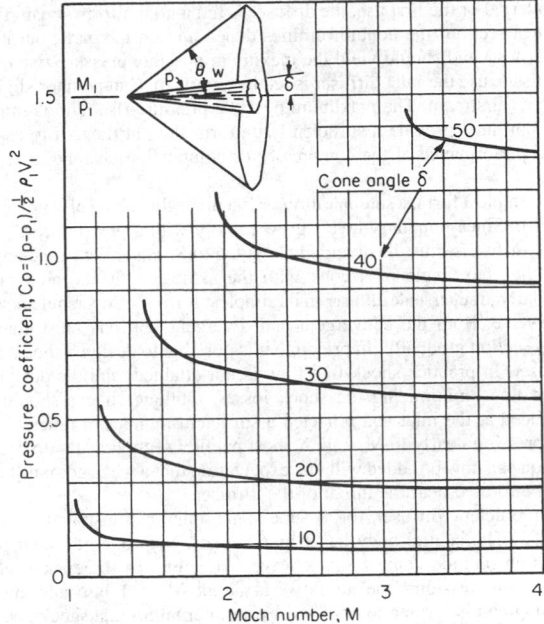

Fig. 11.4.36 Pressure calculations for supersonic flow around cones.

For **rocket nozzles** and other thrust devices the gain in thrust obtained by making the jet uniform and parallel at complete expansion must be balanced against the loss of thrust caused by the friction on the wall of the greater length of nozzle required. A simple conical diverging sec-

Table 11.4.7 Typical Nozzle Ordinates

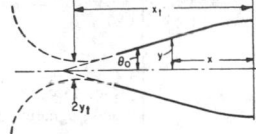

M	1.99		2.42		2.82		3.24		3.62		4.04	
θ_0	7°		9°		12°		13°		14°		15°	
	x	y	x	y	x	y	x	y	x	y	x	y
	0	7.50	0	7.50	0	7.50	0	7.50	0	7.50	0	7.50
	4.38	7.42	4.52	7.42	5.62	7.40	4.94	7.41	5.52	7.40	5.94	7.40
	8.36	7.28	8.69	7.28	9.70	7.26	9.43	7.26	10.39	7.23	11.20	7.21
	11.97	7.09	12.47	7.08	13.38	7.07	13.47	7.05	14.89	7.00	16.10	6.96
	15.19	6.88	15.91	6.84	16.64	6.84	17.15	6.79	18.75	6.73	20.22	6.67
	18.06	6.62	19.04	6.56	19.67	6.57	20.41	6.50	22.24	6.42	23.94	6.34
	20.65	6.35	21.80	6.27	22.35	6.29	23.41	6.19	25.39	6.09	27.17	6.00
	21.84	6.20	24.34	5.96	24.80	5.99	26.03	5.87	28.08	5.76	30.11	5.64
			26.57	5.65	26.95	5.69	28.42	5.53	30.56	5.41	32.69	5.28
	x_t	4.48	27.48	5.49	28.90	5.37	30.46	5.21	32.73	5.07	34.95	4.92
					30.70	5.07	32.35	4.87	34.66	4.73	36.96	4.57
			x_t	3.06	32.20	4.77	33.99	4.56	36.35	4.40	38.69	4.23
					32.90	4.62	35.44	4.25	37.87	4.08	40.26	3.90
							36.13	4.09	39.20	3.77	41.62	3.58
					x_t	2.10			39.80	3.62	42.79	3.29
											43.32	3.15
							x_t	1.41				
									x_t	0.988		
											x_t	0.673

SOURCE: Puckett, Supersonic Nozzle Design, *Jour. Applied Mechanics*, **13**, no. 4, 1948.

tion, cut off experimentally for maximum thrust, is generally used (see also Sec. 11.5 "Jet Propulsion and Aircraft Propellers").

For transonic and supersonic flow, **diffusers** are used for the recovery of kinetic energy. They follow the test sections of supersonic wind tunnels and are used as inlets on high-speed planes and missiles for ram recovery. For the first use, the diffuser is fed a nonuniform stream from the test section (the nonuniformities depending on the particular body under test) and should yield the maximum possible pressure-rise ratio. In missile use the inlet diffuser is fed by a uniform (but perhaps slightly yawed) airstream. The maximum possible pressure-rise ratio is important but must provide a sufficiently uniform flow at the exit to assure good performance of the compressor or combustion chamber that follows.

In simplest form, a **subsonic diffuser** is a diverging channel, a nozzle in reverse. Since boundary layers grow rapidly with a pressure rise, subsonic diffusers must be diverged slowly, 6 to 8° equivalent cone angle, i.e., the apex angle of a cone with the same length and area ratio. Similarly, a **supersonic diffuser** in its simplest form is a supersonic nozzle in reverse. Both the convergent and divergent portions must change cross section gradually. In principle it is possible to design a shock-free diffuser. In practice shock-free flow is not attained, and the design is based upon minimizing the shock losses. Oblique shocks should be produced at the inlet and reflected a sufficient number of times to get compression nearly to M = 1. A short parallel section and a divergent section can now be added with the expectation that a weak normal shock will be formed near the throat of the diffuser.

An efficient diffuser for a supersonic inlet is illustrated in Fig. 11.4.37. The central body has stepped cones, each one of which produces an oblique conical shock wave. After two or three such weak shock compressions, the air flows at about M = 1 into an annular opening and is further compressed by an internal normal shock and by subsonic diffusion. *NACA-TM* 1140 describes this diffuser. A ratio of pressure after diffusion to the total pressure in the atmosphere of as high as 0.6 is obtained with such diffusers at a Mach number of 3. The indications are that higher efficiencies are obtainable by careful design.

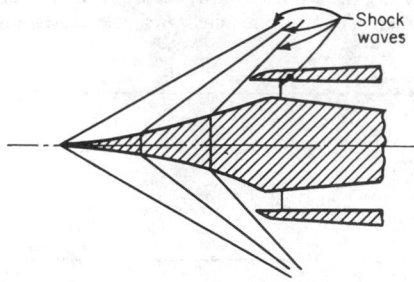

Fig. 11.4.37 Oblique-shock diffuser for ram recovery on a supersonic-plane air intake (Oswatitsch diffuser).

For a **supersonic wind tunnel,** the best way to attain the maximum pressure recovery at a wide range of operating conditions is to make the diffuser throat variable. The ratio of diffuser-exit static pressure to the diffuser-inlet (test-section outlet) total pressure is given in Table 11.4.8. These pressure recoveries are attained by the proper adjustment of the throat section of a variable diffuser on a supersonic wind-tunnel nozzle.

Table 11.4.8

M	1	1.5	2	3	4
p_e/p_0	0.83	0.69	0.50	0.23	0.10

A **supersonic wind tunnel** consists of a compressor or compressor system including precoolers or aftercoolers, a supply section, a supersonic nozzle, a test section with balance and other measuring equipment, a diffuser, and sufficient ducting to connect the parts. The **minimum pressure ratio** required from supply section to diffuser exit is given in Table

11.4.8. Any pressure ratio greater than this is satisfactory. The extra pressure ratio is automatically wasted by additional shock waves that appear in the diffuser. The compression ratio required of the compressor system must be greater than that of Table 11.4.8 by at least an amount sufficient to take care of the pressure drop in the ducting and valves. The latter losses are estimated by the usual hydraulic formulas.

After selecting a compressor system capable of supplying the required maximum pressure ratio, the test section area is computed from

$$A = 1.73 \frac{Q}{V_{s0}} \frac{A}{A^*} \frac{p_e}{p_0} \quad \text{ft}^2 \quad (11.4.12)$$

where Q is the inlet volume capacity of the compressors, ft³/s; V_{s0} is the speed of sound in the supply section, ft/s; A/A^* is the area ratio given in Table 11.4.5 as a function of M; p_e/p_0 is the pressure ratio given in Table 11.4.8 as a function of M.

The nozzle itself is designed for uniform parallel airflow in the test section. Such designs are given in Table 11.4.7, which covers only the part of the nozzle between the exit section and the maximum expansion angle. Since for each Mach number a different nozzle is required, the nozzle must be flexible or the tunnel so arranged that fixed nozzles can be readily interchanged.

The Mach number of a test is set by the nozzle selection, and the Reynolds number is set by the inlet conditions and size of model. The Reynolds number is computed from

$$\text{Re} = \text{Re}_0 D(p_0/14.7)(540/T_0)^{1.268} \quad (11.4.13)$$

where Re_0 is the Reynolds number per inch of model size for atmospheric temperature and pressure, as given by Fig. 11.4.38, D is model diam, in; p_0 is the stagnation pressure, lb/in²; T_0 is the stagnation temperature, °R.

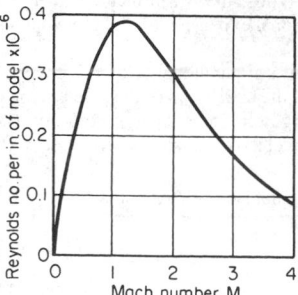

Fig. 11.4.38 Reynolds number–Mach number relation (for a fixed model size and stagnation condition).

For a closed-circuit tunnel, the Reynolds number can be varied independently of the Mach number by adjusting the mass of air in the system, thus changing p_0.

Intermittent-wind tunnels for testing at high speeds do not require the large and expensive compressors associated with continuous-flow tunnels. They use either a large vacuum tank or a large pressure tank (often in the form of a sphere) to produce a pressure differential across the test section. Such tunnels may have steady flow for only a few seconds, but by careful instrumentation, sufficient data may be obtained in this time. A **shock tube** may be used as an intermittent-wind tunnel as well as to study shock waves and their interactions; it is essentially a long tube of constant or varying cross section separated into two parts by a frangible diaphragm. High pressure exists on one side; by rupturing the diaphragm, a shock wave moves into the gas with the lower pressure. After the shock wave a region of steady flow exists for a few milliseconds. Very high stagnation temperatures can be created in a shock tube, which is not the case in a wind tunnel, so that it is useful for studying hypersonic flow phenomena.

Wind-tunnel force measurements are subject to errors caused by the model support strut. Wall interference is small at high Mach numbers for which the reflected model head wave returns well behind the model.

Near M = 1, the wall interference becomes very large. In fact, the tunnel chokes at Mach numbers given in Table 11.4.5, at

$$\frac{A}{A^*} = \cfrac{1}{1 - \cfrac{\text{area of model projected on test-section cross section}}{\text{area of test-section cross section}}}$$

$$(11.4.14)$$

There are two choking points: one subsonic and one supersonic. Between these two Mach numbers, it is impossible to test in the tunnel. As these Mach numbers are approached, the tunnel wall interference becomes very large.

For tests in this range of Mach number, specially constructed **transonic wind tunnels** with perforated or slotted walls have been built. The object here is to produce a mean flow velocity through the walls that comes close to that which would have existed there had the body been moving at that speed in the open. Often auxiliary blowers are needed to produce the necessary suction on the walls and to reinject this air into the tunnel circuit in or after the diffuser section.

Drag is difficult to predict precisely from wind-tunnel measurements, especially in the transonic regime. **Flight tests** of rocket-boosted models which coast through the range of Mach numbers of interest are used to obtain better drag estimates; from telemetered data and radar or optical sighting the deceleration can be determined, which in turn yields the drag.

LINEARIZED SMALL-DISTURBANCE THEORY

As the speed of a body is increased from a low subsonic value (Fig. 11.4.39a), the local Mach number becomes unity somewhere in the fluid along the surface of the body, i.e., the lower critical Mach number is reached. Above this there is a small range of transonic speed in which a supersonic region exists (Fig. 11.4.39b). Shock waves appear in this region attached to the sides of the body (Fig. 11.4.39c) and grow with increasing speed. At still higher transonic speeds a detached shock wave appears ahead of the body, and the earlier side shocks either disappear or move to the rear (Fig. 11.4.39d). Finally, for a sharp-nosed body, the head wave moves back and becomes attached (Fig. 11.4.39e). The flow is now generally supersonic everywhere, and the transonic regime is replaced by the supersonic regime. With the appearance of shock waves there occurs a considerable alteration of the pressure distribution, and the center of pressure on airfoil sections moves from the one-fourth chord point back toward the one-half chord point. There is an asso-

ciated increase of drag and, often, flow separation at the base of the shock.

The redistribution of pressures and the motion of shock waves over the wing surfaces through the transonic regime demands special consideration in the design of control surfaces so that they do not become ineffective by separation or inoperative by excessive loading.

If a body is slender (i.e., planes tangent to its surface at any point make small angles with the flight direction), the disturbance velocities caused by this body will be small compared with the flight speed and, excluding the hypersonic regime, small compared with the speed of sound. This permits the use of linearized small-disturbance theory to predict the approximate flow past the body. This theory relates the steady flow past the body at subsonic speeds to the incompressible flow past a distorted version of this body (the "generalized Prandtl-Glauert rule"). To find the velocity components in the **subsonic** flow about any slender body, first determine the velocity components, u, v, w (in the x, y, z directions, respectively), in the incompressible flow, at the same stream speed, about a stretched shape whose streamwise (x-axis direction) dimensions are $1/\beta$ times as great ($\beta = \sqrt{1 - M^2}$). The desired velocity components are then $\beta^{-2}u$, $\beta^{-1}v$, $\beta^{-1}w$ at corresponding points of the stretched and unstretched bodies. For **thin airfoil sections** this theory predicts

$$C_L = 2\pi\alpha/\sqrt{1 - M^2} \qquad \alpha \text{ in radians} \qquad (11.4.15)$$

where $L = C_L q S$ and $q = (\gamma/2)p_0 M^2$. For finite-span wings and for bodies, no such simple relations exist. However, an approximation to the overall **lift coefficient** for thin flat wings of rectangular platform is (see Fig. 11.4.41):

$$C_L = 2\pi A\alpha/[2 + \sqrt{4 + (\beta A)^2}] \qquad (11.4.16)$$

and the drag is made up of the induced drag due to lift and the skin-friction drag (see above, "Airfoils").

$$C_D = (C_L/\pi A)(1 + 0.01\beta A) + C_{DF} \qquad (11.4.17)$$

At supersonic speeds, the Prandtl-Glauert rule is also applicable if we replace β by $\lambda = \sqrt{M^2 - 1}$ and relate flow to flow past a stretched (or compressed) body at $M = \sqrt{2}$, where $\lambda = 1$. For **thin airflow sections** (see Fig. 11.4.40) this theory predicts

$$C_L = 4\alpha/\sqrt{M^2 - 1} \qquad \alpha \text{ in radians} \qquad (11.4.18)$$

$$C_D = \frac{2}{\lambda} \int_0^1 \left[\left(\frac{dy_u}{dx}\right)^2 + \left(\frac{dy_l}{dx}\right)^2 \right] d\left(\frac{x}{c}\right) + \frac{4a^2}{\lambda} + C_{DF}$$

$$(11.4.19)$$

where dy_u/dx and dy_l/dx are the slopes of the upper and lower surfaces of the airfoil, respectively, and C_{DF} is again the skin-friction drag coefficient. Note that there is a drag due to thickness and a drag due to lift at supersonic speeds for an airfoil section where there is none at subsonic speeds. For symmetric double-wedge airfoil sections, the thickness drag coefficient becomes $(4/\lambda)(t/c)^2$, and for symmetric biconvex airfoil sections it is $(16/3\lambda)(t/c)^2$, where t is the maximum thickness.

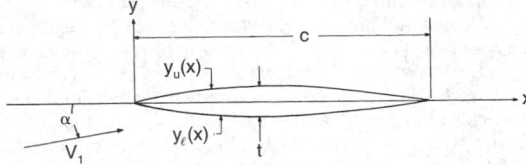

Fig. 11.4.40 Supersonic airfoil section.

Within the acoustic approximation, a disturbance at a point in supersonic flow can affect only the points in the downstream **Mach cone,** i.e., a conical region with apex at the point, axis parallel to the stream direction, and semicone angle equal to the Mach angle (see Fig. 11.4.32). Thus a rectangular wing with constant airfoil section has two-dimensional flow on all parts of the wing except the points within the tip Mach

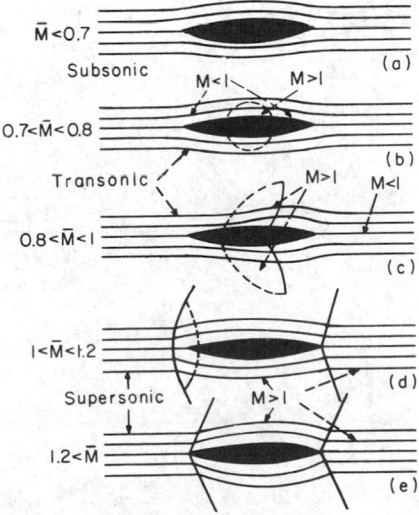

Fig. 11.4.39 Regions of flow about an airfoil (Mach numbers are approximate).

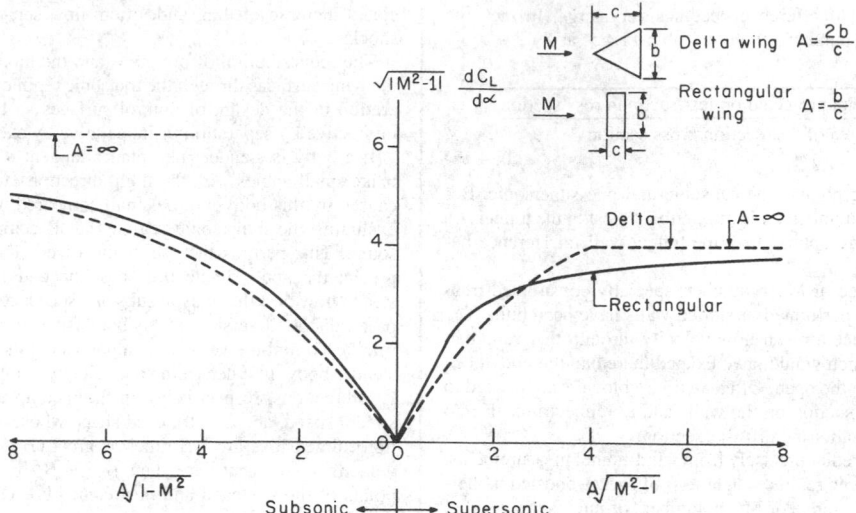

Fig. 11.4.41 Lift-coefficient curve slope for rectangular and delta wings (according to linearized theory).

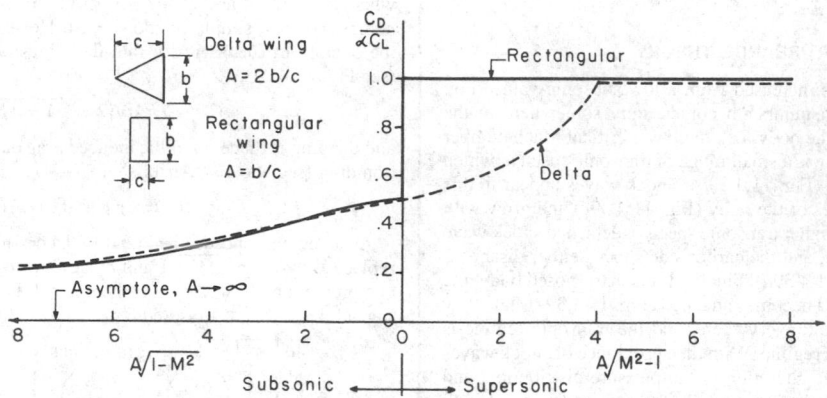

Fig. 11.4.42 Drag due to lift for rectangular and delta wings (according to linearized theory).

cones. At Mach numbers near unity these tip Mach cones cover nearly the entire wing, whereas at high Mach numbers they cover only a small part. Figure 11.4.41 shows the lift-curve coefficient slope for flat **rectangular** and **delta plan-form wings** at subsonic and supersonic speeds; these predictions of acoustic theory are inaccurate for high-aspect-ratio wings in the transonic regime, but elsewhere compare favorably with experiment. Figure 11.4.42 shows the drag due to lift for the same wings. The reduction in drag due to lift of the delta wing relative to the rectangular wing at moderate supersonic speeds, predicted by acoustic theory and shown in Fig. 11.4.42, is due to the fact that the delta-wing leading edge is **swept back,** and the component of velocity normal to the leading edge is subsonic. This creates a leading-edge suction which reduces the drag. This drag reduction is only partially realized in practice; a wing with a rounded leading edge realizes more of this reduction than a wing with a sharp leading edge. Figure 11.4.43 shows the thickness drag of the rectangular and delta wings, which occurs only at supersonic speeds. Note again that the leading-edge sweepback of the delta wing helps to reduce this drag at moderate supersonic speeds.

The lift of a slender **axially symmetric body** at small angle of attack is nearly independent of Mach number and is given approximately by

$$C_L = 2\alpha \qquad (11.4.20)$$

where the lift coefficient is based on the cross-sectional area of the base. The drag for $\alpha = 0$ at subsonic speeds is made up of skin friction and

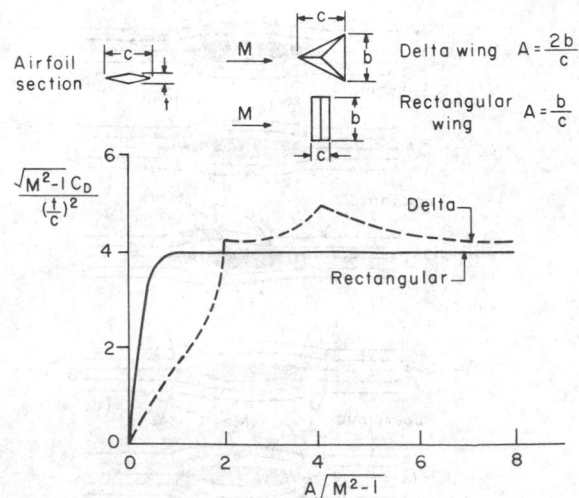

Fig. 11.4.43 Supersonic thickness drag coefficient for rectangular and delta wings with a symmetrical double-wedge airfoil section (according to linearized theory).

base drag; a dead-air region exists just behind the blunt base, and the pressure here is below ambient, causing a rearward suction which is the base drag. At supersonic speeds a wave drag is added, which represents the energy dissipated in shock waves from the nose. Figure 11.4.44 shows a typical drag-coefficient curve for a body of revolution where the **fineness ratio** (ratio of length to diameter) is 12.2. The skin-friction drag coefficient based on wetted area is the same as that of a flat plate, within experimental error.

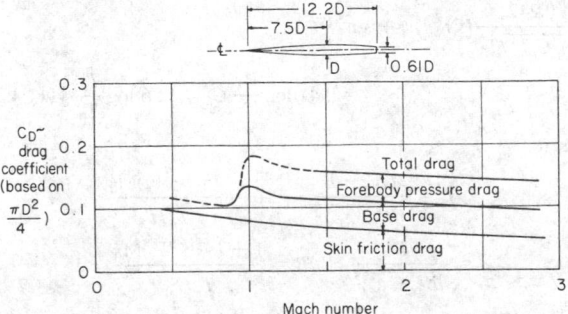

Fig. 11.4.44 Drag coefficients for parabolic-arc (*NACA-RM* 10) body (calculations from experimental data for a 30,000-ft altitude, $D = 12$ in, from *NACR-TR* 1160 and 1161, 1954).

At transonic and moderate supersonic speeds, the wave drag of a **wing body combination** can be effectively reduced by making the cross-sectional area distribution (including wings) a smooth curve when plotted versus fuselage station. This is the **area rule;** its application results in a decided indentation in the fuselage contour at the wing juncture. **Lift interference** effects also occur, especially when the body diameter is not small compared with the wing span. If the wing is attached to a cylindrical body, and if the wing-alone lift coefficient would have been C_L, then the lift carried on the body is $K_B C_L$, and the lift carried on the wing is $K_W C_L$. Figure 11.4.45 shows the variation of K_B and K_W with body diameter to wing-span ratio.

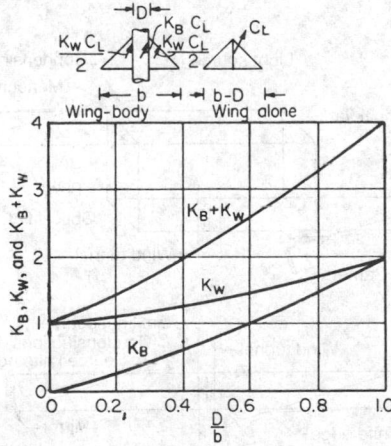

Fig. 11.4.45 Wing-body lift-interface factors.

The **effect of compressibility on skin-friction drag** is slight at subsonic speeds, but at supersonic speeds, a significant reduction in skin-friction coefficient occurs. Figure 11.4.46 shows the turbulent-boundary-layer mean skin-friction coefficient for a cone as a function of wall temperature and Mach number. An important effect at high speed is **aerodynamic heating.**

Figure 11.4.47 illustrates the velocity profile behind the shock wave of a body traveling at hypersonic speed. The shock-wave front repre-

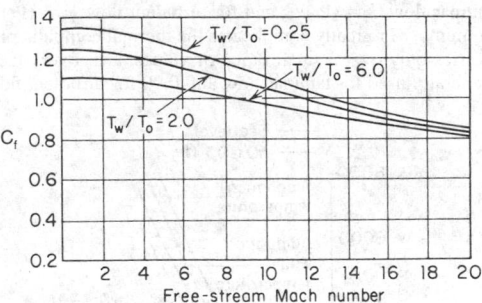

Fig. 11.4.46 Laminar skin-friction coefficient for a cone.

sents an area of high-temperature gas which radiates energy to the body, but boundary-layer convective heating is usually the major contributor. Behind this front is shown the velocity gradient in the boundary layer. The decrease in velocity in the boundary layer is brought about by the forces of interaction between fluid particles and the body (viscosity). This change in velocity is accompanied by a change in temperature and is dependent on the characteristics of the boundary layer. For example, heat transfer from a turbulent boundary layer may be of an order of magnitude greater than for laminar flow.

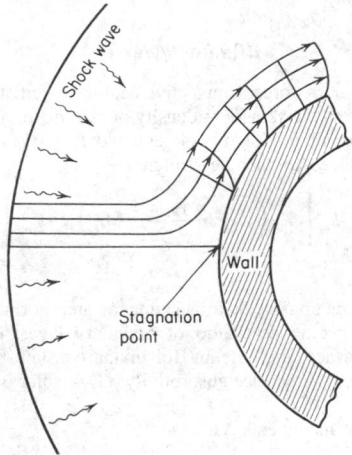

Fig. 11.4.47 Velocity profile behind a shock wave.

If the gas is brought to rest instantaneously, the total energy in the flow is converted to heat and the temperature of the air will rise. The resulting temperature is known as the **stagnation temperature** (see stagnation point, Fig. 11.4.47).

$$T_S = T_\infty \left(1 + \frac{\gamma - 1}{2} M^2 \right) \qquad (11.4.21)$$

T_∞ is the ambient temperature of the gas at infinity, and γ is the specific-heat ratio of the gas. For undissociated air, $\gamma = 1.4$.

In general, this simple, one-dimensional relationship between velocity and temperature does not hold for temperature in the boundary layer. The laminae of the boundary layer are not insulated from each other, and there is cross conduction. This is associated with the **Prandtl number** Pr, which is defined as

$$\Pr = C_p \mu / k_r \qquad (11.4.22)$$

where k_r = thermal conductivity of fluid, C_p = specific heat of fluid at constant pressure, and μ = absolute viscosity of fluid.

Defining the **recovery factor** r as the ratio of the rise in the idealistic wall and stagnation temperature over the free-stream temperature,

$$r = (T_{aw} - T_\infty)/(T_s - T_\infty) = (\Pr)^n \qquad (11.4.23)$$

For laminar flow, $r = (\text{Pr})^{1/2}$; and for turbulent flow, $r = (\text{Pr})^{1/3}$.

Prandtl number Pr greatly complicates the thermal computations, but since it varies only over a small range of values, a recovery factor of 0.85 is generally used for laminar flow and 0.90 for turbulent flow.

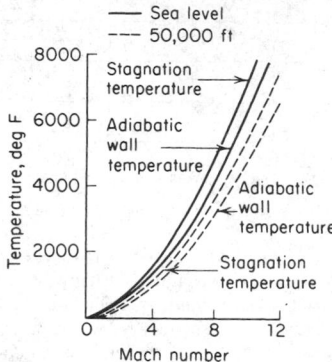

Fig. 11.4.48 Variation of stagnation and adiabatic wall temperature with Mach number.

For a thermally thin wall, the rate of change of the surface temperature is a function of the rate of total heat input and the surface's ability to absorb the heat.

$$dT_w/dt = q_T/wcb \qquad (11.4.24)$$

where t = time, q_T = forced convective heating + radiation heating − heat radiation from the skin, w = density of skin material, c = specific heat of skin material, b = skin thickness, and T_w = skin temperature.

The heat balance may then be written

$$wcb\left(\frac{dT_w}{dt}\right) = k_c\left[T_0\left(1 + r\frac{\gamma_B - 1}{2}M_0^2\right) - T_w\right] + \alpha G_s A_p - \epsilon\sigma T_w^4 \quad (11.4.25)$$

where A_p = correction factor to account for area normal to radiation source, γ_B = specific heat ratio of boundary layer, ϵ = radiative emissivity of surface, σ = Stefan-Boltzmann constant [17.3×10^{-10} Btu/($h \cdot ft^2 \cdot °R^4$), α = surface absorptivity, G_s = solar irradiation Btu/($h \cdot ft^2$).

For small time increments Δt,

$$\frac{dT_w}{dt} = \frac{T_{w2} - T_{w1}}{\Delta t}$$

Then

$$T_{w2} = T_{w1} + \frac{\Delta t}{wcb}\left\{h_c\left[T_0\left(1 + \frac{\gamma_B - 1}{2}M_0^2\right) - T_w\right] + \alpha G_s A_p - T_w^4\right\} \quad (11.4.26)$$

The local heat-transfer coefficient h_c is defined as

$$h_c = \frac{k_r}{r^{4/3}} C_p C_f \rho_0 Vg \qquad (11.4.27)$$

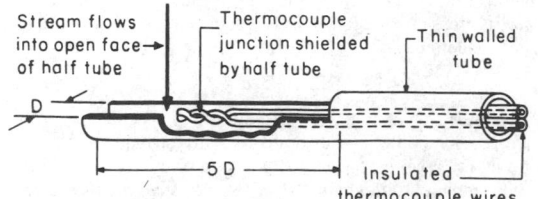

Fig. 11.4.49 Stagnation-temperature probe (recovery factor = 0.98).

where C_p = specific heat of air; C_f = local skin-friction coefficient; ρ_0 = density outside of boundary layer; V = velocity outside of boundary layer; g = acceleration of gravity.

For a cone, $k_r = 1,800$ (laminar Re $< 2 \times 10^6$) and $k_r = 1,800r^{4/3}$ (turbulent).

For a flat-plate transition, Reynolds number is 1×10^6.

Figure 11.4.46 gives laminar-skin-friction coefficient for a cone. For a flat plate, multiply C_f by $\sqrt{3/2}$. For turbulent skin-friction coefficient for a cone,

$$\frac{0.242}{\sqrt{A^2 C_f T_w/T_0}}(\sin^{-1}\psi + \sin^{-1}\theta)$$

$$= 0.41\log\frac{\text{Re}}{2}C_f - 1.26\log\frac{T_w}{T_0} \quad (11.4.28)$$

where

$$\psi = \frac{2A^2 - B}{\sqrt{B^2 + 4A^2}} \qquad A^2 = \frac{(\gamma_0 - \frac{1}{2})M_0^2}{T_w/T_0}$$

$$\theta = \frac{B}{\sqrt{B^2 + 4A^2}} \qquad B = \frac{1 + (\gamma_0 - \frac{1}{2})M_0^2}{T_w/T_0} - 1$$

For a flat plate, use Re instead of Re/2 in Eq. (11.4.28).

Figure 11.4.48 shows data on stagnation and adiabatic wall temperatures.

The primary **measurements in aerodynamics** are pressure measurements. A well-aligned pitot tube with an impact-pressure hole at its nose and static-pressure holes 10 diameters or more back from the nose will accurately measure the **impact pressure** and the **static pressure** of a uniform gas stream. Up to sonic speed the impact pressure is identical with stagnation pressure, but at supersonic speeds, a detached shock wave forms ahead of the probe, through which there is drop in stagnation pressure; the portion of the shock wave just ahead of the probe is normal, so that the equation for water flow in layer form over horizontal tubes (see "Transmission of Heat by Conduction and Convection" in Sec. 4) gives the relation of the measured impact pressure to the isentropic stagnation pressure.

Force measurements of total lift, drag, side force, pitching moment, yawing moment, and rolling moment on models are made on wind-

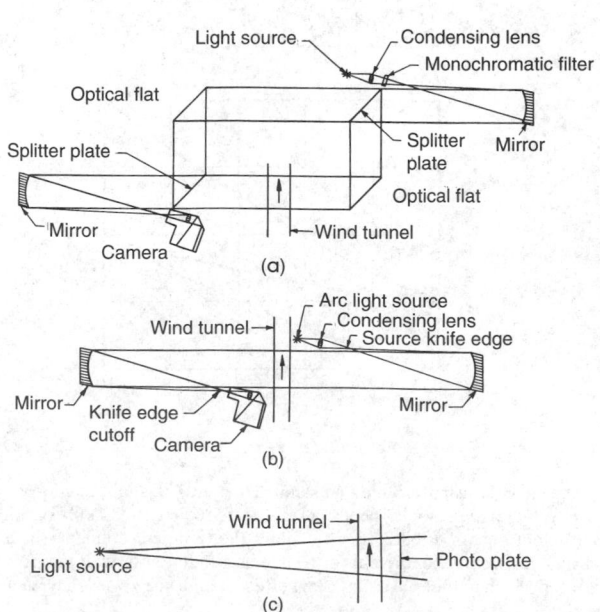

Fig. 11.4.50 Optical systems for observing high-speed flow phenomena. (a) Interferometer; (b) Schlieren (two-mirror) system; (c) shadowgraph.

tunnel balances just as at low speed. Small internal-strain-gage balances are often used to minimize strut interference.

An open thermocouple is unreliable for determining **stagnation temperature** if it is in a stream of high velocity. Figure 11.4.49 shows a simple temperature probe in which the fluid is decelerated adiabatically before its temperature is measured. The recovery factor used in equations, for this probe, accurately aligned to the stream, is 0.98.

Optical measurements in high-speed flow depend on the variation of index of refraction with gas density. This variation is given by

$$n = 1 + k\rho \qquad (11.4.29)$$

where $k = 0.116$ ft³/slug for air. A Mach-Zehnder **interferometer** (Fig. 11.4.50a), is capable of giving accurate density information for two-dimensional and axially symmetric flows. The **Schlieren** optical system (Fig. 11.4.50b) is sensitive to density gradients and is the most commonly used system to determine location of shock waves and regions of compression or expansion. The **shadowgraph** optical system is the simplest system (Fig. 11.4.50c) and is sensitive to the second space derivative of the density.

11.5 JET PROPULSION AND AIRCRAFT PROPELLERS
by Sanford Fleeter

REFERENCES: Zucrow, "Aircraft and Missile Propulsion," Vol. 2, Wiley. Hill and Peterson, "Mechanics and Thermodynamics of Propulsion," Addison-Wesley. Sutton, "Rocket Propulsion Elements," Wiley. Shorr and Zaehringer, "Solid Propellant Technology," Wiley. Forestor and Kuskevics, "Ion Propulsion," AIAA Selected Reprints. Theodorsen, "Theory of Propellers," McGraw-Hill. Smith, "Propellers for High Speed Flight," Princeton. Zucrow, "Principles of Jet Propulsion," Wiley. "Aircraft Propeller Handbook," Depts. of the Air Force, Navy, and Commerce, U.S. Government Printing Office, 1956. Kerrebrock, "Aircraft Engines and Gas Turbines," MIT Press. Oates, ed., "The Aerothermodynamics of Aircraft Gas Turbine Engines," AFAPL-TR-78-52, 1978. Fleeter, "Aeroelasticity for Turbomachine Applications," *AIAA J. of Aircraft,* **16**, no. 5, 1979. Bathie, "Fundamentals of Gas Turbines," Wiley. Hagar and Vrabel, "Advanced Turboprop Project," *NASA SP 495,* 1988. "Aeronautical Technologies for the Twenty-First Century," National Academy Press. Mattingly, Heiser, and Daley, "Aircraft Engine Design," AIAA.

Newton's reaction principle, based on the second and third laws of motion, is the theoretical basis for all methods of propelling a body either in (or on) a fluid medium or in space. Thus, the aircraft propeller, the ships screw, and the jet propulsion of aircraft, missiles, and boats are all examples of the application of the reaction principle to the propulsion of vehicles. Furthermore, all jet engines belong to that class of power plants known as **reaction engines.**

Newton's second law states that a change in motion is proportional to the force applied; i.e. a force proportional to the rate of change of the velocity results whenever a mass is accelerated: $F = ma$. This can be expressed in a more convenient form as $F = d(mv)/dt$, which makes it clear that the application of the reaction principle involves the time rate of increase of the momentum (mv) of the body. It should be noted that the same force results from either providing a small acceleration to a large mass or a large acceleration to a small mass.

Newton's third law defines the fundamental principle underlying all means of propulsion. It states that for every force acting on a body there is an equal and opposite reaction force.

The application of this reaction principle for propulsion involves the indirect effect of increasing the momentum of a mass of fluid in one direction and then utilizing the reaction force for propulsion of the vehicle in the opposite direction. Thus, the reaction to the time rate of increase of the momentum of that fluid, called the **propulsive fluid,** creates a force, termed the **thrust,** which acts in the direction of motion desired for the propelled vehicle. The known devices for achieving the propulsion of bodies differ only in the methods and mechanisms for achieving the time rate of increase in the momentum of the propulsive fluid or matter.

The function of a propeller is to convert the power output of an engine into useful thrust. To do this it accelerates a mass of air in the direction opposite to the direction of flight, generating thrust via Newton's reaction principle. Figure 11.5.1 illustrates schematically, in the relative coordinate system for steady flow, the operating principle of the **ideal aircraft propeller.** Power is supplied to the propeller, assumed to be

equivalent to an **actuator disk,** which imparts only an axial acceleration to the air flowing through it. The rotation of the actuator disk produces a **slipstream** composed of the entire mass of air flowing in the axial direction through the area of the actuator disk, i.e., the circle swept by the propeller. Atmospheric air enters the slipstream with the flight speed V_0 and mass flow rate m_a. It leaves the slipstream with the **wake velocity** w. The thrust is given in Eq. (11.5.5). Propulsion systems utilizing propellers will be termed **propeller propulsion.** (For detailed discussions of the airplane propeller, see the section "Aircraft Propellers.")

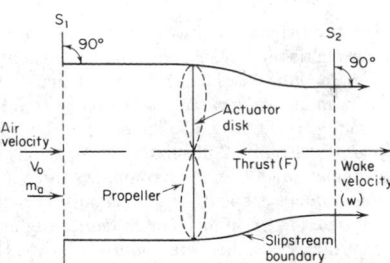

Fig. 11.5.1 Ideal propeller in the relative coordinate system.

Jet propulsion differs from propeller propulsion in that the propulsive fluid (or matter), instead of being caused to flow around the propelled body, is ejected from within the propelled body in the form of one or more high-speed jets of fluid or particles. In fact, the jet propulsion engine is basically designed to accelerate a large stream of fluid (or matter) and to expel it at a high velocity. There are a number of ways of accomplishing this, but in all instances the resultant reaction or thrust exerted on the engine is proportional to the time rate of increase of the momentum of the fluid. Thus, jet engines produce thrust in an analogous manner to the propeller and engine combination. While the propeller gives a small acceleration to a large mass of air, the jet engine gives a larger acceleration to a smaller mass of air.

Theoretically, there are no restrictions on either the type of matter, called the **propellant,** for forming a high-velocity jet or the means for producing the **propulsive jet.** The selection of the most suitable propellant and the most appropriate jet propulsion engine is dictated by the specific mission for the propelled vehicle. For example, in the case of the jet propulsion of a boat, termed **hydraulic jet propulsion,** the propulsive jet is formed from the water on which the boat moves. For the practical propulsion of bodies through either the atmosphere or in space, however, only two types of propulsive jets are suitable:

1. For propulsion within the atmosphere, there is the jet formed by expanding a highly heated, compressed gas containing atmospheric air as either a major or a sole constituent. Such an engine is called either an

airbreathing or a **thermal-jet engine**. If the heating is accomplished by burning a fuel in the air, the engine is a **chemical thermal-jet engine**. If the air is heated by direct or indirect heat exchange with a nuclear-energy source, the engine is termed a **nuclear thermal-jet engine**.

2. For propulsion both within and beyond the atmosphere of earth, there is the exhaust jet containing no atmospheric air. Such an exhaust jet is termed a **rocket jet,** and any matter used for creating the jet is called a **propellant**. A rocket may consist of a stream of gases, solids, liquids, ions, electrons, or a plasma. The assembly of all the equipment required for producing the rocket jet constitutes a **rocket engine**.

Modern airbreathing engines may be segregated into two principal types: (1) **ramjet engines,** and (2) **turbojet engines**. The turbojet engines are of two types: (1) the **simple turbojet engine,** and (2) the **turbofan,** or **bypass, engine**.

Rocket engines can be classified by the form of energy used for achieving the desired jet velocity. The three principal types of rocket engines are (1) **chemical rocket engines,** (2) **nuclear heat-transfer rocket engines,** and (3) **electric rocket engines**.

The propulsive element of a jet-propulsion engine, irrespective of type, is the **exhaust nozzle** or orifice. If the exhaust jet is gaseous, the assembly comprising all the other components of the jet-propulsion engine constitutes a gas generator for supplying highly heated, high-pressure gases to the exhaust nozzle.

These classifications of jet-propulsion engines apply to the basic types of engines. It is possible to have combinations of the different types of thermal-jet engines and also combinations of thermal-jet engines with rocket engines; only the principal types are discussed here.

ESSENTIAL FEATURES OF AIRBREATHING OR THERMAL-JET ENGINES

In the subsequent discussions a *relative coordinate system* is employed wherein the atmospheric air flows toward the propulsion system with the flight speed V_0, and the gases leave the propulsion system with the velocity w, *relative to the walls of the propulsion system*. Furthermore, steady-state operating conditions are assumed.

Ramjet Engine Figure 11.5.2 illustrates schematically the essential features of the simple ramjet engine. It comprises three major components: a diffusion or inlet system, $(0-2)$; a combustion chamber $(2-7)$; and an exhaust nozzle $(7-9)$. In this simple ramjet engine, apart from the necessary control devices, there are no moving parts. However, for accelerating the vehicle or for operating at different Mach numbers, a variable geometry diffuser and exhaust may be required.

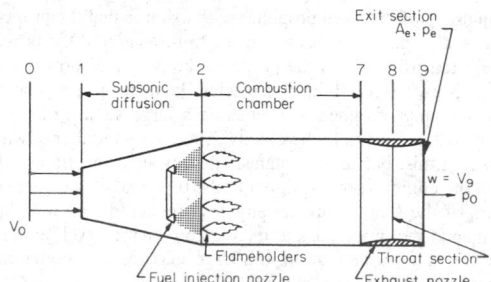

Fig. 11.5.2 Ramjet engine.

The operating principle of the ramjet engine is as follows. The free-stream air flowing toward the engine with a velocity V_0 and Mach number M_0 is decelerated by the inlet diffuser system so that it arrives at the entrance to the combustion chamber with a low Mach number, on the order of $M_2 = 0.2$. When the inlet-flow Mach number is supersonic, this diffuser system consists of a supersonic diffuser followed by a subsonic diffuser. The supersonic diffuser decelerates the inlet flow to approximately unity Mach numbers at the entrance to the subsonic diffuser; this deceleration is accompanied by the formation of shock waves and by an increase in the pressure of the air (diffusion). In the subsonic

diffuser the air is further diffused so that it arrives at the entrance to the combustion chamber with the required low Mach number. If P_0 is the total pressure of the free-stream air having the Mach number M_0 and P_2 that for the air entering the combustion system with the Mach number M_2, then it is desirable that P_2/P_0 be a maximum.

In the combustion chamber a fuel is burned in the air, thereby raising the total temperature of the gases entering the exhaust nozzle (see Sec. 7) to approximately $T_7 = 4260°R$ (2366 K). Most generally a liquid hydrocarbon fuel is used, but experiments have been conducted with solid fuels, liquid hydrogen, liquid methane, and "slurries" of metallic fuels in a liquid fuel. The combustion process is not quite isobaric because of the pressure drops in the combustion chamber, because of the increase in the momentum of the working fluid due to heat addition, and because of friction. The hot gases are discharged to the atmosphere, after expanding in the exhaust nozzle, with the relative velocity $w = V_9$.

Since the ramjet engine can function only if there is a ram pressure rise at the entrance to the combustion chamber, it is not self-operating at zero flight speed. It must, therefore, be accelerated to a flight speed which permits the engine to develop sufficient thrust for accelerating the vehicle it propels to the design flight Mach number. Consequently, a ramjet-propelled missile, for example, must either be launched by dropping it from an airplane or be **boosted** to the required flight speed by means of **launching,** or **bootster,** rockets. It appears that the most appropriate flight regime for the ramjet is at low to moderate supersonic Mach numbers, between approximately $M_0 = 2$ and 5. The lower limit is associated with the necessary ram pressure rise while the upper limit is set by the problem of either cooling or protecting the outer skin of the engine body.

Scramjet Low-cost access to space is one of the major reasons for the interest in **hypersonic propulsion** (speeds above Mach 5). A vehicle to provide access to space must be capable of reaching very high Mach numbers, since orbital velocity (17,000 mi/h) is about Mach 24. The ramjet engine, which must slow the entering air to a low subsonic velocity, cannot be used much above Mach 5 or 6. The solution is to decelerate the flow in the engine to a supersonic Mach number lower than the flight Mach number but greater than the local speed of sound. This propulsion device is the **supersonic combustion ramjet,** or **scramjet,** which is capable, in principle, of operation to Mach 24. As shown schematically in Fig. 11.5.3, in contrast to the ramjet, the scramjet en-

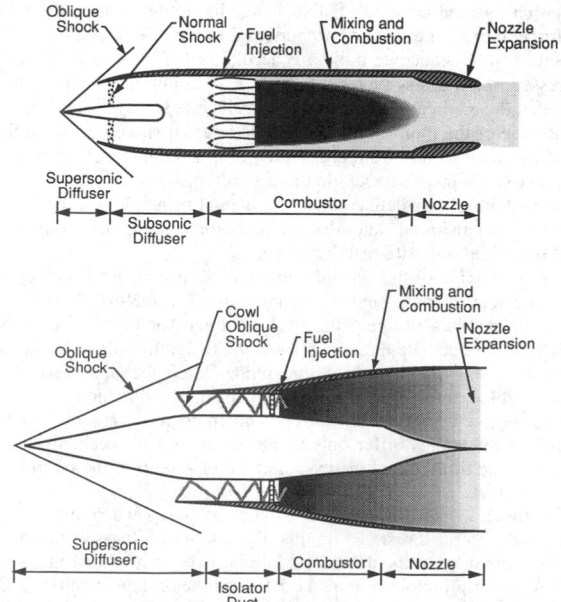

Fig. 11.5.3 Ramjet and scramjet flow configurations.

gine requires that fuel be added to the air, mixed, and burned, all at supersonic velocities, which is a significant technical challenge.

The advantage derived from supersonic combustion of the liquid fuel (usually liquid hydrogen) is that the diffuser of the scramjet engine is required to decelerate the air entering the engine from M_0 to only approximately $M_2 = 0.35\ M_0$ instead of the $M_2 = 0.2$, which is essential with subsonic combustion. This elimination of the subsonic diffusion increases the diffuser efficiency, reduces the static pressure in the combustor (and, therefore, the engine weight and heat transfer rate), and increases the velocity in the combustor (thereby decreasing engine frontal area). It is the combination of the above effects that makes ramjet propulsion at high flight Mach numbers (> 7) feasible.

Pulsejets may be started and operated at considerably lower speeds than ramjets. A pulsejet is a ramjet with an air inlet which is provided with a set of shutters loaded to remain in the closed position. After the pulsejet engine is launched, ram air pressure forces the shutters to open, fuel is injected into the combustion chamber, and is burned. As soon as the pressure in the combustion chamber equals the ram air pressure, the shutters close. The gases produced by combustion are forced out the jet nozzle by the pressure that has built up within the chamber. When the pressure in the combustion chamber falls off, the shutters open again, admitting more air, and the cycle repeats at a high rate.

Simple Turbojet Engine   Figure 11.5.4 illustrates schematically the principal features of a simple turbojet engine, which is basically a gas-turbine engine equipped with a propulsive nozzle and diffuser. Atmospheric air enters the engine and is partially compressed in the diffusion system, and further compressed to a much higher pressure by the air compressor, which may be of either the axial-flow or centrifugal type. The highly compressed air then flows to a combustion chamber wherein sufficient fuel is burned to raise the total temperature of the gases entering the turbine to approximately $T_4 = 2160°R$ (1200 K) for an uncooled turbine. The maximum allowable value for T_4 is limited by metallurgical and stress considerations; it is desirable, however, that T_4 be as high as possible. The combustion process is approximately isobaric. The highly heated air, containing approximately 25 percent of combustion products, expands in the turbine, which is directly connected to the air compressor, and in so doing furnishes the power for driving the air compressor. From the turbine the gases pass through a tailpipe which may be equipped with an **afterburner**. The gases are expanded in a suitably shaped exhaust nozzle and ejected to the atmosphere in the form of a high-speed jet.

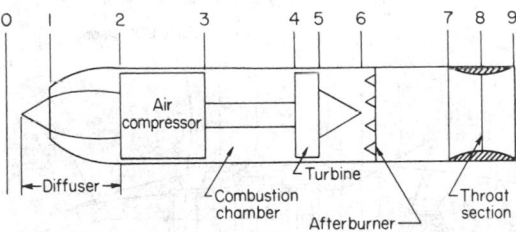

Fig. 11.5.4   Simple turbojet engine.

Like the ramjet engine, the turbojet engine is a continuous-flow engine. It has an advantage over the ramjet engine in that its functioning does not depend upon the ram pressure of the entering air, although the amount of ram pressure recovered does affect its overall economy and performance. The turbojet is the only airbreathing jet engine that has been utilized as the sole propulsion means for piloted aircraft. It appears to be eminently suited for propelling aircraft at speeds above 500 mi/h (805 km/h). As the design flight speed is increased, the ram pressure increases rapidly, and the characteristics of the turbojet engine tend to change over to those of the ramjet engine. Consequently, its top speed appears to be limited to that flight speed where it becomes more advantageous to employ the ramjet engine. It appears that for speeds above approximately 2,000 mi/h (3,219 km/h) it is advantageous to use some form of a ramjet engine.

The thrust of the simple turbojet engine increases rapidly with T_4, because increasing T_4 increases the jet velocity V_j. Actually, V_j increases faster than the corresponding increase in T_4. It is also an inherent characteristic of the gas-turbine engine which produces shaft power, called a **turboshaft** engine, that its useful power increases proportionally faster than a corresponding increase in its turbine inlet temperature T_4. Because of the decrease in strength of turbine materials with increase in temperature, the turbine blades, stators, and disks require cooling at $T_4 > 2160°R$ (1200 K) approximately. T_4 values are currently in the 2860 to $3260°R$ (1589 to 1811 K) range, and are being pushed toward the $4,600°R$ (2556 K) **stoichiometric** limit of JP4 fuel. The cooling air is bled from the compressor at the appropriate stage (or stages) and used to cool the stator blades or rotor blades by convective, film, or transpiration heat transfer. Up to 10 percent of the compressor air may be bled for turbine cooling, and this air is "lost" for turbine work for that blade row where it is used for cooling. Consequently "trade" studies must be made to "weigh" the increased complexity of the engine and the turbine work loss due to air bleed against the increased engine performance associated with increased T_4.

As in the case of any gas-turbine power plant, the efficiencies of the components of the turbojet engine have an influence on its performance characteristics, but its performance is not nearly as sensitive to changes in the efficiency of its component machines as is a gas turbine which delivers shaft power. (See Sec. 9.)

As noted previously, two types of compressors are currently employed, the axial-flow compressor and the centrifugal compressor. Irrespective of the type, the objectives are similar. The compressor must be reliable, compact, easy to manufacture, and have a small frontal area. Because of the limited air induction capacity of the centrifugal compressor, also called the radial compressor, engines for developing thrusts above 7,000 lb (31 kN) at static sea level employ axial-flow compressors (see Sec. 14).

The turbojet engine exhibits a rather flat thrust-versus-speed curve. Because the ratio of the takeoff thrust to thrust in flight is small, certain operational problems exist at takeoff. Since the exhaust gases from the turbine contain considerable excess air, the jet velocity, and consequently the thrust, can be increased by burning additional fuel in the tailpipe upstream from the exhaust nozzle. By employing "**tailpipe burning,**" or "**afterburning,**" as it is called, the thrust can be increased by 35 percent and at 500 mi/h, in a tactical emergency, by approximately 60 percent. With afterburning, the temperature of the gases entering the nozzle T_7, is of the order of $3800°R$ (2110 K).

Turbofan, or Bypass, Engine   For a fixed turbine inlet temperature, the jet velocity from a simple turbojet engine propelling an airplane at subsonic speed is relatively constant. The propulsive efficiency depends on the ratio of the flight speed to the jet velocity and increases as the ratio increases. On the other hand, the thrust depends on the difference between the jet velocity and the flight speed; the larger the difference, the larger the thrust per unit mass of air induced into the engine. By reducing the jet velocity and simultaneously increasing the mass rate of airflow through the engine, the **propulsive efficiency** can be increased without decreasing the thrust. To accomplish this, a **turbofan** or **bypass engine** is required.

Figure 11.5.5 illustrates schematically the components of a turbofan engine. There are two turbines, a low-pressure turbine (LPT) and a high-pressure turbine (HPT); one drives the air compressor of the hot-gas generator and the other drives the fan. Air enters the fan at the rate of $\dot{m}_{aF}$ and is ejected through the nozzle of area A_{7F} with the jet velocity $V_{jF} < V_j$, where V_j is the jet velocity attained by the air flowing through the hot-gas generator with the mass flow rate $\dot{m}_{a1}$; the hot-gas generator is basically a turbojet engine.

A key parameter is the **bypass ratio,** defined as the ratio of the air mass flow rate through the fan bypass duct to that through the hot-gas generator ($\dot{m}_{aF}/\dot{m}_{a1}$). Bypass ratios of about 5 to 10 are in use, with values of about 5 utilized in modern large commercial transport engines, for example, the Pratt & Whitney 4000 and the General Electric CF-6. The hot-gas generator is basically a turbojet engine. The fuel is added to $\dot{m}_{a1}$ at the rate $\dot{m}_f$. The hot-gas stream ($\dot{m}_{a1} + \dot{m}_f$) is discharged to the atmosphere through A_7 with the velocity V_j. Both types of turbofan engine produce an "overall" jet velocity V_{jTF} which is smaller than that for a turbojet engine operating with the same P_3/P_2 and T_4. The arrangements shown in Fig. 11.5.5 are for subsonic propulsion. Fuel can, of course, be burned in the fan air m_{aF} for increasing the thrusts of the

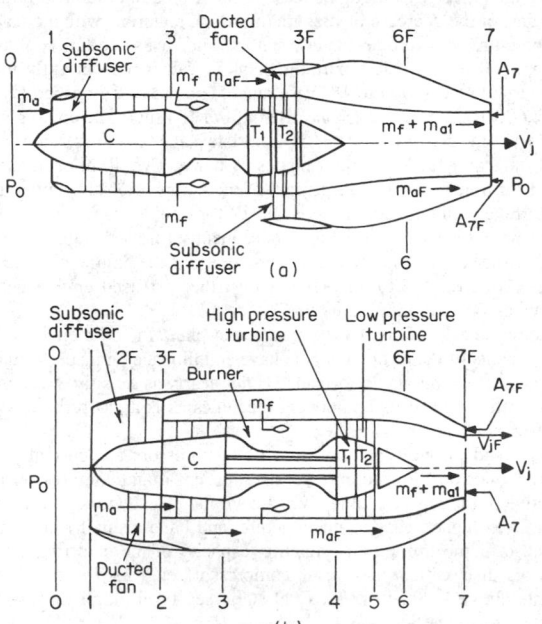

Fig. 11.5.5 Two different schematic arrangements of components of turbofan engines (subsonic flight). (*a*) Aft-fan turbofan engine; (*b*) ducted-fan turbofan engine.

engines. Practically all the newer commercial passenger aircraft are propelled by turbofan engines, and most of the older jet aircraft have been refurbished with new turbofan engines. The advantage of the lower effective jet velocity V_{jTF} is twofold: (1) It increases the propulsive efficiency η_p by reducing $v = V_{jTF}/V_0$ and, consequently, raises the value of η_0. (2) The reduced jet velocity reduces the jet noise; the latter increases with approximately the eighth power of the jet speed.

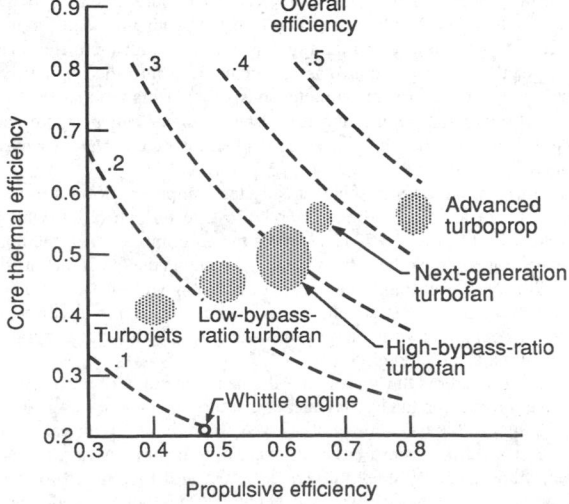

Fig. 11.5.6 Efficiency trends.

The next generation of large transport engines will be conventional high-bypass engines, but will have a larger thrust capability and will offer further improvements in specific fuel consumption, achieved by using even higher overall pressure ratios (over 40 : 1) and turbine inlet temperature (over 1850 K or 2870°F) combined with even more efficient components (Fig. 11.5.6). Beyond these conventional high-bypass engines, will be the **ultrahigh-bypass**-ratio engine in either a ducted or open rotor form (Fig. 11.5.7). The ducted ultrahigh-bypass-ratio engine features a counterrotating, reversible-pitch fan powered by a geared core engine with an overall 80 : 1 cycle pressure ratio. It is likely to be used on the larger aircraft, since it does not restrain the aircraft configuration or flight speed and will provide a 35 percent improvement in fuel burned over the most advanced current subsonic high-bypass engines.

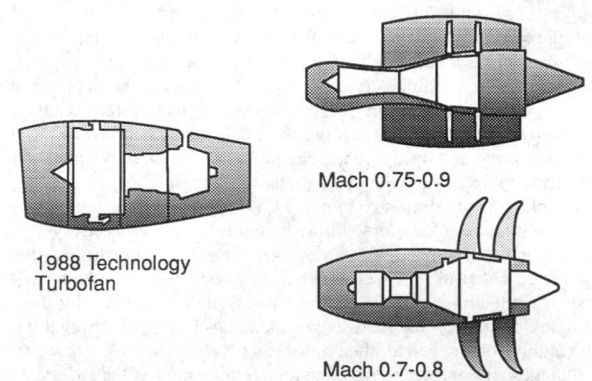

Fig. 11.5.7 Ultrahigh-bypass-ratio engine configurations.

A highly loaded, multiblade swept variable-pitch propeller, the **propfan** or **advanced turboprop,** could be combined with the latest turbine engine technology. The resulting advanced turboprop would offer a potential fuel saving of 50 percent over an equivalent-technology turbofan engine operating at competitive speeds and altitudes because of the turboprop's much higher installed efficiency. To eliminate the gearbox development for a 20,000 shaft horsepower engine, the propfans of the **unducted fan (UDF)** engine were directly driven with counterrotating turbine stages, a unique concept (Fig. 11.5.8). The projected specific

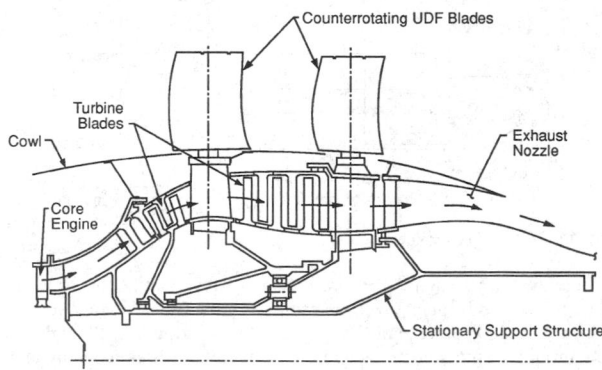

Fig. 11.5.8 UDF counterrotating turbine configuration.

fuel consumption at cruise for the resulting gearless, counterrotating UDF engine was 30 percent lower than that of the most modern turbofan engines, and about 50 percent lower than that of engines presently in use on 150-passenger airplanes. The tremendous fuel-saving potential of the turboprop has already been demonstrated. If economic conditions change, this concept could replace conventional high-bypass-ratio turbofans on small aircraft and military transports.

ESSENTIAL FEATURES OF ROCKET ENGINES

Figure 11.5.9 is a block-type diagram illustrating the essential features of a rocket engine, which comprises three main components: (A) a supply of propellant material contained in the rocket-propelled vehicle, (B) a propellant feed and metering system, and (C) a **thrust chamber,** also called a **rocket motor, thrustor,** or **accelerator.** In any rocket engine, energy must be added to the propellant as it flows through the thrustor.

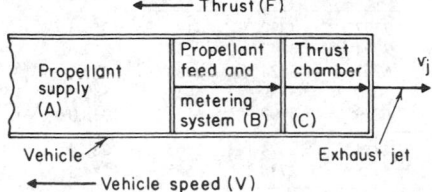

Fig. 11.5.9 Essential features of a rocket engine.

In Figure 11.5.9 propellant material from the supply is metered and fed to the thrust chamber, where energy is added to it. As a consequence of the energy addition, the propellant is discharged from the thrustor with the jet velocity V_j. The thrustor F acts in the direction opposite to that for the jet velocity V_j.

Chemical Rocket Engines

All chemical rocket engines have two common characteristics: (1) They utilize chemical reactions in a thrust chamber to produce a high-pressure, high-temperature gas at the entrance to a converging-diverging exhaust nozzle; (2) the hot propellant gas expands in flowing through the exhaust nozzle, and the expansion process converts a portion of the thermal energy, released by the chemical reaction, into the kinetic energy associated with a high-velocity gaseous-exhaust jet. Chemical rocket engines may be grouped into (1) liquid-bipropellant rocket engines, (2) liquid-monopropellant rocket engines, and (3) solid-propellant rocket engines.

Liquid-Bipropellant Rocket Engine Figure 11.5.10 illustrates the essential features of a liquid-bipropellant rocket engine employing **turbopumps** for feeding two propellants, an **oxidizer** and a **fuel,** to a rocket motor. The motor comprises (1) an injector, (2) a combustion chamber, and (3) a converging-diverging exhaust nozzle. The liquid propellants are fed under pressure, through the injector, into the combustion chamber, where they react chemically to produce large volumes of

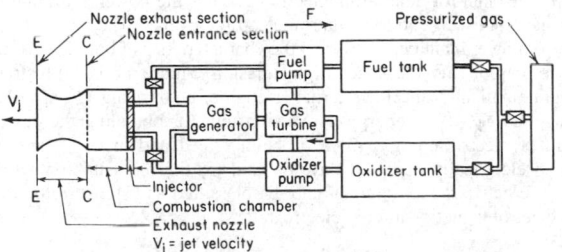

Fig. 11.5.10 Essential features of a liquid-bipropellant rocket engine.

high-temperature, high-pressure gases. For a given propellant combination, the **combustion temperature** T_c depends primarily on the oxidizer-fuel ratio (by weight), termed the **mixture ratio,** and to a lesser extent upon the combustion static pressure p_c. When the mass rate of flow of the liquid propellants equals that of the exhaust gases, the combustion pressure remains constant—the mode of operation that is usually desired.

If the bipropellants react chemically when their liquid streams come in contact with each other, they are said to be **hypergolic.** Propellants which are not hypergolic are said to be **diergolic,** and some form of ignition system is required to initiate combustion.

Except in those cases where the operating duration of the rocket motor is very short or where the combustion temperature is low, means must be provided for protecting the interior walls of the motor. The two most common methods are ablative cooling and regenerative cooling. In **ablative cooling** the inner surfaces of the thrust chamber are covered with an ablative material which vaporizes, thereby providing some cooling. Ordinarily, the material leaves a "char" which acts as a high-temperature insulating material. For high-performance engines which are to operate for relatively long periods, regenerative cooling is employed. In **regenerative cooling,** one of the propellants is circulated around the walls before injection into the motor. In some engines the regenerative cooling is combined with local liquid film cooling at critical areas of the thrust chamber. See Table 11.5.1 for bipropellant combinations.

Refer to Figure 11.5.10. By removing the oxidizer tank, the oxidizer pump, and the plumbing associated with the liquid oxidizer, one obtains the essential elements of a liquid-monopropellant rocket engine. Such an engine does not require a liquid oxidizer to cause the monopropellant to decompose and release its thermochemical energy.

There are basically three groups of monopropellants: (1) liquids which contain the fuel and oxidizer in the same molecule, e.g., hydrogen peroxide (H_2O_2) or nitromethane (CH_3NO_2); (2) liquids which contain either the oxidizer or the fuel constituent in an unstable molecular arrangement, e.g., hydrazine (N_2H_4); and (3) synthetic mixtures of liquid fuels and oxidizers. The most important liquid monopropellants are hydrazine and hydrogen peroxide (up to 98% H_2O_2). Hydrazine can be decomposed by a suitable metal catalyst, which is ordinarily packed in a portion of the thrust chamber at the injector end. The decomposition of hydrazine yields gases at a temperature of 2260°R (1265 K); the gases are a mixture of hydrogen and ammonia. Hydrogen peroxide can be readily decomposed either thermally, chemically, or catalytically. The most favored method is catalytically; a series of silver screens coated with samarium oxide is tightly packed in a decomposition chamber located at the injector end of the thrust chamber. At $P_c = 300$ psia, the decomposition of hydrogen peroxide (98%) yields gases having a temperature of 2260°R (1256 K).

The specific impulse obtainable for a liquid monopropellant is considerably smaller than that obtainable from liquid-bipropellant systems. Consequently, monopropellants are used for such auxiliary purposes as thrust vernier control, attitude reaction controls for space vehicles and missiles, and for gas generation.

Solid-propellant Rocket Engine Figure 11.5.11 illustrates schematically a solid-propellant rocket engine employing an **internal-burning case-bonded grain;** the latter burns radially outward at a substantially

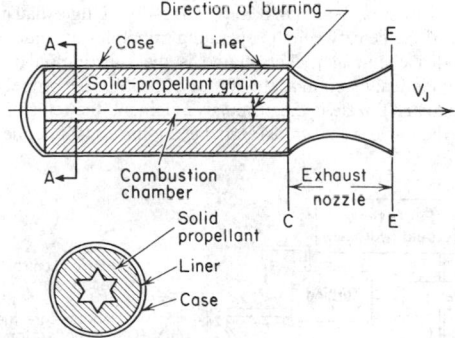

Fig. 11.5.11 Essential features of an internal-burning case-bonded solid-propellant rocket engine.

constant rate. A solid propellant contains both its fuel and the requisite oxidizer. If the fuel and oxidizer are contained in the molecules forming the solid propellant, the propellant is termed a **double-base propellant.** Those solid propellants wherein a solid fuel and a solid oxidizer form an intimate mechanical mixture are termed either **composite or heterogeneous** solid propellants. The chemical reaction of a solid propellant is

Table 11.5.1   Calculated Values of Specific Impulse for Liquid Bipropellent Systems[a]

$[P_c = 1,000 \text{ lbf/in}^2 \ (6,896 \times 10^3 \text{ N/m}^2)$ to $P_e = 1 \text{ atm}]$

Shifting equilibrium; isentropic expansion; adiabatic combustion; one-dimensional flow

| Oxidizer | Fuel | $\dot{m}_o/\dot{m}_f$ | ρ | Temperature T_c | | c^* | I (s) |
				°R	K		
Chlorine trifluoride (CTF)	Hydrazine	2.80	1.50	6553	3640	5961	293.1
(ClF$_3$)	MMH[b]	2.70	1.41	5858	3254	5670	286.0
	Pentaborane (B$_5$H$_9$)	7.05	1.47	7466	4148	5724	289.0
Fluorine (F$_2$)	Ammonia	3.30	1.12	7797	4332	7183	359.5
	Hydrazine	2.30	1.31	8004	4447	7257	364.0
	Hydrogen	7.70	0.45	6902	3834	8380	411.1
	Methane	4.32	1.02	7000	3889	6652	343.8
	RP-1	2.62	1.21	6839	3799	6153	318.0
Hydrogen peroxide	Hydrazine	2.00	1.26	4814	2674	5765	287.4
(100% H$_2$O$_2$)	MMH	3.44	1.26	4928	2738	5665	284.8
IRFNA[c]	Hydine[d]	3.17	1.26	5198	2888	5367	270.3
	MMH	2.57	1.24	5192	2885	5749	275.5
Nitrogen tetroxide	Hydrazine	1.30	1.22	5406	3002	5871	292.2
(N$_2$O$_4$)	Aerozine-50[e]	2.00	1.21	5610	3117	5740	289.2
	MMH	2.15	1.20	5653	3141	5730	288.7
Oxygen (LOX) (O$_2$)	Hydrazine	0.91	1.07	5667	3148	6208	312.8
	Hydrogen	4.00	0.28	4910	2728	7892	291.2
	Methane	3.35	0.82	6002	3333	6080	310.8
	Pentaborane	2.21	0.91	7136	3964	6194	318.1
	RP-1	2.60	1.02	6164	3424	5895	300.0

[a] Based on "Theoretical Performance of Rocket Propellant Combinations," Rocketdyne Corporation, Canoga Park, CA.
[b] MMH: Monomethyl hydrazine (N$_2$H$_3$CH$_3$).
[c] IRFNA: Inhibited red fuming nitric acid; 84.4% (HNO$_3$), 14% (N$_2$O$_4$), 1% (H$_2$O), 0.6% (HF).
[d] Hydine: 60% UDMH[f], 40% Deta[g].
[e] Aerozine-50: 50% UDMH[f], 50% hydrazine (N$_2$H$_4$).
[f] UDMII: Unsymmetrical dimethylhydrazine (CH$_3$)$_2$N$_2$H$_4$.
[g] DETA: Diethylenetriamine (C$_2$H$_{13}$N$_3$).

initiated by an igniter. In general, the configuration of a propellant grain can be designed so that the area of the burning surface of the propellant varies to give a prescribed thrust-versus-time curve.

Nuclear Heat-transfer Rocket Engine

Figure 11.5.12 illustrates schematically a nuclear heat-transfer rocket engine employing a **solid-core reactor.** The heat generated by the fissions of the uranium nucleus is utilized for heating a gaseous propellant, such as hydrogen, to a high temperature of 4000°R (2200 K) approx at the entrance section of the exhaust nozzle. The hot gas is ejected to the surroundings after expansion in a converging-diverging exhaust nozzle. The basic difference between the operating principles of a nuclear heat-transfer and a chemical rocket engine is the substitution of nuclear fission for chemical reaction as the source of heat for the propellant gas. Since the exhaust nozzle is the propulsive element, the remaining components of the nuclear heat-transfer rocket engine constitute the hot-gas generator.

The propellant in a nuclear heat-transfer rocket engine functions basically as a fluid for cooling the solid-core nuclear reactor. Its reaction is not based on energy considerations alone, but on such properties as specific heat, latent heat, molecular weight, and liquid density and on certain practical considerations.

Electric Rocket Engines

Several different types of electric rocket engines have been conceived; the principal types are described below. All of them require some form of power plant for generating electricity. Thus the power plant may be nuclear, solar-cell batteries, thermoelectric, or other. The choice of the type of power plant depends upon the characteristics of both the electric rocket engine and the space-flight mission. Figure 11.5.13 illustrates diagrammatically an electric rocket engine, comprising (1) a **nuclear power source,** (2) an energy-conversion unit for obtaining the desired form of electric energy, (3) a propellant feed and metering system, (4) an electrically operated thrustor, and (5) the requisite control devices. Electric rocket engines may be classified as (1) electrothermal, (2) electromagnetic, and (3) electrostatic.

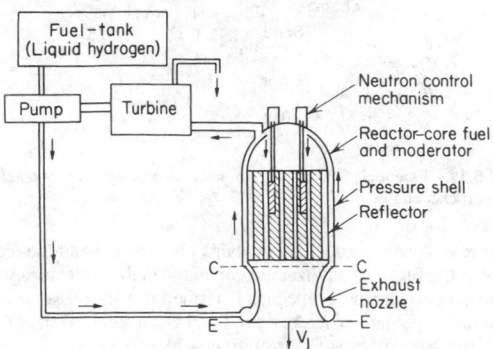

Fig. 11.5.12   Schematic arrangement of a nuclear heat-transfer rocket engine.

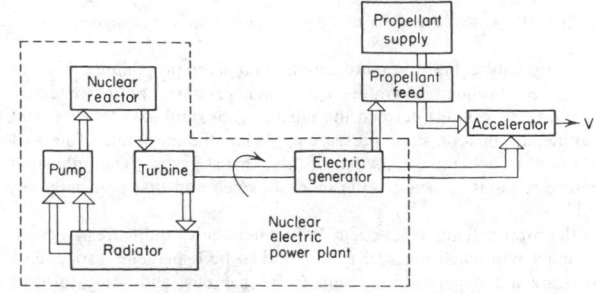

Fig. 11.5.13   Essential features of an electric rocket engine.

Electrothermal Rocket Engine Figure 11.5.14a illustrates the type of engine which uses electric power for heating gaseous propellant to a high temperature before ejecting it through a converging-diverging exhaust nozzle. If an electric arc is employed for heating the propellant, the engine is called a **thermal arc-jet rocket engine**.

Electromagnetic Rocket Engine Figure 11.5.14b illustrates an electromagnetic, or plasma, rocket engine. There is a wide variety of such engines, but all of them utilize the same operating principle. A plasma (a neutral ionized conducting gas) is accelerated by means of its interaction with either a stationary or a varying magnetic field. Basically, a plasma engine differs from a conventional electric motor by the substitution of a conducting plasma for a moving armature.

Electrostatic Rocket Engine Figure 11.5.14c illustrates schematically the electrostatic, or ion, rocket engine, comprising (1) an electric power plant; (2) a propellant supply; (3) ionization apparatus; (4) an ion accelerator; and (5) an electron emitter for **neutralizing** the ion beam ejected from the accelerator. Its operating principle is based on utilizing electrostatic fields for accelerating and ejecting electrically charged particles with extremely large velocities. The overall objective is to transform thermal energy into the kinetic energy associated with an extremely high-velocity stream of electrically neutral particles ejected from one or more thrustors. **Mercury ion engines** are currently being tested. The engine provides thrust by a rapid, controlled discharge of ions created by the ionization of atoms of the mercury fuel supply. Solar cells convert sunlight into electricity, which is then used to bombard the mercury with free electrons. This causes one electron to be stripped from each mercury atom, resulting in ionized mercury atoms. The ions mix with other electrons to form a plasma. A magnetic field set up between two metal screens at the aft end of the engine draws the positively charged ions from the plasma and accelerates them at very high exit speeds out the back of the engine. This creates a high-velocity exit beam producing low-level but extremely long-term thrust.

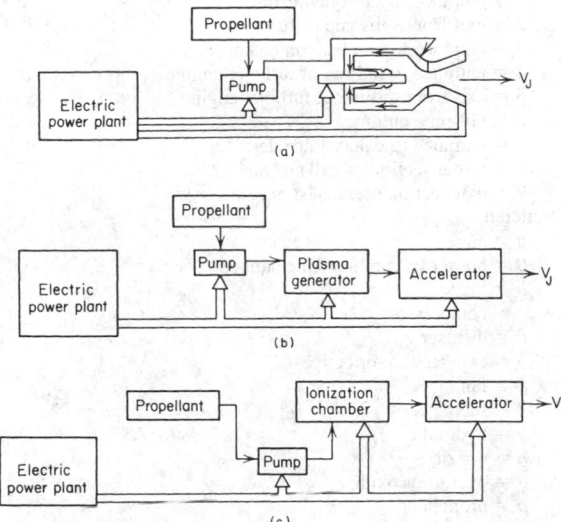

Fig. 11.5.14 Electric rocket engine. (a) Electrothermal rocket engine. (b) electromagnetic rocket engine. (c) electrostatic (ion) engine.

All electric rocket engines are low-thrust devices. The interest in such engines stems from the fact that chemical-rocket technology, for example, the space shuttle, is now so advanced that it has become feasible to place heavy *payloads*, such as a satellite equipped with an electric rocket engine, into an earth orbit. An electric rocket engine which is virtually inoperable terrestrially can be positioned in space and can then operate effectively because of the absence of aerodynamic drag and strong gravitational fields. Consequently, if a vehicle equipped with an electric rocket engine is placed in an earth orbit, the low-thrust electric rocket engine, under the conditions in space, will accelerate it to a large vehicle velocity if the operating time for the engine is sufficiently long.

In a chemical rocket engine, the energy for propulsion, as well as the mass ejected through the exhaust nozzle, is provided by the propellants, but the energy which can be added to a unit mass of the propellant gas is a fixed quantity, limited by the nature of the chemical bonds of the reacting materials. For that reason, chemical rocket engines are said to be **energy-limited rocket engines**. Although a nuclear heat-transfer rocket engine is also energy-limited, the limitation is imposed by the amount of energy that can be added per unit mass of propellant without exceeding the maximum allowable temperature for the materials employed for the solid-core reactor.

In an electric rocket engine, on the other hand, the energy added to the propellant is furnished by an electric power plant. The power available for heating the propellant is limited by the maximum power output of the electric power plant, accordingly, electric rocket engines are **power-limited**.

Notation

a = acoustic speed
a_0 = acoustic speed in free-stream air
a_2 = burning rate constant for solid propellant
A_e = cross-sectional area of exit section of exhaust nozzle
A_p = area of burning surface for solid propellant
A_t = cross-sectional area of throat of exhaust nozzle
c_p = specific heat at constant pressure
c_v = specific heat at constant volume
$c^* = P_c A_t / \dot{m}$ = characteristic velocity for rocket motor
C_d = discharge coefficient for exhaust nozzle
C_F = thrust coefficient
C_{Fg} = gross-thrust coefficient for ramjet engine
C_{Fn} = net thrust coefficient
D or $\mathcal{D}$ = drag
$D_i = \dot{m}_a V_0$ = ram drag for airbreathing or thermal-jet engine
d = diameter
d_p = diameter of propeller
E_{in} = rate at which energy is supplied to propulsion system
E_f = calorific value of fuel
E_p = calorific value of rocket propellants
$f = \dot{m}_f / \dot{m}_a$ = fuel-air ratio
$f' = f / \eta_B$ = fuel-air ratio for ideal combustion chamber
F = force or thrust
$F_i = m_e w - m_i V_0 + (p_e - p_0) A_e$ = thrust due to internal flow
$F_j = m_e V_j$ = jet thrust
$F_p = (p_e - p_0) A_e$ = pressure thrust
$F_g = F_i$ = gross thrust for ramjet engine
$F_g SFC$ = gross-thrust specific-fuel consumption for ramjet engine
g_c = correction factor defined by Newton's second law of motion
h = static specific enthalpy
Δh_n = enthalpy change for exhaust nozzle
H = total (stagnation) specific enthalpy
ΔH_c = lower heating value of fuel
I = specific impulse
$I_a = F / g \dot{m}_a$ = air specific impulse
J = mechanical equivalent of heat; advance ratio, propeller
$k = c_p / c_v$ = specific heat ratio
L or $\mathcal{L}$ = lift
$\mathcal{M}$ = mass
m = mass rate of flow
$\dot{m}_a$ = mass rate of air consumption for thermal-jet engine
$\dot{m}_e$ = mass rate of flow of gas leaving propulsion system
$\dot{m}_f$ = mass rate of fuel consumption
$\dot{m}_i$ = mass rate of flow of gas into propulsion system
$\dot{m}_o$ = mass rate of oxidizer consumption for rocket engine
$\dot{m}_p = \dot{m}_o + \dot{m}_f$ = mass rate of propellant consumption for rocket engine

$\bar{m}$ = molecular weight

M_e = momentum of gases leaving propulsion system in unit time

M_i = momentum of gases entering propulsion system in unit time

$M_0 = V_0/a_0$ = Mach number of free-stream air (flight speed)

n = revolutions per unit time

p = absolute static pressure; pitch of propeller blade

p_c = absolute static pressure of gases in combustion chamber

p_e = absolute static pressure in exit section of exhaust nozzle

p_0 = absolute static pressure of free-stream ambient air

P = absolute total (stagnation) pressure

P_c = absolute total pressure at entrance to exhaust nozzle

$P_0 = p_0 \left(1 + \dfrac{k-1}{2} M_0^2 \right)^{k/(k-1)}$ = absolute total pressure of free-stream air

$\mathscr{P}$ = propulsion power

$\mathscr{P}_L$ = leaving loss

$\mathscr{P}_T$ = thrust power

$q = \rho V^2/2$ = dynamic pressure

$q_0 = \rho_0 V_0^2/2 = k_0 p_0 M_0^2/2$ = dynamic pressure of free-stream air

Q = torque

Q_i = heat supplied to actual combustor

Q_i' = heat supplied to ideal combustor (no losses)

r_0 = linear burning rate for solid propellant

$R = R_u/\bar{m}$ = gas constant

R_u = universal gas constant = $\bar{m} R$

t = absolute static temperature

t_p = temperature of solid propellant prior to ignition

T = absolute total (stagnation) temperature

T_c = absolute total temperature of gas entering exhaust nozzle of rocket motor

T_2 = absolute total temperature at entrance to air compressor (see Fig. 11.5.4)

T_3 = absolute total temperature at exit section of air compressor (see Fig. 11.5.4)

T_3' = absolute total temperature at exit section of ideal compressor (see Fig. 11.5.4) operating between same pressure limits as actual compressor

T_4 = absolute total temperature at entrance to turbine (see Fig. 11.5.4)

T_5 = absolute total temperature at exit from turbine (see Fig. 11.5.4)

T_5' = absolute total temperature at exit from ideal turbine

$u = \pi n d_p$ = propeller tip speed

V = velocity

V_F = forward speed of propeller

V_j = effective jet velocity

V_0 = velocity of free-stream air (flight speed)

w = velocity of exit gases relative to walls of exhaust nozzle or velocity of air in ultimate wake of propeller

$\dot{W}$ = weight rate of flow

$\dot{W}_o$ = weight rate of flow of oxidizer

W_p = weight rate of flow of propellants

Greek

$\alpha = T_4/t_0 = \alpha_d \alpha_1$ = cycle temperature ratio; angle of attack

$\alpha_d = T_2/t_0$ = diffusion temperature ratio

$\alpha_1 = T_4/T_2$

β = bypass ratio for turbofans; helix angle for propeller blade

$\delta = P/P_{std}$ corrected pressure

η = efficiency

$\eta_B = f'/f$ = efficiency of combustion for thermal-jet engine

$\eta_c = (T_3' - T_2)/(T_3 - T_2)$ = isentropic efficiency of compressor

η_d = isentropic efficiency of diffuser

$\eta_n = \varphi^2$ = isentropic efficiency of exhaust nozzle

$\eta_o = \mathscr{P}_T/E_{in}$ = overall efficiency of propulsion system

$\eta_p = \mathscr{P}_T/(\mathscr{P}_T + \mathscr{P}_L)$ = ideal propulsive efficiency

$\eta_t = (T_4 - T_5)/(T_4 - T_5')$ = isentropic efficiency of turbine

$\eta_{th} = \mathscr{P}/E_{in}$ = thermal efficiency of propulsion engine

$\lambda = gI_a/\sqrt{2gJc_p t_0}$ = thrust parameter for turbojet engine; $\frac{1}{2} + \frac{1}{2} \cos \phi$ = divergence coefficient for exhaust nozzle for rocket engines

$\nu = V_0/w$ = speed ratio

ρ = density

$\bar{\rho}$ = mean density

$\Omega = \sqrt{k} \left(\dfrac{2}{k+1} \right)^{(k-1)/(2k-1)}$

ω = angular velocity of propeller shaft

ω' = rate of rotation of slipstream at propeller

ω'' = rate of rotation of slipstream in ultimate slipstream

Φ = fan velocity coefficient = V_F/u

ϕ = semiangle of exhaust-nozzle divergence; effective helix angle

$\varphi = \sqrt{\eta_n}$ = velocity coefficient for exhaust nozzle

ρ = density

σ = solidity of propeller

$\Theta = (P_3/p_0)^{(k-1)/k} = \Theta_d \Theta_c$ = cycle pressure ratio parameter

$\Theta_c = (P_3/P_2)^{(k-1)/k}$ = compressor pressure-ratio parameter

$\Theta_d = (P_2/p_0)^{(k-1)/k}$ = diffuser pressure-ratio parameter

$\Theta_n = (P_7/p_0)^{(k-1)/k} = (P_7/p_9)^{(k-1)/k}$ = nozzle pressure-ratio parameter

$\Theta_t = (P_4/P_5)^{(k-1)/k}$ = turbine pressure-ratio parameter

$\theta = T/T_{std}$ = corrected temperature

Subscripts

(a) numbered

0 = free stream

1 = entrance to subsonic diffuser

2 = exit from subsonic diffuser

3 = entrance to combustion chamber

4 = entrance to turbine of turbojet engine

5 = exit from turbine of turbojet engine

6 = tail-pipe entrance

7 = entrance to exhaust nozzle

8 = throat section of exhaust nozzle

9 = exit section of exhaust nozzle

(b) lettered

a = air

B = burner or combustion chamber

b = blade

c = compressor

d = diffuser

e = exit section; effective

F = fan

f = fuel

h = hydraulic

n = nozzle

o = overall or oxidizer

p = propellant

P = propulsive

std = standard

t = turbine or throat, as specified in text

Statement on Units The dynamic equation employed in the following sections are written for *consistent sets of units,* i.e., for sets in which 1 *unit of force* = 1 *unit of mass* × 1 *unit of acceleration.* Consequently, the gravitational correction factor g_c in Newton's equation $F = (1/g_c)ma$ (see notation) has the numerical value unity and is omitted from the dynamic equations. When the equations are used for calculation purposes, g_c should be included and its appropriate value employed.

THRUST EQUATIONS FOR JET-PROPULSION ENGINES

Refer to Fig. 11.5.15, which illustrates schematically a rotationally symmetrical arbitrary propulsion engine immersed in a uniform flow field. Because of the reactions between the fluid flowing through the engine, called the **internal flow**, and the interior surfaces wetted by the internal flow a resultant **axial force** is produced, that is, a force collinear with the longitudinal axis of the engine. If an axial force acts in the **forward direction**, employing a relative coordinate system, it is called a **thrust**; if it acts in the **backward direction** it is called a **drag**. Similarly, the resultant axial force due to the **external flow**, the flow passing over the external surfaces of the propulsion system, is a thrust or drag depending upon whether it acts in the forward direction or the backward direction.

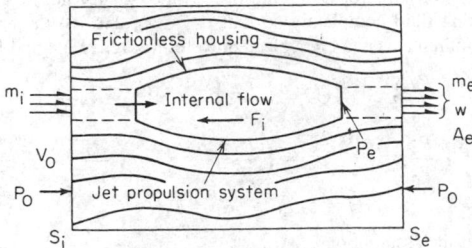

Fig. 11.5.15 Generalized jet-propulsion system.

Application of the momentum equation of fluid mechanics to the generalized propulsion system illustrated in Fig. 11.5.15 gives the following equation for the thrust F_i due to the **internal flow**. Thus, if it is assumed that the external flow is frictionless, there is no change in the rate of momentum for the external flow between S_i and S_e. Hence, the thrust $F = F_i$ is due entirely to the internal flow, and

$$F = F_i = m_e w - \dot{m}_i V_0 + (p_e - p_0)A_e \qquad (11.5.1)$$

In Eq. (11.5.1), $\dot{m}_e w = F_j$ = **jet thrust**, $\dot{m}_i V_0 = D_i$ = **ram drag**, and $(p_e - p_0)A_e = F_p$ = **pressure thrust**. Hence,

$$F_i = F_j - D_i + F_p \qquad (11.5.2)$$

It is convenient to introduce a fictitious **effective jet velocity** V_j, which is defined by

$$F = \dot{m}_e V_j - \dot{m}_i V_0 = \dot{m}_e w - \dot{m}_i V_0 + (p_e - p_0)A_e \quad (11.5.3)$$

If the gases are expanded completely, in the exhaust nozzle, then $w = V_j$ and $F_p = 0$. No appreciable error is introduced, in general, if it is assumed that $F_p = 0$. For **thermal jet engines**, $\dot{m}_i = \dot{m}_a$ (see notation), and $\dot{m}_e = \dot{m}_a + \dot{m}_f$. Let $f = \dot{m}_f/\dot{m}_a$, $v = V_0/V_j$; then

$$F_i = \left(\frac{1+f}{v} - 1\right) \dot{m}_a V_0 + (p_e - p_0)A_e \qquad (11.5.4)$$

In the case of uncooled turbojet engines, $\dot{m}_f$ is not significantly different from the fraction of $\dot{m}_a$ utilized for cooling the bearings and turbine disk. Consequently, no significant error is introduced by assuming that $\dot{m}_e = \dot{m}_a + \dot{m}_f \approx \dot{m}_a = \dot{m}$. Hence, for a simple **turbojet engine**, one may write

$$F_e = \dot{m}_a(V_j - V_0) = \dot{m}V_0(1/v - 1) \qquad (11.5.5)$$

Equation (11.5.5) is also the thrust equation for an **ideal propeller**; in that case V_j is the **wake velocity** for the air leaving its slipstream. The trust equation for a hydraulic jet propulsion system has the same form as Eq. (11.5.5). Let V_0 denote the speed of a boat, and $\dot{m}_w$ the mass rate of flow of the water entering the hydraulic pump and discharged by the exit nozzle with the velocity $w = V_j$ relative to the boat. Hence, for hydraulic jet propulsion

$$F = \dot{m}_s(V_j - V_0) = \dot{m}_w V_0(1/v - 1) \qquad (11.5.6)$$

where $v = V_0/V_j$.

In the case of **rocket engines**, since they do not consume atmospheric air, $\dot{m}_a = 0$, and the flow of gas out of the rocket motor, under steady-state conditions, is equal to $\dot{m}_p - \dot{m}_o + \dot{m}_f$ (see notation).

The effective jet velocity V_j is larger than V_e = **exit velocity** if $p_e > p_0$; that is, if the gases are **underexpanded**. The effective jet velocity is a useful criterion because it can be determined accurately from the measured values of F_i and $\dot{m}_p$ obtained from a static firing test of the rocket motor.

The ratio $F_i/\dot{m}_p$ is denoted by I and called either the **specific thrust** or the **specific impulse**. Hence

$$I = F/\dot{m}_p = V_j/g_c \qquad (11.5.7)$$

Although the dimensions of specific impulse are force/(mass)(s), it is conventional to state its units as *seconds*.

Equation (11.5.5) for the thrust of a **turbojet engine**, when expressed in terms of the effective jet velocity V_j, becomes

$$F = \dot{m}_a(V_j - V_0) \qquad (11.5.8)$$

POWER AND EFFICIENCY RELATIONSHIPS

In a jet-propulsion engine the **propulsion element** is the **exhaust nozzle**, and the rate at which energy is supplied to it is called the **propulsion power**, denoted by $\mathcal{P}$. The rate at which the propulsion system does useful work is termed the **thrust power** $\mathcal{P}_T$ and is given by

$$\mathcal{P}_T = FV_0 \qquad (11.5.9)$$

Assume that $p_e = p_0$, and that the only energy loss in a propulsion system is the **leaving loss** $\mathcal{P}_L = \dot{m}(V_j - V_0)^2/2$, that is, the kinetic energy associated with the jet gases discharged from the system; then the propulsive power is given by

$$\mathcal{P} = \mathcal{P}_L + \mathcal{P}_T \qquad (11.5.10)$$

The **ideal propulsive efficiency** is defined, in general, by

$$\eta_P = \mathcal{P}_T/(\mathcal{P}_T + \mathcal{P}_L) = \text{thrust power/propulsion power} \quad (11.5.11)$$

For a turbojet engine and hydraulic jet propulsion,

$$\eta_P = 2v/(1 + v) \qquad (11.5.12)$$

For a chemical rocket engine,

$$\eta_P = 2v/(2 + v^2) \qquad (11.5.13)$$

The propulsive efficiency η_P is of more or less academic interest. Of more importance is the overall efficiency η_o, which for airbreathing and rocket engines is defined by

$$\eta_o = \eta_{th}\eta_P \qquad (11.5.14)$$

where

$$\eta_{th} = \mathcal{P}/E_{in} = \text{thermal efficiency of system} \qquad (11.5.15)$$

and E_{in} is the rate at which energy is supplied to the propulsion system.

The overall efficiency of a hydraulic jet-propulsion system is given by

$$\eta_o = \eta_{th}\eta_h\eta_p = \eta_h\eta_{th}[2v/(1 + v)] \qquad (11.5.16)$$

where η_{th} is the thermal efficiency of the power plant which drives the water pump, and η_h is the hydraulic efficiency of the water pump. To achieve a reasonable fuel consumption rate, η_h must have a larger value.

For an airbreathing jet engine, $E_{in} = \dot{m}_f(E_f + V_0^2/2)$, and

$$\eta_o = \frac{2v}{1 + v} \frac{\mathcal{P}}{\dot{m}_f(E_f + V_0^2/2)} \qquad (11.5.17)$$

The ratio $\dot{m}_f/F$ is called the **thrust specific-fuel consumption** (*TSFC*) and is measured in mass of fuel per hour per unit of thrust. Hence,

$$TSFC = \dot{m}_f/F = f\dot{m}_a/F \qquad (11.5.18)$$

For a rocket engine, $E_{in} = \dot{m}_p(E_p + V_0^2/2)$, so that

$$\eta_o = \frac{2\nu}{1 + \nu^2} \frac{\mathscr{P}}{\dot{m}_p(E_p + V_0^2/2)} \qquad (11.5.19)$$

PERFORMANCE CHARACTERISTICS OF AIRBREATHING JET ENGINES

The Ramjet Engine In ramjet technology the thrust due to the internal flow F_i is called the **gross thrust** and denoted by F_g. Refer to Fig. 11.5.2 and assume $\dot{m}_0 \approx \dot{m}_9 \approx \dot{m}_a$. Then

$$F_g = F_i \approx \dot{m}_a(V_j - V_0) \qquad (11.5.20)$$

In level unaccelerated flight F_g is equal to the external drag of the propelled vehicle and the ramjet body.

It is customary to express the thrust capabilities of the engine in terms of the **gross-thrust coefficient** C_{Fg}. If A_m is the maximum cross-sectional area of the ramjet engine, $q_0 = \rho_0 V_0^2/2 = k_0 p_0 M_0^2/2 = $ the **dynamic pressure** of the free-stream air, then

$$C_{Fg} = \frac{F_g}{q_0 A_m} = \frac{2F_g}{A_m k_0 p_0 M_0^2} \qquad (11.5.21)$$

In terms of the Mach numbers M_0 and M_9,

$$C_{Fg} = \frac{2A_9/A_m}{k_0 M_0^2}\left(\frac{P_9}{p_0}\frac{1 + k_9 M_9^2}{P_9/p_9} - 1\right) - 2\frac{A_0}{A_m} \qquad (11.5.22)$$

In a fixed-geometry engine, M_9 depends upon the total temperature T_7, the total pressure P_7, the fuel-air ratio f, the nozzle area ratio A_e/A_t, and the efficiency of the nozzle η_n. For estimating purposes, $k_0 = 1.4$ and $k_9 = 1.28$, when the engine burns a liquid-hydrocarbon fuel. The manner in which C_{Fg} varies with altitude M_0 and fuel-air ratio f is shown schematically in Fig. 11.5.16.

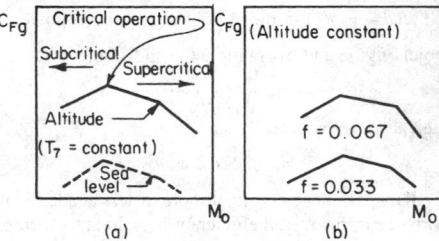

Fig. 11.5.16 Effect of altitude and fuel-air ratio on the gross-thrust coefficient of a fixed-geometry ramjet engine. (a) Effect of altitude; (b) effect of fuel-air ratio.

If $\alpha = T_7/t_0$ = cycle temperature ratio, and k_B is the mean value of k for the combustion gases, the rate at which heat is supplied to the engine is

$$Q_i = Q_i'/\eta_B = \dot{m}_a c_{pB} t_0 (\alpha - T_2)/\eta_B \qquad (11.5.23)$$

It is readily shown that

$$Q_i = \frac{A_0 p_0 V_0}{\eta_B J}\frac{k_B}{k_B - 1}\left[\alpha - \left(1 + \frac{k_0 - 1}{2}M_0^2\right)\right] \qquad (11.5.24)$$

The **gross-thrust specific-fuel consumption** $(F_g SFC)$ is, by definition,

$$F_g SFC = \dot{m}_f/F_g \qquad (11.5.25)$$

The principal sources of loss are aerothermodynamic in nature and cause a decrease in total pressure between stations 0 and 7. For estimating purposes, assuming $M_0 = 2.0$ and $T_6 = 3800°R$ (2111 K), the total pressures across different sections of the engine may be assumed to be approximately those tabulated below.

Part of engine	Total pressure ratio
Supersonic diffuser (0–1)	$P_1/P_0 = 0.92$
Subsonic diffuser (1–2)	$P_2/P_1 = 0.90$
Flameholders (2–7)	$P_7/P_2 = 0.97$
Combustion chamber (2–7)	$P_7/P_2 = 0.92$
Exhaust nozzle (7–9)	$P_9/P_7 = 0.97$

The Simple Turbojet Engine A good insight into the design performance characteristics of the turbojet engine is obtained conveniently by making the following assumptions: (1) The mass rate of flow of working fluid is identical at all stations in the engine; (2) the thermodynamic properties of the working fluid are those for air; (3) the air is a perfect gas and its specific heats are constants; (4) there are no pressure losses due to friction or heat addition; (5) the exhaust nozzle expands the working fluid completely so that $p_9 = p_0$; and (6) the auxiliary power requirements can be neglected. Refer to Fig. 11.5.4 and let

$$\Theta = \left(\frac{P_3}{p_0}\right)^{(k-1)/k} \quad \Theta_d = \left(\frac{P_2}{p_0}\right)^{(k-1)/k} \quad \Theta_c = \left(\frac{P_3}{P_2}\right)^{(k-1)/k}$$

$$\Theta_t = \left(\frac{P_4}{P_5}\right)^{(k-1)/k} \quad \Theta_n = \left(\frac{P_7}{p_0}\right)^{(k-1)/k} = \left(\frac{P_7}{p_9}\right)^{(k-1)/k}$$

$$\alpha = T_4/t_0 \qquad \alpha_d = T_2/t_0 \qquad \alpha_1 = T_4/T_2$$

In view of the assumptions,

$$\Theta = \Theta_d \Theta_c = \Theta_t \Theta_n \qquad (11.5.26)$$

and

$$\alpha = \alpha_a \alpha_1 \qquad (11.5.27)$$

The diffuser pressure-ratio parameter Θ_d is given by

$$\Theta_d = 1 + \eta_d[(k - 1)/2]M_0^2 \qquad (11.5.28)$$

where η_d = isentropic efficiency of diffuser (0.75 to 0.90 for well-designed systems).

The turbine pressure-ratio parameter Θ_t is given by

$$1/\Theta_t = 1 - [(\Theta_c - 1)/\alpha_1 \eta_c \eta_t] \qquad (11.5.29)$$

where η_t = the isentropic efficiency of the turbine (0.90 to 0.95).

Heat supplied, per unit mass of air, is

$$Q_1 = (c_p/\eta_B)(T_4 - T_3)$$

or

$$Q_i = \frac{c_p t_0}{\eta_B \eta_c}\alpha_d\left(\frac{\alpha}{\alpha_d}\eta_c - \eta_c - \Theta_c + 1\right) \qquad (11.5.30)$$

where η_c is the isentropic efficiency of the air compressor (0.85 to 0.90), and η_B is the efficiency of the burner (0.95 to 0.99).

The enthalpy change for the exhaust nozzle Δh_n is given by

$$\frac{\Delta h_n}{c_p t_0} = \frac{\eta_n}{\eta_c}[\alpha\eta_c - \alpha_d(\Theta_c - 1)]$$

$$\times \left\{1 - \frac{\alpha_1 \eta_c \eta_t}{\Theta_c[1 + \eta_d(\alpha_d - 1)][\alpha_1 \eta_c \eta_t - (\Theta_c - 1)]}\right\} \qquad (11.5.31)$$

The **specific thrust**, also called the **air specific impulse**, is

$$I_a = F_i/\dot{m}_a g = (\sqrt{2\Delta h_n} - V_0)/\dot{m}_a \qquad (11.5.32)$$

The overall efficiency η_o is given by

$$\eta_o = \eta_{th}\eta_p = \frac{I_a V_0}{J Q_i}$$

$$= 2\eta_B\eta_c \frac{\lambda M_0\left(\frac{k-1}{2}\right)^{1/2}}{\alpha\eta_c - \left(1 + \frac{k-1}{2}M_0^2\right)(\eta_c + \Theta_c - 1)} \qquad (11.5.33)$$

where η_B ranges from 0.95 to 0.99 and λ is the dimensionless turbojet specific thrust parameter.

The *TSFC* for a turbojet engine is given by

$$TSFC = \frac{\dot{m}_f}{F} \qquad (11.5.34)$$

It can be shown by dimensional analysis, if the effects of Reynolds numbers are neglected, that the variables entering into the performance of a given turbojet engine may be grouped as indicated in Table 11.5.2.

Table 11.5.2

Nondimensional group	Uncorrected form	Corrected form
Flight speed	$V_0/\sqrt{t_0}$	$V_0/\sqrt{\theta}$
Rotational speed	$N/\sqrt{T}$	$N/\sqrt{\theta}$
Air flow rate	$\dot{W}_a\sqrt{T}/D^2P$	$\dot{W}_a\sqrt{\theta}/\delta$
Thrust	F/D^2P	F/δ
Fuel flow rate	$\dot{W}_f J\Delta H_c/D^2P\sqrt{T}$	$\dot{W}_f/\delta\sqrt{\theta}$

$\theta = T/T_{\text{std}} = T/519\ (T/288) =$ corrected temperature (exact value for T_{std} is 518.699°R).
$\delta = P/p_{\text{std}} = P/14.7\ (P/1.013 \times 10^5) =$ corrected pressure.

Figure 11.5.17 is a design-point chart presenting λ and η_o as functions of α, P_3/P_2, and Θ_c for the subsonic performance of simple turbojet engines. The curves apply to propulsion at 30,000 ft. (9.14 km) altitude for engines having characteristic data indicated in the figure. Curve A illustrates the effect of pressure ratio P_3/P_2 or Θ_c for engines operating with $\alpha = 5.0$. It is seen that increasing the compressor pressure ratio increases η_o but it approaches a maximum value at P_3/P_2

approximately equal to 19. Curve B applies to the design point of engines having a fixed compressor pressure ratio of $10:1$ but having different values of α, that is, turbine inlet temperature. It shows that increasing α much above $\alpha = 5$ reduces the overall efficiency and consequently *TSFC*. Curve C presents the design-point characteristics for engines operating with $\alpha = 5$ but having different compressor pressure ratios. The maximum value of λ is obtained at a relatively low pressure ratio: approximately $P_3/P_2 = 5$. Curve D presents the design-point characteristics of engines having a fixed compressor pressure ratio of $10:1$ but different cycle temperature ratios. Increasing α gives significant increases in λ. Hence, large values of turbine inlet temperature T_4 offer the potential for large thrusts per unit of frontal area for such turbojet engines.

Figure 11.5.18 presents the design-point characteristics for engines having $\eta_c = 0.85$, $\eta_t = 0.90$, $\eta_r = 0.93$, and $\eta_B = 0.99$. Different values of diffuser efficiency are presented for flight Mach numbers $M_0 = 0.8$ ($\eta_d = 0.91$), $M_0 = 1.60$ ($\eta_d = 0.90$) and $M_0 = 2.4$ ($\eta_d = 0.88$). Two different classes of engines are considered, those with $P_3/P_2 = 4.0$ and those with $P_3/P_2 = 12.0$. The dimensionless thrust λ is plotted as a function of α with M_0 as a parameter. It is evident that increasing λ, that is, the turbine inlet temperature T_4, offers the advantage of a large increase in thrust per unit area of the engine under all the conditions considered in Fig. 11.5.18.

Figure 11.5.19 presents the effect of flight speed on the design-point performance, at 30,000 ft (9.14 km) altitude, for engines having $\eta_c = 0.85$, $\eta_t = 0.90$, $\eta_B = 0.95$, $T_4 = 2000°R$ (1111 K) and burning a fuel having a heating value $\Delta H_c = 18,700$ Btu/lbm (10,380 kcal/kgm). The magnitude of the compressor pressure ratio P_3/P_2 approaches unity when V_0 is approximately 1,500 mi/h, indicating that at higher speeds the compressor and turbine are superfluous; i.e., the propulsion requirements would be met more adequately by a ramjet engine. Raising T_4 tends to delay the aforementioned condition to a high value of V_0.

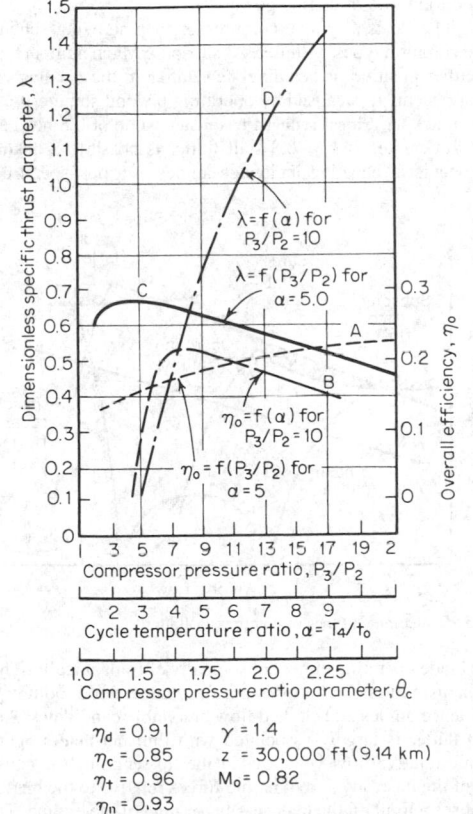

$\eta_d = 0.91$ $\gamma = 1.4$
$\eta_c = 0.85$ $z = 30,000$ ft (9.14 km)
$\eta_t = 0.96$ $M_0 = 0.82$
$\eta_n = 0.93$

Fig. 11.5.17 Dimensionless thrust parameter λ and overall efficiency η_0 as functions of the compressor pressure ratio (simple turbojet engine).

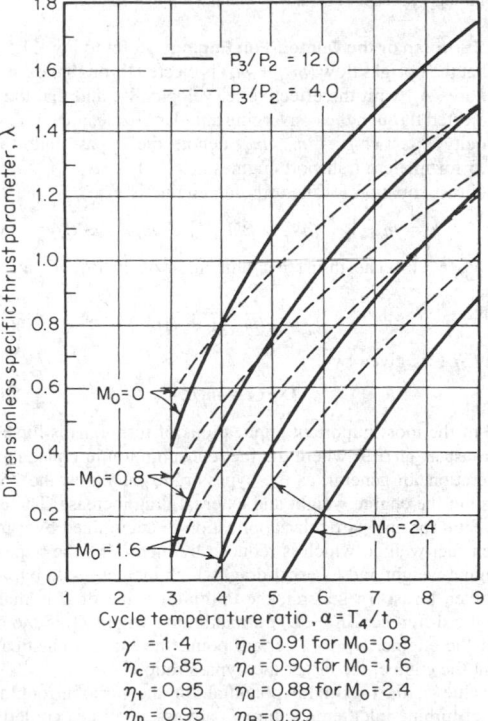

$\gamma = 1.4$ $\eta_d = 0.91$ for $M_0 = 0.8$
$\eta_c = 0.85$ $\eta_d = 0.90$ for $M_0 = 1.6$
$\eta_t = 0.95$ $\eta_d = 0.88$ for $M_0 = 2.4$
$\eta_n = 0.93$ $\eta_B = 0.99$

Fig. 11.5.18 Dimensionless thrust parameter λ for a simple turbojet engine and its overall efficiency η_o as functions of the cycle temperature ratio α, with the flight Mach number as a parameter.

It is a characteristic of turbojet engines, since $\eta_o = \eta_{th}\eta_p$, that for a given M_0 increasing η_{th} causes Δh_n to increase, and hence the jet velocity V_j. As a consequence, η_p is decreased. It is this characteristic which causes $\eta_o = f(\alpha)$, curve B of Fig. 11.5.17, to be rather flat for a wide range of values of $\alpha = T_4/t_0$.

The curves of Figs. 11.5.17 to 11.5.19 are design-point performance curves; each point on a curve is the design point for a different turbojet engine. The performance curves for a specific engine are either computed from experimental data pertaining to its components operating over a wide range of conditions or obtained from testing the complete turbojet engine. It is customary to present the data in standardized form.

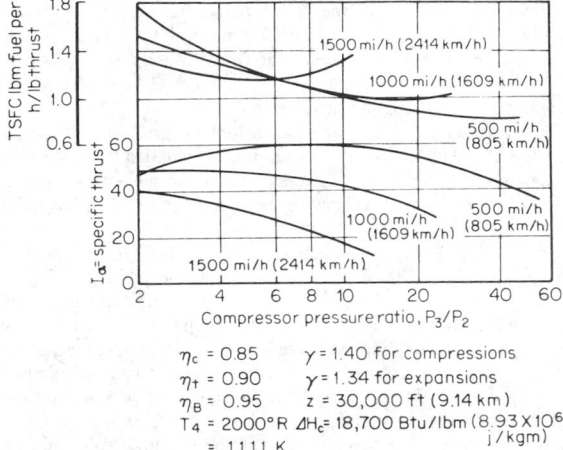

$$\eta_c = 0.85 \qquad \gamma = 1.40 \text{ for compressions}$$
$$\eta_t = 0.90 \qquad \gamma = 1.34 \text{ for expansions}$$
$$\eta_B = 0.95 \qquad z = 30{,}000 \text{ ft } (9.14 \text{ km})$$
$$T_4 = 2000°R \;\; \Delta H_c = 18{,}700 \text{ Btu/lbm } (8.93 \times 10^6 \text{ j/kgm})$$
$$= 1111 \text{ K}$$

Fig. 11.5.19 *TSFC* and specific thrust I_a functions of the compressor pressure ratio, with flight speed as a parameter (simple turbojet engine).

The Turbofan or the Ducted-Fan Engine Refer to Fig. 11.5.5b. Assume that the hot-gas flow $(\dot{m}_{a1} + \dot{m}_f)$ is ejected through the converging nozzle, area A_t, with the effective jet velocity V_j, and that the airflow $\dot{m}_{aF}$ is ejected through the converging annular nozzle, area A_{7F}, with the jet velocity V_{jF}. Let $\beta = \dot{m}_{aF}/\dot{m}_{a1}$ denote the bypass ratio, approximately 5 for modern transport engines.

The thrust equation for the turbofan engine is

$$F = \dot{m}_{a1}(1 + f)V_j + \beta\dot{m}_{a1}V_{jF} - \dot{m}_{a1}(1 + \beta)V_0 \quad (11.5.35)$$

where $V_{jF} < V_j$. The thrust per unit mass of airflow, or air specific impulse, is

$$I_a = F/\dot{m}_{a1}(1 + \beta) \quad (11.5.36)$$

The *TSFC* is given by

$$TSFC = \dot{m}_f/F \quad (11.5.37)$$

One of the most important applications of turbofans is for **high-subsonic transport aircraft** where the fuel consumption in cruise is a major consideration. In general, as the bypass ratio increases, the *TSFC* decreases but the engine weight and external drag increase. The **optimum bypass ratio** for any particular application is determined by a trade-off between fuel weight, which is reduced by increasing the bypass ratio, and engine weight and external drag, which increase with bypass ratio for a given thrust. In general, the optimum value of β is that which makes the difference $(F - D_e)$ a maximum, where D_e is the external drag of the engine. Up to a certain point, the more sophisticated and efficient the engine, the larger the bypass ratio.

To achieve the maximum potential for the turbofan engine, high values of turbine inlet temperature T_4 are essential; i.e., cooled turbine stator and rotor blades must be developed. There is an upper limit to T_4; above it the *TSFC* begins to increase. As T_4 is increased, however, the optimum value of β must also be increased to avoid excessive leaving losses in the propulsive jet.

Mechanical Design

Blades New technologies being developed to achieve higher thrust-to-weight ratios in advanced gas turbine engines also cause higher vibratory blade row responses and stresses that delay engine development and may adversely affect engine reliability. In particular, the design trend toward higher stage loadings and higher specific flow are being attained through increased tip speeds, lower radius ratios, lower aspect ratios, and fewer stages. The resultant axial-flow compressor blade designs utilize thin, low-aspect-ratio blading with corresponding high steady-state stresses. Also, the mechanical damping is considerably reduced in newer rotor designs, particularly those with integral blade-disk configurations (**blisks**) and in those without shrouds. These are similar trends in turbine blading, but they are further complicated by increased inlet and cooling temperatures that have resulted in thin-walled, complex cooling passage designs. As a result, advanced axial-flow blade designs, both compressors and turbines, feature low-aspect-ratio blading which affect the **structural integrity** of the blading. These problems can be classified into two general cateories: (1) **flutter** and (2) **forced response**. A fundamental parameter common to both categories is the **reduced frequency**, $k = \omega b/V$, where b is the blade semichord, ω the frequency of vibration, and V the relative free-stream velocity. For engine applications, the frequency corresponds generally to one of the lower-blade or coupled-blade disk **natural frequencies**.

Under certain conditions, a blade row operating in a completely uniform flow can enter into a self-excited vibrational **instability** termed **flutter**. The vibration is sustained by the extraction of energy from the uniform flow during each vibratory cycle. The outstanding feature of flutter is that **high stresses** exist in the blading, leading to short-term, high-cycle, **fatigue failures**. Its importance is evidenced by the fact that if a portion of a single blade fails due to flutter, the result may be instantaneous and total loss of engine power.

Figure 11.5.20 is a typical compressor map showing schematically the more common types of flutter. At subsonic Mach numbers, up to 0.8 to 0.9, either positive or negative **stall flutter** in the **bending** or **torsion** mode may occur. It is caused by operating beyond some critical airfoil angle of attack. A **critical reduced frequency** value of 0.8 has been cited for stall flutter; i.e., if $k < 0.8$ stall flutter is possible. **Choking flutter** usually occurs at negative incidence angles at a part-speed condition

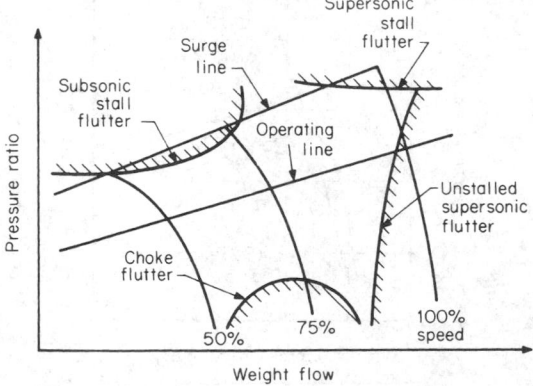

Fig. 11.5.20 Common types of compressor flutter.

with the blade operating either subsonically or transonically. The physical mechanism of choke flutter is not fully understood; both high negative incidence angles and choked flow are viable candidates. **Supersonic unstalled flutter** is largely associated with thin **fan blades** operating at supersonic relative flow velocities at the tip section. It is caused by a lagging of the unsteady aerodynamic forces relative to the blade motion and imposes a limit on the **high-speed operation** of the engine. The stress encountered during this type of flutter can be catastrophically large, decreasing rapidly as a constant speed line is traversed toward higher pressure ratios. High-speed operation near the surge line can lead to

supersonic stall flutter. Because this flutter is associated with **high pressure ratios,** there can be strong shocks present within the blade passages, and in some circumstances, ahead of the blade row.

Destructive aerodynamic **forced responses** have been noted in all engine blading. These failure-level vibratory responses occur when a periodic forcing functions, with frequency equal to a natural blade **resonant frequency,** acts on a blade row. Responses of sufficient magnitude to fail blades have been generated by a variety of sources including upstream and downstream blade rows, flow distortion, bleeds, and mechanical sources.

The importance of forced response lies in the **multiple sources** of excitation for any rotating system. The rotor speeds at which forced responses occur are predicted with "Campbell" or **speed-frequency** diagrams. These display the **natural frequency** of each blade mode versus rotor speed and, at the same time, the forcing function frequency (or **engine order** E lines) versus rotor speed, as indicated schematically in Fig. 11.5.21. These E lines represent the loci of available **excitation energy** at any rotational speed for 1, 2, 3, etc. excitations per revolution. Wherever these curves intersect, a potential source of destructive forced response exists. Not all intersections can be avoided. Design practice is

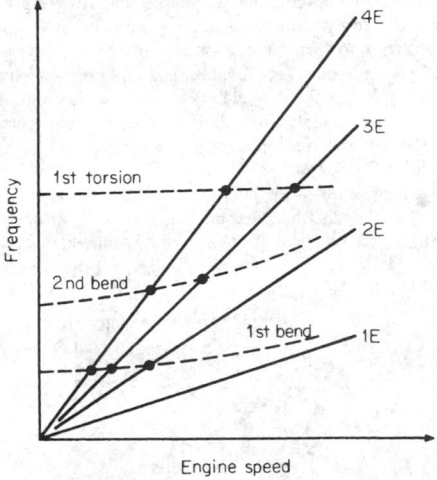

Fig. 11.5.21 Schematic speed-frequency diagram.

to eliminate the lower-order excitations from the operating range whenever possible. This is because the sources of most forced response energy usually result from these lower-order resonances, particularly the 2E line.

Environmental Issues

The increasing worldwide concern for the environment has led to two technical issues becoming significant: (1) airport noise and (2) the Earth's ozone layer.

Airport noise is now a major design criterion for commercial aircraft because of **Federal Aviation Rule, Part 36 (FAR-36),** which sets maximum **takeoff, landing,** and **sideline** noise levels for certification of new turbofan-powered aircraft. Three measuring stations are referred to in FAR-36: (1) under the approach path 1 mi before touchdown, (2) under the takeoff path 3.5 mi from the start of the take-off roll, and (3) along the sides of the runway at a distance of 0.35 mi for four-engined aircraft. Since the power settings and aircraft height are variables, the noise at the approach and takeoff stations are dependent on the engines, the aircraft performance, and operational procedures. The sideline station is more representative of the intrinsic takeoff noise characteristics of the engine, since the engine is at full throttle and the station is at a fixed distance from the aircraft.

The **subsonic turbofan** radiates noise both forward and backward and also produces jet noise from both the fan-jet and the primary jet. How-

ever, because a low jet velocity gives good cruise propulsive efficiency, the jet noise can be reduced. Thus, for the subsonic high-bypass turbofan, **fan noise** is the most critical problem, both on approach and takeoff. For the supersonic turbojet engine, approach noise is not a major problem because it has a long inlet that can be **choked** on approach to suppress fan noise. The jet velocity at takeoff is high if the engine is optimized for cruise performance. However, one environmental barrier issue for the second-generation supersonic transport, the **high-speed civil transport (HSCT),** is airport noise during the takeoff and landing. The prime contributor to an HSCT noise signature is the **jet exhaust,** and thus an approach to quieting the jet exhaust without seriously impacting nozzle aerodynamic performance, size, or weight must be developed.

With regard to aircraft propulsion, depletion of the Earth's **ozone layer** is attributed to the engine exhaust products, specifically the **nitric oxide (NO$_x$)** levels. Thus, combustor designs must evolve so as to produce ultralow levels of NO$_x$ while at the same time not compromising combustion efficiency and operability.

CRITERIA OF ROCKET-MOTOR PERFORMANCE

In most applications of rocket motors the objective is to produce a large thrust for a limited period. The criteria of rocket-motor performance are, therefore, related to thrust-producing capabilities and operating duration. If $\dot{W}_p$ denotes the rate at which the rocket motor consumes propellants, P_c the total pressure at the entrance to the exhaust nozzle, and C_w the **weight flow coefficient,** then

$$\dot{W}_p = C_w P_c A_t \qquad (11.5.38)$$

Ordinarily, the values of C_w as a function of P_c and **mixture ratio** $\dot{W}_o/\dot{W}_f$ are determined experimentally. If no experimental data are available, the values of C_w can be calculated with good accuracy from the calculated thermodynamic properties of the combustion gases. If T_c is the total temperature, and P_c the total pressure of the combustion gases at the entrance section of the nozzle, then, for steady operating conditions, $W_p = W_g$,

$$\dot{W}_g = C_d A_t P_c \Omega \sqrt{\frac{g}{RT_c}} \qquad (11.5.39)$$

where

$$\Omega = \left(\frac{2}{k+1}\right)^{(k-1)/[2(k-1)]} \sqrt{k} \qquad (11.5.40)$$

An equation similar to Eq. (11.5.38) can be written for the thrust developed by the rocket motor. Thus

$$F = C_F P_c A_t \qquad (11.5.41)$$

The values of the **thrust coefficient** C_F for specific propellant combinations are determined experimentally as functions of P_c and the ratio $\dot{W}_o/\dot{W}_f$. When no experimental values are available, C_F can be calculated from the following equation. Therefore

$$C_F = C_d \lambda \phi \Omega \sqrt{\frac{2k}{k-1} Z_t} + \left(\frac{p_e}{P_c} - \frac{p_0}{P_c}\right) \frac{A_e}{A_t} \qquad (11.5.42)$$

where

$$Z_t = 1 - \left(\frac{p_e}{P_c}\right)^{(k-1)/k}$$

Equation (11.5.42) shows that C_F is independent of the combustion temperature T_c and the molecular weight of the combustion gases. The nozzle area ratio A_e/A_t is given by the formula

$$\frac{A_e}{A_t} = \frac{(P_c/p_e)^{1/k}}{\left(\frac{k+1}{2}\right)^{1/(k-1)} \sqrt{[(k+1)/(k-1)]Z_t}} \qquad (11.5.43)$$

Figure 11.5.22 presents C_F calculated by means of the preceding equations, as a function of the area ratio A_e/A_t, with P_c/p_0 as a parameter, for gases having $k = 1.28$; the calculations assumed $\lambda = 0.983$ and $C_d = \varphi = 1.0$. The curves do not take into account the separation phenomena which occur when A_e/A_t is larger than that required to expand the gases to p_0.

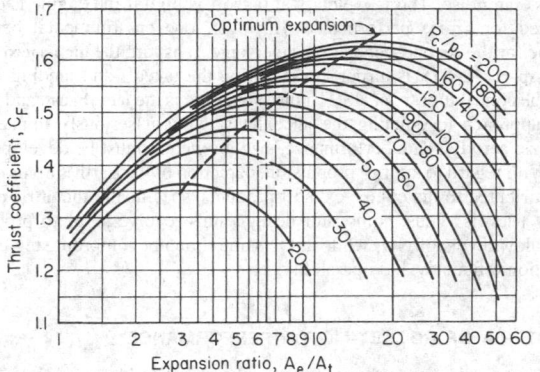

Fig. 11.5.22 Calculated thrust coefficient versus expansion ratio for different pressure ratios (based on $k = 1.28$, $\lambda = 0.9830$, $C_d = \varphi = 1.0$).

The specific impulse I is defined by Eq. (11.5.8) and can be shown to be given by

$$I = F/\dot{m}_p = C_F/C_w = V_j/g_c \qquad (11.5.44)$$

A criterion which is employed for stating the performance of rocket motor is the characteristic velocity c^*. By definition

$$c^* = V_j/C_F = P_cA_t/\dot{m} = I/C_F \qquad (11.5.45)$$

The values of c^* for a given propellant combination are determined experimentally. It should be noted that the value of c^* is independent of the thrust. Basically, c^* measures the effectiveness with which the thermochemical energy of the propellants is converted into effective jet kinetic energy. When no experimental data are available the value of c^* can be estimated quite closely from

$$c^* = \sqrt{RT_c}/\Omega \qquad (11.5.46)$$

Table 11.5.1 presents values of specific impulse for some possible rocket-propellant combinations.

Solid-propellant rocket motors may be segregated into (1) **double-base powders** and (2) **composite, or heterogeneous, propellants.** Double-base powders are gelatinized colloidal mixtures of nitroglycerin and cellulose to which certain stabilizers have been added. Heterogeneous, or composite, propellants are physical mixtures of a solid oxidizer in powder form and some form of solid fuel, such as a plastic or rubberlike material.

Solid-propellant rocket motor technology has made tremendous strides. It is possible to produce solid propellants with a wide range of linear burning rates and with good physical properties. It has been demonstrated experimentally that large thrust rocket motors with chamber diameters of approximately 22 ft (6.7 m) are feasible. As a result, the solid-propellant rocket motor has displaced the liquid rocket engine from a large number of applications, which heretofore were considered the province of the liquid rocket engine. In military applications, there is a strong tendency to supplant liquid rocket engines with solid rocket engines. Only in such areas as the large thrust launching rockets for boosting manned vehicles into space have the liquid rocket engines reigned supreme; the space shuttle is equipped with liquid-bipropellant rocket engines, but it is boosted with auxiliary solid-propellant rocket motors.

The following considerations are what make solid-propellant rocket motors attractive: (1) They are simpler in construction than liquid rocket engines; (2) they are easier to handle in the field; (3) they have instant readiness for use; (4) they have very good storage properties; (5) they have a larger average density $\overline{\rho}_P$ than most liquid-propellant combinations; (6) they have considerably fewer parts; and (7) their reliability in practice has been very good. Furthermore, once the performance characteristics and physical properties of a solid propellant have been established, it is a relatively straightforward engineering problem to design and develop solid-propellant rocket motors of widely different thrust levels and burning times t_R.

Rocket motors which burn double-base propellants find the greatest use in weaponry, e.g., artillery rockets, antitank weapons, etc.

Rocket motors burning heterogeneous, also called **composite,** solid propellants are used for propelling all kinds of military missiles and for sounding rockets. Modern composite propellants have three basic types of ingredients: (1) a fuel which is an organic polymer, called the **binder;** (2) a finely powdered oxidizer, ordinarily ammonium perchlorate (NH_4NO_3); and (3) various additives, for catalyzing the combustion process, for increasing the propellant density, for increasing the specific impulse (e.g., powdered aluminum), for improving the physical properties of the propellant, and the like. After the ingredients have been thoroughly mixed, the resulting viscous fluid is poured, usually under vacuum to prevent the formation of voids, directly into the chamber of the rocket motor which contains a suitable mandril for producing the desired configuration for the propellant grain. The propellant is cured by polymerization to form an elastomeric material. Some of the more common binders are butadiene copolymers and polyurethane.

Modern composite solid propellants have densities ranging from 1.60 to 1.70, depending upon the formulation, and specific impulse values of approximately 245 s, based on a combustion pressure of 1000 psia and expansion to 14.7 psia.

In the case of solid-propellant rockets the rate of propellant consumption $\dot{m}_p$ is related to the **linear burning rate** for the propellant r_0, that is, the rate at which the burning surface of the propellant recedes normal

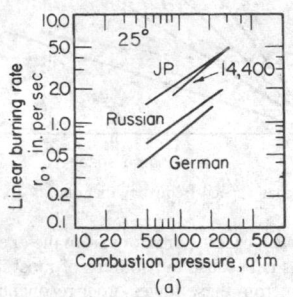

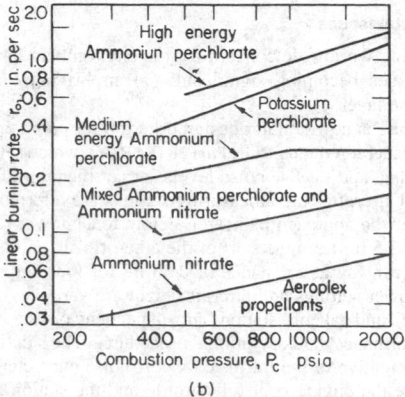

Fig. 11.5.23 (a) Linear burning rate as a function of combustion pressure for double-base propellants; (b) burning-rate characteristics of several heterogeneous propellants at 60°F (15.5°C). (*Aerojet-General Corp.*)

to itself as it burns. For practical purposes it can be assumed that the linear burning rate r_0 is given by

$$r_0 = a_2 p_c^n \qquad (11.5.47)$$

where a_2 and n are constants which are determined by experiment.

If ρ_p denotes the density of the solid propellant and A_p the area of the burning surface, then the propellant consumption rate $\dot{m}_p$ is given by

$$\dot{m}_p = A_p \rho_p r_0 = a_2 A_p \rho_p p_c^n \qquad (11.5.48)$$

Figure 11.5.23 presents the linear burning rate as a function of the combustion pressure for several propellants.

The linear burning rate r_0 is influenced by the temperature of the solid propellant t_p prior to its ignition. Low values of t_p reduce the burning rate, and vice versa. Consequently, the temperature of the propellant must be given in presenting data on linear burning rates. Figure 11.5.24 presents r_0 as a function of p_c for $t_p = -40, 60,$ and $140°F$ for a composite propellant manufactured by the Aerojet-General Corp., Azusa, Calif.

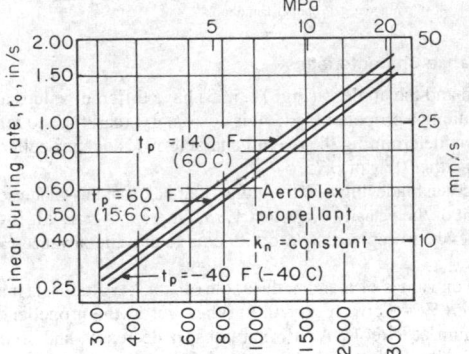

Fig. 11.5.24 Effect of propellant temperature on the linear burning rate of a heterogeneous propellant. *(Aerojet-General Corp.)*

AIRCRAFT PROPELLERS

The function of the aircraft propeller (or airscrew) is to convert the torque delivered to it by an engine into the thrust for propelling an airplane. If the airplane is in steady, level flight, the propeller thrust and the airplane drag equal each other. If the airplane is climbing, the thrust must overcome the drag plus the weight component of the airplane.

The propeller may be the sole thrust-producing element, but it can also be employed in conjunction with another thrust-producing device, such as the exhaust jet of a **turboprop** or **gas-turbine engine**. It may also be employed as a **windmilling** device for producing thrust.

The conventional propeller consists of two or more equally spaced radial blades, which are rotated at a substantially uniform angular velocity. At any arbitrary radius, the section of a blade has the shape of an airfoil, but as the hub is approached, the sections become more nearly circular because of considerations of strength rather than aerodynamic performance. The portions of the blades in the vicinity of the hub contribute, at best, only a small portion of the thrust developed by the propeller blades.

Each section of the propeller blade experiences the aerodynamic reactions of an airfoil of like shape moving through the air in a similar manner. Furthermore, at each radius, the section of one blade forms with the corresponding sections on the other blades a series of similar airfoils, which follow one another as the propeller rotates.

The torque of the rotating propeller imparts a rotational motion to the air flowing through it and, furthermore, causes the pressure immediately behind the propeller to increase while that in front of it is reduced; i.e., the air is sucked toward the front of the propeller and pushed away behind it. A slipstream is created as illustrated in Fig. 11.5.1. The ratio of the cross-sectional area of the slipstream at any station to that of the actuator disk is termed the **slipstream contraction.**

The thrust produced by the propeller follows the relationships developed from the momentum theorem of fluid mechanics for jet engines as presented in Fig. 11.5.15 and Eq. (11.5.5). The thrust equation for the airbreathing engine (simplified) and the propeller is

$$F = \dot{m}(w - V_0) = \rho A V_0 (w - V_0) \qquad (11.5.49)$$

In the case of a propeller the mass flow m is large and the velocity rise across the propeller is small, as compared with a jet engine of the same thrust. However, the thrust of propellers falls off rapidly as air compressibility effects on the propeller become large (flight speeds of 300 to 500 mi/h).

Propeller efficiency is defined as the useful work divided by the energy supplied to the propeller:

$$\eta_p = \frac{P_T}{P} = \frac{FV_0}{P} = \frac{m(v - V_0)V_0}{m/2(w^2 - V_0^2)} = \frac{2v}{1 + v} \qquad (11.5.50)$$

where $v = V_0/w$.

Since w is only slightly larger than V_0, the ideal efficiency is of the order of 0.95, and actual efficiencies are in the range of 0.5 to 0.9.

Propeller Theory

Axial-Momentum Theory   The axial-momentum theory of the propeller is basically a special case of the generalized jet-propulsion system presented in Fig. 11.5.15. It is only a first approximation of the action of the propeller and cannot be employed for design purposes. It neglects such factors as the drag of the blades, energy losses due to slipstream rotation, blade interference, and air compressibility effects. Because of those losses, an actual propeller requires power to rotate at zero thrust, and at zero thrust its propulsive efficiency is, of course, zero.

Blade-Element Theory   This theory, called **strip theory,** takes into account the profile losses of the blade sections. It is the theory most commonly employed in designing a propeller blade and for assessing the off-design performance characteristics of the propeller.

Each section of the propeller is considered to be a rotating airfoil. It is assumed that the radial flow of air may be neglected, so that the flow over a blade section is two-dimensional. Figure 11.5.25 illustrates diagrammatically the velocity vectors pertinent to a blade section of length dr, located at an arbitrary radius r. The projection of the axis of rotation is OO', the plane of rotation is OC, the blade angle is β, the angle of attack for the air flowing toward the blade (in the relative coordinate system) with the velocity V_0 is α, the tangential velocity of the blade element is $u = 2\pi n r$, and the **pitch** or **advance angle** is ϕ. From Fig. 11.5.25.

$$\alpha = \beta - \phi \qquad (11.5.51)$$

and

$$\phi = \tan^{-1}(V_0/\pi n d) \qquad (11.5.52)$$

Figure 11.5.26 illustrates the forces acting on the blade element. If b denotes the width of the blade under consideration and C_L and C_D are

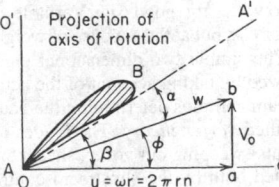

Fig. 11.5.25 Vector diagram for a blade element of a propeller.

the lift and drag coefficients of the blade (airfoil) section, then the thrust force dF acting on the blade element is given by

$$dF = \tfrac{1}{2} \rho V_0^2 b \, dr C_L \frac{\cos(\phi + \gamma)}{\cos \gamma \sin^2 \phi} \qquad (11.5.53)$$

The corresponding torque force dQ is given by

$$dQ = \frac{1}{2}\rho V_0^2 b \, dr C_L \frac{\sin(\phi + \gamma)}{\cos\gamma \sin^2\phi} \quad (11.5.54)$$

where

$$\tan\gamma = d\mathscr{D}/d\mathscr{L} = C_D/C_L = \mathscr{D}/\mathscr{L} \quad (11.5.55)$$

where $\mathscr{D}$ and $\mathscr{L}$ denote drag and lift, respectively.

The propulsion efficiency of the blade element, termed the **blading efficiency**, is defined by

$$\eta_b = \frac{V_0 \, dF}{u \, dQ} = \frac{\tan\phi}{\tan(\phi + \gamma)} = \frac{\mathscr{L}/\mathscr{D} - \tan\phi}{\mathscr{L}/\mathscr{D} + \cot\phi} \quad (11.5.56)$$

The value of ϕ which makes η_b a maximum is termed the **optimum advance angle** ϕ_{opt}. Thus

$$\phi_{opt} = \frac{\pi}{4} - \frac{\gamma}{2} = 45° - \frac{57.3}{2(\mathscr{L}/\mathscr{D})} \quad (11.5.57)$$

The maximum blade efficiency is accordingly

$$(\eta_b)_{max} = \frac{2\gamma - 1}{2\gamma + 1} = \frac{2(\mathscr{L}/\mathscr{D}) - 1}{2(\mathscr{L}/\mathscr{D}) + 1} \quad (11.5.58)$$

A force diagram similar to Fig. 11.5.26 can be constructed for each element of the propeller blade, taking into account the variation in ϕ with the radius. The resultant forces acting on the propeller are obtained by summing (integrating) those forces acting on each blade element.

The integrated lift coefficient is

$$C_{Li} = \int_{r_L/R}^{r/R = 1.0} C_{Lr} \left(\frac{r}{R}\right)^3 d\left(\frac{r}{R}\right) \quad (11.5.59)$$

where C_{Lr} = propeller-blade-section lift coefficient and r/R = fraction of propeller-tip radius.

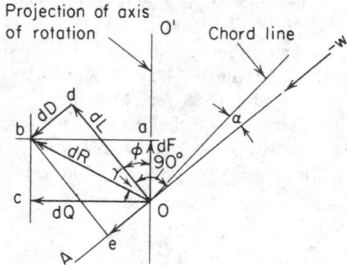

Fig. 11.5.26 Aerodynamic forces acting on a blade element.

The integrated lift coefficient is in the range 0.15 to 0.70. There is a dilemma in the design and selection of blade sections since high static thrust requires high C_{Li} which gives relatively low η_p at cruise. Cruise conditions require low C_{Li} for good η_p. Variable camber propellers permit optimization of C_{Li}, but at the expense of weight and complexity.

Vortex Theory The simple two-dimensional blade-element theory can be further improved by taking account of the actual velocity distribution in the slipstream and thus determining the actual loading on the blading by the so-called *vortex theory*. The blades are represented by bound vortex filaments sweeping out vortex sheets to infinity. The vortex sheets are distorted both by the slipstream contraction and by the profile drag of the blades. If such distortions are neglected, the induced velocity in the wake can be calculated by assuming the vortex sheets to be tubular. Thus, in Fig. 11.5.25, the induced axial velocity becomes the arithmetic mean of the axial velocities at far distances upstream and downstream of the rotor disk; and the induced rotational velocity at the disk becomes $\omega - 0.5\omega'$, where ω' is the rotation of the slipstream in

the plane of the propeller. Hence, the angle of attack α for a rotor element is given by

$$\alpha = \beta - \tan^{-1}[V_F/r(\omega - 0.5\omega')] \quad (11.5.60)$$

where V_F is the induced axial velocity in the plane of the propeller.

The addition of the vortex effects improves the representation of the propeller by the blade sections. There are, however, three-dimensional flow effects which are not included, and the following coefficients have been suggested to include this effect (which reduces the lift coefficient).

Lift Curve Slope Correction Factors for Three-Dimensional Airfoils

Radial station (ratio r/R)	Correction factor
(0.2)(0.3)(0.45)(0.6)(0.7)	0.85
0.8	0.80
0.9	0.70
0.95	0.65

SOURCE: Departments of the Air Force, Navy, and Commerce, "Propeller Handbook."

Performance Characteristics

The pitch and the angle ϕ (Fig. 11.5.25) have different values at different radii along a propeller blade. It is customary, therefore, to refer to all parameters determining the overall characteristics of a propeller to their values at either $0.7r$ or $0.75r$.

For a given blade angle β, the angle of attack α at any blade section is a function of the **velocity coefficient** V_0/u, where $u = \pi n d_p$; it is customary, however, to replace V_0/u by its equivalent, the **advance ratio** $J = V_0/n d_p$.

For given values of β and n, the angle of attack α attains its maximum value when $V_0 = 0$, with $\phi = 0$. In other words, the propeller thrust F is maximum at takeoff. As V_0 increases, α decreases and so does the thrust. Finally, a forward speed is reached for which the thrust is zero.

Increasing $J = V_0/n d_p$, for a constant blade angle, causes the propeller to operate successively as a fan, propeller, brake, and windmill. Most of the operation, however, is conducted in the propeller state. At takeoff, the propeller is in the fan state. Windmilling must be avoided because it may overspeed the engine and damage it.

The lift coefficient C_L is a linear function of α up to the stalling angle, and C_D is a quadratic function of α. Furthermore, the lift $\mathscr{L}$ is zero at a negative value for α. Figure 11.5.27 presents the $\mathscr{L}/\mathscr{D}$ ratios of typical propeller sections. Figure 11.5.28 presents the blade (section) efficiency as a function of the pitch angle ϕ for different values of $\mathscr{L}/\mathscr{D}$. Propellers normally operate at $\mathscr{L}/\mathscr{D}$ ratios of 20 or more. Supersonic propellers operate with $\mathscr{L}/\mathscr{D}$ ratios of approximately 10, and from Fig. 11.5.27 it is apparent that they must operate close to the optimum pitch angle ϕ to obtain usable blade efficiencies.

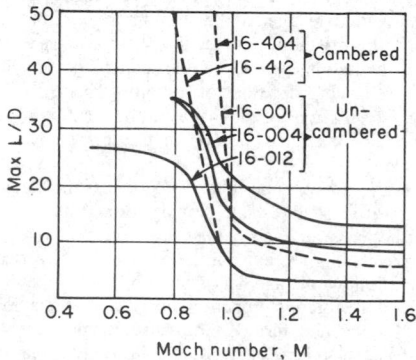

Fig. 11.5.27 Maximum lift-drag ratios for representative sections.

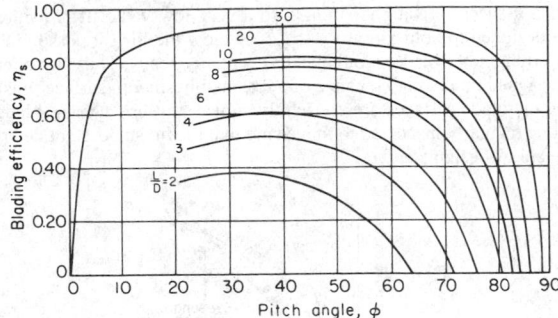

Fig. 11.5.28 Blade-element efficiency versus pitch angle.

Propeller Coefficients It can be shown, neglecting the compressibility of the air, that

$$f(V_0, n, d_p, \rho, F) = 0 \qquad (11.5.61)$$

By dimensional analysis, the following coefficients are obtained for expressing the performances of propellers having the same geometry:

$$F = \rho n^2 d_p^4 C_F \qquad Q = \rho n^2 d_p^5 C_Q \qquad \mathcal{P} = \rho n^3 d_p^5 C_P \quad (11.5.62)$$

C_F, C_Q, and C_P are termed the **thrust, torque, and power coefficients,** respectively. They are independent of the size of the propeller. Consequently, tests of small-scale models can be employed for obtaining the values of F, Q, and $\mathcal{P}$ for geometrically similar full-scale propellers. Hence, the ideal propulsive efficiency of a propeller is given by

$$\eta_p = \frac{F V_0}{\mathcal{P}} = \frac{C_F}{C_P} \frac{V_0}{d_p n} = J \frac{C_F}{C_P} \qquad (11.5.63)$$

Other useful coefficients are the **speed-power coefficient** C_S and the **torque-speed coefficient** C_{QS}. Thus,

$$C_S = V_0 5 \sqrt{\rho / \mathcal{P} n^2} \qquad C_{QS} = V_0 \sqrt{\rho d_p^3 / Q} \qquad (11.5.64)$$

A commonly employed factor related to the geometry of a propeller is the **solidity** σ. By definition,

$$\sigma = c/p \qquad (11.5.65)$$

The solidity varies from radius to radius; at a given radius, σ is proportional to the power-absorption capacity of the annulus of blade elements.

The **activity factor** of a blade AF is an arbitrary measure of the blade solidity and hence of its ability to absorb power. It takes the form

$$AF = \frac{100{,}000}{16} \int_{r_h/R}^{r/R = 1.0} \frac{c}{d_p} \left(\frac{r}{R} \right)^3 d \left(\frac{r}{R} \right) \qquad (11.5.66)$$

where $R = d_p/2$, r = radius at any section, and r_h = spinner radius.

Performance

Static Thrust From the simple axial-momentum theory, the static thrust F_0 of an actuator disk of diameter d_p is given by

$$F_0 = (\pi \rho / 2)^{1/3} (\mathcal{P} d_p)^{2/3} = 10.4 \text{bhp} \times d_p^{2/3} \quad \text{(at sea level)} \quad (11.5.67)$$

The **thrust horsepower** (thp) is accordingly

$$\text{thp} = \frac{F_0}{\text{bhp}} = (38{,}400/N d_p)(1/C_p^{1/3}) \qquad (11.5.68)$$

Equations (11.5.67) and (11.5.68) assume uniform axial velocity, no rotation of the air, and no profile losses. Consequently, they may be in error by as much as 50 percent. Hesse suggests the use of the equations

$$\eta_{p \text{ static}} = \sqrt{\frac{2}{\pi}} \frac{C_F^{3/2}}{C_P} = 0.798 \frac{C_F^{3/2}}{C_P} \qquad (11.5.69)$$

and

$$F = 10.42[d_p (\eta_{p \text{ static}}) \text{shp}]^{2/3} (\rho/\rho_0)^{1/3} \qquad (11.5.70)$$

together with data computed for a specific propeller, as illustrated in Fig. 11.5.29. The static thrust for a given power input increases with the diameter and the number of blades. Controllable pitch propellers selected for good overall performance will develop 3 to 4 lb of thrust per shp.

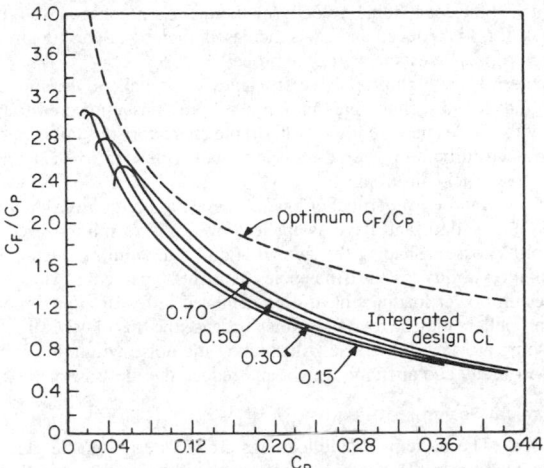

Fig. 11.5.29 Static-thrust performance of a propeller; four-bladed, 100 activity propellers of various integrated design C_L's. (*From Hamilton Standard Division, United Aircraft Corporation.*)

At flight conditions the performance may be obtained from generalized charts presenting C_P as a function of the advance ratio J, with the propeller efficiency and blade pitch angle at 75 percent radius, $\theta_{3/4}$, as parameters. Figure 11.5.30 is such a chart.

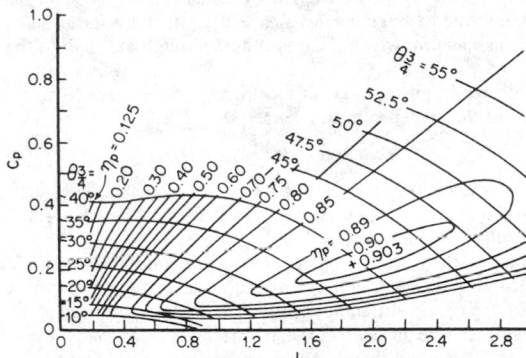

Fig. 11.5.30 Propeller performance—efficiency in forward flight; four-bladed, 100 activity factor, 0.500 integrated design C_L propeller. (*From Hamilton Standard Division, United Aircraft Corporation.*)

Reverse Thrust By operating the propeller blades at large negative angles of attack, reversed thrust can be developed. In that condition, the blades are stalled, and as in the case of static thrust, the forces acting on the blades cannot be calculated. In practice, the reverse-pitch stop is set by trial and error so that the propeller can absorb the rated power at zero speed, i.e., at $V_0 = 0$. With fixed-pitch reversing, the power absorbed when there is forward speed will be smaller than the rated value, but the reversed thrust is usually substantially larger than the static forward thrust corresponding to the rated power. The landing roll of an aircraft with reversed thrust plus brakes is approximately 45 percent of that with brakes alone.

Compressibility Effects As the relative velocity between an airfoil and the free-stream air approaches the sonic speed, the lift coefficient

C_L decreases rapidly, and simultaneously there is a large increase in the drag coefficient C_D. The propeller is subject to those effects. Since the acoustic speed decreases from sea level to the isothermal altitude, the effects appear at lower tip speeds at high altitude. Furthermore, the effect is more pronounced at higher tip speeds. Shock waves are formed at the leading edge of a propeller blade when its pitch-line velocity approaches the local sonic velocity of the air. As a consequence, the coefficient C_F is reduced and C_P is increased, thereby adversely affecting the propulsive efficiency of the propeller.

Because of the high propulsive efficiency of propellers, the application of advanced technologies may provide **turboprop**-powered transports with cruise speeds and other desirable characteristics comparable to advanced turbofan-powered transports, but with a 15 to 20 percent reduction in fuel consumption.

Advanced turboprop propeller blades operate at high power loading, are extremely thin, and have swept leading edges to minimize compressibility losses. Shaping the spinner and nacelle minimizes choking and compressibility losses. Efficiencies of about 70 percent at Mach 0.8 and design power loadings have been achieved in research propellers. Such propellers may be relatively noisy because the tips may be slightly supersonic at Mach 0.8 cruise. Attenuating the noise within the cabin wall will likely add airframe weight and reduce the fuel savings.

Mechanical Design

Blades The principal blade loadings are (1) steady tensile, due to centrifugal forces; (2) steady bending, due to the aerodynamic thrust and torque forces; and (3) vibratory bending, due to cyclic variations in airloads and other excitations originating in the engine. The most serious and limiting stresses usually result from the vibratory loadings. The principal vibratory loading results from the cyclic variation in angle of attack of the blades when the axis of rotation is pitched or yawed relative to airstream. When the axis is pitched up, such as when the aircraft is in a high-angle-of-attack climb, the blades are at a higher angle of attack on the downstroke than when on the upstroke. This results in a once per revolution, $1E$, variation in aerodynamic loading on the blades, usually referred to as $1E$ **aerodynamic excitation**, and a steady **side force** and **yawing moment**, which are transmitted through the shaft to the aircraft.

The degree of pitch or yaw of the propeller axis is usually measured in terms of the **excitation factor**, defined as

$$\text{Excitation factor} = \psi(V_i/400)^2 \qquad (11.5.71)$$

where ψ = angle of pitch or yaw, deg; V_i = indicated flight velocity, mi/h. Another factor sometimes used is Aq, where $A = \psi°$, $q = 12\rho V_i^2$, Aq = excitation factor × 410.

Because of the restoring moment of the **centrifugal loads** on the blade elements when the blade deflects in bending, centrifugal forces have the effect of apparently stiffening the blade in **bending**. In Fig. 11.5.31, the bending natural frequency increase with rpm. At some rpm, the bending natural frequency will come into resonance with the $1E$ excitation, at which point small loadings will be greatly magnified (limited only by the damping present in the blade).

Propellers are designed so that the $1E$ resonant speed is above any expected operating speed. However, there will always be some magnification of the vibratory loads (due to proximity to resonant speed), resulting in disproportionately high vibratory stresses. As indicated in Fig. 11.5.31, blades normally pass through a $2E$ **resonance** in coming up to speed (usually not critical because it is a transient condition) and may pass through or be close to resonance with other modes.

The **natural frequencies** of the rotating propeller blade can be computed with good accuracy, but since the excitation varies from aircraft to aircraft and is usually not well defined, vibration surveys must be made of all new installations. These often result in the definition of ranges at which continuous operation should be avoided. **Allowable vibratory stresses** are limited by the fatigue strength of the material. Unlike the steady-stress limits, **fatigue strength** is not a unique function of the material but is a function of surface finish as well. A sharp notch in the

surface reduces the fatigue strength to a very low value. If propellers were designed to hold vibratory stresses below the lowest possible fatigue strength, the blades would be unacceptably heavy. Thus propellers are designed on the maintenance of a reasonably smooth surface finish. Since the blade surfaces are constantly subject to **store damage**, blades must be regularly inspected to make sure that the finish does not deteriorate below design standards.

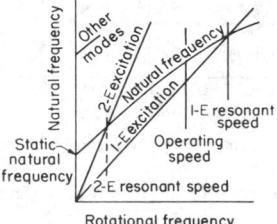

Rotational frequency

Fig. 11.5.31 Relation between blade natural frequency and major excitations.

Propeller blades are stalled when operating at low forward speeds and when reversing (see Performance, above). Under these conditions torsion-mode **stall flutter** can occur.

In general, $1E$ **considerations** control the design of the **inboard** portions of a blade, and **stall flutter** the **outboard** sections.

There are five types of **blade construction** in common use today: **solid aluminum; fabricated steel; one-piece steel; monocoque, fiberglass construction,** supported on a steel shank; and a construction with a **hollow steel central spar and a lightweight fiberglass cover** that forms the aerodynamic contour of the blade. As in any beam designed on the basis of bending loads, it is desirable to concentrate the structural material at the maximum radii, as in the flanges of a simple I beam, and omit it from the center, where it carries no load. This is accomplished, with steel blades, by making them hollow. The structural material is concentrated in the outer fibers, or surface. In general, steel blades weigh about 80 to 85 percent of the equivalent solid aluminum blade. The advantage of steel increases with size. The blades normally constitute from one-third to one-half of the total propeller weight.

The **hub** retains the blades and contains the pitch-change motor. A **pitch-change mechanism** is provided so that the blades can be positioned at the proper angle of attack α to absorb the desired power at the desired rpm, regardless of flight speed. If the pitch is unchanged, as in Fig. 11.5.32, the angle of attack is excessive at low speed. This prevents the engine from reaching rated speed and delivering rated power. The centrifugal loads on the blade elements tend to rotate the blades toward flat pitch (see Fig. 11.5.33). The pitch-change motor acts against this moment by applying a pitch-increasing moment. The addition of **counterweights**, shown dotted in Fig. 11.5.33, reduces the moment and the size of the pitch-change motor. Counterweights, however, add to the radial

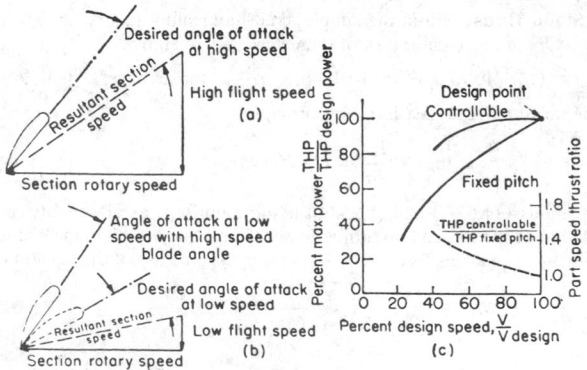

Fig. 11.5.32 Effect of flight speed on the desired blade angle.

centrifugal load of the blades, necessitating a heavier retention, and consequently do not necessarily reduce the net weight of the propeller.

Most pitch-change mechanisms employ a **hydraulic piston,** mounted on the hub, with feed through the propeller shaft and with rotation of the blades by means of gears or links. **Electric motors** and **direct mechanical drives** have been used successfully. In the past, the weight of all systems, when fully developed, has turned out to be about the same.

Action of the pitch-change motor is limited by **low-** and **high-pitch stops.** These prevent the blades from assuming negative angles with reverse thrust. On reciprocating engines, the low-pitch stop is set to allow the propeller to absorb approximately rated power at zero forward speed. A lower setting results in higher windmilling drags in gliding flight, which could be dangerous if carried too far. In reversing propellers the low-pitch stop is automatically removed when the controls are set for reverse pitch. The high-pitch stop prevents the propeller from reversing through the positive range. In feathering propellers it is set at the angle which gives zero aerodynamic moment about the shaft.

Propeller feathering is provided in order to reduce the drag of a dead engine. To accomplish it, the blade-angle range is extended to 90°, approximately. Auxiliary pitch-change power is provided to complete the feathering as the engine slows down and to unfeather when the engine is stationary.

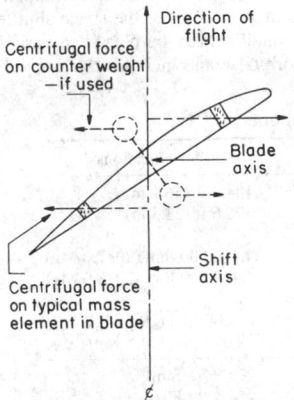

Fig. 11.5.33 Twisting effect of centrifugal loads on blades.

Reversing is provided in order to give propeller braking on the landing roll. It is accomplished by removing the flight low-pitch stop and driving the blades to a high negative **reverse-pitch angle.** The principal mechanical problem is to provide a mechanism which has the least possible chance of going into reverse inadvertently.

The **propeller control** is that portion of the system which regulates the blade angle. In light aircraft, this may consist merely of a mechanism which sets a given blade angle corresponding to a given position of the cockpit control. In such cases the propeller acts as a fixed-pitch propeller except when activated by the pilot. In most cases, the propeller control regulates blade angle so as to maintain a preset rotary rpm, irrespective of flight speed or shaft power. The **basic elements** of this system normally consist of a spring-balanced, engine-drive flyweight delivering an error signal which directs the servo, or pitch-change motor, to increase or decrease blade angle (Fig. 11.5.34). The control loop is closed by rpm feedback through the main engine. This basic system may be embellished with anticipating and delay devices to maintain speed in rapid throttle movements, synchronizing to coordinate two or more engines, and with overriding features to activate the reversing and feathering sequences. (See also Sec. 16.)

Propeller **controls** are fundamentally the same, with several added secondary features. First, because the engine is capable of a very rapid increase or decrease in torque and because of the high gear ratio between the propeller and engine shaft, anticipating features are necessary to prevent overspeeds in the propeller from producing large overspeeds and possible failure in the turbine. Second, because of the high idle rpm needed to maintain the pressure ratio in the compressor, the low-pitch stop of the turboprop engine propeller must be set substantially lower than for a piston engine. This is not true of a free-turbine drive and is only partially true of a split-compressor engine. In the event of an **engine failure,** the propeller, sensing the loss in rpm, will govern down to the low-pitch stop. This may lead to large, often uncontrollable, negative thrust. To avoid this possibility, an engine-shaft torque signal is put into the control, a negative torque automatically activating the feathering system. This portion of the system is known as the **ENT (emergency negative-thrust)** system. Third, because of the high idle rpm characteristic of the turboprop engine, and the resulting low blade angles at low power in the ground-taxi regime, provision must be made to cut out the governing system and substitute direct blade-angle regulation (known as **beta control**). This is required because the propeller will not govern under these conditions as the torque-to-blade-angle relationship goes to zero. The propeller control of a turboprop engine is normally mechanically coordinated directly to the engine throttle, leaving only a single control, with so-called power lever, in the aircraft.

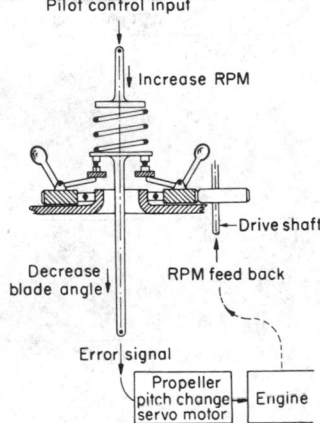

Fig. 11.5.34 Propeller-speed governor.

The shaft on which the propeller is mounted is subjected to the steady-thrust load, vibratory loads due to the propeller side, and **gyroscopic forces** (the $1E$ moment, acting on the blades, is reactionless in propellers with three or more blades). In large aircraft, the shaft size is usually determined by the propeller side force. In aircraft used for aerobatics, gyroscopic loads may become limiting. The **gyroscopic moment** on the propeller shaft is given by $2\omega_1 \cdot \omega_2(I_1 - I_2)$, where ω_1 = angular velocity of maneuver, ω_2 = angular velocity of shaft, I_1 = polar moment of inertia of propeller about shaft, I_2 = moment of inertia of propeller about diameter in plane of rotation. (See also Sec. 3.)

11.6 ASTRONAUTICS
by Aaron Cohen

REFERENCES: "Progress in Astronautics and Aeronautics," AIAA series, Academic Press. "Advances in Astronautical Sciences," American Astronautical Society series. "Handbook of the British Astronautical Association." Koelle, "Handbook of Astronautical Engineering," McGraw-Hill. Purser, Faget, and Smith, "Manned Spacecraft: Engineering Design and Operation," Fairchild Publications. Herrick, "Astrodynamics," Vol. 1, Van Nostrand Reinhold. Cohen, "Space Shuttle," Yearbook of Science and Technology, 1979, McGraw-Hill.

SPACE FLIGHT
by Aaron Cohen

The science of astronautics deals with the design, construction, and operation of craft capable of flight through interplanetary or interstellar space. In addition to orbits around, and trajectories between, such bodies as stars, planets, and planetoids, the ascents from and descents to the surface of such bodies are considered to be part of the field.

The purposes of such craft are numerous and include the transportation of people and cargo, the transmission or relaying of signals, and the carrying of instruments for the measurement of the characteristics of (1) the environment through which the craft flies, (2) the surface of the celestial body over which the craft flies, or (3) astronomical objects or phenomena. The commercial applications of space are now being realized. Many communication, weather, and earth-sensing satellites have been deployed. The space shuttle which provides a method to go to and from low earth orbits offers an expanded use of space.

Processing of materials in a microgravity environment is proving to be very reliable. In microgravity the effects of thermal convection and gravity are eliminated, hence purer materials can be manufactured. Containerless processing can also be accomplished which will lead to additional applications in microgravity materials technology. The space shuttle allows payloads up to 32,000 lb to be placed in earth orbit at altitudes of 150 to 200 nautical miles (nmi) and inclinations of 28.5° to 90°. A description of the space shuttle is contained in Table 11.6.1. Using expendable launch vehicles or the space shuttle and propulsion-assistance modules, satellites can be placed in various orbits. A polar orbit is one whose orbital plane includes the earth's axis and permits

Table 11.6.1   Specifications of the Space Shuttle System and Its Components

Component	Characteristic	Specifications
Overall system	Length	184.2 ft (56.1 m)
	Height	76.6 ft (23.3 m)
	System weight	
	Due-east launch	4,490,800 lbm (2037.0 Mg)
	104° launch azimuth	4,449,000 lbm (2018.0 Mg)
	Payload weight	
	Due-east launch	65,000 lbm (29.5 Mg)
	104° launch azimuth	32,000 lbm (14.5 Mg)
External tank	Diameter	27.8 ft (8.5 m)*
	Length	154.4 ft (47.1 m)
	Weight	
	Launch	1,648,000 lbm (747.6 Mg)
	Inert	70,990 lbm (32.2 Mg)
Solid rocket booster (SRB)	Diameter	12.2 ft (3.7 m)
	Length	149.1 ft (45.4 m)
	Weight (each)	
	Launch	1,285,100 lbm (582.9 Mg)
	Inert	176,300 lbm (80.0 Mg)
	Launch thrust (each)	2,700,000 lbf (12.0 MN)
Separation motors (each SRB), four aft, four forward	Thrust (each)	22,000 lbf (97.9 kN)
Orbiter	Length	121.5 ft (37.0 m)
	Wingspan	78.1 ft (23.8 m)
	Taxi height	57 ft (17.4 m)
	Weight	
	Inert	161,300 lbm (73.2 Mg)
	Landing	
	With payload	203,000 lbm (92.1 Mg)
	Without payload	173,000 lbm (78.5 Mg)
	Cross range	1,100 nmi (2037.2 km)
Payload bay	Length	60 ft (18.3 m)
	Diameter	15 ft (4.6 m)
Main engine (three)	Vacuum thrust (each)	470,000 lbf (2090.7 kN)
Orbital maneuvering system engines (two)	Vacuum thrust (each)	6,000 lbf (26.7 kN)
Reaction control system engines 38 engines	Vacuum thrust (each)	870 lbf (3870 N)
6 vernier rockets	Vacuum thrust (each)	24 lbf (111.2 N)

* Includes spray-on foam insulation.

line-of-sight contact between the spacecraft and every point on earth on a periodic basis. Synchronous orbits have angular rates equal to the rotational rate of the earth about its axis, providing a spacecraft locus of positions at nearly constant longitude. Orbital-parameter variation provides a wide variety of other orbits, permitting the selection of characteristics most suitable for the specific flight objective, such as communication relays, meteorological observations, earth surface monitoring, or interplanetary trajectories.

ASTRONOMICAL CONSTANTS OF THE SOLAR SYSTEM
by Michael B. Duke
NASA

REFERENCES: "Handbook of the British Astronautical Association." Blanco and McCuskey, "Basic Physics of the Solar System," Addison-Wesley. Clarke, Constants and Related Data for Use in Trajectory Calculations, *Tech Rep. 32-604*, Jet Propulsion Lab., Pasadena. Francis, Constants for an Earth-Moon Transit, *Lockheed Rep. LAC/421571*. Krause, On a Consistent System of Astrodynamic Constants, NASA-TN D-1642. Makemson, Baker, and Westrom, Analysis and Standardization of Astrodynamic Constants, *Jour. Astro. Sci.,* Spring 1961. Townsend, "Orbital Flight Handbook," Martin Co. Space and Planetary Environment Criteria Guidelines for Use in Space Vehicle Development, 1982 Revision (Vols. I and II), *NASA Technical Memoranda 82478* and *82501*.

A greater number of astronomical constants is used in astrodynamics and astronautics than in classic astronomy because astronautics is a kind of experimental astronomy and its missions vary so greatly. A higher accuracy is also necessary because the astrodynamic missions include departure, landing, and flyby maneuvers. Some of the constants, given in relative and astronomical units in celestial mechanics, must be known in absolute units for astrodynamic purposes. Tables 11.6.2 to 11.6.4 provide the latest observed data for standardization of astrodynamic computations. See references for additional constants.

Table 11.6.2   General, Terrestrial, and Lunar Constants

1. Ephemeris second: $1 \sec_E = 1/31{,}556{,}925.9747$ of tropical year at 1900.0.
2. Mean solar day (culmination period of the mean sun):

$$1^d = 1^{d*}(.002\ 737\ 909\ 265 + 0^{d*}.589 \times 10^{-10}T = 24^h03^m56^s.555\ 360\ 50 + 0^{s*}.050\ 89 \times 10^{-4}T$$
$$= 1^{rot}.002\ 737\ 811\ 891 - 0^{rot}.001\ 4 \times 10^{-10}T = (1 \pm 10^{-8})^d_E$$

3. Mean sidereal day or mean equinoctial day (culmination period of the vernal equinox):

$$1^{d*} = 0^d.997\ 269\ 566\ 414 \qquad 0^d.587 \times 10^{-10}T = 23^h56^m04^s.090\ 538\ 17 - 0^s.050\ 716\ 8 \times 10^{-4}T$$
$$= 0^{rot}.999\ 999\ 902\ 892 - 0^{rot}.589 \times 10^{-10}T$$

4. Mean stellar day or mean period of the earth's rotation (culmination period of an equatorial star without proper motion):

$$1^d_{st} = 1^{rot} = 1^{d*}.000\ 000\ 097\ 108 + 0^{d*}.589 \times 10^{-10}T$$
$$= 24^h00^m*.008\ 390\ 13 + 0^s*.050\ 89 \times 10^{-4}T$$
$$= 0^d.997\ 269\ 663\ 257 + 0^d.001\ 4 \times 10^{-10}T = 23^h56^m04^s.098\ 905\ 40 + 0^s.001\ 21 \times 10^{-5}T$$

5. Tropical year (equinox to equinox):

$$P_{trop} = (365.242\ 198\ 78 - 0.000\ 006\ 138\ T)^d_E = 365^d_E05^h_E48^m_E45^s_E.530\ T$$

6. Sidereal year (fixed star to fixed star):

$$P_{sid} = (365.256\ 360\ 42 + 0.000\ 000\ 11\ T)^d_E = 365^d_E06^h_E09^m_E09^s_E.54 + 0^s_E.010\ T$$

7. Synodic month (new moon to new moon):

$$P_{syn} = (29.530\ 588\ 2 - 0.000\ 000\ 2\ T)^d = 29^d12^h44^m02^s.78 - 0^s.017\ T$$

8. Tropical month (equinox to equinox):

$$P_{trop} = (27.321\ 581\ 7 - 0.000\ 000\ 2\ T)^d - 27^d07^h43^m04^s.7 - 0^s.017\ T$$

9. Sidereal month (fixed star to fixed star):

$$P_{sid} = (27.321\ 661\ 0 - 0.000\ 000\ 2\ T)^d = 27^d07^h43^m11^s.47 - 0^s.017\ T$$

10. Astronomical unit (mean earth-sun distance): au $= 149{,}598{,}700 \pm 400$ km.
11. Light year: ly $= (9.460\ 530 \pm 0.000\ 003) \times 10^{12}$ km $= 63{,}239.39 \pm 0.15$ au.
12. Parsec: pc $= 206{,}264.806\ 247$ au $= (3.085\ 695 \pm 0.000\ 008) \times 10^{13}$ km.
13. Semimajor axis of the earth's orbit: $a{\oplus} = 1.000\ 000\ 236$ au $= 149{,}598{,}700 \pm 400$ km.
14. Mean orbital speed: $v{\oplus} = 29{,}784.90 \pm 0.08$ m/sec.
15. Mass of the earth: $M{\oplus} = (5.9761 \pm 0.004\ 3) \times 10^{24}$ kg.
16. Equatorial radius of the earth: $R_e{\oplus} = 6{,}378\ 170 \pm 20$ m.
17. Flattening (oblateness, ellipticity): $f{\oplus} = (R_e - R_p)/R_e = 0.003\ 352\ 55 = 1{:}(298.28 \pm 0.05)$.
18. Acceleration of gravity at the earth's surface:

$$g = g_e(1 + \beta \sin^2 \phi + \gamma \sin^2 2\phi) = 9.780\ 315(1 + 0.005\ 302\ 74 \sin^2 \phi - 0.000\ 005\ 9 \sin^2 2\phi)\ \text{m/sec}^2$$

19. Moments of inertia of the earth:

$$A = (0.329\ 681_4 \pm 0.000\ 11)\ M \oplus R_e^2 = (0.801\ 50 \pm 0.000\ 85) \times 10^{38}\ \text{kg-m}^2$$
$$C = (0.330\ 763_9 \pm 0.000\ 11)\ M \oplus R_e^2 = (0.804\ 13 \pm 0.000\ 85) \times 10^{38}\ \text{kg-m}^2$$

20. Circular and escape velocities from the earth's surface at the equator.

$$v_{cir} = 7{,}905.39 \pm 0.06\ \text{m/sec} \qquad v_{esc} = v_{cir}2^{1/2} = 11{,}179.91 \pm 0.08\ \text{m/sec}$$

21. Mean observed distance of the perturbed moon from the earth:

$$\bar{r} \ {\mathbb{C}} = 384{,}401.0 \pm 1.0\ \text{km} = (60.268\ 23 \pm 0.000\ 35)\ R_e\oplus = 0.002\ 569\ 548 \pm 0.000\ 000\ 014\ \text{au}$$

22. Semimajor axis of the moon's orbit: $a\ {\mathbb{C}} = 1.000\ 907\ 681 \bar{r}\ {\mathbb{C}} = 384\ 749.9 \pm 1.0$ km.
23. Mean orbital velocity: $v\ {\mathbb{C}} = 1{,}024.089 \pm 0.003$ m/sec.
24. Mass of the moon: $M\ {\mathbb{C}} = (7.353\ 4 \pm 0.007\ 5) \times 10^{22}$ kg $= M \oplus{:}(81.270 \pm 0.024)$.
25. Semiaxes of the moon ellipsoid: $a = 1{,}738{,}780 \pm 186$ m; $b = 1{,}738{,}452 \pm 209$ m; $c = 1{,}737{,}688 \pm 188$ m.
    The time T is in Julian centuries of 36,525 days from 1900 Jan. 0.5 U.T. (universal time). The constants are given with probable errors (pe).

Table 11.6.3   Planetary Orbit Data

Planet	Number of satellites	Semimajor axis, au	Sidereal period, year	Synodic period, days	Mean daily motion, s/day	Eccentricity, e	Inclination to ecliptic, i, deg	Mean ascending node Ω, deg	Longitude perihelion, ω, deg	Orbital speed, km/s	Planetary escape speed, km/s	Gravity at surface, cm/s²
Mercury ☿		0.387	0.2411	115.88	14,732.42	0.206	7.004	47.857	76.833	47.8	4.3	370
Venus ♀		0.723	0.6152	583.92	5767.2	0.007	3.394	76.320	131.008	35.0	10.3	887
Earth ⊕	1	1.0000	1.0000		3548.19	0.017			102.253	29.8	11.2	982
Mars ♂	2	1.524	1.8822	779.94	1886.52	0.093	1.850	49.250	335.322	24.2	5.0	372
Jupiter ♃	16	5.203	11.86	398.88	299.13	0.048	1.305	99.44	12.72	13.1	61	2599
Saturn ♄	15	9.54	29.46	378.09	120.46	0.056	2.490	113.307	92.264	9.7	39.4	1290
Uranus ♅	5	19.2	84.0	369.66	42.23	0.50	0.773	73.796	170.011	6.8	21.2	830
Neptune ♆	2	30.1	164.8	367.49	21.53	0.009	1.774	131.340	44.274	5.4	23.8	1380
Pluto ♇	1	39.7	247.7	366.74	14.29	0.249	17.170	109.886	112.986	4.7	~1.3	

Table 11.6.4 Physical Data of Sun, Moon, and Planets

Name	Equatorial radius, km	Polar radius, km	Mass, kg	Mean density, g/cm³	Visual albedo	Rotational period	Inclination of equator to orbit, deg
Sun	696,000		1.99×10^{30}	1.42		~ 27 days	7.25
Moon	1,738		7.35×10^{22}	3.34	0.067	27.322 days	6.65
Mercury	2,439		3.30×10^{23}	5.44	0.235	58.646 days	< 1
Venus	6,051		4.87×10^{24}	5.24	0.80	243.0 days	2.2
Earth	6,378	6,356	5.98×10^{24}	5.52	0.39	23.93 h	23.45
Mars	3,397	3,375	6.42×10^{23}	3.93	0.1–0.4	24.62 h	25
Jupiter	71,398	66,793	1.89×10^{27}	1.32	0.51	9.92 h	3.1
Saturn	60,330	55,020	5.68×10^{26}	0.71	0.45	10.66 h	26.7
Uranus	26,145	25,518	8.73×10^{25}	1.2	0.56	11–24 h	97.9
Neptune	24,700	24,060	1.03×10^{26}	1.67	0.57	~ 18 h	28.8
Pluto	1,600		$1.4–2 \times 10^{22}$	0.55–1.75	0.25–0.5	?	?

DYNAMIC ENVIRONMENTS
by Michael B. Duke
NASA

REFERENCES: Harris and Crede, "Shock and Vibration Handbook," McGraw-Hill. Beranek, "Noise Reduction," McGraw-Hill. Crandall, "Random Vibration," Technology Press, MIT. Neugebauer, The Space Environment, *Tech. Rep. 34-229*, Jet Propulsion Lab., Pasadena, 1960. Hart, Effects of Outer Space Environment Important to Simulation of Space Vehicles, *AST Tech Rep. 61-201*. Cornell Aeronautical Lab. Barret, Techniques for Predicting Localized Vibratory Environments of Rocket Vehicles, *NASA TN D-1836*. Wilhold, Guest, and Jones, A Technique for Predicting Far Field Acoustic Environments Due to a Moving Rocket Sound Source, *NASA TN D-1832*. Eldred, Roberts, and White, Structural Vibration in Space Vehicles, *WADD Tech. Rep. 61–62*. Bolt, Beranek, and Newman, Exterior Sound and Vibration Fields of a Saturn Vehicle during Static Firing and during Launching, *U.S.A. Ord. Rep. 764*.

The forces required to propel a space vehicle from the launch pad are tremendous. The dynamic pressures generated in the atmosphere by large rocket engines are exceeded only by nuclear blasts. Slow initial ascent from the launch pad is followed by rapid acceleration and high g loading, and acoustic and aerodynamic forces drive every point of the vehicle surface. The space-vehicle velocity quickly becomes supersonic and continues to hypersonic velocities. Gradually, as the space vehicle leaves the atmosphere, the aerodynamic forces recede. Then suddenly the rocket thrust decays and is followed by the ignition of another rocket, which quickly develops thrust and continues to accelerate the space vehicle. Finally, the space vehicle becomes weightless. In completing a mission, a space vehicle is subjected to the additional environmental extremes of planetary landings, escape, and earth reentry.

These environments represent basic criteria for space-vehicle design. For operational reliability, the most important are the shock, vibration, and acoustic environments which are present to varying magnitudes in all operational phases and which constitute major engineering problems.

It is not possible to have original environmental data prior to fabrication and testing of a particular structure. Vibration, shock, and acoustic noise adversely affect structural integrity and vehicle reliability. These environments must be considered prior to design and fabrication and then again for design verification through ground testing. Precise vibration and acoustic environmental predictions are complicated because of structural nonlinearities, random forcing functions, and multiple-degree-of-freedom systems. Current environmental-prediction techniques consist mainly of extrapolating measured data obtained from existing vehicles. Acoustic environments produced during ground test are important factors in evaluating ground-facility locations and design, personnel safety, and public relations.

A source of sound common to all jet and rocket-engine propulsion systems is the turbulent exhaust flow. This high-velocity flow produces pressure fluctuations, referred to as noise, and has adverse effects on the vehicle and its operations. Also, far-field uncontrolled areas may be subjected to intense acoustic-sound-pressure levels which may require personnel protection or sound-suppression devices. (See also Sec. 12.)

The sound-generation mechanism is presented in empirical and analytical form, utilizing measured sound pressures and the known engine operational parameters. This combination of empirical terms and analytical reasoning provides a generalized relationship to predict sound-pressure levels for the vehicle structure and also for far-field areas. It has been shown by Lighthill (*Proc. Royal Soc.*, 1961) that acoustic-power generation of a subsonic jet is proportional to $\rho V^8 D^2/a_0^3$, where ρ is the density of the ambient medium, a_0 is the speed of sound in the ambient medium, V is the jet or rocket exit exhaust velocity, and D is a characteristic dimension of the engine. Attempts have been made to extend this relationship to allow supersonic-jet acoustic-power predictions to be made, but no one scheme is accepted for broad use.

From acoustic-data-gathering programs utilizing rocket engines, various trends have been noted which facilitate acoustic predictions. For rocket engines of up to 5×10^9 W mechanical power, the acoustic efficiency shows a nonlinear rate of increase. For greater mechanical power, the acoustic efficiency approaches a value slightly higher than 0.5 percent (Fig. 11.6.1).

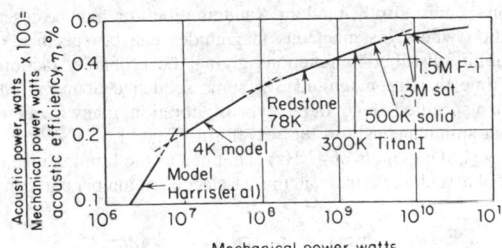

Fig. 11.6.1 Acoustic-efficiency trends.

Turbulent-exhaust-flow sound-source mechanisms also exhibit directional characteristics. Figure 11.6.2 shows data from a Saturn rocket vehicle launch with the microphone flush-mounted on the vehicle skin at approximately 100 ft above the nozzle plane.

During the period between on-pad firing and post-lifting-off or mainstage flight, the exhaust-flow direction changes. The directivity has also shifted 90° to retain its inherent relationship to the exhaust stream. The sound-pressure-level variation during a small time interval—between

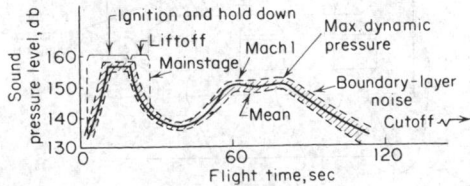

Fig. 11.6.2 Vehicle acoustic environment.

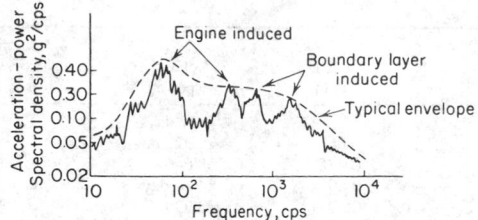

Fig. 11.6.3 Random-vibration spectrum of a space-vehicle structure.

on-pad condition and shortly after liftoff—depends on the changes which have occurred in the directivity and the jet-exhaust-velocity vector (Fig. 11.6.2). The change in sound-pressure level is 20 dB approximately in this case and is indicative of the significance of a sound source's directional properties, which depend on velocity and temperature of the jet.

The acoustic environments produced by a given conventional engine can be empirically described by that engine's flow parameters and geometry. It is difficult to predict the acoustic environments for nonconventional engines such as the plug-nozzle and expansion-deflection engines.

The turbulent-boundary-layer problem of high-velocity vehicle flights is difficult to evaluate in all but the simplest cases. Only for attached, homogeneous, subsonic flows can the boundary-layer noise be estimated with confidence.

The vibratory environment on rocket vehicles consists of total-vehicle bending vibrations, in which the vehicle is considered a nonuniform beam, with low-frequency responses (0 to 20 Hz) and localized vibrations with frequencies up to thousands of hertz. (See also Sec. 5.)

The response of a simple structure to a dynamic forcing function may be described by the universal equation of motion given as $M\ddot{x} + c\dot{x} + Kx = F(t)$ or $F(t)x(t) = (M - K\omega^2) - j(c/\omega)$ and is a complex function containing real and imaginary quantities. In the case of rocket-vehicle vibrations, $F(t)$ is a random forcing phenomenon. Thus, the motion responses must also be random. Random vibration may be described as vibration whose instantaneous magnitudes can be specified only by probability-distribution functions giving the probable fraction of the total time that the magnitude, or some sequence of magnitudes, lies within a specified range. In this type of vibration, many frequencies are present simultaneously in the waveform (Fig. 11.6.3). Power spectral density (PSD; in units of g^2/Hz) is defined as the limiting mean-square value of a random variable, in this case acceleration per unit bandwidth; i.e.,

$$PSD = \lim_{\Delta f \to 0} \overline{g}^2/\Delta f = d\overline{g}^2/df$$

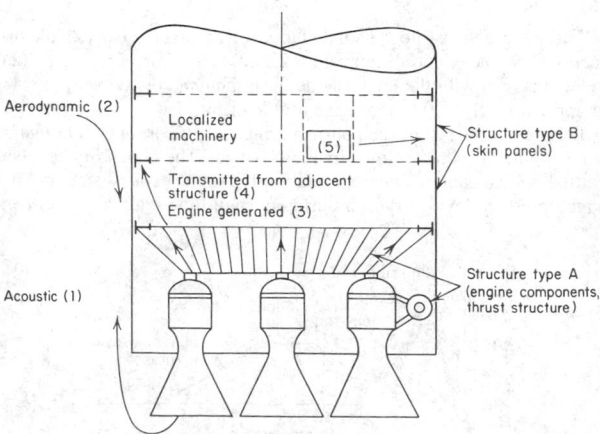

Fig. 11.6.4 Sources of vibration excitement.

The sources of vibration excitation are (1) acoustic pressures generated by rocket-engine operation; (2) aerodynamic pressures created by boundary-layer fluctuations; (3) mechanically induced vibration from rocket-engine pulsation, which is transmitted throughout the vehicle structure; (4) vibration transmitted from adjacent structure; and (5) localized machinery. Since the mean square acceleration is proportional to power, the total vibrational power at any point on the vehicle may be expressed as

$$Pv = Pv(1) + Pv(2) + Pv(3) + Pv(4) + Pv(5)$$

This is illustrated in Fig. 11.6.4 for an arbitrary vehicle. In many instances, the structure is principally excited by only one of these sources and the remaining can be considered negligible (Fig. 11.6.5). The vibration characteristics of structure B, taken from a skin panel, closely correspond to the trends of the acoustic pressures during flight. However, structure A, taken from a rocket-engine component, exhibits stationary trends, indicating no susceptibility to the impinging acoustic field. Thus, for this structure, it can be assumed that only $Pv(3)$ is significant, whereas for structure B, only $Pv(1)$ is the principal exciting source during the on-pad phase and only $Pv(2)$ is significant during the maximum dynamic-pressure phase. This also holds true for captive firings of the vehicle, in which only $Pv(1)$ is significant for structure B and $Pv(3)$ for structure A.

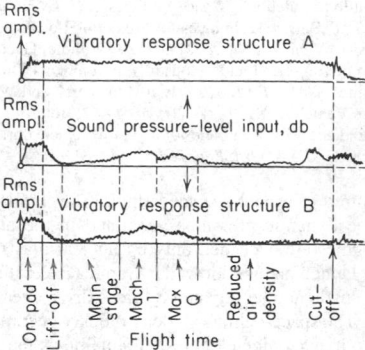

Fig. 11.6.5 Typical structural responses to a random-vibration environment.

SPACE-VEHICLE TRAJECTORIES, FLIGHT MECHANICS, AND PERFORMANCE
by O. Elnan, W. R. Perry, J. W. Russell, A. G. Kromis, and D. W. Fellenz
University of Cincinnati

REFERENCES: Russell, Dugan, and Stewart, "Astronomy," Ginn. Sutton, "Rocket Propulsion Elements," Wiley. White, "Flight Performance Handbook for Powered Flight Operations," Wiley. "U.S. Standard Atmosphere, 1962," NASA, USAF, U.S. Weather Bureau. Space Flight Handbooks, *NASA SP*-33, *SP*-34, and *SP*-35. Gazley, Deceleration and Heating of a Body Entering a Planetary Atmosphere from Space, "Visits in Astronautics," Vol. I, Pergamon. Chapman, An Approximate Analytical Method for Studying Entry into Planetary Atmospheres, *NACA-TN* 4276, May 1958. Chapman, An Analysis of the Corridor and Guidance Requirements for Super-circular Entry into Planetary Atmospheres, *NASA-TN D*-136, Sept 1959. Loh, "Dynamics and Thermodynamics of Planetary Entry," Prentice-Hall. Grant, Importance of the Variation of Drag with Lift in Minimization of Satellite Entry Acceleration, *NASA-TN D*-120, Oct. 1959. Lees, Hartwig, and Cohen, Use of Aerodynamic Lift during Entry into the Earth's Atmosphere, *Jour. ARS,* **29,** Sept 1959. Gervais and Johnson, Abort during Manned Ascent into Space, *AAS,* Jan. 1962.

Notation

A = reference area
A_z = launch azimuth
B = ballistic coefficient
C_L = lift coefficient

e = eccentricity
f = stage-mass fraction (WP/W_A)
H = total energy
H_∞ = hyperbolic excess energy
I = specific impulse
Δ_i = plane-change angle
L = lift force
m = mass
n = mean motion ($2\pi/p_{sid}$)
p = ambient atmospheric pressure
q = dynamic pressure
R = earth's equatorial radius
S = wing area
T = transfer time, time, absolute temperature
t_{pp} = time after perigee passage
V_c = injection velocity
V = velocity
V_g = velocity loss to gravity
W = weight
W_L = payload weight
W_P = propellant weight
Y = distance
α_Ω = right ascension of the ascending node
$\gamma = \theta - 90°$ = flight-path angle used in reentry analysis
ν = true anomaly
ω = earth's rotational velocity
ψ = central angle
$\sigma = \rho/\rho_0$ = relative atmospheric density
Υ = vernal equinox
A_e = nozzle exit area
a = semimajor axis
C_D = drag coefficient
D = drag force
E = eccentric anomaly
F = thrust
g = acceleration due to gravity
H_{esc} = escape energy
h = altitude
i = inclination
J_2 = oblateness coefficient of the earth's potential (second-zonal harmonic)
M = molecular weight, mean anomaly
nm = nautical miles
P = period of revolution
P_e = nozzle exit pressure
q_s = stagnation-point heat-transfer rate
r = radius
s = range
t = time
V_∞ = hyperbolic excess velocity
V_{id} = ideal velocity
ΔV = impulsive velocity
W_A = stage weight ($W_O - W_L$)
W_O = gross weight
X = distance
α = angle of attack, right ascension
β = exponent of density-altitude function
λ = longitude
τ = hour angle
ω = argument of perigee
ϕ' = latitude
ρ = atmospheric density
θ = flight-path angle

Subscripts

a = apogee
e = entry
f = final
o = sea level conditions at 45° latitude
s = space fixed
vac = vacuum
circ = circular orbital condition
esc = escape
i = initial
p = perigee
M = mean
SL = sea level
sid = sidereal

ORBITAL MECHANICS
by O. Elnan and W. R. Perry
University of Cincinnati

The motion of the planets about the sun, as well as that of a satellite in its orbit about a planet, is governed by the inverse-square force law for attracting bodies. In those cases where the mass of the orbiting body is small relative to the central attracting body, it can be neglected. This simplifies the analysis of the orbital motion. The shape of the orbit is always a conic section, i.e., ellipse, parabola, or hyperbola, with the central attracting body at one of the foci. Parabolic and hyperbolic orbits are open, terminate at infinity, and represent cases where the orbiting body escapes the central force field of the attracting body.

The **laws of Kepler** for satellite orbits are (1) the radius vector to the satellite from the central body sweeps over equal areas in equal times; (2) the orbit is an ellipse, with the central attracting body at one of its foci; and (3) the square of the period of the satellite is proportional to the cube of the semimajor axis of the orbit.

Six **orbital elements** are required to describe the orbit in the orbit plane and the orientation of the plane in inertial space. The three elements that define the orbit are the semimajor axis a, eccentricity e, and period of revolution P. The orientation of the orbit (Fig. 11.6.6) is defined by the

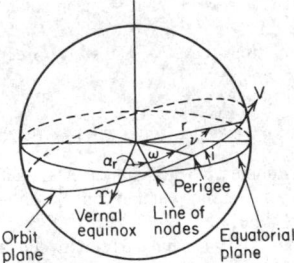

Fig. 11.6.6 Elements of satellite orbit around a planet.

right ascension of the ascending node α_Ω, inclination of the orbit plane to the earth's equatorial plane i, and the argument of perigee ω, which is the central angle measured in the orbit plane from ascending node to perigee. The following equations compute these orbital elements and other orbital parameters.

$$a = (r_a + r_p)/2 = \mu/V_a V_p = r_a/(1 + e) = r_p/(1 - e)$$
$$e = (r_a/a) - 1 = (V_p - V_a)/(V_p + V_a) = (r_p V_p^2/\mu) - 1$$
$$r = a(1 - e^2)/(1 + e \cos \nu)$$
$$r_a = a(1 + e) = r_p(1 + e)/(1 - e)$$
$$r_p = a(1 - e) = r_a(1 - e)/(1 + e)$$
$$V = [\mu(2/r - 1/a)]^{0.5}$$
$$P = 2\pi(a^3\mu)^{0.5} \quad \text{(for circular orbits, } r = a\text{)}$$
$$V_a = V_p(1 - e)/(1 + e) = [\mu(1 - e)/a(1 + e)]^{0.5}$$
$$V_p = V_a(1 + e)/(1 - e) = [\mu(1 + e)/a(1 - e)]^{0.5}$$
$$\nu = \cos^{-1}\{[2r_a r_p - r(r_a + r_p)]/r(r_a - r_p)\}$$
$$t_{pp} = (a^3/\mu)^{0.5}\{\cos^{-1}[(a - r)/ae] - e[1 - (a - r)^2/(ae)^2]^{0.5}\}$$
$$M = nt_{pp} = E - e \sin E$$
$$r = a(1 - e \cos E)$$
$$V_{esc} = (2\mu/r)^{0.5}$$

The elements may be determined from known injection conditions. For example,

$$1/a = V_i^2\mu - 2/r_i$$
$$e^2 = \cos^2 V_i(r_iV_i^2/\mu - 1)^2 + \sin^2 V_i$$

General precession in longitude of vernal equinox:

$$x = 50''.2575 + 0''.0222T$$

where T is in Julian centuries of 36,525 mean solar days reckoned from 1900 Jan. 0.5 UT.

Measurement of Time

The concept of time is based on the position of celestial bodies with respect to the observer's meridian as measured along the geocentric celestial equator. The mean sun and the vernal equinox (Υ) are used to define mean solar time and sidereal time. The mean sun, rather than the apparent (true) sun, is used since the latter's irregular motion from month to month gives a variation in the length of a day. The mean sun moves with uniform speed along the equator at a rate equal to the average of the true sun's angular motion during the year.

One sidereal day is the interval between two successive transits of the vernal equinox across the observer's meridian. A mean solar day is the interval between two successive transits of the mean sun. Local civil time (LCT) is defined to be the hour angle of the mean sun plus 12 h. Local sidereal time is the hour angle of the vernal equinox plus 12 h. For an observer at the Greenwich meridian, the local civil time is Greenwich mean time (GMT) or universal time (UT). The earth is divided into 24 time zones, each 15° of longitude wide. Each zone keeps the time of the standard meridian in the middle of the zone. (See Fig. 11.6.7.)

Time $T = \tau + 12$ h
Local sidereal time LST $= \tau_\Upsilon + 12$ h
Greenwich sidereal time GST $=$ LST $+ \lambda(w)$
Local civil time LCT $= \tau_M + 12$ h $=$ mean solar time
Equation of time ET $= \tau - \tau_M = \alpha_M - \alpha$
Greenwich mean time GMT $=$ UT $= \tau_M$(Greenwich) $+ 12$ h $=$ universal time

$$\text{LCT} = \text{UT} - \lambda(w)$$
$$\text{GMT} = \text{ZT} \pm \lambda°/15$$

The "American Ephemeris and Nautical Almanac" gives the equation of time for every day of the year and conversions of mean solar time to and from sidereal time.

Hohmann Transfer   This is a maneuver between two coplanar orbits where the elliptical transfer orbit is tangent at its perigee to the lower orbit and tangent at its apogee to the higher orbit. To transfer from a low circular orbit to a higher one, the impulsive velocity required is the difference between the velocities in the circular orbits and the velocities in the corresponding points on the transfer orbit. Using the general velocity equation above, the transfer velocities are $\Delta V_i = \{ \mu[2/r_i -$

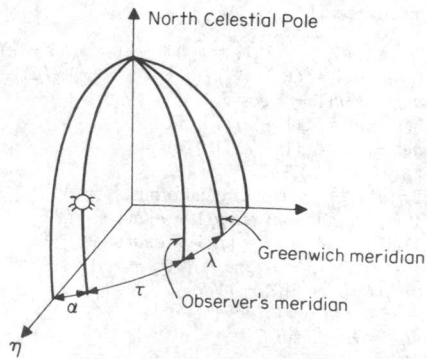

Fig. 11.6.7   Time-zone geometry.

$2/(r_i + r_f)]\}^{0.5} - (\mu/r_i)^{0.5}$ and $\Delta V_f = (\mu/r_f)^{0.5} - \{ \mu[2/r_f - 2/(r_i + r_f)]\}^{0.5}$, where V_i is the orbiting velocity in low orbit and V_f is the orbiting velocity in the high orbit.

Orbital Lifetime   The satellite orbit will gradually decay to lower altitudes because of the drag effects of the atmosphere. This drag force has the form $D = \frac{1}{2}C_DA_pV^2$, where values of the drag coefficient C_D may range from 2.0 to 2.5. The atmospheric density, as a function of altitude, can be used for computing an estimated orbital lifetime of a satellite (Fig. 11.6.8). (See NASA, "U.S. Standard Atmosphere" and "Space Flight Handbooks.")

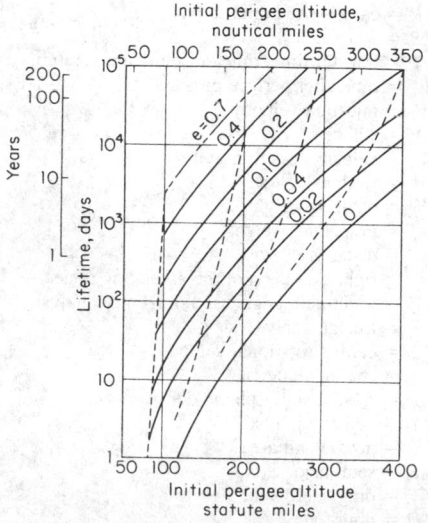

Fig. 11.6.8   Satellite lifetimes in elliptic orbits. $\rho =$ ARDC, 1959; $B = C_DA/2m = 1.0$ ft²/slug; $e =$ initial eccentricity; --- = decay histories.

Perturbations of Satellite Orbits   There are many secular and periodic perturbations of satellite orbits due to the effects of the sun, moon, and some planets. The most significant perturbation on close-earth satellite orbits is caused by the oblateness of the earth. Its greatest effects are the precession of the orbit along the equator (nodal regression) and the rotation of the orbit in the orbit plane (advance of perigee). The nodal-regression rate is given to first-order approximation by

$$\dot\alpha_\Omega = -\tfrac{3}{2}(\mu/a^3)^{0.5}J_2[R/a(1 - e^2)]^2 \cos i$$

The mean motion of the perigee is

$$\dot\omega = J_2n(2 - (\tfrac{5}{2}) \sin^2i)/a^2(1 - e^2)^2$$

LUNAR- AND INTERPLANETARY-FLIGHT MECHANICS
by J. W. Russell
University of Cincinnati

The extension of flight mechanics to the areas of lunar and interplanetary flight must include the effects of the sun, moon, and planets on the transfer trajectories. For most preliminary performance calculations, the **"sphere-of-influence,"** or **"patched-conic,"** method—whereby the multibody force field is treated as a series of central force fields—provides sufficient accuracy. By this method, the trajectories in the central force field are calculated to the "sphere of influence" of each body as standard Keplerian mechanics, and the velocity and position of the extremals are then matched to give a continuous trajectory. After the missions have been finalized, the precision trajectories are obtained by numerically integrating the equations of motion which include all the perturbative elements. It is necessary to know the exact positions of the sun, moon, and planets in their respective orbits.

Lunar-Flight Mechanics   The moon moves about the earth in an orbit having an eccentricity of 0.055 and an inclination to the ecliptic of about 5.145°. The sun causes a precession of the lunar orbit about the ecliptic, making the inclination of the lunar plane to the earth's equatorial plane oscillate between 18.5 and 28.5° over a period of 18.5 years approx. To compute precision earth-moon trajectories, it is necessary to know the precise launch time to be able to include the perturbative effects of the sun, moon, and planets. The data presented are based on the moon being at its mean distance from earth and neglect the perturbation of the sun and planets; they are therefore only to be considered as representative data. The injection velocity (see Gazley) at 100 nmi for earth-moon trajectories and the impulsive velocity for braking into a 100-nmi orbit about the moon are shown in Fig. 11.6.9 as a function of transfer time. Additional energy is required to offset losses due to gravity and aerodynamic drag as well as to provide plane-change and launch-window capabilities.

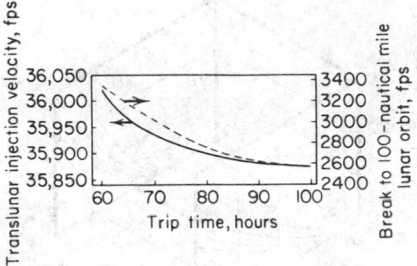

Fig. 11.6.9   Lunar-mission velocity requirements.

Interplanetary-Flight Mechanics   The fact that planetary orbits about the sun are not coplanar greatly restricts the number of feasible interplanetary trajectories. The plane of the interplanetary trajectory must include the position of the departure planet at departure, the sun, and the position of the target planet at arrival. The necessary plane change can be prohibitive even though the relative inclinations of the planetary orbits are small. The impulsive velocity required to effect a plane change for a spacecraft departing earth can be approximated by $\Delta V = 195{,}350 \sin (\Delta_i/2)$, where Δ_i is the amount of plane change required and ΔV is in ft/s.

For interplanetary flight, the "ideal" total energy that must be imparted to the spacecraft is the energy required to escape the gravitational field of the departure planet plus the energy required to change path

about the sun so as to arrive at the target planet at the desired position and time. The energy required to escape the gravitational attraction of a planet can be determined by Keplerian mechanics to be $H_{\text{esc}} = \mu/r$, where μ is the gaussian gravitational constant and r is the distance from the center of the planet. After escaping the departure planet, the velocity must be altered in both magnitude and direction—in order to arrive at the target planet at the chosen time—by supplying additional energy, $H_\infty = \frac{1}{2}V_\infty^2$.

For determination of vehicle size necessary to inject the spacecraft into the interplanetary trajectory, it is convenient to express the total energy $H = H_{\text{esc}} + H_\infty$ in terms of the required injection velocity

$$V_c = \sqrt{2(H_{\text{esc}} + H_\infty)} \qquad \text{or} \qquad V_c \sqrt{(2\mu/r) + V_\infty^2}$$

Hyperbolic excess velocities V_∞ for earth-to-Mars missions are shown in Table 11.6.5 as a function of date. These velocities are near optimum for the trip time and year given but do not represent absolute-optimum trajectories. Mars has a cyclic period with respect to earth of 17 years approx; hence the energy requirements for Mars missions from earth also follow approximate 17-year cycles.

Figure 11.6.10 shows how the hyperbolic excess velocities of Table 11.6.5 can be converted to stage characteristic velocities for a vehicle leaving a 100-nmi earth parking orbit. These are simple cases. In many recent planetary missions, use has been made of planetary flybys to gain velocity and aim the space probe. These "gravity assists" significantly increase the capability of interplanetary flight.

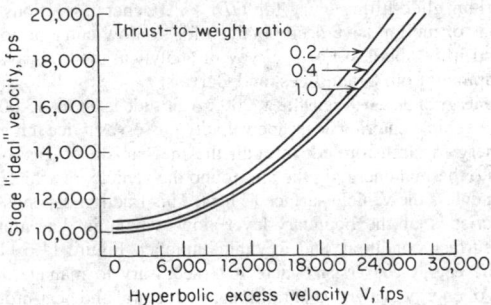

Fig. 11.6.10   Stage-velocity for earth departure from a 100-nmi parking orbit (specific impulse = 450 s).

ATMOSPHERIC ENTRY
by D. W. Fellenz
University of Cincinnati

A vehicle approaching a planet possesses a considerable amount of energy. The entry vehicle must be designed to dissipate this energy without exceeding its limits with respect to maximum decelerations or heating.

The trajectory parameters of an entering vehicle are determined largely by its initial trajectory conditions (suborbital, orbital, superorbital), by the ratio of gas-dynamic forces acting upon it, and by its mass (ballistic factor, lift-drag ratio) and the type of atmosphere it is entering.

Planetary atmospheres (Table 11.6.6) can be assumed, as a first ap-

Table 11.6.5   Hyperbolic-Excess-Velocity Requirements (ft/s) for Typical Mars Missions

Trip time, days	Opposition year				
	1982	1984	1986	1988	1990
100	27,241	23,012	18,070	15,569	19,955
150	15,940	13,088	11,437	10,520	13,792
200	12,268	10,021	11,486	9,308	12,463
250	10,510	12,688	13,410	10,480	15,784

SOURCE: Gammal, "Space Flight Handbooks."

Table 11.6.6   Planetary Atmospheres

Planet	V_{esc}, ft/s	Gases	M, gmol^{-1}	T, K	β^{-1}, ft	$\beta\tau$	ρ_{SL}, lb/ft^3
Venus	34,300	2% N_2 98% CO_2	40	250–350	2×10^4	1,006	1.0
Earth	36,800	78% N_2 21% O_2	29	240	2.35×10^4	880	0.0765
Mars	16,900	95% N_2 3% CO_2	28	130–300	6×10^4	132	0.0062
Jupiter	195,000	89% H_2 11% He	2.2	100–200	6×10^4	3,600	

proximation, to have exponential density-altitude distributions: $\sigma = \rho/\rho_{SL} = e^{-\beta h}$, where

$$\beta = -(1/\rho)(d\rho/dh) = Mg/RT$$

For more exacting analyses, empirical atmospheric characteristics (e.g., U.S. Standard Atmosphere) have to be used.

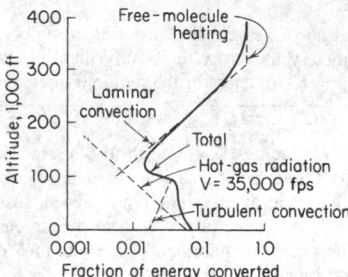

Fig. 11.6.11 Energy conversion during reentry. *(After Gazley in Knoelle, "Handbook of Astronautical Engineering.")*

The trajectory of the vehicle in flight-path fixed notation with the assumption of a nonrating atmosphere is described as $dV/dt = g \sin \gamma - (C_D A/m)(\rho/2)V^2$ and $(V/\cos \gamma)(d\gamma/dt) = g - (V^2/r) - (C_L A/m)(\rho/2)(V^2/\cos \gamma)$. Solutions exist for direct ($\gamma > 5°$) ballistic entry and for equilibrium glide-lifting entry for $L/D > 1$. General solutions of the equations of motion have been obtained for shallow entry of both ballistic and lifting bodies. For a survey of analytical methods available, see Chapman, Loh, Grant, Lees, and Gervais.

The **energy of an earth satellite** at 200 mi altitude is about 13,000 Btu/lb, and a vehicle entering at escape velocity possesses twice this energy. This energy is transformed, through the mechanism of gas-dynamic drag, into thermal energy in the air around the vehicle, of which only a fraction enters the vehicle surface as heat. This fraction depends on the characteristics of the boundary-layer flow, which is determined by shape, surface condition, and Reynolds number. Figure 11.6.11 illustrates the **energy conversion** where it is necessary to manage a given amount of energy in a way that minimizes structural and heat-protection weights, operational restraints, and cost. It would be most desirable if this energy could be dissipated at a constant rate, but with constant vehicle parameters, decelerations vary proportionally to ρV^2 and heating rates proportionally to ρV^3. The selection of a heat-protection system depends on the type of entry flown. Lifting entry from orbit results in relatively long flight times (in the order of ½ h) as compared with 10 min in the case of a steep ballistic entry. **Radiative-heat-transfer sys-**tems favor low heating rates over long time periods. **Ablative systems** favor short heat pulses. In fact, longer soaking periods may melt the ablation coating without getting the benefit of heat absorption through multiple-phase changes. A more uniform dissipation of energy can be achieved by modulation of vehicle parameters.

Determination of an **entry-vehicle configuration** is a process of iteration. The entry-flight profile and the entry and recovery procedures are mainly determined by whether experiments or passengers are carried. The external shape is determined by requirements for hypersonic glide capability and subsonic handling and landing characteristics and also by the relations between aerodynamic shape, heat input, and structural-materials characteristics. Intermediate results are fed back into the evaluation of performance and operational effectiveness of the total transportation system.

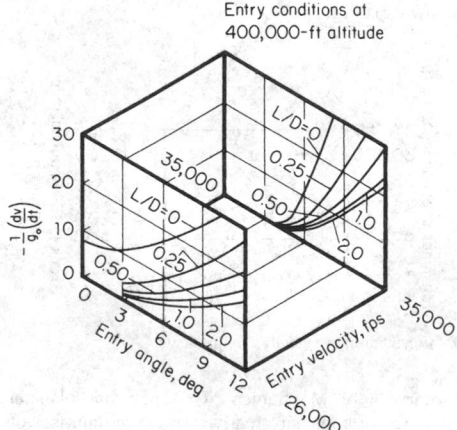

Fig. 11.6.13 Peak decelerations for entry at constant L/D.

If reentry capability from any point of the ascent trajectory is desired for a manned vehicle, vehicle constraints and trajectory requirements must be compatible. **Performance-optimized trajectories** encompass combinations of relatively small velocities and large flight-path angles, which would result in considerable decelerations. In such cases, either reshaping the ascent trajectory or adding velocity at the time of abort, resulting in lower entry-flight-path angles, is effective (see Fig. 11.6.12). Lower flight-path angles during ascent depress the trajectory and increase drag losses. **Atmospheric entry is initiated** by changing the vehicle-velocity vector so that the virtual perigee of the descent ellipse comes to lie inside the atmosphere. The retrovelocity requirement for entry from low orbit is between 250 and 500 ft/s, depending on the range desired from deorbit to landing. For nonlifting entry, maximum deceleration and heat input into the vehicle are largely a function of entry angle, with the ballistic factor $W/C_D A$ determining the alti-

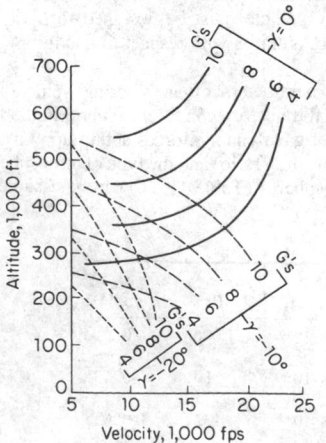

Fig. 11.6.12 Suborbital entry of a lifting vehicle. $W/C_D A = 28$ lb/ft; $L/D = 0.7$.

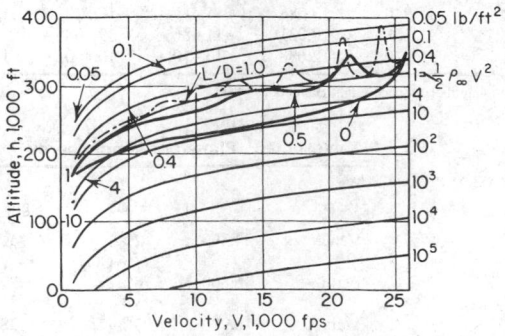

Fig. 11.6.14 Glide reentry, dynamic pressure. $W/C_D A = 1$ lb/ft²; $\gamma = 2°$; $\frac{1}{2}\rho_\infty V = (W/C_D A)(\frac{1}{2}\rho_\infty V^2)$; $h = h_1 - 23,500 \ln (W/C_D A)$; $V = V_1$.

Table 11.6.7

L/D	Lateral, nmi	Longitudinal, nmi
0.5	$\approx$ 200	$\approx$ 2,000
1	$\approx$ 600	$\approx$ 4,000
1.5	$\approx$ 1,200	$\approx$ 6,000
2	$\approx$ 2,000	$\approx$ 9,000

* Assumptions: $\gamma_e = 1°$; $b_e = 400,000$ ft; $V_e = 26,000$ ft/s.

tude at which the maximum decelerations and heating rates occur. Deceleration can be readily determined through the approximate relation $-(1/g)(dv/dt) \approx q/(W/C_D A)$. Decelerations and temperatures are drastically reduced with increasing lift-drag ratio (Fig. 11.6.13) and decreasing $W/C_D A$ (Figs. 11.6.14 and 11.6.15). This effect is particularly beneficial at steeper entry angles. The combination of longer flight times with lower heating rates, however, may actually result in a larger total heat input into the vehicle.

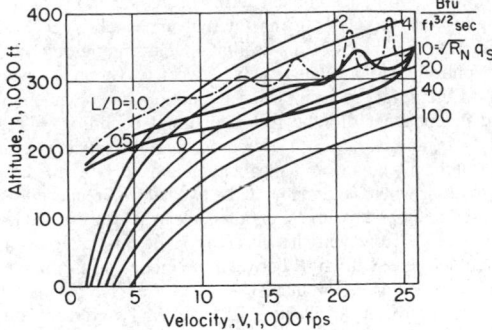

Fig. 11.6.15 Glide reentry, stagnation-heating rate. $W/C_D A = 1$ lb/ft^2; $\gamma = 2°$; $q_s = q_{s1} W/C_D A$; $h = h_1 - 23,500 \ln (W/C_D A)$; $V = V_1$.

The **influence of lift** on the reduction of decelerations is greatest for the step up to $L/D \approx 1$. The influence of $W/C_D A$ on decelerations disappears beyond $L/D \approx 0.6$. Higher, hypersonic L/D ratios serve to improve maneuverability. For entry from low orbit, the landing area selection (footprint) can be increased (see Table 11.6.7).

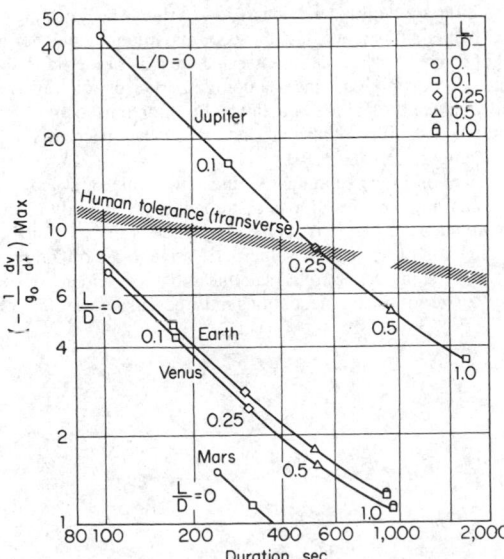

Fig. 11.6.16 Comparison of decelerations and duration for entry into various atmospheres from decaying orbits. *(After Chapman.)*

For entry at parabolic or hyperbolic speeds, it is necessary to dissipate sufficient energy so that the planet can capture the vehicle. Vehicle mass and aerodynamic characteristics determine the minimum allowable entry angle, the skip limit. The vehicle must be steered between the skip limit and the angle for maximum tolerable deceleration and heating. (See Figs. 11.6.16 and 11.6.17.)

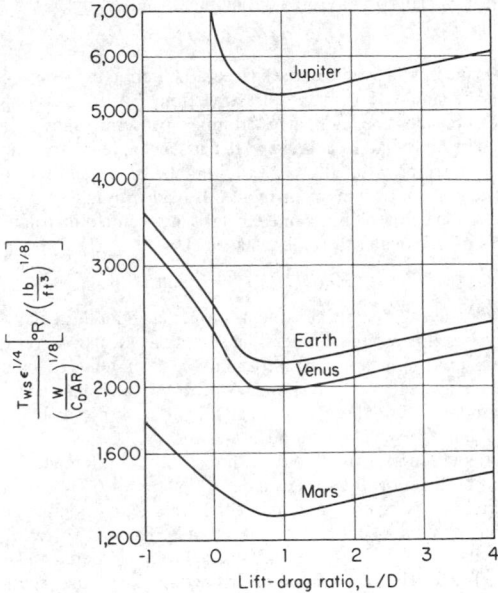

Fig. 11.6.17 Maximum surface temperature for entry into various planets from decaying orbits. *(After Chapman.)*

ATTITUDE DYNAMICS, STABILIZATION, AND CONTROL OF SPACECRAFT
by M. R. M. Crespo da Silva
University of Cincinnati

REFERENCES: Crespo da Silva, Attitude Stability of a Gravity-stabilized Gyrostat Satellite, *Celestial Mech.*, 2, 1970; Non-linear Resonant Attitude Motions in Gravity-stabilized Gyrostat Satellites, *Int. Jour. Non-linear Mech.*, 7, 1972; On the Equivalence Between Two Types of Vehicles with Rotos, *Jour. Brit. Int. Soc.*, 25, 1972. Thomson, Attitude Dynamics of Satellites, in Huang and Johnson (eds.), "Developments in Mechanics," Vol. 3, 1965. Kane, Attitude Stability of Earth Pointing Satellites, *AIAA Jour.*, 3, 1965. Kane and Mingori, Effect of a Rotor on the Attitude Stability of a Satellite in a Circular Orbit, *AIAA Jour.*, 3, 1965. Lange, The Drag-free Satellite, *AIAA Jour.*, 2, 1964. Fleming and DeBra, Stability of Gravity-stabilized Drag-free Satellites, *AIAA Jour.*, 9, 1971. Kendrick (ed.), TWR Space Book, 3d ed., 1967. Greensite, "Analysis and Design of Space Vehicle Flight Control Systems," Spartan Books, 1970.

A spacecraft is required to maintain a certain angular orientation, or attitude, in space in order to perform its mission adequately. For example, within a certain tolerance, it may be required to point a face toward the earth for communications and observation purposes, as well as a solar panel toward the sun for power generation. A vehicle in space is subject to several external forces which produce a moment about its center of mass tending to change its attitude. The environmental moments that can act on a spacecraft can be due to solar radiation pressure, aerodynamic forces, magnetic forces, and the gravity-gradient effect. The relative importance of each of the above moments depends on the spacecraft design and on how close it is to a central attracting body. Most often, the effect on the vehicle's attitude of the first three moments mentioned above is undesirable, although they have been used occasionally for attitude control (e.g., Mariner IV, Tiros, OAO). Micrometeorite impacts can also produce a deleterious effect on the vehicle's attitude.

The most common ways to stabilize a vehicle's attitude in space are the gravity-gradient and the spin-stabilization methods. When a spacecraft is subject to the gravitational force of a central attracting body E, its mass element near the center of the gravity field will be subject to a greater force than that acting on the mass elements farther from E. This creates a moment about the center of mass, C, of the spacecraft, which depends on the inertia properties of the vehicle and also on the distance r from E to C. This gravity-gradient moment is given by

$$\mathbf{M} = (3GM_e/r^3)\hat{\mathbf{r}} \times (\mathbf{I} \cdot \hat{\mathbf{r}})$$

where $\mathbf{I}$ is the inertia dyadic of the spacecraft, referred to its mass center, and $\hat{\mathbf{r}}$ is the unit vector in the direction from E to C. As a specific example, consider the elongated orbiting spacecraft shown in Fig. 11.6.18. The body-fixed x axis (yaw) departs an angle θ_3 from the local vertical, and the body-fixed z axis (pitch) remains normal to the orbital plane. The gravitational moment about C is readily found to be as given in the equation below (it is assumed that the maximum dimension of the vehicle is much smaller than the distance r).

$$\mathbf{M} = (3GM_e/2r^3)(I_x - I_y)(\sin 2\theta_3)\hat{\mathbf{z}}$$

(I_x and I_y are the spacecraft's moments of inertia about the x and y axes, respectively.) It is seen that unless the spacecraft's x axis is vertical or horizontal, this moment is nonzero, and for $I_x < I_y$ it tends to force the x axis of the vehicle to oscillate about the local vertical. Thus, the gravity-gradient moment provides a passive, and therefore very reliable, means for stabilizing the vehicle's attitude.

The orientation of a rigid body with respect to a reference frame is described by a set of three Euler angles. Figures 11.6.19 and 11.6.20 show a rigid body in orbit around a central attracting point E and the coordinate systems used to describe its attitude. The unit vectors $\hat{\mathbf{a}}_1$, $\hat{\mathbf{a}}_2$, and $\hat{\mathbf{a}}_3$, with origin at the spacecraft's center of mass, define the orbital reference frame. The vector $\hat{\mathbf{a}}_1$ points in the direction of the vector from E to C; $\hat{\mathbf{a}}_3$ in the direction of the orbital angular velocity (whose magnitude is denoted by n); and $\hat{\mathbf{a}}_2$ is such that $\hat{\mathbf{a}}_2 = \hat{\mathbf{a}}_3 \times \hat{\mathbf{a}}_1$. The unit vectors $\hat{\mathbf{x}}$, $\hat{\mathbf{y}}$, and $\hat{\mathbf{z}}$ in Figs. 11.6.19 and 11.6.20 are directed along the three principal axes of inertia of the spacecraft. The attitude of the vehicle with respect to the orbiting frame is defined by the three successive rotations θ_1 (yaw), θ_2 (roll), and θ_3 (pitch) as shown in Fig. 11.6.20.

Using the above Euler angles, the gravity-gradient moment (about C) acting on the vehicle is given as

$$\mathbf{M} = \tfrac{3}{2}n^2[\hat{\mathbf{x}}(I_y - I_z) \sin 2\theta_2 \sin \theta_3 \\ + \hat{\mathbf{y}}(I_x - I_z) \sin 2\theta_2 \cos \theta_3 + \hat{\mathbf{z}}(I_x - I_y) \cos^2 \theta_2 \sin 2\theta_3]$$

If the spacecraft is subject to an external (other than the gravitational attraction of E) torque $\mathbf{T} = T_x\hat{\mathbf{x}} + T_y\hat{\mathbf{y}} + T_z\hat{\mathbf{z}}$ about its center of mass C, and if its attitude deviations θ_1, θ_2, and θ_3 remain small, its linearized equations of motion are obtained as given in the following equations.

$$I_x\ddot{\theta}_1 + n\dot{\theta}_2(I_z - I_y - I_x) + n^2\theta_1(I_z - I_y) = T_x$$
$$I_y\ddot{\theta}_2 + n\dot{\theta}_1(I_x - I_z + I_y) - 4n^2\theta_2(I_x - I_z) = T_y$$
$$I_z\ddot{\theta}_3 + 3n^2(I_y - I_x)\theta_3 = T_z$$

Care must be taken in the design of a gravity-gradient stabilized satellite in order to guarantee that $I_x/I_z < I_y/I_z < 1$ for the attitude motions to be stable in the presence of the least amount of damping. For augmenting the gravity-gradient moments, booms are added to the satellite in order

to make its inertia ellipsoid thinly shaped. It is interesting to note that theoretically stable motions can also be obtained when the axis of minimum moment of inertia is normal to the orbital plane. However, this orientation is unstable if damping is present, and this is what occurs in practice since all actual vehicles have parts that can move relative to each other, such as an antenna.

It is seen from the equations above that for infinitesimal oscillations the pitch motion of the satellite is uncoupled from its coupled roll-yaw motions. The natural frequency ω_p of the infinitesimal pitch oscillations is given by

$$\omega_p = n\sqrt{3(I_y - I_x)/I_z}$$

and those of the coupled roll-yaw oscillations are given by the roots of the polynomial $\omega^4 - a_2n^2\omega^2 - a_0n^4 = 0$, where

$$a_2 = 1 - [3 + (I_z - I_y)/I_x](I_x - I_z)/I_y$$
$$a_0 = 4(I_z - I_y)(I_x - I_z)/(I_xI_y)$$

For noninfinitesimal oscillations the pitch motion is coupled to the roll-yaw motion through nonlinear terms in the equations of motion. This coupling may give rise to an internal undesirable energy interchange between the modes of the oscillation, causing the roll-yaw motion, if uncontrolled, to oscillate slowly between two bounds. The upper bound of this nonlinear resonant roll-yaw motion may be much greater than (and independent of) the roll-yaw initial condition. For given values of I_x, I_y, and I_z, it can be decreased only by reducing the initial conditions of the pitch motion. This phenomenon can be excited if $\omega_p \approx 2\omega_1$ or $\omega_p \approx 2\omega_2$, where ω_1 and ω_2 are the two natural frequencies of the roll-yaw oscillations. The same phenomenon can also be excited if $\omega_p \approx \omega_1$ (or $\omega_p \approx \omega_2$). However, the observation of this latter resonance requires a much finer "tuning" between the modes of oscillation, and therefore it is of lesser importance.

As the altitude of the orbit increases, the gravity-gradient moment decreases rapidly. For precise pointing systems, spin stabilization is used. Historically, spin stabilization was the first method used in space and is still the most commonly employed today. Simplicity and reliability are the advantages of this method when only two-axis stabilization is needed. By increasing the spin rate, the vehicle can be made very stiff in resisting disturbance moments.

A rigid body spins in a stable manner about either its axis of maximum or of minimum moment of inertia. However, in the presence of energy dissipation, even if it is infinitesimal, a spin about the axis of minimum moment of inertia leads to an unstable roll-yaw motion. Spinning satellites are built in such a way as to achieve symmetry about the spin axis which, in practice, is the axis of maximum moment of inertia. A spin of a rigid body about its intermediate axis of inertia leads to an unstable roll-yaw motion. Since damping can destroy roll-yaw stability, an individual analysis of the equations of motion must be performed to guarantee the stability of the motions when the spacecraft houses a specific damper.

For planet-orbiting spinning satellites (for which the gravitational moment can now be viewed as a disturbance) the stability of the roll-yaw motion depends not only on the magnitude of the spin rate but also on the direction of its angular velocity (relative to the orbital reference frame) due to spin. Denoting the satellite's angular velocity due to spin by $\mathbf{s} = s\hat{\mathbf{z}}$, the following inequalities must be satisfied for the roll-yaw

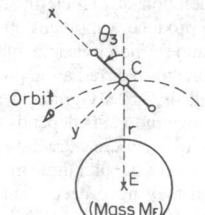

Fig. 11.6.18 Dumbell satellite.

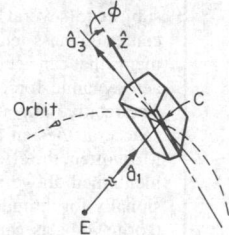

Fig. 11.6.19 Orbiting satellite.

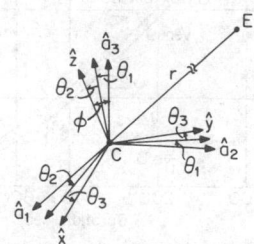

Fig. 11.6.20 Coordinate frames and Euler angles.

motions of a spinning symmetric ($I_x = I_y$) satellite in a circular orbit to be stable in the presence of an infinitesimal amount of damping:

$$(1 + s/n)\frac{I_z}{I_x} - 1 > 0$$

$$(4 + s/n)\frac{I_z}{I_x} - 4 > 0$$

Very often, a spinning pitch wheel is added to a satellite to provide additional stiffness for resisting motions out of the orbital plane. Also, the inclusion of such a wheel provides more freedom to the spacecraft designer when specifying the range of the inertia ratios to guarantee that the attitude motions of the vehicle are stable.

Let us assume that a pitch wheel with axial moment of inertia I_w is connected to the spacecraft shown in Fig. 11.6.19. It is assumed that the wheel is driven by a motor at a constant angular velocity $\mathbf{s} = s\hat{\mathbf{z}}$ relative to the main body of the spacecraft. If the attitude deviations of the main body with respect to the orbiting reference frame remain small, the linearized equations for the attitude motion of the vehicle are now given as

$$I_x\ddot{\theta}_1 + n\dot{\theta}_2(I_z - I_y - I_x + \beta I_z) + n^2\theta_1(I_z - I_y + \beta I_z) = T_x$$

$$I_y\ddot{\theta}_2 + n\dot{\theta}_1(I_x - I_z + I_y - \beta I_z) - n^2\theta_2[4(I_x - I_z) - \beta I_z] = T_y$$

$$I_z\ddot{\theta}_3 + 3n^2(I_y - I_x)\theta_3 = T_z$$

In these equations, the parameter β is defined as $\beta = I_w s/I_z n$, and I_x, I_y, I_z refer to the principal moments of inertia of the entire vehicle (rotor included).

Figure 11.6.21 shows a "stability chart" for this spacecraft when its center of mass, C, is describing a circular orbit around the central attracting point E. In this figure, a cut along the plane $\beta = 0$ produces the stability conditions for a gravity-gradient satellite, whereas a cut along the plane $K_1 + K_2 = 0$ (that is, $I_x = I_y$) gives rise to the inequalities for a spin-stabilized vehicle. Note that a rotor's spin rate with opposite sign to the orbital angular speed ($s < 0$) may destabilize the roll-yaw motions. Nonlinear resonances, as previously discussed, can also be excited in this vehicle. However, since the natural frequencies of the roll-yaw motions now also depend on the magnitude and sign of the internal angular momentum due to the rotor, the resonance lines are displaced in the region shown in Fig. 11.6.21 when the spin rate s of the rotor is changed to a different constant. Therefore, this phenomenon, when excited, can be avoided simply by changing the parameter β.

For a spacecraft to perform its mission adequately during a long period of time, an attitude control system to maintain the vehicle's attitude within specified limits is necessary. Typically, this is done by means of a set of mass expulsion jets mounted on the vehicle, which are actuated by the output of a controller that receives information from the attitude sensors.

Drag-Free Satellites For scientific applications that require the use of a satellite that follows a purely gravitational orbit, a translational control system is incorporated into the vehicle. Typical applications include satellite geodesy and navigation where accurate ephemeris prediction is needed. A drag-free satellite consists of a vehicle that contains a spherical cavity which houses an unsupported spherical proof mass. Sensors in the satellite detect the relative position of the proof mass inside the cavity and activate a set of thrustors placed in the vehicle, forcing it to follow the proof mass without touching it. Since the proof mass is shielded by the satellite from nongravitational forces, it follows a purely gravitational orbit.

METALLIC MATERIALS FOR AEROSPACE APPLICATIONS
by Stephen D. Antolovich
Revised by Robert L. Johnston
NASA

REFERENCES: MIL-HDBK-5C, "Metallic Materials and Elements for Aerospace Vehicle Structures," Sept. 1976. Wolf and Brown (eds.), "Aerospace Structural Metals Handbook," 1978, Mechanical Properties Data Center, Traverse City, Michigan. (This handbook gives information on mechanical, physical, and chemical properties; fabrication; availability; etc. in five published volumes encompassing ferrous, nonferrous, light metal alloys, and nonferrous heat-resistant alloys.) "Damage Tolerant Design Handbook," MCIC-HB-01, Metals and Ceramics Information Center, Battelle Columbus Laboratories, 1975, parts 1 and 2. (These two documents provide the most up-to-date and complete compilation of fracture-mechanics data for alloy steels, stainless steels, aluminum, and titanium alloys.) "Metals Handbook," 8th ed., 11 vols., American Society for Metals: Vol. I, Properties and Selection of Metals, 1961; Vol. II, Heat Treating, Cleaning and Finishing, 1964; Vol. III, Machinery, 1967; Vol. IV, Forming, 1969; Vol. V, Forging and Casting, 1970; Vol. VI, Welding and Brazing, 1971; Vol. VII, Atlas of Microstructures, 1972, Vol. VIII, Metallography, Structure and Phase Diagrams, 1973; Vol. IX, Fractography and Atlas of Fractographs, 1974; Vol. X, Failure Analysis and Prevention, 1975; Vol. XI, Nondestructive Testing and Quality Control, 1976. "Metals Handbook," 9th ed., 5 vols., American Society for Metals: Vol. I, Properties and Selection: Nonferrous Alloys and Pure Metals, 1979; Vol. III, Properties and Selection: Stainless Steels, Tool Materials and Special Purpose Metals, 1980; Vol. IV, Heat Treating; Vol. V, Surface Cleaning, Finishing and Coating. "Structural Alloys Handbook," Mechanical Properties Data Center, Belfour Stulen, Inc., Traverse City, Michigan, 1978. "Titanium Alloys Handbook," MCIC-HB-05, Metals and Ceramics Information Center, Battelle Columbus Laboratories, 1972.

A wide variety of materials are utilized in aerospace applications where performance requirements place extreme demands on the materials. Some of the more important engineering properties to be considered in the selection of materials are (1) strength-to-weight ratio, (2) density, (3) modulus of elasticity, (4) strength and toughness at operating temperature, (5) resistance to fatigue damage, and (6) environmental effects on strength, toughness, and fatigue properties. Factors such as weldability, formability, castability, quality control, and cost must be considered.

Although it is desirable to maximize the strength-to-weight ratio to achieve design goals, materials with high strength-to-weight ratios generally are sensitive to the presence of small cracks and may exhibit catastrophic brittle failure at stresses below the nominal design stress. The problem of premature brittle fracture is accentuated as the temperature decreases and the yield and ultimate strength increase. For these reasons it is imperative that the tendency toward brittle failure be given major consideration for low-temperature applications. For example, liquid oxygen, a common oxidizer, boils at $-297°F$ ($-183°C$) whereas liquid hydrogen, an important fuel, boils at $-423°F$ ($-253°C$), and the problem of brittle failure is predominant. The fracture-mechanics approach has gained wide acceptance, and nondestructive inspection (NDI) characterization of flaws allows safe operating stresses to be calculated. A considerable amount of fracture-mechanics data has been collected and evaluated ("Damage Tolerant Design Handbook"). Stress corrosion cracking and corrosion fatigue become important for repetitive-use components such as aircraft airframes, landing gear, fan shafts, fan and compressor blades, and disks, in addition to brittle fracture problems such as fatigue crack formation and propagation.

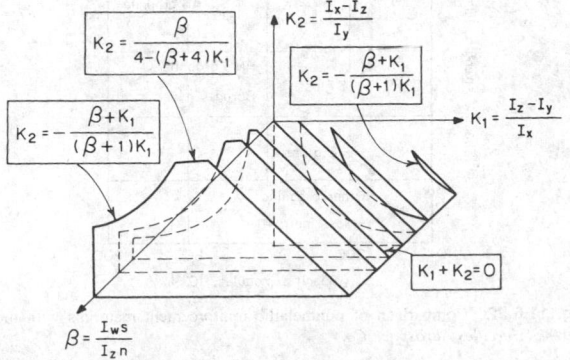

Fig. 11.6.21 Stable region of the vehicle parameter space.

Table 11.6.8 Effective Fiber Properties as Measured in Composites

Fiber	Modulus, 10^6 lb/in^2	Tensile strength, ksi	Tensile failure strain, in/in	Specific gravity	Density, lb/in^3
Boron (4 mil)	60	460	0.8	2.60	0.094
Graphite					
(Thornel 300)	34	470	1.1	1.74	0.063
Graphite pitch	120	350	0.4	2.18,	0.079
Kevlar 49	19	400	1.8	1.45	0.052
Kevlar 29	12	400	3.8	1.45	0.052
E glass	10.5	250–300	2.4	2.54	0.092
S glass	12.5	450	3.6	2.48	0.090

The presence of flaws must be assumed (they may occur either during processing or fabrication) and design must be based on **fatigue crack propagation** (FCP) properties (Sec. 5). FCP of many materials follows an equation of the form

$$da/dN = R(\Delta K)^n \qquad (11.6.1)$$

where a refers to crack length, N is the number of cycles, K is the fluctuation in the stress-intensity parameter, and R and n are material constants. Equation (11.6.1) can be integrated to determine the cyclic life of a component provided that the stress-intensity parameter is known (Sec. 5). Although fatigue-crack initiation is generally retarded by having smooth surfaces, shot-peening, and other surface treatments, FCP of commercial aerospace alloys is relatively insensitive to metallurgical treatment and, to a reasonable approximation, whole classes of materials (i.e., aluminum alloys, high-strength steels, etc.) can be represented within the same scatter band.

It should be noted that increases in the plane-strain fracture toughness K_{IC} result in increased stress, through its linear relationship to the stress-intensity parameter, which in turn causes an exponential increase in the FCP rate and may introduce a fatigue problem where previously there was none.

The effect of environment on materials used for aerospace problems is usually to aggravate the FCP rate. It should be emphasized that fatigue loads have a synergistic effect on the environmental component of cracking and that the crack growth rate during corrosion fatigue is rarely the sum of the FCP and stress-corrosion cracking rates. In weldments of castings or wrought material, tests are needed to simulate anticipated use conditions and to prove weld soundness.

Materials in the turbine section of jet engines are subjected to deleterious gases such as O_2 and SO_2, high temperatures and stresses, and cyclic loads. Consequently the most significant metallurgical factors to consider are microstructural stability, resistance to oxidation and sulfidization, creep resistance, and resistance to low cycle fatigue. Nickel-base superalloys are generally used in the turbine section of the engine for both disks and blades. (See "Aerospace Structural Metals Handbook," Vols. II and IIA, for specific properties.) Although the Ni-base superalloys generally exhibit excellent strength and oxidation resistance at temperatures up to 1800°F (982.22°C), the interactive effects of sustained and cyclic loads are difficult to handle because, at high temperature, plastic deformation of these materials is time-dependent. An approach based on strain range partitioning is described in ASTM STP 520, 1973, pp. 744–782. The utility of this approach is that frequency and temperature dependence are incorporated into the analysis by the way in which the strain is partitioned and reasonable envelopes of the expected life can be calculated.

STRUCTURAL COMPOSITES
by Ivan K. Spiker
NASA

REFERENCES: MIL-HDBK-17, "Plastics for Aerospace Vehicles," Sept. 1973. Advanced Composites Design Guide, Vols. I–V, 3d ed., Sept. 1976.

Table 11.6.9 Matrix Temperature Limits

Material	Maximum use temperature, °F
Polyester	250
Polysulfone	250
Epoxy	350
Polyarylsulfone	400
Phenolic	450
Silicone	450
Polyimide	650
Aluminum	700
Magnesium	700
Glass	2500
Carbon* (W/C)	3200

* With oxidation-resistance.

Composite materials are fibrous materials embedded in a matrix to provide stiffnesses and strengths that neither of the components alone exhibits. There are three general classes of composites: fibers in organic matrices, fibers in metal matrices, and fibers in ceramic matrices.

The selection of a fiber is dictated by the specific strength and/or modulus of the fiber. Table 11.6.8 lists typical properties of the commonly used fibers for structural composites.

The parameter used for selecting the matrix material is normally the maximum use temperature; however, other environmental conditions such as chemical compatibility may dictate the selection of the matrix material. Table 11.6.9 lists matrix materials with their recommended temperature limits.

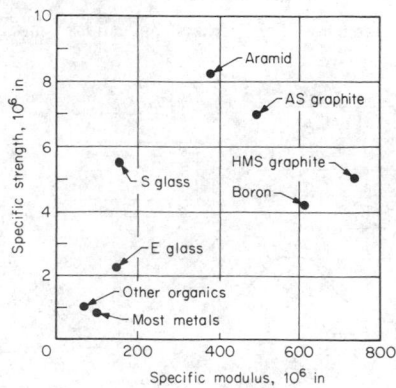

Fig. 11.6.22 Comparison of nonmetallic reinforcement materials with most metals. *(Hercules Aerospace Co.)*

A highly useful feature of composites is the ability to mold parts with complex shapes and multiple contours. A unique feature of designing with graphic composites is the ability to tailor the unidirectional coefficient of thermal expansion from -0.2×10^{-6} in/in · °F to values comparable to those of metals.

Fig. 11.6.22 compares composites with homogeneous organics and metals based on their specific strengths and moduli. A comparison of the fatigue properties of unidirectional composites to an aluminum alloy is depicted in Fig. 11.6.23.

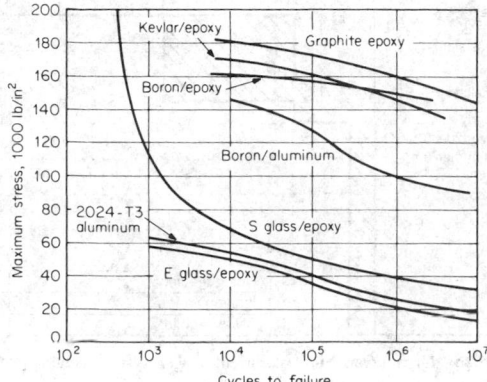

Fig. 11.6.23 Fatigue behavior of unidirectional composites and aluminum. *(Hercules Aerospace Co.)*

STRESS CORROSION CRACKING
by Samuel V. Glorioso
NASA

Stress corrosion cracking is caused by the interaction of sustained tensile or shear stresses and an aggressive environment which results in delayed cracking at stresses well below the yield strength of the material. The occurrence of cracking is dependent upon the duration and magnitude of the applied stress as well as the aggressiveness of the environment.

Sustained stresses can result from residual stresses, applied loads, or both. Residual stresses occur as a result of manufacturing and assembly

Table 11.6.10   Alloys with Low Resistance to Stress Corrosion Cracking

Alloy	Condition
Carbon steel (1000 series)	Above 200 ksi UTS
Low-alloy steel (4130, 4340, D6AC, etc.)	Above 200 ksi UTS
H-11 steel	Above 200 ksi UTS
400C stainless steel	All
AM 350 stainless steel	Below SCT 1000
AM 355 stainless steel	Below SCT 1000
Custom 455 stainless steel	Below H1000
PH 15-7 Mo stainless steel	All except CH900
17-7 PH stainless steel	All except CH900
2011 aluminum	T3, T4
2014 aluminum	All
2024 aluminum	T3, T4
7075 aluminum	T6
7175 aluminum	T6
Cartridge brass, 70% copper, 30% zinc	50% cold-rolled

operations with forming, forging, extrusion, welding, rolling, and assembly fit-up stress as frequent offenders.

Cracking occurs most rapidly for stresses in the material's short-transverse grain direction. Fracture can occur at stresses as low as 8,000 lb/in² for some aluminum alloys exposed to a moist air environment.

Conditions leading to stress corrosion include the exposure of copper to ammonia in condensers and heat exchangers and of stainless steels exposed to chloride-containing solutions. Table 11.6.10 lists some materials which are known to be susceptible. Table 11.6.11 includes materials which are considered to be resistant to stress corrosion cracking in

Table 11.6.11   Alloys with High Resistance to Stress Corrosion Cracking

Alloy	Condition
Carbon steel (1000 series)	Below 180 ksi UTS
Low-alloy steel (4130, 4340, etc.)	Below 180 ksi UTS
300-series stainless steel	All
A286 stainless steel	All
15-5 PA stainless steel	H1000 and above
Beryllium	
Hastelloy X	All
Incoloy 901	All
Inconel 600	Annealed
Inconel 718	All
Inconel x750	All
Monel K-500	All
1000 series aluminum	All
2024 bar, rod aluminum	T8
3000 series aluminum	All
5000 series aluminum	All
6000 series aluminum	All
7075	T73
99.9% copper	
Brass, 85% copper, 15% zinc	

moist salt air. Both tables were condensed from more extensive tables in NASA Specification MSFC-SPEC-522A entitled ''Design Criteria for Controlling Stress Corrosion Cracking,'' which also contains a discussion of material selection and additional conditions under which stress corrosion can occur.

MATERIALS FOR USE IN HIGH-PRESSURE OXYGEN SYSTEMS
by Robert L. Johnston
NASA

Oxygen is relatively reactive at ambient conditions and extremely reactive at high pressures. The design of high-pressure oxygen systems requires special consideration of materials and designs. The types of potential ignition sources that could be present in even a simple component are quite varied. Sources of ignition include electrical failures, particle impact in high-flow regions, pneumatic shock, adiabatic compression, fretting or galling, and Helmholtz resonance in blind passages. Other sources of ignition energy are also possible. A comprehensive review of potential design problems is available in NASA Reference Publication 1113, A.C. Bond et al., ''Design Guide for High Pressure Oxygen Systems.''

Materials currently used in high-pressure oxygen systems range from ignition-resistant materials like Monel 400 to materials of widely varying ignitability like butyl rubber and the silicones. The range of ignitability of various materials is shown in Figs. 11.6.24 and 11.6.25.

While material selection alone cannot preclude ignition, proper choices can markedly reduce the probability of ignition and limit propa-

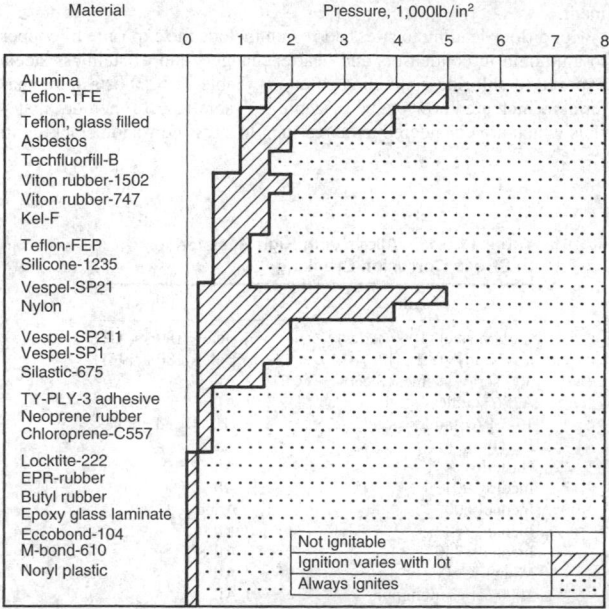

Fig. 11.6.24 Ignition variability of current oxygen-system nonmetallic materials.

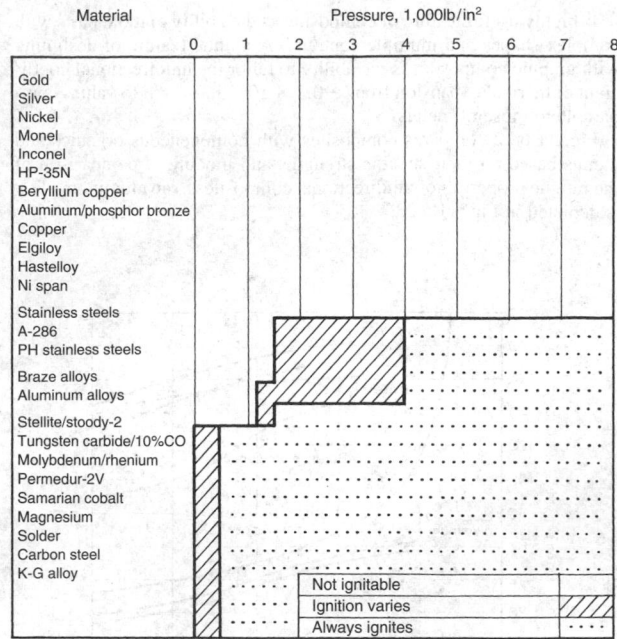

Fig. 11.6.25 Ignition variability of current oxygen-system metallic materials.

gation. Selecting materials with small exothermic heats of combustion will reduce the possibility of propagation. Materials with high heats of combustion (stainless steels) or very high heats of combustion (aluminum) should be avoided. A summary of heats of combustion, as well as other design properties of a few currently used materials, is shown in Table 11.6.12.

The materials listed in Table 11.6.13 have superior resistance to ignition and fire propagation in high-pressure oxygen systems. Monel alloys are available in the necessary range of hardnesses. Springs can be made of Elgiloy. Sapphire poppet balls should replace tungsten carbide or steel balls because sapphire has a lower level of reactivity (and therefore is less combustible) than either tungsten carbide or steel and is more resistant than tungsten carbide to breakup under mechanical impact in an oxygen environment.

Titanium and its alloys, normally attractive as materials for pressure vessels, cannot be used for oxygen vessels because they are impact-sensitive in oxygen. Inconel is a good choice for vessels for high-pressure oxygen.

In no case should an alloy be used at an oxygen pressure above which it can be ignited by particle impact. This criterion would limit the use of aluminum alloys and stainless steels to pressures below 800 lb/in^2 to allow some margin for error in test results and impact predictions.

SPACE ENVIRONMENT
by L. J. Leger and Michael B. Duke
NASA

Materials in the space environment are exposed to vacuum, to temperature extremes, to radiation from the sun, and to the rarified earth atmosphere. Except for thermovacuum conditions, the most damaging aspects of the environment are solar-generated radiation at geosynchronous altitudes and bombardment of surfaces by atomic oxygen (the major constituent of the low-altitude earth atmosphere) in low earth orbit. For long-lived spacecraft impact on surfaces by meteoroids and space debris must be considered.

The vacuum of space may cause the evaporation of a material, or a volatile component of the material, and of the absorbed gases on surfaces. The evaporation rate of a pure material can be calculated by

$$G = \sqrt{\frac{M}{T}} \frac{P}{17.14}$$

where G = evaporation rate, g/(cm^2 · s); M = molecular weight; T = absolute temperature, K; and P = vapor pressure, mmHg at temperature T. (See Dushman, "Scientific Foundations of Vacuum Technique," Wiley.) This simple formula is not applicable to heterogeneous materials or even to a pure substance, such as a plasticizer in an elastomer, which is removed from a matrix of another substance. In this case, other factors, such as migration rate, influence the rate of loss. Although the evaporation of a component of a material may not reduce the effectiveness of the material, e.g., the plasticizer in the insulation of an electrical

Table 11.6.12 Physical Properties of Typical Oxygen System Metallic Materials

Property	Aluminum alloys	Stainless steel	Inconel 718	Monel 400
Density, lb/in^3	0.10	0.28	0.30	0.31
Tensile, ultimate, lb/in^2	40,000	140,000	180,000	145,000
Maximum use temperature, °F	350	950	1,200	1,000
Heat of combustion, Btu/lb	130,000	33,500	15,100	14,200
Maximum use pressure, lb/in^2	1,250	1,250	8,000	8,000

Table 11.6.13   Recommended Materials

Application	Material
Component bodies	Monel
	Inconel 718
Tubing and fittings	Monel
	Inconel 718
Internal parts	Monel
	Inconel 718
	Beryllium copper
Springs	Beryllium copper
	Elgiloy
Valve seats	Gold or silver over Monel or Inconel 718
Valve balls	Sapphire
Lubricants	Batch- or lot-tested Braycote
	3L-38RP
	Batch- or lot-tested Everlube 812
O-seals and backup rings	Batch- or lot-tested Viton
	Batch- or lot-tested Teflon
Pressure vessels	Inconel

conductor, the deposition of the vapor on a colder surface may be intolerable. Metals do not usually evaporate in space at modest temperatures, but organic materials, including elastomers, plastics, coatings, adhesives, and lubricants, must be of very high molecular weight to avoid evaporation. A standard screening test (ASTM E595) has been developed to determine the acceptability of materials from an outgassing standpoint. This screening test addresses only contamination and, therefore, functional characteristic changes must be evaluated separately.

Spacecraft are exposed to electromagnetic radiation and particulate contamination in both low and geosynchronous altitudes; however, the large doses (5×10^{11} ergs/g per 30 years) of high-energy electrons and protons are experienced only at the higher altitudes. Metals and ionic compounds are relatively resistant to space-indigenous radiation. However, semiconductors are sensitive to permanent radiation damage, and other electrical materials are subject to permanent or transient damage. Organic materials are also susceptible to degradation by both electromagnetic and particle radiation, especially in vacuum. Organic polymers of high molecular weight may have such low vapor pressure that their evaporation in vacuum at reasonable temperatures is not significant. However, radiation, which may produce chain scission, yields fragments of reduced molecular weight and increased vapor pressure, which will result in the loss of the mass at the same temperature which did not affect the nonirradiated material (Table 11.6.14). Note the almost complete degradation of the tensile strength of neoprene and Buna N as a result of irradiation in vacuum. On the other hand, Viton A appears to be satisfactory for this environment on the basis of increased tensile strength; however, the decreased elongation would be an important consideration in the application of this material as a seal, particularly as a dynamic seal. Furthermore, Viton A is not suitable for very low temperature applications, another component sometimes encountered in the space environment. Geosynchronous radiation may change thermal expansion characteristics of organic composite materials and this effect must, therefore, be considered for large space structures.

The better known organic liquid lubricants are subject to evaporation in space because of relatively high vapor pressures. A limited number of liquid film lubricants with sufficiently low vapor pressures are available; however, dry film lubricants are preferred. The most widely used dry film utilizes MoS_2 as the lubricating agent in either silicate or organic binders.

Moving electrical-contact surfaces such as brushes, slip rings, and make-break switches require either reliable isolation from the vacuum environment or special selection of materials, especially where long-time operation is involved. Composites of heavy-metal sulfides with silver or copper are promising possibilities for this application.

Spacecraft surfaces exposed in the low earth orbital environment experience 10^{14} to 10^{15} atomic oxygen impacts/(cm$^2 \cdot$ s) when facing into the velocity vector. For even relatively short missions (7 days) total fluence can be as much as 3×10^{20} atoms/(cm^2). This exposure leads to oxidation of the surface of some metals, the most active of which is silver, and to removal of a significant amount of organic material (approx 10 μm), also by oxidation. The extent of reaction is dependent upon many parameters, including atmospheric density, vehicle velocity, attitude, and total exposure time. In turn, atmospheric density is dependent upon altitude and solar activity. Fully fluorinated polymers such as Teflon and silicone-based compounds appear to be more stable than other organic films by a factor of 10. This effect must be taken into account for long-duration flights at altitudes up to 600 km and for short-term low-altitude flights.

Meteoroids derived from comets, asteroids, and possibly larger satellites and planets are present in significant numbers in the inner solar system. These particles range in size from less than 10^{-12} g (cosmic dust) to asteroidal dimensions. In the vicinity of the earth, these particles travel with velocities on the order of 10 to 30 km/s, which makes even millimeter-sized particles potentially damaging to space structures. The current flux of meteoroids in the 1-mm size range is about 0.02/(m$^2 \cdot$ yr); at 1 cm, the flux is about 3 orders of magnitude lower. Some regions of space, for example, the asteroid belt or the vicinity of Saturn's rings, contain significantly higher numbers of solid particles.

The spectral energy distribution for the sun resembles a Planck curve with an effective temperature of 5,800 K, with most of the solar energy lying between 150 nm and 10 nm with a maximum near 450 nmr. The radiant energy of the sun at 1 au is 1.37×10^3 W/m^2. The sun's electromagnetic radiation at wavelengths shorter than visible includes significant fluxes of hard x-rays and gamma rays. The emission of the harder radiation varies with the solar cycle.

The sun also emits charged particles with velocities in the range of 500 to 700 km/s (solar wind) with fluxes on the order of 4×10^3 protons/(cm$^2 \cdot$ s). During solar flares, much higher energy particles are

Table 11.6.14   Effects of Vacuum and Radiation on Elastomers

Material	Test	Pressure, mmHg	Temp, °F	Radiation ergs, g^{-1}/°C	Tensile strength, lb/in^2	Elongation, percent
Neoprene	Air	760	80	0	3,135	426
	Vacuum	1×10^{-5}	80	0	3,350	405
	Air and radiation	760	80	1.9×10^9	2,769	265
	Vac and radiation	5×10^{-6}	80	1.9×10^9	191	218
Buna N	Air	760	80	0	2,630	685
	Vacuum	1×10^{-5}	80	0	2,640	700
	Air and radiation	760	80	1.9×10^9	2,175	390
	Vac and radiation	5×10^{-6}	80	1.7×10^9	203	450
Viton A	Air	760	80	0	1,343	172
	Vacuum	1×10^{-5}	80	0	1,168	238
	Air and radiation	760	80	2×10^{10}	2,629	36
	Vac and radiation	5×10^{-7}	109	1.6×10^{10}	1,830	31

emitted, ranging from 100 keV to 100 MeV energies, with typical proton fluxes at 1 au of 10 to 100 $cm^{-2} \cdot ster^{-1} \cdot s^{-1}$ at 10 MeV to less than 10 $cm^{-2} \cdot ster^{-1} \cdot s^{-1}$ at 30 MeV and higher. Total protons in a flare event are on the order of 10^7 to $10^{10}/cm^2$. Atomic nuclei ranging from protons to uranium nuclei are emitted in solar flares. Galactic cosmic rays, from outside the solar system, with 100-MeV to BeV energies are also observed.

In the vicinity of the earth, interactions between solar radiation and the earth's magnetic field lead to the phenomenon known as the Van Allen radiation belts, in which intense concentrations of radiation are observed in the form principally of protons and electrons, with energies on the order of a few electron volts. These lead to charging problems for spacecraft and structures embedded in the plasma. Beyond the earth's magnetic field, charging can occur, but is not a significant problem. In the vicinity of the earth, variations of charged-particle concentrations occur diurnally and with geographical variation of the magnetic field.

The vacuum of space in the inner solar system is limited by the presence of the solar wind, with the interplanetary pressure being about 10^{-13} mbar. Interaction of solar radiation with the gases at low earth orbital altitudes has been observed to produce ionized oxygen. This species and other ionized molecules may exist which can interact with spacecraft surfaces. In the local vicinity of spacecraft hardware, outgassing characteristics are generally the major determinant of the local pressure and the molecular composition.

Since the beginning of the space age, numerous artificial objects have been placed into earth orbit. Many of these remain in orbit, as does debris from the explosion or disaggregation of various spacecraft. This debris at about 1,000 km (the most intensely contaminated region) now approaches the natural meteoroid flux in the millimeter size range, and is predicted to grow in the future as additional space vehicles are launched into earth orbit. In the micrometer size range, numerous aluminum oxide particles resulting from the firing of upper-stage solid rocket motors are present and may be significantly more abundant than natural meteoroids. Thus, artificial debris from space flights is rapidly accumulating to levels which also present significant impact probabilities. There is widespread acceptance of the dual-wall technique of protecting a craft against particle impact damage. The outer wall, called the bumper, is detached from the load-carrying structural wall. The bumper dissipates some of the kinetic energy of the particle, but primarily it serves to fragment the particle into a fine spray so that the impact energy is spread over a larger area of the structural wall.

SPACE-VEHICLE STRUCTURES
by Thomas L. Moser and Orvis E. Pigg
NASA

REFERENCES: MIL-HDBK-5D, ''Metallic Materials and Elements for Aerospace Vehicle Structures.'' Advanced Composites Design Guide, Sept. 1976, Library Accession no. AD916679 through AD916683. Ashton, Halpin, Pertit, ''Primer on Composite Materials Analysis,'' Technomic Pur. Co. Aeronautical Structures Manual (Vols. I–III), NASA TMX-73305. Bruhn, ''Analysis and Design of Flight Vehicle Structures,'' Tri-State Offset Co., Cincinnati. Barton, ''Fundamentals of Aircraft Structures,'' Prentice-Hall, 1948. Roark, ''Formulas for Stress and Strain,'' McGraw-Hill. ''Nastran Theoretical Manual,'' Computer Software Management and Information Control (COSMIC), University of Georgia. Perry, ''Aircraft Structures,'' McGraw-Hill. Gathwood, ''Thermal Stress,'' McGraw-Hill. Zienkiewicz, ''The Finite Filement,'' McGraw-Hill. MIL-STD-810D, ''Environmental Test Methods and Engineering Guidelines.'' Chapman, ''Heat Transfer,'' Macmillan.

Space vehicles include a large variety of vehicle types, like their predecessors, earth vehicles. Airplanes were the first generation of space vehicles, and our definition of space vehicle has changed as the airplane approached the limits of the earth's atmosphere. Space, in this discussion, is the region beyond the earth's atmosphere.

Many space vehicles are hybrids since they are required to operate in the earth's atmosphere and beyond. The space shuttle orbiter is such a hybrid space vehicle. The Apollo command and service module is also a hybrid, but the Apollo lunar module, which primarily functioned outside the earth's atmosphere, was a space vehicle.

The design procedure for any vehicle structure is basically the same:
1. Establish the design requirements and criteria
2. Define the loads and environments
3. Perform the design and analysis
4. Iterate the previous two steps (as required) because of the interrelation between configuration and loads
5. Verify the design
There are conditions and requirements which are unique for hybrid and space-vehicle structures.

Minimum weight and high reliability are significant but opposing requirements for space-vehicle structures which must be balanced to achieve an efficient, reliable structure design. Depending on the mission of the particular space vehicle, the ratio between the weight of the earth launch vehicle and the payload (e.g., interplanetary probe) can be as high as 400 to 1. The large weight leverages require that the design be as efficient as possible. Structural materials, therefore, must be stressed (or worked) as close to the capabilities as possible. Like aircraft structures, space-vehicle structures should be designed for a zero margin of safety (MS):

$$MS = \frac{\text{(material ultimate allowable)}}{FS \text{ (limit stress)}} - 1$$

1. *Material ultimate allowable.* The stress at which the material will fail.
2. *FS (factor of safety).* This factor usually ranges between 1.25 and 1.50, for space structures, depending on the maturity of the design, the behavior of the material, the confidence in the loads, etc.
3. *Limit stress.* The maximum (reasonable) stress that the structural component would be expected to encounter.
Failure can be based on rupture, collapse, or yield as dictated by the functional requirements of the component. High reliability is necessary for most space vehicle structures since a failure is usually catastrophic or nonrepairable. Total structural reliability is composed of many of the design elements—loads, materials, allowables, load paths, factor of safety, analysis techniques, etc. If the reliability is specified and the statistical parameters are known for each design element, then the process is straightforward. A total structural reliability is seldom specified and good engineering practice is used for the solution of the reliability of each design element. Commonly used to assure structural reliability are (1) material allowables equivalent to A values from MIL-Handbook-5A for safe-life structure and B values for fail-safe structures; (2) limit loads which correspond to events and conditions which would not be exceeded more often than 3 times in 1,000 occurrences; (3) factor of safety equal to 1.4 for crew-carrying space vehicles and 1.25 for those without crews; and (4) verification of the structural design by test demonstration. A good structural design is one which has the correct balance between minimum weight and high reliability.

Space vehicles are usually manufactured and assembled on earth and transported to space. The launch environments of acceleration, vibration, aerodynamic pressure, rapid pressure change, and aerodynamic heating coupled with the space environments of extremely low pressure, wide ranges in temperature, meteroids, and radiation dictate that the design engineers accurately define the magnitudes of these environments and the correct time-compatible combinations of these environments if realistic loads are to be determined. Many natural environments have been defined and are documented in the references. The induced environments, which are a function of the system and/or structural characteristics (e.g., dynamic pressure, vibration, etc.), must be determined by analysis, test, or extrapolation from similar space vehicles. Quantifying the induced loads for the structural designers is one of the major consumers of time and money in any aerospace program and one of the most important. Load determination can be a very complex task. For further guidance the reader is directed to the references.

For a space vehicle which is not exposed to the launch loads, the load determination is greatly simplified. In the absence of an appreciable atmosphere and gravitational accelerations the predominant external force on a space-vehicle structure is that of the propulsion and attitude control systems and the resulting vibration loads. Even though not induced by an external force, thermal stresses and/or deflections can be a major design driver. Space vehicles commonly experience larger thermal gradients because of the time- or attitude-related exposure to solar radiation.

The thermal effects of stress and deflections of the structure require that temperature distributions be defined consistent with the strength and deflection requirements. From a strength integrity standpoint, temperature distribution on a space vehicle can be as important as the pressure distribution on an aircraft. The significance and sensitivity of thermal effects are greatly reduced by the use of composite materials for the structure. The materials can be layered to achieve, within limits, the desired thermal coefficient of expansion and thereby reduce the thermal effects.

Space vehicles which return to the earth must withstand the extremely high temperatures of atmospheric reentry. This produces dynamic temperature distributions which must be combined with the aerodynamic pressures and vehicle dynamic loads.

Innovative, clever, and efficient structural designs minimize the effects of loads associated with launch for a vehicle which is to remain in space. This is accomplished by attaching the space vehicle to the booster rocket to minimize scar weight and to reduce the aerodynamic loads by the use of shrouds, fairings, or controlling the attitude during atmospheric flight.

Efficiencies of weight-critical structures have been greatly improved by more accurately quantifying the loads and stress distributions in complex structural configurations. **Finite element methods** (FEM) of analysis and large capacity, high-speed digital computers have enabled the increased fidelity of analyses. In this process, the structure is modified with finite elements (beams, bars, plates, or solids) by breaking the structure into a group of nodal points and connecting the nodal points with the finite elements which best represent the actual structure. Loads are applied at the nodal points or on the elements which then distribute the loads to the nodal points. Constraints are also applied to the nodal points to support or fix the structure being analyzed. The finite-element program then assembles all the data into a number of linear simultaneous equations and solves for the unknown displacements and then for internal loads and stresses, as required. Many finite-element programs have been developed to perform a variety of structural analyses including static, buckling, vibration, and response.

The loads in a structure depend on the characteristics of the structure and the characteristics of the structure are largely dictated by the loads. This functional dependency between loads and structures requires an iterative design process.

Classically, when structures are worked to stresses near the capability of the materials, e.g., airplane structures, the design is verified by test demonstration. This method of verification was common before high-fidelity structural mathematical models were employed and is still used when practical to apply the design loads and environments to a dedicated test article to verify the design. Design deficiencies are then corrected depending on the importance and the manufacturing schedule.

Space-vehicle structures often do not lend themselves to the classical test approach because of the multitude of loads and environments which must be applied, because testing in earth's gravity precludes realistic loadings, and because of the size and configuration. Verification by analyses with complementary component tests and tests of selected load conditions are common.

VIBRATION OF STRUCTURES
by Lawrence H. Sobel

University of Cincinnati

REFERENCES: Crandall and McCalley, Numerical Methods of Analysis, chap. 28, in Harris and Crede (eds.), "Shock and Vibration Handbook," Vol. 2, McGraw-Hill. Hurty and Rubinstein, "Dynamics of Structures," Prentice-Hall. Meirovitch, "Analytical Methods in Vibrations," Macmillan. Przemieniecki, "Theory of Matrix Structural Analysis," McGraw-Hill. Wilkinson, "The Algebraic Eigenvalue Problem," Oxford.

As a result of the vast improvements in large-scale, high-speed digital computers, numerical methods are being used at an exponentially increasing rate to analyze complex structural vibration problems. The numerical methods most widely used in structural mechanics are the finite element method and the finite difference method. In the finite element method, the actual structure is represented by a finite collection, or assemblage, of structural components whereas, in the finite difference method, spatial derivatives in the differential equations governing the motion of the structure are approximated by finite difference quotients. With either method, the continuous structure with an infinite number of degrees of freedom is, in effect, approximated by a discrete system with a finite number of degrees of freedom (unknowns). The linear, undamped free and forced vibration analysis of such a discrete system is discussed herein. The analysis is most conveniently carried out with the aid of matrix algebra, which is well suited for theoretical and computational purposes. The following discussion will be brief, and the reader is referred to the cited references for more comprehensive treatments and examples.

Equation of Motion

Consider a conservative discrete system undergoing small motion about a state of equilibrium (neutral or stable). Let $q_1(t)$, . . . , $q_n(t)$, where t is time, be the minimum number of independent coordinates (linear or rotational) that completely define the general dynamical configuration of the system and that are compatible with any geometrical constraints imposed on the system. Then the n coordinates q_1, . . . , q_n are called generalized coordinates, and the system is said to have n degrees of freedom. Let $\{Q(t)\}$ be the vector of generalized forces, and let $\{q(t)\}$ be the vector of generalized displacements, $\{q(t)\} = \{q_1(t), . . . , q_n(t)\}$. Corresponding elements of the generalized force and displacement vectors are to be conjugate in the energy sense, which simply means that if one of the elements of $\{q\}$ is a rotation (for instance), then the corresponding element of $\{Q\}$ must be a moment, so that their product represents work or energy. Methods of computing $\{Q\}$ are discussed in the references (e.g., Meirovitch). The matrix equation governing the small undamped motion of the system is given by

$$[M]\{\ddot{q}\} + [K]\{q\} = \{Q\} \qquad (11.6.2)$$

In this equation a dot denotes single differentiation in time, and $[M]$ and $[K]$ are the mass and stiffness matrices, respectively. $[M]$ and $[K]$ are real and constant matrices, which are assumed to be symmetric. Such symmetry will arise whenever an energy approach (e.g., Lagrange's equations) is employed to derive the equation of motion. The mass matrix is assumed always to be positive definite (and hence nonsingular, or can be made to be positive definite; see Przemieniecki).

Free-Vibration Analysis

The following equation governing the free vibration motion of the conservative system is obtained from Eq. (11.6.2) with $\{Q\} = \{0\} : [M]\{\ddot{q}\} + [K]\{q\} = \{0\}$. To obtain a solution of this equation, we note that, for a conservative system, periodic motion may be possible. In particular, let us assume a harmonic solution for $\{q\}$ in the form $\{q\} = \{A\} \cos(\omega t - \alpha)$. That is, we inquire whether it is possible for all coordinates to vibrate harmonically with the same angular frequency ω and the same phase angle α. This means that all points of the system will reach their extreme positions at precisely the same instant of time and pass through the equilibrium position at the same time. Substitution of this trial solution into the free vibration equation yields the eigenvector equation $[K]\{A\} = \omega^2[M]\{A\}$. In this equation $\{A\}$ is called the eigenvector and ω^2 is the eigenvalue (the square of a natural frequency). This homogeneous equation always admits the trivial solution $\{A\} = \{0\}$.

Nontrivial solutions are possible provided that ω^2 takes on certain discrete values, which are obtained from the requirement that the determinant of the coefficient matrix of $\{A\}$ vanishes. This yields the eigenequation (frequency equation) $\Delta(\omega^2) = |[K] - \omega^2[M]| = 0$. It is an nth-degree polynomial in ω^2, and its n roots are denoted by ω_r^2, $r = 1, . . . , n$. The eigenvector $\{A\}_r$, $r = 1, . . . , n$, corresponding to each eigenvalue is determined from $[K]\{\mathbf{A}\}_r = \omega_r^2[M]\{\mathbf{A}\}_r$. However, the homogeneous nature of this equation precludes the possibility of obtaining the explicit values of all the elements of $\{\mathbf{A}\}_r$. It is possible to determine only their relative values or ratios in terms of one of the components. Thus, we see that for each eigenvalue ω_r^2 there is a corresponding eigenvector $\{\mathbf{A}\}_r$, which has a unique shape (based on the relative values of its components) but which is determined only to within a scalar multiplicative factor that may be regarded as being the amplitude of the shape. This unique shape is called the mode shape, and it can be thought of as being the position of the system when it is in its extreme position, and hence momentarily stationary, as in a static problem. Each frequency along with its corresponding mode shape is said to define a so-called natural mode of vibration. Methods of determining these natural modes are presented in the references (e.g., Wilkinson). If the initial conditions are prescribed in just the right way, it is possible to excite just one of the natural modes. However, for arbitrary initial conditions all modes of vibration are excited. Thus, the general solution of the free-vibration problem is given by a linear combination of the natural modes, i.e.,

$$\{\mathbf{q}\} = \sum_{r=1}^{n} C_r\{\mathbf{A}\}_r \cos\left(\omega_r t - \alpha_r\right)$$

The $2n$ arbitrary constants C_r and α_r are determined through specification of the $2n$ initial conditions $\{\mathbf{q}(0)\}$, and $\{\dot{\mathbf{q}}(0)\}$. An explicit representation for $\{\mathbf{q}\}$ in terms of $\{\mathbf{q}(0)\}$ and $\{\dot{\mathbf{q}}(0)\}$ is given later.

Some of the properties of the natural modes of vibration will now be listed. Unless stated otherwise, the following properties are all consequences of the assumptions that $[M]$ and $[K]$ are real and symmetric and that $[M]$ is positive definite: (1) the n eigenvalues and eigenvectors are real. (2) If $[K]$ is positive definite (corresponding to vibration about a stable state of equilibrium), all eigenvalues are positive. (3) If $[K]$ is positive semidefinite (corresponding to vibration about a neutral state of equilibrium), there is at least one zero eigenvalue. All zero eigenvalues correspond to rigid-body modes. (4) Eigenvectors $\{\mathbf{A}\}_r$, $\{\mathbf{A}\}_s$ corresponding to different eigenvalues ω_r^2, ω_s^2 ($\omega_r^2 \neq \omega_s^2$; $r \neq s$; r, $s = 1, \ldots, n$) are orthogonal by pairs with respect to the mass and stiffness matrices (weighting matrices), that is, $\{\mathbf{A}\}_r^T[M]\{\mathbf{A}\}_s = 0$ and $\{\mathbf{A}\}_r^T[K]\{\mathbf{A}\}_s = 0$ for $r \neq s$; r, $s = 1, \ldots, n$. Since the eigenvectors are orthogonal, and hence independent, they form a basis of the n-dimensional vector space. Thus, any vector in the space, such as the solution vector $\{\mathbf{q}\}$, can be represented as a linear combination of the eigenvectors (base vectors). This theorem (expansion theorem) is of vital importance for the response problem to be considered next.

Response Analysis

The response of the system due to arbitrary deterministic excitations in the form of initial displacements $\{\mathbf{q}(0)\}$, initial velocities (impulses) $\{\dot{\mathbf{q}}(0)\}$, or forcing functions $\{\mathbf{Q}(t)\}$ may be obtained by means of the above expansion theorem. That is, the displacement solution of the response problem is expanded with respect to the eigenvectors of the corresponding free-vibration problem. The time-dependent coefficients in this expansion are the so-called normal coordinates of the system. The equations of motion when expressed in terms of the n normal coordinates are uncoupled, and each one of the equations is mathematically identical in form to the equation governing the motion of a simple one-degree-of-freedom spring-mass system. Hence solutions of these equations are readily obtained for the normal coordinates. The normal coordinates are then inserted into the eigenvector expansion to obtain the following explicit solution for displacement response $\{\mathbf{q}(t)\}$ (Przemieniecki) for undamped constrained or unconstrained (free) structures:

From this equation, it is seen that the displacement response is expressed directly in terms of the arbitrary excitations and the free-vibration frequencies and eigenvectors. The eigenvectors have been separated into two types denoted by $\{\mathbf{A}\}_{ro}$ and $\{\mathbf{A}\}_{re}$. The "rigid-body" eigenvectors $\{\mathbf{A}\}_{ro}$ correspond to the m rigid-body modes (if there are any) for which $\omega_r = 0$, $r = 1, \ldots, m$. The "elastic" eigenvectors $\{\mathbf{A}\}_{re}$ correspond to the remaining $n - m$ eigenvectors, for which $\omega_r \neq 0$, $r = m + 1, \ldots, n$. The general solution given by Eq. (11.6.3) is very useful provided that the second integral (Duhamel's integral) can be evaluated. (The first integral is easier to evaluate.) Closed-form solutions of this integral for some of the simpler types of elements of $\{\mathbf{Q}(t)\}$, such as step functions, ramps, etc., are available (see Przemieniecki). Solutions of Duhamel's integral for more complicated forcing functions can be obtained through superposition of solutions for the simpler cases or, of course, from direct numerical integration.

The displacement response obtained from Eq. (11.6.3) may be employed to obtain other response variables of interest, such as velocities, accelerations, stresses, strains, etc. Velocities and accelerations are obtained from successive time differentiations of Eq. (11.6.3). The method used to evaluate the stresses depends on the type of spatial discretization employed to approximate the structural continuum. For example, difference quotients are used in the finite difference approach. Note that the rigid-body components do not contribute to the stress or strain response and hence the first summation in Eq. (11.6.3) can be omitted for computation of these responses.

SPACE PROPULSION
by Henry O. Pohl
NASA

The propulsion system provides the maneuverability, speed, and range of a space vehicle. Modern propulsion systems have diverse forms, however, they all convert an energy source into a controlled high velocity stream of particles to produce force. Among the many possible energy sources, five are considered useful for space-vehicle propulsion:
1. Chemical reaction
2. Solar energy
3. Nuclear energy
4. Electrical energy
5. Stored energy (cold gas)

Accordingly, the various propulsion devices are categorized into chemical energy propulsion, nuclear energy propulsion, stored energy propulsion, solar energy propulsion and electrical energy propulsion.

Chemical propulsion can be further categorized into liquid propulsion or solid propulsion systems. Many combinations of fuels and oxidizers are available. Likewise, electrical energy propulsion can be subdivided into systems using resistive heating, arc heating, plasma, or ion propulsion. Nuclear energy and solar energy are external energy sources used to heat a gas (usually hydrogen).

All space propulsion systems, with the exception of ion propulsion, use nozzle expansion and acceleration of a heated gas as the mechanism for imparting momentum to a vehicle. The fundamental relationships of space propulsion systems are commonly expressed by a number of parameters as follows: The basic performance equation can be derived from the classical newtonian equation $F = MA$. It is defined as

$$F = \left(\frac{\dot{W}}{g}\right) V_e + A_e(P_e - P_a)$$

where F = thrust, lbf; $\dot{W}$ = propellant weight flow rate; g = gravitational acceleration; V_e = exhaust velocity; A_e = nozzle exit area; P_e = pressure at nozzle exit; and P_a = ambient external pressure.

$$\{\mathbf{q}(t)\} = \sum_{r=1}^{m} \frac{\{\mathbf{A}\}_{ro}\{\mathbf{A}\}_{ro}^T}{\{\mathbf{A}\}_{ro}^T[M]\{\mathbf{A}\}_{ro}} \left([M](\{\mathbf{q}(0)\} + t\{\mathbf{q}(0)\}) + \int_{\tau 2=0}^{\tau 2=t} \int_{\tau 1=0}^{\tau 1=\tau 2} \{\mathbf{Q}(\tau_1)\}\, d\tau_1\, d\tau_2 \right)$$

$$+ \sum_{r=m+1}^{n} \frac{\{\mathbf{A}\}_{re}\{\mathbf{A}\}_{re}^T}{\{\mathbf{A}\}_{re}^T[M]\{\mathbf{A}\}_{re}} \left([M]\left(\{\mathbf{q}(0)\} \cos \omega_r t + \frac{1}{\omega_r}\{\dot{\mathbf{q}}(0)\} \sin \omega_r t \right) + \frac{1}{\omega_r}\int_{\tau=0}^{\tau=t} \{\mathbf{Q}(\tau)\} \sin \omega_r(t - \tau)\, d\tau \right) \quad (11.6.3)$$

Table 11.6.15   Specific Impulse for Representative Space Propulsion Systems

Engine type	Working fluid	Specific impulse
Chemical (liquid)		
Monopropellant	Hydrogen peroxide-	110–140
	hydrazine	220–245
Bipropellant	O_2-H_2	440–480
	O_2-hydrocarbon	340–380
	N_2O_4-Monomethyl	300–340
	hydrazine	
Chemical (solid)	Fuel and oxidizer	260–300
Nuclear	H_2	600–1000
Solar heating	H_2	400–800
Arc jet	H_2	400–2000
Cold gas	N_2	50–60
Ion	Cesium	5000–25,000

The specific impulse is one of the more important parameters used in rocket design. It is defined as $I_s = F/W$, where I_s = specific impulse in seconds, F = thrust, and W = propellant weight flow rate per second of operation. I_s increases in engines using gas expansion in a nozzle as the pressure and temperature increase and/or the molecular weight of the gas decreases. The specific impulse values for representative space propulsion systems are shown in Table 11.6.15.

The total impulse $I_t = \int F(t)\, dt$ is useful to determine candidate rocket firing duration and thrust combinations required to satisfy a given change in momentum, where F = thrust and t = time.

Mass fraction is the ratio of propellant to total weight of the vehicle and is defined as $\gamma = (W_i - W_f)/W_i$, where γ = mass fraction; W_i = total weight of the vehicle including the propellant, structure, engine, etc., at the start of operation; and W_f = total weight of the vehicle less the weight of propellant consumed.

In the simplified case of a vehicle operating free of atmospheric drag and gravitational attractions, the velocity increment obtained from a propulsion system is defined as $\Delta V = I_s g \ln (W_i/W_f)$, where ΔV = velocity increment obtained; W_i = total weight of vehicle including propellant, structure, engine, etc., at start of operation; and W_f = total weight of vehicle less weight of propellant consumed during operations. This equation brings together the mission parameters, vehicle parameters and rocket-engine parameters.

Figure 11.6.26 graphically shows the effect specific impulse and mass fraction have on the ability of a propulsion system to change the velocity of a payload in space.

Chemical Propulsion   Only chemical energy has found wide acceptance for use in space propulsion systems. Modern materials and current technologies have made it possible to operate rockets at relatively high pressures. When these rockets are fitted with very large expansion-ratio nozzles (nozzle exits 40 to 200 times the area of the throat) the force produced per unit of propellant consumed is greatly increased when operating in the vacuum of space.

Solid-propellant rocket motors are used for moderate total impulse requirements, in spite of their low specific impulse, provided that the total impulse requirements can be precisely set before the rocket is built. The low specific impulse for these applications is largely offset by the high mass fraction (ratio of propellant weight to total propulsion system weight including propellant weight). They are simple to start, have good storage characteristics and multiple velocity requirements can be accommodated by staging, i.e., using a separate motor for each burn.

Liquid-propellant systems are generally used for moderate to high total impulse requirements, multiple burns, velocity-controlled thrust terminations, variable thrust requirements or for applications where the total impulse requirement is likely to change after the rocket is built or launched.

Bipropellant systems are the most commonly desired systems to satisfy moderate to high total impulse requirements. A fuel and oxidizer are injected into a combustion chamber where they burn, producing large volumes of high-temperature gases. Oxygen and hydrogen produce the

highest specific impulse of any in-service chemical rocket (440 to 460 s). These systems also have good mass fractions (0.82 to 0.85). Oxygen and hydrogen, as stored in the rocket propellant tanks, are cryogenic, with boiling points of $-296.7°F$ and $-442°F$ respectively. Storing these propellants for long periods of time requires thermally

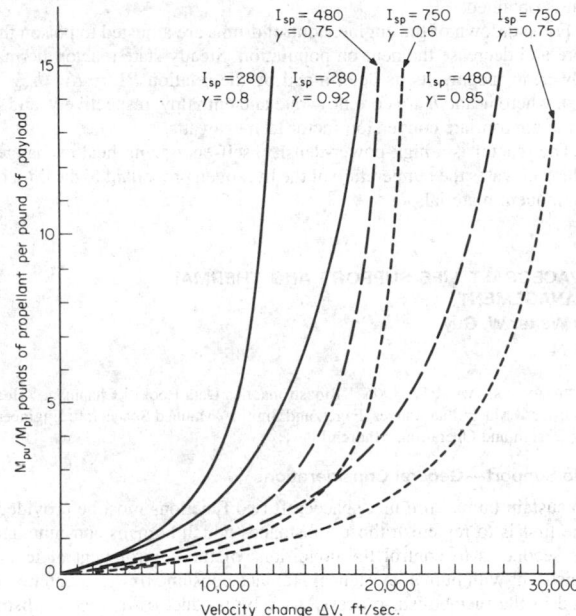

Fig. 11.6.26   Performance curves for a range of specific impulse and mass fractions.

insulated propellant containers and specially designed thermodynamic vents or refrigeration systems to control boil-off losses. The low density of hydrogen (4.4 lb/ft³) requires the use of bulky containers. Because of these characteristics, oxygen/hydrogen rocket systems can best satisfy moderate to high total-impulse requirements. Where designs are constrained by volume, very long durations of time between firings, many short burns, or precise impulse requirements, nitrogen tetroxide and monomethylhydrazine are often preferred, if the total impulse requirements are moderate.

Monopropellant systems generally use a single propellant that decomposes in the presence of a catalyst. These are usually low total-impulse systems used to provide attitude control of a spacecraft or vernier velocity corrections. For attitude control, small thrustors are arranged in clusters around the perimeter of the vehicle. These small thrustors may, in the course of a single mission, produce over 1,000,000 impulses, to keep the vehicle oriented or pointed in the desired direction.

Solar-heating propulsion would use a solar collector to heat the working fluid which is exhausted through a conventional nozzle. The concentrated solar energy heats the rocket propellant directly through the heat exchanger. The heated propellant is then exhausted through the nozzle to produce thrust. Many design problems make the practical development of the solar-heating rocket quite difficult. For example, the solar collector must be pointed toward the sun at all times and the available solar energy varies inversely with the square of the distance from the sun. Using hydrogen as a working fluid, this type system would have a specific impulse of 400 to 800 s and could satisfy very high total impulse requirements at low thrust levels.

Nuclear propulsion is similar to solar-heating propulsion except that the energy source is replaced by a nuclear reactor. A propellant is injected into the reactor heat exchanger, where the propellant is heated under pressure and expanded through the nozzle. Liquid hydrogen is the best choice for the nuclear rocket's working fluid because its low molecular weight produces the highest exhaust velocity for a given nozzle-inlet temperature. For example, for a given entrance nozzle pressure of

43 atm and temperatures of 1650, 3300, and 4950 K the following specific impulses are obtainable: 625, 890, and 1,216 s respectively. The nuclear rocket engine is started by adjusting the reactor neutron-control drums to increase the neutron population. Propellant flow is initiated at a low reactor-power level and is increased in proportion to the increasing neutron population until the design steady-state reactor power output is obtained.

For shutdown of the engine, control drums are adjusted to poison the core and decrease the neutron population. Steady-state reactor thermal power, in megawatts, is determined by the relation $P1 = Kw (h_{out} - h_{in})$, where w and h are core flow rate and enthalpy, respectively, and K is the appropriate conversion factor to megawatts.

The reactor is a high-power-density, self-energizing heat exchanger which elevates the temperature of the hydrogen propellant to the limit of component materials.

SPACECRAFT LIFE SUPPORT AND THERMAL MANAGEMENT
by Walter W. Guy

NASA

REFERENCES: *NASA SP-3006*, "Bioastronautics Data Book." Chapman, "Heat Transfer," Macmillan. Purser, Faget, and Smith, "Manned Spacecraft: Engineering Design and Operation," Fairchild.

Life Support—General Considerations

To sustain human life in a spacecraft two functions must be provided. The first is to replenish those substances which humans consume and the second is to control the conditions of the environment at levels consistent with human existence. Humans consume oxygen, water, and food in the metabolism process. As a by-product of this metabolism, they generate CO_2, respired and perspired water, urine, and feces (Table 11.6.16). The provisioning of man's necessities (and the elimination of his wastes) is a primary function of life support. However, an equally important function is the maintenance of the environment in which he lives.

There are three aspects of environmental control—ambient pressure, atmospheric composition, and thermal condition of the environs. In a typical spacecraft design the ambient pressure is maintained generally in the range between ⅓ atm and sea-level pressure, with the oxygen partial pressure varying in accordance with Fig. 11.6.27 in order to provide the appropriate partial pressure of oxygen in the lungs for breathing.

The other constituents of concern in the atmosphere are the diluent, CO_2, water vapor, and trace amounts of noxious and toxic gases. The percent of the diluent in the atmosphere (or indeed whether a diluent is present at all) is selectable after proper consideration of the design parameters. For instance, short-duration spacecraft can take advantage of human adaptability and use atmospheric pressures and compositions significantly different from those at sea level, with the attendant advantage of system design simplicity. The Mercury, Gemini, and Apollo spacecrafts utilized 5-lb/in² pure oxygen as the basic cabin atmosphere. Skylab used an oxygen-nitrogen mixture although at a reduced pressure, while the shuttle is designed for an essentially sea-level equivalent at-

mosphere. When selecting the pressure and composition, it is important to remember that although a two-gas system adds complexity, it can eliminate the concerns of (1) flammability in an oxygen-rich environment, (2) oxygen toxicity of the crew, and (3) decreased atmospheric cooling potential for electronic equipment in a reduced-pressure cabin. In the selection of a diluent, nitrogen should be considered first, since its properties and effects on humans and equipment are well understood. However, other diluents, such as helium, which have characteristics attractive for a special application, are available.

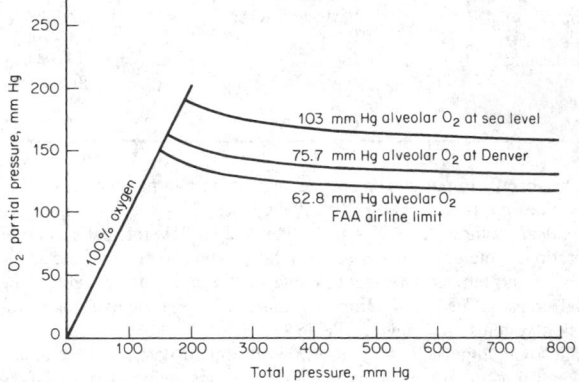

Fig. 11.6.27 Atmospheric oxygen pressure requirements for humans.

The other atmospheric constituents—CO_2, water vapor, and trace gases—must be controlled within acceptable ranges for comfort and well-being. Although the partial pressure of CO_2 is less than half a millimeter of mercury in an ambient earth environment, humans are generally insensitive to CO_2 levels up to 1 percent of the atmosphere (or 7.6 mmHg) unless mission durations increase to more than a month. As spacecraft missions are further extended, the partial pressure of CO_2 should approach the earth ambient level to ensure physiological acceptance. Atmospheric water vapor (or humidity) should be maintained high enough to prevent drying of the skin, eyes, and mucous membranes, but low enough to facilitate evaporation of body perspiration, maximizing crew comfort. The generally accepted range for design is 40 to 70 percent relative humidity. In the closed environment of the spacecraft, cabin noxious and toxic gases are of particular concern. It should be noted that 8-h/day industrial exposure limits are not appropriate for the continuous exposure which results from a sealed spacecraft cabin. Therefore, materials which can offgas should be screened and limited in the crew compartment. The small quantities which cannot be completely eliminated, as well as certain undesirable products of metabo-

Table 11.6.16 Metabolic Balance for Humans

Consumed		Produced	
Oxygen	1.84	Carbon dioxide	2.20
Drinking water	5.18	Perspiration	2.00
Food (⅔ H₂O)	3.98	Respiration	2.00
		Urine	4.53
		Feces	0.27
	11.00		11.00

NOTE: Metabolic rate = 11,200 Btu/human-day and RQ = 0.87; RQ (respiratory quotient) = vol CO_2 produced/vol O_2 consumed. All values are in units of lb/human-day.

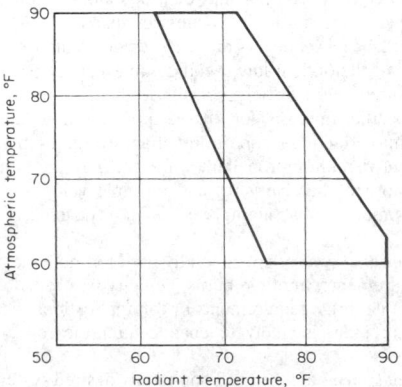

Fig. 11.6.28 Thermal comfort envelope for humans. (Note: metabolic rate 300–600 Btu/h; air velocity 15–45 ft/min; pressure 14.7 lb/in².)

lism, must be actively removed by the life-support system to maintain an acceptable atmosphere for the crew.

The third of the environmental control aspects—atmospheric thermal condition—is subject to several design variables, all of which must be taken into account. Thermal comfort is a function of the temperature and relative humidity of the surrounding atmosphere, the velocity of any ventilation present, and the temperature and radiant properties of the cabin walls and equipment. The interrelationship of these parameters is shown in Fig. 11.6.28 for a reasonable range of flight clothing ensembles. Because of the effective insulations available to the vehicle designers, the problem of thermal control in the spacecraft cabin is generally one of heat removal. Utilizing increased ventilation rates to enhance local film coefficients is one means of improving thermal comfort within the cabin. However, the higher power penalties associated with gas circulation blowers, as compared to liquid pumps, should suggest using the spacecraft thermal control liquid circuit in reducing the temperature of the radiant environment as potentially a lower-power alternative. However, if the walls or equipment are chilled below the dew point, condensation will occur.

As an additional benefit, maintaining an acceptable thermal environment through radiative means reduces the psychological annoyance of a drafty, noisy environment. Ventilation rates between 15 and 45 ft/min are generally considered comfortable, and exceeding this range should be avoided.

Life Support—Subsystems Selection

In designing the life-support system for a spacecraft, subsystems concepts must be selected to accomplish each of the required functions. This process of selection involves many design criteria. Size, weight, power consumption, reliability of operation, process efficiency, service and maintenance requirements, logistics dependency, and environmental sensitivity may each be very important or inconsequential depending on the spacecraft application. In general, for life support subsystems, the dominant selection criterion is mission duration. Various subsystem choices for accomplishing the life support functions are shown in Fig. 11.6.29.

To illustrate the relationship between mission duration and subsystem selection, the various concepts for removal of CO_2 from the atmosphere shown in Fig. 11.6.29b will be discussed. The simplest method of controlling the level of CO_2 in the spacecraft cabin is merely to purge a small amount of the cabin atmosphere overboard. This concept is simple and reliable and requires almost no subsystem hardware, although it does incur a fairly significant penalty for atmospheric makeup. For missions such as Alan Sheppard's first Mercury flight, this concept is probably optimum. However, even a minor increase in mission duration will result in a sizable weight penalty to the spacecraft. By adding a simple chemical absorption bed to the spacecraft ventilation loop (with a chemical such as lithium hydroxide as the absorbent), the expendable penalty for CO_2 removal can be reduced to the weight of the chemical canisters consumed in the absorption process. This concept was used on the Apollo spacecraft. The penalty for this type of concept also becomes prohibitive if mission durations are further extended. For missions such as Skylab, the expendable chemical absorbent can be replaced by a regenerable adsorbent. This change eliminates the lithium hydroxide weight penalty but incurs a small atmospheric ullage loss during the vacuum regeneration of the adsorbent. In addition, the regenerable system is more complex, requiring redundant beds (one for adsorption while the alternate bed desorbs), and mechanical-timing circuits and switching valves. This result is typical in that reducing expendable requirements is generally at the expense of system complexity.

The three CO_2 removal concepts discussed above all prevent the subsequent use of the collected CO_2 after removal from the atmosphere. Should the CO_2 be required for reclaiming the metabolic oxygen (such as will be the case for a space station design), the regenerable concepts can be adapted to accommodate retrieval of the CO_2 during desorption. However, this enhancement further complicates the subsystem design. There is an alternative CO_2 removal concept available which utilizes a fuel-cell-type reaction to concentrate the CO_2 in the atmosphere through

electrochemical means. This process is mechanically simple (as compared to the adsorption-desorption system) but does require more electrical power. However, it has an integration advantage in that it provides the CO_2 collected from the atmosphere mixed with hydrogen, which makes its effluent a natural feedstock for either one of several CO_2 reduction subsystems (which will be required for oxygen recovery on a space station). Although Figure 11.6.29 identifies the various life-support subsystem choices for candidate spacecraft designs of different mission durations, this information should be used only as a guide. The crew size as well as other design parameters can skew the range of applicability for each subsystem (e.g., a large crew can render expendable dependent subsystems inappropriate even for relatively short missions).

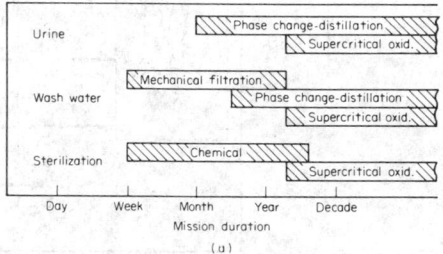

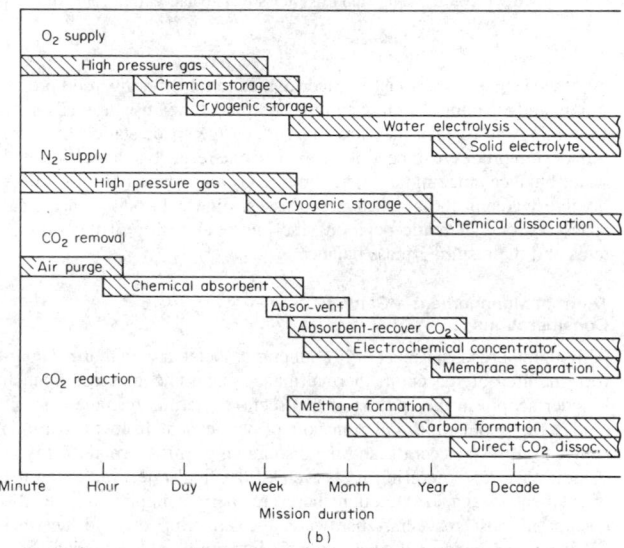

Fig. 11.6.29   Concepts to accomplish the life-support function. (a) Water recovery; (b) atmospheric revitalization.

There is another major aspect of life support system design for spacecraft (in addition to subsystem selection) and that is process integration and mass balance. For example, a space-station-class life-support system design must accommodate wastewater with a wide range of contamination levels—from humidity condensate which includes only minor airborne contamination to various wash and rinse waters (including personal hygiene, shower, clothes washing, and dish washing) and urine. The degree of contamination generally influences the reclamation process selection; therefore, multiple contaminated sources tend to result in multiple subsystem processes. For example, highly contaminated water, such as urine, is most reliably reclaimed through a distillation process since evaporation and condensation can be more effectively utilized to separate dissolved and particulate material from the water, while less contaminated wash waters can be effectively reclaimed through osmotic and mechanical filtration techniques at significantly less power penalties in watts/lb H_2O recovered. Regardless of the tech-

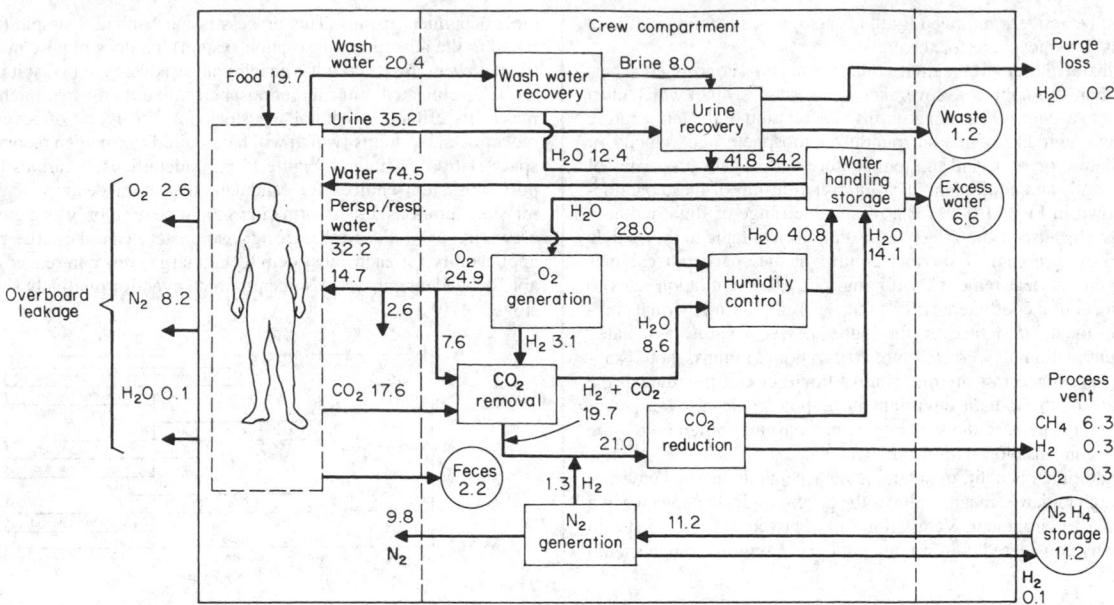

Fig. 11.6.30 Life-support systems mass balance. (All values are in pounds per day for an eight-person crew.)

niques selected, water must be made available for drinking, food preparation, and cleaning the crew and the equipment they use, as well as for generating the oxygen they breathe. In fact, the space station life-support design process can be looked upon, in one respect, as a multifaceted water balance since the dissociation of CO_2 and generation of O_2 both involve water in the respective processes. Figure 11.6.30 is a typical space station schematic depicting the interrelationship of the subsystems and the resulting mass balance.

Thermal Management—General Considerations

Although there is an aspect of life support associated with thermal comfort, the area of spacecraft thermal management encompasses a much broader scope. In the context of this section, thermal management includes the acquisition and transport of waste heat from the various sources on the spacecraft, and the disbursing of this thermal energy at vehicle locations requiring it, as well as the final rejection of the net waste heat to space. Although there are thermal implications to the design of most spacecraft hardware, the Life Support and Thermal Management section only deals with the integration of the residual thermal requirements of the various vehicle subsystems to provide an energy-efficient spacecraft design (i.e., one which utilizes waste heat to the maximum and therefore minimizes requirements for the vehicle heat rejection system).

Thermal management systems are typically divided into two parts. The first, associated with the spacecraft cabin, has design requirements associated with maintaining the crew environment and equipment located in the cabin at acceptable temperatures, and in addition has special requirements imposed on its design because of the location of the system in the pressurized compartment and its proximity to humans. These additional requirements are usually dominant. Any cooling medium which is used in the crew compartment must have acceptable toxicological and flammability characteristics in order to not impose a hazard to the crew in the event of malfunction or leakage. Although there are thermal control fluids which have low toxicity potential, reasonable flammability characteristics and adequate thermal transport properties, none are better than water.

The second aspect of the thermal management system involves those spacecraft systems located outside the cabin. The dominant design criteria in thermal control fluid selection for these systems are thermal

transport efficiency and low-temperature characteristics. The thermal transport efficiency is important since the size, weight, and power requirements of the pumping and distribution system depend on this parameter. However, in the environs of deep space the potential exists for very low temperatures. Therefore, for maximum system operating flexibility, the characteristics of the thermal transport fluid must be such that operation of the system can continue even in a cold environment; thus, low-temperature properties become a fluid selection constraint. Although the range of acceptable fluids is broader here, several fluids in the Freon family meet these requirements. For single-phase, pumped-fluid systems, the temperature levels of the heat sources are important in determining the optimum sequence for locating equipment in the thermal circuit. The equipment with the lowest-temperature heat sink requirements are accommodated by placing them immediately after the heat rejection devices, while equipment which can accept a higher-temperature heat sink are located downstream. Also, those subsystems which require heating must be located at a point in the circuit which can provide the thermal energy at the required temperature level.

Typically spacecraft heat sources include cabin-atmosphere heat exchangers, the avionics system as well as other electronic black boxes, power generation and storage equipment, etc., while heat sinks include propulsion system components, various mechanical equipment requiring intermittent operation, dormant fluid systems which have freezing potential, etc. The ultimate elimination of the net thermal energy (i.e., sources minus sinks) is by rejection to space. This is accomplished through expendables or radiation.

Recent developments in thermal management technology have resulted in the consideration of a two-phase acquisition, transport, and rejection system. This concept uses the heat of evaporation and condensation of the transport fluid to acquire and reject thermal energy, reducing by several orders of magnitude the amount of fluid required to be pumped to the various heat sources and sinks within the spacecraft. This concept has the added advantage of operating nearly isothermally, permitting the idea of a thermal "bus" to be included in future spacecraft design, with indiscriminate sequencing of the heat sources and sinks within the vehicle.

Thermal Management Subsystem Selection

The subsystem selection process associated with the thermal management area depends largely on the magnitude of thermal energy being

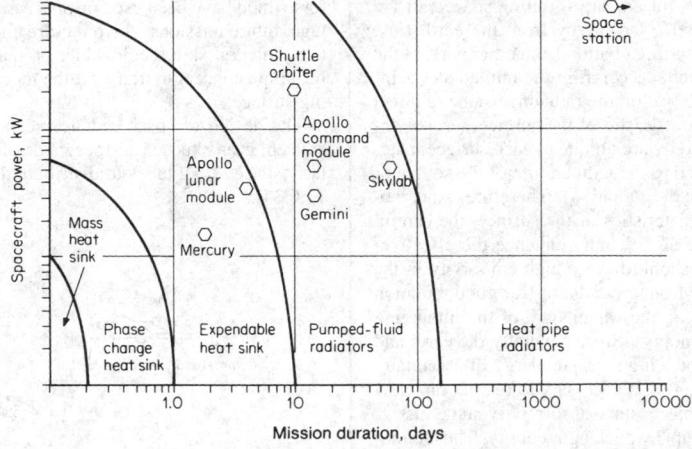

Fig. 11.6.31 Range of application of various heat-rejection methods.

managed and the duration of the mission. Figure 11.6.31 graphically depicts the areas of applicability for various types of heat-rejection devices as a function of these parameters.

Small heat-rejection requirements can sometimes be met by direct heat sinking to the structure (or a fluid reservoir). Often, passive heat sinks need to be supplemented with fusible heat sinks (to take advantage of the latent heat of fusion) or evaporative heat sinks (to take advantage of the latent heat of evaporation). Of the three types of heat sinks, the evaporative heat sink is the only one which has been used as the primary vehicle heat-rejection device in the manned spaceflight program to date. Both the Mercury spacecraft and the lunar module utilized a water evaporative heat sink in this manner. However, the Apollo spacecraft and the shuttle have also used evaporative heat sinks and Skylab used a fusible heat sink as supplemental heat rejection devices.

Since space is the ultimate heat sink for any nonexpendable heat rejection, devices which allow the thermal energy which has been collected in spacecraft coolant circuits to be radiated to space become the mainstay of the thermal management system. These "radiators" take many forms. The simplest design involves routing the cooling circuits to the external skin of the spacecraft so that the thermal energy can be radiated to deep space. Because of spacecraft configuration restrictions, only partial use can be made of the external skin of the spacecraft, and the amount of thermal energy which can be rejected by this technique is thereby limited. Although the Gemini and Apollo spacecraft success-

fully used integral skin radiators, the unique characteristics of the outer surface of the orbiter required a different technique be adopted for its radiator design. Accordingly, the radiators were designed to be deployed from inside of the payload bay doors. This not only provides good exposure to space, but also, in the case of the forward panels, allows two-sided exposure of the panels, thereby doubling the active radiating area.

Heat rejected Q_{rej}, according to the Stefan-Boltzmann law, is given by the following equation:

$$Q_{\text{rej}} = \varepsilon \gamma A T^4$$

where γ = constant, 0.174×10^{-8} Btu/(h $\cdot$ ft^2 $\cdot$ °R^4); ε = emittance property of surface (termed emissivity), A = area available for radiation; and T = radiating temperature. From this equation it is obvious that a radiator with both sides exposed will reject twice as much heat to space as one of the same dimensions which is integral with the skin of the vehicle. A space radiator can absorb as well as reject thermal energy. The net heat rejected Q_{net} from a radiator is given by the following equation:

$$Q_{\text{net}} = Q_{\text{rej}} - \sum_{N}^{A=1} Q_{\text{abs}}$$

The energy absorbed by the radiator, Q_{abs}, can come from various sources (see Fig. 11.6.32).

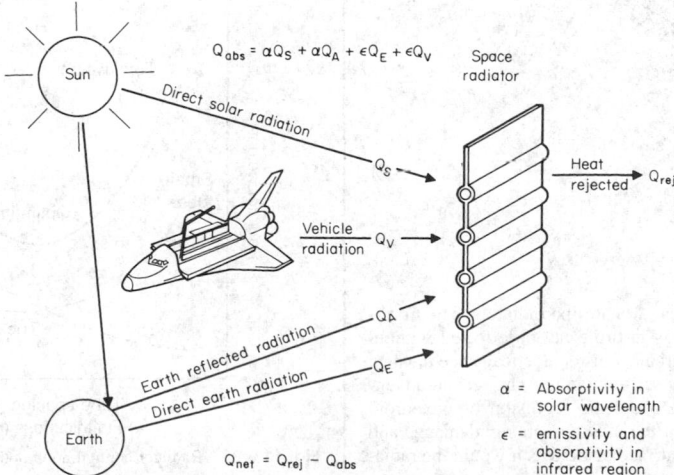

Fig. 11.6.32 Space radiator energy balance.

The most common heat sources for an earth-orbiting spacecraft are direct solar impingement, reflected solar energy from the earth (i.e., albedo), direct radiation from the earth, radiation from other parts of the spacecraft, and radiation from other spacecraft in the immediate vicinity. These energy sources are absorbed on the radiator surface in direct proportion to the absorptance characteristic of the surface, termed absorptivity. Emissivity and absorptivity are unique to each surface material but are variables directly linked to the temperature of the source of the radiation (i.e., wavelength of the radiation). Therefore, since absorptivity and emissivity are characteristics of the surface, the careful selection of a surface coating can significantly enhance the effectiveness of the radiator. The coating should have a high emissivity in the wavelength of the rejected energy from the radiator (for good radiation efficiency) and a low absorptivity in the wavelength of the impinging radiation (for minimal absorbed energy). Since generally the earth and the surrounding spacecraft are producing thermal energy of essentially the same wavelength as the radiator, it is necessary to restrict the view of the radiating surface to these objects since absorptivity and emissivity of a surface are equal for the same wavelength energy. Thus, radiators which can be located so that they are not obstructed from viewing deep space by elements of the spacecraft or other spacecraft in the vicinity will be much more effective in rejecting waste heat. The sun, however, emits thermal energy of a dramatically different wavelength and, therefore, coatings can be selected which absorb less than 10 percent of the solar energy while radiating nearly 90 percent of the energy at the emitting temperature of the radiator.

For applications which require large radiating surface areas as compared to the vehicle's size, integral skin radiators offer insufficient area and thus space radiators which use deployment schemes must be considered. However, as radiators become larger the mechanisms become very complicated. To circumvent the mechanical problems of deployment, a new radiator concept has been developed which allows in-orbit construction. This concept utilizes a single-tube heat pipe as the primary element (see Fig. 11.6.33).

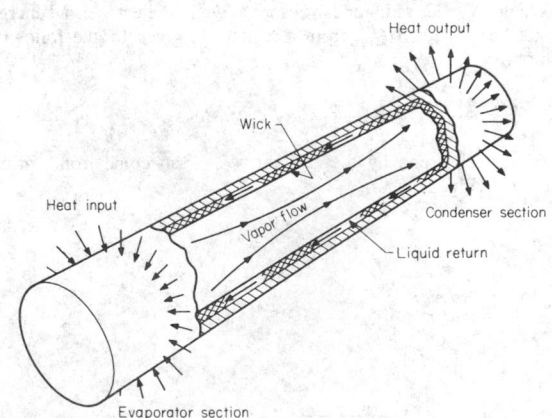

Fig. 11.6.33 Basic heat-pipe operation.

There are two distinct advantages of heat pipe radiators. The first is that assembly of large surface areas in-orbit can be effected without breaking into the thermal control circuit. This is practical since the heat pipe radiator element is completely self-contained. The second advantage is that in-orbit damage from accidental collision or meteoroid puncture affects only that portion of the radiator actually damaged and does not drain the entire radiating surface of fluid as it would the radia-

tors which have been used on spacecraft to date. Because of this advantage, future missions which require either of the three types of radiators (i.e., integral skin, deployable, or space constructible) will utilize the heat pipe concept to transport the thermal energy to the exposed radiating surface.

The design of space radiators can take many forms, but the basic concept is an exposed surface made of a conducting material with regularly spaced fluid tubes to distribute the heat over the surface (see Fig. 11.6.34).

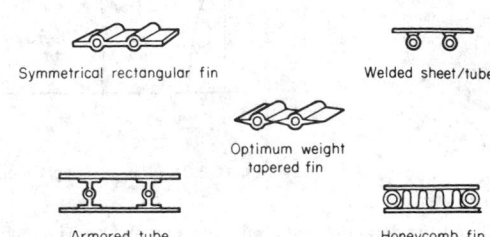

Fig. 11.6.34 Typical radiator tube/fin configurations.

The selection of the material and configuration of the radiating surface (i.e., honeycomb, box structure, thin sheet, etc.) determine its thermal conductance. This conductance, when coupled with tube spacing and the internal heat transfer characteristics within the tube and fluid system, determines the temperature gradient on the radiating surface

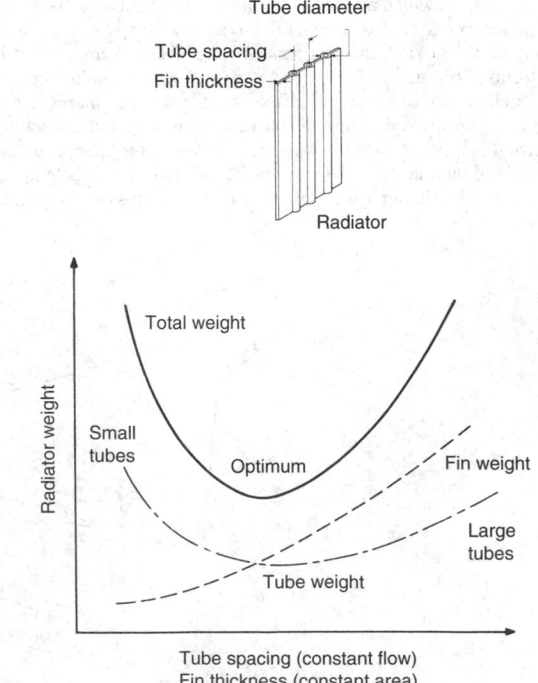

Fig. 11.6.35 Radiator configuration optimization.

between adjacent tubes. This so-called "fin effectiveness" represents a loss in radiating potential for any particular radiator design; therefore, these parameters should be selected carefully to optimize the radiator design within the constraints of each application. Figure 11.6.35 demonstrates the effect of the configuration parameters of the radiator on panel weight.

DOCKING OF TWO FREE-FLYING SPACECRAFT
by Siamak Ghofranian and Matthew S. Schmidt

Rockwell Aerospace

REFERENCES: Ghofranian, Schmidt, Briscoe, and Shliesing, "Space Shuttle Docking to MIR, Mission-1"; Ghofranian, Schmidt, Briscoe, and Shliesing, "Simulation of Shuttle/MIR Docking"; Ghofranian, Schmidt, Lin, Khalessi, and Razi, "Probabilistic Analysis of Docking Mechanism Induced Loads for MIR/Shuttle Mission"; *35th Structures, Structural Dynamics, and Materials Conf.* AIAA/ASME/ASCE/AHS/ASC. Haug, "Computer Aided Kinematics and Dynamics of Mechanical Systems," vol 1: "Basic Methods," Allyn and Bacon. Greenwood, "Principles of Dynamics," Prentice-Hall.

The information in this section covers the mechanics of docking two free-flying unconstrained vehicles in space. Different methods of mating are addressed with the emphasis applied to docking. The content does not encompass all facets of the procedure, but it does bring forth the physics of the operation which make the process an engineering challenge.

Types of Mating

The understanding of mating is a precursor to preliminary design. Different techniques of mating free-flying bodies are often discussed when planning the rendezvous of two vehicles in space. The mating of two structures can be performed in three ways: docking, berthing or a combination of both. **Docking** relies on the transfer of vehicle momentum to provide the means to force the docking ports into alignment. If a significant amount of momentum is required for capture, jet firings can be used to introduce the impulse into the system as needed. **Berthing** relies on a secondary system, such as a robotic arm to enforce the docking ports into alignment. The active vehicle grapples the other with a robot manipulator and guides the structural interfaces together. If manipulator usage is attempted, specific design factors need to be considered: arm configuration, gearbox stiffness, linkage flexibility, and controllability.

Mating of two structures in space is by every means considered a multibody dynamic event. Docking involves a significant level of momentum transfer, while berthing involves relatively little. The dynamics result from the deliberate impact of the two vehicles. As docking occurs the system responds with rigid-body motion and mechanical motion of the docking device. Depending on the relative states between the vehicles and the mode in which the maneuver is attempted, the response of the system and the physics vary considerably.

For discussion purposes the two docking vehicles can be considered as a target vehicle (passive) and a docking vehicle (active). The passive vehicle maintains attitude control as the active vehicle is navigated along an approach corridor. The active vehicle holds an attitude control mode with translational control authority. As the corridor narrows, sighting aids, range, and range-rate feedback on both vehicles are used to guide the docking ports into contact.

As the vehicles approach, a closing velocity exists. From a design standpoint the operation of a docking system involves various mechanism operational phases: mechanism deployment, interface contact and capture, attenuation of relative motion, mechanism retraction, and structural lockup. The **deployment phase** is where the mechanism is driven from its stowed position to its ready-to-dock position. Vehicle interface contact and capture are initiated when the docking ports of each vehicle are maneuvered into contact. Ideally, a positive closing velocity should be maintained from initial contact through capture. Once an opening velocity develops, the probability of a failed capture

increases. As the vehicles interact the mechanism interfaces are indexed into alignment as the system responds to the impact loads. The compliance of the system occurs in two forms: mechanism movement and rigid body vehicle motion.

When the docking interfaces interact, the system responds in six degrees of freedom and the docking system is stroked as the mechanism absorbs the relative misalignments. Once the interfaces are fully aligned, capture latches lock the opposing interfaces together, and the vehicles are "soft-docked." After capture, residual energy in the system allows relative motion between the structures to continue. Depending on the location of the docking system relative to the vehicle center of masses, the motion may dissipate due to the inherent hysteresis in the docking system; otherwise, separate energy-dissipation devices may be necessary. After attenuation the vehicles are likely to be misaligned due to the new relative equilibrium position of the spacecraft. Separate procedures or devices can be used to bring the structural interfaces into alignment.

The **retraction process** is where the active mechanism is driven back to its stowed position. Once they are fully retracted, the structural interfaces are latched at hard interfaces, and the vehicles are "hard-docked."

The mass of each vehicle is an essential part of the docking system design. The momentum of the docking vehicle is the medium which provides the force required by the docking mechanism to comply with the relative misalignments between the vehicles at the docking interface. The operation can be complicated if a vehicle center of mass is offset from the longitudinal axis of the docking system. As a result, the vehicle effective mass at the docking interface is reduced. Furthermore, the center of mass offset induces relative rotation after contact which complicates the capture and attenuation process. Ideally two vehicles should dock such that the velocity vector and longitudinal axis of the docking system have minimum offset from the center of mass. The relative rotation of the vehicles after contact is known as *jackknifing*.

The docking mechanism center-of-mass offset and the vehicle size significantly influence the design of the docking interfaces, attenuation components, docking aids, piloting procedures, and capture-assist methods. The relative vehicle misalignments and rates at the moment the docking interfaces first come into contact (initial conditions) should be known in order to size the mechanism attenuation components, to establish piloting procedures and training criteria, and to study the mechanics of the system before, during, and after docking.

The initial conditions result from vehicle as-built tolerances, thermal and dynamic distortions, variation in piloting performance, and control system tolerances. The misalignments are composed of lateral misalignments in the plane of the docking interface and angular misalignments about thee axes. The rates are composed of translational velocities along three axes and angular velocities about three axes.

The mass and motion of the system are the basic ingredients for providing the energy necessary to join the interfaces. Ideally the interfaces should capture before an opening velocity develops between the docking interfaces. Adequate momentum at the docking port is required to actuate the mechanism. The impulse introduced to the system during the impact must be less than that required to reduce the relative interface velocity to zero. If the relative interface velocity goes to zero before capture, the vehicles will separate. For low closing velocities, a small axial impulse due to initial contact can reduce the relative interface velocity to zero and cause a "no capture" condition.

A drive system along the longitudinal axis of the docking mechanism or reaction control jets increases the probability of capture. It extends the docking system toward the target vehicle once initial contact is made and thus improves capture performance for a docking vehicle with a low closing velocity. However, a drawback of a drive system is that the reaction has the tendency to push the vehicles apart. The action of jet thrusting has the opposite effect; it tends to push the vehicles together. If the momentum required to provide a low possibility of no capture becomes operationally prohibitive, thrusting can be used as a substitute and applied as necessary.

As energy introduced into the system increases, whether in the form of vehicle momentum or jets pulsing, a tradeoff exists between acceptable internal vehicle loads, allowable component stroke limits, vehicle jackknifing, and piloting capability.

Vehicle Mass and Inertia Effects

Concentrated spherical masses colliding in two dimensions can be used to describe the dynamic effects of an impact. The same class of examples are consistently used in beginning physics textbooks to describe the macroscopic concepts of impulse, momentum, and energy. The same theory should be applied to understand the requirements of mating two vehicles and perform conceptual and preliminary design studies. A single-degree-of-freedom mechanical equivalent exhibiting simple harmonic motion is a useful example for understanding the mechanics.

Consider two spacecraft in a two-dimensional plane on a planned course of docking without any misalignments and with some closing velocity. If only the docking-axis degree of freedom is considered, the effective mass of each body, M_1 and M_2, can be computed from

$$M_1 \text{ (or } M_2) = M - (ML)^2/(I + ML^2)$$

where M is the vehicle mass, I is the vehicle mass moment of inertia about the center of mass, and L is the mass center offset from the docking axis. In this case the process of capture and attenuation can be thought of as a perfectly plastic impact of two bodies. The masses dock and stick together. After impact the two bodies continue their trajectory with some final velocity. The combined effective mass of the two bodies represented as a single degree of freedom can be computed from $M_{eff} = M_1 M_2/(M_1 + M_2)$. Since momentum is conserved during a perfectly plastic impact, kinetic energy loss during docking can be computed from $\frac{1}{2} M_{eff} v^2$, where v is the capture velocity. Since at capture we know the effective mass of the system and the capture velocity, the unknown parameter becomes the work required to bring the system to rest.

Design Parameters

The energy loss during docking is equal to the work done, $F \times d$, by the mechanism-attenuation components. If a hysteresis brake is used to bring the system to rest, then a trade can be performed on the force-distance relationship by using

$$\tfrac{1}{2} M_{eff} v^2 = Fd$$

where F represents the force necessary to bring the system to rest over distance d. Using the hysteresis brake as a starting point provides the average about which other devices would oscillate. For an underdamped spring/damper attenuation system, the relative motion and maximum force can be computed from

$$x = v/\omega_d \times e^{-\zeta \omega} \sin \omega_d t \quad \text{and} \quad F = Kx + C\,dx/dt$$

respectively. In this representation,

$$\omega_d = \omega(1 - \zeta^2)^{1/2} \quad \text{and} \quad \omega^2 = K/M_{eff}$$

where ζ represents the damping factor and K and C represent the attenuation component characteristics. As can be seen, minimal knowledge of the basic system characteristics can be used to design the type and arrangement of the attenuation components. During the design of the attenuation components the following components may prove useful: preloaded centering springs, staged springs, ON/OFF dampers, and variable-level hysteresis brakes. The material in this section is only a first look at understanding the design of a docking system. As more detail is developed, the effects of relative vehicle motion, interface contact dynamics, mechanism performance, and vehicle flexibility need to be simulated, which goes beyond the scope of this section. For a comprehensive background, the reader should refer to the cited references.

11.7   PIPELINE TRANSMISSION
by G. David Bounds

REFERENCES: AGA publication "Gas Facts." Huntington, "Natural Gas and Gasoline," McGraw-Hill. Leeston, Crichton, Jacobs, "The Dynamic Natural Gas Industry," University of Oklahoma Press. Lester, Hydraulics for Pipeliners, *Oildom.* Bell, "Petroleum Transportation Handbook," McGraw-Hill. Bureau of Mines, "Mineral Yearbook," Vol. II, Fuels. "The Transportation of Solids in Steel Pipelines," Colorado School of Mines Research Foundation, Inc. Echterhoff, "Pipeline Economics Revisited," PCGA, 1981. *Petroleum Supply Monthly,* Department of Energy, Energy Information Administration, March 1993–February 1994. "US Interstate Pipelines Begin 1993 on Upbeat," *Oil and Gas Journal,* Nov. 22, 1993.

NATURAL GAS

General   According to AGA Estimates for 1991, approximately 65 percent of the U.S. natural-gas reserves occur in seven states (Texas, New Mexico, Oklahoma, Louisiana, Wyoming, Alaska, and Kansas) and only about 40 percent of the total natural gas produced there is consumed within those states. Within these seven states alone, estimated production in 1991 exceeded 16.6 trillion ft^3 (472.6 billion m^3). Total exports, interstate and international, from these states in 1991 were estimated at approximately 13.1 trillion ft^3 (371.8 billion m^3), while imports were approximately 4.5 trillion ft^3 (127.4 billion m^3). Offshore production is becoming increasingly significant. In 1991, approximately 18 percent of the total proven U.S. natural-gas reserves were located offshore.

Implementation of FERC Order 636 has had a significant impact on natural-gas transmission pipelines. Pipeline companies which built and operated gathering systems as needed to support contractual obligations are now restricted from including costs for these activities with the costs of buying, transporting, and selling gas, due to the mandate that gathering be provided as a separate service. The swing in market demand has magnified the need for storage facilities in both gathering and market areas to provide the flexibility required to operate efficiently in an open-access market. Open-access marketing has replaced bundled sales services to meet the instantaneous needs of buyers at competitive prices. When open-access marketing was first tried in the early 1980s, competitive bidding resulted in curtailment of gas flow to approximately 18 trillion ft^3 per year through a national pipeline grid that had moved 24 trillion ft^3 per year in the 1970s. Hubs are now being established to provide access to natural gas capacity release so that pipeline capacity may be more effectively utilized than ever before. Electronic bulletin boards (EBBs) are being refined to operate on an ever-expanding information superhighway to provide open access to gas supplies and markets.

The need to transport natural gas from the producing regions to consuming areas throughout the country has generated a burgeoning network of transmission pipelines. In 1950, there were 314,000 mi (505,000 km) of gas pipelines in this country. By 1991, with natural gas supplying 24 percent of the nation's energy requirements, this mileage had increased to 1,225,000 mi (1,972,000 km), an increase of 390 percent in 40 years. Between the years of 1984 and 1986, the total length of gas transmission pipeline in the United States declined by 2,500 mi (4,000 km). The significant drop in demand for pipeline steel pushed many domestic pipe mills into insolvency. Other pipe mills specialized in production of a single type of longitudinal seam, such as seamless

(SMLS), double-submerged arc weld (DSAW), or electric resistance weld (ERW) pipe. Domestic pipeline installation rebounded in 1987, with the installation of approximately 2,800 mi (4,500 km) of large-diameter steel pipeline.

Modern gas pipelines consist of all diameters from 2 in to 42 in in all grades of specified minimum yield strengths (SMYS) from 35,000 to 70,000 lb/in^2. The type of longitudinal seams available are SMLS in sizes ranging from 2 in to 20 in, ERW in sizes ranging from 2 in to 24 in, and DSAW in sizes ranging from 16 in to 42 in. Higher yield strengths are usually more cost-effective in sizes larger than 10 in.

Although the trend has been toward the use of larger pipe diameters and pipe steel of greater tensile strength, the economics of pipe production and coating and the ease of handling, including transportation as well as installation, has led to greater utilization of 36-in-diameter (0.91-m) pipe with a specified minimum yield strength of 65,000 lb/in^2 (4,570 kg/cm^2). In 1992, 36-in-diameter pipe accounted for approximately 36 percent of the total steel used in construction of large transmission pipelines. Although significant mileage of 42-in-diameter (1.07-m) pipeline has been placed in gas service, no new 42-in-diameter pipeline construction was recorded by the American Gas Association (AGA) in 1992.

Main line valves are usually ball or gate and are of the full-opening variety to allow for the passage of cleaning pigs and inspection tools. Ball, gate, and plug valves are installed on piping other than main line piping.

Pipe bends are usually made on site by a contractor; however, on short-distance pipelines, or in cases where a large degree of bend is required, shop-fabricated bends are produced by induction heating.

Repairs to a pipeline segment according to current regulations of the Department of Transportation (DOT) are to either replace the segment, or use welded steel sleeves. A new method currently being tested by both the DOT and industry is a polymer composite reinforcement wrap (Clock Spring) which is wrapped around the repair area. For this type of repair, welding is not required.

The rapidly growing network of large diameter pipelines, along with increasing pressure capabilities, has brought forth a considerable rise in prime mover and compressor requirements. Natural-gas pipelines in the United States have experienced an increase from about 6 million installed hp (4.5 million kW) in 1960 to 14 million hp (10.6 million kW) by the end of 1991, an increase of 233 percent in 30 years. Natural-gas engines, natural-gas turbines, and electric motors are most generally employed as prime movers on gas pipelines. The gas engines generally drive reciprocating compressors; gas turbines and electric-motor drivers are usually connected to centrifugal compressors. The industry trend is toward the use of larger, centrifugal compressor units because of the inherent economic advantages. Prime movers in general use have increased in size from the early 1950s to the early 1990s as follows: gas engines from 2,500 to 13,500 hp (1,865 to 10,070 kW), gas turbines from 1,000 to 50,000 hp (745 to 37,310 kW), and electric-motor drivers from 2,500 to 20,000 hp (1,865 to 14,920 kW). Modified aircraft-type jet engines are also employed as prime movers driving centrifugal compressors and range from 800 to 50,000 hp (600 to 37,310 kW).

Much progress has been made in automated operations of all types of pipelines. Valves, measuring and regulating stations, and other facilities are operated remotely or automatically. Factory-packaged compressor-station assemblies, most of which are highly automated, are now commonplace. Whole compressor stations with thousands of horsepower are operated unattended by the use of coded dispatching systems handled by high-speed communications.

Automatic supervision and operation of pipeline systems are becoming almost commonplace in today's climate of microprocessor-based remote terminal units (RTUs). These ultra-high-speed data-gathering systems can be used to gather operational data, do calculations, provide proportional integral derivative (PID) process control, and undertake other auxiliary functions. Data may be communicated to a central dispatch office for real-time control. A system of several hundred locations may be scanned and reported out in a few seconds.

Natural gas measurement stations usually consist of pressure control valves, gas measurement, by either AGA dimensioned orifice(s) or by turbine meters, and flow control valves. Overpressure protection may also be required. If a large temperature drop is anticipated across any of the pressure-reducing valves, then a radiant heater or indirect-fired heater may be required. A rough estimate of temperature drop is 7°F per 100 lb/in^2 of pressure drop. Electronic flow measurement has become commonplace, meeting the requirements of current regulations. Microprocessors calculate, remotely monitor, and control flow in real time using the latest calculation methods of the AGA and American Petroleum Institute (API) reports. The quality of the gas is also measured by one of several means. For small-volume stations a continuous gas sampler accumulates a small volume for laboratory testing. Medium volume stations will use a sampler and/or a gravitometer, while large-volume stations will use a gas chromatograph.

Flow   From the thermodynamic energy balance equations for the flow of compressible fluids (Sec. 4), the general flow formula is derived and expressed in *BuMines Monograph 6* (1935) as

$$Q = K \frac{T_0}{P_0} \left[\frac{(P_1^2 - P_2^2)d^5}{GT_f L_f} \right]^{1/2} \tag{11.7.1}$$

where Q is the rate of flow, ft^3/day; T_0 is the temperature base, °R; P_0 is the pressure base, psia; P_1 is the initial gas pressure, psia; P_2 is the final gas pressure, psia; d is the internal pipe diameter, in; G is the specific gravity of the gas, which is dimensionless (for air, $G = 1.0$); T_f is the gas flowing temperature, °R; and L_f is the length of the pipeline, mi. The general flow equation for constant-volume (steady-state), isothermal, horizontal flow of natural gas is given by *U.S. BuMines Monograph 9* (1956) as

$$Q = 38.7744 \frac{T_b}{P_b} \left[\frac{(P_1^2 - P_2^2)D^5}{fZGTL} \right]^{0.5} \tag{11.7.2}$$

where Q = rate of flow, ft^3/day; T_b = temperature base, °R; P_b = pressure base, psia; P_1 = inlet pressure, psia; P_2 = outlet pressure, psia; D = internal pipe diameter, in; f = friction factor, dimensionless; Z = average gas compressibility, dimensionless; G = gas gravity (for air, $G = 1.00$); T = average flowing temperature, °R; L = length of pipe, mi. Monograph 9 also gives a good discussion of friction factors and elevation corrections.

Other flow equations may be derived from the general equation by substitution of the proper expression for the friction factor. This applies to the Panhandle A equation which has the form

$$Q = 435.87 \left(\frac{T_b}{P_b} \right)^{1.0788} \left(\frac{P_1^2 - P_2^2}{G^{0.8539}TL} \right)^{0.5394} D^{2.6182} E \tag{11.7.3}$$

where E = pipeline flow efficiency, a dimensionless decimal fraction (design values of 0.88 to 0.96 are common). In the Panhandle A equation, the compressibility is included in the pipeline efficiency, and the other symbols are as previously defined. This formula may be solved graphically as described by C. W. Marvin, *Oil and Gas Journal*, Sept. 20, 1954.

An AGA paper, Steady Flow in Gas Pipelines (1965), gives a good general review of constant-volume gas flow equations, both with and without elevation corrections. Equations and test data are included for both the partially turbulent (Reynolds number dependent) flow regime and the fully turbulent flow regime using effective internal pipe roughness. In this reference the drag factor is introduced to account for pressure losses for fittings such as bends and valves and is applied in the partially turbulent flow regime.

When the volume of gas flowing varies with time and position (transient or unsteady-state flow), the simultaneous momentum and mass balance equations are usually solved by use of numerical methods and with the aid of computers. An AGA computer program, **Pipetran**, is available for this purpose, as are several related texts on transient flow. A paper entitled Gas Transportation System Modeling appears in the 1972 *AGA Operating Proceedings*. Gas flowing in a pipeline undergoes a number of transient, or unsteady-state, flow conditions caused by such things as prime movers starting and stopping, valves opening and closing, and flows either into or out of the pipeline. DOS-based computer programs, such as **TransFlow** (Gregg Engineering) and **GASUS** (Stoner & Associates) and others, allow the engineer to model simple as well as complex pipeline situations. These programs can be used to test proposed system expansions such as loops or additional compressor stations, to perform capacity studies, and to evaluate the effects of the transients mentioned above.

Compression The theoretical horsepower requirements for a station compressing natural gas may be calculated by polytropic compression formulas (see Sec. 4). The change of state that takes place in almost all reciprocating compressors is close to polytropic. Heat of compression is taken away by the jacket cooling and by radiation, and a small amount of heat is added by piston-ring friction. Compression by centrifugal compressors in even closer to polytropic (see Sec. 14).

For general design, the horsepower requirements for a station compressing natural gas with reciprocating compressors can be calculated as follows:

$$\text{Reciprocating station horsepower} = \text{hp}Z_s \frac{T_s}{520} \quad (11.7.4)$$

The value for hp can be taken from Fig. 11.7.1, typical manufacturer's curve; T_s is suction temperature, °R; and Z_s can be taken from Fig. 11.7.2.

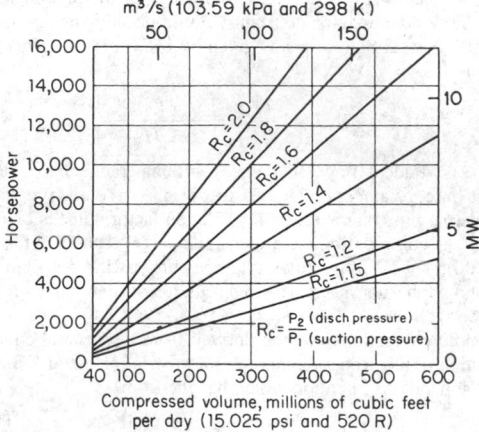

Fig. 11.7.1 Reciprocating-compressor horsepower (MW) graph. $R_c = P_2/P_1 =$ discharge pressure, lb/in² abs ÷ suction pressure, lb/in² abs.

Similarly, the horsepower requirements for a station compressing natural gas with centrifugal compressors can be calculated as follows:

$$\text{Centrifugal station horsepower} = \frac{\text{hp}Z_s}{E_c} \frac{T_s}{520} \quad (11.7.5)$$

The value for hp can be taken from Fig. 11.7.3, typical centrifugal horsepower curve; E_c is the centrifugal-compressor shaft efficiency. (Design shaft efficiencies of 80 to 85 percent are common.)

Compression ratios across a single machine normally do not exceed 3.0 because of outlet temperature limitations. For greater ratios, compressors can be arranged in series operation with an interstage gas cooler placed between the individual units. Multiple units can also be placed in parallel operation in order to increase flexibility and provide for routine maintenance.

Design The pipe and fittings for gas transmission lines are manufactured in accordance with the specifications of the API, ANSI, ASTM, and MSS (see also Sec. 8). Interstate pipelines are installed and operated in conformance with regulations of the U.S. Department of Transportation (DOT). The ANSI B31.8 Code and the *Gas Measure-*

ment Committee Report 3 of the AGA provide supplemental design guidelines. The equation for pipeline design pressure, as dictated by the DOT in Part 192 of Title 49, Code of Federal Regulations, is

$$P = \frac{2St}{D} FET \quad (11.7.6)$$

where $P =$ design pressure, psig; $S =$ specified minimum yield strength, lb/in²; $t =$ nominal wall thickness, in; $D =$ nominal outside

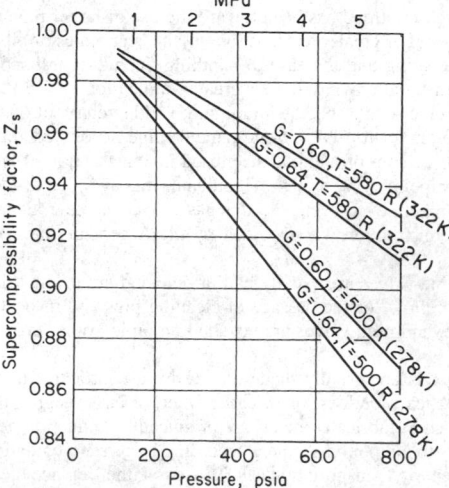

Fig. 11.7.2 Supercompressibility factor Z for natural gas. (Natural gas, 1 percent N_2.)

diameter, in; $F =$ design factor based on population density in the area through which the line passes, dimensionless; $E =$ longitudinal joint factor, dimensionless; and $T =$ temperature derating factor (applies above 250°F only), dimensionless. For offshore lines and cross-country pipelines in areas where the population density is very low, the factor F is 0.72. For seamless, double-submerged arc weld or electric resistance weld pipe with an E value of 1.00 and an operating temperature below 250°F in these low-population density areas, the equation can be written

$$P = 1.44St/D \quad (11.7.7)$$

The selection of the diameter, steel strength, and wall thickness of a line and the determination of the optimum station spacing and sizing are

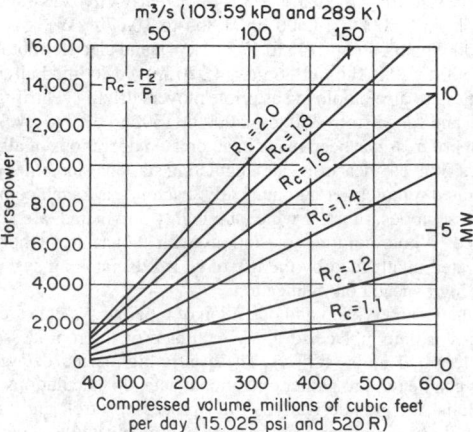

Fig. 11.7.3 Centrifugal-compressor horsepower (MW) graph. $R_c = P_2/P_1 =$ discharge pressure, lb/in² abs ÷ suction pressure, lb/in² abs.

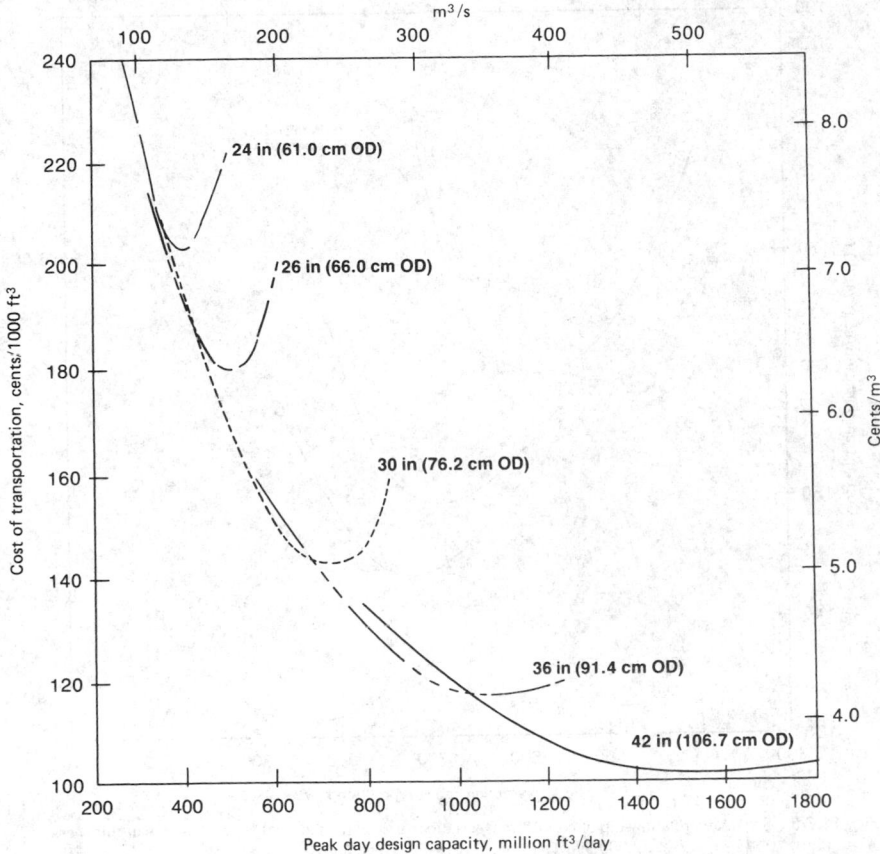

Fig. 11.7.4 Schematic relationship of cost of transportation versus design capacity for a given length of line with various pipeline diameters at equal design pressures.

based on transportation economics. Figures 11.7.4 through 11.7.6 show schematically methods of analysis commonly used to relate design factors to the cost of transportation. A perspective of various elements of the cost of transportation for a long-distance, large-diameter, fully developed pipeline system is presented in Fig. 11.7.7.

The cost of installed pipelines varies greatly with the location, the type of terrain, the design pressure, the total length to be constructed, and many other factors. In general, figures from $23,000 to $38,000 per inch (per 2.54 cm) of OD represent the range of approximate costs per mile (per 1.61 km) of installed pipeline. Offshore pipeline installed costs vary with such items as length, water depth, bottom conditions, pipeline crossings, depth of bury, and many other factors. These costs range from $38,000 to $54,000 per inch (per 2.54 cm) of OD per mile (per 1.61 km) for a typical offshore pipeline. Compressor-station installed costs vary widely with the amount and type of horsepower to be installed and with the location, the type of construction, and weather conditions. The installed costs of gas-turbine stations range from $900 to $1,600 per installed hp ($1,206 to $2,145 per kW); gas-engine stations, from $1,000 to $1,800 per hp ($1,340 to $2,413 per kW); and electric-motor stations, from $600 to $1,500 per hp ($800 to $2,000 per kW).

After a pipeline system is constructed and fully utilized, expansion of delivery capacity can be accomplished by adding pipeline loops (connecting segments of line parallel to the original line), by adding additional horsepower at compressor stations, or by a combination of these two methods.

CRUDE OIL AND OIL PRODUCTS

General   Approximately 91 percent of the U.S. proven liquid hydrocarbon reserves are situated in the seven states of Alaska, California, Louisiana, New Mexico, Oklahoma, Texas, and Wyoming or in the marine areas off their coasts. Offshore production is becoming increasingly significant. The refinement of multiphase pumping systems has enhanced crude oil recovery from formerly marginal offshore fields. In 1991, approximately 11 percent of the total proven crude oil reserves were located in federal waters offshore of California, Louisiana, and Texas. The major markets in which petroleum products are consumed are remote from the proven reserves. About one-half of all petroleum consumed in the United States is imported and must be moved from ports to the centers of consumption. Eighty-nine percent of all movements of petroleum and petroleum products is made by pipelines. The interstate pipelines alone transport over 11.4 billion bbl (1.82 billion m³) of petroleum a year. There are about 163,000 mi (262,300 km) of operating interstate liquid-petroleum trunk lines in the United States.

Centrifugal pumps have been used extensively. Generally, these units are multistaged, installed in series or parallel, depending on the application. Many operators install smaller multiple units instead of larger single units to improve operating efficiency and flexibility.

Reciprocating pumps are installed in parallel; i.e., a common suction and a common discharge header are utilized by all pumps. There are instances where rotary, gear, and vane-type pumps have been employed to fulfill specific requirements.

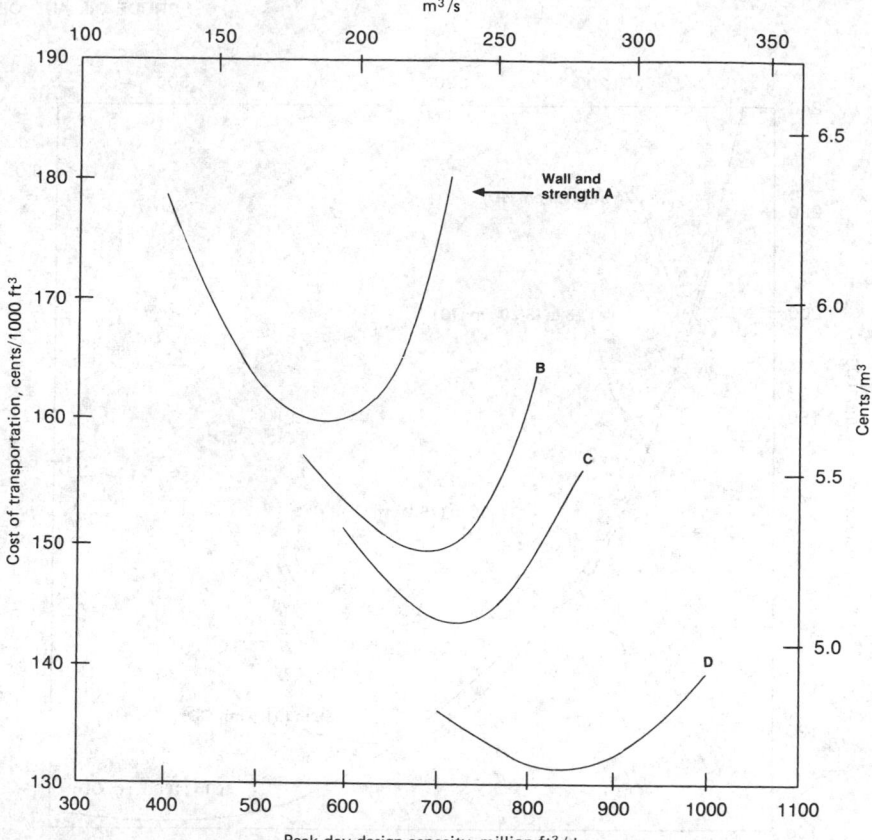

Fig. 11.7.5 Schematic relationship of cost of transportation versus pipeline steel strength and wall thickness.
A = 30 in × 0.312 in, WT × 52,000 lb/in^2 min. yield (779-lb/in^2 operating pressure).
B = 30 in × 0.344 in, WT × 56,000 lb/in^2 min. yield (923-lb/in^2 operating pressure).
C = 30 in × 0.350 in, WT × 60,000 lb/in^2 min. yield (1,008-lb/in^2 operating pressure).
D = 30 in × 0.385 in, WT × 65,000 lb/in^2 min. yield (1,200-lb/in^2 operating pressure).

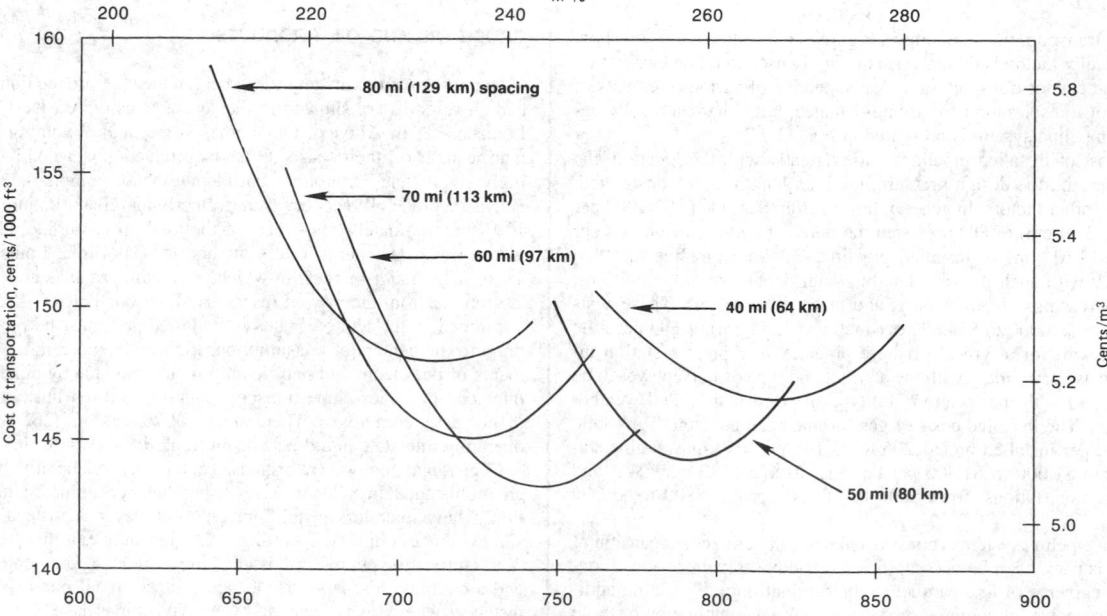

Fig. 11.7.6 Schematic relationship of cost of transportation versus design capacity for a given pipeline diameter and length, with various spacings of compressor stations.

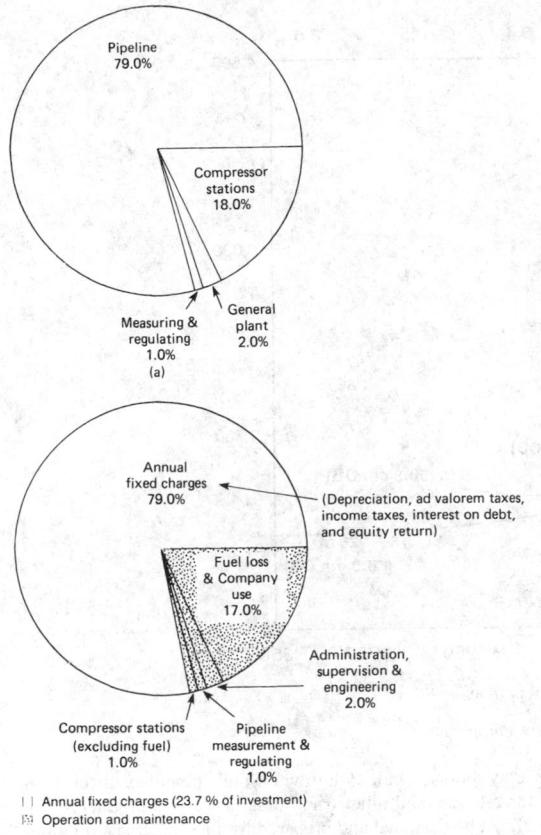

Fig. 11.7.7 Analysis of (a) investment and (b) cost of transportation.

Electrically driven centrifugal pumps are now widely employed for the liquid pipeline industry because they are environmentally friendly and because of their compactness, low initial costs, and ease of control. Centrifugal pumps driven by gas-turbine engines have also come into routine use.

Positive-displacement and turbine meters are in general use on products systems (see Sec. 16). Totalizers for multimeter installations and remote transmission of readings are an industry standard. Greater application of meters with crude-oil gathering facilities is being made as automatic-custody units are installed.

Flow The **Darcy formula,**

$$h_f = fLV^2/D2g \qquad (11.7.8)$$

where h_f = friction loss, ft; f = friction factor (empirical values), dimensionless; L = length of pipe, ft; D = ID, ft; V = velocity flow, ft/s; and g = 32.2 ft/s², is in general use by engineers making crude-oil pipeline calculations. The Darcy formula, when stated in a form utilizing conventional pipeline units, is

$$P = 34.87fB^2S/d^5 \qquad (11.7.9)$$

where P = friction press drop, lb/(in² · mi); f = friction factor (empirical values), dimensionless; B = flow rate, bbl/h (42 gal/bbl); S = specific gravity of the oil, dimensionless; d = ID, in.

The **Williams and Hazen formula,** which has wide acceptance for product pipeline calculations, can be stated in a form employing conventional pipeline terms as follows:

$$P = \frac{2340B^{1.852}S}{C^{1.852}d^{4.870}} \qquad (11.7.10)$$

Table 11.7.1

Pump-station installed hp, total	Cost per installed hp (1hp = 0.746 kW)
250–500	$760–800
500–750	$560–600
750–1,000	$440–480
1,000–2,000	$310–350
Above 2,000	$275

where P = friction press drop, lb/in² per 1,000 ft of pipe length; B = flow rate, bbl/h; S = specific gravity of the oil product; C = friction factor, dimensionless; and d = ID, in. The friction factor C includes the effect of viscosity and differs with each product (C for gasoline is 150; for no. 2 furnace oil, 130; for kerosene, 134).

The brake horsepower for pumping oil is $RSh/3960E$ or $BP/2450E$, where R = flow rate, gal/min; S = specific gravity of the oil; h = pump head, ft; E = pump efficiency, decimal fraction; B = flow rate, bbl/h; and P = pump differential pressure, lb/in².

Design The pipe and fittings for oil transmission lines are manufactured in accordance with specifications of the API and are fabricated, installed, and operated in the United States in conformance with Federal Regulations, DOT, Part 195 (see also Sec. 8). Many factors influence the working pressure of an oil line, and the ANSI B31.4 code for pressure piping is used as a guide in these matters. A study of the hydraulic gradient, land profile, and static-head conditions is made in conjunction with the selection of the main-line pipe (Sec. 3).

Determining the pipe size and station spacing to transport a given oil or oil product at a specified rate of flow and to provide the lowest cost of transportation is a complex matter. Usually the basic approach is to prepare (1) a series of pipeline cost estimates covering a range of pipe diameters and wall thicknesses and (2) a series of station cost estimates for evaluation of the effect of station spacing. By applying capital charges to the system cost estimates and estimating the system operating costs, a series of transportation-cost curves similar to those in Fig. 11.7.8 can be drawn.

Estimates of $20,000 to $40,000 per in (per 2.54 cm) of OD represent the cost per mile (1.61 km) of installed trunk pipeline. The tabulation of pump-station investment costs for automated electric-motor, centrifugal-type installations presented in Table 11.7.1 can be used for preliminary evaluations.

Careful design and good operating practice must be followed to minimize contamination due to interface mixing in oil-product pipelines. When the throughput capacity of an oil line becomes fully utilized, additional capacity can be obtained by installing a parallel pipeline or a partial-loop line. A partial loop is a parallel line which runs for only part of the distance between stations. In the case of product lines, careful attention must be paid to the design of the facilities where the loop line is tied into the original pipeline to prevent commingling of products.

SOLIDS

General Solids may be transported by pipeline by a number of different methods. In *capsule pipelining,* dry solids are placed in a circular capsule which is then pumped down the pipeline. This method can have economic and other advantages in many cases, especially where water is scarce, because the water is never contaminated and can be returned easily in a closed loop.

Another method being developed for the transport of coal is by *coal logs.* Pulverized coal is mixed with a binder (usually petroleum-based) to form a solid log, designed to resist water absorption, that can be transported down the pipeline intact. No dewatering is required with this method of coal transportation and because the binder is combustible it enhances burning.

The most prevalent method of transporting solids by pipeline is by slurry. A slurry is formed by mixing the solid in pulverized form with a true liquid, usually water, to a consistency which can be pumped down the pipeline as a liquid. Nonvolatile gases, such as carbon dioxide, may

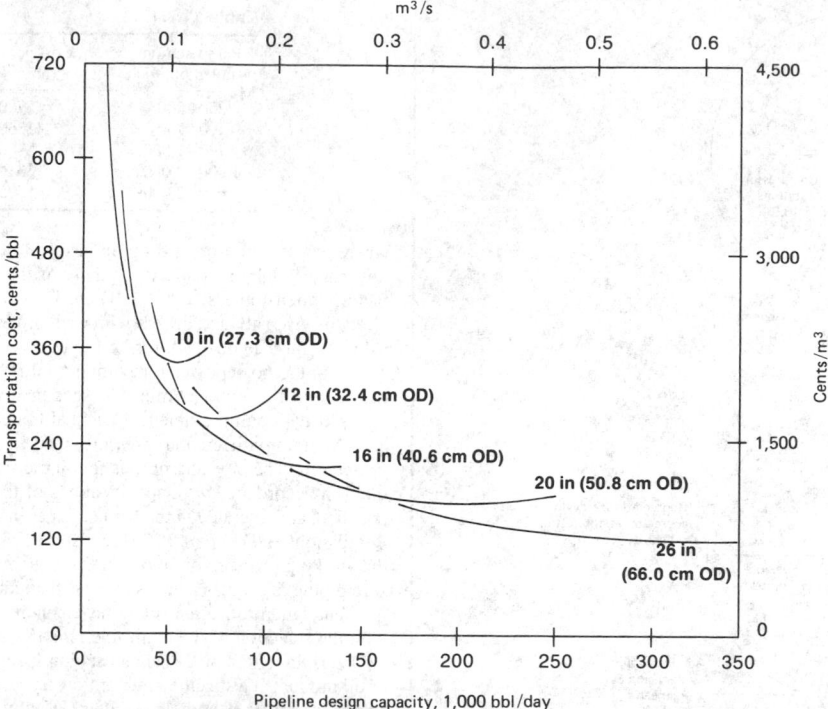

Fig. 11.7.8   Schematic relationship of pipeline capacity to cost of transportation.

also be used as the carrying fluid. In some instances, the carrying fluid may also be marketed after separation from the solid.

Slurry Transport   Some of the solids being transported in slurry form considerable distances by pipeline are coal, coal refuse, gilsonite, phosphate rock, tin ore, nickel ore, copper ore, gold ore, kaolin, lime-stone, clay, borax, sand, and gravel. Solids pipelines differ from pipelines for oil, gas, and other true fluids in that the product to be transported must be designed and prepared for pipeline transportation. After much research in the field of slurry hydrotransportation, methods for predicting the energy requirements for pipeline systems have vastly

Table 11.7.2   Summary of Selected Slurry Pipelines

Location	Length,* mi	Diameter,† in	Capacity,‡ millions of tons/year	Concentration, % by weight
Coal slurry				
Black Mesa, Arizona	273	18	4.8	45–50
Consolidation, Ohio	108	10	1.3	50
Belovo–Novosibirsk, Russia	155	20	3.0	60
Limestone slurry				
Rugby, England	57	10	1.7	50–60
Calaveras, California	17	7	1.5	70
Gladstone, Australia	18	8	1.8	62–64
Trinidad	6	8	0.6	60
Copper concentrate slurry				
Bougainville, New Guinea	17	6	1.0	56–70
KBI, Turkey	38	5	1.0	45
Pinto Valley, Arizona	11	4	0.4	55–65
China	7	5	0.6	60
Iron concentrate slurry				
Savage River, Tasmania	53	9	2.3	55–60
Samarco, Brazil	245	20	12.0	60–70
Kudremukh, India	44	18	7.5	60–70
La Perla–Hercules, Mexico	53/183	8/14	4.5	60–68
Phosphate concentrate slurry				
Valep, Brazil	74	9	2.0	60–65
Chevron Resources, U.S.A.	95	10	2.9	53–60
Goiàsfertil, Brazil	9	6	0.9	62–65
Kaolin clay slurry				
Huber, U.S.A.	21	6	0.7	40
Cornwall, England	25	10	2.9	55

* 1 mi = 1.61 km.
† 1 in = 25.4 mm.
‡ 1 ton = 907 kg.

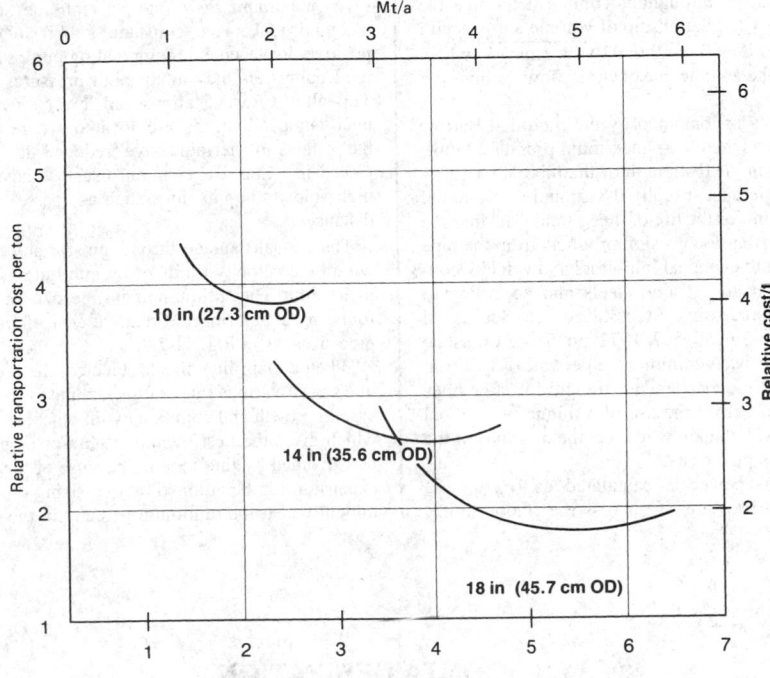

Fig. 11.7.9 System relative transportation cost for various main-line pipe sizes.

improved and slurries are transported today at much higher concentrations than 10 years ago. In the early 1980s, slurry designs ranging from 45 to 58 percent concentration, by weight, were common. Today, ultrafine coal-water mixtures with concentrations, by weight, ranging up to 75 percent, or more, are used. Dispersants and chemical additives, such as **Carbogel** and **Densecoal** are used to reduce drag and facilitate pumping. These additives significantly reduce the operating costs of slurry pipelines. Additional savings are realized by direct burning of these high-concentration coal-water mixtures through a nozzle, since dewatering is not required.

Current trends in the construction of solids pipelines and those proposed for construction indicate that future transportation of minerals and other commodities will bear more consideration. Some of the commercially successful slurry pipelines throughout the world are listed in Table 11.7.2.

Advances in slurry rheology studies and technology related to homogeneous and heterogeneous slurry transport have made the potential of slurry pipelining unlimited. Immediate advantages for slurry pipelines exist where the solids to be transported are remotely located; however, each potential slurry transport project possesses some other advantages. Materials must satisfy the following conditions to be transported by long-distance pipelines: (1) the largest particle size must be limited to that which can readily pass through commercially available pumps, pipes, and other equipment; (2) solids must mix and separate easily from the carrying fluids or be consumable in the slurry state; (3) material degradation must be negligible or beneficial to pipelining and utilization; (4) material must not react with the carrying fluid or become contaminated in the pipeline; (5) the solids-liquid mixture must not be excessively corrosive to pipe, pumps, and equipment; (6) solids must not be so abrasive as to cause excessive wear at carrying velocities.

Coal Slurry Uncleaned coal can be transported by pipeline, but a more stable and economic long-distance pipeline operation can be achieved with clean coal. The use of clean coal allows a variety of supply sources; produces a slurry with a lower, more uniform friction-head loss; reduces the pipe-wall wear; and increases the system capacity. Cleaning pipeline coal costs considerably less than cleaning coals

for other forms of transportation since drying can be eliminated after the cleaning operation.

Coal-slurry preparation normally consists of a wet grinding process for reducing clean coal to the proper particle size and size-range distribution. The final slurry-water concentration can be adjusted before it enters the pipeline with quality control of the slurry holding tanks. The cost of slurry preparation is only slightly affected by changes in concentration or size consist. The normal slurry-preparation-plant practice is to hold the concentration within 1 percent of design to provide a slurry with a uniform pressure drop. Continued research has made it possible to predict accurately the anticipated pressure drop in a pipeline by using a laboratory and computer correlation.

Optimum characteristics for a slurry are (1) maximum concentration, (2) low rate of settling, and (3) minimum friction-head loss. Variables which can be manipulated to control and optimize slurry design are particle shape, size distribution, concentration, and ash content. Established computer techniques are used to evaluate these variables.

Design Because of the many slurry variations, there is no single formula available by which the friction-head loss can be readily determined for all slurry applications. A general approach is described by Wasp, Thompson, and Snoek (The Era of Slurry Pipelines, *Chemtech.*, Sept. 1971, pp. 552-562), and by Wasp, Kenny, and Gandhi in their book, "Solid-Liquid Flow Slurry Pipeline Transportation," giving calculation methods which have been developed for predicting slurry pressure drops. The described methods are quite accurate and can be used for preliminary design. Additional calculation procedures and methods for specifying slurry size consist (solid size-range distribution) and velocity have been developed by continuing research and may be found in the following books: Shook and Roco, "Slurry Flow Principles and Practice," Butterworth Heinemann, 1991, and Wilson, Addie, and Clift, "Slurry Transport Using Centrifugal Pumps," Elsevier, 1992. Slurry hydraulics should be verified with bench-scale tests and compared with commercial operating data reference points or pilot-plant operations before constructing a full-scale long-distance pipeline.

Slurry velocities are generally limited to a minimum critical velocity to prevent solids from settling to the bottom of the pipe and a maximum velocity to prevent excessive friction-head loss and pipe-wall wear.

These velocity limitations and the throughput volume determine the diameter of the mainline pipe. The relationship of volume and velocity to the pipe diameter is expressed as ID = $0.6392 \sqrt{(\text{gal/min})/V}$, where the ID is in inches, gal/min is the volume rate of slurry flow (gallons per minute), and V is flow velocity, ft/s.

Pipe-wall thickness in inches is computed by the modified Barlow formula as $[(PD/2SF) + t_c]/t_m$, where P = maximum pressure, lb/in^2; D = pipe OD, in; S = minimum yield strength of the pipe, lb/in^2; F = safety factor; t_m = pipe mill tolerance, usually 0.875; and t_c = internal erosion/corrosion allowances, in, for the life of the system. The internal metal loss results from the erosion of corrosion products from the pipe wall and thus can be limited with chemical inhibitors and velocity control. This metal loss occurs in all pipeline steels and according to Cowper, Thompson, et al., Processing Steps: Keys to Successful Slurry-Pipeline Systems, *Chem. Eng.*, Feb. 7, 1972, pp. 58–67, must be evaluated on a case-by-case basis. A minimum pipe wall of 0.250 in (0.635 cm) is often used to provide mechanical strength for large lines even where pressures and wear permit the use of a thinner wall. Coal pipelines employ a graduated wall thickness to meet the demands of the hydraulic gradient, which saves pipe costs.

Pump-station hydraulic horsepower is calculated as $(P_d - P_s)(Q)/1,716$ where $(P_d - P_s)$ = pressure rise across the station lb/in^2, Q = maximum flow rate through the station in gallons per minute, and 1,716 is a constant, gal/(ft·in^2). Stations usually employ positive-displacement reciprocating pumps of the type used for oil-field mud pumps, as they attain high pressures with low slurry velocities. This allows fewer stations and lower overall costs than with centrifugal pumps. Stations are located where the friction-head loss and the effects of terrain have reduced the pressure head to 100 ft (30.48 m). Centrifugal pump usage is advantageous in low-pressure, high-velocity applications such as coarse coal transportation for short distances.

The annual tonnage throughput for any diameter line can be varied within the velocity limits or by intermittent operation near the low velocity limit. The optimum main-line size for high-load-factor systems is found by comparing the relative cost of transportation for a series of pipe sizes, as in Fig. 11.7.9.

When a main-line size is selected, the cost of transportation for that line size at various rates of throughput is weighed against the projected system growth and compared with other line sizes to determine the size which gives the best overall system economics. Coal pipelines cannot be expanded by line looping because of the velocity limitations. Some expansion can be attained by oversizing the line and operating it intermittently near the minimum velocity in the earlier years of life.

11.8 CONTAINERIZATION

(Staff Contribution)

REFERENCE: Muller, "Intermodal Freight Transportation," 2d ed., Eno Foundation for Transportation.

Shipping of freight has always suffered from pilferage and other losses associated with the handling of cargo and its repeated exposure to weather conditions and to human access. In addition, the extremely labor-intensive loading and unloading practices of the past became more and more expensive. In the late 1920s and early 1930s the forerunners of present-day container service were offered. They varied from containers on trucks to services in which railroad cars were loaded onto rubber-tired trailers and delivered to the customer's door with no intermediate handling of the cargo. Starting in 1956, containerization as we know it today came into being.

The immense savings in shipping costs resulted in the rapid growth of containerization throughout the world in the 1960s and 1970s and in international standardization. The time savings obtainable through the use of containers is particularly noticeable in maritime service. It is estimated that a container-handling ship can achieve a 4 : 1 sea-to-port time ratio compared to 1 : 1 for a conventional break-bulk ship.

CONTAINER SPECIFICATIONS

There are many **standards** which apply to containers, among which are ISO, ANSI, the International Container Bureau (BIC), and the American Bureau of Shipping (ABS).

Types Ninety percent of containers are dry, nonrefrigerated units with double doors on one end. Other types include refrigerated and/or insulated units, ventilated units, tank containers, and frames with folding ends or corner posts for carrying large items such as automobiles.

Condensation can be a problem inside a closed container. While a good unit can protect against outside elements, it cannot protect against condensation inside the container unless special steps are taken. Condensation typically arises from the moisture in the product being shipped; temperature changes during the trip; or moisture from dunnage, packing, or the container itself.

Construction materials usually consist of a steel, aluminum, or fiberglass outer cover on an internal steel frame. Containers are designed to be stacked up to six units high. Since containers are designed with oceangoing shipment in mind where the container, and those stacked on top of it, will be rising and falling with the movement of the ship, the additional acceleration forces have to be taken into account. This is usually done by rating the maximum stacking load using a g force of 1.8.

Figure 11.8.1 shows a typical 20-ft container. At each of the eight corners there are special fittings which facilitate the lifting, stacking, interconnecting, and tying down of units. These fittings have specially shaped slots and holes which enable standard crane and sling hooks and clevises to be used as well as special hooks and twist-locks.

Sizes The most widely used containers are 20 ft (6.1 m) or 40 ft (12.2 m) long by 8 ft (2.5 m) wide by 8 ft (2.5 m) high. The usual height of 102 in (2.6 m) includes an 8-in-deep (200-mm) underframe. This frame often has two, or sometimes four, transverse slots for forklift forks. Standard lengths are: 20, 35, 40, 45, and 48 ft (6.1, 10.7, 12.2, 13.7, and 14.6 m).

Typical specifications for a 20-ft (6.1-m) container are: maximum gross weight of 52,910 lb (24,000 kg), tare of 5,070 lb (2,300 kg), payload of 47,840 lb (21,700 kg), cubic capacity of 1,171 ft^3 (33.2 m^3), and allowable stacked weight for a g force of 1.8 of 423,280 lb (192,000 kg).

Typical specifications for a 40-ft (12.2-m) container are: maximum gross weight of 67,200 lb (30,480 kg), tare of 8,223 lb (3,730 kg), payload of 58,977 lb (26,750 kg), allowable stacked weight for a g force of 1.8 of 304,170 lb (182,880 kg), cubic capacity of 2,300 ft^3 (65 m^3). Thus a 40-ft (12.2-m) container has roughly the same volume as a 40-ft (12.2-m) highway semitrailer. A typical 40-ft (12.2-m) dry-cargo container with an aluminum outer skin weighs about 6,000 lb (2,720 kg).

Specifications for a typical 40-ft (12.2-m) **refrigerated** container are: maximum gross weight of 67,200 lb (30,480 kg), tare of 8,770 lb (3,980 kg), payload of 58,430 lb (26,500 kg), allowable stacked weight for a g force of 1.8 of 423,307 lb (192,000 kg), cubic capacity of 2,076 ft^3 (58.8 m^3). The refrigeration system specifications are: minimum inside temperature of 0°F (-18°C), maximum outside temperature of 100°F (38°C). The refrigeration system is powered through cables connected to the loading site, or to a generator mounted in the bed of the container chassis or trailer.

Some companies have a sufficiently large volume of shipping to have containers differing from these specifications. For example, the refrigerated containers used to ship Chiquita bananas are 43 ft (13.1 m) long. Their specifications are: maximum gross weight of 67,200 lb (30,480 kg), tare of 8,270 lb (3,750 kg), payload of 58,930 lb (26,730 kg), allowable stacked weight for a g force of 1.8 is 335,980 lb (152,400 kg), and a cubic capacity of 2,324 ft³ (65.8 m³). The height of the floor above the bottom of the underframe is 11 in (280 mm). The refrigeration rating is 16,721 Btu/h (4,214 kcal/h). Electric power supplied is three-phase at 460 V, 60 Hz (or 380 V, 50 Hz). The overall heat-transfer rate is 7,965 Btu/h (2,007 kcal/h).

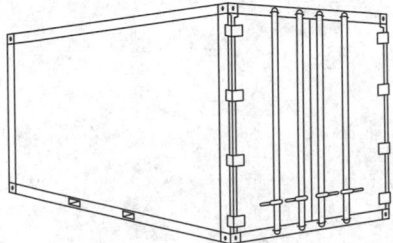

Fig. 11.8.1   Typical 20-ft container.

ROAD WEIGHT LIMITS

The gross weight of a typical 40-ft (12.2-m) dry-cargo container is about 67,000 lb (30,000 kg). The special-purpose semitrailer for carrying containers on the highway is called a **container chassis** and weighs about 10,000 lb (4,500 kg). The towing tractor weighs about 16,000 lb (7,256 kg). Many U.S. highways do not allow such heavy loads. Typical allowable axle loads are 12,000 lb (5,442 kg) on the front wheels, 34,000 lb (15,420 kg) on the tractor rear wheels, and 34,000 lb (15,420 kg) on the chassis wheels. So if the container is to be forwarded by truck in the United States, the load must be limited to 54,000 lb (24,500 kg) and, on many secondary roads, even less. Similar constraints exist elsewhere in the world as well. As with any closed shipping space, depending on the density of the material being shipped, the weight or the volume may be the controlling factor.

CONTAINER FLEETS

As of 1988 about half of the world's containers were owned by container-leasing companies. Most of the remainder were owned by carriers. About four percent were owned by shippers. Container shipment figures and fleet sizes are usually give in TEUs (twenty-foot equivalent units). A **TEU** is the equivalent of a 20-ft (6.1-m) by 8-ft (2.5-m) by 8-ft (2.5-m) unit.

Containers are easy to make. As a result the competition among manufacturers is intense. The low-cost manufacturers are in Japan, Taiwan, the Peoples Republic of China, and South Korea. Prices as low as $2,000 for a 20-ft (6.1-m) container were current in 1984.

CONTAINER TERMINALS

A typical marine container terminal may have 50 acres (20 hectares) of yard space. It can load and unload 150 to 200 ships a year, and in so doing handle 60,000 to 80,000 containers. Under such circumstances the requirements for record keeping and fast, efficient handling are very severe. Specialized container-handling cranes and carriers have been developed to facilitate safe, fast, damage-free handling.

Handling Equipment   By its design and purpose, the container is intended to go from shipper to receiver without any access to the contents. The container with its load may, however, travel by truck trailer from the shipper to the railroad, by rail to a marine terminal, by ship across the ocean, and again by rail and by truck trailer to the receiver. While the container is on the truck trailer, rail car, or ship, little or no handling needs to be done. When a container changes from one mode of transportation to another, at intermodal or transfer points, the container requires handling.

The general transfer tasks of **container-handling machinery** are: ship to/from rail, ship to/from truck, and rail to/from truck. Since a ship may have a capacity of up to 4,400 20-ft containers, extensive yard storage space has to be available. A 50-car double-stacked railroad train can handle up to 200 containers. Container-handling equipment varies from a simple crane with some slings or a large forklift truck at low-volume sites to special-purpose gantry and boom cranes and straddle carriers at high-volume sites.

Terminal designs come in two basic types: decked and wheeled systems. With a **wheeled** system, the containers are stored on their highway trailers. They are transferred to/from the loading crane by terminal tractors, which are also called *yardhorses*. In a **decked** system, the containers are removed from their highway trailers, rail cars, or ships and stacked in rows on the paved yard surface. When the containers are to be loaded, they are picked up at the stacked location, if possible by straddle carriers or mobile cranes, and loaded onto outgoing highway trailers, railcars, or ships. When the full area of the container storage yard is not accessible to the cranes, containers may be loaded onto yard trailers and transported from their stack location to the cranes. An equivalent procedure is also used for incoming containers where they must be transferred by yard trailer from the cranes to their yard storage location.

Section **12**

Building Construction and Equipment

BY

VINCENT M. ALTAMURO   *President, VMA, Inc.*
AINE M. BRAZIL   *Vice President, Thornton-Tomasetti/Engineers*
WILLIAM L. GAMBLE   *Professor of Civil Engineering, University of Illinois at Urbana-Champaign*
NORMAN GOLDBERG   *Consulting Engineer*
ABRAHAM ABRAMOWITZ   *Consulting Engineer; Professor of Electrical Engineering, Emeritus,
   The City College, The City University of New York*
BENSON CARLIN   *President, O.E.M. Medical, Inc.*

12.1 INDUSTRIAL PLANTS
by Vincent M. Altamuro

REFERENCES: Hodson, "Maynard's Industrial Engineering Handbook," 4th ed., McGraw-Hill. Cedarleaf, "Plant Layout and Flow Improvement," McGraw-Hill. Immer, "Materials Handling," McGraw-Hill. Rosaler, "Standard Handbook of Plant Engineering," 2d ed., McGraw-Hill. Merritt and Ricketts, "Building Design and Construction Handbook," McGraw-Hill.

PURPOSES

Industrial plants serve many **functions**. They can:

1. Protect people, products, and equipment from the weather.
2. Preserve and conserve energy.
3. Condition the inside environment to be suitable for the processes, products, and people engaged therein.
4. Protect the outside environment from any fumes, dust, noise, or other contaminants their processes produce.
5. Provide physical security for their contents.
6. Block access, visual and/or physical, of those not authorized to see inside or enter them.
7. Provide the stable, strong, and smooth platform or surface required for operations.
8. Be the frameworks for the distribution networks of the services needed—electric power, lights, fuels, compressed air, gases, steam, air conditioning, fire protection, water, drainage piping, communications.
9. Support the cranes, hoists, racks, and other lifting and holding equipment attached to their superstructures.
10. Be integral parts of equipment (such as drying ovens) by having one or more walls serve as sides of them, and the like.
11. Be safe, pleasant, and efficient places to work for employees, impress visitors, and reflect positively on the company.
12. Fit the surroundings, blend in, be aesthetically attractive, make architectural statements.

This list of design criteria can be expanded to include functions other than those cited and which may be specific to a given proposed plant.

After a plant's expected present and future activities and contents are defined, it is designed to provide the desired functions, and then built to provide spaces for operating personnel and to house equipment and services. An existing plant considered for purchase or lease must be selected so that it will provide its user a competitive advantage. Generally, the decision to own or rent a plant depends on factors such as the expected life of the project, prudent use of capital, the possible need for early occupancy, the availability of a rentable building that matches the requirements, and the possibility of either a very rapid growth or complete failure of the enterprise. Some **commercial real estate** developers will erect a building to meet the specifications of the future operators of the industrial plant and then lease it to them under a long-term contract, often with an option to buy it at the end of the lease. A new structure to be designed and built must have its specifications clearly established so that it will serve its intended use and fit its surroundings.

The creation of an **industrial plant** involves the commitment of sizable resources for many years. The plant is therefore a valuable asset, both present and future. Its design and construction must be timely and within budget, and once undertaken, the project must, at most, be subject to minor modifications only, and preferably none. Once the structure is in place, it is very difficult or not economically feasible to move or change the plant. Even if not truly irreversible, design decisions can be corrected or changed only with great difficulty and added expense. There is often little difference in the cost of designing a plant and configuring its equipment and services layout one way versus another. The difference arises in the operating expenses and efficiencies, i.e., the wrong way versus the best way.

When designing a plant, one should estimate how long it is expected to last—both economically and physically. When making this determination, one accounts for the expected lives of the products manufactured, the processes and manufacturing methods used, the machines used, and the extant technology, as well as the duration of the market, the supply of labor, and so forth. Plant life and production life should coincide to the maximum extent possible. Major difficulties arise when a plant becomes obsolete and may warrant either abandonment or, at best, major modification for conversion. Often the manufacturing methods employed in production have so far outgrown the original methods that the conversion of an existing plant can not be justified at all.

Obsolescence can be deferred by making the factory versatile, adaptable, and flexible. Possible changes in conditions can be anticipated and the ability to alter the plant can be included in the original design. An industrial plant is said to be *versatile* if it can do more than one thing, if its equipment can switch over and make more than one variation of the product, or if its output can be raised or lowered easily to match demand. It is said to be *adaptable* if it can make other related products with only programming and tooling changes to its equipment, without the need to alter their layout. It has *flexibility* to the degree that it is easy to move its services, machines, and other equipment to change the layout and flow paths, and thereby accommodate different products and changing product mixes. Such a building will have a minimum of permanent walls, barriers, and other fixed features. A plant can be made to be general-purpose or special-purpose, or more one than the other. A general-purpose plant is more generic and can be used for a wide variety of purposes. It is usually easier and less expensive to design, build, and convert to new use and is more salable when no longer needed. A special-purpose plant is designed for a specific task. It is more efficient and can make a lower-cost product, as long as the specifications stay within narrow limits. With modular machines, quick-change tooling, and programmable automation, such as robots, it is possible to gain the advantages of both general- and special-purpose construction in one plant. It can be built with the versatility, adaptability, and flexibility to be configured one way for one set of requirements, then reconfigured when conditions change. This enables the manufacturer to produce the wide variety of products that consumers want, and still make them with the lower per unit costs of a special-purpose plant. Both economy of scale and economy of scope are possible in the same facility.

There are trends in the design of some products which are causing plants to be constructed differently, in addition to the demand for the benefits of versatility, adaptability, and flexibility cited above. Some products are getting smaller: computers, electronic circuits, TV cameras, and the like. For other products, greater manufacturing precision and freedom from processing contamination require the establishment of ultraclean facilities such as clean rooms. These and other trends are causing some plants to be built with the ratio of manufacturing space to administrative space different from that in the past. In some plants with large engineering, design, test, quality control, research and development, documentation, clerical, and other departments, the prior ratios have been reversed, so that manufacturing and warehousing spaces constitute a rather small percent of the total plant.

In addition to the basic building structure and its supporting services, an industrial plant comprises people, raw materials, piece parts, work in process, finished products, warehouse stock, machines and supporting equipment, systems and services, office furniture and files, computers, supplies, and a host of other things. While each is an individual entity, the plant must be designed so that all can work together in an **integrated and balanced system**. For a new facility, the best way to achieve this end is to design it in a progressive manner, going from the general to the specific, with each step based on prior decisions, until a set of detailed specifications is established. The sequence of design decisions is not linear, wherein one is finalized before the next one starts. Rather, it is more like a series of loops, with the output of one feeding back and possibly modifying prior decisions. For example, plant size could influence the site chosen, which then could influence the amount of air conditioning and other services needed, all of which, in turn, require space that could change the prior estimate of plant size. Almost all decisions have an impact on one or more interrelated plant specifications. The design of the product influences the degree of automation, which sets in motion a chain of decisions, starting with the number of employees required, down through the number of toilets, the size of the cafeteria, medical department, parking lot, personnel department, etc., and ultimately even placement of emergency exit doors.

PLANT DESIGN

An industrial plant must fulfill its intended functions efficiently and economically. Its design must consider and account for the basic operating conditions to be served. A detailed discussion along those lines follows.

Prerequisite Data, Decisions, and Documentation

The decision to build or buy an individual plant is made by the company's senior executives. It involves the commitment of large sums of money and is properly based on major considerations: need for additional production capacity; introduction of a new product; entering new markets; availability of new and/or better technology and machinery; building a new plant to replace an existing inefficient one or refurbishing and tailoring the existing plant for new production processes and cycles; relocation to a different area, especially if closer proximity to markets and availability of specialized labor are involved; making products in house which were previously bought for resale. Studies, analyses and projections provide input data to help determine the proper site, size, shape, and specifications of a plant. Economic analyses, technological forecasts, and market surveys are used to justify the investment in a plant and to calculate the approximate amount of money required, break-even point, payback period, and profitability. Detailed product design and engineering documentation includes drawings, parts lists, test points, inspection standards, and specifications. Product variations, models, sizes, and options are defined. Sales forecast data addresses expected unit volumes, seasonality, peaks and valleys, growth projections, and other patterns. In addition to projections, constraints on the plant must also be understood at the outset. These may include location restrictions, budgetary limitations, timing deadlines, degree of mechanization, the type of building and its appearance, limits on its effluents, exhaust fumes, and noise, and other effects on the neighborhood and environment.

Activities and Contents

The determination of the activities and contents of a plant is facilitated by a series of analyses and management decisions. Make/buy decisions determine which items or component piece parts are to be made in the plant and which are to be purchased and stored until needed. Some purchased items are used as received, others need work, (painting, plating, cutting to size). Parts purchased on a just-in-time basis will reduce the amount of in-process storage space required.

Processes are classified as **continuous** (refineries, distilleries, paper mills, chemical and plastic resin production); **repetitive** (automobiles, air conditioners, appliances, computers, telephones, toys); or **intermittent,** custom job shop, or to order (elevators, ships, airliners). Some plants include combinations of these processes when they make several types of products. In such mixed, or **hybrid,** situations, one product and processing method usually dominates, but there are cases where a plant set up for mass production work has a special-order shop to make small quantities of variations of the basic product, such as a vending machine that accepts only foreign coins. The manufacturing methods used in

these processes may include casting, shearing, bending, drawing, forming, welding, machining, assembly, etc. (see Sec. 13). **Engineering documentation** used to facilitate manufacturing analysis includes operations analyses, flow process charts, precedence diagrams, "gozinto" charts, bills of materials, and exploded views of subassemblies and final assemblies. With computer-aided design (CAD) and other graphics, these documents can often be constructed, updated, and stored in a common database.

These documents help define the operations, their sequence and interrelationships, inspection points, storage points, and the points in the process where materials and parts join others to form subassemblies and the finished product. (See Hodson, "Maynard's Industrial Engineering Handbook.") Special characteristics of the operations may be noted on these documents, such as "Very noisy operation," "Piece parts can be stacked, but after assembly, they cannot," and the like. Product movement is determined. People/tools/things must move in an industrial plant. Workers and their tools go to the stationary work in process, as in shipbuilding; the worker and the work go to stationary tools, as in a general machine shop; the work in process goes to the workers and the tools, as in an assembly line. Other combinations of the relative movement of workers, tools, and work in progress are used. (See Cederleaf, "Plant Layout and Flow Improvement.")

An industrial plant should be designed to facilitate, not hamper, the **smooth movement of people, material, and information**. The efficiency of flow is also influenced by the layout of the equipment and other contents of the building. A multilevel building implies movement between the levels, which usually takes more time, energy, and expense than having the same activities on one level. Some operations are performed better in high structures, where raw material is elevated and then gravity-fed down through the lower levels as it evolves into a finished product. A plant built into the side of a hill receives material directly into an upper level without the need to elevate it. In most cases, a single-level plant affords the opportunity for the most efficient flow of materials and product. Even a single-level plant should have all its floor at the same elevation so that material moving from one area to another need not go up or down steps, ramps, or inclines. Features of the building such as toilets and utilities should be located where they will not interfere with the most efficient plant layout, and will not have to be moved in relayouts or expansions.

To the maximum extent possible, the **material flow** through an industrial plant should be smooth, straight, unidirectional, and coupled (the output of one machine should be the input to another, without a large cushion of inventory between them); require the least rehandling; be at constant speed over the shortest direct route, with the least energy and expense; and be always directed toward the shipping dock. Raw materials and piece parts should move continuously through the plant as they are converted progressively by a combination of people and equipment into finished products of the desired quantity and quality. Materials must flow with few interruptions, side excursions, or stops, so that they are in the plant for the shortest time possible and thereby facilitate expeditious product completion. Where possible, movements should be combined with operations, such as having a mobile robot or an automatic guided vehicle (AGV) work on (inspect, sort, mark, pack) the item while transporting it. An analysis of the plant's activities, including a list of its contents and an analysis of their relative movements, results in a rational determination of how they must flow through it. This must be done for materials (raw materials, purchased parts, work in process, finished goods, scrap, and supplies), people, data, and services.

Space and Size Calculations

Annual sales forecasts plotted by the month sometimes show peaks and valleys due to **seasonal variations** in demand. A plant with the capacity to produce the highest month's demand has much of its capacity idle or underutilized in the other months. The ideal plant would have a constant output every month, with long production runs of each item to minimize tool changes and setup times. This is not always practicable, for either very high inventory buildups or shortages could result. The compromise is the calculation of a production schedule that is more level than the sales forecast, builds as little inventory as possible without risking shortages, and allows for rejects, equipment breakdowns, vacations, bad weather, and other interruptions and inefficiencies. This will result in a plant with less production equipment and space but with more inventory space (and associated material handling equipment) to accumulate finished product to meet shipping demands in response to sales. Obviously, the warehouse and other storage areas must accommodate the maximum amount of inventory expected to be stored.

The proper size of a plant is determined by **calculating the total space required** (immediate and future) by all its contents: material at all stages of production, people (employees and visitors), equipment (production, materials handling, support, and services) and the ancilliary spaces required (working room, aisles, lobbies). At first, only the **general types of equipment are specified** (spot welders, overhead chain conveyors, forklift trucks, etc.). Later, **specific items are selected** and listed by manufacturer, model number, capacity, speed, size, weight, power, and other requirements. (See Sec. 10.) Finally, one model may be substituted for another to gain a particular advantage. A simplified typical equipment selection calculation follows:

1. Define the operation: Joining.
2. Decide the method: Welding.
3. Note personnel available: Semiskilled.
4. Determine general machine type: Spot welder.
5. Calculate required production output:

Spot Welds Required

	Product			
	A	B	C	Total
Number of spots	120	60	80	
Units per year	10,000	40,000	20,000	
Spots per year, millions	1.2	2.4	1.6	5.2
Spots per day 250 days/year)				20,800
Spots per hour (8 h/day)				2,600
Spots per minute (60 min/h)				44

6. Select a particular candidate machine, considering type of metal, thickness of metal, diameter of spot. From catalog, choose manufacturer (Hobart) and model (series 1500 rocker arm SW-V).

7. Calculate capacity of one machine in minutes per spot.

Operating time	0.03
Material handling time	0.06
Setup time allocated	0.01
Minutes per spot	0.10

8. Calculate the number of machines needed.

$$N = \frac{TP}{60\,HU}$$

where N = number of machines required; T = standard time to perform the operation, minutes; P = production needed per day, operations; H = standard working hours per day; U = use factor—up-time of the machine, its utilization (percent of time it is producing), or its efficiency. Example:

$$N = \frac{0.10 \times 20,800}{60 \times 8 \times 0.80} = 5.4 \text{ machines}$$

9. Round off to next higher number: 5.4 becomes 6 machines.
10. Alternatively, return to catalog to see of there is a faster model. If so, 5.4 could become 5 of a different model machine.
11. Record specifications of selected machine on an equipment card or in a computer file. Appropriate notations should be made therein, such as:

Crane must go over this machine.
Must be near outside wall for venting.
Will require supplemental lighting.
Requires a special foundation.

Requires a drain for water.
Allow room on side for gear changes.
Allow room for largest-size sheet-metal parts.

12. Use data for all machines needed to help establish specifications for equippage of the plant:

Sum of space required (including space for workers, material, aisles, services) becomes size of that department or function

Sum of utilities and supplies (electricity, water, steam, compressed air, oxygen, nitrogen, acetylene, coolants, lubricants, air conditioning) addresses requirements for them.

Sum of weights affects floor loading capacities and specifications. For some parameters, the extreme value is important in establishing the building specifications; i.e., the height of the tallest machine may set the required clear height inside the plant, locally or throughout. Equipment mounted on columns or roof beams will influence their size. Specific equipment may require special foundations or subfloor access pits.

Sums of the prices of the machines and the required tooling, the number of workers, their required skills, and their wages and benefits provide information regarding the investment and operating expenses for the plant.

The same procedure is used to determine the **material handling equipment,** its required space, and its influence on the plant's specifications. (See Immer, "Materials Handling.") Equipment may be capable of moving materials horizontally, vertically, or in both directions; some have a fixed path of travel, while the paths of others can be varied. The general types of equipment that move materials horizontally in a fixed path include conveyors, monorails, and carts pulled by trucks, dragged by chains in floor troughs, or that follow buried wires. Conveyors are overhead or floor-level type. Overhead conveyors clear the floors of some congestion but add to the loads carried by the building's columns and beams. Other types of conveyors are installed at floor level or at working height, and include belt conveyors and powered or unpowered roller conveyors. The general types of material handling equipment that travel in variable horizontal paths include hand trucks, powered trucks, pallet trucks and automatic guided vehicles. Those that travel in fixed vertical paths include elevators, skip hoists, chutes, and lift tables. Those that travel in both horizontal and vertical fixed paths include traveling bridge cranes, gantry cranes, jib cranes, and pneumatic tubes. Those that travel in both horizontal and vertical variable paths include robots and forklift trucks. Forklift trucks may be powered by either battery, gasoline, or liquefied petroleum (LP) gas. Those powered by batteries will require the installation of charging facilities within the plant, and those with internal combustion engines will require safe fuel-storage space.

The installation of each of the above-listed material handling equipment types will influence the plant's specifications in some way. In computing loads on the building's structural members, all static and dynamic loads arising from material handling equipment must be factored into their design, as applicable.

With respect to **raw material,** its form, weight, size, temperature, ruggedness, and other qualities are considered, as are the expected distances to be moved, speed of transit, number of trips, and amount carried per trip. For **warehouses,** the important considerations are cubic space and density of use for various package sizes, weights, and stacking patterns.

The same basic approach is used for **support functions** and **plant services.** The space required for people can be estimated by listing them by functions, jobs, and categories (male versus female, plant versus office, department, location, shift), then allocating space to each. For example, office floor space allocation in ft^2 (m^2) per person may be: plant manager, from 150 to 300 (13.9 to 27.9) depending on size and type of plant; assistant plant manager, 125 to 250 (11.6 to 23.2); department heads, 100 to 200 (9.3 to 18.6); section heads, engineers, specialists, 100 to 150 (9.3 to 13.9); general personnel, 50 to 100 (4.6 to 9.3). When accounting for all employees in an industrial plant, the space per person can range from about 200 to 4,000 ft^2 (18.6 to 370 m^2), depending on the type of industry and the degree of automation employed. An estimate of the male/female ratio often must be made early in the planning stage, for many localities will issue a building permit only if adequate sanitary and rest facilities are included for each gender.

The space required for each department and function is compiled and added to show the approximate total plant size. See Fig. 12.1.1, which also shows the percent of the total plant floor space allotted to functional subtotals. The size of the plant should not be deemed final without including the manner in which to allow for anticipated growth and/or plant expansion. Anticipated growth, if realistic, must enter into the decisions reached as to the initial size of the plant to be built. When dealing with the initial plant size vis-à-vis any anticipated future expansion, some questions usually posed are: Should it be large enough to accommodate expected growth, knowing that it will be too big initially? Should it meet only present needs, then be expanded as and if required? Should a midsize compromise be made? A plant too big for present needs will cost more to build and operate and may place the firm in an uncompetitive position. Expenses for taxes, insurance, heating, air conditioning, etc. for a partially vacant plant are almost the same as for one fully occupied. On the other hand, a plant built just to satisfy present requirements of a growing business can quickly become cramped and inefficient and may lead to rental of external storage and warehouse space, with all the added expense and loss of control entailed thereby. Any future expansion of the original plant will not only be disruptive to operations, but also will cost significantly more than building it bigger in the first place. The correct choice is made by analyzing all options and comparing expected costs and profits and the realistic probability of growth.

If a building is to be expanded or upgraded easily and economically after it is built, that capability must be included at the outset. This is done by deciding in advance the direction and degree of possible future expansion, knowing that if the business grows, not all sections of the plant may have to be expanded to the same degree. After the basic form of the building shape and envelope have been addressed, attention is then directed to other matters: location of the building on the property; materials of construction for internal walls that may have to be rearranged; location of spaces and equipment that will be difficult or impossible to move in the future, such as toilets, shipping/receiving docks, transformers, heavy machinery, and bridge cranes.

Superstructure members, pipes, monorails, ducts, and so on can be terminated at the proposed expansion side of the plant, with suitable end caps or terminations which may be removed easily and activated quickly when required. Installations of major utilities are prudently sized to handle future expansion. If the initial design strategy includes anticipation of vertical expansion via mezzanines, balconies, or additional stories, foundations, footings, superstructure, and other structural elements must be designed with that expansion in mind. To accommodate that requirement at some later date will prove to be prohibitively expensive and will disrupt ongoing operations to a degree unimaginable. By the same token, if future operations require a contraction of active space in the proposed plant, space would be available, at worst, for sublease to others. Regardless of whether possible expansion or contraction is contemplated, the initial plant design should consider both possibilities and seek to have either occur with minimum disruption of production.

Plant Emplacement and Site Selection

The emplacement of a plant is defined by its geographical location, site, position, and orientation, in that order. Geographical location is determined by the country, region, area, state, county, city, and municipality in which the plant is situated. The site is that particular plot of land which can be identified by street address or block and lot numbers in that geographical location. The position and orientation speak to the exact placement and compass heading of the building on the site.

Typical considerations entering each stage of the placement decision, from broad to specific, follow.

Location: Political and economic environment. Stability of currency and government. Market potential. Raw materials supply. Availability and cost of labor. Construction or rental costs. Environmental regulations. Climate. Applicable laws within the governing jurisdiction. Port, river, rail, highway, airport quality. Utilities and their relative costs. Taxes. Incentives offered. Local commercial services. Housing, schools, hospitals, shops, libraries, and other community attractions. Crime rate, police and fire protection. General ambiance of the locality.

Site: Availability. Price. Incentives offered. Building and zoning codes. Near highway entrance/exit, on railroad spur, waterway, and major road. Public transportation for employees. Availability of water, sewer, and other utilities. Topography, elevation, soil properties, subsurface conditions, drainage, flood risk, earthquake faults. Neighbors. Visibility.

Position: Placement for possible expansion with sides of building that may be extended facing an open area or parking lot and the sides not to be extended close to the property line. Proximity to railroad tracks, road, utilities or other fixed features. Positions of buildings on neighboring property.

Orientation: Turned toward or away from winter's wind and summer's sun, as desired, considering the frequently open large shipping and receiving doors. Needs for heating, air conditioning, natural light, color matching, and the like that would be affected by compass heading.

Department or function	Area	
	ft²	m²
Main receiving	5,700	530
Incoming inspection	900	84
Main stockroom	9,600	892
Large-component storage	6,000	557
Coil steel storage	1,840	171
Steel coil line	2,360	219
Sheet metal fabrication	18,000	1,672
Welding	3,000	279
Sheet-steel storage	13,800	1,282
Copper tubing storage	600	56
Tubing fabrication	1,200	111
Cleaning and painting	9,600	892
Insulating	1,200	111
Electrical subassembly	1,200	111
Final assembly	33,600	3,121
Finished goods storage	40,200	3,735
Wood storage and fabrication	2,400	222
Shipping	9,600	892
Maintenance and tool room	600	56
Print shop	600	56
Plant offices	1,200	111
Engineering laboratory	12,000	1,115
Reproduction room	420	39
Canteen/eating area	3,600	334
Medical/first aid	330	31
Offices and lobby	21,600	2,007
Toilets	1,650	154
Total	202,800	18,840

Function	% of total plant	ft²	m²
All receiving, incoming inspection, storage, and shipping	44.2	89,440	8,309
All production	35.3	71,360	6,629
Engineering lab and repro	6.2	12,420	1,154
Support areas: Maintenance and tool, print, first aid, canteen, toilets, plant offices	4.0	7,980	741
General offices and lobby	10.3	21,600	2,007
Total	100.0	202,800	18,840

Fig. 12.1.1 Space required by department, subtotals by function, and for the entire facility for a plant designed and built to manufacture roof-mounted commercial air conditions. (*Source: VMA, Inc.*)

Proximity to competitors may be desirable if the locale has developed into a well-recognized center for that industry, where are found skilled people, and a wide variety of suppliers and supporting services (testing laboratories, local centers of higher learning with faculty available for consultation on an ad hoc basis). Without these conditions, and especially if the product is heavy and expensive to ship, it is desirable to locate close to market areas and distant from competitors.

The amount of land required depends on plant size, the number of employees and their need for parking space, the probability of storing some raw materials and finished goods outdoors, the need to turn large trucks around on the property, the possibility of future expansion of the plant, zoning, and the possible desire to create an open park-like ambiance. Total land area of from 4 to 8 times the plant size is usually adequate; the lower end of the range applies in built-up areas, and the higher end applies in suburban areas. In some locations the land can cost more than the plant.

Before placement of a plant becomes final, all conditions and extremes should be simulated, including winter storms, rainy seasons, floods, several consecutive days of 100°F heat, power outages, labor disruptions, and the like. It is unlikely that the perfect site will be found, thus sought-after attributes must be ranked in order of importance. As an incentive to get new industry, some governments will build roads and schools and reroute buses if necessary. Test borings are often taken at a candidate site before a commitment is made to buy or lease it to ensure that it is suitable. A plot plan (see Fig. 12.1.2) can be made with an approximation of the planned plant on the candidate site indicating land size, topography, drainage, position on the plot, anticipated expansion, compass orientation, location of power lines, railroad tracks, roads, rivers, open fields, neighbors, etc.

Configuration

The configuration of a plant is determined ideally by the optimum layout of its contents. Compromises are usually made, however, often resulting in a conventionally shaped building. A building with several extended branches is very expensive. For a given floor area, a square building requires less total wall length than other quadrilateral shapes, but rectangular buildings are the most common compromise between cost and efficiency. For the lease or purchase of an existing building, the best approach is to design the best layout and then seek an existing building within the desired area in which that layout may be accommodated. For a new building, the layout is set down before the location is selected and then adjusted for the particular features of the site. A layout prepared after the site has been selected can be tailored to fit the site, all the while maintaining the desired features of that layout.

Before the configuration and layout become final, certain broad and tentative choices about the type of building should be made. These include general-purpose versus special-purpose, multilevel versus single level, use of mezzanines and balconies, with windows or windowless, etc. Setting aside considerations of material flow, a multilevel building enjoys the advantages, on a unit floor area basis, of using less land and costing less to build. The columns, however, will generally be spaced closer together, thereby presenting an impediment to desired flow paths and a reduction in options for overall layout. A single-level building, on the other hand, can more easily support heavier floor loads by virtue of its concrete slab on grade floor construction.

The matter of shipping/receiving floor level vis-à-vis truck bed level must be resolved satisfactorily, keeping in mind that loading and unloading ought to be effected with small hoists or forklifts for maximum efficiency. To that end, loading floor level can be built to be flush with

Plot Plan

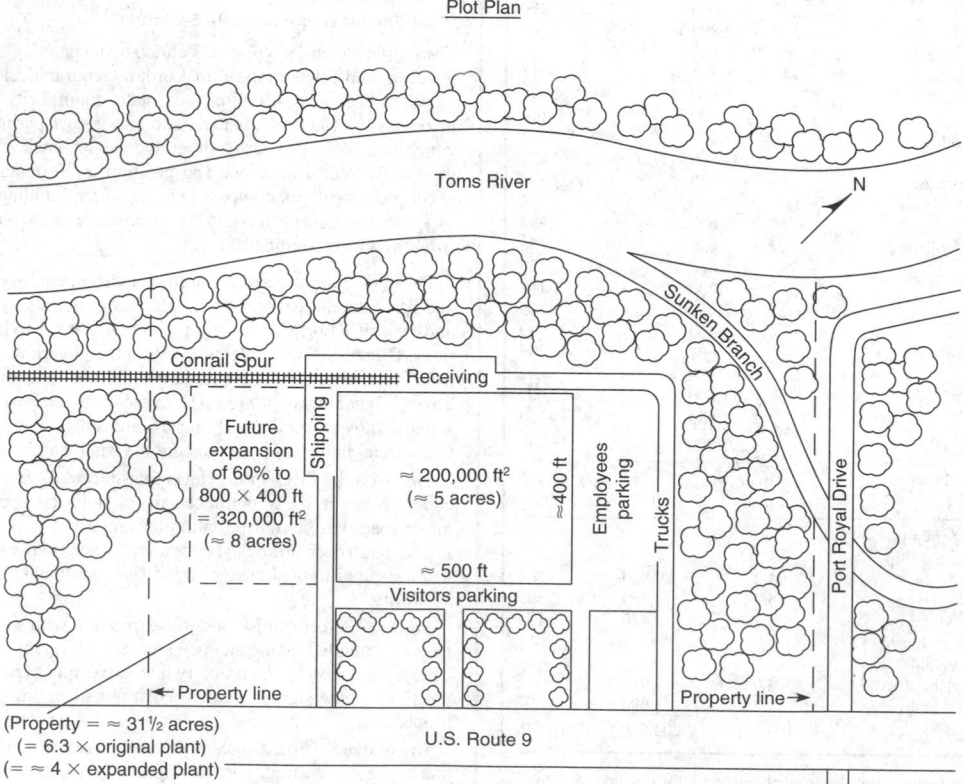

Fig. 12.1.2 Plot plan of an industrial plant. *(Source: VMA, Inc.)*

the bed height of most of the anticipated truck traffic, or the truck parking apron can be inclined downward to align the two. If those cases when trucks with other bed heights have to be accommodated, small demountable ramps can be employed, the same being built sufficiently rigid to permit forklift traverse.

Balconies and mezzanines may be added to gain more storage space, to locate offices with a view of the production areas below, for raw materials or subassembly work that can be gravity-fed and drop-delivered to the operations below, and for the placement of service equipment (hot-water tanks, compressors, air conditioners, and the like). Basements may be included for the placement of heavy equipment, pumps, compressors, furnaces, the lower portion of very tall machines, supplies, and employee parking.

Windows in an industrial plant lower lighting bills, permit truer color matching in natural light, aid cooling and ventilation when opened, may provide a means of egress in case of fire, and may lower fire insurance rates. The advantages of a windowless plant are easy control of the intensity and direction of light (elimination of glare, contrasts, shadows, diffusion, and changing direction); (sometimes) less expensive construction; and less heat transfer, dust and dirt infiltration, maintenance expense to wash and repair glass, and worker distraction; better security; lower theft insurance rates. It is generally easier and more economical to design a windowless building. The absence of fenestration allows more flexibility in interior layouts by virtue of uninterrupted wall space. The absence of low windows, in particular, obviates the need for blinds or drapes to keep the interior private from the passing viewer. Elimination of an enticing target to vandals is not to be dismissed lightly; often damage to machinery and equipment, as well as injury to personnel, results from flying missiles launched at and through windows.

The relative positions and detailed layouts of each department's machines, support equipment, services, and offices are based on analyses of their functions, contents, activities, operations, flow, relationships, and frequency of contacts. The layouts may be product- or process-oriented, or a combination of the two. In a product-oriented layout, machines are arranged in the sequence that the production process requires. This permits the product to advance in a direct path, such as on an assembly line, and with smooth material flow. In a process-oriented layout, the machines are grouped by type, such as a welding or drilling. This requires that the products requiring those operations be brought to that area. It is used where products vary and the output of each is low, such as in a job shop, and allows production to continue even when a machine breaks down.

Another early decision involves the preferred placement of the receiving and shipping departments and the related basic pattern of material flow through the plant. If it is preferred to have the material enter at one end of the plant and exit at the other (Fig. 12.1.3a and b), then separate receiving and shipping facilities, equipment, and personnel will be required. Capital investment and operating economies ensue from combining these functions, with shared personnel, equipment, supervision, and space, but then the basic material flow will loop back to allow the finished product to exit from the same location where raw materials entered (Fig. 12.1.3c).

The drawing showing the size, shape and position of each department or area of an industrial plant is called a **block diagram**. It is developed by constructing a series of **analysis documents**. These include the frequency of relationship chart, proximity preference matrix, relative position block arrangement, sized block arrangement, initial block diagram, refined block diagram, and final block diagram (or simply block diagram). The **frequency of relationship chart** (also called a **from/to chart**) is constructed from an analysis of the engineering and manufacturing documentation and the activities of the plant. It shows the frequency and magnitude of movement, flow, and contacts between the entries. It may be weighted to include the importance of high-priority factors. The **proximity preference matrix** (also called a **relationship chart**) is constructed the same way. It shows (Fig. 12.1.4) which functions, departments, equipment and people should be close to each other, how close, and why, and which should be away from which other and why. It is constructed with a diamond-shaped box at the intersection of every two

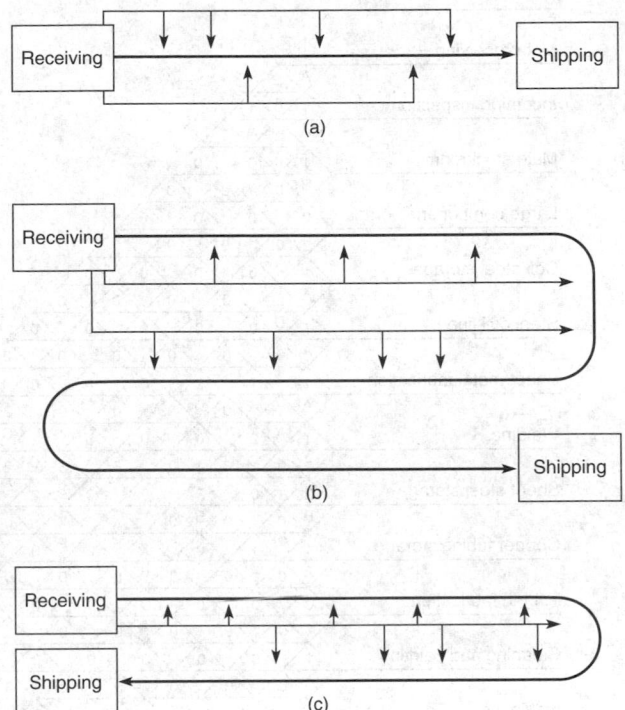

Fig. 12.1.3 Three different relative positions of a plant's receiving and shipping areas. *(Source: VMA, Inc.)*

entries. The notations made in each box show the importance of their need to be either close or separated by a number or code. The reasons can also be shown by entering a code in the lower half of the box, as shown in Fig. 12.1.4. The matrix may be made at the function or department level and, later, at the individual machine or person level. The objective of the matrix and chart is to lay out the plant so that things are as close to (or as far from) other things to satisfy the criteria established, and ensure that those with the highest number of contacts are so located to minimize the time, distance, and energy required.

When this is done, a **relative position block arrangement** (Fig. 12.1.5) may be made; this shows the various arrangements possible by shifting around pieces representing the departments. The arrangement selected is the one that best satisfies the relationships (in decreasing order of importance) and degrees of contacts, as previously determined. The size of each model (or computer graphic representation) is the same because only the relative positions of the departments or areas is of interest at this stage of the design. Assigning a different color or background pattern to each adds to clarity and, when carried through to final and detailed drawings, helps visualize quickly the totals of separated areas; e.g., the total of inventory storage areas spread throughout the plant can be understood quickly if all are shaded the same color on the drawings. The **sized block arrangement diagram** (Fig. 12.1.6) is the relative position block arrangement with the size of each area scaled (but still square) to represent its square footage as determined by its expected contents and room for expansion (unless the expansion is to be handled by extending the building, in which case the department should be placed where expansion is proposed. The **initial block diagram** converts the square representation of each department into a shape that is more suitable for its contents and activities, but of the same square footage. The **refined block diagram** (Fig. 12.1.7) adjusts the initial shapes to effect a compromise between the advantages of having them be the best operational shape and those of fitting them in an economical rectangular building. L-, T-, U-, H-, and E-shaped buildings are often the result of such configuration compromises.

After the main aisles are drawn, as straight as possible, to serve as

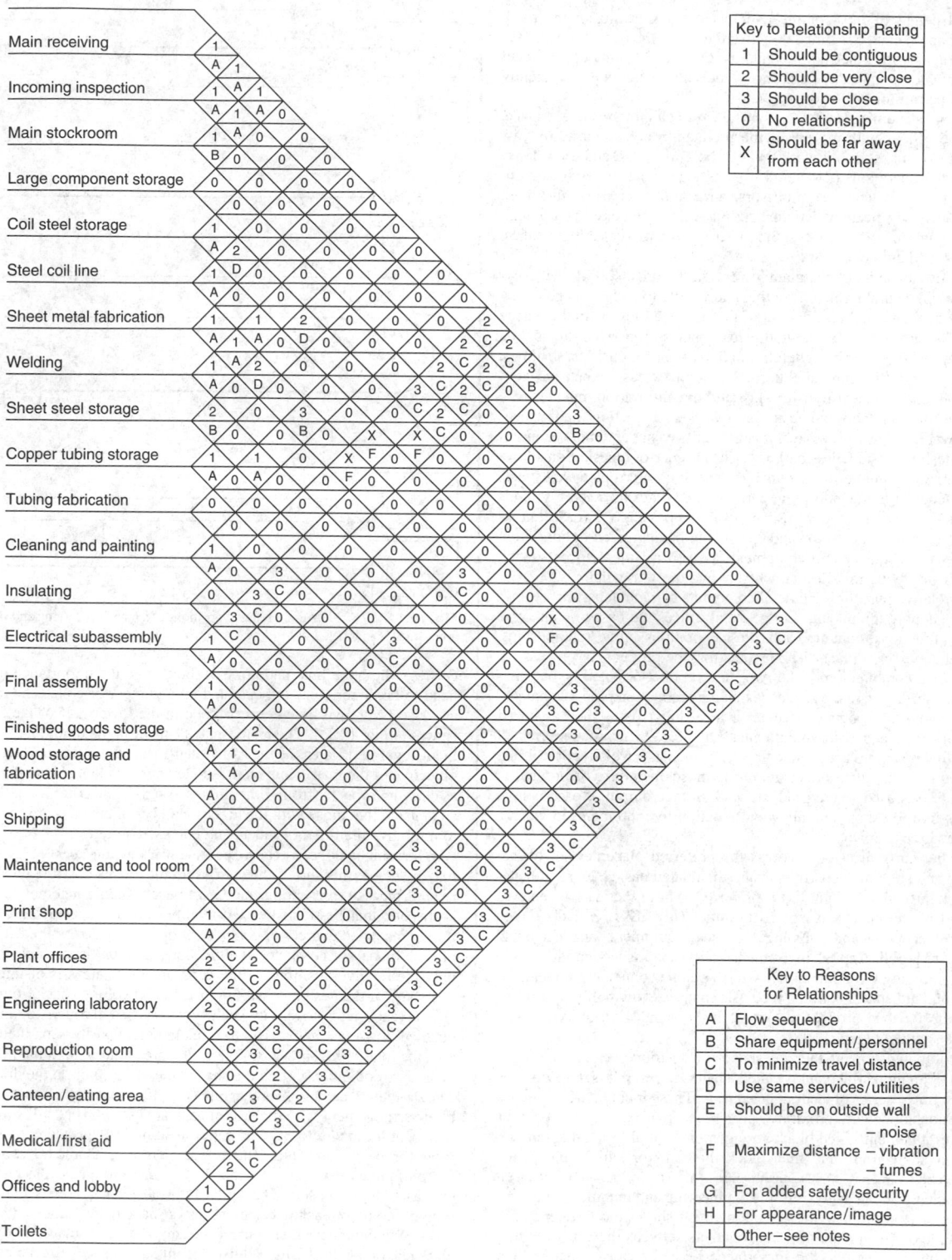

Key to Relationship Rating

1	Should be contiguous
2	Should be very close
3	Should be close
0	No relationship
X	Should be far away from each other

Key to Reasons for Relationships

A	Flow sequence
B	Share equipment/personnel
C	To minimize travel distance
D	Use same services/utilities
E	Should be on outside wall
F	Maximize distance – noise – vibration – fumes
G	For added safety/security
H	For appearance/image
I	Other–see notes

Fig. 12.1.4 A proximity preference matrix listing the departments, functions, and areas, and ranking how near or far each should be relative to the others, and why. *(Source: VMA, Inc.)*

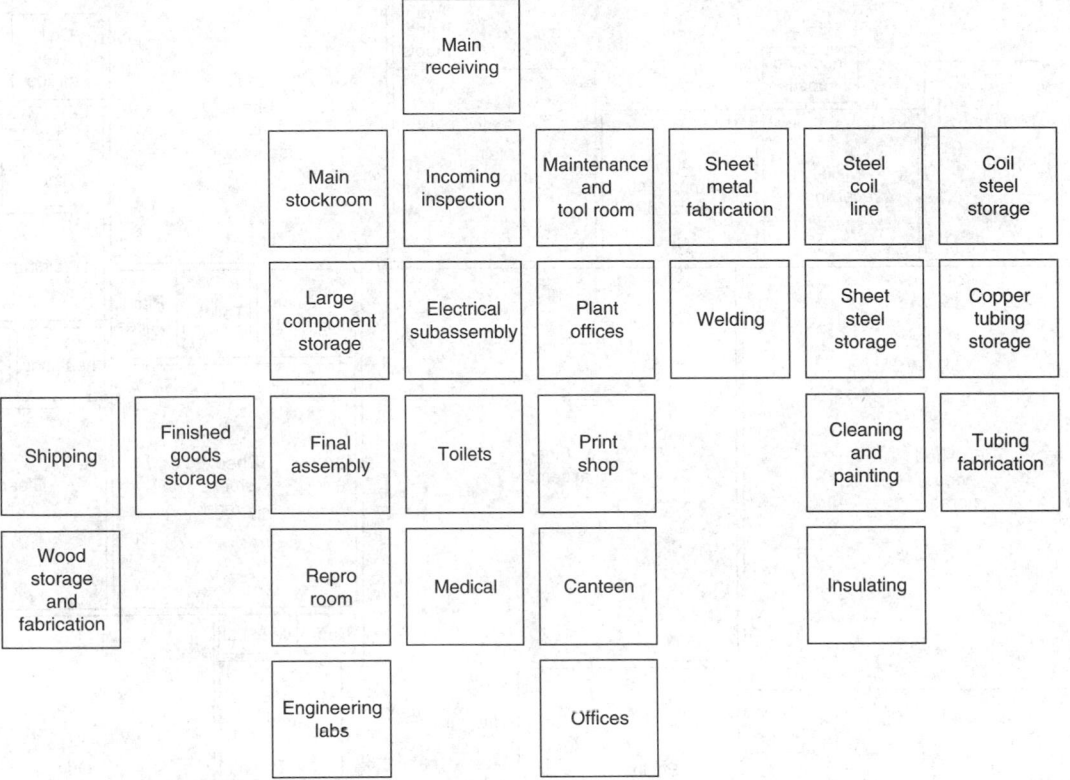

Fig. 12.1.5 A relative position block arrangement that attempts to satisfy and optimize the dictates of the prior proximity preference matrix. (*Source: VMA, Inc.*)

dividers between departments, the size and shape of each area can be fine-tuned, and a **final block diagram**, or simply **block diagram**, is drawn, keeping the same color code scheme used throughout. Detailed layouts are constructed for each department and function to fit within the spaces allocated in the block diagram. When assembled onto one drawing, the block diagrams become the detailed layout for the entire plant.

There are several aids to constructing both block diagrams and detailed layouts. These range from scaled templates to three-dimensional models of machines and equipment. They are available as plain blocks and as highly detailed plastic or cast-metal models. Computer-based optimization programs are also available. Some computer-aided design packages contain libraries of plant components and equipment that can be selected, positioned, rotated, and moved on the screen until a satisfactory layout is achieved. See Figs. 12.1.8 and 12.1.9 for two- and three-dimensional, respectively, computer-generated detailed layout drawings. Arrows are added to these diagrams to show flow paths. Copies of these drawings can be annotated with dimensions and machinery and furniture descriptors and given to vendors and contractors for them to submit bids to supply and install the items. They are also kept on file for use in future maintenance and/or revisions.

Features not expected to be moved in the future or that will become permanent parts of the building should be located first, e.g., stairways, doors, toilets, transformers, steam boilers, fuel tanks, pumps, compressors, piping, and permanent walls. Accurately dimensioned definitive drawings must be made to guide installers of equipment and services.

In addition to the layouts of the production areas, the support functions must also be planned. Support groups assist the production departments; they include research and development, testing laboratories, engineering, production planning and control, quality control, machine shop, sales and advertising, technical literature, purchasing, data processing, accounting and finance, files storage, personnel, medical, administration and management, general office, supply storage, reception lobby, conference rooms, training rooms, library, mail room, copying and reproduction, cafeteria, vending machines, water fountains, washrooms, toilets, lockers, custodial and maintenance, security, and so on. Adequate space must be provided for the people working in these areas and their workstations must be designed with ergonomics, lighting, comfort, and safety in mind.

Many industrial plants have the supervisors' offices situated on the factory floor for better contact with and control of their people. Likewise, offices for quality control inspectors, industrial engineers, and manufacturing or process engineers often are located in the production areas. Such offices may be built with the building or may be purchased and installed as preengineered, prefabricated units. General and administrative offices usually are distant from the plant's machinery and equipment, whose noise and vibrations would otherwise affect the performance of the occupants therein and their ancillary equipment (computers, for example).

Office furniture may be arranged in military style (orderly straight rows), the open plan method (fewer and movable partitions), landscaped (free form with plants and curved dividers), individual offices (glass, wallboard, steel, or masonry walls), or any combination of these. In those arrangements wherein the partitions and dividers are classified as furniture instead of parts of the building, they may be depreciated over a much shorter period of time than the permanent structure. Space is allocated on the basis of position in the organization chart, with the senior executives getting windows (if any) and the most senior getting a large corner office. Floor coverings, ranging from tile to carpeting, again depends on position, as do the amount and type of furniture.

The plant's reception lobby should measure approximately 160 ft² (14.9 m²) if seating for four persons is required, and at least 200 ft² (18.6 m²) for 10 visitors. Add 60 to 100 ft² (5.6 to 9.3 m²) if a recep-

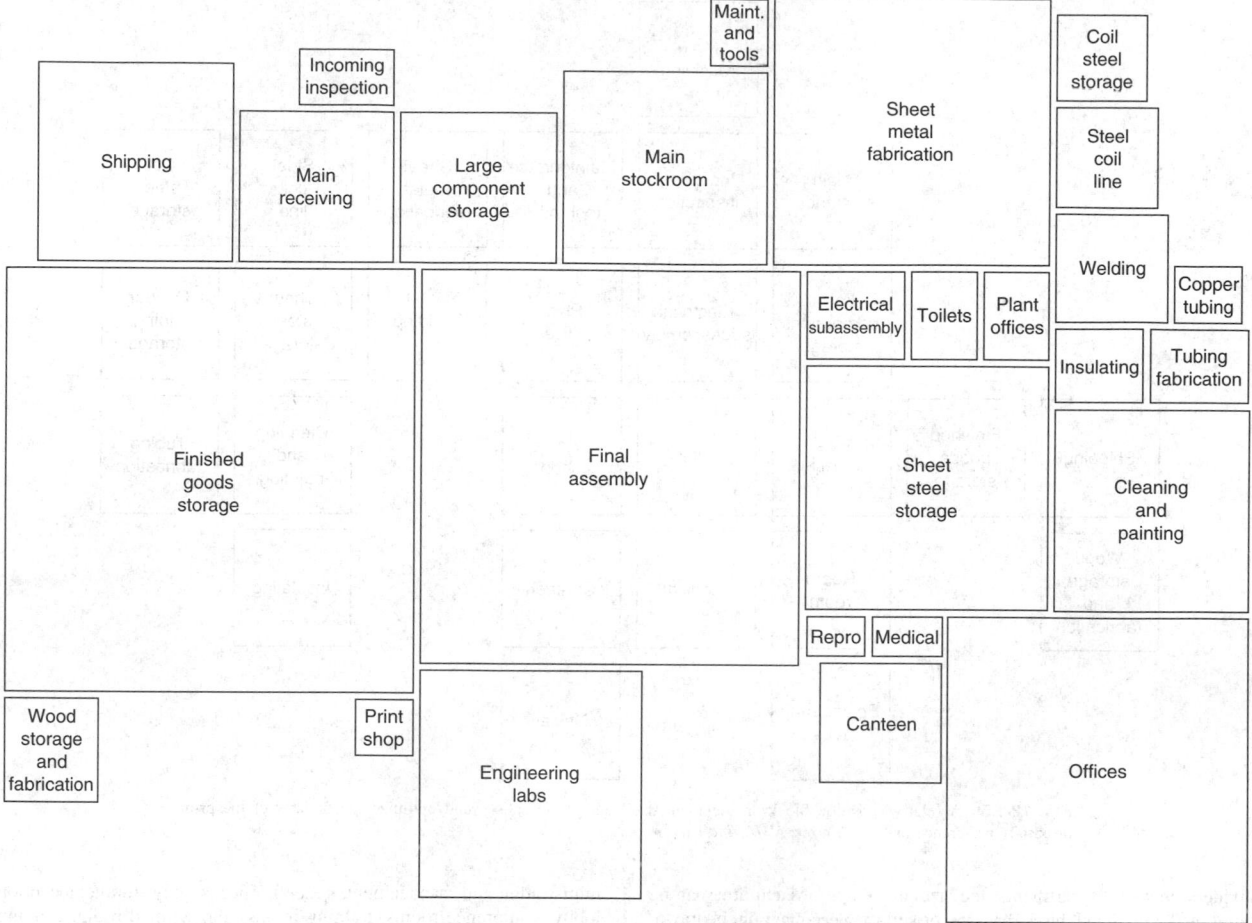

Fig. 12.1.6   A sized block arrangement that converts the prior relative position block arrangement into one that shows the required size of each department, function, and area, while maintaining the preferred relative position of each. *(Source: VMA, Inc.)*

tionist is to be seated there. Cloak rooms require 6 ft² (0.6 m²) per 10 garments.

If a cafeteria is included, 20 ft² per person of expected occupancy should be provided if it is equipped with tables and chairs, plus space required by any vending machines. Conference and meeting rooms should provide about 20 ft² (1.9 m²) per person of expected attendance, and training rooms with theater-type seating should be 400 ft² (37.2 m²) for groups of 20, 600, (55.8) for 40, and 1000 (92.9) for 80. A plant library will range from 400 to 1000 ft² (37.2 to 92.9 m²), depending on its contents. Record storage requires 6 ft² (0.6 m²) per file cabinet. Storage space must be provided for stationery and supplies. Slop sink and mop closets should be 12 to 15 ft² (1.1 to 1.4 m²). Facilities should be placed as close as possible to those who will use them; those for universal use must be located conveniently. In very large plants, spaces and facilities for universal use must be supplied in multiples, and include toilets, clothes closets, lockers, time clocks, emergency exit doors, copy machines, vending machines, and drinking fountains. Aisle locations and widths are critical elements in the management of internal traffic, and are based on: use only for pedestrians or by material handling vehicles as well, in which case load widths must also be included as a design parameter; whether traffic is one-way or two-way, with loaded vehicles passing each other; traversing vehicular traffic only or with dropoff and/or pickup points along the route; and provision of turnaround space for vehicles (forklifts and the like) or restrictions on ma-

neuvering within aisles. For efficient traffic flow, the configuration that will work best most often is one with one main aisle and a number of smaller feeder aisles, with straight, well-marked aisles intersecting at right angles. Aisles located at exterior walls will result in loss of storage space, and generally will be remote from the central area meant to be served. While necessary and functional, aisle space can occupy from 15 to 30 percent of the plant's total floor space. Planning for aisle space must defer to any applicable labor laws or local ordinances.

Services

An industrial plant's services are those utilities that power and supply the production and support functions. They include electric power and backup emergency power; heating, ventilating, and air conditioning (HVAC); water for processes, drinking, toilets, and cleaning; smoke, fume, and fire detection and fire fighting; natural gas and/or fuel oil, process liquids and gases, compressed air, steam; battery chargers for electric forklift trucks, AGVs, and mobile robots; communication networks and links for telephone, facsimile, computers, and other database nodes; surveillance and security systems; and waste, scrap, and effluent disposal drains and piping. The design of these requires not only the specification of the equipment, but also the layout of the distribution networks, with drops to where needed and points of interface to the apparatus served.

Flexibility is increased and unsightly and dangerous wires are elimi-

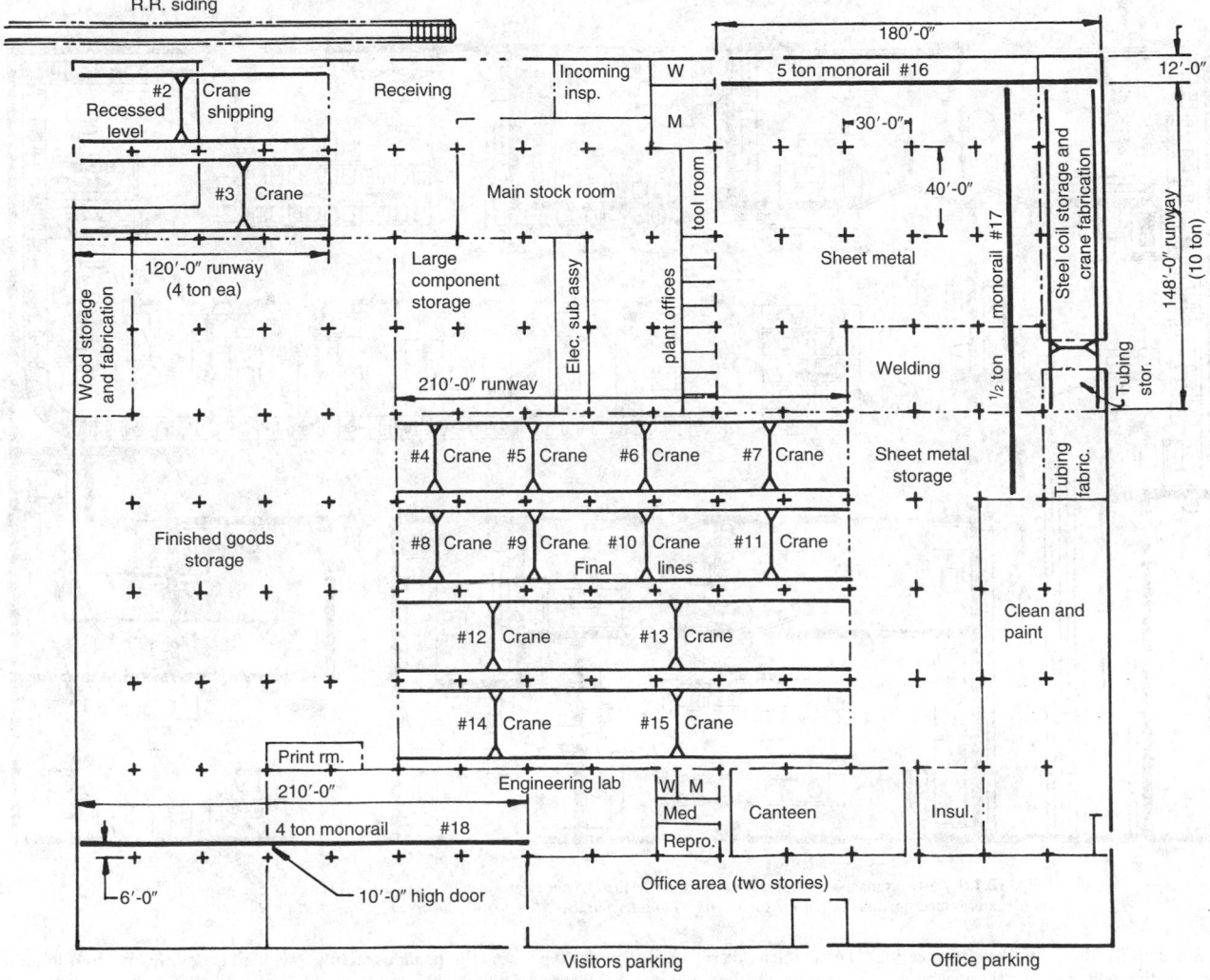

Fig. 12.1.7 A refined block diagram that converts the prior sized block arrangement onto one that makes the shape of each department, function, and area be such that, together, they fit into a building of a more conventional shape, while maintaining the size and relative position of each. *(Source: VMA, Inc.)*

nated by installing buried electric raceways in the floor before concrete is poured. Wires channeled thus are connected to floor-mounted equipment through access caps; changes in machine layout or additions to the complement of machinery are facilitated via connections into the raceways. Definitive, current records of buried raceways document exact locations of raceways and any modifications made thereto over periods of time, and while they are archival in nature, they serve to prevent confusion and guesswork. Other raceways and wire conduits are dispersed through the plant to accommodate electric convenience outlets, communication equipment, and similar services.

Electric power is usually transmitted over the utility's lines at between 22,000 and 115,000 V. The plant usually includes transformers to reduce voltage to 2,300–13,000 V. Most building codes require that these transformers be installed outside the building. The next level of voltage reduction, down to 120–480 V, is provided by transformers usually located inside the building and dispersed to strategic locations to provide balanced service with minimum-length runs of service connections. Alternatively, the plant may arrange for the local utility to supply electric power already stepped down to the levels needed; 240/120-V single phase, three-wire; 208/120-V three-phase, four-wire; 480/277-V

three-phase, four wire. The last is more economical for motors and industrial lighting.

Natural light entering the plant through windows and skylights is erratic and difficult to control. All areas of the plant must be illuminated adequately for the activities conducted therein. Section 12.5 lists typical illumination levels. Density of light, illuminance, is designated in foot-candles, fc (lumens per square foot) or lux, lx (lumens per square meter). One fc = 10.76 lux. In most cases, tasks requiring illuminance more than 100 to 150 fc (1080 to 1600 lux) also require directed supplemental illumination. No area should be illuminated at a level less than 20 percent of nor more than 5 times that of adjacent areas because eyes have trouble adjusting rapidly to drastic differences in illuminance. Proper illumination is also a function of the type and form of the lighting fixtures. Luminaires are selected to shed direct, indirect, or diffused light. Lamp types include incandescent, fluorescent, mercury vapor, metal halide (multivapor), and high-pressure sodium vapor. The number and type of lamps per luminaire, height, spacing, and percent reflectance of the floor, walls, and ceiling are contributing factors. The plant should have a portion of its lights attached to an emergency power source that switches on when the regular power fails. (See Rosaler,

Scale: 1/8" = 1'- 0"

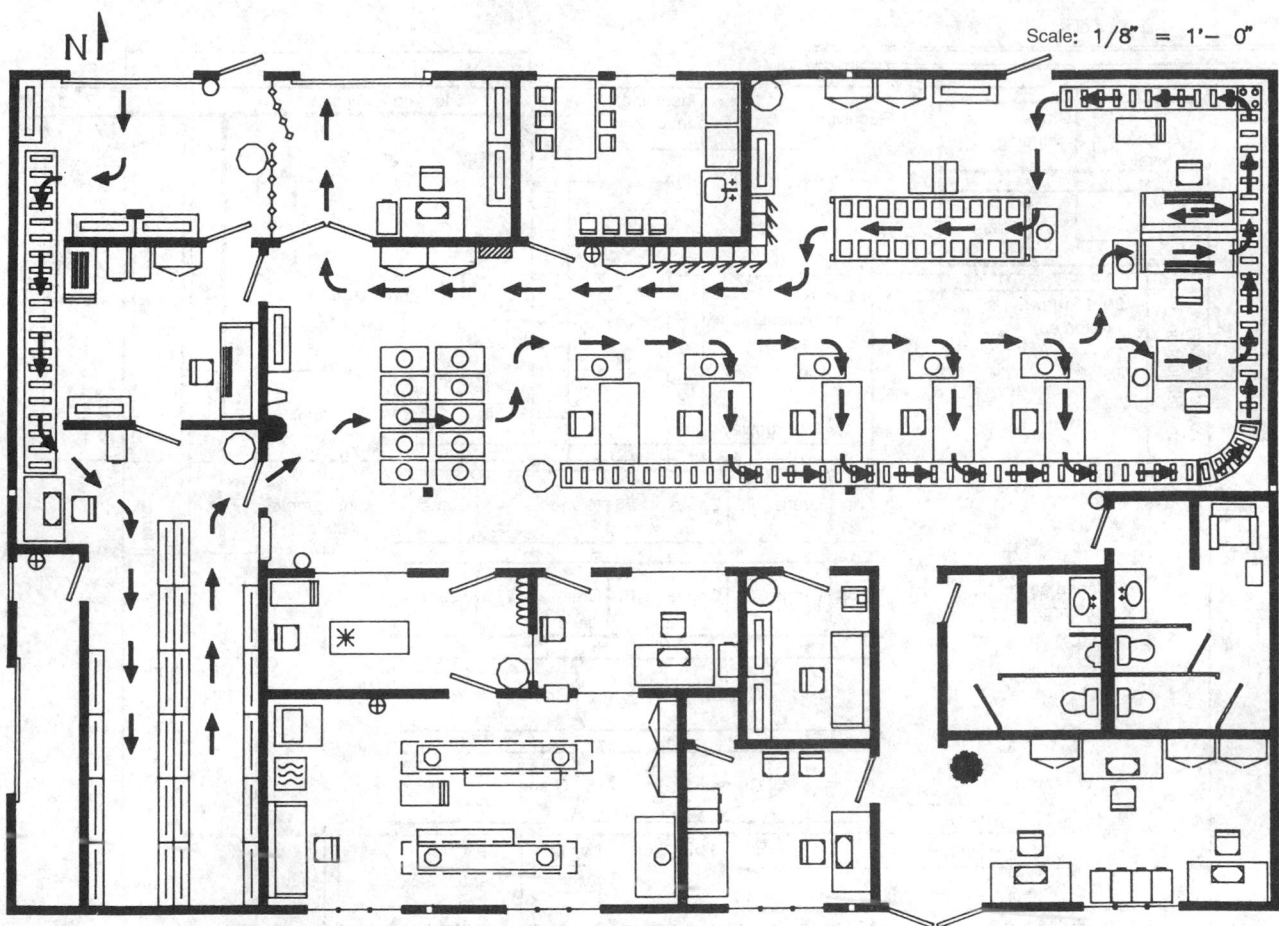

Fig. 12.1.8 A two-dimensional computer-generated detailed layout of one area of an industrial plant, showing furniture, fixtures, and flow path. *(Source: Cederleaf, "Plant Layout and Flow Improvement," McGraw-Hill.)*

"Standard Handbook of Plant Engineering.") A stationary diesel-, gas-, or gasoline-powered electric generator is usually installed to provide emergency electric power. The available fault current at all points in the electric distribution system should be determined so that protective devices can be installed to interrupt it. Selected circuits should have uninterruptible power supplies (UPSs), isolators, and regulators to protect against outages, voltage surges, sags, spikes, frequency drifts, and electrical noise. Exit signs and a clear path to the exits should be capable of being energized by emergency power, from either a standby generator or batteries.

Water, water distribution, and fixtures are essential for many plant processes, including paint booth "waterfalls"; for cooling machines, welders, and the like; for adding water as an ingredient in some products; in toilets, showers, cafeterias, and drinking fountains; for sprinkler systems and fire fighting; and for custodial work and landscaping maintenance. If large amounts of hot and/or cold water are required by the process, boilers and/or chillers are provided, along with associated piping and pumps for fuel and water distribution. The number of toilet fixtures as required by building codes is based on the number and sex of expected building occupants. Dispensers that chill and/or heat water are useful to prepare beverages and soups. Water consumption per person per 8-hour shift in personnel facilities ranges from 30 to 80 gal (114 to 303 L).

Sprinklers are installed according to the recommendations of the National Fire Protection Association (NFPA). Automatic sprinklers can be the wet type, wherein water is always in the pipes up to each head, or the dry type, wherein the pipes are filled with air under pressure to restrain water until the fusible link in the head melts and releases the air pressure, allowing water to flow.

The dry type is used outdoors and in unheated areas where water could freeze. Sprinkler heads are either standard or deluge type. Standard heads have a fusible element in each head that is melted by the heat of a fire, thereby opening the head and releasing water. Deluge heads do not have fusible elements, but a deluge valve which is opened in response to a signal from any of several heat sensors situated in the protected area. In the standard type, only the heads whose elements are melted release water, whereas deluge heads act simultaneously. A deluge system is more likely to extinguish sparks at the periphery of a fire, but it also may ruin materials that are doused unnecessarily with water. A preaction system includes sensors and an alarm which gives plant personnel a chance to deal with the fire before sprinklers actuate and douse valuable merchandise. The alarm may also be wired to signal the local fire department. The sprinkler heads must be spaced in accordance with the prevailing code: generally one head per 200 ft² (18.6 m²) for low-hazard areas, 90 ft² (8.4 m²) for high-hazard areas, and about 120 ft² (11.2 m²) for general manufacturing; about 8 to 12 ft (2.44 to 3.66 m) between heads and height of the highest head not over 15 ft (4.57 m) are typical. (See also Sec. 18.3.) Drains should be installed to carry away sprinkler water. The supply of water for all of the above listed needs must be adequate and reliable. For fire fighting, the greatest fire hazard and the size of the water supply required to deal with it must be estimated. The expected flow from all hoses and sprinklers, the static

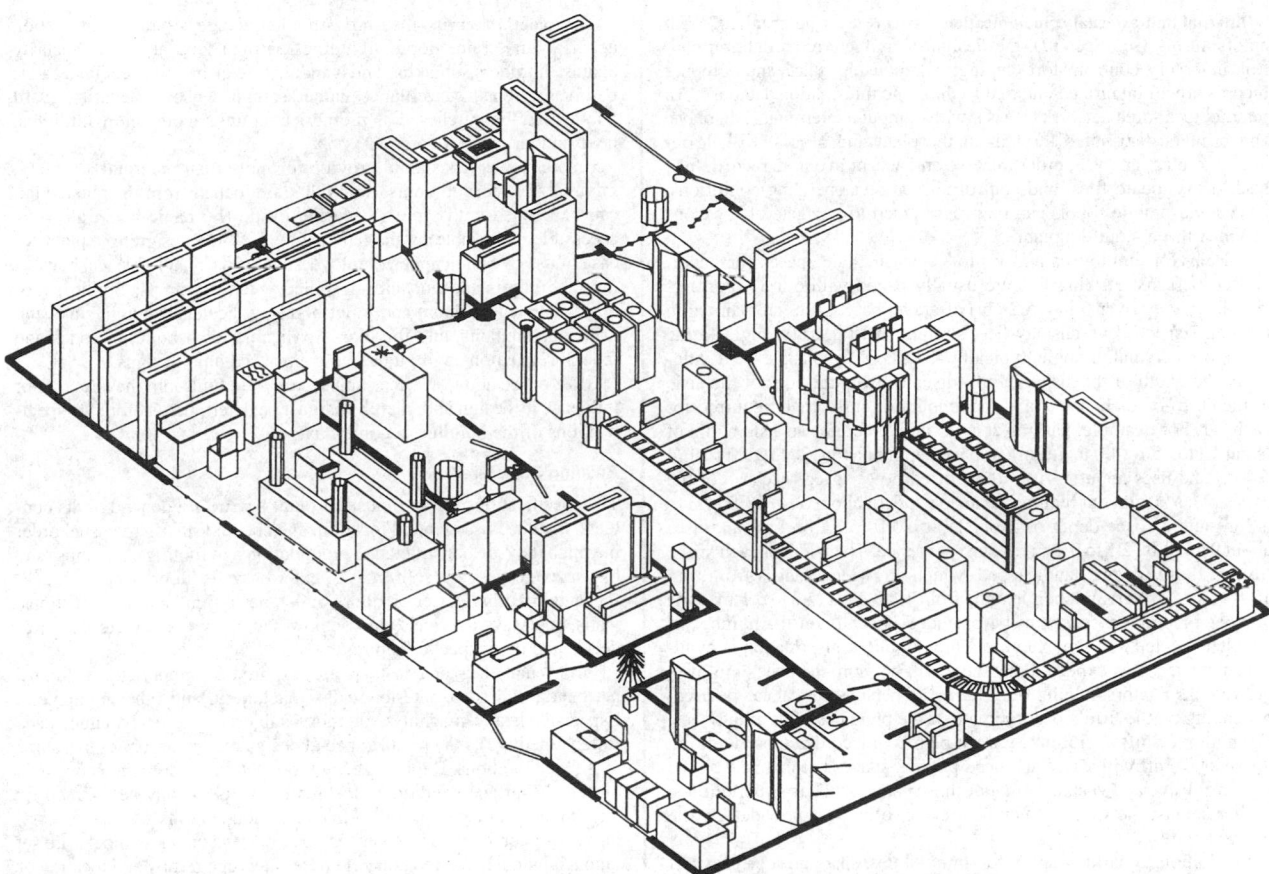

Fig. 12.1.9 A three-dimensional computer-generated detailed layout of the same area shown in Fig. 12.1.8. *(Source: Cederleaf, "Plant Layout & Flow Improvement," McGraw-Hill.)*

pressure, and the minimum flow available at a given residual pressure must be considered. Again, codes and insurance company requirements control. Sources of water include city mains, gravity tanks on towers, reservoirs on roofs or in decorative ponds that are part of the landscaping, wells, nearby lakes, and rivers.

The amount of **heat** required for personnel comfort depends on the plant's location, which in turn, influences the types and capacities of heaters selected. Generally, factory areas can be kept a little cooler, about 65°F than the 72°F recommended for office areas. Heating systems include warm air, hot water, steam, electric, and radiant. Warm air requires a furnace to heat the air and ducts and blowers to distribute it. Circulating hot water and steam heat also require boilers and furnaces and a network of distribution pipes and radiators. Electric space heaters require fixed wiring and local outlets. Unit radiant heats may be fueled by gas, steam, hot water, or electricity and are placed above doors and work areas, oriented to direct heat where it is required. Piping can be embedded in the concrete slab on grade, and in other floors, walls, and ceilings to provide radiant heat from circulating hot water.

Ventilation, the introduction of fresh outside air, the exhaust of stale inside air, or merely the movement of otherwise still inside air, can be effected naturally or mechanically. Natural ventilation requires windows, skylights, louvers, or other openings. Mechanical ventilation requires fans and blowers (and sometimes ducts) to draw air in, circulate it, and exhaust it. The number of cubic feet per minute (cubic metres per minute) of air per person required by people working in an office is about 10 (0.3), and between 25 and 50 (0.7 to 1.4) for those working on the factory floor. The total amount of air needed and the number of air changes per hour determine the number and sizes of fans, motors, and

ducts. One change per hour is too little, with no discernible difference in air quality; 50 to 60 changes of air per hour are too many, and the resulting high-velocity drafts cause sensations of chilling and accompanying discomfort. Minimally, 5 changes of air per hour are required.

Air conditioning adds to employee comfort, increases their productivity, and is essential for the manufacture of certain products. The calculation of the expected heat gains required to specify an **air-conditioning** installation is similar to the procedure for calculating heat loss in the design of a heating system. For air conditioning, the required cooling load is calculated in Btu per hour and converted to required tons of refrigeration by dividing by 12,000. (See Sec. 12.4.) The tonnage required would be one basis for the selection of the type of water-cooled or evaporative condensers, compressors, air handling units, motors, pumps, ducts, registers, circuitry, panels, and controls to be installed. Packaged air-conditioning units mounted on window sills, floors, ceilings, or roofs are often used. (See Merritt and Ricketts, "Building Design and Construction Handbook.")

The design and installation of a **compressed-air system** includes the summation of the volume and pressure of air required at all plant locations. The pipe diameters and total pressure drops between compressor and the most remote point of delivery must be determined. Pressure drop is a function of pipe and hose friction and the requirements of air-powered devices. Piping may be fixed, by and large, but the number and types of air-powered devices will change from time to time as production processes are changed or rearranged. Thus, there is some variability to be expected in the calculations to determine compressor discharge volume and pressure. Suitable allowances for unknown quantities are usually factored into the final design selection.

Internal and external communications systems must be installed. With autofacturing (see Sec. 17), the design, manufacture, production control, inventory control, storekeeping, warehousing, sales, and shipping records are all integrated and tied to the same information database. An internal local-area network (LAN), with computers, terminals, displays, and printers distributed throughout the plant and offices, will be required to plan, analyze, order, receive and accept material, record, feed back data, update files, and control and correct operating conditions. AGVs and mobile robots can be programmed to navigate off beacons installed throughout the plant.

Parking for employees and visitors must include spaces for handicapped persons. Parking lots are usually paved with 3 in (7.6 cm) of asphaltic cover over 6 in (15.2 cm) of gravel base. Lines are painted to designate spaces, which may be "straight in" (at a 90° angle to the curb), or at a smaller angle, typically 45° to 60°. The angle used influences the width of the aisles, depth (distance from the edge of the aisle to the curb) of each space, and the amount of curb length required for each car. For example, parking at a 45° angle requires an aisle width of about 13 to 15 ft (4.0 to 4.6 m), a space depth of about 20 to 21 ft (6.1 to 6.4 m) and uses about 13 ft (4.0 m) of curb per space (because of the overlaps); a 90° layout requires an aisle width of about 24 to 27 ft (7.3 to 8.2 m) and a space depth of about 19 to 20 ft (5.8 to 6.1 m) and uses about 9 to 10 ft (2.7 to 3.1 m) of curb per space. The width of each space (except those for the handicapped, which are wider) ranges from 9 to 10 ft (2.7 to 3.1 m) and their lengths from 19 to 20 ft (5.8 to 6.1 m). The number of parking spaces to be provided depends on the number of people expected and the extent to which public transportation is available and used. It is expected that there will be more than one employee per vehicle. Factors ranging from 1.2 to 2.5 persons per car can be used, depending on the firm's best estimate of the practices of its employees. For multiple-shift operations, it must be recognized that those working the second shift will arrive and need parking spaces before those on the first shift leave and vacate them. For those using public transportation, a shelter against the weather may be erected; often it is provided by the bus company.

In addition to parking space for automobiles, **space must be provided for trucks** that bring material to the plant and carry products away. Sizes of trucks, truck tractors, and their trailers vary widely, as do their turning radii. Many trucks will drive in frontward, maneuver to turn around, and back into the loading docks and platforms. Space must be provided for them to do this, even when there are other trucks present. The minimum size apron (measured by the distance from the outermost obstruction, whether it be a part of the building, the front of another truck already in the loading dock, or anything else that is in the way), required to maneuver a tractor-trailer into or out of a loading position, in one maneuver and with no driver error, varies with the length of the tractor-trailer expected and the width of the position to be provided. For example:

Tractor-trailer length	Position width	Minimum apron size
35 ft (10.7 m)	10 ft (3.1 m)	46 ft (14.0 m)
	12 ft (3.7 m)	43 ft (13.1 m)
	14 ft (4.3 m)	39 ft (11.9 m)
40 ft (12.2 m)	10 ft (3.1 m)	48 ft (14.7 m)
	12 ft (3.7 m)	44 ft (13.4 m)
	14 ft (4.3 m)	42 ft (12.8 m)
45 ft (13.7 m)	10 ft (3.1 m)	57 ft (17.4 m)
	12 ft (3.7 m)	49 ft (15.0 m)
	14 ft (4.3 m)	48 ft (14.6 m)

Truck heights range from about 8 ft (2.4 m) for a panel truck to around 13.5 ft (4.1 m) for double-axle semis and others. Doorways must be sized accordingly. Seals and pads may be added around the doors to fill the gap between the back of trucks and the building. Besides conventional rollup doors, plastic strips and air curtains may be used to provide a partial barrier between the plant's environment and the weather.

The **plant's security** measures should include structural protection against wind, rain, flood, lightning, earthquakes, and fire. Security against intruders, breakins and vandalism can include fencing, gates, perimeter lights, surveillance cameras, and sensors. Security guard booths may be purchased as prefabricated units, complete with lights, heat, and air conditioners.

Emergency egress should provide for more than one route to safety, minimal distances to doors, doors that are locked from the outside but can be opened with a push from the inside (by panic bars), all to be accessible and usable by the handicapped. When designing a plant for ease of use by handicapped employees and visitors, all physical barriers must be eliminated and aids installed, such as ramps, braille signs, wheelchair-width doors and toilet booths, wheelchair-height sinks and drinking fountains, and the like. Applicable local building codes and federal regulations will guide and control the final designs.

Most of the above services and systems are built into the structure or influence its design and, therefore, must be determined before the specifications of the building become final.

Building Structure

The specifications for an industrial plant's **structure** depend on its contents, intended use, general location, and specific site. In the usual order in which they are constructed are foundation and footings, columns (and bay sizes), beams and roof trusses, exterior walls, floor slab, roof decking and covering, exterior doors and fenestration, interior partitions, walls, doorways, services (electric power, compressed air, etc.), interior decor, and other special features.

Firm foundations and footings are required to support the columns, peripheral and some interior walls, machinery, and other equipment. Especially heavy machines, or those subject to large dynamic loads (often repetitive), may require special design that incorporates measures to isolate vibrations. Depth, size, and foundation reinforcement depends on the subsurface conditions. Installation of piles may be dictated by unsuitable load-bearing soil. Almost all foundations consist of cast-in-place concrete; in rare instances, load-bearing concrete blocks are set onto a previously poured concrete base. Site topography and location of the building may require the construction of retaining walls; these may be cast concrete or assembled with precast concrete sections which may serve also to provide a decorative treatment. The subgrade on which the concrete floor slabs will be poured must be well-compacted and made level. Often, the floor slab is not cast until heavy equipment (used to erect superstructure) is removed from the site.

Steel superstructure members are designed to provide the clear spans delineated by the selected **bay sizes**. They also support intermediate floors (if any); roof-mounted equipment; material handling equipment such as monorails; and lighting fixtures, piping, ductwork, and other utilities. The design must account for all dead and live loads, usually in accordance with applicable local building codes and other accepted structural codes. (See Secs. 5, 6, and 12.)

The sizes and locations of **columns and beams** establish the bay sizes. The longer the span, the larger the bay size, but this reflects back into construction with heavier columns, deeper beams and trusses, and concomitant higher costs. Inasmuch as larger bay sizes permit greater freedom in plant layouts and ease material handling, the added expense of large bay sizes is often worth it. Accordingly, the trend in plant design is to incorporate large bays. While any bay size is feasible, a cost/benefit tradeoff may set reasonable limits. Typical bay sizes are 30 × 40 ft (9 × 12 m), 40 × 60 (12 × 18) and 40 × 80 (12 × 24). There are warehouses whose bay size is dictated by the requirement to accommodate stacks of standard pallets with minimum waste space. Aircraft manufacturing plants and commercial hangars enclose enormous column-free cavernous spaces; lengths of 300 to 400 ft (91 to 122 m) are the norm in those applications.

When columns must support heavy traveling bridge cranes, a second row of columns is incorporated into the main columns to provide support for rails and to effect a stiffer structural configuration.

Beams are set at the top of columns and establish the clear height of the building interior; 15, 20, and 25 ft (4.5, 6.1, and 7.6 m) are typical in manufacturing areas, and 30 to 40 ft and more (9.1 to 12.2 m) are employed in warehouse space. When a two-story office is part of a single-level plant (Fig. 12.1.7) the overall height of the building is often set to the second-story height to permit a simple flat roof. Factory clear heights should be at least the height of the highest machine, with a generous margin added; for large products, the clear height is often designed to be twice the height of the largest product. Anything suspended from the beams or girders reduces the clear height. Deeper open roof trusses permit some piping, wiring, and other services to be woven through them and light fixtures placed between them, thereby not reducing clear working space. Roof decking and roofing is affixed to roof trusses or beams.

Exterior walls can be load bearing (support one end of a beam) or nonbearing. If nonbearing, all beam loads are transferred to columns spaced at intervals along the building perimeter. Walls are constructed of masonry [brick or concrete block masonry units (CMUs), sometimes incorporating glass block]; metal panels, usually integrally stiffened; wood for special applications (storage of highway deicing salts); natural stone or stone veneer/precast-concrete panels; or stone veneer on masonry backup walls. Where conditions allow, poured concrete walls are cast on the flat (at grade), then tilted up and secured into place; this technique is attractive especially for warehouses with repetitive wall construction devoid of windows and other openings. Metal panels are usually galvanized steel, with or without paint, or painted aluminum.

Factory floors may be required to sustain live loads from less than 100 pounds per square foot (psf) (0.5 kPa) to over 2,000 psf (96 kPa). A general-purpose light-assembly floor ordinarily will have no less than 100 psf (0.5 kPa) floor load capacity. A special purpose plant floor may be built with floor load capacities which vary from place to place in accordance with the requirements for different load-carrying capacity. Certainly, in the extreme, to build in a maximum floor loading capacity of, say, 2,000 psf (96 kPa) throughout a plant when there is minimal likelihood for that requirement other than in discrete areas, would not be cost-effective. Concentrated live loads (most often from wheeled material handling vehicles) are accounted for in the design of the floor slabs as required. The final design of the plant floor will meld the above factors with other requirements so that the sum of all dead and live loads can be safely sustained in the several areas of the floor.

Virtually all slab on-grade concrete floors are cast in place; in rare cases, precast concrete sections are used. Wooden plank flooring is rarely used on new construction. Upper floors are most often constructed with precast concrete planks overlaid with a cement mortar topping coat, or employ ribbed decking. When upper-floor loading capacities require it, those floors can also be cast in place; in those instances, concrete is cast into ribbed metal decking which acts as wet form work and remains in place. End grain wood blocks may be set atop a concrete floor to mitigate against unavoidable spillage of liquids. (See Sec. 12.2.)

The load-carrying capacity of concrete slabs on grade is ultimately limited not only by the strength of the concrete itself, but also by the ability of the subgrade material to resist deformation. The proper design of the floor, especially at grade, will account for all the parameters and result in a floor which will safely sustain design floor loading. Automatic guided vehicle systems and mobile robots (see Sec. 10) require level and smooth floors. The floor surface may be roughened slightly with a float to reduce slipping hazards, or have a steel trowel finish to accommodate AGV and mobile robot navigation. If it is contemplated that products or equipment will be moved by raising and floating them on films of air, the floor must be crack-free. If a slight floor slope can be tolerated from an operational point of view, that will ease cleaning by allowing water to flow to drains. The slab's surface may be coated, treated, or tiled for protection, comfort, ease of maintenance, and/or aesthetics. Expansion joints around column bases will inhibit propagation of small cracks resulting from differential settlement between column footings and floor slabs.

At this stage, the building looks like the one shown in Fig. 12.1.10.

A **flat roof** is always slightly pitched to drain water toward scuppers, gutters, and downspouts. It is constructed with ribbed metal decking or precast concrete planks, topped with a vapor barrier, insulation, a topping surface, flashing, sealants or caulking, and roof drains. The covering of a **built-up roof (BUR)** comprises three to five layers of roofing felt (fiberglass, polyester fabric, or other organic base material), each layer mopped with hot bitumen (asphalt or coal tar pitch). The wearing surface of roofing may include a cap sheet with fine mineral aggregate, reflective coating, or stone aggregate. When extensive roof traffic is anticipated, pavers or wood walkways are installed. A recent successful roof surface is composed of a synthetic rubber sheet, seamless except at lapped, adhesively bonded joints. It has proved to be a superior product, providing long, carefree life.

The seal between roofing and parapet walls or curbed openings is effected with **flashing** bent and cemented in place with bituminous cement. The flashing material may be galvanized steel aluminum, copper, stainless steel, or a membrane material.

Flat roofs lend themselves to ponding water to help cool the plant. Roof ponds (and water sprays) can reduce interior temperatures by 10 to 15°F (6 to 9°C) in the summer. Roof ponds can serve as a backup source of water for fire fighting. If water is ponded on a roof, roof construction must be handled expressly for that purpose. There are structural implications which must be considered as well.

Doors and ramps should be made 2 ft (0.6 m) larger than the largest equipment or material expected and large enough to allow ingress of fire fighting apparatus. It is often convenient to place very large pieces of equipment such as molding machines and presses inside the building envelope before the walls are completed. After exterior walls and roof are in place, the building interior is sufficiently secure for assembling materials for the remaining construction and to receiving smaller equipment.

Interior walls and **partitions** (masonry, wood, wallboard, and metal) follow, after which office flooring and framing are installed (Fig. 12.1.11). The remainder of the construction involves completion of wiring, lighting, and other services; application of paint or wall coverings, floor coverings, and decor; and so on.

Other Considerations

The **applicable standards, regulations, and procedures** of the 1971 Williams-Steiger Occupational Safety and Health Act (OSHA), as amended, must be followed in the construction and operation of the building and the design and use of the plant's equipment; likewise, the regulations of the Environmental Protection Agency (EPA), as amended, and the design standards of the Americans With Disabilities Act (ADA), as amended, must be followed.

Color can serve several purposes in an industrial plant. As an aid to safety, it can be used to identify contents of pipes, dangerous areas, aisles, moving parts of equipment, emergency switches, and fire fighting and first aid equipment. It also can be used to improve illumination, conserve energy, reduce employee fatigue and influence on morale positively.

OSHA, American National Standards Institute (ANSI) specification Z53.1, and specific safety regulations require the use of specific **identity colors** in certain applications. As a general guide, yellow or wide yellow-and-black bands are used to indicate the need for caution and to highlight physical hazards that persons might trip on, strike against, or fall into, such as edges of platforms, low beams, stairways, and trafficked aisles along which equipment moves. Orange designates dangerous parts of machines or energized equipment that may cut, crush, shock, or otherwise injure, such as the inside of gear boxes, gear covers, and exposed edges of gears, cutting devices, and power jaws. Red is the basic color for the identification of fire protection equipment and apparatus, danger and stop signs, fire alarm boxes, fire exit signs, and sprinkler piping. Green designates safety items, such as first aid kits, gas masks, and safety deluge showers, or their locations. Blue designates caution and is limited to warning against starting, use of, or movement of equipment under repair, and utilizes tags, flags, or painted barriers. Purple designates radiation hazards. Black, white, or a combination of both designated traffic control and housekeeping markings, such as aisleways, drinking fountains, and directional signs.

Fig. 12.1.10 A view of a plant during construction. Shown are its columns, roof trusses, and exterior masonry walls. The reinforced-concrete floor slab is being poured over compacted soil. *(Source: VMA, Inc.)*

There are also color conventions for the identification of fluids conducted in pipes. Instead of being painted their entire length, pipes are painted with colored bands and labels at intervals along the lines, at valves, or where pipes pass through walls. ANSI Standard A13.1-1975 specifies the following colors for pipe identification: red for fire protection; yellow for dangerous materials; blue for protective materials; green and white, black, gray, or aluminum for safe materials.

The **noises** created during the operation of a plant can be dealt with at their **sources, along their paths,** and at their **receivers.** At the sources, noise is minimized or eliminated by: changing the process; replacing (with quieter) equipment; modifying equipment by redesign or component changes; moving to another location; installing mufflers and shock mounts; using isolation pads; and shielding and enclosing equipment with material having a high sound transmission loss (STL) value. STL is a measure of the reduction in sound pressure as it passes through a material. Along the paths, attenuation of noise is enhanced by increasing the distance between sources and receivers; introducing discontinuities in the transmission path, such as barriers and baffles which interrupt direct transmission; installing acoustical barriers with an STL of at least 24 dB and with sufficient absorption to prevent reflection of the noise back to its source; and by placing sound-absorbent material on surfaces along its path, such as acoustical ceilings and floor and wall coverings. At the receivers, enclosures, workstation partitions of sound-absorbing material, earplugs, and "white-noise" generators can be used. (See Secs. 12.6 and 18.2.)

Provisions must be made for **waste disposal,** including: disposal of refuse and garbage, treatment of process fluids and solids, and waste and recyclables recovery, all in accordance with applicable EPA and other regulations.

Among many other things, the plant design process must address matters such as fuel storage, storm and surface drainage, snow removal facilities and procedures (if the local climate warrants), design for ease of maintenance of the building as well as the grounds and associated landscaping, the articulation of the architect's idiom, energy efficiency in the materials of construction and the equipment installed to service the plant, and diligence in compliance with environmental impact studies (if required).

After the plant is designed, but before a commitment is made to build it, its operations should be simulated to bring forth and correct any errors or omissions. In addition to testing its specifications for normal conditions, extremes (severe storms, power outages, truckers' strikes, etc.) should also be evaluated. This simulation is similar to that done before the plant's site is made final, but comes after the structure and layout have been designed. A sensitivity analysis should be made to see the effect on operations of changes in inputs and operating conditions. Depending on these analyses and best guesses as to the future, adjustments may have to be made to ensure that the plant will operate effectively and economically under all reasonable conditions, and that it has no features that will harm people or do damage to the products or the environment.

Fig. 12.1.11 A view of framing for the first- and second-story offices of a plant, with a roof beam and the steelwork supporting the second floor visible. *(Source: VMA, Inc.)*

CONTRACT PROCEDURES

Cost Estimates

Cost estimates fall into three general classes:

1. Preliminary estimates made usually from sketch drawings and brief outline specifications to determine the approximate total cost of a project.

2. Comparative estimates made usually during the progress of design to determine the relative cost of two or more alternative arrangements of equipment, type of building, type of floor framing, and the like.

3. Detail estimates made from final plans and specifications and based on a careful quantity survey of each component part of the work.

Primary quantity estimates are usually more accurate when comparison with comparable projects is lacking or not feasible. The procedure entails computations based on quantity takeoffs from preliminary plans and specifications, and can include the gross area of exterior walls, interior partitions, floors, and roof. Each is multiplied by known (or estimated) unit cost factors. Other items, such as number of electrical outlets and number of plumbing fixtures and sprinklers, are estimated, and their cost computed. Equipment costs can be based on preliminary quotations from manufacturers.

The following items should be included in preliminary estimates (or in other more detailed estimates which may follow as plans develop to the final stage): land cost; fees to real estate brokers, lawyers, architects, engineers, and contractors; interest during construction; building per-

mits; taxes, including local sales or use taxes; demolition of existing structures including removal of old foundations; yard work including leveling, drainage, fencing, roads, walks, landscaping, yard lighting, and parking spaces; transportation facilities including railroad tracks, wharves, and docks; power supply and source; water supply and source; sewer and industrial waste disposal. A judicious **contingency factor** is included to provide for unforeseen conditions that may arise during the development of the project. It may amount to 5 to 15 percent (or more) depending on the character and accuracy of the estimate, whether the proposed design and construction will follow established procedures or incorporate new state-of-the-art features, and the exact purpose for which the estimate is made.

Especially for complex new or altered construction, estimates may be updated continually as a matter of course. This will serve to monitor the evolving cost of the project and the time constraints placed on construction.

Working Drawings, Specifications, and Contracts

The technical staff of a given industry has special knowledge of trade practices, process requirements, operating conditions, and other fundamentals affecting successful operation in that field and is best fitted to determine the basic factors of process design and plant expansion. Unless the company is extremely large, it is unlikely that they will have the specialized staff or that their own staff will have the time available to undertake the complete layout. The efficient transformation of these requirements into completed construction usually requires experience of a different nature. Therefore, it is generally advisable and economical to employ engineers or architects who specialize in this particular field.

Three general procedures are in common use. First, the employment on a percentage or fixed-fee basis of an engineering and construction organization skilled in the industrial field to prepare the necessary working drawings and specifications, purchase equipment and materials, and execute the work. This has become known as **design/build**. For the duration of the project, such an organization becomes a part of the owner's organization, working under the latter's direction and cooperating closely with the owner's technical staff in the development of the design and in the purchase and installation of equipment and materials. This procedure permits construction work to start as soon as basic arrangements and costs have been determined, but before the time required to complete all working drawings and specifications. This is often termed **fast tracking**. Such a program will result in the earliest possible completion consistent with economical construction.

A second procedure is to employ engineers or architects with wide experience in the industrial field to prepare working drawings and specifications and then to obtain **competitive lump-sum bids** and award separate contracts for each or a combination of several subdivisions of the work, such as foundations, structural steel, and brickwork. This method provides direct competition restricted to units of like character and permits intelligent consideration of the bids received. Provision must be made for proper coordination of these separate contracts by experienced and skilled field supervision. In recent years, a **construction manager** has been hired for this purpose. This procedure requires more time than that first described, since all work of a given class should be completely designed before bids for that subdivision are sought. Where construction conditions are uncertain or hazardous, the first method is likely to be more economical. A combination of the first method for uncertain conditions and the second method for the balance may prove most advantageous at times.

A third procedure is to employ engineers or architects to complete all plans and specifications and then **award a lump-sum contract** for the entire work, or one for all building work, and one or more supplementary major contracts to furnish and install equipment. This method is particularly useful where there are no serious complications or hazards affecting construction operations, but it requires considerably more time, since most of the working drawings and specifications must be complete before construction is started. It has the significant advantage, however, of fixing the total cost within narrow limits before the work starts, provided the contracts cover the complete scope of work intended and no major changes ensue.

CONSTRUCTION

The design, construction, occupancy, and operation of an industrial plant is a complex endeavor involving many people and organizations. **Project control tools and techniques** such as the critical path method (CPM), the program evaluation and review technique (PERT), bar charts, dated start/stop schedules, and daily ''do lists'' and ''hot lists'' of late items are all helpful in keeping construction on time and within budget.

Checkpoints and milestones, with appropriate feedback as the project progresses, are established to monitor progress. Bailout points are established in the event the project must be aborted sometime after the beginning of construction. On the other hand, contingency plans should be in hand and include planned reassignment of financial and personnel resources to keep the project on schedule. Figure 12.1.12 shows the scheduled and actual dates of completion of the phases noted for a small plant depicted in several of the previous figures.

Some portion of the organization (often the construction manager's office), is charged with the delicate task of coordinating contractors and the multiplicity of trades at the job site, and as part of its function, it will generate ''punch lists'' of deficient and/or incomplete work which must be completed to the owner's satisfaction before payment therefore is approved.

	Scheduled	Actual
Project planning	March	March
Basic layout	April	April
Structural design	April	April
Equipment specifications	April	April
Construction contract	May	May
Ground breaking	June	June
Detailed office layout	June	June
Detailed plant layout	July	July
Equipment purchases	August	August
Furniture purchases	September	September
Organization firmed	September	September
Steelwork completed	October	October
Masonry completed	October	October
Roof completed	November	November
Staffing and training	November	November
Systems designed	November	November
Forms designed	December	December
Offices completed	December	December
Equipment received	December	December
Building occupied	December	December

Fig. 12.1.12 The project schedule and completion dates for the single-level 202,800 ft^2 industrial plant used as an example in Figs. 12.1.1 to 12.1.7, 12.1.10, and 12.1.11. (*Source: VMA, Inc.*)

12.2 STRUCTURAL DESIGN OF BUILDINGS
by Aine M. Brazil

REFERENCES: ''Manual of Steel Construction—Allowable Stress Design,'' American Institute of Steel Construction. ''National Design Specification for Wood Construction,'' American Forest and Paper Association. ''Design Values for Wood Construction,'' American Forest and Paper Association. ''Uniform Building Code,'' International Conference of Building Officials. ASCE 7-93, ''Minimum Design Loads for Buildings and Other Structures,'' American Society of Civil Engineers. Blodgett, ''Design of Welded Structures,'' J. F. Lincoln Arc Welding Foundation. ''Masonry Designers Guide,'' The Masonry Society.

LOADS AND FORCES

Buildings and other structures should be designed and constructed to support safely all loads of both permanent and transient nature, without exceeding the allowable stresses for the specified materials of construction. Dead loads are defined as the weight of all permanent construction; live loads are those loads produced by the use or occupancy of the building; environmental loads include the effects of wind, snow, rain, and earthquakes.

Live loads on floors are generally regulated by the building codes in cities or states. For areas not regulated, the following values will serve as a guide for live loads in lb/ft^2 (kPa): rooms for habitation, 40 (1.92); offices, 50 (2.39); halls with fixed seats, 60 (2.87); corridors, halls, and other spaces where a crowd may assemble, 100 (4.76); light manufacturing or storage, 125 (5.98); heavy manufacturing or storage, 250 (11.95); foundries, warehouses 200 to 300 (9.58 to 14.37). Floor decks and beams that support only a small floor area must also be designed for any local concentrations of load that may come upon them. Girders, columns, and members that support large floor areas, except in buildings such as warehouses where the full load may extend over the whole area, may often be designed for live loads progressively reduced as the supported area becomes greater. Where live loads, such as cranes and machinery, produce **impact** or **vibration**, static loads should be increased as follows: elevator machinery, 100 percent; reciprocating machinery, 50 percent; others, 25 percent.

Roof live loads should be taken as a minimum of 20 lb per horizontal ft^2 (957.6 kPa) for essentially flat roofs (rise less than 4 in/ft), varying linearly with increasing slope to 12 lb/ft^2 for steep slopes (rise greater than 12 inches per foot). Reductions may also be made on the basis of tributary area greater than 200 ft^2 (18.58 m^2) of the member under consideration, to a maximum of 40 percent for tributary areas greater than 600 ft^2. The minimum roof live load shall be 12 lb/ft^2 after all reduction factors have been applied. Where **snow loads** occur, appropriate design values should be based on the local building codes.

Dead loads are due to the weight of the structure, partitions, finishes, and all permanent equipment not included in the live load. The weights of common building materials used in floors and roofs are given below (see also Sec. 6).

Material	Weight, lb/ft^2
Asphalt and felt, 4-ply	3
Corrugated asbestos board	5
Glass, corrugated wire	5–6
Glass, sheet, ⅛ in thick	2
Lead, ⅛ in thick	8
Plaster ceiling (suspended)	10
Acoustical tile	1–2
Sheet metal	1–2
Shingles, wood	3
Light weight-concrete over metal deck	30–45
Sheathing, 1 in wood	3
Skylight, ³⁄₁₆ to ¼ in, glass and frame	6–8
Slate, ³⁄₁₆ to ½ in thick	8–20
Tar and gravel, 5-ply	6
Tar and slag, 5-ply	5
Roof tiles, plain, ⅝ in thick	20

NOTE: lb/ft^2 × 0.04788 = kPa.

Earthquake Effects Two distinct methods of designing structures for seismic loads are currently accepted. The more familiar method, employed by the Uniform Building Code (UBC), yields equivalent loading for use with the allowable stress design approach. The National Earthquake Hazard Reduction Program (NEHRP) has developed Recommended Provisions for the Development of Seismic Regulations for New Buildings, which is based on the ultimate strength design approach. The NEHRP approach is the basis for model codes, such as BOCA (Building Officials and Code Administrators International, Inc.) National Building Code.

Following the UBC design approach, the static force procedure represents the earthquake effects as equivalent static lateral forces applied at each floor level. The total design lateral force (called **base shear**) is calculated by the formula

$$V = \frac{ZIC}{R_w} W \qquad \text{where} \qquad C = \frac{1.25\,S}{T^{2/3}} \leqq 2.75$$

V = design base shear; Z = factor representing the degree of regional seismicity, ranging from 0.4 for seismically active areas with proximity to certain earthquake faults (Zone 4) to 0.075 for areas of low seismicity (Zone 1); I = importance factor (1.25 for essential facilities, 1.0 for most others); R_w = coefficient representing the type of lateral-force-resisting system of the building, ranging from 4 for a heavy timber bearing wall system (where bracing carries gravity loads as well as lateral loads) to 12 for buildings with highly ductile systems, such as special moment-resisting frames (this coefficient is a measure of the past earthquake resistance of various structural systems); T = the fundamental period of vibration, seconds, of the building in the direction under consideration (this coefficient represents the acceleration effects of the dynamic response of the structure); S = coefficient representing possible amplification effects of soil-structure interaction and is taken as 1.5 unless a lower value is substantiated by soils data; and W = the total dead load (including partitions) plus snow loads over 30 lb/ft^2 (1.436 Pa) and 25 percent of any storage or warehouse live loads.

The **fundamental period** T, used to calculate the seismic coefficient C, may be determined by a rational analysis of the structural properties and deformation characteristics of the structure, or it may be estimated by the formula $T = C_t(h_n)^{3/4}$, where C_t = 0.035 (0.0853) for steel moment-resisting frames, 0.030 (0.0731) for reinforced-concrete moment-resisting frames and eccentrically braced frames, and 0.02 (0.0488) for all other buildings.

The **base shear** V is considered to be distributed over the height of the structure according to the formula

$$F_x = \frac{(V - F_t)w_x h_x}{\Sigma w_i h_i}$$

where $F_t = 0.07TV$ is the lateral force at the top and F_x is the lateral force at any level x; w_x is the weight assigned to level x; and h_x is the height of level x above the base.

For stiff, low-rise buildings, such as one- to three-story, steel-braced frame or concrete shear wall structures, it is common practice to compute the design base shear using the maximum values for the coefficients C. Thus, the design base shear for a two-story concrete shear wall building in Zone 4 might be taken as

$$V = ZICW/R_w = (0.4 \times 1.0 \times 2.75 \times W)/8 = 0.1375W.010$$

Where floors are **rigid diaphragms,** such as concrete fill over metal deck or concrete slabs, lateral forces are distributed to the vertical-resistive elements on the basis of their relative stiffness. Where floors or roofs are flexible diaphragms, such as some metal deck with nonstructural fill, plywood, or timber planking, lateral forces are distributed to the resistive elements on the basis of tributary area. Where a rigid diaphragm exists, a torsional moment, equal to the story shear multiplied by the greater of the real eccentricity between the center of mass and the center of rigidity of the resistive elements or 5 percent of the maximum building dimension at that level, is applied to the diaphragm around a vertical axis through the center of rigidity of the resistive elements. Direct shears in the elements are increased by those induced by the torque when additive but are unaltered when subtractive in order to arrive at design lateral seismic loads to the resistive element.

Earthquake forces on portions of structures, such as walls, partitions, parapets, stacks, appendages, or equipment, are calculated by the formula $F_p = ZI_pC_pW_p$, where F_p = the equivalent lateral static force acting at the center of mass of the element; Z and I_p = coefficients previously defined (although I_p for life safety equipment may be greater than that for the parent structure); C_p = horizontal force factor and is taken as 2.0 for cantilever chimneys, stacks, and parapets as well as ornamental appendages and as 0.75 for other elements such as walls, partitions, ceilings, penthouses, and rigidly mounted equipment; and W_p = weight of element. For flexibly mounted equipment C_p may conservatively be taken as 2.0 or may be determined by rational analysis considering the dynamic properties of both the equipment and the structure which supports it. Seismic loads are applied to walls and partitions normal to their surface and to other elements in any horizontal direction at the center of mass.

Wind Pressures on Structures   Every building and component of buildings should be designed to resist wind effects, determined by taking into consideration the geographic location, exposure, and both the shape and height of the structure. For structures sensitive to dynamic effects, such as buildings with a height-to-width ratio greater than 5, structures sensitive to wind-excited oscillations, such as vortex shedding or icing, and tall buildings [height greater than 400 ft (121.9 m)] special consideration should be given to design for wind effects and procedures used should be in accordance with approved national standards. Wind loads should not be reduced for the shading effects of adjacent buildings.

Wind pressure on walls of buildings should be assumed to be a minimum of 15 lb/ft^2 (0.72 kPa) on surfaces less than 50 ft above the ground. For buildings in exposed locations and in locations with high wind velocity (over 70 mi/h or 112.5 km/h), pressures should be calculated based on the following procedure.

Design wind pressure may be determined by the following formula, which is based on the Uniform Building Code: $P = C_eC_qq_sI_w$, where P = design wind pressure; C_e = coefficient which varies with height, exposure, and gust factor (refer to Table 12.2.1); C_q = pressure coefficient (refer to Table 12.2.2), q_s = wind stagnation pressure at standard height of 33 ft (refer to Table 12.2.3); I_w = importance factor (essential or hazardous facilities, 1.15; other structures, 1.0). The **basic wind speed** is the fastest wind speed at 33 ft (10 m) above the ground of terrain Exposure C and associated with an annual probability of occurrence of 0.02 [varies from 70 to 100 mi/h (112 to 161 km/h)]. The exposure category defines the characteristics of ground surface irregularities at the specific site: Exposure B has terrain with numerous closely spaced obstructions having the size of single-family dwellings or larger; Exposure C is flat, open terrain with scattered obstructions having a height of less than 30 ft; Exposure D is flat, unobstructed areas exposed to wind flowing over large bodies of water.

For exceptionally tall, slender or flexible buildings, it is recommended that a wind tunnel test be performed on a model of the building.

Table 12.2.1   Combined Height, Exposure, and Gust Factor Coefficient C_e*

Height above average level of adjoining ground, ft†	Exposure D	Exposure C	Exposure B
0–15	1.39	1.06	0.62
20	1.45	1.13	0.67
25	1.50	1.19	0.72
30	1.54	1.23	0.76
40	1.62	1.31	0.84
60	1.73	1.43	0.95
80	1.81	1.53	1.04
100	1.88	1.61	1.13
120	1.93	1.67	1.20
160	2.02	1.79	1.31
200	2.10	1.87	1.42
300	2.23	2.05	1.63
400	2.34	2.19	1.80

* Values for intermediate heights above 15 ft (4.6 m) may be interpolated.
† Multiply by 0.305 for meters.

Table 12.2.2 Pressure Coefficients C_q

Structure or part thereof	Description	C_q factor

1. Primary frames and systems	**Method 1** (normal force method)
	Walls:
	    Windward wall
	    Leeward wall
	Roofs[a]:
	    Wind perpendicular to ridge
	    Leeward roof or flat roof
	    Windward roof
	        Less than 2:12 (16.7%)
	        Slope 2:12 (16.7%) to less than 9:12 (75%)
	
	        Slope 9:12 (75%) to 12:12 (100%)
	        Slope > 12:12 (100%)
	    Wind parallel to ridge and flat roofs
		
		0.8 inward
		0.5 outward
		
		0.7 outward
		
		0.7 outward
		0.9 outward or
		    0.3 inward
		0.4 inward
		0.7 inward
		0.7 outward

	Method 2 (projected area method)
	On vertical projected area
	    Structures 40 feet (12 192 mm) or less in height
	    Structures over 40 feet (12 192 mm) in height
	On horizontal projected area[a]
		1.3 horizontal any direction
		1.4 horizontal any direction
		0.7 upward

2. Elements and components not in areas of discontinuity[b]	Wall elements
	    All structures
	    Enclosed and unenclosed structures
	    Partially enclosed structures
	    Parapets walls
		
		1.2 inward
		1.2 outward
		1.6 outward
		1.3 inward or outward

	Roof elements[c] Enclosed and unenclosed structures Slope < 7:12 (58.3%) Slope 7:12 (58.3%) to 12:12 (100%) Partially enclosed structures Slope < 2:12 (16.7%) Slope 2:12 (16.7%) to 7:12 (58.3%) Slope > 7:12 (58.3%) to 12:12 (100%)	 1.3 outward 1.3 outward or inward 1.7 outward 1.6 outward or 0.8 inward 1.7 outward or inward

3. Elements and components in areas of discontinuities[b,d,e]	Wall corners[f]
	
	Roof eaves, rakes or ridges without overhangs[f]
	    Slope < 2:12 (16.7%)
	    Slope 2:12 (16.7%) to 7:12 (58.3%)
	    Slope > 7:12 (58.3%) to 12:12 (100%)
	For slopes less than 2:12 (16.7%)
	    Overhangs at roof eaves, rakes or ridges, and canopies
		1.5 outward or 1.2 inward
		
		2.3 upward
		2.6 outward
		1.6 outward
		
		0.5 added to values above

4. Chimneys, tanks, and solid towers	Square or rectangular Hexagonal or octagonal Round or elliptical	1.4 any direction 1.1 any direction 0.8 any direction

5. Open frame towers[g,h]	Square and rectangular
	    Diagonal
	    Normal
	Triangular
		
		4.0
		3.6
		3.2

6. Tower accessories (such as ladders, conduit, lights and elevators)	Cylindrical members
	    2 inches (51 mm) or less in diameter
	    Over 2 inches (51 mm) in diameter
	Flat or angular members
		
		1.0
		0.8
		1.3

7. Signs, flagpoles, light-poles, minor structures[h]		1.4 any direction

[a] For one story or the top story of multistory partially enclosed structures, an additional value of 0.5 shall be added to the outward C_q. The most critical combination shall be used for design.

[b] C_q values listed are for 10-ft^2 (0.93-m^2) tributary areas. For tributary areas of 100 ft^2 (9.29 m^2), the value of 0.3 may be subtracted from C_q, except for areas at discontinuities with slopes less than 7 units vertical in 12 units horizontal (58.3% slope) where the value of 0.8 may be subtracted from C_q. Interpolation may be used for tributary areas between 10 and 100 square feet (0.93 m^2 and 9.29 m^2). For tributary areas greater than 1,000 ft^2 (92.9 m^2), use primary frame values.

[c] For slopes greater than 12 units vertical in 12 units horizontal (100% slope), use wall element values.

[d] Local pressures shall apply over a distance from the discontinuity of 10 ft (3.05 m) or 0.1 times the least width of the structure, whichever is smaller.

[e] Discontinuities at wall corners or roof ridges are defined as discontinuous breaks in the surface where the included interior angle measures 170° or less.

[f] Load is to be applied on either side of discontinuity but not simultaneously on both sides.

[g] Wind pressures shall be applied to the total normal projected area of all elements on one face. The forces shall be assumed to act parallel to the wind direction.

[h] Factors for cylindrical elements are two-thirds those for flat or angular elements.

Table 12.2.3   Wind Stagnation Pressure q_s at Standard Height of 33 ft

Basic wind speed, mi/h*	70	80	90	100	110	120	130
Pressure q_s, lb/ft^{2†}	12.6	16.4	20.8	25.6	31.0	36.9	43.3

* Multiply by 1.61 for kilometers per hour.
† Multiply by 0.048 for kilonewtons per square meter.

Boundary-layer wind tunnels, which simulate the variation of wind speed with height and gusting, are used to estimate the design wind loading.

The effect of the sudden application of **gust loads** has sometimes been blamed for peculiar failures due to wind. In most cases, these failures can be explained from the pressure distributions in a steady wind. If a relatively flexible structure such as a radio tower, chimney, or skyscraper with a natural period of 1 to 5 s is set into vibration, the stresses in the structure may be increased over those calculated from a static-load analysis. Provision should be made for this effect by increasing the static design wind. A rational analysis should be performed to calculate the magnitude of this increase, which will depend on the flexibility of the structure.

Even a steady wind may give rise to periodic forces which may build up into large vibrations and lead to failure of the structure when the frequency of the exciting force coincides with one of the natural frequencies of vibration of the structure. The periodic exciting force may be due to the separation of a system of **Kármán vortices** (Fig. 12.2.1) in the wake of the body. The exciting frequency n in cycles per second is related to d, the dimension of the body normal to the wind velocity V, by the equation $nd/V = C$, where $C \approx 0.207$ for circular cylinders and $C \approx 0.18$ for rectangular plates (Blenk, Fuchs, and Liebers, Measurements of Vortex Frequencies, *Luftfahrt-Forsch.*, 1935, p. 38). Dangerous vibrations related to the ''flutter'' of airplane wings may arise on bridges and similar flat bodies. These self-induced vibrations may be caused (1) by a negative slope of the curve of lift against angle of attack (Den Hartog, *op. cit.*) or (2) by a dynamic instability which arises when a body having two or more degrees of freedom (such as bending and torsion) moves in such a manner as to extract energy out of the air stream. The first of these vibrations will occur at one of the natural frequencies of the structure. The second type of vibration will occur at a frequency intermediate between the natural frequencies of the structure.

Fig. 12.2.1   Von Kármán vortices.

The wind forces and the **pressure distribution** over a structure corresponding to a design wind V can be determined by **model testing** in a wind tunnel. Extrapolation from model to full scale is based on the fact that at every other point on the body, the pressure p is proportional to the stagnation pressure q (see Sec. 11) and thus the ratio $p/q =$ constant for a fixed point on the body, as the scale of the model or the velocity of the wind is changed. Since the principal component of the wind force is due to the pressures, the force F acting on the surface S is

$$F \approx p_{avg}S = (p/q)_{avg}qS$$

and denoting $(p/q)_{avg}$ by a normal force coefficient or **shape factor** C_N

$$F = C_N \tfrac{1}{2}\rho V^2 S = C_N qS$$

The shape factors so obtained apply to full scale for structures with sharp edges whose principal resistance is due to the pressure forces. For bodies that do not have any sharp edges perpendicular to the flow, such as spheres or streamlined bodies, the factor C_N is not constant. It depends upon the Reynolds number (see Sec. 3). For such bodies, the law for variation of the shape factor C_N must be determined experimentally before safe predictions of full-scale forces can be made from model measurements.

Radio Towers and Other Framed Structures   For open frame towers, by using round structural members instead of flat and angular sections, a substantial reduction in wind force can be effected. Refer to note h in Table 12.2.2.

Load Combinations   Methods of combining types of loading vary with the governing local codes. Dead loads are usually considered to act all the time in combination with either full or reduced live, wind, earthquake, and temperature loads. The reductions in live loads are based on the improbability of fully loaded tributary areas, generally when the tributary area exceeds 150 ft². Most codes consider that wind and earthquake loads need not be taken to act simultaneously. There are two basic design methods in use: allowable stress or working stress design and limit states or ultimate strength design. Whereas allowable stress design compares the actual working stresses with an allowable value, the limit states approach determines the adequacy of each element by comparing the ultimate strength of the element with the factored design loads. The following basic load combinations are applicable for allowable stress design: dead plus live (floor and roof); dead plus live plus wind; dead plus live plus earthquake. Most codes permit a one-third increase in stresses for load combinations considering either wind or earthquake effects.

DESIGN OF STRUCTURAL MEMBERS

Members are usually proportioned so that stresses do not exceed allowable **working stresses** which are based on the strength of the material and, in the case of compressive stresses, on the stiffness of the element under compression. Internal forces and moments in **simple beams,** columns, and pin-connected truss bars are obtained by means of the equations of static equilibrium. **Continuous beams,** rigid frames, and other members characterized by practically rigid joints require for analysis additional equations derived from consideration of deflections and rotations.

Design may also be on the basis of the **ultimate strength** of members, the factor of safety being embodied in stipulated increases in the design loads. In steel-frame construction, the procedures of **plastic design** determine points where the material may be allowed to yield, forming *plastic hinges,* and the resulting redistribution of internal forces permits a more efficient use of the material.

Floors and Roofs

Except in reinforced-concrete flat-slab construction, floors and roofs generally consist of flat decks supported upon beams, girders, or trusses. The decks may usually be considered a series of beamlike strips spanning between beams and themselves designed as beams. The design of a beam consists chiefly in proportioning its cross section to resist the maximum bending and shear and providing adequate connections at its supports, without exceeding the unit stresses allowed in the materials (see Sec. 5) and limiting maximum live load deflection at midspan to $\frac{1}{360}$ of the span.

Up to spans of 20 to 30 ft (6.1 to 9.1 m), either wood or steel beams of uniform section are generally more economical than trusses, while for spans above 50 to 70 ft (15.2 to 21.3 m), trusses are usually more economical. Between these limits, the line of economy is not well-defined. Conditions that favor the use of trusses are as follows: (1) identical trusses are repeated many times, (2) the height of the building need not be increased for the greater depth of the truss (3) fire protection of wood or metal is not required.

Roof trusses often have their top chords sloped with the roof. Common trusses for steeply pitched roofs are shown in Figs. 12.2.2 to 12.2.6, the top chord panels equal in each truss. The members shown by heavy lines are in compression under ordinary loads, those in light lines in tension. The trusses of Figs. 12.2.6 to 12.2.9 are adapted for either steel or wood. In wooden trusses, the tensile web members may be steel rods with plates, nuts, and threaded ends. The truss of Fig. 12.2.6 is usually made of steel.

The forces in any member of these trusses under a vertical load uniformly distributed may be found by multiplying the coefficients in Tables 12.2.4 to 12.2.8 by the panel load P on the truss. For other slopes, types of trusses, or loads, see "General Procedure" below. Trusses for flat roofs are commonly of one of the types shown in Figs. 12.2.7 to 12.2.13 except that the top chords conform to the slope of the roof.

Floor trusses normally have parallel chords. Common types are shown in Figs. 12.2.7 to 12.2.13 in which heavy lines indicate members in compression, light lines in tension, and dash lines members with only

nominal stress, under equal vertical panel loads. The panel lengths l in each truss are equal. The stress in each member is written next to the member in the figure, in terms of the panel load and the lengths of members. For a truss like one of the figures turned upside down, the stresses in the chords and diagonals remain the same in magnitude but

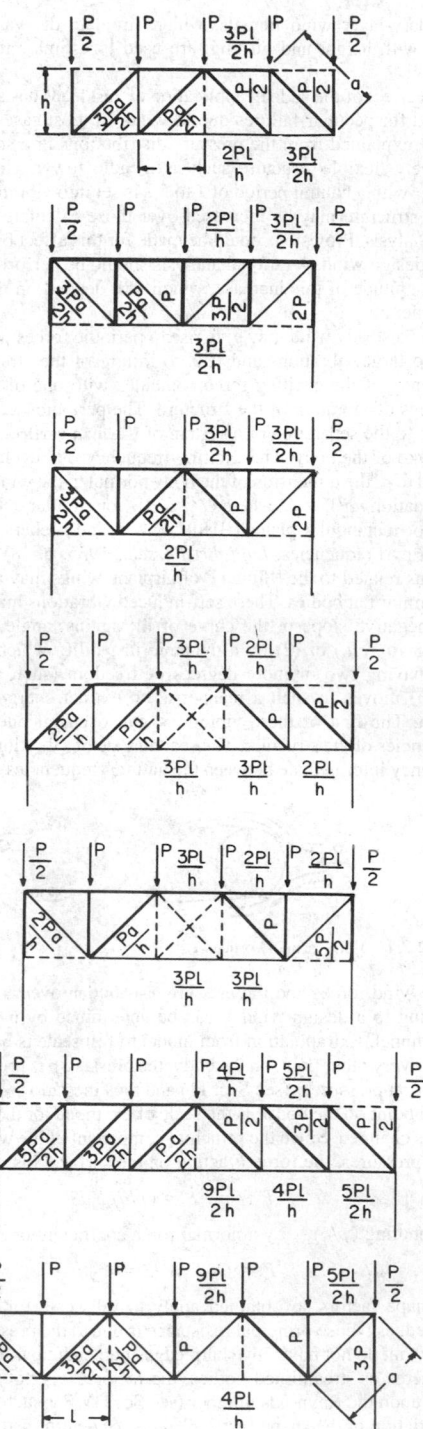

Fig. 12.2.2 to 12.2.6 Types of steep roof trusses.

Fig. 12.2.7 to 12.2.13 Types of floor and roof trusses.

Table 12.2.4 Coefficients for Truss Shown in Fig. 12.2.2

	Coefficients of *P* for force in				
Pitch *h*/*L*	*AD*	*BD*	*CE*	*DE*	*EF*
⅓	2.25	2.71	1.80	0.90	1.00
0.288*	2.60	3.00	2.00	1.00	1.00
¼	3.00	3.35	2.24	1.12	1.00
⅕	3.75	4.03	2.69	1.35	1.00

Table 12.2.5 Coefficients for Truss Shown in Fig. 12.2.3

	Coefficients of *P* for force in								
Pitch *h*/*L*	*AE*	*AG*	*BE*	*CF*	*DH*	*EF*	*FG*	*GH*	*HJ*
⅓	3.75	3.00	4.50	2.7	3.60	0.83	0.72	1.25	2.00
0.288*	4.33	3.46	5.00	3.0	4.00	0.88	0.73	1.32	2.00
¼	5.00	4.00	5.59	3.35	4.47	0.94	0.75	1.42	2.00
⅕	6.25	5.00	6.74	4.04	5.39	1.07	0.79	1.60	2.00

Table 12.2.6 Coefficients for Truss Shown in Fig. 12.2.4

	Coefficients of *P* for force in					
Pitch *h*/*L*	*AD*	*AF*	*BD*	*CE*	*DE*	*EF*
⅓	2.25	1.50	2.70	2.15	0.83	0.75
0.288*	2.60	1.73	3.00	2.50	0.87	0.87
¼	3.00	2.00	3.35	2.91	0.89	1.00
⅕	3.75	2.50	4.04	3.67	0.93	1.25

Table 12.2.7 Coefficients for Truss Shown in Fig. 12.2.5

	Coefficients of *P* for force in							
Pitch *h*/*L*	*AE*	*AH*	*BE*	*CF*	*DG*	*EF*	*FG*	*GH*
⅓	3.75	2.25	4.51	3.91	4.51	0.83	1.42	2.50
0.288*	4.33	2.60	5.00	4.33	5.00	0.88	1.45	2.65
¼	5.00	3.00	5.59	4.84	5.59	0.94	1.49	2.83
⅕	6.25	3.75	6.73	5.83	6.73	1.07	1.57	3.20

Table 12.2.8 Coefficients for Truss Shown in Fig. 12.2.6

	Coefficients of *P* for force in											
Pitch *h*/*L*	*AF*	*AH*	*AM*	*BF*	*CG*	*DK*	*EL*	*FG* *KL*	*GH* *JK*	*HJ*	*JM*	*LM*
⅓	5.25	4.50	3.00	6.31	5.75	5.20	4.65	0.83	0.75	1.66	1.50	2.25
0.288*	6.06	5.20	3.46	7.00	6.50	6.00	5.50	0.87	0.87	1.73	1.73	2.60
¼	7.00	6.00	4.00	7.83	7.38	6.93	6.48	0.89	1.00	1.79	2.00	3.00
⅕	8.75	7.50	5.00	9.42	9.05	8.68	8.31	0.93	1.25	1.86	2.50	3.75

* 30 slope.

reversed in sign. Forces in verticals must be computed (compare Figs. 12.2.7 and 12.2.8). For other loads and other types of trusses see ''General Procedure'' below.

Weights of Trusses The approximate weight in pounds of a wooden roof truss may be taken as $W = LS(L/25 + L^2/6{,}000)$, where L is the span and S the spacing of trusses in feet. The approximate weight in pounds of a steel roof truss may be taken as $W = \frac{1}{5}LS(\sqrt{L} + \frac{1}{8}L)$.

Choice of Roof Trusses

WOODEN TRUSSES. For pitched roofs with spans up to 20 ft (6.1 m) the simple king-post truss (Fig. 12.2.2) may be used. For spans up to 40 ft (12.2 m) the trusses of Figs. 12.2.2 and Fig. 12.2.4 are good. For spans up to 60 ft (18.3 m) the trusses of Figs. 12.2.3 and 12.2.5 are good. The number of panels rarely exceeds eight or the panel length, 10 ft (3m). For flat roofs, the Howe truss (Figs. 12.2.7, 12.2.10, and 12.2.12) is built of wood with steel rods for verticals; the depth is one-eighth to one-twelfth the span. Wooden trusses are usually spaced 10 to 15 ft (3.0 to 4.6 m) on centers. Wood is rarely used in roof trusses with span over 60 ft (18.3 m) long. Steel trusses for pitched roofs may well take the form shown in Figs. 12.2.3 to 12.2.6 for spans up to 100 ft (30.5 m). For flat roofs, the Warren truss (Figs. 12.2.9, 12.2.11, and 12.2.13) is usually constructed in steel; the depth of the trusses ranges from one-eighth to one-twelfth the span, with trusses spaced from 15 to 25 ft (4.6 to 7.6 m) on centers.

Stresses in Trusses

An ideal truss is a framework consisting of straight bars or members connected at their ends by frictionless ball-and-socket joints. The external forces are applied only at these ball-and-socket joints. Internal forces and stresses in such straight bars are axial, either tension or compression, without bending. Since frictionless ball-and-socket joints are impossible, and the ends of bars are often bolted or welded, the ideal truss is never realized. For purposes of analysis, the primary stresses, which are always axial, are determined on the assumption that the truss under consideration conforms to the ideal. Secondary stresses are additional stresses, generally flexural or bending, brought about by all the factors that make the actual truss different from the ideal. In the following discussion, only primary forces and stresses will be considered.

Analytical Solution of Trusses

General Procedure After all external forces (loads and reactions) have been determined, the internal force or stress in any member is found (1) by taking a section, making an imaginary cut through the members of the truss, including the one whose stress is to be found, so as to separate the truss into two parts; (2) by isolating either of these parts; (3) by replacing each bar cut by a force, representing the force in the bar; and (4) by applying the equations of statics to the part isolated.

The various ways in which sections can be taken and the equations used to determine the forces are illustrated by a solution of the truss in Fig. 12.2.14.

A section 1–1 (Fig. 12.2.14) may be taken around a joint, L_0. Isolate the forces inside the section (Fig. 12.2.15a). Assume the unknown forces to be tension. Since the forces are concurrent and coplanar, two independent equations of statics will establish equilibrium. These may be either $\Sigma x = 0$ and $\Sigma y = 0$ or $\Sigma M = 0$ taken about two axes perpendicular to the plane of the forces and passing through two points selected so that neither point is the intersection of the forces, or so that the line joining the two points is not coincident with either of the unknown forces.

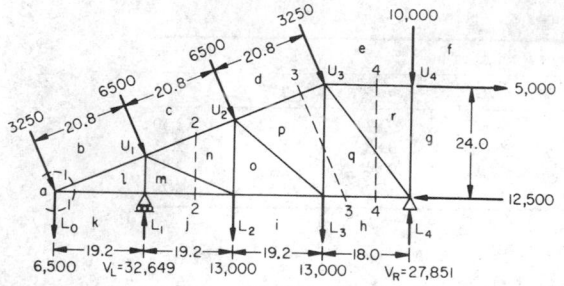

Fig. 12.2.14

Using the first set of equations and taking components of all forces along horizontal and vertical axes; e.g., the horizontal component of the 3,250 lb force is 1,250 and the vertical component is 3,000 lb.

$$\Sigma x = 0 = s_{lk} + s_{bl}(19.2/20.8) + 1,250$$
$$\Sigma y = 0 = s_{bl}(8/20.8) - 9,500$$

From these equations, $s_{lk} = -24,050$ and $s_{bl} = 24,700$.

The minus sign indicates that the force acts opposite to the assumed direction. If all the unknown forces are assumed to be in tension, then a plus sign in the result indicates that the force is tension and a minus sign indicates compression.

Hence, s_{lk} is 24,050 lb compression and s_{bl} is 24,700 lb tension.

Using the $\Sigma M = 0$ twice,

$$\Sigma M \text{ about } U_1 = 0$$
$$= -s_{lk} \times 8 - 1250 \times 8 - 9,500 \times 19.2$$

from which $S_{lk} = 24,050$ lb compression.

$$\Sigma M \text{ about } L_1 = 0 = s_{bl}(8/20.8)19.2 - 9,500 \times 19.2$$

from which $s_{bl} = 24,700$ tension.

Instead of taking a section around a joint, a cut may be made vertically or inclined, cutting a number of bars such as Fig. 12.2.15b or Fig. 12.2.15c. If only three members are cut, and they are neither concurrent nor parallel, the forces can be found by taking moments of all forces on either side of the section about axes passing through the intersections of any two members.

Considering the part to the left of section 2–2, the forces in the three members (Fig. 12.2.15b) may be determined by taking moments about L_0, U_1, and L_2 of all forces acting on the part on either side of the section, e.g., on the left of the section because it has the fewer forces:

$$\Sigma M \text{ about } L_0 = s_{mn}(8/20.8)38.4 + 2,500$$
$$\times 8 - (32,649 - 6,000)19.2$$
$$\Sigma M \text{ about } U_1 = 0 = -s_{mj} \times 8 - 1,250 \times 8 - 9,500 \times 19.2$$
$$\Sigma M \text{ about } L_2 = 0 = s_{cn}(19.2/20.8)16 + 2,500$$
$$\times 8 - 9,500 \times 38.4 + 26,649 \times 19.2$$

from which $s_{mn} = 33,290$ tension, $s_{mj} = 24,050$ compression, and $s_{cn} = 11,298$ compression.

This method is sometimes called the "method of moments."

Considering the part to the right of section 3–3, the forces in the three members (Fig. 12.2.15c) may be determined by taking moments about L_0, U_3, L_3 of all the forces acting on part or either side of the section, e.g., on the right of the section because it has the fewer forces:

$$\Sigma M \text{ about } L_0 = 0 = s_{pq} \times 57.6 + 6,250 \times 24 + 3,000$$
$$\times 57.6 - (27,851 - 10,000) \times 75.6$$
$$\Sigma M \text{ about } L_3 = 0 = -s_{dp}(19.2/20.8)24 + 6,250 \times 24$$
$$- 17,851 \times 18$$
$$\Sigma M \text{ about } U_3 = 0 = s_{qh} \times 24 + 12,500 \times 24 - 17,851 \times 18$$

from which $s_{pq} = 17,825$ tension, $s_{dp} = 7,733$ compression, and $s_{qh} = 888$ tension.

Considering the part to the right of section 4–4, the forces in the three members (Fig. 12.2.15d) may be found by taking moments about axes where two unknowns intersect, e.g., about U_3 and L_4. Since two unknown forces are parallel, their lines of action do not intersect. However, the equation $\Sigma y = 0$ will enable one to find the force in the member which is not parallel to the other two:

$$\Sigma M \text{ about } U_3 = 0 = s_{qh} \times 24 - (27,851 - 10,000)18 + 12,500 \times 24$$
$$\Sigma M \text{ about } L_4 = 0 = -s_{er} \times 24 + 5,000 \times 24$$
$$\Sigma y = 0 = s_{rq}(24/30) + 27,851 - 10,000$$

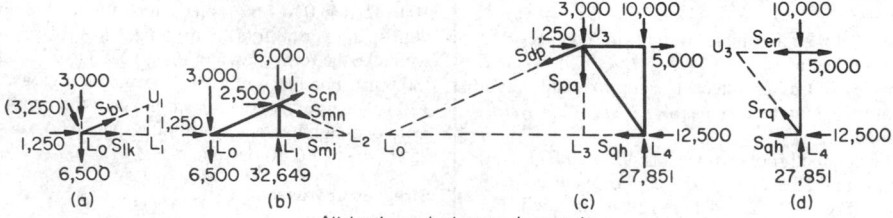

All loads and stresses in pounds

Fig. 12.2.15 Resolution of forces for truss shown in Fig. 12.2.14.

from which s_{qh} = 888 tension, s_{er} = 5,000 tension, and s_{rq} = 22,314 compression.

Columns and Walls

Vertical elements in building construction consist of columns, posts, or partitions that transmit concentrated loads; walls or partitions that transmit linearly distributed loads (and may, if so designed, transmit lateral loads from story to story); rigid frames that transmit lateral as well as vertical loads; and braced frames that, ideally, transmit only lateral loads from story to story.

Columns may be of timber, steel, or reinforced concrete and should be proportioned for the allowable stresses permitted for the material used and for the flexibility of the column. Care must be taken in the framing of beams and girders to avoid or to provide for the additional stresses due to connections which transfer loads to columns with large eccentricities. For instance, a beam framing to a flange of a steel column, with a seat or web connection, will produce an eccentric moment in the column equal to $Rd/2$, where R is the beam reaction and d the depth of the column section. Columns not adequately restrained against deflection of the top may be subject to considerable additional moment because of the resulting eccentricity of otherwise axial loads.

Rigid frames consist of columns and beams welded, bolted, or otherwise connected so as to produce continuity at the joints and permit the entire frame to behave as a unit, capable of resisting both vertical and lateral loads. Advantages of rigid frames are the ease and simplicity of erection, increased headroom, and more open floor plans free of braced frames. Rigid frames composed of rolled sections are commonly used for spans up to 100 ft (30.5 m) in length. Built-up members have been utilized on spans to 250 ft (76.2 m). Welded fabrication offers particular advantages for frames utilizing variable-depth members and on parabolic-shape roofs. The distribution of moments in the statically indeterminate rigid frame is effected by the relationship of the column height to span and roof rise to column height, as well as the relative stiffness of the various members. The solution for the moments in the frame is obtained from the usual equations of statics plus one or more additional equations pertaining to the elastic deformations of the frame under load.

Bearing walls, which are intended to transmit gravity loads from story to story, are designed as vertical elements of unit width such that the combined effects of axial and/or bending stresses do not exceed their allowable values according to the formula $fa/Fa + fb/Fb \leq 1.0$, where f is the actual stress in the axial or bending mode, and F is the allowable stress in the appropriate mode.

Shear walls, either concrete or masonry, which are intended to transmit earthquake, wind, or other lateral loads from story to story parallel to the plane of the wall, are designed as cantilevered vertical shear beams. Loads are distributed among the various wall elements, created by door and window openings, on the basis of their relative rigidities. Unless special conditions indicate otherwise, the assumption is made, in calculating individual stiffnesses, that the wall is fixed against rotation, in the plane of the wall, at the bottom. The individual wall elements are then designed to resist both their share of the wall shear and the moments it induces as well as any vertical loads due to bearing wall action and/or overall wall overturning. Individual wall elements, then, will have special vertical reinforcing at each side (trim reinforcement) to resist bending moments. Diagonal trim reinforcements, 96 bar diameters in total length, where possible, are placed at corners of openings in concrete walls to inhibit cracking. Trim reinforcing should consist of a minimum of two ⅝-in-diameter (1.6-cm) bars (No. 5 bars).

Uplift and downward forces at opposite ends of the shear walls, due to the tendency of the wall to overturn in its own plane when acted upon by lateral forces, must be resisted by vertical reinforcement at those locations. Special boundary elements are required for shear walls in areas of high seismicity (UBC Zones 3 and 4). Only minimum tributary dead loads may be counted upon to help resist uplift while maximum dead plus live loads should be assumed acting simultaneously with downward overturning loads. **Load reversal,** due to wind or seismic loads, must be considered.

Shear walls should be positioned throughout the plan area of the building in such a manner as to distribute the forces uniformly among the individual walls with approximately one-half of the wall area oriented in each of the two major directions of the building. At least one major wall should be positioned, if possible, near each exterior face of the building. Shear walls should be continuous, where possible, from top to bottom of the building; avoid offsetting walls from floor to floor, and particularly avoid discontinuing walls from one level to the level below. Positioning shear walls near major floor openings that would preclude proper anchorage of floor (or roof) to wall should be avoided so that diaphragm forces in the plane of the floor may have an adequate force path into the wall

Where **rigid diaphragms** (such as concrete slabs or metal deck with concrete fill) exist, shears are distributed to the individual walls on the basis of their relative rigidity taking into account the additional forces due to the larger of either accidental or actual horizontal torsion resulting from the eccentricity of the building seismic shear (located at the center of mass) or eccentricity of applied wind load with respect to the center of rigidity of the walls. Where **flexible diaphragms** (such as metal deck with nonstructural fill or plywood) exist, shears are distributed to walls on the basis of tributary area. When diaphragms are semirigid, based on type, construction, or spacing between supporting elements, shears could be distributed by both methods (relative rigidity and tributary area) with individual walls designed to resist the larger of the two shears.

Braced frames, which are intended to transmit lateral loads from floor to floor, are designed as trusses that are loaded horizontally instead of vertically and are principally used in steel construction. Lateral loads are distributed to them on the same basis as to shear walls; therefore, the same general rules apply to their ideal placement. Bracing of individual bays may take the form of X-bracing, single diagonal bracing, or K-bracing. Where K-braces are used, the additional bracing forces associated with vertical floor loads to the braced beam must be considered. Alternative floor framing to avoid large beam loads to the braced beam should be considered where possible.

X-bracing may be designed so that either the tension diagonal takes all the lateral load or so that the tension and compression diagonals share the load. When the tension diagonal resists all the lateral load, the minimum slenderness ratio (kl/r) of the member should not exceed 300; the additional axial load in the column to which the top of the brace is connected, the connection load at each end of the brace and the horizontal footing load at the bottom of the brace, is double that obtained when braces share the load. In X-bracing systems where the tension and compression members share the lateral load, maximum slenderness ratios should be limited to 200 (the members may be considered to brace each other about axes that are normal to the plane of the bay), but connection and individual horizontal foundation loads are decreased. With either method of design, it is important to follow the path of the force from its origin through the structure until it is transferred to the ground. With any bracing system, it is wise to sketch to scale the major connection details prior to finalization of calculations; the desirable intersection of member axes is often more difficult to achieve than a single line drawing would indicate. Connection of braces to columns at intermediate points between floors should be avoided because plastic hinges in columns lead to structural instability.

Stud walls consist of wooden studs with one or more lines of horizontal bridging. The allowable vertical load on the wall is a function of the maximum permissible load on each stud as a column and the spacing of the studs.

Corrugated or flat sheet steel or aluminum, are commonly employed for walls of industrial or mill buildings. These sheets are usually supported on steel girts framing horizontally between columns and supported from heavier eave struts by one or more lines of vertical steel sag rods.

Reinforced-concrete or masonry walls should be designed so that the allowable bending and/or axial stresses are not exceeded, but the minimum thicknesses of such walls should not be less than the following:

Material	Max ratio, unsupported height or length to thickness	Nominal min thickness, in*
Reinforced concrete	25	6
Plain concrete	22	7
Reinforced brick masonry	25	6
Grouted brick masonry	20	6
Plain solid masonry	20	8
Hollow-unit masonry	18	8
Stone masonry (ashlar)	14	16
Interior nonbearing concrete or masonry (reinforced)	48	2
Interior nonbearing concrete or masonry (unreinforced)	36	2

* × 25.4 = mm.

Foundations

Bearing Pressure of Soils The bearing pressure which may be allowed on soil may vary over a large range. For important structures, the nature of the underlying soil should be ascertained by borings or test pits. If the soil consists of medium or soft clay, a settlement analysis based on consolidation tests of undisturbed soil samples from the foundation strata is necessary. Structures founded upon mud, soft clay, silt, peat, or artificial filling will almost certainly settle, and no foundation for a permanent structure should rest on or above such material without adequate provision for the resulting settlement. Table 12.2.9 gives a general classification of soils and typical safe pressures which they may support.

These values approximate the pressures allowed by the building law in most cities. Actual allowable bearing pressures should be based on the quality and engineering characteristics of the material obtained from analysis of borings, standard penetration tests, and rock cores. Other laboratory tests, when the cost thereof is warranted by the magnitude of the project, may show higher values to be safe. The foundation for a building housing heavy vibrating machinery such as steam hammers, heavy punches, and shears should receive some allowance for possible compression and rearrangement of soil due to the vibrations transmitted through it. The foundation for a tall chimney should be designed with a comparatively low pressure upon the soil, because of the disastrous results which might occur from local settlement.

Footings The purpose of footings is to spread the concentrated loads of building walls and columns over an area of soil so that the unit pressure will come within allowable limits. Footings are usually constructed of concrete, placed in open excavations with or without sheeting and bracing. In the past, stone, where available in quantity and the proper quality, has been used economically for residential building footings.

Table 12.2.9 Safe Bearing of Soils

Nature of soil	Safe bearing capacity	
	tons/ft²	MPa
Solid ledge of hard rock, such as granite, trap, etc.	25–100	2.40–9.56
Sound shale and other medium rock, requiring blasting for removal	10–15	0.96–1.43
Hardpan, cemented sand, and gravel, difficult to remove by picking	8–10	0.76–0.96
Soft rock, disintegrated ledge; in natural ledge, difficult to remove by picking	5–10	0.48–0.96
Compact sand and gravel, requiring picking for removal	4–6	0.38–0.58
Hard clay, requiring picking for removal	4–5	0.38–0.48
Gravel, coarse sand, in natural thick beds	4–5	0.38–0.48
Loose, medium, and coarse sand; fine compact sand	1.5–4	0.15–0.38
Medium clay, stiff but capable of being spaded	2–4	0.20–0.38
Fine loose sand	1–2	0.10–0.20

Allowable bearing pressures for granular materials are typically determined as 10 percent of the standard penetration resistances (N values) obtained in the field from standard penetration tests (STPs) and are in tons/ft².

Concrete footings may be either plain or reinforced. Plain concrete footings are generally limited to one- or two-story residential buildings. The center of pressure in the wall or column should always pass through the center of the footing. Where columns, because of fixity, impose a bending moment on the footing, the maximum soil pressures, due to the combined axial and moment components, shall be positive throughout the area of the footing (i.e., no net uplift) and shall be less than the allowable values. Footings in ground exposed to freezing should be carried below the possible penetration of frost.

Deep foundations are required when a suitable bearing soil is deep below the surface, typically more than 10 to 15 ft. (3.0 to 4.6 m). Sometimes deep foundations are necessary even if the suitable bearing materials are shallower, but groundwater conditions make excavation for footings difficult. The most common types of deep foundations include caissons or drilled piers and piles. Where the bearing soil is clay stiff enough to stand with undercutting, and the material immediately above it is peat or silt, the **open-caisson method** may be economical. In this method, cylindrical steel casings 3 ft and more in diameter are sunk as excavation proceeds, the casings having successively smaller diameters. At the bottom of the shaft thus formed, the soil is undercut to obtain sufficient bearing area. The shaft and the enlargement at the base are then filled with concrete, the cylinders sometimes being withdrawn as the concrete is placed. The open-caisson method cannot be used where groundwater flows too freely into the excavation. Where large foundations under very heavy buildings must be carried to great depth to reach rock and hardpan, particularly where groundwater flows freely, the **pneumatic-caisson method** is used.

Drilled-in piers are formed by drilling with special power augers up to 5 ft (1.5 m) diameter or greater. The holes are drilled to the desired bearing level with or without metal casings, depending on the soil conditions. Belling of the bottom may also be performed mechanically from the surface. In poor soils, or where groundwater is present, the hole may be retained with bentonite clay slurry which is displaced as concrete is placed in the caisson by the **tremie method.**

Pile Foundations Piles for foundations may be of wood, concrete, steel, or combinations thereof. Wood piles are generally dressed and, if required, treated offsite; concrete piles may be prepared offsite or cast in place; steel piles are mill-rolled to section. They can be driven, jacked, jetted, screwed, bored, or excavated. Wood piles are best suited for loads in the range of 15 to 30 tons (133.4 to 177.9 KN) per pile and lengths of 20 to 45 feet (6.6 to 15 m). They are difficult to splice and may be driven untreated when located entirely below the permanent water table; otherwise piles treated with creosote should be used to prevent decay. Wood piles should be straight and not less than 6 in in diameter under the bark at the tip. Concrete piles are less destructible, and hence are adaptable to many conditions, including driving in dense gravels, and can be up to approximately 120 ft (40 m) long. Concrete piles are divided into two classes: those poured in place and those precast, cured, and driven. Cast-in-place piles are made by driving a mandrel into the ground and filling the resulting hole with concrete. In one well-known pile of this type (Raymond), a thin sheet-steel corrugated shell is fitted over a tapered mandrel before driving. This shell, which is left in the ground when the mandrel is removed, is filled with concrete. Prestressing of precast-concrete piles provides greater resistance to handling and driving stresses. With a concrete pile, 25 to 60 tons (222.4 to 533.8 KN) or more per pile are carried. Structural steel H columns and steel pipe with capacities of 40 to 300 tons (350 to 1800 kN) per pile have been driven. Experience has shown that corrosion is seldom a practical problem in natural soils, but if otherwise, increasing the steel section to allow for corrosion is a common solution.

Methods of Driving Piles The drop hammer and the steam hammer are usually employed in driving piles. The steam hammer, with its comparatively light blows delivered in rapid succession, is of advantage in a plastic soil, the speed with which the blows are delivered preventing the readjustment of the soil. It is also of advantage in soft soils where the driving is easy, but a light hammer may fail to drive a heavy pile satisfactorily. The water jet is sometimes used in sandy soils. Water supplied under pressure at the point of the pile through a pipe or hose run alongside it erodes the soil, allowing the pile to settle into place. To have full capacity, jetted piles should be driven after jetting is terminated particularly if the pile is to resist uplift loads.

Determination of Safe Loads for Piles   Piles may obtain their supporting capacity from friction on the sides or from bearing at the point. In the latter case, the bearing capacity may be limited by the strength of the pile, considered as a column, to which, however, the surrounding soil affords some lateral support. In the former case, no precise determination of the bearing capacity can be made. Many formulas have been developed for determining the safe bearing capacity in terms of the weight of the hammer, the fall, and the penetration of the pile per blow, the most generally accepted of which is that known as the *Engineering News* formula: $R = 2wh/(s + 1.0)$ for drop hammers, $R = 2wh/(s + 0.1)$ for single-acting steam hammers, $R = 2E/(s + 0.1)$ for double-acting steam hammers, where R = safe load, lb; w = weight of hammer, lb; h = fall of hammer, ft; s = penetration of last blow, in; E = rated energy, ft·lb per blow. This formula and similar ones are based on the determination of the energy in the falling hammer, and from this the pressure which it must exert on the top of the pile. The *Engineering News* formula is currently used typically for timber piles with capacities not exceeding 30 tons. It assumes a factor of safety of 6. It is a wise practice to drive index piles and determine their bearing capacity through pile load tests typically carried to twice the service load of the pile before proceeding with the final design of important structures to be supported on piles. The design then is based on the safe service load capacity so determined, and the piles are driven to the same penetration resistance to which the successfully tested index piles where driven by the same driving hammer.

Spacing of Piles   Wood piles are preferably spaced not closer than 2½ ft (0.76 m), and concrete piles 3 ft (0.91 m) on centers. If driven closer than this, one pile is liable to force another up. Piles in a group must not cause excessive pressure in soil below their tips. The efficiency, or supporting value of friction piles when driven in groups, by the Converse-Labarre method, is

$$\frac{1 - d/s[(n - 1)m + (m - 1)n]}{90mn}$$

where d = pile diameter, in (cm); s = spacing center to center of piles, in (cm); m = number of rows; n = number of piles in a row.

Capping of Piles   Piles are usually capped with concrete; wood piles sometimes with pressure-treated timber. Concrete is the most usual material and the most satisfactory for the reason that it gets a full bearing on all piles. The piles should be embedded 4 to 6 in (10 to 15 cm) in the concrete.

Retaining Walls   A wall used to sustain the pressure of earth behind it is called a retaining wall. Retaining walls which depend for their stability upon the weight of the masonry are classed as gravity walls. Such walls built on firm soil will usually be stable when they have the following proportions: top of fill level, back vertical, base, 0.4 height; top of fill level, back battered, base, 0.5 height; top of fill steeply inclined, back vertical, base, 0.5 height; top of fill steeply inclined, back battered, base, 0.6 height. An additional factor of safety is obtained by building the face on a batter. Care should be taken in the design of a wall that the allowable soil pressure is not exceeded and that drainage is provided for the back of the wall. The foundations of retaining walls should be placed below the level of frost penetration. Retaining walls of reinforced concrete are made thin, with a broad base, and the wall either cantilevered from the base or braced with buttresses or counterforts.

It is impossible to derive formulas for the earth pressure on the back of the wall which will take account of all the actual conditions. Assuming the earth to be a loose, homogeneous, granular mass, and the coefficient of friction to be independent of the pressure, Rankine deduced the following formula for a wall with vertical back:

$$P = (\tfrac{1}{2}wh^2 + vh) \cos d(\cos d$$
$$- \sqrt{\cos^2 d - \cos^2 a})/(\cos d + \sqrt{\cos^2 d - \cos^2 a})$$

the center of pressure being at a height $\tfrac{1}{3}H(wh + 3v)/(wh + 2v)$ above the base, where P = earth pressure per lin ft (m) of the wall, lb (kg); h = height of the wall, ft (m); w = weight of earth per ft³ (m³), lb (kg); v = weight of superposed load per ft² (m²) of surface, lb (kg); d = the angle with the horizontal of the earth surface behind the wall; and a = angle of internal friction of the earth, deg. (For sands, a = 30 to 38°.) The direction of the pressure is parallel to the earth's surface. The retaining wall should have sufficient thickness at the base so that the resultant of the earth pressure P combined with the weight of the wall falls well within the base. If this resultant falls at the outside edge of the middle third, the maximum vertical pressure on the foundation (at the outer edge of the base) will be equal to $2W/T$ lb/ft² (Pa), where W is the total vertical pressure on the base of 1 ft (m) length of wall and T the thickness of the wall at the base, ft (m).

In the design of walls of buildings which must withstand earth pressure and low independent walls, where refinement is not necessary, the earth pressure is frequently assumed to be that of a fluid weighing 40 to 45 lb/ft³ (641 to 721 kg/m³).

MASONRY CONSTRUCTION

The term **masonry** applies to assemblages consisting of fired-clay, concrete, or stone units, mortar, grout, and steel reinforcement (if required). The choice of materials and their proportions is based on the required strength (compressive, flexural, and shear), fire rating, acoustics, durability, and aesthetics. The strength and durability of masonry depends on the size, shape, and quality of the unit, the type of mortar, and the workmanship. Resulting bond strength is affected by the initial rate of absorption, texture and cleanliness of the masonry units.

Brick   Common red bricks are made of clay burned in a kiln. Quality characteristics are hardness and density. Light-colored brick is apt to be soft and porous. Brick for masonry exposed to the weather or where strength is desired should have a crushing strength of not less than 2,500 lb/in² (17.3 MPa) and should absorb not over 20 percent of water by weight, after 5-h immersion in boiling water (see Sec. 6).

Mortar   Mortar is the bonding agent that holds the individual units together as an assembly. All mortars contain cement, sand, lime, and water in varying proportions. Mortars are classified as portland-cement-lime mortar or masonry cement mortar. Proportions for Type M portland-cement-lime mortar are 1 part portland cement, ¼ part lime, and 3 parts sand, by volume; for masonry cement mortar, 1 part masonry cement, 1 part portland cement, and 6 parts sand. Portland-cement-lime mortar should be used for all structural brickwork (see Sec. 6).

Laying and Bonding   Brick should be laid in a full bed of mortar and shoved laterally into place to secure solid bearing and a bed of even thickness and to fill the vertical joints. Clay brick should be thoroughly wet before laying, except in freezing weather. Concrete bricks/blocks should not be wetted before placement. Brick laid with long dimension parallel to the face of the work are called stretchers, perpendicular to the face, headers. Bats (half-brick) should not be used except where necessary to make corners or to form patterns on the face of the wall. Each continuous vertical section of a wall one masonry unit thick is called a **wythe.** Multiwythe walls, of brick or block faced with brick, are now commonly tied together with ties or anchors between each wythe of brick and block. Such ties are typically provided every other course. When the wythes are adequately bonded or tied; the wall may be considered as a composite wall for strength purposes. Walls may also be tied together longitudinally by overlapping stretchers in successive courses. Transverse bond is obtained by making every sixth course headers, the headers themselves overlapping in successive courses in the interior of thick walls. Variations in the arrangement of headers are often used in the face of walls for appearance. The area of cross section of full-length headers should not be less than one-twelfth the face of the

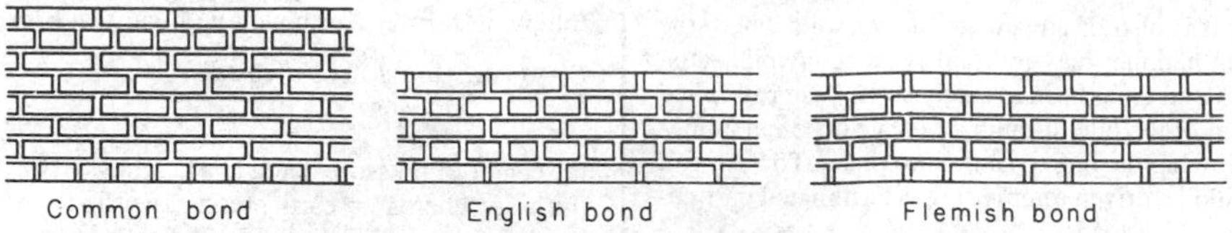

Common bond   English bond   Flemish bond

Fig. 12.2.16 Bonds used in bricklaying.

wall, in bonding each pair of transverse courses of brick. Three examples of bond are shown in Fig. 12.2.16.

Arches over windows and doorways are laid in concentric rings of headers on edge, with radial joints. The radius of the arch should be 1 to 1¼ times the width of the opening.

Lateral Support Brick walls should be supported laterally by bonding to transverse walls or buttresses, or by anchoring to floors, at intervals not exceeding 20 times the thickness. Floors and anchors must be capable of transmitting wind pressure and earthquake forces, acting outward, to transverse walls or other adequate supports, and thus to the ground. The height of piers between lateral supports should not exceed 12 times their least dimension.

Concrete Masonry Units (CMUs) Hollow or solid concrete blocks are used in building walls, often faced with brick or stone on the exterior walls. Standard block sizes are based on a brick module, nominally 4 in high.

Allowable compressive stresses in masonry are related to the masonry unit strength and the type of mortar as follows:

Kind of Masonry	Type M or S mortar		Type N mortar	
	lb/in²	MPa	lb/in²	MPa
Stone ashlar	360	2.48	320	2.20
Rubble	120	0.83	100	0.69
Clay brick (2500 lb/in²)	160	1.10	140	0.97
Hollow CMU (1000 lb/in²)	75	0.52	70	0.48
Solid CMU (2000 lb/in²)	160	1.10	140	0.97

Minimum thickness of load-bearing masonry walls shall be 6 in for up to one story and 8 in for more than one story.

Reinforced Masonry The design of masonry with vertical reinforcing bars placed and fully grouted in some of the cells provides increased capacity to resist flexure and axial loads. In addition, in areas where seismic design is required, due to the brittle nature of unreinforced masonry, all masonry must be reinforced to prevent a brittle failure mode. The design principles for reinforced masonry are similar to the design for reinforced concrete.

Reinforced Concrete (See Sec. 12.3.)

TIMBER CONSTRUCTION

Floors The framing of wooden floors may be divided into two general types: joist construction and solid, or mill, construction. The first consists of **joists** 2 to 6 in wide, of the necessary depth, and spaced about 12 to 16 in (30 to 40 cm) on centers. The wall ends should rest on and be anchored to walls and the interior ends carried by a line of girders on columns. These joists should be securely cross-bridged not over 8 ft (2.4 m) apart in each span to prevent twisting and to assist in distributing concentrated loads. Solid blocking should be provided at ends and at each point of support. The floor is formed of a thickness of rough boarding on which the finish flooring is laid. **Solid** or **mill-construction floors** are designed to do away with the small pockets which exist in joist construction and thus reduce the fire hazard. They are generally framed with beams spaced 8 to 12 ft (2.4 to 3.6 m) on centers and spanning 18 to 25 ft (5.5 to 7.6 m). The wall ends of beams rest on and are anchored to the wall, and the interior ends are carried on columns and tied together to form a continuous tie across the building. Ends of timbers in masonry walls should have metal bearing plates and ½ in space at sides and end for ventilation, to prevent rot. The ends should be beveled and the anchors placed low to avoid overturning the wall if the beams drop in a fire. In all cases, care should be taken to provide sufficient bearing at the points of support so that the allowable intensity of compression across the grain is not exceeded. In case it is desirable to omit columns, or the floor load requires a closer spacing of beams, girders are run lengthwise of the building over the columns to take the beams, the ends of which are hung in hangers or stirrup irons and tied together, over or through the girders. This is called **intermediate framing**. Steel beams are sometimes used in place of wooden beams in this type of construction, in which case a wooden strip is bolted to the top flange of the beam to take the nailing of the plank, or the plank is laid directly on top of the beam and secured by spikes driven from below and clinched over the flange. The floor is formed of 3 or 4 in (7.5 or 10 cm) plank grooved in each edge, put together with splines and securely spiked to beams. On

Table 12.2.10 Properties of Plank and Solid Laminated Floors
(b = breadth = 12 in, f = fiber stress)

Nominal thickness or depth in (1)	Actual thickness d, in (S4S) (2)	Area of section $A = bd$, in² (3)	Moment of inertia $I = bd^3/12$, in⁴ (4)	Section modulus $S = bd^2/6$, in³ (5)	Safe load, lb/ft² on 1-ft span*		Coef of deflection, uniform load‡	
					$f = 1,000$ lb/in²† (6)	$f = 1,600$ (7)	$E = 1,000,000$ (8)	$E = 1,760,000$ (9)
1	¾	9.00	0.422	1.13	753	1,205	53.4	93.98
1½	1¼	15.00	1.95	3.13	2,085	3,336	11.51	20.26
2	1½	18.00	3.38	4.50	3,000	4,800	6.66	11.72
2½	2	24.00	8.00	8.00	5,334	8,534	2.82	4.96
3	2½	30.00	15.60	12.50	8,334	13,334	1.441	2.54
4	3½	42.00	42.9	24.5	16,334	26,134	0.524	0.922
5	4½	54.00	91.1	40.5	27,000	43,200	0.247	0.1404
6	5½	66.00	166.4	60.5	40,300	64,500	0.1348	0.0765
8	7½	90.00	422	112.5	75,000	120,000	0.0533	0.0303
10	9½	114.00	857	180.5	120,400	192,500	0.0263	0.0149
12	11½	138.00	1,521	264.5	176,400	282,000	0.0148	0.0084

NOTE: 1 in = 2.54 cm; 1 lb = 4.45 N; 1 lb/in² = 6.89 kPa.
* Divide tabular value by square of span in feet.
† For other fiber stress f, multiply tabular value by $f/1,000$.
‡ For deflection in, multiply coefficient by load, lb/ft², and by fourth power of span in ft, and divide by 1,000,000. For other modulus of elasticity E, multiply coefficient of col. 8 by 1,000,000, and divide by E.

top of the plank is laid flooring, with a layer of sheathing paper between. In case the floor loads require an excessive thickness of plank, or in localities where heavy plank is not easily obtainable, the floor is built up of 3 × 6 in (7.5 × 15 cm), or other sized pieces, placed on edge, and securely nailed together.

The roofs of buildings of joist and mill construction are framed in a manner similar to the floors of each type and should be securely anchored to the walls and columns. In case columns are not desired in the top story, steel beams or trusses of either steel or wood are used. For spans up to 35 ft (10.7 m), trussed beams can often be used to advantage.

For **unit stresses in timber,** see Sec. 6. For unit stresses in wooden columns, see Table 12.2.12. Table 12.2.10 gives the properties of mill floors made of dressed plank, and of laminated floors made of planks of edge, laid close.

Timber Beams

Properties of Timber Beams   Table 12.2.11 presents those properties of wooden timbers most useful in computing their strength and deflection as beams. (The "nominal size" of a timber is indicated by the breadth and depth of the section in inches. The "actual size" indicates the size of the dressed timber, according to National Lumber

Table 12.2.11   Properties of Wooden Beams (Surfaced Size)

Nominal size, in (1)	Actual size $b \times d$, in, dressed (S4S) size (2)	Area of section bd, in² (3)	Weight at 40 lb/ft³, lb/ft (4)	Moment of inertia $I = bd^3/12$, in⁴ (5)	Section modulus $S = bd^2/6$, in³ (6)	Max safe uniform load, lb, based on Bending on 1 ft span,* $f = 1,000$ lb/in² (7)	Shear at 100† lb/ in² (8)	Coef‡ of deflection, uniform load $E = 1,000,000$ (9)
2 × 4	1½ × 3½	5.25	1.46	5.36	3.06	2,040	700	4.20
3 × 4	2½ × 3½	8.75	2.43	8.93	5.10	3,400	1,166	2.52
4 × 4	3½ × 3½	12.25	3.40	12.51	7.15	4,760	1,632	1.80
2 × 6	1½ × 5½	8.25	2.29	20.8	7.56	5,040	1,100	1.082
3 × 6	2½ × 5½	13.75	3.82	34.7	12.60	8,390	1,835	0.648
4 × 6	3½ × 5½	19.25	5.35	48.5	17.65	11,760	2,570	0.464
6 × 6	5½ × 5½	30.3	8.40	76.3	27.7	18,490	4,040	0.295
2 × 8	1½ × 7¼	10.87	3.02	47.6	13.14	8,760	1,445	0.473
3 × 8	2½ × 7¼	18.12	5.04	79.4	21.9	14,600	2,410	0.284
4 × 8	3½ × 7¼	25.4	7.05	111.1	30.7	20,500	3,380	0.202
6 × 8	5½ × 7½	41.3	11.4	193	51.6	34,400	5,500	0.1162
8 × 8	7½ × 7½	56.3	15.6	264	70.3	46,900	7,500	0.0852
2 × 10	1½ × 9¼	13.87	3.85	98.9	21.4	14,290	1,850	0.227
3 × 10	2½ × 9¼	23.1	6.42	164.9	35.7	23,700	3,080	0.1364
4 × 10	3½ × 9¼	32.4	8.93	231	49.9	33,300	4,310	0.0974
6 × 10	5½ × 9½	52.3	14.5	393	82.7	55,200	6,970	0.0573
8 × 10	7½ × 9½	71.3	19.8	536	113	75,200	9,500	0.0421
10 × 10	9½ × 9½	90.3	25.0	679	143	95,300	12,030	0.0332
2 × 12	1½ × 11¼	16.87	4.69	178	31.6	21,100	2,250	0.1264
3 × 12	2½ × 11¼	28.1	7.81	297	52.7	35,100	3,750	0.0757
4 × 12	3½ × 11¼	39.4	10.94	415	73.9	49,300	5,250	0.0543
6 × 12	5½ × 11½	63.3	17.5	697	121	80,800	8,430	0.0323
8 × 12	7½ × 11½	86.3	23.9	951	165	110,200	11,510	0.0237
10 × 12	9½ × 11½	109.3	30.3	1,204	209	139,600	14,570	0.01864
12 × 12	11½ × 11½	132.3	36.7	1,458	253	169,000	17,620	0.01543
4 × 14	3½ × 13¼	46.4	12.88	678	102.4	68,300	6,180	0.0332
6 × 14	5½ × 13½	74.3	20.6	1,128	167	111,400	9,900	0.01987
8 × 14	7½ × 13½	101.3	28.0	1,538	228	152,000	13,500	0.01462
10 × 14	9½ × 13½	128.3	35.6	1,948	289	192,400	17,120	0.01153
12 × 14	11½ × 13½	155.3	43.1	2,360	349	233,000	20,700	0.00953
14 × 14	13½ × 13½	182.3	50.6	2,770	410	273,000	24,300	0.00812
6 × 16	5½ × 15½	85.3	23.6	1,707	220	146,800	11,380	0.01315
8 × 16	7½ × 15½	116.3	32.0	2,330	300	200,000	15,530	0.00967
10 × 16	9½ × 15½	147.3	40.9	2,950	380	254,000	19,610	0.00762
12 × 16	11½ × 15½	178.3	49.5	3,570	460	307,800	23,800	0.00630
14 × 16	13½ × 15½	209	58.1	4,190	541	360,000	27,900	0.00539
16 × 16	15½ × 15½	240	66.7	4,810	621	414,000	32,000	0.00468
8 × 18	7½ × 17½	131.3	36.4	3,350	383	255,000	17,500	0.00672
10 × 18	9½ × 17½	166.3	46.1	4,240	485	323,000	22,200	0.00531
12 × 18	11½ × 17½	201	55.9	5,140	587	391,000	26,800	0.00438
14 × 18	13½ × 17½	236	65.6	6,030	689	459,000	31,500	0.00373
16 × 18	15½ × 17½	271	75.3	6,920	791	528,000	36,200	0.00325
18 × 18	17½ × 17½	306	85.0	7,820	893	595,000	40,800	0.00288
12 × 20	11½ × 19½	224	62.3	7,110	729	485,000	29,900	0.00316
20 × 20	19½ × 19½	380	106	12,050	1,236	824,000	50,700	0.00187
24 × 24	23½ × 23½	552	153	25,400	2,160	1,440,000	73,400	0.000888
26 × 26	25½ × 25½	650	180.6	35,200	2,760	1,840,000	86,700	0.000639
28 × 28	27½ × 27½	756	210	47,700	3,470	2,320,000	100,600	0.000472
30 × 30	29½ × 29½	870	242	63,100	4,280	2,850,000	116,000	0.000356

NOTE: 1 in = 2.54 cm; 1 ft = 0.305 m; 1 lb = 4.45 N; 1 lb/in² = 6.89 kPa.
* For total safe uniform load, pounds, on beam of span L, feet, divide tabular value by L. For fiber stress f other than 1,000 lb/in² multiply by f and divide by 1,000.
† For shearing stress other than 100 lb/in², multiply by stress and divide by 100.
‡ For deflection, inches, multiply coefficient by total load, pounds, and by cube of span, feet, and divide by 1,000,000. For other modulus of elasticity E, multiply coefficient by 1,000,000 and divide by E.

Manufacturers Assoc. The moment of inertia and section modulus are with the neutral axis perpendicular to the depth at the center. The **safe bending moment** in inch-pounds for a given beam is determined from the section modulus S by multiplying the tabular value by the allowable fiber stress. To select a beam to withstand safely a given bending moment, divide the bending moment in inch-pounds by the allowable fiber stress, and choose a beam whose section modulus S is equal to or larger than the quotient thus obtained. For formulas for computing bending moments, see Sec. 5.2. Note that the allowable fiber stress must be modified by adjustment factors: C_D, load duration factor (0.9 for permanent loads); C_M, wet service factor (approximately 0.8 for moisture content greater than 16 percent); C_t, temperature factor (usually 1.0 for temperatures less than 100°F); C_F, size factor (1.0 for members up to 5 in wide by 12 in deep); and other factors which apply to laminated, curved, round, and/or flat use. (See also Sec. 6.7, ''Properties of Lumber Products.'')

Maximum loads in Table 12.2.11, cols. 7 and 8, are for uniform loading. Use half the values of col. 7 for a single load concentrated at midspan; for other loadings compute the bending moment and use the section modulus, col. 6. The values of col. 8 apply to all symmetrical loadings. For unsymmetrical loading, compute the maximum shear, which must not exceed one-half the tabular value.

The **coefficients of deflection** listed in Table 12.2.11 can be used to deduce deflection as indicated in the footnotes to the table. Coefficients of deflection under concentrated loads applied at the middle of the span may be obtained by multiplying the values in the table by 1.6. The results are only approximate, as the modulus of elasticity varies with the moisture content of the wood.

The deflection due to live load of beams intended to carry plastered ceilings should not exceed $\frac{1}{360}$ of the span.

A convenient rule may be derived by assuming that the modulus of elasticity is 1,000 times the allowable fiber stress, which applies to all woods with sufficient accuracy for the purpose. Beams loaded uniformly to capacity in bending will then deflect $\frac{1}{360}$ of the span when the depth in inches is 0.90 times the span in feet; and beams with central concentration, when the depth is 0.72 times the span in the same units. For such beams, the deflection in inches is, for uniform load, $0.03L^2/d$; for central concentration, $0.024L^2/d$, where L is the span, ft and d the depth, in. Variation in type of loading affects this result comparatively little.

Timber Columns

Timber columns may be either square or round and should have metal bases, usually galvanized steel, to cut off moisture and prevent lateral displacement. For supporting beams, they should have caps which, at roofs, may be of steel, or wood designed for bearing across the grain. At intermediate floors, caps should be of steel, although in some cases hardwood bolsters may be used. Except when caps or beams are of steel, columns should run down and rest directly on the baseplate. Table 12.2.12 gives **working unit stresses for wood columns** recommended where the building laws do not prescribe lower stresses. Use actual, not nominal, dimension of timbers. The column capacity $P_{col} = F_c'A_{net}$, where F_c' is interpolated from Table 12.2.12, and A_{net} is the net cross-sectional area of the column. The values for F_c' in Table 12.2.12 are generally conservative and are based on the factors cited in the table footnotes. In the event $E < 1,000F_c$, the value of F_c' must be computed to include the value of C_P as follows:

$$F_c' = F_c C_D C_M C_t C_F C_P$$

(Note that actual design stress $f_c \leq F_c'$.)

$$C_p = \text{column stability factor} = \frac{1 + (F_{cE}/F_c^*)}{2c}$$
$$- \sqrt{\left[\frac{1 + F_{cE}/F_c^*}{2c}\right]^2 - \frac{F_{cE}/F_c^*}{c}}$$
$$F_c^* = F_c C_D C_M C_t C_F$$

$$F_{cE} = \frac{K_{cE}E}{(l_e/d)^2} = \text{critical buckling design stress in compression parallel to the grain for a given wood species and geometrical configuration of column}$$

where F_c = tabulated allowable design value for compression parallel to the grain for the species (Sec. 6.7, Tables 6.7.6 and 6.7.7); C_D = load duration factor (Sec. 6.7, ''Properties of Lumber Products''); C_M = wet use factor (Sec. 6.7, Table 6.7.8); C_t = temperature factor (Sec. 6.7, Table 6.7.9); C_F = size factor (Sec. 6.7, footnotes to Table 6.7.8); K_{cE} = 0.3 for visually and mechanically graded lumber and 0.418 for glulam members; l_e = effective length of column; d = least dimension of the column cross section; E = modulus of elasticity for the species (Sec. 6.7, Tables 6.7.6 and 6.7.7); c = constant for the type member: 0.8 for sawn lumber, 0.85 for timber piles, 0.9 for glulam members.

Glued Laminated Timber Structural glued laminated timber, commonly called glulam, refers to members which are fabricated by pressure gluing selected wood laminations of either ¾ or 1½ in (19 or 38 mm) surfaced thickness. The grain of all the laminations is approximately parallel longitudinally, with exterior laminations being of generally higher-quality wood since bending stresses are greater at the outer fibers. Curved and tapered structural members are available with the recommended minimum radii of curvature being 9 ft 4 in (2.84 m) for ¾-in laminations and 27 ft 6 in (8.4 m) for 1½-in lamination thickness. Laminations should be parallel to the tension face of members; sawn tapered cuts are permitted on the compression face.

Available net (surfaced) widths of members in inches are 2¼, 3⅛, 5⅛, 8¾, 10¾, 12¼, and 14¼; depths are determined by stress requirements. Economical spans (see ''Timber Construction Manual,'' American Institute of Timber Construction) for roof framing range from 10 to 100 ft (3 to 30 m) for simple spans. Floor framing, which is designed for much heavier live loads, economically spans from 6 to 40 ft (1.8 to 12 m) for simple beams and from 25 to 40 ft (7.5 to 12 m) for continuous beams.

Glued laminated members are generally fabricated from either Douglas fir and larch, Douglas fir (coast region), southern pine, or California redwood, depending on availability. Allowable design stresses depend on whether the condition of use is wet (moisture content in service of 16 percent or more) or dry (as in most covered structures), the species and grade of wood to be used, the manner of loading, and the number of laminations as well as the usual factors for duration of loading. The cumulative reduction factors described above also apply to glulam beams. Additional factors including C_v, volume factor; C_{fu}, flat use factor; and C_c, curvature factor, also must be applied to glued laminated beams. Refer to the National Design Specification for Wood Construction for further information on glued laminated timber design. (See Sec. 6.7.)

Table 12.2.12 Values of F_c', Working Stresses for Square or Rectangular Timber Columns, lb/in²
(Compression parallel to grain.)*

F_c	l_e/d, in/in												
	10	15	20	25	30	35	40	45	50	55†	60†	70†	80†
1000	680	615	508	389	293	225	176	141	116	96	81	60	46
1300	884	799	660	505	381	292	229	184	150	125	106	78	60
1600	1088	984	812	622	469	359	282	226	185	154	130	96	74
1900	1292	1168	964	739	557	427	335	268	220	183	154	114	88

* Values of F_c' in the table are based on $C_D = 0.9$ (long-duration, permanent, 50-year loading); $C_M = 0.8$ (moisture content > 16%); $C_t = 1.0$ (operating temperature < 100°F); $C_F = 1.0$ (cross section up to 5 in wide × 12 in deep); $K_{cE} = 0.3$; $E \geq 1,000F_c$. If $E < 1,000F_c$, see text for procedure to compute F_c'; $c = 0.8$.
† Columns should be limited to $l_e/d = 50$, except for individual members in stud walls, which should be limited to $l_e/d = 80$.

A summary of allowable unit stresses may be found in Sec. 6.7 for glued laminated timber.

Connections

Bolted Joints   Compression may be transmitted by merely butting the timbers, with splice pieces bolted to the sides to keep alignment and resist incidental bending and shear. The same detail (Fig. 12.2.17)

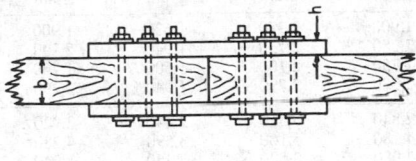

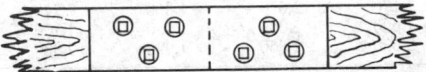

Fig. 12.2.17   Bolted splices for timber framing.

serves in tension, but the entire stress must then be transmitted through the bolts and splice pieces. If of wood, these should have a thickness h equal to $\frac{1}{2}b$. In light, unimportant work, splice pieces may be spiked. Table 12.2.13 gives the allowable load in pounds for one bolt loaded at both ends (double shear) when h is at least equal to $\frac{1}{2}b$. When steel side plates are used for side members, the tabulated loads may be increased 25 percent for parallel-to-grain loading, but no increase should be made for perpendicular-to-grain loads. When a joint consists of two members (single shear), one-half the tabulated load for a piece twice the thickness of the thinner member applies. The safe load for bolts loaded at an angle θ with the grain of the wood is given by the formula $N = PQ/(P \sin^2 \theta + Q \cos^2 \theta)$, where N = allowable load per bolt in a direction at inclination θ with the direction of the grain, lb; P = allowable load per bolt in compression parallel to the grain, lb; and Q = allowable load per bolt in compression perpendicular to the grain, lb.

The size, arrangement, and **spacing of bolts** must be such that tension on the net section of the timber through the bolt holes and shear along the grain do not exceed allowable values. Bolts should be at least 7 diameters from the end of the timber for softwoods and 5 diameters for hardwoods and spaced at least 4 diameters on center parallel to the grain. Crossbolting, to prevent splitting the timber end, is sometimes desirable.

The efficiency of bolted timber connections may be greatly increased by the use of **ring connectors**. Split rings and shear plates are fitted into circular grooves, concentric with the bolt, in the contact surfaces, and transmit shear stresses across the joint. Grooves for split rings and shear plates are cut with a special tool, while toothed rings are usually seated

by drawing together the timbers with high-strength bolts. Allowable loads for these various connectors are given in the "Design Values for Wood Construction," published by the National Lumber Manufacturers Assoc. Selected values are given in Table 12.2.14.

The **holding power of wire nails** is as follows ("Design Values for Wood Construction"): The resistance to withdrawal is proportional to the length of embedment, to the diameter of the nail (where the wood does not split), and to $G^{2.5}$, where G is the ovendry specific gravity of the wood (see Sec. 6 for G values of various species). The safe resistance to withdrawal of common wire nails driven into the side grain of seasoned wood is given by Table 12.2.15. Nails withdrawn from green wood have generally slightly higher resistance, but nails driven into green wood may lose much of their resistance when the wood seasons; the allowable withdrawal load should be one-fourth of that given in Table 12.2.15. Cement and other coatings on nails may add materially to their resistance in softwoods. Drilling lead holes slightly smaller than the nail adds somewhat to the resistance and reduces danger of splitting. The structural design should be such that nails are not loaded in withdrawal from end grain.

The safe **lateral resistance** of common wire nails driven in side grain to the specified penetrations is given in Table 12.2.15 and is proportional to $D^{1.5}$ where D is the diameter, in. These values are for seasoned wood and should be reduced 25 percent for woods which will remain wet or will be loaded before seasoning. For nails driven into end grain, values should be reduced one-third.

Common wire **spikes** are larger for their lengths than nails. Their resistance to withdrawal and lateral resistance are given by the same formulas as for nails, but greater precautions need to be taken to avoid splitting.

The **resistance of wood screws to withdrawal** from side grain of seasoned wood is given by the formula $P = 2,850G^2D$, where P = the allowable load on the screw, lb/in penetration of the threaded portion; G = specific gravity of ovendry wood; D = diameter of screw, in. Wood screws should not be designed to be loaded in withdrawal from end grain.

The **allowable safe lateral resistance** of wood screws embedded 7 diameters in the side grain of seasoned wood is given by the formula $P = KD^2$, where P is the lateral resistance per screw, lb; D is the diameter, in; and K is 4,800 for oak (red and white), 3,960 for Douglas fir (coast region) and southern pine, and 3,240 for cypress (southern) and Douglas fir (inland region).

The following rules should be observed: (1) the size of the lead hole in soft (hard) woods should be about 70 (90) percent of the core or root diameter of the screw; (2) lubricants such as soap may be used without great loss in holding power; (3) long, slender screws are preferable generally, but in hardwood too slender screws may reach the limit of their tensile strength; (4) in the screws themselves, holding power is favored by thin sharp threads, rough unpolished surface, full diameter under the head, and shallow slots.

Table 12.2.13   Allowable Load in Pounds on One Bolt Loaded at Both Ends (Double Shear)
(For additional values and for conditions other than normal, see "Design Values for Wood Construction")

Length of bolt in main member, in	Diam of bolt, in	Douglas fir-larch		California redwood (open grain)		Oak, red and white		Western spruce, pine, fir, cedars	
		Parallel to grain	Perpendicular to grain	Parallel to grain	Perpendicular to grain	Parallel to grain	Perpendicular to grain	Parallel to grain	Perpendicular to grain
1½	½	1,050	470	780	310	1,410	730	760	290
	¾	1,580	590	1,170	370	2,110	890	1,140	360
	1	2,100	680	1,560	440	2,810	1,020	1,520	420
2½	½	1,230	730	990	510	1,530	960	980	490
	¾	2,400	980	1,950	620	2,890	1,480	1,900	600
	1	3,500	1,130	2,590	730	4,690	1,700	2,530	700
3½	½	1,230	730	990	580	1,530	960	980	560
	¾	2,400	1,170	2,010	740	2,890	1,770	1,990	720
	1	4,090	1,350	3,110	870	4,820	2,040	3,040	840
5½	¾	2,400	1,170	2,010	740	2,890	1,770	1,990	720
	1	3,180	1,260	2,690	810	3,780	1,920	2,660	790
	1¼	4,090	1,350	3,110	870	4,820	2,040	3,040	840

NOTE: 1 in = 2.54 cm; 1 lb = 4.45 N.

Table 12.2.14 Allowable Load in Pounds for One-Connector Unit in Single Shear*
(For additional values and for conditions other than normal, see "Design Values for Wood Construction")

Connector unit (diam)	Number of faces of piece with connectors on the same bolt	Net thickness of lumber, in	Min. edge distances, in	Group A Douglas fir-larch and southern pine (dense), oak, red and white ∥ to grain	⊥ to grain	Group B Douglas fir-larch, southern pine (med. grain) ∥ to grain	⊥ to grain	Group C California redwood (close) grain, western hemlock, southern cypress ∥ to grain	⊥ to grain
2½-in split ring, ½-in bolt	1	1 min,	1¾	2,630	1,900	2,270	1,620	1,900	1,350
		1½ or more		3,160	2,280	2,730	1,940	2,290	1,620
	2	1½ min,		2,430	1,750	2,100	1,500	1,760	1,250
		2 or more		3,160	2,280	2,730	1,940	2,290	1,620
4-in split ring, ¾-in bolt	1	1 min,	2¾	4,090	2,840	3,510	2,440	2,920	2,040
		1½ or more		6,020	4,180	5,160	3,590	4,280	2,990
	2	1½ min,		4,110	2,980	3,520	2,450	2,940	2,040
		3 or more		6,140	4,270	5,260	3,660	4,380	3,050
2⅝-in shear plate, ¾-in bolt†	1	1½ min	1¾	3,110	2,170	2,670	1,860	2,220	1,550
	2	1½ min,		2,420	1,690	2,080	1,450	1,730	1,210
		2½ or more		3,330	2,320	2,860	1,990	2,380	1,650
4-in shear plate, ¾-in or ⅞-in bolt†	1	1½ min,	2¾	4,370	3,040	3,750	2,620	3,130	2,170
		1¾ or more		5,090	3,540	4,360	3,040	3,640	2,530
	2	1¾ min,		3,390	2,360	2,910	2,020	2,420	1,680
		2½		4,310	3,000	3,690	2,550	3,080	2,140
		3½ or more		5,030	3,500	4,320	3,000	3,600	2,510

NOTE: 1 m = 2.54 cm; 1 lb = 4.45 N.
* One connector unit consists of one split ring with its bolt in single shear or two shear plates back to back in the contact faces of a timber-to-timber joint with their bolt in single shear.
† Allowable loads for all loadings, except wind, should not exceed 2,900 lb for 2⅝-in shear plates; 4,400 and 6,000 lb for 4-in shear plates with ¾- and ⅞-in bolts, respectively; multiply values by 1.33 for wind loading.

Table 12.2.15 Allowable Loads in Pounds for Common Nails in Side Grain* of Seasoned Wood

Type of load	Specific gravity G		Size of nail									
		d	6	8	10	12	16	20	30	40	50	60
		Length, in	2	2½	3	3¼	3½	4	4½	5	5½	6
		Diam, in	0.113	0.131	0.148	0.148	0.162	0.192	0.207	0.225	0.244	0.263
Withdrawal load per in penetration	0.31		9	10	12	12	13	15	16	18	20	21
	0.40		16	18	20	20	22	27	28	31	33	35
	0.44		20	23	26	26	29	34	37	40	43	46
	0.47		24	27	31	31	34	40	43	47	51	55
	0.51		29	34	38	38	42	49	53	58	63	68
	0.55		34	41	46	46	50	59	64	70	76	81
	0.67		57	66	75	75	82	97	105	114	124	133
Lateral load*†	0.60–0.75		78	97	116	116	132	171	191	218	249	276
	0.50–0.55		63	78	94	94	107	139	154	176	202	223
	0.42–0.50		51	64	77	77	88	113	126	144	165	182
	0.31–0.41		41	51	62	62	70	91	101	116	132	146

NOTE: 1 in = 2.54 cm; 1 lb = 4.45 N.
* The allowable lateral load for nails driven in end grain is two-thirds the values shown above.
† The minimum penetration for full lateral resistance for the four groups listed is 10, 11, 13, and 14 diam from higher to lower specific gravities, respectively. Reduce by interpolation for lesser penetration; minimum penetration is one-third the above.

Table 12.2.16 Allowable Lateral Loads in Pounds on Lag Bolts or Lag Screws

Side member	Length of bolt, in	Diam of bolt at shank, in	Overdry specific gravity of species 0.60–0.75 ∥	⊥	0.51–0.55 ∥	⊥	0.42–0.50 ∥	⊥	0.31–0.41 ∥	⊥
1½-in wood	4	¼	200	190	170	170	130	120	100	100
	4	½	390	250	290	190	210	140	170	110
	6	⅜	480	370	420	320	360	280	290	220
	6	⅝	860	510	710	430	510	310	410	250
2½-in wood	6	½	620	410	470	310	340	220	270	180
	6	1	1,040	520	790	390	560	280	450	230
	8	¾	1,430	790	1,080	600	780	430	620	340
	8	1	1,800	900	1,360	680	970	490	780	390
½-in metal	3	¼	240	185	210	160	155	120	125	100
	3	½	550	285	415	215	295	155	240	125
	6	½	1,100	570	945	490	770	400	615	320
	6	¾	1,970	865	1,480	650	1,060	460	850	370
	10	⅞	3,420	1,420	2,960	1,230	2,340	970	1,890	785
	12	1	4,520	1,810	3,900	1,560	3,290	1,320	2,630	1,050
	16	1¼	7,120	2,850	6,150	2,460	5,500	2,200	4,520	1,810

NOTE: 1 in = 2.54 cm; 1 lb = 4.45 N.

The allowable **withdrawal load of lag screws** in side grain is given by the formula $p = 1,800D^{3/4}G^{3/2}$, allowable load per inch of penetration of threaded portion of lag screw into member receiving the point, lb; D = shank diameter of lag screw, in; G = specific gravity of ovendry wood. Use of lag screws loaded in withdrawal from end grain should be avoided. The allowable load in such case should not exceed 75 percent of that for side grain (see also Sec. 8).

The allowable **lateral resistance of lag screws** for parallel-to-grain loading with screws in side grain is proportional to D^2 and is dependent on species and type of side member. Selected values are given in Table 12.2.16 for one lag screw in single shear in a two-member joint.

Lead holes for lag screws (approximately 75 percent of shank diameter) should be prebored for the threaded portion. Lead holes for the shank should be of the same diameter and length as that of the unthreaded shank. Soap or other lubricant should be used to facilitate insertion and to prevent damage to the screw. Where steel-plate side pieces are used, the allowable loads given by the formula for parallel-to-grain loading may be increased by 25 percent.

The ultimate **withdrawal load per linear inch of penetration of a round drift bolt or pin** from side grain when driven into a prebored hole having a diameter $\frac{1}{8}$ in less than that of the bolt diameter may be determined from the formula $p = 6,000G^2D$, where p = ultimate withdrawal load of penetration, lb/lin in; G = specific gravity of ovendry wood; D = diameter of drift bolt, in. A safety factor of about 5 is suggested for general use. The allowable load in lateral resistance for a drift bolt should ordinarily be taken as less than that for a common bolt.

STEEL CONSTRUCTION

(Note. In the design of steel structures, 1,000 lb is frequently designated as a kilopound or "kip," and a stress of 1 kip per square inch is designated as 1 ksi.)

Structural steel design was based only on the allowable stress design (ASD) approach until the introduction of the load and resistance factor design (LRFD) technique in the mid 1980s. The LRFD approach is an ultimate strength design approach, similar to that adopted by the American Concrete Institute for concrete design. Both ASD and LRFD are accepted in current codes. The ASD method is still the most commonly used for design.

Specifications The following are in part condensed excerpts from the Specifications of the American Institute of Steel Construction.

Material Ordinary steel for rolled shapes, plates, and bars is typically specified by ASTM A36, with a yield stress of 36,000 lb/in² (248.2 MPa). However, advances in mill production methods have resulted in most steels satisfying the higher strength requirements of ASTM A572, Grade 50, leading to increased use of higher-strength steel at little or no cost premium. Other higher-strength steels used in structures are A440, A441, A588, and A242. Steel materials for pipe and tube, specified by ASTM A53 (welded-seam pipe) and A500 (cold-formed), have yield strengths of 33,000 to 50,000 lb/in² (227.5 to 344.7 MPa).

Ordinary unfinished machine bolts are specified by A307. Bolts used for structural steel connections are typically high-strength bolts specified by A325 or A490. Riveting is no longer used, but may often be encountered in older structures. The most common rivets were A502, Grade 1.

Allowable Stresses* in A36 Steel

	lb/in²	MPa
Tension F_t:		
On gross section	22,000	151.6
On net section, except at pinholes	29,000	200
On net section, at pinholes	16,000	110.2
Compression F_c: See Table 12.2.17		
Bending tension and compression on extreme fibers F_b:		
Basic stress, reduced in certain cases	22,000	151.6
Compact, adequately braced beams	24,000	165.3
Rectangular bearing plates	27,000	186.0
Shear F_v: Web of beams, gross section	14,500	99.9

* Allowable stresses may be increased by one-third when produced by wind or seismic loading alone or when combined with design dead and live loads.

Allowable Stresses* in Riveted and Bolted Connections

	lb/in²	MPa
Bearing: A36 steel		
Pins in reamed, drilled, or bored holes	32,400	223.3
Bolts and rivets	69,000	475.7
Roller, lb/lin in (N/lin cm)	760 × diam (in)	1131 × diam (cm)
Shear: bearing-type connections†		
A502, grade 1 hot-driven rivets	17,500	120.6
A307 bolts	10,000	68.9
A325 bolts when threading is excluded from shear planes (std. holes)	30,000	206.8
A325 bolts when threading is not excluded from shear planes (std. holes)	21,000	144.7
Shear: friction-type connections† (with threads included or excluded from shear plane)		
A325 bolts in standard holes	17,500	120.6
A325 bolts in oversized or short slotted holes	15,000	103.4
A325 bolts in long slotted holes	12,000	82.6
Tension:		
A502, grade 1, hot-driven rivets	23,000	158.5
A307 bolts	20,000	137.9
A325 bolts	44,000	303.3
Bending in pins of A36 steel	27,000	186.1

* Allowable stresses are based on nominal body area of fasteners unless indicated.
† Rivets or bolts may not share loads with welds on bearing-type connections but may do so in friction-type connections.

Table 12.2.17   Allowable Stress, in ksi, for Compression Members of A36 Steel

Main and secondary members, Kl/r not more than 120						Main members, Kl/r, 121–200			
$\dfrac{Kl}{r}$	F_a	$\dfrac{Kl}{r}$	F_a	$\dfrac{Kl}{r}$	F_a	$\dfrac{Kl}{r}$	F_a	$\dfrac{Kl}{r}$	F_a
1	21.56	41	19.11	81	15.24	121	10.14	161	5.76
5	21.39	45	18.78	85	14.79	125	9.55	165	5.49
10	21.16	50	18.35	90	14.20	130	8.84	170	5.17
15	20.89	55	17.90	95	13.60	135	8.19	175	4.88
20	20.60	60	17.43	100	12.98	140	7.62	180	4.61
25	20.28	65	16.94	105	12.33	145	7.10	185	4.36
30	19.94	70	16.43	110	11.67	150	6.64	190	4.14
35	19.58	75	15.90	115	10.99	155	6.22	195	3.93
40	19.19	80	15.36	120	10.28	160	5.83	200	3.73

NOTE: 1 ksi = 6.89 MPa.

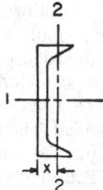

Table 12.2.18 American Standard Channels (C Shapes)

Depth of channel, in	Weight per ft, lb	Area of section, in²	Width of flange, in	Thickness of web, in	Axis 1–1			Axis 2–2		V*	R*
					I, in⁴	r, in	S, in³	r, in	x, in	1,000 lb	
C 15	50.0	14.64	3.716	0.716	404	5.24	53.6	0.87	0.80	156	121
	40.0	11.70	3.520	0.520	349	5.44	46.2	0.89	0.78	113	88
	33.9	9.90	3.400	0.400	315	5.62	41.7	0.91	0.79	87	55
C 12	30.0	8.79	3.170	0.510	162	4.28	26.9	0.77	0.68	89	76
	25.0	7.32	3.047	0.387	144	4.43	23.9	0.79	0.68	67	55
	20.7	6.03	2.940	0.280	129	4.61	21.4	0.81	0.70	49	30
C 10	30.0	8.80	3.033	0.673	103.0	3.42	20.6	0.67	0.65	98	96
	25.0	7.33	2.886	0.526	90.7	3.52	18.1	0.68	0.62	76	75
	20.0	5.86	2.739	0.379	78.5	3.66	15.7	0.70	0.61	55	54
	15.3	4.47	2.600	0.240	66.9	3.87	13.4	0.72	0.64	35	22
C 9	20.0	5.86	2.648	0.448	60.6	3.22	13.5	0.65	0.59	58	62
	15.0	4.39	2.485	0.285	50.7	3.40	11.3	0.67	0.59	37	33
	13.4	3.89	2.430	0.230	47.3	3.49	10.5	0.67	0.61	30	22
C 8	18.75	5.49	2.527	0.487	43.7	2.82	10.9	0.60	0.57	56	
	13.75	4.02	2.343	0.303	35.8	2.99	9.0	0.62	0.56	35	
	11.5	3.36	2.260	0.220	32.3	3.10	8.1	0.63	0.58	26	
C 7	14.75	4.32	2.299	0.419	27.1	2.51	7.7	0.57	0.53	43	
	12.25	3.58	2.194	0.314	24.1	2.59	6.9	0.58	0.53	32	
	9.8	2.85	2.090	0.210	21.1	2.72	6.0	0.59	0.55	21	
C 6	13.0	3.81	2.157	0.437	17.3	2.13	5.8	0.53	0.52	38	
	10.5	3.07	2.034	0.314	15.1	2.22	5.0	0.53	0.50	27.3	
	8.2	2.39	1.920	0.200	13.0	2.34	4.3	0.54	0.52	17.4	
C 5	9.0	2.63	1.885	0.325	9.0	1.83	3.5	0.49	0.48	23.6	
	6.7	1.95	1.750	0.190	7.5	1.95	3.0	0.50	0.49	13.8	
C 4	7.25	2.12	1.720	0.320	4.5	1.47	2.3	0.46	0.46	18.6	
	5.4	1.56	1.580	0.180	3.8	1.56	1.9	0.45	0.46	10.4	
C 3	6.0	1.75	1.596	0.356	2.1	1.08	1.4	0.42	0.46	15.5	
	5.0	1.46	1.498	0.258	1.8	1.12	1.2	0.41	0.44	11.2	
	4.1	1.19	1.410	0.170	1.6	1.17	1.1	0.41	0.44	7.4	

NOTE: 1 in = 2.54 cm; 1 ft = 0.305 m; 1 lb = 4.45 N.
* V amd R values are for channels of A36 steel.

Proportion of Parts

Most simple beams, columns, and truss members are proportioned to limit the actual stresses to the allowable stresses stipulated above. Other stability or serviceability criteria may control the design. **Deflection** may govern in such members as cantilevers and lightly loaded roof beams. **Buckling**, rather than strength, may govern the design of compression members. The slenderness ratio Kl/r, where Kl is the effective length of the member and r is its radius of gyration, should be limited to 200 in compression members, and L/r limited to 300 in tension members. Kl should not be less than the actual unbraced length l in columns of a frame which depends on its bending stiffness for lateral stiffness. **Width-thickness ratios** are specified for projecting elements under compression. Repeated fluctuations in stress leading to fatigue may be a controlling factor. Rules are given for **combined stresses** of tension, compression, bending, and shear.

Tension members should be proportioned for the gross and net section, deducting for bolt or rivet holes ⅛ in (0.3 cm) larger than the nominal diameter of the fastener.

Columns and other **compression members** subject to eccentric load or to axial load and bending are governed by special rules. A long-established rule is that $f_a/F_a + f_b/F_b$ should be equal to or less than unity, where f_a is the axial stress, f_b the bending stress, and F_a and F_b are the corresponding allowable stresses if axial or bending stress alone exist. This is still considered valid when f_a/F_a is less than 0.15. Joints shall be fully spliced, except that where reversal of stress is not expected and the joint is laterally supported, the ends of the members may be milled to plane parallel surfaces normal to the stresses and abutted with sufficient splicing to hold the connected members accurately in place. Column bases should be milled on top for the column bearing, except for rolled-steel bearing plates 4 in (10 cm) or less in thickness.

Beams and girders, of rolled section or built-up, should in general be sized such that the bending moment M divided by the section modulus S is less than the allowable bending stress F_b. For built-up sections a rule of thumb is, for A36 steel, flanges in compression should have a thickness of ¹⁄₁₆ the projecting half width, and webs should have a thickness of ¹⁄₃₂₀ the maximum clear distance between flanges. Web stiffeners should be provided at points of high concentrated loads; additional web stiffeners are required in plate girders. Splices in the webs of plate girders should be made by plates on both sides of the web. When two or more rolled beams or channels are used side by side to form a beam, they should be connected at separators spaced no more than 5 ft (1.52 m); beams deeper than 12 in (30 cm) are to have at least two bolts to each separator.

The **lateral force on crane runways** due to the effect of moving crane trolleys may be assumed as 20 percent of the sum of the weights of the lifted loads and of the crane trolley (but exclusive of the other parts of the crane) applied at the top of the rail, one-half on each side of the runway, and shall be considered as acting in either direction normal to the runway rail. The **longitudinal force** may be assumed as 10 percent of the maximum wheel reactions of the crane applied at the top of the rail.

Bolted or riveted connections carrying calculated stress, except lacing and sag bars, should be designed to support not less than 6,000 lb (27.0 kN). Rivets or high-strength bolts are preferred in all places, and both are implied in these paragraphs wherever "bolting" is mentioned; unfinished bolts, A307, may be used in the shop or in field connections of small unimportant structures of secondary members, bracing, and beams.

Members in tension or compression, meeting at a joint, shall have their lines of center of gravity pass through a support point, if practicable; if not, provision shall be made for the eccentricity. A group of bolts transmitting stress to a member shall have its center of gravity in the line of the stress, if practicable; if not, the group shall be designed for the resulting eccentricity. Where stress is transmitted from one member to another by bolts through a loose filler greater than $\frac{1}{4}$ in in thickness, except in slip critical connections using high-strength bolts, the filler shall be extended beyond the connected member and the extension secured by enough bolts or sufficient welding to distribute the total stress in the member uniformly over the combined sections of the member and the filler.

Most bolted connections transmit shearing forces by developing the shearing or bearing values of the bolts, but bolts in certain connections, such as shelf angles and brackets, are required to transmit tension forces.

Bolts shall be proportioned by the nominal diameter. Rivets and A307 bolts whose grip exceeds 5 diam shall be allowed 1 percent less safe stress for each $\frac{1}{16}$ in (0.16 cm) excess length. The minimum distance between centers of bolt holes shall be $2\frac{2}{3}$ diam of the bolt; but preferably not less than 3 diam.

The minimum distance from the center of any bolt hole to a sheared edge shall be $2\frac{1}{4}$ in (5.7 cm) for $1\frac{1}{4}$ in (32 mm) bolts, 2 in (5.1 cm) for $1\frac{1}{8}$ in (28 mm) bolts, $1\frac{3}{4}$ in (4.4 cm) for 1 in (25 mm) bolts, $1\frac{1}{2}$ in (3.8 cm) for $\frac{7}{8}$ in (22 mm) bolts, $1\frac{1}{4}$ in (3.2 cm) for $\frac{3}{4}$ in (19 mm) bolts, $1\frac{1}{8}$ in (2.8 cm) for $\frac{5}{8}$ in (16 mm) bolts, and $\frac{7}{8}$ in (2.22 cm) for $\frac{1}{2}$ in (13 mm) bolts. The distance from any edge shall not exceed 12 times the thickness of the plate and shall not exceed 6 in (15 cm).

Design of Members

Properties of Standard Structural Shapes Tables 12.2.18 to 12.2.26 give the properties of American Standard channels and I beams, wide-flange beams and columns, angles, and tees. In these tables, I = moment of inertia, r = radius of gyration, S = section modulus, x = distance from gravity axis to face, V = max web shear in kips, and R = max end reaction on $3\frac{1}{2}$-in (9-mm) seat, based on crippling of web, in kips. R values are omitted where web crippling does not govern.

A great variety of tees is produced by shearing or gas-cutting stan-

Table 12.2.19   American Standard I Beams (S Shapes)

Depth of beam, in	Weight per ft, lb	Area of section, in²	Width of flange, in	Thickness of web, in	Neutral axis perpendicular to web at center			Neutral axis coincident with center line of web			V*	R*
					I, in⁴	r, in	S, in³	I, in⁴	r, in	S, in³	1,000 lb	
S 24	120.0	35.13	8.048	0.798	3010.8	9.26	250.9	84.9	1.56	21.1	278	162
	105.9	30.98	7.875	0.625	2811.5	9.53	234.3	78.9	1.60	20.0	218	123
	100.0	29.25	7.247	0.747	2371.8	9.05	197.6	48.4	1.29	13.4	260	140
	90.0	26.30	7.124	0.624	2230.1	9.21	185.8	45.5	1.32	12.8	217	117
	79.9	23.33	7.000	0.500	2087.2	9.46	173.9	42.9	1.36	12.2	174	80
S 20	95.0	27.74	7.200	0.800	1599.7	7.59	160.0	50.5	1.35	14.0	232	150
	85.0	24.80	7.053	0.653	1501.7	7.78	150.2	47.0	1.38	13.3	189	124
	75.0	21.90	6.391	0.641	1263.5	7.60	126.3	30.1	1.17	9.4	186	114
	65.4	19.08	6.250	0.500	1169.5	7.83	116.9	27.9	1.21	8.9	145	83
S 18	70.0	20.46	6.251	0.711	917.5	6.70	101.9	24.5	1.09	7.8	186	123
	54.7	15.94	6.000	0.460	795.5	7.07	88.4	21.2	1.15	7.1	120	70
S 15	50.0	14.59	5.640	0.550	481.1	5.74	64.2	16.0	1.05	5.7	120	91
	42.9	12.49	5.500	0.410	441.8	5.95	58.9	14.6	1.08	5.3	89	58
S 12	50.0	14.57	5.477	0.687	301.6	4.55	50.3	16.0	1.05	5.8	120	116
	40.8	11.84	5.250	0.460	268.9	4.77	44.8	13.8	1.08	5.3	80	78
	35.0	10.20	5.078	0.428	227.0	4.72	37.8	10.0	0.99	3.9	74	66
	31.8	9.26	5.000	0.350	215.8	4.83	36.0	9.5	1.01	3.8	61	45
S 10	35.0	10.22	4.944	0.594	145.8	3.78	29.2	8.5	0.91	3.4	86	89
	25.4	7.38	4.660	0.310	122.1	4.07	24.4	6.9	0.97	3.0	45	38
S 8	23.0	6.71	4.171	0.441	64.2	3.09	16.0	4.4	0.81	2.1	51	
	18.4	5.34	4.000	0.270	56.9	3.26	14.2	3.8	0.84	1.9	31	
S 7	20.0	5.83	3.860	0.450	41.9	2.68	12.0	3.1	0.74	1.6	46	
	15.3	4.43	3.660	0.250	36.2	2.86	10.4	2.7	0.78	1.5	25	
S 6	17.25	5.02	3.565	0.465	26.0	2.28	8.7	2.3	0.68	1.3	40.5	
	12.5	3.61	3.330	0.230	21.8	2.46	7.3	1.8	0.72	1.1	20	
S 5	14.75	4.29	3.284	0.494	15.0	1.87	6.0	1.7	0.63	1.0	35.8	
	10.0	2.87	3.000	0.210	12.1	2.05	4.8	1.2	0.65	0.82	15.2	
S 4	9.5	2.76	2.796	0.326	6.7	1.56	3.3	0.91	0.58	0.65	18.9	
	7.7	2.21	2.660	0.190	6.0	1.64	3.0	0.77	0.59	0.58	11.0	
S 3	7.5	2.17	2.509	0.349	2.9	1.15	1.9	0.59	0.52	0.47	15.2	
	5.7	1.64	2.330	0.170	2.5	1.23	1.7	0.46	0.53	0.40	7.4	

NOTE: 1 in = 2.54 cm; 1 ft = 0.305 m; 1 lb = 4.45 N.
Lightweight beams of each depth are usual stock sizes.
* V and R values are for beams of A36 steel.

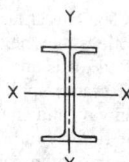

Table 12.2.20   Properties of Wide-Flange Beams and Columns (W Shapes)

Nominal size, in	Weight per ft, lb†	Area of section, in²	Depth of section, in	Flange Width, in	Flange Thickness, in	Web thickness, in	Neutral axis perpendicular to web at center I, in⁴	S, in³	r, in	Neutral axis parallel to web at center I, in⁴	S, in³	r, in	V, 1,000 lb*	R, 1,000 lb*
W 36	300	88.3	36.74	16.66	1.68	0.945	20,300	1,110	15.2	1,300	156	3.83	500	237
	280	82.4	36.52	16.6	1.57	0.885	18,900	1,030	15.1	1,200	144	3.81	465	215
	260	76.5	36.26	16.55	1.44	0.84	17,300	953	15	1,090	132	3.78	439	198
	245	72.1	36.08	16.51	1.35	0.8	16,100	895	15	1,010	123	3.75	416	186
	230	67.6	35.9	16.47	1.26	0.76	15,000	837	14.9	940	114	3.73	393	170
	194	57	36.49	12.12	1.26	0.765	12,100	664	14.6	375	61.9	2.56	402	163
	182	53.6	36.33	12.08	1.18	0.725	11,300	623	14.5	347	57.6	2.55	379	152
	170	50	36.17	12.03	1.1	0.68	10,500	580	14.5	320	53.2	2.53	354	137
	160	47	36.01	12	1.02	0.65	9,750	542	14.4	295	49.1	2.5	337	124
	150	44.2	35.85	11.98	0.94	0.625	9,040	504	14.3	270	45.1	2.47	323	113
W 33	241	70.9	34.18	15.86	1.4	0.83	14,200	829	14.1	932	118	3.63	409	177
	221	65	33.93	15.81	1.275	0.775	12,800	757	14.1	840	106	3.59	379	159
	201	59.1	33.68	15.75	1.15	0.715	11,500	684	14	749	95.2	3.56	347	142
	152	44.7	33.49	11.57	1.055	0.635	8,160	487	13.5	273	47.2	2.47	306	122
	141	41.6	33.3	11.54	0.96	0.605	7,450	448	13.4	246	42.7	2.43	290	109
	130	38.3	33.09	11.51	0.855	0.58	6,710	406	13.2	218	37.9	2.39	276	98
W 30	211	62	30.94	15.11	1.315	0.775	10,300	663	12.9	757	100	3.49	345	162
	191	56.1	30.68	15.04	1.185	0.71	9,170	598	12.8	673	89.5	3.46	314	141
	173	50.8	30.44	14.99	1.065	0.655	8,200	539	12.7	598	79.8	3.43	287	128
	148	43.5	30.67	10.48	1.18	0.65	6,680	436	12.4	227	43.3	2.28	287	131
	132	38.9	30.31	10.55	1	0.615	5,770	380	12.2	196	37.2	2.25	268	115
	124	36.5	30.17	10.52	0.93	0.585	5,360	355	12.1	181	34.4	2.23	254	103
	116	34.2	30.01	10.5	0.85	0.565	4,930	329	12	164	31.3	2.19	244	95
	108	31.7	29.83	10.48	0.76	0.545	4,470	299	11.9	146	27.9	2.15	234	87
W 27	178	52.3	27.81	14.09	1.19	0.725	6,990	502	11.6	555	78.8	3.26	290	141
	161	47.4	27.59	14.02	1.08	0.66	6,280	455	11.5	497	70.9	3.24	262	126
	146	42.9	27.38	13.97	0.975	0.605	5,630	411	11.4	443	63.5	3.21	239	111
	114	33.5	27.29	10.07	0.93	0.57	4,090	299	11	159	31.5	2.18	224	100
	102	30	27.09	10.02	0.83	0.515	3,620	267	11	139	27.8	2.15	201	82
	94	27.7	26.92	9.99	0.745	0.49	3,270	243	10.9	124	24.8	2.12	190	73
W 24	162	47.7	25	12.96	1.22	0.705	5,170	414	10.4	443	68.4	3.05	254	143
	146	43	24.74	12.9	1.09	0.65	4,580	371	10.3	391	60.5	3.01	232	126
	131	38.5	24.48	12.86	0.96	0.605	4,020	329	10.2	340	53	2.97	213	113
	117	34.4	24.26	12.8	0.85	0.55	3,540	291	10.1	297	46.5	2.94	192	94
	104	30.6	24.06	12.75	0.75	0.5	3,100	258	10.1	259	40.7	2.91	173	77
	103	30.3	24.53	9	0.98	0.55	3,000	245	9.96	119	26.5	1.99	194	97
	94	27.7	24.31	9.065	0.875	0.515	2,700	222	9.87	109	24	1.98	180	84
	84	24.7	24.1	9.02	0.77	0.47	2,370	196	9.79	94.4	20.9	1.95	163	70
	76	22.4	23.92	8.99	0.68	0.44	2,100	176	9.69	82.5	18.4	1.92	152	60
W 21	147	43.2	22.06	12.51	1.15	0.72	3,630	329	9.17	376	60.1	2.95	229	140
	132	38.8	21.83	12.44	1.035	0.65	3,220	295	9.12	333	53.5	2.93	204	124
	122	35.9	21.68	12.39	0.96	0.6	2,960	273	9.09	305	49.2	2.92	187	110
	93	27.3	21.62	8.42	0.93	0.58	2,070	192	8.7	92.9	22.1	1.84	181	106
	83	24.3	21.43	8.355	0.835	0.515	1,830	171	8.67	81.4	19.5	1.83	159	85
	73	21.5	21.24	8.295	0.74	0.455	1,600	151	8.64	70.6	17	1.81	139	67
	68	20	21.13	8.27	0.685	0.43	1,480	140	8.6	64.7	15.7	1.8	131	59
	62	18.3	20.99	8.24	0.615	0.4	1,330	127	8.54	57.5	13.9	1.77	121	51
W 18	119	35.1	18.97	11.27	1.06	0.655	2,190	231	7.9	253	44.9	2.69	179	123
	106	31.1	18.73	11.2	0.94	0.59	1,910	204	7.84	220	39.4	2.66	159	106
	97	28.5	18.59	11.15	0.87	0.535	1,750	188	7.82	201	36.1	2.65	143	94
	86	25.3	18.39	11.09	0.77	0.48	1,530	166	7.77	175	31.6	2.63	127	76
	76	22.3	18.21	11.04	0.68	0.425	1,330	146	7.73	152	27.6	2.61	111	60

NOTE: 1 in = 2.54 cm; 1 ft = 0.305 m; 1 lb = 4.45 N.
Flanges of wide-flange beams and columns are not tapered, have constant thickness.
Sections without values of V and R are used chiefly for columns.
Lightweight beams for each nominal size, and beams with depth in even inches, are most usually stocked.
Designation of wide-flange beams is made by giving nominal depth and weight, thus W8 × 40.
* V and R values are for beams of A36 steel.
† Some sections listed are no longer rolled but may be encountered in existing construction. Others currently rolled are not listed. Refer to producers' catalogs for sections currently available.

Table 12.2.20   Properties of Wide-Flange Beams and Columns (W Shapes)   (Continued)

Nominal size, in	Weight per ft, lb†	Area of section, in²	Depth of section, in	Flange Width, in	Flange Thickness, in	Web thickness, in	Neutral axis perpendicular to web at center I, in⁴	S, in³	r, in	Neutral axis parallel to web at center I, in⁴	S, in³	r, in	V, 1,000 lb*	R, 1,000 lb*
W 18	71	20.8	18.47	7.635	0.81	0.495	1,170	127	7.5	60.3	15.8	1.7	132	81
	65	19.1	18.35	7.59	0.75	0.45	1,070	117	7.49	54.8	14.4	1.69	119	67
	60	17.6	18.24	7.555	0.695	0.415	984	108	7.47	50.1	13.3	1.69	109	58
	55	16.2	18.11	7.53	0.63	0.39	890	98.3	7.41	44.9	11.9	1.67	102	51
	50	14.7	17.99	7.495	0.57	0.355	800	88.9	7.38	40.1	10.7	1.65	92	42
W 16	100	29.4	16.97	10.43	0.985	0.585	1,490	175	7.1	186	35.7	2.51	143	107
	89	26.2	16.75	10.37	0.875	0.525	1,300	155	7.05	163	31.4	2.49	127	92
	77	22.6	16.52	10.3	0.76	0.455	1,110	134	7	138	26.9	2.47	108	71
	67	19.7	16.33	10.24	0.665	0.395	954	117	6.96	119	23.2	2.46	93	53
	57	16.8	16.43	7.12	0.715	0.43	758	92.2	6.72	43.1	12.1	1.6	102	63
	50	14.7	16.26	7.07	0.63	0.38	659	81	6.68	37.2	10.5	1.59	89	49
	45	13.3	16.13	7.035	0.565	0.345	586	72.7	6.65	32.8	9.34	1.57	80	41
	40	11.8	16.01	6.995	0.505	0.305	518	64.7	6.63	28.9	8.25	1.57	70	32
	36	10.6	15.86	6.985	0.43	0.295	448	56.5	6.51	24.5	7	1.52	67	29
W 14	426	125	18.67	16.7	3.035	1.875	6,600	707	7.26	2,360	283	4.34	—	—
	398	117	18.29	16.59	2.845	1.77	6,000	656	7.16	2,170	262	4.31	—	—
	370	109	17.92	16.48	2.66	1.655	5,440	607	7.07	1,990	241	4.27	—	—
	342	101	17.54	16.36	2.47	1.54	4,900	559	6.98	1,810	221	4.24	—	—
	311	91.4	17.12	16.23	2.26	1.41	4,330	506	6.88	1,610	199	4.2	—	—
	283	83.3	16.74	16.11	2.07	1.29	3,840	459	6.79	1,440	179	4.17	—	—
	257	75.6	16.38	16	1.89	1.175	3,400	415	6.71	1,290	161	4.13	—	—
	233	68.5	16.04	15.89	1.72	1.07	3,010	375	6.63	1,150	145	4.1	—	—
	211	62	15.72	15.8	1.56	0.98	2,660	338	6.55	1,030	130	4.07	—	—
	193	56.8	15.48	15.71	1.44	0.89	2,400	310	6.5	931	119	4.05	—	—
	176	51.8	15.22	15.65	1.31	0.83	2,140	281	6.43	838	107	4.02	—	—
	159	46.7	14.98	15.57	1.19	0.745	1,900	254	6.38	748	96.2	4	161	145
	145	42.7	14.78	15.5	1.09	0.68	1,710	232	6.33	677	87.3	3.98	145	127
	132	38.8	14.66	14.73	1.03	0.645	1,530	209	6.28	548	74.5	3.76	136	118
	120	35.3	14.48	14.67	0.94	0.59	1,380	190	6.24	495	67.5	3.74	123	106
	109	32	14.32	14.61	0.86	0.525	1,240	173	6.22	447	61.2	3.73	108	92
	99	29.1	14.16	14.57	0.78	0.485	1,110	157	6.17	402	55.2	3.71	99	82
	90	26.5	14.02	14.52	0.71	0.44	999	143	6.14	362	49.9	3.7	89	69
	82	24.1	14.31	10.13	0.855	0.51	882	123	6.05	148	29.3	2.48	105	92
	74	21.8	14.17	10.07	0.785	0.45	796	112	6.04	134	26.6	2.48	92	72
	68	20	14.04	10.04	0.72	0.415	723	103	6.01	121	24.2	2.46	84	61
	61	17.9	13.89	9.995	0.645	0.375	640	92.2	5.98	107	21.5	2.45	75	50
	53	15.6	13.92	8.06	0.66	0.37	541	77.8	5.89	57.7	14.3	1.92	74	49
	48	14.1	13.79	8.03	0.595	0.34	485	70.3	5.85	51.4	12.8	1.91	68	41
	43	12.6	13.66	7.995	0.53	0.305	428	62.7	5.82	45.2	11.3	1.89	60	33
	38	11.2	14.1	6.77	0.515	0.31	385	54.6	5.87	26.7	7.88	1.55	63	34
	34	10	13.98	6.745	0.455	0.285	340	48.6	5.83	23.3	6.91	1.53	57	29
	30	8.85	13.84	6.73	0.385	0.27	291	42	5.73	19.6	5.82	1.49	54	26
W 12	190	55.8	14.38	12.67	1.735	1.06	1,890	263	5.82	589	93	3.25	—	—
	152	44.7	13.71	12.48	1.4	0.87	1,430	209	5.66	454	72.8	3.19	—	—
	136	39.9	13.41	12.4	1.25	0.79	1,240	186	5.58	398	64.2	3.16	—	—
	120	35.3	13.12	12.32	1.105	0.71	1,070	163	5.51	345	56	3.13	134	136
	106	31.2	12.89	12.22	0.99	0.61	933	145	5.47	301	49.3	3.11	113	112
	96	28.2	12.71	12.16	0.9	0.55	833	131	5.44	270	44.4	3.09	101	99
	87	25.6	12.53	12.13	0.81	0.515	740	118	5.38	241	39.7	3.07	93	89
	79	23.2	12.38	12.08	0.735	0.47	662	107	5.34	216	35.8	3.05	84	79
	72	21.1	12.25	12.04	0.67	0.43	597	97.4	5.31	195	32.4	3.04	76	68
	65	19.1	12.12	12	0.605	0.39	533	87.9	5.28	174	29.1	3.02	68	56
	58	17	12.19	10.01	0.64	0.36	475	78	5.28	107	21.4	2.51	63	48
	53	15.6	12.06	9.995	0.575	0.345	425	70.6	5.23	95.8	19.2	2.48	60	44
	50	14.7	12.19	8.08	0.64	0.37	394	64.7	5.18	56.3	13.9	1.96	65	51
	45	13.2	12.06	8.045	0.575	0.335	350	58.1	5.15	50	12.4	1.94	58	42
	40	11.8	11.94	8.005	0.515	0.295	310	51.9	5.13	44.1	11	1.93	51	32

NOTE: 1 in = 2.54 cm; 1 ft = 0.305 m; 1 lb = 4.45 N.
Flanges of wide-flange beams and columns are not tapered, have constant thickness.
Sections without values of V and R are used chiefly for columns.
Lightweight beams for each nominal size, and beams with depth in even inches, are most usually stocked.
Designation of wide-flange beams is made by giving nominal depth and weight, thus W8 × 40.
* V and R values are for beams of A36 steel.
† Some sections listed are no longer rolled but may be encountered in existing construction. Others currently rolled are not listed. Refer to producers' catalogs for sections currently available.

Table 12.2.20 Properties of Wide-Flange Beams and Columns (W Shapes) (Continued)

Nominal size, in	Weight per ft, lb†	Area of section, in²	Depth of section, in	Flange Width, in	Flange Thickness, in	Web thickness, in	Neutral axis perpendicular to web at center I, in⁴	S, in³	r, in	Neutral axis parallel to web at center I, in⁴	S, in³	r, in	V, 1,000 lb*	R, 1,000 lb*
W 12	35	10.3	12.5	6.56	0.52	0.3	285	45.6	5.25	24.5	7.47	1.54	54	33
	30	8.79	12.34	6.52	0.44	0.26	238	38.6	5.21	20.3	6.24	1.52	46	25
	26	7.65	12.22	6.49	0.38	0.23	204	33.4	5.17	17.3	5.34	1.51	40	20
W 10	112	32.9	11.36	10.42	1.25	0.755	716	126	4.66	236	45.3	2.68	—	—
	100	29.4	11.1	10.34	1.12	0.68	623	112	4.6	207	40	2.65	—	—
	88	25.9	10.84	10.27	0.99	0.605	534	98.5	4.54	179	34.8	2.63	—	—
	77	22.6	10.6	10.19	0.87	0.53	455	85.9	4.49	154	30.1	2.6	—	—
	68	20	10.4	10.13	0.77	0.47	394	75.7	4.44	134	26.4	2.59	70	78
	60	17.6	10.22	10.08	0.68	0.42	341	66.7	4.39	116	23	2.57	62	68
	54	15.8	10.09	10.03	0.615	0.37	303	60	4.37	103	20.6	2.56	54	54
	49	14.4	9.98	10	0.56	0.34	272	54.6	4.35	93.4	18.7	2.54	49	45
	45	13.3	10.1	8.02	0.62	0.35	248	49.1	4.32	53.4	13.3	2.01	51	48
	39	11.5	9.92	7.985	0.53	0.315	209	42.1	4.27	45	11.3	1.98	45	39
	33	9.71	9.73	7.96	0.435	0.29	170	35	4.19	36.6	9.2	1.94	41	33
	30	8.84	10.47	5.81	0.51	0.3	170	32.4	4.38	16.7	5.75	1.37	45	35
	26	7.61	10.33	5.77	0.44	0.26	144	27.9	4.35	14.1	4.89	1.36	39	26
	22	6.49	10.17	5.75	0.36	0.24	118	23.2	4.27	11.4	3.97	1.33	35	22
W 8	67	19.7	9	8.28	0.935	0.57	272	60.4	3.72	88.6	21.4	2.12	—	—
	58	17.1	8.75	8.22	0.81	0.51	228	52	3.65	75.1	18.3	2.1	—	—
	48	14.1	8.5	8.11	0.685	0.4	184	43.3	3.61	60.9	15	2.08	49	61
	40	11.7	8.25	8.07	0.56	0.36	146	35.5	3.53	49.1	12.2	2.04	43	53
	35	10.3	8.12	8.02	0.495	0.31	127	31.2	3.51	42.6	10.6	2.03	36	41
	31	9.13	8	7.995	0.435	0.285	110	27.5	3.47	37.1	9.27	2.02	33	35
	28	8.25	8.06	6.535	0.465	0.285	98	24.3	3.45	21.7	6.63	1.62	33	34
	24	7.08	7.93	6.495	0.4	0.245	82.8	20.9	3.42	18.3	5.63	1.61	28	26
	21	6.16	8.28	5.27	0.4	0.25	75.3	18.2	3.49	9.77	3.71	1.26	30	26

NOTE: 1 in = 2.54 cm; 1 ft = 0.305 m; 1 lb = 4.45 N.

Flanges of wide-flange beams and columns are not tapered, have constant thickness.

Sections without values of V and R are used chiefly for columns.

Lightweight beams for each nominal size, and beams with depth in even inches, are most usually stocked.

Designation of wide-flange beams is made by giving nominal depth and weight, thus W8 × 40.

* V and R values are for beams of A36 steel.

† Some sections listed are no longer rolled but may be encountered in existing construction. Others currently rolled are not listed. Refer to producers' catalogs for sections currently available.

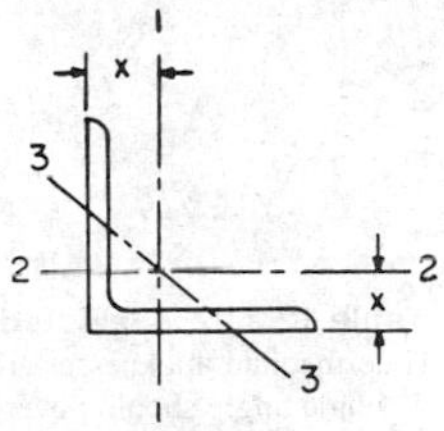

Table 12.2.21 Selected Standard Angles (L Shapes), Equal Legs
(One to three intermediate thicknesses in each size group are available, varying by 1/16 in)
A single angle should never be used as a beam. Two angles, bolted at frequent intervals, may be used.

Size, in	Weight per ft, lb	Areas of section, in²	Axis 1-1 and axis 2-2				Axis 3-3, r min, in	Net areas after deducting holes for ⅞-in rivets	
			I, in⁴	r, in	S, in³	x, in		1 hole	2 holes
8 × 8 × 1⅛	56.9	16.73	98.0	2.42	17.5	2.41	1.56	15.60	14.48
1	51.0	15.00	89.0	2.44	15.8	2.37	1.56	14.00	13.00
⅞	45.0	13.23	79.6	2.45	14.0	2.32	1.57	12.36	11.48
¾	38.9	11.44	69.7	2.47	12.2	2.28	1.57	10.69	9.94
⅝	32.7	9.61	59.4	2.49	10.3	2.23	1.58	8.98	8.36
½	26.4	7.75	48.6	2.50	8.4	2.19	1.59	7.25	6.75
6 × 6 × 1	37.4	11.00	35.5	1.80	8.6	1.86	1.17	10.00	9.00
⅞	33.1	9.73	31.9	1.81	7.6	1.82	1.17	8.86	7.98
¾	28.7	8.44	28.2	1.83	6.7	1.78	1.17	7.69	6.94
⅝	24.2	7.11	24.2	1.84	5.7	1.73	1.18	6.48	5.86
½	19.6	5.75	19.9	1.86	4.6	1.68	1.18	5.25	4.75
⅜	14.9	4.36	15.4	1.88	3.5	1.64	1.19	3.98	3.61
5 × 5 × ⅞	27.2	7.98	17.8	1.49	5.2	1.57	0.97	7.10	6.23
¾	23.6	6.94	15.7	1.51	4.5	1.52	0.97	6.19	5.44
⅝	20.0	5.86	13.6	1.52	3.9	1.48	0.98	5.24	4.61
½	16.2	4.75	11.3	1.54	3.2	1.43	0.98	4.25	3.75
⅜	12.3	3.61	8.7	1.56	2.4	1.39	0.99	3.24	2.86
4 × 4 × ¾	18.5	5.44	7.7	1.19	2.8	1.27	0.78	4.69	3.94
⅝	15.7	4.61	6.7	1.20	2.4	1.23	0.78	3.98	3.36
½	12.8	3.75	5.6	1.22	2.0	1.18	0.78	3.25	2.75
⅜	9.8	2.86	4.4	1.23	1.5	1.14	0.79	2.48	2.11
¼	6.6	1.94	3.0	1.25	1.1	1.09	0.80	1.70	1.45
3½ × 3½ × ½	11.1	3.25	3.6	1.06	1.5	1.06	0.68	2.75	2.25
⅜	8.5	2.48	2.9	1.07	1.2	1.01	0.69	2.10	1.73
¼	5.8	1.69	2.0	1.09	0.79	0.97	0.69	1.44	1.19
3 × 3 × ½	9.4	2.75	2.2	0.90	1.1	0.93	0.58		
⅜	7.2	2.11	1.8	0.91	0.83	0.89	0.58		
¼	4.9	1.44	1.2	0.93	0.58	0.84	0.59		
2½ × 2½ × ½	7.7	2.25	1.2	0.74	0.72	0.81	0.49		
⅜	5.9	1.73	0.98	0.75	0.57	0.76	0.49		
¼	4.1	1.19	0.70	0.77	0.39	0.72	0.49		
2 × 2 × ⅜	4.7	1.36	0.48	0.59	0.35	0.64	0.39		
¼	3.19	0.94	0.35	0.61	0.25	0.59	0.39		
⅛	1.65	0.48	0.19	0.63	0.13	0.55	0.40		
1¾ × 1¾ × ¼	2.77	0.81	0.23	0.53	0.19	0.53	0.34		
⅛	1.44	0.42	0.13	0.55	0.10	0.48	0.35		
1½ × 1½ × ¼	2.34	0.69	0.14	0.45	0.13	0.47	0.29		
⅛	1.23	0.36	0.08	0.47	0.07	0.42	0.30		
1¼ × 1¼ × ¼	1.92	0.56	0.08	0.37	0.09	0.40	0.24		
⅛	1.01	0.30	0.04	0.38	0.05	0.36	0.25		
1 × 1 × ¼	1.49	0.44	0.04	0.29	0.06	0.34	0.20		
⅛	0.80	0.23	0.02	0.30	0.03	0.30	0.20		

NOTE: 1 in = 2.5 cm; 1 ft = 0.305 m; 1 lb = 4.45 N.

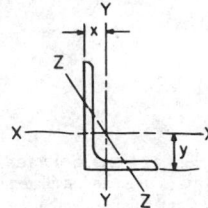

Table 12.2.22   Selected Standard Angles (L Shapes) Unequal Legs

(Intermediate thicknesses are available in each size group, varying by $\frac{1}{16}$ in among the thinner angles)
A single angle should never be used as a beam. Two angles, bolted at frequent intervals, may be used.

Size, in	Thickness, in	Weight per ft, lb	Area of section, in²	Axis X-X				Axis Y-Y				Axis Z-Z	Net areas after deducting holes for ⅞-in rivets	
				I, in⁴	S, in³	r, in	y, in	I, in⁴	S, in³	r, in	y, in	r, in	1 hole	2 holes
8 × 6	1	44.2	13.00	80.8	15.1	2.49	2.65	38.8	8.9	1.73	1.65	1.28	12.00	11.00
	¾	33.8	9.94	63.4	11.7	2.53	2.56	30.7	6.9	1.76	1.56	1.29	9.19	8.44
	½	23.0	6.75	44.3	8.0	2.56	2.47	21.7	4.8	1.79	1.47	1.30	6.25	5.75
	⁷⁄₁₆	20.2	5.93	39.2	7.1	2.57	2.45	19.3	4.2	1.80	1.45	1.31	5.49	5.06
8 × 4	1	37.4	11.00	69.6	14.1	2.52	3.05	11.6	3.9	1.03	1.05	0.85	10.00	9.00
	¾	28.7	8.44	54.9	10.9	2.55	2.95	9.4	3.1	1.05	0.95	0.85	7.69	6.94
	½	19.6	5.75	38.5	7.5	2.59	2.86	6.7	2.2	1.08	0.86	0.86	5.25	4.75
	⁷⁄₁₆	17.2	5.06	34.1	6.6	2.60	2.83	6.0	1.9	1.09	0.83	0.87	4.62	4.18
7 × 4	⅞	30.2	8.86	42.9	9.7	2.20	2.55	10.2	3.5	1.07	1.05	0.86	7.98	7.11
	¾	26.2	7.69	37.8	8.4	2.22	2.51	9.1	3.0	1.09	1.01	0.86	6.94	6.19
	½	17.9	5.25	26.7	5.8	2.25	2.42	6.5	2.1	1.11	0.92	0.87	4.75	4.25
	⅜	13.6	3.98	20.6	4.4	2.27	2.37	5.1	1.6	1.13	0.87	0.88	3.62	3.24
6 × 4	⅞	27.2	7.98	27.7	7.2	1.86	2.12	9.8	3.4	1.11	1.12	0.86	7.10	6.23
	¾	23.6	6.94	24.5	6.3	1.88	2.08	8.7	3.0	1.12	1.08	0.86	6.19	5.44
	½	16.2	4.75	17.4	4.3	1.91	1.99	6.3	2.1	1.15	0.99	0.87	4.25	3.75
	⅜	12.3	3.61	13.5	3.3	1.93	1.94	4.9	1.6	1.17	0.94	0.88	3.24	2.86
	⁵⁄₁₆	10.3	3.03	11.4	2.8	1.94	1.92	4.2	1.4	1.17	0.92	0.88	2.72	2.40
6 × 3½	½	15.3	4.50	16.6	4.2	1.92	2.08	4.3	1.6	0.97	0.83	0.76	4.00	3.50
	⅜	11.7	3.42	12.9	3.2	1.94	2.04	3.3	1.2	0.99	0.79	0.77	3.04	2.67
	¼	7.9	2.31	8.9	2.2	1.96	1.99	2.3	0.85	1.01	0.74	0.78	2.06	1.81
5 × 3½	¾	19.8	5.81	13.9	4.3	1.55	1.75	5.6	2.2	0.98	1.00	0.75	5.06	4.31
	½	13.6	4.00	10.0	3.0	1.58	1.66	4.1	1.6	1.01	0.91	0.75	3.50	3.00
	¼	7.0	2.06	5.4	1.6	1.61	1.56	2.2	0.83	1.04	0.81	0.76	1.81	1.56
5 × 3	½	12.8	3.75	9.5	2.9	1.59	1.75	2.6	1.1	0.83	0.75	0.65	3.25	2.75
	⅜	9.8	2.86	7.4	2.2	1.61	1.70	2.0	0.89	0.84	0.70	0.65	2.48	2.11
	¼	6.6	1.94	5.1	1.5	1.62	1.66	1.4	0.61	0.86	0.66	0.66	1.69	1.44
4 × 3½	⅝	14.7	4.30	6.4	2.4	1.22	1.29	4.5	1.8	1.03	1.04	0.72	3.68	3.05
	½	11.9	3.50	5.3	1.9	1.23	1.25	3.8	1.5	1.04	1.00	0.72	3.00	2.50
	⅜	9.1	2.67	4.2	1.5	1.25	1.21	3.0	1.2	1.06	0.96	0.73	2.30	1.92
	¼	6.2	1.81	2.9	1.0	1.27	1.16	2.1	0.81	1.07	0.91	0.73	1.56	1.31
4 × 3	⅝	13.6	3.98	6.0	2.3	1.23	1.37	2.9	1.4	0.85	0.87	0.64	3.36	2.73
	½	11.1	3.25	5.1	1.9	1.25	1.33	2.4	1.1	0.86	0.83	0.64	2.75	2.25
	¼	5.8	1.69	2.8	1.0	1.28	1.24	1.4	0.60	0.90	0.74	0.65	1.44	1.19
3½ × 3	½	10.2	3.00	3.5	1.5	1.07	1.13	2.3	1.1	0.88	0.88	0.62	2.50	
	¼	5.4	1.56	1.9	0.78	1.11	1.04	1.3	0.59	0.91	0.79	0.63	1.31	
3½ × 2½	½	9.4	2.75	3.2	1.4	1.09	1.20	1.4	0.76	0.70	0.70	0.53	2.25	
	¼	4.9	1.44	1.8	0.75	1.12	1.11	0.78	0.41	0.74	0.61	0.54	1.19	
3 × 2½	½	8.5	2.50	2.1	1.0	0.91	1.00	1.3	0.74	0.72	0.75	0.52		
	⅜	6.6	1.92	1.7	0.81	0.93	0.96	1.0	0.58	0.74	0.71	0.52		
	¼	4.5	1.31	1.2	0.56	0.95	0.91	0.74	0.40	0.75	0.66	0.53		
3 × 2	½	7.7	2.25	1.9	1.0	0.92	1.08	0.67	0.47	0.55	0.58	0.43		
	³⁄₁₆	3.07	0.90	0.84	0.41	0.97	0.97	0.31	0.20	0.58	0.47	0.44		
2½ × 2	⅜	5.3	1.55	0.91	0.55	0.77	0.83	0.51	0.36	0.58	0.58	0.42		
	³⁄₁₆	2.75	0.81	0.51	0.29	0.79	0.76	0.29	0.20	0.60	0.51	0.43		
2 × 1½	¼	2.77	0.81	0.32	0.24	0.62	0.66	0.15	0.14	0.43	0.41	0.32		
	⅛	1.44	0.42	0.17	0.13	0.64	0.62	0.09	0.08	0.45	0.37	0.33		
1¾ × 1¼	¼	2.34	0.69	0.20	0.18	0.54	0.60	0.09	0.10	0.35	0.35	0.27		
	⅛	1.23	0.36	0.11	0.09	0.56	0.56	0.05	0.05	0.37	0.31	0.27		

NOTE: 1 in = 2.54 cm; 1 ft = 0.305 m; 1 lb = 4.45 N.

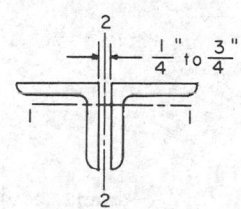

Table 12.2.23 Radii of Gyration for Two Angles, Unequal Legs

Single angle			Two angles	Radii of gyration, in							
				Long legs vertical				Short legs vertical			
					Axis 2-2				Axis 2-2		
Size, in	Weight per ft, lb	Area, in²		Axis 1-1	In contact	⅜ in apart	¾ in apart	Axis 1-1	In contact	⅜ in apart	¾ in apart
8 × 6 × 1	44.2	26.00		2.49	2.39	2.52	2.66	1.73	3.64	3.78	3.92
¾	33.8	19.9		2.53	2.35	2.48	2.62	1.76	3.60	3.74	3.88
8 × 4 × 1	37.4	22.00		2.52	1.47	1.61	1.76	1.03	3.95	4.10	4.25
½	19.6	11.50		2.59	1.38	1.51	1.64	1.08	3.86	4.00	4.14
7 × 4 × ¾	26.2	15.4		2.22	1.48	1.62	1.76	1.09	3.35	3.48	3.64
⅜	13.6	7.96		2.27	1.43	1.55	1.68	1.13	3.28	3.42	3.56
6 × 4 × ¾	23.6	13.9		1.88	1.55	1.69	1.83	1.12	2.80	2.94	3.09
⅜	12.3	7.22		1.93	1.50	1.62	1.76	1.17	2.74	2.87	3.02
5 × 3½ × ¾	19.8	11.62		1.55	1.40	1.54	1.69	0.98	2.34	2.48	2.63
5/16	8.7	5.12		1.61	1.33	1.45	1.59	1.03	2.26	2.38	2.53
4 × 3½ × ½	11.9	7.00		1.23	1.44	1.58	1.72	1.04	1.76	1.89	2.04
5/16	7.7	4.50		1.26	1.42	1.55	1.69	1.07	1.73	1.86	2.00
4 × 3 × ½	11.1	6.50		1.25	1.20	1.33	1.48	0.86	1.82	1.96	2.11
¼	5.8	3.38		1.28	1.16	1.29	1.43	0.90	1.78	1.92	2.06
3½ × 3 × ⅜	7.9	4.59		1.09	1.22	1.36	1.50	0.90	1.53	1.67	1.82
¼	5.4	3.12		1.11	1.21	1.34	1.48	0.91	1.52	1.65	1.80
3 × 2½ × ⅜	6.6	3.84		0.93	1.02	1.16	1.31	0.74	1.34	1.48	1.63
¼	4.5	2.62		0.95	1.00	1.13	1.28	0.75	1.31	1.45	1.60
2½ × 2 × ⅜	5.3	3.10		0.77	0.82	0.96	1.11	0.58	1.13	1.27	1.43
¼	3.6	2.12		0.78	0.80	0.94	1.09	0.59	1.11	1.25	1.40

NOTE: 1 in = 2.54 cm; 1 ft = 0.305 m; 1 lb = 4.45 N.

dard beams (S shapes) or wide-flange sections (W shapes) length-wise at midheight of the web, making two similar shapes of T section. Table 12.2.25 lists a selection of such tees.

For additional data regarding structural shapes, their strengths as beams and columns, and means of making connections, see ''AISC Manual of Steel Construction.''

Welding The main advantage of assembling and connecting steel frames by welding is the reduction in the amount of metal used. The saving in metal is achieved by (1) elimination of bolt holes which reduce the net section of tension members, (2) simplification of details, and (3) elimination of splice plate and gusset plate material. (See also Sec. 13.3.)

Allowable stresses in welds depend on the type of weld, the manner of loading, and the relative strengths of the weld metal and the base metal. For **complete-penetration groove welds** in which the full edge thickness of the thinner part to be joined is beveled in preparation for welding, allowable stresses due to tension or compression normal to the effective area or parallel to the axis of the weld are the same as the base metal, and the allowable shear stress on the effective area is 0.3 times the nominal tensile strength of the weld metal (limited by 0.4 times yield stress in the base metal). The effective area for a complete-penetration groove weld is the width of the part joined times the thickness of the thinner part.

For **partial-penetration groove welds,** allowable stresses due to compression normal to the effective area and tension or compression parallel to the axis of the weld are the same as the base metal. For tension normal to the effective area, the allowable stress is 0.3 times the nominal tensile strength of the weld metal (limited by 0.6 times the yield stress of the base metal); for shear parallel to the axis of the weld, the allowable stress on the effective area is 0.3 times the nominal tensile strength of weld metal (limited by 0.4 times yield stress of the base metal). The effective thickness of partial-penetration groove welds depends on the welding process, welding position, and the included angle of the groove but may be safely taken as the depth of the chamfer less ⅛ in.

For **fillet welds,** allowable shear stresses on the effective area are taken as 0.3 times nominal tensile strength of the weld (limited by 0.4 times yield stress of the base metal), and for tension or compression parallel to the axis of the weld, allowable stresses are the same as the base metal. Fillet welds are not to be loaded in tension or compression normal to the effective area; they transfer loads between members in shear only. The effective area of a fillet weld is the overall length of the full-size weld times the shortest distance from the root to the face (normally leg size × sin 45°). Allowable shear in a fillet weld is taken as 930 lb/in of length (163 N/mm) for each 1/16 in (1.59 mm) of leg size. Fillet welds should be limited to ½ in (12.5 mm) leg size in normal construction and to the thickness of the material up to ¼ in and thickness less 1/16 in for thicker material. The minimum-size fillet weld for ¼-in-thick material should

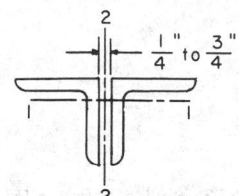

Table 12.2.24 Radii of Gyration for Two Angles, Equal Legs

Single angle		Two angles	Radii of gyration, in			
				Axis 2-2		
Size, in	Weight per ft, lb	Area, in^2	Axis 1-1	In contact	⅜ in apart	¾ in apart
8 × 8 × 1⅛	56.9	33.46	2.42	3.42	3.55	3.69
½	26.4	15.50	2.50	3.33	3.45	3.59
6 × 6 × 1	37.4	22.00	1.80	2.59	2.72	2.87
⅜	14.9	8.72	1.88	2.49	2.62	2.76
5 × 5 × ⅞	27.2	15.96	1.49	2.17	2.31	2.45
⅜	12.3	7.22	1.56	2.09	2.22	2.35
4 × 4 × ¾	18.5	10.88	1.19	1.74	1.88	2.03
¼	6.6	3.88	1.25	1.66	1.79	1.93
3½ × 3½ × ⅜	8.5	4.97	1.07	1.48	1.61	1.75
¼	5.8	3.38	1.09	1.46	1.59	1.73
3 × 3 × ½	9.4	5.50	0.90	1.29	1.43	1.58
¼	4.9	2.88	0.93	1.25	1.38	1.53
2½ × 2½ × ⅜	5.9	3.47	0.75	1.07	1.21	1.36
¼	4.1	2.38	0.77	1.05	1.19	1.34
2 × 2 × ⅜	4.7	2.72	0.59	0.87	1.02	1.18
¼	3.19	1.88	0.61	0.85	0.99	1.14

NOTE: 1 in = 2.54 cm; 1 ft = 0.305 m; 1 lb = 4.45 N.

be ⅛ in; for over ¼- to ½-in material, 3/16 in; for over ½-in to ¾-in material, ¼ in; and for over ¾-in-thick material, 5/16 in. Typical details of welded connections are indicated in Fig. 12.2.18.

Safe Loads for Steel Beams To determine the safe load uniformly distributed, as limited by bending, for a structural steel beam on a given span, apply the formula $W = 8F_bs/l$, where W is the total load, lb.; F_b is the allowable fiber stress (24,000 lb/in^2 or any other); S is the section

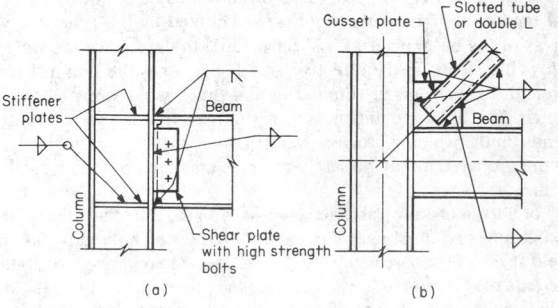

Fig. 12.2.18 Welded connections. (*a*) Moment connection; (*b*) bracing connection.

modulus for the beam in question, given in Tables 12.2.18 to 12.2.26; and l is the span, in. (This formula may also be used with equivalent metric units.) The safe load concentrated at midspan is one-half this amount. For other safe loads, note that F_bS is the safe resistance to bending in inch-pounds (or newton-meters) afforded by the beam.

Compute the load, of whatever type or distribution, which will produce a maximum bending moment equal to safe moment of resistance (see Sec. 5 for bending-moment formulas).

To select a beam to support a given load, compute the maximum bending moment in inch-pounds, divide by the allowable fiber stress F_b, and refer to the table for a beam having a section modulus which is not smaller than the quotient.

Formulas for the safe loads and deflections of beams with various methods of support and of loading are given in Sec. 5.

Short beams should be investigated for crippling of the web. In the tables are given the safe end reactions for beams of A36 steel resting on a seat 3½ in (9 cm) long along the axis of the beam. Short beams should also be investigated for shear, by dividing the maximum shear, in pounds, by the area of the web, excluding the flanges.

Single angles used as beams and loaded in the plane of axis *X-X* or *Y-Y* tend to deflect laterally as well as in the plane of the loads. Unless this is prevented, as by pairing the angles back to back and securing them together, the unit fiber stress due to bending may be as much as 40 percent above that computed by dividing the bending moment by S for the axis perpendicular to the plane of the loads. The relation $f = M/S$ does not hold for single angles, and Z bars, which are unsymmetrical about both axes.

Deflection of I Beams and Other Structural Shapes Table 12.2.27 gives **coefficients of deflection** for steel shapes under uniformly distributed loads, and is based on the formula; deflection in inches = $30fL^2/Ed$, the table giving the values of $30fL^2/E$. (f = fiber stress, lb/in^2, L = span, ft; d = depth of section, in; E = modulus of elasticity = 29,000,000 lb/in^2.)

To find the deflection in inches of a section symmetrical about the neutral axis, such as a beam, channel, etc., divide the coefficient in the table corresponding to given span and fiber stress by the depth of the section in inches.

To find the deflection in inches of a section which is not symmetrical about the neutral axis but which is symmetrical about an axis at right angles thereto, such as a tee or pair of angles, divide the coefficient corresponding to given span and fiber stress by twice the distance of extreme fiber from neutral axis obtained from table of elements of sections.

To find the deflection in inches of a section for any other fiber stress than those given, multiply this fiber stress by either of the coefficients in the table for the given span and divide by the fiber stress corresponding to the coefficients used.

I beams and channels loaded to a fiber stress of 24,000 lb/in^2 (165.5 MPa) will not deflect in excess of $\frac{1}{360}$ of the span (allowed for plastered ceilings) if the depth in inches (cm) is not less than 0.74 (6.21) times the span in feet (m) for uniform loads and 0.60 (5.00) times the span for central concentration.

Beam Supports Steel beams are supported at the ends generally (1) by means of web connections to girders and columns, (2) by resting on structural-steel seats, or (3) by resting on masonry. Limiting values of end reactions of the second type, for seats 3½ in (9 cm) long, are given in Tables 12.2.18 to 12.2.20. Standard AISC web connections of the first type are called *framed beam* connections and are designated by the number of rows of bolts. Examples of connections are given in Fig. 12.2.19. These connections may be specified as "Standard 3 row, 4 row, etc., connections." Connections must always be designated and detailed for the calculated design reaction.

The capacity of web connections is governed by the shearing of the fastener, or the bearing of the fastener on the web or on the material to which the beam is connected, or by the strength of the connecting angles. The supporting values of standard framed beam connections, using ⅞ in fasteners in members of A36 material, are given in Table 12.2.28. For fasteners in webs thicker than 0.34 in use the values in the column headed Double Shear; for thinner webs, bearing limits the value, and the coefficients for web bearing are to be used. For ¾-in fasteners, multiply tabular bearing values by $\frac{6}{7}$ and shear values by $\frac{36}{49}$. Fasteners connecting the outstanding legs to the supporting metal are in double shear if two beams are framed opposite or in single shear if a beam is connected on one side only. If the supporting material of A36 steel is thinner than 0.34 in in double-shear connections or thinner than 0.17 in in single-shear connections, the capacity is limited by bearing. The value of any ⅞-in fastener in bearing on A36

material is 60,900t, where t is the thickness of the plate. The value of ⅞ in A502, grade 1 rivets or A325 HS bolts (slip-critical connections) is 9,020 lb in single shear and 20,400 lb in double shear. The corresponding values for A307 unfinished bolts are 6,000 lb and 12,000 lb, respectively.

Cast-iron columns were often used in the past instead of wood, to save space, in the lower stories of heavy buildings. Their use is now obsolete, but they are occasionally encountered in repair and alteration work to older buildings. The ratio of length to least radius of gyration l/r should not exceed 70, and the average unit stress under axial compression should not exceed $9,000 - 40l/r$ lb/in^2.

Steel joists consisting of lightweight rolled sections, thin for their height, or open-web trussed members fabricated by welding or otherwise, are used with economy in buildings where spans are long and loads are light, and where a plaster ceiling affords sufficient fire protection. They are rarely used in industrial buildings, except to support roof loads.

Steel pipe is often used for columns under light loads. Table 12.2.29 gives the safe loads on standard size pipes (ASTM A501 or A53, grade B) used as columns. For extra-strong and double extra-strong pipe used as columns, the safe loads will increase approximately in the same proportion as the weight per foot. (See Sec. 8.7)

Structural steel tubing (ASTM A500, grade B), in square or rectangular cross section with $F_y = 46$ Ksi (317.1 MPa), is also used for columns, bracing members, and unbraced members subjected to large torsional loads. The closed box shape makes tube sections especially suited for resisting torsional loads.

Corrugated metal deck and siding is used for roofs and walls, respectively, to span between purlins for roof loads or between girts for wind loads. The decking is sized to resist the bending caused by these loads. Roof decking is often also used as a diaphragm to transfer wind or seismic loads to the lateral bracing system below. Load tables specifying safe loads for different spans are available from metal deck manufacturers.

The spacing of purlins on roofs and girts on wall is usually 4 to 6 ft. Numbers 20 and 22, U.S. Standard gage, are generally used for roofing; No. 24 for siding.

Fire Resistance The resistance to fire of building materials has been tested extensively by various agencies. Table 12.2.30 gives the fire-resistance rating of a few of the common building materials and methods of construction as established by the Uniform Building Code from standard fire tests.

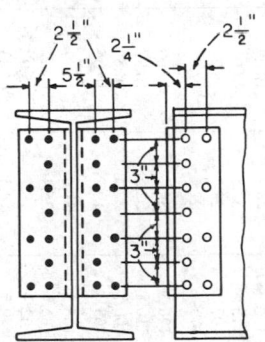
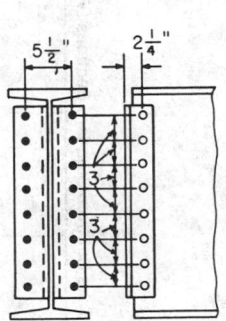
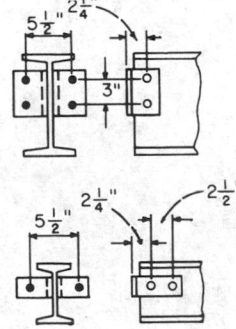

Fig. 12.2.19 Framed beam connections.

Table 12.2.25   Tees Cut from Standard Sections (WT and ST Shapes)*

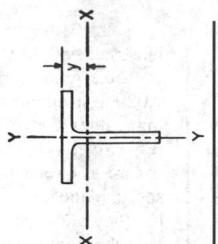

Nominal depth, in	Weight per ft, lb	Area, in²	Depth, of tee, in	Flange Width in	Avg thickness, in	Stem thickness, in	Axis X-X I, in⁴	S, in³	r, in	y, in	Axis Y-Y I, in⁴	S, in³	r, in
WT18	150	44.1	18.37	16.66	1.68	0.95	1,230	86.1	5.27	4.13	648	77.8	3.83
	115	33.8	17.95	16.47	1.26	0.76	934	67.0	5.25	4.01	470	57.1	3.73
	97	28.5	18.25	12.12	1.26	0.77	901	67.0	5.62	4.80	187	30.9	2.56
	75	22.1	17.93	11.98	0.94	0.63	698	53.1	5.62	4.78	135	22.5	2.47
WT16.5	120.5	35.4	17.09	15.86	1.40	0.83	875	65.0	4.96	3.85	466	58.8	3.63
	110.5	32.5	16.97	15.81	1.28	0.78	799	60.8	4.96	3.81	420	53.2	3.59
	100.5	29.5	16.84	15.75	1.15	0.72	725	55.5	4.95	3.70	375	47.6	3.56
	65	19.2	16.55	11.51	0.86	0.58	513	42.1	5.18	4.36	109	18.9	2.39
WT15	106.5	31.0	15.47	15.11	1.32	0.78	610	50.5	4.43	3.40	378	50.1	3.49
	86.5	25.4	15.22	14.99	1.07	0.66	497	41.7	4.42	3.31	299	39.9	3.43
	62	18.2	15.09	10.52	0.93	0.59	396	35.3	4.66	3.90	90.4	17.2	2.23
	49.5	14.5	14.83	10.45	0.67	0.52	322	30.0	4.71	4.09	63.9	12.2	2.10
WT113.5	97	28.5	14.06	14.04	1.34	0.75	444	40.3	3.95	3.03	309	44.1	3.29
	80.5	23.7	13.80	14.02	1.08	0.66	372	34.4	3.96	2.99	248	35.4	3.24
	57	16.8	13.65	10.07	0.93	0.57	289	28.3	4.15	3.42	79.4	15.8	2.18
	42	12.4	13.36	9.96	0.64	0.46	216	21.9	4.18	3.48	52.8	10.6	2.07
WT12	81	23.9	12.50	12.96	1.27	0.71	293	29.9	3.50	2.70	221	34.2	3.05
	52	15.3	12.03	12.75	0.75	0.50	189	20.0	3.51	2.59	130	20.3	2.91
	42	12.4	12.05	9.02	0.77	0.47	166	18.3	3.67	2.97	47.2	10.5	1.95
	27.5	8.1	11.79	7.01	0.51	0.40	117	14.1	3.80	3.50	14.5	4.15	1.34

WT10.5	41.5	12.2	10.72	8.36	0.84	0.52	127	15.7	3.22	2.66	40.7	9.75	1.83
	34	10.0	10.57	8.27	0.69	0.43	103	12.9	3.20	2.59	32.4	7.83	1.80
	28.5	8.37	10.53	6.56	0.65	0.41	90.4	11.8	3.29	2.85	15.3	4.67	1.35
	22	6.49	10.33	6.50	0.45	0.35	71.1	9.7	3.31	2.98	10.3	3.18	1.26
WT9	65	19.1	9.63	11.16	1.20	0.67	127	16.7	2.50	2.02	139	24.9	2.70
	53	15.6	9.37	11.20	0.94	0.59	104	14.1	2.59	1.97	110	19.7	2.66
	30	8.82	9.12	7.56	0.70	0.42	64.7	9.3	2.71	2.16	25	6.63	1.69
	20	5.88	8.95	6.02	0.53	0.32	44.8	6.73	2.76	2.29	9.6	3.17	1.27
WT8	38.5	11.3	8.26	10.30	0.76	0.46	56.9	8.59	2.24	1.63	69.2	13.4	2.47
	25	7.37	8.13	7.07	0.63	0.38	42.3	6.78	2.40	1.89	18.6	5.26	1.59
	20	5.89	8.01	7.00	0.51	0.31	33.1	5.35	2.37	1.81	14.4	4.12	1.57
	15.5	4.56	7.94	5.23	0.44	0.28	27.4	4.64	2.45	2.02	6.20	2.24	1.17
WT7	155.5	45.7	8.56	16.23	2.26	1.41	176	26.7	1.96	1.97	807	99.4	4.20
	60	17.7	7.24	14.67	0.94	0.59	51.7	8.61	1.71	1.24	247	33.7	3.74
	41	12.0	7.16	10.13	0.89	0.51	41.2	7.14	1.85	1.39	74.2	14.6	2.48
	17	5.0	6.99	6.75	0.46	0.29	20.9	3.83	2.04	1.53	11.7	3.45	1.53
WT6	95	27.9	7.19	12.67	1.74	1.06	79.0	14.2	1.68	1.62	295	46.5	3.25
	48	14.1	6.36	12.16	0.90	0.55	32.0	6.12	1.51	1.13	135	22.2	3.09
	32.5	9.54	6.06	12.00	0.61	0.39	20.6	4.06	1.47	0.99	87.2	14.5	3.02
	20	5.89	5.97	8.01	0.52	0.30	14.4	2.95	1.57	1.08	22.0	5.51	1.93
WT5	44	12.9	5.42	10.27	0.99	0.61	20.8	4.77	1.27	1.06	89.3	17.4	2.63
	30	8.82	5.11	10.08	0.68	0.42	12.9	3.04	1.21	0.88	58.1	11.5	2.57
	22.5	5.73	4.96	7.99	0.53	0.32	8.84	2.16	1.24	0.88	22.5	5.64	1.98
	15	4.42	5.24	5.81	0.51	0.30	9.28	2.24	1.45	1.1	8.35	2.87	1.37
WT4	33.5	9.84	4.50	8.28	0.94	0.57	10.9	3.05	1.05	0.94	44.3	10.7	2.12
	20	5.87	4.13	8.07	0.56	0.36	5.73	1.69	0.99	0.74	24.5	6.08	2.04
	14	4.12	4.03	6.34	0.47	0.29	4.22	1.28	1.01	0.73	10.8	3.31	1.62
	7.5	2.22	4.06	4.02	0.32	0.25	3.28	1.07	1.22	1.0	1.7	0.85	0.87
ST 9	35	10.3	9.00	6.25	0.69	0.71	84.7	14.0	2.87	2.94	12.1	3.86	1.08
6	25	7.35	6.00	5.48	0.66	0.69	25.2	6.05	1.85	1.84	7.85	2.87	1.03
4	11.5	3.38	4.00	4.17	0.43	0.44	5.03	1.77	1.22	1.15	2.15	1.03	0.80
3	6.25	1.83	3.00	3.33	0.36	0.23	1.27	0.56	0.83	0.69	0.91	0.55	0.71

NOTE: 1 in = 2.54 cm; 1 ft = 0.305 m; 1 lb = 4.45 N.
* The availability of WT sections listed is governed by the basic W sections from which they are cut. See footnote under Table 12.2.20.

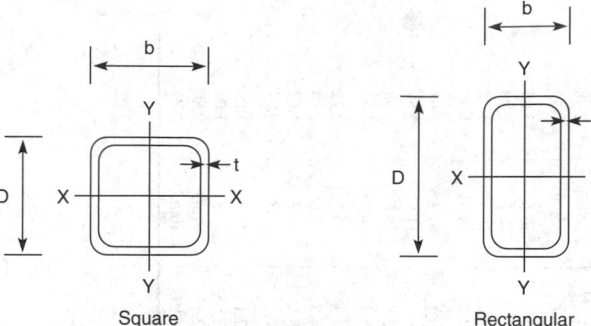

Square   Rectangular

Table 12.2.26   Properties of Square and Rectangular Tubing (TS Sections)*

Nominal Size D, in $\times b$, in	t, in	Weight, lb/ft	Area, in^2	I_{xx}, in^4	S_{xx}, in^3	r_{xx}, in	I_{yy}, in^4	S_{yy}, in^3	r_{yy}, in
TS 12 × 12	0.5	76.07	22.4	485	80.9	4.66	485	80.9	4.66
12	0.375	58.1	17.1	380	63.4	4.72	380	63.4	4.72
TS 10 × 10	0.5	62.46	18.4	271	54.2	3.84	271	54.2	3.84
10	0.375	47.9	14.1	214	42.9	3.9	214	42.9	3.9
10	0.25	32.63	9.59	151	30.1	3.96	151	30.1	3.96
TS 8 × 8	0.5	48.85	14.4	131	32.9	3.03	131	32.9	3.03
8	0.375	37.69	11.1	106	26.4	3.09	106	26.4	3.09
8	0.25	25.82	7.59	75.1	18.8	3.15	75.1	18.8	3.15
TS 6 × 6	0.5	35.24	10.4	50.5	16.8	2.21	50.5	16.8	2.21
6	0.375	27.48	8.08	41.6	13.9	2.27	41.6	13.9	2.27
6	0.25	19.02	5.59	30.3	10.1	2.33	30.3	10.1	2.33
6	0.1875	14.53	4.27	23.8	7.93	2.36	23.8	7.93	2.36
TS 5 × 5	0.5	28.43	8.36	27.0	10.8	1.80	27.0	10.8	1.80
	0.375	22.37	6.58	22.8	9.11	1.86	22.8	9.11	1.86
	0.25	15.62	4.59	16.9	6.78	1.92	16.9	6.78	1.92
	0.1875	11.97	3.52	13.4	5.36	1.95	13.4	5.36	1.95
TS 4 × 4	0.5	21.63	6.36	12.3	6.13	1.39	12.3	6.13	1.39
4	0.375	17.27	5.08	10.7	5.35	1.45	10.7	5.35	1.45
4	0.25	12.21	3.59	8.22	4.11	1.51	8.22	4.11	1.51
4	0.1875	9.42	2.77	6.59	3.3	1.54	6.59	3.3	1.54
TS 3 × 3	0.375	10.58	3.11	3.58	2.39	1.07	3.58	2.39	1.07
	0.25	8.81	2.59	3.16	2.10	1.10	3.16	2.10	1.10
	0.1875	6.87	2.02	2.60	1.73	1.13	2.60	1.73	1.13
TS 2 × 2	0.3125	6.32	1.86	0.815	0.815	0.662	0.815	0.815	0.662
2	0.25	5.41	1.59	0.766	0.766	0.694	0.766	0.766	0.694
2	0.1875	4.32	1.27	0.668	0.668	0.726	0.668	0.668	0.726
TS 20 × 12	0.5	103.3	30.4	1,650	165	7.37	750	125	4.97
12	0.375	78.52	23.1	1,280	128	7.45	583	97.2	5.03
8	0.5	89.68	26.4	1,270	127	6.94	300	75.1	3.38
8	0.375	68.31	20.1	988	98.8	7.02	236	59.1	3.43
8	0.3125	57.36	16.9	838	83.8	7.05	202	50.4	3.46
TS16 × 12	0.5	89.68	26.4	962	120	6.04	618	103	4.84
12	0.375	68.31	20.1	748	93.5	6.11	482	80.3	4.9
8	0.5	76.07	22.4	722	90.2	5.68	244	61	3.3
8	0.375	58.1	17.1	565	70.6	5.75	193	48.2	3.36
8	0.3125	48.86	14.4	481	60.1	5.79	165	41.2	3.39
TS 12 × 6	0.625	67.82	19.9	337	56.2	4.11	112	37.2	2.37
6	0.5	55.66	16.4	287	47.8	4.19	96	32	2.42
6	0.375	42.79	12.6	228	38.1	4.26	77.2	25.7	2.48
6	0.25	29.23	8.59	161	26.9	4.33	55.2	18.4	2.53
6	0.1875	22.18	6.52	124	20.7	4.37	42.8	14.3	2.56
TS 12 × 4	0.625	59.32	17.4	257	42.8	3.84	41.8	20.9	1.55
4	0.5	48.85	14.4	221	36.8	3.92	36.9	18.5	1.6
4	0.375	37.69	11.1	178	29.6	4.01	30.5	15.2	1.66
4	0.25	25.82	7.59	127	21.1	4.09	22.3	11.1	1.71
4	0.1875	19.63	5.77	98.2	16.4	4.13	17.5	8.75	1.74

* On special order, TS sections currently are available in sizes up to 30 × 30 and 30 × 24.

Table 12.2.26 Properties of Square and Rectangular Tubing (TS Sections)* *(Continued)*

Nominal Size D, in $\times$ b, in	t, in	Weight, lb/ft	Area of metal in^2	I_{xx}, in^4	S_{xx}, in^3	r_{xx}, in	I_{yy}, in^4	S_{yy}, in^3	r_{yy}, in
TS 10 $\times$ 4	0.5	42.05	12.4	136	27.1	3.31	30.8	15.4	1.58
4	0.375	32.58	9.58	110	22	3.39	25.5	12.8	1.63
4	0.25	22.42	6.59	79.3	15.9	3.47	18.8	9.39	1.69
TS 8 $\times$ 6	0.5	42.05	12.4	103	25.8	2.89	65.7	21.9	2.31
6	0.375	32.58	9.58	83.7	20.9	2.96	53.5	17.8	2.36
6	0.25	22.42	6.59	60.1	15	3.02	38.6	12.9	2.42
4	0.625	42.3	12.4	85.1	21.3	2.62	27.4	13.7	1.49
4	0.5	35.24	10.4	75.1	18.8	2.69	24.6	12.3	1.54
4	0.375	27.48	8.08	61.9	15.5	2.77	20.6	10.3	1.6
4	0.25	19.02	5.59	45.1	11.3	2.84	15.3	7.63	1.65
2	0.375	22.37	6.58	40.1	10	2.47	3.85	3.85	0.765
2	0.25	15.62	4.59	30.1	7.52	2.56	3.08	3.08	0.819
TS 6 $\times$ 4	0.5	28.43	8.36	35.3	11.8	2.06	18.4	9.21	1.48
4	0.375	22.37	6.58	29.7	9.9	2.13	15.6	7.82	1.54
4	0.25	15.62	4.59	22.1	7.36	2.19	11.7	5.87	1.6
4	0.1875	11.97	3.52	17.4	5.81	2.23	9.32	4.66	1.63
2	0.375	17.27	5.08	17.8	5.94	1.87	2.84	2.84	0.748
2	0.25	12.21	3.59	13.8	4.6	1.96	2.31	2.31	0.802
TS 4 $\times$ 3	0.25	10.51	3.09	6.45	3.23	1.45	4.1	2.74	1.15
3	0.1875	8.15	2.39	5.23	2.62	1.48	3.34	2.23	1.18
2	0.375	12.17	3.58	5.75	2.87	1.27	1.83	1.83	0.715
2	0.25	8.81	2.59	4.69	2.35	1.35	1.54	1.54	0.77
2	0.1875	6.87	2.02	3.87	1.93	1.38	1.29	1.29	0.798
TS 3 $\times$ 2	0.25	7.11	2.09	2.21	1.47	1.03	1.15	1.15	0.742
2	0.1875	5.59	1.64	1.86	1.24	1.06	0.977	0.977	0.771
2	0.125	3.9	1.15	1.38	0.92	1.1	0.733	0.733	0.8

* On special order, TS sections currently are available in sizes up to 30 $\times$ 30 and 30 $\times$ 24.

Table 12.2.27 Coefficients of Deflection for Steel Beams under Uniformly Distributed Loads

Span, ft	Fiber stress, lb/in^2		Span, ft	Fiber stress, lb/in^2		Span, ft	Fiber stress, lb/in^2		Span, ft	Fiber stress, lb/in^2	
	24,000	10,000		24,000	10,000		24,000	10,000		24,000	10,000
1	0.026	0.011	14	4.87	2.029	27	18.1	7.54	39	37.7	15.7
2	0.098	0.041	15	5.59	2.328	28	19.5	8.12	40	39.8	16.6
3	0.223	0.093	16	6.36	2.648	29	20.9	8.71	41	41.8	17.4
4	0.398	0.166	17	7.18	2.990	30	22.4	9.32	42	43.9	18.3
5	0.621	0.259	18	8.04	3.35	31	23.9	9.94	43	45.8	19.1
6	0.892	0.372	19	8.97	3.74	32	25.4	10.60	44	48.0	20.0
7	1.23	0.507	20	9.93	4.14	33	27.0	11.27	45	50.4	21.0
8	1.59	0.662	21	10.9	4.56	34	28.7	11.96	46	52.6	21.9
9	2.01	0.838	22	12.1	5.01	35	30.5	12.7	47	54.7	22.8
10	2.48	1.034	23	13.1	5.47	36	32.2	13.4	48	57.1	23.8
11	3.00	1.251	24	14.3	5.96	37	34.1	14.2	49	59.5	24.8
12	3.58	1.489	25	15.6	6.47	38	35.8	14.9	50	62.2	25.9
13	4.20	1.748	26	16.8	7.00						

NOTE: For a load concentrated at midspan, use $\frac{4}{5}$ of the coefficient given. 1 ft = 0.305 m; 1 lb/in^2 = 6.89 kPa.

Table 12.2.28 Values of Standard Framed-Beam Connections

($\frac{7}{8}$-in A325 HS bolts in standard holes,* A36 members)

AISC designation	Two angles thickness $\times$ length	Shear 1,000 lb	Bearing on beam web (t), 1,000 lb
10 rows	$\frac{5}{16} \times 2'5\frac{1}{2}''$	204	609t
9 rows	$\frac{5}{16} \times 2'2\frac{1}{2}''$	184	548t
8 rows	$\frac{5}{16} \times 1'11\frac{1}{2}''$	164	487t
7 rows	$\frac{5}{16} \times 1'8\frac{1}{2}''$	143	426t
6 rows	$\frac{5}{16} \times 1'5\frac{1}{2}''$	123	365t
5 rows	$\frac{5}{16} \times 1'2\frac{1}{2}''$	102	304t
4 rows	$\frac{5}{16} \times 0'11\frac{1}{2}''$	81.8	243t
3 rows	$\frac{5}{16} \times 0'8\frac{1}{2}''$	61.3	182t
2 rows	$\frac{5}{16} \times 0'5\frac{1}{2}''$	40.9	121t

NOTE: 1 in = 2.54 cm; 1 lb = 4.45 N.

* Values indicated are for slip-critical connections or bearing type where threads are not excluded from the shear plane. For bearing-type connections where threads are excluded from the shear plane, shear values may be increased by 1.47.

† If the web of the supporting beam is thinner than 0.17 in (0.42 in if beams frame on both sides), bearing must also be investigated.

Table 12.2.29   Safe Axial Loads for Standard Pipe Columns, kips
(Stress according to AISC specification for A501 pipe*)

Nominal pipe size, in	Outside diam, in	Wall thickness, in	Effective length of column K1, ft											
			6	7	8	9	10	11	12	14	16	18	20
3	3.500	0.216	38	36	34	31	28	<u>25</u>	<u>22</u>	16	12	10	
3½	4.000	0.226	48	46	44	41	38	35	32	<u>25</u>	19	15	12
4	4.500	0.237	59	57	54	52	49	46	43	<u>36</u>	<u>29</u>	23	19
5	5.563	0.258	83	81	78	76	73	71	68	61	<u>55</u>	47	<u>39</u>
6	6.625	0.280	110	108	106	103	101	98	95	89	82	<u>75</u>	<u>67</u>
8	8.625	0.322	171	168	166	163	161	158	155	149	142	135	127
10	10.750	0.365	246	243	241	238	235	232	229	223	216	209	201
12	12.750	0.375	303	301	299	296	293	291	288	282	275	268	261

NOTE: 1 in = 2.54 cm; 1 ft = 0.305 m. For dimensions of standard pipe see Sec. 8. Safe loads above underscore lines are for values of K1/r more than 120 but not over 200.
* Yield stress is 36 ksi (248.2 MPa). Pipe ordered to ASTM A53, type E or S grade B, or to API standard 5L grade B will have a yield point of 35 ksi (241.3 MPa) and may be designed at stresses allowed for A501 pipe.

Table 12.2.30   Selected Fire-Resistance Ratings

Type	Details of construction	Rating	Type	Details of construction	Rating
Reinforced-concrete beams and girders	Grade A concrete, 1½ in clear to reinforcement	4 h	Wood joists	Wood floor; 1 in tongue-and-groove subfloor and 1 in finish floor with asbestos paper between. Ceiling of ⅝ in Underwriters' Laboratories listed wallboard	1 h
	Grade B concrete, 1½ in clear to reinforcement	3 h			
Steel beams, girders, and trusses	2½ in cover to steel	4 h	Brick walls	Solid walls, unplastered, with no combustible members framed in wall:	
	1 in gypsum-perlite plaster on metal lath, 1¼ in clear of steel	3 h		8 in nominal	4 h
	Ceiling of 1⅛ in gypsum-perlite plaster on metal lath with 2½ in min air space between lath and structural members	4 h		4 in nominal	1 h
			Concrete masonry units	8 in Underwriters' Laboratories listed concrete blocks, laid as specified in Underwriters' Laboratories listing	4 h
Reinforced concrete columns	Grade A concrete 1½ in clear to reinforcement; 12-in columns or larger	4 h		4 in Underwriters' Laboratories listed concrete blocks; laid as specified in Underwriters' Laboratories listing	3 h
	Grade B concrete 2 in clear to reinforcement; 12-in columns or larger	4 h			
Steel columns, 8 × 8 in or larger	Concrete (siliceous gravel):		Steel-stud partitions	¾ in gypsum-perlite plaster both sides on metal lath	2 h
	2½ in clear to steel	4 h		⅝ in gypsum wallboard on 3⅝-in steel studs; attached with 6 d nails; joints taped and cemented	2 h
	2 in clear to steel	3 h			
	1 in clear to steel	2 h			
	1⅛ in gypsum-perlite plaster on metal lath spaced from flanges with 1¼-in steel furring channels	4 h	Wooden-stud partitions	Exterior walls: one side covered with ½ in gypsum sheathing and wood siding; other side faced with ½ in gypsum-perlite plaster on ⅜-in perforated gypsum lath	1 h
	⅞ in portland-cement plaster on metal lath over ¾-in channels	1 h			
Reinforced concrete slabs	5 in concrete (expanded clay, shale, slate, or slag) 1 in clear to reinforcement	4 h		Interior Walls: 2 × 4 in studs with ⅝ in gypsum wallboard on each side	1 h
	6½ in concrete (all other aggregate) 1 in clear to reinforcement	4 h	Plain or reinforced concrete walls	Solid, unplastered:	
				7 in thick	4 h
				6½ in thick	3 h
Heavy-timber floors	3 in tongue-and-groove plank floor with 1 in finish flooring	1 h		5 in thick Grade B Grade A	2 h
				3½ in thick	1 h

NOTE: 1 in = 2.54 cm.
Grade A concrete is made with aggregates such as limestone, calcareous gravel, trap rock, slag, expanded clay, shale, or slate or any other aggregates possessing equivalent fire-resistance properties.
Grade B concrete is all concrete other than Grade A concrete and includes concrete made with aggregates conaining more than 40 percent quartz, cherts, or flint.

12.3 REINFORCED CONCRETE DESIGN AND CONSTRUCTION

by William L. Gamble

REFERENCES: Breen, Jirsa and Ferguson, "Reinforced Concrete Fundamentals," Wiley. Winter and Nilson, "Design of Concrete Structures," McGraw-Hill. Lin and Burns, "Design of Prestressed Concrete Structures," Wiley. Park and Gamble, "Reinforced Concrete Slabs," Wiley. "Building Code Requirements for Reinforced Concrete (318–95)," "Commentary on Building Code Requirements for Reinforced Concrete," "Formwork for Concrete, SP-4," and "Manual of Standard Practice for Detailing Reinforced Concrete Structures," American Concrete Institute. "Standard Specifications for Highway Bridges," American Association of State Highway and Transportation Officials (AASHTO). "Minimum Design Loads for Buildings and Other Structures (ASCE 7)," American Society of Civil Engineers. "Uniform Building Code (UBC)," International Conference of Building Officials.

The design, theory, and notation of this chapter are in general accord with the 1995 Building Code Requirements for Reinforced Concrete of the American Concrete Institute, though many detailed provisions have been omitted.

Standard Notation

Load Factors

D = dead load of structure or force caused by dead load
E = earthquake load or force
L = live load of structure or force caused by live load
W = wind load or force
U = required strength of structure to resist design ultimate loads
ϕ = understrength or capacity reduction factor

Beams and General Notation

a = depth of compression zone, using approximate method
A_b = area of individual reinforcing bar, in^2
A_{ps} = area of prestressing steel
A_s = area of tension reinforcement
A_s' = area of compression reinforcement
A_v = area of steel in one stirrup
b = width of compression face of beam
b_w = width of stem of T beam
$c = k_u d$ = depth to neutral axis at ultimate, from compression face
d = effective depth of beam, compression face to centroid of tension steel
d' = depth of compression steel, from compression face
d_b = diam of individual reinforcing bar, in
E_c = Young's modulus of concrete
E_s = Young's modulus of steel
f_c' = compressive strength of concrete from tests of 6 × 12 in cylinders, lb/in^2
$\sqrt{f_c'}$ = measure of shear and tensile strength of concrete, lb/in^2, i.e., if $f_c' = 4,900$ lb/in^2, then $\sqrt{f_c'} = 70$ lb/in^2
f_{pu} = ultimate stress of prestressing steel, lb/in^2
f_y = yield stress of reinforcing steel, lb/in^2
h = overall height or thickness of member
k_u = ratio of ultimate neutral axis depth to effective depth
l_d = development length of reinforcing bar, in
s = spacing of shear reinforcement
M_u = ultimate moment of section or required ultimate moment
v_c = shear stress in concrete
V_c = shear force resisted by concrete
V_s = shear force resisted by web reinforcement
V_u = shear strength of section or required ultimate shear
β_1 = factor relating neutral axis position to depth of equivalent approximate stress block (see Fig. 12.3.2)
ε_{su} = reinforcement strain at time of failure of member
ε_y = yield strain of reinforcement
ρ = tension reinforcement ratio = A_s/bd

ρ' = compression reinforcement ratio = A_s'/bd
ρ_{bal} = balanced reinforcement ratio

Columns

$A_c = A_g - A_s$ = net area of concrete in cross section
A_{core} = area within spiral
A_g = gross area of column
A_s = area of steel in column
e = eccentricity of axial load on column
$M = Pe$ = applied bending moment
P = axial thrust
P_0 = failure load of short column under concentric load
ρ_g = gross steel ratio = A_s/A_g
ρ_s = spiral steel ratio = volume of spiral steel/volume of core

Floor Systems

b_0 = effective shear perimeter around column in flat plate, flat slab, or footing
c_1 = width of supporting column or capital, in direction of span being considered
c_2 = width of supporting column, in transverse direction
I_b = moment of inertia of beam
K_b = flexural stiffness of beam, moment per unit rotation
K_c = flexural stiffness of columns at joint, moment per unit rotation
K_s = flexural stiffness of slab of width l_2, moment per unit rotation
l_1 = span, center to center of supports, in direction considered
l_2 = span, center to center of supports, in transverse direction
$l_n = l_1 - c_1$ = clear span in direction considered
M_0 = static moment
w = distributed design load, including load factors
α_1 = EI of beam in direction 1/EI of slab of width l_2

Footings

A_1 = loaded area
A_2 = area of same shape and concentric with A_1
f_b = ultimate bearing stress on concrete
β = ratio of long side/short side of footing

Walls

l_c = unbraced height of wall

Prestressed Concrete

A_{ps} = area of prestressed reinforcement
A_t = area of anchorage-zone reinforcement
f_{ci}' = compressive strength of concrete at time of prestressing
f_{ps} = stress in reinforcement at time of failure of member
f_{pu} = ultimate stress of prestressing reinforcement
$f_{se} = f_{si} - \Delta f_s$ = stress in reinforcement at service load
f_{si} = reinforcement stress at time of tensioning steel
Δf_s = loss of prestress from initial tensioning value
$\rho_p = A_{ps}/bd$

MATERIALS

Reinforced concrete is a combination of concrete and steel acting as a unit because of bond between the two materials. Concrete has a high compressive strength but a relatively low tensile strength. Beams of plain concrete fail by tension at very low stresses, but if properly reinforced by embedment of steel in their tensile regions, they may be loaded to utilize the much higher compressive strength of the concrete. Reinforced concrete structures are practically monolithic, are more rigid than steel structures of the same strength, and are inherently fire-resistant. Reinforcement in the concrete also controls cracking caused by temperature changes and shrinkage.

Prestressed concrete is a form of reinforced concrete in which initial stresses opposite those caused by the applied loads are induced by tensioning high-strength steel embedded in the concrete. Members may be *pretensioned,* in which the steel is tensioned and the concrete then cast around it, or *posttensioned,* in which the concrete is cast and cured, after which steel placed in ducts through the concrete is tensioned.

Concrete For reinforced concrete work, only high-quality portland cement concrete may be used, and the aggregates must be carefully selected. The proportions are governed by the required strength, durability, economy, and the quality of the aggregates. Concretes for building construction normally have compressive strengths of 3,000 to 5,000 lb/in² (20 to 35 MPa), except that concrete for columns may be considerably stronger, with 12,000 to 14,000 lb/in² common in some geographic areas. Most concrete for prestressed members will have 5,000 lb/in² or higher compressive strength. The higher-strength concretes require thorough quality control if the strengths are to be consistently obtained. Generally, lean, harsh mixes should be avoided because the bond with the steel will be poor, permeability will be high, and durability of the concrete may be poor. A consistency of concrete that will flow sluggishly but not so wet as to produce segregation of the materials when transported must be used for all reinforced concrete work in order to embed the steel and completely fill the molds or forms. The use of vibration is almost mandatory, and enables the use of stiffer, more economical, concretes than would otherwise be possible (see also Sec. 6.9).

Steel Reinforcing steel may be deformed or plain bars, welded-wire fabric, or high-strength wire and strand for prestressed concrete. Bars with deformations on their surfaces are designed to produce mechanical bond and greater adhesion between the concrete and steel, and are used almost universally in the United States. Welded-wire fabric is suitable in many cases for slabs and walls, and may result in cost savings through labor savings. Fabric is made with both smooth and deformed wires, and the deformed fabric may have some advantage in terms of better crack control. Deformed reinforcing bars having minimum yield stresses from 40,000 to 75,000 lb/in² (275 to 517 MPa) are currently manufactured. The higher-strength steels have the advantage of allowing higher working stresses, but their ductility may be less and it may be difficult to cold-bend them successfully, especially in the larger sizes. The chemical makeup of most reinforcing steels is such that it is not readily weldable without special techniques, including careful preheating and controlled cooling. Furthermore, the variation of the material from batch to batch is so great that separate procedures must be devised for each batch. Tack welding in assembling bar cages can be particularly troublesome because of the stress raisers introduced. A weldable steel is produced with the specification ASTM A-706, but it is not widely available. Prestressing steel is heat-treated high-carbon steel, and 7-wire strand will have a breaking stress of 250,000 or 270,000 lb/in² (1,720 or 1,860 MPa). The usual steel is Grade 270, low-relaxation strand. The strength of solid wire for prestressing is slightly less. ASTM specifications cover the various steels, and all steel used as reinforcement should comply with the appropriate specification.

Reinforcing Steel Sizes The sizes of reinforcing bars have been standardized, and the designation numbers are approximately the bar diameter, in ⅛-in units. Table 12.3.1 gives the nominal diameter, cross-sectional area, and perimeter for each bar size. The areas of the four most common 7-wire prestressing strands are given below:

	Area, in²	
Diam, in	Grade 250	Grade 270
⁷⁄₁₆	0.109	0.115
½	0.144	0.153

Table 12.3.1 Dimensions of Deformed Bars

No.	Diam, in	Area, in²	Perimeter, in
2*	0.250	0.05	0.786
3	0.375	0.11	1.178
4	0.500	0.20	1.571
5	0.625	0.31	1.963
6	0.750	0.44	2.356
7	0.875	0.60	2.749
8	1.000	0.79	3.142
9	1.128	1.00	3.544
10	1.270	1.27	3.990
11	1.410	1.56	4.430
14	1.693	2.25	5.32
18	2.257	4.00	7.09

* No. 2 bars are obtainable in plain rounds only.

Plain and deformed wires are used as reinforcement, usually in the form of welded mats. Plain wires are made in sizes W0.5 through W31, where the number designates the cross-sectional area of the wire in hundredths of square inches. Deformed wires are made in sizes D1 through D31, and the number has the same meaning. Not all sizes are made by all manufacturers, and local suppliers should be consulted about availabilities of sizes. The formerly used wire gage numbers are no longer used for size specifications.

Moduli of Elasticity For concrete the modulus of elasticity E_c may be taken as $w^{1.5}33\sqrt{f'_c}$ in lb/in², where w is the weight of the concrete between 90 and 155 lb/ft³. Normal weight concrete may be assumed to weigh 145 lb/ft³. For steel the modulus of elasticity may be taken as 29,000,000 lb/in² (200 GPa), except for prestressing steel for which the modulus shall be determined by tests or supplied by the manufacturer, but is usually about 28,000,000 lb/in².

The **modular ratio** $n = E_s/E_c$ is of importance in designing reinforced concrete. It may be taken as the nearest whole number but not less than six. The value of n for lightweight concrete may be taken as the same for normal weight concrete of the same strength, except in calculations for deflections.

Protection of Reinforcement Reinforcement, for both regular and prestressed concrete, must be protected by the concrete so as to prevent corrosion. The amount of cover needed for various degrees of exposure is as follows:

Member and Exposure	Cover, in
Concrete surface deposited against the ground	3
Concrete surface to come in contact with the ground after casting:	
Reinforcement larger than No. 5	2
Reinforcement smaller than No. 5	1½
Beams and girders not exposed to weather:	
Main steel	1½
Stirrups and ties	1
Joists, slabs, and walls not exposed to weather	¾
Column spirals and ties	1½

The clear cover and clear bar spacings should ordinarily exceed ⁴⁄₃ times the maximum sized aggregate used in the concrete.

The amount of protection recommended is a minimum, and when corrosive environments or other severe exposure occurs, the cover should be increased. The concrete in the cover should be made as impermeable as possible. Fire-resistance requirements may also control the cover requirements.

LOADS

The dead and live loads are combined in determining the cross sections of the members. The dead load includes the weight of the structure, all finishing materials, and usually the installed equipment. The live load includes the contents and ordinarily refers to the movable items.

The **live load** (pounds per square foot) to be used for design depends upon the loadings that will occur in the particular structure as well as on the requirements of the local building code. The Boston Building Code illustrates good practice and is as follows for floor loads:

Heavy manufacturing, sidewalks, heavy storage, truck garages, 250; public garages, intermediate manufacturing, and hangars, 150; stores, heavy merchandise, light storage, 125; armories, assembly halls, gymnasiums, grandstands, public portions of hotels, theaters, and public buildings, corridors and fire escapes from public assembly buildings, light merchandising stores, stairs, first and basement floors of office buildings, theater stages, 100; upper floors of public buildings, office portions of public buildings, stairs, corridors and fire escapes except from public assembly buildings, theater and assembly halls with fixed seats, light manufacturing, locker rooms, stables, 75; church auditoriums, 60; office buildings above first floor including corridors, classrooms with fixed seats, 50; residence buildings and residence portions of hotels, apartment houses, clubs, hospitals, educational and religious institutions, 40. [Note: 100 lb/ft² = 4.79 kPa.]

Live loads affecting structural members supporting considerable tributary floor areas are sometimes reduced in recognition of the low probability that the entire area will be loaded to the full design load at the same time. Roofs are commonly designed to support live loads of 20 to 30 lb/ft². Wind loads are commonly from 15 to 30 lb/ft² (higher in areas subject to hurricanes) of vertical projection. In the case of heavy moving loads, allowance should be made for impact by increasing the live load by 25 to 100 percent.

Dead loads include the weight of both structure and finishing materials, and the weights of some typical wall, floor, and ceiling materials are as follows:

Description	Weight, lb/ft²
Granolithic finish, per in of thickness	12
⅞-in hardwood, 1⅛-in plank intermediate floor, and tar base	16
3-in wood block in coal-tar pitch	10
Lightweight concrete fill, 2 in thick	14
Plaster on concrete, tile, or concrete block, two coats	5
Plaster on lath	10
6-in concrete block wall	25
Suspended ceiling	12

SEISMIC LOADINGS

Earthquakes induce forces in structures because of inertial forces which resist the ground motions. These forces have damaged or destroyed large numbers of structures throughout the world. The art and science of designing to resist seismic effects are still evolving, and may be thought of in several steps.

1. Determine the ground motion to be expected at the site of interest. This usually requires consulting a map in ASCE-7 or the prevailing local building code, such as the Uniform Building Code (UBC). Several steps may be required, including finding the expected acceleration of the bedrock, followed by an accounting for the soil types between the bedrock and the structure. There is an implicit or explicit assumption of a probability that this ground motion will not be exceeded in some particular time interval, such as 50 or 100 years.

2. Determine the effects of the ground accelerations on the structure which is being analyzed. This will require consideration of the natural period of vibration of the structure and of structural characteristics such as damping and ductility, and more elaborate analyses will be required for large and important structures than for smaller, less crucial cases. Member forces, moment, shear, thrust, torsion, and deformations such as story drift and member end rotations will result from this analysis. The major building codes all have specific requirements and provisions for this analysis, and the UBC is the most widely used in seismically active regions. In most cases, an equivalent static horizontal load will be applied to the structure, with the force in the range of 10 percent or less to as much as 30 percent of the mass of the structure.

3. Design and detail members and connections for the imposed forces. Chapter 21 of the ACI Code has many requirements for the detailing of the reinforcement in the members and joints. When compared with designs for static loadings which include only dead, live, and wind loads, seismic designs will have much more lateral reinforcement in columns, and often much heavier shear reinforcement (especially near the joints), and will usually have extra reinforcement within the joint regions. Much of this added reinforcement is referred to as **confinement reinforcement**, and the intent is to utilize the fact that triaxially confined concrete may be able to sustain much greater strains before failure than unconfined concrete. Obviously, steps 2 and 3 are parts of an iterative process which converges to an acceptable structural design solution.

The current seismic design philosophy includes the assumption that it is not economically possible to design structures to resist extremely large earthquakes without damage. It is expected that structures will resist smaller earthquakes, such as might be expected several times during the life of a structure, with minimal damage, while a very large earthquake which might be expected to occur once during a structure's design life would cause significant damage but would leave the structure still standing. The assignment of an importance factor for buildings is one consequence of this design philosophy. Hospitals, school buildings, and fire stations, for example, would be designed to a higher seismic standard than would a three-story apartment building.

LOAD FACTORS FOR REINFORCED CONCRETE

The current ACI Building Code for Reinforced Concrete is based on a **strength design** concept, in which the strength of a member or cross section is the basis of design. This approach has been adopted because of the difficulty in assigning reasonable and consistent allowable stresses to the concrete and steel. Factors of safety are expressed in terms of *overload* and *understrength* factors. The overload factors, reflecting the uncertainty of the applied loads, are expressed as No wind or earthquake loads:

$$U = 1.4\,D + 1.7\,L \qquad (12.3.1)$$

With wind acting, use the larger of (12.3.2) or (12.3.3):

$$U = 0.75\,(1.4\,D + 1.7\,L + 1.7\,W) \qquad (12.3.2)$$
$$U = 0.9\,D + 1.3\,W \qquad (12.3.3)$$

No section may be weaker than required by Eq. (12.3.1). In case earthquake loadings are considered, 1.1 E is substituted for W in the above equations. Liquids are treated as dead loads, and other provisions are made for earth-pressure loadings. The understrength factors reflect the ductility of failure in the mode considered, and the consequences of a failure on the rest of the structure. These are expressed as ϕ factors with the following values:

Bending and tension	$\phi = 0.90$
Shear and torsion	$\phi = 0.85$
Spiral columns	$\phi = 0.75$
Tied columns and bearing	$\phi = 0.70$

These factors are used to reduce the computed ideal strengths of members, reflecting possible weaknesses related to materials, dimensions, and workmanship.

In addition to strength requirements, serviceability checks must be made to ensure freedom from excessive cracking, deflections, vibrations, etc., at working loads. Prestressed-concrete members must also satisfy a set of allowable stresses at this load level.

Deflections are usually controlled by specifying minimum member depths. The values of $l/16$ and $l/21$, for beams which are simply supported and continuous at both ends, respectively, are typical, unless an investigation is conducted to show that shallower members will result in acceptable deflections. Crack control is obtained by careful distribution of the steel through the tensile zone. In this regard, several small bars are superior to one large bar.

The approach to design is not a "limit design" or "plastic design" concept, as the ultimate forces are derived from elastic analyses of structures, using maximums for each critical section. The plastic col-

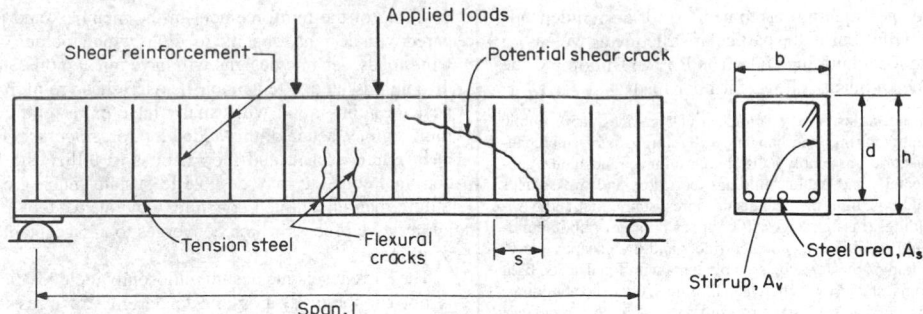

Fig. 12.3.1 Typical arrangement of reinforcement in beams.

lapse loads and mechanisms currently used in the "plastic design" of some steel structures in the United States are not considered.

Bridges may be designed using the same general concepts, but different overload and understrength factors are used, and the serviceability requirements include checks on the fatigue strength.

REINFORCED CONCRETE BEAMS

Concrete beams are reinforced to resist both flexural and shear forces. A typical reinforcement scheme for a simply supported beam is shown in Fig. 12.3.1.

As the load on a reinforced concrete beam is increased, vertical tension cracks appear in the maximum moment regions, and gradually grow in length, width, and number. By the time the working load is reached, some of the cracks extend to the neutral axis and the contribution of the tensile strength of the concrete to the flexural capacity of the beam has become very small. As the load is increased further, the reinforcement eventually yields. This load is nearly the maximum the member can sustain. Further attempts at loading produce large increases in deflection and crack widths with very small increases in load and a gradual reduction in the remaining concrete compression area. When a limiting concrete strain of about 0.003 is reached at the compression face of the beam, the concrete starts crushing and the capacity of the beam starts dropping. For static design purposes, the achievement of the 0.003 strain is usually regarded as the end of the useful life of the beam.

Within limits to be checked later, at the time of flexural failure of a beam the stress in the steel is equal to the yield stress, which greatly simplifies the analysis of the member. The strain and stress distributions in a beam are shown in Fig. 12.3.2, at failure.

It is assumed that plane sections remain plane, that adequate bond exists, that the stress-strain relationships for concrete and steel are known, and that tension in the concrete has a negligible influence.

The **flexural strength** of a cross section may be written as, including the understrength factor and rounding the terms in the parentheses to one significant figure:

$$M_u = \phi A_s f_y d (1 - 0.4\, k_u)$$
$$= \phi A_s f_y d (1 - 0.6 \rho f_y / f_c') \qquad (12.3.4)$$

where $\rho = A_s/bd$ = reinforcement ratio. The reinforcement ratio will ordinarily be between 0.005 and 0.02.

A satisfactory approximate stress distribution is shown in Fig. 12.3.2d. Since $T = C = 0.85 f_c' ba$, then $a = A_s f_y / 0.85 f_c' b$ and the flexural capacity is

$$M_u = \phi A_s f_y (d - a/2) \qquad (12.3.5)$$

It must be demonstrated for each case that the stress in the reinforcement has reached f_y, and the simplest approach is to show that $\varepsilon_{su} \geq \varepsilon_y$. From Fig. 12.3.2b, $\varepsilon_{su} = 0.003 (1 - k_u)/k_u$. From equilibrium (see Fig. 12.3.2c), $k_u = A_s f_y / 0.85\, \beta_1 f_c' bd = \rho f_y / 0.7 f_c'$. The limiting case for the validity of Eqs. (12.3.4) and (12.3.5) is a balanced condition in which the yield strain in the steel and the 0.003 compressive strain in the concrete are reached simultaneously, and this can be found using Fig. 12.3.2 and assuming $\varepsilon_{su} = \varepsilon_y$. It can then be shown that

$$\rho_{bal} = \frac{0.85 \beta_1 f_c'}{f_y} \frac{0.003}{0.003 + \varepsilon_y} \qquad (12.3.6)$$

Values of ρ_{bal} are plotted vs. f_c' in Fig. 12.3.3 for three values of f_y. The

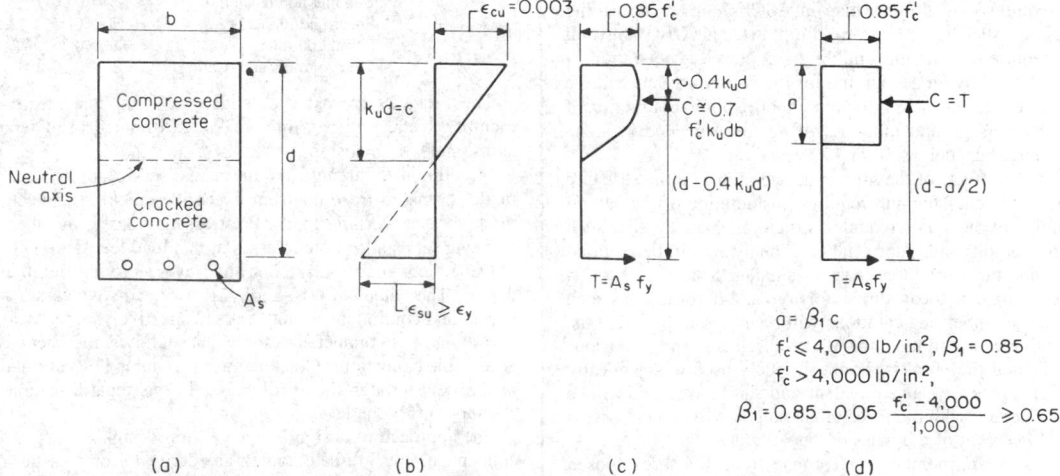

Fig. 12.3.2 Stress and strain distribution in reinforced concrete beam at failure. (*a*) Section; (*b*) strains; (*c*) stresses; (*d*) approximate stresses.

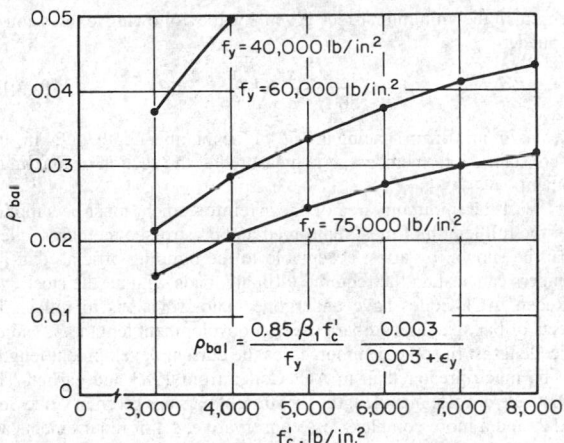

Fig. 12.3.3 Balanced steel ratio as a function of f_y and f_c'.

steel ratio should not exceed $0.75 \, \rho_{bal}$, in order to ensure that at least some yielding, with resultant large deflections, occurs before a member fails.

Beams may contain both compression and tension reinforcement, especially when the beam must be kept as small as possible or when the long-term deflections must be minimized. The forces at ultimate are shown in Fig. 12.3.4, and the moment may be calculated as

$$M_u = \phi[A_s' f_y(d - d') + (A_s - A_s')f_y(d - a/2)] \quad (12.3.7)$$

where $a = (A_s - A_s')f_y/0.85f_c'b$. The net steel ratio $\rho - \rho' = (A_s - A_s')/bd$ should not exceed 0.75 of the balanced value given by Eq. (12.3.6).

Many reinforced concrete beams are flanged sections, or T beams, by virtue of having a slab cast monolithically with the beam, and such a cross section is shown in Fig. 12.3.5. It will be found that the neutral axis lies within the flange in most instances, and if $k_u d \leq t$, then the beam is treated as a rectangular beam of width b. This is checked using $k_u d = A_s f_y/0.85 \, \beta_1 f_c' b$, developed from equilibrium (Fig. 12.3.2). If $k_u d > t$, the flexural capacity is computed by

$$M_u = \phi[(A_s - A_{sw})f_y(d - a/2) + A_{sw}f_y(d - t/2)] \quad (12.3.8)$$

where $A_{sw} = (b - b_w)t0.85f_c'/f_y$; and $a = (A_s - A_{sw})f_y/0.85f_c'b_w$.

Continuous T beams are treated as rectangular beams of width b_w in the negative moment regions where the lower surface of the beam is in compression. Most continuous beams will have compression steel in the negative moment regions, as some bottom steel is always continued into the support regions.

Both T beams and beams with compression steel may be thought of in terms of dividing the beam into two components—one a rectangular beam containing part of the tension steel and the other a couple with the rest of the tension steel at the bottom and either the compression steel or the T-beam flanges at the top. Both components of the beam must satisfy horizontal equilibrium.

The reinforcement ratio should not be less than $\rho_{min} = 3\sqrt{f_c'}/f_y \geq 200/f_y$. This is to ensure that the ultimate moment is somewhat

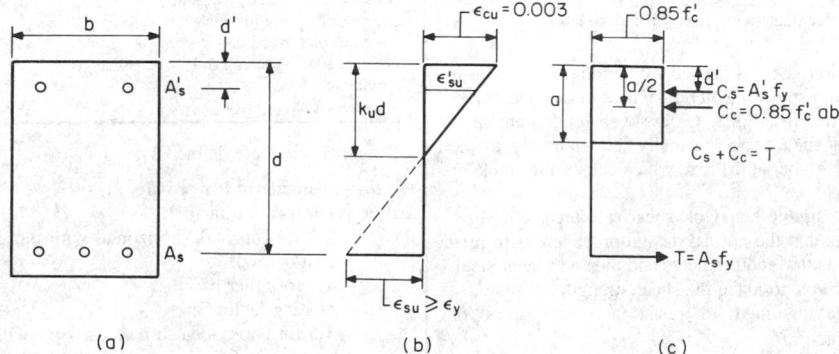

Fig. 12.3.4 Stress and strain distributions in a doubly reinforced concrete beam. (*a*) Section; (*b*) strains; (*c*) approximate stresses.

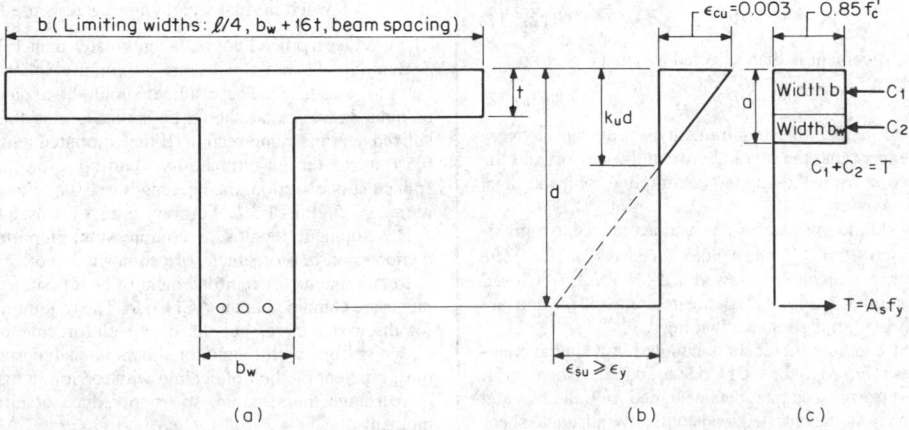

Fig. 12.3.5 Strain and stress distribution in a T beam. (*a*) Section; (*b*) strains; (*c*) approximate stresses.

greater than the initial cracking moment. For T beams, $\rho = A_s/b_w d$ for purposes of the minimum steel requirement.

In the selection of a cross section, it is convenient to transform Eq. (12.3.4) by substituting pbd for A_s and rearranging to obtain

$$M_u/\phi bd^2 = \rho f_y (1 - 0.6\rho f_y/f'_c) \qquad (12.3.9)$$

With given materials f_y and f'_c selection of a ρ value enables calculation of the required bd^2, from which a cross section may be selected. For convenience, values of $M_u/\phi bd^2$, are plotted against ρf_y in Fig. 12.3.6, for several values of f'_c.

The strengths of prestressed-concrete beams are treated much the same as reinforced concrete beams, and the differences will be noted later.

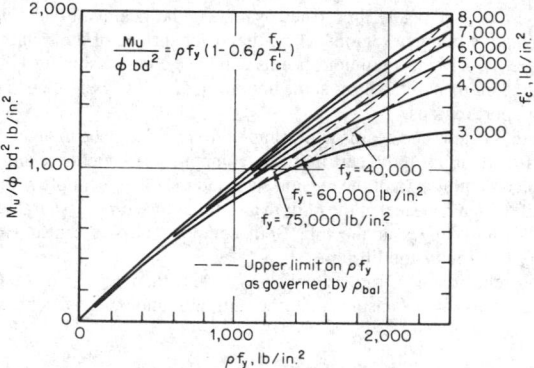

Fig. 12.3.6 Design factors for single reinforced concrete beams.

In addition to the tension stresses caused by bending forces, **shear forces** cause inclined tension stresses which may lead to inclined cracking such as is sketched in Fig. 12.3.1. Unless web reinforcement is present, the formation of such a crack usually leads directly to the complete collapse of the member at the load which caused the crack, or only slightly higher.

As a result of this undesirable behavior, shear reinforcement is required in all major beams, and the normal design procedure is to proportion the beam for the flexural requirements and then add shear steel, usually in the form of stirrups, to make the shear strength adequate.

The concrete can be assumed to resist a shear stress of $v_c = \phi 2\sqrt{f'_c}$, or a shear force of

$$V_c = \phi 2\sqrt{f'_c}\, bd \qquad (12.3.10)$$

(or $V_c = \phi 2\sqrt{f'_c}\, b_w d$ for T beams). The shear reinforcement must resist the force in excess of V_c, so that

$$V_u - V_c = V_s \qquad (12.3.11)$$

The area of shear reinforcement is then selected to satisfy

$$V_s = \phi A_v f_y d/s \qquad (12.3.12)$$

Shear reinforcement to satisfy this requirement is provided at every section of the beam, except that the region between the support and the section at the distance d from the support is supplied with the steel required at d from the support.

The stirrup spacing should not exceed $d/2$, and is reduced to a maximum of $d/4$ if $V_u > \phi 6\sqrt{f'_c}\, bd$. V_u must not exceed $\phi 10\sqrt{f'_c}\, bd$. The minimum area of shear reinforcement allowed is $A_v = 50 b_w s/f_y$. Closed stirrups of the form shown in Fig. 12.3.1 are recommended, and are essential in areas subjected to earthquake loadings.

Shear in prestressed-concrete members is handled in a similar manner, and the designer is referred to the ACI Code for details. However, it will be found that most prestressed members designed for buildings and which do not support major concentrated loads will have adequate shear

strength if the minimum steel given by the following expression is supplied:

$$A_v = \frac{A_{ps}}{80} \frac{f_{pu}}{f_y} \frac{s}{d} \sqrt{\frac{d}{b_w}} \quad \text{in}^2 \qquad (12.3.13)$$

The maximum stirrup spacing is 0.75 of the member depth, or 24 in, and for constant-depth members, d is measured at the section of maximum moment.

At least the minimum area of shear reinforcement must be supplied over the full length of most reinforced and prestressed members unless it can be shown, by a test acceptable to the building official, that the members can sustain the required ultimate loads without the steel.

Recent ACI Codes have undergone major revisions to reflect the effects of bar spacing and bar cover on development length, l_d, and on splice lengths. In some common cases the current development lengths will be much greater than in ACI Codes from 1983 and earlier. The 1995 Code development lengths for straight bars are given in Code Sec. 12.2.2, and a more complex, less conservative, set of requirements are in Sec. 12.2.3. When the available anchorage lengths are less than l_d, hooks are often used. The conservative Sec. 12.2.2 requirements may be summarized as follows:

$$\frac{l_d}{d_b} = K \frac{f_y \alpha \beta \lambda}{\sqrt{f'_c}}$$

in which K is as follows:

	No. 6 and smaller bars and deformed wire	No. 7 and larger bars
1. Clear spacing of bars being developed or spliced $\geq d_b$ and clear cover $\geq d_b$ with minimum ties or stirrups or	$\frac{1}{25}$	$\frac{1}{20}$
2. Clear spacing $\geq 2d_b$ and clear cover $\geq d_b$	$\frac{1}{25}$	$\frac{1}{20}$
3. Other cases	$\frac{3}{50}$	$\frac{3}{40}$

The other terms are defined as:

d_b = diameter of bar or wire, in
α = bar location factor
$\quad$ = 1.3 for top bars (horizontal with more than 12 in of concrete cast below)
$\quad$ = 1.0 for other bars
β = coating factor
$\quad$ = 1.5 for epoxy-coated bars or wires with cover $\leq 3d_b$, or clear spacing $\leq 6d_b$
$\quad$ = 1.2 for all other epoxy-coated bars or wires
$\quad$ = 1.0 for uncoated reinforcement
λ = lightweight concrete factor
$\quad$ = 1.3 when light-weight aggregate concrete is used
$\quad$ = 1.0 when normal weight aggregate is used

The product $\alpha\beta$ need not be taken greater than 1.7.

Bars No. 11 and smaller are commonly spliced by simply lapping them for a distance. These splices should be avoided in regions of high computed stress, and should be spread out so that not many bars are spliced near the same section. If the computed tensile stress is less than $0.5 f_y$ at the design ultimate load and not more than half the bars are spliced at one section, the lap length is $1.0 l_d$. For other cases the minimum lap length is $1.3 l_d$. For lower stress levels, a lap of l_d is sufficient.

Development lengths in compression, important in columns and compression reinforcement, are somewhat shorter.

Requirements for reinforcement of beams subjected to torsional moments are contained in the ACI Code. The requirements are too complex for discussion here, but the basic reinforcement scheme consists of closed stirrups with longitudinal bars in each corner of the stirrup. The most efficient method of dealing with torsion in many instances will be to rearrange the structure so as to reduce or eliminate the torsional moments.

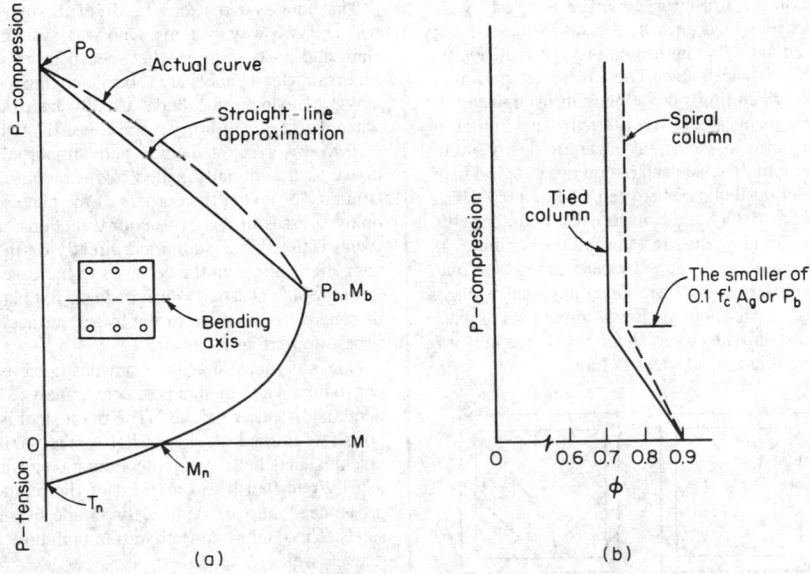

Fig. 12.3.7 Column capacity diagrams. (*a*) Typical moment-thrust interaction diagram; (*b*) variation of ϕ with P.

REINFORCED CONCRETE COLUMNS

Reinforced concrete compression members are proportioned taking into account the applied thrust, the bending moments, and the relationship between length and thickness for the member. The strength of a cross section or a short column can conveniently be shown with the aid of a *moment-thrust interaction diagram* such as is shown in Fig. 12.3.7*a*, which is drawn without considering the ϕ factor. The variation in ϕ with thrust is shown in Fig. 12.3.7*b* for the same column.

The load P_0 is the strength of a short column under a concentric load, and is expressed as

$$P_0 = 0.85\,A_c f'_c + A_s f_y \qquad (12.3.14)$$

The contribution of the concrete is slightly less than the cylinder strength because of differences in workmanship, curing, and position of casting. M_u is simply the strength in flexure, as was discussed for beams. The $M_b - P_b$ point is the "balance point," at which simultaneous crushing of the concrete and yielding of the reinforcement occur. Failure is initiated by crushing of the concrete at loads higher than P_b, and by yielding of the reinforcement in tension at lower loads. Only the reinforcement contributes to the tensile capacity of $T_u = A_s f_y$.

The balance-point moment and thrust are found using the strain and stress distributions shown in Fig. 12.3.8, for a symmetrical section with steel in two faces. From equilibrium,

$$P_b = C_s + C_c - T \qquad (12.3.15)$$

Summing moments about the centroidal axis gives

$$M_b = A_s f_y (d - d')/2 + 0.85 \beta_1 f'_c k_u db (h/2 - \beta_1 k_u d/2) \quad (12.3.16)$$

This assumes that the compression steel has yielded, and this will be true except for very small members or members with exceptionally large cover over the steel.

The portion of the interaction diagram below P_b may be constructed in a point-by-point manner. For any value of $\varepsilon_s > \varepsilon_y$, or for any value of $k_u d < k_u d_{bal}$, the strain and stress distributions are defined. Once the forces are defined, the moments and thrusts are computed using the same formulas as for the balance point. A modification will have to be made when the compression-steel stress is less than f_y.

Care must be used in the selection of the axis about which moments are to be summed, especially if the section is not symmetrical or symmetrically reinforced, since the force system is not a pure couple. The most important thing is to remain consistent, and to be certain that the internal and external moments are summed about the same axis.

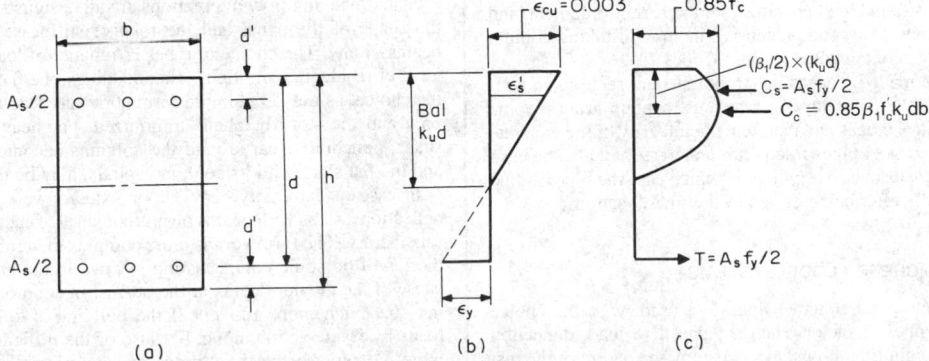

Fig. 12.3.8 Stress and strain distribution in a column at balanced moment and thrust. (*a*) Section; (*b*) strains; (*c*) stresses.

In practice, either the moment-thrust curve can be reduced by the appropriate ϕ factor, or the computed ultimate M and P increased by dividing by ϕ. If the required $M - P$ point lies on or slightly inside the interaction curve, the design is acceptable. The maximum permitted factored thrust is $0.80\ \phi P_0$, which limits the applied thrust capacity in cases where the computed moments are relatively small. This limitation is intended to provide resistance to accidental eccentricities. Column reinforcement consists of longitudinal bars and lateral ties or spiral bars. The ratio of longitudinal steel ρ_g should be between 0.01 and 0.08. Ties are usually No. 3 or No. 4 bars which are bent to enclose the longitudinal bars, and are spaced at not more than 16 longitudinal bar diam, 48 tie diam, or the least thickness of the column. Ties are arranged to bind each corner bar and alternate intermediate bars. Several typical arrangements are shown in Fig. 12.3.9. Ties hold the longitudinal bars in place during construction, and may provide some shear resistance and improve the behavior of the column at loads near failure.

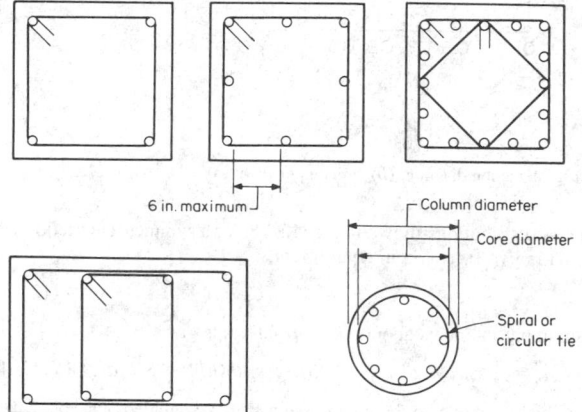

Fig. 12.3.9   Typical arrangement of steel in columns.

Spiral bars serve to confine the concrete core of the section as well as holding the longitudinal bars. The minimum amount of spiral steel, if advantage is to be taken of the higher ϕ factor for spiral columns, is

$$\rho_s = 0.45(A_g/A_{core} - 1)f'_c/f_y \qquad (12.3.17)$$

where f_y is the yield stress of the spiral, but not more than 60,000 lb/in². The clear spacing of the turns of the spiral must be between 1 and 3 in.

The strengths of compression members may be reduced below the cross-sectional strengths by length effects. Most columns in unbraced frames (frames in which the columns resist all horizontal forces) will have some strength reduction because of length effects, and most columns in braced frames will not, but this depends on the precise details of the length, width, restraint by other frame members, reinforcement, and amount of creep expected. A comprehensive method of taking column length into account is contained in the ACI Code.

Columns are also made by encasing structural steel sections in concrete, in which case the covering concrete must contain at least some steel in order to control cracks and maintain the integrity of the concrete in case of fire. Heavy steel pipes filled with concrete may also be used as columns. These are also used as piling, in which case the pipe is filled with concrete, usually after being driven to the final location.

REINFORCED CONCRETE FLOOR SYSTEMS

Several types of reinforced concrete floors are used, with the choice depending on a number of factors such as span, live load, deflection limits, cost, story-height limitations, local custom, the nature of the rest of the structural frame or system, and the probability of future alterations.

The floor systems may be divided, somewhat arbitrarily, into one-way and two-way systems. One-way systems include solid and hollow slabs and joists spanning between parallel supporting beams or walls. Floors in which panels are subdivided into a grid by subbeams spanning between main girders have usually been designed as one-way slabs when the grid length is several times the width.

Two-way systems include slabs supported on all four sides on beams or walls, traditionally called *two-way slabs*. Slabs supported only on columns located at the corners of the panels also carry loads by developing stresses in the two major directions, and are usually termed *flat plates* if the slab is supported directly on the columns and *flat slabs* if there are capitals on the columns to increase the effective support size. A waffle slab is usually designed as a flat plate or flat slab, with pockets of concrete omitted from the lower surface, and the slab appears as a series of crossing joists.

One-way slabs and joists are designed as beams. A 12-in or other convenient width of slab is selected, analyzed as an isolated beam, and a depth and steel are picked. The main steel is perpendicular to the supports. Additional steel, parallel to the supports, is placed to control cracking and help distribute minor concentrated loads. This steel is usually one-fourth to one-third of the main steel, but not less than a gross steel ratio of 0.0018 for Grade 60 and 0.002 for lower-grade steels. These minimums govern in both directions, and in two-way systems as well.

One-way joists, such as that shown in Fig. 12.3.10, are usually cast with reusable sheet metal or fiberglass forms owned or rented by the contractor. Standard form sizes range from about 20 to 36 in wide and 8 to 20 in deep. The web thicknesses are made to suit the shear and fireproofing requirements, and special tapered end sections may be available to increase the web widths in the regions of high shear near the supports. Joists are exempted from the requirement that web steel be supplied regardless of the concrete shear stress. The allowable shear stress for the concrete is 1.1 times that for beams. The top slabs usually range from 2.5 to 4.5 in thick, and are reinforced to span from rib to rib. The joists are essentially designed as isolated T beams, and may be supported on girders or walls. Joist systems are suitable for reasonably long spans and heavy loads, and have low dead weights for the effective depths attainable.

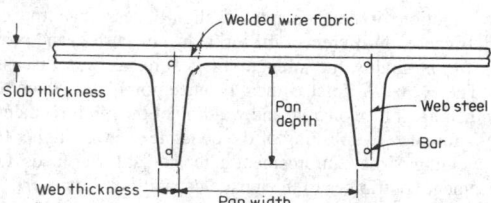

Fig. 12.3.10   Cross section of concrete floor joist.

Slabs spanning in two directions are all designed taking into account the shape of the panel and the relative stiffnesses of the supporting beams, if any. The choice of types is a matter of loadings and economics. For residential and light office loadings, the flat plate is frequently the choice, as the very simple formwork may lead to substantial economy and the story height is minimized. For heavier loads or longer spans, punching shear around the columns becomes a limiting factor, and the flat slab, with its column capitals, may be the most suitable.

In case of extremely heavy floor loads or very stringent limits on deflections, slabs with beams on all four sides of each panel will be most satisfactory. The formwork is more complex than for the other slabs, but there will be some compensating savings in the amounts of steel because of the greater depths of the beams. In addition, the two-way slab may be much more efficient if the building is to resist major lateral loads by frame action alone, because of the difficulties in transferring large moments between columns and flat plates or slabs. The design procedure is the same for slabs with and without beams. The basic steps, for each direction of span in each panel, are:

1. Compute static moment M_0.
2. Distribute M_0 to positive and negative moment sections.
3. Distribute section moments to column and middle strips and beams.

Most buildings slabs are designed for uniformly distributed loads, and the *static moment,* defined as the absolute sum of the midspan positive plus average negative moments, is

$$M_0 = wl_2(l_n)^2/8 \qquad (12.3.18)$$

The dimensions are illustrated in Fig. 12.3.11.

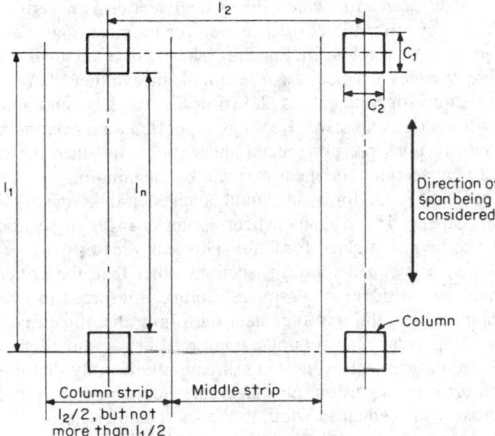

Fig. 12.3.11 Arrangement of typical slab panel.

Relatively simple rules for the distribution of the static moment to various parts of the panel exist as long as the structure meets several simple limits:

1. Minimum of three spans in each direction.
2. Panel length no more than twice panel width.
3. Successive spans differ by not more than one-third the longer span.
4. Columns on a rectangular grid, or offset no more than 10 percent of the span.
5. Live load not more than two times the dead load.
6. If beams are used, they are used on all four sides of each panel and are approximately the same size, except that spandrel beams only are acceptable.

The following is for slabs meeting these restrictions. Information on other cases is contained in the ACI Code.

The positive-negative moment distribution for interior spans is

$$+M = 0.35\,M_0 \qquad (12.3.19)$$
$$-M = 0.65\,M_0 \qquad (12.3.20)$$

For end spans, the stiffness of the exterior support must be taken into account. Table 12.3.2 gives the fractions of M_0 to be assigned to the three critical sections for five common cases.

Once the section moments have been determined, they are distributed to the column and middle strips and beams, taking into account the

panel shape and the beam stiffness. The beam relative stiffness coefficient α_1 is calculated using an effective beam cross section as shown in Fig. 12.3.12, and the full width of the slab panel l_2. The locations of the column and middle strips are shown in Fig. 12.3.11.

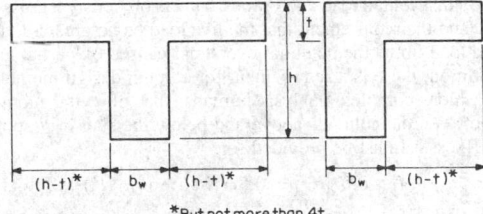

*But not more than 4t

Fig. 12.3.12 Beam sections for calculation of I_b and α_1.

The interior negative and positive moments are distributed to the column strips in the proportions shown in Fig. 12.3.13, with the remainder of the moment going to the middle strip. Linear interpolations are made for intermediate beam stiffnesses, but in most instances where there are beams, they will be found to have

$$\alpha_1 l_2/l_1 \geq 1.0$$

At the exterior supports, the distribution of the negative moments is a complex function of the flexural and torsional stiffnesses of the beams

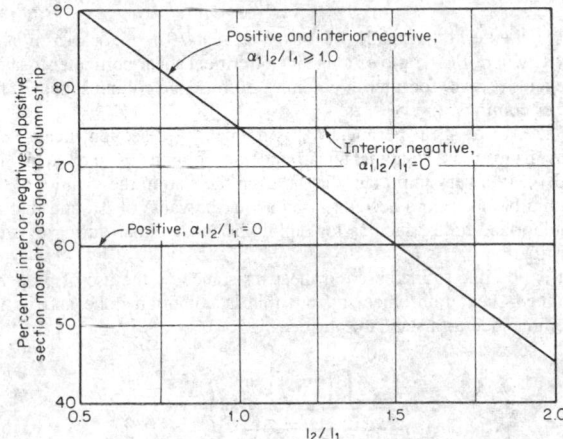

Fig. 12.3.13 Percentage of interior negative and positive moments assigned to a column strip.

and of the panel shape, but satisfactory designs can usually be achieved by assigning all the negative moment to the column strip and detailing the edge beam or edge strip of the slab for torsion, by using closed stirrups at relatively small spacings, and by placing bars parallel to the edge of the structure in each corner of the stirrups. At fully restrained edges, use the distribution for an interior negative moment section.

Table 12.3.2 Moments in End Spans

| | Exterior edge unrestrained (1) | Slab with beams between all supports (2) | Slab without beams between interior supports | | Exterior edge fully restrained (5) |
			Without edge beam (3)	With edge beam (4)	
Interior negative factored moment	0.75	0.70	0.70	0.70	0.65
Positive factored moment	0.63	0.57	0.52	0.50	0.35
Exterior negative factored moment	0	0.16	0.26	0.30	0.65

Beam moments are found by dividing the column strip moment between beam and slab. If $\alpha_2 l_2/l_1 \geq 1.0$, the beam moment is 85 percent of the column strip moment. This moment is reduced linearly to zero as $\alpha_1 l_2/l_1$ approaches zero.

This design method assumes that all panels are loaded with the same uniformly distributed load at all times. This is obviously a gross simplification, and the requirement that the live load be no greater than twice the dead load limits the potential overstress caused by partial loadings.

In addition, there is a requirement for column design moments, and unless a more complete analysis is made, the following moment, divided between the columns above and below the slab in proportion to their stiffnesses, must be provided for:

$$M = 0.07[(w_d + 0.5w_l)l_2 l_n^2 - w_d' l_2' (l_n')^2] \qquad (12.3.21)$$

The loads w_d and w_l are the distributed dead and live loads including the overload factors. The terms w_d', l_2', and l_n' are for the shorter of the two spans meeting at the column considered.

The shear strength of slab structures must always be checked, and shear stresses often govern the thickness of beamless slabs, especially flat plates.

If there are beams with $\alpha_1 l_2/l_1 \geq 1.0$, all shear is assigned to the beams, and stirrups are provided to make the shear capacity adequate, as was described in earlier coverage on beams. The beam shear is linearly reduced to zero as $\alpha_1 l_2/l_1$ is reduced to zero.

For the case of no beams, punching shear around the columns becomes a controlling factor. In this case the average shear stress, calculated as

$$v_u = V_u/b_0 d \qquad (12.3.22)$$

must not exceed the smaller of $\phi 4 \sqrt{f_c'}$, $\phi(2 + 4/\beta_c) \sqrt{f_c'}$, or $\phi(\alpha_s d/b_0 + 2)\sqrt{f_c'}$, where β_c = ratio of long side of critical shear perimeter to short side and $\alpha_s = 40$ for interior columns, 30 for edge columns, and 20 for corner columns.

The critical shear perimeter b_0 is defined by a section located $d/2$ away from and extending all around the column, as shown in Fig. 12.3.14. It is very important that holes in the slab in the vicinity of the column be taken into account in reducing the value of b_0, and that no unauthorized holes, such as for piping, be made either during or after construction.

It is possible to increase the shear resistance by the use of properly designed shear reinforcement, but this is not often done and is not recommended as a standard practice.

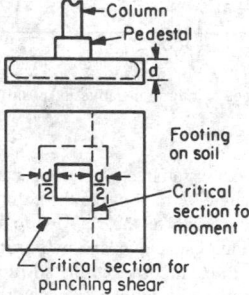

Fig. 12.3.14 Two-way reinforced concrete footing.

Transfer of moments between columns and slabs sets up shear and torsional stresses which must also be considered in the analysis of the shear strength.

FOOTINGS

Footings (Fig. 12.3.14) may be classified as wall footings, isolated column footings, and combined column footings. The bending moments, shears, and bond stresses in such footings should be determined by the principles of statics on the basis of assumed or known soil-pressure distribution over the area of the footing. The bending moment on any projecting portion of a footing may be computed as the moment of the forces acting on the area to one side of a vertical plane through the critical section.

The critical section for bending in a concrete footing supporting a concrete column, pedestal, or wall should be taken at the face of the column, pedestal, or wall. For footings under metallic column bases or under masonry walls where bond with the footings is reduced to the friction value, the critical section is assumed midway between the middle and edge of the base or wall.

Shear stresses must be considered on two sections. The footing may act as a wide beam, for which the critical section is a vertical plane located d away from the critical section for moment, and the stresses must satisfy those for a beam. Punching shear will often govern, and the critical section lies at a distance $d/2$ from the face of the column or other critical section for moment, as shown in Fig. 12.3.14, and as was the case for flat plates and slabs. Footings supported on a small number of high-capacity piles present special shear problems since the conventional critical sections for shear may not be meaningful.

The critical section for bond should be taken at the same plane as for bending. Other vertical planes where abrupt changes of section occur should also be investigated for bond and shear stresses.

In sloped or stepped footings, sections other than the critical ones may require consideration. A square footing, reinforced in two directions, should have the reinforcement uniformly distributed across the entire width. Rectangular footings, reinforced in two directions, should have the reinforcement in the long direction uniformly distributed; in the short direction a portion, Eq. (12.3.23), should be uniformly distributed across a strip equal in width to the short side and centered on the structural element supported and the remainder distributed uniformly in the outer portions. The amount included in the center strip may be computed as follows:

Reinforcement in center strip

$$= \frac{2 \,(\text{total reinforcement in short direction})}{\beta + 1} \qquad (12.3.23)$$

where β is the ratio of the long side to the short side.

Combined Footings Footings supporting two or more columns may be designed with sufficient accuracy by assuming uniform soil pressure and applying the laws of statics. The footing shape must be such that the center of gravity coincides with the center of gravity of the superimposed loads; otherwise unequal settlement may occur. The longitudinal and diagonal tension reinforcement should be designed by the ordinary rules of beam design. Lateral reinforcement should be designed as for isolated footings and should preferably be concentrated in bands under and near the columns proportionate in area to the column loads. The lateral reinforcement at each column should be uniformly distributed within a width centered on the column and should not be greater than the width of the column plus twice the effective depth of the footing.

Spread or raft foundation, consisting of a slab extending over the entire area under the columns or of a slab supported by beams, may be considered as loaded by a uniform upward reaction of the ground. The principle of design is exactly the same as that applied to a floor system, except that the load acts upward instead of downward.

Concrete piles of various types are widely used for foundations as they have larger carrying capacity and greater durability under many conditions of exposure than wooden piles. Precast piles are designed as columns with allowance for driving and handling stresses. Cast-in-place piles are constructed either by driving a steel shell and filling it with concrete or by filling the hole formed by a shell as it is withdrawn. Another method forms a bulb at the bottom by means of a ram which forces the concrete into the ground. The design load or capacity of cast-in-place concrete piles is largely empirical, being based on load-test data. The concrete for precast piles is usually over 5,000 lb/in² strength and for cast-in-place piles, over 3,000 lb/in² strength.

Dowels and Bearing Plates The stress in the longitudinal reinforcement of concrete columns should be transferred to the footing by means of dowels, equal in number and area to the column rods and of sufficient length to transfer the stress as in a lap splice in the column.

Bearing stresses in concrete, under design ultimate loads, should not exceed the following values:

$$\text{Entire surface loaded: } f_b = 0.85 \; \phi f_c' \qquad (12.3.24)$$
$$\text{Part of surface loaded: } f_b = 0.85 \; \phi f_c' \sqrt{A_2/A_1} \qquad (12.3.25)$$

where A_1 = loaded area; and A_2 = surface area of same shape and concentric with A_1. $\sqrt{A_2/A_1}$ should not exceed 2.0.

WALLS AND PARTITIONS

Reinforced concrete is well suited to the construction of walls, especially where they have to withstand heavy pressures, such as the retaining walls of a cellar or basement, walls for coal pockets, silos, reservoirs, or grain elevators. Such walls must be designed for flexural shear and bond stresses as well as stability against overturning, sliding, and soil pressure. Drainage should be provided for by weep holes or drains. Partitions may be built of solid concrete 4 to 6 in thick, reinforced to control temperature and shrinkage cracks. Reinforced concrete walls need to be anchored by reinforcement to adjacent structural members. All walls must be reinforced for temperature with steel placed horizontally and vertically.

The horizontal reinforcement shall not be less than 0.25 percent and the vertical reinforcement not less than 0.15 percent of the area of the reinforced section of the wall when bars are used and three-fourths of these amounts when welded fabric is used. Adequate reinforcement must be provided around all openings for windows and doors.

Retaining walls of reinforced concrete are used to resist the pressures of earth, water, and other retained materials and are usually of T or L shape. The base must be so proportioned that there is sufficient resistance to sliding and overturning and that the safe bearing strength of the soil is not exceeded. The dimensions of the concrete section and the position and amount of steel reinforcement are determined by the moments and shears at critical vertical and horizontal sections at the junction of the wall and the base. Particular attention should be given to drainage to prevent excessive water pressure behind walls retaining earth or other materials. Walls retaining water, such as tanks, should have steel tensile stresses limited to 12,000 lb/in² unless special consideration is given to controlling cracks and should have ample reinforcement to provide for effects caused by shrinkage of the concrete and temperature change.

Bearing Walls The allowable compressive force for reinforced concrete bearing walls subject to concentric loads can be computed as follows:

$$P_u = 0.55 \; \phi f_c' A_g [1 - (l_c/40h)^2] \qquad (12.3.26)$$

For the case of concentrated loads, the effective width for computational purposes can be considered as the width of the bearing plus four times the wall thickness but not greater than the distance between loads. The wall thickness should be at least ¹⁄₂₅ of the unsupported height or width, whichever is smaller. For the upper 15 feet, bearing walls must be at least 6 in thick and increase at least 1 in in thickness for each successive 25 feet downward, except that walls of a two-story dwelling need to be only 6 in thick over the entire height, provided that the strength is adequate.

PRESTRESSED CONCRETE

Prestressed concrete members have initial internal stresses, set up by highly stressed steel tendons embedded in the concrete, which are generally opposite those caused by applied loads. Prestressed members are constructed in one of two ways: (1) *Pretensioned* members are factory precast products, made by tensioning steel tendons between abutments and then casting concrete directly around the steel. After the concrete has reached sufficient strength, the steel is cut and the force transferred to the concrete by bond. (2) *Posttensioned* members, either factory or site cast, contain steel in ducts cast in the concrete. After the concrete has cured, the steel is tensioned and mechanically anchored against the concrete. The ducts are preferably pumped full of grout after tensioning to provide bond and corrosion protection, or the tendons may be coated with corrosion inhibitors.

Very high strength steel is used for prestressing in order to overcome the losses of steel stress due to creep and shrinkage of concrete, and as a result of the strength, relatively small amounts of steel are required. The concrete for pretensioned members will usually be at least 5,000 lb/in² compressive strength, and at least 4,000 lb/in² for posttensioned members.

Because of the initial stress conditions, prestressed concrete members are generally crack-free at working loads and consequently are quite suitable for water-containing structures. Circular tanks and pipes are posttensioned by wrapping them with highly stressed wires, using specialized equipment.

Losses of prestress occur with time owing to creep and shrinkage of concrete and relaxation of steel stress. Pretensioned members also have an initial elastic shortening loss accompanying transfer of force to the concrete, and posttensioned members have losses due to friction between ducts and tendons and anchor set. These losses must be taken into account in the design of members.

Prestressed concrete members are checked for both strength and stresses at working loads. Because of the built-in stresses, the condition of dead load only may also govern. The flexural strength is computed using the same equations as for reinforced concrete which were developed earlier. The steel does not have a well-defined yield stress, and the steel stress at failure can be predicted from the following expressions:

Bonded tendons, low relaxation steel and no compression steel:

$$f_{ps} = f_{pu} \left(1 - \frac{0.28}{\beta_1} \rho_p \frac{f_{pu}}{f_c'} \right) \qquad (12.3.27)$$

Unbonded, $l/h \leq 35$:

$$f_{ps} = f_{se} + 10{,}000 + f_c'/100 \, \rho_p \qquad (12.3.28)$$

but not more than f_{se} or $f_{py} + 60{,}000$.

The stresses are used directly in Eqs. (12.3.4), (12.3.7), or (12.3.8), as appropriate, substituting f_{ps} for f_y and A_{ps} for A_s.

Allowable Stresses at Working Load

Steel:

Maximum jacking stress, but not to exceed recommendation by steel or anchorage manufacturer	$0.8 f_{pu}$
Immediately after transfer or posttensioning	$0.7 f_{pu}$

Concrete:

Temporary stresses immediately after prestressing Compression	$0.6 f_{ci}'$
Tension in areas without reinforcement	$3\sqrt{f_{ci}'}$
Design load stresses (after losses) Compression	$0.45 f_c'$
Tension in precompressed tensile zones	$6\sqrt{f_c'}$

The allowable tension may be increased to $12\sqrt{f_c'}$ if it is demonstrated, by a comprehensive analysis taking cracking into account, that the short- and long-term deflections will be satisfactory.

The final steel stress, at working loads, will usually be 30,000 to 45,000 lb/in² less than the initial stress for pretensioned members. Posttensioned members will have slightly lower losses. Losses, from initial tensioning values, for pretensioned members may be predicted satisfactorily using the following expressions from the AASHTO, Bridge Specifications, Sec. 9.16.2;

$$\Delta f_s = SH + ES + CR_c + CR_s = \text{prestress loss} \qquad \text{lb/in}^2 \quad (12.3.29)$$

where SH = shrinkage loss = $17{,}000 - 150 \, RH$; ES = elastic shortening loss $= (E_s/E_c)f_{cir}$; $CR_c =$ creep loss $= 12f_{cir} - 7f_{cds}$; $CR_s =$ relaxation loss $= 5,000 - 0.1ES - 0.05(SH + CR_c)$, for low relaxation strands; $f_{cir} =$ concrete stress at level of center of gravity of steel (cgs) at section considered, due to initial prestressing force and dead load; $f_{cds} =$ change in concrete stress at cgs due to superimposed composite or noncomposite dead load; and $RH =$ relative humidity, percent. The loss calculations are carried out for each critical moment section. The average annual relative humidity of the service environment should be used in the SH calculation.

The shear reinforcement requirements for simple cases are covered in an earlier section. In addition to the shear steel, a few stirrups or ties should be placed transverse to the member axis as close to the ends as possible, to control potential splitting cracking. The area of steel, from the AASHTO Bridge Specification, should be $A_t = 0.04 f_{si}A_{ps}/20,000$. In posttensioned beams, end blocks will often have to be used to provide space for anchorage bearing plates.

The minimum clear spacing between strands in pretensioned members is three times the strand diameter near the ends of the beam, but many plants are set up to handle only 2-in spacings. Strands may be closer together in central positions of members, which will help in maximizing member effective depths and steel eccentricity.

Few precast, pretensioned members are solid rectangular sections, and single and double T beams and hollow floor slab units are used extensively in buildings. Hollow box beams and I-section beams are used extensively in bridges and in buildings with heavy design loads. Square piling with the prestressing strands arranged in a circular pattern is widely used. Because of the large number of possible sections, it is necessary to check availability of any particular section with local producers before designing any precast structure.

PRECAST CONCRETE

Precast slabs, beams, walls, and partitions as well as piles, retaining wall units, light standards, railroad crossings, and bridge slabs are being increasingly used because of the saving in time and labor cost. Such units vary in size from small slabs for use in floors or residences to large frames for industrial buildings. The small units, such as roof slabs, are cast in steel forms at central plants. Some of the larger units, such as bridge or highway slabs and wall units, are cast in wood forms at or near the place of use. A method by which wall or partition slabs are cast so that they are simply tilted into position has found wide use in housing and industrial construction. Another special adaptation is the method of casting complete floors on top of each other, then lifting into position vertically at the columns.

Particular attention must be given in the design of precast units to reduction in weight and to details to minimize the cost of erection and installation. Reduction in weight is obtained by the use of lightweight aggregates, high-strength concrete, and hollow units. Precast reinforced concrete units are seldom designed for concrete strengths of less than 4,000 lb/in². They are often combined with cast-in-place concrete so as to obtain the advantages of continuity. The combination of precast beams with cast-in-place slabs gives the advantage of T-beam action. Wall units are tied together by interlocking joints or by bolts. Care must be taken in shipping and handling to avoid damage to the precast units, and the design must take care of the stresses that come from such causes. All lifting devices built into the units should be designed for 100 percent impact. All units must be identified as to proper location and orientation in the structure.

Because precast units are made under conditions which allow good control of dimensions, certain restrictions can be relaxed that must be observed for cast-in-place concrete. Cover over the reinforcement for members not exposed to freezing need not be more than the nominal diameter of the steel but not less than 5/8 in. The maximum size of the coarse aggregate can be as large as one-third of the smaller dimension of the member. Precast wall panels are not limited to the minimum thickness requirements for cast-in-place walls.

To reduce the number of connections, precast units should generally be cast as large as can be properly handled. However, some joints will

be needed to transfer moments, torsion, shear, and axial loads from one member to another. The integrity of the structure depends on the adequacy of the design of the various joints and connections. They may be made by use of bolts and pins or clips and keys, by welding the reinforcement or steel insert, or by a number of other methods limited only by the ingenuity of the designer. The connections should not be the weak links in the structure. Thought as to their location will avoid many problems.

JOINTS

Contraction and expansion joints may be needed at intervals in a structure to help care for movement due to temperature changes and shrinkage. Joints at 20- to 30-ft intervals provide good crack control. A weakened plane, formed in the tension side of the member by a slot ¼ in wide and ½ in deep, will induce the formation of contraction cracks at selected points. Structures over 200 ft in length should have special consideration given to contraction provisions.

Construction joints are necessary in most structures because all sections cannot be cast continuously. They should be made at points of minimum shearing stress and reinforced across the joint with a steel area of not less than 0.5 percent of the area of the section cut. Provision must be made for the transfer of shear and other forces through the construction joint. Joints in columns should be made at the underside of the floor members, haunches, T beams, and column capitals.

The hardened concrete at a joint should be properly prepared for bonding with the new concrete by being cleaned, roughened, and wetted. On this surface, a coat of neat cement grout or other bonding agent should be applied just before depositing the new concrete.

FORMS

Forms are usually built of wood or metal but in special cases may be made of plastic or fiberglass reinforced plastic. Wooden forms may be the most economical unless the construction allows for the repeated use of the same forms. Plywood and compressed wood fiber sheets, specially treated to make them waterproof, are frequently used for form faces where good surfaces are required. Forms must be designed so that they can be easily erected, removed, and reerected. The usual order of removing forms is (1) column sides, (2) joists, (3) girder and beam sides, (4) slab bottoms, and (5) girder and beam bottoms. Column forms are held together by clamps made of wood or steel, the spacing of which is smallest at the bottom and increases with the decrease in pressure. Beam forms consist of the bottom and two sides held together by clamps or cleats and supported by posts. Slab forms consist of boards or other form material supported by joists spaced 2 or 3 ft apart or other means. The joists either rest on a horizontal joist bearer fastened to the clamps of the beam or girder or are supported by stringers, or posts, or both.

Special consideration must be given to forms for prestressed concrete members. For pretensioned members the form must be constructed such that it will permit movement of the member during release of the prestressing force. For posttensioned members, the form should provide a minimum of resistance to shortening of the member. It is also necessary to consider the deflection of the members due to the stressing force.

Design in Formwork The formwork is an appreciable portion of the cost of most concrete structures. Any efforts, however, to reduce the cost of the forms must not go beyond the point of safe design to prevent failures which would in themselves raise the cost of construction.

All forms must conform to the dimensions and shape of the members and must be sufficiently tight to prevent leakage of the mortar. They must be properly braced and tied together to maintain their position and shape during the construction procedure.

The formwork must support all the vertical and lateral loads that may be applied until these loads can be carried by the concrete structure. Loads on the form include the weight of the forms, reinforcing steel, fresh concrete, and various construction live loads. The construction live load varies with conditions but is often assumed to be 75 lb/ft² of floor area. The formwork should also be designed to resist lateral loads produced by wind and movement of construction equipment. Most frequently the steel and concrete will not be placed in a symmetrical pattern and frequently large impact loads will occur. Because of the many varied conditions, it is frequently impossible to determine with any great precision the loads which the form must carry. The designer must therefore make safe assumptions by which the forms can be designed such that failure will not result.

Lateral pressures in forms for walls and columns are influenced by a number of factors: weight of concrete, height of placing, vibration, temperature, size and shape of form, amount and distribution of reinforcing steel, and several other variables. Formulas have been suggested for computing safe lateral pressures to be used in form design. However, because of limited test data, they are not generally accepted by all engineers.

Form Liners Absorbent form liners are occasionally applied to the surface of forms to extract the water from the surface of the concrete, eliminate air and water voids, and produce a concrete of uniform appearance with surfaces which are superior in durability and resistance to abrasion.

The vacuum process, whereby water is absorbed from the concrete through a special form liner made of two layers of screen or wire mesh covered by a layer of cloth, has a similar effect and if properly used reduces the water content of the concrete to a depth of several inches.

Neoprene and other types of rubber have been successfully used as liners in precasting work in which a number of units are made from one form. Rubber is particularly suited for patterned work.

Plastic form liners make it easy to obtain a textured surface or a glossy smooth surface. Generally speaking, plastic liners are easily cleaned and if not too thin are suitable for a number of reuses.

Removal of Forms The time that forms should remain in place depends on the character of the members and weather conditions. The strength of concrete must be ascertained before removing the forms. Unless special precautions are taken, concrete should not be placed below 40°F. Fresh concrete should never be subjected to temperatures below freezing. As an approximate guide for the minimum time for form removal, the following rules, which assume moist curing at not less than 70°F for the first 24 h, may be observed.

WALLS IN MASS WORK. In summer, 1 day; in cold weather, 3 days.

THIN WALLS. In summer, 1 day; in cold weather, 5 days.

COLUMNS. In summer, 1 day; in cold weather, 4 days, provided girders are shored to prevent appreciable weight reaching the columns.

SLABS UP TO 7-FT SPAN. In summer, 4 days; in cold weather, 2 weeks.

BEAMS AND GIRDER SIDES. In summer, 1 day; in cold weather, 5 days.

BEAMS AND GIRDERS AND LONG-SPAN SLABS. In summer, 7 days; in cold weather, 2 weeks.

CONDUITS. 2 or 3 days, provided there is not a heavy fill upon them.

ARCHES. If a small size, 1 week; large arches with a heavy dead load, 3 weeks.

Forms for prestressed members may be removed when sufficient prestressing has been applied to enable them to carry their dead loads and the expected construction loads.

EVALUATION OF EXISTING CONCRETE STRUCTURES

This section is intended to give some guidance to the persons responsible for inspections of structures, whether these are done routinely, or before changing a loading or use, or during remodeling, or when something suspicious is found. Serious problems clearly need to be evaluated by a structural engineer experienced in such evaluations, testing, and renovation; such problems should not be evaluated by one who is primarily a structural designer.

The following are some signs of distress. Any cracks wider than about 0.02 in (0.5 mm) are potentially serious, particularly if there are more than a few. If the cracks are inclined like the shear crack shown in Fig. 12.3.1, they are an additional warning sign and should be investigated promptly. At interior supports and other locations where the beams are continuous with the columns, these cracks may also start at the top surface. Any cracking parallel to the member axis is potentially serious.

Rust stains and streaks from cracks indicate corrosion problems. Corrosion, especially from salt, disrupts the concrete surrounding the steel long before the bar areas are significantly reduced, because rust occupies much more volume than the steel it replaces. This rusting causes internal cracking, which can often be detected by tapping on the concrete surface with a hammer (1-lb size), which produces a distinctive "hollow" sound. "Stalactites" growing from the bottoms of members may be either salt or lime, but both indicate water-penetration problems and the need for waterproofing work.

Concrete exposed to freezing and thawing or to attack by some chemicals may crumble and disintegrate, leading to loss of section area and protective cover on the reinforcement. Large deflections and/or slopes in members may be indicative of impending distress or disaster, but sometimes members were not built very straight, and in such cases the warning sign is actually a *change* in the conditions.

Any reinforced concrete building designed before about 1956 has a potential weakness in the shear strength of the beams and girders because of a deficiency in the codes of that era, and major changes in loading and seemingly minor signs of distress should be investigated carefully. This problem may also exist in highway bridges designed before 1974. Prestressed members should not have this problem.

Repairs are made by injecting epoxy into cracks, by surface patching, by chipping away significant volumes of concrete and replacing it, and by other means. The repair materials range from normal concretes to highly modified concrete and polymer materials.

12.4 AIR CONDITIONING, HEATING, AND VENTILATING
By Norman Goldberg

REFERENCES: ASHRAE Handbooks "Fundamentals," "Systems and Equipment," and "Applications." Stamper and Koral, "Handbook of Air Conditioning, Heating and Ventilating," Industrial Press. Carrier, "Handbook of Air Conditioning Design," McGraw-Hill.

Air conditioning is the process of treating air to meet the requirements of a conditioned space by controlling its temperature, humidity, cleanliness, and distribution. This section presents standards, basic data, and physical laws for use in the design of air conditioning and related heating and ventilating systems.

COMFORT INDEXES

The human body generates heat and disipates that heat to the surrounding air by sensible flow and the evaporationg of moisture.

Effective Temperature Effective temperature (ET) combines the effect of ambient temperature and humidity into a single index.

ASHRAE Comfort Chart The American Society of Heating, Refrigeration, and Air Conditioning Engineers (ASHRAE) comfort envelope shown in Fig. 12.4.1 is based on clothing and activity. Comfort varies with skin temperature and skin wettedness.

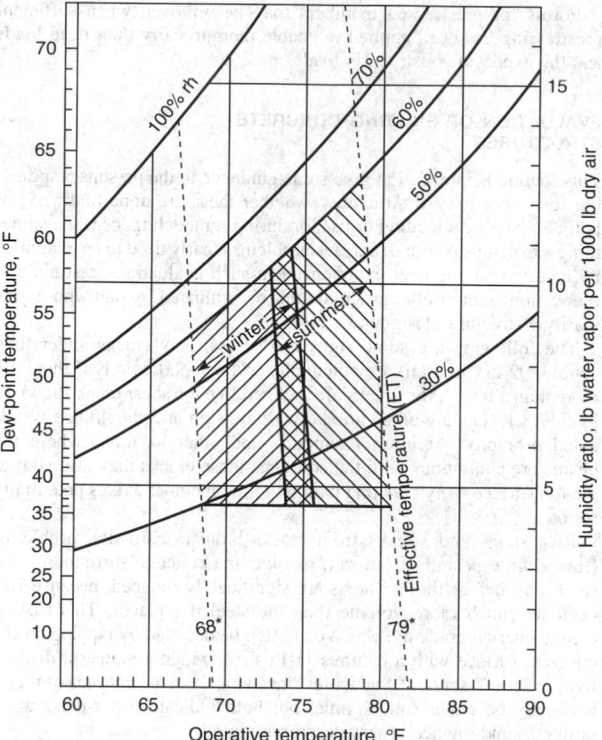

Fig. 12.4.1 Standard effective temperatures and ASHRAE comfort zones. (ASHRAE "Handbook of Fundamentals," 1993.)

Temperature-Humidity Index

The term temperature-humidity index (THI) is used to describe the combined effects of temperature and humidity on comfort experienced by people. Since individual reactions can vary considerably from person to person, this quantity should be considered as a guide rather than as an absolute. However, relatively few people will feel discomfort when the THI is 70 or below. By the time it reaches 75, about half the people will be uncomfortable. An index of 80 in a work area may cause a decrease

in workers' efficiency. To compute THI, any one of the following formulas may be used:

$$\begin{aligned} \text{THI} &= 0.4(t_d + t_w) + 15 \\ &= 0.55 t_d + 0.2 t_{dp} + 17.5 \\ &= t_d - (0.55 - 0.55\ \text{RH})(t_d - 58) \end{aligned} \quad (12.4.1)$$

where t_d = dry-bulb temperature, °F; t_w = wet-bulb temperature, °F; t_{dp} = dew-point temperature, °F; and RH = relative humidity, expressed as a decimal. To convert to degrees Celsius, $t_c = \frac{5}{9}(t_f - 32)$.

Wind-Chill Index

The wind-chill index attempts to describe how much heat the body will lose under certain conditions of wind and temperature. It is determined empirically by an equation which is used to describe the rate of heat loss from a litre cylinder of water at 33°C (91.4°F) as a function of ambient temperature and wind velocity. The formulas for the wind-chill index (WCI) are:

$$\text{WCI} = (10.45 - V + 10\sqrt{V})(33 - t_a) \quad \text{kcal/m}^2/\text{h} \quad (12.4.2)$$

where V = metres per second, t_a = °C.

$$\text{WCI} = (10.45 - 0.447V + 6.6854\sqrt{V})(91.4 - t_a) \quad (12.4.3)$$

where V = miles per hour, t_a = °F.

Instead of using the WCI to express the severity of a cold environment, meteorologists use an index derived from the WCI called the *equivalent wind chill temperature*. This is the ambient temperature that would produce, in a calm wind (defined for this application as 4 mi/h), the same WCI as the actual combination of air temperature and wind velocity. Equivalent wind chill temperature $t_{eq,wc}$ can be calculated by:

$$t_{eq,wc} = -0.0818\ (\text{WCI}) + 91.4 \quad (12.4.4)$$

where $t_{eq,wc}$ is expressed as a temperature, °F. (See Table 12.4.1.)

INDOOR DESIGN CONDITIONS

Indoor design conditions are determined by the application, usually either the comfort of the people occupying the space or the control of conditions within that space to facilitate a process.

Comfort Conditions Typical indoor design conditions for summer and winter comfort are shown in Table 12.4.2.

Conditions for Process Design These vary widely, depending on the process. See the manufacturer's requirements and/or the ASHRAE Handbook "Applications."

Table 12.4.1 Equivalent Wind Chill Temperature of Cold Environments*

Wind Speed, mi/h	Actual Thermometer Reading, °F											
	50	40	30	20	10	0	−10	−20	−30	−40	−50	−60
	Equivalent Chill Temperature, °F											
0	50	40	30	20	10	0	−10	−20	−30	−40	−50	−60
5	48	37	27	16	6	−5	−15	−26	−36	−47	−57	−68
10	40	28	16	3	−9	−21	−34	−46	−58	−71	−83	−95
15	36	22	9	−5	−18	−32	−45	−59	−72	−86	−99	−113
20	32	18	4	−11	−25	−39	−53	−68	−82	−96	−110	−125
25	30	15	0	−15	−30	−44	−59	−74	−89	−104	−119	−134
30	28	13	−3	−18	−33	−48	−64	−79	−94	−110	−125	−140
35	27	11	−4	−20	−36	−51	−67	−83	−98	−114	−129	−145
40†	26	10	−6	−22	−38	−53	−69	−85	−101	−117	−133	−148

Little danger: In less than 5 h, with dry skin. Maximum danger from false sense of security. (WCI less than 1400)	**Increasing danger:** Danger of freezing exposed flesh within one minute. (WCI between 1400 and 2000)	**Great danger:** Flesh may freeze within 30 seconds. (WCI greater than 2000)

* Cooling power of environment expressed as an equivalent temperature under calm conditions.
† Winds greater than 43 mi/h have little added chilling effect.
SOURCE: U.S. Army Research Institute of Environmental Medicine.

Table 12.4.2   Recommended Inside Design Conditions

Type of application	Summer			Winter				
				With humidification			Without humidification	
	Dry bulb, °F (°C)	Rel. hum., %	Temp. swing, °F (°C)*	Dry bulb, °F (°C)	Rel. hum., %	Temp. swing, °F (°C)‡	Dry bulb, °F	Temp. swing, °F (°C)‡
General comfort (apartment, house, hotel, office, hospital, school, etc.)	74–76 (23–24)	50–45	2 (1.1)	68–72 (20–22)	35–30	3 to 4 (1.6–2.2)	70–74	4 (2.2)
Retail shops (short-term occupancy) (bank, barber or beauty shop, department store, supermarket, etc.)	74–76 (23–24)	50–45	2 (1.1)	68–72 (20–22)	35–30†	3 to 4 (1.6–2.2)	70–74	4 (2.2)
Low-sensible-heat-factor applications (high latent load) (auditorium, church, bar, restaurant, kitchen, etc.)	74–76 (23–24)	55–50	2 (1.1)	68–72 (20–22)	40–35	2 to 3 (1.1–1.7)	70–74	4 (2.2)
Factor comfort (assembly areas, machining rooms, etc.)	76–78 (24–26)	55–45	2 (1.1)	68–70 (18–21)	35–30	4 to 6 (2.2–3.3)	65–70	6 (3.3)

* Temperature swing is above the thermostat setting at peak summer load conditions.
† Winter humidification in retail clothing stores is recommended to maintain the quality texture of goods.
‡ Temperature swing is below the thermostat setting at peak winter load conditions (no lights, people, or sun).
°C = ⅝(°F − 32)
SOURCE: Adapted from Carrier Corp. "System Design Manual," with permission.

OUTDOOR DESIGN CONDITIONS

Outdoor design conditions are based primarily on the geographical location but they are affected to some extent by the criticality of the application and by the economics of maintaining indoor conditions at specified design levels during unusual outdoor weather conditions.

Outdoor design conditions for comfort applications are usually not the most severe conditions that have been experienced in a locality, and recommended levels may be exceeded for an average of 5 percent of the hours during the cooling season.

Frequently used winter outside dry-bulb design temperature are shown in Fig. 12.4.2. Summer outdoor design dry-bulb, wet-bulb, and dew-point temperatures for comfort applications are shown in Figs. 12.4.3 to 12.4.5, respectively.

Design of critical air-conditioning systems and selection of atmospheric evaporative cooling equipment are normally based on more severe outdoor conditions than for comfort applications. Figure 12.4.6 shows wet-bulb temperatures in the United States that are exceeded 1 percent of the time during the cooling season.

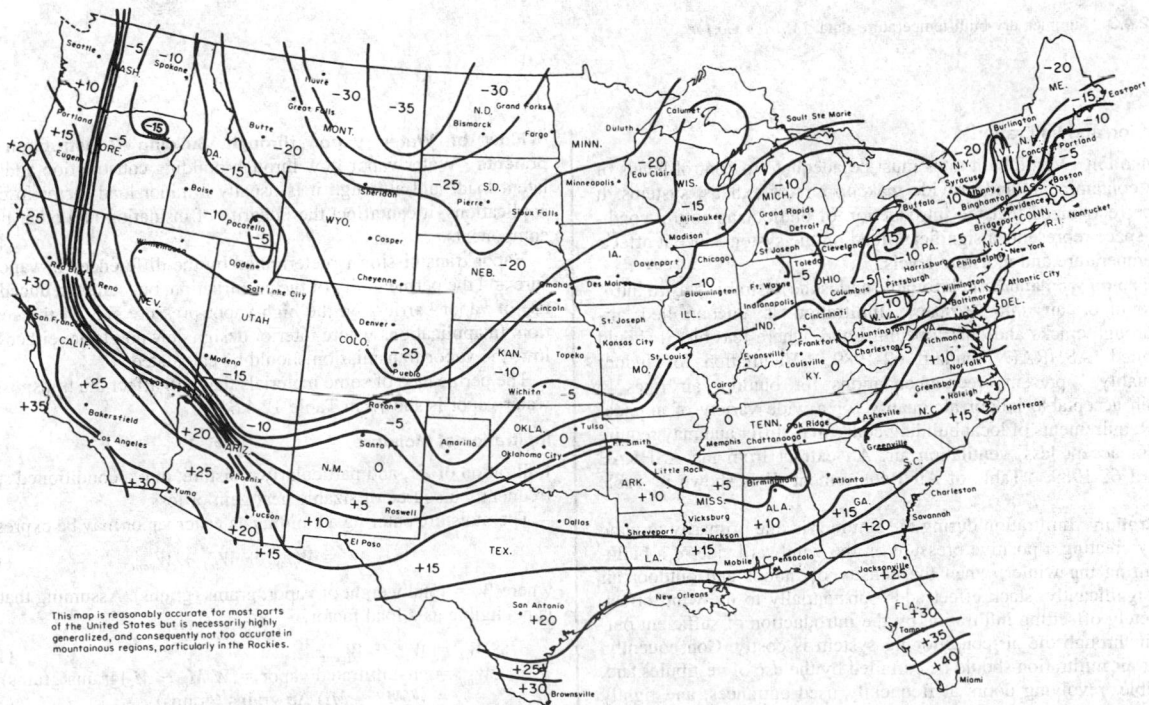

Fig. 12.4.2   Isotherms of winter outdoor temperatures, °F. [*E. Stamper and L. Koral (eds.), "Handbook of Air Conditioning, Heating and Ventilating," 3d ed., 1979, Industrial Press.*]

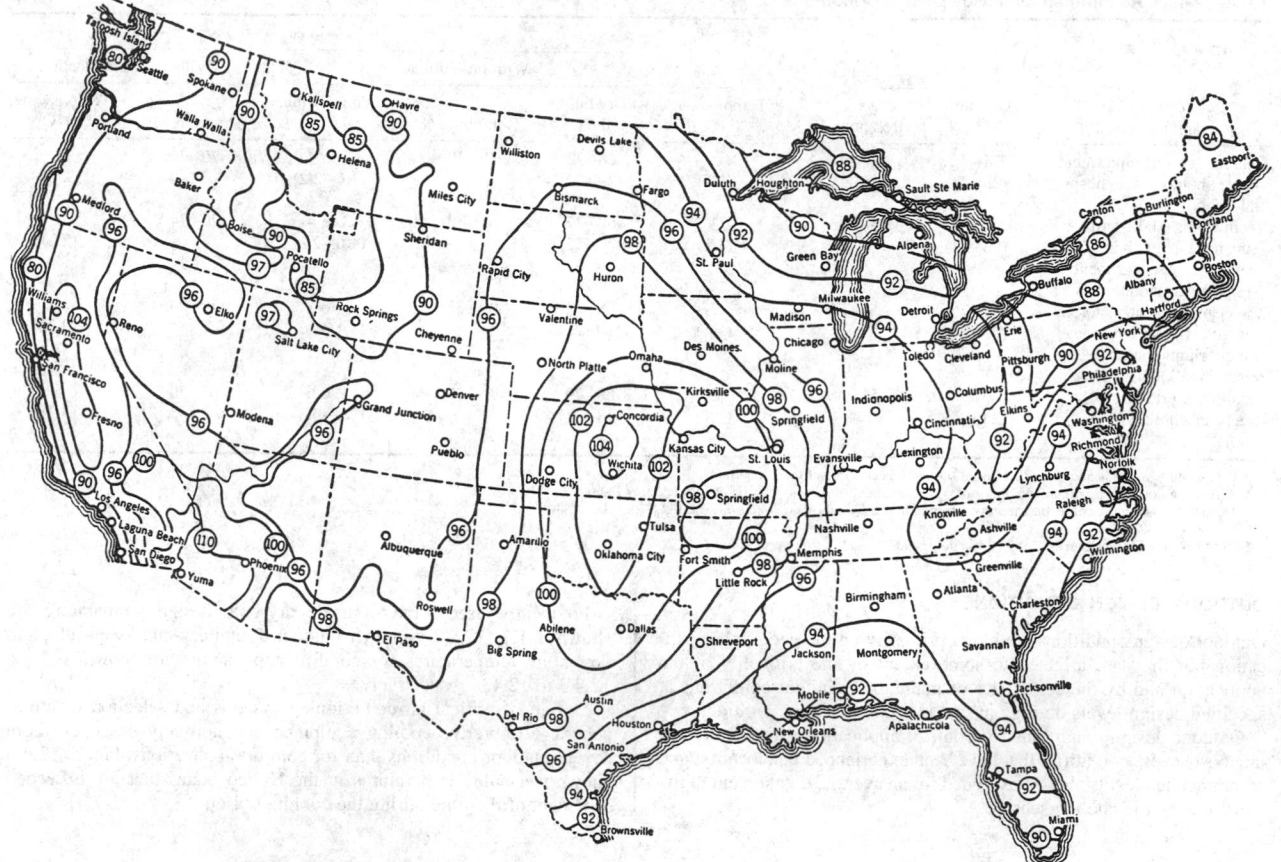

Fig. 12.4.3 Summer dry-bulb temperature data. *(Marley Co.)*

Ventilation and Infiltration

Ventilation   Ventilation rates must be adequate to control levels of indoor contaminants and provide makeup air for exhaust systems. It must be recognized that the introduction of outdoor air into a conditioned space represents a significant load on the system since it affects both temperature and moisture levels.

Mechanical ventilation is controllable and is the desirable way to introduce outdoor air into a space. Infiltration is uncontrolled airflow through cracks and other openings and, where feasible, it should be limited. ASHRAE Standard 62-1989, ''Ventilation for Indoor Air Quality,'' presents recommendations for outdoor air rates to maintain acceptable indoor air quality for a wide variety of applications. Requirements of local building codes may differ and may require more, or accept less, ventilation air. An extract from the ASHRAE Standard 62-1989, ''Table of Air Requirements,'' is shown in Table 12.4.3.

Infiltration   Infiltration during the summer is due primarily to wind velocity creating a positive pressure on the windward side of a building. During the winter, when the density of indoor and outdoor air differ significantly, stack effect adds substantially to the wind effect. Completely offsetting infiltration by the introduction of sufficient outdoor air through the air-conditioning system is costly. Consequently, outdoor air infiltration should be curtailed by the use of vestibules and, if feasible, revolving doors at frequently used entrances, and tightly fitting windows and/or storm sash throughout. See Tables 12.4.4 and 12.4.5.

Flow of Water Vapor through Building Construction Components

Water vapor flow through building construction adds to the latent load, and although it is usually a minor load factor in comfort applications, it can affect the integrity of insulation and other building components.

Vapor transmission is determined by the differences in vapor pressure and the permeability of the construction, but can be reduced by the use of vapor barriers on the high-vapor-pressure side of the construction. In applications where interior design dew points are either high or low, the vapor transmission should be calculated.

The permeance of some materials that will affect the transmission of water vapor is shown in Table 12.4.6.

Infiltration of Moisture

Infiltration of air, and particularly moisture, into a conditioned space is frequently a source of sizable heat gain or loss.

The moisture entering a building as water vapor may be expressed as

$$W_t = W_{trans} + W_{inf} + W_{vent} \qquad (12.4.5)$$

where W_t = total weight of vapor, grains (grams). Assuming that W_{trans} is negligible as a load factor,

$$W_1 = W_{inf} + W_{vent} \qquad (12.4.6)$$

W_{inf} = air-infiltrated vapor = $W(M_o - M_i)$grains(grams)

= $W(M_o - M_i)\,\Delta p$ grains (grams) $\qquad (12.4.7)$

W_{vent} = ventilation air vapor

= $W(M_o - M_i)$ grains (grams) $\qquad (12.4.8)$

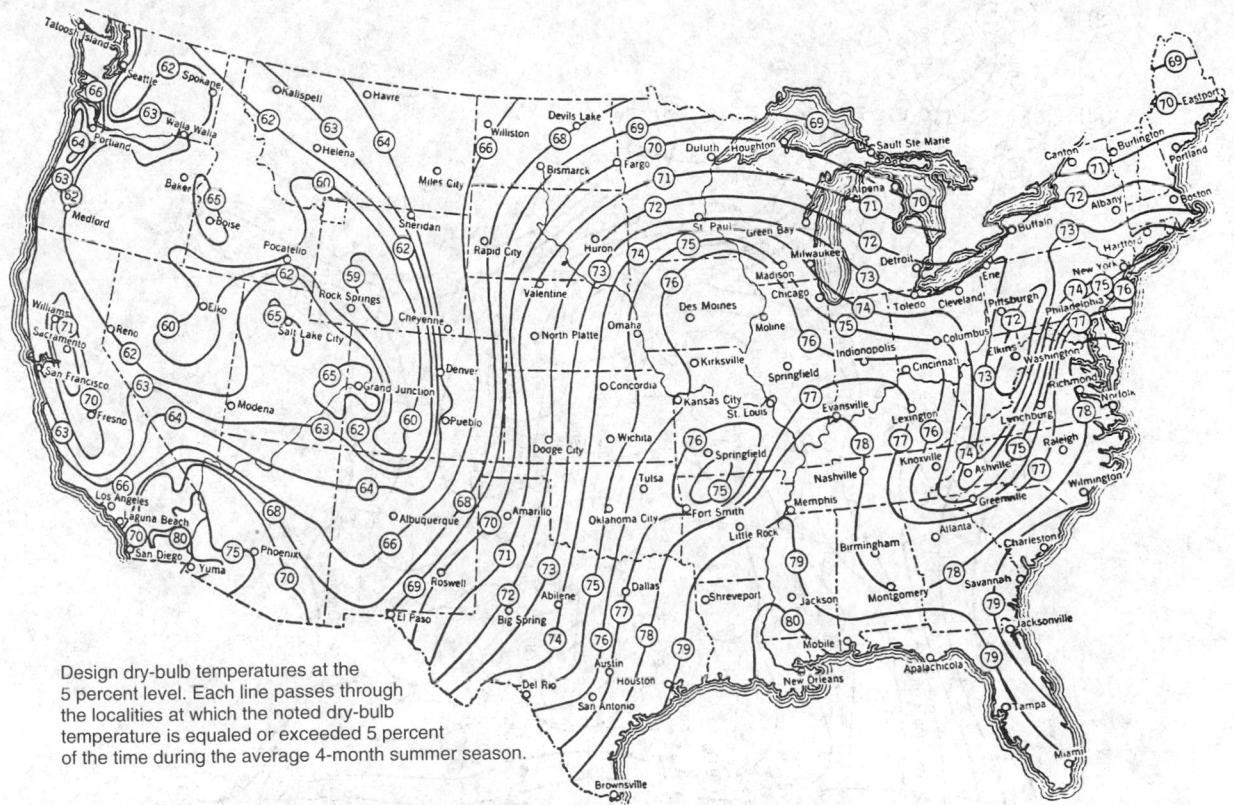

Design dry-bulb temperatures at the
5 percent level. Each line passes through
the localities at which the noted dry-bulb
temperature is equaled or exceeded 5 percent
of the time during the average 4-month summer season.

Fig. 12.4.4 Summer wet-bulb temperature data. *(Marley Co.)*

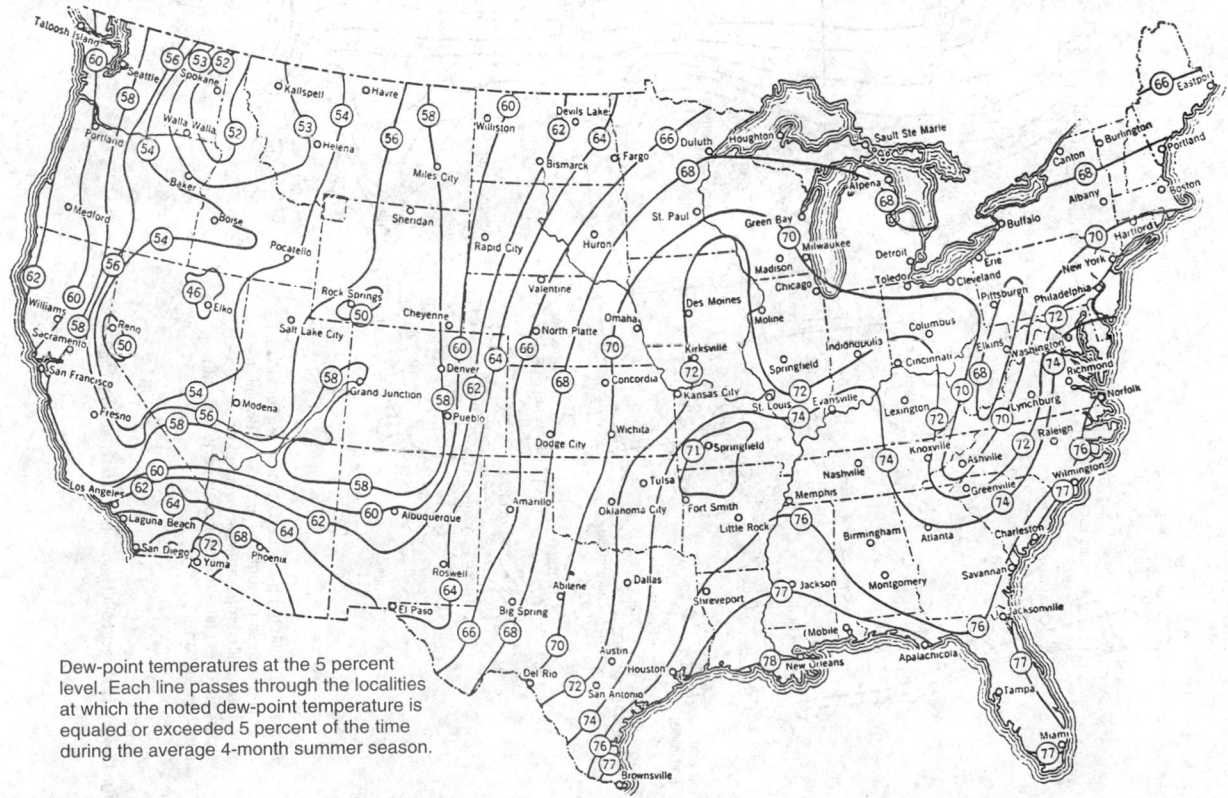

Dew-point temperatures at the 5 percent
level. Each line passes through the localities
at which the noted dew-point temperature is
equaled or exceeded 5 percent of the time
during the average 4-month summer season.

Fig. 12.4.5 Summer dew-point temperature data. *(Marley Co.)*

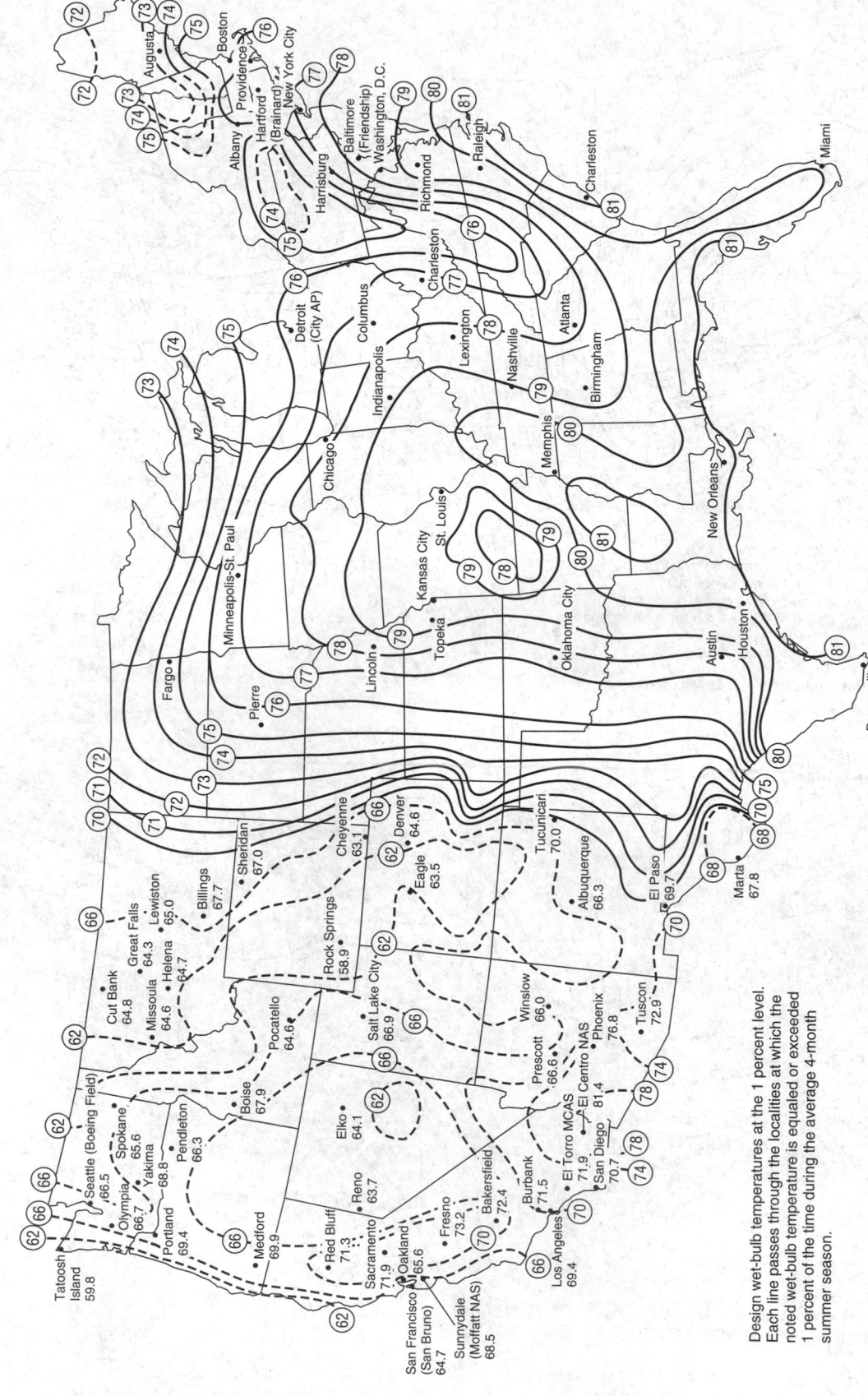

Design wet-bulb temperatures at the 1 percent level.
Each line passes through the localities at which the
noted wet-bulb temperature is equaled or exceeded
1 percent of the time during the average 4-month
summer season.

Fig. 12.4.6 High wet-bulb temperature data. *(Fluor Products Co.)*

Table 12.4.3   Outdoor Air Requirements for Ventilation*

Application	Estimated maximum† occupancy, persons/1000 ft² or 100 m²	Outdoor air requirements cfm/person	L/s · person	cfm/ft²	L/s · m²	Comments
		Commercial facilities				
Hotels, motels, resorts, dormitories						
Bedrooms				30	15	Per room.
Living rooms				30	15	Per room.
Baths				35	18	Per room. Installed capacity for intermittent use.
Lobbies	30	15	8			
Conference rooms	50	20	10			
Assembly rooms	120	15	8			
Dormitory sleeping areas	20	15	8			See also food and beverage services, merchandising, barber and beauty shops, garages.
Gambling casinos	120	30	15			Supplementary smoke-removal equipment may be required.
Offices						
Office space	7	20	10			Some office equipment may require local exhaust.
Reception areas	60	15	8			
Telecommunication centers and data entry areas	60	20	10			
Conference rooms	50	20	10			Supplementary smoke-removal equipment may be required.
Public spaces						
Corridors and utilites				0.05	0.25	
Public restrooms, cfm/wc or cfm/urinal		50	25			Normally supplied by transfer air.
Locker and dressing rooms				0.5	2.5	Local mechanical exhaust with no recirculation recommended.
Smoking lounge	70	60	30			
Food and beverage service						
Dining rooms	70	20	10			⎫ Supplementary smoke-removal equipment may be required.
Cafeteria, fast food	100	20	10			
Bars, cocktail lounges	100	30	15			⎭
Kitchens (cooking)	20	15	8			Exhaust min. 1.5 cfm/ft²
Retail stores, sales floors, and show room floors						
Basement and street	30			0.30	1.50	
Upper floors	20			0.20	1.00	
Storage rooms	15			0.15	0.75	
Dressing rooms				0.20	1.00	
Malls and arcades	20			0.20	1.00	
Shipping and receiving	10			0.15	0.75	
Warehouses	5			0.05	0.25	
Smoking lounge	70	60	30			Normally supplied by transfer air, local mechanical exhaust; exhaust with no recirculation recommended.
Sports and Amusement						
Spectator areas	150	15	8			When internal combustion engines are operated for maintenance of playing surfaces, increased ventilation rates may be required.
Game rooms	70	25	13			
Ice arenas (playing areas)				0.50	2.50	
Swimming pools (pool and deck area)				0.50	2.50	Higher values may be required for humidity control.
Playing floors (gymnasium)	30	20	10			
Ballrooms and discos	100	25	13			
Bowling alleys (seating areas)	70	25	13			
Theaters						Specifcal ventilation will be needed to eliminate special stage effects (e.g., dry ice vapors, mists, etc.)
Ticket booths	60	20	10			
Lobbies	150	20	10			
Auditorium	150	15	8			
Stages, studios	70	15	8			
Miscellaneous						
Photo studios	10	15	8			
Darkrooms	10			0.50	2.50	
Pharmacy	20	15	8			
Bank vaults	5	15	8			
Duplicating, printing				0.50	2.50	Installed equipment must incorporate positive exhaust and control (as required) of undesirable contaminants (toxic or otherwise).

Table 12.4.3   Outdoor Air Requirements for Ventilation*   (*Continued*)

Application	Estimated maximum† occupancy, persons/1000 ft² or 100 m²	Outdoor air requirements				Comments
		cfm/person	L/s·person	cfm/ft²	L/s·m²	
		Institutional Facilities				
Education						
Classroom	50	15	8			
Laboratories	30	20	10			Specifal contaminant control systems
Training shop	30	20	10			may be required for processes or
Music rooms	50	15	8			functions including laboratory animal
Libraries	20	15	8			occupancy.
Locker rooms				0.50	2.50	
Corridors				0.10	0.50	
Auditoriums	150	15	8			
Smoking lounges	70	60	30			Normally supplied by transfer air. Local mechanical exhaust with no recirculation recommended.
Hospitals, nursing and conva- lescent homes						
Patient rooms	10	25	13			Special requirements or codes and
Medical procedure	20	15	8			pressure relationships may determine
Operating rooms	20	30	15			minimum ventilation rates and filter
Recovery and ICU	20	15	8			efficiency. Procedures generating contaminants may require higher rates.
Autopsy rooms				0.50	2.50	Air shall not be recirculated into other spaces.
Physical Therapy	20	15	8			
Correctional facilities						
Cells	20	20	10			
Dining halls	100	15	8			
Guard stations	40	15	8			

* This table prescribes supply rates of acceptable outdoor air required for acceptable indoor air quality. These values have been chosen to control CO_2 and other contaminants with an adequate margin of safety and to account for health variations among people, varied activity levels, and a moderate amount of smoking. Rationale for CO_2 control is presented in Appendix D of Standard 62-1989.
† Net occupable space.
SOURCE: Abstracted from ASHRAE Standard 62-1989, with permission.

Table 12.4.4   Infiltration through Doors—Winter*
15 mi/h (24 km/h) wind velocity†; doors on one or adjacent windward sides‡

	ft³/min per ft² area (m³/min per m²)§									
	Infrequent use		1 and 2 story buildings		Average use					
					Tall buildings, ft (m)					
	ft³/min per ft²	m³/min per m²	ft³/min per ft²	m³/min per m²	50	(15.25 m)	100	(30.5 m)	200	(61 m)
Revolving door	1.6	0.48	10.5	3.15	12.6	3.78	14.2	4.26	17.3	5.19
Glass door [³⁄₁₆ in (5 mm) crack]	9.0	2.70	30.0	9.00	36.0	10.80	40.5	12.15	49.5	14.85
Wood door 3 × 7 ft (0.9 × 2 m)	2.0	0.60	13.0	3.90	15.5	4.65	17.5	5.25	21.5	6.45
Garage and shipping-room door	4.0	1.2	9.0	2.70						

SOURCE: Carrier Corporation "System Design Manual," Part 1, Load Estimating, 1970.
* All values are based on the wind blowing directly at the window or door. When the prevailing wind direction is oblique to the window or doors, multiply the values by 0.60 and use the total window and door areas on the windward side(s).
† Based on a wind velocity of 15 mi/h (24 km.p.a). For design wind velocities different from the base, multiply the table values by the ratio of velocities.
‡ Stack effect in tall buildings may also cause infiltration on the leeward side. To evaluate this, determine the equivalent velocity V_i and subtract the design velocity V. The equivalent velocity is

$$V_i = \sqrt{V^2 - 1.75a} \text{ (upper section)}$$
$$= \sqrt{V^2 - 1.75b} \text{ (lower section)}$$

where a and b are the distances above and below the midheight of the building, respectively, in feet.
Multiple the table values by the ratio $(V_i - V)/15$ for one-half of the windows and doors on the leeward side of the building. (Use values under one- and two-story building for doors on leeward side of tall buildings.)
§ Doors on opposite sides increase values 25%.

Table 12.4.5   Infiltration Through Windows and Doors—Crack Method—Summer–Winter*
Double-hung windows, unlocked on windward side.

	cfm per linear foot of crack											
	Wind velocity, mi/h											
	5		10		15		20		25		30	
Type of double-hung window	No W strip	W strip	No W strip	W strip	No W strip	W strip	No W strip	W strip	No W strip	W strip	No W strip	W strip
Wood sash												
Average Window	0.12	0.07	0.35	0.22	0.65	0.40	0.98	0.60	1.33	0.82	1.73	1.05
Poorly Fitted Window	0.45	0.10	1.15	0.32	1.85	0.57	2.60	0.85	3.30	1.18	4.20	1.53
Poorly Fitted—with Storm Sash	0.23	0.05	0.57	0.16	0.93	0.29	1.30	0.43	1.60	0.59	2.10	0.76
Metal Sash	0.33	0.10	0.78	0.32	1.23	0.53	1.73	0.77	2.3	1.00	2.8	1.27

SOURCE: Carrier "Handbook of Air Conditioning Design," McGraw-Hill, 1965.
* Infiltration caused by stack effect must be calculated separately during the winter.
† No allowance has been made for usage.

Table 12.4.6   Water Vapor Transmission Through Various Materials

Description of material or construction	Permeance, Btu/(h · 100 ft²) (gr/lb diff) latent heat
Packaging materials	
Cellophane, moisture-proof	0.01–0.25
Glassine (1 ply waxed or 3 ply plain)	0.0015–0.006
Kraft paper soaked with paraffin wax, 4.5 lb per 100 ft²	1.4–3.1
Pliofilm	0.01–0.025
Paint films	
2 coats aluminum paint, estimated	0.05–0.2
2 coats asphalt paint, estimated	0.05–0.1
2 coats lead and oil paint, estimated	0.1–0.6
2 coats water emulsion, estimated	5.0–8.0
Papers	
Duplex or asphalt laminae (untreated)	
30-30-30, 3.1 lb per 100 ft²	0.15–0.27
30-60-30, 4.2 lb per 100 ft²	0.051–0.091
Kraft paper	
1 sheet	8.1
2 sheets	5.1
Aluminum foil on one side of sheet	0.016
Aluminum foil on both sides of sheet	0.012
Sheathing paper	
Asphalt impregnated and coated, 7 lb per 100 ft²	0.02–0.10
Slaters felt, 6 lb per 100 ft², 50% saturated with tar	1.4
Roofing Felt, saturated and coated with asphalt	
25 lb/ft²	0.015
50 lb/ft²	0.011
Tin sheet with 4 holes 1/16-in diameter	0.17
Crack 12 in long by 1/32 in wide (approximate from above)	5.2

Painted surfaces: Two coats of a good vapor seal paint on a smooth surface give a fair vapor barrier. More surface treatment is required on a rough surface than on a smooth surface. Data indicates that either asphalt or aluminum paint is good for vapor seals.

Aluminum foil on paper: This material should also be applied over a smooth surface and joints lapped and sealed with asphalt. The vapor barrier should always be placed on the side of the wall having the higher vapor pressure if condemnation of moisture in wall is possible.

Application: The heat gain due to water vapor transmission through walls may be neglected for the normal air-conditioning or refrigeration job. This latent gain should be considered for air-conditioning jobs where there is a great vapor pressure difference between the room and the outside, particularly when the dew point inside must be low. Note that moisture gain due to infiltration usually is of much greater magnitude than moisture transmission through building structures.

Conversion factors: To convert above table values to:
grain/(hr) (sq ft) (inch mercury vapor pressure difference), multiply by 9.8. grain/(hr) (sq ft) (pounds per sq inch vapor pressure difference), multiply by 20.0.
To convert Btu latent heat to grains, multiply by 7000/1060 = 6.6.
SOURCE: Carrier "Air Conditioning Design Manual," 1965.

Table 12.4.7 Overall Heat-Transfer Coefficient _U_

Outside air 15 mi/h wind, inside still air.

Example	Construction	Btu/(h·ft²·°F)	W/(m²·K)
Frame walls	Wood siding, insulation board, air space, gypsum board	0.19	1.079
Frame partition	Gypsum board, air space, gypsum board	0.34	1.931
Frame construction ceilings and floors	Linoleum or tile, felt, plywood, wood subfloor, air space, metal lath, plaster	0.23	1.306
Pitched roofs	Asphalt shingles, building paper, wood sheathing, air space, gypsum, lath, plaster	0.28	1.590
Masonry wall	Face brick 4 in, common brick 4 in	0.48	2.725
	Face brick 4 in, common brick 4 in, air space, gypsum lath, plaster	0.29	1.647
Masonry partition	Cement block (cinder aggregate), plaster on both sides	0.31	1.760
Flat masonry roof	Built-up roofing, roof insulation 1 in, concrete slab 4 in, air space, metal lath, plaster	0.18	1.022

SOURCE: ASHRAE ''Handbook of Fundamentals,'' 1981.

Table 12.4.8 Coefficient Heat Transmission _U_ for Windows and Skylights

Air-to-air heat transfer, Btu/(h·ft²·°F) and W/(m²·K); outside air 0°F, 15 mi/h (24 km/h) wind, no solar radiation; inside still air.

	Vertical glass sheets				Horizontal glass sheets			
	Outdoor exposure		Indoor exposure		Outdoor exposure		Indoor exposure	
	Btu/(h·ft²·°F)	W/(m²·K)	Btu/(h·ft²·°F)	W/(m²·K)	Btu/(h·ft²·°F)	W/(m²·K)	Btu/(h·ft²·°F)	W/(m²·K)
Common window glass, single sheet	1.13	6.416	0.75	4.259	1.22	6.927	0.96	5.450
Common window glass, two sheets, 1-in air space (2.54 cm)	0.53	3.009	0.45	2.555	0.63	3.577	0.56	3.180

SOURCE: ASHRAE ''Handbook of Fundamentals,'' 1981.

Table 12.4.9 Coefficient of Heat Transmission _U_ for Wood Doors

	Single		With glass storm door	
Construction	Btu/(h·ft²·°F)	(m²·K)	Btu/(h·ft²·°F)	W/(m²·K)
1 in-thick solid door (²⁵⁄₃₂ in) (2 cm)	0.64	3.634	0.37	2.101
2 in-thick solid door (1⅝ in) (4 cm)	0.43	2.442	0.28	1.590
Door containing wood or glass panels	0.85	4.826	0.39	2.214

SOURCE: ASHRAE ''Handbook of Fundamentals,'' 1981.

Table 12.4.10 Surface Conductances and Resistances for Air

Position of surface	Direction of heat flow	Nonreflective ε = 0.90				Surface emissivity, reflective ε = 0.20				Reflective ε = 0.05			
		C	C'	R	R'	C	C'	R	R'	C	C'	R	R'
Still air:													
Horizontal	Upward	1.63	9.26	0.61	0.108	0.91	5.17	1.10	0.193	0.76	4.32	1.32	0.232
Vertical	Horizontal	1.46	8.29	0.68	0.121	0.74	4.20	1.35	0.238	0.59	3.35	1.70	0.299
Horizontal	Downward	1.08	6.76	0.92	0.148	0.37	2.10	2.70	0.476	0.22	1.25	4.55	0.800
Moving air (any position):													
15 mi/s (24 km/h) wind (for winter)	Any	6.00	34.07	0.17	0.029								
7½ mi/h (12 km/h) wind (for summer)	Any	4.00	22.71	0.25	0.044								

SOURCE: Adapted from ASHRAE ''Handbook of Fundamentals,'' 1993.
NOTES: C = conductance, Btu/(h·ft²·°F temp. diff.)
    C' = conductance W/(m²·K).
    R = resistance = 1/C.
    R' = resistance = 1/C'.

Table 12.4.11   Determination of *U* value Resulting from Addition of Insulation to Any Given Building Section

| Given building section property*,† | | | | Added R‡,§ (R') | | | | | | | | | | |
| U | U' | R | R' | R = 4   R' = 0.70 | | R = 6   R' = 1.05 | | R = 8   R' = 1.41 | | R = 12   R' = 2.13 | | R = 16   R' = 2.86 | | R = 20   R' = 3.57 | | R = 24   R' = 4.17 | |
				U	U'	U	U'	U	U'	U	U'	U	U'	U	U'	U	U'
1.00	5.68	1.00	0.18	0.20	1.13	0.14	0.79	0.11	0.62	0.08	0.45	0.06	0.34	0.05	0.28	0.04	0.23
0.90	5.11	1.11	0.20	0.20	1.13	0.14	0.79	0.11	0.62	0.08	0.45	0.06	0.34	0.05	0.28	0.04	0.23
0.80	4.54	1.25	0.22	0.19	1.08	0.14	0.79	0.11	0.62	0.08	0.45	0.06	0.34	0.05	0.28	0.04	0.23
0.70	3.97	1.43	0.25	0.19	1.08	0.13	0.74	0.11	0.62	0.07	0.40	0.06	0.34	0.05	0.28	0.04	0.23
0.60	3.41	1.67	0.29	0.19	1.08	0.13	0.74	0.10	0.57	0.07	0.40	0.06	0.34	0.05	0.28	0.04	0.23
0.50	2.84	2.00	0.35	0.18	1.02	0.13	0.74	0.10	0.57	0.07	0.40	0.06	0.34	0.05	0.28	0.04	0.23
0.40	2.27	2.50	0.44	0.16	0.91	0.12	0.68	0.10	0.57	0.07	0.40	0.05	0.28	0.05	0.28	0.04	0.23
0.30	1.70	3.33	0.59	0.14	0.79	0.11	0.62	0.09	0.51	0.07	0.40	0.05	0.28	0.04	0.23	0.04	0.23
0.20	1.14	5.00	0.88	0.11	0.62	0.09	0.51	0.08	0.45	0.06	0.34	0.05	0.28	0.04	0.23	0.03	0.17
0.10	0.57	10.00	1.75	0.06	0.34	0.06	0.34	0.06	0.34	0.05	0.28	0.04	0.23	0.04	0.23	0.03	0.17
0.08	0.45	12.50	2.22	0.06	0.34	0.05	0.28	0.05	0.28	0.04	0.23	0.04	0.23	0.03	0.17	0.03	0.17

SOURCE: Abstracted from ASHRAE "Handbook of Fundamentals," 1981, with permission.
NOTE: U and R expressed in U.S. customary units. U' and R' in SI units.
* For U and R values not shown in table, interpolate as necessary.
† Enter column 1 with U or R of the design building section.
‡ Under appropriate column heading for Added R, find U value of resulting design section.
§ If the insulation occupies a previously considered air space, an adjustment must be made in the given building section R value.

where Δp = vapor-pressure difference through flow path, in Hg (N/m²); W = weight of air, lb (kg); M_o = moisture content of outside air, g/lb (g/kg); M_i = moisture content of inside air, g/lb (g/kg).

For a given locality, the moisture content of the outdoor air may be determined from Fig. 12.4.5 (summer dew-point temperature data) and the appropriate psychrometric chart.

Conductance Coefficients   Tables 12.4.7 to 12.4.9 list conductance heat-transfer coefficients U for various building components. Table 12.4.10 shows conductance heat transfer coefficients for reflective and nonreflective surfaces. Table 12.4.11 facilitates the computation of the enhancement of the U value by the addition of insulation. Table 12.4.12 presents an example of calculation of U for an assembly of various building components.

AIR CONDITIONING

Cooling Load

Cooling load Q_1, the total simultaneous cooling load, expressed in Btu/hour or (watts):

$$Q_t = Q_{ext} + Q_{int} + Q_{outside\ air} \qquad (12.4.9)$$

$$Q_{ext} = \text{external heat gains}$$
$$= Q_{transmission} + Q_{-solar} \qquad (12.4.10)$$

For walls and roofs

$$Q_{tr-sol} = AU\,(Sa\,\Delta t) \qquad \text{Btu/h (watts)} \qquad (12.4.11)$$

Table 12.4.12   Example of Calculation of Coefficient of Transmission *U* for Composite Building Components
U and R expressed in U.S. customary units, U' and R' in SI units

Example 1. Coefficients of transmission U of flat masonry roofs with built-up roofing

Construction (heat flow up)	Resistance R	R'
1. Outside surface (15 mi/h wind 24 km/h)	0.17	0.03
2. Built-up roofing ⅜ in (9 mm)	0.33	0.06
3. Roof insulation (none)		
4. Concrete slab (lt. wt. agg.) (2 in) (50 mm)	2.22	0.39
5. Corrugated metal	0	0
6. Air space	0.85	0.15
7. Metal lath and ¾ in (19 mm) plaster (lt. wt. agg.)	0.47	0.08
8. Inside surface (still air)	0.61	0.11
Total resistance	4.65	0.82
$U = 1/R = 1/4.65 =$	0.22	
$U' = 1/R' = 1/0.82 =$		1.22

Example 2. Coefficients of transmission U of masonry walls

Construction (heat flow up)	Resistance R	R'
1. Outside surface (15 mi/h wind) (24 km/h)	0.17	0.03
2. Face brick (4 in) (100 mm)	0.44	0.08
3. Cement mortar (½ in) (12.5 mm)	0.10	0.02
4. Concrete block (cinder agg.) (8 in) (200 mm)	1.72	0.30
5. Air space (reflective)	2.80	0.49
6. Gypsum wallboard, foil back (½ in) (12.5 mm)	0.45	0.08
7. Inside surface (still air)	0.68	0.12
Total resistance	6.36	1.12
$U = 1/R = 1/6.36 =$	0.36	
$U' = 1/R' = 1/1.12 =$		0.89

SOURCE: Abstracted from ASHRAE "Handbook of Fundamentals," 1981, with permission.

Table 12.4.13 Total Equivalent Temperature Differentials for Calculating Heat Gain through Sunlit Walls

North latitude wall facing	8 D	8 L	10 D	10 L	12 D	12 L	2 D	2 L	4 D	4 L	6 D	6 L	8 D	8 L	10 D	10 L	12 D	12 L	South latitude wall facing
	A.M.								P.M.										
	Exterior color of wall: D = dark, L = light																		
Group A																			
NE	27	16	31	18	26	17	24	17	24	18	23	17	20	15	17	13	15	11	SE
E	32	18	41	24	37	22	29	20	28	20	26	19	23	16	20	14	18	13	E
SE	25	15	36	21	38	23	33	21	28	20	26	18	22	16	19	14	18	12	NE
S	14	9	20	13	28	18	33	22	31	21	25	18	20	15	17	13	15	11	N
SW	17	11	20	13	24	16	34	22	42	27	41	26	28	19	20	14	18	12	NW
W	17	11	20	13	24	16	30	20	42	27	48	30	33	22	22	15	19	13	W
NW	14	9	17	11	21	14	23	17	31	21	38	25	28	19	18	13	16	11	SW
N	14	9	15	10	18	12	20	15	21	16	21	16	18	14	14	11	12	9	S
Group B																			
NE	12	7	27	14	31	17	30	19	31	21	30	22	27	20	21	17	16	13	SE
E	14	8	34	18	45	24	43	25	39	25	35	24	30	22	23	18	17	14	E
SE	9	5	25	13	39	21	44	26	41	26	37	25	31	23	24	18	17	14	NE
S	4	3	7	4	18	11	32	19	41	26	39	27	33	24	25	19	18	15	N
SW	5	3	7	4	11	7	23	15	41	26	54	34	51	33	38	25	26	19	NW
W	6	4	7	4	11	4	18	12	35	23	55	34	59	37	43	28	30	20	W
NW	5	3	6	4	11	7	17	12	26	18	41	27	47	31	36	24	25	18	SW
N	6	4	9	5	12	8	18	12	22	17	25	20	27	21	22	17	16	14	S
Group C																			
NE	9	6	19	10	26	15	28	17	29	18	29	20	28	20	24	19	20	16	SE
E	10	7	22	12	36	19	40	23	39	23	36	24	33	23	28	20	22	17	E
SE	8	6	16	9	29	16	38	21	39	24	37	24	34	23	28	21	23	17	NE
S	7	5	7	4	12	7	22	14	32	20	36	24	34	24	29	21	23	17	N
SW	9	6	8	5	10	6	16	10	28	18	42	26	48	30	42	28	35	22	NW
W	10	7	9	5	10	6	14	9	24	16	40	25	52	32	47	30	37	24	W
NW	8	6	8	5	9	6	13	9	19	14	30	20	40	27	38	26	30	21	SW
N	7	5	8	5	10	7	14	9	18	13	22	16	25	19	23	18	19	16	S
Group D																			
NE	8	5	19	10	28	15	29	17	30	19	30	21	28	21	24	19	19	16	SE
E	9	6	23	12	38	20	42	24	40	24	37	24	33	23	27	20	21	17	E
SE	7	5	16	9	30	16	40	22	41	25	38	25	34	24	28	21	22	17	NE
S	5	4	6	4	12	7	23	14	34	21	38	25	35	24	29	21	23	17	N
SW	8	5	7	4	9	6	16	10	30	19	44	28	51	32	43	28	33	22	NW
W	8	6	7	5	9	6	14	9	25	16	42	27	55	34	49	31	37	25	W
NW	7	5	7	4	9	6	13	9	20	14	31	21	42	28	40	27	31	21	SW
N	6	4	8	5	10	6	14	10	19	14	23	17	25	19	24	19	19	16	S
Group E																			
NE	10	6	23	12	30	16	30	18	30	20	30	21	28	21	23	18	18	14	SE
E	11	6	28	15	42	22	43	24	39	24	36	24	32	23	25	19	19	15	E
SE	8	5	20	11	35	19	42	24	41	25	38	25	33	23	26	20	20	16	NE
S	4	3	6	4	15	9	28	17	38	24	39	26	34	24	27	20	20	16	N
SW	6	4	7	4	10	6	19	12	35	22	49	31	52	33	41	27	30	21	NW
W	7	5	7	4	10	6	16	11	30	20	48	31	57	36	47	30	34	23	W
NW	6	4	6	4	10	6	15	10	23	16	36	24	45	30	38	26	28	20	SW
N	6	4	8	5	11	7	16	11	21	15	24	18	26	20	23	18	18	15	S
Group F																			
NE	9	7	14	9	21	12	25	15	27	17	29	19	28	20	26	19	23	17	SE
E	10	8	17	10	28	15	35	19	37	22	37	23	35	23	31	22	26	19	E
SE	10	7	13	8	22	12	31	17	36	21	37	23	35	23	32	22	27	19	NE
S	9	7	7	5	10	6	17	10	26	16	32	20	33	22	31	22	27	19	N
SW	12	9	10	6	9	6	13	8	22	14	33	21	42	27	42	27	37	25	NW
W	14	9	11	7	10	6	12	8	19	12	31	20	43	27	46	29	41	27	W
NW	12	8	9	6	9	6	11	8	16	11	24	16	33	22	36	24	33	23	SW
N	8	7	8	6	9	6	12	8	15	11	19	14	22	17	23	18	21	17	S

SOURCE: Adapted from Carrier "System Design Manual," 1970.
NOTE: To convert temperature differentials in °F to °C, multiply by 5/9.

where Sa Δt = sol-air equivalent temperature differential, °F (°C). Tables 12.4.13 and 12.4.14.

Heat Transmission

Heat gain through exterior walls and roofs is highly variable and is caused by a combination of solar heat and the temperature difference between indoor and outdoor air. Total heat flow is affected by the type and weight of the construction, exposure, sun time, latitude, and design conditions. The equivalent temperature difference from Tables 12.4.13 and 12.4.14 combined with the appropriate heat-transfer coefficients U for the construction, Tables 12.4.15 and 12.4.16, permits estimating the total heat flow through walls and roofs.

Heat flow through interior construction is essentially constant and may be determined by using the actual temperatures on either side of the construction and the appropriate U values.

Solar Heat Gain

Heat gain through glass is a combination of the heat gains by conduction and solar radiation. Both factors are highly variable and one or both are affected by the type of glass, shading, day of the year, time of day, location and orientation angle to the sun, and outside and inside temperatures.

For glass:

$$Q_{glass} = Q_u + Q_{solar} \tag{12.4.12}$$
$$= A_1 U(t_o - t_i) + A_2(MSHGF)(CLF)(SC) \tag{12.4.13}$$

where A_1 = area of total glass, ft² (m²); A_2 = area of sunlit glass, ft² (m²); U = heat-transfer coefficient (Table 12.4.16); t_0 = outside design temperature, °F (°C) (Fig. 12.4.4); t_i = inside design temperature, °F (°C) (Table 12.4.2); MSHGF = maximum solar heat gain factor (Table 12.4.17); CLF = cooling load factor (Tables 12.4.18, 12.4.19, and 12.4.20); SC = shading coefficient (Tables 12.4.21, 12.4.22, 12.4.23, and 12.4.24).

Note that the data in Table 12.4.17 is for vertical and horizontal glass; the data in Tables 12.4.18 to 12.4.24 addresses the effect of heat storage within the space.

Internal Heat Gains

Internal heat gains are primarily from lighting, equipment, and occupants. They are independent of outdoor temperature and solar effect and include both sensible and latent heat.

$$\begin{aligned} Q_{int} &= \text{internal heat gains} \\ &= Q_{people} + Q_{lighting} + Q_{equipment} \end{aligned} \tag{12.4.14}$$

In some applications such as general office space, internal heat gains are relatively constant; in others where equipment cycles and/or occupancy is not constant, these loads will vary widely.

Heat gains from people vary with the activity and the proportion of men and women. The proportion of sensible latent heat varies with the temperature of the space. Table 12.4.25 lists adjusted heat gains from people for different activities and typical applications.

Lighting heat gains are based on the total electrical wattage of the operating lights, including ballasts, converted to BTU/hr. Note that under some circumstances the entire electrical input to the lights may not immediately become part of the cooling load. This depends on the type and arrangement of the lights, type of air supply and return, space furnishings, the thermal characteristics of the space, and length of time the lights will be on. Unless all of these can be forseen with a high degree of certainty, and lighting represents a significant portion of the cooling load, most designers will use the full electrical input of the operating lights as the lighting load.

Representative heat gains at various lighting levels for the most common lighting fixtures are shown in Fig. 12.4.7. Curves A, B, and C are drawn for fixtures with newer electronic ballasts and high-efficiency lamps. For a given footcandle level of illumination, they use about 40

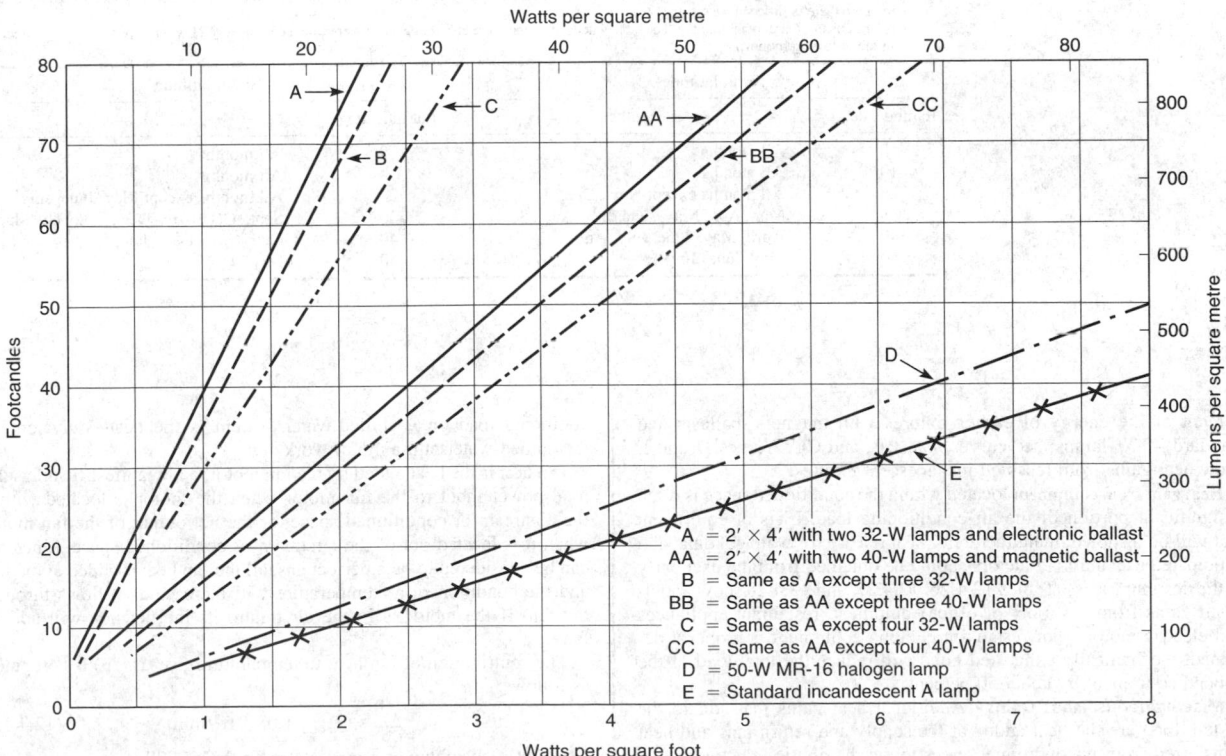

Fig. 12.4.7   Heat gain from typical lighting fixtures. *(Courtesy: S. Ricketts.)*

Table 12.4.14   Total Equivalent Temperature Differentials for Calculating Heat Gain through Flat Roofs

Description of roof construction*·†	Wt., lb/ft²	kg/m²	U value Btu/(h·ft²·°F)	W/(m²·K)
2 in (50 mm) insulation + steel siding	7.8	38.06	0.125	0.710
2 in (50 mm) insulation + 1 in (25 mm) wood‡	8.5	41.48	0.122	0.693
2 in (50 mm) insulation + 2.5 in (62.5 mm) wood‡	13.1	63.93	0.117	0.664
2 in (50 mm) insulation + 4 in (100 mm) wood‡	17.8	88.86	0.113	0.642
2 in (50 mm) insulation + 2 in (50 mm) h.w. concrete	28.8	140.54	0.122	0.693
4 in (100 mm) l.w. concrete	17.8	86.86	0.213	1.209
2 in (50 mm) insulation + 4 in (100 mm) h.w. concrete	52.1	254.25	0.120	0.681
2 in (50 mm) insulation + 6 in (150 mm) h.w. concrete	75.4	367.95	0.117	0.664

* Includes outside surface resistance, ½ in (12.5 mm) membrane and ⅜ in (9 mm) felt on the top and inside surface resistance on the bottom.
† Dark roof (D) = 0.30; light roof (L) = 0.15.
‡ Nominal thickness of wood.

Explanation: Total heat transmission from solar radiation and temperature difference between outdoor and room air, Btu/(h)(ft²)(W/m²) of roof area = equivalent temperature differential from above table × heat-transmission coefficient for summer, Btu/(h·ft²·°F) [W/(m²·K)].

Application. These values may be used for all normal air conditioning estimated; usually without correction (except as noted below in latitude 0 to 50° north or south when the load is calculated for the hottest weather.

Corrections. The values in the table were calculated for an inside temperature of 75°F (24°C) and an outdoor maximum temperature of 95°F (35°C) with an outdoor daily range of 21°F (12°C). The table remains approximately correct for other outdoor maximum (93 to 102°F) (34 to 39°C) and other outdoor daily ranges (16 to 34°F) (9 to 19°C) provided the outdoor daily average temperature remains approximately 85°F (30°C). If the room air temperature is different from 75°F (24°C) and/or the outdoor daily average temperature is different from 85°F (30°C), the following rules can be applied:

1. For room air temperature less than 75°F (24°C) add the difference between 75°F (24°C) and room air temperature; if greater than 75°F (24°C), subtract the difference.
2. For outdoor daily average temperature less than 85°F (30°C) subtract the difference between 85°F (30°C) and the daily average temperature, if greater than 85°F (30°C), add the difference.

Attics or other spaces between the roof and ceiling. If the ceiling is insulated and a fan is used for positive ventilation in the space between the ceiling and roof, the total temperature differential for calculating the room load may be decreased by 25 percent. If the attic space contains a return duct or other air plenum, care should be taken in determining the portion of the heat gain that reaches the ceiling.

Light color. Credit should not be taken for light-colored roofs except where the permanence of light color is established by experience as in rural areas or where there is little smoke.

For solar transmission in other months. The table values of temperature differentials that were calculated for July 21 will be approximately correct for a roof in the following months.

North latitude Latitude, deg	Months	South latitude Latitude, deg	Months
0	All months	0	All months
10	All months	10	All months
20	All months except Nov., Dec., Jan.	20	All months except May, June, July
30	Mar., Apr., May, June, July, Aug., Sept.	30	Sept., Oct., Nov., Dec., Jan., Feb., March
40	April, May, June, July, Aug.	40	Oct., Nov., Dec., Jan., Feb.
50	May, June, July	50	Nov., Dec., Jan.

percent of the energy of earlier fixtures with magnetic ballasts and standard 40-W lamps; see curves AA, BB, and CC. Curves D and E show heat gains from recessed incandescent fixtures.

Heat gain from equipment located within the conditioned space is often a significant portion of the air-conditioning load. Lists of equipment that will be installed, manufacturers' heat release (or wattage) data, and anticipated use and load factors should be obtained from the user early in the design process. Tables 12.4.26, 12.4.27, and 12.4.28 show anticipated gains from a variety of office equipment, restaurant appliances and electric motors. For restaurant cooking equipment, a properly designed, mechanically exhausted hood will reduce the total load of the hooded equipment by about 50 percent.

Miscellaneous Heat Gains Additional heat gains that add to the system load are the heat added at the supply and return fans and heat transferred from unconditioned areas to the distribution ductwork. An additional load on a chilled-water system is the heat equivalent of the chilled-water pump motor work.

Fan heat is the heat added by the fan motor, and manifests itself as the total power input to the fan motor when the motor is located within the airstream or conditioned space, or the net output of the fan motor when it is located out of the airstream or conditioned space. Since air can be considered to be a perfect gas, all the fan heat is added at the fan and the conditioned air temperature will increase as it flows through the fan. If the motor is in the airstream, its inefficiency will add to fan heat.

The contribution of fan heat to temperature rise Δt (°F) is computed as follows:

$$\Delta t = Q_{motor}/1.08 \quad \text{ft}^3/\text{min} \qquad (12.4.15)$$

where Q_{motor} is obtained from data in Table 12.4.28.

				Sun time															
		A.M.						P.M.											
8		10		12		2		4		6		8		10		12			
D	L	D	L	D	L	D	L	D	L	D	L	D	L	D	L	D	L	λ	δ
								Light construction roofs exposed to sun											
24	8	61	29	88	46	96	53	81	46	48	30	10	8	2	2	−3	−3	0.99	1
8	0	41	18	72	36	90	48	88	49	65	38	30	19	9	7	1	0	0.93	2
1	−2	19	6	43	20	65	33	76	41	72	40	53	31	33	20	18	11	0.68	4
								Medium construction roofs exposed to sun											
6	1	13	4	28	12	45	22	58	30	63	34	56	31	43	25	32	18	0.48	5
2	−2	23	9	49	23	70	36	79	43	71	40	49	29	28	17	15	9	0.73	3
1	−3	28	11	59	28	82	43	88	48	74	42	44	27	19	12	6	4	0.87	3
								Heavy construction roofs exposed to sun											
7	2	15	6	30	13	46	23	58	30	61	33	54	30	41	23	31	17	0.45	5
15	7	17	7	25	11	36	17	46	23	51	27	50	27	43	24	36	20	0.30	6

Table 12.4.15 Descriptions of Wall Construction

Group	Components	Weight lb/ft²	Weight kg/m²	U value Btu/(h·ft²·°F)	U value W/(m²·K)
A	1 in (2.5 cm) stucco + 4 in (10 cm) l.w. concrete block + air space	28.6	139.63	0.267	1.516
	1 in (2.5 cm) stucco + air space + 2 in (5 cm) insulation	16.3	79.58	0.106	0.601
B	1 in (2.5 cm) stucco + 4 in (10 cm) common brick	55.9	272.90	0.393	2.232
	1 in (2.5 cm) stucco + 4 in (10 cm) l.w. concrete	62.5	305.13	0.481	2.731
C	4 in (10 cm) face brick + 4 in (10 cm) l.w. concrete block + 1 in (2.5 cm) insulation	62.5	305.13	0.158	0.897
	1 in (2.5 cm) stucco + 4 in (10 cm) h.w. concrete + 2 in (5 cm) insulation	62.9	307.08	0.114	0.647
D	1 in (2.5 cm) stucco + 8 in (20 cm) l.w. concrete block + 1 in (2.5 cm) insulation	41.4	202.11	0.141	0.801
	1 in (2.5 cm) stucco + 2 in (5 cm) insulation + 4 in (10 cm) h.w. concrete block	36.6	178.68	0.111	0.630
E	4 in (10 cm) face brick + 3 in (10 cm) l.w. concrete block	62.2	303.66	0.333	1.891
	1 in (2.5 cm) stucco + 8 in (20 cm) h.w. concrete block	56.6	276.32	0.349	1.982
F	4 in (10 cm) face brick + 4 in (10 cm) common brick	89.5	436.94	0.360	2.044
	4 in (10 cm) face brick + 2 in (5 cm) insulation + 4 in (10 cm) l.w. concrete block	62.5	305.13	0.103	0.585

Table 12.4.16   Coefficients of Transmission U of Windows, Skylights, and Light Transmitting Partitions
These values are for heat transfer from air to air, Btu/(h · ft² · °F).

Part A—Vertical panels (exterior windows, sliding patio doors, and partitions)
—flat glass, glass block, and plastic sheet

Part B—Horizontal panels (skylights)—flat glass, glass block, and plastic domes

Description	Exterior[a]		Interior	Description	Exterior[a]		Interior[l]
	Winter	Summer			Winter[i]	Summer[j]	
Flat glass[b]				Flat glass[e]			
Single glass	1.10	1.04	0.73	Single glass	1.23	0.83	0.96
Insulating glass—double[c]				Insulating glass—doubles[c]			
0.1875-in airspace[d]	0.62	0.65	0.51	0.1875-in airspace[d]	0.70	0.57	0.62
0.25-in airspace[d]	0.58	0.61	0.49	0.25-in airspace[d]	0.65	0.54	0.59
0.5-in airspace[e]	0.49	0.56	0.46	0.5-in airspace[c]	0.59	0.49	0.56
0.5-in airspace, low-emittance coating[f]				0.5-in airspace, low-emittance coating[f]			
e = 0.20	0.32	0.38	0.32	e = 0.20	0.48	0.36	0.39
e = 0.40	0.38	0.45	0.38	e = 0.40	0.52	0.42	0.45
e = 0.60	0.43	0.51	0.42	e = 0.60	0.56	0.46	0.50
Insulating glass—triple[c]				Glass block[h]			
0.25-in airspaces[d]	0.39	0.44	0.38	11 × 11 × 3 in thick with cavity divider	0.53	0.35	0.44
0.5-in airspaces[g]	0.31	0.39	0.30	12 × 12 × 4 in thick with cavity divider	0.51	0.34	0.42
Storm windows							
1-in to 4-in airspace[d]	0.50	0.50	0.44				
Plastic sheet				Plastic domes[k]			
Single glazed				Single-walled	1.15	0.80	—
0.125-in thick	1.06	0.98	—	Double-walled	0.70	0.46	—
0.25-in thick	0.96	0.89	—				
0.5-in thick	0.81	0.76	—				

Part C—adjustment factors for various window and sliding patio door types
(multiply U values in parts A and B by these factors)

Description	Single glass	Double or triple glass	Storm windows
Insulating unit—double[c]			
0.25-in airspace[d]	0.55	0.56	—
0.5-in airspace[c]	0.43	0.45	—
Glass block[h]			
6 × 6 × 4 in thick	0.60	0.57	0.46
8 × 8 × 4 in thick	0.56	0.54	0.44
—with cavity divider	0.48	0.46	0.38
12 × 12 × 4 in thick	0.52	0.50	0.41
—with cavity divider	0.44	0.42	0.36
12 × 12 × 2 in thick	0.60	0.57	0.46

Part C (continued):

Description	Single glass	Double or triple glass	Storm windows
Windows			
All Glass[l]	1.00	1.00	1.00
Wood sash—80% glass	0.90	0.95	0.90
Wood sash—60% glass	0.80	0.85	0.80
Metal sash—80% glass	1.00	1.20[m]	1.20[m]
Sliding patio doors			
Wood frame	0.95	1.00	—
Metal frame	1.00	1.10[m]	—

SOURCE: Abstracted from ASHRAE "Handbook of Fundamentals," 1977, with permission.
[a] See Part C for adjustment for various window and sliding patio door types.
[b] Emittance of uncooled glass surface = 0.84.
[c] Double and triple refer to the number of lights of glass.
[d] 0.125-in glass.
[e] 0.25-in glass.
[f] Coating on either glass surface facing airspace; all other glass surfaces uncoated.
[g] Window design: 0.25-in glass−0.125-in glass−0.25-in glass.
[h] Dimensions are nominal.
[i] For heat flow up.
[j] For heat flow down.
[k] Based on area of opening, not total surface area.
[l] Refers to windows with negligible opaque area.
[m] Values will be less than these when metal sash and frame incorporate thermal breaks. In some thermal break designs, U values will be equal to or less than those for the glass. Window manufacturers should be consulted for specific data.

Table 12.4.17   Maximum Solar Heat Gain Factor (MSHGF), Btu/h · ft², for Sunlit Glass

0°N Lat

	N	NNE/ NNW	NE/ NW	ENE/ WNW	E/ W	ESE/ WSW	SE/ SW	SSE/ SSW	S	Hor.
Jan.	34	34	88	177	234	254	235	182	118	296
Feb.	36	39	132	205	245	247	210	141	67	306
Mar.	38	87	170	223	242	223	170	87	38	303
Apr.	71	134	193	224	221	184	118	38	37	284
May	113	164	203	218	201	154	80	37	37	265
June	129	173	206	212	191	140	66	37	37	255
July	115	164	201	213	195	149	77	38	38	260
Aug.	75	134	187	216	212	175	112	39	38	276
Sep.	40	84	163	213	231	213	163	84	40	293
Oct.	37	40	129	199	236	238	202	135	66	299
Nov.	35	35	88	175	230	250	230	179	117	293
Dec.	34	34	71	164	226	253	240	196	138	288

8°N Lat

	N	NNE/ NNW	NE/ NW	ENE/ WNW	E/ W	ESE/ WSW	SE/ SW	SSE/ SSW	S	Hor.
Jan.	32	32	71	163	224	250	242	203	162	275
Feb.	34	34	114	193	239	248	219	165	110	294
Mar.	37	67	156	215	241	230	184	110	55	300
Apr.	44	117	184	221	225	195	134	53	39	289
May	74	146	198	220	209	167	97	39	38	277
June	90	155	200	217	200	141	82	39	39	269
July	77	145	195	215	204	162	93	40	39	272
Aug.	47	117	179	214	216	186	128	51	41	282
Sep.	38	66	149	205	230	219	176	107	56	290
Oct.	35	35	112	187	231	239	211	160	108	288
Nov.	33	33	71	161	220	245	233	200	160	273
Dec.	31	31	55	149	215	246	247	215	179	265

16°N Lat

	N	NNE/ NNW	NE/ NW	ENE/ WNW	E/ W	ESE/ WSW	SE/ SW	SSE/ SSW	S	Hor.
Jan.	30	30	55	147	210	244	251	223	199	248
Feb.	33	33	96	180	231	247	233	188	154	275
Mar.	35	53	140	205	239	235	197	138	93	291
Apr.	39	99	172	215	227	204	150	77	45	289
May	52	132	189	218	215	179	115	45	41	282
June	66	142	194	217	207	167	99	41	41	277
July	55	132	187	214	210	174	111	44	42	277
Aug.	41	100	168	209	219	196	143	74	46	282
Sep.	36	50	134	196	227	224	191	134	93	282
Oct.	33	33	95	174	223	237	225	183	150	270
Nov.	30	30	55	145	206	241	247	220	196	246
Dec.	29	29	41	132	198	241	254	233	212	234

24°N Lat

	N	NNE/ NNW	NE/ NW	ENE/ WNW	E/ W	ESE/ WSW	SE/ SW	SSE/ SSW	S	Hor.
Jan.	27	27	41	128	190	240	253	241	227	214
Feb.	30	30	80	165	220	244	243	213	192	249
Mar.	34	45	124	195	234	237	214	168	137	275
Apr.	37	88	159	209	228	212	169	107	75	283
May	43	117	178	214	218	190	132	67	46	282
June	55	127	184	214	212	179	117	55	43	279
July	45	116	176	210	213	185	129	65	46	278
Aug.	38	87	156	203	220	204	162	103	72	277
Sep.	35	42	119	185	222	225	206	163	134	266
Oct.	31	31	79	159	211	237	235	207	187	244
Nov.	27	27	42	126	187	236	249	237	224	213
Dec.	26	26	29	112	180	234	247	247	237	199

32°N Lat

	N	NNE/ NNW	NE/ NW	ENE/ WNW	E/ W	ESE/ WSW	SE/ SW	SSE/ SSW	S	Hor.
Jan.	24	24	29	105	175	229	249	250	246	176
Feb.	27	27	65	149	205	242	248	232	221	217
Mar.	32	37	107	183	227	237	227	195	176	252
Apr.	36	80	146	200	227	219	187	141	115	271
May	38	111	170	208	220	199	155	99	74	277
June	44	122	176	208	214	189	139	83	60	276
July	40	111	167	204	215	194	150	96	72	273
Aug.	37	79	141	195	219	210	181	136	111	265
Sep.	33	35	103	173	215	227	218	189	171	244
Oct.	28	28	63	143	195	234	239	225	215	213
Nov.	24	24	29	103	173	225	245	246	243	175
Dec.	22	22	22	84	162	218	246	252	252	158

40°N Lat

	N (Shade)	NNE/ NNW	NE/ NW	ENE/ WNW	E/ W	ESE/ WSW	SE/ SW	SSE/ SSW	S	Hor.
Jan.	20	20	20	74	154	205	241	252	254	133
Feb.	24	24	50	129	186	234	246	244	241	180
Mar.	29	29	93	169	218	238	236	216	206	223
Apr.	34	71	140	190	224	223	203	170	154	252
May	37	102	165	202	220	208	175	133	113	265
June	48	113	172	205	216	199	161	116	95	267
July	38	102	163	198	216	203	170	129	109	262
Aug.	35	71	135	185	216	214	196	165	149	247
Sep.	30	30	87	160	203	227	226	209	200	215
Oct.	25	25	49	123	180	225	238	236	234	177
Nov.	20	20	20	71	151	201	237	248	250	132
Dec.	18	18	18	60	135	188	232	249	253	113

48°N Lat

	N (Shade)	NNE/ NNW	NE/ NW	ENE/ WNW	E/ W	ESE/ WSW	SE/ SW	SSE/ SSW	S	Hor.
Jan.	15	15	15	53	118	175	216	239	245	85
Feb.	20	20	36	103	168	216	242	249	250	138
Mar.	26	26	80	154	204	234	239	232	228	188
Apr.	31	61	132	180	219	225	215	194	186	226
May	35	97	158	200	218	214	192	163	150	247
June	46	110	165	204	215	206	180	148	134	252
July	37	96	156	196	214	209	187	158	146	244
Aug.	33	61	128	174	211	216	208	188	180	223
Sep.	27	27	72	144	191	223	228	223	230	182
Oct.	21	21	35	96	161	207	213	241	242	136
Nov.	15	15	15	52	115	172	212	234	240	85
Dec.	13	13	13	36	91	156	195	224	233	65

56°N Lat

	N (Shade)	NNE/ NNW	NE/ NW	ENE/ WNW	E/ W	ESE/ WSW	SE/ SW	SSE/ SSW	S	Hor.
Jan.	10	10	10	21	74	126	169	194	205	40
Feb.	16	16	21	71	139	184	223	239	244	91
Mar.	22	22	65	136	185	224	238	241	241	149
Apr.	28	58	123	173	211	223	223	213	210	195
May	36	99	149	195	215	218	206	187	181	222
June	53	111	160	199	213	213	196	174	168	231
July	37	98	147	192	211	214	201	183	177	221
Aug.	30	56	119	165	203	216	215	206	203	193
Sep.	23	23	58	126	171	211	227	230	231	144
Oct.	16	16	20	68	132	176	213	229	234	91
Nov.	10	10	10	21	72	122	165	190	200	40
Dec.	7	7	7	7	47	92	135	159	171	23

60°N Lat

	N (Shade)	NNE/ NNW	NE/ NW	ENE/ WNW	E/ W	ESE/ WSW	SE/ SW	SSE/ SSW	S	Hor.
Jan.	7	7	7	7	46	88	130	152	164	21
Feb.	13	13	13	58	118	168	204	225	231	68
Mar.	20	20	56	125	173	215	234	241	242	128
Apr.	27	59	118	168	206	222	225	220	218	178
May	43	98	149	192	212	220	211	198	194	208
June	58	110	162	197	213	215	202	186	181	217
July	44	97	147	189	208	215	206	193	190	207
Aug.	28	57	114	161	199	214	217	213	211	176
Sep.	21	21	50	145	160	202	222	229	231	123
Oct.	14	14	14	56	111	159	193	215	221	67
Nov.	7	7	7	7	45	86	127	148	160	22
Dec.	4	4	4	4	16	51	76	100	107	9

64°N Lat

	N (Shade)	NNE/ NNW	NE/ NW	ENE/ WNW	E/ W	ESE/ WSW	SE/ SW	SSE/ SSW	S	Hor.
Jan.	3	3	3	3	15	45	67	89	96	8
Feb.	11	11	11	43	89	144	177	202	210	45
Mar.	18	18	47	113	159	203	226	236	239	105
Apr.	25	59	113	163	201	219	225	225	224	160
May	48	97	150	189	211	220	215	207	204	192
June	62	114	162	193	213	216	208	196	193	203
July	49	96	148	186	207	215	211	202	200	192
Aug.	27	58	109	157	193	211	217	217	217	159
Sep.	19	19	43	103	148	189	213	224	227	101
Oct.	11	11	11	40	83	135	167	191	199	46
Nov.	4	4	4	4	15	44	66	87	93	8
Dec.	0	0	0	0	1	5	11	14	15	1

SOURCE: Abstracted from ASHRAE "Handbook of Fundamentals," 1989, with permission.

Table 12.4.18   Cooling Load Factors (CLF) for Glass without Interior Shading, in North Latitude Spaces Having Carpeted Floors

Dir.	Room mass	Solar time																							
		0100	0200	0300	0400	0500	0600	0700	0800	0900	1000	1100	1200	1300	1400	1500	1600	1700	1800	1900	2000	2100	2200	2300	2400
N	L	0.00	0.00	0.00	0.00	0.01	0.64	0.73	0.74	0.81	0.88	0.95	0.98	0.98	0.94	0.88	0.79	0.79	0.55	0.31	0.12	0.04	0.02	0.01	0.00
	M	0.03	0.02	0.02	0.02	0.02	0.64	0.69	0.69	0.77	0.84	0.91	0.94	0.95	0.91	0.86	0.79	0.79	0.56	0.32	0.16	0.10	0.07	0.05	0.04
	H	0.10	0.09	0.08	0.07	0.07	0.62	0.64	0.64	0.71	0.77	0.83	0.87	0.88	0.85	0.81	0.75	0.76	0.55	0.34	0.22	0.17	0.15	0.13	0.11
NE	L	0.00	0.00	0.00	0.00	0.01	0.51	0.83	0.88	0.72	0.47	0.33	0.27	0.24	0.23	0.20	0.18	0.14	0.09	0.03	0.01	0.00	0.00	0.00	0.00
	M	0.01	0.01	0.01	0.01	0.01	0.50	0.78	0.82	0.67	0.44	0.32	0.28	0.26	0.24	0.22	0.19	0.15	0.11	0.05	0.03	0.02	0.02	0.01	0.01
	H	0.03	0.03	0.03	0.02	0.03	0.47	0.71	0.72	0.59	0.40	0.30	0.27	0.26	0.25	0.23	0.20	0.17	0.13	0.08	0.06	0.05	0.05	0.04	0.04
E	L	0.01	0.00	0.00	0.00	0.00	0.42	0.76	0.91	0.90	0.75	0.51	0.30	0.22	0.18	0.16	0.13	0.11	0.07	0.02	0.01	0.00	0.00	0.00	0.00
	M	0.01	0.01	0.00	0.00	0.01	0.41	0.72	0.86	0.84	0.71	0.48	0.30	0.24	0.21	0.18	0.16	0.13	0.09	0.04	0.03	0.02	0.01	0.01	0.00
	H	0.03	0.03	0.03	0.02	0.02	0.39	0.66	0.76	0.74	0.63	0.43	0.29	0.24	0.22	0.20	0.18	0.15	0.12	0.08	0.06	0.05	0.05	0.04	0.04
SE	L	0.00	0.00	0.00	0.00	0.00	0.27	0.58	0.81	0.93	0.93	0.81	0.59	0.37	0.27	0.21	0.18	0.14	0.09	0.03	0.01	0.00	0.00	0.00	0.00
	M	0.01	0.01	0.01	0.00	0.01	0.26	0.55	0.77	0.88	0.87	0.76	0.56	0.37	0.29	0.24	0.20	0.16	0.11	0.05	0.04	0.03	0.02	0.02	0.01
	H	0.04	0.04	0.03	0.03	0.03	0.26	0.51	0.69	0.78	0.78	0.68	0.51	0.35	0.29	0.25	0.22	0.19	0.15	0.09	0.08	0.07	0.06	0.05	0.05
S	L	0.01	0.00	0.00	0.01	0.00	0.07	0.15	0.23	0.39	0.62	0.82	0.94	0.93	0.80	0.59	0.38	0.26	0.16	0.06	0.02	0.01	0.00	0.00	0.00
	M	0.01	0.01	0.01	0.01	0.01	0.07	0.14	0.22	0.38	0.59	0.78	0.88	0.88	0.76	0.57	0.38	0.28	0.18	0.09	0.06	0.04	0.03	0.02	0.02
	H	0.05	0.05	0.04	0.04	0.03	0.09	0.15	0.21	0.35	0.54	0.70	0.79	0.79	0.69	0.52	0.37	0.29	0.21	0.13	0.10	0.09	0.08	0.07	0.06
SW	L	0.00	0.00	0.00	0.00	0.00	0.04	0.09	0.13	0.16	0.19	0.23	0.39	0.62	0.82	0.94	0.94	0.81	0.54	0.19	0.07	0.03	0.01	0.00	0.00
	M	0.02	0.02	0.01	0.01	0.01	0.05	0.09	0.13	0.16	0.19	0.22	0.38	0.60	0.78	0.89	0.89	0.77	0.52	0.20	0.10	0.07	0.05	0.04	0.03
	H	0.07	0.06	0.05	0.05	0.04	0.07	0.11	0.14	0.16	0.18	0.21	0.35	0.55	0.71	0.80	0.79	0.69	0.48	0.20	0.14	0.11	0.10	0.08	0.07
W	L	0.00	0.00	0.00	0.00	0.00	0.03	0.07	0.10	0.13	0.15	0.16	0.18	0.31	0.55	0.78	0.92	0.93	0.73	0.25	0.10	0.04	0.01	0.00	0.00
	M	0.02	0.02	0.01	0.01	0.01	0.04	0.07	0.10	0.13	0.14	0.16	0.17	0.30	0.53	0.74	0.87	0.88	0.69	0.24	0.12	0.07	0.05	0.04	0.03
	H	0.06	0.05	0.05	0.04	0.04	0.06	0.09	0.11	0.13	0.15	0.16	0.17	0.28	0.49	0.67	0.78	0.79	0.62	0.23	0.14	0.11	0.09	0.08	0.07
NW	L	0.00	0.00	0.00	0.00	0.00	0.04	0.10	0.14	0.17	0.20	0.22	0.23	0.24	0.31	0.53	0.78	0.92	0.81	0.28	0.10	0.07	0.05	0.04	0.03
	M	0.02	0.02	0.01	0.01	0.01	0.06	0.11	0.13	0.17	0.19	0.21	0.22	0.23	0.30	0.52	0.75	0.88	0.77	0.26	0.12	0.10	0.09	0.08	0.07
	H	0.06	0.05	0.05	0.04	0.04	0.08	0.10	0.14	0.17	0.19	0.20	0.21	0.22	0.28	0.48	0.68	0.79	0.69	0.23	0.14	0.10	0.09	0.08	0.08
Hor.	L	0.02	0.02	0.01	0.01	0.01	0.08	0.24	0.45	0.64	0.80	0.91	0.97	0.97	0.91	0.80	0.64	0.44	0.23	0.08	0.03	0.01	0.00	0.00	0.00
	M	0.02	0.02	0.01	0.01	0.01	0.08	0.24	0.43	0.60	0.75	0.86	0.92	0.92	0.87	0.77	0.63	0.45	0.26	0.12	0.07	0.05	0.04	0.03	0.02
	H	0.07	0.06	0.05	0.05	0.04	0.11	0.25	0.41	0.56	0.68	0.77	0.83	0.83	0.80	0.71	0.59	0.44	0.28	0.17	0.13	0.11	0.10	0.09	0.08

SOURCE: Abstracted from ASHRAE "Handbook of Fundamentals," 1989, with permission.
Values for nominal 15 ft × 15 ft × 10 ft high space, with ceiling, and 50% or less glass in exposed surface at listed orientation.
L = lightweight construction.
M = mediumweight construction.
H = heavyweight construction.

Table 12.4.19 Cooling Load Factors (CLF) for Glass without Interior Shading, in North Latitude Spaces Having Uncarpeted Floors

Dir.	Room mass	0100	0200	0300	0400	0500	0600	0700	0800	0900	1000	1100	1200	1300	1400	1500	1600	1700	1800	1900	2000	2100	2200	2300	2400
N	L	0.00	0.00	0.00	0.00	0.01	0.64	0.73	0.74	0.81	0.88	0.95	0.98	0.98	0.94	0.88	0.79	0.79	0.55	0.31	0.12	0.04	0.02	0.01	0.00
	M	0.12	0.09	0.07	0.06	0.05	0.33	0.45	0.53	0.61	0.69	0.76	0.82	0.85	0.86	0.85	0.81	0.80	0.70	0.60	0.43	0.32	0.24	0.19	0.15
	H	0.24	0.21	0.19	0.18	0.16	0.43	0.48	0.51	0.56	0.61	0.66	0.71	0.73	0.74	0.73	0.71	0.71	0.62	0.52	0.42	0.36	0.32	0.29	0.26
NE	L	0.00	0.00	0.00	0.00	0.01	0.51	0.83	0.88	0.72	0.47	0.33	0.27	0.24	0.23	0.20	0.18	0.14	0.09	0.03	0.01	0.00	0.00	0.00	0.00
	M	0.03	0.02	0.02	0.02	0.02	0.24	0.45	0.57	0.58	0.49	0.41	0.36	0.32	0.29	0.27	0.24	0.21	0.17	0.13	0.10	0.07	0.06	0.05	0.04
	H	0.08	0.07	0.07	0.06	0.06	0.27	0.43	0.49	0.45	0.37	0.32	0.29	0.28	0.27	0.26	0.24	0.22	0.19	0.16	0.14	0.12	0.11	0.10	0.09
E	L	0.00	0.00	0.00	0.00	0.00	0.42	0.76	0.91	0.90	0.75	0.51	0.30	0.22	0.18	0.16	0.13	0.11	0.07	0.02	0.01	0.00	0.00	0.00	0.00
	M	0.03	0.02	0.02	0.02	0.01	0.20	0.41	0.57	0.65	0.64	0.55	0.44	0.36	0.31	0.26	0.23	0.19	0.16	0.12	0.09	0.07	0.06	0.04	0.04
	H	0.08	0.08	0.07	0.06	0.06	0.24	0.40	0.50	0.53	0.50	0.41	0.33	0.30	0.28	0.26	0.24	0.22	0.19	0.16	0.14	0.13	0.11	0.10	0.09
SE	L	0.00	0.00	0.00	0.00	0.00	0.27	0.58	0.81	0.93	0.93	0.81	0.59	0.37	0.27	0.21	0.18	0.14	0.09	0.03	0.01	0.00	0.00	0.00	0.00
	M	0.04	0.03	0.02	0.02	0.02	0.13	0.31	0.48	0.62	0.69	0.69	0.61	0.50	0.41	0.35	0.30	0.25	0.20	0.15	0.12	0.09	0.07	0.06	0.05
	H	0.10	0.09	0.08	0.08	0.07	0.18	0.32	0.45	0.53	0.56	0.54	0.47	0.39	0.35	0.32	0.29	0.26	0.23	0.19	0.17	0.15	0.14	0.12	0.11
S	L	0.00	0.00	0.00	0.00	0.00	0.07	0.15	0.23	0.39	0.62	0.82	0.94	0.93	0.80	0.59	0.38	0.26	0.16	0.06	0.02	0.01	0.00	0.00	0.00
	M	0.05	0.04	0.04	0.03	0.02	0.05	0.09	0.14	0.24	0.38	0.53	0.65	0.72	0.71	0.63	0.52	0.42	0.33	0.24	0.18	0.14	0.11	0.09	0.07
	H	0.13	0.12	0.10	0.09	0.09	0.11	0.14	0.17	0.25	0.36	0.47	0.55	0.58	0.56	0.49	0.41	0.36	0.30	0.25	0.21	0.19	0.17	0.16	0.14
SW	L	0.00	0.00	0.00	0.00	0.00	0.04	0.09	0.13	0.16	0.19	0.23	0.39	0.62	0.82	0.94	0.94	0.81	0.54	0.19	0.07	0.03	0.01	0.00	0.00
	M	0.08	0.07	0.05	0.04	0.03	0.05	0.07	0.09	0.12	0.15	0.17	0.26	0.40	0.54	0.66	0.73	0.72	0.61	0.43	0.31	0.23	0.17	0.13	0.10
	H	0.15	0.14	0.12	0.11	0.10	0.11	0.12	0.14	0.15	0.17	0.18	0.26	0.37	0.48	0.56	0.59	0.57	0.47	0.33	0.27	0.23	0.21	0.19	0.17
W	L	0.00	0.00	0.00	0.00	0.00	0.03	0.06	0.08	0.10	0.12	0.13	0.15	0.18	0.33	0.45	0.63	0.71	0.73	0.25	0.10	0.04	0.01	0.01	0.00
	M	0.08	0.07	0.05	0.04	0.03	0.05	0.07	0.09	0.11	0.13	0.15	0.16	0.21	0.31	0.45	0.53	0.64	0.67	0.46	0.33	0.24	0.18	0.14	0.11
	H	0.14	0.13	0.12	0.11	0.10	0.10	0.11	0.12	0.13	0.14	0.15	0.16	0.21	0.31	0.45	0.54	0.58	0.52	0.46	0.33	0.24	0.19	0.18	0.16
NW	L	0.00	0.00	0.00	0.00	0.00	0.04	0.10	0.12	0.13	0.14	0.16	0.17	0.20	0.23	0.26	0.33	0.46	0.81	0.33	0.10	0.04	0.02	0.01	0.00
	M	0.08	0.06	0.05	0.04	0.03	0.05	0.07	0.10	0.13	0.15	0.16	0.19	0.20	0.24	0.33	0.51	0.64	0.66	0.46	0.32	0.23	0.17	0.13	0.10
	H	0.13	0.12	0.11	0.10	0.09	0.10	0.12	0.13	0.15	0.16	0.17	0.18	0.19	0.23	0.33	0.46	0.55	0.53	0.33	0.25	0.21	0.18	0.16	0.15
Hor.	L	0.00	0.00	0.00	0.00	0.00	0.08	0.25	0.45	0.64	0.80	0.91	0.97	0.97	0.91	0.80	0.64	0.44	0.23	0.08	0.03	0.01	0.00	0.00	0.00
	M	0.07	0.06	0.05	0.04	0.03	0.06	0.14	0.26	0.40	0.53	0.64	0.73	0.78	0.80	0.77	0.70	0.59	0.45	0.33	0.24	0.19	0.14	0.11	0.09
	H	0.16	0.15	0.13	0.12	0.11	0.13	0.20	0.29	0.39	0.48	0.56	0.61	0.65	0.65	0.63	0.57	0.49	0.40	0.32	0.28	0.25	0.22	0.20	0.18

Solar time

SOURCE: Abstracted from ASHRAE "Handbook of Fundamentals," 1989, with permission.
Values for nominal 15 ft × 15 ft × 10 ft high space, with ceiling, and 50% or less glass in exposed surface at listed orientation.
L = lightweight construction.
M = mediumweight construction.
H = heavyweight construction.

Table 12.4.20 Cooling Load Factors (CLF) for Glass with Interior Shading, North Latitudes (All Room Constructions)

Fenestration facing	Solar time																							
	0100	0200	0300	0400	0500	0600	0700	0800	0900	1000	1100	1200	1300	1400	1500	1600	1700	1800	1900	2000	2100	2200	2300	2400
N	0.08	0.07	0.06	0.06	0.07	0.73	0.66	0.65	0.73	0.80	0.86	0.89	0.89	0.86	0.82	0.75	0.78	0.91	0.24	0.18	0.15	0.13	0.11	0.10
NNE	0.03	0.03	0.02	0.02	0.03	0.64	0.77	0.62	0.42	0.37	0.37	0.37	0.36	0.35	0.32	0.28	0.23	0.17	0.08	0.07	0.06	0.05	0.04	0.04
NE	0.03	0.02	0.02	0.02	0.02	0.56	0.76	0.74	0.58	0.37	0.29	0.27	0.26	0.24	0.22	0.20	0.16	0.12	0.06	0.05	0.04	0.04	0.03	0.03
ENE	0.03	0.02	0.02	0.02	0.02	0.52	0.76	0.80	0.71	0.52	0.31	0.26	0.24	0.22	0.20	0.18	0.15	0.11	0.06	0.05	0.04	0.04	0.03	0.03
E	0.03	0.02	0.02	0.02	0.02	0.47	0.72	0.80	0.76	0.62	0.41	0.27	0.24	0.22	0.21	0.19	0.15	0.12	0.07	0.06	0.05	0.04	0.04	0.03
ESE	0.03	0.03	0.02	0.02	0.02	0.41	0.67	0.79	0.80	0.72	0.54	0.34	0.27	0.24	0.21	0.19	0.15	0.12	0.07	0.06	0.05	0.04	0.04	0.03
SE	0.03	0.03	0.02	0.02	0.02	0.30	0.57	0.74	0.81	0.79	0.68	0.49	0.33	0.28	0.25	0.22	0.18	0.13	0.08	0.07	0.06	0.05	0.04	0.04
SSE	0.04	0.03	0.03	0.03	0.03	0.12	0.31	0.54	0.72	0.81	0.81	0.71	0.54	0.38	0.32	0.27	0.22	0.16	0.09	0.08	0.07	0.06	0.05	0.04
S	0.04	0.03	0.03	0.03	0.03	0.09	0.16	0.23	0.38	0.58	0.75	0.83	0.80	0.68	0.50	0.35	0.27	0.19	0.11	0.09	0.08	0.07	0.06	0.05
SSW	0.05	0.04	0.04	0.03	0.03	0.09	0.14	0.18	0.22	0.27	0.43	0.63	0.78	0.84	0.80	0.66	0.46	0.25	0.13	0.11	0.09	0.08	0.07	0.06
SW	0.05	0.05	0.04	0.04	0.03	0.07	0.11	0.14	0.16	0.19	0.22	0.38	0.59	0.75	0.83	0.81	0.69	0.45	0.16	0.12	0.10	0.09	0.07	0.06
WSW	0.05	0.05	0.04	0.04	0.03	0.07	0.10	0.12	0.14	0.16	0.17	0.23	0.44	0.64	0.78	0.84	0.78	0.55	0.16	0.12	0.10	0.09	0.07	0.06
W	0.05	0.05	0.04	0.04	0.03	0.06	0.09	0.11	0.13	0.15	0.16	0.17	0.31	0.53	0.72	0.82	0.81	0.61	0.16	0.12	0.10	0.08	0.07	0.06
WNW	0.05	0.05	0.04	0.03	0.03	0.07	0.10	0.12	0.14	0.16	0.17	0.18	0.22	0.43	0.65	0.80	0.84	0.66	0.16	0.12	0.10	0.08	0.07	0.06
NW	0.05	0.04	0.04	0.03	0.03	0.07	0.11	0.14	0.17	0.19	0.20	0.21	0.22	0.30	0.52	0.73	0.82	0.76	0.17	0.12	0.10	0.08	0.07	0.06
NNW	0.05	0.05	0.04	0.03	0.03	0.11	0.17	0.22	0.26	0.30	0.32	0.33	0.34	0.34	0.39	0.61	0.82	0.76	0.17	0.12	0.10	0.08	0.07	0.06
Hor.	0.06	0.05	0.04	0.04	0.03	0.12	0.27	0.44	0.59	0.72	0.81	0.85	0.85	0.81	0.71	0.58	0.42	0.25	0.14	0.12	0.10	0.08	0.07	0.06

SOURCE: Abstracted from ASHRAE "Handbook of Fundamentals," 1989, with permission.

Table 12.4.21   Shading Coefficients for Single Glass and Insulating Glass*

			Shading coefficient	
Type of glass	Nominal thickness,[†] in (mm)	Solar trans.[†]	$b_0 = 4.0$	$b_0 = 3.0$
Regular sheet	3/32, 1/8 (2, 3)	0.87	1.00	1.00
Regular plate/float	1/4 (6)	0.80	0.95	0.97
	3/8 (9)	0.75	0.91	0.93
	1/2 (12)	0.71	0.88	0.91
Gray sheet	1/8 (3)	0.59	0.78	0.80
	3/16 (5)	0.74	0.90	0.92
	7/32 (6)	0.45	0.66	0.70
	7/32 (6)	0.71	0.88	0.90
	1/4	0.67	0.86	0.88
Heat-absorbing	3/16 (5)	0.52	0.72	0.75
plate/float§	1/4 (6)	0.47	0.70	0.74
	3/8 (9)	0.33	0.56	0.61
	1/2 (12)	0.24	0.50	0.57

Insulating glass*

		Solar trans.[†]		Shading coefficient	
Type of glass	Nominal thickness, ‡ in (mm)	Outer pane	Inner pane	$b_0 = 4.0$	$b_0 = 3.0$
Regular sheet out, regular sheet in	3/32, 1/8 (2, 3)	0.87	0.87	0.90	0.90
Regular plate/float out, regular plate/float in	1/4 (6)	0.80	0.80	0.83	0.83
Heat-absorbing plate/float out, regular plate/float in	1/4 (6)	0.46	0.80	0.56	0.58

SOURCE: Abstracted from ASHRAE ''Handbook of Fundamentals,'' 1993, with permission.
* Refers to factory-fabricated units with 3/16 (5), 1/4 (6), or 1/2 (12) in (mm) airspace or to prime windows plus storm windows.
† Refer to manufacturer's literature for values.
‡ Thickness of each pane of glass, not thickness of assembled unit.
§ Refers to gray-, bronze-, and green-tinted heat-absorbing plate/float glass.

Table 12.4.22   Shading Coefficients for Single Glass with Indoor Shading by Venetian Blinds and Roller Shades

			Type of shading				
			Venetian blinds		Roller shade		
					Opaque		Translucent
Type of glass	Nominal thickness,* in (mm)	Solar trans.[†]	Medium	Light	Dark	White	Light
Regular sheet	3/32–1/4 (2–6)	0.87–0.80					
Regular plate/float	1/4–1/2 (6–12)	0.80–0.71					
Regular pattern	1/8–9/32 (3–7)	0.87–0.79	0.64	0.55	0.59	0.25	0.39
Heat-absorbing pattern	1/8 (3)						
Gray sheet	3/16, 7/32 (5, 6)	0.74, 0.71					
Heat-absorbing plate/float‡	3/16, 1/4 (5, 6)	0.46					
Heat-absorbing pattern	3/16, 1/4 (5, 6)		0.57	0.53	0.45	0.30	0.36
Gray sheet	1/8, 7/32 (3, 5)	0.59, 0.45					
Heat-absorbing plate/float or pattern		0.44–0.30	0.54	0.52	0.40	0.28	0.32
Heat-absorbing plate/float‡	3/8 (10)	0.34					
Heat-absorbing plate or pattern		0.29–0.15 0.24	0.42	0.40	0.36	0.28	0.31
Reflective coated glass SC§ = 0.30			0.25	0.23			
0.40			0.33	0.29			
0.50			0.42	0.38			
0.60			0.50	0.44			

SOURCE: Abstracted from ASHRAE ''Handbook of Fundamentals,'' 1993, with permission.
* Refer to manufacturer's literature for values.
† For vertical blinds with opaque white and beige louvers in the tightly closed position. SC is 0.25 and 0.29 when used with glass of 0.71 to 0.80 transmittance.
‡ Refers to gray-, bronze-, and green-tinted heat-absorbing plate/float glass.
§ Shading coefficient for glass with no shading device.

Table 12.4.23   Shading Coefficients for Insulating Glass* with Indoor Shading by Venetian Blinds and Roller Shades

Type of glass	Nominal thickness, each light in (mm)	Solar trans†		Type of shading				
		Outer pane	Inner pane	Venetial blinds‡		Roller shade		
						Opaque		Translucent
				Medium	Light	Dark	White	Light
Regular sheet out, regular sheet in	³/₃₂, ⅛ (2, 3)	0.87	0.87	0.57	0.51	0.60	0.25	0.37
Regular plate/float out, Regular plate/float in	¼ (6)	0.80	0.80					
Heat absorbing plate/float§ out, regular plate/float in	¼ (6)	0.46	0.80	0.39	0.36	0.40	0.22	0.30
Reflective coated glass SC¶ = 0.20				0.19	0.18			
0.30				0.27	0.26			
0.40				0.34	0.33			

SOURCE: Abstracted from ASHRAE "Handbook of Fundamentals," 1993, with permission.
* Refers to factors-fabricated units with ³/₁₆ (2), ¼ (3), or ½ (12), in (mm) airspace, or to prime windows plus storm windows.
† Refer to manufacturer's literature for exact values.
‡ For vertical blinds with opaque white or beige louvers, tightly closed, SC is approximately the same as for opaque white roller shades.
§ Refers to bronze- or green-tinted heat-absorbing plate/flat glass.
¶ Shading coefficient for glass with no shading device.

Table 12.4.24   Shading Coefficients for Double Glazing with Between-Glass Shading

| Type of glass | Nominal thickness, each pane | Solar trans.* | | Description of airspace | Type of shading | | |
| | | Outer pane | Inner pane | | Venetian blinds | | Louvered sun screen |
					Light	Medium	
Regular sheet out, regular sheet in	³/₃₂, ⅛ (2, 3)	0.87	0.87	Shade in contact with glass or shade separated from glass by air space	0.33	0.36	0.43
Regular plate out, regular plate in	¼ (6)	0.80	0.80	Shade in contact with glass void-filled with plastic			0.49
Heat-absorbing, plate/float† out, Regular plate in	¼ (6)	0.46	0.80	Shade in contact with glass or shade separated from glass by air space	0.28	0.30	0.37
				Shade in contact with glass voids filled with plastic			0.41

SOURCE: Abstracted from ASHRAE "Handbook of Fundamentals," 1993, with permission.
* Refer to manufacturer's literature for exact values.
† Refers to gray-, bronze-, and green-tinted heat-absorbing plate/float glass.

Table 12.4.25   Rates of Heat Gain from Occupants of Conditioned Spaces*†

Degree of activity	Typical application	Total heat adult male, Btu/h	Total heat adjusted,‡ Btu/h	Sensible heat, Btu/h	Latent heat, Btu/h
Seated at theater	Theater—Matinee	390	330	225	105
Seated at theater	Theater—Evening	390	350	245	105
Seated, very light work	Offices, hotels, apartments	450	400	245	155
Moderately active office work	Offices, hotels, apartments	475	450	250	200
Standing, light work; walking	Department store, retail store	550	450	250	200
Walking; standing	Drug store, bank	550	500	250	250
Sedentary work	Restaurant§	490	550	275	275
Light bench work	Factory	800	750	275	475
Moderate dancing	Dance hall	900	850	305	545
Walking 3 mi/h; light machine work	Factory	1,000	1,000	375	625
Bowling¶	Bowling alley	1,500	1,450	580	870
Heavy work	Factory	1,500	1,450	580	870
Heavy machine work; lifting	Factory	1,600	1,600	635	965
Athletics	Gymnasium	2,000	1,800	710	1,090

SOURCE: Abstracted from ASHRAE "Handbook of Fundamentals," 1989, with permission.
* Tabulated values are based on 75°F room dry-bulb temperature. For 80°F room dry-bulb, the total heat remains the same, but the sensible heat values should be decreased by approximately 20%, and the latent heat values increased accordingly.
† All values are rounded to nearest 5 Btu/h.
‡ *Adjusted heat gain* is based on normal percentage of men, women, and children for the application listed, with the postulate that the gain from an adult female is 85% of that for an adult male, and that the gain from a child is 75% of that for an adult male.
§ Adjusted total heat gain for *Sedentary work, Restaurant,* includes 60 Btu/h for food per individual (30 Btu/h sensible and 30 Btu/h latent).
¶ For *Bowling,* figure one person per alley actually bowling, and all others as sitting (400 Btu/h) or standing and walking slowly (550 Btu/h).

Table 12.4.26   Rate of Heat Gain from Selected Office Equipment

Appliance	Size	Maximum input rating, Btu/h	Standby input rating, Btu/h	Recommended rate of heat gain, Btu/h
Cheek processing workstation	12 pockets	16,400	8,410	8,410
Computer devices				
Card puncher		2,730 to 6,140	2,200 to 4,800	2,200 to 4,800
Card reader		7,510	5,200	5,200
Communication/transmission		6,140 to 15,700	5,600 to 9,600	5,600 to 9,600
Disk drives/mass storage		3,410 to 34,100	3,412 to 22,420	3,412 to 22,420
Magnetic ink reader		3,280 to 16,000	2,600 to 14,400	2,600 to 14,400
Microcomputer	16 to 640 Kbyte*	340 to 2,050	300 to 1,800	300 to 1,800
Minicomputer		7,500 to 15,000	7,500 to 15,000	7,500 to 15,000
Optical reader		10,240 to 20,470	8,000 to 17,000	8,000 to 17,000
Plotters		256	128	214
Printers				
Letter quality	30 to 45 char/min	1,200	600	1,000
Line, high speed	5,000 or more lines/min	4,300 to 18,100	2,160 to 9,040	2,500 to 13,000
Line, low speed	300 to 600 lines/min	1,540	770	1,280
Tape drives		4,090 to 22,200	3,500 to 15,000	3,500 to 15,000
Terminal		310 to 680	270 to 600	270 to 600
Copiers/Duplicators				
Blue print		3,930 to 42,700	1,710 to 17,100	3,930 to 42,700
Copiers (large)	30 to 67* copies/min	5,800 to 22,500	3,070	5,800 to 22,500
Copiers (small)	6 to 30* copies/min	1,570 to 5,800	1,020 to 3,070	1,570 to 5,800
Feeder		100	—	100
Microfilm printer		1,540	—	1,540
Sorter/collator		200 to 2,050	—	200 to 2,050
Electronic equipment				
Cassette recorders/players		200	—	200
Receiver/tuner		340	—	340
Signal analyzer		90 to 2,220	—	90 to 2,220
Mail processing				
Folding machine		430	—	270
Inserting machine	3,600 to 6,800 pieces/h	2,050 to 11,300	—	1,330 to 7,340
Labeling machine	1,500 to 30,000 pieces/h	2,050 to 22,500	—	1,330 to 14,700
Postage meter		780	—	510
Word processors/typewriters				
Letter-quality printer	30 to 45 char/min	1,200	600	1,000
Phototypesetter		5,890	—	5,180
Typewriter		270	—	230
Word processor		340 to 2,050	—	300 to 1,800
Vending machines				
Cigarette		250	51 to 85	250
Cold food/beverage		3,920 to 6,550	—	1,960 to 3,280
Hot beverage		5,890	—	2,940
Snack		820 to 940	—	820 to 940
Miscellaneous				
Bar-code printer		1,500	—	1,260
Cash register		200	—	160
Coffee maker	10 cups	5,120	—	3,580 sensible 1,540 latent
Microfiche reader		290	—	290
Microfilm reader		1,770	—	1,770
Microfilm reader/printer		3,920	—	3,920
Microwave oven	1 ft³	2,050	—	1,360
Paper shredder		850 to 10,240	—	680 to 8,250
Water cooler	32 qt/h	2,390	—	5,970

SOURCE: Abstracted from ASHRAE "Handbook of Fundamentals," 1993, with permission.
* Input is not proportional to capacity.

Heat transfer to the distribution duct depends on the duct area exposed, the amount of insulation on the duct, and the temperature differential between the air in the duct and the surroundings.

Outside Air

Outside air loads depend on the amount of outside air intake and its condition. Table 12.4.3 shows recommended outdoor air ventilation rates for different occupancies. Tables 12.4.4 and 12.4.5 can be used to determine the quantity of outside air that will infiltrate because of wind effects. In all cases, sufficient outside air should be provided to balance mechanical exhaust systems.

$$Q_{outside\ air} = q_{sensible} + q_{latent} \quad \text{Btu/h (watts)}$$
$$= AU(t_o - t_i) \text{ ft}^3/\text{min} \quad (12.4.16a)$$
$$+ 0.68 (M_o - M_i) \text{ ft}^3/\text{min} \quad \text{Btu/h}$$

where M_o = outside moisture content at design wet bulb, °F, gr/lb; and M_i = inside moisture content at design relative humidity, gr/lb. A comparable equation in SI units is

$$Q_{outside\ air} = 1.2(t_o' - t_i') \text{ m}^3/\text{s}$$
$$+ 42.6 (M_o' - M_i') \text{ m}^3/\text{s} \quad \text{kW} \quad (12.4.16b)$$

Table 12.4.27   Heat Gain from Restaurant Appliances
Not hooded,* gas burning and steam heated.

Appliance	Type of control	Miscellaneous data	Manufacturer's max. rating, Btu/h	Maintaining rate, Btu/h	Recommended heat gain from average use		
					Sensible heat, Btu/h	Latent heat, Btu/h	Total heat, Btu/h
Gas burning							
Coffee brewer ½ gal	Man.	Combination brewer and warmer	3,400		1,350	350	1,700
Warmer, ½ gal	Man.		500	500	400	100	500
Coffee brewer unit with tank		4 brewers and 4½-gal tank			7,200	1,800	9,000
Coffee urn, 3 gal	Auto.	Black finish	3,200	3,900	2,900	2,900	5,800
Coffee urn, 3 gal	Auto.	Nickel plated		3,400	2,500	2,500	5,000
Coffee urn, 5 gal	Auto.	Nickel plated		4,700	3,900	3,900	7,800
Food warmer, values per sq. ft top surface	Man.	Water bath type	2,000	900	850	450	1,300
Fry kettle, 15 lb fat	Auto.	Frying area 10 × 10	14,250	3,000	4,200	2,800	7,000
Fry kettle, 28 lb fat	Auto.	Frying area 11 × 16	24,000	4,500	7,200	4,800	12,000
Grill, Broil-O-Grill Top burner Bottom burner	Man.	Insulated 22,000 Btu/h 15,500 Btu/h	37,000		14,400	3,600	18,000
Stoves, short order—closed top; values per sq. ft top surface	Man.	Ring-type burners 12,000 to 22,000 Btu each	14,000		4,200	4,200	8,400
Stoves, short order—closed top; values per sq. ft top surface	Man.	Ring-type burners 10,000 to 12,000 Btu each	11,000		3,300	3,300	6,600
Toaster, continuous	Auto.	2 slices wide, 360 slices/h	12,000	10,000	7,700	3,300	11,000
Steam heated							
Coffee urn, 3 gal	Auto.	Black finish			2,900	1,900	4,800
3 gal	Auto.	Nickel plated			2,400	1,600	4,000
5 gal	Auto.	Nickel plated			3,400	2,300	5,700
Coffee urn, 3 gal	Man.	Black finish			3,100	3,100	6,200
3 gal	Man.	Nickel plated			2,600	2,600	5,200
5 gal	Man.	Nickel plated			3,700	3,700	7,400
Food warmer, per sq. ft top surface	Auto.				400	500	900
Food warmer, per sq. ft top surface	Man.				450	1,150	1,500
Coffee brewer, ½ gal	Man.		2,240	306	900	220	1,120
Warmer, ½ gal	Man.		306	306	230	90	320
4 coffee brewing units with 4½-gal tank	Auto.	Water heater, 2,000 W Brewers, 2,960 W	16,900		4,800	1,200	6,000
Coffee urn, 3 gal	Man.	Black finish	11,900	3,000	2,600	1,700	4,300
3 gal	Auto.	Nickel plated	15,300	2,600	2,200	1,500	3,700
5 gal	Auto.	Nickel plated	17,000	3,600	3,400	2,300	5,700
Doughnut machine	Auto.	Exhaust system to outdoors, ½-hp motor	16,000		5,000		5,000
Egg boiler	Man.	Med. ht., 550 W Low ht., 275 W	3,740		1,200	800	2,000
Food warmer with plate warmer, per sq. ft top surface	Auto.	Insulated, separate heating unit for each pot, plate warmer in base	1,350	500	350	350	700
Food warmer without plate warmer, per sq. ft top surface	Auto.	Insulated, separate heating unit for each pot	1,020	400	200	350	550
Fry kettle, 11½ lb fat	Auto.		8,840	1,100	1,600	2,400	4,000
Fry kettle, 25 lb fat	Auto.	Frying area 12 × 14 in	23,800	2,000	3,800	5,700	9,500
Griddle, frying	Auto.	Frying top 18 × 14 in	8,000	2,800	3,100	1,700	4,800
Grille, meat	Auto.	Cooking area 10 × 12 in	10,200	1,900	3,900	2,100	6,000
Grille, sandwich	Auto.	Grill area 12 × 12 in	5,600	1,900	2,700	700	3,400
Roll warmer	Auto.	One drawer	1,500	400	1,100	100	1,200
Toaster, continuous	Auto.	2 slices wide, 360 slices/h	7,500	5,000	5,100	1,300	6,400
Toaster, continuous	Auto.	4 slices wide, 720 slices/h	10,200	6,000	6,100	2,600	8,700
Toaster, pop-up	Auto.	2 slices	4,150	1,000	2,450	450	2,900
Waffle iron	Auto.	One waffle 7-in diam.	2,480	600	1,100	750	1,850
Waffle iron for ice cream sandwich	Auto.	12 cakes, each 2½ × 3¾ in	7,500	1,500	3,100	2,100	5,200

SOURCE: Carrier manual.
* If properly designed positive exhaust hood is used, multiple recommended value by 0.50.

Table 12.4.28   Heat Gain from Typical Electric Motors

Motor nameplate or rated horsepower	Motor type	Nominal r/min	Full load motor efficiency, %	Location of motor and driven equipment with respect to conditioned space or airstream		
				A Motor in, driven equipment in, Btu/h	B Motor out, driven equipment in, Btu/h	C Motor in, driven equipment out, Btu/h
0.05	Shaded pole	1,500	35	360	130	240
0.08	Shaded pole	1,500	35	580	200	380
0.125	Shaded pole	1,500	35	900	320	590
0.16	Shaded pole	1,500	35	1,160	400	760
0.25	Split phase	1,750	54	1,180	640	540
0.33	Split phase	1,750	56	1,500	840	660
0.50	Split phase	1,750	60	2,120	1,270	850
0.75	3-phase	1,750	72	2,650	1,900	740
1	3-phase	1,750	75	3,390	2,550	850
1.5	3-phase	1,750	77	4,960	3,820	1,140
2	3-phase	1,750	79	6,440	5,090	1,350
3	3-phase	1,750	81	9,430	7,640	1,790
5	3-phase	1,750	82	15,500	12,700	2,790
7.5	3-phase	1,750	84	22,700	19,100	3,640
10	3-phase	1,750	85	29,900	24,500	4,490
15	3-phase	1,750	86	44,400	38,200	6,210
20	3-phase	1,750	87	58,500	50,900	7,610
25	3-phase	1,750	88	72,300	63,600	8,680
30	3-phase	1,750	89	85,700	76,300	9,440
40	3-phase	1,750	89	114,000	102,000	12,600
50	3-phase	1,750	89	143,000	127,000	15,700
60	3-phase	1,750	89	172,000	153,000	18,900
75	3-phase	1,750	90	212,000	191,000	21,200
100	3-phase	1,750	90	283,000	255,000	28,300
125	3-phase	1,750	90	353,000	318,000	35,300
150	3-phase	1,750	91	420,000	382,000	37,800
200	3-phase	1,750	91	569,000	509,000	50,300
250	3-phase	1,750	91	699,000	636,000	62,900

SOURCE: Adapted from ASHRAE "Handbook of Fundamentals," 1993, with permission.

where M'_o = outside moisture content at design wet bulb, °C, g/kg; and M'_i = inside moisture content at design relative humidity, g/kg. t'_o, t'_i are expressed in °C.

Moisture Load

Moisture as a cooling load is a latent heat gain expressed in BTU per hour and is computed with the following equations:

$$Q_t = Q_{inf} + Q_{vent} \tag{12.4.17}$$
$$Q_{inf} = 0.68 \times \text{ft}^3/\text{min} \times (M_o - M_i) \quad \text{Btu/h} \tag{12.4.18a}$$
$$Q_{vent} = 0.68 \times \text{ft}^3/\text{min} \times (M_o - M_i) \quad \text{Btu/h} \tag{12.4.19a}$$

where $0.68 = 60/13.5 \times 1,076/7,000$; 60 = min/h; 13.5 = specific volume of moist air at 70°F dB and 50% RH; $1,076$ = average heat removal required to condense 1 lb of water vapor from the room air; and $7,000$ = grains per pound.

Comparable equations in SI units are:

$$Q_{inf} = 42.6 \ \text{m}^3/\text{s} \ (M'_o - M'_i) \quad \text{kW} \tag{12.4.18b}$$
$$Q_{vent} = 42.6 \ \text{m}^3/\text{s} \ (M'_o - M'_i) \quad \text{kW} \tag{12.4.19b}$$

M'_o = moisture content of outside air, g/kg; and M'_i = moisture content of inside air, g/kg.

HEATING

Heating Load

$$Q_t = Q_{tr} + Q_{inf} + Q_{vent} \tag{12.4.20a}$$

where Q_1 = total heating load, Btu/h; Q_{tr} = transmission load = $AU(t_i - t_o)$, Btu/h; Q_{inf} = infiltration load = 1.08 (ft³/min)($t_i - t_o$), Btu/h; Q_{vent} = ventilation load = 1.08 (ft³/min)($t_i - t_o$), Btu/h; A = area through which heat flow occurs, ft²; U = overall heat-transfer

coefficient = $1/R_t$ Btu/(h)(ft²)(°F); t_o = outside-air design dry-bulb temperature, °F (Fig. 12.4.1); t_i = inside design dry-bulb temperature, °F (Table 12.4.1); R_t = thermal resistance, °F/(Btu·h·ft²) = $R_1 + R_2 + R_3 + \cdots + R_n$; and $R_1, R_2, \ldots$ = resistances to heat flow to the individual components of a composite construction. See Fig. 12.4.2 and Table 12.4.2.

A comparable equation in SI units is

$$Q_1 = Q_{tr} + Q_{inf} + Q_{vent} \tag{12.4.20b}$$

where Q_t = total heating load, kW; $Q_{tr} = AU(t_i - t_o)$, kW; $Q_{inf} = 1.2$ (m³/s)($t_i - t_o$), kW; $Q_{vent} = 1.2$ (m³/s)($t_i - t_o$), kW; A = area through which heat flow occurs, m²; U = overall heat-transfer coefficient = $1/R_t$, W/(m²·K); t_o = outside-air design dry-bulb temperature, °C; t_i = inside design dry-bulb temperature, °C; m³/s = cubic meters per second, air; and R_t = thermal resistance, K/(W·m²).

PSYCHROMETRICS

The psychrometric chart can be used to simplify the analysis of air-conditioning processes. The processes can be shown simply and visualized clearly, and the chart lends itself to quick graphical computations. In addition, the state of any air–water vapor mixture can be fixed on the psychrometric chart with reasonable accuracy by means of any two psychrometric characteristics. Usually dry-bulb temperature and wet-bulb temperature are used because they are the simplest to measure.

Figures 12.4.8 and 12.4.9 are psychrometric charts for normal temperatures in inch-pound and SI units, respectively.

Psychrometric charts at different temperatures and barometric pressures are useful for problems falling outside the ranges shown in Figs. 12.4.8 and 12.4.9. A collection of different psychrometric charts, in both USCS and SI units, for low, normal, and high temperatures, at sea

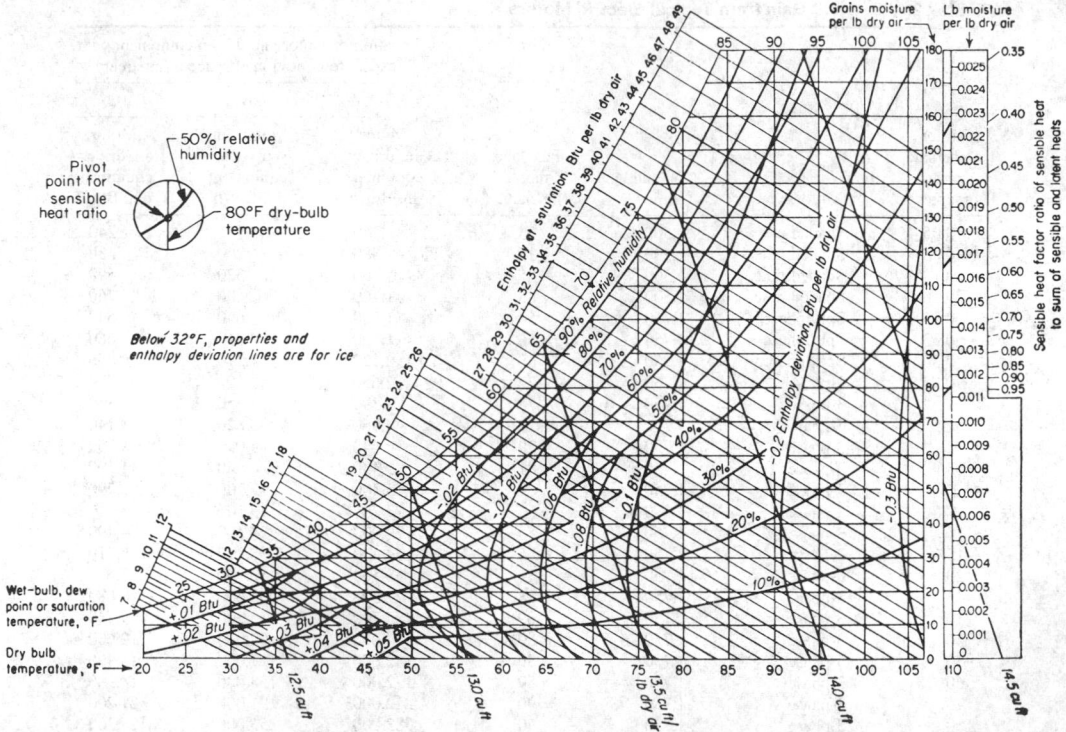

Fig. 12.4.8 Psychrometric chart for normal temperatures in USCS units. (*Carrier Corp.*)

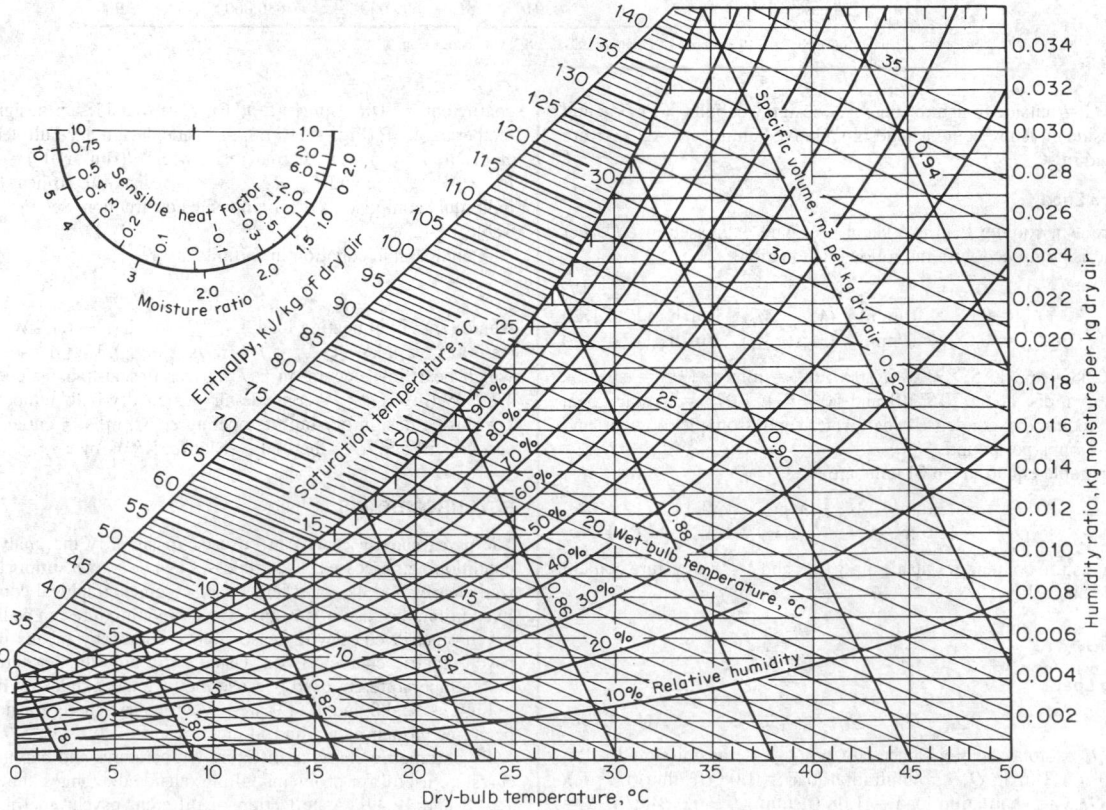

Fig. 12.4.9 Psychrometric chart for normal temperatures in SI units. (*Adapted from material published by Business News Publishing Co., 1975.*)

level and at four elevations above sea level, are available from ASHRAE, the Carrier Corp. of Syracuse, NY, and from other equipment manufacturers.

Figure 12.4.10 shows a psychrometric chart simplified in skeleton form.

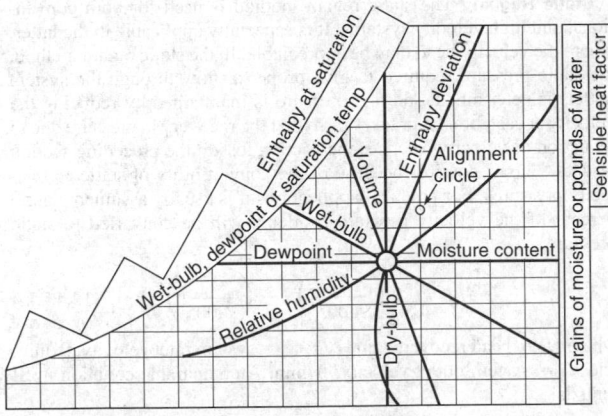

Fig. 12.4.10   Skeleton psychrometric chart.

Terminology

Dry-bulb temperature: The temperature of air as registered by an ordinary thermometer.

Wet-bulb temperature: The temperature registered by a thermometer whose bulb is covered by a wetted wick and exposed to a current of rapidly moving air.

Dew-point temperature: The temperature at which condensation of moisture begins when the air is cooled.

Relative humidity: Ratio of the actual water vapor pressure of the air to the saturated water vapor pressure of the air at the same temperature.

Specific humidity or moisture content: The weight of water vapor in grains (or pounds) of moisture per pound of dry air.

Enthalpy: A thermal property indicating the quantity of heat in the air above an arbitrary datum, in Btu per pound of dry air. The datum for dry air is 0°F and, for the moisture content, 32°F water.

Enthalpy deviation: Enthalpy should be corrected by the enthalpy deviation due to the air not being in the saturated state. On normal air conditioning estimates it is omitted.

Specific volume: The cubic feet of the mixture per pound of dry air.

Sensible heat factor: The ratio of sensible to total heat.

Alignment circle: Used in conjunction with the sensible heat factor to plot the various air-conditioning process lines.

Pounds of dry air: The basis for all psychrometric calculations. Remains constant during all psychrometric processes.

The dry-bulb, wet-bulb, and dew-point temperatures and the relative humidity are so related that, if two properties are known, all other properties shown may then be determined. When **air is saturated,** dry-bulb, wet-bulb, and dew-point temperatures are all equal.

Room sensible heat factor is the ratio of room sensible heat to the total of room sensible and room latent heat.

A line drawn between the room design condition at the slope of the sensible heat factor is the **condition line** and it will intersect the saturation line at the apparatus dew point. See Fig. 12.4.11. In order to maintain design conditions, supply air must be provided to the space at a temperature and moisture level that, when plotted, must fall on the condition line. If air is provided at conditions of apparatus dew point, the minimum quantity of conditioned air would be required to obtain design conditions.

Supply Air Temperature   The room cooling load is the sum of external and internal sensible and latent heat gains plus the difference in

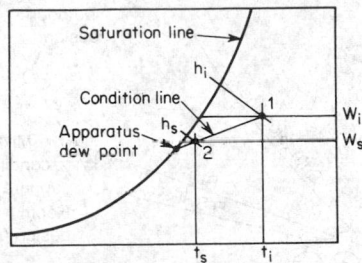

Fig. 12.4.11   Apparatus dew point and condition line.

enthalpy between outside and room air for that portion of outside air which does not contact the cooling-coil surfaces. The percentage of air that passes through a cooling coil untreated is the numerical value of the *coil bypass factor;* e.g., a bypass factor of 20 percent represents a cooling-coil saturation efficiency of 80 percent.

The ratio of room sensible heat gains to total room sensible and latent heat gains is the **room sensible heat ratio** (RSHR).

$$\text{RSHR} = \frac{Q_{rs}}{Q_{rs} + Q_{rl}} \qquad (12.4.21)$$

It represents the ratio of sensible cooling capacity to the total cooling capacity required of the supply air to satisfy room conditions. It is used to plot the slope of the *room-condition line* on a psychometric chart for the determination of the *apparatus dew point* (ADP). See Fig. 12.4.11.

The actual supply air temperature and off coil wet-bulb temperature will depend on the bypass characteristic of the selected cooling coil (Fig. 12.4.11).

Supply Air Rate   The rate of supply air required is expressed by

$$Q_{sa} = Q_{rs}/1.08(t_r - t_s) \qquad \text{ft}^3/\text{min} \qquad (12.4.22a)$$

In SI units:

$$Q'_{sa} = Q'_{rs}/1.2(t_r - t_s) \qquad \text{m}^3/\text{s} \qquad (12.4.22b)$$

where Q_{sa} = supply air, ft³/min (m³/s); Q'_{rs} = room sensible heat, kW; t_r = room design temperature, °F (K); t_s = supply air temperature, °F (K).

The bypass factor is used to express the effectiveness of a cooling coil. It is based on the artificial premise that a portion of the air that enters a cooling coil is cooled completely to saturation, that is, to the apparatus dew point, and the remaining air is not cooled at all. The actual condition of the air that leaves the coil is then represented to be a mixture of the fully conditioned and completely unconditioned air. The ratio of unconditioned air to total air is the **bypass factor**.

The bypass factor is a function of the type and amount of coil surface and the time available for contact as the air passes over the coil. Table 12.4.29 gives approximate bypass factors for various finned coil surfaces and air velocities. A larger bypass factor requires the circulation of more conditioned air; where outside air is introduced, the bypassed air increases the room air load.

Table 12.4.29   Typical Bypass Factors for Unsprayed Coils

Depth of coils (rows)	5-in spacing	
	8 fins/in	14 fins/in
	Face velocity, ft/min	
	300–700	300–700
2	0.42–0.55	0.22–0.38
3	0.27–0.40	0.10–0.23
4	0.19–0.30	0.05–0.14
5	0.12–0.25	0.02–0.09
6	0.08–0.18	0.01–0.06
8	0.03–0.08	

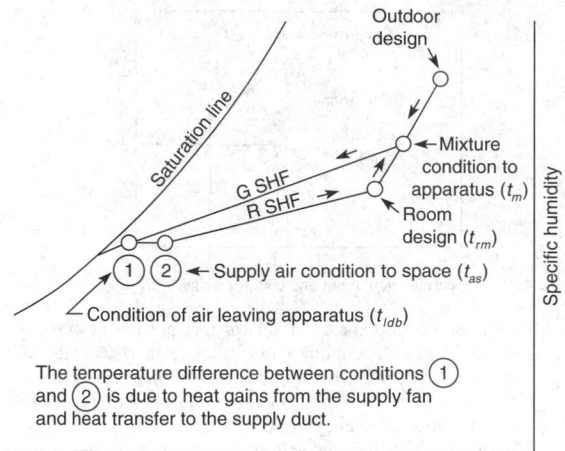

Fig. 12.4.12 Typical cooling process.

Typical Air-Conditioning Cooling Process Figure 12.4.12 depicts a typical cooling process adjusted for fan heating and duct losses.

DUCT DESIGN

Air Velocity Supply and return air ducts and apparatus are sized on the basis of air quantity and are kept within the limitations of allowable friction losses, velocity, and noise. Conveying air at low velocity will be quieter and require less pressure, less fan power and less carefully constructed ductwork than conveying air at medium or high velocity. Lower velocities are preferred where space is available.

Pressure Loss in Duct Systems Pressure losses in duct systems are due to friction of the air in contact with the sides of the duct and dynamic losses caused by changes of duct shape or direction and by obstructions to flow.

$$\text{Friction } H_f = f \frac{L}{D} \left(\frac{V}{4,005} \right)^2 \qquad (12.4.23a)$$

where H_f = head loss due to friction, in H_2O: L = length of duct, ft; D = diameter of duct, ft; V = velocity of air, ft/min; and f = nondimensional friction coefficient. A comparable equation in SI units is

$$H_f = f \frac{L}{d} \frac{(V)^2}{2} \rho \qquad (12.4.23b)$$

where H_f = head loss due to friction, N/m²; L = length of duct, m; D = diameter of duct, m; V = velocity of air, m/s; f = nondimensional friction coefficient; and ρ = density of air, kg/m³.

$$\text{Dynamic losses } H_v = C \left(\frac{V}{4,005} \right)^2 \qquad (12.4.24a)$$

where H_v = velocity-head loss, in H_2O; C = experimentally determined constant; $V = Q/A$ = air velocity, ft/min; Q = airflow rate, ft³/min; and A = cross-sectional area of duct, ft². A comparable equation in SI units is

$$H_v' = C' \frac{(V')^2}{2} \qquad (12.4.24b)$$

where H_v' = velocity-head loss, N/m²; V' = air velocity, m/s; and = density of air, kg/m³. When the equation for H_v' is written in the form above, $C' = C$.

Duct Design Methods Two methods are frequently used: **equal friction** and **static regain**.

Equal Friction The equal friction method is applicable primarily to systems using low or moderate velocities, where the velocity head is not

an important factor. A friction drop per 100 ft of length is chosen, and the duct mains and branches are all sized on the basis of this friction drop. This will invariably result in higher velocities in the mains, where they can be tolerated, and low velocities in the branches, where they are desirable. Limitations are established for friction loss per 100 feet and for velocities in the mains and branches.

Static Regain The static regain method is used for both conventional and high velocity systems. It is especially applicable in the latter, where the velocity head may be appreciable. In the static regain method, the static pressure required to give proper airflow through the system outlets is determined, and this pressure is maintained by reducing the velocity at each branch or takeoff, so that the recovery in pressure due to reduction of velocity balances the friction loss in the preceding section of duct. This is possible because of the convertibility of static and velocity pressures. For practical applications it is usually assumed that 50 percent of the velocity pressure available will be converted to static pressure.

$$H_R = 0.5 \left(\frac{V_1}{4,005} \right)^2 - \left(\frac{V_2}{4,005} \right)^2 \qquad (12.4.25a)$$

where H_R = head recovered, in H_2O; V_1 = system inlet velocity, ft/min; and V_2 = system outlet velocity, ft/min. A comparable equation in SI units is

$$H_R' = \frac{0.5(V_1')^2}{2} - \frac{(V_2')^2}{2} \qquad (12.4.25b)$$

where $H_{R'}$ = head recovered, N/m²; V_1' = system inlet velocity, m/s; and V_2' = system outlet velocity, m/s.

Figure 12.4.13 may be used to determine directly the friction loss per

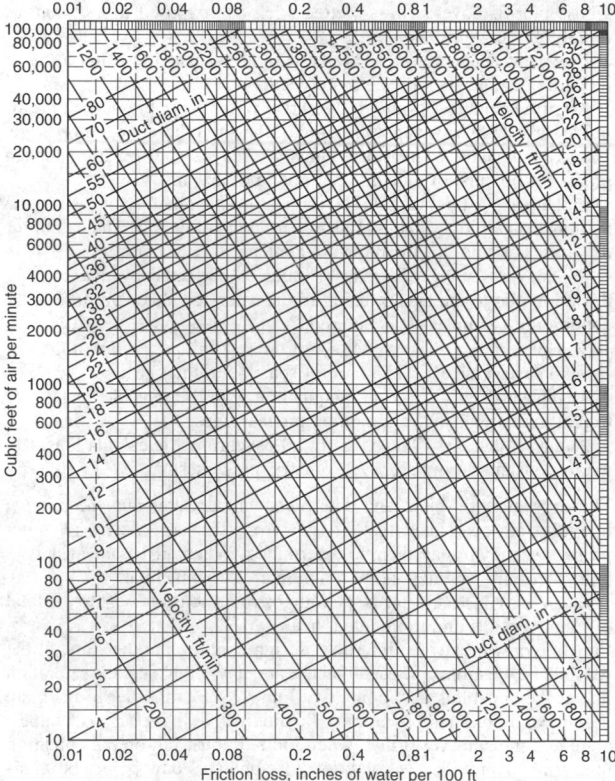

Fig. 12.4.13 Friction loss for usual conditions in round ducts. (This chart applies to smooth, round, galvanized steel ducts, and is based on air at 70°F and 29.96 in Hg absolute pressure. For air of different density, the friction may be assumed to vary directly with the density.)

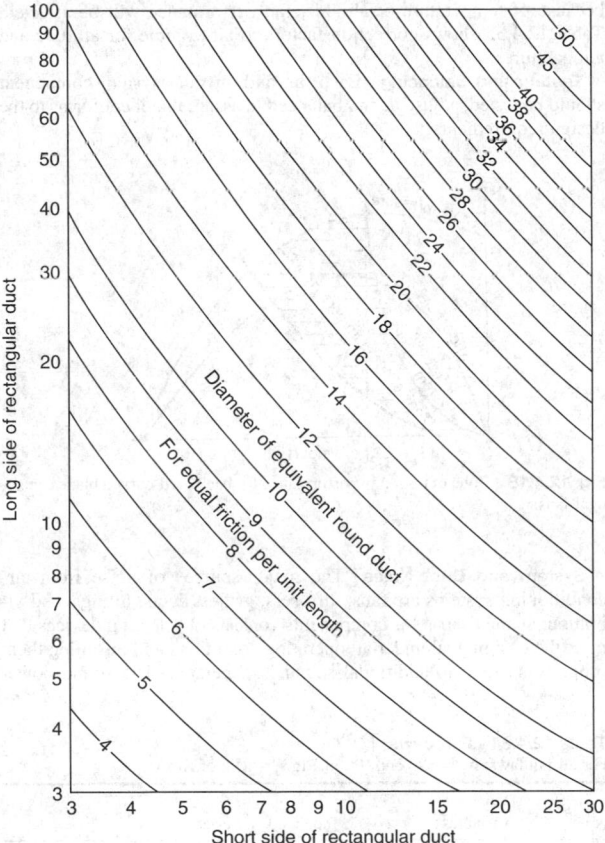

Fig. 12.4.14 Equivalent diameters of round ducts. Use with rectangular ducts, with capacity constant. *(Buffalo Forge.)*

100 ft for round ducts. Figure 12.4.14 provides a conversion from round to rectangular ductwork with the same air capacity. Table 12.4.30 presents a range of friction loss per 100 ft and velocity limits for low- and high-velocity air systems. Typical velocities through some air system components are shown in Table 12.4.31.

Table 12.4.30   Duct System Pressure Losses and Design Velocities

	Pressure loss, in H₂O, gage, per 100 ft	Maximum velocity, ft/min
Low-velocity systems		
Noise critical	0.05–0.07	1,400
Usual design	0.08–0.1	1,800
Straight runs without takeoffs	0.08–0.12	2,400
High-velocity systems		
Usual design	0.4–0.6	2,600
Straight runs without takeoffs	0.6	3,400

Air Distribution

Air outlets are designed to control air motion, noise level, and temperature gradients caused by the introduction of air into a conditioned space and the removal of air from that space.

Supply Outlets   Supply outlets (and diffusers) are usually located at near ceiling height with grilles or registers high on the sidewall. Supply outlets may be located below windows or in the floor as long as they satisfy the criteria for the space and are unobstructed. Factors which usually affect the selection of supply outlets are noise, location of outlet,

Table 12.4.31   Typical Velocities through Air System Components

	Residences		Nonresidential buildings	
Designation	ft/min	m/min	ft/min	m/min
Outdoor-air intakes*	300	80	500	150
Filters*	250	75	300	90
Heating coils*	450	140	500	150
Cooling coils*	450	140	500	150
Air washers	500	150	500	150

SOURCE: Adapted from ASHRAE ''Handbook of Fundamentals,'' 1993, with permission.

 * These velocities are for total face area, not the net free area; other velocities in the table are for net free area.

temperature of supply air, and area of diffusion. Manufacturers' performance data for supply outlets should be used as the basis for selection.

Return Outlets   Selection of return outlets (and registers or grilles) is usually governed by face velocity to limit discomfort to occupants from drafts and/or noise. See Table 12.4.32.

Table 12.4.32   Recommended Return Face Velocities

	Velocity over gross area	
Intake location	ft/min	m/min
Above occupied zone	800 up	245 up
Within occupied zone, not near seats	600–800	185–245
Within occupied zone, near seats	400–600	120–185
Door or wall louvers	200–300	60–90
Undercutting of doors (through undercut area)	200–300	60–90

SOURCE: Adapted from ASHRAE, ''Handbook of Fundamentals,'' 1993, with permission.

FANS

Supply and **return fans** must convey the conditioned air through the system efficiently and with a minimum of disturbing noise and vibrations.

Fans produce pressure by increasing airflow velocity. Velocity change is the result of tangential and radial velocity components in the case of centrifugal fans, and of axial and tangential velocity components in the case of axial-flow fans.

Centrifugal fans are most frequently used, but **axial-flow fans** have been applied successfully. Figures 12.4.15 to 12.4.17 show typical performance curves for forward-curved and backward-curved centrifugal fans

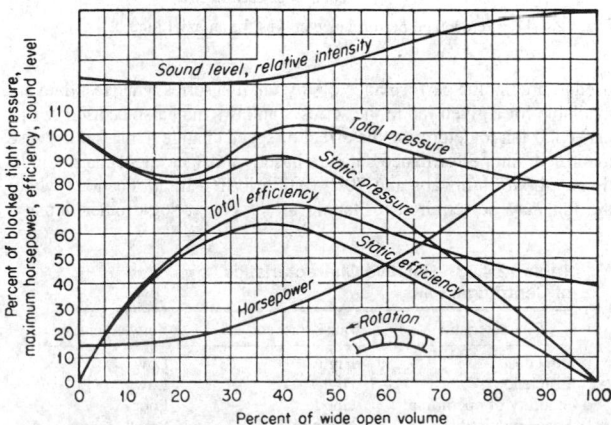

Fig. 12.4.15 Percentage performance curves for a forward-curved blade centrifugal fan.

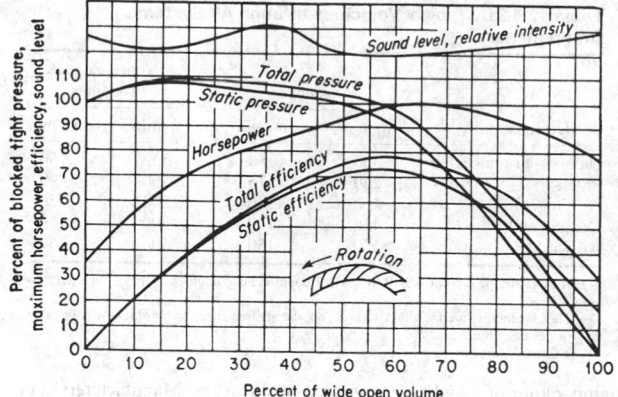

Fig. 12.4.16 Percentage performance curves for a backward-curved blade centrifugal fan.

and axial-flow fans. Table 12.4.33 summarizes the relative characteristics of centrifugal fans. Figure 12.4.18 shows the zone of optimum performance for a backward-curved-blade centrifugal fan.

FAN LAWS

Fan laws in Table 12.4.34 relate the performance variables for any dynamically similar series of fans. (See also Sec. 14.5.) They can be

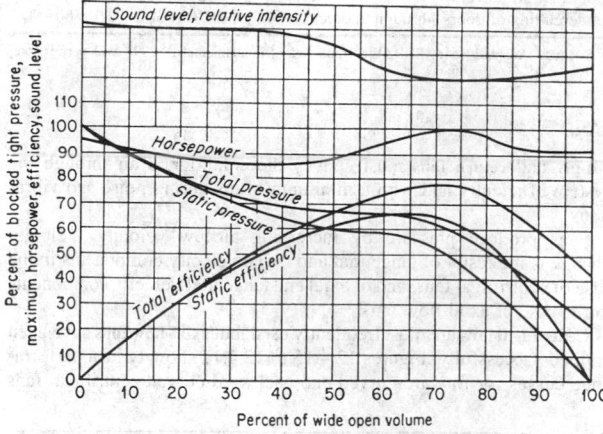

Fig. 12.4.17 Percentage performance curves for an axial fan.

used to predict the performance of any fan in a series when test data are available for a given fan in the series. Fan laws may also be used with a particular fan to determine the effect of speed change. Unless otherwise identified, fan performance data are based on dry air at standard conditions: 14.696 lb/in² abs and 70°F (0.075 lb/ft³). In applications where the fan may be required to handle air or gas at some other density,

Table 12.4.33 Relative Characteristics of Centrifugal Fans

Characteristic	Backward curved	Forward curved
First cost	High	Low
Efficiency	High	Low
Stability of operation	Good	Poor
Space required	Medium	Small
Tip speed	High	Low

because of temperature or altitude, fan performance will be affected. Table 12.4.35 shows correction factors to air volume for altitude and temperature.

Testing and Balancing Each fan and air duct system component should be tested, adjusted, and balanced to assure that it conforms to the design requirements.

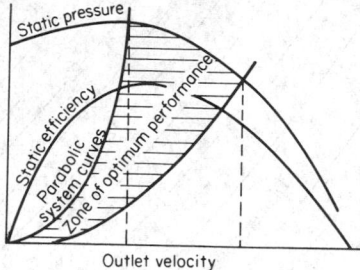

Fig. 12.4.18 Zone of optimum performance for backward-curved blade centrifugal fans.

System and Duct Noise The major sources of noise from air-conditioning systems are fans, diffusers, grilles, ducts, fittings, and vibration. **Sound control** for components consists of selecting devices that meet the design goal under all operating conditions and installing them properly so that no additional sound is generated. The sound power

Table 12.4.34 Fan Laws*†

For all fan laws: $N_{t1} = N_{t2}$ and (Pt. of Rtg.)$_1$ = (Pt. of Rtg.)$_2$

No.	Dependent variables	Independent variables		
1a	$Q_1 = Q_2$	$\times \left(\dfrac{D_1}{D_2}\right)^3 \times \dfrac{N_1}{N_2}$		$\times 1$
1b	Press.$_1$ = Press.$_2$‡	$\times \left(\dfrac{D_1}{D_2}\right)^2 \times \left(\dfrac{N_1}{N_2}\right)^2$		$\times \dfrac{\rho_1}{\rho_2}$
1c	$H_1 = H_2$	$\times \left(\dfrac{D_1}{D_2}\right)^5 \times \left(\dfrac{N_1}{N_2}\right)^3$		$\times \dfrac{\rho_1}{\rho_2}$
2a	$Q_1 = Q_2$	$\times \left(\dfrac{D_1}{D_2}\right)^2 \times \left(\dfrac{\text{Press.}_1}{\text{Press.}_2}\right)^{1/2}$	$\times \left(\dfrac{\rho_2}{\rho_1}\right)^{1/2}$
2b	$N_1 = N_2$	$\times \left(\dfrac{D_2}{D_1}\right) \times \left(\dfrac{\text{Press.}_1}{\text{Press.}_2}\right)^{1/2}$	$\times \left(\dfrac{\rho_2}{\rho_1}\right)^{1/2}$
2c	$H_1 = H_2$	$\times \left(\dfrac{D_1}{D_2}\right)^2 \times \left(\dfrac{\text{Press.}_1}{\text{Press.}_2}\right)^{3/2}$	$\times \left(\dfrac{\rho_2}{\rho_1}\right)^{1/2}$
3a	$N_1 = N_2$	$\times \left(\dfrac{D_2}{D_1}\right)^3 \times \dfrac{Q_1}{Q_2}$		$\times 1$
3b	Press.$_1$ = Press.$_2$	$\times \left(\dfrac{D_2}{D_1}\right)^4 \times \left(\dfrac{Q_1}{Q_2}\right)^2$		$\times \dfrac{\rho_1}{\rho_2}$
3c	$H_1 = H_2$	$\times \left(\dfrac{D_2}{D_1}\right)^4 \times \left(\dfrac{Q_1}{Q_2}\right)^3$		$\times \dfrac{\rho_1}{\rho_2}$

NOTE: D = fan size, N = revolutions per minute, ρ = gas density, Q = volume flow rate, P = pressure, and H = horsepower.
* The subscript 1 denotes that the variable is for the fan under consideration.
† The subscript 2 denotes that the variable is for the tested fan.
‡ P_{total} or P_{static}.

Table 12.4.35 Air Volume Correction Factors for Altitude and Temperature

Altitude, ft (m) above sea level	0	1,000 (305)	2,000 (610)	3,000 (915)	4,000 (1,220)	5,000 (1,525)	6,000 (1,880)	7,000 (2,135)	8,000 (2,440)
Barometric pressure, in Hg (kN/m²)	29.92 (101.3)	28.86 (97.7)	27.82 (94.2)	26.81 (90.8)	25.84 (87.5)	24.89 (84.3)	23.98 (81.2)	23.09 (78.2)	22.2 (75.2)
Air temp, °F (°C)					Correction factors				
70 (21.1)	1.040	1.003	0.967	0.932	0.898	0.865	0.833	0.803	0.772
100 (38)	0.984	0.948	0.915	0.882	0.850	0.818	0.788	0.759	0.731
150 (66)	0.904	0.872	0.840	0.801	0.781	0.752	0.724	0.698	0.672
200 (93)	0.835	0.805	0.777	0.749	0.722	0.694	0.668	0.645	0.620
250 (121)	0.777	0.749	0.722	0.696	0.671	0.647	0.622	0.599	0.577
300 (149)	0.725	0.699	0.674	0.649	0.628	0.603	0.580	0.560	0.538
350 (177)	0.680	0.656	0.632	0.609	0.588	0.566	0.545	0.525	0.505
400 (204)	0.641	0.618	0.596	0.574	0.553	0.533	0.512	0.495	0.476
450 (232)	0.605	0.583	0.564	0.543	0.523	0.503	0.485	0.467	0.450
500 (260)	0.574	0.553	0.534	0.515	0.496	0.477	0.460	0.443	0.426
550 (288)	0.546	0.526	0.508	0.490	0.472	0.454	0.438	0.421	0.406
660 (316)	0.520	0.501	0.484	0.466	0.449	0.433	0.416	0.401	0.387
650 (343)	0.496	0.478	0.462	0.444	0.428	0.413	0.397	0.383	0.368
700 (371)	0.475	0.458	0.442	0.426	0.411	0.395	0.381	0.367	0.354

SOURCE: "Bulletin 3576-B, Correction Factors for Temperature and Altitude," Buffalo Forge Co., Buffalo, NY.

NOTE: Equivalent $= \dfrac{\text{ft}^3 \text{ min at actual conditions}}{\text{correction factor}}$.

output of a fan is determined by the type of fan, airflow, and pressure. Sound control in the duct system requires proper duct layout, sizing, and provision for installing duct noise attenuators, if required. The noise generated in a system increases with both duct velocity and sys-

tem pressure. The ASHRAE Handbook "Applications" and other related sources present methods for sound control in air-conditioning systems.

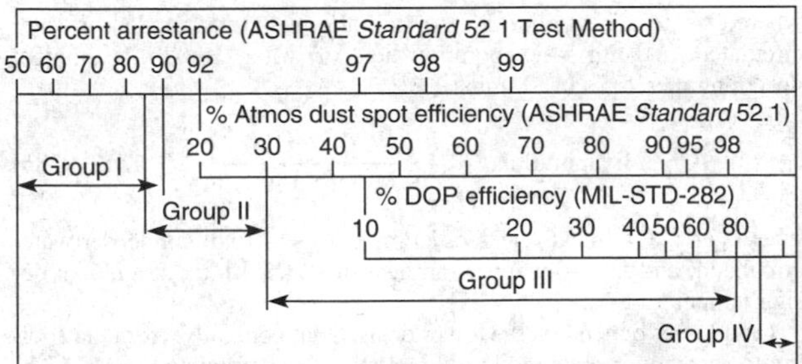

Group I
Panel-type filters of spun glass, open cell foams, expanded metal and screens, synthetics, textile denier woven and nonwoven, or animal hair.

Group II
Pleated panel-type filters of fine denier nonwoven synthetic and synthetic-natural fiber blends, or all natural fiber.

Group III
Extended surface supported and nonsupported filters of fine glass fibers, fine electret synthetic fibers, or wet-laid paper of cellulose-glass or all-glass fibers.

Group IV
Extended-area pleated HEPA-type filters of wet-laid ultra-fine glass fiber paper. Biological grade air filters are generally 95% DOP efficiency; HEPA filters are 99.97 and 99.99%; and ULPA filters are 99.999%.

Notes:
1. Group numbers have no significance other than their use in this figure.
2. Correlation between the test methods shown are approximations for general guidance only.

Fig. 12.4.19 Comparative performance of viscous impingement and dry media filters. *(ASHRAE "Handbook of Fundamentals," 1993.)*

FILTRATION

Filters are provided in air-conditioning systems to assure that the conditioned air will satisfy space cleanliness criteria and to protect air treatment equipment from fouling due to dirt accumulation. Figure 12.4.19 describes types of dry media filters and shows their range of performance. In addition, it shows an approximate comparison of the filtration efficiencies obtained by the use of three different testing methods.

The filters designated as Group I are usually used in least critical applications and as a prefilter to extend the operating life of more costly, higher-efficiency filters. Filters designated as Groups II and III are used in good-quality comfort and process air-conditioning applications. The very high efficiency HEPA and ULPA filters are costly and result in high pressure losses. They are applied in clean rooms and for nuclear and toxic particulate filtration.

HEAT REJECTION APPARATUS

All the heat removed from a conditioned space and the heat equivalent of the work performed by the refrigeration equipment must be removed through the cooling system and dissipated via the refrigeration condenser. Table 12.4.36 lists approximate refrigeration equipment energy consumption values.

In most systems the heat is rejected to the atmosphere, but in some cases, water-cooled refrigeration condensers are used to transfer the rejected heat to surface water or to well water. Some examples of the most common systems used to transfer rejected heat to the atmosphere are shown in Fig. 12.4.20. The **air-cooled condenser and evaporative condenser** are arranged with the refrigeration condenser in the outdoor airstream. Refrigerant is piped directly with no intervening heat transfer. The **air-cooled condenser** is the simplest arrangement, but the condenser must operate at temperatures sufficiently high so that condensing takes place well above the outdoor ambient dry-bulb temperature. The **evaporative condenser** adds a spray system to permit condensing above the outdoor ambient wet-bulb temperature.

Closed-Circuit Coolers For larger systems and where extensive piping of refrigerant is either not desired or otherwise precluded, a water-cooled condenser is utilized. The rejected heat is transferred to water piped to an outdoor radiator—a closed-circuit dry cooler or evaporatively cooled device. The **closed-circuit dry cooler**, or **radiator**, is a simple indirect cooling arrangement, but can operate at temperatures higher than the air-cooled condenser because of the additional heat-transfer process involved.

Table 12.4.36 Refrigeration Energy Consumption per 12,000 Btu/h Cooling Capacity

	Electric consumption, kW/ton cooling capacity	
	Type of heat rejection, kW/ton	
Item	Mechanical-draft cooling tower	Air-cooled condensers
Room-air conditioners		1.2–2.45*
Refrigeration compressors (motor-driven):		
Reciprocating (20 to 100 tons)	1.03–0.86	1.27–1.09
Centrifugal-hermetic	0.94–0.74	
Centrifugal-external drive (350 to 1,250 tons)	1.02–0.81	
Auxiliary equipment:		
Pump, cooling water	0.15	
Blower for air conditioner	0.10	0.10
Blower for rejection air	0.15	0.15

	Steam-consumption rate lb/h/ton (kg/h/ton) cooling capacity	
	Steam rate	
Item	lb/h/ton	kg/h/ton
Centrifugal compressors (steam-driven):		
Condensing (27 in vacuum)	14.5	6.6
Noncondensing (14.7 psig)	30–35	13.6–15.9
Absorption chillers (100 to 1,200 ton) (low-pressure steam)	18–18.5	8.2–8.4
Absorption chillers (high-pressure steam)	11–11.5	5–5.2

	Gas-consumption rate, Btu/ton (J/ton) cooling capacity	
	Gate rate	
Item	Btu/ton	kJ/ton
Refrigeration compressors (natural-gas-engine-driven)	8,500	8,970

* Depending on the EER (energy efficiency ratio) rating for the different manufacturers.

The dry cooler water circuit operating with antifreeze permits year-round operation.

The **closed-circuit evaporative cooler** has an indirect condenser water circuit, but uses a spray system and evaporative cooling to reduce the head pressure. In both the dry cooler and closed-circuit cooler there is no contact of either condenser water or antifreeze with the outdoor atmosphere; often this proves highly desirable.

Cooling Towers In large systems and systems where maximum overall efficiency is important, a portion of the condenser water is evaporated into the atmosphere to cool the remaining condenser water. The water lost by evaporation must be made up, usually from a domestic water source. Figure 12.4.20 shows two cooling tower configurations. The **counterflow** arrangement requires a spray system and is usually taller than the crossflow type. **Crossflow** towers have simpler gravity distribution systems.

All cooling towers and all evaporative spray systems that operate during subfreezing weather must be protected against freezing. A number of methods are available to add heat to the piping circuits and the basins or sumps.

Condenser Water System For electrically driven refrigeration compressors a temperature differential of 10°F (5.5°C) may be assumed, and for steam-driven equipment a temperature differential of 20°F (11°C) is usual. In the latter case the refrigeration and steam con-

densers are piping in series with a temperature rise of approximately 10°F (5.5°C) each.

Condenser water, gal/min

$$= \frac{\text{Total heat rejected, Btu/h}}{500 \times \text{temperature differential, °F}} \quad (12.4.26a)$$

In SI units,

Condenser flow, m³/s

$$= \frac{\text{Total heat rejected, kW}}{4190 \times \text{temperature differential, °C}} \quad (12.4.26b)$$

Atmospheric Evaporative Cooling Equipment The lowest temperature to which water can be cooled in atmospheric cooling equipment is the wet-bulb temperature of the ambient air.

Water-Cooling Effectiveness in Percent

$$E = \frac{(\text{hot-water temp.} - \text{cold-water temp.}) \times 100}{\text{hot-water temp.} - \text{wet-bulb temp. entering}} \quad (12.4.27)$$

The cold-water temperature must be chosen to place the requirement within the effectiveness range of the equipment used.

Makeup Water for Atmospheric Cooling Equipment Makeup water is introduced to replace losses due to evaporation, drift, and blowdown. If all water were cooled by evaporation, the loss by evaporation for the usual 10°F (5.5°C) cooling range would be:

$$\text{Evaporation, \%} = \frac{Q \times 100}{8.3 \times \text{gal/min} \times \text{hfg}} \quad (12.4.28a)$$

where Q = total heat rejected. Btu/h; gal/min = total condenser water circulated; and hfg = evaporation heat of water, Btu/lb, at ambient design temperature. In SI units:

$$\text{Evaporation, \%} = \frac{Q' \times 100}{1,000 \times \text{m}^3/\text{s} \times \text{hfg}} \quad (12.4.28b)$$

where Q' = total heat rejected, kJ/min; m³/s = total condenser water circulated; and hfg = evaporation heat of water, kJ/kg, at ambient design temperature.

Drift losses depend on the tower design, but generally, from the cooling tower, they are limited to 0.2 percent of the circulated rate.

The makeup water replacing losses due to evaporation, drift, and blowdown introduces dissolved solids into the system.

To prevent excessive concentration, a portion of the circulating water is wasted. The quantity of blowdown depends on the original quantity of dissolved solids in the makeup water and the permissible concentration.

For most open systems, chemical water treatment is used to inhibit corrosion and minimize blowdown.

Water Distribution, Chilled Water Systems' Temperature Differential Design temperature differentials for chilled-water systems are determined by the need to supply chilled water at an average temperature sufficiently low to provide adequate cooling and dehumidification and the desire to minimize the costs of piping, pumping energy, and refrigeration. These differentials normally range from 10°F (5.6°C) for average systems to as much as 15°F (8.3°C) or more for very large systems with widely separated loads.

Volume flow rate, gal/min

$$= \frac{(\text{total load} + \text{piping heat gains} + \text{pump heat}), \text{Btu/h}}{500 \times \text{temperature differential, °F}} \quad (12.4.29a)$$

A comparable equation in SI units is

Volume flow rate, m³/s

$$= \frac{(\text{total load} + \text{piping heat gains} + \text{pump heat}), \text{kW}}{4,190 \times \text{temperature differential, °C}} \quad (12.4.29b)$$

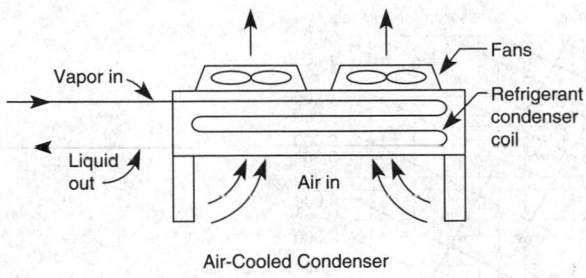

Air-Cooled Condenser

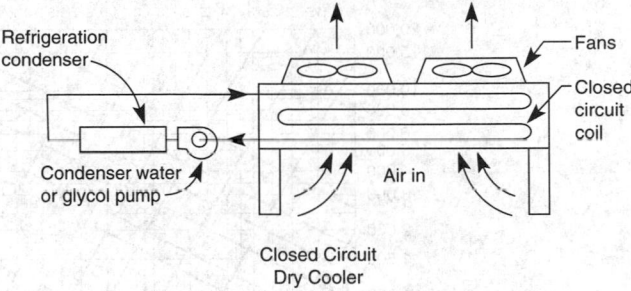

Closed Circuit
Dry Cooler

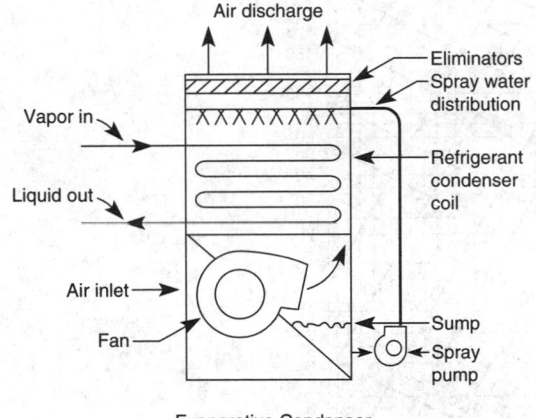

Evaporative Condenser

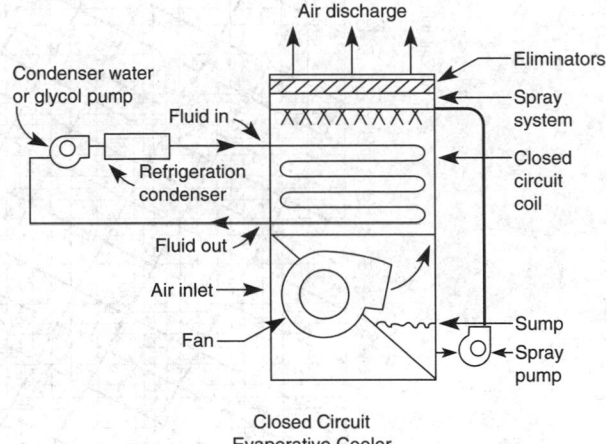

Closed Circuit
Evaporative Cooler

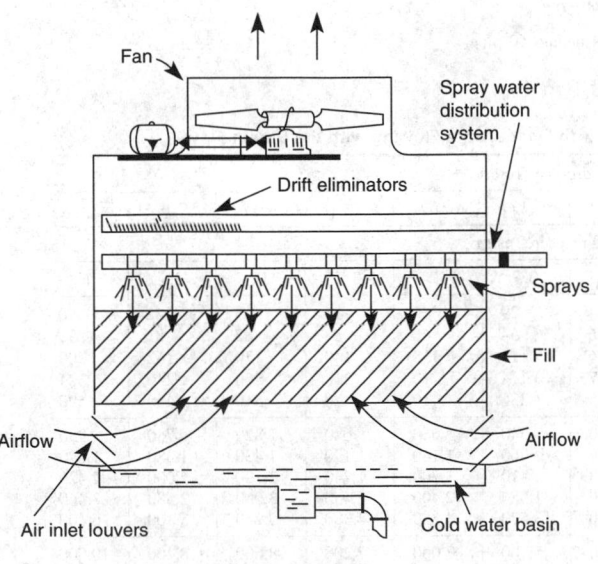

Counterflow
Cooling Tower

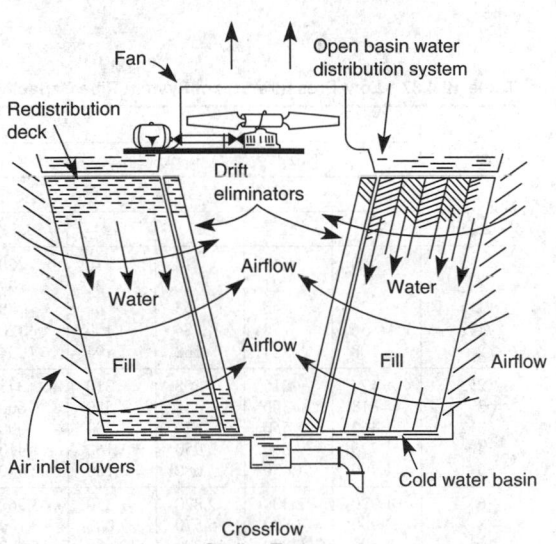

Crossflow
Cooling Tower

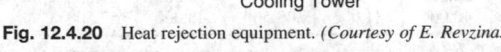

Fig. 12.4.20 Heat rejection equipment. *(Courtesy of E. Revzina.)*

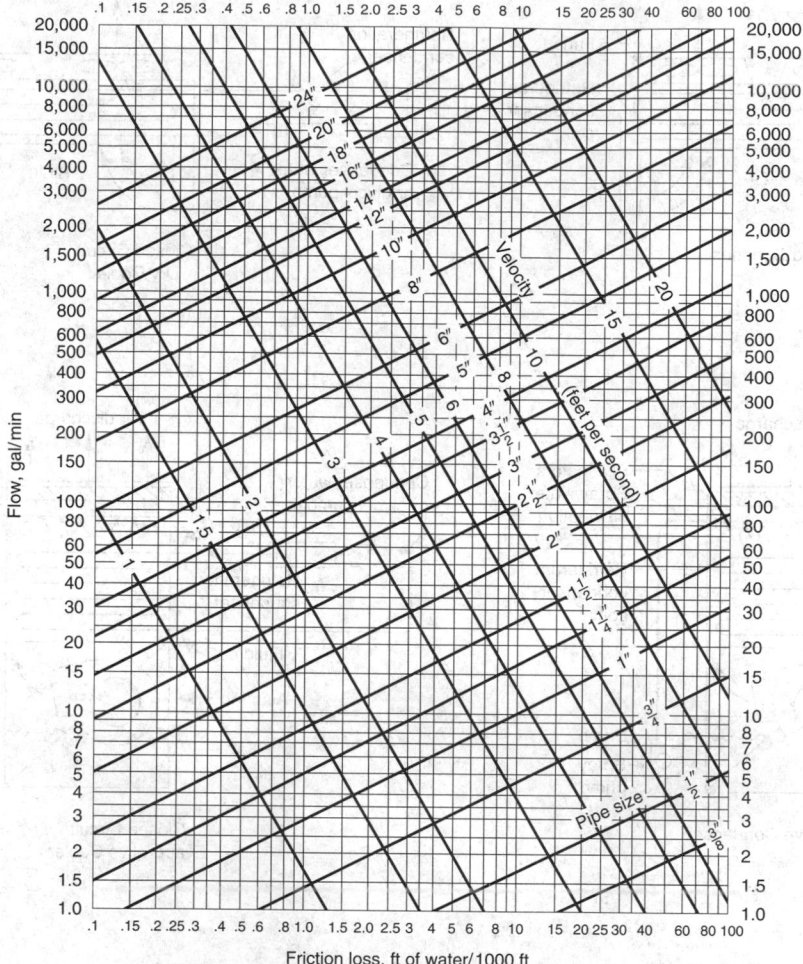

Fig. 12.4.21 Friction loss for open piping systems using schedule 40 pipe.

Table 12.4.37 Low-Pressure Steam System Pipe Capacities (lb/h), with Condensate Flowing with the Steam Flow

Nom. pipe size, in	Pressure drop per 100 ft											
	1/16 psi (1 oz)		1/8 psi (2 oz)		1/4 psi (4 oz)		1/2 psi (8 oz)		3/4 psi (12 oz)		1 psi	
	Saturated pressure (psig)											
	3.5	12	3.5	12	3.5	12	3.5	12	3.5	12	3.5	12
3/4	9	11	14	16	20	24	29	35	36	43	42	50
1	17	21	26	31	37	46	54	66	68	82	81	95
1¼	36	45	53	66	78	96	111	138	140	170	162	200
1½	56	70	84	100	120	147	174	210	218	260	246	304
2	108	134	162	194	234	285	336	410	420	510	480	590
2½	174	215	258	310	378	460	540	660	680	820	780	950
3	318	380	465	550	660	810	960	1,160	1,190	1,430	1,380	1,670
3½	462	550	670	800	990	1,218	1,410	1,700	1,740	2,100	2,000	2,420
4	726	800	950	1,160	1,410	1,690	1,980	2,400	2,450	3,000	2,880	3,460
5	1,700	1,430	1,680	2,100	2,440	3,000	3,570	4,250	4,380	5,250	5,100	6,100
6	1,920	2,300	2,820	3,350	3,960	4,850	5,700	7,000	7,200	8,600	8,400	10,000
8	3,900	4,800	5,570	7,000	8,100	10,000	11,400	14,300	14,500	17,700	16,500	20,500
10	7,200	8,800	10,200	12,600	15,000	18,200	21,000	26,000	26,200	32,000	30,000	37,000
12	11,400	13,700	16,500	19,500	23,400	28,400	33,000	40,000	41,000	49,500	48,000	57,500

NOTE: The weight flow rates at 3.5 psig can be used to cover saturated pressures from 1–6 psig and the weight flow rates at 12 psig can be used to cover saturated pressures from 8–16 psig with an error not exceeding 8%.

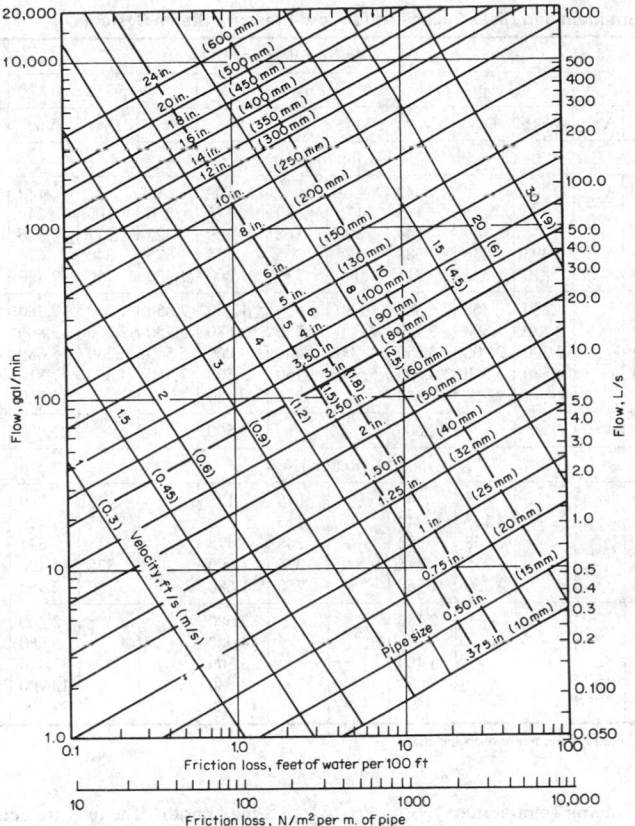

Fig. 12.4.22 Friction loss for closed piping systems using schedule 40 pipe.

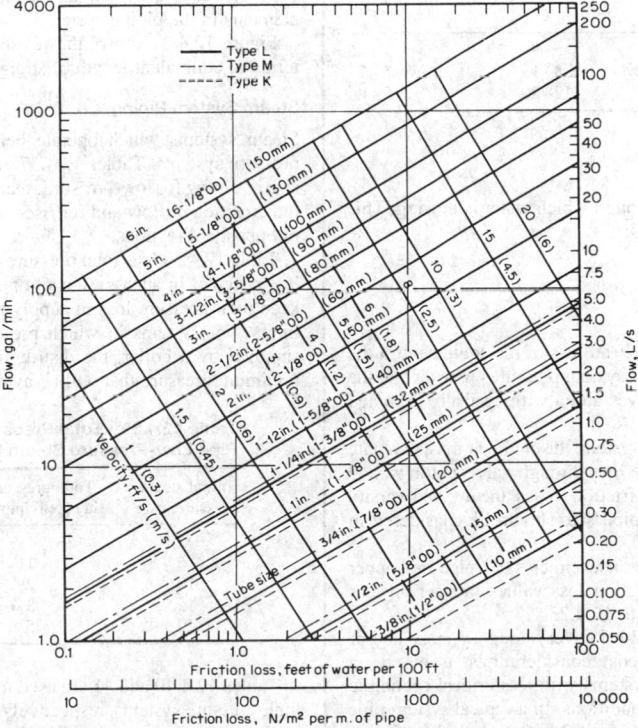

Fig. 12.4.23 Friction loss for closed and open piping systems using copper tubing.

Table 12.4.38   Return Main and Riser Capacities for Low-Pressure Steam Systems

Pipe size, in	1/32 psi (1/2 oz)			1/24 psi (2/3 oz)			1/16 psi (1 oz)			1/8 psi (2 oz)			1/4 psi (4 oz)		
	Wet*	Dry	Vac	Wet*	Dry	Vac	Wet*	Dry	Vac	Wet*	Dry	Vac	Wet*	Dry	Vac
Return mains															
3/4						42			100			142			200
1	125	62		145	71	143	175	80	175	250	103	249	350	115	350
1¼	213	130		248	149	244	300	168	300	425	217	426	600	241	600
1½	338	206		393	236	388	475	265	475	675	340	674	950	378	950
2	700	470		810	535	815	1,000	575	1,000	1,400	740	1,420	2,000	825	2,000
2½	1,180	760		1,580	868	1,360	1,680	950	1,680	2,350	1,230	2,380	3,350	1,360	3,350
3	1,880	1,460		2,130	1,560	2,180	2,680	1,750	2,680	3,750	2,250	3,800	5,350	2,500	5,350
3½	2,750	1,970		3,300	2,200	3,250	4,000	2,500	4,000	5,500	3,230	5,680	8,000	3,580	8,000
4	3,880	2,930		4,580	3,350	4,500	5,500	3,750	5,500	7,750	4,830	7,810	11,000	5,380	11,000
5						7,880			9,680			13,700			19,400
6						12,600			15,500			22,000			31,000
Return risers															
3/4		48			48	143		48	175		48	249		48	350
1		113			113	244		113	300		113	426		113	600
1¼		248			248	388		248	475		248	674		248	950
1½		375			375	815		375	1,000		375	1,420		375	2,000
2		750			750	1,360		750	1,680		750	2,380		750	3,350
2½						2,180			2,680			3,800			5,350
3						3,250			4,000			5,680			8,000
3½						4,480			5,500			7,810			11,000
4						7,880			9,680			13,700			19,400
5						12,600			15,500			22,000			31,000

* Vacuum values may be used for wet return risers and mains.

Water Distribution, Chilled-Water Systems Temperature Differential

	Temp rise	
Application	°C	°F
Close-coupled system on one floor	5–8	3–4.5
Two- or three-story building	8–11	4.5–6
Multistory building	12–20	6.5–11

Water Piping Design

There is a **friction loss** in any pipe through which water is flowing. This loss depends on the following factors:

1. Water velocity
2. Pipe diameter
3. Interior surface roughness
4. Pipe length

System pressure has no effect on the head loss of the equipment in the system. However, higher than normal system pressures may dictate the use of heavier pipe, fittings, and valves along with specially designed equipment.

To properly design a water piping system, the engineer must evaluate not only the pipe friction loss but the loss through valves, fittings, and other equipment. In addition to these friction losses, the use of diversity in reducing the water quantity and pipe size should be considered in designing the water piping system.

Most air conditioning applications use either steel pipe or copper tubing in the piping system. For friction loss values in steel pipe or copper tubing, refer to Figs. 12.4.21 to 12.4.23.

Pipe size determination, especially for very long systems and/or unusual construction or operating cost considerations, is based on an economic analysis, but smaller systems with less critical economic factors are often based on average conditions. In a typical system pipe sizes are selected for a pressure loss of from 3 to 6 ft per 100 ft of pipe.

Pipe Length   The total friction loss in a water piping system is the sum of losses in:

1. The total straight lengths of pipe
2. The equivalent lengths of pipe due to fittings, valves, and other elements in the piping system.

Figure 12.4.24 shows the resistance of some valves and fittings in terms of equivalent length of straight pipe.

Steam System Piping

Steam systems which operate below 15 psig are classified as low-pressure systems. Tables 12.4.37 and 38 may be used to determine the size of piping for low-pressure steam and condensate systems pitched in the direction of flow and for risers in which steam and condensate flow in opposite directions.

Table 12.4.39 lists total pressure drop limitations for long-length piping systems. In such systems, pipe sizes may have to be increased to keep total pressure loss in supply and return piping within the system limits. For systems in which pressure may vary, i.e., when steam is supplied by a boiler, the distribution pressure must be based on the minimum pressure that will be available.

Table 12.4.39   Total Pressure Drop for Two-Pipe Low-Pressure Steam Piping Systems

Initial steam pressure, psig	Total pressure drop in supply piping, psi	Total pressure drop in return piping, psi
2	½	½
5	1¼	1¼
10	2½	2½
15	3¾	3¾
20	5	5

Tables 12.4.40 and 41 are used to size piping for use in medium- and high-pressure systems, respectively, and assume piping is installed with proper pitch and adequate provisions to remove condensate.

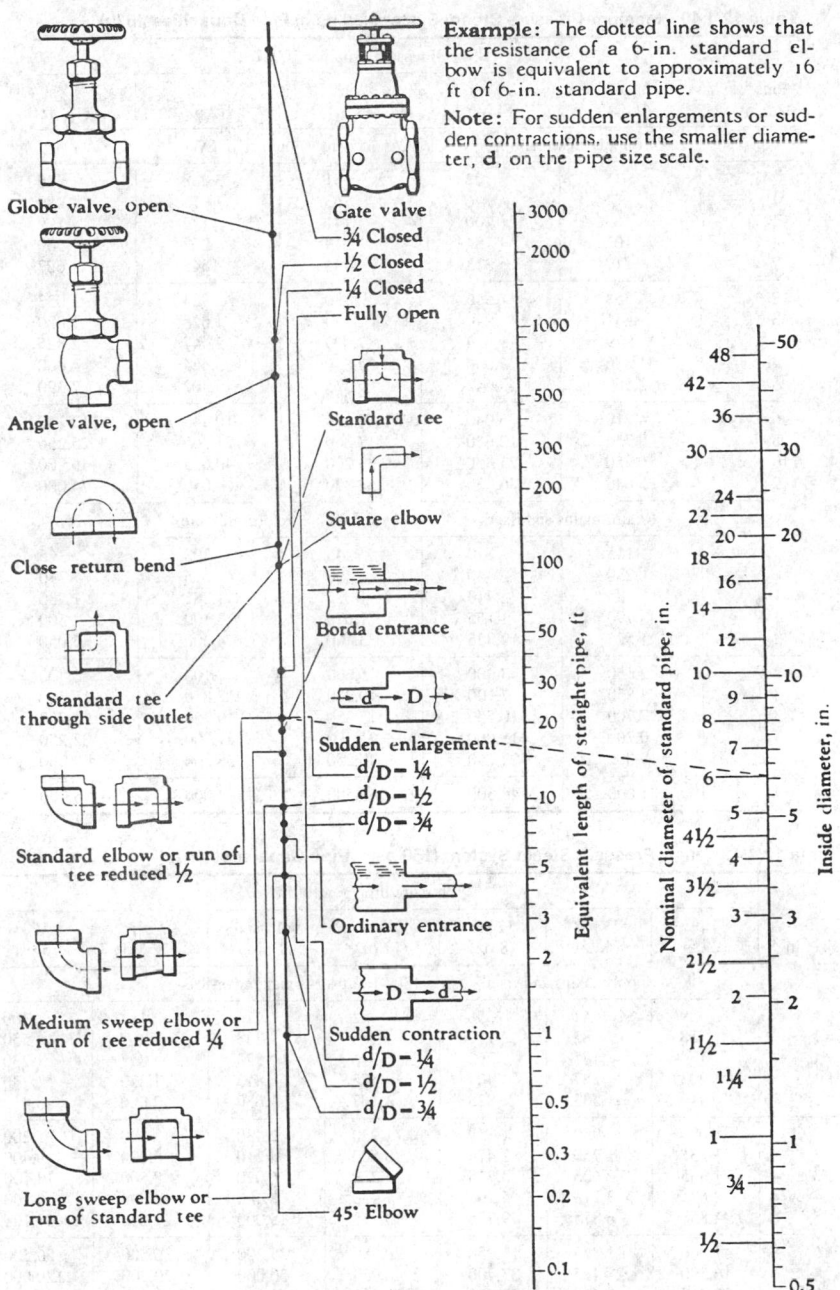

Example: The dotted line shows that the resistance of a 6-in. standard elbow is equivalent to approximately 16 ft of 6-in. standard pipe.

Note: For sudden enlargements or sudden contractions, use the smaller diameter, **d**, on the pipe size scale.

Fig. 12.4.24   Resistance of valves and fittings in terms of equivalent length of straight pipe. *(Crane Co.)*

Table 12.4.40   Medium-Pressure Steam System (30 psig) Pipe Capacities (lb/h)

Pipe size, in	Pressure drop per 100 ft				
	⅛ psi (2 oz)	¼ psi (4 oz)	½ psi (8 oz)	¾ psi (12 oz)	1 psi (16 oz)
Supply mains and risers		25–35 psig—max. error 8%			
¾	15	22	31	38	45
1	31	46	63	77	89
1¼	69	100	141	172	199
1½	107	154	219	267	309
2	217	313	444	543	627
2½	358	516	730	924	1,033
3	651	940	1,330	1,628	1,880
3½	979	1,414	2,000	2,447	2,825
4	1,386	2,000	2,830	3,464	4,000
5	2,560	3,642	5,225	6,402	7,390
6	4,210	6,030	8,590	10,240	12,140
8	8,750	12,640	17,860	21,865	25,250
10	16,250	23,450	33,200	40,625	46,900
12	25,640	36,930	52,320	64,050	74,000
Return mains and risers		0–4 psig—max. return pressure			
¾	115	170	245	308	365
1	230	340	490	615	730
1¼	485	710	1,025	1,285	1,530
1½	790	1,155	1,670	2,100	2,500
2	1,575	2,355	3,400	4,300	5,050
2½	2,650	3,900	5,600	7,100	8,400
3	4,850	7,100	10,250	12,850	15,300
3½	7,200	10,550	15,250	19,150	22,750
4	10,200	15,000	21,600	27,000	32,250
5	19,000	27,750	40,250	55,500	60,000
4	31,000	45,500	65,500	83,000	98,000

Table 12.4.41   High-Pressure Steam System (150 psig) Pipe Capacities (lb/h)

Pipe size, in	Pressure drop per 100 ft						
	⅛ psi (2 oz)	¼ psi (4 oz)	½ psi (8 oz)	¾ psi (12 oz)	1 psi (16 oz)	2 psi (32 oz)	5 psi
Supply mains and risers		130–180 psig—max, error 8%					
¾	29	41	58	82	116	184	300
1	58	82	117	165	233	369	550
1¼	130	185	262	370	523	827	1,230
1½	203	287	407	575	813	1,230	1,730
2	412	583	825	1,167	1,650	2,000	3,410
2½	683	959	1,359	1,920	2,430	3,300	5,200
3	1,237	1,750	2,476	3,500	4,210	6,000	9,400
3½	1,855	2,626	3,715	5,250	6,020	8,500	13,100
4	2,625	3,718	5,260	7,430	8,400	12,300	19,200
5	4,858	6,875	9,725	13,750	15,000	21,200	33,100
6	7,960	11,275	15,950	22,550	25,200	36,500	56,500
8	16,590	23,475	33,200	46,950	50,000	70,200	120,000
10	30,820	43,430	61,700	77,250	90,000	130,000	210,000
12	48,600	68,750	97,250	123,000	155,000	200,000	320,000
Return mains and risers		1–20 psig—max, return pressure					
¾	156	232	360	465	560	890	
1	313	462	690	910	1,120	1,780	
1¼	650	960	1,500	1,950	2,330	3,700	
1½	1,070	1,580	2,460	3,160	3,800	6,100	
2	2,160	3,300	4,950	6,400	7,700	12,300	
2½	3,600	5,350	8,200	10,700	12,800	20,400	
3	6,500	9,600	15,000	19,500	23,300	37,200	
3½	9,600	14,400	22,300	28,700	34,500	55,000	
4	13,700	20,500	31,600	40,500	49,200	78,500	
5	25,600	38,100	58,500	76,000	91,500	146,000	
6	42,000	62,500	96,000	125,000	150,000	238,000	

12.5 ILLUMINATION

by Abraham Abramowitz

REFERENCES: Amick, "Fluorescent Lighting Manual," McGraw-Hill. "IES Lighting Handbook" (1981). Design publications of General Electric Co., North American Philips Co. (successor to Westinghouse Electric Co. Lamp Division), and GTE-Sylvania.

BASIC UNITS

Candela, cd (formerly candle) is the unit of luminous intensity of a light source. One candela is defined as the luminous intensity in a given direction, of a source that emits monochromatic radiation of frequency 540×10^{12} Hz (approximately 555 nm) and of which the radiant intensity in that direction $\frac{1}{683}$ W per steradian (W/sr).

Lumen, lm, is the unit of luminous flux ϕ. It is equal to the flux on a unit surface all points of which are one unit distant from a uniform point source of one candela. Such a point source emits 4π lumens. (For an additional definition of lumen, see the following material on vision.)

Illuminance E is the density of luminous flux on a surface. If the foot is taken as the unit of length and the flux is uniformly distributed over the surface, the density in **lumens per square foot** is called **footcandles, fc;** in SI units **lumens per square metre, lux (lx),** is used. (One footcandle equals 10.76 lux.) In order to make the units comparable, dekalux (10 lux) is frequently used.

The term *illumination* is frequently used for the word *illuminance*. Modern practice reserves *illumination* for the process of lighting and *illuminance* for the result.

Luminance is the luminance intensity of any surface in a given direction per unit of projected area of the surface as viewed from that direction. The unit of luminance is candela/in²; in SI units cd/m² is used. (1 cd/in² = 1,550 cd/m².) In general, a luminous surface will have a different luminance when viewed from different angles. An important exception is a **perfectly diffuse reflecting (lambertian) surface** which has a constant luminance regardless of the viewing angle. If such a surface has a luminance of 1 cd/in², it emits 452 lm/ft². **Footlamberts, fL,** in lumens per square foot, is the unit of luminance applied to this case. While this conversion applies only to the perfectly diffuse case, it is frequently used in all cases. Thus, a perfectly diffuse surface with a luminance of 1 cd/in² is said to have a luminance of 452 fL. In practice the average lumens emitted per square foot of surface is taken to be the footlamberts. This conversion practice is deprecated.

Subjective brightness is the subjective attribute of any light sensation giving rise to the whole scale of qualities of becoming bright, light, brilliant, dim, or dark. Unfortunately, the term "brightness" often is used when referring to luminance.

NOTE. The above definitions are adapted from the "IES Lighting Handbook."

Absorption, reflection, and transmission are the general processes by which incident light flux interacts with a medium. **Absorption** is the process whereby incident flux is dissipated. **Reflection** is the process by which the incident flux leaves a surface or medium from the incident side.

NOTE. Reflection may occur as from a mirror (specular reflection), or it may be reflected at angles different from that of the incident flux to incident plane (diffuse reflection), or it may be a combination of the two types of reflection.

Transmission is the process by which incident flux leaves a surface or medium on a side other than the incident side. If the light ray is reduced only in intensity, the transmission is called *regular*. If the ray emerges in all directions, transmission is called *diffuse*. Both modes may exist in combination.

The incident flux ϕ_i equals the flux absorbed ϕ_a, reflected ϕ_r, and transmitted ϕ_t. That is,

$$\phi_i = \phi_a + \phi_r + \phi_t$$

Dividing this equation by ϕ_i, we obtain

$$1 = \phi_a/\phi_i + \phi_r/\phi_i + \phi_t/\phi_i$$

or

$$1 = \alpha + \rho + \tau$$

α is the *absorptance*, ρ is the *reflectance*, and τ is the *transmittance*. In each case, the incident flux may be restricted to a single wavelength, a particular direction, and a given solid angle. These must be specified.

The **wavelength** of electromagnetic radiation is measured in metres. For the frequencies involved in illumination, the wavelength is given in nanometres, nm, equal to 10^{-9} m, and micrometres, μm, equal to 10^{-6} m.

VISION

Most engineering designs, (bridges, structures, roads etc.) are based on strength and are not concerned with the way the human organism reacts. The **response of the eye** is central to illuminating engineering. The **lens** of the **eye** focuses an image on the **retina**. Here a photochemical process takes place which sends nerve impulses to the brain via the optic nerve. The amount of light entering the eye is controlled by the **pupil**. The normal eye automatically accommodates itself to focus on an object, while the pupil adjusts itself to allow for a high or low level of object luminance. The sensors in the eye are known as **rods** and **cones**. The cones are clustered in a small central part of the retina called the **fovea.** They transmit a sharp image to the brain and give color response. Outside the fovea the rods predominate. They give neither a sharp image nor a color response. When the luminance of the visual field is 0.01 fL or lower, as at night, seeing is due to the rods only and is called **scotopic vision.** At higher levels, with the cones primarily involved, seeing is called **photopic vision.** There is an intermediate region called **mesopic vision.**

The response of the eye to colors of different wavelengths is given in Fig. 12.5.1. Note the shift in maximum response at lower luminance levels called the "Purkinje shift." Note that these curves are relative ones, and that the two peaks do not correspond to the same levels of illumination. The **luminous efficacy** (lumen output per radiated watt) is 683 lm/W at the wavelength of maximum photopic response 555 nm. For white light, radiation which has the characteristic of an equal energy spectrum with all the energy in the visual region, it is approximately 220 lm/W.

Spectral Lumen If the response curve of the eye for photopic vision, versus λ in nanometers, is expressed as $k(\lambda)$, and the spectral power function of the source in watts per nanometer is taken to be $Q_e(\lambda)$, then the luminous flux is given by the equation

$$\phi_{\text{lumens}} = 683 \int_{380}^{780} k(\lambda) \, Q_e(\lambda) \, d(\lambda) \tag{12.5.1}$$

LIGHT METERS

Early light meters compared the luminance of a diffuse highly reflecting surface with that obtained from a calibrated standard. The most common light meter in use today is similar to a photographic exposure meter. A photovoltaic cell is directly connected to a sensitive microammeter calibrated in footcandles (or dekalux). The best meters (called color-corrected) have a response similar to that of the eye in photopic vision. Special shapes are used on the cover to avoid total reflection of

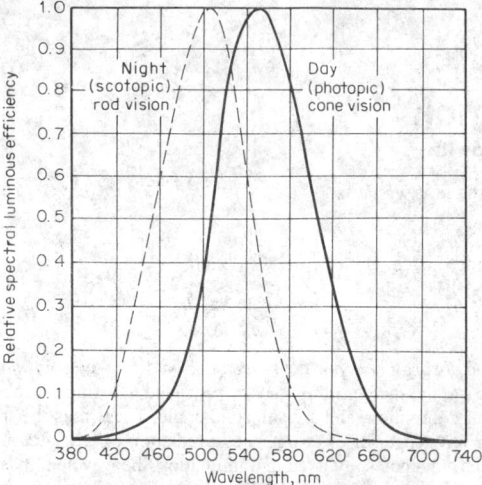

Fig. 12.5.1 Relative spectral luminous efficiency curves for photopic and scotopic vision, showing the Purkinje shift on the wavelength of maximum efficiency. Note the wavelength of the visual region of the electromagnetic spectrum. (*IESNA Lighting Handbook, 5th ed. This material has been modified from its original version and is not reflective of its original form as recognized by the IESNA.*)

light from the glass surface of the cell. Such meters are said to be cosine law corrected. The microammeter is frequently replaced with an electronic amplifier using an analog or digital readout.

LIGHT SOURCES

The original and still major source of light is the sun. Next came fire, derived from candles, oil, and gas lamps. With the discovery of electricity came arc lamps, gas-discharge lamps, and hot-filament lamps. "Flame" or hot sources give a continuous spectrum. Gas-discharge devices such as neon lamps and mercury-arc lamps give discrete, or line, spectra. The lines may be modified in various ways: by pressure broadening, use of phosphor coatings (to convert ultraviolet radiation into visible light), and using a mixture of gases. The continuous spectra of phosphors have colors which depend upon the mixture used. Light-emitting diodes, LED, consisting of a layer of two different semiconductors, are in use for display purposes.

Color Temperature and Luminance

In general, three quantities are required to specify the color of a light and its luminous level. However, an approximate designation is used by specifying the temperature of a hot (black-body) emitter whose color almost matches that of the light. The **color temperature** of daylight is about 6000 K and that of tungsten lamps about 2300 to 3300 K.

Table 12.5.1 Approximate Luminances of Various Light Sources (IES)

Light source	Approximate average luminance	
	cd/in²	kcd/m²
Clear sky	5.16	8
Candle flame (sperm)	6.45	10
60-W inside frosted bulb	77.4	120
60-W "white bulb"	19.35	30
Fluorescent lamp, cool white, T-12 bulb, medium loading	5.3	8.2
High-intensity mercury-arc type H33, 2.5 atm	968	1,500
Clear glass neon tube 15 mm, 60 mA	1.03	1.6

Different light sources have markedly different luminances as shown in Table 12.5.1. "Large" sources have low luminances, while "small" sources have high luminances.

Lamps

Electric lamps are the principal source of artificial light in common use. They convert electrical energy into light or radiant energy.

An **incandescent-filament lamp** contains a filament which is heated by the current passing through it. The filament is enclosed in a glass bulb which has a base suitable to connect the lamp to an electrical socket. To prevent oxidation of the filament at elevated temperature, the bulb is evacuated of air or filled with an inert gas. The bulb also serves to control the light from the incandescent filament, which is essentially a point source. High luminance of the source is typically reduced by acid etching to frost the inside surface of the bulb. Silica coating will also provide additional diffusion and can alter the color of the light emitted. Portions of the bulb's interior can be covered with reflecting material to give a predetermined direction to the emitted light. Chemical tinting of clear glass bulbs provides a variety of colors. Whenever the color that is normally produced by an incandescent filament is changed, the filtering process removes from the radiated light the energy of all wavelengths except those necessary to produce the desired color. This subtractive method of color alteration is less efficacious than the generation of light of varying colors by gaseous discharge.

Sizes and **shapes** of lamp bulbs are designated by a letter code followed by a numeral; the letter indicates the shape (Fig. 12.5.2), and the number indicates the diameter of the bulb in eighths of an inch. Thus a T-12 lamp has a tubular shape and is 1⅛ or 1½ in in diameter.

Fig. 12.5.2 Typical filament lamp shapes: S, straight; F, flame; G, globe; A, general service; T, tubular; PS, pear shape; PAR, parabolic; R, reflector.

Incandescent lamps are available with several **types of bases** (Fig. 12.5.3). Most general-service lamps have medium screw bases; larger or smaller screw bases are used depending on lamp wattage. Bipost and prefocus bases accurately position the filament, as in optical projection systems. Bipost lamps also serve where ruggedness and greater heat dissipation are required.

Fig. 12.5.3 Typical incandescent lamp bases.

Incandescent-lamp filaments are generally constructed of tungsten. Tungsten has a high melting point and a low vapor pressure, which permits high operating temperatures without evaporation: the higher the operating temperature, the higher the efficacy (lumens per watt) and the shorter the life. Filament evaporation throughout the life of the lamp causes blackening of the bulb and thinning of the filament with consequent lower light output. Argon-nitrogen gas filling reduces the rate of evaporation. Figure 12.5.4 shows steps in lamp manufacture.

Tungsten filaments are also placed in compact quartz tubes filled with a halogen atmosphere where the tungsten halide lighting source continuously returns evaporated tungsten particles to the filament. The inside walls do not blacken, and light output remains fairly constant throughout the life of the lamp.

For exposed outdoor lamp signs, **gas-filled incandescent lamps** are often used, especially when high-speed animation is to be depicted by

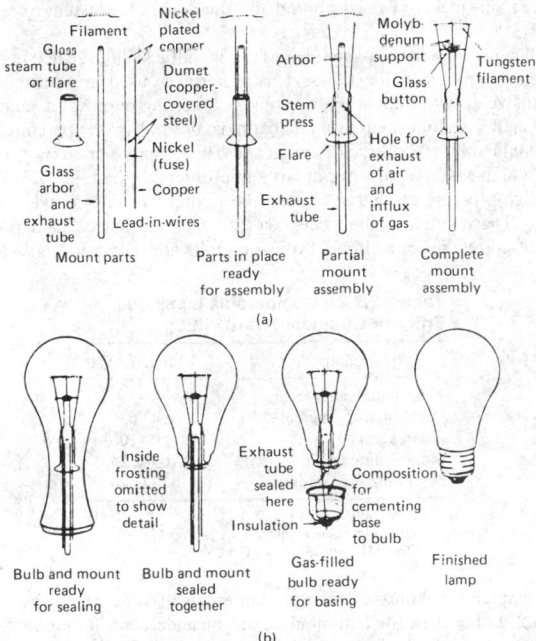

(a)

(b)

Fig. 12.5.4 Steps in the manufacture of a typical filament incandescent lamp. (*a*) Assembly of inner structure; (*b*) final assembly. (*IESNA. This material has been modified from its original version and is not reflective of its original form as recognized by the IESNA.*)

the on-off sequencing of the lamps. The lamp envelope is filled with a gas which enhances the speed with which the filament is heated and cooled. The very fast quenching of the filament without the usual decay trail of light permits a sharply defined illusion of motion.

A **fluorescent lamp** consists of a glass tube coated on the inside with phosphor powders which fluoresce when excited by ultraviolet light; filament electrodes are mounted in end seals connected to base pins (Fig. 12.5.5). The tube is filled with rare gases (such as argon, neon, and

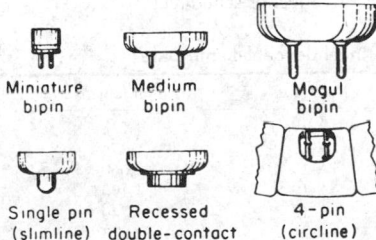

Fig. 12.5.5 Fluorescent lamp bases.

krypton) and a drop of mercury (Fig. 12.5.6), and operates at a relatively low pressure. In operation, electrons are emitted from the hot electrodes. These electrons are accelerated by the voltage across the tube until they collide with mercury atoms, causing them to be ionized and excited. When the mercury atom returns to its normal state, mercury spectral lines in both the visible and the ultraviolet region are generated. The low pressure enhances ultraviolet radiation. The ultraviolet radia-

tion excites the phosphor coating to luminance. The resulting light output is not only much higher than that obtained from the mercury lines alone but also results in a continuous spectrum with colors dependent upon the phosphors used.

As with all gas-discharge devices, these lamps have negative voltampere characteristics. Unless the voltage difference between the applied voltage and the lamp operating voltage is absorbed in some way, damaging currents will result. A reactor is used in series with the lamp. It may be capacitive or inductive (many turns of wire on an iron core). The supply voltage should be at least twice the lamp operating voltage. Where this is not the case, the supply voltage (up to 277 V is used) is stepped up by an autotransformer. The necessary reactance is frequently part of the transformer leakage inductance.

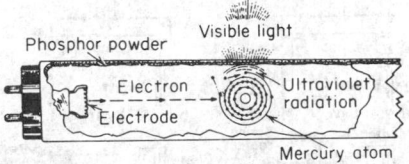

Fig. 12.5.6 Fluorescent lamp operation. (*Westinghouse.*)

For starting purposes, voltages higher than twice lamp operating voltages may be used. For minimum line current a power-factor-correcting capacitor is used and assembled with the autotransformer. A capacitor is put across the lamp to reduce radio-frequency interference with nearby radio receivers. All these elements are placed inside a case filled with a potting compound. The assembly is called a **ballast.** The object of the compound is to reduce noise from the core lamination vibrations and to improve heat dissipation. Built-in thermal protectors, which deenergize the ballast when dangerous temperatures are reached, are now required in the United States. Ballast manufacturers rate their units by noise levels. Some ballasts use solid-state, high-frequency generators.

To start a fluorescent lamp, electron emission from the electrodes must be induced. Two methods are generally employed: (1) the filament electrodes are heated by passing current through them; (2) a high voltage, sufficient to start an electric discharge in the lamp, is applied across it. Once a discharge starts, mercury-ion bombardment keeps the filaments at a hot electron-emitting temperature. Lamps are designed for either type of operation. The first group is further divided into preheat lamps and rapid-start lamps. Some lamps can be used for both types of circuits. The lamp current is carried primarily by electrons emitted from the filaments. The end or weakening of electron emission is an important cause of the end of lamp life.

Preheat circuits contain starters (Fig. 12.5.7) which are switches, closed when power is first applied, permitting current to flow and preheat the electrodes. After a predetermined period of time, the starter switch opens, throwing a potential across the lamp which starts the discharge. Lamps used for preheat circuits have bipin bases (Fig. 12.5.5).

Instant-start circuits have ballasts which apply sufficient voltage across the lamp to induce current flow without preheating the electrodes. **Slimline lamps** are the principal instant-start type. They have single-pin bases because no preheating is required, are available in sizes up to 8 ft in length, and can have varying lumen outputs dependent upon the ballast current rating and wattage. Because of their high starting voltage, they generally employ spring-loaded push-pull lampholders which disconnect the ballast circuit unless the lamps are properly seated in position.

Instant-start lamps are sometimes available with bipin bases similar to those used in preheat lamps. In these instances the lead wires from the pins are connected together inside the lamp. These lamps, marked "instant start," are not interchangeable with rapid-start equipment. Cold-cathode lamps employ the instant-start principle, have cylindrical iron electrodes, and tend to be less efficacious at shorter lengths because of high wattage losses at the electrodes. They are limited to low current densities because the electrodes operate at temperatures below that necessary for thermionic emission. Cold-cathode lamps, whose operation is not affected by dimming or flashing, have long life and are generally used for custom-built shapes and patterns that require bending, such as for electric signs.

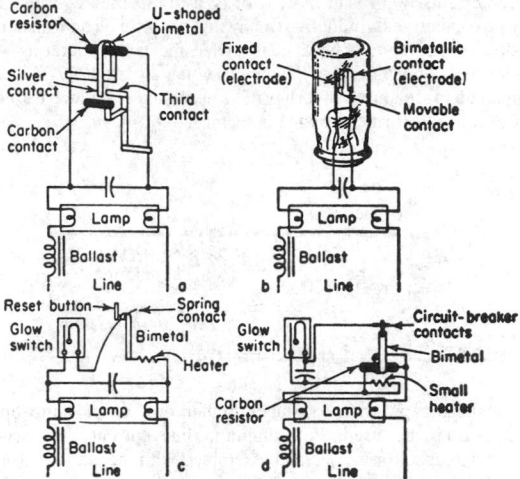

Fig. 12.5.7 Starter switches for preheat cathode circuits. *(IESNA)*

Rapid-start circuit ballasts have separate windings for the electrodes which are immediately and continuously heated when the circuit is energized. This rapid heating causes sufficient ionization in the lamp for a discharge to start from the voltage of the main ballast windings. Two-lamp rapid-start ballasts are of the series sequence type, in which the lamps start in sequence and, when fully lighted, operate in series.

A new type **screw-in fluorescent lamp** with built-in ballast can be used in a standard medium-screw socket. These lamps consume less power than incandescent lamps for the same luminance; accordingly, albeit their first cost is significantly higher, they are expected to prove to be more economical by virtue of their reduced power consumption and much longer life. The verdict of the consuming public is yet to come as significant numbers of them make their way into household and commercial applications.

Fluorescent lamps used in low-ambient-temperature applications, as in outdoor signs, are of the **high-output** (**HO**) type, and require special high-output ballasts to permit the lamps to maintain their luminance at lower operating temperatures.

Typical fluorescent lamp circuits are shown in Fig. 12.5.8.

High-intensity-discharge lamps consist of tubes in which electric arcs in a variety of materials are produced. Outer glass jackets provide thermal insulation in order to maintain the arc tube temperature. The temperature and amount of material is controlled so that the discharge operates in a vapor pressure of several atmospheres. This results in enhancing the radiation in the visible region.

Mercury-vapor lamps consist of mercury-argon-filled quartz tubes surrounded by a nitrogen-filled glass jacket. Clear lamps radiate the visible mercury lines (bluish green). Ultraviolet radiation is absorbed to some extent by the outer jackets. The color of the light and the lumen output is improved by coating the inside of the outer jackets with a phosphor. When excited by the ultraviolet radiation of the arc, the phosphors add light in the red part of the spectrum to the output. The resulting lamps are called white, color improved, or deluxe white. The lamps start by a discharge in argon between an electrode and a starting electrode (see Fig. 12.5.9). As the mercury vaporizes, the pressure builds up and the discharge transfers to a mercury discharge. This takes several minutes. After shutdown, the lamps cannot be restarted until the inner tube pressure drops so that an argon discharge can start.

Metal halide (multivapor) lamps use small quantities of sodium, thallium, scandium, dysprosium, and indium iodides in addition to the usual mercury-argon mix. Color is improved and output substantially increased over high-intensity-discharge lamps using mercury alone. While the construction is similar to mercury lamps, a bimetal switch is built into the lamp to short out the starting resistor after the lamps start. A vacuum jacket is used around the quartz discharge tube (see Fig. 12.5.9).

High-pressure sodium-vapor lamps use metallic sodium sealed in translucent aluminum oxide tubes. This material is used to withstand the corrosive effect of hot sodium vapor. For starting purposes a xenon fill gas and a sodium-mercury amalgam is used. Arc temperatures are maintained by an outer vacuum jacket. The lamp is started by generating a high-voltage pulse for about a microsecond (see Fig. 12.5.9).

High-pressure discharge lamps, like fluorescent lamps, require ballasts. These provide the necessary voltage, reactances, and power-factor-correcting capacitors. Typical circuits are shown in Fig. 12.5.8.

Table 12.5.2 Comparable Luminous Efficacies (lumens/watt)* (IES)

Lamp	Lumens/watt
Tungsten incandescent	8–33
High-intensity mercury†	24–63‡
Fluorescent†	19–100‡
Metal halide (multivapor)†	69–125
High-pressure sodium†	73–140

* Constantly being improved.
† Ballast losses not included.
‡ Depends upon lamp size, type, and color.

Comparative lamp efficacies (lumens/watt) are given in Table 12.5.2. Lamp data for commonly used incandescent, fluorescent, and high-intensity-discharge lamps are listed in Tables 12.5.3, 12.5.4, and 12.5.5.

Luminaires

Luminaires are generally categorized as **industrial, commercial,** or **residential.** Use within these categories usually determines the quality and ruggedness of materials of construction. Generally speaking, style, ornament, and in most cases low cost are prime considerations for residential fixtures. Industrial fixtures require low maintenance, low operating cost, efficiency, and durability. Commercial fixtures combine the elements of all of these and place heavy emphasis on visual comfort.

Luminaires are classified by the International Commission on Illumination (ICI) in accordance with the percentages of total luminaire output emitted above and below the horizontal (Fig. 12.5.10). Industrial fixtures usually are direct or semidirect.

Table 12.5.3 Incandescent-Lamp Data

Watts	Bulb size	Initial lumens	Rated life, h
25	A-19	230	2,500
40	A-19	474	1,500
60	A-19	1,060	1,000
75	A-19	1,190	750
100	A-19	1,740	750
150	A-21	2,873	750
200	A-23	4,000	750
300	PS-30	6,130	750
500	PS-35	10,675	1,000
750	PS-52	16,935	1,000
1,000	PS-52	23,510	1,000

For general-service lamps 115-, 120-, and 125-V service, inside frosted.
NOTE: Lamps are constantly being improved. The latest manufacturer's data should be used for accuracy.
SOURCE: ''IESNA Handbook,'' 8th ed., 1993, reprinted with permission. (This material has been modified from its original version and is not reflective of its original form as recognized by the IESNA.)

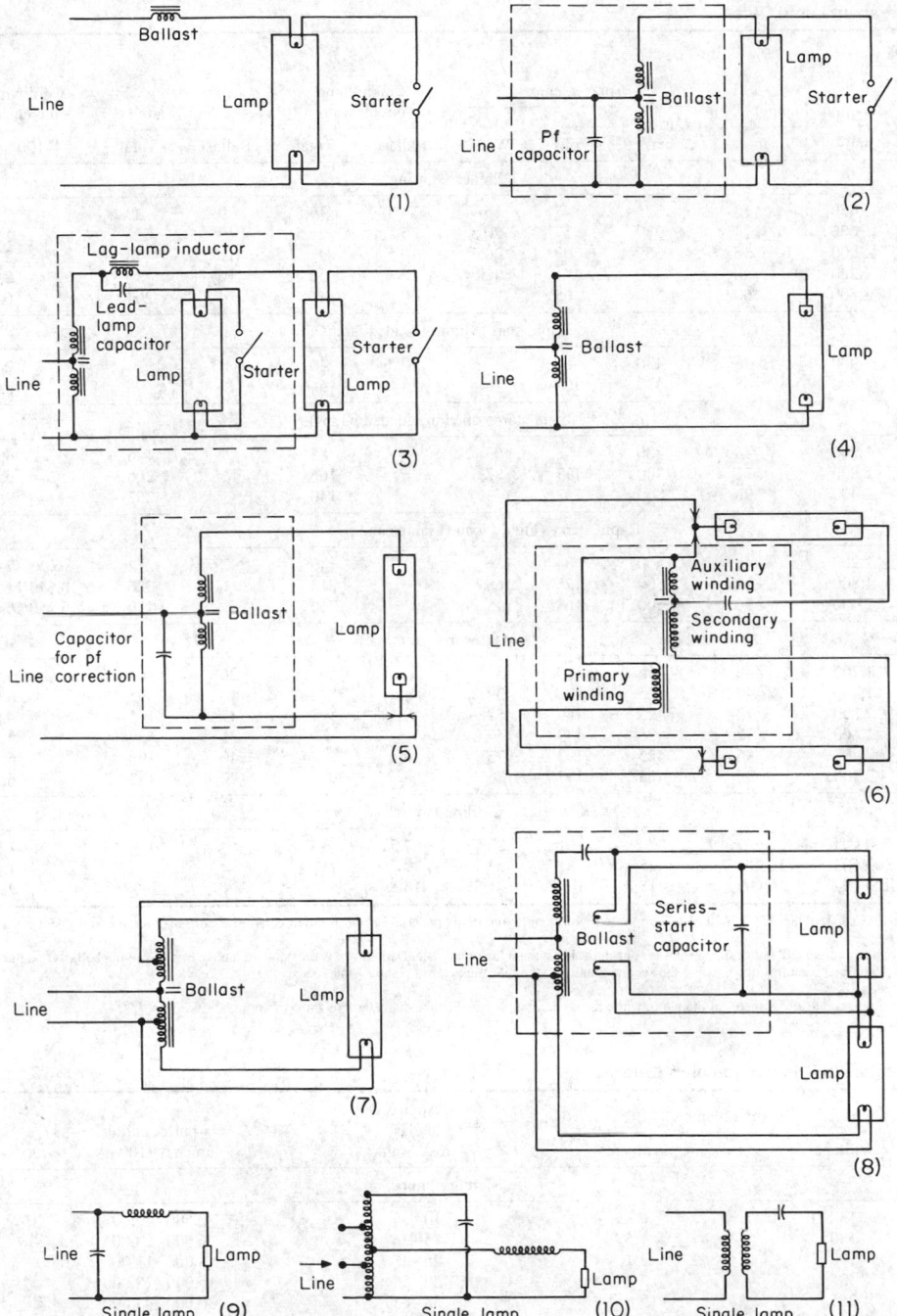

Fig. 12.5.8 Typical circuits for fluorescent and high-intensity discharge lamps. (1) Basic preheat circuit; (2) preheat circuit with autotransformer to step up voltage and capacitor to correct power factor; (3) leadlag preheat; (4) basic instant-start circuit; (5) instant-start circuit showing disconnect lampholder; (6) typical series instant-start circuit; (7) basic rapid-start circuit; (8) two-lamp series lead circuit; (9) mercury reactor ballast circuit; (10) mercury autotransformer circuit; (11) mercury stabilizing ballast circuit. *(GE.)*

Table 12.5.4   Fluorescent Lamp Data*

Lamp designation	Nominal length		Approx lamp			Single-lamp circuit watts		Two-lamp circuit watts		Cool white lumens at	Rated average life, h
	mm	in	Current (ma)	Volts	Watts	Ballast	Total	Ballast	Total	100 h	3 h burning/start
Preheat starting†											
F15T12	450	18	325	47	15	4.5	19.5	9	39	830	9,000
F20T12	600	24	380	57	20.5	5	25.5	10	51	1,283	9,000
F30T8	900	36	355	99	30.5	10.5	41	17	78	2,330	7,500
F40T12	1,200	48	430	101	40	12	52	16	96	2,150	15,000
F90T12	1,500	60	1,500	65	90	20	110	24	204	6,025	9,000
Rapid start† (lightly loaded lamps)											
F30T12	900	36	430	81	33.5	10.5	44			2,210	18,000
F40T12	1,200	48	430	101	41	13	54	13	95	3,150	21,000
Rapid start† (medium loaded lamps)											
F48T12	1,200	48	800	78	63		85		146	4,300	12,000
F72T12	1,800	72	800	117	87		106		200	6,650	12,000
F96T12	2,400	96	800	153	113		140		252	9,150	12,000
Rapid start† (highly loaded lamps and power grove§)											
F48T12/48PG17	1,200	48	1,500	84	116		146		252	6,900/7,450	9,000
F72T12/72PG17	1,800	72	1,500	125	168		213		326	10,640/11,500	9,000
F96T12/96PG17	2,400	96	1,500	163	215		260		450	15,250/16,000	9,000
Instant start† (slimline)											
F48T12 lead/lag	1,200	48	425	100	39			26	104	3,000	7,500–12,000
F48T12 series	1,200	48	425	100	39			17	95	3,000	7,500–12,000
F72T12 lead/lag	1,800	72	425	149	57			47	161	4,585	7,500–12,000
F72T12 series	1,800	72	425	149	57			25	139	4,585	7,500–12,000
F96T12 lead/lag	2,400	96	425	197	75			40	190	6,300	12,000
F96T12 series	2,400	96	425	197	75			22	172	6,300	12,000
Circline lamps¶											
C8T9	200 OD	8¼ OD	370	61	22.5	7.5	30			1,065	12,000
C12T9	300 OD	12 OD	425	81	33	9	42			1,870	12,000
C16T9	400 OD	16 OD	415	108	41.5	16.5	58			2,580	12,000

SOURCE: Adapted from "IESNA Handbook," 8th ed., 1993, reprinted with permission. (This material has been modified from its original version and is not reflective of its original form as recognized by the IESNA.)
* Lamps are continuously being improved. For design purposes consult the latest manufacturers' data. Data shown is for standard lamps. Energy-saving ballasts and fluorescent lamps are available.
† The first number is the "nominal" lamp wattage, while the second number is the tube diameter in eighths of an inch.
§ General Electric Co. trademark.
¶ The first number is the nominal outside diameter of the lamp, while the second number is the tube diameter in eighths of an inch.

Table 12.5.5   High-Intensity-Discharge Lamp Data*

Watt	Nominal lamp		Approx ballast loss, watts	Approx initial lumens† (100 h)	Life, h
	Voltage	Amperes			
Mercury lamps					
100	130	0.85	10–35	2,500–4,400	24,000+
175	130	1.5	15–35	6,000–8,600	24,000+
250	130	2.1	25–35	8,000–13,000	24,000+
400	135	3.2	20–55	15,000–23,000	24,000+
700	265	2.8	35–65	36,000–43,000	24,000+
1,000	265	4.0	40–90	43,000–63,000	24,000+
Metal-halide lamps					
175	130	1.4	35	12,000–14,000	7,500
400	135	3.2	60	31,000–40,000	15,000–20,000
1,000	250	4.3	50–100	105,000–125,000	10,000–12,000
High-pressure sodium-vapor lamps					
250	100	3.0	55–60	25,000–30,000	24,000
400	100	4.7	65–75	47,500–50,000	24,000

SOURCE: Abstracted from "IES Lighting Handbook" and General Electric Co. data.
* Lamps are continuously being improved. For design purposes, consult the latest manufacturers' data.
† Depending upon ballast used, lamps may have outputs which change with burning position.

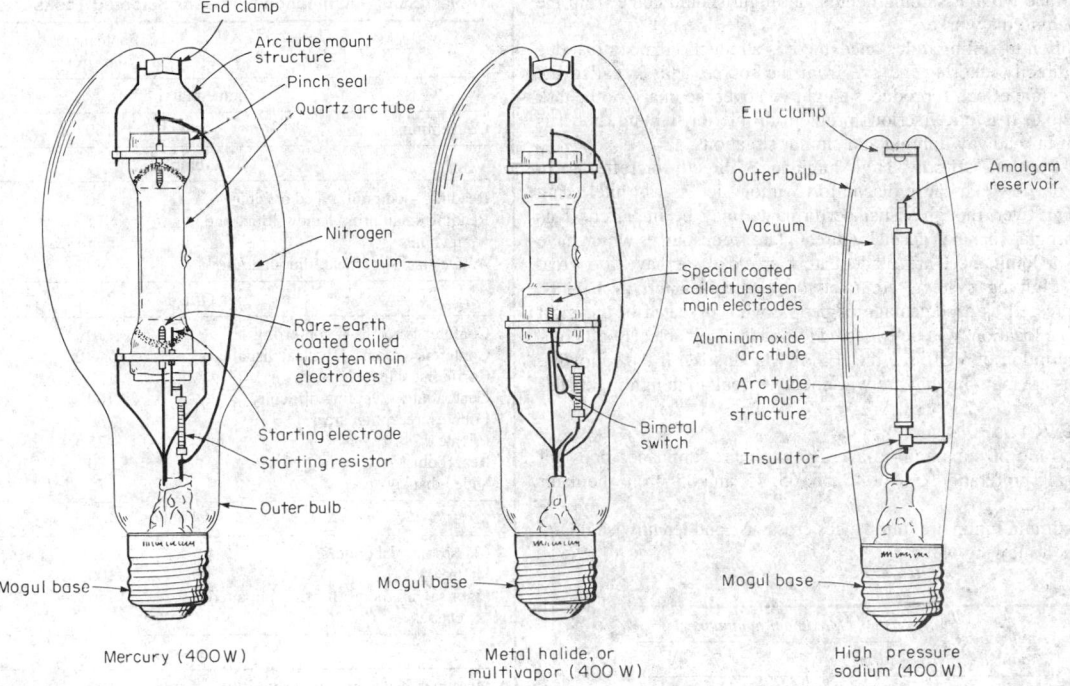

Fig. 12.5.9 High-output lamps. *(GE.)*

Luminaires control the source of light so that it can be better used for a given seeing task. Materials used in luminaires are designed to reflect, refract, diffuse, or obscure light.

Reflectors are commonly made of specular alzak aluminum, glass, baked-enamel steel, and porcelain-enamel steel.

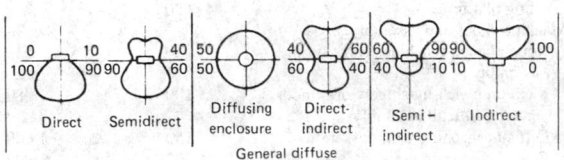

Fig. 12.5.10 Light distribution curves by ICI classifications of luminaires. Upward and downward components are in percentage ranges. *(IES.)*

Lenses with prismatic patterns refract light sources to disperse the rays or to direct them most effectively. Lenses are of glass, acrylic plastic (Plexiglas), or polystyrene. **Glass** is generally superior for incandescent fixtures because of its heat resistance. Fluorescent lamps with **acrylic** lenses have a life at least twice that of **polystyrene** because of their color stability. **Translucent** glass and plastic are used in bottom lenses, louvers, and side panels to diffuse light and to obscure light sources. Baked-enamel steel **louvers,** in egg-crate, concentric ring, or cellular configurations are widely used to shield light sources from normal viewing angles.

Fixture bodies, trims, and lens frames are commonly constructed of steel, electrogalvanized, and/or treated with a rust-inhibiting coating, and painted with several coats of baked enamel. Stamped, spun, cast, and die cast aluminum are also used.

PRESCRIBING ILLUMINATION

The object of a **lighting design** is to provide sufficient illuminance for a given seeing task without introducing discomfort. Sufficient light is not difficult to obtain with modern light sources, but unless properly placed and controlled, uncomfortable, glaring light will result.

A given task has a size, luminance contrast, luminance, and color. The luminance of a perfectly diffusing reflecting surface is given by

$$L = E\rho \qquad (12.5.2)$$

where E is illuminance in footcandles, ρ its reflectance (ratio of reflected light flux to incident flux), and luminance L is in footlamberts. The contrast C between two adjacent areas is given by

$$C = (L_b - L_o)/L_b \qquad (12.5.3)$$

where L_b is the luminance of the larger or background area and L_o is the object or task luminance for a given illuminance E. Substituting Eq. (12.5.2) into Eq. (12.5.3),

$$C = \frac{E\rho_b - E\rho_o}{E\rho_b} = (\rho_b - \rho_o)/\rho_b \qquad (12.5.4)$$

Thus, contrast is basically a function of the task-to-background reflectance. Most surfaces are not perfectly diffuse and require the use of the luminance factor β (ratio of actual luminance for a given viewing angle to the illumination under actual conditions) instead of ρ. The contrast which can just be seen or detected (minimum perceptible contrast) is a function of background luminance for a given task.

Unfortunately most visual tasks have mirrorlike (specular) or semi-mirrorlike surfaces. This results in reducing the work contrast. Where a highly luminous object such as an incandescent-lamp filament is reflected from a polished surface, "reflected glare" results. Reflections of a large luminous area by matte or semimatte material results in "veiling glare." For work on a horizontal surface, the line of sight from the eye for most people is from 0 to 40° from the vertical, with a peak angle of about 25°. If light from the source is directly reflected into the eyes, contrast and resulting visibility are greatly reduced. When a highly luminous source is directly reflected into the eye, it is possible to completely obliterate a task. Veiling reflections are controlled by proper luminaire design and placement of fixtures. Some designs keep flux out of the 0 to 40° zone. As far as possible the work area should have a matte surface without shining details, and light should come from the side or behind the worker.

In addition to veiling reflectance, there is a reduction in contrast due to light directly entering the eye from the source. This is called the disability glare effect. It produces a light veil over the image of the task on the retina. It is not a serious problem in interior lighting, but it is important in roadway lighting and similar situations.

Visual-Comfort Criteria High luminances directly or reflected in the field of view can cause discomfort without necessarily interfering with seeing even though visual performance may be impaired. This discomfort glare can be caused by direct glare from sources which have too high a luminance, are inadequately shielded, or have too great an area. Lighting systems are rated by a **visual-comfort probability, VCP,** expressed as a percentage of people who, if seated in the most undesirable location, will be expected to find it acceptable. (For a complete description of VCP, see the IESNA Handbook.) If the following conditions are met, direct glare will not be a problem in lighting installations:

1. The VCP is 70 or more.
2. The ratio of maximum-to-average luminaire luminance does not exceed 5:1 (preferably 3:1) at 45, 55, 65, 75, and 85° from the nadir crosswise and lengthwise.
3. Maximum luminaire luminances crosswise and lengthwise do not exceed the following values:

Angle above nadir, deg	Maximum luminance	
	cd/m²	fL
45	7,710	2,250
55	5,500	1,605
65	3,860	1,125
75	2,570	750
85	1,695	495

Design of Interior Lighting Systems

Lighting is as much an art as a science. While many studies have been made on what constitutes adequate lighting along with proper quality, the effect to be achieved depends upon the designer. In this section emphasis will be primarily on achieving adequate illumination.

The design approach is to consider the space to be lighted and the task to be performed. An illuminance is then selected. A suitable luminaire is picked, and calculations are made in order to determine the number and layout of the fixtures. The overall quality is then checked. If unsatisfactory, a new layout is made. An economic study is made to check costs. If these are too high, new layouts are studied until all design restraints are met.

Selection of Illuminance Levels

From 1958, the Illuminating Engineering Society (IES) published single-value illuminance levels. Their latest publication, the 1993 "IESNA Lighting Handbook," gives a range of values which permits lighting designers to tailor lighting systems to specific needs. This flexibility permits levels to be adjusted for (1) the visual task; (2) the age of the observers; (3) the need for speed and/or accuracy for visual performance; (4) the reflectance of the task. An illumination-level guide for selected tasks is given in Table 12.5.6. The data are based on an assumption of average conditions for people, tasks, and visual performance requirements. For other conditions see the 1993 "IESNA Lighting Handbook."

Room, Furniture, and Equipment Finishes

The color and finish of rooms, furniture, and equipment are important in the overall lighting design. Best results are obtained when the lighting designer coordinates his or her work with the architect, interior decorator, or plant designer.

Table 12.5.6 Illuminance Guide for Selected Tasks

	Footcandles (lm/ft)	Lux (lm/m²)
Commercial drafting		
Conventional	150*	1,600*
Libraries		
Reading good print, typed originals	30	320
Reading small print, handwriting, photocopies	75	800
Active stacks (vertical, illumination)	30	320
Offices		
Conference rooms—conferring	30	320
Conference rooms—typical visual tasks	75–100	800–1,080
Corridors, stairs, elevators	20	220
General tasks, varying difficulty	100	1,080
Lobbies, reception areas	30	320
Private	75	880
Rest rooms	30	320
Video display areas	75	800
School		
Classrooms, laboratories	75	800
Shops	100	1,080
Sight-saving rooms, hearing-impaired classes	150	1,600
Store		
Mass merchandizing, high activity	100	1,080
Self-service	200	2,200
Circulation, low activity	30	320
Feature displays, low, medium, high activity	150,* 300,* 500*	1,600,* 3,200,* 5,400*
Industrial		
Garages		
Repair	75	800
Active traffic areas	15	160
Loading platform	20	220
Machine shops and assembly areas		
Rough bench-machine work, simple assembly	50	540
Medium bench-machine work, moderately difficult assembly	100*	1,080*
Difficult machine work, assembly	150*	1,600*
Fine bench-machine work, assembly	300*	3,200*
Receiving and shipping	30	320
Warehouse storage rooms		
Active large items	15	160
Active small items, labels	30	320
Inactive	5	54
Outdoor areas		
Storage yards		
Active	20	220
Inactive	1	11
Parking areas		
Open, high activity	2	22
Open, medium activity	1	11
Covered parking, pedestrian areas	5	54
Covered night entrance	5	54
Covered day entrance	50	54

SOURCE: Adapted from General Electric Co. design data.
* Requires supplementary lighting. Care should be taken that the supplementary lighting does not introduce direct and reflected glare.

A color scheme should be selected to give light reflectance values as follows:

Area or unit	Percent reflectance
Ceilings	70–90
Floors	20–40*
Walls, draperies	40–60†
Bench top, desks, machine, and equipment	25–45

* In storage areas, keep reflectance of aisle floors as high as possible in order to reflect light onto the lower shelves. This should also be done where the underside of objects has to be seen.

† These values should be 30 to 40 where video display terminals (VDTs) are used to avoid veiling reflections in the VDT faces.

The color and finish of a space and equipment therein sets the psychological feel of the space. For example, the trend is away from drab finishes on machinery and dark gray filing cabinets. Colors such as yellow, orange, red, and light gray seem to advance toward the eye. They tend to make large spaces feel smaller. Receding colors such as violet, blue, blue green, and dark grays make small spaces feel larger. Some colors are used for safety purposes. Various areas are painted to designate safe or hazardous locations in a fashion similar to piping identification discussed in Sec. 8. These colors have been carefully standardized in ANSI Z53.1-1979.

Designating Color

In order to be able to obtain designed values of a lighting system, it is necessary to be able to specify the exact color wanted. Many methods have been devised for so doing. One method uses carefully controlled sets of colored chips, each one of which has a particular designation. The desired color is matched against these chips. The **Munsell system** uses *scales of hue* (the actual basic color such as red), *value* (a 10-step scale ranging from black through grays to white), and *chroma* (the amount of gray mixed in with the color). This system is used by many manufacturers to designate their colors. The **Ostwald system** describes color in terms of *color content, white content,* and *black content.* The **Inter-Society Color Council–National Bureau of Standards** (ISCC-NBS) **method** designates one-inch square samples with names. For color designation by the **ICI method,** a spectroradiometric curve of the source is determined together with a spectrophotometric curve of the reflecting or transmitting surface. By mathematical manipulation using spectral tristimulus values, chromaticity coordinates are obtained. See the IESNA Handbook for details. Chromaticity coordinates are extensively used for fluorescent lamp colors. These coordinates can be measured directly by photoelectric colorimeters. They are designed with filter photocell responses to be close to each of the ICI tristimulus values. Built-in logic circuitry results in direct reading of the chromaticity. Incandescent and vapor-type lamp colors are specified by color temperature.

LIGHTING DESIGN

Interior lighting is designed by the **lumen** method. This takes into account the interreflections of light inside a room. The average illuminance on the work plane equals the incident luminous flux ϕ divided by the area, or $E = \phi/A$. Lumens reaching the work plane is equal to lamp lumens multiplied by the **coefficient of utilization** CU. This factor is a function of room size, shape, and finish, mounting height of fixture, and type of luminaire used. The lumens ϕ_L initially available from the lamps may be reduced by ambient temperature, lower voltage, and the ballast used. As time goes by the room surfaces and luminaires become dirty, which further reduces the illuminance. In addition, lamp output falls, and some of them burn out. The total effect of all these factors is expressed by the *light-loss factor* LLF. The maintained illuminance E_m is the initial illumination times the LLF, or

$$E_m = (\phi_L \times CU \times LLF)/A \qquad (12.5.5)$$

The required maintained illuminance is selected from Table 12.5.6 or from the more extensive data in the IESNA Handbook. A fixture and lamp is selected, and Eq. (12.5.5) is solved for the necessary lamp flux ϕ_L. The number of luminaires N is found by dividing the total lamp lumens ϕ_L by the lumens per fixture ϕ_F. A trial layout is then made. A simple layout keeps spacing between units equal to twice the distance between fixtures and wall. Spacing is checked against the maximum allowable luminaire spacing from manufacturers' data to ensure uniform illumination. However, this criterion results in inadequate lighting near the walls. In order to light desks and benches along the walls, a spacing of 2½ ft from the luminaire center to the wall is used. The ends of fluorescent luminaire rows should be 6 to 12 in from the walls with a maximum distance of 2 ft.

Wall and ceiling cavity luminances can be obtained by using luminance coefficients (LC) for the fixtures (see the IESNA Handbook). For interior areas, maximum luminance ratios should be 3 : 1 or 1 : 3 between tasks and immediate surround, and 10 : 1 or 1 : 10 between tasks and remote surfaces. To ensure eye comfort, the visual-comfort probability (VCP) is investigated.

The **coefficient of utilization** is found by using the **zonal-cavity** method. In this method effects of the room proportion, luminaire suspension lengths, and work-plane height on the CU are found by dividing the room into three cavities as shown in Fig. 12.5.11. For each cavity a cavity ratio is calculated:

$$\text{Cavity ratio} = \frac{5h\ (\text{room length} + \text{room width})}{(\text{room length}) \times (\text{room width})} \qquad (12.5.6)$$

where $h = h_{RC}$ for the room cavity ratio RCR; $= h_{CC}$ for the ceiling cavity ratio CCR; and $= h_{FC}$ for the floor cavity ratio FCR.

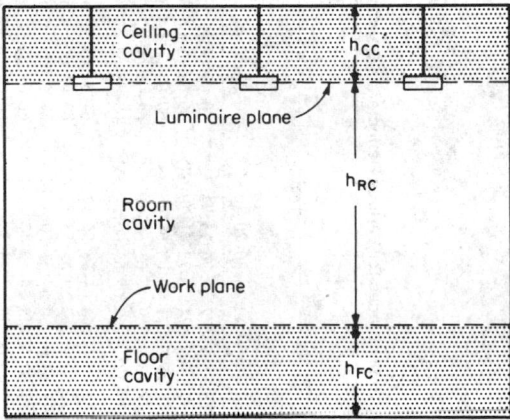

Fig. 12.5.11   The three cavities used in the zonal cavity method.

Table 12.5.7 is used to obtain a single effective ceiling cavity reflectance ρ_{CC} and a single effective floor cavity reflectance ρ_{FC}. For surface-mounted and recessed luminaires, CCR = 0 and the ceiling reflectance is used as ρ_{CC}. Figure 12.5.12 gives CU for selected fixtures. In using Fig. 12.5.12, interpolation may be necessary. Additional fixture data are given in the IES Handbook. Fixture manufacturers furnish such data for their units. Those data should be used for the best accuracy. If the effective floor cavity reflectance ρ_{FC} differs from 20 percent, an adjustment is made by using Table 12.5.8.

For simplicity in calculating the light-loss factor, the effects of ambient temperature, luminaire voltage variation, ballasts, and burnouts will be neglected. Room-surface dirt depreciation factors are shown in Fig. 12.5.13; luminaire dirt depreciation factors are in Fig. 12.5.14. The importance of frequent cleaning is evident. Categories are given for each fixture in Fig. 12.5.12. Lamp lumen depreciation (LLD) depends upon when lamps are replaced before complete burnout. If replacement

Table 12.5.7   Percent Effective Ceiling or Floor Cavity Reflectances for Various Reflectance Combinations (IES)

% base reflectance:	90										80										70										60										50									
% wall reflectance: / Cavity ratio	90	80	70	60	50	40	30	20	10	0	90	80	70	60	50	40	30	20	10	0	90	80	70	60	50	40	30	20	10	0	90	80	70	60	50	40	30	20	10	0	90	80	70	60	50	40	30	20	10	0
0.2	89	88	88	87	86	85	85	84	84	82	79	78	78	77	77	76	76	75	74	72	70	69	68	68	67	67	66	66	65	64	60	59	59	59	58	58	57	56	55	53	50	50	49	49	48	48	47	46	46	44
0.4	88	87	86	85	84	83	81	80	79	76	79	77	76	75	74	73	72	71	70	68	69	68	66	65	64	63	63	62	61	58	60	59	58	57	56	55	55	54	53	50	50	49	48	48	47	46	46	45	44	42
0.6	87	86	84	82	80	79	78	76	74	73	78	76	75	73	71	70	68	66	65	63	69	67	65	64	62	60	59	58	57	54	60	58	57	56	55	53	51	51	50	46	50	48	47	46	45	44	43	42	41	38
0.8	87	85	82	80	77	75	73	71	69	67	78	75	73	71	69	67	65	63	61	57	68	66	64	62	60	58	56	55	53	50	59	57	56	55	54	51	48	47	46	43	50	48	47	45	44	42	40	39	38	36
1.0	86	83	80	77	75	72	69	66	64	62	77	74	72	69	67	65	62	60	57	55	68	65	62	60	58	55	53	52	50	47	59	57	55	53	51	49	47	44	43	41	50	48	46	44	43	41	38	37	36	34
1.2	85	82	78	75	72	69	66	63	60	57	76	73	70	67	64	61	58	55	53	51	67	64	61	59	57	54	50	48	46	44	59	56	54	51	49	46	44	42	40	38	50	47	45	43	41	39	36	35	34	29
1.4	85	80	77	73	69	65	62	59	57	52	76	72	68	65	62	59	55	53	50	48	67	63	60	58	55	51	47	45	44	41	59	56	53	49	47	44	41	39	38	36	50	47	45	42	40	38	35	34	32	27
1.6	84	79	75	71	67	63	59	56	53	50	75	71	67	63	60	57	53	50	47	44	67	62	59	56	53	47	45	43	41	38	59	55	52	48	45	43	41	37	35	33	50	47	44	41	39	36	33	32	30	26
1.8	83	78	73	69	64	60	56	53	50	48	75	70	66	62	58	54	50	47	44	41	66	61	58	54	51	46	44	42	40	35	58	55	51	47	44	40	37	35	33	31	50	46	43	40	38	35	31	30	28	25
2.0	83	77	72	67	62	56	53	50	47	43	74	69	64	60	56	52	48	45	41	38	66	60	56	52	49	45	42	40	38	33	58	54	50	46	43	39	35	33	31	29	50	46	43	40	37	34	30	28	26	24
2.2	82	76	70	65	59	54	50	47	44	40	74	68	63	58	54	49	45	42	38	35	66	60	55	51	48	43	38	36	34	32	58	53	49	45	42	37	34	31	29	28	50	46	42	38	36	33	29	27	24	22
2.4	82	75	69	64	58	53	48	45	41	37	73	67	61	56	52	47	43	40	36	33	65	60	54	50	46	41	37	35	33	30	58	53	48	44	41	36	32	30	27	26	50	46	42	37	35	31	27	25	23	21
2.6	81	74	67	62	56	51	46	42	38	35	73	66	60	55	50	45	41	38	34	31	65	59	54	49	45	40	35	33	30	28	58	53	47	43	39	35	31	28	26	24	50	46	41	37	34	30	26	23	21	20
2.8	81	73	66	60	54	49	44	40	36	34	73	65	59	53	48	43	39	36	32	29	65	59	53	48	43	38	33	30	28	26	58	53	47	43	38	34	30	27	25	22	50	46	41	36	33	30	25	22	19	17
3.0	80	72	64	58	52	47	42	38	34	30	72	65	58	52	47	42	37	34	30	27	64	58	52	47	42	37	32	29	27	24	57	52	46	42	37	33	28	25	23	20	50	45	40	36	32	28	24	21	19	17
3.2	79	71	63	56	50	45	40	36	32	28	72	65	57	51	45	40	35	33	28	25	64	58	51	46	40	36	31	28	25	23	57	51	45	41	36	31	27	23	22	18	50	44	39	35	31	27	23	20	18	16
3.4	79	70	62	54	48	43	38	34	30	27	71	64	56	49	44	39	34	32	27	24	64	57	50	45	39	35	29	27	24	22	57	51	45	40	35	30	26	23	20	17	50	44	39	35	30	26	22	19	17	15
3.6	78	69	61	53	47	42	36	32	28	25	71	63	54	48	43	38	32	30	25	23	63	56	49	44	38	33	28	25	22	20	57	50	44	39	34	29	25	21	18	16	50	44	39	34	29	25	21	18	16	14
3.8	78	69	60	51	45	40	35	31	27	23	70	62	53	47	41	36	31	28	24	22	63	56	49	43	37	32	27	24	21	19	57	50	43	38	33	29	24	21	19	15	50	44	38	33	28	25	21	17	15	13
4.0	77	69	58	51	44	39	33	29	25	22	70	61	53	46	40	35	30	26	22	20	63	55	48	42	36	31	26	22	19	17	57	49	42	37	32	28	23	20	18	14	50	44	38	33	28	24	20	17	15	12
4.2	77	62	57	50	43	37	32	28	24	21	69	60	52	45	39	34	29	25	21	18	62	55	47	41	35	30	25	22	19	16	56	49	42	37	32	27	22	19	17	14	50	43	37	32	28	24	20	17	14	12
4.4	76	61	56	49	42	36	31	27	23	20	69	60	51	44	38	33	28	24	20	17	62	54	46	40	34	29	24	21	18	15	56	49	41	36	31	27	22	19	16	13	50	43	37	32	27	23	19	16	13	11
4.6	76	60	55	47	40	35	30	26	22	19	69	59	50	43	37	32	27	23	19	15	62	53	45	39	33	28	24	20	17	14	56	49	41	35	30	26	21	18	16	13	50	43	36	31	26	22	18	15	13	10
4.8	75	59	54	46	39	34	28	25	21	18	68	58	49	42	36	31	26	22	18	14	62	53	45	38	32	27	23	20	16	13	56	48	41	34	29	25	21	18	15	12	50	43	36	31	26	22	18	15	12	09
5.0	75	59	53	45	38	33	28	24	20	16	68	58	48	41	35	30	25	21	18	14	61	52	44	36	31	26	22	19	16	12	56	48	40	34	28	24	20	17	14	11	50	42	35	30	25	21	17	14	12	09
6.0	73	61	49	41	34	29	24	20	16	11	66	55	44	38	31	27	22	19	16	11	60	51	41	35	28	24	19	16	13	09	55	45	37	31	25	21	17	14	11	07	50	42	34	29	23	19	15	13	10	06
7.0	70	58	45	38	30	27	23	21	18	14	64	53	41	35	28	24	19	16	12	09	58	48	38	32	26	22	17	14	11	06	54	43	35	30	23	20	15	12	09	05	49	41	32	27	21	18	14	11	08	05
8.0	68	55	42	35	27	23	18	15	12	06	62	50	38	32	26	21	17	14	11	05	57	46	35	29	23	19	15	13	10	05	53	42	33	28	22	18	14	11	08	04	49	40	30	25	19	16	12	10	07	03
9.0	66	52	38	31	25	21	16	14	11	05	61	49	36	30	23	19	14	12	09	04	56	45	33	27	21	18	14	12	09	04	52	40	31	26	20	16	12	10	07	03	48	39	29	24	18	15	11	09	07	03
10.0	65	51	36	30	22	19	15	11	09	04	59	46	33	27	21	18	14	11	08	03	55	43	31	25	19	16	12	10	08	03	51	39	29	24	18	15	11	09	07	02	47	37	27	22	17	14	10	08	06	02

* Ceiling, floor, or floor of cavity.

Table 12.5.7 Percent Effective Ceiling or Floor Cavity Reflectance for Various Reflectance Combinations (Continued)

% base* reflectance / % wall reflectance

Cavity ratio	40										30										20										10										0									
% wall reflectance	90	80	70	60	50	40	30	20	10	0	90	80	70	60	50	40	30	20	10	0	90	80	70	60	50	40	30	20	10	0	90	80	70	60	50	40	30	20	10	0	90	80	70	60	50	40	30	20	10	0
0.2	40	40	39	39	39	38	38	37	36	36	31	31	30	30	30	29	29	28	28	27	21	20	20	20	20	20	19	19	19	17	11	11	11	11	10	10	10	09	09	09	02	02	02	01	01	01	01	01	01	00
0.4	41	40	39	39	38	37	36	35	34	34	31	31	30	30	29	28	28	27	26	25	22	21	20	20	19	19	19	18	18	16	12	11	11	11	11	10	10	09	09	08	04	03	03	03	02	02	02	01	01	00
0.6	41	40	39	38	38	37	36	34	33	31	32	31	30	29	28	27	26	26	25	23	22	21	21	20	19	19	18	17	17	15	13	13	12	11	11	10	10	09	08	08	05	05	04	03	03	02	02	02	01	01
0.8	41	40	38	37	36	35	33	32	31	29	32	31	30	29	28	26	25	23	22	22	24	22	21	20	20	19	18	17	17	15	15	14	13	12	11	11	10	09	08	07	07	06	05	04	03	03	02	02	01	01
1.0	42	40	38	37	35	33	32	31	29	27	33	32	30	29	27	25	24	23	22	20	25	23	22	20	19	18	17	16	15	13	16	14	13	12	12	11	10	09	08	07	08	07	06	05	04	03	03	02	01	01
1.2	42	40	38	36	34	32	30	29	27	25	33	32	30	28	27	25	23	22	21	19	25	23	22	19	17	16	16	14	13	12	17	15	14	12	11	11	10	09	07	06	10	08	07	06	05	04	03	02	01	01
1.4	42	39	37	35	33	31	29	27	25	23	34	32	30	28	26	24	22	21	19	18	26	24	22	20	18	17	15	14	13	12	18	16	14	13	12	11	09	08	07	06	11	09	08	07	06	04	03	02	01	01
1.6	42	39	37	35	32	30	27	25	23	22	34	33	29	27	25	23	22	20	18	17	26	24	22	20	18	16	14	13	12	11	19	17	15	13	12	11	09	08	07	06	12	10	09	07	06	05	03	02	01	01
1.8	42	39	36	34	31	29	26	24	22	21	35	33	29	27	25	23	21	19	17	16	27	24	22	20	17	16	14	13	11	10	19	17	15	13	12	11	09	08	06	05	13	11	09	08	07	05	03	02	01	01
2.0	42	39	36	34	31	28	25	23	21	19	35	33	29	26	24	22	20	18	16	14	28	25	23	20	18	16	14	13	11	09	20	18	16	14	13	11	09	08	06	05	14	12	10	09	07	05	04	03	01	01
2.2	42	39	36	33	30	27	24	22	19	18	36	32	29	26	24	22	19	17	15	13	28	25	23	20	18	16	14	12	10	09	21	19	16	14	13	11	09	07	06	05	15	13	11	09	07	06	04	03	02	01
2.4	43	39	35	33	29	27	24	21	18	17	36	32	29	26	24	22	19	16	14	12	29	26	23	20	18	16	14	12	10	08	22	19	17	15	13	11	09	07	06	05	16	13	11	09	08	06	04	03	01	01
2.6	43	39	35	32	29	26	23	20	17	15	36	32	28	25	23	21	18	16	14	12	29	26	23	20	18	16	14	12	10	08	23	20	17	15	13	11	09	07	06	04	17	14	12	10	08	06	05	03	02	01
2.8	43	39	35	32	28	25	22	19	16	14	37	33	29	25	23	21	17	15	13	11	30	27	23	20	17	15	13	11	09	07	23	20	18	16	13	11	09	07	05	03	17	15	13	10	08	07	05	03	02	01
3.0	43	39	35	31	27	24	21	18	16	13	37	33	29	25	22	20	17	15	12	10	30	27	23	20	17	15	13	11	09	07	24	21	18	16	13	11	09	07	05	03	18	16	13	11	09	07	05	03	02	01
3.2	43	39	35	31	27	23	20	17	15	13	37	33	29	25	22	19	16	14	12	10	31	27	23	20	17	15	12	11	09	06	25	21	18	16	13	11	09	05	05	03	19	16	14	11	09	07	05	03	02	01
3.4	43	39	34	30	26	23	20	17	14	12	37	33	29	24	21	19	16	14	11	09	31	27	23	20	17	15	12	10	08	06	26	22	18	16	13	11	09	07	05	03	20	17	14	12	09	07	05	03	02	01
3.6	44	39	34	30	26	22	19	16	14	11	38	33	29	24	21	18	15	13	10	08	32	27	23	20	17	15	12	10	08	05	26	22	19	16	14	11	09	06	04	02	20	17	15	12	10	08	05	04	02	01
3.8	44	38	33	29	25	22	18	16	13	10	38	33	28	24	21	18	15	13	10	07	32	28	23	20	17	14	12	10	07	05	27	23	19	17	14	11	09	06	04	02	21	18	15	12	10	08	05	04	02	01
4.0	44	38	33	29	25	21	18	15	12	10	38	33	28	24	20	18	14	12	09	07	33	28	23	20	17	14	11	09	07	05	27	23	20	17	14	11	09	06	04	02	22	18	15	13	10	08	05	04	02	01
4.2	44	38	33	29	24	21	17	15	12	10	38	33	28	24	20	17	14	12	09	07	33	28	23	20	17	14	11	09	07	04	28	24	20	17	14	11	09	06	04	02	22	19	16	13	10	08	06	04	02	01
4.4	44	38	33	28	24	21	17	14	12	09	39	33	28	24	20	17	14	11	09	06	34	28	24	20	17	14	11	09	07	04	28	24	20	17	14	11	08	06	04	02	23	19	16	13	10	08	06	04	02	01
4.6	44	38	32	28	23	19	16	14	11	08	39	33	27	23	19	17	13	11	08	05	34	29	24	20	17	14	11	08	06	04	29	25	20	17	14	11	08	06	04	02	23	20	17	13	11	08	06	04	02	01
4.8	44	38	32	27	22	19	16	13	10	08	39	33	27	23	19	16	13	10	08	05	35	29	24	20	16	14	10	08	06	04	29	25	20	17	14	11	08	06	04	02	24	20	17	14	11	08	06	04	02	01
5.0	45	38	31	27	22	19	15	13	10	07	39	33	27	23	19	16	13	10	08	05	35	29	24	20	16	13	10	08	06	04	30	25	20	17	14	11	08	06	04	02	25	21	17	14	11	08	06	04	02	01
6.0	44	37	30	25	20	17	13	11	08	05	39	33	26	22	18	15	11	09	06	04	36	30	24	19	16	13	10	08	05	02	31	26	21	17	14	11	08	06	03	01	27	23	18	15	12	09	06	04	02	01
7.0	44	36	29	24	19	16	12	10	07	04	40	33	25	21	17	14	10	08	05	03	36	30	23	19	15	12	09	07	04	02	32	27	21	17	13	11	08	06	03	01	28	24	19	15	12	09	06	04	02	01
8.0	44	35	28	23	18	15	11	09	06	03	40	33	25	21	16	13	09	07	04	02	37	30	23	19	14	11	09	06	03	01	33	27	21	17	13	10	07	05	03	01	30	25	20	15	12	09	06	04	02	01
9.0	44	35	28	23	16	12	10	08	05	02	40	33	24	20	15	12	09	07	04	01	37	29	23	18	14	11	08	06	03	01	34	28	21	17	13	10	07	05	02	01	31	25	20	15	12	09	06	04	02	01
10.0	43	34	25	20	15	12	08	07	05	02	40	32	24	19	14	11	08	06	03	01	37	29	22	18	13	10	07	05	03	01	34	28	21	17	12	10	07	05	02	01	31	25	20	15	12	09	06	04	02	01

* Ceiling, floor, or floor of cavity.

				ρcc* →	80			70			50			30			10			0
Typical luminaire	Maint. cat.	Max S/MH guide§	ρw† → RCR‡ ↓		50	30	10	50	30	10	50	30	10	50	30	10	50	30	10	0
					Coefficients of utilization for 20% effective floor cavity reflectance (ρFC = 20)															
Porcelain-enamel ventilated standard dome with incandescent lamp (0% ↑, 83½% ↓)	IV	1.3	0		.99	.99	.99	.97	.97	.97	.93	.93	.93	.89	.89	.89	.85	.85	.85	.83
			1		.87	.84	.81	.85	.82	.79	.82	.79	.77	.79	.76	.74	.76	.74	.72	.71
			2		.76	.70	.65	.74	.69	.65	.71	.67	.63	.69	.65	.62	.66	.63	.60	.59
			3		.66	.59	.54	.65	.59	.53	.62	.57	.53	.60	.56	.52	.58	.54	.51	.49
			4		.58	.51	.45	.57	.50	.45	.55	.49	.44	.53	.48	.44	.51	.47	.43	.41
			5		.52	.44	.39	.51	.44	.38	.49	.43	.38	.47	.42	.37	.46	.41	.37	.35
			6		.46	.39	.33	.46	.38	.33	.44	.38	.33	.43	.37	.33	.41	.36	.32	.31
			7		.42	.34	.29	.41	.34	.29	.40	.33	.29	.39	.33	.29	.38	.32	.28	.27
			8		.38	.31	.26	.37	.31	.26	.36	.30	.26	.35	.30	.25	.34	.29	.25	.24
			9		.35	.28	.23	.34	.28	.23	.34	.27	.23	.33	.27	.23	.32	.26	.23	.21
			10		.32	.25	.21	.32	.25	.21	.31	.25	.21	.30	.24	.21	.29	.24	.21	.19
EAR-38 lamp above 2″ (51 mm) diameter aperture (increase efficiency to 54½% for 3″ (76 mm) diameter aperture (0% ↑, 44½% ↓)	IV	0.7	0		.52	.52	.52	.51	.51	.51	.48	.48	.48	.46	.46	.46	.45	.45	.45	.44
			1		.49	.48	.47	.48	.47	.46	.46	.45	.45	.44	.44	.43	.43	.43	.42	.41
			2		.46	.44	.43	.45	.44	.43	.44	.43	.42	.43	.42	.41	.41	.41	.40	.39
			3		.43	.41	.40	.43	.41	.40	.42	.40	.39	.41	.39	.38	.40	.39	.38	.37
			4		.41	.39	.37	.41	.39	.37	.40	.38	.37	.39	.37	.36	.38	.37	.36	.35
			5		.39	.37	.35	.39	.37	.35	.38	.36	.35	.37	.36	.34	.36	.35	.34	.34
			6		.37	.35	.33	.37	.35	.33	.36	.34	.33	.35	.34	.33	.35	.34	.32	.32
			7		.35	.33	.31	.35	.33	.31	.34	.33	.31	.34	.32	.31	.33	.32	.31	.30
			8		.34	.31	.30	.33	.31	.30	.33	.31	.30	.32	.31	.29	.32	.31	.29	.29
			9		.32	.30	.28	.32	.30	.28	.31	.30	.28	.31	.29	.28	.31	.29	.28	.28
			10		.31	.28	.27	.31	.28	.27	.30	.28	.27	.30	.28	.27	.30	.28	.27	.26
Enclosed reflector with an incandescent lamp (0% ↑, 71½% ↓)	V	1.4	0		.85	.85	.85	.83	.83	.83	.80	.80	.80	.76	.76	.76	.73	.73	.73	.72
			1		.77	.75	.73	.76	.74	.72	.73	.71	.69	.70	.69	.67	.67	.66	.65	.64
			2		.70	.66	.63	.68	.65	.62	.66	.63	.60	.64	.61	.59	.61	.60	.58	.56
			3		.63	.58	.54	.62	.57	.54	.60	.56	.53	.58	.54	.52	.56	.53	.51	.50
			4		.56	.51	.47	.56	.51	.47	.54	.50	.46	.52	.49	.46	.51	.48	.45	.44
			5		.51	.46	.42	.50	.45	.41	.49	.44	.41	.48	.44	.40	.46	.43	.40	.39
			6		.46	.41	.37	.46	.41	.37	.45	.40	.36	.43	.39	.36	.42	.39	.36	.34
			7		.42	.37	.33	.42	.37	.33	.41	.36	.33	.40	.36	.32	.39	.35	.32	.31
			8		.39	.33	.30	.38	.33	.29	.37	.33	.29	.37	.32	.29	.36	.32	.29	.28
			9		.36	.30	.27	.35	.30	.27	.35	.30	.27	.34	.30	.26	.33	.29	.26	.25
			10		.33	.28	.24	.33	.28	.24	.32	.27	.24	.31	.27	.24	.31	.27	.24	.23
Diffuse aluminum reflector with 35″ CW shielding (17% ↑, 66% ↓)	II	1.5/1.3	0		.95	.95	.95	.91	.91	.91	.83	.83	.83	.76	.76	.76	.69	.69	.69	.66
			1		.85	.82	.79	.81	.79	.76	.75	.73	.71	.69	.67	.66	.63	.62	.61	.58
			2		.75	.71	.67	.72	.68	.65	.67	.63	.61	.62	.59	.57	.57	.55	.53	.51
			3		.67	.61	.57	.65	.59	.55	.60	.56	.52	.55	.52	.49	.51	.49	.46	.44
			4		.60	.54	.49	.58	.52	.48	.54	.49	.45	.50	.46	.43	.46	.43	.41	.39
			5		.54	.47	.43	.52	.46	.42	.49	.43	.40	.45	.41	.38	.42	.39	.36	.34
			6		.49	.42	.37	.47	.41	.37	.44	.39	.35	.41	.37	.33	.38	.35	.32	.30
			7		.44	.38	.33	.43	.37	.32	.40	.35	.31	.38	.33	.30	.35	.31	.28	.27
			8		.40	.34	.29	.39	.33	.29	.37	.31	.28	.34	.30	.27	.32	.28	.26	.24
			9		.37	.31	.26	.36	.30	.26	.34	.29	.25	.32	.27	.24	.30	.26	.23	.21
			10		.34	.28	.24	.33	.27	.23	.31	.26	.23	.29	.25	.22	.28	.24	.21	.19
Diffuse aluminum reflector with 35″CW × 35″LW shielding (17% ↑, 56½% ↓)	II	1.5/1.1	0		.91	.91	.91	.86	.86	.86	.77	.77	.77	.68	.68	.68	.61	.61	.61	.57
			1		.80	.77	.75	.76	.74	.71	.69	.67	.65	.62	.60	.59	.55	.54	.53	.50
			2		.71	.67	.63	.68	.64	.60	.61	.58	.55	.55	.53	.51	.50	.48	.46	.43
			3		.63	.58	.53	.60	.55	.51	.55	.51	.47	.50	.46	.44	.45	.42	.40	.38
			4		.57	.51	.46	.54	.49	.44	.49	.45	.41	.45	.41	.38	.41	.38	.35	.33
			5		.51	.45	.40	.49	.43	.39	.45	.40	.36	.41	.37	.34	.37	.34	.31	.29
			6		.46	.40	.35	.44	.38	.34	.41	.36	.32	.37	.33	.30	.34	.30	.28	.26
			7		.42	.36	.31	.40	.35	.30	.37	.32	.29	.34	.30	.27	.31	.28	.25	.23
			8		.38	.32	.28	.37	.31	.27	.34	.29	.26	.31	.27	.24	.29	.25	.23	.21
			9		.35	.29	.25	.34	.28	.25	.31	.27	.23	.29	.25	.22	.27	.23	.21	.19
			10		.33	.27	.23	.31	.26	.22	.29	.24	.21	.27	.23	.20	.25	.21	.19	.17
2 lamp, surface-mounted, bare lamp unit—photometry with 18-in-wide panel above luminaire (lamps on 6-in centers) (9½% ↑, 78% ↓)	I	1.3	0		1.02	1.02	1.02	.99	.99	.99	.92	.92	.92	.86	.86	.86	.81	.81	.81	.78
			1		.85	.80	.76	.82	.78	.74	.76	.73	.70	.71	.68	.66	.67	.64	.62	.60
			2		.72	.65	.59	.70	.63	.58	.65	.60	.55	.61	.56	.52	.57	.53	.50	.47
			3		.63	.55	.48	.60	.53	.47	.56	.50	.45	.53	.47	.43	.49	.45	.41	.38
			4		.55	.46	.40	.53	.45	.39	.50	.43	.37	.46	.41	.36	.43	.38	.34	.32
			5		.49	.40	.34	.47	.39	.33	.44	.37	.32	.41	.35	.31	.39	.34	.29	.27
			6		.43	.35	.29	.42	.34	.29	.40	.33	.28	.37	.31	.27	.35	.30	.26	.23
			7		.39	.31	.25	.38	.30	.25	.36	.29	.24	.34	.28	.23	.32	.26	.22	.20
			8		.36	.28	.22	.35	.27	.22	.33	.26	.21	.31	.25	.21	.29	.24	.20	.18
			9		.33	.25	.20	.32	.25	.20	.30	.24	.19	.28	.23	.18	.27	.22	.18	.16
			10		.30	.23	.18	.29	.22	.18	.28	.21	.17	.26	.21	.17	.25	.20	.16	.14

Fig. 12.5.12 Coefficients of utilization for typical luminaires. (*Abstracted from IESNA Handbook, 8th ed., 1993. Reprinted with permission. This material has been modified from its original version and is not reflective of its original form as recognized by the IESNA.*)

Coefficients of utilization for 20% effective floor cavity reflectance ($\rho FC = 20$). ρ_{CC}* = percent effective ceiling cavity reflectance (column groups 80, 70, 50, 30, 10, 0). ρ_w† = percent wall reflectance (sub-columns 50, 30, 10).

Typical luminaire	Maint. cat.	Max S/MH guide§	RCR‡	80 / 50	80 / 30	80 / 10	70 / 50	70 / 30	70 / 10	50 / 50	50 / 30	50 / 10	30 / 50	30 / 30	30 / 10	10 / 50	10 / 30	10 / 10	0
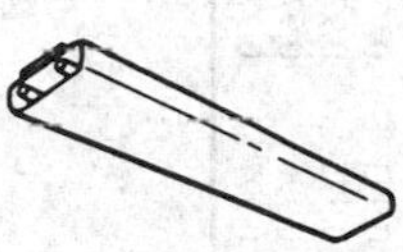 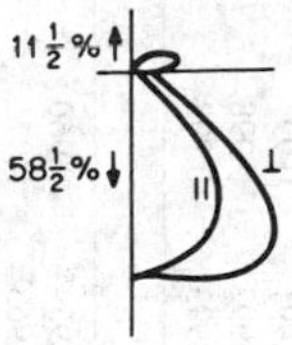 2 lamp prismatic wraparound—multiply by 0.95 for 4 amps (11½%↑, 58½%↓)	V	1.5/1.2	0	.81	.81	.81	.78	.78	.78	.72	.72	.72	.66	.66	.66	.61	.61	.61	.59
			1	.71	.68	.66	.68	.66	.63	.63	.61	59	.58	.57	.56	.54	.53	.52	.50
			2	.63	.58	.55	.60	.56	.53	.56	.53	.50	.52	.50	.47	.48	.46	.45	.43
			3	.56	.50	.46	.54	.49	.45	.50	.46	.43	.47	.43	.41	.43	.41	.39	.37
			4	.50	.44	.40	.48	.43	.39	.45	.40	.37	.42	.38	.35	.39	.36	.34	.32
			5	.45	.39	.34	.43	.38	.34	.40	.36	.32	.38	.34	.31	.35	.32	.30	.28
			6	.40	.34	.30	.39	.34	.30	.37	.32	.28	.34	.30	.27	.32	.29	.26	.25
			7	.37	.31	.27	.35	.30	.26	.33	.29	.25	.31	.27	.24	.30	.26	.23	.22
			8	.33	.28	.24	.32	.27	.23	.30	.26	.23	.29	.25	.22	.27	.24	.21	.20
			9	.31	.25	.21	.30	.25	.21	.28	.24	.20	.26	.23	.20	.25	.22	.19	.18
			10	.28	.23	.19	.27	.22	.19	.26	.21	.18	.24	.21	.18	.22	.20	.17	.16
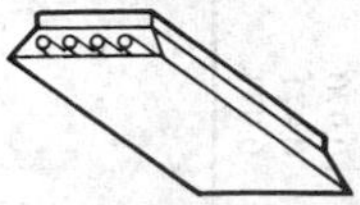 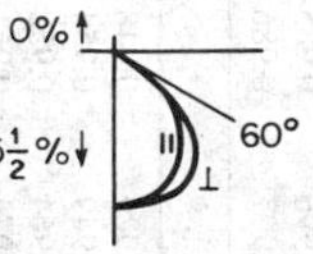 4 lamp, 2' (610 mm) wide unit with sharp cutoff (high angle—low luminance) flat prismatic lens—× 1.05 for 3 lamps, × 0.9 for 2 lamps (0%↑, 65½%↓, 60°)	V	1.4/1.3	0	.71	.71	.71	.70	.70	.70	.66	.66	.66	.64	.64	.64	.61	.61	.61	.60
			1	.64	.62	.60	.63	.61	.60	.60	.59	.58	.58	.57	.56	.56	.55	.54	.53
			2	.57	.54	.51	.56	.53	.51	.54	.52	.50	.52	.50	.48	.51	.49	.47	.46
			3	.51	.47	.44	.50	.46	.43	.49	.45	.43	.47	.44	.42	.46	.43	.41	.40
			4	.46	.41	.38	.45	.41	.37	.44	.40	.37	.42	.39	.36	.41	.38	.36	.35
			5	.41	.36	.33	.40	.36	.32	.39	.35	.32	.38	.35	.32	.37	.34	.31	.30
			6	.37	.32	.28	.36	.32	.28	.35	.31	.28	.34	.31	.28	.34	.30	.28	.27
			7	.33	.29	.25	.33	.28	.25	.32	.28	.25	.31	.27	.25	.30	.27	.24	.23
			8	.30	.26	.22	.30	.25	.22	.29	.25	.22	.28	.25	.22	.28	.24	.22	.21
			9	.28	.23	.20	.27	.23	.20	.27	.23	.20	.26	.22	.20	.25	.22	.19	.18
			10	.25	.21	.18	.25	.21	.18	.25	.20	.18	.24	.20	.18	.23	.20	.18	.17
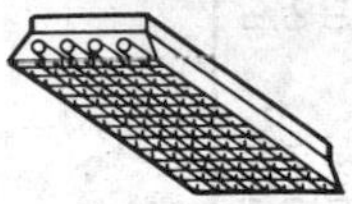 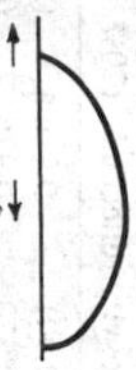 4 lamp, 2' (610 mm) wide troffer with 45° white metal louver—× 1.05 for 3 lamps, × 0.9 for 2 lamps (0%↑, 46%↓)	IV	0.9	0	.55	.55	.55	.54	.54	.54	.51	.51	.51	.49	.49	.49	.47	.47	.47	.46
			1	.49	.48	.46	.48	.47	.46	.46	.45	.44	.45	.44	.43	.43	.42	.42	.41
			2	.44	.42	.40	.43	.41	.39	.42	.40	.38	.40	.39	.37	.39	.38	.37	.36
			3	.40	.37	.34	.39	.36	.34	.38	.36	.33	.37	.35	.33	.36	.34	.32	.32
			4	.36	.33	.30	.36	.33	.30	.35	.32	.30	.34	.31	.29	.33	.31	.29	.28
			5	.33	.30	.27	.33	.29	.27	.32	.29	.27	.31	.28	.26	.30	.28	.26	.25
			6	.30	.27	.24	.30	.27	.24	.29	.26	.24	.29	.26	.24	.28	.25	.24	.23
			7	.28	.25	.22	.28	.24	.22	.27	.24	.22	.26	.24	.22	.26	.23	.22	.21
			8	.26	.23	.20	.26	.22	.20	.25	.22	.20	.25	.22	.20	.24	.22	.20	.19
			9	.24	.21	.19	.24	.21	.19	.23	.20	.18	.23	.20	.18	.23	.20	.18	.18
			10	.23	.19	.17	.22	.19	.17	.22	.19	.17	.22	.19	.17	.21	.19	.17	.16
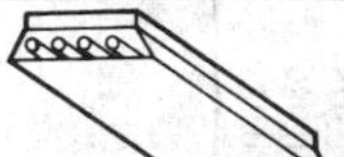  Bilateral batwing distribution—4 lamp, 2' (610 mm) wide fluorescent unit with flat prismatic lens and overlay—× 1.05 for 3 lamps, × 0.9 for 2 lamps (0%↑, 48%↓, 45°)	V	N.A.	0	.57	.57	.57	.56	.56	.56	.53	.53	.53	.51	.51	.51	.49	.49	.49	.48
			1	.50	.48	.46	.49	.47	.45	.47	.45	.44	.45	.43	.42	.43	.42	.41	.40
			2	.43	.40	.37	.42	.39	.36	.40	.38	.35	.39	.37	.35	.37	.36	.34	.33
			3	.37	.33	.30	.37	.33	.30	.35	.32	.29	.34	.31	.29	.33	.30	.28	.27
			4	.33	.28	.25	.32	.28	.25	.31	.27	.24	.30	.27	.24	.29	.26	.24	.23
			5	.29	.24	.21	.28	.24	.21	.27	.24	.21	.26	.23	.20	.25	.23	.20	.19
			6	.26	.21	.18	.25	.21	.18	.24	.21	.18	.24	.20	.18	.23	.20	.17	.16
			7	.23	.19	.16	.23	.18	.15	.22	.18	.15	.21	.18	.15	.21	.17	.15	.14
			8	.21	.17	.14	.21	.16	.14	.20	.16	.13	.19	.16	.13	.19	.16	.13	.12
			9	.19	.15	.12	.19	.15	.12	.18	.14	.12	.18	.14	.12	.17	.14	.12	.11
			10	.17	.13	.11	.17	.13	.11	.17	.13	.11	.16	.13	.11	.16	.13	.10	.10
Diffusing plastic or glass (ρ_{cc} from below ~65%). 1. Ceiling efficiency ~60%; diffuser transmittance ~50%; diffuser reflectance ~40%. Cavity with minimum obstructions and painted with 80% reflectance paints—use $pc = 70$. 2. For lower reflectance paint or obstructions—use $pc = 50$			1				.60	.58	.56	.58	.56	.54							
			2				.53	.49	.45	.51	.47	.43							
			3				.47	.42	.37	.45	.41	.36							
			4				.41	.36	.32	.39	.35	.31							
			5				.37	.31	.27	.35	.30	.26							
			6				.33	.27	.23	.31	.26	.23							
			7				.29	.24	.20	.28	.23	.20							
			8				.26	.21	.18	.25	.20	.17							
			9				.23	.19	.15	.23	.18	.15							
			10				.21	.17	.13	.21	.16	.13							

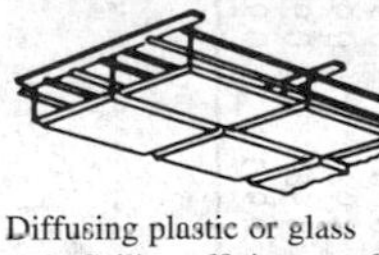

* ρ_{cc} = percent effective ceiling cavity reflectance.
† ρ_w = percent wall reflectance.
‡ RCR = room cavity ratio.
§ Maximum S/MH guide—ratio of maximum luminaire spacing to mounting or ceiling height above work plane.

Fig. 12.5.12 (Continued)

Table 12.5.8 Multiplying Factors for Other than 20 Percent Effective Floor Cavity Reflectance (IES)

Room cavity ratio	80				70				50			30			10		
% wall reflectance ρ_W:	70	50	30	10	70	50	30	10	50	30	10	50	30	10	50	30	10
For 30% effective floor cavity reflectance (20% = 1.00)																	
1	1.092	1.082	1.075	1.068	1.077	1.070	1.064	1.059	1.049	1.044	1.040	1.028	1.026	1.023	1.012	1.010	1.008
2	1.079	1.066	1.055	1.047	1.068	1.057	1.048	1.039	1.041	1.033	1.027	1.026	1.021	1.017	1.013	1.010	1.006
3	1.070	1.054	1.042	1.033	1.061	1.048	1.037	1.028	1.034	1.027	1.020	1.024	1.017	1.012	1.014	1.009	1.005
4	1.062	1.045	1.033	1.024	1.055	1.040	1.029	1.021	1.030	1.022	1.015	1.022	1.015	1.010	1.014	1.009	1.004
5	1.056	1.038	1.026	1.018	1.050	1.034	1.024	1.015	1.027	1.018	1.012	1.020	1.013	1.008	1.014	1.009	1.004
6	1.052	1.033	1.021	1.014	1.047	1.030	1.020	1.012	1.024	1.015	1.009	1.019	1.012	1.006	1.014	1.008	1.003
7	1.047	1.029	1.018	1.011	1.043	1.026	1.017	1.009	1.022	1.013	1.007	1.018	1.010	1.005	1.013	1.008	1.003
8	1.044	1.026	1.015	1.009	1.040	1.024	1.015	1.007	1.020	1.012	1.006	1.017	1.009	1.004	1.013	1.007	1.003
9	1.040	1.024	1.014	1.007	1.037	1.022	1.014	1.006	1.019	1.011	1.005	1.016	1.009	1.004	1.013	1.007	1.002
10	1.037	1.022	1.012	1.006	1.034	1.020	1.012	1.005	1.017	1.010	1.004	1.015	1.009	1.003	1.013	1.007	1.002
For 10% effective floor cavity reflectance (20% = 1.00)																	
1	0.923	0.929	0.935	0.940	0.933	0.939	0.943	0.948	0.956	0.960	0.963	0.973	0.976	0.979	0.989	0.991	0.993
2	0.931	0.942	0.950	0.958	0.940	0.949	0.957	0.963	0.962	0.968	0.974	0.976	0.980	0.985	0.988	0.991	0.995
3	0.939	0.951	0.961	0.969	0.945	0.957	0.966	0.973	0.967	0.975	0.981	0.978	0.983	0.988	0.987	0.992	0.996
4	0.944	0.958	0.969	0.978	0.950	0.963	0.973	0.980	0.972	0.980	0.986	0.980	0.986	0.991	0.987	0.992	0.996
5	0.949	0.964	0.976	0.983	0.954	0.968	0.978	0.985	0.975	0.983	0.989	0.981	0.988	0.993	0.987	0.993	0.997
6	0.953	0.969	0.980	0.986	0.958	0.972	0.982	0.989	0.977	0.985	0.992	0.982	0.989	0.995	0.987	0.993	0.997
7	0.957	0.973	0.983	0.991	0.961	0.975	0.985	0.991	0.979	0.987	0.994	0.983	0.990	0.996	0.987	0.993	0.998
8	0.960	0.976	0.986	0.993	0.963	0.977	0.987	0.993	0.981	0.988	0.995	0.984	0.991	0.997	0.988	0.994	0.998
9	0.963	0.978	0.987	0.994	0.965	0.979	0.989	0.994	0.983	0.990	0.996	0.985	0.992	0.998	0.988	0.994	0.999
10	0.965	0.980	0.989	0.995	0.967	0.981	0.990	0.995	0.984	0.991	0.997	0.986	0.993	0.998	0.988	0.994	0.999
For 0% effective floor cavity reflectance (20% = 1.00)																	
1	0.859	0.870	0.879	0.886	0.873	0.884	0.893	0.901	0.916	0.923	0.929	0.948	0.954	0.960	0.979	0.983	0.987
2	0.871	0.887	0.903	0.919	0.886	0.902	0.916	0.928	0.926	0.938	0.949	0.954	0.963	0.971	0.978	0.983	0.991
3	0.882	0.904	0.915	0.942	0.898	0.918	0.934	0.947	0.936	0.950	0.964	0.958	0.969	0.979	0.976	0.984	0.993
4	0.893	0.919	0.941	0.958	0.908	0.930	0.948	0.961	0.945	0.961	0.974	0.961	0.974	0.984	0.975	0.985	0.994
5	0.903	0.931	0.953	0.969	0.914	0.939	0.958	0.970	0.951	0.967	0.980	0.964	0.977	0.988	0.975	0.985	0.995
6	0.911	0.940	0.961	0.976	0.920	0.945	0.965	0.977	0.955	0.972	0.985	0.966	0.979	0.991	0.975	0.986	0.996
7	0.917	0.947	0.967	0.981	0.924	0.950	0.970	0.982	0.959	0.975	0.988	0.968	0.981	0.993	0.975	0.987	0.997
8	0.922	0.953	0.971	0.985	0.929	0.955	0.975	0.986	0.963	0.978	0.991	0.970	0.983	0.995	0.976	0.988	0.998
9	0.928	0.958	0.975	0.988	0.933	0.959	0.980	0.989	0.966	0.980	0.993	0.971	0.985	0.996	0.976	0.988	0.998
10	0.933	0.962	0.979	0.991	0.937	0.963	0.983	0.992	0.969	0.982	0.995	0.973	0.987	0.997	0.977	0.989	0.999

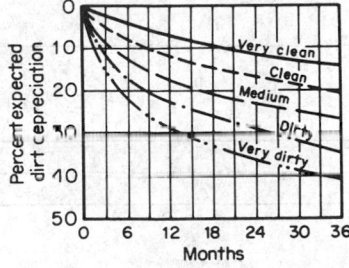

% expected dirt depreciation:	Luminaire distribution type																			
	Direct				Semidirect				Direct-indirect				Semi-indirect				Indirect			
Room cavity ratio	10	20	30	40	10	20	30	40	10	20	30	40	10	20	30	40	10	20	30	40
1	.98	.96	.94	.92	.97	.92	.89	.84	.94	.87	.80	.76	.94	.87	.80	.73	.90	.80	.70	.60
2	.98	.96	.94	.92	.96	.92	.88	.83	.94	.87	.80	.75	.94	.87	.79	.72	.90	.80	.69	.59
3	.98	.95	.93	.90	.96	.91	.87	.82	.94	.86	.79	.74	.94	.86	.78	.71	.90	.79	.68	.58
4	.97	.95	.92	.90	.95	.90	.85	.80	.93	.86	.79	.73	.94	.86	.78	.70	.89	.78	.67	.56
5	.97	.94	.91	.89	.94	.90	.84	.79	.93	.86	.78	.72	.93	.86	.77	.69	.89	.78	.66	.55
6	.97	.94	.91	.88	.94	.89	.83	.78	.93	.85	.78	.71	.93	.85	.76	.68	.89	.77	.66	.54
7	.97	.94	.90	.87	.93	.88	.82	.77	.93	.84	.77	.70	.93	.84	.76	.68	.89	.76	.65	.53
8	.96	.93	.89	.86	.93	.87	.81	.75	.93	.84	.76	.69	.93	.84	.76	.68	.88	.76	.64	.52
9	.96	.92	.88	.85	.93	.87	.80	.74	.93	.84	.76	.68	.93	.84	.75	.67	.88	.75	.63	.51
10	.96	.92	.87	.83	.93	.86	.79	.72	.93	.84	.75	.67	.92	.83	.75	.67	.88	.75	.62	.50

Fig. 12.5.13 Room surface dirt depreciation factors. *(IES)*

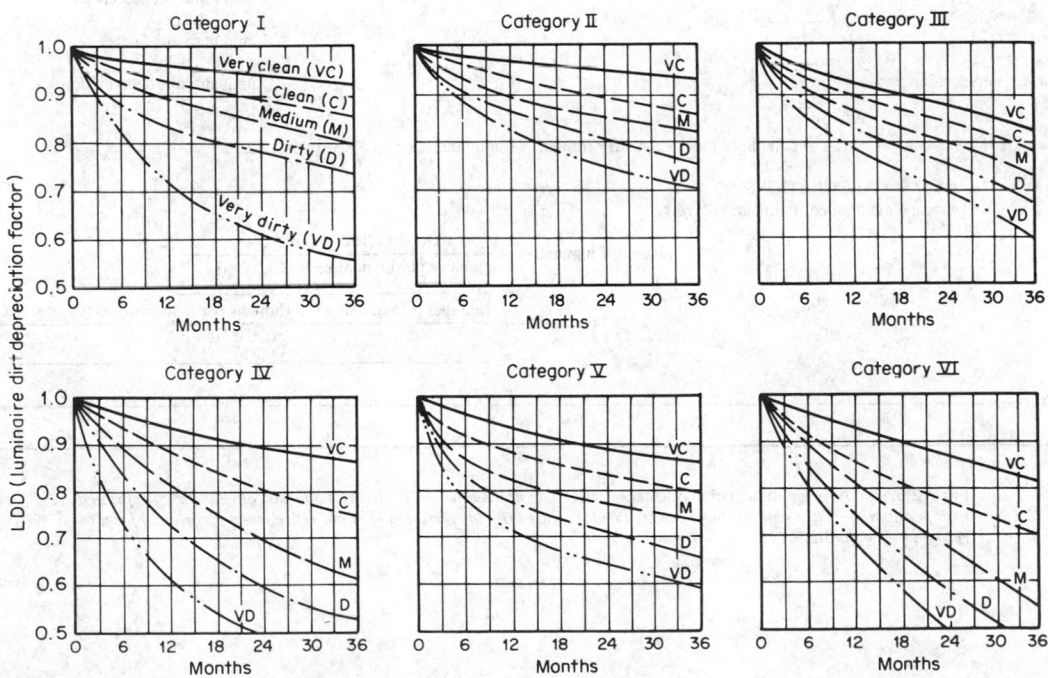

Fig. 12.5.14 Luminaire dirt depreciation (LLD) factors for six luminaire categories (I to VI) and for five degrees of dirtiness. *(IES.)*

GENERAL INFORMATION

Project identification: _____
(Give name of area and/or building and room number)

Average maintained illumination for design: ____ footcandles or

Luminaire data: ____ lux

   Manufacturer: _____

   Catalog number: _____

Lamp data:

   Type and color: _____

   Number per luminaire: _____

   Total lumens per luminaire: _____

SELECTION OF COEFFICIENT OF UTILIZATION

Step 1: Fill in sketch at right.

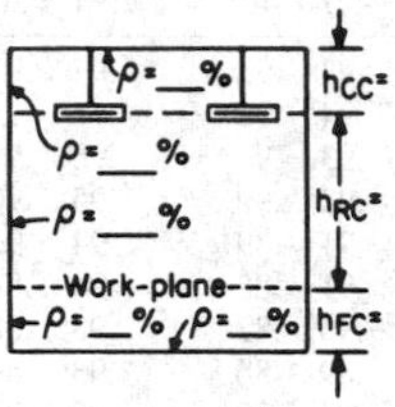

Step 2: Determine cavity ratios [Eq. (12.5.6)]

      Room cavity ratio RCR = _____
      Ceiling cavity ratio CCR: = _____
      Floor cavity ratio FCR = _____

room length: _____

room width: _____

Step 3: Obtain effective ceiling cavity reflectance ρ_{CC} from Table 12.5.7 ρ_{CC} = _____

Step 4: Obtain effective floor cavity reflectance ρ_{FC} from Table 12.5.7 ρ_{FC} = _____

Step 5: Obtain coefficient of utilization CU from manufacturer's data (or Fig. 12.5.12 and Table 12.5.8)   CU = _____

SELECTION OF LIGHT-LOSS FACTORS

Nonrecoverable		Recoverable	
Luminaire ambient temperature		Room surface dirt depreciation	
		RSDD	____
Voltage to luminaire	____	Lamp lumen depreciation	
		LLD	____
Ballast factor	____	Lamp burnouts factor	
		LBO	____
Luminaire surface depreciation		Luminaire dirt depreciation	
	____	LDD	____

Total light loss factor, LLF (product of individual factors above) = _____

CALCULATIONS
(Average maintained illumination level)

$$\text{Number of luminaires} = \frac{(\text{footcandles}) \times (\text{area*, ft}^2)}{(\text{lumens per luminaire}) \times (\text{CU}) \times (\text{LLF})} =$$

$$\text{Footcandles} = \frac{(\text{number of luminaires}) \times (\text{lumens per luminaire}) \times (\text{CU}) \times (\text{LLF})}{(\text{area*, ft}^2)} =$$

Calculated by: _____ Date: _____

---

*If lux is used, area is in m².

Fig. 12.5.15 Average illumination calculation sheet. *[Abstracted from "IESNA Handbook," 8th ed., 1993; reprinted with permission. (This material has been modified from its original version and is not reflective of its original form as recognized by the IESNA.)]*

at 30 percent rated life is used, the LLD for incandescent lamps varies from 78 to 90 percent, with an average about 87 percent. For fluorescent lamps the LLD varies from 67 to 91, with an average about 82 percent. For better values consult the IES Handbook or manufacturers' data.

A design summary sheet is given in Fig. 12.5.15.

While the lumen method is the accepted procedure for calculating interior lighting levels, it is often necessary to have a quick approximation of the quantity of lighting equipment needed to satisfy an illumination-level specification. There are several rules of thumb which serve that purpose.

Spacing Method Table 12.5.9 indicates approximate *base*-maintained footcandle levels according to fixture spacing. For levels other than the base quantity, the level is changed inversely and proportionally to a change in spacing—i.e., doubling the spacing (in one direction) halves the level; halving the spacing doubles the level. Doubling the spacing in both directions reduces the level to one-fourth.

Lumens-per-Square-Foot Method (see footnote for Table 12.5.10);

Less accurate than the previous method, this one permits a fairly reasonable calculation by using the lamp lumens as given in the GE Lamp Catalog and substituting in the formula:

$$\text{Footcandle} = \frac{\text{total lamp lumens per fixture}}{2 \times \text{area per fixture}}$$

Or, for a given footcandle level, transposing the formula to determine area per fixture:

$$\text{Area per fixture} = \frac{\text{total lamp lumens per fixture}}{2 \times \text{footcandles}}$$

The "2" in the denominator assumes loss of half the lumens in fixture utilization, lamp depreciation and dirt accumulation. (*Source: General Electric.*)

Watts-per-Square-Foot Method Table 12.5.10 shows another way to arrive at a quick approximation.

Point Method of Design

If uniformity of lighting is to be investigated, or if outdoor lighting is to be designed, the point method is used. Manufacturers furnish candlepower distribution curves for their fixtures. An average curve is given for symmetrical fixtures while curves in various planes are given for asymmetrical ones. The basic equation for calculating the illumination from such curves is

$$E_h = (I_\theta \cos \theta)/D^2$$
$$= I_\theta H/D^3 \qquad (12.5.7)$$

Table 12.5.10   Watts-per-Square-Foot Method*

A convenient, popular, quick approximation

Lamp	Per 100 fc
Lucalox	1.6 W/ft²
Multi-vapor	2.5 W/ft²
Fluorescent	3.0 W/ft²
Mercury-vapor	3.5 W/ft²
Incandescent (reflector lamp)	8.5 W/ft²

SOURCE: General Electric Co.

* Both Table 12.5.9 and lumen-per-square-foot method are based upon *large* industrial areas, where room width (W) is six times the fixture mounting height (MH). For medium-sized areas (where W = 3MH), reduce footcandles and increase wattage 15 percent. For small areas (where W = MH), reduce footcandles and increase wattage 50 percent. Lucalox fixtures and incandescent reflector lamps tend to be more efficient and have higher lumen-maintenance characteristics; thus, for these sources increase the footcandle value about 25 percent.

where E_h is the illumination on the horizontal plane, I_θ the candlepower in the given direction, and D the distance of the luminaire to the point P. See Fig. 12.5.16.

$$E_v = (I_\theta \sin \theta) D^2$$
$$= I_\theta R / D^3$$

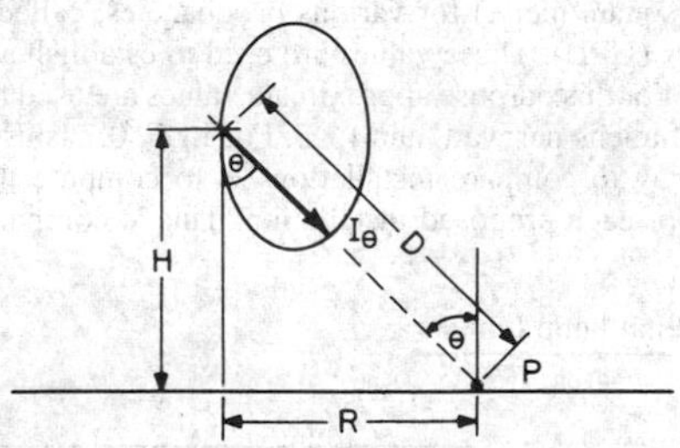

Fig. 12.5.16 Footcandle calculation diagram.

Table 12.5.9   Approximate Footcandle Levels According to Fixture Spacing*

Lighting system		Spacings†				
Lamp	Watts	10 × 10 ft	15 × 15 ft	20 × 20 ft	25 × 25 ft	30 × 30 ft
Lucalox	70	35	15	10	—	—
	100	55	25	15	10	—
	150	95	45	25	15	10
	250	180	80	45	30	20
	400	300	135	75	50	35
	1,000	—	—	210	135	95
Multivapor	175	85	35	20	15	10
	400	200	90	50	35	25
	1,000	—	300	165	105	75

Continuous rows of 2-lamp fixtures on spacing† of:					
Fluorescent (cool white)	6 ft	8 ft	10 ft	12 ft	15 ft
40W Rapid Start	120	90	70	60	50
75W Slimline	120	90	70	60	50
110W High Output	185	140	110	90	75
215W Power Groove	300	225	180	150	120

SOURCE: General Electric Co.

* See footnote for Table 12.5.10.

† Spacings assumed within maximums established by fixture manufacturer.

For vertical surfaces

$$E_v = (I_\theta \sin \theta)D^2 \\ = I_\theta R/D^3 \qquad (12.5.8)$$

Nomograms and graphical solutions are available for Eqs. (12.5.7) and (12.5.8).

THE ECONOMICS OF LIGHTING INSTALLATIONS

The cost of lighting is computed by summing the annual cost of energy; relamping; cost of labor for cleaning, relamping, and servicing; interest; and depreciation.

Different lighting systems can be evaluated by comparing the costs per million lumen hours per luminaire. This is given by Eq. (12.5.9).

$$U = \frac{10}{Q \times D} \left[\frac{(P + h)}{L} + W \times R + \frac{(F + M)}{H} \right] \qquad (12.5.9)$$

where U = unit cost of light in dollars per million lumen-hours; Q = mean lamp lumens; D = luminaire dirt depreciation (average between cleanings); P = price of lamp in cents; h = cost (in cents) to replace one lamp; L = average rated lamp life in thousands of hours; W = mean luminaire input watts (lamps plus ballast); R = energy cost in cents per kilowatthour; F = fixed or owning costs in cents per luminaire-year; M = cleaning costs in cents per luminaire-year; and H = annual hours of operation in thousands of hours.

In the above equations the area can be expressed in square metres. These equations are general and basic. With the advent of energy conservation, it has become important to use electric power effectively and efficiently. The IES, in its 1981 Handbook, established a watts per square foot (square metre) for various occupancies, called a *base unit power density* (UPD). These values are used to establish a power limit for a facility. For this purpose approximate values are used for CU, lamp plus ballast, lumens per watt, and LLF. For LLF, 0.7 is used.

Another way to compare installations is to compute the watts per square foot for each proposed installation. This is computed by either method:

$$\text{watts/ft}^2 = \frac{\text{total lamp lumens}}{\text{area, ft}^2} \\ \times \frac{1}{\text{lumens/watt of lamp and ballast}} \qquad (12.5.10)$$

$$\text{watts/ft}^2 = \frac{\text{designed illuminance}}{\text{CU} \times \text{LLF}} \\ \times \frac{1}{\text{lumens/watt of lamp and ballast}} \qquad (12.5.11)$$

Typical values are shown in Table 12.5.10.

DIMMING SYSTEMS

Dimming systems are used in theaters, auditoriums, ballrooms, etc. Originally, power-consuming rheostats were used. These have been replaced by continuously variable autotransformers, variable reactors, silicon controlled rectifiers (SCR) and triacs. The development of controlled solid-state devices has resulted in small, reliable dimmers which can be readily programmed. Only incandescent and cold-cathode lamps can be dimmed easily. Fluorescent lamps require special ballasts which keep the electrodes hot at all times. Dimmers have been developed for high-intensity discharge (H.I.D.) lamps.

HEAT FROM LIGHTING

Lighting installations are a substantial source of heat, have long been a factor in the design of air-conditioning (cooling) systems, and are increasingly significant in the design of heating systems. The heating effect for 1 W is 3.413 Btu/h. Approximate wattage data for lighting systems at various lighting levels can be calculated by using watts per-square foot calculated from Eq. (12.5.10) or (12.5.11). Heat generated is delivered to surrounding areas in several ways, with energy distribution for fluorescent and incandescent lamps as illustrated in Fig. 12.5.17. With the prevalent high lighting intensities of modern buildings, it is essential to control the heat generated by a lighting system. Substantial portions of the energy which is not radiated into the room may be conducted away from the luminaire by an air stream or by water flowing through a coil attached to the luminaire. In the heating season, this heat

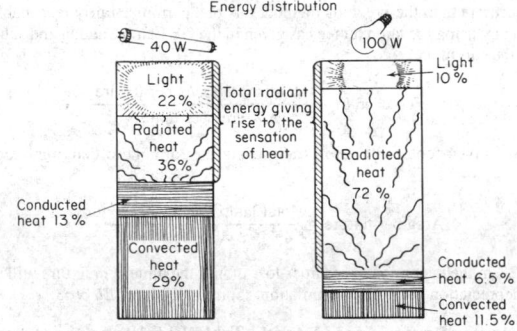

Fig. 12.5.17 Energy distribution of lamps.

energy is delivered to the perimeter of the building for effective space warming. In the cooling season, the heat is rejected to the exterior, thus reducing the load on the cooling system. Air-handling luminaires (Fig. 12.5.18) are receiving wide acceptance.

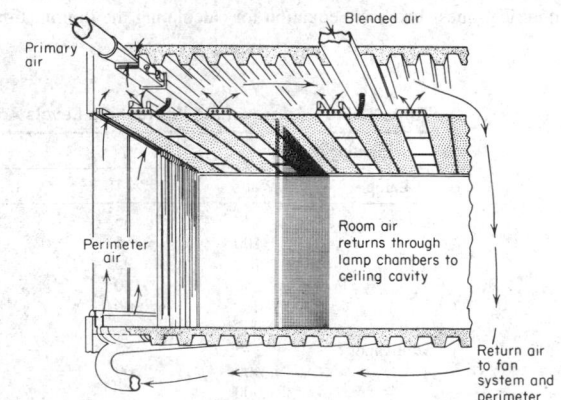

Fig. 12.5.18 Typical air handling system. (*Barber-Colman.*)

12.6 SOUND, NOISE, AND ULTRASONICS

by Benson Carlin and expanded by staff

REFERENCES: ANSI 51.1 Acoustical Terminology. *J Acoust. Soc. Am.* 1929 et seq. Beranek, "Acoustics," McGraw-Hill. Carlin, "Ultrasonics," McGraw-Hill. Morse and Ingard, "Theoretical Acoustics." Harris, "Handbook of Noise Control," McGraw-Hill. Mason, "Physical Acoustics," Academic Press.

DEFINITIONS

Sound is an alteration in pressure, stress, particle displacement, and particle velocity, which is propagated in an elastic material. It is longitudinal in gases but may also be transverse (shear), surface, or other types in elastic media which can support such energy. It may be reflected, diffracted, or refracted at boundaries and under suitable conditions may be changed from one form to another. In longitudinal waves, the molecules move in the direction of wave motion, in the others at right angles to it. Waves may also be plane or circular depending on the source.

In general, the speed of sound V in a medium with mass density ρ is a function of its elastic properties. In solids, $V = \sqrt{E/\rho}$, where E is Young's modulus. In liquids, $V = \sqrt{E'/\rho}$, where E' is the bulk modulus of the liquid (see Table 3.3.2). In gases, the velocity is independent of the pressure, because the elasticity changes to compensate for the density changes; the general equation is $V = \sqrt{kp/\rho}$, where k is the ratio of specific heats and p is the pressure. The velocity in air at 68°F is 1,126 ft/s (33,160 cm/s) and increases by 0.1 percent per °F. In liquids, empirical formulas are easier to use than theoretical ones to predict actual velocities, since velocity varies in a complex way with temperature, pressure, and other factors. With sea water, a standard velocity of 5,100 ft/s (150,000 cm/s) may be used. The velocity of sound in liquids and solids is usually much higher than in gases (see Table 12.6.1 and Kinsler, "Fundamentals of Acoustics," Wiley).

The **frequency** of a sound is the number of periods (cycles) occurring in unit time, customarily expressed as cycles per sec (cps) or "hertz" (Hz); kilocycles per sec, kc = 10^3 cps (kHz); megacycles per sec, Mc = 10^6 cps (MHz). Sound frequencies are usually defined as 20 to 20,000 cps (**audible**), higher (**ultrasonic**), and lower (**infrasonic**). Frequencies as high as the thousand-megacycle range (GHz) are now generated (see Table 12.6.2).

The relation between frequency f and wavelength λ is $V = \lambda f$. In air, at 1,126 cps, the wavelength is 1 ft. In nature, the waves may be simple sinusoidal, complex, or explosive (shock) depending on the source. The first is of course rare.

Attenuation of sound depends on the media of propagation and the frequency and is caused by absorption, spreading, and scattering. At audible frequencies in air attenuation is small except for the spreading of the energy over wide areas as the sound waves are propagated. By this means the intensity drops according to the inverse square law. However, in other media, the absorption, scattering, or other characteristic may be predominant.

The sound **intensity** is the average rate of sound energy transmitted through a unit area normal to the wave direction at the point considered. This is a definition of power and may be expressed in watts per sq metre. It is usual, however, to express power in **decibels,** dB, which is a term used to give the relative magnitude of two powers by comparing the one under consideration to a standard. The sound-pressure level in decibels, dB, is defined as twenty times the logarithm to the base 10 of the ratio of sound pressure to the reference sound pressure. All values are for air at 20°C and atmospheric pressure. Pressure measurements in air use a pressure reference (rms) of 0.0002 dyne/cm^2; 1 dyne/cm^2 is used underwater.

Intensity references for air are 10^{-16} w/cm^2 [equivalent to a pressure (rms) of 0.0002 dyne/cm^2, and 0.02 erg/cm^{2}s, equivalent to a pressure of 1 dyne/cm^2]. Since the references are equivalent (i.e., the reference pressure corresponds to the reference intensity in this particular case), numerical results are identical for plane waves using either expression $IL = 10 \log (I/I_0)$ or $PL = 20 \log (P_e/P_0)$, where IL and PL are the intensity and pressure levels, I_0 and P_0 are the reference intensity and pressure, P_e is the effective pressure, and I is the intensity in question.

Table 12.6.2 Sound Spectrum

Frequency	Action
20–40 cps	Thunder
128 cps	Average speech (male)
250–2,740 cps	Telephone bandwidth
90–5,000 cps	Radio broadcast
15 cps–15 kc	Limits of average human hearing
10–90 kc	Ultrasonic cleaning
15–50 kc	Ultrasonic depth sounding, sonar
20 kc	Ultrasonic bulgar alarm, control apparatus, door opening
30 kc	Highest frequency obtained by friction
40 kc	Highest frequency of Hartmann generator
48 kc	Bat cries
90 kc	Top limit of tuning fork
100 kc	Highest frequency of Galton whistle
500–15,000 kc	Ultrasonic pulse-echo testing
1,000 kc	Medical therapy
1,500–30,000 kc	Ultrasonic delay lines
15,000 kc	Radar trainer

Table 12.6.1 Velocity of Sound

Material	Sound velocity, ft/s	Density, lb/ft^3	Density × velocity, lb/(ft$^2 \cdot$ s)
Aluminum	16,740	168	2.82×10^6
Brass	11,480	530	6.08×10^6
Copper	11,670	555	6.47×10^6
Iron and soft steel	16,410	486	7.98×10^6
Lead	4,026	1125	4.54×10^6
Brick	11,980	125	1.5×10^6
Cork	1,640	15	0.025×10^6
Wood	10,000–15,000	30–50	0.3×10^6–0.75×10^6
Water	4,794	62.4	0.299×10^6
Air, dry, CO$_2$ free, 32°F	1,088.5	0.0808	88.0
Hydrogen	4,165	0.00560	23.3
Water vapor, 212°F	1,564	0.0372	58.2

NOTE: Approximate values from Smithsonian Tables.

When making measurements with pressure or velocity microphones, it is the pressure level or velocity level which is measured and the relationship between the measurement and the intensity is unknown except in the special cases indicated.

Decibels do not add numerically as linear figures do; i.e., 70 dB + 70 dB = 73 dB since doubling power results in a 3-dB increase in sound pressure. Figure 12.6.1 shows how to add decibels within 14 dB of each other. If the difference is greater between two readings, ignore the weaker one.

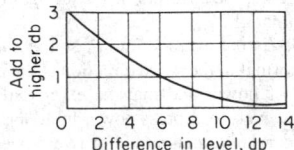

Fig. 12.6.1 Chart for addition of decibels.

Specific Acoustic Impedance The relationship between the pressure and the associated particle velocity at a point in a medium is called the specific acoustic impedance; its unit is the kilogram per meter second or mks rayl. The magnitude ρ_c is called the characteristic impedance of the medium, or the radiation resistance. This applies in the case of plane waves. The ρ_c of material is one of its most useful acoustic characteristics, since by means of it the amount of energy reflected at boundaries may be computed, horns may be analyzed according to the acoustic resistance at throat, and other calculations analogous to those made in electrical design may be carried out.

THE PRODUCTION AND RECEPTION OF SOUNDS

REFERENCES: Rinsler and Frey, ''Fundamentals of Acoustics,'' Wiley. Mason, ''Electromechanical Transducers and Wave Filters,'' Van Nostrand. Olsen, ''Elements of Acoustical Engineering,'' Van Nostrand.

Transducer A device for converting energy from one form to another, e.g., from electrical to acoustic or vice versa, is called a transducer. Among these are loudspeakers, microphones, hydrophones, and piezoelectric and magnetostrictive transducers.

Loudspeakers are usually classified as direct-radiator or horn type. The direct-radiator type consists of a cone, a magnet, a voice coil moving in the magnetic field, a vibrating diaphragm coupled to the cone, and suitable supports. The attachment of a horn improves the impedance match between the speaker and the air since it is essentially an acoustic transformer. The dimensions and flare of the horn contribute to its matching ability.

One of the more common types of horn is exponential, although straight and other types are also possible. In a similar manner, mechanical transformers may be used to concentrate the energy of ultrasonic transducers. In such forms they operate to concentrate rather than to spread the energy. The operation of the speaker may be variously influenced by its enclosure, by the baffle which separates the front from the back radiation, or by its resonances.

Microphones (for gases) and **hydrophones** (for liquids) are transducers for converting mechanical to electrical energy. They may be piezoelectric, electromagnetic, magnetostrictive, or capacitive. The variation in electrical output is proportional to the effect of the acoustic field on the characteristics. Ultrasonic transducers may be any of the above types but are usually crystal (piezoelectric) or magnetostrictive. Among the common piezoelectric materials are quartz, barium titanate, lithium sulfate, ADP (ammonium dihydrogen phosphate), and rochelle salt. In sonar and high-power industrial systems, mosaics of crystals are used; in low-power, high-frequency systems, a single crystal is usual.

Whistles and **sirens** may also be used to produce intense sound fields in gases and liquids. These are devices which produce sound by passing a fluid over an obstacle, thereby creating turbulence in the fluid. When the obstacle is an edge, these are referred to as edge or E tones; when an

orifice, as jet tones. Organ pipes, whistles, and nozzles for spraying are devices of this class. Frequencies up to 100,000 cps are possible, although 30,000 cps is the approximate limit at which appreciable power can be generated. Resonators may be placed in the sound field to reinforce it and to stabilize the frequency. These take the form of small pipes tuned to the approximate frequency. Common types of whistles are the Hartmann and Galton (for gases) and the jet edge (for liquids).

Sirens are devices in which a revolving disk with holes in it interrupts a jet from a nearby tube. Compressed air, steam, and water have been used. Frequencies up to 30 kc may be produced at efficiencies of 50 percent approximately; a 1-hp motor produces between 300 and 1,000 W (see also Jones, *J Acoust. Soc. Am.*, 1946).

Transducers are generally driven by electronic generators, motor generators, or air compressors. As receivers, they activate amplifiers or indicating devices.

The Perception of Sound The average young observer perceives sound between 20 and 20,000 cps. High-frequency response deteriorates with advancing age. The ear responds to a wide range of intensities; e.g., between 500 and 5,000 cps, the ratio of tolerated intensities is about 10^{12}. The minimum intensity perceived varies with frequency. Figure 12.6.2 shows the audible frequency and intensity range for a standard listener, where the lowest curve represents the threshold of hearing and the top one the beginning of sensation in the ear. These curves show the pressure levels required for a given tone to sound as loud as the corresponding reference tone of 1,000 cps (see also Fletcher and Munson, *J. Acoust. Soc. Am.*, 1933).

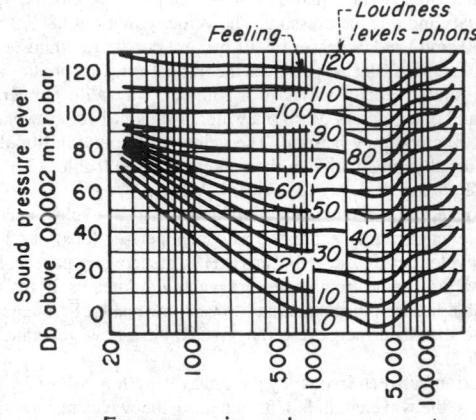

Fig. 12.6.2 Loudness contours.

Loudness is a subjective rather than a purely physical attribute. To provide a qualitative basis, the loudness level in **phons** is defined as the pressure level in decibels of a pure 1,000 cps tone which a typical observer judges to sound as loud as the sound in question. Observers can experimentally judge the loudness of pure or complex tones. However, this does not means that the apparent level is proportional to its level in phons; i.e., a level of 10 phons is not twice as loud as one of 5 phons. An additional expression, **sones,** defined as the loudness of a 1,000 cps tone at 40 dB intensity, is necessary to compare various loudness. The relationship between sones and phons is shown in Fig. 12.6.3.

Quality is a subjective attribute of sound in which equally loud sounds may be distinguished as to kind. Basically, differences in quality arise from differences in the distribution of energy in different parts of the frequency spectrum. In *music* this takes the form of the energy relationship of fundamental and harmonics; in *noise* it is random. These differences affect the sensation of loudness of noise and the psychological annoyance it produces. Shrill, high-pitched, and irregular sounds are usually judged less pleasant than low-pitched and regular sounds.

Among terms used to define quality are **pitch,** determined by frequency (mostly), together with intensity and waveshape (unit is the mel); **timbre,** determined by wave shape (mostly), together with intensity and frequency (see Seashore, "Psychology of Music," Dover).

Masking describes the ability of one sound to make the ear incapable of perceiving a second one. See ANSI 53.20, Psychoacoustical Terminology. It is measured by the shift in the threshold of audibility of the masked sound in decibels. The partial deafening of the ear by the masking effect of noise affords a direct quantitative measure of the interfering effect of the noise. If the masking is measured at several frequencies throughout the audible ranges, the overall pressure level of the sound can be computed. For any given frequency, the masking is expressed in decibels relative to the unmasked threshold for a **critical bandwidth** centered on the frequency of the masked tone. This critical bandwidth is that beyond which an increase in the passband has little effect on the masking of a pure tone at its center frequency. Critical bandwidths vary from 40 cycles at 100 cps to 200 cycles at 6,000 cps.

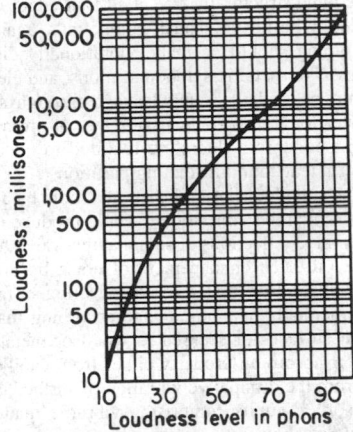

Fig. 12.6.3   Relation between loudness and loudness level.

Noise is an undesired sound. It implies an unwanted disturbance in a useful band which interferes with the useful information. Any elastic structure may produce noise when set into vibration. Generally the motion is an unwanted concomitant of some desired function, e.g., the vibration of a machine tool.

The term has a connotation of unpleasantness in quality or loudness. Typical examples are gear noises, 60-cycle hum, motor traffic, hammer blows, pneumatic-tool operations, and hissing of gases in an orifice. Noise measurements are usually made with a **sound-level meter,** comprising a microphone, attenuator, amplifier, frequency-weighing networks, and an indicator or recorder. See ANSI 51.4, Sound Level Meters. A **sound analyzer** indicates sound pressure as a function of frequency. In some cases electrical filters may be included which permit measurement in certain restricted frequency ranges. By using these instruments the components of the heterogeneous noise may be identified, and this helps to correlate it with its production. Once located, techniques for elimination flow and the results may be used to compute masking. Individual frequencies may be identified by beating against a known source (**beat-frequency oscillator**) until a null is observed.

Contact transducers (vibration pickups) may be used to locate sources of noise, such as in partitions and machine parts. They may be piezoelectric or magnetic and may be used in place of the microphone or the sound-level meter; hydrophones may also be used.

Where the frequency distribution of the noise is significant, an analyzer may be used since the level meter tells nothing about frequency distribution. Analyzers are manufactured in various forms, e.g., the octave-band analyzer, the impact-noise analyzer, and the wave analyzer, each of which uses a different method of finding out which frequency components are present.

Typical sound levels are shown in Table 12.6.3.

Table 12.6.3   Typical Sound Levels

	Decibels	
	120	Threshold of feeling
		Thunder, artillery
Deafening	100	Nearby riveter
		Elevated train
	100	Boiler factory
		Loud street noise
Very loud	90	Noisy factory
		Truck unmuffled
	80	Police whistle
		Noisy office
Loud	70	Average street noise
		Average radio
	60	Average factory
		Noisy home
Moderate	50	Average office
		Average conversation
	40	Quiet radio
		Quiet home or private office
Faint	30	Average auditorium
		Quiet conversation
	20	Rustle of leaves
		Whisper
Very faint	10	Soundproof room
		Threshold of audibility
	0	

NOISE CONTROL

REFERENCES: General Radio Co., "Handbook of Noise Measurement." Harris, "Handbook of Noise Control," McGraw-Hill. Bolt, "Handbook of Acoustic Noise Control." Faulkner, "Handbook of Industrial Noise Control," Industrial Press.

Noise control may be carried out at several stages: (1) at the source, by design changes or by quieting procedures, (2) during transmission, by attention to the path by which it is propagated to the listener, and (3) by quieting at the listening position. It may also be controlled architecturally as by the careful placement of necessarily noisy rooms in a building.

The Source   By inspection and test procedures, a noise is tracked to its source. In some cases, design procedures which attempt to reduce the vibration or to prevent its radiation may be used. This may require the redesign of elements, such as cams, gears, housings, or provisions for cushioning. Viscous damping materials, e.g., putty or tar, may be applied to the vibratory surfaces in the form of nonhardening plastic mixtures. A machine may be isolated by sections or shock mounts to prevent transmission of vibration from one section to another. Absorbing materials may be placed on walls to absorb sound after it has been radiated.

Transmission Isolation   If the vibrations of noisy machinery cannot be suppressed at the source, their transmission to the listener should be impeded. For the higher frequencies constituting noise, the most effective isolation method is the introduction of elastic discontinuities in the structure transmitting the noise (measured by the difference between density-velocity products as given in Table 12.6.1). The discontinuities may be obtained by the use of felt, cork, rubber, or springs in machinery mountings, or by the introduction of alternate lead and cork sheeting at masonry junctions. The isolation treatment should be applied as close to the source as possible in order to eliminate sound radiation from the structures transmitting the vibrations. Where this is not possible, the listening space itself may be isolated. Thus **quiet rooms,** constructed especially for noise measurements, are usually built as separate structures isolated from the main building.

Filtration   Some problems of noise transmission through air lend themselves to solution by methods of filtration. Typical examples are the transmission of sound in ventilating ducts and the noise production at engine exhaust pipes. In each of these cases, the steady flow of gas must not be impeded, but the alternating flow, representing sound transmission, must be effectively suppressed.

For ventilating ducts, an acceptable degree of noise suppression may be obtained by lining the ducts (on at least two nonopposite walls) with an efficient sound absorbent for a distance of 10 to 15 ft from both the inlet and the outlet. Where the length of duct available is insufficient, or where additional noise suppression is required, baffles, covered with absorbing material, may be introduced in the duct. A plenum chamber, used to serve several ducts, should be lined with sound absorbents. If the air velocities are high, it may be necessary to introduce additional baffles at bends in the ducts to avoid noise production through turbulence.

Exhaust mufflers are usually modifications of the elementary low-pass acoustical filter, comprising a through tube to which closed cavities are coupled through small holes at intervals along the tube. Mufflers of this type find application other than in internal combustion engine exhaust systems, and often are termed *passive mufflers*. Typical structures of this type (Fig. 12.6.4) produce little increase in back pressure and considerable attenuation of sound waves having frequencies above a cutoff frequency determined by the size of the holes and cavities. Porous packing, such as steel wool, in the side cavities or studied irregularity in the size and spacing of the cavities will increase the uniformity of noise suppression, whereas increasing the number of side cavities

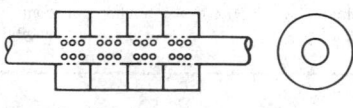

Sectional view

Fig. 12.6.4 Section through a passive exhaust muffler.

and the length of the muffler will increase the amount of suppression. Baffles in the tailpipe or irregular obstructions producing devious flow paths, e.g., the stone-filled pit for stationary engine exhausts, produce muffling action at the expense of appreciable increase in exhaust back pressure.

Shielding of airborne noise must be done by sound-opaque screens large in comparison with the wavelength of the sounds whose transmission they are to impede. This is seldom possible in building interiors except by utilization of building partitions as screens. Sound is transmitted through such partitions principally by minute flexure of the wall as a whole in response to the incident sound pressure on the noisy side, with consequent reradiation on the quiet side. Reduction of sound transmission is obtained by increasing the mass per unit area of the partition, by constructing the partition of material comprising a hollow air-filled or fibrous-filled sandwich with a pattern of holes in one of the surfaces, or by the use of double partitions, vibrationally isolated.

Sound-transmission loss is usually greater for high frequencies than for low and is measured by comparing the average sound level on each side of the partition under standardized conditions. Average values of

transmission loss, for frequencies from 125 to 4,000 cps, for typical partitions, are shown in Table 12.6.4. In any specific case, a more exact measure of the effectiveness of an insulating partition can be obtained by direct comparison of the transmission-loss vs. frequency curve for the partition and the intensity vs. frequency curve for the noise. Additional data on transmission loss for a wide variety of building materials and structures are available in the NIST publications TRBM-44; BMS 17 and Supplements 1 and 2.

In general, double partitions (including floated floor constructions) provide greater transmission loss than equally heavy concrete, masonry, or brick walls but, except for special designs, less transmission loss than equally thick masonry walls. Double walls must be constructed carefully to avoid loss of vibration isolation through mechanical bridging between the opposite surfaces. Sound-absorbing fillers (e.g., mineral wool) are usually detrimental to sound insulation if in contact with both interior surfaces, and a single bridging nail may alter significantly the insulating efficiency. For maximum effectiveness, one of the wall surfaces should be hung structurally free at all four edges with the boundary cracks sealed, with felt or asphalt compounds, against sound leakage. Through piping should be made vibrationally discontinuous by introducing canvas or metallic sylphon sections, and clearance holes at the walls should be sealed. Sound leakage through small clearance cracks contributes to the low transmission loss of ordinary doors. Special self-sealing soundproof doors are required to maintain the effectiveness of an efficient sound-insulating partition.

Quieting The sound level established in a room by a noise source is higher than that which the same source would produce in free space on account of successive reflections of sound at the walls. It is the function of **quieting** to avoid such enhancement of noise by providing a high degree of sound absorption at all interior reflecting surfaces exposed to the noise. Commercially available sound-absorbing materials may be cemented to flat surfaces or secured to wood or metal furring strips. They derive their absorbing property either from capillary porosity of the surface or from the dissipative vibration of surface layers. Hanging ''functional absorber'' units comprising vibratile matte surfaces, enclosing a volume of about 1 ft³, can be used where surface absorbents cannot be installed conveniently. The effectiveness of sound absorbents varies with frequency, usually being greater for high and intermediate than for low frequencies. It may be measured by determining the **absorption coefficient**, defined as the fraction of sound energy diffusely incident on the material that is not reflected, or by determining the **specific acoustic impedance** of the material. The measured absorption coefficient is not a property of the material alone, but depends partly on the size and mounting of the test sample and the size and shape of the test chamber; thus comparison of the coefficients for different materials should be based only on measurements made under identical conditions. Such measurements on a wide variety of materials have been made available by the Acoustical Materials Assoc. (Chicago, Ill.), although it is to be expected that the absorption coefficients effective in various practical applications may differ somewhat from the published values.

Table 12.6.4 Sound-Transmission Loss in Building Partitions

Wall	Thickness, in	Weight lb/ft²	Transmission loss, dB
Wood	0.2	0.45	18.5
Plate glass	0.25	3.2	27.0
Hollow gypsum tile, unplastered	3	11.1	27.2
Brick wall, unplastered	. . . .	22.0	33
Brick wall, plastered	6	46	43
Brick wall, plastered	10.5	93	49
Double wall; metal lath, ½-in gypsum plaster, on staggered 2 × 4 in wood studs	7.5	19.8	44
Double 3-in hollow gypsum tile, unplastered, 3-in airspace	9	22.0	42.6
1-in Thermax nailed over building paper to 3-in Thermax laid up in mortar, ½-in plaster on both sides	5	15	47
Double 2-in solid-gypsum tile, unplastered, completely isolated structurally by separate foundations, 4-in airspace	8	20.4	59

SOURCE: Based on Sabine, ''Acoustics and Architecture,'' McGraw-Hill.

Table 12.6.5 Sound-Absorption Coefficients

Maker, material thickness	Absorption coef at indicated frequencies 128	256	512	1,024	2,048	4,096	Noise-reduction coef	Weight, lb/ft²	Surface	AMA test No.
Armstrong Cork Co.										
Cushiontone A, 3/4 in.	0.10	0.28	0.66	0.91	0.82	0.69	0.65	1.05	484 holes per sq ft, 3/16 in diam, 5/8 in deep. Painted by mfr	47–28
Travertone, 3/4 in.	0.06	0.23	0.78	0.97	0.84	0.80	0.70	1.20	Fissured, painted	48–58
The Celotex Corp.										
Acousti-Celotex Type C-9, 3/4 in	0.11	0.23	0.80	0.93	0.58	0.50	0.65	0.96	441 holes per sq ft, 3/16 in diam. Any paint	46–132
Type M-1, 5/8 in	0.07	0.21	0.64	0.86	0.93	0.83	0.65	1.31	676 holes per sq ft, 5/32 in diam. Any paint	46–12
Q-T Ductliner, in	0.21	0.42	0.71	0.86	0.79	0.75	0.70	1.3	Unpainted	A48–10
Johns-Manville Corp.										
Sanacoustic, KK, pad plus metal facing and pad supports 1 9/16 in	0.25	0.58	0.96	0.97	0.85	0.72	0.85	Pad 1.28	4,608 holes per sq ft, 0.068 in diam. Enameled metal pan backed with wool pad	46–88
Fibretone, 13/16 in	0.14	0.37	0.69	0.80	0.76	0.73	0.65	1.17	484 holes per sq ft, 3/16 in diam. Any paint	46–124
Airacoustic, 1 in	0.29	0.31	0.70	0.82	0.79	0.80	0.70	1.50	Unpainted	46–71
National Gypsum Co.										
Acoustifibre, 5/8 in	0.10	0.16	0.62	0.97	0.81	0.73	0.65	0.56	441 holes per sq ft, 3/16 in diam, 5/16 in deep. Painted	46–137
Owens-Corning Fiberglas Corp.										
Fiberglas Acoustical Tile, plain type, 3/4 in	0.04	0.20	0.63	0.91	0.82	0.82	0.65	0.69	Painted	A48–99
United States Gypsum Co.,										
Acoustone F, 11/16 in	0.08	0.25	0.76	0.84	0.78	0.73	0.65	1.35	Fissured, painted	46–50
Brick wall, painted	0.012		0.017		0.023		0.02			
Concrete wall or floor	0.01		0.015		0.02		0.02			
Wood floor	0.05		0.03		0.03		0.03			
Cork or rubber tile on concrete			0.03–0.08				0.05			
Glass	0.035		0.027		0.02		0.02			

NOTE: This tabulation is based on *Bull.* XI (1949), Acoustical Materials Assoc. All samples were cemented to plasterboard for test, except that the Sanacoustical unit is attached to wood furring with special clips, and the duct linings are laid on 24 gage sheet iron, nailed to 1 × 3-in wood furring, 24-in O.C.

For ordinary noise quieting, the average of absorption coefficients measured at frequencies of 250, 500, 1,000, and 2,000 cps, called the **noise-reduction coefficient,** may be used. Typical values of this coefficient for representative materials are given in Table 12.6.5. In making quantitative estimates of noise reduction, the **total sound absorption** of the room boundaries may be computed by multiplying the noise-reduction coefficient of each different material present by the total exposed area of that material and summing up the resulting products. The noise reduction is then given by

$$\text{Noise reduction in decibels} = 10 \log \frac{\text{total absorption after treatment}}{\text{total absorption before treatment}}$$

When the frequency spectrum of the offending noise is known, greater precision in calculation of total absorption is obtained by replacing the noise-reduction coefficient by the absorption coefficient measured at the frequency of maximum loudness level from the noise source. Subjective judgments of the loudness reduction obtained by quieting can be estimated by using the noise reduction in decibels in connection with the loudness chart of Fig. 1.2.6.3.

In general, the larger the area of absorbing material introduced and the higher its noise-reduction coefficient, the more effectively the noise is reduced. No amount of quieting treatment can reduce the level of the noise received directly from the source. If full coverage of walls and ceiling is not possible, distribution of the material in several small patches is more effective than the same total area of material concentrated in one location. Similarly, the same area of material is more effective when applied to nonopposite walls and ceiling than when concentrated on either of these areas, and more effective when located near the edges and corners of a given area than when located in the center. Recent advances in noise control have resulted from advances made in signal processing chips and in the significant reduction in the cost of those components. These electronic components are built into electronic equipment, the whole of which constitutes an active noise control system, and which is based on the principle that a sound signal can be canceled by an almost identical sound signal produced 180° out of phase with the first one. This concept of an electronic noise muffler has been successfully adapted to control offensive noise in a variety of industrial applications. Further developments are expected along these lines.

APPLICATIONS

REFERENCES: Carlin, ''Ultrasonics,'' McGraw-Hill. Bergmann, ''Ultrasonics,'' S. Hirzel Verlag. ANSI Z24.18, Ultrasonic Therapeutic Equipment.

Industrial Applications Acoustic waves of high powers used in industrial applications are generally called **sonic** (or **ultrasonic,** when greater than about 20,000 cps). High-intensity sonic waves, in the 10,000 cps to the megacycle range (10^6 cps), are applied to many industrial processes. The effects seem to be a function of cavitation, heating, particle acceleration, short wavelength, and other characteristics of the waves. Application categories are (1) high amplitude and (2) low amplitude.

High-amplitude waves are used in operations such as cleaning, welding, drilling, emulsification, soldering, atomization, chemical and biological applications, medical therapy, and sonar. The energy may be continuous, pulsed, or modulated, in various ways.

Low-amplitude waves are used in operations such as material testing, burglar alarms, delay lines, or medical diagnoses. Any waveshape may be used; basically, a physical characteristic of the waves, such as velocity, is measured.

Cavitation may be defined as the formation and collapse of gas- or vapor-filled bubbles. Most significant industrial applications, e.g., cleaning, take place during the vaporous phase. The amount and force of cavitation are affected by the character of the liquid and the gas in it. The bubble collapse generates powerful local forces which cause the desired action. Levels of power in the liquid of approximately 3 W/cm² are required for intense cavitation and are dependent upon factors such as the liquid, temperature, and external pressure; powers as low as 0.3 W/cm² in water will produce a threshold cavitation (see also Briggs et al., *J. Acoust. Soc. Am.*, 1947).

High-Amplitude Applications

(See Fig. 12.6.5.)

Cleaning An ultrasonically agitated bath will erosively clean dirt from immersed articles. A high-power generator, usually electronic, produces sonic energy which is impressed on a transducer to drive the bath. Barium titanate transducers prevail, but magnetostrictive designs may be used. 10,000 to 90,000 cps are commonly used, with the lower frequencies generally more effective. Generator tuning may be manual or by feedback from the transducer controls. Power levels of 5 ± W/cm² are commonly used; size (depth) of the tank may affect the output; 50 ± W/gal is a rough empirical relationship; cavitation must exist for effec-

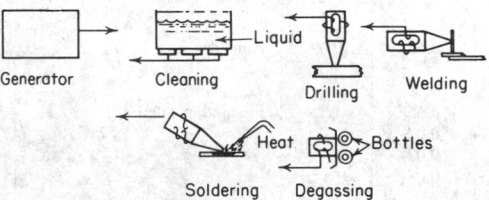

Fig. 12.6.5 High-amplitude systems.

tive, speedy cleaning. A proper cleaning solution must be used, i.e., one which supports intense cavitation and also cleans (e.g., water solutions of alkalines or acids, or solvents like hydrofluoroether); temperatures between 120 and 160°F prevail. Among items commonly cleaned are jewelry (lost-wax castings), eyeglass frames, lenses, metal parts, and watches and clocks.

An alternative form of ultrasonic cleaning can be performed by the introduction of an ultrasonic probe in a cleaning bath. The probe can be made to focus a large amount of energy over a small area or volume of the cleaning solution in which the workpiece is submerged. This method fosters removal of tenacious contaminents from normally inaccessible locations.

Foaming of Beverages Air content, which determines the life of a carbonated beverage, is reduced by foaming the bottles or cans before capping. Magnetostrictive transducers at 20,000 cps and 250 W, in contact with the containers, produce enough foam from the CO_2 to expel the air. The ultrasonic power requirement is basically small, but the losses in the coupling dictate the use of large generators. Similar techniques and apparatus may be used for the removal of gases from materials and for chemical effects such as the acceleration of iodine reactions and oxidation.

Soldering Some materials, such as aluminum, oxidize when exposed to air so that soldering is not possible. However, an ultrasonically driven solder bath will cause wetting of the material with solder and tinning of the surface. Magnetostrictive units are indicated because of the temperature requirements, with external heating and high tin-content solders; pots, as well as irons, may be constructed. Applications include aluminum wire, foil capacitors, and the filling of holes in castings (see also Sec. 13).

Welding Similar equipment (between 100 and 5,000 W output) may be used to weld thin metal or thin-to-thick sections. Such units apply the ultrasonics in a shear direction with respect to the parts to be

welded. The process depends primarily on the sonic energy, the clamping force, and the amount of external heating. Either spot or lap welds are possible, but in all cases one of the sections must be thin (see also Sec. 13 and "AWS Welding Handbook").

Drilling is effected with the same sort of apparatus, but the force is longitudinal rather than shear. An abrasive is flowed over the tool head and is driven by the cavitation against the part to be drilled, causing the material to erode away. Any shape may be obtained in this manner. The head is usually mounted on an apparatus similar to a milling machine. Tolerances depend principally on the physical rigidity of the system and on grit size. Applications include hard materials such as ceramics, jewels, and glasses. The same apparatus has been applied to dental drilling, but the time required and the necessity for use of a slurry have kept it from greater acceptance. However, the method is widely applied to cleaning the surface of the teeth. The technique has been applied to forming lesions in the brain and spinal column of human beings, using a focused beam of sound at 3 ± Mc rather than a velocity transformer.

A **whistle** may be operated in a gas for agglomeration or foam settling and in a liquid for emulsification. Ultrasonic fields introduce additional forces on particles suspended in a gas, causing them to come together. Materials such as smoke, dust, and fog have been experimentally treated in this way. Generally, the whistle drives a resonant cavity, generating about 150 dB intensity (power output of 150 ± W) at 10 to 20 kc. Atomization of a liquid may be effected by introducing the liquid into a strong sonic field, either passing it directly over or through the whistle. The use of waves has been reported for drying solids, such as sugar, for atomically driven sound beacons under the sea, and for emulsification of liquid rubber and other materials.

Sonar Underwater signaling and detection are among the older applications of ultrasonics and comprise the active (pulse-echo, Doppler) and passive (listening) systems. The principles of operation are similar to those of pulse-echo testing in the active case (see Albers, "Underwater Acoustics Handbook," Penn State Univ. Press).

Among the **miscellaneous applications** of high-power ultrasonics is metal treatment, atomization of oil for burners and of cleaning solutions used in the manufacture of delicate electronic devices such as circuit boards, ultrasonic diathermy for bursitis, humidification, and ultrasonic neurosonic surgery.

Testing Materials (See Sec. 5 and McMaster, "Nondestructive Testing Handbook," Ronald.) The most common industrial use of low-power ultrasonic waves is for testing materials. The technique basically depends on the ability of a discontinuity in a material to reflect part of the energy hitting it. Various types of ultrasonic waves such as longitudinal, shear, or surface may be used. Transducers are usually crystal, such as barium titanate or quartz, and measure from ¼ to 1 in. diam (Fig. 12.6.6).

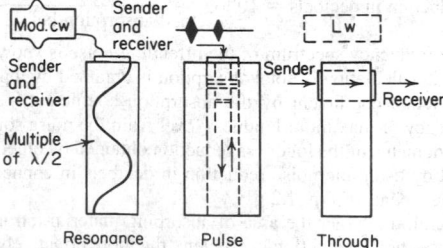

Fig. 12.6.6 Low-amplitude systems.

The basic types of ultrasonic systems are (1) pulse-echo, (2) through-transmission, and (3) resonance. The **pulse-echo system** uses a pulse ranging in length (time) from a fraction of a microsecond to several microseconds and an amplitude from 50 to 250 V radio frequency across the transducer. The pulse travels in the material and is reflected by an interface; the time of travel is measured. The pulse-echo method has also been applied to medical mapping of the body interior and for tracing brain centerline displacement and heart valve action.

The **through-transmission system** places continuous pulsed or modulated waves on a transducer coupled to one side of a part with pickup on the other side. If a flaw interrupts, the waves do not penetrate the part.

The **resonance system** uses a single transducer and varies the frequency applied to it. Within the applied frequencies is one whose wavelength is related to the thickness of the part in such a way that less power is required of the driving system. This condition is indicated, and since the wavelength within the material is known, the thickness is determined. The most common applications of the resonance system are (1) for thickness measurement when one side only is available, and (2) for finding laminations in thin sections and lack of bond (see also Sec. 5).

The **sonic burglar alarm** depends on the Doppler shift caused in a sonic field by a moving object. The unit operates at 20 kc and will find minimum objects of 0.03 ft² in an enclosure of 100 ft². Magnetostrictive transducers are used coupled to diaphragms; they are not unlike radio speaker systems in appearance.

Liquid-Level Sensors   Ultrasonic devices may be used to measure the level of liquid in a tank either by the pulse-echo technique or by indicating the transducer lead, i.e., a transducer driven by an oscillator where the reaction of the liquid load causes a change in the driving current.

Sonic microscopes are devices in which a beam of sound illuminates a part; the shadows of the field are scanned on a cathode-ray tube (usually constructed of barium titanate) by a flying spot.

SAFETY

Under the Occupational Safety and Health Act (OSHA), definitions have been made which legally define levels either as safe or as hazardous. Moreover, noise is now recognized as a pollutant, both as a nuisance and as the cause of hearing impairment.

General noise levels in the environment have already been defined (Fig. 12.6.2). There is some evidence that noise may cause ailments such as anxiety and heart disorders.

Protection from noise is required when sound levels exceed those in Table 12.6.6 (Table G-16 of the Act), when measured on the A scale at slow response on a standard sound-level meter (except for certain alarms, etc., as provided in the Act). When several successive exposures occur, they are combined (see paragraph 1910.95, reference above).

Conversion from octave-band analysis levels to A-weighted levels may be made from the figure in the Act. This figure has been also adopted by some local or state laws (Fig. 12.6.7).

When the environmental noise is greater than specified in the Act, protection must be provided.

However, when the noise is intermittent, if the peaks occur within 1 s or less, it is considered continuous. When protective equipment is required, it must be provided by a trained person and periodic checks made of the effectiveness.

In addition to the levels, time of exposure is also involved, as shown in Table 12.6.6.

OSHA also requires, when 90 dBA is exceeded, "a continuing effective hearing conservation program shall be administered," i.e., consisting of periodic hearing checks and noise surveys.

The type of facility required for these tests is spelled out in the reference above.

Table 12.6.6   Permissible Noise Exposures

Duration per day, h	Sound level, dBA slow response
8	90
6	92
4	95
3	97
2	100
1½	102
1	105
½	110
¼ or less	115

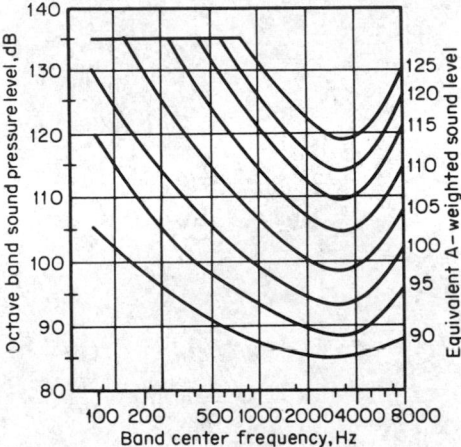

Fig. 12.6.7 Equivalent sound-level contours. Octave-band sound-pressure levels may be converted to the equivalent A-weighted sound level by plotting them on this graph and noting the A-weighted sound level corresponding to the point of highest penetration into the sound-level contours. This equivalent A-weighted sound level, which may differ from the actual A-weighted sound level of the noise, is used to determine exposure limits.

Manufacturing Processes

BY

MICHAEL K. MADSEN *Manager, Industrial Products Engineering, Neenah Foundry Co.*

RAJIV SHIVPURI *Professor of Industrial, Welding, and Systems Engineering, Ohio State University.*

OMER W. BLODGETT *Senior Design Consultant, Lincoln Electric Co.*

DUANE K. MILLER *Welding Design Engineer, Lincoln Electric Co.*

SEROPE KALPAKJIAN *Professor of Mechanical and Materials Engineering, Illinois Institute of Technology.*

THOMAS W. WOLFF *Instructor, Retired, Mechanical Engineering Dept., The City College, The City University of New York.*

RICHARD W. PERKINS *Professor of Mechanical, Aerospace, and Manufacturing Engineering, Syracuse University.*

13.1 FOUNDRY PRACTICE AND EQUIPMENT

by Michael K. Madsen
Expanded by Staff

REFERENCES: Publications of the American Foundrymen's Society: "Cast Metals Handbook," Alloy Cast Irons Handbook," "Copper-base Alloys Foundry Practice," Foundry Sand Handbook." "Steel Castings Handbook," Steel Founders' Society of America. Current publications of ASM International. Current publications of the suppliers of nonferrous metals relating to the casting of those metals; i.e., Aluminum Corp. of America, Reynolds Metal Co., Dow Chemical Co., INCO Alloys International, Inc., RMI Titanium Co., and Copper Development Assn. Publications of the International Lead and Zinc Research Organization (ILZRO).

BASIC STEPS IN MAKING SAND CASTINGS

The basic steps involved in making sand castings are:

1. **Patternmaking.** Patterns are required to make molds. The mold is made by packing molding sand around the pattern. The mold is usually made in two parts so that the pattern can be withdrawn. In horizontal molding, the top half is called the **cope**, and the bottom half is called the **drag**. In vertical molding, the leading half of the mold is called the **swing**, and the back half is called the **ram**. When the pattern is withdrawn from the molding material (sand or other), the imprint of the pattern provides the cavity when the mold parts are brought together. The mold cavity, together with any internal cores (see below) as required, is ultimately filled with molten metal to form the casting.

2. If the casting is to be hollow, additional patterns, referred to as **core boxes,** are needed to shape the sand forms, or **cores,** that are placed in the mold cavity to form the interior surfaces of castings. Thus the void between the mold and core eventually becomes the casting.

3. **Molding** is the operation necessary to prepare a mold for receiving the metal. It consists of ramming sand around the pattern placed in a support, or **flask,** removing the pattern, setting cores in place, cutting the **feeding system** to direct the metal if this feeding system is not a part of the pattern, removing the pattern, and closing the mold.

4. **Melting** and **pouring** are the processes of preparing molten metal of the proper composition and temperature and pouring this into the mold from transfer **ladles.**

5. **Cleaning** is all the operations required to remove the **gates** and **risers** that constitute the feeding system and to remove the adhering sand, scale, and other foreign material that must be removed before the casting is ready for shipment or other processing. Inspection follows, to check for defects in the casting as well as to ensure that the casting has the dimensions specified on the drawing and/or specifications. Inspection for internal defects may be quite involved, depending on the quality specified for the casting (see Sec. 5.4.). The inspected and accepted casting sometimes is used as is, but often it is subject to further processing which may include heat treatment, painting, other surface treatment (e.g., hot-dip galvanizing), and machining. Final operations may include electrodeposited plated metals for either cosmetic or operational requirements.

PATTERNS

Since patterns are the forms for the castings, the casting can be no better than the patterns from which it is made. Where close tolerances or smooth casting finishes are desired, it is particularly important that patterns be carefully designed, constructed, and finished.

Patterns serve a variety of functions, the more important being (1) to shape the mold cavity to produce castings, (2) to accommodate the characteristics of the metal cast, (3) to provide accurate dimensions, (4) to provide a means of getting liquid metal into the mold (gating system), and (5) to provide a means to support cores.

Usual allowances built into the pattern to ensure dimensional accuracy include the following: (1) **Draft,** the taper on the vertical walls of the casting which is necessary to extract the pattern from the mold without disturbing the mold walls. (2) **Shrinkage allowance,** a correction to compensate for the solidification shrinkage of the metal and its contraction during cooling. These allowances vary with the type of metal and size of casting. Typical allowances for cast iron are $\frac{1}{10}$ to $\frac{5}{32}$ in/ft; for steel, $\frac{1}{8}$ to $\frac{1}{4}$ in/ft; and for aluminum, $\frac{1}{16}$ to $\frac{5}{32}$ in/ft. A designer should consult appropriate references (AFS, "Cast Metals Handbook"; ASM, "Casting Design Handbook"; "Design of Ferrous Castings") or the foundry. These allowances also include a size tolerance for the process so that the casting is dimensionally correct. (See also Secs. 6.1, 6.3, and 6.4.) Table 13.1.1 lists additional data for some commonly cast metals. (3) **Machine finish allowance** is necessary if machining operations are to be used so that stock is provided for machining. Tabulated data are available in the references cited for shrinkage allowances. (4) If a casting is prone to distortion, a pattern may be intentionally distorted to compensate. This is a **distortion allowance.**

Patterns vary in complexity, depending on the size and number of castings required. **Loose patterns** are single prototypes of the casting and are used only when a few castings are needed. They are usually constructed of wood, but metal, plaster, plastics, urethanes, or other suitable material may be used. With advancements in solids modeling utilizing computers, CAD/CAM systems, and laser technology, rapid prototyping is possible and lends itself to the manufacture of protype patterns from a number of materials, including dense wax paper, or via stereolithographic processes wherein a laser-actuated polymerized plastic becomes the actual pattern or a prototype for a pattern or a series of patterns. The gating system for feeding the casting is cut into the sand by hand. Some loose patterns may be split into two parts to facilitate molding.

Gated patterns incorporate a gating system along with the pattern to eliminate hand cutting.

Match-plate patterns have the cope and drag portions of the pattern mounted on opposite sides of a wooden or metal plate, and are designed to speed up the molding process. Gating systems are also usually attached. These patterns are generally used with some type of molding machine and are recommended where a large number of castings are required.

For fairly large castings or where an increase in production rate is desired, the patterns can be mounted on separate pattern plates, which are referred to as **cope- and drag-pattern** plates. They are utilized in horizontal or vertical machines. In horizontal molding machines, the pattern plates may be used on separate machines by different workers,

Table 13.1.1 Average Linear Shrinkage of Castings

Bar iron, rolled	1:55	Cast iron	1:96	Steel, puddled	1:72
Bell metal	1:65	Gun metal	1:134	Steel, wrought	1:64
Bismuth	1:265	Iron, fine grained	1:72	Tin	1:128
Brass	1:65	Lead	1:92	Zinc, cast	1:624
Bronze	1:63	Steel castings	1:50	8 Cu + 1 Sn (by weight)	1:13

and then combined into completed molds on the molding floor prior to pouring. In vertical machines, the pattern plates are used on the same machine, with the flaskless mold portions pushed out one behind the other. Vertical machines result in faster production rates and provide an economic edge in overall casting costs.

Special Patterns and Devices   For extremely large castings, **skeleton patterns** may be employed. Large molds of a symmetric nature may be made for forming the sand mold by **sweeps**, which provide the contour of the casting through the movement of a template around an axis.

Follow boards are used to support irregularly shaped, loose patterns which require an irregular parting line between cope and drag. A **master pattern** is used as an original to make up a number of similar patterns that will be used directly in the foundry.

MOLDING PROCESSES AND MATERIALS

(See Table 13.1.2)

Molding Processes

Green Sand   Most castings are made in **green sand**, i.e., sand bonded with clay or bentonite and properly tempered with water to give it **green strength**. Miscellaneous additions may be used for special properties. This method is adaptable to high production of small- or medium-sized castings because the mold can be poured immediately after forming, and the sand can be reused and reprocessed after the casting has solidified.

Dry Sand Molds   These molds are made with green sand but are baked prior to use. The surface is usually given a refractory wash before baking to prevent erosion and to produce a better surface finish. Somewhat the same effects are obtained if the mold is allowed to **air-dry** by leaving it open for a period of time before pouring, or it is **skin-dried** by using a torch, infrared lamps, or heating elements directed at the mold cavity surface.

Core molding makes use of assembled cores to construct the mold. The sand is prepared by mixing with oil, or cereal, forming in core boxes, and baking. This process is used where the intricacy of the casting requires it.

Carbon Dioxide Process Molds   These molds are made in a manner similar to the green sand process but use sand bonded with sodium silicate. When the mold is finished, carbon dioxide gas is passed through the sand to produce a very hard mold with many of the advantages of dry sand and core molds but requiring no baking.

Floor and Pit Molding   When large castings are to be produced, these may be cast either directly on the floor of the foundry or in pits in the floor which serve as the flask. **Loam molding** is a variation of floor molding in which molding material composed of 50 percent sand and 50 percent clay (approx) is troweled onto a brickwork surface and brought to dimension by use of patterns, sweeps, or templates.

Shell Molding   Sand castings having close dimensional tolerances, and smooth finish can be produced by a process using a synthetic resin binder. The sand and resin mixture is dumped onto a preheated metal pattern, which causes the resin in the mixture to set as a thin shell over the pattern. When the shell has reached the proper thickness, the excess sand is removed by rotating the pattern to dump out the sand. The remaining shell is then cured on the pattern and subsequently removed by stripping it off, using mold release pins which have been properly spaced and that are mechanically or hydraulically made to protrude through the pattern. Mating shell halves are bonded, suitably backed by loose sand or other material, and then ready for metal to be poured. Current practice using shell molds has produced castings in excess of 1,000 lb, but often the castings weigh much less.

Plaster Molds   Plaster or plaster-bonded molds are used for casting certain aluminum or copper base alloys. Dimensional accuracy and excellent surface finish make this a useful process for making rubber tire molds, match plates, etc.

A variation of this method of molding is the **Antioch process**, using mixtures of 50 percent silica sand, 40 percent gypsum cement, 8 percent talc, and small amounts of sodium silicate, portland cement, and magnesium oxide. These dry ingredients are mixed with water and poured over the mold. After the mixture is poured, the mold is steam-treated in an autoclave and then allowed to set in air before drying in an oven. When the mold has cooled it is ready for pouring. Tolerances of ± 0.005 in (± 0.13 mm) on small castings and ± 0.015 in (± 0.38 mm) on large castings are obtained by this process.

A problem presented by plaster molds lies in inadequate permeability in the mold material consistent with the desired smooth mold cavity surface. A closely related process, the **Shaw process**, provides a solution. In this process, a refractory aggregate is mixed with a gelling agent and then poured over the pattern. Initial set of the mixture results in a rubbery consistency which allows it to be stripped from the pattern but

Table 13.1.2   Design and Cost Features of Basic Casting Methods

Design and cost features	Process					
	Sand casting	Shell-mold casting	Permanent-mold casting	Plaster-mold casting	Investment casting	Die casting
Choice of materials	Wide—ferrous and nonferrous	Wide—except for low-carbon steels	Restricted—brass, bronze, aluminum, some gray iron	Narrow—brass, bronze, aluminum	Wide—includes hard-to-forge and machine materials	Narrow—zinc, aluminum, brass, magnesium
Complexity	Considerable	Moderate	Moderate	Considerable	Greatest	Considerable
Size range	Great	Limited	Moderate	Moderate	Moderate	Moderate
Minimum section, in	3/32	1/16	0.100	0.010	0.010	0.025
Tolerances, in ft*	1/16–1/8	1/32–3/32	1/32–7/64	1/32–5/64	0.003–0.006	1/32–1/16
Surface smoothness, μin, rms	250–300	150–200	90–125	90–125	90–125	60–125
Design feature remarks	Basic casting method of industry	Considered to be good low-cost casting method	Production economics with substantial quantities	Little finishing required	Best for parts too complicated for other casting methods	Most economical where applicable
Tool and die costs	Low	Low to moderate	Medium	Medium	Low to moderate	High
Optimum lot size	Wide—range from few pieces to huge quantities	More required than sand castings	Best when requirements are in thousands	From one to several hundred	Wide—but best for small quantities	Substantial quantities required
Direct labor costs	High	Moderate	Moderate	High	Very high	Low to medium
Finishing costs	High	Low	Low to moderate	Low	Low	Low
Scrap costs	Moderate	Low	Low	Low	Low	Low

* Closer at extra cost.
SOURCE: Cook, "Engineered Castings," McGraw-Hill.

which is sufficiently strong to return to the shape it had when on the pattern. The mold is then ignited to burn off the volatile content in the set gel and baked at very high heat. This last step results in a hard, rigid mold containing microscopic cracks. The permeability of the completed mold is enhanced by the presence of the so-called microcrazes, while the mold retains the high-quality definition of the mold surface.

Two facts are inherent in the nature of sand molds: First, there may be one or few castings required of a given piece, yet even then an expensive wood pattern is required. Second, the requirement of removal of the pattern from the mold may involve some very intricate pattern construction. These conditions may be alleviated entirely by the use of the **full mold process,** wherein a foamed polystyrene pattern is used. Indeed, the foamed pattern may be made complete with a gating and runner system, and it can incorporate the elimination of draft allowance. In actual practice, the pattern is left in place in the mold and is instantly vaporized when hot metal is poured. The hot metal which vaporized the foam fills the mold cavity to the shape occupied previously by the foam pattern. This process is ideal for casting runs of one or a few pieces, but it can be applied to production quantities by mass-producing the foam patterns. There is extra expense for the equipment to make the destructible foam patterns, but often the economics of the total casting process is quite favorable when compared with resorting to a reusable pattern. There are particular instances when the extreme complexity of a casting can make a hand-carved foam pattern financially attractive.

The **lost-wax, investment, or precision-casting, process** permits the accurate casting of highly alloyed steels and of nonferrous alloys which are impossible to forge and difficult to machine. The procedure consists of making an accurate metal die into which the wax or plastic patterns are cast. The patterns are assembled on a sprue and the assembly sprayed, brushed, or dipped in a slurry of a fine-grained, highly refractory aggregate, and a proprietary bonding agent composed chiefly of ethyl silicate. This mixture is then allowed to set. The pattern is coated repeatedly with coarser slurries until a shell of the aggregate is produced around the pattern. The molds are allowed to stand until the aggregate has set, after which they are heated in an oven in an inverted position so that the wax will run out. After the wax is removed, the molds are baked in a preheat furnace. The molds may then be supported with loose sand and poured in any conventional manner.

There have been attempts in the past to use frozen mercury as a pattern. While mercury is a viable pattern material and can be salvaged totally for reuse, the inherent hazards of handling raw mercury have mitigated against its continued use to make patterns for investment castings.

All dimensions can be held to a tolerance of ± 0.005 in (± 0.13 mm) with some critical dimensions held to 0.002 in (0.05 mm). Most castings produced by this process are relatively small.

Faithful reproduction and accurate tolerances can also be attained by the **Shaw process;** see above. It combines advantages of dimensional control of precision molds with the ease of production of conventional molding. The process makes use of wood or metal patterns and a refractory mold bonded with an ethyl silicate base material. Since the mold is rubbery when stripped from the pattern, some back draft is permissible.

In the **cement-sand process** portland cement is used as the sand binder. A typical mixture has 11 percent portland cement, 89 percent silica sand, and water $4\frac{1}{2}$ to 7 percent of the total sand and cement. New sand is used for facing the mold and is backed with ground-up sand which has been rebonded. Cores are made of the same material. The molds and cores must airdry 24 to 72 h before pouring. The process can be used for either ferrous or nonferrous castings. This molding mixture practically eliminates the generation of gases, forms a hard surface which resists the erosive action of the metal, and produces castings with good surfaces and accurate dimensions. This process is seldom used, and then only for specific castings wherein the preparation of this type of mold outweighs many of its disadvantages.

Permanent-Mold Casting Methods

In the **permanent-mold casting method,** fluid metal is poured by hand into metal molds and around metal cores without external pressure. The molds are mechanically clamped together. Of necessity, the complexity of the cores must be minimal, inasmuch as they must be withdrawn for reuse from the finished casting. Likewise, the shape of the molds must be relatively simple, free of reentrant sections and the like, or else the mold itself will have to be made in sections, with attendant complexity.

Metals suitable for this type of casting are lead, zinc, aluminum and magnesium alloys, certain bronzes, and cast iron.

For making iron castings of this type, a number of metal-mold units are usually mounted on a turntable. The individual operations, such as coating the mold, placing the cores, closing the mold, pouring, opening the mold, and ejection of the casting, are performed as each mold passes certain stations. The molds are preheated before the first casting is poured. The process produces castings having a dense, fine-grained structure, free from shrink holes or blowholes. The tool changes are relatively low, and better surface and closer tolerances are obtained than with the sand-cast method. It does not maintain tolerances as close or sections as thin as the die-casting or the plaster-casting methods.

Yellow brasses, which are high in zinc, should not be cast by the permanent-mold process because the zinc oxide fouls the molds or dies.

The **semipermanent mold casting method** differs from the permanent mold casting in that sand cores are used, in some places, instead of metal cores. The same metals may be cast by this method. This process is used where cored openings are so irregular in shape, or so undercut, that metal cores would be too costly or too difficult to handle. The structure of the metal cast around the sand cores is like that of a sand casting. The advantages of permanent mold casting in tolerances, density, appearance, etc., exist only in the section cast against the metal mold.

Graphite molds may be used as short-run permanent molds since they are easier to machine to shape and can be used for higher-melting point alloys, e.g., steel. The molds are softer, however, and more susceptible to erosive damage. Steel railroad wheels may be made in these molds and can be cast by filling the mold by **low-pressure casting** methods.

In the **slush casting** process, the cast metal is allowed partially to solidify next to the mold walls to produce a thin section, after which the excess liquid metal is poured out of the permanent mold.

In **centrifugal casting** the metal is under centrifugal force, developed by rotating the mold at high speed. This process, used in the manufacture of bronze, steel, and iron castings, has the advantage of producing sound castings with a minimum of risers. In **true centrifugal castings** the metal is poured directly into a mold which is rotated on its own axis. Obviously, the shapes cast by this method must have external and internal geometries which are surfaces of revolution. The external cast surface is defined by the internal surface of the water-cooled mold; the internal surface of the casting results from the effective core of air which exists while the mold is spun and until the metal solidifies sufficiently to retain its cast shape. Currently, all cast-iron pipe intended for service under pressure (e.g., water mains) is centrifugally cast. The process is extended to other metals falling under the rubric of **tubular goods.**

In **pressure casting,** for asymmetrical castings which cannot be spun around their own axes, the mold cavities are arranged around a common sprue located on the neutral axis of the mold. The molds used in the centrifugal-casting process may be metal cores or dry sand, depending on the type of casting and the metal cast.

Die Casting machines consist of a basin holding molten metal, a metallic mold or die, and a metal transferring device which automatically withdraws molten metal from the basin and forces it under pressure into the die. Two forms of die casting machines are in general use. Lead, tin, and zinc alloys containing aluminum are handled in **piston machines**. Aluminum alloys and pure zinc, or zinc alloys free from aluminum, rapidly attack the iron in the piston and cylinder and require a different type of casting machine. The pressures in a piston machine range from a few hundred to thousands of lb/in^2.

The **gooseneck machine** has a cast-iron gooseneck which dips the molten metal out of the melting pot and transfers it to the die. The pressure is applied to the molten metal by compressed air after the gooseneck is brought in contact with the die. This machine, developed primarily for aluminum alloys, is sometimes used for zinc-aluminum alloys, especially for large castings, but, owing to the lower pressure, the casting is likely to be less dense than when made in the piston machine. It is seldom used for magnesium alloys.

In **cold chamber machines** the molten-metal reservoir is separated from the casting machine, and just enough metal for one casting is ladled by hand into a small chamber, from which it is forced into the die under high pressure. The pressures, quite high, ranging from the low thousands to in excess of 10,000 lb/in^2, are produced by a hydraulic system connected to the piston in the hot metal chamber. The alloy is kept so close to its melting temperature that it is in a slushlike condition. The process is applicable to aluminum alloys, magnesium alloys, zinc alloys, and even higher-melting-point alloys like brasses and bronzes, since the pouring well, cylinder, and piston are exposed to the high temperature for only a short time.

All metal mold external pressure castings have close tolerances, sharp outlines and contours, fine smooth surface, and high rate of production, with low labor cost. They have a hard skin and a soft core, resulting from the rapid chilling effect of the cold metal mold.

The dies usually consist of two blocks of steel, each containing a part of the cavity, which are locked together while the casting is being made and drawn apart when it is ready for ejection. One-half of the die (next to the ejector nozzle) is stationary; the other half moves on a carriage. The dies are preheated before using and are either air- or water-cooled to maintain the desired operating temperature. Die life varies with the alloy and dimensional tolerances required. Retractable and removable metal cores are used to form internal surfaces. Inserts can be cast into the piece by placing them on locating pins in the die.

A wide range of sizes and shapes can be made by these processes, including threaded pieces and gears. Holes can be accurately located. The process is best suited to large-quantity production.

A historic application of the process was for typesetting machines such as the linotype. Although now they are obsolescent and rarely found in service, for a long time the end products of typesetting machines were a prime example of a high-quality die-cast metal product.

MOLDING EQUIPMENT AND MECHANIZATION

Flasks may be filled with sand by hand shoveling, gravity feed from overhead hoppers, continuous belt feeding from a bin, sand slingers, and, for large molds, by an overhead crane equipped with a grab bucket.

Hand ramming is the simplest method of compacting sand. To increase the rate, pneumatic rammers are used. The method is slow, the sand is rammed in layers, and it is difficult to gain uniform density.

More uniform results and higher production rates are obtained by **squeezing machines**. Hand-operated squeezers were limited to small molds and are obsolete; **air-operated machines** permit an increase in the allowable size of molds as well as in the production rate. These machines are suitable for shallow molds. Squeezer molding machines produce greatest sand density at the top of the flask and softest near the parting line of pattern. Air-operated machines are also applied in vertical molding processes using flaskless molds. Horizontal impact molding sends shock waves through the sand to pack the grains tightly.

In **jolt molding machines** the pattern is placed on a platen attached to the top of an air cylinder. After the table is raised, a quick-release port opens, and the piston, platen, and mold drop free against the top of the cylinder or striking pads. The impact packs the sand. The densities produced by this machine are greatest next to the parting line of the pattern and softest near the top of the flask. This procedure can be used for any flask that can be rammed on a molding machine. As a separate unit, it is used primarily for medium and large work. Where plain jolt machines are used on large work, it is usual to ram the top of the flask manually with an air hammer.

Jolt squeeze machines use both the jolt and the squeeze procedures. The platen is mounted on two air cylinders: a small cylinder to jolt and a large one to squeeze the mold. They are widely used for small and medium work, and with match-plate or gated patterns. Pattern-stripping devices can be incorporated with jolt or squeezer machines to permit mechanical removal of the pattern. Pattern removal can also be accomplished by using jolt-rockover-draw or jolt-squeeze-rollover-draw machines.

The **sand slinger** is the most widely applicable type of ramming machine. It consists of an impeller mounted on the end of a double-jointed arm which is fed with sand by belt conveyors mounted on the arm. The impeller rotating at high speed gives sufficient velocity to the sand to ram it in the mold by impact. The head may be directed to all parts of the flask manually on the larger machines and may be automatically controlled on smaller units used for the high-speed production of small molds.

Vibrators are used on all pattern-drawing machines to free the pattern from the grip of the sand before drawing. Their use reduces mold damage to a minimum when the pattern is removed, and has the additional advantage of producing castings of more uniform size than can be secured by hand rapping the pattern. Pattern damage is also kept to a minimum. Vibrators are usually air-operated, but some electrically operated types are in use.

Flasks generally consist of two parts: the upper section, called the **cope**, and the bottom section, the **drag**. When more than two parts are used, the intermediate sections are called **cheeks**. Flasks are classified as tight, snap, and slip. **Tight flasks** are those in which the flask remains until the metal is poured. **Snap flasks** are hinged on one corner and have a locking device on the diagonally opposite corner. In use, these flasks are removed as soon as the mold is closed. **Slip flasks** are of solid construction tapered from top to bottom on all four sides so that they can be removed as soon as the mold is closed. Snap or slip flasks permit the molder to make any number of molds with one flask. Before pouring snap- or slip-flask molds, a wood or metal pouring jacket is placed around the mold and a weight set on the top to keep the cope from lifting. The cope and drag sections on all flasks are maintained in proper alignment by flask pins and guides.

Tight flasks can be made in any size and are fabricated of wood, rolled steel, cast steel, cast iron, magnesium, or aluminum. Wood, aluminum, and magnesium are used only for small- and medium-sized flasks. Snap and slip flasks are made of wood, aluminum, or magnesium, and are generally used for molds not over 20 by 20 in (500 mm by 500 mm).

Mechanization of Sand Preparation

In addition to the various types of molding machines, the modern foundry makes use of a variety of equipment to handle the sand and castings.

Sand Preparation and Handling Sand is prepared in **mullers**, which serve to mix the sand, bonding agent, and water. **Aerators** are used in conjunction to loosen the sand to make it more amenable to molding. **Sand cutters** that operate over a heap on the foundry floor may be used instead of mullers. Delivery of the sand to the molding floor may be by means of dump or scoop trucks or by belt conveyors. At the molding floor the molds may be placed on the floor or delivered by conveyors to a pouring station. After pouring, the castings are removed from the flasks and adhering sand at a **shakeout** station. This may be a mechanically operated jolting device that shakes the loose sand from flask and casting. The used sand, in turn, is returned to the storage bins by belt conveyor or other means. Small castings may be poured by using **stack-molding** methods. In this case, each flask has a drag cavity molded in its upper surface and a cope cavity in its lower surface. These are stacked one on the other to a suitable height and poured from a common sprue.

There is an almost infinite variety of equipment and methods available to the foundry, ranging from simple, work-saving devices to completely mechanized units, including completely automatic molding machines. Because of this wide selection available, the degree to which a foundry can be mechanized depends almost entirely on the economics of the operations, rather than the availability or lack of availability of a particular piece of equipment.

MOLDING SAND

Molding sand consists of silica grains held together by some bonding material, usually clay or bentonite.

Grain size greatly influences the surface finish of a casting. The proper grain size is determined by the size of the casting, the quality of surface required, and the surface tension of the molten metal. The grain size should be approximately uniform when maximum permeability is desired.

Naturally bonded sands are mixtures of silica and clay as taken from the pits. Modification may be necessary to produce a satisfactory mixture. This type of sand is used in gray iron, ductile iron, malleable iron, and nonferrous foundries (except magnesium).

Synthetically bonded sands are produced by combining clay-free silica sand with clay or bentonite. These sands can be compounded to suit foundry requirements. They are more uniform than naturally bonded sands but require more careful mixing and control. Steel foundries, gray iron and malleable iron foundries, and magnesium foundries use this type of sand.

Special additives may be used in addition to the basic sand, clay, and water. These include cereals, ground pitch, sea coal, gilsonite, fuel oil, wood flour, silica flour, iron oxide, pearlite, molasses, dextrin, and proprietary materials. These all serve the purpose of altering specific properties of the sand to give desired results.

The properties of the sand that are of major interest to the foundry worker are **permeability,** or the venting power, of the sand; **green compressive strength; green shear strength; deformation,** or the sand movement under a given load; **dry compressive strength;** and **hot strength,** i.e., strength at elevated temperatures. Several auxiliary tests are often made, including moisture content, clay content, and grain-size determination.

The foundry engineer or metallurgist who usually is entrusted with the control of the sand properties makes the adjustments required to keep it in good condition.

Facing sands, for giving better surface to the casting, are used for gray iron, malleable iron, steel, and magnesium castings. The iron sands usually contain **sea coal,** a finely ground coal which keeps the sand from adhering to the casting by generating a gas film when in contact with the hot metal. Steel facings contain silica flour or other very fine highly refractory material to form a dense surface which the metal cannot readily penetrate.

Mold washes are coatings applied to the mold or core surface to improve the finish of the casting. They are applied either wet or dry. The usual practice is to brush or spray the wet mold washes and to brush or rub on the dry ones. Graphite or silica flour mixed with clay and molasses water is frequently used. The washes are mixed usually with water-base or alcohol-base solvent solutions that require oven drying time, during which not only does the wash set, but also the excess moisture is removed from the washed coating.

Core Sands and Core Binders

Green sand cores are made from standard molding-sand mixtures, sometimes strengthened by adding a binder, such as dextrin, which hardens the surface. Cores of this type are very fragile and are usually made with an arbor or wires on the inside to facilitate handling. Their collapsibility is useful to prevent hot tearing of the casting.

Dry sand cores are made from silica sand and a binder (usually oil) which hardens under the action of heat. The amount of oil used should be the minimum which will produce the necessary core strength.

Core binders are either organic, such as core oil, which are destroyed under heat, or inorganic, which are not destroyed.

Organic Binders   The main organic binder is **core oil.** Pure linseed oil is used extensively as one of the basic ingredients in blended-oil core binders. These consist primarily of linseed oil, resin, and a thinner, such as high-grade kerosene. They have good wetting properties, good workability, and better oxidation characteristics than straight linseed oil.

Corn flour produces good green strength and dry strength when used in conjunction with oil. Cores made with this binder are quick drying in the oven and burn out rapidly and completely in the mold.

Dextrin produces a hard surface and weak center because of the migration of dextrin and water to the surface. Used with oil, it produces a hard smooth surface but does not produce a green bond as good as that with corn flour.

Commercial **protein binders,** such as gelatin, casein, and glues, im-

prove flowability of the sand, have high binding power, rapid drying, fair resistance to moisture, and low burning-out point, with only a small volume of gas evolved on burning. They are used where high collapsibility of the core is essential.

Other binders include paper-mill by-products, which absorb moisture readily, have high dry strength, low green strength, high gas ratio, and high binding power for clay materials.

Coal tar pitch and **petroleum pitch** flow with heat and freeze around the grains on cooling. These compounds have low moisture absorption rates and are used extensively for large iron cores. They can be used effectively with impure sands.

Wood and **gum rosin, plastic resins,** and rosin by-products are used to produce collapsibility in cores. They must be well ground. They tend to cake in hot weather, and large amounts are required to get desired strength.

Plastics of the **urea-** and **phenol-formaldehyde** groups and **furan resins** are being used for core binders. They have the advantage of low-temperature baking, collapse readily, and produce only small amounts of gas. These can be used in **dielectric baking ovens** or in the **shell molding, hot box, or air setting** processes for making cores.

Inorganic binders include fire clay, southern bentonite, western bentonite, and iron oxide.

Cores can also be made by mixing sand with sodium silicate. When this mixture is in the core box, it is infiltrated with CO_2, which causes the core to harden. This is called the CO_2 **process.**

Core-Making Methods

Cores are made by the methods employed for sand molds. In addition, **core blowers** and **extrusion machines** are used.

Core blowers force sand into the core box by compressed air at about 100 lb/in². They can be used for making all types of small- and medium-sized cores. The cores produced are very uniform, and high production rates are achieved.

Screw feed machines are used largely for plain cylindrical cores of uniform cross section. The core sand is extruded through a die onto a core plate. The use of these machines is limited to the production of stock cores, which are cut to the desired length after baking.

Core Ovens   Core oven walls are constructed of inner and outer layers of sheet metal separated by rock wool or Fiberglas insulation and with interlocked joints. Combustion chambers are refractory-lined, and the hot gases are circulated by fans. They are designed for operating at temperatures suitable for the constituents in the core body. Time at baking temperatures will, likewise, vary with the composition of the core.

Core driers are light skeleton cast iron or aluminum boxes, the internal shape of which conforms closely to the cope portion of the core. They are used to support, during baking, cores which cannot be placed on a flat plate.

Chaplets are metallic pieces inserted into the mold cavity which support the core. Long unsupported cores will be subject to flotation force as the molten metal fills the mold and may break if the resulting flexural stresses are excessive. Likewise, the liquid forces imposed on cores as metal flows through the mold cavity may cause cores to shift. The chaplets interposed within the mold cavity are placed to alleviate these conditions. They are generally made of the same material as that being cast; they melt and blend with the metal as cast, and they remain solid long enough for the liquid forces to equilibrate through the mold cavity.

CASTING ALLOYS

In general, the types of alloys that can be produced as wrought metals can also be prepared as castings. Certain alloys, however, cannot be forged or rolled and can only be used as cast.

Ferrous Alloys

Steel Castings   (See Sec. 6.3.)   Steel castings may be classified as:

1. Low carbon (C < 0.20 percent). These are relatively soft and not readily heat-treatable.
2. Medium carbon (0.20 percent < C < 0.50 percent). These cast-

ings are somewhat harder and amenable to strengthening by heat treatment.

3. High carbon (C > 0.50 percent). These steels are used where maximum hardness and wear resistance are desired.

In addition to the classification based on carbon content, which determines the maximum hardness obtainable in steel, the castings can be also classified as **low alloy** content (≤ 8 percent) or **high alloy** content (> 8 percent).

Low-alloy steels behave essentially as plain carbon steels but have a higher **hardenability**, which is a measure of ability to be hardened by heat treatment. High-alloy steels are designed to produce some specific property, like corrosion resistance, heat resistance, wear resistance, or some other special property.

Malleable Iron Castings   The carbon content of malleable iron ranges from about 2.00 to 2.80 percent and may reach as high as 3.30 percent if the iron is melted in a cupola. Silicon ranging from 0.90 to 1.80 percent is an additional alloying element required to aid the annealing of the iron. As cast, this iron is hard and brittle and is rendered soft and malleable by a long heat-treating or annealing cycle. (See also Sec. 6.3)

Gray Iron Castings   Gray iron is an alloy of iron, carbon, and silicon, containing a higher percentage of these last two elements than found in malleable iron. Much of the carbon is present in the elemental form as graphite. Other elements present include manganese, phosphorus, and sulfur. Because the properties are controlled by proper proportioning of the carbon and silicon and by the cooling rate of the casting, it is usually sold on the basis of specified properties rather than composition. The carbon content will usually range between 3.00 and 4.00 percent and the silicon will be between 1.00 and 3.00 percent, the higher values of carbon being used with the lower silicon values (usually), and vice versa. As evidence of the fact that gray iron should not be considered as a material having a single set of properties, the ASTM and AFS codify gray cast iron in several classes, with accompanying ranges of tensile strengths available. The high strengths are obtained by proper adjustment of the carbon and silicon contents or by alloying. (See also Sec. 6.3.)

An important variation of gray iron is **nodular iron**, or **ductile iron**, in which the graphite appears as nodules rather than as flakes. This iron is prepared by treating the metal in the ladle with additives that usually include magnesium in alloy form. Nodular iron can exceed 100,000 lb/in^2 (690 MN/m^2) as cast and is much more ductile than gray iron, measuring about 2 to 5 percent elongation at these higher strengths, and even higher percentages if the strength is lower. (See Sec. 6.3.)

Nonferrous Alloys

Aluminum-Base Castings   Aluminum is alloyed with copper, silicon, magnesium, zinc, nickel, and other elements to produce a wide variety of casting alloys having specific characteristics of foundry properties, mechanical properties, machinability, and/or corrosion resistance. Alloys are produced for use in sand casting, permanent mold casting, or die casting. Some alloys are heat-treatable using **solution** and **age-hardening** treatments. (See also Sec. 6.4.)

Copper-Base Alloys   The alloying elements used with copper include zinc (brasses), tin (bronzes), nickel (nickel bronze), aluminum (aluminum bronze), silicon (silicon bronze), and beryllium (beryllium bronze). The brasses and tin bronzes may contain lead for machinability. Various combinations of zinc and tin, or of tin or zinc with other elements, are also available. With the exception of some of the aluminum bronzes and beryllium bronze, most of the copper-base alloys cannot be hardened by heat treatment. (See also Sec. 6.4.)

Special Casting Alloys   Other metals cast in the foundry include **magnesium-base alloys** for light weight, **nickel-base alloys** for high-temperature applications, **titanium-base alloys** for strength-to-weight ratio, etc. The magnesium-base alloys require special precautions during melting and pouring to avoid burning. (See Sec. 6.4.)

MELTING AND HEAT TREATING FURNACES

There are several types of **melting furnaces** used in conjunction with metal casting. Foundry furnaces used in melting practice for ferrous castings are predominantly **electric arc** (direct and indirect), **induction**, and **crucible** for small operations. For cast iron, **cupolas** are still employed, although in ever-decreasing quantities. The previous widespread use of open-hearth furnaces is now relegated to isolated foundries and is essentially obsolete. In general, ferrous foundries' melting practice has become based largely on electric-powered furnaces. **Duplexing** operations are still employed, usually in the form of cupola/induction furnace, or cupola/electric arc furnace.

In nonferrous foundries, electric arc, induction, and crucible furnaces predominate. There are some residual installations which use air furnaces, but they are obsolete and found only in some of the older, small foundries which cater to unique clients.

Vacuum melting and metal refining were fostered by the need for extremely pure metals for high-temperature, high-strength applications (e.g., gas-turbine blades). Vacuum melting is accomplished in a furnace located in an evacuated chamber; the source of heat is most often an electric arc and sometimes induction coils. Gases entrained in the melt are removed, the absence of air prevents oxidation of the base metals, and a high degree of metal purity is retained in the molten metal and in the casting ultimately made from that vacuum-melted metal. The mold is also enclosed in the same evacuated chamber.

The vacuum melting and casting process is very expensive because of the nature of the equipment required, and quantities of metal handled are relatively small. The economics of the overall process are justified by the design requirements for highest-quality castings for ultimately very demanding service.

Annealing and **heat-treating furnaces** used to process castings are the type usually found in industrial practice. (See Secs. 7.3. and 7.5.)

CLEANING AND INSPECTION

Tumbling barrels consist of a power-driven drum in which the castings are tumbled in contact with hard iron stars or balls. Their impact removes the sand and scale.

In **air-blast cleaning units**, compressed air forces silica sand or chilled iron shot into violent contact with the castings, which are tumbled in a barrel, rotated on a table, or passed between multiple orifices on a conveyor. Large rooms are sometimes utilized, with an operator directing the nozzle. These machines are equipped with hoppers and elevators to return the sand or shot to the magazine. Dust-collecting systems are required.

In **centrifugal-blast cleaning units**, a rotating impeller is used to impart the necessary velocity to the chilled iron shot or abrasive grit. The velocities are not so high as with air, but the volume of abrasive is much greater. The construction is otherwise similar to the air blast machine.

Water in large volume at pressures of 250 to 600 lb/in^2 is used to remove sand and cores from medium and large castings.

High-pressure water and sand cleaning (**Hydroblast**) employs high-pressure water mixed with molding sand which has been washed off the casting. A sand classifier is incorporated in the sand reclamation system.

Pneumatic chipping hammers may be used to clean large castings where the sand is badly burned on and for deep pockets.

Removal of Gates and Risers and Finishing Castings   The following tabulation shows the most generally used methods for removing gates and risers (marked R) and for finishing (marked F).

Steel: Oxyacetylene (R) hand hammer or sledge (R), grinders (F), chipping hammer, (F), and machining (F).

Cast iron: Chipping hammer (R, F), hand hammer or sledge (R), abrasive cutoff (R), power saw (R), and grinders (F).

Malleable iron: Hand hammer or sledge (R), grinders (F), shear (F), and machining (F).

Brass and bronze: Chipping hammer (R, F), shear (R, F), hand hammer or sledge (R), abrasive cutoff (R), power saw (R), belt sanders (F), grinders (F), and machining (F).

Aluminum: Chipping hammer (R), shear (R), hand hammer or sledge (R), power saw (R), grinders (F), and belt sander (F).

Magnesium: Band saw (R), machining (F), and flexible-shaft machines with steel burr cutters (F).

Casting Inspection

(See Sec. 5.4)

Castings are inspected for dimensional accuracy, hardness, surface finish, physical properties, internal soundness, and cracks. For **hardness** and for **physical properties,** see Sec. 6.

Internal soundness is checked by cutting or breaking up pilot castings or by nondestructive testing using X-ray, gamma ray, etc.

Destructive testing tells only the condition of the piece tested and does not ensure that other pieces not tested will be sound. It is the most commonly used procedure at the present time.

X-ray, gamma ray, and other methods have made possible the nondestructive checking of castings to determine internal soundness on all castings produced. Shrinks, cracks, tears, and gas holes can be determined and repairs made before the castings are shipped.

Magnetic powder tests (Magnaflux) are used to locate structural discontinuities in iron and steel except austenitic steels, but they are not applicable to most nonferrous metals or their alloys. The method is most useful for the location of surface discontinuities, but it may indicate subsurface defects if the magnetizing force is sufficient to produce a leakage field at the surface.

In this test a magnetic flux is induced in ferromagnetic material. Any abrupt discontinuity in its path results in a local flux leakage field. If finely divided particles of ferromagnetic material are brought into the vicinity, they offer a low reluctance path to the leakage field and take a position that outlines approximately its effective boundaries. The casting to be inspected is magnetized and its surface dusted with the magnetic powder. A low velocity air stream blows the excess powder off and leaves the defect outlined by the powder particles. The powder may be applied while the magnetizing current is flowing (**continuous method**) or after the current is off (**residual method**). It may be applied dry or suspended in a light petroleum distillate similar to kerosene. Expert interpretation of the tests is necessary.

CASTING DESIGN

Design for the best utilization of metal in the cast form requires a knowledge of metal solidification characteristics, foundry practices, and the metallurgy of the metal being used. Metals exhibit certain peculiarities in the formation of solid metal during freezing and also undergo shrinkage in the liquid state during the freezing process and after freezing, and the casting must be designed to take these factors into consideration. Knowledge concerning the freezing process will also be of assistance in determining the fluidity of the metal, its resistance to **hot tearing,** and its tendency to evolve dissolved gases. For economy in production, casting design should take into consideration those factors in molding and coring that will lead to the simplest procedures. Elimination of expensive cores, irregular parting lines, and deep drafts in the casting can often be accomplished with a slight modification of the original design. Combination of the foregoing factors with the selection of the right metal for the job is important in casting design. Consultation between the design engineer and personnel at the foundry will result in well-designed castings and cost-effective foundry procedures. Initial guidance may be had from the several references cited and from updated professional literature, which abounds in the technical journals. Trade literature, as represented by the publications issued by the various generic associations, will be useful in assessing potential problems with specific casting designs. Generally, time is well spent in these endeavors before an actual design concept is reduced to a set of dimensional drawings and/or specifications.

13.2 PLASTIC WORKING OF METALS
by Rajiv Shivpuri

(See also Secs. 5 and 6)

REFERENCES: Crane, "Plastic Working of Metals and Power Press Operations," Wiley. Woodworth, "Punches, Dies and Tools for Manufacturing in Presses," Henley. Jones, "Die Design and Die Making Practice," Industrial Press. Stanley, "Punches and Dies," McGraw-Hill. DeGarmo, "Materials and Processes in Manufacturing," Macmillan. "Modern Plastics Encyclopedia and Engineers Handbook," Plastics Catalogue Corp., New York. "The Tool Engineers Handbook," McGraw-Hill. Bridgman, "Large Plastic Flow and Fracture," McGraw-Hill. "Cold Working of Metals," ASM. Pearson, "The Extrusion of Metals," Wiley.

STRUCTURE

Yieldable structural forces between the particles composing a material to be worked are the key to its behavior. Simple internal structures contain only a single element, as pure copper, silver, or iron. Relatively more difficult to work are the solid solutions in which one element tends to distribute uniformly in the structural pattern of another. Thus silver and gold form a continuous series of solid-solution alloys as their proportions vary. Next are alloys in which strongly bonded molecular groups dispersed through or along the grain boundaries of softer metals offer increasing resistance to working, as does iron carbide (Fe_3C) in solution in iron.

Bonding forces are supplied by electric fields characteristic of individual atoms. These forces in turn are subject to modification by temperature as energy is added, increasing electron activity.

The **particles** which constitute an atom are so small that most of its volume is empty space. For a similar energy state, there is some rough uniformity in the outside size of atoms. In general, therefore, the more complex elements have their larger number of particles more densely packed and so are heavier. For each element, the energy pattern of its electric charges in motion determines the field characteristics of that atom and which of the orderly arrangements it will seek to assume with relation to others like it in the orderly crystalline form.

Space lattice is the term used to describe the orderly arrangement of rows and layers of atoms in the crystalline form. This orderly state is also described as balanced, unstrained, or **annealed.** The working or deforming of materials distorts the orderly arrangement, unbalancing the forces between atoms. Cubic patterns or space lattices characterize the more ductile or workable materials. Hexagonal and more complex patterns tend to be more brittle or more rigid. Flaws, irregularities, or distortions, with corresponding unbalanced strains among adjacent atoms, may occur in the pattern or along grain boundaries. **Slip-plane** movements in working to new shapes tend to slide the once orderly layers of atoms within the grain-boundary limitations of individual crystals. Such sliding movement tends to take place at 45° to the direction of the applied load because much higher stresses are required to pull atoms directly apart or to push them straight together.

Chemical combinations, in liquid or solid solutions, or molecular compounds depend upon relative field patterns of elements or upon actual displacement of one or more electrons from the outer orbit of a donor element to the outer orbit of a receptor element. Thus the molecules of hard iron carbide, Fe_3C, may be held in solid solution in soft pure iron (ferrite) in increasing proportions up to 0.83 percent of carbon in iron, which is described as **pearlite.** Zinc may occupy solid-solution positions in the copper space lattice up to about 45 percent, the range of the ductile red and yellow brasses.

Thermal Changes Adding heat (energy) increases electron activity and therefore also the mobility of the atom. Probability of brittle failure at low temperatures usually becomes less as temperature increases. Transition temperatures from one state to another differ for different elements. Thermal transitions therefore become more complex as such differing elements are combined in alloys and compounds. As temperatures rise, a **stress-relieving** range is reached at which the most severely strained atoms are able to ease themselves around into less strained positions. At somewhat higher temperatures, **annealing** or **recrystallization** of worked or distorted structure takes place. Old grain boundaries disappear and small new grains begin to grow, aligning nearby atoms into their orderly lattice pattern. The more severely the material has been worked, the lower is the temperature at which recrystallization begins. Grain growth is more rapid at higher temperatures. In working materials above their recrystallization range, as in forging, the relief of interatomic strains becomes more nearly spontaneous as the temperature is increased. **Creep** takes place when materials are under some stress above the recrystallization range, and the thermal mobility permits individual atoms to ease around to relieve that stress with an accompanying gradual change of shape. Thus a wax candle droops due to gravity on a hot day. Lead, which recrystallizes below room temperature, will creep when used for roofing or spouting. Steels in rockets and jet engines begin to creep around 1,300 to 1,500°F (704 to 815°C). Creep is more rapid as the temperature rises farther above the recrystallization range.

PLASTICITY

Plasticity is that property of materials which commends them to the mass-production techniques of pressure-forming desired shapes. It is understood more easily if several types of plasticity are considered.

Crystoplastic describes materials, notably metals, which can be worked in the stable crystalling state, below the recrystallization range. Metals which crystallize in the cubic patterns have a wider plastic range than those of hexagonal pattern. Alloying narrows the range and increases the resistance to working. Tensile or compressive testing of an annealed specimen can be used to show the plastic range which lies between the initial yield point and the point of ultimate tensile or compressive failure.

The **plastic range,** as of an annealed metal, is illustrated in Fig. 13.2.1. Changing values of **true stress** are determined by dividing the applied load at any instant by the cross-section area at that instant. As material is worked, a progressive increase in elastic limit and yield point registers the slip-plane movement or work hardening which has taken place and the consequent reduction in residual plasticity. This changing yield point or resistance, shown in Fig. 13.2.1, is divided roughly into three characteristic ranges. The contour of the lower range can be varied by nonuniformity of grain sizes or by small displacements resulting from prior direction of working. Random large, soft grains yield locally under slight displacement, with resulting **surface markings**, described as *orange peel, alligator skin,* or *stretcher strain markings.* These can be

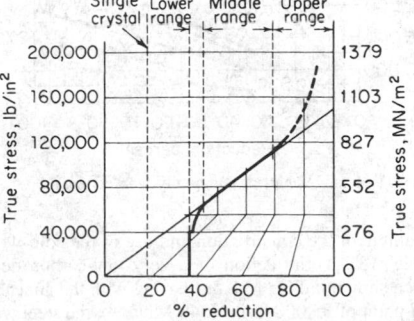

Fig. 13.2.1 Three ranges of crystoplastic work hardening of a low-carbon steel. *(ASME, 1954, W. S. Wagner, E. W. Bliss Co.)*

prevented by preparatory roller leveling, which gives protection in the case of steel for perhaps a day, or by a 3 to 5 percent temper pass of cold-rolling, which may stress relieve in perhaps 3 months, permitting recurrent trouble. The middle range covers most drawing and forming operations. Its upper limit is the point of normal tensile failure. The upper range requires that metal be worked primarily in compression to inhibit the start of tensile fracture. Severe extrusion, spring-temper rolling, and music-wire drawing use this range.

Dispersion hardening of metal alloys by heat treatment (see Fig. 13.2.2) reduces the plastic range and increases the resistance to work hardening. Figure 13.2.2 also shows the common methods of plotting change of true stress against **percentage of reduction**—e.g., reduction of thickness in rolling or compressive working, of area in wire drawing, ironing, or tensile testing, or of diameter in cup drawing or reducing operations—and against **true strain,** which is the natural logarithm of change of area, for convenience in higher mathematics.

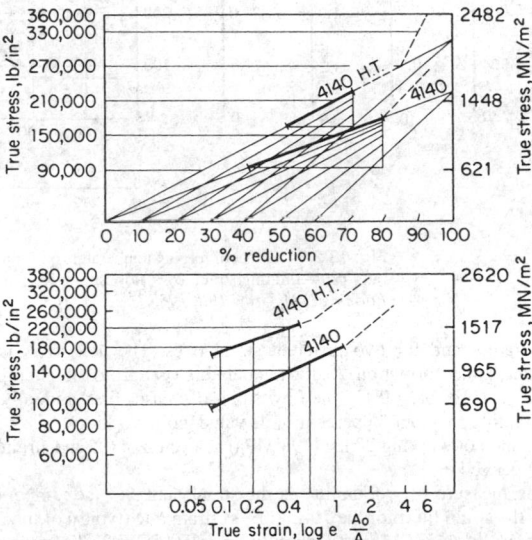

Fig. 13.2.2 High-range plasticity (dotted) of an SAE 4140 steel, showing the effect of dispersion hardening. Two plotting methods. *(ASME, 1958, Crane and Wagner, E. W. Bliss Co.)*

For metals, thermoplastic working is usually described as **hot working**, except for tin and lead, which recrystallize below room temperature. Hot-worked samples may be etched to show **flow lines,** which are usually made up of old-grain boundaries. Where these show, recrystallization has not yet taken place, and some work hardening is retained to improve physical properties. Zinc and magnesium, which are typical of the hexagonal-structure metals, take only small amounts of cold working but can be drawn or otherwise worked severely at rather moderate temperatures [Zn, 200 to 400°F (90 to 200°C); Mg, 500 to 700°F (260 to 400°C)]. Note that, although hexagonal-pattern metals are less easily worked than cubic-pattern metals, they are for that same reason structurally more rigid for a similar relative weight. Advantageous forging temperatures change with alloy composition: copper, 1,800 to 1,900°F (980 to 1,040°C); red brass, Cu 70, Zn 30, 1,600 to 1,700°F (870 to 930°C); yellow brass, Cu 60, Zn 40, 1,200 to 1,500°F (650 to 815°C). See Sec. 6 for general physical properties of metals.

Substantially pure iron shows an increasing elastic limit and decreasing plasticity with increasing amounts of work hardening by cold-rolling. The rate at which such work hardening takes place is greatly increased, and the remaining plasticity reduced, as alloying becomes more complex.

In **steels,** the mechanical working range is conventionally divided into cold, warm, and hot working. Figure 13.2.3 is a plot of flow stress, limit strain, scale factor, and dimensional error for different values of forging

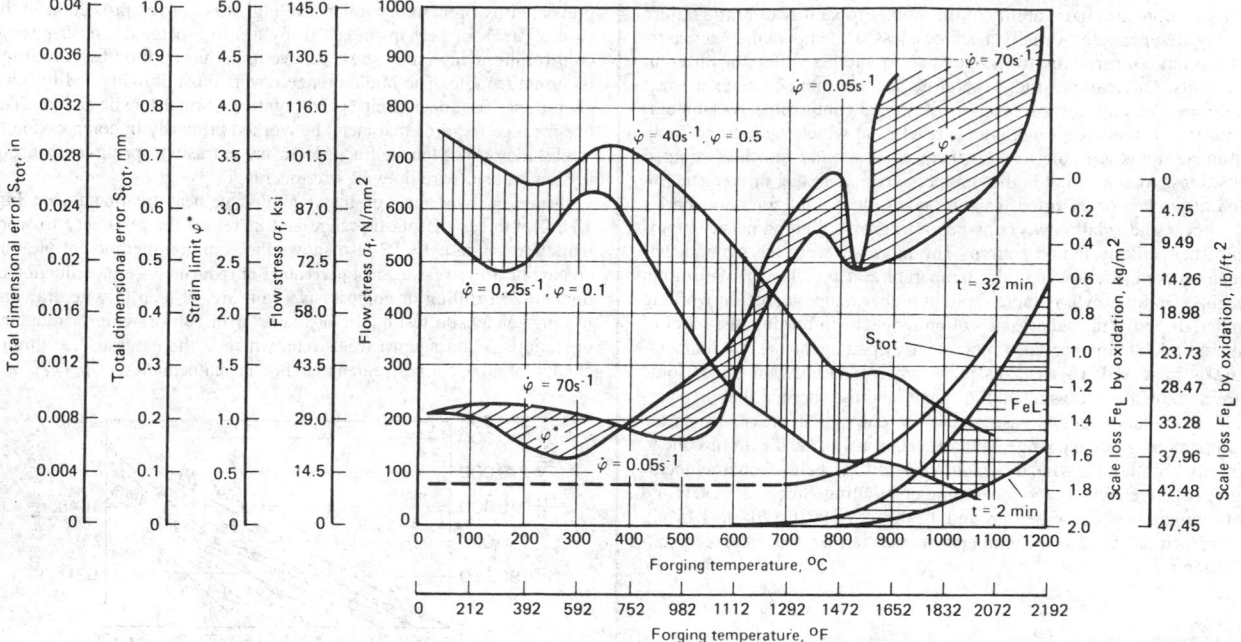

Fig. 13.2.3 Effect of forging temperature on forgeability and material properties. Material: AISI 1015 steel. ϕ = strain rate; ϕ^* = limiting strain; σ_f = flow stress; S_{tot} = dimensional error; Fe_L = scale loss. (*K. Lange, "Handbook of Metal Forming," McGraw-Hill, 1985.*)

temperature and for two different strain rates. The **flow stress** is the resistance to deformation. As the temperature rises from room temperature to 2,072°F (1,100°C), the flow stress decreases first gradually and then rapidly to about 25 percent of its value [cold working 114 ksi (786 MPa) and hot working 28 ksi (193 MPa) at a strain of 0.5 and strain rate of 40 per second].

One measure of workability is the **strain limit**. As the temperature rises, the strain limit for the 70-in (in · s) strain rate (typical of mechanical press forging) decreases slightly up to 500°C (932°F), rises until 750°C (1,382°F), drops rapidly at 800°C (1,472°F) (often called *blue brittleness*), and beyond 850°C (1,562°F) increases rapidly to hot forging temperature of 1,100°C (2,012°F). Therefore, substantial advantages of low material resistance (low tool pressures and press loads) and excellent workability (large flow without material failure) can be realized in the hot-working range. Hot-working temperatures, however, also mean poor dimensional tolerance (total dimensional error), poor surface finish, and material loss due to scale buildup. Forging temperatures above 1,300°C (2,372°F) can lead to **hot shortness** manifested by melting at the grain boundaries.

PLASTIC-WORKING TECHNIQUES

In the **metalworking** operations, as distinguished from metal cutting, material is forced to move into new shapes by plastic flow. **Hot-working** is carried on above the recovery temperature, and spontaneous recovery, or annealing, occurs about as fast as the properties of the material are altered by the deformation. This process is limited by the chilling of the material in the tools, scaling of the material, and the life of the tools at the required temperatures. **Cold-working** is carried on at room temperature and may be applied to most of the common metals. Since, in most cases, no recovery occurs at this temperature, the properties of the metal are altered in the direction of increasing strength and brittleness throughout the working process, and there is consequently a limit to which cold-working may be carried without danger of fracture.

A convenient way of representing the action of the common metals when cold-worked consists of plotting the actual stress in the material against the percentage reduction in thickness. Within the accuracy required for shop use, the relationship is linear, as in Fig. 13.2.4. The

lower limit of stress shown is the yield point at the softest temper, or anneal, commercially available, and the upper limit is the limit of tensile action, or the stress at which fracture, rather than flow, occurs. This latter value does not correspond to the commercially quoted "tensile strength" of the metal, but rather to the "true tensile strength," which is the stress that exists at the reduced section of a tensile specimen at fracture and which is higher than the nominal value in inverse proportion to the reduction of area of the material.

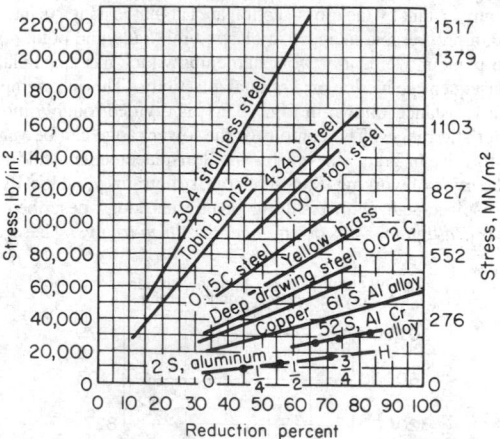

Fig. 13.2.4 Plastic range chart of commonly worked metals.

As an example of the construction and use of the cold-working plots shown in Fig. 13.2.4, the action of a very-low-carbon deep-drawing steel has been shown in Fig. 13.2.5. Starting with the annealed material with a yield point of 35,000 lb/in² (240 MN/m²), the steel was drawn to successive reductions of thickness up to about 58 percent, and the corresponding stresses plotted as the heavy straight line. The entire graph was then extrapolated to 100 percent reduction, giving the **modu-**

lus of strain hardening as indicated, and to zero stress so that all materials might be plotted on the same graph. Lines of equal reduction are slanting lines through the point marking the modulus of strain hardening at theoretical 100 percent reduction. Starting at any initial condition of previous cold work on the heavy line, a percentage reduction from this condition will be indicated by a horizontal traverse to the slanting reduction line of corresponding magnitude and the resulting increase in stress by the vertical traverse from this point to the heavy line.

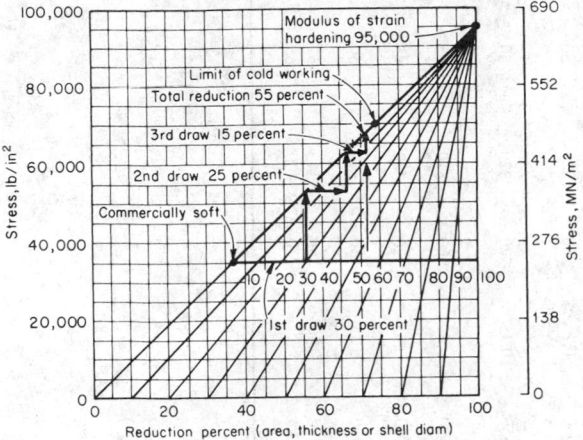

Fig. 13.2.5 Graphical solution of a metalworking problem.

The traverse shown involved three draws from the annealed condition of 30, 25, and 15 percent each, and resulting stresses of 53,000, 63,000, and 68,000 lb/in² (365, 434, and 469 MN/m²). After the initial 30 percent reduction, the next 25 percent uses $(1.00 - 0.30) \times 0.25$, or 17.5 percent more of the cold-working range; the next 15 percent reduction

uses $(1.00 - 0.30 - 0.175) \times 0.15$, or about 8 percent of the original range, totaling $30 + 17.5 + 8 = 55.5$ percent. This may be compared with the test value percent reduction in area for the particular material. The same result might have been obtained, die operation permitting, by a single reduction of 55 percent, as shown. Any appreciable reduction beyond this point would come dangerously close to the limit of plastic flow, and consequently an anneal is called for before any further work is done on the piece.

Figure 13.2.6 shows the approximate **true stress** vs. **true strain** plot of common plastic range values, for comparison with Fig. 13.2.4. In metal forming, a convenient way of representing the resistance of metal to deformation and flow is the **flow stress** σ, also known as the **logarithmic stress** or **true stress**. For most metals, flow stress is a function of the amount of deformation at cold-working temperatures (strain $\bar{\varepsilon}$) and the deformation rate at hot-working temperatures (strain rate $\dot{\bar{\varepsilon}}$). This relationship is often given as a power-law curve; $\sigma = K\bar{\varepsilon}^n$ for cold forming and $\sigma = C\dot{\bar{\varepsilon}}^m$ for hot forming. For commonly used materials, the values of the strength coefficients K and C and hardening coefficients n and m are given in Tables 13.2.1a and b.

A practical manufacturing method of judging relative plasticity is to compute the ratio of initial yield point to the ultimate tensile strength as developed in the tensile test. Thus a General Motors research memo listed steel with a 0.51 **yield/tensile ratio** [22,000 lb/in² (152 MN/m²) yp/43,000 lb/in² (296 MN/m²) ultimate tensile strength] as being suitable for really severe draws of exposed parts. When the ratio reaches about 0.75, the steel should be used only for flat parts or possibly those with a bend of not more than 90°. The higher ratios obviously represent a narrowing range of workability or residual plasticity.

ROLLING OPERATIONS

Rolling of sheets, coils, bars, and shapes is a primary process using plastic ranges both above and below recrystallization to prepare metals for further working or for fabrication. Metal squeezed in the bit area of

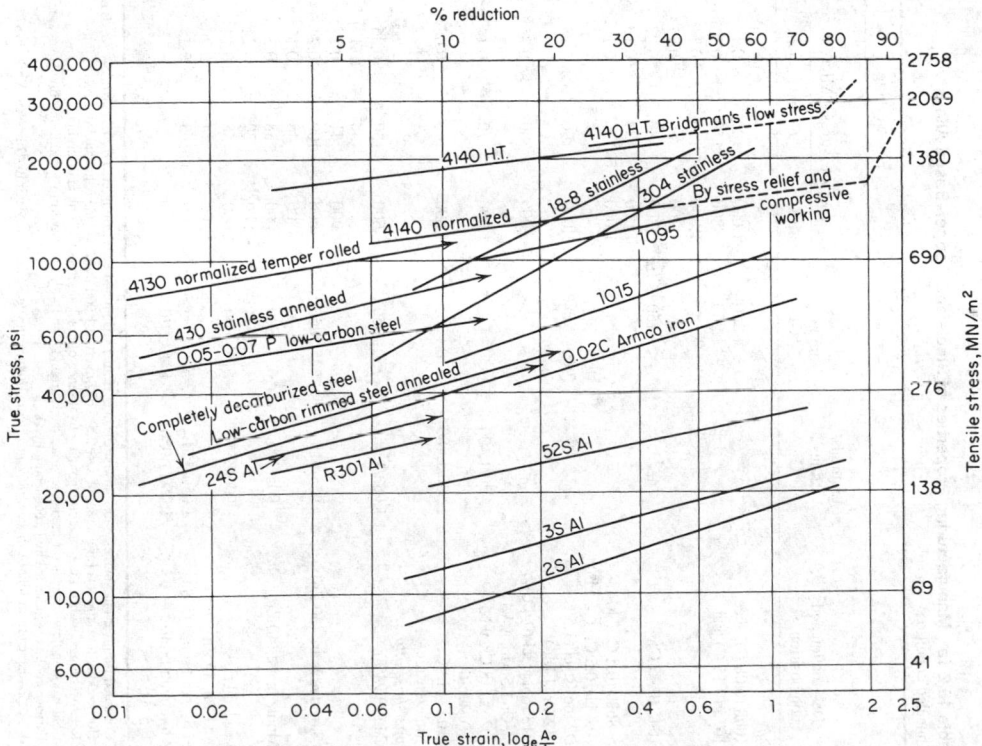

Fig. 13.2.6 True stress vs. true strain curves for typical metals. *(Crane and Hauf, E. W. Bliss. Co.)*

Table 13.2.1a Manufacturing Properties of Steels and Copper-Based Alloys*
(Annealed condition)

Designation and composition, %	Liquidus/ solidus, °C	Usual temp., °C	at °C	C	m	Workability‡	K	n	$\sigma_{0.2}$, MPa	TS, MPa	Elongation, %	q R.A., %	Annealing temp.,¶ °C
		(Hot-working)	(Flow stress,† MPa)				(Flow stress,§ MPa)		(Cold-working)				
Steels:													
1008 (0.08 C), sheet		<1,250	1,000	100	0.1	A	600	0.25	180	320	40	70	850–900 (F)
1015 (0.15 C), bar		<1,250	800	150	0.1	A	620	0.18	300	450	35	70	850–900 (F)
			1,000	120	0.1								
			1,200	50	0.17								
1045 (0.45 C)		<1,150	800	180	0.07	A	950	0.12	410	700	22	45	790–870 (F)
			1,000	120	0.13								
			1,000	120	0.1								
~8620 (0.2 C, 1 Mn 0.4 Ni, 0.5 Cr, 0.4 Mo)		900–1,080	1,000	190	0.13	A	1,300	0.3	350	620	30	60	880 (F)
D2 tool-steel (1.5 C, 12 Cr, 1 Mo)						B							
H13 tool steel (0.4 C, 5 Cr 1.5 Mo, 1 V)			1,000	80	0.26	B							
302 ss (18 Cr, 9 Ni) (austenitic)	1,420/1,400	930–1,200	1,000	170	0.1	B	1,300	0.3	250	600	55	65	1,010–1,120 (Q)
410 ss (13 Cr) (martensitic)	1,530/1,480	870–1,150	1,000	140	0.08	C	960	0.1	280	520	30	65	650–800
Copper-base alloys:													
Cu (99.94%)	1,083/1,065	750–950	600	130	0.06	A	450	0.33	70	220	50	78	375–650
			900	(48)	(0.17)								
				41	0.2								
Cartridge brass (30 Zn)	955/915	725–850	600	100	0.24	A	500	0.41	100	310	65	75	425–750
			800	48	0.15								
Muntz metal (40 Zn)	905/900	625–800	600	38	0.3	A	800	0.5	120	380	45	70	425–600
			800	20	0.24								
Leaded brass (1 Pb, 39 Zn)	900/855	625–800	600	58	0.14	A	800	0.33	130	340	50	55	425–600
			800	14	0.20								
Phosphor bronze (5 Sn)	1,050/950		700	160	0.35	C	720	0.46	150	340	57		480–675
Aluminum bronze (5 Al)	1,060/1,050	815–870				A			170	400	65		425–850

* Compiled from various sources; most flow stress data from T. Altan and F. W. Boulger, *Trans. ASME, Ser. B, J. Eng. Ind.* **95**, 1973, p. 1009.
† Hot-working flow stress is for a strain of $\varepsilon = 0.5$. To convert to 1,000 lb/in², divide calculated stresses by 7.
‡ Relative ratings, with A the best, corresponding to absence of cracking in hot rolling and forging.
§ Cold-working flow stress is for moderate strain rates, around $\varepsilon = 1\ \text{s}^{-1}$. To convert to 1,000 lb/in², divide stresses by 7.
¶ Furnace cooling is indicated by F, quenching by Q.
SOURCE: Adapted from John A. Schey, *Introduction to Manufacturing Processes*, McGraw-Hill, New York, 1987.

Table 13.2.1b Manufacturing Properties of Various Nonferrous Alloys[a]
(Annealed condition, except 6061-T6)

		Hot-working					Cold-working						
Designation and composition, %	Liquidus/ solidus, °C	Usual temp., °C	Flow stress,[b] MPa — at °C	C	m	Workability[f]	Flow stress,[c] MPa — K	n	$\sigma_{0.2}$, MPa	TS,[d] MPa	Elongation,[d] %	R.A.[q] %	Annealing temp.,[e] °C
Light metals:													
1100 Al (99%)	657/643	250–550	300; 500	60; 14	0.08; 0.22	A	140	0.25	35	90	35		340
Mn alloy (1 Mn)	649/648	290–540	400	35	0.13	A			100	130	14		370
~2017 Al (3.5 Cu, 0.5 Mg, 0.5 Mn)	635/510	260–480	400; 500	90; 36	0.12; 0.12	B	380	0.15	100	180	20	65	415 (F)
5052 Al (2.5 Mg)	650/590	260–510	480; 400	35; 50	0.13; 0.16	A	210	0.13	90	190	25		340
6061-0 (1 Mg, 0.6 Si, 0.3 Cu)	652/582	300–550	500	37	0.17	A	220	0.16	55	125	25		415 (F)
6061-T6	NA[g]	NA	NA	NA	NA	NA	450	0.03	275	310	8	45	415
~7075 Al (6 Zn, 2 Mg, 1 Cu)	640/475	250–455	450	40	0.13	B	400	0.17	100	230	16		
Low-melting metals:													
Sn (99.8%)	232	100–200	100	10	0.1	A				15	45	100	150
Pb (99.7%)	327	20–200	75	260	0.1	A				12	35	100	20–200
Zn (0.08% Pb)	417	120–275	225	40	0.1	A				130/170	65/50		100
High-temperature alloys:													
Ni (99.4 Ni + Co)	1,446/1,435	650–1250	1,150	~140	0.2	A			140	440	45	65	650–760
Hastelloy X (47 Ni, 9 Mo, 22 Cr, 18 Fe, 1.5 Co, 0.6 W)	1,290	980–1200				C			360	770	42		1,175
Ti (99%)	1,660	750–1,000	600; 900	200; 38	0.11; 0.25	C; A			480	620	20		590–730
Ti–6 Al–4 V	1,660/1,600	790–1,000	600; 900	550; 140	0.08; 0.4	C; A			900	950	12		700–825
Zirconium	1,852	600–1,000	900	50	0.25	A			210	340	35		500–800
Uranium (99.8%)	1,132	~700	700	110	0.1	A			190	380	4	10	

[a] Empty spaces indicate unavailability of data. Compiled from various sources; most flow stress data from T. Altan and F. W. Boulger, *Trans. ASME, Ser. B. J. Eng. Ind.* **95**, 1973, p. 1009.

[b] Hot-working flow stress is for a strain of $\varepsilon = 0.5$. To convert to 1,000 lb/in², divide calculated stresses by 7.

[c] Cold-working flow stress is for moderate strain rates, around $\dot\varepsilon = 1\ s^{-1}$. To convert to 1,000 lb/in², divide stresses by 7.

[d] Where two values are given, the first is longitudinal, the second transverse.

[e] Furnace cooling is indicated by F.

[f] Relative ratings, with A the best, corresponding to absence of cracking in hot rolling and forging.

[g] NA = Not applicable to the −T6 temper.

SOURCE: Adapted from John A. Schey, *Introduction to Manufacturing Processes*, McGraw-Hill, New York, 1987.

the rolls moves out lengthwise with very little spreading in width. This compressive working above the yield point of the metal may be aided in some cases by maintaining a substantial tensile strain in the direction of rolling.

A cast or forged billet or slab is preheated for the preliminary breakdown stage of rolling, although considerable progress has been made in continuous casting, in which the molten metal is poured continuously into a mold in which the metal is cooled progressively until it solidifies (albeit still at high temperature), whence it is drawn off as a quasi-continuous billet and fed directly into the first roll pass of the rolling mill. The increased speed of operation and production and the increased efficiency of energy consumption are obvious. Most new mills, especially minimills, have incorporated continuous casting as the normal method of operation. A reversing hot mill may achieve 5,000 percent elongation of an original billet in a series of manual or automatic passes. Alternatively, the billet may pass progressively through, say, 10 hot mills in rapid succession. Such a production setup requires precise control so that each mill stand will run enough faster than the previous one to make up for the elongation of the metal that has taken place. Hot-rolled steel may be sold for many purposes with the black mill scale on it. Alternatively, it may be acid-pickled to remove the scale and treated with oil or lime for corrosion protection. To prevent scale from forming in hot-rolling, a nonoxidizing atmosphere may be maintained in the mill area, a highly special plant design.

Pack rolling of a number of sheets stacked together provides means of retaining enough heat to hot-roll thin sheets, as for high-silicon electric steels.

Cold-rolling is practical in production of thin coil stock with the more ductile metals. The number of passes or amount of reduction between anneals is determined by the rate of work hardening of the metal. Successive stands of cold-rolling help to retain heat generated in working. Tension provided by mill reels and between stands helps to increase the practical reduction per step. Bright annealing in a controlled atmosphere avoids surface pockmarks, which are difficult to get out. For high-finish stock, the rolls must be maintained with equal finish.

Protective coating is best exemplified by high-speed tinplate mills in which coil stock passes continuously through the necessary series of cleaning, plating, and heating steps. Zinc and other metals are also applied by plating but not on the same scale. **Clad** sheets (high-strength aluminum alloys with pure aluminum surface for protection against electrolytic oxidation) are produced by rolling together; e.g., an alloy-aluminum billet is hot-rolled together with plates of pure aluminum above and below it through a series of reducing passes, with precautions to ensure clean adhesion.

On the other hand, prevention of adhesion, as by a separating film, is essential in the final stages of **foil rolling**, where two coils may have to be rolled together. Such foil may then be **laminated** with suitable adhesive to paper backing materials for wrapping purposes. (See also Sec. 6.)

Shape-rolling of structural shapes and rails is usually a **hot** operation with roll-pass contours designed to distribute the displacement of metal in a series of steps dictated largely by experience. **Contour rolling** of relatively thin stock into tubular, channel, interlocking, or varied special cross sections is usually done cold in a series of roll stands for lengthwise bending and setting operations. There is also a wide range of simple bead-rolling, flange-rolling, and seam-rolling operations in relatively thin materials, especially in connection with the production of barrels, drums, and other containers.

Oscillating or **segmental rolling** probably developed first in the manually fed contour rolling of agricultural implements. In some cases, the suitably contoured pair of roll inserts or roll dies oscillates before the operator, to form hot or cold metal. In other cases, the rolls rotate constantly, toward the operator. The working contour takes only a portion of the circumference, so that a substantial clearance angle leaves a space between the rolls. This permits the operator to insert the blank to the tong grip between the rolls and against a fixed gage at the back. Then, as rotation continues, the roll dies grip and form the blank, moving it back to the operator. This process is sometimes automated; such units as **tube-reducing mills** oscillate an entire rolling-mill assembly and

feed the work over a mandrel and into the contoured rolls, advancing it and possibly turning it between reciprocating strokes of the roll stands for cold reduction, improved concentricity, and, if desired, the tapering or forming of special sections.

Spinning operations (Fig. 13.2.7) apply a rolling-point pressure to relatively limited-lot production of cup, cone, and disk shapes, from floor lamps and TV tube housings to car wheels and large tank ends. Where substantial metal thickness is required, powerful machines and hydraulic servo controls may be used. Some of the large, heavy sections and difficult metals are spun hot.

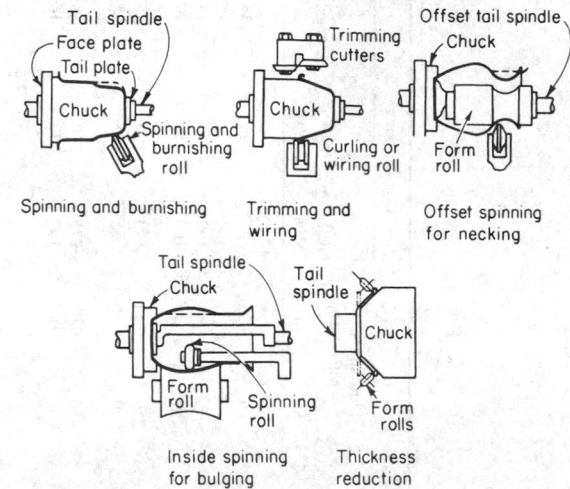

Fig. 13.2.7 Spinning operations.

Rolling operations are distinguished by the relatively rapid and continuous application of working pressure along a limited line of contact. In determining the working area, consider the lineal dimension (width of coil), the bit or reduction in thickness, and the roll-face deflection, which tends to increase the contact area. Approximations of rolling-mill load and power requirements have been worked out in literature of the AISE and ASME.

SHEARING

The shearing group of operations includes such **power press operations** as blanking, piercing, perforating, shaving, broaching, trimming, slitting, and parting. Shearing operations traverse the entire plastic range of metals to the point of failure.

The maximum pressure P, in pounds, required in shearing operations is given by the equation $P = \pi DtS = LtS$, where s is the resistance of the material to shearing, lb/in²; t is the thickness of the material, in; L is the length of cut, in, which is the circumference of a round blank πD or the periphery of a rectangular or irregular blank. Approximate values of s are given in Table 13.2.2.

Shear (Fig. 13.2.8) is the advance of that portion of the shearing edge which first comes in contact with the material to be sheared over the last portion to establish contact, measured in the direction of motion. It should be a function of the thickness t. Shear reduces the maximum pressure because, instead of shearing the whole length of cut at once, the shearing action takes place progressively, shearing on only a portion of the length at any instant. The maximum pressure for any case where the shear is equal to or greater than t is given by $P_{max} = P_{av} t/\text{shear}$, where P_{av} is the average value of the pressure on a punch, with shear $= t$, from the time it strikes the metal to the time it leaves.

Distortion results from shearing at an angle (Fig. 13.2.8) and accordingly, in blanking, where the blank should be flat, the punch should be flat, and the shear should be on the die. Conversely, in hole punching, where the scrap is punched out, the die should be flat and the shear

Table 13.2.2   Approximate Resistance to Shearing in Dies

Material	Annealed state		Hard, cold-worked	
	Resistance to shearing, lb/in²*	Penetration to fracture, percent	Resistance to shearing, lb/in²*	Penetration to fracture, percent
Lead	3,500	50	Anneals at room temperature	
Tin	5,000	40	Anneals at room temperature	
Aluminum 2S, 3S	9,000–11,000	60	13,000–16,000	30
Aluminum 52S, 61S, 62S	12,000–18,000	· · ·	24,000–30,000	
Aluminum 75S	22,000	· · ·	46,000	
Zinc	14,000	50	19,000	25
Copper	22,000	55	28,000	30
Brass	33,000–35,000	50–55	52,000	25–30
Bronze 90–10	· · ·	· · ·	40,000	
Tobin bronze	36,000	25	42,000	
Steel 0.10C	35,000	50	43,000	38
Steel 0.20C	44,000	40	55,000	28
Steel 0.30C	52,000	33	67,000	22
Steel 0.40C	62,000	27	78,000	17
Steel 0.60C	80,000	20	102,000	9
Steel 0.80C	97,000	15	127,000	5
Steel 1.00C	115,000	10	150,000	2
Stainless steel	57,000	39		
Silicon steel	65,000	30		
Nickel	35,000	55		

*1,000 lb/in² = 6.895 MN/m².

NOTE: Available test data do not agree closely. The above table is subject to verification with closer control of metal analysis, rolling and annealing conditions, die clearances. In dinking dies, steel-rule dies, hollow cutters, etc., cutting-edge resistance is substantially independent of thickness: cotton glove cloth (stack, 2 or 3 in thick), 240 lb/in; kraft paper (stack tested, 0.20 in thick), 385 lb/in, celluloid [1½₂ in thick, warmed in water to 120 to 150°F (49 to 66°C)], 300 lb/in.

should be on the punch. Where there are a number of punches, the effect of shear may be obtained by stepping the punches.

Crowding results during the plastic deformation period, before the fracture occurs, in any shearing operation. Accordingly, when small delicate punches are close to a large punch, they should be stepped shorter than the large punch by at least a third of the metal thickness.

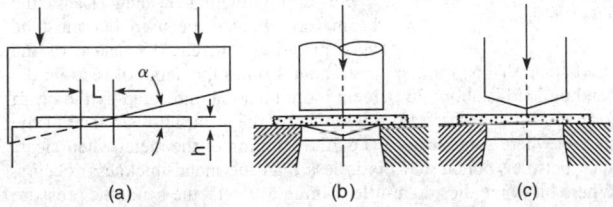

α = shearing angle
t = h = thickness of sheet
L = shearing length

Fig. 13.2.8 Shearing forces can be reduced by providing a rake or shear on (a) the blades in a guillotine, (b) the die in blanking, (c) the punch in piercing. (*J. Schey, "Introduction to Manufacturing Processes," McGraw-Hill, 1987.*)

Clearance between the punch and die is required for a clean cut and durability. An old rule of thumb places the clearance all around the punch at 8 to 10 percent of the metal thickness for soft metal and up to 12 percent for hard metal. Actually, hard metal requires less clearance for a clean fracture than soft, but it will stand more. In some cases, with delicate punches, clearance is as high as 25 percent. Where the hole diameter is important, the punch should be the desired diameter and the clearance should be added to the die diameter. Conversely, where the blank size is important, the die and blank dimensions are the same and clearance is deducted from the punch dimensions.

The **work per stroke** may be approximated as the product of the maximum pressure and the metal thickness, although it is only about 20 to 80 percent of that product, depending upon the clearance and ductility of the metal. Reducing the clearance causes secondary fractures and increases the work done. With sufficient clearance for a clean fracture, the

work is a little less than the product of the maximum pressure, the metal thickness, and the percentage reduction in thickness at which the fracture occurs. Approximate values for this are given in Table 13.2.2. The **power required** may be obtained from the work per stroke plus a 10 to 20 percent friction allowance.

Shaving A sheared edge may be squared up roughly by shaving once, allowing for the shaving of mild sheet steel about 10 percent of the metal thickness. This allowance may be increased somewhat for thinner material and should be decreased for thicker and softer material. In making several cuts, the amount removed is reduced each time. For extremely fine finish a round-edged burnishing die or punch, say 0.001 or 0.0015 in tight, may be used. Aluminum parts may be blanked (as for impact extrusion) with a fine finish by putting a 30° bevel, approx one-third the metal thickness on the die opening, with a near metal-to-metal fit on the punch and die, and pushing the blank through the highly polished die.

Squaring shears for sheet or plate may have their blades arranged in either of the ways shown in Fig. 13.2.9. The square-edged blades in Fig. 13.2.9a may be reversed to give four cutting edges before they are reground. Single-edged blades, as shown in Fig. 13.2.9b, may have a clearance angle on the side where the blades pass, to reduce the working friction. They may also be ground at an angle or rake, on the face which comes in contact with the metal. This reduces the bending and conse-

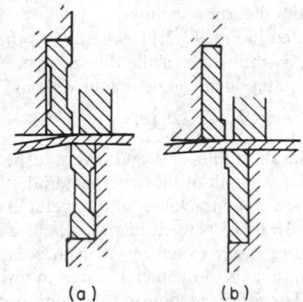

Fig. 13.2.9 Squaring shears.

quent distortion at the edge. Either type of blade distorts also in the other direction owing to the angle of shear on the length of the blades (see Fig. 13.2.8). Cutting speed is 3 to 30 ft/min (1 to 10 m/min), depending in part upon the thickness of the material.

Circular cutters for slitters and circle shears may also be square-edged (on most slitters) or knife-edged (on circle shears). According to one rule, their diameter should be not less than 70 times the metal thickness. Cutting speeds vary from 50 to 200 ft/min (17 to 65 m/min), depending largely upon metal thickness (inverse proportion).

Knife-edge hollow cutters working against end-grain maple blocks represent an old practice in cutting leather, rubber, and cloth in multiple thicknesses. **Steel-rule dies,** made up of knife-edge hard-steel strip economically mounted against a steel plate in a wood matrix with rubber strippers and cutting against hard saw-steel plates, extend the practice to corrugated-carton production and even some limited-lot metal cutting.

Higher precision is often required in **finish shearing** operations on sheet material. For ease of subsequent operations and assembly, the cut edges should be clean (acceptable burr heights and good surface finish) and perpendicular to the sheet surface. The processes include precision or fine blanking, negative clearance blanking, counterblanking, and shaving, as shown in Fig. 13.2.10. By these methods either the plastic behavior of material is suppressed, or the plastically deformed material is removed.

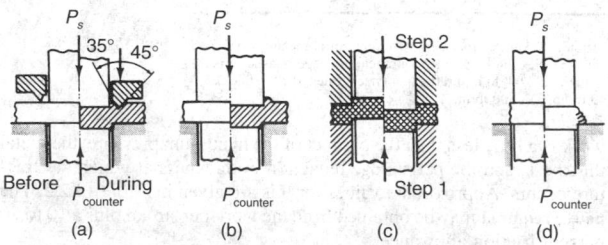

Fig. 13.2.10 Parts with finished edges can be produced by (a) precision blanking, (b) negative-clearance blanking, (c) counterblanking, (d) shaving a previously sheared part. (J. Schey, "Introduction to Manufacturing Processes," McGraw-Hill, 1987.)

BENDING

The bending group of operations is performed in **presses** (variety), **brakes** (metal furniture, cornices, roofing), **bulldozers** (heavy rolled sections), **multiple-roll forming machines** (molding, etc.), **draw benches** (door trim, molding, etc.), **forming rolls** (cylinders), and **roll straighteners** (strips, sheets, plates).

Spring back, due to the elasticity of the metal and amount of the bend, may be compensated for by overbending or largely prevented by striking the metal at the radius with a **coining** (i.e., squeezing, as in production of coins) pressure sufficient to set up compressive stresses to counterbalance surface tensile stresses. A very narrow bead may be used to localize the pinch where needed and minimize danger to the press in squeezing on a large area. Under such conditions, good sharp bends in V dies have been obtained with two to four times the pressure required to shear the metal across the same section.

These are illustrated in Fig. 13.2.11, where P_b is the **bending load** on the press brake, W_b is the width of the die support, and $P_{counter}$ is the counterload. The bending load can be obtained from

$$P_b = wt^2(\text{UTS})/W_b$$

where t and w are the sheet thickness and width, respectively, and UTS is the ultimate tensile strength of the sheet material.

Bending Allowance The thickness of the metal over a small radius or a sharp corner is 10 or 15 percent less than before bending because the metal moves more easily in tension than in compression. For the same reason the neutral axis of the metal moves in toward the center of the corner radius. Therefore, in figuring the length of blank L to be allowed for the bend up to an inside radius r of two or three times the

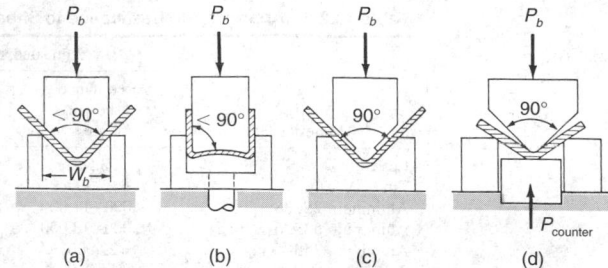

Fig. 13.2.11 Springback may be neutralized or eliminated by (a), (b) overbending; (c) plastic deformation at the end of the stroke; (d) subjecting the bend zone to compression during bending. [Part (d) after V. Kupka, T. Nakagawa, and H. Tyamoto, CIRP 22:73–74 (1973).] (Source: J. Schey, "Introduction to Manufacturing Processes," McGraw-Hill, 1987.)

metal thickness t, the length may be figured closely as along a neutral line at $0.4t$ out from the inside radius. Thus, with reference to Fig. 13.2.12, for any angle a in deg and other dimensions in inches, $L = (r + 0.4t)2\pi a/360 = (r + 0.4t)a/57.3$.

The factor $0.4t$, which locates the neutral axis, is subject to some variation (say 0.35 to $0.45t$) according to radius, condition of metal, and angle. In figuring allowances for sharp bends, note that the metal builds up on the compression side of the corner. Therefore, in locating the neutral axis, consider an inside radius r of about $0.05t$ as a minimum.

Roll straighteners work on the principle of bending the metal beyond its elastic limit in one direction over rolls small enough in diameter, in proportion to the metal thickness, to give a permanent set, and then

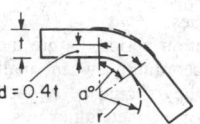

Fig. 13.2.12 Bending allowance.

taking that bend out by repeatedly reversing it in direction and reducing it in amount. Metal is also straightened by gripping and stretching it beyond its elastic limit and by hammering; the results of the latter operation depend entirely upon the skill of the operator.

For approximating **bending loads,** the beam formula may be used but must be very materially increased because of the short spans. Thus, for a span of about 4 times the depth of section, the bending load is about 50 percent more than that indicated by the beam formula. It increases from this to nearly the shearing resistance of the section where some **ironing** (i.e., the thinning of the metal when clearance between punch and die is less than the metal thickness) occurs. Where hit-home dies do a little coining to "set" the bend, the pressure may range from two or three times the shearing resistance, with striking beads and proper care, up to very much higher figures.

The work to roll-bend a sheet or plate t in thick with a volume of V in³, into curved shape of radius r in, is given as $W = CS(t/r)V/48$ ft · lb, in which S is the tensile strength and C is an experience factor between 1.4 and 2.

The **equipment for bending** consists of mechanical presses for short bends, press brakes (mechanical and hydraulic) for long bends, and roll formers for continuous production of profiles. The bends are achieved by bending between tools, wiping motion around a die corner, or bending between a set of rolls. These bending actions are illustrated in Fig. 13.2.13. Complex shapes are formed by repeated bending in simple tooling or by passing the sheet through a series of rolls which progressively bend it into the desired profile. Roll forming is economical for continuous forming for large production volume. Press brakes can be computer-controlled with synchronized feeding and bending as well as spring-back compression.

DRAWING

Drawing includes operations in which metal is pulled or drawn, in suitable containing tools, from flat sheets or blanks into cylindrical cups or rectangular or irregular shapes, deep or shallow. It also includes reduc-

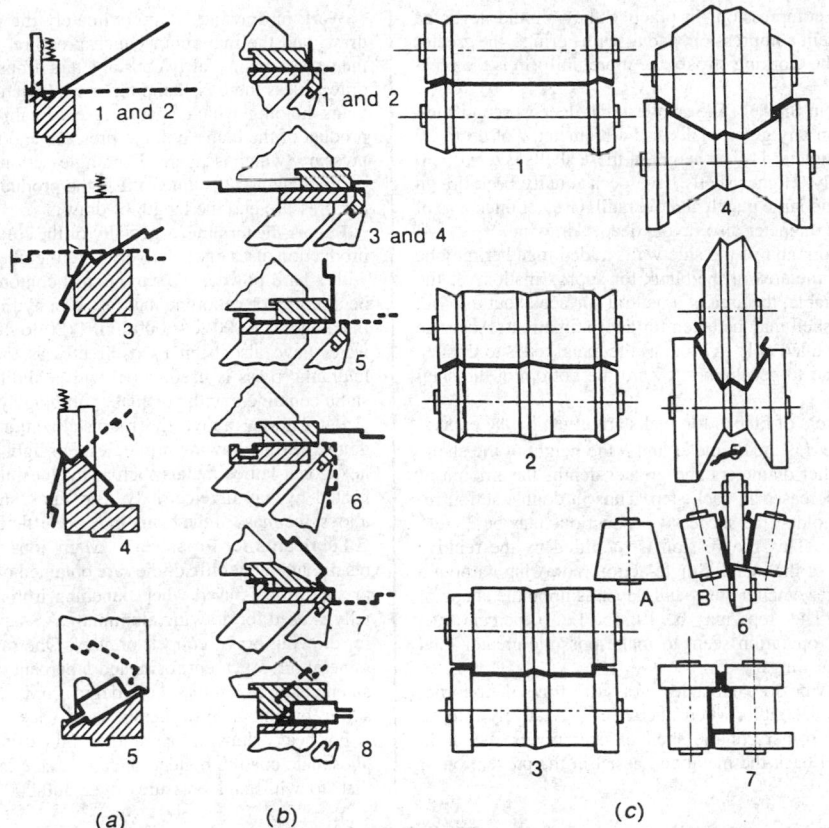

Fig. 13.2.13 Complex profiles formed by a sequence of operations on (*a*) press brakes, (*b*) wiping dies, (*c*) profile rolling. (*After Oehler; Biegen, Hanser Verlag, Munich, 1963.*) Source: J. Schey, *"Introduction to Manufacturing Processes,"* McGraw-Hill, 1987.)

ing operations on shells, tube, wire, etc., in which the metal being drawn is pulled through dies to reduce the diameter or size of the shape. All drawing and reducing operations, by an applied tensile stress in the material, set up circumferential compressive stresses which crowd the metal into the desired shape. The relation of the shape or diameter before drawing to the shape or diameter after drawing determines the magnitude of the stresses. Excessive draws or reductions cause thinning or tearing out near the bottom of a shell. Severe cold-drawing operations require very ductile material and, in consequence of the amount of plastic deformation, harden the metal rapidly and necessitate annealing to restore the ductility for further working.

The **pressure used in drawing** is limited to the load to shear the bottom of the shell out, except in cases where the side wall is ironed thinner, when wall friction makes somewhat higher loads possible. It is less than this limit for round shells which are shorter than the limiting height and also for rectangular shells. Drawing occurs only around the corner radii of rectangular shells, the straight sides being merely free bending.

A **holding pressure** is required in most initial drawing and some re-drawing, to prevent the formation of wrinkles due to the circumferential compressive stresses. Where the blank is relatively thin compared with its diameter, the blank-holding pressure for round work is likely to vary up to about one-third of the drawing pressure. For material heavy enough to provide sufficient internal resistance to wrinkling, no pressure is required. Where a drawn shape is very shallow, the metal must be stretched beyond its elastic limit in order to hold its shape, making it necessary to use higher blank-holding loads, often in excess of the drawing pressure. To grip the edges sufficiently to do this, it is often advisable to use **draw beads** on the blank-holding surfaces if sufficient pressure is available to form these beads.

In sheet/deep-drawing practice, the punch force P can be approximated by

$$P = \pi D_p t_0 (\text{UTS})(D_0/D_p - 0.7)$$

where t_0 is the blank thickness and D_0 and D_p are the diameters of the blank and the punch.

The blank holder pressure for avoiding defects such as wrinkling of bottom/wall tear-out is kept at 0.7 to 1.0 percent of the sum of the yield and the UTS of the material. Punch/die clearances are chosen to be 7 to 14 percent greater than the sheet blank thickness t_0. The die corner radii are chosen to avoid fracture at the die corner from puckering or wrinkling. Recommended values of D_0/t_0 for deep-drawn cups are 6 to 15 for cups without flange and 12 to 30 for cups with flange. These values will be smaller for relatively thick sheets and larger for very thin blank thickness. For deeper-drawn cups, they may be redrawn or reverse-drawn, the latter process taking advantage of strain softening on reverse drawing. When the material has marked strain-hardening propensities, it may be necessary to subject it to an intermediate annealing process to restore some of its ductility and to allow progression of the draws to proceed.

Some shells, which are very thick or very shallow compared with their diameter, do not require a blank holder. Blank-holding pressure may be obtained through toggle, crank, or cam mechanisms built into the machine or by means of air cylinders, spring-pressure attachments, or rubber bumpers under the bolster plate. The length of car springs should be about 18 in/in (18 cm/cm) of draw to give a fairly uniform drawing pressure and long life. The use of car springs has been largely superseded by hydraulic and pneumatic cushions. Rubber bumpers may be figured on a basis of about 7.5 lb/in² (50 kN/m²) of cross-sectional area per 1 percent of compression. In practice they should never be loaded beyond 20 percent compression, and as with springs, the greater the length relative to the working stroke, the more uniform is the pressure.

Dimensions of Drawn Shells The smallest and deepest round shell that can be drawn from any given blank has a diameter d of 65 to 50 percent of the blank diameter D. The height of these shells is $h = 0.35d$ to $0.75d$, approximately. Higher shells have occasionally been drawn with ductile material and large punch and die radii. Greater thickness of material relative to the diameter also favors deeper drawing.

The area of the bottom and of the side walls added together may be considered as equal to the area of the blank for approximations. If the punch radius is appreciable, the area of a neutral surface about $0.4t$ out from the inside of the shell may be taken for approximations. Accurate blank sizes may be obtained only by trial, as the metal tends to thicken toward the top edge and to get thinner toward the bottom of the shell wall in drawing.

Approximate diameters of blanks for shells are given by the expression $\sqrt{d^2 + 4dh}$, where d is the diameter and h is the height of the shell.

In **redrawing** to smaller diameters and greater depths the amount of reduction is usually decreased in each step. Thus in double-action redrawing with a blank holder, the successive reductions may be 25, 20, 16, 13, 10 percent, etc. This progression is modified by the relative thickness and ductility of the metal. Single-action redrawing without a blank holder necessitates smaller steps and depends upon the shape of the dies and punches. The steps may be 19, 15, 12, 10 percent, etc. Smaller reductions per operation seem to make possible greater total reductions between annealings.

Rectangular shells may be drawn to a depth of 4 to 6 times their corner radius. It is sometimes desirable, where the sheet is relatively thin, to use draw beads at the corners of the shell or near reverse bends in irregular shapes to hold back the metal and assist in the prevention of wrinkles.

Work in drawing is approximately the product of the length of the draw, and the maximum punch pressure, as the load rises quickly to the peak, remains fairly constant, and drops off sharply at the end of the draw unless there is stamping or wall friction. To this, add the work of blank holding which, in the case of cam and toggle pressure, is the product of the blank-holding pressure and the spring of the press at the pressure (which is small). For single-action presses with spring, rubber, or air-drawing attachments it is the product of the average blank-holding pressure and the length of draw.

Rubber-die forming, especially of the softer metals and for limited-lot production, uses one relatively hard member of metal, plaster, or plastic with a hard powder filler to control contour. The mating member may be a rubber or neoprene mattress or a hydraulically inflatable bag, confined at 3,000 to 7,000 lb/in² (20 to 48 MN/m²). Babbitt, oil, and water have also been used directly as the mobile member. A large hydraulic press is used, often with a sliding table or tables, and even static containers with adequate pumping systems.

Hot drawing above the recrystallization range applies single- and double-action drawing principles. For light gages of plastics, paper, and hexagonal-lattice metals such as magnesium, dies and punches may be heated by gas or electricity. For thick steel plate and heat-treatable alloys, the mass of the blank may be sufficient to hold the heat required.

Lubricants for Presswork Many jobs may be done dry, but better results and longer life of dies are obtained by the use of a lubricant. Lard or sperm oil is used when punching iron, steel, or copper. Petroleum jelly is used for drawing aluminum. A soap solution is commonly used for drawing brass, copper, or steel. One manufacturer uses 90 percent mineral oil, 5 percent rosin, and 5 percent oleic acid for light work and an emulsion of a mineral oil, degras, and a pigment consisting of chalk, sulfur, or lithopone for heavy work. (See also Sec. 6.)

For heavy drawing operations and extrusion, steels may have a zinc phosphate coating bonded on, and a zinc or sodium stearate bonded to that, to withstand pressures over 300,000 lb/in² (2,070 MN/m²). An

Table 13.2.3 Typical Lubricants* and Friction Coefficients in Plastic Deformation

Workplace material	Working	Forging Lubricant	μ	Extrusion† lubricant	Wire drawing Lubricant	μ	Rolling Lubricant	μ	Sheet metalworking Lubricant	μ
Sn, Pb, Zn alloys		FO-MO	0.05	FO or soap	FO	0.05	FA-MO or MO-EM	0.05 0.1	FO-MO	0.05
Mg alloys	Hot or warm	GR and/or MoS₂	0.1–0.2	None			MO-FA-EM	0.2	GR in MO or dry soap	0.1–0.2
Al alloys	Hot	GR or MoS₂	0.1–0.2	None			MO-FA-EM	0.2		
	Cold	FA-MO or dry soap	0.1 0.1	Lanolin or soap on PH	FA-MO-EM, FA-MO	0.1 0.03	1–5% FA in MO(1–3)	0.03	FO, lanolin, or FA-MO-EM	0.05–0.1
Cu alloys	Hot	GR	0.1–0.2	None (or GR)			MO-EM	0.2		
	Cold	Dry soap, wax, or tallow	0.1	Dry soap or wax or tallow	FO-soap-EM, MO	0.1 0.03	MO-EM	0.1	FO-soap-EM or FO-soap	0.05–0.1
Steels	Hot	GR	0.1–0.2	GL (100–300), GR			None or GR-EM	ST‡ 0.2	GR	0.2
	Cold	EP-MO or soap, on PH	0.1 0.05	Soap on PH	Dry soap or soap on PH	0.05 0.03	10% FO-EM	0.05	EP-MO, EM, soap, or polymer	0.05–0.1
Stainless steel, Ni and alloys	Hot	GR	0.1–0.2	GL (100–300)			None	ST‡	GR	0.2
	Cold	CL-MO or soap on PH	0.1 0.05	CL-MO or soap on PH	Soap on PH or CL-MO	0.03 0.05	FO-CL-EM or CL-MO	0.1 0.05	CL-MO, soap, or polymer	0.1
Ti alloys	Hot	GL or GR	0.2	GL (100–300)					GR, GL	0.2
	Cold	Soap or MO	0.1	Soap on PH	Polymer	0.1	MO	0.1	Soap or polymer	0.1

* Some more frequently used lubricants (hyphenation indicates that several components are used in the lubricant):
  CL = chlorinated paraffin
  EM = emulsion; listed lubricating ingredients are finely distributed in water
  EP = "extreme-pressure" compounds (containing S, Cl, and P)
  FA = fatty acids and alcohols, e.g., oleic acid, stearic acid, stearyl alcohol
  FO = fatty oils, e.g., palm oil and synthetic palm oil
  GL = glass (viscosity at working temperature in units of poise)
  GR = graphite; usually in a water-base carrier fluid
  MO = mineral oil (viscosity in parentheses, in units of centipoise at 40°C).
  PH = phosphate (or similar) surface conversion, providing keying of lubricant
† Friction coefficients are misleading for extrusion and therefore are not quoted here.
‡ The symbol ST indicates sticking friction.
SOURCE: John A. Schey, *Introduction to Manufacturing Processes*, McGraw-Hill, New York, 1987.

anodized coating for aluminum may be used as a host for the lubricant. It is reported that such a chemical treatment plus a lacquer or plastic coating and a lubricant is effective for severe ironing operations.

The problem is to prevent local pressure welding from starting as galling or pickup, with resulting scratching, by maintaining a fluid film separation between metal surfaces. At moderate pressures, almost any viscous liquid lubricant will do the job. Rust protection and easy removal of the lubricant are often major factors in the choice.

Lubricants for metalworking are often classified based on their interface friction coefficient, which depends on the workpiece material and the lubricant being used. The **interface friction coefficient** μ is often defined as the ratio of the friction force to the normal force at the interface. Typical values of μ for different workpiece-lubricant pairs are included in Table 13.2.3.

Shock-wave forming for limited lots is developing in several ways. **Explosive forming**, especially for large-area drawn or formed shapes, usually requires one metal contour-control die immersed in a large container of fluid, or even in a lake or pond. Explosives manufacturers have developed means of computing the charge and the distance that it should be suspended above the blank to be formed. The space back of the blank in the die has to be evacuated. A blank-holding ring to minimize wrinkle formation in the flange area is bolted very tightly to the die, with an O-ring seal to prevent leakage.

Electrohydraulic forming is similar to explosive forming except that the shock wave is imparted electrically from a large battery of capacitors. **Magnetic forming** uses the same source of power but does not require a fluid medium. A flexible pancake coil delivers the magnetic shock pulse.

usually the least severe of the squeezing group. Tolerances are ordinarily closer than for the milling operations which are supplanted. When extremely close tolerances are required, say plus or minus one-thousandth of an inch (0.025 mm), arrange substantial size blocks to take half or two-thirds of the total load. These take up uniformly the bearing-oil films and any slight deflection of the bed and bolster and minimize the error in springback due to variation in thickness, hardness, and area of the rough forging or casting. The usual amount left for squeezing is $\frac{1}{32}$ to $\frac{1}{16}$ in (0.8 to 1.6 mm). Presses may be selected for this service on a basis of 60 to 80 tons/in² (830 to 1,100 MN/m²), although 100 tons/in² (1,380 MN/m²) is more often used in the automobile trade for reserve capacities. When figuring from experimental results obtained in testing machines, the recorded loads are usually doubled in selecting a press, in order to allow for the difference among the speed of the machines, the positive action, and a safety margin.

Swaging or **cold forging** involves squeezing of the blank to an appreciably different shape. Success in performing such operations on steel usually depends upon squeezing a relatively small area with freedom to flow without restraint. Dies for this work must usually be substantially backed up with hardened steel plates. The edge of the blank after coining is usually ragged and must be trimmed for appearance.

Hot forging is similar in certain respects to the above but permits much greater movement of metal. Hot forging may be done in drop hammers, percussion presses, power presses, or forging machines, when dies are used, or in steam hammers, helve hammers, or hydraulic presses, on plain anvils.

The pressure exerted by hydraulic presses and steam hammer for jobbing work should be about as follows:

Ingot diam, in*	5	8	12	16	24	36	48	60	72
Press, tons†	100	200	400	600	1,000	1,500	2,000	3,000	4,000
Hammer, tons†	½	1	3	5	10	20	40	80	120

* 1 in = 2.54 cm.
† 1 ton = 8.9 kN.

BULK FORMING

The squeezing group of operations are those in which the metal is worked in compression. Resultant tensile strains occur, however; in cases where the metal is thin compared with its area and there is an appreciable movement of the metal, there results a pyramiding of pressure toward the center of the die which may prove serious. The metal is incompressible (beyond about 1 percent), and consequently, to reduce the thickness of any volume of metal in the center of the blank, its area must be increased, which involves spreading or stretching all the metal around it. The surrounding metal acts like shrunk bands and offers a resistance increasing toward the center and often many times the compressive resistance of the material.

Squeezing operations and particularly the squeezing of steel are practically the severest of all press operations. They may be divided into four general classifications according to severity, although in every group there will be found examples of working to the limit of what the die steels will stand, which may be taken at about 100 tons/in². The severer operations, such as cold bottom extrusion and wall extrusion, are limited to the softer metals. Squeezing operations ordinarily require pressure through a very short distance, the pressure starting at the compressive yield strength of the material over the surface being squeezed and rising to a maximum at bottom stroke. This maximum is greatest when the metal is thin compared with its area or when the die is entirely closed as for coining. Care must be taken, on all squeezing operations, in the setting of presses and avoidance of double blanks or extra-heavy blanks as the presses must be stiff. In squeezing solidly across bottom center the mechanical advantage is such that a small difference in thickness or setup can make a very large difference in pressure exerted. For this reason high-speed self-contained hydraulic presses, with automatic pressure-control and size blocks, are now finding favor for some of this work.

Sizing, or the flattening or surfacing of parts of forgings or castings, is

Drop hammers are rated according to weight of ram. For carbon steel they may be selected on a basis of 50 to 55 lb of ram weight per square inch of projected area (3.5 to 3.9 kg/cm²) of the forging, including as much of the flash as is squeezed. This allowance should be increased to 60 lb/in² (4.2 kg/cm²) for 0.20 carbon steel, 70 lb/in² (4.9 kg/cm²) for 0.30 carbon steel, and up to about 130 lb/in² (9.2 kg/cm²) for tungsten steel.

In figuring the **forging pressure**, multiply the projected area of the forging, including the portion of the flash that is squeezed, by approximately one-third of the cold compressive strength of the material. Another method gives the forging pressure at three to four times the compressive strength of the material at forging temperatures times the projected area, for presses; or at ten times the compressive strength at forging temperatures times the projected area, for hammers. The pressure builds up to a rather high figure at bottom stroke owing to the cooling of the metal particularly in the flash and to the small amount of relief for excess metal which the flash allows.

For **brass press forgings** a good mixture is about Cu, 59; Zn, 39; Pb, 2 percent forged at 1,300 to 1,400°F (700 to 760°C). The power curve in press forging rises sharply, from the compressive strength at forging heat times the projected area of the slug to three or more times that quantity at bottom stroke. A large flash area assists in driving the metal into deep die recess.

In **heading operations**, hot or cold, the length of wire or rod that can be gathered into a head, without side restraint, in a single operation, is limited to three times the diameter. In coining and then heading large heads, cold, wire of about 0.08 carbon must be used to avoid excessive strain-hardening.

Forging Dies Drop-forge dies are usually of steel or steel castings. A good all-around grade of **steel** is a 0.60 percent carbon open-hearth. Dies of this steel will forge mild steel, copper, and tool steel satisfactorily if the number of forgings required is not too large. For a large number of tool-steel forgings, tool-steel dies of 0.80 to 0.90 percent carbon may be used and for extreme conditions, 3½ percent nickel steel.

Die blocks of alloy steels have special value for the production of drop forgings in large quantities. Widely used die materials and their recommended hardness are listed in Table 13.2.4. For a typical hot-working steel, the relationship between the hardness and the UTS is

HRC	UTS, ksi (MPa)
30	140 (960)
40	185 (1,250)
50	250 (1,700)
60	350 (2,400)

The pressure on dies can be kept as high as 80 percent of the above values. For higher tool pressures, carbide (tungsten carbide is the most common) die material is used. Carbides can withstand very high compressive pressures but have poor tensile properties. Consequently, carbide dies and inserts are always kept under compression, often by the use of shrink rings. These are some design guidelines for punches and dies:

1. *Long punch.* The punch pressure should be kept below the buckling stress σ_b

$$\sigma_b = \sigma_y \left[1 - \left(\frac{4\sigma_y}{\pi^2 E} \right) \left(\frac{L_p}{D_p} \right)^2 \right]$$

where σ_y is the yield stress of the punch material, E the elastic modulus 30,000 ksi (210 GPa) for steel and 50,000 ksi (350 GPa) for tungsten carbide, L_p is the punch length and D_p is the punch diameter.

2. *Short punch.* The failure mode is plastic upset. Therefore, the punch pressure should be kept less than the yield stress of the punch

Table 13.2.4 Typical Die Materials for Deformation Processes*

Process	Die material† and HRC for working			
	Al, Mg, and Cu alloys		Steels and Ni alloys	
Hot forging	6G	30–40	6G	35–45
	H12	48–50	H12	40–56
Hot extrusion	H12	46–50	H12	43–47
Cold extrusion:				
Die	W1, A2	56–58	A2, D2	58–60
	D2	58–60	WC	
Punch	A2, D2	58–60	A2, M2	64–65
Shape drawing	O1	60–62	M2	62–65
	WC		WC	
Cold rolling	O1	55–65	O1, M2	56–65
Blanking	Zn alloy		As for Al, and	
	W1	62–66	M2	60–66
	O1	57–62	WC	
	A2	57–62		
	D2	58–64		
Deep drawing	W1	60–62	As for Al, and	
	O1	57–62	M2	60–65
	A2	57–62	WC	
	D2	58–64		
Press forming	Epoxy/metal powder		As for Al	
	Zn alloy			
	Mild steel			
	Cast iron			
	O1, A2, D2			

* Compiled from "Metals Handbook," 9th ed., vol. 3, American Society for Metals, Metals Park, Ohio, 1980.
† Die materials mentioned first are for lighter duties, shorter runs. Tool steel compositions, percent (representative members of classes):
  6 G (prehardened die steel): 0.5 C, 0.8 Mn, 0.25 Si, 1 Cr, 0.45 Mo, 0.1 V
  H12 (hot-working die steel): 0.35 C, 5 Cr, 1.5 Mo, 1.5 W, 0.4 V
  W1 (water-hardening steel): 0.6–1.4 C
  O1 (oil-hardening steel): 0.9 C, 1 Mn, 0.5 Cr
  A2 (air-hardening steel): 1 C, 5 Cr, 1 Mo
  D2 (cold-working steel): 1.5 C, 12 Cr, 1 Mo
  M2 (Mo high-speed steel): 0.85 C, 4 Cr, 5 Mo, 6.25 W, 2 V
  WC (tungsten carbide)
  SOURCE: John A. Schey, *Introduction to Manufacturing Processes*, McGraw-Hill, New York, 1987.

material [180 ksi (1,200 MPa) for steel and 500 ksi (3,000 MPa) for tungsten carbide].

3. *Flat platen.* Flat platens fail by plastic yielding. A common formula for calculating acceptable maximum platen pressure p is

$$p = \sigma_y \text{ (diameter of platen/diameter of workpiece)}$$

4. *Die cavity.* The die pressures in a deep cavity (impression) are much higher than in a flat platen (die), resulting in die burst-out. A more conservative value of 150 ksi (1,000 MPa) is used for acceptable die pressure. If shrink rings are used, higher values of 250 ksi (1,700 MPa) for steels and 400 ksi (2,700 MPa) for tungsten carbide can be used for the inner insert.

For large massive dies or for intermittent service, chrome-nickel-molybdenum alloys are preferred. In closed die work, where the dies must dissipate considerable heat, the tungsten steels are preferred, with resulting increase in wear resistance but decrease in toughness.

For very large pieces with deep impressions, **cast-steel** dies are sometimes used. For large dies likely to spring in hardening, 0.85 **carbon steel** high in manganese is sometimes used unhardened.

Good **die-block proportions** for width and depth are as follows:

Width, in (cm)	8 (20)	10 (24)	12 (30)	14 (36)
Depth, in (cm)	6 (15)	7 (18)	7 (18)	7 or 8 (18 or 20)

For ordinary work, 1½ in (4 cm) of metal between impression and edge of block is sufficient.

Dimensions of **dovetailed die shanks:** for hammers up to 1,200 lb (550 kg) 4 in (10 cm) wide and 1⅛ in (3 cm) deep, with sides dovetailed at angles of 6° with the vertical; for hammers from 1,200 to 3,000 lb (550 to 1,360 kg) in size, 6 in (15 cm) wide and 1½ in (4 cm) deep, with 6° angles.

The **minimum draft for the impressions** is 7°, although for parts difficult to draw this may be increased up to 15°. It is not uncommon to have several drafts in the same impression.

Open-hearth and tool-steel **dies** are **hardened** by heating in a carbonizing box packed in charcoal and dipping face downward over a jet of brine. The jet is allowed to strike into the impression, thus freeing the face of steam and producing uniform hardness. After hardening, they are drawn in an oil bath to a temperature of 500 to 550°F (260 to 290°C).

The forging production per pair of dies is largely affected by the size and shape of the impression, the material forged, the material in the dies, the quality of heating of stock to be forged, and the care exercised in use. It may vary from a few hundred pieces to 50,000 or more. A normal **life** for a pair of dies under average conditions may be 20,000 pieces.

Coining, Stamping, and Embossing The metal is well confined in closed dies in which it is forced to flow to fill the shape. The U.S. Mint gives the following pressures; silver quarter, 100 tons/in² (1,380 MN/m²); nickel (0.25 Ni, 0.75 Cu), 90 tons/in² (1,240 MN/m²); copper cent, 40 tons/in² (550 MN/m²). In stamping designs, lettering, etc., in sheet metal the thickness is so little compared with the area that there is practically no relief for excess pressure. Where sharp designs are required, as in stamping panels, the dies should be arranged to strike on a narrow line [say ¹⁄₂₂ in (0.8 mm)] around the outline. If a sharp design is not obtained, it is often best to correct deflection in the machine by shimming or more substantial backing. Increasing the pressure only aggravates the condition and may break the press. General practice for light overall stamping is to allow 5 to 10 tons/in² (68 to 132 MN/m²) of area that is to be stamped, except in areas where the yield point must be exceeded.

Extrusion is the severest of the squeezing processes. The metal is forced to flow rapidly through an orifice, being otherwise confined and subject largely to the laws of hydraulics, with allowances for restraint of flow and for work hardening. Power-press **impact extrusion** began with tin and lead collapsible tubes. It has been extended to the backward and forward extrusion of aluminum, brass, and copper in pressure ranges of

30 to 60 or more tons/in² (413 to 825 MN/m²), and mild steel at pressures up to 165 tons/in² (2,275 MN/m²). Hot impact extrusion of steel, as in projectile piercing, ranges from about 25 to 50 tons/in² (345 to 690 MN/m²). **Forward extrusion** of long tubes, rods, and shapes usually performed hot in hydraulic presses has been extended from the softer metals to the extrusion of steels. Most work is done horizontally because of the lengths of the extrusions. Some vertical mechanical-press equipment is used in hot extrusion of steel tubing.

EQUIPMENT FOR WORKING METALS

The **mass production** industries use an extremely wide variety of machines to force materials to flow plastically into desired shapes (as compared with the more gradual methods of obtaining shapes by cutting away surplus material in machine tools). The application of working pressure may use hydraulic, pneumatic, mechanical, or electric means to apply pressing, hammering, or rolling forces. Mechanically and hydraulically actuated devices cover much the same range. In general, the mechanical equipment is faster, easier to maintain, and more efficient to operate by reason of energy-storing flywheels. The hydraulic equipment is more flexible and more easily adjusted to limited lots in pressure, positions, and strokes. Mechanical handling or feeding devices incorporated in or serving many of these more or less specialized machines further extend their productivity.

Power presses consist of a frame or substantial construction with devices for holding the dies or tools and a moving member or slide for actuating one portion of the dies. This slide usually receives its movement from a crankshaft furnished with a clutch for intermittent operation and a flywheel to supply the sudden power requirement. **Hydraulic presses** have no crankshaft, clutch, or flywheel but employ rams actuated by pumps.

The crankshaft is ordinarily the limiting factor in the pressure capacity of the machine and accordingly is often taken as the basis for tonnage ratings. There is no uniform basis for this rating, owing to variations in shaft proportions and materials and in the different relative severity of various press operations. The following valuation is tentative and is based on the shaft diameter in the main bearings. The bending strength is figured at a section through the center of the crankpin and the combined bending and torsional strength at the inside ends of the main bearings, taking the bending fulcrum at a distance out from these points equal to one-third the length of the main bearings. In the case of double-crank presses and twin-drive arrangements, the relative proportion of the torsional load must be varied to suit, but, except in the cases of long strokes, it is usually small. The working strength is based upon a stress in the extreme fibers of 28,000 lb/in² (193 MN/m²). The limit bearing capacities are taken approximately at 5,000 lb/in² (35 MN/m²) on the crankpins and 2,500 lb/in² (18 MN/m²) average over the main bearings for ordinary steel on cast-iron press bearings with proper grooving. On the knuckle-joint-type presses with hardened tool-steel bearing surfaces and flood lubrication, the bearings will take up to about 30,000 lb/in² (207 MN/m²). On eccentric-type shafts where the main bearings support right up to the oversize pin on each side, the limiting factor is the bearing load. The shaft is practically in shear, so that it has a considerable overload capacity (about $7d^2$ tons). In Table 13.2.5, uniform-diameter single crankshafts are those in which for manufacturing reasons the diameter is the same at the crankpin and at the main bearings. Other crank shafts have an oversize crankpin to balance the bending load at the center with the combined bending and torsional load at the side. The strength of the shaft is figured at midstroke, and the stroke and tonnage capacity are given in terms of the diameter d at the main bearings. Where the working load comes on only near the bottom stroke, the shaft press capacity may be figured as if the stroke were shorter in proportion.

Table 13.2.5 gives the rated capacities of a series of power presses as a function of the shaft diameter.

The speed of operation of the press depends upon the energy requirement and the crankpin velocity. The latter determines the velocity of impact on the tools. In blanking, the blow varies directly with the con-

Table 13.2.5   Power-Press Shaft Capacities

Type of press crankshaft	Max stroke, in*	Capacity tons†
Single crank, single drive, uniform diameter	d‡	$2.8d^2$
Single crank, single drive, oversize crankpin	d	$3.5d^2$
Single crank, single drive, oversize crankpin	$2d$	$2.2d^2$
Single crank, single drive, oversize crankpin	$3d$	$1.6d^2$
Single crank, twin drive, oversize crankpin	$2d$	$3.5d^2$
Single crank, twin drive, oversize crankpin	$3d$	$2.7d^2$
Double crank, single drive, oversize crankpin	$0.75d$	$5.5d^2$
Double crank, single drive, oversize crankpin	d	$4.4d^2$
Double crank, single drive, oversize crankpin	$2d$	$2.5d^2$
Double crank, single drive, oversize crankpin	$3d$	$1.7d^2$
Double crank, twin drive, oversize crankpin	$1.5d$	$5.5d^2$
Double crank, twin drive, oversize crankpin	$3d$	$3.2d^2$
Single eccentric, single or twin drive	$0.5d$	$4.3d^2$

* 1 in = 2.54 cm.
† 1 ton = 8.9 kN.
‡ d = shaft diameter in main bearing.
Source: F. W. Bliss Co.

tact speed and the thickness and hardness of the material. In drawing operations the variation depends upon contact speed, ductility of material, lubrication, etc.

The energy required per stroke is practically the product of the average load and the working distance, plus friction allowance, assumed at about 16 percent. On short-stroke operations, such as blanking, the working energy is supplied almost entirely by slowing down the flywheel; motor or belt pull serves merely to return the flywheel to speed during the large part of the cycle in which no work is done. In drawing operations, the working period is considerable, and in many cases the belt takes the largest part of the working load. In this case, add to the available flywheel energy, the work done by the belt. This amounts, for example, to 70 lb/in of width of the belt (123 N/cm), multiplied by the ratio of the belt velocity to the crankpin velocity, multiplied by the length of the working stroke on the crank circle in feet. The maximum flywheel slowdown has been assumed as up to 10 percent for continuous operation and up to 20 percent for intermittent operation. The following formula is based upon average press-flywheel proportions and a slowdown of 10 percent. The result may be doubled for 20 percent slowdown.

The flywheel capacity per stroke at 10 percent slowdown in inch-tons equals $WD^2N^2/5,260,000,000$, where W is the weight of the flywheel, lb, D is the diameter, in, and N is rotation speed, r/min. (See Sec. 8.2.)

The difference between nongeared and geared presses is only in speed of operation and the relatively greater flywheel capacity.

Press frames are designed for stiffness and usually have considerable excess strength. Good practice is to figure cast-iron sections for a stress of about 2,000 to 3,000 lb/in² (13.8 to 20.6 MN/m²). **C-frame presses** are subjected to an appreciable arc spring amount ordinarily of between 0.0005 and 0.002 in/ton (1.5 and 6 mm/MN), because the center of gravity of the frame section is a considerable distance back of the working centerline of the press. **Straight-sided presses** eliminate that portion of the spring or deflection which is on an arc. **Built-up frame presses** are held together with steel tie rods shrunk in under an initial tension in excess of the working load so that they minimize stretch in that portion of the press.

Power presses are built in a very wide variety of styles and sizes with shafts ranging from 1- to 21-in (2.5- to 53-cm) diam. Over a large part of this range they are built with C frames for convenience, straight-sided frames for heavier and thinner work, eccentric shafts for heavy forgings and stampings, double crankshafts for wide jobs, four-point presses for large panel work, underdrive presses in high-production plants where repairs to presses would interfere with flow of production, and knuckle-join presses for intensely high pressures at the very bottom of the stroke. All these are classified as single-action presses and are used for most of the operations previously discussed.

Double-action presses combine the functions of blank holding with drawing. In the smaller sizes, such presses have cams mounted on the cheeks of the crankshaft to actuate the outer or blank-holding slide. In larger machines, toggle mechanisms are provided to actuate the outer slide, with the advantage that the blank-holding load is taken on the frame instead of the crankshaft. Both of these types afford a considerable power saving over single-action presses equipped with drawing attachments, because the latter must add the blank-holding pressure to the working load for the full depth of the draw.

Types of presses include **foot presses,** in which the pendulum type has the lowest mechanical advantage and the longest stroke; the **lever type,** which has higher mechanical advantage and shorter strokes; **toggle** or **knuckle type,** which has the highest mechanical advantage and works through the shortest stroke with considerable advantage obtainable from the use of tie rods on fine stamping or embossing work; long-stroke rack and pinion-driven presses; triple-action drawing presses; cam-actuated presses; etc.

Screw presses consist of a conventional frame and a slide which is forced down by a steep pitch screw on the upper end of which is a flywheel or weight bar. Hand-operated machines are used for die testing and for small production stamping, embossing, forming, and other work requiring more power than foot presses. Power-driven screw presses are built with a friction drive for the flywheel and automatic control to limit the stroke. Such presses are built in comparatively large sizes and used to a considerable extent for press forging. They lack the accuracy and speed of power presses built for this work but have a safety factor which power presses have not, in that their action is not positive. In this they closely resemble a drop hammer, although their motion is slower. The energy available for work in these presses is $\frac{1}{2}I_f v^2 + \frac{1}{2}I_s v^2$, in which I_f is the moment of inertia of the flywheel, I_s is the moment of inertia of the spindle, and v is the angular velocity of both.

Self-contained fast-acting **hydraulic presses** are being increasingly used. Equipped with motor-driven variable-displacement oil hydraulic pumps, the speed and pressure of the operating ram or rams are under instant and automatic control; this is particularly advantageous for deep drawing operations. The punch can be brought into initial contact with the work without shock and moved with a uniform controlled velocity through the drawing portion of the cycle. The drawing of stainless steels and alloy aluminums (in which the control of drawing speed is vital), as well as the hot drawing of magnesium, is best done on hydraulic presses.

The hydraulic press is used on the rubber pad, or *Guerin,* process of blanking or forming metals, in which a laminated-rubber pad replaces one half—usually the female half—of a die. In forming aluminum the practice has developed of using inexpensive dies of soft metal, vulcanized fiber, plastic, wood, or plaster; and cast dies in industries which, like the aircraft industry, require short runs on many different sizes of shapes and parts.

The older accumulator type of hydraulic-press construction is still used for hot extrusion and some forging work.

Drop Hammers or Presses Small belt-lift and board-lift drop presses were used for variety of sheet-metal operations on hardware, cutlery, silverware, etc. Board-lift drop hammers with heads weighing up to 5,000 lb (22 kN) were widely used in the production of steel drop forgings. The energy available for work is the product of the weight of the ram and the length of fall (see below). The following table gives the shaft diameters of trimming presses (for removing the flash) ordinarily used with drop hammers for trimming the same range of forgings:

Helve hammers are usually belt-driven and carry the hammer face or swage on the end of a beam. The belt is provided with a tightening device, treadle-controlled, permitting the operator to regulate the number and speed of the blows. They were **used for general** and duplicate **forging,** welding, plating drawing, swaging, collaring, spindle making, etc.

Strap hammers carry the hammer slung from a strap, usually of leather. The control and operation are the same as for the helve type. They were **adapted for general work.**

Board Drop Hammer Let W = work of blow, ft · lb; H = weight of hammer and die, lb; g = acceleration due to gravity (l = 32.3 ft/s²); h = actual hammer stroke, ft; and v = terminal velocity of hammer, ft/s. Then, if the hammer and die fall of their own weight, $W = (H/g)v^2/2 = Hh = 0.015 Hv^2$ (approx).

Although now obsolete and found only in a few locations, the hammers described above largely have been supplanted by steam or air (pneumatic) hammers. These latter utilize fluid pressure inside a piston to raise and propel the hammer with greater striking velocity, allowing for better control and permitting a higher production rate.

Steam hammers may be divided into **three classes.** In the first, the hammer is lifted by steam and drops of its own weight; in the second, steam is admitted above the piston and through its expansion increases the force of the blow; in the third, live steam is admitted above the piston throughout the stroke and the force of the blow is from combined weight of the falling hammer and the pressure of the steam.

In the **first class,** that of the single-acting steam hammer, part of the force of the steam is used to accelerate the lifting motion, continuing after the steam has been cut off as long as the pressure under the piston $> H$. These hammers are **used** only **for very large work.** The **weight of the hammer** ranges from 25 to 125 tons (220 to 1,110 kN). The disadvantage of this type lies in the fact that the height of the clearance under the piston is directly dependent upon the thickness of the piece of work.

In the **second class,** in which steam is admitted above the piston and allowed to expand, there is an economy in steam consumption, a greater acceleration of the hammer head, and a large number of blows per unit of time.

In the **third class,** live steam is admitted above the piston. Here the **weight of the hammer head** varies from 1 to 25 tons (9 to 1,100 kN). The control is such that the weight of the hammer itself is available for light blows, and for heavier blows steam is admitted above the piston.

In comparatively small hammers where the head weighs 150 to 2,000 lb (670 to 8,900 N), up to 350 **blows per min** can be obtained, depending on the length of the stroke and the tightness of the stuffing box. These hammers are provided with an automatic reversing gear. The **number and force of the blows** can be regulated by throttling the steam, changing the center position of the operating valve, and changing the back lash in the operating gear. The frame of these hammers is usually C-shaped.

The **anvil** in small hammers is usually a single casting. In large hammers it is usually divided into upper and lower anvil blocks. In smaller hammers the anvil is connected with the shears and upper part of the machine. In the larger hammers, however, this is not the case, as the concussions tend to injure the hammer mechanism. For good practice the **weight of the anvil** ($= Q$) for hammers used in forging iron is at least eight times the weight of the hammer head; for forging steel, at least twelve times.

Drop-Hammer Ram Weights and Suitable Trimming-Press Shaft Diameters

Ram weight, lb*	600	800	1,000	1,200	1,500	2,000	2,500	3,000	5,000	8,000
Press shaft diam. in†	½	4	4½	5	5½	6	6½	7	8	9

* 1,000 lb = 4,450 N.
† 1 in = 2.54 cm.
SOURCE: F. W. Bliss Co.

The pressure Q_1 exerted by the anvil block on the surface which it supports is assumed to be as follows: for blooming hammers, $Q_1 = (30 \text{ to } 60)hH + Q$; for billet-forging hammers, $Q_1 = (60 \text{ to } 95)h + Q$; for hammers for steel forging, $Q_1 = (95 \text{ to } 125)hH + Q$.

Pneumatic Hammers A self-contained type of pneumatic forge hammer (the Bêché) has an air-operated ram with an air-compressing cylinder integral with the frame. The ram is raised by admitting compressed air beneath the ram piston; at the same time a partial vacuum is caused above it. The ram is forced down by a reversal of this action.

The **terminal velocity** (velocity at ram-workpiece contact) of the steam, pneumatic, or hydraulic assisted hammer can be calculated as follows:

$$v = \left[2hg \left(1 + \frac{A p_m}{H} \right) \right]^{0.5}$$

where A is the area of the piston and p_m the mean pressure in the drive cylinder. The hammer energy is often calculated by dividing the energy required for plastic work by the mechanical efficiency of the hammer.

Hydraulic presses are energized by pressurized liquid, usually oil. They can deliver high tonnage but are slow. Due to large die chilling (heat loss to dies) present in hydraulic press forging, they are not usually used for hot forging, except in isothermal forging, where slow speed and large die-workpiece contact times are not a major limitation. They are specially suited to sheet metal forming operations, where slower ram speeds produce lower impact loads, speeds can be varied during the stroke, and multiple actions can be obtained for blank holder and die cushion operation.

Hammers and presses are often selected based on their characteristics such as energy, ram mass (t_{up}), force or tonnage, ram speed (stroking rate), stroke length, bed area, and mechanical efficiency. The characteristics for various hammers and presses are summarized in Table 13.2.6.

Rotary motion is used **for working sheet metal** in a variety of machines, including bending rolls (three rolls); rolling straighteners with five, seven, or more rolls; roll forming machines, in which a series of rolls in successive pairs are used to bend the strip material step by step to some desired shape; a series of two-spindle and multiple-spindle machines used for rolling beads, threads, knurls, flanges, and trimming or curling the edges of drawn shells of cylinders; seaming machines for double seaming, crimping, curling, and other operations in the production of tin cans, pieced tinware, etc.; and spinning machines for spinning, burnishing, trimming, curling, shape forming, and thickness reduction. Various production spinning operations and tool arrangements are shown in Fig. 13.2.7.

Plate-Straightening Machines The **horsepower required** for plate-straightening machines operating on steel plate is shown in Table 13.2.7.

Power required for angle-iron-straightening machines: for 4-in angles, 12 hp; for 6-in, 18 hp; for 8-in, 25 hp.

Power or hydraulic presses are used to straighten **large rolled sections**. The presses make 20 to 30 strokes per min, and the amount of flexure is regulated by inserting wedges or pieces of flat iron. The beams are supported on rolls so they can be easily handled. The **power required for presses** of this kind is as follows:

Depth of girder, in*	4	6	8	10	12	16	20	24
Horsepower (approx)†	3	4	7	11	13	19	23	35

* 1 in = 2.54 cm.
† 1 hp = 0.746 kW.

Horizontal plate-bending machines consist of two stationary rolls and a third vertical adjustable upper roll which can be fitted obliquely for taper bending and is held in bearings with spherical seats. The diameter of the rolls can be determined approximately from the equation $r^2 = bt$, in which r is the radius of the roll, b the width of the plate or sheet, and t its thickness, all in inches.

The **power requirements** of horizontal plate and sheet bending machines are shown in Table 13.2.8.

Vertical plate-bending machines have a hydraulically operated piston which moves an upper and a lower pair of rolls between inclined surfaces of the stationary upright and the crosshead. The bending is done piece by piece against a second stationary upright. Heavy ship plates are rigidly clamped down and bent by a roll operated by two hydraulic pistons. For angular bends or for the production of warped surfaces, the pistons can be operated independently or together. In vertical machines, angles and other rolled shapes are bent between suitably shaped rolls. Pipes are filled with sand to prevent flattening when being bent. For

Table 13.2.6   Characteristics of Hammers and Presses*

Equipment type	Energy,† kN · m	Ram mass, kg	Force,‡ kN	Speed, m/s	Strokes per min	Stroke, m	Bed area, m × m	Mechanical efficiency
Hammers								
Mechanical	0.5–40	30–5,000		4–5	350–35	0.1–1.6	0.1 × 0.1 to 0.4 × 0.6	0.2–0.5
Steam and air	20–600	75–17,000 (25,000)		3–8	300–20	0.5–1.2	0.3 × 0.4 to (1.2 × 1.8)	0.05–0.3
Counterblow	5–200 (1,250)			3–5	60–7		0.3 × 0.4 to (1.8 × 5)	0.2–0.7
Herf	15–750			8–20	<2			0.2–0.6
Presses								
Hydraulic, forging			100–80,000 (800,000)	<0.5	30–5	0.3–1 (3)	0.5 × 0.5 to (3.5 × 8)	0.1–0.6
Hydraulic, sheet metalworking			10–40,000	<0.5	130–20	0.1–1	0.2 × 0.2 to 2 × 6	0.5–0.7
Hydraulic, extrusion			1,000–50,000 (200,000)	<0.5	<2	0.8–5	0.06–0.6 diam. container	0.5–0.7
Mechanical, forging			10–80,000	<0.5	130–10	0.1–1	0.2 × 0.2 to 2 × 3	0.2–0.7
Horizontal upsetter			500–30,000 (1–9 in diam)	<1	90–15	0.05–0.4	0.2 × 0.2 to 0.8 × 1	0.2–0.7
Mechanical, sheet metalworking			10–20,000	<1	180–10	0.1–0.8	0.2 × 0.2 to 2 × 6	0.3–0.7
Screw			100–80,000	<1	35–6	0.2–0.8	0.2 × 0.3 to 0.8 × 1	0.2–0.7

* From a number of sources, chiefly A. Geleji, *Forge Equipment, Rolling Mills and Accessories*, Akademiai Kiado, Budapest, 1967.
† Multiply number in column by 100 to get m · kg, by 0.73 to get 10^3 lbf · ft.
‡ Divide number by −10 to get tons. Numbers in parentheses indicate the largest sizes, available in only a few places in the world.
SOURCE: John A. Schey, *Introduction to Manufacturing Processes*, McGraw-Hill, New York, 1987.

Table 13.2.7   Power Requirement for Plate-Straightening Machines (Steel)

Thickness of plate, in*	0.25	0.4	0.6	0.8	1.0	1.2	1.4	1.6
Width of plate, in*	48.0	52.0	60.0	72.0	88.0	102.0	120.0	140.0
Diameter of rolls, in*	5.0	8.0	10.0	12.0	13.0	14.0	15.0	16.0
Horsepower (approx)†	6.0	8.0	12.0	20.0	30.0	55.0	90.0	130.0

* 1 in = 2.54 cm.
† 1 hp = 0.746 kW.

Table 13.2.8   Power Requirement for Plate and Sheet Bending Machines (Steel)

Thickness of plate, in*	0.5	0.6	0.8	1	1.2
Horsepower for plate 120 in wide†	10.0	12.0	18.0	27.0	40.0
Horsepower for plate 240 in wide†	30.0	30.0	40.0	55.0	75.0

* 1 in = 2.54 cm.
† 1 hp = 0.746 kW.

some work, pipes are bent hot between suitable forms operated by hydraulic pressure.

The **rotary swaging machines** for tapering, closing in, and reducing tubes, rods, and hollow articles is essentially a cage carrying a number of rollers and revolving at high speed; e.g., 14 rolls in a cage revolving at 600 r/min will strike 8,400 blows per min on the work.

A rapid succession of light blows is applied to a considerable variety of commercial **riveting** operations such as pneumatic riveting. Another method of riveting, described as spinning, involves rotating small rollers rapidly over the top of the rivet and at the same time applying pressure. Neither of these methods involves pressures as intense as those used in riveting by direct pressure, either hot or cold. Power presses and C-frame riveters, employing hydraulic pressure or air pressure of 80 to 100 lb/in² (550 to 690 kN/m²), are designed to apply 150,000 lb/in² (1,035 MN/m²) on the cross section of the body of the rivet for hot-working and 300,000 lb/in² (2,070 MN/m²) for cold-working. The pieces joined should be pressed together by a pressure 0.3 to 0.4 times that used in riveting.

13.3   WELDING AND CUTTING
by Omer W. Blodgett and Duane K. Miller

REFERENCES: From the American Welding Society (AWS): "Welding Handbook" (six volumes); "Structural Welding Code"; "Filler Metal Specifications," "Welding Terms and Definitions"; "Brazing Manual"; "Thermal Spraying"; "Welding of Chromium-Molybdenum Steels." From the Lincoln Arc Welding Foundation: Blodgett, "Design of Weldments"; Blodgett, "Design of Welded Structures." "Procedure Handbook of Welding," The Lincoln Electric Co. "Safety in Welding and Cutting," ANSI. "Welding and Fabrication Data Book," Penton Pub. Co. AISI, "Steel Construction Manual — ASD," 9th ed., and "Steel Construction Manual—LRFD," 2d ed.

INTRODUCTION

Welded connections and assemblies represent a very large group of fabricated steel components, and only a portion of the aspects of their design and fabrication is treated here. The welding process itself is complex, involving heat and liquid-metal transfer, chemical reactions, and the gradual formation of the welded joint through liquid-metal deposition and subsequent cooling into the solid state, with attendant metallurgical transformations. Some of these items are treated in greater detail in the references and other extensive professional literature, as well as in Secs. 6.2, 6.3, and 13.1.

The material in this section will provide the engineer with an overview of the most important aspects of welded design. In order that the resulting welded fabrication be of adequate strength, stiffness, and utility, the designer will often collaborate with engineers who are experts in the broad area of design and fabrication of weldments.

ARC WELDING

Arc welding is one of several **fusion processes** for joining metal. By the generation of intense heat, the juncture of two metal pieces is melted and mixed—directly or, more often, with an intermediate molten **filler metal**. Upon cooling and solidification, the resulting welded joint metallurgically bonds the former separate pieces into a continuous structural assembly (a **weldment**) whose strength properties are basically those of the individual pieces before welding.

In arc welding, the intense heat needed to melt metal is produced by an **electric arc**. The arc forms between the workpieces and an electrode that is either manually or mechanically moved along the joint; conversely, the work may be moved under a stationary electrode. The **electrode** generally is a specially prepared rod or wire that not only conducts electric current and sustains the arc, but also melts and supplies **filler metal** to the joint; this constitutes a **consumable electrode. Carbon** or **tungsten electrodes** may be used, in which case the electrode serves only to conduct electric current and to sustain the arc between tip and workpiece, and it is not consumed; with these electrodes, any filler metal required is supplied by rod or wire introduced into the region of the arc and melted there. Filler metal applied separately, rather than via a consumable electrode, does not carry electric current.

Most steel welding operations are performed with consumable electrodes.

Welding Process Fundamentals

Heat and Filler Metal   An ac or dc power source fitted with necessary controls is connected by a work cable to the workpiece and by a "hot" cable to an electrode holder of some type, which, in turn, is electrically connected to the welding electrode (Fig. 13.3.1). When the circuit is energized, the flow of electric current through the electrode heats the electrode by virtue of its electric resistance. When the electrode tip is touched to the workpiece and then withdrawn to leave a gap between the electrode and workpiece, the arc jumping the short gap presents a further path of high electric resistance, resulting in the generation of an extremely high temperature in the region of the sustained arc. The temperature reaches about 6,500°F, which is more than adequate to melt most metals. The heat of the arc melts both base and filler metals, the latter being supplied via a consumable electrode or separately. The puddle of molten metal produced is called a **weld pool,** which solidifies as the electrode and arc move along the joint being welded. The resulting weldment is metallurgically bonded as the liquid metal cools, fuses, solidifies, and cools. In addition to serving its main function of supplying heat, the arc is subject to adjustment and/or control to vary the proper transfer of molten metal to the weld pool, remove surface films in the weld region, and foster gas-slag reactions or other beneficial metallurgical changes.

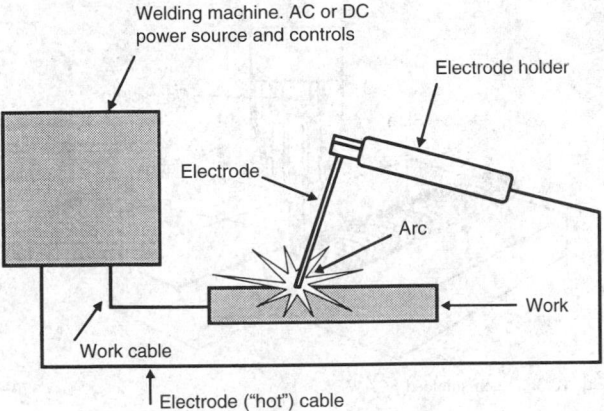

Fig. 13.3.1   Typical welding circuit.

Filler metal composition is generally different from that of the weld metal, which is composed of the solidified mix of both filler and base metals.

Shielding and Fluxing   High-temperature molten metal in the weld pool will react with oxygen and nitrogen in ambient air. These gases will remain dissolved in the liquid metal, but their solubility significantly decreases as the metal cools and solidifies. The decreased solubility causes the gases to come out of solution, and if they are trapped in the metal as it solidifies, **cavities**, termed **porosities**, are left behind. This is always undesirable, but it can be acceptable to a limited degree depending on the specification governing the welding.

Smaller amounts of these gases, particularly nitrogen, may remain dissolved in the weld metal, resulting in drastic reduction in the physical properties of otherwise excellent weld metal. **Notch toughness** is seriously degraded by nitrogen inclusions. Accordingly, the molten metal must be shielded from harmful atmospheric gas contaminants. This is accomplished by gas shielding or slag shielding or both.

Gas shielding is provided either by an external supply of gas, such as carbon dioxide, or by gas generated when the electrode flux heats up. **Slag shielding** results when the flux ingredients are melted and leave behind a slag to cover the weld pool, to act as a barrier to contact between the weld pool and ambient air. At times, both types of shielding are utilized.

In addition to its primary purpose to protect the molten metal, the shielding gas will significantly affect arc behavior. The shielding gas may be mixed with small amounts of other gases (as many as three others) to improve arc stability, puddle (weld pool) fluidity, and other welding operating characteristics.

In the case of shielded-metal arc welding (SMAW), the "stick" electrode is covered with an extruded coating of flux. The arc heat melts the flux and generates a gaseous shield to keep air away from the molten metal, and at the same time the flux ingredients react with deleterious substances, such as surface oxides on the base metal, and chemically combine with those contaminants, creating a **slag** which floats to the surface of the weld pool. That slag crusts over the newly solidified hot metal, minimizes contact between air and hot metal while the metal cools, and thereby inhibits the formation of surface oxides on the newly deposited weld metal, or weld bead. When the temperature of the weld bead decreases, the slag, which has a glassy consistency, is chipped off to reveal the bright surface of the newly deposited metal. Minimal surface oxidation will take place at lower temperatures, inasmuch as oxidation rates are greatly diminished as ambient conditions are approached.

Fluxing action also aids in wetting the interface between the base metal and the molten metal in the weld pool edge, thereby enhancing uniformity and appearance of the weld bead.

Process Selection Criteria

Economic factors generally dictate which welding process to use for a particular application. It is impossible to state which process will always deliver the most economical welds, because the variables involved are significant in both number and diversity. The variables include, but are not limited to, steel (or other base metal) type, joint type, section thickness, production quantity, joint access, position in which the welding is to be performed, equipment availability, availability of qualified and skilled welders, and whether the welding will be done in the field or in the shop.

Shielded Metal Arc Welding

The **SMAW** process (Fig. 13.3.2), commonly known as **stick welding,** or **manual welding,** is the most popular and widespread welding process. It is versatile, relatively simple to do, and very flexible in being applied. To those casually acquainted with welding, **arc welding** usually means **shielded-metal arc welding.** SMAW is used in the shop and in the field for fabrication, erection, maintenance, and repairs. Because of the relative inefficiency of the process, it is seldom used for fabrication of major structures. SMAW has earned a reputation for providing high-quality welds in a dependable fashion. It is, however, inherently slower and more costly than other methods of welding.

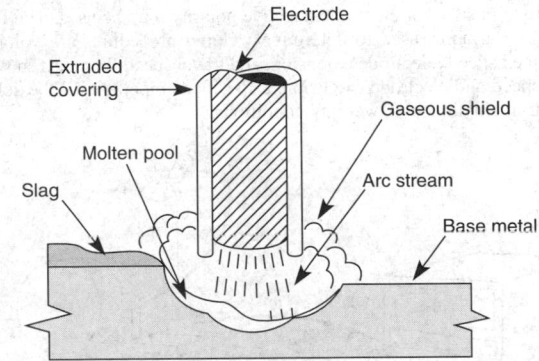

Fig. 13.3.2   SMAW process.

SMAW may utilize either direct current (dc) or alternating current (ac). Generally speaking, direct current is used for smaller electrodes, usually less than 3/16-in diameter. Larger electrodes utilize alternating current to eliminate undesirable arc blow conditions.

Electrodes used with alternating current must be designed specifically to operate in this mode, in which current changes direction 120 times per second with 60-Hz power. All ac electrodes will operate acceptably on direct current. The opposite is not always true.

Flux Cored Arc Welding (FCAW)

In **FCAW,** the arc is maintained between a continuous tubular metal electrode and the weld pool. The tubular electrode is filled with flux and a combination of materials that may include metallic powder(s). FCAW may be done automatically or semiautomatically. FCAW has become the workhorse in fabrication shops practicing semiautomatic welding. Production welds that are short, change direction, are difficult to access, must be done out of position (e.g., vertical or overhead), or are part of a short production run generally will be made with semiautomatic FCAW.

When the application lends itself to automatic welding, most fabricators will select the submerged arc process (see material under

"SAW"). Flux cored arc welding may be used in the automatic mode, but the intensity of arc rays from a high-current flux cored arc, as well as a significant volume of smoke, makes alternatives such as submerged arc more desirable.

Advantages of FCAW FCAW offers two distinct advantages over SMAW. First, the electrode is continuous and eliminates the built-in starts and stops that are inevitable with SMAW using stick electrodes. An economic advantage accrues from the increased operating factor; in addition, the reduced number of arc starts and stops largely eliminates potential sources of weld discontinuities. Second, increased amperages can be used with FCAW. With SMAW, there is a practical limit to the amount of current that can be used. The covered electrodes are 9 to 18 in long, and if the current is too high, electric resistance heating within the unused length of electrode will become so great that the coating ingredients may overheat and "break down." With continuous flux cored electrodes, the tubular electrode is passed through a contact tip, where electric current is transferred to the electrode. The short distance from the contact tip to the end of the electrode, known as electrode extension or "**stickout**," inhibits heat buildup due to electric resistance. This electrode extension distance is typically 1 in for flux cored electrodes, although it may be as much as 2 or 3 in in some circumstances.

Smaller-diameter flux cored electrodes are suitable for all-position welding. Larger electrodes, using higher electric currents, usually are restricted to use in the flat and horizontal positions. Although the equipment required for FCAW is more expensive and more complicated than that for SMAW, most fabricators find FCAW much more economical than SMAW.

FCAW Equipment and Procedures Like all wire-fed welding processes, FCAW requires a power source, wire feeder, and gun and cable assembly (Fig. 13.3.3). The power supply is a dc source, although either electrode positive or electrode negative polarity may be used. The four primary variables used to determine welding procedures are voltage, wire feed speed, electrode extension, and travel speed. For a given wire feed speed and electrode extension, a specified amperage will be delivered to maintain stable welding conditions.

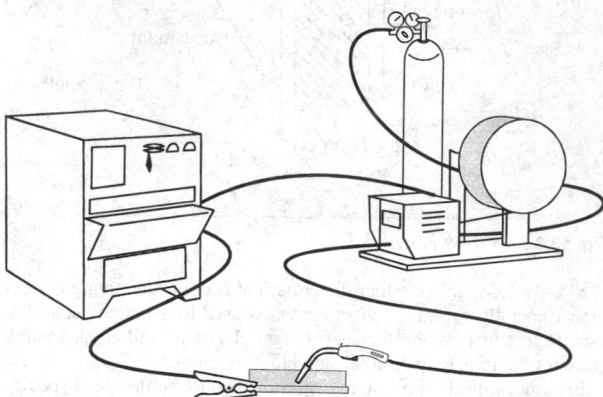

Fig. 13.3.3 FCAW and GMAW equipment.

As wire feed speed is increased, amperage will be increased. On some equipment, the wire feed speed control is called the amperage control, which, despite its name, is just a rheostat that regulates the speed of the dc motor driving the electrode through the gun. The most accurate way, however, to establish welding procedures is to refer to the **wire feed speed (WFS)**, since electrode extension, polarity, and electrode diameter will also affect amperage. For a fixed wire feed speed, a shorter electrical stick-out will result in higher amperages. If procedures are set based on the wire feed speed, the resulting amperage verifies that proper electrode extensions are being used. If amperage is used to set welding procedures, an inaccurate electrode extension may go undetected.

Self-Shielded and Gas-Shielded FCAW Within the category of FCAW, there are two specific subsets: **self-shielded flux core arc welding (FCAW-ss)** (Fig. 13.3.4) and **gas-shielded flux core arc welding (FCAW-g)** (Fig. 13.3.5). Self-shielded flux cored electrodes require no external shielding gas. The entire shielding system results from the flux ingredients contained in the tubular electrode. The gas-shielded variety of flux cored electrode utilizes, in addition to the flux core, an externally supplied shielding gas. Often, CO_2 is used, although other mixtures may be used.

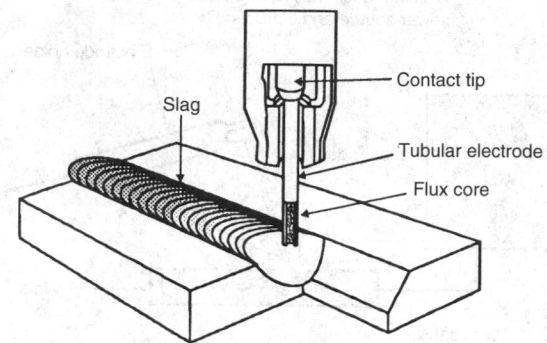

Fig. 13.3.4 Self shielded FCAW.

Both these subsets of FCAW are capable of delivering weld deposits featuring consistency, high quality, and excellent mechanical properties. Self-shielded flux cored electrodes are ideal for field welding operations, for since no externally supplied shielding gas is required, the process may be used in high winds without adversely affecting the

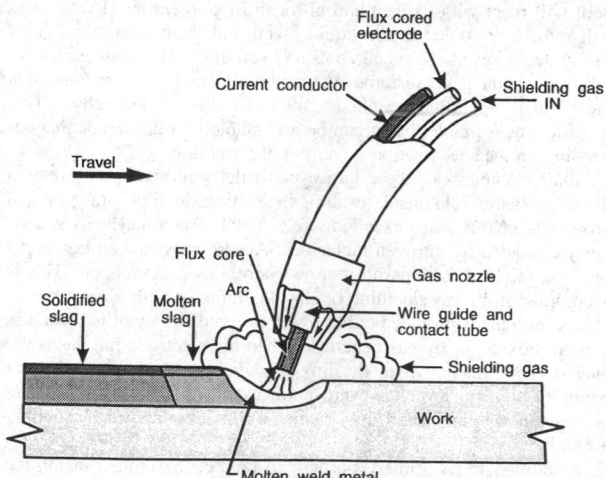

Fig. 13.3.5 Gas shielded FCAW.

quality of the weld metal deposited. With any gas-shielded processes, wind shields must be erected to preclude wind interference with the gas shield. Many fabricators with large shops have found that self-shielded flux core welding offers advantages when the shop door can be left open or fans are used to improve ventilation.

Gas-shielded flux cored electrodes tend to be more versatile than self-shielded flux cored electrodes and, in general, provide better arc action. Operator acceptance is usually higher. The gas shield must be protected from winds and drafts, but this is not difficult for most shop fabrication. Weld appearance is very good, and quality is outstanding. Higher-strength gas-shielded FCAW electrodes are available, but current practice limits self-shielded FCAW deposits to a strength of 80 ksi or less.

Submerged Arc Welding (SAW)

Submerged arc welding differs from other arc welding processes in that a blanket of fusible granular flux is used to shield the arc and molten metal (Fig. 13.3.6). The arc is struck between the workpiece and a bare-wire electrode, the tip of which is submerged in the flux. The arc is completely covered by the flux and it is not visible; thus the weld is made without the flash, spatter, and sparks that characterize the open-arc processes. The flux used develops very little smoke or visible fumes.

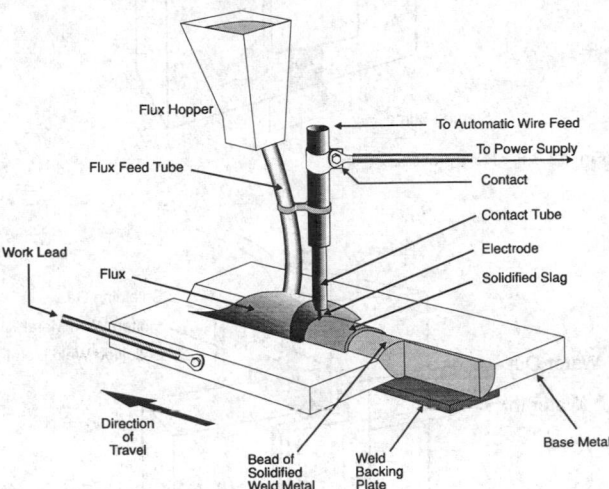

Fig. 13.3.6 SAW process.

Typically, the process is operated fully automatically, although semiautomatic operation is possible. The electrode is fed mechanically to the welding gun, head, or heads. In semiautomatic welding, the welder moves the gun, usually equipped with a flux-feeding device, along the joint.

Flux may be fed by gravity flow from a small hopper atop the torch and then through a nozzle concentric with the electrode, or through a nozzle tube connected to an air-pressurized flux tank. Flux may also be applied in advance of the welding operation or ahead of the arc from a hopper run along the joint. Many fully automatic installations are equipped with a vacuum system to capture unfused flux left after welding; the captured, unused flux is recycled for use once more.

During welding, arc heat melts some of the flux along with the tip of the electrode. The electrode tip and the welding zone are always shielded by molten flux and a cover layer of unfused flux. The electrode is kept a short distance above the workpiece. As the electrode progresses along the joint, the lighter molten flux rises above the molten metal to form slag. The weld metal, having a higher melting (freezing) point, solidifies while the slag above it is still molten. The slag then freezes over the newly solidified weld metal, continuing to protect the metal from contamination while it is very hot and reactive with atmospheric oxygen and nitrogen. Upon cooling and removal of any unmelted flux, the slag is removed from the weld.

Advantages of SAW High currents can be used in SAW, and extremely high heat input can be developed. Because the current is applied to the electrode a short distance above the arc, relatively high amperages can be used on small-diameter electrodes. The resulting extremely high current densities on relatively small-cross-section electrodes permit high rates of metal deposition.

The insulating flux blanket above the arc prevents rapid escape of heat and concentrates it in the welding zone. Not only are the electrode and base metal melted rapidly, but also fusion is deep into the base metal. Deep penetration allows the use of small welding grooves, thus minimizing the amount of filler metal to be deposited and permitting fast welding speeds. Fast welding, in turn, minimizes the total heat input

to the assembly and thus tends to limit problems of heat distortion. Even relatively thick joints can be welded in one pass with SAW.

Versatility of SAW SAW can be applied in more ways than other arc welding processes. A single electrode may be used, as is done with other wire feed processes, but it is possible to use two or more electrodes in submerged arc welding. Two electrodes may be used in parallel, sometimes called **twin arc welding,** employing a single power source and one wire drive. In **multiple-electrode SAW,** up to five electrodes can be used thus, but most often, two or three arc sources are used with separate power supplies and wire drives. In this case, the lead electrode usually operates on direct current while the trailing electrodes operate on alternating current.

Gas Metal Arc Welding (GMAW)

Gas metal arc welding utilizes the same equipment as FCAW (Figs. 13.3.3 and 13.3.7); indeed, the two are similar. The major differences are: (1) GMAW uses a solid or metal cored electrode, and (2) GMAW leaves no residual slag.

GMAW may be referred to as **metal inert gas (MIG),** solid wire and gas, **miniwire** or **microwire welding.** The shielding gas may be carbon

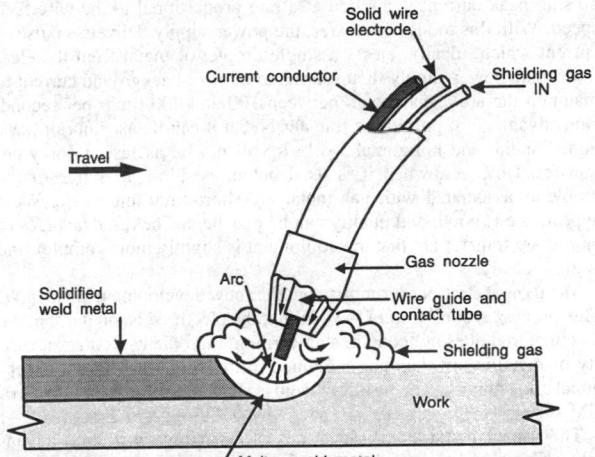

Fig. 13.3.7 GMAW welding process.

dioxide or blends of argon with CO_2 or oxygen, or both. GMAW is usually applied in one of four ways: short arc transfer, globular transfer, spray arc transfer, and pulsed arc transfer.

Short arc transfer is ideal for welding thin-gage materials, but generally is unsuitable for bridge fabrication purposes. In this mode of transfer, a small electrode, usually of 0.035- to 0.045-in diameter, is fed at a moderate wire feed speed at relatively low voltages. The electrode contacts the workpiece, resulting in a short circuit. The arc is actually quenched at this point, and very high current will flow through the electrode, causing it to heat and melt. A small amount of filler metal is transferred to the welding done at this time.

The cycle will repeat itself when the electrode short-circuits to the work again; this occurs between 60 and 200 times per second, creating a characteristic buzz. This mode of transfer is ideal for sheet metal, but results in significant fusion problems if applied to thick sections, when **cold lap** or **cold casting** results from failure of the filler metal to fuse to the base metal. This is unacceptable since the welded connection will have virtually no strength. Caution must be exercised if the short arc transfer mode is applied to thick sections.

Spray arc transfer is characterized by high wire feed speeds at relatively high voltages. A fine spray of molten filler metal drops, all smaller in diameter than the electrode, is ejected from the electrode toward the work. Unlike with short arc transfer, the arc in spray transfer is maintained continuously. High-quality welds with particularly good appearance are obtained. The shielding gas used in spray arc transfer is composed of at least 80 percent argon, with the balance either carbon dioxide or oxygen. Typical mixtures would include 90-10 argon-CO_2, and 95-5 argon-oxygen. Relatively high arc voltages are used with spray arc transfer. Gas metal spray arc transfer welds have excellent appearance and evidence good fusion. However, due to the intensity of the arc, spray arc transfer is restricted to applications in the flat and horizontal positions.

Globular transfer is a mode of gas metal arc welding that results when high concentrations of carbon dioxide are used. Carbon dioxide is not an inert gas; rather, it is active. Therefore, GMAW that uses CO_2 may be referred to as **MAG,** for **metal active gas.** With high concentrations of CO_2 in the shielding gas, the arc no longer behaves in a spraylike fashion, but ejects large globs of metal from the end of the electrode. This mode of transfer, while resulting in deep penetration, generates relatively high levels of spatter, and weld appearance can be poor. Like the spray mode, it is restricted to the flat and horizontal positions. Globular transfer may be preferred over spray arc transfer because of the low cost of CO_2 shielding gas and the lower level of heat experienced by the operator.

Pulsed arc transfer is a relatively new development in GMAW. In this mode, a background current is applied continuously to the electrode. A pulsing peak current is applied at a rate proportional to the wire feed speed. With this mode of transfer, the power supply delivers a pulse of current which, ideally, ejects a single droplet of metal from the electrode. The power supply then returns to a lower background current to maintain the arc. This occurs between 100 and 400 times per second. One advantage of pulsed arc transfer is that it can be used out of position. For flat and horizontal work, it will not be as fast as spray arc transfer. However, when it is used out of position, it is free of the problems associated with gas metal arc short-circuiting mode. Weld appearance is good, and quality can be excellent. The disadvantages of pulsed arc transfer are that the equipment is slightly more complex and is more costly.

Metal cored electrodes comprise another new development in GMAW. This process is similar to FCAW in that the electrode is tubular, but the core material does not contain slag-forming ingredients. Rather, a variety of metallic powders are contained in the core, resulting in exceptional alloy control. The resulting weld is slag-free, as are other forms of GMAW.

The use of metal cored electrodes offers many fabrication advantages. Compared to spray arc transfer, metal cored electrodes require less amperage to obtain the same deposition rates. They are better able to handle mill scale and other surface contaminants. When used out-of-position, they offer greater resistance to the cold lapping phenomenon so common with short arc transfer. Finally, metal cored electrodes permit the use of amperages higher than may be practical with solid electrodes, resulting in higher metal deposition rates.

The weld properties obtained from metal cored electrode deposits can be excellent, and their appearance is very good. Filler metal manufacturers are able to control the composition of the core ingredients, so that mechanical properties obtained from metal cored deposits can be more consistent than those obtained with solid electrodes.

Electroslag/Electrogas Welding (ESW/EGW)

Electroslag and **electrogas welding** (Figs. 13.3.8 and 13.3.9) are closely related processes that allow high deposition welding in the vertical plane. Properly applied, these processes offer tremendous savings over alternative, out-of-position methods and, in many cases, savings over flat-position welding. Although the two processes have similar applications and mechanical setup, there are fundamental differences in the arc characteristics.

Electroslag and electrogas are mechanically similar in that both utilize **copper dams,** or shoes, that are applied to either side of a square-edged butt joint. An electrode or multiple electrodes are fed into the joint. Usually, a starting sump is applied for the beginning of the weld. As the electrode is fed into the joint, a puddle is established that progresses vertically. The water-cooled copper dams chill the weld metal and prevent its escape from the joint. The weld is completed in one pass.

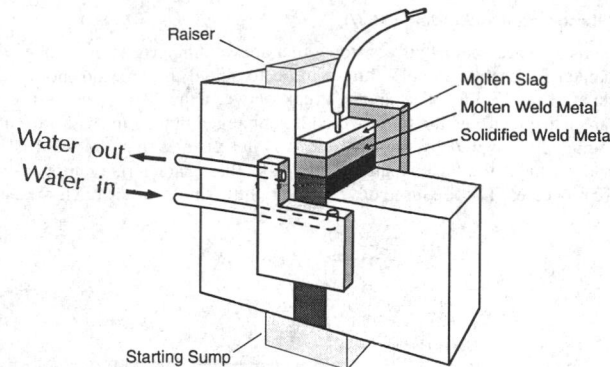

Fig. 13.3.8 ESW process.

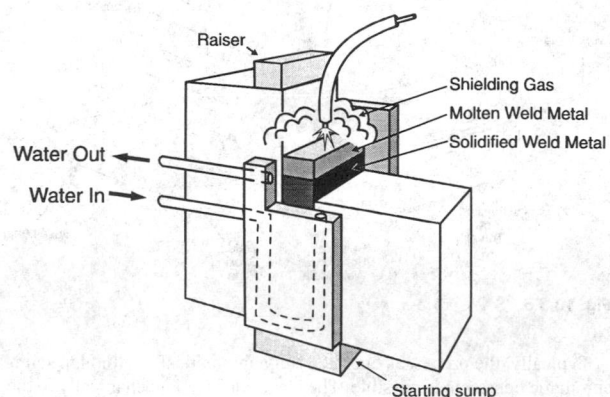

Fig. 13.3.9 EGW process.

Gas Tungsten Arc Welding (GTAW)

The **gas tungsten arc welding** process (Fig. 13.3.10), colloquially called **TIG welding,** uses a nonconsumable tungsten electrode. An arc is established between the tungsten electrode and the workpiece, resulting in heating of the base metal. If required, a filler metal is used. The weld area is shielded with an inert gas, usually argon or helium. GTAW is ideally suited to weld nonferrous materials such as stainless steel and aluminum, and is very effective for joining thin sections.

Highly skilled welders are required for GTAW, but the resulting weld quality can be excellent. The process is often used to weld exotic materials. Critical repair welds as well as root passes in pressure piping are typical applications.

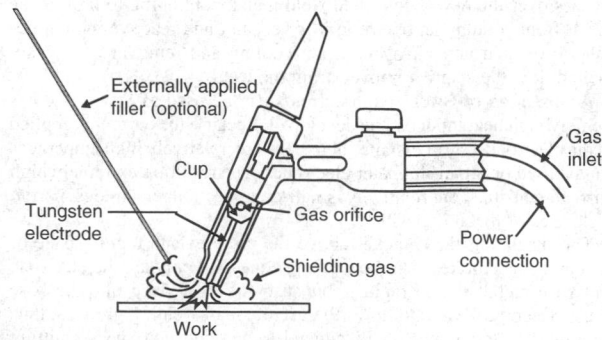

Fig. 13.3.10 Gas tungsten arc welding (GTAW).

Plasma Arc Welding (PAW)

Plasma arc welding is an arc welding process using a constricted arc to generate very high, localized heating. PAW may utilize either a transferred or a nontransferred arc. In the **transferred arc mode**, the arc occurs between the electrode and the workpiece, much as in GTAW, the primary difference being the constriction afforded by the secondary gases and torch design. With the **nontransferred arc** mode, arcing is contained within the torch between a tungsten electrode and a surrounding nozzle.

The constricted arc results in higher localized arc energies than are experienced with GTAW, resulting in faster welding speeds. Applications for PAW are similar to those for GTAW. The only significant disadvantage of PAW is the equipment cost, which is higher than that for GTAW.

Most PAW is done with the transferred arc mode, although this mode utilizes a nontransferred arc for the first step of operation. An arc and plasma are initially established between the electrode and the nozzle. When the torch is properly located, a switching system will redirect the arc toward the workpiece. Since the arc and plasma are already established, transferring the arc to the workpiece is easily accomplished and highly reliable. For this reason, PAW is often preferred for automated applications.

GAS WELDING AND BRAZING

The heat for **gas welding** is supplied by burning a mixture of oxygen and a suitable combustible gas. The gases are mixed in a torch which controls the welding flame.

Acetylene is almost universally used as the combustible gas because of its high flame temperature. This temperature, about 6,000°F (3,315°C), is so far above the melting point of all commercial metals that it provides a means for the rapid localized melting essential in welding. The oxyacetylene flame is also used in cutting ferrous metals.

A **neutral flame** is one in which the fuel gas and oxygen combine completely, leaving no excess of either fuel gas or oxygen. The neutral flame has an inside portion, consisting of a brilliant cone $1/16$ to $3/4$ in (1.6 to 19.1 mm) long, surrounded by a faintly luminous envelope flame. When fuel gas is in excess, the flame consists of three easily recognizable zones: a sharply defined inner cone, an intermediate cone of whitish color, and the bluish outer envelope. The length of the intermediate cone is a measure of the amount of excess fuel gas. This **flame is reducing,** or **carburizing**.

When oxygen is in excess in the mixture, the flame resembles the neutral flame, but the inner cone is shorter, is "necked in" on the sides, is not so sharply defined, and acquires a purplish tinge. A slightly **oxidizing flame** may be used in braze welding and bronze surfacing, and a more strongly oxidizing flame is sometimes used in gas-welding brass, bronze, and copper. A disadvantage of a strongly oxidizing flame is that it can oxidize the surface of the base metal and thereby prevent fusion of the filler metal to the base metal.

In **braze welding**, coalescence is produced by heating above 840°F (450°C) and by using a nonferrous filler metal having a melting point below that of the base metals. Braze welding with brass (bronze) rods is used extensively on cast iron, steel, copper, brass, etc. Since it operates at temperatures lower than base metal melting points, it is used where control of distortion is necessary or lower base metal temperatures during welding are desired. Braze-welded joints on mild steel, made with rods of classifications RCuZn-B and RBCuZn-D, will show transverse tensile values of 60,000 to 70,000 lb/in² (414 to 483 MPa). Joint designs for braze welding are similar to those used for gas and arc welding.

In braze welding it is necessary to remove rust, grease, scale, etc., and to use a suitable flux to dissolve oxides and clean the metal. Sometimes rods are used with a flux coating applied to the outside. Additional flux may or may not be required, notwithstanding the flux coating on the rods. The parts are heated to red heat [1,150 to 1,350°F (621 to 732°C)], and the rod is introduced into the heated zone. The rod melts first and "tins" the surfaces, following which additional filler metal is added.

Welding rods for oxyacetylene braze welding are usually of the copper-zinc (60 Cu-40 Zn) analysis. Additions of tin, manganese, iron, nickel, and silicon are made to improve the mechanical properties and usability of the rods.

Brazing is another one of the general groups of welding processes, consisting of the torch, furnace, induction, dip, and resistance brazing. Brazing may be used to join almost all metals and combinations of dissimilar metals, but some combinations of dissimilar metals are not compatible (e.g., aluminum or magnesium to other metals). In brazing, coalescence is produced by heating above 840°F (450°C) but below the melting point of the metals being joined. The nonferrous filler metal used has a melting point below that of the base metal, and the filler metal is distributed in the closely fitted lap or butt joints by capillary attraction. Clean joints are essential for satisfactory brazing. The use of a flux or controlled atmosphere to ensure surface cleanliness is necessary. Filler metal may be hand-held and fed into the joint (face feeding), or preplaced as rings, washers, shims, slugs, etc.

Brazing with the silver-alloy filler metals previously was known as **silver soldering** and **hard soldering. Braze welding** should not be confused with brazing. Braze welding is a method of welding employing a filler metal which melts below the welding points of the base metals joined, but the filler metal *is not* distributed in the joint by capillary attraction. (See also Sec. 6.)

Torch brazing uses acetylene, propane, or other fuel gas, burned with oxygen or air. The combination employed is governed by the brazing temperature range of the filler metal, which is usually above its liquidus. Flux with a melting point appropriate to the brazing temperature range and the filler metal is essential.

Furnace brazing employs the heat of a gas-fired, electric, or other type of furnace to raise the parts to brazing temperature. Fluxes may be used, although reducing or inert atmospheres are more common since they eliminate postbraze cleaning necessary with fluxes.

Induction brazing utilizes a high-frequency current to generate the necessary heat in the part by induction. Distortion in the brazed joint can be controlled by current frequency and other factors. Fluxes or gaseous atmospheres must be used in induction bearing.

Dip brazing involves the immersion of the parts in a molten bath. The bath may be either molten brazing filler metal or molten salts, which most often are brazing flux. The former is limited to small parts such as electrical connections; the latter is capable of handling assemblies weighing several hundred pounds. The particular merit of dip brazing is that the entire joint is completed all at one time.

Resistance brazing utilizes standard resistance-welding machines to supply the heat. Fluxes or atmospheres must be used, with flux predominating. Standard spot or projection welders may be used. Pressures are lower than those for conventional resistance welding.

RESISTANCE WELDING

In resistance welding, coalescence is produced by the heat obtained from the electric resistance of the workpiece to the flow of electric current in a circuit of which the workpiece is a part, and by the application of pressure. The specific processes include **resistance spot welding, resistance seam welding, and projection welding**. Figure 13.3.11 shows diagrammatic outlines of the processes.

The resistance of the welding circuit should be a maximum at the interface of the parts to be joined, and the heat generated there must reach a value high enough to cause localized fusion under pressure.

Electrodes are of copper alloyed with such metals as molybdenum and tungsten, with high electrical conductivity, good thermal conductivity, and sufficient mechanical strength to withstand the high pressures to which they are subjected. The electrodes are water-cooled. The resistance at the surfaces of contact between the work and the electrodes must be kept low. This may be accomplished by using smooth, clean work surfaces and a high electrode pressure.

In **resistance spot welding** (Fig. 13.3.11), the parts are lapped and held in place under pressure. The size and shape of the electrodes control the size and shape of the welds, which are usually circular.

Designing for spot welding involves six elements: tip size, edge distance, contacting overlap, spot spacing, spot weld shear strength, and electrode clearance. For mild steel, the diameter of the tip face, in terms of sheet thickness t, may be taken as $0.1 + 2t$ for thin material, and as $\sqrt{t}$ for thicker material; all dimensions in inches. Edge distance should be sufficient to provide enough metal around the weld to retain it when in the molten condition. Contacting overlap is generally taken as the diameter of the weld nugget plus twice the minimum edge distance. Spot spacing must be sufficient to ensure that the welding current will not shunt through the previously made weld.

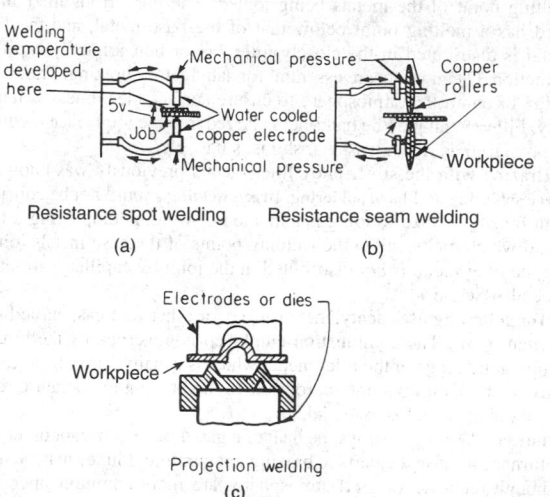

Fig. 13.3.11 (*a*) Resistance spot; (*b*) resistance seam; (*c*) projection welding.

Resistance spot welding machines vary from small, manually operated units to large, elaborately instrumented units designed to produce high-quality welds, as on aircraft parts. Portable gun-type machines are available for use where the assemblies are too large to be transported to a fixed machine. Spot welds may be made singly or in multiples, the latter generally made on special purpose machines. Spacing of electrodes is important to avoid excessive shunting of welding current.

The **resistance seam welding process** (Fig. 13.3.11) produces a series of spot welds made by circular or wheel type electrodes. The weld may be a series of closely spaced individual spot welds, overlapping spot welds, or a continuous weld nugget. The weld shape for individual welds is rectangular, continuous welds are about 80 percent of the width of the roll electrode face.

A **mash weld** is a seam weld in which the finished weld is only slightly thicker than the sheets, and the lap disappears. It is limited to thicknesses of about 16 gage and an overlap of 1½ times the sheet thickness. Operating the machine at reduced speed, with increased pressure and noninterrupted current, a strong quality weld may be secured that will be 10 to 25 percent thicker than the sheets. The process is applicable to mild steel but has limited use on stainless steel; it cannot be used on nonferrous metals. A modification of this technique employs a straight butt joint. This produces a slight depression at the weld, but the strength is satisfactory on some applications, e.g., for the production of some electric-welded pipe and tubing.

Cleanliness of sheets is of even more importance in seam welding than in spot welding. Best results are secured with cold-rolled steel, wiped clean of oil; the next best with pickled hot-rolled steel. Grinding or polishing is sometimes performed, but not sand- or shot-blasting.

In **projection welding** (Fig. 13.3.11), the heat for welding is derived from the localization of resistance at predetermined points by means of projections, embossments, or the intersections of elements of the assembly. The projections may be made by stamping or machining. The process is essentially the same as spot welding, and the projections seem to concentrate the current and pressure. Welds may be made singly or in multiple with somewhat less difficulty than is encountered in spot welding. When made in multiple, all welds may be made simulta-

neously. The advantages of projection welding are (1) the heat balance for difficult assemblies is readily secured, (2) the results are generally more uniform, (3) a closer spacing of welds is possible, and (4) electrode life is increased. Sometimes it is possible to projection-weld joints that could not be welded by other means.

OTHER WELDING PROCESSES

Electron Beam Welding (EBW)

In **electron beam welding,** coalescence of metals is achieved by heat generated by a concentrated beam of high-velocity electrons impinging on the surfaces to be joined. Electrons have a very small mass and carry a negative charge. An electron beam gun, consisting of an emitter, a bias electrode, and an anode, is used to create and accelerate the beam of electrons. Auxiliary components such as beam alignment, focus, and deflection coils may be used with the electron beam gun; the entire assembly is referred to as the electron beam gun column.

The advantages of the process arise from the extremely high energy density in the focused beam which produces deep, narrow welds at high speed, with minimum distortion and other deleterious heat effects. These welds show superior strength compared with those made utilizing other welding processes for a given material. Major applications are with metals and alloys highly reactive to gases in the atmosphere or those volatilized from the base metal being welded.

A disadvantage of the process lies in the necessity for providing precision parts and fixtures so that the beam can be precisely aligned with the joint to ensure complete fusion. Gapped joints are not normally welded because of fixture complexity and the extreme difficulty of manipulating filler metal into the tiny, rapidly moving, weld puddle under high vacuum. When no filler metal is employed, it is common to use the keyhole technique. Here, the electron beam makes a hole entirely through the base metal, which is subsequently filled with melted base metal as the beam leaves the area. Other disadvantages of the process arise from the cost, complexity, and skills required to operate and maintain the equipment, and the safety precautions necessary to protect operating personnel from the X-rays generated during the operation.

Laser Beam Welding and Cutting

By using a laser, energy from a primary source (electrical, chemical, thermal, optical, or nuclear) is converted to a beam of coherent electromagnetic radiation at ultraviolet, visible, or infrared frequency. Because high-energy laser beams are coherent, they can be highly concentrated to provide the high energy density needed for welding, cutting, and heat-treating metals.

As applied to welding, **pulsed** and continuously operating solid-state **lasers** and lasers that produce **continuous-wave (cw) energy** have been developed to the point that multikilowatt laser beam equipment based on CO_2 is capable of full-penetration, single-pass welding of steel to ¾-in thickness.

Lasers do not require a vacuum in which to operate, so that they offer many of the advantages of electron beam welding but at considerably lower equipment cost and higher production rates. Deep, narrow welds are produced at high speeds and low total heat input, thus duplicating the excellent weld properties and minimal heat effects obtained from electron beam welding in some applications. The application of lasers to metals—for cutting or welding—coupled with computerized control systems, allows their use for complex shapes and contours.

Solid-State Welding

Solid-state welding encompasses a group of processes in which the weld is effected by bringing clean metal surfaces into intimate contact under certain specific conditions. In **friction welding,** one part is rotated at high speed with respect to the other, under pressure. The parts are heated, but not to the melting point of the metal. Rotation is stopped at the critical moment of welding. Base metal properties across the joint show little change because the process is so rapid. **Ultrasonic welding** employs mechanical vibrations at ultrasonic frequencies plus pressure to effect the intimate contact between faying surfaces needed to produce a weld. (See also Sec. 12.) The welding tool is essentially a transducer that converts electric frequencies to ultra-high-frequency mechanical vibrations. By applying the tip of the tool, or anvil, to a small area in the external surface of two lapped parts, the vibrations and pressure are transmitted to the faying surfaces. Foils, thin-gage sheets, or fine wires can be spot- or seam-welded to each other or to heavier parts.

Explosion Welding

Explosion welding utilizes extremely high pressures to join metals, often with significantly different properties. For example, it may used to clad a metal substrate, such as steel, with a protective layer of a dissimilar metal, such as aluminum. Since the materials do not melt, two metals with significantly different melting points can be successfully welded by explosion welding. The force and speed of the explosion are directed to cause a series of progressive shock waves that deform the faying surfaces at the moment of impact. A magnified section of the joint reveals a true weld with an interlocking waveshape and, usually, some alloying.

THERMAL CUTTING PROCESSES

Oxyfuel Cutting (OFC)   Oxyfuel cutting (Fig. 13.3.12) is used to cut steels and to prepare bevel and vee grooves. In this process, the metal is heated to its ignition temperature, or kindling point, by a series of preheat flames. After this temperature is attained, a high-velocity stream of pure oxygen is introduced, which causes oxidation or "burning" to occur. The force of the oxygen steam blows the oxides out of the joint, resulting in a clean cut. The oxidation process also generates additional thermal energy, which is radially conducted into the surrounding steel, increasing the temperature of the steel ahead of the cut. The next portion of the steel is raised to the kindling temperature, and the cut proceeds.

Carbon and low-alloy steels are easily cut by the oxyfuel process. Alloy steels can be cut, but with greater difficulty than mild steel. The level of difficulty is a function of the alloy content. When the alloy content reaches the levels found in stainless steels, oxyfuel cutting cannot be used unless the process is modified by injecting flux or iron-rich powders into the oxygen stream. Aluminum cannot be cut with the oxyfuel process. Oxyfuel cutting is commonly regarded as the most economical way to cut steel plates greater than ½ in thick.

A variety of **fuel gases** may be used for oxyfuel cutting, with the choice largely dependent on local economics; they include natural gas, propane, acetylene, and a variety of proprietary gases offering unique advantages. Because of its role in the primary cutting stream, oxygen is always used as a second gas. In addition, some oxygen is mixed with the fuel gas in proportions designed to ensure proper combustion.

Plasma Arc Cutting (PAC)   The plasma arc cutting process (Fig. 13.3.13) was developed initially to cut materials that do not permit the use of the oxyfuel process: stainless steel and aluminum. It was found, however, that plasma arc cutting offered economic advantages when applied to thinner sections of mild steel, especially those less than 1 in thick. Higher travel speed is possible with plasma arc cutting, and the volume of heated base material is reduced, minimizing metallurgical changes as well as reducing distortion.

PAC is a thermal and mechanical process. To utilize PAC, the material is heated until molten and expelled from the cut with a high-velocity stream of compressed gas. Unlike oxyfuel cutting, the process does not rely on oxidation. Because high amounts of energy are introduced

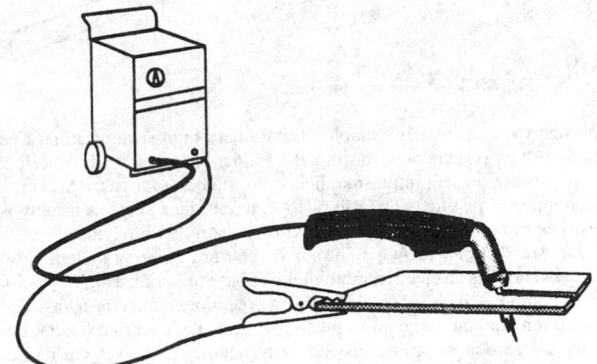

Fig. 13.3.13   Plasma arc cutting process.

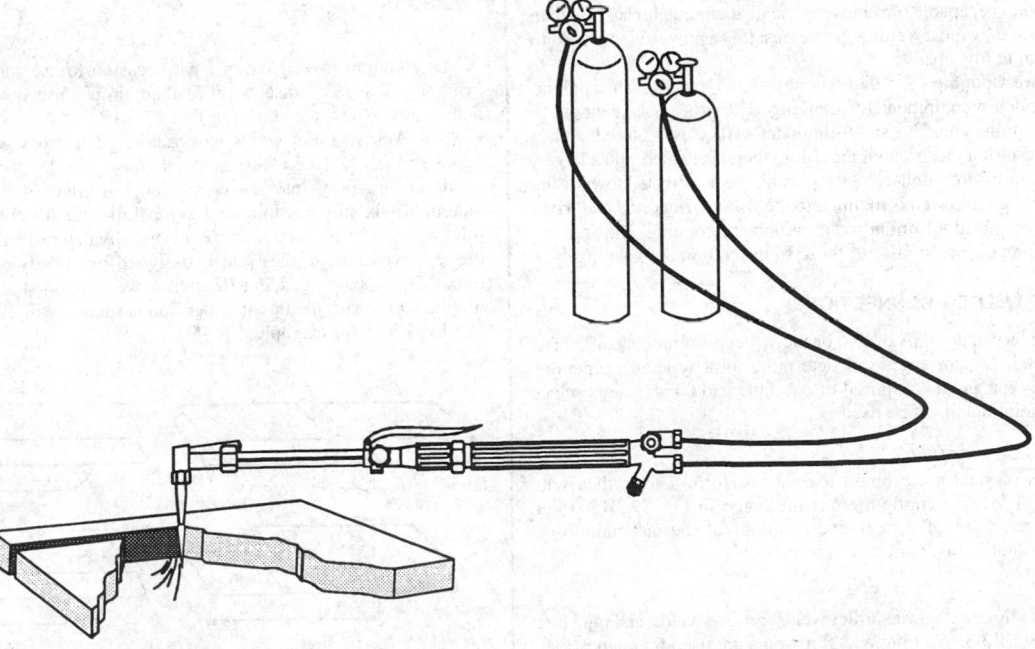

Fig. 13.3.12   Oxyfuel cutting.

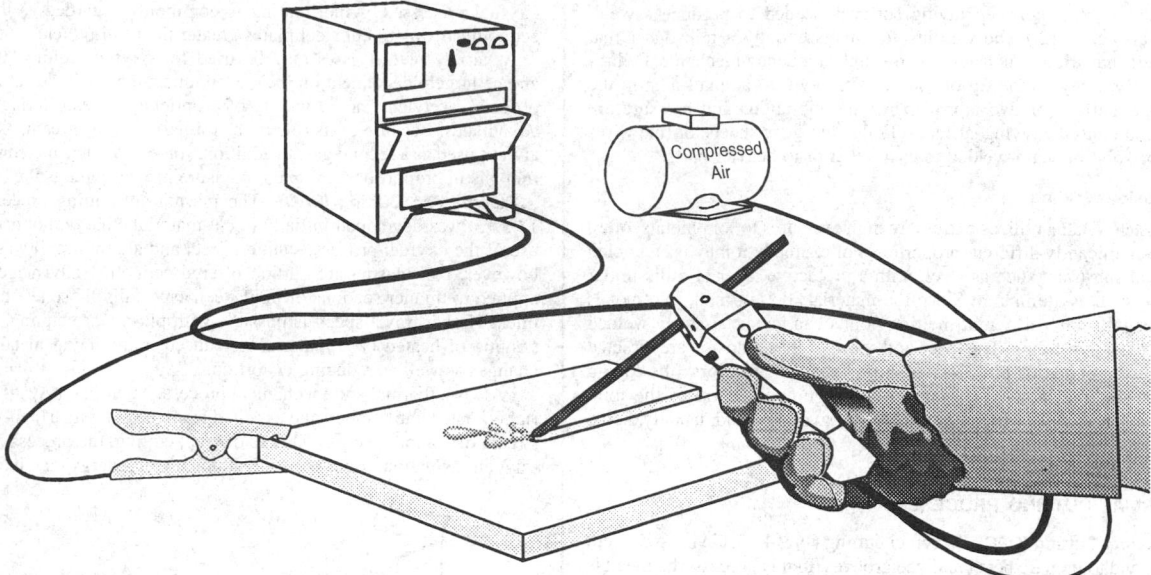

Fig. 13.3.14 Air arc gouging.

through the arc, PAC is capable of extremely high-speed cutting. The thermal energy generated during the oxidation process with oxyfuel cutting is not present in plasma; hence, for thicker sections, PAC is not economically justified. The use of PAC to cut thick sections usually is restricted to materials that do not oxidize readily with oxyfuel.

Air Arc Gouging (AAG) The air carbon arc gouging system (Fig. 13.3.14) utilizes an electric arc to melt the base material; a high-velocity jet of compressed air subsequently blows the molten material away. The air carbon gouging torch looks much like a manual electrode holder, but it uses a carbon electrode instead of a metallic electrode. Current is conducted through the base material to heat it. A valve in the torch handle permits compressed air to flow through two air ports. As the air hits the molten material, a combination of oxidation and expulsion of metal takes place, leaving a smooth cavity behind. The air carbon arc gouging system is capable of removing metal at a much higher rate than can be deposited by most welding processes. It is a powerful tool used to remove metal at low cost.

Plasma Arc Gouging A relatively new development is the application of plasma arc equipment for gouging. The process is identical to plasma arc cutting, but the small-diameter orifice is replaced with a larger one, resulting in a broader arc. More metal is heated, and a larger, broader stream of hot, high-velocity plasma gas is directed toward the workpiece. When the torch is inclined to the work surface, the metal can be removed in a fashion similar to air carbon arc gouging. The applications of the process are similar to those of air carbon arc gouging.

DESIGN OF WELDED CONNECTIONS

A **welded connection** consists of two or more pieces of base metal joined by weld metal. Design engineers determine joint type and generally specify weld type and the required throat dimension. Fabricators select the specific joint details to be used.

Joint Types

When pieces of steel are brought together to form a **joint,** they will assume one of the five configurations presented in Fig. 13.3.15. Joint types are descriptions of the relative positions of the materials to be joined and do not imply a specific type of weld.

Weld Types

Welds fall into three categories: fillet welds, groove welds, and plug and slot welds (Fig. 13.3.16). Plug and slot welds are used for connections that transfer small loads.

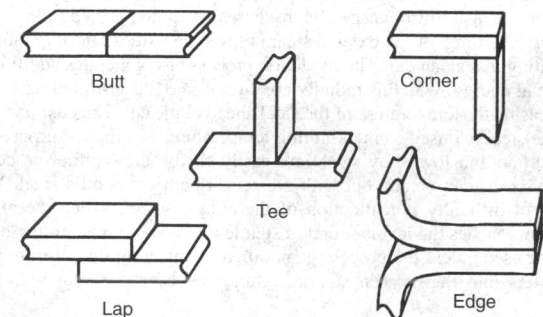

Fig. 13.3.15 Joint types.

Many engineers will see or have occasion to use standard welding symbols. A detailed discussion of their proper use is found in AWS documents. A few are shown in Fig. 13.3.17.

Fillet Welds Fillet welds have a triangular cross section and are applied to the surface of the materials they join. By themselves, fillet welds do not fully fuse the cross-sectional areas of parts they join, although it is still possible to develop full-strength connections with fillet welds. The size of a fillet weld is usually determined by measuring the leg, even though the weld is designed by specifying the required throat. For equal-legged, flat-faced fillet welds applied to plates that are oriented 90° apart, the throat dimension is found by multiplying the leg size by 0.707 (for example, sin 45°).

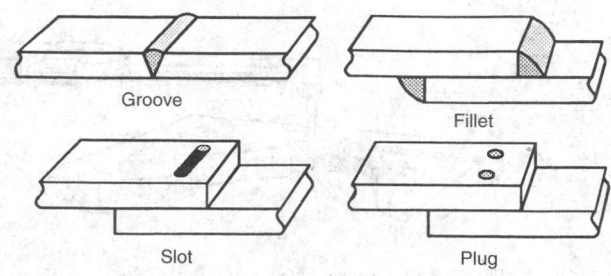

Fig. 13.3.16 Major weld types.

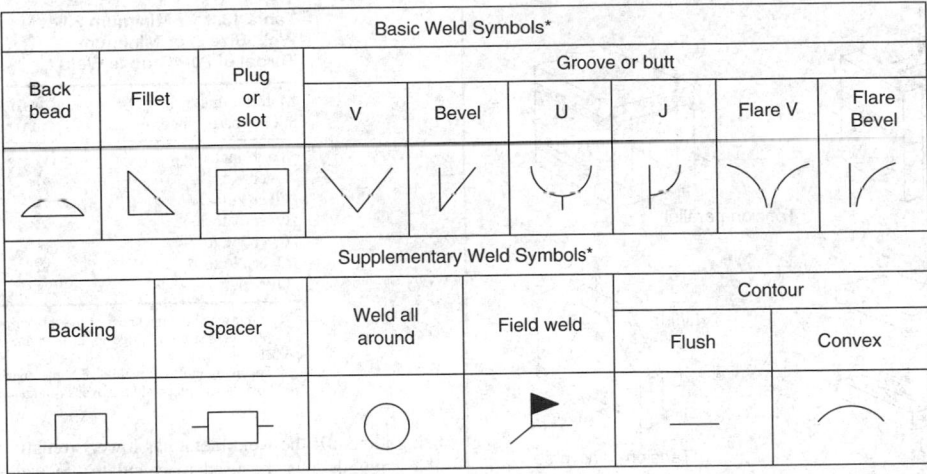

Basic Weld Symbols*								
Back bead	Fillet	Plug or slot	Groove or butt					
			V	Bevel	U	J	Flare V	Flare Bevel

Supplementary Weld Symbols*					
Backing	Spacer	Weld all around	Field weld	Contour	
				Flush	Convex

* See AWS publications for a complete listing.

Fig. 13.3.17   Some welding symbols commonly used. *(AWS)*

Groove Welds   Groove welds comprise two subcategories: complete joint penetration (CJP) groove welds and partial joint penetration (PJP) groove welds (Fig. 13.3.18). By definition, CJP groove welds have a throat dimension equal to the thickness of the material they join; a PJP groove weld is one with a throat dimension less than the thickness of the materials joined.

An **effective throat** is associated with a PJP groove weld. This term is used to differentiate between the depth of groove preparation and the probable depth of fusion that will be achieved. The **effective throat** on a PJP groove weld is abbreviated by E. The required **depth of groove preparation** is designated by a capital S. Since the designer may not

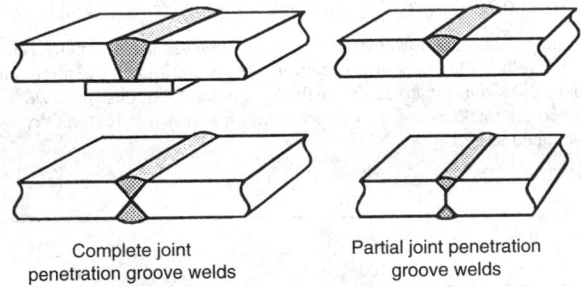

Complete joint penetration groove welds

Partial joint penetration groove welds

Fig. 13.3.18   Types of groove welds.

know which welding process a fabricator will select, it is necessary only to specify the dimension for E. The fabricator then selects the welding process, determines the position of welding, and applies the appropriate S dimension, which will be shown on the shop drawings. In most cases, both the S and E dimensions will appear on the welding symbols of shop drawings, with the effective throat dimension shown in parentheses.

Sizing of Welds

Overwelding is one of the major factors of welding cost. Specifying the correct size of weld is the first step in obtaining low-cost welding. It is important, then, to have a simple method to figure the proper amount of weld to provide adequate strength for all types of connections.

In terms of their application, welds fall into two general types: primary and secondary. **Primary welds** are critical welds that directly transfer the full applied load at the point at which they are located. These welds must develop the full strength of the members they join. Complete joint penetration groove welds are often used for these connections. **Secondary welds** are those that merely hold the parts together to

form a built-up member. The forces on these welds are relatively low, and fillet welds are generally utilized in these connections.

Filler Metal Strength   **Filler metal strength** may be classified as matching, undermatching, or overmatching. **Matching filler metal** has the same, or slightly higher, minimum specified yield and tensile strength as the base metal. CJP groove welds in tension require the use of matching weld metal—otherwise, the strength of the welded connection will be lower than that of the base metal. **Undermatching filler metal** deposits welds of a strength lower than that of the base metal. Undermatching filler metal may be deposited in fillet welds and PJP groove welds as long as the designer specifies a throat size that will compensate for the reduction in weld metal strength. An **overmatching filler metal** deposits weld metal that is stronger than the base metal; this is undesirable unless, for practical reasons, lower-strength filler metal is unavailable for the application. When overmatching filler metal is used, if the weld is stressed to its maximum allowable level, the base metal can be overstressed, resulting in failure in the fusion zone. Designers must ensure that connection strength, including the fusion zone, meets the application requirements.

In welding high-strength steel, it is generally desirable to utilize undermatching filler metal for secondary welds. High-strength steel may require additional preheat and greater care in welding because there is an increased tendency to crack, especially if the joint is restrained. Undermatching filler metals such as E70 are the easiest to use and are preferred, provided the weld is sized to impart sufficient strength to the joint.

Allowable Strength of Welds under Steady Loads   A structure, or **weldment,** is as strong as its weakest point, and "allowable" weld strengths are specified by the American Welding Society (AWS), the American Institute of Steel Construction (AISC), and various other professional organizations to ensure that a weld will deliver the mechanical properties of the members being joined. Allowable weld strengths are designated for various types of welds for steady and fatigue loads.

CJP groove welds are considered **full-strength welds,** since they are capable of transferring the equivalent capacity of the members they join. In calculations, such welds are allowed the same stress as the plate, provided the proper strength level of weld metal is used (e.g., matching filler metal). In such CJP welds, the mechanical properties of the weld metal must at least match those of the base metal. If the plates joined are of different strengths, the weld metal strength must at least match the strength of the weaker plate.

Figure 13.3.19 illustrates representative applications of PJP groove welds widely used in the economical welding of very heavy plates. PJP groove welds in heavy material will usually result in savings in weld metal and welding time, while providing the required joint strength. The

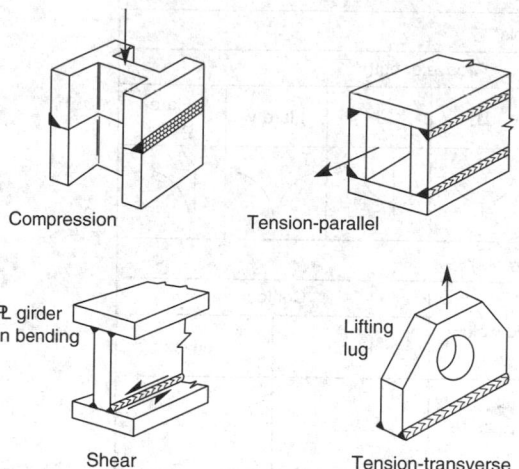

Fig. 13.3.19 Applications of partial joint penetration (PJP) groove welds.

Table 13.3.1 Minimum Fillet Weld Size ω or Minimum Throat of PJP Groove Weld t_e

Material thickness of thicker part joined, in	ω or t_e, in
*To ¼ incl.	⅛
Over ¼ to ½	3/16
Over ½ to ¾	¼
†Over ¾ to 1½	5/16
Over 1½ to 2¼	⅜
Over 2¼ to 6	½
Over 6	⅝

Not to exceed the thickness of the thinner part.
* Minimum size for bridge application does not go below 3/16 in.
† For minimum fillet weld size, table does not go above 5/16-in fillet weld for over ¾-in material.

faster cooling and increased restraint, however, justify establishment of a minimum effective throat t_e (see Table 13.3.1).

Other factors must be considered in determining the allowable stress on the throat of a PJP groove weld. Joint configuration is one. If a V, J, or U groove is specified, it is assumed that the welder can easily reach the bottom of the joint, and the effective weld throat t_e equals the depth of the groove. If a bevel groove with an included angle of 45° or less is specified and SMAW is used, ⅛ in is deducted from the depth of the prepared groove in defining the effective throat. This does not apply to the SAW process because of its deeper penetration capabilities. In the case of GMAW or FCAW, the ⅛-in reduction in throat only applies to bevel grooves with an included angle of 45° or less in the vertical or overhead position.

Weld metal subjected to compression in any direction or to tension parallel to the axis of the weld should have the same allowable strength as the base metal. Matching weld metal must be used for compression, but is not necessary for tension parallel loading.

The existence of tension forces transverse to the axis of the weld or shear in any direction requires the use of weld metal allowable strengths that are the same as those used for fillet welds. The selected weld metal may have mechanical properties higher or lower than those of the metal

being joined. If the weld metal has lower strength, however, its allowable strength must be used to calculate the weld size or maximum allowable weld stress. For higher-strength weld metal, the weld allowable strength may not exceed the shear allowable strength of the base metal.

The AWS has established the **allowable shear value** for weld metal in a fillet or PJP bevel groove weld as

$$\tau = 0.30 \times \text{electrode min. spec. tensile strength} = 0.30 \times \text{EXX}$$

and has proved it valid from a series of fillet weld tests conducted by a special Task Committee of AISC and AWS.

Table 13.3.2 lists the allowable shear values for various weld metal strength levels and the more common fillet weld sizes. These values are for equal-leg fillet welds where the effective throat t_e equals $0.707 \times$ leg size ω. With the table, one can calculate the allowable unit force per lineal inch f for a weld size made with a particular electrode type. For example, the allowable unit force per lineal inch f for a ½-in fillet weld made with an E70 electrode is

$$f = 0.707\omega\tau = 0.707(\text{½ in})(0.30)(70 \text{ ksi}) = 7.42 \text{ kips/in}$$

The minimum allowable sizes for fillet welds are given in Table 13.3.1. When materials of different thickness are joined, the minimum fillet weld size is governed by the thicker material; but this size need not exceed the thickness of the thinner material unless it is required by the calculated stress.

Table 13.3.2 Allowable Loads for Various Size Fillet Welds

	Strength level of weld metal (EXX)						
	60*	70*	80	90*	100	110*	120
	Allowable shear stress on throat, ksi (1,000 lb/in²), of fillet weld or PJP weld						
$\tau =$	18.0	21.0	24.0	27.0	30.0	33.0	36.0
	Allowable unit force on fillet weld, kips/lin in						
$f =$	12.73ω	14.85ω	16.97ω	19.09ω	21.21ω	23.33ω	25.45ω
Leg size ω, in	Allowable unit force for various sizes of fillet welds, kips/lin in						
1	12.73	14.85	16.97	19.09	21.21	23.33	25.45
⅞	11.14	12.99	14.85	16.70	18.57	20.41	22.27
¾	9.55	11.14	12.73	14.32	15.92	17.50	19.09
⅝	7.96	9.28	10.61	11.93	13.27	14.58	15.91
½	6.37	7.42	8.48	9.54	10.61	11.67	12.73
7/16	5.57	6.50	7.42	8.35	9.28	10.21	11.14
⅜	4.77	5.57	6.36	7.16	7.95	8.75	9.54
5/16	3.98	4.64	5.30	5.97	6.63	7.29	7.95
¼	3.18	3.71	4.24	4.77	5.30	5.83	6.36
3/16	2.39	2.78	3.18	3.58	3.98	4.38	4.77
⅛	1.59	1.86	2.12	2.39	2.65	2.92	3.18
1/16	0.795	0.930	1.06	1.19	1.33	1.46	1.59

* Fillet welds actually tested by the joint AISC-AWS Task Committee.

Connections under Simple Loads   For a **simple tensile, compressive or shear load,** the imposed load is divided by weld length to obtain applied force, *f*, in pounds per lineal inch of weld. From this force, the proper leg size of the fillet weld or throat size of groove weld is found.

For **primary welds in butt joints,** groove welds must be made through the entire plate, in other words, **100 percent penetration.** Since a butt joint with a properly made CJP groove has a strength equal to or greater than that of the plate, there is no need to calculate the stress in the weld or to attempt to determine its size. It is necessary only to utilize matching filler metal.

With fillet welds, it is possible to have a weld that is either too large or too small; therefore, it is necessary to be able to determine the proper weld size.

Parallel fillet welds have forces applied parallel to their axis, and the throat is stressed only in shear. For an equal-legged fillet, the maximum shear stress occurs on the 45° throat.

Transverse fillet welds have forces applied transversely, or at right angles to their axis, and the throat is stressed by combined shear and normal (tensile or compressive) stresses. For an equal-legged fillet weld, the maximum shear stress occurs on the 67½° throat, and the maximum normal stress occurs on the 22½° throat.

Connections Subject to Horizontal Shear   A weld joining the flange of a beam to its web is stressed in horizontal shear (Fig. 13.3.20). A designer may be accustomed to specifying a certain size fillet weld for a given plate thickness (e.g., leg size about three-fourths of the plate thickness) in order that the weld develop full plate strength. This particular joint between flange and web is an exception to this rule. In order to

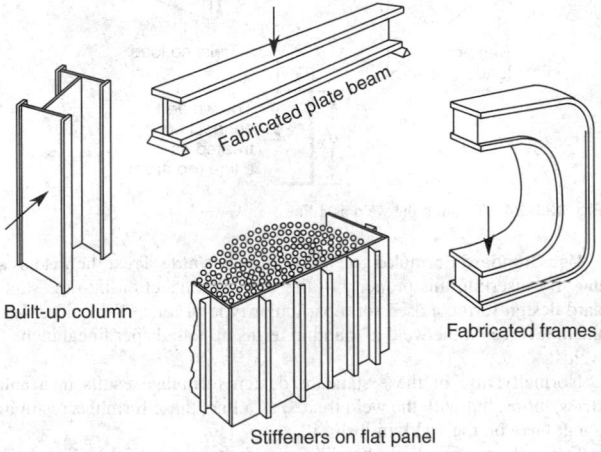

Fig. 13.3.20   Examples of welds stressed in horizontal shear.

prevent web buckling, a lower allowable shear stress is usually used, which results in the requirement for a thicker web. The welds are in an area next to the flange where there is no buckling problem, and no reduction in allowable load is applied to them. From a design standpoint, these welds may be very small; their actual size sometimes is determined by the minimum size allowed by the thickness of the flange plate, in order to ensure the proper slow cooling rate of the weld on the heavier plate.

General Rules about Horizontal Shear   Aside from joining the flanges and web of a beam, or transmitting any unusually high force between the flange and web at right angles to the assembly (e.g., bearing supports, lifting lugs), the weld between flange and web serves to transmit the horizontal shear forces; the weld size is determined by the magnitude of the shear forces. In the analysis of a beam, a shear diagram is useful to depict the amount and location of welding required between the flange and web (Fig. 13.3.21).

Figure 13.3.21 shows that (1) members with applied transverse loads are subject to bending moments; (2) changes in bending moments cause

horizontal shear forces; and (3) horizontal shear forces require welds to transmit them between the flange and web of the beam.

NOTE: (1) Shear forces occur only when the bending moment is changing. (2) It is quite possible for portions of a beam to have little or no shear—i.e., the middle portions of the beams 1 and 2, within which the bending moment is constant. (3) When there is a difference in shear along the length of the beam, the shear forces are usually greatest at the ends of the beam (see beam 3), so that when web stiffeners are used, they are welded continuously when placed at the ends and welded intermittently when placed elsewhere along the length of the beam. (4) Fixing beam ends will alter the moment diagram to reduce the maximum moment; i.e., the bending moment is lower in the middle, but is now introduced at the ends. For the uniform loading configuration in beam 3, irrespective of the end conditions and their effect on bending moments and their location, the shear diagram will remain unchanged, and the amount of welding between flange and web will remain the same.

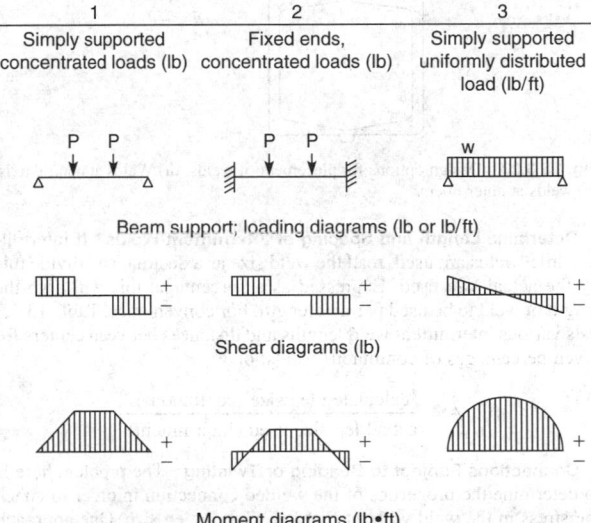

Fig. 13.3.21   Shear and moment diagrams.

Application of Rules to Find Weld Size   Horizontal shear forces acting on the weld joining flange and web (Fig. 13.3.22) may be found from the following formula:

$$f = \frac{Vay}{In} \quad \text{lb/lin in}$$

where *f* = force on weld, lb/lin in; *V* = total shear on section at a given position along beam, lb; *a* = area of flange held by weld, in²; *y* = distance between center of gravity of flange area and neutral axis of whole section, in; *I* = moment of inertia of whole section, in⁴; and *n* = number of welds joining flange to web.

Locate Welds at Point of Minimum Stress   In Fig. 13.3.23*a*, shear force is high because the weld lies on the neutral axis of the section,

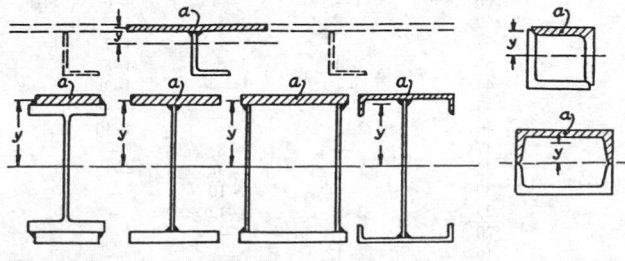

Fig. 13.3.22   Area of flange held by weld.

where the horizontal shear force is maximum. In Fig. 3.3.23*b*, the shear force is resisted by the channel webs, not the welds. In this last case, the shear formula above does not enter into consideration; for the configuration in Fig. 13.3.23*b*, full-penetration welds are not required.

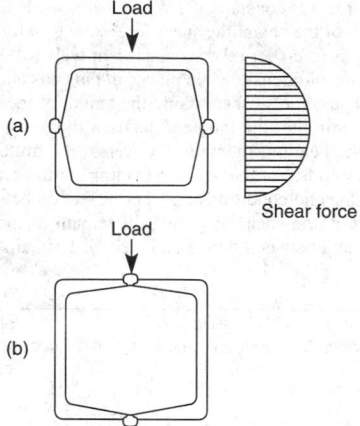

Fig. 13.3.23 Design options for placement of welds. (*a*) Welds at neutral axis; (*b*) welds at outer fibers.

Determine Length and Spacing of Intermittent Welds If intermittent fillet welds are used, read the weld size as a decimal and divide this by the actual size used. Expressed as a percentage, this will give the length of weld to be used per unit length. For convenience, Table 13.3.3 lists various intermittent weld lengths and distances between centers for given percentages of continuous welds; or

$$\% = \frac{\text{calculated leg size (continuous)}}{\text{actual leg size used (intermittent)}}$$

Connections Subject to Bending or Twisting The problem here is to determine the properties of the welded connection in order to check the stress in the weld without first knowing its leg size. One approach suggests assuming a certain weld leg size and then calculating the stress in the weld to see if it is over- or understressed. If the result is too far off, the assumed weld leg size is adjusted, and the calculations repeated. This iterative method has the following disadvantages:

1. A decision must be made as to throat section size to be used to determine the property of the weld. Usually some objection can be raised to any throat section chosen.

2. The resulting stresses must be combined, and for several types of loading, this can become rather complicated.

Table 13.3.3 Length and Spacing of Intermittent Welds

Continuous weld, %	Length of intermittent welds and distance between centers, in		
75	—	3–4	—
66	—	—	4–6
60	—	3–5	—
57	—	—	4–7
50	2–4	3–6	4–8
44	—	—	4–9
43	—	3–7	—
40	2–5	—	4–10
37	—	3–8	—
33	2–6	3–9	4–12
30	—	3–10	—
25	2–8	3–12	—
20	2–10	—	—
16	2–12	—	—

Proposed Method The following is a simple method used to determine the correct amount of welding required to provide adequate strength for either a bending or a torsion load. In this method, the weld is treated as a line, having no area but having a definite length and cross section. This method offers the following advantages:

1. It is not necessary to consider throat areas.

2. Properties of the weld are easily found from a table without knowledge of weld leg size.

3. Forces are considered per unit length of weld, rather than converted to stresses. This facilitates dealing with combined-stress problems.

4. Actual values of welds are given as force per unit length of weld instead of unit stress on throat of weld.

Visualize the welded connection as a line (or lines), following the same outline as the connection but having no cross sectional area. In Fig. 13.3.24, the desired area of the welded connection A_w can be represented by just the length of the weld. The stress on the weld cannot be determined unless the weld size is assumed; but by following the proposed procedure which treats the weld as a line, the solution is more direct, is much simpler, and becomes basically one of determining the force on the weld(s).

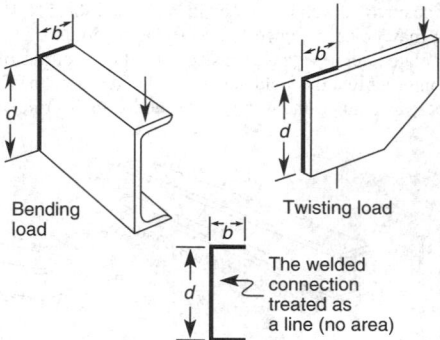

Fig. 13.3.24 Treating the weld as a line.

Use Standard Formulas to Find Force on Weld Treat the **weld as a line**. By inserting this property of the welded connection into the standard design formula used for a particular type of load (Table 13.3.4*a*), the unit force on the weld is found in terms of pounds per lineal inch of weld.

Normally, use of these standard design formulas results in a **unit stress**, lb/in², but with the weld treated as a line, these formulas result in a **unit force** on the weld, in lb/lin in.

For problems involving bending or twisting loads, Table 13.3.4*c* is used. It contains the section modulus S_w and polar moment of inertia J_w of some 13 typical welded connections with the weld treated as a line. For any given connection, two dimensions are needed: width *b* and depth *d*. Section modulus S_w is used for welds subjected to bending; polar moment of inertia J_w for welds subjected to twisting. Section modulus S_w in Table 13.3.4*c* is shown for symmetric and unsymmetric connections. For unsymmetric connections, S_w values listed differentiate between top and bottom, and the forces derived therefrom are specific to location, depending on the value of S_w used.

When one is applying more than one load to a welded connection, they are combined vectorially, but must occur at the same location on the welded joint.

Use Allowable Strength of Weld to Find Weld Size Weld size is obtained by dividing the resulting unit force on the weld by the allowable strength of the particular type of weld used, obtained from Table 13.3.5 (steady loads) or Table 13.3.6 (fatigue loads). For a joint which has only a transverse load applied to the weld (either fillet or butt weld), the allowable transverse load may be used from the applicable table. If part of the load is applied parallel (even if there are transverse loads in addition), the allowable parallel load must be used.

Applying the System to Any Welded Connection

1. Find the position on the welded connection where the combination of forces will be maximum. There may be more than one which must be considered.

2. Find the value of each of the forces on the welded connection at this point. Use Table 13.3.4a for the standard design formula to find the force on the weld. Use Table 13.3.4c to find the property of the weld treated as a line.

3. Combine (vectorially) all the forces on the weld at this point.

4. Determine the required weld size by dividing this value (step 3) by the allowable force in Table 13.3.5 or 13.3.6.

Sample Calculations Using This System The example in Fig. 13.3.25 illustrates the application of this procedure.

Summary

The application of the following guidelines will ensure effective welded connections:

1. Properly select weld type.

2. Use CJP groove welds only where loading criteria mandate.

3. Consider the cost of joint preparation vs. welding time when you select groove weld details.

4. Double-sided joints reduce the amount of weld metal required. Verify welder access to both sides and that the double-sided welds will not require overhead welding.

5. Use intermittent fillet welds where continuous welds are not required.

6. On corner joints, prepare the thinner member.

7. Strive to obtain good fit-up and do not overweld.

8. Orient welds and joints to facilitate flat and horizontal welding wherever possible.

9. Use the minimum amount of filler metal possible in a given joint.

10. Always ensure adequate access for the welder, welding apparatus, and inspector.

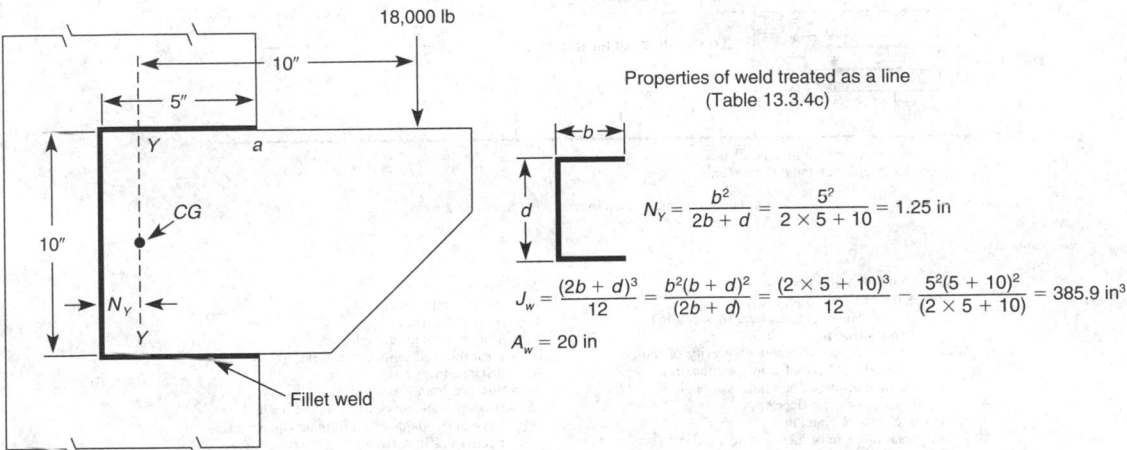

Step 1

18,000 lb

10″

5″

Y

a

CG

10″

N_Y

Y

Fillet weld

Properties of weld treated as a line
(Table 13.3.4c)

$$N_Y = \frac{b^2}{2b + d} = \frac{5^2}{2 \times 5 + 10} = 1.25 \text{ in}$$

$$J_w = \frac{(2b + d)^3}{12} = \frac{b^2(b + d)^2}{(2b + d)} = \frac{(2 \times 5 + 10)^3}{12} - \frac{5^2(5 + 10)^2}{(2 \times 5 + 10)} = 385.9 \text{ in}^3$$

$A_w = 20 \text{ in}$

Step 2 (Table 13.3.4a)

Twisting (horizontal component)

$$f_h = \frac{T C_h}{J_w} = \frac{(180,000)(5)}{(385.9)} = 2340 \text{ lb/in}$$

Twisting (vertical component)

$$f_v = \frac{T C_v}{J_w} = \frac{(180,000)(3.75)}{(385.9)} = 1750 \text{ lb/in}$$

Vertical shear

$$f_s = \frac{P}{A_w} = \frac{180,000}{20} = 900 \text{ lb/in}$$

Step 3

$$f_R = \sqrt{f_v^2 + f_h^2} = \sqrt{.2340^2 + 2650^2} = 3540 \text{ lb/in}$$

$f_h = 2340$

$f_R = 3540$

$f_v = 1750$

$f_s = 900$

2650

Step 4 (Assume E60 electrodes; Table 13.3.2)

$$\omega = \frac{3540}{12,730} = 0.278 \text{ in or } \tfrac{5}{16} \text{ in leg fillet weld}$$

Fig. 13.3.25 Sample problem: Find the fillet weld size required for the connection shown.

Table 13.3.4 Treating a Weld as a Line

Type of loading		Standard design formula	Treating the weld as a line
		Stress lb/in^2	Force, lb/in
Primary welds transmit entire load at this point			
	Tension or compression	$\sigma = \dfrac{P}{A}$	$f = \dfrac{P}{A_w}$
	Vertical shear	$\sigma = \dfrac{V}{A}$	$f = \dfrac{V}{A_w}$
	Bending	$\sigma = \dfrac{M}{S}$	$f = \dfrac{M}{S_w}$
	Twisting	$\sigma = \dfrac{TC}{J}$	$f = \dfrac{TC}{J_w}$
Secondary welds hold section together—low stress			
	Horizontal shear	$\tau = \dfrac{VA_y}{It}$	$f = \dfrac{VA_y}{In}$
	Torsion horizontal shear*	$\tau = \dfrac{T}{2At}$	$f = \dfrac{T}{2A}$

A = area contained within median line.
* Applies to closed tubular section only.

(*a*) Design formulas used to determine forces on a weld

b = width of connection, in
d = depth of connection, in
A = area of flange material held by welds in horizontal shear, in^2
y = distance between center of gravity of flange material and N.A. of whole section, in
I = moment of inertia of whole section, in.^{4}
C = distance of outer fiber, in
t = thickness of plate, in
J = polar moment of inertia of section, in.^{4}
P = tensile or compressive load, lb
V = vertical shear load, lb

M = bending moment, in·lb
T = twisting moment, in·lb.
A_w = length of weld, in
S_w = section modulus of weld, in^2
J_w = polar moment of inertia of weld, in^3
N_x = distance from x axis to face
N_y = distance from y axis to face
S = stress in standard design formula, lb/in^2
f = force in standard design formula when weld is treated as a line, lb/in
n = number of welds

(*b*) Definition of terms

Allowable Fatigue Strength of Welds

The performance of a weld under **cyclic stress** is an important consideration, and applicable specifications have been developed following extensive research by the American Institute of Steel Construction (AISC). Although sound weld metal has about the same fatigue strength as unwelded metal, the change in section induced by the weld may lower the fatigue strength of the welded joint. In the case of a CJP groove weld, reinforcement, any undercut, lack of penetration, or a crack will act as a notch; the notch, in turn, is a **stress raiser** which results in reduced fatigue strength. A fillet weld used in lap or tee joints provides an abrupt change in section; that geometry introduces a stress raiser and results in reduced fatigue strength.

The initial AISC research was directed toward bridge structure components; Table 13.3.6 illustrates a few such combinations. Similar details arise with other classes of fabricated metal products subjected to repeated loading, such as presses, transportation equipment, and material handling devices. The principles underlying fatigue performance are relatively independent of a particular application, and these data shown can be applied to the design of weldments other than for bridge construction.

Table 13.3.6 is abstracted from an extensive tabulation in the AISC "Manual of Steel Construction," 9th ed. The table also lists the variation of allowable range of stress vs. number of stress cycles for cyclic loading. A detailed discussion of the solution of fatigue-loaded welded joints is beyond the scope of this section. The reader is referred to the basic reference cited above and to the references at the head of this section in pursuing the procedures recommended to solve problems involving welded assemblies subjected to fatigue loads.

Table 13.3.4 Treating a Weld as a Line (Continued)

Outline of welded joint b = width d = depth	Bending (about horizontal axis $x - x$)	Twisting
	$S_w = \dfrac{d^2}{6}$ in^2	$J_w = \dfrac{d^3}{12}$ in^3
	$S_w = \dfrac{d^2}{3}$	$J_w - \dfrac{d(3b^2 + d^2)}{6}$
	$S_w = bd$	$J_w = \dfrac{b^3 + 3bd^2}{6}$
 $Ny = \dfrac{b^2}{2(b+d)}$ $Nx = \dfrac{d^2}{2(b+d)}$	$S_w = \dfrac{4bd + d^2}{6} = \dfrac{d^2(4b + d)}{6(2b + d)}$ top bottom	$J_w = \dfrac{(b + d)^4 - 6b^2d^2}{12(b + d)}$
 $Ny = \dfrac{b^2}{2\,b+d}$	$S_w = bd + \dfrac{d^2}{6}$	$J_w = \dfrac{(2b + d)^3}{12} - \dfrac{b^2(b + d)^2}{2b + d}$
$Nx = \dfrac{d^2}{b+2d}$	$S_w = \dfrac{2bd + d^2}{3} = \dfrac{d^2(2b + d)}{3(b + d)}$ top bottom	$J_w = \dfrac{(b + 2d)^3}{12} - \dfrac{d^2(b + d)^2}{b + 2d}$
	$S_w = bd + \dfrac{d^2}{3}$	$J_w = \dfrac{(b + d)^3}{6}$
 $Ny = \dfrac{d^2}{b+2d}$	$S_w = \dfrac{2bd + d^2}{3} = \dfrac{d^2(2b + d)}{3(b + d)}$ top bottom	$J_w = \dfrac{(b + 2d)^3}{12} - \dfrac{d^2(b + d)^2}{b + 2d}$
 $Ny = \dfrac{d^2}{2(b+d)}$	$S_w = \dfrac{4bd + d^2}{3} = \dfrac{4bd^2 + d^3}{6b + 3d}$ top bottom	$J_w = \dfrac{d^3(4b + d)}{6(b + d)} + \dfrac{b^3}{6}$
	$S_w = bd + \dfrac{d^2}{3}$	$J_w = \dfrac{b^3 + 3bd^2 + d^3}{6}$
	$S_w = 2bd + \dfrac{d^2}{3}$	$J_w = \dfrac{2b^3 + 6bd^2 + d^3}{6}$
	$S_w = \dfrac{\pi d^2}{4}$	$J_w = \dfrac{\pi d^3}{4}$
	$I_w = \dfrac{\pi_d}{2}\left(D^2 + \dfrac{d^2}{2}\right)$ $S_w = \dfrac{I_w}{c}$ where $c = \dfrac{\sqrt{D^2 + d^2}}{2}$	

(c) Properties of welded connections

Table 13.3.5a Allowable Stresses on Weld Metal

Type of weld stress	Permissible stress*	Required strength level†‡
Complete penetration groove welds		
Tension normal to effective throat	Same as base metal	Matching weld metal must be used. See table below.
Compression normal to effective throat	Same as base metal	Weld metal with a strength level equal to or one classification (10 ksi) less than matching weld metal may be used.
Tension or compression parallel to axis of weld	Same as base metal	Weld metal with a strength level equal to or less than matching weld metal may be used.
Shear on effective throat	0.30 × nominal tensile strength of weld metal (ksi) except stress on base metal shall not exceed 0.40 × yield stress of base metal	
Partial penetration groove welds		
Compression normal to effective throat	Designed not to bear—0.50 × nominal tensile strength of weld metal (ksi) except stress on base metal shall not exceed 0.60 × yield stress of base metal. Designed to bear. Same as base metal	Weld metal with a strength level equal to or less than matching weld metal may be used.
Tension or compression parallel to axis of weld§	Same as base metal	
Shear parallel to axis of weld	0.30 × nominal tensile strength of weld metal (ksi) except stress on base metal shall not exceed 0.40 × yield stress of base metal	
Tension normal to effective throat¶	0.30 × nominal tensile strength of weld metal (ksi) except stress on base metal shall not exceed 0.60 × yield stress of base metal	
Fillet welds§		
Stress on effective throat, regardless of direction of application of load	0.30 × nominal tensile strength of weld metal (ksi) except stress on base metal shall not exceed 0.40 × yield stress of base metal	Weld metal with a strength level equal to or less than matching weld metal may be used
Tension or compression parallel to axis of weld	Same as base metal	
Plug and slot welds		
Shear parallel to faying surfaces	0.30 × nominal tensile strength of weld metal (ksi) except stress on base metal shall not exceed 0.40 × yield stress of base metal	Weld metal with a strength level equal to or less than matching weld metal may be used

* For matching weld metal, see AISC Table 1.17.2 or AWS Table 4.1.1 or table below.
† Weld metal, one strength level (10 ksi) stronger than matching weld metal may be used when using alloy weld metal on A242 or A588 steel to match corrosion resistance or coloring characteristics (Note 3 of Table 4.1.4 or AWS D1.1).
‡ Fillet welds and partial penetration groove welds joining the component elements of built-up members (ex. flange to web welds) may be designed without regard to the axial tensile or compressive stress applied to them.
§ Cannot be used in tension normal to their axis under fatigue loading (AWS 2.5). AWS Bridge prohibits their use on any butt joint (9.12.1.1), or any splice in a tension or compression member (9.17), or splice in beams or girders (9.21), however, are allowed on corner joints parallel to axial force of components of built-up members (9.12.1.2 (2)). Cannot be used in girder splices (AISC 1.10.8).
¶ AWS D1.1 Section 9 Bridges—reduce above permissible stress allowables of weld by 10%.
SOURCE: Abstracted from AISC and AWS data, by permission. Footnotes refer to basic AWS documents as indicated.

Table 13.3.5b Matching Filler and Base Metals*

Weld metal	60 or 70	70	80	100	110
Type of steel	A36; A53, Gr. B; A106, Gr. B; A131, Gr. A, B, C, CS, D, E; A139, Gr. B; A381, Gr. Y35; A500, Gr. A, B; A501; A516, Gr. 55, 60; A524, Gr. I, II; A529; A570, Gr. D,E; A573, Gr. 65; A709, Gr. 36; API 5L, Gr. B; API 5LX Gr. 42; ABS, Gr. A, B, D, CS, DS, E	A131, Gr. AH32, DH32, EH32, AH36, DH36, EH36; A242; A441; A516, Gr. 65; 70; A537, Class 17; A572, Gr. 42, 45, 50, 55; A588 (4 in and under); A595, Gr. A, B, C; A606; A607, Gr. 45, 50, 55; A618; A633, Gr. A, B, C, D (2½ in and under); A709, Gr. 50, 50W; API 2H; ABS Gr. AH32, DH32, EH32, AH36, DH36, EH36.	A572, Gr. 60, 65; A537, Class 2; A63, Gr. E	A514 (over 2½ in (63 mm)]; A709, Gr. 100, 100W [2½ to 4 in (63 to 102 mm)]	A514 [2½ in (63 mm) and under]; A517; A709, Gr. 100, 100W [2½ in (63 mm) and under]

* Abstracted from AISC and AWS data, by permission

Figure 13.3.26 is a **modified Goodman diagram** for a CJP groove butt weld with weld reinforcement left on. The category is C, and the life is 500,000 to 2 million cycles (see Table 13.3.6). The vertical axis shows maximum stress σ_{max}, and the horizontal axis shows minimum stress σ_{min}, either positive or negative. A steady load is represented by the 45° line to the right, and a complete reversal by the 45° line to the left. The region to the right of the vertical line ($K = 0$) represents tensile loading. The fatigue formulas apply to welded butt joints in plates or other joined members. The region to the left of this line represents cycles going into compressive loading.

The fatigue formulas are meant to reduce the allowable stress as cyclic loads are encountered. The resulting allowable fatigue stress

Table 13.3.6   AISC Fatigue Allowable Stresses for Cyclic Loading

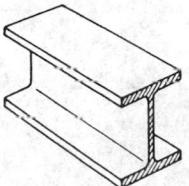

Base metal–no attachments–rolled or clean surfaces

(A)

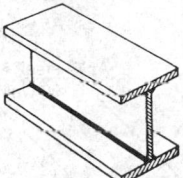

Base metal–built-up plates or shapes–connected by continuous complete penetration groove welds or fillet welds–without attachments. Note: don't use this as a fatigue allowable for the fillet weld to transfer a load. See (F) for that case.

(B)

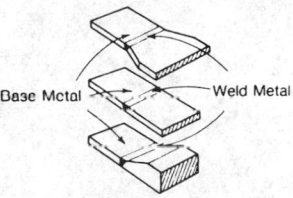

Base metal and weld metal at full penetration groove welds–changes thickness or width not to exceed a slope of 1 in 2½ (22°). Ground flush and inspected by radiography or ultrasound (B).

For A514 steel (B′)

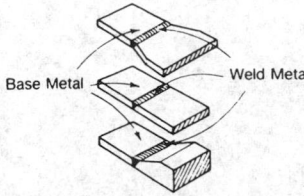

Base metal and weld metal at full penetration groove welds–changes in thickness or width not to exceed a slope of 1 in 2½ (22°) Weld reinforcement not removed inspected by radiography or ultrasound.

(C)

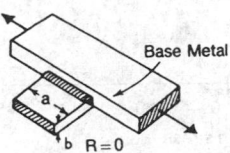

Longitudinal loading
Base metal–full penetration groove weld
Weld termination ground smooth
Weld reinforcement not removed
Not necessarily equal thickness

$2 < a < 12$ b or 4 in	(D)
$a > 12b$ or 4 in when $b \pm 1$ in	(E)
$a > 12b$ or 4 in when $b > 1$ in	(E′)

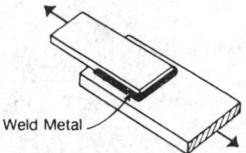

Weld metal of continuous or intermittent longitudinal or transverse fillet welds.

(F)

	Allowable Stress Range, σ_{sr} Ksi			
Category	20,000 to 100,000 ~	100,000 to 500,000 ~	500,000 to 2×10^6 ~	Over 2×10^6 ~
A	63	37	24	24
B	49	29	18	16
B′	39	23	15	12
C	35	21	13	10 (Note 1)
D	28	16	10	7
E	22	13	8	5
E′	16	9	6	3
F	15	12	9	8

Note 1: Flexural stress range of 12 ksi permitted at toe of stiffener welds on flanges.

Allowable fatigue stress:

$$\sigma_{max} = \frac{\upsilon_{sr}}{1-K} \quad \text{for normal stress } \sigma \qquad \tau_{max} = \frac{\tau_{sr}}{1-K} \quad \text{for shear stress } \tau$$

but shall not exceed steady allowables

$$\sigma_{max} \text{ or } \tau_{max} = \text{maximum allowable fatigue stress}$$
$$\sigma_{sr} \text{ or } \tau_{sr} = \text{allowable range of stress from table above}$$
$$K = \frac{\sigma_{min}}{\sigma_{max}} = \frac{M_{min}}{M_{max}} = \frac{F_{min}}{F_{max}} = \frac{\tau_{min}}{\tau_{max}} = \frac{V_{min}}{V_{max}}$$

where S = shear, T = tension, R = reversal, M = stress in metal, and W = stress in weld.

SOURCE: Abstracted from AISC "Manual of Steel Construction," 9th ed., by permission.

should not exceed the usual steady load allowable stress. For this reason, these fatigue curves are cut off with horizontal lines representing the steady load allowable stress for the particular type of steel used. In Fig. 13.3.26, A36 steel with E60 or E70 weld metal is cut off at 22 ksi; A441 steel with E70 weld metal at 30 ksi; and A514 with E110 weld metal at either 54 or 60 ksi, depending on plate thickness.

Figure 13.3.27 is a modification of Fig. 13.3.26, where the horizontal axis represents the range K of the cyclic stress. Here, two additional strength levels of weld metal have been added—E80 and E90—along with equivalent strength levels of steel. Note that for a small range in stress of $K = 0.6$ to 1.0, higher-strength welds and steels show increased allowable fatigue stress. However, as the stress range increases—lower values of K—the increase is not as great, and below $K = +0.35$ all combinations of weld and steel strengths exhibit the same allowable fatigue stress.

Figure 13.3.28 represents the same welded joint as in Fig. 13.3.27,

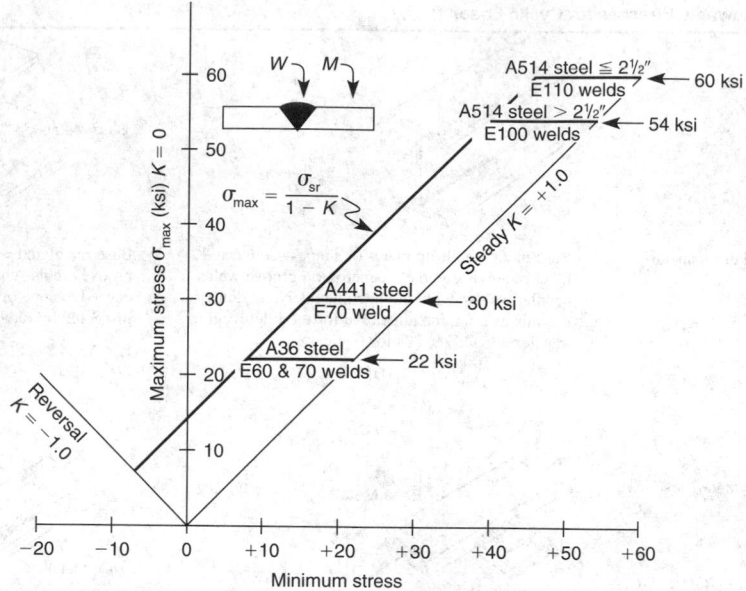

Fig. 13.3.26 Modified Goodman diagram for butt weld. [Butt weld and plate, weld reinforcement left on. Category C; 500,000 to 2,000,000 cycles (see Table 13.3.6).]

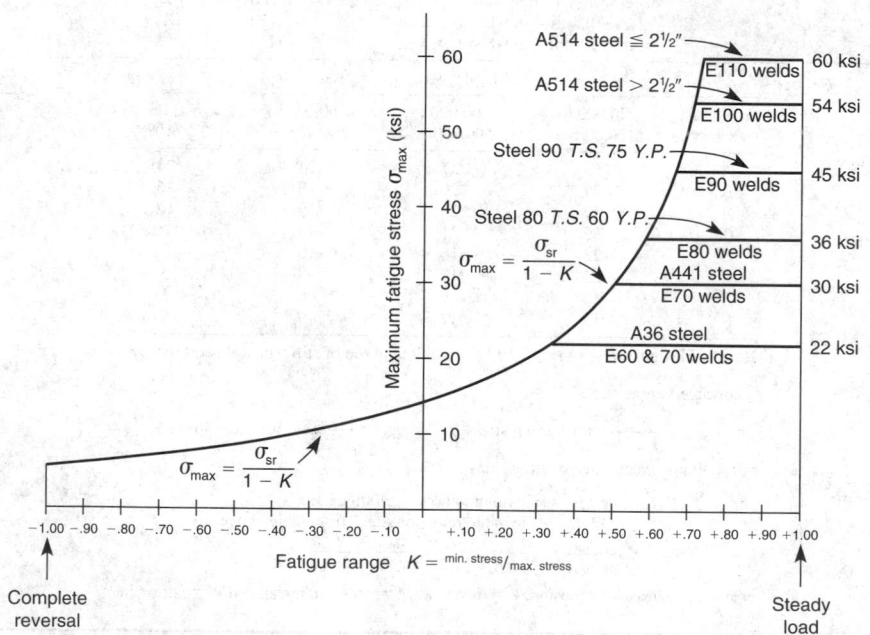

Fig. 13.3.27 Fatigue allowable for groove weld. [Butt weld and plate, weld reinforcement left on. Category C; 500,000 to 2,000,000 cycles (see Table 13.3.6).]

but with a lower life of 20,000 to 100,000 cycles. Here, the higher-strength welds and steels have higher allowable fatigue stresses and over a wider range. A conclusion can be drawn that the wider the range of cycling, the less useful the application of a high-strength steel. When there is a complete stress reversal, there is not much advantage in using a high-strength steel.

BASE METALS FOR WELDING

Introduction

When one is considering welding them, the nature of the base metals must be understood and recognized, i.e., their chemical composition, mechanical properties, and metallurgical structure. Cognizance of the

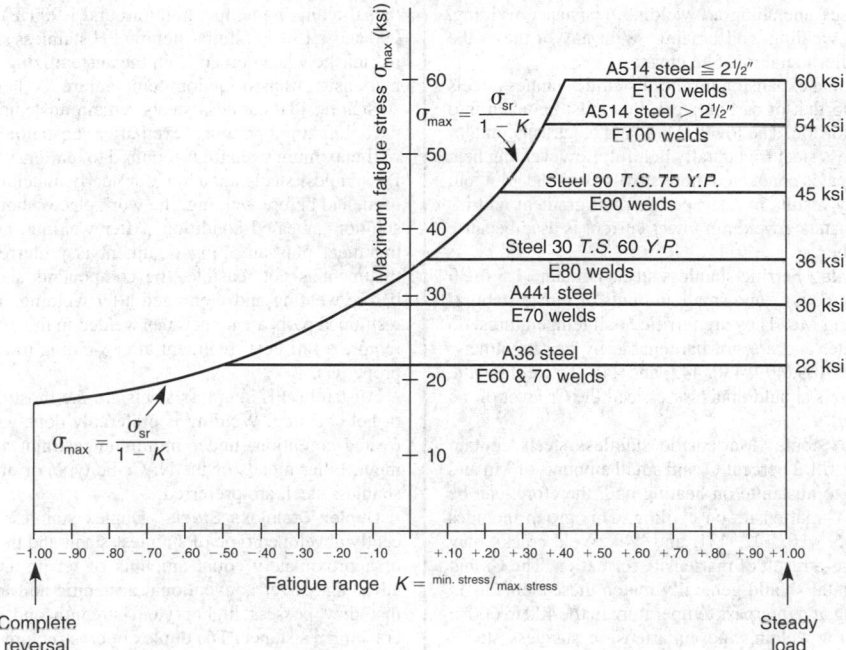

Fig. 13.3.28 Fatigue allowable for groove weld. [Butt weld and plate, weld reinforcement left on. Category C; 20,000 to 100,000 cycles (see Table 13.3.6).]

mechanical properties of the base metal will guide the designer to ensure that the weld metal deposited will have properties equal to those of the base metal; knowledge of the chemical composition of the base metal will affect the selection of the filler metal and/or electrode; finally, the metallurgical structure of the base metal as it comes to the welding operation (hot-worked, cold-worked, quenched, tempered, annealed, etc.) will affect the weldability of the metal and, if it is weldable, the degree to which the final properties are as dictated by design requirements. Welding specifications may address these matters, and base metal suppliers can provide additional data as to the weldability of the metal.

In some cases, the identity of the base metal is absolutely not known. To proceed to weld such metal may prove disastrous. Identification may be aided by some general characteristics which may be self-evident: carbon steel (oxide coating) vs. stainless steel (unoxidized); brush-finished aluminum (lightweight) vs. brush-finished Monel metal (heavy); etc. Ultimately, it may become necessary to subject the unknown metal to chemical, mechanical, and other types of laboratory tests to ascertain its exact nature.

Steel

Low-Carbon Steels   (Carbon up to 0.30 percent)   Steels in this class are readily welded by most arc and gas processes. Preheating is unnecessary unless parts are very heavy or welding is performed below 32°F (0°C). Torch-heating steel in the vicinity of welding to 70°F (21°C) offsets low temperatures. Postheating is necessary only for important structures such as boilers, pressure vessels, and piping. GTAW is usable only on killed steels; rimmed steels produce porous, weak welds. Resistance welding is readily accomplished if carbon is below 0.20 percent; higher carbon requires heat-treatment to slow the cooling rate and avoid hardness. Brazing with BAg, BCu, and BCuZn filler metals is very successful.

Medium-Carbon Steels   (Carbon from 0.30 to 0.45 percent)   This class of steel may be welded by the arc, resistance, and gas processes. As the rapid cooling of the metal in the welded zone produces a harder structure, it is desirable to hold the carbon as near 0.30 percent as possible. These hard areas are proportionately more brittle and difficult to machine. The cooling rate may be diminished and hardness decreased by preheating the metal to be welded above 300°F (149°C) and preferably to 500°F (260°C). The degree of preheating depends on the thickness of the section. Subsequent heating of the welded zone to 1,100 to 1,200°F (593 to 649°C) will restore ductility and relieve thermal strains. Brazing may also be used, as noted for low-carbon steels above.

High-Carbon Steels   (Carbon from 0.45 to 0.80 percent)   These steels are rarely welded except in special cases. The tendency for the metal heated above the critical range to become brittle is more pronounced than with lower- or medium-carbon steels. Thorough preheating of metal, in and near the welded zone, to a minimum of 500°F is essential. Subsequent annealing at 1,350 to 1,450°F (732 to 788°C) is also desirable. Brazing is often used with these steels, and is combined with the heat treatment cycle.

Low-Alloy Steels   The weldability of low-alloy steels is dependent upon the analysis and the hardenability, those exhibiting low hardenability being welded with relative ease, whereas those of high hardenability requiring preheating and postheating. Sections of ¼ in (6.4 mm) or less may be welded with mild-steel filler metal and may provide joint strength approximating base metal strength by virtue of alloy pickup in the weld metal and weld reinforcement. Higher-strength alloys require filler metals with mechanical properties matching the base metal. Special alloys with creep-resistant or corrosion-resistant properties must be welded with filler metals of the same chemical analysis. Low-hydrogen-type electrodes (either mild- or alloy-steel analyses) permit the welding of alloy steels, minimizing the occurrence of under-head cracking.

Stainless Steel

Stainless steel is an iron-base alloy containing upward of 11 percent chromium. A thin, dense surface film of chromium oxide which forms on stainless steel imparts superior corrosion resistance; its passivated nature inhibits scaling and prevents further oxidation, hence the appellation "stainless." (See Sec. 6.2.)

There are five types of stainless steels, and depending on the amount and kind of alloying additions present, they range from fully austenitic to fully ferritic. Most stainless steels have good weldability and may be welded by many processes, including arc welding, resistance welding, electron and laser beam welding, and brazing. With any of these, the joint surfaces and any filler metal must be clean.

The coefficient of thermal expansion for the austenitic stainless steels is 50 percent greater than that of carbon steel; this must be taken into account to minimize distortion. The low thermal and electrical conductivity of austenitic stainless steel is generally helpful. Low welding heat is required because the heat is conducted more slowly from the joint, but low thermal conductivity results in a steeper thermal gradient and increases distortion. In resistance welding, lower current is used because electric resistivity is higher.

Ferritic Stainless Steels Ferritic stainless steels contain 11.5 to 30 percent Cr, up to 0.20 percent C, and small amounts of ferrite stabilizers, such as Al, Nb, Ti, and Mo. They are ferritic at all temperatures, do not transform to austenite, and are not hardenable by heat treatment. This group includes types 405, 409, 430, 442, and 446. To weld ferritic stainless steels, filler metals should match or exceed the Cr level of the base metal.

Martensitic Stainless Steels Martensitic stainless steels contain 11.4 to 18 percent Cr, up to 1.2 percent C, and small amounts of Mn and Ni. They will transform to austenite on heating and, therefore, can be hardened by formation of martensite on cooling. This group includes types 403, 410, 414, 416, 420, 422, 431, and 440. Weld cracks may appear on cooled welds as a result of martensite formation. The Cr and C content of the filler metal should generally match these elements in the base metal. Preheating and interpass temperature in the 400 to 600°F range is recommended for welding most martensitic stainless steels. Steels with over 0.20 percent C often require a postweld heat treatment to avoid weld cracking.

Austenitic Stainless Steels Austenitic stainless steels contain 16 to 26 percent Cr, 10 to 24 percent Ni and Mn, up to 0.40 percent C, and small amounts of Mo, Ti, Nb, and Ta. The balance between Cr and Ni + Mn is normally adjusted to provide a microstructure of 90 to 100 percent austenite. These alloys have good strength and high toughness over a wide temperature range, and they resist oxidation to over 1,000°F. This group includes types 302, 304, 310, 316, 321, and 347. Filler metals for these alloys should generally match the base metal, but for most alloys should also provide a microstructure with some ferrite to avoid hot cracking. Two problems are associated with welding austenitic stainless steels: sensitization of the weld-heat-affected zone and hot cracking of weld metal.

Sensitization is caused by chromium carbide precipitation at the austenitic grain boundaries in the heat-affected zone when the base metal is heated to 800 to 1,600°F. Chromium carbide precipitates remove chromium from solution in the vicinity of the grain boundaries, and this condition leads to intergranular corrosion. The problem can be alleviated by using low-carbon stainless-steel base metal (types 302L, 316L, etc.) and low-carbon filler metal. Alternately, there are **stabilized stainless-steel** base metals and filler metals available which contain alloying elements that react preferentially with carbon, thereby not depleting the chromium content in solid solution and keeping it available for corrosion resistance. Type 321 contains titanium and type 347 contains niobium and tantalum, all of which are stronger carbide formers than chromium.

Hot cracking is caused by low-melting-point metallic compounds of sulfur and phosphorus which penetrate grain boundaries. When present in the weld metal or heat-affected zone, they will penetrate grain boundaries and cause cracks to appear as the weld cools and shrinkage stresses develop. Hot cracking can be prevented by adjusting the composition of the base metal and filler metal to obtain a microstructure with a small amount of ferrite in the austenite matrix. The ferrite provides ferrite-austenite boundaries which control the sulfur and phosphorus compounds and thereby prevent hot cracking.

Precipitation-Hardening Stainless Steels Precipitation-hardening (PH) stainless steels contain alloying elements such as aluminum which permit hardening by a solution and aging heat treatment. There are three categories of PH stainless steels: martensitic, semiaustenitic, and austenitic. Martensitic PH stainless steels are hardened by quenching from

the austenitizing temperature (around 1,900°F) and then aging between 900 and 1,150°F. Semiaustenitic PH stainless steels do not transform to martensite when cooled from the austenitizing temperature because the martensite transformation temperature is below room temperature. Austenitic PH stainless steels remain austenitic after quenching from the solution temperature, even after substantial amounts of cold work.

If maximum strength is required to martensitic PH and semiaustenitic PH stainless steels, matching, or nearly matching, filler metal should be used, and before welding, the work pieces should be in the annealed or solution-annealed condition. After welding, a complete solution heat treatment plus an aging treatment is preferred. If postweld solution treatment is not feasible, the components should be solution-treated before welding and then aged after welding. Thick sections of highly restrained parts are sometimes welded in the overaged condition. These require a full heat treatment after welding to attain maximum strength properties.

Austenitic PH stainless steels are the most difficult to weld because of hot cracking. Welding is preferably done with the parts in solution-treated condition, under minimum restraint and with minimum heat input. Filler metals of the Ni-Cr-Fe type, or of conventional austenitic stainless steel, are preferred.

Duplex Stainless Steels Duplex stainless steels are the most recently developed type of stainless steel, and they have a microstructure of approximately equal amounts of ferrite and austenite. They have advantages over conventional austenitic and ferritic stainless steels in that they possess higher yield strength and greater stress corrosion cracking resistance. The duplex microstructure is attained in steels containing 21 to 25 percent Cr and 5 to 7 percent Ni by hot-working at 1,832 to 1,922°F, followed by water quenching. Weld metal of this composition will be mainly ferritic because the deposit will solidify as ferrite and will transform only partly to austenite without hot working or annealing. Since hot-working or annealing most weld deposits is not feasible, the metal composition filler is generally modified by adding Ni (to 8 to 10 percent); this results in increased amounts of austenite in the as-welded microstructure.

Cast Iron

Even though cast iron has a high carbon content and is a relatively brittle and rigid material, welding can be performed successfully if proper precautions are taken. Optimum conditions for welding include the following: (1) A weld groove large enough to permit manipulation of the electrode or the welding torch and rod. The groove must be clean and free of oil, grease, and any foreign material. (2) Adequate preheat, depending on the welding process used, the type of cast iron, and the size and shape of the casting. Preheat temperature must be maintained throughout the welding operation. (3) Welding heat input sufficient for a good weld but not enough to superheat the weld metal; i.e., welding temperature should be kept as low as practicable. (4) Slow cooling after welding. Gray iron may be enclosed in insulation, lime, or vermiculite. Other irons may require postheat treatment immediately after welding to restore mechanical properties. ESt and ENiFe identify electrodes of steel and of a nickel-iron alloy. Many different welding processes have been used to weld cast iron, the most common being manual shielded metal-arc welding, gas welding, and braze welding.

Aluminum and Aluminum Alloys

(See Sec. 6.4)

The **properties** that distinguish the **aluminum alloys** from other metals determine which welding processes can be used and which particular procedures must be followed for best results. Among the welding processes that can be used, choice is further dictated by the requirements of the end product and by economic considerations.

Physical properties of aluminum alloys that most significantly affect all welding procedures include low melting-point range, approx 900 to 1,215°F (482 to 657°C), high thermal conductivity (about two to four times that of mild steel), high rate of thermal expansion (about twice that of mild steel), and high electrical conductivity (about 3 to 5 times that of mild steel). Interpreted in terms of welding, this means that, when compared with mild steel, much higher welding speeds are demanded, greater care must be exercised to avoid distortion, and for arc and resistance welding, much higher current densities are required.

Aluminum alloys are not quench-hardenable. However, weld cracking may result from excessive shrinkage stresses due to the high rate of thermal contraction. To offset this tendency, welding procedures, where possible, require a fast weld cycle and a narrow-weld zone, e.g., a highly concentrated heat source with deep penetration, moving at a high rate of speed. Shrinkage stresses can also be reduced by using a filler metal of lower melting point than the base metal. The filler metal ER4043 is often used for this purpose.

Welding procedures also call for the removal of the thin, tough, transparent film of aluminum oxide that forms on and protects the surface of these alloys. The oxide has a melting point of about 3,700°F (2,038°C) and can therefore exist as a solid in the molten weld. Removal may be by chemical reduction or by mechanical means such as machining, filing, rubbing with steel wool, or brushing with a stainless-steel wire brush. Most aluminum is welded with GTAW or GMAW. GTAW usually uses alternating current, with argon as the shielding gas. The power supply must deliver high current with balanced wave characteristics, or deliver high-frequency current. With helium, weld penetration is deeper, and higher welding speeds are possible. Most welding, however, is done using argon because it allows for better control and permits the welder to see the weld pool more easily.

GMAW employs direct current, electrode positive in a shielding gas that may be argon, helium, or a mixture of the two. In this process, the welding arc is formed by the filler metal, which serves as the electrode. Since the filler metal is fed from a coil as it melts in the arc, some arc instability may arise. For this reason, the process does not have the same precision as the GTAW process for welding very thin gages. However, it is more economical for welding thicker sections because of its higher deposition rates.

Copper and Copper Alloys

In welding **commercially pure copper,** it is important to select the correct type. Electrolytic, or ''tough-pitch,'' copper contains a small percentage of copper oxide, which at welding heat leads to oxide embrittlement. For welded assemblies it is recommended that deoxidized, or oxygen-free, copper be used and that welding rods, when needed, be of the same analysis. The preferred processes for welding copper are GTAW and GMAW; manual SMAW can also be used. It is also welded by oxyacetylene method and braze-welded; brazing with brazing filler metals conforming to BAg, BCuP, and RBCuZn-A classifications is also employed. The high heat conductivity of copper requires special consideration in welding; generally higher welding heats are necessary together with concurrent supplementary heating. (See also Sec. 6.4.)

Copper alloys are extensively welded in industry. The specific procedures employed are dependent upon the analysis, and reference should be made to the AWS Welding Handbook. Filler metals for welding copper and its alloys are covered in AWS specifications.

SAFETY

Welding is safe when sufficient measures are taken to protect the welder from potential hazards. When these measures are overlooked or ignored, welders can be subject to electric shock; overexposure to radiation, fumes, and gases; and fires and explosion. Any of these can be fatal. Everyone associated with welding operations should be aware of the potential hazards and help ensure that safe practices are employed. Infractions must be reported to the appropriate responsible authority.

NOTE: Oxygen is incorrectly called **air** in some fabricating shops. Air from the atmosphere contains only 21 percent oxygen and obviously is different from the 100 percent pure oxygen used for cutting. The unintentional confusion of oxygen with air has resulted in fatal accidents. When compressed oxygen is inadvertently used to power air tools, e.g., an explosion can result. While most people recognize that fuel gases are dangerous, the case can be made that oxygen requires even more careful handling.

Information about welding safety is available from American Welding Society, P.O. Box 351040, Miami, FL 33135.

13.4   MATERIAL REMOVAL PROCESSES AND MACHINE TOOLS
by Serope Kalpakjian

REFERENCES: ''Machining,'' ASM Handbook, vol. 16, ASM International. ''Machining,'' Tool and Manufacturing Engineers Handbook, vol. 1, Society of Manufacturing Engineers. ''Machining Data Handbook,'' 3d ed., Machinability Data Center. ''Machinery's Handbook,'' Industrial Press. ''Metal Cutting Tool Handbook,'' 7th ed., United States Cutting Tool Institute. Boothroyd and Knight, ''Fundamentals of Machining and Machine Tools,'' 2d ed., Marcel Dekker. Borkowski and Szymanski, ''Uses of Abrasives and Abrasive Tools,'' Ellis Horwood. Byers (ed.), ''Metalworking Fluids,'' Marcel Dekker. Endoy, ''Gear Hobbing, Shaping and Shaving,'' Society of Manufacturing Engineers. Farago, ''Abrasive Methods Engineering,'' 2 vols., Industrial Press. Gibbs and Crandell, ''An Introduction to CNC Machining and Programming,'' Industrial Press. Kalpakjian, ''Manufacturing Engineering and Technology,'' 3d ed., Addison-Wesley. King and Hahn, ''Handbook of Modern Grinding Technology,'' Chapman & Hall/Methuen. Krar and Ratterman, ''Superabrasives: Grinding and Machining with CBN and Diamond,'' McGraw-Hill. Lynch, ''Computer Numerical Control for Machining,'' McGraw-Hill. Malkin, ''Grinding Technology: Theory and Applications of Machining with Abrasives,'' Wiley. Maroney, ''A Guide to Metal and Plastic Finishing,'' Industrial Press. McGeough, ''Advanced Methods of Machining,'' Chapman & Hall. Meyers and Slattery, ''Basic Machining Reference Handbook,'' Industrial Press. Nachtman and Kalpakjian, ''Lubricants and Lubrication in Metalworking Operations,'' Marcel Dekker. Polywka and Gabrel, ''Programming of Computer Numerically Controlled Machines,'' Industrial Press. Shaw, ''Metal Cutting Principles,'' Oxford. Thyer, ''Computer Numerical Control of Machine Tools,'' 2d ed., Industrial Press. Trent, ''Metal Cutting,'' 4th ed., Butterworth Heinemann. Walsh, ''McGraw-Hill Machining and Metalworking Handbook.'' Weck, ''Handbook of Machine Tools,'' 4 vols., Wiley. Stephenson and Agapiou, ''Metal Cutting: Theory and Practice,'' Marcel Dekker. DeVries, ''Analysis of Material Removal Processes,'' Springer. ''Tool Materials,'' ASM Specialty Handbook, ASM International. Juneja, ''Fundamentals of Metal Cutting and Machine Tools,'' McGraw-Hill. Salmon, ''Modern Grinding Process Technology,'' McGraw-Hill. Benedict, ''Nontraditional Manufacturing Processes,'' Marcel Dekker. Webster et al., ''Abrasive Processes: Theory, Technology, and Practice,'' Marcel Dekker.

INTRODUCTION

Material removal processes, which include machining, cutting, grinding, and various nonmechanical chipless processes, are desirable or even necessary for the following basic reasons: (1) Closer dimensional tolerances, surface roughness, or surface-finish characteristics may be required than are available by casting, forming, powder metallurgy, and other shaping processes; and (2) part geometries may be too complex or too expensive to be manufactured by other processes. However, material removal processes inevitably waste material in the form of chips, production rates may be low, and unless carried out properly, the processes can have detrimental effects on the surface properties and performance of parts.

Traditional material removal processes consist of turning, boring, drilling, reaming, threading, milling, shaping, planing, and broaching, as well as abrasive processes such as grinding, ultrasonic machining, lapping, and honing. Nontraditional processes include electrical and chemical means of material removal, as well as the use of abrasive jets, water jets, laser beams, and electron beams. This section describes the principles of these operations, the processing parameters involved, and the characteristics of the machine tools employed.

BASIC MECHANICS OF METAL CUTTING

The basic mechanics of chip-type machining processes (Fig. 13.4.1) are shown, in simplest two-dimensional form, in Fig. 13.4.2. A tool with a certain **rake angle** α (positive as shown) and **relief angle** moves along the surface of the workpiece at a depth t_1. The material ahead of the tool is sheared continuously along the **shear plane**, which makes an angle of ϕ

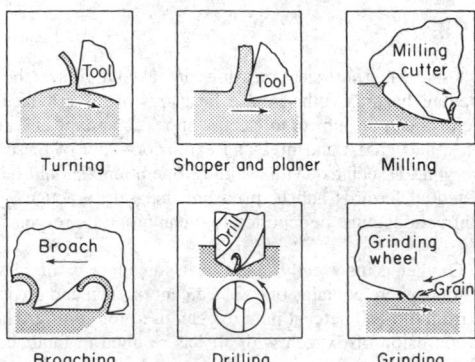

Fig. 13.4.1 Examples of chip-type machining operations.

with the surface of the workpiece. This angle is called the **shear angle** and, together with the rake angle, determines the chip thickness t_2. The ratio of t_1 to t_2 is called the **cutting ratio** r. The relationship between the shear angle, the rake angle, and the cutting ratio is given by the equation $\tan \phi = r \cos \alpha / (1 - r \sin \alpha)$. It can readily be seen that the shear angle is important in that it controls the thickness of the chip. This, in turn, has great influence on cutting performance. The **shear strain** that the material undergoes is given by the equation $\gamma = \cot \phi + \tan (\phi - \alpha)$. Shear strains in metal cutting are usually less than 5.

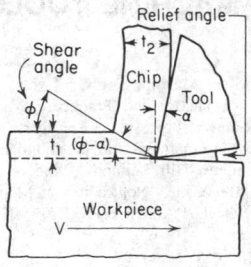

Fig. 13.4.2 Basic mechanics of metal cutting process.

Investigations have shown that the shear plane may be neither a plane nor a narrow zone, as assumed in simple analysis. Various formulas have been developed which define the shear angle in terms of such factors as the rake angle and the friction angle β. (See Fig. 13.4.3.)

Because of the large shear strains that the chip undergoes, it becomes hard and brittle. In most cases, the chip curls away from the tool. Among possible factors contributing to chip curl are nonuniform normal stress distribution on the shear plane, strain hardening, and thermal effects.

Regardless of the type of machining operation, some basic types of chips or combinations of these are found in practice (Fig. 13.4.4):

Continuous chips are formed by continuous deformation of the work-

piece material ahead of the tool, followed by smooth flow of the chip along the tool face. These chips ordinarily are obtained in cutting ductile materials at high speeds.

Discontinuous chips consist of segments which are produced by fracture of the metal ahead of the tool. The segments may be either loosely

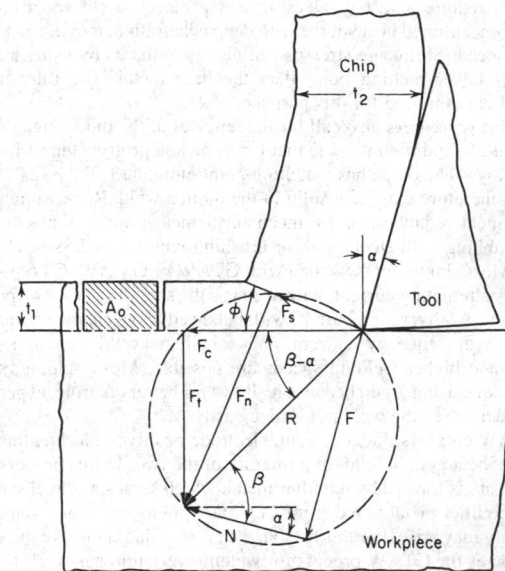

Fig. 13.4.3 Force system in metal cutting process.

connected to each other or unconnected. Such chips are most often found in the machining of brittle materials or in cutting ductile materials at very low speeds or low or negative rake angles.

Inhomogeneous (serrated) chips consist of regions of large and small strain. Such chips are characteristic of metals with low thermal conductivity or metals whose yield strength decreases sharply with temperature. Chips from titanium alloys frequently are of this type.

Built-up edge chips consist of a mass of metal which adheres to the tool face while the chip itself flows continuously along the face. This type of chip is often encountered in machining operations at low speeds and is associated with high adhesion between chip and tool and causes poor surface finish.

The **forces** acting on the cutting tool are shown in Fig. 13.4.3. The resultant force R has two components, F_c and F_t. The cutting force F_c in the direction of tool travel determines the amount of work done in cutting. The thrust force F_t does no work but, together with F_c, produces deflections of the tool. The resultant force also has two components on the shear plane: F_s is the force required to shear the metal along the shear plane, and F_n is the normal force on this plane. Two other force components also exist on the face of the tool: the friction force F and the normal force N.

Whereas the cutting force F_c is always in the direction shown in Fig. 13.4.3, the thrust force F_t may be in the opposite direction to that shown in the figure. This occurs when both the rake angle and the depth of cut are large, and friction is low.

From the geometry of Fig. 13.4.3, the following relationships can be derived: The **coefficient of friction** at the tool-chip interface is given by $\mu = (F_t + F_c \tan \alpha) / (F_c - F_t \tan \alpha)$. The **friction force** along the tool is $F = F_t \cos \alpha + F_c \sin \alpha$. The **shear stress** in the shear plane is $\tau = (F_c \sin \phi \cos \phi - F_t \sin^2 \phi) / A_0$, where A_0 is the cross-sectional area that is being cut from the workpiece.

The coefficient of friction on the tool face is a complex but important factor in cutting performance; it can be reduced by such means as the use of an effective cutting fluid, higher cutting speed, improved tool material and condition, or chemical additives in the workpiece material.

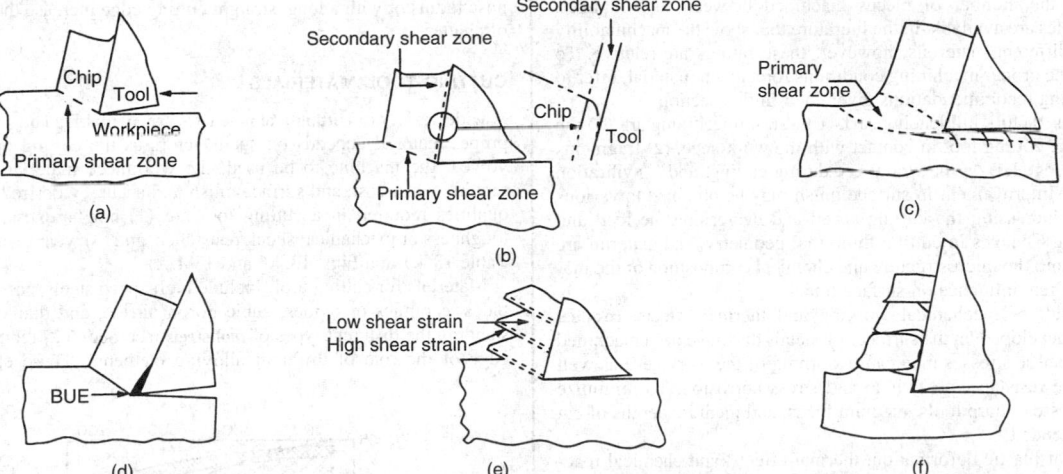

Fig. 13.4.4 Basic types of chips produced in metal cutting: (a) continuous chip with narrow, straight primary shear zone; (b) secondary shear zone at the tool-chip interface; (c) continuous chip with large primary shear zone; (d) continuous chip with built-up edge; (e) segmented or nonhomogeneous chip, (f) discontinuous chip. (*Source: After M. C. Shaw.*)

The net **power** consumed at the tool is $P = F_c V$. Since F_c is a function of tool geometry, workpiece material, and process variables, it is difficult reliably to calculate its value in a particular machining operation. Depending on workpiece material and the condition of the tool, **unit power** requirements in machining range between 0.2 hp·min/in³ (0.55 W·s/mm³) of metal removal for aluminum and magnesium alloys, to 3.5 for high-strength alloys. The power consumed is the product of unit power and rate of metal removal: $P = (\text{unit power})(\text{vol/min})$.

The power consumed in cutting is transformed mostly to **heat**. Most of the heat is carried away by the chip, and the remainder is divided between the tool and the workpiece. An increase in cutting speed or feed will increase the proportion of the heat transferred to the chip. It has been observed that, in turning, the average interface **temperature** between the tool and the chip increases with cutting speed and feed, while the influence of the depth of cut on temperature has been found to be limited. Interface temperatures to the range of 1,500 to 2,000°F (800 to 1,100°C) have been measured in metal cutting. Generally the use of a cutting fluid removes heat and thus avoids temperature buildup on the cutting edge.

In cutting metal at high speeds, the chips may become very hot and cause safety hazards because of long spirals which whirl around and become entangled with the tooling. In such cases, **chip breakers** are introduced on the tool geometry, which curl the chips and cause them to break into short sections. Chip breakers can be produced on the face of the cutting tool or insert, or are separate pieces clamped on top of the tool or insert.

A factor of great significance in metal cutting is **tool wear**. Many factors determine the type and rate at which wear occurs on the tool. The major critical variables that affect wear are tool temperature, type and hardness of tool material, grade and condition of workpiece, abrasiveness of the microconstituents in the workpiece material, tool geometry, feed speed, and cutting fluid. The type of wear pattern that develops depends on the relative role of these variables.

Tool wear can be classified as (1) flank wear (Fig. 13.4.5); (2) crater wear on the tool face; (3) localized wear, such as the rounding of the cutting edge; (4) chipping or thermal softening and plastic flow of the cutting edge; (5) concentrated wear resulting in a deep groove at the edge of a turning tool, known as **wear notch**.

In general, the wear on the flank or relief side of the tool is the most dependable guide for **tool life**. A wear land of 0.060 in (1.5 mm) on high-speed steel tools and 0.015 or 0.030 in (0.4 or 0.8 mm) for carbide tools is usually used as the endpoint. The cutting speed is the variable

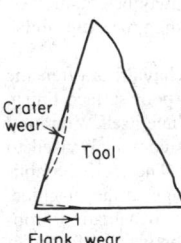

Fig. 13.4.5 Types of tool wear in cutting.

which has the greatest influence on tool life. The relationship between tool life and cutting speed is given by the Taylor equation $VT^n = C$, where V is the cutting speed; T is the actual cutting time to develop a certain wear land, min; C is a constant whose value depends on workpiece material and process variables, numerically equal to the cutting speed that gives a tool life of 1 min; and n is the exponent whose value depends on workpiece material and other process variables. The recommended cutting speed for a high-speed steel tool is generally the one which produces a 60- to 120-min tool life. With carbide tools, a 30- to 60-min tool life may be satisfactory. Values of n range from 0.08 to 0.2 for high-speed steels, from 0.2 to 0.5 for carbides, and from 0.5 to 0.7 for ceramic tools.

When one is using tool-life equations, caution should be exercised in extrapolation of the curves beyond the operating region for which they are derived. In a log-log plot, tool life curves may be linear over a short cutting-speed range but are rarely linear over a wide range of cutting speeds. In spite of the considerable data obtained to date, no simple formulas can be given for quantitative relationships between tool life and various process variables for a wide range of materials and conditions.

An important aspect of machining on computer-controlled equipment is **tool-condition monitoring** while the machine is in operation with little or no supervision by an operator. Most state-of-the-art machine controls are now equipped with tool-condition monitoring systems. Two common techniques involve the use of (1) transducers that are installed on the tool holder and continually monitor torque and forces and (2) acoustic emission through a piezoelectric transducer. In both methods the signals are analyzed and interpreted automatically for tool wear or chipping, and corrective actions are taken before any significant damage is done to the workpiece.

A term commonly used in machining and comprising most of the items discussed above is **machinability**. This is best defined in terms of (1) tool life, (2) power requirement, and (3) surface integrity. Thus, a good machinability rating would indicate a combination of long tool life, low power requirement, and a good surface. However, it is difficult to develop quantitative relationships between these variables. Tool life is considered as the important factor and, in production, is usually expressed as the number of pieces machined between tool changes. Various tables are available in the literature that show the machinability rating for different materials; however, these ratings are relative. To determine the proper machining conditions for a given material, refer to the machining recommendations given later in this section.

The major factors influencing **surface finish** in machining are (1) the profile of the cutting tool in contact with the workpiece, (2) fragments of built-up edge left on the workpiece during cutting, and (3) vibration and chatter. Improvement in surface finish may be obtained to various degrees by increasing the cutting speed and decreasing the feed and depth of cut. Changes in cutting fluid, tool geometry, and material are also important; the microstructure and chemical composition of the material have great influence on surface finish.

As a result of mechanical working and thermal effects, **residual stresses** are developed on the surfaces of metals that have been machined or ground. These stresses may cause warping of the workpiece as well as affect the resistance to fatigue and stress corrosion. To minimize residual stresses, sharp tools, medium feeds, and medium depths of cut are recommended.

Because of plastic deformation, thermal effects, and chemical reactions during material removal processes, alterations of machined surfaces may take place which can seriously affect the **surface integrity** of a part. Typical detrimental effects may be lowering of the fatigue strength of the part, distortion, changes in stress-corrosion properties, burns, cracks, and residual stresses. Because of its great importance to manufacturing technology, particularly in critical aerospace components, surface integrity is now a recognized and rapidly developing subject. Surface control generally results in decreasing production rate and increasing costs. Improvements in surface integrity may be obtained by post-processing techniques such as polishing, sanding, peening, finish machining, and fine grinding.

Vibration in machine tools, a very complex behavior, is often the cause of premature tool failure or short tool life, poor surface finish, damage to the workpiece, and damage to the machine itself. Vibration may be **forced** or **self-excited**. The term **chatter** is commonly used to designate self-excited vibrations in machine tools. The excited amplitudes are usually very high and may cause damage to the machine. Although there is no complete solution to all types of vibration problems, certain measures may be taken. If the vibration is being forced, it may be possible to remove or isolate the forcing element from the machine. In cases where the forcing frequency is near a natural frequency, either the forcing frequency or the natural frequency may be raised or lowered. Damping will also greatly reduce the amplitude. Self-excited vibrations are generally controlled by increasing the stiffness and damping. (See also Sec. 5.)

Good machining practice requires a rigid setup. The machine tool must be capable of providing the **stiffness** required for the machining conditions used. If a rigid setup is not available, the depth of cut must be reduced. Excessive tool overhang should be avoided, and in milling, cutters should be mounted as close to the spindle as possible. The length of end mills and drills should be kept to a minimum. Tools with large

nose radius or with a long, straight cutting edge increase the possibility of chatter.

CUTTING-TOOL MATERIALS

A wide variety of **cutting-tool materials** are available. The selection of a proper material depends on such factors as the cutting operation involved, the machine to be used, the workpiece material, production requirements, cost, and surface finish and accuracy desired. The major qualities required in a cutting tool are (1) hot hardness, (2) impact toughness or mechanical shock resistance, and (3) wear resistance. (See Table 13.4.1 and Figs. 13.4.6 and 13.4.7.)

Materials for cutting tools include high-speed steels, cast alloys, carbides, ceramics or oxides, cubic boron nitride, and diamond. Understanding the different types of **tool steels** (see Sec. 6.2) requires knowledge of the role of different alloying elements. These elements are

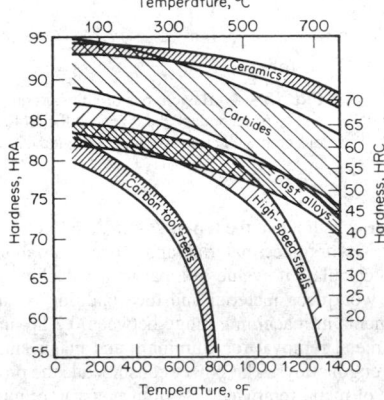

Fig. 13.4.6 Hardness of tool materials as a function of temperature.

added to (1) obtain greater hardness and wear resistance, (2) obtain greater impact toughness, (3) impart hot hardness to the steel such that its hardness is maintained at high cutting temperatures, and (4) decrease distortion and warpage during heat treating.

Carbon forms a carbide with iron, making it respond to hardening and thus increasing the hardness, strength, and wear resistance. The carbon content of tool steels ranges from 0.6 to 1.4 percent. **Chromium** is added to increase wear resistance and toughness; the content ranges from 0.25 to 4.5 percent. **Cobalt** is commonly used in high-speed steels to increase hot hardness so that tools may be used at higher cutting speeds and still maintain hardness and sharp cutting edges; the content ranges from 5 to 12 percent. **Molybdenum** is a strong carbide-forming element and increases strength, wear resistance, and hot hardness. It is always used in conjunction with other alloying elements, and its content ranges to 10 percent. **Tungsten** promotes hot hardness and strength; content ranges

Table 13.4.1 Characteristics of Cutting-Tool Materials

	High-speed steels	Cast cobalt alloys	Carbides	Coated carbides	Ceramics	Polycrystalline cubic boron nitride	Diamond
Hot hardness	←	increasing					→
Toughness	←	increasing					
Impact strength	←	increasing					
Wear resistance	←	increasing					→
Chipping resistance	←	increasing					
Cutting speed	←	increasing					→
Thermal shock resistance	←	increasing					
Tool material cost	←	increasing					→

NOTE: These tool materials have a wide range of compositions and properties; thus overlapping characteristics exist in many categories of tool materials.

SOURCE: After R Komandurr (ed.), ''Advances in Hard Material Tool Technology,'' Carnegie Press, Pittsburgh, PA.

from 1.25 to 20 percent. **Vanadium** increases hot hardness and abrasion resistance; in high-speed steels, it ranges from 1 to 5 percent.

High-speed steels are the most highly alloyed group among tool steels and maintain their hardness, strength, and cutting edge. With suitable procedures and equipment, they can be fully hardened with little danger of distortion or cracking. High-speed steel tools are widely used in operations using form tools, drilling, reaming, end-milling, broaching, tapping, and tooling for screw machines.

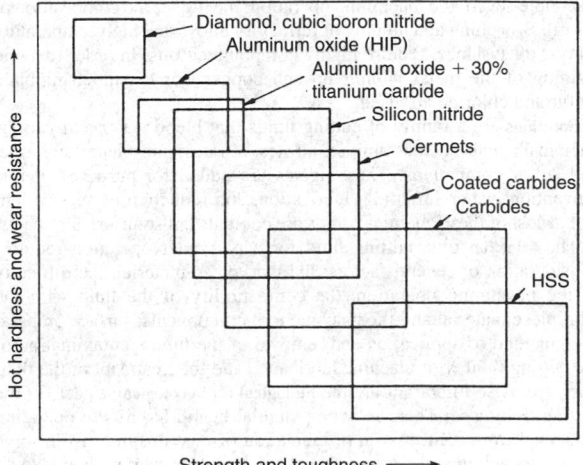

Fig. 13.4.7 Ranges of properties of various groups of tool materials.

Cast alloys maintain high hardness at high temperatures and have good wear resistance. Cast-alloy tools, which are cast and ground into any desired shape, are composed of cobalt (38 to 53 percent), chromium (30 to 33 percent), and tungsten (10 to 20 percent). These alloys are recommended for deep roughing operations at relatively high speeds and feeds. Cutting fluids are not necessary and are usually used only to obtain a special surface finish.

Carbides have metal carbides as key ingredients and are manufactured by powder-metallurgy techniques. They have the following properties which make them very effective cutting-tool materials: (1) high

hardness over a wide range of temperatures; (2) high elastic modulus, 2 to 3 times that of steel; (3) no plastic flow even at very high stresses; (4) low thermal expansion; and (5) high thermal conductivity. Carbides are used in the form of inserts or tips which are clamped or brazed to a steel shank. Because of the difference in coefficients of expansion, brazing should be done carefully. The mechanically fastened tool tips are called **inserts** (Fig. 13.4.8); they are available in different shapes, such as square, triangular, circular, and various special shapes.

There are three general groups of carbides in use: (1) tungsten carbide with cobalt as a binder, used in machining cast irons and nonferrous abrasive metals; (2) tungsten carbide with cobalt as a binder, plus a solid solution of WC-TiC-TaC-NbC, for use in machining steels; and

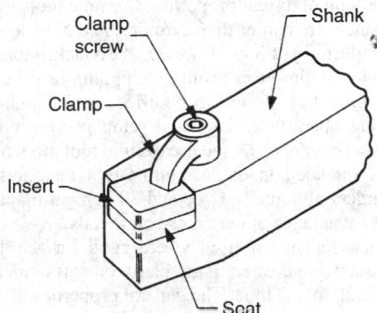

Fig. 13.4.8 Insert clamped to shank of a toolholder.

(3) titanium carbide with nickel and molybdenum as a binder, for use where cutting temperatures are high because of high cutting speeds or the high strength of the workpiece material. Carbides are classified by ISO and ANSI, as shown in Table 13.4.2 which includes recommendations for a variety of workpiece materials and cutting conditions. (See also Sec. 6.4.)

Coated carbides consist of conventional carbide inserts that are coated with a thin layer of titanium nitride, titanium carbide, or aluminum oxide. The coating provides additional wear resistance while maintaining the strength and toughness of the carbide tool. Coatings are also applied to high-speed steel tools, particularly drills and taps. The desirable properties of individual coatings can be combined and optimized

Table 13.4.2 Classification of Tungsten Carbides According to Machining Applications

ISO standard	ANSI classification no. (grade)	Materials to be machined	Machining operation	Type of carbide	Characteristics of	
					Cut	Carbide
K30-K40	C-1	Cast iron, nonferrous metals, and nonmetallic materials requiring abrasion resistance	Roughing	Wear-resistant grades; generally straight WC-Co with varying grain sizes	Increasing cutting speed ↑	Increasing hardness and wear resistance ↑
K20	C-2		General purpose			
K10	C-3		Light finishing			
K01	C-4		Precision finishing		↓ Increasing feed rate	↓ Increasing strength and binder content
P30-P50	C-5	Steels and steel alloys requiring crater and deformation resistance	Roughing	Crater-resistant grades; various WC–Co compositions with TiC and/or TaC alloys	Increasing cutting speed ↑	Increasing hardness and wear resistance ↑
P20	C-6		General purpose			
P10	C-7		Light finishing			
P01	C-8		Precision finishing		↓ Increasing feed rate	↓ Increasing strength and binder content

NOTE: The ISO and ANSI comparisons are approximate.

by using **multiphase coatings**. Carbide tools are now available with, e.g., a layer of titanium carbide over the carbide substrate, followed by aluminum oxide and then titanium nitride. Various alternating layers of coatings are also used, each layer being on the order of 80 to 400 μin (2 to 10 μm) thick. New developments in coatings include diamond, titanium carbonitride, chromium carbide, zirconium nitride, and hafnium nitride.

Stiffness is of great importance when using carbide tools. Light feeds, low speeds, and chatter are deleterious. No cutting fluid is needed, but if one is used for cooling, it should be applied in large quantities and continuously to prevent heating and quenching.

Ceramic, or **oxide,** inserts consist primarily of fine aluminum oxide grains which have been bonded together. Minor additions of other elements help to obtain optimum properties. Ceramic tools have very high abrasion resistance, are harder than carbides, and have less tendency to weld to metals during cutting. However, they lack impact toughness, and premature tool failure can result by chipping or general breakage. Ceramic tools have been found to be effective for high-speed, uninterrupted turning operations. Tool and setup geometry is important. Tool failures can be reduced by the use of rigid tool mountings and rigid machine tools. Included in oxide, cutting-tool materials are **cermets** (such as 70 percent aluminum oxide and 30 percent titanium carbide), combining the advantages of ceramics and metals.

Polycrystalline **diamond** is used where good surface finish and dimensional accuracy are desired, particularly on soft nonferrous materials that are difficult to machine. The general properties of diamonds are extreme hardness, low thermal expansion, high heat conductivity, and a very low coefficient of friction. The polycrystalline diamond is bonded to a carbide substrate. Single-crystal diamond is also used as a cutting tool to produce extremely fine surface finish on nonferrous alloys, such as copper-base mirrors.

Next to diamond, **cubic boron nitride (CBN)** is the hardest material presently available. Polycrystalline CBN is bonded to a carbide substrate and used as a cutting tool. The CBN layer provides very high wear resistance and edge strength. It is chemically inert to iron and nickel at elevated temperatures; thus it is particularly suitable for machining high-temperature alloys and various ferrous alloys. Both diamond and CBN are also used as abrasives in grinding operations.

CUTTING FLUIDS

Cutting fluids, frequently referred to as lubricants or coolants, comprise those liquids and gases which are applied to the cutting zone to facilitate the cutting operation. A cutting fluid is used (1) to keep the tool cool and prevent it from being heated to a temperature at which the hardness and resistance to abrasion are reduced; (2) to keep the workpiece cool, thus preventing it from being machined in a warped shape to inaccurate final dimensions; (3) through lubrication to reduce the power consumption, wear on the tool, and generation of heat; (4) to provide a good finish on the workpiece; (5) to aid in providing a satisfactory chip formation; (6) to wash away the chips (this is particularly desirable in deep-hole drilling, hacksawing, milling, and grinding); and (7) to prevent corrosion of the workpiece and machine tool.

Classification Cutting fluids may be classified as follows: (1) air blast, (2) emulsions, (3) oils, and (4) solutions. Cutting fluids are also classified as light-, medium-, and heavy-duty; light-duty fluids are for general-purpose machining.

Induced **air blast** may be used with internal and surface grinding and polishing operations. Its main purpose is to remove the small chips or dust, although some cooling is also obtained, especially in machining of plastics.

Emulsions consist of a soluble oil emulsified with water in the ratio of 1 part oil to 10 to 100 parts water, depending upon the type of product and the operation. Emulsions have surface-active or **extreme-pressure** additives to reduce friction and provide an effective lubricant film under high pressure at the tool-chip interface during machining. Emulsions are low-cost cutting fluids and are used for practically all types of cutting and grinding when machining all types of metals. The more concen-

trated mixtures of oil and water, such as 1 : 10, are used for broaching, threading, and gear cutting. For most operations, a solution of 1 part soluble oil to 20 parts water is satisfactory.

A variety of **oils** are used for metal cutting. They are used where lubrication rather than cooling is essential or on high-grade finishing cuts, although sometimes superior finishes are obtained with emulsions. Oils generally used in machining are mineral oils with the following compositions: (1) straight mineral oil, (2) with fat, (3) with fat and sulfur, (4) with fat and chlorine, and (5) with fat, sulfur, and chlorine. The more severe the machining operation, the higher the composition of the oil. Broaching and tapping of refractory alloys and high-temperature alloys, for instance, require highly compounded oils. In order to avoid staining of the metal, aluminum and copper, for example, inhibited sulfur and chlorine are used.

Solutions are a family of cutting fluids that blend water and various chemical agents such as amines, nitrites, nitrates, phosphates, chlorine, and sulfur compounds. These agents are added for purposes of rust prevention, water softening, lubrication, and reduction of surface tension. Most of these chemical fluids are coolants but some are lubricants.

The **selection** of a cutting fluid for a particular operation requires consideration of several factors: the workpiece material, the difficulty of the machining operation, the compatibility of the fluid with the workpiece material and the machine tool components, surface preparation, method of application and removal of the fluid, contamination of the cutting fluid with machine lubricants, and the treatment of the fluid after use. Also important are the **biological** and **ecological** aspects of the cutting fluid used. There may be potential health hazards to operating personnel from contact with or inhalation of mist or fumes from some fluids. Recycling and waste disposal are also important problems to be considered.

Methods of Application The most common method is **flood cooling** in quantities such as 3 to 5 gal/min (about 10 to 20 L/min) for single-point tools and up to 60 gal/min (230 L/min) per cutter for multiple-tooth cutters. Whenever possible, multiple nozzles should be used. In **mist cooling** a small jet equipment is used to disperse water-base fluids as very fine droplets in a carrier that is generally air at pressures 10 to 80 lb/in^2 (70 to 550 kPa). Mist cooling has a number of advantages, such as providing high-velocity fluids to the working areas, better visibility, and improving tool life in certain instances. The disadvantages are that venting is required and also the cooling capability is rather limited.

MACHINE TOOLS

The general types of **machine tools** are lathes; turret lathes; screw, boring, drilling, reaming, threading, milling, and gear-cutting machines; planers and shapers; broaching, cutting-off, grinding, and polishing machines. Each of these is subdivided into many types and sizes. General items common to all machine tools are discussed first, and individual machining processes and equipment are treated later in this section.

Automation is the application of special equipment to control and perform manufacturing processes with little or no manual effort. It is applied to the manufacturing of all types of goods and processes, from the raw material to the finished product. Automation involves many activities, such as handling, processing, assembly, inspecting, and packaging. Its primary objective is to lower manufacturing cost through controlled production and quality, lower labor cost, reduced damage to work by handling, higher degree of safety for personnel, and economy of floor space. Automation may be partial, such as gaging in cylindrical grinding, or it may be complete.

The conditions which play a role in decisions concerning automation are rising production costs, high percentage of rejects, lagging output, scarcity of skilled labor, hazardous working conditions, and work requiring repetitive operation. Factors which must be carefully studied before deciding on automation are high initial cost of equipment, maintenance problems, and type of product (See also Sec. 16.)

Mass production with modern machine tools has been achieved through the development of self-contained **power-head** production units and the development of **transfer** mechanisms. Power-head units, consisting of a frame, electric driving motor, gearbox, tool spindles, etc., are available for many types of machining operations. Transfer mechanisms move the workpieces from station to station by various methods. Transfer-type machines can be arranged in several configurations, such as a straight line or a U pattern. Various types of machine tools for mass production can be built from components; this is known as the **building-block** principle. Such a system combines flexibility and adaptability with high productivity. (See **machining centers.**)

Numerical control (NC), which is a method of controlling the motions of machine components by numbers, was first applied to machine tools in the 1950s. Numerically controlled machine tools are classified according to the type of cutting operation. For instance, in drilling and boring machines, the positioning and the cutting take place sequentially (point to point), whereas in die-sinking machines, positioning and cutting take place simultaneously. The latter are often described as **continuous-path** machines, and since they require more exacting specifications, they give rise to more complex problems. Machines now perform over a very wide range of cutting conditions without requiring adjustment to eliminate chatter, and to improve accuracy. Complex contours can be machined which would be almost impossible by any other method. A large variety of programming systems has been developed.

The control system in NC machines has been converted to computer control with various software. In **computer numerical control (CNC),** a microcomputer is a part of the control panel of the machine tool. The advantages of computer numerical control are ease of operation, simpler programming, greater accuracy, versatility, and lower maintenance costs.

Although numerical control of machine tools has many advantages such as high productivity and flexibility, it has certain limitations. Among these are high initial cost of equipment and the need for trained personnel and special maintenance.

Further developments in machine tools are **machining centers.** This is a machine equipped with as many as 200 tools and with an automatic tool changer (Fig. 13.4.9). It is designed to perform various operations on different surfaces of the workpiece, which is placed on a pallet capable of as much as five-axis movement (three linear and two rotational). Machining centers, which may be vertical or horizontal spindle, have flexibility and versatility that other machine tools do not have, and thus they have become the first choice in machine selection in modern manufacturing plants and shops. They have the capability of tool and part checking, tool-condition monitoring, in-process and postprocess gaging, and inspection of machined surfaces. **Universal machining centers** are the latest development, and they have both vertical and horizontal spindles. **Turning centers** are a further development of computer-controlled lathes and have great flexibility. Many centers are now constructed on a **modular** basis, so that various accessories and peripheral equipment can be installed and modified depending on the type of product to be machined.

An approach to optimize machining operations is **adaptive control.** While the material is being machined, the system senses operating conditions such as forces, tool-tip temperature, rate of tool wear, and surface finish, and converts these data into feed and speed control that enables the machine to cut under optimum conditions for maximum productivity. Combined with numerical controls and computers, adaptive controls are expected to result in increased efficiency of metalworking operations.

With the advent of sophisticated computers and various software, modern manufacturing has evolved into **computer-integrated manufacturing (CIM).** This system involves the coordinated participation of computers in all phases of manufacturing. **Computer-aided design** combined with **computer-aided manufacturing (CAD/CAM),** results in a much higher productivity, better accuracy and efficiency, and reduction in design effort and prototype development. CIM also involves the management of the factory, inventory, and labor, and it integrates all these activities, eventually leading to untended factories.

The highest level of sophistication is reached with a **flexible manufacturing system (FMS).** Such a system is made of **manufacturing cells** and an automatic materials-handling system interfaced with a central computer. The manufacturing cell is a system in which CNC machines are used to make a specific part or parts with similar shape. The workstations, i.e., several machine tools, are placed around an **industrial robot** which automatically loads, unloads, and transfers the parts. FMS has the capability to optimize each step of the total manufacturing operation, resulting in the highest possible level of efficiency and productivity.

The proper design of **machine-tool structures** requires analysis of such factors as form and materials of structures, stresses, weight, and manufacturing and performance considerations. The best approach to obtain the ultimate in machine-tool accuracy is to employ both improvements in structural stiffness and compensation of deflections by use of special controls. The C-frame structure has been used extensively in the past because it provides ready accessibility to the working area of the machine. With the advent of computer control, the box-type frame with its considerably improved static stiffness becomes practical since the need for manual access to the working area is greatly reduced. The use of a box-type structure with thin walls can provide low weight for a given stiffness. The light-weight-design principle offers high dynamic stiffness by providing a high natural frequency of the structure through combining high static stiffness with low weight rather than through the use of large mass. (Dynamic stiffness is the stiffness exhibited by the system when subjected to dynamic excitation where the elastic, the damping, and the inertia properties of the structure are involved; it is a frequency-dependent quantity.)

TURNING

Turning is a machining operation for all types of metallic and nonmetallic materials and is capable of producing circular parts with straight or various profiles. The cutting tools may be single-point or form tools. The most common machine tool used is a **lathe;** modern lathes are computer-controlled and can achieve high production rates with little labor. The basic operation is shown in Fig. 13.4.10, where the workpiece is held in a chuck and rotates at N r/min; a cutting tool moves along the length of the piece at a feed f (in/r or mm/r) and removes material at a radial depth d, reducing the diameter from D_0 to D_f.

Lathes are generally considered to be the oldest member of machine tools, having been first developed in the late eighteenth century. The most common lathe is called an engine lathe because it was one of the first machines driven by Watt's steam engine. The basic lathe has the following main parts: bed, headstock, tailstock, and carriage. The types of lathes available for a variety of applications may be listed as follows: engine lathes, bench lathes, horizontal turret lathes, vertical lathes, and

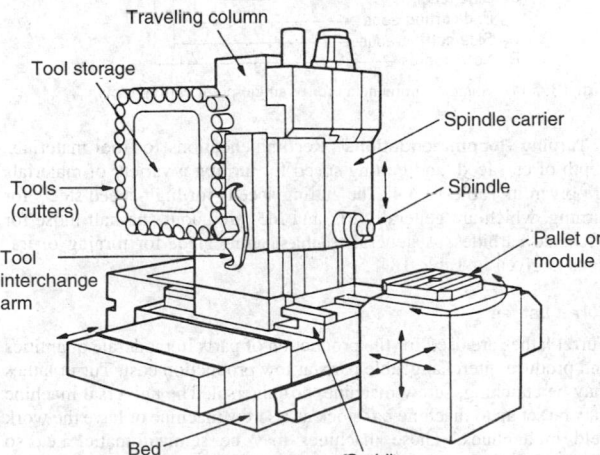

Fig. 13.4.9 Schematic of a horizontal spindle machining center, equipped with an automatic tool changer. Tool magazines can store 200 different cutting tools. (*Source: Courtesy of Cincinnati Milacron, Inc.*)

automatics. A great variety of lathes and attachments are available within each category, also depending on the production rate required.

It is common practice to specify the size of an engine lathe by giving the **swing** (diameter) and the **distance between centers** when the tailstock is flush with the end of the bed. The maximum swing over the ways is usually greater than the nominal swing. The **length of the bed** is given frequently to specify the overall length of the bed. A lathe size is indicated thus: 14 in (356 mm) (swing) by 30 in (762 mm) (between centers) by 6 ft (1,830 mm) (length of bed). Lathes are made for light-, medium-, or heavy-duty work.

Geometric progression is used extensively in designing machine-tool feeds and speeds. Feeds in geometric progression are used on cylindrical grinders, boring mills, milling machines, drilling machines, etc.; but for screw-cutting lathes, the power feeds for thread cutting and turning must be in proportion to the pitch of threads to be cut.

All geared-head lathes, which are single-pulley (belt-driven or arranged for direct-motor drive through short, flat, or V belts, gears, or silent chain), increase the power of the drive and provide a means for obtaining 8, 12, 16, or 24 spindle speeds. The teeth may be of the spur, helical, or herringbone type and may be ground or lapped after hardening.

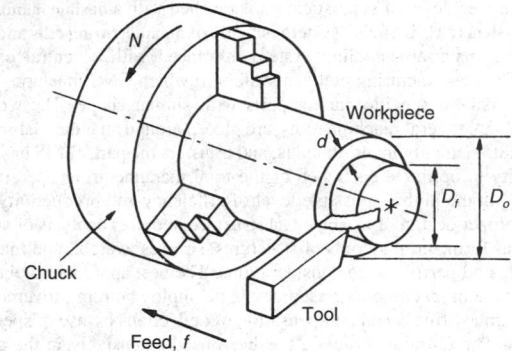

Fig. 13.4.10 A turning operation on a round workpiece held in a three-jaw chuck.

Variable speeds are obtained by driving with adjustable-speed dc shunt-wound motors with stepped field-resistance control or by electronics or motor-generator system to give speed variation in infinite steps. AC motors driving through infinitely variable speed transmissions of the mechanical or hydraulic type are also in general use.

Modern lathes, many of which are now computer-controlled (**turning centers**), are built with the speed capacity, stiffness, and strength capable of taking full advantage of new and stronger tool materials. The main drive-motor capacity of lathes ranges from fractional to more than 200 hp (150 kW). Speed preselectors, which give speed as a function of work diameter, are introduced, and variable-speed drives using dc motors with panel control are standard on many lathes. Lathes with contour facing, turning, and boring attachments are also available.

Tool Shapes for Turning

The standard **nomenclature** for single-point tools, such as those used on lathes, planers, and shapers, is shown in Fig. 13.4.11. Each tool consists of a shank and point. The point of a single-point tool may be formed by grinding on the end of the shank; it may be forged on the end of the shank and subsequently ground; a tip or insert may be clamped or brazed to the end of the shank (see Fig. 13.4.8). The **best tool shape** for each material and each operation depends on many factors. For specific information and recommendations, the various sources listed in the References should be consulted. See also Table 13.4.3.

Positive **rake angles** improve the cutting operation with regard to forces and deflection; however, a high positive rake angle may result in early failure of the cutting edge. Positive rake angles are generally used in lower-strength materials. For higher-strength materials, negative rake

angles may be used. **Back rake** usually controls the direction of chip flow and is of less importance than the **side rake**. The purpose of **relief angles** is to avoid interference and rubbing between the workpiece and tool flank surfaces. In general, they should be small for high-strength materials and larger for softer materials. Excessive relief angles may weaken the tool. The **side cutting-edge angle** influences the length of chip contact and the true feed. This angle is often limited by the workpiece geometry, e.g., the shoulder contour. Large angles are apt to cause tool chatter. Small **end cutting-edge angles** may create excessive force normal to the workpiece, and large angles may weaken the tool point. The purpose of the **nose radius** is to give a smooth surface finish and to obtain longer tool life by increasing the strength of the cutting edge. The nose radius should be tangent to the cutting-edge angles. A large nose radius gives a stronger tool and may be used for roughing cuts; however, large radii may lead to tool chatter. A small nose radius reduces forces and is therefore preferred on thin or slender workpieces.

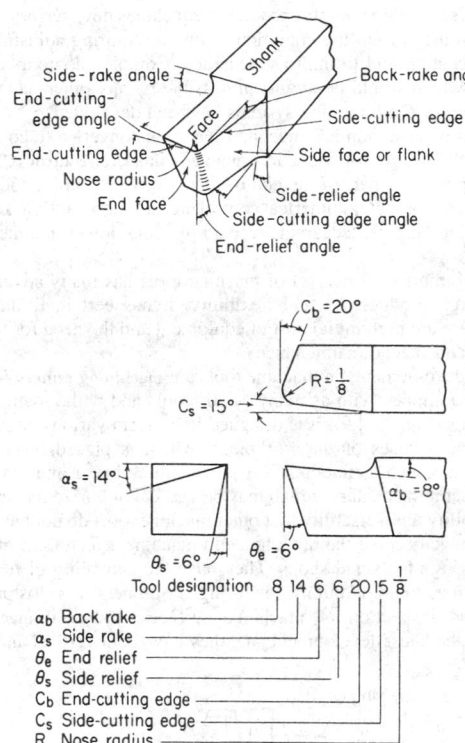

Fig. 13.4.11 Standard nomenclature for single-point cutting tools.

Turning Recommendations Recommendations for tool materials, depth of cut, feed, and cutting speed for turning a variety of materials are given in Table 13.4.4. The cutting speeds for high-speed steels for turning, which are generally M2 and M3, are about one-half those for uncoated carbides. A general **troubleshooting guide** for turning operations is given in Table 13.4.5.

Turret Lathes

Turret lathes are used for the production of parts in moderate quantities and produce interchangeable parts at low production cost. Turret lathes may be chucking, screw machine, or universal. The universal machine may be set up to machine bar stock as a screw machine or have the work held in a chuck. These machines may be semiautomatic, i.e., so arranged that after a piece is chucked and the machine started, it will complete the machining cycle automatically and come to a stop. They may be horizontal or vertical and single- or multiple-spindle; many of these lathes are now computer-controlled and have a variety of features.

Table 13.4.3 Recommend Tool Geometry for Turning (degrees)

	High-speed steel and cast-alloy tools					Carbide tools (inserts)				
Material	Back rake	Side rake	End relief	Side relief	Side and end cutting edge	Back rake	Side rake	End relief	Side relief	Side and end cutting edge
Aluminum alloys	20	15	12	10	5	0	5	5	5	15
Magnesium alloys	20	15	12	10	5	0	5	5	5	15
Copper alloys	5	10	8	8	5	0	5	5	5	15
Steels	10	12	5	5	15	−5	−5	5	5	15
Stainless steels, ferritic	5	8	5	5	15	0	5	5	5	15
Stainless steels, austenitic	0	10	5	5	15	0	5	5	5	15
Stainless steels, martensitic	0	10	5	5	15	−5	−5	5	5	15
High-temperature alloys	0	10	5	5	15	5	0	5	5	45
Refractory alloys	0	20	5	5	5	—	—	5	5	15
Titanium alloys	0	5	5	5	15	−5	−5	5	5	5
Cast irons	5	10	5	5	15	−5	−5	5	5	15
Thermoplastics	0	0	20–30	15–20	10	0	0	20–30	15–20	10
Thermosetting plastics	0	0	20–30	15–20	10	0	15	5	5	15

SOURCE: ''Matchining Data Handbook,'' published by the Machinability Data Center, Metcut Research Associates, Inc.

Table 13.4.4 General Recommendations for Turning Operations

		General-purpose starting conditions			Range for roughing and finishing		
Workpiece material	Cutting tool	Depth of cut, mm (in)	Feed mm/r (in/r)	Cutting speed, m/min (ft/min)	Depth of cut, mm (in)	Feed mm/r (in/r)	Cutting speed, m/min (ft/min)
Low-C and free-machining steels	Uncoated carbide	1.5–6.3 (0.06–0.25)	0.35 (0.014)	90 (300)	0.5–7.6 (0.02–0.30)	0.15–1.1 (0.006–0.045)	60–135 (200–450)
	Ceramic-coated carbide	1.5–6.3 (0.06–0.25)	0.35 (0.014)	245–275 (800–900)	0.5–7.6 (0.02–0.30)	0.15–1.1 (0.006–0.045)	120–425 (400–1,400)
	Triple-coated carbide	1.5–6.3 (0.06–0.25)	0.35 (0.014)	185–200 (600–650)	0.5–7.6 (0.02–0.30)	0.15–1.1 (0.006–0.045)	90–245 (300–800)
	TiN-coated carbide	1.5–6.3 (0.06–0.25)	0.35 (0.014)	105–150 (350–500)	0.5–7.6 (0.02–0.30)	0.15–1.1 (0.006–0.045)	60–230 (200–750)
	Al_2O_3 ceramic	1.5–6.3 (0.06–0.25)	0.25 (0.010)	395–440 (1,300–1,450)	0.5–7.6 (0.02–0.30)	0.15–1.1 (0.006–0.045)	365–550 (1,200–1,800)
	Cermet	1.5–6.3 (0.06–0.25)	0.30 (0.012)	215–290 (700–950)	0.5–7.6 (0.02–0.30)	0.15–1.1 (0.006–0.045)	105–455 (350–1,500)
Medium- and high-C steels	Uncoated carbide	1.2–4.0 (0.05–0.20)	0.30 (0.012)	75 (250)	2.5–7.6 (0.10–0.30)	0.15–0.75 (0.006–0.03)	45–120 (150–400)
	Ceramic-coated carbide	1.2–4.0 (0.05–0.20)	0.30 (0.012)	185–230 (600–750)	2.5–7.6 (0.10–0.30)	0.15–0.75 (0.006–0.03)	120–410 (400–1,350)
	Triple-coated carbide	1.2–4.0 (0.050–0.20)	0.30 (0.012)	120–150 (400–500)	2.5–7.6 (0.10–0.30)	0.15–0.75 (0.006–0.03)	75–215 (250–700)
	TiN-coated carbide	1.2–4.0 (0.05–0.20)	0.30 (0.012)	90–200 (300–650)	2.5–7.6 (0.10–0.30)	0.15–0.75 (0.006–0.03)	45–215 (150–700)
	Al_2O_3 ceramic	1.2–4.0 (0.05–0.20)	0.25 (0.010)	335 (1,100)	2.5–7.6 (0.10–0.30)	0.15–0.75 (0.006–0.03)	245–455 (800–1,500)
	Cermet	1.2–4.0 (0.05–0.20)	0.25 (0.010)	170–245 (550–800)	2.5–7.6 (0.10–0.30)	0.15–0.75 (0.006–0.03)	105–305 (350–1,000)
Cast iron, gray	Uncoated carbide	1.25–6.3 (0.05–0.25)	0.32 (0.013)	90 (300)	0.4–12.7 (0.015–0.5)	0.1–0.75 (0.004–0.03)	75–185 (250–600)
	Ceramic-coated carbide	1.25–6.3 (0.05–0.25)	0.32 (0.013)	200 (650)	0.4–12.7 (0.015–0.5)	0.1–0.75 (0.004–0.03)	120–365 (400–1,200)
	TiN-coated carbide	1.25–6.3 (0.05–0.25)	0.32 (0.013)	90–135 (300–450)	0.4–12.7 (0.015–0.5)	0.1–0.75 (0.004–0.03)	60–215 (200–700)
	Al_2O_3 ceramic	1.25–6.3 (0.05–0.25)	0.25 (0.010)	455–490 (1,500–1,600)	0.4–12.7 (0.015–0.5)	0.1–0.75 (0.004–0.03)	365–855 (1,200–2,800)
	SiN ceramic	1.25–6.3 (0.05–0.25)	0.32 (0.013)	730 (2,400)	0.4–12.7 (0.015–0.5)	0.1–0.75 (0.004–0.03)	200–990 (650–3,250)
Stainless steel, austenitic	Triple-coated carbide	1.5–4.4 (0.06–0.175)	0.35 (0.014)	150 (500)	0.5–12.7 (0.02–0.5)	0.08–0.75 (0.003–0.03)	75–230 (250–750)
	TiN-coated carbide	1.5–4.4 (0.06–0.175)	0.35 (0.014)	85–160 (275–525)	0.5–12.7 (0.02–0.5)	0.08–0.75 (0.003–0.03)	55–200 (175–650)
	Cermet	1.5–4.4 (0.06–0.175)	0.30 (0.012)	185–215 (600–700)	0.5–12.7 (0.02–0.5)	0.08–0.75 (0.003–0.03)	105–290 (350–950)
High-temperature alloys, nickel base	Uncoated carbide	2.5 (0.10)	0.15 (0.006)	25–45 (75–150)	0.25–6.3 (0.01–0.25)	0.1–0.3 (0.004–0.012)	15–30 (50–100)
	Ceramic-coated carbide	2.5 (0.10)	0.15 (0.006)	45 (150)	0.25–6.3 (0.01–0.25)	0.1–0.3 (0.004–0.012)	20–60 (65–200)

Table 13.4.4 General Recommendations for Turning Operations (*Continued*)

Workpiece material	Cutting tool	General-purpose starting conditions			Range for roughing and finishing		
		Depth of cut, mm (in)	Feed mm/r (in/r)	Cutting speed, m/min (ft/min)	Depth of cut, mm (in)	Feed mm/r (in/r)	Cutting speed, m/min (ft/min)
High-temperature alloys, nickel base (*cont.*)	TiN-coated carbide	2.5 (0.10)	0.15 (0.006)	30–55 (95–175)	0.25–6.3 (0.01–0.25)	0.1–0.3 (0.004–0.012)	20–85 (60–275)
	Al₂O₃ ceramic	2.5 (0.10)	0.15 (0.006)	260 (850)	0.25–6.3 (0.01–0.25)	0.1–0.3 (0.004–0.012)	185–395 (600–1,300)
	SiN ceramic	2.5 (0.10)	0.15 (0.006)	215 (700)	0.25–6.3 (0.01–0.25)	0.1–0.3 (0.004–0.012)	90–215 (300–700)
	Polycrystalline CBN	2.5 (0.10)	0.15 (0.006)	150 (500)	0.25–6.3 (0.01–0.25)	0.1–0.3 (0.004–0.012)	120–185 (400–600)
Titanium alloys	Uncoated carbide	1.0–3.8 (0.04–0.15)	0.15 (0.006)	35–60 (120–200)	0.25–6.3 (0.01–0.25)	0.1–0.4 (0.004–0.015)	10–75 (30–250)
	TiN-coated carbide	1.0–3.8 (0.04–0.15)	0.15 (0.006)	30–60 (100–200)	0.25–6.3 (0.01–0.25)	0.1–0.4 (0.004–0.015)	10–100 (30–325)
Aluminum alloys, free-machining	Uncoated carbide	1.5–5.0 (0.06–0.20)	0.45 (0.018)	490 (1,600)	0.25–8.8 (0.01–0.35)	0.08–0.62 (0.003–0.025)	200–670 (650–2,000)
	TiN-coated carbide	1.5–5.0 (0.06–0.20)	0.45 (0.018)	550 (1,800)	0.25–8.8 (0.01–0.35)	0.08–0.62 (0.003–0.025)	60–915 (200–3,000)
	Cermet	1.5–5.0 (0.06–0.20)	0.45 (0.018)	490 (1,600)	0.25–8.8 (0.01–0.35)	0.08–0.62 (0.003–0.025)	215–795 (700–2,600)
	Polycrystalline diamond	1.5–5.0 (0.06–0.20)	0.45 (0.018)	760 (2,500)	0.25–8.8 (0.01–0.35)	0.08–0.62 (0.003–0.025)	305–3,050 (1,000–10,000)
High-silicon	Polycrystalline diamond	1.5–5.0 (0.06–0.20)	0.45 (0.018)	530 (1,700)	0.25–8.8 (0.01–0.35)	0.08–0.62 (0.003–0.025)	365–915 (1,200–3,000)
Copper alloys	Uncoated carbide	1.5–5.0 (0.06–0.20)	0.25 (0.010)	260 (850)	0.4–7.5 (0.015–0.3)	0.15–0.75 (0.006–0.03)	105–535 (350–1,750)
	Ceramic-coated carbide	1.5–5.0 (0.06–0.20)	0.25 (0.010)	365 (1,200)	0.4–7.5 (0.015–0.3)	0.15–0.75 (0.006–0.03)	215–670 (700–2,200)
	Triple-coated carbide	1.5–5.0 (0.06–0.20)	0.25 (0.010)	215 (700)	0.4–7.5 (0.015–0.3)	0.15–0.75 (0.006–0.03)	90–305 (300–1,000)
	TiN-coated carbide	1.5–5.0 (0.06–0.20)	0.25 (0.010)	90–275 (300–900)	0.4–7.5 (0.015–0.3)	0.15–0.75 (0.006–0.03)	45–455 (150–1,500)
	Cermet	1.5–5.0 (0.06–0.20)	0.25 (0.010)	245–425 (800–1,400)	0.4–7.5 (0.015–0.3)	0.15–0.75 (0.006–0.03)	200–610 (650–2,000)
	Polycrystalline diamond	1.5–5.0 (0.06–0.20)	0.25 (0.010)	520 (1,700)	0.4–7.5 (0.015–0.3)	0.15–0.75 (0.006–0.03)	275–915 (900–3,000)
Tungsten alloys	Uncoated carbide	2.5 (0.10)	0.2 (0.008)	75 (250)	0.25–5.0 (0.01–0.2)	0.12–0.45 (0.005–0.018)	55–120 (175–400)
	TiN-coated carbide	2.5 (0.10)	0.2 (0.008)	85 (275)	0.25–5.0 (0.01–0.2)	0.12–0.45 (0.005–0.018)	60–150 (200–500)
Thermoplastics and thermosets	TiN-coated carbide	1.2 (0.05)	0.12 (0.005)	170 (550)	0.12–5.0 (0.005–0.20)	0.08–0.35 (0.003–0.015)	90–230 (300–750)
	Polycrystalline diamond	1.2 (0.05)	0.12 (0.005)	395 (1,300)	0.12–5.0 (0.005–0.20)	0.08–1.35 (0.003–0.015)	150–730 (500–2,400)
Composites, graphite-reinforced	TiN-coated carbide	1.9 (0.075)	0.2 (0.008)	200 (650)	0.12–6.3 (0.005–0.25)	0.12–1.5 (0.005–0.06)	105–290 (350–950)
	Polycrystalline diamond	1.9 (0.075)	0.2 (0.008)	760 (2,500)	0.12–6.3 (0.005–0.25)	0.12–1.5 (0.005–0.06)	550–1,310 (1,800–4,300)

NOTE: Cutting speeds for high-speed-steel tools are about one-half those for uncoated carbides.
SOURCE: Based on data from Kennametal Inc.

Table 13.4.5 General Troubleshooting Guide for Turning Operations

Problem	Probable causes
Tool breakage	Tool material lacks toughness; improper tool angles; machine tool lacks stiffness; worn bearings and machine components; cutting parameters too high
Excessive tool wear	Cutting parameters too high; improper tool material; ineffective cutting fluid; improper tool angles
Rough surface finish	Built-up edge on tool; feed too high; tool too sharp, chipped, or worn; vibration and chatter
Dimensional variability	Lack of stiffness; excessive temperature rise; tool wear
Tool chatter	Lack of stiffness; workpiece not supported rigidly; excessive tool overhang

The basic principle of the turret lathe is that, with standard tools, setups can be made quickly so that combined, multiple, and successive cuts can be made on a part. By **combined** cuts, tools on the cross slide operate simultaneously with those on the turret, e.g., facing from the cross slide and boring from the turret. **Multiple** cuts permit two or more tools to operate from either or both the cross slide or turret. By **successive** cuts, one tool may follow another to rough or finish a surface; e.g., a hole may be drilled, bored, and reamed at one chucking. In the tool-slide machine only roughing cuts, such as turn and face, can be made in one machine. A second machine similarly tooled must be available to make the finishing cuts.

Ram-type turret lathes have the turret mounted on a ram which slides in a separate base. The base is clamped at a position along the bed to suit a long or short workpiece. A cross slide can be used so that combined cuts can be taken from the turret and the cross slide at the same time. Turret and cross slide can be equipped with manual or power feed. The short stroke of the turret slide limits this machine to comparatively short light work, in both small and quantity-lot production.

Saddle-type turret lathes have the turret mounted on a saddle which slides directly on the bed. Hence, the length of stroke is limited only by the length of bed. A separate square-turret carriage with longitudinal and transverse movement can be mounted between the head and the hex-turret saddle so that combined cuts from both stations at one time are possible. The saddle type of turret lathe generally has a large hollow vertically faced turret for accurate alignment of the tools.

Screw Machines

When turret lathes are set up for bar stock, they are often called **screw machines.** Turret lathes that are adaptable only to bar-stock work are constructed for light work. As with turret lathes, they have spring collets for holding the bars during machining and friction fingers or rolls to feed the bar stock forward. Some bar-feeding devices are operated by hand and others semiautomatically.

Automatic screw machines may be classified as single-spindle or multiple-spindle. Single-spindle machines rotate the bar stock from which the part is to be made. The tools are carried on a turret and on cross slides or on a circular drum and on cross slides. Multiple-spindle machines have four, five, six, or eight spindles, each carrying a bar of the material from which the piece is to be made. Capacities range from ⅓ to 6 in (3 to 150 mm) diam of bar stock.

Feeds of forming tools vary with the width of the cut. The wider the forming tool and the smaller the diameter of stock, the smaller the feed. On multiple-spindle machines, where many tools are working simultaneously, the feeds should be such as to reduce the actual cutting time to a minimum. Often only one or two tools in a set are working up to capacity, as far as actual speed and feed are concerned.

BORING

Boring is a machining process for producing internal straight cylindrical surfaces or profiles, with process characteristics and tooling similar to those for turning operations.

Boring machines are of two general types, horizontal and vertical, and are frequently referred to as horizontal boring machines and vertical boring and turning mills. A classification of boring machines comprises horizontal boring, drilling, and milling machines; vertical boring and turning mills; vertical multispindle cylinder boring mills; vertical cylinder boring mills; vertical turret boring mills (vertical turret lathes); carwheel boring mills; diamond or precision boring machines (vertical and horizontal); and jig borers.

The **horizontal type** is made for both precision work and general manufacturing. It is particularly adapted for work not conveniently revolved, for milling, slotting, drilling, tapping, boring, and reaming long holes, and for making interchangeable parts that must be produced without jigs and fixtures. The machine is universal and has a wide range of speeds and feeds, for a face-mill operation may be followed by one with a small-diameter drill or end mill.

Vertical boring mills are adapted to a wide range of faceplate work that can be revolved. The advantage lies in the ease of fastening a workpiece to the horizontal table, which resembles a four-jaw independent chuck with extra radial T slots, and in the lessened effect of centrifugal forces arising from unsymmetrically balanced workpieces.

A **jig-boring machine** has a single-spindle sliding head mounted over a table adjustable longitudinally and transversely by lead screws which roughly locate the work under the spindle. Precision setting of the table may be obtained with end measuring rods, or it may depend only on the accuracy of the lead screw. These machines, made in various sizes, are used for accurately finishing holes and surfaces in definite relation to one another. They may use drills, rose or fluted reamers, or single-point boring tools. The latter are held in an adjustable **boring head** by which the tool can be moved eccentrically to change the diameter of the hole.

Precision-boring machines may have one or more spindles operating at high speeds for the purpose of boring to accurate dimensions such surfaces as wrist-pin holes in pistons and connecting-rod bushings.

Boring Recommendations Boring recommendations for tool materials, depth of cut, feed, and cutting speed are generally the same as those for turning operations (see Table 13.4.4). However, tool deflections, chatter, and dimensional accuracy can be significant problems because the boring bar has to reach the full length to be machined and space within the workpiece may be limited. Boring bars have been designed to dampen vibrations and reduce chatter during machining.

DRILLING

Drilling is a commonly employed hole-making process that uses a **drill** as a cutting tool for producing round holes of various sizes and depths. Drilled holes may be subjected to additional operations for better surface finish and dimensional accuracy, such as reaming and honing, described later in this section.

Drilling machines are intended for drilling holes, tapping, counterboring, reaming, and general boring operations. They may be classified into a large variety of types.

Sizes of Drilling Machines **Vertical drilling machines** are usually designated by a dimension which roughly indicates the diameter of the largest circle that can be drilled at its center under the machine. This dimensioning, however, does not hold for all makes of machines. The sizes begin with about 6 and continue to 50 in. Heavy-duty drill presses of the vertical type, with all-geared speed and feed drive, are constructed with a box-type column instead of the older cylindrical column.

The size of a **radial drill** is designated by the length of the arm. This represents the radius of a piece which can be drilled in the center.

Twist drills (Fig. 13.4.12) are the most common tools used in drilling and are made in many sizes and lengths. For years they have been grouped according to numbered sizes, 1 to 80, inclusive, corresponding approximately to Stub's steel wire gage; some by lettered sizes A to Z, inclusive; some by fractional inches from 1⁄64 up, and the group of millimeter sizes.

Straight-shank twist drills of fractional size and various lengths range from 1⁄64 in diam to 1¼ in by 1⁄64 in increments; to 1½ in by 1⁄32 in; and to 2 in by 1⁄16 in. **Taper-shank drills** range from ⅛ in diam to 1¾ in by 1⁄64 increments; to 2¼ in by 1⁄32 in; and to 3½ in by 1⁄16 in. Larger drills are made by various drill manufacturers. Drills are also available in metric dimensions.

Tolerances have been set on the various features of all drills so that the products of different manufacturers will be interchangeable in the user's plants.

Twist drills are decreased in diameter from point to shank (back taper) to prevent binding. If the web is increased gradually in thickness from point to shank to increase the strength, it is customary to reduce the helix angle as it approaches the shank. The shape of the groove is important, the one that gives a straight cutting edge and allows a full curl to the chip being the best. The **helix angles** of the flutes vary from 10 to 45°. The standard **point angle** is 118°. There are a number of **drill grinders** on the market designed to give the proper angles. The point may be ground either in the **standard** or the **crankshaft** geometry. The drill geometry for high-speed steel twist drills for a variety of workpiece materials is given in Table 13.4.6.

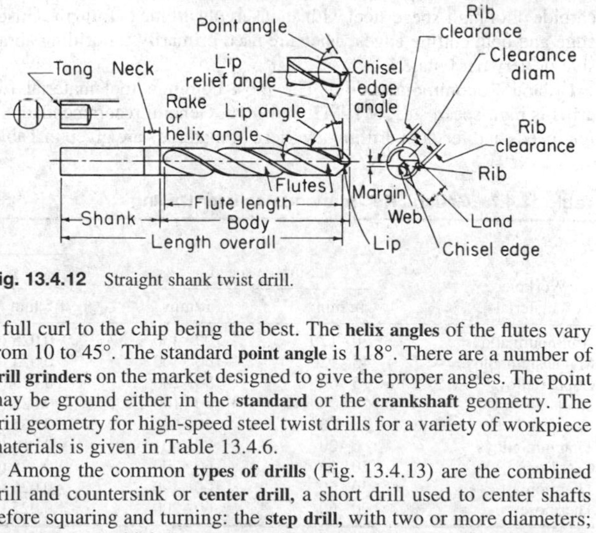

Fig. 13.4.12 Straight shank twist drill.

Among the common **types of drills** (Fig. 13.4.13) are the combined drill and countersink or **center drill,** a short drill used to center shafts before squaring and turning: the **step drill,** with two or more diameters; the **spade drill** which has a removable tip or bit clamped in a holder on the drill shank, used for large and deep holes; the **trepanning tool** used to

Table 13.4.6   Recommended Drill Geometry for High-Speed Steel Twist Drills

Material	Point angle, deg	Lip relief angle, deg	Chisel edge angle, deg	Helix angle, deg	Point grind
Aluminum alloys	90–118	12–15	125–135	24–48	Standard
Magnesium alloys	70–118	12–15	120–135	30–45	Standard
Copper alloys	118	12–15	125–135	10–30	Standard
Steels	118	10–15	125–135	24–32	Standard
High strength steels	118–135	7–10	125–135	24–32	Crankshaft
Stainless steels, low-strength	118	10–12	125–135	24–32	Standard
Stainless steels, high-strength	118–135	7–10	120–130	24–32	Crankshaft
High-temperature alloys	118–135	9–12	125–135	15–30	Crankshaft
Refractory alloys	118	7–10	125–135	24–32	Standard
Titanium alloys	118–135	7–10	125–135	15–32	Crankshaft
Cast irons	118	8–12	125–135	24–32	Standard
Plastics	60–90	7	120–135	29	Standard

SOURCE: "Machining Data Handbook," published by the Machinability Data Center, Metcut Research Associates Inc.

cut a core from a piece of metal instead of reducing all the metal removed to chips; the **gun drill,** run at a high speed under a light feed, and used to drill small long holes; the **core drill** used to bore out cored holes; the **oil-hole drill,** having holes or tubes in its body through which oil is forced to the cutting lips; **three-** and **four-fluted drills,** used to enlarge holes after a leader hole has been cored, punched, or drilled with a

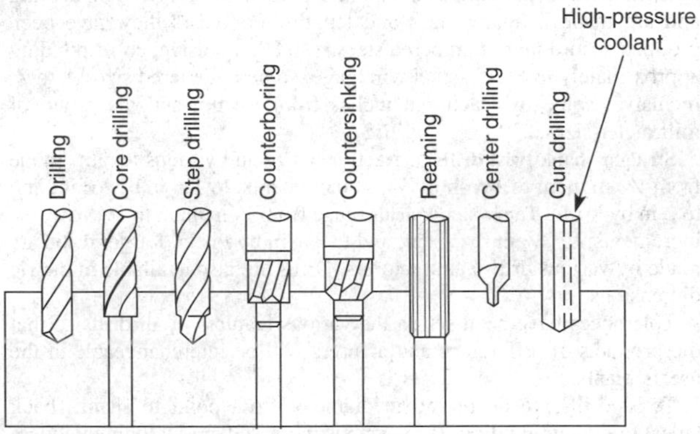

Fig. 13.4.13   Various types of drills, and drilling and reaming operations.

two-fluted drill; **twisted drills** made from flat high-speed steel or drop-forged to desired shape and then twisted. Drills are also made of solid carbide or of high-speed steel with an insert of carbide to form the chisel edge and both cutting edges. They are used primarily for drilling abrasive or very hard materials.

Drilling Recommendations   The most common tool material for drills is high-speed steel M1, M7, and M10. General recommendations for speeds and feeds in drilling a variety of materials are given in Table

13.4.7. Hole depth is also a factor in selecting drilling parameters. A general **troubleshooting guide** for drilling is given in Table 13.4.8.

Table 13.4.8   General Troubleshooting Guide for Drilling Operations

Problem	Probable causes
Drill breakage	Dull drill; drill seizing in hole because of chips clogging flutes; feed too high; lip relief angle too small
Excessive drill wear	Cutting speed too high; ineffective cutting fluid; rake angle too high; drill burned and strength lost when sharpened
Tapered hole	Drill misaligned or bent; lips not equal; web not central
Oversize hole	Same as above; machine spindle loose; chisel edge not central; side pressure on workpiece
Poor hole surface finish	Dull drill; ineffective cutting fluid; welding of workpiece material on drill margin; improperly ground drill; improper alignment

REAMING

A **reamer** is a multiple-cutting edge tool used to enlarge or finish holes, and to provide accurate dimensions as well as good finish. Reamers are of two types: (1) rose and (2) fluted.

The **rose reamer** is a heavy-bodied tool with end cutting edges. It is used to remove considerable metal and to true up a hole preparatory to flute reaming. It is similar to the three- and four-fluted drills. Wide cylindrical lands are provided back of the flute edges.

Fluted reamers cut principally on the periphery and remove only 0.004 to 0.008 in (0.1 to 0.2 mm) on the bore. Very narrow cylindrical margins are provided back of the flute edges, 0.012 to 0.015 in (0.3 to

Table 13.4.7   General Recommendations for Drilling

| Workpiece material | Surface speed | | Feed, mm/r (in/r) | | r/min | |
| | | | Drill diameter | | | |
	m/min	ft/min	1.5 mm (0.060 in)	12.5 mm (0.5 in)	1.5 mm	12.5 mm
Aluminum alloys	30–120	100–400	0.025 (0.001)	0.30 (0.012)	6,400–25,000	800–3,000
Magnesium alloys	45–120	150–400	0.025 (0.001)	0.30 (0.012)	9,600–25,000	1,100–3,000
Copper alloys	15–60	50–200	0.025 (0.001)	0.25 (0.010)	3,200–12,000	400–1,500
Steels	20–30	60–100	0.025 (0.001)	0.30 (0.012)	4,300–6,400	500–800
Stainless steels	10–20	40–60	0.025 (0.001)	0.18 (0.007)	2,100–4,300	250–500
Titanium alloys	6–20	20–60	0.010 (0.0004)	0.15 (0.006)	1,300–4,300	150–500
Cast irons	20–60	60–200	0.025 (0.001)	0.30 (0.012)	4,300–12,000	500–1,500
Thermoplastics	30–60	100–200	0.025 (0.001)	0.13 (0.005)	6,400–12,000	800–1,500
Thermosets	20–60	60–200	0.025 (0.001)	0.10 (0.004)	4,300–12,000	500–1,500

NOTE: As hole depth increases, speeds and feeds should be reduced. Selection of speeds and feeds also depends on the specific surface finish required.

0.4 mm) wide for machine-finish reaming and 0.004 to 0.006 in (0.1 to 0.15 mm) for hand reaming, to provide free cutting of the edges due to the slight body taper and also to pilot the reamer in the hole. The hole to be flute- or finish-reamed should be true. A rake of 5° is recommended for most reaming operations. A reamer may be straight or helically fluted. The latter provides much smoother cutting and gives a better finish.

Expansion reamers permit a slight expansion by a wedge so that the reamer may be resharpened to its normal size or for job shop use; they provide slight variations in size. **Adjustable reamers** have means of adjusting inserted blades so that a definite size can be maintained through numerous grindings and fully worn blades can be replaced with new ones. **Shell reamers** constitute the cutting portion of the tool which fits interchangeably on arbors to make many sizes available or to make replacement of worn-out shells less costly. Reamers float in their holding fixtures to ensure alignment, or they should be piloted in guide bushings above and below the work. They may also be held rigidly, such as in the tailstock of a lathe.

The **speed** of high-speed steel reamers should be two-thirds to three-quarters and **feeds** usually are two or three times that of the corresponding drill size. The most common tool materials for reamers are M1, M2, and M7 high-speed steels and C-2 carbide.

THREADING

Threads may be formed on the outside or inside of a cylinder or cone (1) with single-point threading tools, (2) with threading chasers, (3) with taps, (4) with dies, (5) by thread milling, (6) by thread rolling, and (7) by grinding. There are numerous types of taps, such as hand, machine screw, pipe, and combined pipe tap and drill. Small taps usually have no radial relief. They may be made in two, three, or four flutes. Large taps may have still more flutes.

The **feed** of a tap depends upon the lead of the screw thread. The **cutting speed** depends upon numerous factors: Hard tough materials, great length of hole, taper taps, and full-depth thread reduce the speed; long chamfer, fine pitches, and a cutting fluid applied in quantity increase the speed. Taps are cut or formed by grinding. The ground-thread taps may operate at much higher speeds than the cut taps. Speeds may range from 3 ft/min (1 m/min) for high-strength steels to 150 ft/min (45 m/min) for aluminum and magnesium alloys. Common high-speed steel tool materials for taps are M1, M7, and M10.

Threading dies, used to produce external threads, may be solid, adjustable, spring-adjustable, or self-opening die heads. Replacement chasers are used in die heads and may be of the fixed or self-opening type. These chasers may be of the radial type, hobbed or milled; of the tangental type; or of the circular type. Emulsions and oils are satisfactory for most threading operations.

MILLING

Milling is one of the most versatile machining processes and is capable of producing a variety of shapes involving flat surfaces, slots, and contours (Fig. 13.4.14).

Milling machines use cutters with multiple teeth in contrast with the single-point tools of the lathe and planer. The workpiece is generally fed past the cutter perpendicular to the cutter axis. Milling usually is face or peripheral cutting.

Standard spindle noses and arbors for milling machines provide interchangeability of arbors and face-milling cutters, regardless of make or size of machine. The taper of the spindle end and arbor is 3½ in/ft, to make them **self-releasing**. The retention of the shank is dependent upon a positive locking device, such as screws or draw-in bolt. When unlocked, these tapers release themselves.

Milling-machine classification is based on design, operation, or purpose. **Knee-and-column** type milling machines have the table and saddle supported on the vertically adjustable knee gibbed to the face of the column. The table is fed longitudinally on the saddle, and the latter transversely on the knee to give three feeding motions.

Knee-type machines are made with horizontal or vertical spindles. The **horizontal** universal machines have a swiveling table for cutting helices. The plain machines are used for jobbing or production work, the universal for toolroom work. **Vertical** milling machines with fixed or sliding heads are otherwise similar to the horizontal type. They are used for face or end milling and are frequently provided with a rotary table for making cylindrical surfaces.

The **fixed-bed** machines have a spindle mounted in a head dovetailed to and sliding on the face of the column. The table rests directly on the bed. They are simple and rigidly built and are used primarily for high-production work. These machines are usually provided with work-holding fixtures and may be constructed as plain or multiple-spindle machines, simple or duplex.

Rotary-type millers usually have a rotating table on which fixtures carrying the workpiece are mounted. The cutter spindles are mounted over the edge of the table past which the workpiece is fed.

Drum-type millers consist of a drum carrying the work and rotating on a horizontal axis. Both rotary and drum types are mass-production machines, there being no idle time, for the drums rotate continuously while the parts, loaded on one side of the machine, pass first the roughing and then the finishing cutters and are replaced when they return to the loading position.

In **planetary** milling machines the workpiece is stationary on the bed or clamped to the tailstock while the cutter rotates. Plain and formed internal or external surfaces and threads are produced by inserting the cutter into the bore to be milled, feeding it to depth radially, making a sweeping cut about the bore, and withdrawing first radially and then axially.

Planar-type millers are used only on the heaviest work. They are used to machine a number of surfaces on a particular part or group of parts arranged in series in fixtures on the table.

Thread millers are used to cut threads and worms. A single formed cutter may be used or all the threads may be cut at one time by a multiple-thread cutter.

Milling Cutters

Milling cutters are made in a wide variety of shapes and sizes. The nomenclature of tooth parts and angles is standardized as in

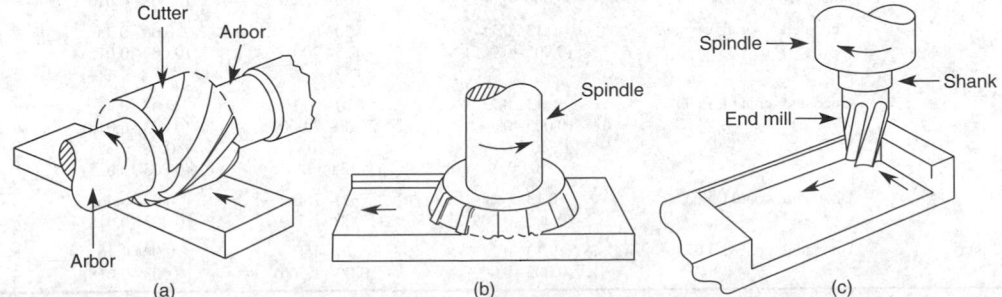

Fig. 13.4.14   Basic types of milling cutters and operations. (*a*) Slab (peripheral) milling; (*b*) face milling; (*c*) end milling.

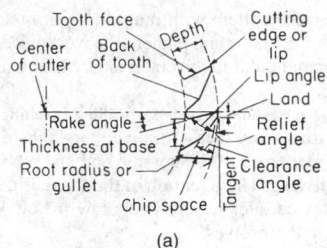

(a)

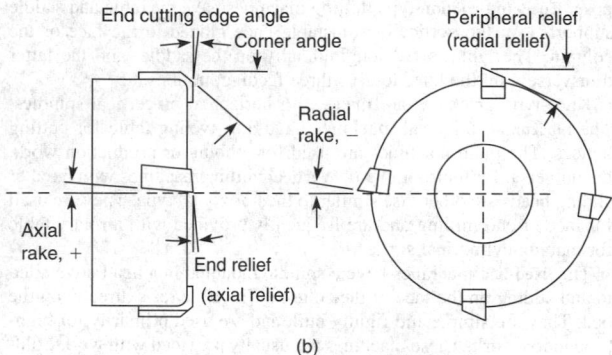

(b)

Fig. 13.4.15   (a) Plain milling cutter teeth; (b) face milling cutter.

Table 13.4.10   General Troubleshooting Guide for Milling Operations

Problem	Probable causes
Tool breakage	Tool material lacks toughness; improper tool angles; cutting parameters too high
Tool wear excessive	Cutting parameters too high; improper tool material; improper tool angles; improper cutting fluid
Rough surface finish	Feed too high; spindle speed too low; too few teeth on cutter; tool chipped or worn; built-up edge; vibration and chatter
Tolerances too broad	Lack of spindle stiffness; excessive temperature rise; dull tool; chips clogging cutter
Workpiece surface burnished	Dull tool; depth of cut too low; radial relief angle too small
Back striking	Dull cutting tools; cutter spindle tilt; negative tool angles
Chatter marks	Insufficient stiffness of system; external vibrations; feed, depth, and width of cut too large
Burr formation	Dull cutting edges or too much honing; incorrect angle of entry or exit; feed and depth of cut too high; incorrect insert geometry
Breakout	Lead angle too low; incorrect cutting edge geometry; incorrect angle of entry or exit; feed and depth of cut too high

Table 13.4.9   General Recommendations for Milling Operations

Workpiece material	Cutting tool	General-purpose starting conditions		Range of conditions	
		Feed, mm/tooth (in/tooth)	Speed, m/min (ft/min)	Feed, mm/tooth (in/tooth)	Speed, m/min (ft/min)
Low-C and free-machining steels	Uncoated carbide, coated carbide, cermets	0.13–0.20 (0.005–0.008)	120–180 (400–600)	0.085–0.38 (0.003–0.015)	90–425 (300–1,400)
Alloy steels					
Soft	Uncoated, coated, cermets	0.10–0.18 (0.004–0.007)	90–170 (300–550)	0.08–0.30 (0.003–0.012)	60–370 (200–1,200)
Hard	Cermets, PCBN	0.10–0.15 (0.004–0.006)	180–210 (600–700)	0.08–0.25 (0.003–0.010)	75–460 (250–1,500)
Cast iron, gray					
Soft	Uncoated, coated, cermets, SiN	0.10–0.20 (0.004–0.008)	120–760 (400–2,500)	0.08–0.38 (0.003–0.015)	90–1,370 (300–4,500)
Hard	Cermets, SiN, PCBN	0.10–0.20 (0.004–0.008)	120–210 (400–700)	0.08–0.38 (0.003–0.015)	90–460 (300–1,500)
Stainless steel, austenitic	Uncoated, coated, cermets	0.13–0.18 (0.005–0.007)	120–370 (400–1,200)	0.08–0.38 (0.003–0.015)	90–500 (300–1,800)
High-temperature alloys, nickel base	Uncoated, coated, cermets, SiN, PCBN	0.10–0.18 (0.004–0.007)	30–370 (100–1,200)	0.08–0.38 (0.003–0.015)	30–550 (90–1,800)
Titanium alloys	Uncoated, coated, cermets	0.13–0.15 (0.005–0.006)	50–60 (175–200)	0.08–0.38 (0.003–0.015)	40–140 (125–450)
Aluminum alloys					
Free-machining	Uncoated, coated, PCD	0.13–0.23 (0.005–0.009)	610–900 (2,000–3,000)	0.08–0.46 (0.003–0.018)	300–3,000 (1,000–10,000)
High-silicon	PCD	0.13 (0.005)	610 (2,000)	0.08–0.38 (0.003–0.015)	370–910 (1,200–3,000)
Copper alloys	Uncoated, coated, PCD	0.13–0.23 (0.005–0.009)	300–760 (1,000–2,500)	0.08–0.46 (0.003–0.018)	90–1,070 (300–3,500)
Thermoplastics and thermosets	Uncoated, coated, PCD	0.13–0.23 (0.005–0.009)	270–460 (900–1,500)	0.08–0.46 (0.003–0.018)	90–1,370 (300–4,500)

NOTE: Depths of cut, d, usually are in the range of 1–8 mm (0.04–0.3 in). PCBN: polycrystalline cubic boron nitride; PCD: polycrystalline diamond.
SOURCE: Based on data from Kennametal Inc.

Fig. 13.4.15. Milling cutters may be classified in various ways, such as purpose or use of the cutters (Woodruff keyseat cutters, T-slot cutters, gear cutters, etc.); construction characteristics (solid cutters, carbide-tipped cutters, etc.); method of mounting (arbor type, shank type, etc.); and relief of teeth. The latter has two categories: profile cutters which produce flat, curved, or irregular surfaces, with the cutter teeth sharpened on the land; and formed cutters which are sharpened on the face to retain true cross-sectional form of the cutter.

Two kinds of milling are generally considered to represent all forms of milling processes: **peripheral (slab)** and **face** milling. In the peripheral-milling process the axis of the cutter is parallel to the surface milled, whereas in face milling, the cutter axis is generally at a right angle to the surface. The peripheral-milling process is also divided into two types: **conventional (up) milling** and **climb (down) milling**. Each has its advantages, and the choice depends on a number of factors such as the type and condition of the equipment, tool life, surface finish, and machining parameters.

Milling Recommendations Recommendations for tool materials, feed per tooth, and cutting speed for milling a variety of materials are given in Table 13.4.9. The cutting speeds for high-speed steels for turning, which are generally M2 and M7, are about one-half those for uncoated carbides. A general **troubleshooting guide** for milling operations is given in Table 13.4.10.

GEAR MANUFACTURING

(See also Sec. 8)

Gear Cutting Most gear-cutting processes can be classified as either **forming** or **generating**. In a forming process, the shape of the tool is reproduced on the workpiece; in a generating process, the shape produced on the workpiece depends on both the shape of the tool and the relative motion between the tool and the workpiece during the cutting operation. A soft live center on a lathe can be formed by means of a broad, flat-form tool fed at right angles to the lathe spindle, or generated by a single-point tool fed at the point angle in the compound rest. In general, a generating process is more accurate than a forming process.

In the **form cutting** of gears, the tool has the shape of the space between the teeth. For this reason, form cutting will produce precise tooth profiles only when the cutter is accurately made and the tooth space is of constant width, such as on spur and helical gears. A form cutter may cut or finish one of or all the spaces in one pass. Single-space cutters may be disk-type or end-mill-type milling cutters. In all single-space operations, the gear blank must be retracted and indexed, i.e., rotated one tooth space, between each pass.

Single-space form milling with disk-type cutters is particularly suitable for gears with large teeth, because, as far as metal removal is concerned, the cutting action of a milling cutter is more efficient than that of the tools used for generating. Form milling of spur gears is done on machines that retract and index the gear blank automatically.

For the same tooth size (pitch), the shape (profile) of the teeth on an involute gear depends on the number of teeth on the gear. Most gears have active profiles that are wholly, partially, or approximately involute, and, consequently, accurate form cutting would require a different cutter for each number of teeth. In most cases, satisfactory results can be obtained by using the eight cutters for each pitch that are commercially available. Each cutter is designed to cut a range of tooth numbers; the

no. 1 cutter, for example, cuts from 135 teeth to a rack, and the no. 8 cuts 12 and 13 teeth. (See Table 13.4.11.)

Multiple-space form cutting is done with a broach or with a patented Shear Speed toolhead. The broach is pushed down into or over the gear blank, usually on a hydraulic press, and all the spaces can be cut in one pass.

The Shear-Speed toolhead contains three main parts: a housing, a member with radial slots in which a tool for each of the tooth spaces on the gear being cut can slide, and a movable, double cone-shaped guiding unit that controls the radial movement (feed) of the tools. With the toolhead stationary, the part is reciprocated past the cutting tools; each tool is fed radially a predetermined amount each stroke until the full tooth depth is cut. To avoid drag, the tools are retracted on each return (noncutting) stroke of the gear.

Gears always operate in pairs and are usually required to have contacting profiles of such a shape that the ratio of the speeds of the pair remains constant at all times; such gears are said to be **conjugate** to one another.

In a **gear generating machine**, the generating tool can be considered as one of the gears in a conjugate pair and the gear blank as the other gear. The correct relative motion between the tool arbor and the blank arbor is obtained by means of a train of indexing gears within the machine.

One of the most valuable properties of the involute as a gear-tooth profile is that if a cutter is made in the form of an involute gear of a given pitch and any number of teeth, it can generate all gears of all tooth numbers of the same pitch and they will all be conjugate to one another. The generating tool may be a pinion-shaped cutter, a rack-shaped (straight) cutter, or a hob, which is essentially a series of racks wrapped around a cylinder in a helical, screwlike form.

On a **gear shaper**, the generating tool is a pinion-shaped cutter that rotates slowly at the proper speed as if in mesh with the blank; the cutting action is produced by a reciprocation of the cutter parallel to the work axis. These machines can cut spur and helical gears, both internal and external; they can also cut continuous-tooth helical (herringbone) gears and are particularly suitable for cluster gears, or gears that are close to a shoulder.

On a **rack shaper** the generating tool is a segment of a rack that moves perpendicular to the axis of the blank while the blank rotates about a fixed axis at the speed corresponding to conjugate action between the rack and the blank; the cutting action is produced by a reciprocation of the cutter parallel to the axis of the blank. Since it is impracticable to have more than 6 to 12 teeth on a rack cutter, the cutter must be disengaged from the blank at suitable intervals and returned to the starting point, the blank meanwhile remaining fixed. These machines can cut both spur and helical external gears.

A gear-cutting **hob** is basically a worm, or screw, made into a generating tool by cutting a series of longitudinal slots or "gashes" to form teeth; to form cutting edges, the teeth are "backed off," or relieved, in a lathe equipped with a backing-off attachment. A hob may have one, two, or three threads; on involute hobs with a single thread, the generating portion of the hob-tooth profile usually has straight sides (like an involute rack tooth) in a section taken at right angles to the thread.

In addition to the conjugate rotary motions of the hob and workpiece, the hob must be fed parallel to the workpiece axis for a distance greater than the face width of the gear. The feed, per revolution of the work-

Table 13.4.11

No. of cutter	1	2	3	4	5	6	7	8
No. of teeth	135−∞	55−134	35−54	27−34	21−26	17−20	14−16	12 and 13

For more accurate gears, 15 cutters are available

No. of cutter	1	1½	2	2½	3	3½	4	4½
No. of teeth	135−∞	80−134	55−79	42−54	35−41	30−34	26−29	23−25

No. of cutter	5	5½	6	6½	7	7½	8
No. of teeth	21 and 22	19 and 20	17 and 18	15 and 16	14	13	12

piece, is produced by the feed gears, and its magnitude depends on the material, pitch, and finish desired; the feed gears are independent of the indexing gears. The hobbing process is continuous until all the teeth are cut.

The same machines and the same hobs that are used for cutting spur gears can be used for helical gears; it is only necessary to tip the hob axis so that the hob and gear pitch helices are tangent to one another and to correlate the indexing and feed gears so that the blank and the hob are advanced or retarded with respect to each other by the amount required to produce the helical teeth. Some hobbing machines have a differential gear mechanism that permits the indexing gears to be selected as for spur gears and the feed gearing to be chosen independently.

The threads of worms are usually cut with a disk-type milling cutter on a thread-milling machine and finished, after hardening, by grinding. Worm gears are usually cut with a hob on the machines used for hobbing spur and helical gears. Except for the gashes, the relief on the teeth, and an allowance for grinding, the hob is a counterpart of the worm. The hob and workpiece axes are inclined to one another at the shaft angle of the worm and gear set, usually 90°. The hob may be fed in to full depth in a radial (to the blank) direction or parallel to the hob axis.

Although it is possible to approximate the true shape of the teeth on a straight bevel gear by taking two or three cuts with a form cutter on a milling machine, this method, because of the taper of the teeth, is obviously unsuited for the rapid production of accurate teeth. Most straight bevel gears are roughed out in one cut with a form cutter on machines that index automatically and then finished to the proper shape on a generator.

The generating method used for straight bevel gears is analogous to the rack-generating method used for spur gears. Instead of using a rack with several complete teeth, however, the cutter has only one straight cutting edge that moves, during generation, in the plane of the tooth of a basic crown gear conjugate to the gear being generated. A crown gear is the rack among bevel gears; its pitch surface is a plane, and its teeth have straight sides.

The generating cutter moves back and forth across the face of the bevel gear like the tool on a shaper; the "generating roll" is obtained by rotating the gear slowly relative to the tool. In practice two tools are used, one for each side of a tooth; after each tooth has been generated, the gear must be retracted and indexed to the next tooth.

The machines used for cutting spiral bevel gears operate on essentially the same principle as those used for straight bevel gears; only the cutter is different. The spiral cutter is basically a disk that has a number of straight-sided cutting blades protruding from its periphery on one side to form the rim of a cup. The machines have means for indexing, retracting, and producing a generating roll; by disconnecting the roll gears, spiral bevel gears can be form cut.

Gear Shaving For improving the surface finish and profile accuracy of cut spur and helical gears (internal and external), gear shaving, a free-cutting gear finishing operation that removes small amounts of metal from the working surfaces of the teeth, is employed. The teeth on the shaving cutter, which may be in the form of a pinion (spur or helical) or a rack, have a series of sharp-edged rectangular grooves running from tip to root. The intersection of the grooves with the tooth profiles creates cutting edges; when the cutter and the workpiece, in tight mesh, are caused to move relative to one another along the teeth, the cutting edges remove metal from the teeth of the work gear. Usually the cutter drives the workpiece, which is free to rotate and is traversed past the cutter parallel to the workpiece axis. Shaving requires less time than grinding, but ordinarily it cannot be used on gears harder than approximately 400 HB (42 HRC).

Gear Grinding Machines for the grinding of spur and helical gears utilize either a forming or a generating process. For form grinding, a disk-type grinding wheel is dressed to the proper shape by a diamond held on a special dressing attachment; for each number of teeth a special index plate, with V-type notches on its periphery, is required. When grinding helical gears, means for producing a helical motion of the blank must be provided.

For grinding-generating, the grinding wheel may be a disk-type, double-conical wheel with an axial section equivalent to the basic

rack of the gear system. A master gear, similar to the gear being ground, is attached to the workpiece arbor and meshes with a master rack; the generating roll is created by rolling the master gear in the stationary rack.

Spiral bevel and hypoid gears can be ground on the machines on which they are generated. The grinding wheel has the shape of a flaring cup with a double-conical rim having a cross section equivalent to the surface that is the envelope of the rotary cutter blades.

Gear Rolling The cold-rolling process is used for the finishing of spur and helical gears for automatic transmissions and power tools; in some cases it has replaced gear shaving. It differs from cutting in that the metal is not removed in the form of chips but is displaced under heavy pressure.

There are two main types of cold-rolling machines, namely, those employing dies in the form of racks or gears that operate in a parallel-axis relationship with the blank and those employing worm-type dies that operate on axes at approximately 90° to the workpiece axis. The dies, under pressure, create the tooth profiles by the plastic deformation of the blank.

When racks are used, the process resembles thread rolling; with gear-type dies the blank can turn freely on a shaft between two dies, one mounted on a fixed head and the other on a movable head. The dies have the same number of teeth and are connected by gears to run in the same direction at the same speed. In operation, the movable die presses the blank into contact with the fixed die, and a conjugate profile is generated on the blank. On some of these machines the blank can be fed axially, and gears can be rolled in bar form to any convenient length.

On machines employing worm-type dies, the two dies are diametrically opposed on the blank and rotate in opposite directions. The speeds of the blank and the dies are synchronized by change gears, like the blank and the hob on a hobbing machine; the blank is fed axially between the dies.

Although gears have been cold-rolled from solid blanks, most commercial rolling is done as a finishing operation on fine-pitch, precut gears.

PLANING AND SHAPING

Planers are used to rough and finish large flat surfaces, although arcs and special forms can be made with proper tools and attachments. Surfaces to be finished by scraping, such as ways and long dovetails and, particularly, parts of machine tools, are, with few exceptions, planed. With fixtures to arrange parts in parallel and series, quantities of small parts can be produced economically on planers.

Shapers are used for miscellaneous planing, surfacing, notching, key seating, and production of flat surfaces on flat parts. They are virtually obsolete, but some are still in use. The tool is held in a holder supported on a clapper on the end of a ram which is reciprocated hydraulically or by crank and rocker arm, in a straight line. A table carrying the vise and the workpiece feeds transversely on each return stroke.

BROACHING

Broaching is a production process whereby a cutter, called a **broach,** is used to finish internal or external surfaces such as holes of circular, square, or irregular section, keyways, the teeth of internal gears, multiple spline holes, and flat surfaces. In broaching, the action of the broach itself serves as a clamping medium so that in many cases the operation may be completed in the time ordinarily taken to chuck the piece. The expense of making a broach may be large. However, the cost of maintaining its size is usually not excessive, and very little scrap work results. Broaching round holes gives greater accuracy and better finish than reaming, but since the broach may be guided only by the workpiece it is cutting, the hole may not be accurate with respect to previously machined surfaces. Where such accuracy is required, it is better practice to broach first and then turn other surfaces with the workpiece mounted on a mandrel. The broach is usually long and is provided with many teeth so graded in size that each takes a small chip when the tool is pulled or pushed through the previously prepared leader hole or past the surface.

The main features of the broach are the pitch, degree of taper or increase in height of each successive tooth, relief, tooth depth, and rake.

The **pitch** of the teeth, i.e., the distance from one tooth to the next, depends upon tooth strength, length of cut, shape and size of chips, etc. The pitch should be as coarse as possible to provide ample chip clearance, but at least two teeth should be in contact with the workpiece at all times. The formula $p = 0.35 \sqrt{l}$ may be used, where p is pitch of the roughing teeth and l the length of hole or surface, in. An average pitch for small broaches is $\frac{1}{8}$ to $\frac{1}{4}$ in (3.175 to 6.35 mm) and for large ones $\frac{1}{2}$ to 1 in (12.7 to 25.4 mm). Where the hole or other surface to be broached is short, the teeth are often cut on an angle or helix, so as to give more continuous cutting action by having at least two teeth cutting simultaneously.

The degree of **taper**, or increase in size per tooth, depends largely on the hardness or toughness of the material to be broached and the finish desired. The degree of taper or feed for broaching cast iron is approximately double that for steel. Usually the first few teeth coming in contact with the workpiece are undersize but of uniform taper to take the greatest feeds per tooth, but as the finished size is approached, the teeth take smaller and smaller feeds with several teeth at the finishing end of nearly zero taper. In some cases, for soft metals and even cast iron, the large end is left plain or with rounded lands a trifle larger than the last cutting tooth so as to burnish the surface. For medium-sized broaches, the taper per tooth is 0.001 to 0.003 in (0.025 to 0.076 mm). Large broaches remove 0.005 to 0.010 in (0.127 to 0.254 mm) per tooth or even more. The teeth are given a **front rake** angle of 5 to 15° to give a curl to the chip, provide a cleaner cut surface, and reduce the power consumption. The **land** back of the cutting edge, which may be $\frac{1}{64}$ to $\frac{1}{16}$ in (0.4 to 1.6 mm) wide, usually is provided with a land relief varying from 1 to 3° with a clearance of 30 to 45°.

The heavier the feed per tooth or the longer the surface being broached, the greater must be the **chip clearance** or space between successive teeth for the chips to accumulate. The root should be a smooth curve.

Broaches are generally made of M2 or M7 high-speed steel; carbide is also used for the teeth of large broaches. Broaches of complicated shape are likely to warp during the heat-treating process. For this reason, in hardening, they are often heated in a vertical cylindrical furnace and quenched by being hung in an air blast furnished from small holes along the side of pipes placed vertically about the broach.

Push broaches are usually shorter than pull broaches, being 6 to 14 in (150 to 350 mm) long, depending on their diameter and the amount of metal to be removed. In many cases, for accuracy, four to six broaches of the push type constitute a set used in sequence to finish the surface being broached. Push broaches usually have a large cross-sectional area so as to be sufficiently rigid. With **pull broaches,** pulling tends to straighten the hole, whereas pushing permits the broaches to follow any irregularity of the leader hole. Pull broaches are attached to the crosshead of the broaching machine by means of a key slot and key, by a threaded connection, or by a head that fits into an automatic broach puller. The threaded connection is used where the broach is not removed from the drawing head while the workpiece is placed over the cutter, as in cutting a keyway. In enlarging holes, however, the small end of the pull broach must first be extended through the reamed, drilled, or cored hole and then fixed in the drawing head before being pulled through the workpiece.

Broaching Machines Push broaching is done on machines of the press type with a sort of fixture for holding the workpiece and broach or on presses operated by power. They are usually vertical and may be driven hydraulically or by screw, rack, or crank. The pull type of broach may be either vertical or horizontal. The ram may be driven hydraulically or by screw, rack, or crank. Both are made in the duplex- and multiple-head type.

Processing Parameters for Broaching Cutting speeds for broaching may range from 5 ft/min (1.5 m/min) for high-strength materials to as high as 30 to 50 ft/min (9 to 15 m/min) for aluminum and magnesium alloys. The most common tool materials are M2 and M7 high-speed steels. An emulsion is often used for broaching for general work, but oils may also be used.

CUTTING OFF

Cutting off involves parting or slotting bars, tubes, plate, or sheet by various means. The machines come in various types such as a lathe (using a single-point cutting tool), hacksaws, band saws, circular saws, friction saws, and thin abrasive wheels. Cutting off may also be carried out by shearing and cropping, as well as using flames and laser beams.

In **power hacksaws,** the frame in which the blade is strained is reciprocated above the workpiece which is held in a vise on the bed. The cutting feed is effected by weighting the frame, with 12 to 50 lb (55 to 225 N) of force from small to large machines; adding weights or spring tension giving up to 180 lb (800 N); providing a positive screw feed or a friction screw feed; and by a hydraulic feed mechanism giving forces up to 300 lb (1.34 kN) between the blade and workpiece. With high-speed steel blades, cutting speeds range from about 30 strokes per minute for high-strength materials to 180 strokes per minute for carbon steels.

Hacksaw **blades** for hand frames are made 8, 10, and 12 in long, $\frac{7}{16}$ to $\frac{9}{16}$ in wide, and 0.025 in thick. Number of teeth per inch for cutting soft steel or cast iron, 14; tool steel and angle iron, 18; brass, copper, and heavy tubing, 24; sheet metal and thin tube, 32.

Blades for power hacksaws are made of alloy steels and of high-speed steels. Each length is made in two or more widths. The coarsest teeth should be used on large workpieces and with heavy feeds.

Band saws, vertical, horizontal, and universal, are used for cutting off. The kerf or width of cut is small with a consequently small loss in expensive material. The teeth of band saws, like those of hacksaws, are set with the regular alternate type, one bent to the right and the next to the left; or with the alternate and center set, in which one tooth is bent to the right, the second to the left, and the third straight in the center. With high-speed steel saw blades, band speeds range from about 30 ft/min (10 m/min) for high-temperature alloys to about 400 (120) for carbon steels. For aluminum and magnesium alloys the speed ranges up to about 1,300 ft/min (400 m/min) with high-carbon blades. The band speed should be decreased as the workpiece thickness increases.

Friction sawing machines are used largely for cutting off structural shapes. Peripheral speeds of about 20,000 ft/min (6,000 m/min) are used. The wheels may be plain on the periphery, V-notched, or with milled square notches.

Abrasive cutoff wheels are made of thin resinoid or rubber bonded abrasives. The wheels operate at surface speeds of 12,000 to 16,000 ft/min (3,600 to 4,800 m/min). These wheels are used for cutting off tubes, shapes, and hardened high-speed steel.

Circular saws are made in a wide variety of styles and sizes. Circular saws may have teeth of several shapes, as radial face teeth for small fine-tooth saws; radial face teeth with a land for fine-tooth saws; alternate bevel-edged teeth to break up the chips, with every other tooth beveled 45° on each side with the next tooth plain; alternate side-beveled teeth; and one tooth beveled on the right, the next on the left, and the third beveled on both sides, each leaving slightly overlapping flats on the periphery of the teeth. The most common high-speed steels for circular saws are M2 and M7; some saws have inserted carbide teeth.

ABRASIVE PROCESSES

(See also Sec. 6)

Abrasive processes consist of a variety of operations in which the tool is made of an abrasive material, the most common examples of which are grinding (using wheels, known as **bonded abrasives**), honing, and lapping. An abrasive is a small, nonmetallic hard particle having sharp cutting edges and an irregular shape. Abrasive processes, which can be performed on a wide variety of metallic and nonmetallic materials, remove material in the form of tiny chips and produce surface finishes

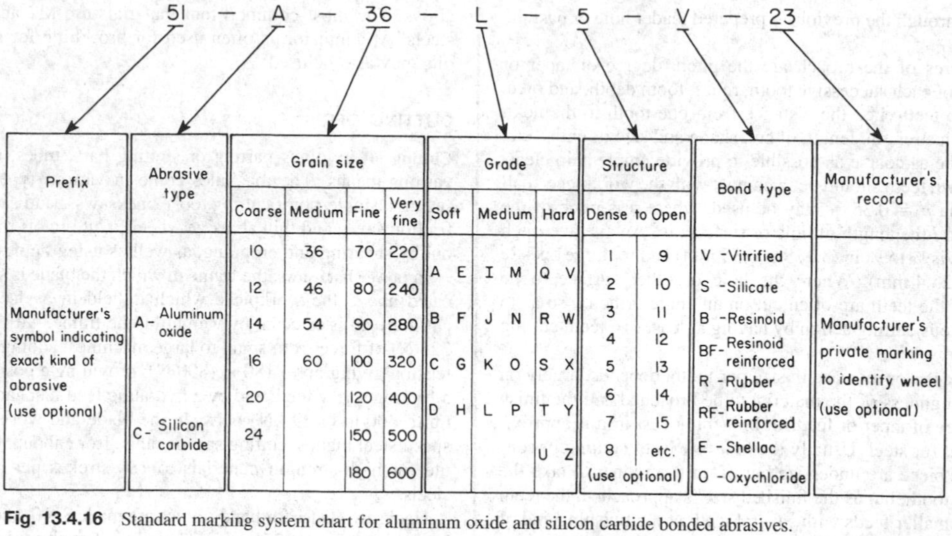

51 – A – 36 – L – 5 – V – 23

Prefix	Abrasive type	Coarse	Medium	Fine	Very fine	Soft	Medium	Hard	Structure (Dense to Open)	Bond type	Manufacturer's record
Manufacturer's symbol indicating exact kind of abrasive (use optional)	A – Aluminum oxide	10	36	70	220	A	E I M	Q V	1 9	V – Vitrified	Manufacturer's private marking to identify wheel (use optional)
		12	46	80	240	B	F J N	R W	2 10	S – Silicate	
		14	54	90	280	C	G K O	S X	3 11	B – Resinoid	
		16	60	100	320	D	H L P	T Y	4 12	BF – Resinoid reinforced	
	C – Silicon carbide	20		120	400			U Z	5 13	R – Rubber	
		24		150	500				6 14	RF – Rubber reinforced	
				180	600				7 15	E – Shellac	
									8 etc. (use optional)	O – Oxychloride	

Fig. 13.4.16 Standard marking system chart for aluminum oxide and silicon carbide bonded abrasives.

and dimensional accuracies that are generally not obtainable through other machining or manufacturing processes.

Grinding wheels have characteristics influenced by (1) type of abrasive; (2) grain size; (3) grade; (4) structure; and (5) type of bond. (See Figs. 13.4.16 and 13.4.17.)

Selection of Abrasive Although a number of natural abrasives are available, such as emery, corundum, quartz, garnet, and diamond, the most commonly used abrasives in grinding wheels are **aluminum oxide** and **silicon carbide,** the former being more commonly used than the latter. Aluminum oxide is softer than silicon carbide, and because of its friability and low attritious wear it is suitable for most applications. Silicon carbide is used for grinding aluminum, magnesium, titanium, copper, tungsten, and rubber. It is also used for grinding very hard and brittle materials such as carbides, ceramics, and stones. **Diamond** and **cubic boron nitride** grains are used to grind very hard materials and are known as **superabrasives.**

Selection of **grain size** depends on the rate of material removal desired and the surface finish. Coarse grains are used for fast removal of stock; fine grain for low removal rates and for fine finish. Coarse grains are also used for ductile materials and a finer grain for hard and brittle materials.

The **grade** of a grinding wheel is a measure of the strength of its bond. The force that acts on the grain in grinding depends on process variables (such as speeds, depth of cut, etc.) and the strength of the work material. Thus a greater force on the grain will increase the possibility of dislodging the grain; if the bond is too strong, the grain will tend to get dull, and if it is too weak then wheel wear will be high. If glazing occurs, the wheel is **acting hard;** reducing the wheel speed or increasing the work speed or the depth of cut causes the wheel to **act softer.** If the wheel breaks down too rapidly, reversing this procedure will make the wheel act harder. Harder wheels are generally recommended for soft work materials, and vice versa.

A variety of **bond** types are used in grinding wheels; these are generally categorized as **organic** and **inorganic.** Organic bonds are materials such as resin, rubber, shellac, and other similar bonding agents. Inorganic materials are glass, clay, porcelain, sodium silicate, magnesium

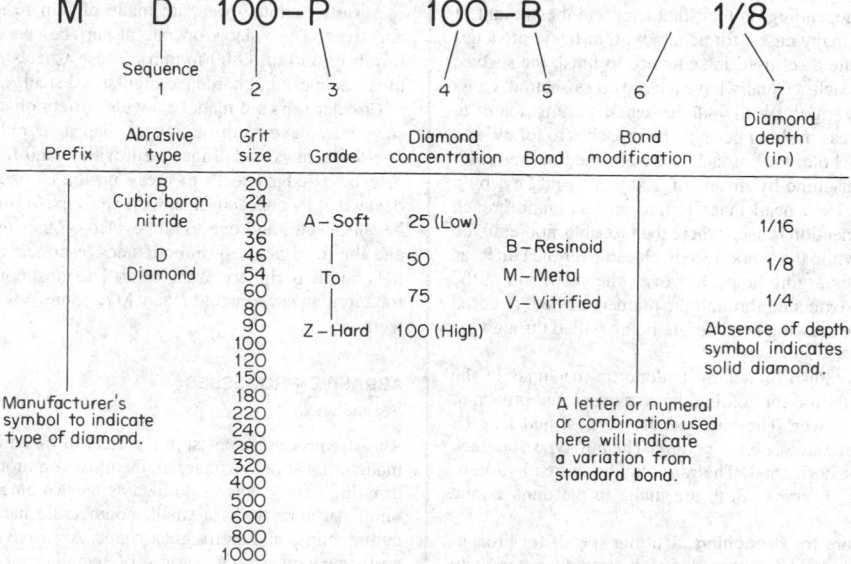

Fig. 13.4.17 Standard marking system chart for diamond tool and cubic boron nitride (CBN) bonded abrasives.

oxychloride, and metal. The most common bond type is the vitrified bond which is composed of clay, glass, porcelain, or related ceramic materials. This type of bond is brittle and produces wheels that are rigid, porous, and resistant to oil and water. The most flexible bond is rubber which is used in making very thin, flexible wheels. Wheels subjected to bending strains should be made with organic bonds. In selecting a bonding agent, attention should be paid to its sensitivity to temperature, stresses, and grinding fluids, particularly over a period of time. The term **reinforced** as applied to grinding wheels indicates a class of organic wheels which contain one or more layers of strengthening fabric or filament, such as fiberglass. This term does not cover wheels with reinforcing elements such as steel rings, steel cup backs, or wire or tape winding. Fiberglass and filament reinforcing increases the ability of wheels to withstand operational forces when cracked.

The **structure** of a wheel is important in two aspects: It supplies a clearance for the chip, and it determines the number of cutting points on the wheel.

In addition to wheel characteristics, grinding wheels come in a very large variety of shapes and dimensions. They are classified as **types,** such as type 1-straight wheels, type 4-taper side wheels, type 12-dish wheels, etc.

The **grinding ratio** is defined as the ratio of the volume of material removed to the volume of wheel wear. The grinding ratio depends on parameters such as the type of wheel, workpiece speed, wheel speed, cross-feed, down-feed, and the grinding fluid used. Values ranging from a low of 2 to over 200 have been observed in practice. A high grinding ratio, however, may not necessarily result in the best surface integrity of the part.

Wheel Speeds Depending on the type of wheel and the type and strength of bond, wheel speeds for standard applications range between 4,500 and 16,000 surface ft/min (1,400 and 4,800 m/min). The lowest speeds are for low-strength, inorganic bonds whereas the highest speeds are for high-strength organic bonds. The majority of surface grinding operations are carried out at speeds from 5,500 to 6,500 ft/min (1,750 to 2,000 m/min). A recent trend is toward high-efficiency grinding where wheel speeds from 12,000 to 18,000 ft/min (3,600 to 5,500 m/min) are employed.

It has been found that, by increasing the wheel speed, the rate of material removal can be increased, thus making the process more economical. This, of course, requires special grinding wheels to withstand the high stresses. Design changes or improvements involve items such as a composite wheel with a vitrified bond on the outside and a resinoid bond toward the center of the wheel; elimination of the central hole of the wheel by providing small bolt holes; and clamping of wheel segments instead of using a one-piece wheel. Grinding machines for such high-speed applications have requirements such as rigidity, high work and wheel speeds, high power, and special provisions for safety.

Workpiece speeds depend on the size and type of workpiece material and on whether it is rigid enough to hold its shape. In surface grinding, table speeds generally range from 50 to 100 ft/min (15 to 30 m/min); for cylindrical grinding, work speeds from 70 to 100 ft/min (20 to 30 m/min), and for internal grinding they generally range from 75 to 200 ft/min (20 to 60 m/min).

Cross-feed depends on the width of the wheel. In roughing, the workpiece should travel past the wheel ¾ to ⅞ of the width of the wheel for each revolution of the work. As the workpiece travels past the wheel with a helical motion, the preceding rule allows a slight overlap. In finishing, a finer feed is used, generally ⅒ to ¼ of the width of the wheel for each revolution of the workpiece.

Depth of Cut In the roughing operation, the depth of cut should be all the wheel will stand. This varies with the hardness of the material and the diameter of the workpiece. In the finishing operation, the depth of cut is always small: 0.0005 to 0.001 in (0.013 to 0.025 mm). Good results as regards finish are obtained by letting the wheel run over the workpiece several times without cross-feeding.

Grinding Allowances From 0.005 to 0.040 in (0.13 to 1 mm) is generally removed from the diameter in rough grinding in a cylindrical machine. For finishing, 0.002 to 0.010 in (0.05 to 0.25 mm) is common.

Workpieces can be finished by grinding to a tolerance of 0.0002 in (0.005 mm) and a surface roughness of $50+$ μin R_q (1.2 μm).

In situations where grinding leaves unfavorable surface **residual stresses,** the technique of **gentle** or **low-stress** grinding may be employed. This generally consists of removing a layer of about 0.010 in (0.25 mm) at depths of cut of 0.0002 to 0.0005 in (0.005 to 0.013 mm) with wheel speeds that are lower than the conventional 5,500 to 6,500 ft/min.

Truing and Dressing In **truing,** a diamond supported in the end of a soft steel rod held rigidly in the machine is passed over the face of the wheel two or three times to remove just enough material to give the wheel its true geometric shape. **Dressing** is a more severe operation of removing the dull or loaded surface of the wheel. Abrasive sticks or wheels or steel star wheels are pressed against and moved over the wheel face.

Safety If not stored, handled, and used properly, a grinding wheel can be a very dangerous tool. Because of its mass and high rotational speed, a grinding wheel has considerable energy and, if it fractures, it can cause serious injury and even death to the operator or to personnel nearby. A safety code B7.1 entitled ''The Use, Care, and Protection of Abrasive Wheels'' is available from ANSI; other safety literature is available from the Grinding Wheel Institute and from the National Safety Council.

The salient features of safety in the use of grinding wheels may be listed as follows: Wheels should be stored and handled carefully; a wheel that has been dropped should not be used. Before it is mounted on the machine, a wheel should be visually inspected for possible cracks; a simple ''ring'' test may be employed whereby the wheel is tapped gently with a light nonmetallic implement and if it sounds cracked, it should not be used. The wheel should be mounted properly with the required blotters and flanges, and the mounting nut tightened not excessively. The label on the wheel should be read carefully for maximum operating speed and other instructions. An appropriate guard should always be used with the machine, whether portable or stationary.

Newly mounted wheels should be allowed to run idle at the operating speed for at least 1 min before grinding. The operator should always wear safety glasses and should not stand directly in front of a grinding wheel when a grinder is started. If a grinding fluid is used, it should be turned off first before stopping the wheel to avoid creating an out-of-balance condition. Because for each type of operation and workpiece material there usually is a specific type of wheel recommended, the operator must make sure that the appropriate wheel has been selected.

Finishing Operations

Polishing is an operation by which scratches or tool marks or, in some instances, rough surfaces left after forging, rolling, or similar operations are removed. It is not a precision operation. The nature of the polishing process has been debated for a long time. Two mechanisms appear to play a role: One is fine-scale abrasion, and the other is softening of surface layers. In addition to removal of material by the abrasive particles, the high temperatures generated because of friction soften the asperities of the surface of the workpiece, resulting in a smeared surface layer. Furthermore, chemical reactions may also take place in polishing whereby surface irregularities are removed by chemical attack.

Polishing is usually done in stages. The first stage is rough polishing, using abrasive grain sizes of about 36 to 80, followed by a second stage, using an abrasive size range of 80 to 120, a third stage of size 150 and finer, etc., with a final stage of buffing. For the first two steps the polishing wheels are used dry. For finishing, the wheels are first worn down a little and then coated with tallow, oil, beeswax, or similar substances. This step is partly polishing and partly buffing, as additional abrasive is often added in cake form with the grease. The cutting action is freer, and the life of the wheel is prolonged by making the wheel surface flexible. Buffing wheels are also used for the finishing step when tallow, etc., containing coarse or fine abrasive grains is periodically rubbed against the wheel.

Polishing wheels consisting of wooden disks faced with leather, turned to fit the form of the piece to be polished, are used for flat surfaces or on work where it is necessary to maintain square edges. A large variety of other types of wheels are in common use. Compress wheels are used extensively and are strong, durable, and easily kept in balance. They consist of a steel center the rim of which holds a laminated surface of leather, canvas, linen, felt, rubber, etc., of various degrees of pliability. Wheels of solid leather disks of walrus hide, buffalo hide, sheepskin, or bull's neck hide, or of soft materials such as felt, canvas, and muslin, built up of disks either loose, stitched, or glued, depending on the resiliency or pliability required, are used extensively for polishing as well as buffing. **Belts** of cloth or leather are often charged with abrasive for polishing flat or other workpieces. Wire brushes may be used with no abrasive for a final operation to give a satin finish to nonferrous metals.

For most polishing operations speeds range from 5,000 to 7,500 surface ft/min (1,500 to 2,250 m/min). The higher range is for high-strength steels and stainless steels. Excessively high speeds may cause burning of the workpiece and glazing.

Buffing is a form of finish polishing in which the surface finish is improved; very little material is removed. The powdered abrasives are applied to the surface of the wheel by pressing a mixture of abrasive and tallow or wax against the face for a few seconds. The abrasive is replenished periodically. The wheels are made of a soft pliable material, such as soft leather, felt, linen, or muslin, and rotated at high speed.

Buffing generally comprises two stages: **cutting down** and **coloring**. Cutting down makes a surface smoother by removing scratches and other marks from previous operations. Coloring further refines the surface and brings out maximum luster. A variety of buffing compounds are available: Tripoli (an amorphous silica), aluminum oxide, chromium oxide, soft silica, rouge (iron oxide), pumice, lime compounds, emery, and crocus. In cutting down nonferrous metals, Tripoli is used; and for steels and stainless steels, aluminum oxide is the common abrasive. For coloring, soft silica, rouge, and chromium oxide are the more common compounds used. Buffing speeds range from 6,000 to 10,000 surface ft/min (1,800 to 3,000 m/min); the higher speeds are for steels, although the speed may be as high at 12,000 surface ft/min (3,600 m/min) for coloring brass and copper.

Lapping is a process of producing extremely smooth and accurate surfaces by rubbing the surface which is to be lapped against a mating form which is called a **lap**. The lap may either be charged with a fine abrasive and moistened with oil or grease, or the fine abrasive may be introduced with the oil. If a part is to be lapped to a final accurate dimension, a mating form of a softer material such as soft close-grained cast iron, copper, brass, or lead is made up. Aluminum oxide, silicon carbide, and diamond grits are used for lapping. Lapping requires considerable time. No more than 0.0002 to 0.0005 in (0.005 to 0.013 mm) should be left for removal by this method. Surface plates, rings, and plugs are common forms of laps. For most applications grit sizes range between 100 and 800, depending on the finish desired. For most efficient lapping, speeds generally range from 300 to 800 surface ft/min (150 to 240 m/min) with pressures of 1 to 3 lb/in^2 (7 to 21 kPa) for soft materials and up to 10 lb/in^2 (70 kPa) for harder materials.

Honing is an operation similar to lapping. Instead of a metal lap charged with fine abrasive grains, a honing stone made of fine abrasives is used. Small stones of various cross-sectional shapes and lengths are manufactured for honing the edges of cutting tools. Automobile cylinders are honed for fine finish and accurate dimensions. This honing usually follows a light-finish reaming operation or a precision-boring operation using diamonds or carbide tools. The tool consists of several honing stones adjustable at a given radius or forced outward by springs or a wedge forced mechanically or hydraulically and is given a reciprocating (25 to 40 per min) and a rotating motion (about 300 r/min) in the cylinder which is flooded with kerosine.

Hones operate at speeds of 50 to 200 surface ft/min (15 to 60 m/min) and use universal joints to allow the tool to center itself in the workpiece. The automatic pressure-cycle control of hone expansion, in which the pressure is reduced in steps as the final finish is reached, removes metal 10 times as fast as with the spring-expanded hone. Rotational and reciprocating movements are provided to give an uneven ratio and thus prevent an abrasive grain from ever traversing its own path twice.

Superfinishing is a honing process. Formed honing stones bear against the workpiece previously finished to 0.0005 in (0.013 mm) or at the most to 0.001 in (0.025 mm) by a very light pressure which gradually increases to several pounds per square inch (1 lb/in^2 = 0.0069 MPa) of stone area in proportion to the development of the increased area of contact between the workpiece and stone. The workpiece or tool rotates and where possible is reciprocated slowly over the surface which may be finished in a matter of 20 s to a surface quality of 1 to 3 μin (0.025 to 0.075 μm). Superfinishing is applied to many types of workpieces such as crankshaft pins and bearings, cylinder bores, pistons, valve stems, cams, and other metallic moving parts.

Sand blasting consists of particles of sand, powdered quartz, chilled-iron shot, emery, or other hard granular material blown by a jet of compressed air or of steam against a hard surface which it is desired to abrade. It is commonly used for cleaning metal castings, frosting smooth surfaces, etc. Portions of the surface which are not to be abraded can be protected by coating with a soft material such as wax, lead, or rubber.

Grinders

Grinding machines may be classified as to purpose and type as follows: for **rough removal of stock**, the swinging-frame, portable, flexible shaft, two-wheel stand, and disk; **cutting off** or parting, the cutting-off machine; **surface finishing**, band polisher, two-wheel combination, two-wheel polishing machine, two-wheel buffing machine, and semiautomatic polishing and buffing machine; **precision grinding**, tool post, cylindrical (plain and universal), crankshaft, centerless, internal, and surface (reciprocating table with horizontal or vertical wheel spindle, and rotary table with horizontal or vertical wheel spindle); **special form** grinders, gear or worm, ball-bearing balls, cams, and threads; and **tool and cutter** grinders for single-point tools, drills, and milling cutters, reamers, taps, dies, knives, etc.

Grinding equipment of all types (many with computer controls) has been improved during the past few years so as to be more rigid, provide more power to the grinding wheel, and provide automatic cycling, loading, clamping, wheel dressing, and automatic feedback.

Centerless grinders are used to good advantage where large numbers of relatively small pieces must be ground and where the ground surface has no exact relation to any other surface except as a whole; the work is carried on a support between two abrasive wheels, one a normal grinding wheel, the second a rubber-bonded wheel, rotating at about ½oth the grinding speed, and is tilted 3 to 8° to cause the work to rotate and feed past the grinding wheel (see also Sec. 6).

Cylinder grinders are a special type for grinding the inside of cylinders of engines; one form has a planetary motion for the grinding spindle. The **cylindrical grinder** is a companion machine to the engine lathe; shafts, cylinders, rods, studs, and a wide variety of other cylindrical parts are first roughed out on the lathe, then finished accurately to size by the cylindrical grinder. The work is carried on centers, rotated slowly, and traversed past the face of a grinding wheel.

Universal grinders are cylindrical machines arranged with a swiveling table so that both straight and taper internal and external work can be ground. **Drill grinders** are provided with rests so mounted that by a simple swinging motion, correct cutting angles are produced automatically on the lips of drills; a cupped wheel is usually employed. **Internal grinders** are used for finishing the holes in bushings, rolls, sleeves, cutters, and the like; spindle speeds from 15,000 to 30,000 r/min are common.

Horizontal surface grinders range from small capacity, used mainly in tool making or small production work, to large sizes used for production work.

Vertical surface grinders are used for producing flat surfaces on production work. **Vertical** and **horizontal disk grinders** are used for surfacing. Grinding machines are used for **cutting off** steel, especially tubes, structural shapes, and hard metals. A thin resinoid or rubber-bonded wheel is used, with aluminum oxide abrasive for all types of steel, aluminum, brass, bronze, nickel, Monel, and Stellite; silicon carbide for cast iron, copper, carbon, glass, stone, plastics, and other nonmetallic materials; and diamond for cemented carbides and ceramics.

Belt grinders use a coated abrasive belt running between pulleys. Belt grinding is generally considered to be a roughing process, but finer finishes may be obtained by using finer grain size. Belt speeds generally range from 2,000 to 10,000 ft/min (600 to 3,000 m/min) with grain sizes ranging between 24 and 320, depending on the workpiece material and the surface finish desired. The process has the advantage of high-speed material removal and is applied to flat as well as irregular surfaces.

Although grinding is generally regarded as a finishing operation, it is possible to increase the rate of stock removal whereby the process becomes, in certain instances, competitive with milling. This type of grinding operation is usually called **creep-feed grinding**. It uses equipment such as reciprocating table or vertical-spindle rotary table surface grinders with capacities up to 300 hp (220 kW). The normal stock removal may range up to ¼ in (6.4 mm) with wheel speeds between 3,400 and 5,000 surface ft/min (1,000 and 1,500 m/min).

A good ground surface with **surface roughness** of 50 to 200 μin (1 to 5 μm) is sufficient for many purposes and is a basic requirement for further finishing operations, such as lapping, honing, and superfinishing. The size and depth of the scratches can be varied considerably by the selection of the wheel.

MACHINING AND GRINDING OF PLASTICS

The low strength of thermoplastics permits high cutting speeds and feeds, but the low heat conductivity and greater resilience require increased reliefs and less rake in order to avoid undersize cutting. Hard and sharp tools should be used. Plastics are usually abrasive and cause the tools to wear or become dull rapidly. Dull tools generate heat and cause the tools to cut to shallow depths. The depth of cut should be small. When high production justifies the cost, diamond turning and boring tools are used. Diamond tools maintain sharp cutting edges and produce an excellent machined surface. They are particularly advantageous when a more abrasive plastic such as reinforced plastic is machined.

A cutting fluid, such as a small blast of air, a stream of water, or a cutting fluid, improves the **turning** and cutting of plastics as it prevents the heating of the tool and causes the chips to remain brittle and to break rather than become sticky and gummy. A zero or slightly negative back rake and a relief angle of 8 to 12° should be used. For thermoplastics cutting speeds generally range from 250 to 400 ft/min (75 to 120 m/min) and for thermosetting plastics from 400 to 1,000 ft/min (120 to 300 m/min). Recommended tool materials are M2 and T5 high-speed steels and C-2 carbide.

In **milling** plastics, speeds of 400 to 1,000 ft/min (120 to 300 m/min) should be used, with angles similar to those on a single-point tool. From 0 to 10° negative rake may be used. Good results have been obtained by hobbing plastic gears with carbide-tipped hobs. Recommended tool materials are M2 and M7 high-speed steels and C-2 carbide.

In **drilling**, speeds range from 150 to 400 ft/min (45 to 120 m/min), and the recommended drill geometry is given in Table 13.4.6. Tool materials are M1, M7, and M10 high-speed steel. Usually the drill cuts undersize; drills 0.002 to 0.003 in (0.05 to 0.075 mm) oversize should be used.

In **sawing** plastics, either precision or buttress tooth forms may be used, with a pitch ranging from 3 to 14 teeth/in (1.2 to 5.5 teeth/cm), the thicker the material the lower the number of teeth per unit length of saw. Cutting speeds for thermoplastics range from 1,000 to 4,000 ft/min (300 to 1,200 m/min) and for thermosetting plastics from about 3,000 to 5,500 ft/min (900 to 1,700 m/min), with the higher speeds for thinner stock. High-carbon-steel blades are recommended. An air blast is helpful in preventing the chips from sticking to the saw. Abrasive saws operating at 3,500 to 6,000 ft/min (1,000 to 1,800 m/min) are also used for cutting off bars and forms.

Plastics are tapped and threaded with standard tools. Ground M1, M7, or M10 high-speed steel **taps** with large polished flutes are recommended. **Tapping speeds** are usually from 25 to 50 ft/min (8 to 15 m/min); water serves as a good cutting fluid as it keeps the material brittle and prevents sticking in the flutes. **Thread cutting** is generally accomplished with tools similar to those used on brass.

Reaming is best accomplished in production by using tools of the expansion or adjustable type with relatively low speeds but high feeds. Less material should be removed in reaming plastics than in reaming other materials.

Polishing and **buffing** are done on many types of plastics. Polishing is done with special compounds containing wax or a fine abrasive. Buffing wheels for plastics should have loose stitching. Vinyl plastics can be buffed and polished with fabric wheels of standard types, using light pressures.

Thermoplastics and thermosets can be **ground** with relative ease, usually by using silicon carbide wheels. As in machining, temperature rise should be minimized.

MACHINING AND GRINDING OF CERAMICS

The technology of machining and grinding of ceramics, as well as composite materials, has advanced rapidly, resulting in good surface characteristics and product integrity. Ceramics can be machined with carbide, high-speed steel, or diamond tools, although care should be exercised because of the brittle nature of ceramics and the resulting possible surface damage. **Machinable ceramics** have been developed which minimize machining problems. Grinding of ceramics is usually done with diamond wheels.

OTHER MATERIAL REMOVAL PROCESSES

In addition to the mechanical methods of material removal described above, there are a number of other important processes which may be preferred over conventional methods. Among the important factors to be considered are the hardness of the workpiece material, the shape of the part, its sensitivity to thermal damage, residual stresses, tolerances, and economics. Some of these processes produce a heat-affected layer on the surface; improvements in surface integrity may be obtained by postprocessing techniques such as polishing or peening. Almost all machines are now computer-controlled.

Electric-discharge machining (EDM) is based on the principle of erosion of metals by spark discharges. Figure 13.4.18 gives a schematic diagram of this process. The spark is a transient electric discharge through the space between two charged electrodes, which are the tool and the workpiece. The discharge occurs when the potential difference between the tool and the workpiece is large enough to cause a breakdown in the medium (which is called the **dielectric fluid** and is usually a hydrocarbon) and to procure an electrically conductive spark channel. The breakdown potential is usually established by connecting the two electrodes to the terminals of a capacitor charged from a power source. The spacing between the tool and workpiece is critical; therefore, the feed is controlled by servomechanisms. The dielectric fluid has the additional functions of providing a cooling medium and carrying away particles produced by the electric discharge. The discharge is repeated rapidly, and each time a minute amount of workpiece material is removed.

The rate of metal removal depends mostly on the average current in the discharge circuit; it is also a function of the electrode characteristics, the electrical parameters, and the nature of the dielectric fluid. In practice, this rate is normally varied by changing the number of discharges per second or the energy per discharge. Rates of metal removal may range from 0.01 to 25 in³/h (0.17 to 410 cm³/h), depending on surface

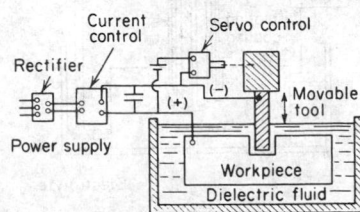

Fig. 13.4.18   Schematic diagram of the electric-discharge machining process.

finish and tolerance requirements. In general, higher rates produce rougher surfaces. Surface finishes may range from 1,000 μin R_q (25 μm) in roughing cuts to less than 25 μin (0.6 μm) in finishing cuts.

The response of materials to this process depends mostly on their thermal properties. Thermal capacity and conductivity, latent heats of melting and vaporization are important. Hardness and strength do not necessarily have significant effect on metal removal rates. The process is applicable to all materials which are sufficiently good conductors of electricity. The tool has great influence on permissible removal rates. It is usually made of graphite, copper-tungsten, or copper alloys. Tools have been made by casting, extruding, machining, powder metallurgy, and other techniques and are made in any desired shape. Tool wear is an important consideration, and in order to control tolerances and minimize cost, the ratio of tool material removed to workpiece material removed should be low. This ratio varies with different tool and workpiece material combinations and with operating conditions. Therefore, a particular tool material may not be best for all workpieces. Tolerances as low as 0.0001 to 0.0005 in (0.0025 to 0.0127 mm) can be held with slow metal removal rates. In machining some steels, tool wear can be minimized by reversing the polarity and using copper tools. This is known as "no-wear EDM."

The electric-discharge machining process has numerous applications, such as machining cavities and dies, cutting small-diameter holes, blanking parts from sheets, cutting off rods of materials with poor machinability, and flat or form grinding. It is also applied to sharpening tools, cutters, and broaches. The process can be used to generate almost any geometry if a suitable tool can be fabricated and brought into close proximity to the workpiece.

Thick plates may be cut with **wire EDM**. A slowly moving wire travels a prescribed path along the workpiece and cuts the metal with the sparks acting like saw teeth. The wire, usually about 0.01 in (0.25 mm) in diameter, is made of brass, copper, or tungsten and is generally used only once. The process is also used in making tools and dies from hard materials, provided that they are electrically conducting.

Electric-discharge grinding (EDG) is similar to the electric-discharge machining process with the exception that the electrode is in the form of a grinding wheel. Removal rates are up to 1.5 in³/h (25 cm³/h) with practical tolerances on the order of 0.001 in (0.025 mm). A graphite or brass electrode wheel is operated around 100 to 600 surface ft/min (30 to 180 m/min) to minimize splashing of the dielectric fluid. Typical applications of this process are in grinding of carbide tools and dies, thin slots in hard materials, and production grinding of intricate forms.

The **electrochemical machining (ECM)** process (Fig. 13.4.19) uses electrolytes which dissolve the reaction products formed on the workpiece by electrochemical action; it is similar to a reverse electroplating process. The electrolyte is pumped at high velocities through the tool. A gap of 0.005 to 0.020 in (0.13 to 0.5 mm) is maintained. A dc power supply maintains very high current densities between the tool and the workpiece. In most applications, a current density of 1,000 to 5,000 A is required per in² of active cutting area. The rate of metal removal is proportional to the amount of current passing between the tool and the workpiece. Removal rates up to 1 in³/min (16 cm³/min) can be obtained

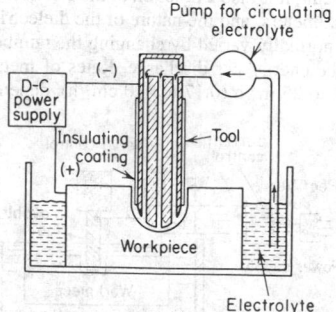

Fig. 13.4.19 Schematic diagram of the electrochemical machining process.

with a 10,000-A power supply. The penetration rate is proportional to the current density for a given workpiece material.

The process leaves a burr-free surface. It is also a cold machining process and does no thermal damage to the surface of the workpiece. Electrodes are normally made of brass or copper; stainless steel, titanium, sintered copper-tungsten, aluminum, and graphite have also been used. The electrolyte is usually a sodium chloride solution up to 2.5 lb/gal (300 g/L); other solutions and proprietary mixtures are also available. The amount of overcut, defined as the difference between hole diameter and tool diameter, depends upon cutting conditions. For production applications, the average overcut is around 0.015 in (0.4 mm). The rate of penetration is up to 0.750 in/min (20 mm/min).

Very good surface finishes may be obtained with this process. However, sharp square corners or sharp corners and flat bottoms cannot be machined to high accuracies. The process is applied mainly to round or odd-shaped holes with straight parallel sides. It is also applied to cases where conventional methods produce burrs which are costly to remove. The process is particularly economical for materials with a hardness above 400 HB.

The **electrochemical grinding (ECG)** process (Fig. 13.4.20) is a combination of electrochemical machining and abrasive cutting where most of the metal removal results from the electrolytic action. The process consists of a rotating cathode, a neutral electrolyte, and abrasive particles in contact with the workpiece. The equipment is similar to a conventional grinding machine except for the electrical accessories. The cath-

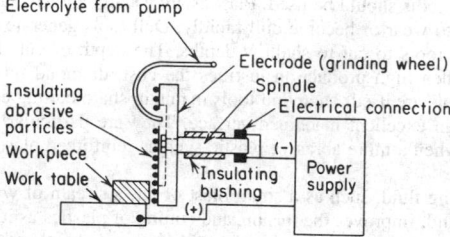

Fig. 13.4.20 Schematic diagram of the electrochemical grinding process.

ode usually consists of a metal-bonded diamond or aluminum oxide wheel. An important function of the abrasive grains is to maintain a space for the electrolyte between the wheel and workpiece.

Surface finish, precision, and metal-removal rate are influenced by the composition of the electrolyte. Aqueous solutions of sodium silicate, borax, sodium nitrate, and sodium nitrite are commonly used as electrolytes. The process is primarily used for tool and cutter sharpening and for machining of high-strength materials.

A combination of the electric-discharge and electrochemical methods of material removal is known as **electrochemical discharge grinding (ECDG)**. The electrode is a pure graphite rotating wheel which electrochemically grinds the workpiece. The intermittent spark discharges remove oxide films that form as a result of electrolytic action. The equipment is similar to that for electrochemical grinding. Typical applications include machining of fragile parts and resharpening or form grinding of carbides and tools such as milling cutters.

In **chemical machining (CM)** material is removed by chemical or electrochemical dissolution of preferentially exposed surfaces of the workpiece. Selective attack on different areas is controlled by masking or by partial immersion. There are two processes involved: chemical **milling** and chemical **blanking**. Milling applications produce shallow cavities for overall weight reduction, and are also used to make tapered sheets, plates, or extrusions. Masking with paint or tapes is common. Masking materials may be elastomers (such as butyl rubber, neoprene, and styrene-butadiene) or plastics (such as polyvinyl chloride, polystyrene, and polyethylene). Typical blanking applications are decorative panels, printed-circuit etching, and thin stampings. Etchants are solutions of sodium hydroxide for aluminum, and solutions of hydrochloric and nitric acids for steel.

Ultrasonic machining (USM) is a process in which a tool is given a high-frequency, low-amplitude oscillation, which, in turn, transmits a high velocity to fine abrasive particles that are present between the tool and the workpiece. Minute particles of the workpiece are chipped away on each stroke. Aluminum oxide, boron carbide, or silicone carbide grains are used in a water slurry (usually 50 percent by volume), which also carries away the debris. Grain size ranges from 200 to 1,000 (see Sec. 6 and Figs. 13.4.16 and 13.4.17).

The equipment consists of an electronic oscillator, a transducer (Fig. 13.4.21), a connecting cone or toolholder, and the tool. The oscillatory motion is obtained most conveniently by magnetostriction, at approximately 20,000 Hz and a stroke of 0.002 to 0.005 in (0.05 to 0.13 mm). The tool material is normally cold-rolled steel or stainless steel and is brazed, soldered, or fastened mechanically to the transducer through a

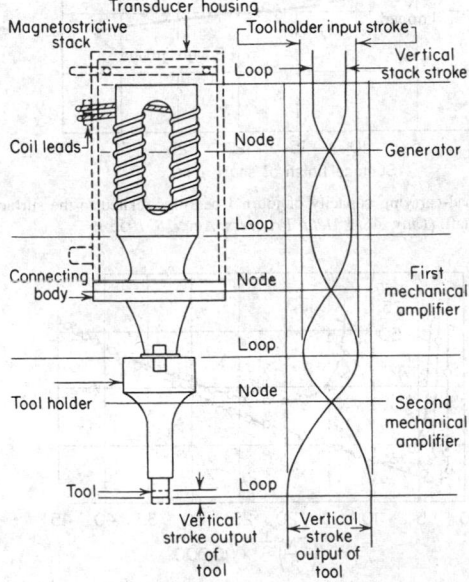

Fig. 13.4.21 Schematic diagram of a transducer used in the ultrasonic machining (USM) process.

toolholder. The tool is ordinarily 0.003 to 0.004 in (0.075 to 0.1 mm) smaller than the cavity it produces. Tolerances of 0.0005 in (0.013 mm) or better can be obtained with fine abrasives. For best results, roughing cuts should be followed with one or more finishing operations with finer grits. The ultrasonic machining process is used in drilling holes, engraving, cavity sinking, slicing, broaching, etc. It is best suited to materials which are hard and brittle, such as ceramics, carbides, borides, ferrites, glass, precious stones, and hardened steels.

In **abrasive-jet machining (AJM)**, material is removed by fine abrasive particles (aluminum oxide or silicon carbide) carried in a high-velocity stream of air, nitrogen, or carbon dioxide. The gas pressure ranges up to 120 lb/in^2 (800 kPa), providing a nozzle velocity of up to 1,000 ft/s (300 m/s). Nozzles are made of tungsten carbide or sapphire. Typical applications are in drilling, sawing, slotting, and deburring of hard, brittle materials such as glass.

In **water jet machining (WJM)**, water is ejected from a nozzle at pressures as high as 200,000 lb/in^2 (1,400 MPa) and acts as a saw. The process is suitable for cutting and deburring of a variety of materials such as polymers, paper, and brick in thicknesses ranging from 0.03 to 1 in (0.8 to 25 mm) or more. The cut can be started at any location, wetting is minimal, and no deformation of the rest of the piece takes place. Abrasives can be added to the water stream to increase material removal rate, and this is known as **abrasive water jet machining (AWJM)**.

In **laser-beam machining (LBM)**, material is removed by converting electric energy into a narrow beam of light and focusing it on the workpiece. The high energy density of the beam is capable of melting and vaporizing all materials, and consequently, there is a thin heat-affected zone. The most commonly used laser types are CO_2 (pulsed or continuous-wave) and Nd : YAG. Typical applications include cutting a variety of metallic and nonmetallic materials, drilling (as small as 0.0002 in or 0.005 mm in diameter), and marking. The efficiency of cutting increases with decreasing thermal conductivity and reflectivity of the material. Because of the inherent flexibility of the process, programmable and computer-controlled laser cutting is now becoming important, particularly in cutting profiles and multiple holes of various shapes and sizes on large sheets. Cutting speeds may range up to 25 ft/min (7.5 m/min).

The **electron-beam machining (EBM)** process removes material by focusing high-velocity electrons on the workpiece. Unlike lasers, this process is carried out in a vacuum chamber and is used for drilling small holes in all materials including ceramics, scribing, and cutting slots.

13.5 SURFACE TEXTURE DESIGNATION, PRODUCTION, AND CONTROL
by Thomas W. Wolff

REFERENCES: American National Standards Institute, "Surface Texture," ANSI/ASME B 46.1-1985, and "Surface Texture Symbols," ANSI Y 14.36-1978. Broadston, "Control of Surface Quality," Surface Checking Gage Co., Hollywood, CA. ASME, "Metals Engineering Design Handbook," McGraw-Hill SME, "Tool Engineers Handbook," McGraw-Hill.

Rapid changes in the complexity and precision requirements of mechanical products since 1945 have created a need for improved methods of determining, designating, producing, and controlling the surface texture of manufactured parts. Although standards are aimed at standardizing methods for measuring by using stylus probes and electronic transducers for surface quality control, other descriptive specifications are sometimes required, i.e., interferometric light bands, peak-to-valley by optical sectioning, light reflectance by commercial glossmeters, etc. Other parameters are used by highly industrialized foreign countries to solve their surface specification problems. These include the high-spot counter and bearing area meter of England (Talysurf); the total peak-to-

valley, or R_1, of Germany (Perthen); and the R or average amplitude of surface deviations of France. In the United States, peak counting is used in the sheet-steel industry, instrumentation is available (Bendix), and a standard for specification, SAE J-911, exists.

Surface texture control should be considered for many reasons, among them being the following:

1. Advancements in the technology of metal-cutting tools and machinery have made the production of higher-quality surfaces possible.

2. Products are now being designed that depend upon proper quality control of critical surfaces for their successful operation as well as for long, troublefree performance in service.

3. Artisans who knew the function and finish requirements for all the parts they made have been replaced, in most cases, by machine operators who are not qualified to determine the proper texture requirements for critical surfaces. The latter must depend, instead, on drawing specifications.

4. Remote manufacture and the necessity for controlling costs have made it preferable that finish requirements for all the critical surfaces of a part be specified on the drawing.

5. The design engineer, who best understands the overall function of a part and all its surfaces, should be able to determine the requirement for surface texture control where applicable and to use a satisfactory standardized method for providing this information on the drawing for use by manufacturing departments.

6. Manufacturing personnel should know what processes are able to produce surfaces within specifications and should be able to verify that the production techniques in use are under control.

7. Quality control personnel should be able to check conformance to surface texture specifications if product quality is to be maintained and product performance and reputation ensured.

8. Too much control may be worse than too little; hence, overuse of available techniques may hinder rather than assist, there being no payoff in producing surfaces that are more expensive than required to ensure product performance to establish standards.

DESIGN CRITERIA

Surfaces produced by various processes exhibit distinct differences in texture. These differences make it possible for honed, lapped, polished, turned, milled, or ground surfaces to be easily identified. As a result of its unique character, the surface texture produced by any given process can be readily compared with other surfaces produced by the same process through the simple means of comparing the average size of its irregularities, using applicable standards and modern measurement methods. It is then possible to predict and control its performance with considerable certainty by limiting the range of the average size of its characteristic surface irregularities. Surface texture standards make this control possible.

Variations in the texture of a critical surface of a part influence its ability to resist wear and fatigue; to assist or destroy effective lubrication; to increase or decrease its friction and/or abrasive action on other parts, and to resist corrosion, as well as affect many other properties that may be critical under certain conditions.

Clay has shown that the load-carrying capacity of nitrided shafts of varying degrees of roughness, all running at 1,500 r/min in diamond-turned lead-bronze bushings finished to 20 μin (0.50 μm), varies as shown in Fig. 13.5.1. The effects of roughness values on the friction between a flat slider on a well-lubricated rotating disk are shown in Fig. 13.5.2.

Surface texture control should be a normal design consideration under the following conditions:

1. For those parts whose roughness must be held within closely controlled limits for optimum performance. In such cases, even the process may have to be specified. Automobile engine cylinder walls, which should be finished to about 13 μin (0.32 μm) and have a circumferential (ground) or an angular (honed) lay, are an example. If too rough, excessive wear occurs, if too smooth, piston rings will not seat properly, lubrication is poor, and surfaces will seize or gall.

2. Some parts, such as antifriction bearings, cannot be made too smooth for their function. In these cases, the designer must optimize the tradeoff between the added costs of production and various benefits derived from added performance, such as higher reliability and market value.

3. There are some parts where surfaces must be made as smooth as possible for optimum performance regardless of cost, such as gages, gage blocks, lenses, and carbon pressure seals.

4. In some cases, the nature of the most satisfactory finishing process may dictate the surface texture requirements to attain production efficiency, uniformity, and control even though the individual performance of the part itself may not be dependent on the quality of the controlled surface. Hardened steel bushings, e.g., which must be ground to close tolerance for press fit into housings, could have outside surfaces well beyond the roughness range specified and still perform their function satisfactorily.

5. For parts which the shop, with unjustified pride, has traditionally finished to greater perfection than is necessary, the use of proper surface texture designations will encourage rougher surfaces on exterior and other surfaces that do not need to be finely finished. Significant cost reductions will accrue thereby.

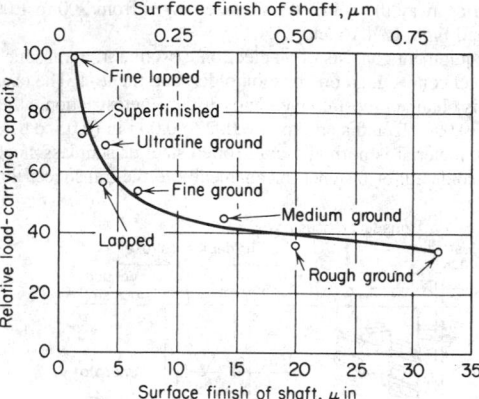

Fig. 13.5.1 Load-carrying capacity of journal bearings related to the surface roughness of a shaft. (*Clay, ASM Metal Progress, Aug. 15, 1955.*)

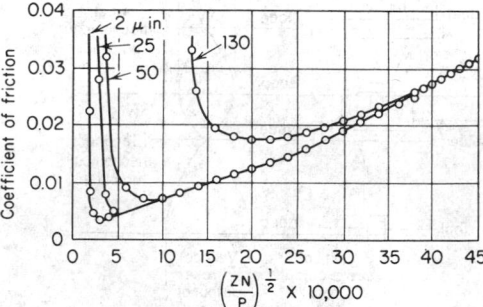

Fig. 13.5.2 Effect of surface texture on friction with hydrodynamic lubrication using a flat slider on a rotating disk. Z = oil viscosity, cP; N = rubbing speed, ft/min; P = load, lb/in^2.

It is the designer's responsibility to decide which surfaces of a given part are critical to its design function and which are not. This decision should be based upon a full knowledge of the part's function as well as of the performance of various surface textures that might be specified. From both a design and an economic standpoint, it may be just as unsound to specify too smooth a surface as to make it too rough—or to control it at all if not necessary. Wherever normal shop practice will produce acceptable surfaces, as in drilling, tapping, and threading, or in keyways, slots, and other purely functional surfaces, unnecessary surface texture control will add costs which should be avoided.

Whereas each specialized field of endeavor has its own traditional criteria for determining which surface finishes are optimum for adequate performance, Table 13.5.1 provides some common examples for design review, and Table 13.5.6 provides data on the surface texture ranges that can be obtained from normal production processes.

DESIGNATION STANDARDS, SYMBOLS, AND CONVENTIONS

The precise definition and measurement of surface texture irregularities of machined surfaces are almost impossible because the irregularities are very complex in shape and character and, being so small, do not lend themselves to direct measurement. Although both their shape and length may affect their properties, control of their average height and direction usually provides sufficient control of their performance. The standards

Table 13.5.1 Typical Surface Texture Design Requirements

(250 μin)	6.3	Clearance surfaces Rough machine parts	(16 μin)	0.40	Motor shafts Gear teeth (heavy loads) Spline shafts
(125 μin)	3.2	Mating surfaces (static) Chased and cut threads Clutch-disk faces Surfaces for soft gaskets			O-ring grooves (static) Antifriction bearing bores and faces Camshaft lobes Compressor-blade airfoils Journals for elastomer lip seals
(63 μin)	1.60	Piston-pin bores Brake drums Cylinder block, top Gear locating faces Gear shafts and bores Ratchet and pawl teeth Milled threads Rolling surfaces Gearbox faces Piston crowns Turbine-blade dovetails	(13 μin) (8 μin) (4 μin)	0.32 0.20 0.10	Engine cylinder bores Piston outside diameters Crankshaft bearings Jet-engine stator blades Valve-tappet cam faces Hydraulic-cylinder bores Lapped antifriction bearings Ball-bearing races Piston pins Hydraulic piston rods Carbon-seal mating surfaces
(32 μin)	0.80	Broached holes Bronze journal bearings Gear teeth Slideways and gibs Press-fit parts Piston-rod bushings Antifriction bearing seats Sealing surfaces for hydraulic tube fittings	(2 μin) (1 μin)	0.050 0.025	Shop-gage faces Comparator anvils Bearing balls Gages and mirrors Micrometer anvils

do not specify the surface texture suitable for any particular application, nor the means by which it may be produced or measured. Neither are the standards concerned with other surface qualities such as appearance, luster, color, hardness, microstructure, or corrosion and wear resistance, any of which may be a governing design consideration.

The standards provide definitions of the terms used in delineating critical surface-texture qualities and a series of symbols and conventions suitable for their designation and control. In the discussion which follows, the reference standards used are "Surface Texture" (ANSI/ASME B46.1-1985) and "Surface Texture Symbols" (ANSI Y14.36-1978).

The basic ANSI symbol for designating surface texture is the checkmark with horizontal extension shown in Fig. 13.5.3. The symbol with the triangle at the base indicates a requirement for a machining allowance, in preference to the old *f* symbol. Another, with the small circle in the base, prohibits machining; hence surfaces must be produced without the removal of material by processes such as cast, forged, hot- or cold-finished, die-cast, sintered- or injection-molded, to name a few. The surface-texture requirement may be shown at A; the machining allowance at B; the process may be indicated above the line at C; the roughness width cutoff (sampling length) at D, and the lay at E. The ANSI symbol provides places for the insertion of numbers to specify a wide variety of texture characteristics, as shown in Table 13.5.2.

Control of **roughness,** the finely spaced surface texture irregularities resulting from the manufacturing process or the cutting action of tools

or abrasive grains, is the most important function accomplished through the use of these standards, because roughness, in general, has a greater effect on performance than any other surface quality. The roughness-height index value is a number which equals the arithmetic average deviation of the minute surface irregularities from a hypothetical perfect surface, expressed in either millionths of an inch (microinches, μin, 0.000001 in) or in micrometres, μm, if drawing dimensions are in metric, SI units. For control purposes, roughness-height values are taken from Table 13.5.3, with those in boldface type given preference.

The term *roughness cutoff,* a characteristic of tracer-point measuring instruments, is used to limit the length of trace within which the asperities of the surface must lie for consideration as roughness. Asperity spacings greater than roughness cutoff are then considered as waviness.

Waviness refers to the secondary irregularities upon which roughness is superimposed, which are of significantly longer wavelength and are usually caused by machine or work deflections, tool or workpiece vibration, heat treatment, or warping. Waviness can be measured by a dial indicator or a profile recording instrument from which roughness has been filtered out. It is rated as maximum peak-to-valley distance and is indicated by the preferred values of Table 13.5.4. For fine waviness control, techniques involving contact-area determination in percent (90, 75, 50 percent preferred) may be required. Waviness control by interferometric methods is also common, where notes, such as "Flat within XX helium light bands," may be used. Dimensions may be determined from the precision length table (see Sec. 1).

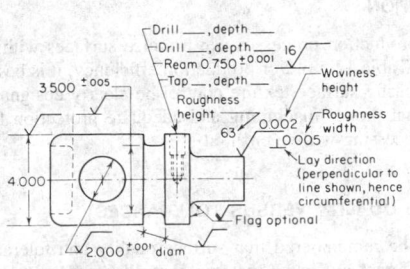
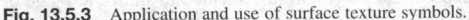
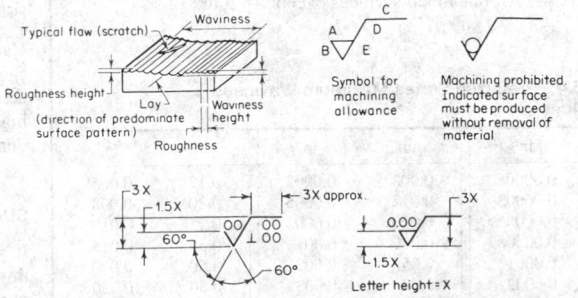

Fig. 13.5.3 Application and use of surface texture symbols.

Table 13.5.2 Application of Surface Texture Values to Surface Symbols

(63) 1.6	Roughness average rating is placed at the left of the long leg. The specification of only one rating shall indicate the maximum value and any lesser value shall be acceptable. Specify in micrometres (microinches).
(63) 1.6 (32) 0.8	The specification of maximum value and minimum value roughness average ratings indicates permissible range of value rating. Specify in micrometres (microinches).
(32) 0.8 0.05	Maximum waviness height rating is placed above the horizontal extension. Any lesser rating shall be acceptable. Specify in millimetres (inches).
(32) 0.8 0.05 − 100	Maximum waviness spacing rating is placed above the horizontal extension and to the right of the waviness height rating. Any lesser rating shall be acceptable. Specify in millimetres (inches).
(63) 1.6 3.5	Machining is required to produce the surface. The basic amount of stock provided for machining is specified at the left of the short leg of the symbol. Specify in millimetres (inches).
(63) 1.6	Removal of material by machining is prohibited.
(32) 0.8 ⊥	Lay designation is indicated by the lay symbol placed at the right of the long leg.
(32) 0.8 2.5 (0.100)	Roughness sampling length or cutoff rating is placed below the horizontal extension. When no value is shown, 0.80 mm is assumed. Specify in millimetres (inches).
(32) 0.8 ⊥0.5	Where required, maximum roughness spacing shall be placed at the right of the lay symbol. Any lesser rating shall be acceptable. Specify in millimetres (inches).

Table 13.5.3 Preferred Series Roughness Average Values R_a, μm and μin

μm	μin	μm	μin	μm	μin	μm	μin	μm	μin
0.012	0.5	0.125	5	0.50	20	2.00	80	8.0	320
0.025	1	0.15	6	0.63	25	2.50	100	10.0	400
0.050	**2**	**0.20**	**8**	**0.80**	**32**	**3.20**	**125**	**12.5**	**500**
0.075	3	0.25	10	1.00	40	4.0	160	15.0	600
0.10	**4**	0.32	13	1.25	50	5.0	200	20.0	800
		0.40	**16**	**1.60**	**63**	**6.3**	**250**	**25.0**	**1,000**

Lay refers to the direction of the predominant visible surface roughness pattern. It can be controlled by use of the approved symbols given in Table 13.5.5, which indicate desired lay direction with respect to the boundary line of the surface upon which the symbol is placed.

Flaws are imperfections in a surface that occur only at infrequent intervals. They are usually caused by nonuniformity of the material, or they result from damage to the surface subsequent to processing, such as scratches, dents, pits, and cracks. Flaws should not be considered in surface texture measurements, as the standards do not consider or classify them. Acceptance or rejection of parts having flaws is strictly a matter of judgment based upon whether the flaw will compromise the intended function of the part.

To call attention to the fact that surface texture values are specified on any given drawing, a note and typical symbol may be used as follows:

√ Surface texture per ANSI B46.1

Values for nondesignated surfaces can be limited by the note

xx
√ All machined surfaces except as noted

Table 13.5.4 Preferred Series Maximum Waviness Height Values

mm	in	mm	in	mm	in
0.0005	0.00002	0.008	0.0003	0.12	0.005
0.0008	0.00003	0.012	0.0005	0.20	0.008
0.0012	0.00005	0.020	0.0008	0.25	0.010
0.0020	0.00008	0.025	0.001	0.38	0.015
0.0025	0.0001	0.05	0.002	0.50	0.020
0.005	0.0002	0.08	0.003	0.80	0.030

MEASUREMENT

Two general methods exist to measure surface texture: **profile methods** and **area methods**. Profile methods measure the contour of the surface in a plane usually perpendicular to the surface. Area methods measure an area of a surface and produce results that depend on area-averaged properties.

Another categorization is by **contact methods** and **noncontact methods**. Contact methods include stylus methods (tracer-point analysis) and capacitance methods. Noncontact methods include light section microscopy, optical reflection measurements, and interferometry.

Replicas of typical standard machined surfaces provide less accurate but often adequate reference and control of rougher surfaces with R_a over 16 μin.

The United States and 25 other countries have adopted the **roughness average R_a** as the standard measure of surface roughness. (See ANSI/ASME B46.1-1985.)

PRODUCTION

Various production processes can produce surfaces within the ranges shown in Table 13.5.6. For production efficiency, it is best that critical areas requiring surface texture control be clearly designated on drawings so that proper machining and adequate protection from damage during processing will be ensured.

SURFACE QUALITY VERSUS TOLERANCES

It should be remembered that surface quality and tolerances are distinctly different attributes that are controlled for completely separate purposes. **Tolerances** are established to limit the range of the size of a

Table 13.5.5 Lay Symbols

Lay symbol	Interpretation	Example showing direction of tool marks
=	Lay approximately parallel to the line representing the surface to which the symbol is applied	
⊥	Lay approximately perpendicular to the line representing the surface to which the symbol is applied	
X	Lay angular in both directions to line representing the surface to which symbol is applied	
M	Lay multidirectional	
C	Lay approximately circular relative to the center of the surface to which the symbol is applied	
R	Lay approximately radial relative to the center of the surface to which the symbol is applied	
P	Lay particulate, nondirectional, or protuberant	

part at the time of manufacture, as measured with gages, micrometres, or other traditional measuring devices having anvils that make contact with the part. **Surface quality** controls, on the other hand, serve to limit the minute surface irregularities or asperities that are formed by the manufacturing process. These lie under the gage anvils during measurement and **do not use up tolerances.**

Table 13.5.6 Surface-Roughness Ranges of Production Processes

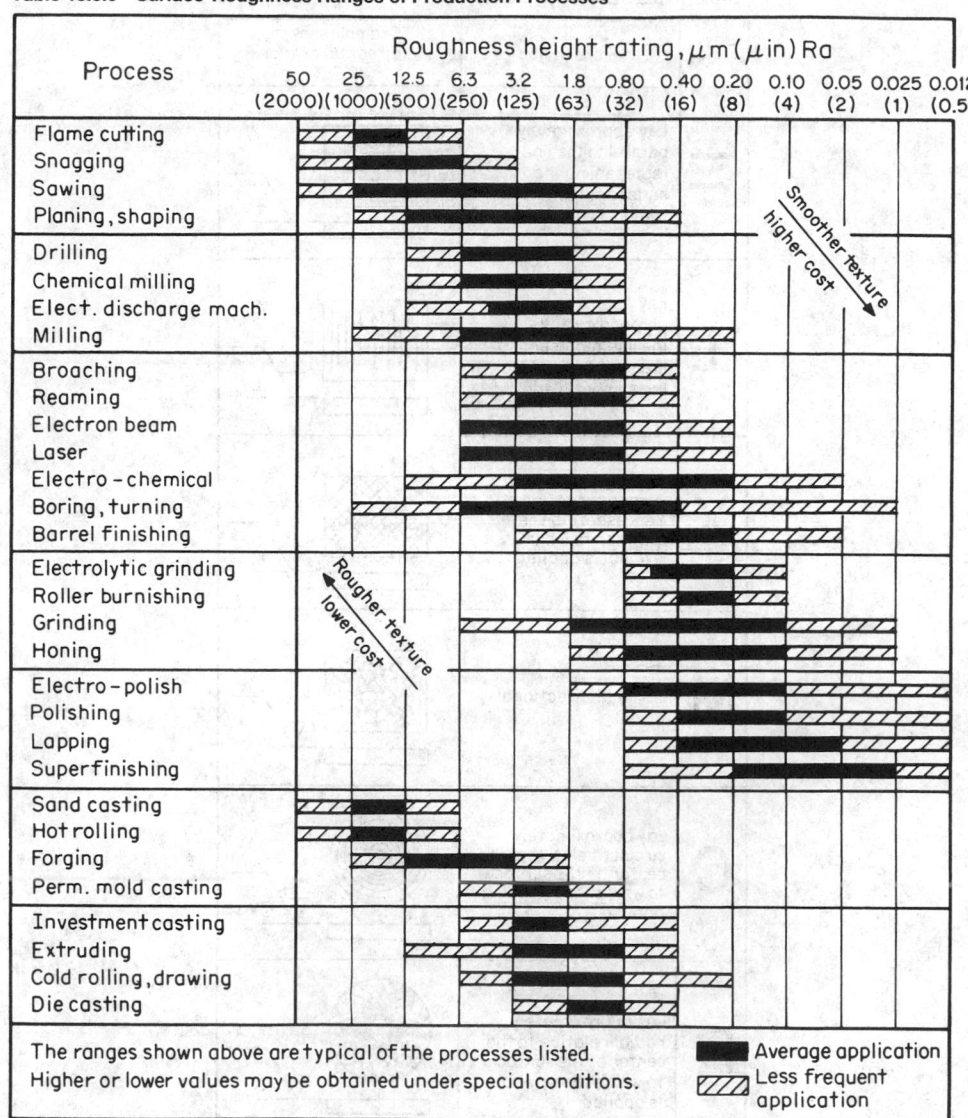

The ranges shown above are typical of the processes listed.
Higher or lower values may be obtained under special conditions.

■ Average application
▨ Less frequent application

13.6 WOODCUTTING TOOLS AND MACHINES
by Richard W. Perkins

REFERENCES: Davis, Machining and Related Characteristics of United States Hardwoods, *USDA Tech. Bull.* 1267. Harris, "A Handbook of Woodcutting," Her Majesty's Stationery Office, London. Koch, "Wood Machining Processes," Ronald Press. Kollmann, Wood Machining, in Kollmann and Côté, "Principles of Wood Science and Technology," chap. 9, Springer-Verlag.

SAWING

Sawing machines are classified according to basic machine design, i.e., band saw, gang saw, chain saw, circular saw. Saws are designated as ripsaws if they are designed to cut along the grain or **crosscut** saws if they are designed to cut across the grain. A **combination** saw is designed to cut reasonably well along the grain, across the grain, or along a direction at an angle to the grain (**miter**). Sawing machines are often further classified according to the specific operation for which they are used, e.g., **headsaw** (the primary log-breakdown saw in a sawmill), **resaw** (saw for ripping cants into boards), **edger** (saw for edging boards in a sawmill), **variety saw** (general-purpose saw for use in furniture plants), **scroll saw** (general-purpose narrow-band saw for use in furniture plants).

The thickness of the saw blade is designated in terms of the Birmingham wire gage (BWG) (see Sec. 6). Large-diameter [40 to 60 in (1.02 to 1.52 mm)] circular-saw blades are tapered so that they are thicker at the center than at the rim. Typical headsaw blades range in thickness from 5 to 6 BWG [0.203 to 0.220 in (5.16 to 5.59 mm)] for use in heavy-duty applications to 8 to 9 BWG [0.148 to 0.165 in (3.76 to 4.19 mm)] for lighter operations. Small-diameter [6 to 30 in (152 to 762 mm)] circular saws are generally flat-ground and range from 10 to 18 BWG [0.049 to 0.134 in (1.24 to 3.40 mm)] in thickness. Band-saw and gang-saw blades are flat-ground and are generally thinner than circular-saw blades designed for similar applications. For example, typical wide-band-saw blades for sawmill use range from 11 to 16 BWG [0.065 to 0.120 in (1.65 to 3.05 mm)] in thickness. The thickness of a band-saw blade is determined by the cutting load and the diameter of the band wheel. Gang-saw blades are generally somewhat thicker than band-saw blades for similar operations. Narrow-band-saw blades for use on scroll band saws range in thickness from 20 to 25 BWG [0.020 to 0.035 in (0.51 to 0.89 mm)] and range in width from ⅛ to about 1¾ in (3.17 to 44.5 mm) depending upon the curvature of cuts to be made.

The considerable amount of heat generated at the cutting edge results in compressive stresses in the rim of the saw blade of sufficient magnitude to cause mechanical instability of the saw blades. Circular-saw blades and wide-band-saw blades are commonly prestressed (or **tensioned**) to reduce the possibility of buckling. Small circular-saw blades for use on power-feed rip-saws and crosscut saws are frequently provided with **expansion slots** for the same purpose.

The **shape of the cutting portion of the sawtooth** is determined by specifying the hook, face bevel, top bevel, and clearance angles. The optimum tooth shape depends primarily upon cutting direction, moisture content, and density of the workpiece material. Sawteeth are, in general, designed in such a way that the portion of the cutting edge which is required to cut across the fiber direction is provided with the maximum effective rake angle consistent with tool strength and wear considerations. Ripsaws are designed with a hook angle between some 46° for inserted-tooth circular headsaws used to cut green material and 10° for solid-tooth saws cutting dense material at low moisture content. Ripsaws generally have zero face bevel and top bevel angle; however, spring-set ripsaws sometimes are provided with a moderate top bevel angle (5 to 15°). The hook angle for crosscut saws ranges from positive 10° to negative 30°. These saws are generally designed with both top and face bevel angles of 5 to 15°; however, in some cases top and face bevel angles as high as 45° are employed. A compromise design is used for combination saws which embodies the features of both ripsaws and crosscut saws in order to provide a tool which can cut reasonably well in all directions. The clearance angle should be maintained at the smallest possible value in order to provide for maximum tooth strength. For ripsawing applications, the clearance angle should be about 12 to 15°. The minimum satisfactory clearance angle is determined by the nature of the work material, not from kinematical considerations of the motion of the tool through the work. In some cases of cutoff, combination, and narrow-band-saw designs where the tooth pitch is relatively small, much larger clearance angles are used in order to provide the necessary gullet volume.

A certain amount of clearance between the saw blade and the generated surface (**side clearance** or **set**) is necessary to prevent frictional heating of the saw blade. In the case of solid-tooth circular saws and band or gang saws, the side clearance is generally provided either by deflecting alternate teeth (**spring-setting**) or by spreading the cutting edge (**swage-setting**). The amount of side clearance depends upon density, moisture content, and size of the saw blade. In most cases, satisfactory results are obtained if the side clearance S is determined from the formula S [in (mm)] = $A/2[f(g-5) - f(g)]$, where g = gage number (BWG) of the saw blade, $f(n)$ = dimension in inches (mm) corresponding to the gage number n, and A has values from Table 13.6.1. Certain specialty circular saws such as planer, smooth-trimmer, and miter saws are hollowground to provide side clearance. Inserted-tooth saws, carbide-tipped saws, and chain-saw teeth are designed so that sufficient side clearance is provided for the life of the tool; consequently, the setting of such saws is unnecessary.

Table 13.6.1 Values of *A* for Computing Side Clearance

	Workpiece material			
	Specific gravity less than 0.45		Specific gravity greater than 0.55	
Saw type	Air dry	Green	Air dry	Green
Circular rip and combination	0.90	1.00	0.85	0.95
Glue-joint ripsaw	0.80	—	0.60	—
Circular crosscut	0.95	1.05	0.90	1.00
Wide-band saw	0.55	0.65	0.30	0.40
Narrow-band saw	0.65	—	0.55	—

The **tooth speed** for sawing operations ranges from 3,000 to 17,000 ft/min (15 to 86 m/s) approx. Large tooth speeds are in general desirable in order to permit maximum work rates. The upper limit of permissible tooth speed depends in most cases on machine design considerations and not on considerations of wear or surface quality as in the case of metal cutting. Exceptionally high tooth speeds may result in charring of the work material, which is machined at slow feed rates.

In many sawing applications, **surface quality** is not of prime importance since the sawed surfaces are subsequently machined, e.g., by planing, shaping, sanding; therefore, it is desirable to operate the saw at the largest feed per tooth consistent with gullet overloading. Large values of feed per tooth result in lower amounts of work required per unit volume of material cut and in lower amounts of wear per unit tool travel. Large-diameter circular saws, wide band saws, and gang saws for ripping green material are generally designed so that the feed per tooth should be about 0.08 to 0.12 in (2.03 to 3.05 mm). Small-diameter circular saws are designed so that the feed per tooth ranges from 0.03 in (0.76 mm) for dense hardwoods to 0.05 in (1.27 mm) for low-density softwoods. Narrow-band saws are generally operated at somewhat smaller values of feed per tooth, e.g., 0.005 to 0.04 in (0.13 to 1.02 mm). Smaller values of feed per tooth are necessary for applications where surface quality is of prime importance, e.g., glue-joint ripsawing and variety-saw operations. The degree of gullet loading is measured by the **gullet-feed index (GFI),** which is computed as the feed per tooth times the depth of face divided by the gullet area. The maximum GFI depends primarily upon species, moisture content, and cutting direction. It is generally conceded that the maximum GFI for ripsawing lies between 0.3 for high-density, low-moisture-content material and 0.4 for low-density, high-moisture-content material. For specific information, see Telford, *For. Prod. Res. Soc. Proc.,* 1949.

Saws vary considerably in design of the **gullet shape.** The primary design considerations are gullet area and tooth strength; however, special design shapes are often required for certain classes of workpiece material, e.g., for ripping frozen wood.

Materials Saw blades and the sawteeth of solid-tooth saws are generally made of a nickel tool steel. The bits for inserted-tooth saws were historically plain carbon tool steel; however, high-speed steel bits or bits with a cast-alloy inlay (e.g., Stellite) are sometimes used in applications where metal or gravel will not be encountered. Small-diameter circular saws of virtually all designs are made with cemented-carbide tips. This type is almost imperative in applications where highly abrasive material is cut, namely, in plywood and particleboard operations.

Sawing Power References: Endersby, The Performance of Circular Plate Ripsaws, *For. Prod. Res. Bull.* 27, Her Majesty's Stationery Office, London, 1953. Johnston, Experimental Cut-off Saw, *For. Prod. Jour.,* June 1962. Oehrli, Research in Cross-cutting with Power Saw Chain Teeth, *For. Prod. Jour.,* Jan. 1960. Telford, Energy Requirements for Insert-point Circular Headsaws, *Proc. For. Prod. Res. Soc.,* 1949.

An approximate relation for computing the power P, ft·lb/min (W), required to saw is

$$P = kvb(A + Bt_a)/p$$

where k is the kerf, in (m); v is the tooth speed, ft/min (m/s); p is the tooth pitch, in (m); A and B are constants for a given sawing operation,

Table 13.6.2   Constants for Sawing-Power Estimation

Species	Material		Tool		Constants			
	Specific gravity	Moisture content, %	Angles[e]	Sawing situation	A, lb/in	A, N/m	B, lb/in² × 10⁻³	B, N/m² × 10⁻⁶
Beech, European[a]	0.72	12	20, 0, 12	SS, R	27.8	4,869	5.760	39.71
Birch, yellow[b]	0.55	FSP	− 30, 10, 10	SS, CC	19.7	3,450	4.100	28.27
Elm, wych[a]	0.67	12	20, 0, 12	SS, R	23.2	4,063	4.840	33.37
Maple, sugar[c]	0.63	FSP	41, 0, 0	IT, R	85.6	15,991	2.995	20.65
Maple, sugar[c,f]	0.63	FSP	41, 0, 0	IT, R	48.0	8,406	4.400	30.34
Pine, northern white[c]	0.34	FSP	41, 0, 0	IT, R	27.1	4,746	1.675	11.55
Pine, northern white[c,f]	0.34	FSP	41, 0, 0	IT, R	28.2	4,939	2.085	14.38
Pine, northern white[b]	0.34	FSP	− 30, 10, 10	SS, CC	0.0	0	3.300	22.75
Pine, ponderosa[d]	0.38–0.40	15–40	28, 25, 0	OFT, R	29.3	5,131	1.700	11.72
Pine, ponderosa[d]	0.38–0.40	15–40	28, 25, 0	OFT, CC	0.0	0	2.120	14.62
Poplar (*P. serotina*)[a]	0.48	12	20, 0, 12	SS, R	18.6	3,257	3.290	22.68
Redwood, California[a]	0.37	12	20, 0, 12	SS, R	15.0	2,627	2.260	15.58
Spruce, white[b]	0.32	FSP	− 30, 10, 10	SS, CC	0.0	0	4.680	32.27

[a] Endersby.
[b] Johnston.
[c] Hoyle, unpublished report, N.Y. State College of Forestry, Syracuse, NY, 1958.
[d] Oehrli.
[e] The numbers represent hook angle, face bevel angle, and top bevel in degrees.
[f] Cutting performed on frozen material.
NOTE: FSP = moisture content greater than the fiber saturation point; CC = crosscut; IT = insert-tooth; OFT = offset-tooth; R = rip; SS = spring-set.

lb/in (N/m) and lb/in² (N/m²), respectively; and t_a is the average chip thickness, in (m). The average chip thickness is computed from the relation $t_a = \gamma f_t \times d/b$, where f_t is the feed per tooth; d is the depth of face; b is the length of the cut path through the workpiece; and γ has the value 1 except for saws with spring-set or offset teeth, in which case γ has the value 2. The constants A and B depend primarily upon cutting direction (ripsawing, crosscutting), moisture content below the fiber-saturation point and specific gravity of the workpiece material, and tooth shape. The values of A and B (see Table 13.6.2) depend to some degree upon the depth of face, saw diameter, gullet shape, gullet-feed index, saw speed, and whether the tool motion is linear or rotary; however, the effect of these variables can generally be neglected for purposes of approximation.

Computers have now been utilized in sawmills where raw logs are first processed into rough-cut lumber. With suitable software, the mill operator can input key dimensions of the log and receive the cutting pattern which provides a mix of cross sections of lumber so as to maximize the yield from the log. The saving in waste is sizable, and this technique is especially attractive in view of the decreasing availability of large-caliper old stand timber, together with the cost of same.

PLANING AND MOLDING

Machinery   Planing and molding machines employ a rotating cutterhead to generate a smooth, defect-free surface by cutting in a direction approximately along the grain. A **surfacer** (or **planer**) is designed to machine boards or panels to uniform thickness. A **facer** (or **facing planer**) is designed to generate a flat (plane) surface on the wide faces of boards. The **edge jointer** is intended to perform the same task on the edges of boards in preparation for edge-gluing into panels. A **planer-matcher** is a heavy-duty machine designed to plane rough boards to uniform width and thickness in one operation. This machine is commonly used for dressing dimension lumber and producing millwork. The **molder** is a high-production machine for use in furniture plants to generate parts of uniform cross-sectional shape.

Recommended Operating Conditions   It is of prime importance to adjust the operating conditions and knife geometry so that the machining defects are reduced to a satisfactory level. The most commonly encountered defects are torn (chipped) grain, fuzzy grain, raised and loosened grain, and chip marks. **Torn grain** is caused by the wood splitting ahead of the cutting edge and below the generated surface. It is generally associated with large cutting angle, large chip thickness, low moisture content, and low workpiece material density. The **fuzzy-grain** defect is characterized by small groups of wood fibers which stand up above the generated surface. This defect is caused by incomplete sever-

ing of the wood by the cutting edge and is generally associated with small cutting angles, dull knives, low-density species, high moisture content, and (often) the presence of reaction wood. The **raised-grain** defect is characterized by an uneven surface where one portion of the annual ring is raised above the remaining part. **Loosened grain** is similar to raised grain; however, loosened grain is characterized by a separation of the early wood from the late wood which is readily discernible to the naked eye. The raised- and loosened-grain defects are attributed to the crushing of springwood cells as the knife passes over the surface. (Edge-grain material may exhibit a defect similar to the raised-grain defect if machining is performed at a markedly different moisture content from that encountered at some later time.) Raised and loosened grains are associated with dull knives, excessive jointing of knives [the jointing land should not exceed ¹⁄₃₂ in (0.79 mm)], and high moisture content of workpiece material. **Chip marks** are caused by chips which are forced by the knife into the generated surface as the knife enters the workpiece material. Chip marks are associated with inadequate exhaust, low moisture content, and species (e.g., birch, Douglas fir, and maple have a marked propensity toward the chip-mark defect).

Depth of cut is an important variable with respect to surface quality, particularly in the case of species which are quite prone to the torn-grain defect (e.g., hard maple, Douglas fir, southern yellow pine). In most cases, the depth of cut should be less than ¹⁄₁₆ in (1.59 mm). The number of **marks per inch** (**marks per metre**) (reciprocal of the feed per cutter) is an important variable in all cases; however, it is most important in those cases for which the torn-grain defect is highly probable. The marks per inch (marks per meter) should be between 8 and 12 (315 and 472) for rough planing operations and from 12 to 16 (472 to 630) for finishing cuts. Slightly higher values may be necessary for refractory species or for situations where knots or curly grain are present. It is seldom necessary to exceed a value of 20 marks per inch (787 marks per metre). The **clearance angle** should in all cases exceed a value of 10°. When it is desired to hone or joint the knives between sharpenings, a value of about 20° should be used. The optimum **cutting angle** lies between 20 and 30° for most planing situations; however, in the case of interlocked or wavy grain, low moisture content, or species with a marked tendency toward the torn-grain defect, it may be necessary to reduce the cutting angle to 10 or 15°.

BORING

Machinery   The typical general-purpose wood-boring machine has a single vertical spindle and is a hand-feed machine. Production machines are often of the vertical, multiple-spindle, adjustable-gang type or the horizontal type with two adjustable, independently driven spindles. The former type is commonly employed in furniture plants for boring holes in the faces of parts, and the latter type is commonly used for boring dowel holes in the edges and ends of parts.

Tool Design   A wide variety of tool designs is available for specialized boring tasks; however, the most commonly used tools are the taper-head drill, the spur machine drill, and the machine bit. The **taper-head** drill is a twist drill with a point angle of 60 to 90°, lip clearance angle of 15 to 20°, chisel-edge angle of 125 to 135°, and helix angle of 20 to 40°. Taper-head drills are used for drilling screw holes and for boring dowel holes along the grain. The **spur machine** drill is equivalent to a twist drill having a point angle of 180° with the addition of a pyramidal point (instead of a web) and spurs at the circumference. These drills are designed with a helix angle of 20 to 40° and a clearance angle of 15 to 20°. The **machine bit** has a specially formed head which determines the configuration of the spurs. It also has a point. Machine bits are designed with a helix angle of 40 to 60°, cutting angle of 20 to 40°, and clearance angle of 15 to 20°. Machine bits are designed with spurs contiguous to the cutting edges (**double-spur machine bit**), with spurs removed from the vicinity of the cutting edges (**extension-lip machine bit**), and with the outlining portion of the spurs removed (**flat-cut machine bit**).

The purpose of the spurs is to aid in severing wood fibers across their axes, thereby increasing hole-wall smoothness when boring across the grain. Therefore, drills or bits with spurs (double-spur machine drill and bit) are intended for boring across the grain, whereas drills or bits without spurs (taper-head drill, flat-cut machine bit) are intended for boring along the grain or at an angle to the grain.

Taper-head and spur machine drills can be sharpened until they become too short for further use; however, machine bits and other bit styles which have specially formed heads can only be sharpened a limited number of times before the spur and cutting-face configuration is significantly altered. Since most wood-boring tools are sharpened by filing the clearance face, it is important to ensure that sufficient clearance is maintained. The clearance angle should be at least 5° greater than the angle whose tangent (function) is the feed per revolution divided by the circumference of the drill point.

Recommended Operating Conditions   The most common defects are tearing of fibers from the end-grain portions of the hole surface and charring of hole surfaces. Rough hole surfaces are most often encountered in low-density and ring-porous species. This defect can generally be reduced to a satisfactory level by controlling the chip thickness. Charring is commonly a problem in high-density species. It can be avoided by maintaining the peripheral speed of the tool below a level which depends upon density and moisture content and by maintaining the chip thickness at a satisfactory level. Large chip thickness may result in excessive tool temperature and therefore rapid tool wear; however, large chip thickness is seldom a cause of hole charring. The following recommendations pertain to the use of spur-type drills or bits for boring material at about 6 percent moisture content across the grain. For species having a specific gravity less than 0.45, the chip thickness should be between 0.015 and 0.030 in (0.38 and 0.76 mm), and the peripheral speed of the tool should not exceed 900 ft/min (4.57 m/s). For material of specific gravity between 0.45 and 0.65, satisfactory results can be obtained with values of chip thickness between 0.015 and 0.045 in (0.38 and 1.14 mm) and with peripheral speeds less than 700 ft/min (3.56 m/s). For material of specific gravity greater than 0.65, the chip thickness should lie between 0.015 and 0.030 in (0.38 and 0.76 mm) and the peripheral speed should not exceed 500 ft/min (2.54 m/s). Somewhat higher values of chip thickness and peripheral

speed can be employed when the moisture content of the material is higher.

SANDING

(See Secs. 6.7 and 6.8)

Machinery   Machines for production sanding of parts having flat surfaces are multiple-drum sanders, automatic-stroke sanders, and wide-belt sanders. **Multiple-drum** sanders are of the endless-bed or roll-feed type and have from two to six drums. The drum at the infeed end is fitted with a relatively coarse abrasive (40 to 100 grit), takes a relatively heavy cut [0.010 to 0.015 in (0.25 to 0.38 mm)], and operates at a relatively slow surface speed [3,000 to 3,500 ft/min (15.24 to 17.78 m/s)]. The drum at the outfeed end has a relatively fine abrasive paper (60 to 150 grit), takes a relatively light cut [about 0.005 in (0.13 mm)], and operates at a somewhat higher surface speed [4,000 to 5,000 ft/min (20.3 to 25.4 m/s)]. **Automatic-stroke** sanders employ a narrow abrasive belt and a reciprocating shoe which forces the abrasive belt against the work material. This machine is commonly employed in furniture plants for the final white-sanding operation prior to finish coating. The automatic-stroke sander has a relatively low rate of material removal (about one-tenth to one-third of the rate for the final drum of a multiple-drum sander) and is operated with a belt speed of 3,000 to 7,500 ft/min (15.2 to 38.1 m/s). **Wide-belt** sanders are commonly used in board plants (plywood, particle board, hardboard). They have the advantage of higher production rates and somewhat greater accuracy than multiple-drum sanders [e.g., feed rates up to 100 ft/min (0.51 m/s) as opposed to about 35 ft/min (0.18 m/s)]. Wide-belt sanders operate at surface speeds of approximately 5,000 ft/min (25.4 m/s) and are capable of operating at depths of cut of 0.006 to 0.020 in (0.15 to 0.51 mm) depending upon workpiece material density.

Abrasive Tools   The abrasive tool consists of a **backing** to carry the **abrasive** and an **adhesive** coat to fix the abrasive to the backing. Backings are constructed of paper, cloth, or vulcanized fiber or consist of a cloth-paper combination. The adhesive coating (see also Sec. 6) is made up of two coatings; the first coat (**make coat**) acts to join the abrasive material to the backing, and the second coat (**size coat**) acts to provide the necessary support for the abrasive particles. Coating materials are generally animal glues, urea resins, or phenolic resins. The choice of material for the make and size coats depends upon the required flexibility of the tool and the work rate required of the tool. Abrasive materials (see also Sec. 6) for woodworking applications are garnet, aluminum oxide, and silicon carbide. **Garnet** is the most commonly used abrasive mineral because of its low cost and acceptable working qualities for low-work-rate situations. It is generally used for sheet goods, for sanding softwoods with all types of machines, and for sanding where the belt is loaded up (as opposed to worn out). **Aluminum oxide** abrasive is used extensively for sanding hardwoods, particleboard, and hardboard. **Silicon carbide** abrasive is used for sanding and polishing between coating operations and for machine sanding of particleboard and hardboard. Silicon carbide is also frequently used for the sanding of softwoods where the removal of raised fibers is a problem. The size of the abrasive particles is specified by the mesh number (the approximate number of openings per inch in the screen through which the particles will pass). (See Commercial Std.CS217-59, ''Grading of Abrasive Grain on Coated Abrasive Products,'' U.S. Government Printing Office.) Mesh numbers range from about 600 to 12. Size may also be designated by an older system of symbols which range from 10/0 (mesh no. 400) through 0 (mesh no. 80) to 4½ (mesh no. 12). Some general recommendations for common white wood sanding operations are presented in Table 13.6.3.

Table 13.6.3 Recommendations for Common Whitewood Sanding Operations

		Abrasive									
		Backing		Adhesive		Mineral			Grit size		
		Material	Weight	Make coat	Size coat	1st drum	2d drum	3d drum	1st drum	2d drum	3d drum
Multiple drum	Softwood	Paper	E	Glue	Resin	G	G	G or S	50	80	100
	Hardwood	Paper	E	Glue	Resin	A	A	A	60	100	120
	Particleboard	Paper	E	Glue	Resin	G	G	G or A	40	60	80
		Fiber	0.020	Resin	Resin	S	S	S	40	60	80
	Hardboard	Paper	E	Glue	Resin	S	S	S	60	80	120
		Fiber	0.020	Resin	Resin	S	S	S	60	80	120
Wide-belt	Softwood	Paper	E	Glue	Resin	G	or	S		80–220*	
		Cloth	X	Resin	Resin	G	or	S		80–220*	
	Hardwood	Paper	E	Glue	Resin		A			80–220*	
		Cloth	X	Resin	Resin		A			80–220*	
	Burnishing	Paper	E	Glue	Glue		A			280–400	
	Particleboard	Cloth	X	Resin	Resin		S			24–150*	
	Hardboard	Cloth	X	Resin	Resin		S			100–150	
Stroke sanding	Softwood	Paper	E	Glue	Resin		G			80; 120†	
	Hardwood	Paper	E	Glue	Resin		A			100; 150–180†	
		Cloth	X	Glue	Resin		A			100; 150–180†	
	Particleboard	Paper	E	Glue	Resin	A	or	S		80; 120†	
		Cloth	X	Resin	Resin	A	or	S		80; 120†	
	Hardboard	Paper	E	Glue	Resin	A	or	S		100; 150†	
		Cloth	X	Resin	Resin	A	or	S		100; 150†	
Edge sanding	Softwood	Cloth	X	Glue	Resin		G			60; 100†	
	Hardwood	Cloth	X	Glue	Resin		A			60–150*	
		Cloth	X	Resin	Resin		A			60–150*	
Mold sanding	Softwood	Cloth	J	Glue	Glue		G			80–120*	
		Cloth	J	Glue	Resin		G			80–120*	
	Hardwood	Cloth	J	Glue	Glue	G	or	A		80–120*	
		Cloth	J	Glue	Resin	G	or	A		80–120*	

* May be single- or muiltiple-grit operation.
† First number for cutting-down operations, second number for finishing operations.
SOURCE: Graham, *Furniture Production,* July and Aug., 1961; and Martin, *Wood Working Digest,* Sept. 1961.
NOTE: G = garnet; A = aluminum oxide; S = silicon oxide.

Fans, Pumps, and Compressors

BY

T. L. HENSHAW *Consulting Engineer, Battle Creek, MI.*
IGOR J. KARASSIK *Late Senior Consulting Engineer, Ingersoll-Dresser Pump Co.*
JAMES L. BOWMAN *Senior Engineering Consultant, Rotary-Reciprocating Compressor Division, Ingersoll-Rand Co.*
BENJAMIN B. DAYTON *Consulting Physicist, East Flat Rock, NC.*
ROBERT JORGENSEN *Engineering Consultant.*

14.1 DISPLACEMENT PUMPS

by T. L. Henshaw

REFERENCES: Buse, "Positive-Displacement Pumps," Marks' Handbook, 9th ed., McGraw-Hill. Henshaw, "Reciprocating Pumps," Van Nostrand Reinhold. "Hydraulic Institute Standards," 14 ed., Hydraulic Institute. Chesney, Water Injection—Pump Development, paper 68-Pet-11, ASME. Lees, Designing for High Pressures, paper 93-DE-9, ASME. "Positive Displacement Pumps—Reciprocating," API Standard 674, American Petroleum Institute.

NOTE: Much of the text and some of the illustrations are from the author's book: Henshaw, "Reciprocating Pumps," Van Nostrand Reinhold, 1987.

GENERAL

A **displacement pump** (also called **positive-displacement,** or just **p-d**) is a pump which imparts energy to the pumpage (the material pumped) by trapping a fixed volume at suction (inlet) conditions, compressing it to discharge pressure, then pushing it into the discharge (outlet) line. A displacement pump does not rely on velocity to achieve pumping action, as does a centrifugal pump or ejector.

Displacement pumps fall into two major classes: reciprocating and rotary, as illustrated in Fig. 14.1.1.

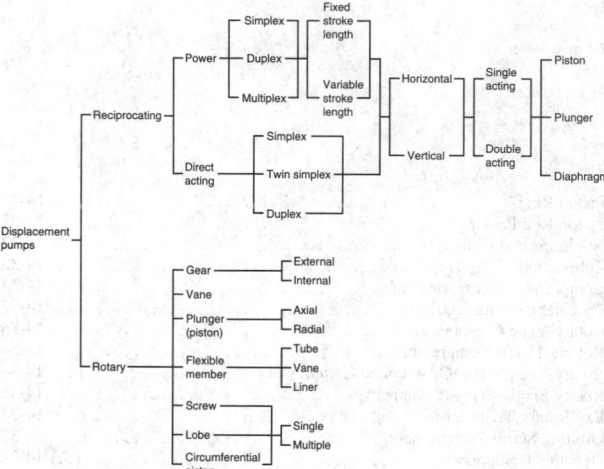

Fig. 14.1.1 Classification diagram of displacement pumps.

Uses and Applications

Displacement pumps serve primarily in applications of low capacity and high pressure, those mostly beyond the capabilities of centrifugal pumps. Some of these services could be performed by centrifugals, but not without an increase in power requirements and/or maintenance. Because displacement pumps achieve high pressures with low pumpage velocities, they are well-suited for abrasive-slurry and high-viscosity services. A reciprocating pump must have special fittings to be suitable for most slurries.

Net Positive Suction Head

Net positive suction head (NPSH), also called **net positive inlet pressure (NPIP)** and **net inlet pressure (NIP),** is the difference between suction pressure and vapor pressure, at the pump suction nozzle, when the pump is running. In a reciprocating pump, NPSH is required to push the suction valve from its seat and to overcome the friction losses and acceleration head within the pump liquid end. In a rotary pump, NPSH

is required to push the pumpage into the cavities created by the pumping elements.

If sufficient NPSH is not provided by the system, the pumpage will begin to flash (boil) as it flows into the pump. The vapor will cause a deterioration of pump performance. As the vapor flows into regions of higher pressure in the pump, it condenses back to a liquid. This collapse of the vapor bubbles occurs with significant impact, impinging upon metal surfaces with enough energy to break out small pieces of the metal. This is called **cavitation** damage. The shock created by the bubble collapse may be severe enough to crack a fluid cylinder or break a crankshaft.

All pumps require the system to provide some NPSH. The NPSH provided by the system is called NPSHA (the A is for available). The NPSH required by the pump is called NPSHR. For displacement pumps, NPSHR is normally expressed in pressure units (lb/in^2 or kPa).

Since water usually contains dissolved air, the vapor pressure of the solution is higher than for deaerated water, but this is often overlooked when NPSH calculations are performed. The Hydraulic Institute recommends an NPSH margin of 3 lb/in^2 (20 kPa) for power pumps in systems where the pumpage has been exposed to a gas other than the liquid's own vapor. A liquid (such as propane) at its bubble point in the suction vessel requires no such margin.

RECIPROCATING PUMPS

A **reciprocating pump** is a displacement pump which reciprocates the pumping element (piston, plunger, or diaphragm). The capacity of a reciprocating pump is proportional to its speed, and is relatively independent of discharge pressure.

A **power pump** is one that reciprocates the pumping element with a crankshaft or camshaft (see Figs. 14.1.2 and 14.1.3). It requires a driver which has a rotating shaft, such as a motor, engine, or turbine. A constant-speed power pump will deliver essentially the same capacity at any pressure within the capability of the driver and the strength of the pump.

A **direct-acting** pump is a reciprocating pump driven by a fluid which has a differential pressure (see Fig. 14.1.4). The motive fluid pushes on

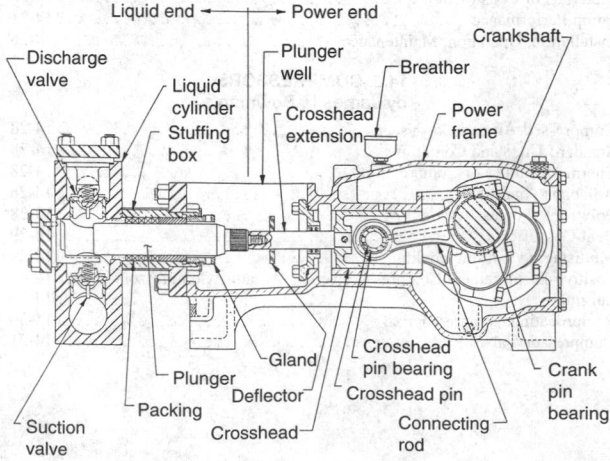

Fig. 14.1.2 Horizontal power pump.

The Pumping Cycle

As the pumping element reverses from the discharge stroke to the suction stroke, liquid within the pumping chambers expands, and the pressure drops. Since most liquids are relatively incompressible, little movement of the element is required to reduce the pressure. When the pressure drops sufficiently below suction pressure, the differential pressure (suction pressure minus chamber pressure) pushes the suction valve open. This occurs when the element is moving slowly, so that the valve opens slowly and smoothly as the velocity of the element increases. Liquid then flows through the valve and follows the element on its suction stroke. As the element decelerates near the end of the stroke, the suction valve returns to its seat. When the element stops, the suction valve closes.

The pumping element then reverses and starts its discharge stroke. The liquid trapped in the pumping chamber is compressed until chamber pressure exceeds discharge pressure by an amount sufficient to push the discharge valve from its seat. The action of the discharge valve, during the discharge stroke, is then the same as that of the suction valve during the suction stroke.

Power Pumps

The drive end of a power pump is called a **power end** (see Fig. 14.1.2). Its function is to convert rotary motion from a driver into reciprocating motion for the liquid end. The main component of the power end is the **power frame,** which supports all other power end parts and, usually, the liquid end. The second major component of the power end is the **crankshaft** (sometimes a camshaft). The crankshaft transmits power from the driver to the connecting rods.

The **connecting rod** is driven by a throw (journal) of the crankshaft on one end, and drives a crosshead on the other. The crankshaft moves in pure rotating motion, the crosshead in pure reciprocating motion. The connecting rod is the link between the two. **Main bearings** support the shaft in the power frame.

The **crosshead** transmits the force from the connecting rod into the pumping element. It is fastened directly to a plunger or piston rod, or to a rod called an **extension, stub,** or **pony rod.** The other end of this rod is fastened to the pumping element.

Power pump overall efficiencies normally range from 85 to 94 percent, higher than any other type of pump.

Selection Graphs Different manufacturers produce these graphs in different configurations. All graphs consist of straight lines since displacement is directly proportional to speed. Figure 14.1.14 shows one such graph for a triplex power pump. Displacement is plotted against crankshaft speed for the different plunger diameters available in the pump. The pressures listed are those at which each plunger size would operate to load the power end to its rated value (4,500 lbf or 20,000 N).

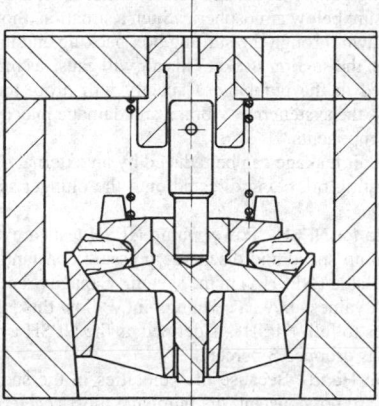

Fig. 14.1.10 Reciprocating pump elastomeric-insert valve. *(TRW-Mission Manufacturing Co.)*

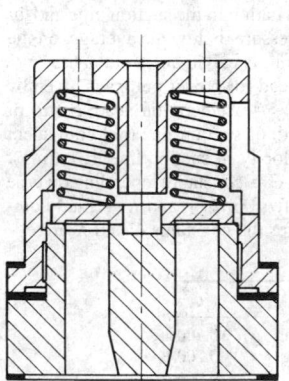

Fig. 14.1.11 Reciprocating pump double-ported disk valve.

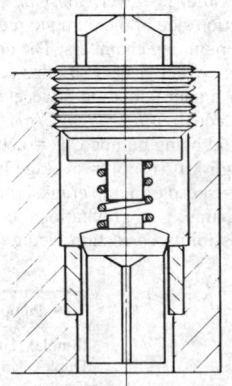

Fig. 14.1.12 Reciprocating pump conical-faced (bevel-seat) wing-guided valve.

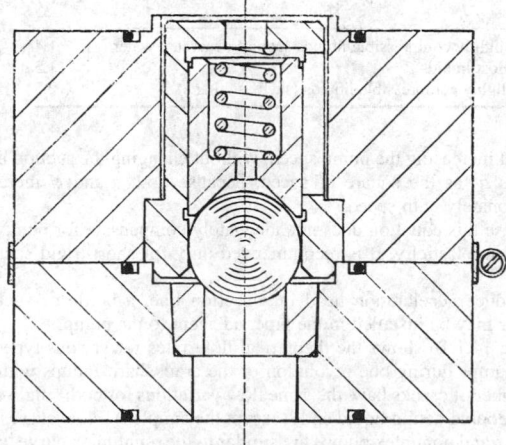

Fig. 14.1.13 Reciprocating pump ball valve. *(Ingersoll-Rand Company.)*

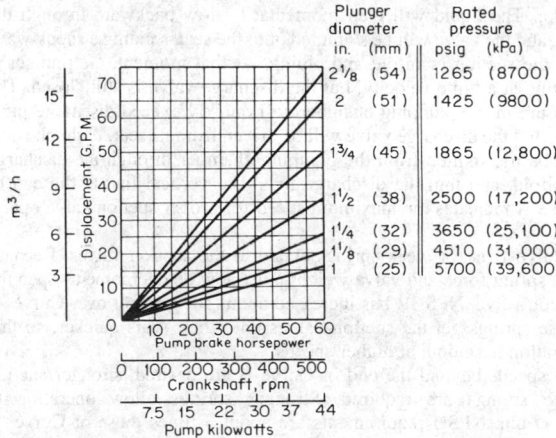

Fig. 14.1.14 Power pump selection graph. *(Ingersoll-Rand Company.)*

NPSH Curves Figure 14.1.15 provides insight into power pump NPSHR characteristics, valve action, and the effect of valve springs on pump performance. These curves are for a 3-in (76-mm) stroke, horizontal, triplex plunger power pump with 2¼-in-diameter (57-mm) plungers, and with suction valves that operate vertically. The valves are wing-guided, and have a diameter approximately equal to the plunger diameter.

Because the axis of the suction valve is vertical, the valve can operate without a spring if the pump speed is kept low. Curve A represents the NPSH requirements with no springs on the suction valves. NPSHR at 100 r/min is only 0.7 lb/in² (5 kPa) (1.6 ft or 0.5 m of water), less than for most centrifugals.

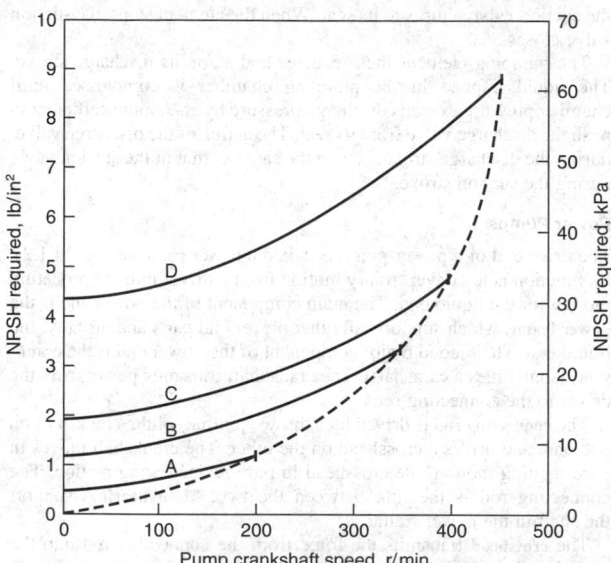

A = no suction valve springs (std. disch.)
B = weak suction valve springs (std. disch.)
C = standard suction valve springs (std. disch.)
D = heavy valve springs—suction and discharge

Fig. 14.1.15 NPSH curves for a 3-in- (75-mm-) stroke triplex power pump.

The speed of the pump in this configuration is limited by the ability of the suction valve to keep up with the plunger. Since there is no spring to push the valve back onto its seat, gravity is the only force tending to close the valve against the entering fluid. If the pump is run too fast, the valve will still be off its seat when the plunger reverses to the discharge stroke. The liquid will then momentarily flow backward through the seat, and the valve will be slammed onto the seat, sending a shock wave into the suction manifold and piping. At that moment, the plunger is moving at a finite velocity, but the discharge valve is still closed. The pressure in the pumping chamber will quickly exceed discharge pressure, and the discharge valve will be driven from its seat. A shock wave will be transmitted from the pumping chamber through the discharge manifold, and into the discharge line. The vertical line at the end of Curve A indicates the maximum speed for proper suction valve operation.

Curve B is for weak springs added to the suction valves. Because both spring force and valve weight must now be overcome to open the suction valve, NPSHR has increased about 100 percent over Curve A. These springs get the suction valves onto their seats quicker, so that operation is smooth at higher speeds.

If speeds beyond the end of Curve B are desired, stronger suction valve springs are required. Stronger springs allow operation to 400 r/min. NPSH requirements are about 3 times those of Curve A, ranging from about 2 to 5 lb/in² (14 to 35 kPa).

Curves A through C represent a pump equipped with the same (standard) discharge valve spring. Only the suction spring has been changed.

If operation is required at speeds exceeding the limits of Curve C, extra-strong springs are required on both suction and discharge valves. As shown by Curve D, NPSHR is about double that required for the standard springs, ranging from about 4.5 to 9 lb/in² (31 to 62 kPa).

The dashed line passing through the ends of the NPSH curves illus-

trates the dramatic increase of NPSHR as pump speed is increased. When it is necessary to change suction springs, NPSHR increases approximately as the **square** of speed.

To minimize the problem of dissolved air, the NPSH tests that produced these curves were performed with water at, or near, its boiling point in the suction vessel.

Reciprocating pumps, under the correct conditions, can operate with a suction pressure below atmospheric. Such a situation, though, can lead to air being drawn through the stuffing box packing and into the pumping chamber on the suction stroke. This air will cause as many problems as air entrained in the pumpage. Capacity will drop, the pump may operate noisily, the system may vibrate, and damage may occur to pump and system components.

This inward air leakage can be reduced by an external sealing liquid, such as lubricating oil, being directed onto the plunger surface or into the packing.

Test Criteria for NPSH Power pump NPSH tests are performed by holding the pump speed and discharge pressure constant, and varying the NPSH available (NPSHA) in the system. Capacity remains constant for all NPSHA values above a certain point. Below this NPSHA value, capacity begins to fall. NPSHR is defined as the NPSH available when the capacity has dropped 3 percent.

Acceleration Head Because the velocities in the suction and discharge piping are not constant, the pumpage must accelerate a number of times for each revolution of the crankshaft. Since the liquid has mass and, therefore, inertia, energy is required to produce the acceleration. This energy is returned to the system upon deceleration. Sufficient pressure, however, must be provided to accelerate the liquid on the suction side of the pump to prevent cavitation in the suction pipe and/or the pumping chambers. The drop in pressure, below the average, on the suction side, caused by this acceleration, is called **acceleration head**.

An approximation of acceleration head for power pumps is given by the equation $h_a = LvNC/gK$, where h_a = acceleration head, ft (m) of liquid being pumped; L = actual length of suction line (not equivalent length), ft (m); v = average liquid velocity in suction line, ft/s (m/s); N = speed of pump crankshaft, r/min; C = constant, depending on type of pump; g = gravitational constant, ft/s² (m/s²); and K = fluid compressibility correction factor. Values for the constants C and K are:

Pump type	C
Simplex, single-acting	0.400
Duplex, single-acting	0.200
Triplex	0.066
Quintuplex	0.040
Septuplex	0.028
Nonuplex	0.022

Fluid	K
Slightly compressible liquids such as deaerated water	1.4
Most liquids	1.5
Highly compressible liquids (such as ethane)	2.5

Note that increasing the pump speed, without changing the suction line, increases h_a as the square of speed, because both v and N increase proportionally with speed.

Because this equation does not adequately compensate for pumpage and system elasticity, it is recommended only for short, rigid suction lines.

To reduce acceleration head in a suction line, a bottle or suction stabilizer may be installed in the pipe, adjacent to the pump.

Figure 14.1.16 shows the theoretical flow rates for various types of power pumps during one revolution of the crankshaft. Pumps with an odd number of cranks have the same flow variations for both single-acting and double-acting liquid ends (except the simplex). Flow curves for septuplex and nonuplex pumps are similar to the quintuplex curve, with more but smaller variations, as indicated in the tabulation. The values tabulated will vary slightly with variations in the ratio of connecting rod length to stroke length. These curves indicate flow-rate variations, not **pressure pulsations**. Pressure pulsations are a product of the acceleration created by the pump and the mass of the pumpage in the system.

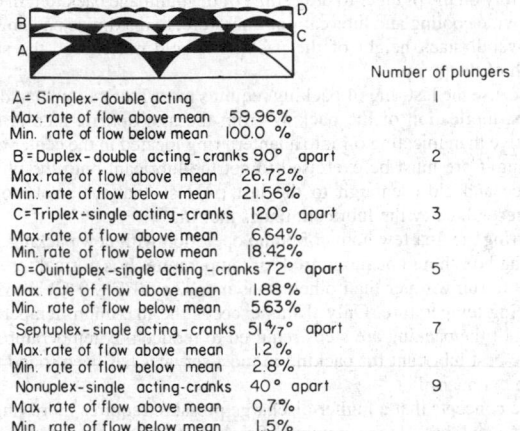

	Number of plungers
A = Simplex - double acting	1
Max. rate of flow above mean 59.96%	
Min. rate of flow below mean 100.0 %	
B = Duplex - double acting - cranks 90° apart	2
Max. rate of flow above mean 26.72%	
Min. rate of flow below mean 21.56%	
C = Triplex - single acting - cranks 120° apart	3
Max. rate of flow above mean 6.64%	
Min. rate of flow below mean 18.42%	
D = Quintuplex - single acting - cranks 72° apart	5
Max. rate of flow above mean 1.88%	
Min. rate of flow below mean 5.63%	
Septuplex - single acting - cranks 51 4/7° apart	7
Max. rate of flow above mean 1.2%	
Min. rate of flow below mean 2.8%	
Nonuplex - single acting - cranks 40° apart	9
Max. rate of flow above mean 0.7%	
Min. rate of flow below mean 1.5%	

Fig. 14.1.16 Flow-rate variations in reciprocating power pumps.

Actual flow rates may vary considerably from these curves, particularly with high-speed pumps, where valve closing and opening may lag the crankshaft rotation as much as 15°, and at high pressures, where liquid compression may cause a further lag of 5° to 10° (or more for a large clearance volume).

Power Pump Speeds Probably the most controversial factor in the selection of power pumps is the maximum allowable speed. Some manufacturers, for example, offer 3-in- (76-mm-) stroke triplex pumps for operation at 500 r/min. One builder of portable water-jetting units routinely operates this size pump at 600 r/min. Most users of this size pump, in continuous-duty applications, prefer to operate it no faster than about 400 r/min.

Top speeds quoted by one manufacturer for the 3-in-stroke triplex increased over a period of years from 150 to 520 r/min. Although improvements in power end and liquid end construction made the pump capable of operating at higher speeds, albeit with shorter packing life, a major obstacle to higher speeds is system design. Initially, some thought that a smaller pump, running faster to achieve the same capacity, would produce less pulsation in the suction and discharge pipes. Unfortunately, the opposite is true.

The velocity variation in the piping for a triplex pump is 25 percent, regardless of the pump size or speed (see Fig. 14.1.16). For the same capacity, a smaller pump, running faster, will produce the same maximum and minimum flow rates and more pulses per second. Since the acceleration head is proportional to frequency, doubling the pump speed doubles the acceleration head, thereby reducing the NPSH available from the system. Also, doubling the speed requires a much stiffer valve spring, thereby significantly increasing the NPSHR of the pump. If the NPSHA drops below the NPSHR, cavitation and pounding will occur.

The best solution for excessive acceleration is an effective dampener in the suction line; but such devices often are less effective at higher frequencies, due to the inertia of their moving parts and inertia of the liquid inside them, which must oscillate for them to function.

Through field observations, the author developed a set of **maximum recommended speeds** for plunger-type power pumps in continuous-duty services. These speeds, as adopted by API 674, are shown in Fig. 14.1.17, along with the Hydraulic Institute basic speeds. Intermittent and cyclic operation is often satisfactory above these speeds. Lower speeds may be dictated by factors such as pump construction, system design, low NPSHA, entrained gas, high temperature, entrained solids, high viscosity, and requirements for a low sound level.

The **minimum speed** of a power pump is determined by its ability to provide sufficient lubrication to all bearing surfaces in the power end. Some units can run satisfactorily at 20 r/min. Others must be maintained at 100 r/min or more.

Plunger load (also called **frame load** and **rod load**) is the force transmitted to the power end by one plunger. For a single-acting pump, the discharge plunger load is calculated by multiplying discharge pressure by the cross-sectional area of one plunger. Suction plunger load is the suction pressure multiplied by the plunger area.

A power pump is rated by the maximum discharge plunger load that the power end is capable of absorbing when the suction pressure is zero. Some units are rated for continuous operation, some for intermittent operation, and some are rated both ways.

Torque Characteristics For fixed suction and discharge pressures, a power pump requires an **average** input torque that is independent of speed (except for increases at very low and very high speeds). The torque actually has a variation that mirrors the discharge flow-velocity curve. A power pump will require the same torque at one-half or one-quarter rated speed and, therefore, will require one-half or one-quarter rated power, respectively. Figure 14.1.18 illustrates the variation of average torque with speed for a typical triplex power pump.

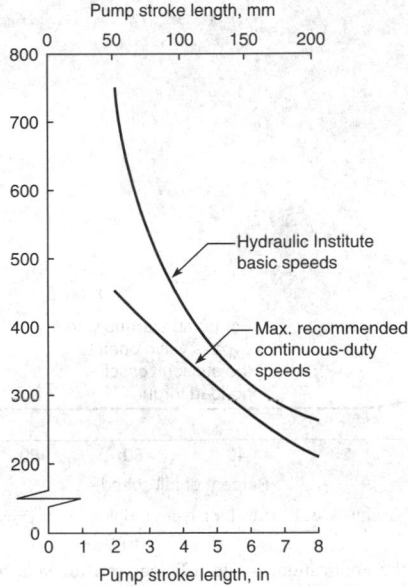

Fig. 14.1.17 Maximum recommended speeds for plunger-type power pumps in continuous-duty services.

The curve for full-load torque in Fig. 14.1.18 is for starting against full discharge pressure. Breakaway torque is about 150 percent of average full-load running torque. As speed is increased, and proper lubrication is established in the power end and packing, torque drops to the full-load, full-speed value, and is thereafter constant up to full speed.

For a start-up that is easier on equipment, pump discharge is piped back to the suction vessel, making the discharge pressure near suction pressure. The no-load curve in Fig. 14.1.18 illustrates the resulting torque requirements imposed by the pump on the drive train. Breakaway torque is approximately 25 percent of full-load torque. This will vary, depending on the type of packing and bearings in the pump and the length of time the pump has been idle. As speed increases, the torque drops to less than 10 percent of the full-load torque.

Packing The component normally requiring the most maintenance on reciprocating pumps is packing. Although the life of packing in a power pump is typically about 2,500 h (3 months), some installations, with special stuffing box arrangements, have experienced lives of more than 18,000 h (2 years), at discharge pressures to 4,000 lb/in² (28 MPa).

Short packing life can result from any of the following conditions: (1) misalignment of plunger (or rod) with stuffing box; (2) worn plunger, rod, stuffing-box bushings, or stuffing-box bore; (3) improper

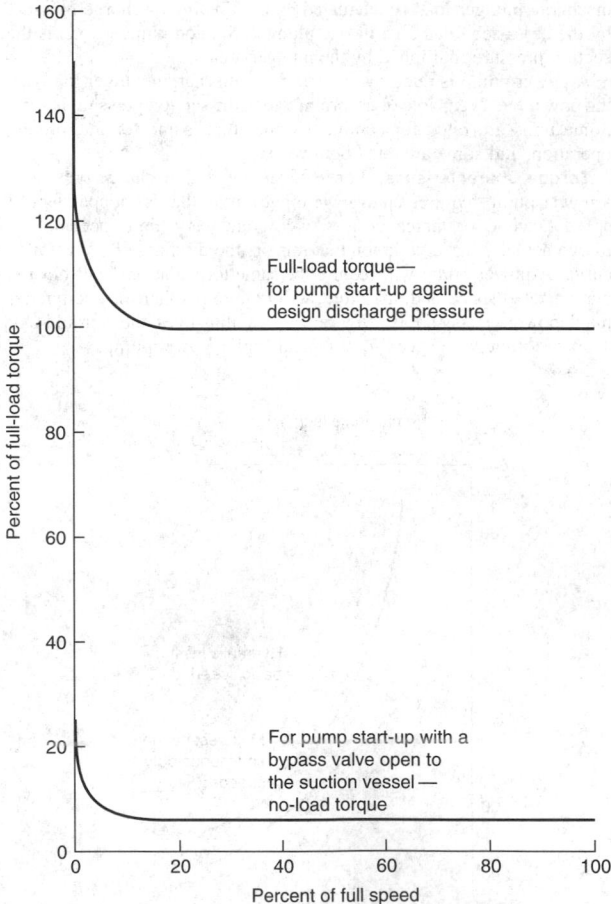

Fig. 14.1.18 Torque-speed curves for a typical triplex power pump.

packing for the application; (4) insufficient or excessive lubrication; (5) packing gland too tight or too loose; (6) excessive speed or pressure; (7) high or low temperature of pumpage; (8) excessive friction (too much packing); (9) packing running dry (pumping chamber gas-bound); (10) shock conditions caused by entrained gas or cavitation, broken or weak valve springs, or system problems; (11) solids from the pumpage, environment, or lubricant; (12) improper packing installation or break-in; and (13) icing caused by volatile liquids that refrigerate and form ice crystals on leakage to atmosphere, or by pumpage at temperatures below 32°F (0°C). As is evident from this list, short packing life can indicate problems elsewhere in the pump or system.

To achieve a low leakage rate, the clearance between the plunger (or rod) and packing must be essentially zero. This requires that the sealing rings be relatively soft and pliant. Because the packing is pliant, it tends to flow into the stuffing box clearances, especially between the plunger and follower bushing. If this bushing does not provide an effective barrier, the packing will extrude, and leakage will increase.

A set of square or V-type packing rings will experience a pressure gradient, during operation, as indicated in Fig. 14.1.7. The last ring of packing, adjacent to the gland-follower bushing, will experience the largest axial loading, resulting in greater deformation, tighter sealing

and, therefore, the largest pressure drop. The gap between the plunger and the follower must be small enough to prevent packing extrusion. Most packing failures originate at this critical sealing point.

Because this last ring of packing is the most critical, does the most sealing, and generates the most friction, it requires more lubrication than the others. In a nonlubricated arrangement (Fig. 14.1.7), this ring must rely on the plunger to drag some of the pumpage back to it in order to provide cooling and lubrication. Therefore, to maximize packing life, the overall stack height of the packing should not exceed the stroke length of the pump.

Because the last ring of packing requires more lubrication than do the others, lubrication of the packing from the atmospheric side is more effective than injecting oil into a lantern ring located in the center of the packing. Care must be exercised to get the lubricant onto the plunger surface and close enough to the last ring, so that the stroke of the plunger will carry the lubricant under the ring.

During the first few hours of pump operation with new packing, each stuffing box should be monitored for temperature. It is normal for some boxes to run warmer than others—as much as 50°F (30°C) above the pumping temperature. Only if this exceeds the maximum temperature rating of the packing are steps required to reduce box temperature.

The best lubricant for packing in most services has been found to be steam cylinder oil.

The concepts that a higher discharge pressure requires more rings of packing and that a larger number of rings lasts longer are true for long-stroke, low-speed machines, but have been disproven in a number of moderate- to high-speed power-pump applications. Unless they are profusely lubricated, the larger number of rings create additional frictional heat and wipe lubricant from the plunger surface, thus depriving some rings of lubrication. On numerous saltwater injection pumps operating at pressures above 4,000 lb/in² (28 MPa), Chesney reported that packing life was only two weeks with twelve rings of packing in each stuffing box. With three rings in each box, packing life was approximately 6 months.

Plungers Next to packing, the plunger is the component of a power pump that typically requires most frequent replacement. The high speed of the plunger and the friction load of the packing tend to wear the plunger surface. For longer life, plungers are sometimes hardened, although it is more common to apply a hard coating. Such coatings are of chrome, various ceramics, nickel-based alloys, or cobalt-based alloys. Desired characteristics of the coatings include hardness, smoothness, high bond strength, low porosity, corrosion resistance, and low cost. No one coating optimizes all characteristics.

Ceramic coatings are harder than the metals, but are brittle, porous, and sometimes lower in bond strength. Porosity contributes to shorter packing life. Mixing of hard particles, such as tungsten carbide, into the less-hard nickel or cobalt alloys has resulted in longer plunger life at the expense of shorter packing life.

Stuffing Boxes Various stuffing box designs (various lubrication and bleed-off arrangements to minimize leakage and extend packing life) are shown in Fig. 14.1.19.

The most significant advance in packing arrangements in recent years has been spring loading. Spring loading is applied most to V-ring (chevron) packing but also works well with square packing. The spring must be located on the pressure side of the packing. Springs of various types can be used, including single-coil, multiple coil, wave-washer, belleville, and thick-rubber-washer. The force provided by the spring is small compared to the force imposed on the packing by discharge pressure. The major functions of the spring are to provide a small preload to help set the packing, and to hold all bushings and packing in place during operation.

Spring loading of packing has the following advantages:

Requires no adjustment of the gland. The gland is tightened until it bottoms, then is locked. This removes one of the biggest variables in packing life: personnel skill.

Alloys expansion. If the packing expands during the initial break-in, the spring allows for the expansion.

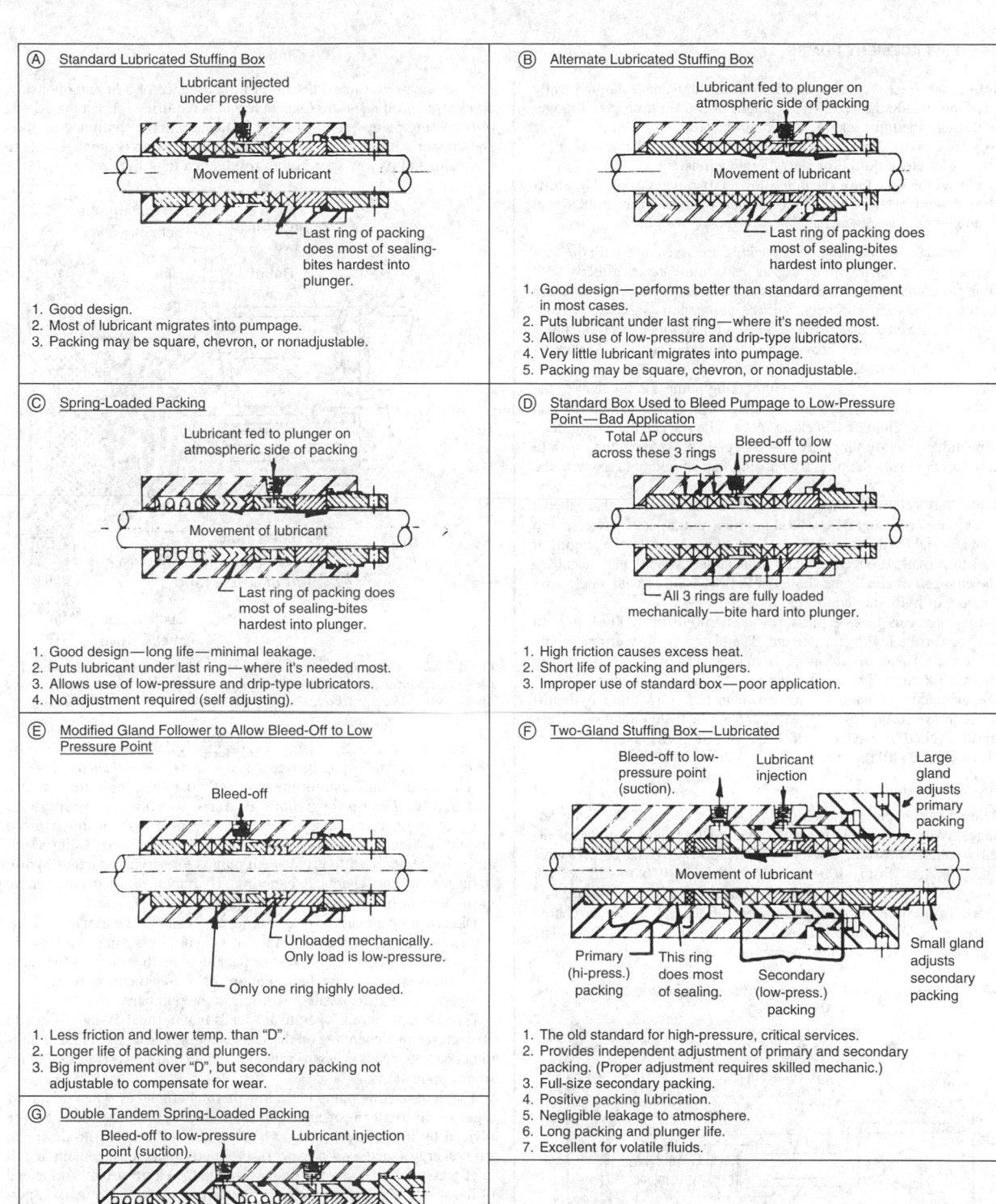

(A) Standard Lubricated Stuffing Box

Lubricant injected under pressure

Movement of lubricant

Last ring of packing does most of sealing-bites hardest into plunger.

1. Good design.
2. Most of lubricant migrates into pumpage.
3. Packing may be square, chevron, or nonadjustable.

(B) Alternate Lubricated Stuffing Box

Lubricant fed to plunger on atmospheric side of packing

Movement of lubricant

Last ring of packing does most of sealing-bites hardest into plunger.

1. Good design—performs better than standard arrangement in most cases.
2. Puts lubricant under last ring—where it's needed most.
3. Allows use of low-pressure and drip-type lubricators.
4. Very little lubricant migrates into pumpage.
5. Packing may be square, chevron, or nonadjustable.

(C) Spring-Loaded Packing

Lubricant fed to plunger on atmospheric side of packing

Movement of lubricant

Last ring of packing does most of sealing-bites hardest into plunger.

1. Good design—long life—minimal leakage.
2. Puts lubricant under last ring—where it's needed most.
3. Allows use of low-pressure and drip-type lubricators.
4. No adjustment required (self adjusting).

(D) Standard Box Used to Bleed Pumpage to Low-Pressure Point—Bad Application

Total ΔP occurs across these 3 rings

Bleed-off to low pressure point

All 3 rings are fully loaded mechanically—bite hard into plunger.

1. High friction causes excess heat.
2. Short life of packing and plungers.
3. Improper use of standard box—poor application.

(E) Modified Gland Follower to Allow Bleed-Off to Low Pressure Point

Bleed-off

Unloaded mechanically. Only load is low-pressure.

Only one ring highly loaded.

1. Less friction and lower temp. than "D".
2. Longer life of packing and plungers.
3. Big improvement over "D", but secondary packing not adjustable to compensate for wear.

(F) Two-Gland Stuffing Box—Lubricated

Bleed-off to low-pressure point (suction).

Lubricant injection

Large gland adjusts primary packing

Movement of lubricant

Primary (hi-press.) packing

This ring does most of sealing.

Secondary (low-press.) packing

Small gland adjusts secondary packing

1. The old standard for high-pressure, critical services.
2. Provides independent adjustment of primary and secondary packing. (Proper adjustment requires skilled mechanic.)
3. Full-size secondary packing.
4. Positive packing lubrication.
5. Negligible leakage to atmosphere.
6. Long packing and plunger life.
7. Excellent for volatile fluids.

(G) Double Tandem Spring-Loaded Packing

Bleed-off to low-pressure point (suction).

Lubricant injection

Movement of lubricant

1. The best design available for most high-pressure, critical services.
2. Combines best features of two-gland box and spring-loaded packing—negligible leakage, long life, and the packing is self-adjusting.

Fig. 14.1.19 Various stuffing box designs for reciprocating pumps.

Adjusts for wear. As the packing wears, adjustment automatically occurs from inside the box. The problem of transmitting gland movement through the top packing ring is eliminated.

Provides a cavity. The spring cavity provides an annular space for the injection of a clean liquid for slurry applications.

Eliminates the need for a gland, if allowed by pump design. The stuffing box assembly (if a separate component) can be disassembled and reassembled on a workbench.

Disadvantages of spring-loading packing are associated with the cavity created by the spring. Since this cavity communicates directly with the pumping chamber, the additional clearance volume will cause a reduction in volumetric efficiency if the pumpage is sufficiently compressible. This cavity also provides a place for vapors to accumulate. If the pump design does not provide for venting this space, a further reduction in volumetric efficiency may occur.

Spring-loaded packing is the reciprocating pump's equivalent to the mechanical seal for rotating shafts. Leakage is low, life is extended, and adjustments are eliminated. Packing sets can be stacked in tandem (they must be independently supported) for a stepped pressure reduction, or to capture leakage from the primary packing that should not escape to the environment.

Controlled-Volume Pumps   A **controlled-volume** pump (also called a **metering, proportioning,** or **chemical-injection** pump) is a power pump with an adjustable stroke length (or adjustable *effective* stroke length). It is used to provide an accurate, and adjustable, capacity. The capacity can be changed manually, as shown in Fig. 14.1.20, or with an electric, pneumatic, or hydraulic controller.

Such pumps can develop pressures to 30,000 lb/in² (200 MPa) with capacity controlled with ± 1 percent. The piston or plunger may come in direct contact with the pumpage, or may be separated from the pumpage by a diaphragm. The diaphragm may be flat, tubular, or conical. It can be actuated mechanically, as shown in Fig. 14.1.20, or hydraulically, as shown in Fig. 14.1.21. Mechanically actuated diaphragms are generally limited to capacities of 25 gal/h (100 L/h) and pressures of 250 lb/in² (1,700 kPa).

Direct-Acting Pumps

The direct-acting pump (see Fig. 14.1.4) has some of the same advantages as the power pump, plus others. These units are also well-suited for high-pressure, low-flow applications. Discharge pressures normally range from 100 to 5,000 lb/in² (0.7 to 35 MPa), but may exceed 10,000 lb/in² (70 MPa). Capacity is proportional to speed from stall to maximum speed, affected little by discharge pressure. Speed is controlled by controlling the motive fluid. The unit is normally self-priming, particularly the low-clearance-volume type.

Direct-acting pumps are negligibly affected by hostile environments,

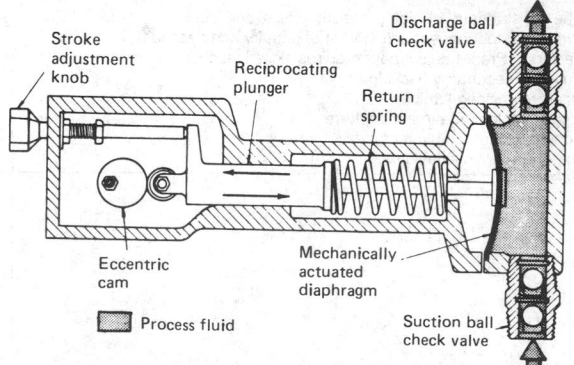

Fig. 14.1.20 Controlled-volume pump with mechanically actuated diaphragm and manual-adjustment stroke length. *(Reprinted from James P. Poynton, "Metering Pumps: Selection and Application," Marcel Dekker, Inc., Copyright 1983. Used by permission.)*

such as corrosive fumes, because of the absence of a bearing housing, crankcase, or oil reservoir (except for units requiring lubricators). Some direct-acting pumps, inundated by floodwater, have continued to operate without adverse effects. The direct-acting pump is quiet, simple to maintain, and its low speed often results in a long life.

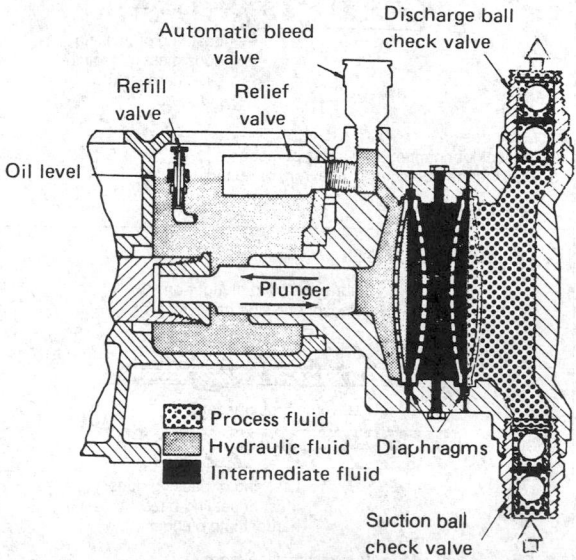

Fig. 14.1.21 Controlled-volume pump with hydraulically actuated double diaphragms. *(Reprinted from James P. Poynton, "Metering Pumps: Selection and Application," Marcel Dekker, Inc., Copyright 1983. Used by permission.)*

Since the direct-acting pump was originally designed to be driven by steam (initially, most pumps were), it was known as a **steam pump**—not for the pumped fluid, but for the motive fluid. Other fluids are now also used to drive direct-acting pumps. Fuel gas, which would otherwise be throttled through a pressure regulator for plant use, is often piped through a direct-acting pump to provide "free" pumping. Compressed air is frequently used to drive small pumps for services such as hydrostatic testing and chemical metering. Hydraulic oil is used to drive **intensifiers** for high-pressure water-jetting.

Direct-acting air-driven diaphragm pumps handle a variety of liquids at pressures up to the pressure of the driving air. This pump, as shown in Fig. 14.1.22, has no stuffing box or packing, and therefore has no leakage, unless a diaphragm fails. Failure of a diaphragm may result in pumpage in the exhaust air and/or air in the pumpage.

Direct-acting pumps operate at speeds that normally range from 0 to 50 cycles/min, depending on the stroke length. The ability of these units to operate at speeds down to zero (i.e., stall conditions) is desirable for some applications.

The direct-acting pump has a low thermal efficiency when driven by a gas (such as steam or air). The mechanical efficiency (output force divided by input force) is high; but because the unit has no device to store energy (such as a flywheel), the motive gas must remain at full inlet pressure in the drive cylinder through the entire stroke. At the end of the stroke, the gas expands to exhaust pressure, but no work is performed during this expansion. Hence, the thermal energy of the gas is lost to friction. Steam pumps consume about 100 lb/h of steam for each hydraulic horsepower (hhp) (60 kg/h · kW) developed on the liquid end. When natural gas or air is the motive fluid, consumption is about 3,500 std. ft³/h · hhp (130 m³/h · kW).

The **drive end** (or steam end, or gas end) of a direct-acting pump converts the differential pressure of the motive fluid to reciprocating motion for the liquid end. It is similar in construction to the liquid end, containing a piston (or diaphragm) and valving. The major difference is that the valve is mechanically actuated by a control system which senses the location of the drive piston, causing the valve to reverse the flow of the motive fluid when the drive piston reaches the end of its stroke.

The main component of the drive end is the drive cylinder. This cylinder forms the major portion of the pressure boundary, and supports the other drive end parts.

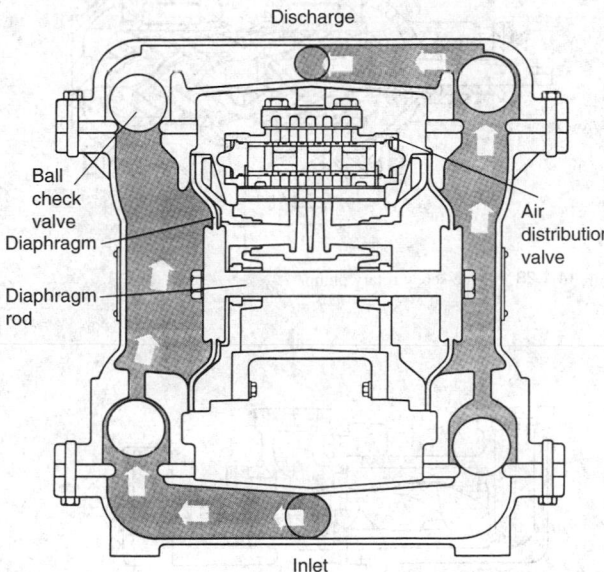

Fig. 14.1.22   Direct-acting, air-driven diaphragm pump. *(Used by permission of Warren Rupp, Inc., a unit of IDEX Corp., Mansfield, OH.)*

Mechanical Efficiency—Direct-Acting Pumps   For a direct-acting pump, mechanical efficiency is the ratio of the force transmitted to the liquid by the piston (or plunger or diaphragm) to the force transmitted to the drive piston (or diaphragm) by the motive fluid. For double-acting

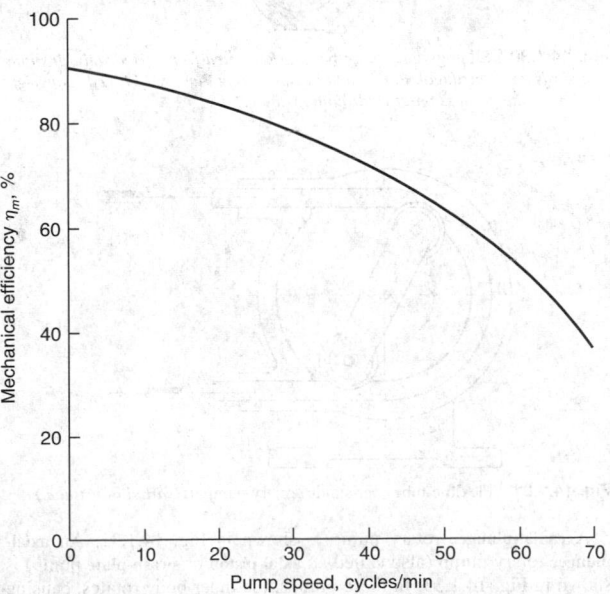

Fig. 14.1.23   Mechanical efficiency of a direct-acting pump.

pumps, the differential pressures are used at both ends of the pump, and since fluid friction losses in valves and ports must be charged to the pump, the pressures are as measured at the inlet and outlet ports. In equation form:

$$\eta_m = \frac{A_L \, \Delta p_L}{A_{dr} \, \Delta p_{dr}} \qquad (14.1.1)$$

where η_m = mechanical efficiency of pump, A_L = area of liquid piston or plunger, Δp_L = differential pressure across liquid end, A_{dr} = area of drive piston, and Δp_{dr} = differential pressure across drive end.

The area of the piston rod is usually small relative to the piston area, and is often ignored. However, it must be considered when the rod area becomes a significant portion of the piston area.

The mechanical efficiency of a typical direct-acting pump is shown in Fig. 14.1.23. Note that η_m rises as speed is reduced. It is this characteristic that allows the unit to be controlled by throttling the motive fluid. The reduction in the available differential driving pressure forces the pump to operate more efficiently, i.e., at a lower speed.

ROTARY PUMPS

Rotary pumps are displacement pumps which have rotating pumping elements, such as gears, lobes, screws, vanes, or rollers. They do not contain inlet and outlet check valves, as do reciprocating pumps. They are built for capacities from a fraction of a gallon per minute (cubic meter per hour) (for domestic oil burners) to about 10,000 gal/min (2000 m³/h) (for marine cargo service). Though used for pressures up to 5,000 lb/in² (35 MPa), their particular field is for pressures of 25 to 500 lb/in² (170 to 3,500 kPa).

Most rotary pumps rely on close running clearances to prevent the pumpage from leaking from the discharge side back to the suction side of the pump. Because of the close clearances, the pumpage must be clean. Most rotary pumps are noted for their ability to handle viscous liquids, and many actually *require* a viscous liquid to achieve peak performance. Viscosity affects the mechanical efficiency, volumetric efficiency, and NPSHR of a rotary pump.

The **rotor** is the pumping element of the rotary pump, and is usually the feature by which the pump is classified.

Gears   The teeth of the gear trap and displace the pumpage. If the teeth are on the outside of the gear, as shown in Fig. 14.1.24, it is called an **external gear**. As the teeth disengage, suction pressure pushes liquid into the cavities between the teeth. The teeth then carry the liquid around to the discharge side of the pump. As the teeth engage, the fluid

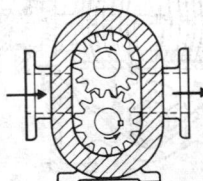

Fig. 14.1.24   External-gear rotary pump. *(Reprinted from "Hydraulic Institute Standards for Centrifugal, Rotary and Reciprocating Pumps," 14th ed., copyright 1983, with the permission of Hydraulic Institute.)*

is pushed into the discharge line. In the most common type of gear pump, one of the pumping gears drives the other pumping gear. This direct contact between the teeth requires the pumpage to be a good lubricant. In some units, though, the two pumping gears are driven by gears (called **timing gears**) mounted on the two shafts external to the pumpage. With this arrangement, the two pumping gears do not contact each other, and the unit is more suitable for liquids with low lubricity. This arrangement does, however, require a larger, more complex pump.

If the teeth are on the inside of a ring, as shown in Fig. 14.1.25, it is called an **internal gear**. In this unit, the larger internal-tooth gear drives the smaller external-tooth gear. Both gears transfer liquid from the suction to the discharge.

Both external- and internal-gear rotary pumps are used in lubrication systems of engines, compressors, and larger pumps.

Lobe pumps are similar in construction and pumping action to external gear pumps, but one lobe does not drive the other. They do not even

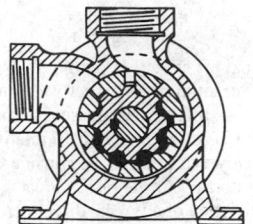

Fig. 14.1.25   Internal-gear rotary pump.

touch each other, and therefore must have their shafts independently driven by external timing gears. The lobes are often made of elastomers, and operate at low speeds. These units are used to transfer delicate items such as cherries and other foods—and even live fish. Figure 14.1.26 shows a rotary pump with three-lobe rotors.

Screw pumps are constructed with one, two, or three screws. The single-screw pump is more commonly called a progressing-cavity pump, and is discussed separately below. The two-screw pump, as illustrated by Fig. 14.1.27, is used to handle liquids with viscosities to greater than 1,000,000 SSU (200,000 mm²/s), with capacities to about 10,000 gal/min (2000 m³/h), pressures to 2500 lb/in² (17 MPa), and temperatures to 600°F (300°C). It is particularly suited for high-viscosity liquids.

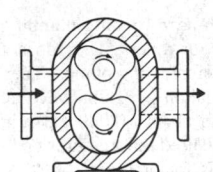

Fig. 14.1.26   Three-lobe rotary pump. *(Reprinted from "Hydraulic Institute Standards for Centrifugal Rotary and Reciprocating Pumps," 14th ed., copyright 1983, with the permission of Hydraulic Institute.)*

The three-screw rotary pump, as illustrated in Fig. 14.1.28, is a high-speed pump (up to 6,000 r/min) used primarily for lubrication systems on turbines, compressors, and centrifugal pumps.

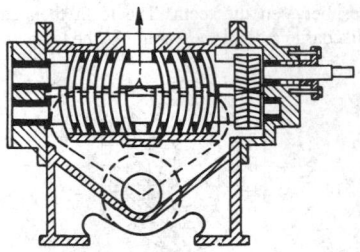

Fig. 14.1.27   Two-screw rotary pump.

A **progressing-cavity** pump is illustrated in Fig. 14.1.29. The rotor is made of polished steel, and in the shape of a single-lead screw. The stator, usually made of elastomer, is in the shape of a double-lead screw. As the rotor turns, its axis prescribes a circle. This characteristic requires the rotor to be driven by an internal universal joint (or equivalent). The pumpage moves through the pump in a spiral. The rubber stator makes this unit well-suited for abrasive services.

Figure 14.1.30 illustrates a **sliding vane** rotary pump. The single rotor contains multiple vanes which slide in radial slots. The rotor and casing are eccentric. The vanes maintain contact with the casing by centrifugal force and pressure. These units are typically available for capacities to 1000 gal/min (200 m³/h), viscosities to 500,000 SSU (100,000 mm²/s), and pressures to 125 lb/in² (900 kPa). Some sliding-vane pumps are suitable for low-lubricity liquids such as light hydrocarbons.

Flexible Member   Some rotary pumps are built with flexible vanes, liners, and tubes. The flexible tube pump, also called a **peristaltic** pump, is shown in Fig. 14.1.31.

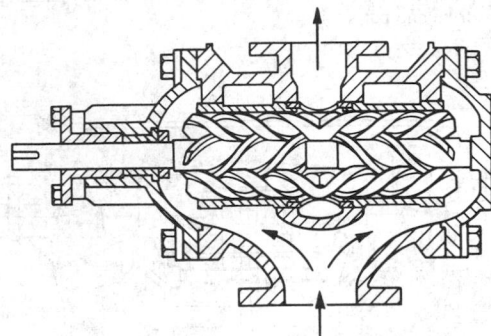

Fig. 14.1.28   Three-screw rotary pump.

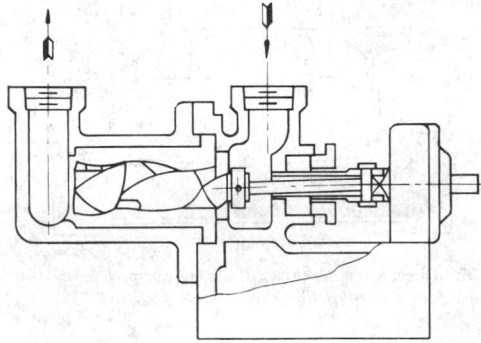

Fig. 14.1.29   Progressing-cavity (single-screw) rotary pump. *(Hydraulic Institute.)*

Fig. 14.1.30   Sliding-vane rotary pump. *(Reprinted from "Hydraulic Institute Standards for Centrifugal, Rotary and Reciprocating Pumps," 14th ed., copyright 1983, with the permission of Hydraulic Institute.)*

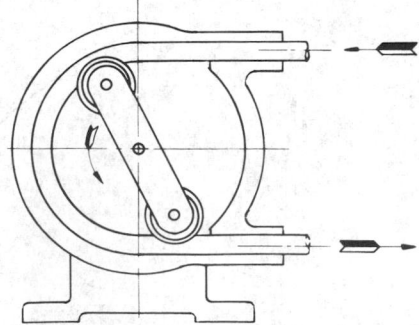

Fig. 14.1.31   Flexible-tube (peristaltic) rotary pump. *(Hydraulic Institute.)*

A **radial-plunger** rotary pump is shown in Fig. 14.1.32. An **axial-plunger** rotary pump (also called an **axial-piston** or **swash-plate** pump) is shown in Fig. 14.1.33. In these pumps, the inner body rotates, causing each plunger to alternately accept pumpage from the inlet and deliver it to the discharge port. In the radial pump, the length of stroke of each plunger is established by the eccentricity of the rotor. In the axial pump, the stroke length is established by the angle of the plate or shaft. On some pumps, the eccentricity (or angle) is adjustable, providing variable

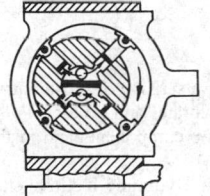
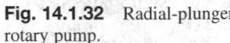

Fig. 14.1.32 Radial-plunger rotary pump.

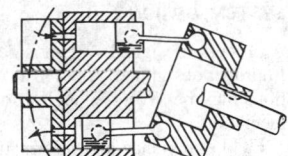

Fig. 14.1.33 Axial-plunger (swash-plate) rotary pump.

displacement. Rotary plunger pumps are used in hydraulic systems to provide power to hydraulic motors and cylinders.

Circumferential-Piston Pump Although sometimes considered a lobe pump, this unit (shown in Fig. 14.1.34) differs from the lobe unit in that there is no close clearance between the two rotors. Close clearance does exist, though, between each rotor and adjacent stationary parts.

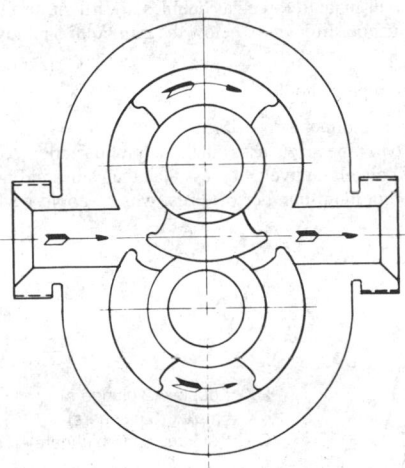

Fig. 14.1.34 Circumferential-piston rotary pump. *(Hydraulic Institute.)*

The **casing** (also called a **body** or **housing**) is the main pressure-retaining component of a rotary pump. It supports the rotor, contains the suction and discharge nozzles, and usually forms a portion of the boundaries of the pumping chambers.

The **bearings** of a rotary pump may be internal or external. If internal, they are lubricated by the pumpage. If the pumpage is not suitable for bearing lubrication, the bearings must be external. External bearings require two pumpage seals on each shaft. Some rotary pumps contain four seals.

Figure 14.1.35 shows typical **rotary pump performance curves**. Capacity, mechanical efficiency, and power are plotted against pressure. The difference between displacement and capacity is called **slip**. Power input is the sum of power output and power losses (due to friction and slip).

As defined by the Hydraulic Institute, the NPSH required by a rotary pump is that value of NPSHA when cavitation noise, a sharp drop in capacity, or a 5 percent reduction in capacity occurs, whichever occurs first. The definition is not as precise as the reciprocating pump 3 percent capacity drop.

Increased Running Clearances An increase in the clearances between the rotating parts, and/or between the rotating and stationary parts, of a rotary pump will cause the slip to increase. An increase in slip

causes the volumetric efficiency to drop, the mechanical efficiency to drop, and the NPSHR to rise. If the unit is running at a constant speed, the lower volumetric efficiency (lower capacity) will result in a lower discharge pressure if the system head varies with capacity.

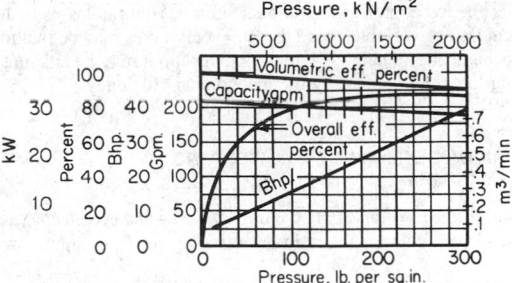

Fig. 14.1.35 Rotary pump performance curves for a herringbone gear pump at 600 r/min with a 400-SSU (90-mm²/s) liquid.

VOLUMETRIC EFFICIENCY

Volumetric efficiency η_v of a displacement pump is the ratio of capacity Q to displacement D: $\eta_v = Q/D$. Capacity is the volume flow rate in the suction pipe. Displacement is the volume swept by the pumping elements per unit time. For a single-acting triplex plunger pump, displacement may be calculated as follows: $D = 3(\pi/4)d^2LN$, where D = displacement; d = diameter of plunger; L = stroke length; and N = crankshaft speed, r/min.

To predict the capacity of a reciprocating pump, in the selection process, it is necessary to predict the volumetric efficiency. Volumetric efficiency may be calculated with the following equation:

$$\eta_v = 1 - \frac{S}{D'} - \frac{C}{D'}\left(\frac{v_s}{v_d} - 1\right) \qquad (14.1.2)$$

where η_v = volumetric efficiency; S = slip through both suction and discharge valves, typically about 3 percent of D'; D' = displacement of one pumping element during one stroke; C = clearance volume (dead space) in the pumping chamber (when the pumping element is at the end of the discharge stroke); v_s = specific volume of pumpage at suction conditions; and v_d = specific volume of the pumpage at discharge conditions.

Although volumetric efficiency has an effect on mechanical efficiency, in reciprocating pumps the two do not necessarily move in unison. It is possible to have a high η_m with a low η_v (due to liquid compressibility), or a low η_m with a high η_v (such as would occur in a high-suction-pressure, low-differential-pressure application).

The same equation can be applied to rotary pumps, but usually $v_s = v_d$, so that Eq. (14.1.2) reduces to $\eta_v = 1 - S/D'$. For a rotary pump, S = leakage of pumpage from the discharge side of the pump to the suction side. Because rotary pumps have no check valves, they must rely on close clearances to restrict slip. Slip is a function of the differential pressure, the viscosity of the pumpage, and the clearance between the parts. Slip in a rotary pump is usually a higher fraction of displacement than in a reciprocating pump.

POWER OUTPUT AND INPUT

The **power output** (also called **hydraulic power, hydraulic horsepower,** and **water horsepower**) of any pump is the usable power it imparts to the pumpage. It is the product of capacity and differential pressure.

In U.S. units, $P_o = Q \times \Delta p/1{,}714 = Q(p_d - p_s)/1{,}714$, where P_o = power output, hp; Q = pump capacity, gal/min; Δp = pump differential pressure, lb/in²; p_d = pump discharge pressure, lb/in²; and p_s = pump suction pressure, lb/in².

In metric units, $P_o = Q \times \Delta p/3{,}600 = Q(p_d - p_s)/3{,}600$, where P_o = power output, kW; Q = pump capacity, m³/h; Δp = pump differential pressure kPa; p_d = pump discharge pressure, kPa; and p_s = pump suction pressure, kPa.

The **power input** (sometimes called **brake horsepower**) is the power required to drive power and rotary pumps. Normally, the power input includes the losses in an integral gear unit, but not the losses in a separate gear unit or variable-speed drive. V-belt losses may be included. If the mechanical efficiency is known, power input may be calculated by dividing the power output by the mechanical efficiency.

MECHANICAL EFFICIENCY—POWER AND ROTARY PUMPS

Mechanical efficiency (also called **pump efficiency** and **overall efficiency**) of a power-driven pump is the ratio of output power P_o to input power P_i: $\eta_m = P_o/P_i$.

As shown in Fig. 14.1.36, mechanical efficiency drops as the frame load is reduced, because the power output falls faster than friction losses, and becomes a smaller part of the power input. Mechanical efficiency is zero when the differential pressure is zero.

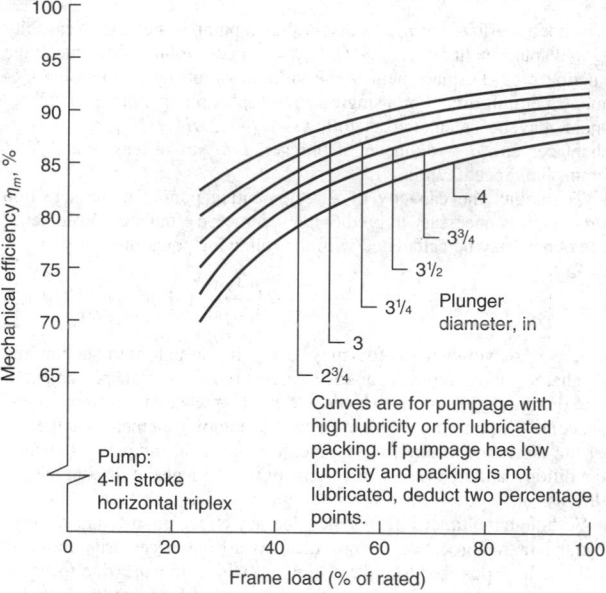

Fig. 14.1.36 Mechanical (overall) efficiency of a triplex power pump.

PULSATION DAMPENERS

A **pulsation dampener** is a device which reduces the pressure pulsations created, in a suction or discharge pipe, by the interaction of a pump and a system. The dampener performs this function by reducing the velocity variations of the pumpage in the pipe.

Most reciprocating pumps, and many rotary pumps, create a variation in the velocity of the pumpage in the piping. Unless the suction and discharge vessels are very close to the pump, these velocity variations will be converted by the system into pressure variations. To reduce these pulses, a small vessel can be installed in the suction and/or discharge line. Such a vessel mounted in the discharge line is called a **pulsation dampener**. If the suction pressure is high enough, this same type of device can be used in the suction line, but a special dampener, called a **suction stabilizer** will perform the dual function of reducing pulses and separating free gas from the pumpage.

For optimum effectiveness, each dampener must be installed close to the pipe and adjacent to the pump. There must be minimal restriction

between the dampener and pipe. For a low suction pressure, a simple air-filled standpipe is often adequate. For the discharge side, and for high suction pressures, the dampener often contains a diaphragm or bladder that is precharged with nitrogen to about 70 percent of system pressure.

SYSTEM DESIGN

A properly designed system is necessary for a satisfactory installation. Improper design will result in a system that vibrates and is noisy. Pulsations may be severe enough to damage pump components and instrumentation.

Field experience and information from the Hydraulic Institute standards are condensed below into a list of general design guidelines for the (1) suction vessel, (2) suction piping, and (3) discharge piping. The asterisks in the following lists show features that *do not* apply to rotary pumps that produce a uniform flow rate (constant velocity).

The suction vessel should:

Be large enough to provide sufficient retention time to allow free gas to rise to the liquid surface.

Have the feed and return lines enter below the minimum liquid level.

Include a vortex breaker over the outlet (pump suction) line and/or provide sufficient submergence to preclude vortex formation.

Contain a weir plate to force gas bubbles toward the surface. The top of the weir must be sufficiently below the minimum tank level to avoid disturbance.

The suction piping should:

Be as short and direct as possible.

Be one or two pipe sizes larger than the pump suction connection.

Have an average liquid velocity less than the values in Fig. 14.1.37.*

Contain a minimum number of turns, which should be long-radius elbows or laterals.*

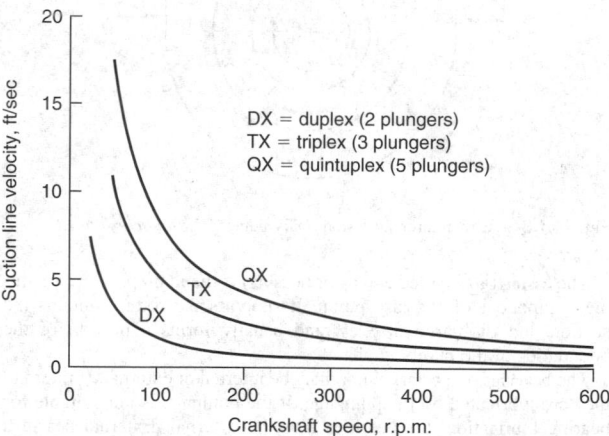

Fig. 14.1.37 Suction line velocities for constant acceleration head, for reciprocating power pumps.

Be designed to preclude the collection of vapor in the piping. (There should be no high points, unless vented. The reducer at the pump should be the eccentric type, installed with the straight side up.)

Be designed so that NPSHA, allowing for acceleration head, exceeds NPSHR.

Include a suitable suction stabilizer/gas separator located in the suction pipe adjacent to the pump liquid end if the acceleration head is excessive or the pumpage contains free gas.

Contain a full-opening block valve so that flow to the pump is not restricted.

Not include a strainer or filter unless regular maintenance is assured. (The starved condition resulting from a plugged strainer can cause more damage to a pump than solids.)

The discharge piping should:

Be one or two pipe sizes larger than the pump discharge connection.

Have an average velocity less than 3 times the maximum suction line velocity.

Contain a minimum number of turns, using long-radius elbows and laterals where practical.

Include a suitable pulsation dampener (or provisions for adding one) adjacent to the pump liquid end.*

Contain a relief valve, sized to pass full pump capacity at a pressure that does not exceed 110 percent of cracking pressure (the opening pressure of the relief valve). The discharge from the relief valve should be piped back to the suction vessel so that gases liberated through the valve are not fed back into the pump.

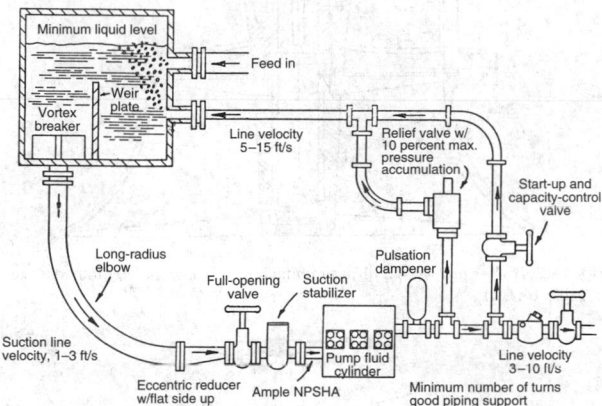

Fig. 14.1.38   A good system for a reciprocating pump.

Contain a bypass line and valve so that the pump may be started against negligible discharge pressure.

Contain a check valve to prevent the imposition of system pressure on the pump during start-up.

The features of a good system are illustrated in Fig. 14.1.38.

Remedies for Low NPSHA/High NPSHR

In designing the suction system for a displacement pump, if NPSHA is found to be less than NPSHR, a remedy may be found in the following list:

Increase the diameter of the suction line.

Reduce the length of the suction line by providing a more direct route, or by moving the pump closer to the suction vessel.

Install a suction bottle or stabilizer adjacent to the pump liquid end. A fabricated bottle has often been successfully used at pressures below about 50 lb/in² (300 kPa), but maintenance of a liquid level is required.

A section of rubber hose in the suction line, adjacent to the pump, will often reduce the acceleration head.

Elevate the suction vessel or the level of liquid in the suction vessel.

Reduce temperature of pumpage. (If the pumpage is at bubble point in the vessel, it must be cooled after it leaves the vessel.)

Reduce speed of the pump, or select a larger pump running slower.

At a lower speed, it may be possible to operate a reciprocating pump with light suction valve springs, or none at all.)

If the above steps are insufficient, impractical, or impossible, a booster pump must be provided. A booster for a displacement pump is normally a centrifugal, although direct-acting reciprocating and rotary pumps are sometimes used. The NPSHR of the booster must be less than that available from the system. The head of the booster should exceed the main pump NPSHR plus suction-line losses by about 20 percent. The booster should be installed near the suction vessel, and a pulsation bottle or stabilizer should be installed in the suction line, adjacent to the main pump, to protect the booster from pulsating flow.

14.2 CENTRIFUGAL AND AXIAL-FLOW PUMPS
by Igor J. Karassik

REFERENCES: (1) Theory: Stepanoff, "Centrifugal and Axial Flow Pumps," Wiley. Wislicenus, "Fluid Mechanics of Turbomachinery," McGraw-Hill. Pfleiderer, "Die Kreiselpumpen," Springer. Spannhake, "Centrifugal Pumps, Turbines and Propellers," Technology Press. (2) Practice: Hicks, "Pump Selection and Application," and "Pump Operation and Maintenance," McGraw-Hill. Standards of the Hydraulic Institute. Karassik, "Centrifugal Pump Clinic," Marcel Dekker, Inc.

NOMENCLATURE AND MECHANICAL DESIGN

Pumps covered in this section fall into three general classes: (1) **centrifugal** or **radial-flow**, (2) **mixed-flow**, and (3) **axial-flow** or propeller pumps. The essential elements of a centrifugal pump are (1) the **rotating element,** consisting of the shaft and the impeller, and (2) the **stationary element,** consisting of the casing, stuffing boxes, and bearings (see Fig. 14.2.1). Other parts, such as wearing rings and shaft sleeves, are generally added to produce better-operating, more economical machines as warranted by the various services on which the pumps are to be used. Names recommended by the Hydraulic Institute for various parts are given in Table 14.2.1.

In a centrifugal pump the liquid is forced, by atmospheric or other pressure, into a set of rotating vanes which constitute an impeller discharging the liquid at a higher pressure and a higher velocity at its periphery. The major portion of the velocity energy is then converted into pressure energy by means of a **volute** (Fig. 14.2.2) or by a set of stationary **diffusion vanes** (Fig. 14.2.3) surrounding the impeller periph-

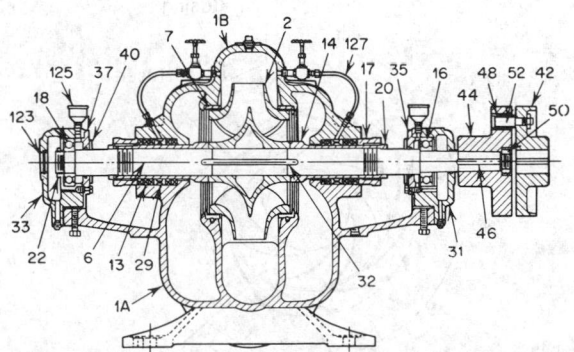

Fig. 14.2.1   Horizontal single-stage double-suction volute pump. (Numbers refer to parts listed in Table 14.2.1.)

Table 14.2.1 Recommended Names of Centrifugal-Pump Parts
(These parts are called out in Figs. 14.2.1, 14.2.4, and 14.2.5)

Item no.	Name of part	Item no.	Name of part
1	Casing	33	Bearing housing (outboard)
1A	Casing (lower half)	35	Bearing cover (inboard)
1B	Casing (upper half)	36	Propeller key
2	Impeller	37	Bearing cover (outboard)
4	Propeller	39	Bearing bushing
6	Pump shaft	40	Deflector
7	Casing ring	42	Coupling (driver half)
8	Impeller ring	44	Coupling (pump half)
9	Suction cover	46	Coupling key
11	Stuffing-box cover	48	Coupling bushing
13	Packing	50	Coupling lock nut
14	Shaft sleeve	52	Coupling pin
15	Discharge bowl	59	Handhole cover
16	Bearing (inboard)	68	Shaft collar
17	Gland	72	Thrust collar
18	Bearing (outboard)	78	Bearing spacer
19	Frame	85	Shaft-enclosing tube
20	Shaft-sleeve nut	89	Seal
22	Bearing lock nut	91	Suction bowl
24	Impeller nut	101	Column pipe
25	Suction-head ring	103	Connector bearing
27	Stuffing-box-cover ring	123	Bearing end cover
29	Lantern ring	125	Grease (oil) cup
31	Bearing housing (inboard)	127	Seal piping (tubing)
32	Impeller key		

ery. Pumps with volute casings are called **volute pumps;** those with diffusion vanes are called **diffuser pumps.** The latter were once commonly called turbine pumps, but this term has recently been more selectively applied to vertical deep-well centrifugal diffuser pumps, now called **vertical turbine pumps.**

Centrifugal pumps are divided into other categories, several of which relate to the impeller. First, impellers are classified according to the major direction of flow in reference to the axis of rotation. Centrifugal pumps may have (1) **radial-flow** impellers (Figs. 14.2.1, 14.2.2, and 14.2.5), (2) **axial-flow** impellers (Fig. 14.2.4), and (3) **mixed-flow** impellers, which combine radial- and axial-flow principles (Fig. 14.2.7).

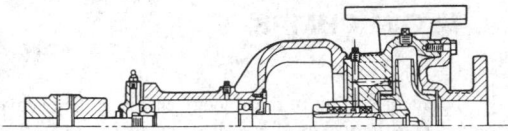

Fig. 14.2.2 Single-stage end-suction volute pump.

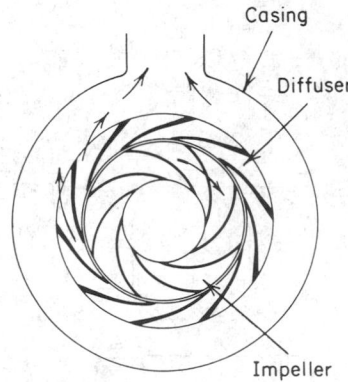

Fig. 14.2.3 Diffuser-type pump.

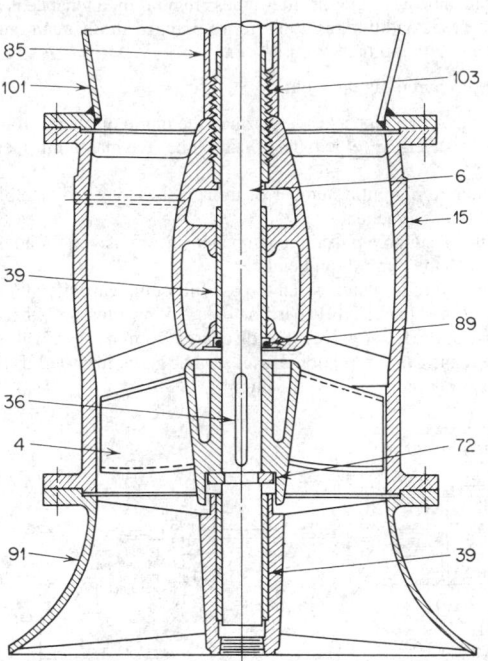

Fig. 14.2.4 Vertical wet-pit diffuser pump bowl. (Numbers refer to parts listed in Table 14.2.1.)

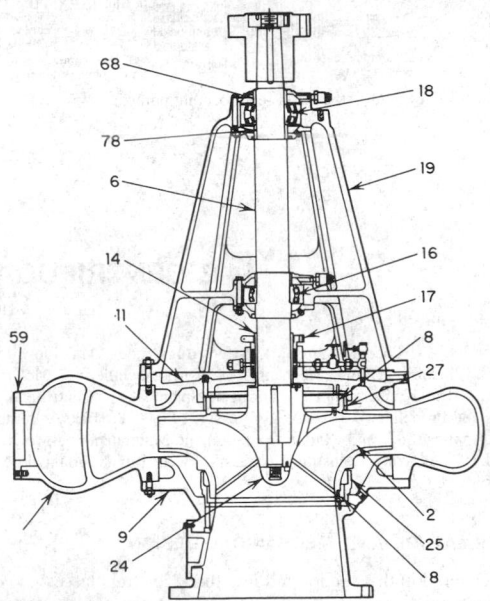

Fig. 14.2.5 Vertical end-suction pump with a double-volute casing. (Numbers refer to parts listed in Table 14.2.1.)

Impellers are further classified according to the flow arrangement into (1) **single-suction,** with a single inlet on one side, and (2) **double-suction,** with water flowing to the impeller symmetrically from both sides. They are categorized according to their mechanical construction into (1) **closed,** with shrouds or sidewalls enclosing the waterways, (2) **open,** with no shrouds, and (3) **semiopen** or semiclosed.

If the pump is one in which the head is developed by a single impeller, it is called a **single-stage pump.** When two or more impellers operating in series are used, the unit is called a **multistage pump.** The mechanical design of the casing provides the added classification of **axially split** (Fig. 14.2.1) or **radially split** (Figs. 14.2.2 and 14.2.5), and the axis of rotation determines whether the pump is horizontal-shaft, vertical-shaft, or (occasionally) inclined-shaft. Usually these are referred to simply as **horizontal** or **vertical** units.

Horizontal centrifugal pumps are classified still further according to suction-nozzle location into (1) **end-suction,** (2) **side-suction,** (3) **bottom-suction,** and (4) **top-suction.**

Some pumps operate with the liquid conducted to and from the unit by piping. Other pumps, usually vertical types, are submerged in their suction supply. Vertical pumps are therefore called either **dry-pit** or **wet-pit** types. If the wet-pit pumps are axial-flow, mixed-flow, or vertical-turbine types, the liquid is discharged up through the supporting drop or column pipe to a discharge point either above or below the supporting floor. These pumps are consequently designated as **above-ground discharge** or **below-ground discharge** units.

Casings

The pressure acting on the impeller in a single-volute pump-casing design is nearly uniform when the pump is operated at or near its design capacity. At other capacities, the pressures around the impeller are not uniform, causing a radial reaction (or radial thrust) which can substantially increase the pump-shaft deflection. When it becomes impractical to counteract this radial thrust through the use of a heavier shaft and heavier bearings, a **double-** or **twin-volute** design (Fig. 14.2.5) may be used.

End-suction single-stage pumps are made of one-piece solid casings. At least one side of the casing must have an opening with a cover so that the impeller can be assembled in the pump. If the cover is on the suction side, it becomes the casing sidewall and contains the suction opening (Fig. 14.2.2). This is called the **suction cover** or **casing suction head.** Other designs are made with stuffing-box covers (Fig. 14.2.6), and still others have both casing suction covers and stuffing-box covers (Figs. 14.2.5 and 14.2.6).

In the inexpensive open-impeller pump, the impeller rotates within close clearance of the pump casing (Fig. 14.2.6). If the intended service is more severe, a side plate is mounted within the casing to provide a renewable close-clearance guide to the liquid flowing through the open impeller.

The discharge nozzle of end-suction single-stage horizontal pumps is usually in a top-vertical position (Fig. 14.2.2). However, other nozzle positions may be obtained, such as top-horizontal, bottom-horizontal, or bottom-vertical discharge. Practically all double-suction axially split casing pumps have a side-discharge nozzle and either a side- or a bottom-suction nozzle. Single-stage bottom-suction pumps are rarely made in sizes below 10 in discharge-nozzle diameter.

Both axially split (Fig. 14.2.8) and radially split (Fig. 14.2.9) casings are used for multistage pumps. The choice between the two designs is

dictated by the discharge pressure, with 1,300 to 2,000 lb/in² (90 to 140 kg/cm²) forming the approximate limit between the two. Radially split casings are normally designed as double casings; the working parts of the pump are enclosed in an inner casing, which is then inserted into a second, or outer, casing. The space between the two casings is maintained at the discharge pressure of the last pump stage.

Impellers and Wearing Rings

In addition to being classified with reference to the suction flow into the impeller, the basic flow component, and their mechanical features, impellers are also classified with reference to their profile and to their head-capacity characteristics at a given speed. This last relationship will be described later, in the discussion of **specific speed.**

Many impellers are designed for specific applications. Special nonclogging impellers with blunt edges and large waterways are used for sewage which ordinarily contains rags or stringy material. Impellers designed for handling paper-pulp stock are fully open and nonclogging and have screw conveyor vanes which project into the suction nozzle.

Wearing rings provide an easily and economically renewable leakage joint between the impeller and casing. A leakage joint without renewable parts is used only in very small, inexpensive pumps. The stationary ring is called (1) **casing ring** if mounted in the casing, (2) **suction-cover ring** or **suction-head ring** if mounted in a suction cover or head, and (3) **stuffing-box cover ring** if mounted in the stuffing-box cover. A renewable part for the impeller wearing surface is called the **impeller ring.** Pumps with both stationary and rotating rings are said to have **double-ring** construction.

There are various types of wearing-ring designs, and selection of the most desirable type depends on the liquid being handled, the pressure differential across the leakage joint, the surface speed, and the particular pump design. In general, centrifugal-pump designers use the ring construction they have found suitable for each particular pump service. The most common ring constructions are the **flat type** (Fig. 14.2.2) and the **L type** (Figs. 14.2.1 and 14.2.7).

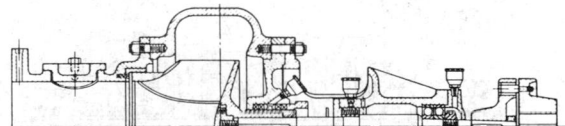

Fig. 14.2.7 End-suction pump with removable suction and stuffing-box heads.

Axial Thrust in Single-Stage and Multistage Pumps

Axial hydraulic thrust is the summation of unbalanced impeller forces acting in the axial direction. Theoretically, a double-suction impeller is in hydraulic balance, with the pressures on one side equal to and counterbalancing the pressures on the other. In practice, some slight unbalance may exist, and even double-suction pumps are provided with thrust bearings.

The single-suction radial-flow impeller is subject to axial thrust because a portion of the front wall is exposed to suction pressure, with a greater back-wall surface subject to discharge pressure. In addition, an overhung single-suction impeller with a single stuffing box is subject to an axial force equivalent to the product of the shaft area through the stuffing box and the difference between suction and atmospheric pressure. This force acts toward the impeller suction when the suction pressure is less than the atmospheric and in the opposite direction when it is higher than the atmospheric.

To eliminate the axial thrust of a single-suction impeller, a pump can be provided with both front and back wearing rings (Figs. 14.2.2 and 14.2.5). Pressure approximately equal to the suction pressure is maintained in a chamber located on the inner side of the back wearing ring by providing so-called **balancing holes** through the impeller. Leakage past the back wearing ring is returned into the suction area through these holes. In large pumps, a piped connection usually replaces the balancing holes.

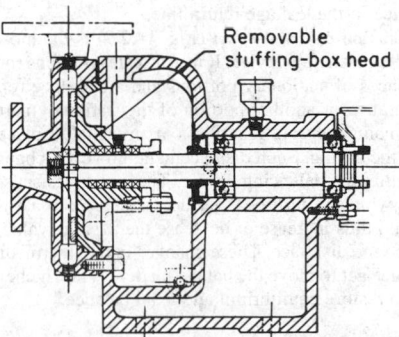

Removable stuffing-box head

Fig. 14.2.6 End-suction pump with removable stuffing-box head.

Most multistage pumps are built with single-suction impellers. To balance the axial thrust of these impellers, two arrangements are used: (1) The impellers all face in the same direction and are mounted in the ascending order of the stages. The axial thrust is balanced by a hydraulic balancing device (Figs. 14.2.8 and 14.2.9). (2) An even number of single-suction impellers is used, one half of these facing in a direction opposite to the second half (Fig. 14.2.10). This mounting of single-suction impellers back to back is frequently called **opposed impellers.**

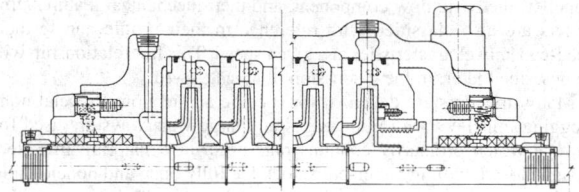

Fig. 14.2.8 Multistage pump with single-suction impellers facing in one direction and with hydraulic balancing device.

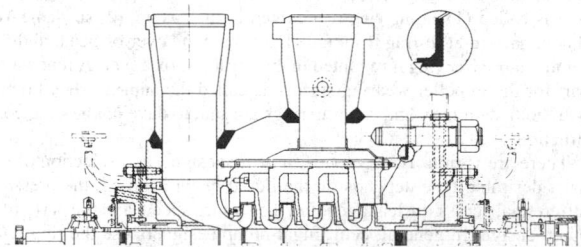

Fig. 14.2.9 Double-casing multistage pump with radially split inner casing.

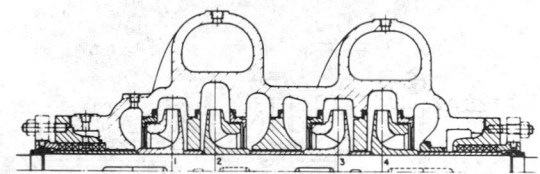

Fig. 14.2.10 Four-stage pump with opposed impellers.

Hydraulic balancing devices may take the form of (1) a **balancing drum,** (2) a **balancing disk,** or (3) a combination of these two. The **balancing drum** is illustrated in Fig. 14.2.11. The balancing chamber at the back of the last-stage impeller is separated from the pump interior by a drum mounted on the shaft. The drum is separated by a small radial clearance from the stationary portion of the balancing device, called the **balancing**

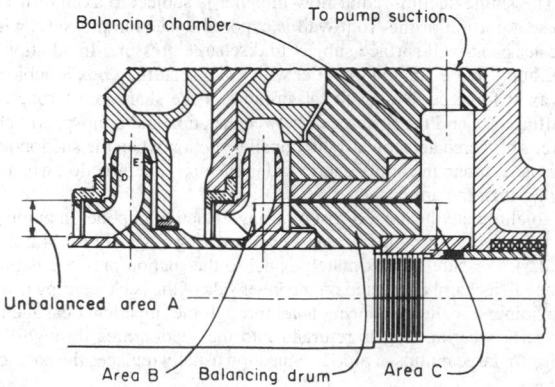

Fig. 14.2.11 Balancing drum.

drum head, which is fixed to the pump casing. The balancing chamber is connected either to the pump suction or to the vessel from which the pump takes its suction. The forces acting on the balancing drum are (1) toward the discharge end—the discharge pressure multiplied by the front balancing area (area *B*) of the drum; (2) toward the suction end— the back pressure in the balancing chamber multiplied by the back balancing area (area *C*) of the drum. The first force is greater than the second, thereby counterbalancing the axial thrust exerted upon the single-suction impellers. The drum diameter can be selected to balance the axial thrust completely or to balance 90 to 95 percent of this thrust, depending on whether a slight thrust load in a specific direction on the thrust bearing is desirable.

The operation of the simple **balancing disk** is illustrated in Fig. 14.2.12. The rotating disk is separated from the balancing-disk head by a small axial clearance. The leakage through this clearance flows into the balancing chamber and from there either to the pump suction or to the suction vessel. The back of the balancing disk is subject to the balancing-chamber back pressure, whereas the disk face experiences a range of pressures. These vary from discharge pressure at its smallest diameter to back pressure at its periphery. The inner and outer disk diameters are chosen so that the difference between the total force act-

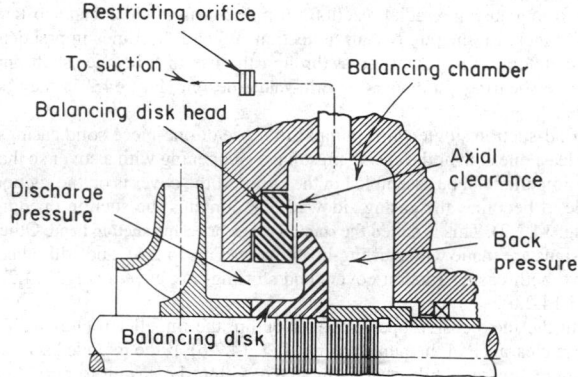

Fig. 14.2.12 Simple balancing disk.

ing on the disk face and that acting on its back will balance the impeller axial thrust. If the axial thrust of the impellers should exceed the thrust acting on the disk during operation, the latter is moved toward the disk head, reducing the axial clearance. The amount of leakage through this clearance is reduced so that the friction losses in the leakage return line are also reduced, lowering the back pressure in the balancing chamber. This automatically increases the pressure difference acting on the disk and moves it away from the disk head, increasing the clearance. Now the pressure builds up in the balancing chamber, and the disk is again moved toward the disk head until an equilibrium is reached. To ensure proper balancing-disk operation, the change in back pressure must be of an appreciable magnitude. This is accomplished by introducing a restricting orifice in the leakage return line.

The **combination disk and drum** (Fig. 14.2.9) is the most commonly used hydraulic balancing device. It incorporates portions rotating within radial clearances of stationary portions and a disk face rotating within an axial clearance of another portion of the stationary part. The radial clearance remains constant regardless of any axial displacement of the rotor within the casing. Such displacement, however, changes the axial clearance within the balancing device. These changes cause changes in the leakage, which in turn change the pressure drop across the radial clearances and thus increase or decrease the average value of the pressure acting on the disk face. These changes in the intermediate pressure on the disk face act to move the balancing device in whichever direction is required to restore equilibrium and axial balance.

Shafts and Shaft Sleeves

Pump-shaft diameters are usually larger than actually needed to transmit the torque because their size is dictated by the maximum permissible or desirable shaft deflection. This deflection is itself chosen to prevent possible contact at the wearing surfaces while maintaining reasonable clearances that will not affect pump efficiency too unfavorably. The first **critical speed** of a shaft is related to its deflection. It follows that a shaft design permitting a deflection of, for instance, 0.005 to 0.006 in (0.13 to 0.15 mm) will have a first critical speed of 2,400 to 2,650 r/min. This is the reason for using rigid shafts (operating below their first critical speed) for pumps that operate at 1,750 r/min or lower. Multistage pumps operating at 3,600 r/min or higher use shafts of equal stiffness (for the same purpose of avoiding wearing-ring contact). However, their corresponding critical speed is about 25 to 40 percent less than their operating speed. This margin is sufficient to avoid any danger to the operation caused by critical-speed effect.

Pump shafts are usually protected from erosion, corrosion, and wear at the stuffing boxes and leakage joints and in the waterways by renewable **sleeves**. The most common shaft-sleeve function is that of protecting the shaft from wear at a stuffing box. Shaft sleeves serving other functions are given specific names to indicate their purpose. For example, a shaft sleeve used between two multistage impellers in conjunction with an interstage bushing to form an interstage leakage joint is called an **interstage** or **distance** sleeve.

Stuffing Boxes

Stuffing boxes have the primary function of protecting the pump against leakage at the point where the shaft passes out through the pump casing. If the pump handles a suction lift and the pressure at the interior stuffing-box end is below atmospheric, the stuffing-box function is to prevent air leakage into the pump. If this pressure is above atmospheric, the function is to prevent liquid leakage out of the pump. The stuffing box takes the form of a cylindrical recess that accommodates a number of rings of **packing** around the shaft or shaft sleeve. If sealing the box is desired, a **lantern ring** or **seal cage** is used to separate the rings of packing into approximately equal sections. The packing is compressed to give the desired fit on the shaft or sleeve by a gland that can be adjusted in an axial direction.

Water or some other sealing fluid can be introduced under pressure into the space provided by the seal cage, causing flow of sealing fluid in both axial directions. This is useful for pumps handling flammable or chemically active and dangerous liquids since it prevents outflow of the pumped liquid.

When a pump handles clean, cool water, stuffing-box seals are usually connected to the pump discharge or, in multistage pumps, to an intermediate stage. An **independent supply of sealing water** should be provided if any of the following conditions exist: (1) a suction lift in excess of 15 ft (4.5 m); (2) a discharge pressure under 10 lb/in² (0.7 kg/cm²); (3) hot water handled without adequate cooling (except for boiler feed pumps, in which seal cages are not used); (4) muddy, sandy, or gritty water handled; (5) for all hot-well pumps; (6) no leakage to atmosphere permitted of the liquid handled. When sealing water is taken from the pump discharge, an external connection is generally made to the seal cage through small-diameter piping (Fig. 14.2.1), or an internal-passage connection is made within the pump itself (Fig. 14.2.2).

High temperatures or pressures complicate the problem of maintaining stuffing-box packing. Pumps in these more difficult services are usually provided with jacketed, **water-cooled stuffing boxes**. If the pressure ahead of the stuffing box makes it impractical to pack the stuffing box satisfactorily, a pressure-reducing breakdown or **labyrinth** may be located ahead of the box, with the leakage past the pressure-reducing breakdown being returned to some point of lower pressure in the pumping cycle.

Basically, **stuffing-box packing** is a pressure-breakdown device that is sufficiently plastic to be adjusted for proper operation. The most common types are asbestos packing and metallic packing, the latter being composed of flexible metallic strands or foil with graphite or oil lubricant and with either an asbestos or a plastic core. Other types of packing used may be hemp, cord, braided, duck fabric, chevron, etc. Packing is supplied either in continuous coils of square cross section or in preformed, die-molded rings.

Mechanical seals are used in centrifugal pumps when it becomes im-practical to use conventional packing with radial sealing surfaces. The sealing surfaces of a mechanical seal are located in a plane perpendicular to the pump shaft and consist of two highly polished surfaces running adjacently, one surface being connected to the shaft and the other to the stationary portion of the pump. These surfaces are held essentially in contact by a spring, the axial clearance between the surfaces being provided by a thin film of liquid. The flow of liquid may be only a drop every few minutes or even a haze of escaping vapor.

There are two basic seal arrangements: (1) internal assembly and (2) external assembly. Two mechanical seals may be mounted inside a stuffing box to make a **double seal assembly** (Fig. 14.2.13.). Such an arrangement is used for pumps handling toxic or highly inflammable liquids. A clear, filtered, and generally inert sealing liquid is injected between two seals at a pressure slightly in excess of the pressure in the pump ahead of the seal.

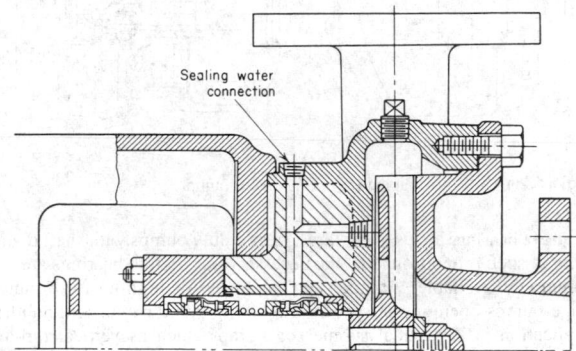

Fig. 14.2.13 Single-stage end-suction pump with double mechanical seal.

Mechanical seals can reduce the escape of toxic or flammable liquids very effectively, but lately we have seen constantly more stringent regulations by both federal and local authorities. This has led to a greater use of hermetic, or so-called *sealless* pumps. Two solutions have emerged to meet this requirement. The first is the canned-motor pump in which a portion of the pumped liquid is permitted to flow into the motor. The liquid is isolated from contact with the windings, insulation, and cores of rotor and stator by metal "cans." The second solution is the use of magnetic couplings. These consist of two sets of magnets, the outer (or driver) magnets, and the inner (or driven) magnets. A motor is connected to the outer portion of the coupling and the pump impeller is connected to the inner portion. A containment shell separates the two magnets and hermetically seals the pump proper. This latter solution is finding preference because it permits the use of standard motors, making pumps more readily maintainable in the field, while canned-motor pumps usually have to be returned to the manufacturer's plant for rebuilding.

For some power-plant services, **condensate-injection sealing** is superior to either conventional packing or mechanical seals. A labyrinth breakdown bushing is substituted for the conventional packing, and the pump-shaft sleeve runs within this bushing with a small radial clearance. Cold condensate at a pressure in excess of the internal pump pressure is introduced centrally in this breakdown bushing (Fig. 14.2.9). A small portion of the injection water flows inwardly into the pump proper; the remainder flows out into a collecting chamber vented to the atmosphere and is piped back to the condenser.

Bearings

(See also Sec. 8.)

All types of **bearings** are used in centrifugal pumps. Even the same basic design of pump is often made with two or more different bearings, required by varying service conditions. Two external bearings are usually used for the double-suction single-stage general-service pump, one on either side of the casing. In horizontal pumps with bearings on each end, the **inboard bearing** is the one between the casing and the coupling and the **outboard bearing** is located at the opposite end. Pumps with overhung impellers have both bearings on the same side of the casing; the bearing nearest the impeller is the inboard bearing, and the one farthest away the outboard bearing.

Ball bearings are the most common antifriction bearings used on centrifugal pumps. Roller bearings are used less often, although the spherical roller bearing is used frequently for large shaft sizes. Ball bearings used in centrifugal pumps are usually grease-lubricated, although some services use oil lubrication.

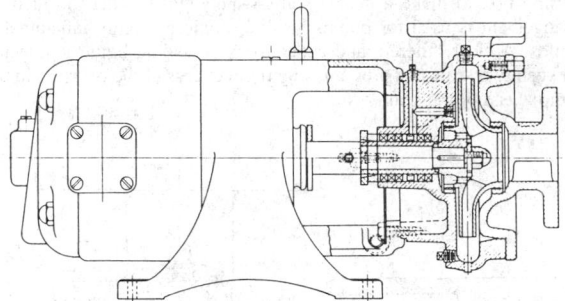

Fig. 14.2.14 Close-coupled (motor-mounted) pump.

Sleeve bearings are used for large, heavy-duty pumps with shaft diameters of such proportions that the necessary antifriction bearings are not commonly available. Another application is for high-pressure multistage pumps operating at speeds of 3,600 to 9,000 r/min. Still another application is in vertical submerged pumps, such as vertical turbine pumps, in which the bearings are subject to a water contact. Most sleeve bearings are oil-lubricated.

Thrust bearings used in combination with sleeve bearings are generally Kingsbury or Kingsbury-type bearings.

Couplings

Centrifugal pumps are connected to their drivers through **couplings** of one sort or another, except for close-coupled units (Fig. 14.2.14), in which the impeller is mounted on an extension of the shaft of the driver. Couplings used with centrifugal pumps can be either rigid (of the clamp or compressor type) or **flexible** (pin-and-buffer, gear, grid, or flexible-disk type). (See also Sec. 8.)

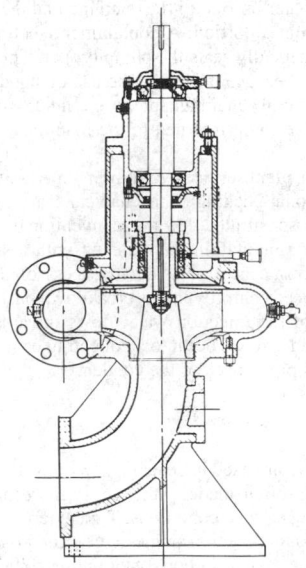

Fig. 14.2.15 Small vertical sewage pump with intermediate shafting.

Pump Mounting

It is desirable that pumps and their drivers be removable from their mountings. Consequently, they are usually bolted and doweled to machined surfaces that, in turn, are firmly connected to the foundations. These machined surfaces are usually part of a common **bedplate** on which the pump and its driver have been prealigned. Bedplates are made of either cast iron or structural steel. Cast-iron or steel **soleplates** are customarily used for vertical dry-pit pumps and also for some of the larger horizontal units.

Vertical Pumps

Dry-pit pumps with external bearings include most sewage pumps, most medium and large drainage and irrigation pumps for medium and high head, many large condenser-circulating and water-supply pumps, and many marine pumps. Some vertical dry-pit pumps are basically horizontal designs with minor modifications to adapt them for vertical-shaft drive. Other applications, such as small- and medium-sized sewage pumps, employ a purely vertical design. Most of these sewage pumps have elbow suction nozzles (Fig. 14.2.15) containing a handhole to provide easy access to the impeller. Although the driving motors are frequently mounted right on top of the pump casing, the use of vertical-shaft design permits mounting the motor at an elevation sufficiently above the pump to prevent accidental flooding. For such applications, the pump and its driver are separated by a length of shafting, which may require steady bearings between the two units.

Vertical wet-pit centrifugal pumps can be classified as (1) vertical turbine pumps, (2) propeller or modified propeller pumps, (3) sewage pumps, (4) volute pumps, and (5) sump pumps. The first of these is the most common type. **Vertical turbine pumps** (Fig. 14.2.16) are built with either closed or semiopen impellers and with either enclosed or open line shafting. The bowl assembly consists of the suction head, the impeller or impellers, the discharge bowl, the intermediate bowl or bowls, the discharge case, the various bearings, the shaft, and miscellaneous parts such as keys and impeller locking devices. The column-pipe assembly consists of the column pipe itself, the shafting above the bowl assembly, the shaft bearings, and the cover pipe or bearing retainers. The pump is suspended from the driving head, which consists of the discharge elbow, the motor or driver support, and either the stuffing box (in open-shaft construction) or the assembly for providing tension on and the introduction of lubricant to the cover pipe.

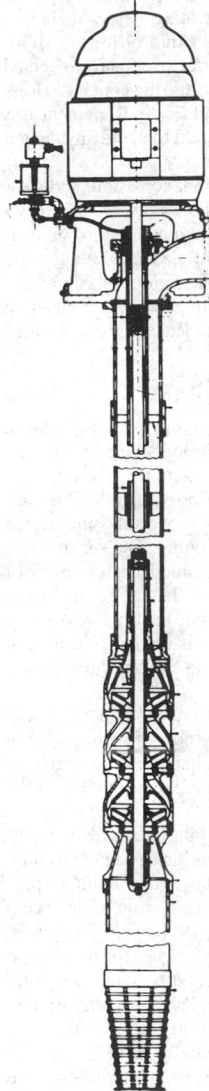

Fig. 14.2.16 Vertical turbine pump with closed impellers and oil-lubricated enclosed shafting.

MATERIALS OF CONSTRUCTION

Centrifugal pumps can be fabricated of almost any of the known common metals or metal alloys, as well as of porcelain, glass, and even synthetics. A listing of materials commonly recommended for various

Table 14.2.2 Materials for Various Fittings
(Materials for bearing housings, bearings, and other parts are not usually affected by the liquid handled)

Part	Standard fitting	All-iron fitting	All-bronze fitting
Casing	Cast iron	Cast iron	Bronze
Suction head	Cast iron	Cast iron	Bronze
Impeller	Bronze	Cast iron	Bronze
Impeller ring	Bronze	Cast iron or steel	Bronze
Casing ring	Bronze	Cast iron	Bronze
Diffuser	Cast iron or bronze	Cast iron	Bronze
Stage piece	Cast iron or bronze	Cast iron	Bronze
Shaft, with sleeve	Steel	Steel	Steel, bronze, or Monel
Without sleeve	Stainless steel or steel	Stainless steel or steel	Bronze or Monel
Shaft sleeve	Bronze	Steel or stainless steel	Bronze
Gland	Bronze	Cast iron	Bronze

liquids can be found in the Standards of the Hydraulic Institute. Table 14.2.2 indicates the materials most commonly used for various pump parts.

PUMP PERFORMANCE

The performance of a centrifugal pump is generally described in terms of the following of its **characteristics:** (1) rate of flow, or **capacity** Q, expressed in units of volume per unit of time, most frequently ft³/s, gal/min, or m³/h (1 ft³/s = 449 gal/min; 1 m³/h = 4.403 gal/min); (2) increase of energy content in the fluid pumped, or **head** H, expressed in units of energy per unit of mass, usually ft·lb/lb or, more simply, ft, or m; (3) input **power** P, expressed in units of work per unit of time, bhp; (4) **efficiency** η, the ratio of useful work performed to power input, (5) **rotative speed** N, in r/min.

Since the parameters indicated are all mutually interdependent, it is customary to represent the performance of a centrifugal pump by means of **characteristic curves** similar to that shown in Fig. 14.2.17. While it is possible, within certain limits, for the pump designer to regulate the shape of these curves to suit the needs of a particular application, this is essentially a function of design capacity, head, and speed and hence for

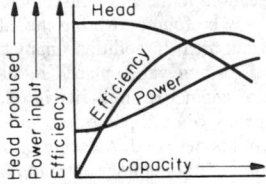

Fig. 14.2.17 Typical characteristic curves at constant speed.

a standard production unit is determined by experiment and is not subject to modification. For any given capacity on such a characteristic curve, the relationship between performance characteristics is expressed by the equation $\eta = \gamma QH/3{,}960P$, where η is efficiency expressed as a decimal, γ is the specific gravity of the fluid pumped, Q is in gal/min, H in ft, and P in bhp. In metric units, $\eta = \gamma QH/270P$, where Q is in m³/h and H in meters.

Inasmuch as the actual performance of centrifugal pumps is determined largely by experimental means, it is highly desirable to be able to use the results of past tests as a basis for predicting the performance of future designs. To this end, an interesting and widely used characteristic number known as **specific speed**, $N_s = N\sqrt{Q}/H^{3/4}$, has been developed. In this expression, values of N, Q, and H are all for the point of best efficiency. While not dimensionless, N_s is generally expressed simply as a number since its practical application is such that units are of no consequence except for their influence on the absolute magnitude of the number itself. Specific speed is of interest to both the pump designer and the pump user essentially in two ways: (1) All geometrically similar pumps, regardless of their size, will have identical specific speeds (but

all pumps of the same specific speed will not necessarily be geometrically similar). (2) Within reasonable limits, pump geometry and performance can be predicted as a function of N_s and Q. For N in r/min, Q in gal/min, and H in ft, the practical range of N_s is approximately 500 to 15,000. In metric units Q is in m³/s and H is in meters, so that this range

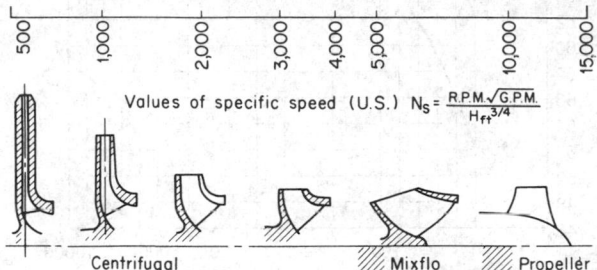

Fig. 14.2.18 Approximate relative impeller shapes versus specific speed.

becomes approximately 10 to 300 ($N_{sm} = N_s/51.66$) The shape of the profiles for impellers varies over the range of specific speeds, as shown in Fig. 14.2.18. The commercially attainable efficiencies also vary with the specific speed (Fig. 14.2.19). It must be remembered that these efficiencies are attainable at the capacity at best efficiency. The actual specified capacity may differ from this best-efficiency flow. In addition, the general shape of the pump-characteristic curves will vary widely from one end of this range to the other, as illustrated by Figs. 14.2.24 and 14.2.25.

Pump Theory

The basic purpose of a centrifugal pump in any fluid-handling system is to add energy to the fluid, and since it is a dynamic machine, the pump depends entirely on changes in velocity relationships to provide the energy. While the measurable evidence of energy addition is in most cases largely in the form of static pressure, this is partially the result of velocity reductions and constraints occurring in the diffuser or casing and to this extent represents a conversion from the velocity energy produced by the impeller. Thus, any discussion of centrifugal-pump theory generally becomes a discussion of velocities occurring at various points within the pump.

The true velocity relationships existing within a pump are extremely complex and, to a substantial degree, still unknown in their ultimate detail; but for practical purposes, a one-dimensional analysis serves to illustrate the basic concepts and, indeed, has served as the basis of design for virtually all centrifugal pumps ever built.

For radial and mixed-flow impellers ($500 \leq N_s \leq 7{,}500$ or $10 \leq N_{sm} \leq 150$) the velocities at the inlet and outlet of the impeller are shown by the vector diagrams in Fig. 14.2.20. The head produced by such an impeller is represented by

$$H = \eta_H(U_2 V_{u2} - U_1 V_{u1})/g \qquad (14.2.1)$$

where H = head, ft (m); U = circumferential velocity of impeller at radius being considered, ft/s (m/s), V_u = average value of the circumferential component of absolute fluid velocity, ft/s (m/s), and g = 32.174 ft/s² (9.80 m/s²). Subscript 1 refers to the impeller inlet section, and subscript 2 to the impeller discharge section. The coefficient η_H is

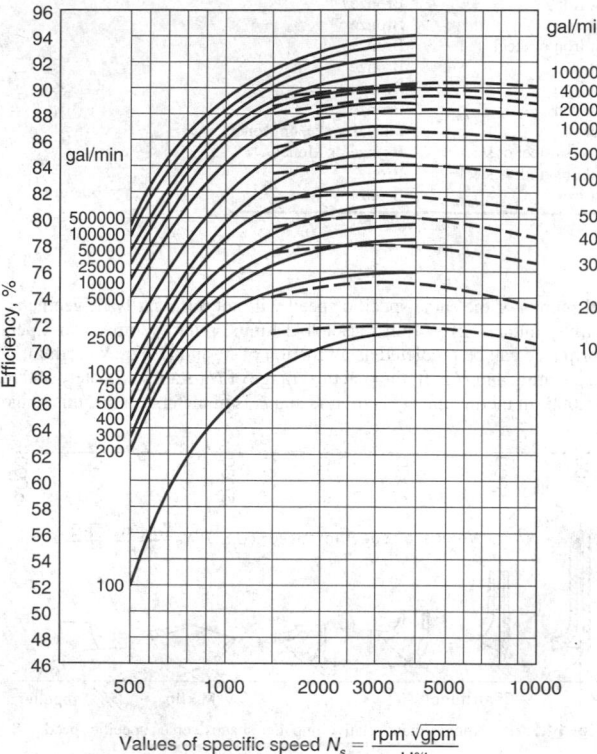

Fig. 14.2.19 Efficiencies of single-stage end-suction and double-suction centrifugal pumps (solid lines) and wet pit pumps (dashed lines).

the hydraulic efficiency of the rotating-vane system and, for the range of specific speeds indicated above, will generally fall between 0.85 and 0.95. This hydraulic efficiency is considerably higher than pump efficiency η since it does not include mechanical losses due to bearing or packing friction, volumetric losses due to internal wearing-ring clearances, impeller-disk friction, or fluid-friction losses due to velocity conversion or boundary-layer considerations ahead of or following the impeller. For pumps in this specific-speed range, the hydraulic losses $1 - \eta_H$ will generally be between one-quarter and three-quarters of total pump losses $1 - \eta$.

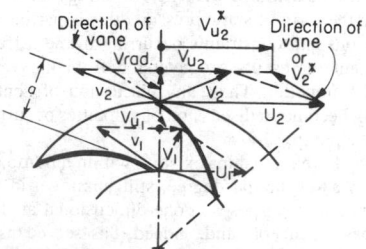

Fig. 14.2.20 Velocity diagram of a radial-flow impeller.

For pumps arranged with an axial inlet to the impeller (such as that shown in Fig. 14.2.5), it is generally assumed that the entering flow will

have no rotational component, and V_{u1} is therefore zero. Equation 14.2.1 can thus be reduced to

$$H = \eta_H U_2 V_{u2}/g \qquad (14.2.2)$$

In practice, Eq. (14.2.2) will provide a close approximation to total head for any pump up to a specific speed of 2,000 ($N_{sm} \approx 40$) since the term $U_1 V_{u1}$ in Eq. (14.2.1) is very small compared with $U_2 V_{u2}$.

It should also be noted in Fig. 14.2.20 that the relative velocity v_2 does not coincide in direction with the vane angle at the impeller discharge. The angular difference between v_2 and the direction of the vane is due to the irrotational nature of the flow between the vanes. The effect of this difference can be taken into account as the vector difference $V_{u2}* - V_{u2} = Na/229$, where a is the shortest distance in inches taken in the radial plane between the discharge tip of any vane and the upper surface of the following vane. ($V_{u2}* - V_{u2} = Na/19,100$, where a is in mm and velocities are in m/s.)

The foregoing relationships provide the basis for determining impeller diameters and vane angles required to produce a given total head requirement at a specified rotative speed. It is equally necessary, of course, to provide in the design for handling a specified flow volume, and this is readily accomplished by providing the necessary cross-sectional area A between vanes to pass the required flow at velocities previously determined. A useful relationship for this purpose, in light of the units commonly employed in pump design, is $Q = AV/0.321$, where Q is in gal/min, V in ft/s, and A in in² ($Q = AV/278$, where Q is in m³/h, V in m/s, and A in mm²).

Upon leaving the impeller, the liquid pumped enters either (1) a system of diffusing vanes surrounded by an outer casing or (2) directly into a casing designed to contain the fluid and control its velocities. Where diffusing vanes are used, they are designed on the basis of velocity relationships very similar to those employed in impeller design but with the objective being to slow the fluid down to convert velocity energy to pressure energy and, further, to reduce frictional losses in the discharge system following the diffuser. Where the impeller discharges directly into the casing, this component of the machine is most frequently designed in the form of a volute to provide constant velocity all around the impeller periphery up to the point of entry into the discharge nozzle. From this point, commonly called the **casing throat,** to the discharge flange or to the inlet of a succeeding stage, the velocity is gradually reduced. Special circumstances related to pump design or application often result in modifications to the constant-velocity design of such a casing, and variations may be found covering the entire range from constant velocity to constant area. In addition, many casings are now designed with one or more spiral vanes placed in such a way as to approximate a condition of geometric similarity in relation to the impeller, which is advantageous when a pump is operated at capacities other than those for which it is designed. A casing of this nature represents an effort by the designer to obtain an optimum balance between the desirable geometric similarity of the diffuser discharge and the manufacturing simplicity and generally high efficiency of the volute-type casing. The most common form of such a casing is the twin-volute type discussed under Nomenclature and Mechanical Design (see Fig. 14.2.1 and Table 14.2.1).

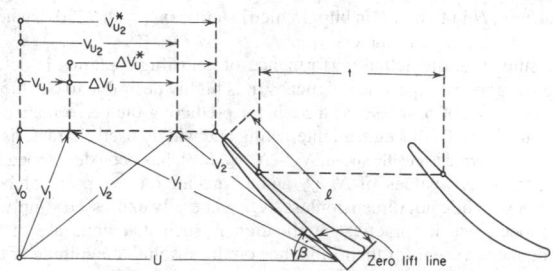

Fig. 14.2.21 Velocity diagram of an axial-flow vane system.

For **axial-flow impellers** ($7{,}500 \leq N_s \leq 15{,}000$ or $150 \leq N_{sm} \leq 300$) velocity relationships can be approximated in a manner similar to that used for lower-specific-speed pumps, but refinement of these approximations is approached in a somewhat different manner, largely because of the considerable body of knowledge available in the form of airfoil data which can be applied. Velocity diagrams for pumps of this type are shown in Fig. 14.2.21.

Considering a cylindrical stream tube intersecting the vanes of an axial-flow impeller, we can rewrite Eq. (14.2.1) in the form

$$H = \eta_H U \, \Delta V_u / g \qquad (14.2.3)$$

where ΔV_u represents the increase in the tangential component of the absolute velocity as the fluid passes through the impeller. For pumps in this specific-speed range, η_H will generally fall between 0.80 and 0.90 and $1 - \eta_H$ will generally be between one-half and three-quarters of $1 - \eta$.

As in the case of radial-flow impellers, the relative velocity v_2 at the impeller exit does not coincide with the vane angle. In this case, the necessary correction can be applied by means of the expression

$$\Delta V_u = 2\Delta V_u^* / [(t/l)(2/\pi K)(1/\sin \beta) + 1] \qquad (14.2.4)$$

where t is vane spacing, l is vane length, β is the discharge vane angle, and K is the coefficient determined from Fig. 14.2.22, which provides for the fact that the impeller blades are, in effect, arranged in a continuous lattice.

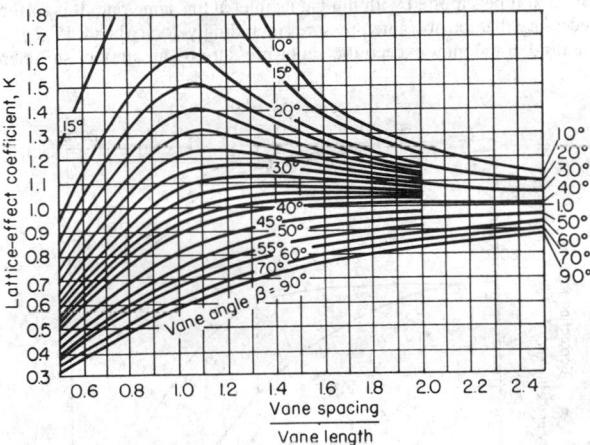

Fig. 14.2.22   Lattice-effect coefficient. *(Weinig.)*

In the design of axial-flow pumps, it is generally assumed that the head developed by the blade elements within all the cylindrical stream tubes between the impeller hub and its outer diameter will be the same. For this condition to be achieved, it will be evident from Eq. (14.2.3) that since U will vary directly with the radius, ΔV_u must vary inversely with the radius. Thus vane camber (or curvature) will be greater near the hub than at the periphery. It is further generally assumed that the axial velocity is constant throughout the impeller, and to satisfy this condition, the blade angles will be greater at the hub than at the periphery, giving rise to the twist of the vane.

By designing the vane element at the hub, the periphery, and any reasonable number of radial stations between, it is possible to define the vane over its entire surface to produce the total head desired. The design capacity can be simply established from the constant axial velocity and the annular area between the hub and the periphery.

Axial-flow impellers will almost invariably discharge into a vaned diffuser designed primarily to convert into pressure the tangential component of the absolute velocity leaving the impeller, thus producing a uniform, nonrotating velocity profile at the pump discharge or at the entrance to any succeeding stage.

Pump Application

In applying any centrifugal pump to a practical fluid-handling system, the engineer generally has two variables in the pump which may be used to effect a match between it and the system, namely, speed of operation and impeller diameter. (In axial-flow pumps, impeller diameter cannot conveniently be changed, but similar results can be accomplished within a limited range by reducing vane length.) To take advantage of these variables, it is necessary to understand the similarity relationships which govern pump behavior.

The effect of change in speed can be most readily explained by referring back to Eq. (14.2.1). For a fixed impeller diameter, it will immediately be evident that any increase in rotative speed will result in a directly proportional increase in the peripheral velocities of the impeller at both inlet and outlet, U_1 and U_2, respectively. Furthermore, since the directions of both the relative and absolute velocities v and V, respectively, are controlled by the vane angles, it follows that both the inlet and outlet velocity diagrams remain geometrically similar, in other words, as pump speed changes, all fluid velocities change in direct proportion. Thus the effect of a speed change on pump capacity can be represented for two speeds, N_1 and N_2 by

$$Q_1/Q_2 = N_1/N_2$$

Referring again to Eq. (14.2.1), it will be evident that the change in head will be proportional to the square of the change in speed, resulting in

$$H_1/H_2 = N_1^2/N_2^2$$

Also, since P is proportional to QH, then

$$P_1/P_2 = Q_1 H_1 / Q_2 H_2 = N_1^3/N_2^3$$

The effect of a reduction in impeller diameter is for practical purposes, identical to that of a reduction in speed, and the above equations can all be rewritten in the same form, substituting D_1 and D_2 for N_1 and N_2. In this case, however, there is somewhat less latitude for change, the practical limit for cut-down being approximately 25 to 30 percent in the case of low-specific-speed impellers and decreasing from this value as specific speed increases.

From the foregoing it should be noted that running a pump at speeds far below its normal rated speed would be uneconomical since the pump would obviously be overdesigned for the service. Conversely, the same pump could not be run at speeds far above its rating since it would soon exceed its horsepower capability. Thus the normal approach to pump application calls for selection of a unit which will meet the system requirements at or near the pump's maximum rated speed and preferably as close to maximum impeller diameter as possible.

A further aspect of the similarity laws, of interest primarily to pump designers, is the matter of geometrically similar pumps, or as they are commonly called, **factors of each other.** Assuming two such pumps with a size ratio f_1/f_2, then **at the same rotative speed** $H_1/H_2 = f_1^2/f_2^2$, which is the same as for a change in speed of the same ratio. In the case of capacity, however, we have, in addition to the linear increase due to velocity change, a further increase due to the change in the cross-sectional area of all fluid channels which is proportional to the square of the size factor, resulting in the relationship $Q_1 Q_2 = f_1^3/f_2^3$. The ratio of horsepower requirements therefore becomes $P_1/P_2 = f_1^5/f_2^5$.

Lines of geometrically similar pumps are more frequently designed, however, for constant head than for constant speed, and to have this condition apply for the size ratio indicated in the preceding paragraph, it is necessary that the speeds vary in exact inverse proportion to size ratio. Thus, **for constant head,** $N_2/N_1 = f_1/f_2$ and

$$H_1/H_2 = 1 = (f_1/f_2)^2(N_2/N_1)^2$$
$$Q_1/Q_2 = (f_1/f_2)^3(N_2/N_1) = (f_1/f_2)^2$$
$$P_1/P_1 = (H_1/H_2)(Q_1/Q_2) = (f_1/f_2)^2$$

To simplify selection of pumps, it is customary to plot, for any given speed, performance curves to show pump characteristics over the available range of impeller diameters rather than at the single diameter which would be implicit in a curve of the type shown in Fig. 14.2.17. A typical rating curve of this nature is shown in Fig. 14.2.23. Such rating curves will vary widely in their nature with changes in N_s, and an understanding of these variations is essential to the selection of properly sized pump drivers and, in many cases, proper design of discharge piping and/or the determination of limiting ranges of pump operation. Some

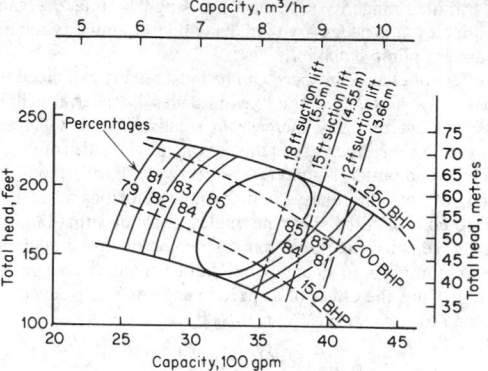

Fig. 14.2.23 Rating curve of 10-in double-suction single-stage pump.

idea of the diversity of characteristics which is available can be obtained by a comparison between Figs. 14.2.24 and 14.2.25, both of which are plotted entirely in percentages of design values occurring at the point of best efficiency.

In the case of the lower-specific-speed pump shown in Fig. 14.2.24, a

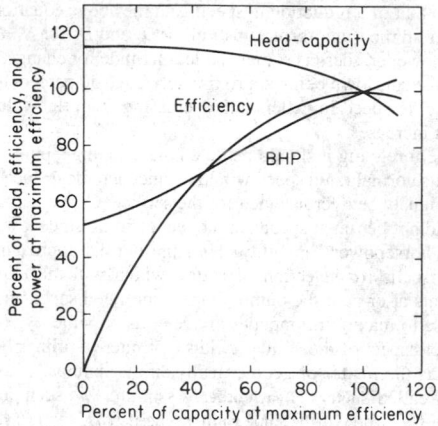

Fig. 14.2.24 Type characteristics for $N_s = 1,550$ single-suction impeller ($N_{sm} = 30$).

driving motor selected for the rated condition would be more than adequate at lower capacities, and discharge piping would be subjected to only modest overpressures. On the other hand, if the pump shown in Fig. 14.2.25 is to be operated at reduced capacities, then both motor size and discharge piping must be designed for the minimum flows to be encountered.

Selection of drivers, as well as pumps, is also influenced by **properties of the fluid** handled. Pump horsepower varies directly with the specific gravity of the liquid and is influenced in a more complex manner by viscosity. The effect of the latter is illustrated in Fig. 14.2.26. For the example shown on the chart, the pump in question is rated for 750 gal/min (170 m³/h) at 100-ft (30-m) head. When handling a fluid having a viscosity of 1,000 SSU (220 cS), its capacity is reduced to 95 percent of that on water over its complete range of operation. Its head is reduced to 96, 94, 92, and 89 percent of its head on cold water at 60, 80, 100, and 120 percent of these reduced capacities, and its efficiency is

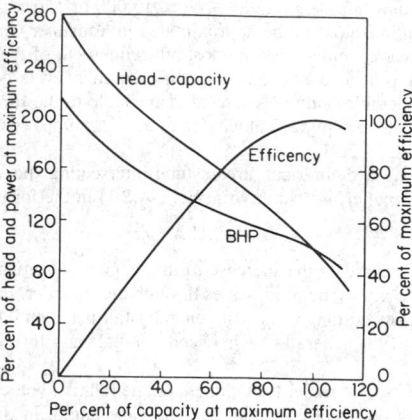

Fig. 14.2.25 Type characteristics for $N_s = 10,000$ single-suction impeller ($N_{sm} = 194$).

reduced to 63.5 percent of its efficiency on water over its reduced capacity range. Thus, if the pump had a rated efficiency on cold water of 70 percent, it would now deliver 712.5 gal/min (161.8 m³/h) at 92-ft (28.0-m), head with an efficiency of 44.5 percent.

In order for a pump to deliver its rated output, it is obviously necessary that it be supplied with fluid at its inlet at the same rate. It is further necessary that the absolute pressure (including velocity head, $V^2/2g$) of the fluid at the inlet exceed the vapor pressure by an amount sufficient

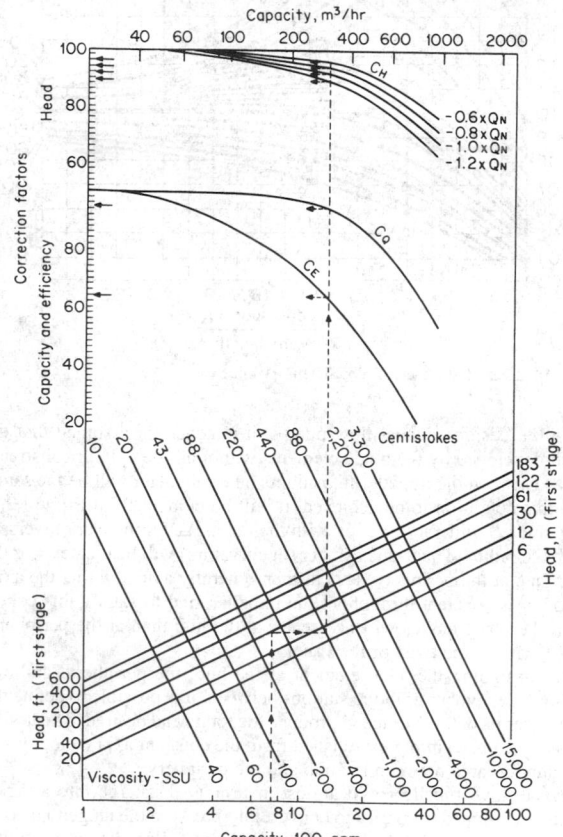

Fig. 14.2.26 Performance-correction chart for effect of viscosity. (*Hydraulic Institute.*)

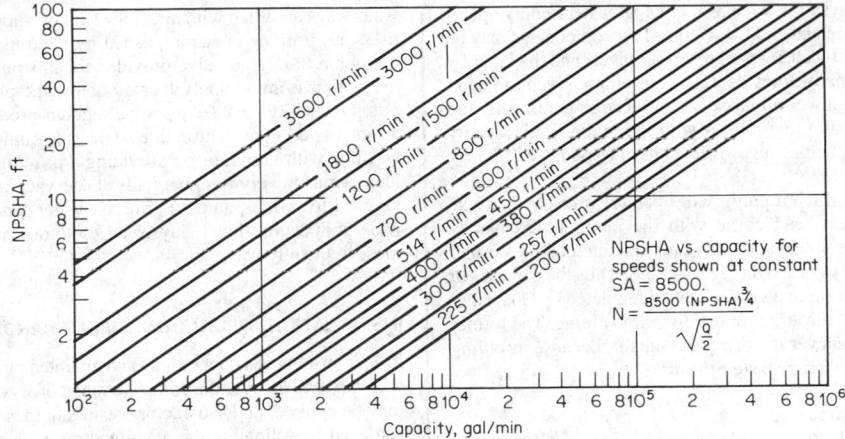

Fig. 14.2.27 Recommended maximum operating speeds for double-suction pumps. *(Hydraulic Institute Standards.)*

to overcome (1) any entrance or frictional losses between the point of entry into the pump and the impeller, and (2) the shock losses occurring at the impeller inlet. This gives rise to the definition of **net positive suction head (NPSH)** which is the absolute pressure at the pump inlet expressed in feet of liquid, plus velocity head, minus the vapor pressure of the fluid at pumping temperature, and corrected to the elevation of the pump centerline in the case of horizontal pumps or to the entrance to the first-stage impeller for vertical pumps.

NPSH required is determined by the pump manufacturer and is a function of both pump speed and pump capacity. **NPSH available** represents the energy level of the fluid over the vapor pressure at the pump inlet and is determined entirely by the system preceding the pump. Unless NPSH available at least equals NPSH required at any condition of operation, some of the fluid will vaporize in the pump inlet and bubbles of vapor will be carried into the impeller. These bubbles will collapse violently at some point downstream of the pump inlet (usually at some point within the impeller) and produce very sharp, crackling noises, frequently accompanied by physical damage of adjacent metal surfaces. This phenomenon is known as **cavitation** and is generally highly undesirable.

A term similar to that used for pump specific speed has been developed for pump inlet characteristics and has been identified as **suction specific speed**, $S = N \sqrt{Q}/H_{sv}^{3/4}$, where H_{sv} is NPSH. For double-suction impellers, $S = N \sqrt{Q/2}/H_{sv}^{3/4}$. The value S, like N_s, is expressed simply as a number and, for general-purpose pumps, does not generally exceed 10,000, where N is in r/min, Q is in gal/min, and H_{sv} is in ft. When Q is in m³/s and H_{sv} is in meters, the equivalent value S_m is 194.

Within the limits of NPSH capabilities, it is desirable to select a pump of the highest specific speed since this will produce the highest pump speed and consequently use the smallest pump. As a convenience in accomplishing this, the Hydraulic Institute has published two charts indicating the recommended NPSH as a function of pump capacity and operating speed and, conversely, the maximum recommended operating speed for different capacities if the NPSH value is known or first selected. These charts are based on a recommended value of suction specific speed of 8,500 ($S_m = 165$). The chart for double-suction pumps is reproduced in Fig. 14.2.27.

Recirculation A better understanding has been recently reached of the flow pattern in a pump impeller when the pump is operated at capacities below that at its best efficiency. All pumps exhibit the phenomenon of recirculation which is a flow reversal at the inlet or at the discharge tips of the impeller. The capacities at which flow recirculation occurs at the suction or discharge are not necessarily coincidental. It is now possible to predict the flow patterns that must exist to produce this flow reversal. Depending on the size and speed of the pump, the effects of recirculation can be very damaging not only to the operation but also

to the life of the impeller and casing. The symptoms associated with recirculation and a method for calculating the capacity at which it takes place at both the suction and discharge of the impeller are given in Fraser, "Recirculation in Centrifugal Pumps," ASME Winter Annual Meeting, Nov. 16, 1981, Washington, D.C.

Just as the head-capacity characteristic of a pump is represented by a curve indicating reducing head developed with increasing capacity, the head-capacity requirements of the system served by the pump can be depicted by a curve showing increased head requirements for increased flow, i.e., a **system-head curve** (Fig. 14.2.28). The capacity at which a pump will operate in this system is that at which the pump head-capacity curve intersects the system-head curve.

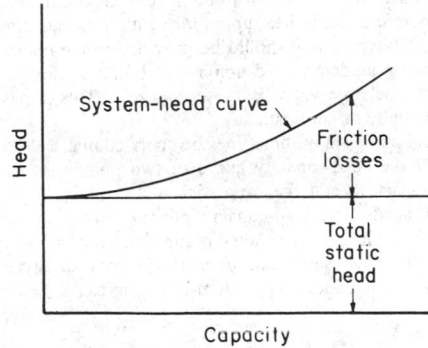

Fig. 14.2.28 System-head curve.

Every system-head curve consists of (1) a total static head (which may in some cases be equal to zero) plus (2) friction losses. The total static head $H_{stat} = H_d - H_s + P_d - P_s$, where H_d and H_s are the elevations at the system's termination and origin, respectively, and P_d and P_s are the gage pressures at the same points. In the special case of a continuous freshwater system where $P_d = P_s = $ zero, i.e., atmospheric, it must be remembered that siphon recovery cannot exceed 1 atm (34 ft less vapor pressure at sea level) and that recovery will not occur if the siphon leg is broken. Thus, any system over this height will have a total static-head component at least equal to the difference between its highest elevation and H_s minus the static-head equivalent of 1 atm plus the vapor pressure, even where $H_d = H_s$.

The **friction losses** H_{lr} are the sum of the line losses H_1, fitting losses H_f, and changes in velocity head across the system, H_v. (See Sec. 3 for calculation methods.) H_v is simply the velocity-head difference between the system's termination and origin; $H_v = V_d^2/2g - V_s^2/2g$.

In complex pumping systems, such as municipal water-supply operations, both the total static-head and the friction-loss components may be variable, e.g., the former by changes in reservoir levels and the latter by the particular combination of lines being served at any given moment. This results in upper and lower limits of system-head requirements, and the intersection of these limiting curves with the pump head-capacity curve will define the capacity range over which the pump will be required to operate.

Since the capacity at which a pump will operate corresponds to the intersection of the system-head curve with the pump head-capacity curve, any changes in pump capacity can only be obtained by varying one or the other of these two curves (Fig. 14.2.29). Thus the capacity of a centrifugal pump operating in a system can be regulated by (1) changing the pump speed or (2) throttling in the discharge piping. The former method is preferable whenever the driver permits it, because throttling always involves an appreciable waste of power.

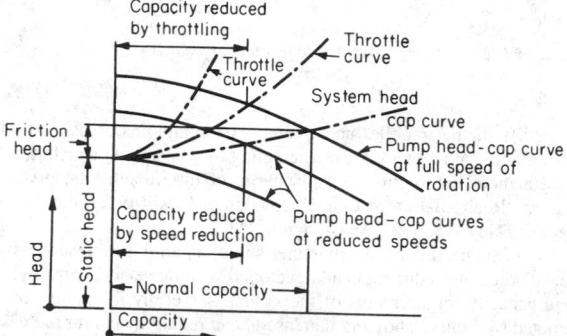

Fig. 14.2.29 Pump operation in a system.

Pump capacity should never be permitted to be reduced to zero because the fluid within the pump would have to absorb the entire power input and therefore would heat up rapidly with injurious effects on the pump. A small bypass line should be provided in the pump discharge line to permit a predetermined amount of liquid to flow through the pump if the discharge valve is closed entirely. This bypass may be operated manually or automatically.

Centrifugal pumps may sometimes be operated in **parallel** or in **series**. To construct the head-capacity curve of two pumps in parallel, it is merely necessary to add the capacities of the individual pumps for various total heads. The head-capacity curve of two pumps in series is constructed by adding the individual pump heads for various capacities. Figure 14.2.30 shows series and parallel operation of two centrifugal pumps with both flat and steep system-head curves.

Priming

A centrifugal pump is primed when the waterways of the pump are filled with the liquid to be pumped. When first put into service, the

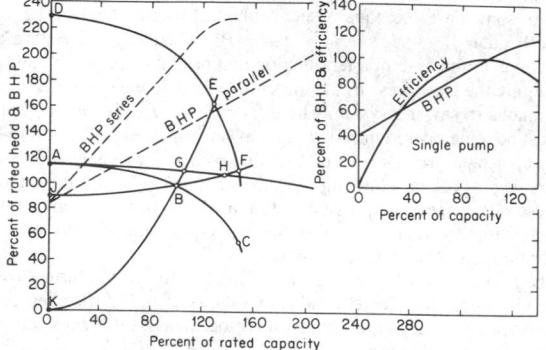

Fig. 14.2.30 Series and parallel operation of centrifugal pumps.

waterways are filled with air. If the suction supply is above atmospheric pressure, priming is accomplished by venting the entrapped air out of the pump through a valve provided for this purpose. If the pump takes its suction from a supply located below the pump itself, the air in the pump must be evacuated by some vacuum-producing device, by placing a foot valve in the suction line so that the pump and suction piping can be filled with liquid, or by providing a priming chamber in the suction line. Almost every commercially made vacuum-producing device can be used to prime pumps. Formerly, water- and steam-jet primers had wide application, but today electric-motor-driven vacuum pumps are most frequently used.

INSTALLATION, OPERATION, MAINTENANCE

Proper installation, operation, and maintenance of centrifugal pumps will vary widely over the complete range of services to which the pumps may be applied, and satisfactory results in these areas can only be fully achieved by following the manufacturer's instructions for the size and type of unit involved. There are, however, certain general considerations which should be observed and which will seldom need to be modified under any circumstances.

In general, the location selected for installation should be as close to the source of the fluid as possible, consistent with the requirement that adequate space be made available to provide accessibility for operation, inspection, and maintenance. The pumping unit should be mounted on a foundation of sufficient size and rigidity to support the unit itself plus the weight of the fluid it will contain during operation and to maintain accurate alignment. Piping should be independently supported and anchored to avoid imposing stresses on the pump, and suction piping in particular must be designed to minimize friction losses and to present a uniform velocity profile at the pump inlet. Suction and discharge (and/or check) valves must be suitable for the pressures involved and, in the case of large units, may also require independent support. If the pump will be required to operate against a suction lift, a suitable priming system must be installed, and where it is to be provided with a head-on suction, it will often be necessary to provide a venting arrangement. Care must also be exercised to ensure that all auxiliary connections for sealing, cooling, flushing, and drainage are made as required for the particular unit being installed.

Prior to initial operation of any centrifugal pump, it is necessary to make sure that the driver is connected to provide proper direction of rotation, that any shaft couplings between separate components of the entire unit are aligned within the manufacturer's stated limits, and that all bearings are provided with the proper amounts and grades of lubricants. The normal starting sequence will then be as follows: (1) open valves in all auxiliary sealing, cooling, flushing, and bypass lines; (2) open suction valve; (3) close discharge valve for low-specific-speed pumps where no check valve is installed after the pump, or open discharge valve for high-specific-speed pumps or wherever a discharge check valve has been provided; (4) prime or vent as necessary; (5) energize the driver; and (6) open discharge valve if it was previously closed in step 3.

Following start-up and until proper operation has been adequately established, it is desirable to monitor bearing temperature, stuffing-box leakage, and other outward symptoms of the unit's behavior. Securing of the pump is accomplished by a reversal of the start-up sequence, encompassing steps 6, 5, 3, and 1, in that order.

Until the early 1970s there were only four factors to consider when setting an acceptable minimum flow for centrifugal pumps:

Higher radial thrust developed by single-volute pumps at reduced flows.

Temperature rise in the liquid pumped as capacity is reduced.

Desire to avoid overload of drivers on high-specific-speed axial-flow pumps, which have a rising power consumption as capacity is reduced.

For pumps handling liquids with significant amounts of dissolved or entrained air or gas, there is the need to maintain sufficiently high velocities in the pump casing to wash out this air or gas with the liquid.

Since then, the phenomenon of internal recirculation in the impeller has been discovered. The recirculation occurs at flows less than the capacity at best efficiency. Problems include pressure pulsations at the suction and discharge and rapid deterioration of the pump and casing. The unfavorable effects of this internal recirculation have brought a fifth factor into the picture when setting minimum flows. This is discussed more fully above under "Recirculation."

Each of the effects mentioned above may dictate a different minimum operating capacity. Obviously, the final decision must be based on the greatest of the individual minimums. The internal recirculation usually sets the recommended minimum. Assuming that the pump is properly furnished with the necessary instrumentation, such as flowmeters, pressure gages with sufficient sensitivity to show pulsations, and vibration- and noise-monitoring equipment, an experienced test engineer should be able to pinpoint the onset of internal recirculation.

If the flow demand of the service in which the pump is installed is expected to fall below the minimum permissible flow, some means must be provided to accommodate the difference between the minimum permissible flow and that required by the service. This is accomplished by installing a bypass in the discharge line from the pump, located on the pump side of the check and gate valves and leading to some lower pressure point in the installation where excess heat absorbed through operation at low flows may be dissipated. A valve is located in this bypass line. For small installations, the bypass valve is operated as fully open or fully closed. For large installations, a modulating valve is usually used to bypass the difference between the required flow and the minimum permissible.

In the matter of pump maintenance, a generally accepted cardinal rule is that as long as operation continues normal, the unit should be left alone. Thus, except in special circumstances, periodic overhauls are not recommended. The amount and degree of maintenance likely to be required are influenced primarily by the nature of the service to which the pump is applied, and maintenance practices must therefore be determined largely by the user as a result of personal experience.

14.3 COMPRESSORS

James L. Bowman

REFERENCES: Chlumsky, "Reciprocating and Rotary Compressors," E&FN Spons Ltd., Loomis, "Compressed Air and Gas Data," 3rd ed., Ingersoll-Rand. Rollins, "Compressed Air and Gas Handbook," 5th ed., Compressed Air and Gas Institute. Sheppard, "Principles of Turbomachinery," Macmillan. Stepanoff, "Turboblowers," Wiley. Koppers Research and Engineering Staff, "Engineers Handbook of Piston Rings, Seal Rings, Mechanical Shaft Seals," 8th ed., Koppers Company, Inc. Lundberg and Glanvall, "A Comparison of SRM and Globoid Type Screw Compressors at Full Load," Proceedings of the 1978 Purdue Compressor Technology Conference, Purdue University, W. Lafayette, IN. Von Nimitz, "Pulsation and Vibration Control Requirements in the Design of Reciprocating Compressor and Pump Installations," Proceedings of the 1982 Purdue Compressor Technology Conference, Purdue University, W. Lafayette, IN. Mechanics, Simulation and Design of Compressor Valves, Gas Passages and Pulsation Mufflers, short course, July 12–14, 1992, Purdue Univeristy, West Lafayette, IN.

Notation

a = speed of sound
A = piston area
A_v = valve flow area
c_p = constant pressure specific heat
c_v = constant volume specific heat
C = clearance volume (decimal)
d = depth of ring section, diameter
D = diameter
e = eccentricity
E = modulus of elasticity (lb/in^2)
f = valve resistance, in velocity heads
g = ring free gap less end clearance (in)
g_c = gravitational constant
ghp = gas horsepower
h = enthalpy
H_p = polytropic head (ft·lb/lbm or kJ/kg)
k = ratio of specific heats
K = area constant (decimal)
L = length, leakage coefficient, connecting rod length
L_{eq} = equivalent length of pipe
m = mass
MW = molecular weight
n = polytropic coefficient
N_m = number of lobes on main rotor
n_s = number of stages

n_v = number of vanes
N = rotational speed
N_s = specific speed
P = pressure
Q = heat flow, volume flow at inlet conditions
r = compression ratio (P_2/P_1)
r_s = pressure ratio per stage
r_t = pressure ratio across entire compressor
r_δ = ratio of cylinder pressure at bottom dead center to inlet pressure
R = gas constant [for air 53.34 ft·lbf/(lb$_m$·R)], crank radius
s = entropy, stress
sg = specific gravity relative to air
T = absolute temperature, thrust
T_v = vane thickness
U = piston velocity, peripheral velocity
V = volume
V_i = built-in volume ratio
$\overline{V}$ = velocity
W = work, mass flow
Z = compressibility factor

Greek

α = ratio of piston area over valve flow area
β = blade angle, degrees
λ = wave length
Δ = pressure loss, differential pressure (lb/in^2)
η_c = adiabatic efficiency
η_p = polytropic efficiency
η_v = volumetric efficiency
μ = pressure coefficient (gH_p/U^2)
ϕ = flow coefficient (Q/ND^3)
ω = angular speed (radians/second)

Subscripts

1 = inlet conditions
2 = discharge conditions
A = axial
R = radial

Compressed-Air and Gas Usage

Compressed-air is used for machine and tool operation, drilling, painting, soot blowing, pneumatic conveying, food processing, instrument operations, and in situ operations (e.g., underground combustion). Pressures range from 25 psig (172 kPa) to 60,000 psig (413,790 kPa). The largest usage is at 90 to 110 psig which is the normal plant air-pressure range.

Gas compressors are used for refrigeration, air conditioning, heating, pipeline conveying, natural gas gathering, catalytic cracking, polymerization, and in other chemical processes.

Standard Units and Conditions

In the ISO system, the standard unit of pressure for compressors is the kilopascal (kPa). In some countries this is the only unit which can appear on the compressor pressure gages by law. In Europe, the European Committee of Manufacturers of Compressors, Vacuum Pumps and Pneumatic Tools (PNEUROP) and, in the United States, the Compressed Air and Gas Institute (CAGI) prefer the bar as the standard unit of pressure. PNEUROP and CAGI have selected as standard conditions 1 bar (14.5 lb/in²) (100 kPa), 20°C (68°F), and 0 percent relative humidity. The unit of flow in the ISO system is m³/s. Other units still in common usage are m³/h, m³/min, and L/s. In the United States the most commonly used units are ft³/min (cfm) and ft³/h (cfh). Power is normally expressed in kilowatts (PNEUROP) and horsepower (CAGI).

Thermodynamics of Compression

Most compressors are analyzed using the ideal-gas law and assuming constant specific heat. Real-gas deviations are handled by applying a compressibility factor (also called *super compressibility*). The ideal-gas law will give satisfactory results for nonhydrocarbon gases for pressures to approximately 1,000 psig (6,900 kPa) at normal temperatures. Most hydrocarbon gases and refrigerants deviate strongly from the ideal-gas laws even at moderate pressures. For these cases, thermodynamics property tables, Mollier Charts, or compressibility charts should be used. Unfortunately these are not available for many gases of industrial importance. In this case a generalized compressibility chart is used. Compressibility charts for various gases can be found in Loomis, "Compressed Air and Gas Data," 3d ed., Ingersoll-Rand, and Rollins, "Compressed Air and Gas Handbook," 5th ed., Compressed Air and Gas Institute.

Adiabatic Analysis

The equation of state for an ideal gas is:

$$PV = mRT \qquad (14.3.1)$$

The first law of thermodynamics for a steady-flow process is (per unit mass):

$$Q = (h_2 - h_1) + \frac{\overline{V}_2^2 - \overline{V}_1^2}{2g_c} + W \qquad (14.3.2)$$

Neglecting the kinetic energy of the gas and assuming constant specific heat, for a reversible adiabatic process, Eq. (14.3.2) becomes:

$$W = c_p(T_1 - T_2') \qquad (14.3.3)$$

Substituting $(P_2/P_1)^{(k-1)/k}$ for T_2'/T_1 and $k/(k-1)$ for c_p/R, Eq. (14.3.3) becomes:

$$W = P_1V_1 \frac{k}{1-k}\left[\left(\frac{P_2}{P_1}\right)^{(k-1)/k} - 1\right] \qquad (14.3.4)$$

The same results can be obtained by evaluating $W = \int V \, dp$ on an idealized P-V diagram (Fig. 14.3.1).

The isentropic efficiency (ratio of isentropic work to actual work) is given by:

$$\eta_c = T_1 \frac{(P_2/P_1)^{(k-1)/k} - 1}{T_2 - T_1} \qquad (14.3.5)$$

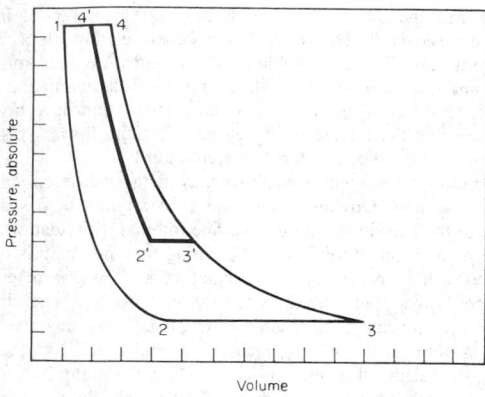

Fig. 14.3.1 Idealized indicator diagram for a positive-displacement compressor.

Polytropic Process

As an alternative to the adiabatic analysis, a polytropic process can be defined as

$$PV^n = C \qquad (14.3.6)$$

All the adiabatic equations apply if $(n-1)/n$ is substituted for $(k-1)/k$ where

$$\frac{n-1}{n} = \frac{k-1}{k}\frac{1}{\eta_p}$$

If the inlet and discharge temperatures and pressures are known, n can be calculated from:

$$n = \frac{1}{1 - \ln(T_2/T_1)/\ln(P_2/P_1)} \qquad (14.3.7)$$

where n is directly related to the heat transferred during the process by

$$n = \frac{C_p(T_2 - T_1) - Q}{C_v(T_2 - T_1) - Q} \qquad (14.3.8)$$

The reversible work done along the polytropic path is called the **polytropic work.**

The polytropic efficiency η_p (ratio of polytropic work to actual work) is given by:

$$\eta_p = \frac{\dfrac{k-1}{k}\ln(P_2/P_1)}{\ln(T_2/T_1)} \qquad (14.3.9)$$

The polytropic efficiency and the isentropic efficiency can be related by:

$$\eta_c = \frac{r^{(k-1)/k} - 1}{r^{(k-1)/(k\eta_p)} - 1} \qquad (14.3.10)$$

Equation (14.3.10) is strictly valid only for ideal gases with constant specific heats. However, because the real-gas errors occur in both the numerator and denominator, the equation gives reasonably accurate results for real gases which deviate moderately from ideal gases.

Positive-displacement compressors are almost universally analyzed using the adiabatic model. Dynamic compressors are analyzed using both the adiabatic and polytropic models with the trend toward the polytropic model. The advantage of the polytropic model is that the discharge temperature is immediately available and η_p is essentially constant for different gases. The polytropic efficiency is independent of the thermodynamic state of the gas. Also, the total polytropic head is the sum of the polytropic heads for the stages. This is not the case for isentropic heads. The advantage of the isentropic model is that the isentropic work can be read immediately from thermodynamic tables or charts. Comparing adiabatic efficiencies at different pressure ratios is not a valid procedure.

Real-Gas Effects

Deviations from the ideal-gas law are accounted for by introducing a compressibility factor

$$Z = \frac{PV}{RT} \qquad (14.3.11)$$

The isentropic work of compression for a real gas becomes:

$$W = P_1 V_1 \frac{k}{1-k} \left[\left(\frac{P_2}{P_1} \right)^{(k-1)/k} - 1 \right] \frac{Z_1 + Z_2}{2Z_1} \qquad (14.3.12)$$

where Z_1 and Z_2 are the compressibilities at conditions 1 and 2. Compressibility will also affect the volume flow of the compressor because of the reexpansion of the clearance volume gas.

Mutistaging and Intercooling

Temperature rise and mechanical stresses limit the maximum pressure differential across a single stage of any compressor type. The pressure rise across a dynamic compressor stage is further limited by the available polytropic head which the stage can develop. The volumetric efficiency of a reciprocating compressor decreases with increasing pressure ratio placing a practical limit on the maximum ratio per stage.

Multistaging is used to overcome these limitations and to save power. With perfect intercooling and no pressure losses between stages, theoretically the minimum power is obtained when

$$r_s = \sqrt[n_s]{r_t} \qquad (14.3.13)$$

where n_s = the number of stages; r_s = the pressure ratio per stage; and r_t = the ratio across the compressor.

For every 10°F of imperfect intercooling, the horsepower increases approximately 1 percent. Perfect intercooling is achieved when the temperature of the gas leaving the intercooler equals the temperature of the gas at the compressor inlet. The maximum theoretical work that can be saved by perfect intercooling is:

$$W = \left(\frac{k}{1-k} P_1 V_1 \right) \left[\left(\frac{P_2}{P_1} \right)^{(k-1)/k} \right.$$
$$\left. - n_s \left(\frac{P_2}{P_1} \right)^{(k-1)/n_s k} + n_s - 1 \right] \qquad (14.3.14)$$

On the ideal indicator diagram (Fig. 14.3.1), this is area $3'4\ 4'2'$.

In practice, perfect intercooling is seldom obtained. Normal practice is to cool the interstage gas to within 15°F to 20°F of the inlet gas. For water-cooled machines, the coldest water available should be used for intercooling.

Positive-Displacement Compressors Versus Dynamic Compressors

Positive-displacement compressors collect a fixed volume of gas within a chamber and compress it by reducing the chamber volume. The ideal work for such a process is given by Eq. (14.3.4). The work could also be obtained from the area enclosed by an ideal indicator diagram (Fig. 14.3.1). The actual horsepower, taking into account real-gas deviations, is

$$hp = \frac{144\, P_1 V_1 k}{33,000(1-k)} \left[\left(\frac{P_2}{P_1} \right)^{(k-1)/k} - 1 \right]$$
$$\times \frac{1}{\eta_c} \left(\frac{Z_1 + Z_2}{2Z_1} \right) \qquad (14.3.15)$$

with pressures in lbf/in² and volumes in ft³/min. With the volume rate V in m³/s and the pressures in pascals (dropping the 144/33,000), the power will be in watts.

For a given volume flow, the ideal power is affected by inlet pressure, the k value of the gas, and the compression ratio. The ideal power is not affected by the inlet temperature of the gas or its molecular weight.

The actual power is increased due to losses through the intake and discharge valves or ports as can be seen on the indicator diagram Fig. 14.3.2.

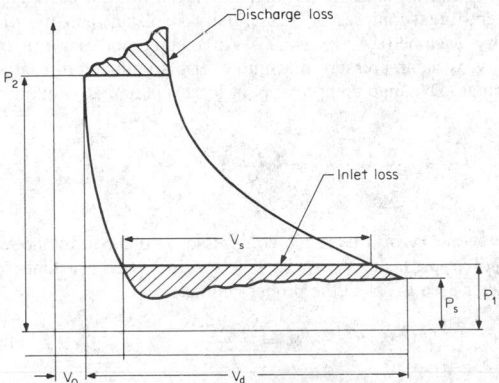

Fig. 14.3.2   Typical indicator diagram showing intake and discharge losses.

Additional losses are caused by turbulence within the compression chamber, preheating of the inlet gas, and leakage from the compression chamber. The isentropic compression efficiency (ratio of isentropic work to actual compression work) can be determined from the indicator diagram or calculated from Eq. (14.3.5) provided that there is no liquid injection. In addition to the thermodynamic losses, the mechanical losses must be added to the power requirement.

Volumetric efficiency is defined for positive-displacement compressors as

$$\eta_v = \frac{\text{actual delivery at intake conditions}}{\text{compressor displacement}}$$

For multistaging compressors, only the first-stage displacement is used.

Positive-displacement compressors are essentially constant-volume, variable-pressure machines as shown in Fig. 14.3.3. For fixed inlet and discharge pressures and varying speed, positive-displacement compressors are essentially constant-torque machines.

Dynamic compressors operate by transferring momentum to the gas via a high-speed rotor. The process is steady flow with the work given by Eq. (14.3.4). Head is defined as the energy per unit mass of the fluid. Head is imparted to the fluid through the change in magnitude and radius of the velocity components of the fluid as it passes through the rotor. The head developed by the rotor is

$$H = (U_1 V_{U1} - U_2 V_{U2})/g_c \qquad (14.3.16a)$$

Where U_1 and U_2 are the linear velocity of the rotor at radii 1 and 2. V_{U1} and V_{U2} are the fluid tangential velocity at radii 1 and 2.

The polytropic head required for a given compression ratio is

$$H_p = \frac{Z_{ave} R T_1 [r^{(n-1)/n} - 1]}{(n-1)/n} \qquad (14.3.16b)$$

The gas horsepower on a CFM basis is

$$ghp = Q_1 P_1 [r^{(n-1)/n} - 1] \frac{n}{229 \eta_p (n-1)} \left(\frac{Z_1 + Z_2}{2Z_1} \right) \qquad (14.3.17)$$

and on a mass flow basis:

$$ghp = \frac{W H_p}{33,000 \eta_p} \qquad (14.3.18)$$

With H_p in J/kg and W in kg/s, the power in kilowatts is

$$kW = H_p \left(\frac{W}{\eta_p} \right) \qquad (14.3.19)$$

Mechanical losses must be added to the gas power to obtain the shaft power. Volumetric efficiency is not defined for dynamic compressors.

For a fixed impeller design and a fixed speed, the energy that is transferred to a unit mass of fluid as it passes through the impeller (i.e., the head) is constant. Pressure rise and power vary directly with inlet gas density independent of the cause of the density change. An increase in k will cause the pressure rise to decrease but will not affect the power.

A centrifugal compressor is essentially a constant-pressure, variable-capacity machine (see Fig. 14.3.3). An axial compressor is a constant-capacity, variable-pressure machine over a significant discharge pressure range. Dynamic compressors at least qualitatively follow the fan laws.

$$\frac{Q_1}{Q_2} = \frac{N_1}{N_2} \qquad \frac{H_{p1}}{H_{p2}} = \left(\frac{N_1}{N_2}\right)^2 \qquad \frac{ghp_1}{ghp_2} = \left(\frac{N_1}{N_2}\right)^3$$

The geometry of a dynamic compressor is dictated by the required flow, polytropic head, and speed. The specific speed is a dimensionless parameter used to select the proper geometry.

$$N_s = \frac{N Q^{(1/2)}}{H_p^{(3/4)}} \qquad (14.3.20)$$

For specific speeds from 400 to 900, radial impellers are used. Between 800 and 1,400, mixed-flow impellers are preferred. For specific speeds above 1,400, axial impellers are normally used.

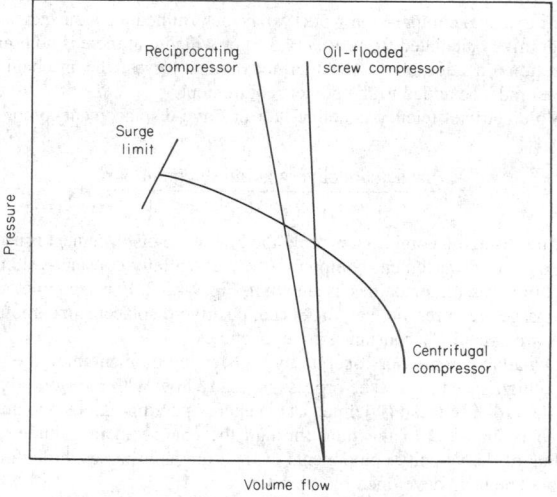

Fig. 14.3.3 Pressure capacity curves for different compressor types.

Surging

All dynamic compressors have a minimum flow point called the **surge limit** below which the operation of the machine is unstable. The surge limit is a function of the compressor type, design pressure ratio, gas properties, inlet temperature, blade angle, and speed. Operation at or below the surge limit must be avoided. The existence of the surge limit makes it essential to know the demand curve for the particular installation involved.

Reciprocating Compressors

Reciprocating compressors as shown in Figs. 14.3.4 and 14.3.5 range in size from fractional cfm to 15,000 cfm (25,485 m³/h) with discharge pressures as high as 60,000 psig (413,790 kPa). The majority of applications fall in the pressure range of 10 to 300 psig (690 to 2,069 kPa) and capacities less than 2,500 cfm (4,250 m³/h). Single-acting compressors (which compress gas on one side of the piston only) have their widest application below 50 hp (37 kW). Larger compressors are usually double-acting (both sides of the piston are used to compress gas).

Most plant air systems operate at 90 to 110 psig. Reciprocating and oil-flooded rotary-screw compressors share this market about equally. Above 200 psig, reciprocating compressors dominate.

Single-stage compressors of 100 hp (75 kW) operating at 125 psig were once common. Modern practice is to use multistaging with intercooling for pressures above 80 psig and sizes as low as 10 hp (7.5 kW). The intercooling saves power, and the reduced discharge temperatures improve safety and prolong compressor life.

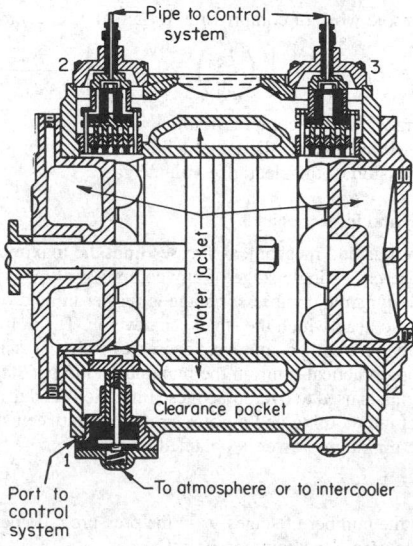

Fig. 14.3.4 Cross section of gas-compressor cylinder, showing water jackets, clearance pocket with pneumatic control, and suction-valve lift devices.

The capacity of a reciprocating compressor is determined by the first-stage displacement multiplied by the volumetric efficiency.

$$\eta_v = (1 + C)(r_\delta)^{(1/k)} - \frac{C(r)^{(1/k)}}{(Z_2/Z_1)} - L \qquad (14.3.21)$$

where $r_\delta = P_s/P_1$; $r = P_2/P_1$; and $L =$ a leakage coefficient $= 0.05$ for lubricated compressors and 0.10 for nonlubricated. $P_s =$ the pressure that exists in the cylinder when the piston is at bottom dead center. C is the clearance volume as a decimal fraction and includes valve passage volumes up to the sealing elements (point $P_2 V_c$ in Fig. 14.3.2). Clearance volume for normal service is 5 to 15 percent. High-pressure machines have as small a clearance volume as practical whereas pipeline compressors may have 100 percent clearance. The indicator volumetric efficiency from Fig. 14.3.2 is V_s/V_d. In general, this will be greater than the value obtained by Eq. (14.3.21).

The compression efficiency varies between 0.85 and .95. The mechanical efficiency falls in the range 0.88 to 0.95. The overall efficiency is the product of the compression and mechanical efficiencies.

The discharge temperature can be calculated from Eq. (14.3.22):

$$T_2 = T_1 \left[1 + \frac{(P_2/P_1)^{(k-1)/k} - 1}{\eta_c} \right] \qquad (14.3.22)$$

If the inlet pressure is varied while holding the discharge pressure constant, the horsepower of the compressor will pass through a maximum value. This can be seen by substituting displacement times η_v for V_1 in the horsepower equation (14.3.12). For correct sizing of the driver, it is necessary to know the range of inlet pressures as well as discharge pressure. Interstage pressures are selected approximately by Eq. (14.3.13) with empirical adjustments made for interstage piping, intercooler, and valve losses. The loss of volume due to condensation in the intercoolers should be accounted for.

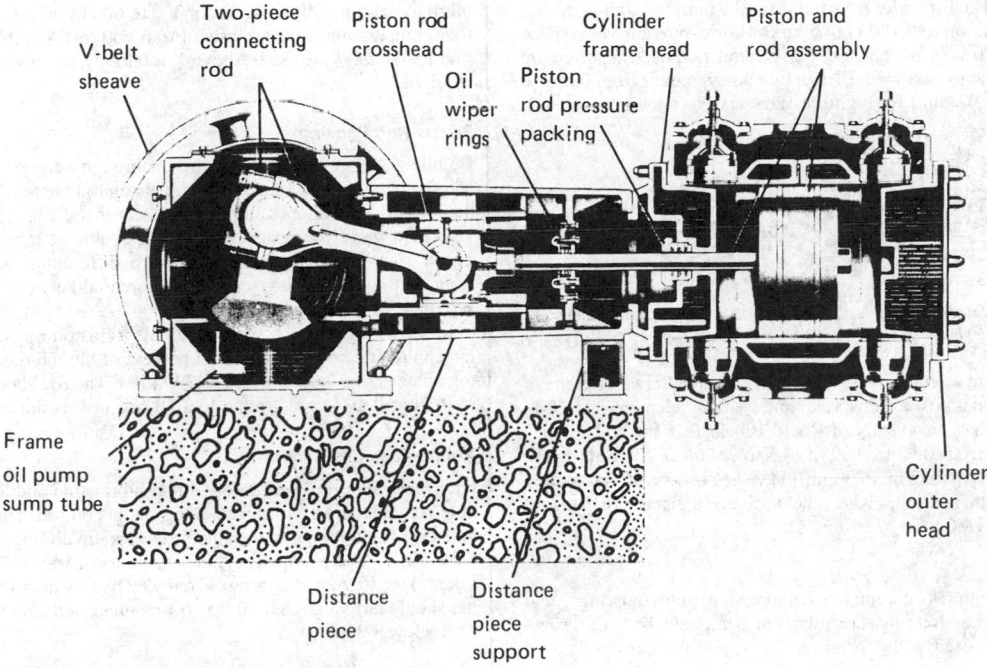

Fig. 14.3.5 Single-stage double-acting reciprocating compressor.

Valve losses can be approximated by:

$$\Delta = 1.26 \times 10^{-6} \alpha f U_{\text{ave}}^2 \text{sg} \left(\frac{P}{Z}\right) \text{ lbs/in}^2 \qquad (14.3.23)$$

where α = ratio of piston area to valve flow area; f = valve resistance in velocity heads ($\cong 4$); sg = specific gravity relative to air; and U_{ave} = average piston speed, ft/s.

The piston velocity as a function of time is:

$$U = R\omega \left(\sin \omega t + \frac{R}{2L} \sin 2\omega t\right) \qquad (14.3.24)$$

where R = crank radius; L = connecting rod length; and ω = crank speed (rad/s).

If there were no sealing elements in the valves, the velocity through the valves would be:

$$U_v = \left(\frac{A}{A_v}\right) U$$

This pulsating velocity is amplified by the valve sealing elements so that pulsating flow is inherent in reciprocating compressors. The full wavelength for a double-acting cylinder is $\lambda = 60 \, a/2N$, where a is the speed of sound. For a single-acting cylinder, it is two times this value. Pipes with total equivalent lengths equal to multiples of $\lambda/4$ should be avoided.

For a double-acting cylinder, the major problems occur at $\lambda/4$ and $3\lambda/4$ although resonance can occur at the other multiples of $\lambda/4$. Equivalent length L_{eq} is given by

$$L_{\text{eq}} = \frac{\lambda}{360} \tan^{-1} \left(\frac{2\pi V}{S\lambda}\right) \qquad (14.3.25)$$

where V = equivalent cylinder and passage volume; and S = cross-sectional flow area of the pipe.

If it is not feasible to avoid the designated lengths, a volume bottle or an orifice can be installed in the piping system. If an orifice is used, it should be ½ the pipe diameter and located at a node (one node occurs at the open end of the pipe). A volume bottle should be located as close to the cylinder as possible. Von Nimitz ("Pulsation and Vibration Control Requirements in the Design of Reciprocating Compressor and Pump Installations," Proceedings of the 1982 Purdue Compressor Technology Conference, Purdue University) recommends the following equations for sizing:

$$V_s = 4\text{PD} \left(\frac{kT_s}{\text{MW}}\right)^{1/4} \qquad V_d = \frac{V_s}{R^{1/k}} \qquad (14.3.26)$$

PD is the total double-acting displacement volume of all cylinders to be manifolded to the surge bottle (ft³). Subscript s refers to suction, d to discharge. R is the compression ratio. Temperatures are in degrees Rankine.

Compressor Valves

The effective valve area is defined as the product of the element lift and the sum of the valve-seat peripheries or strip edges, less the guide and end contacting surfaces. The plate and strip-type valves (Figs. 14.3.6 and 14.3.7) are used in air and low-pressure service. The concentric-

Fig. 14.3.6 Plate valve, illustrating floating element, cushion spring, and seat. *(Ingersoll-Rand.)*

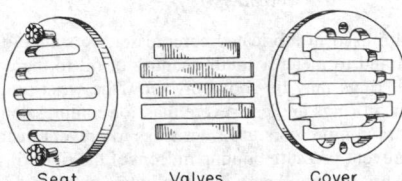

Fig. 14.3.7 Feather valve, illustrating seat, flexing elements, and valve cover. *(Worthington.)*

disk type (Fig. 14.3.8) is used in high-pressure, chemical, and natural-gas operations. The lift varies from 0.035 in (0.9 mm) for high-pressure, high-speed operation to 0.180 in (4.6 mm) for low-pressure, low-speed operation, with 0.100 in (2.5 mm) for general purposes. Springs are used to reduce the impact load. Element thickness ranges from 0.050 to 0.125 in (1.3 to 3.2 mm) for high-pressure service. The thickness, lift,

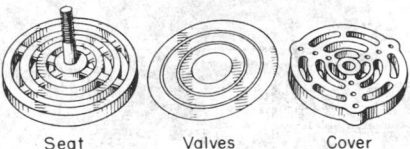

Seat   Valves   Cover

Fig. 14.3.8 Concentric-disk valve, illustrating seat, floating elements, and valve cover. (*Chicago Pneumatic.*)

and spring load are selected by trial and error. The feather valve element (Fig. 14.3.7) functions as a spring and sealing element. An 8-in (203-mm) strip has an average lift of 0.100 in (2.5 mm). The strip thickness varies from 0.02 to 0.09 in (0.5 to 2.0 mm). Multiple nylon poppet valves [lift = 0.25 in (6.3 mm)] have been successfully applied to pipeline and other services where the discharge temperature does not exceed 250°F (121°C).

Piston Rings

Reciprocating compressors employ compression rings, oil rings, and rider rings. Compression rings are always present, but whether oil rings or rider rings are used will depend on the type of compressor and its service.

The standard compression ring is the one-piece square-cut or angle-cut ring although two-piece and segmental rings are sometimes used. The one-piece ring usually has a rectangular cross section, but occasionally the cylinder contact face is crowned to prevent edge loading and to improve lubrication. Cast-iron rings are still widely used but are being displaced in many applications by filled TFE rings. Carbon-filled TFE rings are preferred for process compressors. For air compressors, carbon, glass, and bronze fillers are used. TFE rings can be used from − 450 to 500°F (− 268 to 260°C). Bronze-filled TFE is good for higher-temperature operations because of its higher thermal conductivity. Carbon-filled TFE is preferred for oil-free service. Many compressors that handle gases with corrosive components (e.g., H_2S) use piston rings made of cloth-reinforced thermoset laminates with molybdenum disulfide added to improve the lubricity. Polyamide, polyimide, and poly(amide-imide) rings work well with marginal or no lubrication.

A one-piece ring must be stretched over the piston to install it in the piston groove. The maximum stress occurs 180° from the joint and is given by Koppers Research and Engineering Staff (''Engineers Handbook of Piston Rings, Seal Rings, Mechanical Shaft Seals,'' 8th ed., Koppers Company, Inc.) as:

$$S_o = \frac{0.424\ E(8d - g)/d}{(D/d - 1)^2} \quad \text{lb/in}^2 \qquad (14.3.27)$$

When closed to the cylinder diameter, the ring is stressed to supply the initial seating pressure. The maximum stress (180° from the joint) is given by the Koppers ''Engineers Handbook'' as:

$$S_c = \frac{0.482\ E\ (g/d)}{(D/d - 1)^2} \quad \text{lb/in}^2 \qquad (14.3.28)$$

Oil rings are used in lubricated compressors to scrape oil from the cylinder wall and to meter a small quantity of oil to the compression rings. The oil rings may be above or below the wrist pin. Operating conditions for oil rings are less severe than for compression rings.

Nonlubricant compressors use rider rings to prevent piston cylinder contact. To prevent pressure buildup in front of the rider rings, longitudinal grooves are cut into the rings. Usually two or more rider rings are required.

Below 300 lb/in² differential, two compression rings and one or two oil rings are normally used. From 300 to 600 lb/in² differential, three to four compression rings are used. From 600 to 1,500 lb/in² differential, four to five rings are used. Above 1,500 lb/in², six or more rings may be required.

Piston-Rod Packing

Double-acting compressors are driven through a crosshead (Fig. 14.3.5) and require a rod packing to seal the piston rod connecting the piston to the crosshead. The operating conditions for rod packings are more severe than those for piston rings. In a double-acting compressor, the piston rings seal against the pressure differential across the stage whereas the rod packing seals the differential between discharge pressure and atmospheric pressure.

Rod packings use the same materials as piston rings. Metallic packing can tolerate rod wear of 0.15 percent of the rod diameter, with less tolerance above 2,000 lb/in² (13,790 kPa). The rod should be hardened to Rockwell 40 C and ground to a 10-rms or less finish.

Nonlubricated Cylinders

NL-1 class does not permit any direct lubrication and tolerates the minute contamination introduced along the piston rod. This can be eliminated by an additional spacer and rod wiper, qualifying the cylinder as a class NL-2. The pistons for this service use various Teflon rings and riders. The life of carbon rings is reduced by low humidities. The cylinder should have a ground 10- to 16-rms finish and a minimum hardness of Rockwell 20 C.

Lubrication

Lubricants serve to (1) prevent wear by providing a supporting film between the rubbing surfaces, (2) seal close clearances, (3) protect against corrosion, and (4) transmit heat of friction and minute wear particles away from points of contact. Industrial cylinders and packings require force-feed lubricators. Viscosity is the best index of suitability: general service at pressures below 500 lb/in² (3,448 kPa) requires 400 SSU (86 centistokes) at 100°F (38°C) (SAE 40); 700 SSU (151 centistokes) for 2,000 lb/in² (13,793 kPa); and 1,000 SSU (216 centistokes) (SAE 50) for 8,000 lb/in² (55,000 kPa) operations. These oils have very high viscosity at 40°F (45°C), which will result in poor lubrication. Good winterizing practice provides continuous circulation of warm jacket water and an immersed electric heater in the oil reservoir. Crankcase oils should include a foam inhibitor, sludge dispersant, and rust inhibitor. Phosphorous extreme-pressure agents are added to high-pressure lubricants to avoid wear and scuff damage. Castor-bean and rapeseed oils are additives resistant to solvent action of condensing hydrocarbons.

Cellulube, Houghto-safe, Pydraul AC, Anderol, and Fluorolubes are used to resist the hazard of exothermic reaction in air compression. The crankcase lubricant consumption for 10 plants over a score of years averaged 20,000 bhp/(h)(gal). Table 14.3.1 shows lubricant requirements for general-process and natural-gas operations.

Table 14.3.1   Lubrication Required for Compressor Cylinders

Range of cylinder diam, in	Film thickness, μm	10^3 hp/(h) per gal	Pints per cylinder per day per 100 hp of load	Oil drops/ min
24–36	0.40	13	1.5	13
15–23	0.45	19	1.0	9
10–14	0.55	24	0.8	7
7–9	0.60	32	0.6	5
4–7	1.20	24	0.8	7
3–5	1.40	27	0.7	6

NOTE: Based on 12,500 drops per pint of SAE 40 oil at 75°F with vacuum-type lubricator. Reduced drop count to roughly one-half for pressure-type lubricator. Example: 500 bhp, 20-in cylinder, requires 5 pint/day and feed of 45 drops/min. Each installation will vary on lubrication requirements. Manufacturers' recommendations on oil type and quantity should be followed.

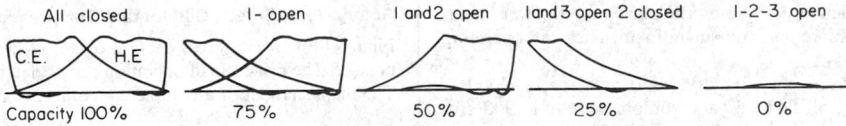

Fig. 14.3.9   Cards from a compressor with combination clearance and suction-valve bypass control.

Compressor Accessories

Most state laws and safe practice require a relief valve ahead of the first stop valve in every positive-displacement compressor. It is set to release at 1.25 times normal discharge or at the maximum working pressure of the cylinder, whichever is lower. The relief-valve piping system sometimes includes a manual vent valve and/or a bypass valve to the suction to facilitate start-up and shutdown operations. Quick line-sizing equations are (1) line connection, $d/1.75$; (2) bypass, $d/4.5$; (3) vent, $d/6.3$; (4) relief-valve port, $d/9$.

Volume flow is controlled by variable-speed drivers; steam engines can operate at 20 percent of rated speed and gas engines at 60 percent, and electric-motor speed can be varied by means of eddy-current and hydraulic couplings and special wound rotors with rheostats, which are both costly and inefficient. Unloading can be applied by means of valve-lift and clearance pockets, as shown in Figs. 14.3.4 and 14.3.9. Utility air plants are throttled with suction unloaders, as shown in Fig. 14.3.10. Suction throttling and bypass controls are used in process operations.

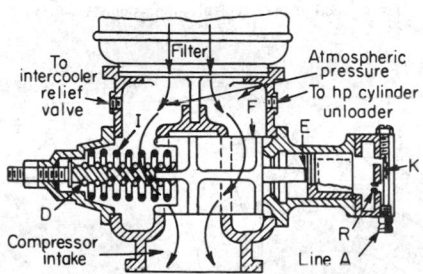

Fig. 14.3.10   Air suction unloader. Piston E moves to closed position as line A pressure reaches design point.

Cylinder Cooling

Single-acting compressors, especially smaller ones, are usually air cooled by fins cast into the cylinder. Cooling air is drawn across the fins by a fan. Double-acting compressors usually have cylinders that are water jacketed.

The principal purpose of a jacket-water system is to normalize cylinder-casting strains. It has some effect on compressor efficiency due to reduced preheating of the air. This effect is reduced as the compressor speed is increased. The heat rejection to the jacket water is approximated by:

$$H_j = 4(t_{ag} - t_{aw}) + 100 \text{ Btu/(bhp·h)} \qquad (14.3.29)$$

where t_{ag} and t_{aw} represent the average gas and water temperature, respectively, usually 155 and 140°F, respectively, and a rejection of 160 Btu/(bhp·h). Where the temperature rise is between 2 and 5°F, the jacket-water requirement is 160 bhp/(500 × 3.2 × t_w), or 0.1 gal/(min·bhp). The use of cold water in jacket-water systems causes condensation on the cylinder walls, washing of lubricant, and excessive ring wear.

Rotary-Vane Compressors

Figure 14.3.11 is a section through a typical rotary-vane compressor. The displacement can be calculated using the following formula.

$$V_{th} = 2eL(\pi D - T_v N_v)N \qquad (14.3.30)$$

Rotary-vane compressors have a built-in volume ratio V_i. This means that they will compress the gas within the compression chamber to $P_1(V_i)^n$ before opening the compression chamber to the discharge port. P_1 is the inlet pressure, and V_i is the ratio of the volume of the compression chamber at the inlet cutoff point to the volume at the point just prior

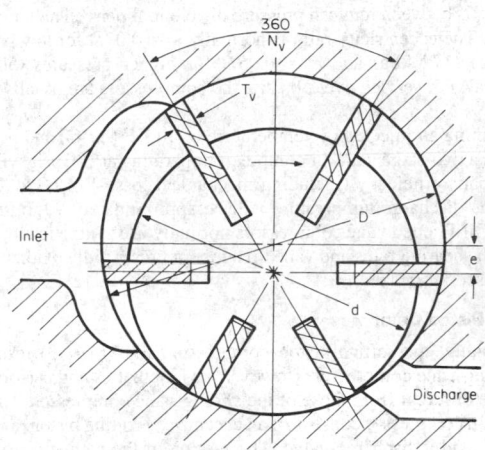

Fig. 14.3.11   Section through a typical vane compressor with N_v vanes.

to the opening of the discharge port. If the operating-pressure ratio does not match the built-in ratio, overcompression or undercompression will take place as indicated in Fig. 14.3.12. For this case the ideal work should be calculated as:

$$W_{ideal} = \frac{P_1 V_1}{(1 - n)} [V_i^{n-1} - n] - P_2 \frac{V_1}{V_i} \qquad (14.3.31)$$

The ratio of this work over the ideal work required for a perfectly matched compressor can be read from Fig. 14.3.16. For example, the theoretical work for an air compressor with $V_i = 2$ operating at $P_2 = 0.8P_i$ theoretically will require 2.5 percent more work than it would if it were designed so that the built-in ratio matched the operating ratio.

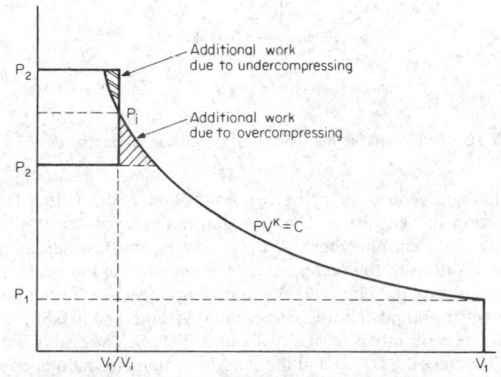

Fig. 14.3.12   Idealized P-V diagram for a compressor with a built-in volume ratio (also referred to as a *built-in pressure ratio*).

However, normal practice is to base efficiencies on a power calculated as though the compressor built-in ratio matches the operating ratio.

Rotary-vane compressors may be dry, lubricated, or oil-flooded. Prior to 1960 two-stage, oil-flooded vane compressors up to 900 cfm (1,530 m³/h) compressing to 150 psig (10.3 bar) (1,034 kPa) were used in portable compressors. They are now usually limited to less then 100 cfm (170 m³/h) air compressors and to small refrigeration compressors and boosters.

Vane materials may be phenolic (micarta, polyamid-imides, polyimides, Ryton) or metal (aluminum-silicon alloy).

The major limitations of vane compressors are imposed by vane-tip friction, vane-bending stress, and vane-length limitations. Maximum vane tip speeds are approximately 65 ft/s (20 m/s). Normally 6 to 8 vanes are used although as many as 20 vanes have been used. The trade-off is between reduced pressure differential per cell and increased friction. The eccentricity ratio is normally $e = 0.07D$ for low pressure (25 psig) (172 kPa) and $e = 0.05D$ for higher pressures (50 psig) (345 kPa). For service over 50 psig, the compressors are usually multi-staged.

Assuming an injected oil temperature of 140°F (60°C) and no intercooling, a two-stage vane air compressor operating at 100 psig will have an adiabatic efficiency (including mechanical losses) of 60 to 72 percent. The discharge temperature will be approximately 90 to 95°C or 200°F. Oil-flooded vane compressors normally use synthetic lubricants. The high shearing rates and vane tip temperatures rapidly oxidize petroleum based lubricants.

Rolling-Piston Compressors

Superficially, the rolling-piston compressor (Fig. 14.3.13) resembles the sliding-vane compressor. However, its kinematics and performance are quite different from those of the sliding-vane compressor. The rolling-piston compressor consists of an eccentric, a rolling piston, a vane, a cylinder, and a discharge valve. The motion of the rolling piston with

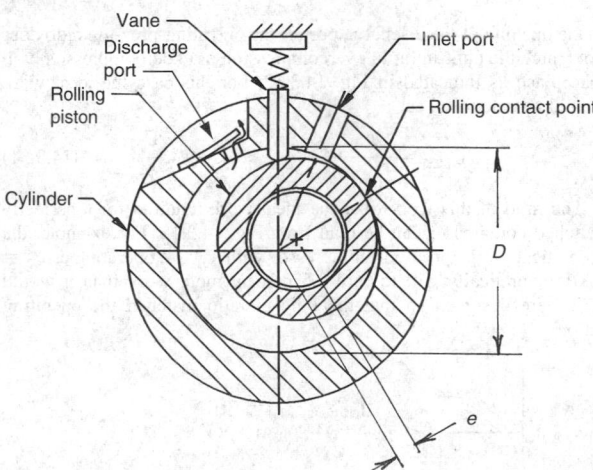

Fig. 14.3.13 Section through a typical rolling-piston compressor.

respect to the eccentric and cylinder is controlled by the various friction forces action on the piston. Compression and suction occur simultaneously in different chambers. A cycle covers two revolutions of the eccentric. Rolling-piston compressors are usually used in refrigeration systems, typically 0.25 hp (180 W) to 0.75 hp (550 W). Typical values for volumetric and power efficiencies are 0.90 and 0.84 to 0.86, respectively. The eccentricity is normally about $0.09 \times D$. The piston diameter is approximately $0.81 \times D$ and the length-to-diameter ratio is approximately 0.5. The compressor displacement can be calculated from Eq. (14.3.30) with $N_v = 1$.

Rotary, Twin-Screw, Oil-Flooded Compressors

Figure 14.3.14 is a cross section of an oil-flooded, rotary-screw compressor. The principle of operation can be determined from Fig. 14.3.15. The displacement of a twin-screw compressor is

$$V_{th} = KD^3 \frac{L}{D} C_{th} N \qquad (14.3.32)$$

where $K = N_m(A_{mg} + A_{fg})D^2$; D = the rotor diameter; L = the rotor length; N_m = the number of lobes on the main rotor; A_{mg} = the area of one groove on the main rotor; and A_{fg} = the area of one groove on the secondary rotor. C_{th} is an overlap constant (between 0.95 and 1.0 for most designs) which corrects for the overlapping of the filling and compressing processes.

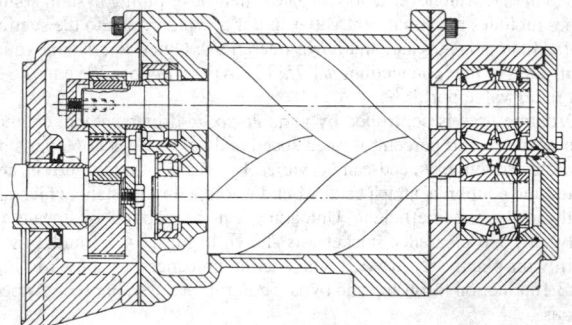

Fig. 14.3.14 Cross section of a typical single-stage, oil-flooded, twin-screw compressor.

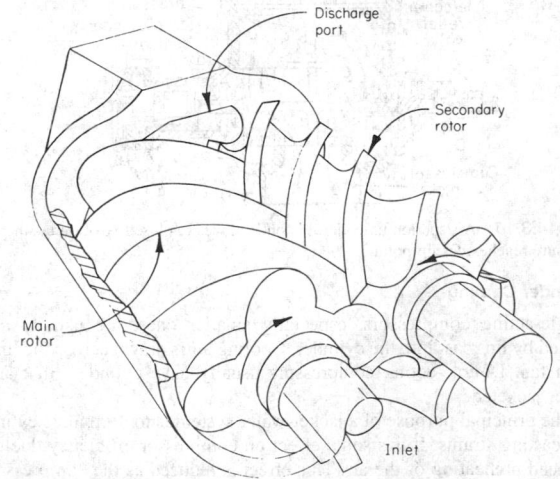

Fig. 14.3.15 Operating principle of a twin-screw compressor.

C_{th} depends on the wrap angle, the profile, and the number of lobes. Lobe combinations of (4 and 5), (4 and 6), (5 and 6), and (5 and 7) are used. The most common design uses the (4 and 6) combination. Typical values for C_{th} and K are given in Table 14.3.2. A complete cycle (filling and compressing) takes place over approximately 750° of main rotor rotation. There are four cycles per revolution of the male rotor. The discharge process for adjacent grooves overlap for approximately 40° giving an inherently smooth discharge process.

Rotary-screw compressors have a built-in volume ratio V_i (typically 4.4 for single-stage air compressors and 1.9 to 3.5 for refrigeration compressors). For an inlet pressure of P_1, the compressor will have a built-in pressure of $P_1 V_i^n$. The effect on the P-V diagram and on the theoretical power can be found from Fig. 14.3.12 and Fig. 14.3.16. In practice, the increased discharge port losses associated with higher built-in volume ratios (smaller port area) skew the results in Fig. 14.3.16. In practice, it is more efficient, at least in air compressors, to have the built-in ratio slightly lower than the theoretical ratio. Single-stage, twin-screw air compressors rarely have built-in pressure ratios

Table 14.3.2 Typical Constants for Different Lobe Combinations (300° Male Wrap)

N_m/N_f	3/4	4/5	4/6	5/6	5/7
C_{th}	0.954	0.973	0.969	0.984	0.980
K	0.530	0.439	0.507	0.454	0.468

greater than 9, and most are designed with ratios of 7 to 8. An air compressor designed for 100 psig can operate at 175 psig with a loss in adiabatic efficiency of approximately 5 and 4 percent loss of volumetric efficiency.

Adiabatic efficiency (including mechanical losses in the air end) vary from 70 percent for small sizes (80 mm) to 88 percent for large (300 mm) slow-running compressors. An average value for a midsize compressor is 0.78 to 0.75 based on an injection temperature of in-

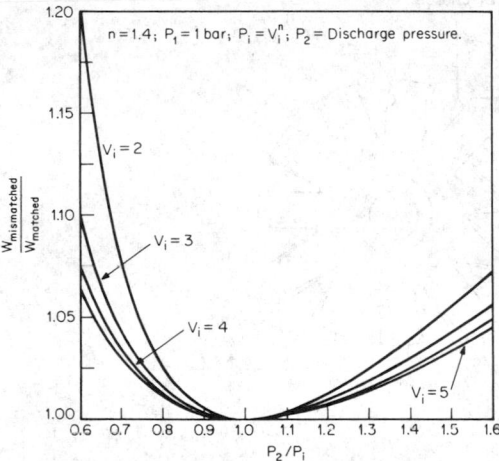

Fig. 14.3.16 The effect of the built-in volume ratio on theoretical work of compression for a compressor operating off the design point.

let temperature plus 50°F. Accuracy in manufacturing the rotors is critical to achieving good efficiency. The heat absorbed by the cooling oil can be estimated by:

$$Q_h = \text{power}\left[1 - \frac{\eta_c(T_2 - T_1)}{T_1(r^{(k-1)/k} - 1)}\right] \qquad (14.3.33)$$

where power = the shaft power to the compressor. T_2 is normally 70 to 90°F over T_1.

Tip speeds range from 15 to 50 m/s. Rotor sizes range from 40 to 400 mm diameters with L/D from 1 to 2. Oil-flooded screw compressors dominate the portable-compressor market and share the plant-air market about equally with the reciprocating compressor up to about 3,000 cfm. In the refrigeration industry, the screw compressor finds its widest application in the 300 to 500 kW range.

In heat-pump applications, the screw compressor's zero-clearance volume is an advantage over reciprocating compressors. Reciprocating compressors lose volumetric efficiency as the outside temperature drops (the compression ratio increases) requiring supplemental heat to compensate for the lost capacity. Refrigeration screw compressors are now available with a secondary suction port known as an **economizer connection.** This allows for two-temperature-level operation. At full-load operation, the system COP is improved significantly. As the compressor is unloaded, the improvement is lost.

Capacity can be controlled by on-line/off-line operation, inlet throttling, slide-valve or turn-valve operation, or speed regulation. Stationary air-compressors are normally controlled by on-line/off-line operation or inlet throttling with unload below 50 percent capacity.

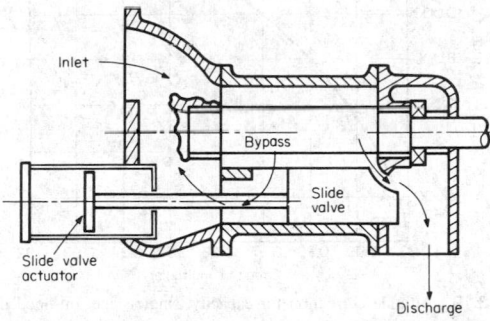

Fig. 14.3.17 Principle of slide-valve operation. As the valve is moved toward the discharge end of the compressor, the point of inlet cutoff is delayed.

The air-oil sump is usually blown down to atmospheric pressure at unload.

Portable compressors are normally controlled by a combination of inlet throttling and speed regulation; the air-oil sumps are not blown down.

Refrigeration compressors are controlled by inlet throttling or more often by a slide valve or turn valve. A compressor with a slide valve is shown in Fig. 14.3.17. The effect on power consumption for each type of capacity control can be found from Fig. 14.3.18.

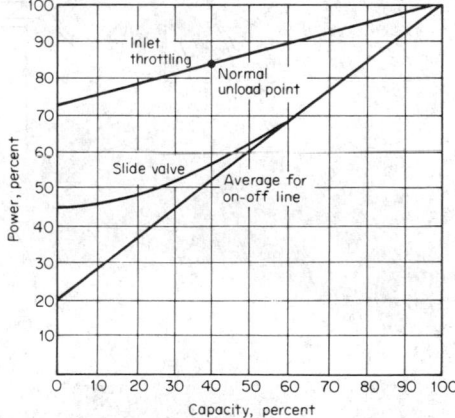

Fig. 14.3.18 Comparison of different capacity control methods for oil-flooded screw compressors operating at constant speed.

Oil-flooded, twin-screw compressors have been operated successfully on petroleum-based oils, automatic transmission fluids, synthetic diesters, polyalphaolefins, synthesized hydrocarbons, and polyglycols. The trend is to synthetic lubricants.

Regardless of the lubricant selected, the discharge temperature for an air compressor must be kept high enough to prevent condensation in the lubricant during part-load operation. This is usually accomplished by a thermal valve set to maintain the minimum injection temperature at 140 to 165°F (60 to 76°C). Standard oil separators limit oil carryover to 5 ppm or less. Optional treatment can reduce the levels further.

The capacity loss of an oil-flooded, twin-screw compressor with altitude is quite small compared to a normal reciprocating compressor. The approximate capacity correction for altitude can be read from Fig. 14.3.19.

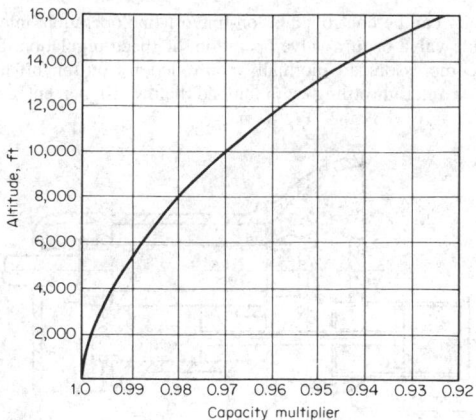

Fig. 14.3.19 Altitude correction for capacity, single-stage, oil-flooded screw compressor.

Rotary Single-Screw Compressors

The operating principle of the single-screw compressor can be determined from Fig. 14.3.20. Lundberg and Glanvall give the displacement as:

$$\beta = \arcsin \frac{B - D/2}{R}$$

$$\Omega = \pi - 2\beta$$

$$C = 2R \sin \frac{\Omega}{2}$$

$$\gamma = \frac{\alpha\Omega}{\xi}$$

$$t = R - \frac{B - D/2}{\sin(\beta + \gamma)}$$

$$u = B - \frac{D}{2} + \frac{t}{2} \sin(\beta + \gamma)$$

$$V = \sum_{\alpha = \alpha 1}^{\xi} AU \, \Delta\alpha$$

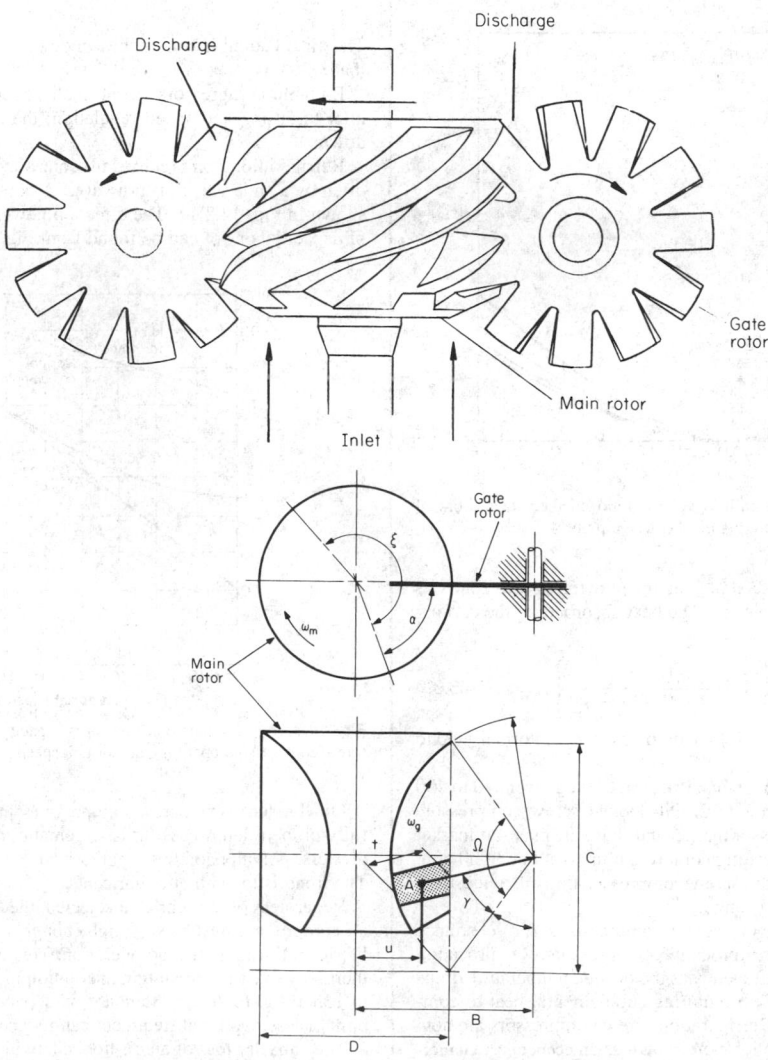

Fig. 14.3.20 Principle of operation and geometry of a single-screw compressor. *(Lundberg and Glanvall, "A Comparison of SRM and Globoid Type Screw Compressors at Full Load," Proceedings of the 1978 Purdue Compressor Technology Conference, Purdue University, West Lafayette, IN.)*

The rotary single-screw compressor is an oil-flooded compressor with a built-in volume ratio. The compressions for each star wheel are almost in phase so the gross effect as far as the discharge piping is that there are six discharges per revolution of the main rotor.

The single-screw compressor is used in the refrigeration industry where the low bearing load on the main rotor is an advantage. It is also being used as an air compressor in the smaller sizes.

Adiabatic efficiency of a single-screw compressor will be 2 to 5 percent lower than that of a comparable twin-screw compressor. The effect of off design operation on theoretical power can be found from Fig. 14.3.16. Capacity control is identical to the twin-screw compressor.

Dry Rotary Twin-Screw Compressors

A dry rotary, twin-screw compressor has timing gears to separate the rotors which may or may not be coated. The typical compressor is water-jacketed. Seals prevent gas leakage along the shafts and seal the oil to the bearings and timing gears. Sleeve bearings are most common, but some units have antifriction bearings. The displacement is the same as for the oil-flooded screw. Tip speeds vary from 80 to 120 m/s. Maximum compression ratio per stage is 4.5 based on $k = 1.4$. Sizes range from about 400 to 20,000 cfm (680 to 34,000 m³/h). Pressures range

from 15 to 180 psig (1,240 kPa). Two-stage compressors are almost always intercooled. With intercooling to 15°F they approach adiabatic efficiency (based on zero intercooling) including mechanical-loss ranges from 72 to 80 percent. While oil-flooded screw compressors can be controlled to very low capacities by inlet throttling, dry screw compressors cannot. The resulting pressure ratio causes excessive temperature increase across the compressor.

Orbiting Scroll Compressors

The primary elements of the orbiting scroll compressor are the fixed scroll, orbiting scroll, Oldham ring, crankshaft, and crankcase. Typically the two scrolls are involutes of circles phased 180° apart. The orbiting scroll is driven by the crankshaft and prevented from rotating via the Oldham ring (Fig. 14.3.21). As the scroll orbits, the pockets formed between the flanks of the scrolls move toward the center reducing in size until the built-in volume ratio is achieved. At this point two pockets start discharging simultaneously through the discharge port. The filling, compression, and discharging process usually takes 900 to 1,100 degrees of shaft rotation, depending on the number of wraps. The compressor displacement is a function of the involute generating radius, the starting angle, the thickness angle, and the scroll height. The crank

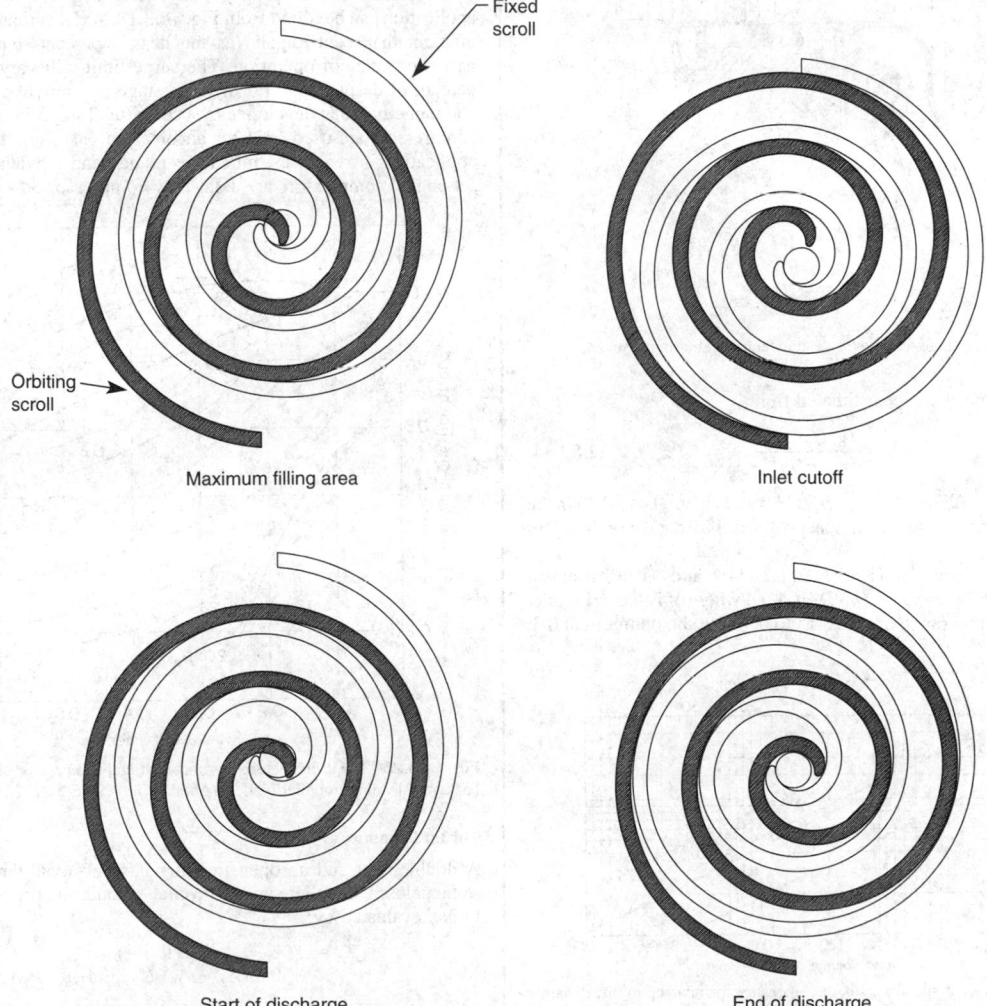

Fixed scroll

Orbiting scroll

Maximum filling area

Inlet cutoff

Start of discharge

End of discharge

Fig. 14.3.21   Operating principle of an orbiting scroll compressor (moving scroll is orbiting clockwise).

radius is equal to 0.5 times the involute pitch minus the scroll thickness. The scroll compressor is used primarily as an air-conditioning or heat-pump compressor in the 5 to 50 hp (3.7 to 37 kW) range.

Dynamic Compressors

A typical centrifugal compressor is shown in Fig. 14.3.22. The optimum range of operation can be found from Fig. 14.3.23. An estimate of the polytropic efficiency can be obtained from Fig. 14.3.24 or calculated from

$$\eta_p = 0.014 \ln (Q) + 0.600 \qquad (14.3.34)$$

where Q is in cubic feet per minute.

For normal industrial compressors, $H_p = 10,000$ ft·lb/lb (30 kJ/kg) per stage although much higher values can be obtained.

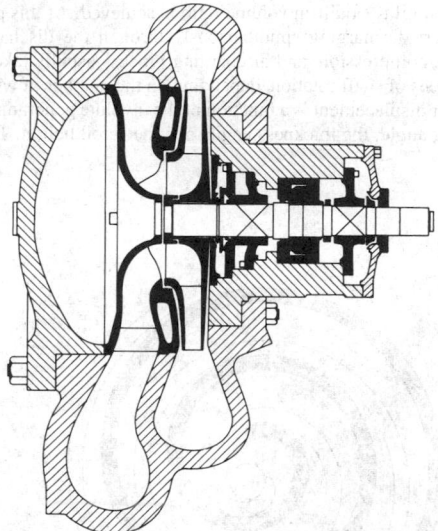

Fig. 14.3.22 Single-stage, cantilever-design centrifugal compressor.

The required speed can be estimated from:

$$N = \frac{(g_c H_p/\mu)^{0.5}}{\pi D} \qquad (14.3.34)$$

The pressure coefficient ($\mu = g_c H_p/U^2$) varies from 0.4 to 0.7 (in the high-efficiency range). In the absence of specific data, use 0.55 for estimating.

The diameter is determined by the required flow and can be estimated from the flow coefficient $\varphi = Q/ND^3 \propto V/U$, where V is the axial inlet velocity. The normal range of φ is 0.15 to 0.4 with the optimum at 0.3.

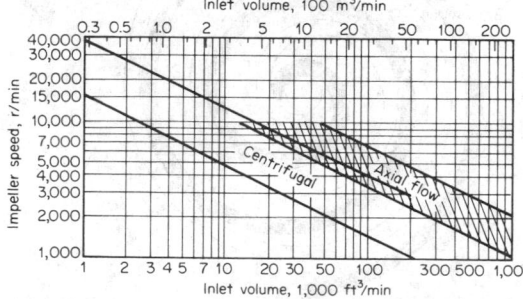

Fig. 14.3.23 Range of speed versus inlet capacity for near-optimum dynamic compressor efficiency. (*Loomis, "Compressed Air and Gas Data," 3d ed., Ingersoll-Rand, pp. 4–28.*)

The slope of the μ-φ curve (and the $H_p \cdot Q$ curve) is strongly dependent on the backward lean of the impeller blades as shown in Fig. 14.3.25. $\beta_2 = 90°$ is a radial blade. Normal closed industrial impeller blades will have an angle of 65 to 55°. It should be noted that an increase in density of the gas will flatten the slope.

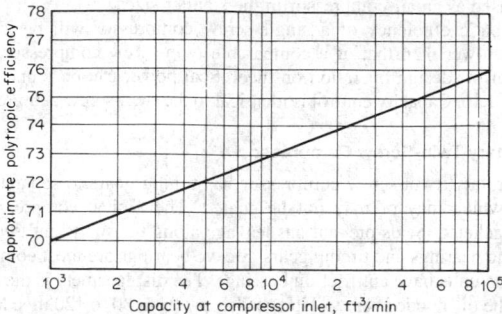

Fig. 14.3.24 Approximate polytropic efficiency versus compressor inlet capacity. (*Rollins, "Compressed Air and Gas Handbook," 4th ed., Compressed Air and Gas Institute, pp. 3–71.*)

A typical axial compressor is shown in Fig. 14.3.26. The range of application can be found from Fig. 14.3.18. Axial compressors are more efficient than centrifugals (as much as 5 percent) but have a much narrower range of operation. The surge limit will vary from 85 to 95 percent of design flow. For a given stage, the polytropic head will be about one-half that developed by a centrifugal stage, i.e., $\mu = 0.3$. Axial compressors are used only for air or clean gas. They find their major application in refineries, butadiene plants, and ethylene oxide plants. Jet-engine compressors are axial-flow compressors.

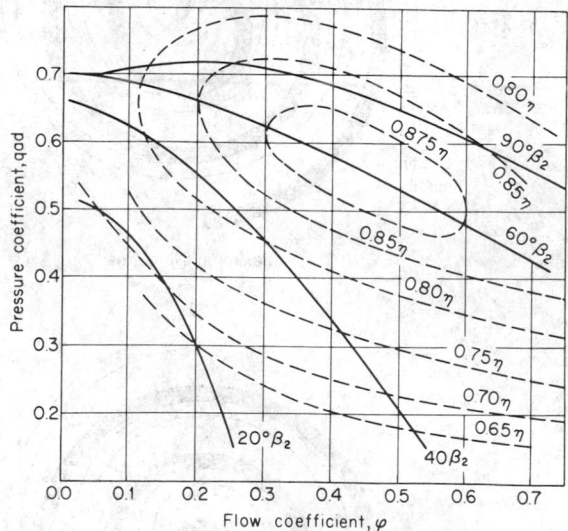

Fig. 14.3.25 Basic characteristics of centrifugal compressor, showing effects of backward-leaning impeller blades. (*After Balje, ASME 51-F-12.*)

Thrust Pressures

A double-inlet and an open impeller have no axial thrust, whereas a semienclosed impeller, with a frontal shroud, can impose substantial thrust, evaluated by:

$$T_A = Y(A_1 - A_2) \, \Delta \, (\text{lb/in}^2)$$
$$+ 0.8 \, (A_2 - A_3) \, \Delta \, (\text{lb/in}^2) - P_1 A_3$$

where A_1 = full impeller area; A_2 = impeller area less the area of the inlet-eye seal; A_3 = shaft area through the seal, all in in^2; and

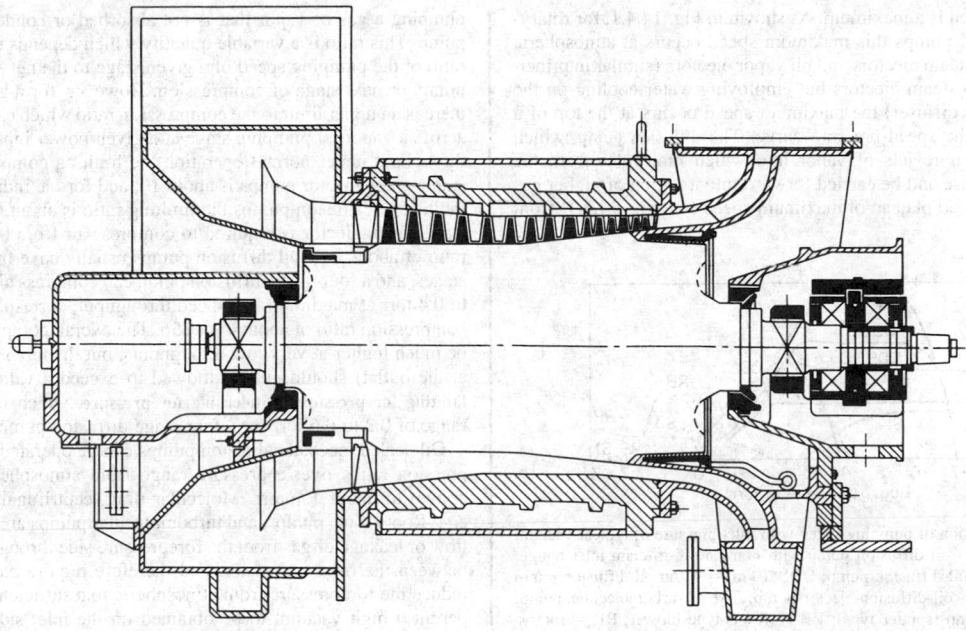

Fig. 14.3.26 Multistage axial compressor.

Δ (lb/in^2) = pressure differential per stage. Y is the percentage of Δ (lb/in^2) acting in back of a single-stage disk: (1) Without a frontal shroud and a plain back, $Y = 0.35$; (2) with 0.060-in back ribs, $Y = 0.28$; (3) with ¼-in equalizer holes at half radius and about 2 in apart, $Y = 0.22$; and (4) with deep scallops, $Y = 0.18$. For fully shrouded impellers, as used in multiple stage, $Y = 1.00$. These thrusts are balanced by opposing impeller inlet ends or, more generally, by a balancing drum wherein an equal and opposite thrust is created, equal to the

product of the compressor Δ (lb/in^2) and the area of the drum. Stepanoff ("Turboblowers," Wiley) shows that radial impellers create a radial thrust, caused by the uneven volute-pressure distribution, which is:

$$T_R = 0.36 \Delta \text{ (lb/in}^2\text{) } db$$

where d = the impeller diameter; and b = the width, including the shrouds, both in inches.

14.4   HIGH-VACUUM PUMPS
by Benjamin B. Dayton

REFERENCES: Dushman, "Scientific Foundations of High Vacuum Technique," 2d ed., Lafferty (ed.), Wiley. Roth, "Vacuum Technology," North-Holland. O'Hanlon, "A User's Guide to Vacuum Technology," Wiley. Power, "High Vacuum Pumping Equipment," Reinhold. Van Atta, "Vacuum Science and Engineering," McGraw-Hill. LaPelle, "Practical Vacuum Systems," McGraw-Hill. Holland, Steckelmacher, and Yarwood, "Vacuum Manual," E. & F. N. Spon. M. H. Hablanian, "High-Vacuum Technology," Marcel Dekker.

The pressure of gas in a chamber at a given temperature can be reduced by allowing the gas to escape through a port into a vacuum pump, or by a pumping system comprising two or more pumps in series, which compresses the gas and discharges it into the atmosphere, or by allowing the gas (or vapor) to condense on a sufficiently cold surface or to react with a chemically active surface exposed within the chamber or within an appendage to the chamber. The movement of the gas toward the pump or trapping surface can be accelerated by heating the walls of the chamber or by ionizing the gas and applying an electric field. **Rotary mechanical pumps** and **steam ejectors** which can compress the gas from about 10^{-3} torr up to atmospheric pressure are described earlier in Secs. 9 and 14. Pumps and trapping techniques required to produce pressures in the "high-vacuum region" (below 10^{-3} torr) are described below.

Units   The term *pascal* (abbreviated Pa and equal to one newton per square meter) is recommended as the SI unit for the pressure of the gas in vacuum systems, but the unit *torr* (equal to 133.322 Pa) is still widely used in equipment catalogs and published articles. The ISO also permits the use of bar (10^5 Pa) and millibar (mbar), which are convenient units for vacuum technology (one torr equals 1.333 millibar). **Pumping speed** is normally expressed in litres per second (L/s) or cubic feet per minute (ft^3/min); one litre per second equals 2.12 cubic feet per minute.

SELECTION OF PUMPS

The type of vacuum pump selected depends primarily on the lowest pressure to be attained and the possible effects of vapor contamination. The lowest pressure that the pump itself can attain without vapor traps is called the **ultimate pressure** of the pump. The **size** of the pump depends either on the time to "pump-down" from atmospheric pressure or on the pressure to be maintained in the presence of a given gas load during the process cycle and the number of pumps which can be installed in parallel. The gas load is usually expressed in terms of **throughput** defined as the product of the static pressure and the volumetric flow rate across a given section at a given temperature. Size of pump is rated in terms of volumetric pumping speed (L/s or ft^3/min) at the inlet pressure for which the speed is a maximum. As shown in Fig. 14.4.1, for rotary-piston mechanical pumps this maximum speed occurs at atmospheric pressure, but for steam ejectors and oil vapor ejectors (similar in principle and design to steam ejectors but employing water cooling on the venturi section or diffuser) the maximum speed occurs at the top of a narrow peak on the speed-pressure curve. The **diffusion pump,** which employs one or more jets of vapor into which molecules from the chamber can diffuse and be carried forward into a region of higher gas pressure, has a broad plateau of maximum speed (see Fig. 14.4.1) from

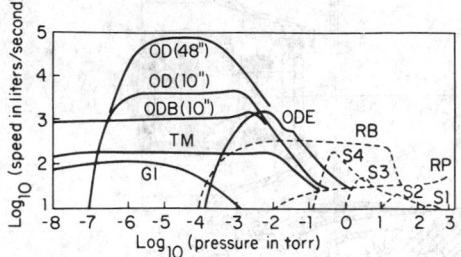

Fig. 14.4.1 Variation of pumping speed with inlet pressure of typical vacuum pumps. OD(48 in) = oil diffusion pump with nominal 48-in-diam inlet flange; OD(10 in) = 10-in oil-diffusion pump; ODB(10 in) = 10-in oil-diffusion pump with baffle; ODE = oil-diffusion-ejector pump; TM = turbomolecular pump; GI = getter-ion pump (sputter type); RB = Roots-type blower; RP = rotary-piston, oil-sealed mechanical pump; S1, S2, S3, and S4 = matching stages in series of four-stage steam ejector.

about 10^{-3} torr (1 mtorr) to pressures 20 times the ultimate pressure, the latter being approximately equal to the vapor pressure at room temperature of the fluid used to form the vapor jet. Placing a cooled baffle between the diffusion pump and the vacuum chamber reduces the net pumping speed but allows the attainment of lower pressures by condensing the pump-fluid vapor. The turbomolecular pump, employing many turbine blades rotating between slotted stator disks, also has a broad pumping range up to about 0.01 torr, as shown by curve TM in Fig. 14.4.1.

TYPES AND SIZES

The number, types, and sizes of vacuum pumps in series required to compress gas from the chamber up to atmospheric pressure usually depends on the compression ratio per pumping stage at the maximum throughput during the process cycle. For **ultrahigh vacuum** systems (pressures less than 10^{-9} torr) the number of pumps or stages required to reach the lowest pressures may depend on the rate of back leakage or back diffusion through the pumps at the minimum throughput. **Compression ratio** is defined as the ratio of the outlet (exhaust or discharge) pressure to the inlet (or intake) pressure at a given throughput when

pumping a gas or vapor that is not absorbed or condensed within the pump. This ratio is a variable quantity which depends in general on the ratio of the pumping speed of a given stage to the net speed of the fore pump, or next stage of compression. However, for a given throughput there is an upper limit to the compression ratio which can be maintained across a vapor jet pumping stage at a given power input. At maximum throughput under normal operation the limiting compression ratio for single-stage ejector pumps is about 10, and for the individual stages of multistage diffusion pumps the limiting ratio is about 4. Thus, a three-stage steam ejector is required to compress air from 0.8 to 760 torr, a ratio of about 10^3. Oil diffusion pumps usually have from three to five stages, and a four-stage diffusion pump can compress air from 1.5 mtorr to 0.4 torr at maximum full-speed throughput, corresponding to a total compression ratio of about 4^4 or 256. The overall compression ratio can be much higher at very low throughputs, but the **forepressure** (pressure at the outlet) should not be allowed to exceed a value known as the **limiting forepressure** or **tolerable forepressure**, which is usually in the range of 0.2 to 0.6 torr for a four-stage diffusion pump.

Oil-sealed mechanical piston pumps can be operated with high compression ratios over a pressure range from atmospheric pressure (or above) to about 1 mtorr. Molecular drag, centrifugal, rotary blower, e.g., Roots-type, pumps, and turbomolecular pumps are limited by back flow or leakage of gas from the forepressure side through the clearances between the rotor and stator, and therefore require backing pumps to reduce the forepressure from atmospheric to a sufficiently low value to permit a high vacuum to be obtained on the inlet side. Oil vapor-jet pumps cannot discharge gas directly to the atmosphere because of limitations on the vapor pressure which can be generated without excessive thermal decomposition of the oil, and backing pumps must therefore be used to reduce their forepressures to values less than the vapor pressure of the oil at a safe temperature in the boiler. Some gas molecules (particularly hydrogen and helium) may succeed in diffusing back through a vapor jet from the forevacuum to the high vacuum, and for ultrahigh vacuum systems this back diffusion must be reduced by using a sufficient number of pumping stages and vapor jets with high vapor density and velocity.

Table 14.4.1 lists types, sizes, and operating characteristics of typical commercially available pumps. The maximum (standard measured) speed (in litres per second) of oil diffusion pumps is about $18D^2$ for three-stage models and $28D^2$ for four-stage models, where D is the true inlet diameter (in inches). The larger pumps are usually water-cooled, and about 80 percent of the heat generated must be carried away by the cooling water, the remainder being lost by convection, radiation, and conduction to the surroundings. The exit temperature of the cooling water should usually not exceed 40°C, and the required rate of flow in cubic centimeters per minute of water having an inlet temperature of 20°C and an exit temperature of 30°C is about numerically equal to the power input in watts. This rule of thumb applies to either diffusion pumps or water-cooled mechanical pumps where the horsepower of the motor is converted to watts by multiplying by 746.

Table 14.4.1   Types and Available Sizes of High-Vacuum Pumps

Type	Nominal inlet ID, in	Maximum air speed L/s	Starting pressure, torr	Ultimate pressure, torr	Power input, kW
Rotary piston	0.5–8	0.1–400	(atm)	$\sim 10^{-3}$	0.2–30
Blower (Roots)	2–24	40–13,000	30–100	$\sim 10^{-3}$	0.5–70
Turbodrag (hybrid)	2–12	60–2,000*	5–10	$\sim 10^{-8}$	0.1–2.6
Turbomolecular	2–26	50–10,000*	~ 0.1	$< 10^{-9}$	0.4–6
Diffusion (vapor)	2–48	100–90,000	~ 0.5	$< 10^{-9}$	0.3–30
Diffusion ejector	12–24	4,000–12,500	2–5	$\sim 10^{-4}$	6–23
Getter-ion (sputter)	0.3–36	1–50,000	~ 0.02	$< 10^{-10}$	0.1–50
Sorption	1–2	1–80	(atm)	$\sim 10^{-3}$	0.4–2
Cryopump	4–48	400–57,000*	$\sim 10^{-3}$	$< 10^{-10}$	1–8
Nonevaporable getter	3–8	150–1600†	$\sim 10^{-4}$	$< 10^{-10}$	0.2–1.8

* Nitrogen speed.
† Hydrogen speed.
1 torr = 133.32 Pa.

Diffusion Pumps Figure 14.4.2 shows in cross section a typical vacuum system with a diffusion pump A having three stages (annular nozzles forming vapor jets). The operating fluid, which is usually a low-vapor-pressure organic or silicone oil of molecular weight ranging from 350 to 500, is located in the bottom of the metal casing and is

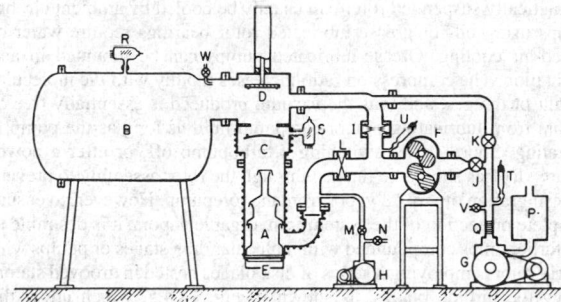

Fig. 14.4.2 Typical high-vacuum system. A = diffusion pump; B = chamber; C = baffle; D = valve; E = pump heater; F = Roots-type blower; G = roughing (mechanical) pump; H = holding (mechanical) pump; I to N = valves; P = oil reservoir and separator; Q = bellows; R, S = ion-gage tubes; T, U = Pirani-gage tubes; V, W = air-inlet valves.

heated by an external electrical heating unit E to about 200°C to generate the vapor. Mercury can be used as the pump fluid, but efficient refrigerated traps are required to keep the mercury vapor out of the vacuum chamber B. Oil diffusion pumps require some form of purging or purification of the oil during use to eliminate dissolved gases and volatile decomposition products. **Fractionating oil diffusion pumps** purify the oil by circulating it through a series of boilers, or boiler compartments, feeding vapor through separate chimneys to the various nozzles in a multistage pump. The volatile impurities are ejected with the vapor feeding the stages nearest the fore vacuum, and the purged oil of lowest vapor pressure supplies the top nozzle from which vapor molecules scattered back out of the inlet port create a partial pressure of oil vapor in the high vacuum which limits the ultimate pressure obtainable without cold traps [Hickman, *J. Appl. Phys.* **11**, 303 (1940)].

Getter-ion pumps or **sputter-ion pumps** employ chemically active metal layers which are continuously or intermittently deposited on the wall of the pump by either thermal evaporation or sputtering and which chemisorb oxygen, nitrogen, water vapor, and other active gases while the inert gases such as helium, neon, and argon are "cleaned up" by ionizing them in an electric discharge and drawing the positive ions to the wall where the neutralized ions are buried by fresh deposits of metal. These pumps require a roughing pump to reduce the pressure to less than about 20 mtorr at which point the active metal (usually titanium) can be evaporated or sputtered at the required rate, but after they begin operation no backing pump is required since all of the gas is trapped at the wall. When isolated by valves from the roughing pump, the getter-ion pumps form an enclosure sealed to the vacuum system so that a power failure cannot result in leakage of atmospheric air or vapors from a forepump into the system. Sputter-ion pumps can operate continuously for more than 1 year at pressures below 10^{-6} torr.

Cryopumps consist of one or more exposed surfaces refrigerated to a temperature usually below 100 K, at which certain gases will be condensed and form a layer having an equilibrium vapor pressure below a specified limit. Figure 14.4.3 shows a cryopump with plate arrays cooled to 15 K and 80 K directly by a mechanical cryocooler, usually based on the Gifford-McMahon cycle (Sec. 19.2) and charged with helium gas. Activated charcoal is fixed inside the 15 K array panels. Cooldown time is about 1 to 3 h. For vacuum systems to be cycled to atmospheric pressure, a high-vacuum valve should be placed over the cryopump inlet and closed when venting the system. The condition for again opening the valve is that the product of system volume and pressure in torr-litres should be slightly below the manufacturer's rated crossover limit (ranging from about 40 to 500 depending on the pump

size). When the cryopanels become saturated with thick condensed layers and the charcoal needs reactivation, the inlet valve is closed and the cryocooler switched off while the pump is allowed to warm up and purged with warm dry nitrogen. A pressure-relief valve is incorporated in the cryopump to protect against too-rapid evaporation of the condensed gases, and since hydrogen gas may be released first there should be no object present which might ignite the gas mixture. A plate cooled to 20 K, by circulating helium gas from a refrigeration unit through coils attached to the plate, will condense N_2, CO_2, CO, H_2O, O_2, Ar, and Xe to maintain partial pressures of these gases less than 10^{-10} torr.

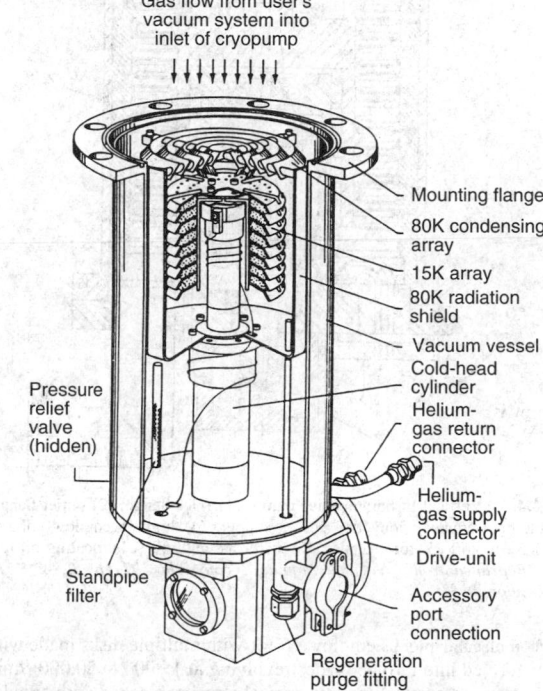

Gas flow from user's vacuum system into inlet of cryopump

Mounting flange
80K condensing array
15K array
80K radiation shield
Vacuum vessel
Cold-head cylinder
Helium-gas return connector
Helium-gas supply connector
Drive-unit
Accessory port connection
Regeneration purge fitting
Standpipe filter
Pressure relief valve (hidden)

Fig. 14.4.3 Cryopump with two-stage mechanical refrigeration unit. (*Courtesy CTI-CRYOGENICS.*)

Hydrogen, helium, and neon are not adequately condensed at 20 K but may be "cryotrapped" in a deposit of H_2O and other gases condensing on the plate. Cryopumping is used in space simulation chambers to create the necessary low pressures and to act as a heat sink comparable to "cold black space." In this application the 20-K plates are shielded by liquid-nitrogen-cooled panels, and the whole array is shaped to cover the inside wall of the vacuum chamber.

Sorption pumps employ a sorbent such as activated charcoal or synthetic zeolite (Molecular Sieve) cooled by liquid nitrogen or other refrigerant. They can be used to rough a system down from atmospheric pressure to a few millitorr at which getter-ion pumps may begin operation, or an additional preconditioned sorption pump can be valved in to reduce the pressure to 10^{-5} torr or less.

Nonevaporable getter pumps employing cartridges or strips of zirconium-aluminum alloy or zirconium-vanadium-iron alloy mounted on flanges with central heating elements can be directly inserted in the wall of vacuum systems or enclosed in a separate housing attached to the system. These pump all active gases, especially hydrogen, which is the common residual gas in ultrahigh-vacuum systems. When very high pumping speed is required, the getter is heated above room temperature to promote diffusion of gas into the bulk material. The Zr-Al alloy must first be activated by heating for about 45 min to 700°C, but the Zr-V-Fe alloy needs to be heated to only about 450°C. Starting pressure is less than 10^{-4} torr and ultimate pressure for active gases can be less than

10^{-10} torr. Regeneration of the getter by heating is periodically required. The pumping of hydrogen is reversible by controlling the temperature, and an application is the pumping, storing, and releasing of hydrogen isotopes.

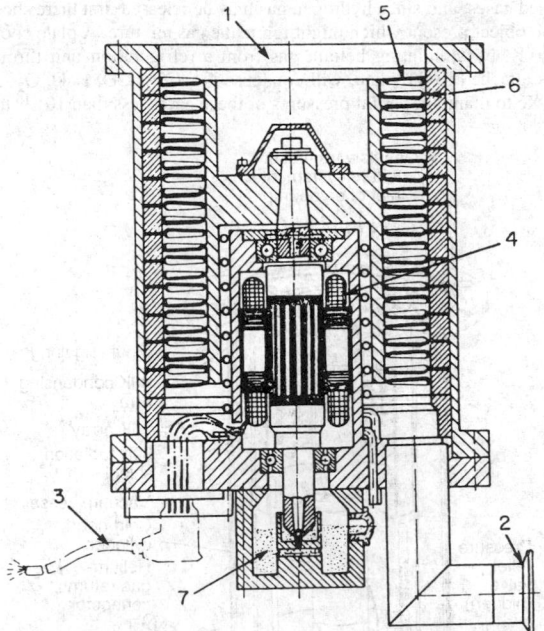

Fig. 14.4.4 Vertical turbomolecular pump. (1) Inlet flange; (2) outlet flange; (3) cable to frequency-converter power supply; (4) high-speed, low-voltage, three-phase motor; (5) rotor blades; (6) stator assembly; (7) lubricating oil reservoir. *(Reprinted from "Vacuum Manual," copyright 1974 by E. & F. N. Spon, London.)*

Turbomolecular pumps employ a rotor with multiple disks made with slots or formed into turbine blades revolving at 15,000 to 50,000 r/min (depending on diameter) in a cylindrical housing stacked with annular stator disks fitting between the rotor disks and also provided with slots or blades inclined at a reverse angle. In the horizontal-axis, dual-disk

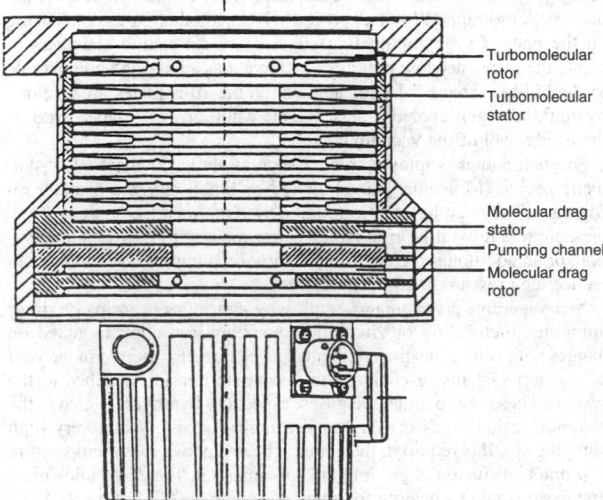

Fig. 14.4.5 Hybrid turbomolecular pump with molecular drag stages. *(Reprinted with permission from the Journal of Vacuum Science and Technology, vol. 11, p. 1616, 1993.)*

assembly turbopump (U.S. Patent 2,918,208), the gas enters through a large central port and then flows to the left or right through separate disk assemblies parallel to the axis. In the vertical single-disk assembly turbopump (Fig. 14.4.4), the gas enters through a protective screen in a large inlet port (1) and is impelled downward parallel to the axis by rotor blades (5) with tip speeds of about 300 m/s. Turbopumps with a magnetically suspended rotor can usually be cooled by ambient air, but pumps using oil- or grease-lubricated rotor bearings require water or forced-air cooling. Grease-lubricated pumps can be mounted in any orientation. The compression ratio increases rapidly with the molecular weight of the gas, and thus the vacuum produced is essentially free of vapors from lubricating oils or forepump fluid as long as the pump is operating. Shortly after switching a turbopump off, or after a power failure, dry gas should be flushed through the rotor assembly to prevent back migration of any oil vapor from the forepump. However, to ensure complete protection of the system from organic vapors, it is desirable to use turbopumps compounded with molecular drag stages or pumps with hybrid rotors employing a series of disk blades nested in grooved stators below the turbine blades, as shown in Fig. 14.4.5, which allow the operating exit pressure to be extended up to 10 torr so that oil-vapor-free "dry" mechanical backing pumps can be used. (M. H. Hablanian, *J. Vac. Sci. Technol. A*, **11,** p. 1614, 1993.)

INSTALLATION

In a typical assembly of pumps for a high vacuum system (Fig. 14.4.2) the diffusion pump A must be located close to the chamber B and should always have a water-cooled or refrigerated baffle C over the pump inlet so that vapor scattered back from the jets (backstreaming) will be condensed and returned as liquid to the pump. The pipe or manifold, which may include a valve D, connecting the chamber and the baffled pump should be of diameter equal to or larger than the inlet of the pump, and the length of the passage between the baffled pump and the chamber should preferably be not more than about 3 times the mean diameter of the passage. A space at least 6 in high should be allowed below the diffusion pump boiler E for easy servicing of the heaters, draining the pump fluid, or removal of the pump from the system.

Some mechanical pumps have appreciable vibration so that they should be firmly anchored or mounted on vibration damping pads, and flexible metal bellows Q should be installed in the forevacuum line. These bellows also aid in aligning the pumps during assembly. It is not necessary to have the mechanical pumps near the chamber because the resistance to gas flow of moderate lengths of pipe having the same diameter as the inlet of these pumps is not sufficiently large to create a serious pressure drop at mean pressures above 200 mtorr. Below 200 mtorr the pressure drop along the pipe is larger, but usually the diffusion pump operation is not affected unless the limiting forepressure (about 100 to 600 mtorr) is exceeded at the pump outlet.

VAPOR CONTAMINATION

The vapors evolved from vacuum systems may be condensed within mechanical pumps or the intercondensers of steam-ejector systems. If the condensed vapor is the same as the working fluid in the pump, such as water vapor condensed in "wet" vacuum pumps, no harm is done, provided that the fluid level is maintained at the optimum working value. If the condensed vapor can be separated from the working fluid by centrifuging, evaporation, filtering, or settling, suitable separating means can be installed. In certain types of rotary pumps the Gaede **gas-ballast** principle may be employed to avoid condensation by admitting air at a certain point in the compression cycle. If separation cannot be satisfactorily accomplished during operation of the pump, the vapor should be condensed in cold traps or liquid absorption columns before reaching the pump.

Baffles and **traps** are often required to prevent the backstreaming or migration of the pump operating fluid or sealing fluid into the high-vacuum chamber. Unless cooled baffles and traps or sorption traps for oil vapor are included in the connecting line, the ultimate pressure in the vacuum chamber will usually not be less than the vapor pressure of the pump fluid at ambient temperature. While a water-cooled "optically tight" baffle installed above the inlet of a diffusion pump will condense the backstreaming pump-fluid vapor and return most of the fluid to the pump, some fluid will reevaporate from such a baffle and migrate back into the vacuum chamber. To reduce the partial pressure of pump-fluid vapor below the vapor pressure at the water cooling temperature it is necessary to add a cold trap or baffle refrigerated to temperatures usually below the pour point of the fluid. These cold traps must be periodically warmed and the condensate removed to avoid inefficient cooling of the exposed surfaces during operation.

FLOW OF GASES AT LOW PRESSURE

The pipe line between the high-vacuum pump and the vacuum chamber limits the volumetric flow so that the **net pumping speed** as measured by a vacuum gage located in the chamber is given by $S_n = S_0U/(S_0 + U)$, where S_0 is the measured speed of the pump at its inlet and U is the **conductance** of the pipe defined by $U = Q/(P_n - P_0)$, where Q is the throughput, P_n is the pressure in the chamber, and P_0 is the pressure near the inlet of the pump as measured by a gage installed in a similar manner to that used to determine the pump speed S_0. When there is no loss or gain of gas within the pipe line, $Q = S_nP_n = S_0P_0 = U(P_n - P_0)$.

The conductance of a pipe depends on the geometry and the **Knudsen number** K (defined as the ratio of the mean free path of the gas molecules to the mean diameter of the cross section) as well as the direction and velocity of the molecules entering the pipe. For $K > 1$ the conductance for air in litres per second at 25°C of a circular pipe of length L (feet) and inside diameter D (inches) connecting a high-vacuum pump to a chamber of diameter greater than $3D$, including the "entrance correction" at the chamber, but neglecting the "exit correction" at the pump which depends on the inlet diameter and other factors, may be calculated from $U = 6.6D^3/(L + 0.11D)$. The effect of right-angle bends in the pipe for "molecular flow" $K > 1$ and for $L > D/3$ may be approximated by adding $0.05D$ to L for each bend (where L is in feet and D in inches). A single right-angle bend in a short pipe ($L < D/3$) has practically no effect on the conductance as computed for a straight pipe of the same length along the centerline [Davis, *J. Appl. Phys,* **5**, 358 (1954)].

When $K < 0.01$, the conductance in litres per second for air at 20°C of a long circular pipe length L (feet) and diameter D (inches) may be calculated from $U = 0.25D^4\overline{P}/L$, where $\overline{P}$ is the mean pressure in the pipe in millitorr (micrometers of Hg). For $0.01 < K < 1$ the conductance for air at 20°C of a long tube can be estimated from

$$U = \frac{0.25D^4\overline{P}}{L} + \left(\frac{1 + 0.65D\overline{P}}{1 + 0.80D\overline{P}}\right)\left(\frac{6.6D^3}{L}\right)$$

The size of the primary pump, which acts as a "roughing pump" to pump the chamber down from atmospheric pressure to a pressure at which a diffusion pump or other high-vacuum pump can operate, may depend on the peak gas load during the process as well as the desired pump-down time. However, the size indicated by the peak load condition is frequently much smaller than the size required to meet the specified pump-down or roughing time. In this case, if the process cycle is much longer than the pump-down time, it is advisable to use two fore-pumps (primary pumps), a large one (G in Fig. 14.4.2) for roughing down and a smaller one (H in Fig. 14.4.2) for holding the vapor pumps during the roughing period and backing them during the processing period. Roots-type blowers are useful for shortening the roughing time for large chambers in the range below 20 torr and for handling unusually large bursts of gas which occur in some processes (F in Fig. 14.4.2).

An oil-sealed rotary mechanical pump is normally used as the primary pump, and the roughing time t_r (in minutes) required to evacuate a chamber of volume V (cubic feet) from atmospheric pressure to about 0.7 torr at which high-vacuum pumps begin to operate can be estimated

from $t_r = 10V/C$, where C is the rated speed (cubic feet per minute) at atmospheric pressure of the rotary pump. Since the speed of the high-vacuum pumps is usually of the order of 100 times that of the forepump, the pressure should drop quickly as soon as the high-vacuum pumps "take hold," but below 10^{-4} torr the pressure may begin to decrease more slowly because of the out-gassing of the materials exposed inside the vacuum system.

For most of the materials exposed in high-vacuum systems the **outgassing rate** can be assumed proportional to the exposed area A_m, although for some very porous materials the rate is more proportional to the bulk or mass. Except for evaporation from pure liquid or solid phases, the outgassing rate normally decreases with time when the temperature is constant. For many industrial-type vacuum systems it has been found that the pressure P in the chamber decreases approximately according to $P = P_u + K_1A_m/S_nt^\alpha$ where P_u is the ultimate pressure, K_1 is a constant which may be considered equal to the average outgassing rate per unit area after 1 h of pumping, A_m is the exposed area, t is the total pumping time in hours, and α is an exponent which is usually nearly constant for the first few hours. For rough calculations in typical metal vacuum systems it may be assumed that $\alpha = 1$, but for systems containing large amounts of elastomers or plastics α may be closer to 0.5 [Dayton, *Trans. 6th Natl. Vacuum Symp.,* pp. 101–119, Pergamon Press, Oxford, 1960; Kraus, ibid., pp. 204–205].

When the system is first evacuated after prolonged exposure to the atmosphere, most of the outgassing load for the first 10 to 100 h is usually water vapor, and for unbaked metal systems the numerical value of K_1 is of the order of 10^{-4} when A_m is in square feet and S_n in litres per second. For systems containing large amounts of elastomers or plastics K_1 is more of the order of 10^{-3} to 10^{-2}. Systems constructed entirely of metal and glass may be heated to temperatures as high as 500°C to accelerate the outgassing, the average rate increasing by approximately a factor of 10 for each 100°C increase in temperature. In order to reach 10^{-6} torr in 10 h in an unconditioned vacuum chamber, the net pumping speed (in litres per second) should be 10 to 100 times the exposed area (in square feet).

The time t_v (in minutes) to vent a vacuum chamber of volume V (cubic feet) from pressures less than 1 mtorr to atmospheric pressure through a standard globe valve of nominal diameter D (inches) is given approximately by $t_v = V/100D^2$. The oil in the boiler of oil diffusion pumps should be cooled to below 100°C before opening the pump to atmospheric pressure.

APPLICATIONS OF HIGH-VACUUM PUMPS

Steam ejectors, Roots-type blowers, diffusion-ejector oil vapor pumps, and rotary piston pumps are used to produce pressures in the 10^{-1} to 10^{-4} torr range for distillation of plasticizers, fat-soluble vitamins, and certain other organic chemicals; for dehydration of frozen foods, animal tissue, blood plasma, serum, and antibiotics; for refining, degassing, and casting metals in vacuum furnaces; for vacuum sintering of cermets and powder metallurgy parts; for vacuum annealing of special alloys; and for pumping down vacuum chambers to a pressure at which high vacuum pumps can begin operation. Diffusion pumps, getter-ion pumps, and turbomolecular pumps are used to evacuate radio, television, and X-ray tubes and vacuum-insulated containers before sealing off; to produce pressures in the 10^{-3} to 10^{-6} torr range in chambers for coating injection-molded plastic parts, rolls of plastic sheet, glass plates, and other substrates with aluminum and other metal films; for coating optical parts with antireflection films or layers producing dichroic mirrors and interference filters; for depositing metal and semiconductor films to produce microelectronic circuits; to maintain pressures of 10^{-4} to 10^{-7} torr in mass spectrometers, electron microscopes, synchrocyclotrons, betatrons, and other devices for accelerating or separating charged particles.

14.5 FANS
by Robert Jorgensen

REFERENCES: Jorgensen, "Fan Engineering," 8th ed., Buffalo Forge Company. "Laboratory Methods of Testing Fans for Rating," ANSI/ASHRAE Standard 51-85 and ANSI/AMCA Standard 210-85. "Fans and Systems," AMCA Publication 201. "Test Code for Sound Rating," AMCA Standard 300-86. "Standards Handbook," AMCA Publication 99-83. "ASME Performance Test Codes, Code on Fans," ANSI/ASME PTC 11-1984.

Symbols

A = area, ft² (m²)
b = barometric pressure, in Hg (Pa)
D = diameter, ft (m)
D_s = specific diameter, ft (m)
H = fan power input, hp (W)
H_0 = fan power output, hp (W)
h = head, ft (m)
K_p = compressibility factor, dimensionless
L_p = sound pressure level, dB
L_w = sound power level, dB
L_{ws} = specific sound power level, dB
$\log$ = logarithm to base 10
M = molecular weight, dimensionless
N = speed of rotation, r/min
N_s = specific speed, r/min
n = number or polytropic exponent
p_s = fan static pressure, in wg (Pa)
p_t = fan total pressure, in wg (Pa)
p_v = fan velocity pressure, in wg (Pa)
p_{sx} = static pressure at plane x, in wg (Pa)
P_{tx} = total pressure at plane x, in wg (Pa)
P_{vx} = velocity pressure at plane x, in wg (Pa)
$\dot{Q}$ = fan capacity, ft³/min (m³/s)
$\dot{Q}_x$ = volumetric flow rate at plane x, ft³/min (m³/s)
T = absolute temperature, °R (K)
W = power, W
x = factor used to determine K_p, dimensionless
z = factor used to determine K_p, dimensionless
γ = isentropic exponent, dimensionless
η_s = fan static efficiency, per unit
η_t = fan total efficiency, per unit
ρ = fan air density, lbm/ft³ (kg/m³)
ρ_x = air density at plane x, lbm/ft³ (kg/m³)

Subscripts

1 = fan inlet plane
2 = fan outlet plane
a = absolute
b = basic known conditions
c = calculated condition
r = reading
x = plane 1, 2, 3, or other

FAN TYPES AND NOMENCLATURE

Any device which produces a current of air may be called a fan. This discussion will be limited to fans which have a rotating impeller to produce the flow and a stationary casing to guide the flow into and out of the impeller (see Fig. 14.5.1). The form of the casing or the impeller may vary widely.

Fan Classification

One of the characteristics by which fans are classified is the nature of the flow through the blade passages of the impeller. Axial flow, radial flow, mixed flow, and cross flow are all possible in fan impellers. Certain fan names result from these classifications. Other fan names derive from other characteristics.

Propeller fans, tube-axial fans, and vane-axial fans all utilize **axial-flow impellers,** but their casings differ. Propeller fans may be mounted in a ring or panel. Tube-axial fans and vane-axial fans both use tubular casings, but for vane-axial fans they are equipped with stationary guide vanes. A great deal of the energy transferred to the air in axial-flow machines is in kinetic form. Some of this kinetic energy can be transformed into pressure energy by straightening the swirl, e.g., with vanes, or by reducing the exit velocity, e.g., with a diffuser. Propeller fans effect very little transformation and hence have very low pressure-producing capability. Vane-axial fans can be equipped for maximum transformation as well as high transfer of energy and hence have high pressure-producing potential depending on tip speed and blade angles. High hub ratios promote high energy transfer.

Centrifugal fans and tubular centrifugal fans both utilize **radial-flow impellers.** Centrifugal fans usually employ a volute or scroll-type casing, the flow entering the casing axially and leaving tangentially. Tubular centrifugals use tubular casings so that the flow both entering and leaving the casing is axial. A considerable portion of the energy transferred to the air in a radial-flow machine is due to centrifugal action; hence the name centrifugal fan. Since centrifugal action varies with blade depth, the pressure-producing capability of radial-flow fans will vary with this factor as well as tip speed and blade angles.

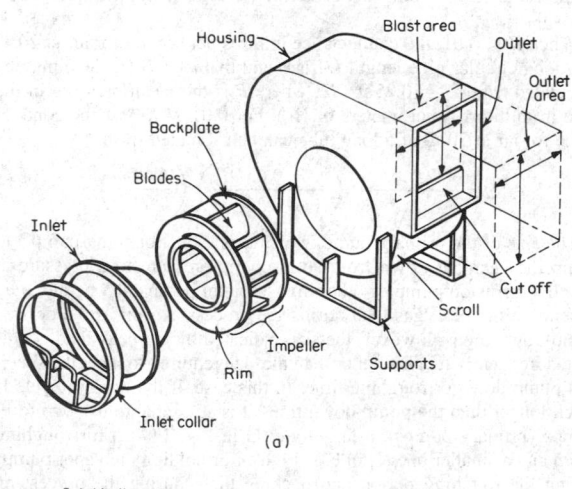

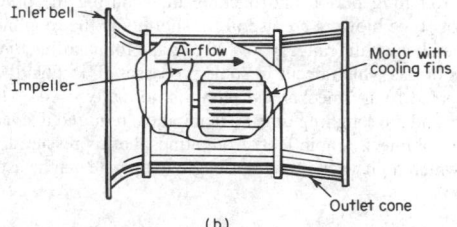

Fig. 14.5.1 Elements of fans and preferred terminology. (*a*) Centrifugal fans; (*b*) axial fans. (*Based on ACMA Publ. 201.*)

Mixed-flow impellers can be used in either axial or scroll-type casings. They are characterized as mixed flow because both axial and radial flow take place in the blading. Mixed-flow impellers used in axial-flow casings have a hub similar to a pure axial-flow impeller, but the inlet portion of the blading extends down over the face of the hub, thereby giving some radial guidance. Mixed-flow impellers used in scroll-cased ~~on ~~ blades which give most of the axial guidance in the inlet ~~ of the radial guidance in the discharge portion. ~~ the air passes through the blading twice, en- ~~ through the tip, passing across the im- ~~ ~~ as are designed to provide this ~~ ~~ own as tangential fans or depends

Fan Details

Propeller fans and other **axial fans** may use ~~ blade.low and depends sections or blades of uniform thickness. Blading may be fixed, adjustable at standstill, or variable in operation. Propeller fans have very small hubs. Hub-to-tip diameter ratios ranging from 0.4 to 0.7 are common in vane-axial fans. The larger the hub, the more important it is to have an inner cylinder approximately the hub size located downstream of the impeller. The guide vanes of a vane-axial fan are located in the annular space between the tubular casing and the inner cylinder. Diffusers are generally used between the fan and the discharge ductwork.

Centrifugal fans use various types of blading. Forward-curved blades are shallow and curved so that both the tip and the heel point in the direction of rotation. Radial and radial-tip blades both are radial at the tip, but the latter are curved at the heel to point in the direction of rotation. Backward-curved and backward-inclined blades point in the direction opposite rotation at the tip and in the direction of rotation at the heel. All the above blades are of uniform thickness and are designed for radial flow. Airfoil blades have backward-curved chord lines so that the leading edge of the airfoil is at the heel pointing forward and the trailing edge at the tip pointing backward with respect to rotation. Impellers for all blade shapes are usually shrouded and may have single or double inlets. Blade widths are related to the inlet-to-tip-diameter ratio. Tip angles may vary widely, but heel angles should be set to minimize entrance losses. Scroll casings may be fitted with a streamlined inlet bell, an inlet cone, or simply a collar.

Tubular centrifugals may be designed for backward-curve, air-foil or mixed-flow impellers. An inlet bell and discharge guide vanes are required for good performance.

Cross-flow fans utilize impellers with blading similar to that of a forward-curved centrifugal, but the end shrouds have no inlet holes. Blade-length-to-tip-diameter ratios are limited only by structural considerations.

Power roof ventilators may use either axial- or radial-flow impellers. The casings will include either a propeller fan mounting ring or a tubular casing to guide the flow for an axial-flow impeller. If a radial-flow impeller is used, an inlet bell is required, but the scroll case may be replaced by the ventilator hood.

All fan types may have direct or indirect drive arrangements. Various standards including arrangement numbers have been adopted by AMCA. Arr. 1 signifies an overhung impeller on a fan shaft with two bearings on a base; Arr. 3 signifies an impeller on a fan shaft between bearings; Arr. 4 is for overhung impeller on a motor shaft; Arr. 7 is Arr. 3 with a motor base; Arr. 8 is Arr. 1 with an extended base for motor; Arr. 9 is Arr. 1 with motor mounted on side of unit; Arr. 10 is Arr. 1 with motor mounted inside of base. Rotation is specified as cw or ccw when viewed from the drive side. All fans can be equipped with variable-speed drives, variable inlet vanes, or dampers, but controllable-pitch axials do not need speed or vane control.

FAN PERFORMANCE AND TESTING

The conventional terms used to describe fan performance in the United States are defined blow.

Fan Air Density Air density is the mass per unit volume of the air. The density of a perfect gas is a function of its molecular weight, temperature, and pressure as indicated by

$$\rho_x = \frac{M}{386.7} \frac{529.7}{T_x} \frac{b_x}{29.92}$$

where the subscript x indicates the plane of the measurements. For dry air this reduces to

$$\rho_x = 1.325 b_x / T_x$$

This expression will usually be accurate enough even when moist air is involved. Fan air density is the density of the air corresponding to the total pressure and total temperature at the fan inlet

$$\rho = \rho_1$$

Fan Capacity Volume flow rate is usually determined from pressure measurements, e.g., a velocity pressure traverse taken with a Pitot tube or the pressure drop acrossa flowmeter. The average velocity pressure for a Pitot traverse is

$$p_{vx} = (\Sigma \sqrt{p_{vxr}} / n)^2$$

where subscript x indicates the plane of the measurements, subscript r indicates a reading at one station, n is the number of stations, and Σ is the summation sign. The corresponding capacity is:

$$\dot{Q}_x = 1{,}097 A_x \sqrt{p_{vx} / \rho_x}$$

For a flow meter

$$\dot{Q}_x = 1{,}097 CA_x Y \sqrt{\Delta p / \rho_x} / F$$

where C = coefficient of discharge of meter, Y = expansion factor for gas, Δp = measured pressure drop, and F = velocity-of-approach factor for the meter installation. Fan capacity is the volumetric flow rate at fan air density

$$\dot{Q} = \dot{Q}_x \rho_x / \rho$$

Fan Total Pressure Fan total pressure is the difference between the total pressure at the fan outlet and the total pressure at the fan inlet

$$p_t = p_{t2} - p_{t1}$$

When the fan draws directly from the atmosphere.

$$p_{t1} = 0$$

When the fan discharges directly to the atmosphere,

$$p_{t2} = p_{v2}$$

If either side of the fan is connected to ductwork, etc., and the measuring plane is remote, the measured values should be corrected for the approximate pressure drop

$$p_{t2} = p_{tx} + \Delta p_{2-x} \qquad p_{t1} = p_{tx} + \Delta p_{x-1}$$

Fan Velocity Pressure Fan velocity pressure is the pressure corresponding to the average velocity at the fan outlet

$$p_v = \left(\frac{\dot{Q}_2 / A_2}{1{,}097} \right)^2 \rho_2$$

Fan Static Pressure Fan static pressure is the difference between the fan total pressure and the fan velocity pressure. Therefore, fan static pressure is the difference between the static pressure at the fan outlet and the total pressure at the fan inlet

$$p_s = p_t - p_v = p_{s2} - p_{t1}$$

Fan Speed Fan speed is the rotative speed of the impeller.

Compressibility Factor The compressibility factor is the ratio of the fan total pressure p'_t that would be developed with an incompressible fluid to the fan total pressure p_t that is developed with a compressible fluid, all other conditions being equal:

$$K_p = \frac{p'_t}{p_t} = \frac{[n/(n-1)] [(p_{t2a}/p_{t1a})^{(n-1)/n} - 1]}{(p_{t2a}/p_{t1a}) - 1}$$

Compressibility factor can be determined from test measurements using

$$x = p_t/p_{t1a}$$
$$z = [(\gamma - 1)/\gamma]\, 6356H/\dot{Q}p_{t1a}$$
$$K_p = [z \log (1 + x)]/[x \log (1 + z)]$$

Fan Power Output Fan power output is the product of fan capacity and fan total pressure and compressibility factor

$$H_o = \dot{Q}p_t K_p/6{,}356$$

Fan Power Input Fan power input is the power required to drive the fan and any elements in the drive train which are considered a part of the fan. Power input can be calculated from appropriate measurements for a dynamometer, torque meter, or calibrated motor.

Fan Total Efficiency Fan total efficiency is the ratio of the fan power output to the fan power input

$$\eta_t = \dot{Q}p_t K_p/6{,}356H$$

Fan Static Efficiency Fan static efficiency is the fan total efficiency multiplied by the ratio of fan static pressure to fan total pressure.

$$\eta_s = \eta_t p_s/p_t$$

Fan Sound Power Level Fan sound power level is 10 times the logarithm (base 10) of the ratio of the actual sound power in watts to 10^{-12} watts,

$$L_w = 10 \log (W/10^{-12})$$

Fan sound power level can be calculated from sound pressure-level measurements in a known acoustical environment. The standard laboratory method of testing is to use a calibrated sound source to calibrate a semireverberant room. In-duct test methods are being developed. The total sound power level of a fan is usually assumed to be 3 dB higher than either the inlet or outlet component. The casing component varies with construction but will usually range from 15 to 30 dB less than the total.

Head The difference between head and pressure is important in fan engineering. Both are measures of the energy in the air. Head is energy per unit weight and can be expressed in ft·lb/lb, which is often abbreviated to ft (of fluid flowing). Pressure is energy per unit volume and can be expressed in ft·lb/ft³, which simplifies to lb/ft² or force per unit area. The use of the inch water gage (in wg) is a convenience in fan engineering reflecting the usual methods of measurement. It sounds like a head measurement but is actually a pressure measurement corresponding to 5.192 lb/ft², the pressure exerted by a column of water 1 inch high. Pressures can be converted into heads and vice versa:

$$p = \rho h/5.192 \qquad h = 5.192p/\rho$$

For instance, for air at 0.075 lbm/ft³ density, 1 in wg corresponds to 69.4 ft of air. That is, a column of air at that density would exert a pressure of 5.192 lb/ft². Lighter air would exert less pressure for a given head. For a given pressure, the head would be higher with lighter air. Although it is not widely used in the United States, head is commonly used in a number of European countries.

Fans, like other turbomachines, can be considered constant-head-constant-capacity (volumetric) machines. This means that a fan will develop the same head at a given capacity regardless of the fluid handled, all other conditions being equal. Of course, this also means that a fan will develop a pressure proportional to the density at a given capacity, all other conditions being equal.

All the preceding equations are based on the U.S. Customary Units listed under Symbols. If **SI units** are to be used, certain numerical coefficients will have to be modified. For instance, substitute $\sqrt{2}$ for 1,097 in the flow equations when using Pa for pressure, kg/m³ for density, and m³/s for capacity. Similarly, substitute 1.0 for 6,356 in the power equations when using Pa for pressure, m³/s for capacity, and W for power.

Common metric practice in fan engineering leads to other numerical coefficients. When mm wg is used for pressure, m³/s for capacity, kW for power, and kg/m³ for density, substitute 4.424 for 1,097 in the flow equations and 102.2 for 6,356 in the power equations.

The preceding discussion utilized what may be termed the **volume-flow-rate-pressure** approach to expressing fan performance. An alternative, the **mass-flow-rate-specific-energy** approach is equally valid but not generally used in the United States. Refer to ASME PTC-11 for additional details.

FAN AND SYSTEM PERFORMANCE CHARACTERISTICS

The performance characteristics of a fan are best described by
The conventional method of graphing fan performance
of curves with capacity as abscissa and all right, i.e., the energy re-
System characteristics can be plo

System Char ... flow through the system (which can be expressed as pressure or head) varies approximately as the square of the flow. In some cases, the system characteristics will not pass through the origin because the energy required to produce flow through an element of the system may be controlled at a particular value, e.g., with venturi scrubbers. In some cases, the system characteristic will not be parabolic because the flow through an element of the system is laminar rather than turbulent, e.g., in some types of filters. Whatever the case, the system designer should establish the characteristics by determining the energy requirements at various flow rates. The energy requirement (pressure drops or head losses) for each element can be determined by reference to handbook or manufacturer's literature or by test.

The true measure of the energy requirement for a system element is the total pressure drop or the total head loss. Only if the entrance velocity for the element equals the exit velocity will the change in static pressure equal the total pressure drop. There are some advantages in using static pressure change, but the system designer is usually well advised to use total pressure drops to avoid errors in fan selection. The sum of the total pressure losses for elements on the inlet side of the fan will equal $-p_{t1}$. This should include energy losses at the entrance to the system but not the energy to accelerate the air to the velocity at the fan inlet, which is chargeable to the fan. The sum of the total pressure losses for elements on the discharge side of the fan will equal p_{t2}. This should include the kinetic energy of the stream issuing from the system. The total system requirement will be the arithmetic sum of all the appropriate losses or the algebraic difference $p_{t2} - p_{t1}$, which is also p_t.

A system characteristic curve based on total pressure is plotted in Fig. 14.5.2. A characteristic based on static pressure is also shown. The latter recognizes the definition of fan static pressure so that the only difference is the velocity pressure corresponding to the fan outlet velocity. This system could operate at any capacity provided a fan delivered the exact pressure to match the energy requirements shown on the system curve for that capacity. The advantages of plotting the system characteristics on the fan graph will become evident in the following discussions.

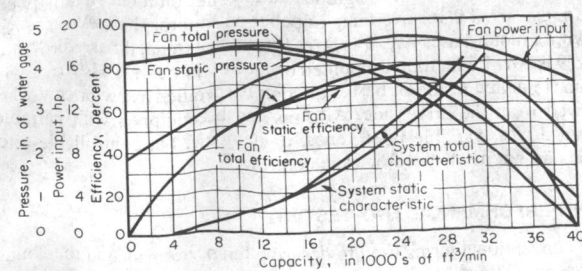

Fig. 14.5.2 Fan and system characteristics.

Fan Characteristics

The constant-speed performance characteristics of a fan are illustrated in Fig. 14.5.2. These characteristics are for a particular size and type of fan operating at a particular speed and handling air of a particular density. The fan can operate at any capacity from zero to the maximum shown, but when applied on a particular system the fan will operate only at the intersection of the system characteristics with the appropriate fan pressure characteristic. For the case illustrated, the fan will operate at $Q = 27,300$ ft^3/min and $p_t = 3.4$ in wg or $p_s = 3$ in wg, requiring $H = 18.3$ hp at the speed and density for which the curves were drawn. The static efficiency at this point of operation is 73 percent, and the total efficiency is 80 percent. If the system characteristic had been lower, it would have intersected the fan characteristic at a higher capacity and the fan would have delivered more air. Contrariwise, if the system characteristic had been higher, the capacity of the fan would have been less. Capacity reduction can be accomplished, in fact, by creating additional resistance, as with an outlet damper.

Figure 14.5.3 illustrates the characteristics of a fan with damper control, variable inlet vane control, and variable speed control. These particular characteristics are for a backward-curved centrifugal fan, but the general principles apply to all fans. Outlet dampers do not affect the flow to the fan and therefore can alter fan performance only by adding resistance to the system and producing a new intersection. Point 1 is for wide-open dampers. Points 2 and 3 are for progressively closed dampers. Note that some power reduction accompanies the capacity reduction for this particular fan. Variable-inlet vanes produce inlet whirl, which reduces pressure-producing capability. Point 4 is for wide-open inlet vanes and corresponds to point 1. Points 5 and 6 are for progressively closed vanes. Note that the power reduction at reduced capacity is better for vanes than for dampers. Variable speed is the most efficient means of capacity control. Point 7 is for full speed and points 8 and 9 for progressively reduced speed. Note the improved power savings over other methods. Variable speed also has advantages in terms of lower noise and reduced erosion potential but is generally at a disadvantage regarding first cost.

Figure 14.5.4 illustrates the characteristics of a fan with variable pitch control. These characteristics are for an axial-flow fan, but comparisons to the centrifugal-fan ratings in Fig. 14.5.3 can be made. Point 10 corresponds to points 1, 4, or 7. Reduced ratings are shown at points 11 and 12 obtained by reducing the pitch. Power savings can be almost as good as with speed control. Notice that in all methods of capacity control,

operation is at the intersection of a system characteristic with a fan characteristic. With variable vanes, variable speed, and variable pitch, the fan characteristic is modified. With damper control, the effect could be considered a change in fan characteristics, but it generally makes more sense to consider it a change in system characteristics.

Fan-System Matching

It has already been observed in the discussions of both system and fan characteristics that the point of operation for **one fan** on a particular system will be at the intersection of their characteristics. Stated another way, the energy required by the system must be provided by the fan exactly. If the fan delivers too much or too little energy, the capacity will be more or less than desired. The effects of utilizing dampers, variable speed, variable-inlet vanes, and variable pitch were illustrated, but in all cases the fan was preselected. The more general case is the one involving the selection or design of a fan to do a particular job. The design of a fan from an aerodynamicist's point of view is beyond the scope of this discussion. Fortunately, most fan problems can be solved by selecting a fan from the many standard lines available commercially. Again, the crux of the matter is to match fan capability with system requirements.

Most fan manufacturers have several standard lines of fans. Each line may consist of various sizes all resembling each other. If the blade angles and proportions are the same, the line is said to be **homologous.** There is only one fan size in a line of homologous fans that will operate at the maximum efficiency point on a given system. If the fan is too big, operation will be at some point to the left of peak efficiency on a standard characteristic plot. If the fan is too small, operation will be to the right of peak. In either case it will be off peak. A slightly undersized fan is often preferred for reasons of cost and stability.

Selecting and rating a fan from a catalog is a matter of fan-system matching. When it is recognized that many of the catalog sizes may be able to provide the capacity and pressure, selection becomes a matter of trying various sizes and comparing speed, power, cost, etc. The choice is a matter of evaluation.

There are various reasons for using **more than one fan** on a system. Supply and exhaust fans are used in ventilation to avoid excessive pressure build-up in the space being served. Forced and induced-draft fans are used to maintain a specified draft over the fire. Two fans may fit the available space better than one larger fan. Capacity control by various fan combinations may be more economical than other control methods.

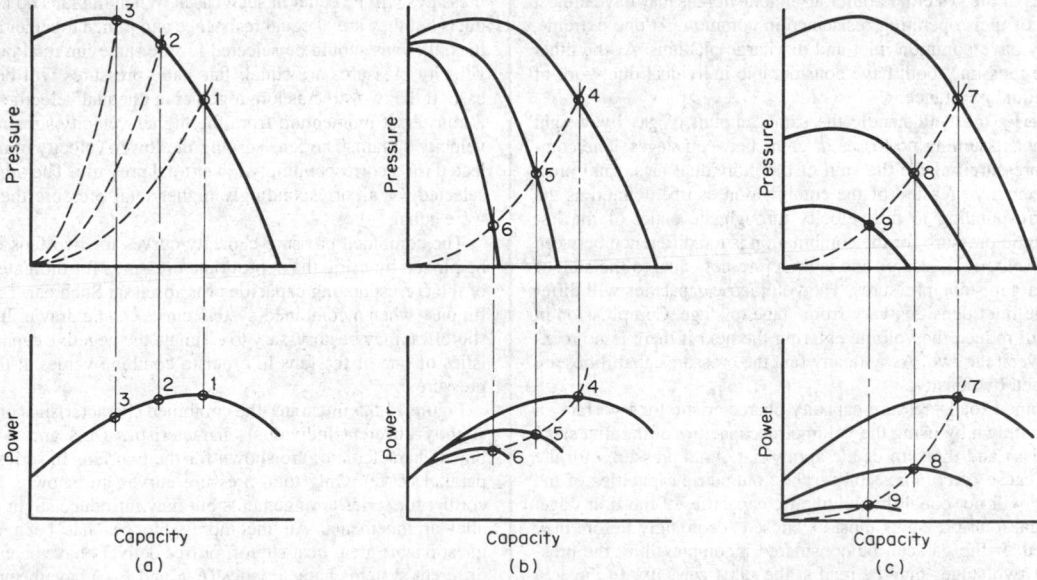

Fig. 14.5.3   Fan characteristics with (a) damper control, (b) variable-inlet vane control, (c) variable speed control.

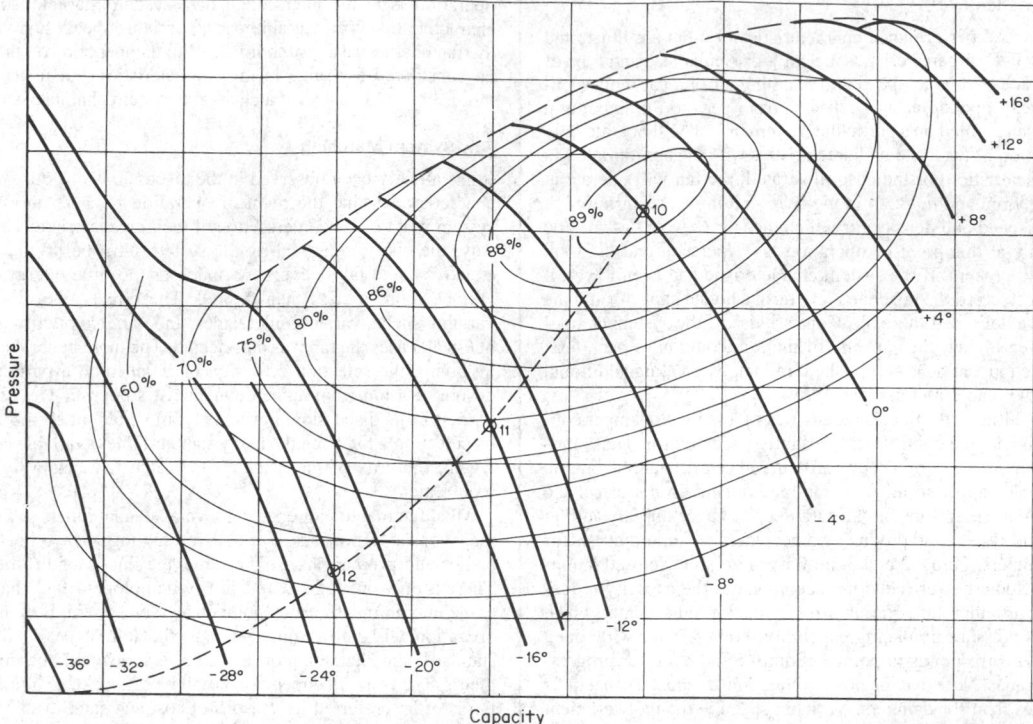

Fig. 14.5.4   Characteristics of an axial-flow fan with variable-pitch control.

Multistage arrangements may be necessary when pressure requirements exceed the capabilities of a single-stage fan. Standby fans are frequently required to ensure continuous operation.

When two fans are used, they may be located quite remote from each other or they may be close enough to share shaft and bearings or even casings. Double-width, double-inlet fans are essentially two fans in parallel in a common housing. Multistage blowers are, in effect, two or more fans in series in the same casing. Fans may also be in series but at opposite ends of the system. Parallel-arrangement fans may have almost any amount of their operating resistance in common. At one extreme, the fans may have common inlet and discharge plenums. At the other extreme, the fans may both have considerable individual ductwork of equal or unequal resistance.

Fans in **series** must all handle the same amount of gas by weight measurements, assuming no losses or gains between stages. The combined total pressure will be the sum of the individual fans, total pressures. The velocity pressure of the combination can be defined as the pressure corresponding to the velocity through the outlet of the last stage. The static pressure for the combination is the difference between its total and velocity pressures and is therefore not equal to the sum of the individual fan static pressures. The volumetric capacities will differ whenever the inlet densities vary from stage to stage. Compression in one stage will reduce the volume entering the next if there is no reexpansion between the two. As with any fan, the pressure capabilities are also influenced by density.

The combined total pressure-capacity characteristic for two fans in series can be drawn by using the volumetric capacities of the first stage for the abscissa and the sum of the appropriate total pressures for the ordinate. Because of compressibility, the volumetric capacities of the second stage will not equal the volumetric capacities of the first stage. The individual total pressures must be chosen accordingly before they are combined. If the gas can be considered incompressible, the pressures for the two stages may be read at the same capacity. In the area near free delivery, it may be necessary to estimate the negative pressure

characteristics of one of the fans in order to combine values at the appropriate capacity.

Fans in **parallel** must all develop sufficient pressure to overcome the losses in any individual ductwork, etc., as well as the losses in the common portions of the system. When such fans have no individual ductwork but discharge into a common plenum, their individual velocity pressures are lost and the fans should be selected to produce the same fan static pressures. If fan velocity pressures are equal, the fan total pressures will be equal in such cases. When the fans do have individual ducts but they are of equal resistance and joined together at equal velocities, the fans should be selected for the same fan total pressures. If fan velocity pressures are equal, fan static pressures will be equal in this case. If the two streams join together at unequal velocities, there will be a transfer of momentum from the higher-velocity stream to the lower-velocity stream. The fans serving the lower-velocity branch can be selected for a correspondingly lower total pressure. The other fan must be selected for a correspondingly higher total pressure than if velocities were equal.

The combined pressure-capacity curves for two fans in parallel can be plotted by using the appropriate pressures for ordinates and the sum of the corresponding capacities for abscissa. Such curves are meaningful only when a combined-system curve can be drawn. In the area near shutoff, it may be necessary to estimate the negative capacity characteristics of one of the fans in order to combine values at the appropriate pressure.

Figure 14.5.5 illustrates the combined characteristics of two fans with slightly different individual characteristics (*A-A* and *B-B*). The combined characteristics are shown for the two fans in series (*C-C*) and in parallel (*D-D*). Only total-pressure curves are shown. This is always correct for series arrangements but may introduce slight errors for parallel arrangements. An incompressible gas has been assumed. The questionable areas near shutoff or free delivery have been omitted. Two different system characteristics (*E-E* and *F-F*) have been drawn on the chart. With the two fans in series, operation will be at point *EC* or *FC* if

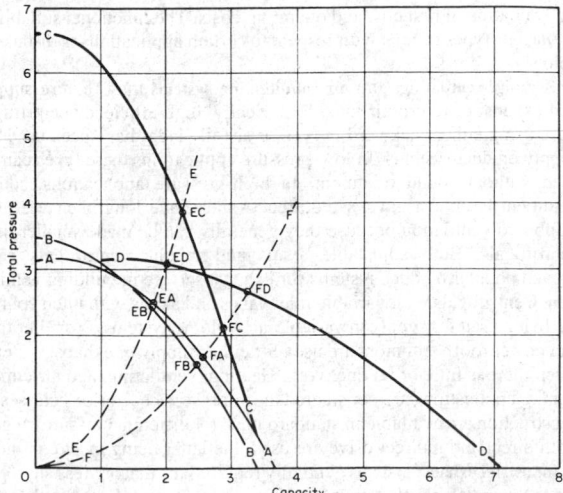

Fig. 14.5.5 Combined characteristics of fans in series and parallel.

the fan is on system E or F, respectively. Parallel arrangement will lead to operation at ED or FD. Single-fan operation would be at the point indicated by the intersection of the appropriate fan and system curves provided the effect of an inoperative second fan is negligible. Some sort of bypass is required around an inoperative fan in series whereas an inoperative fan in parallel need only be dampered shut. Parallel operation yields a higher capacity than series operation on system F, but the reverse is true on system E. For the type of fan and system characteristics drawn there is only one possible point of operation for any arrangement.

Fan Laws

The fan laws are based on the experimentally demonstrable fact that any two members of a homologous series of fans have performance curves which are homologous. At the same point of rating, i.e., at similarly situated points of operation on their characteristic curves, efficiencies are equal and other variables are interrelated according to the fan laws. If size and speed are considered independent variables and if compressibility effects are ignored, the fan laws can be written as follows:

$$\eta_{tc} = \eta_{tb}$$
$$\dot{Q}_c = \dot{Q}_b (D_c/D_b)^3 (N_c/N_b)$$
$$p_{tc} = p_{tb}(D_c/D_b)^2(N_c/N_b)^2(\rho_c/\rho_b)$$
$$H_c = H_b(D_c/D_b)^5(N_c/N_b)^3(\rho_c/\rho_b)$$
$$L_{wc} = L_{wb} + 70 \log (D_c/D_b) + 50 \log (N_c/N_b)$$
$$+ 20 \log (\rho_c/\rho_b)$$

The above laws are useful, but they are dangerous if misapplied. The calculated fan must have the same point of rating as the known fan. When in doubt, it is best to reselect the fan rather than attempt to use the fan laws.

The fan designer utilizes the fan laws in various ways. Some of the more useful relationships in addition to those above derive from considering fan capacity and fan total pressure as the independent variables. This leads to specific diameter, specific speed, and specific sound power level:

$$D_s = D(p_t/\rho)^{1/4}/\dot{Q}^{1/2}$$
$$N_s = N\dot{Q}^{1/2}/(p_t/\rho)^{3/4}$$
$$L_{ws} = L_w - 10 \log \dot{Q} - 20 \log p_t$$

D_s, N_s, and L_{ws} are the diameter, speed, and sound power level of a homologous fan which will deliver 1 ft³/min at 1 in wg at the same point of rating as Q and p_t for D and N. D_s and N_s can be used to advantage in fan selection. They can also be used by a designer to determine how

well a line will fit in with other lines. This is illustrated in Fig. 14.5.6. Each segment represents a particular fan line. Note the trends for each kind of fan. Incidentally, some fan engineers utilize different formulas for specific diameter and specific speed

$$D_{se} = Dp_{te}^{1/4}/\dot{Q}^{1/2} \qquad N_{se} = N\dot{Q}^{1/2}/p_{te}^{3/4}$$

where p_{te} is the equivalent total pressure based on standard air. This makes $D_{se} = D_s \times 1.911$ and $N_{se} = N_s \times 6.978$.

Specific sound power level is useful in predicting noise levels as well as in comparing fan designs. There appears to be a lower limit of L_{ws} in the vicinity of forty-five dB for the more efficient types ranging to 70 dB or more for cruder designs. Actual sound power levels can be figured from

$$L_w = L_{ws} + 10 \log \dot{Q} + 20 \log p_t$$

Another useful parameter which derives from the fan laws is orifice ratio

$$R_o = \dot{Q}/D^2(p_t/\rho)^{1/2}$$

This ratio can be plotted on a characteristic curve for a known fan. If the ratio is determined for a calculated homologous fan, the point of rating can be established by inspection. Other ratios can be used in the same manner including p_v/p_t, p_t/Q^2, D_s, and N_s.

Stability Considerations

The flow through a system and its fan will normally be steady. If the fluctuations occasioned by a temporary disturbance are quickly damped out, the fan system may be described as having a **stable** operating characteristic. If the unsteady flow continues after the disturbance is removed, the operating characteristic is **unstable**.

To ensure stable operation the slopes of the pressure-capacity curves for the fan and system should be of opposite sign. Almost all systems have a positive slope; i.e., the pressure requirement or resistance increases with capacity. Therefore, for stable operation the fan curve should have a negative slope. Such is the case at or above the design capacity.

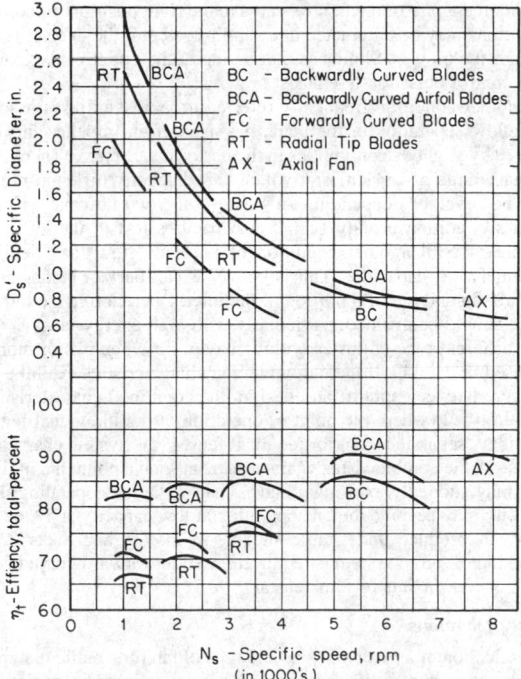

Fig. 14.5.6 Specific diameter and efficiency versus specific speed for single-inlet fan types.

When the slopes of the fan and system characteristics are of opposite sign, any system disturbance tending to produce a temporary decrease in flow is nullified by the increase in fan pressure. When the slopes are of the same sign, any tendency to decrease flow is strengthened by the resulting decrease in fan pressure. When fan and system curves coincide over a range of capacities, the operating characteristics are extremely unstable. Even if the curves exactly coincide at only one point, the flow may vary over a considerable range.

There may or may not be any obvious indication of unstable operation. The pressure and power fluctuations that accompany unsteady flow may be so small and rapid that they cannot be detected by any but the most sensitive instruments. Less rapid fluctuations may be detected on the ordinary instruments used in fan testing. The changes in noise which occur with each change in flow rate are easily detected by ear as individual beats if the beat frequency is below about 10 Hz. In any event, the overall noise level will be higher with unsteady flow than with steady flow.

The conditions which accompany unsteady flow are variously described as pulsations, hunting, surging, or pumping. Since these conditions occur only when the operating point is to the left of maximum pressure on the fan curve, this peak is frequently referred to as the **surge point** or **pumping limit**.

Pulsation can be prevented by rating the fan to the right of the surge point. Fans are usually selected on this basis, but it is sometimes necessary to control the volume delivered to the value below that at the surge point. This may lead to pulsation, particularly if the fan pressure exceeds 10 in wg.

If the required capacity is less than that at the pumping limit, pulsation can be prevented in various ways, all of which in effect provide a negatively sloping fan curve at the actual operating point. To accomplish this effect the required pressure must be less than the fan capabilities at the required capacity. One method is to bleed sufficient air for actual operation to be beyond the pumping limit. Other possible methods are the use of pitch, speed, or vane control for volume reduction. In any of these cases, the point of operation on the new fan curve must be to the right of the new surge point. Although in the section on Capacity Control dampers were considered a part of the system, they may also be considered a part of the fan if located in the right position. Accordingly, pulsations may be eliminated in a supply system if the damper is on the inlet of the blower. Similarly, dampering at the outlet of the exhauster may control pulsation in exhaust systems.

Another condition frequently referred to as instability is associated with flow separation in the blade passages of an impeller and is evidenced by slight discontinuities in the performance curve. There may be a small range of capacities at which two distinctly different pressures may be developed depending on which of the two flow patterns exists. Such a condition usually occurs at capacities just to the left of peak efficiency. Still another condition of unsteady flow may develop at extremely low capacities. This is known as **blowback** or **puffing** because air puffs in and out of a portion of the inlet. Operation in the blowback range should be avoided, particularly with high-energy fans.

A different type of unsteady flow may occur when two or more fans are used in parallel. If the individual fan characteristics exhibit a dip in pressure between shutoff and design, the combined characteristic will contain points where the point of operation of the individual fans may be widely separated even for identical fans. If the system characteristic intersects the combined-fan characteristic at such a point, the individual fans may suddenly exchange loads. That is, the fan operating at high capacity may become the one operating at low capacity and vice versa. This can produce undesirable shocks on motors and ducts. Careful matching of fan to system is required for either forward-curved centrifugals or most axials for this reason.

Fan Applications

The selection of a particular size and type of fan for a particular application involves considerations of aerodynamic, economic, and functional suitability. Many of the factors involved in aerodynamic suitability have been discussed above. Determination of economic suitability requires an evaluation of first cost and operating costs. The functional suitability of various types of fans with respect to certain applications is discussed below.

Heating, ventilating, and **air-conditioning systems** may require supply and exhaust or return air fans. Historically, high-efficiency centrifugal fans, using either backward-curved or airfoil blades have been used for supply on duct systems. In low-pressure applications these types can be used without sound treatment. In high-pressure applications, sound treatment is almost always required. Axials have long been used for shipboard ventilation because they generally can be made smaller than centrifugals. Both adjustable axials and tubular centrifugals have proved popular on duct systems for exhaust service in building ventilation. Centrifugals with variable inlet vanes and axials with pitch control are being used for supply in variable-air-volume systems. Propeller fans and power roof ventilators are used for either supply or exhaust systems when there is little or no ductwork. Heating, ventilating, and air-conditioning applications are considered clean-air service. Various classes of construction are available in standard lines for different pressure ranges. Both direct and indirect drive are used, the latter being most common.

Industrial exhaust systems generally require fans that are less susceptible to the unbalance that may result from dirty-gas applications than the clean-air fans used for heating, ventilating, and air conditioning. Simple, rugged, industrial exhausters are favored for applications up to 200 hp. They have a few radial blades and relatively low efficiency. Most are V-belt-driven. Extra-heavy construction may be required where significant material passes through the fan.

Process air requirements can be met with either centrifugal or axial fans; the latter may be used in single or double stage. The higher pressure ratings are usually provided by a single-stage centrifugal fan with radial blades, known as a pressure blower. These units are generally direct connected to the driver. They not only compete with centrifugal compressors but resemble them.

Large industrial process and **pollution-control systems** involving more than about 200 hp are generally satisfied with a somewhat more sophisticated fan than an industrial exhauster or pressure blower. Centrifugal fans with radial-tip blades are frequently used on the more severe service. For the less severe requirements, backward-curved or airfoil blades may be used. Rugged fixed-pitch axials have also been used. Industrial fans are usually equipped with inlet boxes and independently mounted bearings and are usually direct-driven. Journal bearings are usually preferred. Inlet-box damper control can be used to approximate the power saving available from variable-inlet vane control. Variable-speed hydraulic couplings may be economically justified in some cases. Special methods or special construction may be required to provide protection against corrosion or erosion. These fans tend to take on the name of the application such as sintering fan, scrubber exhaust fan, etc.

Mechanical draft systems may utilize any of the fan types described in connection with the above applications. Ventilating fans, industrial exhausters, and pressure blowers have been used for forced draft, induced draft, and primary air service on small steam-generating units. The large generating units are generally equipped with the most efficient fans available consistent with the erosion-corrosion potential of the gas being handled.

Both centrifugals and axials are used for forced and induced draft. Axials have predominated in Europe, and there is a growing trend throughout the rest of the world toward axials. Centrifugals have been used almost exclusively in the United States until very recently.

Forced-draft centrifugals invariably have airfoil blade impellers. Induced-draft centrifugals may have airfoil-blade impellers, but for scrubber exhaust, radial-tip blades are more common. This is due, in part, to the high pressures required for scrubber operation and in part to the erosion-corrosion potential downstream of a scrubber. Gas recirculating fans are usually of the radial-tip design. Forced-draft control may be by variable speed but is more likely to be variable vanes on centrifugal fans. Variable inlet vanes or inlet-box dampers may be used to control induced-draft fans. All large centrifugals are direct-connected and have independent pedestal-mounted bearings. Journal bearings are almost always used.

Forced-draft axials are likely to be of the variable-pitch full-airfoil-section design. Hydraulic systems are almost always used for pitch control, but pneumatic and mechanical systems have been tried. Bearings are usually of the antifriction type. Fixed-pitch axials are usually used for induced-draft duty. Control is by variable-inlet vanes. Either journal or antifriction bearings may be used.

Other Systems

Fans are incorporated in many different kinds of machines. Electronic equipment may require cooling fans to prevent hot spots. Driers use fans to circulate air to carry heat to, and moisture away from, the product. Air-support structures require fans to inflate them and maintain the supporting pressure. Ground-effect machines use fans to provide the lift pressure. Air conditioners and other heat exchangers incorporate fans. Aerodynamic, economic, and functional considerations will dictate the type and size of fan to be used.

Tunnel ventilation can be achieved by using either axial or centrifugal fans. Transverse ventilation utilizes supply and exhaust fans connected to the tunnel by ductwork. These fans are rated in the manner described above. Another method utilizes specially tested and rated axial fans called **jet fans**. The fans, which are placed in the tunnel, increase the momentum of some of the air flowing with the traffic. This air induces additional flow.

Section **15**

Electrical and Electronics Engineering

BY

C. JAMES ERICKSON *Retired Principal Consultant, E. I. du Pont de Nemours & Co., Inc.*
CHARLES D. POTTS *Retired Project Engineer, E. I. du Pont de Nemours & Co., Inc.*
BYRON M. JONES *Consulting Engineer, Assistant Professor of Electrical Engineering,
University of Wisconsin — Platteville.*

15.1 ELECTRICAL ENGINEERING
by C. James Erickson
revised by Charles D. Potts

REFERENCES: Knowlton, "Standard Handbook for Electrical Engineers," McGraw-Hill. Pender and Del Mar, "Electrical Engineers' Handbook," Wiley. Dawes, "Course in Electrical Engineering," Vols. I and II, McGraw-Hill. Gray, "Principles and Practice of Electrical Engineering," McGraw-Hill. Laws, "Electrical Measurements," McGraw-Hill. Karapetoff-Dennison, "Experimental Electrical Engineering and Manual for Electrical Testing," Wiley. Langsdorf, "Principles of Direct-current Machines," McGraw-Hill. Hehre and Harness, "Electric Circuits and Machinery," Vols. I and II, Wiley. Timbie-Higbie, "Alternating Current Electricity and Its Application to Industry," Wiley. Lawrence, "Principles of Alternating-current Machinery," McGraw-Hill. Puchstein and Lloyd, "Alternating-current Machinery," Wiley. Lovell, "Generating Stations," McGraw-Hill. Underhill, "Coils and Magnet Wire" and "Magnets," McGraw-Hill. Abbott, "National Electrical Code Handbook," McGraw-Hill. Dyke, "Automobile and Gasoline Engine Encyclopedia," The Goodheart-Wilcox Co., Inc. Fink and Carrol, "Standard Handbook for Electrical Engineers," McGraw-Hill.

ELECTRICAL AND MAGNETIC UNITS

System of Units The International System of Units (SI) is being adopted universally. The SI system has its roots in the metre, kilogram, second (mks) system of units. Since a centimeter, gram, second (**cgs**) system has been widely used, and will still be used in some instances, Tables 15.1.1 and 15.1.2 are provided for conversion between the two systems. Basic SI units are metre, kilogram (mass), second, ampere, kelvin, mole (quantity), and candela (luminous intensity). Other SI units are derived from these basic units.

Electrical Units
(See Table 15.1.1.)

Current (I, i) The SI unit of current is the **ampere,** which is equal to one-tenth the absolute unit of current (**abampere**). The abampere of current is defined as follows: if 0.01 metre (1 centimetre) of a circuit is bent into an arc of 0.01 metre (1 centimetre) radius, the current is 1 abampere if the magnetic field intensity at the center is 0.01257 ampere per metre (1 oersted), provided the remainder of the circuit produces no magnetic effect at the center of the arc. One **international ampere** (9.99835 amperes) (dc) will deposit 0.001118 gram per second of silver from a standard silver solution.

Quantity (Q) The **coulomb** is the quantity of electricity transported in one second by a current of one ampere.

Potential Difference or Electromotive Force (V, E, emf) The **volt** is the difference of electric potential between two points of a conductor carrying a constant current of one ampere, when the power dissipated between these points is equal to one watt.

Resistance (R, r) The **ohm** is the electrical resistance between two points of a conductor when a constant difference of potential of one volt, applied between these two points, produces in this conductor a current of one ampere, this conductor not being the source of any electromotive force.

Resistivity (ρ) The resistivity of a material is the dc resistance between the opposite parallel faces of a portion of the material having unit length and unit cross section.

Conductance (G, g) The **siemens** is the electrical conductance of a conductor in which a current of one ampere is produced by an electric potential difference of one volt. One siemens is the reciprocal of one ohm.

Conductivity (γ) The conductivity of a material is the dc conductance between the opposite parallel faces of a portion of the material having unit length and unit cross section.

Capacitance (C) is that property of a system of conductors and dielectrics which permits the storage of electricity when potential difference exists between the conductors. Its value is expressed as a ratio of a quantity of electricity to a potential difference. A capacitance value is always positive. The farad is the capacitance of a capacitor between the plates of which there appears a difference of potential of one volt when it is charged by a quantity of electricity equal to one coulomb.

Permittivity or dielectric constant (ε_0) is the electrostatic energy stored per unit volume of a vacuum for unit potential gradient. The permittivity of a vacuum or free space is 8.85×10^{-12} farads per metre.

Relative Permittivity or Dielectric Constant (ε_r) is the ratio of electrostatic energy stored per unit volume of a dielectric for a unit potential gradient to the permittivity (ε_0) of a vacuum. The relative permittivity is a number.

Self-inductance (L) is the property of an electric circuit which determines, for a given rate of change of current in the circuit, the emf induced in the same circuit. Thus $e_1 = -L di_1/dt$, where e_1 and i_1 are in the same circuit and L is the coefficient of self-inductance.

The **henry** is the inductance of a closed circuit in which an electromotive force of one volt is produced when the electric current varies uniformly at a rate of one ampere per second.

Mutual inductance (M) is the common property of two associated electric circuits which determines, for a given rate of change of current in one of the circuits, the emf induced in the other. Thus $e_1 = -M di_2/dt$ and $e_2 = -M di_1/dt$, where e_1 and i_1 are in circuit 1; e_2 and i_2 are in circuit 2; and M is the mutual inductance.

The **henry** is the mutual inductance of two separate circuits in which an electromotive force of one volt is produced in one circuit when the electric current in the other circuit varies uniformly at a rate of one ampere per second.

If M is the mutual inductance of two circuits and k is the coefficient of coupling, i.e., the proportion of flux produced by one circuit which links the other, then $M = k(L_1 L_2)^{1/2}$, where L_1 and L_2 are the respective **self-inductances** of the two circuits.

Energy (J) in a system is measured by the amount of work which a system is capable of doing. The **joule** is the work done when the point of application of a force of one newton is displaced a distance of one metre in the direction of the force.

Power (W) is the time rate of transferring or transforming energy. The **watt** is the power which gives rise to the production of energy at the rate of one joule per second.

Active power (P) at the points of entry of a single-phase, two-wire circuit or of a polyphase circuit is the time average of the values of the instantaneous power at the points of entry, the average being taken over a complete cycle of the alternating current. The value of active power is given in **watts** when the rms currents are in amperes and the rms potential differences are in volts. For sinusoidal emf and current, $P = EI \cos \theta$, where E and I are the rms values of volts and currents, and θ is the phase difference of E and I.

Reactive power (Q) at the points of entry of a single-phase, two-wire circuit, or for the special case of a sinusoidal current and sinusoidal potential difference of the same frequency, is equal to the product obtained by multiplying the rms value of the current by the rms value of the potential difference and by the sine of the angular phase difference by which the current leads or lags the potential difference. $Q = EI \sin \theta$. The unit of Q is the **var** (volt-ampere-reactive). One kilovar = 10^3 var.

Apparent power (EI) at the points of entry of a single-phase, two-wire circuit is equal to the product of the rms current in one conductor multiplied by the rms potential difference between the two points of entry. Apparent power = EI.

Power factor (pf) is the ratio of power to apparent power. pf = P/EI =

Table 15.1.1 Electrical Units

Quantity	Symbol	Equation	SI unit	SI unit symbol	CGS unit	Ratio of magnitude of SI to cgs unit
Current	I, i	$I = E/R; I = E/Z; I = Q/t$	Ampere	A	Abampere	10^{-1}
Quantity	Q, q	$Q = it; Q = CE$	Coulomb	C	Abcoulomb	10^{-1}
Electromotive force	E, e	$E = IR; E = W/Q$	Volt	V	Abvolt	10^8
Resistance	R, r	$R = E/I; R = \rho l/A$	Ohm	Ω	Abohm	10^9
Resistivity	ρ	$\rho = RA/l$	Ohm-metre	$\Omega \cdot m$	Abohm-cm	10^{11}
Conductance	G, g	$G = \gamma A/l; G = A/\rho l$	Siemens	S	Abmho	10^{-9}
Conductivity	γ	$\gamma = 1/\rho; \gamma = l/RA$	Siemens/meter	S/m	Abmho/cm	10^{-11}
Capacitance	C	$C = Q/E$	Farad	F	Abfarad*	10^{-9}
Permittivity	ε		Farads/meter	F/m	Stat farad*/cm	8.85×10^{-12}
Relative permittivity	ε_r	$\varepsilon_r = \varepsilon/\varepsilon_0$	Numerical		Numerical	1
Self-inductance	L	$L = -N(d\phi/dt)$	Henry	H	Abhenry	10^9
Mutual inductance	M	$M = K(L_1 L_2)^{1/2}$	Henry	H	Abhenry	10^9
Energy	J	$J = eit$	Joule	J	Erg	10^7
	kWh	kWh = kW/3600; 3.6 MJ	Kilowatthour	kWh		36×10^{12}
Active power	W	$W = J/t; W = EI \cos \theta$	Watt	W	Abwatt	10^7
Reactive power	jQ	$Q = EI \sin \theta$	Var	var	Abvar	10^7
Apparent power	VA	$VA = EI$	Volt-ampere	VA		
Power factor	pf	$pf = W/VA; pf = W/(W + jQ)$				1
Reactance, inductive	X_L	$X_L = 2\pi fL$	Ohm	Ω	Abohm	10^9
Reactance, capacitive	X_C	$X_C = 1/(2\pi fC)$	Ohm	Ω	Abohm	10^9
Impedence	Z	$Z = E/I; Z = R + j(X_L - X_C)$	Ohm	Ω	Abohm	10^9
Conductance	G	$G = R/Z^2$	Siemens	S	Abmho	10^{-9}
Susceptance	B	$B = X/Z^2$	Siemens	S	Abmho	10^{-9}
Admittance	Y	$Y = I/E; Y = G + jB$	Siemens	S	Abmho	10^{-9}
Frequency	f	$f = 1/T$	Hertz	Hz	Cps, Hz	1
Period	T	$T = 1/f$	Second	s	Second	1
Time constant	T	$L/R; RC$	Second	s	Second	1
Angular velocity	ω	$\omega = 2\pi f$	Radians/second	rad/s	Radians/second	1

* 1 Abfarad (EMU Units) = 9×10^{-20} stat farads (ESU units).

$\cos \theta$, where θ is the phase difference between E and I, both assumed to be sinusoidal.

The **reactance** (X) of a portion of a circuit for a sinusoidal current and potential difference of the same frequency is the product of the sine of the angular phase difference between the current and potential difference times the ratio of the rms potential difference to the rms current, there being no source of power in the portion of the circuit under consideration. $X = (E/I) \sin \theta = 2\pi fL$ ohms, where f is the frequency, and L the inductance in henries; or $X = 1/2\pi fC$ ohms, where C is the capacitance in farads.

The **impedance** (Z) of a portion of an electric circuit to a completely specified periodic current and potential difference is the ratio of the rms value of the potential difference between the terminals to the rms value of the current, there being no source of power in the portion under consideration. $Z = E/I$ ohms.

Admittance (Y) is the reciprocal of impedance. $Y = I/E$ siemens.

The **susceptance** (B) of a portion of a circuit for a sinusoidal current and potential difference of the same frequency is the product of the sine of the angular phase difference between the current and the potential difference times the ratio of the rms current to the rms potential difference, there being no source of power in the portion of the circuit under consideration. $B = (I/E) \sin \theta$.

Magnetic Units

(See Table 15.1.2.)

Magnetic flux (Φ, ϕ) is the magnetic flow that exists in any magnetic circuit.

The **weber** is the magnetic flux which, linking a circuit of one turn, produces in it an electromotive force of one volt as it is reduced to zero at a uniform rate in one second.

Magnetic flux density (β) is the ratio of the flux in any cross section to the area of that cross section, the cross section being taken normal to the direction of flux.

The **tesla** is the magnetic flux density given by a magnetic flux of one weber per square metre.

Unit magnetic pole, when concentrated at a point and placed one metre apart in a vacuum from a second unit magnetic pole, will repel or attract the second unit pole with a force of one **newton.**

The **weber** is the magnetic flux produced by a unit pole.

Table 15.1.2 Magnetic Units

Quantity	Symbol	Equation*	SI unit	SI unit symbol	CGS unit	Ratio of magnitude of SI to cgs unit
Magnetic flux	Φ, ϕ	$\phi = F/R$	Weber	wb	Maxwell	10^8
Magnetic flux density	β	$\beta = \phi/A$	Tesla	T	Gauss	10^4
Pole strength	Q_m	$Q_m = F/\beta; Q_m = Fl/NI\mu_0\mu_r$	Ampere-turns-metre	A·m		
			Unit pole		Unit pole	0.7958×10^7
Magnetomotive force	$\mathcal{F}$	$\mathcal{F} = NI$	Ampere-turns	A	Gilbert	1.257
Magnetic field intensity	H	$H = \mathcal{F}/l$	Ampere-turns per metre	A/m	Oersted	0.01257
Permeability air	μ_0	$\mu_0 = \beta/H$	Henry per metre	H/m	Gilbert per oersted	1.257×10^{-6}
Relative permeability	μ_r	$\mu_r = \mu/\mu_0$	Numeric		Numeric	1
Reluctivity	γ	$\gamma = 1/\mu_r$	Numeric		Numeric	1
Permeance	P	$P = \mu_0\mu_r A/l$	Henry	H		7.96×10^7
Reluctance	R	$R = l/\mu_0\mu_r A$	1/Henry	1/H		1.257×10^{-8}

* l = length in metres; A = area in square metres; F = force in newtons; N = number of turns.

Magnetomotive force ($\mathscr{F}$, mmf) produces magnetic flux and corresponds to electromotive force in an electric circuit.

The **ampere** (turn) is the unit of mmf.

Magnetic field intensity (H) at a point is the vector quantity which is measured by a mechanical force which is exerted on a unit pole placed at the point in a vacuum.

An **ampere per metre** is the unit of field intensity.

Permeability (μ) is the ratio of unit magnetic flux density to unit magnetic field intensity in air (B/H). The permeability of air is 1.257×10^{-6} **henry per metre.**

Relative permeability (μ_r) is the ratio of the magnetic flux in any element of a medium to the flux that would exist if that element were replaced with air, the magnetomotive force (mmf) acting on the element remaining unchanged ($\mu_r = \mu/\mu_0$).

The relative permeability is a number.

Permeance (P) of a portion of a magnetic circuit bounded by two equipotential surfaces, and by a third surface at every point of which there is a tangent having the direction of the magnetic induction, is the ratio of the flux through any cross section to the magnetic potential difference between the surfaces when taken within the portion under consideration. The equation for the permeance of the medium as defined above is $P = \mu_0 \mu_r A/l$. Permeance is the **reciprocal of reluctance.**

Reluctivity (γ) of a medium is the reciprocal of its permeability.

Reluctance (R) is the **reciprocal of permeance.** It is the resistance to magnetic flow. In a homogeneous medium of uniform cross section, reluctance is equal to the length divided by the product of the area and permeability, the length and area being expressed in metre units. $R = l/A\mu_0\mu_r$, where $\mu_0 = 1.257 \times 10^{-6}$.

CONDUCTORS AND RESISTANCE

Resistivity, or **specific resistance,** is the resistance of a sample of the material having both a length and cross section of unity. The two most common resistivity samples are the centimetre cube and the cir mil·ft.

If l is the length of a conductor of uniform cross section a, then its resistance is

$$R = \rho l/a \qquad (15.1.1)$$

where ρ is the resistivity. With a cir mil·ft ρ is the resistance of a cir mil·ft and a is the cross section, cir mils. Since $v = la$ is the volume of a conductor,

$$R = \rho l^2/v = \rho v/a^2 \qquad (15.1.2)$$

A **circular mil** is a unit of area equal to that of a circle whose diameter is 1 mil (0.001 in). It is the unit of area which is used almost entirely in this country for wires and cables. To obtain the cir mils of a solid cylindrical conductor, square its diameter expressed in mils. For example, the diameter of 000 AWG solid copper wire is 410 mils and its cross section is $(410)^2 = 168{,}100$ cir mils. The diameter in mils of a solid cylindrical conductor is the square root of its cross section expressed in cir mils.

A **cir mil·ft** is a conductor having a length of 1 ft and a uniform cross section of 1 cir mil. In terms of the copper standard the resistance of a cir mil·ft of copper at 20°C is 10.371 Ω. As a first approximation 10 Ω may frequently be used.

At 60°C a cir mil·in of copper has a resistance of 1.0 Ω. This is a very convenient unit of resistivity for magnet coils since the resistance is merely the length of copper in inches divided by its cross section in cir mils.

Temperature Coefficient of Resistance The resistance of the pure metals increases with temperature. The resistance at any temperature $t°C$ is

$$R = R_0(1 + \alpha t) \qquad (15.1.3)$$

where R_0 is the resistance at 20°C and α is the **temperature coefficient of resistance.** For copper, $\alpha = 0.00393$.

With any initial temperature t_1, the resistance at temperature $t°C$ is

$$R = R_1[1 + \alpha_1(t - t_1)] \qquad (15.1.4)$$

Table 15.1.3 Properties of Metals and Alloys
(See Table 15.1.27 for properties of resistor alloys)

Metals	Resistivity, 20°C		Temperature coefficient of resistance at 20°C
	$\mu\Omega\cdot cm$	$\Omega\cdot$cir mil/ft	
Aluminum	2.828	17.01	0.00403
Antimony	42.1	251.0	0.0036
Bismuth	111.0	668.0	0.004
Brass	6.21	37.0	0.0015
Carbon: amorphous	3,800–4,100	....	(−)
Retort (graphite)	720–812*	....	(−)
Copper (drawn)	1.724	10.37	0.00393
Gold	2.44	14.7	0.0034
Iron: electrolytic	10.1	59.9	0.0064
Cast	75.2–98.8	448–588	
Wire	97.8	588	
Lead	22.0	132	0.00387
Molybdenum	5.78	34.8	
Monel metal	43.5	262	0.0019
Mercury	96.8	576	0.00089
Nickel	8.54	50.8	0.0041
Platinum	10.72	63.8	0.003
Platinum silver, 2Ag + 1Pt	24.6†	148.0	0.00031
Silver	1.628	9.8	0.0038
Steel: soft	15.9	95.8	0.0016
Glass hard	45.7	275	
Silicon (4 percent)	51.18	308	
Transformer	11.09	66.7	
Trolley wire	12.7	76.4	
Tin	11.63	70	0.0042
Tungsten	5.51	33.2	0.005
Zinc	5.97	35.58	0.0037

NOTE: Max working temperature: Cu, 260°C; Ni, 600°C; Pt, 1,500°C.
* Furnace electrodes, 3,000°C.
† 0°C.

where R_1 is the resistance at temperature $t_1°C$ and α_1 is the temperature coefficient of resistance at temperature t_1 [see Eq. (15.1.5)].

For any initial temperature t_1 the value of α_1 is

$$\alpha_1 = 1/(234.5 + t_1) \qquad (15.1.5)$$

Inferred Absolute Zero   Between 100 and 0°C the resistance of copper decreases at a rate which is practically uniform and which if continued would give a resistance of zero at $-234.5°C$ (an easy number to remember). If the resistance at $t_1°C$ is R_1 and the resistance at $t_2°C$ is R_2, then

$$R_2/R_1 = (234.5 + t_2)/(234.5 + t_1) \qquad (15.1.6)$$

EXAMPLE.   The resistance of a copper coil at 25°C is 4.26 Ω. Determine its resistance at 45°C. Using Eq. (15.1.4) and $\alpha_1 = 1/(234.5 + 25) = 0.00385$, $R = 4.26[1 + 0.00385(45 - 25)] = 4.59$ Ω. Using Eq. (15.1.6) $R = 4.26(234.5 + 45)/(234.5 + 25) = 4.26 \times 1.077 = 4.59$ Ω.

The inferred absolute zero for aluminum is $-228°C$.

Materials   The materials generally used for the transmission and distribution of electrical energy are copper, aluminum, and sometimes iron and steel. For resistors and heaters, iron, steel, commercial alloys, and carbon are most used.

Copper is the most widely used electrical conductor. It has high conductivity, relatively low cost, good resistance to oxidation, is readily soldered, and has good mechanical characteristics such as tensile strength, toughness, and ductility. Its tensile strength together with its low linear temperature coefficient of expansion are desirable characteristics in its use for overhead transmission lines. The international copper standard for 100 percent conductivity annealed copper is a density of 8.89 g/cm³ (0.321 lb/in³) and resistivity is given in Table 15.1.3. ASTM specifications for minimum conductivities of copper wire are as follows:

Conductor diam, in	Soft or annealed	Medium hard drawn	Hard drawn
0.040–0.324	98.16%	96.60%	96.16%
0.325–0.460	98.16%	97.66%	97.16%

Table 15.1.4   Working Table, Standard Annealed Copper Wire, Solid
[American Wire Gage (B & S)]

Gage no.	Diam, mils	Cross section cir mils	Cross section in²	Ω per 1,000 ft 25°C (=77°F)	Ω per 1,000 ft 65°C (=149°F)	Ω/mi at 25°C (=77°F)	Weight per 1,000 ft, lb
0000	460.0	212,000	0.166	0.0500	0.0577	0.264	641.0
000	410.0	168,000	0.132	0.0630	0.0727	0.333	508.0
00	365.0	133,000	0.105	0.0795	0.0917	0.420	403.0
0	325.0	106,000	0.0829	0.100	0.116	0.528	319.0
1	289.0	83,700	0.0657	0.126	0.146	0.665	253.0
2	258.0	66,400	0.0521	0.159	0.184	0.839	201.0
3	229.0	52,600	0.0413	0.201	0.232	1.061	159.0
4	204.0	41,700	0.0328	0.253	0.292	1.335	126.0
5	182.0	33,100	0.0260	0.319	0.369	1.685	100.0
6	162.0	26,300	0.0206	0.403	0.465	2.13	79.5
7	144.0	20,800	0.0164	0.508	0.586	2.68	63.0
8	128.0	16,500	0.0130	0.641	0.739	3.38	50.0
9	114.0	13,100	0.0103	0.808	0.932	4.27	39.6
10	102.0	10,400	0.00815	1.02	1.18	5.38	31.4
11	91.0	8,230	0.00647	1.28	1.48	6.75	24.9
12	81.0	6,530	0.00513	1.62	1.87	8.55	19.8
13	72.0	5,180	0.00407	2.04	2.36	10.77	15.7
14	64.0	4,110	0.00323	2.58	2.97	13.62	12.4
15	57.0	3,260	0.00256	3.25	3.75	17.16	9.86
16	51.0	2,580	0.00203	4.09	4.73	21.6	7.82
17	45.0	2,050	0.00161	5.16	5.96	27.2	6.20
18	40.0	1,620	0.00128	6.51	7.51	34.4	4.92
19	36.0	1,290	0.00101	8.21	9.48	43.3	3.90
20	32.0	1,020	0.000802	10.4	11.9	54.9	3.09
21	28.5	810	0.000636	13.1	15.1	69.1	2.45
22	25.3	642	0.000505	16.5	19.0	87.1	1.94
23	22.6	509	0.000400	20.8	24.0	109.8	1.54
24	20.1	404	0.000317	26.2	30.2	138.3	1.22
25	17.9	320	0.000252	33.0	38.1	174.1	0.970
26	15.9	254	0.000200	41.6	48.0	220	0.769
27	14.2	202	0.000158	52.5	60.6	277	0.610
28	12.6	160	0.000126	66.2	76.4	350	0.484
29	11.3	127	0.0000995	83.4	96.3	440	0.384
30	10.0	101	0.0000789	105	121	554	0.304
31	8.9	79.7	0.0000626	133	153	702	0.241
32	8.0	63.2	0.0000496	167	193	882	0.191
33	7.1	50.1	0.0000394	211	243	1,114	0.152
34	6.3	39.8	0.0000312	266	307	1,404	0.120
35	5.6	31.5	0.0000248	335	387	1,769	0.0954
36	5.0	25.0	0.0000196	423	488	2,230	0.0757
37	4.5	19.8	0.0000156	533	616	2,810	0.0600
38	4.0	15.7	0.0000123	673	776	3,550	0.0476
39	3.5	12.5	0.0000098	848	979	4,480	0.0377
40	3.1	9.9	0.0000078	1,070	1,230	5,650	0.0200

Aluminum is used to considerable extent for high-voltage transmission lines, because its weight is one-half that of copper for the same conductance. Moreover, the greater diameter reduces corona loss. As it has 1.4 times the linear temperature coefficient of expansion, changes in sag with temperature are greater. Because of its lower melting point, spans may fail more readily with arc-overs. In aluminum cable steel-reinforced (ACSR), the center strand is a steel cable, which gives added tensile strength. Aluminum is used occasionally for bus bars because of its large heat-dissipating surface for a given conductance. The greater cross section for a given conductance requires a greater volume of insulation for a given voltage. When the ratio of the cost of aluminum to the cost of copper becomes economically favorable, aluminum is often used for insulated wires and cables. The international aluminum standard for 62 percent conductivity aluminum is a density of 2.70 g/cm³ (0.0976 lb/in³) and resistivity as given in Table 15.1.3.

Steel, either **galvanized** or **copper-covered** ("copperweld"), is used for high-voltage transmission spans where tensile strength is more important than high conductance. Steel is also used for third rails.

Copper alloys and **bronzes** are of increasing importance as electrical conductors. They have lower electrical conductivity but greater tensile strength and are resistant to corrosion. **Hitenso, Calsum bronzes, Signal bronze, Phono-electric,** and **Everdur** are bronzes containing phosphorus, silicon, manganese, or zinc. Their conductivities vary from 20 to 85 percent of 100 percent conductivity copper, and they have tensile strengths up to 130,000 lb/in², about twice that of hard-drawn copper. Such alloys were frequently used for trolley wires. Copper alloys having lower conductivity are usually classified as resistor materials.

In Table 15.1.3 are given the electrical properties of some of the pure metals and alloys.

American Wire Gage (AWG) The AWG (formerly Brown & Sharpe gage) is based on a constant ratio between diameters of successive gage numbers (Table 15.1.4). The ratio of any diameter to the next smaller is 1.123, and the corresponding ratio of cross sections is $(1.123)^2 = 1.261$, or 1¼ approximately. $(1.123)^6$ is 2.0050, so that diameters differing by 6 gage numbers have a ratio of approximately 2; cross sections differing by 3 gage numbers also have a ratio of approximately 2. The ratio of cross sections differing by 2 numbers is $(1.261)^2 = 1.590$, or 1.6 approximately. The ratio of cross sections differing by 10 numbers is approximately 10. The gage ordinarily extends from no. 40 to 0000 (4/0). Wires larger than 0000 must be stranded, and their cross section is given in cir mils.

The diameter of no. 10 wire is 102.0 mils. As an approximation this may be considered as being 100 mils; the cross section is 10,000 cir mils; the resistance is 1 Ω per 1,000 ft; and the weight of 1,000 ft is $31.4(10\pi)$ lb. Also the weight of 1,000 ft of no. 2 is 200 lb. These facts give many short cuts in estimating resistances and weights of various gage numbers.

Lay Cables In order to obtain sufficient flexibility, wires larger than 0000 are stranded, and they are designated by their circular mils (Table 15.1.5). Smaller wires may be stranded also since sizes as small as no. 4 when insulated are usually too stiff for easy handling. Lay cables are made up geometrically as shown in Fig. 15.1.1. Six strands will just fit around the single central conductor; the number of strands in each succeeding layer increases by 6. The number of strands that can thus be laid up are 1, 7, 19, 37, 61, 91, 127, etc. In order to obtain sufficient flexibility with large cables, the strands themselves frequently consist of stranded cable.

Fig. 15.1.1 Makeup of a 19-strand cable.

The **resistance of cables** is readily computed from Eq. (15.1.1), using the cir mil ft as the unit of resistivity.

EXAMPLE. Determine the resistance of 3,500 ft of 800,000 cir mil cable at 20°C. *Answer:* ρ (of a cir mil·ft) = 10.37. $R = 10.37 \times 3,500/800,000 = 0.0454\ \Omega$.

$\rho = 10\ \Omega/\text{cir mil}\cdot\text{ft}$ is often sufficiently accurate for practical purposes.

Table 15.1.5   Bare Concentric Lay Cables of Standard Annealed Copper

AWG no.	cir mils	Ω per 1,000 ft		Weight per 1,000 ft, lb	Standard concentric standing		
		25°C (=77°F)	65°C (=149°F)		No. of wires	Diam of wires, mils	Outside diam, mils
	2,000,000	0.00539	0.00622	6,180	127	125.5	1,631
	1,700,000	0.00634	0.00732	5,250	127	115.7	1,504
	1,500,000	0.00719	0.00830	4,630	91	128.4	1,412
	1,200,000	0.00899	0.0104	3,710	91	114.8	1,263
	1,000,000	0.0108	0.0124	3,090	61	128.0	1,152
	900,000	0.0120	0.0138	2,780	61	121.5	1,093
	850,000	0.0127	0.0146	2,620	61	118.0	1,062
	750,000	0.0144	0.0166	2,320	61	110.9	998
	650,000	0.0166	0.0192	2,010	61	103.2	929
	600,000	0.0180	0.0207	1,850	61	99.2	893
	550,000	0.0196	0.0226	1,700	61	95.0	855
	500,000	0.0216	0.0249	1,540	37	116.2	814
	450,000	0.0240	0.0277	1,390	37	110.3	772
	400,000	0.0270	0.0311	1,240	37	104.0	728
	350,000	0.0308	0.0356	1,080	37	97.3	681
	300,000	0.0360	0.0415	926	37	90.0	630
	250,000	0.0431	0.0498	772	37	82.2	575
0000	212,000	0.0509	0.0587	653	19	105.5	528
000	168,000	0.0642	0.0741	518	19	94.0	470
00	133,000	0.0811	0.0936	411	19	83.7	418
0	106,000	0.102	0.117	326	19	74.5	373
1	83,700	0.129	0.149	258	19	66.4	332
2	66,400	0.162	0.187	205	7	97.4	292
3	52,600	0.205	0.237	163	7	86.7	260
4	41,700	0.259	0.299	129	7	77.2	232

NOTE: See Table 15.1.21 for the carrying capacity of wires.
SOURCE: From *NBS Cir.* 31.

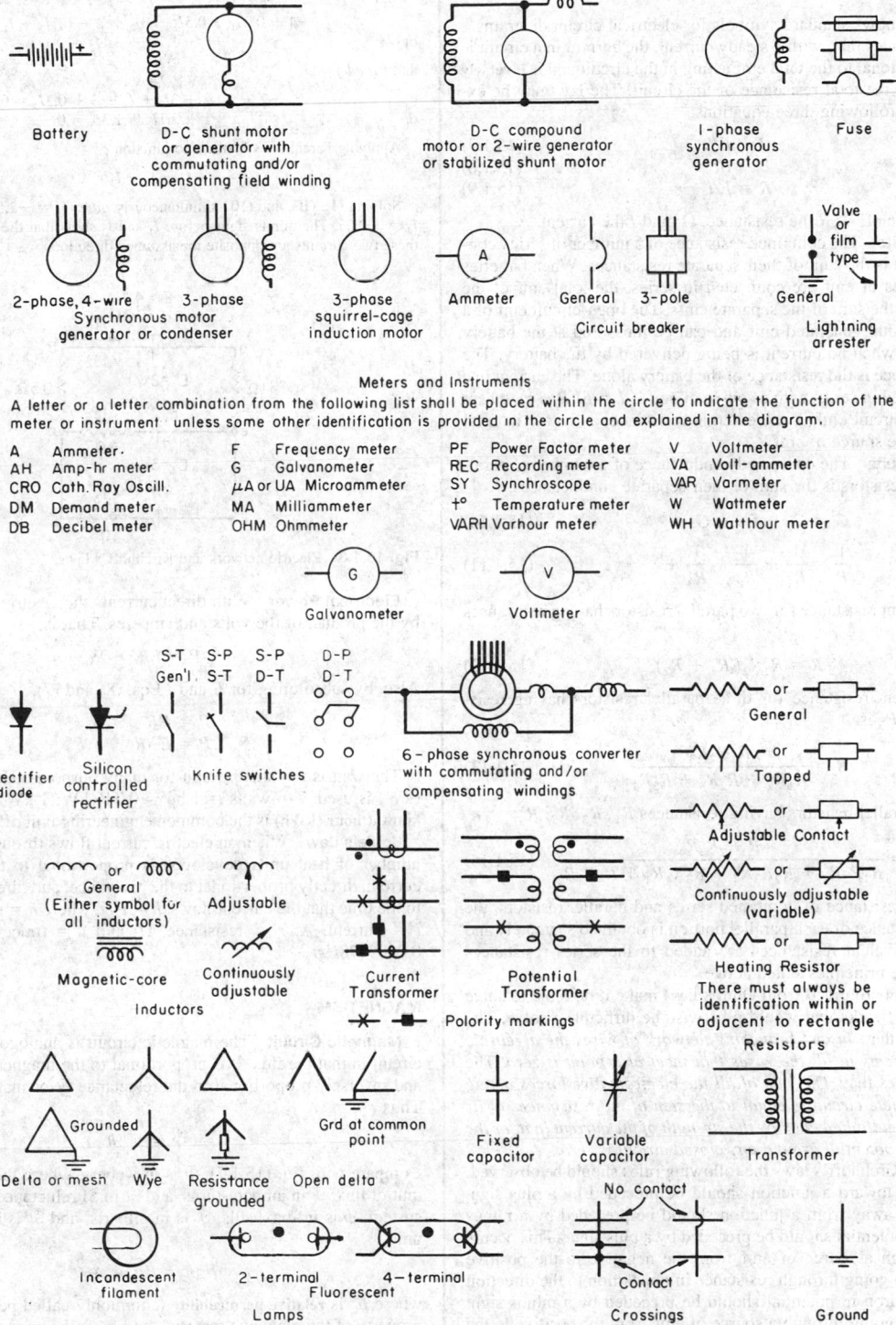

Meters and Instruments

A letter or a letter combination from the following list shall be placed within the circle to indicate the function of the meter or instrument unless some other identification is provided in the circle and explained in the diagram.

A	Ammeter·	F	Frequency meter	PF	Power Factor meter	V	Voltmeter
AH	Amp-hr meter	G	Galvanometer	REC	Recording meter	VA	Volt-ammeter
CRO	Cath. Ray Oscill.	μA or UA	Microammeter	SY	Synchroscope	VAR	Varmeter
DM	Demand meter	MA	Milliammeter	†°	Temperature meter	W	Wattmeter
DB	Decibel meter	OHM	Ohmmeter	VARH	Varhour meter	WH	Watthour meter

Fig. 15.1.2 Diagrammatic symbols for electrical machinery and apparatus. *(American Standard, "Graphic Symbols for Electrical and Electronic Diagrams," ANS/IEEE, 315, 1975.)*

ELECTRICAL CIRCUITS

Figure 15.1.2 shows standard symbols for electrical circuit diagrams.

Ohm's law states that, with a steady current, the current in a circuit is **directly** proportional to the **total** emf acting in the circuit and is **inversely** proportional to the total resistance of the circuit. The law may be expressed by the following three equations:

$$I = E/R \qquad (15.1.7)$$
$$E = IR \qquad (15.1.8)$$
$$R = E/I \qquad (15.1.9)$$

where E is the emf, V; R the resistance, Ω; and I the current, A.

Series Circuits The combined resistance of a number of series-connected resistors is the sum of their separate resistances. When batteries or other sources of emf are connected in series, the total emf of the combination is the sum of the separate emfs. The open-circuit emf of a battery is the total generated emf and can be measured at the battery terminals only when no current is being delivered by the battery. The internal resistance is the resistance of the battery alone. The current in a circuit connected in series with a source of emf is $I = E/(R + r)$, where E is the open-circuit emf, R the external resistance, and r the internal resistance of the source of emf.

Parallel Circuits The combined conductance of a number of parallel-connected resistors is the sum of their separate conductances.

$$G = G_1 + G_2 + G_3 + \cdots \qquad (15.1.10)$$
$$\frac{1}{R} = \frac{1}{R_1} + \frac{1}{R_2} + \frac{1}{R_3} + \cdots \qquad (15.1.11)$$

The equivalent resistance for two parallel resistors having resistances R_1, R_2 is

$$R = R_1R_2/(R_1 + R_2) \qquad (15.1.12)$$

The equivalent resistance for three parallel resistors having resistances R_1, R_2, R_3 is

$$R = \frac{R_1R_2R_3}{R_1R_2 + R_2R_3 + R_3R_1} \qquad (15.1.13)$$

and for four parallel resistors having resistances R_1, R_2, R_3, R_4

$$R = \frac{R_1R_2R_3R_4}{R_1R_2R_3 + R_2R_3R_4 + R_3R_4R_1 + R_4R_1R_2} \qquad (15.1.14)$$

To obtain the resistance of combined series and parallel resistors, the equivalent resistance of each parallel portion is obtained separately and then these equivalent resistances are added to the series resistances according to the principles stated above.

Kirchhoff's laws (derived from Ohm's law) make it possible to solve many circuit networks that would otherwise be difficult to solve. The first law states that: *In any branching network of wires the algebraic sum of the currents in all the wires that meet at a point is zero.* The second law states that: *The sum of all the electromotive forces acting around a complete circuit is equal to the sum of the resistances of its separate parts multiplied each by the strength of the current in it, or the total change of potential around any closed circuit is zero.*

In applying Kirchhoff's laws the following rules should be observed. Currents going toward a junction should be preceded by a plus sign. Currents going away from a junction should be preceded by a minus sign. A rise in potential should be preceded by a plus sign. (This occurs in going through a source of emf from the negative to the positive terminal, and in going through resistance in opposition to the direction of current.) A drop in potential should be preceded by a minus sign. (This occurs in going through a source of emf from the positive to the negative terminal and in going through resistance in conjunction with the current.)

The application of Kirchhoff's laws is illustrated by the following example.

EXAMPLE. Determine the three currents I_1, I_2, and I_3 in the circuit network (Fig. 15.1.3). The arrows show the assumed directions of the three currents.

Applying Kirchhoff's second law to circuit *abcdea*,

$$+4 + 0.2I_1 + 0.5I_1 - 3I_2 + 2 - 0.1I_2 + I_1 = 0$$
or
$$+6 + 1.7I_1 - 3.1I_2 = 0 \qquad (I)$$

and for *edcfge*.

$$-2 + 0.1I_2 + 3I_2 + I_3 + 3 + 0.3I_3 = 0$$
or
$$+1 + 3.1I_2 + 1.3I_3 = 0 \qquad (II)$$

Applying Kirchhoff's first law to junction *c*,

$$-I_1 - I_2 + I_3 = 0 \qquad (III)$$

Solving (I), (II), and (III) simultaneously gives $I_1 = -2.56$, $I_2 = +0.53$, and $I_3 = -2.03$. The minus signs before I_1 and I_3 show that the actual directions of these two currents are opposite the assumed directions.

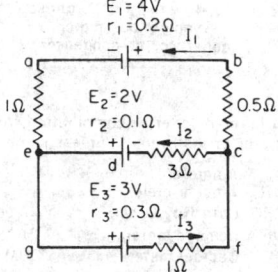

Fig. 15.1.3 Electric network and Kirchhoff's laws.

Electrical Power With direct currents the electrical power is given by the product of the volts and amperes. That is,

$$P = EI \qquad \text{W} \qquad (15.1.15)$$

Also, by substituting for E and I Eqs. (8) and (7),

$$P = I^2R \qquad \text{W} \qquad (15.1.16)$$
$$P = E^2/R \qquad \text{W} \qquad (15.1.17)$$

The watt is too small a unit for many purposes. Hence, the **kilowatt** (kW) is used. 746 watts = 1 hp = 0.746 kW; 1 kW = 1.340 hp. The **kilowatthour (kWh)** is the common engineering unit of electrical energy.

Joule's Law When an electric current flows through resistance, the number of heat units developed is proportional to the square of the current, directly proportional to the resistance, and directly proportional to the time that the current flows. $h = i^2rt$, where h = number of joules; i = current, A; r = resistance, Ω; and t = time, s. h (in Btu) = $0.0009478i^2rt$.

MAGNETISM

Magnetic Circuit The magnetic circuit is analogous to the electric circuit in that the flux Φ is proportional to the magnetomotive force $\mathscr{F}$ and inversely proportional to the reluctance $\mathscr{R}$ or magnetic resistance. Thus

$$\Phi = \mathscr{F}/\mathscr{R} \qquad (15.1.18)$$

Compare with Eq. (15.1.7). Φ is in webers, where the weber is the SI unit of flux, $\mathscr{F}$ in ampere-turns, and $\mathscr{R}$ in SI reluctance units. In the cgs system, ϕ is in maxwells, $\mathscr{F}$ is in gilberts, and $\mathscr{R}$ is in cgs reluctance units.

$$\mathscr{R} = l/\mu_r\mu_v A \qquad (15.1.19)$$

where μ_r is **relative permeability** (commonly called permeability, μ), a property of the magnetic material, and μ_v is the permeability of evacuated space = $4\pi \times 10^{-7}$, and A is in square metres. In the cgs system $\mu_v = 1$

$$\mathscr{R} = \frac{l}{\mu_r(4\pi \times 10^{-7})A} = \frac{l}{\mu_r(1.257 \times 10^{-6})A} \qquad (15.1.20)$$

l is in metres and A in square metres.

The unit of **flux density** in the SI system is the tesla, which is equal to the number of webers per square metre taken perpendicular to their direction. One ampere-turn between opposite faces of a metre cube of a magnetic medium produces μ_r tesla. For air, $\mu_r = 4\pi \times 10^{-7}$. In the cgs system the unit of flux density is gauss = 10^4T (see Table 15.1.2).

Magnetic-circuit calculations cannot be made with the same degree of accuracy as electric-circuit calculations because of several factors. The cross-sectional dimensions of the magnetic circuit are large relative to its length; magnetic paths are irregular, and their geometry can only be approximated as with the air gap of electric machines, which usually have slots on one or both sides of the gap.

Magnetic flux cannot be confined to definite magnetic paths, but a considerable proportion usually takes paths external to the circuit giving magnetic leakage (see Fig. 15.1.7). The relative permeability of iron varies over wide ranges with the flux density and with the previous magnetic condition (see Fig. 15.1.5). These variations of relative permeability cannot be expressed by any simple equation. Although the foregoing factors prevent the obtaining of extremely high accuracy in magnetic calculations, yet, with experience, it is possible to design magnetic circuits with a precision that is satisfactory for all practical purposes.

The **magnetomotive force** $\mathcal{F}$ in Eq. (15.1.18) is expressed in ampere-turns = NI, where N is the number of turns linked with the circuit and I is the current, A. The unit of reluctance is the reluctance of a 1-m cube of air. The total reluctance is proportional to the length and inversely proportional to the cross-sectional area of the magnetic circuit, which is analogous to electrical resistance. Hence the reluctance of any given path of uniform cross section A is $l/A\mu$, where l = length of path, cm; A = its cross section, cm^2; and μ = permeability. Reluctances in series are added to obtain their combined reluctance. Ohm's law of the magnetic circuit becomes

$$\Phi = \frac{NI}{l_1/A_1\mu_1 + l_2/A_2\mu_2 + l_3/A_3\mu_3 \cdots} \text{ Mx} \qquad (15.1.21)$$

where l_1, A_1, μ_1, etc., are the lengths, cross sections, and relative permeabilities of each series part of the circuit.

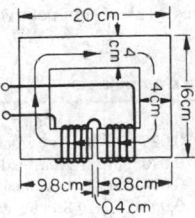

Fig. 15.1.4   Magnetic circuit.

EXAMPLE.   In Fig. 15.1.4 is shown a magnetic circuit of cast steel with a 0.4-cm air gap. The cross section of the core is 4 cm square. There are 425 turns wound on the core and the current is 10 A. The relative permeability of the steel at the operating flux density is 1,100. Assume that the path of the flux is as shown, the average path at the corners being quarter circles. Neglect fringing at the air gap and any leakage. Determine the flux and the flux density.

Using the SI system, the length of the iron is 0.522 m, the length of the air gap is 0.004 m, and the cross section of the iron and air gap is 0.0016 m^2.

$$\Phi = \frac{425 \times 10}{\dfrac{0.522}{1,100 \times 4\pi \times 10^{-7} \times 0.0016} + \dfrac{0.004}{4\pi \times 10^{-7} \times 0.0016}}$$
$$= 0.00191 \text{ Wb}$$

Using the cgs system, the length of the magnetic path in the iron = 12 + 8 + 8 + 5.8 + 5.8 + 4π = 52.2 cm. From Eq. (15.1.21),

$$\Phi = \frac{0.4\pi \times 425 \times 10}{[52.2/(16 \times 1,100)] + (0.4/16)} = 191,000 \text{ Mx}$$

$$B = \frac{191,000}{16} = 11,940 \text{ G}$$

Magnetization and Permeability Curves   The magnetic permeability of air is a constant and is taken as unity. The relative permeability of iron and other magnetic substances varies with the flux density. In Fig. 15.1.5 is shown a magnetization curve for cast steel in which the flux density B in tesla is plotted as a function of the field intensity, amperes

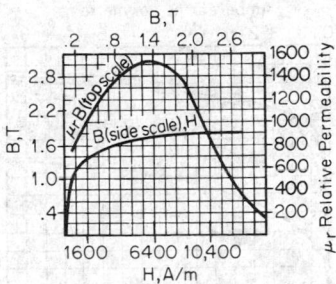

Fig. 15.1.5   Magnetization and relative-permeability curves for cast steel.

per metre, H. Also the relative permeability $\mu_r = B/H$ is plotted as a function of the flux density B. Note the wide range over which the relative permeability varies. No satisfactory equation has been found to express the relation between magnetizing force and flux density and between relative permeability and flux density. If an attempt is made to solve Eq. (15.1.21) for flux, the factors μ_1, μ_2, etc., are unknown since they are functions of the flux density, which is being determined. The simplest method is one of trial and error, i.e., a value of flux, and the corresponding permeability, is first assumed, the equation solved for the flux, and if the computed flux differs widely from the assumed flux, a second approximation is made, etc. In nearly all magnetic designs either the flux or flux density is the independent variable, and it is required to find the necessary ampere-turns to produce them. Let the flux $\Phi = BA$ where B is the flux density, G. Then

$$\Phi = BA = 0.4\pi NI(l/A\mu_r)$$
$$NI = Bl/\mu_0\mu_r = 0.796Bl/\mu_r \times 10^6 \qquad (15.1.22)$$

and

Equation (15.1.22) shows that the necessary ampere-turns are proportional to the **flux density** and the length of path and are inversely proportional to the relative permeability.

With air and nonmagnetic substances μ_r [Eq. (15.1.22)] becomes unity, and

$$NI = 0.796Bl \times 10^6 \qquad (15.1.23)$$

in metre units. With inch units

$$NI = 0.313B'l' \qquad (15.1.24)$$

where B' is the flux density, Mx/in^2; and l' the length of the magnetic path, in.

EXAMPLE.   The average flux density in the air gap of a generator is 40,000 Mx/in^2, and the effective length of the gap is 0.2 in. How many ampere-turns per pole are necessary for the gap?

$$NI = 0.313 \times 40,000 \times 0.2 = 2,500$$

Since the relation of μ_r to flux density B in Eq. (15.1.22) is not simple, the relation of ampere-turns per unit length of magnetic circuit to flux density is ordinarily shown graphically. Typical curves of this character are shown in Fig. 15.1.6, inch units being used although scales of tesla, and ampere turns per metre are also given. To determine the number of ampere-turns necessary to produce a given total flux in a magnetic circuit composed of several parts in series having various lengths, cross sections, and relative permeabilities, determine the flux density if the cross section is fixed, or otherwise choose a cross section to give a suitable flux density. From the magnetization curve obtain the ampere-turns necessary to drive this **flux density** through a unit length of the portion of the circuit considered and multiply by the length. Add together the ampere-turns required for each series part of the magnetic circuit to obtain the total ampere-turns necessary to give the assumed flux.

It is desirable to operate magnetic circuits at as high flux densities as is practicable in order to reduce the amount of iron and copper. The air gaps of dynamos are operated at average densities of 40,000 to

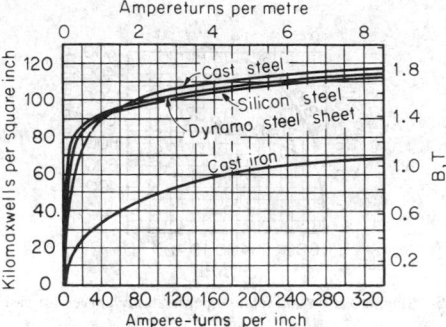

Fig. 15.1.6   Typical magnetization curves.

50,000 Mx/in². Higher densities increase the exciting ampere-turns and tooth losses. At 45,000 Mx/in² the flux density in the teeth may be as high as 120,000 to 130,000 Mx/in². The flux densities in transformer cores are limited as a rule by the permissible losses. At 60 Hz and with silicon steel the maximum density is 60,000 to 70,000 Mx/in², at 25 Hz the density may run as high as 75,000 to 90,000 Mx/in². With laminated cores, the net iron is approximately 0.9 the gross cross section.

Magnetic Leakage   It is impossible to confine all magnetic flux to any desired path since there is no known insulator of magnetic flux. Figure 15.1.7 shows the magnetic circuit of a modern four-pole dy-

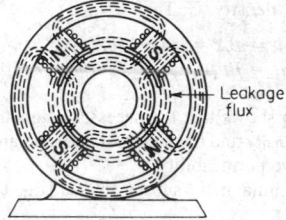

Fig. 15.1.7   Magnetic circuit of a four-pole dynamo with leakage flux.

namo. A considerable proportion of the useful magnetic flux leaks between the pole shoes and cores, rather than across the air gap. The ratio of the maximum flux, which exists in the field cores, to the useful flux, i.e., the flux that crosses the air gap, is the **coefficient of leakage.** This coefficient must always be greater than unity and in carefully designed dynamos may be as low as 1.15. It is frequently as high as 1.30. Although the geometry of the leakage-flux paths is not simple, the leakage flux may be determined by approximations with a fair degree of accuracy.

Magnetic Hysteresis   The magnetization curves shown in Figs. 15.1.5 and 15.1.6 are called **normal curves.** They are taken with the magnetizing force continuously increased from zero. If at any point the magnetizing force be decreased, a greater value of flux density for any given magnetizing force will result. The effect of carrying iron through a complete cycle of magnetization, both positive and negative, is shown in Fig. 15.1.8.

The curve OKB, taken with increasing values of magnetizing force per centimeter H, is the **normal induction** curve. If after the magnetizing force has reached the value OA, it is decreased, the magnetic flux density B will decrease in accordance with curve BCD, between A and O the values being much greater than those given by the normal curve, i.e., the flux density lags the magnetizing force. At zero magnetizing force, the flux density is OC, call the **remanence.** A negative magnetizing force

OD, called the **coercive force,** is required to bring the flux density to zero. If the magnetizing force is increased negatively to OA', the flux density will be given by the curve DE. If the magnetizing force is increased positively from A' to A, the flux density will be given by the curve EFGB, which is similar to the curve BCDE. OF is the negative remanence and OG again is the coercive force. The complete curve is called a

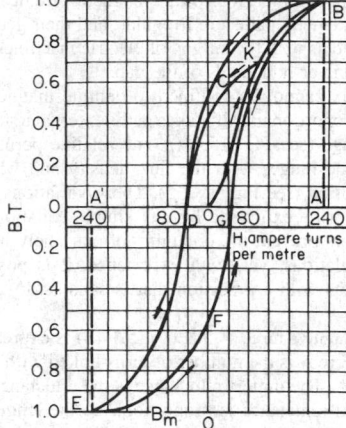

Fig. 15.1.8   Hysteresis loop for dynamo steel.

hysteresis loop. When the normal curve reaches the point K, if the magnetizing force is then decreased, another hysteresis loop, a portion of which is shown at KL, will be obtained. It is seen that the flux density lags the magnetizing force throughout.

The energy dissipated per cycle is proportional to the area of the loop and is equal to $(1/4\pi)\int H\,dB$ ergs/(Hz)(cm³). For moderately high densities the energy loss per cycle varies according to the **Steinmetz law**

$$W = 10\eta B_m^{1.6} \qquad \text{W} \cdot \text{s/m}^3 \qquad (15.1.25)$$

where B_m is the maximum value of the flux density, T (Fig. 15.1.8). Table 15.1.6 gives values of the Steinmetz coefficient η for common magnetic steels.

Table 15.1.6   Steinmetz Coefficients

Hard tungsten steel	0.058	Ordinary sheet iron	0.004
Hard cast steel	0.025	Pure iron	0.003
Forged steel	0.020	Annealed iron sheet	0.002
Cast iron	0.013	Best annealed sheet	0.001
Electrolytic iron	0.009	Silicon steel sheet	0.00046
Soft machine steel	0.009	Permalloy	0.0001
Annealed cast steel	0.008		

A permanent increase in the hysteresis constant occurs if the temperature of operation remains for some time above 80°C. This phenomenon is known as **aging** and may be much reduced by proper annealing of the iron. Silicon steels containing about 3 percent silicon have a lower hysteresis loss, somewhat larger eddy-current loss, and are practically nonaging.

Eddy-current losses, also known as Foucault-current losses, occur in iron subjected to cyclic magnetization. Eddy-current losses are reduced by laminating the iron, which subdivides the emf and increases greatly the length of path of the parasitic currents. Eddy currents also have a screening effect, which tends to prevent the flux penetrating the iron. Hence laminating also allows the full cross section of the iron to be utilized unless the frequency is too high.

Eddy-current loss in sheets is given by

$$P_e = (\pi t f B_m)^2/6\rho 10^{16} \qquad \text{W/cm}^3 \qquad (15.1.26)$$

where t = thickness, cm; f = frequency, Hz; B_m = the maximum flux density, G; ρ = the resistivity, $\Omega \cdot$ cm.

Relations of Direction of Magnetic Flux to Current Direction The direction of the magnetizing force of a current is at right angles to its direction of flow. Magnetic lines about a cylindrical conductor carrying current exist in circular planes concentric with and normal to the conductor. This is illustrated in Fig. 15.1.9a. The ⊕ sign, corresponding to the feathered end of the arrow, indicates a direction of current away from the observer; a ⊙ sign, corresponding to the tip of an arrow, indicates a direction of current toward the observer.

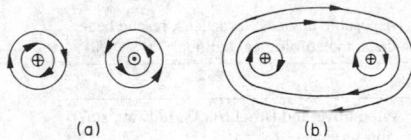

Fig. 15.1.9 Currents in (a) opposite directions, (b) in the same direction.

Corkscrew Rule The direction of the current and that of the resulting magnetic field are related to each other as the forward travel of a corkscrew and the direction in which it is rotated.

Hand Rule Grasp the conductor in the right hand with the thumb pointing in the direction of the current. The fingers will then point in the direction of the lines of flux.

The applications of these rules are illustrated in Fig. 15.1.9. If the currents in parallel conductors are in opposite directions (Fig. 15.1.9a), the conductors tend to move apart; if the currents in parallel conductors are in the same direction (Fig. 15.1.9b), the conductors tend to come together. The magnetic lines act like stretched rubber bands and, in attempting to contract, tend to pull the two conductors together.

The relation of the direction of current in a solenoid helix to the direction of flux is shown in Fig. 15.1.10. Figure 15.1.11 shows the

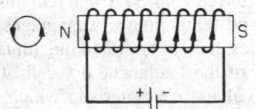

Fig. 15.1.10 Direction of current and poles in a solenoid.

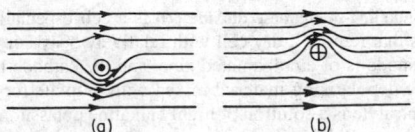

Fig. 15.1.11 Effect of a current on a uniform magnetic field.

effect on a uniform field of placing a conductor carrying current in that field and normal to it. In (a) the direction of the current is toward the observer. By applying the corkscrew rule it is seen that the current weakens the field immediately above it and strengthens the field immediately below it. The reverse is true in (b), where the direction of the current is away from the observer.

Figure 15.1.11 is illustrative of the force developed on a conductor carrying current in a magnetic field. In (a) the conductor will tend to move upward owing to the stretching of the magnetic lines beneath it. Similarly, the conductor in (b) will tend to move downward. This principle is the basis of motor action. (See also "Magnets.")

BATTERIES

In an **electric cell,** or **battery,** chemical energy is converted into electrical energy. The word *battery* may be used for a single cell or for an assembly of cells connected in series or parallel. A battery utilizes the potential difference which exists between different elements. When two different elements are immersed in electrolyte an emf exists tending to send current within the cell from the negative pole, which is the more

highly electropositive, to the positive pole. The **poles,** or **electrodes** of a battery form the junction with the external circuit.

If the external circuit is closed, current flows from the battery at the **positive electrode,** or **anode,** and enters the battery at the **negative electrode,** or **cathode.**

In a **primary battery** the chemically reacting parts **require renewal;** in a **secondary battery,** the electrochemical processes **are reversible** to a high degree and the chemically reacting parts are restored after partial or complete discharge by reversing the direction of current through the battery. See Table 15.1.7 for a summary of battery types and applications.

Electromotive force of a battery is the total potential difference existing between the electrodes on open circuit. When current flows, the potential difference across the terminal drops because of the resistance drop within the cell and because of **polarization.**

Polarization When current flows in a battery, hydrogen is deposited on the cathode. This produces two effects, both of which reduce the terminal voltage of the battery. The hydrogen in contact with the cathode constitutes a hydrogen battery which opposes the emf of the battery; the hydrogen bubbles reduce the contact area of the electrolyte with the cathode, thus increasing the battery resistance. The most satisfactory method of reducing polarization is to have present at the cathode some compound that supplies negative ions to combine with the positive hydrogen ions at the plate. In the Leclanché cell, manganese peroxide in contact with the carbon cathode serves as a depolarizer, its oxygen ion combining with the hydrogen ion to form water.

If E is the emf of the cell, E_p the emf of polarization, r the internal resistance, V the terminal voltage, when current I flows, then

$$V = (E - E_p) - Ir \tag{15.1.27}$$

Primary Batteries

Dry Cells A dry cell is one in which the electrolyte exists in the form of a jelly, is absorbed in a porous medium, or is otherwise restrained from flowing from its intended position, such a cell being completely portable and the electrolyte nonspillable. The Leclanché cell consists of a cylindrical zinc container which serves as the negative electrode and is lined with specially prepared paper, or some similar absorbent material, to prevent the mixture of carbon and manganese dioxide, which is tamped tightly around the positive carbon electrode, from coming in contact with the zinc. The absorbent lining and the mixture are moistened with a solution of zinc chloride and sal ammoniac. In smaller cells (Fig. 15.1.12) the manganese-carbon mixture is often molded into a cylinder around the carbon electrode, the whole is then set into the zinc cup, and the space between the molded mixture and the zinc is filled with electrolyte made into a paste in such a manner that it can be solidified by either standing or heating. The top of the cell is closed with a sealing compound, and the cell is placed in a cardboard container. The emf of a dry cell when new is 1.4 to 1.6 V.

In **block assembly** the dry cells, especially in the smaller sizes, are assembled in series and sealed in blocks of insulating compound with only two terminals and, sometimes, intermediate taps brought out. This type of battery is used for radio B and C batteries. Another construction is to build the battery up of layers in somewhat the manner of the old voltaic pile. Each cell consists of a layer of zinc, a layer of treated paper, and a flat cake of the manganese-carbon mixture. The cells are separated by layers of a special material which conducts electricity but which is impervious to electrolyte. A sufficient number of such cells are built up to give the required voltage and the whole battery is sealed into the carton.

Leclanché cells are generally available in sizes ranging from small, thin penlight batteries to large assemblies of cells in series or parallel for special high-voltage or high-current applications.

The **efficiency** of a standard-size dry battery depends on the rate at which it is discharged. Up to a certain rate the lower the discharge rate, the greater the efficiency. Above this rate the efficiency decreases (see *Natl. Bur. Stand. Circ.* 79, p. 39).

When used efficiently, a 6-in dry cell will give over 30 A·h of ser-

Table 15.1.7   Battery Types and Applications

Battery type	Cell type	Nominal cell voltage	Capacity, wH/kg	Applications
		Primary		
Leclanché (zinc-carbon)	Dry	1.5	22–44	Flashlights, emergency lights, radios
Zinc-mercury (Ruben)	Dry	1.34	90–110	Medical, marine, space, laboratory, and emergency devices
Zinc-alkaline-manganese dioxide	Dry	1.5	66	Models, cameras, shavers, lights
Silver or cuprous chloride-magnesium	Wet		55–120	Disposable devices: torpedoes, rescue beacons, meteorological balloons
		Secondary		
Lead-acid	Wet	2		Automotive, industrial trucks, railway, station service
Lead-calcium	Wet	2		Standby
Edison (nickel-iron)	Wet	1.2		Industrial trucks; boat and train lights
Nickel-cadmium	Wet	1.2	28	Engine starting, emergency lighting, station service
Silver oxide-cadmium	Wet	1.4	45–65	Space
Silver-zinc	Wet	1.55	90–155	Models, photographic equipment, missiles

vice. As ordinarily used, however, the dry cell give no more than 8 to 10 A·h of service and at times even less. The 1¼ by 2¼ in flashlight battery is usually employed with a lamp taking 0.25 to 0.35 A. Under these conditions 3 A·h or thereabouts may be expected if the battery is used for not more than an hour or so a day. The so-called ''heavy-duty'' radio battery will give about 8 to 10 A·h when efficiently used.

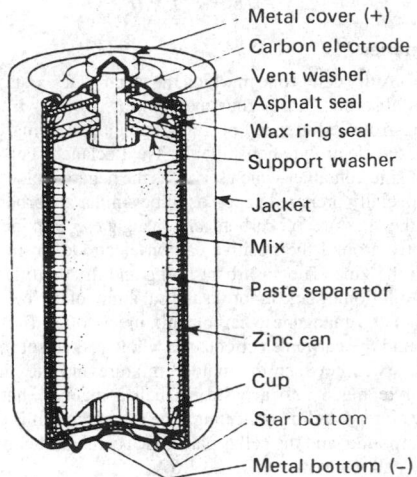

Fig. 15.1.12   Cross section of a standard round zinc-carbon cell. *(From "Standard Handbook for Electrical Engineers," Fink and Carrol, McGraw-Hill, NY, copyright 1968.)*

For the best results 6-in dry cells should not be used for current drains of over 0.5 A except for very short periods of time. Flashlight batteries should not be used for higher than the preceding current drain, and heavy-duty radio batteries will give best results if the current drain is kept below 25 mA.

Dry cells should be stored in a cool, dry place. Extreme heat during storage will shorten their life. The cell will not be injured by being frozen but will be as good as new after being brought back to normal temperature. In extreme cold weather dry cells may not give more than half of their normal service. At a temperature of about − 30°F they freeze solid and give neither voltage nor current.

The amperage of a dry cell by definition is the current that it will give when it is short-circuited (at about 70°F) through an ammeter which with its leads has a resistance of 0.01 Ω.

The **Ruben cell** (Ruben, Balanced Alkaline Dry Cells, *Trans. Electrochem. Soc.*, **92**, 1947) was developed jointly by the Ruben Laboratories and P. R. Mallory & Company during World War II for the operation of radar equipment and other electronic devices which require a high ratio of ampere-hour capacity to the volume of the cell at higher current densities than were considered practicable for the Leclanché type. The anode is of amalgamated zinc, and the cathode is a mercuric oxide depolarizing material intimately mixed with graphite in order to reduce its electrical resistivity. The electrolyte is a solution of potassium hydroxide (KOH) containing potassium zincate. The cell is made in three forms as shown in Fig. 15.1.13, the wound-anode type (*a*), the button type (*b*), and the cylindrical type (*c*).

The no-load emf of the cell is 1.34 V and remains essentially constant irrespective of time and temperature. Advantages of the cell are long shelf life, which enables them to be stored indefinitely; long service life, about four times that of the Leclanché dry cell of equivalent volume; small weight; a flat voltage characteristic which is advantageous for electronic uses in which the characteristics of tubes vary widely with voltage; adaptability to operating at high temperatures without deterioration; high resistance to shock.

The **zinc-alkaline-manganese dioxide cell** is a cell especially useful in applications that require a dry cell with relatively heavy or continuous drain. The anode is of amalgamated zinc, and the cathode is a manganese dioxide depolarizing material mixed with graphite for conductivity. The electrolyte is a solution of highly alkaline potassium hydroxide immobilized in cellulosic-type separators. These cells are available in standard-size cylindrical construction and wafer (flat) construction for cassette and tape recorder applications.

Wet Cells   The **silver** or **cuprous chloride-magnesium cell** is a one-shot battery with a life of days after the electrolyte is added. A wet cell may be stored for years in a dry state. The cathode is either compacted copper chloride and graphite or sheet silver chloride, while the cathode is a thin magnesium sheet. The electrolyte is a solution of sodium chloride. The silver chloride cells are more expensive and are available in more and larger ratings.

The **Weston cell** is a primary cell used as a standard of emf. It consists of a glass H tube in the bottom of one leg of which is mercury which forms the cathode; in the bottom of the other leg is cadmium amalgam forming the anode. The electrolytes consist of mercurous sulfate and cadmium sulfate. There are two forms of the Weston cell: the saturated or normal cell, and the unsaturated cell. In the normal cell the electrolyte is saturated. This is the official standard since it is more permanent than the unsaturated type and can be reproduced with far greater accuracy. When carefully made, the emfs of cells agree within a few parts per million. There is, however, a small temperature coefficient. Although the unsaturated cell is not so reliable as the normal cell and must be standardized, it has a negligible temperature coefficient and is more convenient for general use. The manufacturers recommend that the temperature be not less than 4°C and not more than 40°C and the current

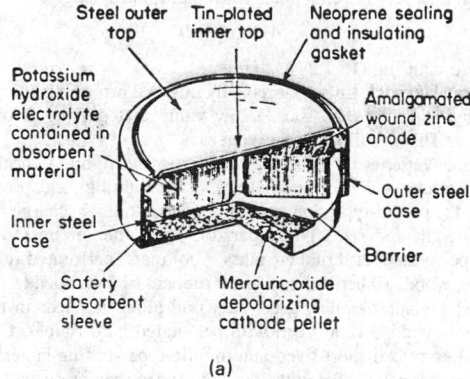

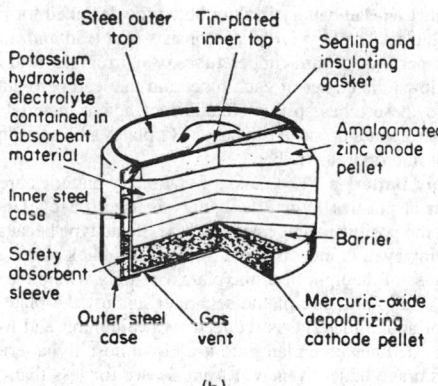

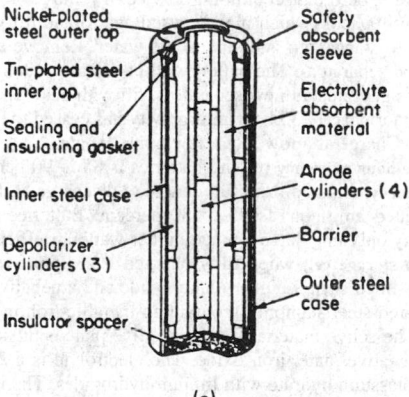

Fig. 15.1.13   Ruben cells. (*a*) Wound-anode; (*b*) button; (*c*) cylindrical. (*From "Standard Handbook for Electrical Engineers," Fink and Carrol, McGraw-Hill, NY, copyright 1968.*)

should not exceed 0.0001 A. The emf is between 1.0185 and 1.0190 V. Since no appreciable current can be taken from the cell, a null method must be used to utilize its emf.

Storage (Secondary) Batteries

In a **storage battery** the electrolytic action must be **reversible** to a high degree. There are three types of storage batteries; the lead-lead-acid type, the nickel-iron-alkaline type (**Edison** battery), and the nickel-cadmium-alkali type (**Nicad**). In addition, there are various specialized types of cells for scientific and military purposes, and there is continuous development work in the search for higher capacities.

In the manufacture of the **lead-lead-acid cells** there are three general types of plates, or electrodes. In the **Planté type** the active material is

electrically formed of pure lead by repeated reversals of the charging current. In the **Faure,** or **pasted-plate,** type, the positive and negative plates are formed by applying a paste, largely of lead oxides (PbO_2, Pb_3O_4), to lead-antimony or lead calcium supporting grids. A current is passed through the plates while they are immersed in weak sulfuric acid, the positive plates being connected as anodes and the negative ones as cathodes. The paste on the positive plates is converted into lead peroxide while that on the negative plate is reduced to spongy lead. The **tubular plate** (iron-clad) type has lead-alloy rods surrounded by perforated dielectric tubes with powdered-lead oxides packed between the rod and tube for the positive plate.

In order to obtain high capacity per unit weight it is necessary to expose a large plate area to the action of the acid. This is done in the Planté plate by "ploughing" with sharp steel disks, and by using corrugated helical inserts as active positive material (Manchester plate). In the pasted plate a large area of the material is necessarily exposed to the action of the acid.

The chemical reactions in a lead cell may be expressed by the following equation, based on the double sulfation theory:

$$\underset{\substack{\text{positive} \\ \text{plate}}}{PbO_2} + \underset{\substack{\text{negative} \\ \text{plate}}}{Pb} + \underset{\substack{\text{sulfuric} \\ \text{acid}}}{2H_2SO_4} \overset{\text{charge}}{\underset{\text{discharge}}{=}} \underset{\substack{\text{positive and} \\ \text{negative plates}}}{2PbSO_4} + \underset{\text{water}}{2H_2O}$$

Between the extremes of complete charge and discharge, complex combinations of lead and sulfate are formed. After complete discharge a hard insoluble sulfate forms slowly on the plates, and this is reducible only by slow charging. This sulfation is objectionable and should be avoided.

Specific Gravity   Water is formed with discharge and sulfuric acid is formed on charge, consequently the specific gravity must decrease on discharge and increase on charge. The variation of the specific gravity for a stationary battery is shown in Fig. 15.1.14. With starting and

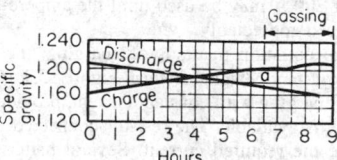

Fig. 15.1.14   Variations of specific gravity in a stationary battery.

vehicle batteries it is necessary to operate the electrolyte from between 1.280 to 1.300 when fully charged to as low as 1.100 when completely discharged. The condition of charge of a battery can be determined by its specific gravity.

Battery electrolyte may be made from concentrated sulfuric acid (oil of vitriol, sp gr 1.84) by **pouring the acid into the water** in the following proportions:

Parts Water to 1 Part Acid

Specific gravity	1.200	1.210	1.240	1.280
Volume	4.3	4.0	3.4	2.75
Weight	2.4	2.2	1.9	1.5

Freezing Temperature of Sulfuric Acid and Water Mixtures

Specific gravity	1.180	1.200	1.240	1.280
Freezing temp, °F	−6	−16	−51	−90

Voltage   The emf of a lead cell when fully charged and idle is 2.05 to 2.10 V. Discharge lowers the voltage in proportion to the current. When charging at constant current and normal rate, the terminal voltage gradually increases from 2.14 to 2.3 V, then increases rapidly to between 2.5

and 2.6 V (Fig. 15.1.15). This latter interval is known as the **gassing period.** When this period is reached, the charging rate should be reduced in order to avoid waste of power and unnecessary erosion of the plates.

Practically all batteries have a **normal rating** based on the 8-h rate of

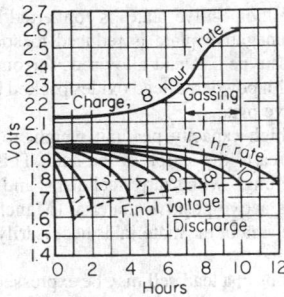

Fig. 15.1.15 Voltage curves on charge and discharge for a lead cell.

discharge. Thus a 320 A·h battery would have a normal rate of 40 A. The ampere-hour capacity of batteries falls off rapidly with increase in discharge rate.

Effect of Discharge Rate on Battery Capacity

Discharge rate, h	8	5	3	1	1/3	1/10
Percentage of rated capacity,						
Planté type	100	88	75	55.8	37	19.5
Pasted type	100	93	83	63	41	25.5

The following rule may be observed in **charging a lead battery.** The charging rate in amperes should be less than the number of ampere-hours out of the battery. For example, if 200 A·h are out of a battery, a charging rate of 200 A may be used until the ampere-hours out of the battery are reduced appreciably.

There are two common methods of charging: the **constant-current method** and the **constant-potential method.** Figure 15.1.16a shows a common method of charging with constant current, provided a low-voltage dc power supply is available. The resistor connected in series may be adjusted to give the required current. Several batteries may be connected in series. Figure 15.1.16b shows a more common method, using a copper oxide or silicon rectifier, since ac power supply is more common than dc. The rectifier disks, mounted in a stack, are bridge-connected, the directions of rectification being indicated. The polarity of the two wires can readily be determined by means of a dc voltmeter.

The constant-potential method is to be preferred since the rate automatically tapers off as the cell approaches the charged condition. Without resistance the terminal voltage should be 2.3 V per cell, but it is preferable to use 2.4 to 2.5 V per cell with low resistance in series.

When a battery is being charged, its terminal voltage

$$V = E + Ir \qquad (15.1.28)$$

Compare with Eq. (15.1.27).

When a battery is fully charged, any rate will produce gassing, but the rate may be reduced to such a low value that gassing is practically harmless. This is called the **finishing rate.**

Portable batteries for automobile starting and lighting, airplanes, industrial trucks, electric locomotives, train lighting, and power boats employ the pasted-type plates because of their high discharge rates for a given weight and size. The separators are either of treated grooved wood; perforated hard rubber; glass-wool mats; perforated rubber, and grooved wood; ribbed microporous rubber. In low-priced short-lived batteries for automobiles, grooved wood alone is used; in the better types, the wood is reinforced with perforated hard rubber. Containers for the low-priced short-lived automobile-type starting batteries are of asphaltic compound; for other portable types they are usually of hard rubber.

The **Exide iron-clad** battery is a portable type designed for propelling electric vehicles. The positive plate consists of a lead-antimony frame supporting perforated hard-rubber tubes. An irregular lead-antimony core runs down the center of each tube, and the lead peroxide paste is packed into these tubes so that shedding of active material from the positive plate cannot occur. Pasted negative plates are used. The separators are flat microporous rubber.

Stationary Batteries The tanks of stationary batteries are made of hard rubber or plastics. When the battery is used for regulating or cycling duty, the positive plates may be of the Planté type because of their long life. However, in most modern installations thick pasted plates are used. Because of the tight fit of the plate assembly within the container and the resulting pressure of the separator against the plate surfaces, shedding of active material is reduced to a minimum and long life is obtained. Pasted negative plates are used in almost all batteries.

A lead storage battery **removed from service** for less than 9 months should be charged once a month if possible; if not, it should be given a heavy overcharge before discontinuing service. If removed for a longer period, siphon off acid (which may be used again) and fill with fresh water. Allow to stand 15 h and siphon off water. Remove and throw away the wood separators. The battery will now stand indefinitely. To put in service again, install new separators, fill with acid (sp gr 1.210) and charge at normal rate 35 h or until gravity has ceased to rise over a period of 5 h. Charge at a low rate a few hours longer.

The **ampere-hour efficiency** of lead batteries is 85 to 90 percent. The **watthour efficiency** obtained from full charge to discharge at the normal rate and at rated amp-hour is 75 to 80 percent. Batteries which do regulating duty only may have a much higher watthour efficiency.

The **Edison storage cell** when fully charged has a positive plate of nickel pencils filled with a higher nickel oxide and a negative plate of flat nickel-plated-steel stampings containing metallic iron in finely divided form. The active material for the positive plate is nickel hydrate and for the negative plate, iron oxide. The electrolyte is a 21 percent solution of potassium hydrate with lithium hydroxides. The initial emf is about 1.4 V and the average emf about 1.1 V throughout discharge. In

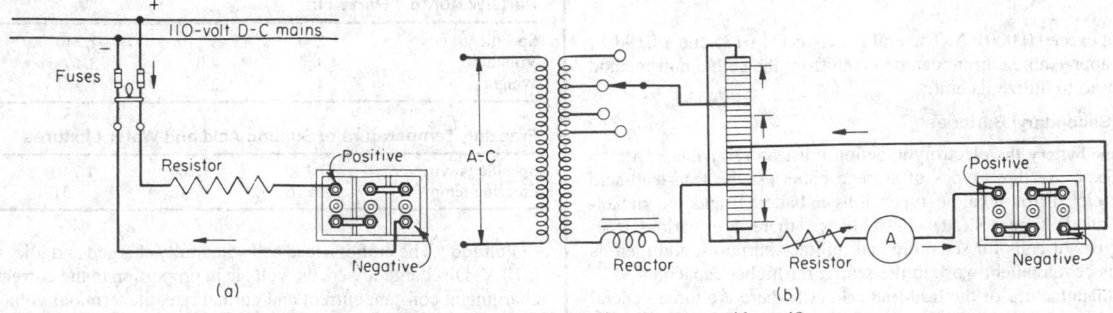

Fig. 15.1.16 Connections for charging a storage battery from (a) 110-V dc mains, (b) copper oxide rectifier.

Fig. 15.1.17 are shown typical voltage characteristics on charge and discharge for an Edison cell. On account of the higher internal resistance of the cell the battery is not so efficient from the energy standpoint as the lead cell. The jar is welded nickel-plated steel. The battery is compact and extremely light and strong and for these reasons is particularly adapted for propelling electric vehicles and for boat- and train-lighting systems. The battery is rugged, and since there is no opportunity for the growth of active material on the plates or flaking of active material, the battery has long life.

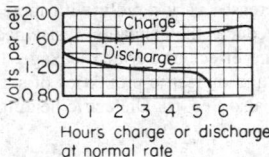

Fig. 15.1.17 Voltage during charge and discharge of an Edison cell.

Nickel-Cadmium-Alkali (Nicad) Battery The positive active material is nickelic (black) hydroxide mixed with graphite to give it high conductivity. The negative active material is cadmium oxide. Both materials are used in powdered form and are contained within flat perforated steel pockets. These pockets are locked into steel plates, the positive and negative being alike in construction. All steel parts are nickel-plated. A complete plate group consists of a number of positive and negative plates assembled on bolts and terminal posts common to plates of the same polarity. The separators are thin strips of polystyrene, and all other battery insulation is also polystyrene. The entire plate assembly is contained within a welded-steel tank. The electrolyte is potassium hydroxide (KOH), specific gravity 1.210 at 72°F (22°C); it does not enter into any chemical reactions with the electrode materials, and its specific gravity remains constant during charge and discharge, neglecting any slight change due to the small amount of gassing. On charge, the voltage is 1.4 to 1.5 V until near the end when it rises to 1.8 V. On discharge, the voltage is nearly constant at 1.2 V.

Nicad batteries are strong mechanically and are not damaged by overcharge; they hold their charge over long periods of idleness, the active material cannot flake off, the internal resistance is low, there is no corrosion, and the battery has an indefinitely long life. It is a general-purpose battery.

In the **Sonotone** nickel-cadmium battery the positive plates are nickel oxide when the battery is charged, and the negative plates are metallic cadmium. On discharge the positive plates are reduced to a state of lower oxidation, and the negative plates regain oxygen. The electrolyte is a 30 percent solution of potassium hydroxide, the specific gravity of which is 1.29 at room temperature. The case is a transparent plastic. The terminal voltage at the normal discharge rate is 1.2 V per cell.

Rechargeable batteries, exemplified by Gould Nicad cells (Alkaline Battery Division, Gould National Batteries, Inc.), are hermetically sealed nickel-cadmium cells that contain no free alkaline electrolyte. Since there is no spillage or leakage, they can operate in any position, have long life, and require no maintenance or servicing, and their weight is small for their output. They are thus well adapted to power many types of cordless appliances such as tools, hedge shears, cameras, dictating equipment, electric razors, radios, and television sets. The electrodes consist of a plaque of microporous sintered nickel having an extremely high surface area. The electrochemical reactions differ from those of the conventional vented-type alkaline battery, a type which at the end of a charge liberates both oxygen and hydrogen gases as well as electrolytic fumes that must be vented through a valve in the top of the cell. In the sealed nickel-cadmium cell, the negative electrode (at the time that the cell is sealed) never becomes fully charged, and the evolution of hydrogen is completely suppressed. On charging, when the positive electrode has reached its full capacity, the oxygen which has evolved is channeled through the porous separator to the negative electrode and oxidizes the finely divided cadmium of the microporous plate to cadmium hydroxide, which at the same time is reduced to metallic cadmium. The cells are constructed in three different forms: the button

type, the cylindrical type, and the prismatic type. Their ratings range from 20 mA·h to 23 A·h. Their average discharge voltage is 1.22 V, and they require 14 h of charge at the normal rate (one-tenth A·h rating), which for a 3.5 A·h cell is 0.35 A.

Precautions in the **care of storage batteries:** An ammeter should not be connected directly across the terminals to test the condition of a cell; a battery should not be left to stand in a discharged condition; a flame should not be brought in the vicinity of a battery that is being charged; the battery should not be allowed to become heated when charging; water should never be added to the concentrated acid—always acid to the water; acid should never be equalized except when the battery is in a charged condition; a battery should never be exposed to the influence of external heat; voltmeter tests should be made when the current is flowing; batteries should always be kept clean. To replace acid lost through slopping, use a solution of 2 parts concentrated sulfuric acid in 5 parts water by weight, unless a hydrometer is at hand to enable the solution to be made up according to the specifications of the makers of the cell.

DIELECTRIC CIRCUITS

Dynamic and Static Electricity Electricity in motion such as an electric current is dynamic electricity; electricity at rest is static electricity. The two are identical physically. Since static electricity is frequently produced at high voltage and small quantity, the two are frequently considered as being two different types of electricity.

Capacitors

Capacitors (formerly condensers) Two conducting bodies, or electrodes, separated by a dielectric constitute a capacitor. If a positive charge is placed on one electrode of a capacitor, an equal negative charge is induced on the other. The medium between the capacitor plates is called a **dielectric.** The dielectric properties of a medium relate to its ability to conduct **dielectric lines.** This is in distinction to its **insulating** properties which relate to its property to conduct **electric current.** For example, air is an excellent insulator but ruptures dielectrically at low voltage. It is not a good dielectric so far as breakdown strength is concerned.

With capacitors

$$Q = CE \qquad (15.1.29)$$
$$C = Q/E \qquad (15.1.30)$$
$$E = Q/C \qquad (15.1.31)$$

where Q = quantity, C; C = capacitance, F; and E = voltage. The unit of capacitance in the practical system is the **farad.** The farad is too large a unit for practical purposes, so that either the **microfarad** (μF) or the **picofarad** (pF) are used. However, in voltage, current, and energy relations the capacitance must be expressed in farads.

The energy stored in a capacitor is

$$W = \tfrac{1}{2}QE = \tfrac{1}{2}CE^2 = \tfrac{1}{2}Q^2/C \qquad J \qquad (15.1.32)$$

Capacitance of Capacitors The capacitance of a **parallel-electrode** capacitor (Fig. 15.1.18) is

$$C = \varepsilon_r A/(4\pi d \times 9 \times 10^3) \qquad \mu F \qquad (15.1.33)$$

where ε_r = relative capacitivity; A = area of one electrode, m²; and d = distance between electrodes, m.

Fig. 15.1.18 Parallel-electrode capacitor.

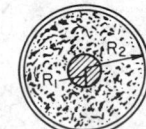

Fig. 15.1.19 Coaxial-cylinder capacitor.

The capacitance of **coaxial cylindrical** capacitors (Fig. 15.1.19) is

$$C = 0.2171\varepsilon_r l/[9 \times 10^5 \log (R_2/R_1)] \qquad \mu F \qquad (15.1.34)$$

where ε_r is the relative capacitivity and l the length, m. Also

$$C = 0.03882\varepsilon_r/\log(R_2/R_1) \qquad \mu\text{F/mi} \qquad (15.1.35)$$

Equation (15.1.35) is useful in that it is applicable to cables.

The capacitance of two **parallel cylindrical conductors** D m between centers and having radii of r m is

$$C = 0.01941/\log(D/r) \qquad \mu\text{F/mi} \qquad (15.1.36)$$

In practice, the capacitance to neutral or to an infinite conducting plane midway between the conductors and perpendicular to their plane is usually used. The capacitance to neutral is

$$C = 0.03882/\log(D/r) \qquad \mu\text{F/mi} \qquad (15.1.37)$$

Equations (15.1.36) and (15.1.37) are used for calculating the capacitance of overhead transmission lines. When computing charging current, use voltage between lines in (15.1.36) and to neutral in (15.1.37).

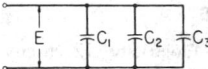

Fig. 15.1.20 Capacitances in parallel.

Capacitances in Parallel The equivalent capacitance of capacitances in parallel (Fig. 15.1.20) is

$$C = C_1 + C_2 + C_3 \qquad (15.1.38)$$

Capacitances in parallel are all across the same voltage. If the voltage is E, then the total quantity $Q = CE$ and $Q_1 = C_1E$, etc.

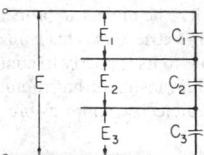

Fig. 15.1.21 Capacitances in series.

Capacitances in Series The equivalent capacitance C of capacitances in series (Fig. 15.1.21) is found as follows:

$$1/C = 1/C_1 + 1/C_2 + 1/C_3 \qquad (15.1.39)$$

If the capacitances are not leaky, the charge Q is the same on each. $Q = CE$, $E_1 = Q/C_1$, $E_2 = Q/C_2$, etc.

Insulators and Dielectrics Insulating materials are applied to electric circuits to prevent the leakage of current. Insulating materials used with high voltage must not only have a high resistance to leakage current, but must also be able to resist dielectric puncture; i.e., in addition to being a good insulator, the material must be a good dielectric. Insulation resistance is usually expressed in $M\Omega$ and the resistivity given in $M\Omega \cdot$ cm. The dielectric strength is usually given in terms of voltage gradient, common units being V/mil, V/mm, and kV/cm. Insulation resistance decreases very rapidly with increase in temperature. Absorbed moisture reduces the insulation resistance, and moisture and humidity have a large effect on surface leakage. In Table 15.1.8 are given the insulating and dielectric properties of several common insulating materials (see also Sec. 6). Dielectric heating of materials is described in Sec. 7.

TRANSIENTS

Induced EMF If a flux ϕ webers linking N turns of conductor changes, an emf

$$e = -N(d\phi/dt) \qquad \text{V} \qquad (15.1.40)$$

is induced.

Self-inductance Let a flux ϕ link N turns. The linkages of the circuit are $N\phi$ weber-turns. If the permeability of the circuit is assumed constant, the number of these linkages per ampere is the **self-inductance** or **inductance** of the circuit. The unit of inductance is the **henry.** The inductance is

$$L = N\phi/(i) \qquad \text{H} \qquad (15.1.41)$$

If the permeability changes with the current

$$L = N(d\phi/di) \qquad \text{H} \qquad (15.1.42)$$

The energy stored in the magnetic field

$$W = \frac{1}{2}Li^2 \qquad \text{J} \qquad (15.1.43)$$

EMF of Self-induction If Eq. (15.1.41) is written $Li = N\phi$ and differentiated with respect to the time t, $L(di/dt) = N(d\phi/dt)$ and from Eq. (15.1.40)

$$e = -L(di/dt) \qquad \text{V} \qquad (15.1.44)$$

e is the emf of self-induction. If a rate of change of current of 1 A/s induces an emf of 1 V, the inductance is then 1 H.

Table 15.1.8 Electrical Properties of Insulating Materials

Material	Volume resistivity, $M\Omega \cdot$ cm	Dielectric constant, 60Hz	Dielectric strength	
			V/mil	V/mm
Asbestos board (ebonized)	10^7		55	2×10^3
Bakelite	$5-30 \times 10^{11}$	4.5–5.5	450–1,400	$(17-55) \times 10^3$
Epoxy	10^{14}	3.5–5	300–400	$(12-16) \times 10^3$
Fluorocarbons:				
Fluorinated ethylene propylene	10^{18}	2.1	500	20×10^3
Polytetrafluoroethylene	10^{18}	2.1	400	16×10^3
Glass	17×10^9	5.4–9.9	760–3,800	$(3-15) \times 10^4$
Magnesium oxide		2.2	300–700	$(12-27) \times 10^3$
Mica	$10^{14}-10^{17}$	4.5–7.5	1,000–4,000	$(4-16) \times 10^4$
Nylon	$10^{14}-10^{17}$	4–7.6	300–400	$(12-16) \times 10^3$
Neoprene		7.5	600	23.5×10^3
Oils:				
Mineral	21×10^6	2–4.7	300–400	$(12-16) \times 10^3$
Paraffin	10^{15}	2.41	410–550	$(16-22) \times 10^3$
Paper		1.7–2.6	110–230	$(4-9) \times 10^3$
Paper, treated		2.5–4	500–750	$(20-30) \times 10^3$
Phenolic (glass filled)	$10^{12}-10^{13}$	5–9	140–400	$(5.5-16) \times 10^3$
Polyethylene	$10^{15}-10^{18}$	2.3	450–1,000	$(17-40) \times 10^3$
Polyimide	$10^{16}-10^{17}$	3.5	400	16×10^3
Polyvinyl chloride (flexible)	$10^{11}-10^{15}$	5–9	300–1,000	$(12-40) \times 10^3$
Porcelain	3×10^8	5.7–6.8	240–300	$(9.5-12) \times 10^3$
Rubber	$10^{14}-10^{16}$	2–3.5	500–700	$(20-27) \times 10^3$
Rubber (butyl)	10^{18}	2.1		

Current in Inductive Circuit   If a circuit containing resistance R and inductance L in series is connected across a steady voltage E, the voltage E must supply the iR drop in the circuit and at the same time overcome the emf of self-induction. That is $E = Ri + L\,di/dt$. A solution of this differential equation gives

$$i = (E/R)(1 - e^{-Rt/L}) \quad \text{A} \quad (15.1.45)$$

where e is the base of the natural system of logarithms.

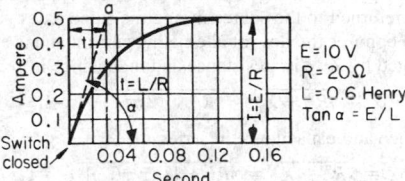

Fig. 15.1.22   Rise of current in an inductive circuit.

Figure 15.1.22 shows this equation plotted when $E = 10$ V, $R = 20\ \Omega$, $L = 0.6$ H. It is to be noted that inductance causes the current to rise slowly to its Ohm's law value, $I_0 = E/R = \tfrac{10}{20} = 0.5$ A. When $t = L/R$, the current has reached 63.2 percent of its Ohm's law value. L/R is the **time constant** of the circuit. In the foregoing circuit, the time constant $L/R = 0.6/20 = 0.03$ s. The initial rate of rise of current is $\tan \alpha = E/L$. If current continued at this rate, it would reach $\alpha = E/R$ in L/R s $[(E/L) \times (L/R) = E/R]$.

If a circuit containing inductance and resistance in series is short-circuited when the current is I_0, the equation of current becomes

$$i = I_0 e^{-Rt/L} \quad \text{A} \quad (15.1.46)$$

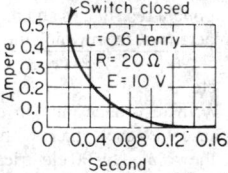

Fig. 15.1.23   Decay of current in an inductive circuit.

Figure 15.1.23 shows this equation plotted when $I_0 = 0.5$ A, $R = 20\ \Omega$, $L = 0.6$ H. It is seen that inductance opposes the decay of current. Inductance always opposes change of current.

Mutual Inductance   If two circuits having inductances L_1 and L_2 henrys are so related to each other geometrically that any portion of the flux produced by the current in one circuit links the other circuit, the two circuits possess **mutual inductance.** It follows that a change of current in one circuit causes an emf to be induced in the other. Let ε_2 be induced in circuit 2 by a change di_1/dt in circuit 1. Then

$$e_2 = -M\,di_1/dt \quad \text{V} \quad (15.1.47)$$

M is the mutual inductance of the two circuits.

$$M = k\sqrt{L_1 L_2} \quad (15.1.48)$$

where k is the **coefficient of coupling** of the two circuits, or the proportion of the flux in one circuit which links the other. Also a change of current di_2/dt in circuit 2 induces an emf e_1 in circuit 1, $e_1 = -M\,di_2/dt$.

The stored energy is

$$W = \tfrac{1}{2}L_1 I_1^2 + \tfrac{1}{2}L_2 I_2^2 + M I_1 I_2 \quad \text{J} \quad (15.1.49)$$

where I_1 and I_2 are the currents in circuits 1 and 2.

Current in Capacitive Circuit   If capacitance C farads and resistance R ohms are connected in series across the steady voltage E, the current is

$$i = (E/R)e^{-t/CR} \quad \text{A} \quad (15.1.50)$$

If a capacitor charged to voltage E is discharged through resistance R, the current is

$$i = -(E/R)e^{-t/CR} \quad \text{A} \quad (15.1.51)$$

Except for sign, these two equations are identical and are of the same form as Eq. (15.1.46).

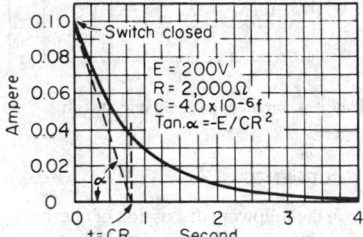

Fig. 15.1.24   Transient current to a capacitor.

In Fig. 15.1.24 is shown the transient current to a capacitor in series with a resistor when $E = 200$ V, $C = 4.0\ \mu$F, $R = 2$ kΩ. When $t = CR$, the current has reached $1/e = 0.368$ its initial value. CR is the **time constant** of the circuit. The initial rate of decrease of current is $\tan \alpha = -E/CR^2$. If the current continued at this rate it would reach zero when the time is CR s. If, in its fully charged condition, the capacitor of Fig. 15.1.24 is discharged through the resistor R, the curve will be the negative of that shown in Fig. 15.1.24.

Resistance, Inductance, and Capacitance in Series   If a circuit having resistance, inductance, and capacitance in series is connected across a source of steady voltage, a transient condition results. If $R > \sqrt{4L/C}$, the circuit is nonoscillatory or overdamped.

The current is

$$i = \frac{EC}{\sqrt{R^2 C^2 - 4LC}} \left(e^{(-\alpha+\beta)t} - e^{(-\alpha-\beta)t}\right) \quad \text{A} \quad (15.1.52)$$

where $\alpha = R/2L$ and $\beta = (\sqrt{R^2 C^2 - 4LC})/2LC$.

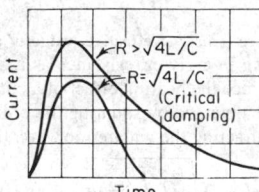

Fig. 15.1.25   Transient current in nonoscillatory circuits.

In Fig. 15.1.25 is shown the curve corresponding to Eq. (15.1.52). When $R = \sqrt{4L/C}$, the system is **critically damped** and the transient dies out rapidly without oscillation. The current is

$$i = (E/L)te^{-Rt/2L} \quad \text{A} \quad (15.1.53)$$

Figure 15.1.25 shows also the curve corresponding to Eq. (15.1.53).

If $R < \sqrt{4L/C}$, the transient is oscillatory, being a logarithmically damped sine wave. The current is

$$i = \frac{2EC}{\sqrt{4LC - R^2 C^2}}\, e^{-Rt/2L} \sin \frac{\sqrt{4LC - R^2 C^2}}{2LC}\, t \quad \text{A} \quad (15.1.54)$$

The transient oscillates at a frequency very nearly equal to $1/(2\pi\sqrt{LC})$ Hz. This is the **natural frequency** of the circuit.

In Fig. 15.1.26 is shown the curve corresponding to Eq. (15.1.54). If the capacitor, after being charged to E V, is discharged into the foregoing series circuits, the currents are given by Eqs. (15.1.52) to (15.1.54) multiplied by -1. Equations (15.1.52) to (15.1.54) are the same types obtained with dynamic mechanical systems with friction, mass, and elasticity.

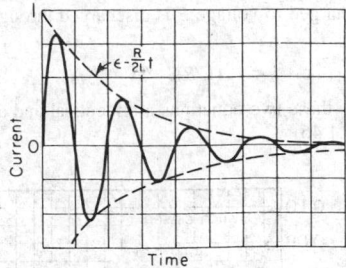

Fig. 15.1.26 Transient current in an oscillatory circuit.

ALTERNATING CURRENTS

Sine Waves In the following discussion of alternating currents, sine waves of voltage and current will be assumed. That is, $e = E_m \sin \omega t$ and $i = I_m \sin(\omega t - \theta)$, where E_m and I_m are maximum values of voltage and current; ω, the angular velocity, in rad/s, is equal to $2\pi f$, where f is the frequency; θ is the angle of phase difference.

Cycle; Frequency When any given armature coil has passed a pair of poles, the emf or current has gone through 360 electrical degrees, or 1 **cycle**. An **alternation** is one-half cycle. The **frequency** of a synchronous machine in cycles per second (hertz) is

$$f = NP/120 \qquad \text{Hz} \qquad (15.1.55)$$

where N is the speed in r/min and P the number of poles. In the United States and Canada the frequency of 60 Hz is almost universal for general lighting and power. For the ac power supply to dc transit systems, and for railroad electrification, a frequency of 25 Hz is used in many installations. In most of Europe and Latin America the frequency of 50 Hz is in general use. In aircraft the frequency of 400 Hz has become standard.

Static inverters make it possible to obtain high and variable frequencies to drive motors at greater than the 3,600 r/min limitation on 60-Hz circuits, and to vary speeds. The textile industry has small motors operating at 12,000 r/min (200 Hz), and larger motors have been run at 6,000 r/min (100 Hz). Many large mainframe computers have been powered at 400 Hz.

The **root-mean-square (rms)**, or **effective**, value of a current wave produces the same heating in a given resistance as a direct current of the same ampere value. Since the heating effect of a current is proportional to $i^2 r$, the rms value is obtained by squaring the ordinates, finding their average value, and extracting the square root, i.e., the rms value is

$$I = \sqrt{1/T \int_0^T i^2 \, dt} \qquad \text{A} \qquad (15.1.56)$$

where T is the time of a cycle. The rms value I of a sine wave equals $(1/\sqrt{2})I_m = 0.707 I_m$.

Average Value of a Wave The average value of a sine wave over a complete cycle is zero. For a half cycle the average is $(2/\pi)I_m$, or $0.637 I_m$, where I_m is the maximum value of the sine wave. The average value is of importance only occasionally. A dc measuring instrument gives the average value of a pulsating wave. The average value is of use (1) when the effects of the current are proportional to the number of coulombs, as in electrolytic work and (2) when converting alternating to direct current.

Form Factor The form factor of a wave is the ratio of rms value to average value. For a sine wave this is $\pi/(2\sqrt{2}) = 1.11$. This factor is important in that it enters equations for induced emf.

Inductive reactance, $2\pi fL$ or ωL, opposes an alternating current in inductance L. It is expressed in Ω. Reactance is usually denoted by the symbol X. Inductive reactance is denoted by X_L.

The **current** in an inductive reactance X_L when connected across the voltage E is

$$I = E/X_L = E/(2\pi fL) \qquad \text{A} \qquad (15.1.57)$$

This current lags the voltage by 90 electrical degrees. Inductance absorbs no energy. The energy stored in the magnetic field during each half cycle is returned to the source during the same half cycle.

Capacitive reactance is $1/(2\pi fC) = 1/\omega C$ and is denoted by X_C, where C is in F. If C is given in μF, $X_C = 10^6 2\pi fC$. The current in a capacitive reactance X_C when connected across voltage E is

$$I = E/X_C = 2\pi fCE \qquad \text{A} \qquad (15.1.58)$$

This current leads the voltage by 90 electrical degrees. Pure capacitance absorbs no energy. The energy stored in the dielectric field during each half cycle is returned to the source during the same half cycle.

Impedance opposes the flow of alternating current and is expressed in Ω. It is denoted by Z. With resistance and inductance in series

$$Z = \sqrt{R^2 + X_L^2} = \sqrt{R^2 + (2\pi fL)^2} \qquad \Omega \qquad (15.1.59)$$

With resistance and capacitance in series

$$Z = \sqrt{R^2 + X_C^2} = \sqrt{R^2 + [1/(2\pi fC)]^2} \qquad \Omega \qquad (15.1.60)$$

With resistance, inductance, and capacitance in series

$$Z = \sqrt{R^2 + (X_L - X_C)^2}$$
$$= \sqrt{R^2 + [2\pi fL - 1/(2\pi fC)]^2} \qquad \Omega \qquad (15.1.61)$$

The current is

$$I = E/\sqrt{R^2 + [2\pi fL - 1/(2\pi fC)]^2} \qquad \text{A} \qquad (15.1.62)$$

Phasor or Vector Representation Sine waves of voltage and current can be represented by phasors, these phasors being proportional in magnitude to the waves that they represent. The angle between two phasors is also equal to the time angle existing between the two waves that they represent.

Phasors may be combined as forces are combined in mechanics. Both graphical methods and the methods of complex algebra are used. Impedances and also admittances may be similarly combined, either graphically or symbolically. The usual method is to resolve series impedances into their component resistances and reactances, then combine all resistances and all reactances, from which the resultant impedance is obtained. Thus $Z_1 + Z_2 = \sqrt{(r_1 + r_2)^2 + (x_1 + x_2)^2}$, where r_1 and x_1 are the components of Z_1, etc.

Phase Difference With resistance only in the circuit, the current and the voltage are in phase with each other; with inductance only in the circuit, the current lags the voltage by 90 electrical degrees; with capacitance only in the circuit, the current leads the voltage by 90 electrical degrees.

With resistance and inductance in series, the voltage leads the current by angle θ where $\tan \theta = X_L/R$. With resistance and capacitance in series, the voltage lags the current by angle θ where $\tan \theta = -X_C/R$.

With resistance, inductance, and capacitance in series, the voltage may lag, lead, or be in phase with the current.

$$\tan \theta = (X_L - X_C)/R = (2\pi fL - 1/2\pi fC)/R \qquad (15.1.63)$$

If $X_L > X_C$ the voltage leads; if $X_L < X_C$ the voltage lags; if $X_L = X_C$ the current and voltage are in phase and the circuit is in resonance.

Power Factor In ac circuits the power $P = I^2 R$ where I is the current and R the effective resistance (see below). Also the power

$$P = EI \cos \theta \qquad \text{W} \qquad (15.1.64)$$

where θ is the phase angle between E and I. Cos θ is the **power factor** (pf) of the circuit. It can never exceed unity and is usually less than unity.

$$\cos \theta = P/EI \qquad (15.1.65)$$

P is often called the true power. The product EI is the volt-amp $(V \cdot A)$ and is often called the apparent power.

Active or **energy** current is the projection of the total current on the voltage phasor. $I_e = I \cos \theta$. Power $= EI_e$.

Reactive, quadrature, or **wattless current** $I_q = I \sin \theta$ and is the component of the current that contributes no power but increases the $I^2 R$ losses of the system. In power systems it should ordinarily be made low.

The **vars** (volt-amp-reactive) are equal to the product of the voltage and reactive current. **Vars** = EI_q. **Kilovars** = $EI_q/1,000$.

Effective Resistance   When alternating current flows in a circuit, the losses are ordinarily greater than are given by the losses in the ohmic resistance alone. For example, alternating current tends to flow near the surface of conductors (skin effect). If iron is associated with the circuit, eddy-current and hysteresis losses result. These power losses may be accounted for by increasing the ohmic resistance to a value R, where R is the **effective resistance**, $R = P/I^2$. Since the iron losses vary as $I^{1.8}$ to I^2, little error results from this assumption.

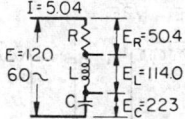

Fig. 15.1.27   Resistor, inductor, and capacitor in series.

SOLUTION OF SERIES-CIRCUIT PROBLEM.   Let a resistor R of 10 Ω, an inductor L of 0.06 H, and a capacitor C of 60 μF be connected in series across 120-V 60-Hz mains (Fig. 15.1.27). Determine (1) the impedance, (2) the current, (3) the voltage across the resistance, the inductance, the capacitance, (4) the power factor, (5) the power, (6) the angle of phase difference.
   (1) $\omega = 2\pi 60 = 377$. $X_L = 0.06 \times 377 = 22.6$ Ω; $X_C = 1/(377 \times 0.000060) = 44.2$ Ω; $Z = \sqrt{(10)^2 + (22.6 - 44.2)^2} = 23.8$ Ω; (2) $I = 120/23.8 = 5.04$ A; (3) $E_R = IR = 5.04 \times 10 = 50.4$ V; $E_L = IX_L = 5.04 \times 22.6 = 114.0$ V; $E_C = IX_C = 5.04 \times 44.2 = 223$ V; (4) $\tan \theta = (X_L - X_C)/R = -21.6/10 = -2.16$, $\theta = -65.2°$, $\cos \theta = $ pf $= 0.420$; (5) $P = 120 \times 5.04 \times 0.420 = 254$ W; $P = I^2R = (5.04)^2 \times 10 = 254$ W (check); (6) From (4) $\theta = -65.2°$. Voltage lags. The phasor diagram to scale of this circuit is shown in Fig. 15.1.28. Since the current is common for all elements of the circuit, its phasor is laid horizontally along the axis of reference.

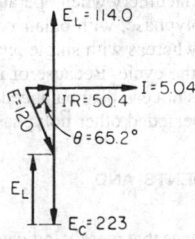

Fig. 15.1.28   Phasor diagram for a series circuit.

Resonance   If the voltage E and the resistance R [Eq. (15.1.62)] are fixed, the maximum value of current occurs when $2\pi fL - 1/2\pi fC = 0$. The circuit so far as its terminals are concerned behaves like a noninductive resistor. The current $I = E/R$, the power $P = EI$, and the power factor is unity.
   The voltage across the inductor and the voltage across the capacitor are opposite and equal and may be many times greater than the circuit voltage. The frequency

$$f = 1/(2\pi \sqrt{LC}) \qquad \text{Hz} \qquad (15.1.66)$$

is the **natural frequency** of the circuit and is the frequency at which it will oscillate if the circuit is not acted upon by some external frequency. This is the principle of radio sending and receiving circuits. Resonant conditions of this type should be avoided in power circuits, as the piling up of voltage may endanger apparatus and insulation.

EXAMPLE.   For what value of the inductance in the circuit (Fig. 15.1.27) will the circuit be in resonance, and what is the voltage across the inductor and capacitor under these conditions?
   From Eq. (15.1.66) $L = 1/(2\pi f)^2C = 0.1173$ H. $I = E/R = 120/10 = 12$ A. $L\omega I = I/C\omega = 0.1173 \times 377 \times 12 = 530$ V. This voltage is over four times the line voltage.

Parallel Circuits   Parallel circuits are used for nearly all power distribution. With several series circuits in parallel it is merely necessary to find the current in each and add all the current phasors vectorially to find the total current. Parallel circuits may be solved analytically.
   A series circuit has resistance r_1 and **inductive reactance** x_1. The **conductance** is

$$g_1 = r_1/(r_1^2 + x_1^2) = r_1/Z_1^2 \qquad \text{S} \qquad (15.1.67)$$

and the **susceptance** is

$$b_1 = x_1/(r_1^2 + x_1^2) = x_1/Z_1^2 \qquad \text{S} \qquad (15.1.68)$$

Conductance is not the reciprocal of resistance unless the reactance is zero; susceptance is not the reciprocal of reactance unless the resistance is zero. With inductive reactance the susceptance is **negative**; with capacitive reactance the susceptance is **positive**.
   If a second circuit has resistance r_2 and **capacitive reactance** x_2 in series, $g_2 = r_2/(r_2^2 + x_2^2) = r_2/Z_2^2$; $b_2 = x_2/(r_2^2 + x_2^2) = x_2/Z_2^2$. The total conductance $G = g_1 + g_2$; the total susceptance $B = -b_1 + b_2$. The **admittance** is

$$Y = \sqrt{G^2 + B^2} = 1/Z \qquad \text{S} \qquad (15.1.69)$$

The energy current is EG; the reactive current is EB; the power is

$$P = E^2G \qquad \text{W} \qquad (15.1.70)$$
$$\text{vars} = E^2B \qquad \text{W} \qquad (15.1.71)$$

The power factor is

$$pf = G/Y \qquad (15.1.72)$$

Also the following relations hold:

$$r = g/(g^2 + b^2) = g/Y^2 \qquad \Omega \qquad (15.1.73)$$
$$x = b/(g^2 + b^2) = b/Y^2 \qquad \Omega \qquad (15.1.74)$$

SOLUTION OF A PARALLEL-CIRCUIT PROBLEM.   In the parallel circuit of Fig. 15.1.29 it is desired to find the joint impedance, the total current, the power in each branch, the total power, and the power factor, when $E = 100$, $f = 60$, $R_1 = 2$ Ω, $R_2 = 4$ Ω, $L_1 = 0.00795$ H, $X_1 = 2\pi fL_1 = 3$ Ω, $C_2 = 1,326$ μF, $X_2 = 1/2\pi fC_2 = 2$ Ω, $Z_1 = \sqrt{2^2 + 3^2} = 3.6$ Ω, and $Y_1 = 1/3.6 = 0.278$ S.
   Solution: $g_1 = R_1/(R_1^2 + X_1^2) = 2/13 = 0.154$; $b_1 = -3/13 = -0.231$; $Z_2 = \sqrt{16 + 4} = 4.47$; $Y_2 = 1/4.47 = 0.224$; $g_2 = R_2/(R_2^2 + X_2^2) = 4/(16 + 4) = 0.2$ S; $b_2 = 2/20 = 0.1$ S; $G = g_1 + g_2 = 0.154 + 0.2 = 0.354$ S; $B = b_1 + b_2 = -0.231 + 0.1 = -0.131$ S; $Y = \sqrt{G^2 + B^2} = \sqrt{0.354^2 + (-0.131)^2} = 0.377$ S, and joint impedance $Z = 1/0.377 = 2.65$ Ω. Phase angle $\theta = \tan^{-1}(-0.131/0.354) = -20.3°$. $I = EY = 100 \times 0.377 = 37.7$ A; $P_1 = E^2g_1 = 100^2 \times 0.154 = 1,540$ W; $P_2 = E^2g_2 = 100^2 \times 0.2 = 2,000$ W; total power = $E^2G = 100^2 \times 0.354 = 3,540$ W. Power factor = $\cos \theta = 3,540/(100 \times 37.7) = 93.8$ percent.

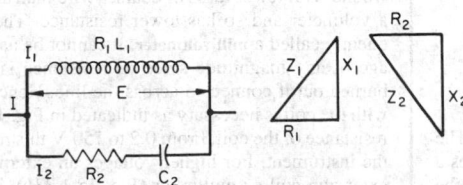
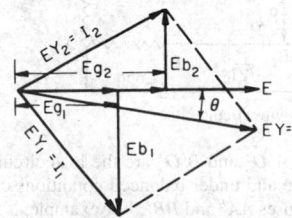

Fig. 15.1.29   Parallel circuit and phasor diagrams.

With parallel circuits, unity power factor is obtained when the algebraic sum of the quadrature currents is zero. That is, $b_1 + b_2 + b_3 \cdots = 0$.

Three-Phase Circuits Ac generators are usually wound with three armature circuits which are spaced 120 electrical degrees apart on the armature. Hence these coils generate emfs 120 electrical degrees apart. The coils are connected either in Y (star) or in Δ (mesh) as shown in Fig. 15.1.30. Whether Y- or Δ-connected, with a balanced load, the three

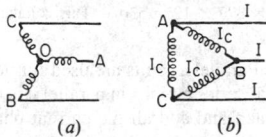

Fig. 15.1.30 Three-phase connections. (*a*) Y connection; (*b*) Δ connection.

coil emfs E_c and the three coil currents I_c are equal. In the Y connection the line and coil currents are equal, but the line emfs E_{AB}, E_{BC}, E_{CA} are $\sqrt{3}$ times in magnitude the coil emfs E_{OA}, E_{OB}, E_{OC}, since each is the phasor difference of two coil emfs. In the delta connection the line and coil emfs are equal, but I, the line current, is $\sqrt{3} I_c$, the coil current, i.e., it is the phasor difference of the currents in the two coils connected to the line. The power of a coil is $E_c I_c \cos \theta$, so that the total power is $3E_c I_c \cos \theta$. If θ is the angle between coil current and coil voltage, the angle between line current and line voltage will be $30° \pm \theta$. In terms of line current and emf, the power is $\sqrt{3} EI \cos \theta$. A fourth or neutral conductor connected to O is frequently used with the Y connection. The neutral point O is frequently grounded in transmission and distribution circuits. The coil emfs are assumed to be sine waves. Under these conditions they balance, so that in the delta connection the sum of the two coil emfs at each instant is balanced by the third coil emf. Even though the third, ninth, fifteenth . . . harmonics, $3(2n + 1)f$, where $n = 0$ or an integer, exist in the coil emfs, they cannot appear between the three external line conductors of the three-phase Y-connected circuit. In the delta circuit, the same harmonics $3(2n + 1)f$ cause local currents to circulate around the mesh. This may cause a very appreciable heating. In a three-phase system the power

$$P = \sqrt{3} \ EI \cos \theta \qquad \text{W} \qquad (15.1.75)$$

the power factor is

$$P/\sqrt{3} \ EI \qquad (15.1.76)$$

and the kV $\cdot$ A

$$\sqrt{3} \ EI/1{,}000 \qquad (15.1.77)$$

where E and I are line voltages and currents.

Two-Phase Circuits Two-phase generators have two windings spaced 90 electrical degrees apart on the armature. These windings generate emfs differing in time phase by 90°. The two windings may be independent and power transmitted to the receiver though the two single-phase circuits are entirely insulated from each other. The two circuits may be combined into a two-phase three-wire circuit such as is shown in Fig. 15.1.31, where OA and OB are the generator circuits (or

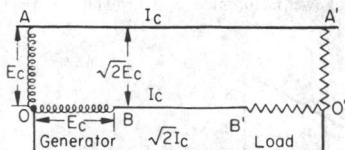

Fig. 15.1.31 Two-phase, three-wire circuit.

transformer secondaries) and $A'O'$ and $B'O'$ are the load circuits. The wire OO' is the common wire and under balanced conditions carries a current $\sqrt{2}$ times the current wires AA' and BB'. For example, if I_c is the coil current, $\sqrt{2} \ I_c$ will be the value of the current in the common con-

ductor OO'. If E_c is the voltage across OA or OB, $\sqrt{2} \ E_c$ will be the voltage across AB. The power of a two-phase circuit is twice the power in either coil if the load is balanced. Normally, the voltages OA and OB are equal, and the current is the same in both coils. Owing to nonsymmetry and the high degree of unbalancing of this system even under balanced loads, it is not used at the present time for transmission and is little used for distribution.

Four-Phase Circuit A four-phase or quarter-phase circuit is shown in Fig. 15.1.32. The windings AC and BD may be independent or con-

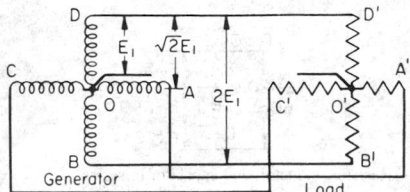

Fig. 15.1.32 Four-phase or quarter-phase circuit.

nected at O. The voltages AC and BD are 90 electrical degrees apart as in two-phase circuits. If a neutral wire O-O' is added, three different voltages can be obtained. Let E_1 = voltage between O-A, O-B, O-C, O-D. Voltages between A-B, B-C, C-D, D-A = $\sqrt{2} \ E_1$. Voltages between A-C, B-D = $2E_1$. Because of this multiplicity of voltages and the fact that polyphase power apparatus and lamps may be connected at the same time, this system is still used to some extent in distribution.

Advantages of Polyphase Power The advantages of polyphase power over single-phase power are as follows. The output of synchronous generators and most other rotating machinery is from 60 to 90 percent greater when operated polyphase than when operated single phase; pulsating fluxes and corresponding iron losses which occur in many common types of machinery when operated single phase are negligible when operated polyphase; with balanced polyphase loads polyphase power is constant whereas with single phase the power fluctuates over wide limits during the cycle. Because of its minimum number of wires and the fact that it is not easily unbalanced, the three-phase system has for the most part superseded other polyphase systems.

ELECTRICAL INSTRUMENTS AND MEASUREMENTS

Electrical measuring devices that merely indicate, such as ammeters and voltmeters, are called **instruments;** devices that totalize with time such as watthour meters and ampere-hour meters are called **meters.** (See also Sec. 16.) Most types of electrical instruments are available with digital read out.

DC Instruments Direct current and voltage are both measured with an indicating instrument based on the principle of the D'Arsonval galvanometer. A coil with steel pivots and turning in jewel bearings is mounted in a magnetic field produced by permanent magnets. The motion is restrained by two small flat coiled springs, which also serve to conduct the current to the coil. The deflections of the coil are read with a light aluminum pointer attached to the coil and moving over a graduated scale. The same instrument may be used for either current or voltage, but the method of connecting in circuit is different in the two cases. Usually, however, the coil of an instrument to be used as an ammeter is wound with fewer turns of coarser wire than an instrument to be used as a voltmeter and so has lower resistance. The instrument itself is frequently called a **millivoltmeter.** It cannot be used alone to measure voltage of any magnitude since its resistance is so low that it would be burned out if connected across the line. Hence a resistance r' in series with the coil is necessary as indicated in Fig. 15.1.33a in which r_c is the resistance of the coil. From 0.2 to 750 V this resistance is usually within the instrument. For higher voltages an external resistance R, called an extension coil or multiplier (Fig. 15.1.33b), is necessary. Let e be the reading of the instrument, in volts (Fig. 15.1.33b), r the internal resistance of the instrument, including r' and r_c in Eq. (15.1.33a), R the resistance of the multiplier. Then the total voltage is

$$E = e(R + r)/r \qquad (15.1.78)$$

It is clear that by using suitable values of R a voltmeter can be made to have several scales.

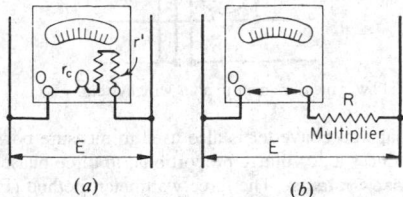

Fig. 15.1.33 Voltmeters. (a) Internal resistance; (b) with multiplier.

Instruments themselves can only carry currents of the magnitudes of 0.01 to 0.06 A. To measure larger values of current the instrument is provided with a shunt R (Fig. 15.1.34). The current divides inversely as the resistances r and R of the instrument and the shunt. A low resistance r' within the instrument is connected in series with the coil. This permits some adjustment to the deflection so that the instrument can be

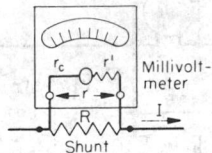

Fig. 15.1.34 Millivoltmeter with shunt.

adapted to its shunt. Usually most of the current flows through the shunt, and the current in the instrument is negligible in comparison. Up to 50 and 75 A the shunt can be incorporated within the instrument. For larger currents it is usually necessary to have the shunt external to the instrument and connect the instrument to the potential terminals of the shunt by means of leads. Any given instrument may have any number of ranges by providing it with a sufficient number of shunts. The range of the usual instrument of this type is approximately 50 mV. Although the same instrument may be used for voltmeters or ammeters, the moving coils of voltmeters are usually wound with more turns of finer wire. They take approximately 0.01 A so that their resistance is approximately 100 Ω/V. Instruments used as ammeters alone operate with 0.01 to 0.06 A.

Permanent-magnet moving-coil instruments may be used to measure unidirectional pulsating currents or voltages and in such cases will indicate the average value of the periodically varying current or voltage.

AC Instruments   Instruments generally used for alternating currents may be divided into five types: **electrodynamometer, iron-vane, thermocouple, rectifier,** and **electronic.** Instruments of the **electrodynamometer** type, the most precise, operate on the principle of one coil carrying current, turning in the magnetic field produced by a second coil carrying current taken from the same circuit. If these circuits or coils are connected in series, the torque exerted on the moving system for a given relative position of the coil system is proportional to the square of the current and is not dependent on the direction of the current. Consequently, the instrument will have a compressed scale at the lower end and will usually have only the upper two-thirds of the scale range useful for accurate measurement. Instruments of this type ordinarily require 0.04 to 0.08 A or more in the moving-coil circuit for full-scale deflection. They read the rms value of the alternating or pulsating current. The wattmeter operates on the electrodynamometer principle. The fixed coil, however, is energized by the current of the circuit, and the moving coil is connected across the potential in series with high resistance. Unless shielded magnetically the foregoing instruments will not, in

general, indicate so accurately on direct as on alternating current because of the effects of external stray magnetic fields. Also reversed readings should be taken. **Iron vane** instruments consist of a fixed coil which actuates magnetically a light movable iron vane mounted on a spindle; they are rugged, inexpensive, and may be had in ranges of 30 to 750 V and 0.05 to 100 A. They measure rms values and tend to have compressed scales as in the case of electrodynamometer instruments.

The compressed part of the scale may, however, be extended by changing the shape of the vanes. Such instruments operate with direct current and are accurate to within 1 percent or so. AC instruments of the **induction type** (Westinghouse Electric Corp.) must be used on ac circuits of the frequency for which they have been designed. They are rugged and relatively inexpensive and are used principally for switchboards where a long-scale range and a strong deflecting torque are of particular advantage. **Thermocouple instruments** operate on the **Seebeck effect.** The current to be measured is conducted through a heater wire, and a thermojunction is either in thermal contact with the heater or is very close to it. The emf developed in the thermojunction is measured by a permanent-magnet dc type of instrument. By controlling the shape of the air gap, a nearly uniform scale is obtained. This type of instrument is well adapted to the measurement of high-frequency currents or voltages, and since it operates on the heating effect of current, it is convenient as a transfer instrument between direct current and alternating current.

In the **rectifier-type instrument** the ac voltage or current is rectified, usually by means of a small copper oxide or a selenium-type rectifier, connected in a bridge circuit to give full-wave rectification (Fig. 15.1.35). The rectified current is measured with a dc permanent-mag-

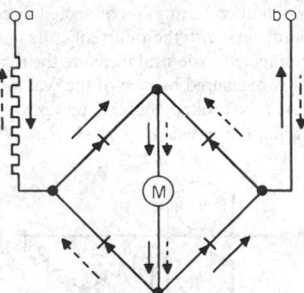

Fig. 15.1.35 Rectifier-type instrument.

net-type instrument M. The instrument measures the **average** value of the half waves that have been rectified, and with the sine waves, the average value is 0.9 the rms value. The scale is calibrated to indicate rms values. With nonsinusoidal waves the ratio of average to rms may vary considerably from 0.9 so that the instrument may be in error up to ± 5 percent from this cause. This type of instrument is widely used in the measurement of high-frequency voltages and currents. **Electronic voltmeters** operate on the principle of the amplification which can be obtained with a transistor. Since the emf to be measured is applied to the base, the instruments take practically no current and hence are adapted to measure potential differences which would change radically were any appreciable current taken by the measuring device. This type of instrument can measure voltages from a few tenths of a volt to several hundred volts, and with a potential divider, up to thousands of volts. They are also adapted to frequencies up to 100 MHz.

Particular care must be used in selecting instruments for measuring the nonsinusoidal waves of rectifier and controlled rectifier circuits. The **electrodynamic, iron vane,** and **thermocouple** instruments will read rms values. The **rectifier** instrument will read average values, while the **electronic** instrument may read either rms or average value, depending on the type.

Power Measurement in Single-Phase Circuits   Wattmeters are not rated primarily in W, but in A and V. For example, with a low power factor the current and voltage coils may be overloaded and yet the needle be well on the scale. The current coil may be carrying several times its rated current, and yet the instrument reads zero because the potential circuit is not closed, etc. Hence it is desirable to use both an ammeter and a voltmeter in conjunction with a wattmeter when measuring power (Fig. 15.1.36a). The instruments themselves consume appreciable power, and correction is often necessary unless these losses are negligible compared with the power being measured. For example, in Fig. 15.1.36a, the wattmeter measures the $I^2 R$ loss in its own current coil and in the ammeter (1 to 2 W each), as well as the loss in the

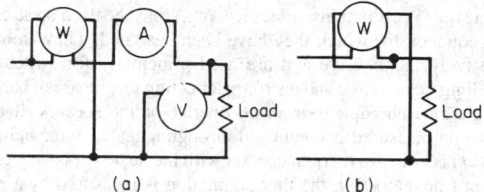

Fig. 15.1.36 Connections of instruments to single-phase load.

voltmeter ($= E^2/R$ where R is the resistance of the voltmeter). The losses in the ammeter and voltmeter may be eliminated by short-circuiting the ammeter and disconnecting the voltmeter when reading the wattmeter. If the wattmeter is connected as shown in Fig. 15.1.36b, it measures the power taken by its own potential coil (E^2/R_p) which at 110 V is 5 to 7 W. (R_p is the resistance of the potential circuit.) Frequently correction must be made for this power.

Power Measurement in Polyphase Circuits; Three-Wattmeter Method Let ao, bo, and co be any Y-connected three-phase load (Fig. 15.1.37). Three wattmeters with their current coils in each line and their potential circuits connected to neutral measure the total power, since the power in each load is measured by one of the wattmeters. The connection oo' may, however, be broken, and the total power is still the sum of the three readings; i.e., the power $P = P_1 + P_2 + P_3$. This method is

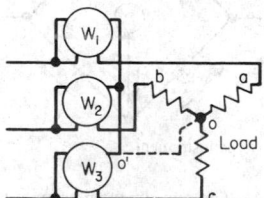

Fig. 15.1.37 Three-wattmeter method.

applicable to any system of n wires. The current coil of one wattmeter is connected in each of the n wires. The potential circuit of each wattmeter is connected between its own phase wire and a junction in common with all the other potential circuits. The wattmeters must be connected symmetrically, and the readings of any that read negative must be given the negative sign.

In the general case any system of n wires requires at least $n - 1$ wattmeters to measure the power correctly. The $n - 1$ wattmeters are connected in series with $n - 1$ wires. The potential circuit of each is connected between its own phase wire and the wire in which no wattmeter is connected (Fig. 15.1.38).

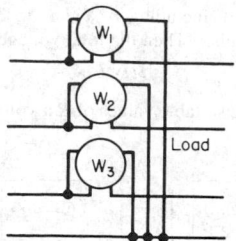

Fig. 15.1.38 Power measurement in an n-wire system.

The thermal watt converter is also used to measure power. This instrument produces a dc voltage proportional to three-phase ac power.

Three-Phase Systems The three-wattmeter method (Fig. 15.1.37) is applicable to any three-phase system. It is commonly used with the three-phase four-wire system. If the loads are balanced, $P_1 = P_2 = P_3$ and the power $P = 3P_1$.

The **two-wattmeter** method is most commonly used with three-phase three-wire systems (Fig. 15.1.39). The current coils may be connected in *any* two wires, the potential circuits being connected to the third. It will be recognized that this is adapting the method of Fig. 15.1.38 to three wires. With balanced loads the readings of the wattmeters are

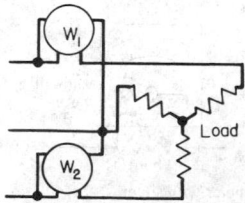

Fig. 15.1.39 Two-wattmeter method.

$P_1 = Ei \cos (30° + \theta)$, $P_2 = Ei \cos (30° - \theta)$, and $P = P_2 \pm P_1$. θ is the angle of phase difference between coil voltage and current. Since

$$P_1/P_2 = \cos (30° + \theta)/\cos (30° - \theta) \qquad (15.1.79)$$

the power factor is a function of P_1/P_2. Table 15.1.9 gives values of power factor for different ratios of P_1/P_2.

$$P = P_2 + P_1 \text{ when } \theta < 60°.$$

When $\theta = 60°$, pf $= \cos 60° = 0.5$, $P_1 = \cos (30° + 60°) = 0$, $P = P_2$. When $\theta > 60°$, pf < 0.5, $P = P_2 - P_1$. Also,

$$\tan \theta = \sqrt{3} (P_2 - P_1)/(P_2 + P_1) \qquad (15.1.80)$$

In a **polyphase wattmeter** the two single-phase wattmeter elements are combined to act on a single spindle. Hence the adding and subtracting of the individual readings are done automatically. The total power is indicated on one scale. This type of instrument is almost always used on switchboards. The connections of a portable type are shown in Fig. 15.1.40.

In the foregoing instrument connections, Y-connected loads are shown. These methods are equally applicable to delta-connected loads. The two-wattmeter method (Fig. 15.1.39) is obviously adapted to the two-phase three-wire system (Fig. 15.1.31).

Table 15.1.9 Ratio P_1/P_2 and Power Factor

P_1/P_2	Power factor	P_1/P_2	Power factor	P_1/P_2	Power factor	P_1/P_2	Power factor
+ 1.0	1.000	+ 0.4	0.804	− 0.1	0.427	− 0.6	0.142
+ 0.9	0.996	+ 0.3	0.732	− 0.2	0.360	− 0.7	0.102
+ 0.8	0.982	+ 0.2	0.656	− 0.3	0.296	− 0.8	0.064
+ 0.7	0.956	+ 0.1	0.576	− 0.4	0.240	− 0.9	0.030
+ 0.6	0.918	0.0	0.50	− 0.5	0.188	− 1.0	0.000
+ 0.5	0.866						

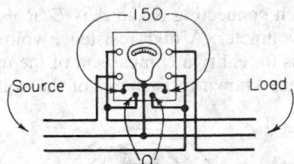

Fig. 15.1.40 Connections for a polyphase wattmeter in a three-phase circuit.

Measurement of Energy

Watthour meters record the energy taken by a circuit over some interval of time. Correct registration occurs if the angular velocity of the rotating element at every instant is proportional to the power. The method of accomplishing this with dc meters is illustrated in Fig. 15.1.41. The meter is in reality a small motor. The field coils FF are in series with the line. The armature A is connected across the line, usually in series with a resistor R. The movable field coil F' is in series with the armature A and serves to compensate for friction. C is a small commutator, either of

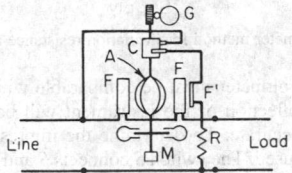

Fig. 15.1.41 DC watthour meter.

copper or of silver, and the two small brushes are usually of silver. An aluminum disk, rotating between the poles of permanent magnets M, acts as a magnetic brake the retarding torque of which is proportional to the angular velocity of the disk. A small worm and the gears G actuate the recording dials.

The following relation, or an equivalent, holds with most types of meter. With each revolution of the disk, K Wh are recorded, where K is the **meter constant** found usually on the disk. It follows that the average watts P over any period of time t sec is

$$P = 3,600KN/t \qquad (15.1.81)$$

where N is the revolutions of the disk during that period. Hence, the meter may be calibrated by connecting standardized instruments to measure the average power taken by the load and by counting the revolutions N for t s. Near full load, if the meter registers fast, the magnets M should be moved outward radially; if it registers slow, the magnets should be moved inward. If the meter registers fast at light (5 to 10 percent) load, the starting coil F' should be moved further away from the armature; if it registers slow, F' should be moved nearer the armature. A meter should not register more than 1.5 percent fast or slow, and with calibrated standards it can be made to register to within 1 percent of correct.

The **induction watthour meter** is used with alternating current. Although the dc meter registers correctly with alternating current, it is more expensive than the induction type, the commutator and brushes may cause trouble, and at low power factors compensation is necessary. In the induction watthour meter the driving torque is developed in the aluminum disk by the joint action of the alternating magnetic flux produced by the potential circuit and by the load current. The driving torque and the retarding torque are both developed in the same aluminum disk, hence no commutator and brushes are necessary. The rotating element is very light, and hence the friction torque is small. Equation (15.1.81) applies to this type of meter. When calibrating, the average power W for t s is determined with a calibrated wattmeter. The friction compensation is made at light loads by changing the position of a small hollow stamping with respect to the potential lug. The meter should also be adjusted at low power factor (0.5 is customary). If the meter is slow with lagging current, resistance should be cut out of the compensating circuit; if slow with leading current, resistance should be inserted.

Power-Factor Measurement The usual method of determining power factor is by the use of voltmeter, ammeter, and wattmeter. The wattmeter gives the watts of the circuit, and the product of the voltmeter reading and the ammeter reading gives the volt-amperes. The power factor is the ratio of the two [see Eqs. (15.1.65) and (15.1.76)]. Also single-phase and three-phase power-factor indicators, which can be connected directly in circuit, are on the market.

Instrument Transformers

With voltages higher than 600 V, and even at 600 V, it becomes dangerous and inaccurate to connect instruments and meters directly into power lines. It is also difficult to make potential instruments for voltages in excess of 600 V and ammeters in excess of 60-A ratings. To insulate such instruments from high voltage and at the same time to permit the use of low-range instruments, instrument transformers are used. **Potential transformers** are identical with power transformers except that their volt-ampere rating is low, being 40 to 500 W. Their primaries are wound for line voltage and their secondaries for 110 V. **Current transformers** are designed to go in series with the line, and the rated secondary current is 5 A. The secondary of a current transformer **should always be closed** when current is flowing; it should never be allowed to become open circuited under these conditions. When open-circuited the voltage across the secondary becomes so high as to be dangerous and the flux becomes so large in magnitude that the transformer overheats. Semiconductors that break down at safe voltages and short current-transformer secondaries are available to ensure that the secondary is closed. The secondaries of both potential and current transformers should be well grounded at one point (Figs. 15.1.42 and 15.1.43). Instrument transformers introduce slight errors because of small variations in their ratio with load. Also there is slight phase displacement in both current and potential transformers. The readings of the instruments must be multiplied by the instrument transformer ratios. The scales of switchboard instruments are usually calibrated to take these ratios into account.

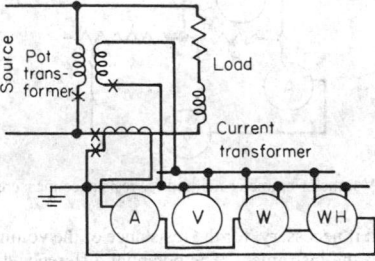

Fig. 15.1.42 Single-phase connections of instruments with transformers.

Figure 15.1.42 shows the use of instrument transformers to measure the voltage, current, power, and kilowatthours of a single-phase load. Figure 15.1.43 shows the connections that would be used to measure the voltage, current, and power of a 26,400-V 600-A three-phase load.

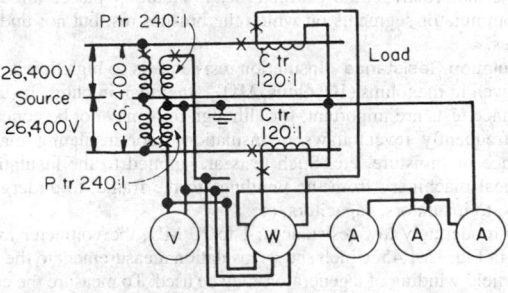

Fig. 15.1.43 Three-phase connections of instruments and instrument transformers.

Measurement of High Voltages Potential transformers such as those shown in Figs. 15.1.42 and 15.1.43 may be used even for very high voltages, but for voltages above 132 kV they become so large and expensive that they are used only sparingly. A convenient method used with testing transformers is the employment of a voltmeter coil, which consists of a coil of a few turns interwoven in the high-voltage winding and insulated from it. The voltage ratio is the ratio of the turns in the high-voltage winding to those in the voltmeter coil. A capacitance voltage divider consists of two or more capacitors connected in series across the high voltage to be measured. A high-impedance voltmeter, such as an electronic one, is connected across the capacitor at the grounded end. The high voltage $V = V_m C_m / C$ V, where V_m is the voltmeter reading, C the capacitance (in μF) of the entire divider, and C_m the equivalent capacitance (in μF) of the capacitor at the grounded end. A bushing potential device consists of a high-voltage-transformer bushing having a capacitance tap brought out from one of the metallic electrodes within the bushing which is near ground potential. This device is obviously a capacitance voltage divider. For testing, **sphere gaps** are used for the very high voltages. Calibration data for sphere gaps are given in the ANSI/IEEE Std. 4-1978 Standard Techniques for High Voltage Testing. Even when it is not being used for the measurement of voltage, it is frequently advisable to connect a sphere gap in parallel with the specimen being tested so as to prevent overvoltages. The gap is set to a slightly higher voltage than that which is desired.

Measurement of Resistance

Voltmeter-Ammeter Method A common method of measuring resistance, known as the voltmeter-ammeter or fall-in-potential method, makes use of an ammeter and a voltmeter. In Fig. 15.1.44, the resistance to be measured is R. The current in the resistor R is I A, which is measured by the ammeter A in series. The drop in potential across the resistor R is measured by the voltmeter V. The current shunted by the voltmeter is so small that it may generally be neglected. A correction

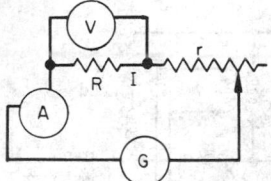

Fig. 15.1.44 Voltmeter-ammeter method for resistance measurement.

may be applied if necessary, for the resistance of the voltmeter is generally given with the instrument. The potential difference divided by the current gives the resistance included between the voltmeter leads. As a check, determinations are generally made with several values of current, which may be varied by means of the controlling resistor r. If the resistance to be measured is that of the armature of a dc machine and the voltmeter leads are placed on the brush holders, the resistance determined will include that of the brush contacts. To measure the resistance of the armature alone, the voltmeter leads should be placed directly on the commutator segments on which the brushes rest but not under the brushes.

Insulation Resistance Insulation resistance is so high that it is usually given in megohms (10^6 ohms, MΩ) rather than in ohms. Insulation resistance tests are important, for although they may not be conclusive they frequently reveal flaws in insulation, poor insulating material, presence of moisture, etc. Such tests are applied to the insulation of electrical machinery from the windings to the frame, to underground cables, to insulators, capacitors, etc.

For moderately low resistances, 1 to 10 MΩ, the voltmeter method given in Fig. 15.1.45, which shows insulation measurement to the frame of the field winding of a generator, may be used. To measure the current when a voltage E is impressed across the resistor R, a high-reading voltmeter V is connected in series with R. The current under this condi-

tion with the switch connecting S and A is $E/(R + r)$, where r is the resistance of the voltmeter. A high-resistance voltmeter is necessary, since the method is in reality a comparison of the unknown insulation resistance R with the known resistance r of the voltmeter. Hence, the

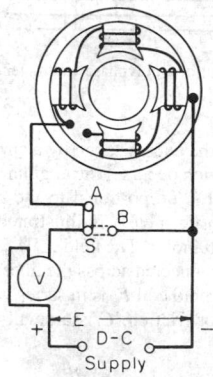

Fig. 15.1.45 Voltmeter method for insulation resistance measurement.

resistance of the voltmeter must be comparable with the unknown resistance, or the deflection of the instrument will be so small that the results will be inaccurate. To determine the impressed voltage E, the same voltmeter is used. The switch S connects S and B for this purpose. With these two readings, the unknown resistance is

$$R = r(E - e)/e \qquad (15.1.82)$$

where e is the deflection of the voltmeter when in series with the resistance to be measured as when S is at A. If a special voltmeter, having a resistance of 100 kΩ per 150 V, is available, a resistance of the order of 2 to 3 MΩ may be measured very accurately.

When the insulation resistance is too high to be measured with a voltmeter, a sensitive galvanometer may be used. The connections for measuring the insulation resistance of a cable are shown in Fig. 15.1.46. The battery should have an emf of at least 100 V. Radio B batteries are convenient for this purpose. The method involves comparing the unknown resistance with a standard 0.1 MΩ. To calibrate the galvanometer the cable is short-circuited (dotted line) and the switch S is thrown to

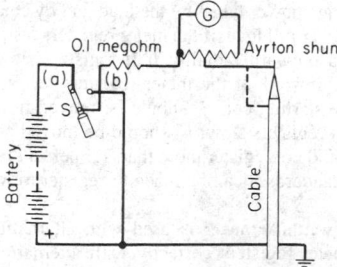

Fig. 15.1.46 Measurement of insulation resistance with a galvanometer.

position (a). Let the galvanometer deflection be D_1 and the reading of the Ayrton shunt S_1. The short circuit is then removed. The 0.1 MΩ is left in circuit since it is usually negligible in comparison with the unknown resistance X. Let the reading of the galvanometer now be D_2 and the reading of the shunt S_2. Then

$$X = 0.1 \, S_2 D_1 / S_1 D_2 \qquad \text{M}\Omega \qquad (15.1.83)$$

When the switch S is thrown to position (b), the cable is short-circuited through the 0.1 MΩ and becomes discharged.

The **Megger** insulation tester is an instrument that indicates insulation resistance directly on a scale. It consists of a small hand or motor-driven generator which generates 500 V, 1,000 V, 2,500 V, or 5,000 V. A clutch slips when the voltage exceeds the rated value. The current through the unknown resistance flows through a moving element consisting of two coils fastened rigidly together, but which move in different portions of the magnetic field. A pointer attached to the spindle of the moving element indicates the insulation resistance directly. These instruments have a range up to 10,000 MΩ and are very convenient where portability and convenience are desirable.

The insulation resistance of electrical machinery may be of doubtful significance as far as dielectric strength is concerned. It varies widely with temperature, humidity, and cleanliness of the parts. When the insulation resistance falls below the prescribed value, it can (in most cases of good design) be brought to the required standard by cleaning and drying the machine. Hence it may be useful in determining whether or not the insulation is in proper condition for a dielectric test. IEEE Std. 62-1978 specifies minimum values of insulation resistance in MΩ = (rated voltage)/(rating in kW + 1,000). If the operating voltage is higher than the rated voltage, the operating voltage should be used. The rule specifies that a dc voltage of 500 be used in testing. If not, the voltage should be specified.

Wheatstone Bridge   Resistors from a fraction of an Ω to 100 kΩ and more may be measured with a high degree of precision with the Wheatstone bridge (Fig. 15.1.47). The bridge consists of four resistors $ABCX$ connected as shown. X is the unknown resistance; A and B are ratio arms, the resistance units of which are in even decimal Ω as 1, 10, 100, etc. C is the rheostat arm. A battery or low-voltage source of direct current is connected across ab. A galvanometer G of moderate sensitivity is connected across cd. The values of A and B are so chosen that three

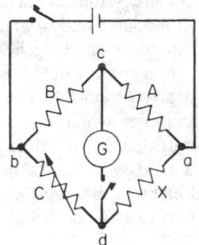

Fig. 15.1.47   Wheatstone bridge.

or four significant figures in the value of C are obtained. As a first approximation it is well to make A and B equal. When the bridge is in balance,

$$X/C = A/B \qquad (15.1.84)$$

The positions of the battery and galvanometer are interchangeable. There are many modifications of the bridge which adapt it to measurements of very low resistances and also to ac measurements.

Kelvin Double Bridge   The simple Wheatstone bridge is not adapted to measuring very low resistances since the contact resistances of the test specimen become comparable with the specimen resistance. This error is avoided in the Kelvin double bridge, the diagram of which is shown in Fig. 15.1.48. The specimen X, which may be a short length of copper wire or bus bar, is connected in series with an adjustable calibrated resistor R whose resistance is comparable with that of the specimen. The arms A and B of the bridge are ratio arms usually with decimal values of 1, 10, 100 Ω. One terminal of the galvanometer is connected to X and R by means of two resistors a and b. If these resistors are set so that $a/b = A/B$, the contact resistance r between X and R is eliminated in the measurement. The contact resistances at c and d have no effect since at balance the galvanometer current is zero. The contact resistances at f and e need only be negligible compared with the resistances of arms A and B both of which are reasonably high. By means of the variable resistor Rh the value of current, as indicated by ammeter A, may be adjusted to give the necessary sensitivity. When the bridge is in balance,

$$X/R = A/B \qquad (15.1.85)$$

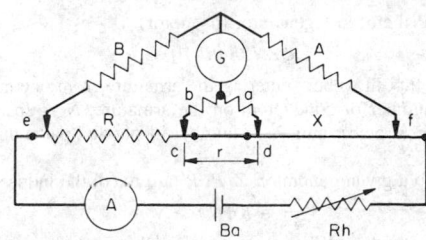

Fig. 15.1.48   Kelvin double bridge.

Potentiometer   The principle of the potentiometer is shown in Fig. 15.1.49. ab is a slide wire, and bc consists of a number of equal individual resistors between contacts. A battery Ba the emf of which is approximately 2 V supplies current to this wire through the adjustable rheostat R. A slider m makes contact with ab, and a contactor m' connects with the contacts in bc. A galvanometer G is in series with the wire connecting to m. By means of the double-throw double-pole switch Sw, either

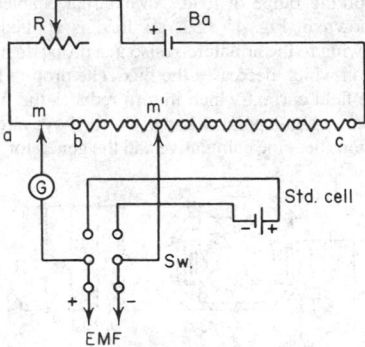

Fig. 15.1.49   Potentiometer principle.

the standard cell or the unknown emf (EMF) may be connected to mm' through the galvanometer G. The potentiometer is standardized by throwing Sw to the standard-cell side, setting mm' so that their positions on ab and bc correspond to the emf of the standard cell. The rheostat R is then adjusted until G reads zero. (In commercial potentiometers a dial which may be set directly to the emf of the standard cell is usually provided.) The unknown emf is measured by throwing Sw to EMF and adjusting m and m' until G reads zero. The advantage of this method of measuring emf is that when the potentiometer is in balance no current is taken from either the standard cell or the source of emf. Potentiometers seldom exceed 1.6 V in range. To measure voltage in excess of this, a **volt box** which acts as a multiplier is used. To measure current, the voltage drop across a standard resistor of suitable value is measured with the potentiometer. For example, with 50 A a 0.01-Ω standard resistance gives a voltage drop of 0.5 V which is well within the range of the potentiometer.

Potentiometers of low range are used extensively with thermocouple pyrometers. Figure 15.1.49 merely illustrates the principle of the potentiometer. There are many modifications, conveniences, etc., not shown in Fig. 15.1.49.

DC GENERATORS

All electrical machines are comprised of a magnetic circuit of iron (or steel) and an electric circuit of copper. In a generator the armature conductors are rotated so that they cut the magnetic flux coming from and entering the field poles. In the dc generator (except the unipolar type) the emf induced in the individual conductors is alternating, but this is rectified by the commutator and brushes, so that the current to the external circuit is unidirectional.

The induced emf in a generator (or motor)

$$E = \phi ZNP/60P'10^8 \quad \text{V} \qquad (15.1.86)$$

where ϕ = flux in webers entering the armature from one north pole; Z = total number of conductors on the armature; N = speed, r/min; P = number of poles; and P' = number of parallel paths through the armature.

Since with a given generator, Z, P, P' are fixed, the induced emf

$$E = K\phi N \quad \text{V} \qquad (15.1.87)$$

where K is a constant. When the armature delivers current, the terminal volts are

$$V = E - I_a R_a \qquad (15.1.88)$$

where I_a is the armature current and R_a the armature resistance including the brush and contact resistance, which vary somewhat.

There are three standard types of dc generators: the **shunt generator,** the **series generator,** and the **compound generator.**

Shunt Generator The field of the shunt generator in series with its rheostat is connected directly across the armature as shown in Fig. 15.1.50. This machine maintains approximately constant terminal voltage over its working range of load. An external characteristic of the generator is shown in Fig. 15.1.51. As load is applied the terminal voltage drops owing to the armature-resistance drop [Eq. (15.1.88)] and armature reaction which decreases the flux. The drop in terminal voltage reduces the field current which in turn reduces the flux, hence the induced emf, etc. At some point B, usually well above rated current, the foregoing reactions become cumulative and the generator starts to break

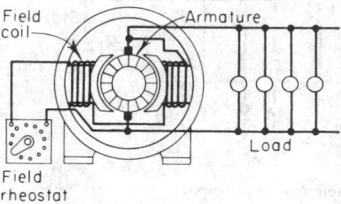

Fig. 15.1.50 Shunt generator.

down. The current reaches a maximum value and then decreases to nearly zero at short circuit. With large machines, point B is well above rated current, the operating range being between O and A. The voltage may be maintained constant by means of the field rheostat. Automatic regulators which operate through field resistance are frequently used to maintain constant voltage.

Shunt generators are used in systems which are all tied together where their stability when in parallel is an advantage. If a generator fails to build up, (1) the load may be connected; (2) the field resistance may be too high; (3) the field circuit may be open; (4) the residual magnetism may be insufficient; (5) the field connection may be reversed.

Series Generator In the series generator (Fig. 15.1.52) the entire load current flows through the field winding, which consists of relatively few turns of wire of sufficient size to carry the entire load current without undue heating. The field excitation, and hence the terminal voltage, depends on the magnitude of the load current. The generator

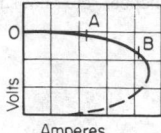

Fig. 15.1.51 Shunt generator characteristic.

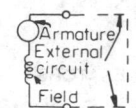

Fig. 15.1.52 Series generator.

supplies an essentially constant current and for years was used to supply series arc lamps for street lighting requiring direct current. Except for some special applications, the series generator is now obsolete.

Compound-Wound Generators By the addition of a series winding to a shunt generator the terminal voltage may be automatically maintained very nearly constant, or, by properly proportioning the series turns, the terminal voltage may be made to increase with load to compensate for loss of voltage in the line, so that approximately constant voltage is maintained at the load. If the shunt field is connected outside the series field (Fig. 15.1.53), the machine is **long shunt;** if the shunt field is connected inside the series field, i.e., directly to the armature terminals, it is **short shunt.** So far as the operating characteristic is concerned, it makes little difference which way a machine is connected. Table 15.1.10 gives performance characteristics.

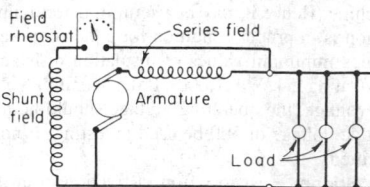

Fig. 15.1.53 Compound-wound dc generator.

Compound-wound generators are chiefly used for small isolated plants and for generators supplying a purely motor load subject to rapid fluctuations such as in railway work. When first putting a compound generator in service, the shunt field must be so connected that the machine builds up. The series field is then connected so that it aids the shunt field. Figure 15.1.54 gives the characteristics of an overcompounded 200-kW 600-V compound-wound generator.

Amplidynes The amplidyne is a dc generator in which a small amount of power supplied to a control field controls the generator output, the response being nearly proportional to the control field input. The amplidyne is a dc amplifier which can supply large amounts of power. The amplifier operates on the principle of armature reaction. In Fig. 15.1.55, NN and SS are the conventional north and south poles of a dc generator with central cavities. BB are the usual brushes placed at right angles to the pole axes of NN and SS. A control winding CC of

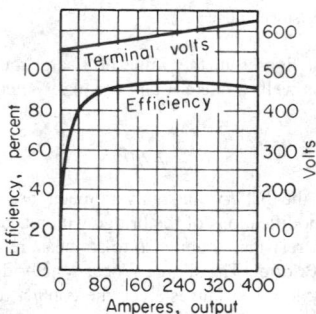

Fig. 15.1.54 Characteristics of a 200-kW compound-wound dc generator.

Table 15.1.10 Approximate Test Performance of Compound-Wound DC Generators with Commutating Poles

kW	Speed, r/min	Volts	Amperes	Efficiencies, percent		
				¼ load	½ load	Full load
5	1,750	125	40	77.0	80.5	82.0
10	1,750	125	80	80.0	83.0	85.0
25	1,750	125	200	84.0	86.5	88.0
50	1,750	125	400	83.0	86.0	88.0
100	1,750	125	800	87.0	88.5	90.0
200	1,750	125	1,600	88.0	90.5	91.0
400	1,750	250	1,600	91.7	91.9	91.7
1,000	1,750	250	4,000	92.2	92.6	92.1

SOURCE: Westinghouse Electric Corp.

small rating, as low as 100 W, is wound on the field poles. In Fig. 15.1.55, for simplicity, the control winding is shown as being wound on one pole only. The brushes BB are short-circuited, so that a small excitation mmf in the control field produces a large short-circuit current along the brush axis BB. This large short-circuit current produces a large armature-reaction flux AA along brush axis BB. The armature rotating in this field produces a large voltage along the brush axis B'B'. The load or working current is taken from brushes B'B' as shown. In Fig. 15.1.55 the working current only is shown by the crosses and dots in the circles. The short-circuit current would be shown by crosses in the conductors to the left of brushes BB and by dots in the conductors to the right of brushes BB.

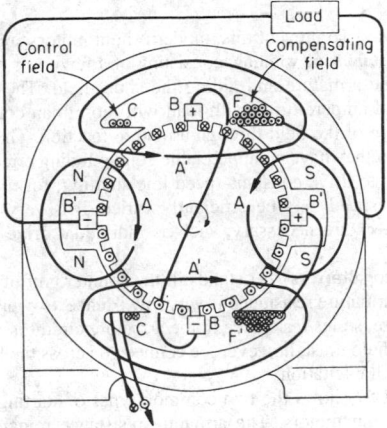

Fig. 15.1.55 Amplidyne.

A small current in the control winding produces a high output voltage and current as a result of the large short-circuit current in brushes BB.

In order that the brushes B'B' shall not be short-circuiting conductors which are cutting the flux of poles NN and SS, cavities are cut in these poles. Also the load current from brushes B'B' produces an armature reaction mmf in opposition to flux A'A' produced by the control field CC. Were this mmf not compensated, the flux A'A' and the output of the machine would no longer be determined entirely by the control field. Hence there is a compensating field FF' in series with the armature, which neutralizes the armature-reaction mmf which the load current produces. For simplicity the compensating field is shown on one field pole only.

The amplidyne is capable of controlling and regulating speed, voltage, current, and power with accurate and rapid response. The amplification is from 10,000 to 250,000 times in machines rated from 1 to 50 kW. Amplidynes are frequently used in connection with **selsyns** and are employed for gun and turret control and for accurate controls in many industrial power applications.

Parallel Operation of Shunt Generators It is desirable to operate generators in parallel so that the station capacity can be adapted to the load. Shunt generators, because of their drooping characteristics (Fig. 15.1.51), are inherently stable when in parallel. To connect shunt generators in parallel it is necessary that the switches be so connected that like poles are connected to the same bus bars when the switches are closed. Assume one generator to be in operation; to connect another generator in parallel with it, the incoming generator is first brought up to speed and its terminal voltage adjusted to a value slightly greater than the bus-bar voltage. This generator may then be connected in parallel with the other without difficulty. The proper division of load between them is adjusted by means of the field rheostats and is maintained automatically if the machines have similar voltage-regulation characteristics.

Parallel Operation of Compound Generators As a rule, compound generators have either flat or rising voltage characteristics. Therefore, when connected in parallel, they are inherently unstable. Stability may, however, be obtained by using an equalizer connection, Fig. 15.1.56,

which connects the terminals of the generator at the junctions of the series fields. This connection is of low resistance so that any increase of current divides proportionately between the series fields of the two machines. The equalizer switch (E.S.) should be closed first and opened last, if possible. In practice, the equalizer switch is often one blade of a

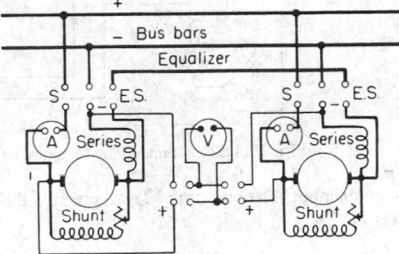

Fig. 15.1.56 Connections for compound-wound generators operating in parallel.

three-pole switch, the other two being the bus switch S, as in Fig. 15.1.56. When compound generators are used on a three-wire system, two series fields—one at each armature terminal—and two equalizers are necessary. It is possible to operate any number of compound generators in parallel provided their characteristics are not too different and the equalizer connection is used.

DC MOTORS

Motors operate on the principle that a conductor carrying current in a magnetic field tends to move at right angles to that field (see Fig. 15.1.11). The ordinary dc generator will operate entirely satisfactorily as a motor and will have the same rating. The conductors of the motor rotate in a magnetic field and therefore must generate an emf just as does the generator. The induced emf

$$E = K\phi N \qquad (15.1.89)$$

where K = constant, ϕ flux entering the armature from one north pole, and N = r/min [see Eq. (15.1.87)]. This emf is in opposition to the terminal voltage and tends to oppose current entering the armature. Its value is

$$E = V - I_a R_a \qquad (15.1.90)$$

where V = terminal voltage, I_a = armature current, and R_a = armature resistance [compare with Eq. (15.1.88)]. From Eq. (15.1.89) it is seen that the speed

$$N = K_s E/\phi \qquad (15.1.91)$$

when $K_s = 1/K$. This is the fundamental speed equation for a motor. By substituting in Eq. (15.1.90)

$$N = K_s(V - I_a R_a)/\phi \qquad (15.1.92)$$

which is the general equation for the speed of a motor.

The internal or electromagnetic torque developed by an armature is proportional to the flux and to the armature current; i.e.,

$$T_t = K_t \phi I_a \qquad (15.1.93)$$

when K_t is a constant. The torque at the pulley is slightly less than the internal torque by the torque necessary to overcome the rotational losses, such as friction, windage, eddy-current and hysteresis losses in the armature iron and in the pole faces.

The total mechanical power developed internally

$$P_m = EI_a \quad W \quad (= EI_a/746 \quad hp) \qquad (15.1.94)$$

The internal torque thus becomes

$$T = EI_a 33,000/(2\pi \times 746N) = 7.04 EI_a/N \qquad (15.1.95)$$

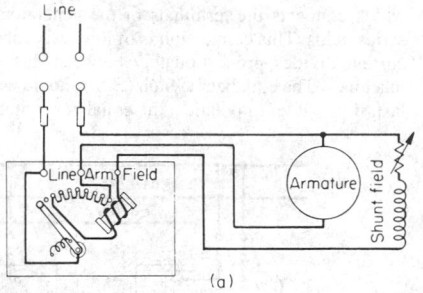
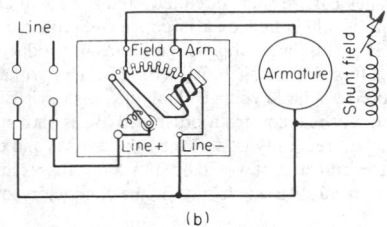

Fig. 15.1.57 Connections for shunt dc motors and starters. (*a*) Three-point box; (*b*) four-point box.

Let VI be the motor input. The output is $VI\eta$ where η is the efficiency. The horsepower is

$$P_H = VI\eta/746 \qquad (15.1.96)$$

and the torque is

$$T = 33,000P_H/2\pi N = 5,260P_H/N \qquad \text{lb} \cdot \text{ft} \qquad (15.1.97)$$

where N is r/min.

Shunt Motor In the shunt motor (Fig. 15.1.57) the flux is substantially constant and $I_a R_a$ is 2 to 6 percent of V. Hence from Eq. (15.1.92), the speed varies only slightly with load (Fig. 15.1.58), so that the motor is adapted to work requiring constant speed. The speed regulation of constant-speed motors is defined by ANSI/IEEE Std. 100-1992 Standard Dictionary of Electrical and Electronic terms as follows:

The speed regulation of a constant-speed direct-current motor is the change in speed when the load is reduced gradually from the rated value to zero with constant applied voltage and field rheostat setting expressed as a percent of speed at rated load.

In Fig. 15.1.61 the speed regulation under each condition is $100(ac - bc)/bc$ (Fig. 15.1.58a). Also from Eq. (15.1.93) it is seen that the torque is practically proportional to the armature current (see Fig. 15.1.58b). The motor is able to develop full-load torque and more on starting, but

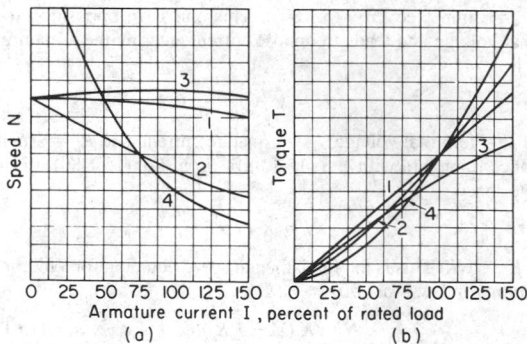

Fig. 15.1.58 Speed and torque characteristics of dc motors. (1) Shunt motor (2) cumulative compound motor; (3) differential compound motor; (4) series motor.

the ordinary starter is not designed to carry the current necessary for starting under load. If a motor is to be started under load, the starter should be provided with resistors adapted to carry the required current without overheating. A controller is also adapted for starting duty under load.

Commutating poles have so improved commutation in dc machines that it is possible to use a much shorter air gap than formerly. Since, with the shorter air gap, fewer field ampere-turns are required, the armature becomes magnetically strong with respect to the field. Hence, a sudden overload might weaken the field through armature reaction, thus causing an increase in speed; the effect may become cumulative and the

motor run away. To prevent this, modern shunt motors are usually provided with a **stabilizing winding,** consisting of a few turns of the field in series with the armature and aiding the shunt field. The resulting increase of field ampere-turns with load will more than compensate for any weakening of the field through armature reaction. The series turns are so few that they have no appreciable compounding effect. The shunt motor is used to drive constant-speed line shafting, for machine tools, etc. Since its speed may be efficiently varied, it is very useful when **adjustable speeds** are necessary, such as individual drive for machine tools.

Shunt-Motor Starters At standstill the counter emf of the motor is zero and the armature resistance is very low. Hence, except in motors of very small size, series resistance in the armature circuit is necessary on starting. The field must, however, be connected across the line so that it may obtain full excitation.

Figure 15.1.57 shows the two common types of starting boxes used for starting shunt motors. The armature resistance remains in circuit only during starting. In the **three-point box** (Fig. 15.1.57a) the starting lever is held, against the force of a spring, in the running position, by an electromagnet in series with the field circuit, so that, if the field circuit is interrupted or the line voltage becomes too low, the lever is released and the armature circuit is opened automatically. In the **four-point starting box** the electromagnet is connected directly across the line, as shown in Fig. 15.1.57b. In this type the arm is released instantly upon failure of the line voltage. In the three-point type some time elapses before the field current drops enough to effect the release. Some starting rheostats are provided with an overload device so that the circuit is automatically interrupted if too large a current is taken by the armature. The four-point box is used where a wide speed range is obtained by means of the field rheostat. The electromagnet is not then affected by changes in field current.

In large motors and in many small motors, automatic starters are widely used. The advantages of the automatic starter are that the current is held between certain maximum and minimum values so that the circuit does not become opened by too rapid starting as may occur with manual operation; the acceleration is smooth and nearly uniform. Since workers can stop and start a motor merely by the pushing of a button, there results considerable saving by the shutting down of the motor when it is not needed. Automatic controls are essential to elevator motors so that smooth rapid acceleration with frequent starting and stopping may be obtained. Also automatic starting is very necessary with multiple-unit operation of electric-railway cars and with rolling-mill motors which are continually subjected to rapid acceleration, stopping, and reversing.

Series Motor In the series motor the armature and field are in series. Hence, if saturation is neglected, the flux is proportional to the current and the torque [Eq. (15.1.93)] varies as the current squared. Therefore any increase in current will produce a much greater proportionate increase in torque (see Fig. 15.1.58b). This makes the motor particularly well adapted to traction work, cranes, hoists, fork-lift trucks, and other types of work which require large starting torques. A study of Eq. (15.1.92) shows that with increase in current the numerator changes only slightly, whereas the change in the denominator is nearly proportional to the change in current. Hence the speed of the series motor is practically inversely proportional to the current. With overloads the speed drops to very low values (see Fig. 15.1.58a). With decrease in load the speed approaches infinity, theoretically. Hence the series motor should always be connected to its load by a direct drive, such as gears, so that it cannot reach unsafe speeds (see Speed Control of Motors). A series-motor starting box with no-voltage release is shown in Fig. 15.1.59.

Differential Compound Motors   The cumulative compound winding of a generator becomes a differential compound winding when the machine is used as a motor. Its speed may be made more nearly constant than that of a shunt motor, or, if desired, it may be adjusted to increase with increasing load.

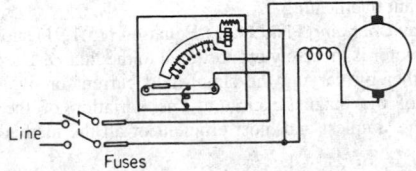

Fig. 15.1.59   Series motor starter, no-voltage release.

The speed as a function of armature current is shown in Fig. 15.1.58a and the torque as a function of armature current is shown in Fig. 15.1.58b.

Since the speed of the shunt motor is sufficiently constant for most purposes and the differential motor tends toward instability, particularly in starting and on overloads, the differential motor is little used.

Cumulative compound motors develop a more rapid increase in torque with load than shunt motors (Fig. 15.1.58b); on the other hand, they have much poorer speed regulation (Fig. 15.1.58a). Hence they are used where larger starting torque than that developed by the shunt motor is necessary, as in some industrial drives. They are particularly useful where large and intermittent increases of torque occur as in drives for shears, punches, rolling mills, etc. In addition to the sudden increase in torque which the motor develops with sudden applications of load, the fact that it slows down rapidly and hence causes the rotating parts to give up some of their kinetic energy is another important advantage in that it reduces the peaks on the power plant. Performance data for compound motors are given in Table 15.1.11.

Commutation   The brushes on the commutator of either a motor or generator should be set in such a position that the induced emf in the armature coils undergoing commutation and hence short-circuited by the brushes, is zero. In practice, this condition can at best be only approximately realized. Frequently conditions are such that it is far from being realized. At no load, the brushes should be set in a position corresponding to the geometrical neutral of the machine, for under these conditions the induced emf in the coils short-circuited by the brushes is zero. As load is applied, two factors cause sparking under the brushes. The mmf of the armature, or **armature reaction**, distorts the flux; when the current in the coils undergoing commutation reverses, an emf of self-induction $-L \, di/dt$ tends to prolong the current flow which produces sparking. In a generator, armature reaction distorts the flux in the direction of rotation and the brushes should be advanced. In order to neutralize the emf of self-induction the brushes should be set a little ahead of the neutral plane so that the emf induced in the short-circuited coils by the cutting of the flux at the fringe of the next pole is opposite to this emf of self-induction. In a motor the brushes are correspondingly moved backward in the direction opposite rotation.

Theoretically, the brushes should be shifted with every change in load. However, practically all dc generators and motors now have **commutating poles** (or **interpoles**) and with these the brushes can remain in the no-load neutral plane, and good commutation can be obtained over the entire range of load. Commutating poles are small poles between the main poles (Fig. 15.1.60) and are excited by a winding in series with the armature. Their function is to neutralize the flux distortion in the **neutral plane** caused by armature reaction and also to supply a flux that will cause an emf to be induced in the conductors undergoing commutation, opposite and equal to the emf of self-induction. Since armature reaction

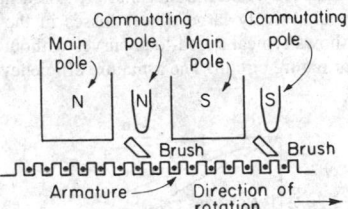

Fig. 15.1.60   Commutating poles in motor.

and the emf of self-induction are both proportional to the armature current, saturation being neglected, they are neutralized theoretically at every load. Commutating poles have made possible dc generators and motors of very much higher voltage, greater speeds, and larger kW ratings than would otherwise be possible.

Occasionally, the commutating poles may be connected incorrectly. In a motor, passing from an N main pole in the direction of rotation of the armature, an N commutating pole should be encountered as shown in Fig. 15.1.60. In a generator under these conditions an S commutating pole should be encountered. The test can easily be made with a compass. If poor commutation is caused by too strong interpoles, the winding may be shunted. If the poles are too weak and the shunting cannot be reduced, they may be strengthened by inserting sheet-iron shims between the pole and the yoke thus reducing the air gap.

Although the emfs induced in the coils undergoing commutation are relatively small, the resistance of the coils themselves is low so that unless further resistance is introduced, the short-circuit currents would be large. Hence, with the exception of certain low-voltage generators, carbon brushes that have relatively large contact resistance are almost always used. Moreover, the graphite in the brushes has a lubricating action, and the usual carbon brush does not score the commutator.

Table 15.1.11   Test Performance of Compound-Wound DC Motors

| Power, hp | Speed, r/min | 115 V | | 230 V | | 550 V | |
		Current, A	Full-load efficiency, %	Current, A	Full-load efficiency, %	Current, A	Full-load efficiency, %
1	1,750	8.4	78	4.3	79	1.86	73.0
2	1,750	16.0	80	8.0	81	3.21	82.0
5	1,750	40.0	82	20.0	83	8.40	81.0
10	1,750	75.0	85.6	37.5	85	15.4	86.5
25	1,750	182.0	87.3	91.7	87.5	38.1	88.5
50	850	—	—	180.0	89	73.1	90.0
100	850	—	—	350.0	90.5	149.0	91.0
200	1,750	—	—	700.0	91	295.0	92.0

SOURCE: Westinghouse Electric Corp.

Speed Control of Motors

Shunt Motors In Eq. (15.1.91) the speed of a shunt motor $N = K_s E/\phi$, where K_s is a constant involving the design of the motor such as conductors on armature surface and number of poles. Obviously, in order to change the speed of a motor, without changing its construction, two factors may be varied, the counter emf E and the flux ϕ.

Armature-Resistance Control The counter emf $E = V - I_a R_a$, where V is the terminal voltage, assumed constant. R_a must be small so that the armature heating can be maintained within permissible limits. Under these conditions the speed change with load is small. By inserting an external resistor, however, into the armature circuit the counter emf E may be made to decrease rapidly with increase in load; that is, $E = V - I_a(R_a + R)$ [see Eq. (15.1.90)] where R is the resistance of the external resistor. The resistor R must be inserted in the **armature** circuit only. The advantages of this method are its simplicity, the full torque of the motor is developed at any speed, and the method introduces no commutating difficulties. Its disadvantages are the increased speed regulation with change of load (Fig. 15.1.61), the low efficiency, particularly at the lower speeds, and the fact that provision must be made to dissipate the comparatively large power losses in the series resistor. Figure 15.1.61 shows typical speed-load curves without and with series resistors in the armature circuit. The armature efficiency is nearly equal

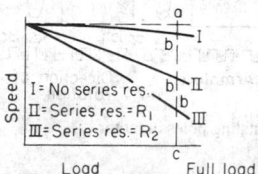

Fig. 15.1.61 Speed-load characteristics with armature resistance control.

to the ratio of the operating speed to the no-load speed. Hence at 25 percent speed the armature efficiency is practically 25 percent. Frequently the controlling and starting resistors are one, and the device is called a **controller**. Starting rheostats themselves are not designed to carry the armature current continuously and must not be used as controllers. The armature-resistance method of speed control is frequently used to regulate the speed of ventilating fans where the power demand diminishes rapidly with decrease in speed.

Control by Changing Impressed Voltage From Eq. (15.1.92) it is evident that the speed of a motor may be changed if V is changed by connecting the armature across different voltages. Speed control by this method is accomplished by having mains (usually four), which are maintained at different voltages, available at the motor.

The shunt field of the motor is generally permanently connected to one pair of mains, and the armature circuit is provided with a controller by means of which the operator can readily connect the armature to any pair of mains. Such a system gives a series of distinct and widely separated speeds and generally necessitates the use of field-resistance control, in combination, to obtain intermediate speeds. This method, known as the **multivoltage method,** has the disadvantage that the system is expensive, for it requires several generating machines, a somewhat complicated switchboard, and a number of service wires. The system is used somewhat in machine shops and is extensively used for dc elevator starting and speed control.

In the **Ward Leonard system,** the variable voltage is obtained from a separately excited generator whose armature terminals are connected directed to the armature terminals of the working motor. The generator is driven at essentially constant speed by a dc shunt motor if the power supply is direct current, or by an induction motor or a synchronous motor if the power supply is alternating current. The field circuit of the generator and that of the motor are connected across a constant-voltage dc supply. The terminal voltage of the generator, and hence the voltage applied to the armature of the motor, is varied by changing the generator-field current with a field rheostat. The rheostat has a wide range of resistance so that the speed of the motor may be varied smoothly from 0

to 100 percent. Since three machines are involved the system is costly, somewhat complicated, and has low power efficiency. However, because the system is flexible and the speed can be smoothly varied over wide ranges, it has been used in many applications, such as elevators, mine hoists, large printing presses, paper machines, and electric locomotives.

The **Ward-Leonard system** has been largely replaced by a **static converter system** where a silicon-controlled-rectifier bridge is used to convert three-phase alternating current, or single-phase on smaller drives, to dc voltage that can be smoothly varied by **phase-angle firing** of the rectifiers from full voltage to zero. This system is smaller, lighter, and less expensive than the motor-generator system. Care must be taken to assure that the dc motor will accept the **harmonics** present in the dc output without overheating.

Control by Changing Field Flux Equation (15.1.91) shows that the speed of a motor is inversely proportional to the flux ϕ. The flux can be changed either by varying the shunt-field current or by varying the reluctance of the magnetic circuit. The variation of the **shunt-field current** is the simplest and most efficient of all the methods of speed control.

With the ordinary motor, speed variation of 1.5 to 1.0 is obtainable with this method. If attempt is made to obtain greater ratios, severe sparking at the brushes results, owing to the field distortion caused by the armature mmf becoming large in comparison with the weakened field of the motor. Speed ratios of 5 : 1 and higher are, however, obtainable with motors which have commutating poles. Since the field current is a small proportion of the total current (1 to 3 percent), the rheostat losses in the field circuit are always small. This method is efficient. Also for any given speed adjustment the speed regulation is excellent, which is another advantage. Because of its simplicity, efficiency, and excellent speed regulation, the control of speed by means of the field current is by far the most common method. Output power remains constant when the field is weakened, so output torque varies inversely with motor speed.

Speed Control of Series Motors The series motor is fundamentally a variable-speed motor, the speed varying widely from light load to full load and more (see Fig. 15.1.58a). From Eq. (15.1.92) the speed for any value of ϕ, or current, can be changed by varying the impressed voltage. Hence the speed can be controlled by inserting resistance in series with the motor. This method, which is practically the same as the armature-resistance control method for shunt motors, has the same objections of low efficiency and poor regulation with fluctuating loads. It is extensively used in controlling the speed of hoist and crane motors.

The **series-parallel** system of series-motor speed control is almost universally used in electric traction. At least two motors are necessary. The two motors are first connected in series with each other and with the starting resistor. The starting resistor is gradually cut out and, since each motor then operates at half line voltage, the speed of each is approximately half speed. Both motors take the same current, and each can develop full torque. This condition of operation is efficient since there is no external resistance in circuit. When the controller is moved to the next position, the motors are connected in parallel with each other and each in series with starting resistors. Full speed of the motors is obtained by gradually cutting out these resistors. Connecting the two motors in series on starting reduces the current to one-half the value that would be required for a given torque were both motors connected in parallel on starting. The power taken from the trolley is halved, and an intermediate running speed is efficiently obtained.

In the **multiple-unit** method of speed control which is used for electric railway trains, the starting contactors, reverser, etc., for each car are located under that car. The relays operating these control devices are actuated by energy taken from the train line consisting usually of seven wires. The train line runs the entire length of the train, the connections between the individual cars being made through the couplers. The train line is energized by the action of the motorman operating any one of the small master controllers which are located in each car. Hence corresponding relays, contactors, etc., in every car all operate simultaneously. High accelerations may be reached with this system because of the large tractive effort exerted by the wheels on every car.

SYNCHRONOUS GENERATORS

The synchronous generator is the only type of ac generator now in general use at power stations.

Construction In the usual synchronous generator the armature or stator is the stationary member. This construction has many advantages. It is possible to make the slots any reasonable depth, since the tooth necks increase in cross section with increase in depth of slot; this is not true of the rotor. The large slot section which is thus obtainable gives ample space for copper and insulation. The conductors from the armature to the bus bars can be insulated throughout their entire lengths, since no rotating or sliding contacts are necessary. The insulation in a stationary member does not deteriorate as rapidly as that on a rotating member, for it is not subjected to centrifugal force or to any considerable vibration.

The **rotating member** is ordinarily the field. There are two general types of field construction: the **salient-pole type** and the **cylindrical,** or **nonsalient-pole, type.** The salient-pole type is used almost entirely for slow and moderate-speed generators since this construction is the least expensive and permits ample space for the field ampere-turns.

It is not practicable to employ salient poles in high-speed turboalternators because of the excessive windage and the difficulty of obtaining sufficient mechanical strength. The **cylindrical type** consists of a cylindrical steel forging with radial slots in which the field copper, usually in strip form, is placed. The fields are ordinarily excited at low voltage, 125 and 250 V, the current being conducted to the rotating member by means of slip rings and brushes. An ac generator armature to supply **field voltage** can be mounted on the generator shaft and supply dc to the motor field through a static rectifier bridge, also mounted on the shaft, eliminating all slip rings and brushes. The field power is ordinarily only 1.5 percent and less of the rated power of the machine (see Table 15.1.12).

Classes of Synchronous Generators Synchronous generators may be divided into three general classes: (1) the slow-speed engine-driven type; (2) the moderate-speed waterwheel-driven type; and (3) the high-speed turbine-driven type. In (1) a hollow box frame is used as the stator support, and the field consists of a spider to which a larger number of salient poles are attached, usually bolted. The speed seldom exceeds 75 to 90 r/min, although it may run as high as 150 r/min. Waterwheel generators also have salient poles which are usually dovetailed to a cylindrical spider consisting of steel plates riveted together. Their speeds range from 80 to 900 r/min and sometimes higher, although the 9,000-kVA Keokuk synchronous generators rotate at only 58 r/min, operating at a very low head. The speed rating of direct-connected waterwheel generators decreases with decrease in head. It is desirable to operate synchronous generators at the highest permissible speed since the weight and costs diminish with increase in speed. Waterwheel-driven generators must be able to run at double speed, as a precaution against accident, should the governor fail to shut the gate sufficiently rapidly in case the circuit breakers open or should the governing mechanism become inoperative.

Turbine-driven generators operate at speeds of 720 to 3,600 r/min. Direct-connected exciters, belt-driven exciters from the generator shaft, and separately driven exciters are used. In large stations separately driven (usually motor) exciters may supply the excitation energy to excitation bus bars. Steam-driven exciters and storage batteries are frequently held in reserve. With slow-speed synchronous generators, the belt-driven exciter is frequently used because it can be driven at higher speed, thus reducing the cost.

Table 15.1.12 Performance Data for Synchronous Generators

					Efficiencies, %			Approx. net weight, lb
kVA	Poles	Speed, r/min	Excitation, kW		½ load	¾ load	Full load	
25	4	1,800	0.8		81.5	85.7	87.6	900
93.8	8	900	2		87	89.5	90.9	2,700
250	12	600	5		90	91.3	92.2	6,000
500	18	400	8		91.7	92.6	93.2	10,000
1,000	24	300	14.5		92.6	93.4	93.9	16,100
3,125	48	150	40		93.4	94.2	94.6	52,000

80% pf, 3 phase, 60 Hz, 240 to 2,400 V, horizontal-coupled or belted-type engine

Industrial-size turbine generators, direct-connected type, 80% pf, 3 phase, 60 Hz, air-cooled

			Excitation		Efficiency, %			Volume of air, ft³/min		Approx. weight, including exciter, lb
kVA	Poles	Speed, r/min	kW	V	½ load	¾ load	Full load		Voltage	
1,875	2	3,600	18	125	95.3	96.1	96.3	3,500	480–6,900	21,900
2,500	2	3,600	22	125	95.3	96.1	96.3	5,000	2,400–6,900	22,600
3,125	2	3,600	24	125	95.3	96.3	96.5	5,500	2,400–6,900	25,100
3,750	2	3,600	24	125–250	95.3	96.3	96.6	6,500	2,400–6,900	27,900
5,000	2	3,600	29	125–250	95.3	96.3	96.6	11,000	2,400–6,900	40,100
6,250	2	3,600	38	125–250	95.3	96.3	96.7	12,000	2,400–13,800	43,300
7,500	2	3,600	42	125–250	95.5	96.5	96.9	15,000	2,400–13,800	45,000
9,375	2	3,600	47	125–250	95.5	96.5	96.9	16,500	2,400–13,800	61,200

Central-station-size turbine generators, direct-connected type, 85% pf, 3 phase, 60 Hz, 11,500 to 14,400 V

			Excitation		Efficiency, %			Volume of air, ft³/min		Approx. weight, including exciter, lb
kVA	Poles	Speed, r/min	kW	V	½ load	¾ load	Full load		Ventilation	
13,529	2	3,600	70	250	96.3	97.1	97.3	22,000	Air-cooled	116,700
17,647	2	3,600	100	250	97.7	97.9	97.9	22,000	H₂-cooled	115,700
23,529	2	3,600	115	250	98.0	98.2	98.2	25,000	H₂-cooled	143,600
35,294	2	3,600	145	250	98.1	98.3	98.3	34,000	H₂-cooled	194,800
47,058	2	3,600	155	250	98.3	98.5	98.5	42,000	H₂-cooled	237,200
70,588	2	3,600	200	250	98.4	98.7	98.7	50,000	H₂-cooled	302,500

SOURCE: Westinghouse Electric Corp.

Synchronous-Generator Design At the present time single-phase generators are seldom built. For single-phase service two phases of a standard three-phase Y-connected generator are used. A single-phase load or unbalanced three-phase load produces flux pulsations in the magnetic circuits of synchronous generators, which increase the iron losses and introduce harmonics into the emf wave. Two-phase windings consist of two similar single-phase windings displaced 90 electrical space degrees on the armature and ordinarily occupying all the slots on the armature. The most common type of winding is the three-phase lap-wound two-layer type of winding. In three-phase windings three windings are spaced 120 electrical space degrees apart, the individual phase belts being spaced 60° apart. Usually, all the slots on the armature are occupied. Standard voltages are 550, 1,100, 2,200, 6,600, 13,200, and 20,000 V. It is much more difficult to insulate for 20,000 V than it is for the lower voltages. However, if the power is to be transmitted at this voltage, its use would be justified by the saving of transformers. In machines of moderate and larger ratings it is common to generate at 6,600 and 13,200 V if transformers must be used. The higher voltage is preferable, particularly for the higher ratings, because it reduces the cross section of the connecting leads and bus bars.

The **standard frequency** in the United States for lighting and power systems is 60 Hz; the few former 50-Hz systems have practically all been converted to 60 Hz. The frequency of 25 Hz is commonly used in street-railway and subway systems to supply power to the synchronous converters and other ac-dc conversion apparatus; it is also commonly used in railroad electrification, particularly for single-phase series-motor locomotives (see Sec. 11). At 25 Hz incandescent lamps have noticeable flicker. In European (and most other) countries 50 Hz is standard. The frequency of a synchronous machine

$$f = P \times r/\text{min}/120 \qquad \text{Hz} \qquad (15.1.98)$$

where P is the number of poles. Synchronous generators are rated in kVA rather than in kW, since heating, which determines the rating, is dependent only on the current and is independent of power factor. If the kilowatt rating is specified, the power factor should also be specified.

Induced EMF The induced emf per phase in synchronous generator is

$$E = 2.22 k_b k_p \Phi f Z \qquad \text{V/phase} \qquad (15.1.99)$$

where k_b = breadth factor or belt factor (usually 0.9 to 1.0), which depends on the number of slots per pole per phase, 0.958 for three-phase, four slots per pole per phase; k_p = pitch factor = 1.0 for full pitch, 0.966 for ⅚ pitch; Φ = total flux Wb, entering armature from one north pole and is assumed to be sinusoidally distributed along the air gap; f = frequency; and Z = number of series conductors per phase.

Synchronous generators usually are Y-connected. The advantages are that for a given line voltage the voltage per phase is $1/\sqrt{3}$ that of the delta-connected winding; third-harmonic currents and their multiples cannot circulate in the winding as with a delta-connected winding; third-harmonic emfs and their multiples cannot exist in the line emfs; a neutral point is available for grounding.

Regulation The terminal voltage of synchronous generator at constant frequency and field excitation depends not only on the current load but on the power factor as well. This is illustrated in Fig. 15.1.62, which shows the voltage-current characteristics of a synchronous generator

with lagging current, leading current and in-phase current (pf = 1.00). With leading current the voltage may actually rise with increase in load; the rate of voltage decrease with load becomes greater as the lag of the current increases. The regulation of a synchronous generator is defined by the ANSI/IEEE Std. 100-1992 Standard Dictionary of Electrical and Electronic Terms as follows:

> The voltage regulation of a synchronous generator is the rise in voltage with constant field current, when, with the synchronous generator operated at rated voltage and rated speed, the specified load at the specified power factor is reduced to zero, expressed as a percent of rated voltage.

For example, in Fig. 15.1.62 the regulation under each condition is

$$100(ac - bc)/bc \qquad (15.1.100)$$

With leading current the regulation may be negative.

Three factors affect the regulation of synchronous generators; the **effective armature resistance**, the **armature leakage reactance,** and the **armature reaction**. With alternating current the armature loss is greater than the value obtained by multiplying the square of the armature current by the ohmic resistance. This is due to hysteresis and eddy-current losses in the iron adjacent to the conductor and to the alternating flux-producing losses in the conductors themselves. Also the current is not distributed uniformly over conductors in the slot, but the current density tends to be greatest in the top of the slot. These factors all have the effect of increasing the resistance. The ratio of effective to ohmic resistance varies from 1.2 to 1.5. The **armature leakage reactance** is due to the flux produced by the armature current linking the conductors in the slots and also the end connections.

The armature mmf reacts on the field to change the value of the flux. With a single-phase generator and with an unbalanced load on a polyphase generator, the armature mmf is pulsating and causes iron losses in the field structure. With polyphase machines under a constant balanced load, the armature mmf is practically constant in magnitude and fixed in its relation to the field poles. Its direction with relation to the field-pole axis is determined by the power factor of the load.

A component of current in phase with the no-load induced emf, or the excitation emf, merely distorts the field by strengthening the trailing pole tip and weakening the leading pole tip. A component of current lagging the excitation emf by 90° weakens the field without distortion. A component of current leading the excitation emf by 90° strengthens the field without distortion. Ordinarily, both cross magnetization and one of the other components are acting simultaneously.

The foregoing effects are called **armature reaction**. Frequently the effects of armature reactance and armature reaction can be combined into a single quantity.

It is difficult to determine the regulation of synchronous generator by actual loading, even when in service, owing to the difficulty of obtaining, controlling, and absorbing the large balanced loads. Hence methods of **predetermining regulation without actually loading** the machine are used.

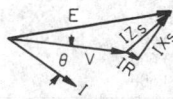

Fig. 15.1.63 Phasor diagram for synchronous impedance method.

Synchronous Impedance Method Both armature reactance and armature reaction have the same effect on the terminal voltage. In the synchronous impedance method the generator is considered as having no armature reaction, but the armature reactance is increased a sufficient amount to account for the effect of armature reaction. The phasor diagram for a current I lagging the terminal voltage V by an angle θ is shown in Fig. 15.1.63. In a polyphase generator the phasor diagram is applicable to one phase, a balanced load almost always being assumed.

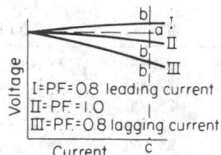

I=PF=0.8 leading current
II=PF=1.0
III=PF=0.8 lagging current

Fig. 15.1.62 Synchronous generator characteristics.

The power factor of the load is cos θ; IR is the effective armature resistance drop and is parallel to I; IX_s is the synchronous reactance drop and is at right angles to I and leading it by 90°. IX_s includes both the reactance drop and the drop in voltage due to armature reaction. That part of IX_s which replaces armature reaction is in reality a fictitious quantity. The synchronous impedance drop is given by IZ_s. The no-load or open-circuit (excitation) voltage

$$E = \sqrt{(V\cos\theta + IR)^2 + (V\sin\theta \pm IX_s)^2} \quad \text{V} \quad (15.1.101)$$

All quantities are per phase. The negative sign is used with leading current.

$$\text{Regulation} = 100(E - V)/V \quad (15.1.102)$$

With leading current E may be less than V and a negative regulation results.

The synchronous impedance is determined from an open-circuit and a short-circuit test, made with a weak field. The voltage E' on open circuit is divided by the current I' on short circuit for the same value of field current.

$$Z_s = E'/I' \qquad X_s = \sqrt{Z_s^2 - R^2} \qquad \Omega \quad (15.1.103)$$

R is so small compared with X_s that for all practical purposes $X_s = Z_s$. R may be determined by measuring the ohmic resistance per phase and multiplying by 1.4 to 1.5 to obtain the effective resistance. This value of R and the value of X_s obtained from Eq. (15.1.103) may then be substituted in Eq. (15.1.101) to obtain E at the specified load and power factor.

Since the synchronous reactance is determined at low saturation of the iron and used at high saturation, the method gives regulations that are too large; hence it is called the **pessimistic method**.

MMF Method In the mmf method the generator is considered as having no armature reactance but the armature reaction is increased by an amount sufficient to include the effect of reactance. That part of armature reaction which replaces the effect of armature reactance is in reality a fictitious quantity. To obtain the data necessary for computing the regulation, the generator is short-circuited and the field adjusted to give rated current in the armature. The corresponding value of field current I_a is read. The field is then adjusted to give voltage E' equal to rated terminal voltage + IR drop ($= V + IR$, as phasors, Fig. 15.1.64) on open circuit and the field current I' read.

I_a is 180° from the current phasor I, and I' leads E' by 90° (Fig. 15.1.64). The angle between I' and I_a is $90 - \theta + \phi$, but since ϕ is small, it can usually be neglected. The phasor sum of I_a and I' is I_o. The

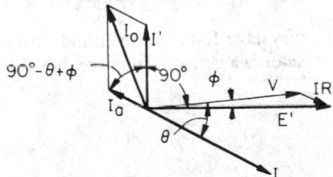

Fig. 15.1.64 Phasor diagram for the mmf method.

open-circuit voltage E corresponding to I_o is the no-load voltage and can be found on the saturation curve. The regulation is then found from Eq. (15.1.102). This method gives a value of regulation less than the actual value and hence is called the **optimistic method**. The actual regulation lies somewhere between the values obtained by the two methods but is more nearly equal to the value obtained by the mmf method.

ANSI Method The ANSI method (American Standard 50, Rotating Electrical Machinery) which has become the accepted standard for the predetermination of synchronous generator operation, eliminates in large measure the errors due to saturation which are inherent in the synchronous impedance and mmf methods. In Fig. 15.1.65a is shown the saturation curve OAF of the generator. The axis OP is not only the field-current axis but also the axis of the current phasor I as well. V the terminal voltage is drawn θ deg from I or OP, where θ is the power

factor angle. The effective resistance drop IR and the **leakage reactance** drop IX are drawn parallel and perpendicular to the current phasor. E_a, the phasor sum of V, IR, and IX, is the internal **induced** emf. Arcs are swung with O as the center and V and E_a as radii to intercept the axis of ordinates at B and C. OK, tangent to the straight portion of the saturation

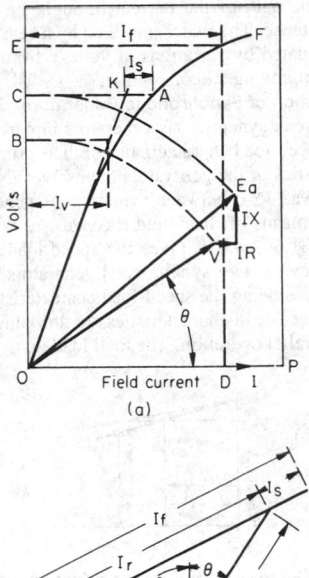

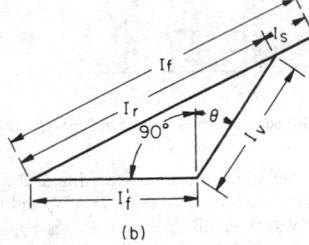

Fig. 15.1.65 ANSI method of synchronous generator regulation.

curve, is the **air-gap line**. If there is no saturation, I_v is the field current necessary to produce V, and CK is the field current necessary to produce E_a. The field current I_s is the increase in field current necessary to take into account the saturation corresponding to E_a.

The corresponding phasor diagram to a larger scale is shown in Fig. 15.1.65b. I'_f, the field current necessary to produce rated current at short circuit, corresponding to I_a (Fig. 15.1.64), is drawn horizontally. The field current I_v is drawn at an angle θ to the right of a perpendicular erected at the right-hand end of I'_f. I_r is the resultant of I'_f and I_v. I_s is added to I_r giving I_f the resultant field current. The no-load emf E is found on the saturation curve, Fig. 15.5.65a, corresponding to $I_f = OD$.

Excitation is commonly supplied by a small dc generator driven from the generator shaft. On account of commutation, except in the smaller sizes, the dc generator cannot be driven at 3,600 r/min, the usual speed for turbine generators, and belt or gear drives are necessary. The use of the silicon rectifier has made possible simpler means of excitation as well as voltage regulation. In one system the exciter consists of a small rotating-armature synchronous generator (which can run at high speed) mounted directly on the main generator shaft. The three-phase armature current is rectified by six silicon rectifiers and is conducted directly to the main generator field without any sliding contacts. The main generator field current is controlled by the current to the stationary field of the exciter generator. In another system there is no rotating exciter, the generator excitation being supplied directly from the generator terminals, the 13,800 V, three-phase, being stepped down to 115 V, three-phase, by small transformers and rectified by silicon rectifiers. Voltage regulation is obtained by saturable reactors actuated by potential transformers connected across the generator terminals.

Most regulators such as the following operate through the field of the exciter. In the **Tirrill regulator** the field resistance of the exciter is short-circuited temporarily by contacts when the bus-bar voltage drops. Actually, the contacts are vibrating continuously, the time that they are closed depending on the value of the bus-bar voltage. The General Electric Co. manufactures a direct-acting regulator in which the regulating rheostat is part of the regulator itself. The rheostat consists of stacks of graphite plates, each plate being pivoted at the center. Tilting the plates changes the path of the current through the rheostat and thus changes the resistance. The plates are tilted by a sensitive torque armature which is actuated by variations of voltage from the normal value (for regulators employing silicon rectifiers).

Parallel Operation of Synchronous Generators The kilowatt division of load between synchronous generators in parallel is determined entirely by the speed-load characteristics of their prime movers and not by the characteristics of the generators themselves. No appreciable adjustment of kilowatt load between synchronous generators in parallel can be made by means of their field rheostats, as with dc generators. Consider Fig. 15.1.66, which gives the speed-load characteristics in terms of frequency, of two synchronous generators, no. 1 and no. 2, these characteristics being the speed-load characteristics of their prime movers. These speed-load characteristics are drooping, which is necessary for stable parallel operation. The total load on the two machines is

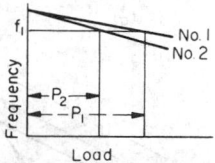

Fig. 15.1.66 Speed-load characteristics of synchronous generators in parallel.

$P_1 + P_2$ kW. Both machines must be operating at the same frequency F_1. Hence generator 1 must be delivering P_1 kW, and generator 2 must be delivering P_2 kW (the small generator losses being neglected). If, under the foregoing conditions, the field of either machine is strengthened, it cannot deliver a greater kilowatt load, for its prime mover can deliver more power only by dropping its speed. This is impossible, for both generators must operate always at the same frequency f_1. For any fixed total power load, the division of kilowatt load between synchronous generators can be changed only by modifying in some manner the speed-load characteristics of their prime movers, such, for example, as changing the tension in the governor spring. Synchronous generators in parallel are of themselves in stable equilibrium. If the driving torque of one machine is increased, the resulting electrical reactions between the machines cause a circulating current to flow between machines. This current puts more electrical load on the machine whose driving torque is increased and tends to produce motor action in the other machines. In an extreme case, the driving torque of one prime mover may be removed entirely, and its generator will operate as a synchronous motor, driving the prime mover mechanically.

Variations in driving torques cause currents to circulate between synchronous generators, transferring power which tends to keep the generators in synchronism. If the power transfer takes the form of recurring pulsations, it is called **hunting**, which may be reduced by building heavy copper grids called **amortisseurs**, or **damper windings**, into the pole faces. Turbine- and waterwheel-driven synchronous generators are much better adapted to parallel operation than are synchronous generators which are driven by reciprocating engines, because of their uniformity of torque.

Increasing the field current of synchronous generators in parallel with others causes it to deliver a greater lagging component of current. Since the character of the load determines the total current delivered by the system, the lagging components of current delivered by the other generators must decrease and may even become leading components. Likewise if the field of one generator is weakened, it delivers a greater leading component of current and the other machines deliver compo-

nents of current which are more lagging. These leading and lagging currents do not affect appreciably the division of **kilowatt** load between the synchronous generators. They do, however, cause unnecessary heating in their armatures. The fields of all synchronous generators should be so adjusted that the heating due to the quadrature components of currents is a minimum. With two generators having equal armature resistances, this occurs when both deliver equal quadrature currents.

Armature reactance in the armature of machines in parallel is desirable. If not too great, it stabilizes their operation by producing the synchronizing action. Synchronous generators with too little reactance are sensitive, and if connected in parallel with slight phase displacement or inequality of voltage, considerable disturbance results. Armature reactance also reduces the current on short circuit, particularly during the first few cycles when the short-circuit current is a maximum. Frequently, external power-limiting reactances are connected in series to protect the generators and equipment from injury that would result from the tremendous short-circuit currents. For these reasons, poor regulation in large synchronous generators is frequently considered to be an advantage rather than a disadvantage.

Ground Resistors Most power systems operate with a grounded neutral. When the station generators deliver current directly to the system (without intervening transformers), it is customary to ground the neutral (of the Y-connected windings) or one generator in a station; this is usually done through a grounding resistor of from 2 to 6 Ω. If the neutral of more than one generator is grounded, third-harmonic (and multiples thereof) currents can circulate between the generators. The ground resistor reduces the short-circuit currents when faults to ground occur, and hence reduces the violence of the short circuit as well as the duty of the circuit breakers. Grounding reactors are sometimes used but have limited application owing to the danger of high voltages resulting from resonant conditions.

INDUCTION GENERATORS

The **induction generator** is an induction motor driven above synchronous speed. The rotor conductors cut the rotating field in a direction to convert shaft mechanical power to electrical power. Load increases as speed increases, so the generator is self-regulating and can be used without governor control. On short circuits the induction generator will deliver current to the fault for only a few cycles because, unlike the synchronous generator, it is not self-exciting. Since it is not self-excited, an induction generator must always be used in parallel with an electrical system where there are some synchronous machines, or capacitor banks, to deliver lagging current (VARs) to the induction generator for excitation.

Induction generators have found favor in industrial **cogeneration** applications and as **wind-driven** generators where they provide a small part of the total load.

CELLS

Fuel cells convert chemical energy of fuel and oxygen directly to electrical energy. **Solar cells** convert solar radiation to electrical energy. At present, these conversion methods are not economically competitive with historic generating techniques, but they have found applications in isolated areas, such as microwave relaying stations, satellites, and residential lighting, where power requirements are small and costs for transmitting electrical power from more conventional sources is prohibitive. As solar technology continues to improve, many other applications will be evaluated.

TRANSFORMERS

Transformer Theory The transformer is a device that transfers energy from one electric circuit to another without change of frequency and usually, but not always, with a change in voltage. The energy is transferred through the medium of a magnetic field: it is supplied to the transformer through a primary winding and is delivered by means of a secondary winding. Both windings link the same magnetic circuit. With no load on the secondary, a small current, called the exciting current, flows in the primary and produces the alternating flux. This flux links both primary and secondary windings and induces the same volts per turn in each. With a sine wave the emf is

$$E = 4.44\Phi_m nf \quad \text{V} \qquad (15.1.104)$$

where Φ_m = maximum instantaneous flux in webers, n = turns on either winding, and f = frequency. Equation (15.1.104) may also be written

$$E = 4.44B_m Anf \quad \text{V} \qquad (15.1.105)$$

B_m = maximum instantaneous flux density in iron and A = net cross section of iron. If B_m is in T, A is in m^2; if B_m is in Mx/in^2, A is in in^2.

In English units, Eq. (15.1.104) becomes

$$E = 4.44\Phi_m nf\, 10^{-8} \quad \text{V} \qquad (15.1.104a)$$

where Φ_M is in maxwells; Eq. (15.1.105) becomes

$$E = 4.44B_m Anf\, 10^{-8} \quad \text{V} \qquad (15.1.105a)$$

where B_m is in Mx/in^2 and A is in in^2.

B_m is practically fixed. In large transformers with silicon steel it varies between 60,000 and 75,000 Mx/in^2 at 60 Hz and between 75,000 and 90,000 Mx/in^2 at 25 Hz. It is desirable to operate the iron at as high density as possible in order to minimize the weight of iron and copper. On the other hand, with too high densities the eddy-current and hysteresis losses become too great, and with low frequency the exciting current may become excessive. It follows from Eq. (15.1.104) that

$$E_1/E_2 = n_1/n_2 \qquad (15.1.106)$$

where E_1 and E_2 are the primary and secondary emfs and n_1 and n_2 are the primary and secondary turns. Since the impedance drops in ordinary transformers are small, the terminal voltages of primary and secondary are also practically proportional to their number of turns. As the change in secondary terminal voltage in the ordinary constant-potential transformer over its range of operation is small (1.5 to 3 percent), the flux must remain substantially constant. Therefore, the added ampere-turns produced by any secondary load must be balanced by opposite and equal primary ampere-turns. Since the exciting current is small compared with the load current (1.5 to 5 percent) and the two are usually out of phase, the exciting current may ordinarily be neglected. Hence,

$$n_1 I_1 = n_2 I_2 \qquad (15.1.107)$$
$$I_1/I_2 = n_2/n_1 \qquad (15.1.108)$$

where I_1 and I_2 are the primary and secondary currents.

When load is applied to the secondary of a transformer, the secondary ampere-turns reduce the flux slightly. This reduces the counter emf of the primary, permitting more current to enter and thus supply the increased power demanded by the secondary.

Both primary and secondary windings must necessarily have resistance. All the flux produced by the primary does not link the secondary; the counter ampere-turns of the secondary produce some flux which does not link the primary. These **leakage fluxes** produce reactance in each winding. The combined effect of the resistance and reactance produces an impedance drop in each winding when current flows. These impedance drops produce a slight drop in the secondary terminal voltage with load.

Transformer Testing Transformer regulation and losses are so small that it is far more accurate to compute the regulation and efficiency than to determine them by actual measurement. The necessary measurements and computations are comparatively simple, and little power is involved in making the tests. In the **open-circuit** test, the power input to either winding is measured at its rated voltage. Usually it is more convenient to make this test on the low-voltage winding, particularly if it is rated at 110, 220, or 550 V. The open-circuit power practically all goes to supply the core losses, consisting of eddy-current and hysteresis losses. Let this value of power be P_0. The eddy-current loss varies as the square of the voltage and frequency; the hysteresis loss varies as the 1.6 power of the voltage, and directly as the frequency. In

the **short-circuit** test one winding is short-circuited, and the current in the other is adjusted to near its rated value. The voltage V_c, the current I_1, and the power input P_c are measured. When one winding of a transformer is short-circuited, the voltage across the other winding is 3 to 4 percent of rated value when rated current flows. Since a voltage range of from 110 to 250 V is best adapted to measuring instruments, that winding whose rated voltage, multiplied by 0.03 or 0.04, is closest to this voltage range should be used for making the short-circuit test, the other winding being short-circuited. Practically all the power on short circuit goes to supply the copper loss of primary and secondary. If the measurements are made on the primary,

$$R_{01} = P_c/I_1^2 \qquad (15.1.109)$$
$$Z_{01} = V_c/I_1 \qquad (15.1.110)$$
$$X_{01} = \sqrt{Z_{01}^2 - R_{01}^2} \qquad (15.1.111)$$

where R_{01}, Z_{01}, and X_{01} are the equivalent resistance, impedance, and reactance referred to the primary. Also $R_{02} = R_{01}(n_2/n_1)^2$; $Z_{02} = Z_{01}(n_2/n_1)^2$; $X_{02} = X_{01}(n_2/n_1)^2$, these quantities being the equivalent resistance, impedance, and reactance referred to the secondary. If the dc resistances R_1 and R_2 of the primary and secondary are measured,

$$R_{01} = R_1 + (n_1/n_2)^2 R_2 \qquad (15.1.112)$$
$$R_{02} = R_2 + (n_2/n_1)^2 R_1 \qquad (15.1.113)$$

The ac or effective resistances, determined from Eq. (15.1.109), are usually 10 to 15 percent greater than these values.

Regulation The regulation may be computed from the foregoing data as follows:

$$V_1' = \sqrt{(V_1 \cos\theta + I_1 R_{01})^2 + (V_1 \sin\theta \pm I_1 X_{01})^2} \qquad (15.1.114)$$
$$\text{Regulation} = 100(V_1' - V_1)/V_1 \qquad (15.1.115)$$

V_1 = rated primary terminal voltage; $\cos\theta$ = load power factor; I_1 = rated primary current; R_{01} = equivalent resistance referred to primary [from Eq. (15.1.109)]; X_{01} = equivalent reactance referred to primary. The + sign is used with lagging current and the − sign with leading current. Equations (15.1.114) and (15.1.115) are equally applicable to the secondary if the subscripts are changed.

Efficiency The only two losses in a constant-potential transformer are the core loss in W, P_0, which is practically independent of load, and P_c the copper loss in W, which varies as the load current squared. The efficiency for any current I_1 is

$$\eta = V_1 I_1 \cos\theta/(V_1 I_1 \cos\theta + P_0 + I_1^2 R_{01}) \qquad (15.1.116)$$

Equation (15.1.116) applies equally well to the secondary if the subscripts are changed. The maximum efficiency occurs when the core and copper losses are equal.

All-Day Efficiency Since transformers must usually be on the line 24 h/day, part of which time the load may be very light, the all-day efficiency is important. This is equal to the total energy or watthour output divided by the total energy or watthour input for the 24 h. That is,

$$\eta = \frac{(V_1 I_1 \cos\theta_1)t_1 + \cdots}{(V_1 I_1 \cos\theta_1)t_1 + \cdots + (I_1^2 R_{01})t_1 + \cdots + 24 P_0} \qquad (15.1.117)$$

where t_1 = time in hours that load $V_1 I_1 \cos\theta_1$ is being delivered, etc.

Polyphase Transformer Connections Three-phase transformer banks may be connected Δ-Δ, Δ-Y, Y-Y, and Y-Δ. The Δ-Δ connection is very common, particularly at the lower voltages, and has the important advantage that the bank will operate V-connected if one transformer is disabled. The Δ-Y connection is advantageous for stepping up to high voltages since the secondary of the transformers need be wound only for 58 percent ($1/\sqrt{3}$) of the line voltage; it is also necessary when a four-wire three-phase system is obtained from a three-wire three-phase system since "a floating neutral" on the secondary cannot occur. This connection has found increased application in step-down distribution at 600 V and lower because of the relative ease of applying sensitive fault protection. The Y-Y system may be used for stepping up voltage. It should not be used for obtaining a three-phase four-wire system from a three-phase three-wire system, because of the "floating neutral" on the

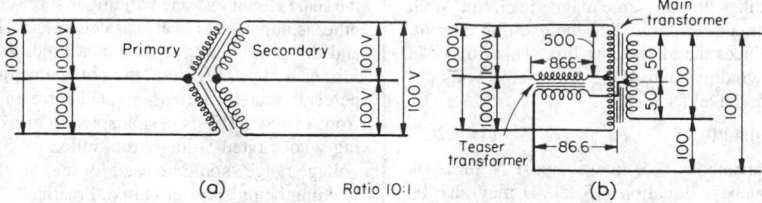

Fig. 15.1.67 Transformer connections for transforming moderate amounts of three-phase power. (a) Open delta connection; (b) T connection.

secondary and the resulting high degree of unbalance of the secondary voltages. The Y-Δ system may be used to step down high voltages, the reverse of the Δ-Y connection. Y-connected windings also preclude third harmonic, and their multiples, circulating currents in the transformer windings. In the Δ-Y and Y-Δ systems the ratio of line voltage is obviously not that of the individual transformers. Because of different phase displacement between primaries and secondaries, a Δ-Δ bank cannot be connected in parallel (on both sides) with a Δ-Y bank, etc., even if they both have the correct voltage ratios between lines.

Three-phase transformers combine the magnetic circuits of three single-phase transformers so that they have parts in common. A material saving in cost, in weight, and in space results, the greatest saving occurring in the core and oil. The advantages of three-phase transformers are often outweighed by their lack of flexibility. The failure of a single phase shuts down the entire transformer. With three single units, one unit may be readily replaced with a single spare. The primaries of single-phase transformers may be connected in Y or Δ at will and the secondaries properly phased. The primaries, as well as the secondaries of three-phase transformers, must be phased.

For the transformation of moderate amounts of power from three-phase to three-phase, two transformers employing either the **open delta** or **T connection** (Fig. 15.1.67) may be used. With each connection the ratio of line voltages is the same as the transformer ratios. In the figure, ratios 10:1 are shown. In the T connection the primary and the secondary of the main transformer must be provided with a center tap to which one end of the teaser transformer is connected. The ratings of these systems are only 58 percent of the rating of the system using three similar transformers, one for each phase. Owing to dissymmetry, the terminal voltages become somewhat unbalanced even with a balanced load.

To transform from two- to three-phase or the reverse, the **T connection** of Fig. 15.1.68 is used. To make the secondary voltages symmetrical a tap (called a **Scott tap**) is brought out at 86.6 percent ($\sqrt{3}/2$) of the primary winding of the teaser transformer as shown in Fig. 15.1.68. With balanced no-load voltages the voltages become slightly unbalanced even under a symmetrical load, owing to unequal phase differences in the individual coils. The three-phase neutral O is one-third the winding of the teaser transformer from the junction. In Fig. 15.1.68a the transformation is from three-phase to a two-phase three-wire system. In Fig. 15.1.68b the transformation is from three-phase to a four-phase, five-wire system. The voltages are given on the basis of 100-V primaries with 1:1 transformer ratios.

An **autotransformer,** also called **compensator,** consists essentially of a single winding linking a magnetic circuit. Part of the energy is transformed, and the remainder flows through conductively. Suitable taps are provided so that, if the primary voltage is applied to two of the taps, a voltage may be taken from any other two taps. The ratio of voltages is equal practically to the ratio of the turns between their taps. An autotransformer should be installed only when the ratio of transformation is not large. The ratio of power transformed to total power is $1 - n$, where n is the ratio of low-voltage to high-voltage emf. This gives the saving over the ordinary transformer and is greatest when the ratio is not far from unity. Figure 15.1.69a shows 100 kW being changed from 3,300 to 2,300 V; 30.3 kW only are being actually transformed, and the remainder of the power flows through conductively. Figure 15.1.69b shows how an ordinary 10:1, 10-kW lighting transformer may be connected to boost 110 kW 10 percent in voltage. In Fig. 15.1.69b, however, the 230-V secondary must be insulated for 2,300 V to the core and

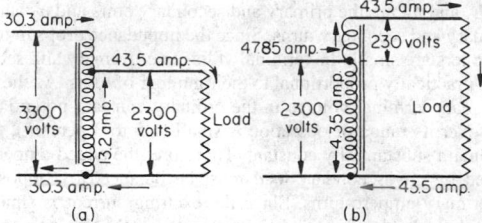

Fig. 15.1.69 Autotransformer.

ground. The voltage may likewise be reduced by reversing the 230-V coil. An autotransformer should never be used when it is desired to keep dangerous primary potentials from the secondary. It is used for starting induction motors (Fig. 15.1.71) and for a number of similar purposes.

Data on Transformers Single-phase 55° self-cooled oil-insulated transformers for 2,300-V primaries, 230/115-V secondaries, and in sizes from 5 to 200 kVA for 60(25) Hz have efficiencies from one-half to full load of about 98 (97 to 98.7) percent and regulation of 1.5 (1.1 to 2.1) percent with pf = 1, and 3.5 (2.7 to 4.1) percent with pf = 0.8. Power transformers with 13,200-V primaries and 2,300-V secondaries in sizes from 667 to 5,000 kVA and for both 60 and 25 Hz have efficiencies from one-half to full load of about 99.0 percent and regulation of about 1.0 (4.2) percent with pf = 1 (0.8).

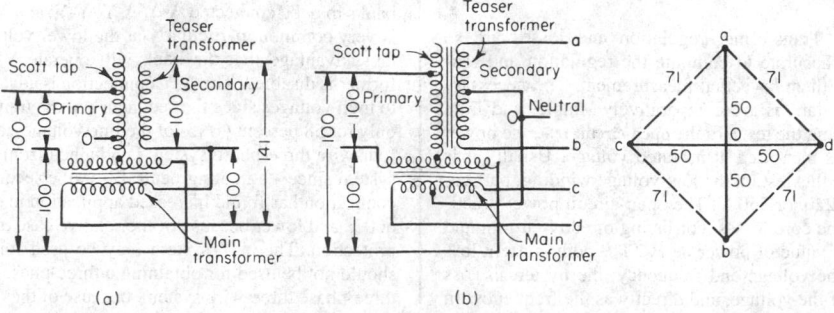

Fig. 15.1.68 Connections for transforming from three-phase to two- and four-phase power.

AC MOTORS

Polyphase Induction Motor The polyphase induction motor is the most common type of motor used. It consists ordinarily of a stator which is wound in the same manner as the synchronous-generator stator. If two-phase current is supplied to a two-phase winding or three-phase current to a three-phase winding, a rotating magnetic field is produced in the air gap. The number of poles which this field has is the same as the number of poles that a synchronous generator employing the same stator winding would have. The speed of the rotating field, or the **synchronous speed,**

$$N = 120f/P \quad \text{r/min} \quad (15.1.118)$$

where f = frequency and P = number of poles.

There are two general types of rotors. The **squirrel-cage type** consists of heavy copper or aluminum bars short-circuited by end rings, or the bars and end rings may be an integral aluminum casting. The **wound rotor** has a polyphase winding of the same number of poles as the stator, and the terminals are brought out to slip rings so that external resistance may be introduced. The rotor conductors must be cut by the rotating field, hence the rotor cannot run at synchronous speed but must slip. The per unit **slip** is

$$s = (N - N_2)/N \quad (15.1.119)$$

where N_2 = the rotor speed, r/min. The rotor frequency

$$f_2 = sf \quad (15.1.120)$$

The torque is proportional to the air-gap flux and the components of rotor current in space phase with it. The rotor currents tend to lag the emfs producing them, because of the rotor-leakage reactance. From Eq. (15.1.120) the rotor frequency and hence the rotor reactance ($X_2 = 2\pi f_2 L_2$) are low when the motor is running near synchronous speed, so that there is a large component of rotor current in space phase with the flux. With large values of slip the increased rotor frequency increases the rotor reactance and hence the lag of the rotor currents behind their emfs, and therefore considerable space-phase difference between these currents and the flux develops. Consequently even with large values of current the torque may be small. The torque of the induction motor increases with slip until it reaches a maximum value called the **breakdown torque,** after which the torque decreases (see Fig. 15.1.72). The breakdown torque varies as the square of the voltage, inversely as the stator impedance and rotor reactance, and is independent of the rotor resistance.

The squirrel-cage motor develops moderate torque on starting ($s = 1.0$) even though the current may be three to seven times rated current. For any value of slip the torque of the induction motor varies as the square of the voltage. The torque of the squirrel-cage motor which, on starting, is only moderate may be reduced in the larger motors because of starting voltage drop from inrush and possible necessity of applying reduced-voltage starting.

Polyphase squirrel-cage motors are used mostly for constant-speed work; however, recent developments in variable-frequency pulse-width modulation (PWM) ac drives have seen these motors used in variable-speed applications where the ratio of maximum to minimum speed does not exceed 4 : 1. They are used widely on account of their rugged construction and the absence of moving electrical contacts, which makes them suitable for operation when exposed to flammable dust or gas. General-purpose squirrel-cage motors have starting torques of 100 to 250 percent of full load torque at rated voltage. The highest torques occur at the higher rated speeds. The locked rotor currents vary between four and seven times full-load current. In the **double-squirrel-cage** type of motor there is a high-resistance winding in the top of the rotor slots and a low-resistance winding in the bottom of the slots. The low-resistance winding is made to have a high leakage reactance, either by separating the windings with a magnetic bridge, Fig. 15.1.70a, or by making the slot very narrow in the area between the two windings, Fig. 15.1.70b. On starting, because of the high reactance of the low-resistance winding, most of the rotor current will flow in the high-resistance winding, giving the motor a large starting torque. As the rotor approaches the low

value of slip at which it normally operates, the rotor frequency and hence the rotor reactance become low and most of the rotor current now flows in the low-resistance winding. Cage bars can be shaped so that one winding gives comparable results. See Fig. 15.1.70c. The rotor operates with a low value of slip. The high starting torque of the high-resistance motor and the excellent constant-speed operating characteristics of the low-resistance squirrel-cage rotor are combined in one motor.

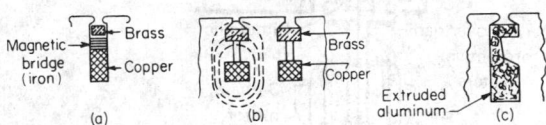

Fig. 15.1.70 Types of slots for squirrel-cage windings.

The **single shaped** cage bar is more economical, and almost any shape for required characteristics can be extruded from aluminum. Single bars also eliminate the problems of differential expansion with double cage bars.

Nameplates of polyphase integral-hp squirrel-cage induction motors carry a *code* letter and a *design* letter. These provide information about motor characteristics, the former on locked rotor or starting inrush current (see Table 15.1.26) and the latter on torque characteristics. National Electrical Manufacturers Association standards publication No. MG1-1978 defines four design letters: A, B, C, and D. In all cases the motors are designed for full voltage starting. Locked rotor current and torque, pull-up torque and breakdown torque are tabulated according to horsepower and speed. Designs A, B, and C have full load slips less than 5 percent and design D more than 5 percent. The nature of the various designs can be understood by reference to the full voltage values for a 100-hp, 1800-r/min motor which follow:

	Design			
	A	B	C	D
Locked rotor torque	125*	125*	200*	275*
Pull-up torque	100†	100†	140†	. . .
Breakdown torque	200†	200†	190†	. . .
Locked rotor current	. . .	600*	600*	600*
Full load slip (%)	5§	5§	5§	5‡

NOTE: All quantities, except slip, are percent of full load value.
* Upper limit.
† Not less than.
‡ Greater than.
§ Less than.

Starting It is desirable to start induction motors by direct connection across the line, since reduced voltage starters are expensive and almost always reduce the starting torque. The capacity of the distribution system dictates when reduced voltage starting must be used to limit voltage dips on the system. On stiff industrial systems 25,000-hp motors have been successfully started **across-the-line.**

In Fig. 15.1.71a is shown an **"across-the-line" starter** which may be operated from different push-button stations. The START push button closes the solenoid circuit between phases C and A through three bimetallic strips in series. This energizes solenoid S, which attracts armature D, which in turn closes the starting switch and the auxiliary blade G. This blade keeps the solenoid circuit closed when the START push button is released. Pressing the STOP push button opens the solenoid circuit, permitting the starting switch to open. A prolonged heavy overload raises the temperature of the heaters by an amount that will cause at least one of the bimetallic strips to open the solenoid circuit, releasing the starting switch.

A common method of applying reduced-voltage start is to use a **compensator** or **autotransformer** or **autostarter** (Fig. 15.1.71b). When the switch is in the starting position, the three windings AB of the three-phase autotransformer are connected in Y across the line and the motor terminals are connected to the taps which supply reduced voltage. When the switch is in the running position, the starter is entirely disconnected

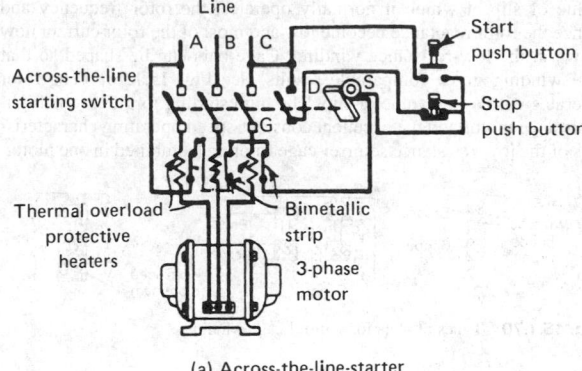

(a) Across-the-line-starter

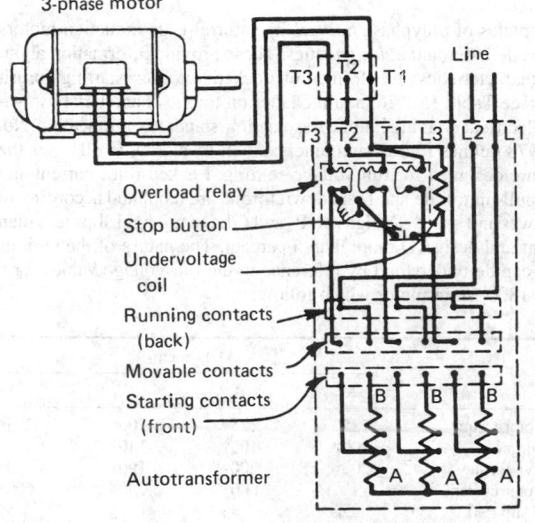

(b) Autostarter

Fig. 15.1.71 Starters for squirrel-cage induction motor. (*a*) Across-the-line starter; (*b*) autostarter.

from the line. In modern practice, motors are protected by thermal or magnetic overload relays (Fig. 15.1.71) which operate to trip the circuit breaker. Since a time element is involved in the operation of such relays, they do not respond to large starting currents, because of their short duration. To limit the current to as low a value as possible, the lowest taps that will give the motor sufficient voltage to supply the required starting torque should be used. As the torque of an induction motor varies as the square of the voltage, the compensator produces a very low starting torque.

Resistors in series with the stator may also be used to start squirrel-cage motors. They are inserted in each phase and are gradually cut out as the motor comes up to speed. The resistors are generally made of wire-type resistor units or of graphite disks enclosed within heat-resisting porcelain-lined iron tubes. The disadvantage of resistors is that if the motor is started slowly the resistor becomes very hot and may burn out. Resistor starters are less expensive than autotransformers. Their application is to motors that start with light loads at infrequent intervals.

A **phase-controlled,** silicon-controlled rectifier (SCR) may be used to limit the motor-starting current to any value that will provide sufficient starting torque by reducing voltage to the motors. The SCR can also be used to start and stop the motors. A positive opening device such as a contactor or circuit breaker should be used in series with the SCR to

stop the motor in case of a failed SCR, which will normally fail shorted and apply full voltage to the motor. The SCR starter has been combined with a microprocessor to provide gradual motor terminal voltage increases during an adjustable acceleration period. This is commonly called *soft start.* Other features include the ability to limit motor starting current and save energy when the motor is lightly loaded.

By **introducing resistance into the rotor circuit** through slip rings, the rotor currents may be brought nearly into phase with the air-gap flux and, at the same time, any value of torque up to maximum torque obtained. As the rotor develops speed, resistance may be cut out until there is no external resistance in the rotor circuit. The speed may also be controlled by inserting resistance in the rotor circuit. However, like the armature-resistance method of speed control with shunt motors, this method is also inefficient and gives poor speed regulation. Figure 15.1.72 shows graphically the effect on the torque of applying reduced voltage (curves *b*, *c*) and of inserting resistance in the rotor circuit (curve *d*). As shown by curves *b* and *c*, the torque for any given slip is proportional to the square of the line voltage. The effect of introducing resistance into the rotor circuit is shown by curve *d*. The point of maximum torque is shifted toward higher values of slip. The maximum

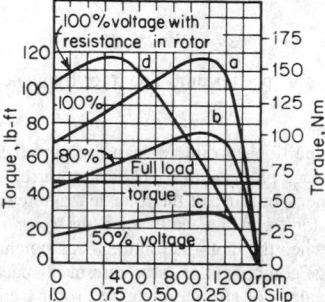

Fig. 15.1.72 Speed-torque curve for 10-hp, 60-Hz, 1,140-r/min induction motor.

torque at starting (slip = 1.0) occurs when the rotor resistance is equal to the rotor reactance at standstill. The wound-rotor motor is used where large starting torque is necessary as in railway work, hoists, and cranes. It has better starting characteristics than the squirrel-cage motor, but, because of the necessarily higher resistance of the rotor, it has greater slip even with the rotor resistance all cut out. Obviously, the wound rotor, controller, and external resistance make it more expensive than the squirrel-cage type. See Table 15.1.13 for performance data.

One **disadvantage of induction motors** is that they take lagging current, and the power factor at half load and less is low. The speed- and torque-load characteristics of induction motors are almost identical with those of the shunt motor. The speed decreases slightly to full load, the slip being from 10 percent in small motors to 2 percent in very large motors. The torque is almost proportional to the load nearly up to the breakdown torque. The power factor is 0.8 to 0.9 at full load. The direction of rotation of any three-phase motor may be reversed by interchanging any two stator wires.

Speed Control of Induction Motors The induction motor inherently is a constant-speed motor. From Eqs. (15.1.118) and (15.1.119) the rotor speed is

$$N_2 = 120f(1 - s)/P \qquad (15.1.121)$$

The speed can be changed only by changing the frequency, poles, or slip. In some applications where the motors constitute the only load on the generators, as with electric propulsion of ships, their speed may be changed by changing the frequency. Even then the range is limited, for both turbines and generators must operate near their rated speeds for good efficiency. By employing two distinct windings or by reconnecting a single winding by switching it is possible to change the number of poles. Complications prevent more than two speeds being readily obtained in this manner. Elevator motors frequently have two distinct

Table 15.1.13 Drip-Proof Motors

3 phase, 230 V, 60 Hz, 1,750 r/min, squirrel cage

hp	Weight, lb	Current, A	Power factor, %			Efficiency, %*		
			½ load	¾ load	Full load	½ load	¾ load	Full load
1	40	3.8	45.3	64.8	66.2	63.8	71.4	75.5
2	45	6.0	54.3	67.6	76.5	75.2	79.9	81.5
5	65	14.2	61.3	73.7	80.7	77.0	80.9	81.5
10	110	26.0	66.3	77.7	83.1	83.5	86.0	86.5
20	190	53.6	69.5	78.4	80.9	85.0	87.0	86.5
40	475	101.2	69.8	79.4	83.6	86.8	88.3	88.5
100	830	230.0	83.7	88.2	89.0	92.2	92.4	91.7
200	1,270	446.0	82.5	87.8	89.2	93.5	94.2	94.1

3 phase, 230 V, 60 Hz, 1,750 r/min, wound rotor

5	155	15.6	50.7	64.1	74.0	77.6	81.0	81.4
10	220	27.0	66.8	77.5	82.5	84.3	85.4	84.8
25	495	60.0	79.2	86.6	89.4	88.1	88.6	87.8
50	650	122.0	72.4	82.3	86.8	86.7	87.9	87.7
100	945	234.0	78.4	86.2	89.4	90.0	90.4	89.8
200	2,000	446.0	79.8	87.6	90.8	90.5	92.1	92.5

3 phase, 2,300 V, 60 Hz, 1,750 r/min, squirrel cage

300	2,300	70.4	76.2	83.4	86.0	90.8	92.5	92.8
700	3,380	155.0	85.5	89.4	90.0	91.7	93.4	93.7
1,000	4,345	221.0	86.2	89.5	90.0	92.1	93.8	94.1

3 phase, 2,300 V, 60 Hz, 1,750 r/min, wound rotor

300	3,900	68	84.4	86.2	89.9	90.8	92.5	92.8
700	5,750	154	82.7	87.7	90.9	91.7	93.4	93.7
1,000	8,450	218	82.9	87.9	87.9	92.1	93.8	94.1

* High-efficiency motors are available at premium cost.
SOURCE: Westinghouse Electric Corp.

windings. Another objection to changing the number of poles is the fact that the design is a compromise, and sacrifices of desirable characteristics usually are necessary at both speeds.

The change of slip by introducing resistance into the rotor circuit has been discussed under the wound-rotor motor. It is also possible to introduce an **inverter** into the rotor circuit and convert the slip frequency power to line frequency power and return it to the distribution system.

Inverters are also used to convert line frequencies to variable frequencies to operate squirrel-cage motors at almost any speed up to their mechanical limitations. Inverters also reduce the starting stresses on a motor.

Polyphase voltages should be evenly **balanced** to prevent phase current **unbalance**. If voltages are not balanced, the motor must be de-rated in accordance with National Electrical Manufacturers Association (NEMA), Publication No. MG-1-14.34.

Approximately 5 percent voltage unbalance would cause about 25 percent increase in temperature rise at full load. The input current unbalance at full load would probably be 6 to 10 times the input voltage unbalance.

Single phasing, one phase open, is the ultimate unbalance and will cause overheating and burnout if the motor is not disconnected from the line.

Single-Phase Induction Motor Single-phase induction motors are usually made in fractional horsepower ratings, but they are listed by NEMA in integral ratings up to 10 hp. They have relatively high rotor resistances and can operate in the single-phase mode without overheating. Single-phase induction motors are not self-starting.

However, the single-phase motor runs in the direction in which it is started. There are several methods of starting single-phase induction motors. Short-circuited turns, or **shading coils,** may be placed around the pole tips which retard the time phase of the flux in the pole tip, and thus a weak torque in the direction of rotation is produced. A high-resistance starting winding, displaced 90 electrical degrees from the main winding, produces poles between the main poles and so provides a rotating field which is weak but is sufficient to start the motor. This is called the

split-phase method. In order to minimize overheating this winding is ordinarily cut out by a centrifugal device when the armature reaches speed. In the larger motors a repulsion-motor start is used. The rotor is wound like an ordinary dc armature with a commutator, but with short-circuited brushes pressing on it axially rather than radially. The motor starts as a repulsion motor, developing high torque. When it nears its synchronous speed, a centrifugal device pushes the brushes away from the commutator, and at the same time causes the segments to be short-circuited, thus converting the motor into a single-phase induction motor.

Capacitor Motors Instead of splitting the phase by means of a high-resistance winding, it has become almost universal practice to connect a capacitor in series with the auxiliary winding (which is displaced 90 electrical degrees from the main winding). With capacitance, it is possible to make the flux produced by the auxiliary winding lead that produced by the main field winding by 90° so that a true two-phase rotating field results and good starting torque develops. However, the 90° phase relation between the two fields is obtainable at only one value of speed (as at starting), and the phase relation changes as the motor comes up to speed. Frequently the auxiliary winding is disconnected either by a centrifugal switch or a relay as the motor approaches full speed, in which case the motor is called a **capacitor-start** motor. With proper design the auxiliary winding may be left in circuit permanently (frequently with additional capacitance introduced). This improves both the power factor and torque characteristics. Such a motor is called **permanent-split capacitor motor.**

Phase Converter If a polyphase induction motor is operating single-phase, polyphase emfs are generated in its stator by the combination of stator and rotor fluxes. Such a machine can be utilized, therefore, for converting single-phase power into polyphase power and, when so used, is called a **phase converter.** Unless corrective means are utilized, the polyphase emfs at the machine terminals are somewhat unbalanced. The power input, being single-phase and at a power factor less than unity, not only fluctuates but is negative for two periods during each cycle. The power output being polyphase is steady, or nearly so. The cyclic differences between the power output and the power input are accounted for in the kinetic energy stored in the rotating mass of the armature. The armature accelerates and decelerates, but only slightly, in accordance with the difference between output and input. The phase converter is used principally on railway locomotives, since a single trolley wire can be used to deliver single-phase power to the locomotive, and the converter can deliver three-phase power to the three-phase wound-rotor driving motors.

AC Commutator Motors Inherently simple ac motors are not adapted to high starting torques and variable speed. There are a number of types of commutating motor that have been developed to meet the requirement of high starting torque and adjustable speed, particularly with single phase. These usually have been accompanied by compensating windings, centrifugal switches, etc., in order to overcome low power factors and commutation difficulties. With proper compensation, commutator motors may be designed to operate at a power factor of nearly unity or even to take leading current.

One of the simplest of the **single-phase commutator motors** is the ac series railway motor such as is used on the erstwhile New York, New Haven, and Hartford Railroad. It is based on the principle that the torque of the dc series motor is in the same direction irrespective of the polarity of its line terminals. This type of motor must be used on low frequency, not over 25 Hz, and is much heavier and more costly than an equivalent dc motor. The torque and speed curves are almost identical with those of the dc series motor. Unlike most ac apparatus the power factor is highest at light load and decreased with increasing load. Such motors operate with direct current even better than with alternating current. For example, the New Haven locomotives also operate from the 600-V dc third-rail system (two motors in series) from the New York City line (238th St.) into Grand Central Station. (See also Sec. 11.)

On account of difficulties inherent in ac operation such as commutation and high reactance drops in the windings, it is economical to construct and operate such motors only in sizes adaptable to locomotives, the ratings being of the order of 300 to 400 hp. **Universal motors** are small simple series motors, usually of fractional horsepower, and will operate on either direct or alternating current, even at 60 Hz. They are used for vacuum cleaners, electric drills, and small utility purposes.

Synchronous Motor Just as dc shunt generators operate as motors, a synchronous generator, connected across a suitable ac power supply, will operate as a motor and deliver mechanical power. Each conductor on the stator must be passed by a pole of alternate polarity every half cycle so that at constant frequency the rpm of the motor is constant and is equal to

$$N = 120f/P \qquad \text{r/min} \qquad (15.1.122)$$

and the speed is independent of the load.

There are two types of synchronous motors in general use: the **slip-ring** type and the **brushless** type. The motor field current is transmitted to the motor by brushes and slip rings on the slip-ring type. On the brushless type it is generated by a shaft-mounted exciter and rectified and controlled by shaft-mounted static devices. Eliminating the slip rings is advantageous in dirty or hazardous areas.

The synchronous motor has the desirable characteristic that its power factor can be varied over a wide range merely by changing the field excitation. With a weak field the motor takes a lagging current. If the load is kept constant and the excitation increased, the current decreases (Fig. 15.1.73) and the phase difference between voltage and current becomes less until the current is in phase with the voltage and the power factor is unity. The current is then at its minimum value such as I_0, and the corresponding field current is called the **normal excitation**. Further increase in field current causes the armature current to lead and the power factor to decrease. Thus **underexcitation** causes the current to **lag**; **overexcitation** causes the current to **lead**. The effect of varying the field current at constant values of load is shown by the V curves (Fig. 15.1.73). Unity power factor occurs at the minimum value of armature current, corresponding to normal excitation. The power factor for any point such as P is I_0/I_1, leading current. Because of its adjustable power factor, the motor is frequently run light merely to improve power factor

or to control the voltage at some part of a power system. When so used the motor is called a **synchronous condenser**. The motor may, however, deliver mechanical power and at the same time take either leading or lagging current.

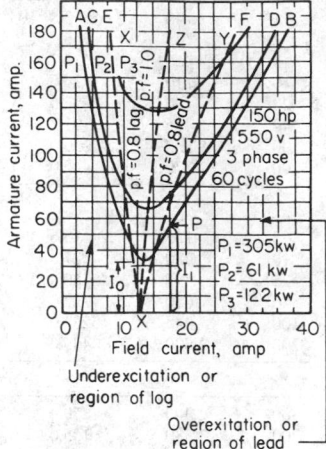

Fig. 15.1.73 V curves of a synchronous motor.

Synchronous motors are used to drive **centrifugal** and **axial compressors**, usually through speed increasers, **pumps, fans,** and other high-horsepower applications where **constant speed, efficiency,** and **power factor** correction are important. **Low-speed** synchronous motors, under 600 r/min, sometimes called **engine** type, are used in driving **reciprocating compressors** and in **ball mills** and in other slow-speed applications. Their low length-to-diameter ratio, because of the need for many poles, gives them a high moment of inertia which is helpful in smoothing the **pulsating torques** of these loads.

If the motor **field current** is separately supported by a **battery** or **constant voltage transformer,** the synchronous motor will maintain speed on a lower voltage dip than will an induction motor because torque is proportional to voltage rather than voltage squared. However, if a synchronous motor drops **out of step,** it will normally not have the ability to reaccelerate the load, unless the driven equipment is automatically unloaded.

The synchronous motor is usually not used in smaller sizes since both the motors and its controls are more expensive than induction motors, and the ability of a small motor to supply VARs to correct power factor is limited.

If situated near an inductive load the motor may be overexcited, and its leading current will neutralize entirely or in part the lagging quadrature current of the load. This reduces the I^2R loss in the transmission lines and also increases the kilowatt ratings of the system apparatus. The **synchronous condenser and motor** can also be used to control voltage and to stabilize power lines. If the condenser or motor is overexcited, its leading current flowing through the line reactance causes a rise in voltage at the motor; if it is underexcited, the lagging current flowing through the line reactance causes a drop in voltage at the motor. Thus within limits it becomes possible to control the voltage at the end of a transmission line by regulating the fields of synchronous condensers or motors. Long 220-kV lines and the 287-kV Hoover Dam–Los Angeles line require several thousand kVa in synchronous condensers floating at their load ends merely for voltage control. If the load becomes small, the voltage would rise to very high values if the synchronous condensers were not underexcited, thus maintaining nearly constant voltage. See Table 15.1.14 for characteristics.

A **salient-pole synchronous motor** may be started as an induction motor. In **laminated-pole** machines conducting bars of copper, copper alloy, or aluminum, **damper** or **amortisseur windings** are inserted in the pole face and short-circuited at the ends, exactly as a squirrel-cage winding in the

Table 15.1.14 Performance Data for Coupled Synchronous Motors

Power, hp	Poles	Speed, r/min	Current, A	Excitation, kW	Efficiencies, %			Weight, lb
					½ load	¾ load	Full load	
Unity power factor, 3 phase, 60 Hz, 2,300 V								
500	4	1,800	100	3	94.5	95.2	95.3	5,000
2,000	4	1,800	385	9	96.5	97.1	97.2	15,000
5,000	4	1,800	960	13	96.5	97.3	97.5	27,000
10,000	6	1,200	1,912	40	97.5	97.9	98.0	45,000
500	18	400	99.3	5	92.9	93.9	94.3	7,150
1,000	24	300	197	8.4	93.7	94.6	95.0	15,650
4,000	48	150	781	25	94.9	95.6	95.6	54,000
80% power factor, 3 phase, 60 Hz, 2,300 V								
500	4	1,800	127	4.5	93.3	94.0	94.1	6,500
2,000	4	1,800	486	13	95.5	96.1	96.2	24,000
5,000	4	1,800	1,212	21	95.5	96.3	96.5	37,000
10,000	6	1,200	2,405	50	96.8	97.3	97.4	70,000
500	18	400	125	7.2	92.4	93.4	93.6	9,500
1,000	24	300	248	11.6	93.3	94.2	94.4	17,500
4,000	48	150	982	40	94.6	95.3	95.5	11,500

SOURCE: Westinghouse Electric Corp.

induction motor is connected. The bars can be designed only for starting purposes since they carry no current at synchronous speed and have no effect on efficiency. In solid-pole motors a block of steel is bolted to the pole and performs the current-carrying function of the damper winding in the laminated-pole motor. At times the pole faces are interconnected to minimize starting-pulsating torques. When the synchronous motor reaches 95 to 98 percent speed as an induction motor, the motor field is applied by a **timer** or **slip frequency** control circuit, and the motor pulls into step at 100 percent speed. While accelerating, the motor field is connected to resistances to minimize induced voltages and currents.

Two-pole motors are built as **turbine** type or **round-rotor motors** for mechanical strength and do not have the thermal capacity or space for starting windings, so they must be started by supplementary means.

One such supplementary starting means is the use of a variable-frequency source, either a **variable-speed generator** or more commonly a static **converter-inverter**. The motor is brought up to speed in synchronism with a slowly increasing frequency. One common application is the starting of the large motor-generators used in **pump-storage** utility systems.

Variable frequency may be used to start salient-pole machines also. Requirements for a start without high torques and pulsations or high voltage drops on small electrical systems may dictate the use of something other than full voltage starting.

The **synchronous reluctance motor** is similar to an induction machine with salient poles machined in the rotors. Under light loads the motor will synchronize on reluctance torque and lock in step with the rotating field at synchronous speed. These motors are used in small sizes with variable-frequency inverters for speed control in the paper and textile industry.

The **synchronous-induction motor** is fundamentally a wound-rotor slip-ring induction motor with an air gap greater than normal, and the rotor slots are larger and fewer. On starting, resistance is inserted in the rotor circuit to produce high torque, and this is cut out as the speed increases. As synchronism is approached, the rotor windings are connected to a dc power source and the motor operates synchronously.

Timing or **clock motors** operate synchronously from ac power systems. Figure 15.1.74a illustrates the Warren Telechron motor which operates on the hysteresis principle. The stator consists of a laminated element with an exciting coil, and each pole piece is divided, a short-circuited shading turn being placed on each of the half poles so formed. The rotor consists of two or more hard-steel disks of the shape shown, mounted on a small shaft. The shaded poles produce a 3,600 r/min rotating magnetic field (at 60 Hz), and because of hysteresis loss, the disk follows the field just as the rotor of an induction motor does. When the rotor approaches the synchronous speed of 3,600 r/min, the rotating magnetic field takes

a path along the two rotor bars and locks the rotor in with it. The rotor and the necessary train of reducing gears rotate in oil sealed in a small metal can. Figure 15.1.74b shows a subsynchronous motor. Six squirrel-cage bars are inserted in six slots of a solid cylindrical iron rotor, and the spaces between the slots form six salient poles. The motor, because of the squirrel cage, starts as an induction motor, attempting to attain the speed of the rotating field, or 3,600 r/min (at 60 Hz). However, when the rotor reaches 1,200 r/min, one-third synchronous speed, the salient poles of the rotor lock in with the poles of the stator and hold the rotor at 1,200 r/min.

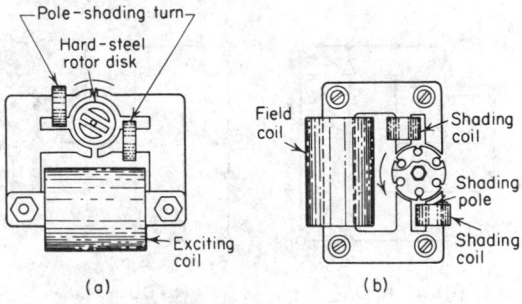

Fig. 15.1.74 Synchronous motors for timing. (a) Warren Telechron motor; (b) Holtz induction reluctance subsynchronous motor.

AC-DC CONVERSION

Static Rectifiers Silicon devices, and to a lesser extent gas tubes, are the primary means of ac to dc or dc to ac conversion in modern installations. They are advantageous when compared to synchronous converters or motor generators because of efficiency, cost, size, weight, and reliability. Various bridge configurations for single-phase and three-phase applications are shown in Fig. 15.1.75a. Table 15.1.15 shows the relative outputs of rectifier circuits. The use of two three-phase bridges fed from an ac source consisting of a three-winding transformer with both a Δ and Y secondary winding so that output voltages are 30° out of phase will reduce dc ripple to approximately 1 percent.

The use of **silicon-controlled rectifiers** (SCRs) to replace rectifiers in the various bridge configurations allows the output voltages to be varied from rated output voltage to zero. The output voltage wave will not be a sine wave but a series of square waves, which may not be suitable for some applications. Dc to ac conversions are shown in Fig. 15.1.75b.

A new technology of ac-to-dc conversion commonly called *switch*

Table 15.1.15 **Relationships for AC-DC Conversion Static Devices**

Device description	Voltages, %		Currents, %		Ripple, %
	E_{ac}	E_{dc}	I_{ac}	I_{dc}	
1ϕ, half wave	100	45	100	100	121
1ϕ, full wave	100	90	100	90	48
3ϕ, full wave	100	135	100	123	4.2

mode power supplies (SMPSs) is found in almost all microprocessor-based modern electronic equipment for dc supplies between 3 and 15 V. The supply voltage (120 V ac) is rectified by a single-phase full-wave bridge circuit. See Fig. 15.1.75. The output is stored in a capacitor. A switcher will then switch the dc voltage from the capacitor on and off at a high frequency, usually between 10 and 100 kHz. These high-frequency pulses are stepped down in voltage by a transformer and rectified by diodes. The diodes' output is filtered to the dc supply required. These SMPSs are small in size, have higher efficiency, and are lower in cost. They do create harmonic and power quality problems that have to be addressed.

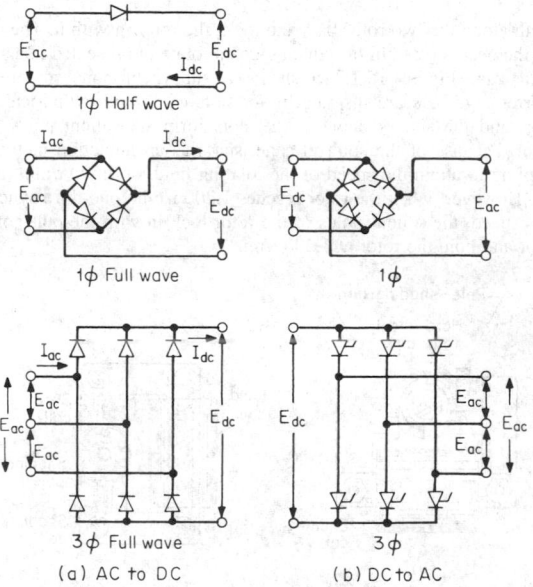

Fig. 15.1.75 AC-DC conversion with static devices.

SYNCHRONOUS CONVERTERS

The synchronous converter is essentially a dc generator with slip rings connected by taps to equidistant points in the armature winding. Alternating current may also be taken from and delivered to the armature. The machine may be single-phase, in which case there are two slip rings and two slip-ring taps per pair of poles; it may be three-phase, in which case there are three slip rings and three slip-ring taps per pair of poles, etc. Converters are usually used to convert alternating to direct current, in which case they are said to be operating **direct;** they may equally well convert direct to alternating current, in which case they are said to be operating **inverted.** A converter will operate satisfactorily as a dc motor, a synchronous motor, a dc generator, a synchronous generator, or it may deliver direct and alternating current simultaneously, when it is called a **double-current generator.**

The rating of a converter increases very rapidly with increase in the number of phases owing, in part, to better utilization of the armature copper and also because of more uniform distribution of armature heating.

Because of the materially increased rating, converters are nearly all operated six-phase. The rating decreases rapidly with decrease in power factor, and hence the converter should operate near unity power factor. The diametrical ac voltage is the ac voltage between two slip-ring taps 180 electrical degrees apart. With a two-pole closed winding, i.e., a winding that closes on itself when the winding is completed, the diametrical ac voltage is the voltage between any two slip-ring taps diametrically opposite each other.

With a sine-voltage wave, the dc voltage is the peak of the diametrical ac voltage wave. The voltage relations for sine waves are as follows: dc volts, 141; single phase, diametrical, 100; three-phase, 87; four-phase, diametrical, 100; four phase, adjacent taps, 71; six phase, diametrical, 100; six phase, adjacent taps, 50. These relations are obtained from the sides of polygons inscribed in a circle having a diameter of 100 V, as shown in Fig. 15.1.76.

Selsyns The word **selsyn** is an abbreviation of self-synchronizing and is applied to devices which are connected electrically, and in which an angular displacement of the rotating member of one device produces an equal angular displacement in the rotating member of the second

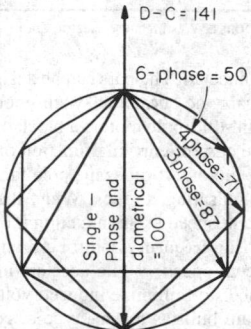

Fig. 15.1.76 EMF relations in a converter.

device. There are several types of selsyns and they may be dc or ac, single-phase or polyphase. A simple and common type is shown in Fig. 15.1.77. The two stators S_1, S_2 are phase-wound stators, identical electrically with synchronous-generator or induction-motor stators. For simplicity Gramme-ring windings are shown in Fig. 15.1.77. The two stators are connected three-phase and in parallel. There are also two bobbin-type rotors R_1, R_2, with single-phase windings, each connected to a single-phase supply such as 115 V, 60 Hz. When R_1 and R_2 are in the same angular positions, the emfs induced in the two stators by the ac flux of the rotors are equal and opposite, there are no interchange currents between stators, and the system is in equilibrium. However, if the angular displacement of R_1, for example, is changed, the magnitudes of the emfs induced in the stator winding of S_1 are correspondingly changed. The emfs of the two stators then become unbalanced, currents flow from S_1 to S_2, producing torque on R_2. When R_2 attains the same angular position as R_1, the emfs in the two rotors again become equal and opposite, and the system is again in equilibrium.

If there is torque load on either rotor, a resultant current is necessary to sustain the torque, so that there must be an angular displacement between rotors. However, by the use of an auxiliary selsyn a current may be fed into the system which is proportional to the angular difference of the two rotors. This current will continue until the error is corrected. This is called **feedback.** There may be a master selsyn, controlling several secondary units.

Selsyns are used for position indicators, e.g., in bridge–engine-room signal systems. They are also widely used for fire control so that from any desired position all the turrets and guns on battleships can be turned and elevated simultaneously through any desired angle with a high degree of accuracy. The selsyn itself rarely has sufficient power to perform these operations, but it actuates control through power multipliers such as amplidynes.

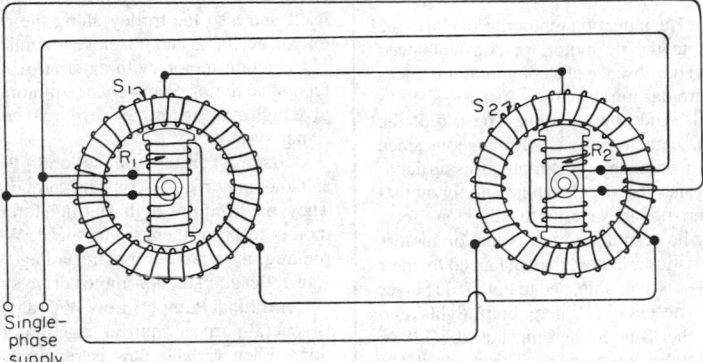

Fig. 15.1.77   Selsyn system.

RATING OF ELECTRICAL APPARATUS

The **rating** of electrical apparatus is almost always determined by the maximum temperature at which the materials in the machine, especially the insulation and lubricant, may be operated for long periods without deterioration. It is permissible, as far as temperature is concerned, to overload the apparatus so long as the safe temperature is not exceeded. The ANSI/IEEE Standard 100-1992 classifies **insulating materials** in seven different **classes:**

1. *Class 90 insulation.* Materials or combinations of materials such as cotton, silk, and paper without impregnation which will have suitable thermal endurance if operated continually at 90°C.

2. *Class 105 insulation.* Materials or combinations of materials such as cotton, silk, and paper when suitably impregnated or coated or when immersed in a dielectric liquid. This class has sufficient thermal endurance at 105°C.

3. *Class 130 insulation.* Materials or combinations of materials such as mica, glass fiber, asbestos, etc., with suitable bonding substances. This class has sufficient thermal endurance at 130°C.

4. *Class 155 insulation.* Same materials as class 130 but with bonding substances suitable for continuous operation at 155°C.

5. *Class 180 insulation.* Materials or combinations of materials such as silicone elastomer, mica, glass fiber, asbestos, etc., with suitable bonding substances such as appropriate silicone resins. This class has sufficient thermal life at 180°C.

6. *Class 220 insulation.* Materials suitable for continuous operation at 220°C.

7. *Class over-220 insulation.* Materials consisting entirely of mica, porcelain, glass, quartz, and similar inorganic materials which have suitable thermal life at temperatures over 220°C.

NOTE: In all cases, other materials or combinations of materials other than those mentioned above may be used in a given class if from experience or accepted tests they can be shown to have comparable thermal life. It is common practice also to specify **insulation systems in electrical machinery** by letter. For example, integral horsepower ac motors may have a maximum temperature rise in the winding (determined by winding resistance) or 60°C for **class A** insulation, 80°C for **class B**, 105°C for **class F**, and 125°C for **class H**—all based on a 40°C ambient.

The recommended methods of measurement are: (1) the thermometer method is preferred for uninsulated windings, exposed metal parts, gases and liquids, or surface methods generally; thermocouples are preferred for rapidly changing surface temperatures; (2) the applied-thermocouple method is suitable for making surface temperature measurements when it is desired to measure the temperature of surfaces that are accessible to thermocouples but not to liquid-in-glass thermometers; (3) the contact-thermocouple method is suitable for measuring temperatures of bare metal surfaces such as those of commutator bars and slip rings; (4) the resistance method is suitable for insulated windings, except for windings of such low resistance that measurements

cannot be accurately made due to uncontrollable resistance in contacts or where it is impracticable to make connections to obtain measurements before an undesirable drop in temperature occurs; (5) the embedded-detector method is suitable for interior measurements at designated locations as specified in the standards for certain kinds of equipment, such as large rotating machines.

Efficiency of Electrical Motors

Methods of determining efficiency are by **direct measurement** or by **segregated losses.** Methods are outlined in Standard Test Procedure for Polyphase Induction Motors and Generators, ANSI/IEEE Std. 112-1991; Standard Test Code for DC Machines, IEEE Std. 113-1985, Test Procedure for Single-Phase Induction Motors, ANSI/IEEE Std. 114-1982; and Test Procedures for Synchronous Machines, IEEE Std. 115-1983.

Direct measurements can be made by using calibrated **motors, generators,** or **dynamometers** for input to generators and output from motors, and precision electrical motors for input to motors and output from generators.

$$\text{Efficiencies} = \frac{\text{output}}{\text{input}} \qquad (15.1.123)$$

The segregated **losses** in motors are classified as follows: (1) Stator I^2R (shunt and series field I^2R for dc); (2) rotor I^2R (armature I^2R for dc); (3) core loss; (4) stray-load loss; (5) friction and windage loss; (6) brush-contact loss (wound rotor and dc); (7) brush-friction loss (wound rotor and dc); (8) exciter loss (synchronous and dc); and (9) ventilating loss (dc). Losses are calculated separately and totaled. Measure the electrical output of the generator; then

$$\text{Efficiency} = \frac{\text{output}}{\text{output} + \text{losses}} \qquad (15.1.124a)$$

Measure the electrical input of the motors; then

$$\text{Efficiency} = \frac{\text{input} - \text{losses}}{\text{input}} \qquad (15.1.124b)$$

When testing dc motors, compensation should be made for the **harmonics** associated with rectified ac used to provide the variable dc voltage to the motors. Instrumentation should be chosen to accurately reflect the rms value of currents.

Temperature rise under full-load conditions may be measured by tests as outlined in the IEEE Standards referred to above. Methods of loading are: (1) Load motor with **dynamometer** or **generator** of similar capacity and run until temperatures stabilize; (2) load generator with **motor-generator** set or **plant** load and run until temperature stabilizes; (3) alternately apply **dual frequences** to motor until it reaches rated temperature; (4) synchronous motor may be operated as synchronous **condenser** at no load with **zero power factor** at rated current, voltage, and frequency until temperatures stabilize.

Industrial Applications of Motors

Alternating or Direct Current The induction motor, particularly the squirrel-cage type, is preferable to the dc motor for constant-speed work, for the initial cost is less and the absence of a commutator reduces maintenance. Also there is less fire hazard in many industries, such as sawmills, flour mills, textile mills, and powder mills. The use of the induction motor in such places as cement mills is advantageous since with dc motors the grit makes the maintenance of commutators difficult.

For variable-speed work like cranes, hoists, elevators, and for adjustable speeds, the dc motor characteristics are superior to induction-motor characteristics. Even then, it may be desirable to use induction motors since their less desirable characteristics are more than balanced by their simplicity and the fact that ac power is available, and to obtain dc power conversion apparatus is usually necessary. Where both lights and motors are to be supplied from the same ac system, the 208/120-V four-wire three-phase system is now in common use. This gives 208 V three-phase for the motors, and 120 V to neutral for the lights.

Full-load speed, temperature rise, efficiency, and power factor as well as breakdown torque and starting torque have long been parameters of concern in the application and purchase of motors. Another qualification is service factor. The service factor of an alternating current motor is a multiplier applicable to the horsepower rating. When so applied, the result is a permissible horsepower loading under the conditions specified for the service factor. When operated at service factor load with 1.15 or higher service factor, the permissible temperature rise by resistance is as follows: class A insulation 70°C; class B, 90°C; and class F, 115°C.

Special enclosures, fittings, seals, ventilation systems, electromagnetic design, etc., are required when the motor is to be operated under unusual service conditions, such as exposure to (1) combustible, explosive, abrasive, or conducting dusts, (2) lint or very dirty conditions where the accumulation of dirt might impede the ventilation, (3) chemical fumes or flammable or explosive gases, (4) nuclear radiation, (5) steam, salt laden air, or oil vapor, (6) damp or very dry locations, radiant heat, vermin infestation, or atmosphere conducive to the growth of fungus, (7) abnormal shock, vibration, or external mechanical loading, (8) abnormal axial thrust or side forces on the motor shaft, (9) excessive departure from rated voltage, (10) deviation factors of the line voltage exceeding 10 percent, (11) line voltage unbalance exceeding 1 percent, (12) situations where low noise levels are required, (13) speeds higher than the highest rated speed, (14) operation in a poorly ventilated room, in a pit, or in an inclined attitude, (15) torsional impact loads, repeated abnormal overloads, reversing or electric braking, (16) operation at standstill with any winding continuously energized, and (17) operation with extremely low structureborne and airborne noise. For dc machines, a further unusual service condition occurs when the average load is less than 50 percent over a 24-h period or the continuous load is less than 50 percent over a 4-h period.

The standard direction of rotation for all nonreversing dc motors, ac single-phase motors, synchronous motors, and universal motors is counterclockwise when facing the end of the machine opposite the drive end. For dc and ac generators, the rotation is clockwise.

Further information may be found in Publication No. MG-1 of the National Electrical Manufacturers Association.

It must be recognized that heat is conducted by electrical conductors. Windings in motors operating in a 40°C ambient at class F temperature rises are running at temperatures 90°C higher than the maximum allowable temperature (75°C) of cable ordinarily used in interior wiring. Heat conducted by the motor leads in such a situation could cause a failure of the branch circuit cable in the terminal box. See Tables 15.1.21 and 15.1.22.

ELECTRIC DRIVES

Cranes and Hoists The dc series motor is best adapted to cranes and hoists. When the load is heavy the motor slows down automatically and develops increased torque thus reducing the peaks on the electrical system. With light loads, the speed increases rapidly, thus giving a lively crane. The series motor is also well adapted to moving the bridge itself and also the trolley along the bridge. Where alternating current only is available and it is not economical to convert it, the slip-ring type of induction motor, with external-resistance speed control, is the best type of ac motor. Squirrel-cage motors with high resistance end rings to give high starting torque (design D) are used (design D motors; also see Ilgner system).

Constant-Torque Applications **Piston pumps, mills, extruders,** and **agitators** may require constant torque over their complete speed range. They may require high starting torque design C or D squirrel-cage motors to bring them up to speed. Where speed is to be varied while running, a variable armature voltage **dc motor** or a **variable-frequency** squirrel-cage induction-motor drive system may be used.

Centrifugal Pumps Low WK^2 and low starting torques make design B general-purpose squirrel-cage motors the preference for this application. When **variable flow** is required, the use of a **variable-frequency** power supply to vary motor speed will be energy efficient when compared to changing flow by **control-valve** closure to increase head.

Centrifugal Fans High WK^2 may require high starting torque design C or D squirrel-cage motors to bring the fan up to speed in a reasonable period of time. When **variable flow** is required, the use of a **variable-frequency** power supply or a multispeed motor to vary fan speed will be energy efficient when compared to closing **louvers**. For large fans, **synchronous-motor** drives may be considered for high efficiency and improved power factor.

Axial or Centrifugal Compressors For smaller compressors, say, up to 100 hp, the squirrel-cage induction motor is the drive of choice. When the WK^2 is high, a design C or D high-torque motor may be required. For larger compressors, the synchronous motor is more efficient and improves power factor. Where **variable flow** is required, the **variable-frequency** power supply to vary motor speed is more efficient than controlling by **valve** and in some applications may eliminate a **gearbox** by allowing the motor to run at compressor operating speed.

Pulsating-Torque Applications **Reciprocating compressors, rock crushers,** and **hammer mills** experience widely varying torque pulsations during each revolution. They usually have a **flywheel** to store energy, so a high-torque, high-slip design D motor will accelerate the high WK^2 rapidly and allow energy recovery from the flywheel when high torque is demanded. On larger drives a slow-speed, engine-type synchronous motor can be directly connected. The motor itself supplies significant WK^2 to smooth out the torque and current pulsations of the system.

SWITCHBOARDS

Switchboards may, in general, be divided into four classes: direct-control panel type; remote mechanical-control panel type; direct-control truck type; electrically operated. With **direct-control panel-type boards** the switches, rheostats, bus bars, meters, and other apparatus are mounted on or near the board and the switches and rheostats are operated directly, or by operating handles if they are mounted in back of the board. The voltages, for both direct current and alternating current, are usually limited to 600 V and less but may operate up to 2,500 V ac if oil circuit breakers are used. Such panels are not recommended for capacities greater than 3,000 kVA. **Remote mechanical-control panel-type boards** are ac switchboards with the bus bars and connections removed from the panels and mounted separately away from the load. The oil circuit breakers are operated by levers and rods. This type of board is designed for heavier duty than the direct-control type and is used up to 25,000 kVA. **Direct-control truck-type switchboards** for 15,000 V or less consist of equipment enclosed in steel compartments completely assembled by the manufacturers. The high-voltage parts are enclosed, and the equipment is interlocked to prevent mistakes in operation. This equipment is designed for low- and medium-capacity plants and auxiliary power in large generating stations. **Electrically operated switchboards** employ solenoid or motor-operated circuit breakers, rheostats, etc., controlled by small switches mounted on the panels. This makes it possible to locate the high-voltage and other equipment independently of the location of switchboard.

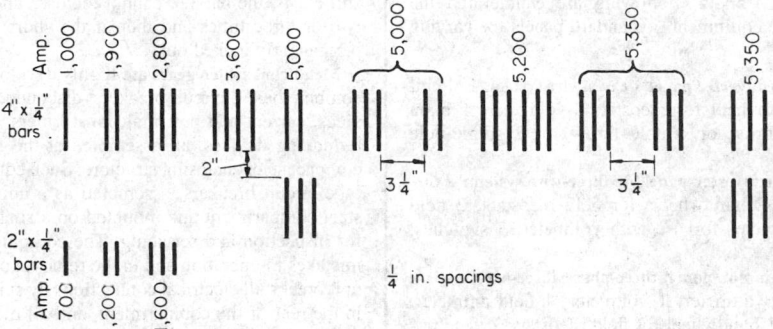

Fig. 15.1.78 Current-carrying capacity of copper bus bars.

In all large stations the switching equipment and buses are always mounted entirely either in separate buildings or in outdoor enclosures. Such equipment is termed **bus structures** and is electrically operated from the main control board.

Marble has high dielectric qualities and was formerly used exclusively for the panels. It is now used occasionally where its appearance is desired for architectural purposes. Slate is used extensively and is finished in black enamel, marine, and natural black. Ebony asbestos is also used frequently, is lighter than marble or slate, has high dielectric strength and insulation resistivity, and can be readily cut, drilled, and machined. Steel panels, usually ⅛ in thick, are light, economical in construction and erection, and at the present time are favored over other types.

Switchboards should be erected at least 3 to 4 ft from the wall. Switchboard frames and structures should be grounded. The only exceptions are effectively insulated frames of single-polarity dc switchboards. For low-potential work, the conductors on the rear of the switchboard are usually made up of flat copper strip, known as **bus-bar** copper. The size required is based upon a current density of about 1,000 A/in². Figure 15.1.78 gives the approximate continuous dc carrying capacity of copper bus bars for different arrangements and spacings for 35°C temperature rise.

Switchboards must be individually adapted for each specific electrical system. Space permits the showing of the diagrams of only three boards each for a typical electrical system (Fig. 15.1.79). Aluminum bus bars are also frequently used.

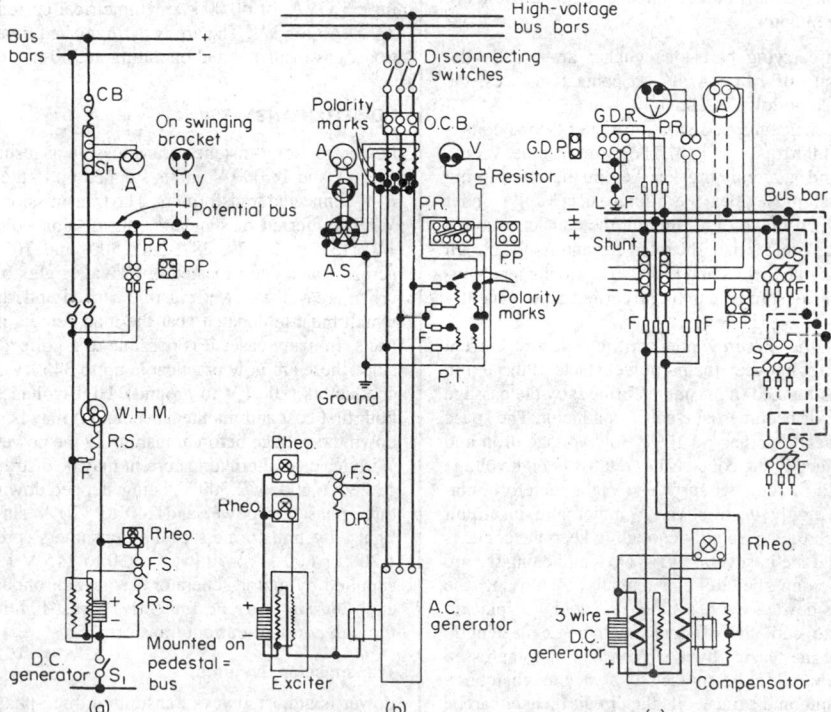

Fig. 15.1.79 Switchboard wiring diagrams for generators. (*a*) 125-V or 250-V dc generator; (*b*) three-phase, synchronous generator and exciter for a small or isolated plant; (*c*) three-wire dc generator for a small or isolated plant. A, ammeter; AS, three-way ammeter switch; CB, circuit breaker, CT, current transformer; DR, ground detector receptacle; L, ground detector lamp; OC, overload coil; OCB, oil circuit breaker; PP, potential ring; PR, potential receptacle; PT, potential transformer; Rheo, rheostat; RS, resistor; S, switch; Sh, shunt; V, voltmeter; WHM, watthour meter.

Equipment of Standard Panels Following are enumerated the various parts required in the equipment of standard panels for varying services:

Generator or synchronous-converter panel, dc two-wire system: 1 circuit breaker; 1 ammeter; 1 handwheel for rheostat; 1 voltmeter; 1 main switch (three-pole single throw or double throw) or 2 single-pole switches.

Generator or synchronous-converter panel, dc three-wire system: 2 circuit breakers; 2 ammeters; 2 handwheels for field rheostats; 2 field switches; 2 potential receptacles for use with voltmeter; 3 switches; 1 four-point starting switch.

Generator or synchronous-motor panel, three-phase three-wire system: 3 ammeters; 1 three-phase wattmeter; 1 voltmeter; 1 field ammeter; 1 double-pole field switch; 1 handwheel for field rheostats; 1 synchronizing receptacle (four-point); 1 potential receptacle (eight-point); 1 field rheostat; 1 triple-pole oil switch; 1 power-factor indicator; 1 synchronizer; 2 series transformers; 1 governor control switch.

Synchronous-converter panel, three-phase: 1 ammeter; 1 power-factor indicator; 1 synchronizing receptacle; 1 triple-pole oil circuit breaker; 2 current transformers; 1 potential transformer; 1 watthour meter (polyphase); 1 governor control switch.

Induction motor panel, three-phase: 1 ammeter; series transformers; 1 oil switch.

Feeder panel, dc, two-wire and three-wire: 1 single-pole circuit breaker; 1 ammeter; 2 single-pole main switches; potential receptacles (1 four-point for two-wire panel; 1 four-point and 1 eight-point for three-wire panel).

Feeder panel, three-wire, three-phase and single-phase: 3 ammeters; 1 automatic oil switch (three-pole for three-phase, two-pole for single-phase); 2 series transformers; 1 shunt transformer; 1 wattmeter; 1 voltmeter; 1 watthour meter; 1 handwheel for control of potential regulator.

Exciter panel (for 1 or 2 exciters): 1 ammeter (2 for 2 exciters); 1 field rheostat (2 for 2 exciters); 1 four-point receptacle (2 for 2 exciters); 1 equalizing rheostat for regulator.

Switches The current-carrying parts of switches are usually designed for a current density of 1,000 A/in². At contact surfaces, the current density should be kept down to about 50 A/in².

Circuit Breakers Switches equipped with a tripping device constitutes an elementary load interrupter switch. The difference between a load interrupter switch and a circuit breaker lies in the interrupting capacity. A circuit breaker must open the circuit successfully under short circuit conditions when the current through the contacts may be several orders of magnitude greater than the rated current. As the circuit is being opened, the device must withstand the accompanying mechanical forces and the heat of the ensuing arc until the current is permanently reduced to zero.

The opening of a metallic circuit while carrying electric current causes an electric arc to form between the parting contacts. If the action takes place in air, the air is ionized (a plasma is formed) by the passage of current. When ionized, air becomes an electric conductor. The space between the parting contacts thus has relatively low voltage drop and the region close to the surface of the contacts has relatively high voltage drop. The thermal input to the contact surfaces (VI) is therefore relatively large and can be highly destructive. A major aim in circuit breaker design is to quench the arc rapidly enough to keep the contacts in a reusable state. This is done in several ways: (1) lengthening the arc mechanically, (2) lengthening the arc magnetically by driving the current-carrying plasma sideways with a magnetic field, (3) placing barriers in the arc path to cool the plasma and increase its length, (4) displacing and cooling the plasma by means of a jet of compressed air or inert gas, and (5) separating the contacts in a vacuum chamber.

By a combination of shunt and series coils the circuit breaker can be made to trip when the energy reverses. Circuit breakers may trip unnecessarily when the difficulty has been immediately cleared by a local breaker or fuse. In order that service shall not be thus interrupted unnecessarily, **automatically reclosing breakers** are used. After tripping, an automatic mechanism operates to reclose the breaker. If the short circuit still exists, the breaker cannot reclose. The breaker attempts to reclose two or three times and then if the short circuit still exists it remains permanently locked out.

Metal-clad switch gears are highly developed pieces of equipment that combine buses, circuit breakers, disconnecting devices, controlling devices, current and potential transformers, instruments, meters, and interlocking devices, all assembled at the factory as a single unit in a compact steel enclosing structure. Such equipment may comprise truck-type circuit breakers, assembled as a unit, each housed in a separate steel compartment and mounted on a small truck to facilitate removal for inspection and servicing. The equipment is interlocked to prevent mistakes in operation and in the removal of the unit; the removal of the unit breaks all electrical connections by suitable disconnecting switches in the rear of the compartment, and all metal parts are grounded. This design provides compactness, simplicity, ease of inspection, and safety to the operator.

High-voltage circuit breakers can be oil type, in which the contacts open under **oil, air-blast** type, in which the arc is extinguished by a powerful blast of air directed through an orifice across the arc and into an arc chute, H_2S type, or vacuum contact type. The tripping of high-voltage circuit breakers is initiated by an abnormal current acting through the secondary of a current transformer on an inverse-time relay in which the time of closing the relay contacts is an inverse time function of the current; i.e., the greater the current the shorter the time of closing. The breaker is tripped by a dc tripping coil, the dc circuit being closed by the relay contacts. Modern circuit breakers should open the circuit within 3 cycles from the time of the closing of the relay contacts.

Vacuum circuit breakers have received wide acceptance in all fields in recent years, both for indoor work and for outdoor applications. Indoor breakers are available up to 40 kV and interrupting capacities up to 2.5 GVA. Outdoor breakers are available in ratings up to that of the EHV (extra-high voltage 765-kV three-pole breaker capable of interrupting 55 GVA, or 40,000-A symmetrical current. Its operating rating is 3,000 A, 765 kV. The arc is extinguished in a vacuum. Switching stations, gas-insulated and operating at 550 kV, are also in use.

POWER TRANSMISSION

Power for long-distance transmission is usually generated at 6,600, 13,200, and 18,000 V and is stepped up to the transmission voltage by Δ-Y-connected transformers. The transmission voltage is roughly 1,000 V/mi. Preferred or standard transmission voltages are 22, 33, 44, 66, 110, 132, 154, 220, 287, 330, 500, and 765 kV. High-voltage lines across country are located on private rights of way. When they reach urban areas, the power must be carried underground to the substations which must be located near the load centers in the thickly settled districts. In many cases it is possible to go directly to underground cables since these are now practicable up to 345 kV between three-phase line conductors (200 kV to ground). High-voltage cables are expensive in both first cost and maintenance, and it may be more economical to step down the voltage before transmitting the power by underground cables. Within a city, alternating current may be distributed from a substation at 13,200, 6,600, or 2,300 V, being stepped down to 600, 480, and 240 V, three-phase for power and 240 to 120 V single-phase three-wire for lights, by transformers at the consumers' premises. **Direct current** at 1,200 or 600 V for railways, 230 to 115 V for lighting and power, is supplied by motor-generator sets, synchronous converters, and rectifiers. **Constant current** for series street-lighting systems is obtained through constant-current transformers.

Transmission Systems

Power is almost always transmitted three-phase. The following fundamental relations apply to any transmission system. The weight of conductor required to transmit power by any given system with a given percentage power loss varies directly with the power, directly as the square of the distance, and inversely as the square of the voltage. The cross-sectional area of the conductors with a given percentage power

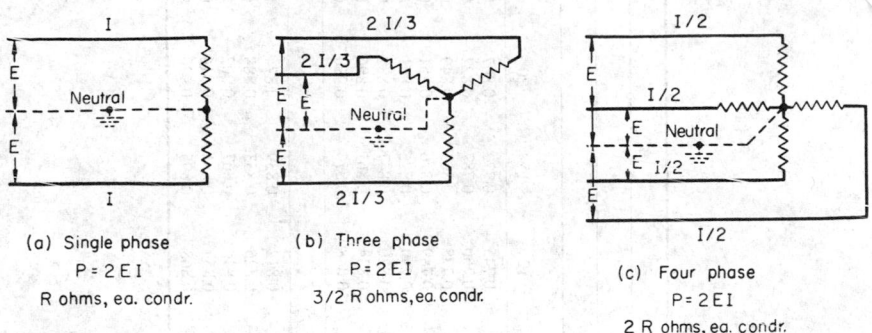

Fig. 15.1.80 Three equivalent symmetrical transmission, or distribution, systems. (*a*) Single-phase; (*b*) three-phase; (*c*) four-phase.

loss varies directly with the power, directly with the distance, and inversely as the square of the voltage.

For two systems of the same length transmitting the same power at different voltages and with the same power loss for both systems, the cross-sectional area and weight of the conductors will vary inversely as the square of the voltages. The foregoing relations between the cross section or weight of the conductor and transmission distance and voltage hold for all systems, whether dc, single-phase, three-phase, or four-phase. With the power, distance, and power loss fixed, all symmetrical systems having equal voltages to neutral require equal weights of conductor. Thus, the three symmetrical systems shown in Fig. 15.1.80 all deliver the same power, have the same power loss and equal voltages to neutral, and the transmission distances are all assumed to be equal. They all require the same weight of conductor since the weights are inversely proportional to all resistances. (No actual neutral conductor is used.) The respective power losses are (1) $2I^2R$ W; (2) $3(2I/3)^2(3R/2) = 2I^2R$ W; (3) $4(I/2)^2(2R) = 2I^2R$ W, which are all equal.

Size of Transmission Conductor Kelvin's law states, "The most economical area of conductor is that for which the annual cost of energy wasted is equal to the interest on that portion of the capital outlay which can be considered proportional to the weight of copper used." In Fig. 15.1.81 are shown the annual interest cost, the annual cost of I^2R loss, and the total cost as functions of circular mils cross section for both typical overhead conductors and three-conductor cables. Note that the total-cost curves have very flat minimums, and usually other factors such as the character of the load and the voltage regulation, are taken into consideration.

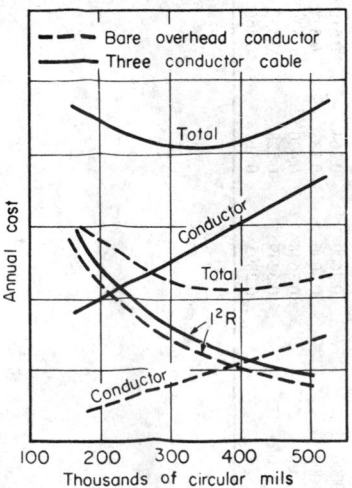

Fig. 15.1.81 Most economical sizes of overhead and underground conductors.

In addition to resistance, overhead power lines have inductive reactance to alternating currents. The inductive reactance

$$X = 2\pi f\{80 + 741.1 \log [(D - r)/r]\} \, 10^{-6}$$
$$\Omega/\text{conductor mile} \quad (15.1.126)$$

where f = frequency, D = distance between centers of conductors (in), and r their radius (in). Table 15.1.16 gives the inductive reactance per mile at 60 Hz and the resistance of stranded and solid copper conductor. (See Table 15.1.20.)

Any **symmetrical system** having n conductors can be divided into n equal single-phase systems, each consisting of one wire and a return circuit of zero impedance and each having as its voltage the system voltage to neutral.

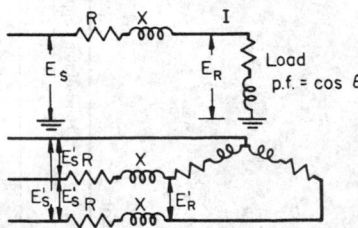

Fig. 15.1.82 Three-phase power system.

Figure 15.1.82 shows a symmetrical three-phase system, with one phase detached. The load or received voltage between line conductors is E'_R so that the receiver voltage to *neutral* is $E_R = E'_R/\sqrt{3}$ V. The current is I A, the load power factor is $\cos \theta$, and the line resistance and reactance are R and X Ω per wire, and the sending-end voltage is E_S. The phasor diagram is shown in Fig. 15.1.83 (compare with Fig. 15.1.63). Its solution is

$$E_S = \sqrt{(E_R \cos \theta + IR)^2 + (E_R \sin \theta + IX)^2} \quad (15.1.127)$$

[see Eq. (15.1.101)].

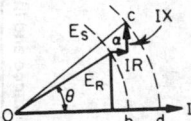

Fig. 15.1.83 Phasor diagram for a power line.

Figure 15.1.84 (Mershon diagram) shows the right-hand portion of Fig. 15.1.83 plotted to large scale, the arc 00 corresponding to the arc *ab* (Fig. 15.1.83). The abscissa 0 (Fig. 15.1.84) corresponds to point *b* (Fig. 15.1.83) and is the load voltage E_R taken as 100 percent. The concentric circular arcs 0–40 are given in percentage of E_R. To find the sending-

Table 15.1.16 Resistance and Inductive Reactance per Single Conductor

Size, cir mils or AWG	No. of strands	OD, in	Resistance, Ω/mi	60 Hz Spacing, ft												
				1	2	3	4	5	6	7	8	10	12	15	20	30
Hard-drawn copper, stranded																
500,000	37	0.814	0.1130	0.443	0.527	0.576	0.611	0.638	0.660	0.679	0.695	0.722	0.745	0.772	0.807	0.856
400,000	19	0.725	0.1426	0.458	0.542	0.591	0.626	0.653	0.675	0.694	0.710	0.737	0.760	0.787	0.822	0.871
300,000	19	0.628	0.1900	0.476	0.560	0.609	0.644	0.671	0.693	0.712	0.728	0.755	0.778	0.805	0.840	0.889
250,000	19	0.574	0.2278	0.487	0.571	0.620	0.655	0.682	0.704	0.723	0.739	0.766	0.789	0.816	0.851	0.900
0000	19	0.528	0.2690	0.497	0.581	0.630	0.665	0.692	0.714	0.733	0.749	0.776	0.799	0.826	0.861	0.917
000	7	0.464	0.339	0.518	0.602	0.651	0.686	0.713	0.735	0.754	0.770	0.797	0.820	0.847	0.882	0.931
00	7	0.414	0.428	0.532	0.616	0.665	0.700	0.727	0.749	0.768	0.784	0.811	0.834	0.861	0.896	0.945
0	7	0.368	0.538	0.546	0.630	0.679	0.714	0.741	0.763	0.782	0.798	0.825	0.848	0.875	0.910	0.959
Hard-drawn copper, solid																
0000	—	0.4600	0.264	0.510	0.594	0.643	0.678	0.705	0.727	0.746	0.762	0.789	0.812	0.839	0.874	0.923
000	—	0.4096	0.333	0.524	0.608	0.657	0.692	0.719	0.741	0.760	0.776	0.803	0.826	0.853	0.888	0.937
00	—	0.3648	0.420	0.538	0.622	0.671	0.706	0.733	0.755	0.774	0.790	0.817	0.840	0.867	0.902	0.951
0	—	0.3249	0.528	0.552	0.636	0.685	0.720	0.747	0.769	0.788	0.804	0.831	0.854	0.881	0.916	0.965
1	—	0.2893	0.665	0.566	0.650	0.699	0.734	0.761	0.783	0.802	0.818	0.845	0.868	0.895	0.930	0.979

end voltage E_S for any power factor $\cos\theta$, compute first the resistance drop IR and the reactance drop IX in percentage of E_R. Then follow the ordinate corresponding to the load power factor to the inner arc 00 (a, Fig. 15.1.83). Lay off the percentage IR drop horizontally to the right, and the percentage IX drop vertically upward. The arc at which the IX drop terminates (c, Fig. 15.1.83) when added to 100 percent gives the sending-end voltage E_S in percent of the load voltage E_R.

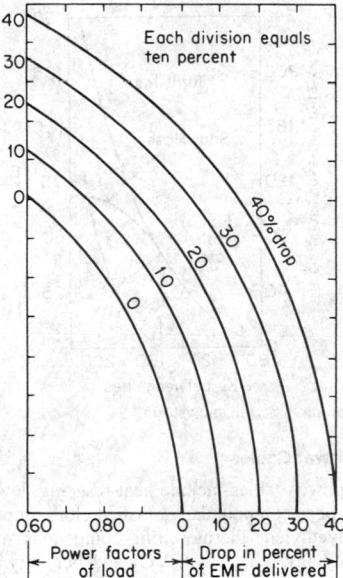

Fig. 15.1.84   Mershon diagram for determining voltage drop in power lines.

EXAMPLE.   Let it be desired to transmit 20,000 kW three-phase 80 percent power factor lagging current, a distance of 60 mi. The voltage at the receiving end is 66,000 V, 60 Hz and the line loss must not exceed 10 percent of the power delivered. The conductor spacing must be 7 ft (84 in). Determine the sending-end voltage and the actual efficiency. $I = 20{,}000{,}000/(66{,}000 \times 0.80 \times \sqrt{3}) = 218.8$ A. $3 \times 218.8^2 \times R' = 0.10 \times 20{,}000{,}000$. $R' = 13.9$ $\Omega = 0.232$ Ω/mi. By

referring to Table 13.1.16, 250,000 cir mils copper having a resistance of 0.2278 Ω/mi may be used. The total resistance $R = 60 \times 0.2278 = 13.67$ Ω. The reactance $X = 60 \times 0.723 = 43.38$ Ω. The volts to neutral at the load, $E_R = 66{,}000/\sqrt{3} = 38{,}100$ V. $\cos\theta = 0.80$; $\sin\theta = 0.60$. Using Eq. (15.1.127), $E_S = \{[(38{,}100 \times 0.80) + (218.8 \times 13.67)]^2 + [(38{,}100 \times 0.60) + (218.8 \times 43.38)]^2\}^{1/2} = 46{,}500$ V to neutral or $\sqrt{3} \times 46{,}500 = 80{,}500$ between lines at the sending end. The line loss is $3(218.8)^2 \times 13.67 = 1{,}963$ kW. The efficiency $\eta = 20{,}000/(20{,}000 + 1{,}963) = 0.911$, or 91.1 percent. This same line is solved by means of the Mershon diagram as follows. Let $E_R = 38{,}100$ V $= 100$ percent. $IR = 218.8 \times 13.67 = 2{,}991$ V $= 7.85$ percent. $IX = 218.8 \times 43.38 = 9{,}490$ V $= 24.9$ percent. Follow the 0.80 power-factor ordinate (Fig. 15.1.84) to its intersection with the arc 00; from this point go 7.85 percent horizontally to the right and then 24.9 percent vertically. (These percentages are measured on the horizontal scale.) This last distance terminates on the 22.5 percent arc. The sending-end voltage to neutral is then $1.225 \times 38{,}100 = 46{,}500$ V, so that the sending-end voltage between line conductors is $E'_S = 46{,}500 \sqrt{3} = 80{,}530$ V.

In Table 15.1.16 the spacing is the distance between the centers of the two conductors of a single-phase system or the distance between the centers of each pair of conductors of a three-phase system if they are equally spaced. If they are not equally spaced, the geometric mean distance (GMD) is used, where $GMD = \sqrt[3]{D_1 D_2 D_3}$ (Fig. 15.1.85a). With the flat horizontal spacing shown in Fig. 15.1.85b, $GMD = \sqrt[3]{2D^3} = 1.26D$.

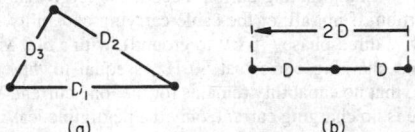

Fig. 15.1.85   Unequal spacing of three-phase conductors. (a) $GMD = (D_1 D_2 D_3)^{1/3}$; ($b$) flat horizontal spacing; $GMD = 1.26$ D.

In addition to copper, aluminum cable steel-reinforced (ACSR), Table 15.1.17, is used for transmission conductor. For the same resistance it is lighter than copper, and with high voltages the larger diameter reduces corona loss.

Until 1966, 345 kV was the highest operating voltage in the United States. The first 500 kV system put into operation (1966) was a 350-mi transmission loop of the Virginia Electric and Power Company; the longest transmission distance was 170 mi. The towers, about 94 ft high,

Table 15.1.17   Properties of Aluminum Cable Steel-Reinforced (ACSR)

| Cir mils or AWG | | No. of wires | | | Cross section, in² | | Total | Ω/mi of single conductor at 25°C | | | | |
| | | | | | | | | | 200 A | | 600 A | |
Aluminum	Copper equivalent	Aluminum	Steel	OD, in	Aluminum	Total	lb/mi	0 A dc	25 Hz	60 Hz	25 Hz	60 Hz
1,590,000	1,000,000	54	19	1.545	1.249	1.4071	10,777	0.0587	0.0589	0.0594	0.0592	0.0607
1,431,000	900,000	54	19	1.465	1.124	1.2664	9,699	0.0652	0.0654	0.0659	0.0657	0.0671
1,272,000	800,000	54	19	1.382	0.9990	1.1256	8,621	0.0734	0.0736	0.0742	0.0738	0.0752
1,192,500	750,000	54	19	1.338	0.9366	1.0553	8,082	0.0783	0.0785	0.0791	0.0787	0.0801
1,113,000	700,000	54	19	1.293	0.8741	0.9850	7,544	0.0839	0.0841	0.0848	0.0843	0.0857
1,033,500	650,000	54	7	1.246	0.8117	0.9170	7,019	0.0903	0.0906	0.0913	0.0908	0.0922
954,000	600,000	54	7	1.196	0.7493	0.8464	6,479	0.0979	0.0980	0.0985	0.0983	0.0997
874,500	550,000	54	7	1.146	0.6868	0.7759	5,940	0.107	0.107	0.108	0.107	0.109
795,000	500,000	26	7	1.108	0.6244	0.7261	5,770	0.117	0.117	0.117	0.117	0.117
715,500	450,000	54	7	1.036	0.5620	0.6348	4,859	0.131	0.131	0.133	0.131	0.133
636,000	400,000	54	7	0.977	0.4995	0.5642	4,319	0.147	0.147	0.149	0.147	0.149
556,500	350,000	26	7	0.927	0.4371	0.5083	4,039	0.168	0.168	0.168	0.168	0.168
477,000	300,000	26	7	0.858	0.3746	0.4357	3,462	0.196	0.196	0.196	0.196	0.196
397,500	250,000	26	7	0.783	0.3122	0.3630	2,885	0.235	0.235	0.235	0.235	0.235
336,400	0000	26	7	0.721	0.2642	0.3073	2,442	0.278	0.278	0.278	0.278	0.278
266,800	000	26	7	0.642	0.2095	0.2367	1,936	0.350	0.350	0.350	0.350	0.350
0000	00	6	1	0.563	0.1662	0.1939	1,542	0.441	0.443	0.446	0.447	0.464
000	0	6	1	0.502	0.1318	0.1537	1,223	0.556	0.557	0.561	0.562	0.579
00	1	6	1	0.447	0.1045	0.1219	970	0.702	0.703	0.707	0.706	0.718
0	2	6	1	0.398	0.0829	0.0967	769	0.885	0.885	0.889	0.887	0.893

SOURCE: Aluminum Co. of America.

are of corrosion-resistant steel, and the conductors are 61-strand cables of aluminum alloy, rather than the usual aluminum cable with a steel core (ACSR). The conductor diameter is 1.65 in with two ''bundled'' conductors per phase and 18-in spacing. The standard span is 1,600 ft, the conductor spacing is flat with 30-ft spacing between phase-conductor centers, and the minimum clearance to ground is 34 to 39 ft. To maintain a minimum clearance of 11 ft to the towers and 30 ft spacing between phases, vee insulator strings, each consisting of twenty-four 10-in disks, are used with each phase. The highest EHV system in North American is the 765-kV system in the midwest region of the United States. A dc transmission line on the west coast of the United States operating at ± 450 kV is transmitting power in bulk more than 800 miles.

High-voltage dc transmission has a greater potential for savings and a greater ability to transmit large blocks of power longer distances than has three-phase transmission. For the same crest voltage there is a saving of 50 percent in the weight of the conductor. Because of the power stability limit due to inductive and capacitive effects (inherent with ac transmission), the ability to transmit large blocks of power long distances has not kept pace with power developments, even at the present highest ac transmission voltage of 765 kV. With direct current there is no such power stability limit.

Where cables are necessary, as under water, the capacitive charging current may, with alternating current become so large that it absorbs a large proportion, if not all, of the cable-carrying capability. For example, at 132 kV, three-phase (76 kV to ground), with a 500 MCM cable, at 36 mi, the charging current at 60 Hz is equal to the entire cable capability so that no capability remains for the load current. With direct current there is no charging current, only the negligible leakage current, and there are no ac dielectric losses. Furthermore, the dc voltage at which a given cable can operate is twice the ac voltage.

The high dc transmission voltage is obtained by converting the ac power voltage to direct current by means of mercury-arc rectifiers; at the receiving end of the line the dc voltage is inverted back to a power-frequency voltage by means of mercury-arc inverters. The maximum dc transmission voltage in use in the United States, as of 1994, is 500 kV.

Alternating direct to alternating current is nonsynchronous transmission of electric power. It can be overhead or under the surface. In the early 1970s the problem of bulk electric-power transmission over high-voltage transmission lines above the surface developed the insistent discussion of land use and environmental cost.

While the maximum ac transmission voltage in use in the United States (1994) is 765 kV, ultrahigh voltage (UHV) is under consideration. Transmission voltages of 1,200 to 2,550 kV are being tested presently. The right-of-way requirements for power transmission are significantly reduced at higher voltages. For example, in one study the transmission of 7,500 MVA at 345 kV ac was found to require 14 circuits on a corridor 725 ft (221.5 m) wide, whereas a single 1,200-kV ac circuit of 7,500-MVA capacity would need a corridor 310 ft (91.5 m) wide.

Continuing research and development efforts may result in the development of more economic high-voltage underground transmission links. Sufficient bulk power transmission capability would permit **power wheeling**, i.e., the use of generating capacity to the east and west to serve a given locality as the earth revolves and the area of peak demand glides across the countryside.

Corona is a reddish-blue electrical discharge which occurs when the voltage-gradient in air exceeds 30 kV peak, 21.1 kV rms, at 76 cm pressure. This electrical discharge is caused by ionization of the air and becomes more or less concentrated at irregularities on the conductor surface and on the outer strands of stranded conductors. Corona is accompanied by a hissing sound; it produces ozone and, in the presence of moisture, nitrous acid. On high-voltage lines corona produces a substantial power loss, corrosion of the conductors, and radio and television interference. The fair-weather loss increases as the square of the voltage above a critical value e_0 and is greatly increased by fog, smoke, rainstorms, sleet, and snow (see Fig. 15.1.86). To reduce corona, the diameter of high-voltage conductors is increased to values much greater than

would be required for the necessary conductance cross section. This is accomplished by the use of hollow, segmented conductors and by the use of aluminum cable, steel-reinforced (ACSR), which often has inner layers of jute to increase the diameter. In extra-high-voltage lines (400 kV and greater), corona is reduced by the use of bundled conductors in which each phase consists of two or three conductors spaced about 16 in (0.41 m) from one another.

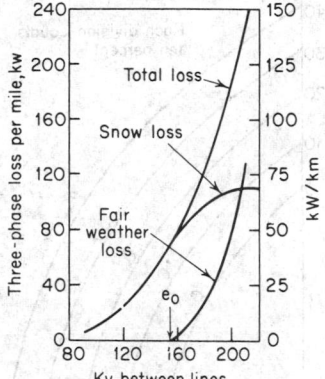

Fig. 15.1.86 Corona loss with snowstorm.

Underground Power Cables

Insulations for power cables include heat-resisting, low-water-absorptive synthetic rubber compounds, varnished cloth, impregnated paper, cross-linked polyethylene thermosetting compounds, and thermoplastics such as polyvinyl chloride (PVC) and polyethylene (PE) compounds (see Sec. 6).

Properly chosen **rubber-insulated cables** may be used in wet locations with a nonmetallic jacket for protective covering instead of a metallic sheath. Commonly used jackets are flame-resisting, such as neoprene and PVC. Such cables are relatively light in weight, easy to train in ducts and manholes, and easily spliced. When distribution voltages exceed 2,000 V phase to phase, an ozone-resisting type of compound is required. Such rubber insulation may be used in cables carrying up to 28,000 V between lines in three-phase grounded systems. The insulation wall will be thicker than with varnished cloth, polyethylene, ethylene propylene rubber, or paper.

Varnished-cloth cables are made by applying varnish-treated closely woven cloth in the form of tapes, helically, to the metallic conductor. Simultaneously a viscous compound is applied between layers which fills in any voids at laps in the taping and imparts flexibility when the cable is bent by permitting movement of one tape upon another. This type of insulation has higher dielectric loss than impregnated paper but is suitable for the transmission of power up to 28,000 V between phases over short distances. Such insulated cables may be used in dry locations with flame-resisting fibrous braid, reinforced neoprene tape, or PVC jacket and are often further protected with an interlocked metallic tape armor; but in wet locations these cables should be protected by a continuous metallic sheath such as lead or aluminum. Since varnish-cloth-insulated cable has high ozone resistance, heat resistance, and impulse strength, it is well adapted for station or powerhouse wiring or for any service where the temperature is high or where there are sudden increases in voltage for short periods. Since the varnish is not affected by mineral oils, such cables make excellent leads for transformers and oil switches.

PVC is readily available in several fast, bright colors and is often chosen for color-coded multiconductor control cables. It has inherent flame and oil resistance, and as single conductor wire and cable with the proper wall thickness for a particular application, it usually does not need any outside protective covering. On account of its high dielectric constant and high power factor, its use is limited to low voltages, i.e., under 1,000 V, except for series lighting circuits.

Table 15.1.18 Ampacities of Insulated Cables in Underground Raceways
[Based on conductor temperature of 90°C, ambient earth temperature of 20°C, 100 percent load factor, thermal resistance (RHO) of 90, and three circuits in group]

Conductor size, AWG or MCM	Copper				Aluminum			
	3-1/C cables per raceway		1-3/C cable per raceway		3-1/C cables per raceway		1-3/C cable per raceway	
	2,001–5,000-V ampacity	5,001–35,000-V ampacity	2,001–5,000-V ampacity	5,001–35,000-V ampacity	2,001–5,000-V ampacity	5,001–35,000-V ampacity	2,001–5,000-V ampacity	5,001–35,000-V ampacity
8	56		53		44		41	
6	73	77	69	75	57	60	54	59
4	95	99	89	97	74	77	70	75
2	125	130	115	125	96	100	90	100
1	140	145	135	140	110	110	105	110
1/0	160	165	150	160	125	125	120	125
2/0	185	185	170	185	145	145	135	140
3/0	210	210	195	205	160	165	155	160
4/0	235	240	225	230	185	185	175	180
250	260	260	245	255	205	200	190	200
350	315	310	295	305	245	245	230	240
500	375	370	355	360	295	290	280	285
750	460	440	430	430	370	355	345	350
1,000	525	495	485	485	425	405	400	400

NOTE: This is a general table. For other temperatures and installation conditions, see NFPA 70, 1993, National Electrical Code, Tables 77 through 80 and associated notes.

Polyethylene, because of its excellent electrical characteristics, first found use when it was adapted especially for high-frequency cables used in radio and radar circuits; for certain telephone, communication, and signal cables; and for submarine cables. Submarine telephone cables with built-in repeaters laid first in the Atlantic Ocean and then in the Pacific are insulated with polyethylene. Because of polyethylene's thermal characteristics, the standard maximum conductor operating temperature is 75°C. It is commonly used for power cables (including large use for underground residential distribution), with transmissions up to 15,000 V. Successful installations have been in service at 46 kV and some at 69 kV. The upper limit has not been reached, inasmuch as work is in progress on higher-voltage polyethylene power cables as a result of advancements in the art of compounding.

Cross-linked polyethylene is another insulation which is gaining in favor in the process field. For power cable insulations, the cross-linking process is most commonly obtained chemically. It converts polyethylene from a thermoplastic into a thermosetting material; the result is a compound with a unique combination of properties, including resistance to heat and oxidation, thus permitting an increase in maximum conductor operating temperature to 90°C. The service record with this compound has been good at voltages which have been gradually increased to 35 kV. Another thermosetting material, ethylene propylene rubber (EPR), has found wide acceptance in the 5- to 35-kV range.

Impregnated-paper insulation is used for very-high-voltage cables whose range has been extended to 345 kV. To eliminate the detrimental effects of moisture and to maintain proper impregnation of the paper, such cables must have a continuous metallic sheath such as lead or aluminum or be enclosed within a steel pipe; the operation of the cable depends absolutely on the integrity of that enclosure. In three-conductor belted-type cables the individual insulated conductors are surrounded by a belt or wall of impregnated paper over which the lead sheath is applied. When all three conductors are within one sheath, their inductive effects practically neutralize one another and eddy-current loss in the sheath is negligible. In the type-H cable, each of the individual conductors is surrounded with a perforated metallic covering, either aluminum foil backed with a paper tape or thin perforated metal tapes wound over the paper. All three conductors are then enclosed within the metal sheath. The metallic coverings being grounded electrically, each conductor acts as a single-conductor cable. This construction eliminates "tangential" stresses within the insulation and reduces pockets or voids. When paper tapes are wound on the conductor, impregnated with an oil or a petrolatum compound, and covered with a lead sheath, they are called **solid type.**

Three-conductor cables are now operating at 33,000 V, and single-conductor cables at 66,000 V between phases (38,000 V to ground). In New York and Chicago, special hollow-conductor oil-filled single-conductor cables are operating successfully at 132,000 V (76,000 V to ground). In France, cables are operating at 345 kV between conductors.

Other methods of installing underground cables are to draw them into steel pipes, usually without the sheaths, and to fill the pipes with oil under pressure (**oilstatic**) or **nitrogen** under 200 lb pressure. The ordinary medium-high-voltage underground cables are usually drawn into duct lines. With a straight run and ample clearance the length of cable between manholes may reach 600 to 1,000 ft. Ordinarily, the distance is more nearly 400 to 500 ft. With bends of small radius the distance must be further reduced.

Cable ratings are based on the permissible operating temperatures of the insulation and environmental installation conditions. See Table 15.1.18 for ampacities.

POWER DISTRIBUTION

Distribution Systems The choice of the system of power distribution is determined by the type of power that is available and by the nature of the load. To transmit a given power over a given distance with a given power loss (I^2R), the weight of conductor varies inversely as the square of the voltage. Incandescent lamps will not operate economically at voltages much higher than 120 V; the most suitable voltages for dc motors are 230, 500, and 550 V; for ac motors, standard voltages are 230, 460, and 575 V, three-phase. When power for lighting is to be distributed in a district where the consumers are relatively far apart, alternating current is used, being distributed at high voltage (2,400, 4,160, 4,800, 6,900, and 13,800 V) and transformed at the consumer's premises, or by transformers on poles or located in manholes or vaults under the street or sidewalks, to 240/120 V three-wire for lighting and domestic customers, and to 208, 240, 480, and 600 volts, three-phase, for power.

The first central station power systems were built with dc generation and distribution. The economical transmission distance was short. Densely populated, downtown areas of cities were therefore the first sections to be served. Growth of electric service in the United States was phenomenal in the last two decades of the nineteenth century. After 1895 when ac generation was selected for the development of power from Niagara Falls, the expansion of dc distribution diminished. The economics overwhelmingly favored the new ac system.

Direct current service is still available in small pockets in some cities. In those cases, ac power is generated, transmitted, and distributed. The conversion to dc takes place in rectifiers installed in manholes near the load or in the building to be served. Some dc customers resist the change to ac service because of their need for motor speed control. Elevator and printing press drives and some cloth-cutting knives are examples of such needs. (See Low-Voltage AC Network.)

Series Circuits These **constant-current** circuits were widely used for **street lighting**. The voltage was automatically adjusted to match the number of lamps in series and maintain a constant current. With the advent of **HID lamps** and individual **photocells** on street lighting fixtures which require parallel circuitry, this system fell into disuse and is rarely seen.

Parallel Circuits **Power** is usually distributed at **constant potential**, and all the devices or receivers in the circuit are connected in parallel, giving a constant-potential system, Fig. 15.1.87a. If conductors of constant cross section are used and all the loads, L_1, L_2, etc., are operating, there will be a greater voltage IR drop per unit length of wire in the portion of the circuit AB and CD than in the other portions; also the voltage will not be the same for the different lamps but will decrease along the mains with distance from the generating end.

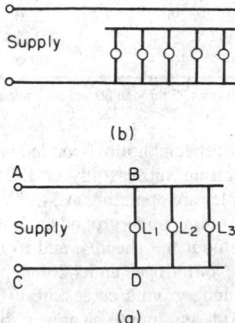

Fig. 15.1.87 (a) Parallel circuit; (b) loop circuit.

Loop Circuits A more nearly equal voltage for each load is obtained in the loop system, Fig. 15.1.87b. The electrical distance from one generator terminal to the other through any receiver is the same as that through any other receiver, and the voltage at the receivers may be maintained more nearly equal, but at the expense of additional conductor material.

Series-Parallel Circuit For incandescent lamps the power must be at low voltage (115 V) and the voltage variations must be small. If the transmission distance is considerable or the loads are large, a large or perhaps prohibitive investment in conductor material would be necessary. In some special cases, lamps may be operated in groups of two in series as shown in Fig. 15.1.88. The transmitting voltage is thus doubled, and, for a given number of lamps, the current is halved, the permissible voltage drop (IR) in conductors doubled, the conductor resistance quadrupled, the weight of conductor material thus being reduced to 25 percent of that necessary for simple parallel operation.

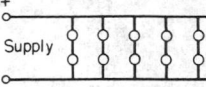

Fig. 15.1.88 Series-parallel system.

Three-Wire System In the series-parallel system the loads must be used in pairs and both units of the pair must have the same power rating. To overcome these objections and at the same time to obtain the economy in conductor material of operating at higher voltage, the three-wire system is used. It consists merely of adding a third wire or neutral to the system of Fig. 15.1.88 as shown in Fig. 15.1.89.

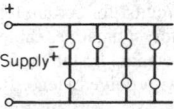

Fig. 15.1.89 Three-wire system.

If the neutral wire is of the same cross section as the two outer wires, this system requires only 37.5 percent of the copper required by an equivalent two-wire system. Since the neutral ordinarily carries less current than the outers, it is usually smaller and the ratio of copper to that of the two-wire system is even less than 37.5 percent (see Table 15.1.19).

When the loads on each half of the system are equal, there will be no

Table 15.1.19 Resistance and 60-Hz Reactance for Wires with Small Spacings, Ω, at 20°C
(See also Table 15.1.16.)

AWG and size of wire, cir mils	Resistance in 1,000 ft of line (2,000 ft of wire), copper	Reactance in 1,000 ft of line (2,000 ft of wire) at 60 Hz for the distance given in inches between centers of conductors										
		½	1	2	3	4	5	6	9	12	18	24
14– 4,107	5.06	0.138	0.178	0.218	0.220	0.233	0.244	0.252	0.271	0.284	0.302	
12– 6,530	3.18	0.127	0.159	0.190	0.210	0.223	0.233	0.241	0.260	0.273	0.292	
10– 10,380	2.00	0.116	0.148	0.180	0.199	0.212	0.223	0.231	0.249	0.262	0.281	
8– 16,510	1.26	0.106	0.138	0.169	0.188	0.201	0.212	0.220	0.238	0.252	0.270	0.284
6– 26,250	0.790	0.095	0.127	0.158	0.178	0.190	0.201	0.209	0.228	0.241	0.260	0.272
4– 41,740	0.498	0.085	0.117	0.149	0.167	0.180	0.190	0.199	0.217	0.230	0.249	0.262
2– 66,370	0.312	0.074	0.106	0.138	0.156	0.169	0.180	0.188	0.206	0.220	0.238	0.252
1– 83,690	0.248	0.068	0.101	0.132	0.151	0.164	0.174	0.183	0.201	0.214	0.233	0.246
0– 105,500	0.196	0.063	0.095	0.127	0.145	0.159	0.169	0.177	0.196	0.209	0.228	0.241
00– 133,100	0.156	0.057	0.090	0.121	0.140	0.153	0.164	0.172	0.190	0.204	0.222	0.236
000– 167,800	0.122	0.052	0.085	0.116	0.135	0.148	0.158	0.167	0.185	0.199	0.217	0.230
0000– 211,600	0.098	0.046	0.079	0.111	0.130	0.143	0.153	0.161	0.180	0.193	0.212	0.225
250,000	0.085	—	0.075	0.106	0.125	0.139	0.148	0.157	0.175	0.189	0.207	0.220
300,000	0.075	—	0.071	0.103	0.120	0.134	0.144	0.153	0.171	0.185	0.203	0.217
350,000	0.061	—	0.067	0.099	0.188	0.128	0.141	0.149	0.168	0.182	0.200	0.213
400,000	0.052	—	0.064	0.096	0.114	0.127	0.138	0.146	0.165	0.178	0.197	0.209
500,000	0.042	—	—	0.090	0.109	0.122	0.133	0.141	0.160	0.172	0.192	0.202
600,000	0.035	—	—	0.087	0.106	0.118	0.128	0.137	0.155	0.169	0.187	0.200
700,000	0.030	—	—	0.083	0.102	0.114	0.125	0.133	0.152	0.165	0.184	0.197
800,000	0.026	—	—	0.080	0.099	0.112	0.122	0.130	0.148	0.162	0.181	0.194
900,000	0.024	—	—	0.077	0.096	0.109	0.119	0.127	0.146	0.159	0.178	0.191
1,000,000	0.022	—	—	0.075	0.094	0.106	0.117	0.125	0.144	0.158	0.176	0.188

NOTE: For other frequencies the reactance will be in direct proportion to the frequency.

current in the middle or neutral wire, and the condition is the same as that shown in Fig. 15.1.88. When the loads on the two sides are unequal, there will be a current in the neutral wire equal to the difference of the currents in the outside wires.

AC Three-Wire Distribution   Practically all energy for lighting and small motor work is distributed at 1,150, 2,300, or 4,160 V ac to transformers which step down the voltage to 240 and 120 V for three-wire domestic and lighting systems as well as 208, 240, 480, and 600 volts, three-phase, for power. For the three-wire systems the transformers are so designed that the secondary or low-voltage winding will deliver power at 240 V, and the middle or neutral wire is obtained by connecting to the center or midpoint of this winding (see Figs. 15.1.90 and 15.1.91).

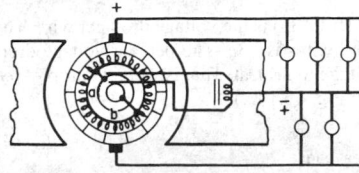

Fig. 15.1.90   Three-wire generator.

Disturbances of utility power have been present since the inception of the electric utility industry; however, in the past equipment was much more forgiving and less affected. Present electronic data processing equipment is microprocessor-based and consequently requires a higher quality of ac power. Although the quality of utility power in the United States is very good, sensitive electronic equipment still needs to be protected against the adverse and damaging effects of transients, surges, and other power-system aberrations.

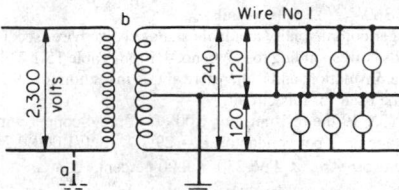

Fig. 15.1.91   Three-wire 230/115-V ac system.

This has led to the development of technologies which condition power for use by this sensitive equipment. Power-enhancing devices (surge suppressors, isolation transformers, and line voltage regulators) modify the incoming waveform to mitigate some aberrations. Power-synthesizing devices (motor-generator sets, magnetic synthesizers, and uninterruptible power supplies) use the incoming power as a source of energy to generate a new, completely isolated waveform. Each type has advantages and limitations.

Grounding   The neutral wire of the secondary circuit of the transformer should be grounded on the pole (or in the manhole) and at the service switch in the building supplied. If, as a result of a lightning stroke or a fault in the transformer insulation, the transformer primary circuit becomes grounded at *a* (Fig. 15.1.91) and the transformer insulation between primary and secondary windings is broken down at *b* and if there were no permanent ground connection in the secondary neutral wire, the potential of wire 1 would be raised 2,300 V above ground potential. This constitutes a very serious hazard to life for persons coming in contact with the 120 V system. The National Electrical Code requires the use of a ground wire not smaller than 8 AWG copper. With the neutral grounded (Fig. 15.1.91), voltages to ground on the secondary system cannot exceed 120 V. (See National Electrical Code 1991, Art. 250.)

Feeders and Mains   Where power is supplied to a large district, improved voltage regulation is obtained by having **centers of distribution**. Power is supplied from the station bus at high voltage to the centers of

distribution by large cables known as **feeders**. Power is distributed from the distribution centers to the consumers through the mains and transformed to a usable voltage at the user's site. As there are no loads connected to the feeders between the generating station and centers of distribution, the voltage at the latter points may be maintained constant. **Pilot wires** from the centers of distribution often run back to the station, allowing the operator or automatic controls to maintain constant voltage at the centers of distribution. This system provides a means of maintaining very close **voltage regulation** at the consumer's premises.

A common and economical method of supplying business and thickly settled districts with high load densities is to employ a 208/120-V, three-phase, four-wire **low-voltage ac network**. The network operates with 208 V between outer wires giving 120 V to neutral (Fig. 15.1.92). Motors are connected across the three outer wires operating at 208 V, three-phase. Lamp loads are connected between outer wires and the grounded neutral. The network is supplied directly from 13,800-V feeders by 13,800/208-V three-phase transformer units, usually located in manholes, vaults, or outdoor enclosures. This system thus eliminates the necessity for transformation in the substation. A large number of such units feed the network, so that the secondaries are all in parallel. Each transformer is provided with an overload reverse-energy circuit breaker (network protector), so that a feeder and its transformer are isolated if trouble develops in either. This system is flexible since units can be easily added or removed in accordance with the rapid changes in local loads that occur particularly in downtown business districts.

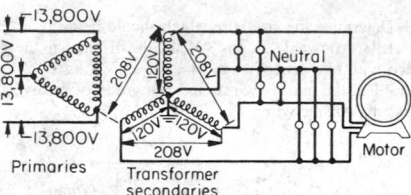

Fig. 15.1.92   208/120-V secondary network (single unit) showing voltages.

Voltage Drops   In ac distribution systems the voltage drop from transformer to consumer in lighting mains should not exceed 2 percent in first-class systems, so that the lamps along the mains can all operate at nearly the same voltage and the annoying flicker of lamps may not occur with the switching of appliances. This may require a much larger conductor than the most economical size. In transmission lines and in feeders where there are no intermediate loads and where means of regulating the voltage are provided, the drop is not limited to the low values that are necessary with mains and the matter of economy may be given consideration.

WIRING CALCULATIONS

These calculations can be used for dc, and for ac if the reactance can be neglected. The determination of the proper size of conductor is influenced by a number of factors. Except for short distances, the minimum size of conductor shown in Table 15.1.21, which is based on the maximum permissible current for each type of insulation, cannot be used; the size of conductor must be larger so that the voltage drop IR shall not be too great. With branch circuits supplying an incandescent-lamp load, this drop should not be more than a small percentage of the voltage between wires. The National Electrical Code 1993 requires that conductors for feeders, i.e., from the service equipment to the final branch circuit overcurrent device, be sized to prevent (1) a voltage drop of more than 3 percent at the farthest outlet of power, heating, and lighting loads or combinations thereof, and (2) a maximum voltage drop on combined feeders and branch circuits to the farthest outlet of more than 5 percent.

The resistance of 1 cir mil·ft of commercial copper may be taken as 10.8 Ω. The resistance of a copper conductor may be expressed as $R = 10.8 l/A$, where l = length, ft and A = area, cir mils. If the length is expressed in terms of the transmission distance d (since the two wires are usually run parallel), the voltage drop IR to the end of the circuit is

$$e = 21.6Id/A \qquad (15.1.128)$$

and the size of conductor in circular mils necessary to give the permissible voltage drop e is

$$A = 21.6Id/e \qquad (15.1.129)$$

If e is expressed as a percentage x of the voltage E between conductors, then

$$A = 2,160Id/xE \qquad (15.1.130)$$

EXAMPLE. Determine the size of conductor to supply power to a 10-hp, 220-V dc motor 500 ft from the switchboard with 5 V drop. Assume a motor efficiency of 86 percent. The motor will then require a current of $(10 \times 746)/(0.86 \times 220) = 39.4$ A. From Eq. (15.1.129), $A = 21.6 \times 39.4 \times 500/5 = 85,100$ cir mils. The next largest wire is no. 0 AWG.

The calculation of the size conductor for three-wire circuits is made in practically the same manner. With a balanced circuit there is no current in the neutral wire, and the current in each outside wire will be equal to one-half the sum of the currents taken by all the receiving devices connected between neutral and outside wires plus the sum of the currents taken by the receivers connected between the outside wires. Using this total current and neglecting the neutral wire, make calculations for the size of the outside wires by means of Eq. (15.1.129). The neutral wire should have the same cross section as the outside wires in interior wiring.

EXAMPLE. Determine the size wire which should be used for the three-wire main of Fig. 15.1.93. Allowable drop is 3 V and the distance to the load center 40 ft; circuit loaded with two groups of receivers each taking 60 A connected between the neutral and the outside wires, and one group of receivers taking 20 A connected across the outside wires.

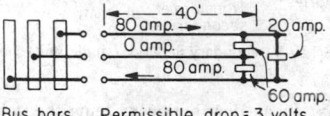

Bus bars   Permissible drop = 3 volts

Fig. 15.1.93   Three-wire 230/115-V main.

Solution: load $= (60 + 60)/2 + 20 = 80$ A. Substituting in Eq. (15.1.129), cir mils $= 21.6Id/e = 21.6 \times 80 \times 40/3 = 23,030$ cir mils.

From Tables 15.1.19 and 15.1.21, no. 6 wire, which has a cross section of 26,250 cir mils, is the next size larger. This size of wire would satisfy the voltage-drop requirements, but rubber-insulated no. 6 has a safe carrying capacity of but 55 A. The current in the circuit is 80 A. Therefore, rubber-insulated wire no. 3, which has a carrying capacity of 80 A, should be used. The neutral wire should be the same size as the outside wires.

See also examples in the National Electrical Code 1993.

Wiring calculations for ac circuits require some consideration of power factor, reactance, and skin effect. Skin effect becomes pronounced only when very large conductors are used for alternating current. For interior wiring, conductors larger than 700,000 cir mils should not be used, and many prefer not to use conductors larger than 300,000 cir mils. Should the required copper cross section exceed these values, a number of conductors may be operated in parallel.

For voltages under 5,000 the effect of line capacitance may be neglected. With ordinary single-phase interior wiring, where the effect of the line reactance may be neglected and where the power factor of the load (incandescent lamps) is nearly 100 percent, the calculations are made the same as for dc circuits. Three-wire ac circuits of ordinary length with incandescent lamp loads are also determined in the same manner. When the load is other than incandescent lamps, it is necessary to know the power factor of the load in order to make calculations. When the exact power factor cannot be accurately determined, the following approximate values may be used: incandescent lamps, 0.95 to 1.00; lamps and motors, 0.75 to 0.85; motors 0.5 to 0.80. Equation

(15.1.131) gives the value of current in a single-phase circuit. See also Table 15.1.25.

$$I = (P \times 1,000)/(E \times \text{pf}) \qquad (15.1.131)$$

where I = current, A; P = kW; E = load voltage; and pf = power factor of the load. The size of conductor is then determined by substituting this value of I in Eq. (15.1.129) or (15.1.130).

For three-phase three-wire ac circuits the current per wire

$$I = 1,000P/\sqrt{3}E\text{pf} = 580P/E\text{pf} \qquad (15.1.132)$$

Computations are usually made of voltage drop **per wire** (see Fig. 15.1.94). Hence, if reactance can be neglected, the conductor cross section in cir mils is one-half that given by Eq. (15.1.129). That is,

$$A = 10.8Id/e \qquad \text{cir mils} \qquad (15.1.133)$$

where e in Eq. (15.1.133) is the voltage drop per wire. The voltage drop between any two wires is $\sqrt{3}e$. The percent voltage drop should be in terms of the voltage to **neutral**. That is, percent drop $= [e/(E/\sqrt{3})]100 = [\sqrt{3}e/E]100$ (see Fig. 15.1.82).

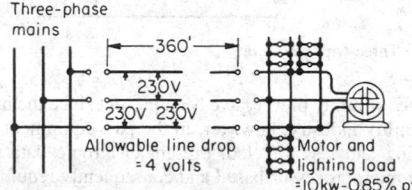

Fig. 15.1.94   Three-phase lamp and induction motor load.

EXAMPLE. In Fig. 15.1.94, load 10 kW; voltage of circuit 230; power factor 0.85; distance 360 ft; allowable drop per wire 4 V. Substituting in Eq. (15.1.132) $I = (580 \times 10)/(230 \times 0.85) = 29.7$ A. Substituting in Eq. (15.1.133), $A = 10.8 \times 29.7 \times 360/4 = 28,900$ cir mils.

The next larger commercially available standard-size wire (see Table 15.1.19) is 41,700 cir mils corresponding to AWG no. 4. From Table 15.1.21 this will carry 70 A with rubber insulation, and is therefore ample in section for 29.7 A. Three no. 4 wires would be used for this circuit.

From Table 15.1.19 the resistance of 1,000 ft of no. 4 copper wire is 0.249 Ω. Hence, the voltage drop per conductor, $e = 29.7 \times (360/1,000)0.249 = 2.66$ V. Percent voltage drop $= \sqrt{3} \times 2.66/230 = 2.00$ percent.

Where all the wires of a circuit, two wires for a single-phase circuit, four wires for a four-phase circuit (see Fig. 15.1.32 and Fig. 15.1.80c), and three wires for a three-phase circuit, are carried in the same conduit or where the wires are separated less than 1 in between centers, the effect of line (inductive) reactance may ordinarily be neglected. Where circuit conductors are large and widely separated from one another and the circuits are long, the inductive reactance may increase the voltage drop by a considerable amount over that due to resistance alone. Such problems are treated using IR and IX phasors. Line reactance decreases somewhat as the size of wire increases and decreases as the distance between wires decreases.

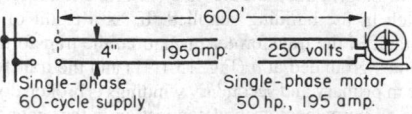

Fig. 15.1.95   Single-phase induction motor load on branch circuit.

EXAMPLE. Determine the size of wire necessary for the branch to the 50-hp, 60-Hz, 250-V single-phase induction motor of Fig. 15.1.95. The name-plate rating of the motor is 195 A, and its full-load power factor is 0.85. The wires are run open and separated 4 in; length of circuit, 600 ft. Assume the line drop must not exceed 7 percent, or $0.07 \times 250 = 17.5$ V. The point made by this example is emphasized by the assumption of an outsize motor.

Solution. To ascertain approximately the size of conductor, substitute in Eq. (15.1.129) giving cir mils $= 21.6 \times 195 \times 600/17.5 = 144,400$. Referring to

Table 15.1.19, the next larger size wire is no. 000 or 167,800 cir mils. This size would be ample if there were no line reactance. In order to allow for reactance drop, a larger conductor is selected and the corresponding voltage drop determined. Inasmuch as this is a motor branch, the code rules require that the carrying capacity be sufficient for a 25 percent overload. Therefore the conductor should be capable of carrying $195 \times 1.25 = 244$ A. From Table 15.1.21, a 350,000-cir mil conductor rubber-insulated cable would be required to carry 244 A. Resistance drop (see Table 15.1.19), $IR = 195 \times 0.061 \times 0.6 = 7.14$ V. $7.14/250 = 2.86$ percent. From Table 15.1.19, $X = 0.128 \times 0.6 = 0.0768$ Ω. $IX = 195 \times 0.768 = 14.98$ V. $14.98/250 = 5.99$ percent. Using the Mershon diagram (Fig. 15.1.84), follow the ordinate corresponding to power factor, 0.85, until it intersects the smallest circle. From this point, lay off horizontally the percentage resistance drop, 2.86. From this last point, lay off vertically the percentage reactance drop 5.99. This last point lies about on the 6.0 percent circle, showing that with 195 amp the difference between the sending-end and receiving-end voltages is $0.06 \times 250 = 15.0$ V, which is within the specified limits.

Also Eq. (15.1.127) may be used. $\cos \theta = 0.85$; $\sin \theta = 0.527$.

$$E_s = \{[(250 \times 0.85) + 7.14]^2 + [(250$$
$$\times 0.527) + 14.98]^2\}^{1/2} = 264.3 \text{ V}$$
$$264.3/250 = 105.7 \text{ percent}$$

In the calculation of three-phase three-wire circuits where line reactance must be considered, the method found above under Power Transmission may be used. The system is considered as being three single-phase systems having a ground return the resistance and inductance of which are zero, and the voltages are equal to the line voltages divided by $\sqrt{3}$. When the three conductors are spaced unequally, the value of GMD given in Fig. 15.1.85 should be used in Tables 15.1.16 and 15.1.19. (When the value of resistance or reactance per 1,000 ft of **conductor** is desired, the values in Table 15.1.19 should be divided by 2.)

The National Electrical Code of 1993 specifies that the size of conductors for branch circuits should be such that the voltage drop will not exceed 3 percent to the farthest outlet for power, heating, lighting, or combination thereof, requiring further that the total voltage drop for feeders and branch circuits should not exceed 5 percent overall. For examples of calculations for interior wiring, see National Electrical Code of 1993 (Chap. 9).

INTERIOR WIRING

Interior wiring requirements are based, for the most part, on the **National Electrical Code** (NEC), which has been adopted by the National Fire Protection Association, American National Standards Institute (ANSI), and the Occupational Safety and Health Act (OSHA).

The Occupational Safety and Health Act of 1970 (OSHA) made the National Electrical Code a national standard. Conformance with the NEC became a requirement in most commercial, industrial, agricultural, etc., establishments in the United States. Some localities may not accept NEC standards. In those cases, local rules must be followed.

NEC authority starts at the point where the connections are made to the conductor of the **service drop** (overhead) or **lateral** (underground) from the electricity supply system. The **service equipment** must have a rating not less than the load to be carried (computed according to NEC methods). Service equipment is defined as the necessary equipment, such as circuit breakers or fused switches and accompanying accessories. This equipment must be located near the point of entrance of supply conductors to a building or other structure or an otherwise defined area. Service equipment is intended to be the main control and means of cutoff of the supply.

Service-entrance conductors connect the electricity supply to the service equipment. Service-entrance conductors running along the exterior or entering a building or other structure may be installed (1) as separate conductors, (2) in approved cables, (3) as cable bus, or (4) enclosed in rigid conduit. Also, for voltages less than 600 V, the conductors may be installed in electrical metallic tubing, wireways, auxiliary gutters, or busways. Service-entrance cables which are exposed to physical damage from awnings, swinging signs, coal chutes, etc., must be of the protected type or be protected by conduit, electrical metallic tubing, etc. Service heads must be raintight. Thermoplastic or rubber insulation is

required in overhead services. A grounded conductor may be bare. If exposed to the weather or embedded in masonry, raceways must be raintight and arranged to drain. Underground service raceway or duct entering from an underground distribution system must be sealed with a suitable compound (spare ducts, also).

NEC rules permit **multiple service** to a building for various reasons, such as: (1) fire pumps, (2) emergency light and power, (3) multiple occupancy, (4) when the calculated load is greater than 3,000 A, (5) when the building extends over a large area, and (6) where different voltages, frequencies, number of phases, or classes of use are required.

Ordinary service drops (overhead) and lateral (underground) must be large enough to carry the load but not smaller than no. 8 copper or no. 6 aluminum. As an exception, for installations to supply only limited loads of a single branch circuit, such as small polyphase power, etc., service drops must not be smaller than no. 12 hard-drawn copper or equivalent, and service laterals must be not smaller than no. 12 copper or no. 10 aluminum.

The phrase **large enough to carry the load** requires elaboration. The various conductors of public-utility electric-supply systems are sized according to the calculations and decisions of the personnel of the specific public utility supplying the service drop or lateral. At the load end of the drop or lateral, the NEC rules apply, and from that point on into the consumer's premises, NEC rules are the governing authority. There is a discontinuity at this point in the calculation of combined load demand for electricity and allowable current (ampacity) of conductors, cables, etc. This discontinuity in calculations results from the fact that the utility company operates locally, whereas the NEC is a set of national standards and therefore cannot readily allow for regional differences in electrical coincident demand, ambient temperature, etc. The NEC's aim is the assurance of an electrically "safe" human environment. This will be fostered by following the NEC rules.

Service-entrance cables are conductor assemblies which bear the type codes **SE** (for overhead services) and **USE** (for underground services). Under specified conditions, these cables may also be used for interior feeder and branch-circuit wiring.

The service-entrance equipment must have the **capability of safely interrupting** the current resulting from a short circuit at its terminals. **Available short-circuit current** is the term given to the maximum current that the power system can deliver through a given circuit to any negligible-impedance short circuit applied at a given point. (This value can be in terms of symmetrical or asymmetrical, momentary or clearing current, as specified.)

In most instances, the available short-circuit current is limited by the impedance of the last transformer in the supply system. **Large power users,** however, must become aware of changes in the electricity supply system which, because of growth of system capacity or any other reason, would increase the short-circuit current available to their service-entrance equipment. If this current is too great, explosive failure can result.

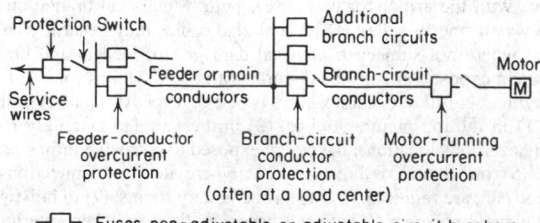

Fig. 15.1.96 Motor and wiring protection.

Kilowatthour and sometimes demand-metering equipment are connected to the service-entrance conductors. Proceeding toward the utilization equipment, the power-supply system fans out into feeders and branch circuits (see Fig. 15.1.96). Each of the feeders, i.e., a run of untapped conductor or cable, is connected to the supply through a switch and fuses or a circuit breaker. At a point, usually near that portion of the electrical loads which are to be supplied, a **panel box** or perhaps a **load-center assembly** of switching and/or control equipment is installed. From this panel box or load-center assembly, circuits radiate; i.e., circuits are installed to extend into the area being served to connect electrical machinery or devices or make available electric receptacles connected to the source of electric power.

Each feeder and each branch circuit will have its own **over-current protection** and **disconnect** means in the form of a fuse and switch combination or a circuit breaker.

There is a provision in the NEC 1993 rules for the following types of feeder and branch circuit wiring:

1. Open Wiring on Insulators (NEC 1993, Art. 320). This wiring method uses approved cleats, knobs, tubes, and flexible tubing for the protection and support of insulated conductors run on or in buildings and not concealed by the building structure. It is permitted only in industrial or agricultural establishments.

2. Concealed Knob-and-Tube Work (NEC 1993, Art. 324). Concealed knob-and-tube work may be used in the hollow spaces of walls and ceilings. It may be used only for the extension of existing facilities.

3. Flat Conductor Cable, Type FCC (NEC 1993, Art. 328). Type FCC cable may be installed under carpet squares. It may not be used outdoors or in wet locations, in corrosive or hazardous areas, or in residential, school, or hospital buildings.

4. Mineral-Insulated Metal-Sheathed Cable, Type MI (NEC 1993, Art. 330). Type MI cable contains one or more electrical conductors insulated with a highly compressed refractory mineral insulation and enclosed in a liquid- and gastight metallic tube sheath. Appropriate approved fittings must be used with it. It may be used for services, feeders, and branch circuits either exposed or concealed and dry or wet. It may be used in Class I, II, or III hazardous locations. It may be used for under-plaster extensions and embedded in plaster finish or brick or other masonry. It may be used where exposed to weather or continuous moisture, for underground runs and embedded in masonry, concrete or fill, in buildings in the course of construction or where exposed to oil, gasoline, or other conditions. If the environment would cause destruction of the sheath, it must be protected by suitable materials.

5. Power and Control Tray Cable (NEC 1993, Art. 340). Type TC cable is a factory assembly of two or more insulated conductors with or without associated bare or covered-grounding conductors under a nonmetallic sheath, approved for installation in cable trays, in raceways, or where supported by messenger wire.

6. Metal-Clad Cable, Type MC and AC Series (NEC 1993, Art. 333 and 334). These are metal-clad cables, i.e., an assembly of insulated conductors in a flexible metal enclosure. Type MC are power cables and in the range up to 600 V are made in conductor sizes of no. 4 and larger for copper and no. 2 and larger for aluminum. Type AC are branch and feeder cables with armor of flexible metal tape. All AC types except ACL have an internal bonding strip of copper or aluminum in intimate contact with the armor for its entire length. Metal-clad branch circuit cable was formerly called BX. Metal-clad cables may generally be installed where not subject to physical damage, for feeders and branch circuits in exposed or concealed work, with qualifications for wet locations, direct burial in concrete, etc. The use of Type AC cable is prohibited (1) in motion-picture studios, (2) in theaters and assembly halls, (3) in hazardous locations, (4) where exposed to corrosive fumes or vapors, (5) on cranes or hoists except where flexible connections to motors, etc., are required, (6) in storage-battery rooms, (7) in hoistways or on elevators except (i) between risers and limit switches, interlocks, operating buttons, and similar devices in hoistways and in escalators and moving walkways and (ii) short runs on elevator cars, where free from oil, and if securely fastened in place, or (8) in commercial garages in hazardous locations. Type ACL (lead-covered) shall not be used for direct burial in the earth.

7. Nonmetallic-Sheathed Cable, Types NM and NMC (NEC 1993, Art. 336). These are assemblies of two more insulated conductors (nos. 14 through 2 for copper, nos. 12 through 2 for aluminum) having an outer sheath of moisture-resistant, flame-retardant, nonmetallic ma-terial. In addition to the insulated conductors, the cable may have an approved size of uninsulated or bare conductor for grounding purposes only. The outer covering of NMC cable is flame retardant and corrosion resistant. The use of this type of cable, commonly called **Nomex,** is permitted in one- or two-family dwellings, multifamily dwellings, and other structures provided that such structures do not exceed three floors above grade.

8. Shielded Nonmetallic-Sheathed Cable, Type SNM (NEC 1993, Art. 337). Type SNM is a factory-assembled cable consisting of two or more insulated conductors (nos. 14 through 2 copper and nos. 12 through 2 aluminum) in an extruded core of moisture-resistant material, covered with an overlapping spiral metal tape and wire shield and jacketed with an extruded moisture-, flame-, corrosion-, fungus-, and sunlight-resistant material. This cable is to be used (1) under appropriate ambient-temperature conditions and (2) in continuous rigid-cable support or in raceways. It can be used in some hazardous locations as defined by the NEC.

9. Service Entrance Cable, Types SE and USE (NEC 1993, Art. 338). These cables, containing one or more individually insulated conductors, are primarily used for electric services. Type SE has a flame-retardant, moisture-resistant covering and is not required to have built-in protection against mechanical abuse. Type USE is recognized for use underground. It has a moisture-resistant covering, but not necessarily a flame-retardant one. Like the SE cable, USE cable is not required to have inherent protection against mechanical abuse. Under specified conditions, SE and USE cables can be used for feeders and branch circuits.

10. Underground Feeder and Branch-Circuit Cable, Type UF (NEC 1993, Art. 339). This cable is made in sizes 14 through 4/0, and the insulated conductors are Types TW, RHW, and others approved for the purpose. As in the NM cable, the UF type may contain an approved size of uninsulated or bare conductor for grounding purposes only. The outer jacket of this cable shall be flame-retardant, moisture-resistant, fungus-resistant, corrosive-resistant, and suitable for direct burial in the ground.

11. Other Installation Practices. The NEC details rules for nonmetallic circuit extensions and underplaster extensions. It also provides detailed rules for installation of electrical wiring in (a) rigid metal conduit (which may be used for all atmospheric conditions and locations with due regard to corrosion protection and choice of fittings), (b) rigid nonmetallic conduit (which is essentially corrosion-proof), in electrical metallic tubing (which is lighter-weight than rigid metal conduit), (c) flexible metal conduit, (d) liquidtight flexible metal conduit, (e) surface raceways, (f) underfloor raceways, (g) multioutlet assemblies, (h) cellular metal floor raceways, (i) structural raceways, (j) cellular concrete floor raceways, (k) wireways (sheet-metal troughs with hinged or removable covers), (l) flat, Type FC, cable assemblies installed in a surface metal raceway (Type FC cable contains three or four no. 10 special stranded copper wires), (m) busways, and (n) cable-bus. Busways and cable-bus installations are permitted for exposed work only.

In all installation work, only approved outlets, switch and junction boxes, fittings, terminal strips, and dead-end caps shall be used, and they are to be used in an approved fashion (see NEC 1993, Art. 370).

Table 15.1.20   Wire Table for Standard Annealed Copper at 20°C in SI Units

AWG size	Diameter, mm	kgf/km	m/Ω	Area, mm²
14	1.628	18.50	120.7	2.08
12	2.053	29.42	191.9	3.31
10	2.588	46.77	305.1	5.261
8	3.264	74.37	485.2	8.367
6	4.115	118.2	771.5	13.30
4	5.189	188.0	1227	21.15
2	6.544	299.0	1951	33.62
1	7.348	377.0	2460	42.41
0	8.252	475.4	3102	53.49
00	9.266	599.5	3911	67.43
000	10.40	755.9	4932	85.01
0000	11.68	935.2	6219	107.2

Table 15.1.20 relates AWG wire sizes to metric units. Table 15.1.21 lists the allowable ampacities of copper and aluminum conductors. Table 15.1.22 lists various conductor insulation systems approved by the 1993 NEC for conductors used in interior wiring. Dimensions and allowable fill of conduit and tubing are listed in Table 15.1.23. Table 15.1.24 lists the cross-sectional area of various insulated conductors. For installations not covered by the tables presented here, review the 1993 NEC. Estimated full-load currents of motors can be taken from Table 15.1.25.

Switching Arrangements Small quick-break switches must be set in or on a metal box or fitting and may be of the push, tumbler, or rotary type. The following types of switches are used to control lighting circuits: (1) single-pole, (2) double-pole, (3) three-point or three-way, (4) four-way, in combination with three-way switches to control lights from three or more stations, (5) electrolier.

In all metallic protecting systems, such as conduit, armored cable, or metal raceways, joints and splices in conductors must be made only in junction boxes or other proper fittings; therefore, these fittings can be located only in accessible places and never concealed in partitions. Splices or joints in the wire must never be in the conduit piping, raceway, or metallic tubing itself, for the splices may become a source of trouble as a result of corrosion or grounding if water should enter the conduit.

All conductors of an ac system must be placed in the same metallic casing so that their resultant magnetic field is nearly zero. If this is not done, eddy currents are set up causing heating and excessive loss. With single conductors in a casing, an excessive reactance drop may result.

Service wires are the conductors that bring the electric power into a building and should enter the building as near as possible to the service switch, so that when the switch is open all the electrical conductors and equipment inside the building will be dead. The service wires must be rubber- or thermoplastic-covered from the point of support on the outside of the building to the service switch or cutout and must be no. 6 wire or larger except for installations consisting of two-wire branch circuits where no. 8 wire may be permitted. A minimum of 100-A three-wire service is recommended for all single-family residences.

Table 15.1.21 Allowable Ampacities of Insulated Conductors
(Not more than three conductors in conduit. Based on ambient air temperature of 40°C.)

	Temperature rating of insulation, °C					
	Copper			Aluminum		
	60	75	90	60	75	90
Conductor size: AWG or MCM	Types TW, UF	Types RH, RHW, THW, THWN, XHHW, USE, ZW	Types SA, AVB, FEP, FEPB, THHN, RHH, XHHW*	Types TW, UF	Types RH, RHW THW, THWN, XHHW, USE	Types SA, AVB, THHN, RHH, XHHW*
14	18†	22†	25†	—	—	—
12	23†	28†	32†	18†	22†	26†
10	29†	37†	42†	23†	29†	34†
8	36	48	55	28	37	43
6	50	64	75	37	50	58
4	65	83	97	50	65	76
3	76	98	114	59	76	89
2	87	112	130	68	87	102
1	104	134	156	81	104	122
0	119	153	179	93	119	139
00	135	175	204	106	137	159
000	160	207	242	125	162	189
0000	184	238	278	144	186	217
250	210	271	317	165	213	249
300	232	300	351	183	236	276
350	254	328	384	201	259	303
400	274	354	415	218	281	329
500	314	407	477	252	326	381
600	345	448	525	280	362	424
700	376	489	574	308	399	467
750	392	509	598	322	417	488
800	403	524	616	334	432	506
900	426	555	653	357	463	542
1,000	499	585	689	380	493	578
Ambient temp., °C	For ambient temperatures other than 40°C, multiply the ampacities shown above by the appropriate factors shown below.					
21–35	1.32	1.20	1.14	1.32	1.20	1.14
26–30	1.22	1.13	1.10	1.22	1.13	1.10
31–35	1.12	1.07	1.05	1.12	1.07	1.05
36–40	1.00	1.00	1.00	1.00	1.00	1.00
41–45	0.87	0.93	0.95	0.87	0.93	0.95
46–50	0.71	0.85	0.89	0.71	0.85	0.89
51–55	0.50	0.76	0.84	0.50	0.76	0.84
56–60	—	0.65	0.77	—	0.65	0.77
61–70	—	0.38	0.63	—	0.38	0.63
71–80	—	—	0.45	—	—	0.45

* For dry locations only, rated 75°C for wet locations.

† Overcurrent protection shall not exceed 15 A for no. 14 copper and no. 12 aluminum, 20 A for no. 12 copper, 25 A for no. 10 aluminum, and 30 A for no. 10 copper.

NOTE: This is a general table. For other installation conditions, see NFPA 70-1994 National Electric Code®, Article 310.

SOURCE: NEC® 1993, Table 310-16. Reprinted with permission from NFPA 70-1984, National Electrical Code®, Copyright © 1992, National Fire Protection Association, Quincy, Massachusetts 02269. This reprinted material is not the complete and official position of the NFPA on the referenced subject, which is represented only by the standard in its entirety.

Table 15.1.22   Conductor Type and Application

Type letter	Insulation, trade name (see NEC 1984, Art. 310 for complete information)	Environment	Outer covering[a]
	Max operating temperature = 60°C (140°F)		
TW	Moisture-resistant thermoplastic	Dry and wet	None
TF	Thermoplastic-covered, solid or 7-strand	[c,d]	None
TFF	Thermoplastic-covered, flexible stranding	[c,d]	None
MTW	Moisture-, heat-, and oil-resistant thermoplastic machine-tool wiring (NFPA Stand. 79, NEC 1975, Art. 670)	Wet	None or nylon
UF	Moisture-resistant, underground feeder	Dry and wet	None
	Max operating temperature = 75°C (167°F)		
RH	Heat-resistant rubber	Dry	1,2
RHW	Moisture- and heat-resistant rubber[e]	Dry and wet	1,2
THW	Moisture- and heat-resistant thermoplastic	Dry and wet	None
THWN	Moisture- and heat-resistant thermoplastic	Dry and wet	Nylon
XHHW	Moisture- and heat-resistant cross-linked polymer	Wet	None
RFH-1 and 2	Heat-resistant rubber-covered solid or 7-strand	[b-d]	None
FFH-1 and 2	Heat-resistant rubber-covered flexible stranding	[b-d]	None
UF	Moisture-resistant and heat-resistant	Dry and wet	None
USE	Heat- and moisture-resistant	Dry and wet	4
ZW	Modified ethylene tetrafluoroethylene	Wet	None
	Max operating temperature = 85°C (185°F)		
MI	Mineral-insulated (metal-sheathed)	Dry and wet	Copper
	Max operating temperature = 90°C (194°F)		
RHH	Heat-resistant rubber	Dry	1,2
THHN	Heat-resistant thermoplastic	Dry	Nylon
THW	Moisture- and heat-resistant thermoplastic	[f]	None
XHHW	Moisture- and heat-resistant cross-linked synthetic polymer	Dry	None
FEP	Fluorinated ethylene propylene	Dry	None
FEPB	Fluorinated ethylene propylene	Dry	3
TFN	Heat-resistant thermoplastic covered, solid or 7-strand	[c,d]	Nylon
TFFN	Heat-resistant thermplastic flexible stranding		Nylon
MTW	Moisture-, heat-, and oil-resistant thermoplastic machine-tool wiring (NFPA Stand. 79, NEC 1975, Art. 670)	Dry	None or nylon
SA	Silicone asbestos	Dry	Asbestos or glass
	Max operating temperature = 150°C (302°F)		
Z, ZW	Modified ethylene tetrafluoroethylene	Dry	None
	Max operating temperature = 200°C (392°F)		
FEP, FEPB	Fluorinated ethylene propylene	Dry	None
	Special applications		3
PF, PGF	Fluorinated ethylene propylene	[c,d]	None or glass braid
PFA	Perfluoroalkoxy	Dry	None
SF-2	Silicone rubber, solid or 7-strand	[c,d]	Nonmetallic
	Max operating temperature = 250°C (482°F)		
MI	Mineral-insulated (metal-sheathed), for special applications	Dry and wet	Copper
TFE	Extruded polytetrafluoroethylene, only for leads within apparatus or within raceways connected to apparatus, or as open wiring (silver or nickel-coated copper only)	Dry	None
PFAH	Perfluoroalkoxy (special application)	Dry	None
PTF	Extruded polytetrafluoroethylene, solid or 7-strand (silver or nickel-coated copper only)	[c,d]	None

[a] 1: Moisture-resistant, flame-retardant nonmetallic; 2: outer covering not required when rubber insulation has been specifically approved for the purpose; 3: no. 14-8 glass braid, no. 6-2 asbestos braid; 4: moisture-resistant nonmetallic.
[b] Limited to 300 V.
[c] No. 18 and no. 16 conductor for remote controls, low-energy power, low-voltage power, and signal circuits; NEC 1975 Sec. 725-16, Sec. 760-16.
[d] Fixture wire no. 18-16.
[e] For over 2,000 V, the insulation shall be ozone-resistant.
[f] Special applications within electric discharge lighting equipment. Limited to 1,000 V open-circuit volts or less.
SOURCE: NEC 1993 Tables 310-13 and 402-3. Reprinted with permission from NFPA 70-1993, National Electrical Code®, Copyright © 1992, National Fire Protection Association, Quincy, Massachusetts 02269. This reprinted material is not the complete and official position of the NFPA on the referenced subject, which is represented only by the standard in its entirety.

Table 15.1.23 Dimensions and Allowable Fill of Conduit and Tubing

Trade size, in	ID, in	Area, in²	Allowable fill, in² of conductors (not lead-covered)		
			One conductor, 53% fill	Two conductors, 31% fill	Over two conductors,* 40% fill
½	0.622	0.30	0.16	0.09	0.12
¾	0.824	0.53	0.28	0.16	0.21
1	1.049	0.86	0.46	0.27	0.34
1¼	1.380	1.50	0.80	0.47	0.60
1½	1.610	2.04	1.08	0.63	0.82
2	2.067	3.36	1.78	1.04	1.34
2½	2.469	4.79	2.54	1.43	1.92
3	3.068	7.38	3.91	2.29	2.95
3½	3.548	9.90	5.25	3.07	3.96
4	4.026	12.72	6.74	3.94	5.09
4½	4.506	15.94	8.45	4.94	6.38
5	5.047	20.00	10.60	6.20	8.00
6	6.065	28.89	15.31	8.96	11.56

* For conductor derating with more than three conductors see NEC 1993, Art. 310.
SOURCE: NEC 1993, p. 70-824. Reprinted with permission from NFPA 70-1993, National Electrical Code®, Copyright © 1992, National Fire Protection Association, Quincy, Massachusetts 02269. This reprinted material is not the complete and official position of the NFPA on the referenced subject, which is represented only by the standard in its entirety.

Generally, when the conductors from overhead lines enter a building, the wires are encased in rigid conduit equipped with a weather cap or a service entrance cable (type ASE armored or SE type, unarmored) may be attached directly to the building wall. The inner end of the service enters a metal service cabinet in which the service fuses and switch are located. Service conductors may also terminate at an air-break or oil-immersed switch in a metal case or on a panel board which is accessible to qualified persons only.

All underground service wires must be connected to the interior wiring through a blade of the service switch or circuit breaker and be fused or automatically interrupted at the service switch. A service switch controlling a three-wire dc, a single-phase or four-wire three-phase system having a grounded neutral wire does not need to open that conductor.

The single-line diagram, Fig. 15.1.96, indicates a simplified interior arrangement of circuits and the necessary protection of the conductors and terminal load. Where a reduction is made in the wire size a protective device shall be installed to limit the conductor current to a safe value. Large motors and other terminal loads should also have overcurrent protection.

The maximum permissible fuse ratings and the setting of the protective devices for starting and for running protection of motors are given in Table 15.1.26.

Grounding of direct- and alternating-current systems of 300 V and

---

Table 15.1.24 Nominal Cross-Sectional Area of Rubber-Covered and Plastic-Covered Conductors

Type	Size, AWG					
	18	16	14	12	10	8
	Cross-sectional area, in²					
RFH-2	0.0167	0.0196	(fixture wire)			
SF-2	0.0167	0.0196	0.0230	(fixture wire)		
RH	—	—	0.0230	0.0278	0.0460	0.0854
RHH, RHW	—	—	0.0327	0.0384	0.0460	0.0854
RHH, RHW (without outer covering)	—	—	0.0206	0.0251	0.0311	0.0598
THW	—	—	0.0206	0.0251	0.0311	0.0598
TW, RUH	—	—	0.0135	0.0172	0.0224	0.0471
TF	0.0088	0.0109	—	—	—	—
THWN, THHN	—	—	0.0087	0.0117	0.0184	0.0373
TFN	0.0064	0.0079	—	—	—	—
FEPB, Z, ZF, ZFF	—	—	0.0087	0.0115	0.0159	0.0272
FEP, TFE	—	—	0.0087	0.0115	0.0159	0.0333
PTF	0.0052	0.0066	0.0087	—	—	—
XHHW	—	—	0.0131	0.0167	0.0216	0.0456

Type	Size, AWG								
	6	4	3	2	1	1/0	2/0	3/0	4/0
	Cross-sectional area, in²								
RH, RHH,* RHW*	0.1238	0.1605	0.1817	0.2067	0.2715	0.3107	0.3578	0.4151	0.4840
RUH	0.0819	0.1087	0.1263	0.1473	—	—	—	—	—
TW, THW	0.0819	0.1087	0.1263	0.1473	0.2027	0.2367	0.2781	0.3288	0.3904
TFE	0.0467	0.0669	0.0803	0.0973	0.1385	0.1676	0.1974	0.2436	0.2999
FEPB, ZF, ZFF	0.0716	0.0962	0.1122	0.1316	—	—	—	—	—
FEP	0.0467	0.0669	0.0803	0.0973	—	—	—	—	—
THHN, THWN	0.0519	0.0845	0.0995	0.1182	0.1590	0.1893	0.2265	0.2715	0.3278
XHHW	0.0625	0.0845	0.0995	0.1182	0.1590	0.1893	0.2265	0.2715	0.3278
Z	0.0716	0.0962	0.1122	0.1320	0.1385	0.1676	0.1948	0.2463	0.3000

Type	Size, MCM					
	250	500	750	1000	1500	2000
	Cross-sectional area, in²					
RH, RHH,* RHW*	0.5917	0.9834	1.4082	1.7531	2.5475	3.2079
TW, T, THW	0.4877	0.8316	1.2252	1.5482	2.2748	2.9013
THHN, THWN	0.4026	0.7163	1.0623	1.3623	—	—
XHHW	0.4026	0.7163	1.0936	1.3893	2.0612	2.6590

* RHH and RHW without covering have the same dimension as THW.
SOURCE: NEC 1993, pp. 70-825 and 826. Reprinted with permission from NFPA 70-1993, National Electrical Code®, Copyright © 1992 National Fire Protection Association, Quincy, Massachusetts 02269. This reprinted material is not the complete and official position of the NFPA on the referenced subject, which is represented only by the standard in its entirety.
NOTE: For general branch and feeder circuits the minimum conductor size is 14. Sizes 14 to 8 are solid wire. Sizes 6 and larger are stranded.

Table 15.1.25   Approximate Full-Load Currents of Motors,* A
(See NEC 1993 for more complete information.)

	Three-phase ac motors squirrel-cage and wound-rotor induction types				Synchronous type, unity power factor				Single-phase ac motors		DC motors†	
hp	230 V	460 V	575 V	2,300 V	230 V	460 V	575 V	2,300 V	115 V‡	230 V‡	120 V	240 V
½	2	1	0.8	—	—	—	—	—	9.8	4.9	5.4	2.7
¾	2.8	1.4	1.1	—	—	—	—	—	13.8	6.9	7.6	3.8
1	3.6	1.8	1.4	—	—	—	—	—	16	8	9.5	4.7
1½	5.2	2.6	2.1	—	—	—	—	—	20	10	13.2	6.6
2	6.8	3.4	2.7	—	—	—	—	—	24	12	17	8.5
3	9.6	4.8	3.9	—	—	—	—	—	34	17	25	12.2
5	15.2	7.6	6.1	—	—	—	—	—	56	28	40	20
7½	22	11	9	—	—	—	—	—	80	40	58	29
10	28	14	11	—	—	—	—	—	100	50	76	38
15	42	21	17	—	—	—	—	—	—	—	—	55
20	54	27	22	—	—	—	—	—	—	—	—	72
25	68	34	27	—	53	26	21	—	—	—	—	89
30	80	40	32	—	63	32	26	—	—	—	—	106
40	104	52	41	—	83	41	33	—	—	—	—	140
50	130	65	52	—	104	52	42	—	—	—	—	173
60	154	77	62	16	123	61	49	12	—	—	—	206
75	192	96	77	20	155	78	62	15	—	—	—	255
100	248	124	99	26	202	101	81	20	—	—	—	341
125	312	156	125	31	253	126	101	25	—	—	—	425
150	360	180	144	37	302	151	121	30	—	—	—	506
200	480	240	192	49	400	201	161	40	—	—	—	675

* The values of current are for motors running at speeds customary for belted motors and motors having normal torque characteristics. Use name-plate data for low-speed, high-torque, or multispeed motors. For synchronous motors of 0.8 pf multiply the above amperes by 1.25, at 0.9 pf by 1.1. The motor voltages listed are rated voltages. Respective nominal system voltages would be 220 to 240, 440 to 480, and 550 to 600 V. For full-load currents of 208-V motors, multiply the above amperes by 1.10; for 200-V motors, multiply by 1.15.
  † Ampere values are for motors running at base speed.
  ‡ Rated voltage. Nominal system voltages are 120 and 240.

Table 15.1.26   Maximum Rating or Setting of Motor Branch-Circuit Protective Devices and Starting-Inrush Code Letters

Type of motor	Percent of full-load current				Code letter	kVA/hp with locked rotor
	Non-time delay fuse	Dual-element (time delay) fuse	Instantaneous breaker	Time-limit breaker		
Single-phase, all types, no code letter	300	175	700	250	A	0–3.14
					B	3.15–3.54
AC motors: single-phase, polyphase squirrel-cage, or					C	3.55–3.99
synchronous with full-voltage, resistor, or reactor					D	4.0–4.49
starting:					E	4.5–4.99
No code letter	300	175	700	250	F	5.0–5.59
F–V	300	175	700	250	G	5.6–6.29
B–E	250	175	700	200	H	6.3–7.09
A	150	150	700	150	J	7.1–7.99
AC squirrel-cage or synchronous motors with autotrans-					K	8.0–8.99
former starting:					L	9.0–9.99
Not more than 30 A					M	10.0–11.19
No code letter	250	175	700	200	N	11.2–12.49
More than 30 A					P	12.5–13.99
No code letter	200	175	700	200	R	14.0–15.99
F–V	250	175	700	200	S	16.0–17.99
B–E	200	175	700	200	T	18.0–19.99
A	150	150	700	150	U	20.0–22.39
High reactance squirrel-cage:					V	22.4 and up
Not more than 30 A						
No code letter	250	175	700	250		
More than 30 A						
No code letter	200	175	700	200		
Wound rotor, no code letter	150	150	700	150		
DC motors:						
No more than 50 hp						
No code letter	150	150	250	150		
More than 50 hp						
No code letter	150	150	175	150		
Low-torque, low-speed (450 r/min or lower) synchronous motors which start unloaded	200	200	200	200		

SOURCE: NEC 1993, Tables 430-152 and 430-7(b). Reprinted with permission from NFPA 70-1993, National Electrical Code®, Copyright © 1992, National Fire Protection Association, Quincy, Massachusetts 02269. This reprinted material is not the complete and official position of the NFPA on the referenced subject, which is represented only by the standard in its entirety.

less is usually required. Inside a building the **grounded conductor** (one of the two conductors in two-wire system and the **neutral conductor** in a three-wire or a four-wire system) should have a **white or natural gray covering** throughout to distinguish it from the ungrounded conductors. This identified grounded conductor must not be fused or be opened unless the other conductors are opened simultaneously. **Green wires** only shall be used for grounding electrical equipment, such as motors, as well as conduits, armor, boxes, and such metallic enclosures. Four-wire circuits have **black, white, red, and blue** conductors or alphanumeric identification.

DC systems need be grounded only at the generating stations because the grounded wire is electrically connected to one of the conductors in all the circuits throughout the system. In ac systems, since one section can be insulated from the other by a transformer, each section of 300 V or under is grounded at the individual services. The conductor grounding the ac system should not be less than no. 8 copper wire and must be without a joint or a splice and run from the supply side of the service switch.

The service conduit that protects the service wires on the outside of the building must be grounded by a wire at least as large as no. 8, run directly to ground.

The entire metallic system surrounding the conductors must be at ground potential. It is only necessary to ground the metallic system, including motors and other equipment, at one point, provided that each section makes a good electrical connection with the next.

Since January 1, 1973, **ground-fault circuit interrupters** have been required by the NEC in some areas for personnel protection from line-to-ground electrical shock. Such circuit interrupters are required by the 1984 NEC in branch circuits supplying certain areas in residences, hotels and motels, health-care facilities, marinas and boat yards, mobile homes and recreational vehicles, swimming pools, and construction sites. The 1984 or subsequent code should be carefully reviewed for exact requirements.

Ground-fault circuit protection may be used at other locations and, if so used, will provide additional protection against line-to-ground shock hazard.

Ground fault-circuit interrupters monitor the current in the two conductors of a circuit. These two currents should be equal. If they are not equal, some current is leaking to ground, indicating a line to ground fault. If the difference between the two currents is 5 mA or more, the ground-fault circuit interrupter will automatically disconnect the faulted circuit in about 0.025 s.

Overload Protection   A fuse or circuit breaker must be provided in all ungrounded conductors. Induction motors are usually protected by overload or thermal relays operating as automatic circuit breakers. To protect wiring properly, an automatic cutout must be installed at every point where a change is made in the size of the wire. Fuses or circuit breakers should not be placed either in a grounded line or in a ground wire. (See Fig. 15.1.96.)

Fuses, cutout bases, and **switches** are manufactured and change sizes, as follows: Edison plug (125 V only), 0 to 30 A; spring-clip cartridge (ferrule contact), 0 to 30, 31 to 60 A; knife-blade cartridge type, 61 to 100, 101 to 200, 201 to 400, 401 to 600 A; for 601 A and larger, knife-blade cartridge type with equal-size fuses in parallel may be used except for the protection of a branch motor circuit where a circuit breaker can be installed.

Since the rating of a fuse is only about 90 percent of the current that it will carry indefinitely, and since it may also take a few minutes before the heat due to slightly excessive current would be sufficient to melt the fuse wire and hence open the circuit, insulation may be permanently damaged if the fuses are larger than the current-carrying capacities given by Table 15.1.21.

Demand Calculations for Building Feeder Sizes   The **demand factor** or **demand** is the ratio of maximum demand to the total connected load. This depends on the type of building, whether hotel, theater, factory, etc. The demand factor for any particular class of installation decreases as the floor area increases. Values of demand factors are found in the 1993 NEC Art. 220.

Load centers are panels or cabinets which act as distribution centers and which are supplied by feeder or main conductors, and from which the current to several branch circuits is taken. In each branch circuit there is usually a small combined switch and circuit breaker. Load centers for widely different types of circuits such as single-phase two-wire, single-phase three-wire, and three-phase are readily obtainable from several manufacturers.

RESISTOR MATERIALS

For use in rheostats, electric furnaces, ovens, heaters, and many electrical appliances, a resistor material with high melting point and high resistivity which does not disintegrate or corrode at high temperatures is necessary. These requirements are met by the nickel-chromium and nickel-chromium-iron alloys. For electrical instruments and measuring apparatus, the resistor material should have high resistivity, low temperature coefficient, and, for many uses, low thermoelectric power against copper. The properties of resistor materials are given in Table 15.1.27. Most of these materials are available in ribbon as well as in wire form. Cast-iron and steel wire are efficient and economical resistor materials for many uses, such as power-absorbing rheostats and motor starters and controllers. (See also Sec. 6.)

Advance has a low temperature coefficient and is useful in many types of measuring instrument and precision equipment. Because of its high

Table 15.1.27   Properties of Metals, Alloys, and Resistor Materials

| | | | Resistivity | | | | |
Material	Composition	Sp gr	Microhms-cm at 20°C	Ohms cir-mil-ft at 20°C	Temp coef of resistance per deg C	Temp range, °C	Max safe working temp, °C	Approx melting point, °C
Advance	Cu 0.55; Ni 0.45	8.9	48.4	294	±0.00002	20–100	500	1210
Comet	Ni 0.30; Cr 0.05; Fe 0.65	8.15	95	570	0.00088	20–500	600	1480
Bronze, commercial	Cu; Zn	8.7	4.2	25	0.0020	0–100	—	1040
Hytemco-Balco	Ni 0.50; Fe 0.50	8.46	20	120	0.0045	20–100	600	1425
Kanthal A	Al 0.055; Cr 0.22; Co 0.055; Fe 0.72	7.1	145	870	0.00002	0–500	1330	1510
Magno	Ni 0.955; Mn 0.045	8.75	20	120	0.0036	20–100	400	1435
Manganin	Cu 0.84; Mn 0.12; Ni 0.04	8.19	48.2	290	±0.000015	15–35	100	1020
Monel metal	Ni 0.67; Cu 0.28	8.9	42.6	256	0.0001	0–100	425	1350
Nichrome	Ni 0.60; Fe 0.25; Cr 0.15	8.247	112	675	0.00017	20–100	930	1350
Nichrome V	Ni 0.80; Cr 0.20	8.412	108	650	0.00013	20–100	1100	1400
Nickel, pure	Ni 0.99	8.9	10	60	0.0050	0–100	400	1450
Platinum	Pt	21.45	10.616	63.80	0.003	0–100	1200	1773
Silver	Ag	10.5	1.622	9.755	0.00361	0–100	650	960
Tungsten	W	19.3	5.523	33.22	0.0045	—	*	3410

* Tungsten subject to rapid oxidation in air above 150°C.

thermoelectric power to copper, it is valuable for thermoelements and pyrometers. It is noncorrosive and is used to a large extent in industrial and radio rheostats. **Hytemco** is a nickel-iron alloy characterized by a high temperature coefficient and is used advantageously where self-regulation is required as in immersion heaters and heater pads. **Magno** is a manganese-nickel alloy used in the manufacture of incandescent lamps and radio tubes. **Manganin** is a copper-manganese-nickel alloy which, because of its very low temperature coefficient and its low thermal emf with respect to copper, is very valuable for high-precision electrical measuring apparatus. It is used for the resistance units in bridges, for shunts, multipliers, and similar measuring devices. **Nichrome V** is a nickel-chromium alloy free from iron, is noncorrosive, nonmagnetic, withstands high temperatures, and has high resistivity. It is recommended as material for heating elements in electric furnaces, hot-water heaters, ranges, radiant heaters, and high-grade electrical appliances. **Kanthal** is used for heating applications where higher operating temperatures are required than for Nichrome. Mechanically, it is less workable than Nichrome. **Platinum** is used in specialized heating applications where very high temperatures are required. **Tungsten** may be used in very high temperature ovens with an inert atmosphere. **Pure nickel** is used to satisfy the high requirements in the fabrication of radio tubes, such as the elimination of all gases and impurities in the metal parts. It also has other uses such as in incandescent lamps, for combustion boats, laboratory accessories, and resistance thermometers.

Carbon withstands high temperatures and has high resistance; its temperature coefficient is negative; it will safely carry about 125 A/in². Amorphous carbon has a resistivity of 3,800 and 4,100 $\mu\Omega \cdot$ cm, retort carbon about 720 $\mu\Omega \cdot$ cm, and graphite about 812 $\mu\Omega \cdot$ cm. The properties of any particular kind of carbon depend on the temperature at which it was fired. Carbon for rheostats may best be used in the form of compression rheostats. **Silicon carbide** is used to manufacture heating rods that will safely operate at 1,650°C (3,000°F) surface temperature. It has a negative coefficient of resistance up to 650°C, after which it is positive. It must be mechanically protected because of inherent brittleness.

MAGNETS

A **permanent magnet** is one that retains a considerable amount of magnetism indefinitely. Permanent magnets are used in electrical instruments, telephone receivers, loudspeakers, magnetos, tachometers, magnetic chucks, motors, and for many purposes where a constant magnetic field or a constant source of magnetism is desired. The magnetic material should have high retentivity, a high remanence, and a high coercive force (see Fig. 15.1.8). These properties are usually found with hardened steel and its alloys and also in ceramic permanent magnet materials.

Since permanent magnets must operate on the molecular mmf imparted to them when magnetized, they must necessarily operate on the portion *CDO* of the hysteresis loop (see Fig. 15.1.8). The area *CDO* is proportional to the **stored energy** within the magnet and is a criterion of its usefulness as a permanent-magnetic material. In the left half of Fig. 15.1.97 are given the *B-H* characteristics of several permanent-magnetic materials; these include 5 to 6 percent tungsten steel (curve 1); 3½ percent chrome magnet steel (curve 2); cobalt magnet steel, containing 16 to 36 percent cobalt and 5 to 9 percent chromium and in some alloys tungsten (curve 3); and the carbon-free aluminum-nickel-cobalt-steel alloys called Alnico. There are many grades of Alnico; the characteristics of three of them are shown by curves 4, 5, 6. Their composition is as follows:

Curve	Alnico no.	Composition, percent					
		Al	Ni	Co	Cu	Ti	Fe
4	5	8	14	24	3	. . .	51
5	6	8	14	24	3	1.25	49.75
6	12	6	18	35	. . .	8.0	33

All the Alnicos can be made by the sand or the precision-casting (lost-wax) process, but the most satisfactory method is by the sintering process.

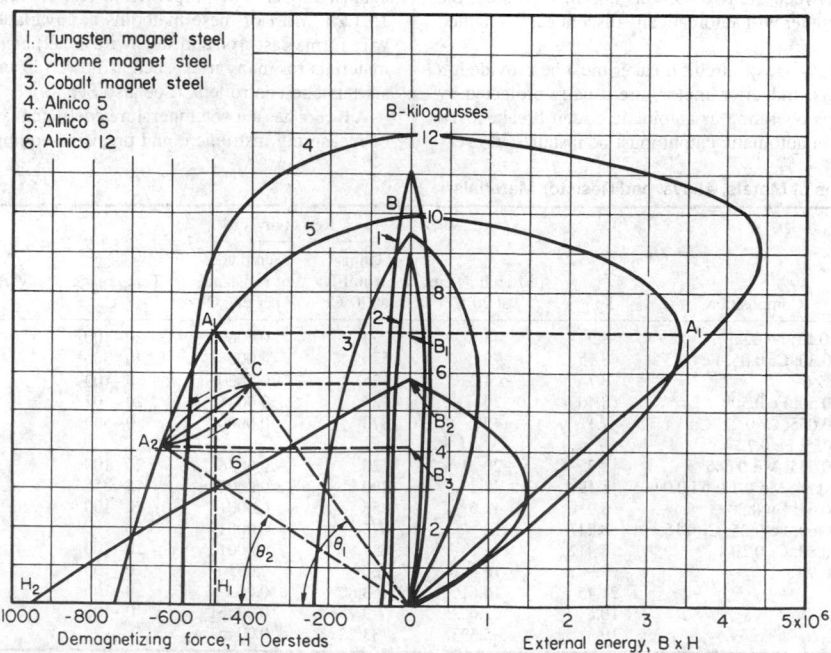

1. Tungsten magnet steel
2. Chrome magnet steel
3. Cobalt magnet steel
4. Alnico 5
5. Alnico 6
6. Alnico 12

B-kilogausses

Demagnetizing force, H Oersteds

External energy, B x H

Fig. 15.1.97 Characteristics of permanent magnet materials.

If the alloys are held in a magnetic field during heat-treatment, a **magnet grain** is established and the magnetic properties in the direction of the field are greatly increased. The alloys are hard, can be formed only by casting or sintering, and cannot be machined except by grinding.

The curves in the right half of Fig. 15.1.97 are "external energy" curves and give the product of B and H. The optimum point of operation is at the point of maximum energy as is indicated at A_1 on curve 5.

Considering curve 5, if the magnetic circuit remained closed, the magnet would operate at point B. To utilize the flux, an air gap must be introduced. The air gap acts as a demagnetizing force, H_1 $(= B_1A_1)$, and the magnet operates at point A_1 on the HB curve. The line OA_1 is called the *air-gap line* and its slope is given by $\tan \theta_1 = B_1/H_1$ where $H_1 = B_g l_g/l_m$ and $B_1/B_g = A_g/A_m$, where B_g = flux density in gap, T; l_g = length of gap, m; l_m = length of magnet, m; A_g, A_m = areas of the gap and magnet, m².

If the air gap is lengthened, the magnet will operate at A_2 corresponding to a lesser flux density B_3 and the new air-gap line is OA_2. If the gap is now closed to its original value, the magnet will not return to operation at point A_1 but will operate at some point C on the line OA_1. If the air gap is varied between the two foregoing values, the magnet will operate along the **minor hysteresis loop** A_2C. Return to point A_1 can be accomplished only by remagnetizing and coming back down the curve from B to A_1.

Alnico magnets corresponding to curves 4 and 5 are best adapted to operation with short air gaps, since the introduction of a long air gap will demagnetize the magnet materially. On the other hand, a magnet with a long air gap will operate most satisfactorily on curve 6 on account of the high coercive force H_2. With change in the length of the air gap, the operation will be essentially along that curve and the magnet will lose little of its original magnetization.

There are several other grades of Alnico with characteristics between curves 4, 5, and 6. Ceramic PM materials have a very large coercive force.

The steels for permanent magnets are cut in strips, heated to a red-hot temperature, and forged into shape, usually in a "bulldozer." If they are to be machined, they are cooled in mica dust to prevent air hardening. They are then ground, tumbled, and tempered. Alnico and ceramic types are cast and then finish-ground.

Permanent magnets are magnetized either by placing them over a bus bar carrying a large direct current, by placing them across the poles of a powerful electromagnet, or by an ampere-turn pulse.

Unless permanent magnets are subjected to artificial aging, they gradually weaken until after a long period they become stabilized usually at from 85 to 90 percent of their initial strength. With magnets for electrical instruments, where a constant field strength is imperative, artificial aging is accomplished by mechanical vibration or by immersion in oil at 250°F for a period of a few hours.

In an **electromagnet** the magnetic field is produced by an electric current. The core is usually made of soft iron or mild steel because, the permeability being higher, a stronger magnetic field may be obtained. Also since the retentivity is low, there is little trouble due to the sticking of armatures when the circuit is opened. Electromagnets may have the form of simple solenoids, iron-clad solenoids, plunger electromagnets, electromagnets with external armatures, and lifting magnets, which are circular in form with a flat holding surface.

A **solenoid** is a winding of insulated conductor and is wound helically; the direction of winding may be either right or left. A **portative electromagnet** is one designed only for holding material brought in contact with it. A **tractive electromagnet** is one designed to exert a force on the load through some distance and thus do work. The **range** of an electromagnet is the distance through which the plunger will perform work when the winding is energized. For long range of operation, the plunger type of tractive magnet is best suited, for the length of core is governed practically by the range of action desired, and the area of the core is determined by the pull. **Solenoid and plunger** is a solenoid provided with a movable iron rod or bar called a plunger. When the coil is energized, the iron rod becomes magnetized and the mutual action of the field in the solenoid on the poles created on the plunger causes the plunger to move within the solenoid. This force becomes zero only when the magnetic centers of the plunger and solenoid coincide. If the load is attached to the plunger, work will be done until the force to be overcome is equal to the force that the solenoid exerts on the plunger. When the iron of the plunger is not saturated, the strength of magnetic field in the solenoid and the induced poles are both proportional to the exciting current, so that the pull varies as the **current squared**. When the plunger becomes highly saturated, the pull varies almost **directly** with the current. The **maximum uniform pull** occurs when the end of the plunger is at the center of the solenoid and is equal to

$$F = CAnI/l \qquad \text{lb} \qquad (15.1.134)$$

where A = cross-sectional area of plunger, in²; n = number of turns; I = current, A; l = length of the solenoid, in; and C = pull, (lb/in²)/(A-turn/in). C depends on the proportions of the coil, the degree of saturation, the length, and the physical and chemical purity of the plunger. Table 15.1.28 gives values of C for several different solenoids.

Curve 1, Fig. 15.1.98, shows the characteristic pull of an open-magnetic circuit solenoid, 12 in long, having 10,000 A-turns or 833 A-turns/in.

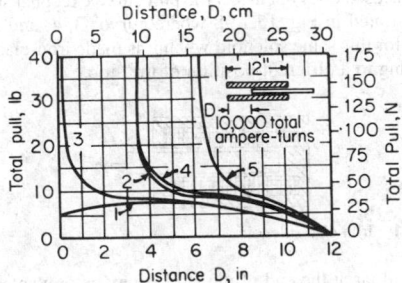

Fig. 15.1.98 Pull on solenoid with plunger. (1) Coil and plunger; (2) coil and plunger with stop; (3) iron-clad coil and plunger; (4) and (5) same as (3) with different lengths of stop.

When a **strong pull** is desired at the end of the stroke, a stop may be used as shown in Fig. 15.1.99. Curve 2, Fig. 15.1.98, shows the pull obtained by adding a stop to the plunger. It will be noted that, except when the end of the plunger is near the stop, the stop adds little to the

Table 15.1.28   Maximum Pull per Square Inch of Core for Solenoids with Open Magnetic Circuit

Length of coil l, in	Length of plunger, in	Core area A, in²	Total ampere-turns $I \times n$	Max pull P, psi	$1,000 \times C$	Length of coil l, in	Length of plunger, in	Core area A, in²	Total ampere-turns $I \times n$	Max pull P, psi	$1,000 \times C$
6	Long	1.0	15,900	22.4	9.0	12	Long	1.0	11,200	8.75	9.4
9	Long	1.0	11,330	11.5	9.1	12	Long	1.0	20,500	16.75	9.8
9	Long	1.0	14,200	14.6	9.2	18	36	1.0	18,200	9.8	9.7
10	10	2.76	40,000	40.2	10.0	18	36	1.0	41,000	22.5	9.8
10	10	2.76	60,000	61.6	10.3	18	18	1.0	18,200	9.8	9.7
10	10	2.76	80,000	80.8	10.1	18	18	1.0	41,000	22.5	9.8

SOURCE: From data by Underhill, *Elec. World*, 45, 1906, pp. 796, 881.

solenoid pull. The pull is made up of two components: one due to the attraction between plunger and winding, the other to the attraction between plunger and stop. The equation for the pull is

$$P = AIn[(In/l_a^2 C_1^2) + (C/l)] \qquad (15.1.135)$$

where A = area of the core, in^2; n = number of turns; l_a = length of gap between core and stop; and C, C_1 = constants. At the beginning of the stroke the second member of the equation is predominant, and at the end

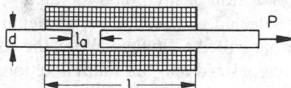

Fig. 15.1.99 Solenoid with stop.

of the stroke the first member represents practically the entire pull. Approximate values of C and C_1 are $C_1 = 2,660$ (for l greater than $10d$), $C = 0.0096$, where d is the diameter of the plunger, in. In SI units

$$P = 1.7512 AnI \left(\frac{2.54nI}{l_a^2 C_1^2} + \frac{C}{l} \right) \qquad N \qquad (15.1.135a)$$

where A is in cm^2, l and l_a in cm, and the pull P in N. All other quantities are unchanged.

The **range of uniform** pull can be extended by the use of conical ends of stop and plunger, as shown in Fig. 15.1.100. A stronger magnet mechanically can be obtained by using an iron-clad solenoid, Fig. 15.1.101, in which an iron return path is provided for the flux. Except for low flux densities and short air gaps the dimensions of the iron

Fig. 15.1.100 Conical plunger and stop.

return path are of no practical importance, and the fact that an iron return path is used does not affect the pull curve except at short air gaps. This is illustrated in Fig. 15.1.98 where curves 3, 4, and 5 are typical pull curves for this same solenoid when it is made iron-clad, each curve corresponding to a different position of the stop.

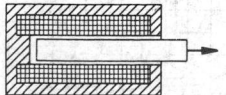

Fig. 15.1.101 Iron-clad solenoid.

Mechanical jar at the end of the stroke may be prevented by leaving the end of the solenoid open. The plunger then comes to equilibrium when its middle is at the middle of the winding, thus providing a magnetic cushion effect. Electromagnets with external armatures are best adapted for **short-range** work, and the best type is the horseshoe magnet. The pull for short-range magnets is expressed by the equation

$$F = B^2A/72,134,000 \qquad lb \qquad (15.1.136)$$

where B = flux density, Mx/in^2, and A = area of the core, in^2.

In SI units

$$F = 397,840B^2A \qquad N \qquad (15.1.136a)$$

where the flux density B is in Wb/m^2; A = area, m^2; and F = force, N.

A greater holding power is obtained if the surfaces of the armature and core are not machined to an absolutely smooth contact surface. If the surface is slightly irregular, the area of contact A is reduced but the flux density B is increased approximately in proportion (if the iron is being operated below saturation) and the pull is increased since it varies as the square of the density B. Nonmagnetic stops should be used if it is desired that the armature may be released readily when the current is interrupted.

Lifting magnets are of the portative type in that their function is merely

to hold the load. The actual lifting is performed by the hoisting apparatus. The magnet is almost toroidal in shape. The coil shield is of manganese steel which is very hard and thus resists wear and is practically nonmagnetic. The holding power is given by Eq. (15.1.136), where A = area of holding surface, in^2. It is difficult to calculate accurately the holding force of a lifting magnet for it depends on the magnetic characteristics of the load, the area of contact, and the manner in which the load is applied.

Rapid action in a magnet can be obtained by reducing the time constant of the winding and by subdividing the metal parts to reduce induced currents which have a demagnetizing effect when the circuit is closed. The movement of the plunger through the winding causes the winding and its bobbin to be cut by a magnetic field; if the bobbin is of metal and not slotted longitudinally, it is a short-circuited turn linked by a changing magnetic field and hence currents are induced in it. These currents oppose the flux and hence reduce the pull during the transient period. They also cause some heating. Where it is found impossible to reduce the time constant sufficiently, an electromagnet designed for a voltage much lower than normal is often used. A resistor is connected in series which is short-circuited during the stroke of the plunger. At the completion of the stroke the plunger automatically opens the short circuit, reducing the current to a value which will not overheat the magnet under continuous operation. The extremely short time of overload produces very rapid action but does not injure the winding. The solenoids on many automatic motor-starting panels are designed in this manner.

When **slow action** is desired, it can be obtained by using solid cores and yoke and by using a heavy metallic spool or bobbin for the winding. A separate winding short-circuited on itself is also used to some extent.

Sparking at switch terminals may be reduced or eliminated by neutralizing the inductance of the winding. This is accomplished by winding a separate short-circuited coil with its wires parallel to those of the active winding. (This method can be used with dc magnets only.) This is not economical, since one-half the winding space is wasted. By connecting a capacitor across the switch terminals, the energy of the inductive discharge on opening the circuit may be absorbed. For the purpose of neutralizing the inductive discharge and causing a quick release, a small **reverse current** may be sent through the coil winding automatically on opening the circuit. Sleeves of tin, aluminum, or copper foil placed over the various layers of the winding absorb energy when the circuit is broken and reduce the energy dissipated at the switch terminals. This scheme can be used for dc magnets only. **Sticking** of the parts of the magnetic circuit due to residual magnetism may be prevented by the use of nonmagnetic stops. In the case of lifting magnets subjected to rough usage and hard blows (as in a steel works), these stops usually consist of plates of manganese steel, which are extremely hard and nonmagnetic.

AC Tractive Magnets Because of the iron losses due to eddy currents, the magnetic circuits of ac electromagnets should be composed of laminated iron or steel. The magnetic circuit of large magnets is usually built up of thin sheets of sheet metal held together by means of suitable clamps. Small cores of circular cross section usually consist of a bundle of soft iron wires. Since the iron losses increase with the flux density, it is not advisable to operate at as high a density as with direct current. The current instead of being limited by the resistance of the winding is now determined almost entirely by the inductive reactance as the resistance is small. With the removal of the load the current rises to high values. The pull of ac magnets is nearly constant irrespective of the length of air gap.

In a **single-phase magnet** the pull varies from zero to a maximum and back to zero twice every cycle, which may cause considerable chattering of the armature against the stop. This may be prevented by the use of a spring or, in the case of a solenoid coil, by allowing the plunger to seek its position of equilibrium in the coil. **Chattering** may also be prevented by the use of a short-circuited winding or shading coil around one tip of the pole piece or by the use of polyphase. In a **two-phase magnet** the pull is constant and equal to the maximum instantaneous pull produced by one phase so long as the voltage is a sine function. In a **three-phase magnet** under the same conditions the pull is constant and equal to 1.5 times the maximum instantaneous pull of one phase. Should the load become greater than the minimum instantaneous pull, there will be chattering as in a single-phase magnet.

Heating of Magnets The lifting capacity of an electromagnet is limited by the permissible current-carrying capacity of the winding, which, in turn, is dependent on the amount of heat energy that the winding can dissipate per unit time without exceeding a given temperature rise. **Enamels, synthetic varnishes, and thermoplastic wire insulations** are available to allow a wide variety of temperature rises.

Design of Exciting Coil Let n = number of turns, l = mean length of turn, in ($l = 2\pi r$, where r is the mean radius, in), A = cross section of wire, cir mils. The resistance of 1 cir mil·ft of copper is practically 12 Ω at 60°C, or 1 Ω/cir mil·in. Hence the resistance, $R = nl/A$ Ω; the current, $I = EA/nl$; the ampere-turns, $nI = EA/l$; the power to be dissipated, $P = E^2A/nl$ W. From the foregoing equations the cross section of wire and the number of turns can be calculated.

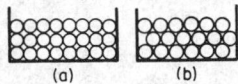

Fig. 15.1.102 Winding space factor.

Space Factor of Winding Space factor of a coil is the ratio of the space occupied by the conductor to the total volume of the coil or winding. Only in the theoretical case of uninsulated square or rectangular conductor may the space factor be 100 percent. For wire of circular section with insulation of negligible thickness, wound as shown in Fig. 15.1.102a, the space factor will be 78.5 percent. When the turns of wire are "bedded," as shown in Fig. 15.1.102b (the case in most windings, particularly with smaller wires), there is a theoretical gain of about 7 percent in space factor. Experiments have shown that in most cases this gain is about neutralized in practice by the flattening out of the insulation of the wire due to the tension used in winding. When wound in a haphazard manner, the space factors of magnet wires vary according to size, substantially as follows:

	Double cotton covered				Single cotton covered			
Size, AWG	0	5	10	15	20	25	30	35
Space factor, %	60	53.8	45.5	35.1	32.2	32	25.7	16

Magnet wire is usually a soft insulated annealed copper wire of high conductivity. It can be obtained in square, rectangular, and circular section, but the round or cylindrical wire is used almost entirely in the smaller sizes. Ribbons are frequently used in the larger sizes. **Aluminum** has been used at times. A number of different varnish or enamel insulating compounds are available in different temperature classes, up to 220°C, the **temperature** rating and thickness for **dielectric strength** being dependent on the usage. Where **mechanical strength** is important and space is not at a premium, textile or paper insulation is used on the wires and later varnish impregnated for high dielectric strength. **Asbestos-covered** magnet wire is used where the temperature is high. It combines

Table 15.1.29   Magnet-Wire Dimensions, Sizes 14 to 44 AWG

	Bare wire diameter, in			Single		Heavy	
AWG	Minimum	Nominal	Maximum	Minimum increase in diameter, in	Maximum overall diameter, in	Minimum increase in diameter, in	Maximum overall diameter, in
14	0.0635	0.0641	0.0644	0.0016	0.0666	0.0032	0.0682
15	0.0565	0.0571	0.0574	0.0015	0.0594	0.0030	0.0609
16	0.0503	0.0508	0.0511	0.0014	0.0531	0.0029	0.0545
17	0.0448	0.0453	0.0455	0.0014	0.0475	0.0028	0.0488
18	0.0399	0.0403	0.0405	0.0013	0.0424	0.0026	0.0437
19	0.0355	0.0359	0.0361	0.0012	0.0379	0.0025	0.0391
20	0.0317	0.0320	0.0322	0.0012	0.0339	0.0023	0.0351
21	0.0282	0.0285	0.0286	0.0011	0.0303	0.0022	0.0314
22	0.0250	0.0253	0.0254	0.0011	0.0270	0.0021	0.0281
23	0.0224	0.0226	0.0227	0.0010	0.0243	0.0020	0.0253
24	0.0199	0.0201	0.0202	0.0010	0.0217	0.0019	0.0227
25	0.0177	0.0179	0.0180	0.0009	0.0194	0.0018	0.0203
26	0.0157	0.0159	0.0160	0.0009	0.0173	0.0017	0.0182
27	0.0141	0.0142	0.0143	0.0008	0.0156	0.0016	0.0164
28	0.0125	0.0126	0.0127	0.0008	0.0140	0.0016	0.0147
29	0.0112	0.0113	0.0114	0.0007	0.0126	0.0015	0.0133
30	0.0099	0.0100	0.0101	0.0007	0.0112	0.0014	0.0119
31	0.0088	0.0089	0.0090	0.0006	0.0100	0.0013	0.0108
32	0.0079	0.0080	0.0081	0.0006	0.0091	0.0012	0.0098
33	0.0070	0.0071	0.0072	0.0005	0.0081	0.0011	0.0088
34	0.0062	0.0063	0.0064	0.0005	0.0072	0.0010	0.0078
35	0.0055	0.0056	0.0057	0.0004	0.0064	0.0009	0.0070
36	0.0049	0.0050	0.0051	0.0004	0.0058	0.0008	0.0063
37	0.0044	0.0045	0.0046	0.0003	0.0052	0.0008	0.0057
38	0.0039	0.0040	0.0041	0.0003	0.0047	0.0007	0.0051
39	0.0034	0.0035	0.0036	0.0002	0.0041	0.0006	0.0045
40	0.0030	0.0031	0.0032	0.0002	0.0037	0.0006	0.0040
41	0.0027	0.0028	0.0029	0.0002	0.0033	0.0005	0.0036
42	0.0024	0.0025	0.0026	0.0002	0.0030	0.0004	0.0032
43	0.0021	0.0022	0.0023	0.0002	0.0026	0.0004	0.0029
44	0.0019	0.0020	0.0021	0.0001	0.0024	0.0004	0.0027

SOURCE: "Standard Handbook for Electrical Engineers," Fink and Carrol, McGraw-Hill, New York, 1968.

resistance to heat and abrasion, can resist mild acids, has good dielectric strength, and is fireproof.

Table 15.1.29 gives the diameters of magnet wire with the different types of insulation. For further data on electrical insulating materials, see Sec. 6.

AUTOMOBILE SYSTEMS

Automobile Ignition Systems

The ignition system in an automobile produces the spark which ignites the combustible mixture in the engine cylinders. This is accomplished by a high-voltage, or high-tension, spark between metal points in a spark plug. (A spark plug is an insulated bushing screwed into the cylinder head.) Spark plugs usually have porcelain insulation, but for some special uses, such as in airplane engines, mica may be used. There are two general sources for the energy necessary for ignition; one is the electrical system of the car which is maintained by the generator and the battery (**battery ignition**), and the other is a **magneto**. Battery ignition systems have traditionally operated electromechanically, using a spark coil, a high-voltage distributor, and low-voltage breaker points. Electronic ignition systems, working from the battery, became standard on U.S. cars in 1975. These vary in complexity from the use of a single transistor to reduce the current through the points to pointless systems triggered by magnetic pulses or interrupted light beams. Capacitor discharge into a pulse transformer is used in some systems to obtain the high voltage needed to fire the spark plugs.

Battery ignition is most widely used since it is simple, reliable, and low in cost, and the electrical system is a part of the car equipment. The high voltage for the spark is obtained from an ignition coil which consists of a primary coil of relatively few turns and a secondary coil of a large number of turns, both coils being wound on a common magnetic core consisting of either thin strips of iron or small iron wires. In a 6-V system the resistance of the primary coil is from 0.9 to 2 Ω and the inductance is from 5 to 10 mH. The number of secondary turns varies from 9,000 to 25,000, and the ratio of primary to secondary turns varies from 1 : 40 to 1 : 100.

The coil operates on the following principle. It stores energy in a magnetic field relatively slowly and then releases it suddenly. The power developed ($p = dw/dt$) is thus relatively large ($w = $ stored energy). The high emf e_2 which is required for the spark is induced by the sudden change in the flux ϕ in the core of the coil when the primary current is suddenly interrupted, $e_2 = -n_2(d\phi/dt)$, where n_2 is the number of secondary turns. For satisfactory ignition, peak voltages from 10 to 20 kV are desirable. Figure 15.1.103 shows the relation between the volts required to produce a spark and pressure with compressed air.

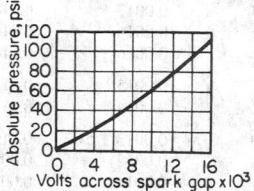

Fig. 15.1.103 Pressure-voltage curve for spark plug.

A battery ignition system for a four-cylinder engine is shown diagrammatically in Fig. 15.1.104. The primary circuit supplied by the battery consists of the primary coil P and a set of contacts, or "points" operated by a four-lobe cam, in series. In order to reduce arcing and burning of the contacts and to produce a sharp break in the current, a capacitor C is connected across the contacts. The contacts, which are of pure tungsten, are operated by a four-lobe cam which is driven at one-half engine speed. A strong spring tends to keep the contacts closed.

In the Delco-Remy distributor (Fig. 15.1.105) two breaker arms are connected in parallel; one coil and one capacitor are used. One set of contacts is open when the other is just breaking but closes a few degrees

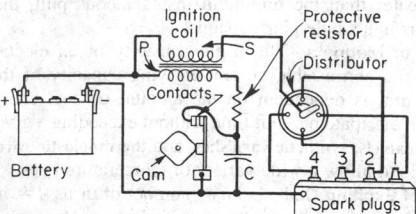

Fig. 15.1.104 Battery-ignition system.

after the break occurs. This closes the primary of the ignition coil immediately after the break and increases the time that the primary of the ignition coil is closed and permits the flux in the iron to reach its full value. The interrupter shown in Fig. 15.1.105 is designed for an eight-cylinder engine.

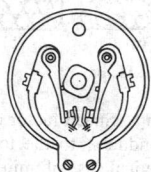

Fig. 15.1.105 Delco-Remy eight-cylinder interrupter.

Electronic ignition systems have no problem operating at the required speed.

The spark should advance with increase in engine speed so as to allow for the time lag in the explosion. To take care of this **automatically** most timers are now equipped with centrifugally operated weights which advance the breaker cam with respect to the engine drive as the speed increases.

Automobile Lighting and Starting Systems

Automobile lighting and starting systems initially operated at 6 V, but at present nearly all cars, except the smaller ones, operate at 12 V because larger engines, particularly V-8s, are now common and require more starting power. With 12 V, for the same power, the starting current is halved, and the effect of resistance in the leads, connections, and brushes is materially reduced. In some systems the positive side of the system is grounded, but more often the negative side is grounded.

A further development is the application of an **ac generator,** or alternator, combined with a rectifier as the generating unit rather than the usual dc generator. One advantage is the elimination of the commutator, made up of segments, which requires some maintenance due to the sparking and wear of the carbon brushes. With the alternator the dc field rotates, the brushes operating on smooth slip rings require almost no maintenance. Also, the system is greatly simplified by the fact that rectifiers are "one-way" devices, and the battery **cannot** deliver current back to the generator when its voltage drops below that of the battery. Thus, no cutout relay, such as is required with dc generators, is necessary. This ac development is the result of the development of reliable, low-cost germanium and silicon semiconductor rectifiers.

Figure 15.1.106 shows a schematic diagram of the Ford system (adapted initially to trucks). The generator stator is wound three-phase Y-connected, and the field is bipolar supplied with direct current through slip rings and brushes. The rectifier diodes are connected full-wave bridge circuit to supply the battery through the ammeter.

Regulator The function of the regulator is to control the generator current so that its value is adapted to the battery voltage which is related to the condition of charge of the battery (see Fig. 15.1.15). Thus, when the battery voltage drops (indicating a lowered condition of charge), the current should be increased, and, conversely, when the battery voltage increases (indicating a high condition of charge), the current should be decreased.

Fig. 15.1.106 Schematic diagram of Ford lighting and starting system.

Neglecting for the moment the starting procedure, when the ignition switch is thrown to the normal "on" position at c, the coil actuating the field relay is connected to the battery + terminal and causes the relay contacts a to close. This energizes the regulator circuits, and, if the two upper voltage regulator contacts b are closed as shown, the rotating field of the alternator is connected directly to the battery + terminal, and the field current is then at its maximum value and produces a high generator voltage and large output current. At the same time the voltage regulator coil in series with the 0.3- and 14-Ω resistors is connected between the battery + terminal and ground. If the voltage of the battery rises owing to its higher condition of charge, the current to the voltage regulator coil increases, causing it to open the two upper contacts at b. Current from the battery now flows through the 0.3-Ω resistor and divides, some going through the 10-Ω resistor and dividing between the field and the 50-Ω resistor, and the remainder going to the voltage regulator coil. The current to the rotating field is thus reduced, causing the alternator output to be reduced. Because of the 0.3 Ω now in circuit, the current to the coil of the voltage regulator is reduced to such a value that it holds the center contact at b in the mid, or open, position. If the battery voltage rises to an even higher value, the regulator coil becomes strong enough to close the lower contacts at b; this short-circuits the field, reducing its current almost to zero, and thus reducing the alternator output to zero. On the other hand when the battery voltage drops, the foregoing sequence is reversed, and the contacts at b operate to increase the current to the alternator field.

As was mentioned earlier, the battery cannot supply current to the alternator because of the "one-way" characteristic of the rectifier. Thus, when the alternator voltage drops below that of the battery and even when the alternator stops running, its current automatically becomes zero. The alternator has a normal rectifier open-circuit voltage of about 14 V and a rating of 20 A.

Starting In most cases, for starting, the ignition key is turned far to the right and held there until the motor starts. Then, when the key is released, the ignition switch contacts assume a normal operating position. Thus, in Fig. 15.1.106, when the ignition-switch contact is in the starting position S, the starter relay coil becomes connected by a lead to the battery + terminal and thus becomes energized, closing the relay contacts. The starter motor is then connected to the battery to crank the engine. At the time that the contact closes it makes contact with a small metal brush e which connects the battery + terminal to the primary

terminal of the ignition coil through the protective resistor R. After the motor starts, the ignition switch contacts spring to the normal operating position C, and the starter relay switch opens, thereby breaking contact with the small brush e. However, when contact C is closed, the ignition coil primary terminal is now connected through leads to the battery + terminal. The interrupter, the ignition coil, and the distributor now operate in the manner described earlier (see Fig. 15.1.104); the system shown in Fig. 15.1.106 is that for a six-cylinder engine.

The connection of **accessories** to the electric system is illustrated in Fig. 15.1.106 for the horn, head and other lights, and temperature and fuel gages.

Magneto Ignition

Principle of Magneto A magneto is an electric generator in which the magnetic flux is provided by one or more permanent magnets. It is a self-contained unit and is used advantageously for ignition where a generator and a battery are not needed to supply power to other accessories. The design of magnetos was radically changed when Alnico, with its very high retentivity (Fig. 15.1.97), became available as a permanent-magnet material. One method of utilizing Alnico magnets is to insert the bar magnets in the frame of the magneto (Fig. 15.1.107). The rotor is a soft-iron bobbin. A primary winding of relatively few turns and a secondary winding of a relatively large number of turns are wound over the laminated yoke Y. The position of the rotor shown in (a) provides a low reluctance path for the magnetic flux of the left-hand mag-

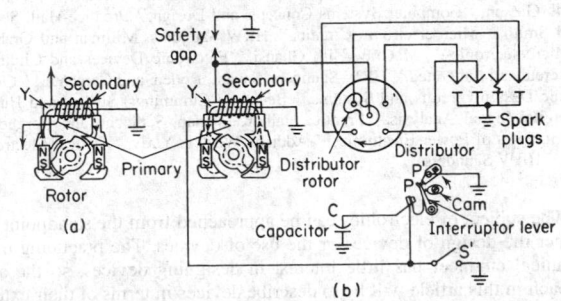

Fig. 15.1.107 Magneto-ignition system.

net and a high reluctance path for the magnetic flux of the right-hand magnet so that the flux goes through the yoke from left to right as shown. When the rotor turns one-eighth of a revolution, it becomes horizontal; obviously, each of the magnets acts in opposition relative to the yoke, and the flux therein becomes zero. In (*b*), which also shows the external electrical connections, the rotor is shown as having turned one-fourth of a revolution, or 90°, from its position in (*a*).

The rotor now provides a low-reluctance path for the right-hand magnet and a high-reluctance path for the left-hand one. Thus the magnetic flux now goes through the yoke from right to left. It follows that in each 90° interval the flux in the yoke undergoes a complete reversal. It will be recognized that this is an **inductor** type of ac generator.

In the diagram in (*b*), one end each of the primary and of the secondary are grounded together. The other end of the primary is connected to an insulated interrupter lever having a contact point *P*. This makes intermittent contact with the grounded contact point *P'*. The contact point *P* is actuated by a cam which is driven by the same shaft as the magneto rotor and is in a definite relation to it. A switch *S'* is provided to ground and thus short-circuit the secondary when it is desired to stop the engine. A capacitor *C* is connected between the point *P* and ground to absorb the energy of the spark which occurs when the contacts open.

With the contacts closed, the primary is short-circuited, and the varying flux in the core produced by the rotation of the rotor induces an alternating current in the winding which in turn produces an alternating flux in the core. With the rotation of the rotor the current in the secondary rises cyclically to maximum values, and at these instants the cam causes the points *PP'* to open suddenly, interrupting the current in the primary and thus causing a sudden collapse of the flux in the core. This induces a high-impulse emf in the secondary which is transmitted to the distributor and thence to the proper spark plug as shown.

On starting, the speed of the magneto may be so low that the emf is not sufficient to produce a hot spark. This difficulty can be met by impulse starting, in which the rotor is driven through a spring. During cranking the rotor is restrained from turning until the engine comes to the proper firing position, at which time the rotor is suddenly released. The energy stored in the spring produces a high, instantaneous, angular velocity to the rotor, resulting in a high emf and hot spark.

Inductor-type magnetos, having a large number of rotor poles and arranged differently from those shown in Fig. 15.1.107, are used for airplane-engine ignition. In another magneto design the rotor is a solid cylindrical Alnico magnet, permanently magnetized with an N and an S pole diametrically opposite. The frame is laminated, and there is a yoke with primary and secondary windings. When the rotor rotates, its N and S poles produce an alternating flux in the yoke which induces a short-circuit current in the primary winding and the method of producing the spark is then the same as with Fig. 15.1.107*b*.

Miscellaneous Automobile Electronics Systems

The use of electronics on automobiles is expanding rapidly. **Microprocessors** are being installed to control many functions that were previously not controlled or poorly controlled.

Air-fuel ratio is controlled from sensing oxygen in the exhaust system. **Timing** is based on crankshaft position, acceleration, and engine temperature. **Air injection** into exhaust and recirculation of exhaust reduces emissions. **Knock** is prevented by stopping untimely detonations. **Diagnosis** of problems is performed. Car leveling and load matching are made by shock absorber adjustments. Wheel **spin** is prevented. **Transmission** controls provide improved efficiency. Operator interfaces are by **cathode ray tube, fluid crystal, vacuum fluorescent displays,** and **speech synthesis. Navigation aids** are in a rudimentary state.

15.2 ELECTRONICS
by Byron M. Jones

REFERENCES: ''Reference Data for Radio Engineers,'' Howard Sams & Co. ''Transistor Manual,'' General Electric Co. ''SCR Manual,'' General Electric Co. ''The Semiconductor Data Book,'' Motorola Inc. Fink, ''Television Engineering,'' McGraw-Hill. ''Industrial Electronics Reference Book,'' Wiley. Mano, ''Digital and Logic Design,'' Prentice-Hall. McNamara, ''Technical Aspects of Data Communication,'' Digital Press. Fletcher, ''An Engineering Approach to Digital Design,'' Prentice-Hall. ''The TTL Data Book,'' Texas Instruments, Inc. ''1988 MOS Products Catalog,'' American Microsystems, Inc. ''1988 Linear Data Book,'' National Semiconductor Corp. ''CMOS Standard Cell Data Book,'' Texas Instruments, Inc. ''Power MOSFET Transistor Data,'' Motorola, Inc. Franco, ''Design with Operational Amplifiers and Analog Integrated Circuits,'' McGraw-Hill. Soclof, ''Applications of Analog Integrated Circuits,'' Prentice-Hall. Gibson, ''Computer Systems Concepts and Design,'' Prentice-Hall. Sedra and Smith, ''Microelectronic Circuits,'' HRW Saunders. Millman and Grabel, ''Microelectronics,'' McGraw-Hill. Ghausi, ''Electronic Devices and Circuits: Discrete and Integrated,'' HRW Saunders. Savant, Roden, and Carpenter, ''Electronic Design Circuits and Systems,'' Benjamin Cummings. Stearns and Hush, ''Digital Signal Analysis,'' Prentice-Hall. Kassakian, Schlecht, and Verghese, ''Principles of Power Electronics,'' Addison Wesley. Yariv, ''Optical Electronics,'' HRW Saunders.

The subject of electronics can be approached from the standpoint of either the design of devices or the use of devices. The practicing mechanical engineer has little interest in designing devices, so the approach in this article will be to describe devices in terms of their external characteristics.

Components

Resistors, capacitors, reactors, and **transformers** are described earlier in this section, along with basic circuit theory. These explanations are equally applicable to electronic circuits and hence are not repeated here. A description of additional components peculiar to electronic circuits follows.

A **rectifier,** or **diode,** is an electronic device which offers unequal resistance to forward and reverse current flow. Figure 15.2.1 shows the schematic symbol for a diode. The arrow beside the diode shows the direction of current flow. Current flow is taken to be the flow of positive

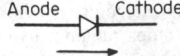

Anode   Cathode

Fig. 15.2.1 Diode schematic symbol.

charges, i.e., the arrow is counter to electron flow. Figure 15.2.2 shows typical forward and reverse volt-ampere characteristics. Notice that the scales for voltage and current are not the same for the first and third quadrants. This has been done so that both the forward and reverse characteristics can be shown on a single plot even though they differ by several orders of magnitude.

Diodes are rated for forward current capacity and reverse voltage breakdown. They are manufactured with maximum current capabilities ranging from 0.05 A to more than 1,000 A. Reverse voltage breakdown varies from 50 V to more than 2,500 V. At rated forward current, the forward voltage drop varies between 0.7 and 1.5 V for silicon diodes. Although other materials are used for special-purpose devices, by far the most common semiconductor material is silicon. With a forward

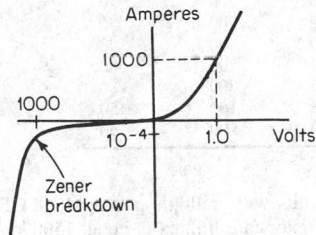

Fig. 15.2.2   Diode forward-reverse characteristic.

current of 1,000 A and a forward voltage drop of 1 V, there would be a power loss in the diode of 1,000 W (more than 1 hp). The basic diode package shown in Fig. 15.2.3 can dissipate about 20 W. To maintain an acceptable temperature in the diode, it is necessary to mount the diode on a **heat sink**. The manufacturer's recommendation should be followed very carefully to ensure good heat transfer and at the same time avoid fracturing the silicon chip inside the diode package.

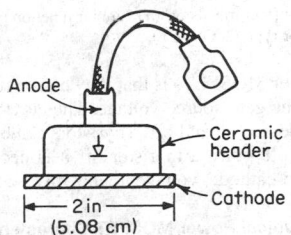

Fig. 15.2.3   Physical diode package.

The selection of **fuses** or **circuit breakers** for the protection of **rectifiers** and rectifier circuitry requires more care than for other electronic devices. Diode failures as a result of circuit faults occur in a fraction of a millisecond. Special semiconductor fuses have been developed specifically for semiconductor circuits. Proper protective circuits must be provided for the protection of not only semiconductors but also the rest of the circuit and nearby personnel. Diodes and diode fuses have a short-circuit rating in amperes-squared-seconds (I^2t). As long as the I^2t rating of the diode exceeds the I^2t rating of its protective fuse, the diode and its associated circuitry will be protected. Circuit breakers may be used to protect diode circuits, but additional line impedance must be provided to limit the current while the circuit breaker clears. Circuit breakers do not interrupt the current when their contacts open. The fault is not cleared until the line voltage reverses at the end of the cycle of the applied voltage. This means that the **clearing time for a circuit breaker** is about ½ cycle of the ac input voltage. Diodes have a 1-cycle overcurrent rating which indicates the fault current the diode can carry for circuit breaker protection schemes. Line inductance is normally provided to limit fault currents for breaker protection. Often this inductance is in the form of leakage reactance in the transformer which supplies power to the diode circuit.

A **thyristor,** often called a **silicon-controlled rectifier (SCR),** is a rectifier which blocks current in both the forward and reverse directions. Conduction of current in the forward direction will occur when the anode is positive with respect to the cathode and when the gate is pulsed positive with respect to the cathode. Once the thyristor has begun to conduct, the gate pulse can return to 0 V or even go negative and the thyristor will continue to pass current. To stop the cathode-to-anode current, it is necessary to reverse the cathode-to-anode voltage. The thyristor will again be able to block both forward and reverse voltages until current

flow is initiated by a gate pulse. The schematic symbol for an SCR is shown in Fig. 15.2.4. The physical packaging of thyristors is similar to that of rectifiers with similar ratings except, of course, that the thyristor must have an additional gate connection.

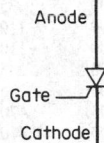

Fig. 15.2.4   Thyristor schematic symbol.

The gate pulse required to fire an SCR is quite small compared with the anode voltage and current. Power gains in the range of 10^6 to 10^9 are easily obtained. In addition, the power loss in the thyristor is very low, compared with the power it controls, so that it is a very efficient power-controlling device. Efficiency in a thyristor power supply is usually 97 to 99 percent. When the thyristor blocks either forward or reverse current, the high voltage drop across the thyristor accompanies low current. When the thyristor is conducting forward current after having been fired by its gate pulse, the high anode current occurs with a forward voltage drop of about 1.5 V. Since high voltage and high current never occur simultaneously, the power dissipation in both the on and off states is low.

The thyristor is rated primarily on the basis of its forward-current capacity and its voltage-blocking capability. Devices are manufactured to have equal forward and reverse voltage-blocking capability. Like diodes, thyristors have I^2t ratings and 1-cycle surge current ratings to allow design of protective circuits. In addition to these ratings, which the SCR shares in common with diodes, the SCR has many additional specifications. Because the thyristor is limited in part by its average current and in part by its rms current, forward-current capacity is a function of the duty cycle to which the device is subjected. Since the thyristor cannot regain its blocking ability until its anode voltage is reversed and remains reversed for a short time, this time must be specified. The time to regain blocking ability after the anode voltage has been reversed is called the **turn-off time.** Specifications are also given for minimum and maximum gate drive. If forward blocking voltage is reapplied too quickly, the SCR may fire with no applied gate voltage pulse. The maximum safe value of rate of reapplied voltage is called the *dv/dt* rating of the SCR. When the gate pulse is applied, current begins to flow in the area immediately adjacent to the gate junction. Rather quickly, the current spreads across the entire cathode-junction area. In some circuits associated with the thyristor an extremely fast rate of rise of current may occur. In this event localized heating of the cathode may occur with a resulting immediate failure or in less extreme cases a slow degradation of the thyristor. The maximum rate of change of current for a thyristor is given by its *di/dt* rating. Design for *di/dt* and *dv/dt* limits is not normally a problem at power-line frequencies of 50 and 60 Hz. These ratings become a design factor at frequencies of 500 Hz and greater. Table 15.2.1 lists typical thyristor characteristics.

A **triac** is a bilateral SCR. It blocks current in either direction until it receives a gate pulse. It can be used to control in ac circuits. Triacs are widely used for light dimmers and for the control of small universal ac motors. The triac must regain its blocking ability as the line voltage crosses through zero. This fact limits the use of triacs to 60 Hz and below.

A **transistor** is a semiconductor amplifier. The schematic symbol for a transistor is shown in Fig. 15.2.5. There are two types of transistors, *p-n-p* and *n-p-n.* Notice that the polarities of voltage applied to these devices are opposite. In many sizes matched *p-n-p* and *n-p-n* devices are available. The most common transistors have a collector dissipation rating of 150 to 600 mW. Collector to base breakdown voltage is 20 to 50 V. The amplification or gain of a transistor occurs because of two facts: (1) A small change in current in the base circuit causes a large

Table 15.2.1 Typical Thyristor Characteristics

| Voltage | Current, A | | | 1-cycle surge, A | di/dt, A/μs | dv/dt, V/μs | Turn-off time, μs |
	rms	avg	I^2t, A²·s				
400	35	20	165	180	100	200	10
1,200	35	20	75	150	100	200	10
400	110	70	4,000	1,000	100	200	40
1,200	110	70	4,000	1,000	100	200	40
400	235	160	32,000	3,500	100	200	80
1,200	235	160	32,000	3,500	75	200	80
400	470	300	120,000	5,500	50	100	150
1,200	470	300	120,000	5,500	50	100	150

change in current in the collector and emitter leads. This current amplification is designated *hfe* on most transistor specification sheets. (2) A small change in base-to-emitter voltage can cause a large change in either the collector-to-base voltage or the collector-to-emitter voltage. Table 15.2.2 shows basic ratings for some typical transistors. There is a great profusion of transistor types so that the choice of type depends upon availability and cost as well as operating characteristics.

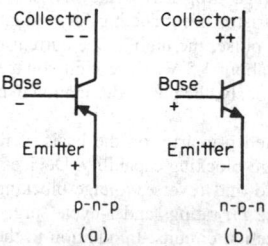

Fig. 15.2.5 Transistor schematic symbol.

The gain of a transistor is independent of frequency over a wide range. At high frequency, the gain falls off. This cutoff frequency may be as low as 20 kHz for audio transistors or as high as 1 GHz for radio-frequency (rf) transistors.

The schematic symbols for the field effect transistor (FET) is shown in Fig. 15.2.6. The flow of current from source to drain is controlled by an electric field established in the device by the voltage applied between the gate and the drain. The effect of this field is to change the resistance of the transistor by altering its internal current path. The FET has an extremely high gate resistance (10^{12} Ω), and as a consequence, it is used for applications requiring high input impedance. Some FETs have been designed for high-frequency characteristics. The two basic constructions used for FETs are **bipolar junctions** and **metal oxide semiconductors**. The schematic symbols for each of these are shown in Fig. 15.2.6*a* and 15.2.6*b*. These are called JFETs and MOSFETs to distinguish between them. JFETs and MOSFETs are used as stand-alone devices and are also widely used in integrated circuits. (See below, this section.)

A MOSFET with higher current capacity is called a **power MOSFET**. Table 15.2.3 shows some typical characteristics for power MOSFETs. Power MOSFETs are somewhat more limited in maximum power than

are thyristors. Circuits with multiple power MOSFETs are limited to about 20 kW. Thyristors are limited to about 1500 kW. As a point of interest, there are electric power applications in the hundreds of megawatts which incorporate massively series and parallel thyristors. An

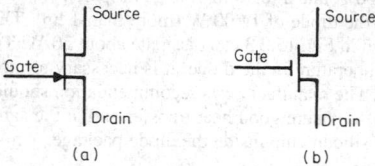

Fig. 15.2.6 Field-effect transistor. (*a*) Bipolar junction type (JFET); (*b*) metal-oxide-semiconductor type (MOSFET).

advantage of power MOSFETs is that they can be turned on and turned off by means of the gate-source voltage. Thus low-power electric control can turn the device on and off. Thyristors can be turned on only by their gate voltage. To turn a thyristor off it is necessary for its high-powered anode-to-cathode voltage to be reversed. This factor adds

Table 15.2.3 Typical Power MOSFET Characteristics

Drain voltage	Device	Drain amperes	Power dissipation, W	Case type
600	MTH6N60	6.0	150	TO-218
400	MTH8N40	8.0	150	TO-218
100	MTH25N10	25	150	TO-218
50	MTH35N05	35	150	TO-218
600	MTP1N60	1.0	75	TO-220
400	MTP2N40	2.0	75	TO-220
100	MTP10N10	10	75	TO-220
200	MTE120N20	120	500	346-01
100	MTE150N10	150	500	346-01
50	MTE200N05	200	500	346-01

Table 15.2.2 Typical Transistor Characteristics

JEDEC number	Type	Collector-emitter volts at breakdown, BV_{CE}	Collector dissipation, P_c(25°C)	Collector current, I_c	Current gain, h_{fe}
2N3904	*n-p-n*	40	310 mW	200 mA	200
2N3906	*p-n-p*	40	310 mW	200 mA	200
2N3055	*n-p-n*	100	115 W	15 A	20
2N6275	*n-p-n*	120	250 W	50 A	30
2N5458	JFET	40	200 mW	9 mA	*
2N5486	JFET	25	200 mW		†

* JFET, current gain is not applicable.
† High-frequency JFET—up to 400 MHz.

complication to many thyristor circuits. Another power device which is similar to the power MOSFET is the **insulated gate-bipolar transistor (IGBT)**. This device is a **Darlington** combination of a MOSFET and a bipolar transistor (see Fig. 15.2.13). A low-power MOSFET first transistor drives the base of a second high-power bipolar transistor. These two transistors are integrated in a single case. The IGBT is applied in high-power devices which use high-frequency switching. IGBTs are available in ratings similar to thyristors (see Table 15.2.1), so they are power devices. The advantage of the IGBT is that it can be switched on and off by means of its gate. The advantage of an IGBT over a power MOSFET is that it can be made with higher power ratings.

The **unijunction** is a special-purpose semiconductor device. It is a pulse generator and widely used to fire thyristors and triacs as well as in timing circuits and waveshaping circuits. The schematic symbol for a unijunction is shown in Fig. 15.2.7. The device is essentially a silicon resistor. This resistor is connected to base 1 and base 2. The emitter is fastened to this resistor about halfway between bases 1 and 2. If a positive voltage is applied to base 2, and if the emitter and base 1 are at zero, the emitter junction is back-biased and no current flows in the emitter. If the emitter voltage is made increasingly positive, the emitter junction will become forward-biased. When this occurs, the resistance between base 1 and base 2 and between base 2 and the emitter suddenly switches to a very low value. This is a regenerative action, so that very fast and very energetic pulses can be generated with this device.

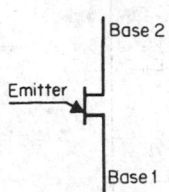

Fig. 15.2.7   Unijunction.

Before the advent of semiconductors, electronic rectifiers and amplifiers were **vacuum tubes** or **gas-filled tubes**. Some use of these devices still remains. If an electrode is heated in a vacuum, it gives up surface electrons. If an electric field is established between this heated electrode and another electrode so that the electrons are attracted to the other electrode, a current will flow through the vacuum. Electrons flow from the heated cathode to the cold anode. If the polarity is reversed, since there are no free electrons around the anode, no current will flow. This, then, is a vacuum-tube rectifier. If a third electrode, called a **control grid,** is placed between the cathode and the anode, the flow of electrons from the cathode to the anode can be controlled. This is a basic vacuum-tube amplifier. Additional grids have been placed between the cathode and anode to further enhance certain characteristics of the vacuum tube. In addition, multiple anodes and cathodes have been enclosed in a single tube for special applications such as radio signal converters.

If an inert gas, such as neon or argon, is introduced into the vacuum, conduction can be initiated from a cold electrode. The breakdown voltage is relatively stable for given gas and gas pressure and is in the range of 50 to 200 V. The **nixie** display tube is such a device. This tube contains 10 cathodes shaped in the form of the numerals from 0 to 9. If one of these cathodes is made negative with respect to the anode in the tube, the gas in the tube glows around that cathode. In this way each of the 10 numerals can be made to glow when the appropriate electrode is energized.

An **ignitron** is a vapor-filled tube. It has a pool of liquid mercury in the bottom of the tube. Air is exhausted from the enclosure, leaving only mercury vapor, which comes from the pool at the bottom. If no current is flowing, this tube will block voltage whether the anode is plus or minus with respect to the mercury-pool cathode. A small rod called an **ignitor** can form a cathode spot on the pool of mercury when it is withdrawn from the pool. The ignitor is pulled out of the pool by an electromagnet. Once the cathode spot has been formed, electrons will continue to flow from the mercury-pool cathode to the anode until the anode-to-cathode voltage is reversed. The operation of an ignitron is very similar to that of a thyristor. The anode and cathode of each device perform similar functions. The ignitor and gate also perform similar functions. The thyristor is capable of operating at much higher frequencies than the ignitron and is much more efficient since the thyristor has 1.5 V

forward drop and the ignitron has 15 V forward drop. The ignitron has an advantage over the thyristor in that it can carry extremely high overload currents without damage. For this reason ignitrons are often used as electronic "crowbars" which discharge electrical energy when a fault occurs in a circuit.

Discrete-Component Circuits

Several common rectifier circuits are shown in Fig. 15.2.8. The waveforms shown in this figure assume no line reactance. The presence of line reactance will make a slight difference in the waveshapes and the conversion factors shown in Fig. 15.2.8. These waveshapes are equally applicable for loads which are pure resistive or resistive and inductive. In a resistive load the current flowing in the load has the same waveshape as the voltage applied to it. For inductive loads, the current waveshape will be smoother than the voltage applied. If the inductance is high enough, the ripple in the current may be indeterminantly small. An approximation of the ripple current can be calculated as follows:

$$I = \frac{E_{dc}\,PCT}{200\,\pi fNL} \qquad (15.2.1)$$

where I = rms ripple current, E_{dc} = dc load voltage, PCT = percent ripple from Fig. 15.2.8, f = line frequency, N = number of cycles of ripple frequency per cycle of line frequency, L = equivalent series inductance in load. Equation (15.2.1) will always give a value of ripple higher than that calculated by more exact means, but this value is normally satisfactory for power-supply design.

Capacitance in the load leads to increased regulation. At light loads, the capacitor will tend to charge up to the peak value of the line voltage and remain there. This means that for either the single full-wave circuit or the single-phase bridge the dc output voltage would be 1.414 times the rms input voltage. As the size of the loading resistor is reduced, or as the size of the parallel load capacitor is reduced, the load voltage will more nearly follow the rectified line voltage and so the dc voltage will approach 0.9 times the rms input voltage for very heavy loads or for very small filter capacitors. One can see then that dc voltage may vary between 1.414 and 0.9 times line voltage due only to waveform changes when **capacitor filtering** is used.

Four different **thyristor rectifier circuits** are shown in Fig. 15.2.9. These circuits are equally suitable for resistive or inductive loads. It will be noted that the half-wave circuit for the thyristor has a rectifier across the load, as in Fig. 15.2.8. This diode is called a **freewheeling diode** because it freewheels and carries inductive load current when the thyristor is not conducting. Without this diode, it would not be possible to build up current in an inductive load. The gate-control circuitry is not shown in Fig. 15.2.9 in order to make the power circuit easier to see. Notice the location of the thyristors and rectifiers in the single-phase full-wave circuit. Constructed this way, the two diodes in series perform the function of a free-wheeling diode. The circuit can be built with a thyristor and rectifier interchanged. This would work for resistive loads but not for inductive loads. For the full three-phase bridge, a freewheeling diode is not required since the carryover from the firing of one SCR to the next does not carry through a large portion of the negative half cycle and therefore current can be built up in an inductive load.

Capacitance must be used with care **in thyristor circuits**. A capacitor directly across any of the circuits in Fig. 15.2.9 will immediately destroy the thyristors. When an SCR is fired directly into a capacitor with no series resistance, the resulting di/dt in the thyristor causes extreme local heating in the device and a resultant failure. A sufficiently high series resistor prevents failure. An inductance in series with a capacitor must also be used with caution. The series inductance may cause the capacitor to "ring up." Under this condition, the voltage across the capacitor can approach twice peak line voltage or 2.828 times rms line voltage.

The advantage of the thyristor circuits shown in Fig. 15.2.9 over the rectifier circuits is, of course, that the thyristor circuits provide variable output voltage. The output of the thyristor circuits depends upon the magnitude of the incoming line voltage and the phase angle at which the

Type	Circuit	Output voltage waveform	E_{dc} (avg)	Ripple fundamental frequency	% ripple	Peak inverse voltage
Half-wave 1φ			0.318 E_M 0.45 E_{ac}	F	121	3.14 E_{dc}
Full-wave 1φ			0.636 E_M 0.9 E_{ac}	2F	48	3.14 E_{dc}
Bridge 1φ			0.636 E_M 0.9 E_{ac}	2F	48	1.57 E_{dc}
Half-wave 3φ			0.827 E_M 1.17 E_{ac}	3F	18	2.09 E_{dc}

E_M = maximum value of e_{ac}
E_{ac} = effective value of e_{ac}
E_{dc} = average value of d-c load voltage
F = line frequency
% ripple = 100 x rms ripple/ E_{dc}

Fig. 15.2.8 Comparison of rectifier circuits.

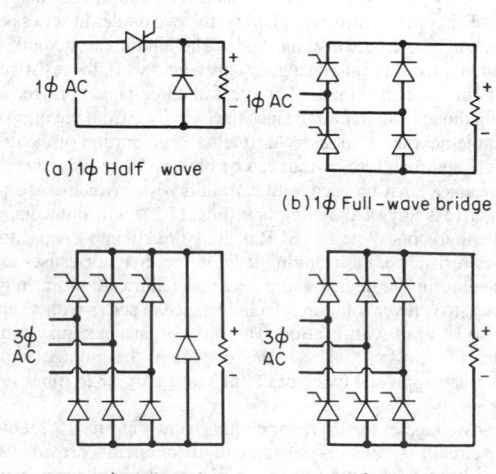

(a) 1φ Half wave

(b) 1φ Full-wave bridge

(c) 3φ Half bridge

(d) 3φ Full bridge

Fig. 15.2.9 Basic thyristor circuits.

thyristors are fired. The control characteristic for the thyristor power supply is determined by the waveshape of the output voltage and also by the phase-shifting scheme used in the firing-control means for the thyristor. Practical and economic power supplies usually have control characteristics with some degree of nonlinearity. A representative characteristic is shown in Fig. 15.2.10. This control characteristic is usually given for nominal line voltage with the tacit understanding that variations in line voltage will cause approximately proportional changes in output voltage.

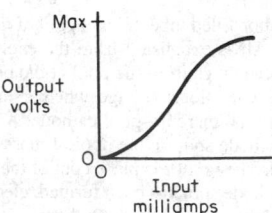

Fig. 15.2.10 Thyristor control characteristic.

Transistor amplifiers can take many different forms. A complete discussion is beyond the scope of this handbook. The circuits described here illustrate basic principles. A basic **single-stage amplifier** is shown in Fig. 15.2.11. The transistor can be cut off by making the input terminal sufficiently negative. It can be saturated by making the input terminal sufficiently positive. In the linear range, the base of an *n-p-n* transistor will be 0.5 to 0.7 V positive with respect to the emitter. The collector voltage will vary from about 0.2 V to V_c (20 V, typically). Note that there is a sign inversion of voltage between the base and the collector; i.e., when the base is made more positive, the collector becomes less positive. The resistors in this circuit serve the following functions. Resistor *R*1 limits the input current to the base of the transistor so that it is not harmed when the input signal overdrives. Resistors *R*2 and *R*3 establish the transistor's operating point with no input signal. Resistors *R*4 and *R*5 determine the voltage gain of the amplifier. Resistor *R*4 also serves to stabilize the zero-signal operating point, as established by resistors *R*2 and *R*3. Usual practice is to design single-stage gains of 10 to 20. Much higher gains are possible to achieve, but low gain levels permit the use of less expensive transistors and increase circuit reliability.

Figure 15.2.12 illustrates a basic **two-stage transistor amplifier** using complementary *n-p-n* and *p-n-p* transistors. Note that the first stage is

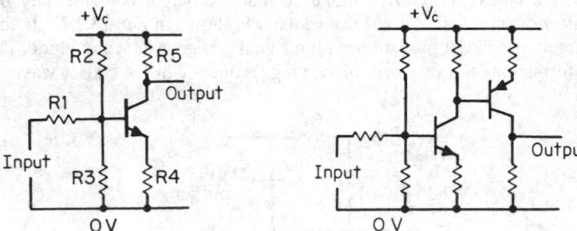

Fig. 15.2.11 Single-stage amplifier.   **Fig. 15.2.12** Two-stage amplifier.

identical to that shown in Fig. 15.2.11. This *n-p-n* stage drives the following *p-n-p* stage. Additional alternate *n-p-n* and *p-n-p* stages can be added until any desired overall amplifier gain is achieved.

Figure 15.2.13 shows the **Darlington** connection of transistors. The amplifier is used to obtain maximum current gain from two transistors.

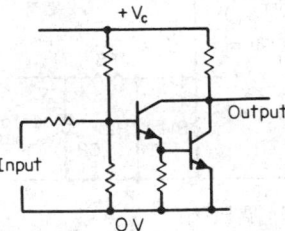

Fig. 15.2.13 Darlington connection.

Assuming a base-to-collector current gain of 50 times for each transistor, this circuit will give an input-to-output current gain of 2,500. This high level of gain is not very stable if the ambient temperature changes, but in many cases this drift is tolerable.

Figure 15.2.14 shows a circuit developed specifically to minimize

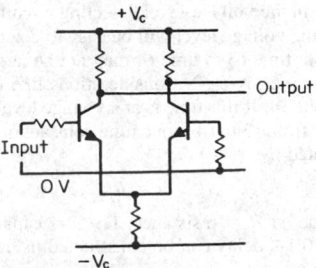

Fig. 15.2.14 Differential amplifier.

temperature drift and drift due to power supply voltage changes. The **differential amplifier** minimizes drift because of the balanced nature of the circuit. Whatever changes in one transistor tend to increase the output are compensated by reverse trends in the second transistor. The input signal does not affect both transistors in compensatory ways, of course, and so it is amplified. One way to look at a differential amplifier is that twice as many transistors are used for each stage of amplification to achieve compensation. For very low drift requirements, matched transistors are available. For the ultimate in differential amplifier performance, two matched transistors are encapsulated in a single unit. **Operational amplifiers** made with discrete components frequently use differential amplifiers to minimize drift and offset. The operational amplifier is a low-drift, high-gain amplifier designed for a wide range of control and instrumentation uses.

Oscillators are circuits which provide a frequency output with no signal input. A portion of the collector signal is fed back to the base of the transistor. This feedback is amplified by the transistor and so maintains a sustained oscillation. The frequency of the oscillation is determined by parallel inductance and capacitance. The oscillatory circuit consisting of an inductance and a capacitance in parallel is called an *LC* **tank circuit**.

This frequency is approximately equal to

$$f = 1/2\pi\sqrt{CL} \qquad (15.2.2)$$

where f = frequency, Hz; C = capacitance, F; L = inductance, H. A 1-MHz oscillator might typically be designed with a 20-μH inductance in parallel with a 0.05-μF capacitor. The exact frequency will vary from the calculated value because of loading effects and stray inductance and capacitance. The **Colpitts** oscillator shown in Fig. 15.2.15 differs from the **Hartley** oscillator shown in Fig. 15.2.16 only in the way energy is fed back to the emitter. The Colpitts oscillator has a capacitive voltage

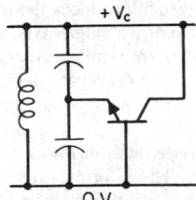   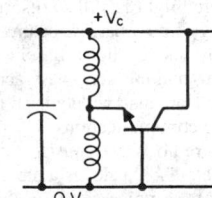

Fig. 15.2.15 Colpitts oscillator.   **Fig. 15.2.16** Hartley oscillator.

divider in the resonant tank. The Hartley oscillator has an inductive voltage divider in the tank. The **crystal oscillator** shown in Fig. 15.2.17 has much greater frequency stability than the circuits in Figs. 15.2.15 and 15.2.16. Frequency stability of 1 part in 10^7 is easily achieved with a crystal-controlled oscillator. If the oscillator is temperature-controlled by mounting it in a small temperature-controlled oven, the frequency

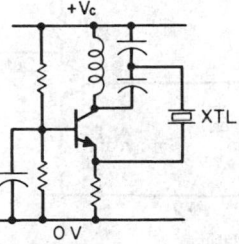

Fig. 15.2.17 Crystal-controlled oscillator.

stability can be increased to 1 part in 10^9. The resonant *LC* tank in the collector circuit is tuned to approximately the crystal frequency. The crystal offers a low impedance at its resonant frequency. This pulls the collector-tank operating frequency to the crystal resonant frequency.

As the desired operating frequency becomes 500 MHz and greater, **resonant cavities** are used as tank circuits instead of discrete capacitors and inductors. A rough guide to the relationship between frequency and resonant-cavity size is the wavelength of the frequency

$$\lambda = 300 \times 10^6/f \qquad (15.2.3)$$

where λ = wavelength, m; 300×10^6 = speed of light, m/s; f = frequency, Hz. The resonant cavities will be smaller than indicated by Eq. (15.2.3) because in general the cavity is either one-half or one-fourth wavelength and also, in general, the electromagnetic wave velocity is less in a cavity than in free space.

The operating principles of these devices are beyond the scope of this article. There are many different kinds of **microwave tubes** including **klystrons, magnetrons,** and **traveling-wave** tubes. All these tubes employ moving electrons to excite a resonant cavity. These devices serve as either oscillators or amplifiers at microwave frequencies.

Lasers operate at a frequency of light of approximately 600 THz. This corresponds to a wavelength of 0.5 μm, or, the more usual measure of visual-light wavelength, 5×10^3 Å. Most laser oscillators are basically a variation of the Fabry-Perot etalon, or interferometer. High-power lasers may be gas, liquid, or ruby-based devices. High-power lasers are used for machining and surveying. A more recent device, invented in 1961, is the semiconductor laser. The semiconductor laser is constructed using gallium arsenide (GaAs) as the semiconductor material. These devices are quite small and can be controlled by means of field-effect transistors. The GaAs laser can be modulated (turned on and off) at a 10 GHz rate, making it ideal for modern fiber-optic communications.

Most light is disorganized insofar as the axis of vibration and the frequency of vibration are concerned. When radiation along different axes is attenuated, as with a polarizing screen, the light is said to be *polarized.* White light contains all visible frequencies. When white light is filtered, the remaining light is colored, or frequency-limited. A single color of light still contains a broad range of frequencies. Polarized, colored light is still so disorganized that it is difficult to focus the light energy into a narrow beam. Laser light is inherently a single-axis, single-frequency light. Lasers can be focused into an extremely narrow beam, making them very accurate cutting tools and surveying devices. Lasers are also widely used for high-speed, wide-frequency-band fiber-optic communications.

A radio wave consists of two parts, a **carrier,** and an information signal. The carrier is a steady high frequency. The information signal may be a voice signal, a video signal, or telemetry information. The carrier wave can be modulated by varying its amplitude or by varying its frequency. **Modulators** are circuits which impress the information signal onto the carrier. A **demodulator** is a circuit in the receiving apparatus which separates the information signal from the carrier. A simple amplitude modulator is shown in Fig. 15.2.18. The transistor is base-driven with the carrier input and emitter driven with the information signal. The modulated carrier wave appears at the collector of the transistor. An

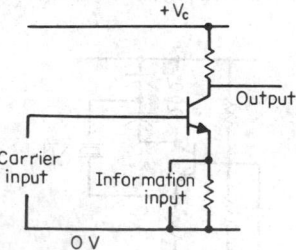

Fig. 15.2.18 AM modulator.

FM modulator is shown in Fig. 15.2.19. The carrier must be changed in frequency in response to the information signal input. This is accomplished by using a saturable ferrite core in the inductance of a Colpitts oscillator which is tuned to the carrier frequency. As the collector current in transistor $T1$ varies with the information signal, the saturation level in the ferrite core changes, which in turn varies the inductance of

the winding in the tank circuit and alters the operating frequency of the oscillator.

The **demodulator** for an AM signal is shown in Fig. 15.2.20. The diode rectifies the carrier plus information signal so that the filtered

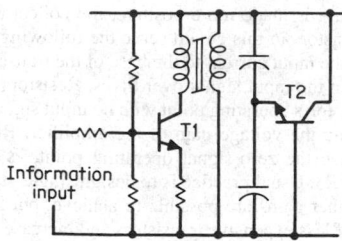

Fig. 15.2.19 FM modulator.

voltage appearing across the capacitor is the information signal. Resistor $R2$ blocks the carrier signal so that the output contains only the information signal. An FM **demodulator** is shown in Fig. 15.2.21. In this circuit, the carrier plus information signal has a constant amplitude. The information is in the form of varying frequency in the carrier wave. If

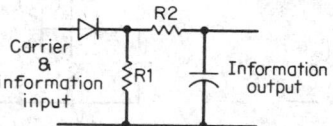

Fig. 15.2.20 AM demodulator.

inductor $L1$ and capacitor $C1$ are tuned to near the carrier frequency but not exactly at resonance, the current through resistor $R1$ will vary as the carrier frequency shifts up and down. This will create an AM signal across resistor $R1$. The diode, resistors $R2$ and $R3$, and capacitor $C2$ demodulate this signal as in the circuit in Fig. 15.2.20.

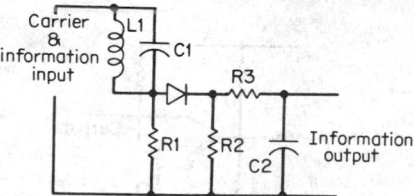

Fig. 15.2.21 FM discriminator.

The waveform of the basic **electronic timing circuit** is shown in Fig. 15.2.22 along with a basic timing circuit. Switch $S1$ is closed from time $t1$ until time $t3$. During this time, the transistor shorts the capacitor and holds the capacitor at 0.2 V. When switch $S1$ is opened at time $t3$, the transistor ceases to conduct and the capacitor charges exponentially due to the current flow through resistor $R1$. Delay time can be measured to any point along this exponential charge. If the time is measured until time $t6$, the timing may vary due to small shifts in supply voltage or slight changes in the voltage-level detecting circuit. If time is measured until time $t4$, the voltage level will be easy to detect, but the obtainable time delay from time $t3$ to time $t4$ may not be large enough compared with the reset time $t1$ to $t2$. Considerations like these usually dictate detecting at time $t4$. If this time is at a voltage level which is 63 percent of V_c, the time from $t3$ to $t4$ is one time constant of $R1$ and C. This time can be calculated by

$$t = RC \qquad (15.2.4)$$

where t = time, s; R = resistance, Ω; C = capacitance, F. A timing circuit with a 0.1-s delay can be constructed using a 0.1-μF capacitor and a 1.0-MΩ resistor.

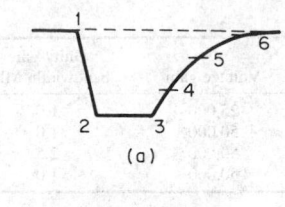

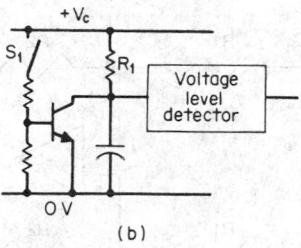

Fig. 15.2.22 Basic timing circuit.

An improved timing circuit is shown in Fig. 15.2.23. In this circuit, the unijunction is used as a level detector, a pulse generator, and a reset means for the capacitor. The transistor is used as a constant current source for charging the timing capacitor. The current through the transistor is determined by resistors $R1$, $R2$, and $R3$. This current is adjustable by means of $R1$. When the charge on the capacitor reaches approximately 50 percent of V_c, the unijunction fires, discharging the capacitor

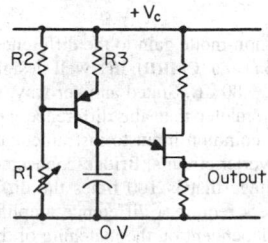

Fig. 15.2.23 Improved timing circuit.

and generating a pulse at the output. The discharged capacitor is then recharged by the transistor, and the cycle continues to repeat. The pulse rate of this circuit can be varied from one pulse per minute to many thousands of pulses per second.

Integrated Circuits

Table 15.2.4 lists some of the more common physical packages for discrete component and integrated semiconductor devices. Although discrete components are still used for electronic design, **integrated circuits (ICs)** are becoming predominant in almost all types of electronic equipment. Dimensions of common dual in-line pin (DIP) integrated-

Table 15.2.4 Semiconductor Physical Packaging

Signal devices	
Plastic	TO92
Metal can	TO5, TO18, TO39
Power devices	
Tab mount	TO127, TO218, TO220
Diamond case	TO3, TO66
Stud mount	
Flat base	
Flat pak (Hockey puck)	
Integrated circuits	
Dip (dual in-line pins)	(See Fig. 15.2.24)
Flat pack	
Chip carrier (50-mil centers)	

circuit devices are shown in Fig. 15.2.24. An IC costs far less than circuits made with discrete components. Integrated circuits can be classified in several different ways. One way to classify them is by complexity. **Small-scale integration (SSI), medium-scale integration (MSI), large-scale integration (LSI),** and **very large scale integration (VLSI)** refer

Number of pins	W	P	L	Approximate height
64	0.8	0.1	3.3	0.24
40	0.6	0.1	2.0	0.2
28	0.6	0.1	1.4	0.2
24	0.6	0.1	1.3	0.2
22	0.4	0.1	1.1	0.2
20	0.3	0.1	1.0	0.2
18	0.3	0.1	0.9	0.2
16	0.3	0.1	0.87	0.2
14	0.3	0.1	0.78	0.2
8	0.3	0.1	0.4	0.2

Fig. 15.2.24 Approximate physical dimensions of dual in-line pin (DIP) integrated circuits. All dimensions are in inches. Dual in-line packages are made in three different constructions—molded plastic, cerdip, and ceramic.

to this kind of classification. The cost and availability of a particular IC are more dependent upon the size of the market for that device than on the level of its internal complexity. For this reason, the classification by circuit complexity is not as meaningful today as it once was. The literature still refers to these classifications, however. For the purpose of this text, ICs will be separated into two broad classes: linear ICs and digital ICs.

The trend in IC development has been toward greatly increased complexity at significantly reduced cost. Present-day ICs are manufactured with internal spacings as low as 0.5 μm. The limitation of the contents of a single device is more often controlled by external connections than by internal space. For this reason, more and more complex combinations of circuits are being interconnected within a single device. There is also a tendency to accomplish functions digitally that were formerly done by analog means. Although these digital circuits are much more complex than their analog counterparts, the cost and reliability of ICs make the resulting digital circuit the preferred design. One can expect these trends will continue based on current technology. One can also anticipate further declines in price versus performance. It has been demonstrated again and again that digital IC designs are much more stable and reliable than analog designs.

For years, the complexity of large-scale integrated circuits doubled each year. This meant that, over a 10-year period, the complexity of a single device increased by 2^{10} times, or over 1000 times, and, over the period from 1960 to 1980 grew from one transistor on a chip to one million transistors on a chip. More recently the complexity increase has fallen off to only 1.7 times, or 1.7^{10}, or over 200 times in a 10-year period. Manufacturers also have developed application-specific integrated circuits (ASICs). These devices allow a circuit designer to design integrated circuits almost as easily as printed board circuits.

Linear Integrated Circuits

The basic building block for many linear ICs is the **operational amplifier**. Table 15.2.5 lists the basic characteristics for a few representative IC operational amplifiers. In most instances, an adequate design for an operational amplifier circuit can be made assuming an "ideal" operational amplifier. For an ideal operational amplifier, one assumes that it

Table 15.2.5 Operational Amplifiers

Type	Purpose	Input bias current, nA	Input res., Ω	Supply voltage, V	Voltage gain	Unity-gain bandwidth, MHz
LM741	General purpose	500	2×10^6	+ 20	25,000	1.0
LM224	Quad gen. purpose	150	2×10^6	3 to 32	50,000	1.0
LM255	FET input	0.1	10^{12}	+ 22	50,000	2.5
LM444A	Quad FET input	0.005	10^{12}	+ 22	50,000	1.0

has infinite gain and no voltage drop across its input terminals. In most designs, feedback is used to limit the gain of each operational amplifier. As long as the resulting closed-loop gain is much less than the open-loop gain of the operational amplifier, this assumption yields results that are within acceptable engineering accuracy. Operational amplifiers use a balanced input circuit which minimizes input voltage offset. Furthermore, specially designed operational amplifiers are available which have extremely low input offset voltage. The input voltage must be kept low because of temperature drift considerations. For these reasons, the assumption of zero input voltage, sometimes called a "virtual ground," is justified. Figure 15.2.25 shows three operational amplifier circuits and the equations which describe their behavior. In this figure S is the **Laplace transform** variable. In these equations the input and output voltages are functions of S. The equations are written in the frequency domain. A simple transformation to steady variable-frequency behavior can be obtained by simply substituting $\cos \omega t$ for the variable S wherever it appears in the equation. By varying ω, the frequency in radians per second, one can obtain the steady-state frequency response of the circuit.

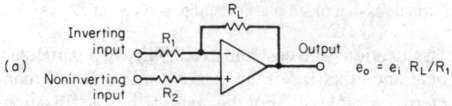

(a)
$$e_o = e_i \, R_L/R_1$$

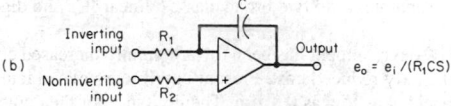

(b)
$$e_o = e_i \, /(R_1CS)$$

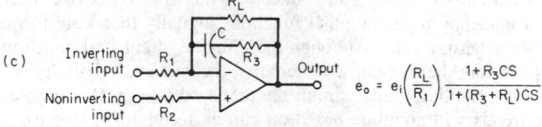

(c)
$$e_o = e_i \left(\frac{R_L}{R_1}\right) \frac{1 + R_3CS}{1 + (R_3 + R_L)CS}$$

Fig. 15.2.25 Operational amplifier circuits.

The operational amplifier circuits shown in Figs. 15.2.25, 15.2.26, 15.2.27, and 15.2.28, show only the signal wires. There are additional connections to a dc power supply, and in some instances, to stabilizing circuits and guard circuits. These connections have been omitted for conceptual clarity.

One of the most useful analog integrated circuits is the **difference amplifier**. This is a balanced input amplifier that is a fundamental component in instrumentation and control applications. The difference amplifier is shown in Fig. 15.2.26. This amplifier maximizes the voltage gain for $V_{\text{output}}/V_{\text{dif}}$, the **difference gain**, and minimizes the voltage gain for $V_{\text{output}}/V_{\text{CM}}$, the **common-mode gain**. There are two resistors in the circuit shown as $R1$ and two resistors shown as $R2$. The resistance values of the two $R1$ resistors are equal, and similarly the resistance values of the two $R2$ resistors are equal. The difference gain is given by

$$\frac{V_{\text{output}}}{V_{\text{dif}}} = -\frac{R2}{R1} \qquad (15.2.5)$$

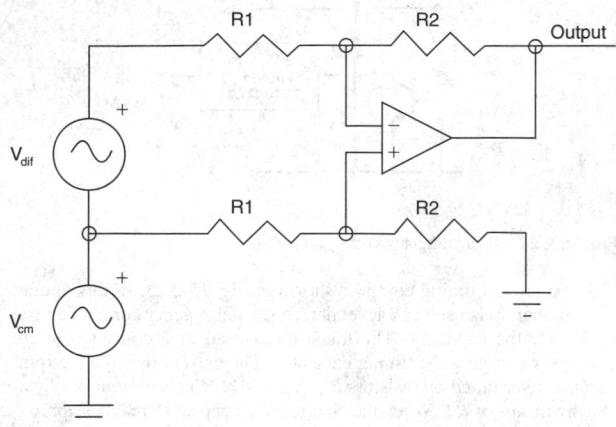

Fig. 15.2.26 Difference amplifier.

The ratio of the common-mode gain to the difference gain is called the **common-mode rejection ratio (CMRR)**. In a well-designed difference amplifier, the CMRR is -80 dB. Stated another way, the common mode gain is 10,000 times smaller than the difference gain. The difference amplifier is the most common input to instrumentation amplifiers and analog-to-digital converter circuits. Bridge-connected transducers have a common-mode voltage that is 100 times the difference voltage, or more. Such transducers require a difference amplifier. The common-mode gain is highly dependent on the matching of the $R1$ resistors and the $R2$ resistors. A 1 percent difference in these resistors causes the CMRR to be degraded to -30 or -40 dB.

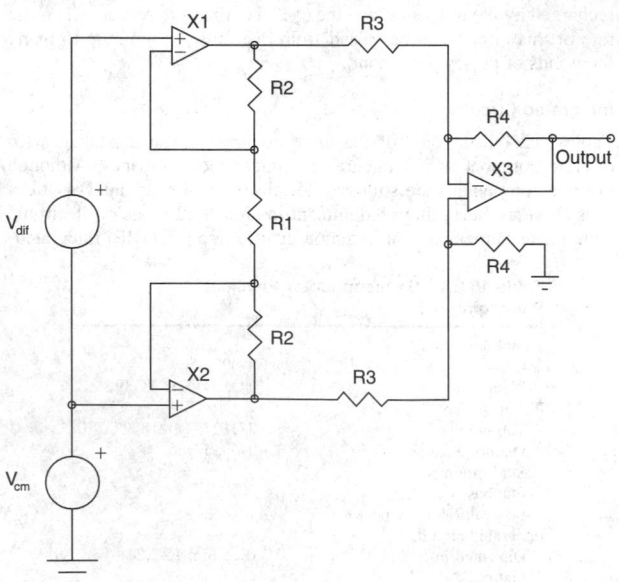

Fig. 15.2.27 Instrumentation amplifier.

An **instrumentation amplifier** is a high-grade difference amplifier. Although there are many implementations of the instrumentation amplifier, the three op-amp circuit is quite common, and will illustrate this device. Figure 15.2.27 shows the three op-amp instrumentation amplifier. This is a two-stage amplifier. The first stage is composed of amplifiers $X1$ and $X2$ and resistors $R1$, $R2$, and $R2$. The second stage is

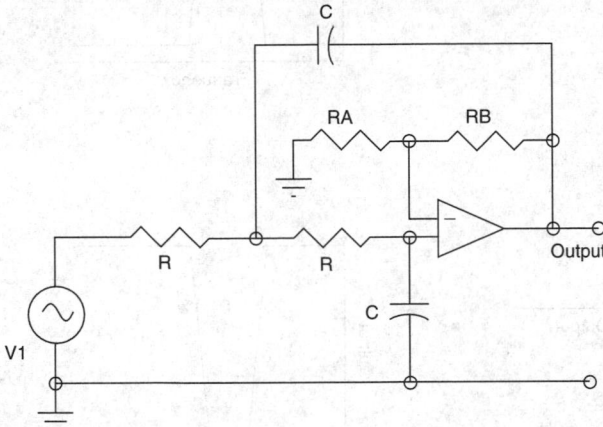

Fig. 15.2.28   Modified Sallen-Key filter circuit.

composed of amplifier $X3$ and resistors $R3$, $R3$, $R4$, and $R4$. The CMRR is usually much lower for the instrumentation amplifier, typically -120 dB. The differential mode gain is much higher for the instrumentation amplifier. For stability and frequency response considerations, the difference gain of a normal difference amplifier is usually 10 or less. For the instrumentation amplifier, the difference gain is typically 1000. In the circuit shown in Fig. 15.2.27, the difference gain of the first stage is given by

$$\frac{V_{\text{internal}}}{V_{\text{dif}}} = \frac{(2 \times R2 + R1)}{R1} \tag{15.2.6}$$

The difference gain of the second is given by

$$\frac{V_{\text{internal}}}{V_{\text{internal}}} = -\frac{R4}{R3} \tag{15.2.7}$$

The overall gain for the amplifier is the product of the individual stage gains. In the typical case the gain of the first stage is 100, and for the second stage is 10, giving an overall gain of 1,000. Integrated circuit instrumentation amplifiers are available which include all of the circuitry in Fig. 15.2.27. The resistors are matched so that the circuit designer does not have to contend with component matching.

A word about cost is in order. In 1994, high-grade operational amplifiers cost $0.50 for four op amps in a single IC chip, or about $0.125 for each amplifier. The instrumentation amplifier is somewhat more expensive, but can be obtained for less than $5.00.

Filtering of electronic signals is often required. A filter passes some frequencies and suppresses others. Filters may be classed as low pass, high pass, band pass, or band stop. Figure 15.2.28 shows a low-pass active filter circuit which is a modification of the Sallen-Key circuit. In this circuit, the two resistors labeled R are matched and the two capacitors labeled C are matched. The frequency response of this circuit is given by its transfer function:

$$\frac{V_{\text{output}}}{V_{\text{input}}} = \frac{1}{S^2 + \dfrac{\omega_N}{Q}S + \omega_N^2} \tag{15.2.8}$$

In this equation, ω_N is the resonant frequency of this circuit, in radians per second, and Q is a quality factor that indicates the amount of

amplitude increase at the resonant frequency. For the circuit shown in Fig. 15.2.28, ω_N is given by

$$\omega_N = \frac{1}{RC} \tag{15.2.9}$$

and Q is given by

$$Q = \frac{1}{2 - (RB/RA)} \tag{15.2.10}$$

The filter shown in Fig. 15.2.28 is a two-pole low-pass filter. Higher-order filters—four pole, six, etc.—can be constructed by cascading sections of two-pole filters. Thus, a six-pole low-pass filter would consist of three circuits of the configuration shown in the figure. The components R, R, C, C, RA, and RB would vary in each filter section.

Continuing with the low-pass filter, there are three common variations in filter response: Butterworth response, Chebychev response, and elliptical response. The Butterworth filter has no ripple in either the pass band or the stop band. The Chebychev filter has equal ripple variations in the pass band, but is flat in the stop band, and the elliptical filter has equal ripple variations in both the pass band and the stop band. There is a transition band of frequencies between the pass band and the stop band. The Butterworth filter has the greatest transition band. The Chebychev response has a sharper cutoff frequency characteristic than the Butterworth response, and the elliptical response has the sharpest transition from pass band to stop band. The circuits shown in Figs. 15.2.27 and 15.2.28 can be used to realize Butterworth and Chebychev filters. The elliptical filter requires a more complex circuit.

The high-pass filter can be formulated by substituting $1/S$ for S in Eq. (15.2.8). The circuit configuration for a high-pass filter is shown in Fig. 15.2.29. Notice that only the position of the Rs and Cs have changed. In general, the values of Rs, Cs, RA, and RB will change for the high-frequency filter, so it would be inappropriate to design a low-pass filter and simply reverse the positions of the Rs and Cs.

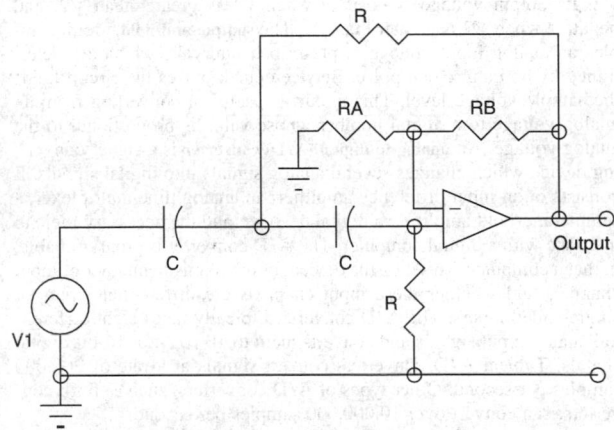

Fig. 15.2.29   High-pass filter section.

Band-pass and band-stop filters can be made by combinations of low-pass and high-pass filter sections. Figure 15.2.30a shows the diagram configuration of filters to realize a band-pass filter. The low-pass filter would be designed for the high-frequency transition, and the high-pass filter would be designed for the low-frequency transition. Figure 15.2.30b shows the block diagram configuration for a band-stop filter. The band-stop filter and the low-pass filter would be designed for the low-frequency transition, and the high-pass filter would be designed for the high-frequency transition. Band-pass and band-stop filters may also be realized by means of special high-Q circuits. A more complete discussion of filter technology can be found in the references at the beginning of this section.

Table 15.2.6 lists some typical linear ICs, most of which contain operational amplifiers with additional circuitry. The **voltage comparator**

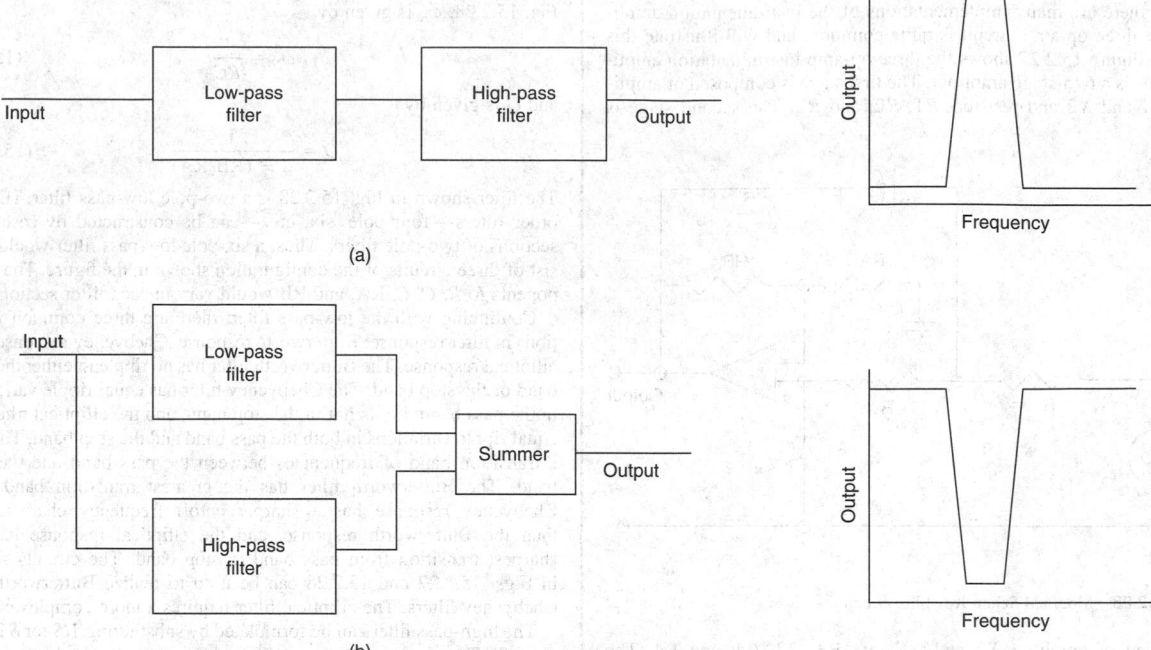

Fig. 15.2.30 (*a*) Band-pass and (*b*) band-stop filters.

is an operational amplifier that compares two input voltages, $V1$ and $V2$. Its output voltage is positive when $V1$ is greater than $V2$, and negative when $V2$ is greater than $V1$. The **sample-and-hold** circuit samples an analog input voltage at prescribed intervals, which are determined by an input clock pulse. Between clock pulses the circuit holds the sample voltage level. This circuit is useful in converting from an analog voltage to a digital number whose value is proportional to the analog voltage. An **analog-to-digital (A/D) converter** is a signal-converting device which changes several analog signals into digital signals. It consists of an input difference amplifier, an analog time multiplexer, a sample-and-hold amplifier, a digital decoder, and the necessary logic to interface with a digital computer. The A/D converter is programmable, in that a computer can set up the device for the requisite number of input channels and whether these input channels are differential inputs or single-ended inputs. The A/D converter typically takes eight differential analog input signals and converts them to 10-, 12-, or 16-bit digital signals. Typical A/D converters convert signals at a rate of 200,000 samples per second. Other types of A/D converters, such as flash converters, can convert over 10,000,000 samples per second.

A companion circuit to the A/D converter is the **digital-to-analog (D/A) converter**, which is used to convert from digital signals to analog signals. This is also a programmable device, but in general the D/A converter is less complex than is the A/D converter. **Voltage-regulators** and **voltage references** are electronic circuits which create precision dc voltage sources. The voltage reference is more precise than the voltage regulator. The **voltage-controlled oscillator** is a circuit which converts from a dc signal to a proportional ac frequency. The output of a voltage-controlled oscillator is usually a rectangular ac wave rather than a sine wave. The **NE555 timer/oscillator** is a general-purpose timer/oscillator which has been integrated into a single IC chip. It can function as a **monostable multivibrator**, a **free-running multivibrator**, or as a **synchronized multivibrator**. It can also be used as a **linear ramp generator**, or for time delay or sequential timing applications.

Table 15.2.7 lists linear ICs that are used in audio, radio, and television circuits. The degree of complexity that can be incorporated in a single device is illustrated by the fact that a complete AM-FM radio circuit is available in a single IC device. The **phase-locked loop** is a device that is widely utilized for accurate frequency control. This device produces an output frequency that is set by a digital input. It is a highly accurate and stable circuit. This circuit is often used to demodulate FM radio waves.

Table 15.2.8 lists linear IC circuits that are used in telecommunications. These circuits include digital circuits within them and/or are used with digital devices. Whether these should be classed as linear ICs or

Table 15.2.6   Linear Integrated-Circuit Devices

Operational amplifier	Voltage comparator
Sample and hold	Digital-to-analog converter
Analog-to-digital converter	Voltage reference
Voltage regulator	NE555 Timer/oscillator
Voltage-controlled oscillator	

Table 15.2.7   Audio, Radio, and Television Integrated-Circuit Devices

Audio amplifier	Tone-volume-balance circuit
Dolby filter circuit	Phase-locked loop (PLL)
Intermediate frequency circuit	AM-FM radio
TV chroma demodulator	Digital tuner
Video-IF amplifier-detector	

Table 15.2.8   Telecommunication Integrated-Circuit Devices

Radio-control transmitter-encoder
Radio-control receiver-decoder
Pulse-code modulator–coder-decoder (PCM CODEC)
Single-chip programmable signal processor
Touch-tone generators
Modulator-demodulator (modem)

digital ICs may be questioned. Several manufacturers include them in their linear device listings and not with their digital devices, and for this reason, they are listed here as linear devices. The radio-control **transmitter-encoder** and **receiver-decoder** provide a means of sending up to four control signals on a single radio-control frequency link. Each of the four channels can be either an on-off channel or a **pulse-width-modulated** (**PWM**) proportional channel. The **pulse-code modulator–coder-decoder** (**PWM CODEC**) is typical of a series of IC devices that have been designed to facilitate the design of digital-switched telephone circuits.

Integrated circuits are normally divided into two classes: linear or analog ICs and digital ICs. There are a few hybrid circuits which have both analog and digital characteristics. Some representative hybrid integrated circuits are listed in Table 15.2.9, and shown in Fig. 15.2.31. In the **superdiode,** there are two modes of operation. When the input signal is positive, the output of the op amp is negative, causing diode $D2$ to conduct and diode $D1$ to block current flow. When the input is negative, diode $D2$ blocks and diode $D1$ conducts. The output of the circuit is zero when the input is negative, and the output is proportional to the input when the input is positive. A normal diode has 0.5 to 1.0 V of forward voltage drop. This circuit is called a super diode because its switching point is at zero volts. There is a voltage inversion from input to output, but this can be reversed by means of an additional op-amp inverter following the super diode. The action of the circuit can be reversed by reversing diodes $D1$ and $D2$. The **limiter** circuit, shown in Fig. 15.2.31b, provides linear operation at low output voltages, either positive or negative, and neither diode is conducting. When the input voltage goes sufficiently negative, diode $D1$ begins to conduct, limiting the negative output of the op-amp. Similarly when the input goes sufficiently positive, diode $D2$ begins to conduct, limiting the positive output of the op-amp. The **Schmitt trigger** circuit, shown in Fig. 15.2.31c, has positive feedback to the op-amp, through resistors $R1$ and $R2$. If the amplifier output is initially negative, the output will not change until the input becomes as negative as the non-inverting terminal of the op-amp. When the input becomes slightly more negative, the op-amp output will suddenly switch from negative to positive due to the positive feedback. The op-amp will remain in this condition until the input becomes sufficiently positive, causing the op-amp to switch to a negative output.

The **dead-band** circuit is similar to the limiter circuit except that for input voltages around zero, there is no output voltage. Above a threshold input voltage, the output voltage is proportional to the input voltage. The logarithmic amplifier relies upon the fact that the voltage drop across a diode causes logarithmically varying current to flow through the diode. This relationship is remarkably true for current changes of 10^6 to 1. The logarithmic amplifiers can have their outputs added together to effect a multiplication of the input signals.

Digital Integrated Circuits

The basic circuit building block for digital ICs is the gate circuit. A gate is a switching amplifier that is designed to be either on or off. (By contrast, an operational amplifier is a proportional amplifier.) For 5-V logic levels, the gate switches to a 0 whenever its input falls below 0.8 V and to a 1 whenever its input exceeds 2.8 V. This arrangement ensures immunity to spurious noise impulses in both the 0 and the 1 state.

Several representative **transistor-transistor-logic** (**TTL**) **gates** are listed in Table 15.2.10. Gates can be combined to form logic devices of two fundamental kinds: combinational and sequential. In **combinational logic,** the output of a device changes whenever its input conditions change. The basic gate exemplifies this behavior.

A number of gates can be interconnected to form a **flip-flop** circuit. This is a bistable circuit that stays in a particular state, a 0 or a 1 state,

until its "clock" input goes to a 1. At this time its output will stay in its present state or change to a new state depending upon its input just prior to the clock pulse. Its output will retain this information until the next time the clock goes to a 1. The flip-flop has memory, because it retains its output from one clock pulse to another. By connecting several flip-flops together, several sequential states can be defined permitting the design of a **sequential logic** circuit.

Table 15.2.11 shows three common flip-flops. The **truth table,** sometimes called a **state table,** shows the specification for the behavior of each circuit. The present output state of the flip-flop is designated $Q(t)$. The next output state is designated $Q(t + 1)$. In addition to the truth

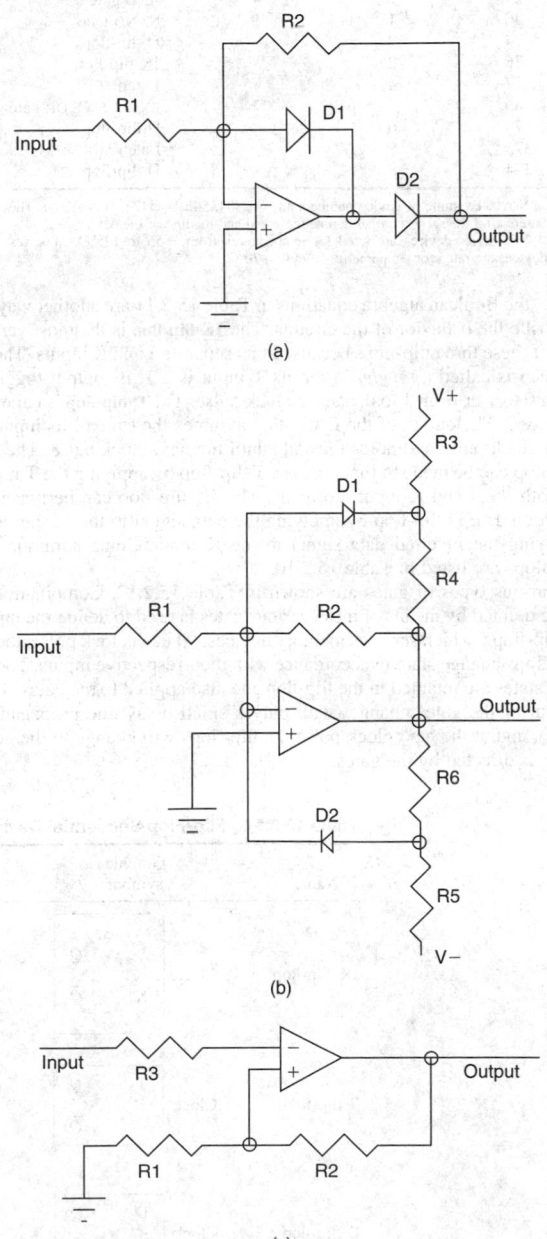

Fig. 15.2.31 Hybrid circuits. (*a*) Superdiode circuit; (*b*) limiter circuit; (*c*) Schmitt trigger circuit.

Table 15.2.9 Hybrid Circuits

Superdiode	Limiters
Schmitt trigger circuits	Dead-band circuits
Switched-capacitor filters	Logarithmic amplifiers

Table 15.2.10 Digital Integrated-Circuit Devices

Type 54/74*	No. circuits per device	No. inputs per device	Function
00	4	2	NAND gate
02	4	2	NOR gate
04	6	1	Inverter
06	6	1	Buffer
08	4	2	AND gate
10	3	3	NAND gate
11	3	3	AND gate
13	2	4	Schmitt trigger
14	6	1	Schmitt trigger
20	2	4	NAND gate
21	2	4	AND gate
30	1	8	NAND gate
74	2		D flip-flop
76	2		JK flip-flop
77	4		Latch
86	4	2	EXCLUSIVE OR gate
174	6		D flip-flop
373	8		Latch
374	8		D flip-flop

NOTE: Example of device numbers are 74LS04, 54L04, 5477, and 74H10. The letters after the series number denote the speed and loading of the device.

* 54 series devices are rated for temperatures from -55 to $125°C$. 74 series devices are rated for temperatures from 0 to $70°C$.

table, the Boolean algebra equations in Table 15.2.11 are another way to describe the behavior of the circuits. The **JK flip-flop** is the most versatile of these three flip-flops because of its separate J and K inputs. The **T flip-flop** is called a *toggle*. When its T input is a 1, its output toggles, from 0 to 1 or from 1 to 0, at each clock pulse. The **D flip-flop** is called a *data cell*. The output of the D flip-flop assumes the state of its input at each clock pulse and holds this data until the next clock pulse. The JK flip-flop can be made to function as a T flip-flop by applying the T input to both the J and K input terminals. The JK flip-flop can be made to function as a D flip-flop by applying the data signal to the J input and applying the inverted data signal to the K input. Some common IC flip-flops are listed in Table 15.2.10.

Various types of gates are shown in Table 15.2.12. Combinational logic defined by means of these various gates is used to define the input to flip-flops, which serve as memory devices. At each clock pulse, these flip-flops change state in accordance with their respective inputs. These new states are retained in the flip-flop and also applied to the gates. The output of the gates change (with only a small delay due propagation time), and at the next clock pulse the flip-flops will change to the next state as directed by the gates.

The gates shown in Table 15.2.12 have only two inputs. As was seen in Table 15.2.10, gates may have as many as eight inputs. In the case of an AND **gate**, all its inputs must be 1 in order for its output to be a 1. For an OR gate, if any of its inputs become a 1, then its output will become a 1. The NAND and NOR gates function in a similar way.

Boolean algebra is the branch of mathematics used to analyze logic circuits. Boolean algebra has two operators: $\cdot$, which indicates an AND operation, and $+$, which indicates an OR operation: The $=$ has the same meaning in Boolean algebra as in ordinary algebra. The symbol for "X not" is X' (or sometimes $\overline{X}$). The identity element for the AND operation is 0; the identity element for the OR operation is 1. The rules for Boolean algebra can be derived from set theory applied to a system in which only two numbers exist, i.e., zero and one. These rules are summarized in Huntington's postulates and DeMorgan's theorem and are listed in Table 15.2.13.

To facilitate the analysis of digital circuits and to aid in the application of the rules given in Table 15.2.13, **Karnaugh maps** are used. Typical two-variable and four-variable Karnaugh maps are shown in Fig. 15.2.32 along with the algebraic expressions represented by each map.

Looking at Fig. 15.2.32b and realizing each of the ones individually would lead to the Boolean expression:

$$OUT = W'X'YZ + WXYZ + WX'YZ + WXYZ' \quad (15.2.11)$$

Realizing this expression directly would require four four-input AND gates, and one four-input OR gates. The reduced expression

$$OUT = WXY + X'YZ \quad (15.2.12)$$

requires only two three-input AND gates and one two-input OR gate, a simpler and less expensive logic circuit.

Some of the more common large-scale integrated circuits are listed in Table 15.2.14. The adder can add two binary numbers together and output the result. For a single bit adder which adds a bit from word A, a bit from word B, and a carry bit, C, the Boolean expressions for the two outputs of the adder are

$$SUM = AB'C' + A'BC' + A'B'C + ABC \quad (15.2.13)$$
$$CO = AB + BC + AC \quad (15.2.14)$$

Adders are available, not for a single bit, but rather for 4 bits, 8 bits, or 16 bits. There is only one carry bit for the whole device. The devices in Table 15.2.14 will be described in words, but the specifications for each of these devices has a truth table as a part of the product description. An arithmetic logic unit (ALU) or accumulator can perform several simple logical or arithmetic operations, such as add, complement, shift left, and shift right. In addition, the set of instructions which command these operations will also be decoded by the ALU. As one might expect, the Boolean expression for a 16-bit or 32-bit ALU can be formidable. A

Table 15.2.11 Flip-Flop Sequential Devices

Name		Graphic symbol	Algebraic function	Truth table			
JK flip-flop	Clock	S / J / < / K / R / Q / $\overline{Q}$	$Q(t + 1) = JQ'(t) + K'Q(t)$	J 0 1 X X	K X X 0 1	Q(t) 0 0 1 1	Q(t+1) 0 1 1 0
T flip-flop	Clock	T / < / Q / $\overline{Q}$	$Q(t + 1) = TQ'(t) + T'Q(t)$	T 0 0 1 1		Q(t) 0 1 0 1	Q(t+1) 0 1 1 0
D flip-flop	Clock	D / < / Q / $\overline{Q}$	$Q(t + 1) = D$	D 0 0 1 1		Q(t) 0 1 0 1	Q(t+1) 0 0 1 1

Table 15.2.12 Combinational Gate Logic

Name	Graphic symbol	Algebraic function	Truth table
AND		$Q = xy$	$x\ y \mid Q$ $0\ 0 \mid 0$ $0\ 1 \mid 0$ $1\ 0 \mid 0$ $1\ 1 \mid 1$
OR		$Q = x + y$	$x\ y \mid Q$ $0\ 0 \mid 0$ $0\ 1 \mid 1$ $1\ 0 \mid 1$ $1\ 1 \mid 1$
Inverter		$Q = x'$	$x \mid Q$ $0 \mid 1$ $1 \mid 0$
Buffer		$Q = x$	$x \mid Q$ $0 \mid 0$ $1 \mid 1$
NAND		$Q = (xy)'$	$x\ y \mid Q$ $0\ 0 \mid 1$ $0\ 1 \mid 1$ $1\ 0 \mid 1$ $1\ 1 \mid 0$
NOR		$Q = (x + y)'$	$x\ y \mid Q$ $0\ 0 \mid 1$ $0\ 1 \mid 0$ $1\ 0 \mid 0$ $1\ 1 \mid 0$
EXCLUSIVE-OR		$Q = xy' + x'y$ $= x + y$	$x\ y \mid Q$ $0\ 0 \mid 0$ $0\ 1 \mid 1$ $1\ 0 \mid 1$ $1\ 1 \mid 0$

parity generator/checker is a circuit that can make a cyclic redundancy code which can be used to check the accurate transmission of data. Shift registers perform only the shifting functions of ALUs. Counters are made in many configurations—up-down counters, binary-coded-decimal counters, Johnson counters, etc. A decoder can best be explained by a simple example. A three- to eight-line decoder will be described. The three input lines can have eight binary variations, 000, 001, 010, 011, 100, 101, 110, and 111. For each of these input conditions one, and only one, of the output lines will be set. The input signal is said to have been decoded. The encoder performs the related operation of encoding from

Table 15.2.13 Rules for Boolean Algebra

$X + 0 = X$	$X \cdot 1 = X$
$X + 1 = 1$	$X \cdot X' = 0$
$X + X' = 1$	$X \cdot X = X$
$(X')' = X$	$X \cdot 0 = 0$
$X + Y = Y + X$	$X \cdot Y = Y \cdot X$
$X + (Y + Z) = (X + Y) + Z$	$X \cdot (YZ) = (X \cdot Y)Z$
$X + X \cdot Y = X$	$X \cdot (X + Y) = X$

DeMorgan's theorem:

$(X + Y)' = X' \cdot Y'$	$(X \cdot Y)' = X' + Y'$

eight to three lines. Depending on which one input line is set, an appropriate output bit pattern is set. Encoders and decoders are available in 2 to 4, 3 to 8, and 4 to 16 line configurations. The multiplexer/demultiplexer performs a similar function. The multiplexer has a data input terminal and control input terminal and several data output terminals. It switches a single train of information, ones and zeros, on an input line, and routes this information to one of many output lines depending on the bit pattern placed on a control input. The demultiplexer performs the reverse operation. Multiplexers/demultiplexers are also configured in 2 to 4, 3 to 8, and 4 to 16 line devices.

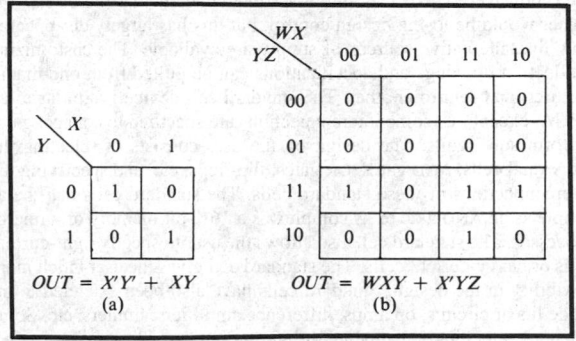

Fig. 15.2.32 Karnaugh maps of typical logic functions.

There are a wide variety of display controllers and drivers. These are devices which interface between digital circuits and displays which humans can use. Light-emitting diodes, liquid crystals, and fluorescent displays have different drive requirements. The digital device may perform in binary, but the display requires decimal. In many cases, the display driver is separate from the display itself, but more recently the display driver has been incorporated in the display.

The complexity of integrated circuits is growing continually. The simple gates and flip-flops discussed so far are still used to some extent. Logically, these gates and flip-flops can be integrated into larger single-chip circuits called *large-scale integrated circuits* (**LSI**). Combinations of these circuits are grouped together on single chips called *very large scale integrated circuits* (**VLSI**). VLSI chips contain 500,000 to more than 2,000,000 gates. VLSI chips have enabled the explosive growth of personal computers, by allowing ever-increasing functionality at lower cost. These circuits are custom-designed VLSI circuits which are developed for large-volume applications. Bridging the gap between LSI and custom VLSI circuits are a series of linear and digital user-designed circuits.

The first of these user-designed circuits is called programmable logic arrays (**PLAs**) or programmable array logic (**PAL**). The PLA grew out of the programmable read-only memory (**PROM**). A programmable read-only memory is a computer memory which can be programmed one time, and then can be read many times. The PROM consists of a series of gates, initially all ones, each of which can be electrically altered to be a zero. A PROM might be logically organized to be 1 bit wide by 64K bits long, or 4 bits by 16K bits, 8 bits by 8K bits. The example just given is a series of 64K-bit PROMs. PROMs are available up to 4M bits on a single chip. The PLA is also field programmable, but has a more limited programming capability than the PROM. The limited capability makes

Table 15.2.14 Large-Scale Digital Integrated Circuits

Adder	Accumulator
Arithmetic logic unit	Parity generator-checker
Shift register	Encoder
Multiplexer	Decoder
Demultiplexer	Counters
Display controller-driver	

it less costly to program. The PLA is adaptable to replacing Boolean functions in both combinational logic and sequential logic circuits. The PLA can replace 10 to 40 gate and flip-flop chips, resulting in a simpler, more reliable, less costly design.

There are two other approaches to application-specific integrated circuits, (**ASICs**). These are field-programmable gate arrays (**FPGAs**) and standard cells. The FPGA is sometimes referred to as a "sea of gates." The chip is designed with a series of identical gates, but without any interconnections between them. The user specifies the interconnections, and thus makes the chip specific to the application. The manufacture of the chip cannot be completed until the interconnections have been specified, however. There was initially some concern that the circuit designer would be losing design control, but this has largely disappeared now that alternative sources of supply are available. The customizing production drawings and specifications can be pulled from one manufacturer, and sent to another. The standard cell device is similar to an FPGA, except that some interconnections are specified to connect gates to form subcircuits. The design by the user consists of selecting the individual cells, AND gates, OR gates, flip-flops, etc. and specifying the interconnections of these standard cells. The standard cells may be as simple as an AND gate or as complex as a 16K-bit memory or a microprocessor. The standard cell also allows the user to specify light-current cells or heavy-current cells. The standard cell gives the user much more flexibility in the design. Standard cells have also been extended to include linear circuits, op amps, difference amplifiers, limiters, etc. Some standard cell chips are suitable for hybrid digital and linear circuits on a single chip.

Two problems arise as more complex circuitry is incorporated in ASICs, and also in microprocessors (which will be discussed in the next section). These devices require more input and output data pins than are available in dual in-line packages (see Fig. 15.2.24), and they generate more heat on the chip. One solution to the pin problem is the pin array package, shown in Fig. 15.2.33. These packages still maintain 0.1-in spacing between pins, but the pin count has been increased to 132 and 208 in the examples shown. The package size has increased from dual in-line packages, and the force necessary to install and remove the

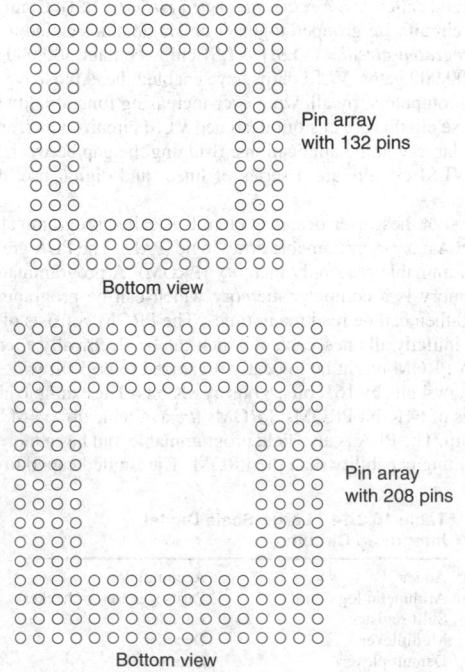

Pin array with 132 pins

Bottom view

Pin array with 208 pins

Bottom view

Fig. 15.2.33 Typical pin grid array packages. (All pins are on a 0.1 in grid.)

package is not appreciably greater than for dual in-line packages. There are additional microcircuit packages at various stages of development and standardization that have up to 500 pins and 0.05-in pin spacing.

Higher-speed and higher-density digital electronics lead to higher heat dissipation in devices. One solution is to provide heat sinks and fans to remove heat. A more recent development is the **heat pipe**. The heat pipe consists of boiling fluid/vapor refrigerant in a closed chamber. The semiconductor boils the fluid to change it to the vapor state. A pipe conducts the vapor to a vapor-to-air heat exchanger, where the vapor is condensed. The heat pipe also provides a return path for the fluid to return to the semiconductor. An attractive feature of the heat pipe is that there is no requirement for a mechanical compressor.

A further problem for complex electronic circuits is proof of design and proof of manufacture testing. To adequately test a device, it is necessary to have access to additional test points in the circuit, beyond those needed to interface in actual operation. With a limited number of output pins and highly complex internal logic, the design of the circuit must include a fairly long test vector that will excite the various circuit functions. In addition, a fairly long results vector will be generated by the circuit. One approach to test design is called **boundary scan** design, described in specification IEEE 1149.1 of the Institute of Electrical and Electronics Engineers. This specification outlines a four- or five-wire test circuit, and the internal circuitry, which allows a test vector to be entered serially onto the chip, and allows a results vector to be serially unloaded from the chip. Internal circuitry is required to accept each bit of the input test vector and to generate each bit of the results vector. The serial transfer is a tradeoff between the time required to enter and retrieve the test vectors and the cost of additional input/output pins for the device. Serial transfer would be unacceptable for normal operations, but is all right for test purposes.

Computer Integrated Circuits

One of the devices which has become feasible as a result of VLSI is the **microprocessor**. A complete computer can be built in a single IC device. In most cases, however, several devices are employed to build a complete computer system. In most computers, the cost of the microprocessor is negligible compared with the total system cost. Not long ago, the **central processing unit (CPU)** was the most expensive part of a computer. The microprocessor provides the total CPU function at a fraction of the earlier cost.

The power of a microprocessor is a function of its clock speed and the size and number of its registers. Clock rates vary from 1 to 200 MHz. Common register sizes are 8, 16, 24, 32, and 64 bits. Most personal computers currently use 32-bit registers. Commonly microprocessors have 12 to 16 registers. Other factors affect processor power and speed. Most microprocessors have a data bus, an address bus, and a set of control and power supply terminals. The number of each of these buses may vary. A small inexpensive microprocessor might have 8 data lines, 16 address lines, and 8 control/power supply lines. A large sophisticated microprocessor might have 64 data lines, 24 address lines, and 20 control/power supply lines. Some microprocessors have 32-bit internal registers and only 16 data bit terminals. The 16 data terminals are time-multiplexed to allow the complete 32-bit word to be entered onto the microprocessor chip in two read cycles. Once in the microprocessor, the 32-bit registers communicate with each other over a data bus which is a full 32 bits wide. This scheme reduces the number of terminals on the microprocessor, and reduces its cost, with a slight timing penalty.

Clock speed is limited by clock skew and parasitic capacitance. There is a finite time required for electric pulses to move along wires and for gates to change state. Typically electric pulses move along wires at about half the speed of light. Gates change state in a few nanoseconds. Parasitic capacitance is the stray capacitance that exists between any circuit and ground. When the gate changes state, this capacitance must be charged. The charging current increases linearly with switching rate and voltage. Because of the small circuit components and short wire lengths, the parasitic capacitance is much smaller for circuits on a chip compared with circuits on printed boards. Modern microprocessors use an on-chip clock rate that is 3 times as fast as the off-chip clock.

The **microcontroller** is a device similar to the microprocessor, in that it is also a CPU. Microcontrollers are configured for controlling applications. In general, they have greater input and output capability. It is very easy to control a single output bit for turning on an electric relay. With a microprocessor, 8 bits are switched together, but with a microcontroller individual bits can be switched. Some microcontrollers have 256 internal registers, rather than the 15 to 20 of a microprocessor. Microcontrollers are often stand-alone single-chip devices, and are a less expensive system than a microcomputer. Typical applications for microcontrollers are microwave oven controllers, washer and dryer controls, and automobile engine controls.

Memory is an important part of a computer system. Memory can be divided into two broad classes: volatile and nonvolatile. Volatile memory forgets what it contains when the power is removed from it. Nonvolatile memory retains its contents when power is removed. Random-access memory (**RAM**) is volatile memory that is sometimes called *main memory*. This is the memory that the CPU uses for most of its operations. RAM is a semiconductor memory. RAM may be dynamic memory or static. In a static RAM the memory elements are flip-flops, and retain their memory for as long as power is applied. Dynamic RAMs are also semiconductor memories, but hold the information as charges on capacitors. This charge will leak off even though the power is still applied to the memory. To retain the memory in a dynamic RAM, it is necessary to continually refresh the memory. This is typically done every 10 ms. Dynamic RAMs are much cheaper and have less loss than static RAMs. Large banks of memory are made with dynamic RAMs. Small memories, or memories with special requirements, are made with static RAMs. Typical dynamic RAMs are 256K, 1M, and 4M bytes. Typical static RAM may be as small as 1K bytes and can be as large as 1M bytes.

Nonvolatile memory can take several forms: ROMs, disk storage, compact disks, and tapes. A read-only memory (ROM) is a set of flip-flops which are programmed to take a particular state. This programming may be installed at the time of manufacture. Some devices are programmable by the user (**PROMs**). A PROM is typically all ones as shipped. The individual memory cells may be programmed to a zero or left a one. Programming memory cell is done by applying a high voltage to a programming pin when that cell's address is set. Normally the ROM or PROM is written once and read many times. An **EPROM** is an erasable, programmable read-only memory. The EPROM is programmed in the same way as a PROM. To erase the memory, the EPROM is placed under a high-intensity ultraviolet light. When the EPROM is erased, all of the memory is lost, so the entire program must be reentered.

Disks and tapes are magnetic storage devices. When power is removed from these devices, the memory is retained due magnetic retentivity. Magnetic memory can only be read (or written) by moving the magnetic medium under a read/write head. **Compact disks** are a form of read-only memory. In the compact disk, the information is stored optically rather than magnetically.

The magnetic tape is a removable nonvolatile memory device. Hard disks are usually not removable. Floppy disks are removable, but for large quantities of data, the tape medium is cheaper and more convenient. Tape is either in a cartridge or on a reel. Newer tape cartridges have large volumes, 50M bytes to 1000M bytes. The trend is that cartridge tape is replacing reel tape as a storage means. Magnetic tape can store computer data for over a year, and with a periodic refreshing can store data indefinitely.

In addition to the CPU and storage devices, other integrated circuits have been designed for computer systems. These are listed in Table 15.2.15. The **tristate buffer** is a switching amplifier which can be a zero, a one, or high impedance. When a computer bus line can be driven by several devices, all of the devices are high impedance except the one which is controlling the bus line at that time. Typically eight tristate buffers are contained in one IC package. The **tristate transceiver** is similar to the tristate buffer except that it allows the bus signal to be received as well as written. The **parallel interface adapter** is a device that permits a parallel output device, such as a printer, to be connected to a computer

Table 15.2.15   Digital Computer Integrated-Circuit Devices

Tristate buffer	Tristate transceiver
Parallel interface adapter	Programmable peripheral interface
Universal asynchronous transmitter-receiver	Floppy/hard disk controller
Analog-to-digital converter	Digital-to-analog converter
Programmable timer	Direct memory access controller
Programmable interrupt controller	

bus. There are several variations of the parallel interface adapter. Some control two, three, four, or more parallel devices. Some have been designed to work with Intel microprocessors, others to work with Motorola microprocessors. The **programmable peripheral interface** is a parallel interface adapter designed for use with Intel computers. The **universal asynchronous receiver-transmitter (UART)** is a device which will allow a serial output device to be connected to a computer bus. The UART has a parallel connection to the computer bus, but provides a serial output to the peripheral device. The UART is programmable, so that it can be set for 7-bit or 8-bit transmission. It can be set for even or odd parity checking, and the transmitting/receiving speed can be set. The **floppy/hard disk controller** is a programmable device which can handle the interfacing requirements for floppy and/or hard disks. Analog-to-digital and digital-to-analog converters allow the digital computer to interface with analog signals. The **programmable timer** is a timing circuit which can be programmed by the computer either to generate an interrupt periodically, or to generate an interrupt after a programmed time. An interrupt is a signal on one of the control lines of the computer bus. The **direct memory access controller (DMA)** is a CPU type of device which can control the computer bus. The direct memory access controller generates a DMA request signal on a computer control line. The main CPU will complete the portion of the instruction that it is executing, and generate a DMA grant signal. The DMA controller then transfers data into or out of main memory or disk memory. Since the CPU is not involved in the memory transfer, it can proceed with the instruction it is executing, as long as that instruction does not require the computer bus that the DMA controller is using. DMA controllers can off-load the CPU, allowing the overall system to have more power. The **programmable interrupt controller** is a general-purpose semiconductor which can be used in various types of computer printed boards to service interrupts.

Computer Applications

The computer is universally used, and a complete discussion of computer usage is beyond the scope of this section. Computer techniques are used in electronic design so extensively that some description of these electronic applications of computers are necessary in order to understand modern electronic design.

Computers can be used as control devices. The computer might be a controlling device to simply turn a pump on and off. On the other hand, a variable displacement pump might incorporate a microcontroller in its feedback control loop in place of pilot hydraulic circuits. Entire processes may be controlled by computers. For example, modern paper machines use computers throughout for control purposes. Computer terminals at various locations show the operation conditions throughout the machine. Entire chemical-producing plants are under computer control.

Some automotive applications include control of air, fuel mixture, and spark timing in an automobile engine. This is typical of an application in which there are several input variables that must be measured, and several controlled outputs. The input variables that must be measured are air temperature, manifold temperature, crankcase temperature, engine speed, fuel flow, etc. The outputs which are controlled by the computer are ignition timing and fuel/air mixture. Automobile brakes, transmissions, and suspensions are automotive subsystems that incorporate microcontrollers for improved automobile performance. Microcomputers may also be used to reduce the complexity of automobile wiring. In such a system, battery power is distributed to all parts of the car along a single wire. Intelligence for the control of lights and accessories flows along a single control wire. Relays are provided at each lamp, switch, etc. to interface between the control and battery wire. Messages along the control wire are time-multiplexed so that all devices share the same control wire.

Testing, data acquisition, and analysis is another use of computers. Modern testing systems make extensive use of computers. Either digital or analog signals can be incorporated into a test system. Analog-to-digital converters convert analog signals into digital signals which the computer can use. Within the computer, a model of the process being tested can be constructed, and monitored by the test data being accumulated. Within the computer, signals can be generated which are not being actually measured. For example, the velocity of a mass may be measured, but the acceleration or the position of the mass may be calculated within the computer. Highly reliable systems can be developed using electronic control and computers, if redundancy is incorporated into the measuring and computer systems. In the aerospace industry, these systems are called **fly-by-wire** control systems. The space shuttle and supersonic fighters are examples of craft using fly-by-wire. In such systems, failure of a computer, a controller, or a sensor must not cause a failure of the craft.

Data acquisition systems give the user much greater insight into the data being taken. Limitations and anomalies of the measured data become much more apparent than with analog data. The need for signal conditioning becomes imperative in a digital data acquisition system. Signal processing is an integral part of any digital data system. Analog signals can be filtered as a part of signal conditioning, but after the signal has been converted to digital data, digital filtering may also be used. Digital filters can be broken into two classes: **finite impulse response (FIR)** and **infinite impulse response (IIR)** filters. As can be seen in Fig 15.2.34, the FIR filter does not have any feedback from a later part

of the filter to a previous part. The IIR filter incorporates feedback from a later part of the filter to an earlier part. The FIR filter normally requires a greater length to accomplish the filtering objective than does the IIR filter. Because the IIR filter incorporates feedback, it is possible for it to be unstable. The FIR filter cannot be unstable. Both FIR and IIR filters employ a clock to generate a fixed delay from point to point in the filter. The clock delays are shown as ΔT in the figure. Each delay is a clock period. FIR filters are often 45 to 75 delays long. IIR filters are usually 8 to 10 delays long. The effective delay for an IIR filter is longer than the actual number of delays, because of the feedback in these filters. Although the infinite response suggests that the filter never completes its response, in a practical sense the IIR filter response is finite. The design approach for FIR filters and IIR filters is quite different. For an FIR filter, the filter coefficients A_0 through A_n are designed to optimally fit a desired filter response curve. The response of this optimal filter often has excessive sidebands, that is, high output in the nonpass frequency range. The sidebands can be reduced by applying a window function, which modifies the filter coefficients. The effect of the window function is to broaden the transition region between the passband and the stopband, and reduce the magnitude of the sidebands. The response equation, or transfer function, for the FIR filter with even symmetry is

$$H(\omega) = \sum_{i=0}^{i=n} A_i \cos(\omega i) \qquad (15.2.15)$$

In this equation, the frequency ω is a continuous variable in radians per second. Although the filter is a discrete device, its frequency response can be viewed as continuous. The digital signal is discrete, and when a signal is converted from digital to analog, this conversion adds its own modification of the signal.

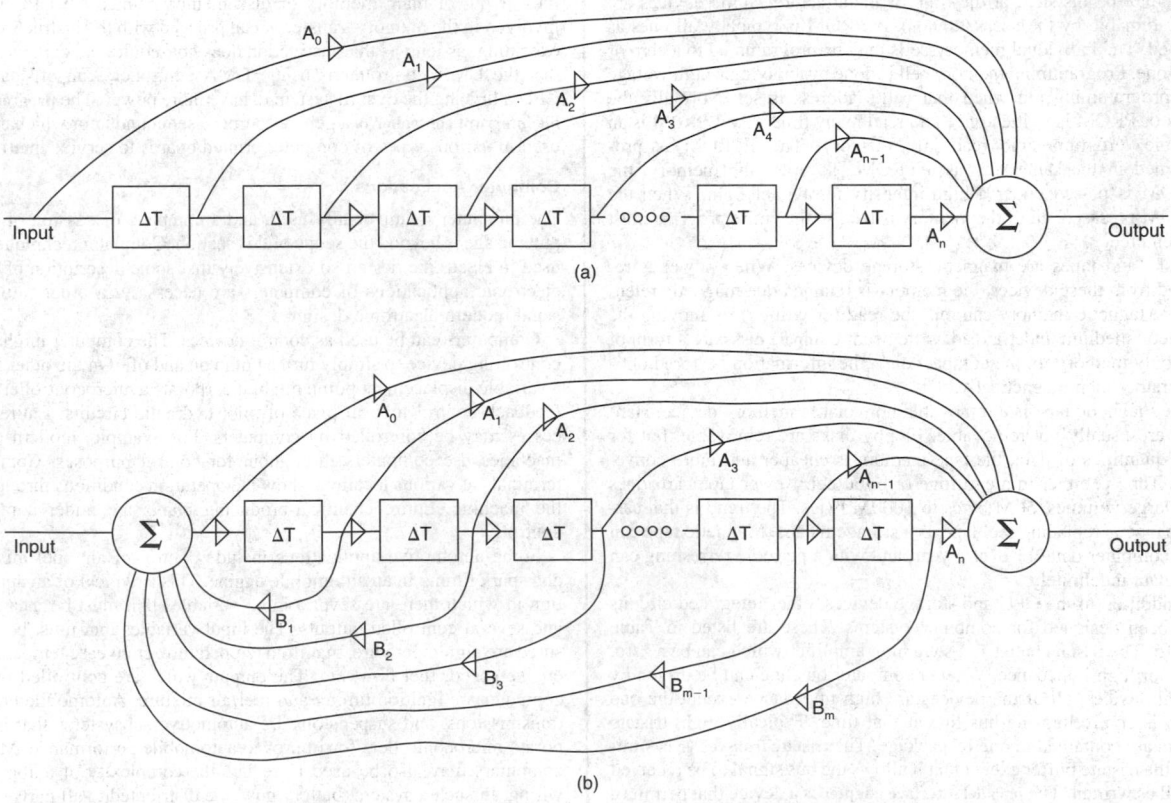

Fig. 15.2.34 (a) Finite impulse response (FIR) and (b) infinite impulse response (IIR) digital filters.

The design approach for an IIR filter is quite different. For an IIR filter, the first step is to design a low-pass analog filter for a particular filter response, such as a Butterworth, Chebychev, or Cauer response. The analog filter can then be converted to a low-pass digital filter using the bilinear transform. Finally, the low-pass digital filter can be converted to a high-pass, band-pass, or band-stop filter. The resulting filter has a transfer function given by

$$H(\omega) = \frac{\sum\limits_{i=0}^{i=n} A_i \cos(\omega i)}{1 - \sum\limits_{i=1}^{i=m} B_i \cos(\omega i)} \qquad (15.2.16)$$

A more complete discussion of digital filter technology can be found in the references at the beginning of this section. Further help in digital filter design can be obtained by using a computer program, such as MATLAB.

Digital filter calculations consist of a series of mathematical operations which follow a consistent pattern. Each calculation is performed by retrieving a number, multiplying it by a coefficient, adding it to another number, and storing the result. This is the operation at each of the summation points in the FIR and the IIR. A microprocessor called a **digital signal processor (DSP)** has been developed for this purpose. DSPs complete a retrieve, multiply, add, and store in a single clock pulse (30 ns), several times faster than a general-purpose microprocessor. DSP microprocessors are widely used in test equipment, telephone equipment, television, and military equipment. Because of the wide usage, DSP microprocessors are quite inexpensive. DSPs were developed in order to increase the bandwidth of digital signal processing. It should be further noted that digital filters can employ parallel processing, in which several DSP chips can operate in parallel to further increase bandwidth.

Digital signal processing is not limited to analog and digital filtering. Digital data may also be improved by using smoothing algorithms, or by removing inconsistent data points, such as outliers. Insight into data content can also be obtained by statistical analysis. Digital signal processing is also used to improve the quality of voice signals, and to enhance video signals.

Digital Communications

The bandwidth required to transmit information depends on the digital pulse rate. A fundamental relationship exists for the bandwidth required for the transmission of digital or analog information. For analog signals, the bandwidth depends on the highest frequency component in the signal. For digital circuits, the bandwidth depends on the maximum pulse rate. The relationship between the two is known as the Nyquist criterion. The minimum pulse rate pr necessary to transmit an analog signal digitally must be twice the highest frequency f:

$$\text{pr} = 2f \qquad (15.2.17)$$

This is a bilateral relationship. The frequency bandwidth BW of a transmission system must be equal to at least half the pulse rate:

$$\text{BW} = \text{pr}/2 \qquad (15.2.18)$$

These conditions are minimum requirements. A transmission system that has greater bandwidth can support slower pulse rates, and similarly, a high-pulse-rate system can be used to pass a low-frequency signal. A higher pulse rate than the Nyquist criterion will approximate an analog signal waveform more accurately, but is not necessary to pass the information. Voice-grade lines have a bandwidth of 4,000 Hz, and will pass 8,000 bits per second (bps). Twisted-pair cable, properly terminated, can pass about 1 MHz or 2 Mbps. Coaxial cable can pass 100 to 200 MHz. To obtain greater bandwidths, light is being used to transmit information, rather than electricity. Inexpensive fiber-optic devices can be used for distances up to several hundred feet. For longer distances, phase-locked lasers and single-mode fiber-optic cable is being employed. Single-mode fiber is quite small and somewhat hard to terminate. Tooling to facilitate connecting single-mode cable is becoming more readily available.

Electric instrumentation has several communication standards which facilitate the connection of measuring systems. A low-speed instrument bus, IGIB, has been developed and described in a standard, IEEE 488. For higher communication rates, a CAMAC crate has been standardized. These standards give physical characteristics, such as connectors and plugs, mounting dimensions, printed board sizes, etc. These standards also define communications protocol, bus timing, and connector pin assignment.

In most computers, the interconnection of computer printed boards is accomplished using a "mother board." The mother board may contain the CPU, semiconductor memory, and several input/output devices. While these devices are somewhat arbitrary, it is essential that the mother board contain a computer bus structure. The computer bus is defined by a standard which defines the type connector, the connector circuit configuration, and the computer communication protocol. There are several agencies which define computer bus standards in the United States and around the world. Some of the more common IEEE bus standards are listed in Table 15.2.16.

Power Electronics

Power electronics can be divided into three major groups: motor drives, power supplies, and power amplifiers. Motor drives can be divided into dc motor drives and ac motor drives. Power supplies include power supplies for electronic circuits, battery back-up systems, electric power applications, and induction/dielectric heating supplies. Power amplifiers cover other power conditioning topics.

The power for dc motor armatures can be derived from thyristor circuits like those shown in Fig. 15.2.9. Single-phase bridge circuits are used for 5-hp drives and smaller. Three-phase bridge circuits are used for drives larger than 5 hp. A single set of six thyristors can supply power for about 300 hp. Above 300 hp, multiple sets of thyristors must be used in parallel. Mill drives have been built with more than 10,000 hp provided by thyristors.

The control of dc motors whether powered by thyristors or by dc generators is accomplished electronically. Control of individual drives can be accomplished by tachometer feedback or by armature voltage feedback. The speed-regulation accuracy for armature feedback is 5 percent; for tachometer feedback speed-regulation accuracy is from 0.1 to 1.0 percent. When two drives must be coordinated with each other, as in a continuous-web processing machine, they can be regulated to control torque, speed, position, draw, or a combination of these parameters. Torque controls can be achieved using dc motor armature current for a feedback signal. Speed-control signals are derived as for single motors. Position or draw control can be accomplished by using selsyn ties or dancer rolls. A **dancer roll** is a weight- or spring-loaded roll that rides on the web. It is free to move up and down, and as it does, a signal is taken from its position to serve as a feedback for the drive regulator before the dancer or after it.

Coordination of the motions of two or more drives requires tracking of the drives in both steady-state and transient conditions. Linearity of the control and feedback signals determine steady-state tracking. Provision must be made for both low-speed and high-speed matching signals. Transient matching requires that signals not only be the right magnitude but also arrive at the right time. An example will serve to illustrate this point. Suppose it is desired to have two drives with tachometer feedback

Table 15.2.16   IEEE Computer Bus Standards

Identifier	Common name	Description
IEEE 488	GPIB	General-purpose instrument bus
IEEE 796	Multibus	General-purpose computer bus
IEEE 961	STD bus	Standard computer bus
IEEE 1014	VMEBUS	European computer bus
IEEE Proposal	VXIBUS	Instrumentation adaptation of VMEBUS
IEEE Proposal	MXIBUS	Extension of VXIBUS

which have a continuous web between them. One way to accomplish this would be to designate one drive as a master and the other as a slave. The tachometer on the master drive would serve as its own feedback signal and as the reference or command signal for the slave drive. The slave drive would have its own feedback from its own tachometer and so its regulator would try to minimize the difference between the two tachometer signals. On a transient basis the master drive will always start before the slave. An alternate and more common arrangement is to provide a common reference for both drives and let each drive receive its command signals at the same instant.

Digital computers are being used on-line in mills and continuous processing industries. DC motors can be controlled by either analog or digital regulators. With the greatly reduced cost of integrated circuits, digital regulators are being increasingly used.

For digital speed control, feedback is taken from a digital tachometer, which generates a pulse frequency proportional to motor speed. An adjustable frequency reference is compared with the feedback, and the difference is applied to an up/down counter. The output of the counter is converted to a dc control signal for the motor controller. In this system the dc drive functions as a form of phase-locked loop, and in fact, a phase-locked loop can be substituted for the up/down counter in digital speed control. For digital position control, a shaft encoder is used for feedback. In a shaft encoder, a single turn of the motor shaft is encoded in a binary code. For example, an eight-track shaft encoder can define 2^8, or 256, parts of a revolution. The output of the shaft encoder is compared with a digitally derived reference code. The difference between the feedback and the reference is converted into a dc signal which controls the motor, and holds the desired position. Sometimes the position encoder is mounted on the driven machine. The accuracy of sensing the machine position is improved. Accurate milling machines have feedback from machine position, rather than from the drive motor. Computers can be used to derive a speed reference or a position reference for several motors, and thus, an entire process line can be controlled. Modern high-speed paper making machines are one example of coordinated computer control of several motors. Graphic terminals, which display the entire paper making machine, are located at various positions allowing several operators to coordinate the operation of the machine.

DC motors can also be supplied from a fixed dc voltage such as a battery or a fixed dc power supply. Inherently, fixed voltage on the field and on the armature of a dc motor implies constant-speed operation. The armature voltage can be controlled by a proportional device such as a transistor. Proportional devices are limited to low-power applications. At low speeds, the dc motor requires low voltage and full current to produce full torque. The losses in the control device are proportional to the product of voltage and current. The full battery voltage is applied to the control device and the motor in series. Thus, the control device has a high voltage drop across it at low motor armature voltages and low motor speeds. To reduce controller losses, a **chopper** may be used. The chopper is thyristor, power MOSFET, or IGBT circuit which can rapidly switch on and off. When the chopper is on, the loss in the control device is low because the voltage drop is small. When the chopper is off, the loss in the control device is small because the current through the device is very small. Large amounts of power can be controlled with low losses by means of a switching controller. For large dc motors, the switching rate of the chopper is made high enough that the inherent inductance of the motor armature maintains a nearly constant motor current.

The speed range of a variable-speed dc motor is limited by its armature resistance, which produces IR drop. For a very small motor, the IR drop can be as much as 20 percent of rated armature voltage. The resistance-to-inductance ratio of small motors is much greater than for large motors. To extend the speed range of small motors, a low-frequency chopper can be used which will cause the armature current to be discontinuous. Discontinuous conduction reduces the effective IR drop for the motor allowing its speed range to be extended.

DC motors have been widely used for variable-speed applications because of their excellent characteristics. AC motors have been used primarily for constant-speed applications. The control schemes described above are equally applicable to ac motors (except of course for armature voltage and armature current feedback). If power circuitry is properly handled, the control of an ac motor is just as flexible and versatile as that of a dc motor.

AC motors can be supplied either from **phase-controlled circuits** or from **inverter circuits**. Phase control uses a simpler electronic circuit than the inverter, but it results in high loss in the ac motor. The phase control switches at line frequency, and so the frequency applied to the motor is constant. This means that the synchronous speed of the induction motor is constant. The difference between the operating speed of the motor and synchronous speed produces losses in the rotor of the motor proportional to the torque demanded of the motor. For this reason, phase-controlled ac motor drives are limited to applications with slight speed change requirements or to loads that require much lower torque at low speed, such as pumps or fans. Inverters are used for ac motor drives which require constant torque operation over a wide speed range. Several different schemes have been used for ac motor inverters. Some of these are **pulse width modulation (PWM), six-step inverters,** and **current link inverters**. Most modern drives are PWM inverters. In a PWM inverter, the ac line voltage is converted into a fixed dc voltage, and then this dc voltage is inverted into a variable frequency and a variable voltage. Figure 15.2.35 shows a block diagram of PWM inverter. Since the motor is an ac motor, both positive and negative voltages must be applied to each motor phase. The motor shown in Fig. 15.2.35 has three phases. Consider the operating conditions at only one of the phases of the motor. To apply a positive voltage to phase A-B of the motor, switches SW1 and SW4 are turned on. To apply a negative voltage to phase A-B, SW2 and SW3 are turned on. To apply zero volts to phase A-B, either SW2 and SW4 or SW1 and SW3 are turned on. A similar operation takes place for phase B-C and phase C-A of the motor. The dc voltage in the PWM inverter is constant. To generate an approximate sine wave of voltage, the inverter must produce a low voltage during part of the cycle and high voltage during another part of the cycle, and it must do this on both the positive half of the sine wave as well as the negative half. On the positive half-cycle, switching occurs alternately between switches SW1 and SW4 and switches SW1 and SW3. To control the speed of the ac motor and keep its losses low, one must vary both the frequency of the motor voltage and its magnitude. The motor requires an approximately constant volts per hertz rate for full torque and low losses. Although this control scheme seems very complicated, PWM inverter drives have been developed which are far more dependable than dc drives. The commutator and brushes required for a dc motor limit its reliability. The motor frequency for an ac motor inverter drive varies from nearly zero to a maximum of 120 Hz typically. To reverse the rotation of the ac motor, the phase sequence applied to the motor is reversed. The actual switching frequency of the inverter is approximately 10,000 Hz, allowing the inverter switch on and off several times during a cycle applied to the motor. The pulse width is varied or modulated, to change the output voltage, hence the name pulse-width-modulated inverter. The inductance of the motor keeps the current variations in the motor from being too great.

Switching power supplies use technology similar to ac motor inverters. In Fig. 15.2.35, SW1 and SW2 are called an *inverter leg*. The inverter leg is fundamental to switching power supplies. The dc chopper described earlier could be made with a single inverter leg. Regulated dc power supplies are also made using an inverter leg. A higher intermediate voltage is produced in an ac-to-dc rectifier. This voltage is then chopped to make a lower output voltage. Feedback is taken from the load point of the power supply to control the pulse width of the inverter leg, or chopper. Precise computer power supplies are often switching power supplies of this type. They maintain precise dc output voltage under changing load conditions and changing supply voltage. Because the power supplies switch, the loss in these power supplies is much lower than for proportionally controlled power supplies.

The switching behavior of the inverter leg is actually more complex than simply turning on one switch and turning off another. Each switching action involves a resonant transient. At each switching point an

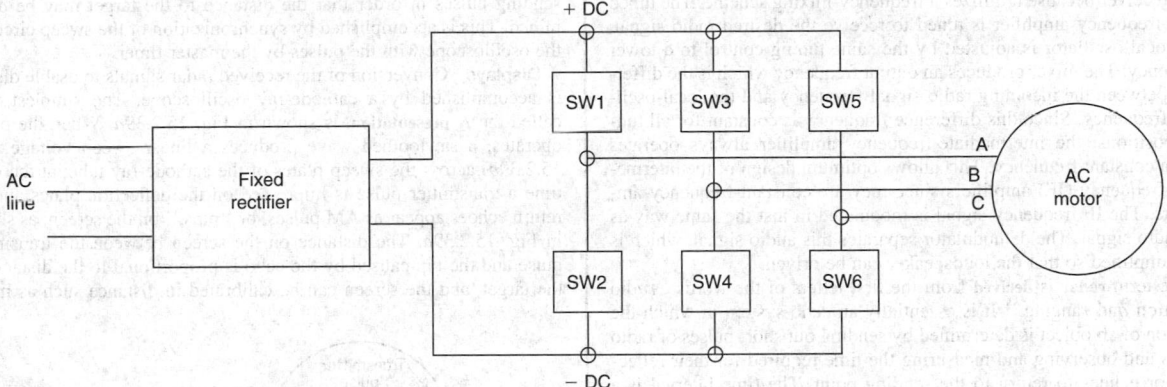

Fig. 15.2.35 Pulse width modulation inverter.

R-L-C circuit rings from one voltage to another. At very high switching rates, a second transient starts before the first transient is over. This leads to a power supply with a parade of transients. Such power supplies are called *resonant converters*. These power supplies may be used for dc-to-dc, dc-to-ac, ac-to-dc, and ac-to-ac conversions. By using the ring-up inherent in resonant circuits, voltage levels can be increased as well as decreased. Induction heating power supplies are typical of these kinds of circuits. Induction heating is produced by the magnetic losses generated in nonmagnetic and magnetic metals. If the frequency and voltage are increased sufficiently, dielectric heating may also be produced in high-frequency power supplies. High-frequency, high-voltage power supplies are also used to produce a corona discharge to treat the surface of plastic materials. Plastics which are normally hydrophobic can be made hydrophilic by corona discharge treatment.

Communications

The Federal Communications Commission (FCC) regulates the use of radio-frequency transmission in the United States. This regulation is necessary to prevent interfering transmissions of radio signals. Some of the frequency allocations are given in Table 15.2.17. The frequency bands are also classified as shown in Table 15.2.18. Very low frequencies are used for long-distance communications across the surface of the earth. Higher frequencies are limited to line-of-sight transmission. Because of bandwidth considerations, high frequencies are used for high-density communication links. Orbiting **satellites** allow the use of high-frequency transmission for long-distance high-density communications.

A **radio transmitter** is shown in Fig. 15.2.36. It consists of four basic parts: an rf oscillator tuned to the carrier frequency, an information-input device (microphone), a modulator to impress the input signal on the carrier, and an antenna to radiate the modulated carrier wave.

A radio wave can be modulated in several ways. It can be amplitude-, frequency-, or phase-modulated. These are examples of analog signal modulation. A radio wave may also be digitally modulated. Examples

Table 15.2.17 Partial Table of Frequency Allocations
(For a complete listing of frequency allocations, see ''Reference Data for Radio Engineers,'' published by Howard Sams & Co.)

Frequency, MHz	Utilization
0.535–1.605	Commercial broadcast band
27.255	Citizen's personal radio
54–72	Television channels 2–4
76–88	Television channels 5–6
88–108	Frequency-modulation broadcasting
174–216	Television channels 7–13
460–470	Citizens' personal radio
470–890	Television channels 14–83

Table 15.2.18 Frequency Bands

Designation	Frequency	Wavelength
VLF, very low frequency	3–30 kHz	100–10 km
LF, low frequency	30–300 kHz	10–1 km
MF, medium frequency	300–3,000 kHz	1,000–100 m
HF, high frequency	3–30 MHz	100–10 m
VHF, very high frequency	30–300 MHz	10–1 m
UHF, ultra-high frequency	300–3,000 MHz	100–10 cm
SHF, super-high frequency	3,000–30,000 MHz	10–1 cm
EHF, extremely high frequency	30,000–300,000 MHz	10–1 mm

NOTE: Wavelength in meters = $300/f$, where f is in megahertz.

of digital modulation include on-off, binary phase-shift keying (BPSK), frequency-shift keying (FSK), quadrature phase-shift keying (QPSK), and quadrature amplitude modulation. Regardless of which method of modulation is used, the frequency content in the signal requires a bandwidth of frequencies to support the transmission. If a 10-kHz signal is to be broadcast, the transmitted radio wave will consist of a center frequency, called the *carrier frequency,* and an upper and lower sideband,

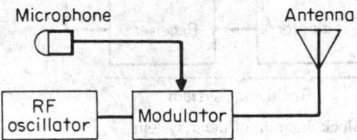

Fig. 15.2.36 Radio transmitter.

for a frequency bandwidth of 2 times 10 kHz, or 20 kHz. It is possible to suppress the upper or lower sideband leading to single-sideband transmission, but this is a special case. Regardless of the modulation scheme, the bandwidth required for signal transmission is the same for a given signal bandwidth. Higher frequencies or higher digital pulse rates require greater bandwidth for transmission.

A **radio receiver** is shown in Fig. 15.2.37. This is called a **superhetero-**

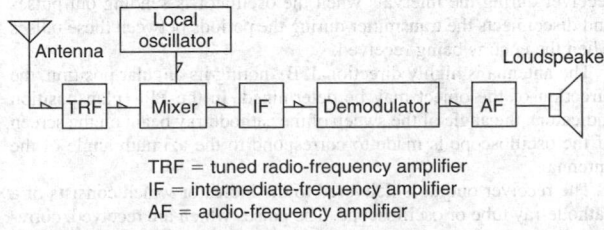

TRF = tuned radio-frequency amplifier
IF = intermediate-frequency amplifier
AF = audio-frequency amplifier

Fig. 15.2.37 Superheterodyne radio receiver.

dyne receiver because it utilizes a frequency-mixing scheme. The tuned radio-frequency amplifier is tuned to receive the desired radio signal. The local oscillator is adjusted by the same tuning control to a lower frequency. The mixer produces an output frequency which is the difference between the incoming radio-signal frequency and the local-oscillator frequency. Since this difference frequency is constant for all tuning positions, the intermediate-frequency amplifier always operates with a constant frequency. This allows optimum design of the intermediate-frequency (IF) amplifiers since they are constant-frequency amplifiers. The IF frequency signal is modulated in just the same way as the radio signal. The demodulator separates this audio signal, which is then amplified so that the loudspeaker can be driven.

The term **radar** is **derived** from the first letters of the words "*radio detection and ranging.*" It is essentially an echo system in which the location of an object is determined by sending out short pulses of radio waves and observing and measuring the time required for their reflections or echoes to return to the sending point. The time interval is a measure of the distance of the object from the transmitter. The velocity of radio waves is the same as the velocity of light, or 984 ft/μs, so that each microsecond interval corresponds to a distance of 492 ft. The direction of an object can be determined by the position of the directional transmitting and receiving antenna. Radio waves penetrate darkness, fog, and clouds, and hence are able to detect objects that otherwise would remain concealed. Radar can be used for the automatic "tracking" of objects such as airplanes.

A block diagram of a radar system is shown in Fig. 15.2.38. The transmitting system consists of an rf oscillator which is controlled by a modulator, or pulser, so that it sends to the antenna intermittent trains of rf waves of relatively high power but of very short duration, corresponding to the pulses received by the modulator. The energy of the oscillator is transmitted through the duplexer and to the antenna through

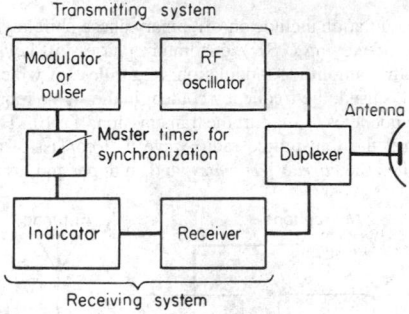

Fig. 15.2.38 Block diagram of radar system.

either coaxial cable or waveguides. The **receiver** is an ordinary heterodyne-type radio receiver which has high sensitivity in the band width corresponding to the frequency of the oscillator. For low frequencies the local oscillator is an ordinary oscillator for frequencies of 2,000 MHz; and higher a reflex **klystron** (hf cavity oscillator) is used. A common intermediate frequency is 30 MHz but 15 and 60 MHz are also frequently used.

In most radar systems the same antenna is used for receiving as for transmitting. This requires the use of a **duplexer** which cuts off the receiver during the intervals when the oscillator is sending out pulses and disconnects the transmitter during the periods between these pulses when the echo is being received.

The antenna is highly directional. By noting its angular position, the direction of the object may be determined. In the PPI (plan position indicator), the angle of the sweep of the cathode-ray beam on the screen of the oscilloscope is made to correspond to the azimuth angle of the antenna.

The receiver output is delivered to the indicator which consists of a cathode-ray tube or oscilloscope. The pulses which are received, corresponding to echoes from the target, must be synchronized with the

sending pulses in order that the distance to the target may be determined. This is accomplished by synchronization of the sweep circuit of the oscilloscope with the pulses by the master timer.

Displays Conversion of the received radar signals to usable display is accomplished by a cathode-ray oscilloscope. The simplest type, called the **A presentation,** is shown in Fig. 15.2.39a. When the pulser operates, a sawtoothed wave produces a linear sweep voltage (Fig. 15.2.39b) across the sweep plates of the cathode-ray tube; at the same time a transmitter pulse is impressed on the deflection plates and the return echoes appear as AM pulses, or "pips," on the screen, as shown in Fig. 15.2.39a. The distance on the screen between the transmitter pulse and the pip caused by the echo is proportional to the distance to the target, and the screen can be calibrated in distance such as miles.

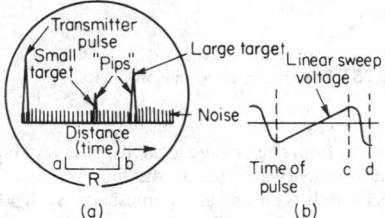

Fig. 15.2.39 Type A presentation.

[The return of the spot to its initial starting position, produced by the sweep interval cd (Fig. 15.2.39b), is so rapid that it is not detectable by the eye.] The direction of the target may be determined by the angular position of the antenna, which can be transmitted to the operator by means of a selsyn. Different objects, such as airplanes, ships, islands, and land approaches, have characteristic pips, and operators become skilled in their interpretation. A bird in flight can be recognized on the screen. Also a portion of the scale such as ab can be segregated and amplified for close study of the characteristics of the pips.

Plan Position Indicator (PPI) In the PPI (Fig. 15.2.40) the direction of a radial sweep of the electron beam is synchronized with the azimuth sweep of the antenna. The sweep of the beam is rotated continuously in synchronism with the antenna, and the received signals intensity-modulate the electron beam as it sweeps from the center of the oscilloscope screen radially outward. In this way the direction and range position of

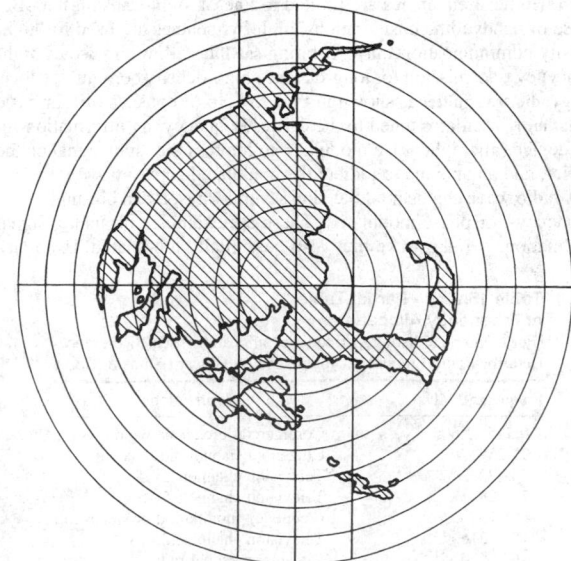

Fig. 15.2.40 Plan position indicator (PPI) of southeastern Massachusetts.

an object can be determined from the pattern on the screen of the oscilloscope, as shown in Fig. 15.2.40.

There are two methods by which the angular direction of the cathode spot is made to correspond with the angular position of the antenna. In one method, used on board ship, two magnetic deflecting coils are rotated around the neck of the tube in synchronism with the antenna, by means of a sclsyn. In the other method, used on aircraft, two fixed magnetic deflecting coils at right angles to each other and placed at the neck of the tube are supplied with current from a small two-phase synchronous generator whose rotor is driven by the antenna. Thus a rotating field, similar to that produced by the stator of an induction motor, is produced by the magnetic deflecting coils. These two rotating fields, although produced by different means, are equivalent and cause the cathode beam to sweep radially in synchronism with the antenna. Circular coordinates spaced radially corresponding to distance are obtained by impressing on the control electrode short positive pulses synchronized with the transmitted pulse but delayed by time values corresponding to the desired distances. These coordinates appear as circles on the screen. Since the time of rotation of the antenna is relatively slow, it is necessary that a persistent screen be used in order that the operator may view the entire pattern. In Fig. 15.2.32 is shown a line drawing of a PPI presentation of Cape Cod, Mass., on a radar screen, taken from an airplane.

The applications of radar to war purposes are well known, such as detecting enemy ships and planes, aiming guns at them, and locating cities, rivers, mountains, and other landmarks in bombing operations. In peacetime, radar is used to navigate ships in darkness and poor visibility by locating navigational aids such as buoys and lighthouses, as well as protruding ledges, islands, and other landmarks. It can be similarly used in air navigation, as well as to operate altimeters for determining the height of the plane above ground. It is also used for aerial mapping.

Sonar is closely related to radar. The difference is that sound is used as the transmitting signal. Since the velocity of sound in water is so much slower than radio waves in free space, the timing for sonar signals is much slower than for radar signals. A 1-s delay of a sonar return echo implies that the target is 2,500 ft away. A 1-s delay of a radar return echo would imply that the target is 193,000 miles away. Although A displays and PPI displays are used with sonar signals, much reliance is placed on hearing. Transmitting a sonar pulse is an excellent way of disclosing your presence to an enemy vessel. Today more reliance is placed on passive detection of enemy vessels, by **signature analysis**. Signature analysis is signal analysis using fast Fourier transforms and filters.

Television is accomplished by systematically scanning a scene or the image of a scene to be reproduced and transmitting at each instant a current or a voltage which is proportional to the light intensity of the elementary area of the scene which at the instant is being scanned. The varying voltage or current is amplified, modulated on a carrier wave, and then transmitted as a radio wave. At the receiver the radio wave enters the antenna, is amplified, and demodulated to give a voltage or a current wave similar to the original wave. This voltage or current wave is then used to control the intensity of a cathode-ray beam which is focused on a fluorescent screen in a cathode-ray reproducing tube. The cathode-ray beam is caused to move over the screen in the same pattern as the scanning beam at the transmitter and in synchronism with it. Thus each small area of the receiver screen is illuminated instantaneously with light intensity corresponding to that of a similarly placed area in the original scene. This process is conducted so rapidly that owing to persistence of vision of the eye, the reproduction of each instantaneous scene appears to be a complete picture and the effect with successive scenes is similar to that produced by the projection of successive frames of a motion picture.

Scanning and Blanking In the United States the ratio of width to height of a standard television picture is 4 : 3, and the picture is composed of 525 lines repeated 30 times a second, this last factor being one-half 60, the prevalent electric power frequency in the United States. The scanning sequence along the individual lines is from left to right and the sequence of the lines is from top to bottom. Also, interlacing is

employed, the general method of which is shown in Fig. 15.2.41. The cathode-ray spot starts at 1 in the upper left-hand corner and is swept rapidly from left to right either by a sawtooth emf wave applied to the sweep plates or by the sawtooth current wave applied to the sweep coils of the tube. When the spot arrives at the right-hand side of the picture, the sawtooth wave of either emf or current in the sweep circuit acts to return the cathode-ray spot rapidly to point 3 at the left-hand side of the picture. However, during this period the cathode-ray is blanked, or entirely eliminated, by the application of a negative potential to the control grid of the tube. At the end of the return period, the blanking effect ceases and the spot appears at point 3, from which it again is swept

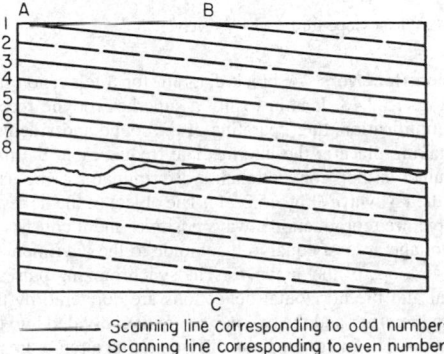

Scanning line corresponding to odd numbers
Scanning line corresponding to even numbers

Fig. 15.2.41   Pattern of interlaced scanning.

across the picture and this process is repeated for 262.5 lines until the spot reaches a midpoint C at the bottom of the picture. It is then carried vertically and rapidly to B, the midpoint of the top of the picture, the beam also being blanked during this period. This process of scanning is then repeated, a second set of lines corresponding to the even numbers 2, 4, 6 being established between the lines designated by the odd numbers. These lines are shown dashed in Fig. 15.2.41. This method or pattern of scanning is called **interlacing**. The two sets of lines taken together produce a frame of 525 lines, which are repeated 30 times each second. However, owing to interlacing, the flicker frequency is 60 Hz which is not noticeable, 50 Hz having been determined as the threshold of flicker noticeable to the average eye. In Fig. 15.2.41, for the sake of clarity, the distances between horizontal lines are greatly exaggerated and no attempt is made to maintain proportions.

Frequency Band In order to obtain the necessary resolution of pictures, television frequencies must be high. In the United States, VHF frequencies from 54 to 88 MHz (omitting 72 to 76 MHz) and 174 to 216 MHz are assigned for television broadcasting. A UHF band of frequencies for commercial television use is also allocated and consists of the frequencies of from 470 to 890 MHz (see also Tables 15.2.17 and 15.2.18).

In order to obtain the 525 lines repeated 30 times per second, a bandwidth of 6 MHz is necessary. The video, or picture, signal with the superimposed scanning and blanking pulses is amplitude-modulated, amplified, and transmitted. The carrier frequency associated with the sound transmitter is 4.5 MHz higher than the video carrier frequency and is frequency-modulated with a maximum frequency deviation of 25 kHz.

In scanning motion-picture films a complication arises because standard film rate is 24 frames per second, while the television rate is 30 frames per second. This difficulty is overcome by scanning the first of two successive film frames twice and the second frame three times at the 60-Hz rate, making the total time for the two frames $\frac{1}{12}$ ($\frac{2}{60}$ + $\frac{3}{60}$) or $\frac{1}{24}$ s average per frame.

Kinescope The kinescope (Fig. 15.2.42) is the terminal tube in which the televised picture is reproduced. It is relatively simple, being not unlike the cathode-ray oscilloscope tube. It has an electric gun operating at 8,000 to 20,000 V which produces an electron beam focused on a fluorescent surface within the front wall of the tube. The picture is viewed at the front wall. The horizontal and vertical deflections of the beam are normally controlled by deflection coils, as shown in Fig. 15.2.42.

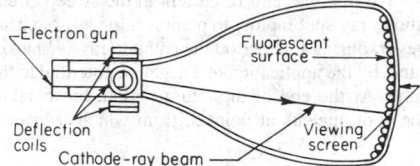

Fig. 15.2.42 Kinescope for television receiver.

Television Receivers A block diagram for a television receiver is given in Fig. 15.2.43. It is in reality a superheterodyne receiver with tuned rf amplification, the separating of the sound and video or picture channels taking place at the intermediate frequency in the mixer. The sound channel is then conventional, a discriminator being used to demodulate the FM wave (Fig. 15.2.21). The object of the dc restorer is to make the picture reproduction always positive, and it consists of applying a dc voltage at least equal in magnitude to the maximum values of the negative loops of the ac waves. The synchronizing pulses for both the vertical and the horizontal deflections are delivered by the dc restorer to an amplifier and the two pulses are then divided into the V and H components. The integrating and differentiating circuits are necessary to separate horizontal and vertical synchronizing signals.

As stated earlier, at any instant the magnitude of the current from the pickup tube varies in accordance with the light intensity of the part of the scene being scanned at that instant. This current is amplified and, together with the sound and synchronizing current, is broadcast and received by the circuit shown in Fig. 15.2.43. The video current is detected by rectification, is amplified, and is then made to control the intensity of the kinescope electron beam. Tubes produce a scanning pattern, identical with that in the pickup tube, and these tubes are triggered by the synchronizing pulses which are transmitted in the broadcast wave. Hence, the original televised scene is reproduced on the fluorescent screen of the kinescope.

Color-television transmission is similar to black-and-white television, and the two signals must be compatible with each other. The kinescope for color TV has three electron guns, one for each primary color. The fluorescent screen has a matrix of three different colors of phosphor and a mask with many small holes in it. The intensity signals for each color are phase-shifted from each other so that the proper phosphors are excited by each electron stream at each mask point over the entire screen. A black-and-white signal does not have the same synchronizing signal as a color signal. The color receiver has circuits which recognize this state and switch it to black and white reception.

Telephone Communications

Telephone communications is an important aspect of electronics because it provides a source of low-cost electronic devices which can be adapted for electronic circuit design. One of the big problems in telephone system design is switching. Each long-distance phone call requires that several circuits be joined together, and thousands of phone calls take place simultaneously. Older telephone exchange buildings had rack upon rack of telephone relays. These are being replaced by semiconductor switching. Older trunk lines were coaxial cable or microwave radio links. These are being replaced by fiber-optic cables. Fiber-optic cables have a tremendous bandwidth. A telephone voice circuit has a bandwidth of 4 kHz. Thousands of these circuits can be routed through one fiber-optic cable. Separating these voice circuits into communication channels by frequency separation and filtering is not feasible. Communication on fiber-optic cables is done digitally, and in packets. Each packet is a few hundred bytes long, and includes a header with an address and other control information, the message, and a trailer with error-correcting codes. The address is a session address to identify where the packet is from and where it is going. Users time-share the fiber-optic cable, and all of this signal processing must be done at a high enough rate to appear to be instantaneous to the subscriber.

Any form of signal modulation can be considered to be data encryption. The information is coded or encrypted onto the transmitting medium. At the transmitting end, a clear signal with no noise is sent, but at the receiving end of the medium, the signal contains the signal plus unwanted noise. Since the coding method is known, and since the noise is known to be random, separation of signal and noise is possible. This process of separating signal from noise will be repeated several times for long-distance communications. Each time, the noise is discarded and only the signal is sent on. Such a scheme of data encryption, or modulation, is also warranted in data telemetry, in order to allow noise to be removed from the desired signal.

Long-distance communications takes several different forms. Historically, cables with hundreds of twisted-pair circuits were used for long-distance communication. These are used for local subscriber service today. The next step was a cable made up of many wideband microwave coaxial cables. The coaxial cables were replaced by microwave radio links. Then, fiber-optic cables replaced microwave links. Wireless communication, using laser-generated signals, is the latest form of high volume data transmission. In a telephone system, all of these systems must coexist and work together.

There is great interest in cellular communications. This is a microwave broadcast system. The cellular communication system is evolving today, and promises to allow mobility of telephone communications.

Global Positioning Information

Navigational aids take several different forms. **Shoran** (short-range navigation) and **loran** (long-range navigation) are methods by which ships or planes can locate their position. A loran station consists of a master

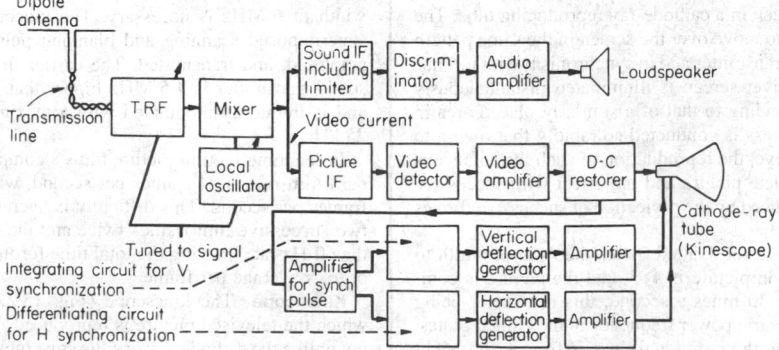

Fig. 15.2.43 Block diagram for television receiver. TRF: tuned radio frequency; IF: intermediate frequency.

transmitter and slave transmitter located about 200 mi apart. A pulse is broadcast from the master transmitter and, when received by the slave transmitter, causes it to also broadcast a pulse. The loran receiver, located on a ship or airplane, receives both the master and the slave pulses. The receiver allows the user to measure the delay between receiving the master and slave pulses. The delay defines a line across the surface of the earth. By picking three loran stations, determining the delay of each of the transmitter pairs, and using loran charts, three lines can be defined, allowing the user to determine the triangle which defines the receiver's position. **Very high frequency omnirange (VOR)** is used for short-range aircraft navigation, often in conjunction with **direction measuring equipment (DME)**. VOR stations are located approximately 100 mi apart. The receiver in the aircraft receives a signal from the VOR transmitter with two pieces of information, which allows the pilot to determine the bearing from the VOR station. The first piece of information is a 30-Hz amplitude-modulated signal which is broadcast from a rotating directional antenna. The antenna rotates 30 times per second, and broadcasts in the 108- to 112-MHz range. The second piece of information is a frequency-modulated signal which varies between 9,480 and 10,440 Hz. This FM signal changes frequency at 30 Hz, and forms a reference for the VOR system. A receiver on the aircraft compares the 30-Hz amplitude-modulated signal and the 30-Hz frequency-modulated reference, and translates the difference in phase between the FM signal and the AM signal to give the airplane its bearing. The VOR also broadcasts station identification information in the 112- to 118-MHz range. Most VOR stations can also broadcast a **distance measuring equipment (DME)**. The aircraft has a DME transmitter which broadcasts a pulse which activates a ground mounted transponder. The aircraft sends an FM signal in the 1,025- to 1,150-MHz range. The DME ground station responds with a delayed signal in the 962- to 1,024-MHz or 1,151- to 1,213-MHz range. Distance is determined by the time delay required for the propagation of the signals. The depressed, straight line distance from the aircraft is measured by the DME equipment. To relieve the pilot of the job of correcting for altitude in the DME signal and to allow the pilot to fly any heading, a computer system called RNAV was introduced. RNAV uses the VOR and DME information, plus wind velocity and airspeed, to calculate the required aircraft heading and the estimated time of arrival at the destination.

Most major airports have an **instrument low approach system (ILS)**. This produces a set of signals in the 108- to 112-MHz region. There a right-hand antenna lobe, relative to the aircraft, which is modulated at 150 Hz, and a left-hand lobe modulated at 90 Hz. In addition, in the 329- to 335-MHz range there is a vertical set of antenna lobes, with the upper lobe modulated at 90 Hz and lower lobe at 150 Hz. The display on the aircraft is an azimuth needle and a glide path needle. The pilot simply flies the aircraft to center the needles to determine a correct landing approach. There are also ILS marker beacons to tell the pilot how close the aircraft is to the airport. Other aircraft navigation systems are based on ground-operated radar systems. Still others are based on aircraft-mounted Doppler radar. Doppler radar is a velocity measuring system.

Satellite positioning systems are the newest development for long-range and short-range navigation. The **global positioning system (GPS)** is a system involving 21 to 24 satellites which broadcast accurate time marks. By knowing which satellite has broadcast the time mark, and measuring its delay to the aircraft, one can determine a spherical surface in space. By determining the delay from four satellites, four spherical surfaces in space can be determined. Worldwide navigation accurate within 50 m in longitude and latitude and within 100 m vertically is being realized commercially. The military capabilities are better than the commercial, in the submeter range. The commercial timing signals are deliberately disturbed with a random timing variation to degrade them. The capability of the military system was demonstrated by the accuracy of "smart" bombs during the Iraq-Kuwait war.

GPS is being used for commercial truck fleets to provide better fleet utilization. GPS is also being used by the Tennessee Valley Authority to monitor power system performance, by determining the phase angle of the 60-Hz voltage wave throughout its system. GPS is also being used by search and rescue teams. GPS receivers for search and rescue cost $300 to $600, and are dropping in price at this time.

Instruments and Controls

BY

O. MULLER-GIRARD *Consulting Engineer, Rochester, NY.*
GREGORY V. MURPHY *Process Control Consultant, DuPont Co.*
W. DAVID TETER *Professor, Department of Civil Engineering, College of Engineering, University of Delaware.*

16.1 INSTRUMENTS
by Otto Muller-Girard

REFERENCES: ASME publications: "Instruments and Apparatus Supplement to Performance Test Codes (PTC 19.1–19.20)"; "Fluid Meters, pt. II, Application." ASTM, "Manual on the Use of Thermocouples in Temperature Measurement," STP 470B. ISA publications: "Standards and Recommended Practices for Instrumentation and Controls," 11 ed. Spitzer, "Flow Measurement." Preston-Thomas, The International Temperature Scale of 1990 (ITS-90), *Metrologia,* **27,** 3–10 (1990), Springer-Verlag. NIST Monograph 175, "Temperature-Electromotive Force Reference Functions and Tables for the Letter-Designated Thermocouple Types Based on the ITS-90," Government Printing Office, April 1993. Schooley, (ed.), "Temperature, Its Measurement and Control in Science and Industry," Vol. 6, Pts. 1 and 2, American Institute of Physics. Time and frequency services offered by the National Institute of Standards and Technology (NIST). Lombardi and Beehler, NIST, paper 37-93. Beckwith, et al., "Mechanical Measurements," Addison-Wesley. Considine, "Encyclopedia of Instrumentation and Control," Krieger reprint. Considine, "Handbook of Applied Instrumentation," McGraw-Hill, Krieger reprint. Dally, et al., "Instrumentation for Engineering Measurements," Wiley. Doebelin, "Measurement Systems, Application and Design," McGraw-Hill. Erikson and Graber, Harris et al., "Shock and Vibration Control Handbook," McGraw-Hill. Holman, "Experimental Methods for Engineers," McGraw-Hill. Jones (ed.), "Instrument Science and Technology, Vol. 1, Measurement of Pressure, Level, Flow and Temperature," Heyden. Lion, "Instrumentation in Scientific Research, Electrical Input Transducers," McGraw-Hill. Sheingold, (ed.), "Transducer Interfacing Handbook," Analog Devices, Inc. Norwood, MA. Snell, "Nuclear Instruments and Their Uses," Wiley. Spink, "Principles and Practice of Flow Meter Engineering," Foxboro Co. Stout, "Basic Electrical Measurements," Prentice-Hall. *Periodicals: Instruments & Control Systems,* monthly, Chilton Co. *InTech,* monthly, ISA. *Measurements & Control,* bimonthly, Measurements and Data Corp., Pittsburgh. *Sensors,* monthly, Helmers Publishing. *Test & Measurement World,* Cahners.

INTRODUCTION TO MEASUREMENT

An **instrument,** as referred to in the following discussion, is a device for determining the value or magnitude of a quantity or variable. The variables of interest are those which help describe or define an object, system, or process. Thus, in a manufacturing operation, product quality is related to measurements of its various dimensions and physical properties such as hardness and surface finish. In an industrial process, measurement and control of temperature, pressure, flow rates, etc., determine quality and efficiency of production.

Measurements may be direct, e.g., using a micrometer to measure a dimension, or indirect, e.g., determining moisture in steam by measuring the temperature in a throttling calorimeter.

Because of physical limitations of the measuring device and the system under study, practical measurements always have some error. The **accuracy** of an instrument is the closeness with which its reading approaches the true value of the variable being measured. Accuracy is commonly expressed as a percentage of measurement span, measurement value, or full-scale value. Span is the difference between the full-scale and the zero scale value. **Uncertainty,** the sum of the errors at work to make the measured value different from the true value, is the accuracy of measurement standards. Uncertainty is expressed in parts per million (ppm) of a measurement value. **Precision** refers to the reproducibility of the measurements, i.e., with a fixed value of the variable, how much successive readings differ from one another. **Sensitivity** is the ratio of output signal or response of the instrument to a change in input or measured variable. **Resolution** relates to the smallest change in measured value to which the instrument will respond.

Error may be classified as systematic or random. Systematic errors are those due to assignable causes. These may be static or dynamic. Static errors are caused by limitations of the measuring device or the physical laws governing its behavior. Dynamic errors are caused by the instrument not responding fast enough to follow the changes in mea-

sured variable. Random errors are those due to causes which cannot be directly established because of random variations in the system.

Standards for measurement are established by the National Institute of Standards and Technology. Secondary standards are prepared by very precise comparison with these primary standards and, in turn, form the basis for calibrating instruments in use. A well-known example is the use of precision gage blocks for the calibration of measuring instruments and machine tools.

There are three essential parts to an instrument: the **sensing element,** the **transmitting means,** and the **output** or **indicating element.** The sensing element responds directly to the measured quantity, producing a related motion, pressure, or electrical signal. This is transmitted by linkage, tubing, wiring, etc., to a device for display, recording, and/or control. Displays include motion of a pointer or pen on a calibrated scale, chart, oscilloscope screen, or direct numerical indication. Recording forms include writing on a chart and storage on magnetic tape or disk. The instrument may be actuated by mechanical, hydraulic, pneumatic, electrical, optical, or other energy medium. Often a combination of several energy modes is employed to obtain the accuracy, sensitivity, or form of output desired.

The transmission of measurements to distant indicators and controls is industrially accomplished by using the standardized electrical current signal of 4 to 20 mA; 4 mA represents the zero scale value and 20 mA the full-scale value of the measurement range. A pressure of 3 to 15 lb/in^2 is commonly used for pneumatic transmission of signals.

COUNTING EVENTS

Event counters are used to measure the number of items passing on a conveyor line, the number of operations of a machine, etc. Coupled with time measurements, they yield measures of average rate or frequency. They find important application, therefore, in inventory control, production analysis, and in the sequencing control of automatic machines.

Choice of the proper counting device depends on the kind of events being counted, the necessary counting speed, and the disposition of the measurement; i.e., whether it is to be indicated remotely, used to actuate a machine, etc. Errors in the total count may be introduced by events being too close together or by too much nonuniformity in the items being counted.

The **mechanical counter** is shown in Fig. 16.1.1. Motion of the event being counted deflects the arm, which through an appropriate linkage advances the count register one unit. Alternatively, motion of the actu-

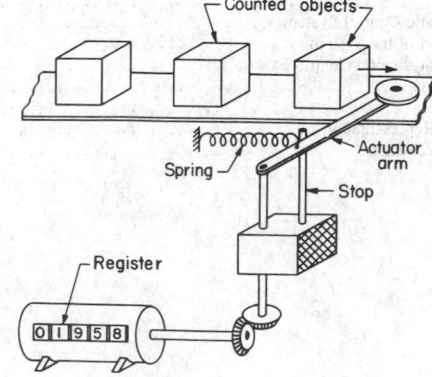

Fig. 16.1.1 Mechanical counter.

ating arm may close an electrical switch which energizes a relay coil to advance the count register one step.

Where there is a desire to avoid contact or close proximity with the object being counted, the photoelectric cell or diode, in conjunction with a lamp, a light-emitting diode (LED), or a laser light source, is employed in the transmitted or reflected light mode (Fig. 16.1.2). A signal to a counter is generated whenever the received light level is altered by the passing objects. Objects may be very small and very high counting speeds may be achieved with electronic counters.

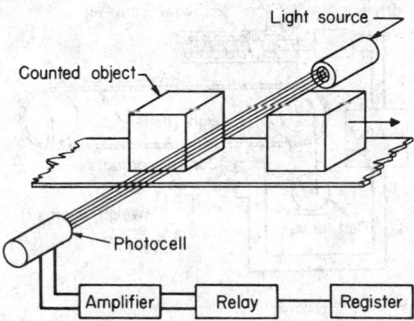

Fig. 16.1.2 Photoelectric counter.

Sensing methods based on electrical capacitance, magnetic, and eddy-current effects are extremely sensitive and fast acting, and are suitable for objects in close proximity to the sensor. The capacitive probe senses dielectrics other than air, such as glass and plastic parts. The magnetic pickup, by induction, responds to the motion of iron and nickel. The eddy-current sensor, by energy absorption, detects nonmagnetic conductors. All are suitable for counting machine operations.

The count is displayed by either a mechanical register as in Fig. 16.1.1, a dial-type register (as on the household watthour meter), or an electronic pulse counter with either number indicators or digital printing output. **Electronic counters** can operate accurately at rates exceeding 1 million counts per second.

TIME AND FREQUENCY MEASUREMENT

Measurement of time is basic to time and motion studies, time program controls, and the measurements of velocity, frequency, and flow rate. (See also Sec. 1.)

Mechanical clocks, chronometers, and stopwatches measure time in terms of the natural oscillation period of a system such as a pendulum, or hairspring balance-wheel combination. The minimum resolution is one-half period. Since this period is somewhat affected by temperature, precise timepieces employ a compensating element to maintain timing accuracies over long periods. Stopwatches may be obtained to read to better than 0.1 s. The major limitation, however, is in the response time of the user.

Electric timers are simple, inexpensive, and readily adaptable to remote-control operations. The majority of these are ac synchronous motors geared in the proper ratio to the indicator. These depend for their accuracy on the frequency of the line voltage. Consequently, care must be exercised in using such devices for precise short-time measurements.

Electronic timers are started and stopped by electrical pulses and hence are not limited by the observer's reaction time. They may be made extremely accurate and capable of measuring to less than 1 μs. These measure time by counting the number of cycles in a high-frequency signal generated internally by means of a quartz crystal. Stopwatch versions read at 0.01 s. Commercial instruments offer one or more functions: counting, measurement of frequency, period, and time intervals. Microprocessor-equipped versions increase versatility.

There are a variety of timing devices designed to indicate or control to a fixed time. These include timers based on the charging time of a condenser (e.g., type 555 integrated circuit), and the flow of oil or other fluid through a restriction.

Timing devices can be **calibrated** by comparison with a standard instrument or by reference to the National Institute of Standards and Technology timed radio signals, carrier frequencies and audio modulation of radio stations WWV and WWVB, Colorado, and WWVH, Hawaii. WWV and WWVH broadcast with carrier frequencies of 2.5, 5, 10, and 15 MHz. WWV also broadcasts on 20 MHz. Broadcasts provide second, minute, and hour marks with once-per-minute time announcements by voice and binary-coded decimal (BCD) signal on a 100-Hz subcarrier. Standard audio frequencies of 440, 500, and 600 Hz are provided. Station WWVB uses a 60-kHz carrier and provides second and minute marks and BCD time and date. Time services are also issued by NIST from geostationary satellites of the National Oceanographic and Atmospheric Administration (NOAA) on frequencies of 468.8375 MHz for the 75° west satellite and 468.825 MHz for the 105° west satellite. Automated Computer Time Service (ACTS) is available to 300- or 1200-baud modems via phone number 303-494-4774. (See also Sec. 1.2.)

Fast-moving, repetitive motions may be timed and studied by the use of stroboscopes which generate brilliant, very brief flashes of light at an adjustable rate.

The frequency of the observed motion is measured by adjusting the stroboscopic frequency until the system appears to stand still. The frequency of the motion is then equal to the stroboscope frequency or an integer multiple of it.

Many other means exist for **measuring vibrational or rotational frequencies.** These include timing a fixed number of rotations or oscillations of the moving member. Contact sensing can be done by an attached switch, or noncontact sensing can be done by magnetic or optical means. The pulses can be counted by an electronic counter or displayed on an oscilloscope or recorder and compared with a known frequency. Also used are reeds which vibrate when the measured oscillation excites their natural frequencies, flyball devices which respond directly to angular velocity, and generator-type tachometers which generate a voltage proportional to the speed.

MASS AND WEIGHT MEASUREMENT

Mass is the measure of the quantity of matter. The fundamental unit is the kilogram. The U.S. customary unit is the pound; 1 lb = 0.4536 kg (see Sec. 1.2, "Measuring Units"). Weight is a measure of the force of gravity acting on a mass (see "Units of Force and Mass" in Sec. 4).

A general equation relating weight W and mass M is $W/g = M/g_c$, where g is the local acceleration of gravity, and g_c = 32.174 lbm · ft/ (lbf) (s²) [(1 kg · m/(N) (s²)] is a property of the unit system. Then $W = Mg/g_c$. The specific weight w and the mass density p are related by $w = pg/g_c$. Masses are conveniently compared by comparing their weights, and masses are often loosely referred to as weights. Indeed, almost all practical measures of mass are based on weight.

Weighing devices fall into two major categories: balances and force-deflection systems. The device may be batch or continuous weighing, automatic or manual. Accuracies are expected to be of the order of 0.1 to better than 0.0001 percent, depending on the type and application of the scale. Calibration is normally performed by use of standard weights (masses) with calibrations traceable to the National Institute of Standards and Technology.

The **equal arm balance** compares the weight of an object with a set of standard weights. The laboratory balance shown in Fig. 16.1.3 is used for extreme precision and sensitivity. A chain poise provides fine adjustment of the final balance weight. The magnetic damper causes the balance to come to equilibrium quickly.

Large weighing scales operate on the same principle; however, the arms are unequal to allow multiplication between the tare and the measured weights. In this group are platform, track, hopper, and tank scales. Here balance is achieved by adjusting the position of one or more balance weights along a beam directly calibrated in weight units. In dial-indicating-type scales, balance is achieved automatically through the deflection of calibrated pendulum weights from the vertical. The deflection is greatly magnified by the pointer-actuating mechanism, providing a direct-reading weight indication on the dial.

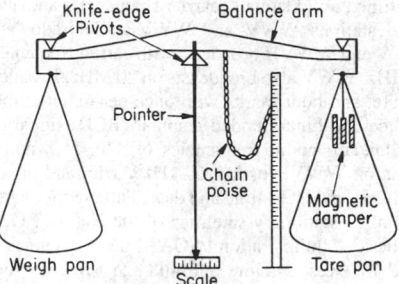

Fig. 16.1.3 Laboratory balance.

Since the deflection of a spring (within its design range) is directly proportional to the applied force, a calibrated spring serves as a simple and inexpensive weighing device. Applications include the **spring scale** and **torsion balance**. These are subject to hysteresis and temperature errors and are not used for precise work.

Other force-sensing elements are adaptable to weight measurement. Strain-gage load cells eliminate pivot maintenance and moving parts and provide an electrical output which can be used for direct recording and control purposes. Pneumatic pressure cells are also used with similar advantages.

In production processes, **continuous and automatic operating scales** are employed. In one type, the balancing weight is positioned by a reversible electric motor. Deflection of the beam makes an electrical contact which drives the motor in the proper direction to restore balance. The final balance position is translated by means of a potentiometer or digital encoding disk into a signal which is used for recording or control purposes.

The **batch-type scale** (Fig. 16.1.4) is adaptable to continuous flow streams of either liquids or solid particles. Material flows from the feed hopper through an adjustable gate into the scale hopper. When the weight in the scale hopper reaches that of the tare, the trip mechanism operates, closing the gate and opening the door. As soon as the scale hopper is empty, the weight of the tare forces the door closed again, resets the trip, and opens the gate to repeat the cycle. The agitator rotates

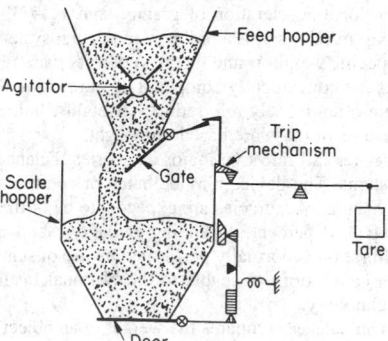

Fig. 16.1.4 Automatic batch-weighing scale.

while the gate is open, to prevent the solids from packing. Also, a "dribble" (partial closing of the gate just before the mechanism trips) is employed to minimize the error from the falling column of material at the instant balance is achieved. Since each dump of the scale represents a fixed weight, a counter yields the total weight of material passing through the scale.

In **continuous weighers,** a section of conveyor belt is balanced on a weigh beam (Fig. 16.1.5). The belt is driven at a constant speed; hence, if the total weight is held constant, the weight rate of material fed through the scale is fixed. Unbalance of the weight beam causes the rate of material flow onto the belt to be changed in the direction of restoring balance. This is accomplished by a mechanical adjustment of the feed gate or by varying the speed of a belt or screw feeder drive.

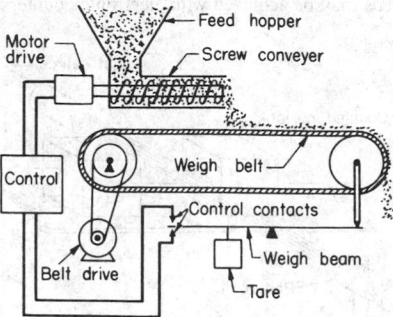

Fig. 16.1.5 Continuous-weighing scale.

If the density of the material is constant, **volume measurements** may be used to determine the mass. Thus, calibrated tanks are frequently used for liquids and vane and screw-type feeders for solids. Though often simpler to apply, these are not generally capable of as high accuracies as are common in weighing.

MEASUREMENT OF LINEAR AND ANGULAR DISPLACEMENT

Displacement-measuring devices are employed to measure dimension, distances between points, and some derived quantities such as velocity, area, etc. These devices fall into two major categories: those based on comparison with a known or reference length and those based on some fixed physical relationship.

The **measurement of angles** is closely related to displacement measurements, and indeed, one is often converted into the other in the process of measurement. The common unit is the degree, which represents $\frac{1}{360}$ of an entire rotation. The radian is used in mathematics and is related to the degree by π rad = 180°; 1 rad = 57.3°. The grad is an angle unit = $\frac{1}{400}$ rotation.

Figure 16.1.6 illustrates some methods of **rotary to linear conversion**. Figure 16.1.6*a* is a simple link and lever, Fig. 16.1.6*b* is a flexible link and sector, and Fig. 16.1.6*c* is a rack-and-pinion mechanism. These can be used to convert in either direction according to the relationship $D = RA/57.3$, where R = mean radius of the rotating element, in; D = displacement, in; and A = rotation, deg. (This equation holds for the link and lever of Fig. 16.1.6*a* only if the angle change from the perpendicular is small.)

Comparative devices are generally of the indicating type and include ruled or graduated devices such as the machinist's scale, folding rule, tape measure, digital caliper (Fig. 16.1.7), digital micrometer (Fig. 16.1.8), etc. These vary widely in their accuracy, resolution, and measuring span, according to their intended application. The manual readings depend for their accuracy on the skill and care of the operator.

The digital caliper and digital micrometer provide increased sensitivity and precision of reading. The stem of the digital caliper carries an embedded encoded distance scale. That scale is read by the slider. The distance so found shows on the digital display. The device is battery-operated and capable of displaying in inches or millimeters. Typical resolution is 0.0005 in or 0.01 mm.

The **digital micrometer,** employing rotation and translation to stretch the effective encoded scale length, provides resolution to 0.0001 in or 0.003 mm.

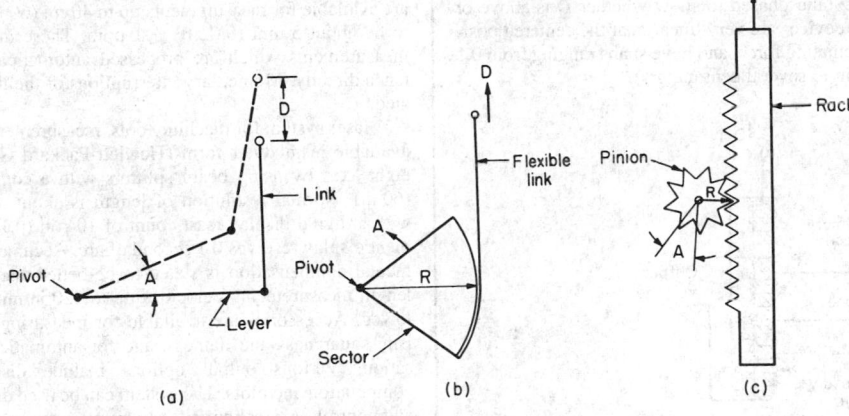

Fig. 16.1.6 Linear-rotary conversion mechanisms.

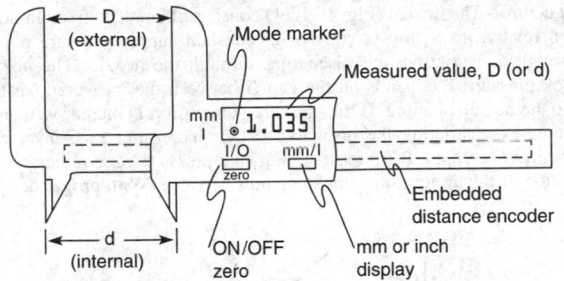

Fig. 16.1.7 Digital caliper.

Dial gages are also used to magnify motion. A rack and pinion (Fig. 16.1.6c) converts linear into rotary motion, and a pointer moves over a calibrated scale.

Various modifications of the above-mentioned devices are available for making special kinds of measurements; e.g., **depth gages** for measuring the depth of a hole or cavity, **inside and outside calipers** (Fig. 16.1.7) for measuring the internal and external dimensions respectively of an object, **protractors** for angular measurement, etc.

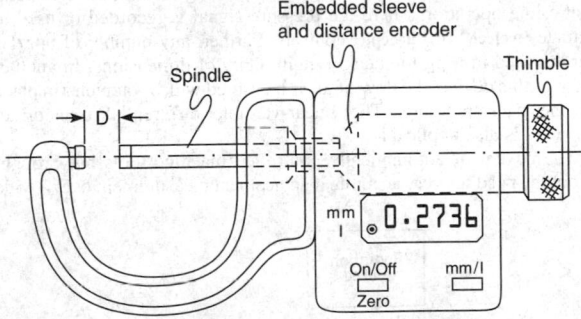

Fig. 16.1.8 Digital micrometer.

For line production and inspection work, **go no-go gages** provide a rapid and accurate means of dimension measurement and control. Since the measured values are fixed, the dependence on the operator's skill is considerably reduced. Such gages can be very complex in form to embrace a multidimensional object. They can also take the more general forms of the **feeler, wire, or thread-gage** sets. Of particular importance are **precision gage blocks**, which are used as standards for calibrating other measuring devices.

Displacement can be measured electrically through its effect on the resistance, inductance, reluctance, or capacitance of an appropriate sensing element.

The **potentiometer** is comparatively inexpensive, accurate, and flexible in application. It consists of a fixed linear resistance over which slides a rotating contact keyed to the input shaft (Fig. 16.1.9). The resistance or voltage (assuming constant voltage across terminals 1 and 3) measured across terminals 1 and 2 is directly proportional to the angle A. For straight-line motion, a mechanism of the type shown in Fig. 16.1.6 converts to rotary motion (or a rectilinear-type potentiometer can be used directly). (See also Sec. 15.) Versions with multiturns, straight-line motion, and special nonlinear resistance vs. motion are available.

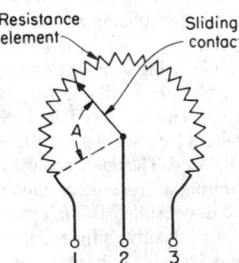

Fig. 16.1.9 Potentiometer.

The **synchro**, the **linear variable differential transformer (LVDT)**, and the **E transformer** are devices in which the input motion changes the inductive coupling between primary and secondary coils. These avoid the limitations of wear, friction, and resolution of the potentiometer, but they require an ac supply and usually an electronic amplifier for the output. (See also Sec. 15.)

The **synchro** is a rotating device which is used to transmit rotary motions to a remote location for indication or control action. It is particularly useful where the rotation is continuous or covers a wide range. They are used in pairs, one transmitter and one receiver. For measurement of difference in angular position, the **control-transmitter** and **control-transformer** synchros generate an electrical error signal useful in control systems. A synchro differential added to the pair serves the same purpose as a gear differential.

The **linear variable differential transformer (LVDT)** consists of a primary and two secondary coils wound around a common core (Fig. 16.1.10). An armature (iron) is free to move vertically along the axis of the coils. An ac voltage is applied to the primary. A voltage is induced in each secondary coil proportional to the relative length of armature linking it with the primary. The secondaries are connected to oppose each other so that when the armature is centered, the output voltage is zero.

When the armature is displaced off center by an amount D, the output will be proportional to D (and phased to show whether D is above or below the center). These devices are very linear near the centered position, require negligible actuating force, and have spans ranging from 0.1 to several inches (0.25 cm to several centimeters).

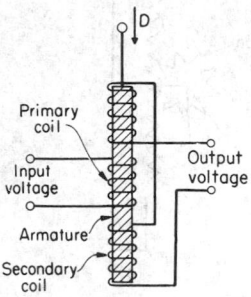

Fig. 16.1.10 Linear variable differential transformer (LVDT).

The **E transformer** is very similar to the above except that the coils are wound around a laminated iron core in the shape of an E (with the primary and secondaries occupying the center and outside legs respectively). The magnetic path is completed through an armature whose motion, either rotary or translational, varies the induced voltage in the secondaries, as in the device of Fig. 16.1.10. This, too, is sensitive to extremely small motions.

A method that is readily applied, if a strain-gage analyzer is handy, is to measure the deflection of a cantilever spring with strain gages bonded to its surface (see Strain Gages, Sec. 5).

The **change of capacitance** with the displacement of the capacitor plates is extremely sensitive and suitable to very small displacements or large rotation. Often, one plate is fixed within the instrument; the other is formed or rotated by the object being measured. The capacitance can be measured by an impedance bridge, by determining the resonant frequency of a tuned circuit or using a relaxation oscillator.

Many optical instruments are available for obtaining precise measurements. The **transit and level** are used in surveying for measuring angles and vertical distances (see Sec. 16.3). A telescope with fine cross hairs permits accurate sighting. The angle scales are generally equipped with verniers. The **measuring microscope** permits measurement of very small displacements and dimensions. The microscope table is equipped with micrometer screws for sensitive adjustment. In addition, templates of scales, angles, etc., are available to permit measurement by comparison. The **optical comparator** projects a magnified shadow image of an object on a screen where it can readily be compared with a reference template.

Light can be used as a standard for the measurement of distance, straightness, and related properties. The wavelength of light in a medium is the velocity of light in vacuum divided by the index of refraction n of the medium. For dry air $n - 1$ is closely proportional to air density and is about 0.000277 at 1 atm and 15°C for 550-nm green light. Since the wavelength changes about $+1$ ppm/°C, and about -0.36 ppm/mmHg, density gradients bend light slightly. A temperature gradient of 1°C/m (0.5°F/ft) will cause a deviation from a tangent line of about 0.05 mm (0.002 in) at 10 m (33 ft).

Optical equipment to establish and test alignment, plumb lines, squareness, and flatness includes **jig transits, alignment telescopes, collimators,** optical squares, mirrors, targets, and scales.

Interference principles can be used for distance measurements. An optical flat placed in close contact with a polished surface and illuminated perpendicular to the surface with a monochromatic light will show interference bands which are contours of constant separation distance between the surfaces. Adjacent bands correspond to separation differences of one-half wavelength. For 550-nm wavelength this is 275 nm (10.8 μin). This test is useful in examining surfaces for flatness and in length comparisons with gage blocks.

Laser beams can be used over great distances. Surveying instruments are available for measurements up to 40 mi (60 km). Accuracy is stated to be about 5 mm (0.02 ft) $+1$ ppm. These instruments take several measurements which are processed automatically to display the distance directly. Momentary interruptions of the light beam can be tolerated.

A laser system for machine tools, measurement tables, and the like is available in modular form (Hewlett-Packard Co.). It can serve up to eight axes by using beam splitters with a combined range of 200 ft (60 m). Normal resolution of length is about one-fourth wavelength, with a digital display least count of 10 μin (0.1 μm). Angle-measurement display resolves 0.1 second of arc. Accuracy with proper environmental compensation is stated to be better than 1 ppm $+$ 1 count in length measurement. Velocities up to 720 in/min (0.3 m/s) can be followed. Accessories are available for measuring straightness, parallelism, squareness and flatness, and for automatic temperature compensation. Various output options include displays and automatic computation and plots. The system can be used directly in measurement and control or to calibrate lead screws and other conventional measuring devices.

Pneumatic gaging finds an important place in line inspection and quality control. The device (Fig. 16.1.11) consists of a nozzle fixed in position relative to a stop or jig. Air at constant supply pressure passes through a restriction and discharges through the nozzle. The nozzle back pressure P depends on the gap G between the measured surface and the nozzle opening. If the measured dimension D increases, then G decreases, restricting the discharge of air, increasing P. Conversely, when D decreases, P decreases. Thus, the pressure gage indicates deviation of the dimension from some normal value. With proper design,

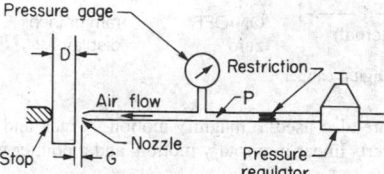

Fig. 16.1.11 Pneumatic gage.

this pressure is directly proportional to the deviation, limited, however, to a few thousandths of an inch span. The device is extremely sensitive [better than 0.0001 in (0.003 mm)], rugged, and, with periodic calibration against a standard, quite accurate. The gage is adaptable to automatic line operation where the pressure signal is recorded or used to actuate "reject" or "accept" controls. Further, any number of nozzles can be used in a jig to check a multiplicity of dimensions. In another form of this device, the flow of air is measured with a rotameter in place of the back pressure. The linear-variable differential transformer (LVDT) is also applicable.

The advent of automatically controlled machine tools has brought about the need for very accurate displacement measurement over a wide

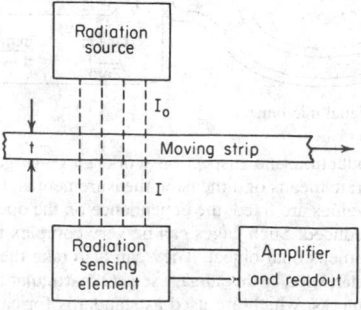

Fig. 16.1.12 Radiation-type thickness gage.

range. Most commonly applied for this purpose is the calibrated **lead screw** which measures linear displacement in terms of its angular rotation. **Digital** systems greatly extend the resolution and accuracy limitations of the lead screw. In these, a uniformly spaced optical or inductive grid is displaced relative to a sensing element. The number of grid lines counted is a direct measure of the displacement (see discussion of lasers above).

Measurement of strip thickness or coating thickness is achieved by **X-ray** or **beta-radiation-** type gages (Fig. 16.1.12). A constant radiation source (X-ray tube or radioisotope) provides an incident intensity I_0; the radiation intensity I after passing through the absorbing material is measured by an appropriate device (scintillation counter, Geiger-Müller tube, etc.). The thickness t is determined by the equation $I = I_0 e^{-kt}$, where k is a constant dependent on the material and the measuring device. The major advantage here is that measurements are continuous and nondestructive and require no contact. The method is extended to measure liquid level and density.

MEASUREMENT OF AREA

Area measurements are made for the purpose of determining surface area of an object or area inside a closed curve relating to some desired physical quantity. Dimensions are expressed as a length squared; e.g., in^2 or m^2. The areas of simple forms are readily obtained by formula. The area of a complex form can be determined by subdividing into simple forms of known area. In addition, various numerical methods are available (see **Simpson's rule**, Sec. 2) for estimating the area under irregular curves.

Area measuring devices include various mechanical, electrical, or electronic **flow integrators** (used with flowmeters) and the **polar planimeter**. The latter consists of two arms pivoted to each other. A tracer at the end of one arm is guided around the boundary curve of the area, causing rotation of a recorder wheel proportional to the area enclosed.

MEASUREMENT OF FLUID VOLUME

For a liquid of known density, volume is a quick and simple means of measuring the amount (or mass) of liquid present. Conversely, measuring the weight and volume of a given quantity of material permits calculation of its density. Volume has the dimensions of length cubed; e.g., cubic metres, cubic feet. The volume of simple forms can be obtained by formula.

A volumetric device is any container which has a known and fixed calibration of volume contained vs. the level of liquid. The device may be calibrated at only one point (**pipette, volumetric flask**) or may be graduated over its entire volume (**burette, graduated cylinder, volumetric tank**). In the case of the tank, a sight glass may be calibrated directly in liquid volume.

Volumetric measure of continuous flow streams is obtained with the **displacement meter**. This is available in various forms: the nutating disk, reciprocating piston, rotating vane, etc. The **nutating-disk meter** (Fig. 16.1.13) is relatively inexpensive and hence is widely used (water meters, etc.). Liquid entering the meter causes the disk to nutate or "roll" as the liquid makes its way around the chamber to the outlet. A pin on the disk causes a counter to rotate, thereby counting the total number of rolls of the disk. Meter accuracy is limited by leakage past the disk and friction. The **piston meter** is like a piston pump operated

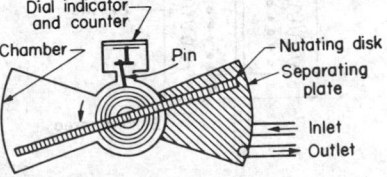

Fig. 16.1.13 Nutating-disk meter.

backward. It is used for more precise measure (available to 0.1 percent accuracy).

Volumetric gas measurement is commonly made with a **bellows meter**. Two bellows are alternately filled and exhausted with the gas. Motion of the bellows actuates a register to indicate the total flow. Various liquid-sealed displacement meters are also available for this purpose.

For precise volume measurements, corrections for temperature must be made (because of expansion of both the material being measured and the volumetric device). In the case of gases, the pressure also must be noted.

FORCE AND TORQUE MEASUREMENT

Force may be measured by the deflection of an elastic element, by balancing against a known force, by the acceleration produced in an object of known mass, or by its effects on the electrical or other properties of a stress-sensitive material. The common unit of force is the pound (newton). **Torque** is the product of a force and the perpendicular distance to the axis of rotation. Thus, torque tends to produce rotational motion and is expressed in units of pound feet (newton metres). Torque can be measured by the angular deflection of an elastic element or, where the moment arm is known, by any of the force measuring methods.

Since weight is the force of gravity acting on a mass, any of the weight-measuring devices already discussed can be used to measure force. Common methods employ the deflection of springs or cantilever beams.

The strain gage is an element whose electrical resistance changes with applied strain (see Sec. 5). Combined with an element of known force-strain, motion-strain, or other input-strain relationship it is a transducer for the corresponding input. The relation of gage-resistance change to input variable can be found by analysis and calibration. Measure of the resistance change can be translated into a measure of the force applied. The gage may be bonded or unbonded. In the bonded case, the gage is cemented to the surface of an elastic member and measures the strain of the member. Since the gage is very sensitive to temperature, the readings must be compensated. For this purpose, four gages are connected in a Wheatstone-bridge circuit such that the temperature effect cancels itself. A four-element unbonded gage is shown in Fig. 16.1.14. Note that as the applied force increases, the tension on two of the elements increases while that on the other two decreases. Gages subject to strain change of the same sign are put in opposite arms of the bridge. The zero adjustment permits balancing the bridge for zero output at any desired input. The e_1 and e_2 terminal pairs may be used interchangeably for the input excitation and the signal output.

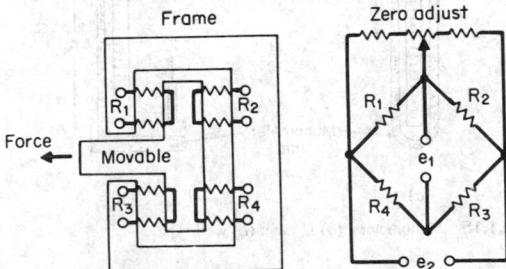

Fig. 16.1.14 Unbonded strain-gage board.

The **piezoelectric** effect is useful in measuring rapidly varying forces because of its high-frequency response and negligible displacement characteristics. Quartz rochelle salt, and barium titanate are common piezoelectric materials. They have the property of varying an output charge in direct proportion to the stress applied. This produces a voltage inversely proportional to the circuit capacitance. Charge leakage produces drifting at a rate depending on the circuit time constant. The voltage must be measured with a device having a very high input resistance. Accuracy is limited because of temperature dependence and some hysteresis effect.

Forces may also be measured with any of the pressure devices described in the next section by balancing against a fluid pressure acting on a fixed area.

PRESSURE AND VACUUM MEASUREMENT

Pressure is defined as the force per unit area exerted by a fluid. Pressure devices normally measure with respect to atmospheric pressure (mean value = 14.7 lb/in^2), $p_a = p_g + 14.7$, where p_a = total or **absolute** pressure and p_g = **gage** pressure, both lb/in^2. Conventionally, gage pressure and vacuum refer to pressures above and below atmospheric, respectively. Common units are lb/in^2, in Hg, ftH$_2$O, kg/cm^2, bars, and mmHg. The mean SI atmosphere is 1.013 bar.

Pressure devices are based on (1) measure of an equivalent height of liquid column; (2) measure of the force exerted on a fixed area; (3) measure of some change in electrical or physical characteristics of the fluid.

The manometer measures pressure according to the relationship $p = wh = \rho gh/g_c$, where h = height of liquid of density ρ and specific weight w (assumed constants) supported by a pressure p. Thus, pressures are often expressed directly in terms of the equivalent height (head) of manometer liquid, e.g., inH$_2$O or inHg. Usual manometer fluids are water or mercury, although other fluids are available for special ranges.

The **U-tube manometer** (Fig. 16.1.15a) expresses the pressure difference $p_1 - p_2$ as the difference in levels h. If p_2 is exposed to the atmosphere, the manometer reads the gage pressure of p_1. If the p_2 tube is evacuated and sealed ($p_2 = 0$), the absolute value of p_1 is indicated. A common modification is the **well-type manometer** (Fig. 16.1.15b). The scale is specially calibrated to take into account changes of level inside the well so that only a single tube reading is required. In particular, Fig. 16.1.15b illustrates the form usually applied to measurement of atmospheric pressure (**mercury barometer**).

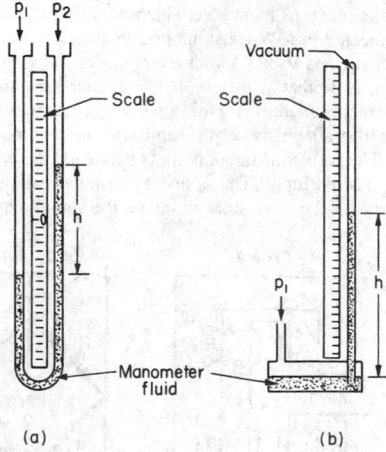

Fig. 16.1.15 Manometers. (*a*) U tube; (*b*) well type.

The sensitivity of readings can be increased by inclining the manometer tubes to the vertical (**inclined manometer**), by use of low-specific-gravity manometer fluids, or by application of optical-magnification or level-sensing devices. Accuracy is influenced by surface-tension effects (reading of the meniscus) and changes in fluid density (due to temperature changes and impurities).

By definition, pressure times the area acted upon equals the force exerted. The pressure may act on a diaphragm, bellows, or other element of fixed area. The force is then measured with any force-measuring device, e.g., spring deflection, strain gage, or weight balance. Very

commonly, the unknown pressure is balanced against an air or hydraulic pressure, which in turn is measured with a gage. By use of unequal-area diaphragms, the pressure can thus be amplified or attenuated as required. Further, it permits isolating the process fluid which may be corrosive, viscous, etc.

The **Bourdon-tube gage** (Fig. 16.1.16) is the most commonly used pressure device. It consists of a flattened tube of spring bronze or steel bent into a circle. Pressure inside the tube tends to straighten it. Since one end of the tube is fixed to the pressure inlet, the other end moves proportionally to the pressure difference existing between the inside and outside of the tube. The motion rotates the pointer through a pinion-and-sector mechanism. For amplification of the motion, the tube may be bent through several turns to form spiral or helical elements as are used in pressure recorders.

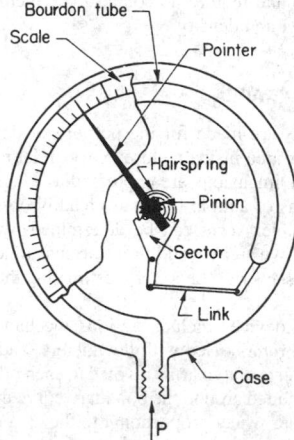

Fig. 16.1.16 Bourdon-tube gage.

In the **diaphragm gage,** the pressure acts on a diaphragm in opposition to a spring or other elastic member. The deflection of the diaphragm is therefore proportional to the pressure. Since the force increases with the area of the diaphragm, very small pressures can be measured by the use of large diaphragms. The diaphragm may be metallic (brass, stainless steel) for strength and corrosion resistance, or nonmetallic (leather, neoprene, silicon, rubber) for high sensitivity and large deflection. With a stiff diaphragm, the total motion must be very small to maintain linearity.

The **bellows gage** (Fig. 16.1.17) is somewhat similar to the diaphragm gage, with the advantage, however, of providing a much wider range of motion. The force acting on the bottom of the bellows is balanced by the deflection of the spring. This motion is transmitted to the output arm, which then actuates a pointer or recorder pen.

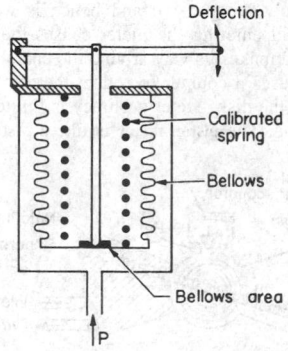

Fig. 16.1.17 Bellows gage.

The motion (or force) of the pressure element can be converted into an electrical signal by use of a differential transformer or strain-gage element or into an air-pressure signal through the action of a nozzle and pilot. The signal is then used for transmission, recording, or control.

The **dead-weight tester** is used as a standard for calibrating gages. Known hydraulic or gas pressures are generated by means of weights loaded on a calibrated piston. The useful range is from 5 to 5,000 lb/in² (0.3 to 350 bar). For low pressures, the water or mercury manometer serves as a reference.

For many applications (fluid flow, liquid level), it is important to measure the **difference between two pressures**. This can be done directly with the manometer. Other pressure devices are available as differential devices where (1) the case is made pressure-tight so that the second pressure can be applied external to the pressure element; (2) two identical pressure elements are mounted so that their outputs oppose each other.

Similar devices to those discussed are used to measure **vacuum,** the only difference being a shift in range or at most a relocation of the zeroing spring. When the vacuum is high (absolute pressure near zero) variations in atmospheric pressure become an important source of error. It is here that absolute-pressure devices are employed.

Any of the differential-pressure elements can be converted to an **absolute-pressure device** by sealing one pressure side to a perfect vacuum. A common instrument for the range 0 to 30 inHg employs two bellows of equal area set back to back. One bellows is completely evacuated and sealed; the other is connected to the measured pressure. The output is a bellows displacement, as in Fig. 16.1.17.

There are many instruments for high-vacuum work (0.001 to 10,000 μm range). These kinds of devices are based on the characteristic properties of gases at low pressures. The **McLeod gage** amplifies the pressure to be measured by compressing the gas a known amount and then measuring its pressure with a mercury manometer. The ratio of initial to final pressure is equal to the ratio of final to initial volume (for common gases). This gage serves as a standard for low pressures.

The **Pirani gage** (Fig. 16.1.18) is based on the change of heat conductivity of a gas with pressure and the change of electrical resistance of a wire with temperature. The wire is electrically heated with a constant current. Its temperature changes with pressure, producing a voltage across the bridge network. The compensating cell corrects for room-temperature changes.

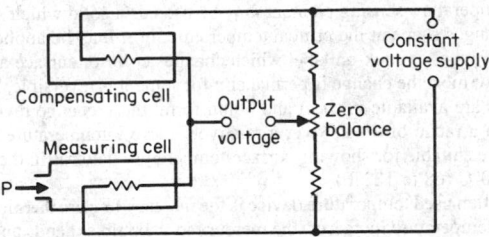

Fig. 16.1.18 Pirani gage.

The **thermocouple gage** is similar to the Pirani gage, except that a thermocouple is used to measure the temperature difference between the resistance elements in the measuring and compensating cells, respectively.

The **ionization gage** measures the ion current generated by bombardment of the molecules of the gas by the electron stream in a triode-type tube. This gage is limited to pressures below 1 μm. It is, however, extremely sensitive.

LIQUID-LEVEL MEASUREMENT

Level instruments are used for determining (or controlling) the height of liquid in a vessel or the location of the interface between two liquids of different specific gravity. In large storage tanks the level is indicated by a **calibrated tape or chain** which is attached to a float riding the liquid surface or by converting the signal reflection time of a radar or ultrasonic beam radiated onto the surface of the liquid into a level indication. For measuring small changes in level, the **fixed displacer** is common (Fig. 16.1.19). The buoyant force is proportional to the volume of displacer submerged and hence changes directly with the level. The force is balanced by the air pressure acting in the bellows, which in turn is generated by the flapper and nozzle. A pressure gage (or recorder) indicates the level.

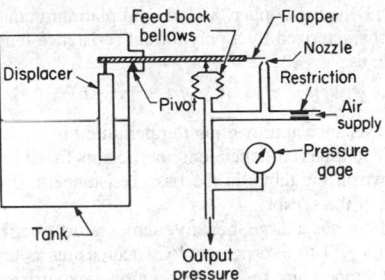

Fig. 16.1.19 Displacer-type level meter.

The level is often measured by means of a **differential-pressure meter** connected to taps in the top and bottom of the tank. As indicated in the discussion on manometers, the pressure difference is the height times the specific weight of the liquid. Where the liquid is corrosive or contains solids, then liquid seals, water purge, or air purge may be used to isolate the meter from the process.

For special applications, the dielectric, conducting, or absorption properties of the liquid can be used. Thus, in one model the liquid rises between two plates of a condenser, producing a **capacitance change** proportional to the change in level, and in another the **radiation** from a small radioactive source is measured. Since the liquid has a high absorption for the rays (compared with the vapor space), the intensity of the measured radiation decreases with the increase in level. An important advantage of this type is that it requires no external connections to the process.

TEMPERATURE MEASUREMENT

The common temperature scales (Fahrenheit and Celsius) are based on the freezing and boiling points of water (see Sec. 4 for discussion of temperature standards, units, and conversion equations).

Temperature is measured in a number of different ways. Some of the more useful are as follows.

1. *Thermal expansion of a gas* (**gas thermometer**). At constant volume, the pressure p of an (ideal) gas is directly proportional to its absolute temperature T. Thus, $p = (p_0/T_0)T$, where p_0 is the pressure at some known temperature T_0.

2. *Thermal expansion of a liquid or solid* (**mercury thermometer, bimetallic element**). Substances tend to expand with temperature. Thus, a change in temperature $t_2 - t_1$ causes a change in length $l_2 - l_1$ or a change in volume $V_2 - V_1$, according to the expressions.

$$l_2 - l_1 = a'(t_2 - t_1)l_1 \quad \text{or} \quad V_2 - V_1 = a'''(t_2 - t_1)V_1$$

where a' and a''' = linear and volumetric coefficients of thermal expansion, respectively (see Sec. 4). For many substances, a' and a''' are reasonably constant over a limited temperature range. For solids, $a''' = 3a'$. For mercury at room temperature, a''' is approximately $0.00018°C^{-1}$ $(0.00010°F^{-1})$.

3. *Vapor pressure of a liquid* (**vapor-bulb thermometer**). The vapor pressure of all liquids increases with temperature. The Clapeyron equation permits calculation of the rate of change of vapor pressure with temperature.

4. *Thermoelectric potential* (**thermocouple**). When two dissimilar metals are brought into intimate contact, a voltage is developed which depends on the temperature of the junction and the particular metals

used. If two such junctions are connected in series with a voltage-measuring device, the measured voltage will be very nearly proportional to the temperature difference of the two junctions.

5. *Variation of electrical resistance* (**resistance thermometer, thermistor**). Electrical conductors experience a change in resistance with temperature which can be measured with a Wheatstone- or Mueller-bridge circuit, or a digital ohmmeter. The platinum resistance thermometer (PRT) can be very stable and is used as the temperature scale interpolation standard from -160 to $660°C$. Commercial resistance temperature detectors (RTD) using copper, nickel, and platinum conductors are in use and are characterized by a polynomial resistance-temperature relationship, such as

$$t = A + B \times R_t + C \times R_t^2 + D \times R_t^3 + E \times R_t^4$$

where R_t = resistance at prevailing temperature t in °C. $A, B, C, D,$ and E are range- and material-dependent coefficients listed in Table 16.1.1. R_0, also shown in the table, is the base resistance at 0°C used in the identification of the sensor.

The thermistor has a large, negative temperature coefficient of resistance, typically -3 to -6 percent/°C, decreasing as temperature increases. The temperature-resistance relation is approximated (to perhaps 0.01° in range 0 to 100°C) by:

$$R_t = \exp\left(A_0 + A_1/t + \frac{A_2}{t^2} + \frac{A_3}{t^3}\right)$$

and

$$\frac{1}{t} = a_0 + a_1 \ln R_t + a_2(\ln R_t)^2 + a_3(\ln R_t)^3$$

with the constants chosen to fit four calibration points. Often a simpler form is given:

$$R = R_0 \exp\left\{\beta\left[\left(\frac{1}{t}\right) - \left(\frac{1}{t_0}\right)\right]\right\}$$

Typically β varies in the range of 3,000 to 5,000 K. The reference temperature t_0 is usually 298 K($= 25°C$, 77°F), and R_0 is the resistance at that temperature. The error may be as small as 0.3°C in the range of 0° to 50°C. Thermistors are available in many forms and sizes for use from -196 to $+450°C$ with various tolerances on interchangeability and matching. (See "Catalog of Thermistors," Thermometrics, Inc.) The AD590 and AD592 integrated circuit (Analog Devices, Inc.) passes a current of 1 $\mu A/°K$ very nearly proportional to absolute temperature. All these sensors are subject to self-heating error.

6. *Change in radiation* (**radiation and optical pyrometers**). A body radiates energy proportional to the fourth power of its absolute temperature. This principle is particularly adaptable to the measurement of very high temperatures where either the total quantity of radiation or its intensity within a narrow wavelength band may be measured. In the former type (radiation pyrometer), the radiation is focused on a heat-sensitive element, e.g., a thermocouple, and its rise in temperature is measured. In the latter type (optical pyrometer) the intensity of the radiation is compared optically with a heated filament. Either the filament brightness is varied by a control calibrated in temperature, or a fixed brightness filament is compared with the source viewed through a calibrated optical wedge.

The infrared thermometer accepts radiation from an object seen in a definite field of view, filters it to select a portion of the infrared spectrum, and focuses it on a sensor such as a blackened thermistor flake, which warms and changes resistance. Electronic amplification and signal processing produce a digital display of temperature. Correct calibration requires consideration of source emissivity, reflection, and transmission from other radiation sources, atmospheric absorption between the source object and the sensor, and compensation for temperature variation at the sensor's immediate surroundings.

Electrical nonconductors generally have fairly high (about 0.95) emissivities, while good conductors (especially smooth, reflective metal surfaces), do not; special calibration or surface conditioning is then needed. Very wide band (0.7 to 20 μm) instruments gather relatively large amounts of energy but include atmospheric absorption bands which reduce the energy received from a distance. The band 8 to 14 μm is substantially free from atmospheric absorption and is popular for general use with source objects in the range 32 to 1,000°F (0 to 540°C). Other bands and two-color instruments are used in some cases. See Bonkowski, Infrared Thermometry, *Measurements and Control*, Feb. 1984, pp. 152–162.

Fiber-optics probes extend the use of radiation methods to hard-to-reach places.

Important relationships used in the design of these instruments are the Wien and Stefan-Boltzmann laws (in modified form):

$$\lambda_m = k_1/T \qquad q = k_2 \varepsilon A(T_2^4 - T_1^4)$$

where λ_m = wavelength of maximum intensity, μm (nm); q = radiant energy flux, Btu/h (W); A = radiation surface, ft^2 (m^2); ε = mean emissivity of the surfaces; T_2, T_1 = absolute temperatures of radiating and receiving surfaces, respectively, °R (K); $k_1 = 5215$ μm. °R (2898 $\mu m \cdot K$); $k_2 = 0.173 \times 10^{-8}$ Btu/(h·ft^2·°R^4) [5.73 × 10^{-8} W/(m^2·K^4)]. The emissivity depends on the material and form of the surfaces involved (see Sec. 4). Radiation sensors with scanning capability can produce maps, photographs, and television displays showing temperature-distribution patterns. They can operate with resolutions to under 1°C and at temperatures below room temperature.

7. *Change in physical or chemical state* (**Seger cones, Tempilsticks**). The temperatures at which substances melt or initiate chemical reaction are often known and reproducible characteristics. Commercial products are available which cover the temperature range from about 120 to 3600°F (50 to 2000°C) in intervals ranging from 3 to 70°F (2 to 40°C). The temperature-sensing element may be used as a solid which softens and changes shape at the critical temperature, or it may be applied as a paint, crayon, or stick-on label which changes color or surface appearance. For most the change is permanent; for some it is reversible. Liquid crystals are available in sheet and liquid form: these change reversibly through a range of colors over a relatively narrow temperature range. They are suitable for showing surface-temperature patterns in the range 20 to 50°C (68 to 122°F).

An often used temperature device is the **mercury-in-glass thermometer**. As the temperature increases, the mercury in the bulb expands and rises through a fine capillary in the graduated thermometer stem. Useful range extends from -30 to 900°F (-35 to 500°C). In many applications of the mercury thermometer, the stem is not exposed to the mea-

Table 16.1.1 Polynomial Coefficients for Resistance Temperature Detectors

Material of conductor	ID R_0, Ω	Useful range, °C	A, °C	B, °C/Ω	C, °C/Ω^2	D, °C/Ω^3	E, °C/Ω^4	Typical accuracy,* °C
Copper 10Ω @25°C	9.042	−70 to 0	−225.64	23.30735	+0.246864	−0.00715		1.5
		0 to 150	−234.69	25.95508				1.5
Nickel	120	−80 to 320	−199.47	1.955336	−0.00266	1.88E − 6		1
Platinum DIN/IEC $\alpha = 0.00385/°C$	100	−200 to 0	−241.86	2.213927	0.002867	−9.8E − 6	1.64E − 8	1
		0 to 850	−236.06	2.215142	0.001455			0.5

* For higher accuracy consult the table or equation furnished by the manufacturer of the specific RTD being used. Temperatures per ITS-90, resistances per SI-90.

sured temperature; hence a correction is required (except where the thermometer has been calibrated for **partial immersion**). Recommended formula for the correction K to be added to the thermometer reading is $K = 0.00009 \, D(t_1 - t_2)$, where D = number of degrees of exposed mercury filament, °F; t_1 = thermometer reading, °F; t_2 = the temperature at about middle of the exposed portion of stem, °F. For Celsius thermometers the constant 0.00009 becomes 0.00016.

For **industrial applications** the thermometer or other sensor is encased in a metal or ceramic protective well and case (Fig. 16.1.20). A threaded union fitting is provided so that the thermometer can be installed in a line or vessel under pressure. Ideally the sensor should have the same temperature as the fluid into which the well is inserted. However, heat

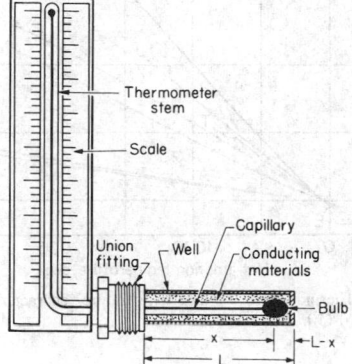

Fig. 16.1.20 Industrial thermometer.

conduction to or from the pipe or vessel wall and radiation heat transfer may also influence the sensor temperature (see ASME PTC 19.3-1974 Temperature Measurement, on well design). An approximation of the conduction error effect is

$$T_{sensor} - T_{fluid} = (T_{wall} - T_{fluid})E$$

For a sensor inserted to a distance $L - x$ from the tip of a well of insertion length L, $E = \cosh[m(L - x)]/\cosh mL$, where $m = (h/kt)^{0.5}$; x and L are in ft (m); h = fluid-to-well conductance, Btu/(h) (ft²)(°F) [J/(h) (m²)(°C)]; k = thermal conductivity of the well-wall material. Btu/(h)(ft)(°F) [J/(h)(m)(°C)]; and t = well-wall thickness, ft (m). Good thermal contact between the sensor and the well wall is assumed. For $(L - x)/L = 0.25$:

mL	1	2	3	4	5	6	7
E	0.67	0.30	0.13	0.057	0.025	0.012	0.005

Radiation effects can be reduced by a polished, low-emissivity surface on the well and by radiation shields around the well. Concern with mercury contamination has made the bimetal thermometer the most commonly used expansion-based temperature measuring device. Differential thermal expansion of a solid is employed in the simple **bimetal** (used in thermostats) and the bimetallic helix (Fig. 16.1.21). The bimetallic element is made by welding together two strips of metal having different coefficients of expansion. A change in temperature then causes the element to bend or twist an amount proportional to the temperature.

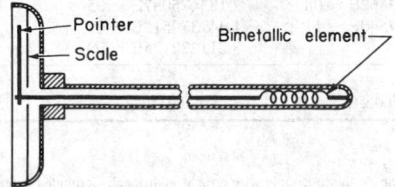

Fig. 16.1.21 Bimetallic temperature gage.

A common bimetallic pair consists of invar (iron-nickel alloy) and brass.

For control or alarm indications at fixed temperatures, thermometers may be equipped with electrical contacts such that when the temperature matches the contact point, an external relay circuit is energized.

A popular industrial-type instrument employs the deflection of a **pressure-spring** to indicate (or record) the temperature (Fig. 16.1.22). The sensing element is a metal bulb containing some specific gas or liquid. The bulb connects with the pressure spring (in the form of a spiral or helix) through a capillary tube which is usually enclosed in a

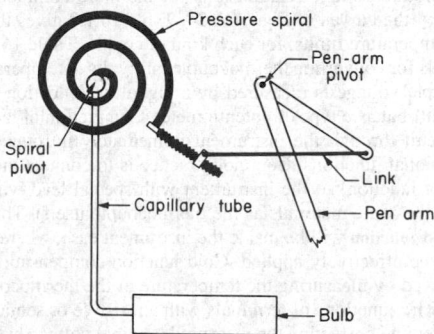

Fig. 16.1.22 Pressure-spring element.

protective sheath or armor. Increasing temperature causes the fluid in the bulb to expand in volume or increase in pressure. This forces the pressure spring to unwind and move the pen or pointer an appropriate distance upscale.

The **bulb fluid** may be mercury (mercury system), nitrogen under pressure (gas system), or a volatile liquid (vapor-pressure system). Mercury and gas systems have linear scales; however, they must be compensated to avoid ambient temperature errors. The capillary may range up to 200 ft in length with, however, considerable reduction in speed of response.

For transmitting temperature readings over any distance (up to 1,000 ft), the **pneumatic transmitter** (Fig. 16.1.23) is better suited than the methods outlined thus far. This instrument has the additional advantages of greater compactness, higher response speeds, and generally better accuracy. The bulb is filled with gas under pressure which acts on the diaphragm. An increase in bulb temperature increases the upward force acting on the main beam, tending to rotate it clockwise. This causes the baffle or flapper to move closer to the nozzle, increasing the nozzle back pressure. This acts on the pilot, producing an increase in output pressure, which increases the force exerted by the feedback bellows. The system returns to equilibrium when the increase in bellows pressure exactly balances the effect of the increased diaphragm pressure. Since the lever ratios are fixed, this results in a direct proportionality between bulb temperature and output air pressure. For precision,

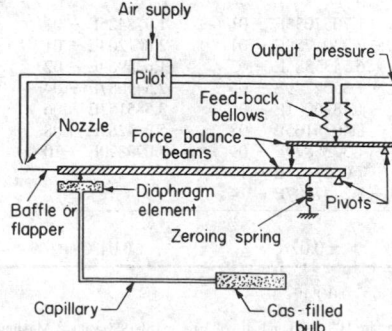

Fig. 16.1.23 Pneumatic temperature transmitter.

compensating elements are built into the instrument to correct for the effects of changes in barometric pressure and ambient temperature.

Electrical systems based on the thermocouple or resistance thermometer are particularly applicable where many different temperatures are to be measured, where transmission distances are large, or where high sensitivity and rapid response are required. The thermocouple is used with high temperatures; the resistance thermometer for low temperatures and high accuracy requirements.

The **choice of thermocouple** depends on the temperature range, desired accuracy, and the nature of the atmosphere to which it is to be exposed. The temperature-voltage relationships for the more common of these are given by the curves of Fig. 16.1.24. Table 16.1.2 gives the recommended temperature limits, for each kind of couple. Table 16.1.3 gives polynomials for converting thermocouple millivolts to temperature. The thermocouple voltage is measured by a digital or deflection millivoltmeter or null-balance type of potentiometer. Completion of the thermocouple circuit through the instrument immediately introduces one or more additional junctions. Common practice is to connect the thermocouple (hot junction) to the instrument with special lead wire (which may be of the same materials as the thermocouple itself). This assures that the **cold junction** will be inside the instrument case, where compensation can be effectively applied. Cold junction compensation is typically achieved by measuring the temperature of the thermocouple wire to copper wire junctions or terminals with a resistive or semiconductor thermometer and correcting the measured terminal voltage by a derived equivalent millivolt cold junction value. Figure 16.1.25 shows a digital temperature indicator with correction for different ANSI types of ther-

mocouple voltage to temperature nonlinearities being stored in and applied to the analog-to-digital converter (A/D) by a read-only memory (ROM) chip.

The **resistance thermometer** employs the same circuitry as described above, with the resistance element (RTD) being placed external to the instrument and the cold junction being omitted (Fig. 16.1.26). Three types of RTD connections are in use: two wire, three wire, and four wire. The two-wire connection makes the measurement sensitive to lead

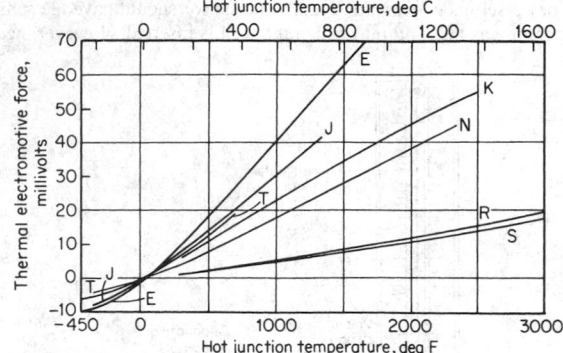

Fig. 16.1.24 Thermocouple voltage-temperature characteristics [reference junction at 32°F (0°C)].

Table 16.1.2   Limits of Error on Standard Wires without Selection*†

ANSI symbol‡	Materials and polarities		°F: −150	−75	32	200	530	600	700	1,000	1,400	2,300	2,700
	Positive	Negative	°C: −101	−59	0	93	277	316	371	538	760	1,260	1,482
T	Cu	Constantan§		2%	1.5°F (0.8°C)		¾%						
E	Ni-Cr	Constantan				3°F (1.7°C)				½%			
J	Fe	Constantan				4°F (2.2°C)		¾%					
K	Ni-Cr	Ni-Al				4°F (2.2°C)		¾%					
N	Ni-Cr-Si	Ni-SiMn				4°F (2.2°C)		¾%					
R	Pt-13% Rh	Pt				3°F (1.5°C)					¼%		
S	Pt-10% Rh	Pt				3°F (1.5°C)					¼%		

* Protect copper from oxidation above 600°F; iron above 900°F. Protect Ni-Al from reducing atmospheres. Protect platinum from nonreducing atmospheres. Type B (Pt-30% Rh versus Pt-6%) is used up to 3,200°F(1,700°C). Its standard error is ½ percent above 1,470°F (800°C).

† Closer tolerances are obtainable by selection and calibration. Consult makers' catalogs. Tungsten-rhenium alloys are in use up to 5,000°F (2,760°C). For cryogenic thermocouples see Sparks et al., *Reference Tables for Low-Temperature Thermocouples. Natl. Bur. Stand. Monogr.* 124.

‡ Individual wires are designated by the ANSI symbol followed by P or N; thus iron is JP.

§ Constantan is 55% Cu, 45% Ni. The nickel-chromium and nickel-aluminum alloys are available as Chromel and Alumel, trademarks of Hoskins Mfg. Co.

Table 16.1.3   Polynomial Coefficients for Converting Thermocouple emf to Temperature*

Range	Type E	Type J	Type K	Type N	Type S	Type T
mV	0 to 76.373	0 to 42.919	0 to 20.644	0 to 47.513	1.874 to 11.95	0 to 20.872
°C	0 to 1000°	0 to 760°	0 to 500°	0 to 1300°	250 to 1200°	0 to 400°
°F	32 to 1832°	32 to 1400°	32 to 932°	32 to 2372°	482 to 2192°	32 to 752°
α_0	0	0	0	0	1.291507177E + 01	0
α_1	1.7057035E + 01	1.978425E + 01	2.508355E + 01	3.8783277E + 01	1.466298863E + 02	2.592800E + 01
α_2	−2.3301759E − 01	−2.001204E − 01	7.860106E − 02	−1.1612344E + 00	−1.534713402E + 01	−7.602961E − 01
α_3	6.5435585E − 03	1.036969E − 02	−2.503131E − 01	6.9525655E − 02	3.145945973E + 00	4.637791E − 02
α_4	−7.3562749E − 05	−2.549687E − 04	8.315270E − 02	−3.0090077E − 03	−4.163257839E − 01	−2.165394E − 03
α_5	−1.7896001E − 06	3.585153E − 06	−1.228034E − 02	8.8311584E − 05	3.187963771E − 02	6.048144E − 05
α_6	8.4036165E − 08	−5.344285E − 08	9.804036E − 04	−1.6213839E − 06	−1.291637500E − 03	−7.293422E − 07
α_7	−1.3735879E − 09	5.099890E − 10	−4.413030E − 05	1.6693362E − 08	2.183475087E − 05	
α_8	1.0629823E − 11		1.057734E − 06	−7.3117540E − 11	−1.447379511E − 07	
α_9	−3.2447087E − 14		−1.052755E − 08		8.211272125E − 09	
Maximum deviation	± 0.02°C	± 0.04°C	− 0.05 to + 0.04°C	± 0.06°C	± 0.01°C	± 0.03°C

* $T(°C) = \sum_{i=0}^{n} a_i \times (mV)^i$

All temperatures are ITS-1990 and all voltages are SI-1990 values. Maximum deviation is that from the ITS-1990 tables; thermocouple wire error is additional. Computed temperature deviates greatly outside of given ranges. Consult source for thermocouple types B and R and for other millivolt ranges.

SOURCE: NIST Monograph 175, April 1993.

wire temperature changes. The three-wire connection, preferred in industrial applications, eliminates the lead wire effect provided the leads are of the same gage and length, and subject to the same environment. The four-wire arrangement makes no demands on the lead wires and is preferred for scientific measurements.

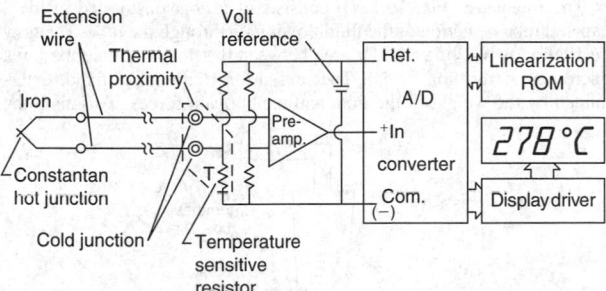

Fig. 16.1.25 Temperature measurement with thermocouple and digital millivoltmeter.

The resistance bulb consists of a copper or platinum wire coil sealed in a protective metal tube. The **thermistor** has a very large temperature coefficient of resistance and may be substituted in low-accuracy, low-cost applications.

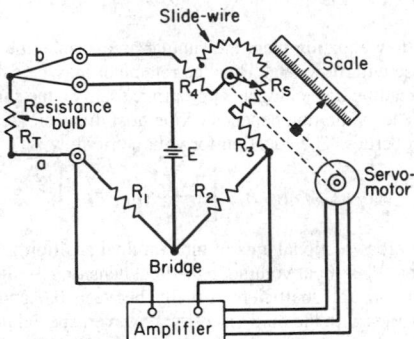

Fig. 16.1.26 Three-wire resistance thermometer with self-balancing potentiometer recorder.

By use of a **selector switch,** any number of temperatures may be measured with the same instrument. The switch connects in order each thermocouple (or resistance bulb) to the potentiometer (or bridge circuit) or digital voltmeter. When balance is achieved, the recorder prints the temperature value, then the switch advances on to the next position.

Optical pyrometers are applied to high-temperature measurement in the range 1000 to 5000°F (540 to 2760°C). One type is shown in Fig. 16.1.27. The surface whose temperature is to be measured (target) is focused by the lens onto the filament of a calibrated tungsten lamp. The light intensity of the filament is kept constant by maintaining a constant

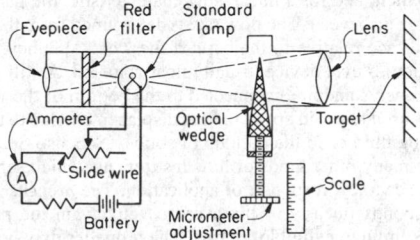

Fig. 16.1.27 Optical pyrometer.

current flow. The intensity of the target image is adjusted by positioning the optical wedge until the image intensity appears exactly equal to that of the filament. A scale attached to the wedge is calibrated directly in temperature. The red filter is employed so that the comparison is made at a specific wavelength (color) of light to make the calibration more reproducible. In another type of optical pyrometer, comparison is made by adjusting the current through the filament of the standard lamp. Here, an ammeter in series is calibrated to read temperature directly. **Automatic operation** may be had by comparing filament with image intensities with a pair of photoelectric cells arranged in a bridge network. A difference in intensity produces a voltage, which is amplified to drive the slide wire or optical wedge in the direction to restore zero difference.

The **radiation pyrometer** is normally applied to temperature measurements above 1000°F. Basically, there is no upper limit; however, the lower limit is determined by the sensitivity and cold-junction compensation of the instrument. It has been used down to almost room temperature. A common type of radiation receiver is shown in Fig. 16.1.28. A lens focuses the radiation onto a thermal sensing element. The temperature rise of this element depends on the total radiation received and the conduction of heat away from the element. The radiation relates to the temperature of the target; the conduction depends on the temperature of the pyrometer housing. In normal applications the latter factor is not very great; however, for improved accuracy a compensating coil is added to the circuit. The sensing element may be a thermopile, vacuum

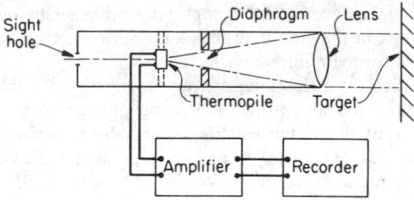

Fig. 16.1.28 Radiation pyrometer.

thermocouple, or bolometer. The **thermopile** consists of a number of thermocouples connected in series, arranged so that all the hot junctions lie in the field of the incoming radiation; all of the cold junctions are in thermal contact with the pyrometer housing so that they remain at ambient temperature. The **vacuum thermocouple** is a single thermocouple whose hot junction is enclosed in an evacuated glass envelope. The **bolometer** consists of a very thin strip of blackened nickel or platinum foil which responds to temperature in the same manner as the resistance thermometer. The **thermal sensing element** is connected to a potentiometer or bridge network of the same type as described for the self-balance thermocouple and resistance-thermometer instruments. Because of the nature of the radiation law, the scale is nonlinear.

Accuracy of the optical- and radiation-type pyrometers depends on:

1. *Emissivity of the surface being sighted on.* For closed furnace applications, blackbody conditions can be assumed (emissivity = 1). For other applications corrections for the actual emissivity of the surface must be made (correction tables are available for each pyrometer model). Multiple color or wavelength sensing is used to reduce sensitivity to hot object emissivity. For measuring hot fluids, a target tube immersed in the fluid provides a target of known emissivity.

2. *Radiation absorption between target and instrument.* Smoke, gases, and glass lenses absorb some of the radiation and reduce the incoming signal. Use of an enclosed (or purged) target tube or direct calibration will correct this.

3. *Focusing of the target on the sensing element.*

MEASUREMENT OF FLUID FLOW RATE

(See also Secs. 3 and 4.)

Flow is expressed in volumetric or mass units per unit time. Thus gases are generally measured in ft³/min (m³/min) or ft³/h (m³/h), steam in lb/h (kg/h), and liquid in gal/min (L/min) or gal/h (L/h). Conversion between volumetric flow Q and mass flow m is given by $m = K\rho Q$, where ρ = density of the fluid and K is a constant depending on the units of m, Q, and ρ. Flow rate can be measured directly by attaching a rate device to a volumetric meter of the types previously described, e.g., a tachometer connected to the rotating shaft of the nutating-disk meter (Fig. 16.1.13).

Flow is most frequently measured by application of the principle of conservation of mechanical energy through conversion of fluid velocity to pressure (head). Thus, if the fluid is forced to change its velocity from V_1 to V_2, its pressure will change from p_1 to p_2 according to the equation (neglecting friction, expansion, and turbulence effects):

$$V_2^2 - V_1^2 = \frac{2g_c}{\rho}(p_1 - p_2) \qquad (16.1.1)$$

where g = acceleration due to gravity, ρ = fluid density, and g_c = 32.184 lbm·ft/(lbf)(s²) [1.0 kg·m/(N)(s²)]. **Caution:** If the flow pulsates, the average value of $p_1 - p_2$ will be greater than that for steady flow of the same average flow.

See "ASME Pipeline Flowmeters," and "Pitot Tubes" in Sec. 3 for coverage of venturi tubes, flow nozzles, compressible flow, orifice meters, ASME orifices, and Pitot tubes.

The tabulation orifice coefficients apply only for **straight pipe** upstream and downstream from the orifice. In most cases, satisfactory results are obtained if there are no fittings closer than 25 pipe diameters upstream and 5 diameters downstream from the orifice. The upstream limitation can be reduced a bit by employing **straightening vanes**. Reciprocating pumps in the line may introduce serious errors and require special efforts for their correction.

A wide variety of differential pressure meters is available for measuring the orifice (or other primary element) pressure drop.

Figure 16.1.29 shows the **diaphragm, or "dry," meter**. The orifice differential acts across a metal or rubber diaphragm, generating a force which tends to rotate the lever clockwise, moving the baffle toward the nozzle. This increases the nozzle back pressure, which acts on the pilot diaphragm to open the air supply port and increase the output pressure.

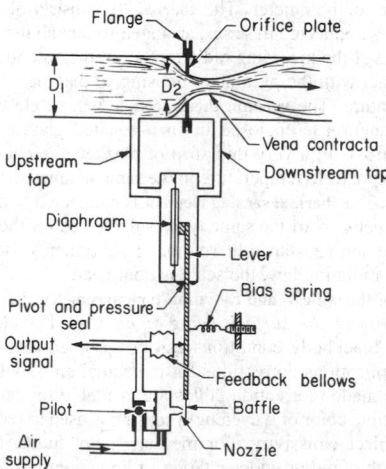

Fig. 16.1.29 Orifice plate and diaphragm-type meter.

This increases the force exerted by the feedback bellows, which generates a force opposing the motion of the main diaphragm. Equilibrium is reached when a change in orifice differential is exactly balanced by a proportionate change in output pressure. Often a damping device in the form of a simple oil dashpot is attached to the lever to reduce output fluctuations.

The flowmeter normally exhibits a square-root flow calibration. Some meters are designed to take out the square root by use of **cams, characterized floats or displacers (Ledoux bell)** or devices which describe a square-root behavior. These methods do not improve accuracy or performance but merely provide the convenience of a linear scale.

The meters described thus far are termed **variable-head** because the pressure drop varies with the flow, orifice ratio being fixed. In contrast, the **variable-area** meter maintains a constant pressure differential but varies the orifice area with flow.

The **rotameter** (Fig. 16.1.30) consists of a float positioned inside a tapered tube by action of the fluid flowing up through the tube. The flow restriction is now the annular area between the float and the tube (area increases as the float rises). The pressure differential is fixed, determined by the weight of the float and the buoyant forces. To satisfy the

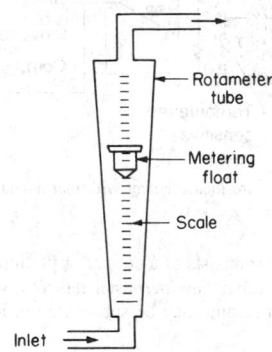

Fig. 16.1.30 Rotameter.

volumetric flow equation then, the annular area (hence the float level) must increase with flow rate. Thus the rotameter may be calibrated for direct flow reading by etching an appropriate scale on the surface of the glass tube. The calibration depends on the float dimensions, tube taper, and fluid properties. The equation for volumetric flow is

$$Q = C_R(A_T - A_F)\left[\frac{2gV_F}{\rho A_F}(\rho_F - \rho)\right]^{1/2}$$

where A_T = cross-sectional area of tube (at float position), A_F = effective float area, V_F = float volume, ρ_F = float density, ρ = fluid density, and C_R = rotameter coefficient (usually between 0.6 and 0.8). The coefficient varies with the fluid viscosity; however, special float designs are available which are relatively insensitive to viscosity effects. Also, fluid density compensation can be obtained.

The **rotameter reading may be transmitted** for recording and control purposes by affixing to the float a stem which connects to an armature or permanent magnet. The armature forms part of an inductance bridge whose signal is amplified electronically to drive a pen-positioning motor. For pneumatic transmission, the magnet provides magnet coupling to a pneumatic motion transmitter external to the rotameter tube. This generates an air pressure proportional to the height of the float.

The **area meter** is similar to the rotameter in operation. Flow area is varied by motion of a piston in a straight cylinder with openings cut into the wall. The piston position is transmitted as above by an armature and inductance bridge circuit.

Primary elements for **flow in open channels** usually employ **weirs** or **open nozzles** to restrict the flow. Weir designs include the rectangular slot; the V notch; and for a linear-flow characteristic, the parabolically shaped weir (Sutro weir). The flow rate is determined from the height of the liquid surface relative to the base of the weir. This height is measured by a liquid-level device, usually float-actuated. A still well (float chamber or open standpipe) connected to the bottom of the weir or the nozzle tap is used to avoid errors in float displacement due to the motion of the flowing fluid or to the buildup of solids. (See also Sec. 3.)

There are many other kinds of flow instruments which serve special purposes of accuracy, response, or application. The **propeller type** (Fig. 16.1.31) responds linearly to the average velocity in the path of the propeller, assuming negligible friction. The propeller may be mechanically geared to a tachometer to indicate flow rate and to a counter to show total quantity flow. The magnetic pickup (Fig. 16.1.31) generates a pulse each time a propeller tip passes. The frequency of pulses (measured by means of appropriate electronic circuitry) is then proportional to the local stream velocity. If the propeller occupies only part of the flow stream, an individual calibration is necessary and the velocity distribution must remain constant. The **turbine** type is similar, but is fabricated as a unit in a short length of pipe with vanes to guide the flow approaching the rotor. Its magnetic pickup permits hermetic sealing. A minimum flow is needed to overcome magnetic cogging and start the rotor turning.

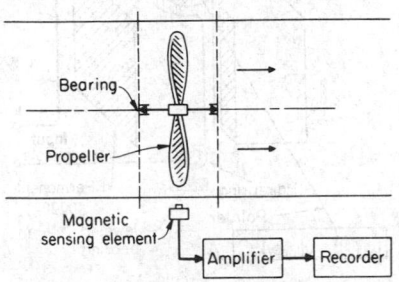

Fig. 16.1.31   Propeller-type flowmeter.

The **metering pump** is an accurately calibrated positive-displacement pump which provides both measurement and control of fluid-flow rate. The pump may be either fixed volumetric displacement-variable speed or constant speed-variable displacement.

For air flow, a **vane-type meter (anemometer)** is often used. A mechanical counter counts the number of revolutions of the vane shaft over a timed interval. Instantaneous airflow readings are more readily obtained with the **hot-wire anemometer**. Here, a resistance wire heated by an electric current is placed in the flow stream. The temperature of the wire depends on the current and the rate at which heat is conducted away from it. This latter factor is related to the thermal properties of the air and its velocity past the wire. Airflow can be measured in terms of (1) the current through the wire to maintain a fixed temperature, (2) temperature of the wire for a fixed current, or (3) temperature rise of the air passing the wire for fixed current. The wire temperature is readily measured in terms of its resistance. The anemometer must be specially calibrated for the application. Lasers have also been applied to anemometer use.

The **electromagnetic flowmeter** has no moving parts and does not require any insertions in the flow stream. It is based on the voltage induced by the flow of charged particles of the fluid past a strong magnetic field. It is suitable for liquids having resistivities of 50 k$\Omega \cdot$ cm or less. The **vortex-shedding meter** has a flow obstruction in the pipe; vortices form behind it at a rate nearly proportional to the volume flow rate. Vortex-formation-rate data give flow rate; a counter gives the integrated flow.

Doppler-effect flowmeters depend on reflection from particles moving with the fluid being metered; the shift in frequency of the reflected wave is proportional to velocity. Two transducers are used side by side, directed so that there is a large component of flow velocity along the sound path. One transmits and one receives.

Transit-time **ultrasonic flowmeters** use one or more pairs of transducers on opposite sides of the pipe, displaced along the length of the pipe. The apparent velocity of sound is $c \pm v$, where c is the speed of sound with no flow, and v is the component of flow velocity in the direction of the sound propagation path. The difference in sound velocity in the two directions is proportional to the flow's velocity component along the sound propagation path. The transit time difference is $2vl/(c^2 + v^2)$, where l is the path length. For $v \ll c$, the factor $(c^2 + v^2)$ is nearly constant. These meters cause no pressure drop and can be applied to pipes up to very large diameters. Multipath meters improve accuracy.

Mass flowmeters measure changes in momentum related to the mass flow rate.

Flowmeters measure rate of flow. To measure the total quantity of fluid flowing during a specified interval of time, the flow rate must be integrated over that interval. The integration may be done manually by estimating from the chart record the hourly flow averages or by measuring the area under the flow curve with a special square-root planimeter. **Mechanical integrators** use a constant-speed motor to rotate a counter. A cam converts the square-root meter reading into a linear displacement such that the fraction of time that the motor is engaged to the counter is proportional to the flow rate, resulting in a counter reading proportional to the integrated flow. **Electrical integrators** are similar in principle to the watthour meter in that the speed of the integrating motor is made proportional to the magnitude of the flow signal (see Sec. 15).

POWER MEASUREMENT

Power is defined as the rate of doing work. Common units are the horsepower and the kilowatt: 1 hp = 33,000 ft$\cdot$lb/min = 0.746 kW. The power input to a rotating machine in hp (W) = $2\pi nT/k$, where n = r/min of the shaft where the torque T is measured in lbf$\cdot$ft (N$\cdot$m), and k = 33,000 ft$\cdot$lbf/hp$\cdot$min [60 N$\cdot$m/(W$\cdot$min)]. The same equation applies to the power output of an engine or motor, where n and T refer to the output shaft. Mechanical power-measuring devices (**dynamometers**) are of two types: (1) those absorbing the power and dissipating it as heat and (2) those transmitting the measured power. As indicated by the above equation, two measurements are involved: shaft speed and torque. The speed is measured directly by means of a tachometer. Torque is usually measured by balancing against weights applied to a fixed lever arm; however, other force measuring methods are also used. In the **transmission dynamometer**, the torque is measured by means of strain-gage elements bonded to the transmission shaft.

There are several kinds of **absorption dynamometers**. The **Prony brake** applies a friction load to the output shaft by means of wood blocks, flexible band, or other friction surface. The **fan brake** absorbs power by "fan" action of rotating plates on surrounding air. The **water brake** acts as an inefficient centrifugal pump to convert mechanical energy into heat. The pump casing is mounted on antifriction bearings so that the developed turning moment can be measured. In the **magnetic-drag** or **eddy-current brake**, rotation of a metal disk in a magnetic field induces eddy currents in the disk which dissipate as heat. The field assembly is mounted in bearings in order to measure the torque.

One type of **Prony brake** is illustrated in Fig. 16.1.32. The torque developed is given by $L(W - W_0)$, where L is the length of the brake arm, ft; W and W_0 are the scale loads with the brake operating and with the brake free, respectively. The brake horsepower then equals $2\pi nL(W - W_0)/33,000$, where n is shaft speed, r/min.

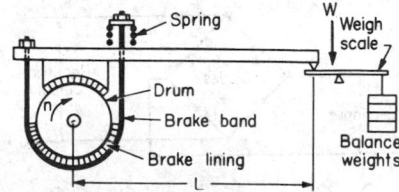

Fig. 16.1.32   Prony brake.

In addition to eddy-current brakes, **electric dynamometers** include calibrated generators and motors and cradle-mounted generators and motors. In calibrated machines, the efficiency is determined over a range of operating conditions and plotted. Mechanical power measurement can then be made by measuring the electrical power input (or output) to the machine. In the electric-cradle dynamometer, the motor or generator stator is mounted in trunnion bearing so that the torque can be measured by suitable scales.

The **engine indicator** is a device for plotting cylinder pressure as a function of piston (or volume) displacement. The resulting p-v diagram (Fig. 16.1.33) provides both a measure of the work done in a reciprocating engine, pump, or compressor and a means for analyzing its performance (see Secs. 4, 9, and 14). If A_d is the area inside the closed curve drawn by the indicator, then the indicated horsepower for the cylinder under test $= KnA_pA_d$ where K is a proportionality factor determined by the scale factors of the indicator diagram, n = engine speed, r/min, A_p = piston area.

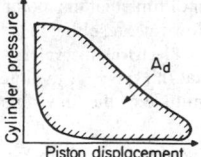

Fig. 16.1.33 Indicator diagram.

Completely mechanical indicators can be used only for low-speed machines. They have largely been superseded by electrical transducers using strain gages, variable capacitance, piezoresistive, and piezoelectric principles which are suitable for **high-speed** as well as low-speed pressure changes (the piezoelectric principle has low-speed limitations). The usual diagram is produced on an oscilloscope display as pressure vs. time, with a marker to indicate some reference event such as spark timing or top dead center. Special transducers can be coupled to a crank or cam shaft to give an electrical signal representing piston motion so that a p-v diagram can be shown on an oscilloscope.

ELECTRICAL MEASUREMENTS

(See also Sec. 15.)

Electrical measurements serve two purposes: (1) to measure the electrical quantities themselves, e.g., line voltage, power consumption, and (2) to measure other physical quantities which have been converted into electrical variables, e.g., temperature measurement in terms of thermocouple voltage.

In general, there is a sharp distinction between ac and dc devices used in measurements. Consequently, it is often desirable to transform an ac signal to an equivalent dc value, and vice versa. An ac signal is converted to dc (rectified) by use of **selenium rectifiers, silicon or germanium diodes, or electron-tube diodes.** Full-wave rectification is accomplished by the **diode bridge,** shown in Fig. 16.1.34. The rectified signal may be passed through one or more low-pass filter stages to smooth the waveform to its average value. Similarly, there are many ways of modulating a dc signal (converting it to alternating current). The most common method used in instrument applications is a solid-state oscillator.

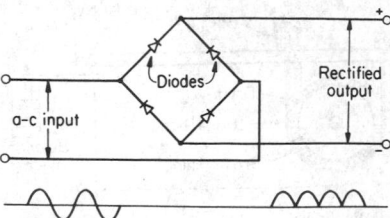

Fig. 16.1.34 Full-wave rectifier.

The **galvanometer** (Fig. 16.1.35), recently supplanted by the direct-reading **digital voltmeter** (DVM), is basic to dc measurement. The input signal is applied across a coil mounted in jeweled bearings or on a taut-band suspension so that it is free to rotate between the poles of a permanent magnet. Current in the coil produces a magnetic moment which tends to rotate the coil. The rotation is limited, however, by the restraining torque of the hairsprings. The resulting deflection of the coil θ is proportional to the current I:

$$\theta = \frac{NBWL}{K} I$$

where N = number of turns in coil; W, L = coil width and length, respectively; B = magnetic field intensity; K = spring constant of the hairsprings. Galvanometer deflection is indicated by a balanced pointer attached to the coil. In very sensitive elements, the pointer is replaced by a mirror reflecting a spot of light onto a ground-glass scale; the bearings and hairspring are replaced by a torsion-wire suspension.

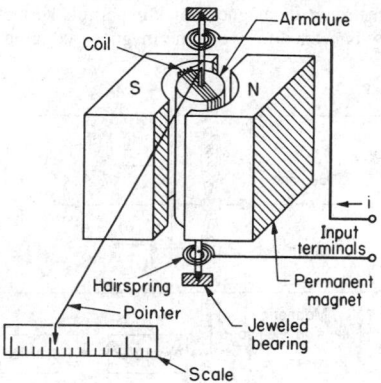

Fig. 16.1.35 D'Arsonval galvanometer.

The galvanometer can be converted into a **dc voltmeter, ammeter, or ohmmeter** by application of Ohm's law, $IR = E$, where I = current, A; E = electrical potential, V; and R = resistance, Ω.

For a **voltmeter,** a fixed resistance R is placed in series with the galvanometer (Fig. 16.1.36a). The current i through the galvanometer is proportional to the applied voltage E: $i = E/(r + R)$, where r = coil resistance. Different voltage ranges are obtained by changing the series resistance.

An **ammeter** is produced by placing the resistance in parallel with the galvanometer or DVM (Fig. 16.1.36b). The current then divides between the galvanometer coil or DVM and the resistor in inverse ratio to their resistance values (r and R, respectively); thus, $i = IR/(r + R)$, where i = current through coil and I = total current to be measured. Different current ranges are obtained by using different shunt resistances.

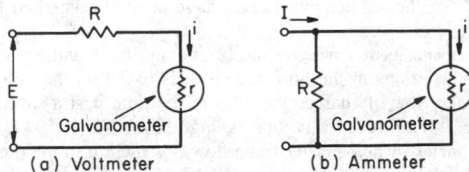

Fig. 16.1.36 (a) Voltmeter; (b) ammeter.

The common ohmmeter consists of a battery, a galvanometer with a shunt rheostat, and resistance in series to total Ri Ω. The shunt is adjusted to give a full-scale (0 Ω) reading with the test terminals shorted. When an unknown resistance R is connected, the deflection is $Ri/(Ri + R)$ fraction of full scale. The scale is calibrated to read R directly. A half-scale deflection indicates $R = Ri$. Alternatively, the galvanometer is connected to read voltage drop across the unknown R while a known current flows through it. This principle is used for low-value resistances and in digital ohmmeters.

Digital instruments are available for all these applications and often offer higher resolution and accuracy with less circuit loading. Fluctuating readings are difficult to follow, however.

Alternating current and voltage must be measured by special means. A dc instrument with a rectifier input is commonly used in applications requiring high input impedance and wide frequency range. For precise measurement at power-line frequencies, the electrodynamic instrument is used. This is similar to the galvanometer except that the permanent magnet is replaced by an electromagnet. The movable coil and field coils are connected in series; hence they respond simultaneously to the same current and voltage alternations. The pointer deflection is proportional to the square of the input signal. The moving-iron-type instruments consist of a soft-iron vane or armature which moves in response to current flowing through a stationary coil. The pointer is attached to the iron to indicate the deflection on a calibrated nonlinear scale. For measuring at very high frequencies, the **thermocouple voltmeter or ammeter** is used. This is based on the heating effect of the current passing through a fixed resistance R. Heat is liberated at the rate of E^2/R or $I^2 R W$.

DC **electrical power** is the product of the current through the load and the voltage across the load. Thus it can be simply measured using a voltmeter and ammeter. AC power is directly indicated by the wattmeter, which is similar to the electrodynamic instrument described above. Here the field coils are connected in series with the load, and the movable coil is connected across the load (to measure its voltage). The deflection of the movable coil is then proportional to the effective load power.

Precise voltage measurement (direct current) can be made by balancing the unknown voltage against a measured fraction of a known reference voltage with a **potentiometer** (Fig. 16.1.9). Balance is indicated by means of a sensitive current detector placed in series with the unknown voltage. The potentiometer is calibrated for angular position vs. fractional voltage output. Accuracies to 0.05 percent are attainable, dependent on the linearity of the potentiometer and the accuracy of the reference source. The reference standard may be a Weston standard cell or a regulated voltage supply (based on diode characteristics). The balance detector may be a galvanometer or electronic amplifier.

Precision resistance and general impedance measurements are made with bridge circuits (Fig. 16.1.37) which are adjusted until no signal is detected by the null detector (bridge is balanced). Then $Z_1 Z_3 = Z_2 Z_4$. The basic Wheatstone bridge is used for resistance measurement where all the impedances (Z's) are resistances (R's). If R_1 is to be measured, $R_1 = R_2 R_4 / R_3$ when balanced. A sensitive galvanometer for the null detector and dc voltage excitation is usual. All R_2, R_3, R_4 must be calibrated, and some adjustable. For general impedance measurement, ac voltage excitation of suitable frequency is used. The null detector may be a sensitive ac meter, oscilloscope, or, for audio frequencies, simple earphones. The basic balance equation is still valid, but it now

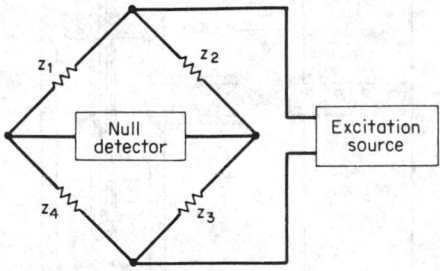

Fig. 16.1.37 Impedance bridge.

requires also that the sum of the phase angles of Z_1 and Z_3 equal the sum of the phase angles of Z_2 and Z_4. As an example, if Z_1 is a capacitor, the bridge can be balanced if Z_2 is a known capacitor while Z_3 and Z_4 are resistances. The phase-angle condition is met, and $Z_1 Z_3 = Z_2 Z_4$ becomes $(\frac{1}{2}\pi f C_1)R_3 = (\frac{1}{2}\pi f C_2)R_4$ and $C_1 = C_2 R_3 / R_4$. Variations on the basic principle include the Kelvin bridge for measurement of low resistance, and the Mueller bridge for platinum resistance thermometers.

Voltage measurement requires a meter of substantially higher impedance than the impedance of the source being measured. The vacuum-tube cathode follower and the field-effect transistor are suitable for high-impedance inputs. The following circuitry may be a simple amplifier to drive a pointer-type meter, or may use a digital technique to produce a digital output and display. Digital counting circuits are capable of great precision and are widely adapted to measurements of time, frequency, voltage, and resistance. Transducers are available to convert

temperature, pressure, flow, length and other variables into signals suitable for these instruments.

The charge amplifier is an example of an operational amplifier application (Fig. 16.1.38). It is used for outputs of piezoelectric transducers in which the output is a charge proportional to input force or other input converted to a force. Several capacitors switchable across the feedback path provide a range of full-scale values. The output is a voltage.

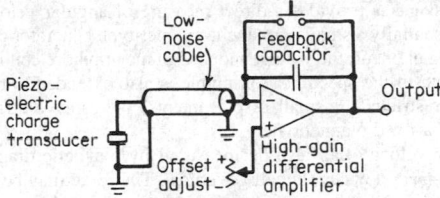

Fig. 16.1.38 Charge amplifier application.

The **cathode-ray oscilloscope** (Fig. 16.1.39) is an extremely useful and versatile device characterized by high input impedance and wide frequency range. An electron beam is focused on the phosphor-coated face of the cathode-ray tube, producing a visible spot of light at the point of impingement. The beam is deflected by applying voltages to vertical and horizontal deflector plates. Thus, the relationship between two varying voltages can be observed by applying them to the vertical and horizontal plates. The horizontal axis is commonly used for a linear time base generated by an internal sawtooth-wave generator. Virtually any desired sweep speed is obtainable as a calibrated sweep. Sweeps which change value part way across the screen are available to provide localized time magnification. As an alternative to the time base, any arbitrary voltage can be applied to drive the horizontal axis. The vertical axis is usually used to display a dependent variable voltage. **Dual-beam** and

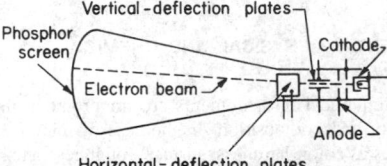

Fig. 16.1.39 Cathode-ray tube.

dual-trace instruments show two waveforms simultaneously. Special long-persistence and storage screens can hold transient waveforms for from seconds to hours. Greater versatility and unique capabilities are afforded by use of **digital-storage** oscilloscope. Each input signal is sampled, digitized, and stored in a first-in–first-out memory. Since a record of the recent signal is in memory when a trigger pulse is received, the timing of the end of storing new data into memory determines how much of the stored signal was before, and how much after, the trigger. Unlike storage screens, the stored signal can be amplified and shifted on the screen for detailed analysis, accompanied by numerical display of voltage and time for any point. Care must be taken that enough samples are taken in any waveform; otherwise aliasing results in a false view of the waveform.

The stored data can be processed mathematically in the oscilloscope or transferred to a computer for further study. Accessories for microcomputers allow them to function as digital oscilloscopes and other specialized tasks.

VELOCITY AND ACCELERATION MEASUREMENT

Velocity or speed is the time rate of change of displacement. Consequently, if the displacement measuring device provides an output signal which is a continuous (and smooth) function of time, the velocity can be measured by **differentiating** this signal either graphically or by use of a differentiating circuit. The accuracy may be very limited by noise (high-frequency fluctuations), however. More commonly, the output of an accelerometer is integrated to yield the velocity of the moving member. **Average speed** over a time interval can be determined by measuring the time required for the moving body to pass two fixed points a known distance apart. Here photoelectric or other rapid sensing devices may be used to trigger the start and stop of the timer. **Rotational speed** may be similarly measured by counting the number of rotations in a fixed time interval.

The **tachometer** provides a direct measure of angular velocity. One form is essentially a small permanent-magnet-type generator coupled to the rotating element; the voltage induced in the armature coil is directly proportional to the speed. The principle is also extended to rectilinear motions (restricted to small displacements) by using a straight coil moving in a fixed magnetic field.

Angular velocity can also be measured by magnetic drag-cup and **centrifugal-force** devices (flyball governor). The force may be balanced against a spring with the resulting deflection calibrated in terms of the shaft speed. Alternatively, the force may be balanced against the air pressure generated by a pneumatic nozzle-baffle assembly (similar to Fig. 16.1.23).

Vibration velocity pickups may use a coil which moves relative to a magnet. The voltage generated in the coil has the same frequency as the vibration and, for sine motion, a magnitude proportional to the product of vibration frequency and amplitude. Vibration **acceleration pickups** commonly use strain-gage, piezoresistive, or piezoelectric elements to sense a force $F = Ma/g_c$. The maximum usable frequency of an accelerometer is about one-fifth of the pickup's natural frequency (see Sec. 3.4). The minimum usable frequency depends on the type of pickup and the associated circuitry. The output of an accelerometer can be integrated to obtain a velocity signal; a velocity signal can be integrated to obtain a displacement signal. The **operational amplifier** is a versatile element which can be connected as an integrator for this use.

Holography is being applied to the study of surface vibration patterns.

MEASUREMENT OF PHYSICAL AND CHEMICAL PROPERTIES

Physical and chemical measurements are important in the control of product quality and composition. In the case of manufactured items, such properties as color, hardness, surface, roughness, etc., are of interest. Color is measured by means of a **colorimeter,** which provides comparison with color standards, or by means of a **spectrophotometer,** which analyzes the color spectrum. The **Brinell and Rockwell testers** measure surface hardness in terms of the depth of penetration of a hardened steel ball or special stylus. Testing machines with **strain-gage** elements provide measurement of the strength and elastic properties of materials. **Profilometers** are used to measure surface characteristics. In one type, the surface contour is magnified optically and the image projected onto a screen or viewer; in another, a stylus is employed to translate the surface irregularities into an electrical signal which may be recorded in the form of a highly magnified profile of the surface or presented as an averaged roughness-factor reading.

For liquids, attributes such as density, viscosity, melting point, boiling point, transparency, etc., are important. Density measurements have already been discussed. **Viscosity** is measured with a **viscosimeter,** of which there are three main types: flow through an orifice or capillary (Saybolt), viscous drag on a cylinder rotating in the fluid (MacMichael), damping of a vibrating reed (Ultrasonic) (see Secs. 3 and 4). **Plasticity** and consistency are related properties which are determined with special apparatus for heating or cooling the material and observing the temperature-time curve. The **photometer, reflectometer, and turbidimeter** are devices for measuring transparency or turbidity of nonopaque liquids and solids.

A variety of properties can be measured for determining chemical composition. **Electrical properties** include pH, conductivity, dielectric constant, oxidation potential, etc. **Physical properties** include density, refractive index, thermal conductivity, vapor pressure, melting and boiling points, etc. Of increasing industrial application are **spectroscopic**

measurements: **infrared absorption spectra, ultraviolet and visible emission spectra, mass spectrometry,** and **gas chromatography**. These are specific to particular types of compounds and molecular configurations and hence are very powerful in the analysis of complex mixtures. As examples, infrared analyzers are in use to measure low-concentration contaminants in engine oils resulting from wear and in hydraulic oils to detect deterioration. **X-ray diffraction** has many applications in the analysis of crystalline solids, metals, and solid solutions.

Of special importance in the realm of composition measurements is the determination of **moisture content**. A common laboratory procedure measures the loss of weight of the oven-dried sample. More rapid methods employ electrical conductance or capacitance measurements, based on the relatively high conductivity and dielectric constant values for ordinary water.

Water vapor in air (humidity) is measured in terms of its physical properties or effects on materials (see also Secs. 4 and 12). (1) The **psychrometer** is based on the cooling effect of water evaporating into the airstream. It consists of two thermal elements exposed to a steady airflow; one is dry, the other is kept moist. See Sec. 4 for psychrometric charts. (2) The **dew-point recorder** measures the temperature at which water just starts to condense out of the air. (3) The **hygrometer** measures the change in length of such humidity-sensitive elements as hair and wood. (4) **Electric sensing elements** employ a wire-wound coil impregnated with a hygroscopic salt (one that maintains an equilibrium between its moisture content and the air humidity) such that the resistance of the coil is related to the humidity.

The **throttling calorimeter** (Fig. 16.1.40) is most commonly used for determining the moisture in steam. A sampling nozzle is located preferably in a vertical section of steampipe far removed from any fittings. Steam enters the calorimeter through a throttling orifice and into a well-insulated expansion chamber. The steam quality x (fraction dry steam) is determined from the equation $x = (h_c - h_f) / h_{fg}$, where h_c is the enthalpy of superheated steam at the temperature and pressure measured

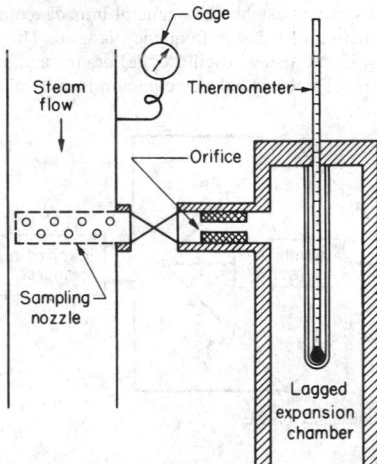

Fig. 16.1.40 Throttling calorimeter.

in the calorimeter; h_f and h_{fg} are, respectively, the liquid enthalpy and the heat of vaporization corresponding to line pressure. The chamber is conveniently exhausted to atmospheric pressure; then only line pressure and temperature of the throttled steam need be measured. The range of the throttling calorimeter is limited to small percentages of moisture; a **separating calorimeter** may be employed for larger moisture contents.

The **Orsat** apparatus is generally used for chemical analysis of flue gases. It consists of a graduated tube or burette designed to receive and measure volumes of gas (at constant temperature). The gas is analyzed for CO_2, O_2, CO, and N_2 by bubbling through appropriate absorbing reagents and measuring the resulting change in volume. The reagents normally employed are KOH solution for CO_2, pyrogallic acid and

KOH mixture for O_2, and cuprous chloride (Cu_2Cl_2) for CO. The final remaining unabsorbed gas is assumed to be N_2. The most common errors in the Orsat analysis are due to **leakage** and poor **sampling**. The former can be checked by simple test; the latter factor can only be minimized by careful sampling procedure. Recommended procedure is the taking of several simultaneous samples from different points in the cross-sectional area of the flue-gas stream, analyzing these separately, and averaging the results.

There are many instruments for **measuring CO_2 (and other gases) automatically.** In one type, the CO_2 is absorbed in KOH, and the change in volume determined automatically. The more common type, however, is based on the difference in **thermal conductivity** of CO_2 compared with air. Two thermal conductivity cells are set into opposing arms of a Wheatstone-bridge circuit. Air is sealed into one cell (reference), and the CO_2-containing gas is passed through the other. The cell contains an electrically heated resistance element; the temperature of the element (and therefore its resistance) depends on the thermal conductivity of the gaseous atmosphere. As a result, the unbalance of the bridge provides a measure of the CO_2 content of the gas sample.

The same principle can be employed for analyzing other constituents of gas mixtures where there is a significant thermal-conductivity difference. A modification of this principle is also used for determining CO or other combustible gases by mixing the gas sample with air or oxygen. The combustible gas then burns on the heated wire of the test cell, producing a temperature rise which is measured as above.

Many other physical properties are employed in the determination of specific components of gaseous mixtures. An interesting example is the **oxygen analyzer,** based on the unique paramagnetic properties of oxygen.

NUCLEAR RADIATION INSTRUMENTS

(See also Sec. 9.)

Nuclear radiation instrumentation is increasing in importance with two main areas of application: (1) measurement and control of radiation variables in nuclear reactor-based processes, such as nuclear power plants and (2) measurement of other physical variables based on radioactive excitation and tracer techniques. The instruments respond in general to electromagnetic radiation in the gamma and perhaps X-ray regions and to beta particles (electrons), neutrons, and alpha particles (helium nuclei).

Gas Ionization Tubes   The **ion chamber, proportional counter,** and **Geiger counter** are common instruments for radiation detection and measurement. These are different applications of the gas-ionization tube distinguished primarily by the amount of applied voltage.

A simple and very common form of the instrument consists of a gas-filled cylinder with a fine wire along the axis forming the anode and the cylinder wall itself (at ground potential) forming the cathode, as shown in Fig. 16.1.41. When a radiation particle enters the tube, its collision with gas molecules causes an ionization consisting of electrons (negatively charged) and positive ions. The electrons move very rapidly toward the positively charged wire; the heavier positive ions move relatively slowly toward the cathode. The above activity is detected by the resulting current flow in the external circuitry.

When the voltage applied across the tube is relatively low, the number of electrons collected at the anode is essentially equal to that produced by the incident radiation. In this voltage range, the device is called an **ion chamber.** The device may be used to count the number of radiation particles when the frequency is low; when the frequency is high, an external integrating circuit yields an output current proportional to the radiation intensity. Since the amplification factor of the ion chamber is low, high-gain electronic amplification of the current signal is necessary.

If the applied voltage is increased, a point is reached where the radiation-produced ions have enough energy to collide with other gas molecules and produce more ions which also enter into collisions so that an "avalanche" of electrons is collected at the anode. Thus, there is a very considerable amplification of the output signal. In this region, the device is called a **proportional counter** and is characterized by the voltage or current pulse being proportional to the energy content of the incident radiation signal.

With still further increase in the applied voltage, a point of saturation is reached wherein the output pulses have a constant amplitude independent of the incident radiation level. The resulting **Geiger counter** is capable of producing output pulses up to 10 V in amplitude, thus greatly reducing the requirements on the external circuitry and instrumentation. This advantage is offset somewhat by a lower maximum counting rate and more limited ability to differentiate among the various types of radiation as compared with the proportional counter.

The **scintillation counter** is based on the excitation of a phosphor by incident radiation to produce light radiation which is in turn detected by a photomultiplier tube to yield an output voltage. The signal output is greatly amplified and nearly proportional to the energy of the initial radiation. The device may be applied to a wide range of radiations, it has a very fast response, and, by choice of phosphor material, it offers a large degree of flexibility in applications.

Applications to the Measurement of Physical Variables   The ready availability of radioactive isotopes of long half-life, such as cobalt 60, make possible a variety of industrial and laboratory measuring techniques based on radiation instruments of the type described above. Most applications are based on (1) radiation absorption, (2) tracer identification, and (3) other properties. These techniques often have the advantages of isolation of the measuring device from the system, access to a variable not observable by conventional means, or measurement without destruction or modification of the system.

In the utilization of **absorption** properties, a radioactive source is separated from the radiation-measuring device by that part of the system to be measured. The measured radiation intensity will depend on the fraction of radiation absorbed, which in turn will depend on the distance traveled through the absorbing medium and the density and nature of the material. Thus, the instrument can be adapted to measuring thickness (see Fig. 16.1.12), coating weight, density, liquid or solids level, or concentration (of certain components).

Tracer techniques are effectively used in measuring flow rates or velocities, residence time distributions, and flow patterns. In flow measurement, a sharp pulse of radioactive material may be injected into the flow stream; with two detectors placed downstream from the injection point and a known distance apart, the velocity of the pulse is readily measured. Alternatively, if a known constant flow rate of tracer is injected into the flow stream, a measure of the radiation downstream is easily converted into a measure of the desired flow rate. Other applications of tracer techniques involve the use of tagged molecules embedded in the process to provide measures of wear, chemical reactions, etc.

Other applications of radiation phenomena include level measurements based on a floating radioactive source, level measurements based on the back-scattering effect of the medium, pressure measurements in the high-vacuum region based on the amount of ionization caused by alpha rays, location of interface in pipeline transmission applications, and certain chemical analysis applications.

INDICATING, RECORDING, AND LOGGING

An important element of measurement is the display of the measured value in a form which the human operator can readily interpret. Two

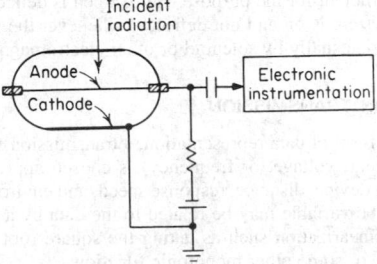

Fig. 16.1.41   Gas-ionization tube.

basic types of display are employed: analog and digital. **Analog** refers to a reading obtained from the motion of a pointer on a scale or the record of a pen moving over a chart. Digital refers to the reading displayed as a number, a series of holes on a punched card, a sequence of pulses on magnetic tape, or dots on a heat sensitive paper surface forming a trace. Further classification relates to indicating and recording functions. The **indicator** consists merely of a pointer moving over a calibrated scale. The scale may be concentric, as in the Bourdon gage (Fig. 16.1.16) or eccentric, as in the flowmeter. There are also digital indicators which directly display or illuminate the specific digits corresponding to the reading. Obviously, use of the indicator is limited to cases where the variable of interest is constant during the measuring period, or at most, changes slowly.

The **recorder** is used where long-term trends or detailed variations with time are of interest, or where the response is too rapid for the human eye to follow. In the common **circular-chart recorder** (Fig. 16.1.42), the pointer is replaced by a pen which writes on a chart rotated by a constant-speed electrical or spring-wound clock. Various chart

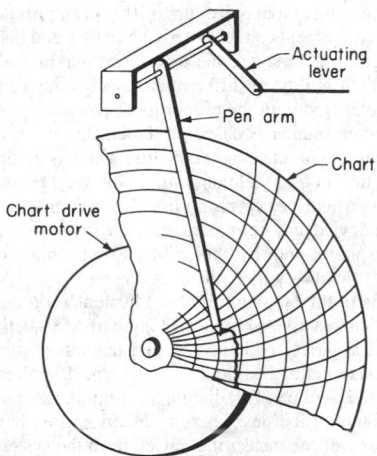

Fig. 16.1.42   Circular-chart recorder.

speeds are available from 1 r/min to 1 every 7 days. Up to four recording pens on a single chart are available (with a print-wheel mechanism, six color-identified records may be had). The **strip-chart recorder** shown in Fig. 16.1.43 is of the type used in electronic potentiometers, where the pen is positioned by a servomotor or a stepper motor as in the case of a digital recorder. A constant-speed motor drives the chart vertically past the pen, which deflects horizontally. **Multipoint recording** is achieved by replacing the pen with a print-wheel assembly. A selector switch

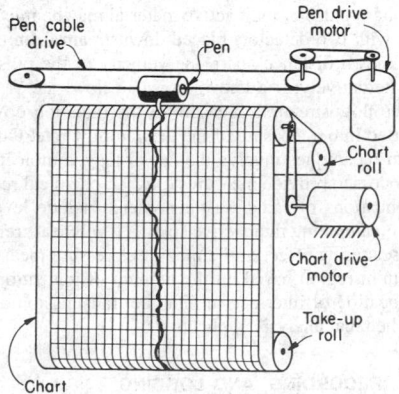

Fig. 16.1.43   Strip-chart recorder.

switches the input signal from one variable to another at the same time that the print wheel switches from one number (or symbol) to another. The record of each variable appears then as a sequence of dots with an identifying numeral. Up to 16 different records may be recorded on a chart (with external switching, as many as 144 records have been applied). Miniature recorders with 3- and 4-in strip charts are gaining favor in process industries because of their compactness and readability. The pen may be pneumatically or electrically actuated. Maximum number of records per chart is two.

For **direct-writing recording** of high-speed phenomena up to about 100 Hz, a pen or stylus can be driven by a galvanometer. The chart is in strip form and is driven at a speed suitable for the resolution needed. Recording may be done with ink and standard chart paper or heated stylus and special heat-sensitive paper. Mirror galvanometers projecting a spot of light onto a moving chart of light-sensitive paper can be used up to several kilohertz. A number of galvanometers can be used side by side to record several signals simultaneously on the same chart.

For higher frequencies a form of **magnetic recording** is common. Analog signals can be recorded by amplitude and frequency modulation. The latter is particularly convenient for playback at reduced speed.

Digital signals can be recorded in magnetic form. They can be recorded to any desired precision by using more bits to represent the data. Resolution is 1 part in 2^n, where n is the number of bits used in straight binary form, less in binary-coded decimal, where 4 bits are used to encode each decimal digit. Digital recording and data transmission have the advantage that in principle error rates can be made as small as desired in the presence of noise by adding more bits which serve as checks in error correcting codes (Raisbeck, "Information Theory: An Introduction for Scientists and Engineers," M.I.T. Press).

Most physical variables are in analog form. Popular standards for the transmission of analog signals include 3- to 15-psig pneumatic signals, direct currents of 4 to 20 or 10 to 50 mA, 0 to 5 and 0 to 10 V. Suitable **signal conditioners** are needed to convert thermocouple outputs and the like to these levels. (Of course, if the instrument is specifically for the particular thermocouple, this conversion is not needed.) This standardization gives greater flexibility in interconnecting signal sources with indicators and recorders. Some transducers and signal conditioners are designed to receive their power over the same two wires used to transmit their output signals.

Often it is necessary to convert from analog to digital form (as for the input to a digital computer) and vice versa. The **analog-digital (A/D)** and **digital-analog (D/A)** converters provide these interfaces. They are available in various conversion speeds and resolutions. Resolution is specified in terms of the number of bits in the digital signal.

Data which have been stored in magnetic form can be recovered at any time by connecting the storage device to an electrically actuated typewriter, printer, or other readout device. Modern **logging systems** have the measurements from hundreds of different points in the process tabulated periodically. These systems may provide such additional features as the printing of deviations from the normal in red and the more frequent scanning of abnormal conditions. Computer elements are also used in conjunction with logging systems to compute derived variables (such as operating efficiency, system losses, etc.) and to apply corrections to measured variables, e.g., temperature and pressure compensation of gas-flow readings.

In quality control and time-motion studies, often a simple **on-off-type recorder** is sufficient for the purpose. Here, a pen is deflected when the machine or system is on and not deflected whenever the system is off. Pen actuation is usually by solenoid or other electromagnetic element.

INFORMATION TRANSMISSION

In the analog form of data representation, a transmission variable (e.g., pressure, current, voltage, or frequency) is chosen appropriate to the data receiving device, distance, response speed, and environmental considerations. The variable may be related to the data by a simple linear function, by linearization such as taking the square root of an orifice pressure drop, or some other monotonic function.

A 3 to 15 (or 3 to 27, 6 to 30) psig air pressure, a 4 to 20 (or 10 to 50) mA dc current, and a 1 to 5 (or 0 to 5, 0 to 10) V dc voltage are in use to represent a data range of 0 to 100 percent. Where the 0 percent data level is transmitted as a nonzero value (e.g., 3 psig, 4 mA), a loss of power or a line break is detectable as an out-of-range condition. A variety of two-wire transmitters are powered by the loop current and they vary this same loop current to transmit data values. They are available for inputs including pressure, differential pressure, thermocouples, RTDs (resistance temperature detectors), strain gages, and pneumatic signals. Other systems use three or four wires, permitting separation of power and output signal. Converters are available to change a signal from one form to another, e.g., a 3- to 15-psig signal is converted to a 4- to 20-mA signal.

In the digital form of data transmission, patterns of binary (two-level) signals are sent in an agreed-upon manner to represent data. Binary coded decimal (BCD) uses 4 information units (bits) to represent each of the digits 0 through 9 (0000, 0001, 0010, 0011, 0100, 0101, 0110, 0111, 1000, 1001). The ASCII (American National Standard Code for Information Interchange) code uses 7 bits (128 different codes) to represent the alphabet, digits, punctuation, and control codes. The PC character set extends the ASCII 7-bit code set by an eighth bit to provide another 128 codes (codes 128 through 255) devoted to foreign characters, mathematical symbols, lines, box, and shading elements. Data are often sent in groups of 8 bits (1 byte). Commonly used 8-bit transmission permits the full ASCII/PC character set to be transmitted.

Two distinct signal levels are used in binary digital transmission. Some values (at the receiving device) are:

Name	Space	Mark
Binary name	0	1
TTL (5 V), V	0.0 to 0.8	2.0 to 5.0
CMOS (3 to 15 V), % of supply volts	0 to 30	70 to 100
RS-232-C, V	+ 3 to + 15	− 3 to − 15
20-mA loop	Current off	Current on
Telephone modem (Bell System 103), Hz tone		
From originator	1,070	1,270
To originator	2,025	2,225

The band between the two levels provides some protection against noise. The 20-mA loop uses a pair of wires for each transmission direction; optoisolators convert the 20-mA current to appropriate voltages at each end while providing electrical isolation.

The EIA Standard RS-232-C specifies an "Interface Between Data Terminal Equipment (DTE) and Data Communication Equipment Employing Serial Binary Data Interchange." Twenty lines are defined; a minimum for two-way systems uses: line 1, protective ground; line 2, transmitted data (DTE to DCE); line 3, received data (DCE to DTE); and line 7, signal ground (common return). When a 9-pin (rather than 25-pin) connector is used, pin 3 is transmitted data and pin 2 is received data. If both devices are the same type, a crossover between wires 2 and 3 is needed. A cable with a 25-pin connector on one end and a 9-pin connector at the other end does not require a crossover.

In asynchronous serial transmission, an example of which is shown in Fig. 16.1.44, the no-signal state is the MARK level. A change to the SPACE level indicates that an ASCII code will start 1-bit-timer later. The start bit is always SPACE. The ASCII code follows, least significant bit first. This example is the letter S, binary form (most significant bit, msb, written first): 1 010 011, or 124 in octal form.

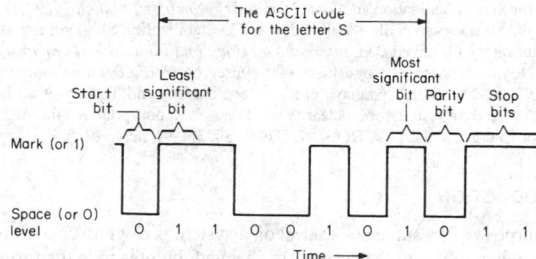

Fig. 16.1.44   ASCII transmission.

The parity bit is optional for error checking. This example uses even parity: the total number of 1s in the ASCII code and the parity bit is even. The bit sequence concludes with one or more stop bits at the MARK level. The minimum number required is set by the receiving device. Standard RS-232-C baud rates (bits per second) include 110, 150, 300, 600, 1,200, 2,400, 4,800, 9,600, 19,200, and 38,400. The EIA Standard RS-422 improves upon the RS-232-C by using transmission lines balanced to ground. This improves noise immunity and increases usable baud rates and transmission distance.

Transmissions are classed as **simplex** (one direction only), **half-duplex** (one direction at a time), and **full-duplex** (capable of simultaneous transmission in both directions). An agreed-upon protocol allows the receiver to signal the sender whether or not it is able to accept data. This may be done by a separate line(s) or by special (XON/XOFF; control Q/control S) ASCII signals on the return path of a full-duplex line.

For parallel transmission, multiple wires carry signals representing all the bits at once. Separate lines indicate when the receiver is ready for new data and when the sender has put new data on the lines. This exchange is called **handshaking**.

The above forms are used for communication between two devices. Where more than two devices are to be interconnected, a network, or bus system, is employed. The IEEE-488 General Purpose Interface Bus, GPIB (based on the Hewlett-Packard HP-1B), uses a parallel bus structure and can interconnect up to 15 devices, at a total connection path length of 20 m. One device acts as a controller at any time. Multiple instruments and control devices may be interconnected using 2 or 4 wire circuits and serial bus standard RS-485. Alternately, the ISA SP-50 protocol and other schemes still in development may be employed to achieve serial multidrop communications over distances of up to 2,500 m (8,200 ft). See also Sec. 2.2, "Computers," and Sec. 15.2, "Electronics."

16.2   AUTOMATIC CONTROLS
by Gregory V. Murphy

REFERENCES: Thaler, "Elements of Servomechanism Theory," McGraw-Hill. Shinskey, "Process Control Systems: Application, Design and Tuning," McGraw-Hill. Kuo, "Automatic Control Systems," Prentice-Hall. Phillips and Nagle, "Digital Control System Analysis and Design," Prentice-Hall. Lewis, "Applied Optimal Control and Estimation: Digital Design and Implementation," Prentice-Hall. Cochin and Plass, "Analysis and Design of Dynamic Systems," HarperCollins. Astrom and Hagglund, "Automatic Tuning of PID Controllers,"

Instrument Society of America. Maciejowski, "Multivariable Feedback Design," Addison-Wesley. Murphy and Bailey, LQG/LTR Control System Design for a Low-Pressure Feedwater Heater Train with Time Delay, *Proc. IECON,* 1990. Murphy and Bailey, Evaluation of Time Delay Requirements for Closed-Loop Stability Using Classical and Modern Methods, *IEEE Southeastern Symp. on System Theory,* 1989. Murphy and Bailey, "LQG/LTR Robust Control System Design for a Low-Pressure Feedwater Heater Train," *Proc. IEEE Southeastcon,*

1990. Kazerooni and Narayanan, "Loop Shaping Design Related to LQG/LTR for SISO Minimum Phase Plants," *IEEE American Control Conf.,* Vol. 1, 1987. Murphy and Bailey, "Robust Control Technique for Nuclear Power Plants," ORNL-10916, March 1989. Birdwell, Crockett, Bodenheimer, and Chang, The CASCADE Final Report: Vol. II, "CASCADE Tools and Knowledge Base," University of Tennessee. Wang and Birdwell, A Nonlinear PID-Type Controller Utilizing Fuzzy Logic, *Proc. Joint IEEE/IFAC Symp. on Controller-Aided Control System Design,* 1994. Upadhyaya and Eryurek, Application of Neural Networks for Sensor Validation and Plant Monitoring, *Nuclear Technology,* **97,** no. 2, Feb. 1992. Vasadevan et al., Stabilization and Destabilization Slugging Behavior in a Laboratory Fluidized Bed, *International Conf. on Fluidized Bed Combustion,* 1995. Doyle and Stein, "Robustness with Observers." *IEEE Trans. Automatic Control, AC-24,* 1979. Upadhaya et al. "Development and Testing of an Integrated Validation System for Nuclear Power Plants," Report prepared for the U.S. Dept. of Energy. Vols. 1–3, DOE/NE/37959-34, 35, 36, Sept. 1989.

INTRODUCTION

The purpose of an **automatic control** on a system is to produce a desired output when inputs to the system are changed. Inputs are in the form of commands, which the output is expected to follow, and disturbances, which the automatic control is expected to minimize. The usual form of an automatic control is a **closed-loop feedback control** which Ahrendt defines as "an operation which, in the presence of a disturbing influence, tends to reduce the difference between the actual state of a system and an arbitrarily varied desired state and which does so on the basis of this difference." The general theories and definitions of automatic control have been developed to aid the designer to meet primarily three basic specifications for the performance of the control system, namely, stability, accuracy, and speed of response.

The **terminology** of automatic control is being constantly updated by the ASME, IEEE, and ISA. Redundant terms, such as *rate, preact,* and *derivative,* for the same controller action are being standardized. Common terminology is still evolving. The introduction of the digital computer as a control device has necessitated the introduction of a whole new subset of terminology. The following terms and definitions have been selected to serve as a reference to a complex area of technology whose breadth crosses several professional disciplines.

Adaptive control system: A control system within which automatic means are used to change the system parameters in a way intended to improve the performance of the system.

Amplification: The ratio of output to input, in a device intended to increase this ratio. A gain greater than 1.

Attenuation: A decrease in signal magnitude between two points, or a gain of less than 1.

Automatic-control system: A system in which deliberate guidance or manipulation is used to achieve a prescribed value of a variable and which operates without human intervention.

Automatic controller: A device, or combination of devices, which measures the value of a variable, quantity, or condition and operates to correct or limit deviation of this measured value from a selected command (set-point) reference.

Bode diagram: A plot of log-gain and phase-angle values on a log-frequency base, for an element, loop, or output transfer function.

Capacitance: A property expressible by the ratio of the time integral of the flow rate of a quantity (heat, electric charge) to or from a storage, divided by the related potential charge.

Command: An input variable established by means external to, and independent of, the automatic-control system, which sets the ideal value of the controlled variable. See *set point.*

Control action: Of a control element or controlling system, the nature of the change of the output affected by the input.

Control action, derivative: That component of control action for which the output is proportional to the rate of change of input.

Control action, floating: A control system in which the rate of change of the manipulated variable is a continuous function of the actuating signal.

Control action, integral (reset): Control action in which the output is proportional to the time integral of the input.

Control action, proportional: Control action in which there is a continuous linear relationship between the output and the input.

Control system, sampling: Control using intermittently observed values of signals such as the feedback signal or the actuating signal.

Damping: The progressive reduction or suppression of the oscillation of a system.

Deviation: Any departure from a desired or expected value or pattern. Steady-state deviation is known as *offset.*

Disturbance: An undesired variable applied to a system which tends to affect adversely the value of the controlled variable.

Error: The difference between the indicated value and the accepted standard value.

Gain: For a linear system or element, the ratio of the change in output to the causal change in input.

Load: The material, force, torque, energy, or power applied to or removed from a system or element.

Nyquist diagram: A polar plot of the loop transfer function.

Nichols diagram: A plot of magnitude and phase contours using ordinates of logarithmic loop gain and abscissas of loop phase angle.

Offset: The steady-state deviation when the command is fixed.

Peak time: The time for the system output to reach its first maximum in responding to a disturbance.

Proportional band: The reciprocal of gain expressed as a percentage.

Resistance: An opposition to flow that results in dissipation of energy and limitation of flow.

Response time: The time for the output of an element or system to change from an initial value to a specified percentage of the steady state.

Rise time: The time for the output of a system to increase from a small specified percentage of the steady-state increment to a large specified percentage of the increment.

Self-regulation: The property of a process or a machine to settle out at equilibrium at a disturbance, without the intervention of a controller.

Set point: A fixed or constant command given to the controller designating the desired value of the controlled variable.

Settling time: The time required, after a disturbance, for the output to enter and remain within a specified narrow band centered on the steady-state value.

Time constant: The time required for the response of a first-order system to reach 63.2 percent of the total change when disturbed by a step function.

Transfer function: A mathematical statement of the influence which a system or element has on a signal or action compared at input and output terminals.

BASIC AUTOMATIC-CONTROL SYSTEM

A **closed-loop control system** consists of a process, a measurement of the controlled variable, and a controller which compares the actual measurement with the desired value and uses the difference between them to automatically adjust one of the inputs to the process. The **physical system** to be controlled can be electrical, thermal, hydraulic, pneumatic, gaseous, mechanical, or described by any other physical or chemical process. Generally, the control system will be designed to meet one of two **objectives.** A servomechanism is designed to follow changes in set point as closely as possible. Many electrical and mechanical control systems

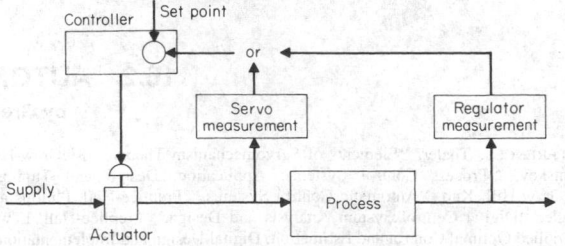

Fig. 16.2.1 Feedback control loop showing operation as servomechanism or regulator.

are servomechanisms. A **regulator** is designed to keep output constant despite changes in load or disturbances. Regulatory controls are widely used on chemical processes. Both objectives are shown in Fig. 16.2.1.

The **control components** can be actuated pneumatically, hydraulically, electronically, or digitally. Only in very few applications does actuation affect controllability. **Actuation** is chosen on the basis of economics.

The **purpose** of the control system must be defined. A large capacity or inertia will make the system sluggish for servo operation but will help to minimize the error for regulator operation.

PROCESS AS PART OF THE SYSTEM

Figure 16.2.1 shows the **process** to be part of the control system either as load on the servo or process to be controlled. Thus the process must be designed as part of the system. The process is **modeled** in terms of its static and dynamic gains in order that it be incorporated into the system diagram. Modeling uses Ohm's and Kirchhoff's laws for electrical systems, Newton's laws for mechanical systems, mass balances for fluid-flow systems, and energy balances for thermal systems.

Consider the **electrical system** in Fig. 16.2.2

$$i = \frac{E_1 - E_2}{R}$$
$$i = C\frac{dE_2}{dt}$$
(16.2.1)

Combining

$$RC\frac{dE_2}{dt} + E_2 = E_1$$
(16.2.2)

where $\left(R\frac{\text{volt-second}}{\text{coulomb}}\right)\left(C\frac{\text{coulomb}}{\text{volt}}\right) = \tau\,\text{s} = \text{time constant}$

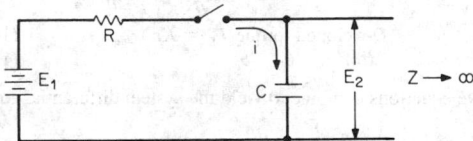

Fig. 16.2.2 Electrical system where current flows upon closing switch.

Consider the **mass balance** of the vessels shown in Figs. 16.2.3 and 16.2.4:

$$\text{Accumulation} = \text{input} - \text{output}$$
$$\frac{d(\rho V)}{dt} = w_1 - w_2 \quad \text{lb/min}$$
(16.23)

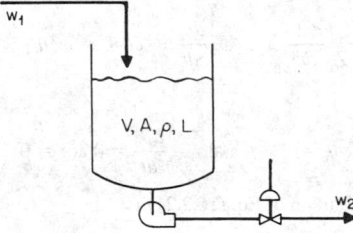

Fig. 16.2.3 Liquid level process.

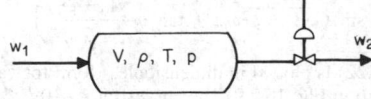

Fig. 16.2.4 Gas pressure process.

For the liquid level process (Fig. 16.2.3):

$$V = A L$$
(16.2.4)

Linearizing

$$w_2 = \left.\frac{w_2}{x}\right|_{ss} x$$

Substituting

$$\rho A\frac{dL}{dt} = w_1 - \left.\frac{w_2}{x}\right|_{ss} x$$
(16.25)

where $\rho A = $ analogous capacitance.
  For the **gas pressure process** (Fig. 16.2.4):

$$V\frac{d\rho}{dt} = w_1 - w_2$$
(16.2.6)

From thermodynamics

$$\left(\frac{\rho}{p}\right)^n = \text{constant}$$
(16.2.7)

Linearizing

$$w_2 = \left.\frac{w_2}{x}\right|_{ss} x + \left.\frac{w_2}{2(p - p_2)}\right|_{ss} p - p_2$$
(16.2.8)

Substituting

$$RC\frac{dp}{dt} + p = \left.\frac{2(p - p_2)}{x}\right|_{ss} x + p_2$$
(16.2.9)

where $\quad C = \dfrac{V}{nRT} \quad R = \dfrac{2(p - p_2)}{w_2} \quad RC = \tau\,\text{min}$

and the terms are defined: $V = $ volume, ft³ (m³); $A = $ cross-sectional area, ft² (m²); $L = $ level, ft (m); $\rho = $ density, lb/ft³ (g/ml); $x = $ valve stem position (normalized 0 to 1); $T = $ temperature, °F (°C); $p = $ pressure, lb/in² (kPa); and $w = $ mass flow, lb/min (kg/min).

The **thermal process** of Fig. 16.2.5 is modeled by a heat balance (Shinskey, "Process Control Systems," McGraw-Hill):

$$Mc\frac{dT}{dt} = wc(T_0 - T) - UA(T - T_w)$$
(16.2.10)

$$\frac{Mc}{wc + UA}\frac{dT}{dt} + T = \frac{wc}{wc + UA}T_0 + \frac{UA}{wc + UA}T_w$$
(16.2.11)

where $\quad C = Mc \quad R = \dfrac{1}{wc + UA} \quad RC = \tau\,\text{min}$

$$RC\frac{dT}{dt} + T = \frac{wc}{wc + UA}T_0 + \frac{UA}{wc + UA}T_w$$
(16.2.12)

The terms are defined: $M = $ weight of process fluid in vessel, lb (kg); $c = $ specific heat, Btu/lb·°F (J/kg·°C); $U = $ overall heat-transfer coefficient, Btu/ft²·min·°F (W/m²·°C); and $A = $ heat-transfer area, ft² (m²).

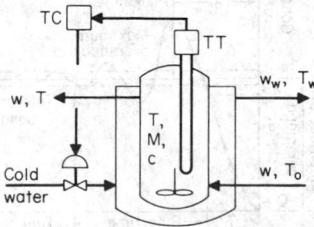

Fig. 16.2.5 Thermal process with heat from vessel being removed by cold water in jacket.

Newton's laws can be applied to the manometer shown in Fig. 16.2.6.

Inertia force = restoring force − flow resistance

$$\frac{Al\rho}{g_c}\frac{d^2h}{dt^2} = A\left(p - 2h\rho\frac{g}{g_c}\right) - RA\frac{dh}{dt} \qquad (16.2.13)$$

Flow resistance for laminar flow is given by the Hagen-Poiseuille equation:

$$R = \frac{\text{driving force}}{\text{rate of transfer}} = \frac{321\mu}{d^2g_c} \qquad (16.2.14)$$

Substituting

$$\frac{1}{2g}\frac{d^2h}{dt^2} + \frac{161\mu}{\rho dg^2}\frac{dh}{dt} + h = h_i \qquad (16.2.15)$$

In standard form

$$\tau_c^2\frac{d^2h}{dt^2} + 2\tau_c\zeta\frac{dh}{dt} + h = h_i \qquad (16.2.16)$$

where τ_c = characteristic response time, min, $= 1/[(60)(2\pi)\omega_n]$ where ω_n = natural frequency, Hz; and ζ = damping coefficient (ratio), dimensionless.

The variables τ_c and ζ are very valuable design aids since they define **system response** and **stability** in terms of system parameters.

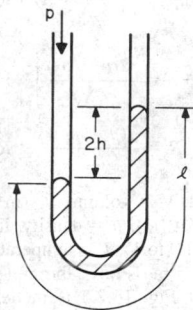

Fig. 16.2.6 Filled manometer measuring pressure P.

TRANSIENT ANALYSIS OF A CONTROL SYSTEM

The stability, accuracy, and speed of response of a control system are determined by analyzing the **steady-state** and the **transient** performance. It is desirable to achieve the steady state in the shortest possible time, while maintaining the output within specified limits. Steady-state performance is evaluated in terms of the accuracy with which the output is controlled for a specified input. The transient performance, i.e., the behavior of the output variable as the system changes from one steady-state condition to another, is evaluated in terms of such quantities as maximum overshoot, rise time, and response time (Fig. 16.2.7).

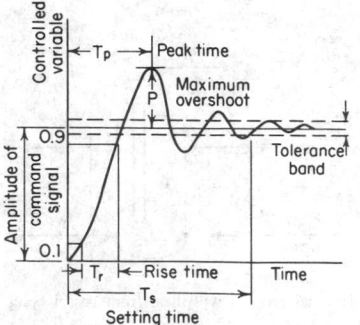

Fig. 16.2.7 System response to a unit step-function command.

Transient-Producing Disturbances A number of factors affect the **quality of control,** among them **disturbances** caused by set-point changes and process-load changes. Both set point and process load may be defined in terms of the setting of the final control element to maintain the controlled variable at the set point. Thus both cause the final control element to reposition. For a purely mechanical system the disturbance may take the form of a vibration, a displacement, a velocity, or an acceleration. A process-load change can be caused by variations in the rate of energy supply or variations in the rate at which work flows through the process. Reference to Fig. 16.2.5 and Eq. (16.2.12) shows disturbances to be variations in inlet process fluid temperature and cooling-water temperature. Further linearization would show variations in process flow and the overall heat-transfer coefficient to also be disturbances.

The Basic Closed-Loop Control To illustrate some characteristics of a basic closed-loop control, consider a mechanical, rotational system composed of a prime mover or motor, a total system inertia J, and a viscous friction f. To control the system's output variable θ_o, a command signal θ_i must be supplied, the output variable measured and compared to the input, and the resulting signal difference used to control the flow of energy to the load. The basic control system is represented schematically in Fig. 16.2.8.

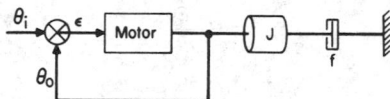

Fig. 16.2.8 A basic closed-loop control system.

The differential equation of this basic system is readily obtained from the idealized equations

$$\text{Load torque } T_L = J\frac{d^2\theta_o}{dt^2} + f\frac{d\theta_o}{dt} \qquad (16.2.17)$$

$$\text{Developed torque } T_D = K\varepsilon \qquad (16.2.18)$$
$$\text{Error } \varepsilon = \theta_i - \theta_o \qquad (16.2.19)$$

The above equations combine to yield the system differential equation:

$$J\frac{d^2\theta_o}{dt^2} + f\frac{d\theta_o}{dt} + K\theta_o = K\theta_i \qquad (16.2.20)$$

Step-Input Response of a Viscous-Damped Control If the control system described in Fig. 16.2.8 by Eq. (16.2.20) is subjected to a step change in the input variable θ_i, a solution $\theta_o = \theta_o(t)$ can be obtained as follows. (1) Let the ratio $\sqrt{K/J}$ be designated by the symbol ω_n and be called the **natural frequency.** (2) Let the quantity $2\sqrt{JK}$ be designated by the symbol f_c and be called the **friction coefficient** required for critical damping. (3) Let f/f_c be designated by the symbol ζ and be called the **damping ratio.** Equation (16.2.20) can then be written as

$$\frac{d^2\theta_o}{dt^2} + 2\zeta\omega_n\frac{d\theta_o}{dt} + \omega_n^2\theta_o = \omega_n^2\theta_i \qquad (16.2.21)$$

For $\theta_i = 1$:

$$\theta_o = 0 \qquad \text{and} \qquad \frac{d\theta_o}{dt} = 0 \text{ at } t = 0$$

The complete solution of Eq. (16.2.21) is

$$\theta_o = 1 - \frac{\exp(-\zeta\omega_n t)}{\sqrt{1 - \zeta^2}}$$

$$\sin\left(\sqrt{1 - \zeta^2}\omega_n t + \tan^{-1}\frac{\sqrt{1 - \zeta^2}}{\zeta}\right) \qquad (16.2.22)$$

Equation (16.2.22) is plotted in dimensionless form for various values of damping ratio in Fig. 16.2.9. The curves for $\zeta = 0.1$, 2, and 1 illustrate the underdamped, overdamped, and critically damped case, where any further decrease in system damping would result in overshoot.

Damping is a property of the system which opposes a change in the output variable.

The immediately apparent features of an observed transient performance are (1) the existence and magnitude of the maximum overshoot, (2) the frequency of the transient oscillation, and (3) the response time.

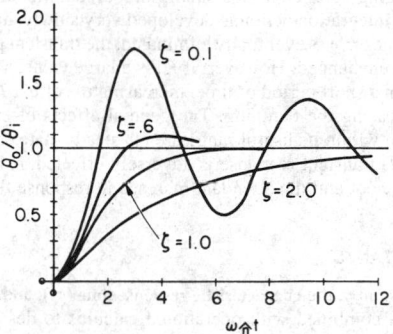

Fig. 16.2.9 Transient response of a second-order viscous-damped control to unit-step input displacement.

Maximum Overshoot When an automatic-control system is underdamped, the output variable overshoots its desired steady-state condition and a transient oscillation occurs. The first overshoot is the greatest, and it is the effect of its amplitude which must concern the control designer. The primary considerations for limiting this maximum overshoot are (1) to avoid damage to the process or machine due to excessive excursions of the controlled variable beyond that specified by the command signal, and (2) to avoid the excessive settling time associated with highly underdamped systems. Obviously, exact quantitative limits cannot generally be specified for the magnitude of this overshoot. However, experience indicates that satisfactory performance can generally be obtained if the overshoot is limited to 30 percent or less.

Transient Frequency An undamped system oscillates about the final steady-state condition with a frequency of oscillation which should be as high as possible in order to minimize the response time. The designer must, however, avoid resonance conditions where the frequency of the transient oscillation is near the natural frequency of the system or its component parts.

Rise Time T_n**, Peak Overshoot** P**, Peak Time** T_P These quantities are related to ζ and ω_n in Figs. 16.2.10 and 16.2.11. Some useful formulas are listed below:

$$\omega_n T_n \approx 1.02 + 0.48\zeta + 1.15\zeta^2 + 0.76\zeta^3 \qquad 0 \le \zeta \le 1$$

$$\omega_n T_s = \left.\begin{array}{ll} 17.6 - 19.2\zeta & 0.2 \le \zeta \le 0.75 \\ -3.8 + 9.4\zeta & 0.75 \le \zeta \le 1 \end{array}\right\} \; 2\% \text{ tolerance band}$$

$$P = \exp\left(\frac{-\pi\zeta}{\sqrt{1-\zeta^2}}\right) \qquad T_p = \frac{\pi}{\omega_n\sqrt{1-\zeta^2}}$$

Although these quantities are defined for a second-order system, they may be useful in the early design states of higher-order systems if the response of the higher-order system is dominated by roots of the characteristic equation near the imaginary axis.

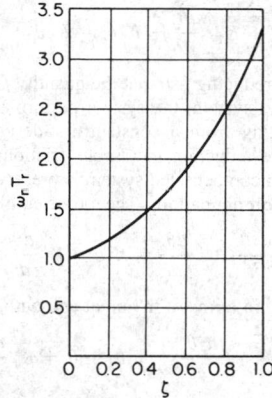

Fig. 16.2.10 Rise time T_r as a function of ζ and ω_n.

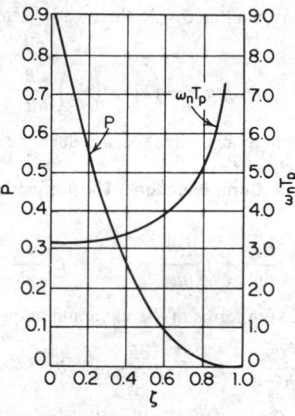

Fig. 16.2.11 Peak overshoot P and peak time T_p as functions of ζ and ω_n.

Derivative and Integral Compensation (Thaler) Four common compensation methods for improving the steady-state performance of a proportional-error control without damaging its transient response are shown in Fig. 16.2.12. They are (1) error derivative compensation, (2) input derivative compensation, (3) output derivative compensation, (4) error integral compensation.

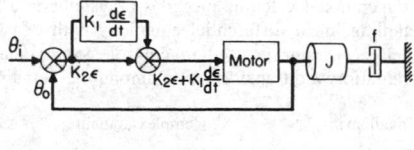

(a)

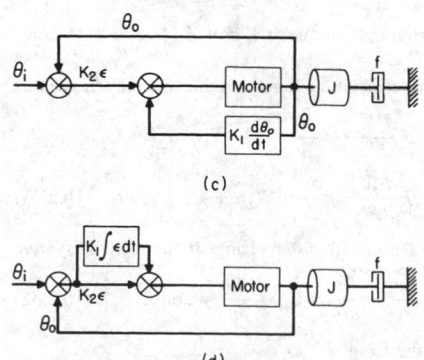

(c)

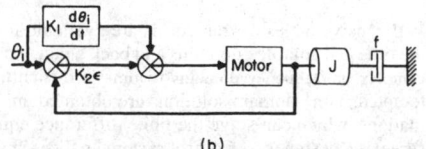

(b)

(d)

Fig. 16.2.12 Derivative and integral compensation of a basic closed-loop system. (a) Error derivative compensation; (b) input derivative compensation; (c) output derivative compensation; (d) error integral compensation. (Thaler.)

Error Derivative Compensation   The torque equilibrium equation is

$$J \frac{d^2\theta_o}{dt^2} + f \frac{d\theta_o}{dt} = K_2\varepsilon + K_1 \frac{d\varepsilon}{dt} \qquad (16.2.23)$$

Writing Eq. (16.2.23) in terms of the input and output variables yields

$$J \frac{d^2\theta_o}{dt^2} + (f + K_1)\frac{d\theta_o}{dt} + K_2\theta_o = K_2\theta_i + K_1 \frac{d\theta_i}{dt} \qquad (16.2.24)$$

By adjusting K_1 and reducing f so that the quantity $f + K_1$ is equal to f in the uncompensated system, the system performance is affected as follows: (1) ε resulting from a constant-first-derivative input is reduced because of the reduction in viscous friction; (2) the transient performance of the uncompensated system is preserved unchanged.

Derivative Input Compensation   The torque equilibrium equation is

$$J \frac{d^2\theta_o}{dt^2} + f \frac{d\theta_o}{dt} = K_2\varepsilon + K_1 \frac{d\theta_i}{dt} \qquad (16.2.25)$$

Writing Eq. (16.2.25) in terms of the input and output variables yields

$$J \frac{d^2\theta_o}{dt^2} + f \frac{d^2\theta_o}{dt} + K_2\theta_o = \theta_i + K_1 \frac{d\theta_i}{dt} \qquad (16.2.26)$$

Examination of Eq. (16.2.26) yields the following information about the compensated system's performance: (1) since the characteristic equation is unchanged from that of the uncompensated system, the transient performance is unaltered; (2) the steady-state solution to Eq. (16.2.26) is

$$\theta_o = \theta_i - \frac{f}{K_2}\left(1 - \frac{K_1}{K_2}\right)\frac{d\theta_i}{dt} \qquad (16.2.27)$$

Therefore the input derivative signal can reduce the steady-state error by adjusting K_1 to equal K_2.

Derivative Output Compensation   The torque equilibrium equation is

$$J \frac{d^2\theta_o}{dt^2} + f \frac{d^2\theta_o}{dt} = K_2\varepsilon \pm K_1 \frac{d\theta_o}{dt} \qquad (16.2.28)$$

Writing Eq. (16.2.28) in terms of the input and output variables yields

$$J \frac{d^2\theta_o}{dt^2} + (f \pm K_1)\frac{d\theta_o}{dt} + K_2\theta_o = K_2\theta_i \qquad (16.2.29)$$

Examination of Eq. (16.2.29) yields the following information about the compensated system's performance. (1) Output derivative feedback produces the same system effect as the viscous friction does. This compensation therefore damps the transient performance. (2) Under conditions where $\theta_i = ct$, the steady-state error is increased.

Error Integral Compensation   Error integral compensation is used where it is necessary to eliminate steady-state errors resulting from input signals with constant first derivatives or under conditions of externally applied loads. The torque equilibrium equation is

$$J \frac{d^2\theta_o}{dt^2} + f \frac{d\theta_o}{dt} \pm \text{external load torque} = K_2\varepsilon + K_1 \int_0^t \varepsilon \, dt \quad (16.2.30)$$

Writing Eq. (16.2.30) in terms of the input variable and the error yields

$$\text{External load torque} + J \frac{d^2\theta_i}{dt^2} + f \frac{d\theta_i}{dt}$$
$$= J \frac{d^2\varepsilon}{dt^2} + f \frac{d\varepsilon}{dt} + K_2\varepsilon + K_1 \int_0^t \varepsilon \, dt \quad (16.2.31)$$

At steady state for $t \gg 0$ and with a step change in the input derivative

$$\frac{d^2\theta_i}{dt^2} = \frac{d\varepsilon}{dt} = \frac{d^2\varepsilon}{dt^2} = 0 \qquad \frac{d\theta_i}{dt} = \text{const} \qquad (16.2.32)$$

Eq. (16.2.31) assumes the form

$$\pm \text{External load torque} + f \frac{d\theta_i}{dt} = K_2\varepsilon + K_1 \int_0^t \varepsilon \, dt \quad (16.2.33)$$

Since the sum of the load torque and the term $f(d\theta_i/dt)$ is finite, $\varepsilon = 0$ for

$$\int_0^\infty \varepsilon \, dt \to \infty \qquad (16.2.34)$$

If the integrating coefficient is a small number, additional torque produced by the integration action is developed very slowly, and, although the steady-state error is eventually eliminated, the transient performance is essentially unchanged. However, if K_1 is a large value, a large torque is produced in a short period of time, increasing the effect T/J ratio, and thereby decreasing the damping. The general effects of error integral compensation within its useful range are (1) steady-state error is eliminated; and (2) transient response is adversely effected, resulting in increased overshoot and the attendant increase in response time.

TIME CONSTANTS

The **time constant** τ, the **characteristic response time** (τ_c), and the **damping coefficient** are combined with operational calculus to design a control system without solving the classical differential equations. Note that the electrical, Eq. (16.2.2), mass, Eq. (16.2.9), and energy, Eq. (16.2.12) processes are all described by analogous first-order differential equations of the form

$$\tau \frac{dc}{dt} + c = Ka \qquad (16.2.35)$$

Solving with initial conditions equal to zero, the **time-domain response** to a step disturbance is

$$c = Ka(1 - e^{t/\tau}) \qquad (16.2.36)$$

Plotting is shown in Fig. 16.2.13.

Figure 16.2.13 shows the **time constant** to be defined by the point at 63.2 percent of the response, while response is essentially complete at three time constants (95 percent).

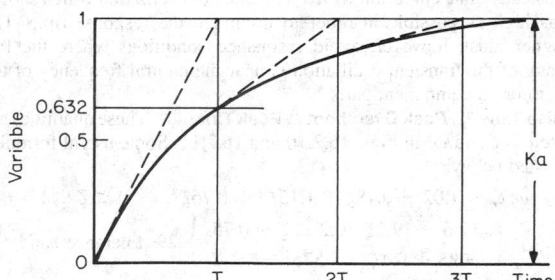

Fig. 16.2.13   Response of first-order system showing time constant relationship.

Operational calculus provides a systematic and simple method for handling linear differential equations with constant coefficients. The **Laplace operator** is used in control system analysis because of the straightforward transformation among the domains of interest:

Time domain	Complex domain	Frequency domain
$\dfrac{d}{dt}$	s	$j\omega$

Extensive tables of transforms are available for converting between the time and complex domains and back again. Transformation from the **complex** or **frequency domains** to **time** is difficult, and it is seldom attempted. Time-domain solutions are obtained only from computer simulations, which can solve the finite difference equations fast enough to serve as a design tool. Control system analysis and design proceed with a number of analytical and graphical techniques which do not require a time-domain solution.

The **root locus method** (Evans), is a graphical technique used in the complex domain which provides substantial insight into the system. It has the weaknesses of handling dead time poorly and of the graph's being very tedious to plot. It is used only when computerized plotting is available.

BLOCK DIAGRAMS

The **physical diagram** of the system is converted to a **block diagram** in order that the different components of the system (all the way from a steam boiler to a thermocouple) can be placed on a common mathematical footing for analysis as a system. The block diagram shows the functional relationship among the parts of the system by depicting the action of the variables in the system. The circle represents an algebraic function of addition. Each system component is represented by a **block** which has one input and one output. The block represents a dynamic function such that the output is a function of time and also of the input. The dynamic function is called a **transfer function**—the ratio of the Laplace transform of the output variable to the input variable with all initial conditions equal to zero. The input and output variables are considered as signals, and the blocks are connected by arrows to show the flow of information in the system.

Rewriting Eq. (16.2.12) for the thermal system,

$$RC\frac{dT}{dt} + T = K_1 T_0 + K_2 T_w \qquad (16.2.37)$$

Transforming

$$(\tau s + 1)T = K_1 T_0 + K_2 T_w \qquad (16.2.38)$$

for which the block diagram is shown in Fig. 16.2.14. Two conditions are specified: (1) the components must be described by linear differential equations (or nonlinear equations linearized by suitable approximations), and (2) each block is unilateral. What occurs in one component may not affect the components preceding.

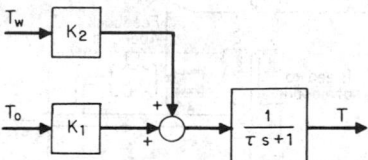

Fig. 16.2.14 Block diagram of thermal process.

Block-Diagram Algebra

The block diagram of a single-loop feedback-control system subjected to a command input $R(s)$ and a disturbance $U(s)$ is shown in Fig. 16.2.15.

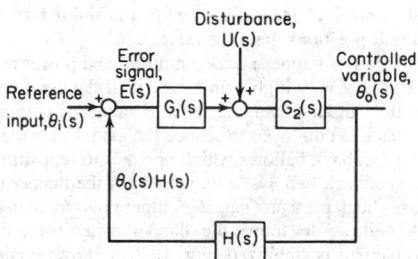

Fig. 16.2.15 Single-loop feedback control system.

When $U(s) = 0$ and the input is a reference change, the system may be reduced as follows:

$$E(s) = -\theta_o(s)H(s)$$
$$\theta_o(s) = E(s)[G_1(s)G_2(s)] \qquad (16.2.39)$$

Therefore
$$\frac{\theta_o(s)}{\theta_i(s)} = \frac{G_1(s)G(s)}{1 + G_1(s)G_2(s)H(s)}$$

When $\theta_i(s) = 0$ and the input is a disturbance, the system may be reduced as follows:

$$E(s) = -\theta_o(s)H(s)$$
$$[E(s)\,G_1(s) + U(s)]G_2(s) = \theta_o(s)$$
$$\frac{\theta_o(s)}{U(s)} = \frac{G_2(s)}{1 + G_1(s)G_2(s)H(s)} \qquad (16.2.40)$$

Closed-loop transfer function Eq. (16.2.40) can be determined by observation from

$$\frac{\theta_o(s)}{U(s)} = \frac{\text{feedforward functions}}{1 + \text{complete-loop functions}}$$

The denominator of Eq. (16.2.40) is the **characteristic equation** which determines **stability**. Most systems are designed on the basis of the characteristic equation since it sets both response time τ_c and damping ζ. The equation is undefined (unstable) if $G_1(s)G_2(s)H(s)$ equals -1. But -1 is a vector of magnitude 1 and a phase of $-180°$. This fact is used to determine stable parameter adjustments in the graphical techniques to be discussed later.

The input disturbance $[U(s)]$ can be any time function in actual operation. The **step input** is widely used for analysis and testing since it is easily implemented; it results in a simple Laplace transform; it is the most severe type of disturbance; and the response to a step change shows the maximum error that could occur.

In many complex control systems, especially in the nonmechanical process-control field, auxiliary feedback paths are provided in order to adjust the system's performance. Figure 16.2.16a illustrates such a condition. In analyzing such a system it is usually best to combine secondary loops into the main control loop to form an equivalent series block and transfer function. The system of Fig. 16.2.16a might be reduced in the following sequence.

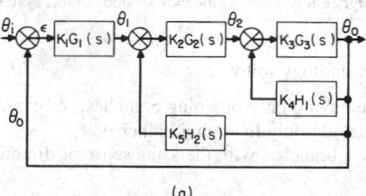

(a)

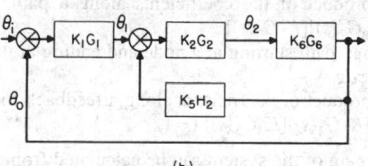

(b)

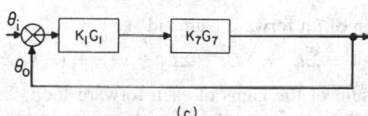

(c)

Fig. 16.2.16 Reduction of a closed-loop control system with multiple secondary loops.

1. Replace $K_3G_3(s)$ and $K_4H_1(s)$ with a single equivalent element

$$\frac{\theta_o}{\theta_2} = \frac{K_3G_3(s)}{1 + K_4H_1(s)K_3G_3(s)} = K_6G_6 \qquad (16.2.41)$$

The result of this first reduction is shown in Fig. 16.2.16b:

2. Figure 16.2.16b can be treated in a similar fashion and a single block used to replace K_2G_2, K_6G_6, and K_5H_2

$$\frac{\theta_o}{\theta_1} = \frac{K_2G_2K_6G_6(s)}{1 + K_5H_2K_2G_2K_6G_6(s)} = K_7G_7 \qquad (16.2.42)$$

The result of this second reduction is shown in Fig. 16.2.16c. The resulting open-loop transfer function is

$$\theta_o/\varepsilon = K_1 G_1 K_7 G_7(s) \qquad (16.2.43)$$

The closed-loop or frequency response function is

$$\frac{\theta_o}{\theta_i} = \frac{K_1 G_1 K_7 G_7(s)}{1 + K_1 G_1 K_7 G_7(s)} \qquad (16.2.44)$$

Equation (16.2.44) can, of course, be expanded to include the terms of the system's secondary loops.

SIGNAL-FLOW REPRESENTATION

An alternate graphical representation of the mathematical relationships is the signal-flow graph. For complicated systems it allows a more compact representation and more rapid reduction techniques than the block diagram.

In Fig. 16.2.17, the nodes represent the variables θ_i, ε, θ_1, ..., θ_o, and the branches the relationships between the nodes, of the system shown in Fig. 16.2.16. For example,

$$\theta_1(s) = \varepsilon K_1 G_1(s) - K_5 H_2(s) \theta_o(s) \qquad (16.2.45a)$$

and

$$\varepsilon = \theta_i - \theta_o \qquad (16.2.45b)$$

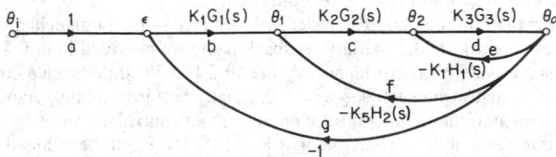

Fig. 16.2.17 Signal flow graph of the closed-loop control system shown in Fig. 16.2.16.

Signal-flow terminology follows:

Source: node having only outgoing branches, for example, θ_i
Sink: node having only incoming branches, θ_o
Path: series of branches with the same sense of direction, for example, *abcd*, *cdf*
Forward path: path originating at a source and ending at a sink, with no node encountered more than once, for example, *abcd*
Path gain: product of the coefficients along a path, for example, $1 [K_1 G_2(s)][K_2 G_2(s)][K_3 G_3(s)]$
Feedback loop: path starting at a node and ending at the same node, for example, *bcdg*
Loop gain: product of coefficients along a feedback loop, for example, $[K_1 G_2(s)][K_2 G_2(s)][K_3 G_3(s)] (-1)$

The overall gain of the system can be calculated from

$$G = \frac{\Sigma_i G_i \Delta_i}{\Delta} \qquad (16.2.46)$$

where G_i = gain of ith forward path and

$$\Delta = 1 - \Sigma L_1 + \Sigma L_2 - \Sigma L_3 + \cdots + (-1)^k \Sigma L_k$$

where ΣL_1 = sum of the gains of each forward loop; ΣL_2 = sum of products of loop gains for nontouching loops (no node is common), taken two at a time; ΣL_3 = sum of products of loop gains for nontouching loops taken three at a time; Δ_i = value of Δ for signal flow graph resulting when ith path is removed.

From Fig. 16.2.17 there is only one forward path, *abcd*.

$$\therefore G_1 = K_1 G_1(s) K_2 G_2(s) K_3 G_3(s)$$

Closed loops are *de*, *cdf*, and *bcdg*, with gains $-K_1 H_1(s) K_3 G_3(s)$, $-K_2 G_2(s) K_3 G_3(s) K_5 H_2(s)$, and $-K_1 G_1(s) K_2 G_2(s) K_3 G_3(s)$.

There are no nontouching closed loops;

$$\therefore \Delta = 1 + K_1 H_1(s) K_3 G_3(s) + K_2 G_2(s) K_3 G_3(s) K_5 H_2(s)$$
$$+ K_1 G_1(s) K_2 G_2(s) K_3 G_3(s)$$

There are no loops remaining if the forward path *abcd* is removed;

$$\therefore \Delta_1 = 1$$

Thus

$$\theta_o/\theta_1 = K_1 G_1(s) K_2 G_2(s) K_3 G_3/[1 + K_1 H_1(s) K_3 G_3(s)$$
$$+ K_2 G_2(s) K_3 G_3(s) K_5 H_2(s)$$
$$+ K_1 G_1(s) K_2 G_2(s) K_3 G_3(s)] \qquad (16.2.47)$$

which is identical with Eq. (16.2.44).

CONTROLLER MECHANISMS

The **controller** modifies the error signal in a desired manner to produce an output pressure which is used to actuate the valve motor. The several controller modes used singly or in combination are (1) the proportional mode in which $P_{out}(t) = K_c E(t)$, (2) the integral mode, in which $P_{out}(t) = 1/T_1 \int E(t)\, dt$, and (3) the rate mode, in which $P_{out}(t) = T_2 dE(t)/dt$. In these expressions $P_{out}(t)$ = controller output pressure, $E(t)$ = input error signal, K_c = proportional gain, $1/T_1$ = reset rate, and T_2 = rate time.

A **pneumatic controller** consists of an **error-detecting mechanism, control modes** made up of proportional (P), integral (I), and derivative (D) actions in almost any combination, and a **pneumatic amplifier** to provide output capacity. The error-detecting mechanism is a differential link, one end of which is positioned by the signal proportional to the controlled variable, and the other end of which is positioned to correspond to the command set point. The proportional action is provided by a flapper nozzle (Fig. 16.2.18), where the flapper is positioned by the error signal. A motion of 0.0015 in by the flapper is sufficient for nearly full output range. Nozzle back pressure is inversely proportional to the distance between nozzle opening and flapper.

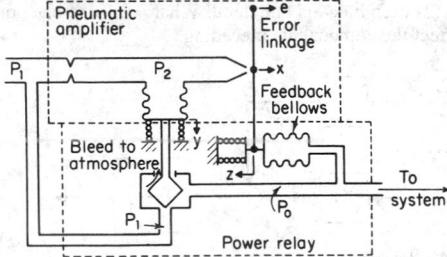

Fig. 16.2.18 Gain reduction of pneumatic amplifier by means of feedback bellows. (*Raven, "Automatic Control Engineering," McGraw-Hill.*)

The controller employs a power-amplifying pilot for providing a larger quantity of air than could be provided through the small restriction shown in Fig. 16.2.18. The nozzle back pressure, instead of operating the final control element directly, is transmitted to a bellows chamber where it positions the pilot valve.

The combination of flapper-nozzle amplifier and power relay shown in Fig. 16.2.18 has a very high gain since small flapper displacements can result in the output traversing the full range of output pressure. Negative feedback is employed to reduce the gain. Controller output is connected to a feedback bellows which operates to reposition the flapper. With the feedback bellows, a movement of the flapper toward the nozzle increases back pressure, causing output pressure to decrease and the feedback bellows to move the flapper away from the nozzle. Thus the mechanism is stabilized. Fig. 16.2.19 shows controller gain being varied by adjusting the point at which the feedback bellows bears on the flapper.

The **rate mode,** called "anticipatory," can take large corrective action when errors are small but have a high rate of change. The mode resists not only departures from the set point but also returns and so provides a stabilizing action. Since the rate mode cannot control to a set point, it is not used alone. When used with the proportional mode, its stabilizing influence (90° phase lead; see Table 16.2.3) may allow an increase in gain K_c and a consequent decrease in steady-state error.

Proportional-derivative (PD) control is obtained by adding an adjustable feedback restriction as shown in Fig. 16.2.20. This results in delayed negative feedback. The restriction delays and reduces the feedback, and, since the feedback is negative, the output pressure is momentarily higher and leads, instead of lags, the error signal.

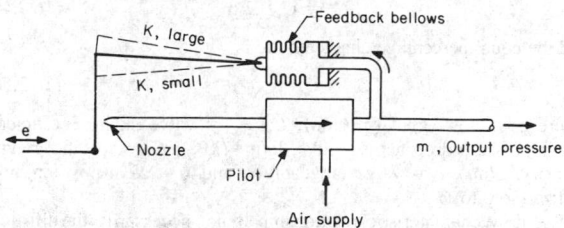

Fig. 16.2.19 Proportional controller with negative feedback. *(Reproduced by permission. Copyright© John Wiley & Sons, Inc. Publishers, 1958. From D. P. Eckman, "Automatic Process Control.")*

Since the proportional mode requires an error signal to change output pressure, set point and load changes in a proportionally controlled system are accompanied by a steady-state error inversely proportional to the gain. For systems which, because of stability considerations, cannot tolerate high gains, the integral mode added to the proportional will

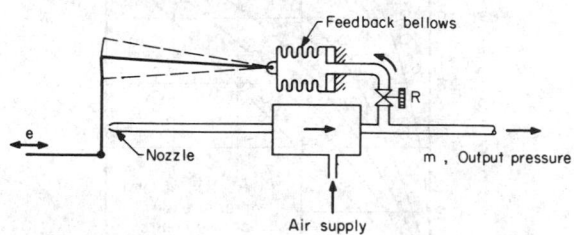

Fig. 16.2.20 Proportional-derivative controller. *(Reproduced by permission. Copyright© John Wiley & Sons, Inc. Publishers, 1958. From D. P. Eckman, "Automatic Process Control.")*

eliminate the steady-state error since the output from this mode is continually varying so long as an error exists. The addition of the integral mode to a proportional controller has an adverse effect on the relative stability of the control because of the 90° phase lag introduced.

Proportional-integral (PI) control is obtained by adding a positive feedback bellows and an adjustable restriction (Fig. 16.2.21). The addition of the positive feedback bellows cancels the gain reduction brought about by the negative feedback bellows at a rate determined by the adjustable restriction.

Electronic controllers which are analogous to the pneumatic controller have been developed. They have the advantages of elimination of time lags; compatability with computers; being less expensive to install (although more expensive to purchase); being more energy efficient; and being immune to low temperatures (water in pneumatic lines freezes).

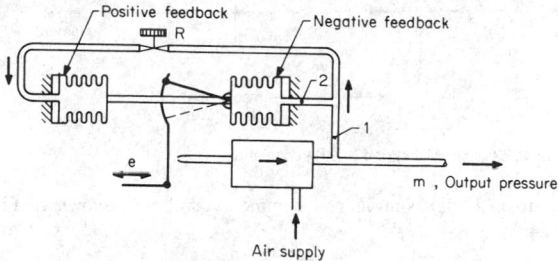

Fig. 16.2.21 Proportional-integral controller. *(Reproduced by permission. Copyright© John Wiley & Sons, Inc. Publishers, 1958. From D. P. Eckman, "Automatic Process Control.")*

The heart of the electronic controller is the **high-gain operational amplifier** with passive elements at the input and in the feedback. Figure 16.2.22 shows the classical control elements, though circuits in commercially available controllers are far more sophisticated. The current drawn by the dc amplifier is negligible, and the amplifier gain is around 1,000, so that the junction point is essentially at ground potential. The reset amplifier has delayed negative feedback, similar to the pneumatic circuit in Fig. 16.2.21. The derivative amplifier has advanced negative feedback, similar to Fig. 16.2.20. Gains are adjusted by changing the ratios of resistors or capacitors, while adjustable time constants are made up of a variable resistor and a capacitor (RC).

Regardless of actuation, the commonly applied **controller modes** are as follows:

Designation	Transform	Symbol
Proportional	K (gain)	P
Reset	$1 + \dfrac{1}{\tau_i s}$	I
Derivative	$\dfrac{\tau_d s + 1}{(t_d/\alpha)s + 1}$	D
Floating	$\dfrac{1}{\tau_F s}$	F

The modes are combined as needed into P, PI, PID, PD, and F. Choice depends on the application and can be made from the Bode diagram. The block diagram for a PID controller is shown in Fig. 16.2.23.

The **derivative mode** can be placed on the measured variable signal (as shown), on the error, or on the controller output. The arrangement of

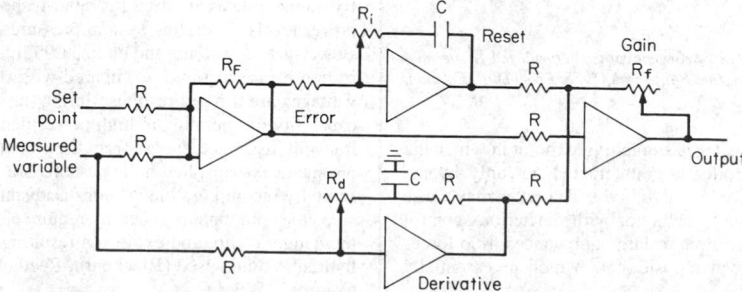

Fig. 16.2.22 Simplified circuit for a typical electronic controller.

Fig. 16.2.23 is preferred for regulation since the derivative does not affect set-point changes, and the derivative acts to reduce overshoot on start-up (Zoss, "Applied Instrumentation in the Process Industries," Vol. IV, Gulf Publishing Co.). Regardless of a sequence, the block

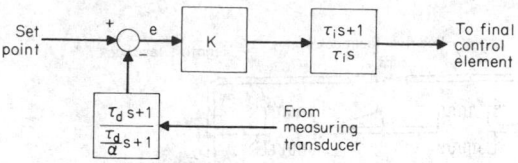

Fig. 16.2.23 Block diagram of PID controller.

diagram of a PID controller is commonly drawn as shown in Fig. 16.2.24.

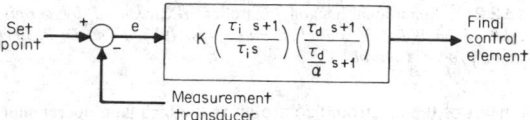

Fig. 16.2.24 Combined controller block diagram.

FINAL CONTROL ELEMENTS

The **final control element** is a mechanism which alters the value of the **manipulated variable** in response to an output signal from the automatic-control device. It will typically receive a signal from the controller and manipulate a flow of material or energy to the process. The final control element can be a control valve, an electrical motor, a servovalve, or a damper. Servovalves are discussed in the next section, "Hydraulic-Control Systems."

The final control element often consists of two parts: first, an **actuator** which translates the controller signal into a command for the manipulating device, and, second, a **mechanism** to adjust the manipulated variable. Figure 16.2.25 shows a control valve made up of an actuator and an adjustable orifice.

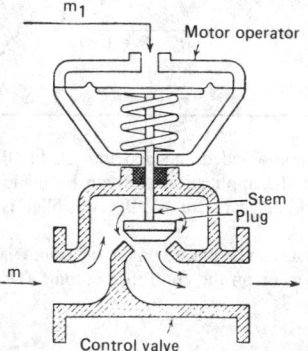

Fig. 16.2.25 Control valve and pneumatic actuator. *(Reproduced by permission. Copyright© John Wiley & Sons, Inc. Publishers, 1958. From D. P. Eckman, "Automatic Process Control.")*

The most commonly used **actuator** is a diaphragm motor in which the output pressure from the controller is counteracted not only by the spring but also by fluid forces in the valve body. The latter may cause serious deviation from linear static behavior with deleterious control effects. High friction at the valve stem or large unbalanced fluid forces at the plug can be overcome by valve positioners, which are essentially proportional controllers.

A **control valve** in liquid service is described by Eq. (16.2.48):

$$q = C_v \sqrt{\frac{\Delta p}{G}} \qquad (16.2.48)$$

And C_v is described by its relationship to stem lift x, which is called the **valve-flow characteristic.**

The linear characteristic is

$$C_v = C_v|_{max} \, x \qquad (16.2.49)$$

and the equal-percentage characteristic is

$$C_v = C_v|_{max} \, (r)^{x-1} \qquad (16.2.50)$$

where q = flow, gpm (m^3/min); $C_v|_{max}$ = valve sizing coefficient; Δp = pressure drop across valve, lb/in^2 (kPa); G = specific gravity (density, g/ml); r = valve rangeability; and x = stem position, normalized to 0 to 1.

The flow characteristic relationship is not necessarily the lift-flow characteristic of the valve when installed since the valve is but one component in a piping system in which pressure drops vary with the flow rate. The change in characteristic as less percentage of total system drop is taken across the control valve is shown in Fig. 16.2.26.

Valve characteristics are generally selected according to their ability to compensate for nonlinearities in the system. Nonlinearities typically represent a change in gain which in turn changes the character of the control response for a given controller setting when load- or set-point

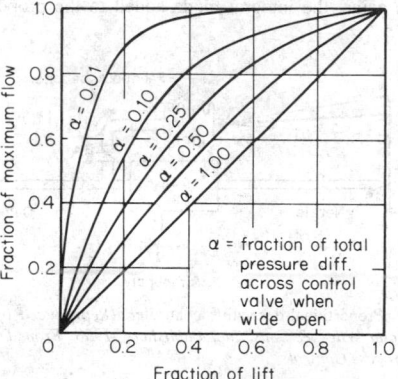

Fig. 16.2.26 Control valve linear flow characteristic. *(Reproduced by permission. Copyright© John Wiley & Sons, Inc. Publishers, 1958. From D. P. Eckman, "Automatic Process Control.")*

changes occur or when there is a variable overall pressure drop across the system. Equal-percentage valves, for example, tend to control over widely varying operating conditions, since the change in flow is always proportional to flow rate. The selection of the proper valve therefore depends on study of the particular system.

HYDRAULIC-CONTROL SYSTEMS

Hydraulic systems are used for rapid-response servomechanisms at high power levels. Operating system pressures are from 50 to 100 lb/in^2 for slower-acting systems and up to 5,000 lb/in^2 where lightweight and fast responses are required. Compared with **electrical systems** the major advantages are a rapid response in the large horsepower ranges and the capability of operating at high power-density levels since the fluid can transmit dissipated energy from the point of generation. Compared with **pneumatic systems,** hydraulic systems are faster because the fluid is essentially incompressible. Major disadvantages are vulnerability to dirt, since the components generally require close machining tolerances, and the danger of fire and explosion resulting from the flammability of the hydraulic fluids used (Blackburn, Reethof, and Shearer, "Fluid Power Control," Wiley).

The direction and volume of flow are controlled by **servo valves** in the system. They may be single-stage (pilot-operated) and mechanically or electrically actuated. A schematic of a spool-type four-way single-stage control piston and inertia load is shown in Fig. 16.2.27. Hydraulic fluid at constant pressure enters at the supply port. With displacement of the spool valve downward, for example, inflow to the top side of the piston

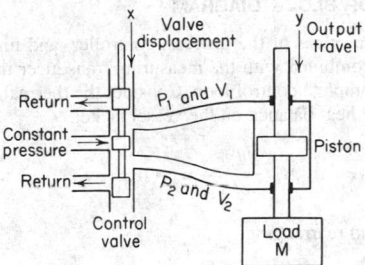

Fig. 16.2.27 Four-way valve-piston circuit. (*Truxal, "Control Engineer's Handbook," McGraw-Hill.*)

moves the piston downward. Because of machining tolerances the spool dimensions are either large (overlapped) or smaller (underlapped) than the port dimension. Underlapped valves permit leakage to the piston in the centered position; overlapped valves result in a dead zone, where motion x results in no flow until a port is opened.

The transfer function of this circuit is given as (Truxal, "Control Engineers' Handbook")

$$\frac{y}{x} = \frac{C_1 \dfrac{1}{1 + \alpha(C_1/C_2)}}{s\left[\dfrac{VM}{2BA^2}\dfrac{1}{1 + \alpha(C_1/C_2)}s^2 + \dfrac{C_1 m/C_2 + V\alpha/2BA^2}{1 + \alpha(C_1/C_2)}s + 1\right]}$$

where C_1 = servo velocity gradient, in/(s)(in), C_2 = servo force gradient, lb/in, α = viscous friction of load and piston, lb/(in)(s), B = bulk modulus of fluid, lb/in² M = mass of load and piston, lb/(in)(s²), A = piston area, in², and V = effective entrained fluid volume, in³ (one-half of total entrained volume between valve and piston).

The velocity of the output is proportional to the input resulting in a **velocity-control** servo. To convert this system to a **position-control** servo, mechanical, hydraulic, or electrical feedback may be employed. A valve-piston position servo with **mechanical feedback** is shown in Fig. 16.2.28. Any difference between the input D and the piston position y causes a motion x, which causes the piston to move in a direction opposite to D, that is, in a direction to reduce x. The lever ratio establishes the relationship between y and D.

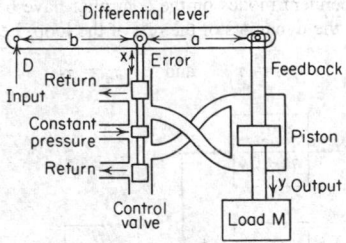

Fig. 16.2.28 Valve-piston position servomechanical feedback. (*Truxal, "Control Engineer's Handbook," McGraw-Hill.*)

Most commercially available servo valves are two stages, permitting electrohydraulic action. The pilot stage can be operated by a low-power, short-travel electrical device, with a concomitant increase in flexibility. A typical pilot-operated servo valve is shown in Fig. 16.2.29. In this case the pilot is a double-flapper valve rather than a spool valve. (In

general, small, accurate low-leakage spool valves are costly.) Upward movement of the flapper by the actuating motor results in increased pressure to the right end of the power spool. Hydraulic feedback occurs because of the increased flow across restrictor a. The power spool moves to the left until the unbalanced pressure is matched by the spring resistance. The disadvantage of this valve is the continual leakage flow through the flapper nozzle, but the torque motor has a low-power requirement and is inexpensive.

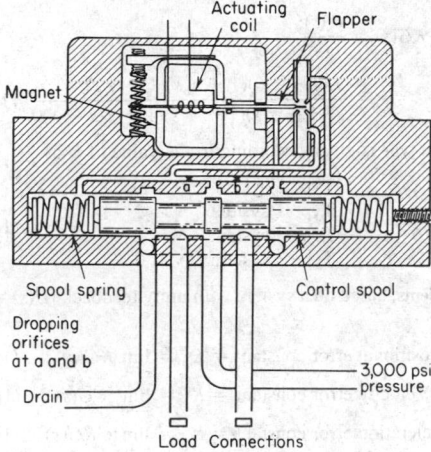

Fig. 16.2.29 Two-stage electrohydraulic servo valve. The first-stage is a four-way flapper valve with a calibrated pressure output driving a second-stage, spring-loaded, four-way spool valve. (*Moog Servocontrols, Inc.*)

The major disadvantages of spool-type valves are (1) high cost, because of high-tolerance requirements between the valve lands, (2) high static friction and inertia, and (3) susceptibility to dirt. The **flapper** valve is less expensive to manufacture than a spool valve of equivalent characteristics and is not so susceptible to damage by dirt particles. In Fig. 16.2.30, P_1, the supply pressure, is constant. Input motion of the flapper toward the nozzle increases P_2 and drives the piston toward the right. The steady-state characteristic P_2 vs. x is shown in Fig. 16.2.31.

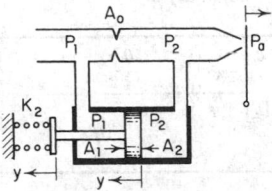

Fig. 16.2.30 Flapper valve. (*Raven.*)

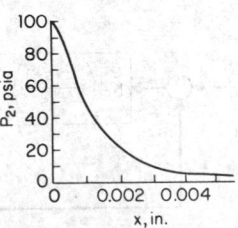

Fig. 16.2.31 Equilibrium curve of P_2 versus x for a flapper valve. (*Raven, "Automatic Control Engineering," McGraw-Hill.*)

STEADY-STATE PERFORMANCE

The steady-state error of a control system can be determined by using the final value theorem (see Laplace Transforms, Sec. 2) in which

$$\lim_{t\to\infty} \theta_o(t) = \lim_{s\to 0} s\,\theta_o(s) \quad \text{provided } \theta_o(t) \text{ is stable}$$

The classification of control systems according to the form of the open-loop transfer function facilitates the determination of the steady-state errors when the system is subjected to various inputs.

The open-loop transfer function $\theta_o(s)/\varepsilon(s) = KG(s)$ may be written

$$
\begin{aligned}
KG(s) &= \frac{\theta_o(s)}{\varepsilon(s)} \\
&= \frac{K(1 + \tau_a s)(1 + \tau_a s)(1 + \tau_c s)\;\cdots}{s^N(1 + \tau_1 s)(1 + \tau_2 s)(1 + \tau_3 s)\;\cdots}
\end{aligned}
\quad (16.2.51a)
$$

The **system type** is given according to the value of N as

$$
\begin{aligned}
\text{Type 0 system} &\quad N = 0 \\
\text{Type 1 system} &\quad N = 1 \\
\text{Type 2 system} &\quad N = 2 \\
\text{Type 3 system} &\quad N = 3
\end{aligned}
\quad (16.2.51b)
$$

Error coefficients, based on a system with unity feedback [$H(s) = 1$], are defined as

$$\text{Positional error constant} = K_o = \lim_{s\to 0} KG(s) \quad (16.2.52a)$$

$$\text{Velocity error constant} = K_v = \lim_{s\to 0} sKG(s) \quad (16.2.52b)$$

$$\text{Acceleration error constant} = K_a = \lim_{s\to 0} s^2 KG(s) \quad (16.2.52c)$$

A summary of error coefficients for systems of different types is given in Table 16.2.1.

Table 16.2.1 Summary of Error Coefficients

N	K_o	K_v	K_a
0	const	0	0
1	∞	const	0
2	∞	∞	const

A summary of the errors for types 0, 1, and 2 systems, when subjected to various inputs, is given in Table 16.2.2.

Table 16.2.2 Summary of Errors

Input error	$\theta_i(t) = A$ ε_0/A	$\theta_i(t) = vt$ ε_v/v	$\theta_i(t) = at^2$ ε_a/a
$N = 0$	$1/(1 + K_0)$	∞	∞
1	0	$1/K_v$	∞
2	0	0	$1/K_a$

The higher the system type, the better is the output able to follow the higher degrees of input. Higher-type systems, however, are more difficult to stabilize, and a compromise must be made between the steady-state error and the settling time of the response.

CLOSED-LOOP BLOCK DIAGRAM

The **transfer functions** of the process, controller, and final control element can be combined with the measuring transducer into a **block diagram** of the complete control loop. Consider the thermal system of Fig. 16.2.5 and the heat balance on the vessel jacket:

$$M_j c \frac{dT_w}{dt} = w_w c(T_{w1} - T_w) \quad (16.2.53)$$

Linearizing and rearranging:

$$\frac{M_j}{w_w} T_w s + T_w = \left[\frac{(T_{w1} - T_w)}{w_w}\right] \left[\frac{w_w}{x}\right] x \quad (16.2.54)$$

where M_j = weight of cooling water in jacket, lb (kg); c = specific heat of water, Btu/lb $\cdot$ °F (J/kg $\cdot$ °C); T = temperature, °F; x = valve stem position (normalized 0 to 1); and w_w = water flow, lb/min (kg/min).

For simplicity, let the jacket time constant M_j/w_w be so small as to be negligible (not typically the case). The block diagram of Fig. 16.2.32 then represents process diagram Fig. 16.2.5. The **gain** K in each case is an incremental change in output divided by an incremental change in input. For example, if the temperature transmitter has a temperature span of 100°F and an output of 4 to 20 mA, the gain would be

$$K_\tau = \frac{20 - 4}{100} = \frac{16}{100} = 0.16 \text{ mA/°F}$$

The **open-loop gain** is defined as the product of all the gains in the control loop:

$$K_o = K_c K_v K_T \left(\frac{T_{w1} - T_w}{x}\right) \left(\frac{UA}{wc + UA}\right)$$

Combining the **closed-loop system** of Fig. 16.2.32 results in the closed-loop equation

$$T = \frac{\left(\dfrac{wc}{wc + UA}\right)\left(\dfrac{1}{\tau s + 1}\right) T_0}{1 + \dfrac{K_o(\tau_i s + 1)(\tau_d s + 1)}{\tau_i s(\tau s + 1)\left(\dfrac{\tau_d}{\alpha}s + 1\right)(\tau_T s + 1)}} \quad (16.2.55)$$

The time-dependent modes on the controller have been designed to compensate for the dynamics of the rest of the loop. Let

$$\tau_i = \tau \quad \text{and} \quad \tau_d = \tau_T$$

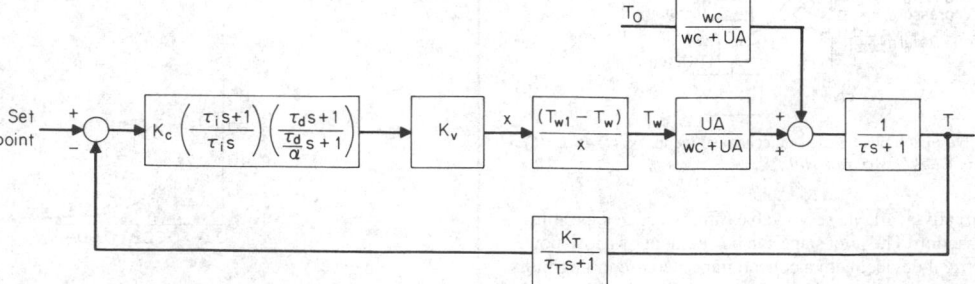

Fig. 16.2.32 Block diagram of a thermal system.

Table 16.2.3   Process Characteristics versus Mode of Control

| Number of process capacities | Process reaction rate | Process time lags | | Load changes | | Suitable mode of control |
		Resistance-capacitance (RC)	Dead time (transportation)	Size	Speed	
Single	Slow	Moderate to large	Small	Any	Any	Two-position. Two-position with differential gap
				Moderate	Slow	Multiposition. Proportional input
Single (self-regulating)	Fast	Small	Small	Any	Slow	Floating modes: Single-speed Multispeed
					Moderate	Proportional-speed floating
Multiple	Slow to moderate	Moderate	Small	Small	Moderate	Proportional position
Multiple	Moderate	Any	Small	Small	Any	Proportional plus rate
Multiple	Any	Any	Small to moderate	Large	Slow to moderate	Proportional plus reset
Multiple	Any	Any	Small	Large	Fast	Proportional plus reset plus rate
Any	Faster than that of the control system	Small or nearly zero	Small to moderate	Any	Any	Wideband proportional plus fast reset

SOURCE: Considine, "Process Instruments and Controls Handbook," McGraw-Hill.

After cancellation, Eq. (16.2.55) becomes

$$T = \frac{\left(\dfrac{wc}{wc + UA}\right) T_0}{(\tau s + 1)\left[1 + \dfrac{K_c}{\tau_i s \left(\dfrac{\tau_d}{\alpha} s + 1\right)}\right]} \qquad (16.2.56)$$

Equation (16.2.56) can be multiplied out to become

$$T = \frac{\left(\dfrac{wc}{wc + UA}\right)(\tau_i s)\left(\dfrac{\tau_d}{\alpha} s + 1\right) T_0}{(\tau s + 1)\left(\dfrac{\tau_i \tau_d}{\alpha} s^2 + \tau_i s + K_c\right)} \qquad (16.2.57)$$

Equation (16.2.57) can be written in the standard form of Eqs. (16.2.16) and (16.2.21):

$$T = \frac{\left(\dfrac{wc}{wc + UA}\right)(\tau_i s)\left(\dfrac{\tau_d}{\alpha} s + 1\right) T_0}{K_c(\tau s + 1)(\tau_c^2 s^2 + 2\tau_c \zeta s + 1)} \qquad (16.2.58)$$

Controller gain K_c can then be chosen to give the best response τ_c with stability ζ. More rigorous techniques for determining controller parameter values are shown under Bode diagrams and Routh's criterion. Table 16.2.3 provides preliminary guidance for the selection of control modes.

FREQUENCY RESPONSE

Although it is the time response of the control system that is of major importance, study of the effect on transient response of changes in system parameters, either in the process or controller, is more conveniently made from a **frequency-response** analysis of the system. The **frequency response** of a system is the steady-state output of the system to input sinusoids of varying frequency. The output for a linear system can be completely described in terms of the amplitude ratio of the output sinusoid to the input sinusoid. The amplitude ratio or gain, and phase, are functions of the frequency of the input sinusoid.

The use of **sinusoidal methods** to analyze and test dynamic systems has gained widespread popularity because system response is obtained easily from the response of the individual elements, no matter how many

elements are included. By contrast, transient analysis is quite tedious, with only three dynamic components in the system, and is too difficult to be worthwhile for four or more components.

Frequency-response analysis provides quantitative information concerning the system on maximum gain K_o for stability, time response τ_c, ω_n, and the controller parameter adjustments τ_i, τ_d. In most cases, this information is adequate for assurance of stability, for comparing alternatives, or for judging the merits of a proposed control system. System parameters are generally obtained from open-loop frequency response, which is the response with the loop broken between the controller and the final control element (Fig. 16.2.32).

Frequency response also lends itself to system identification by testing for systems not readily amenable to mathematical analysis by subjecting the system to input sinusoids of varying frequency.

The shift from the **complex domain** to the **frequency domain** (linear systems, zero initial conditions) is accomplished by simply substituting $j\omega$ for s. Consider Eq. (16.2.38).

$$(\tau s + 1)T = K_1 T_0 \qquad (16.2.59)$$

Substituting

$$\frac{T}{T_0} = \frac{K_1}{\tau j\omega + 1}$$

Multiply by the complex conjugate

$$\frac{T}{T_0} = \frac{K_1}{\tau j\omega + 1}\frac{\tau j\omega - 1}{\tau j\omega - 1}$$

Collecting terms

$$\frac{T}{T_0} = K_1\left[\frac{1}{\tau^2\omega^2 + 1} - j\frac{\tau\omega}{\tau^2\omega^2 + 1}\right] \qquad (16.2.60)$$

The first term is the real part of the solution, and the second term, the imaginary part. These terms define a vector, shown in Fig. 16.2.33, which has magnitude M and phase angle θ as determined either graphically or by the Pythagorean theorem.

$$M = \left[\frac{1}{\tau^2\omega^2 + 1}\right]^{1/2} \qquad (16.2.61)$$

$$\theta = \tan^{-1}\left(\frac{-\zeta\omega}{1}\right) \qquad (16.2.62)$$

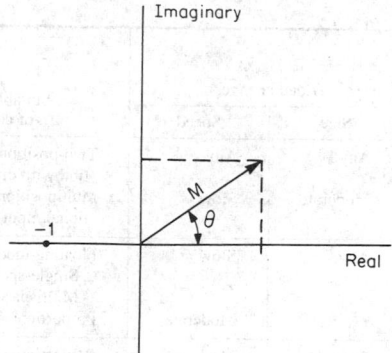

Fig. 16.2.33 Polar plot showing vector.

GRAPHICAL DISPLAY OF FREQUENCY RESPONSE

Sinusoidal response is **plotted** in three different ways: (1) the rectangular plot with the log of the amplitude versus the log of the frequency, called a **Bode plot;** (2) a phase-margin plot with magnitude shown versus a function of phase with frequency as a parameter, called a **Nichols plot;** and (3) a polar plot with magnitude and phase shown in vector form with frequency as a parameter, called a **Nyquist plot.** The Nichols plot is actually a plot of phase margin $(180° - \theta)$ versus frequency and is used to define system performance. The polar plot enables the absolute stability of the system to be determined without the need for obtaining the roots of the characteristic equation. The logarithmic (Bode) plot has the advantage of ease in plotting, especially in design, since the individual effects of cascaded elements can be gaged by superposition. All the graphical procedures use the ''minus 1'' point [discussed with Eq. (16.2.38)] as the criterion for stability (magnitude = 1, phase = $-180°$).

NYQUIST PLOT

The **Nyquist diagram** is a graphical method of determining stability. The diagram is essentially a mapping into the $G(s)$ plane of a contour enclosing the major portion of the right half of the s plane. Figure 16.2.34 shows the Nyquist plot of a typical system.

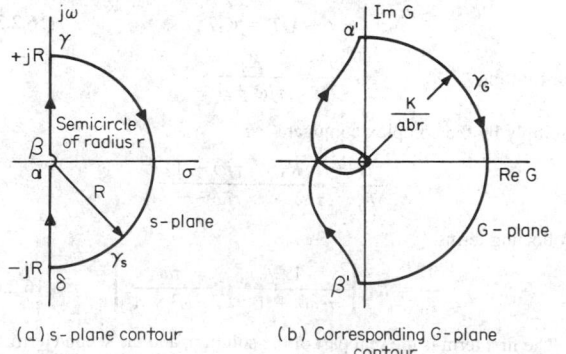

(a) s-plane contour
(b) Corresponding G-plane contour

Fig. 16.2.34 Nyquist diagram for $G(s) = 1/[s(s + a)(s + b)]$. *(Truxal, ''Control System Synthesis,'' McGraw-Hill.)*

The frequency response of a closed-loop system can also be derived from the **polar plot** of the direct transfer function $KG(s)$. Systems with dynamic elements in the feedback can be transformed into direct systems by block-diagram manipulation. In Fig. 16.2.35, the amplitude ratio θ_o/θ_i at any frequency ω is the ratio of the lengths of the vectors $O\beta$

and $\alpha\beta$. The angle formed by the vectors $\alpha\beta$ and $O\beta$ is the phase angle of the frequency response $\angle \theta_o/\theta_i(j\omega)$. The numerator of the transfer function is a constant representing the closed-loop gain. Changing the gain proportionally changes the length of the vector $O\beta$. Figure 16.2.36 shows another typical Nyquist plot.

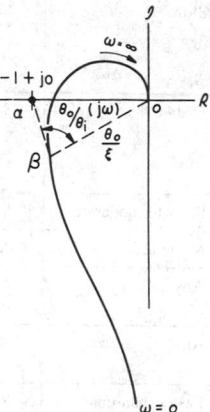

Fig. 16.2.35 Typical $(\theta_o/\varepsilon)/(j\omega)$ plot.

The Nyquist diagram is widely used in the design of electrical systems, but plots are very tedious to make and are generated via computer programs.

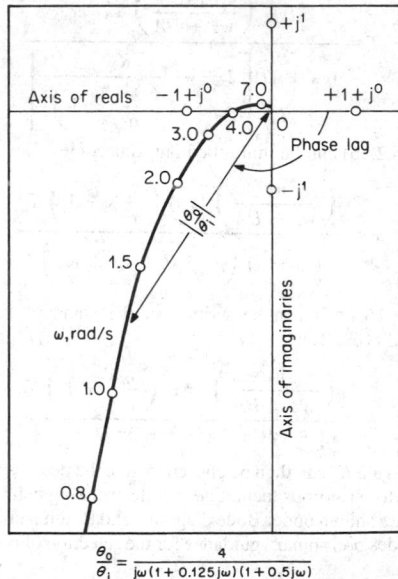

$$\frac{\theta_o}{\theta_i} = \frac{4}{j\omega(1 + 0.125\,j\omega)(1 + 0.5\,j\omega)}$$

Fig. 16.2.36 Typical Nyquist diagram showing loop transfer function $|\theta_o/\theta_i|$. *(ASME, ''Terminology for Automatic Control,'' Pub. ASA C85.1-1963.)*

BODE DIAGRAM

The most common method of presenting the **response data at various frequencies** is to use the log-log plot for amplitude ratios, accompanied by a semilog plot for phase angles. These rectangular plots are called **Bode diagrams,** after H. W. Bode, who did basic work on the theory of feedback amplifiers. A major advantage of this method is that the numerical computation of points on the curve is simplified by the fact that log magnitude versus log frequency can be approximated to engineering accuracy with straight-line approximations. In the **sinusoidal testing** of a system, the system dynamics will be faster than a very low frequency test signal, and the output amplitude has time to recover fully to the input amplitude. Thus the magnitude ratio is 1. Conversely, at high frequency, the system does not have time to reach equilibrium, and output amplitude will decrease. Magnitude ratio goes to a very small value under this condition. Thus the system will show a decrease in magnitude ratio as frequency increases. By similar reasoning, the difference between input and output phase angles becomes progressively more negative as frequency increases.

Consider the equations for magnitude, Eq. (16.2.61), and phase, Eq. (16.2.62). If $\tau_w \ll 1$,

$$M = \left(\frac{1}{1}\right)^{1/2} = 1 \qquad \theta = \tan^{-1}(0) = 0$$

and if $\tau_w \gg 1$,

$$M = \left(\frac{1}{\tau^2\omega^2}\right)^{1/2} = \frac{1}{\tau\omega} \approx 0 \qquad \theta = \tan^{-1}\left(\frac{-\tau_w}{1}\right) \approx -90°$$

The two **straight-line asymptotes** of magnitude, if extended, would intersect at $\tau\omega = 1$. Thus a relationship between frequency and time constant is established. The point at which $\tau\omega = 1$ is called the *corner frequency*, or the *break frequency*. The asymptote past the break frequency is easily plotted from a point at the break frequency $\omega = 1/\tau$ and a point at 10 times the frequency and 0.1 times the magnitude ratio.

The frequency response of systems containing different dynamic elements may be calculated by suitably employing the magnitude and phase characteristics of each element separately. Combined magnitudes are calculated

$$|M_1 M_2| = |M_1| \times |M_2|$$

Or, on the log-log Bode diagram,

$$\log |M_1 M_2| = \log |M_1| + \log |M_2|$$

The logarithmic scale permits **multiplication** by adding vertical distances. For phase

$$\underline{/\theta_1 \theta_2} = \angle\theta_1 + \angle\theta_2$$

Thus **phase addition** is also made by adding vertical distances. Figure 16.2.37 shows the combination of the straight-line asymptotes for the following transfer function:

$$G(s) = \frac{1}{\tau s(\tau s + 1)} \qquad (16.2.63)$$

Table 16.2.4 lists additional frequency-response relationships for system elements. The noninteracting controller, item 8, has the advantage of no interaction among tuning parameters but has the disadvantages of being more difficult to plot on the Bode diagram and of not offering the direct compensation of the controller shown in Fig. 16.2.24.

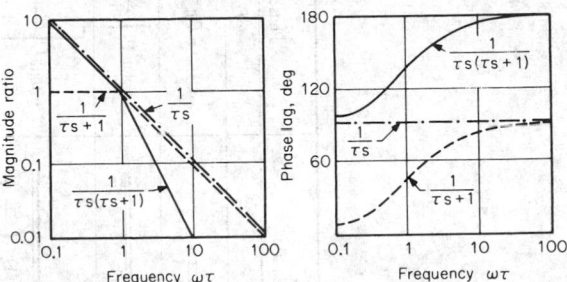

Fig. 16.2.37 Combined transfer functions. *(Reproduced by permission. Copyright© John Wiley & Sons, Inc. Publishers, 1958. From D. P. Eckman, "Automatic Process Control.")*

Figure 16.2.38 shows a typical combined Bode diagram. The ordinate has a scale in decibels in addition to the magnitude ratio scale. Decibels (dB) are sometimes used to provide a linear scale. They are defined

$$dB = -20 \log M$$

The discontinuity at the break frequency due to the meeting of the straight-line asymptotes will not be found in actual test data. Instead, the corner will be rounded, as shown dotted in Fig. 16.2.39. The error due to the approximation is some 3 dB at the break frequency, as shown in the figure.

Quadratic terms (Table 16.2.4) also lend themselves to asymptotic approximation, but the form of the magnitude and phase plots around the break frequency depends on the damping coefficient ζ. The slope of the magnitude ratio attenuation past the break point is twice that of the first-order system.

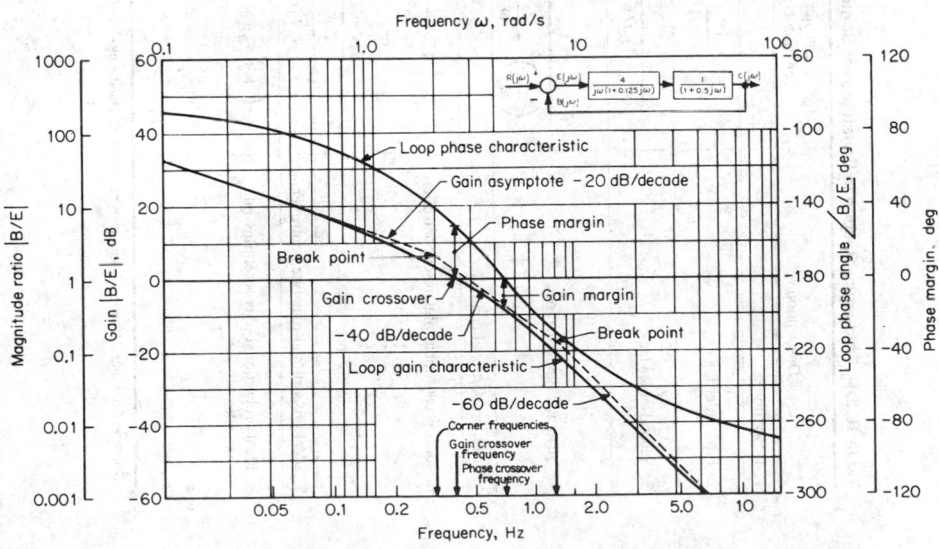

Fig. 16.2.38 Typical Bode diagram showing loop-transfer function *B/E*. (ASME, *"Terminology for Automatic Control,"* Pub. ASA C85.1-1963.)

Table 16.2.4 Frequency-Response Equations for Some Common Control-System Elements

Description	Transfer function $G(s)$	Frequency response $G(j\omega)$	Magnitude ratio	Phase angle
1. Dead time	$\varepsilon^{-T_L s}$	$\varepsilon^{-j\omega T_L}$	1	$-\omega T_L$ radians
2. First-order lag	$\dfrac{1}{Ts+1}$	$\dfrac{1}{j\omega T+1}$	$\dfrac{1}{\sqrt{\omega^2 T^2+1}}$	$-\tan^{-1}(\omega T)$
3. Second-order lag	$\dfrac{1}{(Ts+1)(aTs+1)}$	$\dfrac{1}{-a\omega^2 T^2+j(1+a)\omega T+1}$	$\dfrac{1}{\sqrt{(1-a\omega^2 T^2)^2+(1+a)^2\omega^2 T^2}}$	$-\tan^{-1}\left[\dfrac{(1+a)\omega T}{1-aT^2\omega^2}\right]$
4. Quadratic (underdamped)	$\dfrac{1}{\left(\dfrac{s}{\omega_n}\right)^2+\dfrac{2\zeta}{\omega_n}s+1}$	$\dfrac{1}{-\left(\dfrac{\omega}{\omega_n}\right)^2+j2\zeta\dfrac{\omega}{\omega_n}+1}$	$\dfrac{1}{\sqrt{\left(1-\dfrac{\omega^2}{\omega_n^2}\right)^2+4\zeta^2\left(\dfrac{\omega}{\omega_n}\right)^2}}$	$-\tan^{-1}\left[\dfrac{2\zeta\dfrac{\omega}{\omega_n}}{1-\left(\dfrac{\omega}{\omega_n}\right)^2}\right]$
5. Ideal proportional controller	K	K	K	0
6. Ideal proportional-plus-reset controller $T_i=\dfrac{1}{r}$ r = reset rate	$K\left(1+\dfrac{1}{T_i s}\right)$ or $K\dfrac{T_i s+1}{T_i s}$	$K\left(1+\dfrac{1}{j\omega T_i}\right)$ or $K\dfrac{j\omega T_i+1}{j\omega T_i}$	$K\sqrt{1+\left(\dfrac{1}{\omega T_i}\right)^2}$	$-\tan^{-1}\left(\dfrac{1}{\omega T_i}\right)$
7. Ideal proportional-plus-rate controller	$K(1+T_d s)$	$K(1+j\omega T_d)$	$K\sqrt{1+\omega^2 T_d^2}$	$\tan^{-1}(\omega T_d)$
8. Ideal proportional-plus-reset-plus-rate controller	$K\left(1+T_d s+\dfrac{1}{T_i s}\right)$	$K\left(1+j\omega T_d+\dfrac{1}{j\omega T_i}\right)$ or $K\dfrac{j\omega T_i-\omega^2 T_d T_i+1}{j\omega T_i}$	$K\sqrt{(\omega T_i)^2+(1-\omega^2 T_d T_i)^2}$	$\tan^{-1}\left(\omega T_d-\dfrac{1}{\omega T_i}\right)$

SOURCE: Considine, "Process Instruments and Controls Handbook," McGraw-Hill.

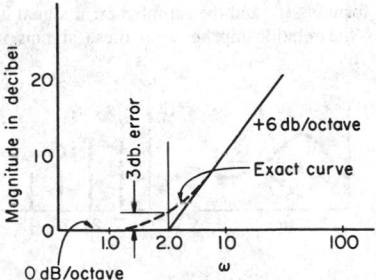

Fig. 16.2.39 Bode plot of term $(j\omega\tau_a + 1)$.

CONTROLLERS ON THE BODE PLOT

Controller modes, gain, reset, and derivative are plotted individually (in the form of Fig. 16.2.24) on the Bode diagram and summed as required. The functions for reset (integral) and derivative modes are moved horizontally on the diagram to determine proper parameter adjustment. The combined magnitude ratio curve for the whole system is moved vertically to determine open-loop gain.

After all the dynamic elements of the system are plotted and combined, reset and derivative are set in accordance with the following **criteria:**

Reset: The reset phase curve is positioned horizontally so that it contributes $10°$ of phase lag where the system phase lag is $170°$.

Derivative: The derivative phase curve is positioned horizontally so that it contributes $60°$ of phase lead where the system phase lag is $180°$.

The magnitude ratio curves are positioned with the phase-angle curves, and parameter values for reset and derivative can be read at the break points.

STABILITY AND PERFORMANCE OF AN AUTOMATIC CONTROL

An automatic-control system is stable if the amplitude of transient oscillations decreases with time and the system reaches a steady state. The stability of a system can be evaluated by examining the roots of the differential equation describing the system. The presence of positive real roots or complex roots with positive real parts dictates an unstable system. Any stability test utilizing the open-loop transfer function or its plot must utilize this fact as the basis of the test.

The Nyquist Stability Criterion The $KG(j\omega)$ locus for a typical single-loop automatic-control system plotted for all positive and negative frequencies is shown in Fig. 16.2.40. The locus for negative values of ω is the mirror image of the positive ω locus in the real axis. To complete the diagram, a semicircle (or full circle if the locus approaches $-\infty$ on the real axis) of infinite radius is assumed to connect in a positive sense, the $+$ locus at $\omega \to 0$ with the negative locus at $\omega \to -0$. If this locus is traced in a positive sense from $\omega \to \infty$ to $\omega \to 0$, around the circle at ∞, and then along the negative-frequency locus the following may be concluded: (1) if the locus **does not** enclose the $-1 + j0$ point, the system is stable; (2) if the locus **does** enclose the $-1 + j0$ point, the system is unstable. The Nyquist criterion can also be applied to the log magnitude of $KG(j\omega)$ and phase-vs.-log ω diagrams. In this method of display, the criterion for stability reduces to the requirement that the log magnitude of $KG(j\omega)$ must cross the 0-dB axis at a frequency less than the frequency at which the phase curve crosses the $-180°$ line. Two stability conditions are illustrated in Fig. 16.2.41. The Nyquist criterion not only provides a simple test for the stability of an automatic-control system but also indicates the degree of stability of the system by indicating the degree to which $KG(j\omega)$ locus avoids the $-1 + j0$ point.

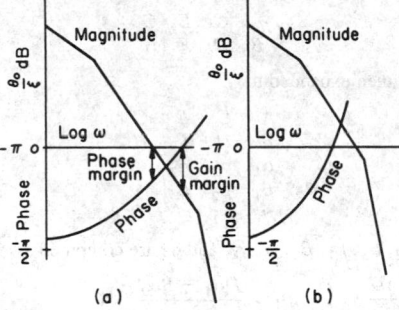

Fig. 16.2.41 Nyquist stability criterion in terms of log magnitude $KG(j\omega)$ diagrams. (a) Stable; (b) unstable. (Porter, "An Introduction to Servomechanism," Wiley.)

The concepts of **phase margin** and **gain margin** are employed to give this quantitative indication of the degree of stability of an automatic-control system. **Phase margin** is defined as the additional negative phase shift necessary to make the phase angle of the transfer function $-180°$ at the frequency where the magnitude of the $KG(j\omega)$ vector is unity. Physically, phase margin can be interpreted as the amount by which the unity KG vector has to be shifted to make a stable system unstable.

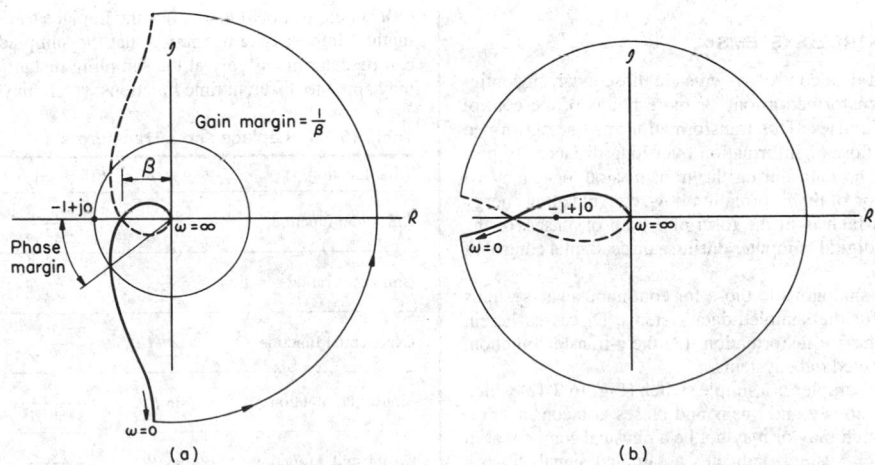

Fig. 16.2.40 Typical $KG(j\omega)$ loci illustrating application of Nyquist's stability criterion. (a) Stable; (b) unstable.

In a similar manner, **gain margin** is defined as the reciprocal of the magnitude of the KG vector at $-180°$. Physically, gain margin is the number by which the gain must be multiplied to put the system to the limit of stability. Thaler suggests satisfactory results can be obtained in most control applications if the phase margin is between 40 and 60° while the gain margin is between 3 and 10 (10 to 20 dB). These values will ensure a small transient overshoot with a single cycle in the transient. The margin concepts are qualitatively illustrated in Figs. 16.2.40 and 16.2.41.

Routh's Stability Criterion The frequency-response equation of a closed-loop automatic control is

$$\frac{\theta_o}{\theta_i} = \frac{KG(j\omega)}{1 + KG(j\omega)} \qquad (16.2.64)$$

The characteristic equation obtained therefrom has the algebraic form

$$A(j\omega)^n + B(j\omega)^{n-1} + C(j\omega)^{n-2} + \cdots = 0 \qquad (16.2.65)$$

The purpose of Routh's method is to determine the existence of roots of this equation which are positive or which are complex with positive real parts and thus identify the resulting instability. To apply the criterion the coefficients are written alternately in two rows as

$$\begin{matrix} A & C & E & G \\ B & D & F & H \end{matrix}$$

This array is then expanded to

$$\begin{matrix} A & C & E & G \\ B & D & F & H \\ \alpha_1 & \alpha_2 & \alpha_3 & \\ \beta_1 & \beta_2 & \beta_3 & \\ \gamma_1 & \gamma_2 & & \end{matrix}$$

where $\alpha_1, \alpha_2, \alpha_3, \beta_1, \beta_2, \beta_3, \gamma_1$ and γ_2 are computed as

$$\alpha_1 = \frac{BC - AD}{B} \qquad \beta_1 = \frac{D\alpha_1 - B\alpha_2}{\alpha_1}$$

$$\alpha_2 = \frac{BE - AF}{B} \qquad \beta_2 = \frac{F\alpha_1 - B\alpha_3}{\alpha_1} \qquad \gamma_1 = \frac{\beta_1\alpha_2 - \alpha_1\beta_2}{\beta_1}$$

$$\alpha_3 = \frac{BG - AH}{B} \qquad \beta_3 = \frac{H\alpha_1 - B_o}{\alpha_1} \qquad \gamma_2 = \frac{\beta_1\alpha_3 - \alpha_1\beta_3}{\beta_1}$$

When the array has been computed, the left-hand column (A, B, α_1, β_1, γ_1) is examined. If the signs of all the numbers in the left-hand column are the same, there are no positive real roots. If there are changes in sign, the number of positive real roots is equal to the number of changes in sign. It should be recognized that this is a test for instability; the absence of sign changes does not guarantee stability.

SAMPLED-DATA CONTROL SYSTEMS

Definition Sampled-data control systems are those in which continuous information is transformed at one or more points of the control system into a series of pulses. This transformation may be performed intentionally, e.g., the flow of information over long distances to preserve the accuracy of the data during the transmission, or it may be inherent in the generation of the information flow, e.g., radiating energy from a radar antenna which is in the form of a train of pulses, or the signals developed by a digital computer during a direct digital control of machine-tool operation.

Methods of analysis analogous to those for continuous-data systems have been developed for the sampled-data systems. Discussed herein are (1) sampling, (2) the z transformation, (3) the z-transfer function, and (4) stability of sampled data systems.

Sampling The ideal sampler is a simple switch (Fig. 16.2.42) which is closed only instantaneously and opens and closes at a constant frequency. The switch, which may or may not be a physical component in a sampled-data feedback system, indicates a sampled signal. Such a sampled-data feedback system is shown in Fig. 16.2.43. The error signal

in continuous form is $\varepsilon(t)$, and the sampled error signal is $\varepsilon^*(t)$. Figure 16.2.42 shows the relationship between these signals in a graphical form.

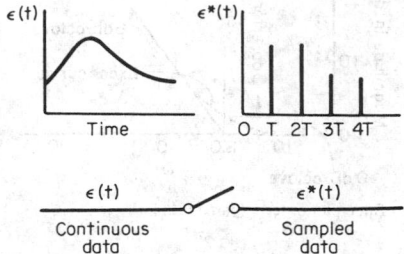

Fig. 16.2.42 Ideal sampler, showing continuous input and sampled output.

z Transformation In the analysis of continuous-data systems, it has been shown that the Laplace transformation can be used to reduce ordinary differential equations to algebraic equations. For sampled-data systems, an operational calculus, the z transform, can be used to simplify the analysis of such systems.

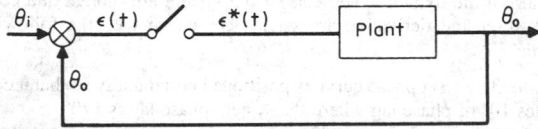

Fig. 16.2.43 Sampled data feedback system.

Consider the sampler as an impulse modulator; i.e., the sampling modulates an infinite train of unit impulses with the continuous-data variable. Then the Laplace transformation can be shown to be

$$F^*(s) = \sum_{n=0}^{\infty} f(nT)e^{-nTs} \qquad (16.2.66)$$

(See A. W. Langill, "Automatic Control Systems Engineering.")
In terms of the z transform, this becomes

$$F^*(s) = F(z) = \sum_{n=0}^{\infty} f(nT)z^{-n} \qquad (16.2.67)$$

where $z^{-n} = e^{-nTs}$.

Tables in Sec. 2 list the Laplace transforms for a number of continuous time functions. Table 16.2.5 lists the z transforms for some of these functions.

It should be noted that unlike the Laplace transforms, the z-transform method imposes the restriction that the sampled-data-system response can be determined only at the sampling instant. The same z transform may apply to different time functions which may have the same value at

Table 16.2.5 Laplace and z Transforms

Time function	$f(t)$	$F(s) = \pounds[f(t)]$	$F^*(z) = F(z)$
Unit ramp function	t	$\dfrac{1}{s^2}$	$\dfrac{Tz}{(z-1)^2}$
Unit acceleration function	$\dfrac{t^2}{2}$	$\dfrac{1}{s^3}$	$\dfrac{1}{2}T^2\dfrac{z(z+1)}{(z-1)^3}$
Exponential function	$e^{-\alpha t}$	$\dfrac{1}{s+\alpha}$	$\dfrac{z}{z-e^{-\alpha T}}$
Sinusoidal function	$\sin \beta t$	$\dfrac{\beta}{s^2+\beta^2}$	$\dfrac{z \sin \beta T}{z^2 - 2z \cos \beta T + 1}$
Cosinusoidal function	$\cos \beta t$	$\dfrac{s}{s^2+\beta^2}$	$\dfrac{z(z - \cos \beta T)}{z^2 - 2z \cos \beta T + 1}$

Table 16.2.6   Typical Block Diagrams of Sampled-Data Control Systems and Their Transforms

System	Laplace transform of the output, $C(s)$	z-transform of the output, $C^*(z)$
1.	$\dfrac{G_1(s)}{1 + G_1 G_2^*(z)} R^*(z)$	$\dfrac{G_1^*(z)}{1 + G_1 G_2^*(z)} R^*(z)$
2.	$\dfrac{G_1(s)}{1 + G_1(s)G_2(s)} R^*(z)$	$\left[\dfrac{G_1(s)}{1 + G_1(s)G_2(s)}\right]^* R^*(z)$
3.		$\dfrac{G_1^*(z)}{1 + G_1^*(z)G_2^*(z)} R^*(z)$
4.	$G_1(s)\left[R(s) - G_2(s)\dfrac{RG_1^*(z)}{1 + G_1 G_2^*(z)} \right]$	$\dfrac{RG_1^*(s)}{1 + G_1 G_2^*(s)}$
5.	$G_1(s)\left[R^*(z) - \dfrac{G_1^*(z)G_2^*(z)\,R^*(z)}{1 + G_1^*(z)G_2^*(z)} \right]$	$\dfrac{G_1^*(z)}{1 + G_1^*(z)G_2^*(z)} R^*(z)$
6.	$G_1(s)\left[R^*(z) - G_2(s)\dfrac{R^*(z)G_1^*(z)}{1 + G_1 G_2^*(z)} \right]$	$\dfrac{G_1^*(z)}{1 + G_1 G_2^*(z)} R^*(z)$
7.	$\dfrac{G_2(s)}{1 + G_1 G_2 G_3^*(z)} RG_1^*(z)$	$\dfrac{G_2^*(z)}{1 + G_1 G_2 G_3^*(z)} RG_1^*(z)$
8.	$\dfrac{G_2(s)}{1 + G_2^*(z) + G_1 G_2^*(z)} RG_1^*(z)$	$\dfrac{G_2^*(z)}{1 + G_2^*(z) + G_1 G_2^*(z)} RG_1^*(z)$

SOURCE: John G. Truxal (ed.), "Control Engineers' Handbook," McGraw-Hill.

the instant of sampling. The z-transform function is not defined, therefore, in a continuous sense, and the inverse z transform is not unique.

z **Transfer Function**   The ratio of the sampled output function of a discrete network to the sampled input function is the z-transfer function. A discrete network is one which has both a sampled input and output. A table of block diagrams of a number of sampled-data control systems with their associated transfer functions is presented in Table 16.2.6.

Stability of Sampled-Data Systems   The stability of sampled-data systems can be demonstrated utilizing frequency-response methods which have been discussed in this section. See "Stability and Performance of an Automatic Control." The Nyquist stability criterion again applies with the same conclusions relative to the -1 point, and the methods of generalizing the open- and closed-loop frequency response plots remain the same.

MODERN CONTROL TECHNIQUES

Many engineers and researchers have been actively pursuing the application and development of advanced control strategies. In recent years this effort has included extensive work in the area of robust control theory. One of the primary attractions of this theory is that it generalizes the **single input–single system output (SISO)** concept of gain and phase margins, and its effect on system sensitivity to multivariable control system.

In this section the design technique for a **linear quadratic gaussian with loop transfer recovery (LQG/LTR)** robust controller design method will be summarized. The concepts required to understand this method will be reviewed. A linearized model of a simple process has been chosen to illustrate this technique. The simple process has three state variables, one input, and one output. Three control system design methods are compared: LQG, a LQG/LTR, and a proportional plus integral controller (PI).

Previous Work and Results

Most applications of modern control techniques to process control have appeared in the form of **optimal control strategies** which were in the form of full-state feedback control with and without state estimation [linear quadratic regulator (LQR)]. The regulator problem, coupled with a state estimation problem, was shown not to have desirable robustness properties. Doyle and Stein (Doyle and Stein, 1979) proposed the multivariable robust design philosophy known as LQG/LTR to eliminate the shortcomings of the LQR and LQG design techniques. The LQG/LTR design technique has been applied to control submersible vehicles, engine speed, helicopters, ship steering, and large, flexible space structures.

The LQG/LTR method requires the plant to be **minimum phase,** which cannot be guaranteed in a plant process. Typically **nonminimum phase** conditions in a plant process are due to time delay. The LQG/LTR design procedure offers no guarantees when a nonminimum-phase plant is used. However recent work develops a method to obtain a robust controller design for plants with time delay (Murphy and Bailey, IEEE 1989 and IECON 1990).

The LQG/LTR synthesis method for minimum phase plants achieves desired loop shapes and maximum robustness properties in the design of feedback control systems. The LQG/LTR design procedure uses **singular value analysis** and design procedures to obtain the desired performance and stability robustness. The use of singular values yields a frequency-domain design and analysis method generally expected by control engineers to determine system robustness. The LQG/LTR synthesis method is applicable to both **multiple-input/multiple-output (MIMO)** and SISO control systems. The performance limitations of this design method are analogous to those of a SISO system.

Definition of Robustness

A system is considered to be robust and have **good robustness properties** if it has a large stability margin, good disturbance attenuation, and/or low sensitivity to parameter variations. The term **stability margin** refers to the gain margin and the phase margin, which are quantitative measures of stability. A proven method of obtaining good robustness properties is the use of a feedback control system, which can be designed to allow for variations in the system dynamics. Some causes of **variation** in system dynamics are

Modeling and data errors in the nominal plant and system.
Changes in environmental conditions, manufacturing tolerances, wear due to aging, and noncritical material failures.
Errors due to calibration, installation, and adjustments.

Feedback control systems with good feedback properties have been synthesized for SISO systems. Classical frequency domain techniques such as Nyquist, root-locus, Bode, and Nichols plots have been used to obtain the feedback control system for the SISO system. These design techniques have allowed the synthesis of feedback control systems yielding **insensitivity** to bounded parameter variations and a large stability margin. The success of the feedback control system for the SISO system has led to the direct extension of the classical frequency domain technique to the design of a multivariable feedback control system. This extension to the multivariable design problem examines an individual feedback loop as the phase and gain margins are varied, while the nominal phase and gain values in the remaining feedback paths are held constant. This technique, however, fails to consider the results of simultaneous variation of gain and phase in all paths, which is a real-world possibility and needs to be considered. A method of obtaining a feedback control system with good robustness properties that takes into consideration simultaneous gain and phase variation is the LQG/LTR.

The LQG/LTR technique can be applied to MIMO systems or SISO systems. The LQG/LTR technique not only has the good robustness properties of the classical frequency domain techniques but also is capable of minimizing the effects of unmodeled high-frequency dynamics, neglected nonlinearities, and a reduced-order model. **Computer control system design tools,** for instance, developed by Integrated Systems Incorporated (ISI) and The MathWorks can be used in synthesizing the LQG/LTR technique and other controller design techniques. The use of computer-aided design tools eliminates the numerical programming burden of programming the complex LQG/LTR algorithm.

In what follows, the needed concepts are defined and the LQG/LTR procedure for development of a **model-based compensator** (MBC) is described. A unity-feedback MBC is selected for the controller structure, because of its similarity to the SISO unity-feedback control system well

known to classical control designers. The MBC closed-loop control system has proven to offer great practical considerations in the design of automatic control systems.

Robustness Concepts of the LQG and LQG/LTR Control Systems

The LQG/LTR design procedure is based on the system configuration of the LQG controller shown in Fig. 16.2.44. The LQG controller consists of a Kalman filter state estimator and a linear quadratic regulator. The Kalman filter state estimator has good robustness properties for plant perturbations at the plant output. The linear quadratic regulator (LQR) has good robustness properties for perturbations at the plant input. Even though its components separately have good robustness properties, the LQG controller is found to have no guaranteed robustness properties at either the input (point 2) or the output (point 3) of the plant.

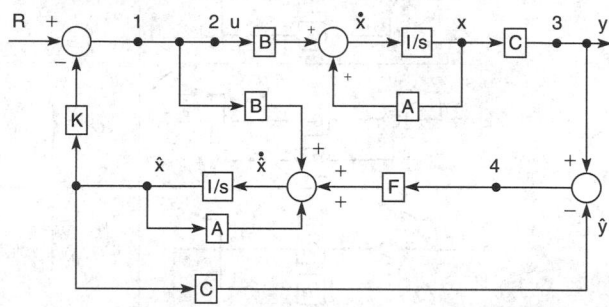

- LQG guaranteed no robustness properties at the input or output of the plant.
- Point 1 has the good robustness properties of the full-state feedback system.
- Point 4 has the good robustness properties of the Kalman filter.
- Point 2 has no guaranteed robustness properties.
- Point 3 has no guaranteed robustness properties.
- The LQG/LTR design method permits recovery of the robustness properties of point 1 at point 2 or the robustness properties of point 4 at point 3.

Fig. 16.2.44 Summary of the robustness properties of the LQG block diagram.

The LQG/LTR design procedure allows us to recover robustness properties at either the input or the output of the plant. If robustness is desired at the input to the plant, first a nominal robust LQR design is made to satisfy the design constraints. Next, an LTR step is made to design a Kalman filter gain that recovers the robustness at the input to the plant of the LQG controller that is approximately that of the nominal LQR design. This implies from Fig. 16.2.44 that the robustness properties at points 1 and 2 are approximately the same.

If robustness is desired at the **output of the plant,** first a nominal robust Kalman filter design is made to satisfy the performance constraints. Next, an LTR step is made to design an LQR gain that recovers the robustness at the output of the plant that is approximately that of the nominal Kalman filter design. This implies from Fig. 16.2.44 that the robustness properties at points 3 and 4 are approximately the same.

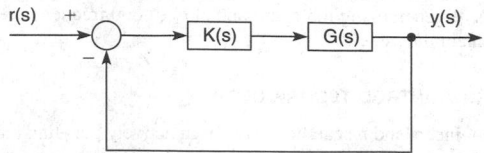

Fig. 16.2.45 Unity feedback model-based compensator (UFMBC).

The block diagram of the unity-feedback MBC is shown in Fig. 16.2.45. This control system structure allows tracking and regulation of a reference input at the output of the plant. The filter and regulator gains used in the controller $R(s)$ are obtained appropriately, depending on whether robustness is desired at the input or the output of the plant. Note that the robustness properties of the **unity-feedback MBC** (UFMBC) are the same as those of the LQG system, since the UFMBC is just an alternative structure of the LQG system.

MATHEMATICS AND CONTROL BACKGROUND

The use of **matrix algebra** is very important in the study of MIMO control system design methods. Some MIMO control system design methods use matrix algebra to decouple the system so that the plant matrix transfer function consists only of individual **decoupled transfer functions**, each represented conventionally as the ratio of two polynomials. This simplification then allows the use of conventional methods such as the Bode plot, Nyquist plot, and Nichols chart as a means of designing a controller for each loop of the MIMO plant. The LQG/LTR control system design method, however, does not require **decoupling** of the plant but rather preserves the coupling between the individual loops of the MIMO plant. Each step of the LQG/LTR design technique therefore requires the use of some matrix algebra, vector and matrix norms, and singular values during the control system design procedure. For instance, matrix algebra is used to derive a suitable representation of the closed-loop system for stability and robustness analysis. Matrix norms and singular values are used to examine the matrix magnitude of the MIMO transfer function. The **matrix magnitude** or **singular values** of a system allows preservation of the coupling between the various loops of the MIMO control system.

The concepts of matrix norm, singular values, controllability, observability, stabilizability, and detectability are very important in applying MIMO design procedures. It is therefore important that these concepts be well understood.

The purpose of this section is to provide **background material** in mathematics and control systems that has proved useful in deriving the procedures and analysis required to implement the control system design procedure.

Matrix Norm and Singular Values

The matrix norm of $\|A\|_2$ is defined for the $m \times n$ matrix A. The matrix A defines a linear transformation from the vector space V, called the domain to a vector space W, which is called the range. Thus the linear transformation

$$A(x) = Ax \tag{16.2.68}$$

transforms a vector x in $V \in C^n$ into a vector $A(x)$ in $W \in C^m$. The matrix norm of $\|A\|_2$ is defined as

$$\|A\|_2 = \max_{x \neq 0} \frac{\|A\|_2}{\|x\|_2} = l_2 \text{ norm} \tag{16.2.69}$$

Since the l_2 **norm** is a useful tool in control system analysis, further insight into its definition is given. The l_2 norm of Eq. (16.2.69) can be written as

$$\|A\|_2 = \max_i [\lambda_i(A^H A)]^{1/2} \quad i = 1, 2, \cdots, n \tag{16.2.70}$$

$$\|A\|_2 = \max_i [\lambda_i(AA^H)]^{1/2} \quad i = 1, 2, \cdots, n \tag{16.2.71}$$

The matrices $A^H A$ and AA^H are Hermitian and positive semidefinite. The eigenvalues of these matrices are real and nonnegative. The definition of the **singular value** of a complex matrix $A \in C^{n \times n}$ is

$$\sigma_i(A) = [\lambda_i(A^H A)]^{1/2}$$
$$= [\lambda_i(AA^H)]^{1/2} \geq 0 \quad i = 1, 2, \cdots, n \tag{16.2.72}$$

If $A \in C^{m \times n}$, which indicates that A is a nonsquare matrix, then

$$\sigma_i(A) = [\lambda_i(A^H A)]^{1/2} = [\lambda_i(AA^H)]^{1/2} \tag{16.2.73}$$

for $1 \leq i \leq k$, where k is the number of singular values $= \min(m, n)$. The maximum and minimum singular values are defined respectively as

$$\bar{\sigma}(A) = \|A\|_2 \tag{16.2.74}$$

and

$$\underline{\sigma}(A) = \max_{x \neq 0} \frac{\|Ax\|}{\|x\|_2} = \frac{1}{\|A^{-1}\|_2} \tag{16.2.75}$$

provided A has an inverse.

Controllability, Observability, Stabilizability, and Detectability

The LQG/LTR control system design method has a certain **requirement** of the plant for which the controller is being designed: the plant must be at least stabilizable and detectable. Therefore it is important that the conditions of controllability, observability, and the less restricted conditions of stabilizability and detectability be understood. A description of these concepts follows.

Given the state equation for the linear time invariant (LTI) system described as $\dot{x}(t) = Ax(t) + Bu(t)$ and $y(t) = Cx(t)$. The matrices A, B, and C are constant. Note that $x \in R^n$, $n \in R^p$, and $y \in R^p$ are respectively the state, input, and output vectors. Variables n and p are the dimensions of the states and the input and output vectors. The input and output vectors are of the same dimension.

A system is said to be controllable if it is possible to transfer the system from any initial state x_0 in state space to any other state x_F using the input in a finite period of time. The system is said to be uncontrollable otherwise. The condition of controllability places **no constraint** on the input in the system.

The algebraic test for controllability of the LTI system state equation uses the controllability matrix

$$Q = [B \ AB \cdots A^{n-1}B] \tag{16.2.76}$$

where n is the **number of states.** The LTI system is controllable if the rank of Q is equal to n. If the rank of Q is less than n, the system is not controllable. An uncontrollable system indicates that some modes or poles are not affected by the control input. Uncontrollable systems with **unstable modes** imply that a controller design is not possible to ensure system stability. If the uncontrollable modes of the uncontrollable system are stable, then the system is stabilizable.

An LTI system is considered **observable** at initial time t_0 if there exists a finite time t_1 greater than t_0 such that for any initial state x_0 in state space using knowledge of the input u and output y over the time interval $t_0 < t < t_1$, the state x_0 can be determined. The system is said to be unobservable otherwise.

The algebraic test for observability of the LTI system uses the observability matrix

$$N = [C^T \ A^T C^T \cdots (A^T)^{n-1} C^T] \tag{16.2.77}$$

The LTI system is observable if the rank of N is equal to n, where n is the number of states. If the rank of N is less than n, the system is unobservable. An unobservable system indicates that not all system modes contribute to the output of the system. If the **unobservable modes** (states) are stable, then the system is considered to be detectable.

EVALUATING MULTIVARIABLE PERFORMANCE AND STABILITY ROBUSTNESS OF A CONTROL SYSTEM USING SINGULAR VALUES

Performance Robustness

The requirements for the control system design are formulated by placing certain restrictions on the singular values of the **return ratio, return difference,** and **inverse return difference** of the control system. The block diagram in Fig. 16.2.46 shows the controller $K(s)$ and plant $G(s)$ with input $d_i(s)$ and output $d_o(s)$ disturbances and sensor noise $n(s)$.

Using Fig. 16.2.46, a **frequency-domain transfer function** representation for the plant output is first obtained. These transfer function representations, used in conjunction with concepts of singular values, allow the formulation of requirements in the frequency domain to obtain a control system with good performance and stability robustness.

In this design, performance robustness is of concern at the output of the plant. Therefore it is required that the plant output transfer function be derived in order to analyze and assist in obtaining a control system design to meet certain performance robustness requirements. Using Fig. 16.2.46, the plant output transfer function is derived and performance

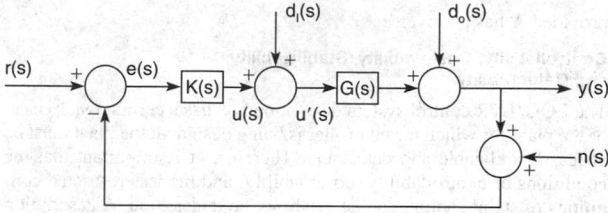

Fig. 16.2.46   Model-based unity feedback MIMO control system.

robustness requirements established. From Fig. 16.2.46 it can be shown that

$$y(s) = d_o(s) + G(s)u'(s) \qquad (16.2.78)$$
$$e(s) = r(s) - y(s) - n(s) \qquad (16.2.79)$$

and
$$u'(s) = d_o(s) + u(s) = d_f(s) + K(s)e(s). \qquad (16.2.80)$$

where $G(s) = \mathbf{C}[s\mathbf{I} - \mathbf{A}]^{-1}\mathbf{B}$ is the plant transfer function and $\mathbf{A}, \mathbf{B},$ and $\mathbf{C}$ are system matrices. The transfer functions $y(s), d_o(s), e(s), r(s), n(s), u(s), d_f(s),$ and $K(s)$ are respectively the output, output disturbance, error, reference, noise, input, input disturbance, and controller transfer functions.

Using Eqs. (16.2.79) and (16.2.80) it can be shown that Eq. (16.2.78) becomes

$$\begin{aligned}
y(s) &= [G(s)K(s)][r(s) - y(s) - n(s)] \\
&\quad + d_o(s) + G(s)d_f(s) \\
&= G(s)K(s)r(s) - G(s)K(s)y(s) - G(s)K(s)n(s) \\
&\quad + d_o(s) + G(s)d_f(s) \qquad (16.2.81)
\end{aligned}$$

Therefore,

$$\begin{aligned}
y(s) &= [\mathbf{I} + G(s)K(s)]^{-1}[G(s)K(s)r(s) - G(s)K(s)n(s) \\
&\quad + d_o(s) + G(s)d_f(s)] \\
&= [\mathbf{I} + G(s)K(s)]^{-1}\{G(s)K(s)[r(s) - n(s)]\} \\
&\quad + [\mathbf{I} + G(s)K(s)]^{-1}[d_o(s) + G(s)d_f(s)] \\
&= y_r(s) + y_{d_f}(s) + y_{d_o}(s) + y_n(s), \qquad (16.2.82)
\end{aligned}$$

where $y_r(s), y_{d_f}(s), y_{d_o}(s),$ and $y_n(s)$ are, respectively, the contributions to the plant output due to the reference input, input disturbance, output disturbance, and noise input to the system. The requirements for good command following, disturbance rejection, and insensitivity to sensor noise are defined by using singular values to characterize the magnitude of MIMO transfer functions. These requirements are also **applicable to SISO systems** (Kazerooni and Narayan; Murphy and Bailey, April 1989).

Command Following   The conditions required to obtain a MIMO control system design with good command following is described. Good command following of the reference input $r(s)$ by the plant output $y(s)$ implies that

$$y(s) \approx r(s) \qquad \forall \; s \in s_R, \qquad (16.2.83)$$

where $s = j\omega$ and $s_R = j\omega_R$ and ω_R is the frequency range of the input $r(s)$. Letting $d_o(s) = d_f(s) = n(s) = 0$ in Eq. (16.2.82) results in

$$\begin{aligned}
y_r(s) &= [\mathbf{I} + G(j\omega)K(j\omega)]^{-1}G(j\omega)K(j\omega)r(j\omega) \qquad \forall \; \omega \in \omega_R \\
&= \{\mathbf{I} + [G(j\omega)K(j\omega)]^{-1}\}^{-1}r(j\omega) \qquad \forall \; \omega \in \omega_R \qquad (16.2.84)
\end{aligned}$$

where $G(j\omega)K(j\omega)$ is defined as the return ratio. Therefore, to obtain good command following, the system **inverse return difference** $\mathbf{I} + [G(j\omega)K(j\omega)]^{-1}$ must have the property such that

$$\mathbf{I} + [G(j\omega)K(j\omega)]^{-1} \approx \mathbf{I} \qquad \forall \; \omega \in \omega_R \qquad (16.2.85)$$

which implies that

$$\underline{\sigma}\,[G(j\omega)K(j\omega)] \gg 1 \qquad \forall \; \omega \in \omega_R \qquad (16.2.86)$$

Thus the magnitude of the minimum singular value of the **return ratio** $\underline{\sigma}[G(j\omega)K(j\omega)]$ must be large over the frequency range of the input to obtain good command following of the control system.

Disturbance Rejection   To minimize the effect of the output disturbance, $y_{d_o}(s)$, on the output signal $y(s)$, Eq. (16.2.82) is used with $n(s) = r(s) = d_f(s) = 0$, which results in

$$y_{d_o}(s) = [\mathbf{I} + G(s)K(s)]^{-1}d_o(s). \qquad (16.2.87)$$

Rejection of this output disturbance at the plant output requires that the magnitude of the **return difference** $\mathbf{I} + G(s)K(s)$ to meet the following condition

$$\underline{\sigma}\,[\mathbf{I} + G(j\omega)K(j\omega)] \gg 1 \qquad \forall \; \omega \in \omega_{d_o} \qquad (16.2.88)$$

which implies

$$\underline{\sigma}\,[G(j\omega)K(j\omega)] \gg 1 \qquad \forall \; \omega \in \omega_{d_o} \qquad (16.2.89)$$

where y_{d_o} is the frequency range of the output disturbance $d_o(s)$.

With $n(s) = r(s) = d_o(s) = 0$ in Eq. (16.2.82), the effect of input disturbance on the plant output is examined. This effect is seen to be

$$y_{d_f}(s) = [\mathbf{I} + G(j\omega)K(j\omega)]^{-1}[G(j\omega)d_f(j\omega)] \qquad \forall \; \omega \in \omega_{d_f} \qquad (16.2.90)$$

where ω_{d_f} is the frequency range of the input disturbance $d_f(s)$. To minimize the effect of $y_{d_f}(s)$ on the plant output requires that the magnitude of the return difference be large (in general, the larger the better). This requirement implies that

$$\underline{\sigma}\,[G(j\omega)K(j\omega)] \gg 1 \qquad \forall \; \omega \in \omega_{d_f} \qquad (16.2.91)$$

to reject the input disturbance. It should be noted that this equation gives only general conditions for rejection. The control engineer must determine how much greater than 1 the return ratio magnitude must be.

Noise Rejection   The requirements to minimize the effects of sensor noise on the plant output must be determined. First letting $n(s) = d_f(s) = d_o(s) = 0$ in Eq. (16.2.82) results in

$$\begin{aligned}
y_n(s) &= [\mathbf{I} + G(s)K(s)]^{-1}[G(s)K(s)]\,n(s) \qquad \forall \; s \in s_n \\
&= \{\mathbf{I} + [G(s)K(s)]^{-1}\}^{-1}n(s) \qquad \forall \; s \in s_n \qquad (16.2.92)
\end{aligned}$$

where $s_n = j\omega_n$ and ω_n is the frequency range of the sensor noise. Thus to minimize the effects of sensor noise on the plant output it is desired to have the magnitude of $y_n(s)$ small over the frequency range of the noise. Thus

$$\overline{\sigma}\,[G(j\omega)K(j\omega)] \gg 1 \qquad \forall \; \omega \in \omega_n \qquad (16.2.93)$$

is required to minimize the effect of noise on the plant output.

Any overlapping of any of the frequency ranges $\omega_R, \omega_{d_f}, \omega_{d_o}, \omega_n$ will require system design specification tradeoffs to be made. This will definitely occur when the lower frequency ranges of ω_R and/or ω_{d_f} and ω_{d_o} overlap the higher frequency range ω_n.

Stability Robustness

All linear control system design methods are based on a linear model of the plant. Because this design model only approximates the actual plant, an error exists between the design model and the actual plant. This error can be evaluated as **absolute** or **relative.** The method of evaluation of this error (model uncertainty) depends on the nature of the differences between the actual design model and the actual plant. The evaluation of an absolute error defines an additive model uncertainty, whereas the evaluation of a relative error defines a multiplicative model uncertainty. The **additive** and multiplicative model uncertainties can be further classified as either structured or unstructured model uncertainties (Murphy and Bailey, ORNL 1989).

A **structured model uncertainty** is usually a low-frequency phenomenon characterized by variations in the parameters of the linear time-invariant plant design model. These parameter variations are caused by changes in the plant operating conditions, wear due to aging, and environmental conditions. A structured uncertainty also indicates that the gain and phase information concerning the uncertainty is known. Additive model uncertainties are generally considered structured uncertainties.

An **unstructured uncertainty** is typically a high-frequency phenomenon in which the only information known about the uncertainty is its magnitude. Unstructured uncertainties usually have the characteristic of becoming significant at high frequencies. Multiplicative uncertainties are characterized as unstructured. Unstructured uncertainties are usually caused by the truncation of high-frequency plant dynamics or modes due to the linearization process, and the neglect of plant dynamics such as actuators and sensors. Some plant dynamics that characterize unstructured uncertainty are flexible-body dynamics, electrical and mechanical resonances, and transport delays. Most linear control-system design methods neglect these high-frequency plant dynamics. The result is that if a high-frequency bandwidth controller is implemented, the high-frequency modes could be excited, resulting in an unstable control system.

Model uncertainty clearly is seen to impose limitations on the achievable performance of a feedback control system design. Hence this section will focus on the limitations imposed by uncertainty, not on the difficult problem of exact representation of the system uncertainty.

Multiplicative Uncertainty Multiplicative uncertainty (perturbation) can be represented at the input or output of the plant, and is called respectively the input multiplicative uncertainty or the **output multiplicative uncertainty.** The model uncertainty due to actuator and sensor dynamics is modeled respectively as an input or output multiplicative perturbation. A reduced-order model of an already **linearized model** results in a multiplicative model uncertainty. **Time delay** at the input or output of a plant yields, respectively, input and output multiplicative uncertainties in the plant. Even though other multiplicative model uncertainties exist among various types of control system design problems, the model uncertainties mentioned above are a common consideration in most control problems.

Output Multiplicative Uncertainty

The output multiplicative uncertainty $\Delta G(s)$ is defined as

$$\Delta G(s) = [G'(s) - G(s)] \, G^{-1}(s) \qquad (16.2.94)$$

where

$$G'(s) = L(s)G(s) \qquad (16.2.95)$$

is the perturbed transfer function, $L(s)$ represents dynamics not included in the nominal plant $G(s)$. Thus resulting in

$$\Delta G(s) = [L(s) - \mathbf{I}] \qquad (16.2.96)$$

A block diagram of a control system with output multiplicative perturbation is shown in Fig. 16.2.47. The conditions of stability for a control system with output multiplicative uncertainty require that

$$\underline{\sigma}\{\mathbf{I} + [G(j\omega)K(j\omega)]^{-1}\} > \overline{\sigma}\,[\Delta G(j\omega)] \qquad \forall\ \omega > 0 \quad (16.2.97)$$

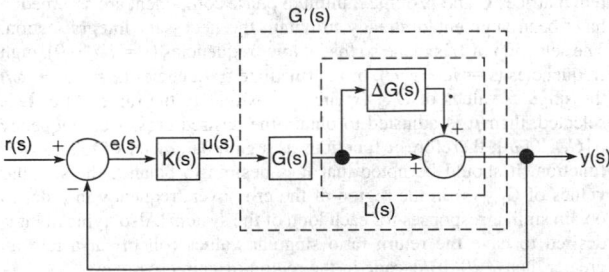

Fig. 16.2.47   Control system with output multiplicative perturbation.

Additive Uncertainty The block diagram of a control system with a plant additive model uncertainty is shown in Fig. 16.2.48. The representation of the additive model uncertainty using an absolute error criterion is defined as

$$\Delta G(s) = G'(s) - G(s) \qquad (16.2.98)$$

where in this case the perturbed transfer function $G'(s)$ is

$$G'(s) - [\mathbf{C} + \Delta\mathbf{C}][s\mathbf{I} - [\Lambda + \Delta\Lambda]]^{-1}[\mathbf{B} + \Delta\mathbf{B}] \quad (16.2.99)$$

and the nominal plant transfer function $G(s)$ is

$$G(s) = \mathbf{C}[s\mathbf{I} - \mathbf{A}]^{-1}\mathbf{B} \qquad (16.2.100)$$

and where $\mathbf{A}$, $\mathbf{B}$, and $\mathbf{C}$ are the nominal system matrices and $\Delta\mathbf{A}$, $\Delta\mathbf{B}$, and $\Delta\mathbf{C}$ are the perturbation matrices that indicate the respective deviations from the nominal system matrices. The condition for stability of the additively perturbed control system requires that

$$\underline{\sigma}[G(j\omega)K(j\omega)] > \overline{\sigma}[\Delta G(j\omega)K(j\omega)] \qquad \forall\ \omega > 0 \quad (16.2.101)$$

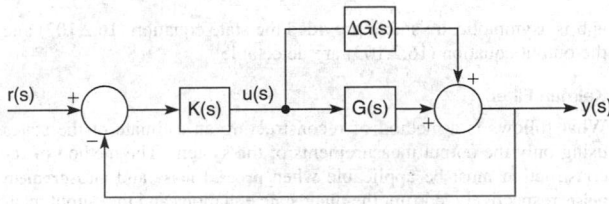

Fig. 16.2.48   Control system with additive perturbation.

Table 16.2.7 contains a summary of the general design requirements presented in this section to obtain a control system with good performance and stability robustness.

REVIEW OF OPTIMAL CONTROL THEORY

Linear Quadratic Regulator

Consider the linear time-invariant state-space system, first-order vector differential equation model

$$\dot{x}(t) = \mathbf{A}x(t) + \mathbf{B}u(t) \qquad (16.2.102)$$
$$y(t) = \mathbf{C}x(t) \qquad (16.2.103)$$

where $\mathbf{A}$, $\mathbf{B}$, and $\mathbf{C}$ are constant system matrices, and $x(t)$, $u(t)$, and $y(t)$ are respectively the system state, input, and output vectors. The system is stabilizable and controllable. The goal of this procedure is to minimize the finite performance index

$$J = \int_0^\infty [x^{\mathrm{T}}(t)\mathbf{Q}x(t) + u^{\mathrm{T}}(t)\mathbf{R}u(t)]\, dt \qquad (16.2.104)$$

$$J = \int_0^\infty [x^{\mathrm{T}}\mathbf{C}^{\mathrm{T}}\mathbf{C}x + u^{\mathrm{T}}\mathbf{R}u]\, dt$$

$$= \int_0^\infty [y^{\mathrm{T}}y + u^{\mathrm{T}}\mathbf{R}u]\, dt$$

where $\mathbf{Q}$ is a symmetric positive semidefinite matrix and $\mathbf{R}$ is positive definite.

Table 16.2.7   General System Design Specifications

System requirements	Range	
Good command-following	$\underline{\sigma}[G(j\omega)K(j\omega)] \gg 0$ dB	$\forall\ \omega \in \omega_R$
Good disturbance rejection	$\underline{\sigma}[G(j\omega)K(j\omega)] \gg 0$ dB	$\forall\ \omega \in \omega_d$
Good immunity to noise	$\overline{\sigma}[G(j\omega)K(j\omega)] \ll 0$ dB	$\forall\ \omega \in \omega_n$
Good system response to high-frequency modeling error*	$\underline{\sigma}\{I + [G(j\omega)K(j\omega)]^{-1}\} > \|\Delta G(j\omega)\|$	$\forall\ \omega > 0$ rad/s
Good insensitivity to parameter variations at low frequencies*	$\underline{\sigma}[G(j\omega)K(j\omega)] > \overline{\sigma}[\Delta G(j\omega)K(j\omega)]$	

* The $\Delta G(j\omega)$ indicated corresponds to the applicable multiplicative or additive model uncertainty.

The **performance index** J, referred to as the quadratic performance index, implies that a control $u(t)$ is sought to facilitate the minimization of J. The weighting matrices $\mathbf{Q}$ and $\mathbf{R}$ are selected to reflect the importance of particular states and control inputs. In general the weighting of the diagonal elements of $\mathbf{Q}$ can be determined by the importance of the state, as observed by the $\mathbf{C}$ matrix of the output equation. The effect of $\mathbf{Q}$ is the ability to control the **transient response** of the LQR. The effect of $\mathbf{R}$ is the ability to **control** the **energy** resulting from u.

Therefore a linear control law that minimizes J is defined as

$$u(t) = -\mathbf{K}x(t) \qquad (16.2.105)$$
$$\text{where} \qquad \mathbf{K} = \mathbf{R}^{-1}\mathbf{B}^{\mathsf{T}}\mathbf{S} \qquad (16.2.106)$$

where, in turn, $\mathbf{S}$ is a constant symmetric positive semidefinite matrix that satisfies the algebraic Riccati equation

$$\mathbf{Q} - \mathbf{S}\mathbf{B}\mathbf{R}^{-1}\mathbf{B}^{\mathsf{T}}\mathbf{S} + \mathbf{A}^{\mathsf{T}}\mathbf{S} + \mathbf{S}\mathbf{A} = 0 \qquad (16.2.107)$$

The closed loop regulator is given by

$$\dot{\mathbf{x}}(t) = [\mathbf{A} - \mathbf{B}\mathbf{K}]x(t) = [\mathbf{A} - \mathbf{B}\mathbf{R}^{-1}\mathbf{B}^{\mathsf{T}}\mathbf{S}]x(t) \qquad (16.2.108)$$

and is asymptotically stable provided the state equation (16.2.102) and the output equation (16.2.103) are detectable.

Kalman Filter

What follows is a method of reconstructing an estimate of the states using only the output measurements of the system. The method of reconstruction must be applicable when **process noise** and **measurement noise**, respectively, corrupt the plant state equations and the output measurement equations.

Let us consider the stochastic linear system state space vector differential equation model

$$\dot{\mathbf{x}}(t) = \mathbf{A}x(t) + \mathbf{B}u(t) + \mathbf{\Gamma}d(t) \qquad (16.2.109)$$
$$y(t) = \mathbf{C}x(t) + n(t) \qquad (16.2.110)$$

where $\mathbf{A}$, $\mathbf{B}$, $\mathbf{C}$, and $\mathbf{\Gamma}$ are constant system matrices. Vectors $s(t)$, $u(t)$, and $y(t)$ are respectively the state, input, and output vectors. Vector $d(t)$ is a process noise random vector and vector $n(t)$ is a measurement noise random vector. Assuming that $d(t)$ and $n(t)$ are zero-mean, uncorrelated, gaussian white noises,

$$E[d(t)]E[n(t)] = 0 \qquad \text{for all } t \qquad (16.2.111)$$
$$E[d(t)d^{\mathsf{T}}(\tau)] = \mathbf{D}_o\delta(t - \tau) \qquad (16.2.112)$$
$$E[n(t)n^{\mathsf{T}}(\tau)] = \mathbf{N}\delta(t - \tau) \qquad \text{for all } t < \tau \qquad (16.2.113)$$
$$E[d(t)n^{\mathsf{T}}(\tau)] = 0. \qquad (16.2.114)$$

where τ is the difference between two points in time, $E[\ ]$ is mean value, δ is an impulse function, and $\mathbf{N}$ and $\mathbf{D}_o$ are constant symmetric, positive definite, and positive semidefinite matrices, respectively. Note that $\mathbf{N}$ and $\mathbf{D}_o$ are constant because it is assumed that $d(t)$ and $n(t)$ are wide-sense stationary.

The **state estimate** $\hat{\mathbf{x}}$ **of the actual measured state** x is obtained from the noisy measurement y. Therefore we can define a state error vector

$$e(t) = x(t) - \hat{\mathbf{x}}(t) \qquad (16.2.115)$$

where we desire to minimize the mean square error

$$\bar{\mathbf{e}} = E[e^{\mathsf{T}}(t)e(t)] \qquad (16.2.116)$$

The estimator can thus be shown to take the form

$$\dot{\hat{\mathbf{x}}} = \mathbf{A}x(t) + \mathbf{B}u(t) + \mathbf{F}[y(t) - \mathbf{C}x(t)] \qquad (16.2.117)$$

where $\mathbf{F}$ is the filter gain which minimizes Eq. (16.2.116); $\mathbf{F}$ is defined as

$$\mathbf{F} = \mathbf{\Sigma}\mathbf{C}^{\mathsf{T}}\mathbf{N}^{-1} \qquad (16.2.118)$$

where $\mathbf{\Sigma}$ is a constant-variance matrix of the error $e(t)$. It is assumed here that $e(t)$ is wide-sense stationary. The matrix $\mathbf{\Sigma}$ is obtained by solving the algebraic variance Riccati equation

$$\mathbf{D} - \mathbf{\Sigma}\mathbf{C}^{\mathsf{T}}\mathbf{N}^{-1}\mathbf{C}\mathbf{\Sigma} + \mathbf{A}\mathbf{\Sigma} + \mathbf{\Sigma}\mathbf{A}^{\mathsf{T}} = 0 \qquad (16.2.119)$$
$$\text{where} \qquad \mathbf{D} = \mathbf{\Gamma}\mathbf{D}_o\mathbf{\Gamma} \qquad (16.2.120)$$

A sufficient condition to obtain a unique and positive definite $\mathbf{\Sigma}$ from Eq. (16.2.117) requires that $[\mathbf{A}, \mathbf{C}]$ be completely observable. However, if the pair $[\mathbf{A}, \mathbf{C}]$ is required to be detectable, then $\mathbf{\Sigma}$ can be positive semidefinite.

The **reconstruction error** $e(t)$ satisfies the differential equation

$$\dot{e}(t) = [\mathbf{A} - \mathbf{F}\mathbf{C}]e(t) + [\mathbf{\Gamma} - \mathbf{F}]\begin{bmatrix} d(t) \\ n(t) \end{bmatrix} \qquad (16.2.121)$$

such that

$$e(t) \rightarrow 0 \text{ as } t \rightarrow \infty \qquad \forall\, t \geq t_0$$

if and only if the **observer** is asymptotically stable. The poles of the observer (filter) are found by using the closed-loop dynamics matrix $[\mathbf{A} - \mathbf{F}\mathbf{C}]$. To obtain asymptotic stability of the filter, it is required that the pair $[\mathbf{A}, \mathbf{\Gamma}]$ be completely controllable. The necessary and sufficient condition of **stabilizability** of the pair $[\mathbf{A}, \mathbf{\Gamma}]$ will ensure **stability** also.

PROCEDURE FOR LQG/LTR COMPENSATOR DESIGN

An LQG/LTR compensator will be designed for a **minimum phase plant (no right half plane zeroes)**. The control system so designed will be guaranteed to be stable and will be robust at the plant output.

First, it is assumed that if **zero steady-state error** is desired in the control system design, the plant inputs will be augmented to obtain the desired **integral action**. In some cases the dynamics of the plant may have poles that are nearly zero, which establishes a **near integral action** in the plant. In such a case the addition of integrals at the plant input causes a much greater than **20-dB/decade roll-off**, thus yielding unreasonable plant dynamics characterized by the return ratio singular-value plots. Therefore caution is required in deciding how or whether to augment the plant dynamics. In the following description of the compensator design it is assumed that the nominal plant transfer function $G(s)$ has been augmented as required by the control engineers' desires.

Step 1 The first step requires selecting the appropriate transfer function $[G_{\text{FOL}}(s)]$ that satisfies the desired **performance and stability robustness** of the final control system. Therefore, the appropriate scalar $\mu > 0$ and a constant matrix $\mathbf{L}$ such that the singular values of

$$G_{\text{FOL}}(s) = (1/\sqrt{\mu})[\mathbf{C}(s\mathbf{I} - \mathbf{A})^{-1}\mathbf{L}] \qquad (16.2.122)$$

approximately meet the desired control system requirements. The system matrices $\mathbf{C}$ and $\mathbf{A}$ of the minimum phase component are assumed to have been augmented already to obtain the necessary integral action. The selection of $\mathbf{L}$ is made so that at low frequencies ($s = j\omega \rightarrow 0$), high frequencies ($s = j\omega \rightarrow j\omega$), or intermediate frequencies ($s = j\omega \rightarrow j\omega_l$) the singular values of $G_{\text{FOL}}(s)$ are approximately the same. After $\mathbf{L}$ is selected, then μ is adjusted to obtain the desired crossover frequency $\sigma[G_{\text{FOL}}(j\omega)]$; therefore μ functions as a gain parameter of the transfer function. It should be noted that it is desired to balance the singular values of $G_{\text{FOL}}(s)$ in the region of the crossover frequency in order to obtain similar responses for each loop of the system. Also, typically it is desired to have the return ratio singular values roll off at a rate no greater than -20 dB/decade in the region of gain crossover.

Step 2 In this step the Kalman filter transfer function is defined as

$$G_{\text{KF}}(s) = \mathbf{C}(s\mathbf{I} - \mathbf{A})^{-1}\mathbf{F} \qquad (16.2.123)$$

where $\mathbf{F}$ is the Kalman filter gain and $\mathbf{C}$ and $\mathbf{A}$ are the system matrices of the linear time-invariant state-space system.

To compute the **filter gain F**, first the following Kalman filter **algebraic Riccati equation (ARE)** is solved

$$\mathbf{L}\mathbf{L}^{\mathsf{T}} - (1/\sqrt{\mu})\mathbf{\Sigma}\mathbf{C}^{\mathsf{T}}\mathbf{C}\mathbf{\Sigma} + \mathbf{A}\mathbf{\Sigma} + \mathbf{\Sigma}\mathbf{A}^{\mathsf{T}} = 0 \qquad (16.2.124)$$

where $\mathbf{L}$ and μ are as obtained in step 1 and $\mathbf{\Sigma}$ (the covariance matrix) is the solution to the equation. The filter gain is thus computed to be

$$\mathbf{F} = (1/\sqrt{\mu})\mathbf{\Sigma}\mathbf{C}^{\mathsf{T}} \qquad (16.2.125)$$

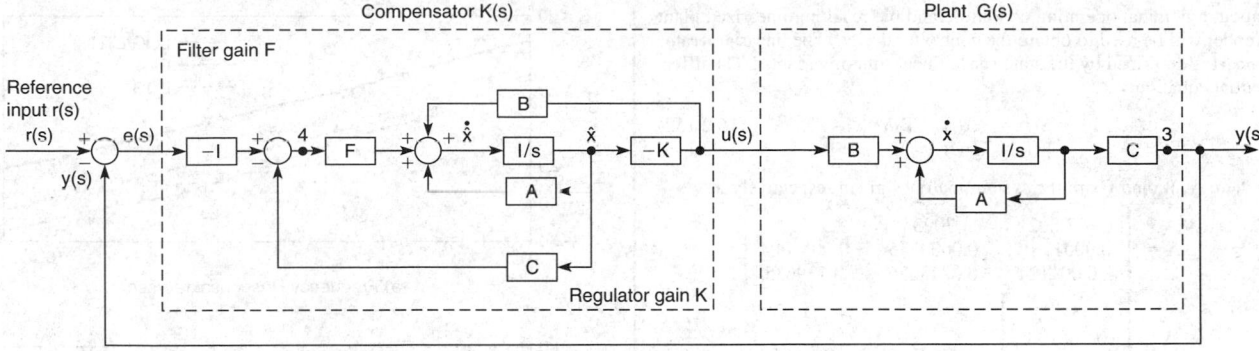

Fig. 16.2.49   Block diagram of unity-feedback LQG/LTR control system.

Thus as seen above, $\mathbf{L}$ and μ are tunable parameters used to produce the necessary filter gain $\mathbf{F}$ to meet the desired system performance and stability robustness constraints. This step completes the LQG/LTR design requirement of designing the target Kalman filter transfer function matrix with the desired properties which will be recovered at the plant output during the loop transfer recovery step. The general block diagram of the LQG/LTR control system is shown in Fig. 16.2.49.

Step 3   The transfer function of the LQG/LTR controller is defined as

$$K(s) = \mathbf{K}[s\mathbf{I} - (\mathbf{A} - \mathbf{BK} - \mathbf{FC})]\mathbf{F} \qquad (16.2.126)$$

where $\mathbf{A}$, $\mathbf{B}$, and $\mathbf{C}$ are the system matrices of Eqs. (16.2.102) and (16.2.103). The filter gain $\mathbf{F}$ is as stated in Eq. (16.2.125). The **regulator gain $\mathbf{K}$** will be computed in this step to achieve loop transfer recovery. The selection of $\mathbf{K}$ will yield a return ratio at the plant output such that

$$G(s)K(s) \to G_{\mathrm{KF}}(s) \qquad (16.2.127)$$

where $G(s)$ is the plant transfer function, and $K(s)$ is the controller transfer function.

The first step in selecting $\mathbf{K}$ requires solving the LQR problem, which has the following ARE:

$$\mathbf{Q}(q) - \mathbf{SBR}_0^{-1}\mathbf{B}^{\mathrm{T}}\mathbf{S} + \mathbf{SA} + \mathbf{A}^{\mathrm{T}}\mathbf{S} = 0 \qquad (16.2.128)$$

where $\mathbf{R}_0 = \mathbf{I}$ is the control weighting matrix and $\mathbf{Q}(q) = \mathbf{Q}_0 + q^2\mathbf{CC}^{\mathrm{T}}$ is the modified state weighting matrix. The scalar q is a free design parameter. The resulting regulator gain $\mathbf{K}$ is computed as

$$\mathbf{K} = \mathbf{B}^{\mathrm{T}}\mathbf{S} \qquad (16.2.129)$$

As $q \to 0$ Eq. (16.2.128) becomes the ARE to the optimal regulator problem. As $q \to \infty$ the LQG/LTR technique guarantees that, since the design model $G(s) = \mathbf{C}(s\mathbf{I} - \mathbf{A})^{-1}\mathbf{B}$ has no nonminimum phase zeroes, then pointwise in s

$$\lim_{q \to \infty} G(s)K(s) \to G_{\mathrm{KF}}(s) \qquad (16.2.130)$$

$$\lim_{q \to \infty} \{\mathbf{C}(s\mathbf{I} - \mathbf{A})^{-1}\mathbf{BK}[s\mathbf{I} - (\mathbf{A} - \mathbf{BK} - \mathbf{FC})]^{-1}\mathbf{F}\} \to$$
$$\mathbf{C}(s\mathbf{I} - \mathbf{A})^{-1}\mathbf{F} \qquad (16.2.131)$$

This completes the LQG/LTR design procedure. Now that the loop transfer recovery step is complete, the singular-value plots of the return ratio, return difference, and inverse return difference must be examined to verify that the desirable loop shape has been obtained.

EXAMPLE CONTROLLER DESIGN FOR A DEAERATOR

In this section, three linear control methods are used to obtain a level-control system design for a deaerator. The **three linear control system design methods** that will be used are proportional plus integral, linear quadratic gaussian, and linear quadratic gaussian with loop transfer recovery. In this investigation, the dual of the procedure developed for the LQG/LTR design at the plant input will be used to obtain a **robust control system design** at the **plant output** (Murphy and Bailey, ORNL 1989).

The **Bode gain** and **phase plots** of the resulting control system design will be presented for each controller. Also, the singular-value plots of the **return ratio, return difference,** and **inverse return difference** for opening the loop at the plant output (point 3 in Fig. 16.2.50) will be presented.

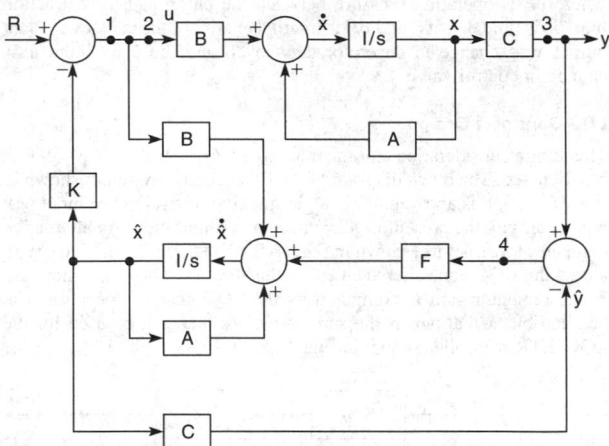

Fig. 16.2.50   Block diagram of LQG system.

The deaerator was chosen because of its **simple mathematical structure;** one input, one output, and three state variables. Thus the model is easy to follow but complex enough to illustrate the design technique.

The Linearized Model

The mathematical model of the deaerator is **nonlinear,** with a process flow diagram as shown in Fig. 16.2.51. This nonlinear plant is **linearized**

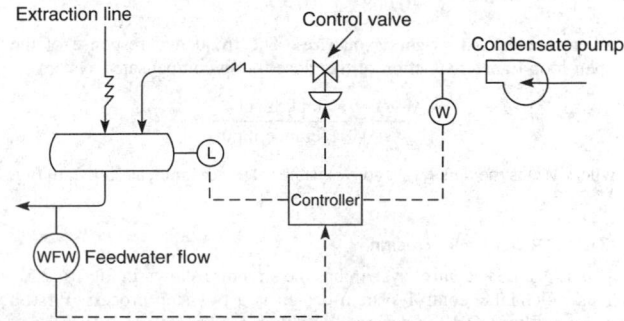

Fig. 16.2.51   General process flow diagram of the deaerator.

about a nominal operating condition, and the resulting linearized plant model will be used to obtain the controller design. The linear **deaerator model** is described by the state space linear time-invariant (LTI) differential equation

$$\dot{x}(t) = \mathbf{A}x(t) + \mathbf{B}u(t) \qquad (16.2.132)$$
$$y(t) = \mathbf{C}x(t) \qquad (16.2.133)$$

where **A, B,** and **C** are the system matrices given respectively as

$$\mathbf{A} = \begin{bmatrix} -53.802 & 1.7093 & 9.92677 \\ 0.0001761 & -0.0009245 & -0.0053004 \\ -0.001124 & -0.0243636 & -0.139635 \end{bmatrix}$$

$$\mathbf{B} = \begin{bmatrix} -9526.25 \\ 0.265148 \\ -1.72463 \end{bmatrix}$$

$$\mathbf{C} = [0.0 \quad 3.2833 \quad 0.06373].$$

The system states are defined as

$$x = \begin{bmatrix} \delta P \\ \delta \rho \\ \delta H \end{bmatrix}$$

where δP is operating pressure between the pump and the extraction line, $\delta \rho$ is fluid density, and δH is internal energy of the tank level. Plant output y = change in deaerator tank level, and the plant input u = change in control valve.

LQG Controller Design

The design considerations for the tracking LQG control system will first be discussed. The block diagram of the LQG control system is shown in Fig. 16.2.50. It is apparent that output tracking of a reference input will not occur with the present LQG controller configuration. An alternative compensator structure, shown in Fig. 16.2.52, is therefore used to obtain a tracking LQG controller. The controller $K(s)$ uses the same filter gain **F** and regulator gain **K** computed by the LQG design procedure. The detailed block diagram is the same structure that will be used for the LQG/LTR controller shown in Fig. 16.2.49.

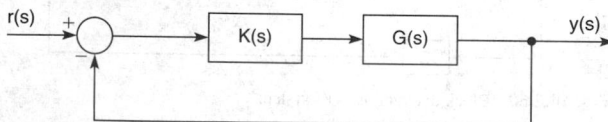

Fig. 16.2.52 Alternative compensator structure.

Using a **control system design package** MATRIX$_X$, the regulator gain **K** and filter gain **F** are computed as

$$\mathbf{K} = 10^5 \times [0.0223 \quad 3.145 \quad -0.1024]$$

and

$$\mathbf{F} = \begin{bmatrix} 0.1962 \\ 9.9998 \\ -0.0083 \end{bmatrix}$$

given **user-defined weighting matrices**. The frequency response of the open-loop transfer function of the closed-loop compensated system

$$\frac{y(s)}{r(s)} = \frac{\text{tank level}}{\text{reference input}}$$

where $y(s)$ is the **tank level** and $r(s)$ is the **reference input,** as shown in Fig. 16.2.53.

LQG/LTR Controller Design

This LQG/LTR control system has the structure shown in Fig. 16.2.52. This LQG/LTR control system design is a two-step process. First, a Kalman filter (KBF) is designed to obtain **good command following** and **disturbance rejection** over a specified low-frequency range. Also, the KBF design is made to meet the required robustness criteria with system

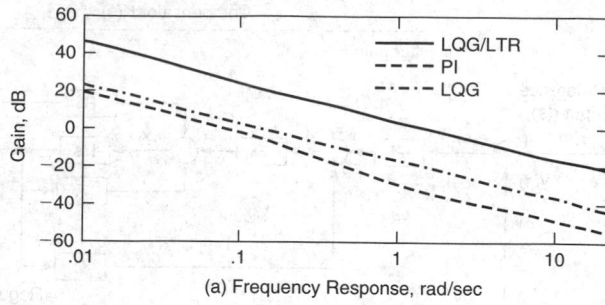

(a) Frequency Response, rad/sec

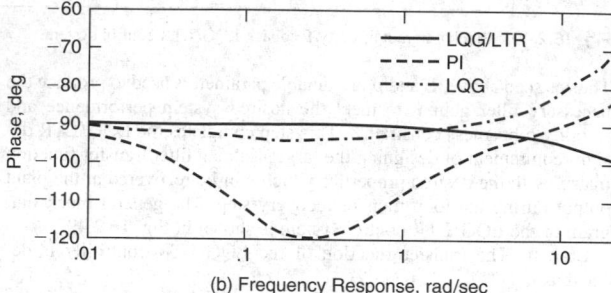

(b) Frequency Response, rad/sec

Fig. 16.2.53 Open-loop frequency response of the PI, LQG, LQG/LTR control system. (*a*) Frequency response of system gain; (*b*) frequency response of system phase.

uncertainty of $\Delta G(s)$ at the system's output. To obtain good command following and disturbance rejection requires that

$$\underline{\sigma}(G_4(s)) \geq 20 \text{ dB} \qquad \forall \ \omega < 0.1 \text{ rad/s} \qquad (16.2.134)$$

where $G_4(s)$ is the loop transfer function at the output of the KBF (point 4 of Fig. 16.2.50). Assuming the **high-frequency uncertainty** $\Delta G(s)$ for this system becomes significant at 5 rad/s, then for robustness

$$\underline{\sigma}\{\mathbf{I} + [G_4(s)]^{-1}\} \geq \|\Delta G(j\omega)\| = 5 \text{ dB}$$
$$\forall \ \omega \geq 5 \text{ rad/s.} \qquad (16.2.135)$$

In this application, the frequency at which system uncertainty becomes significant, 5 rad/s, and the magnitude of the uncertainty, 5 dB at 5 rad/s, are assumed values. This assumption is required because of the lack of information on the high-frequency modeling errors of the process being studied.

The **second step** is to **recover the good robustness properties** of the loop transfer function $G_4(s)$ of point 4 at the output node y (point 3). This will be accomplished by applying the loop transfer recovery (LTR) step at the output node (point 3). The LTR step requires the computation of a regulator gain (**K**) to obtain the robustness properties of the loop transfer function $G_4(s)$ at the output node y.

The tool used to obtain the LQG/LTR design for this system, considering the low- and high-frequency bound requirements is CASCADE, a computer-aided system and control analysis and design environment that synthesizes the LQG/LTR design procedure (Birdwell et al.). Note that more recent computer-aided control system design tools synthesize this design procedure.

The resulting filter gain **F** and regulator **K** gain for this system are respectively

$$\mathbf{F} = \begin{bmatrix} -0.0107 \\ 0.472 \\ -0.02299 \end{bmatrix}$$

and

$$\mathbf{K} = [0.660 \quad 27703.397 \quad 589.4521]$$

The resulting open-loop frequency response of the compensated system is shown in Fig. 16.2.53.

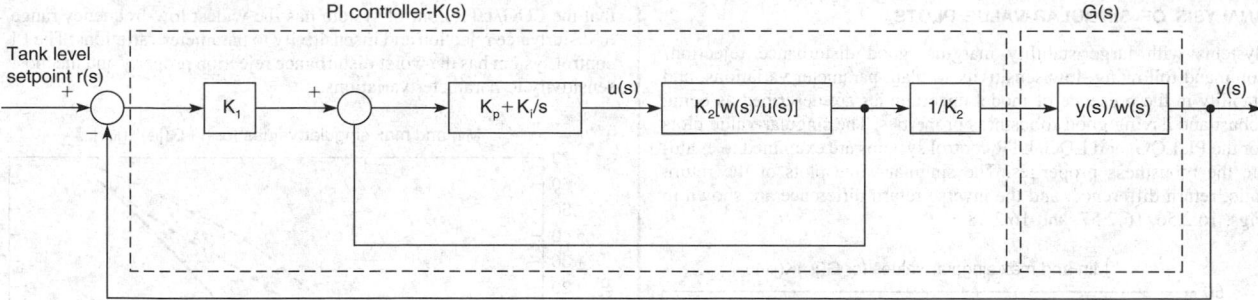

Fig. 16.2.54   Block diagram of a PI control system.

PI Controller Design

The PI controller for this system design takes on a structure similar to the three-element controller prevalent in process control. Therefore the PI control system structure takes the form shown in Fig. 16.2.54. The classical design details of the deaerator are shown in Murphy and Bailey, ORNL 1989.

The transfer functions used in this control structure are defined as

$$\frac{w(s)}{u(s)} = \frac{\text{condensate flow}}{\text{controller output}} \qquad (16.2.136)$$

$$= 4.5227 \times 10^8 \times \frac{s + 0.142}{(s + 0.1405)(s + 53.8015)}$$

and

$$\frac{y(s)}{w(s)} = \frac{\text{tank level}}{\text{condensate}} \qquad (16.2.137)$$

$$= 1.6804 \times 10^{-9} \times \frac{(s + 0.1986)(s + 47.458)}{s(s + 0.142)}$$

The constants K_1 and K_2 are used for unit conversion and defined respectively as 1.0 and 10^{-6}. The terms K_p and K_i are defined respectively as the proportional gain and the integral gain.

Using the practical experience of a prior three-element controller design, the proportional and integral gains are selected to be $K_p = 0.1$ and $K_i = 0.05$. The resulting open-loop frequency response plot of the compensated system $y(s)/r(s)$ is shown in Fig. 16.2.54.

Precompensator Design

The transient responses of the PI, LQG, and LQG/LTR control systems are shown in Fig. 16.2.55. The transient response of the LQG/LTR controller is seen to have a faster time to peak than the LQG and PI transient responses. Considering the practical physical limitations of the plant, it appears unlikely that the required rise time dictated by the transient response of the LQG/LTR control system is possible. The obvious solution would be to redesign the LQG/LTR control system to obtain a slower, more practical response. In the case of the LQG/LTR control system, the possibility of a redesign is eliminated, because the control system was designed to meet certain command-tracking, disturbance-rejection, and stability-robustness requirements.

Therefore, to eliminate this difficulty, a second-order precompensator will be placed in the forward path of the reference input signal. This precompensator will shape the output transient response of the system, since the control system design requires the output to follow the reference input. The precompensator in this investigation will be required to have a time to peak of 30 s with a maximum overshoot of about 3 percent. The requirements are selected to emulate somewhat the transient response of the PI controller.

The transfer function of the second-order precompensator used is as follows:

$$\frac{r(s)}{r_i(s)} = \frac{\omega_n^2}{s^2 + 2\xi\omega_n s + \omega_n^2} \qquad (16.2.138)$$

where ξ is the damping ratio, ω_n is the natural frequency, $r(s)$ is the output of the precompensator, and $r_i(s)$ is the actual reference input. The

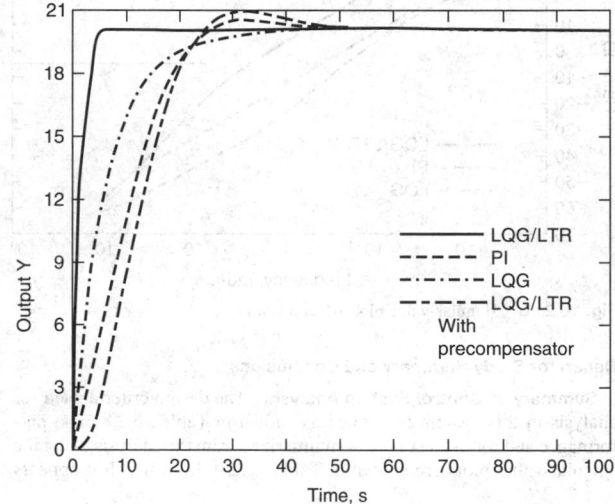

Fig. 16.2.55   PI, LQG, LQG/LTR, LQG/LTR (with precompensator) control system closed-loop transient responses.

precompensator Eq. (16.2.138) can be represented in state equation form as follows:

$$\begin{bmatrix} x_1 \\ x_2 \end{bmatrix} = \begin{bmatrix} 0 & 1 \\ -\omega_n^2 & -\zeta\omega_n \end{bmatrix} \begin{bmatrix} x_1 \\ x_2 \end{bmatrix} + \begin{bmatrix} 0 \\ 1 \end{bmatrix} r_i(t) \qquad (16.2.139)$$

and

$$r(t) = [\omega_n^2 \quad 0] \begin{bmatrix} x_1 \\ x_2 \end{bmatrix} \qquad (16.2.140)$$

where $r_i(t)$ is the time-domain representation of the reference input signal and $r(t)$ is the time-domain signal at the output of the precompensator.

Considering the 3 percent overshoot requirement, then ξ must be 0.7. Using the expression

$$t_{\max} = \frac{\pi}{\omega_n\sqrt{1 - (0.7)^2}} \qquad (16.2.141)$$

where $t_{\max}$ must be 30 s, gives

$$\omega_n = \frac{\pi}{30\sqrt{1 - (0.7)^2}} \qquad (16.2.142)$$

The complete state equation of the precompensator is

$$\begin{bmatrix} x_1 \\ x_2 \end{bmatrix} = \begin{bmatrix} 0 & 1 \\ -0.0215 & -0.2052 \end{bmatrix} \begin{bmatrix} x_1 \\ x_2 \end{bmatrix} + \begin{bmatrix} 0 \\ 1 \end{bmatrix} r_i(t) \qquad (16.2.143)$$

and

$$r(t) = [0.0215 \quad 0] \begin{bmatrix} x_1 \\ x_2 \end{bmatrix} \qquad (16.2.144)$$

The resulting transient response of the LQG/LTR control system using the precompensator is shown in Fig. 16.2.55.

ANALYSIS OF SINGULAR-VALUE PLOTS

Systems with large stability margins, good disturbance rejection/command following, low sensitivity to plant parameter variations, and stability in the presence of model uncertainties are described as being robust and having good robustness properties. The singular-value plots for the PI, LQG, and LQG/LTR control systems are examined to evaluate the robustness properties. The singular-value plots of the return ratio, return difference, and the inverse return difference are shown in Figs. 16.2.56, 16.2.57, and 16.2.58.

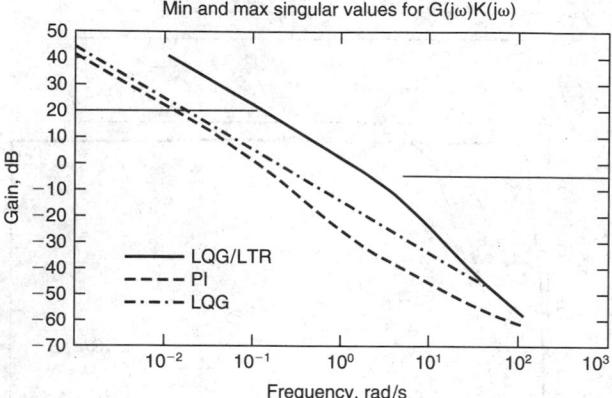

Fig. 16.2.56 Singular-value plots of return ratio.

Deaerator Study Summary and Conclusions

Summary of Control System Analysis The design criteria used for analysis in this system is defined as shown in Table 16.2.8. The performance and robustness results summarized from the analysis using the singular-value plots are shown in Table 16.2.9, from which it appears

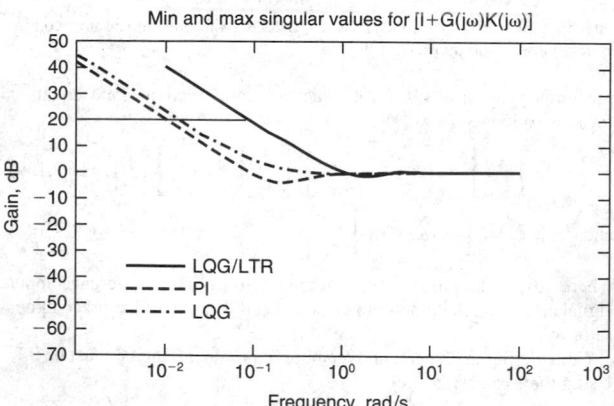

Fig. 16.2.57 Singular-value plots of return difference.

that the LQG/LTR control system has the widest low-frequency range of disturbance rejection and insensitivity to parameter variations. The PI control system has the worst disturbance rejection property and the most sensitivity to parameter variations.

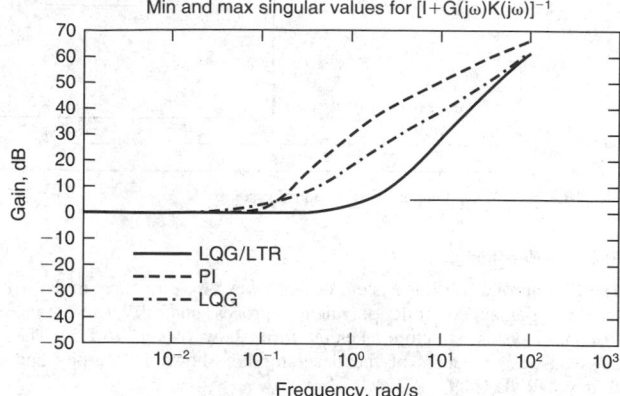

Fig. 16.2.58 Singular-value plots of inverse return difference.

All of the control systems are capable of maintaining a stable robust system in the presence of high-frequency modeling errors, but the PI control system is best. Though all the control systems also have good immunity to noise at frequencies significantly greater than 5 rad/s, the PI control system has the best immunity.

Unlike the other systems, the LQG/LTR control system meets all the design criteria. Its design was obtained in a systematic manner, in contrast to the trial-and-error method used for the LQG control system. The LQG controller design was obtained by shaping the output transient response using the system state weighting matrix, then examining the system's singular-value plots. The PI control system used is simply a three-element control system strategy that has previously been used in the simulation study of the deaerator flow control system.

Conclusions It should be evident that performance characteristics such as a suitable transient response do not imply that the system will have good robustness properties. Obtaining suitable performance characteristics and a stable robust system are two separate goals. Classical unity-feedback design methods have been used for transient response

Table 16.2.8 System Design Specifications

System requirements	Range
Good command-following/ disturbance rejection	$\underline{\sigma}[L(j\omega)] > 20$ dB $\forall\ \omega \leq 0.1$ rad/s
Good system response to high-frequency modeling error	$\underline{\sigma}(I + [L(j\omega)]^{-1}) \geq \|\Delta L\| = 5$ dB $\forall\ \omega \leq 5$ rad/s
Good insensitivity to parameter variations at low frequencies	$\underline{\sigma}[I + L(j\omega)] \geq 26$ dB for low frequencies
Good immunity to noise, $\omega \geq 5$ rad/s	$\overline{\sigma}[L(j\omega)] \ll 0$ dB for high frequencies

Table 16.2.9 Performance and Robustness Results

System properties	PI control system	LQG control system	LQG/LTR control system
Command following/disturbance	$\underline{\sigma}[L(j\omega)] \geq 20$ dB $\forall\ \omega \leq 0.01$ rad/s	$\underline{\sigma}[L(j\omega)] \geq 20$ dB $\forall\ \omega \leq .017$ rad/s	$\underline{\sigma}[L(j\omega)] \geq 20$ dB $\forall\ \omega \leq 0.1$ rad/s
System response to high-frequency modeling error ($\omega \geq 5$ rad/s)	$\underline{\sigma}(I + [L(j\omega)]^{-1})$ ≥ 43.0 dB	$\underline{\sigma}(I + [L(j\omega)]^{-1}) \geq 30$ dB	$\underline{\sigma}(I + [L(j\omega)]^{-1}) \geq 16.7$ dB
Insensitivity to parameter variations at low frequencies	$\underline{\sigma}[I + L(j\omega)] \geq 26$ dB $\forall\ \omega \leq 0.0055$ rad/s	$\underline{\sigma}[I + L(j\omega)] \geq 26$ dB $\forall\ \omega \leq 0.008$ rad/s	$\underline{\sigma}[I + L(j\omega)] \geq 26$ dB $\forall\ \omega \leq 0.055$ rad/s
Immunity to noise ($\omega \geq 5$ rad/s)	$\overline{\sigma}[L(j\omega)] \leq -42.5$ dB	$\overline{\sigma}[L(j\omega)] \leq -30$ dB	$\overline{\sigma}[L(j\omega)] \leq -16$ dB

shaping, with little consideration for robustness properties. The main advantage of a closed-loop feedback system is that good performance and stability robustness properties are obtainable. As has been shown, transient response shaping can be obtained easily using a precompensator. Therefore it appears that the goal of the control system designer is first to design a stable robust system and then use prefiltering to obtain the desired transient response.

Optimal control methods as demonstrated by the LQG control system do not guarantee good robustness properties when applied systematically to meet minimization requirements of a performance index. Methods such as **pole placement** emphasize transient response characteristics without regard to robustness, which could be detrimental to the system integrity in the presence of model uncertainties.

TECHNOLOGY REVIEW

Fuzzy Control

Fuzzy controllers are a popular method for construction of simple control systems from **intuitive knowledge of a process**. They generally take the form of a set of **IF . . . THEN . . .** rules, where the conditional parts of the rules have the structure "Variable i is <fuzzy value 1> AND Variable j is <fuzzy value 2> . . . ." Here the <fuzzy value> refers to a **membership function,** which defines a range of values in which the variable may lie and a number between zero and one for each element of that range specifying the possibility (where certainty is one) that the variable takes that value. Most fuzzy control applications use an **error signal** and the rate of change of an error signal as the two variables to be tested in the conditionals.

The actions of the rules, specified after THEN, take a similar form. Because only a finite collection of membership functions is used, they can be named by mnemonics which convey meaning to the designer. For **example,** a fuzzy controller's rule base might contain the following rules:

IF **temperature-error is high and temperature-error-rate is slow THEN fuel-rate is negative.**

IF **temperature-error is zero and temperature-error-rate is slow THEN fuel-rate is zero.**

Fuzzy controllers are an attractive control design approach because of the ease with which fuzzy rules can be interpreted and can be made to follow a designer's **intuitive knowledge** regarding the control structure. They are inherently nonlinear, so it is easy to incorporate effects which mimic variable gains by adjusting either the rules or the definitions of the membership functions. Fuzzy controller technology is attractive, in part because of its intuitive appeal and its **nonlinear nature,** but also because of the frequent claim that fuzzy rules can be utilized to approximate arbitrary continuous functional relationships. While this is true, in fact it would require substantial complexity to construct a fuzzy rule base to mimic a desired function sufficiently well for most engineers to label it an approximation. Herein also lie the potential liabilities of fuzzy controller technology: there is very little theoretical guidance on how rules and membership functions should be specified. Fuzzy controller design is an **intuitive process;** if intuition fails the designer, one is left with very little beyond trial and error.

One remedy for this situation is the use of machine, or **automated learning technology** to **infer** fuzzy controller **rules** from either collected data or simulation of a process model. This process is inference, rather than deductive reasoning, in that proper operation of the controller is **learning by example,** so if the examples do not adequately describe the process, problems can occur. This inference process, however, is quite similar to the methods of system identification, which also infer model structure from examples, and which is an accepted method of model construction. Machine learning technology enables utilization of more complex controller structures as well, which introduce additional degrees of freedom in the design process but also enhance the capabilities of the controller. One **example** of this approach is the **Fuzzy PID** con-

troller structure (Wang and Birdwell), which utilizes a fuzzy rule base to generate gains for a PID controller. Other similar approaches have been reported in the literature. An advantage of this approach is the ability to establish theoretical results on closed loop stability, which are otherwise lacking for fuzzy controllers.

Signal Validation Technology

Continuous monitoring of instrument channels in a process industry facility serves many purposes during plant operation. In order to achieve the desired operating condition, the system states must be measured accurately. This may be accomplished by implementing a **reliable signal validation procedure** during both normal and transient operations. Such a system would help reduce challenges on control systems, **minimize plant downtime,** and help plan maintenance tasks. Examples of measurements include pressure, temperature, flow, liquid level, electrical parameters, machinery vibration, and many others. The performance of control, safety, and plant monitoring systems depends on the accuracy of signals being used in these systems. Signal validation is defined as the detection, isolation, and characterization of faulty signals. This technique must be applicable to systems with redundant or single sensor configuration.

Various methods have been developed for signal validation in aerospace, power, chemical, metals, and other process industries. Most of the early development was in the aerospace industry. The signal validation techniques vary in complexity depending on the level of information to be extracted. Both **model-based** and **direct data-based techniques** are now available. The following is a list of **techniques** often implemented for **signal validation:**

Consistency checking of redundant sensors
Sequential probability ratio testing for incipient fault detection
Process empirical modeling (static and dynamic) for state estimation
Computational neural networks for state estimation
Kalman filtering technique for state estimation in both linear and nonlinear systems
Time-series modeling techniques for sensor response time estimation and frequency bandwidth monitoring

PC-based signal validation systems (Upadhyaya, 1989) are now available and are being implemented on-line in various industries. The computer software system consists of one or more of the above signal processing modules, with a decision maker that provides sensor status information to the operator.

An example of **signal validation** (Upadhyaya and Eryurek) in a power plant is the monitoring of water level in a steam generator. The objective is to estimate the water level using other related measurements and compare this with the measured level. An empirical model, a neural network model, or the Kalman filtering technique may be used. For example, the inputs to an empirical model consist of steam generator main feedwater flow rate, steam generator pressure, and inlet and outlet temperatures of the primary water through the steam generator. Such a model would be developed during normal plant operation, and is generally referred to as the *training phase.* The signal estimation models should be updated according to the plant operational status. The use of multiple signal validation modules provides a high degree of confidence in the results.

Chaos

Recent advances in dynamical process analysis have revealed that nonlinear interactions in simple deterministic processes can often result in highly complex, aperiodic, sometimes apparently "random" behavior. Processes which exhibit such behavior are said to be exhibit *deterministic chaos.* The term *chaos* is perhaps unfortunate for describing this behavior. It evokes images of purely erratic behavior, but this phenomenon actually involves highly structured patterns. It is now recognized that deterministic chaos dominates many engineering systems of practical interest, and that, in many cases, it may be possible to exploit previously unrecognized deterministic structure for improved understanding and control.

Deterministic chaos and nonlinear dynamics both apply to phenomena that arise in process equipment, motors, machinery, and even control systems as a result of nonlinear components. Since nonlinearities are almost always present to some extent, nonlinear dynamics and chaos are the rule rather than the exception, although they may sometimes occur to such a small degree that they can be safely ignored. In many cases, however, nonlinearities are sufficiently large to cause significant changes in process operation. The effects are typically seen as unstable operation and/or apparent noise that is difficult to diagnose and control with conventional linear methods. These instabilities and noise can cause reduced operating range, poor quality, excessive downtime and maintenance, and, in worst cases, catastrophic process failure.

The apparent erratic behavior in chaotic processes is caused by the extreme sensitivity to initial conditions in the system, introduced by nonlinear components. Though the system is deterministic, this sensitivity and the inherent uncertainty in the initial conditions make long-term predictions impossible. In addition, small external disturbances can cause the process to behave in unexpected ways if this sensitivity has not been considered. Conversely, if one can learn how the process reacts to small perturbations, this inherent sensitivity can be taken advantage of. Once the system dynamics are known, small changes in system parameters can be used to drive the system to a desired state and to keep it there. Thus, great effects can be achieved through minimal input. Various control algorithms have been developed that use this principle to achieve their goals. A recent example from engineering is the control of a slugging fluidized bed by Vasadevan et al.

The following example illustrates the use of chaos analysis and control which involves a laboratory fluidized-bed experiment. Several recent studies have recognized that some modes of fluidization in fluidized beds can be classified as deterministic chaos. These modes involve large scale cyclic motions where the mass of particles move in a piston-like manner up and down in the bed. This mode is usually referred to as *slugging* and is usually considered undesirable because of its inferior transfer properties. Slugging is notoriously difficult to eradicate by conventional control techniques. Vasadevan et al. have shown that the inherent sensitivity to small perturbations can be exploited to alleviate the slugging problem. They did this by adding an extra small nozzle in the bed wall at the bottom of the bed. The small nozzle injected small pulses of process gas into the bed, thus disrupting the "normal" gas flow. The timing of the pulses was shown to be crucial to achieve the desired effect. Different injection timing caused the slugging behavior to be either enhanced or destroyed.

16.3 SURVEYING
by W. David Teter

REFERENCES: Moffitt, Bouchard, "Surveying," HarperCollins. Wolf and Brinker, "Elementary Surveying," Harper & Rowe. Kavanaugh and Bird, "Surveying Principles and Applications," Prentice-Hall. Kissam, "Surveying for Civil Engineers," McGraw-Hill. Anderson and Mikhail, "Introduction to Surveying," McGraw-Hill. McCormac, "Surveying Fundamentals," Prentice-Hall.

INTRODUCTION

Surveying is often defined as the art and science of measurement for location or establishment of position above, on, or beneath the earth's surface. The principles of surveying practice have remained consistent from their earliest inception, but in recent years the equipment and technology have changed rapidly. The emergence of the total station device, electronic distance measurements (EDM), Global Positioning System (GPS), and geographic information systems (GIS) is of significance in comparing modern surveying to past practice.

The information required by a surveyor remains basically the same and consists of the measurement of direction (angle) and distance, both horizontally and vertically. This requirement holds regardless of the type of survey such as land boundary description, topographic, construction, route, or hydrographic.

HORIZONTAL DISTANCE

Most surveying and engineering measurements of distance are horizontal or vertical. Land measurement referenced to a map or plat is reduced to horizontal distance regardless of the manner in which the field measurements are made.

The methods and devices used by the surveyor to measure distance include pacing, odometers, tachometry (stadia), steel tape, and electronic distance measurement. These methods produce expected precision ranging from ± 2 percent for pacing to + 0.0003 percent for EDM. The surveyor must be aware of the necessary precision for the given application.

Tapes

Until the advent of EDM technology, the primary distance-measuring instrument was the steel tape, usually 100 ft or 30 m (or multiples) in length, with 1 ft or 1 dm on the end graduated to read with high precision (0.01 ft or 1 mm). Cloth tapes can be used when low precision is acceptable.

Corrections Measurements with a steel tape are subject to variations that normally must be corrected for when high precision is required. Tapes are manufactured to a standard length at usually 68°F (20°C), for a standard pull of between 10 and 20 lb (4.5 and 9 kg), and supported over the entire length. The specifications for the tape are provided by the manufacturer. If any of the conditions vary during field measurement, corrections for temperature, pull, and sag will have to be made according to the manufacturer's instructions.

Tape Use Most surveying measurements are made with reference to the horizontal, and if the terrain is level, involves little more than laying down the tape in a sequential manner to establish the total distance. Care should be taken to measure in a straight line and to apply constant tension. When measuring on a slope, either of two methods may be employed. In the first method, the tape is held horizontal by raising the low end and employing a plumb bob for location over the point. In cases of steep terrain a technique known as *breaking tape* is used, where only a portion of the total tape length is used. With the second method, the slope distance is measured and converted by trigonometry to the horizontal distance. In Fig. 16.3.1 the horizontal distance $D = S \cos (\alpha)$ where S is the slope distance and α is the slope angle which must be estimated or determined.

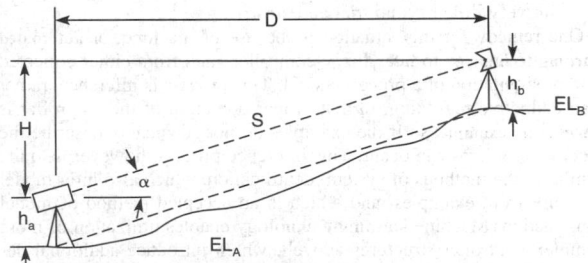

Fig. 16.3.1 Slope-reduction measurements used with EDM devices.

Electronic Distance Measurement

Modern surveying practice employs EDM technology for precise and rapid distance measurement. A representation of an EDM device is shown in Fig 16.3.2. The principle of EDM is based on the comparison of the modulated wavelength of an electromagnetic energy source (beam) to the time required for the beam to travel to and return from a point at an unknown distance. The EDM devices may be of two types: (1) electrooptical devices employing light transmission within or just beyond the visible region or (2) devices transmitting in the microwave spectrum. Electrooptical devices require an active transmitter and a passive reflector, while microwave devices require an active transmitter

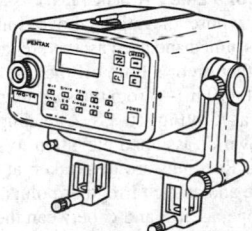

Fig. 16.3.2 Electronic distance meter (EDM). *(PENTAX Corp.)*

and an identical unit for reception and retransmission at the endpoint of the measured line. Ideally the EDM beam would propagate at the velocity of light; however, the actual velocity of propagation is affected by the index of refraction of the atmosphere. As a result, the atmospheric conditions of temperature, pressure, and humidity must be monitored in order to apply corrections to the measurements. Advances in electrooptical technology have made microwave devices, which are highly sensitive to atmospheric relative humidity, nearly obsolete.

Atmospheric Corrections   EDM devices built prior to about 1982 require a manual computation for atmospheric correction; however, the more modern instruments allow for keystroke entry of values for temperature and pressure which are processed and applied automatically as a correction for error to the output reading of distance. The relationships between temperature, pressure, and error are shown in Fig. 16.3.3. Microwave devices require an additional correction for relative humidity. A psychrometer wet-dry bulb temperature difference of 3°F (2°C) causes an error of about 0.001 percent in distance measurement.

Instrument Corrections   Electrooptical devices may also be subject to error attributed to *reflector constant*. This results from the physical

center of the reflector not being coincident with the effective or optical center. The error can range to 30 or 40 mm and is unique, but is usually known, for each reflector. The older EDM devices required this value to be subtracted from each reading, but in more modern devices the reflector constant can be preset into the instrument and the compensation occurs automatically. Finally, microwave EDM devices are subject to error from ground reflection which is known as *ground swing*. Care must be taken to minimize this effect by deploying the devices as high as possible and averaging multiple measurements.

EDM Field Practice   Practitioners have developed many variations in employing EDM devices. Generally the field procedures are governed by the fact that the distance readout from the device is a slope measurement which must be reduced to the horizontal equivalent. Newer EDM devices may allow for automatic reduction by keystroke input of the vertical angle, if known, as would be the case when the EDM device is employed with a theodolite or if the device is a total station (see later in the section). Otherwise the slope reduction can be accomplished by using elevation differences. With reference to Fig. 16.3.1, it is seen that $H = (EL_A + h_a) - (EL_B + h_b)$ and that the horizontal distance $D = (S^2 - H^2)^{1/2}$. EDM devices are sometimes mounted on a conventional theodolite. In such cases the vertical angle as measured by the theodolite must be adjusted to be equivalent to the angle seen by the EDM device. It is common to sight the theodolite at a point below the reflector equal to the vertical distance between the optical centers of the theodolite and the EDM device.

VERTICAL DISTANCE

The acquisition of data for vertical distance is also called **leveling**. Methods for the determination of vertical distance include direct measurement, tachometry or stadia leveling, trigonometric leveling, and differential leveling. Direct measurement is obvious, and stadia leveling is discussed later.

Trigonometric Leveling   This method requires the measurement of the vertical angle and slope distance between two points and is illustrated in Fig. 16.3.1. The vertical distance $H = S \sin (\alpha)$. A precise value for the vertical angle and slope distance will not be obtained if the EDM device is mounted on the theodolite or the instrument heights h_a and h_b are not equal. See "EDM Field Practice," above. Before the emergence of EDM devices, the slope distance was simply taped and the vertical angle measured with a theodolite or transit.

If the elevation of an inaccessible point is required, the EDM reflector cannot be placed there and a procedure employing two setups of a transit or theodolite may be used, as shown in Fig. 16.3.4. Angle $Z_3 = Z_1 - Z_2$, and the application of the law of sines shows that distance $A/\sin Z_2 = C/\sin Z_3$. C is the measured distance between the two instrument setups. The vertical distance $H = A \sin Z_1$. The height of the stack above the instrument point is equal to $H + h_1$, where h_1 is the height of the instrument.

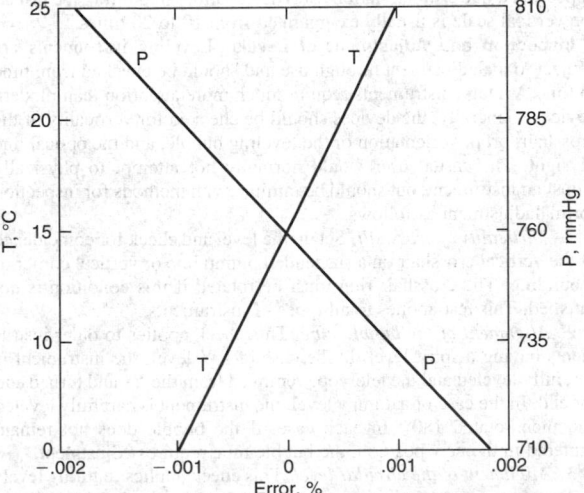

Fig. 16.3.3 Relationship of error to atmospheric pressure and temperature.

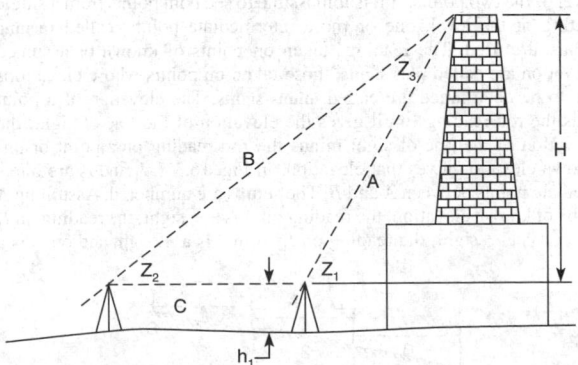

Fig. 16.3.4 Determination of elevation of an inaccessible point.

Differential Leveling This method requires the use of an instrument called a **level**, a device that provides a horizontal line of sight to a rod graduated in feet or metres. Figure 16.3.5 shows a vintage level, and Fig. 16.3.6 shows a modern "automatic" level. The difference is that with the older device, four leveling screws are used to center the spirit-level bubble in two directions so that the instrument is aligned with the

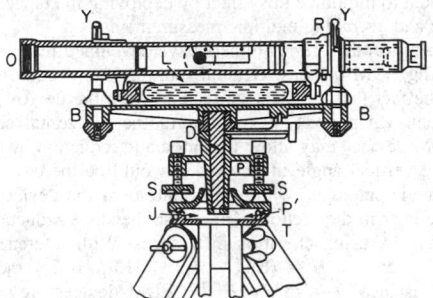

Fig. 16.3.5 Y level.

horizontal. The modern device adjusts its own optics for precise levelness once the three leveling screws have been used to center the rough-leveling/circular level bubble within the scribed circle. The instrument legs should be planted on a firm footing. The eyepiece is checked and adjusted to the user's eye for clear focus on the crosshairs, and the objective-lens focus knob is used to obtain a clear image of the graduated rod.

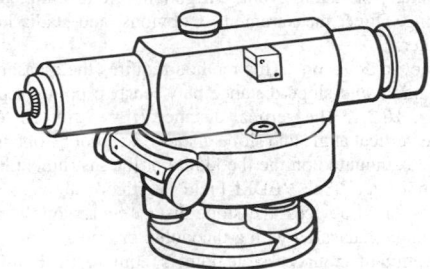

Fig. 16.3.6 Automatic (self-adjusting) level.

To determine the difference in elevation between two points, set the level nearly midway between the points, hold a rod on one, look through the level and see where the line of sight, as defined by the eye and horizontal cross wire, cuts the rod, called **rod reading**. Move the rod to the second point and read. The difference of the readings is the difference in level of the two points. If it is impossible to see both points from a single setting of the level, one or more intermediate points, called **turning points,** are used. The readings taken on points of known or assumed elevation are called **plus sights**, those taken on points whose elevations are to be determined are called **minus sights**. The elevation of a point plus the rod readings on it gives the elevation of the line of sight; the elevation of the line of sight minus the rod reading on a point of unknown elevation gives that elevation. In Fig. 16.3.7, I_1 and I_2 are intermediate points between A and B. The setups are numbered. Assuming A to be of known elevation, the reading on A is a + sight; the reading on I_1 from 1 is a − sight; the reading on I_1 from 2 is a + sight and on I_2 is a

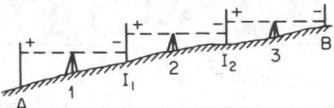

Fig. 16.3.7 Determining elevation difference with intermediate points.

− sight. The algebraic sum of the plus and minus sights is the difference of elevation between A and B. Target rods can usually be read by vernier to thousandths of a foot. In grading work the nearest tenth of a foot is good; in lining shafting the finest possible reading is none too good. It is desirable that the sum of the distances to the plus sights approximately equal the sum of the distances to the minus sights to ensure compensation of errors of adjustment. On a side hill this can be accomplished by zigzagging. When the direction of pointing is changed, the position of the level bubble should be checked. In the case of a vintage instrument, the bubble should be adjusted back to a centered position along the spirit level. An automatic level requires the bubble to be repositioned within the centering circle.

To Make a Profile of a Line A **bench mark** is a point of reasonably permanent character whose elevation above some surface—as sea level—is known or assumed and used as a reference point for elevation. The level is set up either on or a little off the line some distance—not more than about 300 ft (90 m)—from the starting point or a convenient benchmark (BM), as at K in Fig. 16.3.8. A reading is taken on the BM and added to the known or assumed elevation to get the height of the instrument, called HI. Readings are then taken at regular intervals (or **stations**) along the line and at such irregular points as may be necessary to show change of slope, as at B and C between the regular points. The regular points are marked by stakes previously set "on line" at distances of 100 ft (30 m), 50 ft (15 m), or other distance suitable to the

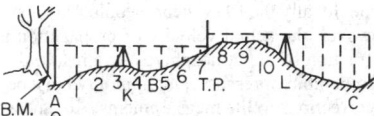

Fig. 16.3.8 Making a profile.

character of the ground and purpose of the work. When the work has proceeded as far as possible—not more than about 300 ft (90 m) from the instrument for good work—a **turning point** (TP) is taken at a regular point or other convenient place, the instrument moved ahead and the operation continued. The first reading on the BM and the first reading on a TP after a new setup are plus sights (+ S); readings to points along the line and the first reading on a TP to be established are minus sights (− S).

The **notes** are taken in the form shown in Fig. 16.3.9. The elevation of a given point, both sights taken on it, and the HI determined from it all appear on a line with its station (Sta) designation. In plotting the **profile**, the vertical scale is usually exaggerated from 10 to 20 times.

Inspection and Adjustment of Levels Leveling instruments are subject to maladjustment through use and should be checked from time to time. Vintage instruments require much more attention than modern devices. Generally, the devices should be checked for verticality of the crosshair, proper orientation of the leveling bubble, and the optical line of sight. The casual user would normally not attempt to physically adjust an instrument, but should be familiar with methods for inspection for maladjustment as follows.

1. *Verticality of crosshair.* Set up the level and check for coincidence of the vertical crosshair on a suspended plumb line or vertical corner of a building. The crosshair ring must be rotated if this condition is not satisfied. This test applies to all types of instruments.

2. *Alignment of the bubble tube.* This check applies to older instruments having a spirit level. In the case of a Y level, the instrument is carefully leveled and the telescope removed from the Ys and turned end for end. In the case of a dumpy level, the instrument is carefully leveled and then rotated 180°. In each case, if the bubble does not remain centered in the new position, the bubble tube requires adjustment.

3. *Alignment of the circular level.* This check applies to tilting levels and modern automatic levels equipped with a circular level consisting of a bubble to be centered within a scribed circle. The instrument is carefully adjusted to center the bubble within the leveling circle and

Left-hand page

	This space for a heading, telling what the work is, who does it, and the date on which it is done.			
Sta.	+ S	H.I.	− S	Elev.
B.M.	6.42	506.42	....	500.0
A = 0........			10.4	496.0
1........			8.2	98.2
2........			6.1	500.3
+30........			5.5	0.9
3........			6.1	0.3
4........			7.9	498.5
B = + 40....			8.4	98.0
5........			7.5	98.9
6........			5.1	501.3
7........			3.2	3.2
+ 10 T.P.....	4.27	509.13	1.56	504.86
8........			2.2	506.9

Right-hand page

This page is for remarks describing B.M.s and T.P.s or other important particulars.

Fig. 16.3.9 Form for surveyor's notes.

then rotated through 180°. If the bubble does not remain centered, adjustment is required.

4. *Line of sight coincident with the optical axis.* This check (commonly called the *two-peg* test) applies to all instruments regardless of construction. The check is performed by setting the instrument midway between two graduated rods and taking readings on both. The instrument is then moved to within the 6 ft of one of the rods, and again readings are taken on both. The difference in readings should agree with the first set. If they do not, an adjustment to raise or lower the crosshair is necessary.

ANGULAR MEASUREMENT

The instruments used for measurement of angle include the transit (Fig. 16.3.10), the optical theodolite, and the electronic total station (Fig. 16.3.11). Angular measurements in both horizontal and vertical planes are obtained with the quantitative value derived in one of three ways depending on the era in which the instrument was manufactured. In principle, the effective manner in which these devices operate is shown in Fig. 16.3.12. The graduated outer circle is aligned at zero with the inner-circle arrow, and an initial point A is sighted on. Then, with the outer circle held fixed by means of clamps on the instrument, the inner circle (with pointer and moving with the telescope) is rotated to a position for sighting on point B. The relative motion between the two circles describes the direction angle which is read to precision with the aid of the scale's vernier. In the actual case, only the transit operates in the mechanical manner described. The operation of an optical theodolite occurs internally with the user viewing internal scales that move relative to each other. Many optical theodolites have scale microscopes or scale micrometers for precise reading. The electronic total station (discussed later) is also dependent on internal optics, but in addition it gives its angle values in the form of a digital display.

Examination of these devices reveals the presence of an *upper clamp* (controlling the movement of the inner circle described before) and a *lower clamp* (controlling the movement of the outer circle). Both clamps have associated with them a *tangent* screw for very fine adjustment. Setting up the instrument involves leveling the device and, unlike

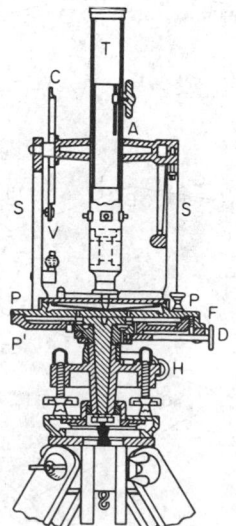

Fig. 16.3.10 Vintage transit.

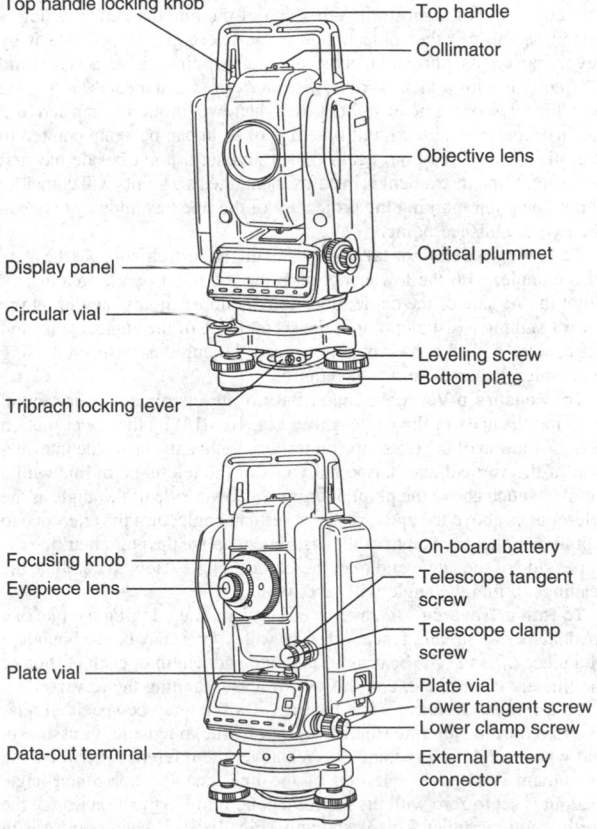

Top handle locking knob — Top handle
— Collimator
— Objective lens
Display panel — — Optical plummet
Circular vial —
Tribrach locking lever — — Leveling screw
— Bottom plate

Focusing knob — — On-board battery
Eyepiece lens — — Telescope tangent screw
— Telescope clamp screw
Plate vial — — Plate vial
— Lower tangent screw
— Lower clamp screw
Data-out terminal — — External battery connector

Fig. 16.3.11 Electronic total station. *(PENTAX Corp.)*

the procedure with a level, locating the instrument exactly over a particular point (the apex of the angle). This is accomplished with a suspended plumb (in the case of older transits) or an optical (line-of-sight) plummet in modern instruments. Prior to any measurement the clamps are manipulated so as to zero the initial angular reading prior to turning the angle to be read. In the case of the digital output total station, the initial angular reading is zeroed by pushing the zero button on the display. A total station, in effect, has only one motion clamp.

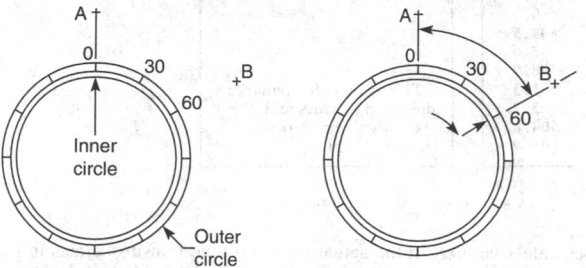

Fig. 16.3.12 Use of graduated circles to measure the angles between points.

Angle Specification Figure 16.3.13 shows several ways in which direction angles are expressed. The angle describing the direction of a line may be expressed as a **bearing,** which is the angle measured east or west from the north or south line and not exceeding 90°, e.g.: N45°E, S78°E, N89°W. Another system for specifying direction is by **azimuth** angle. Azimuths are measured clockwise from north (usually) up to 360°. Still other systems exist where direction is expressed by angle to the right (or left), deflection angles, or interior angles.

To Produce a Straight Line Set up the instrument over one end of the line; with the lower motion clamp and tangent screw bring the telescopic line of sight to the other end of the line marked by a flag, a pencil, a pin, or other object; transit the telescope, i.e., plunge it by revolving on its horizontal axis, and set a point (drive a stake and "center" it with a tack or otherwise) a desired distance ahead in line with the telescopic line of sight; loosen the lower motion clamp and turn the instrument in azimuth until the line of sight can be again pointed to the other end of the line; again transit and set a point beside the first point set. If the instrument is in adjustment, the two points will coincide; if not, the point marking the projection of the line lies midway between the two established points.

To Measure a Horizontal Angle Set up the instrument over the apex of the angle; with the lower motion bring the line of sight to a distant point in one side of the angle; unclamp the upper motion and bring the line of sight to a distant point in the second side of the angle, clamp and set exactly with the tangent screw; read the angle as displayed using scale micrometer or vernier if required.

To Measure a Vertical Angle Set up the instrument over a point marking the apex of the angle A (see Fig. 16.3.14) by the lower motion and the motion of the telescope on its horizontal axis, bring the intersection of the vertical and horizontal wires of the telescope in line with a point as much above the point defining the lower side of the angle as the telescope is above the apex; read the vertical angle, turn the telescope to a point which is the height of the instrument above the point marking the upper side of the angle and read the vertical angle. How to combine the readings to find the angle will be obvious.

To Run a Traverse A traverse is a broken line marking the line of a road, bank of a stream, fence, ridge, or valley, or it may be the boundary of a piece of land. The bearing or azimuth and length of each portion of the line are determined, and this constitutes "running the traverse."

To Establish Bearing The bearing of a line may be specified relative to a north-south line (meridian) that is true, magnetic, or assumed, and whose angle is less than 90°. Whatever the reference meridian, the instrument is set over one end of the line, and the horizontal angle readout is set to zero with the instrument pointing in the direction of the north-south meridian. This is shown in Fig. 16.3.12 when point A is in

the direction of the meridian (north). Using appropriate clamps, the angle is turned by pointing the telescope to the other end of the desired line. In Fig. 16.3.12 the bearing angle would be N60°E. Note that modern electron total stations generally do not have a magnetic needle and, as a result, the bearings or azimuths are typically measured relative to an arbitrary (assumed) meridian.

To establish azimuth the same procedure as described for the determination of bearing may be used. An azimuth may have values up to 360°. When the preceding line of a traverse is used to orient the instrument, the azimuth of this line is known as the **back azimuth** and its value should be set into the instrument as it is pointed along the preceding line

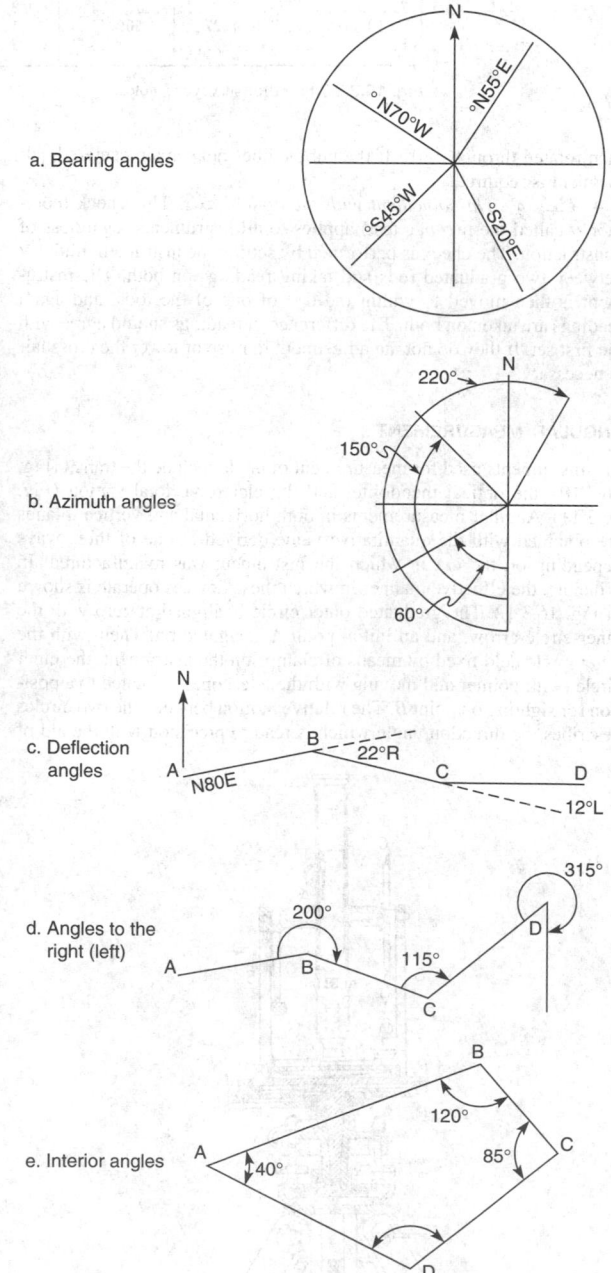

Fig. 16.3.13 Diagram showing various methods of specifying direction angles.

of the traverse. The telescope is turned clockwise to the next line of the traverse to establish the **forward azimuth**.

In traverse work, azimuths and bearings are not usually measured except for occasional checks. Instead, deflection angles or angles to the right from one course to the next are measured. The initial line (course)

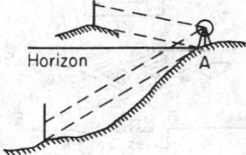

Fig. 16.3.14 Measuring a vertical angle.

is taken as an assumed meridian, and the bearings or azimuths of the other courses with respect to the initial course are calculated. The calculations are derived from the measured deflection angles or angles to the right. If the magnetic or true meridian for the initial course is determined, then the derived bearings or azimuths can likewise be adjusted.

In Fig. 16.3.15, the bearing of a is N40°E, of b is N88° 30′E, of c is S49°20′E = 180° − (40° + 48°30′ + 42°10′), of d is S36°40′W = 86° − 49°20′, or 40° + 48°30′ + 42°10′ + 86° − 180°, of e is N81°20′W = 180° − (36°40′ + 62°). No azimuths are shown in Fig. 16.3.15, but note that the azimuth of a is 40° and the back azimuth of a is 220° (40° + 180°). The azimuths of b and c are 88°30′ and 130°40′ respectively.

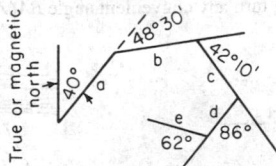

Fig. 16.3.15 Bearings derived from measured angles.

Inspection and Adjustment of the Instrument The transit, optical theodolite, and total station devices are subject to maladjustment through use and should be checked from time to time. Regardless of the type of instrument, the basic relationships that should exist are (1) the plate level-bubble tubes must be horizontal, (2) the line of sight must be perpendicular to the horizontal axis, (3) the line of sight must move in the vertical plane, (4) the bubble (if there is one) of the telescope tube must be centered when the scope is horizontal, and (5) the reading of vertical angles must be zero when the instrument is level and the telescope is level.

All modern devices are subject to the following checks: (1) adjustment of the plate levels, (2) verticality of the crosshair, (3) line of sight perpendicular to the horizontal axis of the telescope, (4) the line of sight must move in a vertical plane, (5) the telescope bubble must be centered (two-peg test), (6) the vertical circle must be indexed, (7) the circular level must be centered, (8) the optical plummet must be vertical, (9) the reflector constant should be verified, (10) the EDM beam axis and the line of sight must be closely coincident, and (11) the vertical and horizontal zero points should be checked.

To Measure Distances with the Stadia In the transit and optical theodolite telescope are two extra horizontal wires so spaced (when fixed by the maker) that they are $\frac{1}{100}$ of the focal length of the objective apart. When looking through the telescope at a rod held in a vertical position, 100 times the rod length S intercepted between the two extra horizontal wires plus an instrumental constant C is the distance D from the center of the instrument to the rod if the line of sight is horizontal, or $D = 100 S + C$ (see Fig. 16.3.16). If the line of sight is inclined by a vertical angle A, as in Fig. 16.3.17, then if S is the space intercepted on the rod and C is the instrumental constant, the distance is given by the formula $D = 100 S \cos^2 A + C \cos A$. For angles less than 5 or 6°, the distance is given with sufficient exactness by $D = 100S$. Although

theory would indicate that distances can thus be determined to within 0.2 ft, in practice it is not well to rely on a precision greater than the nearest foot for distances of 500 ft or less.

Only the oldest transits, known as *external-focusing devices,* have an instrument constant. They can be recognized by noting that the objective lens will physically move as the focus knob is turned. Such devices

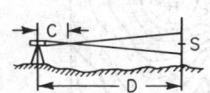

Fig. 16.3.16 Horizontal stadia measurement.

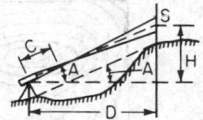

Fig. 16.3.17 Sloping stadia measurement.

have not been manufactured for at least 50 years, and therefore it is most probable that, for stadia applications, the value of $C = 0$ should be used in the equations. Likewise, it is certain that the stadia interval factor K in the following equation will be exactly 100. Stadia surveying falls into the category of low-precision work, but in certain cases where low precision is acceptable, the speed and efficiency of stadia methods is advantageous.

Stadia Leveling This method is related to the previously described trigonometric leveling. The elevation of the instrument is determined by sighting on a point of known elevation with the center crosshair positioned at the height of the instrument (HI) at the setup position. Record the elevation angle A from the setup point to the distant point of known elevation, read the rod intercept S, and apply the following equation to determine the difference in elevation H between the known point and the instrument:

$$H = KS \cos A \sin A + C \sin A$$

Since K is sure to be equal to 100 and that the stadia constant C for most existing instruments will be equal to zero, the value of H is easily determined and the elevation of the instrument HI is known. With known HI, sights can now be performed on other points and the above equation applied to determine the elevation differences between the instrument and the selected points. Note also that the horizontal distance D can be determined by a similar equation:

$$D = \frac{1}{2} KS \sin 2A + C \cos A$$

Stadia Traverse A stadia traverse of low precision can be quickly performed by recording the stadia data for rod intercept S and vertical angle A plus the horizontal direction angle (or deflection angle, angle to the right, or azimuth) for each occupied point defining a traverse. Application of the equation for distance D eliminates the need for laborious taping of a distance. Note that modern surveying with EDM devices makes stadia surveying obsolete.

Topography Low-precision location of topographic details is also quickly performed by stadia methods. From a known instrument setup position, the direction angle to the landmark features, along with the necessary values of rod intercept S and vertical angle A, allows computation of the vertical and horizontal location of the landmark position from the foregoing equations for H and D. If sights are made in a regular pattern as described in the next paragraph, a contour map can be developed.

Contour Maps A contour map is one on which the configuration of the surface is shown by lines of equal elevation called **contour lines**. In Fig. 16.3.18, contour lines varying by 10 ft in elevation are shown. H, H are hill peaks, R, R ravines, S, S saddles or low places in the ridge $HSHSH$. The horizontal distance between adjacent contours shows the distance for a fall or rise of the contour interval—10 ft in the figure. A profile of any line as AB can be made from the contour map as shown in the lower part of the figure. Conversely, a contour map may be made from a series of profiles, properly chosen. Thus, a profile line run along the ridge $HSHSH$ and radiating profile lines from the peaks down the hills and from the saddles down the ravines would give data for projecting points of equal elevation which could be connected for contour lines. This is the best method for making contour maps of very limited areas, such as city squares, or very small parks. If the ground is not too much broken, the small tract is divided into squares and elevations are taken at each square, corner, and between two corners on some lines if necessary to get correct profiles.

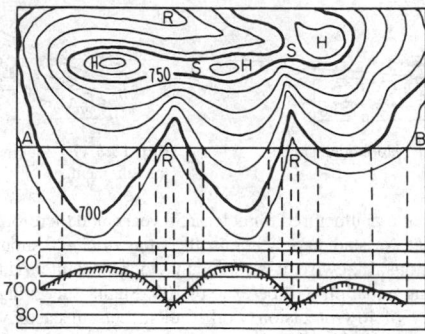

Fig. 16.3.18 Contour map.

SPECIAL PROBLEMS IN SURVEYING AND MENSURATION

Volume of Earth in Foundation and Area Grading The volume of earth removed from a foundation pit or in grading an area can be computed in several ways, of which two follow.

1. The area (Fig. 16.3.19) is divided into squares or rectangles, elevations are taken at each corner before and after grading, and the volumes are computed as a series of prisms. If A is the area (ft^2) of one of the squares or rectangles—all being equal—and b_1, b_2, b_3, b_4 are corner heights (ft) equal to the differences of elevation before and after grading, the subscripts referring to the number of prisms of which b is a corner, then the volume in cubic yards is

$$Q = A(\Sigma h_1 + 2\Sigma h_2 + 3\Sigma h_3 + 4\Sigma h_4) / (4 \times 27)$$

In Fig. 16.3.19 the h's at A_0, D_0, D_3, C_5, and A_5, would be h_1's; those at B_0, C_0, D_1, D_2, C_4, B_5, A_4, A_3, A_2, and A_1 would be h_2's; that at C_3 an h_3; and the rest h_4's. The rectangles or squares should be of such size that their tops and bottoms are practically planes.

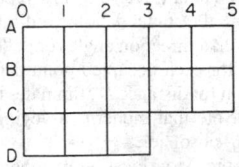

Fig. 16.3.19 Estimating volume of earth by squares.

2. A large-scale profile of each line one way across the area is carefully made, as the A, B, C and D lines of Fig. 16.3.19, the final grade line is drawn on it, and the areas in excavation and embankment are separately measured with a planimeter or by estimation from the drawing. The excavation area of profile A is averaged with that of profile B, and the result multiplied by the distance AB and divided by 27 to reduce to cubic yards. Similarly, the material between B and C is found.

To Pass an Obstacle Four cases are shown in Fig. 16.3.20. If the obstacle is large, as a building, (1) turn right angles at B, C, D and E, making $BC = DE$ when $CD = BE$. All distances should be long enough to ensure sufficiently accurate sighting. (2) At B turn the angle K and measure BC to a convenient point. At C turn left $= 360° - 2K$; measure $CD = BC$. At D turn K for line DE. $BD = 2BC$ cos $(180° - K)$. (3) At B lay off a right angle and measure BC. At C measure any angle to clear object and measure $CD = BC/\cos C$. At D lay off $K = 90° + C$ for the line DE. $BC = BC \times$ tan C. If the obstacle is small, as a tree. (4) at A,

some distance back, turn the small angle a necessary to pass the obstacle and measure AB. At B turn the angle $2a$ and measure $BC = AB$. At C turn the small angle a for the line AC, and transit, or turn the large angle $K = 180° - a$. If a is but a few minutes of arc, $AC = AB + BC$ with

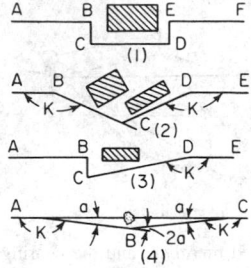

Fig. 16.3.20 Surveying past an obstacle.

sufficient exactness. If only a tape is available, the right-angle method (1) above given may be used, or an equilateral triangle, ABC (Fig. 16.3.21) may be laid out, AC produced a convenient distance to F, the similar triangle DEF laid out, FE produced to H making $FH = AF$, and the similar triangle GHI then laid out for the line GH. $AH = AF$.

To Measure the Distance across a Stream To measure AB, Fig. 16.3.22, B being any established point, tree, stake, or building corner: (1) Set the instrument over A; turn a right angle and measure any distance AC; set over C and measure the angle ACB. $AB = AC$ tan ACB. (2) Set over A, turn any convenient angle BAC' and measure AC';

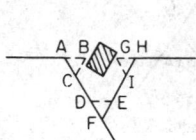

Fig. 16.3.21 Surveying past an obstacle by using an equilateral triangle.

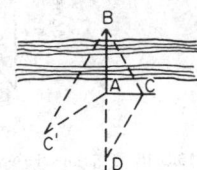

Fig. 16.3.22 Measuring across a stream.

set over C' and measure $AC'B$. Angle $ABC' = 180° - AC'B - BAC'$. $BA = AC' \times$ sin $AC'B$/sin ABC'. (3) Set up on A and produce BA any measured distance to D; establish a convenient point C about opposite A and measure BAC and CAD; set over D and measure ADC; set over C, and measure DCA and ACB; solve ACD for AC, and ABC for AB. For best results the acute angles of either method should lie between 30 and 60°.

To Measure a Visible but Inaccessible Distance (as AB in Fig. 16.3.23) Measure CD. Set the instrument at C and measure angles ACB and BCD; set at D and measure angles CDA and ADB. $CAD = 180° - (ACB + BCD + CDA)$. $AD = CD \times$ sin ACD/sin CAD. $CBD = 180° - (BCD + CDA + ADB)$. $BD = CD \times$ sin BCD/sin CBD. In the triangle ABD, $\frac{1}{2}(B + A) = 90° - \frac{1}{2}D$, where A, B and D are the angles of the triangle; tan $\frac{1}{2}(B - A) = $ cot $\frac{1}{2}D(AD - BD) / (AD + BD)$; $AB = BD$ sin D/sin $A = AD$ sin D/sin$B$.

Random Line On many surveys it is necessary to run a random line from point A to a nonvisible point B which is a known distance away. On the basis of compass bearings, a line such as AB is run. The distances AB and BC are measured, and the angle BAC is found from its calculated tangent (see Fig. 16.3.24).

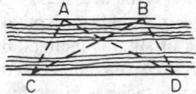

Fig. 16.3.23 Measuring an inaccessible distance.

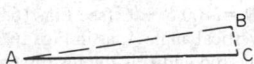

Fig. 16.3.24 Running a random line.

To Stake Out a Simple Horizontal Curve   A simple horizontal curve is composed of a single arc. Usually the curve must be laid out so that it joins two straight lines called tangents, which are marked on the ground by PT (points on tangent). These tangents are run to intersection, thus locating the PI (point of intersection). The plus of the PI and the angle I are measured (see Fig. 16.3.25). With these values and any given value of R (the radius desired for the curve), the data required for staking out the curve can be computed.

$$R = \frac{5,729.58}{D} \qquad L = 100\,\frac{\Delta}{D} \qquad T = R \tan \frac{\Delta}{2}$$

where R = radius, T = tan distance, L = curve length, C = long chord. In a sample computation, assume that Δ = 8°24′ and a 2° curve is required. R = 5,729.58/2 = 2,864.79 ft (873 m); L = (100) (8.40/2) = 420.0 ft (128 m); T = 2,864.79 × 0.07344 = 210.39 ft (64 m). The degree of the curve is always twice as great as the deflection angle for a chord of 100 ft (30 m).

Setting Stakes for Trenching   A common way to give line and grade for trenching (see Fig. 16.3.26) is to set stakes K ft from the center line, driving them so that the near face is the measuring point and the top is some whole inch or tenth of a foot above the bottom grade or grade of the center or top of the pipe to be laid. The top of the pipe barrel is perhaps the better line of reference. If preferred, two stakes can be driven on opposite sides and a board nailed across, on which the center-line is marked and the depth to pipeline given. When only one stake is

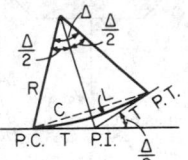

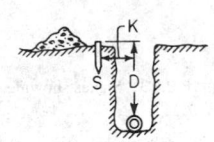

Fig. 16.3.25   Staking out a horizontal curve.

Fig. 16.3.26   Setting stakes for trenching.

used, a graduated pole sliding on one end of a level board at right angles is convenient for workmen and inspectors. On long grades, the grade stakes are set by "shooting in." Two grade stakes are set, one at each end of the grade, the instrument is set over one, its height above grade determined, and a rod reading calculated for the distance stake such as to make the line of sight parallel to the grade line; the line of sight is then set at this rod reading; when the rod is taken to any intermediate stake, the height of instrument above grade less the rod reading will be the height of the top of the stake above grade. If the ground is uniform, the stakes may all be set at the same height above grade by driving them so as to give the same rod readings throughout.

To Reference a Point   The Point P (Fig. 16.3.27), which must be disturbed during construction operations and will be again required as a line point in a railway, pipeline, or other survey, is referenced as follows: (1) Set the instrument over it and set four points, A, B, and C, D on two intersecting lines. When P is again required, the transit is set over B and, with foresight on A, two temporary points close together near P but on opposite sides of the line DC are set; the instrument is then set on D and, with foresight on C, a point is set in the lines DC and BA by setting it in DC under a string stretched between the two temporary points on BA. (2) Points A and E and C and F may be established instead of A, B, C, D. (3) If the ground is fairly level and is not to be much disturbed, only points A and C need be located, and these by simple tape measurement from P. They should be less than a tape length from P. When P is wanted, arcs struck from A and C with the measured distance for radii will give P at their intersection.

Foundations   The corners and lines of a foundation are preserved by setting stakes outside the area to be disturbed, as in Fig. 16.3.28. Cords stretched around nails in the stakes marking the reference points will give the referenced corners at their intersections and the main lines of the building. These corners can be plumbed down to the level desired if

the height of the stakes above grade is given. It is well to nail boards across the stakes at AB, putting nails in the top edge of the board to mark the points A and B, and if the ground permits, to put all the boards at the same level.

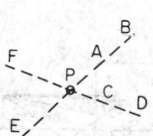

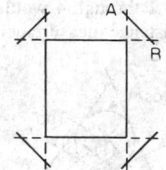

Fig. 16.3.27   Referencing a point which will be disturbed.

Fig. 16.3.28   Reference points for foundation corners.

To Test the Alignment and Level of a Shaft   Having placed the shaft hangers as closely in line as possible by the use of a chalk line, the shaft is finally adjusted for line by hanging plumb lines over one side of the shaft at each hanger and bringing these lines into a line found by stretching a cord or wire or by setting a theodolite at one end and adjusting at each hanger till its plumb line is in the line of sight. The position of the line will be known either on the floor or on the ceiling rafters or beams to which the hangers are attached. If the latter, the instrument may be centered over a point found by plumbing down, and sighted to a plumb line at the farther end.

To level the shaft, an ordinary carpenter's level may be used near each hanger, or, better, a pole with an improvised sliding target may be hung over the shaft at each hanger by a hook in one end. The target is brought to the line of sight of a leveling instrument set preferably about under the middle of the shaft, by adjusting the hanger.

When the hangers are attached to inclined roof rafters, the two extreme hangers can be put in a line at right angles to the vertical planes of the rafters by the use of a square and cord. The other hangers will then be put as nearly as possible without instrumental test in the same line. The shaft being hung, the two extreme hangers, which have been attached to the rafters about midway between their limits of adjustment, are brought to line and level by trial, using an instrument with a well-adjusted telescope bubble, a plumb line, and inverted level rod or target pole. Each intermediate hanger is then tested and may be adjusted by trial.

To Determine the Verticality of a Stack   If the stack is not in use and its top is accessible, a board can be fitted across the top, the center of the opening found, and a plumb line suspended to the bottom, where its deviation from the center will show any leaning. If the stack is in use or its top not accessible and its sides are battered, the following procedure may be followed. Referring to Fig. 16.3.29, set up an instrument at any point T and measure the horizontal angles between vertical planes tangent respectively to both sides of the top and the base and also the angle a to a second point T_1. On a line through T approximately at right angles to the chimney diameter, set the transit at T_1 and perform the same operations as at T_1, measuring also K and the angle b. On the drawing board, lay off K to as large a scale as convenient, and from the plotted T and T_1 lay off the several angles shown in the figure. By trial, draw circumferences tangent to the two quadrilaterals formed by the intersecting tangents of the base and top, respectively. The line joining the centers of these circumferences will be the deviation from the vertical in direction and amount. If the base is square, T and T_1 should be established opposite the middle points of two adjacent sides, as in Fig. 16.3.30.

To Determine Land Area and Boundaries   Two procedures may be followed to gather data to establish and define land area and boundaries. Generally the data is gathered by the traverse method or the radial/radiation-survey method.

The **traverse** method entails the occupation of each end point on the boundary lines for a land area. The length of each boundary is measured (in more recent times with EDM) and the direction is measured by deflection angle or angle to the right. Reference sources explain how direction angles and boundary lengths are converted to **latitudes** and **departures** (north-south and east-west change along a boundary). The latitudes and departures of each course/boundary allow for a simple computation of double meridian distance (DMD), which in turn is directly related by computation to land area. In Fig. 16.3.31 each of the points 1 through 4 would be occupied by the instrument for direction and distance measurement.

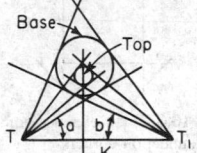
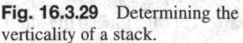
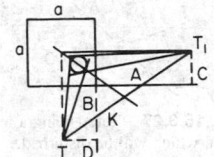

Fig. 16.3.29 Determining the verticality of a stack.

Fig. 16.3.30 Determining the verticality of a stack.

The **radial** method requires, ideally, only a single instrument setup. In Fig. 16.3.31 the setup point might be at A or any boundary corner with known or assumed coordinates and with line of sight available to all other corners. This method relies on gathering data to permit computation of boundary corners relative to the instrument coordinates. The data required is the usual direction angle from the instrument to the point (e.g., angle α in Fig. 16.3.31) and the distance to the point. The references detail the computation of the coordinates of each corner and the computation of land area by the **coordinate method**. The coordinate method used with the radial survey technique lends itself to the use of modern EDM and total station equipment. In very low precision work (approximation) the older methods might be used.

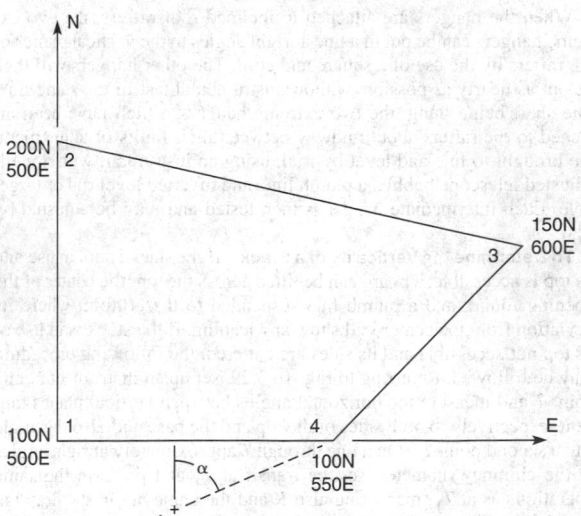

Fig. 16.3.31 Land area survey points.

Stream flow estimates may be obtained by using leveling methods to determine depth along a stream cross section and then measuring current velocity at each depth location with a current meter. As shown in Fig. 16.3.32, the total stream flow Q will be the sum of the products of area A and velocity V for each (say) 10-ft-wide division of the cross section.

To define limits of cut and fill once the route alignment for the roadway has been established (see Fig. 16.3.15), the surveyor uses leveling methods to define transverse ground profiles. From this information and the required elevation of the roadbed, the distance outward from the roadway center is determined, thus defining the limits of cut and fill. The surveyor then implants slope stakes to guide the earthmoving equipment operators. See Fig. 16.3.33.

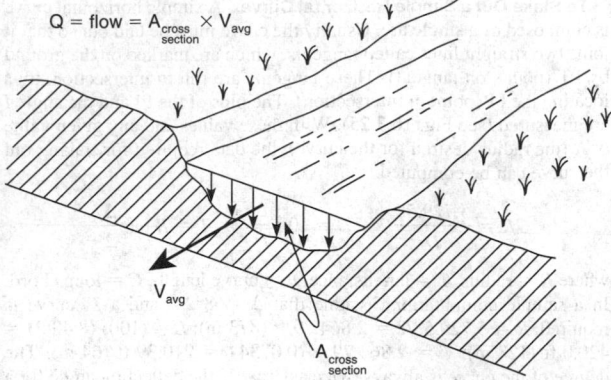

Fig. 16.3.32 Determination of stream flow.

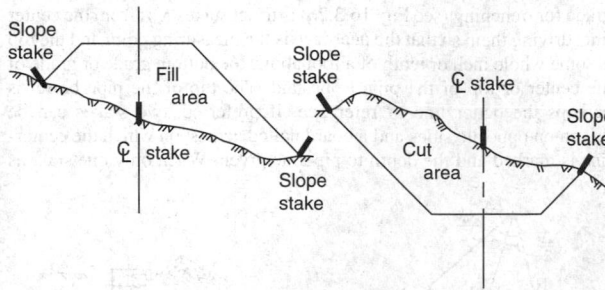

Fig. 16.3.33 Stakes showing limits of roadway cut and fill.

GLOBAL POSITIONING SYSTEM

The GPS concept is based on a system of 24 satellites in six orbital planes with four satellites per orbit. Two elements of data support the underlying principle of GPS surveying. First, radio-frequency signals are transmitted from the satellites and received by GPS receiver units at or near the earth's surface. The signal transmission time is combined with the velocity of propagation (approximately the speed of light) to result in a pseudodistance $d = Vt$, where V is the velocity of propagation and t is the time interval from source to receiver. A second requirement is the precise location of the satellites as published in satellite ephemerides.

The satellite data (ephemeride and time) is transmitted on two frequencies referred to as L1 (1575.42 MHz) and L2 (1227.60 MHz). The transmissions are band-modulated with codes that can be interpreted by the GPS receiver–signal processor. The L1 frequency modulation results in two signals: the precise positioning service (PPS) P code and the C/A or standard positioning service (SPS) code. The L2 frequency contains the P code only. The most recent GPS receivers use dual-frequency receiving and process the C/A code from L1 and the P code from L2.

Single-Receiver Positioning A single point position can be established using one GPS receiver where C/A or SPS code is processed from several satellites (at least four for both horizontal and vertical position). The reliability depends on the uncertainty of the satellite position, which is $+15$ m in the ephemeride data transmitted. This handicap can be circumvented through the application of Loran C technology. The surveyor may also obtain high precision through the use of two receivers and differential positioning. The U.S. government reserves the right to degrade ephemeride data for reasons of national security.

Differential Positioning Multiple GPS receivers may be used to attain submeter accuracy. This is done by placing one receiver at a known location and multiple receivers at locations to be determined. Virtually all error inherent in the transmissions is eliminated or cancels out, since the difference in coordinates from the known position is being determined. By averaging data taken over time and postprocessing, the precision of this technique can approach millimeter precision.

Field Practice A procedure, sometimes called *leapfrogging,* can be used with three GPS receivers (R1, R2, and R3) operating simultaneously (see Fig. 16.3.34). Receiver R1 is placed at a known control point P, while R2 and R3 are placed at L and M. The three units are operated for a minimum of 15 min to obtain time and position data from at least four satellites. The R1 at P and R2 at L are leapfrogged to N and O respectively. This procedure is repeated until all points have been occupied. It is also desirable to include a partial leapfrog to obtain data for the OP line if a closed traverse is wanted.

Fig. 16.3.34 Determining positions by leapfrogging.

A second procedure requires two receivers to be initially located at two known control points. The units are operated for at least 15 min in order to establish data defining the relative location of the two points. Then one of the GPS receivers is moved to subsequent positions until all points have been occupied. As in the previous procedure, signals from a minimum of four satellites must be acquired.

GPS Computing The data for time and position of a satellite acquired by a GPS station is subsequently postprocessed on a computer. The object is to convert the time and satellite-position data to the earth-surface position of the GPS receiver. At least four satellites are needed for a complete (three-dimensional) definition of coordinate location and elevation of a control point. The computer program typically requires input of the ephemerides-coordinate location of the (at least) four satellites, the propagation velocity of the radio transmission from the satellite, and a satellite-clock offset determined by calibration or by receipt of correction data from the satellite.

General Survey Computing The microcomputer is now an integral part of the processing of survey data. Software vendors have developed many programs that perform virtually every computation required by the professional surveyor. Some programs will also produce the graphic output required for almost every survey: site drawings, plat drawings, profile and transverse sections, and topographic maps.

The Total Station The construction of the total station device is such that it performs electronically all the functions of a transit, optical theodolite, level, and tape. It has built-in EDM capability. Thus the total data-gathering requirements are available in a single instrument: horizontal and vertical angles plus the distance from the instrument's optical center to the EDM reflector. A built-in microprocessor converts the slope distance to the desired horizontal distance, and also to vertical distance for leveling requirements. If the height of the instrument's optical center and the height of the reflector are keyed in, the on-board computer can display that actual elevation difference between ground points with correction for earth curvature and atmospheric refraction. A typical total station is shown in Fig. 16.3.11.

Another feature found in some total stations is data-collection capability. This permits data as read by the instrument (angle, distance) to be electronically output to a data collector for transfer to a computer for later processing. A total station operating in conjunction with a remote-processing unit allows a survey to be conducted by a single person. The variety of features found on total station devices is enormous.

The total station changes some of the traditional procedures in surveying practice. An example is the laying out of a route curve as shown in Fig. 16.3.24. The references describe the newer methods.

Section **17**

Industrial Engineering

BY

B. W. NIEBEL *Professor Emeritus of Industrial Engineering, The Pennsylvania State University.*
SCOTT JONES *Professor, Department of Accounting, College of Business and Economics,*
University of Delaware.
ASHLEY C. COCKERILL *Senior Engineer, Motorola Corp.*
VINCENT M. ALTAMURO *President, VMA, Inc., Toms River, NJ.*
ROBERT J. VONDRASEK *Assistant Vice President of Engineering, National Fire Protection Assoc.*
JAMES M. CONNOLLY *Section Head, Projects Department, Jacksonville Electric Authority.*
EZRA S. KRENDEL *Emeritus Professor of Operations Research and Statistics, Wharton School,*
University of Pennsylvania.

17.1 INDUSTRIAL ECONOMICS AND MANAGEMENT
by B. W. Niebel

REFERENCES: Fabrycky and Thuesen, "Economic Decision Analysis," Prentice-Hall. Niebel, "Motion and Time Study," Irwin. Moore, "Manufacturing Management," Irwin. Folts, "Introduction to Industrial Management," McGraw-Hill. Bock and Holstein, "Production Planning and Control," Merrill. Mayer, "Production Management," McGraw-Hill. Maynard, "Handbook of Modern Manufacturing Management," McGraw-Hill. Eary and Johnson, "Process Engineering for Manufacturing," Prentice-Hall. Shamblin and Stevens, "Operations Research—A Fundamental Approach," McGraw-Hill. Fay and Beatty, "The Compensation Sourcebook," Human Resource Development Press.

PLANT ORGANIZATION

Organization generally is recognized as the foundation of management. The term, as it is used in industry and business, means the distribution of the functions of the business to the personnel logically qualified to handle them. It should be noted that the organization should be built around functions rather than individuals.

In the past and to a large extent today, the majority of progressive concerns are organized on a **line-and-staff** basis. The relationships usually are shown on an **organization chart,** which reveals the relationships of the major divisions and departments and the lines of direct authority from superior to subordinate. Lines of authority usually are shown as vertical lines. **Staff authority** frequently is indicated by a dotted line, which distinguishes it from direct authority. This same procedure is usually used to indicate committee relationships. Departments or activities are clearly identified within framed rectangles. The names of individuals responsible for a given department or activity often are included with their job organization titles. Although the organization chart shows the relationship of organization units, it does not clearly define the responsibilities of the individuals and the groups. Thus organization charts must be supplemented with carefully prepared job descriptions for all members of the organization. **Position descriptions** are written definitions of jobs enumerating the duties and responsibilities of each position.

A **line organization** comprises those individuals, groups, and supervising employees concerned directly with the productive operation of the business. The paths of authority are clearly defined, as each individual has but one superior from whom he or she obtains orders and instructions. This superior reports to but one individual, who has complete jurisdiction over his or her operation and supplies necessary technical information. In large- and middle-sized organizations, a pure line-type enterprise cannot exist because of the complexity of our business society.

A **staff organization** involves personnel, departments, or activities that assist the line supervisor in any advisory, service, coordinating, or control capacity. It should be noted that a staff position is a full-time job and is essentially the work of a specialist. Typical staff functions are performed by the company's legal department, controller, and production control. Figure 17.1.1 illustrates a typical line-staff activity.

Committees are used in some instances. A committee is a group of individuals which meets to discuss problems or projects within its area of assigned responsibility in order to arrive at recommendations or decisions. A **committee** operates on a staff basis. Although committees are time-consuming and frequently delay action, their use combines the experience and judgment of several persons, rather than a single individual, in reaching decisions.

The control of organization is the responsibility of two groups of management: (1) **administrative management,** which has the responsibility for determining policy and coordinating sales, finance, production, and distribution, and (2) **production management,** which has the responsibility for executing the policies established by administration.

In building an efficient organization, management should abide by certain principles, namely:

1. Clear separation of the various functions of the business should be established to avoid overlap or conflict in the accomplishment of tasks or in the issuance or reception of orders.

2. Each managerial position should have a definite location within the organization, with a written job specification.

3. There should be a clear distinction between line and staff operation and control.

4. A clear understanding of the authority under each position should prevail.

5. Selection of all personnel should be based on unbiased techniques.

6. A recognized line of authority should prevail from the top of the organization to the bottom, with an equally clear line of responsibility from the bottom to the top.

7. A system of communication should be well established and definitely known—it should be short, yet able to reach rapidly everyone in the organization.

Staff members usually have no authority over any portion of the organization that the staff unit assists. However, the department or division that is being assisted by the staff can make demands upon the staff to provide certain services. There are instances where a control type of staff may be delegated to direct the actions of certain individuals in the organization that they are servicing. When this takes place, the delegated authority may be termed **staff authority;** it is also frequently known as **functional authority** because its scope is determined by the functional specialty of the staff involved.

Many businesses today are finding it fruitful to establish "temporary organizations" in which a team of qualified individuals, reporting to management, is assembled to accomplish a mission, goal, or project, and then this organization is disbanded when the goal is reached.

The term **virtual corporation** is used to identify those combinations of business and industry where technology is used to execute a wide array of temporary alliances in order to grasp specific market opportunities. With business becoming more complex and global, it is highly likely there will be more partnerships emerging among companies and entrepreneurs. Thus today's joint ventures, strategic alliances, and outsourc-

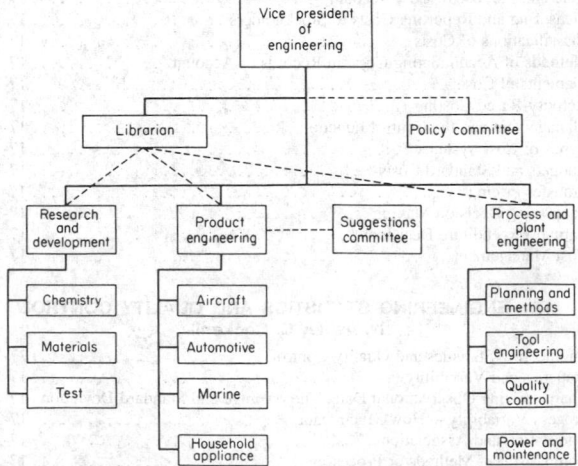

Fig. 17.1.1 Organization chart illustrating the activities reporting to the vice president of engineering.

ing will expand to the virtual corporation utilizing high-speed communication networks. Here, common standards will be used allowing for interchange of design drawings, specifications, and production data.

Thus many businesses and industries are deviating from the classic line-and-staff organization into a more horizontal organizational structure. Each partner in the organization brings its core competence to the effort. Thus, it becomes practical for companies to share skills, costs, and global markets, making it easier to manage a large enterprise since others will be managing components of it for them.

In such organizations there is a strong trend toward **project teams** made up of qualified persons, reporting to management, who are given the authority to do a project (short or long term), complete it, and then be restructured. These project teams frequently are referred to as "temporary organizations." This type of organization characterizes **participative management.** Here position descriptions become less important than functional assignments. An extension of participative management is employee-owned companies in which employees not only assume a production role but also the decision-making shareholder type of role.

Good organization requires that (1) responsibilities be clearly defined; (2) responsibility be coupled with corresponding authority; (3) a change in responsibility be made only after a definite understanding exists to that effect by all persons concerned; (4) no employee be subject to definite orders from more than one source; (5) orders not be given to subordinates over the head of another executive; (6) all criticism be made in a constructive manner and be made privately; (7) promotions, wage changes, and disciplinary action always be approved by the executive immediately superior to the one directly responsible; (8) employees whose work is subject to regular inspection or appraisal be given the facilities to maintain an independent check of the quality of their work.

PROCESS PLANNING

Process planning encompasses selecting the best process to be used in the most advantageous sequence, selecting the specific jigs, fixtures, gages, etc. to be used, and specifying the locating points of the special tools and the speeds, feeds, and depths of cut to be employed.

The processes that take place in transforming a part from chosen raw material to a finished piece include the following.

Basic Processes The first processes used in the sequence that leads to the finished design.

Secondary Processes Those operations required to transform the general form created by the basic process to product specifications. These include:

1. **Critical manufacturing operations** applied to areas of the part where dimensional or surface specifications are sufficiently exacting to require quality control or used for locating the workpiece in relation to other areas or mating parts.

2. **Placement operations** whose method and sequence are determined principally by the nature and occurrence of the critical operations. Placement operations prepare for a critical operation or correct the workpiece to return it to its required geometry or characteristic.

3. **Tie-in operations,** those productive operations whose sequence and method are determined by the geometry to be imposed on the work as it comes out of a basic process or critical operation in order to satisfy the specification of the finished part. Thus, tie-in operations are those secondary productive operations which are necessary to produce the part, but which are not critical.

4. **Protection operations,** those necessary operations that are performed to protect the product from the environment and handling during its progress through the plant and to the customer, and also those operations that control the product's level of quality.

Effective process planning requires the consideration of a large number of manufacturing aspects. Today, the modern computer is able to make the many comparisons and selections in order to arrive at an economic plan that will meet quality and quantity requirements. With computerized planning, considerably less time is required and it can be completed by a technician having less factory experience than needed for manual planning.

PROCESS ANALYSIS

Process analysis is a procedure for studying all productive and nonproductive operations for the purpose of optimizing cost, quality, throughput time, and production output. These four criteria are not mutually exclusive and they are not necessarily negatively correlated. High quality with few if any rejects can result in high production output with low throughput time and cost. All four of these criteria need to be addressed if a facility is going to be a world competitor producing a quality product. It is possible, for example, to have high productivity with efficient equipment producing good quality, but still fall short in the competitive world because of excessive throughput time. The high throughput time will cause poor deliveries and high cost due to excessive in-process inventory resulting from poor planning and scheduling.

In applying process analysis to an existing plant producing a product line, the procedure is first to acquire all information related to the volume of the work that will be directed to the process under study, namely, the expected volume of business, the chance of repeat business, the life of the job, the chance for design changes, and the labor content of the job. This will determine the time and effort to be devoted toward improving the existing process or planning a new process.

Once an estimate is made of quantity, process life, and labor content, then all pertinent factual information should be collected on operations; facilities used for transportation and transportation distances; inspections, inspection facilities, and inspection times; storage, storage facilities, and time spent in storage; vendor operations, together with vendor prices; and all drawings and design specifications. When the information affecting cost and method is gathered, it should be presented in a form suitable for study, e.g., a **flow process chart.** This chart presents graphically and chronologically all manufacturing information. Studies should be made of each event with thought toward improvement. The recommended procedure is to take each step in the present method individually and analyze it with a specific approach toward improvement, considering the key points of analysis. After each element has been analyzed, the process should be reconsidered with thought toward overall improvement. The primary approaches that should be used when analyzing the flow chart include (1) purpose of operation, (2) design of parts, (3) tolerances and specifications, (4) materials, (5) process of manufacture, (6) setup and tools, (7) working conditions, (8) materials handling, (9) plant layout, and (10) principles of motion economy. (See also Sec. 12.1, "Industrial Plants.")

Purpose of Operation Many operations can be eliminated if sufficient study is given the procedural process. Before accepting any operation as necessary, its purpose should be clearly determined and **checklist questions** should be asked to stimulate ideas that may result in eliminating the operation or some component of it. Typical checklist questions are: Can purpose be accomplished better in another way? Can operation be eliminated? Can operation be combined with another? Can operation be performed during idle period of another? Is sequence of operations the best possible?

Design of Parts Design should never be regarded as permanent. Experience has shown that practically every design can be improved. The analyst should consider the existing design to determine if it is possible to make improvements. In general, improvements can be made by (1) simplifying the design through reduction of the number of parts, (2) reducing the number of operations required to produce the design, (3) reducing the length of travel in the manufacture of the design, and (4) utilizing a better material in design.

Tolerances and Specifications These frequently can be liberalized to decrease unit costs without detrimental effects on quality; in other instances, they should be made more rigid to facilitate manufacturing and assembly operations. Tolerances and specifications must be investigated to ensure the use of an optimum process.

Materials Five considerations should be kept in mind relative to both the direct and the indirect material used in the process: (1) finding a less expensive material, (2) finding materials easier to process, (3) using materials more economically, (4) using salvage materials, and (5) using supplies and tools economically.

Process of Manufacture Improvement in the process of manufacture is perhaps the salient point, and possible improvements deserving special consideration include (1) mechanizing manual operations, (2) utilizing more efficient facilities on mechanical operations, (3) operating mechanical facilities more efficiently, and (4) when changing an operation, considering the possible effects on subsequent operations. There are almost always many ways to produce a given design, and better production methods are continually being developed. By systematically questioning and investigating the manufacturing process, more effective methods will be developed.

Setup and tools have such a dominant influence on economics that consideration must include quantity to be produced, chance for repeat business, amount of labor involved, delivery requirements of the customer, and capital needed to develop the setup and provide the tools. Specifically, consideration should be given to reducing the setup time by better planning in production control, designing tooling for the full-capacity utilization of the production facility, and introducing more efficient tooling such as quick-acting clamps and multiple part orientations.

Good working conditions are an integral part of an optimum process as they improve the safety record, reduce absenteeism and tardiness, raise employee morale, improve public relations, and increase production. Consideration should include (1) improved lighting; (2) controlled temperature; (3) adequate ventilation; (4) sound control; (5) promotion of orderliness, cleanliness, and good housekeeping; (6) arrangement for immediate disposal of irritating and harmful dusts, fumes, gases, and fogs; (7) provision of guards at nip points and points of power transmission; (8) installation of personnel-protection equipment; and (9) sponsorship and enforcement of a well-formulated first aid and safety program.

Materials Handling The handling of materials is an essential part of each operation and frequently consumes the major share of the time. Materials handling adds nothing but cost to the product and increased throughput time. It should accordingly be reduced. When analyzing the flow process chart, keep in mind that the best-handled part is the least manually handled part. Whether distances of moves are large or small, points to be considered for reduction of time and energy spent in handling materials are (1) reduction of time spent in picking up material, (2) maximum use of mechanical handling equipment, (3) better use of existing handling facilities, (4) greater care in the handling of materials.

Plant Layout Good process design requires good **plant layout**. This involves development of the workplace so that the location of the equipment introduces low throughput time and maximum economy during the manufacturing process. In general, plant layouts represent one or a combination of (1) product, or straight-line, layouts, and (2) process, or functional, arrangements. In the **straight-line layout**, machinery is located so the flow from one operation to the next is minimized for any product class. To avoid temporary storages between facilities and excess in-processing inventories there needs to be a balance in the number of facilities of each type. **Process, or functional, layout** is the grouping of similar facilities, e.g., all turret lathes in one section, department, or building.

The principal advantage of **product grouping** is lower materials-handling costs since distances moved are minimized. The major disadvantages are:

1. Since a broad variety of occupations are represented in a small area, employee discontent can readily be fostered.
2. Unlike facilities grouped together result in operator training becoming more difficult since no experienced operator on a given facility may be located in the immediate area to train new employees.
3. The problem of finding competent supervisors is increased due to the variety of facilities and jobs to be supervised.
4. Greater initial investment is required because of duplicate service lines such as air, water, gas, oil, and power lines.
5. The arrangement of facilities tends to give a casual observer thought that disorder prevails. Thus it is more difficult to promote good housekeeping.

In general, the disadvantages of product grouping are more than offset by the advantage of low handling cost and lower throughput time.

Process, or functional, layout gives an appearance of neatness and orderliness and, consequently, tends to promote good housekeeping; new workers can be trained more readily, and it is easier to obtain experienced supervision since the requirements of supervising like facilities are not so arduous. The obvious disadvantages of process grouping are the possibilities of long moves and of backtracking on jobs that require a series of operations on diversified facilities. In planning the process, important points to be considered are: (1) For straight-line mass production, material laid aside should be in position for the next operation. (2) For diversified production, the layout should permit short moves and deliveries and the material should be convenient to the operator. (3) For multiple-machine operations, equipment should be grouped around the operator. (4) For efficient stacking, storage areas should be arranged to minimize searching and rehandling. (5) For better worker efficiency, service centers should be located close to production areas. (6) Throughput time is always a major consideration. Scheduling should be well-planned so in-process inventory is kept to minimum levels yet is high enough so that production facilities are not shut down because of lack of material and throughput time is controlled.

Principles of Motion Economy The last of the primary approaches to process design is the analysis of the flow chart for the incorporation of basic principles of motion economy. When studying work performed at any work station, the engineer should ask: (1) Are both hands working at the same time and in opposite, symmetrical directions? (2) Is each hand going through as few motions as possible? (3) Is the workplace arranged so that long reaches are avoided? (4) Are both hands being used effectively, with neither being used as a holding device? In the event that "no" is the answer to any of these questions, then the work station should be altered to incorporate improvements related to motion economy. (See also Sec. 17.4, Methods Engineering.)

JUST-IN-TIME TECHNIQUES, MANUFACTURING RESOURCE PLANNING, AND PRODUCTION CONTROL

Just-in-time techniques, developed by the Japanese as a procedure to control in-process inventories and lead to continual improvement, requires that parts for production be produced or delivered as they are needed. Typically, materials are delivered to the floor when they are expected to be needed according to some planned schedule. Often the difference in time between when parts are thought to be needed and when they are needed is substantial and inventory costs and throughput times can soar.

Manufacturing resource planning (MRP II) procedures recommend that materials be released to the factory floor at that time that the production control system indicates the shop should be ready to receive them.

Production control includes the scheduling of production; the dispatching of materials, tools, and supplies at the required time so that the predicted schedules can be realized; the follow-up of production orders to be sure that proposed schedules are realized; the maintenance of an adequate inventory to meet production requirements at optimum cost; and the maintenance of cost and manufacturing records to establish controls, estimating, and equipment replacement. Consideration must be given to the requirements of the customer, the available capacity, the nature of the work that precedes the production to be scheduled, and the nature of the work that succeeds the current work being scheduled. Centered in the production control effort should be an ongoing analysis to continually focus on any bottlenecks within the plant, since it is here where increases in throughput time take place.

Scheduling may be accomplished with various degrees of refinement. In low-production plants where the total number of hours required per unit of production is large, scheduling may adequately be done by departmental loading; e.g., if a department has a total of 10 direct-labor employees, it has 400 available work hours per week. Every new job is scheduled by departments giving consideration to the average number of available hours within the department. A refinement of this method is to schedule groups of facilities or sections, e.g., to schedule the milling machine section as a group. In high-production plants, detailed facility scheduling frequently is necessary in order to ensure optimum results from all facilities. Thus, with an 8-hour shift, each facility is recorded as having 8 available hours, and work is scheduled to each piece of equipment indicating the time that it should arrive at the work station and the time that the work should be completed.

Scheduling is frequently done on control boards utilizing commercially available devices, such as *Productrol*, *Sched-U-Graph*, and *Visitrol*. These, in effect, are mechanized versions of **Gantt charts**, where schedules are represented by paper strips cut to lengths equivalent to standard times. The strips are placed in the appropriate horizontal position adjacent to the particular order being worked; delays are conspicuously marked by red signals at the delay point. Manual posting to a ledger maintains projected schedules and cumulative loads. The digital computer is successfully used as a scheduling facility.

An adaptation of the Gantt chart, **PERT** (Program Evaluation and Review Techniques), has considerable application to project-oriented scheduling (as opposed to repetitive-type applications). This prognostic management planning and control method graphically portrays the optimum way to attain some predetermined objective, usually in terms of time. The **critical path** (CPM = Critical Path Method) consists of that sequence of events in which delay in the start or completion of any event in the sequence will cause a delay in the project completion.

In using PERT for scheduling, three time estimates are used for each activity, based upon the following questions: (1) What is the earliest time (optimistic) in which you can expect to complete this activity if everything works out ideally? (2) Under average conditions, what would be the most likely time duration for this activity? (3) What is the longest possible time (pessimistic) required to complete this activity if almost everything goes wrong? With these estimates, a probability distribution of the time required to perform the activity can be made (Fig. 17.1.2). The activity is started, and depending on how successfully events take place, the finish will occur somewhere between a and b (most likely close to m). The distribution closely approximates that of the beta distribution and is used as the typical model in PERT. The weighted linear approximation for the expected mean time, using probability theory, is given by

$$t_e = (a + 4m + b)/6$$

With the development of the project plan and the calculation of activity times (time for all jobs between successive nodes in the network, such as the time for "design of rocket ignition system"), a chain of

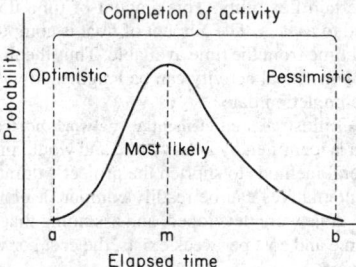

Fig. 17.1.2 Probability distribution of time required to perform an activity.

activities through the project plan can be established which has identical early and late event times; i.e., the completion time of each activity comprising this chain cannot be delayed without delaying project completion. These are the critical events.

Events are represented by nodes and are positions in time representing the start and/or completion of a particular activity. A number is assigned to each event for reference purposes.

Each operation is referred to as an **activity** and is shown as an arc on the diagram. Each arc, or activity, has attached to it a number representing the number of weeks required to complete the activity. Dummy activities, shown as a dotted line, utilize no time or cost and are used to maintain the correct sequence of activities.

The **time to complete** the entire project would correspond to the longest path from the initial node to the final node. In Fig. 17.1.3 the time to complete the project would be the longest path from node 1 to node 12. This longest path is termed **the critical path** since it establishes the minimum project time. There is at least one such chain through any given project. There can, of course, be more than one chain reflecting the minimum time. This is the concept behind the meaning of critical paths. The critical path method (**CPM**) **as compared to PERT** utilizes estimated times rather than the calculation of "most likely" times as previously referred to. Under CPM the analyst frequently will provide two time-cost estimates. One estimate would be for normal operation and the other could be for emergency operation. These two time estimates would reflect the impact of cost on quick-delivery techniques, i.e., the shorter the time the higher the cost, the longer the time the smaller the cost.

It should be evident that those activities that do not lie on the critical

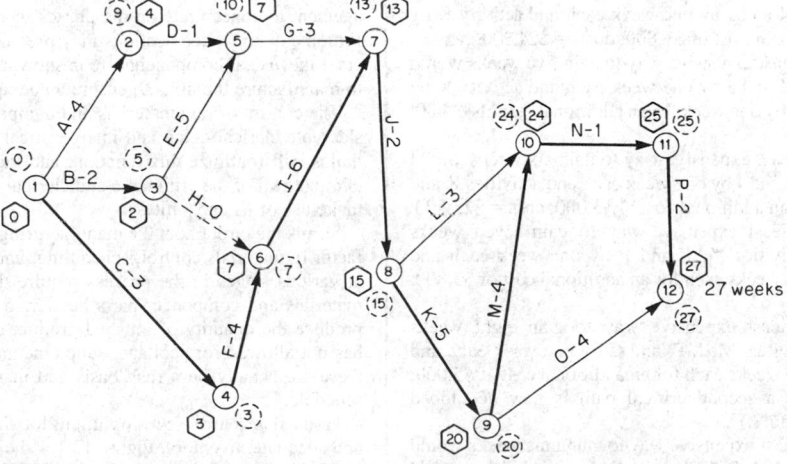

Fig. 17.1.3 Network showing critical path (heavy line). Code numbers within nodes signify events. Connecting lines with directional arrows indicate operations that are dependent on prerequisite operations. Time values on connecting lines represent normal time in weeks. Hexagonals associated with events show the earliest event time. Dotted circles associated with events present the latest event time.

path have a certain flexibility. This amount of time flexibility or freedom is referred to as **float**. The amount of float is computed by subtracting the normal time from the time available. Thus the float is the amount of time that a noncritical activity can be lengthened without increasing the project's completion date.

Figure 17.1.3 illustrates an **elementary network portraying the critical path**. This path is identified by a heavy line and would include 27 weeks. There are several methods to shorten the project's duration. The cost of various time alternatives can be readily computed. For example, if the following cost table were developed, and assuming that a linear relation between the time and cost per week exists, the cost per week to improve delivery is shown.

Activities	Normal		Emergency	
	Weeks	Dollars	Weeks	Dollars
A	4	4000	2	6000
B	2	1200	1	2500
C	3	3600	2	4800
D	1	1000	0.5	1800
E	5	6000	3	8000
F	4	3200	3	5000
G	3	3000	2	5000
H	0	0	0	0
I	6	7200	4	8400
J	2	1600	1	2000
K	5	3000	3	4000
L	3	3000	2	4000
M	4	1600	3	2000
N	1	700	1	700
O	4	4400	2	6000
P	2	1600	1	2400

The **cost of various time alternatives** can be readily computed.

27-week schedule—Normal duration for project; cost = $22,500.

26-week schedule—The least expensive way to gain one week would be to reduce activity M or J for an additional cost of $400; cost = $22,900.

25-week schedule—The least expensive way to gain two weeks would be to reduce activities M and J (one week each) for an additional cost of $800; cost = $23,300.

24-week schedule—The least expensive way to gain three weeks would be to reduce activities M, J, and K (one week each) for an additional cost of $1,300; cost = $23,800.

23-week schedule—The least expensive way to gain four weeks would be to reduce activities M and J by one week each and activity K by two weeks for an additional cost of $1,800; cost = $24,300.

22-week schedule—The least expensive way to gain five weeks would be to reduce activities M and J by one week each and activity K by two weeks and activity I by one week for an additional cost of $2,400; cost = $24,900.

21-week schedule—The least expensive way to gain six weeks would be to reduce activities M and J by one week each and activities K and I by two weeks each for an additional cost of $3,000; cost = $25,500.

20-week schedule—The least expensive way to gain seven weeks would be to reduce activities M, J, and P by one week each and activities K and I by two weeks each for an additional cost of $3,800; cost = $26,300.

19-week schedule—The least expensive way to gain eight weeks would be to reduce activities M, J, P, and C by one week each and activities K and I by two weeks each for an additional cost of $5,000; cost = $27,500. (Note a second critical path is now developed through nodes 1, 3, 5, and 7.)

18-week schedule—The least expensive way to gain nine weeks would be to reduce activities M, J, P, C, E, and F by one week each and activities K and I by two weeks each for an additional cost of $7,800; cost = $30,000. (Note that by shortening time to 18 weeks, we develop a second critical path.)

TOTAL QUALITY CONTROL

Total quality control implies the involvement of all members of an organization who can affect the quality of the output—a product or service. Its goal is to provide defect-free products 100 percent of the time, thus completely meeting the needs of the customer.

ISO 9000 is a quality assurance management system that is rapidly becoming the world standard for quality. The ISO 9000 series standards are a set of four individual, but related, international standards on quality management and quality assurance with one set of application guidelines. The system incorporates a comprehensive review process covering how companies design, produce, install, inspect, package, and market products. As a series of technical standards, ISO 9000 provides a three-way balance between internal audits, corrective action, and corporate management participation leading to the successful implementation of sound quality procedures.

The series of technical standards include four divisions:

ISO 9001 This is the broadest standard covering procedures from purchasing to service of the sold product.

ISO 9002 This is targeted toward standards related to processes and the assignment of subcontractors.

ISO 9003 These technical standards apply to final inspection and test.

ISO 9004 These standards apply to quality management systems.

CONTROL OF MATERIALS

Control of materials is critical to the smooth functioning of a plant. Raw materials and purchased parts must be on hand in the required quantities and at the time needed if production schedules are to be met. Unless management is speculating on raw materials, inventories should be at the lowest practicable level in order to minimize the capital invested and to reduce losses due to obsolescence, design changes, and deterioration. However, some minimum stock is essential if production is not to be delayed by lack of materials. The quantity for ordering replenishment stocks is determined by such factors as the lead time needed by the supplier, the reliability of the sources of supply in meeting promised delivery dates, the value of the materials, the cost of storage, and the risks of obsolescence or deterioration.

In many instances, plant management has the choice of manufacturing the components used in its product or procuring them from outside suppliers. Where suppliers specialize in certain components, they may be able to reach high-volume operations and produce more economically than can the individual users. Procurement from outside suppliers simplifies the manufacturing problem within a plant and permits management to concentrate on the phases where it has critical know-how. Extreme quality specifications may preclude the use of outside suppliers. Likewise, if components are in short supply, the user may be forced to manufacture the units to ensure an adequate supply.

The control of raw materials and component parts may involve considerable clerical detail and many critical decisions. Systems and formulas will routinize this function, and the computer has been able to eliminate all of the arithmetic and clerical activities in those plants that make use of its capability.

Shrinkage throughout the manufacturing process may be a significant factor in materials control, scheduling, and dispatching. Spoilage rates at various stages in the process require that excess quantities of raw materials and component parts be started into the process in order to produce the quantity of finished product desired. If the original order has not allowed for spoilage, supplementary orders will be necessary; these are usually on a rush basis and may seriously disrupt the plant schedule.

Production control seeks optimum lot sizes with minimum total cost and adequate inventory. Figure 17.1.4 shows the time pattern, under an ideal situation, for active inventory. With assumed fixed cost per unit of output (except for starting and storage costs) and with zero minimum inventory, the optimum lot size Q is given by $\sqrt{ah/B}$, where $B =$ factor when storage space is reserved for maximum inventory =

$0.5[1 - (d/r)](2s + ip)$; h = starting cost per lot (planning and setup); a = annual demand; s = annual cost of storage per unit of product; i = required yield on working capital; p = unit cost of production; d = daily demand; and r = daily rate of production during production period.

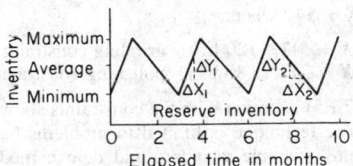

Fig. 17.1.4   Inventory time pattern. $\Delta Y_1/\Delta X_1$ = rate of increase of inventory. $\Delta Y_2/\Delta X_2$ = rate of decrease of inventory.

STRATEGIC ECONOMIC EVALUATIONS

New equipment or facilities may be acquired for a variety of reasons: (1) Existing machines may be so badly worn that they are either beyond repair or excessively costly to maintain. (2) The equipment may be incapable of holding specified quality tolerances. (3) A technical development may introduce a process producing higher-quality products. (4) Changes in the product line may require new kinds of machines. (5) An improved model which reduces operating costs, especially power costs, may come on the market.

A decision to invest in new capital equipment involves the risk that improved models of machines may become available and render the new equipment obsolete before its mechanical life has expired. Aggressive competitors who regularly modernize their plants may force other companies to adopt a similar policy. The **paramount question** on strategic manufacturing expenditures is: Will it pay? Asking this question usually involves the consideration of alternatives. In comparing the economy of alternatives it is important that the engineer understand the concept of return on investment. Let:

i = interest rate per period
n = number of interest periods
C = cash receipts
D = cash payments
P = present worth at the beginning of n periods
S = lump sum of money at the end of n periods
R = an end-of-period payment or receipt in a uniform series continuing for n periods. The present value of the sum of the entire series at interest rate i is equal to P.

Thus, \$1 n years from now = $1/(1 + i)^n$

$$P = \frac{C_0 - D_0}{(1 + i)^0} + \frac{C_1 - D_1}{(1 + i)^1} + \frac{C_2 - D_2}{(1 + i)^2} \cdots \frac{C_n - D_n}{(1 + i)^n}$$

Then:

$S = P(1 + i)^n$ single payment

$P = S \dfrac{1}{(1 + i)^n}$ single payment

$R = S \dfrac{i}{(1 + i)^n - 1}$ uniform series, sinking fund

$R = P \dfrac{i(1 + i)^n}{(1 + i)^n - 1}$ uniform series, capital recovery

$S = R \dfrac{(1 + i)^n - 1}{i}$ uniform series, compound amount

$P = R \dfrac{(1 + i)^n - 1}{i(1 + i)^n}$ uniform series, present worth

For example, a new-type power-lift truck is being contemplated in the receiving department in order to reduce hand labor on a particular product line. The annual cost of this labor and labor extras such as social

security taxes, industrial accident insurance, paid vacations, and various employees' fringe benefits are \$16,800 at present. An alternative proposal is to procure the new power equipment at a first cost of \$20,000. It is expected that this equipment will reduce the annual cost of the hand labor to \$6,900. Annual payments to utilize the new equipment include power (\$700); maintenance (\$2,400); and insurance (\$400).

It has been estimated that the need for this particular operation will continue for at least the next 8 years and that equipment will have no salvage value at the end of the life of this product line. Management requires a 15 percent rate of return before income taxes.

Since the burden of proof is on the proposed investment, its cost will include the 15 percent rate of return in the equivalent uniform annual cost of capital recovery. This computation approach assures that if the new method is adopted the savings will be at least as much before taxes as 15 percent per year.

The following is a **comparison of annual cost** of the present and proposed methods.

The **uniform equivalent cost** of the proposed method shows that nearly \$2,000 per year is saved in addition to the required minimum of 15 percent already included as a cost of investing in the new power-lift truck. Clearly the proposed plan is more economical.

Present Method

Labor + labor extras = \$16,800	
Total annual cost of the present method = \$16,800.00	

Proposed Method

Equivalent uniform annual cost of capital recovery =	
($20,000)(0.15)(1 + 0.15)^8 = 20,000 \times 0.22285 = \$4,457.00	
Labor and labor extras	\$6,900.00
Power	700.00
Maintenance	2,400.00
Insurance	400.00
Total uniform equivalent annual cost of proposed method	\$14,857.00

OPTIMIZATION TECHNIQUES

It is important that modern management be trained in the various **decision-making processes**. Judgment by itself will not always provide the best answer. The various decision-making processes are based on theory of probability, statistical analysis, and engineering economy.

Decision making under **certainty** assumes the states of the product or market are known at a given time. Usually the decision-making strategy under certainty would be based on that alternative that maximizes if we are seeking quality, profit, etc., and that minimizes if we are studying scrap, customer complaints, etc. The several alternatives are compared as to the results for a particular state (quantity, hours of service, anticipated life, etc.). Usually decisions are not made under an assumed certainty. Often **risk** is involved in providing a future state of the market or product. If several possible states of a market prevail, a probability value is assigned to each state. Then a logical decision-making strategy would be to calculate the expected return under each decision alternative and to select the largest value if we are maximizing or the smallest value if we are minimizing. Here

$$E(a) = \sum_{j=1}^{n} p_j c_{ij}$$

where $E(a)$ = expected value of the alternative; p_j = probability of each state of product or market occurring; c_{ij} = outcome for particular alternative i at a state of product or market j.

A different decision-making strategy would be to consider the state of the market that has the greatest chance of occurring. Then the alternative, based upon the **most probable future**, would be that one that is either a maximum or minimum for that particular state.

A third decision-making strategy under risk would be based upon a **level of aspiration**. Here the decision maker assigns an outcome value c_{ij} which represents the consequence she or he is willing to settle for if it is reasonably certain that this consequence or better will be achieved most of the time. This assigned value may be referred to as representing a level of aspiration which can be identified as A. Now the probability for each a_j where the c_{ij} (each decision alternative) is equal or greater to A is determined. The alternative with the greatest $p(c_{ij} \geq A)$ is selected if we are maximizing.

There are other decision-making strategies based on decision under risk and uncertainty. The above examples provide the reader the desirability of considering several alternatives with respect to the different states of the product or market.

Linear Programming

At the heart of management's responsibility is the best or optimum use of limited resources including money, personnel, materials, facilities, and time. Linear programming, a mathematical technique, permits determination of the best use which can be made of available resources. It provides a systematic and efficient procedure which can be used as a guide in decision making.

As an example, imagine the simple problem of a small machine shop that manufactures two models, standard and deluxe. Each standard model requires 2 h of grinding and 4 h of polishing. Each deluxe model requires 5 h of grinding and 2 h of polishing. The manufacturer has three grinders and two polishers; therefore, in a 40-h week there are 120 h of grinding capacity and 80 h of polishing capacity. There is a profit of $3 on each standard model and $4 on each deluxe model and a ready market for both models. The management must decide on: (1) the allocation of the available production capacity to standard and deluxe models and (2) the number of units of each model in order to maximize profit.

To solve this linear programming problem, the symbol X is assigned to the number of standard models and Y to the number of deluxe models. The profit from making X standard models and Y deluxe models is $3X + 4Y$ dollars. The term *profit* refers to the **profit contribution,** also referred to as **contribution margin** or **marginal income.** The profit contribution per unit is the selling price per unit less the unit variable cost. Total contribution is the per-unit contribution multiplied by the number of units.

The restrictions on machine capacity are expressed in this manner: To manufacture one standard unit requires 2 h of grinding time, so that making X standard models uses $2X$ h. Similarly, the production of Y deluxe models uses $5Y$ h of grinding time. With 120 h of grinding time available, the grinding capacity is written as follows: $2X + 5Y \leq 120$ h of grinding capacity per week. The limitation on polishing capacity is expressed as follows: $4X + 2Y \leq 80$ h per week. In summary, the basic information is:

	Grinding time	Polishing time	Profit contribution
Standard model	2 h	4 h	$3
Deluxe model	5 h	2 h	4
Plant capacity	120 h	80 h	

Two basic linear programming techniques, the graphic method and the simplex method, are described and illustrated using the above capacity-allocation–profit-contribution maximization data.

Graphic Method

Operations	Hours available	Hours required per model		Maximum number of models	
		Standard	Deluxe	Standard	Deluxe
Grinding	120	2	5	$\frac{120}{2} = 60$	$\frac{120}{5} = 24$
Polishing	80	4	2	$\frac{80}{4} = 20$	$\frac{80}{2} = 40$

The lowest number in each of the two columns at the extreme right measures the impact of the hours limitations. The company can produce 20 standard models with a profit contribution of $60 (20 × $3) or 24 deluxe models at a profit contribution of $96 (24 × $4). Is there a better solution?

To determine production levels in order to maximize the profit contribution of $3X + $4Y when:

$$2X + 5Y \leq 120 \text{ h} \quad \text{grinding constraint}$$
$$4X + 2Y \leq 80 \text{ h} \quad \text{polishing constraint}$$

a graph (Fig. 17.1.5) is drawn with the constraints shown. The two-dimensional graphic technique is limited to problems having only two variables—in this example, standard and deluxe models. However, more than two constraints can be considered, although this case uses only two, grinding and polishing.

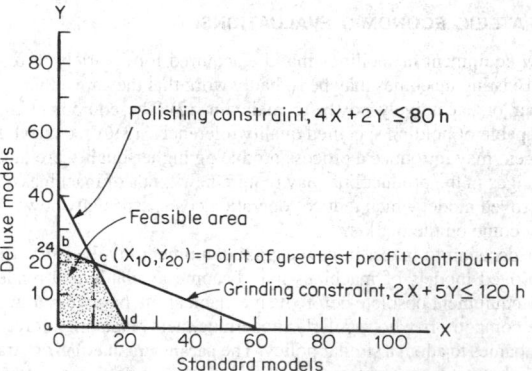

Fig. 17.1.5 Graph depicting feasible solution.

The **constraints** define the solution space when they are sketched on the graph. The solution space, representing the area of feasible solutions, is bounded by the corner points a, b, c, and d on the graph. Any combination of standard and deluxe units that falls within the solution space is a feasible solution. However, the *best* feasible solution, according to mathematical laws, is in this case found at one of the corner points. Consequently, all corner-point variables must be tried to find the combination which maximizes the profit contribution: $3X + $4Y.

Trying values at each of the corner points:

$$a = (X = 0, Y = 0); \$3(0) + \$4(0) = \$ 0 \text{ profit}$$
$$b = (X = 0, Y = 24); \$3(0) + \$4(24) = \$ 96 \text{ profit}$$
$$c = (X = 10, Y = 20); \$3(10) + \$4(20) = \$110 \text{ profit}$$
$$d = (X = 20, Y = 0); \$3(20) + \$4(0) = \$ 60 \text{ profit}$$

Therefore, in order to maximize profit the plant should schedule 10 standard models and 20 deluxe models.

Simplex Method The simplex method is considered one of the basic techniques from which many linear programming techniques are directly or indirectly derived. The method uses an iterative, stepwise process which approaches an optimum solution in order to reach an objective function of maximization (for profit) or minimization (for cost). The pertinent data are recorded in a tabular form known as the **simplex tableau.** The components of the tableau are as follows (see Table 17.1.1):

The **objective row** of the matrix consists of the coefficients of the objective function, which is the profit contribution per unit of each of the products.

The **variable row** has the names of the variables of the problem including slack variables. **Slack variables** S_1 and S_2 are introduced in order to transform the set of inequalities into a set of equations. The use of slack variables involves simply the addition of an arbitrary variable to one side of the inequality, transforming it into an equality. This arbitrary variable is called **slack variable,** since it takes up the slack in the inequality. The simplex method requires the use of equations, in contrast to the inequalities used by the graphic method.

The **problem rows** contain the coefficients of the equations which represent constraints upon the satisfaction of the objective function. Each constraint equation adds an additional problem row.

The **objective column** receives different entries at each iteration, representing the profit per unit of the variables. In this first tableau (the only one illustrated due to space limitations) zeros are listed because they are the coefficients of the slack variables of the objective function. This column indicates that at the very beginning every S_n has a net worth of zero profit.

The **variable column** receives different notations at each iteration by replacement. These notations are the variables used to find the profit contribution of the particular iteration. In this first matrix a situation of no (zero) production is considered. For this reason, zeros are marked in the objective column and the slacks are recorded in the variable column. As the iterations proceed, by replacements, appropriate values and notations will be entered in these two columns, objective and variable.

The **quantity column** shows the constant values of the constraint equations.

Based on the data used in the graphic method and with a knowledge of the basic components of the simplex tableau, the first matrix can now be set up.

Letting X and Y be respectively the number of items of the standard model and the deluxe model that are to be manufactured, the system of inequalities or the set of constraint equations is

$$2X + 5Y \leq 120$$
$$4X + 2Y \leq 80$$

in which both X and Y must be positive values or zero ($X \geq 0$; $Y \geq 0$) for this problem.

The objective function is $3X + 4Y = P$; these two steps were the same for the graphic method.

The set of inequalities used by the graphic method must next be transformed into a set of equations by the use of slack variables. The inequalities rewritten as equalities are

$$2X + 5Y + S_1 = 120$$
$$4X + 2Y + S_2 = 80$$

and the objective function becomes

$$3X + 4Y + 0S_1 + 0S_2 = P \quad \text{to be maximized}$$

The first tableau with the first solution would then appear as shown in Table 17.1.1.

The tableau carries also the first solution which is shown in the **index row**. The index row carries values computed by the following steps:

1. Multiply the values of the quantity column and those columns to the right of the quantity column by the corresponding value, by rows, of the objective column.

2. Add the results of the products by column of the matrix.

3. Subtract the values in the objective row from the results in step 2. For this operation the objective row is assumed to have a zero value in the quantity column. By convention the profit contribution entered in the cell lying in the quantity column and in the index row is zero, a condition valid only for the first tableau; in the subsequent matrices it will be a positive value.

Index row:

Steps 1 and 2:	Step 3:
$120(0) + 80(0) = 0$	$0 - 0 = 0$
$2(0) + 4(0) = 0$	$0 - 3 = -3$
$5(0) + 2(0) = 0$	$0 - 4 = -4$
$1(0) + 0(0) = 0$	$0 - 0 = 0$
$0(0) + 1(0) = 0$	$0 - 0 = 0$

In this first tableau the slack variables were introduced into the product mix, variable column, to find a *feasible* solution to the problem. It can be proven mathematically that beginning with slack variables assures a feasible solution. One possible solution might have S_1 take a value of 120 and S_2 a value of 80. This approach satisfies the constraint equation but is undesirable since the resulting profit is zero.

It is a rule of the simplex method that the optimum solution has not been reached if the index row carries any negative values at the completion of an iteration in a maximization problem. Consequently, this first tableau does not carry the optimum solution since negative values appear in its index row. A second tableau or matrix must now be prepared, step by step, according to the rules of the simplex method.

Duality of Linear Programming Problems and the Problem of Shadow Prices Every linear programming problem has associated with it another linear programming problem called its **dual**. This duality relationship states that for every maximization (or minimization) problem in linear programming, there is a unique, similar problem of minimization (or maximization) involving the same data which describe the original problem. The possibility of solving any linear programming problem by starting from two different points of view offers considerable advantage. The two paired problems are defined as the dual problems because both are formed by the same set of data, although differently arranged in their mathematical presentation. Either can be considered to be the **primal**; consequently the other becomes its dual.

Shadow prices are the values assigned to one unit of capacity and represent economic values per unit of scarce resources involved in the restrictions of a linear programming problem. To maximize or minimize the total value of the total output it is necessary to assign a quantity of unit values to each input. These quantities, as cost coefficients in the dual, take the name of "shadow prices," "accounting prices," "fictitious prices," or "imputed prices" (values). They indicate the amount by which total profits would be increased if the producing division could increase its productive capacity by a unit. The shadow prices, expressed by monetary units (dollars) per unit of each element, represent the least cost of any extra unit of the element under consideration, in other words, a kind of marginal cost. The real use of shadow prices (or values) is for management's evaluation of the manufacturing process.

Queuing Theory

Queuing theory or **waiting-line** theory problems involve the matching of servers, who provide, to randomly arriving customers, services which take random amounts of time. Typical questions addressed by queuing theory studies are: how long does the average customer wait before being waited on and how many servers are needed to assure that only a given fraction of customers waits longer than a given amount of time.

In the typical problem applicable to queuing theory solution, people

Table 17.1.1 First Simplex Tableau and First Solution

		0	3	4	0	0	Objective row
	Mix	Quantity	X	Y	S_1	S_2	Variable row
0	S_1	120	2	5	1	0	Problem rows
0	S_2	80	4	2	0	1	
		0	−3	−4	0	0	Index row
Objective column	Variable column	Quantity column					

(or customers or parts) arrive at a server (or machine) and wait in line (in a queue) until service is rendered. There may be one or more servers. On completion of the service, the person leaves the system. The rate at which people arrive to be serviced is often considered to be a random variable with a Poisson distribution having a parameter λ. The average rate at which services can be provided is also generally a Poisson distribution with a parameter μ. The symbol k is often used to indicate the number of servers.

Monte Carlo

Monte Carlo simulation can be a helpful method in gaining insight to problems where the system under study is too complex to describe or the model which has been developed to represent the system does not lend itself to an analytical solution by other mathematical techniques. Briefly, the method involves building a mathematical model of the system to be studied which calculates results based on the input variables, or **parameters.** In general the variables are of two kinds: decision parameters, which represent variables which the analyst can choose, and stochastic, or random, variables, which may take on a range of values and which the analyst cannot control.

The random variables are selected from specially prepared probability tables which give the probability that the variable will have a particular value. All the random variables must be independent. That means that the probability distribution of each variable is **independent** of the values chosen for the others. If there is any correlation between the random variables, that correlation will have to be built into the system model.

For example, in a model of a business situation where market share is to be calculated, a decision-type variable representing selling price can be selected by the analyst. A variable for the price of the competitive product can be randomly selected. Another random variable for the rate of change of market share can also be randomly selected. The purpose of the model is to use these variables to calculate a market share suitable to those market conditions. The algebra of the model will take the effects of all the variables into account. Since the rate-of-change variable can take many values which cannot be accurately predicted, as can the competitive price variable, many runs will be made with different randomly selected values for the random variables. Consequently a range of probable answers will be obtained. This is usually in the form of a **histogram.** A histogram is a graph, or table, showing values of the output and the probability that those values will occur. The results, when translated into words, are expressed in the typical Monte Carlo form of: if such a price is chosen, the following probability distribution of market shares is to be expected.

WAGE ADMINISTRATION

Workers are compensated for their efforts in two principal ways: by hourly rates and by financial-type incentives. An hourly rate is paid to the worker for the number of hours worked and usually is not dependent upon the quantity or quality of the worker's output. Each worker is assigned a job title depending upon qualifications, experience, and skill. Under a structured system of wages, jobs are grouped into classifications, and a similar range of rates is applied to all jobs in a classification. The bottom of the range is paid to beginners, and periodic increases to the top of the range may be automatic or may depend upon the supervisor's appraisal of the individual's performance.

Job evaluation is a formal system for ranking jobs in classes. Each job is studied in relation to other jobs by analyzing such factors as responsibility, education, mental skill, manual skill, physical effort, and working conditions. A total numerical rating for job comparison is obtained by assigning points for each factor.

Merit rating is a point-scale evaluation of an individual's performance by a supervisor, considering such factors as quantity of output, quality of work, adaptability, dependability, ability to work with others, and attitude. These ratings serve as criteria for pay increases within a job classification.

The general level of hourly rates, or base rates, for a company should be determined with reference to the community level of rates. "Going rates" for the community are obtained from wage surveys conducted by the company, a local trade group, or a government agency. Generally, a company which offers wages noticeably lower than the community rates will attract less-competent and less-permanent employees.

Under **financial-type incentives**—or **piece rates,** as they are commonly called—the worker's compensation is dependent upon his or her rate of output. Ordinarily, there is a minimum hourly guarantee below which pay will not decline. Penalties for substandard work may reduce the worker's pay. In some instances, a maximum for incentive earnings is established. The incentive may be calculated so that only the individual's output affects his pay, or when the individual's output cannot be measured, a **group incentive** may be paid, where the pay of each member of the group is determined by his or her base rate plus the output of the group. Group incentives tend to promote cooperation among workers. The administration of incentive plans requires careful management attention if abuses are to be avoided. Restrictions on output and deteriorated standards may lead to higher unit labor costs rather than the anticipated lower costs.

Gainsharing plans are becoming increasingly popular in labor-intensive industries as an additive means of compensation in piecework systems. Gainsharing measures and rewards employees for those facets of the business they can influence. Thus, gainsharing rewards are based on some measure of productivity differing from profit-sharing which uses a more global measure of profitability.

Improshare gainsharing plans measure performance rather than dollar savings. They are based on engineered standards and are not affected by changes in sales volume, technology, or capital expenditures. Productivity generally is measured on a departmental basis where gains above a predetermined efficiency are shared according to formulas which allocate a portion of the gains to the individual department and a portion to the total plant group and a portion to the company; it is not unusual to share 40 percent of the gains to the department, 20 percent to the plant group, and 40 percent to the company. The theory is to measure and reward each department and also to achieve recognition that "we are all in this together." The base, where gains commence, usually is computed from actual department performances over a period of time such as 1 year. Gainsharing rewards are usually distributed monthly as an additive to piecework plans. In addition to financial participation, gainsharing plans offer:

1. The ability to see the outcome of high performance of work in monetary terms.
2. A means to increase employees' commitment and loyalty.
3. Group rewards that lead to cohesion and peer pressure.
4. An enlargement of jobs that encourages initiative and ingenuity.
5. Expanded communication resulting in the enhancement of teamwork.

A **profit-sharing plan** is a form of group incentive whereby each participating worker receives a periodic bonus in addition to regular pay, provided the company earns a profit. A minimum profit is usually set aside for a return on invested capital, and beyond this amount a percentage of profits goes into a pool to be shared by the employees. Many factors other than worker productivity affect profits, e.g., fluctuations in sales volume, selling prices, and costs of raw materials and purchased parts. To protect the workers against adverse developments outside their control, some plans give the workers a bonus whenever the actual payroll dollars are less than the normal amount expected for a given volume of production. Bonuses may be distributed quarterly or even annually and may consequently be less encouraging than incentives paid weekly and related directly to output.

EMPLOYEE RELATIONS

Increasingly, management deals with collective bargaining units in setting conditions of work, hours of work, wages, seniority, vacations, and the like. The bargaining unit may be affiliated with a national union or may be independent and limited to the employees of a particular plant.

A union contract is usually negotiated annually, although managements have sought longer-term agreements. Some managements have negotiated contracts covering a 3- to 5-year period. Under these contracts, the union frequently reserves the right to reopen the wage clauses annually. Today a more conciliatory atmosphere exists in connection with union-management relations.

Grievance procedures facilitate the processing of minor day-to-day disputes between workers and management. Grievances most commonly occur when the worker:

1. Thinks he or she is unfairly treated in matters of (a) pay rates and/or time standards, (b) promotion, (c) work assignment, (d) distribution of overtime, (e) seniority, or (f) disciplinary action.

2. Believes he or she is handicapped by (a) lack of clear policies or working rules; (b) inadequate supervision; (c) too many bosses; (d) supervisors who play favorites; (e) coworkers who are careless, inefficient, or uncooperative; or (f) lack of opportunity to show his or her ability.

3. Is dissatisfied with (a) general job conveniences (e.g., washrooms), (b) working conditions (e.g., light), (c) equipment and tools, (d) plant or office setup (e.g., working space), or (e) protection against job hazards and accidents.

A union steward acts as the employee's representative in discussing a complaint with the supervisor. If a settlement cannot be reached, the discussion may move to the general superintendent's level; the personnel manager is often involved at the various stages. Ways to reduce employee grievances are:

1. Make employees feel accepted; give them a sense of "belonging."

2. Make employees feel significant; give them recognition as people.

3. Make employees feel safe as to (a) job security, and (b) suitable working conditions.

4. Let employee experience the help of leadership.

5. Increase employees' knowledge about (a) the company, (b) its product(s), (c) their jobs, and (d) their next jobs.

6. Give employees fair and impartial treatment.

7. Give employees a chance to be heard: (a) Ask them for suggestions; acknowledge these suggestions; use where practicable and give credit. (b) Encourage them to discuss their problems and gripes; follow through if and as needed.

8. Aid employees to make their contribution to the solution of their problems.

9. Assist employees to develop pride in their work.

10. Recognize employees' status.

An outside arbitrator may be helpful when all internal grievance procedures have been exhausted.

Selection of workers must give attention to the suitability of applicants as well as their previous training. Interviewing and tests for qualities required on the job are necessary for good placement. **Preemployment testing** has proven to be helpful in developing a faster-learning, more disciplined, more quality-conscious, and more productive work force. General aptitude testing in the areas of general learning, verbal comprehension, numerical skills, spatial perception, form perception, clerical perception, motor coordination, finger dexterity, and manual dexterity provide helpful information in the evaluation of an applicant for any job. For example, the attributes of motor coordination, finger dexterity, and manual dexterity have been found to be fundamental in such jobs as sewing, fine assembly, and precision machining. Induction and instruction of new workers are equally important and should not be a secondary task of a busy supervisor; they may be a responsibility of the personnel manager, a member of the training staff, or of a subsupervisor. Induction includes an orientation to the total plant operations as well as to the duties of the specific job; introduction to the supervisor, subsupervisor, and coworkers; and an explanation of the employee's relations with the personnel department and with employee committees or groups that deal with management.

Promotions and recognition of accomplishment profoundly affect morale and require constant supervision by the personnel manager and the line managers. Training programs are used to prepare workers and apprentices for advancement; formal training programs for supervisors are used to develop the capabilities of members of management. The training sessions may be organized and conducted under the personnel manager, training staff member(s), or outside specialists.

Federal law and state laws in most states prohibit **discrimination** because of race, color, religion, age, sex, or national origin by most businesses and labor organizations. The EEOC (Equal Employment Opportunity Commission) polices these activities.

Quality circles, sometimes referred to as *employee participation groups,* are small intact work groups formed in an organization for the purpose of continuous improvement in product quality, productivity, and overall employee performance. Quality circles can be thought of as a process of organizational development based on a management philosophy of employee involvement leading to continuous improvement. This philosophy includes establishing a work climate that promotes the concept that, for the most part, accomplishments are dependent on development and utilization of the potential of the organization's human resources. To successfully utilize quality circles, it is important that both management and union officials become involved in administering the concept and in identifying training needs. Training that should be provided should extend beyond the quality circle members. In addition, upper-level and middle management as well as line supervision and staff support personnel should receive instruction as to the mission, goals, feedback, rewards, and support of the proposed quality circles. In order for quality circles to be successful, the following skills must be utilized regularly: information and data collection, problem analysis and solution, decision making, group dynamics, and communication. Quality circle programs should include a usable measure, such as statistical quality control, to help identify problems and to keep track of performance.

17.2 COST ACCOUNTING
by Scott Jones

REFERENCES: Horngren and Foster, "Cost Accounting—A Managerial Emphasis," Prentice-Hall. Anthony, Dearden, and Govindarajan, "Management Control Systems," Irwin. Cooper and Kaplan, "The Design of Cost Management Systems—Text, Cases, and Readings," Prentice-Hall.

ROLE AND PURPOSE OF COST ACCOUNTING

Cost measurements and reporting procedures are integral components of management information systems, providing financial measurements of economic value and reports useful to many and varied objectives. In a functional organization, where lines of authority are drawn between engineering, production, marketing, finance, and so on, cost accounting information is primarily used by managers for control in guiding departmental units toward the attainment of specific organizational goals. In team-oriented organizations, authority is distributed to multifunctional teams empowered as a group to make decisions. The focus of cost accounting in this organizational structure is not so much on control, but on supporting decisions through the collection of relevant information for decision making. Cost accounting also provides insight for the attainment of competitive advantage by providing an information set and analytical framework useful for the analysis of product or process design, or service delivery alternatives.

The basic purpose of cost measurements and reporting procedures can be organized into a few fundamental areas. These are: (1) identifying and measuring the economic value to be placed on goods and services for reporting periodic results to external information users (creditors, stockholders, and regulators); (2) providing control frameworks for the implementation of specific organizational objectives; (3) supporting operational and strategic decision making aimed at achieving and sustaining competitive advantage.

MEASURING AND REPORTING COSTS TO STOCKHOLDERS

Accounting in general and cost accounting in particular are most visible to the general public in the role of **external reporting**. In this role, cost accounting is geared toward measuring and reporting periodic results, typically annual, to users outside of the company such as stockholders, creditors, and regulators. The annual report presents to these users management forecasts for the coming period and results of past years operations as reflected in **general-purpose financial statements**. (See also Sec. 17.1, "Industrial Economics and Management.") Performance is usually captured through the presentation of three reports: the income statement, the balance sheet, and the statement of cash flows (see Fig. 17.2.1). These statements are audited by independent certified public accountants who attest that the results reported present a fair picture of the financial position of the company and that prescribed rules have been followed in preparing the reports. The accounting principles and procedures that guide the preparation of those reports are governed by **generally accepted accounting principles** (GAAP).

The underlying principles that guide GAAP financial statements encourage comparability among companies and across time. Therefore, these rules are usually too constraining for reports destined for internal use. External reports such as the balance sheet and income statement are prepared on the **accrual basis**. Under this basis, revenues are recognized (reported to stockholders) when earned, likewise costs are expensed against (matched with) revenue when incurred. This is to be contrasted with the **cash basis**, which recognizes revenues and expenses when cash is collected or paid. For example, if a drill press is purchased for use in the factory, a cash outflow occurs when the item is paid for. The cash basis would recognize the purchase price of the drill press as an expense when the payment is made. On the other hand, the accrual basis of accounting would capitalize the price paid, and this amount would be the cost (or basis) of the asset called *drill press*. In accrual accounting, an asset is something having future economic benefit, and therefore the cost of this asset must be distributed among the periods of time when it is used to generate revenue. The cost of the drill press would be expensed periodically by deducting a small amount of that cost from revenue as the drill press is used over its economic life, which may be several years. This periodic charge is called *depreciation*. To capture the effects that revenue-generating activities have on cash, GAAP financial statements also include the statement of cash flows. The statement of cash flows is not prepared on an accrual basis; rather, it reflects the amount of cash flowing into a company during a period, as well as the cash outflow. The first section of that statement, "Cash flows from operating activities," is essentially an income statement prepared on the cash basis.

Another application of cost accounting measurements for external users involves the preparation of reports such as income tax returns for governmental agencies. Federal, state, and local tax authorities prescribe specific accounting procedures to be applied in determining taxable income. These rules are conceptually similar to general-purpose financial reporting but differ mainly in technical aspects of the computations, which are modified to support whatever public finance goals may exist for a particular period. Whereas GAAP financial statements allow for the analysis of credit and investment opportunities, Internal

Revenue Service regulations are designed to raise revenue, stimulate the economy, or both. Regulations in effect at the time of writing were primarily aimed at reducing the burgeoning federal deficit and hence assigned rather long "useful lives" to depreciable assets; at other times in history useful lives were shortened to stimulate investment and economic growth. An example of how the IRS regulations could differ from GAAP can be illustrated in determining the useful life of a drill press. For GAAP, the drill press would be an asset, and may be estimated as depreciable over a wide range of economic lives. The IRS would also view the drill press to be a depreciable asset, but consider the class life to be 9.5 years if used in the manufacture of automobiles, or 8 years in the manufacture of aerospace products. For practicality, many companies will follow the IRS Code when determining useful life for GAAP, though this practice is generally not required.

CLASSIFICATIONS OF COSTS

The purposes of cost accounting require classifications of costs so that they are recognized (1) by the nature of the item (a natural classification), (2) in their relation to the product, (3) with respect to the accounting period to which they apply, (4) in their tendency to vary with volume or activity, (5) in their relation to departments, (6) for control and analysis, and (7) for planning and decision making.

Direct material and direct labor may be listed among the items which have a **variable** nature. Factory overhead, however, must be carefully examined with regard to items of a variable and a fixed nature. It is impossible to budget and control factory-overhead items successfully without regard to their tendency to be fixed or variable; the division is a necessary prerequisite to successful budgeting and intelligent cost planning and analysis.

In general, **variable expenses** show the following characteristics: (1) variability of total amount in direct proportion to volume, (2) comparatively constant cost per unit or product in the face of changing volume, (3) easy and reasonably accurate assignments to operating departments, and (4) incurrence controllable by the responsible department head.

The characteristics of **fixed expenses** are (1) fixed amount within a relative output range, (2) decrease of fixed cost per unit with increased output, (3) assignment to departments often made by managerial decisions or cost-allocation methods, and (4) control for incurrence resting with top management rather than departmental supervisors. Whether an expense is classified as fixed or variable may well be the result of managerial decisions.

Some **factory overhead** items are semivariable in nature; i.e., they vary with production but not in direct proportion to the volume. For practical purposes, it is desirable to resolve each semivariable expense item into its variable and fixed components.

A factory is generally organized along departmental lines for production purposes. This factory departmentalization is the basis for the important classification and subsequent accumulation of costs by departments to achieve (1) cost control and (2) accurate costing. The departments of a company generally fall into two categories: (1) producing, or productive, departments, and (2) nonproducing, or service, departments. A producing department is one in which manual and machine operations are performed directly upon any part of the product manufactured. A service department is one that is not directly engaged in production but renders a particular type of service for the benefit of other departments. The expense incurred in the operation of service departments represents a part of the total factory overhead that must be absorbed in the cost of the product.

For **product costing**, the factory may be divided into departments, and departments may also be subdivided into cost centers. As a product passes through a cost center or department, it is charged with a share of the indirect expenses on the basis of a departmental factory-overhead rate. For cost-control purposes, budgets are established for departments and cost centers. Actual expenses are compared with budget allowances in order to determine the efficiency of a department and to measure the manager's success in controlling expenses.

Factory overhead, which is charged to a product or a job on the basis

Balance Sheet (Illustrative)
Black Carbon, Inc.
December 31, 19–

Assets

Current Assets:

Cash		$ 5,050,000
Accounts Receivable (net)		6,990,000
Inventories:		
Raw Materials and Supplies	$ 1,000,000	
Work in Process	1,800,000	
Finished Goods	2,900,000	5,700,000
Investments		1,000,000
Deferred Charges		340,000
Total Current Assets		$19,080,000

Property, Plant, and Equipment:

Land		$ 4,000,000	
Buildings and Equipment	$75,500,000		
Less: Allowance for Depreciation	47,300,000	28,200,000	
Total Fixed Assets			32,200,000
Total Assets			$51,280,000

Liabilities

Current Liabilities:

Accounts Payable and Accruals		$ 3,580,000	
Provision for Income Taxes:			
Federal	$ 2,250,000		
State	65,000	2,315,000	
Total Current Liabilities			$ 5,895,000
Long-term Debt			5,300,000
Total Liabilities			$11,195,000

Stockholders Equity:

Common Stock—no par value			
Authorized—2,000,000 shares			
Outstanding—1,190,000 shares		$11,900,000	
Earnings retained in the business		28,185,000	
Total Stockholders' Equity			$40,085,000
Total Liabilities and Stockholders' Equity			$51,280,000

Income Statement (Illustrative)
Black Carbon, Inc.
for the year 19–

Net Sales		$50,087,000
Cost of products:		
Material, Labor, and Overhead (excluding depreciation)	$32,150,000	
Depreciation	5,420,000	37,570,000
Gross Profit		$12,517,000
Less: Selling and Administrative Expenses		3,220,000
Profit from Operations		9,297,000
Other Deductions		305,000
		8,992,000
Other Income		219,000
Income before Federal and State Income Taxes		9,211,000
Less: Provision for Federal and State Income Taxes		4,055,000
Total Net Income		5,156,000
Dividends paid to shareholders		2,200,000
Income retained in the business		$ 2,956,000

Statement of Cash Flows (Illustrative)
Black Carbon, Inc.
December 31, 19–

Cash flows from operating activities		
Cash received from customers	$49,525,000	
Cash paid to suppliers and employees	(40,890,000)	
Cash paid for interest	(1,050,000)	
Cash paid for income taxes	(2,860,000)	
Net cash provided by operating activities		$ 4,725,000
Cash flows from investing activities		
Purchase of equipment	(1,970,000)	
Net cash from investing activities		(1,970,000)
Cash flows from financing activities		
Principal payment on loans	(2,780,000)	
Dividend payments	(2,200,000)	
Net cash used by financing activities		(4,980,000)
Net increase (decrease) in cash		(2,225,000)
Cash at beginning of year		7,275,000
Cash at end of year		$ 5,050,000

Fig. 17.2.1 Examples of the balance sheet, income statement, and statement of cash flows based on the published annual report.

of a predetermined overhead rate, is considered indirect with regard to the product or the job to which the expense is charged. Service-department expenses are prorated to other service departments and/or to the producing departments. The proration is accomplished by using some rational basis such as area occupied or number of workers. The prorated costs are termed **indirect departmental charges.** When all service-department expenses have been prorated to the producing departments, each producing department's total factory overhead will consist of its own direct departmental expense and the indirect (or prorated, or apportioned) charges. This total cost is charged to the product or the job on the basis of the predetermined factory-overhead rate.

A company's cost system provides the data required for establishing **standard costs** and for the preparation and operation of a **budget.**

The **budget** program enlists all members of management in the task of creating a workable and acceptable plan of action, welds the plan into a homogeneous unit, communicates to the managerial levels differences between planned activity and actual performance, and points out unfavorable conditions which need corrective action. The budget not only will help promote coordination of people, clarification of policy, and crystallization of plans, but with successful use will create greater internal harmony and unanimity of purpose among managers and workers.

The established **standard-cost** values for material, labor, and factory overhead form the foundation for the budget. Since standard costs are an invaluable aid in the process of setting prices, it is essential to set these standard costs at realistic levels. The measurement of deviations from established standards or norms is accomplished through the use of variance accounts.

Costs as a basis for planning are estimated costs which may be incurred if any one of several alternative courses of action is adopted. Different types of costs involve varying kinds of consideration in managerial planning and decision making.

METHODS OF ACCUMULATING COSTS IN RECORDS OF ACCOUNT

The balance sheet lists the components of inventory as raw materials, work in process, and finished goods. These accounts reflect the cost of unsold production at various stages of completion. The costs in work in process and finished goods are accumulated or tabulated in the record of accounts according to one of two methods:

1. The Job-Order Cost Method   When orders are placed in the factory for specific jobs or lots of product, which can be identified through all manufacturing processes, a job cost system is appropriate. This method has certain characteristics. A manufacturing order often corresponds to a customer's order, though sometimes a manufacturing order may be for stock. The customer's order may be obtained on the basis of a bid price computed from an estimated cost for the job. The goods in each order are kept physically separate from those of other jobs. The costs of a manufacturing order are entered on a job cost sheet which shows the total cost of the job upon completion of the order. This cost is compared with the estimated cost and with the price which the customer agreed to pay.

2. The Process Cost Method   When production proceeds in a continuous flow, when units of product are not separately identifiable, and when there are no specific jobs or lots of product, a process cost system is appropriate, for it has certain characteristics: work is ordered through the plant for a specific time period until the raw materials on hand have all been processed or until a specified quantity has been produced; goods are sold from the stock of finished goods on hand since a customer's order is not separately processed in the factory; the cost-of-production sheet is a record of the costs incurred in operating the process —or a series of processes—for a period of time. It shows the quantity produced in pounds, tons, gallons, or other units, and the cost per unit is obtained by dividing the total costs of the period by the total units produced. Performance is indicated by comparing the quantity produced and the cost per unit of the current period with similar figures of other periods or with standard cost figures.

ELEMENTS OF COSTS

The main items of costs shown on the income statement are factory costs which include direct materials, direct labor and factory overhead; and selling and administrative expenses. A breakdown of costs is shown in Figure 17.2.2.

Materials   The cost of materials purchased is recorded from purchase invoices. When the materials are used in the factory, an assumption must be made as to **cost flow,** that is, whether to charge them to operations at average prices, at costs based on the first-in, first-out method of costing, or at costs based on the last-in, first-out method of costing. Each method will lead to a different cost figure, depending on how prices change. Each situation must be studied individually to determine which practice will give a maximum of accuracy in cost figures with a minimum of accounting and clerical effort. Once the choice has been made, records must be set up to charge materials to operations based on requisitions. Indirect material is necessary to the completion of the product, but its consumption with regard to the final product is either so small or so complex that it would be futile to treat it as a direct-material item.

Labor   Labor also consists of two categories: direct and indirect. Direct labor, also called **productive labor,** is expended immediately on the materials comprising the finished product. Indirect labor, in contrast to direct labor, cannot be traced specifically to the construction or composition of the finished product. The term includes the labor of supervisors, shop clerks, general helpers, cleaners, and those employees engaged in maintenance work.

Factory Overhead   Indirect materials or factory supplies and indirect labor constitute an important segment of factory overhead. In addition, costs of fuel, power, small tools, depreciation, taxes on real estate, patent amortization, rent, inspection, supervision, social security taxes, health and accident insurance, workers' compensation insurance, and many others fall into this large category. These expenses must be collected and allocated to all jobs or units produced. Many expenses are definitely applicable to a specific department and are easily assigned thereto. Other expenses relate to the entire plant and must be prorated to departments on some suitable basis. For instance, heat might be prorated to departments on the basis of volume of space occupied. The expenses of the service departments are prorated to the producing departments on some basis such as service rendered in the case of a maintenance department or so much dollar payroll processed in the case of a cost department.

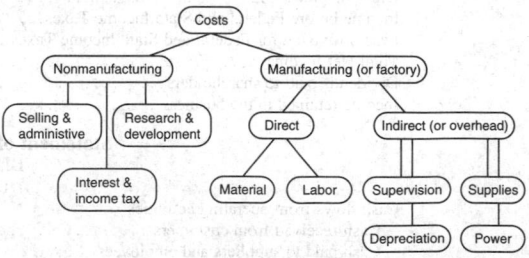

Fig. 17.2.2   Summary diagram of cost relationships.

The charging of factory overhead to jobs or products is accomplished by means of an overhead or burden rate. This rate is essentially a ratio computed to show the relationship of the total burden of a department to some other easily measurable total figure for the department. For example, the total burden cost of a department may be divided by its direct-labor cost to give a percentage-of-direct-labor rate. This percentage applied to the direct-labor cost of a job or a product gives the amount of overhead chargeable thereto. Other common types of burden rates are the labor-hour rate (departmental expenses ÷ total direct-labor hours) and the machine-hour rate (departmental expenses ÷ total machine hours available). Labor rates are most commonly used. When, however, machines perform the greater amount of the work, machine-hour rates give better results. It must be clearly understood that these rates are computed in advance of production, generally at the beginning of the year. They are used throughout the fiscal period unless seasonal fluctuations or unusual changes in expense amounts necessitate the creation of a new rate. The determination of the overhead rate is closely tied up with overhead budgets.

ACTIVITY-BASED COSTING

The method of assigning overhead to products based on labor hours or machine hours is referred to as **volume-based** overhead absorption because the amount assigned will vary strictly with the volume of either labor or machine time consumed. In applications where production costs may not be strictly driven by volume of labor hours, **activity-based** or **transaction** costing is appropriate. This situation typically occurs when there are many options or alternatives available to the customer. Typically, these products are produced in low volumes and have a high degree of complexity. An example would be the option of a premium radio in an automobile. Though the actual purchase cost of that radio would not be overlooked in pricing the automobile, the indirect costs would be overlooked in a volume-based system because the indirect costs associated with the premium radio, such as holding an additional item of inventory, documenting and producing separate receiving orders, added clerical and assembly coordination effort, increased engineering complexity, and so on would be grouped under the general category of overhead. On the other hand, in an activity-based system, indirect costs would be assigned to a unique cost pool, such as shown in Fig. 17.2.3 and 17.2.4, which compare the procedures of volume and

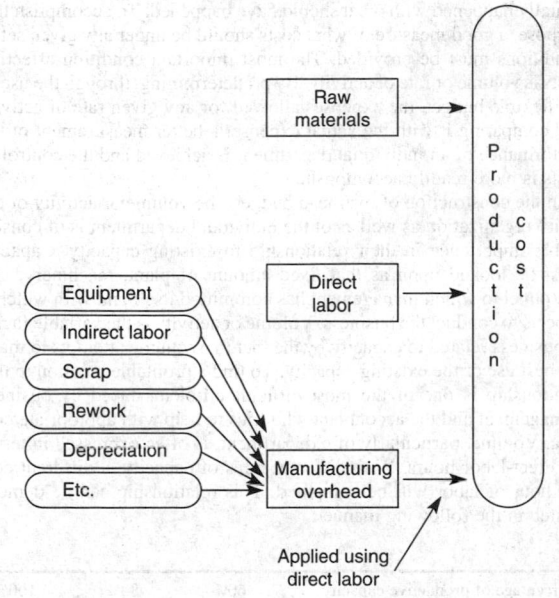

Fig. 17.2.3 Volume-based costing.

activity-based costing. The striking difference is that the overhead cost pool used in volume systems is not present. In activity-based systems, many more cost pools are used and are closely related to some **causal aspect** of the process such as machine setup, receiving orders, or material movement. The costs are assigned to products based on the relative amounts of each cost driver consumed by that product. Therefore, low-volume options such as the premium radio receive a larger share of indirect costs relative to the actual volume of labor used to insert the component. The amount of each cost pool attributed to a product can then be spread over the number of units produced and a more accurate assignment of costs obtained.

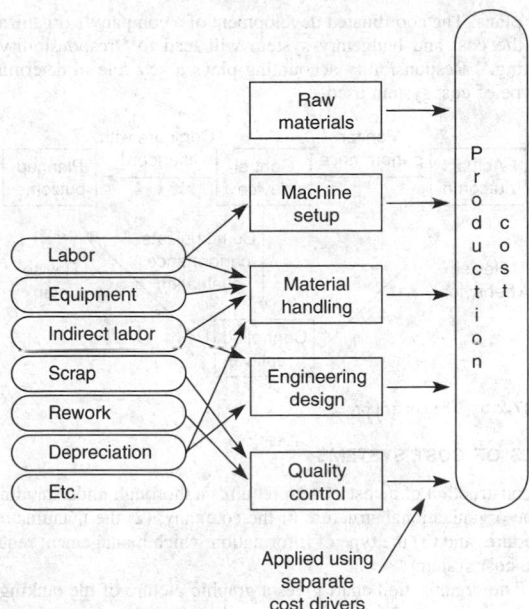

Fig. 17.2.4 Activity-based costing.

Departmental Classification As mentioned above, the establishment of departmental lines is important not only for costing purposes but also for budgetary control purposes. Departmental lines are set up in order to (1) segregate basically different processes of production, (2) secure the smoothest possible flow of production, and (3) establish lines of responsibility for control over production and costs. When the costing methods are designed to fit in with the departmentalization of factory and office, costs can be accumulated within a department with production being on either the job-order or process cost method.

MANAGEMENT AND THE CONTROL FUNCTION

To be successful, management must integrate its own knowledge, skills, and practices with the know-how and experience of those who are entrusted with the task of carrying out company objectives. Management, together with its employees and workers, can achieve its objectives through performance of the three managerial functions: (1) planning and setting objectives, (2) organizing, and (3) controlling.

Planning is a basic function of the management process. Without planning there is no need to organize or control. However, planning must precede doing, and the budget is the most important planning tool of an enterprise.

Organizing is essentially the establishment of the framework within which the required activities are to be performed, together with a list of who should perform them. Creation of an organization requires the establishment of organizational or functional units generally known as departments, divisions, sections, floors, branches, etc.

Controlling is the process or procedure by which management ensures operative performance which corresponds with plans. The control process is pictured diagrammatically in Fig. 17.2.5. Recognition of accounting as an important tool in the controlling phase is evidenced through the role of performance reports in pointing out areas and jobs or tasks which require corrective action. These reports should make possible "management by exception."

The effectiveness of the control of costs depends upon proper communication through control and action reports from the accountant to the various levels of operating management. An organization chart is essential to the development of a cost system and cost reports which parallel the responsibilities of individuals for implementing management plans. The coordinated development of a company's organization with the cost and budgetary system will lead to "responsibility accounting." Responsibility accounting plays a key role in determining the type of cost system used.

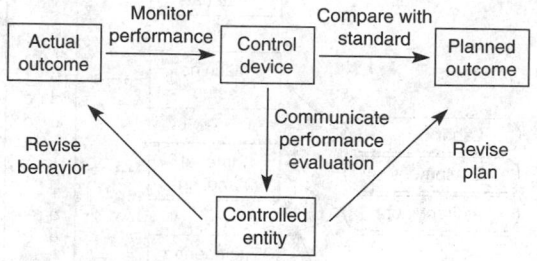

Fig. 17.2.5 The control process.

TYPES OF COST SYSTEMS

The construction of a cost system requires a thorough understanding of (1) the organizational structure of the company, (2) the manufacturing procedure, and (3) the type of information which management requires of the cost system.

1. The organization chart gives a graphic picture of the ranking authority of superintendents, department heads, and managers who are responsible for (a) providing the detailed information needed by the accounting division in order to install a successful system; (b) incurring expenditures in personnel, materials, and other cost elements, which the cost accountant must segregate and report to those in charge. The cost system with its operating accounts must correspond to organizational divisions of authority so that the individual supervisor, department head, or executive can be held "accountable" for the costs incurred in the department.

2. The manufacturing procedure and shop methods lead to a consideration of the type of pay (piece rate, incentive, day rate, etc.); the method of collecting hours worked; the control of inventories; the problem of costing tools, dies, jigs, and machinery; and many other problems connected with the factory.

3. The organizational setup on the one hand and the manufacturing procedure on the other form the background for the design of a cost system that is based on (a) recognition of the various cost elements, (b) departmentalization of factory and office, and (c) the chart of accounts.

Any cost system should be perfected so that it will (1) aid in the control and management of the company; (2) measure the efficiency of personnel, materials, and machines; (3) help in eliminating waste; (4) provide comparison within individual industries; (5) provide a means of valuing inventories; and (6) aid in establishing selling prices. In an organization departmentalized or segmented along product lines, it is often arbitrary to allocate certain indirect costs especially when common facilities or personnel are shared. This is because there is no objective basis to compute a division of common costs. Control methods for evaluating performance often rely on the **segment margin** statement.

Example of a Segment Margin Statement

	Product A	Product B	Total
Sales	$9,000	$11,000	$20,000
Direct material & labor	(4,000)	(5,000)	(9,000)
Contribution margin	5,000	6,000	11,000
Product specific overhead	(1,000)	(2,500)	(3,500)
Segment margin	4,000	3,500	7,500
Common costs			(2,000)
Operating income			$ 5,500

The cost system's value is greatly enhanced when it is interlocked with a **budgetary control system.** When budget figures are based upon standard costs, the greatest benefit will be derived from such a combination.

Basically, two types of cost systems exist: (1) the **actual** (or **historical**) and (2) the **standard** (or **predetermined**). The actual cost system accumulates and summarizes costs as they occur and determines a final product cost after all manufacturing operations have been completed. The job is charged with actual quantities and costs of materials used and labor expended; the overhead or burden is allocated on the basis of some predetermined overhead rate. This predetermined overhead rate shows that even the so-called **actual system** does not entirely live up to its name. Under a standard cost system all costs are predetermined in advance of production. Both the actual (historical) and the standard cost system may be used in connection with either (1) the job-order cost method or (2) the process cost method.

BUDGETS AND STANDARD COSTS

A budget provides management with the information necessary to attain the following major objectives of budgetary control: (1) an organized procedure for planning; (2) a means for coordinating the activities of the various divisions of a business; (3) a basis for cost control. The planning phase provides the means for formalizing and coordinating the plans of the many individuals whose decisions influence the conduct of a business. Sales, production, and expense budgets must be established. Their establishment leads necessarily to the second phase of coordination. Production must be planned in relation to expected sales, materials and labor must be acquired or hired in line with expected production requirements, facilities must be expanded only as foreseeable future needs justify, and finances must be planned in relation to volume of sales and production. The third phase of cost control is predicated on the idea that actual costs will be compared with budgeted costs, thus relating what actually happened with what should have happened. To accomplish this purpose, a good measure of what costs should be under any given set of conditions must be provided. The most important condition affecting costs is volume or rate of activity. By predetermining, through the use of the flexible budget, the expenses allowed for any given rate of activity and comparing it with the actual expense, a better measurement of the performance of an individual department is achieved and the control of costs is more readily accomplished.

In the construction of overhead budgets the volume or activity of the entire organization as well as of the individual department is of considerable importance in their relationship to existing capacity. Capacity must be looked upon as that fixed amount of plant, machinery, and personnel to which management has committed itself and with which it expects to conduct the business. Volume or activity is the variable factor in business related to capacity by the fact that volume attempts to make the best use of the existing capacity. To find a profitable solution to this relationship is one of the most difficult problems faced by business management and the accountant who tries to help with appropriate cost data. Volume, particularly of a department, is often expressed in terms of direct-labor hours. With different rates of capacity, a different cost per hour of labor will be computed. This relationship can be demonstrated in the following manner:

Percentage of productive capacity	60%	80%	100%
Direct-labor hours	600	800	1,000
Factory overhead			
Fixed overhead	$1,200	$1,200	$1,200
Variable overhead	600	800	1,000
Total	$1,800	$2,000	$2,200
Overhead rate per direct-labor hour	$3.00	$2.50	$2.20

The existence of fixed overhead causes a higher rate at lower capacity. It is desirable to select that overhead rate which permits a full recovery of production costs by the end of the business cycle. The above tabulation reveals another important axiom with respect to fixed and variable overhead. Fixed overhead remains constant in total but varies in respect to cost per unit or hour. Variable overhead varies in total but remains fixed in relationship to the unit or hour.

Standard Costs   The budget, as a statement of expected costs, acts as a guidepost which keeps business on a charted course. Standards, however, do not tell what the costs are expected to be but rather what they will be if certain performances are attained. In a well-managed business, costs never exceed the budget. They should constantly approach predetermined standards. The uses of standard costs are of prime importance for (1) controlling and reducing costs, (2) promoting and measuring efficiencies, (3) simplifying the costing procedures, (4) evaluating inventories, (5) calculating and setting selling prices. The success of a standard cost system depends upon the reliability and accuracy of the standards. To be effective, standards should be established for a definite period of time so that control can be exercised and variances from standards computed. Standards are set for materials, labor, and factory overhead. When actual costs differ from standard costs with respect to material and labor, two causes can generally be detected. (1) The price may be higher or lower or the rate paid a worker may be different; the difference is called a material price or a labor rate variance. (2) The quantity of the material used may be more or less than the standard quantity or the hours used by the worker may be more or less. The difference is called material-quantity variance or labor-efficiency variance, respectively. For factory overhead, the computation is somewhat more elaborate. Actual expenses are compared not only with standard expenses but also with budget figures. Various methods are in vogue, resulting in different kinds of overhead variances. Most accountants compute a controllable and a volume variance. The controllable variance deals chiefly with variable expenses and measures the efficiency of the manager's ability to hold costs within the budget allowance. The volume variance portrays fixed overhead with respect to the use or nonuse of existing capacity. It measures the success of management in its ability to fill capacity with sales or production volume. These two variances can be analyzed further into an expenditure and efficiency variance for the controllable variances and into an effectiveness and capacity variance for the volume variance. Such detailed analyses might bring forth additional information which would help management in making decisions. Of absolute importance for any cost system is the fact that the information must reach management promptly, with regularity, and in a report that is analytical, permitting quick comparison with targets and goals. Only in this manner can management, which includes all echelons from the foreman to the president, exercise control over costs and therewith over profits.

TRANSFER PRICING

A **transfer price** refers to the selling price of a good or service when both the buyer and seller are within the same organization. For example, one division of a company may produce a component, such as an engine, and transfer this component to an assembly division. For purposes of control, these organizational units may be treated as **profit centers** (responsible for earning a specified profit or return on investment). Accordingly, the transfer price is a revenue for the seller and a cost to the buyer. Because organizational control is at issue whenever interdivisional transfers are made, companies must often specify a policy to dictate the basis for determining a transfer price. Transfer prices should be based on market prices when available. Most taxing authorities require intercompany transfers to be made at market price as well. To solve situations of suboptimal resource usage (e.g., idle capacity) it is often possible to construct transfer prices based on manufacturing cost plus some allowance for profit. If the producing division is a **cost center** (responsible for controlling costs to achieve a certain budgeted level), in order to promote efficiency transfers are usually made on the basis of **standard cost.**

SUPPORTING DECISION MAKING

The analytical phase of cost accounting has become more important and influential in the last few years. Management must make many decisions, some of a short-range, others of a long-range nature. To base judgment upon good, reliable data and analyses is a major task for controllers and their staffs. Cost analysis comprises such matters as analysis of distribution costs, gross-profit analysis, break-even analysis, profit-volume analysis, differential-cost analysis, direct costing, capital-expenditure analysis, return on capital employed, and price analysis. A detailed discussion of each phase mentioned lies beyond the scope of this section, but a short description is appropriate.

Distribution-cost analysis deals with allocation of selling expenses to territories, customers, channels of distribution, products, and sales representatives. Once so allocated it might be possible to determine the most profitable and the least profitable commodity, product, territory, or customer. Segment margin statements are useful for this analysis. Standards have been introduced recently in these analyses. The Robinson-Patman Act, an amendment to Section 2 of the Clayton Act, gave additional impetus to the analytical phase of distribution costs. This act prohibits pricing the same product at different amounts when the amounts do not reflect actual cost differences (such as distribution or warranty).

Gross-profit analysis attempts to determine the causes for an increase or decrease in the gross profit. Any change in the gross profit is due to one or a combination of the following: (1) changes in the selling price of the products; (2) changes in the volume sold; (3) changes in the types of products sold, called the sales-mix; (4) changes in the cost elements. Cost elements are analyzed through budgetary control methods. Sales figures must be scrutinized to unearth the changes from the contemplated course and therewith from the final profit.

Break-even analysis, generally presented in the form of a break-even chart, constitutes one of the briefest and most easily understood devices for data presentation for policy-making decisions. The name "break-even" implies that point at which the company neither makes a profit nor suffers a loss from the operations of the business. A break-even chart can be defined as a portrayal in graphic form of the relation of production and sales to profit or, more briefly, a graphic variable income statement. The computation of the break-even point can be made by the following formula.

$$\text{Break-even sales volume} = \frac{\text{total fixed expenses}}{1 - \dfrac{\text{total variable expenses}}{\text{total sales volume}}}$$

EXAMPLE. Assume fixed expenses, \$13,800,000; variable expenses, \$27,000,000; total sales volume, \$50,000,000. Computation: Break-even sales volume = \$13,800,000/[1 − (\$27,000,000/\$50,000,000)] = \$30,000,000.

Results can be obtained in chart form (Fig. 17.2.6).

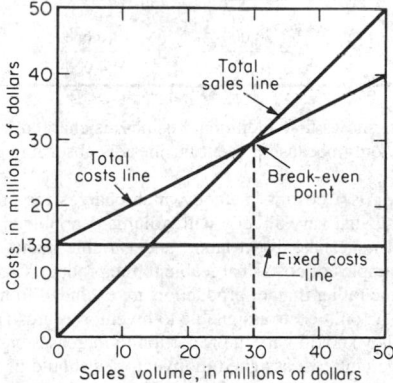

Fig. 17.2.6   Illustrative break-even chart.

Cost-volume-profit analysis deals with the effect that a change of volume, cost, price, and product-mix will have on profits. Managements of many enterprises attempt to stimulate the public to purchase their products by conducting intensive promotion campaigns in radio, press, mail, and television. The customer, however, makes the final decision. What management wants to know is which product or model will yield the most profitable margin; which is the least profitable; what effect a reduction in sales price will have on final profit; what effect a shift in volume or product-mix will have on product costs and profits; what the new break-even point will be under such changing conditions; what the effect of expected increases in wages or other operating costs on profit will be; what the effect will be on costs, profit, and sales volume should there be an expansion of the plant. Cost-volume-profit analysis can also be presented graphically in a so-called *volume-profit-analysis graph.* Using the same data as in the break-even chart, a volume-profit analysis graph takes the form shown in Fig. 17.2.7.

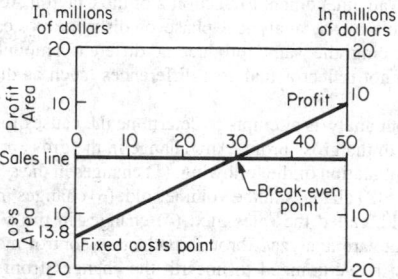

Fig. 17.2.7 Illustrative cost-volume-profit analysis graph.

Differential-cost analysis treats differences, as the title suggests. These differences, also called alternative courses, arise when management wants to know whether or not to take business at a special price, to risk a decline in price of total sales, to sacrifice volume for price, to shut down part of the plant, or to enlarge plant capacity. While accountants generally use the term "differential," economists speak of "marginal" and engineers of "incremental" costs in connection with such a study. As in any of the previously discussed analyses, the classification of costs into their fixed and variable components is absolutely essential. However, while in break-even analysis the emphasis rests upon the fixed expenses, differential-cost studies stress the variable costs. The differential-cost statement presents only the differences in the following manner:

	Present business	Additional business	Total
Sales	$100,000	$10,000	$110,000
Variable costs	60,000	6,000	66,000
Marginal income	40,000	4,000	44,000
Fixed expenses	30,000	none	30,000
Profit	$ 10,000	$ 4,000	$ 14,000

This statement shows that additional business is charged with the variable expenses only because present business is absorbing all fixed expenses.

Direct costing is a costing method which charges the products with only those costs that vary directly with volume. Variable or direct costs such as direct materials, direct labor, and variable manufacturing expenses are examples of costs chargeable to the product. Costs that are a function of time rather than of production are excluded from the cost of the product. The only costs assignable to inventories are variable costs, and because they should vary in proportion to increases or decreases in production, the unit cost assigned to inventories should be uniform.

CAPITAL-EXPENDITURE DECISIONS

The preparation of a capital-expenditure budget must be preceded by an analytical and decision-making process by management. This area of managerial decisions not only is important to the success of the company but also is crucial in case of errors. Financial requirements, present

and anticipated costs, profits, tax considerations, and legal, personnel, and market problems must be studied and reviewed before making the final decision.

Four evaluation techniques are generally accepted as representative tools for decision making: (1) **payback-** or **payout-period method;** (2) **average-return-on-investment method;** (3) **present-value method;** and (4) **discounted-cash-flow method.** None of these methods serves every purpose or every firm. The methods should, however, aid management in exercising judgment and making decisions. Of significance in the evaluation of a capital expenditure is the time value of money which is employed in the present value and the discounted-cash-flow methods. The **present value** means that a dollar received a year hence is not the equivalent of a dollar received today, because the use of money has a value. For this reason, the estimated results of an investment proposal can be stated as a cash equivalent at the present time, i.e., its present value. Present-value tables have been devised to facilitate application of present-value theory.

In the present-value method the discount rate is known or at least predetermined. In the **discounted-cash-flow** (DCF) method the rate is to be calculated and is defined as the rate of discount at which the sum of positive present values equals the sum of negative present values. The DCF method permits management to amortize corporate profits by selecting proposals with the highest rates of return as long as the rates are higher than the company's own **cost of capital** plus management's allowance for risk and uncertainty. Cost of capital represents the expected return for a given level of risk that investors demand for investing their money in a given firm or venture. However, when related to capital-expenditure planning, the cost of capital refers to a specific cost of capital from a particular financing effort to provide funds for a specific project or numerous projects. Such use of the concept connotes the marginal cost of capital point of view and implies linkage of the financing and investment decisions. It is, therefore, not surprising that the cost of capital differs, depending upon the sources. A company could obtain funds from (1) bonds, (2) preferred and common stock, (3) use of retained earnings, and (4) loans from banks. If a company obtains funds by some combination of these sources to achieve or maintain a particular capital structure, then the cost of capital (money) is the weighted average cost of each money source.

Return-on-Capital Concept This aids management in making decisions with respect to proposed capital expenditures. This concept can also be used for (1) measuring operating performance, (2) profit planning and decision making, and (3) product pricing. The return on capital may be expressed as the product of two factors: the percentage of profit to sales and the rate of capital turnover. In the form of an equation, the method appears as

$$\frac{\text{Profit}}{\text{Sales}} = \text{profit margin}$$

$$\times \to = \begin{array}{c}\text{return on}\\\text{capital}\\\text{investment}\end{array}$$

$$\frac{\text{Sales}}{\text{Capital}} = \begin{array}{c}\text{investment}\\\text{turnover}\end{array}$$

Whether for top executive, plant or product manager, plant engineer, sales representative, or accountant, the concept of return on capital employed tends to mesh the interest of the entire organization. An understanding and appreciation of the return-on-capital concept by all employees help in building an organization interested in achieving fair profits and an adequate rate of return.

COST MANAGEMENT

Often, programs of **continuous improvement** require that costs be computed according to the activity-based method. That method facilitates identifying and setting priorities for the elimination of non-value-adding activities. **Non-value-added** activities decrease cycle time efficiency, where cycle time efficiency is the sum of all value-added activity times divided by total cycle time. The engineer may redesign the product using common parts or through process redesign so as to eliminate those activities or cost pools that add to product cost without adding to value. Some examples of these activities are material movement, run setup, and queue time.

Efforts to manage product costs by eliminating non-value-adding activities are frequently the result of a need to attain a specific target cost. Traditionally selling price was determined by adding a required markup to total cost. Global competition has forced producers to accept a market price determined by competitive forces:

Target cost = market selling price − required return on investment

Accordingly, the company that stays in business is the one that can accept this price and still earn a return on investment. **Target costing** is a concerted effort to design, produce, and deliver the product at a cost that will assure long-term survival.

17.3   ENGINEERING STATISTICS AND QUALITY CONTROL

Ashley C. Cockerill

REFERENCES: Brownlee, "Statistical Theory and Methodology in Science and Engineering," Wiley. Conover, Two k-sample slippage tests, *Journal of the American Statistical Association,* **63**: 614–626. Conover, "Practical Nonparametric Statistics," Wiley. Duncan, "Quality Control and Industrial Statistics," Richard D. Irwin. Gibbons, "Nonparametric Statistical Inference," McGraw-Hill. Olmstead, A corner test for association, *Annals of Mathematical Statistics,* **18**: 495–513. Owen, "Handbook of Statistical Tables," Addison-Wesley. Pearson, "Biometrika Tables for Statisticians," vol. 1, 2d ed., Cambridge University Press. Tukey, "A quick compact two sample test to Duckworth's specifications," *Technometrics,* **1**: 31–48. Wilks, "Mathematical Statistics," Wiley.

ENGINEERING STATISTICS AND QUALITY CONTROL

Statistical models and statistical methods play an important role in modern engineering. Phenomena such as turbulence, vibration, and the strength of fiber bundles have statistical models for some of their underlying theories. Engineers now have available to them batteries of computer programs to assist in the analysis of masses of complex data. Many textbooks are needed to cover fully all these models and methods; many are areas of specialization in themselves. On the other hand, every engineer has a need for easy-to-use, self-contained statistical methods to assist in the analysis of data and the planning of tests and experiments. The sections to follow give methods that can be used in many everyday situations, yet they require a minimum of background to use and need little, if any, calculation to obtain good results.

STATISTICS AND VARIABILITY

One of the primary problems in the interpretation of engineering and scientific data is coping with **variability**. Variability is inherent in every physical process; no two individuals in any population are the same. For example, it makes no real sense to speak of the tensile strength of a synthetic fiber manufactured under certain conditions; some of the fibers will be stronger than others. Many factors, including variations of raw materials, operation of equipment, and test conditions themselves, may account for differences. Some factors may be carefully controlled, but variability will be observed in any data taken from the process. Even tightly designed and controlled laboratory experiments will exhibit variability.

Variability or variation is one of the basic concepts of statistics. Statistical methods are aimed at giving objective, quantitative, and reproducible ways of assessing the effects of variability. In particular, they aim to provide measures of the uncertainty in conclusions drawn from observational data that are inherently variable.

A second important concept is that of a **random sample**. To make valid inferences or conclusions from a set of observational data, the data should be able to be considered a random sample. What does this mean? In an operational sense it means that everything we are interested in seeing should have an equal chance of being represented in the observa-tions we obtain. Some examples of what not to do may help. If machine setup is an important contributor to differences, then all observations should not be taken from one setup. If instrumental variation can be important, then measurements on the same item should not be taken successively—time to "forget" the last reading should pass. A random sample of n items in a warehouse is not the first n that you can find. It is the n that is selected by a procedure guaranteed to give each item of interest an equal chance of selection. One should be guided by generalizations of the fact that the apples on top of a basket may not be representative of all apples in the basket.

CHARACTERIZING OBSERVATIONAL DATA: THE AVERAGE AND STANDARD DEVIATION

The two statistics most commonly used to characterize observational data are the **average** and the **standard deviation**. Denote by $x_1, x_2, \ldots, x_n$ the n individual observations in a random sample from some process. Then the average and standard deviation are defined as follows:
Average:

$$\bar{x} = \sum_{i=1}^{n} x_i / n$$

Standard deviation:

$$s = \left[\sum_{i=1}^{n} (x_i - \bar{x})^2 / (n - 1) \right]^{1/2}$$

Clearly, the average gives one number around which the n observations tend to cluster. The standard deviation gives a measure of how the n observations vary or spread about this average. The square of the standard deviation is called the **variance**. If we consider a unit mass at each point x_i, then the variance is equivalent to a moment of inertia about an axis through $\bar{x}$. It is readily seen that for a fixed value of $\bar{x}$, greater spreads from the average will produce larger values of the standard deviation s. The average and the standard deviation can be used jointly to summarize where the observations are concentrated. Tchebysheff's theorem states: A fraction of at least $1 - (1/k^2)$ of the observations lie within k standard deviations of the average. The theorem guarantees lower bounds on the percentage of observations within k (also known as z in some textbooks) standard deviations of the average.

Interval	Lower bound on % of measurements
$\bar{x} - 2s$ to $\bar{x} + 2s$	75%
$\bar{x} - 3s$ to $\bar{x} + 3s$	89%
$\bar{x} - 4s$ to $\bar{x} + 4s$	94%
$\bar{x} - 5s$ to $\bar{x} + 5s$	96%
$\bar{x} - 6s$ to $\bar{x} + 6s$	97%

Table 17.3.1 Average of Range $f_n = \sigma$

Sample size	f_n	Sample size	f_n
2	0.8862	11	0.3152
3	0.5908	12	0.3069
4	0.4857	13	0.2998
5	0.4299	14	0.2935
6	0.3946	15	0.2880
7	0.3698	16	0.2831
8	0.3512	17	0.2787
9	0.3367	18	0.2747
10	0.3249	19	0.2711
		20	0.2677

Since the average and the standard deviation are computed from a sample, they are themselves subject to fluctuation. However, if μ is the long-term average of the process and σ is the long-term standard deviation, then:

$$\text{Average } (\bar{x}) = \mu, \text{ process average}$$
$$\text{Average } (s) = \sigma, \text{ process standard deviation}$$

Furthermore, the intervals $\mu \pm k\sigma$ contain the same percentage of all values, as do the intervals $\bar{x} \pm ks$ for the sample; that is, at least 89 percent of all the long-term values will be contained in the interval $\mu - 3\sigma$ to $\mu + 3\sigma$, etc.

Range Estimate of the Standard Deviation

For $n \leq 20$ it is more convenient to compute the range r to estimate the standard deviation σ. The range is $x_{(n)} - x_{(1)}$, where $x_{(n)}$ is the largest value in a random sample of size n and $x_{(1)}$ is the smallest value. For example, if $n = 10$ and the observations are 310, 309, 312, 316, 314, 303, 306, 308, 302, 305, the range is $r = 316 - 302 = 14$. An estimate of the standard deviation σ is obtained by multiplying r by the factor f_n in Table 17.3.1. The average value of $r \cdot f_n$ is σ. Thus, in the example above, an estimate of σ and a value that can be used for s is $0.3249r = 0.3249(14) = 4.5486$.

PROCESS VARIABILITY—HOW MUCH DATA?

Since the output of all processes is variable, one can make reasonable decisions about the output only if one can obtain a measure of how much variability or spread one can expect to see under normal conditions. Variability cannot be measured accurately with a small amount of data. Methods for assessing how much data are needed are given for two general situations.

Specified Tolerances

A convenient statement about the variability or spread of a process can be based on the smallest and largest values in a random sample of the output. There are no practical limitations on its use. Suppose that we have a random sample of n values from our process. Denote the values by $X_1, X_2, \ldots, X_n$. After obtaining the values we find the smallest, $X_{(1)}$, and the largest, $X_{(n)}$. Now we want to assess what percent of all future values that this process might generate will be covered by $X_{(1)}$ and $X_{(n)}$. In statistics, $X_{(1)}$ and $X_{(n)}$ are called **tolerance limits**. If the process generates bolts and X is the diameter, then the engineering concept of tolerance and the statistical concept of tolerance are seen to be quite similar.

Let p be the percentage of all the process values that on a long-term basis will be between $X_{(1)}$ and $X_{(n)}$. Let P be a lower bound for this percentage p. Now consider the probability statement: Probability ($p \geq P$) $= C$. The quantity C we call **confidence**. Since it is a probability its value is between 0 and 1. As C approaches 1 our confidence in the percentage P increases. The interpretation of P and C can be explained in terms of Table 17.3.2.

Suppose that we take a random sample of size $n = 269$ values from our process output; the smallest value is 10 and the largest is 54. In Table 17.3.2 we see that 269 is the entry for $P = 99$ and $C = 0.75$. This tells us that at least $P = 99$ percent of all future values that this process

Table 17.3.2

P, %	Confidence, C			
	0.995	0.99	0.95	0.75
99.9	7427	6636	4742	2692
99.5	1483	1325	947	538
99	740	661	473	269
98	368	330	235	134
97	245	219	156	89
96	183	163	117	67
95	146	130	93	53
90	71	64	46	26
80	34	31	22	13
75	27	24	18	10

NOTE: Sample size r required to have a confidence C that at least P percent of all future values will be included between the smallest and largest values in a random sample.

will generate will be between 10 and 54, the smallest and largest values seen in the sample of 269. The confidence $C = 0.75 = \frac{3}{4}$ tells us that the chances are 3 out of 4 that our statement is correct. As we increase the sample size n, we increase the chances that our statement is correct. For example, if our sample size had been $n = 473$, then $C = 0.95$ and the chances are 95 in 100 that we are correct in making the statement that at least 99 percent of all process values will be between the 10 and 54 seen in the sample. Similarly, if the sample size had been 740, then the chances of being correct increase to 995 in 1000. If sample size n is decreased sufficiently, the confidence $C = 0.50 = \frac{1}{2}$ indicates that the chances are one in two of being right, *and* one in two of being *wrong*. Therefore, it is important to select n so as to keep C as large as possible. The cost of acquiring the data will determine the upper limit for n.

Further information on tolerance limits can be found in Wilks (1962) and Duncan (1986).

Wear-Out and Life Tests

A special case of coverage occurs if our interest is in a wear-out phenomenon or a life test. For example, suppose we put a number of incandescent light bulbs on test; our interest is in the length of time to failure. Clearly we do not want to wait until all specimens fail to draw a conclusion; it might take an inordinate length of time for the last one to fail. From a practical point of view we would probably be interested in those that fail first anyway. If the sample size is properly chosen, there will be important information as soon as we obtain the first failure.

In a random sample of size n let the failure times be $T_1 \geq T_2 \geq \cdots \geq T_n$. The value T_1 is thus the smallest value in a random sample of size n. Now let q be the percentage of future units that can be expected to fail in a time less than T_1, the smallest value in the sample. As before we can make a probability statement about q. Let Q be an upper bound to q. Then we can compute: Probability ($q \leq Q$) $= C$.

For example, suppose that we put a random sample of 299 items on test and the first one fails in time $T_1 = 151$ h. From Table 17.3.3 we see that 299 is an entry for $Q = 1$ and $C = 0.95$. Thus we can conclude

Table 17.3.3

Q, %	Confidence, C			
	0.995	0.99	0.95	0.75
0.1	5296	4603	2995	1386
1	528	459	299	138
2	263	228	149	69
3	174	152	99	46
4	130	113	74	34
5	104	90	59	28
10	51	44	29	14
15	33	29	19	9
20	24	21	14	7
25	19	17	11	5

NOTE: Sample size r required to have a confidence C that fewer than Q percent of future units will fail in a time shorter than the shortest life in the sample. For a more extensive table of values, see Owen (1962).

that not more than 1 percent of future units should fail in a time less than 151 h. The confidence in the statement is 95 chances in 100 of being correct. Again referring to Table 17.3.3, we see that if $T_1 = 151$ were based on a sample of 528, then the confidence would be increased to 995 chances in 1000. Most importantly, Table 17.3.3 tells us that we need to test a very large sample if we want to have high confidence that only a small percentage of future units will fail in a time less than the smallest observed. The theory behind Table 17.3.3 can be found in Wilks (1962). For a more extensive table of values see Owen (1962). If $Q' = Q/100$, use

$$r = [\log(1 - C)]/[\log(1 - Q')]$$

CORRELATION AND ASSOCIATION

One of the most common problems in the analysis of engineering data is to determine if a correlation or an association exists between two variables X and Y, where the data occur in pairs (X_i, Y_i). The "corner test of association" developed by Olmstead and Tukey (1947) is a quick and simple test to assist in making this determination.

Corner Test

Conditions for Use Each pair (X_i, Y_i) should have been obtained independently; there are no other practical assumptions for its use. Of course, the user should consider the physical process generating the data when interpreting any correlation or association that is determined to exist.

Procedure

1. Make a scatter plot on graph paper of the data pairs (X_i, Y_i), with the usual convention that X is the horizontal scale and Y the vertical.

2. Determine the median X_m of the X_i values. Determine the median Y_m of the Y_i values.

The median splits the data into two parts so that there is an equal number of values above and below the median. Let N denote the total number of points. If N is odd, then N can be written as $2k + 1$ and the median is the $(k + 1)$st value as the values are ordered from the smallest to the largest. If N is even, then N can be written as $2k$. Then the median is taken to be midway between the kth and $(k + 1)$st values.

3. Draw a vertical line through X_m.

4. Draw a horizontal line through Y_m.

5. The lines in (3) and (4) divide the graph into four quadrants. Label the upper right and lower left as plus. Label the upper left and lower right as minus.

6. Begin at the right side of the plot. Count the values, in order of decreasing X, until forced to cross the horizontal median Y_m. Give the count a plus sign if the values are in a plus quadrant, a minus sign if they are in a minus quadrant.

7. Repeat the procedure in (6), moving down from above until you have to cross the vertical median, moving from left to right until you have to cross the horizontal median, and moving up from below until you have to cross the vertical median.

8. Compute the algebraic sum of the four counts obtained in (6) and (7). Denote the sum by T.

Test If $|T| \geq 11$, then there is evidence of correlation between X and Y; $|T|$ is the value of T ignoring the sign.

EXAMPLE. Table 17.3.4 gives 33 pairs of values obtained from samples of a paper product. The X coordinate is proportional to the reciprocal of light transmitted by the sample. The Y coordinate is proportional to tensile strength.
1. The 33 pairs of values are plotted in Fig. 17.3.1.
2. The median of X values is $X_m = 11.94$. The median of Y values is $Y_m = 6.3$.
3 to 5. The medians are shown in Fig. 17.3.1, and the quadrants are labeled.
6 to 7. The counts are as follows:

Right to left: -2 (points at 19.99 and 17.86 on X)
Top to bottom: -4 (points at 16.6, 12.9, 12.5, 11.7 on Y)
Left to right: -2 (points at 0.93 and 3.18 on X)
Bottom to top: -4 (points at 0.1, 1.6, 2.6, 2.7 on Y)

8. The algebraic sum of the counts is -12. Hence $T = -12$. And since $|T| \geq 11$, one can conclude that there is evidence of correlation or association between the variables X and Y.

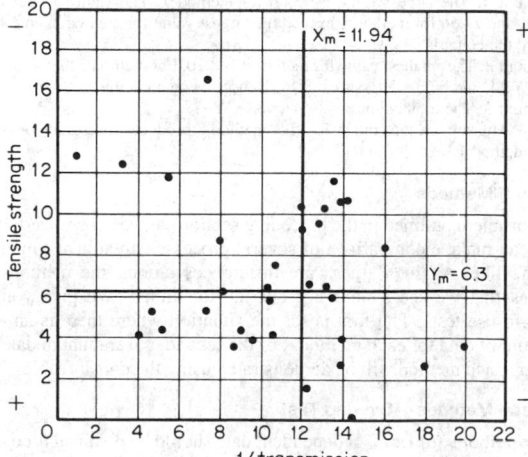

Fig. 17.3.1 Plot of example data used in conjunction with the Corner Test.

COMPARISON OF METHODS OR PROCESSES

A common problem in engineering investigations is that of using experimental or observational data to assess the performance of two processes, two treatment methods, or the like. Often one process or treatment is a standard or the one in current use. The other is an alternative that is a candidate to replace the standard. Sometimes it is cheaper, and one hopes to see no performance difference. Sometimes it is supposed to offer superior performance, and one hopes to see a measurable difference in the variable of interest. In either case we know that the variable of interest will have a distribution of values; and if the two processes are to be measurably different the distribution of values should not overlap too much. For an objective assessment we need to have some way to calibrate the overlap. A quick and easy-to-use test for this purpose is the **outside count test** developed by Tukey (1959).

Two Methods—Outside Count Test

Conditions for Use Given two groups of measurements taken under conditions 1 and 2 (treatments, methods, etc.), we identify the direction

Table 17.3.4

i	X	Y	i	X	Y
1	10.45	4.1	18	9.65	3.8
2	13.81	2.7	19	7.44	5.4
3	12.22	1.6	20	10.70	7.6
4	9.05	4.3	21	13.38	6.0
5	17.86	2.6	22	13.00	10.4
6	14.54	0.1	23	13.90	10.7
7	19.99	3.7	24	11.94	9.4
8	8.73	3.5	25	14.11	10.7
9	4.66	5.3	26	0.93	12.9
10	13.88	3.9	27	3.18	12.5
11	5.10	4.4	28	13.13	6.5
12	3.98	4.1	29	13.45	11.7
13	8.12	6.3	30	12.70	9.6
14	12.26	6.6	31	15.95	8.5
15	10.30	6.5	32	7.30	16.6
16	5.40	11.9	33	7.78	8.8
17	10.39	5.8			

NOTE: Data are on paper samples. X is proportional to reciprocal of light transmission. Y is proportional to tensile strength.

of difference by insisting that the two groups have minimum overlap. Use 1 to denote the group with the smaller number of measurements and let N_1 be the number of measurements for that group. Let N_2 be the number of measurements for the other group. The number of observations for each group should be about the same.

The conditions to be satisfied are:

$$4 \leq N_1 \leq N_2 \leq 30$$
$$N_2 \leq (4N_1/3) + 3$$

Procedure

1. Count the number of values in the one group *exceeding* all values in the other.

2. Count the number of values in the other group *falling below* all those in the other.

3. Sum the two counts in (1) and (2). (It is required that neither count be zero. One group must have the largest value and the other the smallest.)

Test If the sum of the two counts in (3) is 7 or larger, there is sufficient evidence to conclude that the two methods are measurably different.

EXAMPLE. The following data represent the results of a trial of two methods for increasing the wear resistance of a grinding wheel. The data are proportional to wear:

Method 1: 13.06**, 9.52, 9.98, 8.83, 12.78, 9.00, 11.56, 8.10*, 12.21.

Method 2: 8.44, 9.64, 9.94, 7.30, 8.74, 6.30*, 10.78**, 7.24, 9.30, 6.66.

The smallest value for each method is marked with an asterisk; the largest value is marked with two asterisks.

Count 1: The largest value is 13.06 for method 1. The values 13.06, 12.78, 12.21, and 11.56 for method 1 exceed the largest value for method 2, viz., 10.78. Hence the count is 4.

Count 2: The smallest value for method 1 is 8.10. For method 2 the values 7.30, 6.30, 7.24, and 6.66 are less than 8.10. Hence the count is 4.

Count 3: The total count is $4 + 4 = 8 > 7$.

The data support the conclusion that method 2 produces measurably less wear than method 1.

Several Methods

The problem outlined in the preceding section can be generalized so that one can make a comparison of several processes, treatments, methods, or the like. Again, if there are differences among the methods, the values that we see should not overlap too much. We give you two easy-to-use tests. The first is for the situation where there is an equal amount of data for each method. For the second, the amount of data may differ. Each method will be demonstrated using the data in Table 17.3.5.

Several Methods—Overlap Test

Conditions for Use Independent data should be obtained for each of the k methods. *The number of values n should be the same* for each method.

Table 17.3.5

Sample no.	Supplier 1	Supplier 2	Supplier 3	Supplier 4
1	45.37	30.05	41.30	46.21
2	21.68	36.04	31.09	36.01
3	43.91	18.04	24.31	46.28
4	47.76*	32.91	15.64	21.80
5	23.81	41.67	54.85*	28.57
6	19.90	37.40	32.96	48.45
7	44.68	46.67*	45.48	33.49
8	11.81	27.93	45.14	53.07*
9	35.42	45.20	45.49	35.65
10	39.85	29.54	52.82	14.95

NOTE: The data are proportional to the time to failure of a standard cutting tool. Asterisks denote largest value in each group.

Procedure

1. For each of the k methods, determine the *largest* value. Label it with an asterisk.

2. Scan the largest values. Label the group with the *largest largest* value as BIG. Label the group with the *smallest* largest value as SMALL, and its largest value as S.

3. In the group labeled BIG *count* the number of values that are larger than S, the largest value in SMALL. Denote this count by C.

4. Enter Table 17.3.6 for n values of k groups. If C exceeds the tabled value, then the data support a conclusion that the methods are different. Otherwise, they do not.

EXAMPLE. 1. In Table 17.3.5 the largest value for each of the four groups is marked with an asterisk.

2. Group 3 is BIG. Group 2 is SMALL; the largest value in Group 2 is $S = 46.67$.

3. The number of values in Group 3 larger than 46.67 is 2 (52.82, 54.85).

4. Enter Table 17.3.6 with $n = 10$ and $k = 4$. The entry is 5 which is greater than 2. Hence, the data do not support a conclusion that the time to failure for the cutting tools of the four suppliers is measurably different.

Several Methods—Rank Test

Conditions for Use Independent data should be obtained for each of the methods. The amount of data for each method may be different.

Procedure

1. Let n_i be the number of values in Group i.

2. Let $N = \sum_1^k n_i$ be the total number of values.

3. Rank each value from 1 to N beginning with the smallest. (If there are ties among t values, divide the successive ranks equally among them.)

Table 17.3.6 95% Point for k-Sample Problems

n	k = 3	4	5	6	7	8	9	10
5	4	4	4	4	4	4	4	4
6	4	4	4	5	5	5	5	5
7	4	5	5	5	5	5	5	5
8	4	5	5	5	5	5	5	5
9	5	5	5	5	5	5	6	6
10	5	5	5	5	6	6	6	6
12	5	5	5	6	6	6	6	6
14	5	5	6	6	6	6	6	6
16	5	5	6	6	6	6	6	6
18	5	6	6	6	6	6	6	7
20	5	6	6	6	6	6	7	7
25	5	6	6	6	6	7	7	7
30	5	6	6	6	7	7	7	7
40	5	6	6	7	7	7	7	7
>40	6	6	7	7	7	7	8	8

NOTE: k is the number of groups; n is the number of values per group. For other n, k, and percentage points see Conover (1968).

4. Compute r_i, the sum of the ranks for the ith group. [*Note:* For a check $\sum_1^k r_i = N(N+1)/2$.]

Test

1. Compute

$$T = [12/N(N+1)] \left[\sum_1^k (r_i^2/n_i) \right] - 3(N+1)$$

2. Go to Table 17.3.7; find the entry under $k - 1$.

If T exceeds the entry, then the data support the conclusion that the groups are different. Otherwise, they do not.

Table 17.3.7 Chi-Square Distribution

k	w	k	w
1	3.841	16	26.30
2	5.991	17	27.59
3	7.815	18	28.87
4	9.488	19	30.14
5	11.07	20	31.41
6	12.59	22	33.92
7	14.07	24	36.42
8	15.51	26	38.89
9	16.92	28	41.34
10	18.31	30	43.77
11	19.68	40	55.76
12	21.03	50	67.50
13	22.36	60	79.08
14	23.68	70	90.53
15	25.00	80	101.9

NOTE: Entries are $P(W > w) = p = 0.05$. For other values of k and p, see Pearson and Hartley (1962).

EXAMPLE. We again use the data shown in Table 17.3.5. In Table 17.3.8 the numerical values representing times to failure have been replaced by their ranks. To facilitate such ranking it is convenient to order the values in each group from smallest to largest. Then all values are ranked from smallest to largest. In Table 17.3.8 the values have been reordered this way. The ranks are in parentheses.

1. The number of values in each group is 10. Hence $n_i = 10$ for each value of i.
2. The total number of values $N = 40$.
3 and 4. The sum of the ranks r_i is shown for each group. [Note that $\Sigma r_i = 820 = (40)(41)/2$.]

$$\begin{aligned}
T &= [12/N(N+1)][\Sigma(r_i^2/10)] - 3(N+1) \\
&= [12/(40)(41)][170182/10] - 3(41) \\
&= 124.523 - 123. = 1.523
\end{aligned}$$

Now go to Table 17.3.7 and obtain the entry under $k = 4 - 1 = 3$. The entry is 7.815, which is larger than 1.523. Hence, the data do not support a conclusion that the time to failure for the cutting tools for the four suppliers is measurably different.

Table 17.3.8

Supplier*			
1	2	3	4
11.81 (1)	18.04 (4)	15.64 (3)	14.95 (2)
19.90 (5)	27.93 (10)	24.31 (9)	21.80 (7)
21.68 (6)	29.54 (12)	21.09 (14)	28.57 (11)
23.81 (8)	30.05 (13)	32.96 (16)	33.49 (17)
35.42 (18)	32.91 (15)	41.30 (24)	35.65 (19)
39.85 (23)	36.04 (21)	45.14 (28)	36.01 (20)
43.91 (26)	37.40 (22)	45.48 (31)	46.21 (33)
44.68 (27)	41.67 (25)	45.49 (32)	46.28 (34)
45.37 (30)	45.20 (29)	52.82 (38)	48.45 (37)
47.76 (36)	46.67 (35)	54.85 (40)	53.07 (39)
180	186	235	219 r_i
32400	34596	55225	47961 r_i^2

* These are the data of Table 17.3.5 with the values for each supplier listed from smallest to largest. The values in parentheses are the ranks of the time to failure values from smallest to largest.

GO/NO-GO DATA

Quite often the data that we encounter will be **attribute** or **go/no-go data**; that is, we will not have quantitative measurements but only a characterization as to whether an item does or does not have some attribute. For example, if a manufactured part has a specification that it should not be shorter than 2 in, we might construct a template; and if a part is to meet the specification, it should not fit into the template. After inspecting a series of units with the template our data would consist of a tabulation of "gos" and "no-gos"—those that did not meet the specification and those that did.

If the items that are checked for an attribute are obtained by random sampling, the resulting data will follow what is known as the **binomial distribution**. Its standard form is as follows:

p is the long-term fraction of failures
$q = 1 - p$ is the long-term fraction of successes
n is the size of the random sample.

Then the probability that our sample gives x failures and $n - x$ successes is

$$f(x; n, p) = \binom{n}{x} p^x q^{n-x}; x = 0, 1, 2, \ldots, n$$

where

$$\binom{n}{x} = n!/x! \, (n - x)!$$

From $f(x; n, p)$, for a given n and p, we can calculate the probability of x failures in a sample size n. Similarly, by summing terms for different values of x, we can calculate the probability of having more than w failures or fewer than r failures, etc. Here we are not going to try to be so precise; rather we are going to try to show the general picture of the relationship between x, p, and n by the use of examples and the graph in Fig. 17.3.2.

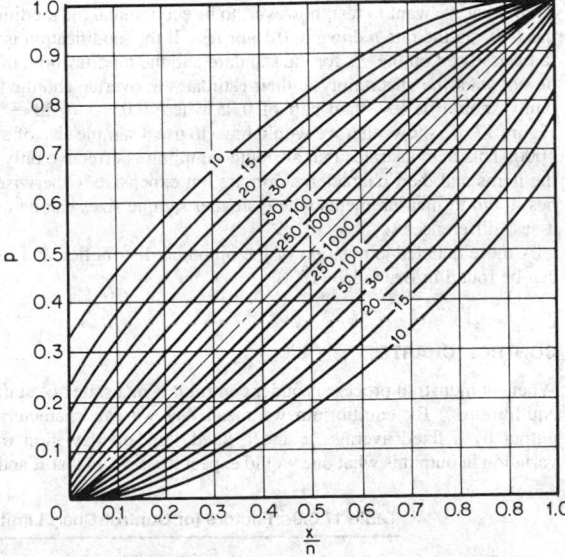

Fig. 17.3.2 Binomial distribution, 95 percent confidence bands. (*Reproduced with the permission of the Biometrika Trust from C. J. Clopper and E. S. Pearson, "The Use of Confidence or Fiducial Limits Illustrated in the Case of the Binomial,"* Biometrika, 26 (1934), p. 410.)

Estimating the Failure Rate

In a manufacturing process a general index of quality is the fraction of items which fail to pass a certain test. Suppose that we take a random sample of size $n = 100$ from the process and observe $x_0 = 10$ failures. Clearly we have met the conditions for a binomial distribution and an estimate of p, the long-term failure rate is $\hat{p} = x_0/n = 10/100 = 0.1$.

However, we would also like to know the accuracy of the estimate. In other words, if we operate the process for a long time under these conditions and obtain a large sample, what might be the value of p? One simple way to assess the estimate of $\hat{p}$ is to find values p_1 and p_2 ($p_1 < p_2$) such that the following probabilities are satisfied for a fixed value of α:

$$\Pr[x \geq x_0 | p_1] = \alpha/2$$
$$\Pr[x \leq x_0 | p_2] = \alpha/2$$

These values are the solutions for p of the two equations.

$$\sum_{x=x_0}^{n} \binom{n}{x} p_1^x (1 - p_1)^{n-x} = \alpha/2$$

$$\sum_{x=0}^{x_0} \binom{n}{x} p_2^x (1 - p_2)^{n-x} = \alpha/2$$

General solutions for these equations for $\alpha = 0.05$ are shown in Fig. 17.3.2. If we go to Fig. 17.3.2 with $x/n = 0.1$ and read where the lines for $n = 100$ intersect the ordinate or p scale, we see that $p_1 = 0.07$ and $p_2 = 0.18$. We can then state that we have reasonable confidence (the probability is 0.95) that the long-term failure rate for the process is between 0.07 and 0.18.

Estimating the Sample Size

It should be evident that Fig. 17.3.2 can also be used to determine how large a sample is needed to estimate a proportion or a percentage with a specific accuracy or tolerance. Suppose that the proportion of interest is assumed to be around $p = 0.20$. Now enter Fig. 17.3.2 with $x/n = 0.20$. From the figure we see that if we take a sample of size 50, our estimate will have a range of about ± 0.10 (actually $-0.10, +0.13$). On the other hand, if the sample size is 250, the estimate will have a range of about ± 0.06.

Often one wants to compare the performance of two processes. As above, suppose that the rate p for our process is 0.20. We have a modification that we want to test; however, to be economical the modification has to bring the rate p down to 0.15 or less. If the modification is going to be assessed on the p's for the standard and the modification, then we do not want the uncertainty in their estimates to overlap and the uncertainty should be less than half of 0.05 where $0.05 = (0.20 - 0.15)$. Figure 17.3.2 shows that we would have to use a sample size of at least 1000. This demonstrates that attribute sampling is effective only when the items and their characterization are not expensive. Otherwise, it is best to go to measurements where smaller sample sizes can be used to assess differences.

A more detailed exposition of the binomial distribution and its uses can be found in Brownlee (1970).

CONTROL CHARTS

When an industrial process is under control it is in a state of "statistical equilibrium." By equilibrium we mean that we can characterize its output by a fixed average μ and a fixed standard deviation σ. The variation in output is what one would expect to see from that μ and σ, as bounded by the values given in Tchebysheff's theorem, let us say. However, if control is lost, we tend to get a greater spread in values. In effect, the average μ or the standard deviation σ is changing because of some cause. The causes of lack of control are manifold—it can be a change in raw materials, tool wear, instrumentaton failure, operator error, etc. The important thing is that one wants to be able to detect when this lack of control occurs and take the appropriate steps to make corrections.

One of the most frequently used statistical tools for analyzing the state of an industrial process is the control chart. The two most commonly used charts are those for the **average** and the **range**. The control chart procedure consists of these steps:

1. Choose a characteristic X which will be used to describe the product coming from the process.

2. At time t_i, take a small number of observations n on the process. The number n should be small enough so that it is reasonable to assume that conditions will not change during the course of obtaining the observations.

3. For each set of n observations, compute the average $\bar{x}_i$ and the range r_i as defined in the section "Characterizing Observational Data."

There are two different control situations of interest.

4a. *Control standards given.* Suppose that from past operation of the process or from the need to meet certain specifications, a goal average μ^* and a goal standard deviation σ^* are specified. Then we set up two charts as follows:

Average chart:	Upper limit line:	$\mu^* + A\sigma^*$
	Central line:	μ^*
	Lower limit line:	$\mu^* - A\sigma^*$
Range chart:	Upper limit line:	$D_2\sigma^*$
	Central line:	$d\sigma^*$
	Lower limit line:	$D_1\sigma^*$

The values of A, d, D_1, and D_2 depend upon n and can be found in Table 17.3.9.

Plot the values of $\bar{x}_i$ and r_i obtained in (3) on the two charts as shown in Fig. 17.3.3. Whenever a value falls outside the limit lines, there is an indication of lack of control, and one is justified in seeking the causes for a change.

4b. *Control, no standards given.* Often one has no prior information about the process μ and σ, and one wants to determine if the process behaves as though it is in statistical equilibrium, and if not, take actions to get it there. In this case one has to determine the central lines for the charts from process data. To do this one first accumulates the data for 25 to 50 time periods as indicated in (2). Then two charts are set up as follows: Let K be the 25 to 50 time periods observed. Compute an overall average $\bar{X} = \sum_{i=1}^{k} \bar{x}_i/K$ and the average range $\bar{R} = \sum_{i=1}^{k} r_i/K$. Set up charts with limits defined from:

Average chart:	Upper limit line:	$\bar{X} + A_2\bar{R}$
	Central line:	$\bar{X}$
	Lower limit line:	$\bar{X} - A_2\bar{R}$
Range chart:	Upper limit line:	$D_4\bar{R}$
	Central line:	$\bar{R}$
	Lower limit line:	$D_3\bar{R}$

Table 17.3.9 Factors for Control-Chart Limits

Sample size n	For averages			For ranges			
	A	A_2	d	D_1	D_2	D_3	D_4
2	2.12	1.88	1.128	0	3.69	0	3.27
3	1.73	1.02	1.693	0	4.36	0	2.57
4	1.50	0.73	2.059	0	4.70	0	2.28
5	1.34	0.58	2.326	0	4.92	0	2.11
6	1.22	0.48	2.534	0	5.08	0	2.00
7	1.13	0.42	2.704	0.21	5.20	0.08	1.92
8	1.06	0.37	2.847	0.39	5.31	0.14	1.86
9	1.00	0.34	2.970	0.55	5.39	0.18	1.82
10	0.95	0.31	3.078	0.69	5.47	0.22	1.78

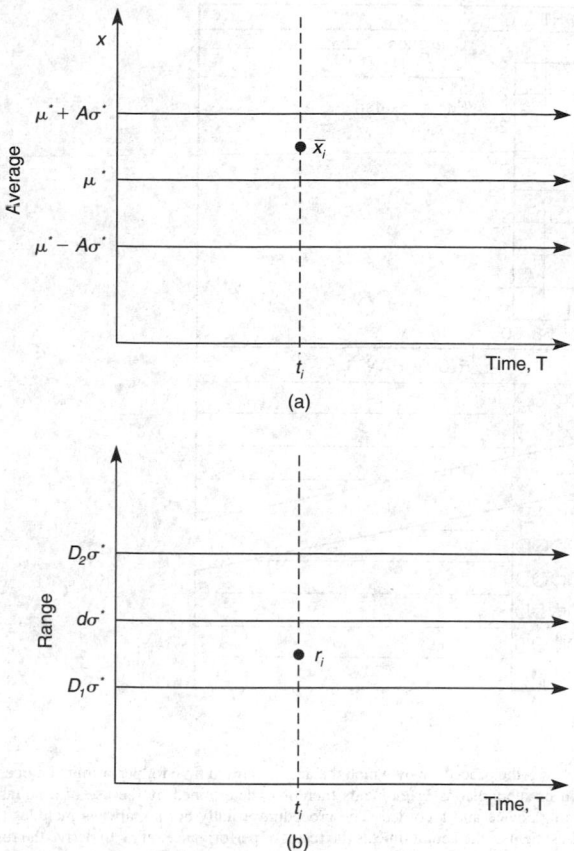

Fig. 17.3.3   Control chart. (*a*) Average chart; (*b*) range chart.

The values of A_2, D_3, and D_4 depend upon n and can be found in Table 17.3.9. The individual $\bar{x}_i$ and r_i are plotted on the charts, and again a value outside the limits is an indication of lack of control and is justification for seeking the cause for a change.

Process Capability Indices

If the process is in statistical control, an estimate for the process standard deviation can be obtained by using

$$\hat{\sigma} = \overline{R}/d$$

In turn, $\hat{\sigma}$ is used to calculate the **process capability indices** C_p and C_{pk}. These two indices compare the actual spread of the data with the desired range (usually as specified by a customer). C_p is used when the actual process average is equal to the goal average. C_{pk} is used when the actual process average is not equal to the goal average. The desired range is called the **specification tolerance** (ST) and is equal to the upper specification limit minus the lower specification limit, namely, $2A\sigma^*$.

C_p is given by the following equation:

$$C_p = \text{ST}/6\hat{\sigma} = 2A\sigma^*/6\hat{\sigma}$$

If $C_p \geq 1$, the process is considered to be capable, which means that most or all of the data stayed within the desired range. If $C_p < 1$, the process is considered to be not capable and requires adjustment. Ideally, one should control the process variability so that $C_p \geq 2$.

The other index, C_{pk}, is given by

$$C_{pk} = C_p(1 - k)$$

where

$$k = 2(\mu^* - \overline{X})/\text{ST} = (\mu^* - \overline{X})/A\sigma^*$$

and

$$\overline{X} = \sum \bar{x}_i/K$$

The $(1 - k)$ factor modifies C_p so as to allow the actual average $\overline{X}$ to be different from the goal average μ^*. Ideally, one should control the process so that $C_{pk} \geq 1.5$. In process capability analysis, the indices should be calculated on a frequent basis, but the trends should be examined only monthly or quarterly in order to be meaningful.

Charts for Go/No-Go Data

The control chart concept can also be used for attribute or go/no-go data. The procedures are, in general, the same as outlined for averages and ranges. Briefly, they are as follows:

1. Select a sample of size n from the process; for best results n should be in the range of 50 to 100.

2. Let x_i denote the number of defective units in the sample of size n at time t_i; then $\hat{p}_i = x_i/n$ is an estimate of the process fraction defective.

3. Set control limits for a standard fraction defective p^* at

$$p^* \pm 3[p^*(1 - p^*)/n]^{1/2}$$

If no standard is given, then take $K = 25$ to 50 samples of size n to get a good estimate of the fraction-defective p. Define $\overline{p} = \Sigma_{i=1}^{K} \hat{p}_i/K$. In this case set control limits at

$$\overline{p} \pm 3[\overline{p}(1 - \overline{p})/n]^{1/2}$$

4. Interpret a $\hat{p}_i$ outside the limits as an indication of a change worthy of investigation.

Further information on control charts can be found in Duncan (1986).

17.4   METHODS ENGINEERING
by Vincent M. Altamuro

REFERENCES: ASME Standard Industrial Engineering Terminology. Barnes, "Motion and Time Study," Wiley. Krick, "Methods Engineering," Wiley. Maynard, Stegemerten, and Schwab, "Methods-Time-Measurement," McGraw-Hill.

SCOPE OF METHODS ENGINEERING

Methods engineering is concerned with the selection, development, and documentation of the methods by which work is to be done. It includes the analysis of input and output conditions, assisting in the choice of the processes to be used, operations and work flow analyses, workplace design, assisting in tool and equipment selection and specifications, ergonomic and human factors considerations, workplace layout, motion analysis and standardization, and the establishment of work time standards. A primary concern of methods engineering is the integration of humans and equipment in the work processes and facilities.

PROCESS ANALYSIS

Process analysis is that step in the conversion of raw materials to a finished product at which decisions are made regarding what methods, machines, tools, inspections and routings are best. In many cases, the product's specifications can be altered slightly, without diminishing its function or quality level, so as to allow processing by a preferred

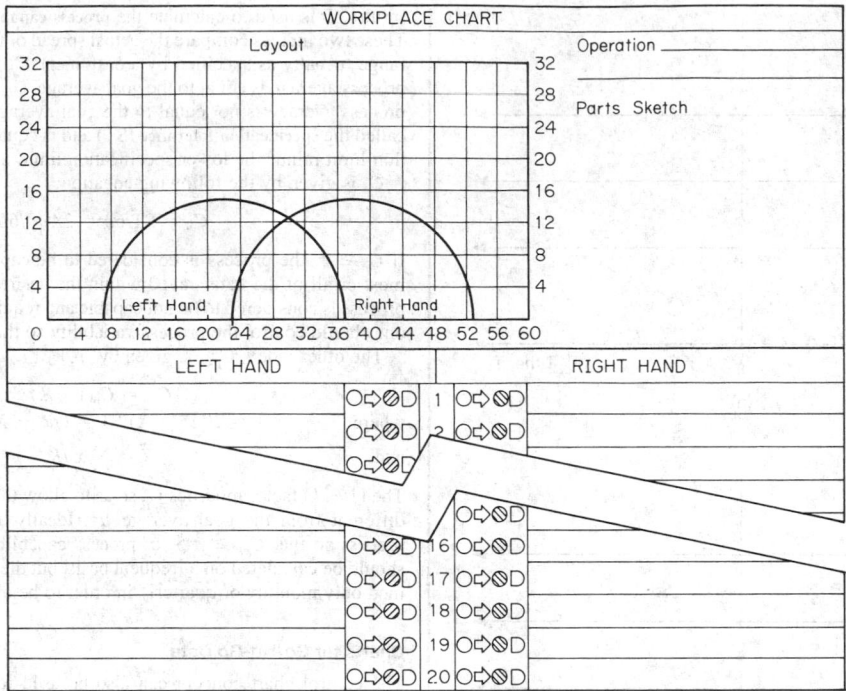

Fig. 17.4.1 Workplace layout chart.

method. For this reason, it is desirable to have the product's designer and the process engineer work together before specifications are finalized.

WORKPLACE DESIGN

Material usually flows through a facility, stopping briefly at stations where additional work is done on it to bring it closer to a finished product. These workstations, or workplaces, must be designed to permit performance of the required operations, to contain all the tooling and equipment needed to fit the capabilities and limitations of the people working at them, to be safe and to interface smoothly with neighboring workplaces.

Human engineering and ergonomic factors must be considered so that all work, tools, and machine activation devices are not only within the comfortable reach of the operator but are designed for safe and efficient operation. A workplace chart (Fig. 17.4.1) which analyzes the required actions of both hands is an aid in workplace design.

METHODS DESIGN

Methods design is the analysis of the various ways a task can be done so as to establish the one best way. It includes *motion analysis*—the study of the actions the operator can use and the advantages and/or disadvantages of each variation—and *standardization of procedure*—the selection and recording of the selected and authorized work methods.

While "time and motion study" is the more commonly used term, it is more correct to use "motion and time study," as the motion study to establish the standard procedure must be done prior to the establishment of a standard time to perform that work.

According to ASME Standard Industrial Engineering Terminology, motion study is defined as

. . . the analysis of the manual and the eye movements occurring in an operation or work cycle for the purpose of eliminating wasted movements and establishing a better sequence and coordination of movements.

In the same publication, time study is defined as

. . . the procedure by which the actual elapsed time for performing an operation or subdivisions or elements thereof is determined by the use of a suitable timing device and recorded. The procedure usually but not always includes the adjustment of the actual time as the result of performance rating to derive the time which should be required to perform the task by a workman working at a standard pace and following a standard method under standard conditions.

Attempts have been made to separate the two functions and to assign each to a specialist. Although motion study deals with method and time study deals with time, the two are nearly inseparable in practical application work. The method determines the time required, and the time determines which of two or more methods is the best. It has, therefore, been found best to have both functions handled by the same individual.

ELEMENTS OF MOTION AND TIME STUDY

Figure 17.4.2 presents graphically the steps which should be taken to make a good motion and time study and shows their relation to each other and the order in which they must be performed.

METHOD DEVELOPMENT

The first step is the development of the method. Starting with the drawing of the product, the operations which must be performed are determined and tools and equipment are specified. In large companies, this is usually done by a specialist called a process engineer. In smaller companies, processing is commonly done by the time study specialist.

Next, the detailed method by which each operation should be performed is developed. The procedures used for this are known as operation analysis and motion study.

OPERATION ANALYSIS

Operation analysis is the procedure employed to study all major factors which affect a given operation. It is used for the purpose of uncovering possibilities of improving the method. The study is made by reviewing the operation with an open mind and asking either of oneself or others

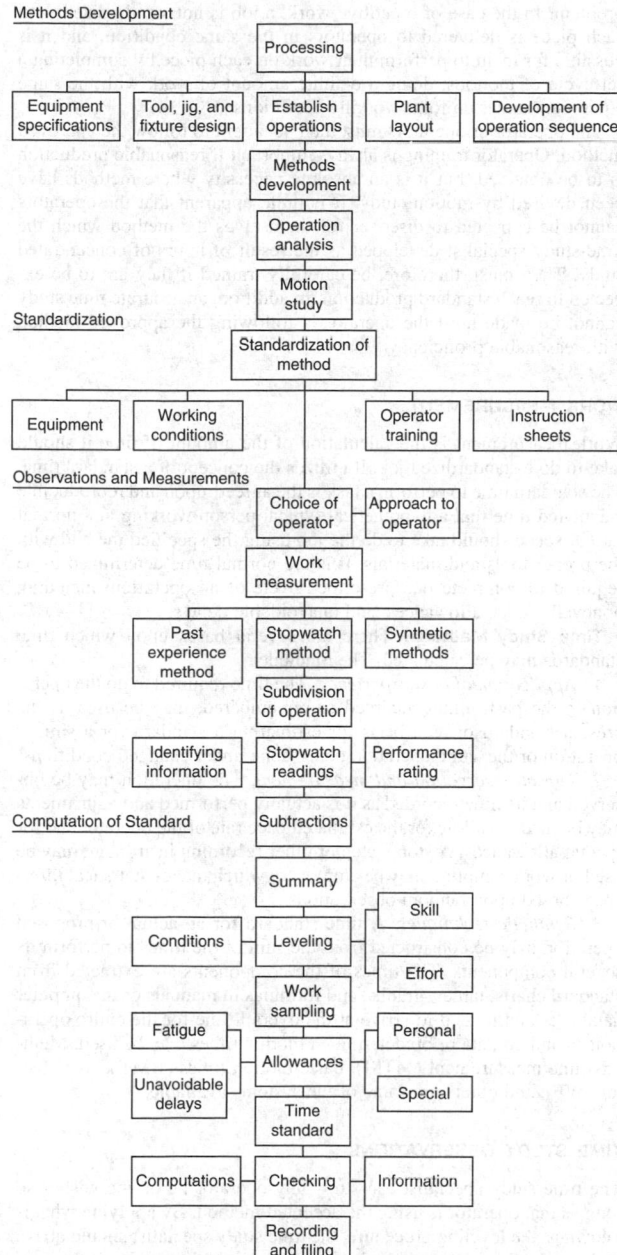

Methods Development

Standardization

Observations and Measurements

Computation of Standard

Fig. 17.4.2 Graphic analysis of the elements of motion and time study.

questions which are likely to lead to methods-improving ideas. If this is done systematically, so that the possibility of overlooking factors which should be considered is minimized, worthwhile improvements are almost certain to result.

The 10 major factors explored during operation analysis, together with typical questions which should be asked about each factor, are as follows:

1. Purpose of operation
   a. Is the result accomplished by the operation necessary?
   b. Can the purpose of the operation be accomplished better in any other way?

2. Design of part
   a. Can motions be eliminated by design changes which will not affect the functioning and other desirable characteristics of the product?
   b. Is the design satisfactory for automated assembly?
3. Complete survey of all operations performed on part
   a. Can the operation being analyzed be eliminated by changing the procedure or the sequence of operations?
   b. Can it be combined with another operation?
4. Inspection requirements
   a. Are tolerance, allowance, finish, and other requirements necessary?
   b. Will changing the requirements of a previous operation make this operation easier to perform?
5. Material
   a. Is the material furnished in a suitable condition for use?
   b. Is material utilized to best advantage during processing?
6. Material handling
   a. Where should incoming and outgoing material be located with respect to the work station?
   b. Can a progressive assembly line be set up?
7. Workplace layout, setup, and tool equipment
   a. Does the workplace layout conform to the principles of motion economy?
   b. Can the work be held in the machine by other means to better advantage?
8. Common possibilities for job improvement
   a. Can "drop delivery" be used?
   b. Can foot-operated mechanisms be used to free the hand for other work?
9. Working conditions
   a. Has safety received due consideration?
   b. Are new workers properly introduced to their surroundings, and are sufficient instructions given them?
10. Method
    a. Is the repetitiveness of the job sufficient to justify more detailed motion study?
    b. Should full automation be considered?

When the method has been developed, conditions are standardized, and the operators are trained to follow the approved method.

At this time, not before, the job is ready for time study. Suitable operators are selected, the purposes of the study are carefully explained to them, and the time study observations are made. During the study, time study specialists rate the performance being given by operators either by judging the skill and effort they are exhibiting or by assessing the speed with which motions are made as compared with what they consider to be a normal working pace. The final step is to compute the standard.

PRINCIPLES OF MOTION STUDY

Operation analysis is a primary analysis which eliminates inefficiencies. Motion study is a secondary analysis which refines the method still further. Motion study may and often does suggest further improvements in the factors considered during operation analyses, such as tools, material handling, design, and workplace layouts. In addition, it studies the human factors as well as the mechanical and sets up operations in conformance with the limitations, both physical and psychological, of those who must perform them.

The technique of motion study rests on the concept originally advanced by Frank B. and Lillian M. Gilbreth that all work is performed by using a relatively few basic operations in varying combinations and sequences. These Gilbreth Basic Elements have also been called "therbligs" and "basic divisions of accomplishment."

The basic elements together with their symbols (for definitions see ASME Industrial Engineering Terminology), grouped in accordance with their effect on accomplishment, are as follows:

Group 1	
Accomplishes	
Reach	R
Move	M
Grasp	G
Position	P
Disengage	D
Release	RL
Examine	E
Do	DO

Group 2	
Retards accomplishment	
Change direction	CD
Preposition	PP
Search	S
Select	SE
Plan	PL
Balancing delay	BD

Group 3	
Does not accomplish	
Hold	H
Avoidable delay	AD
Unavoidable delay	UD
Rest to overcome fatigue	F

Group 1 is the useful group of basic elements or the ones that accomplish work. They do not necessarily accomplish it in the most effective way, however, and a study of these elements will often uncover possibilities for improvement.

Group 2 contains the basic elements that tend to retard accomplishment when present. In most cases, they do this by slowing down the group 1 basic elements. They should be eliminated wherever possible.

Group 3 is the nonaccomplishment group. The greatest improvements in method usually come from the elimination of the group 3 basic elements from the cycle. This is done by rearranging the motion sequence, by providing mechanical holding fixtures, and by improving the workplace layout.

An operation may be analyzed into its basic elements either by observation or by making a micromotion study of a motion picture of the operation.

Methods improvement may be made on any operation by eliminating insofar as possible the group 2 and group 3 basic elements and by arranging the workplace so that the group 1 basic elements are performed in the shortest reasonable time. In doing this, certain laws of motion economy are followed. The following, derived from the laws originally stated by the Gilbreths, are the most important.

1. When both hands begin and complete their motions simultaneously and are not idle during rest periods, maximum performance is approached.

2. When motions of the arms are made simultaneously in opposite directions over symmetrical paths, rhythm and automaticity develop most naturally.

3. The motion sequence which employs the fewest basic elements is the best for performing a given task.

4. When motions are confined to the lowest practical classification, maximum performance and minimum fatigue are approached. Motion classifications are: Class 1, finger motions; Class 2, finger and wrist motions; Class 3, finger, wrist, and forearm motions; Class 4, finger, wrist, forearm, and upper-arm motions; Class 5, finger, wrist forearm, upper-arm, and body motions.

STANDARDIZING THE JOB

When an acceptable method has been devised, equipment, materials, and conditions must be standardized so that the method can always be followed. Information and records describing the standard method must be carefully made and preserved, for experience has shown that, unless this is done, minor variations creep in which may in time cause a major problem. In the case of repetitive work, a job is not standardized until each piece is delivered to operators in the same condition, and it is possible for them to perform their work on each piece by completing a set cycle of motions, doing a definite amount of work with the same equipment under uniform working conditions.

The operator or operators must then be taught to follow the approved method. Operator training is always important if reasonable production is to be obtained, but it is an absolute necessity where methods have been devised by motion study. It is quite apparent that the operators cannot be expected to discover for themselves the method which the time-study specialist developed as the result of hours of concentrated study. They must, therefore, be carefully trained if they are to be expected to reach standard production. In addition, an accurate time study cannot be made until the operator is following the approved method with reasonable proficiency.

WORK MEASUREMENT

Work measurement is the calculation of the amount of time it should take to do a standardized job. It utilizes the concept of a standard time. The standard time to perform a task is the agreed-upon and reproducible calculated time that a hypothetical typical person working at a normal rate of speed should take to do the job using the specified method with the proper tools and materials. It is the normal time determined to be required to complete one prescribed cycle of an operation, including noncyclic tasks, allowances, and unavoidable delays.

Time Study Methods There are several bases upon which time standards may be calculated. They include:

1. *Application of past experience.* The time required to do the operation in the past, either recorded or remembered, may be used as the present standard or as a basis for estimating a standard for a similar operation or the same operation being done under changed conditions.

2. *Direct observation and measurement.* The operation may be observed and its time recorded as it is actually performed and adjustments may be made to allow for the estimated pace rate of the operator and for special allowances. A stop watch or other recording instrument may be used or work sampling may be employed, which makes statistical inferences based upon random observations.

3. *Synthetic techniques.* A time standard for an actual or proposed operation may be constructed from the sum of the times to perform its several components. The times of the components are extracted from standard charts, tables, graphs, and formulas in manuals or in computer databases and totaled to arrive at the overall time for the entire operation. Standard data or predetermined motion times may be used. Methods-time measurement (MTM), basic-motion times (BMT), work factor (WF), and others are some of the systems available.

TIME STUDY OBSERVATIONS

The time study specialist can study any operator he or she wishes so long as that operator is using the accepted method. By applying what is known as the leveling procedure, the time study specialist should arrive at the same final time standard regardless of whether studying the fastest or the slowest worker.

The manner in which the operator is approached at the beginning of the study is important. This is particularly true if the operator is not accustomed to being studied. Time study specialists should be courteous and unassuming and should show a recognition of and a respect for the problems of the operator. They should be frank in their dealings with the operator and should be willing at any time to explain what they are doing and how they do it.

The first step in making time study observations is to subdivide the operation into a number of smaller operations which will be studied and timed separately. These subdivisions are known as elements, or elemental operations. Each element is exactly described in a few well-chosen words, which are recorded on the top of the time study form. Figure 17.4.3 shows how this is done. The beginning and ending points of these elements must be clearly recognizable so that the chances of overlapping watch readings will be minimized.

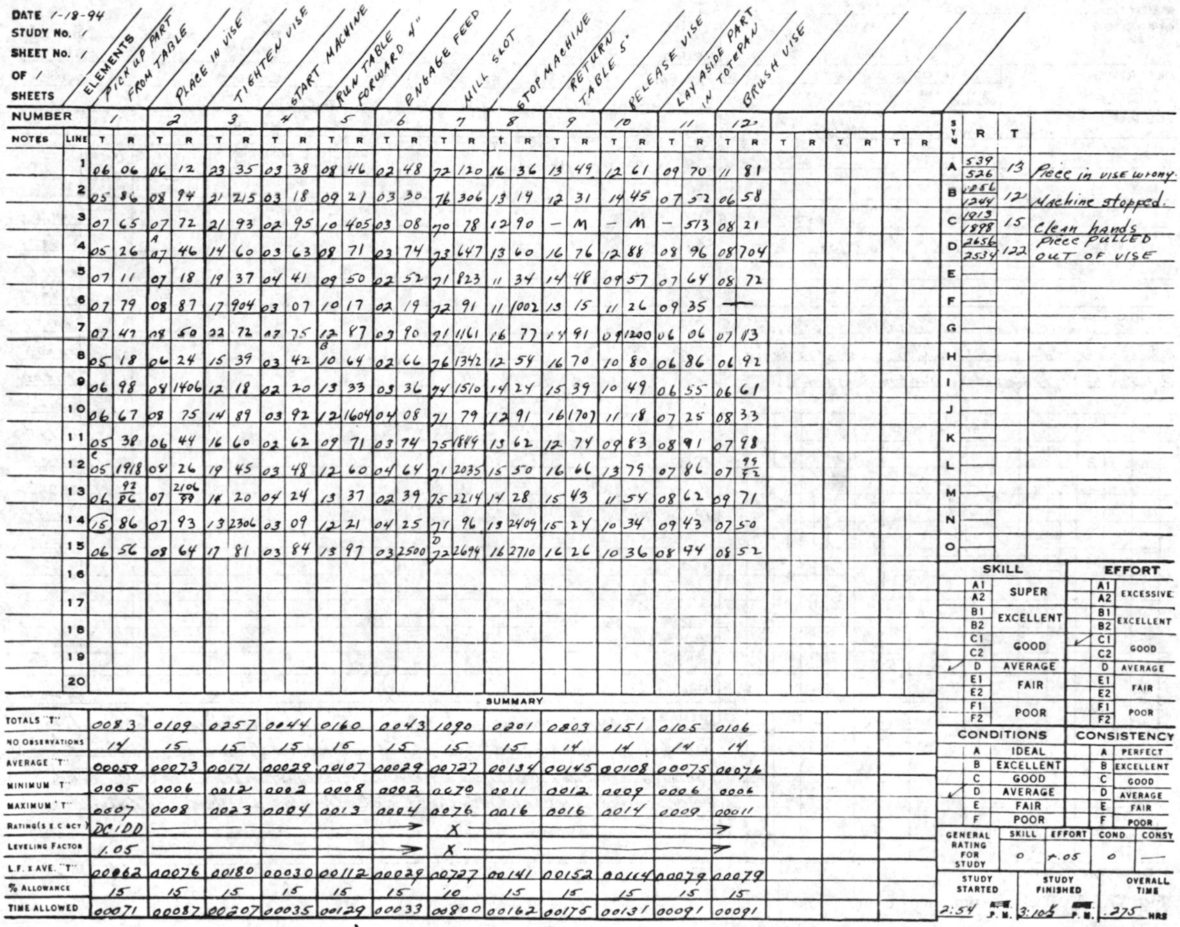

DATE 1-18-94
STUDY No. 1
SHEET No. 1
OF 1
SHEETS

Elements:
1. PICK UP PART FROM TABLE
2. PLACE IN VISE
3. TIGHTEN VISE
4. START MACHINE
5. RUN TABLE FORWARD 4"
6. ENGAGE FEED
7. MILL SLOT
8. STOP MACHINE
9. RETURN TABLE 5"
10. RELEASE VISE
11. LAY ASIDE PART IN TOTE PAN
12. BRUSH VISE

LINE	1 (t/R)	2 (t/R)	3 (t/R)	4 (t/R)	5 (t/R)	6 (t/R)	7 (t/R)	8 (t/R)	9 (t/R)	10 (t/R)	11 (t/R)	12 (t/R)
1	06/06	06/12	23/35	03/38	08/46	02/48	72/120	16/36	13/49	12/61	09/70	11/81
2	05/86	08/94	21/215	03/18	09/21	03/30	76/306	13/19	12/31	14/45	07/52	06/58
3	07/65	07/72	21/93	02/95	10/405	03/08	70/78	12/90	–/M	–/M	–/573	08/21
4	05/26	07/46	14/60	03/63	08/71	03/74	73/647	13/60	16/76	12/88	09/96	08/704
5	07/11	07/18	19/37	04/41	09/50	02/52	71/823	11/34	14/48	09/57	07/64	08/72
6	07/79	08/87	17/904	03/07	10/17	02/19	72/91	11/1002	13/15	11/26	09/35	—
7	07/47	08/50	22/72	07/75	12/87	02/90	71/1161	16/77	14/91	09/1200	06/06	07/113
8	05/18	06/24	15/39	03/42	10/64	02/66	76/1342	12/54	16/70	10/80	06/86	06/92
9	06/98	08/1406	12/18	02/20	13/33	03/36	94/1510	14/24	15/39	10/49	06/55	06/61
10	06/67	08/75	14/89	03/92	12/1604	04/08	71/79	12/91	16/1707	11/18	07/25	08/33
11	05/38	06/44	16/60	02/62	09/71	03/74	75/1844	13/62	12/74	09/83	08/91	07/98
12	05/198	08/26	19/45	03/48	12/60	04/64	71/2035	15/50	16/66	13/79	07/86	07/99
13	06/92 86	07/2106 39	14/20	04/24	13/37	02/39	75/2214	14/28	15/43	11/54	08/62	09/71
14	5/86	07/93	13/2306	03/09	12/21	04/25	71/96	13/2409	15/24	10/34	09/43	07/50
15	06/56	08/64	17/81	03/84	13/97	03/2500	72/2694	16/2710	16/26	10/36	08/44	08/52

SYM	R	T	Note
A	539 / 526	13	Piece in vise wrong.
B	1256 / 1244	12	Machine stopped.
C	1913 / 1898	15	Clean hands
D	2656 / 2534	122	Piece pulled out of vise

SUMMARY

	1	2	3	4	5	6	7	8	9	10	11	12
TOTALS "T"	0083	0109	0257	0044	0160	0043	1090	0201	0803	0151	0105	0106
NO. OBSERVATIONS	14	15	15	15	16	15	15	15	14	14	14	14
AVERAGE "T"	00059	00073	00171	00029	00107	00029	00727	00134	00145	00108	00075	00076
MINIMUM "T"	0005	0006	0012	0002	0009	0002	0670	0011	0012	0009	0006	0006
MAXIMUM "T"	0007	0008	0023	0004	0013	0004	0076	0016	0016	0014	0009	0011
RATING (S E C &CY)	DC100						X					
LEVELING FACTOR	1.05						X					
L.F. x AVE. "T"	00062	00076	00180	00030	00112	00029	00727	00141	00152	00114	00079	00079
% ALLOWANCE	15	15	15	15	15	15	10	15	15	15	15	15
TIME ALLOWED	00071	00087	00207	00035	00129	00033	00800	00162	00175	00131	00091	00091

SKILL / EFFORT:
SKILL		EFFORT	
A1 A2	SUPER	A1 A2	EXCESSIVE
B1 B2	EXCELLENT	B1 B2	EXCELLENT
C1 C2	GOOD	C1 C2	GOOD (✓ C1)
D	AVERAGE (✓)	D	AVERAGE
E1 E2	FAIR	E1 E2	FAIR
F1 F2	POOR	F1 F2	POOR

CONDITIONS / CONSISTENCY:
CONDITIONS		CONSISTENCY	
A	IDEAL	A	PERFECT
B	EXCELLENT	B	EXCELLENT
C	GOOD	C	GOOD
D	AVERAGE (✓)	D	AVERAGE
E	FAIR	E	FAIR
F	POOR	F	POOR

GENERAL RATING FOR STUDY: SKILL 0 | EFFORT +.05 | COND 0 | CONSY

STUDY STARTED 2:54 P.M.   STUDY FINISHED 3:10½ P.M.   OVERALL TIME .275 HRS

Fig. 17.4.3 Face of a time study form.

The timing is done with the aid of a stopwatch, or less frequently, with a special type of "time study machine." There are several types of stopwatches as well as several methods of recording watch readings in common use. The study illustrated by Fig. 17.4.3 was made using a decimal-hour stopwatch that reads directly in ten-thousandths of an hour. The readings were recorded using what is known as the continuous method of recording. In this method, the watch runs continuously from the beginning of the study to the end. Thus every moment of time is accounted for, something that may be important if the correctness of the study is ever questioned. The watch is read at the end of each elemental operation, and the reading is recorded in the "R" column under the proper element description. The elapsed time for each element is later secured by subtracting successive readings. This observation procedure gives results as accurate as any other and more accurate than some.

Occasionally variations from the regular sequence of elemental operations occur. The time study specialist must be prepared to handle such situations when they happen. These variations may be divided into four general classes as follows: (1) elements performed out of order, (2) elements missed by the time study specialist, (3) elements omitted by the operator, (4) foreign elements.

The time study illustrated by Fig. 17.4.3 contains examples of each of these kinds of irregularity. Elements 12 and 1 on lines 12 and 13 were performed out of order. On line 3, the time study specialist missed obtaining the watch readings for elements 9 and 10. On line 6, element 12 was omitted by the operator. Foreign elements A, B, C, and D occurred during regular elements 2, 5, 1, and 7, respectively. A study of these examples will show how the time study specialist handles varia-

tions from the regular sequence of elements which occur during the making of a time study.

A time study to be of value for future use must tell the whole story of a job in such a way that it will be understood by anyone familiar with the time study procedure. This will not be possible unless all identifying and other pertinent information is recorded at the time the study is made. Records should be made to show complete identification of the operator; the part or assembly; the machines, tools, and equipment used; the operation; the department in which the operation was performed; and the conditions existing at the time the study was made. Sketches are generally a desirable part of this description. Figure 17.4.4 shows the information which would be recorded on the reverse side of the time study form illustrated by Fig. 17.4.3.

PERFORMANCE RATING

The objective of a time study is to determine the time which a worker giving average performance will require to do the job under average or normal conditions. It is important to understand that when the time study specialist speaks of average performance, he or she is not referring to the mathematical average of all human beings, or even the average of all persons engaged in a given occupation. Average performance is established by definition and not statistically. It represents the time study specialist's conception of the normal, steady, but unhurried performance which may reasonably be expected from anyone qualified for the work. If sufficient inducement is offered by incentives or otherwise, this performance may be considerably surpassed.

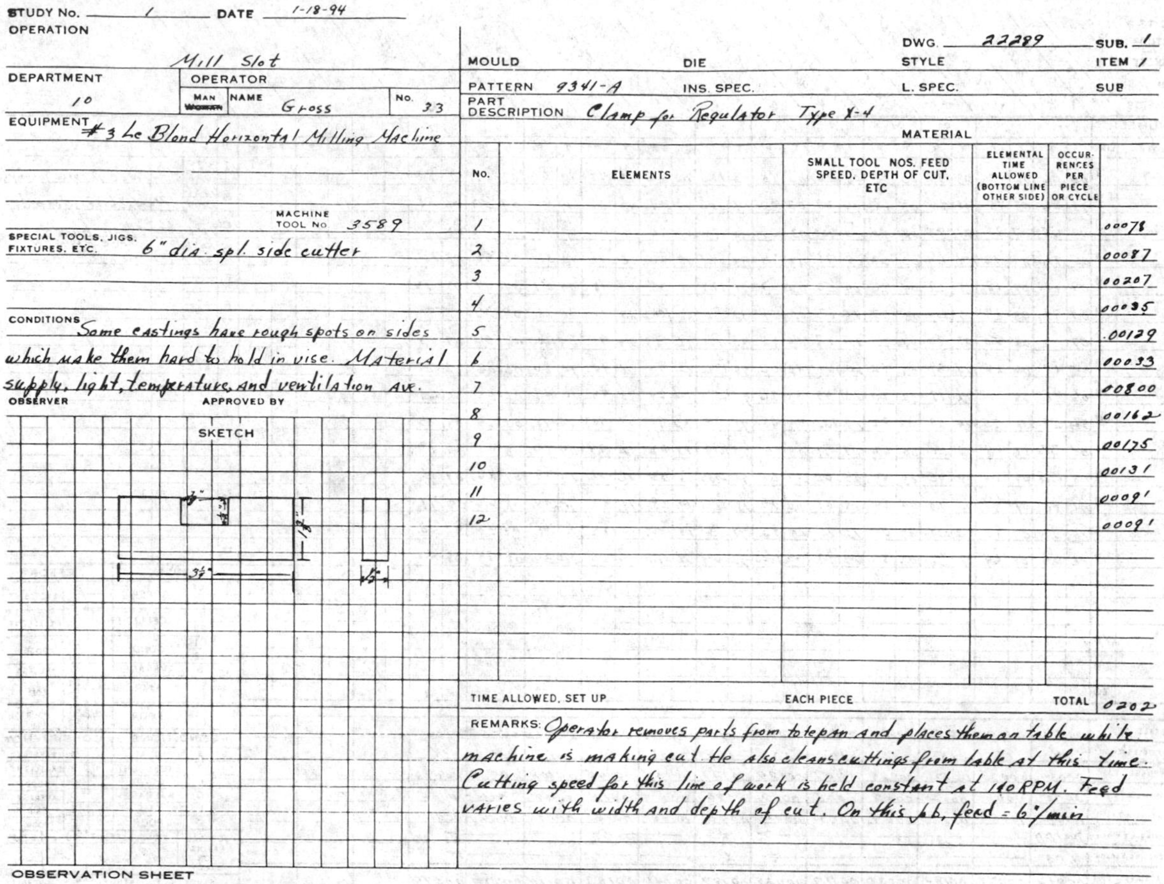

STUDY No. _____ *1* _____ DATE _____ *1-18-94* _____

OPERATION _____ *Mill Slot* _____

DWG. *22289* SUB. *1*
STYLE ____ ITEM *1*

DEPARTMENT *10*

OPERATOR
MAN / WOMAN NAME *Gross* NO. *33*

PATTERN *9341-A* INS. SPEC. ____ L. SPEC. ____ SUB ____
PART DESCRIPTION *Clamp for Regulator Type X-4*
MOULD ____ DIE ____

EQUIPMENT *#3 Le Blond Horizontal Milling Machine*

MATERIAL ____

NO.	ELEMENTS	SMALL TOOL NOS. FEED SPEED, DEPTH OF CUT, ETC	ELEMENTAL TIME ALLOWED (BOTTOM LINE OTHER SIDE)	OCCUR-RENCES PER PIECE OR CYCLE	
1					.00078
2					.00087
3					.00207
4					.00035
5					.00129
6					.00033
7					.00800
8					.00162
9					.00175
10					.00131
11					.00091
12					.00091

MACHINE TOOL No. *3589*

SPECIAL TOOLS, JIGS, FIXTURES, ETC. *6" dia. spl. side cutter*

CONDITIONS *Some castings have rough spots on sides which make them hard to hold in vise. Material supply, light, temperature, and ventilation ave.*

OBSERVER ____ APPROVED BY ____

SKETCH

TIME ALLOWED, SET UP ____ EACH PIECE ____ TOTAL *.0202*

REMARKS *Operator removes parts from totepan and places them on table while machine is making cut. He also cleans cuttings from table at this time. Cutting speed for this line of work is held constant at 140 RPM. Feed varies with width and depth of cut. On this job, feed = 6"/min*

OBSERVATION SHEET

Fig. 17.4.4 Back of a time study form.

If all operators available for study worked at the average performance level, the task of establishing a standard would be easy. It would be necessary merely to average the elapsed elemental times determined from time study and add an allowance for fatigue and personal and unavoidable delays. It is seldom, however, that a performance is observed which is rated throughout as average. Therefore, to establish a standard which represents the time which would be taken had an average performance been observed, it is necessary to use some method of adjusting the recorded elemental times when other than average performance is timed.

One of the well-known methods of doing this is the **leveling procedure**. When properly applied it gives excellent results. It must be correctly understood, of course, and the time study specialist who uses it must be thoroughly trained to apply it correctly.

The procedure recognizes that when the correct method is being followed, skill, effort, and working conditions will affect the level at which the operator works. These factors are judged during the making of the time study. Skill is defined as *proficiency at following a given method.* This is not subject to variation at will by the operator but develops with practice over a period of time. Effort is defined as *the will to work.* It is controllable by the operator within the limits imposed by skill. Conditions are those conditions which affect the operator and not those which affect the method.

Definitions have been established for different degrees of skill, effort, and conditions. Numerical factors have been established by extensive research for each degree of skill, effort, and conditions. These are shown by Fig. 17.4.5. The algebraic sum of these numerical values added to 1.0 gives the leveling factor by which all actual elemental times are multiplied to bring them to the average or normal level. The leveling factor represents in effect the amount in percent which actual performance times are above and below the average performance level.

	Skill			Conditions			Effort	
+0.15	A1	Superskill	+0.06	A	Ideal	+0.13	A1	Excessive
+0.13	A2		+0.04	B	Excellent	+0.12	A2	
+0.11	B1	Excellent	+0.02	C	Good	+0.10	B1	Excellent
+0.08	B2		0.00	D	Average	+0.08	B2	
+0.06	C1	Good	−0.03	E	Fair	+0.05	C1	Good
+0.03	C2		−0.07	F	Poor	+0.02	C2	
0.00	D	Average				0.00	D	Average
−0.05	E1	Fair				−0.04	E1	Fair
−0.10	E2					−0.08	E2	
−0.16	F1	Poor				−0.12	F1	Poor
−0.22	F2					−0.17	F2	

Fig. 17.4.5 Leveling factors for performance rating.

ALLOWANCES FOR FATIGUE AND PERSONAL AND UNAVOIDABLE DELAYS

The leveled elemental time values are net elapsed times adjusted to the average performance level. They do not provide for delays and other legitimate allowances. Something, therefore, must be added to take care of such things as fatigue, and special conditions of the work.

Fatigue allowances vary according to the nature of the work. Flat percentages are determined for each general class of work, such as bench work, machine-tool operation, hard physical labor, and so on. Personal allowances are the same for most classes of work. Unavoidable delay allowances vary with the nature of the work and the conditions under which it is performed. Peculiar conditions surrounding specific jobs sometimes require additional special allowances.

It is apparent, therefore, that the proper allowance factor to use can only be determined by a study of the class of work to which it is to be applied. Allowances are determined either by a series of all-day time studies or by a statistical method known as work sampling, or both. When an allowance factor has once been established, it is then applied to all time studies made on that class of work thereafter.

DEVELOPING THE TIME STANDARD

When time study observations have been completed, a series of calculations are made to develop the time standard. Elapsed times are determined by subtracting successive watch readings. Each subtraction is recorded between the two watch readings that determine its value. Elapsed time is noted in ink to ensure a permanent record and to distinguish it from the watch readings which are usually recorded in pencil. A study of Fig. 17.4.3 will show how subtractions are entered on the time study form and later summarized.

The several elapsed times for each element are next carefully compared and examined for abnormal values. If any are found, they are circled so that they can be distinguished and excluded from the summary.

The remaining elapsed times for each element are added and are averaged by dividing by the number of elapsed time readings. The results are average elapsed times which represent the time taken by the operator during that particular study. These times must be adjusted by multiplying them by a leveling factor to bring them to the average performance level. This factor is determined by the rating of skill, effort, and conditions made during the period of observation.

Each average elapsed time is multiplied by the leveling factor, except when the element is not controlled by the operator. An element that is outside the control of the operator, such as element 7 in Fig. 17.4.3 which is a cut with power feed, should not be leveled, because it is unaffected by the ability of the operator. As long as the proper feed and speed are used, the time for performing this element will be the same whether the worker is an expert or a learner.

If workers were able to work continuously, the leveled time would be the correct value to allow for doing the operation studied, but constant application to the job is nether possible nor desirable. In the course of a day, there are certain to be occasional interruptions and delays, for which due allowance must be made in establishing the final standard. Therefore, each elemental time is increased by an allowance which covers time that will be consumed by personal and unavoidable delays, fatigue, and any special factors that may affect the job.

The numbers and descriptions of the elemental operations together with their allowed time are transcribed on the back of the time study form as shown by Fig. 17.4.4. The number of times an elemental operations occurs on each piece or cycle of the operation is taken into account, and the total time allowed for each element is determined and recorded. The final standard for the operation is the sum of the amounts recorded in the "time-allowed" column. When all computations have been checked and all supporting records have been properly identified and filed, the task of developing the time standard is complete.

TIME FORMULAS AND STANDARD DATA

On repetitive work, time study is a satisfactory tool of work measurement. A single time study may be sufficient to establish a standard which will cover the work of one or more operators for a long period of time.

As quantities become smaller, however, the cost of establishing standards by individual time study increases until at length it becomes prohibitive. In the extreme case, where products are manufactured in quantities of one, it would require at least one time study specialist for each operator if standards were established by detailed time study, and the standards would not be available until after the jobs had been completed.

In order to simplify the task of setting standards on a given class of work and in order to improve the consistency of the standards, standard data are frequently used by time study specialists. A compilation of standard data in its simplest form is merely a list of all the different elements that have occurred during all the time studies made on a given class of work, with representative time values for each element. Every element that differs even slightly from any other element has its own time value.

When a job comes into the shop on which no standard has previously been established, time study specialists analyze the job either mentally or by direct observation and determine the elements required to perform it. They then select time values from the standard data for each element. Their sum gives the standard for the job.

This method, although a decided improvement from a time, cost, and consistency standpoint over individual time study, is capable of further refinement and improvement. On a given class of work, certain elements will be performed—for example, "pick up part"—on every piece produced, while others—such as "secure in steady rest"—will be performed only when a piece has certain characteristics. In some cases, the performance of a certain element will always require the performance of another element, e.g., "start machine" will always require the subsequent performance of the element "stop machine." Then again, the time for performing certain elements—for example, "engage feed"—will be the same regardless of the characteristics of the part being worked upon, while the time for performing certain other elements—like "lay part aside"—will be affected by the size and shape of the part.

Thus it is possible to make certain combinations and groupings which will simplify the task of applying standard data. Time study specialists construct various charts, and tables which they still call standard data, or, in the ultimate refinement, develop time formulas. A time formula is a convenient arrangement of standard data which simplifies their accurate application. Much of the analysis which is necessary when applying standard data is done once and for all at the time the formula is derived. The job characteristics which make the performance of certain elements or groups of elements necessary are determined, and the formula is expressed in terms of these characteristics.

Figure 17.4.3 illustrates a detailed time study made to establish a standard on a simple milling machine operation. The same standard can be derived much more quickly from the following time formula:

$$\text{Curve } A + \text{Table } T = \text{each piece time}$$

Curve A combines the times for the variable elements "pick up part from table," "place in vise," and "lay aside part in totepan" with the times for the constant elements "tighten vise," "start machine," "run table forward 4 in," "engage feed," "stop machine," "release vise," and "brush vise." Table T combines the times for "mill slot" and "return table." The standard time for milling a slot in a brass clamp of any size is computed by determining the variable characteristics of the job from the drawing—in this case, the volume of the clamp and the perimeter of the cut—and adding together the time read from curve A and the time read from Table T.

The amount of time which the use of time formulas will save the time study specialist is readily apparent. It takes a certain amount of time and no little know-how to develop a time formula, but once it is available, the job of establishing accurate standards becomes a simple, fairly routine task. The time required to make and work up a time study will be from 1 to 100 or more hours, depending upon the length of the operation cycle studied. The time required to establish a standard from a time formula will, in the majority of cases, range from 1 to 15 min, depending upon the complexity of the formula and the amount of time required to determine the characteristics of the job. Where all the necessary information may be obtained from the drawing of the part, the standard may generally be computed in less than 5 min.

USES OF TIME STANDARDS

Some of the more common uses of time standards are in connection with
1. Wage incentive plans
2. Plant layout
3. Plant capacity studies
4. Production planning and control
5. Standard costs
6. Budgetary control
7. Cost reduction activities
8. Product design
9. Tool design
10. Top-management controls
11. Equipment selection
12. Bidding for new business
13. Machine loading
14. Effective labor utilization
15. Material-handling studies

Time standards can be established not only for direct labor operations but also for indirect work, such as maintenance and repair, inspection, office and clerical operations, engineering, and management. They can also be set for machinery and equipment, including robots.

17.5   COST OF ELECTRIC POWER
by Robert J. Vondrasek and James M. Connolly

REFERENCES: Department of Energy (DOE), "Electric Plant Cost and Power Production Expenses," 1991. Electric Power Research Institute, "Technical Assessment Guide," 1993. Oak Ridge National Laboratory, "CONCEPT-V, A Computer Code for Conceptual Cost Estimates of Steam Electric Power Plants." "Handy-Whitman Index of Public Utility Construction Costs," Whitman, Requardt and Associates. Grant and Ireson, "Principles of Engineering Economy," Ronald Press.

This section is primarily concerned with the economics of electric power production by central station generating plants. The cost of power generated by stationary thermal and hydroelectric plants makes a significant contribution to the price of electric service provided by interconnected and pooled electric utility systems. For a single, isolated plant, often assigned specific industrial or commercial loads, generation costs can comprise a major portion of, or can establish the total price of, power. Generation costs consist of production expenses (fuel plus operating and maintenance expenses) and fixed charges on investment (cost of investment capital, depreciation or amortization, taxes, and insurance).

Interconnected power supply systems must price power to include transmission costs, distribution costs, and commercial expenses. Power price savings are realized by scheduling the installation and operation of a range of plant types to optimize overall generation costs. Generally, efficient plants burning lowest-cost fuels or operating on river water flows that are available the year round are assigned to continuous baseload operation at or near full capacity. Plants bearing low unit-kilowatt investment costs and lower efficiencies are installed to provide peaking capacity. Operation of these units during short daily periods of peak load limits their energy output and thus minimizes characteristic cost penalties associated with their poorer station efficiency or their requirement for higher-priced gas or distillate fuels. Large interconnected utilities may also achieve price benefits by accommodating the staggered installation of large-size units which carry lower unit-kilowatt investments, by supplementing capacity and reserve needs of neighboring systems, and by the daily and seasonal exchange of off-peak, low-incremental-cost energy. Historically, there have been significant regional differences in the price of power. These differences have reflected such factors as availability of hydroelectric energy, the cost of fossil fuels, labor costs, ownnership type and investment composition, local tax structure, and the opportunities pooling and coordination provide to exploit the economies of scale.

Cost or price is a basic characteristic of power supply. Along with relative abundance, reliability, and high quality of service, low prices have encouraged the widespread use of electric energy that has come to be associated with our national way of life. Industrial use of electric energy is particularly heavy in the electroprocess and metallurgical industries. The price of electricity has a significant effect on the end-product cost in these industries. In many manufacturing and process industries, quality of service in terms of voltage regulation, frequency control, and reliability is a major concern. Uninterrupted supply is of crucial importance in many process industries where a power failure may entail material waste or damage to equipment, in addition to loss of production revenue. Because of the flexibility and convenience of electric power, future increases in industrial, residential, and commercial consumption are anticipated despite deregulation, fuel price increases, environmental considerations, and energy conservation efforts.

Various types of power-generating units are owned and operated by industry and private sector investors to provide power (and sometimes steam) to industrial facilities and often sell excess power to the local utility at the avoided cost of incremental power for the utility. Because of the general small size of these units, often dual purpose (power and steam), and lack of reporting information on their economics, this section does not address cogeneration or "dual-purpose" power plants. However, the calculations and information in this section can assist in deciding the advisability of purchasing versus generating electric power.

CONSTRUCTED PLANT COSTS

A **central station** serving a utility system is designed to meet not only the existing and prospective loads of the system in which it is to function but also the pooling and integration obligations to adjacent systems. Service requirements which establish station size, type, location, and design characteristics ultimately affect cost of delivered power. Selection of plant type and the overall philosophy followed in design must accommodate a combination of objectives which may include high operating efficiency, minimum investment, high reliability and availability, maximum reserve capability margins, rapid load change capability, quick-start capability, or service adaptability as spinning reserve.

For a given plant, the design must account for siting factors such as environmental impacts, subsoil conditions, local meteorology and air quality, quality and quantity of available water supply, access for construction, transmission intertie, fuel delivery and storage, and maintainability. In addition, plant siting and design will be significantly affected by legal restrictions on effluents which may have adverse impacts on the environment. In the case of **nuclear plants,** siting must also consider the proximity of population centers and the size of exclusion areas. Reactor plant design must bear the investments required to control radioactive releases and to provide safeguard systems which protect against accidents. **Fossil-fueled power plants** may be sited adjacent to fuel supplies or in proximity to load centers, thereby increasing transmission costs on the one hand or fuel delivery charges on the other. Depending primarily on climate, plants may be enclosed, semienclosed, or of the outdoor type. Spare auxiliary components can be installed to improve reliability. Increased investments in sophisticated heat cycles and controls, for improved equipment performance, can achieve higher plant efficiencies. **Combustion-turbine combined-cycle plant** outputs are restricted by both increased elevation and high ambient air temperatures. Hydroelectric sites are frequently very distant from load centers and thus require added costs for extensive transmission facilities. Also, **hydroelectric facilities** may provide for flood control, navigation, or recreation as by-products of power production. In such instances, total cost should be properly allocated to the various product elements of the multipurpose project.

An **industrial power plant** provided to meet the requirements of an isolated load entails design considerations and exhibits cost characteristics which differ from those of a central station power plant assigned an integrated role within a connected generating system. Industrial power plants often produce both process steam and electric power. Industrial facilities must often accommodate both base- and peak-load requirements. They may be designed to provide for on-site reserve capacity or spinning reserve capacity. Frequency control and voltage regulation must be viewed as a special problem because of the limited capability of a single plant to meet load changes.

Table 17.5.1 provides typical installed cost data for central station generating plants. The figures represent costs of facilities in place, excluding interest during construction. The cost of land, waste-disposal facilities, fuel in storage, or loaded nuclear fuel is not included. Costs apply to plants completed in 1993. Interest during construction can be estimated by multiplying the simple interest rate per year by the con-

Table 17.5.1   Typical Investments Costs* (1993 Price Level)

Plant description			
Type	Net capacity, MW	Fuel	Total investment cost in $/net kW
Conventional fossil	500	Coal	1,580
		Oil	1,250
		Gas	930
Advanced light-water reactor	600	Nuclear	1,950
Combustion turbine/ combined cycle	300	Gas	590

* Capital investments exclude costs for the following: initial fuel supply, cost of decommissioning for nuclear plants, main transformers, switchyard, transmission facilities, waste disposal, land and land rights, and interest during construction.

struction period in years and dividing by 2 to reflect the carrying costs on the average commitment of capital toward equipment and labor during construction. Escalation effects for plants to be completed beyond this date may be extrapolated in accordance with anticipated cost trends for labor, material, and equipment. Historical cost trends, by region, as experienced in the power industry, which can be helpful in forecasting future costs, may be determined by use of the figures in Table 17.5.2. Escalation can significantly affect plant costs on future projects, especially in view of the 10- to 12-year engineering and construction periods historically experienced for nuclear facilities and the corresponding 5 to 6 years required for fossil-fueled power plants. In addition to rising equipment, construction labor, and material costs, major factors influencing the upward trend in plant costs include increased investment in environmental control systems, an emphasis on improved quality assurance and plant reliability, and a concern for safety, particularly in the nuclear field.

Conventional Steam-Electric Plants   Conventional fossil plant investment costs given in Table 17.5.1 are for 500-MW nominal units which are deemed to be representative of future central station fossil units. Costs will vary from those in the table due to equipment arrangement, pollution control systems, foundations, and cooling-water-system

Table 17.5.2   U.S. Cost Trends of Electric Plant Construction by Region (1973 Index is 100)

	North Atlantic	South Atlantic	North Central	South Central	Plateau	Pacific
Total: Steam generating plants						
Annual average, Jan. 1, 1983	233	236	234	241	244	251
1984	242	242	242	249	252	259
1985	255	254	255	258	260	269
1986	257	253	256	256	259	269
1987	262	256	260	259	262	272
1988	276	270	275	272	277	287
1989	290	280	286	284	285	297
1990	304	293	299	291	300	311
1991	312	297	306	296	303	317
1992	317	298	307	299	306	323
1993	329	308	318	304	317	333
Total: Hydro generating plants						
Annual average, Jan. 1, 1983	212	216	216	228	228	224
1984	221	219	224	232	235	229
1985	235	232	236	237	247	241
1986	240	231	238	234	247	241
1987	245	232	244	236	252	245
1988	253	238	253	236	256	252
1989	264	246	263	242	264	262
1990	275	255	269	247	267	271
1991	276	256	272	245	271	273
1992	279	248	271	239	267	273
1993	291	259	281	250	278	282

SOURCE: "Handy-Whitman Index of Public Utility Construction Costs," compiled and published by Whitman, Requardt & Associates, 2315 Saint Paul Street, Baltimore, MD 21218.

designs dictated by plant site conditions. The variety of cycle arrangements and steam conditions selected also affects plant capital cost. Plant designers try to economically balance investment and operating costs for each plant. Present-day parameters, in the face of current economics, call for a drum boiler, regenerative reheat cycles at initial steam pressures of 2,400 psig with superheat and reheat temperatures of 1,000°F (539°C). Similar temperature levels are employed for 3,500-psig initial steam pressure supercritical-reheat and double-reheat cycles which require higher investment outlays in exchange for efficiency or heat rate improvements of between 5 and 10 percent. Increased investment costs are also caused by more generous boiler-furnace sizing, and larger fuel- and ash-handling, precipitator, and scrubber facilities required for burning poor quality coals. Investment increments are also required to provide partial enclosure of the turbine building and furnace structure and to fully enclose the boiler by providing extended housing to weather-protect duct work and breeching.

Nuclear Plants The construction of nuclear power plant capacity in the United States has been suspended for over a decade. Current construction activities in the nuclear power arena include completing delayed nuclear units and converting existing or partially completed units to fossil fuels. The costs associated with delays and the reengineering required to adapt units to new fuels and thermodynamic cycles is not indicative of the cost associated with committing new nuclear units today. A new generation of smaller light-water reactor nuclear power plants is currently in the design stage. Costs associated with these units, including siting, licensing, and fuel cycle, remain unclear, as they are precommercial. The earliest units will not be offered until the late 1990s. However, they are the basis for nuclear capacity in this section as they are the most likely nuclear units to become commercial in the future. They utilize preengineered designs to reduce construction and licensing time.

In general, light-water reactor nuclear plants require considerably higher investments than do fossil-fueled plants, reflecting the need for leakproof reactor pressure containment structures, radiation shielding, and a host of reactor plant safety-related devices and redundant equipment. Also, light-water reactor plants operate at lower initial steam conditions than do fossil-fueled plants and, because of their poorer turbine cycle efficiency, require larger steam flows and increased equipment sizes at added investment.

Both light-water reactor plants and conventional fossil plants evidence declining unit cost with increasing size. The economy of scale has been demonstrated most strikingly in the nuclear field where orders for units at the 1,200-MW size level had predominated. The utilization of large-size units has been made possible by the interconnection of utility transmission and distribution systems, many of which have grown to levels capable of absorbing large individual units without assuming economic penalties for the reserve equipment required to assure service continuity in the event of unscheduled outages.

Combustion Turbine/Combined-Cycle Plants Packaged combined-cycle plants are offered by a number of vendors in the 100- to 600-MW size range. These plants consist of multiple installations of combustion gas turbines arranged to exhaust to waste-heat steam generators which may be equipped for supplementary firing of fuel. Steam produced is supplied to a conventional nonreheat steam turbine cycle. Advantages of combined cycles are lower unit investment costs, efficient thermal performance, increased flexibility (which allows independent operation of the gas-turbine portion of the plant), shorter installation schedules, reduced cooling-water requirements, and the reduction in sulfur oxide and particulate emissions characteristic of gaseous fuels.

Hydroelectric Plants Hydroelectric generation offers unique advantages. Fuel, a heavy contributor to thermal plant operating costs, is eliminated. Also, hydro facilities last longer than do other plant types; thus they carry lower depreciation rates. They have lower maintenance and operating expenses, eliminate air and thermal discharges, and because of their relatively simple design, exhibit attractive availability and forced-outage rates. Quick-start capability and rapid response to load change ideally suit hydro turbines to spinning reserve and frequency-control assignments.

The constructed cost of a hydroelectric station is strongly site dependent. Overall costs fluctuate significantly with variations in dam costs, intake and discharge system requirements, pondage required to firm up capacity, and with the cost of relocating facilities within the areas inundated by the impoundments. For a given investment in structures, available head and flow quantity may vary considerably, resulting in a wide range of outputs and unit investment costs. Installed plant costs reported by the Department of Energy for hydro plants include a $398-per-kilowatt investment for the 140-MW Keowee Plant in South Carolina, completed by Duke Power Company in 1971. The Northfield Mountain Plant of Western Massachusetts Electric Company, completed in 1973, carries an investment cost of $145/kW owing to a high gross head and smaller pondage volume. Cost prediction for future hydroelectric construction is difficult, particularly in view of the decreasing availability of economical sites and restrictions imposed by concern for the ecological and social consequences of disrupting the natural flow patterns of rivers and streams.

Pumped-Storage Plants Pumped-storage plants involve a special application of hydroelectric generation, allowing the use of off-peak energy supplied at incremental charges by low-operating-cost thermal stations to elevate and store water for the daily generation of energy during peak-load hours. Pumped hydro projects must justify the inefficiencies of storage pumping and hydroelectric reconversion of off-peak thermal plant energy by investment cost savings over competing peaking plants. Installation of a pumped hydro station calls for a suitable high head site which minimizes required water storage and upper and lower reservoir areas and an available makeup source to supply the evaporative losses of the closed hydraulic loop. Despite the added complications of installing both pumping and generating units, or of utilizing reversible motor-generator pump-turbines, costs for pumped hydro stations generally fall below those for conventional hydroelectric stations. Lower installed costs are the result of elimination of dams, extensive pondage, and the siting need for appreciable natural water flow. The Department of Energy reports a 1991 installed cost of $937/kW for the Duke Power Company's 1,065-MW Bad Creek Project. The 1985 Virginia Electric Power Company's Bath County 2101-MW plant carries an investment charge of $803/kW. Differences in gross head, impoundment, and siting make plant cost comparisons difficult.

Geothermal Plants Geothermal generation utilizes the earth's heat by extracting it from steam or hot water found within the earth's crust. Prevalent in geological formations underlying the western United States and the Gulf of Mexico, geothermal energy is predominantly unexploited, but it is receiving increased attention in view of escalating demands on limited worldwide fossil-fuel supplies. Because natural geothermal heat supplants fuel, the atmospheric release of combustion products is eliminated. Nevertheless, noxious gases and chemical residues, usually contained in geothermal steam and hot water, must be treated when geothemral resources are tapped. There is a current lack of significant cost data covering geothermal plants. The major commercial U.S. facility, the Geysers Plant in northern California, began in 1960 as a phased expansion. It uses dry steam at 600°F (316°C). Because boiler and associated fuel-handling facilities are eliminated, investment in these generating plants is considerably less than the cost of comparable fossil-fueled units. However, overall investment chargeable to geothermal facilities includes significant exploration and drilling costs which are site dependent and cannot be accurately predicted without extensive geophysical investigation.

Environmental Considerations Environmental protection has become a dominant factor in the siting and design of new power generating stations. Both stack emissions to the atmosphere and thermal discharges to natural water courses must be significantly reduced in order to meet increasingly stringent environmental criteria. In many cases older plants are being required to reduce emission levels to achieve legislated ambient air-quality standards and to control thermal discharges by the use of closed cooling systems to prevent aquatic thermal pollution.

Control of **air pollution** in fossil-fueled power plants includes the reduction of particulates, sulfur oxides, and nitrogen oxides in flue gas emissions. Particulate collection can be achieved by electrostatic precipitators, baghouse filters, or as part of stack gas scrubbing. Stack gas scrubbing is required for all new coal-fired power plants, regardless of sulfur content of the fuel. Fossil-fueled power plants are believed to be a major contributor to acid rain.

Scrubbers reduce sulfur oxide emissions by contacting flue gas with a sorbate composed of metal (usually sodium, magnesium, or calcium) hydroxides in solution which act as bases to produce sulfate and sulfite precipitates when they contact sulfur oxides in the flue gas. Wet scrubbers contact flue gas with a sorbate in solution. Dry scrubbers evaporate sorbate solution into the gas stream.

Fossil-fueled power plants are required to burn low-sulfur fuels. Restrictions on the use of oil in new power plants and its high cost have virtually eliminated new central station steam power plants designed to utilize oil.

Control of nitrogen oxides NO_x is attained primarily through modifications to flame propagation and the combustion processes.

The 1990 amendments to the federal Clean Air Act and proposed restrictions in the new amendment to the federal Water Pollution Act of 1972 (Clean Water Act) will have cost implications that have not been fully documented. Under the influence of the nitrous oxide (NO_x) emission limitations proposed for Phase I and Phase II of the 1990 amendments to the Clean Air Act, steam generator designs will change to incorporate some form of low-NO_x burners, overfire air dampers, catalyst injection, flue gas recirculation to the burners, and other modifications to meet these new federal standards. New combustion turbine designs are also being developed to meet these new NO_x emission regulations.

In order to avoid plant discharges of waste heat to the aquatic environment, evaporative-type closed-cooling cycles are employed in lieu of once-through cooling system designs. Closed-loop cooling systems in current use employ evaporative **cooling towers** or **cooling ponds**. Use of these systems entail investment penalties consisting of net increases in equipment and facilities cost, and penalties incurred by losses in peak capability due to the lower plant efficiency associated with evaporative cooling systems.

More advanced closed-cooling-loop designs anticipate the use of dry and wet-dry cooling towers. These represent possible alternates to conventional evaporative systems where makeup water is in short supply or where visible vapor plumes or ice formed by vapor discharge present hazards. The penalties for dry-tower cooling are significantly higher than those for conventional evaporative designs. Large, more costly water-to-air heat-transfer surfaces are required, and characteristically higher condensing temperatures result in higher turbine backpressure, severely restricting plant capability.

Although no large-size dry towers are currently in operation in the United States, estimates indicate incremental investment cost penalties for conventional fossil-fueled plants in the range of $150 to $200 per kilowatt for mechanical-draft dry cooling towers and of $300 to $400 per kilowatt for natural draft, dry cooling designs. Similarly equipped nuclear plants bear dry tower investment cost penalties approximately 40 percent higher than the above. Wet-dry towers show promise for practical application, combining the advantages of both wet and dry tower designs. Wet-dry towers incur added investment penalties for closed heat-transfer surface only to the extent necessary to eliminate visible plume and/or to reduce makeup requirements. Investment cost penalties for wet-dry towers fall between those for wet towers and dry towers.

FIXED CHARGES

Costs that are established by the amount of capital investment in plant and which are fixed regardless of production level are termed **fixed charges**. Annual fixed charges are ordinarily expressed as a percentage of investment and include interest or the cost of money, funds applied to amortize investment or to allow for replacement of depreciated plant, and charges covering property taxes and insurance. Additionally, fixed charges may include an interim replacement allowance to cover the replacement cost of plant equipment not expected to last the full life of the plant.

For investor-owned utilities, the **cost of money** employed for utility plant expansion depends upon financial market conditions in general and upon the attitude of investors with regard to a particular utility enterprise or specific project. Funds for investor-owned utility expansion are derived from both the risk capital (equity) and debt capital (bond) markets. Utility bonds command a return of 5 to 7 percent in the current market, while equity capital returns of between 6 and 8 percent are typical. Prevailing return rates for investments in utility plant facili-

Table 17.5.3   Comparison of Interest or Rate of Return and Revenue Requirements for Private and Public Projects

	Investor-owned utility, %	Government-owned utility, %	Industrial-commercial ownership, %
Distribution of investment			
Equity capital (stocks)	40	0	100
Debt capital (bonds)	60	100	0
Total	100	100	100
Rate of return or interest			
On stocks	7.0	0.0	8.0
On bonds	6.0	5.0	0.0
Income subject to federal income tax (FIT)			
Average rate of return	6.4*	5.0	8.0
Deduction for interest	3.6†	5.0	0.0
Net taxable income	2.8	0.0	8.0
Return on equity before 35% FIT			
2.8%/(100 − 35%)	4.3‡		
8.0%/(100 − 35%)			12.3
FIT as percentage of capital			
4.3−2.8%	1.5		
12.3−8.0%			4.3
Summary revenue requirements			
Equity return before FIT	4.3	0.0	12.3
Bond interest	3.6§	5.0	0.0
Total	7.9	5.0	12.3

* Average return is 7% equity return on 40% of investment plus 6% bond interest on 60% of investment.
† Deduction is 6% bond interest on 60% of investment.
‡ Return on equity equals 7% equity return to stockholders on 40% of investment.
§ Return on debt equals 6% bond interest on 60% of investment.

ties are influenced by the rate-setting practices of public utility regulatory agencies and by supply-and-demand factors in the investment market.

Public utility facilities owned by state and municipal government organizations are generally financed by long-term revenue bonds, most of which qualify for tax-free-income status. Interest rates currently fall between 4 and 6 percent and reflect the tax relief on interest income enjoyed by the bondholders.

Generating facilities are often financed by industrial concerns whose primary business is the production of a manufactured product. In these instances, the annual cost of money invested in power facilities will be established by considering alternate investment of the required funds in manufacturing plant. Inasmuch as returns on equity capital invested in manufacturing industries are usually on the order of 10 to 20 percent, the rate of return for industrial power plants will tend to be set at these higher levels.

Corporate federal taxes are levied on equity capital income. Thus, corporate earnings on equity investment must exceed the return paid the investor by an amount sufficient to cover the tax increment. Current federal tax rates of 35 percent apply. Therefore, revenue requirements for a given return to the investor will amount to slightly less than twice the given rate (see Table 17.5.3).

The amortization of debt capital or the provision for **depreciation** over the physical life of utility plant facilities may be effected by several methods. Straight-line depreciation requiring uniform charges in each year over a predetermined period of useful service is commonly applied because of its simplicity. The percentage method of depreciation assumes a constant percentage decrease in the value of capital investment from its value the previous year, thereby resulting in annual depreciation charges which progressively diminish. The sinking-fund method of economic analysis assumes equal annual payments which, when invested at a given interest rate, will accumulate the capital value of facilities less their salvage value, over a predetermined useful service life. Table 17.5.4 illustrates the representative useful life for alternate utility facilities.

Use may be made of interest tables which show, for any rate of interest and any number of years, the equal annual payment (sinking-fund) rate which will amortize an investment and additionally will yield an annual return on investment equal to the interest rate.

$$\text{Equal annual payment} = \frac{i(1 + i)^n}{(1 + i)^n - 1}$$

where n = number of years of life, and i = interest rate or rate of return.

Property taxes and property insurance premiums are normally established as a function of plant investment and thus are properly included as fixed charges. Property taxes vary with the location of installed facilities and with the rates levied by the various governmental authorities having jurisdiction. In general, public power authorities will be free of taxes, although public enterprises often render payments to government in lieu of taxes. Annual property tax rates for private enterprise will amount to perhaps 2 to 4 percent of investment, while property insur-

Table 17.5.4   Representative Useful Life of Alternate Utility Facilities

Facility	Representative useful service life, years
Steam-electric generating plant	30
Hydroelectric plant	50
Combustion turbine-combined cycle	30
Nuclear plant	30
Transmission and distribution plant	40

ance may account for annual costs of between 0.3 and 0.5 percent.

A representative makeup of fixed charges on investment in a conventional steam-electric station having a useful service life of 30 years is shown in various classes of ownership in Table 17.5.5.

OPERATING EXPENSES

Fossil Fuels   Currently, fossil fuels contribute approximately three-fourths of the primary energy consumed by the United States in the production of electric energy. Generating station demands for fossil fuels, particularly coal, are continuing to increase as the electric utility and industrial power markets grow. Price comparisons of fossil fuel are generally made on the basis of delivered cost per million Btu. This cost includes mine-mouth or well-head price, plus the cost of delivery by pipeline or carrier. Price comparisons must recognize that solid and, to a lesser extent, liquid fuels require plant investments and operating expenditures for fuel receipt, storage, handling and processing facilities, and for ash collection and removal. High transportation cost contributions on a Btu basis will be incurred by high-moisture and ash-content coals with low heating values. It is, therefore, advantageous to fire lignite and subbituminous coals at mine-mouth generating plants.

Coal represents our most abundant indigenous energy resource, with enough economically recoverable supplies at current use rates to last well into the next century. About half of the recoverable coal reserves have a sulfur content above 1 percent and are considered high-sulfur coal. Low-sulfur coals are found chiefly in the low-load areas of the mountainous West, and delivered cost at the major markets east of the Mississippi include high transportation charges. Increasingly, coal production is bearing the cost of more rigid enforcement of stringent mine safety regulations, and the charges associated with strip-mine land restoration. Delivered price depends upon transportation economies as may be affected by barging, unit train haulage, or pumping in slurry pipelines. Delivered price levels for coal fuel vary with plant location. Plants conveniently located with respect to eastern coal reserves report delivered-coal prices in the general range of $1.25 to $2.00 per million Btu, depending upon sulfur and ash content. Low-sulfur western subbituminous coals have delivered prices of between $1.00 to $1.75 per million Btu.

The domestic supply of **petroleum** is now outstripped by nationwide demand. The United States has become dependent on overseas sources

Table 17.5.5   Derivation of Fixed-Charge Rate for Conventional Fossil-Fueled Steam-Electric Plant with 30-Year Economic Life

	Investor-owned utility, %	Government-owned utility, %	Industrial-commercial ownership, %
Rate of return or interest*	6.4	5.0	8.0
Amortization or depreciation†	1.2	1.5	0.9
Federal income tax	1.5	0.0	4.3
Local taxes (or payment in lieu of taxes)	2.0	2.0	2.0
Insurance	0.3	0.3	0.3
Total	11.4‡	8.8	15.5

* Interest or return rate is listed in Table 17.5.3.

† Sinking-fund amortization or depreciation rate (at given rate of return or interest), which when added to the rate of return or interest equals the capital recovery factor.

‡ Assuming initial plant investment of $930/kW and 7,500 h/year operation at full load, the annual fixed charges in mills per kilowatthour will be 930 × 0.114 × 1,000/7,500 = 14 mills/kWh.

Table 17.5.6   Operating Expense for Fuel for Representative Heat Rates and Fuel Prices

	Nominal size, MW	Typical heat rate, Btu/kWh	Fuel price, ¢/MBtu	Fuel cost, mills/kWh
Conventional coal fired	500	9,800	200	19.6
Advanced light-water reactor	600	10,700	60	6.4
Combustion turbine/combined cycle	300	8,000	220	17.6

NOTE: Operating fuel cost, mills/kWh = (Btu/kWh) × (¢/Btu) × 10^{-5}.

to meet growing energy demands. The consequences of this trend are a continued unfavorable balance of payments and dependence on foreign oil supplies from politically unstable areas in the Middle East and North Africa. Residual and distillate oil prices have risen because of short supply and pressure by the major oil producers on worldwide market price levels. Blends of low-sulfur oils delivered during 1993 to generating plants along the eastern seaboard were priced upward from $3.50 per million Btu to as high as $4.50 per million Btu for the 0.3 percent sulfur fuel required for firing in some metropolitan areas. During this same period distillate oil commanded a nationwide price ranging from approximately $4.00 to $5.00 per million Btu.

Consumption of **natural gas** as a power fuel is increasing because it is clean burning, convenient to handle, and generally requires smaller and cheaper furnaces. Historically, its limited supply made it best-suited for consumption by residential and commercial users and meeting industrial process needs. Present-day power plant use of natural gas is on the rise, especially in plants located in the gas-producing areas of the south central and western states. Gas fuel prices for 1993 ranged from $2.00 to $3.50 per million Btu. Price levels of natural gas, severely regulated in the past by government controls imposed at the well, have been falling sharply in response to current supply-and-demand factors and deregulation.

Nuclear Fuel   Reactor plant fuel costs present a special case. Actually, the initial core loading which will support operation of a nuclear plant over its early years of life requires a single purchase prior to commissioning of the plant. For comparison with fossil-fuel prices, therefore, nuclear-fuel cycle costs, including first-core investment, periodic charges for reload fuel, and spent-fuel shipment and processing costs, are ordinarily extrapolated at assigned load factors over the life of the plant and are converted to an economic equivalent expressed in dollars per million Btu of released fission heat. Nuclear-fuel costs are not only influenced by ore prices and by fuel fabrication and processing costs, they are also sensitive to investment and uranium-enrichment costs. Future costs are subject to inflationary pressures and cost-saving technology changes such as extended burn-up cycles. Charges of 50 to 70 cents per million Btu are representative of the levelized fuel prices for nuclear plants.

Operating Costs of Fuel   Fuel price contributions to energy-generation costs will reflect start-up and furnace-banking losses and will depend upon plant efficiencies which, at low loads, show considerable departure from the best-point performance achieved at or near full unit loadings. These factors significantly affect the operating expenses of load-following utility system units as well as plants assigned fluctuating demands in manufacturing or industrial service. As an example, calculated performance for a nominal 500-MW, 2,400-psig coal-fired regenerative reheat steam unit shows a best-point heat rate of 9,800 Btu/net kWh at rated output. Load reduction yields heat rates of 10,500 and 12,400 Btu/kWh at loadings of 250 MW and 125 MW, respectively. Typical operating-expense ranges for given fuel prices and estimated full-load heat rates may be determined directly where continuous operation at or near unit rating is assumed (Table 17.5.6).

Operating Labor and Maintenance   In addition to fuel costs, operating expenses include labor costs for plant operation and maintenance, plus charges for operating supplies and maintenance materials, general administrative expenses, and other costs incidental to normal plant operation. Operating labor and maintenance costs vary considerably with unit size, operating regimen, plant-design conditions, type of facility, and the local labor market. Representative figures appear in Table

17.5.7. Nuclear liability insurance costs are included under reactor plant operating expenses. They finance a program of government indemnity, coupled with private insurance, for public damage that could arise from a nuclear incident.

Environmental Controls   Systems and equipment required for air and water pollution abatement generally carry increased fuel and maintenance labor and materials costs. Reductions in plant output resulting from the higher condensing pressures associated with cooling-tower operation or the added auxiliary power for stack gas clean-up systems lower plant efficiency and increase fuel consumption.

OVERALL GENERATION COSTS

The total cost of power generation may now be estimated by reference to the preceding material assuming type of ownership, capital structure, plant type, fuel, and loading regimen. Table 17.5.8 comprises an illustrative tabulation of the factors determining the overall generation cost of an investor-owned nuclear generating facility.

It should be noted that capacity factor, or the ratio of average-actual to peak-capable load carried by a given generating facility, will have a significant effect on generation expenses. In addition to the effects of part-load operation on fuel costs as previously discussed, capacity factor will determine the plant generation which will support fixed charges. High-capacity-factor operation will spread fixed charges over a large number of kilowatt-hours of output, thereby reducing unit generation costs. During its initial life, a thermal plant is usually operated at high-capacity factor. As inevitable obsolescence brings newer and more efficient equipment into service, a unit's baseload position on the utility system load duration curve is relinquished, and capacity factor tends to drop. This decline in capacity factor must be acknowledged in estimating output and generation costs over the life of a given facility.

Most modern electric utilities incorporate computerized systems designed to economically dispatch power generated at each production plant feeding the load. Individual generating-unit loads are assigned in a manner that can be demonstrated to result in minimum overall cost; i.e., at each system power level, load is shared between units so that all operate at the same incremental production cost. Telemetered data reflecting system load and generation is transmitted to central dispatch computers by multiplexing via power line carrier, microwave, or telephone lines. The communication schemes include channels for transmitting load adjustment commands developed by the computer to on-line generating units. Loading instructions account for the unit production efficiencies and transmission losses associated with each dispatch assignment.

Table 17.5.7   Representative Operating and Maintenance Costs (1993 Price Level)

Type of plant	Nominal size, MW	Fuel	Operating and maintenance costs, mills/kWh
Conventional fossil	500	Coal	6–8
Advanced light-water reactor	600	Nuclear	5–9
Combustion turbine/combined cycle	300	Gas	6–10
Conventional hydro	300	—	5–7

NOTE: Unit kilowatt-hour costs include labor, maintenance materials, operating supplies, and incidental expenses. Costs shown assume base-load operation.

Table 17.5.8   Total Cost of Generating Power
(Plant type: 500-MW conventional fossil; plant net heat rate: 9,800 Btu/kWh)

	Generating costs, mills/kWh
Coal fuel @ $2.00/mBtu	19.6
Operating and maintenance	7.0
Fixed charges @ 11.4% per annum	24.0
Total operating costs	50.6

NOTES: Fixed charges are based on assumed initial plant investment of 1,580/kW and 7,500 h/year of operation. Values shown are typical and could vary significantly for individual plants.

TRANSMISSION COSTS

Because of a growing scarcity of urban sites and increasing emphasis on environmental protection and nuclear safety, it has become more and more difficult to site major power stations near centers of load. As a consequence, added cost of transmission along with attendant resistive power losses add significantly to the overall cost of service. Because of the distances involved, these costs are generally greatest for nuclear facilities, mine-mouth stations, and hydroelectric plants where remoteness or remote resources strongly govern siting. Transmission plant investment also reflects a trend toward interconnection of neighboring utilities. Designed to improve service reliability by the pooling of reserves and to effect savings by capacity and energy interchanges, such interties must carry substantial ratings so that emergency power transfers can be accommodated without exceeding system stability limits. Costs for major overhead transmission ties (1,000 MVA and up) are estimated to range between $300,000 and $700,000 per circuit mile depending on terrain and voltage level. Investments are also sensitive to factors of climate and proximity to urban areas that can increase right-of-way costs significantly. Where underground transmission is elected, installed investment costs can be as much as 10 times the cost of overhead lines.

The choice of **transmission voltage** level and whether ac or dc is used for bulk power transfer depends on the amount of power transmitted, the transmission distance involved, and at each voltage level, the cost of line and substation equipment. Voltages generally employed are 230, 345, 500, and 765 kV ac and 5000 + kV dc. Where long distances on the order of 400 mi (650 km) or more are encountered, dc transmission becomes economically attractive. For shorter lines, however, savings in fewer conductors and lighter transmission towers are nullified by the high cost of dc-ac conversion equipment at both line terminals.

POWER PRICES

The price of power delivered by a utility system must account for the production costs at each of its generating stations. As previously noted, these costs depend upon labor rates, fuel prices, and material charges. They reflect investment levels in generating plant and the fixed-charge rates established by funding patterns, type of ownership, and expected equipment service life. Overall system production costs are affected by the investment requirements of specific mixes of generating equipment types, and by the manner in which load is shared by units, i.e., how production is allocated between highly efficient base-load stations and the less-efficient peaking equipment which normally runs for only a few hours each day. Also, important cost reductions are achieved in hydro systems by controlling natural and stored water flows to allow optimized sizing and scheduling of hydroelectric output, thereby reducing needs for thermal peaking capacity and decreasing the generation requirements of high fuel cost fossil plants.

Power prices cover the investment charges, maintenance costs, and capacity and energy losses chargeable to the transmission and distribution plant. They also include the administrative costs incurred to maintain corporate enterprise and the commercial expense of metering and billing.

Generally, power prices must provide a return to cover the average cost of power production throughout a given system. Rate schedules and supply contracts, however, are drawn to reflect the reduced cost of off-peak energy produced by available generating units during periods of low system load. Additionally, prices for high load factor service often recognize the cost reductions effected by spreading fixed charges over increased units of energy output. Large blocks of capacity and energy supplied for industrial use are often priced by establishing an annual charge for capacity which equals the fixed charges on investment in committed generation and transmission plant, plus charges for energy representing the sum of the variable kilowatt-hour production costs for fuel, maintenance, and operation.

Historically, utilities have applied rate schedules which promote consumption by applying progressively lower rates to blocks of increased energy usage. Rationale for such pricing is the savings that load growth can realize through economies of scale, as well as the improved utilization of existing utility plant. However, regulatory pressure, reflecting a policy of minimizing the industry's impact on the environment and the critical need to conserve high-cost imported fuel, has favored a marginal cost-pricing system more nearly reflecting the actual cost of production and transmission of a particular user's supply of power. Rate setting under this conservationist approach calls for flat, rather than reduced, rates as usage increases and for high unit energy charges during peak-load periods.

Facilities designed for the **dual-purposes production of power and process steam** permit investment savings, principally in steam-generation plants, which result in combined production charges falling below the total cost of separate, single-purpose production of power and of steam. Such savings can permit proportionate decreases in the prices ordinarily charged for separate single-purpose production of each of the products, or they may be assigned in total to reduce the price of one or the other by-product. This latter option is often exercised in the case of dual-purpose water product plants arranged for seawater flash evaporation using power-turbine extraction as a process steam source. Where severe shortages of fresh water exist, social considerations favor the total assignment of dual-purpose savings to the water product. Thus, minimal prices for desalted product water are achieved, while dual-purpose power is marketed at prices competing with single-purpose power generation costs. Similarly, assignment of the total savings of dual-purpose power and process steam production to the power product may justify on-site industrial plant power generation in preference to outside purchases of higher-priced utility system power supplies.

Table 17.5.9   Average Revenue per Kilowatt-Hour Sold to Customers for the Total U.S. Electric Utility Industry
(Cents per kilowatt-hour)

Year	Residential	Commercial	Industrial	All
1973	2.35	2.30	1.17	1.86
1974	2.83	2.85	1.55	2.30
1975	3.21	3.23	1.92	2.70
1976	3.45	3.46	2.07	2.89
1977	3.78	3.84	2.33	3.21
1978	4.03	4.10	2.59	3.46
1979	4.43	4.50	2.91	3.82
1980	5.12	5.22	3.44	4.49
1981	5.86	6.00	4.03	5.16
1982	6.44	6.61	4.66	5.79
1983	6.83	6.80	4.68	6.00
1984	7.17	7.14	4.88	6.27
1985	7.39	7.27	5.04	6.47
1986	7.43	7.22	4.99	6.47
1987	7.45	7.10	4.82	6.39
1988	7.49	7.04	4.71	6.36
1989	7.65	7.20	4.79	6.47
1990	7.83	7.33	4.81	6.57
1991	8.05	7.54	4.89	6.76
1992	8.17	7.62	4.89	6.81

SOURCE: Edison Electric Institute.

Where hydro facilities supply power in combination with irrigation, flood control, navigation, or recreational benefits, power costs are largely sensitive to the allocation of investment charges against each of the multipurpose project functions. Should generation be treated as a by-product, power prices can be reduced drastically to reflect equipment operating expenses and the limited fixed charges covering investment in only the generating plant itself.

Cheap fuel, advances in design, the economies of scale, and the economic application of alternate generating unit types produced downward trends in the price of electric service in the 1960s. In the 1970s these trends were reversed by inflationary effects on plant costs, high interest rates, and the increased prices commanded by fossil and nuclear fuels (see Table 17.5.9). A dwindling number of favorable plant sites, licensing delays, and the added costs of environmental impact controls combined to cause further upward pressure on power prices in the 1980s. However, falling interest rates and fuel prices began to slow the rate of increase in the cost of electric power. These stabilizing influences have continued into the 1990s, tempered by stricter environmental regulation costs. Increases in electric rates are expected into the foreseeable future.

17.6 HUMAN FACTORS AND ERGONOMICS
by Ezra S. Krendel

REFERENCES: McRuer, Pilot-Induced Oscillations and Human Dynamic Behavior, *NASA CR-4683*, July 1995. McRuer, Human Dynamics in Man-Machine Systems, *Automatica*, **16**, 1980. McRuer and Krendel, Mathematical Models of Human Pilot Behavior, *AGARDograph no. 188*, 1974. Krendel and McRuer, "A Servomechanism Approach to Skill Development," *J. Franklin Inst.*, **269**, 1960. Wright, "Factors Affecting the Cost of Airplanes," *J. Aeronautical Sciences*, **3**, 1936. Reason, "Human Error," Cambridge University Press. Konz, "Work Design: Industrial Ergonomics," 3d ed., Publishing Horizons. Atkinson, Herrnstein, Lindzey, and Luce (eds.), "Stevens' Handbook of Experimental Psychology," 2d ed., vols. I and II, Wiley. Salvendy (ed.), "Handbook of Human Factors," Wiley. Boff, Kaufman, and Thomas (eds.), "Handbook of Perception and Human Performance," vols. I and II, Wiley. Swain and Guttman, *Handbook of human reliability analysis with emphasis on nuclear power applications*, NUREG/CR-2744, U.S. NRC, 1983.

SCOPE

The goal of human factors and ergonomics is to improve the performance, reliability, and efficiency of systems in which humans operate in concert with machines, man/machine/systems (**MMSs**). This discipline developed rapidly during and after World War II. The life-or-death stakes and the advances in military technology made even minor improvements in military systems highly desirable and major improvements essential. In subsequent years the risks of accidents and the complexity and costs of military hardware, manned space exploration, nuclear power plants, vehicular control, and civil air transportation were behind the continued development of human factors/ergonomics as an engineering discipline. Commercial applications in product design, industrial process control, quality control, health care technology, computers, and office design have expanded this development. Human factors engineering and ergonomics draw on or interact with the theories and data of many diverse disciplines: psychology, physiology, and applied physical anthropology; aeronautical, electrical, industrial, mechanical, and systems engineering; and computer and cognitive sciences.

Humans operate as sensors, information processors, actuators, and decision makers. The models used to describe these operations are determined by the MMS. For an important and extensively studied family of MMSs—the manual control of air, space, ground, or sea vehicles—the range and variability of allowable manual control is constrained, particularly when the system is operating near its stability limits. The repertory of human control dynamics behavior modes, which demonstrates the adaptive skills the operator applies to reorganizing inputs so as to maintain system performance, can then be described with the same mathematics as the inanimate system components. In applications which emphasize signal detection and decision making (as in monitoring or visual search) or estimate human reliability, performance is best described probabilistically. When the human is a significant source of energy (as in heavy industry or intense athletic activities), power output decrements over time are the main interest.

What follows presents hints at this mass of information, with some useful empirical rules, and concludes with an example of closed-loop compensatory control, the first stage in the repertory of human dynamics behavior modes.

PSYCHOMOTOR BEHAVIOR

Psychomotor behavior is the activity of receiving sensory input signals and interpreting and physically responding to them. Humans can receive inputs by vision, hearing, smell, and the cutaneous senses which respond to temperature, mechanical energy, or electrical energy. Kinesthesis and the vestibular sense inform about location and position. Vision followed by hearing are the most important senses for transmitting signals carrying complex information for decisions and for control of MMSs. Signals for warning or alerting need not be complex and can be transmitted by one or a combination of the sensory channels. The choice is determined by the situation and the task being performed by the person or persons to be warned rather than by differences in modality reaction times.

The sense dominating psychomotor behavior, vision, receives light signals by either of two different retinal receptors, cones for photoptic and rods for scotopic vision, which convert optical images to signals sent to the brain. Daylight color vision and detailed visual acuity are photopic. Scotopic vision is black and white with shades of gray. Cones are concentrated in the fovea, a central area whose visual angle subtends 30 minutes of arc, and are found in very small numbers elsewhere on the retina. The distribution of rods extends to the periphery of the retina but not to the fovea. Their concentration is greatest at about 20° from the fovea.

After adaptation in the dark for 5 min, the cones are at their maximum sensitivity to low light levels below which color vision is lost. After 30 min in the dark, the rods reach their maximum sensitivity to low light levels. Adapting from a dark environment to a light one takes about a minute; time to readapt to the dark can be brief if the time has been spent under red light. Red light stimulates cones, but not rods, enabling the rods to retain their previous adaptation to a low light level.

Perception of objects and events results from sensory inputs and their interpretation by the brain. Deliberately devised size and distance illusions trick the brain by manipulating expected cues for size, distance, and perspective. Unintended illusions or misleading perceptions can arise when the brain relies on inappropriate sensations or expectations. The process of perception of the size and distance of an object begins when light from the object passes through the observer's pupil and lens. An image is focused on the retina by the action of the ciliary muscles, which can change the accommodation of the lens. Estimates of an object's apparent distance are influenced by this muscle action, by the visual cues leading to the object, and by the expectations of the viewer for the size of the object. These cues may be inadequate or they may be inconsistent with one another and as a consequence create a perception of distance which differs from reality. As an example, the vision of the driver of an automobile may be accommodated to the dashboard displays at a distance focus of 0.5 m. Refocusing on an unexpected distant object may take as long as 0.4 s. The object may be unfamiliar and visual cues obscured by darkness or fog. Even a briefly held misperception of object size or distance risks an error and an accident.

Complex psychomotor behavior may be constructed by incrementally aggregating a sequence of elemental discrete actions, by continuous closed-loop control, by complex open-loop activities, and by combinations of the above. Choice reaction time (RT) is an elemental discrete action whose duration is a function of the number of alternative choices and their probabilities. An example is a horizontal line of n signal light bulbs, below each of which is a key to be struck as quickly as possible when the corresponding bulb lights up. If $n = 1$, RT would be about 200 ms for a practiced alert operator. As n increases, the operator's uncertainty, $H = \log_2 (n + 1)$, which allows for no response, increases, and RT increases proportionally. This is the *Hick-Hyman law*:

$$RT = a + bH \qquad (17.6.1)$$

In information theory H is a measure of uncertainty or entropy in bits. H can be calculated whether the stimuli are equally likely or whether their probabilities of occurrence differ. The upper limit of H is slightly more than 3 bits. For the example given a is about 200 ms, b is about 150 ms, and H ranges from 0 to a little over 3 bits. An application is the calculation of RT differences among MMS designs for which values of a are likely to be similar.

The movement time (MT) for rapid discrete or repetitive actions, usually by the hands, is proportional to the difficulty of the task. *Fitts' law* defines an index of difficulty I_D in terms of target width W and movement amplitude A:

$$I_D = \log_2 (2A/W) \qquad (17.6.2)$$
$$MT = c + dI_D \qquad (17.6.3)$$

The form of Eq. (17.6.3) is an empirical description of data from a wide range of rapid movements as in operating a key pad or in sorting items into bins. In applications, d is about 100 ms and c, which depends on the movement geometry, is approximately 200 ms. Estimates of MT provide comparisons among operating procedures.

The equations for RT and MT imply a stable level of skilled performance. The learning curves for psychomotor skills which depend on negative feedback are exponential functions of time. For other psychomotor tasks, skill learning follows a power function whose form is the same as the empirical learning curve developed by Wright to predict the unit cost of producing aircraft, and is

$$y = a\, x^{-b} \qquad (17.6.4)$$

where y is the number of direct labor hours to produce the xth unit, a is the hours to produce the first unit, and b is the rate that labor hours decreases with cumulative output. For learning a psychomotor skill, a ranges from 0.2 to 0.6. For production a ranges from 0.1 to 0.5 with an expected value across industries of about 0.3. The empirical power function when fitted with a straight line on log-log coordinates smooths variability and simplifies quantifying production improvements over time. Each doubling of cumulative output results in a reduction of unit cost by a fixed percentage of its beginning cost. A progress ratio p of 80 percent means that after each doubling of the cumulative unit output, the unit cost is 80 percent of its value before doubling.

SKILLS AND ERRORS

The dynamics of psychomotor skill development and its negative, poor performance and the emergence of errors, is a pervasive consideration in MMS design. The highest level of psychomotor skill is attained by a process of successive organization of perceptions (SOP), whereby operators fully familiar with the dynamics of the machine under their control and the appropriate responses to the input signals can reorganize the system, adapt their behavior to create a repertory of special responses, and then select from this repertory the appropriate response for system performance. In closed-loop control this response may be anticipatory and compensate for inherent human control lags. In an open-loop mode the response may be programmed, as in executing the sequential actions in preparing to stop a car on approaching a red light. Skilled performance can degrade precipitously if inappropriate cues contaminate the perceptual organization or cause it to regress; training in cue recognition is essential.

The level of skilled performance achieved and its variability is affected by individual abilities, training, motivation, attention, workload, work scheduling, the social environment of the activity, and impairment due to substance abuse, stressors, or fatigue. By adhering to human factors findings in the selection and positioning of information displays, controls, monitors, keyboards, seating, and accommodating the physical dimensions of the operator, the designer can make sustained desirable performance more likely, and thwarted or unintended responses less likely. Unintended responses such as misreading a display or inadvertently reaching for the wrong switch result in errors characterized as *slips*. The potential for catastrophic accidents in the operation of nuclear power plants, chemical plants, aircraft, and oil tankers emphasizes the need to design against such slips and to be able to estimate the reliability of both the MMS tasks and the total system. By ascribing human error probabilities to slips and other deteriorations in performance and developing detailed event trees for the many complex MMS tasks, Swain and Guttman were able to estimate human reliability in much the same way that the reliability of inanimate devices has been estimated. Although their numerical reliability estimates cannot always be precise, they enable comparisons to be made of the MMS's reliability under alternative physical configurations, training disciplines, and institutional policies.

A different type of error in behavior arises from *mistakes* in the human's thought processes. Mistakes are similar to errors arising from perceptual illusions in their inappropriate reliance on expectations and experiences. When the error is a mistake, the human operator has misinterpreted the situation, and reliability estimates are not feasible. The mental processes leading to mistakes have been developed from detailed reconstructions of the human actor's thought processes and behavior in accidents and near accidents. The basic sources are accident reports, anecdotes, interviews with survivors, and skilled introspection on the part of the analyst. If the mistake is made under time pressure, it is difficult for the operator to reexamine the situation, sort out misconceptions, and examine options in a deliberate manner. Ergonomic design can make mistakes less likely and recovery more likely by reorganizing the information displays. A more direct approach to lessening the incidence of mistaken intentions is to design redundancy into the MMSs and to develop training and personnel selection procedures which emphasize the search for options and discourage perseverative mistaken thought processes.

MANUAL CONTROL

Compensatory closed-loop tracking is that control mode in which the human responds to the error signal alone. Progressively more effective modes attained via SOP enable the human to respond to the input and output signals as well, to preview the input, and finally to respond intermittently open-loop in a dual-mode configuration. Instabilities and oscillations in the compensatory closed-loop task can occur if the open-loop gain is too high. A human operator who has regressed to the compensatory mode may inadvertently generate excessive gain and destabilize the system. Pilot-induced oscillations are undesirable, occasionally catastrophic, sustained aircraft oscillations which occur because the pilot insists on maintaining closed-loop control of what would otherwise be a stable aircraft.

Figure 17.6.1 is a simplified block diagram for compensatory closed-loop control. If the controlled element is an automobile or an aircraft, the dynamics can be approximated by first- and second-order systems in the region of human control. The human operator is a quasi-linear element. For such an element the response to a given input is divided into two parts: describing-function components which correspond to the response of the equivalent linear elements driven by that input, and a "remnant" component which represents the difference between the response of the actual system and a system composed of equivalent linear elements. The describing function and remnant depend explicitly on the task variables (which are inputs, disturbances, and controlled-element dynamics) and on operator-centered environmental and procedural variables. The effects of these variables have been integrated into a set of rules for applying and adjusting human-operator describing functions. These rules are fairly simple when the plant dynamics either are or can be approximated by first- or second-order dynamic systems. The following conditions must apply before using these rules:

1. The forcing functions which act as inputs to the MMS are unpredictable, have low bandwidth (below 1 Hz), and are continuous waveforms.

2. The controlled element is a low-order system or can be so approximated and has no highly resonant modes over the input bandwidth.

3. The display and controls are reasonably well scaled and smooth so as to minimize the effects of thresholds, detents, and friction.

4. The other task demands are sufficiently light to permit the operator to devote the majority of his or her attention to minimizing the displayed error.

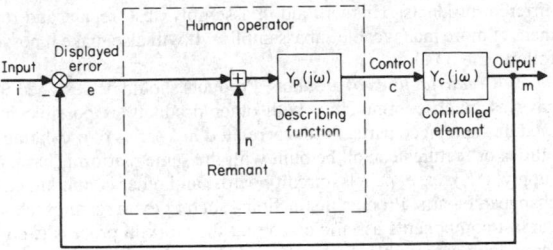

Fig. 17.6.1 Simplified block diagram for manual control.

McRUER'S RULE

Under these conditions (which are commonly met in most operational control tasks), theory, experiments, and practice have shown that the human operator attempts to compensate for the dynamics of the controlled element so that the open-loop characteristics of the human-ma-

chine system acts like an integrator in series with a time delay in the frequency range most critical to system stability and performance. The open loop for Fig. 17.6.1 is $Y_p(j\omega)Y_c(j\omega)$. This behavior, most noticeable in the region of open-loop crossover—e.g., where the open-loop gain is unity—is known as **McRuer's rule** and is summarized in Eq. (17.6.5) where τ is the effective time delay.

$$Y_p(j\omega)Y_c(j\omega) \doteq \frac{\omega_c e^{-j\omega\tau}}{j\omega} \qquad (17.6.5)$$

CROSSOVER FREQUENCY

Two examples where the conditions on controlled-element dynamics readily obtain are for automobile heading control at low to moderate speeds where $Y_c(j\omega) = K_c/j\omega$, and for attitude control of a spacecraft with damper off where $Y_c(j\omega) = K_c/(j\omega)^2$. The term ω_c, is the crossover frequency which is defined at the condition where the open-loop magnitude is unity. This frequency is important in the design of closed-loop control systems because many of the basic qualities of feedback systems relate to it. Thus, in the range of input frequencies from very low up to and including the crossover frequency ω_c, the output of the closed-loop system (m in Fig. 17.6.1) follows the system input (i in Fig. 17.6.1) closely, and the system error (e in Fig. 17.6.1) is reduced. At increasingly higher frequencies beyond crossover, these properties are lost as the error increases and the input and output of the system are no longer similar. For desirable closed-loop performance, the crossover frequency must exceed the largest frequency in the input or in the disturbances to the system for which there is appreciable power. McRuer's rule also permits simple calculations of system stability from the two-parameter model in the vicinity of crossover. Phase margin, $\phi_M = (\pi/2) - \omega_c\tau$. System performance can be estimated by use of this model. The conditions are that the input can be approximated by a nearly flat spectrum of bandwidth ω_i and variance σ_i^2, and that ω_i/ω_c be somewhat less than unity. Under these conditions, the relative mean-square error coherent with the input can be expressed as

$$\frac{\overline{e_i^2}}{\sigma_i^2} \doteq \frac{(\omega_i/\omega_c)^2}{3} \qquad (17.6.6)$$

The total mean-square error will be somewhat larger since it includes an operator-induced remnant as well.

17.7 AUTOMATIC MANUFACTURING
by Vincent M. Altamuro

The production of many products involves both **fabrication** and **assembly** activities. Fabrication is the making of the component piece parts that are later assembled into the final product. Fabrication methods include casting, molding, forging, forming, stamping, and machining. Assembly may be manual or automatic. Automatic manufacturing can range from **semiautomatic** to **fully automatic**, depending on the number of human operations required. Automatic manufacturing installations may also be called **fixed** and **flexible** according to how easily they can be altered to make product variations. These several available modes allow the creation of a mode that draws on all of them—the **autofacturing** mode, that combination of the manual and the specific types of automatic operations which best achieves the quality, operational, and economic objectives sought.

The selection of assembly mode must be consistent with the design of the product, the personnel skills, the equipment available, and other factors. Manual assembly is suitable for low volume output and relatively short production runs with high product-to-product variations.

Fixed automatic assembly is suitable for high-speed, high-volume production of uniform products. Also called *hard, dedicated,* or *conventional* automation, it can produce low-unit-cost items of uniformly high quality once its machinery is perfected and until a product variation is wanted. If a product's assembly cost is the sum of its *setup cost* (the cost of getting everything ready to make it) and its *run cost* (the cost of making it), then manual assembly has a relatively low setup cost but a high run cost, while fixed automatic assembly has the reverse. Flexible automatic assembly is suitable for midrange-volume production with variations in product specifications. Also called *soft* or *programmable* automation, it uses robots and other computer-based equipment that are more easily movable and adjustable so as to reduce setup costs and still enjoy low run costs. It fits between fully manual and fully fixed automatic assembly, as shown in Fig. 17.7.1.

Autofacturing is a production system that is comprised primarily of automated equipment which is configured as several integrated subsystems, using one common database and computer controls to make, test, and transport specifically designed products at high and uniform quality levels meeting flexible specifications with a minimum of human effort. There are many levels of autofacturing from individual cells, or *islands,* all the way up to a complete and integrated system. Most situations are somewhere in between, but progressing toward a total system.

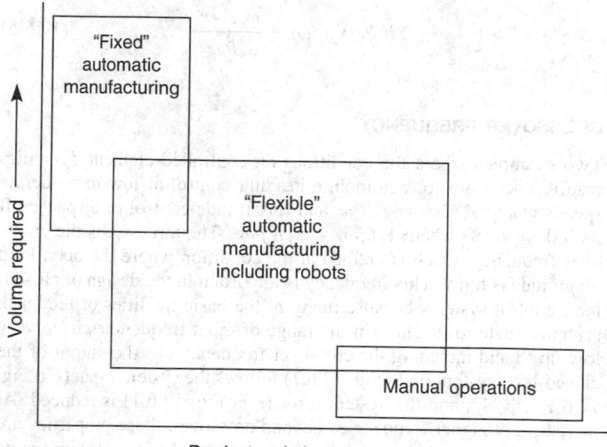

Fig. 17.7.1 The most appropriate assembly mode, as a function of output volume and variations.

Autofacturing is a subset of the overall field of automation. It refers to automated manufacturing activities as distinguished from the many other things called automation, such as data processing, office automation, electronic funds transfers, automated banking, central station telephone operations, airline ticket reservations, and the like.

An autofacturing system is composed of several integrated subsystems. Also, along with other systems such as marketing and distribution systems, general accounting and cost accounting systems, management information systems (MISs), and others, it is a subsystem of the corporation's overall operating system. All subsystems should be interconnected, complementary, in balance, and integrated into one total system. Some of the prerequisites and subsystems of an autofacturing system are listed and described below.

Design for Assembly

After management decisions regarding organization, staffing, and product characteristics have been made, the first engineering step is to design the product for ease and economy of assembly. In designing a product, decisions must be made regarding which portions of it should be assembled by people, hard automation, and programmable devices, since the shape and features of its parts will often be different for each mode.

Human assemblers, for example, have the advantage of three-dimensional vision, color sensitivity, two hands, eye-hand coordination, the facility to jiggle parts together, the ability to back off when things do not go together as they should and look for the reason rather than force them, the ability to pick out patterns which are varied or complex, etc. However, they cannot work in dangerous environments, apply large or exact forces, perform with precision consistently cycle after cycle, or handle very large, very small, or very fragile items, etc.

Hard automation requires the input of a steady, voluminous stream of very uniform parts, all positioned and oriented precisely so that the high-speed machines can operate without jamming or stopping. Machines and instrumentation are better than humans in monitoring and responding rapidly and consistently to many stimuli simultaneously—some beyond the range of human capabilities. They also can be made to do several things at once for long periods of time with little variance or deterioration of performance.

Soft, or programmable, automation permits variations in inputs, as it can sense and adjust for deviations and continue to operate. It is usually faster than manual work but slower than dedicated machines, and its output can permit variations from the standard product so as to achieve marketing and inventory advantages that often more than offset its added cost per unit of output.

In autofacturing, each assembly operation is analyzed to determine whether it is best done manually, with fixed, or with flexible automation. An autofacturing system, then, results in a mixed mode of operation, each with that portion of the product designed to be assembled in the most efficient way available. In it, the actions of conventional machines, automated equipment, human operators, and robots are all interrelated and in balance.

There are over 100 guidelines on how to configure a product and its component piece parts so that they can be assembled easily and economically as possible. The most important of these rules are as follows:

1. *Minimize and simplify.* Reduce the number of piece parts as much as possible. Determine the essential functions of each part. Transfer functions to other parts. Combine parts. Eliminate as many parts as economically feasible. Simplify before automating.

2. *Modularize.* Design the product such that parts are grouped into modules or sections of the end product. Create modules, like building blocks, so that they can be selected and joined in various ways to make different products. This will aid in assembly, test, repair, and replacement of more manageable subassemblies. It will also make it possible to follow rule 3 below.

3. *Create families of products.* Products should be designed so as to have as much commonality with other products as possible in their modules and piece parts. Products related in a series of escalating capabilities or features can all be built with the same platform (base, power supply, etc.), case, panels, circuit boards, and other common and interchangeable parts. Product distinctions, such as extra features, should be the last components assembled so that the work in process is common for as long as possible. Parts that make a product distinctive should be grouped into the same module. Even seemingly different products can be designed to have the same power supply, switches, gears, and other internal components and modules.

4. *Design parts to have as many different uses as possible,* depending on how they are installed and which portion of them is used. Determine whether the extra cost of the added complexity is more than offset by the benefits.

5. *Use the best overall method to make each part.* That is, consider not only its fabrication costs, but the relative costs of handling, subsequent operations, assembly, inspection, scrap, etc.

6. *Design in layers.* Where possible, design the product so that it is made up of layers of components, such that making it requires the adding of one layer on top of another, built up from the base to the cover or outer case. See Fig. 17.7.2.

7. *Assemble in short, straight vertical strokes.* If rule 6, above, is observed, then this should be possible. Avoid the need for lateral, curved, or complex motion paths in assembling components. Minimize the number of directions of assembly. Minimize the amount of lifting, rotating, and other handling of the components, subassemblies, or product. See Fig. 17.7.2.

8. *Present parts appropriately.* Bring parts to the machines, robots, and human operators in the position and orientation best suited to use without the need for rehandling or repositioning. Hold them in trays, on reels, in connected strips, etc. rather than disoriented in a tote box or bin. Parts stamped from strips or molded should be left on their webs as long as possible, for they are orderly. Parts to be fed to assembly stations by vibratory bowl feeders should be designed so as to feed through easily and not jam, snag, tangle, nest, shingle, bridge, or interlock.

9. *Design for symmetry and orientation.* To the degree possible, design parts to be symmetrical so that no matter what position they are in, it is proper for assembly. If perfect symmetry is not possible, strive to have the parts go together in as many ways as possible. Where that is not possible, accentuate the asymmetry and give the parts an orientation feature, so that a person or robot can easily sense it and orient the part correctly. See Fig. 17.7.2.

10. *Make errors impossible.* Design the product and its parts so that the components physically cannot be assembled wrong. If a part has internal or hard-to-sense features, add a feature that can be sensed, even if it has no operational function. Do not have the dimensions of a part be too similar if the part's proper orientation and assembly depend on discerning a small difference.

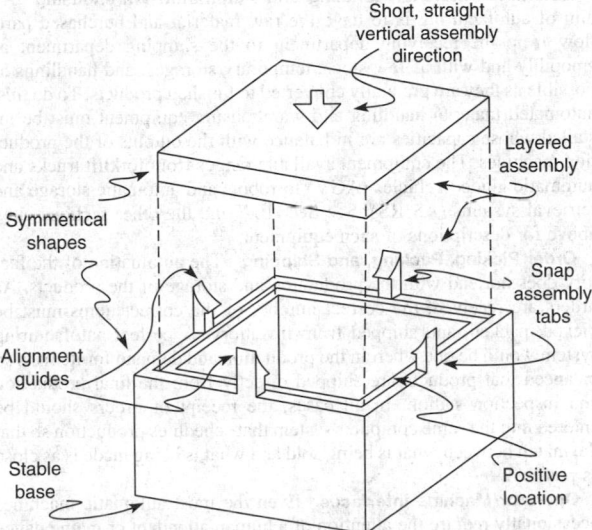

Fig. 17.7.2 Selected design for assembly guidelines.

11. *Give each part a positive location.* Through the use of shoulders, recesses, guides, tabs, and the like, create an exact place where each component is to go. See Fig. 17.7.2.

12. *Minimize fasteners.* Eliminate or reduce to the lowest essential number the product's discrete fasteners. Nuts, bolts, screws, and the like are difficult and expensive to handle and assemble. They should be used only where future disassembly, adjustments, or other needs make them necessary. Very short screws, bolts, and rivets are more difficult to position, as the weight of their heads often makes them fall out of the holes. It is more difficult to place the blade of a driver in screws and bolts with round heads and slotted tops than in those whose tops are flat with closed patterns for the driver's matching blade tips.

13. *Snap in.* Where possible have the parts snap together instead of being held with discrete fasteners. Plastic tabs can be added to a part while it is being molded at negligible extra cost. The operations of placing the part in position for assembly and of assembling it become a single action when it is snapped into place. See Fig. 17.7.2.

14. *Reduce variations.* Where fasteners must be used, have as few variations as possible. Seven or eight different screw sizes can often be reduced to two or three. The added cost of overdesign where a slightly larger size than needed is used is more than made up in purchasing, handling, inventory, record keeping, and tooling savings.

15. *Chamfer.* Assembly can be speeded up and less expensive tooling can be used if holes are chamfered and pins are slightly pointed and rounded. See Fig. 17.7.3.

16. *Avoid problem parts.* Flimsy, amorphous, and long flexible parts such as wire, fabric, insulation, and the like are difficult to handle and should be avoided or provided for in the process. Some coil springs tangle because of their helix and wire diameter or if they are open at their ends.

17. *Help the robot.* Give different parts a common feature so that the same gripper can be used to handle them all. Changing grippers takes time. Extra or complex grippers add to expense. Allow space around components so that the gripper can get in to assemble the part. Design the grippers to fit the contours of the parts or to wrap around them rather than merely hold them with pressure and friction. This will minimize slippage and movement.

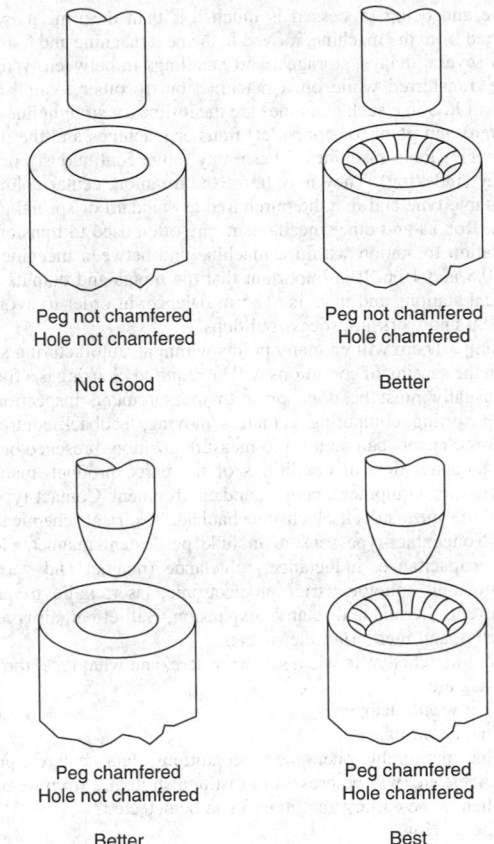

Fig. 17.7.3 The use of chamfers to accommodate slight misalignments.

Autofacturing Subsystems

Material Holding, Feeding, and Metering In the progression from raw materials and purchased parts to the finished product, the first autofacturing system subsystems are those whose functions are to hold, feed, and meter the correct amount of input to the operating stations. In designating the holding system, the material must be described in detail. Then the required holding status must be defined. Is it to be held in bulk form or as discrete items? Is each item unique or the same as all the others? Must each item be segregated from the others or may they all be commingled at this stage? Then, the best-suited holding devices must be specified, e.g., bins, hoppers, reservoirs, reels, spools, bobbins, trays, racks. Subsequent to holding, the material or parts are to be fed into the next stage of the process. Again, a series of determinations must be made. Are parts to be fed individually or en masse, in metered amounts or at any rate, separated or connected, oriented or unaligned, selectively or randomly, pretested or untested, interruptible or nonstop? Then the most suitable feeding and metering devices must be specified, e.g., dispensers, pumps, applicators, vibratory feeders, escapements. See also Sec. 10.1.

Operations The operating stations of the system include those machines, tools, and devices that convert the raw materials to parts or finished products, or bring parts together and assemble them into finished products. It is at these stations that the casting, molding, bending, machining, soldering, welding, riveting, bolting, snapping together, painting, and the like are done. The operators of the machines may be humans or robots or the machines may be designed to operate automatically or under the direction of a computer.

Transferring When more than one operation is required, means of transferring the work in process from one machine or station to the next must be provided. In many companies, the time that material is on a machine and being processed is much less than the time it is being unfixtured from the machine, moved to the next machine and fixtured in it, with several delays, storages, and handlings in between. Work can also be transferred while on a machine, but to other tooling or test stations of it. Some such machines are the in-line (or straight line) pallet (or platen) and plain (or nonpallet) transfer machines and the dial (or turret) rotary table machines. These may move continuously or intermittently (indexing). They may be of the trunnion, center column, or shuttle-table type and may be purchased as standard or special custom designs. Robots and other mechanisms are often used to transfer work from station to station within a machine and between machines. See Secs. 10.5 and 10.6. It is important that the inputs and outputs of the individual stations and machines be in balance, in order to avoid idle equipment and work-in-process buildups.

Sensing There will be many points within an autofacturing system at which the sensing of conditions will be required. Sensing is a function which usually must be done prior to measurement, inspection/test, counting, sorting, computing, actuating/moving, feedback/control, and correction. Sensors can be used to measure position, presence or proximity, characteristics, or conditions of the parts, product, machines, tooling/fixture, equipment, people, and environment. Contact-type sensors include mechanical, electromechanical, electrical, chemical, and air jet. Noncontact-type sensors include permanent magnet, electromagnet, capacitance, inductance, reluctance (magnet and variable), sonic/ultrasonic, photoelectric/optoelectronic, laser, radio frequency, eddy current, thermal and relative expansion, Hall effect, air (pressure and fluidic), and inertial/accelerometer.

In deciding whether to use a sensor, where, and what type, the questions to ask are

1. What *should* happen?
2. What *could* happen?
3. What measurable phenomena, conditions, data, features, properties, changes, etc. will be present to distinguish item 2 from item 1?
4. When is the earliest that item 3 can be detected?
5. Where? How?
6. What is the best sensor or combination of sensors to do this?
7. What should be done with the information sensed?

Measurement, Inspection, and Test In autofacturing, it is essential that the product be inspected while on the machines and while it is being made. The objective is to make only good products. To do this, critical parameters are sensed and measured in real time. If they begin to drift beyond preset statistical limits, error signals are fed back and adjustments in the machine are made automatically to bring the variables back in control before any bad products are made. It is important, therefore, that the product be designed so that there is access to inspection and test points during processing. It also means that the machines should include sensors and inspection stations amid the processing stations.

Equipment Monitoring and Performance Maintenance In addition to the product, all of the machines and equipment must be monitored constantly to assure continued peak performance and to avoid breakdowns. Again, a multitude of sensors, clocks, and counters is a must for the measurement, feedback, correction, and control of the status of the system. For tooling, where wear is certain, automatic tool changers that pull the used tool out and replace it with a new one can be used. An operator or maintenance person then removes the used tool from the device and loads it with another new one so the tool changer is ready again. In some cases, duplicate modular workstations are kept on hand so that a malfunctioning one can be replaced rapidly.

Sorting, Counting, and Marking After the product is made, its characteristics can be sensed and measured and it can be sorted according to whatever decision rules are set. Deflection devices—such as conveyor belt gates, air jets, and the like—can be used to divert each class of product to its own receptacle. They can then be counted (with the data going to the computer) and marked or labeled, as desired.

Wrapping and Packaging The wrapping and packaging of the product is just as much an integrated part of the overall system as is assembly, and should be just as automated. Even the packing cartons should be designed to facilitate automated operations. Instead of conventional flaps that tend to close and block the robot's path in trying to put products in, boxes should have straight sides and a lid so that both the products and the lid can be handled in straight, vertical motions.

Automated Material Handling and Automated Warehousing An aim of autofacturing is to have the raw material and purchased parts flow from the receiving department to the shipping department as smoothly and with as few stops, temporary storages, and handlings as possible as they are gradually converted to finished products. To do this, automated material handling and warehousing equipment must be installed whose capacities are in balance with the outputs of the production machines. The equipment available ranges from forklift trucks and automatic guided vehicles (AGV) to robots and automatic storage and retrieval systems (AS/RS). See Sec. 10.7 and the other sections cited above for descriptions of such equipment.

Order Picking, Packing, and Shipping The automation of the factory does not end with the production and storage of the products. As orders are received, the correct number of the correct items must be picked, packed, and shipped from inventory. A perfect autofacturing system would be one wherein the production and demand for items is so balanced that products are shipped directly from the final production and inspection station. In all cases, the receipt of orders should be entered into the same computer system that schedules production so that the match between what is being sold and what is being made is as close as possible.

Operator/Machine Interfaces Even the most automatic machines occasionally require the attention of a human attendant or maintenance person. The fields of study that examine the human/machine interfaces and the degrees of effort and responses of people at work are called *human engineering, human factors engineering, biomechanics, ergonomics,* and *cybernetics.* While each concentrates on a particular aspect of the subject, they are all concerned with matching the needs and capabilities of humans to the designs of the machines (and total environment) with which they interface. Humans control machines via buttons, knobs, handles, switches, keyboards, voice recognition devices, and the like, and these must all be designed, colored, located, etc. to minimize human effort and maximize effectiveness. Machines output information to humans via lights, sounds, dials, gages, video screens, and the like. These, too, must be designed to maximize response and minimize errors. At all times, the human/machine interface must be designed with safety in mind. See Sec. 17.6.

Communication, Computation, and Control The system's robots and other equipment have a multitude of microprocessors embedded in them. That, and the fact that they also have many sophisticated sensors justifies calling them "smart" machines. They, in turn, are controlled by more powerful microcomputers that report to minicomputers, thence to a powerful central host computer. The host computer does not control each detail of the system; it merely integrates and balances a distributed and hierarchical communication, computation, and control capability that uses the same database and is linked by a local-area network (LAN).

Auxiliary Systems There are several ancillary systems that can be employed to enhance automated manufacturing and the autofacturing system. Some of these are group technology (GT), concurrent engineering or simultaneous engineering, and computer-aided design and computer-aided manufacturing (CAD/CAM) in the product design and prototyping stages; just-in-time (JIT) delivery, material requirements planning (MRP), and manufacturing resources planning (MRP II) in the production planning and inventory control stages; computer-aided testing (CAT) and total quality control (TQC) in the production, inspection, and testing stages; computer-integrated manufacturing (CIM) in the communication, computation, and control stages; and management information systems (MISs) in the reporting and analysis stages.

Section **18**

The Engineering Environment

PASCAL M. RAPIER *Scientist III, Retired, Lawrence Berkeley Laboratory.*
ANDREW C. KLEIN *Associate Professor, Nuclear Engineering, Oregon State University.*
PAUL E. CRAWFORD, ESQ. *Partner, Connolly, Bove, Lodge & Hutz, Wilmington, DE.*
R. ERIC HUTZ, ESQ. *Associate, Connolly, Bove, Lodge & Hutz, Wilmington, DE.*

18.1 ENVIRONMENTAL CONTROL
by Pascal M. Rapier and Andrew C. Klein

REFERENCES: "McGraw-Hill Encyclopedia of Environmental Science." Beckmann, "Health Hazards of Not Going Nuclear," Golem Press, Boulder, CO. Edinger et al., "Heat Exchange and Transport in the Environment," P.B. 240757 NTIS, USDA, Nov. 1974. Corbitt, "Standard Handbook of Environmental Engineering," McGraw-Hill.

PRINCIPLE OF ENVIRONMENTAL CONTROL

Nature has provided two almost inexhaustible sumps for maintaining a steady state environment on earth. The first of these is the 2.7 K background temperature of absolute space, which nature uses for heat rejection to close its heat balances. The second is the oceans, which serve to close the material balances of its cyclic processes by accepting the combined runoffs of the continents. The greatest engineering progress comes when people control their environmental activities so as to take maximum advantage at minimum cost of these sumps and of nature's cyclic processes. This is the **basic principle** upon which the science of **environmental control** is founded. With it, any environmental problem can be modeled on a conveniently small scale and then solved precisely by LeChatelier's controlled equilibria principle for any large-scale operation. When the principle is applied to obtain optimum yield of a desired end product or result, it will help predict the necessary input compositions, pressure, temperature, and concentration. This same principle is applied in other diverse fields of study such as genetics.

NATIONAL POLICY

The National Environmental Policy Act of 1969, Public Law 91-190 (NEPA), concerns air, water, and land quality and conservation of natural resources. Section 2 of NEPA declares a national policy which will encourage productive and enjoyable harmony between people and their environment and which will enrich their understanding of ecological systems and natural resources.

Section 101(b) (2) provides for the "widest range of beneficial uses of the environment without degradation, risk to health or safety, or other undesirable consequences."

Section 101(b) (5) provides for achieving "a balance between population and resource use which will permit high standards of living" (e.g., cost/benefit balancing).

Section 101(b) (6) provides for enhancing the quality of renewable resources and the maximum attainable recycling of depletable resources (e.g., developing breeder reactors to extend nuclear fuel reserves).

Section 102(c) requires that all agencies of the federal government shall file environmental impact statements. These detail in a published document for each case: (a) its adverse environmental effects, (b) its alternatives, (c) its enhancement of long-term environmental productivity, and (c) its irreversible and irretrievable commitments of natural resources.

Legislation most responsive to the NEPA objectives so far is the Magnetic Fusion Energy Act of 1980. Aiming for a working magnetic fusion system by the year 2000, this act set as a goal a controlled nuclear fusion economy by the middle to late twenty-first century. (See Considine, "Energy Technology Handbook," McGraw-Hill.)

AN OVERVIEW OF ECOLOGY
(See also Sec. 9.1)

Ecology is the study of the **biosphere,** a hypothetical system comprising the surface of the earth and all the subsystems necessary to maintain a steady-state life-support system. The **thermal cycle** is open, isenthalpic, and irreversible. Solar energy is utilized, becomes degraded, and is finally rejected into outer space by means of long-wave radiation through the optical window of the earth's atmosphere. Major quasicyclic processes in the biosphere, necessary for life support, are the **hydrological cycle,** the **carbon cycle,** the **nitrogen cycle,** the **potassium cycle,** the **phosphorous cycle,** and the **sulfur cycle.** These are almost closed cycles, which are occasionally interrupted and renewed by techtonic geological upheavals.

In these cycles, vegetation and animal wastes decay, thereby furnishing nutrients to land, air, and water. New green plants grow, animals eat them, and the recycling continues. Ecologically, the bacteria which decompose all these wastes are called *destroyers,* the green plants which employ photosynthesis are called *producers,* and animals which eat the plants are called *consumers.* A hypothetically bounded small portion of the biosphere is called an **ecosystem.** All ecosystems are **open,** as their essential cycles extend into the biosphere.

An ecologically idealized community would be one in which the ecosystem, including the surrounding agricultural area, would operate steady-state, almost closed cycle. Tucson, Arizona, approaches this state, its crops being irrigated by tertiary sewage treatment so that water and all the plant nutrients, are recycled. Minerals entering the ecosystem through the Santa Cruz River and imported fuels are almost counterbalanced by exports of crops from the area and by irrigation runoff into the normally dry Gila River bed. These activities conform to nature's naturally occurring cycles, and maintain the fertility of the soil.

In larger, heavily overpopulated communities, however, people's activities are counter to nature's, and so humanity suffers. The earth is denuded of local vegetation, and crops are imported. All the nitrogen, potassium, sulfur, and phosphorus contained in the agricultural commodities and fuels, instead of being recycled to the farmlands, are burned or handled by the waste disposal systems of the communities and dumped into rivers and harbors, thereby reaching the oceans and being lost to the continents, despite the recycling and soil fixation methods intended by nature to prevent such losses. Severe pollution problems result, mainly from nonpoint sources and storm run-offs which are not amenable to control, no matter how much money is spent. But the major damage done is depletion of wetlands, forests, and farmlands and buildup of atmospheric CO_2 from fossil-fuel use. It is irreversible and irretrievable and will continue until our society finally faces the fact that the changes required to help stem these events must be made on a local as well as planetwide level.

Environmental control seeks to subdue and to utilize nature's ecological cycles in order to serve people's needs, thereby conserving natural energy and mineral resources, and to replenish desirable local flora and fauna populations by agriculture and cultivation to provide adequate food, clothing, and shelter. Environmental control seeks to extend depletable fuel supplies with clean, abundant forms of gravitational, solar, and nuclear energy (fission and fusion). Also, replenishable substitutes for other depletable resources are sought, as well as recovery and recycling means for scarce and irretrievable substances. Environmental control seeks to conserve land and water quality by diversion into adequately controlled drainage canals of concentrated runoffs, e.g., from irrigation, municipalities, industries and mining, so that these nonrecyclable, unsalvageable mixtures are ultimately discharged into the oceans, where they blend inconsequentially with the vastly larger amounts of runoff from the continents, occurring naturally as part of the hydrological cycle. The public standard of living is highest when all these things are done voluntarily by a responsible, cost-conscious citizenry.

The following subsections describe typical, cost-effective, and productive practices among the free nations. These should be considered to the extent permitted by local regulations.

COST/BENEFIT BALANCING AND ISO 9000 REGISTRATION

REFERENCES: Summary of UWAG-EEI Comments on EPA's Proposed Regulations under 304, 306, and 316(a) of Public Law 92-500, the Federal Water Pollution Control Act Amendments of 1972, June 1974. Effluent Limitations Guidelines for Steam Electric Power Generating Point Source Category 40 CFR, Chapter I, Subchapter N, Part 423. Craig, "The No Nonsense Guide to Achieving ISO 9000 Registration," ASME.

Cost/Benefit Calculations

In order to assess the feasibility [under Section 304b (1B) of the FWPCA Amendments of 1972] of any environmental control technology or augment, both the public cost and benefit of its operation must be considered. For demonstrating feasibility, this dimensionless cost/benefit ratio has the limits

$$0 \le C/B \le 1$$

The upper limit of unity means that $1 spent in the incremental public operational costs of the augment will yield $1 of return in public benefits. If an augment is required for which the cost/benefit ratio is greater than unity, it is unfeasible. (It is more feasible to accept the pollution.) If an augment is required for which the cost/benefit ratio is negative, the public benefit is negative and constitutes a public detriment. The more that is spent on such an augment, the worse the environmental degradation becomes.

EXAMPLE.   Once-through vs recirculating cooling systems: A 1,000-MW(e) steam/electric generating plant is operating with once-through cooling on a river. The heat rate is 10,300 Btu/kWh (0.33 thermal efficiency), with about one-half of the heat, or 5,150 Btu/kWh, being discharged into the river. The public benefit of this thermal enrichment is a stable and abundant fish population, which is estimated to offset all public detriments except occasional fish kills caused by emergency shut-downs. These average each year not more than 8-h duration and $45,000 public detriment.

Proper environmental management to provide emergency heating of the water during unexpected outages will avoid the fish kills with an estimated 90 percent confidence factor. A cooling tower retrofit will reduce the thermal discharge by 90 percent but will substitute some undesirable atmospheric, noise, and stream pollution conditions. Neglect these problems, but otherwise compare the feasibility of an open-recirculation cooling tower retrofit with that of proper environmental management.

First alternative: Environmental management of existing plant. The annual public detriment for the fish kills is $0.045/kW(e) of installed capacity. For avoiding these by emergency heating of the river water, let the estimated operating cost to the consumer be $1 per million Btu. The annual environmental management cost for an 8-h outage per year becomes:

$$8 \text{ h} \times \$.00515/\text{kWh} = \$0.0412/\text{kW(e)}$$

The cost/benefit ratio becomes $0.412/$.045 or 0.916, so that 91.6¢ spent will yield $1 return in public benefits. This is a wise investment. For the existing plant the possibility of a fish kill is reduced to one-tenth its original value, and the remaining annual public detriment becomes $0.0045 per kW(e) of installed capacity.

Second alternative: Retrofit of an open recirculating cooling tower. The viability of a cooling tower retrofit must be assessed on the basis of this reduced public detriment for the existing plant. The estimated annual operating cost to the public for a cooling tower augment, based on public operational overhead, a 1970 overall plant investment, and a 3 percent incremental capacity loss, due to the retrofit, is $20/kW(e) of installed capacity. Dividing this by the $0.0045 expected public benefit yield a cost/benefit ratio of 4,444. This means that $4,444 must be spent (and the cost passed on to consumer) for every dollar of public benefit. As this is clearly not in the public interest, it is unfeasible. It is cheaper to replenish the fish supply. Moreover, the 3 percent capacity loss of power generation due to the augment results in a 3 percent wastage of fuel and a 3 percent increase in the total thermal discharge to the environment. Energy, in contrast to the fish population, is a nonreplenishable natural resource. It must be conserved, and it must be considered when assessing the public detriment. Valued at 3 percent of the heat rate, $1 per million Btu and 0.9 load factor, this waste of natural resources will cost $2.435 per kW(e) annually.

Loss of cooling water by evaporation in the cooling tower is also a public detriment that can be conservatively valued at 25¢ per 1,000 gal. At 5,150 Btu/

kWh thermal discharge rate and 0.9 evaporative load factor, this depletion of a natural resource will cost:

$$0.9 \times 5,150 \times \$0.25 \times 24 \times 365 \div 8.33 \times 10^6 = \$1.219/\text{kW(e)}$$

Subtracting these two detriments, amounting to $3.654, from the annual public benefit of $0.0045 for conserving the fish population yields a negative public benefit due to the augment of −$3.65. The overall cost/benefit of this installation therefore becomes

$$\$20/(-\$3.65) = -5.48$$

The negative sign indicates that money spent to improve the environment by this means simply makes matters worse. The expected public benefit is not possible, and therefore the technology is not economically achievable.

Achieving and Practicing ISO 9000 Registration

Both capital and operating costs of a new or reorganized operation or venture can be minimized through the mechanism of ISO 9000 registration. In this way parts, machinery, construction, management, and operating and servicing techniques are standardized locally, and ultimately internationally. The favorable economics that ensue are of immeasurable consequence in a competitive marketplace and contribute to the optimum utilization of labor, facilities, and raw materials. A large portion of the benefits which accrue are related to personnel involved in the business entity.

Direct Water Cooling

For large-scale power generation, natural surface cooling in open water bodies is most cost-effective. The intake and outfall should be so oriented as to minimize possibility of heat recirculation and short-circuiting, which can usually be done by locating the outfall sufficiently down current of the intake.

Surface discharge of the condenser water as a hot, buoyant plume requires the least area of water surface for heat dissipation, yields the largest value for the overall surface heat transfer coefficient K, and hence minimizes both the cost and environmental impact of the facility.

The surface heat-transfer coefficient K combines the heat loss contributions due to longwave radiation, evaporation, and convection into a function which can be reliably estimated by two independent equations (J. E. Edinger, 1974):

$$K = H_s(T_s - T_{dp})^{-1} \tag{18.1.1}$$

$$K = \frac{T_s - 183.16}{20} + (0.47 + B)f(W) \tag{18.1.2}$$

where K = overall surface heat-transfer coefficient, W/(m$^2 \cdot$ K); H_s = gross 24-h average sunshine factor, W/m^2; T_s = water surface temperature, ambient, K; T_{dp} = dew point temperature, K; W = wind speed, m/s; $f(W)$ = wind speed function, $9.2 + 0.46W^2$ (Brady, 1969), W/(m$^2 \cdot$ mmHg); B = slope of vapor pressure curve at $T_m = \frac{1}{2}(T_s - T_{dp})$, mmHg/K.

The derivative of the 1967 ASME steam table saturation line B is given accurately over the temperature range 0 to 50°C by the Rapier log-hyperbolic curve fit

$$B = \frac{5,301.734}{T_m^2} \exp\left(20.94673 - \frac{5,301.734}{T_m}\right)$$

The overall heat transfer coefficient K can be most accurately obtained from Eq. (18.1.1), since all the variables can be directly measured; then the wind speed function $f(W)$ can be determined for the particular site by substitution of K into Eq. (18.1.2). On the other hand, if T_s is not known, both T_s and K can be approximated by iteration between the two equations, after adopting a reasonable wind speed function, such as the Brady parabola, given above.

Once the value of K has been determined and the temperature rise through the condensers has been decided upon, the initial temperature difference or excess above ambient lake surface temperature at the discharge point, and the final temperature difference or excess above ambient at the plant intake, can be readily calculated, since the effective area A_e in square metres from the discharge point up to the intake determines the log-mean temperature difference, as given by the equation

$$\text{LMTD, °C} = \frac{\text{thermal discharge rate, W}}{KA_e}$$

At locations where limited quantities of water are available, some heat recirculation occurs and, particularly during summer months, can cause temperatures in excess of 32°C at the inlet of the plant. When considering such a location, tradeoffs should be made to determine whether it is more economical to accept a higher temperature at the inlet or to install auxiliary cooling means such as towers or spray ponds.

EXAMPLE. During July, a proposed inland lake site has a gross sunshine factor of 387.3, a dew point of 13.9°C, and a wind speed of 4.07 m/s. Iteration yields 26.1°C for the water surface temperature and 31.8 W/(m² · *K*) for the heat-transfer coefficient. If thermally loading the lake will cause a temperature excess of more than 8° at the intake, tradeoffs should be considered here.

CONTROL OF THERMAL DISCHARGES

(See also Sec. 9.5.)

REFERENCES: "Water Resources and Waste Heat," *Power Eng.,* pp. 26–33, June 1973.

The total solar flux reaching the earth is 182 trillion kW; 32.4 trillion kW is utilized in the hydrological cycle, whereby freshwater distilled from the oceans is utilized on the continents by the plant and animal kingdoms and finally drains back into the ocean to complete the cycle. The entire thermal demand estimated for the United States in 1980 was 5 billion kW, very small compared with the solar flux in the earth's biosphere. The temperature of the earth's biosphere is stabilized by reradiating the solar flux as long wave radiation back into absolute space through the "optical window" (7- to 13-μm wavelength) of the earth's atmosphere. For example, if once-through cooling for 25,000 MW(e) power generation were added to Lake Michigan, the average surface temperature rise would be limited to ¼°F by longwave radiation from the surface of the lake.

The Joint Committee on Atomic Energy, 93d Congress (1973), anticipated that by 1980 the total heat dissipated into the environment by the generation and use of energy in the United States will be equivalent to 43.2 million bbl of oil per day. The combined heat loss caused by electric and industrial power generation will be 11.5 million bbl per day, or 26.6 percent, which includes all the point sources. Distributed and diffuse sources, such as automotive, residential, commercial, and industrial heating and cooling systems, which constitute the balance, account for 73.4 percent of the heat dissipation, and create "heat islands" around every major metropolis. About half the heat loss from power generation is presently discharged into bodies of water via nonconsumptive once-through cooling systems and does not contribute to the "heat island." The use of consumptive evaporative cooling for this would add to the atmospheric discharges in the "heat island" and raise the ambient temperatures. By 1978, increased urbanization in the Los Angeles Basin will raise the ambient temperature an estimated 5°F. The surface temperatures of water bodies in the Basin would be raised over 5°F, which exceeds guideline limitations for thermal discharges. By the year 2000, if evaporative cooling towers are required for all the incremental power generating capacity, the New York–Philadelphia "heat island" will suffer a mean ambient temperature increase of about 15°F, and the cooling towers will consume twice the flow of the entire Colorado River Basin, 31,000 ft³/s, or 3,000 times their 1970 consumption of water. Not included is the "greenhouse" effect of excess CO_2 in the atmosphere from the burning of fossil fuels. An estimated trillion tons of CO_2 will have accumulated, and trapped solar radiation will have raised the earth's mean temperature 3.6°F (2 K). (See Plass, Carbon Dioxide and Climate, *Sci. Amer.,* 41–47, July 1959, pp. 41–47. Also pp. 173–175, Sept. 1983.)

For the year 2000, the percentages of total heat discharged to the environment from various human activities have been estimated as shown in Table 18.1.1.

Table 18.1.1 Heat Discharged into Environment; Year 2000 (Estimate)

Thermal discharge source	Coolant	Total heat, %
Residential and commercial	Atmospheric and water	31.75
Industrial	Atmospheric and water	27.00
Transportation (all forms)	Atmospheric	22.20
Electric power generation	Atmospheric and water	14.30
Electric power consumption	Atmosphere	4.75
Total heat into environment		100.00

The data and projections cited above are dated, and certainly the current situation is different from all projections, but only in degree. The data in Table 18.1.1 are useful in that they address a global problem. Increased use of nuclear fuel to power central-station electric generating plants reduces the heat discharged into the atmosphere from exhaust stacks, but there is almost the same amount of rejected heat dissipated to condenser cooling water or to the atmosphere in cooling towers. Likewise, there is much controversy among the polity in the matter of the greenhouse effect, ozone holes, mandated use of electric-driven vehicles, and so on. The solutions offered to ameliorate those problems are not universally acceptable, but recognition of the existence of problems in this regard is not usually contested. In certain quarters the solution prated is for so-called renewable sources of energy which are "benign" (i.e., solar energy), but the technology is not refined yet to make the extraction of solar energy competitive. Much intellectual effort and well-intentioned persuasion must be brought to bear on this and associated type problems as the population grows and the consumption of fossil fuels continues apace.

The largest thermal discharge, 31.75 percent, is primarily attributable to residential and commercial heating and ventilating systems, with air conditioning units making the most undesirable contributions to the environmental "heat island" effect. Cogenerative district heating, better thermal insulation, and reduction of electrical loads (such as by replacing electrical heating units with steam or gas-fired units and atmospheric air conditioners with once-through, water-cooled units) could substantially reduce the undesirable side effects.

The second largest thermal discharge is from industry. Because of keen marketplace competition, thermal processes have been selected to provide the most favorable cost/benefit ratio for the consumer. Regulatory control here, except to prevent malpractices, can do more public harm than good. Industrial/nuclear power complexes, where practicable, will produce the maximum public benefit.

Transportation is the third largest contributor to the "heat island" effect and the one with the worst side effects. Thermal discharge to city streets could be reduced to one-third its present value and petroleum stocks conserved if all-electric transportation could be substituted economically for the internal combustion engine. This would be particularly advantageous in the summertime.

The smallest contributor to the "heat island" effect is electric power generation, which accounts for 7.15 percent of the total atmospheric discharge, with another 7.15 percent thermal discharge to water. Irrigation canals, navigation canals, and artificial waterway complexes designed to include treated municipal effluents could, if implemented in the future, provide sufficient water for a significant number of thermal power generating units, save energy, and reduce the atmospheric discharge near cities by as much as 7.15 percent.

In short, the environmental impact of heat discharged by human activities can be minimized and fuel conserved by cascading processes to get the highest practicable thermodynamic efficiencies and the lowest practicable heat-sink temperatures. Greater efforts must be brought to bear in the matter of curtailment of CO_2 emissions from all sources, inasmuch as they are recognized to have a deleterious effect in polluting the global environment as well as bringing about higher ambient temperatures. Reforestation and/or the maintenance of green spaces and the increased utilization of nonfossil fuels are looked upon with favor in these regards.

Use of Cooling Towers
(See Sec. 9.5.)

REFERENCES: Nomographs for Thermal Pollution Control Systems, EPA 660/2-73-004, Sept. 1973.

In locales where water supplies are inadequate to provide nonconsumptive once-through cooling, evaporative open-recirculating cooling towers become necessary. These consume water at about 2 percent of the recirculation rate, but reduce circulation into rivers or streams by over 95 percent even when employing low concentrating cycles. Nomographs are available from EPA which provide cost comparisons of the cooling systems sampled by Table 18.1.2. Selection of the most feasible cooling system depends on specific site conditions. Economics dictate once-through cooling where possible, wet cooling towers or ponds where water supplies are limited, and dry cooling towers where water is unavailable. All of these will contribute significantly to the "heat island" effect unless specifically designed to provide buoyant plumes that will puncture inversion layers.

Table 18.1.2   Cooling System Augments

Cooling systems	Relative capital cost	Power consumption
Natural draft wet towers	3.2	Low
Mechanical draft wet towers	1	High
Spray ponds	2.3	High
Cooling ponds	1.4	Low
Natural/mechanical draft dry towers	7.3/6.7	Medium/high

Wet cooling towers substitute water pollution for thermal pollution by concentrating blowdown, require chemical treatment, and release heavy-metal corrosion products into the blowdown stream. Their operation requires an optimized water chemistry which balances the acid treatment and cycles of concentration to provide nonaggressive water with a slightly positive Langlier index, and to retard the buildup of scale. The Langlier (saturation) index is the difference between the actual measured pH and the calculated pH_s at saturation with $CaCO_3$. A zero Langlier index balances pH, alkalinity, and calcium hardness, and establishes the threshold for alkaline scale formation. (For further information on this and other water treatments, see Betz, "Handbook of Industrial Water Conditioning," Betz Laboratories.) For environmental reasons, pretreatments based upon zinc, chromate, and phosphate inhibitors should generally be avoided, while posttreatments to remove pollutants and corrosion products from the blowdown may have to be employed. With all constraints considered, the most economical design results from a series of tradeoffs. Dry cooling towers are justified only where water shortages preclude the use of evaporative coolers, as their cost, inefficiency, and environmental impact from abnormal heating of the air are otherwise excessive.

Cooling towers may be required by environmental regulations, and "zero discharge" of blowdown may also be required, with high concentrating cycles and "perpetual storage" of the resultant salty wastes. The salty plume will cause additional difficulties. Blowdown volume can be reduced by concentrating to 10 percent solids using vapor compression. The power requirement for this is 90 kWh/kgal evaporated. Further concentration requires open kettles. (For further information, see Kolflat, Cooling Tower Practices, *Power Eng.*, pp. 32–39, Jan. 1974, and Boles and Levin, Treating Cooling Tower Blowdown, *Power*, pp. 68–76, Aug. 1974.)

WASTEWATER CONTROL

REFERENCES: The Clean Water Act of 1972, as amended through 1981, PL 97-117. Sittig, "Resource Recovery and Recycling Handbook of Industrial Wastes," Noyes Data Corp., Park Ridge, NJ. Lund, "Industrial Pollution Control Handbook," McGraw-Hill. "Pollution Control Technology," Research and Education Association, New York.

Industrial Categorization by the 1972 FWPCA Amendments (Clean Water Act)

The Federal Water Pollution Control Act (FWPCA) Amendments of 1972, Public Law 92-500, preempt all existing federal regulations by authorizing guidelines for "effluent discharges," rather than by controlling "receiving water quality."

Title III, Section 306(b) (1) (A) categorizes sources. "Source" means any building, structure, facility, or installation from which there is or may be the discharge of pollutants for which effluent limitations guidelines and revisions will be developed. These are published periodically in the Federal Register, under the 40 CFR part numbers shown by Table 18.1.3. The limitations were to be achieved by July 1, 1977, for "best practicable control technology currently available" (BPCTCA) and by July 1, 1983, for "best available technology economically achievable" (BATEA).

Both BPCTCA and BATEA guidelines must meet feasibility and achievability criteria under cost/benefit testing and will be modified periodically to reflect current needs and technology.

Pretreatment guidelines have been established by EPA under Section 307(b) to control industries with effluent discharges into municipal systems. Industrial wastes not so regulated become part of the municipal wastewater flow and thus a municipal responsibility.

Permit and Application for Industrial Discharges

The Permit Program for Industrial Discharges under the provisions of the "Refuse Act of 1899" was previously administered by the Army Corps of Engineers. The Clean Water Act of 1972 incorporated this program and transferred control from the Army Corps of Engineers to the EPA.

A Permit Program Application plus an environmental report [see NEPA-102(c) for criteria] is filed by the discharger with the state, which then refers it to the Regional EPA permit section for review. The state draws a permit that includes limitations, compliance schedules, and state monitoring requirements, making the permit forms available to the public 30 days prior to issuing the permit. Hearings on the permit may cause much delay. After this procedure is completed, the permit may be issued for a maximum of 5 years.

Environmental Considerations for Construction Projects

Introduction The environmental considerations which follow should be taken into account by industry when preparing an environmental report for construction projects, and before an effluent discharge permit can be granted. Cost/benefit factors for applicable environmental control technologies and regulations limiting effluent discharges should be obtained and included in the record of public hearings and should lead to rational decision-making based on a total short- and long-term project cost to the public consumer in relation to the worth of anticipated public benefits. (*Note:* Cost/benefit ratios for meeting effluent discharge regulations can be obtained from the comments submitted to EPA by the categories of industries listed by Table 18.1.3. Public costs generally range from 100 to 1,000 times greater than social or environmental benefits. The engineer will also find these reports a valuable source of design information for environmental projects.)

It should be recognized that the considerations identified should be individually tailored to respond to the unique guidelines for effluent discharge encountered on a particular project and in a particular state.

General Construction Projects

1. What site selection criteria will be used? Will they include cost/benefit balancing of all environmental worth factors such as the conservation of land and fuel reserves, air and water quality, noise, impacts on residents of the area, fish, wildlife, and vegetation? Will alternative sites and alternative orientations of the plant on the selected site be considered?

Table 18.1.3 Industries Categorized under 40 CFR, Chapter I, Subchapter N

Industrial groups	Part nos.	Stds. of performance
1. Food and grain processing	405–409	In general, BPCTCA 1977
2. Textiles	410	guidelines limit pH* between 6
3. Feed lots; meat products	412; 432	and 9, BOD,* TSS,* and
4. Chemical; petroleum metal-lurgy	410–424	COD,* to the ppm range, toxic substances, e.g., Ba, Pb, Hg,
5. Power generation	423	Ag, As, Cr^{+6}, F, to nontoxic
6. Leather, glass, and rubber products	425–427	concentration. BATEA 1983 guidelines limit these further on
7. Wood products, paper, and pulp	428–431	nonprocess streams, and seek "zero discharge" of process
8. Oil, gas, coal, minerals and biochemicals	434–440	streams.
9. Tars, asphalt, paint inks, and gums	443–458	
10. Pesticides, chems. and ex-plosives	455–457	
11. Photographic, hospts.	459–460	

* BOD (biochemical oxygen demand); TSS (total suspended solids); COD (chemical oxygen demand); pH (log of the reciprocal hydrogen ion concentration).

2. What disposition will be made of solid and liquid residues (ashes and chemical and industrial wastes)? Does the disposal method include adequate cassetting or neutralization to minimize the danger of soil or water pollution? What steps are planned to contain and reclaim ash dumps and chemical and oil spills to avoid pollution of surface and groundwater by runoff? What provision has been made for the disposal of waste-concentrated brines? If waste disposal into water bodies is planned, what will be the dollar value of its detriment to aquatic life? To what degree will tidal action and currents dilute plant effluents? What disposition will be made of potentially harmful and toxic corrosion products occurring in the plant effluent? Can they be injected into underground aquifers? What provision will be made for controlling the release of waste material into water bodies? If additional units are constructed, what will be the total load of waste materials? What downstream environmental effects can be anticipated with respect to humans, crops, forests, and wildlife? With what cost-effectiveness can any harmful effects be minimized by dilution and toxic materials be discharged below toxic concentrations to merge into the normal background of compositions?

3. What public benefits and detriments will thermal effluents have on the receiving waters? What temperature increase can be anticipated? Is there sufficient water motion in the receiving bodies to dissipate heat effectively? Has the use of cooling towers been explored? What is the probable cost of their public detriment, incurred by the cooling towers' consumptive use of water and energy, by producing fog and undesirable weather conditions through the dissipation of waste heat, and by producing blowdown which may require treatment?

4. What impact will the impoundment for a water supply have in terms of the destruction of agricultural, forest lands, and habitats for fish and wildlife? What measures are planned to mitigate the loss of natural habitats for fish and wildlife? To what degree will archaeological and scenic values be affected? How will the reservoir and downstream flow affect water quality parameters, i.e., temperature, dissolved oxygen, nutrients, nitrogen concentration, hydrogen sulfide, and color?

5. How vulnerable is the facility to surface subsidence, earthquakes, hurricanes, war, insurrection, and other catastrophes? What is the probable annual cost (e.g., liability insurance cost) of the environmental degradation which could be expected from such catastrophes? What preventive and remedial safeguards for downstream inhabitants will be incorporated in plant design and construction? What provisions are made for training plant operators in environmental protection?

6. Have the environmental consequences of transportation, pipeline, and power transmission been considered in site selection? What steps are planned to avoid soil erosion and the silting of streams as pipelines, transmission facilities, and access roads are constructed?

7. What provision has been made for recycling and for industrial by-product development associated with the facility? (See also Sec. 6.10.) What impact will that activity have on the environment?

Process Selection, Recycling, and Ultimate Disposal of Residual Wastes
(See also Sec. 6.10.)

In the design of an industrial installation, segregation of process and nonprocess streams, together with good housekeeping practices and diversion of storm waters will generally result in minimum volumes of water requiring special treatments before being discharged. Wastewater treatment processes involved are: chemical and physical treatments for removal of undesirable waste products; suspended and dissolved solids from process waste streams; removal of corrosion products and carbonaceous wastes (biological treatment) from nonprocess streams; oil-water separations; suspended-solids removal from storm runoffs; good housekeeping and protected stockpiling to prevent contamination of runoffs; and blending of treated effluents and impoundment to meet local end-of-pipe regulations (see Table 18.1.4).

Recycling of consumer and commercial wastes back to their sources (e.g., crankcase oil, glycol, fat, photographic and plating wastes, paper, rags, rubber, metals, and irrigational and fertilizer values of municipal wastes), following the good housekeeping practices of industry, is a very cost-effective method for reducing municipal stream pollution. However, within most industries recycling of salvageable material is almost always practiced to the point of optimum return, so that proposed benefits from further recycling in an effort to obtain "zero discharge" are usually minimal. Recycling, or "closed-cycle" water reuse schemes consume water and produce blow-down streams of concentrated pollutants in contrast to "open-cycle" schemes, whereby solids have traditionally been discharged at normal and below toxic levels, within the assimilative capacity of the receiving bodies of water. The latter, and not the former, conserves water and enhances the naturally occurring ecological cycles—provided, of course, that abnormal toxic substances are not discharged, but are removed by a posttreatment for ultimate disposal.

Descriptions, flow sheets, and costs for a variety of industrial water-

Table 18.1.4 Removal Process Guide for Selected Contaminants

Treatment processes	pH, acidic, alkaline	Suspended and precipitated solids	Bacteria, odor, organic, color	Nutrients (N, P)	Herbicides, pesticides, CN^-	TDS	Oils, solvents	Heavy metals Cr^{6+}
Chemical*	C	B	A	A	A	A, B	B	A, B
Clarification† and settling		G, D, I	D, E, H	H, D		D	F, H, D	D, I
Selected‡ biological			J, K	J, K	K		J	
Desalting§			O, M	L, N		L		L, N
Sludge¶ dewatering		P	P				O, M	P

* A, selected chemical treatment; B, coagulation; C, neutralization.
† D, selected clarification; E, aeration, oxidation; F, flotation, decantation; G, filtration; H, lagooning; I, sedimentation.
‡ J, secondary (e.g., aerobic, anaerobic); K, tertiary (e.g., carbon adsorption; extended aeration); anaerobic dentrification.
§ L, selected desalting and brine blowdown; M, membranes; N, ion exchange; O, distillation.
¶ P, dewatering (e.g., gravity, centrifugal, vacuum filtration; drying).

use and recovery schemes are found in "Water Reuse, Industry's Opportunity" (1973), published by the AIChE.

Water reuse by consumptive processes always results in a stream of concentrated residual wastes for which an ultimate disposal site must be found. Thus, for brine from saline-water conversion processes, there are four major avenues: (1) ocean discharge; (2) underground discharge into saline aquifers; (3) dry lake beds, liner protected; and (4) solidification and use for landfill. Reference to current regulatory and trade publications should be made to determine which of these is permitted for a given situation.

ATMOSPHERIC POLLUTION

REFERENCES: Laws: Current Legislation, *Chem. Eng.*, June 21, 1971. Victor Sussman, New Priorities in Air Pollution Control, *Jour. Air Poll. Control Assoc.*, **21**, no. 4, April 1971, p. 201. Corbitt, "Standard Handbook of Environmental Engineering," McGraw-Hill.

Abstract of Federal Air Pollution Control Legislation

Clean Air Amendments, 1965 Authorized federal control of new motor vehicle emissions. Extended federal control to cover international pollution originating in the United States.

Clean Air Act of November 21, 1967 Under Public Law 90-148 the states are held responsible for establishing local air quality standards consistent with federal criteria.

Clean Air Act Amendments of 1977 Under Public Law 95-95, the EPA can require offending sources to be shut down immediately. Unsatisfactory state enforcement programs can be taken over and run by the EPA. New plants and new additions to old ones must have the best available pollution control technology, and meet emission standards, including zero levels for hazardous substances. A 90 percent reduction from baseline in automotive emissions is required.

The deadline set forth in the act requiring cities to comply with federal standards by 1987 was not possible to meet universally, although some cities did meet the criteria for SO_2 and ozone.

Clean Air Act of 1990 This was the first new air pollution legislation since the 1977 amendments. Briefly, its major requirements are to reduce SO_2 emission by 50 percent and to set emission standards for a number of chemicals which had not been addressed in previous legislation, as amended. The gradual phaseout of use of CFCs is set forth, with complete elimination set for 1996. Airborne pollution respects no national boundaries; accordingly, there has been much cooperation among sovereign nations to solve, among others, the serious problem of acid rain resulting from some forms of pollution.

Some Existing Federal Standards The EPA publishes in the Federal Register national ambient air quality standards covering six pollutants (see Table 18.1.5).

Effect of Energy Crisis on Legislation The energy crisis of the 1970s led Congress in 1974 to consider legislation to postpone the NEPA and to modify the Clean Air Act in order to provide relief from the shortage of petroleum and natural gas. Currently, there are abundant supplies of all types of fossil fuels available in the global market, and at competitive prices. There was no actual postponement of then-extant

Table 18.1.5 Ambient Air Quality Standards

Pollutant	Primary	Secondary, if not same as primary
Particulates, $\mu g/m^3$		
Annual geometric mean	75	60
Max. 24-h conc.*	260	150
Sulfur dioxide, $\mu g/m^3$		
Annual arith. aver.	80 (0.03 ppm)	
Max. 24-h conc.*	365 (0.14 ppm)	
Max. 3-h conc.*		1,300 (0.5 ppm)
Carbon monoxide, mg/m^3		
Max. 8-h conc.*	10 (9 ppm)	
Max. 1-h conc.*	40 (35 ppm)	
Ozone, μ/m^3		
1-h max.*	160 (0.08 ppm)	
Lead particulates, $\mu g/m^3$		
Calendar qtr., aver.	160 (0.24 ppm)	
Nitrogen dioxide, $\mu g/m^3$		
Annual arith. aver.	100 (0.05 ppm)	

* Not to be exceeded more than once a year.
NOTE: All values subject to periodic revision.

legislation, but concerted efforts to abate CO_2 pollution resulting from a fossil-fuel-based economy continue. (See discussion above under "Control of Thermal Discharges.") Electric power generation increases worldwide. Most of it is represented by fossil fuel plants as dictated by local resources available and the state of those local economies. Few unexploited very large water power sites exist, and of those, only a handful are economically viable in the current market. Nuclear-powered plants are being emplaced in many parts of the world, although no new plants have been ordered in the United States for almost 20 years. In some quarters it appears inevitable that the conflicting requirements of pollution abatement, reduction of CO_2 in the atmosphere, and reluctance to exploit nuclear power will ultimately result in the reemergence of nuclear power. The timing is not clear, and it will be a function, among other things, of the successful and timely resolution of the problems attendant disposal of high-level radioactive wastes, depletion of fossil fuel reserves, and possible development of advanced-type nuclear reactors which may prove economically and environmentally viable sometime in the twenty-first century.

It would be prudent to conserve irreplaceable natural resources by utilizing them in the most productive energy, materials, and ecological cycles. Often, alternative methods are at cross purposes with those objectives and result in inadvertent squandering of those resources.

SOURCES AND CONTROL OF AIR POLLUTION

REFERENCES: Magil, Holden, and Ackley, "Air Pollution Handbook," McGraw-Hill. Gibbs, "Clouds and Smoke," Churchill. White, "Industrial Electrostatic Precipitation," Addison-Wesley. Morgensen, "Fan Engineering," Buffalo Forge. Dallavalle, "Micrometrics," Pitman. Stern, "Air Pollution," Academic Press. "Air Pollution Engineering Manual," J. A. Danielson, EPA-Research Triangle Park, NC, May 1973.

Table 18.1.6 Principal Air Pollutants

Gases	Vapors	Particulate matter			
		Dusts	Fumes	Smokes	Mists
Acid gases	Alcohols	Alumina	Metallic halogens	Ash	Acid
Carbon monoxide	Esters	Calcium fluoride	Metallic oxides	Cinders	Chromic
Hydrogen chloride	Hydrocarbons	Cement	Silicon tetrafluoride	Organic compounds	Phosphoric
Hydrogen fluoride	Ketones	Coal		Soot	Sulfuric
Hydrogen sulphide	Mercaptans	Grain		Tobacco	Organic chemicals
Nitrogen oxides		Limestone			Oils
Sulfur dioxide		Metal			Salt spray
Alkaline gases		Ore			
Ammonia		Rock			
		Wood			

Table 18.1.7   Pounds of Contaminants Discharged Daily from Domestic Sources in a Metropolitan Area of 100,000 Persons

| | Pounds per day per 100,000 persons using each category of heating and refuse disposal | | | | | |
| | Fuel—domestic heating | | | Domestic incineration | | |
Principal contaminants	Coal	Oil	Gas	Backyard burning	Household* incinerator	Apartment incinerator
SO_x	42,000	17,000	0.4	180	1	12
NO_x	8,000	6,000	6	90	1,150	30
H_2S	1,000	500	0.1			24
NH_3	2,000	800	0.3	345		24
HCl	2,000	500	0.3			
Aldehydes	2,000	800	1	600	8,400	72
Organics	20,000	4,000	1	42,000	12,000	1,800
Organic acids	30,000	12,000	1	225	1,900	4,800
Solids	200,000	800	0.1	3,400	16,500	4,000
Total of above categories, lb	307,000	42,400	10	46,800	40,000	10,700
Total of above categories, kg	139,100	19,210	4.5	21,200	18,120	4,850

* Add 17 percent for cigarette smoking (1984 estimate).
NOTES: Each column shows estimates of the pollutants released to the atmosphere if the entire population used the noted fuel or method of incineration.
   SO_x—oxides of sulfur—SO_2 and SO_3.
   NO_x—oxides of nitrogen.
   Aldehydes measured as formaldehyde.
   Organic acids measured as acetic acid.
   Total includes only above categories and does not imply total contamination in the ambient atmosphere.
SOURCES: Eliassen, Domestic and Municipal Sources of Air Pollution, *Proc. National Conference on Air Pollution*, 1958.

Domestic and Municipal Sources of Air Pollution

(See Tables 18.1.7 and 18.1.8.)

All **pollutants** are usually classified as gases, vapors, and **particulate matter** (see Table 18.1.6 and Fig. 18.1.1). Particulate matter includes: **Dusts**—a loose term applied to solid particles predominantly larger than colloidal and capable of temporary gas suspension. Dusts do not tend to flocculate except under electrostatic forces; they do not diffuse but settle under the influence of gravity. Dusts result from such operations as crushing, grinding, drilling, screening, and blasting. **Fumes**—the solid particles generated by condensation from the gaseous state, generally after volatilization from melted substances, and often accompanied by a chemical reaction such as oxidation. For example, zinc vapor will evaporate from the surface of heated liquid metal and will then condense upon contact with ambient air to form small, fluffy particles of zinc oxide. Fumes are usually less than 1 μm, although they may coalesce or flocculate to form larger particles. **Smokes**—gas-borne particles (usually less than 0.5 μm) resulting from incomplete combustion of materials such as wood, tobacco, coal, and oil. **Mists** and **plumes**—loose terms applied to dispersions of liquid particles (0.1 to 2.5 μm), the dispersion being of low concentration and the particles **large**.

Low-Emission Automotive Propulsion Systems

(See Secs. 9.6 and 11.1.)

The automobile is a major source of air pollution and a significant contributor to photochemical smog (caused by the products of photoreactions in the atmosphere—organic peroxides, peracids, hydroxy peracids, peroxyacyl nitrate, and other compounds). For every 1,000 gal of gasoline consumed by an automobile engine, there are discharged the following air pollutants, in pounds: carbon monoxide, 3,000; hydrocarbons, 200 to 400; nitrogen oxides, 50 to 150; aldehydes, 5; sulphur compounds, 5 to 10; organic acids, 2; ammonia, 2; solids, 0.3.

Conventional Gasoline Engines   A major thrust aimed at the reduction of air pollution from automotive exhausts resulted in enactment of legislation to increase the fuel economy obtained by passenger cars. Commonly known as the **corporate average fuel economy (CAFE)** program, the legislation sought to increase average fuel economy from 18 mi/gal (1978) to 27.5 mi/gal (1985). Current average fuel economy for passenger cars is slightly above 28 mi/gal (1995).

In the quest to reduce pollutants entrained in automotive exhaust gases, it was also recognized that the alarmingly high levels of lead in the local atmosphere, especially in large urban areas and their surrounds, had to be eliminated. This led to a dual approach: introduction of nonleaded gasoline (with the concomitant phaseout of leaded fuel) and incorporation of catalytic converters as original equipment by passenger car manufacturers. The two matters went hand in hand, for it was known that a practical catalytic converter would be rendered ineffective if it were contaminated by lead (or phosphorus). At this time, all passenger cars (with a few exceptions, e.g., some off-road vehicles on farms) are equipped with catalytic converters. The converter functions to reduce the levels of emission of carbon monoxide (CO), hydrocarbons (HC), and oxides of nitrogen (NO_x). Chemical reactions take place when the exhaust gases pass through the converter, resulting in conversion of the offending pollutants to CO_2, water vapor, and nitrogen. The

Table 18.1.8   Sources of Air and Typical Loss Rates

Class	Aerosols	Gases and vapors	Typical loss rates
Combustion processes	Dust, fume	NO_2, SO_2, CO, organics, acids	0.05–1.5% by weight of fuel
Stationary engines	Fume	NO_2, CO, acids, organics	4–7% by weight of fuel (hydrocarbons)
Petroleum operations	Dust, mist	SO_2, H_2S, NH_3, CO, hydrocarbons, mercaptans	0.25–1.5% by weight of material processed
Chemical processes	Dust, mist, fume, spray	Process-dependent (SO_2, CO, NH_3, acids, organics, solvents, odors, sulfides)	0.5–2% by weight of material processed
Pyro- and electro-metallurgical processes	Dust, fume	SO_2, CO, fluorides, organics	0.5–2% by weight of material processed
Mineral processing	Dust, fume	Process-dependent (SO_2, CO, fluorides, organics)	1–3% by weight of material processed
...od and feed operations	Dust, mist	Odorous materials	0.25–1% by weight of material processed

E: Rose et al., Prevention and Control of Air Pollution by Process Changes and Equipment, W.H.O. Rept. Air Pollution, *Monograph Series* 46-307-343.

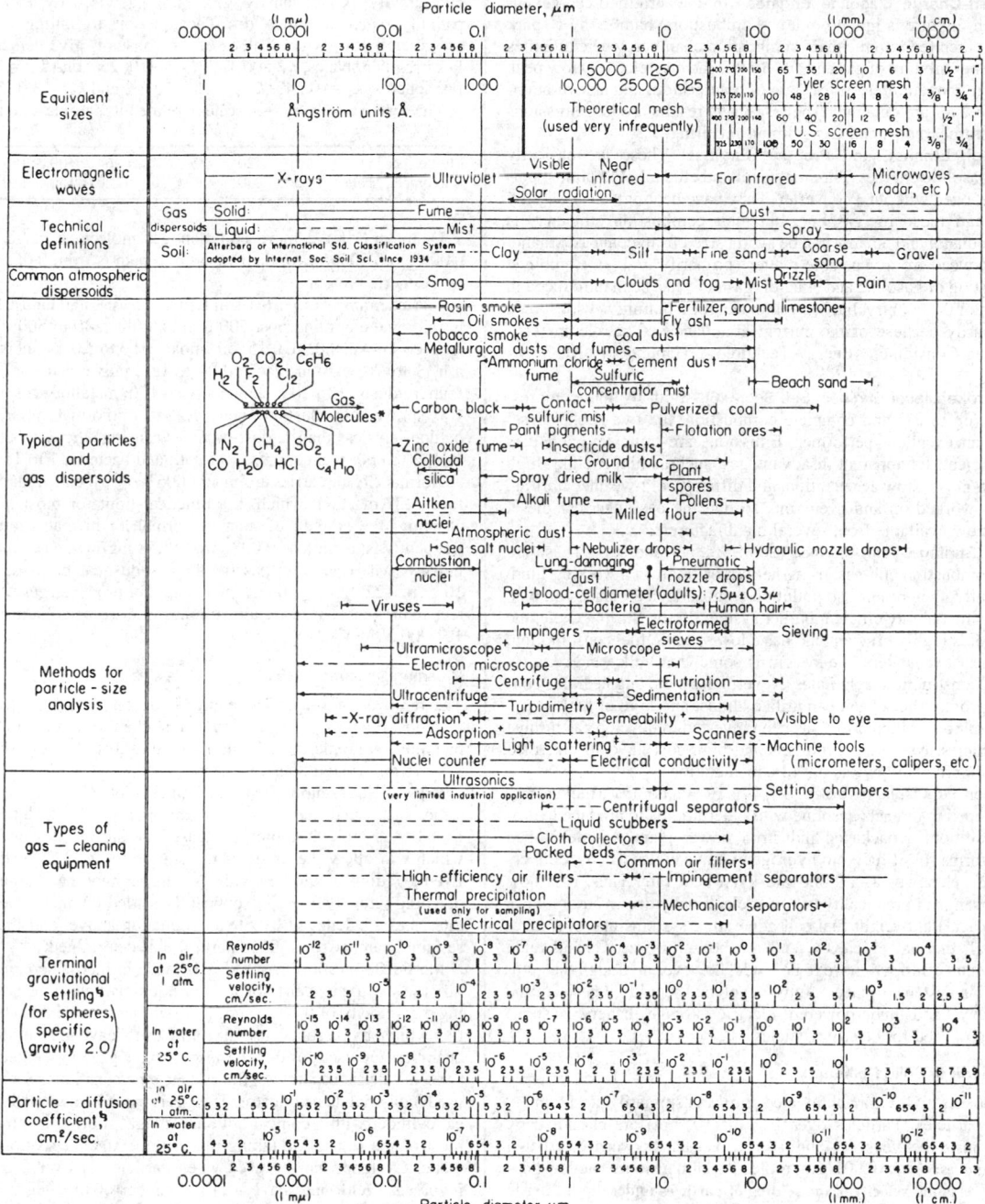

Fig. 18.1.1 Sizes and characteristics of airborne particles. *(From C. E. Lapple, Stanford Res. Inst. Jour., vol. 5, p. 94, Third Quarter 1961. Reproduced by permission.)*

catalytic converter will function properly only if free of contaminants (lead and phosphorus, cited above) and if maintained properly to prevent operation at extremely high temperatures. Continued improvements in design and construction have proved effective in reducing pollutants ascribable to automotive exhaust gases, and have done so at no appreciable loss in fuel economy.

Fuel Injection Gasoline Engines A typical fuel injection engine of about 1,700-cm³ displacement and burning 91 octane fuel, when compared with a conventional two-carburetor engine of the same displacement and using the same gasoline, will demonstrate an improved gasoline economy and substantially lower emissions because of improved combustion.

Stratified-Charge Gasoline Engines In the stratified-charge engine, a rich mixture is ignited in a precombustion chamber by a spark plug and is then fed into the main combustion chamber, where it ignites a much leaner mixture of gas and air. By this means, relatively low peak combustion temperatures, a slow rate of combustion, and an overabundance of free air are obtained. The respective results are low emissions of NO_x, hydrocarbons, and CO, without misfiring.

Methanol, Ethanol, and Propane Engines These are basically modified gasoline engines. Pure methanol and ethanol burn almost pollution-free, even without a converter. They have heating values slightly over one-half those for gasoline, but their combustion efficiency is higher. Methanol and ethanol can be synthesized from coal gas; ethanol is also produced from vegetable matter (primarily grains). Combined fuel consisting of gasoline and ethanol (gasahol) has been introduced in some parts of the country, but has yet to become a definitive widespread fuel, primarily because of the current abundance of petroleum feedstocks. (See Considine, "Energy Technology Handbook," McGraw-Hill.)

Two-Stroke Diesel Engines (See Sec. 9.6) With its high compression ratio (up to 21) and complete combustion, the two-stroke diesel engine is an excellent performer. Emissions are quite low and it is highly efficient. Its apparent disadvantages, such as high weight, high initial cost, noise, slow acceleration, and difficult cold-weather starting, have been worked on and overcome, so that the passenger car diesel engine is now available from several manufacturers.

Philips (Stirling Cycle) Engines (See Secs. 4.1 and 9.1) These are external combustion engines, using helium or freon as a working fluid. Consequently, they require no pollution control devices, since combustion occurs in a highly efficient burner system. In the Stirling cycle, the engine, too, is highly efficient and has a low specific fuel consumption. With its heat exchangers, the system is somewhat bulky and requires expensive construction techniques. Nevertheless, the engine holds considerable promise because it can utilize almost any kind of fuel.

Gas Turbines Much effort has been invested to develop gas turbine driven vehicles. Numerous problems and high cost have all but ruled these highly efficient devices out of consideration.

Hydrogen Fuel Cells (See Sec. 9.1) In 1974 a compact fuel system was developed that reacts gasoline with other liquids to produce hydrogen. In addition, a packaged industrial power plant was developed which re-forms fossil fuel into hydrogen and, using fuel cells, delivers 26 MW of net power at a 9,000 Btu/kWh heat rate. Thus, it is now within the range of practicality to power electrically driven automobiles with compact, lightweight fuel cell generating systems that operate at high overall thermal efficiency, and are pollution-free. Considerable developmental effort will have to be made, however, to design the automobile. This system, if successfully developed, could outperform the huge, heavy, and underpowered electric storage battery systems currently proposed for electric vehicles.

Classification of Solid Pollutants

The **micrometer** (10^{-6} m), abbreviated μm, is customarily used to measure fine particles. Particles over 10 μm (10^{-5} m) are classified by Gibbs as **dusts;** between 0.1 and 10 μm, **clouds;** between 0.001 and 0.1 μm, **smokes;** below 0.001 μm, molecular dimensions. The law governing the **terminal velocity of the settling of particles** under the influence of gravity varies with the size of the particle. In the turbulent region (above 2,000 μm)

$$V_t = Ks^{1/2}p^{1/2}D^{1/2} = k_1\sqrt{SD}$$

In the intermediate region (between 1,000 and 100 μm), the law is $V_n = K's^{2/3}\rho^{-1/3}\eta^{-1/3}D = k_2s^{2/3}D$. In the streamline region (2 to 50 μm), Stokes' law holds: $V_s = K''sD^2\eta^{-1} = k_3sD^2$. From 0.1 to 1.0 μm, Cunningham's correction must be applied to Stokes' law:

$$V_c = V_s(1 + 1.72\lambda/D) = V_s(1 + 0.172/D)$$

Below 0.1 μm, the movement due to molecular shock (brownian motion) exceeds that due to velocity. In these equations, V is the terminal velocity, ft/min; D, particle diameter, μm; p, gas pressure, atm; s, spe-

cific gravity; ρ, gas density, g/cm^3; η, gas viscosity, P; λ, mean free path of the gas molecules, μm. The equations for falling in standard air are given below, where values of the constants give velocities in m/s. For irregular shapes, $k_1 = 0.142$, $k_2 = 0.00259$, and $k_3 = 1.98 (10)^{-5}$. For spheres, $k_1 = 0.243$, $k_2 = 4.11 (10)^{-3}$, and $k_3 = 3.0 (10)^{-5}$.

The **Tyler standard screen scale** is related to particle size as below:

Meshes per in	10	20	35	48	64	100	150	200	325
Micrometer scale	1,650	830	420	300	220	150	110	74	44

Diameters of the more common gas molecules range from about 0.0003 to 0.00045 μm; their mean free path is from 0.06 to 0.2 μm at atmospheric pressure.

The size ranges of **particles** in typical **aerosols** and **industrial dusts** are, in micrometers: raindrops, 500 to 5,000; mist, 40 to 500; fog, 1 to 40; tobacco smoke, 0.01 to 0.15; oil smoke, 0.03 to 1.0; pigments, 1 to 7; fly ash, 3 to 70; carbon black, 0.04 to 0.2; pulverized coal, 10 to 400; foundry dusts, 1 to 200; cement, 10 to 150; metallurgical fumes, 0.1 to 100; sprayed zinc dust (condensed), 2 to 15; normal impurities in quiet outdoor air, less than 1; dust particles causing silicosis, below 10; pollens, 20 to 60; plant spores, 10 to 30; and bacteria, 1 to 15.

Normal **city air** carries around 0.0006 g of suspended matter per cubic foot (1.37 mg/m^3), which is a practical limit for most industrial gas cleaning; the amount of dust in normal air in manufacturing plants frequently is as high as 0.002 gr/ft^3 (4.58 mg/m^3). The amount of dust in **blast-furnace gas** after passing the first dust catcher is of the order of 10 gr/ft^3 (22.9 g/m^3), the same as raw hot producer gas. All dust content figures are based on air volumes at 60°F and 1 atm (15.6°C and 101,000 N/m^2).

Cleaning Apparatus

The removal of suspended matter is accomplished by the following methods. Each particle case must be studied to find the most desirable method. Economic considerations require that cleaning should not be any more thorough than necessary.

Gravity Separation This method is applicable only to the larger-sized suspended particles (100 μm and over). A long horizontal chamber is built, in which the gases are slowed down to a velocity which will allow the particles to settle to the bottom of the chamber. Even gas flow should be maintained throughout the chamber. Hoppers or a drag scraper are used to collect the settled-out material. The settling rate can be calculated by Stokes' equation above, and the size of the settling chamber for a given particle size determined.

Inertia Separation An effort is made in inertia separators to magnify the settling tendency of solid or liquid particles in gases by increasing the velocity of the gas and by providing for rapid changes in direction which, by inertia, cause the particles to leave the gas stream. Baffle chambers are used for this purpose with a zigzag movement of the gas.

Of all inertia separators, the **cyclone** type constitutes the most widely used industrial dust collector. The gas is passed tangentially into a vertical cylinder with a conical bottom. The gas follows a spiral path, with most of the separation taking place in the smaller sections of the cyclone. Cyclones can be used when particles of over 5 μm diam are involved. **Multiclones** have a large number of parallel, small cyclone units. The gas pressure drop is about 4 times the entrance velocity head (1 to 4 in w. g.). The dust content of the cleaned gas is seldom less than 1 gr/ft^3 (2.29 g/m^3).

In the **Rotoclone** apparatus, the impeller is a concave disk with curved blades along which the dust slips and is discharged over the edge of the disk into a ring-shaped dust chamber. The main air stream leaves in a wide scroll on the side of this chamber. The apparatus is usually preceded by a cyclone for removal of coarse dust. By spraying water in the entering air, the precipitation efficiency of dust removal can be materially raised. Particles of minimum size 8 μm are claimed to be removed by the dry apparatus, 1 μm by the wet apparatus. The power for dust separation cannot readily be separated from that for compressing the gases. The power requirement is 3 bhp per 1,000 ft^3/min for an outlet

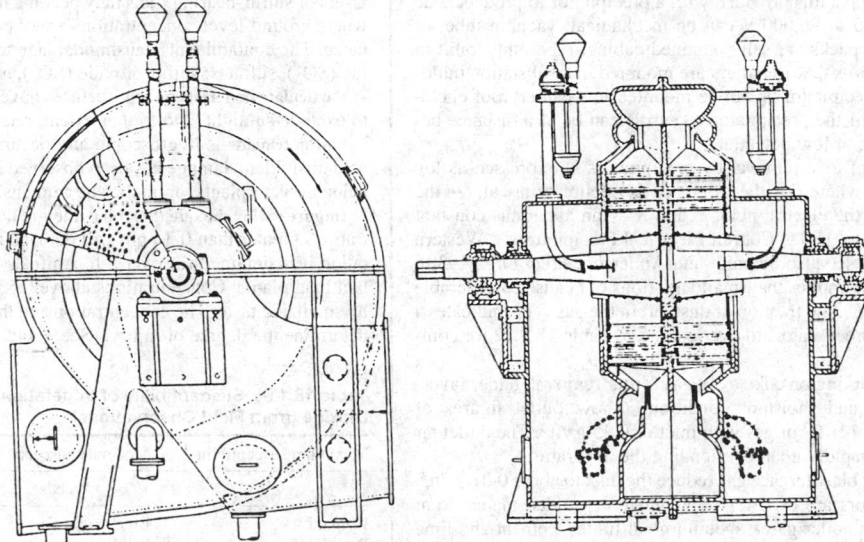

Fig. 18.1.2   Theisen disintegrator air washer.

pressure of 10 in w. g.; in the water-sprayed machine, the required power is at least 25 percent more.

Static spray scrubbers are usually of the tower type, the gas passing upward countercurrently to the descending liquid. Sets of sprays are placed in the top zone, with various materials used in layers to channel and mix the gas and water. Hurdles, cylindrical tiles, and random-packed ceramic tiles or metal spirals are common packing materials. With a gas flow-rate of about 350 ft³/min per ft² of cross-sectional area and a water rate of 25 gal/Mft³ of gas, a cleanliness of 0.1 to 0.3 gr/ft³ can be obtained with blast furnace, coke oven, and producer gas. The pressure drop through a tower scrubber is in the range of 4 to 10 in w. g. The scrubber process is used as the primary cleaning and cooling stages before the cleaning of gases.

Dynamic Spray Scrubbers   Contact between water droplets and suspended matter in gas is improved by mechanical agitation of gas and water. The *Thiesen* disintegrator (Fig. 18.1.2) is an example of a modern type of dynamic spray scrubber. It consists of a substantial cast-iron casing enclosing two stationary and one motor-driven rotor basket. The water is injected by gravity to the center of the perforated cones. Centrifugal force distributes the water over the bar system. The rotation atomizes the water which agglomerates with the dust, wetting it to effect its removal. The disintegrator is always followed by a moisture eliminator which removes the wetted particles from the gas. The range of cleanliness is between 0.02 and 0.005 grains of dust per cubic foot. In addition, any desired pressure increase up to 16 in w. g. can be obtained. The disintegrator is not sensitive to the quality of the incoming gas. It will clean just as well with 0.30 gr/ft³ in the entering gas as with 0.10 gr. It has been installed mostly on blast-furnace gas cleaning systems. In recent years, it has been successful on waste-gas cleaning for the oxygen steelmaking process and electric-furnace waste-gas cleaning systems. A similar dynamic scrubbing effect can be obtained by spraying water into the "eye" of the final exhaust fan and dewatering with a wet cyclone. Water requirements range from 4 to 8 gal per 1,000 ft³/min of gas, and a power demand of from 4 to 10 kW per 1,000 ft³/min of gas.

The other example of dynamic spray scrubbers is the **Pease-Anthony Venturi**. In this system, the gas is forced through a Venturi throat in which the gas is mixed with high-pressure water sprays. A tank after the Venturi is needed to cool and eliminate the moisture. Cleanliness in the range of 0.1 to 0.3 gr/ft³ has been reported.

Filtration is comparatively little used for cleaning fuel gases; it is in extensive use in cleaning air and waste gases. Materials used for gas filtration are ordinarily thickly woven cotton or wool cloth for tempera-

tures up to 250°F; for higher temperatures, metal cloth or woven glass cloth is advisable. The gases filtered should be well above their dew point, as condensation on the filter cloth plugs the pores. If necessary, saturated gas should be reheated. The cloth is frequently made into "bags," tubes of 6 to 12 in in diameter and up to 40 ft long, suspended from a steel framework (baghouse). The gas inlet is at the lower end through a header to which the bags are connected in parallel; the outlet is through a housing surrounding all the bags. At frequent intervals, the operation of all or part of each unit is interrupted while the bags are pulsed, shaken, or reverse-air-cleaned to dislodge the accumulated dust, which drops back into the gas-inlet header and is removed by a conveyor screw. The dust content can be reduced to 0.01 gr/ft³ or less at reasonable cost. The apparatus also is used for the recovery of valuable solids from gases.

Electrical Precipitation   This principle, although involving highly technical physioelectrical phenomena, is generally familiar to all. If suspended particles are placed in a high-voltage electric field, they receive an electric charge and move toward one or the other of the electrodes between which the electric field is established. If one electrode is a pipe or flat plate and the other electrode is a wire axially suspended in the center of the pipe or between two plates, the field is strongest around the wire and weakest near the surface of the pipe or plate. The charged dust particles move from the strong part of the field toward the weak part of the field.

Two types of precipitators are manufactured. **The vertical-round type** is usually built in a cylindrical tank having a header sheet near the inside top. In this header sheet are nested pipes which act as the collecting electrodes. The high-voltage ionizing electrodes are twisted square steel rods, about ¼ in in size, supported from above the header sheet and hung axially in the collecting electrode pipes. Water is introduced above the header sheet and flows over weir rings at the tops of the pipes to form a water film on the inside of the pipes. **The horizontal-plate type** has vertical parallel plates as collecting electrodes. The discharge electrodes are a series of wires suspended equidistantly between the plates, and can be operated as either a "wet" or "dry" unit. When used as a wet precipitator, weir boxes are built on the top of each plate to produce a wetting of both sides of the plate, washing the collected particles down the plate. As a dry precipitator, a rapping device operated on a time cycle shakes the plates, causing the collected particles to fall into a hopper under the plates. Dry collection requires injection of a conductive gas such as SO_3 to work efficiently. (See Marchello, "Control of Air Pollution Sources," Marcel Dekker, p. 182.)

The electrical apparatus provided with a precipitator to produce a dc potential of 30,000 to 90,000 V can be mechanical, vacuum tube, or solid state. Power packs are self-contained cabinet-type units, built in sizes from 2½ to 50 kVA. Rectifiers are mounted in a substation building close to the precipitator or can be mounted in weatherproof enclosures on the roof of the precipitator. Controls can be at a distance because the wiring is of low potential.

The efficiency of dust removal by precipitators is expressed as $\log(1 - E) = t \log K$, where $E =$ the efficiency as fraction removal; $t =$ the time the gas is in the electric field, s; and $K =$ an apparatus constant varying from 0.05 to 0.80 ("Cottrell Electrical Precipitators," Western Precipitator Corp.; see also Schmidt and Anderson, *Elec. Eng.*, 1938). The above function implies that small variations of t cause considerable variations of $(1 - E)$, the fraction of dust left in the gas. This indicates a sensitivity of the precipitators to overload. (See Table 18.1.12 for common applications).

Precipitators working on coke oven gas for tar removal, under favorable temperature and moisture conditions, show outlet figures of 0.005 gr of dry tar per ft³ of gas with inlet of 0.25 gr/ft³. The outlet tar content increases rapidly upon overloading the apparatus.

Precipitators for blast furnace gas reduce the dust to about 0.01 gr/ft³. Electric precipitators are used for removal of suspended matter from gases and air from boiler gases, metallurgical fumes, cement and lime dusts, etc. Precipitators can be built for dust in practically all particle sizes, as well as mist. They cannot collect material in the gaseous or vapor state. Precipitators are not applicable where explosive gas or dust is involved in the presence of air or oxygen in explosive proportions because of the hazard of ignition by an electric spark.

Blast-Furnace Gas Cleaning Modern equipment for this service represents a field in which gas cleaning has been developed to a high degree of efficiency. The method described here is used on most modern blast-furnace gas-cleaning plants.

The **dust catcher** is a large cylindrical steel tank, brick-lined, about 30 to 40 ft in diameter. The dust catcher is fundamentally a settling chamber. Dust is dropped out of the gas by reducing the velocity of the flow. Gas enters at 400°F approx and with 10 to 20 gr of dust per cubic foot. A good dust catcher will clean the gas to 3 gr of dust per cubic foot.

The **primary washer** is a cylindrical steel shell 16 to 21 ft in diameter, with a conical bottom section. Four to five banks of ceramic, round, or cross-partition tile are arranged in the tower. At the top, sprays cover the top bank of tile with an even distribution of water at the rate of 7 to 9 gal/(ft²·min). Size of the tower is determined with a gas rate of 350 ft³/min per ft² of cross-sectional area. The gas enters the washer at an angle of 45° downward, passes up through the tile and out a main at the top of the washer. The water passes down through the tile banks and collects the dust removed from the gas. The water collects in the bottom section and overflows through a pipe arranged with a gas seal. The water then is sent to a thickener for removal of the iron-ore dust. The gas enters the washer at 400°F approx and is cooled to 100°F approx. The dust content of the gas is reduced to about 0.15 to 0.30 gr/ft³, which is satisfactory for entrance into a precipitator or Theisen disintegrator.

Electrostatic precipitators reduce the dust content of the gas to the cleanliness required by its final use. The following are common requirements for blast furnace gas in grains per cubic foot at standard conditions (1 atm, 70°F): hot blast stoves, 0.01; boilers, 0.01; soaking pits, 0.10; coke ovens, 0.005.

Effect of Sulfur Gases and Particulates on Vegetation

REFERENCES: Kamprath, E. J., "Possible Benefits from Sulfur in the Atmosphere." *Combustion,* **44,** no. 4, 1972, pp. 16, 17; Grennard and Ross, Progress Report on Sulfur Dioxide, *Combustion,* **45,** no. 7, 1974, pp. 4–9; Wagner, TVA's Proposed Clean Air Strategy, *Public Utilities Fortnightly,* pp. 34–37, June 6, 1974.

Damage from Sulfur Fumigation Vegetation is damaged by abnormal concentrations (above 0.3 ppm) of sulfur gases at ground level.

Users of sulfur-bearing fuels may become involved with damage claims where ground-level concentrations are of consequent increasing importance. The contaminants causing damage to vegetation are sulfur trioxide (SO_3), sulfates, sulfur dioxide (SO_2), and particulate materials.

Particulate materials on leaf surfaces have no harmful effect other than to exclude sunlight, and to that extent, retard growth.

Sulfur trioxide is hygroscopic and occurs at ground level as sulfuric acid mist. Plant damage has been observed in the field. Sulfur trioxide is deleterious to plants, animals, and humans.

Impure sulfur dioxide (SO_2) has the greatest effect on plants. Concentrations greater than 0.3 ppm will damage the leaves of sensitive vegetation if maintained for over 5 h. Animals and humans are more resistant than plants. Concentrations above 0.5 to 2.0 ppm are recognizable by smell and taste. The concentration and the duration of the fumigation determine the degree of injury. (See Table 18.1.9). Environmental con-

Table 18.1.9 Susceptibility of Vegetation to Injury by Sulfur Dioxide (from Field Observations)

Maximum susceptibility	Moderate susceptibility	Resistant
Weeds		
Plantain	Bracken fern	Goldenrod
Ragweed	Wild carrot	Small-leaved milkweed
Smartweed	Wild grape	Knotweed
Dewberry	Sweet clover	Milkweed
Greenbriar		
Dandelion		
Galinsoga		
Pigweed (redroot)		
Wildflowers		
Greenbriar	Blackberry	Goldenrod
Galinsoga	Witch hazel	Small-leaved milkweed
Dandelion	Huckleberry	Elderberry
	Blueberry	Mountain laurel
	Bracken fern	Yarrow
	Viburnum	Knotweed
	Wild carrot	Joe-Pye weed
	Sweet clover	Milkweed
Farm crops		
Alfalfa	Alsike clover	Potato
Oats	Corn	
Buckwheat	Cabbage	
Barley		
Garden vegetables		
Beet	Huckleberry	Onion
Endive	Blueberry	Cucumber
Bean	Tomato	Corn
Peas	Parsley	Potato
Brussels sprouts	Cauliflower	
Landscape materials		
Hawthorn	Huckleberry	Hazelnut
Sunflower	Blueberry	Mountain laurel
Cosmos	Nasturtium	
Sweet William	Dahlia	
	Gladiolus	
	Willow	
Trees		
White pine	Larch	Red pine
Hemlock	Scotch pine	Blue spruce
Hawthorn	Norway spruce	Black locust
Black birch	Sumac	Black oak
Yellow birch	Ash	Red oak
	Wild cherry	White oak
	Domestic apple	Sugar maple
	Willow	Swamp red maple
		Norway maple

Table 18.1.10   Emission of Sulfur-Bearing Gases and Solids from a 1,000,000-kW (1,000-MW) Fossil Fuel Fired Power Station

Fuel:	Mid-western bit coal	Eastern bit coal	Central Ill. bit coal	High-sulfur fuel oil	Low-sulfur fuel oil
Firing method:	Pulv.-coal round burners	Pulv.-coal round burners	Cyclone burners	Round burners	Round burners
Sulfur content of fuel, %	4.14	1.15	4.65	2.41	0.40
Ash content of fuel, %	18.16	7.59	15.00	0.10	0.10
Sulfur-bearing gases leaving boiler:					
Sulfur dioxide, lb/h	69,000	15,100	97,500	23,200	3,620
Sulfur trioxide, lb/h	3,450	755	4,875	1,160	180
Total sulfur-bearing gases, lb/h	72,450	15,855	102,375	24,360	3,800
Solids in flue gas leaving boiler:					
Sulfur compounds, lb/h	4,600	920	920	2,520	420
Carbon, lb/h	8,280	2,760	1,190	4,200	4,200
Fly ash, lb/h	118,680	38,640	20,250	420	420
Total solids, lb/h	131,560	42,320	22,360	7,140	5,040
Collection efficiency,* %	99+‡	99+‡	99+‡	†	†
Solids in flue gas leaving stack, lb/h	360	360	15	†	†

* Based on clear stack.
† See text. Firebox temperatures are kept below 1,900°F to control NO_2 emissions.
‡ Total efficiency whether combination mechanical and electrostatic units or electrostatic alone, with SO_3 injection.
NOTE: 1 lb/h = $7.55(10)^{-3}$ kg/s.

ditions of temperature, humidity, soil moisture, soil fertility, nutrient supply, light intensity, age of plants, and moist-leaf conditions have a marked influence on the effect of sulfur dioxide on vegetation.

Benefits from Sulfur in the Atmosphere   In normal atmospheric concentrations below 0.3 ppm, as occurs in metropolitan areas 95 percent of the time, sulfur dioxide is an essential plant nutrient. Vegetation utilizes it for growth by absorption from the atmosphere through leaves, and from rain-saturated earth through the roots. Fossil fuels, by virtue of their origin as former vegetation, contain from 0.4 to over 4 percent sulfur (Table 18.1.10). To maintain the natural ecological sulfur cycle for sustaining plant growth, this organic sulfur should be recycled back into the soil. As sulfur in stack gases is brought down by rainfall largely over a radius of about 15 mi from steam-electric plants in rural areas, averaging about 5 lb/(acre·year) (0.55 g/m²), anywhere from 20 to 50 percent of the annual sulfur requirement for crops can be furnished by stack emissions, while the balance must be supplied by fertilizers. Agriculture requires annually 10 to 40 lb/acre (1.11 to 4.44 g/m²) of sulfur, which, in the absence of stack emissions, must be furnished by direct sulfur application for optimum yields of crops.

The sulfur dioxide problem, therefore, is mainly one of preventing excessive ground-level concentrations, whereby a normal plant nutrient can become a menace. Two methods of control are possible when burning fossil fuels: sulfur oxide removal by wet scrubbing, or dispersion by tall stacks with subsequent precipitation by normal rainfall.

Sulfur Oxide Removal Processes   Stack-gas scrubbing processes remove sulfur from the flue gas but also increase the buildup of atmospheric CO_2. In 1973 the seven most promising stack-gas scrubbing processes selected for further study by the Sulfur Oxide Control Technology Assessment Panel were

1. Catalytic oxidation of SO_2 to SO_3, to recover sulfuric acid as a saleable by-product
2. Wet lime/limestone scrubbing, which produces calcium sulfite sludge for disposal
3. Alkali scrubbing without regeneration, which produces 18 percent sodium sulfite brine for disposal
4. Alkali scrubbing with lime regeneration, which produces calcium sulfite sludge for disposal
5. Alkali scrubbing with thermal regeneration, which recovers SO_2 gas by-product, and a sodium sulfite/sulfate brine for disposal
6. Magnesium oxide scrubbing, with off-site sulfuric acid manufacture and regeneration of magnesium oxide by others
7. Alkali scrubbing, plus electrodialytic regeneration to recover SO_2 gas and a purge stream of impure brine

A comparison of these systems is given by Table 18.1.11. The sulfur

Table 18.1.11   Comparison of SO₂ Control Systems

Process	Requirements	Throwaway or recovery	Additional cost of power generation, %	SO₂ control efficiency, %	Additional plant capacity investment, %
SDEL*	Fuel-switching and tall stacks		1–2.35	99+	1–2.8
Low-sulfur fuel (coal or oil)			28–85	Varies	—
Dry fluidized-bed combustors	Limestone (200% of stoichiometric)	Throwaway CaSO₃/CaSO₄	25–38	50–90	25–90
Wet lime/limestone scrubbing	Lime (100–120% stoich.); limestone (120–150% stoich.)	Throwaway CaSO₃/CaSO₄	19–37	60–85	17–30
MgO scrubbing	MgO carbon and fuel (for regeneration and drying)	Recover conc. H₂SO₄	25–50	90	20–35
Catalytic oxidation (add-on)	V₂O₅ catalyst heat	Recover dilute H₂SO₄	25–45	85–90	25–40
Na₂SO₃ scrubbing	Sodium makeup heat	Recover conc. H₂SO₄ or sulfur	25–50	90	25–40
Double alkali	Sodium makeup plus lime (100–130% stoich.)	Throwaway CaSO₃/CaSO₄	17–35	90	15–30

BASIS: 80 percent plant load factor. Yearly fixed charges are 18 percent of capital investment. Cost ranges are for 200- to 1,000-MW plant sizes. Costs of sludge disposal ponds not included.
SOURCES: Burns and Roe data, TVA's Clean Air Strategy. *Public Utilities Fortnightly,* 34: June 6, 1974.

dioxide emissions limitations (SDEL) case, with its low-cost and negligible environmental impact, is technically justifiable with a cost/benefit ratio of 0.19. The other designs are bulky, costly, noisy, and experimental. They consume energy, water, fuel, and chemicals, and require direct sulfur application to surrounding agricultural areas, plus waste sludge disposal. These adverse environmental impacts usually result in negative total public benefits, unjustifiable except under unusually favorable circumstances.

By far the most difficult and costly environmental hazard produced by sulfur removal is the volume of high chemical-oxygen-demand (COD) liquid and solid wastes produced. This contaminated thixotropic sludge is environmentally unacceptable for landfill or for any other conceivable use. Reduction to elemental sulfur for permanent underground storage by an inverse Frasch process would reduce the volume of wastes, but the cost is exorbitant. Moreover, even if sulfur is removed, the buildup of acid gases (mainly CO_2) in the global atmosphere will continue unabated, along with increasing acid rains, as the CO_2 content of the air increases. Substitution of nuclear fuels for fossil fuels will correct the situation.

Sulfur Dioxide Emission Limitation (SDEL) Programs

To avoid the water pollution difficulties and the wet sludge disposal difficulties introduced by wet scrubbing, Tennessee Valley Authority and others have developed various SDEL programs.

SDEL methods employ combinations of five or more sensing, dispersing, and limitation approaches simultaneously.

1. Very tall stacks (1,000 ft) for dispersing SO_2 gases high in the atmosphere, thereby minimizing unacceptable ground-level concentrations and occurrences. (See the Holland equation for stack height, below.)

2. Atmospheric (AFBC) and pressurized (PFBC) fluidized bed combustor units utilizing pulverized coal with limestone or dolomite added to the bed to absorb sulfur dioxide. The bed is cooled internally by steam coils. Ninety percent sulfur removal along with low NO_x is possible. Stack gases are cleaned mechanically and electrostatically. (See Energy Technology Conference VIII, Government Institutes, Rockville, MD, pp. 893–919.)

3. Computers, which quickly correlate meteorological and plant operational data, so as to anticipate and prevent unacceptable ground-level concentrations.

4. Sensors and monitors, which sense and record automatically the local ambient SO_2 concentrations, and pH's, thereby ensuring the effectiveness of the control operations. (Limestone application is made directly to lakes in the area to raise pH.)

5. Emission limitations, accomplished by fuel switching between high- and low-sulfur fuels during normal power plant operations, and by local power generation curtailment during emergency situations, when even switching to low-sulfur fuels might not provide acceptable ground-level concentrations.

6. Daily local plant weather observations and meteorological measurements by means of light aircraft and balloons. These approaches allow high sulfur fuels to be burned most of the time, thereby conserving scarce low sulfur fuels for domestic and other ground level uses not employing SDEL methods.

As opposed to wet scrubbers for SO_2 control, SDEL methods are estimated to require less than one-tenth the total investment cost and one-thirteenth the annual operating cost, exclusive of wet sludge disposal costs.

Atmospheric Dispersion of Pollutants from Stacks

Stacks can provide effective atmospheric dispersion of gaseous and particulate pollutants with acceptable ground-level concentrations. Theoretical and empirical formulas are available to estimate the dispersion of air borne pollutants continuously emitted from stacks. Effective stack height of a plume is given by $H = h_s + h_r$. (See notation below.)

Notation

Symbol	Definition	System of consistent units
b	Atmospheric-dispersal parameter	Dimensionless
d	Stack-exit diameter	m
H	Effective stack height	m
h_r	Rise of plume	m
h_s	Stack height	m
p	Atmospheric-dispersal parameter	Dimensionless
Q	Emission rate of pollutant	m^3/s
Q_h	Heat emission rate	cal/s
q	Atmospheric-dispersal parameter	Dimensionless
u	Mean wind speed	m/s
v	Stack-exit velocity	m/s
X_g	Ground-level concentration of pollutant at a distance x from base of stack	ppm by vol
X_{mg}	Maximum ground-level concentration of pollutant at a distance x_m from base of stack	ppm by vol
x	Downwind distance from emission source in direction of mean wind	m
x_m	Distance from base of stack to maximum ground-level concentration	m
σ_A	Basic diffusion parameter related to azimuth plume angle in radians	m^{1-q}
σ_E	Basic diffusion parameter related to elevation plume angle in radians	m^{1-p}

The Holland formula, recommended for calculating h_r, is $h_r = (1.5vd + 4.09 \times 10^{-5}Q_h)/u$. A ratio of stack height h_s to building height of 2½ to 1 or more is commonly used to avoid entrapment of the plume in the vortex of adjacent buildings and the associated high values of ground-level concentration X_g. Stack-exit velocity v should be 14 to 28 m/s (45 to 90 ft/s) to minimize plume entrapment in stack vortices. The lower the ratio of stack inside to outside diameter at its top, the higher the necessary v.

The **Cramer equation,** recommended for calculating X_g, is $X_g = 10^6 Q/(\pi u x^b \sigma_A \sigma_E) \exp [H^2/(2\sigma_E^2 x^{2p})]$. The equation is for sampling intervals of approximately 20 min, with X_g varying approximately inversely as the one-fifth power of the interval.

Simplified approximate formulas for X_{mg} and x_m are

$$X_{mg} = 2Q \, 10^6 \sigma_E/(\pi e^{uH^{1.9}} \sigma_A) \text{ and } x_m = H/(1.35 \, \sigma_E)$$

Atmospheric dispersal characteristics and hence the values of X_g, X_{mg}, and x_m are influenced by thermal stratification and atmospheric turbulence. A superadiabatic (adiabatic) (subadiabatic) lapse rate, i.e., a

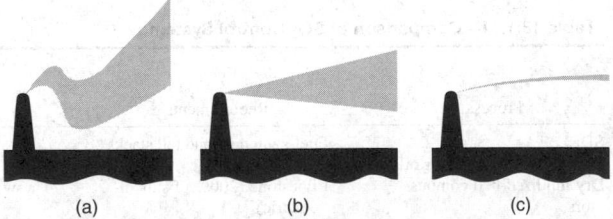

Fig. 18.1.3 Height and plume behavior for (a) unstable, (b) neutral, and (c) stable conditions.

drop in temperature with altitude greater than (equal to) (less than) 5.4°F per 1,000 ft (9.84 K/km) results in an unstable (neutral) (stable) atmosphere with much (moderate) (very little) vertical mixing. (See Fig. 18.1.3.)

The recommended parameters for use in the Cramer equations under the above conditions of thermal stratification are:

Thermal stratification	σ_A, rad	σ_E, rad	b	p	q
Unstable	0.39	0.13	2	1.1	0.9
Neutral	0.21	0.07	1.85	1.0	0.85
Stable	0.12	0.04	1.6	0.8	0.8

The above formulas deal only with steady-state conditions. There are transient meteorological conditions which can result in significant ground-level concentrations of pollutants and which cannot readily be analyzed by computations. These conditions include the breakup of a temperature inversion and the slow, steady buildup of pollutants during stagnations accompanied by deep ground fogs in a valley. For large power plants, high stacks (400 ft and more) (122 m), high stack-exit velocities [45 to 90 ft/s (13.7 to 27.4 m/s)], and adequate exit temperatures [250 to 300°F (394 to 422 K)] are usually effective in avoiding such fumigations.

Stack Emissions

Emission of sulfur-bearing gases and solids from a 1,000-MW fossil-fuel fired power station is shown in Table 18.1.10. (See also Tables 18.1.12 and 18.1.13 for more data.) Reducing the remaining solids in stack emissions from bituminous coal fired (excluding stoker firing) power boilers to 0.02 gr/standard ft³ (45.8 mg/m³) and less results in a clear stack discharge. This requires collection efficiencies of 99+ percent, attainable with commercial equipment. These high collection-efficiency requirements demonstrate the need for optical instruments which will readily determine stack-emission quality. The **Ringelman chart** is frequently used to evaluate stack emission, but it is a crude and inaccurate method. A good indicator is a stack emission invisible to the naked human eye. With oil firing, present practice is to control stack emissions by furnace design, combustion control, use of additives, and multiple

cyclone-type collectors located in the low- or high-temperature zones of the flue-gas system. For future plants of 1,000 MW and larger, use of high-temperature electrostatic precipitators located ahead of the air preheater may be considered.

Air Pollution Control in Various Industries

(See Tables 18.1.12 and 18.1.13.)

Industrial sources of air pollution and typical loss rates are given in Table 18.1.8.

Nonferrous metal smelting (e.g., copper, zinc, lead) and thermal operations (e.g., blast furnace, sintering, converters) are sources of air pollution and of potentially recoverable metallic materials. A large variety of collection equipment has been tried. Experience has proved that precipitators, cloth filters, and gas washers are best.

Paper industry sources of air pollution are (1) recovery furnace gases and (2) lime kiln exhaust gases. Noxious gases require scrubbing. **Recovery-furnace gases** in the sulfate and soda processes of manufacture emit solids [2 to 6 gr/standard ft³ (4.6 to 13.7 g/m³)] which are mainly sodium salts. The small particle size is particularly bothersome because of the large, obscuring plume in the stack tail. Chemically, the solids are destructive to painted and finished surfaces (e.g., automobiles, structures). Valuable constituents are recovered for return to the cycle. Electrostatic precipitators with 95 to 99 percent efficiency are favored for collection and return of solids. Corrosion-resistant materials (glazed tile, special cements) must be used to reduce maintenance. **Lime kilns** are used to calcine calcium carbonate and recover it as CaO for use in the process. The consequent air pollutant is recovered by gas washers and mechanical collectors.

Cement industry sources of air pollutants are (1) clinker kilns and (2) bagging and other mechanical operations. The **kilns** are the major source of air pollution, with large, unattractive exhaust plumes depositing cement-making solids on the landscape and on painted surfaces. Damage

Table 18.1.12 Common Applications of Industrial Precipitators

Industry	Application	Gas-flow range, ft³/min*	Temp. range, °F	Percent weight of dust (below 10)	Usual efficiencies, %	Dust conc. range in ppm† by wt in air	Temp. range, K
Electric power	Fly ash from pulverized-coal-fired boilers	50,000–750,000	270–600	25–75	95–99+	760–9,500	405–589
Portland cement	Dust from kilns	50,000–1,000,000	300–750	35–75	85–99+	950–29,000	422–672
	Dust from dryers	30,000–100,000	125–350	10–60	95–99	1,905–29,000	325–450
	Mill ventilation	2,000–10,000	50–125	35–75	95–99	9,500–98,000	283–325
Steel	Cleaning blast-furnace gas for fuel	20,000–100,000	90–110	100	95–99	38–950	306–317
	Collecting tars from coke-oven gases	50,000–200,000	80–120	100	95–99	190–1,905	300–322
	Collecting fume from open-hearth and electric furnaces	30,000–400,000	300–700	95	90–99	95–5,700	422–644
Nonferrous metals	Fume from kilns, roasters, sintering machines, aluminum pot-lines, etc.	5,000–1,000,000	150–1100	10–100	90–98	5,700–95,000	339–867
Pulp and paper	Soda-fume recovery in kraft pulp mills	50,000–200,000	275–350	99	90–95	950–7,600	408–450
Chemical	Acid mist	2,500–20,000	100–200	100	95–99	38–1,905	311–367
	Cleaning hydrogen, CO₂, SO₂, etc.	5,000–20,000	70–200	100	90–99	19–1,905	295–367
	Separate dust from vaporized phosphorus	2,500–7,500	500–600	30–85	99+	19–1,905	533–589
Petroleum	Powdered catalyst recovery	50,000–150,000	350–550	50–75	90–99.9	190–4,800	450–561
Rock products	Roofing, magnesite, dolomite, etc.	5,000–200,000	100–700	30–45	90–98	950–4,800	311–644
Gas	Tar from gas	2,000–50,000	50–150	100	90–98	19–3,800	283–339
Carbon black	Collecting and agglomerating carbon black	20,000–150,000	300–700	100	10–35	57–9,500	422–644
Gypsum	Dust from kettles, conveyors, etc.	5,000–20,000	250–350	95	90–98	2,850–9,500	394–450

* ft³/min = 4.72(10)⁻⁴ m³/s.
† 1 ppm = 0.525 gr/1,000 ft³ = 2 mg/m³, in air.
SOURCE: Holden and Ackley, "Air Pollution Handbook," McGraw-Hill.

Table 18.1.13   Characteristics of Air- and Gas-Cleaning Devices

General class	Specific type	Device most suitable for	Removable contaminants	Optimum size particle, μm	Limits of gas temperature, °F	Opt. conc., ppm by wt	Lim. gas temp., K
Odor adsorbers	Shallow bed			(Molecular)	0–100	<1.9	256–311
Air washers	Spray chamber		Malodors, gases	>20	40–700		
	Wet cell			>5	40–700	<9.5	278–644
Electro. precip., low-voltage	Two-stage, plate		Lints, dusts, pollens, tobacco smoke	<1	0–250		
	Two-stage, filter	Atmospheric air cleaning		<1	0–180	<1.9	256–394
Air filters, viscous-coated	Throwaway			>5	0–180	<3.81	256–356
	Washable			>5	0–250	<3.81	256–394
Air filters, dry-fiber	5–10 μm			>3	0–180	<1.9	256–356
	2–5 μm			>0.5	0–180	<1.9	256–356
Absolute filters	Paper		Special†	<1	0–1,800	<1.9	256–1,256
Industrial filters	Cloth bag			>0.3	0–180‡	>190	256–356
	Cloth envelope			>0.3	0–180‡	>190	256–356
Electro, precip., high-voltage	Single-stage, plate			<2	0–700	>190	256–644
	Single-stage, pipe			<2	0–700	>190	256–644
Dry inertial collectors	Settling chamber			>50	0–700	>9,520	256–644
	Baffled chamber		Dusts, fumes, smokes, mists	>50	0–700	>9,520	256–644
	Skimming chamber			>20	0–700	>1,905	256–644
	Cyclone			>10	0–700	>1,905	256–644
	Multiple-cyclone			>5	0–700	>1,905	256–644
	Impingement			>10	0–700	>1,905	256–644
	Dynamic	Stack gas cleaning		>10	0–700	>1,905	256–644
Scrubbers§	Cyclone			>10	40–700	>1,905	256–644
	Impingement			>5	40–700	>1,905	256–644
	Dynamic			>10	40–700	>1,905	256–644
	Fog			<2	40–700	>190	256–644
	Pebble bed			>5	40–700	>190	256–644
	Multidynamic			<1	40–700	>190	256–644
	Venturi			<2	40–700	>190	256–644
	Submerged nozzle			>2	40–700	>190	256–644
	Jet			<5	40–700	>190	256–644
Incinerators Afterburners	Direct		Gases, vapors, malodors	Any	2,000	Combustible	1,367
	Catalytic			(Molecular)	1,000	any	811
Gas absorbers	Spray tower			(Molecular)	40–100	>1.9	278–311
	Packed column			(Molecular)	40–100	>1.9	278–311
	Fiber cell			(Molecular)	40–100	>1.9	278–311
Gas adsorbers	Deep bed			(Molecular)	0–100	>1.9	256–311

* Based on std. air @ 0.075 lb/ft³ (1.2 kg/m³).
† Bacteria, radioactive, or highly toxic fumes.
‡ 500°F for glass (553 K), 450°F for Teflon (505 K), 275°F for dacron (408 K), and 240°F for orion (389 K).
§ Reheating of scrubbed stack is necessary to avoid plumes.
SOURCE: Jorgensen, "Fan Engineering," Buffalo Forge Co. Used by permission.

is reduced by the use of electrostatic precipitators with the wet process and cyclones plus cloth filters with the dry process. **Bagging and other mechanical losses** are reduced by use of cloth filters following cyclone separators.

Carbon-black production utilizes electrostatic precipitators, cloth filters, and scrubbers to collect the black from the gas stream and to control air pollution.

Iron-and-steel-industry sources of air pollution are cupolas, oxygen-furnaces, electric-furnaces, and sintering operations. The dispersoids in these operations are mostly oxides of iron in particle sizes below 10 μm and with loadings ranging to 15 gr/standard ft³ (34.3 g/m³). Dust emittance is acceptably controlled generally by use of precipitators and scrubbers for open-hearth and basic oxygen processes, use of cloth filters and scrubbers for electric furnaces, and use of precipitators, scrubbers, and filters for sintering.

Petroleum-refining sources of air pollution are mainly gaseous in nature. Mercaptans are removed by scrubbing. Solid dispersoids from the fluid-cracking processes are removed by electrostatic precipitators and/or cyclone-type collectors.

Miscellaneous Dusts Four industrial dusts have been classified as causing severe respiratory ailments: asbestos, causing asbestosis; crystalline silica, causing silicosis; cotton and cottonseed dust, a powerful allergen causing bronchial asthma and byssonosis; and lead, causing plumbism.

In mining operations and foundries, cyclonic collectors, followed by baghouses, have been found most effective. Likewise, in textiles and in cottonseed-oil pressing, baghouses following the lint-collecting cyclones can remove almost 100 percent of the hazardous pollutant.

Large central incineration plants can produce odors, noxious gases, and particulates (see Sec. 7.4). Odors and noxious gases can be minimized by carefully controlled combustion, using modern traveling-grate continuous systems. Furnace temperatures between 1,600 and 1,900°F (1,150 to 1,300 K) are recommended. Particulates can be minimized by well-managed covered conveying of dust-bearing refuse, incinerator residues and fly ash, and the use of electrostatic precipitators and wet scrubbers to comply with increasingly stringent air quality standards. (See Table 18.1.13.) Gaseous emissions of SO_x, CO, and NO_x do not cause problems with continuous incineration, as their concentrations are small, but HCl and HF from plastics are hazardous. Stack height should be sufficient to minimize ground effects of any residual pollution. Municipal refuse can also be fired into waste heat utility boilers, a fuel conservation measure which will be increasingly important in electric power cogeneration.

RADIOACTIVE WASTE MANAGEMENT

REFERENCES: Cochran and Tsoulfanidis, ''The Nuclear Fuel Cycle: Analysis and Management,'' American Nuclear Society. ''Radiological Health Handbook.'' Nuclear Waste Policy Act of 1982 and its 1987 amendments. Low Level Radioactive Waste Policy Act of 1980 and its 1985 amendments, 10CFR 50 Appendix 1, and 10 CFR 20.

General Considerations

Under existing legislation and authorization the Nuclear Regulatory Commission (NRC) has the primary responsibility for developing and regulating waste management methods and practices, sites of operation, and enforcement of all applicable standards, including those developed adjunctively with EPA. Radioactive wastes are generally divided into five general categories by the Nuclear Waste Policy Act: (1) **high-level waste (HLW)**; (2) **transuranic (TRU) waste**; (3) **low-level waste (LLW)**; (4) **uranium mill tailings**; and (5) **naturally occurring and accelerator-produced radioactive material (NARM)**. HLW is ''the highly radioactive material resulting from the reprocessing of spent nuclear fuel, including liquid waste produced directly in reprocessing, and any solid material derived from such liquid waste that contains fission products in sufficient concentrations, and other highly radioactive material that the NRC, consistent with existing law, determines by rule requires permanent isolation.'' TRU wastes are wastes that contain alpha-particle-emitting radioisotopes with atomic numbers greater than 92 and half-lives longer than 5 years. LLW is defined as material which is ''not HLW, spent nuclear fuel, TRU or by-product material, and material which the NRC, consistent with existing law, classifies as LLW.'' LLW is generated by the nuclear power industry, hospitals, and other medical facilities, universities, and research laboratories. Examples of LLW include contaminated clothing, tools, syringes, cotton wipes, paper, rags, and other trash. Uranium mill tailings also are treated as radioactive wastes and are not transported from their generation point. NARM is not regulated by the NRC, but is regulated by the individual states. Usually, most NARM qualifies as LLW and most often is treated as such.

HLW and TRU wastes are covered under the **Nuclear Waste Policy Act (NWPA)**, which spells out the step-by-step procedures for all tasks related to their disposal. The NWPA details the process to be followed in the selection of repositories for the disposal of HLW, TRU waste, and spent nuclear fuel; the development of interim storage and monitored retrievable storage for HLW and spent fuel; and the responsibilities and obligations of the Department of Energy and the nuclear utilities.

LLW is governed by the **Low Level Radioactive Waste Policy Act (LLRWPA)**. The LLRWPA dictates that every state is responsible for the disposal of all LLW generated within its borders, and assumes that the LLW can be stored safely on a regional basis. To carry out the disposal of LLW, many states have entered into ''compacts'' with other states to site and operate regional disposal facilities. Other states have chosen to develop their own disposal sites. The LLW generated by nuclear power plants usually consists of three types: gases, liquids, and solids.

Gases   From the buildings which contain the fuel-handling system, auxiliary equipment, and waste systems, ventilating air, filtered if necessary, is released through a vent which is monitored for radioactive gases, iodine, and airborne particulates. Automatic controls will shut down the ventilating system, thereby stopping any release of airborne activity if the monitor set points.

Ventilating air for the containment vessel, which holds the reactor, primary pumps, and steam generators, is drawn when it is necessary for inspections or for other reasons to purge the vessel. **Purge air** is filtered through high-efficiency particulate air (HEPA) and through charcoal filters before release. In case monitor set points are exceeded, purging is automatically curtailed.

The third source of gases is the nitrogen cover gas of the volume control system in the PWR, and the steam jet ejector system in the BWR. Krypton, xenon, and iodine are produced within the reactor and mix with the cover gas. To release them, the principle employed is to ''delay and decay'' by compressing them into decay tanks of about 45 days' storage capacity in the PWR case, and into decay pipes of 20- to 60-min storage capacity in the BWR case. All releases are carefully monitored to be within established limits.

Liquids   Liquid radwaste streams from boiling water reactors may be classified into two major systems. In one of these, relatively high-radioactivity, low-conductivity water is reclaimed by filtration and deionization, and is stored for recycling to plant makeup systems. In the other, relatively low-radioactivity, high-conductivity water from floor and laboratory drains, laundry drains, makeup demineralizer regeneration, and equipment decontamination is treated by various means, such as filtration, waste concentration, and deionization, is monitored and recycled for in-plant use. Pressurized water reactors can be characterized by three major liquid streams. The first of these consists of borated, relatively high-radioactivity demineralized water from the primary cooling loop. During the feed-and-bleed operating mode, borated water is removed and replaced with freshwater, which slowly reduces the amount of chemical shim (boric acid) present. A boric acid evaporator makes it possible to recover boric acid and to recycle a large portion of the distillate. The second major stream consists of floor and equipment drains from the containment vessel and the auxiliary building and from the radiochemistry laboratory, which go to a large **waste holdup tank**. These are subsequently transferred to an evaporator, and the distillate therefrom is deionized, passed through a radiation monitor, and then recycled. With both types of reactor, the heavy ends, crud, and dirt concentrate in the evaporator feed tank. This waste is pumped to the radwaste **drumming station**, where it is solidified with Portland cement or chemical grout. The third of the major streams is soapy water from laundry and showers. Because of foaming difficulties, this stream is not always processed through an evaporator, but instead is filtered or put through a reverse osmosis unit, monitored for radioactivity, and if satisfactory, released.

Solids   Solid wastes for BWRs and PWRs are handled similarly. Evaporator and filter sludges are pumped to the radwaste drumming station and solidified with Portland cement or chemical grout in 55-gal (or larger) drums. Spent ion-exchange resins that are not regenerated are sent to an accumulation tank for decay and are then flushed into a shipping cask and dewatered for removal to a burial ground. They may also be solidified. Balable wastes are drummed but, if highly radioactive, are set in concrete. Large, heavy, junk equipment items can sometimes be stored on-site more economically than cutting them up and packaging them for off-site shipment.

Reference to current federal regulations and trade publications must be made to keep abreast of the latest developments.

SOLID AND HAZARDOUS WASTE MANAGEMENT

REFERENCES: ''Pollution Control Technology, Solid Waste Disposal,'' Research & Educational Assn., New York. Sittig, ''Handbook of Toxic and Hazardous Chemicals''; also Sittig, ''Resource Recovery and Recycling Handbook of Industrial Wastes,'' Noyes Data Corp., Park Ridge, NJ. Exner, ''Detoxication of Hazardous Waste,'' Ann Arbor Science Publishers. The Resource Conservation and Recovery Act of 1976 (RCRA) (as amended in 1980), PL 96-482 and CERCLA, PL 96-510 (SW-171). Metry, ''The Handbook of Hazardous Waste Management,'' Technomic Publishing, Westport, CT. Peirce and Vesilind, ''Hazardous Waste Management,'' Ann Arbor Science Publishers, Ann Arbor, MI. Current EPA regulations.

Solid waste includes solids discarded permanently, or temporarily, and materials suspended in air or water. Also forming part of this group are wet solids with insufficient liquid content to be free-flowing. Total solid wastes (excluding sewage) produced currently in the United States

Table 18.1.14   Cost/Benefit Evaluation of Recycled Sewage Sludge as Fertilizer (Denver)*

	Total cost delivery and application, \$/dry ton	Current worth of nutrients, \$/dry ton	Disposal cost savings, \$/dry ton	Total benefit (worth and savings), \$/dry ton	Cost/ benefit
Air-dried (trucked)	\$ 4.00	\$ 8.37	\$11.34	\$19.71	0.20
Liquid (5% solids) (trucked)	14.00	12.20	11.34	23.54	0.595
Liquid (5% solids) (pipelined)	9.00	12.20	11.34	23.54	0.382

* 1975 data.

amount to about 250 million tons/year, or about 1 ton/(person · year), and include all industrial, domestic, commercial, and agricultural wastes. Between 4 and 5 percent of that total constitutes **hazardous wastes** subject to special handling and/or disposal. It is projected that the total solid waste stream will increase to 300 to 350 million tons by early in the twenty-first century. The increasing accumulation of dry waste which results from the high standard of living of modern society will deplete existing economically exploited natural resources and further aggravate the already costly waste disposal problem. Product and energy recovery, therefore, constitutes the preferable direction to follow rather than elaborate means of disposal. There is, however, a growth factor for solid residuals which is caused by environmental control legislation: unpredictable amounts of waste generated by air and water pollution control. Environmental regulations in effect simply transfer the pollution from those media to the solid state, and vice versa.

Federal Environmental Legislation   See RCRA and CERCLA following.

Solid Waste Disposal

Solid waste handling involves two main steps: collection and disposal. **Solid waste conditioning** involves recovering those materials of value already present while conditioning the rest for conversion, recycling, or ultimate disposal. **Solid waste segregation** prior to collection into refuse, combustible wastes, noncombustible wastes, and hazardous waste is always a beneficial preconditioning step.

The principal solid disposal methods used are **incineration, landfill,** and **ocean disposal.** Many mining and processing operations leave a process residue on the premises since disposal cost elsewhere is prohibitive. Examples are red mud from refining bauxite and sulfite sludge from SO_2 stack gas scrubbing processes.

Landfilling requires spreading and compacting the waste and covering it with soil. This system is applicable both to biodegradable and nondegradable types of wastes. Leaching contamination of ground water must be prevented by installing liners. Careful attention must be given to possible subsidence and gas formation in landfilling. Topsoil disposal is limited to biodegradable organic materials.

Ocean dumping takes advantage of the ocean bottom trench configuration and the dispersing effects of ocean currents. Until recently, ocean dumping was the ultimate form of ocean disposal, but the practice has effectively ceased. The cost/benefit of ocean disposal was very favorable compared to alternate methods, but is now highly regulated and, with rare exceptions, the practice is proscribed and illegal. Ecological considerations effectively rule out this form of waste disposal for the foreseeable future.

Incineration is a volume reduction process rather than solid waste disposal, resulting in a residue which still must ultimately be disposed of. Heating values of solid wastes range from 5,000 to 15,000 Btu/lb for ordinary trash and industrial wastes. The growth of incineration is affected by higher fuel costs, construction costs, and added expense for air pollution and hazardous waste control.

Another means of solid waste reduction is **composting,** not widely used because of the limited market for its products, but ultimately may become more widely spread to maintain a rural-urban ecological balance.

Product Recovery

The average solid waste (per 100 lb) produced daily in the United States is roughly 6 lb industrial (from process), 5 lb municipal, and 89 lb mine tailings, smelter slags, dredging spoil, and agricultural wastes. Materials salvaged from solid waste recycled back to their sources not only reduce disposal costs but also conserve farmland values and natural resources. Recovery of metal, paper, glass, plastics, and hazardous substances is an outstanding example of materials recycling.

Materials separation from solid waste may be performed by physical or chemical means. The most common types of equipment used are material separators (varies with application: magnetic, thermal, flotation, size reduction—shredders, crushers, grinders, and pulverizers), materials flow (conveyors, lifters), mixers, and blenders. (See Sec. 10.)

Basic chemical conversion processes for industrial waste are hydrogenation, oxidation, acetylation, cross-linking cellulosic materials. For more detail, see Engdahl, "Solid Waste Processing—A State of the Art Report on Unit Operations and Processes," Public Health Service Publication No. 1856, Supt. of Documents, Washington, DC.

Biochemical processing of wastes from the food and chemical industries is common practice. An example is recovery of B-complex vitamins and animal feed from solid brewery wastes.

Pyrolysis of organic wastes can economically utilize them as supplementary fuels depending upon waste types and waste energy conversion processes. New efforts are being made in that direction to process municipal refuse as well as industrial and hazardous wastes.

Dry sludge production from municipal sewage has been achieved with the development of activated sludge processes using high purity oxygen. These processes reduce the volume of activated sludge by two-thirds and improve its dewatering characteristics.

Table 18.1.14 itemizes the 1975 cost factors at Denver from which cost benefit ratios have been developed. (See Cost/Benefit Balancing.) The cost benefit for air-dried sludge is particularly favorable, inasmuch as for every 20¢ invested, the public consumer receives \$1.00 in benefits (1990 data).

RCRA AND CERCLA

The Resource Conservation and Recovery Act of 1976 (RCRA), as amended in 1980, regulates current and future waste disposal practices, including both municipal and hazardous waste, while the **Comprehensive Environmental Response Compensation and Liability Act (CERCLA),** or "Superfund," cleans up old, hazardous waste sites. Also, see 39 CFR 158, parts 240 and 241. "Thermal Processing and Land Disposal of Solid Waste."

Hazardous Waste Management   RCRA defines *hazardous waste* as a solid waste, or combination of solid wastes, which because of its quantity, concentration, or physical, chemical, or infectious characteristics may:

cause, or significantly contribute to an increase in mortality or an increase in serious irreversible, or incapacitating reversible, illness; or pose a substantial present or potential hazard to human health or the environment when improperly treated, stored, transported, or disposed of, or otherwise managed.

The act requires EPA to promulgate regulations establishing criteria and standards for all aspects of hazardous waste management. The regulations developed by the EPA designate a large number of substances as hazardous and also provide specific characteristics to determine whether a particular waste is hazardous. The characteristics include ignitability, corrosivity, reactivity, and EP toxicity.

The EPA also promulgated regulations establishing standards for hazardous waste treatment, storage, and disposal facilities. Provisions include groundwater protection, development and implementation of contingency plans, closure and postclosure activities, and specific facility performance standards. The standards cover facilities such as tanks, surface impoundments, land treatment, landfills, incineration, underground injection, and physical, chemical, and biological treatment. Other regulations cover "cradle-to-grave" control of the collection, source separation, storage, transportation, processing, treatment, recovery, and disposal of such wastes. Generators of these wastes bear the ultimate legal responsibility and are required to keep specific records for the disposition of their hazardous wastes by means of a manifest system.

18.2   OCCUPATIONAL SAFETY AND HEALTH
by Pascal M. Rapier and Andrew C. Klein

REFERENCES: "General Industry Guide for Applying Safety and Health Standards," 29CFR 1910 OSHA, U.S. Dept. of Labor, 1973, and Amendments. 29CRF 1910.1–1910.309, *Fed. Reg.*, **37**:202, Oct. 18, 1972, and Amendments. "Handbook of Heavy Construction," McGraw-Hill. Labor laws applicable in the jurisdiction where the working enterprise is located.

WORKER'S COMPENSATION LAWS

Worker's compensation laws in this country were introduced in 1911 and have now been enacted in all the states. The principle involved is that the worker injured or disabled in industry should be enabled, through proper medical treatment, to return to wage-earning capacity as promptly as possible and, while incapacitated, should receive compensation in lieu of wages, regardless of fault. The expense of medical treatment and compensation should properly be borne by industry and become a part of the cost of its products. The laws generally provide that workers injured in industry shall be furnished the necessary medical treatment, and, in addition, compensation based on a percentage of their average weekly wages, payable periodically. Dependents of employees killed in industry are likewise compensated.

Many states have supplemented their worker's compensation laws by providing comparable benefits in cases of incapacity or death due to occupational disease. Some states make these provisions applicable only in case of specified occupational diseases; others make them of general application to cases of any disease directly attributable to the employment. At the present time, all states and the District of Columbia provide for compensation benefits in occupational-disease cases either by enlarging the scope of the worker's compensation law, by separate legislative enactment, or by judicial construction.

The enactment of worker's compensation laws has been followed in many jurisdictions by more stringent provisions relating to factory inspections for the prevention of accidents in industry and of occupational disease.

The enactment of worker's compensation and occupational-disease laws has increased materially the cost of insurance to industry. The increased cost and the certainty with which it is applied have put a premium on accident-prevention work. This cost can be materially reduced by the installation of safety devices. Experience has shown that approximately 80 percent of all industrial accidents are preventable.

SAFETY

REFERENCES: "General Industry Guide for Applying Safety and Health Standards," 29CFR 1910 OSHA, U.S. Dept. of Labor, 1973. 29CRF 1910.1–1910.309, *Fed. Reg.*, **37**:202, Oct. 18, 1972. "Accident Prevention Manual for Industrial Operations." National Safety Council. "Manual of Accident Prevention in Construction," Associated General Contractors of America, Inc. American Conference of Governmental Industrial Hygienists, "Industrial Ventilation—A Manual of Recommended Practice." Blake, "Industrial Safety." Prentice-Hall. DeReamer, "Modern Safety Practices," Wiley. Heinrich, "Industrial Accident Prevention," McGraw-Hill. Patty (ed.), "Industrial Hygiene and Toxicology," Interscience. Simonds and Grimaldi, "Safety Management," Irwin. Stubbs (ed.),

The person(s) responsible for overseeing the safety of all operating personnel must be cognizant of the latest laws and regulations pertaining to worker safety, which are changed and/or updated from time to time.

The logical time to install safety devices is when new machines are being built, while general construction work is being done, or when alterations or repairs are being made; results can be accomplished with a minimum of expense and delay at the time plans and specifications are being prepared.

Checking for Safety   To ensure that the question of safety will not be overlooked, it is well to have all plans, specifications, and drawings checked for safety, making special provision for this in each set of specifications and in the title plate of each drawing.

Buildings

(See 29CFR 1910, Subparts D, E, and F, as amended.)

Plant Arrangement   A fundamental factor in effective accident prevention is the provision of ample ground space in the plant site to reduce crowding of buildings, congestion of plant traffic, unusual fire hazards, unsafe yard conditions, etc. One-story buildings have definite advantages in regard to fire hazards, building collapse, and natural lighting.

Where a plant is located near a main line of a railroad, consideration for safe access by employees and vehicular traffic should be given serious attention. Safe passageways from one building to another are also important. Blind corners, doorways opening onto yard railway tracks, etc., should be avoided as much as possible, and, where such conditions must necessarily exist, safety railings, gates, signs, or gongs should be installed to warn pedestrians of danger.

Because of their inherent **hazards of fire and explosion,** the storage, handling, and utilization of *flammable liquids* should be carefully controlled. The National Fire Codes, promulgated by the NFPA, are considered authoritative in establishing minimum safeguards necessary to control these hazards. It is necessary in any event to conform to the requirements of any applicable state and local code.

Buildings in which **dusty operations** are carried on should be designed to present a minimum area of projections, ledges, and resting places for dust accumulations.

Floors, Stairways, Aisles, etc.   Probably the most important factor to be considered in connection with floors, stairways, etc., concerns **slipperiness.** Floors and stairs should be free from projecting nails, boltheads, etc., as noiseless as possible, wear well, and be strong enough to carry safely any static or moving load. The weight of modern industrial machinery and material handling equipment should be carefully considered in checking floor-load calculations. Floors and stairways should be kept clear of unnecessary obstructions over which workers may trip.

Spilling of oil, water, acid, etc., should be prevented to eliminate slipping hazards. Splash guards, drip collectors, etc., can be designed in many instances to reduce slippage. Excessive spillage of dusty materials onto the floor is sometimes taken care of by installing floor gratings beneath which pits or conveyor systems are located to collect the falling material.

Ample aisle space is very important, especially in foundries where workers carry ladles of molten metal and where there is considerable shop traffic involving power-driven trucks, etc. Aisles should be clearly marked off to assist in keeping them clear. One-way traffic is often advantageous.

Stairways should be provided with **handrails** on both sides, and an intermediate middle handrail should be installed on stairways over 88 in in width. Nonslip treads on stairs are desirable. Stairways should be adequately lighted (see Sec. 12.5).

Exits and Fire Escapes (See Life Safety Code NFPA.) As far as practicable, all doors should open outward or with the natural direction of egress; they must not block passageways from other floors or parts of the building.

For factories, not less than two means of exit should be provided on every floor, including basements, of all buildings or sections; these exits should be separated in such a manner that they are not likely to be cut off by a single local fire. The location of stairways around or adjacent to passenger elevators is undesirable unless there is separation by fire walls.

Outside fire escapes are inferior to stairways as means of egress. Where they are used, they should be located on blank walls or arranged so that persons on them will be protected from flames issuing from windows or openings underneath by use of wired glass in standard metal frames, fire doors, etc. Outside fire escapes, to provide maximum protection for persons using them during fire, should be enclosed in noncombustible towers which will protect against weather, smoke, or fire, with access through or over some intermediate balcony or structure to the building proper.

Lighting (See Sec. 12.5) Adequate lighting has a definite bearing on the prevention of accidents. Workrooms should be well lighted to reduce eye strain and the possibility of permanent eye impairment and also to remove any danger of employees falling over obstructions or being caught in machinery in darkened areas.

Ventilation (See Sec. 12.4) A lack of adequate ventilation in a workroom tends to bring on fatigue and reduces the alertness of workers, thus making them more susceptible to accidents. Where injurious dusts or noxious vapors are encountered, it is necessary to provide for their removal by the installation of adequate local exhaust systems (see Sec. 12.4 and ACGIH ''Industrial Ventilation'' and ASA Standard, ''Fundamentals Governing the Design and Operation of Local Exhaust Systems'').

Identification of Piping (See Sec. 8.7) It is desirable that a plan of identifying the contents of various pipelines be adopted so that in case of an emergency it will be possible to determine quickly the service of all pipelines involved.

Mechanical Guarding in Machine Design

(29CFR 1910, Subpart 0, as Amended.)
The most logical time at which to consider the safeguarding of a machine is during its design. In this stage, features of safe operation can be incorporated so that there will be a minimum of specific guarding required on the finished machine. The following points are fundamental considerations concerning accident prevention which should be taken into account in machine design.

Care should be taken in arranging clearances of moving parts to avoid shearing or crushing points in which hands or other parts of operator's body might be caught or injured.

Arrangements should be made so that adjustments, inspections, and hand lubrications can be done safely.

Machines should be so designed that operators are not required to stand in an uncomfortable position, reach over moving parts, or exert themselves in awkward positions.

Machines should be designed so that there will be little danger of the operator tripping over parts of the frame or striking against projecting parts during normal operation movements.

Careful attention should be given to strength of all parts whose failure might result in injury to operator.

All guards, covers, or enclosures should be designed strong enough to prevent the possibility of their giving way and permitting an accident in case the operator should fall or be thrown against them.

Point-of-Operation Guarding The point of operation on a machine is taken to be that zone where the work of the machine is actually performed and where the operator, by manipulating the material being processed, is exposed to a hazard from moving parts of the machine. Guards for point-of-operation protection are placed on the machines as additional equipment. The first requirement for a successful guard of this type is that it shall be convenient and not interfere with the operator's movement or affect the output of the machine. The following statements describe several basic principles which may be utilized in point-of-operation guarding.

Where possible, the **danger point** should be completely **covered** by a barrier or enclosure before the hazardous operation of the machine begins. This may be accomplished for example on a treadle-controlled machine by having the treadle operate the guard which, when in proper location, will then activate the machine.

Operator's hands may be kept out of dangerous positions by installing **starting devices** so located that, to operate the machines, both hands of the operator must be out of the danger zone.

Feeding devices may be used so that material to be processed is placed in a feeding mechanism at a point where there is no exposure to moving parts. Special holders or feeding tongs can also be used to place work in hazardous positions.

Electronic controls, which operate by the interruption of a light beam or other energy source to protect the danger zone, may be used to start and stop machines.

Electrical interlocks on guarding devices may be utilized in operating circuits so that unless a guard is in proper position the circuit is open and no current will flow until the machine is safely protected by the guard being brought into proper position.

Automation has minimized the hazards associated with the manual handling of stock and has eliminated the need for repetitive exposure at the point of operation, but the urgency of making repairs or adjustments introduces the need for special precautions covering maintenance work, such as locking out the power source.

Electric Equipment

(See 29CFR 1910, Subpart S, as amended.)

In considering electrical equipment for industrial establishments, it should be borne in mind that the **hazard** to human life **increases with increase in voltage.** Thus, economy in transmission wiring and copper parts, which is achieved through the use of high voltage motors, may be offset by increased danger of accident. For small motors, lights, and general service inside industrial plants, installations of 110 or 220 V are recommended.

All **switches, fuse boxes, terminals, starting rheostats, motors,** etc., located within 8 ft of a floor or working platform, should be enclosed or guarded in such a manner as will prevent accidental contact with live parts, irrespective of voltage. **Switches** should be arranged so that they can be locked in the open position, to guard against a switch being activated accidentally while workers are at work on the lines or equipment that the switch controls. **Equipment for operation at 550 V and higher** should be isolated from other operating equipment, in separate rooms or enclosures, with provision to lock these enclosures. **All metallic cases, frames,** and **supports** of such equipment should be permanently grounded; foundation bolts should not be depended upon for this purpose, but substantial ground conductors should be used. It is preferable to have these ground wires accessible for inspection. **All low voltage secondary circuits** of 300 V or less (such as light, motor, and meter circuits) should be permanently grounded whenever a neutral point is available for the purpose. Secondary circuits of 250 V or less should be permanently grounded even though a neutral point is not available. Whenever grounding has to be omitted, the frame or casing of the apparatus should be permanently grounded, and all live parts of the secondary circuit should be shielded to prevent accidental contact therewith. It is also desirable to have floors or platforms adjacent to switchboards built of nonconducting material, or suitably insulated. In the absence of such construction, rubber mats may be used.

Portable electric lamps, tools, and machines are subjected to severe operating conditions. Many electric shock fatalities have occurred on electric portables, even with 110-V lighting circuits. Cable used to service portable equipment should be high-grade insulated cord designed for the service, not ordinary twisted lamp cord. Its mechanical attachment to both the portable and the attachment plug or source of power should be designed to prevent sharp bending or chafing that would break down the insulation. All portable lamps should have nonmetallic sockets (such as rubber, plastic, or porcelain). The heavier portable tools and machines cannot practically have their metallic parts insulated from contact by the operator, and protection from shock due to accidental charge thereon must be provided by means of a special grounding wire. Such protective grounding should be in the form of an additional wire in the cable that feeds the portable device. One end should be permanently connected to the machine frame and the other connected to ground through an additional pole in both the attachment plug and the receptacle to which the device is to be attached.

Occupational-Disease Prevention

(See 29CFR 1910, Subpart G, as amended.)

In any industrial operation where a toxic material is being processed in such a manner that those persons engaged in or working near the operation are exposed to appreciable quantities of dusts, fumes, vapor, or gas, it is important that adequate control measures be adopted. The following statements cover the major considerations involved in the application of effective control to industrial occupational disease.

Contaminants (See 29CFR 1910, 93.) The physical and chemical characteristics of a contaminant should be known. In the case of dusts or fumes, the chemical nature, particle size, solubility, etc., should be determined. For gases or vapors, the composition, vapor pressure, flash point, etc., are important factors. In all atmospheric contamination, the quantity of material in the worker's breathing zone must be known before the degree of hazard can be evaluated. The chemical charactristics are important in the selection of materials to be used in the construction of any control equipment where corrosion, etc., might be factors.

A careful investigation should be made to determine accurately the sources from which the contaminant is being produced or from which it is being dispersed. The most common types of dust-producing operations are crushing, screening, grinding, polishing, etc. Dispersion of dust is encountered in practically all dry handling operations of fine materials. Vapors and fumes are produced by chemical processes and reactions and are most commonly found in connection with the use of solvents.

A number of testing instruments and procedures have been developed for determining quantitatively the concentrations or amounts of various toxic material in the atmosphere of a work room (see ACGIH, ''Air Sampling Instruments'' and reports of U.S. Public Health Service and U.S. Bureau of Mines; see also AIHA Hygienic Guide Series and ACGIH, ''Annual Threshold Limits''). In general, the seriousness of any exposure involving a health hazard is directly proportional to the dosage (concentration and length of time of the exposure). Engineering control should be directed to the reduction of these two factors.

Local Ventilation at the Source of Contaminant Removal by means of exhaust hoods, enclosures, etc., so located as to prevent the escape into occupied areas of any appreciable amount of contaminant at its source is the most effective method of control. For details of hood design, piping, collectors, fan characteristics, and for general principles of exhaust design system, see Sec. 12.4. and ANSI Standard, ''Fundamentals Governing the Design and Operation of Local Exhaust Systems''; AFS, ''Engineering Manual for Control of In-plant Environment in Foundries.''

Natural ventilation has a limited application in industrial occupational disease control. It is important with a natural ventilation system to maintain close supervision of adjustment as required by changes in temperature, wind directions, etc. Regular tests of air content should be made to check on the degree of dilution being obtained. This method of control is not recommended for exposures where severe hazards exist owing to extreme toxicity or high concentrations.

Isolation Under some circumstances, the isolation of a hazardous operation, physically, or in point of time, is required. For example, a hazardous operation may be carried on in a separated room in which all contamination can be confined or it may be carried on outside regular working hours when no one except the persons engaged in the operation will be present. By isolating an operation, the number of persons exposed to any accompanying hazard may be reduced to a minimum.

Process Revision The possibility of substituting a less toxic material should be borne in mind; e.g., high-boiling-point distillates for benzol in rubber cements, thinners, etc.; dolomite lime for quartz in foundry parting compounds. It may be possible to reduce the concentrations of objectionable contaminants and the time of exposure by changing handling operations, by substituting mechanical feeds or conveyors for manual operations, or by installing automatic machinery or controls so as to dispense with the presence of an attendant. An automatic temperature or pressure control device, in connection with a chemical process, may reduce materially the time of an operator's exposure as well as prevent the production of excess or undesirable fumes or gases. An enclosed conveyor system handling dry material and automatically weighing it may be utilized to eliminate a severe exposure that would result from hand shoveling.

Physical Hazards In addition to exposures to toxic gases, vapors, dusts, fumes, etc., there are several physical conditions such as abnormal pressures, temperatures, and humidities as well as radiation (including ultraviolet; infrared; X-ray; α, β, and γ rays from radioactive substances; and radiation from handling radioactive isotopes), all of which may, in case of excessive exposure, prove detrimental to the health of exposed persons. Evaluation depends on physical methods of measurement. Protective measures include shielding from radiation and control of exposure time rather than ventilation, as in the case of atmospheric contaminations.

Ambient Noise Control (See also Sec. 12.6) Outer property noise propagation is regulated by EPA, under provisions of the Noise Control Act of 1972; Section 6 applies to new-product noise emission standards; Section 17 is concerned with noise emission regulations for railroads and Section 18 for motor carriers. EPA is directed to publish noise emission guidelines in the *Federal Register*. The regulation for aircraft-airports are promulgated by the Federal Aviation Administration (FAA) consistent with Section 611 of the Federal Aviation Act of 1958 and guidelines proposed by EPA.

Occupational Noise Exposure Mandatory protection of employees against noise is required by 29CFR 1910.95 as follows:

(*a*) Protection against the effects of noise exposure shall be provided when the sound levels exceed those shown in Table 18.2.1 when mea-

Table 18.2.1   Permissible Noise Exposure*

Duration, h/day	Sound level, slow response, dBA
8	90
6	92
4	95
3	97
2	100
1½	102
1	105
½	110
¼ or less	115

* 29CFR 1910.95.

sured on the A scale of a standard sound-level meter at slow response. When noise levels are determined by octave band analysis, the equivalent A-weighted sound level may be determined from Fig. 18.2.1. Octave band sound-pressure levels may be converted to the equivalent A-weighted sound level by plotting them on this graph and noting the A-weighted sound level corresponding to the point of highest penetration into the sound-level contours. This equivalent A-weighted sound level, which may differ from the actual A-weighted sound level of the noise, is used to determine exposure limits from Table 18.2.1.

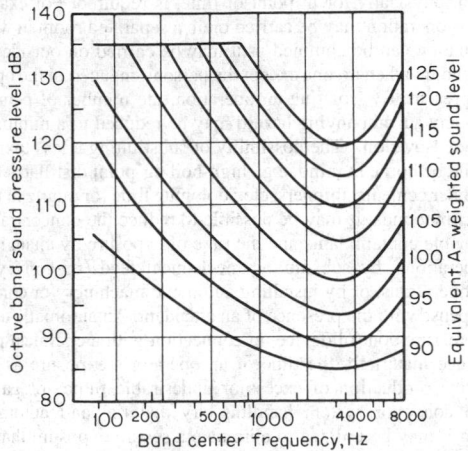

Fig. 18.2.1   Equivalent sound-level contours.

(*b*) (1) When employees are subjected to sound levels exceeding those listed in Table 18.2.1, feasible administrative or engineering controls shall be utilized. If such controls fail to reduce sound levels within the levels of Table 18.2.1, personal protective equipment shall be provided and used to reduce sound levels within the levels of the table. (2) If the variations in noise level involve maxima at intervals of 1 s or less, it is to be considered continuous. (3) In all cases where the sound levels exceed the values shown herein, a continuing, effective hearing conservation program shall be administered.

When the daily noise exposure is composed of two or more periods of noise exposure of different levels, their combined effect should be considered, rather than the individual effect of each. If the sum of the fractions $C_1/T_1 + C_2/T_2 + \cdots + C_n/T_n$ exceeds unity, then the mixed exposure should be considered to exceed the limit value. C_n indicates the total time of exposure at a specified noise level, and T_n indicates the total time of exposure permitted at that level. Exposure to impulsive or impact noise should not exceed 140-dB peak sound-pressure level.

Personal Respiratory Protection (See 29CFR1910, 134, as amended.)   Where it is not practicable to control air contamination in the breathing zone of workers by adequate exhaust ventilation, etc., and it is necessary for personnel to be exposed to harmful amounts of dust, smoke, fumes, or gases or to work in an atmosphere with a deficiency of oxygen, personal respiratory equipment must be provided. Such equipment is not, in general, suited for prolonged daily use because of inherent discomfort and inconvenience to the wearer. For emergency or temporary situations or until effective control of contamination can be developed and applied, personal respiratory protection should be given very careful consideration where harmful exposures are encountered (see "The Respirator Manual," prepared jointly by the American Industrial Hygiene Association and the American Conference of Governmental Industrial Hygienists). The U.S. Bureau of Mines (Schedules 19 and 21) has developed certain standards for approval of respiratory protection devices. The equipment may be (1) **supplied-air respirators** (hose type), devices that supply clean, respirable air to the wearer through a hose line extending from a source outside the contaminated zone; (2) **supplied-oxygen respirators** (self-contained type of oxygen-

breathing apparatus), devices that supply oxygen to the wearer from a source of supply that he or she carries as part of the respirator; or (3) **filter-type respirators,** which may be chemical filter respirators that remove the harmful constituents from air passing through them by chemical reactions, absorption, and adsorption, including ordinary gas masks and chemical respirators; or which may be mechanical-filter respirators that remove the harmful constituents from air by mechanical filtration, including dust, mist, or fume respirators.

Safety Codes and Published Material. OSHA   The American National Standards Institute (ANSI) publishes a series of ANSI Standards. State and local ordinances must be complied with as required.

The National Safety Council publishes a series of Industrial Data Sheets and Safe Practice Pamphlets covering the safeguarding of industrial operations, in addition to literature, posters, films, and other safety educational material. The annual publication "Accident Facts" reports current trends in accident experience on an overall basis.

United States government agencies, such as the Bureau of Mines, the Bureau of Labor Statistics, the National Bureau of Standards, and the Division of Labor Standards publish statistical information concerning industrial-accident occurrence and accident prevention. The National Fire Protection Association (NFPA) promulgates fire codes and publishes information of all types relating to fire protection.

The Williams Steiger Act of 1970 created the **Federal Occupational Safety and Health Administration** (OSHA), which has now codified and published blanket standards for national enforcement in all public safety and health areas. Updated revisions will continue to appear in the Federal Register under Title 29 CFR1910 and 1518, and these standards make all preexisting standards obsolete.

Radiation Health Physics (Nuclear Regulatory Commission, Code of Federal Regulations 10CFR 20; International Commission of Radiological Protection, Pub. 30, Radiological Health Handbook)   **Radiation,** which accompanies all nuclear reactions, can be advantageous or disadvantageous, depending generally on the manner in which contact is made and on the amount of energy involved. In its application to humans, the common terminology includes **radiation absorbed dose (rad)** and **dose equivalent,** measured in SI units as **gray (Gy)** and **sieverts (Sv),** respectively. In an older, though still common, usage, the respective units are designated **rad** and **rem.** Absorbed dose, measured in grays (or rads), is a measure of the energy absorbed from radiation per unit mass of irradiated material. Absorbed dose is applicable to all types of ionizing radiation and irradiating material. The amount of biological damage to human tissue, however, depends on the type of radiation and its energy. Thus, absorbed dose measurements must be modified in order to relate the amount of energy absorbed to the biological effect. Dose equivalent, as measured in sieverts (or rems), accounts for these effects through the use of **radiation weighting factors** (formerly known as **quality factors**), Q. For radiation protection purposes, the **dose equivalent** is defined as the product of the absorbed dose and the radiation weighting factor. The currently accepted weighting factors are 1, for photons and electrons of all energies; 5, for protons with energies greater than 2 meV; 20, for alpha particles, fission fragments, and heavy nuclei; from 5 to 20, for neutrons, depending on the neutron energy.

Exposure to radiation may be divided into two general types: (1) external exposure, which is that resulting from radiation sources external to the body; and (2) internal exposure, which is that resulting from radionuclides within the body. Radiation-protection standards are established for both types of exposure. Limits for external exposure are given in units of sieverts per unit of time. Basic limits currently are based on a control point of 1 year. Local operational rules may specify control limits for shorter exposure times (day, week, month), such limits being based on the basic limits. Radiation-protection standards for controlling internal exposures are based on the same basic limits as those for external exposures. Controls, however, are affected by the use of **annual levels of intake (ALIs)** and **derived air concentration (DAC)** values established for all radionuclides. Many factors (physical, chemical, biological, and physiological) are involved in determining the ALI and DAC of a radionuclide. As such, the ALIs and DACs of the radionuclides have a very wide range of values. All **ALIs** and **DACs** are normally expressed in terms of **bequerels** and **bequerels per cubic metre,** respectively. Generally, ALI and DAC values are based on a 40-h workweek for persons exposed to radiation.

Monitoring is the periodic or continuous determination of the presence and extent of ionizing radiation and radioactive contamination. It provides information for radiation protection and normally is assigned to a health physics group, which establishes schedules, executes or supervises the operation, and evaluates their results. It is divided into these special types: personnel, area, source, surface, air, and water.

Regulations governing exposure to radiation are promulgated in the form of Nuclear Regulation Commission regulations found in 10CFR 20. The current **occupational dose limit** is 0.050 Sv (or 5.0 rem) total effective dose equivalent per year. The total effective dose equivalent includes both external and internal doses. Additional radiation protection limits exist for radiation exposures to individual organs of the body, the lens of eye, and the extremities. Reference to the current full statement of 10CFR 20 is mandatory for a complete understanding of the regulations.

18.3 FIRE PROTECTION
by Pascal M. Rapier and Andrew C. Klein

REFERENCES: ''Fire Protection Handbook,'' ''National Fire Codes,'' ''Fire Inspection Manual,'' National Fire Protection Assn. ''Handbook of Industrial Loss Prevention,'' McGraw-Hill. ''Uniform Building Code 1976,'' International Conference of Building Officials, Building Offic. and Code Admin. International (BOCA). Fire Insurance Assn. (FIA), Factory Mutual (FM), Code of Federal Regulations: 29CFR 1518, July 1976, and 29CFR 1910, July 1976, Jan. 28, 1977, as amended.

IMPORTANCE OF FIRE PROTECTION
(29CFR 1910, Subpart L)

The profitable use or availability for use of facilities is the aim of all business, whether industrial, mercantile, professional, scientific, or educational. Destruction of or damage to the facilities cripples the attainment of that purpose.

Monetary compensation for destroyed property and for resulting lost profits is obtained by the purchase of insurance. Reputation, goodwill, and other intangible factors may be irreparably damaged. Loss of life is not recoverable. Loss by fire is largely an avoidable, nonproductive tax resulting from carelessness, ignorance, apathy, or incompetence. The annual fire loss in the United States (refer to NFPA) has increased steadily since 1964 at about 8 to 12 percent per year, reflecting the population and economic growth and inflation. In recent years, the number of losses has exceeded 2 million. Building fires have made up in excess of 80 percent of the **amount** of loss but only about 40 percent of the **number.**

Fire-Loss Prevention Neither insurance organizations nor legally established regulatory bodies assume the responsibility of industry for conserving its own resources and operating its facilities safely, although they may detect deficiencies or a lack of compliance with regulations and may serve in a consulting capacity. Fire loss prevention is an indispensable element in industry and business. It exists only with top management direction and the support of labor.

The designation **fire protection** usually encompasses the entire field of **prevention of loss by fire,** including both the causes for the occurrence of fires and methods for minimizing their consequences. Included along with fire are other destructive agencies such as explosion, lightning, electric current, wind, earthquake, nuclear excursions, and radioactive contamination. OSHA and building codes mandate minimum standards of protection to prevent injury and loss of life. Higher standards to protect investment are a worthwhile management option.

Fire brigades (29CFR 1910.164) are essential to the development and maintenance of an effective fire protection program at every job site. By prompt, immediate response and notification of the local fire department, every effort should be made to bring the fire quickly under control during the early minutes of an outbreak. The immediate availability of the correct fire protection and suppression equipment is essential.

Fire protection engineering involves the application of sound engineering principles to the reduction of loss by fire and related hazards. The Society of Fire Protection Engineers has established a well-defined scope for fire-protection engineering practice, both in building design and in safe operating practices.

Limitation of Loss With large businesses it is difficult to limit single destructive occurrences to an amount that will not be catastrophic to the owner or to others having a financial interest. Principles and methods for preventing large losses by fire are well known to fire protection engineers.

Sources of Fire Protection Information Much specific, detailed, and technical information is given in the standards, codes, and rules of the NFPA, Factory Mutual Engineering Division, Factory Insurance Assoc., American Insurance Association, ANSI, Underwriters Laboratories Inc., and may also be found in the procedures of many insurance companies and inspection bureaus.

The NFPA ''Handbook of Fire Protection'' and the Factory Mutual ''Handbook of Industrial Loss Prevention'' are especially comprehensive references. The most complete and readily available source of standards is the ''National Fire Codes,'' prepared and published by the NFPA, currently in ten volumes as follows: 1—Flammable Liquids; 2—Gases; 3—Combustible Solids, Dusts, and Explosives; 4—Building Construction and Facilities; 5—Electrical; 6—Sprinklers, Fire Pumps and Water Tanks; 7—Alarms and Special Extinguishing Systems; 8—Portable and Manual Fire-control Equipment; 9—Occupancy Standards and Process Hazards; 10—Transportation. Many of the standards have been adopted by insurance inspection and advisory groups, such as the American Insurance Association, and by Federal, state, and municipal regulatory bodies and have been incorporated into building codes and other regulations. The Occupational Safety and Health Administration (OSHA) U.S. Dept. of Labor, is promulgating blanket national standards for fire protection, which appear in the Federal Register under Title 29, Code of Federal Regulations for the working place, and Part 1518 during construction. Where applicable, these supersede and make obsolete all existing standards.

CONSTRUCTION

Building construction and the protection of buildings and their contents against fire are frequently governed by local building codes and ordinances and by insurance standards. When new construction or changes are planned, the property owner should have the advice of a competent fire protection engineer and should consult local authorities and insurance carriers to avoid delay and the possibility of expensive changes later.

Some construction features which have increasing importance in fire-loss prevention are (1) the trend toward large fire areas which present high values to possible loss and make manual fire fighting difficult; (2) blank wall, air-conditioned, and artificially lighted buildings which interfere with ready access; and (3) mechanization of materials handling, resulting in conveyor systems that interconnect areas, and in larger storage areas with high-piled stock which make fires difficult to extinguish. These cause increased cost of fire protection and emphasize the need for automatic fire control methods, sources of dependable water supplies, and special extinguishing systems. A measure that should be adopted where practical is the use of outdoor, totally enclosed process equipment, with shelters provided only for control rooms, laboratories, and some maintenance functions.

Types of Construction   **Fire-resistive** refers to types that withstand considerable fire without serious damage, such as reinforced concrete or protected steel. **Noncombustible** refers to any construction that contains no elements of burnable material built which may be structurally damaged by fire, such as unprotected metal. **Combustible** means structures entirely of combustible materials or having combustible elements of such character and distribution that a fire can spread and contribute fuel so that severe damage results. Combustible types of construction are frequently subdivided into (1) "heavy timber," also called "plank on timber," "mill" or "slow burning" that has masonry walls with floors and roof of plank on heavy timbers; (2) "ordinary construction" that has masonry walls with floors and roof made of boards on joists and is sometimes called "quick burning"; (3) "wood frame" that has all of its elements of wood, except that the exterior may be surfaced with a noncombustible sheathing.

Many types of construction have composite elements that may include combustible materials. Insulation, acoustic materials, and surface treatments may aid the start and contribute to the rapid spread of fire. A roof of interlocking metal sheets, with asphalt on its upper surface as part of a vapor barrier or as an adhesive for insulation or weather surfacing, can, if initially heated by a local interior exposing fire, furnish gaseous fuel through the joints to produce a spreading damaging fire. Owing to the difficulty and cost of protection, enclosed spaces at roofs, ceilings, in walls, and below floors should be avoided. Important steel structural members and steel supports for heavy equipment that may be exposed to severe fire should have heat-insulating protection. Economic tradeoffs, based on the probability of exposure, should be made.

Fire Performance of Building Materials   The kind and amount of fire protection needed are governed by the fire performance of materials used in construction and by the combustibility of building contents. Completely noncombustible construction avoids the need for sprinkler protection, provided that contents are also certain to be noncombustible. Knowledge of the fire performance of materials, based upon tests made by qualified laboratories, is essential before they are adopted (see Sec. 6.8). Fire performance is in two categories, fire resistance and fire hazard.

Fire resistance measures the susceptibility of materials to damage by exposure to fire and is usually measured as the time period of exposure, without significant damage, to a standard fire-exposure as specified by standard fire tests, such as "ASTM E-119 Fire Tests of Building Construction and Materials," and expressed as a **fire-resistance classification** in hours as ½ h, 2 h, 6 h, etc.

Fire-hazard indicators are also in two generally recognized categories used by the Underwriters Laboratories. The **listing** of specific appliances and materials indicates that the products comply with the Laboratories' test requirements, with particular reference to fire-preventive and fire protective capabilities when the manufacturer's instructions and limitations of use have been followed. The **fire-hazard classification** of materials and products is related mainly to burning characteristics. It has a numerical basis determined by (1) flame spread, (2) fuel contributed, (3) smoke-developed. A fire hazard classification of 0 indicates a material with a fire performance equivalent to that of cement board composition. A value of 100 corresponds to the behavior of red oak lumber. Building codes and other regulations generally prescribe limiting values for fire hazard classifications.

Protection against Exposure Fires   Protection of buildings or other structures against fires in nearby property must sometimes be provided. A practical barrier against a conflagration is the presence of fire-resistive buildings along the exposed side. Usually the most severe exposure is localized, and protection against it may be provided by a blank brick or concrete wall, by wired-glass metal-frame windows, or by open sprinklers alone or in combination. The relative value of various safeguards against exposure fires has been estimated to be roughly blank brick or concrete walls, 100; tin-clad shutters, 60; wired-glass windows, metal frames, glass block (4 in minimum thickness), 40; wired-glass

windows, wooden frames, 20; plain glass windows, 5; additional value of open sprinklers with any of the foregoing, except blank fire walls, 30; open windows, 0.

Horizontal Cutoffs   Provide effective fire walls between important buildings, and subdivide large building areas in order to limit the probable maximum damage from a single fire.

Vertical Cutoffs   Enclose stairs, elevator wells, vertical conveyors, chutes, and other floor openings with adequate fire-resistive walls, having fire doors or their equivalent at openings to prevent the rapid spread of fire and heat upward from floor to floor.

Isolation of Hazards   Cut off hazardous occupancies by fire-resistive partitions or fire walls or, if the degree of hazard warrants, isolate them in separate buildings. Make provision for adequate ventilation and for explosion vents where needed.

Floor Leakage   Provide drained, watertight floors over large values susceptible to water damage.

Storehouses and Vaults   Provide suitable storehouses for large quantities of combustible raw stock, tools, and patterns, and for valuable finished goods. Provide reliable vaults for the storage of business records and valuable drawings.

SAFEGUARDS DURING BUILDING CONSTRUCTION
(20CFR 1518, Subpart F)

A building under construction is more vulnerable to fire than the same building completed and in use with normal fire protection. As construction progresses, concentrations of readily combustible materials appear at new locations from day to day. So do many potential ignition sources—temporary heaters, welders' torches, roofers' tar kettles, and workers' discarded matches or cigarettes. If fire starts, wall and floor openings create draft and help flames to spread rapidly. Many serious construction fires have taken place. Most can be prevented. Guard against cutting and welding hazards by proper supervision of operations. Plan in advance for effective notification of, and coordinated firefighting operations with, any available public fire department. (See 29CFR 1518.24.)

Start installation of the fire-protective equipment—including water supplies, yard mains, and hydrants—as soon as construction of the building starts. Make layouts for sprinklers as soon as building drawings are completed, and let contracts without delay. It is sometimes possible to arrange for limited temporary protection by providing connections for hose from the piping system which furnishes water for construction uses. Have hose and nozzles available as soon as the yard piping, hydrants, and water supplies are ready. Distribute ample hand extinguishing equipment throughout the premises, including contractor's temporary buildings.

Temporary Heating   Arrange any needed heat safely. Use coke as fuel for salamanders, rather than wood trash or rubbish. Oil-fired heaters, if well-designed and properly located, are safer. Unit heaters are still better and are recommended if an adequate steam supply is available. Locate salamanders or oil heaters away from woodwork or tarpaulins. Keep the floor clean and free from combustible material. Flameproofed, water-repellent tarpaulins are available, and their use is advised. The tarpaulin salamander wooden-form combination has caused many serious fires, particularly in reinforced concrete construction work during freezing weather. Scaffolding constructed of metal, or of fire-retardant-treated wood, largely avoids a serious fire hazard. Locate construction sheds a safe distance from a building under erection, never inside.

Temporary Storage   Keep combustible storage out of buildings under construction, as far as possible and at least 10 ft (3.05 m) from structures when stored outside. Under no conditions erect canvas tents or wooden shelters inside. A safe location and good fire protection are important for valuable equipment or machinery, delivered before the building is ready for its installation.

AUTOMATIC SPRINKLERS (29CFR 1910.195)

Advantages of Automatic Sprinklers   Automatic sprinklers are the most dependable and effective means of fire protection. The advantages are (1) automatic sprinklers go into operation soon after a fire starts, and before it has gained dangerous proportions; (2) the automatic sprinklers that open are over, and in the vicinity of, the fire; (3) fires at all locations, including out-of-the-way places, are controlled as effectively as fires easily seen and reached; (4) sprinklers operate where fire fighters could not enter or would be driven out by heat or smoke; (5) sprinklers are always ready; (6) only sprinklers that are needed to control a fire operate, so that water is used to better advantage than from hose streams.

Where Sprinklers are Needed   Automatic sprinklers are needed for complete protection where there is an appreciable amount of combustible material in either building construction or in building contents, e.g., (1) throughout buildings having floors or roofs of sufficient combustibility to contribute to a spreading fire, whether or not the contents are combustible; (2) in wholly noncombustible buildings where contents are combustible, including the storage or use of flammable liquids; (3) in concealed spaces, such as attics and under low roofs, if they contain any combustibles, including heat insulation and exposed electric wiring; (4) in vacant spaces beneath a combustible first floor of a building, unless the floor is completely tight, the space sealed against entry, and no possible sources of ignition are present; (5) in dryers, large ducts, closets, and small offices, unless there is no fuel for a fire in either the construction or the contents; (6) under wide storage shelves or work tables over 4 ft in width having any combustible stock, and under canopies over platforms where combustible materials may be present.

Sprinklers over Electric Equipment   If the construction is fire-resistive or adequately fireproofed, sprinklers can generally be omitted in electric generator rooms and over switchboards. If voltages exceed 600, it is generally better to replace combustible roofs or ceilings with non-combustible construction or to protect all exposed woodwork with metal lath and cement plaster, rather than to install sprinklers. Metal shields or hoods may be used to protect generators, switchboards, or other important electrical equipment from possible water damage.

Automatic-Sprinkler Equipment   The responsibility for the design of automatic sprinkler systems should be given only to experienced and responsible parties. Approval of preliminary layouts and working plans are usually required by municipal or insurance inspection bureaus before installation is started. Sprinkler installation is a well-established trade by itself. Standards for the Installation of Sprinkler Systems are given in complete detail in a publication under that name by the National Fire Protection Assoc.

Dry Pipe Systems   The ordinary wet pipe sprinkler system cannot be used where subject to temperatures below freezing. A dry pipe system must be substituted. In the dry pipe system the sprinkler piping contains air under pressure instead of water. When a sprinkler is opened by fire, the air pressure falls and water is admitted automatically by the operation of a dry pipe valve.

Special Sprinkler Systems   The deluge system is a special type of sprinkler equipment frequently used in hazardous occupancies, such as airplane hangers and storage areas for flammable liquids or materials, where a flash fire could spread before the regular automatic sprinklers became operative, and where the prompt discharge of a large amount of water over a considerable area is needed. The deluge system has sprinkler heads with the fusible element removed. Water is controlled by a quick-opening valve (deluge valve) operated by heat sensitive elements distributed over the area protected. Piping and heads are arranged as in a standard sprinkler system, with larger pipe sizes.

Precision systems are actuated by heat-sensitive devices, smoke detectors, or ionization detectors, which trip a controlling valve and admit water to the otherwise dry sprinkler piping before the regular closed automatic sprinkler heads operate. The system has the advantage of giving an alarm before sprinklers operate. Accidental opening of automatic sprinklers or mechanical breakage of piping does not result in the discharge of water.

Deluge and other systems actuated by heat-sensitive or other special devices introduce additional mechanical equipment and lose the simplicity of the standard automatic sprinkler system. Some additional skilled maintenance is required. Various arrangements for mechanically supervising the condition of such equipment can be provided.

Nonfreeze sprinkler systems are small and are sometimes used in relatively unimportant locations where it is impracticable to provide heat. They consist of special piping connections to wet-pipe systems, with a nonfreezing solution in the exposed piping. Antifreeze solutions (glycols) are acceptable for this purpose. They should be limited to systems with less than 20 heads. It is better to provide heat, to connect to existing dry pipe systems, or to provide small dry pipe valves. Nonfreeze systems are preferable to shutting off the sprinklers and draining the piping during the winter. No portion of an automatic sprinkler system connected to a public water supply should be filled with a nonfreeze liquid without determining that the arrangement will comply with water department health regulations.

Outside sprinklers are used for protection against fire from outside sources. Windows in outside walls present the most frequent problem. Outside sprinklers may be operated either manually or automatically. The effectiveness of these systems depends on the water supply valve opening in time. They can sometimes be adapted to automatic control by a thermostatically operated deluge valve.

Types of Automatic Sprinkler Heads   Approved automatic-sprinkler heads are well standardized as to the rate of water discharge, pattern of discharge, and temperature operating characteristics. The essential elements of an automatic sprinkler are a nozzle, a closing device that is released at a definite temperature, and a deflector that produces the desired distribution pattern. Temperature ratings appropriate for the location must be selected. The ratings for normal atmospheric temperatures, not substantially above 100°F, are 135 to 165°F. Ratings for other ceiling temperature ranges are as follows:

Max ceiling temp		Sprinkler rating	
°F	K	°F	K
101–150	311–338	175–212	352–373
151–225	339–380	250–286	394–414
226–300	381–442	325–360	436–455

Heads of higher rating are obtainable for ambient temperatures up to 500°F. Corrosion-resistant heads, usually wax or plastic coated, are obtainable for use in corrosive atmospheres.

Sprinkler Flow Alarms   The flow of water from a sprinkler system may be made to sound an alarm which gives warning of both a fire and accidental leakage from the system. This alarm may be either a hydraulic gong or an electric bell, or both. If the property has no constant attendance, or if added security is desired, central-station supervisory service is available in many industrial centers. This service transmits water flow alarms to public fire departments and can perform numerous other supervisory functions. Preaction alarms give advance notice of hazard.

Location and Spacing of Automatic Sprinklers   As the purpose of an automatic sprinkler system is to protect both building and contents, the water distribution pattern and the quantity of water applied must be held within close limits. Sprinklers customarily are installed so that the discharge from one head is allowed for each 60 to 130 ft^2 of floor area. The distance between adjacent heads should not exceed 15 ft. Spacing averages 80 to 120 ft^2 per head with 8 to 12 ft between heads.

Recent developments in sprinkler system design for the more severe hazards produced by high piled combustible goods or flammable liquids are leading to standards including a **discharge density** in terms of gallons per minute per square foot of floor area. In this consideration building contents, rather than building construction, are the determining factor. Densities called for commonly have been in a range of 0.15 to 0.3 gal/min per ft^2 of floor area. High-piled storage of combustible materials, such as rolled paper, imposes a severe hazard which may call for a water density of 0.6 gal/min per ft^2, or more. NFPA standards (1973) call for intermediate levels of sprinkler heads in high rack storage areas. Open and deluge sprinklers require hydraulically designed systems based on friction loss instead of standard pipe schedules (see NFPA No. 13, No. 13a, and No. 231c).

The number of heads, to be supplied through pipes of standard sizes, with ordinary hazards and reasonably uniform water discharge, is as follows:

Size of pipe, in	Max. no. of sprinklers	Size of pipe, in	Max. no. of sprinklers
1	2	3½	65
1¼	3	4	100
1½	5	5	160
2	10	6	275
2½	20	8	400
3	40		

Not more than eight sprinklers should be on one branch line on either side of a cross main. For deluge systems, with which all heads are open and must be supplied simultaneously, and for especially hazardous occupancies where a large proportion of the automatic heads are expected to operate, larger pipe sizes are required as follows:

Size of pipe, in	Max. no. of sprinklers	Size of pipe, in	Max. no. of sprinklers
1	1	3	27
1¼	2	3½	40
1½	5	4	55
2	8	5	90
2½	15	6	150

WATER SUPPLIES FOR FIRE PROTECTION
(See Sec. 3.3 for hydraulics, Sec. 8.7 for piping, Sec. 9.6 for engines, Secs. 14.1 and 14.2 for pumps.)

Insurance companies or their inspection and engineering bureaus should be consulted concerning water supply requirements for fire service. In addition to providing water for automatic sprinklers, the supply should be adequate for hydrants and hose streams. Common water supplies are public water systems, gravity tanks or private reservoirs, and fire pumps taking suction from aboveground suction tanks, rivers, or ponds. Two or more independent supplies are usually needed for larger properties. Strong, dependable public water systems alone are adequate only for small plants, of good construction, having safe occupancies, and not dangerously exposed.

The "primary" water supply automatically maintains pressure on the property fire system at all times. The "secondary" supply supplements it as needed. The **primary supply** should not be less than 500 gpm available at a pressure that will be effective for automatic sprinklers at the highest plant elevation. A **secondary water supply** serves several purposes. The most important usually is to provide more water, frequently at higher pressure than is available from the constant primary supply. A secondary supply is also of value in maintaining protection at any time the primary supply may be interrupted due to public supply deficiencies or to the necessity of taking tanks or reservoirs out of service for repairs and maintenance.

High water demands for special hazard protection, especially where large areas must have deluge system protection, can seldom be met by connections to public water systems. Fire pumps and suction reservoirs can give the added volume of water at desirably high pressure for the expected duration of the total demand, which is made up of the anticipated sprinkler demand and an allowance for the number of hose streams likely to be used. Large airplane hangars, large building areas containing flammable liquids, or very extensive properties that could have a general fire, approaching conflagration proportions, can have an estimated total demand of 5,000 gal/min or even more.

Gravity Water Tanks   The smallest gravity tank on a tower considered advisable for fire protection is 25,000 gal, which may be suitable for the protection of small properties. Water from tanks of such limited size should be reserved for automatic sprinklers only. Tanks of 50,000- to 100,000-gal capacity may provide either the primary or secondary

supplies for large properties of moderate hazard. Tanks combining fire service and domestic or industrial supplies in a single structure are allowable, provided that the amount needed for the fire supply cannot be withdrawn into the industrial use system.

Gravity tanks should be at such elevation that the bottom of the tank will be at least 35 ft above the highest sprinkler to be supplied. The bottom of large tanks, intended to provide water for hose streams as well as sprinklers, should be at least 75 ft, and preferably 100 ft, above ground level.

Pressure Tanks   Hydropneumatic steel tanks can be used for primary water supplies for demands of short duration, such as needed to bring automatic fire pumps into full operation. Under some conditions they can be used to facilitate the automatic control of the automatic fire pumps by providing a rapid drop in the control pressure. High cost and limited capacity deter the general use of pressure tanks.

Private Fire Pumps   Well-located fire pumps with ample suction supplies and capable of maintaining high pressure over a long period provide a very satisfactory secondary supply. Centrifugal fire pumps of approved design are in common use. Practically all new installations use centrifugal pumps driven by electric motors, gasoline or diesel engines, or steam turbines. The electric motor driven centrifugal pump is most desirable because of simplicity of control and operation. Gasoline engine drives are used to a limited extent where other reliable sources of power are not available, or as auxiliary units in connection with other pumps.

Adequate suction and reliable power supply are essential for fire pumps. The suction supply should be sufficient to operate the pump at rated capacity for at least 1½ h at the smaller plants, and an inexhaustible suction supply is desirable for good protection of the largest industrial properties.

An automatically controlled, electrically driven centrifugal fire pump is sometimes used as a booster pump when the public water is of good volume but too low in pressure for direct use in sprinkler systems.

Fire pumps are used in capacities of 500, 750, 1,000, 1,500, and occasionally 2,000 or 2,500 gal/min. The 750 and 1,000 gal/min sizes are most common.

Fire Department Connections   For properties where the public water supply is at comparatively low pressure but adequate in quantity, fire department connections to the plant fire-service piping can serve as a valuable auxiliary for pumpers to deliver water into the fire system at high pressures. In city properties, particularly, fire department connections are usually a part of the sprinkler system.

UNDERGROUND WATER PIPING FOR FIRE SERVICE

Complete specifications or standards covering all of the main features of underground piping for fire service are given in NFPA National Fire Codes, vol. 6, "Sprinklers, Fire Pumps, and Water Tanks," and the Factory Mutual "Handbook of Industrial Loss Prevention." The "Handbook of Cast-Iron Pipe," published by the Cast-Iron Research Association, contains much useful information.

Underground Mains   Buried underground piping should be located so that hydrants and control valves are at a safe distance from possible falling building walls. A complete loop system around buildings or groups of buildings, with multiple supplies preferably connected at opposite sides and with prudently located division valves, affords the best hydraulic characteristics for heavy flows and for freedom from impairments. Pipe trenches should allow careful laying and uniform support for the pipe. Foreign materials should be kept out of the pipe. Pipe should be anchored at bends, dead ends, and branch connections. The system should be hydraulically tested for 2 h at a minimum pressure of 200 lb/in² or at least 50 lb/in² in excess of the maximum static pressure if it is to be more than 150 lb/in². Leakage at joints must be less than specified amounts. Underground piping should be thoroughly flushed before it is connected to indoor piping. Pipe should be buried well below the deepest frost penetration as shown by weather charts. In the coldest areas a cover of at least 5 ft is necessary. In the southern states 2½ ft may be adequate, except that under roadways a greater cover may be needed for protection against traffic loads. Clearance to prevent breakage by settlement is needed at foundation walls.

Soil Corrosion Rapid external corrosion of cast iron pipe may be expected if the mains pass under coal piles, through cinder fill, or where pickling liquors, acids, alkalies, or salts penetrate the soil. Cast iron pipe covered with heavy coating of asphalt, or other pipe suitably covered should be used to offset soil corrosion in such locations. Backfill should be of clear sand or gravel. (See Secs. 6.8 and 6.9.)

Type of Pipe Pipe for underground use is usually cast iron or steel. Ordinarily the working water pressure rating does not need to exceed 150 lb/in². Representative specifications for pipe suitable for fire service are: "Specifications for Cast-Iron Pit-Cast Pipe" (ANSI A21.2 or AWWA C102), "Specifications for Cast-Iron Pipe Centrifugally Cast in Sand-Lined Molds" (ANSI 21.8 or AWWA C108); asbestos-cement pipe should conform to "Tentative Standard Specifications for Asbestos-Cement Water Pipe" (AWWA C400) or "Commodity Federal Specification SS-P-55/A."

OUTSIDE HYDRANTS AND HOSE

Location of Hydrants Where space permits, outdoor hydrants and hose provide a supplemental to automatic sprinklers and afford means for fighting fires in combustible yard storage, in railroad cars and vehicles, and small combustible unsprinkled sheds. Recommended space between hydrants varies from 150 to 300 ft, depending upon the type of buildings and the character of outdoor combustibles. Hose for outdoor use, 2½ in in size, is made of woven cotton or modern synthetic fiber and rubber-lined.

STANDPIPES AND INSIDE HOSE
(29CFR 1910.158)

Standpipe systems furnish the best means of obtaining effective fire streams in the upper stories of buildings. They are designed for small hose streams used by the occupants and for large hose streams to be used by public or plant fire departments. Outlets may be designed to supply both. In buildings of unusual height, standpipes are sometimes supplied by a series of fire pumps and tanks at different elevations.

Small hose is of particular value in areas with hazardous occupancies, such as opener, picker, spinning, and woodworking rooms. Hose should be 1½-in cotton, rubber-lined, or unlined linen, with ⅜- or ½-in nozzles. Spray-type nozzles are usually best for the hazards mentioned. The water supply should preferably be taken from a connection independent of the sprinkler piping so that hose streams will be available when the sprinklers are shut off after a fire or during sprinkler repairs or changes. Small hose, for fire service only, in locations of ordinary hazards, may be connected to wet pipe sprinkler systems, but in no case should hose connections be made to sprinkler lines smaller than 2½-in diam. (See Sec. 8.7.)

SPECIAL FORMS OF FIRE PROTECTION

Special types of protection are adapted to the control of unusual hazards, such as flammable liquids. Equipment should be secured from makers specializing in the form of protection required, but such use does not supplant general building protection by automatic sprinklers.

Water Spray A dense strong spray of water from suitably designed nozzles is effective for controlling fires in flammable liquids of moderate hazard, for unusually flammable solid materials, and for surface fires of ordinary combustible materials. Such a system may be most appropriate for the protection of transformers and other oil-filled electric equipment and for systems handling fuel or lubricating oil under pressure. The entire zone to be protected must be within reach of the strong spray. Water pressures of 50 lb/in² prevail with properly designed equipment. (See Sec. 3.3.)

Foam Foams for fire control provide a blanketing action to exclude air and in some cases to give useful insulating effect. Foams are commonly designated as "chemical foam" and "air foam," depending upon their service or method of production. For flammable-liquid fires, foam is available for manual or automatic applications with a selection of characteristics appropriate for a wide variety of conditions. The amount of foam varies from ½ ft³ to several cubic feet per square foot of surface protected. Rate of application and means of distribution largely determine its effectiveness. A sprinkler system, designated as "Foam-water," provides an initial discharge of very fluid foam backed up by a sprinkler discharge of water. Special **alcohol type** foam is needed for application to alcohols, alcohol-type liquids, and organic solvents, all of which seriously break down the commonly used foams. **High expansion** foams are available. They are easily produced in a **high expansion foam generator** by blowing air through a screen wet with a continuous spray of water having a bubble producing additive. The foam is very light and fluid and can be applied to fill completely and quickly a room or other enclosed area of considerable size. One gallon of high-expansion foam-producing liquid can form as much as 1,000 gal of foam. Ordinary foam producing liquids yield only about 10 gal of foam per gal of liquid.

Carbon Dioxide (29CFR 1910.161) For (1) flammable liquids; (2) electric equipment, such as large enclosed electric generators; and (3) hazards where a space-filling effect by an inert atmosphere is needed, or where a nonwetting extinguishing agent is desired. Carbon dioxide for fire extinguishing is available in small cylinders for manual use, in banks of large cylinders, or in refrigerated storage tanks for piped extinguishing systems.

Dry chemical (29CFR 1910.160) extinguishers and extinguishing systems are mainly used for flammable liquids and electrical fires. They are effective also on surface fires in combustible fibers. Multipurpose dry chemical extinguishers have special ingredients which make them suitable for fires in ordinary combustibles.

PORTABLE HAND EXTINGUISHERS
(29 CFR 1910.157)

Hand-operated extinguishers and small hose are effective when employees are on hand to attack fires immediately after discovery. They are frequently used to put out the final vestiges of fires brought under control by automatic sprinklers. Grouped by hazards, the hand extinguishers are selected from four classifications, as shown by Table 18.3.1.

Table 18.3.1 Portable Extinguishers by Hazard Classification

Extinguisher type	Class A general purpose	Class B flammable liquids	Class C electrical fires	Class D combustible metal
Foam	x			
Loaded stream	x	x		
Dry chemical		x	x	
CO₂ (plastic horn)		x	x	
Vaporizing liquid		x	x	
Dry powder				x

SOURCE: NFPA Handbook, vol. 17.

18.4 PATENTS, TRADEMARKS, AND COPYRIGHTS

by Paul E. Crawford and R. Eric Hutz

PATENTS

REFERENCES: Chisum, "Patents," Matthew Bender. Rules of Practice of U.S. Patent and Trademark Office, 37 C.F.R. §1, et seq.

United States

(All statutory references are to the Patent Act of 1952, 35 U.S.C., et seq.)

Types of Patents Granted in the United States United States patents are classified as utility, design, and plant patents with the first category comprising the vast majority of patents granted in the United States. A design patent can be obtained for any "new, original and ornamental design for an article of manufacture" (35 U.S.C. §171). The emphasis in design patents is on the ornamental, rather than functional, aspects of the article of manufacture. A plant patent may be obtained by "whoever invents or discovers and asexually reproduces any distinct and new variety of plant" (35 U.S.C. §161), and gives that person the right to exclude others from asexually reproducing the plant or selling or using the plant so produced (35 U.S.C. §163).

The Patent Grant In return for a full disclosure of an invention to the public in the patent document, the United States will grant an exclusive monopoly to an inventor or her or his assignee for the term of the patent. As a result of U.S. ratification of the GATT treaty, the patent term for utility patents has changed from 17 years from the grant of the patent to 20 years from the initial filing date of the application. This new patent term begins on the date the patent is issued and ends 20 years from the date of initial filing. The new 20-year term from date of filing applies to all patents granted on applications filed on or after June 8, 1995. All patents based on applications filed prior to June 8, 1995, will have a term that is the greater of the 20-year term provided by the new law or 17 years from the grant. The 20-year term can also be extended for a maximum of 5 years for delays in grant of a patent due to interferences, secrecy orders, or successful appeals to the Patent and Trademark Office (PTO) or the courts. However, any extensions are subject to certain limitations. The term for a design patent is 14 years. After the term of the monopoly expires, the public is then free to practice the invention. Alternatively, an inventory may choose not to disclose his or her invention to the public but instead maintain it as a trade secret; but by doing so, an inventor cannot stop others from using his or her invention if someone else independently develops the same invention.

Determining What Is Patentable The basic statutory requirements for patentability are utility of the invention (35 U.S.C. §101), novelty (35 U.S.C. §102), and nonobviousness (35 U.S.C. §103). To meet the utility requirement, an inventor need only disclose in his or her patent an invention which is new, useful, and functional in the field of applied technology as opposed to theoretical or abstract ideas. Also, developments in the nontechnical arts, social sciences, methods of doing business, computer programs per se, and the like are not generally patentable. However, there have been recent developments in the law that have expanded the possibilities for patenting certain types of business systems and some aspects of computer programs. A patent professional should be consulted regarding these developments.

The novelty and nonobviousness requirements of the patent laws are assessed in light of the prior art which generally comprises all written materials antedating the inventor's application for a patent and/or his or her actual date of invention (see section on interferences below for more detail on what constitutes date of invention). Also included in prior art are prior public uses and sales by others of the same or similar processes or products as that of the inventor. A prior public disclosure or offer to sell by the inventor made more than 1 year before the inventor's patent application also constitutes prior art which can prevent the grant of a patent [35 U.S.C. §102(b)]. Thus, it is crucial that patent applications be promptly filed (within a year) after any effort at commercial exploitation of an invention by or on behalf of the inventor. In assessing an invention against the novelty requirement of 35 U.S.C. §102, the courts have held that if the invention is different—without regard to how different—from each item of prior art, it meets that requirement. In other words, if no single piece of prior art discloses the invention, it is considered novel under 35 U.S.C. §102.

Since few inventions—at least not those for which patent protection is sought—are identical to any single item of prior art, the patentability of most patents must be assessed under 35 U.S.C. §103, i.e., whether "the differences between the subject matter sought to be patented and the prior art are such that the subject matter as a whole would have been obvious at the time the invention was made to a person having ordinary skill in the art to which said subject matter pertains." Such assessment requires a factual determination at the time of the invention of the prior art, differences between the prior art and the invention, and the level of skill in the art to which the invention pertains, and a further assessment whether those differences would have been obvious in light of the level of skill of the practitioners in the art. To assist in this determination, the Supreme Court has indicated that courts should look to important background facts such as long-felt need for the invention, failure of others to achieve the invention, and ultimately, any commercial success of the invention [*Graham v. John Deere Company of Kansas City,* 383 U.S. 1, 86 S.Ct. 684 (1966)].

Documentation of an Invention In ongoing research, recordation of developments is essential to prove the date of invention. The date of invention may become important in a PTO proceeding to determine priority of invention as between two inventors. Such a proceeding is called an **interference**. In an interference, the priority of invention must usually be established by contemporaneous, witnessed documentation of the work underlying the invention. A notebook used to record, explain, and date research work is invaluable in an interference to prove an early invention date. The notebook should be signed by the researcher and witnessed by one or more persons who have direct knowledge of, and understand, what was being done during the making of the invention.

Prior to the recent ratification of the GATT and NAFTA treaties, inventive activity occurring in foreign countries could not be relied upon to establish a date of invention in the United States. As a result of GATT and NAFTA, 35 U.S.C. §104 has been amended to provide that evidence of inventive activity in the territory of a World Trade Organization (WTO) member country, Canada, or Mexico be treated the same as inventive activity in the United States. This change in law applies to all applications filed on or after January 1, 1996. However, a date of invention may not be established in a WTO member country (other than the United States, Canada, or Mexico) that is earlier than January 1, 1996, except as provided by statute (35 U.S.C. §119 and §365). Thus, companies or individuals engaging in research in foreign countries may rely on that activity in an interference.

Preparing the Patent Application A recommended first step in preparing an application is to evaluate the prior art and its relation to the invention for which patent protection is to be sought, utilizing the obviousness guidelines of the Supreme Court noted above. The most comprehensive source of prior art is the search room at the PTO in Arlington, Virginia, where over 4 million patents and publications are available to the public and are broken down into thousands of classes and subclasses to facilitate a search. Less comprehensive compilations of U.S. patents can also be found in the libraries of many large cities. Copies of U.S. patents can be obtained from these libraries or directly from the PTO. Search results should be evaluated by a competent patent attorney or agent who can advise whether the invention is patentable over the prior art uncovered and who can also focus on the differences between the invention and prior art when preparing a patent application covering the invention.

If the invention, after careful evaluation, is considered to be patentable, the next step is identification of the proper inventors. In the United States, patent applications are filed in the name of the individual(s) who contributed to all or part of the invention rather than the legal owner of the invention or application, e.g., the inventor's employer. Where two or more persons contributed to all or part of the invention (a joint invention), the respective efforts of these persons must have been made in collaboration with each other and the invention must be the product of their aggregate efforts.

The patent application comprises four essential parts: the specification or detailed description of the invention; claims, particularly pointing out the features which the inventor considers to be his or her invention; a drawing illustrating the invention, where applicable; and an oath or declaration of the inventor(s) stating a belief that he or she believes himself or herself to be the original and first inventor of the invention claimed in the application (35 U.S.C. §§111–115).

In addition to the standard patent application, applicants may elect to file a provisional application. A provisional application only requires the filing of a specification complying with 35 U.S.C. §112, paragraph 1, and drawings where necessary for the understanding of the invention. The application must be made in the name of the inventors, include the appropriate filing fee and a suitable cover sheet. A provisional application does not require any claims, nor does it require an oath or declaration. A provisional application is not examined by the PTO, cannot claim priority of an earlier application, and cannot issue as a patent. In addition, provisional applications will automatically become abandoned 12 months after filing and can become abandoned prior to the expiration of that 12-month-period to respond to a PTO requirement or to pay a required fee. A provisional application may be revived, but its pendency cannot extend beyond 12 months from its filing date. The provisional application is a form of domestic priority system which provides a quick and inexpensive method to establish an early effective filing date which establishes a constructive reduction to practice for any invention described in the application. However, this priority is only effective provided a suitable nonprovisional application is filed prior to the expiration of the provisional application.

Proceedings in the U.S. Patent and Trademark Office After the application is filed with the appropriate fee ($365 to $730 basic fee for utility patents; $150 to $300 for design patent; and $245 to $290 for plant patent), an examiner in the PTO reviews the application to ensure compliance with all statutory requirements as to the form of the application and to independently assess the patentability of the claims in the application in light of prior art. If found to be patentably distinct from the prior art, a patent will be granted upon payment of another fee ($605 to $1,210 for a utility patent; $210 to $420 for a design patent; $305 to $610 for a plant patent). If the examiner concludes the invention claimed in the application is not patentable over prior art, all such claims will be rejected. An administrative appeal to the Patent Office Board of Appeals is available to challenge such rejection, and a further appeal to the federal courts is available from an adverse decision of the Board of Appeals (35 U.S.C. §§134, 141, 145).

To maintain a patent, the owner must also make periodic payments to the PTO before the fourth, eighth, and twelfth anniversaries of the patent grant. These maintenance payments increase from $480 to $960 to $1,450 to $2,900 as the patent ages. Since the above-mentioned fees change annually, all fees should be determined in advance to avoid the consequences of fee underpayment.

Rights of a Patent Owner A patent gives the patent owner (either original applicant or her or his assignee) the right to prevent anyone else from infringing upon his or her patent rights by making, using, offering to sell, selling within the United States, or importing the patented invention into the United States. If someone else does infringe a patent, the patent owner can seek an injunction against further infringement and/or damages for such unauthorized use of his or her invention (35 U.S.C. §§281–284). The federal courts have exclusive jurisdiction over suits relating to patents.

A patent owner may commercially exploit his or her patent by selling (assigning) or licensing it to another. Multiple licenses can be granted under a single patent giving each licensee nonexclusive rights under the patent. Geographic and use limitations in a license are also possible as a means of maximizing the patent owner's benefit from his or her patent.

Reissue and Reexamination of Patent If through inadvertence, accident, or mistake, the original patent was defective or claimed too much or too little, the error may be repaired by a reissue proceeding provided the subject matter needed to correct the mistake appeared in the original patent (35 U.S.C. §251). The term of a reissued patent expires the day the original patent would have expired.

Also, if the patentability of a patent is put in question, either the patent owner or a third party may ask the PTO to reexamine and reevaluate the patent vis-à-vis the prior art. The PTO then proceeds to examine the patent much the same as it would with an original application.

Foreign Countries

International Convention for the Protection of Industrial Property The important provision of the convention is that any person who has duly applied for a patent in one of the contracting states shall have a right of priority for 12 months in making application in the other states.

Patent Cooperation Treaty (PCT) and European Patent Convention (EPC) To simplify the process of obtaining foreign patents, it is now possible to file a single application, e.g., in the United States, and have that application mature into a patent in each country worldwide which is a member of the Patent Cooperation Treaty. Similarly, if the focus of intended patent protection is in Europe, a single application in the European Patent Office located in Berlin, Munich, or the Hague can be issued as a patent in most of the countries on the European continent under the European Patent Convention.

Use of either the PCT or EPC provisions will generally be cost-effective versus separate filings in individual countries, since, for example, translation costs are usually minimized or eliminated. However, as in all such complex matters, knowledgeable patent professional counsel should be consulted.

TRADEMARKS

REFERENCES: Callman, ''Unfair Competition, Trademarks and Monopolies,'' Callaghan and Co., Chicago. McCarthy, ''Trademarks and Unfair Competition,'' Lawyers Cooperative, Bancroft-Whitney. Gilson, ''Trademark Protection and Practice,'' Matthew Bender.

United States

[Section references (§) refer to Lanham Act of 1946, and 15 U.S.C., both section numbers being given.]

What Is a Trademark? The term *trademark* includes any word, name, symbol, or device or any combination thereof adopted and used by manufacturers or merchants to identify their goods and distinguish them from those manufactured or sold by others (§45, §1127). A *service mark* is defined as a mark used in the sale or advertising of services to identify the services of one person and distinguish them from the services of others (§45, §1127).

How Rights in a Mark Are Obtained Unlike a patent, but somewhat like a copyright, ownership of a mark is obtained by use. If use occurs only within a state, a mark may be registered in that state. Forms for registering marks under state law are obtainable from the Secretary of State of each state. If use occurs only in interstate or foreign commerce, a mark may be registered in the U.S. Patent and Trademark Office (PTO) by filing an application in the form prescribed (37 C.F.R. Pt. IV).

After federal registration the symbol ® may be used with the mark. Prior to federal registration, the symbols **TM** for a trademark or **SM** for a service mark may be used to give notice of a claim of common law or state law rights.

A person who has a bona fide, good-faith intention to use a trademark in commerce, but is not actually using the mark, may apply for registration by filing an Intent To Use application in the PTO. Thus, a person who is not actually using a mark may still obtain federal registration provided that person commences using the mark in interstate commerce within a specified period after allowance of the Intent To Use application.

Tests for Registrability and Infringement A mark may be registrable on the Principal Register only if, when applied to the goods or used in connection with the services, it is not likely to cause confusion or mistake or to deceive. The same test is used to determine whether a mark infringes on an earlier mark. Marks that are descriptive, misdescriptive, geographical, or primarily merely a surname may not be registrable on the Principal Register, if at all, until after 5 years of substantially exclusive and continuous use in interstate commerce (§2[f], §1052[f]). Marks not registrable on the Principal Register may be registrable on the Supplemental Register if they are capable of becoming distinctive as to the applicant's goods or services (§23, §1091).

Term of Registration Federal registrations remain in force for 10 years unless canceled. However, unless an affidavit is filed during the sixth year after registration showing that the mark is still in use, the Commissioner of Patents and Trademarks will cancel the registration of the mark (§8, §1058). In addition, federal registrations only remain in force for marks which are actually in use. Nonuse of a registered mark can result in a finding that the mark has been abandoned by its owner. The law currently provides that nonuse of a mark for 2 years or longer is prima facie evidence that the mark has been abandoned. Within the sixth year after registration on the Principal Register a registered mark may become more secure from legal challenge, i.e. ''incontestable'' upon the filing of the required affidavit (§15[3], §1065[b]). During the last 6 months of the registration term an application for renewal may be filed (§9, §1059). The terms of state registrations vary.

Preliminary Search Before adopting a mark, it is advisable to have a search made in the PTO to determine whether the mark under consideration would conflict with any pending application, registrations, or others' use of the mark in connection with any similar goods or services.

Cost of Registering a Mark Government fees and time for payment are fixed by law (§31, §1113) but attorneys' fees vary.

Effect of Federal Registration Federal registration of a mark affords nationwide protection, and once the certificate has been issued, no person can acquire any additional rights superior to those obtained by the federal registrant. Federal registration of a mark establishes federal jurisdiction in an infringement action, can be the basis for treble damages, and is admissible as evidence of trademark rights. Registration on the Principal Register constitutes constructive notice, constitutes prima facie or conclusive evidence of the exclusive right to use the mark in interstate commerce, may become incontestable, and may be recorded with the United States Treasury Department to bar importation of goods bearing an infringing trademark.

Assignment of a Mark A mark can be assigned only in conjunction with the goodwill of the business or that portion thereof with which the mark is associated. Assignments of registered marks or of applications for registration should be recorded in the United States Patent and Trademark Office within 3 months after the date of assignment (§10, §1060).

Foreign Countries

Registration In general, most foreign countries require the registration of a mark in compliance with local requirements for trademark protection. In most countries registration is compulsory and provides the sole basis for protection of a trademark. A few countries afford common-law protection to unregistered marks. Because the laws and regulations in foreign countries regarding trademark registration vary so much, it is not practicable to summarize the requirements for registra-

tion in the space allocated to this note. Anyone interested in foreign protection of a mark should consult an attorney.

COPYRIGHTS

REFERENCES: Nimmer, ''Nimmer on Copyright,'' Matthew Bender. Copyright Office Regulations, 37 C.F.R., Chapter 2.

United States

[Section references (§) refer to Copyright Act of 1976, 17 U.S.C.]

Subject Matter of Copyright Copyright protection subsists in works of authorship in literary works, musical works including any accompanying words, dramatic works, including any accompanying music, pantomimes and choreographic works, pictorial, graphic, and sculptural works, motion pictures and other audiovisual works, and sound recordings, but not in any idea, procedure, process, system, method of operation, concept, principle, or discovery [§102]. However, copyright protection may be obtained for computer programs, including both the source code and object code. In addition, owners of copyrights covering computer programs (including any tape, disk, or other medium embodying such program) generally have the right to authorize or prohibit commercial rental to the public of the originals or copies of their copyrighted work. Noncommercial transfers of lawfully made copies of computer programs by a nonprofit educational institution to another nonprofit educational institution or to faculty, staff, and students does not constitute rental, lease, or lending under the copyright laws.

Method of Registering Copyright After publication (sale, placing on sale, public distribution), registration may be effected by an application to the Register of Copyrights along with the fee ($20.00) and two copies of the writing (§401; 407; 409). The Copyright Office (Register of Copyrights, Library of Congress, Washington, DC 20540) supplies without charge the necessary forms for use when applying for registration of a claim to copyright.

Form of Copyright Notes The notice of copyright required (§401) shall consist either of the word ''copyright,'' the abbreviation ''Copr.,'' or the symbol ©, accompanied by the name of the copyright owner and the year date of first publication.

The copyright notice for phonorecords of sound recordings should contain the symbol ℗, the name of the copyright owner, and the year date of first publication.

Who May Obtain Copyright Only the author or those deriving their rights through the author can rightfully claim copyright protection. Where a copyrighted work is prepared by an employee within the scope of his or her employment (a ''work made for hire''), the employer and not the employee is considered to be the author. The name of the author (creator of the work) is normally used in the copyright notice even though an assignee may file the application for registration.

Duration of Copyright Copyright in a work created on or after January 1, 1978, subsists from its creation and, with certain exceptions, endures for a term consisting of the life of the author and 50 years after the author's death (§302). Copyright in the work created before January 1, 1978, but not theretofore in the public domain or copyrighted, subsists from January 1, 1978, and endures for the term provided by §302. In no case, however, shall the copyright term in such a work expire before December 31, 2002; and, if the work is published on or before December 31, 2002, the term of copyright shall not expire before December 31, 2027 (§303). Any copyright, the first term of which is subsisting on January 1, 1978, shall endure for 28 years from the date it was originally secured, with certain exceptions (§304). All terms of copyright provided by §302–304 run to the end of the calendar year in which they would otherwise expire (§305).

Infringement of Copyright A violation of any of the exclusive rights (§1) by a person not licensed could lead to liability (§101) for infringement. Actions for infringement of copyright are brought in the U.S. district courts.

Consulting Attorneys An attorney should be consulted before publication unless the author or proprietor has previously registered copyrights. A mistake could cause the copyright to be lost if the work is published without proper notice.

Foreign Countries

Universal Copyright Convention Under this convention a U.S. citizen may obtain a copyright in most countries of the world simply by publishing within the United States using the prescribed notice, namely, © accompanied by the name of the copyright proprietor and the year of first publication placed in such manner and location as to give reasonable notice of claim of copyright. While the term ''Copyright'' on the notice is adequate for copyright under U.S. law, only the © is recognized under the convention.

18.5 MISCELLANY
Staff Contribution

The engineering environment may be broadened to include the professional working environment in which engineers conduct their practice. Ultimately, the engineers' endeavors result in a product of some sort, be it a tangible item or drawings, specifications, inspection documents, research reports, and the like. Keeping that in mind, the subjects which follow address briefly some aspects pertinent to the practice of engineering.

PERSONNEL

The engineer often will employ other persons, and as an employer, the engineer is subject to all applicable laws and regulations related to the employer-employee relationship. The administration of personnel matters will involve the engineer directly or will be delegated to a surrogate within the organization. Among the things which relate to personnel are worker's compensation insurance; accident and health insurance; antidiscrimination laws related to hiring, termination, and promotion; general liability insurance; and so on. Depending upon the staffing and the type of work which is performed in the engineer's office, it may prove prudent to carry employer's liability insurance to cover situations which are excluded from worker's compensation and general liability insurance. In these regards, it is wise to seek legal counsel.

If the employees are affiliated with organized labor, collective bargaining will be involved with the one or more unions representing the employees, and it will encompass matters such as pay, fringe benefits, and working conditions.

The objective of good personnel administration is to have the entity run smoothly, with maximum interpersonal harmony and good will so that the organization will prosper.

Some other pertinent details related to personnel administration are discussed briefly in Sec. 17.1.

PROFESSIONAL ENGINEER LICENSING AND REGISTRATION

Many engineers, whether self-employed or employed by others, may seek to obtain a license as a registered professional engineer. Generally, the terms *licensed professional engineer* and *registered professional engineer* are synonymous.

The licensing of professional engineers (and of those in other professions and trades) falls under the aegis of the individual state education departments, which control and administer professional licensing within the state's jurisdiction. In years past, some states granted the license as a professional engineer to a few persons who, although lacking formal credentials (education and otherwise), had a demonstrated record of performance of the duties of a professional engineer. The licenses were granted under ''grandfather'' clauses. This is now very seldom the case.

The usual route to obtain a license as a professional engineer is via a combination of education, experience, and written examination. The successful attainment of an undergraduate degree in engineering will admit the applicant to a first examination, the successful completion of which results in the applicant being granted a certificate as Engineer in Training (EIT). Additional suitable experience for another 4 years admits the EIT into the final written examination for professional engineer. The successful completion of the final examination results in the award of a license as Professional Engineer.

Federal, state, and most local municipalities require that their key engineering personnel be licensed professional engineers. Engineering documents submitted to those bodies, likewise, will require that the signatory thereto be a licensed professional engineer.

There are minor variations from state to state as to acceptable experience, the degree to which academic experience may be acceptable as engineering experience, etc.; but by and large, there is a degree of uniformity among the state education departments in the matter of licensing professional engineers.

Professional engineers active in several state jurisdictions may be required to obtain a license as professional engineer in those different states. To that end, most states have established procedures to grant a license as professional engineer by comity, but usually this will be possible only if the original license was granted on the basis of a written examination. Comity is often initiated through the offices of the National Society of Professional Engineers as a service to its members.

The work bearing the professional engineer's seal is considered creative work and is so addressed in most stated education laws. The work carries an implicit copyright, even though a statement to that effect may not be placed on the documents. Said work is the property of the professional engineer; unauthorized copying of such documents is illegal, and violators are subject to legal action and redress.

The basic intent of formal licensing of professional engineers is to protect the public from incompetent practice. A licensed professional engineer whose competence is called into question is subject to formal investigative procedures, and when the circumstances warrant it, the license may be revoked or suspended.

Detailed information regarding license as a professional engineer can be obtained from the state education department having jurisdiction, and it will set forth the requirements, privileges, and responsibilities inherent in the granting of the license.

PRODUCT LIABILITY

In the current consumer and/or user climate, the matter of product liability has assumed a large role and often impacts upon the engineer's input to certain products. In some cases, the engineer may be held personally responsible and liable for some outcome even though the engineer may have been employed by others. The case reports of litigation involving product liability are voluminous, and it is not surprising that identical circumstances will result in verdicts different from one jurisdiction to another. The laws related to product liability are based principally on two concepts: negligence and strict liability. The details are not addressed here; suffice it to say that the engineer who faces product liability litigation must retain competent legal counsel to prepare and implement an appropriate defense.

PROFESSIONAL LIABILITY INSURANCE

This type of insurance is often carried by engineers in private practice to protect the engineer as well as to provide relief to an injured party, when the latter may result from error(s) and/or omission(s) in the documents prepared by the engineer; hence, the term *errors and omissions insurance,* or E&O insurance. The nature of professional liability insurance is quite complex and is usually handled between the engineer and the insurer, often with the active participation of the engineer's legal counsel. Essentially, in obtaining professional liability insurance, the engineer enlists a broader financial entity (the insurance company) to absorb the major portion of the costs of claims in exchange for a premium paid to the insurance company. Decisions related to professional liability insurance evolve from the management decisions made by the engineer (or the firm), as part of the overall conduct of the engineering practice and consideration of the financial risks incurred thereby.

In many cases, before any work proceeds, the client will require that the engineer provide evidence that professional liability insurance is in effect, and that evidence often is incorporated into the contractual agreement between engineer and client.

Professional liability insurance policies are traditionally of the claims-made type. Claims arising out of work undertaken and performed while the policy is in effect will be covered. Claims arising from that work at a later date will **not** be covered unless the insurance has continued in effect to the time the claim is made. All coverage ceases when the policy is canceled or not renewed.

Insurance premiums depend on several factors: limits of liability, deductibles, type of work performed and the services rendered in connection therewith, experience of engineer and his or her "track record," and the geographic location of the work (there are low-risk states and high-risk states). There are other factors which may be subject to consideration by the insurer.

Section **19**

Refrigeration, Cryogenics, Optics, and Miscellaneous

BY

A. J. RYDZEWSKI   *Consultant, Engineering Department, E. I. du Pont de Nemours & Co., Inc.*
WARREN W. RICE   *Senior Project Engineer, Piedmont Engineering Corp.*
F. J. EDESKUTY   *Los Alamos National Laboratory, retired.*
MICHAEL J. CLARK   *Manager, Optical Tool Engineering and Manufacturing, Bausch & Lomb, Rochester, NY.*

19.1 MECHANICAL REFRIGERATION

by A. J. Rydzewski and Warren W. Rice

REFERENCES: Publications of the ASRE and ASHRAE; "Air-conditioning Refrigerating Data Books"; *Refrigerating Engineering;* ARI Equipment Standards; ASHRAE Guide and Data Book. Merkel and Bosnjakovic, "Diagrammen und Tabellen zur Berechnung der Absorptionskältemaschinen," Springer, Berlin. Jordan and Priester, "Refrigeration and Air Conditioning," Prentice-Hall.

REFRIGERATION MACHINES AND PROCESSES

(For general theory of refrigeration, see Sec. 4; for air-conditioning uses, see Sec. 12.4; for cryogenics, see Sec. 19.2.)

Refrigeration is a term used to describe thermal systems which maintain a process space or material at a temperature less than available from ambient conditions. Heat is transferred from the materials to be cooled into a lower-temperature substance referred to as the **refrigerant.** The refrigerant is cooled by physical processes which pump heat into the ambient environment. These apply several physical phenomena such as phase changes, sensible heating, thermoelectric effects, work extraction by mechanical devices, and endothermic chemical reactions. Stored energy in the form of ice, sublimating solid carbon dioxide, and cold liquids are included with processes described by the word **refrigeration.**

Many refrigeration systems use a continuous cycle. This requires that the refrigerant be recovered for recycling after it has absorbed heat from the product or space to be cooled. The most commonly used system has a recirculating fluid that boils at a low temperature to remove heat. The vapor is compressed to a pressure such that the condensing temperature is elevated. The vapor is cooled and condensed by a heat exchanger which rejects the heat into the environment. The condensed liquid is passed through a restricting valve or orifice. The liquid changes into a mixture of liquid and vapor that is colder than the refrigerated space. This expansion at constant enthalpy is known as the *Joule-Thompson effect.* An expansive turbine can remove heat from a high-pressure vapor by extracting energy in the form of mechanical work. When the exhaust conditions are proper, the material will be a mixture of gas and condensate. This fluid can be used as a refrigerant or stored in liquid form as product. The remaining vapor can be recycled as needed to continue the process.

The equipment used in refrigeration processes is referred to as follows: high-pressure vapor heat exchangers are called *condensers,* low-pressure heat exchangers are called *evaporators,* throttling devices are called *expansion valves* or *capillaries,* and pressure-raising devices are called *compressors.*

With continuous recirculation of the cycle fluid, the low-temperature heat pickup must be pumped, via energy input, to a higher level so as to discharge it to atmospheric temperature. In this sense the system acts as an **energy-transport** device or **heat pump** or, from another viewpoint, as a temperature transformer. This philosophy of energy transport may equally well be applied to more remote systems, such as thermoelectric. It is, of course, a manifestation of the principles of the Second Law, and the effectiveness of the system's operation may be evaluated numerically.

Efficiency ratings of refrigeration equipment are expressed several ways. Some of the more common terms include **coefficient of performance** (COP) and **energy efficiency ratio** (EER). **COP** is defined as refrigerant effect divided by net work input, where the refrigerant effect is the absolute value of the heat transferred from the lower temperature source, and the net work input is the absolute value of heat transferred to the higher temperature sink minus the refrigerant effect. COP can alternatively be defined as the ratio calculated by dividing the total heating capacity in Btu per hour provided by the refrigeration system, including circulating fan heat, but excluding supplementary resistance, by the total electric input in watts × 3.412. This definition applies primarily to heat pumps. **EER** is a ratio calculated by dividing the cooling capacity in Btu per hour by the power input in watts at any set of rating conditions, expressed in Btu/W · h.

Other rating terms used primarily for residential and commercial air-conditioning units and heat pumps include **seasonal energy efficiency ratio** (SEER) and **heating season performance factor** (HSPF). Both adjust efficiency ratings for the variations encountered over a normal heating or cooling season.

The various association and society standards have specific conditions at which equipment is rated. In efficiency comparisons, familiarity with the specific rating method is important. The factors used in the calculation of manufacturer's published data may not be representative of the specific application. Part load conditions must be considered in any evaluation when the unit will not normally be fully loaded. If the load is constant and is known, comparisons should be made at that point. Other important factors include whether the stated efficiency is just the compressor efficiency, or if it includes motor, seal, bearing, and drive losses. It is also recommended that rated operating conditions be adjusted to actual conditions when known. For example, the actual condensing temperatures may be significantly higher or lower than rated conditions depending on the geographic region.

Refrigeration capacity is defined in terms of the "ton."

Units of Refrigeration In the United States, a **standard ton of refrigeration** corresponds to a heat absorption at a rate of 288,000 Btu/day or 200 Btu/min (3.5168 kW). The heat absorption per day is approximately the heat of fusion of 1 ton (907.19 kg) of ice at 32°F (0°C). The **standard rating** of a refrigerating machine, using a condensable vapor, is the number of standard tons of refrigeration it can produce under the following conditions: (1) liquid only enters the expansion valve and vapor only enters the compressor or the absorber of an absorption system; (2) the liquid entering the expansion valve is subcooled 9°F (5°C) and the vapor entering the compressor or absorber is superheated 9°F (5°C), these temperatures to be measured within 10 ft (3.05 m) of the compressor cylinder or absorber; (3) the pressure at the compressor or absorber inlet corresponds to a saturation temperature of 5°F (−15°C); and (4) the pressure at the compressor or absorber outlet corresponds to a saturation temperature of 86°F (30°C).

The **British unit of refrigeration** corresponds to a heat-absorption rate of 237.6 Btu/min (4.175 kW) with inlet and outlet pressures corresponding to saturation temperatures of 23°F (−5°C) and 59°F (15°C), respectively. The standard practice in Europe is to use kilowatts to specify refrigeration loads. Occasionally a unit of refrigeration capacity called the **frigorie** is used. A frigorie is approximately equivalent to 50 Btu/min (0.8786 kW), or one-quarter of a standard ton of refrigeration.

PROPERTIES OF REFRIGERANTS

(For detailed properties, see Sec. 4.)

Refrigerants are the transport fluids which convey the heat energy from the low-temperature level to the high-temperature level, where it can, in terms of heat transfer, give up its heat. In the broad sense, gases involved in liquefaction processes or in gas-compression cycles go through low-temperature phases and hence may be termed "refrigerants," in a way similar to the more conventional vapor-compression fluids.

Refrigerants are **designated by number.** The identifying number of the refrigerant, or the word *Refrigerant* or both, may be used in conjunction with the trade name. For example, Refrigerant 134a can also be referred to as R-134a, HFC-134a, or SUVA-134a. The number designation

Table 19.1.1   Environmental Properties of Refrigerants

Refrigerant and number	Ozone depletion potential	Global warming potential	Toxicity, ppm (v/v)	AEL,* TLV†	Flash point, lower/upper, vol. %
R-11/(CFC-11)	1.0	1.0	1000	TLV	–
R-12/(CFC-12)	1.0	2.8	1000	AEL	–
R-502	0.23	3.75	1000	TLV	–
R-114	0.7	3.9	1000	TLV	–
R-402B (SUVA HP81)	0.03	0.52	1000	AEL	None
HCFC-123	0.02	0.02	30	AEL	None
HFC-134a	0.0	0.26	1000	AEL	None
HFC-125	0.0	0.84	1000	AEL	None
HCFC-124	0.02	0.1	500	AEL	None
R-22/(HCFC-22)	0.05	0.3	1000	TLV	None
HFC-23	0.0	6.0	1000	AEL	None
HFC-32	0.0	0.11	1000	TLV	14/31

* AEL is the recommended time-weighted average concentration of an airborne chemical to which nearly all workers may be exposed during an 8-h and/or 12-h day, 40-h week without adverse effect. Determined by the du Pont Company for compounds that do not have a TLV.

† TLV, established for industrial chemicals by the American Conference of Governmental Industrial Hygienists, is the recommended time-weighted average concentration of an airborne chemical to which nearly all workers may be exposed during an 8-h day, 40-h week without adverse effect.

SOURCE: Data from E. I. du Pont de Nemours & Co., Inc. Bulletin AG-1.

and safety classification of refrigerants is covered in ANSI/ASHRAE Standard 34-1992.

Recent evidence indicates that much of the damage to the atmospheric ozone layer is the result of decomposition of chlorofluorocarbon (CFC) chemicals. An international agreement known as the Montreal Protocol took effect in 1989 and a new Clean Air Act was signed into law in 1990 to limit the production and regulate the use and disposal of chlorofluorocarbons. The chlorofluorocarbons that cause ozone-layer depletion have been designated as CFC-type materials. Other refrigerants that are chlorofluorocarbons but cause little or no ozone destruction are designated as hydrochlorofluorocarbon (HCFC) or hydrofluorocarbon (HFC) refrigerants.

Several terms are used to specify the relative destructiveness of chlorofluorocarbon refrigerants to the ozone layer, and other effects on the environment. The **ozone depletion potential** (ODP) is the ozone-destroying power of a substance measured relative to Refrigerant 11 (CFC-11) (R-11). The **global warming potential** (GWP) is a relative measure of the ability of a substance to cause an increase in the temperature of the atmosphere by absorbing solar and earth radiation that is relative to the effect of Refrigerant 11. A listing of the ozone depletion potential, global warming potential, and 8-hour toxicity limits (AEL or TLV) for halo-carbon refrigerants is given in Table 19.1.1. Table 19.1.2 is a listing of replacement refrigerants for existing CFC compounds.

The **normal boiling point** is an important attribute of refrigerants for vapor compression systems. The desirable characteristic is a positive gage pressure throughout the refrigerating system. This will eliminate contamination of the refrigerant and resulting heat-transfer inefficiencies due to air inleakage. Table 19.1.3 shows some common fluids in order of boiling point: included are halocarbon compounds that offer reduced ozone depletion potential, hydrocarbons, sulfur compounds, and inorganic compounds.

The desirable thermal properties of a refrigerant are (1) convenient evaporation and condensation pressures, (2) high critical and low freezing temperatures, (3) high latent heat of evaporation and high vapor specific heat, (4) low viscosity and high film heat conductivity. Desir-

able practical properties include (1) low cost, (2) chemical and physical inertness under operating conditions, (3) noncorrosiveness toward ordinary construction materials, and (4) low explosive hazard both alone and mixed with air. The refrigerant should be nonpoisonous and nonirritating and should not cause deterioration in the lubricant used. Leakage should be detectable by simple tests, easily performed.

The **specific volume of refrigerant** to be handled determines the size of positive displacement compressors, but with centrifugal compressors a large volume is not objectionable and may be an advantage for small units.

A comparison of various refrigerants based on *ideal performance* is given in Table 19.1.4. The results are for the **standard temperature range** of 5 to 86°F (− 15 to 30°C) and also for certain other ranges. For these calculations, zero piston clearance and expansion through a throttle valve were assumed. The use of an expansion cylinder instead of a valve would have yielded work (available for supplying some of the work of compression), but this is always a negligible quantity, except when the compression pressure approaches the critical pressure, as for carbon dioxide. Theoretical *coefficients of performance* may be obtained by dividing 4.72 by the theoretical horsepower given in the table.

Ammonia (R-717) is used extensively in commercial, industrial, and moderately low-temperature installations. It is toxic, and recent laws regulate the operation, maintenance, and design of ammonia systems that contain 10,000 lb of refrigerant or more. (See Document 29 CFR 1910.119 of the Federal Register for further information.) Ammonia is provided commercially in several purity grades. It is recommended that refrigerant-grade anhydrous ammonia of 99.95 percent purity be used. Moisture contamination from impure ammonia will accumulate in evaporator coils and inhibit performance. Ammonia is corrosive to copper and copper-bearing alloys. It is not miscible to any large extent with lubricating oils. Exposure to it in a confined space at concentrations below 400 ppm is rarely fatal. Ammonia is immediately fatal at 0.5 to 0.6 percent by volume. The presence of ammonia can be detected by burning a sulfur candle in the vicinity of a leak. A white cloud will form if ammonia vapors are present. Hydrochloric acid solutions and wetted

Table 19.1.2   Replacement Refrigerant Compounds

Current refrigerant	Replacement refrigerant	Formula	End uses*
R-11/CFC-11	HCFC-123	$CHCL_2CF_3$	Chillers [+ 25°F (− 3.9°C)]
R-12/CFC-12	HFC-134a	CH_2FCF_3	Medium temperature [− 5°F (− 20.6°C)]
R-13/CFC-13	HFC-23	CHF_3	Low temperature [− 80°F (− 62.2°C)]
R-502	R-402A, R404A	Blend	Ice machines, freezers [− 50°F (− 45.6°C)]
R-22	HCFC-22	$CHClF_2$	General use [− 50°F (− 45.6°C)]

* Lower optimum use temperature.

Table 19.1.3 Selected Refrigerants and Gases in Low-Temperature Applications

Refrigerant	Refrigerant no.	Symbol	Molecular weight	Boiling point, °F	Critical point Temperature, °F	Critical point Pressure, lb/in² abs
Helium	704	He	4.0	−452	−450.0	33
Hydrogen	702	H_2	2.0	−423	−400.0	188
Nitrogen	728	N_2	28.0	−320	−232.0	492
Air	729	—	29.0	−312	−220.3	547
Oxygen	732	O_2	32.0	−297	−182.0	730
Methane	50	CH_4	16.0	−258.9	−115.8	673
Carbon tetrafluoride	14	CF_4	88.0	−198.4	−49.9	542
Ethylene	1,150	C_2H_4	28.0	−155.0	48.8	732
Ethane	170	C_2H_6	30.0	−127.5	90.1	708
Trifluoromethane	23	CHF_3	70.01	−115.66	78.6	701.4
Carbon dioxide	744	CO_2	44.0	−109.3	87.8	1,071
Propane	290	C_3H_8	44.1	−44.2	202	661
Monodichlorodifluoromethane	22	$CHClF_2$	86.5	−41.36	204.8	716
Ammonia	717	NH_3	17.0	−28.0	271.4	1,657
1,1,1,2-tetrafluoroethane	134a	CH_2FCF_3	102	−15	213.9	588.3
Methylchloride	40	CH_3Cl	50.5	−10.8	289.4	969
Isobutane	601	C_4H_{10}	58.1	10.3	272.7	537
2-chloro-1,1,1,2-tetrafluoroethane	124	$CHClFCF_3$	136.5	12.2	252	524.5
Sulfur dioxide	764	SO_2	64.1	14.0	314.8	1,142
Butane	600	C_4H_{10}	58.1	31.3	306	550
Dichloromonofluoromethane	21	$CHCl_2F$	102.9	48.1	353.3	750
2,2-dichloro-1,1,1-trifluoroethane	123	$CHCl_2CF_3$	152.9	82	362.8	533.2
Ethyl-ether	610	$C_4H_{10}O$	74.1	94.3	522.1	381
Methylene chloride	30	CH_2Cl_2	84.9	103.6	480	670
Water	718	H_2O	18.0	212	706.1	3,226

Table 19.1.4 Ideal Performance of Refrigerants for Various Temperature Ranges

Refrigerant and (number)	Operating temperature range, °F	Suction pressure, lb/in² abs	Head pressure, lb/in² abs	Ratio of head to suction pressure	With dry and saturated suction vapor, per ton Weight of vapor, lb/min	Piston displacement, ft³/min	Theoretical hp	Temperature at end of compression, °F	Type of compressor
Air (729)	5–86	14.7	73.5	5.0	7.02	82.3	2.82	277.0	Recip. and exp. cyl
Water (718)	32–86	0.0885	0.6152	6.95	0.1957	647.0	0.618	332	Centrif. or ejector
	32–100	0.0885	0.9492	10.73	0.1985	656.2	0.819	420	
	40–100	0.1217	0.9492	7.80	0.1978	483.4	0.687	366	
Carbon dioxide (CO_2) (744)	5–86	332.0	1,043.0	3.14	3.61	0.960	1.827	160.3	Recip.
Ammonia (NH_3) (717)	5–86	34.27	169.2	4.94	0.421	3.44	0.99	209.8	Recip.
	20–100	48.21	211.9	4.40	0.421	2.49	0.94	212.8	
	40–100	73.32	211.9	2.89	0.427	1.70	0.65	176.0	
SUVA 123 ($CHCl_2CF_3$) (HCFC-123)	5–86	2.303	15.9	6.9	3.26	45.6	0.95*	90.5	Recip.
	20–100	3.48	20.8	5.97	3.33	31.7	0.91	101.	Centrif.
	40–100	5.79	20.8	3.59	3.18	18.8	0.6*	100.	Centrif.
SUVA 134a (CH_2FCF_3) (HFC-134a)	5–86	23.77	111.8	4.7	3.14	6.1	1.04	98.	Centrif.
	20–100	33.13	139.0	4.19	3.28	4.62	1.0*	109.	
	40–100	49.77	139.0	2.79	3.14	3.0	0.68	106.	
SUVA HP 81 blend [R-402 (38/2/60)]	5–86	54.37	200.5	3.68	3.56	3.37	1.07	99.	Screw or rotary or recip.
	20–100	72.12	243.9	3.38	3.78	2.71	1.04	121.5	
	40–100	102.26	243.9	2.385	3.62	1.83	0.68*	114.	
R 22 ($CHClF_2$) (22)	5–86	43.02	174.5	4.06	2.926	3.65	1.03	131.7	Recip.
	20–100	57.98	212.6	3.67	3.023	2.83	0.99	152.7	
	40–100	83.72	212.6	2.54	2.936	1.93	0.68	131.0	
Methylene chloride (CH_2Cl_2) (30)	5–86	1.28	10.07	8.56	1.485	74.0	0.96*	205.1*	Centrif.
	20–100	1.92	13.25	6.90	1.520	47.72	0.91*	157.1*	
	40–100	3.38	13.25	3.92	1.493	27.76	0.63*	167.7*	
Methyl chloride (CH_3Cl) (40)	5–86	21.15	94.70	4.48	1.331	5.95	0.96	178.1	Recip.
	20–100	29.16	116.7	4.00	1.363	4.51	0.90	184.4	
	40–100	43.25	116.7	2.69	1.342	3.07	0.62	157.0	
Sulfur dioxide (SO_2) (764)	5–86	11.81	66.45	5.63	1.415	9.08	0.97	191.4	Recip.
	20–100	17.18	84.52	4.92	1.453	6.52	0.92	193.4	
	40–100	27.10	84.52	3.12	1.444	4.17	0.63	162.6	
Propane (C_3H_8) (290)	5–86	42.1	155.3	3.69	1.653	4.10	1.35*	92.9*	Recip.
	20–100	55.5	187.0	3.37	1.730	3.29	1.32*	103.9*	
	40–100	78.0	187.0	2.40	1.646	2.26	0.90*	101.5*	
Ethane (C_2H_5) (170)	5–86	236.0	675.9	2.87	3.41	1.82	2.180	105	Recip.
Ethyl chloride (C_2H_5Cl) (160)	5–86	4.65	27.10	5.83	1.405	24.0	0.95*	106.3*	Rotary
	20–100	6.80	34.79	5.12	1.425	17.19	0.92*	116.1*	
	40–100	10.79	34.79	3.22	1.375	10.73	0.63*	109.9*	

* Values may be slightly in error.

19-4

Table 19.1.5   Properties of Superheated Ammonia

(v = specific volume in ft^3/lb; h = enthalpy in Btu/lb; s = entropy; h_f and s_f are measured from $-40°$F.)

Pressure, lb/in^2 abs	Temperature of saturated vapor, °F		Temperature of superheated vapor, °F								
			-30	-20	-10	0	10	20	30	40	50
10	-41.34	v	26.58	27.26	27.92	28.58	29.24	29.90	30.55	31.20	31.85
		h	603.2	608.5	613.7	618.9	624.0	629.1	634.2	639.3	644.4
		s	1.4420	1.4542	1.4659	1.4773	1.4884	1.4992	1.5097	1.5200	1.5301
20	-16.64	v			13.74	14.09	14.44	14.78	15.11	15.45	15.78
		h			610.0	615.5	621.0	626.4	631.7	637.0	642.3
		s			1.3784	1.3907	1.4025	1.4138	1.4248	1.4356	1.4460
30	-0.57	v				9.250	9.492	9.731	9.966	10.20	10.43
		h				611.9	617.8	623.5	629.1	634.6	640.1
		s				1.3371	1.3497	1.3618	1.3733	1.3845	1.3953
40	11.66	v						7.203	7.387	7.568	7.746
		h						620.4	626.3	632.1	637.8
		s						1.3231	1.3353	1.3470	1.3583
50	21.67	v							5.838	5.988	6.135
		h							623.4	629.5	635.4
		s							1.3046	1.3169	1.3286

Pressure, lb/in^2 abs	Temperature of saturated vapor, °F		100	120	140	160	180	200	240	280	320
80	44.40	v	4.190	4.371	4.548	4.722	4.893	5.063	5.398	5.730	
		h	658.7	670.4	681.8	693.2	704.4	715.6	738.1	760.7	
		s	1.3199	1.3404	1.3598	1.3784	1.3963	1.4136	1.4467	1.4781	
100	56.05	v	3.304	3.454	3.600	3.743	3.883	4.021	4.294	4.562	
		h	655.2	667.3	679.2	690.8	702.3	713.7	736.5	759.4	
		s	1.2891	1.3104	1.3305	1.3495	1.3678	1.3854	1.4190	1.4507	
120	66.02	v	2.712	2.842	2.967	3.089	3.190	3.326	3.557	3.783	
		h	651.6	664.2	676.5	688.5	700.2	711.8	734.9	758.0	
		s	1.2628	1.2850	1.3058	1.3254	1.3441	1.3620	1.3960	1.4281	
140	74.79	v	2.288	2.404	2.515	2.622	2.727	2.830	3.030	3.227	3.420
		h	647.8	661.1	673.7	686.0	698.0	709.9	733.3	756.7	780.0
		s	1.2396	1.2628	1.2843	1.3045	1.3236	1.3418	1.3763	1.4088	1.4395
160	82.64	v	1.969	2.075	2.175	2.272	2.365	2.457	2.635	2.809	2.980
		h	643.9	657.8	670.9	683.5	695.8	707.9	731.7	755.3	778.9
		s	1.2186	1.2429	1.2652	1.2859	1.3054	1.3240	1.3591	1.3919	1.4229
180	89.78	v	1.720	1.818	1.910	1.999	2.084	2.167	2.328	2.484	2.637
		h	639.9	654.4	668.0	681.0	693.6	705.9	730.1	753.9	777.7
		s	1.1992	1.2247	1.2477	1.2691	1.2891	1.3081	1.3436	1.3768	1.4081
200	96.34	v	1.520	1.612	1.698	1.780	1.859	1.935	2.082	2.225	2.364
		h	635.6	650.9	665.0	678.4	691.3	703.9	728.4	752.5	776.5
		s	1.1809	1.2077	1.2317	1.2537	1.2742	1.2935	1.3296	1.3631	1.3947
220	102.42	v		1.443	1.525	1.601	1.675	1.745	1.881	2.012	2.140
		h		647.3	662.0	675.8	689.1	701.9	726.8	751.1	775.3
		s		1.1917	1.2167	1.2394	1.2604	1.2801	1.3168	1.3507	1.3825
240	108.09	v		1.302	1.380	1.452	1.521	1.587	1.714	1.835	1.954
		h		643.5	658.8	673.1	686.7	699.8	725.1	749.8	774.1
		s		1.1764	1.2025	1.2259	1.2475	1.2677	1.3049	1.3392	1.3712
260	113.42	v		1.182	1.257	1.326	1.391	1.453	1.572	1.686	1.796
		h		639.5	655.6	670.4	684.4	697.7	723.4	748.4	772.9
		s		1.1617	1.1889	1.2132	1.2354	1.2560	1.2938	1.3258	1.3608

Source: Condensed from NBS Circ. 142, 1923.

litmus papers will also react with ammonia to give visible evidence of the presence of vapors. See Tables 19.1.5 and 19.1.6 for thermodynamic properties of R-717.

Carbon dioxide (R-744) was used for a long time as a "safety refrigerant." It is not toxic, and exposure to it is not dangerous unless concentrations are high. Like many other refrigerants, it is heavier than air, and caution must be exercised in confined spaces, as it may displace oxygen and result in asphyxiation. With cooling water at 70°F (21°C), the condensing pressure of CO_2 is high, and since its critical temperature is 87.8°F (31°C), condensation will not occur at water temperatures above this. Specific power consumption is high on CO_2 machines.

Sulfur dioxide (R-764) has rapidly gone out of use. It is extremely corrosive unless absolutely anhydrous. It is extremely irritating in small concentrations and toxic in high concentrations.

Methyl chloride (R-40) (CH_3Cl), an anesthetic in amounts of 5 to 10 percent by volume, has been used in air-cooled units of moderate or small sizes. It is miscible with mineral oils; small amounts of moisture in a methyl chloride system will cause trouble by freezing in expansion valves.

For industrial work, butane (C_4H_{10}), propane (C_3H_8), and, at low temperatures, ethane (C_2H_6) are used. **Dieline** (dichloroethylene, $C_2H_2Cl_2$) and **Carrene** (dichloromethane, CH_2Cl_2) have been used to some extent, usually in centrifugal compressors.

As a result of legislation requiring the discontinuation of production

Table 19.1.6 Properties of Saturated Ammonia

(h_f and s_f are measured from $-40°$F.)

Temp., °F, t	Pressure, lb/in² abs, p	Specific volume, ft³/lb		Enthalpy, Btu			Entropy		
		Sat. liquid, v_f	Sat. vapor, v_g	Sat. liquid, h_f	Vapor-ization, h_{fg}	Sat. vapor, h_g	Sat. liquid, s_f	Vapor-ization, s_{fg}	Sat. vapor, s_g
-40	10.41	0.02322	24.86	0.0	597.6	597.6	0.0000	1.4242	1.4242
-38	11.04	0.02326	23.53	2.1	596.2	598.3	0.0051	1.4142	1.4193
-36	11.71	0.02331	22.27	4.3	594.8	599.1	0.0101	1.4043	1.4144
-34	12.41	0.02335	21.10	6.4	593.5	599.9	0.0151	1.3945	1.4096
-32	13.14	0.02340	20.00	8.5	592.1	600.6	0.0201	1.3847	1.4048
-30	13.90	0.02345	18.97	10.7	590.7	601.4	0.0250	1.3751	1.4001
-28	14.71	0.02349	18.00	12.8	589.3	602.1	0.0300	1.3655	1.3955
-26	15.55	0.02354	17.09	14.9	587.9	602.8	0.0350	1.3559	1.3909
-24	16.42	0.02359	16.24	17.1	586.5	603.6	0.0399	1.3464	1.3863
-22	17.34	0.02364	15.43	19.2	585.1	604.3	0.0448	1.3370	1.3818
-20	18.30	0.02369	14.68	21.4	583.6	605.0	0.0497	1.3277	1.3774
-18	19.30	0.02374	13.97	23.5	582.2	605.7	0.0545	1.3184	1.3729
-16	20.34	0.02378	13.29	25.6	580.8	606.4	0.0594	1.3092	1.3686
-14	21.43	0.02383	12.66	27.8	579.3	607.1	0.0642	1.3001	1.3643
-12	22.56	0.02388	12.06	30.0	577.8	607.8	0.0690	1.2910	1.3600
-10	23.74	0.02393	11.50	32.1	576.4	608.5	0.0738	1.2820	1.3558
-8	24.97	0.02399	10.97	34.3	574.9	609.2	0.0786	1.2730	1.3516
-6	26.26	0.02404	10.47	36.4	573.4	609.8	0.0833	1.2641	1.3474
-4	27.59	0.02409	9.991	38.6	571.9	610.5	0.0880	1.2553	1.3433
-2	28.98	0.02414	9.541	40.7	570.4	611.1	0.0928	1.2465	1.3393
0	30.42	0.02419	9.116	42.9	568.9	611.8	0.0975	1.2377	1.3352
2	31.92	0.02424	8.714	45.1	567.3	612.4	0.1022	1.2290	1.3312
4	33.47	0.02430	8.333	47.2	565.8	613.0	0.1069	1.2204	1.3273
6	35.09	0.02435	7.971	49.4	564.2	613.6	0.1115	1.2119	1.3234
8	36.77	0.02440	7.629	51.6	562.7	614.3	0.1162	1.2033	1.3195
10	38.51	0.02446	7.304	53.8	561.1	614.9	0.1208	1.1949	1.3157
12	40.31	0.02451	6.996	56.0	559.5	615.5	0.1254	1.1864	1.3118
14	42.18	0.02457	6.703	58.2	557.9	616.1	0.1300	1.1781	1.3081
16	44.12	0.02462	6.425	60.3	556.3	616.6	0.1346	1.1697	1.3043
18	46.13	0.02468	6.161	62.5	554.7	617.2	0.1392	1.1614	1.3006
20	48.21	0.02474	5.910	64.7	553.1	617.8	0.1437	1.1532	1.2969
22	50.36	0.02479	5.671	66.9	551.4	618.3	0.1483	1.1450	1.2933
24	52.59	0.02485	5.443	69.1	549.8	618.9	0.1528	1.1369	1.2897
26	54.90	0.02491	5.227	71.3	548.1	619.4	0.1573	1.1288	1.2861
28	57.28	0.02497	5.021	73.5	546.4	619.9	0.1618	1.1207	1.2825
30	59.74	0.02503	4.825	75.7	544.8	620.5	0.1663	1.1127	1.2790
32	62.29	0.02508	4.637	77.9	543.1	621.0	0.1708	1.1047	1.2755
34	64.91	0.02514	4.459	80.1	541.4	621.5	0.1753	1.0968	1.2721
36	67.63	0.02521	4.289	82.3	539.7	622.0	0.1797	1.0889	1.2686
38	70.43	0.02527	4.126	84.6	537.9	622.5	0.1841	1.0811	1.2652
40	73.32	0.02533	3.971	86.8	536.2	623.0	0.1885	1.0733	1.2618
42	76.31	0.02539	3.823	89.0	534.4	623.4	0.1930	1.0655	1.2585
44	79.38	0.02545	3.682	91.2	532.7	623.9	0.1974	1.0578	1.2552
46	82.55	0.02551	3.547	93.5	530.9	624.4	0.2018	1.0501	1.2519
48	85.82	0.02558	3.418	95.7	529.1	624.8	0.2062	1.0424	1.2486
50	89.19	0.02564	3.294	97.9	527.3	625.2	0.2105	1.0348	1.2453
52	92.66	0.02571	3.176	100.2	525.5	625.7	0.2149	1.0272	1.2421
54	96.23	0.02577	3.063	102.4	523.7	626.1	0.2192	1.0197	1.2389
56	99.91	0.02584	2.954	104.7	521.8	626.5	0.2236	1.0121	1.2357
58	103.7	0.02590	2.851	106.9	520.0	626.9	0.2279	1.0046	1.2325
60	107.6	0.02597	2.751	109.2	518.1	627.3	0.2322	0.9972	1.2294
62	111.6	0.02604	2.656	111.5	516.2	627.7	0.2365	0.9897	1.2262
64	115.7	0.02611	2.565	113.7	514.3	628.0	0.2408	0.9823	1.2231
66	120.0	0.02618	2.477	116.0	512.4	628.4	0.2451	0.9750	1.2201
68	124.3	0.02625	2.393	118.3	510.5	628.8	0.2494	0.9676	1.2170
70	128.8	0.02632	2.312	120.5	508.6	629.1	0.2537	0.9603	1.2140
72	133.4	0.02639	2.235	122.8	506.6	629.4	0.2579	0.9531	1.2110
74	138.1	0.02646	2.161	125.1	504.7	629.8	0.2622	0.9458	1.2080
76	143.0	0.02653	2.089	127.4	502.7	630.1	0.2664	0.9386	1.2050
78	147.9	0.02661	2.021	129.7	500.7	630.4	0.2706	0.9314	1.2020

Table 19.1.6   Properties of Saturated Ammonia   (Continued)
(h_f and s_f are measured from $-40°F$.)

Temp., °F, t	Pressure, lb/in² abs, p	Specific volume, ft³/lb		Enthalpy, Btu			Entropy		
		Sat. liquid, v_f	Sat. vapor, v_g	Sat. liquid, h_f	Vapor-ization, h_{fg}	Sat. vapor, h_g	Sat. liquid, s_f	Vapor-ization, s_{fg}	Sat. vapor, s_g
80	153.0	0.02668	1.955	132.0	498.7	630.7	0.2749	0.9242	1.1991
82	158.3	0.02675	1.892	134.3	496.7	631.0	0.2791	0.9171	1.1962
84	163.7	0.02684	1.831	136.6	494.7	631.3	0.2833	0.9100	1.1933
86	169.2	0.02691	1.772	138.9	492.6	631.5	0.2875	0.9029	1.1904
88	174.8	0.02699	1.716	141.2	490.6	631.8	0.2917	0.8958	1.1875
90	180.6	0.02707	1.661	143.5	488.5	632.0	0.2958	0.8888	1.1846
92	186.6	0.02715	1.609	145.8	486.4	632.2	0.3000	0.8818	1.1818
94	192.7	0.02723	1.559	148.2	484.3	632.5	0.3041	0.8748	1.1789
96	198.9	0.02731	1.510	150.5	482.1	632.6	0.3083	0.8678	1.1761
98	205.3	0.02739	1.464	152.9	480.0	632.9	0.3125	0.8608	1.1733
100	211.9	0.02747	1.419	155.2	477.8	633.0	0.3166	0.8539	1.1705
105	228.9	0.02769	1.313	161.1	472.3	633.4	0.3269	0.8366	1.1635
110	247.0	0.02790	1.217	167.0	466.7	633.7	0.3372	0.8194	1.1566
115	266.2	0.02813	1.128	173.0	460.9	633.9	0.3474	0.8023	1.1497
120	286.4	0.02836	1.047	179.0	455.0	634.0	0.3576	0.7851	1.1427

of CFC refrigerants, two new families of refrigerants, hydrochloro-fluorocarbons (HCFC) and hydrofluorocarbons (HFC) have been commercialized to replace them. Current information indicates that these replacement compounds have very low levels of toxicity. They are odorless and nonirritating. With the exception of HFC-32, HFC-143a, and HFC-152a, they are nonflammable. The replacement compounds include HCFC-22 (R-22): chloro-difluoromethane (CHClF$_2$) (see Fig. 19.1.1); HFC-23: trifluoromethane (CHF$_3$); HFC-32: difluoromethane (CH$_2$F$_2$); HCFC-123: 2,2-dichloro-1,1,1-trifluoroethane (CHCl$_2$CF$_3$) (see Fig. 19.1.2); HCFC-124: 2-chloro-1,1,1,2-tetrafluoroethane (CHClFCF$_3$); HFC-125: pentafluoroethane (CHF$_2$CF$_3$); HFC-143a: trifluoroethane (CH$_3$CF$_3$); HFC-134a: 1,1,1,2-tetrafluoroethane (CH$_2$FCF$_3$) (see Fig. 19.1.3); HFC-152a: 1,1-difluoroethane (CH$_3$CHF$_2$); HCFC-124: 2-chloro-1,1,1,2-tetrafluoroethane (CHClFCF$_3$); and HCFC-123: 2,2-dichloro-1,1,1-trifluoroethane (CHCl$_2$CF$_3$). HCFC-22 has been in use for many years, and because of its low ODP, will continue to be used in air-conditioning, commercial, and industrial refrigeration systems. HFC-134a is a medium-pressure refrigerant used as the replacement for R12. HFC-123 is a low-pressure refrigerant used as the replacement for R-11.

Several of the HCFC class of compounds have been blended together to create refrigerants with properties which simulate properties of banned refrigerants. One such blend has been proposed as the replacement for R-502. This blend is designated as R-402a, also known by the brand name SUVA HP80 and others. The thermodynamic properties of this mixture are shown in Fig. 19.1.4. Use caution when using blended refrigerants in centrifugal compressors. The compressor impeller may act to separate the components of the blend by molecular weight.

The HFC and HCFC refrigerants may not be miscible with the petroleum-based lubricants used in CFC systems. Existing systems depend on oil and refrigerant miscibility to return oil from the evaporators and piping back into the compressor lubricating system. Synthetic oils such as polyol ester and polyalkylene glycol must be used with the new refrigerants. In addition, seals, bushings, gaskets, motor insulation (hermetic machines) may also need replacement.

Building codes often require monitoring of refrigerant levels in equipment spaces. In addition to a monitor, an alarm system and exhaust facilities should be installed to prevent exposure above the acceptable exposure limit (AEL). ANSI/ASHRAE 15-1994, "Safety Code for Mechanical Refrigeration," has recommendations covering equipment room safety.

Leak Detection   Detection and repair of refrigerant leaks is an essential part of responsible refrigeration equipment operation. Worker safety, potential environmental damage, and refrigerant cost are the primary reasons for a detection and repair program. Area monitors are useful in determining that a leak has occurred. Hand-held detectors are useful in determining specific leak location. Refrigerants containing chlorine as part of the molecule are readily detected in minute amounts by passing them through a copper gauze kept hot by the essentially colorless flame of burning methyl alcohol. Even traces of such refrigerants in air are readily detectable by the intense green color imparted to the flame by their presence. The alternative refrigerants have either no chlorine or a lower amount of chlorine. They require new types of equipment which do not rely on chlorine detection. Nonselective, halogen-specific, and compound-specific detectors are available. Each type has limitations, and selection must be based on the specific application. For example, the nonspecific-type detector is poorly suited for area monitoring because of its high detection limit and potential for false alarms, but is a reliable, low-cost choice for pinpointing leaks. The halogen-specific detectors can be applied for either leak pinpointing or area monitoring. They are best applied where only one refrigerant is in use. The compound-specific detectors are relatively expensive, but are very selective. They are best applied in environments where several refrigerants or other chemical compounds may be present.

The pressure-temperature relations for the saturated vapors of many of the commercially important refrigerants are given in Fig. 19.1.5. If the temperature at which the refrigeration is desired and the temperature at which heat can be discarded (condenser temperature) are known, the chart is convenient for determining for any chosen refrigerant the pressures which must be maintained. For instance, if the use of methyl chloride is contemplated in an air-cooled cabinet food freezer located in a room where the air temperature may rise to 90°F (32°C), the compressor discharge must be at least 100 psia (690 kPa absolute), and if the cabinet cooling coil is to be held at $-30°F$ ($-34°C$), compressor suction will be 9 psia (62 kPa absolute).

The large volumes required for HCFC-123 and dieline can be handled satisfactorily by centrifugal compressors. When the evaporating pressure is below atmospheric—as in the case of HCFC-123, dieline, ammonia in food freezers, sulfur dioxide, water vapor, and butane—air leaks are likely to occur. The air tends to accumulate in the condenser and the heat-transfer performance is degraded.

Refrigerating systems which operate with a positive pressure in the equipment will leak refrigerant into the atmosphere. Many HFC- and HCFC-type refrigerants are practically odorless. Dangerous concentrations can build up in unventilated enclosed rooms such as freezers should a significant leak occur. Leak detection devices are available commercially to provide warnings of equipment leakage before harmful concentrations are present in occupied spaces. Addition of certain fluorescent compounds into the refrigerant in an operating system will allow identification of equipment leaks. The additives are placed into the

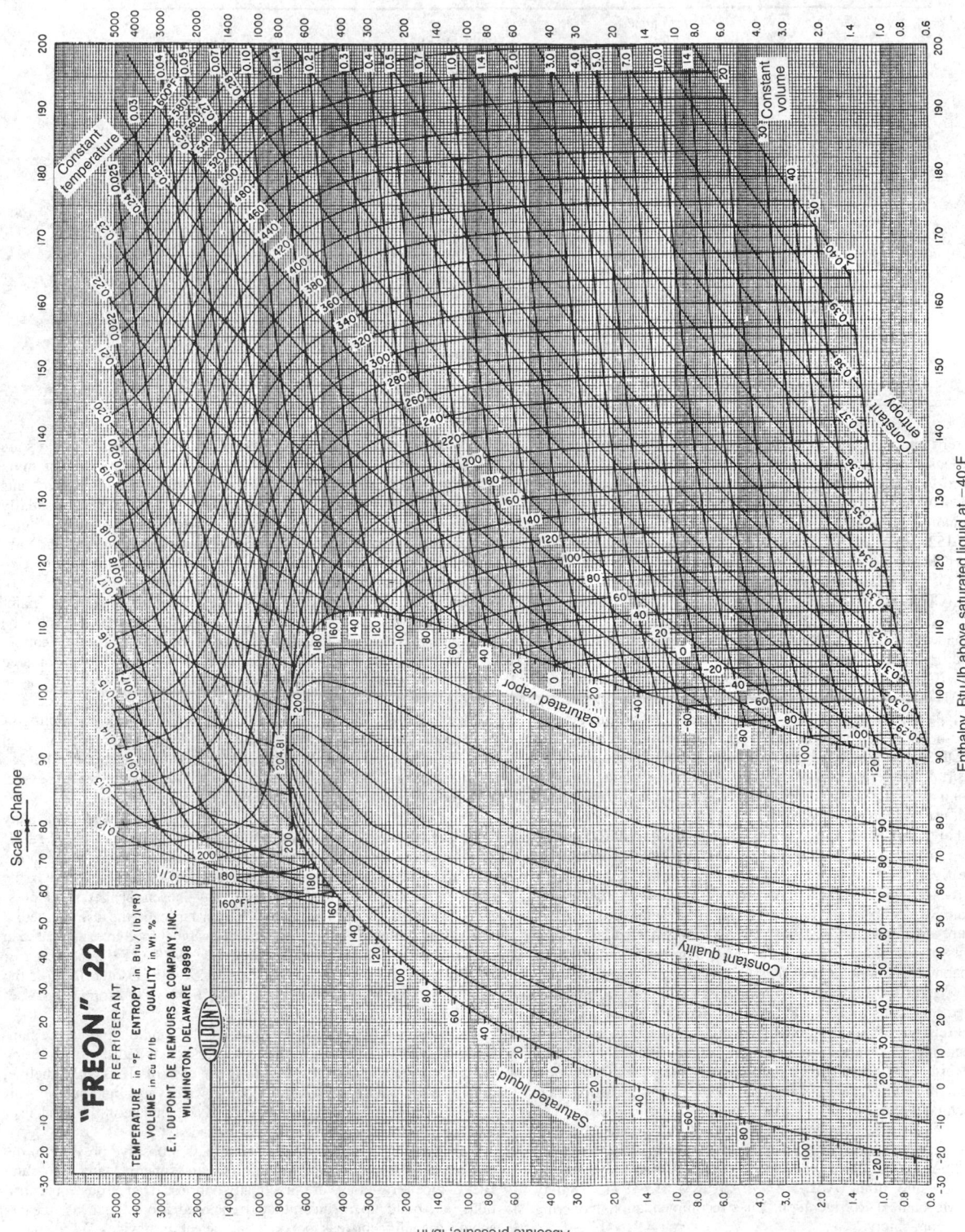

Fig. 19.1.1 Thermodynamic properties of R-22. (*Reprinted with permission of E. I. du Pont de Nemours & Co.*)

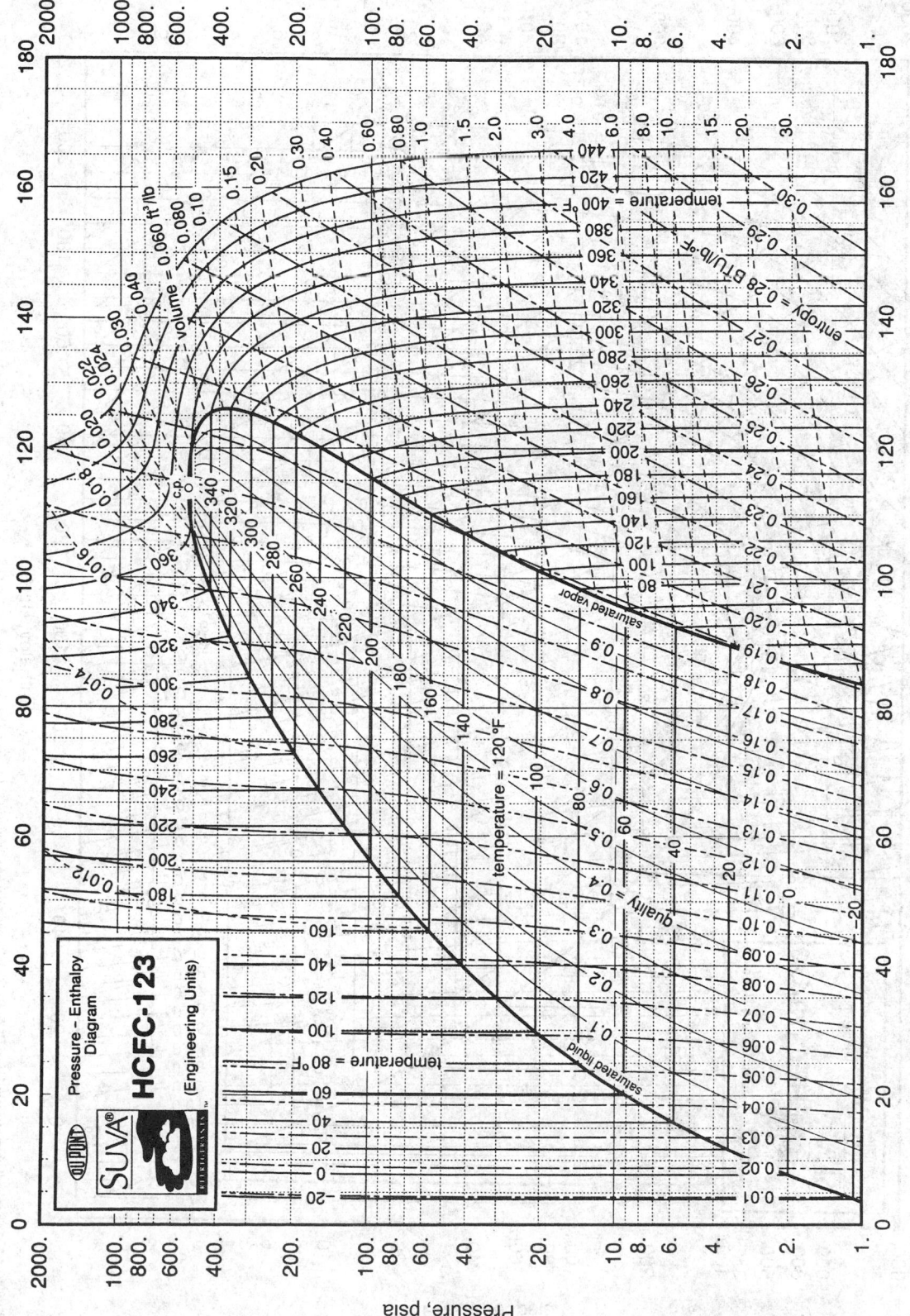

Fig. 19.1.2 Thermodynamic properties of HCFC-123. (*Reprinted with permission of E. I. du Pont de Nemours & Co.*)

19-9

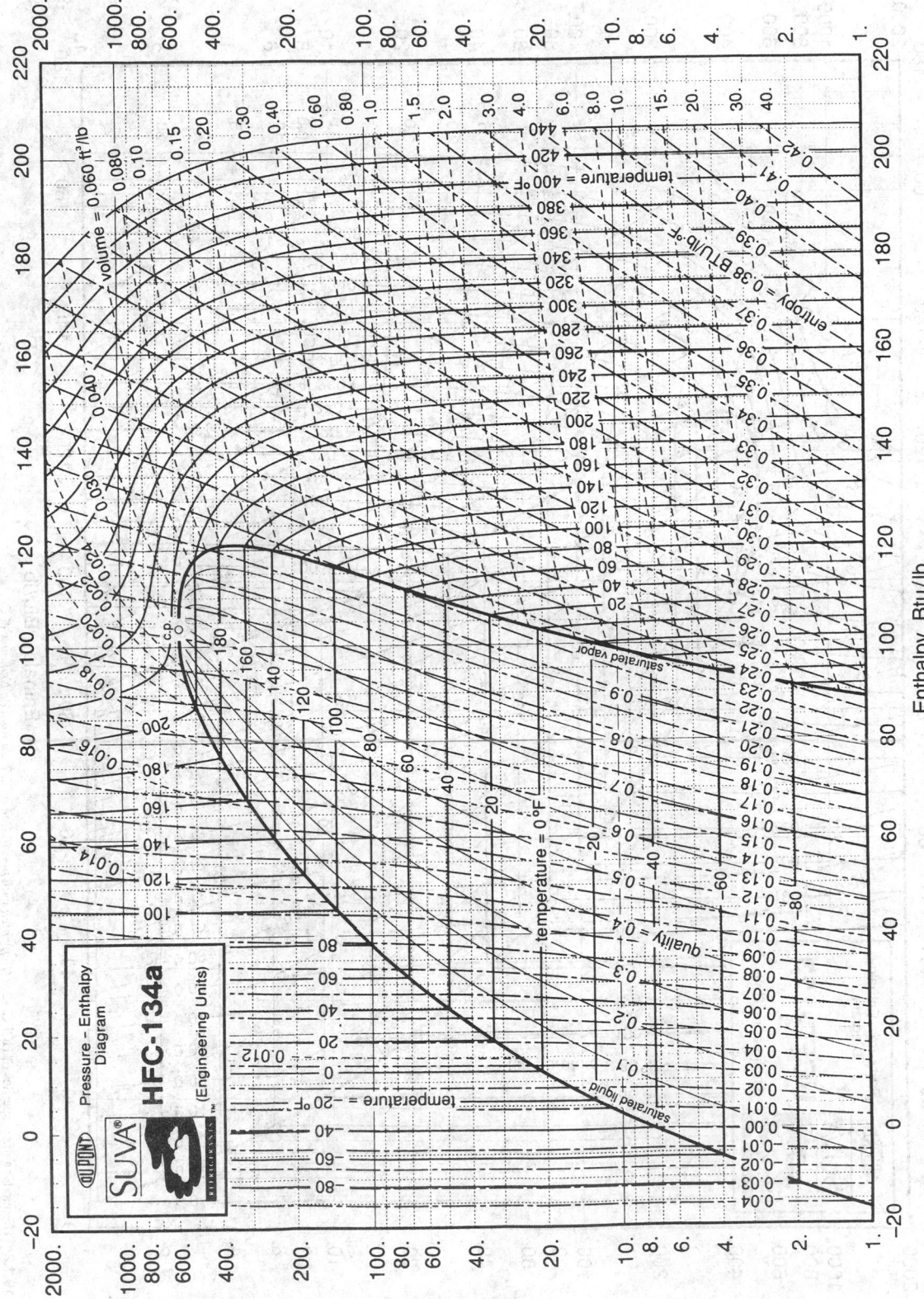

Fig. 19.1.3 Thermodynamic properties of HCFC-134a. (*Reprinted with permission of E. I. du Pont de Nemours & Co.*)

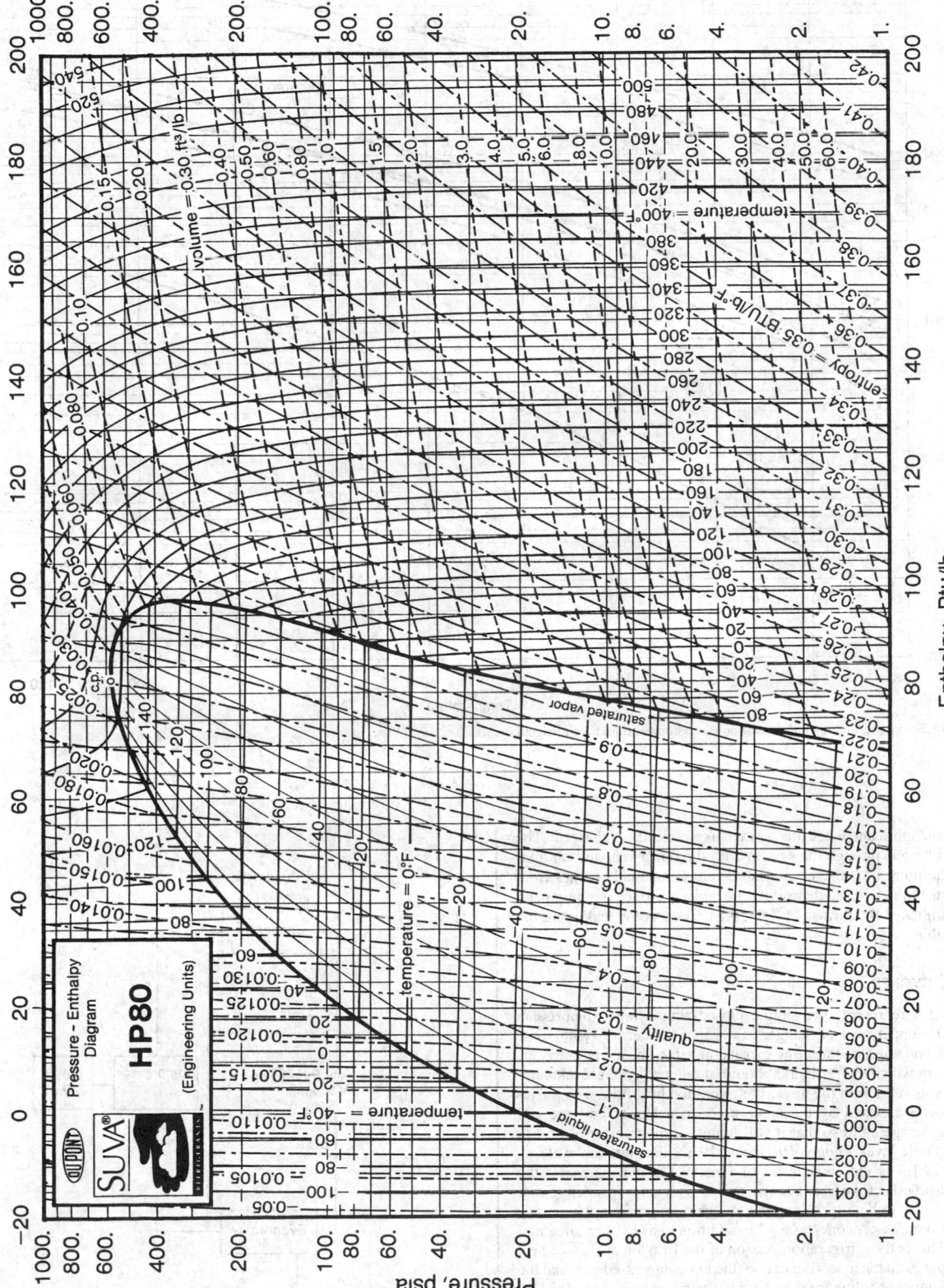

Fig. 19.1.4 Thermodynamic properties of SUVA HP80 (R-402A). *(Reprinted with permission of E. I. du Pont de Nemours & Co.)*

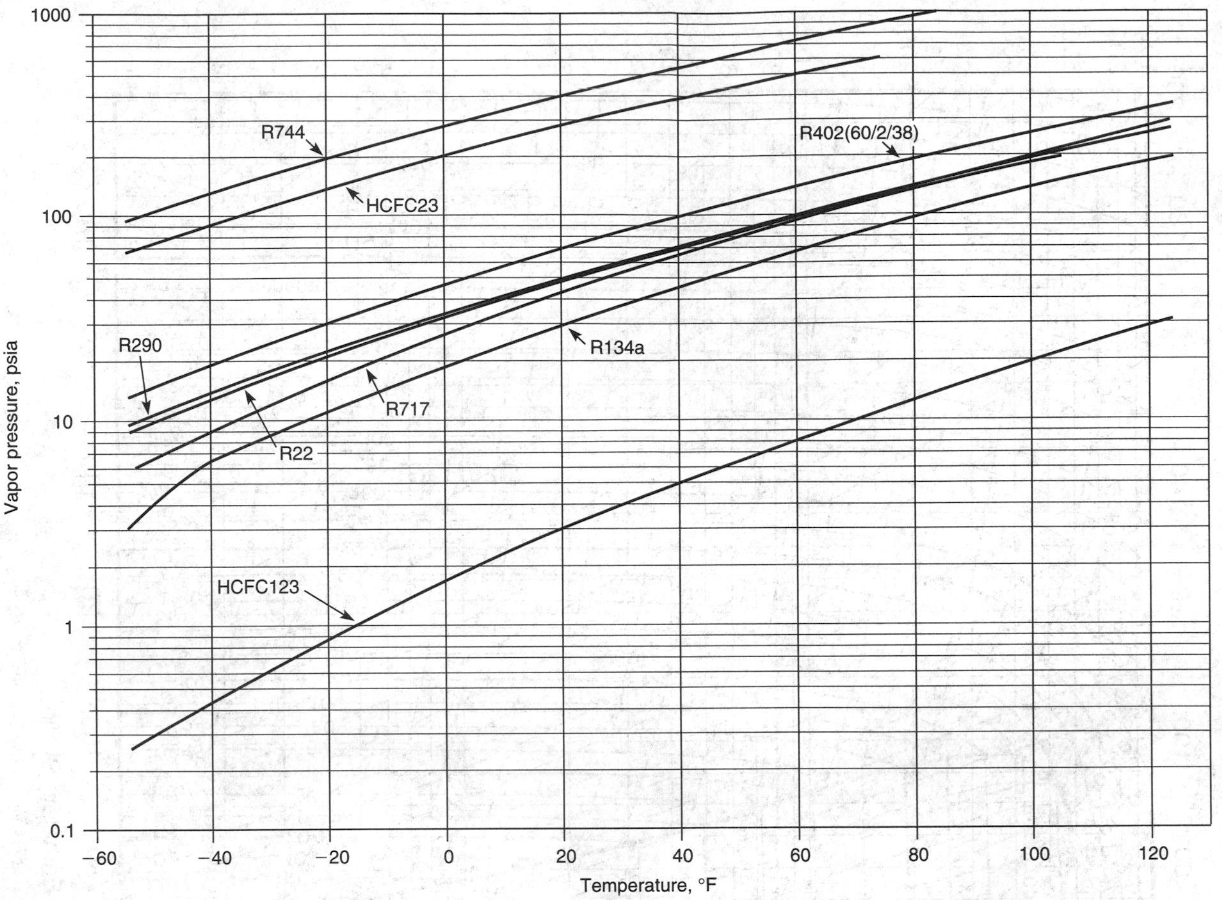

Fig. 19.1.5 Pressure-temperature relations for saturated liquids of refrigerants.

refrigeration lubricant when the system is serviced or charged. The compound travels through the system with the refrigerant and oil mixture. At the point of leakage, the fluorescent compound escapes from the system, and is left as a deposit on the outside surface of equipment. When illuminated by ultraviolet light, the deposit at the leak point becomes visible.

OVERALL CYCLES

Figure 19.1.6 represents the simple closed-circuit **vapor-compression** system. The upper heat exchanger is a vapor condenser (with some superheat), and after the throttling expansion valve, the lower one is the evaporator in which the refrigerant liquid at reduced pressure and temperature evaporates with the inward refrigeration heat flow. The refrigerant vapor is elevated by the compressor to a higher pressure and condensing temperature so that it will liquefy in its transfer of heat to the atmospheric level. Figures 19.1.7 and 19.1.8 illustrate the cycle.

Figure 19.1.9 is the counterpart of Fig. 19.1.6 and represents the closed sensible-heat **gas-compression** cycle, with an expander (either displacement or turbine-type) parallel to a gas compression which may provide a compressor-work "assist" in addition to its exhaust stream of cold gas. This is the refrigeration version of the Brayton cycle (see Sec. 4), the upper exchanger serving to cool the hot compressed gas, and the lower exchanger handling the sensible-heat refrigeration effect. In place of the expander, an expansion valve could be used; this is particularly

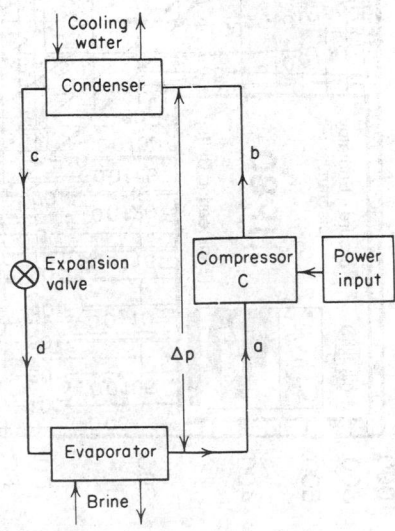

Fig. 19.1.6 Simple closed-circuit vapor-compression system.

effective on very low-temperature open-cycle operations, such as with gas liquefaction processes incorporating secondary "cold-regenerating" heat exchangers.

An extended temperature-entropy diagram (Fig. 19.1.10) illustrates paths for gas-compression sensible-heat cycles as follows: (1) 1-2-3-4 isentropic expander; (2) 1-2-3-4' polytropic expander; (3) 1-2-5-6 throttling (isenthalpic); and (4) 1-2-7-8 throttling (low temperature).

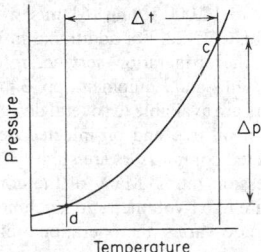

Fig. 19.1.7 Pressure-temperature curve illustrating overall differences.

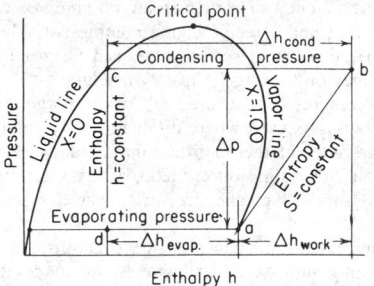

Fig. 19.1.8 Pressure-enthalpy diagram for simple ideal refrigeration cycle.

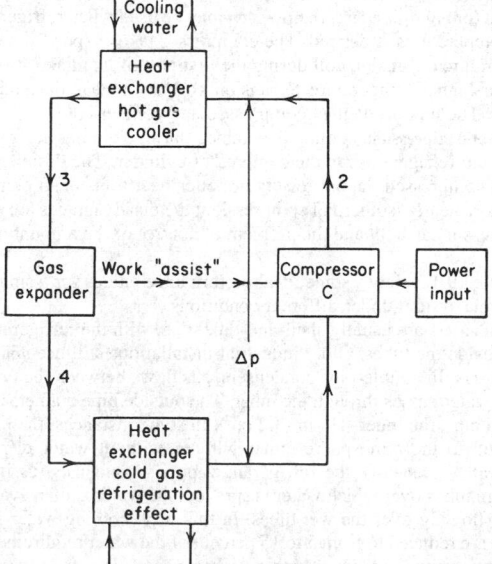

Fig. 19.1.9 Closed sensible-heat gas-compression cycle.

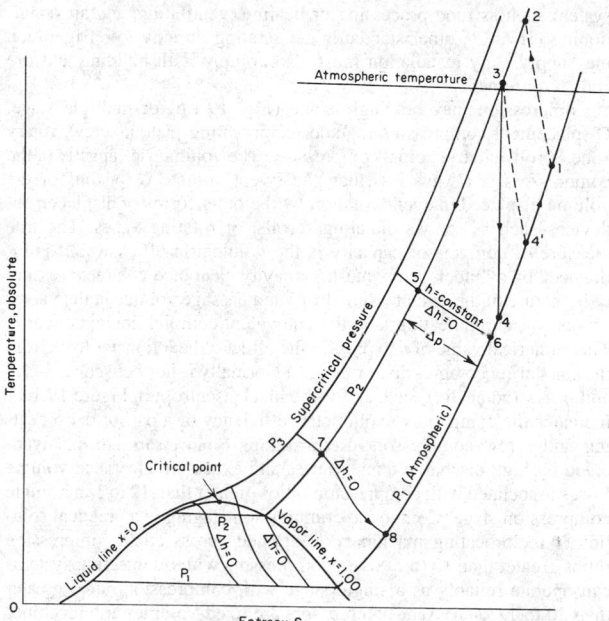

Fig. 19.1.10 Extended temperature-entropy diagram for air illustrating sensible-heat air-compression processes.

COMPONENTS OF COMPRESSION SYSTEMS

A refrigeration system is an energy-transport complex of assorted components. It may be a conventional system in which the same fluid, the refrigerant, recirculates continuously in a closed circuit, or it may entail a partially open system with discharge of some of the processed fluid either as a liquefied gas or as a solidified gas, with replacement makeup, e.g., by air liquefaction and solid carbon dioxide production. Components include vapor and gas compressors; liquid pumps; heat-transfer equipment (gas coolers, intercoolers, aftercoolers, exchangers, economizers); vapor condensers and the counterpart evaporators; liquid coolers and receivers; expanders; control valves and pressure-drop throttling devices (capillaries, refrigerant-mixture separating chambers, stream-mixing chambers); and connecting piping and insulation.

Codes and **standards** have been published by various associations and societies. These cover design, selection, designation, testing, sound ratings, safety, efficiency, and other areas relating to mechanical refrigeration. Publication of a comprehensive list is impractical because of frequent revisions. Several of the organizations include Air Conditioning Contractors of America (ACCA), American National Standards Institute (ANSI), Air-Conditioning and Refrigeration Institute (ARI), American Society of Heating, Refrigerating and Air-Conditioning Engineers, Inc. (ASHRAE), The American Society of Mechanical Engineers (ASME), Canadian Standards Association (CSA), International Institute of Ammonia Refrigeration (IIAR), and Underwriters Laboratories Inc. (UL). A list of current codes and standards can be obtained by contacting the particular organization.

Several of the more general refrigeration equipment standards include: Safety Code for Mechanical Refrigeration, ANSI/ASHRAE 15; Number Designation and Safety Classification of Refrigerants, ANSI/ASHRAE 34; and General Standard on Refrigeration Equipment, CAN/CSA-C22.2.

There are numerous additional standards relating to specific types of refrigeration equipment (air conditioners, heat pumps, chillers) or specific components (compressors, water- or air-cooled condensers, lubricating oil). Because refrigeration equipment requires auxiliaries (electric motors, steam turbines, piping, insulation) and is part of a larger system such as food processing or heating, ventilating, and air conditioning (HVAC), other standards not relating directly to refrigeration may apply. Any installation must also comply with building and fire code requirements.

Compression may be single stage (Fig. 19.1.6) or multiple stage. Displacement compressors include reciprocating piston, screw, rotary vane, scroll, and trochoidal compressors. The volumetric capacity of the compressors is always less than the swept volume of pistons or the volume displaced in cyclic rotation by the other forms of displacement devices such as screws, nutating scrolls, or rotating vanes. The true measure of compressor capacity is the volumetric efficiency. It is influenced by cylinder, screw mesh, or cavity clearance volume; pressure ratio; nature of the gas; pressure drops and passage volume in the valves or vanes; suction heating; and the equivalent compression exponent n. The numerical value of n expresses the effect of heat transfer to or from the gas during compression of the gas. Normally it lies between $n = 1.0$ and $n = k$ (adiabatic), such as for the ideal (isentropic). Figure 19.1.11 indicates the comparative volumetric efficiency of a pair of reciprocating and screw compressors. **Reciprocating compressors** are disadvantaged by high discharge temperatures and excessive clearance volume losses associated with compression ratios greater than 12 to 1 in a single compression stage. Compound compression staging is a practical solution if reciprocating machinery is applied across cycle compression ratios greater than 10 to 1. **Screw** compressors with oil injection systems can operate reliably as a single stage with compression ratios greater than 20 to 1. **Rotary vane** compressors are fixed-volume-ratio machines and can be used across compression ratios of 18 to 1, vane material and lubricating oil permitting; however, in the larger-size machines, a 7 to 1 compression ratio is commonly used in order to maintain efficiency. **Scroll** compressors have been used in commercial practice for systems that have capacities less than 10 tons (35 kW). **Trochoidal** compressors

are often called "Wankel" compressors because, in one form, the rotor is similar to the Wankel internal combustion engine rotor.

Reciprocating-piston compressors may be horizontal or vertical. The pistons can be designed to be either single acting or double acting with compression occurring on either stroke of the piston. Commercial designs have evolved into multiple cylinder machines arranged in a V block similar to an automotive engine. Average piston speed may be as great as 800 ft/min (4 m/s). Design characteristics of the suction and discharge valves installed on the cylinder determine the energy efficiency, and maintenance frequency. Valves most frequently used are the free-floating reed, clamped reed, and ring valve. Gas velocity through the valves is limited to 12,000 ft/min (61 m/s) with R-717 and 9,000 ft/min (46 m/s) with HCFC-22. For additional information on reciprocating compressors in refrigeration service, refer to the ASHRAE Handbook, 1992 Systems and Equipment, pp. 34.4 to 34.9.

Screw compressors are available in several designs, both single screw and twin screw, with oil-free and oil-injected designs in both types. Twin-screw oil-injected compressors are slightly more energy efficient at moderate compression ratios. Many of the commercially available screw compressors are fixed-volume-ratio machines. (See also Sec. 14.) If the compression ratio varies, the operational efficiency of the compressor also varies significantly. The energy reductions that are available by operating with cooler condensing pressures during spring, fall, and winter may not be fully achievable. A movable slide valve stop which will vary the compressor internal volume ratio to achieve optimum energy consumption is now available. Figure 19.1.11 illustrates the adiabatic efficiency of a twin-screw compressor with a movable slide stop control and a reciprocating compressor. Flash intercoolers (Fig. 19.1.13), also called *economizers,* can be provided to improve energy efficiency on the larger-capacity machines.

Air-cooled compressors are used where discharge temperatures are low; water cooling is used where discharge temperatures are high, as with ammonia, and on larger industrial units. **Oil separators** return to the compressor the lubricating oil carried over by the refrigerant vapors. **Rotary compressors** of the vane, eccentric, gear, and screw types are in use.

Kinetic compressors include high-speed centrifugal and axial flow machines, usually multistaged, and jet-entrainment devices. Centrifugal machines are especially adapted to high-volume flow (> 500 ft³/min) (0.24 m³/s). (See also Sec. 14.)

The entire machine-compression operation may be replaced by a secondary absorber-pump-generator system, in which the complex is known as an absorption refrigeration system.

Dual (or **multiple effect**) **compression** may be used when refrigeration at two temperatures is desired. The compressor takes vapor from a lower temperature expansion coil during the first part of its intake stroke, and from a higher temperature expansion coil at or near the end of the stroke. The mixture is then compressed and condensed.

In **wet compression,** cooling is obtained during compression by spraying liquid refrigerant into the compressor cylinder. The desuperheating of the compressed vapors results in better heat transfer in condensers and more nearly isothermal compression; the disadvantages are reduced compressor capacity and the problem of control of the amount of injection.

Table 19.1.7 shows some ideal performance values for a single-state ammonia system under different conditions.

Condensers are usually shell-and-tube type, with the refrigerant passing outside the tubes. Older industrial installations still use **double-pipe condensers**. In double-pipe condensers, gas flows between the two pipes while water passes through the inner. The outside pipe diameter may be 2 in (5 cm); the inner, 1¼ in (3.2 cm). In some instances, the pipes are exposed to the atmosphere either with or without water drip. In an **evaporative condenser** the refrigerant vapor is condensed as it passes through tubes over which water is sprayed; the water is then evaporated by air flowing over the wet tubes. In this way, cooling-water requirements are reduced to from 5 to 15 percent of the water requirements of a nonevaporative condenser with no water reuse. The evaporative condenser combines in a single unit a refrigerant condenser and the atmo-

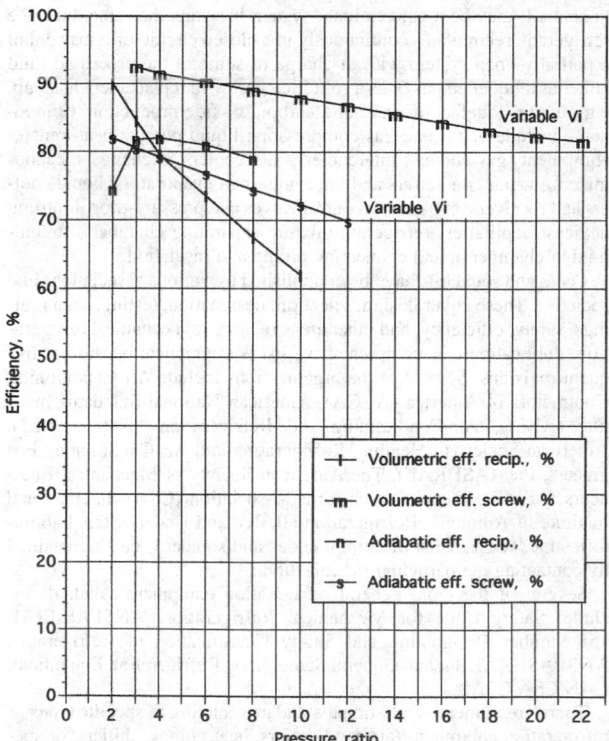

Fig. 19.1.11 Screw and reciprocating compressor volumetric and adiabatic efficiency versus pressure ratio. (*Courtesy of J. W. Pillis, "Development of a Variable Volume Ratio Screw Compressor," Frick Company, 1983.*)

Table 19.1.7   Theoretical Horsepower and Theoretical Volume of Dry Ammonia Gas Pumped per Minute to Produce 1 Ton of Refrigeration

| Suction pressure and temperature | | Condenser pressure, lb/in² gage (temperature, °F) | | | | | | | | | |
| lb/in² gage | °F | 103 (65) | | 127 (75) | | 153 (85) | | 182 (95) | | 215 (105) | |
		hp	ft³/min	hp	ft³/min	hp	ft³/min	hp	ft³/min	hp	ft³/min
4	−20	1.058	5.84	1.205	5.96	1.361	6.09	1.525	6.23	1.691	6.43
6	−15	0.997	5.35	1.145	5.46	1.300	5.58	1.461	5.70	1.546	5.83
9	−10	0.903	4.66	1.045	4.76	1.193	4.86	1.347	4.97	1.435	5.08
13	−5	0.818	4.09	0.954	4.17	1.094	4.25	1.244	4.35	1.321	4.44
16	0	0.735	3.59	0.865	3.66	1.002	3.74	1.147	3.83	1.219	3.91
20	5	0.666	3.20	0.795	3.27	0.928	3.34	1.066	3.41	1.138	3.49
24	10	0.592	2.87	0.726	2.93	0.854	2.99	0.991	3.06	1.060	3.12
28	15	0.541	2.59	0.664	2.65	0.792	2.71	0.922	2.76	0.994	2.82
33	20	0.474	2.31	0.592	2.36	0.715	2.41	0.842	2.46	0.903	2.51
39	25	0.410	2.06	0.523	2.10	0.599	2.15	0.767	2.20	0.829	2.24
45	30	0.351	1.85	0.461	1.89	0.576	1.93	0.694	1.97	0.759	2.01
51	35	0.300	1.70	0.410	1.74	0.521	1.77	0.640	1.81	0.701	1.85

spheric cooling tower or spray pond which is required if the cooling water is to be reused. Air-cooled finned-tube condensers with forced ventilation are widely used on small units, and shell-and-coil or double-tube water-cooled condensers on medium units.

Evaporators may have plain or finned surfaces. Fin density is available from 1 to 14 fins per inch. Fin density of 2 to 4 fins per inch is specified for air-cooling service below 28°F (−2.2°C), because air can flow through the coil while thin layers of frost are forming. Four types of defrosting methods are used: hot gas from the compressor, electrical air heaters, hot water, and room air circulation. Flooded evaporators are operated practically full of refrigerant. The level is controlled by a float valve. Wet-expansion evaporators are operated with a level approaching flooded operation. Dry-expansion (once-through) units operate with an indefinite amount of liquid in the evaporator. Usually a thermal bulb measures the outlet pipe temperature and adjusts a thermostatic expansion valve to maintain 15°F (8.3°C) superheat in the outlet vapor. Flooded or wet operation gives high heat-transfer rates. Pumped recirculation of refrigerant is used to promote heat transfer in finned coil evaporators, and also on some shell and tube evaporators (liquid chillers) by completely wetting the surfaces with a rapidly moving layer of refrigerant. The most commonly used liquid feed ratio is 3 to 1. However, a feed ratio as high as 6 to 1 may be required to accommodate distribution imbalances in the piping system. Combined liquid and vapor **return piping** should be sloped at 0.25 in/ft (2.08 percent) to prevent formation of waves in the liquid-phase layer that can react with the flowing vapors to cause liquid hammer that, in turn, can severely damage piping components. **Receivers** and **recirculating pumps** must be carefully matched to the operating system to prevent pump cavitation, liquid droplet carryover, and gross liquid overflow due to surges of liquid that may occur as a result of equipment operations such as defrosting processes. Specifically, the tank drop leg liquid velocity should be limited to 100 ft/min (0.51 m/s) and the vapor velocity over the liquid level in the tank should also be limited to 100 ft/min (0.51 m/s).

For a low-cost unit operating in limited space under conditions of motion, dry operation is preferred.

Controls are required on the liquid level of the refrigerant and on the temperature of the refrigerated space. The liquid control regulates the flow of refrigerant into the evaporator and also serves as the pressure barrier between the high operating pressure of the condenser and the lower operating pressure of the evaporator. It may take the form of a **capillary tube** between the condenser (high side) and the evaporator (low side), in which case it is nonadjustable. Plugging due to dirt is a common difficulty. Capillary-tube liquid control is largely confined to relatively small units assembled and charged at the factory and particularly for hermetically sealed systems. The **constant-pressure expansion valve**, maintaining a constant evaporator pressure, and the **thermal expansion valve**, maintaining a constant superheat leaving the evaporator, are standard liquid controls for most commercial applications. A **low-side float** liquid control, used with a flooded evaporator operating at evaporator (low) pressure, consists of a float-operated valve to admit liquid refrigerant to the evaporator in accordance with demand so that a constant liquid level is held in it. A **high-side float** liquid control is often used with a single flooded evaporator; the float operating the valve between the evaporator and the condenser is in a float chamber containing liquid refrigerant at the condenser (high-side) pressure. As the liquid level in the float chamber falls, the valve closes, thus preventing hot gaseous refrigerant from passing from the high to the low side as is possible when using a capillary or a low-side float.

Several types of noncondensable-gas removal devices have been developed to remove tramp air from refrigerating systems. The exhaust of the purge system contains a portion of the refrigerant. This refrigerant loss is detrimental to the environment, and the replacement of lost refrigerant is costly. Three types of refrigerant purge systems have been developed. One uses a scavenging condenser, cooled by system refrigerant, followed by a carbon absorption device to remove the last vestiges of refrigerant from the purge vapors. Other types, such as oil-absorption systems and reverse-osmosis systems, are developed but not offered commercially. The efficiency of purge systems is specified by the weight ratio of refrigerant lost to the air that is exhausted. Present-day systems release as little as 0.0047 unit of refrigerant per unit of released air. Systems designed to meet earlier requirements may lose 3 to 7 units of refrigerant per unit of air.

Refrigerant charge in a system must be removed in an environmentally acceptable and economical manner to accomplish occasional repairs. Pump-out systems remove the refrigerant from equipment and store it for reuse. These systems consist of a compressor, air-cooled condenser, and receiver tank. They are available in portable or stationary configurations. The capture and recycle of CFC and HCFC refrigerants are mandated by the 1990 Clean Air Act.

Vapor-Compression Circuits

Compound Low-Temperature Systems Multistaging is advantageous as it reduces operating temperatures by intercooling and power requirements if the fluid superheats considerably on adiabatic compression, as with ammonia; with F-12 refrigerant, propane, and butane, the superheat is negligible (see Figs. 19.1.12 to 19.1.15).

Gas Intercooler Plus Liquid Injection Since gas temperature in the intercooler is limited by cooling water temperature, further favorable reduction of gas temperature, prior to entrance to the second stage, may be accomplished by direct injection, via an expansion valve, of refrigerant liquid (Fig. 19.1.12).

Flash Intercooler With a flash intercooler separator located midway between successive expansion valves, the gas leaving the low-stage (I) compressor passes through the flash intercooler, gets cooled by direct contact with the liquid, and, augmented by the flash vapor and at reduced temperature, enters the second stage (II) (Fig. 19.1.13).

For positive-displacement compressors, selection of flash intercooler temperature is determined by the following relation:

$$P_{\text{intermediate}} = (P_{\text{I suction}} \times P_{\text{II discharge}})^{1/2}$$

where P = pressure in absolute units, lb/in² abs (Pa). This equation is valid for centrifugal machines with modification. The common practice is to limit single-impeller-stage adiabatic head to approximately

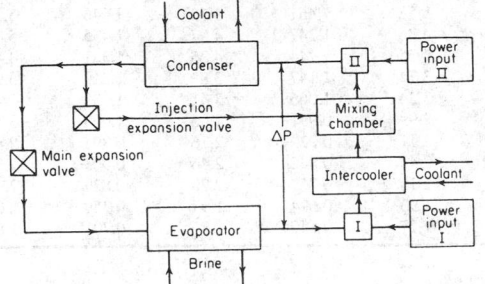

Fig. 19.1.12 Two-stage system with conventional intercooler plus refrigerant injection.

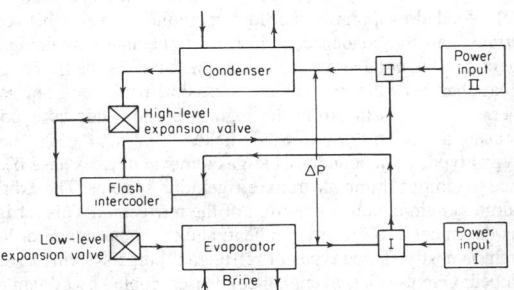

Fig. 19.1.13 Two-stage system with flash intercooler.

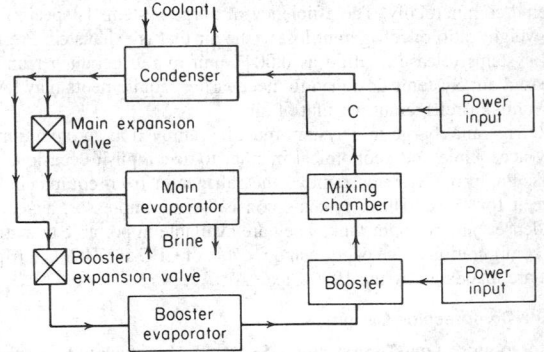

Fig. 19.1.14 Two-stage system incorporating low-temperature booster circuit.

10,000 ft (3,000 m) for low-molecular-weight gas and 6,000 ft (1,800 m) for the heavier HCFC and HFC refrigerants. Economical rotor design may require an interstage pressure that does not match the ideal flash intercooler pressure. Where choice is required, select a pressure slightly greater than the ideal pressure given by the above equation.

Low-Temperature Booster Plus Single Stage For the addition of a second lower-temperature evaporator to an existing single-stage system, a booster for the low-temperature vapor may be employed, its output feeding into the single-stage suction. The booster is a centrifugal or rotary compressor, suitable for handling the large vapor volumes encountered at low temperatures (and pressures); the reciprocating compressor is more advantageous at high pressures (Fig. 19.1.14).

Cascade Systems with Segregated Circuits Here two separate circuits are employed with the low-stage vapor condensed in the second-stage evaporator. Two different refrigerants may be appropriately chosen, a lower-boiling-point refrigerant serving for the first stage. Thus, for low temperatures ($< -50°C$ or $-50°F$), CO_2 may be employed in the lower stage with NH_3 in the second stage condensing the CO_2. Or the same refrigerant may be used in both stages, limiting the migration of the lubricant to the low-temperature realm (Fig. 19.1.15).

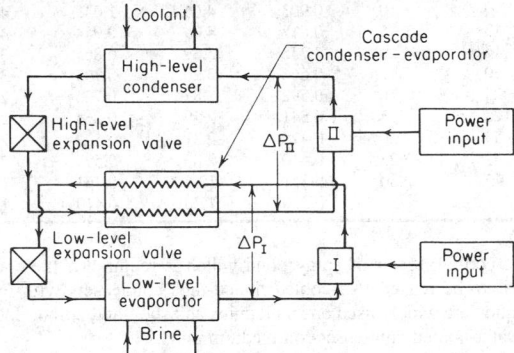

Fig. 19.1.15 Two-stage system in cascade form.

Water-vapor refrigeration is necessarily limited to near-atmospheric temperature levels; it involves large suction volumes at high vacuum and utilizes a nontoxic refrigerant; this type of refrigeration was frequently used in the early days of space air conditioning. It utilizes (1) centrifugal compressors or (2) steam-jet compressors, entraining low-density water vapor from the evaporator. The centrifugal water-vapor compressor, usually multistage, must operate at high speeds (7,000 to 10,000 r/min). Operation at high vacuum in the evaporator requires an air pump at the condenser to eliminate leakage effects.

Steam-jet refrigeration systems are used where cooling to temperatures above 0°C is desired. Applications include industrial air conditioning and cooling of city gas to condense out tar and other objectionable impurities, of gas absorbers to increase efficiency of absorption, of reaction units where heat removal and temperature control are important during chemical transformations, and of wort and mash in the brewing and other fermentation industries particularly in the summer months. Jet refrigeration has been used for cooling passenger trains and is coming into popularity for marine installations, e.g., cooling of banana boats and large passenger vessels. The system is a compression-type refrigerator: it uses water as a refrigerant, a part of which is evaporated to produce cooling of the remainder, steam-jet ejectors to compress the water vapor resulting from evaporation, and a condenser, either the surface type, where refrigerant water is to be reused, or the barometric type, where refrigerant water can be discarded. The advantages of the system are: low installation cost; the absence of moving parts except for small liquid pumps; and safety, since no noxious or toxic refrigerants are used.

For considerable differences in temperatures between condensing-water temperature and refrigerating temperatures (10°C or more), cooling of the refrigerant water by evaporation in two stages operating at different pressures (and temperatures) will usually give better operating economy, but at somewhat higher installation cost.

Figure 19.1.16 shows the variation of refrigerating capacity with variation of the chilled-water temperature, capacity increasing as chilled-water temperature rises.

Household compression machines are designed for continuous automatic operation and for conservation of the charges of refrigerant and oil. These units are almost universally motor-driven air-cooled compressors, the principal exception being the Electrolux-Servel absorption unit. The compressor may be hermetically sealed, with the compressor and motor enclosed in the same casing, or may use a shaft seal, embodying some form of sylphon bellows, with an outside coil spring at the place where the crankshaft passes out of the casing. The compressor may be reciprocating, but rotary types are being increasingly used either with a floating piston ring driven by an eccentric or with sliding blades. Lubrication is by internal forced-feed circulation in most cases. The

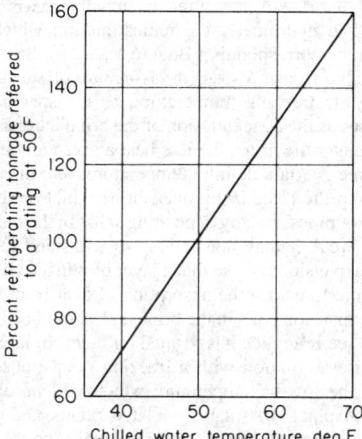

Fig. 19.1.16   Variation of refrigerating capacity with the jet-chilled water temperature in steam refrigeration.

condenser may be of the radiator, coil, or plate type, cooled by natural draft, by a fan on the main motor, or by a separate motor. Refrigerant feed control to the evaporator may be by (1) float feed (flooded system), (2) pressure-actuated diaphragm valve, or most commonly (3) fixed orifice. The fixed orifice may be either a plate orifice or a capillary tube. The evaporator is usually made of aluminum. Controls are generally pressure-operated thermostatic switches. The power consumption of an average home refrigerator (18 ft³) (0.5 m²) is 133 kWh per month. A typical home deep freezer consumes approximately 152 kWh per month.

ABSORPTION SYSTEMS

Absorption refrigeration machines are essentially vapor-compression plants (Fig. 19.1.6) in which the mechanical compressor has been replaced by a thermally activated arrangement (Fig. 19.1.17). The basic element are the **absorber, pump, heat exchanger, throttle valve,** and **generator**.

In an absorption cycle, a secondary fluid, the **absorbent,** is used to absorb the primary fluid, the **refrigerant,** after the refrigerant has left the evaporator. The vaporized refrigerant is converted back to liquid phase in the absorber. Heat released in the absorption process must be rejected to cooling water. The solution of absorbent and refrigerant is then pumped to the generator. A heat exchanger may be used for heat recovery and a corresponding increase in efficiency. In the generator, heat is added, and the more volatile refrigerant is separated from the absorbent through distillation. The refrigerant continues to the condenser, expansion valve, and evaporator, while the absorbent returns to the absorber.

Although there are many refrigerant-absorbent combinations, the most common systems are **ammonia-water** and **water-lithium bromide.** A common variation of the absorption system is the **Electrolux-Servel** process (Platen-Munters Patent). It has had wide application in gas-fired household refrigeration machines and air-conditioning systems. The use of hydrogen in a hermetically sealed system eliminates the use of pumps and other moving parts.

Absorption machines represent only a small percentage of refrigeration systems in operation. This is because they are generally bulky and difficult to operate efficiently. There has been renewed interest in absorption machines with the advent of high energy costs. This is because an absorption machine is an excellent way to utilize waste heat. Exam-

ples of potential applications are varied. One example may be on an exothermic process, or where extensive use of steam turbine with limited turndown results in vented steam especially during warm weather when air conditioning is required. A site where peak electrical demand is set by air-conditioning load and ratcheted demand charges are high may justify absorption refrigeration if adequate steam-generation facilities are already present.

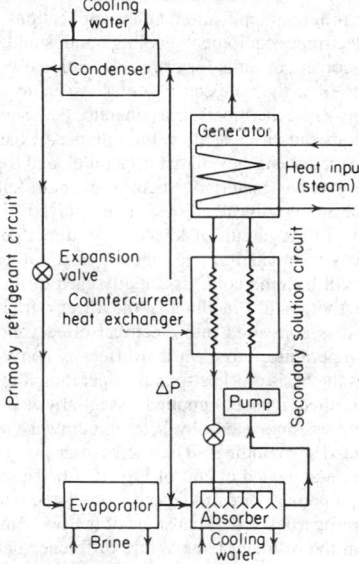

Fig. 19.1.17   Elemental absorption system circuit.

A typical absorption machine requires about 20 lb (9 kg) of steam for 1 ton (3.5 kW) of refrigeration effect at 45°F (7°C).

A more detailed explanation of absorption systems is found in ASHRAE Handbook, 1993 Fundamentals, pp. 1.20 to 1.25.

AIR MACHINES

Compressed-air machines in which air is compressed, cooled, and then adiabatically expanded in an engine while doing work are obsolete for usual industrial refrigeration because of bulk and inefficiency. Their coefficient of performance is less than unity as compared with 4 or 5 for vapor machines. Operation may be closed cycle, in which the air is reused, or open cycle, in which the air is discarded after expansion. Compression ratios of about 3 seem best. In the **dense-air** closed system, the pressure at the compressor intake averages 75 lb/in² gage (560 kPa) and up to 250 lb/in² gage (1,700 kPa) at the outlet. The use of superatmospheric pressure throughout the cycle decreases the size, and the reuse of dried air eliminates the ice formation on engine valves which occurs with open-system operation. Air cycles are used for liquefaction of the so-called permanent gases and to an increasing extent for comfort cooling in high-speed planes where the small lightweight equipment and the safety of the cooling medium are important. High-speed (90,000 to 100,000 r/min) gas turbines are used to extract work from 5 to 100 lb/in² gage compressed air.

THERMOELECTRIC COOLING

Thermoelectric cooling utilizes the **Peltier effect** whereby a temperature differential will occur between two junctions of a closed loop of dissimilar metallic conductors when an electric current is imposed. It is variously called *Peltier effect cooling, thermoelectric refrigeration,* and *electronic cooling.* It is the inverse of the thermocouple operating principle whereby a temperature differential across the junctions of two dissimilar metals produces an electric current.

The capacity of a thermoelectric couple is small. System capacity can be increased by increasing the number of elements used. The temperature produced can be lowered by cascading the systems. Capacities approaching 25 tons (87.9 kW) cooling effect have been achieved, and temperatures as low as $-266°F$ ($-166°C$) have been attained.

Aside from limited practical capacities, the major disadvantage of a thermoelectric cooling system is its poor coefficient of performance. The absence of moving parts, silent operation, ability to function in zero gravity, and lack of pressurized vessels, has given thermoelectric cooling wide space-program application. Other applications include use on submarines, electronic equipment cooling, and small units such as drinking water coolers or units for use in recreational vehicles.

The elements of a thermoelectric cooler are quite simple and are equivalent to those of a thermoelectric generator (see Sec. 9). For development of various output capacities, the number of basic elements is proportionally increased; they all act in parallel, and their outputs add together. **Semiconductor** materials, such as bismuth-telluride-selenide and bismuth-antimony-telluride alloys, are employed.

With an input to the circuit of low-voltage direct current and with heat continuously abstracted at one junction at room temperature, the other junction will become cold. Essentially each of the two junctions becomes an ''activity cell,'' in which at the lower refrigeration temperature level, heat is converted into electrical effect, and at the higher atmospheric temperature, the electrical effect is converted into heat. This establishes the basis of ''heat-pump'' operation: the lower-temperature heat requirement must be supplied, essentially as in the evaporator of a vapor-compression system, by heat withdrawal via heat transfer, from all connected surroundings. The energy-transport circuit thus parallels the machine-activated circuit of Fig. 19.1.6. The amount of heat for a single-component circuit depends on the circulating electric current, on the properties and dimensions of the two conductors (p and n types), and on the resultant temperature difference between the high side room temperature and the low side input heat that is divided into two groups, useless and useful. The **useless heat** is the heat conduction between the high and low temperature through the thermoelements plus the current-generated resistance-loss heat within the elements that flows to the colder junction. The **useful heat** is the refrigeration effect in a regular evaporator. All the heat must in turn be discharged via the Peltier effect at the higher room temperature, as in the condenser of a vapor-compression system. Ultimate performance of the circuit depends on the voltage-generating nature of the two dissimilar conductors and on their electrical resistance, thermal conductivity, and physical dimensions. The different properties are mathematically combined into a **figure of merit, Z**.

METHODS OF APPLYING REFRIGERATION

In **direct expansion systems** the evaporator is placed in the space which is to be cooled; in **indirect systems** a secondary fluid (**brine**) is cooled by contact with the evaporator surface, and the cooled brine goes to the space which is to be refrigerated. Brine systems require 40 to 60 percent more surface than do direct expansion; they have an equalizing effect due to the large heat capacity of cold brine, they are safer (particularly if the refrigerating effect must be carried considerable distance or widely distributed), and they permit closer temperature regulation than is possible with direct expansion. If two temperatures are to be held, the lower may be direct expansion, the higher by brine cooling. Development of better controls and newer piping methods has made direct expansion more attractive than previously.

Brines used for industrial refrigeration are usually aqueous solutions of calcium chloride, ethylene glycol, propylene glycol, or undiluted methylene chloride, and silicone-based alkylated fluids. Calcium chloride should not contain over 0.2 percent magnesia, calculated as magnesium chloride. Calcium chloride brines are recommended down to $-45°F$ ($-43°C$). Brines should be chemically neutral: acidic brines attack ferrous materials, alkaline brines attack zinc, ammonia in brine (resulting from leaks in the ammonia system) is especially harmful to most nonferrous metals. Corrosion by brine is increased by the presence of oxygen, air, or carbon dioxide and by galvanic action between dissimilar metals. Corrosion inhibitors are widely used, a satisfactory one being about 1,600 g of sodium chromate or dichromate per 1 m^3 of calcium chloride brine or 3,200 g per 1 m^3 of sodium chloride brine.

The properties of sodium and calcium chloride brines are given in Tables 19.1.8 to 19.1.10.

The density of brine is measured by a **salinometer** (or **salometer**), which is a simple hydrometer the indications on which are 4 times greater than on the corresponding Baumé scale (see Sec. 1).

It is undesirable to use a **strength of solution** of salt greater than is necessitated by its freezing temperature, as the specific heat (Table 19.1.9) decreases as the concentration of the brine increases, and consequently the stronger the brine, the less heat a given amount of it is able to convey between certain definite temperatures and the more power is required to pump the brine. Moreover, brine which is too strong may cause clogging of pipes, etc., by depositing salt. On the other hand, if the solution is too weak it may not be able to withstand the temperature existing in the expansion coil, so that a layer of thin ice will form around the latter and interfere with the absorption of heat from the brine. The surface of the expansion coils in the brine tank should be inspected from time to time to see if any ice has formed on them. In larger plants, it is customary to use a solution with a freezing point not less than 10°F (5.5°C) below the lowest temperature which will be obtained in the operation of the plant. In smaller isolated plants and where careful supervision is not ensured, it is customary to make the solution as strong as possible without being unstable, usually 1.240 to 1.250 sp gr.

Ethylene glycol brine is used in various strengths from 15 to 50 percent by weight for refrigeration temperatures down to $-1°F$ ($-18°C$). It is toxic, and therefore its usage is not recommended in food or beverage processing equipment where a leak can contaminate the product. A corrosion inhibitor is recommended to protect carbon steel equipment and piping. Specific heat ranges from 0.975 to 0.76 Btu/lb · °F (4,080 to 3,180 J/kg · K) in typical systems. Specific gravity varies from 1.08 to 1.05.

Propylene glycol brine in 15 to 35 percent weight strength is considered nontoxic and is often used in brewing and other food and beverage applications. The brine is more viscous (10 centipoise at $-6.7°C$ for 30 percent weight solution) than ethylene glycol. Specific heat varies from 0.9 to 0.975 Btu/lb · °F. Specific gravity ranges from 1.05 to 1.01. Propylene glycol has often replaced ethylene glycol due to the latter's toxicity.

Methylene chloride brine is often used in systems requiring -20 to $-125°F$ (-30 to $-87°C$) low temperatures. It has low flammability. Hydrolysis and water contamination must be prevented to avoid equipment corrosion. Specific heat varies from 0.269 to 0.273 Btu/lb · °F (1,126 to 1,142 J/kg · K). Specific gravity ranges from 1.4 to 1.5.

Proprietary chemical blends, such as Dowtherm J and Syltherm, are used where the fluid properties are advantageous. Dowtherm J, a mixture of isomers of an alkylated aromatic compound, is usable between the temperatures of 358°F (181°C) and $-120°F$ ($-84°C$). The usual range of service is between 0°F ($-18°C$) and $-70°F$ ($-21°C$). Physical properties vary from 0.412 to 0.390 Btu/lb · °F (0.412 to 0.390 cal/g · °C) for specific heat; 0.894 to 0.926 for specific gravity (ref. 68°F water); 0.079 to 0.084 Btu · ft/h · ft^2 · °F (0.136 to 0.145 W/m · K) for thermal conductivity; and 1.72 to 4.61 cP in the range of usage.

Sodium chloride brine usage is being reduced due to its corrosivity and its relatively high freezing point.

Table 19.1.11 gives a comparison of the performance of the four major brines in a typical shell and tube chiller. This chart does not indicate effects of specific heat, ingredient cost, pumping power requirements, or system investment cost in the selection of brines.

Brine coolers may be of three types: shell-and-tube, shell-and-coil, and double pipe. The shell-and-tube type is the most widely used, the brine flowing through the tubes which are surrounded by the evaporating refrigerant. Tubes may be arranged for multipass operation. The effective heat-transfer surface varies from 8 to 15 ft^2 per ton (0.21 to 0.4 m^2/kW), varying with temperature and brine velocity. A submerged coil in an open brine tank is used for ice making by the can process.

Table 19.1.8 Properties of Calcium Chloride Solutions
(For variation of specific gravity, with temperature, see Table 19.1.9.)

Parts of $CaCl_2$ by weight in 100 parts of the solution	Specific gravity at 60°F	Deg Baumé	Weight per gal, lb	Weight per cu ft, lb	Freezing point, °F	Specific heat at					
						−4°F	14°F	32°F	50°F	68°F	86°F
6	1.050	7.0	8.76	65.52	28.0	—	—	—	—	—	—
8	1.069	9.33	8.926	66.70	24.2	—	—	0.882	0.887	0.892	0.897
10	1.087	11.57	9.076	67.83	21.4	—	—	0.853	0.858	0.863	0.868
12	1.105	13.78	9.227	68.95	18.2	—	—	0.825	0.831	0.836	0.842
14	1.124	15.96	9.377	70.08	14.4	—	—	0.799	0.805	0.811	0.817
16	1.143	18.12	9.536	71.26	9.9	—	0.768	0.775	0.781	0.787	0.792
18	1.162	20.24	9.703	72.51	4.7	—	0.745	0.752	0.759	0.764	0.769
20	1.182	22.32	9.853	73.63	−1.0	—	0.723	0.731	0.738	0.744	0.749
22	1.202	24.38	10.04	75.0	−7.3	0.695	0.704	0.711	0.718	0.724	0.729
24	1.223	26.41	10.21	76.32	−14.1	0.678	0.686	0.693	0.700	0.706	0.712
26	1.244	28.41	10.38	77.56	−22.0	0.663	0.670	0.677	0.683	0.690	0.696
28	1.265	30.39	10.56	78.94	−32.0	0.649	0.656	0.662	0.669	0.675	0.682
30	1.287	32.34	10.75	80.35	−46.0	0.638	0.643	0.648	0.655	0.661	0.668

Table 19.1.9 Specific Gravities of Brines
(To change to lb/ft³ multiply by 62.43; to change to lb/gal multiply by 8.35.)

Parts by weight of salt in 100 parts of brine	Sodium chloride				Calcium chloride				Magnesium chloride			
	Temperature, °F											
	14	32	50	68	14	32	50	68	14	32	50	68
6	—	1.046	1.044	1.041	—	1.053	1.051	1.049	—	1.053	1.051	1.049
8	—	1.061	1.059	1.056	—	1.071	1.069	1.066	—	1.070	1.069	1.067
10	—	1.077	1.074	1.071	—	1.090	1.087	1.084	—	1.089	1.087	1.084
12	—	1.093	1.090	1.086	—	1.108	1.106	1.103	1.108	1.107	1.105	1.102
14	—	1.108	1.105	1.101	—	1.127	1.124	1.121	1.127	1.126	1.123	1.121
16	1.128	1.124	1.121	1.116	1.150	1.147	1.144	1.140	1.147	1.145	1.142	1.139
18	1.144	1.140	1.136	1.132	1.170	1.167	1.163	1.159	1.166	1.164	1.161	1.158
20	1.161	1.157	1.152	1.148	1.190	1.187	1.183	1.179	1.186	1.183	1.181	1.178
22	1.178	1.173	1.169	1.164	1.211	1.208	1.203	1.199	1.206	1.203	1.201	1.197
24	1.195	1.190	1.185	1.180	1.233	1.229	1.224	1.219	1.226	1.224	1.221	1.218

Table 19.1.10 Weight of Commercial Calcium Chloride in Brine

Sp gr	1.10	1.12	1.14	1.16	1.18	1.20	1.22	1.24	1.26	1.28	1.30	1.32
Wt per gal, lb	1.41	1.70	2.00	2.30	2.59	2.90	3.20	3.50	3.83	4.13	4.46	4.78
Wt per cu ft, lb	10.55	12.72	14.96	17.20	19.37	21.69	23.94	26.18	28.62	30.89	33.36	35.75

NOTE: Specific gravity is at 60°F for both brine and water. The weights are of 73 to 75 percent solid calcium chloride per gal of brine at 60°F. For flake 77 to 80 percent calcium chloride multiply the weights given by 0.94.

Table 19.1.11 Comparative Performance of Brines†

Brine	Heat-transfer coefficient, U_o*	Fouling factor		Refrigerant no.	Tube material
		Inner	Outer		
Calcium chloride, 24%	53	0.001	0.002	R717	Carbon steel
Propylene glycol, 40%	26.3	0.0005	0.0	HCFC 22	Copper
Ethylene glycol, 40%	44.7	0.0005	0.0	HCFC 22	Copper
Methylene chloride	111.9	0.0	0.0	HCFC 22	Copper

* Values presented in Btu/h · ft² · °F; to convert to W/m² · K, multiply by 5.678.
† Values calculated using ¾-in diameter tubes; integrally finned; 19 fins per inch; 7 ft/s velocity; brine is inside the tube; brine bulk temperature is 5°F.

Table 19.1.12 Capacity of Multipass Shelf-and-Tube Brine Coolers, Flooded
(Tons refrigeration)

Diameter of shell, in	Length of shell, ft	Velocity of brine through cooler, ft/min							
		200				400			
		Total brine, gal/min	Mean temp diff, brine and ammonia, °F			Total brine, gal/min	Mean temp diff, brine and ammonia, °F		
			7½	12½	17½		5	10	15
26	6	140	7.74	12.9	18.1	290	7.15	14.3	21.2
26	9	140	11.5	19.3	26.0	290	10.7	21.4	32.0
26	12	140	15.5	25.8	36.2	290	14.3	28.6	42.9
34	9	230	18.5	30.7	43.1	440	17.1	34.1	51.1
34	12	230	24.6	41.0	57.4	440	22.6	45.3	68
34	18	230	37.0	62.0	86.2	440	34.1	68.2	101
42	12	510	45.0	74.8	105	970	41.5	83	123
42	18	510	67.4	112	157	970	62.2	124	187

The **double-pipe cooler** is usually of 2-in (50-mm) inner or brine-flow pipe and 3-in (75-mm) outer pipe. The commercial rating is 15 to 20 ft (5 to 6 m) length of coil per ton of refrigeration.

The **shell-and-tube cooler** is used with closed heads and is erected both vertically and horizontally; brine flows through the tubes and ammonia is in the shell. It is made in sizes from 1 to 350 tons with ratings of 8 to 15 ft² effective surface per ton, varying with the temperature and brine velocities; tubes 1 to 2½ in (25 to 63 mm) arranged multipass. This type of cooler has largely displaced all other types in recent installations (Table 19.1.12).

REFRIGERANT PIPING

(See also Sec. 8.)

It is important to the proper operation of a refrigeration system that the piping or mains interconnecting the compressors, condensers, evaporators, and receivers be properly sized. This piping must be considered in three categories, viz., liquid lines, suction lines, and discharge lines. Essentially pipeline sizing is governed by pressure drop, first cost, operating cost, noise, and oil entrainment. Excessive pressure drops penalize compressor efficiencies and may affect control-valve operation adversely. Liquid-line velocities for most refrigerants are in the order of 60 to 400 ft/min (3.3 to 22 m/s), suction lines from 700 to 4,600 ft/min (38 to 250 m/s), and discharge lines from 1,000 to 5,000 ft/min (55 to 275 m/s). Pressure drops vary approximately as the square of the velocity (or tonnage) and directly as the length of the piping. For liquid lines, when evaporator is located above condenser, the following pressure drops in pounds per square inch per foot (kilopascals per meter) of static lift should be allowed: ammonia, 0.26 (5.9); R 12, 0.57 (12.9); R 22, 0.51 (11.5); R 11, 0.64 (14.5). See also Secs. 3 and 4.)

Ammonia Mains For the average installation, Table 19.1.13 shows the maximum tons of refrigeration normally allowed for various sizes of standard pipe, based on 100 ft (30 m) equivalent length of piping (measured length plus allowance for valves and fittings). The gas pressure drops per 100 ft (30 m) equivalent length upon which Table 19.1.13 is based are as follows: suction lines at 5 lb/in² (gage) (34.5 kPa) = 0.25 lb/in² (1.7 kPa); suction lines at 20 lb/in² (gage) (138 kPa) = 0.50 lb/in² (3.4 kPa); suction lines at 45 lb/in² (gage) (310 kPa) = 1 lb/in² (6.9 kPa); and discharge lines = 1 lb/in² (6.9 kPa).

Standard-weight (schedule 40) steel pipe is used for ammonia mains, except for liquid lines 1½ (38 mm) and smaller where extra-strong (schedule 80) pipe is used. Joints may be either screwed, flanged, or welded, but welding is preferred.

R 12 Mains Maximum tons refrigerant for various pressure drops per 100 ft equivalent length of R 12 mains are given in Table 19.1.14. For average conditions, allowable pressure drops in R 12 suction lines are 0.5 to 1.0 lb/in² (3.4 to 6.9 kPa) below 0°F (− 18°C) evaporator temperature; 1.0 to 1.5 lb/in² (6.9 to 10.3 kPa) from 0 to 25°F (− 18 to − 4°C); and 2.0 to 2.5 lb/in² (13.8 to 17.2 kPa) from 25 to 50°F (− 4 to 10°C). Discharge-line pressure drops upon which Table 19.1.14 is based are approximately 1 lb/in² (6.89 kPa) per 100 ft (30 m) equivalent length. Pressure drops of 1 to 4 lb/in² (6.9 to 27.6 kPa) are permissible. Liquid-line pressure drops of 3 to 5 lb/in² (20.7 to 34.5 kPa) per 100 ft (30 m) equivalent length are normal, with 10 lb/in² (69 kPa) usually considered maximum. To facilitate oil return in vertical suction lines with the evaporator located below the compressor, the velocity in the vertical section should be 1,000 ft/min (38 m/s) or more. Suction-line capacities in Table 19.1.14 may be increased 9 percent for 50°F (10°C) evaporator temperature and will decrease approximately 7 percent for each 10°F (5.6°C) below 40°F (4.4°C).

R 22 piping will, in general, be smaller for the same tonnage than R 12 piping. Suction lines will handle about one-third more tonnage for the same pressure drop. Liquid and discharge lines can be the same as for the equivalent tonnage of R 12 or slightly smaller.

Piping may be standard-weight steel pipe, but copper tubing is used in most cases. Medium-weight type L copper tubing is normally used for land installations, and the heavier type K copper tubing is used for marine work. Joints in the copper tubing may be flared compression fittings for tubing ¾ in (18 mm) OD and smaller. However, hard solder (silver-base alloy melting above 1000°F) (540°C) is preferred for most tubing connections. Copper tubing is almost always used for liquid lines. Steel pipe may be used for the larger suction and discharge lines but should be sandblasted on the inside and carefully cleaned before installation.

Piping flexibility is an important consideration, especially in low-temperature systems [0°F(− 18°C) or colder] and where large-diameter or long runs of piping are utilized. The coefficient of thermal expansion is about 9×10^{-6}/°F (16×10^{-6}/°C) for carbon steel pipe. This means a 20-ft (6-m) length will shrink about 0.15 in (3.8 mm) when cooled from 70 to 0°F (21 to − 18°C). If the piping configuration were such to exert a moment on a compressor flange, premature equipment failure could result. Long runs of cold brine lines should be analyzed for proper anchoring, guiding, and support to prevent system failure.

Flexibility is provided by a combination of changes in direction and use of properly positioned anchors, spring hangers, and guides. Refer to Sec. 5 for additional information on piping flexibility.

Table 19.1.13 Maximum Tons Refrigeration for Ammonia Mains
(100 ft equivalent pipe length)

| Pipe size, in | Suction line | | | Discharge line | Liquid line | |
| | Suction pressure, lb/in² gage | | | | Condenser to receiver | Receiver to system |
	5(− 17.2°F)	20(5.5°F)	45(30°F)			
⅜	—	—	—	—	2.5	12.0
½	0.6	1.1	2.0	3.1	6.0	20.0
¾	1.2	2.2	4.1	6.0	14.0	75.0
1	2.2	4.0	7.5	11.4	24.0	137
1¼	4.4	8.0	15.0	22.4	50.0	245
1½	6.4	11.8	21.6	30.9	77.0	400
2	12.1	22.2	42.0	62.0	140	850
2½	19.1	35.5	65.0	97.5	220	1,475
3	31.5	59.0	108	160	375	2,400
3½	46.6	87.5	156	238	540	3,500
4	64.0	118	240	330	740	
5	117	208	385	560	1,320	
6	175	306	600	905	2,030	
8	362	650	1,200	1,810	4,200	
10	640	1,180	2,160	3,200		
12	940	1,850				

SOURCE: From ASRE, Air-Conditioning Refrigerating Data Book, 1955–1956 ed., based on ARI Equipment Standards; see ASHRAE, Handbook, 1994 Refrigeration, for more complete tables.

Table 19.1.14 Maximum Tons Refrigeration for R 12 Lines

Line sizes, in	Suction-line (based on 105°F condensing temp) pressure drop, lb/in² per 100 ft equivalent length at 40°F saturation						Discharge-line condensing temp.		Liquid-line (based on 105°F condensing temp; 25°F evaporating temp) pressure drop, lb/in² per 100 ft equivalent length (type L tubing)			
	½	1	2	3	4	5	115°F	90°F	3	5	10	20
⅜ OD	—	—	—	—	—	—	—	—	0.88	1.14	1.80	2.58
½ OD	0.14	0.20	0.28	0.35	0.41	0.45	—	—	2.89	3.64	5.56	8.50
⅜ IPS	0.17	0.24	0.34	0.42	0.49	0.54						
⅝ OD	0.25	0.35	0.51	0.62	0.73	0.81	1.43	1.15	4.86	6.81	10.2	15.8
½ IPS	0.35	0.45	0.65	0.79	0.93	1.03	1.87	1.50	4.86	6.81	10.2	15.8
⅞ OD	0.55	0.76	1.10	1.34	1.58	1.75	2.97	2.38	10.5	14.1	21.8	33.0
¾ IPS	0.68	0.94	1.35	1.65	1.92	2.12	3.26	2.62	9.73	12.6	18.5	27.0
1⅛ OD	1.26	1.80	2.57	3.17	3.76	4.15	5.05	4.05	21.4	28.2	41.3	60.8
1 IPS	1.43	2.01	2.89	3.54	4.17	4.60	5.29	4.25	21.4	28.2	41.3	60.8
1⅜ OD	2.21	3.12	4.45	5.50	6.38	7.05	7.72	6.19	36.9	48.1	70.5	101
1¼ IPS	2.70	3.82	5.37	6.72	7.68	8.48	9.16	7.35	36.9	48.1	70.5	101
1⅝ OD	3.40	4.78	6.79	8.42	9.77	10.8	10.92	8.75	62.0	80.2	114	160
1½ IPS	4.05	5.75	8.10	10.12	11.6	12.8	12.5	10.0	62.0	80.2	114	160
2⅛ OD	6.12	8.60	12.1	15.1	17.4	19.2	19.2	15.3				
2 IPS	7.66	10.9	15.3	19.2	22.2	24.5	20.6	16.5	124	161	231	328
2⅝ OD	12.0	17.1	24.0	30.1	34.6	38.2	32.2	25.9				
2½ IPS	12.0	17.1	24.0	30.1	34.6	38.2	32.2	25.9	230	297	426	607
3⅛ OD	19.1	27.2	38.2	47.8	55.0	60.7	51.5	39.8				
3 IPS	20.9	29.4	42.3	51.8	60.0	66.2	54.5	43.8	364	469	676	972
3⅝ OD	27.8	39.7	55.7	69.8	80.3	88.7	72.0	57.6				
3½ IPS	30.2	43.2	61.0	76.1	87.0	96.0	78.8	62.3	539	704	1,005	1,430
4⅛ OD	38.6	55.2	78.0	97.3	111	123	95.8	77.1				
4 IPS	40.7	58.6	83.0	103	118	130	101.6	81.6	753	972	1,385	1,945
5 IPS	71.3	100	141	176	203	224	171.5	137.8				
6 IPS	126	183	257	322	366	403	266	214				
8 IPS	211	297	422	523	602	664	461	370				
10 IPS	352	503	712	887	1,024	1,130	725	582				
12 IPS	550	780	1,106	1,373	1,582	1,748	1,041	836				

SOURCE: From ASRE, Air-Conditioning Refrigerating Data Book, 1955–1956 ed., based on ARI Equipment Standards; see ASHRAE, Handbook, 1994 Refrigeration, for more complete tables.

Mains of Other Refrigerants For sizing piping for methyl chloride, sulphur dioxide, carbon dioxide, and other refrigerants, reference may be made to the ASHRAE Handbook, 1985 Fundamentals, Chap. 34.

COLD STORAGE

The first consideration in cold-storage room design is the temperature to be maintained in the refrigerated space. The temperature can generally be divided into three ranges: above 32°F (0°C), 32°F (0°C) and below, and 0°F (− 18°C) and below. Methods of construction and degree of insulation are dependent on the maintained temperature. Other more specialized design considerations are humidity control, ventilation requirements, controlled atmosphere (control of O_2, N_2, or CO_2), fire protection, and material-handling facilities (see Table 19.1.15).

Cold-storage rooms can either be site fabricated or prefabricated. The increasing availability of preinsulated panels with low *U* factors has led to increasing use of prefabricated rooms. Many of these are available as complete packages with the refrigeration system included.

Vapor Barrier Although good insulation is important, a well-designed and installed vapor barrier is the key to maintaining satisfactory insulation performance and life. A poor vapor barrier allows moisture to condense in the insulation, greatly reducing its insulating value. A well-designed system is of little value if poorly installed. Poor workmanship, in fact, accounts for most vapor barrier inadequacies.

Common vapor barrier materials include metal foils, polyethylene, asphalted felts and kraft paper, hot-melt asphalt, and special vapor barrier paints. Vapor barriers must be installed on the warm side of the wall.

Table 19.1.15 Cold-Storage Usage and Transmission Factors

	Room volume, ft³									
	20	50	100	500	1,000	2,000	5,000	10,000	50,000	100,000
Usage load*										
Average	4.68	2.28	1.61	1.21	1.10	0.835	0.403	0.240	0.178	0.173
Heavy	5.51	3.55	2.52	1.87	1.67	1.29	0.625	0.408	0.305	0.295
Storage temp, °F	40 and above		25–40		15–25		0–15		−25–0	
Insulation thickness, in	3		4		5		6		8	
Theoretical *U* factor†	0.086		0.066		0.055		0.046		0.035	
Practical *U* factor‡	0.111		0.083		0.066		0.055		0.046	
Piping *U* factor§	2.5		2.2		2.0		1.8		1.6	

* Btu per cu ft per 24 h per °F temp diff between outside and inside. Usage load includes 10°F or less of product cooling, and infiltration losses, etc., but does not include any freezing of product or fan motor loads when unit coolers are used. Values are for normal conditions and should not be used for unusual product loads.

† Theoretical *U* factor is based on corkboard ($k = 0.30$) with plaster on both sides, Btu per ft² per h per °F temp diff.

‡ Practical *U* factor allows for inefficiency of joints and structural supports in insulation.

§ Piping *U* factor is Btu/ft² outside pipe surface per h per °F temp diff between cooling medium and air (gravity circulation). Allowance has been made for frosting at the lower temperatures.

Table 19.1.16 Product Storage Data

Product	Quick-freeze temp., °F	Storage temp., °F Long	Storage temp., °F Short	Humidity, % R.H.	Specific heat Above freezing	Specific heat Below freezing	Latent heat	Freezing point	Respiration, Btu per lb per day
Apples	−15	30–32	38–42	85–88	0.86	0.45	121.0	28.4	0.75
Asparagus	−30	32	40	85–90	0.95	0.44	134.0	29.8	
Bacon, fresh	—	0–5	36–40	80	0.55	0.31	30.0	25.0	
Bananas	—	56–72	56–72	85–95	0.80	0.42	108.0	28.0	4.18
Beans, green	—	32–34	40–45	85–90	0.92	0.47	128	29.7	3.3
Beans, dried	—	36–40	50–60	70	0.30	0.237	18		
Beef, fresh, fat	−15	30–32	38–42	84	0.60	0.35	79		
Beef, fresh, lean	−15	30–32	38–42	85	0.77	0.40	100		
Beets, topped	—	32–35	45–50	95–98	0.86	0.47	129	31.1	2.0
Blackberries	−15	31–32	42–45	80–85	0.89	0.46	125	28.9	
Broccoli	—	32–35	40–45	90–95	0.92	0.47	130	29.2	
Butter	+15	—	40–45	—	0.64	0.34	15	15.0	
Cabbage	−30	32	45	90–95	0.93	0.47	130	31.2	
Carrots, topped	−30	32	40–45	95–98	0.87	0.45	120	29.6	1.73
Cauliflower		32	40–45	85–90	0.93	0.47	132	30.1	
Celery	−30	31–32	45–50	90–95	0.92	0.48	135	29.7	2.27
Cheese	+15	32–38	39–45	—	0.70				
Cherries	—	31–32	40	80–85	0.87	0.45	120	26.0	6.6
Chocolate coatings	—	45–50	—	—	0.3				
Corn, green	—	31–32	45	85–90	0.80	0.43	108	29.0	4.1
Cranberries	—	36–40	40–45	85–90	0.90	0.46	124	27.3	
Cream	—	34	40–45	—	0.88	0.37	84		
Cucumbers	—	45–50	45–50	80–85	0.97	0.47	137	30.5	
Dates, cured	—	28	55–60	50–60	0.83	0.44	104		
Eggs, fresh	−10	30–31	38–45	—	0.76	0.40	98	31.0	
Eggplants	—	45–50	46–50	85–90	0.94	0.47	132	30.4	
Flowers	—	35–40	—	85–90					
Fish, fresh, iced	−15	25	25–30	—	0.82	0.41	105	30.0	
Fish, dried	—	30–40	—	60–70	0.56	0.34	65		
Furs	—	32–34	40–42	40–60					
Furs, to shock	—	15	15						
Grapefruit	—	32	32	85–90	0.91	0.46	126	28.4	0.5
Grapes	—	30–32	35–40	80–85	0.86	0.44	116	27.0	0.5
Ham, fresh	—	28	36–40	80	0.68	0.38	87		
Honey	—	31–32	45–50	—	0.35	0.26	26		
Ice cream	−20	—	0–10	—	0.5–0.8	0.45	96		
Lard	—	32–34	40–45	80	0.52	0.31	90		
Lemons	—	55–58	—	80–85	0.92	0.46	127	28.1	0.4
Lettuce	—	32	45	90–95	0.96	0.48	136	31.2	8.0
Liver, fresh	—	32–34	36–38	83	0.72	0.42	94		
Lobster, boiled	—	25	36–40	—	0.81	0.42	105		
Maple syrup	—	31–32	45	—	0.24	0.215	7.0		
Meat, brined	—	31–32	40–45	—	0.75	0.36	75.0		
Melons	—	34–40	40–45	75–85	0.92	0.35	115	28.5	1.0
Milk	—	34–36	40–45	—	0.92	0.46	124	31.0	
Mushrooms	—	32–35	55–60	80–85	0.93	0.47	130	30.2	
Mutton	—	32–34	34–42	82	0.81	0.39	96	29.0	
Nut meats	—	32–50	35–40	65–75	0.30	0.24	14	20.0	
Oleomargarine	—	34–36	—	—	0.65	0.34	35	15.0	
Onions	—	32	50–60	70–75	0.91	0.46	120	30.1	1.0
Oranges	—	32–34	50	85–90	0.90	0.46	124	27.9	0.7
Oysters	—	—	32–35	—	0.85	0.45	120.0		
Parsnips	−30	32–34	34–40	90–95	0.82	0.45	120.0	28.9	
Peaches, fresh	—	31–32	50	85–90	0.90	0.46	126	29.4	1.0
Pears, fresh	—	29–31	40	85–90	0.86	0.45	118	28.0	6.6
Peas, green	—	32	40–45	85–90	0.80	0.42	108	30.0	
Peas, dried	—	35–40	50–60	—	0.28	0.22	14		
Peppers	—	32	40–45	85–90	0.94	0.47	—	30.1	2.35
Pineapples, ripe	—	40–45	50	85–90	0.88	0.45	122	29.9	
Plums	—	31–32	40–45	80–85	0.88	0.45	123	28.0	

Table 19.1.16 Product Storage Data (*Continued*)

| Product | Quick-freeze temp, °F | Storage temp., °F | | Humidity, % R.H. | Specific heat | | Latent heat | Freezing point | Respiration, Btu per lb per day |
		Long	Short		Above freezing	Below freezing			
Pork, fresh	—	30	36–40	85	0.60	0.38	66	28.0	
Potatoes, white	−30	36–50	45–60	85–90	0.77	0.44	105	28.9	0.85
Poultry, dressed	−10	28–30	29–32	—	0.80	0.41	99	27	
Pumpkins	—	50–55	55–60	70–75	0.92	0.47	130	30.2	
Quinces	—	31–32	40–45	80–85	0.90	—		28.1	
Raspberries	—	31–32	40–45	80–85	0.89	0.46	125	30.0	3.3
Sardines, canned	—	—	35–40		0.76	0.410	101		
Sausage, fresh	—	31–36	36–40	80	0.89				
Sauerkraut	—	33–36	36–38	85	0.91	0.47	128	26	
Squash	—	50–55	55–60	70–75	0.92	0.47	130	29.3	
Spinach	—	32	45–50	85	0.94	0.48	132	30.8	
Strawberries	−15	31–32	42–45	80–85	0.92	0.48	129	30.0	3.3
Tomatoes, ripe	—	40–50	55–70	85–90	0.95	0.48	135	30.4	0.5
Turnips	—	32	40–45	95–98	0.93	0.40	137	30.5	1.0
Veal	−15	28–30	36–40	—	0.71	0.39	91	29	

Insulation In recent years newer insulating materials have largely supplanted the use of plaster and corkboard in cold-storage room construction. Insulation has generally been used in three forms for this application: board form, panelized, and poured, sprayed, or foamed in place.

Board form insulation includes corkboard, rigid polystyrene, polyisocyanurate, and polyurethane, foamed, and fibrous glass.

Panelized insulation has gained acceptance, particularly in prefabricated installations. The outer metal skin can serve as a vapor barrier, and the panels are factory insulated with polystyrene, polyisocyanurate, polyurethane, or fibrous glass. *U* values as low as 0.035 have been attained in 4-in (100-mm) thick panels.

Poured, sprayed, or foamed in place insulation has its greatest applications on existing installations. It is primarily a urethane material. The application requires special equipment, and precautionary measures must be taken.

Cold Storage Temperatures A great deal of research and experience are required to obtain authoritative information on optimum temperatures and humidities for various products in cold storage. Table 19.1.16 is representative of good practice. The safe storage period depends upon the product and the storage temperature, and operational techniques vary greatly. Modern cold-storage warehouses of the larger concerns are cooled by brine which is furnished at two different temperatures only. The higher temperature for the mild-temperature warehouses is 10 to 12°F (−12 to −11°C), and the low-temperature brine for the freezers is −10 to −12°F (−23 to −24°C). All temperatures above that of the brine are obtained by regulating the amount of brine circulated in any particular set of coils. In the low-temperature warehouses, the piping is arranged for two classes of service: (1) **sharp freezers,** where the goods which are to be frozen are kept in a blast of air whose temperature is −30°F (−34°C) while their temperature is brought down quickly to the holding temperature (say in from 6 to 10 h) after which they are stored in (2) **holding rooms** where the desired temperature is maintained.

The system of cooling which is now being installed in the highest type of warehouses consists of a coil room containing the necessary brine-type or wetted, direct-expansion-type coils, through which the air from the different rooms is circulated by a pressure blower. The inlet and outlet of each room are so arranged that the cooled circulating air will cover the entire room in its transit; this is usually accomplished by having the cold-air inlet in the center of the room and two cold air outlets—one at each end of the room. The piping ratio for the coil rooms in this system, assuming a high-grade insulation with 2 to 4 Btu transmission per ft² per 24 h per °F temp diff, should be 1 ft² of external pipe surface to 15 ft³ of space to be cooled [with brine at −10 to −12°F (−23 to −24°C)] for warehouses carrying temperatures of zero and

below, and 1 ft² of pipe surface to 24 ft³ of space to be cooled [with brine at 10 to 12°F (−12 to −11°C)] in warehouses carrying mild temperatures of from 30 to 40° F (−1 to 4°C). Another system is used in which the blower and coils are supplemented by coils in the rooms. With this arrangement, it is possible to reduce temperatures quickly and hold them with the coils.

The average coil transmission in cold-storage rooms without forced circulation of the air is about 2 Btu/ft² of outside metal surface per h per °F temp diff with horizontal piping and 2.5 Btu with vertical piping. When forced air circulation is used, the transmission rate will increase to 20 Btu or more. In **brine circulation,** the brine, at same compressor back pressure, has a higher temperature than the ammonia, and consequently 1½ times as much pipe or more is used in brine circulation as is direct expansion for a given back pressure.

Piping of Rooms The **size of pipe** usually employed for piping rooms varies from 1 to 2 in (25 to 50 mm) with either brine circulation or direct expansion.

The extra cost of liberal piping allowance will often be offset by the consequent improvement in the efficiency of operation of the compressor. An expansion valve should be provided for every 500-ft (150 m) length of 1-in (25 mm) pipe, every 650 ft (200 m) of 1¼-in (30 mm) pipe, and every 1,000 ft (300 mm) of 2-in (50 mm) pipe when direct expansion is used.

Values of **overall coefficients of heat transfer** in Btu/h · ft² · °F for refrigerating practice are given in Tables 19.1.15 and 19.1.17. (See Sec. 4.)

Brine circulation is generally preferred to direct expansion so as to avoid danger from escaping ammonia or other refrigerant in case the pipes should leak. An advantage of the brine system is that there is always a considerable mass of refrigerated brine which can be drawn on in case the machinery should have to be stopped for any reason. In small plants, the general machinery may be stopped at night and only the brine pump be kept going to distribute the surplus refrigeration which has been accumulated in the brine during the day. Brine piping must consist of two lines, a flow and a return, usually of the same size. Brine storage is seldom used in large plants because of its bulk, its first cost, and the practical inability to store much refrigeration.

The **brine coils** in each cooled space are in parallel across the supply and return pipes. It has been common practice to allow 100 to 120 running ft (30 to 36 m) of **pipe** per circuit for low temperatures, and 400 to 440 (120 to 135 m) for high temperatures. The **tons refrigeration produced by brine** at various temperature differences and rates of pumping may be calculated approximately as follows: tons refrigeration = gal/min × °F range/28.

Air Conditioning of Low-Temperature Storage Rooms Bacterial growth and chemical decomposition are retarded by lowering the tem-

Table 19.1.17 Overall Coefficients of Heat Transfer
(Btu/ft^2 · h · °F)

Can ice-making piping		Brine coolers	
Old-style feed, non-flooded	12–15	Shell and tube	45–100
Flooded	20–40	Double pipe	150–300
High-velocity raceway trunk coils	80–110	Cooling coils	
Ammonia condensers		Boiling refrigerant to air in unit coolers	4–8
Submerged (obsolete)	30–40	Water to air in unit coolers	5–9
Atmospheric, gas entering at top	60–65	Brine to unagitated air	2.2–2.8
Atmospheric, drip or bleeder	125–200	Direct expansion to unagitated air*	1.6–2.5
Flooded	125–150	Water cooler, shell and coil	15–25
Shell and tube	150–300	Water cooler, shell and tube	50–150
Double pipe	150–250	Water cooler, shell and finned tube (Freon)	30–150
Baudelot coolers, counterflow, atmo-		Liquid-ammonia cooler, shell and coil ac-	
spheric type		cumulator	45
Milk coolers	75	Air dehydrator	
Cream coolers	60	Shell and coil (brine in coil) { 1st coil	5.0
Oil coolers	10	{ 2d coil	3.0
Water coolers		Double pipe	6–7
Direct expansion	60–150	Superheat remover, shell and tube	15–25
Flooded	60–200		

* U factor increases to 3.3 at 15°F temp diff with ammonia recirculation systems.
NOTE: Forced circulation of the air increases the coefficient to 1½ to 2½ times the values for still air. One inch of frost decreases the value 25 percent.

perature of perishable goods. Too low a temperature will freeze the goods and may result in spoilage. A holding temperature above the freezing point of the article is best for storage conditions. The maintenance of a proper relative **humidity** in a storage space holding unwrapped goods is as important as the maintenance of the proper temperature. High humidity favors the growth of mold and bacteria. Low humidities rob the product of moisture resulting in losses in value through impaired appearance and lost weight. **Air motion** is of importance in maintaining uniform conditions throughout the storage space. Too high air velocities cause excessive drying, and stagnant air through high humidities will cause mold. An optimum air motion falling between stagnation on one side and excessive drying on the other must be selected.

Low-temperature conditioning for storage purposes is preferably accomplished by the use of **cold-diffuser methods**. The cold diffuser connected to or located in the storage space consists of a fan and cooling means. The cooling means may be either a brine or cold-water spray or a surface-type cooler using brine, or a volatile refrigerant such as ammonia, Freon, etc. Unitary equipment, performing the functions of the cold diffuser, is commercially available. For the requirements for various perishable products, see ASHRAE Handbooks, 1993 Fundamentals and 1995 Applications, 1994 Refrigeration, and Table 19.1.16.

Quick freezing is freezing in 2 h or less; as employed for foods the temperatures may be as low as − 20 to − 50°F (− 29 to − 46°C). The methods of freezing include direct immersion in cold brine or a brine spray with the commodity held in a metal container. Another procedure is the use of a freezing tunnel approximately 50 ft (15 m) long, equipped with a stainless-steel conveyor belt. The commodities to be frozen pass through this tunnel in 15 to 30 min. The speed of the belt is variable, and the heat transfer is accomplished by high-velocity air circulation at a number of points in the tunnel, the circulation being transverse to the belt. The circulating air is cooled by brine sprays or wetted, direct-expansion coils which maintain the required temperature and the high humidity necessary to prevent shrinkage. In another process, packages of the material to be quick-frozen are clamped between steel platens containing evaporating refrigerant. Successful results are obtained with temperatures as high as 0°F (− 18°C). Storage may be at − 5 to − 10°F (− 20 to − 23°C) and transportation at 10°F (− 12°C) or lower.

Cold-storage lockers for individual family use are used in large numbers, particularly in suburban and rural areas. Lockers are rented usually on a yearly basis. A locker plant may have several hundred rental steel lockers, 6-ft^3 (0.17-m^3) capacity each, placed in rooms at about 0°F (− 18°C); a room for cutting and wrapping the meat; and a sharp freezer room held at about − 20°F (− 29°C). Two compressors are frequently specified or a single compressor using automatic control valves on the

separate rooms. Ammonia and R 12, operating under direct expansion, are commonly used.

ICE MAKING

Ice refrigeration is economical for some applications. The use of natural ice is fast disappearing. Manufactured ice is made by several methods. In the can system, galvanized cans containing up to 400 lb (180 kg) of water are immersed in brine for 25 to 40 h for freezing. Air dissolved in the water separates as bubbles causing opacity unless the water is agitated during the early stages of cooling, and any dissolved impurities come out of solution to form a central off-color core in the ice block unless the core is removed and replaced with fresh water just before freezing. The time and investment required for making ice in cans are often uneconomical. In the **Flakice** method (Crosby Field, *Trans. ASME*, May 1951, pp. 347–357) refrigerant is sprayed against the inner wall of a flexible, slowly rotating cylinder which is partially immersed in a tank of water. As the cylinder rotates, a thin layer of ice forms on the outside and then is broken off as further rotation causes the cylinder surface to flex under the action of an internal cam. For a patent history of small ice machines of this type, refer to Crosby Field, *Refrig. Eng.,* **58,** Dec., 1950, p. 1163. In the **PakIce** process (Taylor, *Refrig. Eng.,* **22,** 1931, p. 307) liquid ammonia is evaporated in an annular space formed by two concentric cylinders, the inner being corrugated. Water sprayed on the inner corrugated surface quickly freezes in a thin layer about .25 mm thick and is continually removed by mechanical scrapers. In an **extrusion method** (Watt, *Trans. ASME,* 1949) a slightly tapered circular or rectangular vertical cylinder with large end up is open at the top, but closed by a ram at the bottom. This cylinder has a jacket in which refrigerant is evaporated directly against the cylinder walls. Water enters the cylinder and freezes to a tapered block at the start of the operation. As soon as this occurs, the ram lifts (about 6 mm), shearing the ice block from the cold cylinder walls. Water admitted to this space quickly freezes, the ram again operates and the block of ice is again lifted, and shearing occurs again between the freshly formed ice and the cylinder walls, but not at the fresh ice and old ice interface. In this way, a continuous cylinder of ice is formed by the series of nested ice shells. The installation and upkeep costs are small and so is the space required per ton of the ice produced; a pilot-plant 1-ton (907 kg) unit requires 3 ft^2 (0.3 m^2) of floor space.

Batch ice makers can manufacture up to 75 tons (68 tonnes) of ice per day. These machines use constant-diameter ice-forming tubes to produce ice in shell and toroidal forms. The ice forms on a bank of vertical tubes in either of two configurations: inside the tubes for pellets or outside of the tubes for shell ice. A water distribution system is provided at the top. An ice conveyor and mechanical size-reducing device is mounted below the tubes. The refrigerant is circulated through the annulus of concentric tubes when ice pellets are made, and inside of single tubes when shell ice is made. The refrigerant supply can be designed for flooded operation; however, equipment capacity is reduced. The cycle starts as water passes through the distributors and flows over the surface of the tubes as a film. The refrigerant freezes the water to form a thickening layer of ice as time passes. Unfrozen water drains out of the tube into a sump. Then the water is recirculated, by a pump, up to the distribution system at the top of the tube bank. A purge of water is taken from the sump to remove accumulated salts and other impurities. When harvest is initiated, the refrigerant system is drained, and hot gas from the condenser is introduced into the refrigerant side of the tube and the pressure is raised. This causes the ice to partially melt and release from the tube wall. The cylinder or shell of ice falls by gravity into the conveying and size-reducing machinery. The size of an ice pellet can vary from 0.75 in (19 mm) to 1.375 in (35 mm) outside diameter. Ice clarity can be controlled by flushing the ice with water before and during harvest. The machine is controlled by digital logic microprocessors. Energy consumption when using 65°F (18°C) feedwater varies from 1.94 to 2.33 tons (6.8 to 8.2 kW) refrigeration per ton (909 kg) of ice when using R-717 versus HCFC-22 in a recirculation system. Standard model ice makers can be operated with the refrigerant system either in the flooded condition or with circulating pumped refrigerant in the tubes. Similar units lose 18 percent capacity when operating in a flooded state compared to operating with pumped recirculating refrigerant.

Making clear ice using continuous methods is often accomplished by washing the newly frozen ice surface with cold water. This method can increase water requirements by up to 50 percent and energy usage by 15 percent.

The refrigeration tonnage required for making 1 ton (909 kg) of ice varies from 1.84 with 55°F (12.8°C) inlet water temperature to about 2.0 if the water to be frozen enters as high as 75°F (23.9°C).

SKATING RINKS

Rinks for ice skating, hockey, or curling vary in size from about 400 ft² (37 m²) area to 16,000 ft² (1,486 m²) or more. Construction of the ice floor is important. For a low-cost, efficient unit, brine pipes are laid directly in sand, which is kept wet during freezing to increase heat transfer. As freezing progresses, a layer of ice is built up by spraying water on the surface. The expense of maintaining such a floor in a satisfactory level condition may be considerable, and corrosion of the pipes, from the outside in, in contact with the wet sand may be excessive, particularly during shutdown periods. Moreover, it is not possible to use such a floor for other purposes. When a floor for diversified activities is needed, it is usual to provide a lower layer of insulation, primarily to prevent sweating on ceilings under the floor, surmounted by a concrete slab in which the steel pipes carrying the brine are imbedded and on which the ice layer is formed. This construction protects the pipe from external corrosion, affords fast freezing, and makes possible use of the floor in a few hours for purposes where an ice surface is not desired. Floor pipes can be mild steel or polyethylene and are usually 1 or 1¼ in (25 or 32 mm) spaced 3 to 6 in (75 to 150 mm) on centers; care must be exercised to ensure a uniform distribution of brine. Brine flow may be 10 gal/min per ton of refrigeration, corresponding to about 1 gal/min per 10 to 20 ft² of piping surface. In a few installations, cold brine is sprayed directly against the under side of a steel floor on which the ice layer is frozen. Close control of ice surface temperature is necessary for good skating, usually −3°C ± 0.5. Refrigeration capacity should be from 0.4 to 0.85 ton per 100 ft² (150 to 320 W per m²) of floor surface, although the load may be much higher for open-air rinks or other unusual conditions.

REFRIGERATION IN THE CHEMICAL INDUSTRIES

The use of refrigeration in petroleum and other industries is widespread. In petroleum processing, refrigeration is used (1) to control vapor pressure of highly volatile constituents (methane, ethane, propane, and butane), such control being necessary during distillation, during processing, and for recovery of gasoline fractions from natural gas; (2) to shift the solubility relationship so that undesired constituents such as asphalt and wax in lubricating oils may be removed by precipitation; (3) to produce selective chemical reaction such as occurs when sulphuric acid is used to remove gum-forming constituents from light fuels or when an alkylate fuel fraction is formed by combining a low molecular weight unsaturated with a similar saturated hydrocarbon. Alkylate is a valuable and important constituent of aviation fuels.

For (1), temperature ranging from −35°F (−38°C) or lower to as high as 60°F (15°C) are required, the refrigerant being ammonia, propane, or butane. For the higher temperatures spray-cooled water or jet refrigerating units (see above) are sometimes satisfactory. For (2), it is possible to use propane as a combined solvent and refrigerant, evaporation of the propane causing the required cooling, temperature level [approximately −40°F (−40°C)] being controlled by the pressure held on the equipment. The propane vapor is recompressed, condensed, and reused as in any compression cycle. As much as ⅓ ton of refrigeration per barrel of lubricating oil dewaxed is needed. For (3), temperatures of up to 60°F (−16 to 15°C) are general, and ammonia is a satisfactory and widely used refrigerant.

The chemical industry uses refrigeration to control condensation polymerization processes in the manufacture of synthetic rubbers and plastics. Syntheses of refrigerants and other chemicals require systems to maintain temperatures which range from 5°C down to −85°C in process equipment. Large centralized refrigeration systems which use ethylene glycol and methylene chloride brine are common. Several modern chemical processes use the eventual product as a refrigerant (for example, the manufacture of ammonia). Cascade systems with segregated circuits are used to produce temperatures below −50°C.

Recently imposed regulations limit volatile organic vapor emissions to the atmosphere from storage tank vents and other process systems. The recovery efficiency required of cleanup systems has led to development of refrigerated liquid absorption processes to scrub vaporous effluents from organic chemical storage tanks. Recently installed systems have required liquid absorbent temperatures of −40°F (−40°C). The capacity of these systems varies from 2 tons (7 kW) to 500 tons (1760 kW). The absorbent liquid is often accumulated from the effluent stream over a period of time.

DEEP REFRIGERATION

Dry ice (solid CO_2) is useful as a refrigerating medium in special cases. The main objection to the general use of dry ice is its cost, but the ease in its handling, its low temperature [−110°F (−79°C) at atmospheric pressure], its noncorrosive and nontoxic properties, its high latent heat, and the absence of liquid drip make it desirable. The heat absorbed per pound of solid CO_2 during sublimation at atmospheric pressure and −110°F (−70°C) is 245 Btu approx. The specific heat of the gas at constant pressure is about 0.2.

Carbon dioxide is obtained commercially either by fermentation or by burning. The gas is compressed, usually in three stages, and cooled to atmospheric temperature, thereby forming a liquid at about 1,000 psia (690 kPa); the liquid is then throttled to below 5.3 atm pressure. During throttling part of the CO_2 solidifies and is compressed to form a dense ice. The fraction that returns to the vapor phase is recompressed.

Production of Solid Carbon Dioxide Figure 19.1.18 shows a simple cycle and Fig. 19.1.19 the elemental pressure-enthalpy diagram for the production of solid carbon dioxide. From the high pressure at atmospheric temperature the liquid is throttled (isenthalpic) to atmospheric pressure, the resultant point being within the sublimation solid-vapor zone below the triple-point realm. In this resultant throttling expansion, part of the CO_2 solidifies; it is then removed in the separator from the residual vapor and is compacted to form a dense CO_2 ice. The fraction representing the vapor phase is augmented by fresh makeup gas to continue the production cycle.

Beyond the manufacture of dry ice, operation at deep refrigerations,

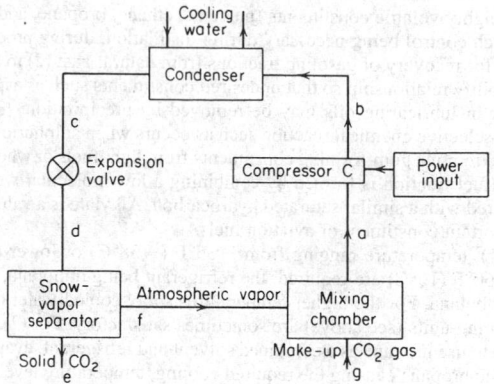

Fig. 19.1.18 Elemental dry-ice production circuit.

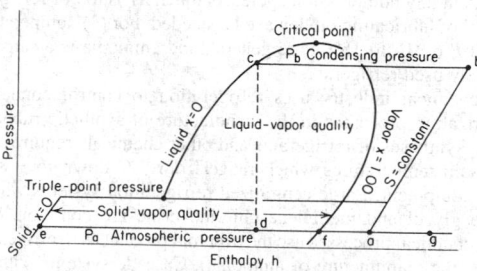

Fig. 19.1.19 Pressure-enthalpy diagram illustrating dry-ice production circuit.

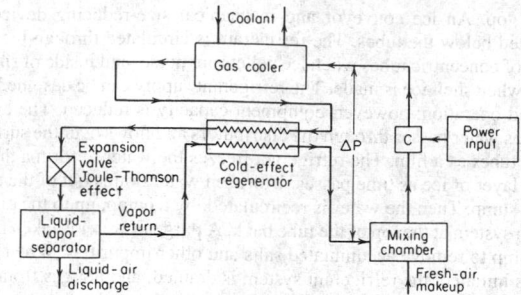

Fig. 19.1.20 Elemental Linde system circuit.

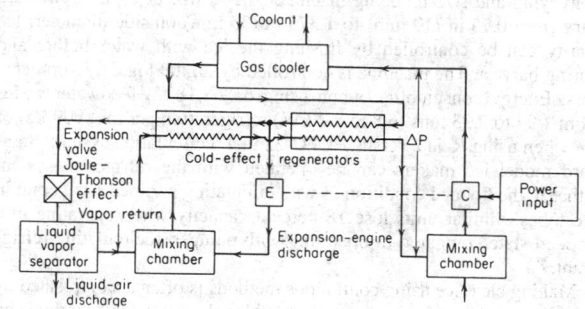

Fig. 19.1.21 Elemental Claude system circuit.

such as air liquefaction [at $-312°F$ ($-191°C$) approx], lead into **cryogenics** (see Sec. 19). Illustrative systems are (1) Linde (throttling) and (2) Claude (expander). In the elemental **Linde system** (Fig. 19.1.20), by virtue of the "cold-effect" regenerator, the process air is reduced in temperature by the return fraction of the air from the separator. The

throttling process thus ends in the "wet zone" at atmospheric pressure with generation of low-temperature liquid air that flows from the separator. In the **Claude system** (Fig. 19.1.21), use is made of an expander (whose work output reduces the net work requirement of the compressor) and of an expansion valve with the Joule-Thomson effect.

19.2 CRYOGENICS
by F. J. Edeskuty

REFERENCES: Timmerhaus (ed.), "Advances in Cryogenic Engineering" (annual), Plenum Press. Vance and Duke, "Applied Cryogenic Engineering," Wiley. Scott, "Cryogenic Engineering," Met Chem Research. Vance (ed.), "Cryogenic Technology," Wiley. Mendelssohn, "Cryophysics," Interscience. McClintock, "Cryogenics," Reinhold. Scott, Denton, and Nicholls (eds.), "Technology and Uses of Liquid Hydrogen," Pergamon. Hoare, Jackson, and Kurti, "Experimental Cryophysics," Butterworth. White, "Experimental Techniques in Low-temperature Physics," Oxford. Johnson (ed.), "Compendium of Properties of Materials at Low Temperature," WADD Technical Report 60-56, U.S. Dept. of Commerce. Durham, McClintock, and Reed, "Cryogenic Materials Data Handbook," PB 171-809, U.S. Dept. of Commerce. Cook (ed.), "Argon, Helium, and the Rare Gases," Interscience. Perry, Chilton, and Kirkpatrick, "Chemical Engineers' Handbook," McGraw-Hill. Barron, "Cryogenic Systems," McGraw-Hill. Bailey, "Advanced Cryogenics," Plenum. Haselden, "Cryogenic Fundamentals," Academic. Williamson and Edeskuty (eds.), "Liquid Cryogens," CRC. Timmerhaus and Flynn, "Cryogenic Process Engineering," Plenum Press. Van Sciver, "Helium Cryogenics," Plenum Press.

Notation

Some symbols in this section differ from the notation of engineering practice but are retained to facilitate use of the prevalent scientific sources of information.

A = area, ft^2 (m^2)
b.c.c. = body centered cubic lattice structure
C = specific heat, Btu/lbm · °R(J/kg · K)
C_p, C_v = specific heats at constant pressure and volume respectively, Btu/lbm · °R(J/kg · K)
f.c.c. = face centered cubic lattic structure
H = enthalpy, Btu/lbm (J/kg)
H = magnetic field, $G(T)$
h.c.p. = hexagonal close-packed lattice structure
κ = thermal conductivity, Btu/h · ft · °R (W/m · K)
σ = electrical conductivity (reciprocal of resistivity) or pair separation for zero potential in Lenard-Jones equation
s = Boltzmann's constant
L = length, ft (m)
LH$_2$ = liquid hydrogen (H$_2$)
LHe = liquid helium (He)
LN$_2$ = liquid nitrogen (N$_2$)
LNG = liquefied natural gas
LO$_2$ = liquid oxygen
n = number of moles

P = pressure, psia or torr (vacuum) (Pa)
R_e = electrical resistance, Ω
T = temperature, °R (K)
V = volume, ft^3 (m^3)
α = thermal linear-expansion coefficient, °R^{-1} (K^{-1}), or accommodation coefficient
η = viscosity, lbm/ft · h (N · s/m^2)
ρ = density, lbm/ft^3 (kg/m^3)
ΔT = temp difference, °R(K)
* = Denotes property divided by quantum parameter (Table 19.2.5) which is not the critical point value.

Cryogenics is the study, production, and utilization of low temperatures. Cryogenic temperatures have been defined so ambiguously that "upper limits" to the cryogenic range from 216 to 396° R (120 K to 220 K) may be found in the literature. In this section the cryogenic range for a property of a given substance is considered to embrace the scale between absolute zero and the temperature above which the property has the expected or normal behavior. Cryogenics thus embraces the unusual and unexpected variations which appear at low temperatures and make extrapolations of properties from ambient to low temperatures unreliable.

Progressively lower temperatures become increasingly difficult to attain in practice. As the working temperature of a refrigerator is lowered, the work required to transfer a given amount of heat increases as demonstrated by the Carnot limitation, to wit, $W = Q[(T_1 - T_2)/T_2]$, where W is the work required to extract the heat Q at a low temperature T_2 and reject it at a higher temperature T_1 (see Sec. 4). The actual work is always greater than this because of inefficiencies of mechanical equipment, thermal losses associated with finite temperature differences in heat exchangers, and heat leaks from the surroundings to the cold equipment.

REFRIGERATION METHODS

(See also Sec. 4 and Sec. 19.1, "Mechanical Refrigeration.")

Cryogenic refrigerators (cryocoolers) may be classified by (1) the functions they perform (e.g., the delivery of liquid cryogens, the separation of mixtures of gases, and the maintenance of spaces at cryogenic temperatures, (2) their refrigerating capacities and (3) the temperatures they reach. Large industrial-sized plants (a) deliver LNG (~ 120 K), LO$_2$ (~ 83 K), LN$_2$ (~ 77 K), LH$_2$ (~ 20 K), and LHe (~ 4 K), (b) separate gaseous mixtures, e.g. the constituents of the atmosphere, H$_2$ from petroleum refinery gases, H$_2$ and CO from coke oven and coal-water gas reactors, and He from natural gas, and (c) provide refrigeration to maintain spaces at low temperatures. For the latter, i.e., (c), a unit has been manufactured and installed that delivers 13 kW of refrigeration at less than 3.8 K.

An important area of refrigerator development that is being commercially exploited now is laboratory-sized cryogenic refrigerators for laboratory research and development at LHe temperatures (< 5 K). These refrigerators are used for many different purposes, refrigerating for example: high-field superconducting electromagnets, LHe bubble chambers for high-energy (nuclear) particle research, experimental superconducting electric generators and motors and superconducting magnets for levitation of railroad trains. Another area of commercial exploitation that is important for the progress of cryogenic physics research is the development of refrigerators for the continuous production of refrigeration at temperatures below 1 K. These include L^{3}He evaporation refrigerators and L^{3}He-L^{4}He dilution refrigerators that reach temperatures of 0.4 and 0.003 K, respectively.

Of various methods of refrigeration, the most commonly used to produce temperatures as low as 1 K are (1) the evaporation of a volatile liquid (referred to as the **cascade** method when applied in several successive stages using progressively lower-boiling liquids), (2) Joule-Thomson (isenthalpic) expansion of a compressed gas, and (3) an adiabatic (isentropic) expansion of a compressed gas in an engine (reciprocating or turbine) or from a bomb or cylinder through a throttling valve

(Simon expansion). Using L^{3}He, method (1) is capable of reaching temperatures as low as ~ 0.4 K. Numerous refrigeration cycles have been devised utilizing various combinations of the above three methods. The final stage of refrigeration in plants that deliver liquid cryogens is generally a Joule-Thomson (isenthalpic) expansion. **Expansion engines** (reciprocating and turbine) are commonly used for producing the refrigeration needed to reach the final stage. Turbine expanders are preferred for the large refrigerators because their efficiencies (70 to 85 percent) are higher than for reciprocating expanders. The efficiencies of the heat exchangers for counterflowing "cold" and "warm" gases are an important factor in the overall refrigerator efficiency. In industrial plants heat exchanger efficiencies reach 98 percent. Their design represents a compromise between the attainment of a high heat-transfer coefficient, on the one hand, and a low resistance to the flow on the other.

Figure 19.2.1 typifies a commonly employed modified Brayton cycle which uses an expansion engine for generating the refrigeration. The theoretical figure of merit (the ratio of the work W_c done by the compressor to the heat Q, transferred to the refrigerant at the low temperature) is for this cycle

$$\frac{W_c}{Q} = \frac{RT_1 \ln (P_1/P_5)}{M(H_4 - H_3)}$$

where M is the molecular weight of the gas whose enthalpy per unit mass is H. The other symbols are defined by reference in Fig. 19.2.1.

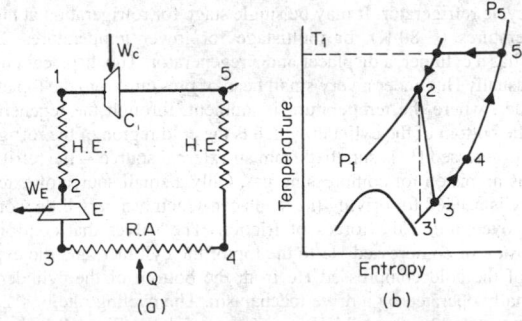

Fig. 19.2.1 (a) Schematic of a modified (isothermal compression) Brayton cycle, and (b) the process path. W_c is the energy (work) to drive the compressor C. Q is the heat absorbed by the refrigerator from the refrigerated area R.A., where the working fluid passes from state 3 to state 4. The heat-exchanger (H.E.) processes are isobaric. The dashed line 2-3' is the ideal isentropic expansion; 2-3 represents the actual expansion in the expander E.

The **Stirling refrigerating engine** and modifications of it are used in a number of commercial makes of laboratory and miniature-sized refrigerators. The Stirling-cycle refrigerator is well suited to (1) the liquefaction of air on a laboratory scale (~ 7 l/h), (2) the recondensation of evaporated liquid cryogens, and (3) the refrigeration of closed spaces where the refrigeration load is not very large. These laboratory scale refrigerators supply ~ 1 kW of refrigeration at ~ 80 K and ~ 2 kW at 160 K. They are used with liquid air fractionating columns for the production of LN$_2$ (~ 7 l/h) and LO$_2$ (~ 5 l/h).

The Stirling-cycle refrigerator consists of a piston for compressing isothermally the working fluid (usually He), and a displacer which can operate in the same cylinder with the piston. The displacer and piston are connected to the same electrically driven shaft but displaced in phase by 90°. The displacer pushes compressed gas isochorically from the warm region where it was compressed, through the regenerator into the cold region where the compressed gas is expanded isothermally doing work on the piston and producing refrigeration. The regenerator consists of a porous mass (packed metal wool of high heat capacity) in which a steep temperature gradient is established between the warm region of compression and the cold region of expansion. The displacer returns the gas after expansion to the region for compression through the regenerator in which it is warmed to the temperature of the isothermal compression. The refrigerant (He) is recycled. The regenerators in the larger refrigerators are usually placed around the outside of the engine cylinders, but in small capacity, lower temperature refrigerators they are put inside the displacers.

In Stirling refrigerators that reach temperatures of 17 K and produce 1 W of refrigeration at 25 K, the compressed He is expanded in two or three stages at temperatures intermediate between room temperature and the lowest temperature reached. There is only a single piston for compression. The expansion in stages allows part of the heat that leaks into the coldest region of a single stage engine to be absorbed and removed at a higher (intermediate) temperature where the efficiency for transferring heat is greater.

The **Vuilleumier refrigerating engine** is a modification of the Stirling refrigerator. It *resembles* two single Stirling engines placed back to back. One of the engines operates as a heat engine and the other as a refrigerator. The lower temperature at which the heat engine discharges its waste heat, is also the top temperature at which the refrigerator engine discharges its waste heat. Hence, the Vuilleumier refrigerator operates at elevated, intermediate and low temperature levels which may be 800, 300 (ambient), and 90 K, respectively. Each engine has a cylinder, a displacer and a regenerator, and the two engines are connected to a common crankshaft but displaced by 90°. Very little external power is needed to operate the two displacers which are operated in sinusoidal motion, displaced 90° in phase. The working fluid (He) is recycled.

The **Gifford-McMahon refrigerator** is another modification of the Stirling-cycle refrigerator. It may be single stage for refrigeration at higher temperatures (~ 80 K), or multistage for lower temperatures. Each stage has a cylinder, a displacer and a regenerator. The displacer pushes gas (usually He) under a very small head of pressure from the top of the cylinder, where the temperature is ambient, through the regenerator, into the bottom of the cylinder which is the cold region of the refrigerator. Compressed He is supplied from an external source—the refrigerator has no piston for compressing gas. Only a small source of external energy is needed for driving the displacer which has very little work to do in overcoming the forces of friction. The valves that control the admission of compressed He to the top of the cylinder, and the expansion of the cold compressed He from the bottom of the cylinder are externally operated by a drive mechanism. The cycling rate is < 2 Hz. The expansion of the cold He is of the adiabatic-Simon-bomb type. Expanded gas is discharged from the refrigerator.

Some applications of cryogenic refrigerators require very high reliability without maintenance. The fewer the working parts, the more likely is the needed reliability to be achieved. The **pulse-tube** and **thermoacoustic refrigerators** are promising candidates in this respect in that they have only one moving part, and that part operates at ambient temperature. Pressure oscillations are introduced into a tube with a closed end and fitted with the proper arrangement of heat exchangers. For the optimum frequency of the pulsation, a temperature gradient is established between the heat exchangers, and this gradient can pump heat from a low temperature to a higher one. A number of different designs have resulted, including single-stage and multistage units, and operating frequencies have ranged from about 1 Hz up to acoustic frequencies. This type of refrigerator, still in the research stage, is capable of reaching low temperatures down to 28 K with a single-stage unit, 26 K with a two-stage unit, and 11.5 K with a three-stage unit, all with ambient temperature heat sinks. Present research is also focused on the use of an acoustically driven pulse-tube refrigerator that would have no moving parts and no close tolerances of the components, thus making it inexpensive and easy to produce.

Magnetic refrigeration was originally proposed in 1925 but has yet to find acceptance for commercial exploitation. Magnetic refrigeration depends upon the ability of paramagnetic materials at low temperature (or ferromagnetic materials near their Curie temperatures) to warm up (or expel heat at constant temperature) with the application of a magnetic field. Conversely, these materials will cool (or absorb heat at constant temperature) upon removal or reduction of the magnetic field. With the use of a paramagnetic material magnetized at a higher temperature of about 1.5 K, temperatures of about 0.001 K have been reached by a

single demagnetization. Although a continuously operating paramagnetic salt refrigerator was marketed in the late 1950s, commercial magnetic refrigerators are not now available. However, magnetic refrigeration is applicable over a wide range of temperatures, and recent research work is leading to continuous application of this process by mechanisms that promise a considerable increase in efficiency over that of more conventional refrigerators.

^{3}He-^{4}He dilution refrigerators for reaching temperatures between 0.01 and 0.3 K and generating continuously as much as 750 ergs/s of refrigeration are commercially available. Refrigeration results from the solution of L^{3}He in L^{4}He, the heat of solution being negative. Figure 19.2.2 is a schematic diagram of a dilution refrigerator. ^{3}He vapor from the "still" at ~ 1 K (the upper operating temperature of the refrigerator) is collected and returned by a pump to the "mixing" chamber where solution takes place. The ^{3}He arrives at the mixing chamber as L^{3}He near the mixing chamber temperature (the lower operating temperature of the refrigerator) after flowing through a heat exchanger counter to the

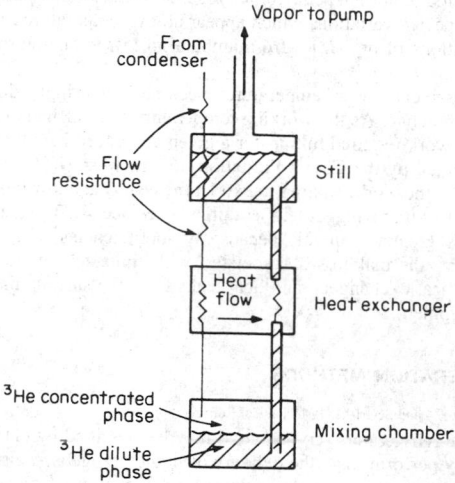

Fig. 19.2.2 Schematic diagram of the mixing chamber and still of a ^{3}He-^{4}He dilution refrigerator.

flow of cold ^{3}He-^{4}He solution on its way to the "still." In the electrically heated still, ^{3}He is evaporated from the solution. The evaporated ^{3}He is withdrawn by the collecting pump (diffusion) that returns the ^{3}He to the mixing chamber. At 1 K the vapor pressure of L^{4}He is negligible. L^{4}He in the still, freed of ^{3}He, returns to the mixing chamber by superfluid flow, building up in the mixing chamber an osmotic pressure that drives ^{3}He atoms towards the still against the forces of viscous flow.

In the mixing chamber there is a two-phase separation of ^{3}He and ^{4}He. A layer of nearly pure L^{3}He of lower density rides on a heavier L^{4}He-rich layer containing ~ 6 percent ^{3}He. At temperatures lower than 0.87 K liquid solutions of ^{3}He and ^{4}He separate into two liquid phases, one ^{3}He-rich and the other ^{4}He-rich, in thermodynamic equilibrium. At 0 K, the ^{3}He-rich phase is 100 percent ^{3}He, whereas the ^{4}He-rich phase (in equilibrium with L^{3}He phase) contains ~ 6 percent ^{3}He.

In the mixing chamber, L^{3}He enters the upper ^{3}He-rich phase; solution of ^{3}He in ^{4}He takes place at the interphase boundary. The lower (denser) ^{4}He-rich phase connects with the still.

Starting at 50 mK (a temperature reached in a ^{3}He-^{4}He dilution refrigerator), temperatures in the 2- to 3-mK range are attainable with a **Pomeranchuk refrigerator**. Refrigeration is generated by compression of a mixture of liquid and solid phases of ^{3}He at pressures in excess of 28.9 atm, the minimum P at which these phases can coexist in equilibrium. The entropy of solid ^{3}He exceeds the entropy of L^{3}He, which is contrary to the normal behavior for other substances. This occurs in ^{3}He because of its nuclear magnetic properties. The ^{3}He nucleus is magnetic and at temperatures in the mK range and above, in solid ^{3}He, the nuclear moments are randomly oriented whereas in L³He, at temperatures lower than ~ 0.3 K, the nuclear moments are paired or partially paired, antiferromagnetically. An adiabatic increase of P converts liquid to solid with a reduction in T, whereas an isothermal increase of P results in an absorption of heat. ³He, condensed, is confined in a container with flexible metal walls in the "cold" region of a ³He-⁴He dilution refrigerator. The ³He container is surrounded with L⁴He which serves as the pressure transmitting fluid.

GAS LIQUEFACTION

A conventional method of gas liquefaction utilizes a gas compressor, a countercurrent heat exchanger, and a throttling valve through which the gas expands isenthalpically (Joule-Thomson) and cools if it has already been cooled below its inversion temperature. After expansion, the cold gas returns through the heat exchanger, in which it exchanges heat with the countercurrent compressed gas flowing to the throttling valve. It leaves the heat exchanger near the temperature of the entering compressed gas. The temperature decreases at the throttling valve until the condensing temperature is reached, and then liquefaction occurs at a rate determined by the rate of refrigeration.

The rate of liquefaction x in pounds per hour is $x = \{[H$ (expanded *gas* out) $- H$ (compressed *gas* in)] $w - q]/[H$ (compressed *gas* in) $- H$ (*liquid* following valve)]}, where the H's are enthalpies in Btu per pound for the final heat exchanger, w is the flow rate in the same units as x, and q is the heat leak in Btu per hour from outside into the heat exchanger. For an ideal heat exchanger the expanded gas leaves at the temperature of the entering compressed gas. In practice, there is a small difference in temperature which represents a small loss of refrigeration. For ideal gases, H is independent of P for a given T, and hence no liquefaction of an ideal gas results. The H's of air, O_2, N_2, A, CO_2, the hydrocarbons, and the normal refrigerants decrease with increasing pressure at ambient T, except at very high P's, and these gases are liquefiable by this kind of isenthalpic (Joule-Thomson) expansion. The H's of H_2 and He at ambient T, however, increase with increasing P, even at low P's, and hence an auxiliary mechanism is required for the liquefaction of these gases. In one method, the stream of compressed gas is split in the liquefier, and a part goes to an engine in which it is cooled by an isentropic expansion with the performance of work. This part of the flow, thus cooled, is sent to a heat exchanger, where it flows counter to the other fraction of compressed gas, cooling it to a temperature below the inversion temperature (see Sec. 4). This flow of compressed gas, thus precooled, is run through another and final heat exchanger with a throttling valve at its lower end, with the result that the quantity x is liquefied.

Another method of precooling H_2 and He below their inversion temperatures makes use of a boiling liquid cryogen through which the compressed gas flows in a heat exchanger. LN_2 is used for precooling H_2, and LH_2 for He.

APPLICATIONS OF CRYOGENICS

Gases, such as O_2, N_2, natural gas, H_2, and He, are liquefied for transportation because of the gain in density at low pressure. Shipments have been made of LH_2 by trailer trucks in quantities of 13×10^3 gal (50 m³) for thousands of miles and by railway tank cars holding as much as 28×10^3 gal (100 m³). LHe is regularly shipped in quantities from 25 gal (0.1 m³) to 10,000 gal (40 m³). Liquid cryogens are used as liquid refrigerants. The aerospace, steel, and bottled-gas industries are the principal users of liquefied cryogens. Important applications as coolants can be found in vacuum technology, electronics, biology, medicine, and metal forming.

In recent years, large superconducting magnets have been undergoing development for a variety of applications including fusion reactors, energy storage, electric-power transmission, high-energy physics experiments, and magnetically levitated trains. Magnets of Nb-Ti are available with fields up to 80 kG (8 T) with large magnetic field volumes (1 m³). Higher field magnets [up to 150 kG (15 T)] of Nb₃Sn are available with smaller magnetic field volumes (10⁻⁴ m³). The Los

Alamos National Laboratory installed a 30-MJ superconducting magnetic energy-storage system in the Bonneville Power Administration's Tacoma, Washington, substation to provide stabilization of utility longline power-transmission systems. The response characteristics of the unit have been demonstrated on the utility's power grid, and the unit has been run successfully for extended periods. Two 100-m long superconducting ac transmission cables have been successfully tested at Brookhaven National Laboratory at 4,100 A and the equivalent of 138 kV three-phase. At the other end of the size spectrum, Josephson junction devices are being considered for applications such as radiation detectors, highly precise and accurate electrical measurements, geophysical instrumentation, medical instrumentation, and computer memory cells.

Cryogenics is also becoming increasingly important in the storage and transport of energy. Large quantities of natural gas are routinely transported as a liquid (LNG). Peak shaving in municipal gas systems is accomplished by storage of LNG. If, due to depletion of fossil fuels, hydrogen becomes an important synthetic portable fuel, cryogenic hydrogen will play an important role as it already does in space applications including the shuttle. Another potential application is aircraft fuel. This may be extended to surface transportation applications. Slush hydrogen, a mixture of liquid and solid hydrogen, is being considered as a fuel for the National Aerospace Plane.

PROPERTIES OF SOLIDS AT LOW TEMPERATURES

(See also Sec. 4.)

Specific heats of solids in general decrease with decreasing T, becoming zero at 0°R (Fig. 19.2.3). The downward approach to $C_p = 0$ at 0° R is interrupted for some substances (principally compounds) by "bumps" on the curve (excess C_p that rises to a maximum and then decreases). Paramagnetic salts undergoing transitions to either a ferromagnetic or an antiferromagnetic state are examples. This excess specific heat of a paramagnetic salt is connected with the effectiveness of the salt for reaching low temperatures by the method of adiabatic demagnetization. There are also transitions in solids from a more orderly to a less orderly arrangement of atoms and molecules in the lattice that give rise to excess C_p. The transition in solid ortho- and normal H_2 below 20°R (11 K) is an example. In Fig. 19.2.3, C_p's are plotted for various materials.

Heat is transferred in dielectrics by lattice vibrations, or waves. In

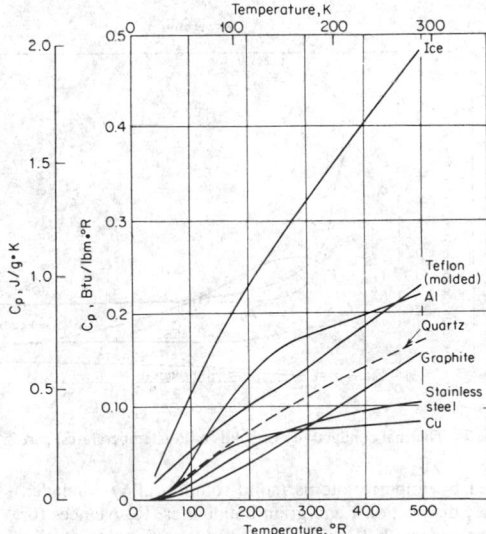

Fig. 19.2.3 Specific heats of solids at low temperatures.

good electrical conductors, heat is transferred principally by the conduction electrons, and thermal and electrical conductivities are related by the Wiedemann-Franz law; $\kappa/\sigma T = \text{const}$. This means that the ratio of the thermal and electrical conductivities at a given T is approximately the same for the good conductors. In the poor electrical conductors, alloys for example, both lattice waves and conduction electrons play important parts in the transfer of heat. The κ's of "pure" dielectrics and "pure" metals rise with increasing T from $\kappa = 0$ at $0°R$ (proportional to T for metals and to T^3 for dielectrics), reach a maximum, normally between 10 and $100°R$ (5 and 55 K), and then decrease to a value approximately independent of T (Fig. 19.2.5 and ice in Fig. 19.2.4). The κ's of alloys (Fig. 19.2.4) are an order of magnitude smaller than for "pure" metals and *do not* exhibit the maxima characteristic of the "pure" metals at low T. Lattice disorder introduced by alloying, even in small amounts, and working a metal, even a pure metal, reduces κ. Annealing, in general, raises κ.

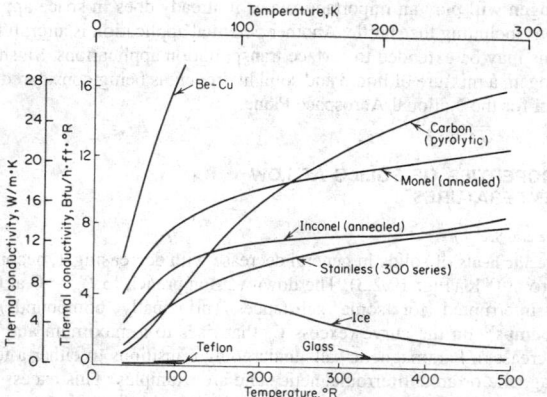

Fig. 19.2.4 Thermal conductivity of solids at low temperatures, part 1.

At $0°R$, an absolutely pure and perfect single crystal of metal would have zero electrical resistance, $R_e = 0$. **Electrical resistance** arises from the scattering of the conduction electrons as they move through the lattice of metal ions under the influence of an externally applied electric field. Scattering arises for two reasons: (1) the amplitudes of the thermal vibrations of the lattice which increase with T at a rate proportional to $\sqrt{C}$ of the metal, and (2) the imperfections in the regularity of the lattice

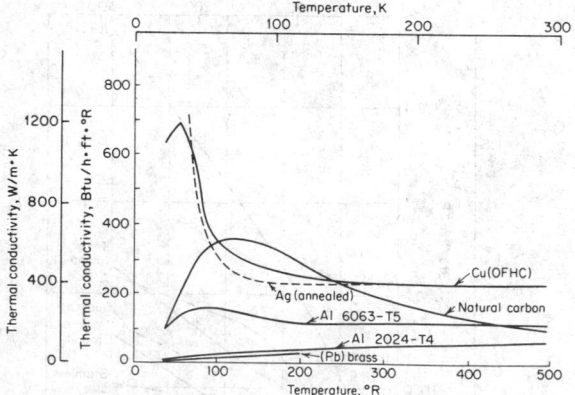

Fig. 19.2.5 Thermal conductivity of solids at low temperatures, part 2.

as caused by impurity atoms (solid-solution alloys included), lattice vacancies, dislocations, and grain boundaries. Resistances for "pure" metals increase with T and are roughly proportional to $C \cdot T$, except at very small T's where they are proportional to T^5 for absolutely pure

crystals. The resistance due to impurities and imperfections is very roughly independent of T (Matthiessen's rule). For very pure metals, R_e is approximately constant below $20°R$ (11 K). Magnetic impurities can give rise to a minimum in resistivity [usually below $36°R$ (20 K)] called the Kondo effect. Resistivity in $\Omega \cdot \text{cm}$ (Fig. 19.2.6) is the R_e of a 1-cm cube.

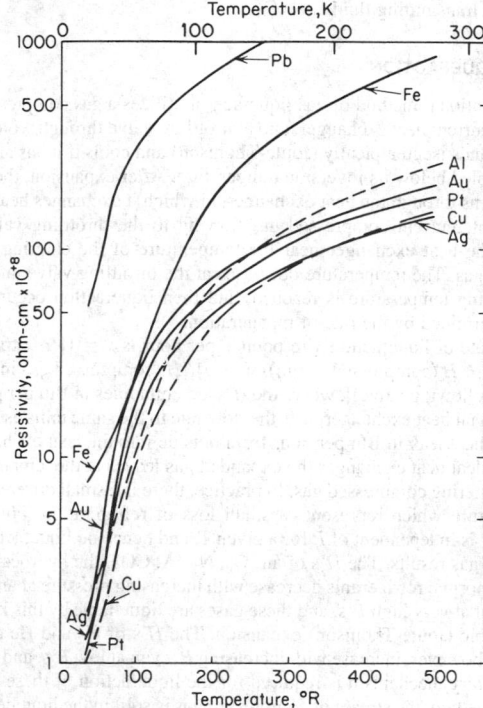

Fig. 19.2.6 Electric resistivity of solids at low temperatures.

Some metals, including elements (none from column 1 of the Mendeleeff table or the ferromagnetic elements), intermetallic compounds, metal alloys, and a number of oxides exhibit the phenomenon of **superconductivity,** characterized by a zero dc electrical resistance from $0°R$ up to a transition temperature T_c, at which normal resistance (value extrapolated from higher T) appears. In pure, single crystals the transition range ΔT, from $R_e = 0$ to the normal value, may be as small as a few millidegrees. Until the discovery of the so-called high-temperature superconductors (HTSCs), the higher T_c in a practical superconductor was that of the compound Nb_3Ge at $41.9°R$ (23.3 K). The search for higher T_c materials continues with T_c's that have already been reported as high as $239°R$ (133 K). Closed circuits consisting entirely of superconductors can support a persistent, resistanceless current without an external source of voltage. A superconducting circuit therefore maintains constant the value of the total magnetic flux that was enclosed by the circuit at the time it entered the superconducting state. Hence, superconducting materials influence the magnetic fields in their environment. For this reason, their use in the construction of equipment and apparatus must sometimes be avoided when expected temperatures could be below their T_c. Lead brasses and some solders, in particular Pb-Sn alloys, become superconductors.

Superconductivity, in materials designated Type 1 superconductor, is characterized by (1) perfect electric conductivity ($R_e = 0$) and (2) the internal magnetic induction $B = 0$ when H (external) is not zero (the Meissner effect). Persistent currents at the surface of the specimen shield the interior from H (external). H parallel to the surface of a specimen is continuous at the surface and falls off exponentially below the surface. The penetration depth of H for perfect (chemically and mechanically) specimens is of the order of 5×10^{-8} m. The penetration depth is larger for alloys, specimens with lattice imperfections, and all superconducting specimens near their superconducting transition temperatures T_c.

The normal-to-superconducting transition temperature T_c is also a function of H (external) as well as of the current being carried by the superconductor. For a Type I superconductor, the effect of H (external) is expressed by $T_c = T_{c0} [1 - (H_c/H_0)]^{1/2}$ where T_{c0} is the value of T_c at essentially zero H, and H_0 is the value of H_c at 0 K. For every $H < H_c$, a further decrease in T_c occurs when the superconductor is carrying a current, the maximum current density for maintaining the superconducting state being denoted J_c. Thus the superconducting state exists when $T < T_c$, $H < H_c$, and $J < J_c$, and the normal, or resistive, state exists when either $T > T_c$, $H > H_c$, or $J > J_c$. Because T_c decreases with increases in H and J, the practical application of a superconductor requires that it be maintained at a temperature considerably below its T_c.

Tables 19.2.1 and 19.2.2 are only representative. The number of superconductors including compounds and alloys runs into the hundreds. Of the chemical elements, 38 are known to become superconductors. Some that are not superconductors at normal pressures have high-pressure allotropic modifications that are superconductors, e.g., Bi, Si, and Ge at pressures of ~ 25, 130, and 120×10^8 Pa, respectively. Many of the superconducting alloys have nonsuperconducting constituents, and there are compounds all of whose constituent elements in their pure state are nonsuperconductors (see Table 19.2.2). It is interesting that good conductors Cu, Ag, and Au do not become superconductors to the lowest temperatures at which they have been tested (~ 0.01 K).

Type I superconductor is the designation given to those superconductors that exhibit a complete Meissner effect for $0 < H < H_c$; i.e., B = 0 until H_c is reached and penetration of the specimen by B becomes complete. The superconducting elemental metals in the pure and mechanically perfect state are Type I superconductors. Superconducting alloys (intermetallic compounds, solid solutions, and mixed phases) and work-hardened (unannealed) metals are Type II superconductors. For Type II superconductors, the Meissner effect is complete only for $0 < H < H_{c1}$ where H_{c1} is smaller than H_c. For $H_{c1} < H < H_{c2}$, the magnetic field penetrates the body of the specimen in the form of current vortices, called *fluxons*, that generate a fixed quantized amount of magnetic flux. Each fluxon is a cylindrical region of circulating current centered on a resistive core of material in which the superconductivity is suppressed. Hence the region defined by $H_{c1} < H < H_{c2}$ is called the *mixed state* of a Type II superconductor. The number of fluxons increases with H from zero at H_{c1} to a density at which complete overlap of the cores occurs along with the restoration of the normal state at H_{c2}. Thus H_{c2} is an upper limit to the superconducting state that increases as T decreases.

The alloys commonly used for high-field superconducting magnets (Nb_3Sn, NbTi, and NbZr, see Table 19.2.2) are Type II superconduc-

Table 19.2.1   Transition Temperature T_0 and Critical Magnetic Fields H_0 of Some Type I Superconductors

Superconductor	T_0, K	H_0, A/m*
Nb	9.25	1.57×10^5
Pb	7.23	6.4×10^4
V	5.31	8.8×10^4
Hg	4.154	3.3×10^4
Sn	3.722	2.4×10^4
In	3.405	2.2×10^4
Al	1.175	8.4×10^3
Mo	0.916	7.2×10^3
Zr	0.53	3.7×10^3
Ti	0.39	8.0×10^3
W	0.0154	9.2×10
Na Bi	2.2	The elements, in pure state at normal pressure, are not superconductors.
Au_2Bi	1.7	
Cu S	1.6	

* To convert to oersteds, divide by 79.57.

Table 19.2.2   Transition Temperatures T_0 and Upper Critical Fields H_{c2} of Some High-Field, Type II Superconductors

Superconductor	T_0, K	H_{c2}, A/m*	T, K, for H_{c2}
$Al_{0.75}Ge_{0.25}Nb_3$	18.5	3.4×10^7	4.2
Nb_3Sn	18.0	1.9×10^7	4.2
$N_{0.93}Nb$	15.9	1.3×10^7	0
GaV_3	14.8	1.9×10^7	0
NbZr	10.8	7.4×10^6	0
$Nb_{0.2}Ti_{0.8}$	7.5	6.4×10^6	4.2
$Ti_{0.6}V_{0.4}$	7.0	8.8×10^6	2

* To convert to oersteds, divide by 79.57.

tors. H_c's for Type I superconductors are small, whereas magnet alloy wires have H_{c2} values that are 100 to 1,000 times higher.

The fluxons in the mixed state of a Type II superconductor transporting a current are acted on by a Lorentz force that is proportional to $\mathbf{i} \times \mathbf{B}$ and acts in a direction perpendicular to $\mathbf{B}$, and to $\mathbf{i}$ (the current intensity). Unless the fluxons are pinned to crystal lattice sites they are propelled across the superconductor under the influence of the Lorentz force. This movement involves the performance of work by the current and results in (1) production of heat within the superconductor and (2) the appearance of electrical resistance to the flow of current and, eventually, the destruction of the superconducting state. Fluxons are pinned by lattice imperfections (chemical and mechanical) and their displacement is resisted until the Lorentz force exceeds a breakaway value, freeing the fluxons to move. Lattice imperfections are introduced in the wires for high-field superconducting magnets to pin fluxons as well as to increase H_{c2}. Currents avoid the normal core of a pinned fluxon and thus retain their resistanceless character.

A current exceeding the critical value destroys the superconducting state. This critical current is determined by the critical Lorentz force (the product of current and magnetic field) at which pinning forces are exceeded. Hence, the critical value of an applied field depends on the current transported by the superconductor and vice versa. In general, the larger the applied field, the smaller is the critical current and vice versa. For superconducting magnets, it is essential that the values of the limiting magnetic field and the limiting current be simultaneously large.

A cryogenic refrigerator operating on the Carnot cycle (ideal) will require the least amount of input work to effect refrigeration (remove heat) at a low temperature and reject the heat at a higher temperature (usually ambient temperature). The ideal ratio of input work to refrigeration produced (both expressed in watts) can be given as $(T_1 - T_2)/T_2$ (the Carnot efficiency) where T_1 represents the temperature at which the heat will be rejected and T_2 the refrigeration temperature. Cryogenic refrigerators built to date do not approach this limit very closely: typically less than 40 percent of Carnot efficiency for the largest refrigerators and progressively less as the size of the refrigerator becomes smaller. For the Types I and II superconductors discussed above, practical application requires operation in the temperature region of 12 K or lower. Thus the lower limit (ideal refrigeration) for the work to heat removal ratio is seen to be 24 for operation between ambient temperature and 12 K. Raising the operating temperature of a superconducting system to 20, 40, or 77 K (the temperature of liquid nitrogen boiling at normal atmospheric pressure) reduces this ratio to 14, 6.5, and 2.9 respectively. Although the actual refrigerator may take 2.5 or more times this ratio, the saving in input energy that can be effected by raising the operating temperature is obvious. The equipment for producing refrigeration at these higher temperatures is also less complicated and less expensive than that needed for producing refrigeration at the lower temperatures required for the operation of the low-temperature superconductors.

In 1986 a true breakthrough was achieved in producing superconductors able to enter the superconducting state at much higher temperatures. The HTSCs are oxide compounds consisting of several elements. They are more like ceramics and have a complicated crystal structure. Many of them exhibit the superconducting phenomenon only along

Cu-O planes and their performance is greatly influenced by the oxygen content of the compound. If the crystals are not properly oriented, the HTSCs have small sections of superconductor connected by rather poorly conducting sections. This has been referred to a "having weak links." Also the ability of these superconductors to conduct useful quantities of current can be severely limited by an imposed magnetic field at high temperatures. At temperatures below 30 K, their extraordinarily high H_{c2}'s allow them to carry high current densities at much higher magnetic fields than is possible with the conventional superconductors. The HTSC's are fabricated as thin films and multifilimentary wires. The necessity for optimum chemistry of the constituents, proper crystal orientation, and the fact that the ceramic-like nature of the superconductor makes it brittle complicate both the manufacture of the wires in useful lengths and the subsequent fabrication of equipment such as coils for motors, generators, etc. A great deal of research is continuing in the attempt to improve the commercial usefulness of the HTSCs. Table 19.2.3 lists several of the HTSCs along with pertinent properties. There is much active research on methods to produce HTSCs, including research on a number of other materials in addition to those listed in Table 19.2.3.

The **coefficient of linear expansion** α is $(1/L)(dL/dT)$. Thermal expansion of asymmetric crystals differs along different crystal axes. For an isotropic solid, the volume or cubical-expansion coefficient $(1/V)(dV/dT) = 3\alpha$. Expansion coefficients do not vary appreciably in the temperature region near ambient temperatures, but they all approach zero at 0°R, and the approach to zero is tangential to the T axis ($d\alpha/dT = 0$ at 0°R.) Some expansion coefficients are negative at low temperatures, e.g., stainless steel, some Invars, and fused quartz. The expansivity of some crystals are negative in some directions even though their volume coefficients are positive. Cold-working may produce differences in expansivity in different directions. Annealing restores isotropy. Figure 19.2.7 shows the relative change in length, the integral of α from T to 528°R (293 K).

Because the cold interiors of cryogenic equipment must at times be

Table 19.2.3 Properties and Status of some HTSCs

HTSC*	T_c, K	Comments
$La_{2-x}Ba_xCuO_4$	35	First HTSC discovered
$YBa_2Cu_3O_7$ (YBCO or 1-2-3)	95	First HTSC with T_c above normal boiling point of liquid nitrogen, $H_{c2} = 30$ T at 77 K; decreased interest due to weak link problem
$HgBa_2Ca_2Cu_3O_{8+\delta}$	133	Highest T_c to date
$Bi_2Sr_2Ca_2Cu_3O_8$ (2223 BSCCO)	110	$H_{c2} = 30$ T at 77 K, $J_c > 2 \times 10^4$ A/cm² at 77 K and $H_{ext} = 0$; industrial pilot-plant-scale production
$Tl_2Ba_2CaCu_2O_8$ (2212 TBCCO)	~115	$J_c = 7 \times 10^6$ A/cm² at 75 K and $H_{ext} = 0$

* These HTSCs are representative of classes of superconductors.

warmed to ambient temperature, provisions have to be made for those changes in dimensions of the interior that results from large changes in T. Even for equipment of ordinary size, these changes can be too large for accommodation within the elastic limits of the materials of construction or of the structure. In such situations, flexibility has to be designed into the structure. For example, bellows and U bends are commonly employed to obtain this flexibility in insulated transfer lines for liquid cryogens.

The **mechanical properties** important for the design and construction of cryogenic equipment are the same as for other temperatures (see also Sec. 5). Frequently, the choice of materials for the construction of cryogenic equipment will depend upon other considerations besides mechanical strength, e.g., lightness (density or weight), thermal conductivity (heat transfer along structural support members), and thermal expansivity (change of dimensions when cycling between ambient and low temperatures). Frequently, mechanical properties at low temperature are significantly different from properties at ambient temperature. This makes room-temperature data unreliable for engineering use at low temperatures.

It is not possible to make generalizations that would not have numerous exceptions about the temperature variations of the mechanical properties. In this discussion, Figs. 19.2.8 to 19.2.10 are only illustrative guides. There is no substitute for test data on a truly representative specimen when designing for the limit of effectiveness of a cryogenic material or structure. Just as the mechanical properties at ambient temperatures are dependent upon the impurities (metallic and nonmetallic), their chemical nature and concentration, the thermal history of the specimen, the amount and kind of working (microstructure and dislocations of the lattice), and the rate of loading and type of stress (uni-, bi-, and triaxial), so also are the changes, quantitative and qualitative, in the mechanical properties when changing the temperature from ambient to low.

Metals, nonmetallic solids (glass), plastics, and elastomers are discussed in order. The metals are classed by their lattice (crystal) symmetry.

The **f.c.c. metals** and their alloys are most commonly used for the construction of cryogenic equipment. Al, Cu, Ni, their alloys, and the austenitic stainless steels of the 18-8 type (300 series) are f.c.c. They do not exhibit an impact (or a notched-tensile) ductile-to-brittle transition at low temperatures. As a general rule, which has some exceptions, the mechanical properties of these metals improve as T is reduced: (1) Young's modulus at 40°R (22 K) is 5 to 20 percent larger than at 530°R (294 K), (2) the yield strength at 40°R (22 K) is considerably greater than the strength at 530°R (294 K). (Cu is an exception; see Fig. 19.2.9), and (3) the fatigue properties at low T are improved. There is a large difference between the yield strength and the ultimate tensile strength of these metals and alloys, especially when they have been annealed. Pb (f.c.c.) and In (face centered tetragonal) are used for low-T deformable gaskets because of their creep properties. Low-T creep data are meager, but the rate of creep decreases with T.

Beta brass (f.c.c.) is ductile down to 7°R, though it is like all alloys of

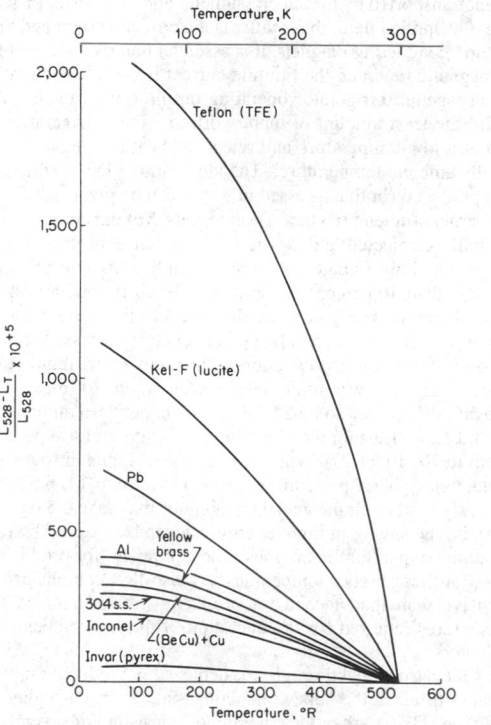

Fig. 19.2.7 Relative change in length from T to 528°R (293 K), equal to $\int_t^{528°R} \alpha \, dt$. Bracketed materials are within ~ 5 percent of the curve shown.

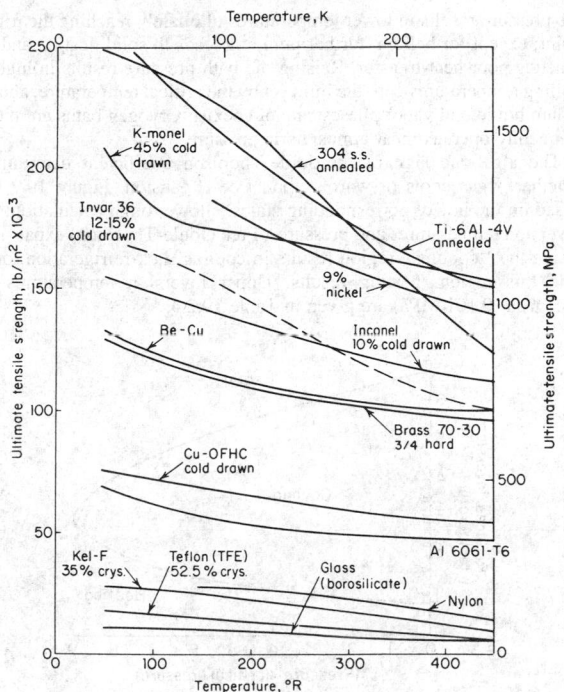

Fig. 19.2.8 Ultimate tensile strength of solids at low temperatures.

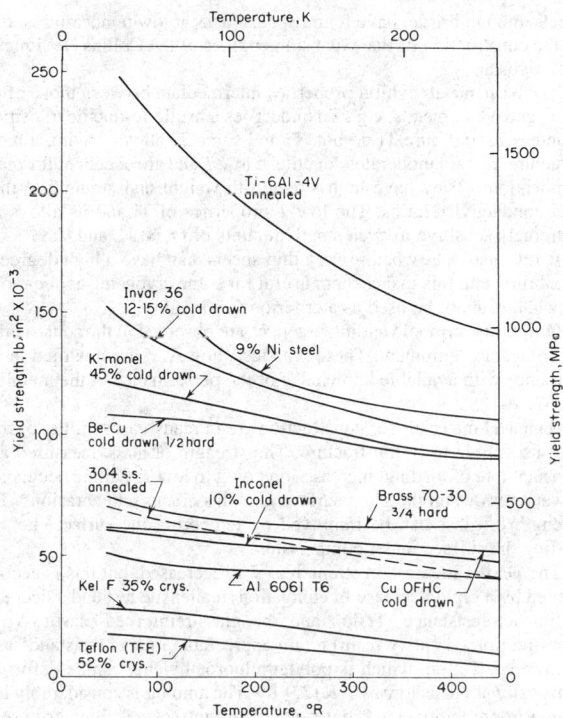

Fig. 19.2.9 Yield strength of solids at low temperatures.

Cu in being less ductile than Cu itself. Thin brass is sometimes porous, and this limits its usefulness for high-vacuum enclosures at low temperatures. Free-machining brass normally contains Pb, which even in low concentrations can render the brass superconducting at LHe temperatures. In the superconducting state, it may then affect the magnetic field in its vicinity (see electrical properties, above).

The b.c.c. **metals and alloys** are normally classed as undesirable. These include Fe, the martensitic steels (low carbon and the 400 series of stainless steels), Mo, and Nb. If not brittle at room temperature, they have a ductile-to-brittle transition at low T. Working can induce the austenite-to-martensite transition in some steels. AISI 301 austenitic

stainless is an example. It remains moderately "tough" at very low T and is a valuable material for construction, but cold-drawing, in excess of 70 percent strain, induces a partial transformation of structure and reduces the elongation and notched-strength ratio to nearly zero. This same type of reduction in toughness is observed in the "heat-treatable" stainless steels. Improved tensile properties can be obtained by cold-working and by heat-treating. These alloys usually have decreased toughness at low temperatures. Alloys of V, Nb and Ta, although not f.c.c., behave well as regards brittleness at low temperatures. These alloys have the advantage of being suitable for use at high T. Carbon

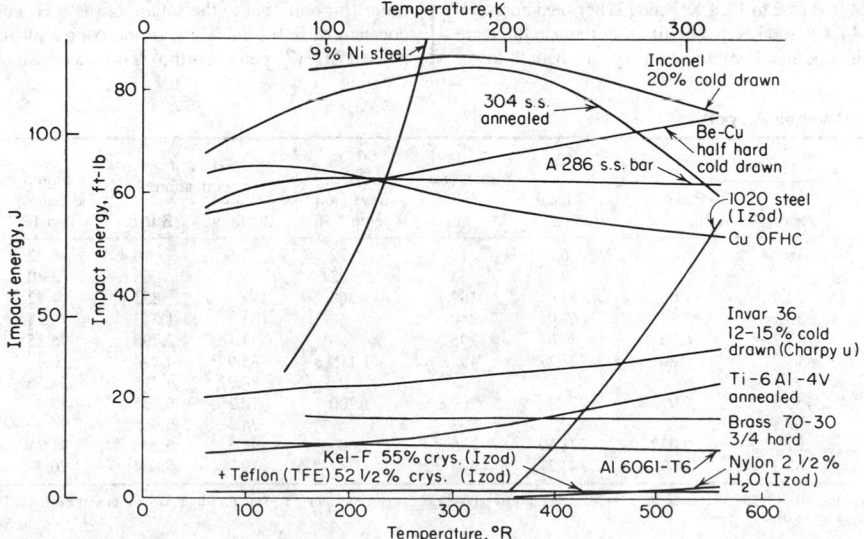

Fig. 19.2.10 Impact energy for solids at low temperatures (Charpy V unless noted). For Kel-F, nylon, and Teflon, the impact energy units are foot-pound per inch of notch width.

steels, though brittle, have found special uses at low temperatures, e.g., in the construction of the expansion engines of the Collins He liquefier and cryostat.

The **h.c.p. metals** exhibit properties intermediate between those of the f.c.c. and b.c.c. metals; e.g., Zn undergoes a brittle-to-ductile transition, whereas Zr and pure Ti do not. Ti and some Ti alloys, having a h.c.p. structure, remain moderately ductile at low T and are excellent for many applications. They have high strength-to-weight and strength-to-thermal-conductivity ratios. The low-T properties of Ti and its alloys are extremely sensitive to even small amounts of O, N, H, and C.

Brittle materials, when in very thin sheets, can have a high degree of flexibility, and this makes them useful for some cryogenic applications. **Flexibility** cannot be used as a criterion for ductility.

Ordinarily, normal welding practices are observed in the construction of cryogenic equipment. These practices, however, are modified in accordance with available knowledge of the performance of the metals at low T.

Nonmetal materials for construction are in many cases brittle, or they are susceptible to brittle fracture. The strength of **glass**, measured at a constant rate of loading, increases on going to low T. Failure occurs at a lower stress when the glass surface contains cracks and abrasions. The strength of glass can be improved by tempering the surface, i.e., by putting the surface under compression.

The **plastics** increase in strength as T is decreased, but this is accompanied by a rapid decrease in elongation in a tensile test and a decrease in impact resistance. Teflon and the glass-reinforced plastics (e.g., glass-reinforced epoxy resin) retain appreciable impact resistance as T is lowered. Teflon, which is polytetrafluoroethylene, can be deformed plastically at T's as low as 7°R (3.9 K). The amount is considerably less than at room temperature, but it is enough to make Teflon very useful for some cryogenic applications. The glass-reinforced epoxies, besides having appreciable impact resistance at low T's, also have high strength-to-weight and strength-to-thermal-conductivity ratios.

All the **elastomers** become brittle at low T. However, elastomers like natural rubber, nitrile rubber, Viton A, and plastics such as Mylar and nylon that become brittle at low T can be used for static seal gaskets when *highly compressed* at room temperature, prior to cooling.

PROPERTIES OF CRYOGENIC FLUIDS (CRYOGENS)

Table 19.2.4 indicates the **temperature ranges** accessible with liquid cryogen baths. There are two inaccessible ranges: (1) between helium and hydrogen [9.36 to 24.9°R (5.2 to 13.8 K)] and (2) between neon and oxygen [80 to 97.9°R (44.4 to 54.4 K)]. The limiting temperatures are set by the critical and triple points. Pumping on a cryogen bath to lower

the pressure results in lower temperatures ultimately reaching the triple point; except for helium, further pumping leads to solidification and in practice poor heat-transfer. Raising the bath pressure results in higher boiling temperatures with the limit set by the critical temperature, above which liquid and vapor phases cannot coexist. Cryogen baths are most frequently operated near atmospheric pressure.

The algebraic sign of the Joule-Thomson coefficient determines whether a gas cools or warms upon free expansion. Figure 19.2.11, based on the law of corresponding states, allows rough calculations of inversion temperatures and pressures. Free (Joule-Thomson) expansion inside the "cooling" region results in cooling, i.e., refrigeration; outside this region, heating results. Upper inversion temperatures at 14.7 psia (0.101 MPa) are given in Table 19.2.4.

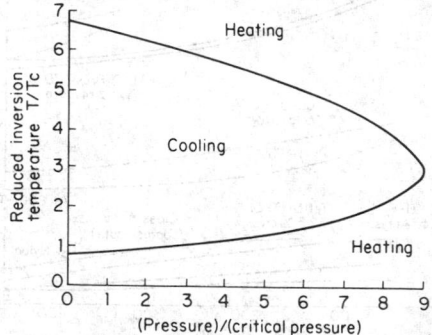

Fig. 19.2.11 Reduced inversion temperature versus reduced pressure.

Volumetric latent heats (Table 19.2.4) decrease with the lower-boiling cryogens and emphasize the importance of the insulation in storing and handling LH$_2$ and LHe.

Two forms of **hydrogen** and **deuterium** exist: **ortho** and **para**. The ortho and para molecules differ in the relative orientation of the nuclear spins of the two atoms composing the diatomic molecule. In the ortho form of H$_2$ the nuclear spins of the two atoms in the molecule are parallel (in the same direction), whereas in the para form the spins are antiparallel. The relative orientations for ortho and para D$_2$ differ from those for H$_2$. The thermodynamic equilibrium composition of ortho and para varieties is temperature dependent as shown in Fig. 19.2.12.

In liquid hydrogen, the uncatalyzed ortho-para reaction is second order (proportional to the square of the o-H$_2$ concentration) with a rate constant of 0.0114/h. Thus, in the course of uncatalyzed liquefaction, normal LH$_2$ (75 percent ortho) is produced. Since the heat of conversion

Table 19.2.4 Common Cryogen Properties

Cryogen	Boiling point, °R	Triple-point Temp., °R	Triple-point Pressure, lb/in² abs	Critical temp., °R	Critical pressure, lb/in² abs	Upper inversion temp.,* °R	Heat of vaporization* Btu/lbm	Heat of vaporization* Btu/ft³	Liquid density,† lbm/ft³	Vapor density,† lbm/ft³	Gas density,‡ lbm/ft³
He³	5.7			6	17	72	3.6	13.4	3.72	1.50	0.0084
He⁴	7.6			9.36	33.2	92	8.8	68.6	7.80	1.04	0.0111
H₂ (equilib.)	36.5	24.9	1.02	59.4	187.7	368	192	853	4.42	0.0837	0.00561
D₂ (normal)	42.6	33.7	2.49	69.0	240		131	1,343	10.25	0.14	0.0112
Ne	48.7	44.2	6.26	80.0	395.3	486	37.1	2,790	75.35	0.58	0.056
N₂	139.2	113.7	1.86	227.27	492.3	1,115	85.9	4,294	50.4	0.28	0.078
Air	(141.8)						88.22	4,813	54.56	0.28	0.0807
A	157.1	150.8	9.97	271.3	709.8	1,300	70.2	6,135	87.4	0.36	0.111
F₂	151	(96.4)§		259.3	808		74.1	6,965	94	0.50	0.106
O₂	162.3	97.9	0.022	278.59	736.3	1,605	91.5	6,538	71.24	0.28	0.089
CH₄	201.1	163.2	1.69	343.27	673		219.2	5,804	26.5	0.115	0.0448

NOTE: To convert °R to K, multiply °R by 0.556; to convert lb/in² to MPa, multiply by 0.00689; to convert Btu/lbm to J/g, multiply by 2.326; to convert Btu/ft³ to J/m³, multiply by 3.73 × 10⁴; to convert lbm/ft³ to kg/m³, multiply by 16.01.

* 14.7 lb/in² abs.
† At normal boiling point.
‡ At 14.7 lb/in² abs and 492°R.
§ Melting point, 14.7 lb/in² abs.

of o-H_2 to p-H_2 at the normal boiling point is 302 Btu/lbm (702 J/g) (greater than the heat of vaporization; see Table 19.2.4), long-term storage of normal LH_2 is impractical. Therefore, in the commercial production of LH_2 a catalyst is used to produce LH_2 with more than 95 percent para which for practical purposes is equivalent to the equilibrium composition (99.8 percent para).

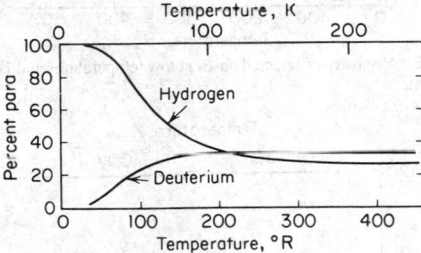

Fig. 19.2.12 Thermodynamic equilibrium composition of ortho and para varieties of H_2 and D_2; percent ortho = 100 − percent para.

For pressures of no more than 150 lb/in² (1.03 MPa) and temperatures at least twice the critical temperature, the ideal gas law ($PV = nRT$) enables one to calculate PVT data with sufficient accuracy for many engineering purposes. For cases in which the ideal gas law is not adequate or in which experimental data are not available, the procedures outlined in Sec. 4 may be used. The theoretical prediction of PVT data for liquids is considerably more difficult than for gases. Consequently no universal equation analogous to the perfect gas law exists for extensive ranges of P and T.

Reduced quantum mechanical correlations of saturated liquid densities, viscosities, and thermal conductivities for several cryogens are shown in Figs. 19.2.13 to 19.2.15. Dashed lines represent areas void of experimental data.

Utilization of Figs. 19.2.13 to 19.2.15 requires knowledge of the

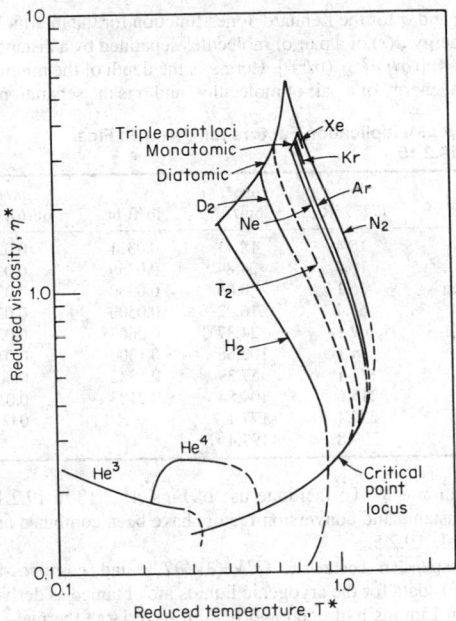

Fig. 19.2.14 Reduced viscosity versus reduced temperature along the saturation curve for cryogenic liquids. Values of T^* are obtained by dividing the desired temperature (°R) by the value of T/T^* given in Table 19.2.5. Reduced viscosity η^* must be multiplied by η/η^* given in Table 19.2.5 to obtain viscosity in lbm/ft · h. Multiply lbm/ft · h by 0.000413 to obtain Pa · s.

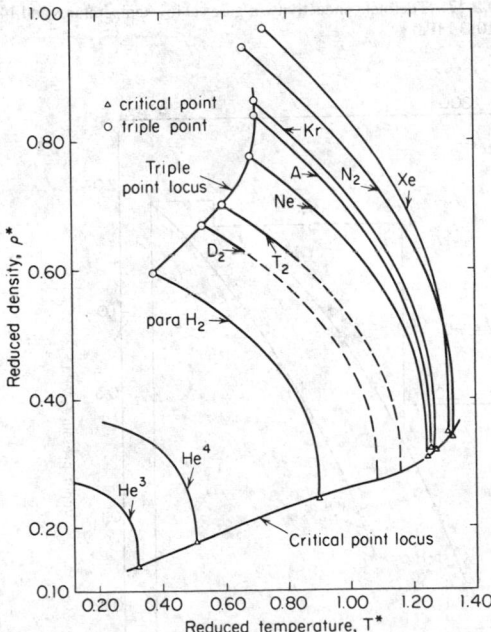

Fig. 19.2.13 Reduced density versus reduced temperature along the saturation curve for cryogenic liquids. Values of T^* are obtained by dividing the desired temperature (°R) by the value of T/T^* given in Table 19.2.5. Reduced density ρ^* must be multiplied by ρ/ρ^* given in Table 19.2.5 to obtain density in lbm/ft³. Multiply lbm/ft³ by 16.01 to obtain kg/m³ and °R by 0.556 to obtain K.

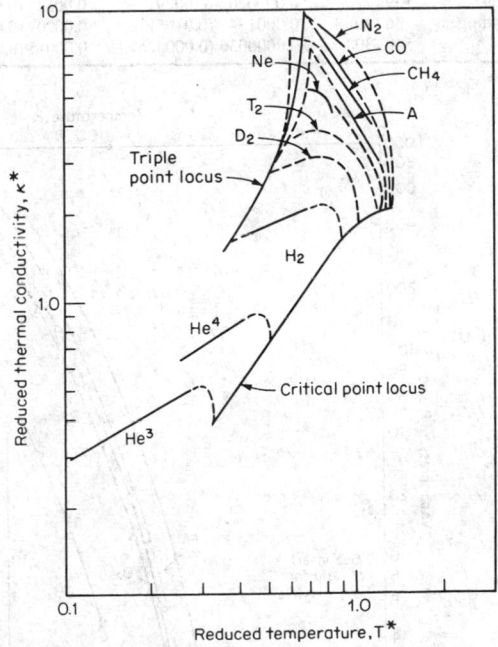

Fig. 19.2.15 Reduced thermal conductivity versus reduced temperature along the saturation curve for cryogenic liquids. Values of T^* are obtained by dividing the desired temperature (°R) by the value of T/T^* given in Table 19.2.5. Reduced thermal conductivity κ^* must be multiplied by κ/κ^* given in Table 19.2.5 to obtain thermal conductivity in Btu/h · ft · °R. Multiply by 1.73 to obtain W/m · K.

constants ε and σ for the Lennard-Jones function for the intermolecular potential energy $\varphi(r)$ of a pair of molecules separated by a distance r, so that $\varphi(r) = 4\varepsilon[(\sigma/r)^{12} - (\sigma/r)^6]$. Here ε is the depth of the minimum in the potential energy of a pair of molecules, and σ is the separation of the

Table 19.2.5 Multiplication Factors for Use with Figs. 19.2.13 to 19.2.15

Substance	T/T^*, °R	ρ/ρ^*, lbm/ft³	η/η^*, lb/ft-hr	κ/κ^*, Btu/(h·ft·°R)
Helium 3	18.4	18.71	0.0311	0.0204
Helium 4	18.4	24.84	0.0359	0.0178
Para-hydrogen	66.1	8.06	0.0360	0.0363
Deuterium	63.4	16.22	0.0500	0.0257
Tritium	62.1	24.37	0.0607	0.0210
Neon	64.1	100.66	0.1300	0.0128
Nitrogen	171.1	57.39	0.1382	0.0098
Argon	215.6	104.84	0.2183	0.0109
Krypton	297.4	177.1		0.0080
Xenon	397.8	197.4		

pair at which $\varphi = 0$. To facilitate use of Figs. 19.2.13 to 19.2.15, all required constants and conversion factors have been combined and are listed in Table 19.2.5.

Thermal-expansion coefficient $(1/V)(\partial V/\partial T)_P$ and **compressibility** $(1/V)(\partial V/\partial P)_T$ data for the cryogenic liquids are obtained as derivatives of PVT data. Liquids hydrogen and helium have large thermal-expansion and compressibility coefficients (see Table 19.2.6).

Table 19.2.6 Compressibility and Thermal-Expansion Coefficients. Coefficients for Liquid Nitrogen, Hydrogen, and Helium at 14.7 lb/in² abs (0.101 MPa)

Cryogen	T, °R (K)	Compressibility, $(1/V)(\partial V/\partial P)_T$, 1/(lb/in² abs) (1/MPa)		Thermal expansion, $(1/V)(\partial V/\partial T)_p$, 1/°R (1/K)	
Helium 4	7 (4)	0.0028	(0.4)	0.085	(0.153)
Nitrogen	139 (77)	0.000020	(0.0029)	0.0032	(0.0058)
Para-hydrogen	36 (20)	0.000134	(0.0195)	0.0090	(0.0162)
Water	540 (300)	0.0000036	(0.000526)	0.00015	(0.00027)

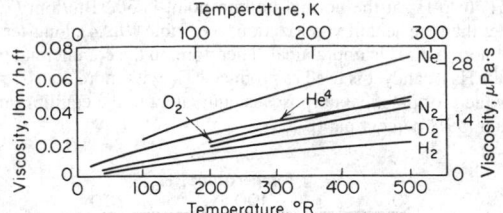

Fig. 19.2.16 Viscosity of selected gases at low temperatures and 14.7 lb/in² abs (0.1013 MPa).

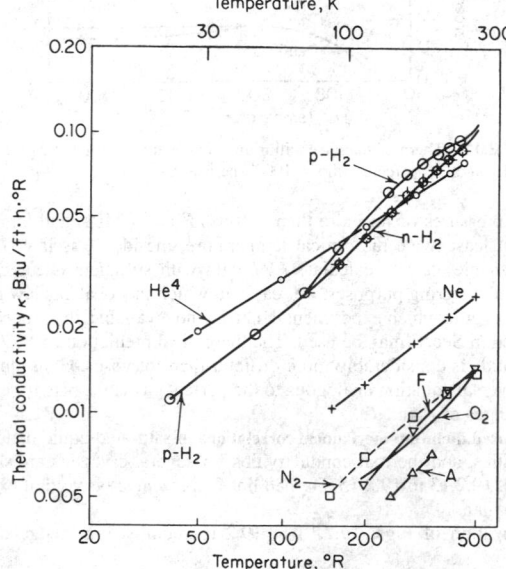

Fig. 19.2.17 Thermal conductivity of gases at low temperatures and 14.7 lb/in² abs (0.1013 MPa).

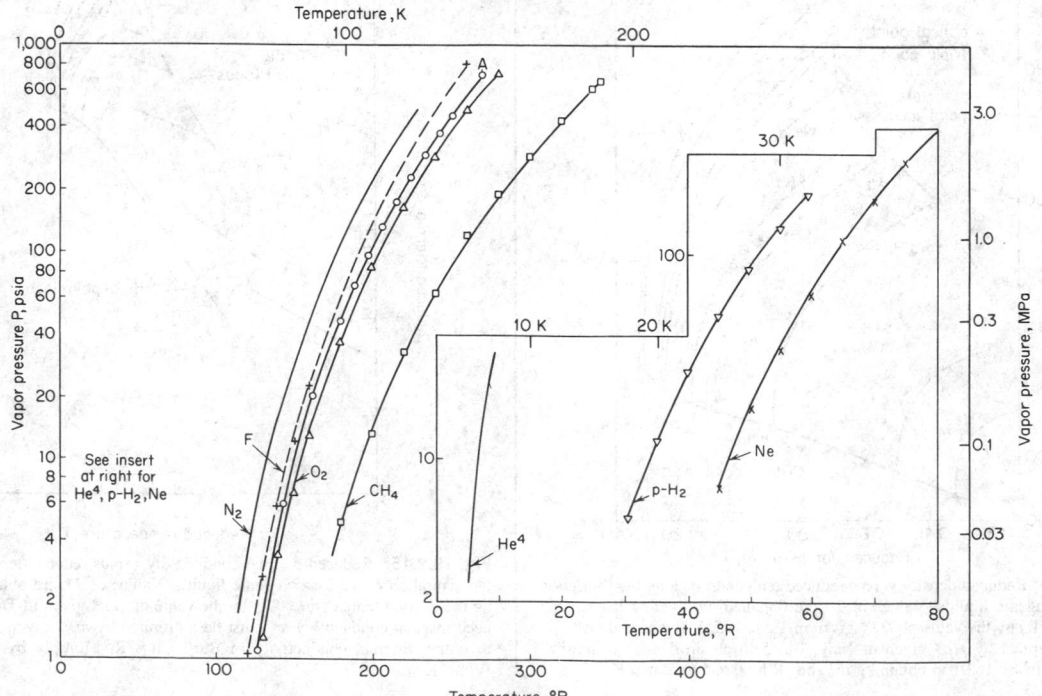

Fig. 19.2.18 Vapor pressures of the common cryogens.

For gases, kinetic theory predicts an increase of **viscosity** with rising temperature as shown in Fig. 19.2.16 ($\eta \sim \sqrt{T}$). Normally, the **thermal conductivities** of liquids decrease with increasing temperature, whereas the thermal conductivities of gases increase. Fig. 19.2.17 presents thermal conductivity for several gases as a function of temperature at 14.7 psia (0.1013 MPa).

Figure 19.2.18 is a graph of **vapor pressures** of the common cryogens.

Figure 19.2.19 presents **specific heats** at constant pressure for several of the cryogenic liquids, while Fig. 19.2.20 presents the specific heats for liquid hydrogen and helium along the saturation curve. Specific heats of gases are given in Fig. 19.2.21.

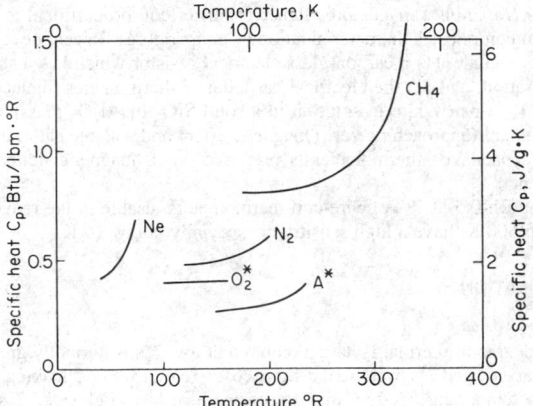

Fig. 19.2.19 Specific heat at constant pressure (C_p) for liquid cryogens at saturation pressure. Asterisk means C_p evaluated along an isobar at 14.7 lb/in² abs (0.1013 MPa) instead of saturation pressure.

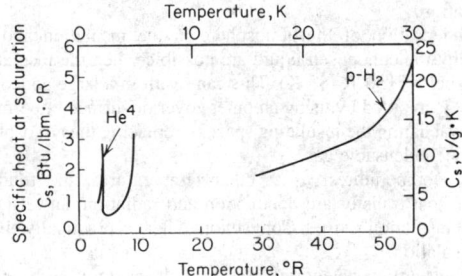

Fig. 19.2.20 Specific heats at saturation C_s along the saturation curve versus temperature for ^{4}He and para hydrogen.

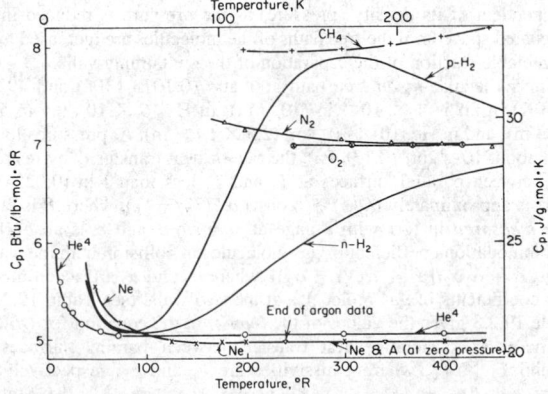

Fig. 19.2.21 Specific heat C_p of cryogen gases at 14.7 lb/in² abs (0.1013 MPa) versus temperature.

INSTRUMENTATION

(See also Secs. 15 and 16.)

The usual instrumentation problems of engineering practice are further complicated by low-temperature problems which require special calibration procedures.

Pressures are measured with normally used apparatus, e.g., bourdon gages, transducers, and manometers. If the measuring device is located outside the cryogenic environment and insulated from it by a thermal barrier, no special problems are introduced. Occasionally response time requirements necessitate the placing of the transducer in the cryogen space. Then the problems of temperature compensation and calibration must be considered.

Level measurements currently utilize several methods. The differential-pressure method can in principle be used with all cryogens. With cryogens of low densities and low heats of vaporization (e.g., H_2 and He) pressure oscillation occurs in the liquid leg. In the case of LH_2 these can be eliminated by the proper design of the liquid leg and the introduction of He gas which does not condense.

Direct weighing is also used to determine "levels," although difficulties are encountered owing to the large tare-to-cryogen weight ratios (as large as 10 to 1 for a full container in the case of H_2) and extraneous loadings introduced by permanently connected piping, frost, and wind. Weighing systems have nevertheless been successfully used on 50,000-gal (190 m³) LH_2 dewars with an accuracy in level changes of 250 gal (1 m³).

With the **capacitance gage,** the sensing element consists of a concentric tube capacitor which can be calibrated to give liquid-level measurements accurate to a few tenths of a percent. An inexpensive, accurate **point-level sensor** is easily fabricated from a carbon resistor, typically $\frac{1}{10}$ W, 1,500 Ω. Its use depends upon the fact that its resistance is a strong function of temperature (Fig. 19.2.22). A small current (50 to 80 mA) is passed through the resistor to heat it. Since heat from the resistor is dissipated more readily in the liquid than in the gas, the

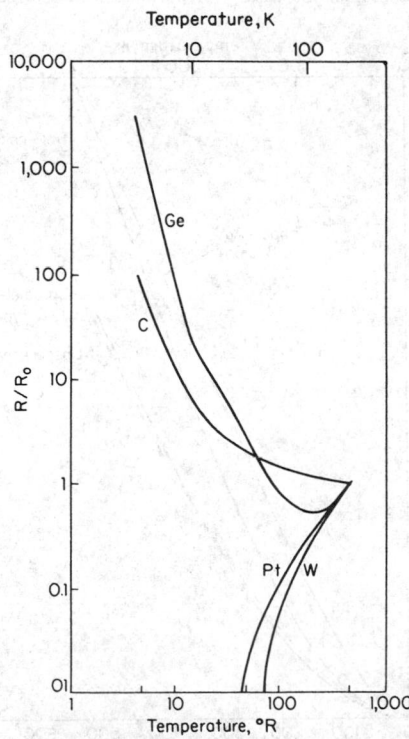

Fig. 19.2.22 Ratio of resistance R to the resistance R_0 at 492°R (273 K) versus temperature for several resistance thermometers. (Values for germanium and carbon are representative only.)

temperature of the resistor undergoes a step increase when the resistor environment changes from liquid to gas, whereas the resistance decreases stepwise.

Flow measurements utilize orifices, venturis, and turbine-type meters. Various devices for direct mass-flow measurement exist, but none are widely used. "Quality" meters have been built to determine the percentage liquid in two-phase flow. The determination of quality is complicated by nonequilibrium of temperature in the gas phase.

Temperature measurements are usually made with thermocouples, resistance thermometers, and vapor-pressure thermometers.

The **vapor-pressure thermometer** depends upon a vapor-pressure-temperature relationship (see Fig. 19.2.18). In general, a small cavity is filled with a gas which has a condensation temperature in the vicinity of the temperature to be measured. If sufficient gas is present, the pressure will be that of the liquid vapor pressure at the coldest part of the measuring system. Maximum speed of response is obtained if the total quantity of gas is minimized to permit only a thin film of liquid to form. In the case of H_2, a catalyst (e.g., iron hydroxide) to promote ortho-para conversion should be included in the bulb for an accurate measurement since the vapor pressure of H_2 is dependent upon its ortho-para composition (see above).

The pressure at the surface of a liquid cryogen in a dewar may, under conditions of thermal equilibrium, be used to measure the temperature of the cryogen and of apparatus immersed in it. Corrections for the pressure of the hydrostatic head of liquid may be required. If using this method, it should be realized that large vertical temperature gradients can exist in an unstirred liquid cryogen in a vessel with good insulation.

Thermocouples are favored for the measurement of temperatures because of their low cost, ease of application, and rapid response. The thermoelectric powers (temperature sensitivities) of the thermocouples commonly used at higher T's decrease with decreasing temperature, and spurious emfs generated in wires of non-uniform composition are troublesome. The proper use of a reference junction at a known fixed temperature close to the temperature being measured is often advantageous for accuracy. Variability in composition of thermocouple alloys makes individual calibrations necessary for accurate results. Figure 19.2.23

gives typical thermocouple emf-versus-temperature curves for some common thermocouples. A thermocouple consisting of gold with 0.07-at % iron versus chromel has been found useful over the entire temperature range from 2 to 300 K. Its sensitivity increases from 11 μV/K at 4 K to 17 μV/K at 77 K to 22 μV/K at 300 K.

Radio carbon resistors (*Allen Bradley* for 2 to 100 K and *Speer Carbon* for 0.01 to 4.2 K) are much used as temperature sensors because of their low cost and high sensitivity. However, the permanence and stability of their calibrations are not very good, and germanium (doped) resistance thermometers are preferred (0.01 to 100 K) for their good reproducibility of calibration, although they are considerably more expensive.

The carbon and germanium resistors (thermometers) in magnetic fields H exhibit a magnetoresistance (Hall) effect, proportional to H^2, which complicates the determination of temperatures. Recent developments include: (1) a carbon-glass electrical resistor which has a stable calibration; and (2) the electrical capacitance thermometer (dielectric: $SrTiO_3$, a perovskite in solution in a solid SiO_2 or Al_2O_3 glass) with very small (approaching zero) magnetic effect and a stable calibration. The capacitance thermometer is preferred when magnetic fields are present.

The Rh-0.5-at % Fe, wire-coil thermometers, usable in the range of 0.1 to 300 K, have a high sensitivity especially below 10 K.

INSULATION

(See also Sec. 6.)

The degree of thermal isolation required at low T's is normally greater than at elevated T's because it is more costly to remove heat leaking into a low-temperature system than to replace heat lost at elevated T's, as demonstrated by the Carnot cycle. Other differences between insulating for low and elevated temperatures are:

1. Condensation of moisture in low-temperature insulation is possible. When this occurs, the conductance of the insulation is significantly increased. This problem is avoided by a vapor barrier on the outside of the insulation.

2. Condensation of the atmosphere in the insulation and surface washing by the condensed liquid are possible when the insulating surfaces are below 150°R (83 K). This can result in a large added transfer of heat. It is avoided by using an outer cover or surface impermeable to air, and evacuating the insulating space or replacing the atmosphere in it with a noncondensable gas.

At low temperatures, as at other temperatures, the fundamental modes of heat transfer are conduction and radiation, and it is against these that insulation is used. Convection of heat is practically eliminated by the insulation.

Low-T insulation categories are (1) high vacuum, with or without multiple radiation shields, (2) powders, (3) rigid foams, and (4) low-conductivity solids, such as balsa wood and corkboard.

1. (i) **Heat is transferred across an evacuated space** by radiation and by conduction through the residual gas. The conductivity of a gas is almost independent of its density (pressure) as the pressure is reduced in an evacuated space until the free paths of the molecules are increased to an appreciable fraction of the separation of the containing walls.

The molecular *mean* free path at 1 atm (0.1013 MPa) and 492°R (273 K) in air is 4×10^{-6} in (10^{-7} m); in H_2, 6×10^{-6} in (1.5×10^{-7} m); and in He, 10×10^{-6} in (2.5×10^{-7} m). At pressures lower than about 10^{-3} torr (0.133 Pa), the rate of heat transfer Q by residual gas between parallel surfaces at T_1 and T_2 less than 1 in (0.0254 m) apart is approximately $(Q/A) \approx (\text{const}) \, \alpha P(T_2 - T_1)$, where P is pressure *measured* in torr with a gage at *ambient T* and α is an *overall* accommodation coefficient of gas molecules in collision with the walls. Thus, $\alpha = \alpha_1\alpha_2/[\alpha_2 + \alpha_1 (1 - \alpha_2)]$, where α_1 and α_2 are accommodation coefficients of gas molecules at the two walls (see Table 19.2.7). Table 19.2.8 gives the values of the *(const)* in the equation for Q/A.

The **rate of radiative heat transfer** between parallel surfaces at T_1 and $T_2 > T_1$, where emissivities are e_1 and e_2, respectively, is $Q = e_1e_2s/[e_2 + (1 - e_2)e_1] \, [T_2^4 - T_1^4]$, where s is the Stefan-Boltzmann constant whose value is 1.712×10^{-9} Btu/ft$^2 \cdot$ h $\cdot$ °R^4

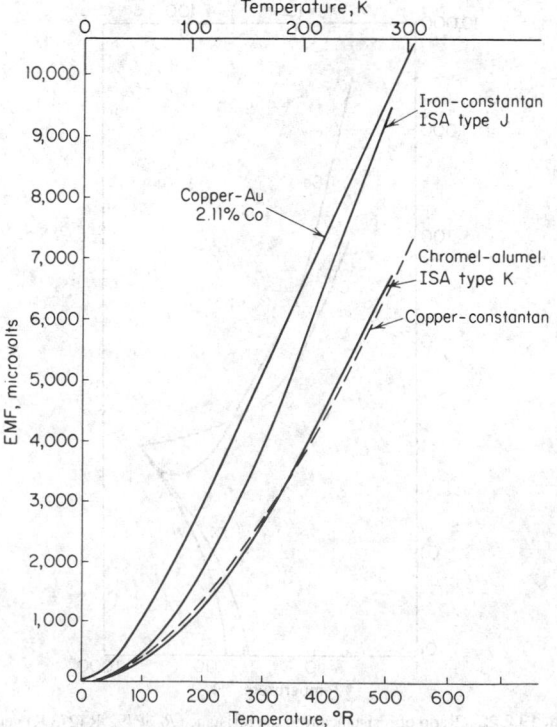

Fig. 19.2.23 EMF versus temperature for several thermocouples.

Table 19.2.7   Approximate Values of Accomodation Coefficients, α

Temp, °R (K)	He	H_2	Air
540 (300)	0.3	0.3	0.8–0.9
140 (78)	0.6	0.5	1
35 (19)	0.6	1	1

Table 19.2.8   Constant in the Gas-Conduction Equation

$Q/A \approx (\text{const}) \, \alpha P(T_2 - T_1)$

Gas	T_2 and T_1, °R (K)	Const, Btu/h · ft² · torr · °R (J/N · s · K)
N_2	<700 (389)	28 (1.19)
O_2	<540 (300)	26 (1.11)
H_2	540 and 140 (300 and 78)	93 (3.96)
H_2	140 and 36 (78 and 20)	70 (2.98)
He	Any	49 (2.09)

NOTE: T_1—inner wall (cold). T_2—outer wall (hot).

$(5.67 \times 10^{-8} \text{ W/m}^2\text{K}^4)$ (see also Sec. 4). Table 19.2.9 contains emissivities for 540°R (300 K) thermal radiation; these for most engineering purposes may be considered equivalent to minimal values of the e's for very carefully prepared and cleaned surfaces. Organic coatings have high emissivities approaching unity (>0.9). Mechanically polished surfaces have higher e's than the minimal values. Handling a surface transfers an absorbing (high e) coating to it.

(ii) **Refrigerated Radiation Shield**   An LN_2 (140°R) (78 K) cooled surface interposed in the insulating vacuum space of an LH_2 or an LHe container reduces the heat transferred by radiation to the LH_2 or LHe by a factor of about 250 when compared with that transferred by a surface of the same emissivity at room temperature without the interposition of the refrigerated shield. Nearly all the heat radiated by the surface at room temperature is absorbed by the LN_2, and this results in evaporation of the relatively inexpensive LN_2.

(iii) **A floating radiation shield** is an opaque layer (e.g., a metal sheet), having surfaces of low emissivity suspended in the vacuum space with a minimum of thermal contacts with surfaces that are warmer or cooler. If the emissivity of the surfaces of the floating shield is the same as the emissivity of the vacuum-space walls, the floating shield decreases the *radiative* heat transfer by a factor of 2. For m floating shields arranged in series between the inner cold and outer warm surfaces, the rate of *radiative* heat transfer becomes Q_m (m floating shields) = $[1/(m + 1)]Q_0$ where Q_0 = heat transfer for $m = 0$.

(iv) **Multilayer (Super) Insulation**   The principle of multiple floating radiation shields has been extended to the use of many thin metal layers (up to 75 layers per inch or 3,000 layers per meter) separated by thin thermal insulation. Best results have been obtained with Al foil separated by glass-fiber paper. Other materials have been used successfully, e.g., nylon nets as separators and aluminized Mylar with no spacer

Table 19.2.9   Minimal Values of Emissivity of Metal Surfaces at Various Temperatures for 540°R (300 K) Thermal Radiation

Surfaces	Surface temp, °R (K) 7 (3.9)	140 (78)	540 (300)
Copper	0.005	0.008	0.018
Silver	0.0044	0.008	0.02
Aluminum	0.011	0.018	0.03
Chromium		0.08	0.08
Nickel		0.022	0.04
Brass	0.018		0.035
Stainless steel 18-8		0.048	0.08
50 Pb-50 Sn solder		0.032	
Glass, paints, carbon			>0.9
Nickel plate on copper		0.033	

material. The *apparent* thermal *conductivity* of Al foil multilayer insulation spaced with glass fiber paper is variable depending on the number of layers of foil, the thickness of the paper, and the thickness and compacting of the insulating layer. For $\Delta T = (540 - 36)$°R or $(300 - 20)$ K, apparent mean conductivities range from 2.0 to 4.0×10^{-5} Btu/h · ft · °R (3.5 to 7×10^{-5} J/s · m · K). For $\Delta T = [540 - 140$°R $(300 - 78)$ K], the values are about one-third larger. Using nylon nets, glass fabric or glass fibers for separating the Al foil increases the conductance from three to seven times. The above conductivities are for the direction normal to the foil surfaces. Lateral conductivities are many thousands of times larger. An advantage of aluminized Mylar is its reduced lateral conductivity. Another advantage is its relatively low density (one-third to one-half of other multilayer insulations). If the residual gas pressure in multilayer insulation is less than 10^{-4} torr, the conductance is practically independent of the residual gas pressure. At 10^{-3} torr (0.133 Pa), the conductance is increased by the order of 50 percent.

2. (i) **Powder insulation** consists of finely divided materials. Conductances of evacuated powders may vary by as much as 100 percent with variations of particle size and *apparent* density (packing) of the powder. The most commonly used powders are perlite, expanded SiO_2 (aerogel), calcium-silicate, diatomaceous earth, and carbon black. The conductance of powder insulation is practically independent of gas pressure down to 10 torr (1.333 kPa), at which pressure it begins to decrease rapidly with decreasing pressure as the free path of the gas molecules becomes comparable with the space between powder particles. The most rapid decrease of conductance occurs between 1 and 10^{-1} torr (133 and 13.3 Pa). At 10^{-3} torr (0.133 Pa) the conductance reaches practically the lower limit. Even at 10^{-2} torr (1.33 Pa) the conductance is quite low. For this reason, good mechanical pumps are adequate for the evacuation of powder insulation (fine mesh filters are needed to protect the pump from being damaged by the powder). When evacuated to 10^{-4} torr (1.33×10^{-2} Pa) or less, the apparent mean thermal conductivity for perlite [540°R (300 K) to 140°R (78 K)] ranges from 0.6 to 1.2 mW/m · K. For nonevacuated perlite filled with He or H_2 an apparent mean thermal conductivity of 115 mW/m · K can be used. With N_2 a typical value is 30 mW/m · K.

The principal function of the powder is to impede the transfer of heat by radiation. Powders are used when radiation would constitute an important heat leak. If an insulating vacuum is bounded by highly reflecting walls, such as clean Cu, Ag, or Al, adding powder may result in an increase of heat transfer because of the paths of solid conduction through the powder.

(ii) **Al powder or flakes added to a powder insulation** reduces the radiative transfer through the insulation. Powdered Al is more effective than flaked Al. The conductance of highly evacuated perlite insulation [$\Delta T = 540 - 140$°R $(300 - 78$ K)] may be reduced 40 percent by the addition of 25 percent of Al powder. A similar addition of Al powder to Santocel and Cab-O-Sil (diatomaceous earth) reduces the conductance by 70 percent. Metal powder additions are most effective with those powders that are partially transparent to infrared radiation of wave lengths longer than 3 μm.

3. **Rigid-Foam insulation**   The foams most used in cryogenic applications are the more or less closed-cell foams. Table 19.2.10 gives thermal conductivities of selected samples of insulating foams. The conductivity of a foam is dependent on the conduction through the intracellular gas, on the transfer by thermal radiation, and on solid conduction. Temperatures below the condensing T of the encased gas condense it and improve the insulation. Gases encased when polymer foams are blown gradually diffuse out of the cells and are replaced by ambient gases. The conductivities of many Freon and CO_2 blown foams increase in time as much as 30 percent because of the diffusion of air into the foam. Conductivities may increase by a factor of 3 or 4 when cells are permeated with H_2 or He. Glass and silica foams appear to be the only ones having fully closed cells. The **structural rigidity** of foams may be used to eliminate the need for mechanical supports. It is possible to use rigid foams without inner liners and outer casings. Polymer foams have relatively high **thermal expansions**, even greater than the same polymer without voids. This makes cracking of the insulation a problem

Table 19.2.10 Apparent Mean Thermal Conductivities of Selected Foams, Balsa Wood, and Corkboard

Material	Density lbm/ft^3 (kg/m^3)	Thermal conductivity,* Btu/h·ft·°R (W/K·m)
Polystyrene foam†	2.4 (38)	0.019 (0.033)
	2.9 (46)	0.015 (0.026)
Epoxy resin foam†	5.0 (80)	0.019 (0.033)
Polyurethane foam†	5.0–8.7 (80–139)	0.019 (0.033)
Rubber foam†	5.0 (80)	0.021 (0.036)
Silica foam†	10.0 (160)	0.032 (0.055)
Glass foam†	8.7 (139)	0.020 (0.035)
Balsa wood, across grain	7.3 (117)	0.027 (0.047)
	8.8 (141)	0.032 (0.055)
	20.0 (320)	0.048 (0.083)
Corkboard, no added binder	5.4 (86)	0.021 (0.036)
	7.0 (112)	0.0225 (0.039)
	10.6 (170)	0.025 (0.043)
	14.0 (224)	0.028 (0.048)
Corkboard with asphalt binder	14.5 (232)	0.027 (0.047)

* Test space pressure = 14.7 lb/in^2 abs (0.101 MPa).
† ΔT = 540 to 140°R (300 to 78 K).

when it is bonded to metal. Sliding expansion joints have been used to overcome this. Mylar film bags (0.001 in or 0.025 mm thick) have been used to hold cryogenic liquids inside rigid-foam insulation and plastic films outside to keep the ambient atmosphere from the insulation.

4. **Balsa wood and corkboard** have been used for insulating cryogens that are not very cold, e.g., liquid methane (201°R or 112 K) and higher-boiling cryogens. The conductivities of these insulators (Table 19.2.10) are considerably higher than for the other types of low-T insulation discussed above. Their use, therefore, is restricted to special applications, as where their ability to support internal structures mechanically is important.

Conduction of heat along structural supports passing through the insulation merits special attention for an adequate realization of the benefits of superior insulation. Ideal materials of construction would have high mechanical strength and low thermal conductivity. In Table 19.2.11, the yield strengths, in tension, of construction materials are compared with their thermal conductivities. This kind of comparison favors the nonmetallic materials in Table 19.2.11. When large quantities of cryogens are handled in the field, metals are commonly used for

structural members because glass and plastics (Teflon excepted) are brittle at low T's.

Minimizing the transfer of heat through the supporting structure offers opportunities for ingenuity and inventiveness in design. The transfer is minimized by lengthening the supporting members. A stack of thin metal disks or washers, utilizing the contact resistance between adjacent disks, will increase thermal resistance without increase in length of support. A stack of 0.0008-in (0.02-mm) stainless steel disks under a compressive stress of 1,000 lb/in^2 (6.895 MPa), in vacuum, has the same thermal resistance as a solid rod 50 times the length of the stack.

Insulation Selection Because of their smaller heats of vaporization and greater cost of production, the lowest-T cryogens (LHe and LH$_2$) ordinarily merit more effective insulation than the higher-boiling cryogens like LO$_2$ and LN$_2$, and these in turn are ordinarily better insulated than other cryogens boiling at still more elevated T's. The choice of insulation ordinarily involves the conditions of use of the equipment, as well as considerations of the tolerable loss of refrigeration and the cost of the insulation and its installation. If equipment is to have intermittent, rather than long, uninterrupted operating periods, the insulation heat capacity and time lag in reaching steady state are important. An insulation with large heat capacity (e.g., powders) will evaporate large quantities of refrigerant and take a long time to reach steady state. Powders are compacted by mechanical shocks and vibrations. Continued mechanical loading will compact both multilayer insulation and powders. Weight and bulkiness of the insulation can be important. The heat transported by the mechanical supports and vents connecting the cold interior with the warm exterior becomes the minimum refrigeration loss, attainable only with perfect insulation of the rest of the system. A large loss of refrigeration through supports and vents ordinarily reduces the attractiveness of costly insulation having the lowest conductance.

SAFETY

Precautions are needed for the safe handling and storage of cryogens because (1) air can contaminate the cryogen, creating a potential explosive, either directly as when air mixes with H$_2$ or CH$_4$ or indirectly by transforming an inert cryogen like LN$_2$ into an oxidant (a potential hazard for combustible insulation), and (2) moisture and air can freeze on cold surfaces, clogging vents and preventing normal operation of valves.

Vent systems should be sized for sufficient exhaust velocities to prevent back-diffusion of air into the cryogen space. Storage at slightly

Table 19.2.11 Comparison of Materials for Support Members

Material	E_y,*¶ kpsi (MPa)	κ,* Btu/h·ft·°R (W/K·m)	E_y/κ kpsi·h·ft·°R/Btu (MN·K/W·m)
Aluminum 2024	55 (379)	47 (81)	1.17 (4.68)
Aluminum 7075	70 (483)	50 (87)	1.4 (5.55)
Copper, ann	12 (83)	274 (474)	0.044 (0.18)
Hastelloy† B	65 (448)	5.4 (9.3)	12 (48)
Hastelloy† C	48 (331)	5.9 (10.2)	8.1 (32)
K Monel‡	100 (690)	9.9 (17.1)	10.1 (40)
Stainless steel 304, ann	35 (241)	5.9 (10.2)	5.9 (24)
Stainless steel (drawn 210,000 lb/in^2)	150 (1034)	5.2 (9.0)	29 (115)
Titanium, pure	85 (586)	21 (36.3)	4.0 (16)
Titanium alloy (4A1–4Mn)	145 (1000)	3.5 (6.1)	41.4 (164)
Dacron§	20 (138)	0.088 (0.15)¶	227 (920)
Mylar§	10 (69)	0.088 (0.15)¶	113 (460)
Nylon§	20 (138)	0.18 (0.31)	111 (445)
Teflon§	2 (14)	0.14 (0.24)	14.3 (58)

* E_y is the yield stress, and κ is the average thermal conductivity between 36 and 540°R (20 and 300 K).
† Haynes Stellite Co.
‡ International Nickel Co.
§ E.I. du Pont de Nemours & Company
¶ Room-temperature value.

elevated pressure is preferable to prevent in-leakage of contaminants. Condensation can also be a problem on poorly insulated lines carrying low-boiling cryogens; frost is not a hazard but the oxygen-enriched condensed air is hazardous in the presence of combustibles. The insulation of liquid vent lines is a necessary precaution. Where this is not possible, appropriately placed guttering can prevent the contact of any condensed air with combustible objects.

Any space, either containing a cryogen or being refrigerated by a cryogen, should be protected with proper **safety-relief mechanisms**, i.e., relief valves or rupture disks. Such space includes less obvious ones such as the insulating vacuum space surrounding a cryogenic fluid or a section of line which can trap liquid between two valves. A heat leak to trapped liquid confined without a vent for escape of vapor may develop pressures sufficient to burst the containing vessel. Gas pressures necessary to maintain liquid density at ambient temperature are for helium, 18,000 lb/in² (124 MPa); for hydrogen, 28,000 lb/in² (193 MPa); and for nitrogen, 43,000 lb/in² (296 MPa).

The **selection of structural** materials calls for consideration of the effects of low temperatures on the properties of those materials. A number of otherwise suitable materials become brittle at low temperature (Fig. 19.2.10). Materials must perform satisfactorily over the complete range from cryogenic to room temperature, with hydrogen embrittlement particularly significant. While hydrogen embrittlement is not believed to be a problem at liquid-hydrogen temperatures, it has been shown to occur at room temperature. The 300-series stainless steels, aluminum alloys and BeCu, among other alloys, are more resistant to hydrogen embrittlement.

In calculating allowable stresses, room temperature yield and tensile strengths should be used because (1) pressure testing of containers at ambient temperature is frequently necessary, and (2) in the case of large vessels, large unknown temperature gradients can exist above the liquid level both in the ullage space and in the walls of the vessel. The 300-series stainless steels are normally austenitic (f.c.c.). However, in some 300-series stainless steels austenite is partially transformed by cold-working and possibly by temperature cycling to martensite. As martensite is brittle at low T's, the ductility of these steels is reduced by this kind of treatment. For this reason, it is recommended as an added safety factor that room temperature strength of the 300 series stainless steels be used for the design of structural members.

Careful **stress analysis** is essential and must include stresses due to (1) thermal contraction of equipment, and (2) radial, axial, and circumferential temperature gradients caused by non-uniform cooling rates. The latter are frequently encountered in the cool-down of cryogenic transfer lines.

Cryogenic equipment should be **purged** before being placed in service to remove unwanted condensables and gases that could form explosive mixtures. Commonly used methods are (1) evacuating and back-filling with purge gas and (2) flowing of purge gas through the system.

The cryogen can be a direct **hazard to personnel**. Cryogen ''burns'' can result from direct contact with either the cryogen or uninsulated equipment containing the cryogen. The large evolution of gases associated with cryogenic spills can result in asphyxiation. Asphyxiation is a hazard in entering a warmed cryogenic vessel. Further safety details are found in the references.

19.3 OPTICS
by Michael J. Clark

REFERENCES: AIP Handbook, McGraw-Hill. Born and Wolf, ''Principles of Optics,'' Pergamon Press. Hecht and Zajak, ''Optics,'' Addison-Wesley. Jenkins and White, ''Fundamentals of Optics,'' McGraw-Hill. Kingslake, ''Lens Design Fundamentals,'' Academic Press. Malacara, ''Optical Shop Testing,'' Wiley. Photonics Handbook, Laurin Publishing Co.

FUNDAMENTAL THEORIES OF LIGHT

Electromagnetic radiation spans a very large spectrum (Fig. 19.3.1). **Visible light** is a small band in the electromagnetic spectrum with wavelengths ranging between 0.390 and 0.770 μm where approximately 0.560 μm (green-yellow) seems the brightest to the average person. Units of measurement for electromagnetic radiation are wavelengths of light. Wavelength is the distance between peaks of a light wave. **Wavelength** λ has a reciprocal relationship to frequency of vibration of a wave, f, and for a traveling light wave that relationship is $\lambda = c/f$, where c is the speed of light. The **speed of light** in a vacuum for electromagnetic waves of all wavelengths is taken as 2.99793×10^8 m/s (for calculations made to four significant digits or less it is practical to use the estimation of 3.0×10^8 m/s for the speed of light). The speed of light is reduced when the light is traveling in any medium. The ratio of the speed of light in a vacuum c to the speed of light in a medium c' is defined as the **index of refraction,** $n = c/c'$. The index of refraction of air at standard conditions (0°C and 760 mmHg) is taken as 1.00292; it is often assumed to be unity. Most optical substances; gases, liquids, and solids (glasses, crystals, etc.) have refractive indexes between 1 and 4 (see AIP Handbook or Photonics Handbook). The index of refraction of most substances (including optical glass) is not constant but varies with the color of light. The **dispersion** ν of glass is normally taken as $\nu = (n_D - 1)/(n_F - n_C)$, where n_C, n_D, and n_F are the indexes of refraction for three spectral lines at 0.6563 μm (red), 0.5893 μm (yellow), and 0.4861 μm (blue), respectively. These lines are obtained from hydrogen

(n_C and n_F) and sodium (n_D) discharge tube light sources. Most optical glasses have dispersions between 20 and 65.

FUNDAMENTALS OF GEOMETRICAL OPTICS

At an interface of two media with different refractive indexes, n and n' (Fig. 19.3.2), light rays at an incident angle I measured from the direction normal to the interface will be partially reflected back into the incident medium and partially refracted into the second medium. Two fundamental laws of physics describe these occurrences: The **law of reflection** states that the angle of incidence I equals the angle of reflection I'', and the angle of reflection is on the opposite side of the normal in the incident medium. The **law of refraction** (also referred to as **Snell's law**) states that the relationship between the incident angle and the refracted angle I' obeys $n \sin I = n' \sin I$. Given a spherical interface on an optical axis (Fig. 19.3.3), the relationship between object distance l, image distance l', and radius of curvature of the spherical interface r, is given by $n'/l' = (n' - n)/r + n/l$. This relationship holds only for rays that are close to the optical axis (paraxial rays).

The **magnification** m for such a surface is $m = h'/h = nl'/n'l$. If the image is inverted with respect to its object, the magnification is a negative number. Fractional magnifications indicate an actual reduction in size. The **focal length** f of a thin lens is the distance from the lens when the object is infinitely far away. The **power** Φ of a lens is taken as $\Phi = n/f$ (see Jenkins and White). The focal lengths of typical commercial lenses for use in air vary between 3 mm for microscope objectives to 30.5 m (100 ft) for large observatory telescope objectives.

The larger the **aperture** of an optical element, the brighter will be the image and the better the resolving power, if the rest of the system (including the eye) is capable of accommodating all of the light. The aperture of telescope objectives and camera projection lenses is usually

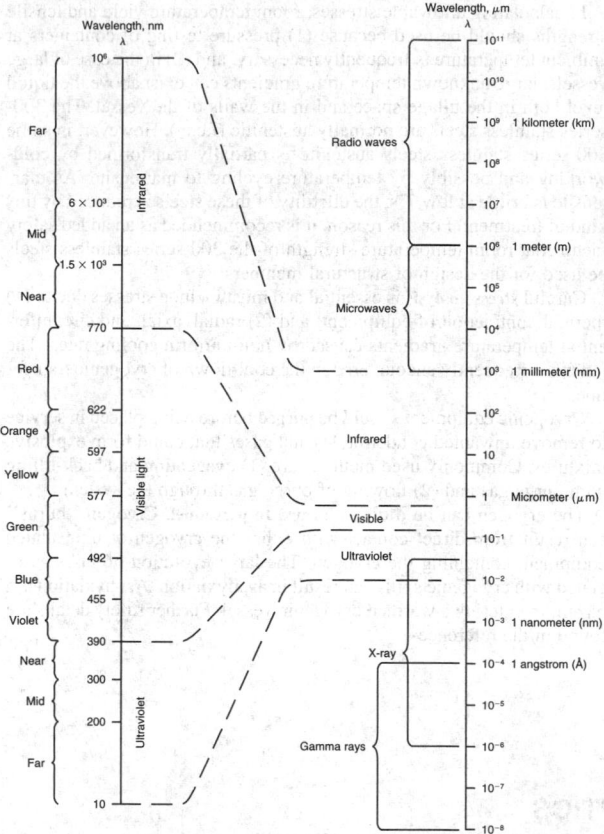

Fig. 19.3.1 Electromagnetic spectrum.

described by the **f number,** which is defined as the ratio of the focal length to the clear diameter. Commercially available lenses usually fall between $f/1.2$ and $f/15$. The aperture of microscope objectives is usually described by the **numerical aperture** NA. The numerical aperture is defined as $n \sin U$, where n is the index of refraction of the object space and U is the half angle of the vertex of the cone of light from the object that passes through the lens. Commercially available lenses usually fall between NA/1.40 and NA/0.08.

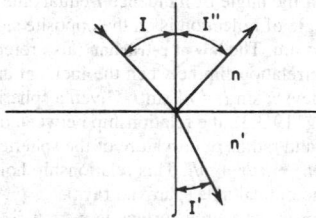

Fig. 19.3.2 Reflection and refraction at a plane surface.

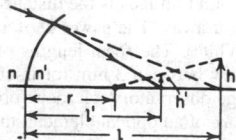

Fig. 19.3.3 Refraction at a spherical surface; all parameters are shown in their positive sense.

Photographic film speeds are usually given as an ASA exposure index. Black-and-white films are available commercially in the ASA range from 25 to 3,000. Color films fall in the slower end of this range. A rule of thumb for exposing films on a bright, sunny day is to set the aperture of the camera at $f/16$ and the shutter speed at 1/ASA exposure index.

OPTICAL DESIGN

The first step in developing a cost-effective optical system is to investigate what components are commercially available that would meet your specific needs. If commercially available components do not sufficiently meet the requirements of the application, then a customized design would need to be developed. The majority of optical designs are performed through the use of computer-aided design (CAD) packages. Originally, early in the 1970s, the optical CAD market was dominated by mainframe packages. Due to recent enhancements in computer hardware technology, the availability of PC-based optical design packages has expanded (see Photonics Handbook for more information on optical design packages). The optical design package is used as a tool by designers to develop an optical system. The designer generally enters initial design parameters and specifications. The CAD software then optimizes the design parameters and outputs a design proposal. Several iterations may be required to develop an appropriate system. When developing an optical design consideration should be made for f number, clear aperture, field of view, spectral range, packaging constraints, environmental parameters, transmittance, and performance specification.

LASERS

The name **laser** is an acronym of light amplification by the stimulated emission of radiation. A laser is a device that generates a highly coherent monochromatic source of intense radiation. The laser contains a cavity that is excited to stimulate emissions of photons that are repeatedly reflected between highly reflective mirrors that are precisely aligned at either end of the cavity. The intensity of the emission increases as it oscillates within the cavity. One mirror is made to partially transmit the intense monochromatic source. The wavelength of the source is determined by the type of material used to make up the cavity. The cavity may be made up of any material, gas, liquid, crystal or glass, whose atoms are capable of being excited from the ground state to a semistable state. Some typical laser applications are bar code scanning, optical discs, optical fiber communication, photolithography, ablation, micromachining, microsurgery, fusion, spectroscopy, interferometry, and distance measurement, to name a few.

INTERFERENCE AND INTERFEROMETRY

Optical **interference** is the combination of two or more waves of light. Constructive interference occurs when the peaks and valleys of the waves align with each other; these waves are said to be *in phase*. If the waves are out of phase they will destructively interfere. For wavefronts with equal amplitudes and wavelengths, and the appropriate phase relationship, destructive interference will cause the waves to cancel each other and therefore create dark bands called **fringes**. Optical surfaces, plano or spherical, may be measured through the relative interference of the wavefront of a test surface with the wavefront of a known highly precise reference surface. **Test glasses** and/or **test/plates** are frequently used as a reference surface in measuring an optical surface. The test glass is brought into contact with the surface under test and placed under a monochromatic source of light so that interference fringes become visible. The interference fringes are then evaluated to determine the accuracy of the surface under test relative to the reference surface (see Photonics Handbook for additional information on fringe interpretation). An **interferometer** is a measurement instrument that utilizes interference theory to determine the accuracy of optical surfaces and/or systems. Interferometers typically generate a precise monochromatic wavefront that is split to generate a test wavefront and reference wavefront. These wavefronts are passed through the test and reference optical systems and then recombined to create interference fringes. There are several different configurations of interferometers with variable applications (see Malacara). Some of the most common configurations are the Fizeau interferometer, Michelson interferometer, and the Twyman-Green interferometer. Many of the commercially available interferometers have incorporated sophisticated computer systems to capture surface data and enhance the analysis of optical surfaces and systems.

Defects of an optical surface or system that cause deviations in the image accuracy are called **aberrations**. There are several primary geometrical aberrations that are evaluated to develop optimum performance of an optical design. **Spherical aberration** is an axial error where rays across the aperture have different focal lengths. **Astigmatism** occurs when light in the tangential plane focuses to a different point than light in the sagittal plane. In the absence of astigmatism, the image is formed on a curved surface referred to as the Petzval surface. The curvature of the Petzval image, or **field curvature**, is 1/(radius of the Petzval surface). **Coma** occurs when rays through the outer portion of a lens produce a higher or lower magnification than the central rays. **Distortion** is a change of magnification as a function of the field of view. There are two **chromatic aberrations**: axial and lateral. Axial chromatic aberration is the effect of different wavelengths of light focusing to a different point on the optical axis due to the effects of dispersion. Lateral color is a change or difference in the off-axis image focus for different wavelengths of light.

FIBER OPTICS

Optical fibers consist of an optically transparent central fiber, surrounded by a less dense cladding. Bundles of such fibers may be used to transmit images over long distances and/or around curves. Optical fibers utilize the principle of total internal reflection where the incident angle is greater than or equal to the critical angle. The **critical angle** Φ_C can be computed by the expression $\sin \Phi_C = n/n'$, where n is the refractive index of the fiber and n' is the refractive index of the cladding surrounding the fiber. To achieve total internal reflection, the index of the fiber material must be greater than the index of the cladding. Optical fibers are very effective in communication applications—video, voice, and data—with expanding applications in industrial facilities and military systems. Significant advantages to optical fibers over electrical wires are electrical isolation, low signal degradation, broad signal bandwidth, high reliability, light weight, and low cost.

19.4   MISCELLANEOUS
(Staff Contribution)

SIZES OF TYPE

The unit of height of a line of printer's type is the "point." One point = $\frac{1}{72}$ in. Six-point type is consequently $\frac{6}{72} = \frac{1}{12}$ in high. The sizes generally employed in books and periodicals are 6-, 7-, 8-, 9-, 10-, 11-, and 12-point. These are shown in Fig. 19.4.1.

6-point	**10-point**
7-point	**11-point**
8-point	**12-point**
9-point	

Fig. 19.4.1 Examples of type sizes.

COPY PREPARATION AND COPY FITTING

It is recommended that manuscript material be printed double-spaced on 8½- by 11-in paper with 1½-in margins at the top and left side, 1-in margins on the bottom and left side, and with each page containing 25 lines, double-spaced. Thus a printed line will be 60 characters long. Nearly all printers can print at 10 characters to the inch and double-spaced (three lines to the inch). For convenience in copyediting, printed copy should be printed double-spaced. One manuscript page so printed is equivalent to about 250 words (assuming that each word averages six characters).

One way to determine approximately how many lines of manuscript copy are equivalent to how many lines of finished book type is to print, to the specifications given above, a number of lines from a book which has been printed in the desired type font. This will show the relationship between the number of manuscript-copy lines and lines set in type.

Nearly all word-processing programs will produce satisfactory manuscript. Authors working directly with a publisher will usually make arrangements to submit their manuscripts in computer-readable form. Some publishers require it. Publishers have very comprehensive format-conversion facilities. Authors working in groups or with intermediate editors, however, face the problem that others of their associates or editors may not have much conversion capability. In such cases, the editors or group leaders will make arrangements to accept such computer-readable copy as they can. In all cases, a printed copy should accompany all computer-readable material.

PROOFREADING

When correcting proof, use ink or pencil of a different color from the marks already on the proof. Put your marks in the right or left margin, whichever is nearer to the word corrected. If there are several corrections in a single line, place them in order from left to right, separated by a slant line (e.g., tr/cap/,). If the same correction is made several times in the same line with no intervening correction, make your correction once in the margin, followed by an appropriate number of slant lines. For instance, if you add an *s* to three words in a line, write in the margin s///. When you wish to insert words, put a caret at the point of insertion and write the additional words in the margin. To delete material without substituting anything, cross it out and put a delete sign in the margin. When you delete words and substitute other words, cross out the unwanted material, use a caret within the line, and write the new material in the margin; the delete sign is then unnecessary. Standard proofreader's marks are shown in Fig. 19.4.2.

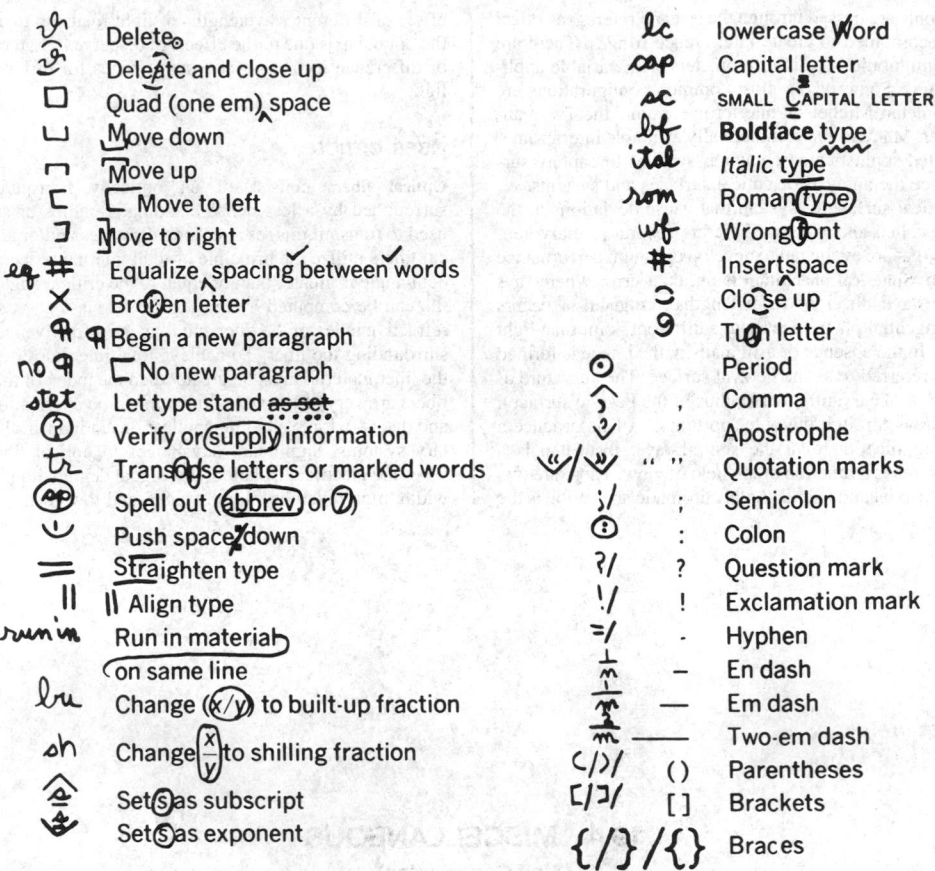

Fig. 19.4.2 Proofreader's marks.

Occasionally you may decide that material you have deleted should be restored. Place a row of dots under the crossed-out material, cross off the delete sign in the margin, and write "stet" (let it stand).

SPECIAL ALPHABETS

The primary alphabets used as symbols are Greek, script, and German. Examples of these are shown in Figs. 19.4.3 through 19.4.6.

ΑΒΓΔΕΖΗΘΙΚΛΜΝΞΟΠΡΣΤΥΦΧΨΩ
αβγδεζηθικλμνξοπρστυφχψω

Fig. 19.4.3 Greek alphabet.

ΑΒΓΔΕΖΗΘΙΚΛΜΝΞΟΠΡΣΤΥΦΧΨΩ
αβγδεζηθικλμνξοπρστυφχψω

Fig. 19.4.4 Boldface Greek alphabet.

Note that the capital Greek letters used by the printer are easily mistaken for the Roman letters A, B, E, Z, H, I, K, M, N, O, P, T, and X and

may not be suitable as symbols. If they are used in a manuscript, they should be identified with a marginal note so they will not be improperly set. The form of many Greek boldface symbols is markedly different from the style of the lightface characters.

Lightface: 𝒜 ℬ 𝒞 𝒟 ℰ ℱ 𝒢 ℋ 𝒥 𝒦 ℒ
𝑀 𝒩 𝒪 𝒫 𝒬 ℛ 𝒮 𝒯 𝒰 𝒱 𝒲 𝒳 𝒴 𝒵
Boldface: 𝓑 𝓔 𝓔 𝓕 𝓚 𝓘 𝓜

Fig. 19.4.5 Script alphabet.

𝔄𝔅ℭ𝔇𝔈𝔉𝔊ℌℑ𝔍𝔎𝔏𝔐𝔑𝔒𝔓𝔔ℜ𝔖𝔗𝔘𝔙𝔚𝔛𝔜ℨ

𝔞𝔟𝔠𝔡𝔢𝔣𝔤𝔥𝔦𝔧𝔨𝔩𝔪𝔫𝔬𝔭𝔮𝔯𝔰𝔱𝔲𝔳𝔴𝔵𝔶𝔷

äüö 𝔄𝔘𝔒 ch ck ß ſſ ſt tz

Fig. 19.4.6 German Fraktur alphabet.

Script and German symbols (Figs. 19.4.5 and 19.4.6) may be indicated by circling in a distinctive color the corresponding English letter.

Index

About the Editors

EUGENE A. AVALLONE, Editor, is Professor of Mechanical Engineering, Emeritus, The City College of the City University of New York. He has been engaged for many years as a consultant to industry and to a number of local and national governmental agencies.

THEODORE BAUMEISTER, III, Editor, is now retired from Du Pont where he was an internal consultant. His specialties are operations research, business decision making, and long-range planning. He has also taught financial modeling in the United States, South America, and the Far East.

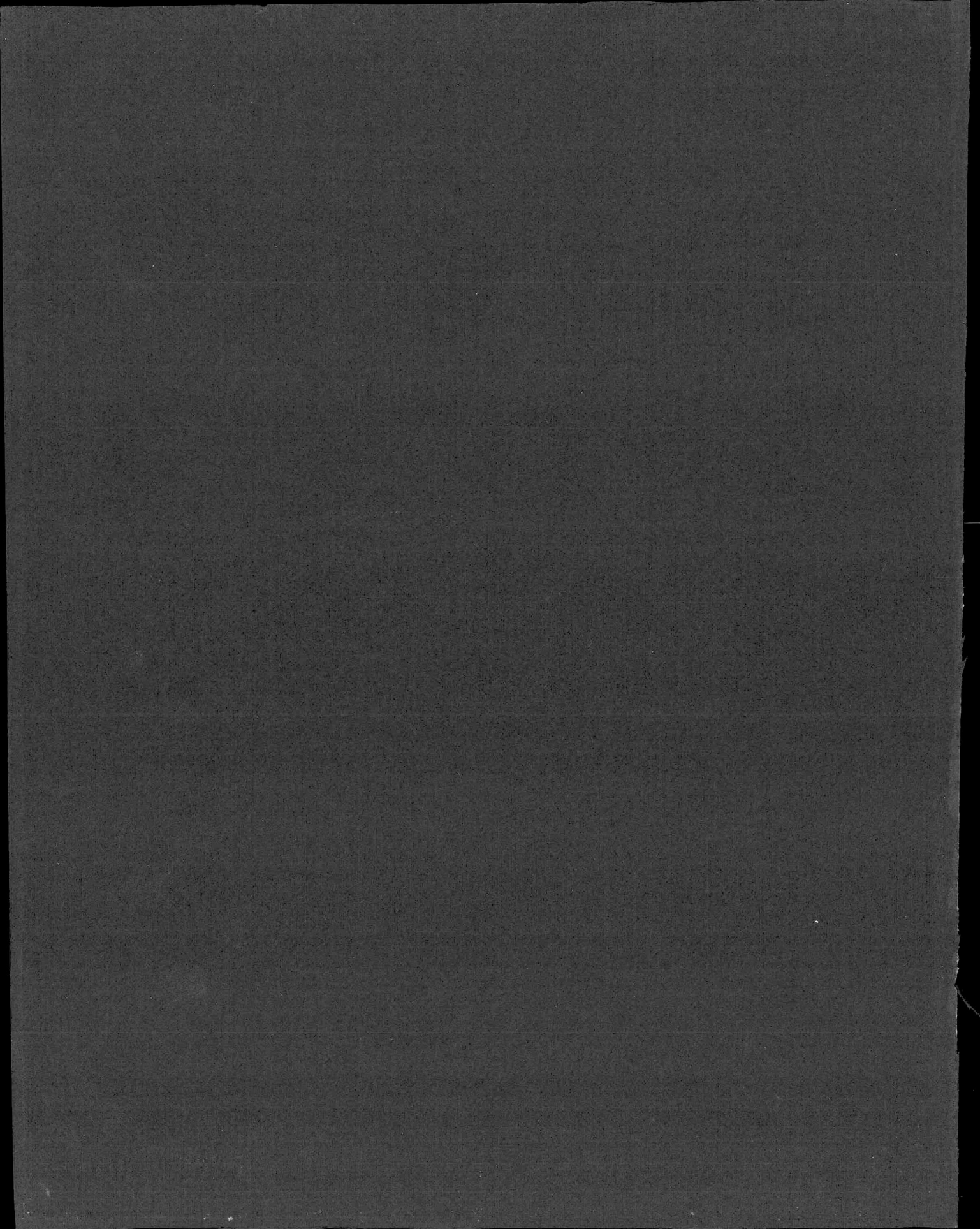